CRC Handbook
of
Chemistry and Physics

87th Edition

CRC Handbook
of
Chemistry and Physics

A Ready-Reference Book of Chemical and Physical Data

HANDBOOK OF CHEMISTRY AND PHYSICS

2006-2007

87th
EDITION

CRC PRESS

Editor-in-Chief

David R. Lide, Ph.D.

Former Director, Standard Reference Data
National Institute of Standards and Technology

Taylor & Francis
Taylor & Francis Group

Boca Raton   London   New York

A CRC title, part of the Taylor & Francis imprint, a member of the
Taylor & Francis Group, the academic division of T&F Informa plc.

PREFACE

This 87th Edition of the *CRC Handbook of Chemistry and Physics* is the second to appear in the larger page size and new type font. Comments from users of the 86th Edition have been generally favorable; the elimination of landscape format for wide tables seems particularly appreciated. However, we always welcome your comments and suggestions.

The overall philosophy of the *Handbook* remains the same, namely, to provide broad coverage of all types of physical science data commonly encountered by scientists and engineers, with as much depth as can be accommodated in a one-volume format. Notwithstanding the growing availability of specialized databases on the Internet, we feel there is still a need for a concise, reliable reference source spanning the full range of the physical sciences and focusing on key data that are frequently needed by R&D professionals, engineers, and students. The *CRC Handbook of Chemistry and Physics*, in its print, CD-ROM, and Web formats, is aimed at serving these needs. The data contained in the *Handbook* have been carefully evaluated by experts in each field; quality control is a high priority and the sources are documented. The annual updates make it possible to add new and better data in a timely fashion. In this way we hope to continue the role of the *Handbook* as a unique reference source.

Among the changes in the 87th Edition are major revisions of four heavily used tables:

- *Physical Constants of Inorganic Compounds* has been completely updated, and the number of compounds has been increased by 20%.
- *Bond Dissociation Energies* has been updated with results from the latest literature, and the coverage has been expanded to include organometallics, low molecular weight biochemical compounds, and positive ions. The total number of chemical bonds covered is now 4193, as compared to 2579 in the 86th Edition.

- *Table of the Isotopes*, the comprehensive listing of the energies and radiation properties of all known isotopes, has been brought up-to-date with results from the literature up to the beginning of 2006. This definitive compilation now includes over 4500 individual isotopes.
- *Scientific Abbreviations and Symbols* has been expanded to about 1100 entries and includes more acronyms from quantum chemistry and abbreviations for chemicals of environmental interest.

Fourteen other tables have been updated. Of particular note are two tables based on very recent IUPAC recommendations: *Standard Atomic Weights (2005)* and *Nomenclature for Inorganic Ions and Ligands*. There is also a new table on *Specific Enthalpies of Solution of Polymers and Copolymers*.

The Editor appreciates suggestions on new topics for the *Handbook* and notification of any errors. Input from users plays a key role in keeping the book up-to-date. Address all comments to Editor-in-Chief, *CRC Handbook of Chemistry and Physics*, Taylor & Francis Group, 6000 Broken Sound Parkway NW, Suite 300, Boca Raton, FL 33487. Comments may also be sent by electronic mail to drlide@post.harvard.edu.

The *CRC Handbook of Chemistry and Physics* is dependent on the efforts of many contributors throughout the world. The list of current contributors follows this Preface. Valuable suggestions have been received from the Editorial Advisory Board and from many users. The assistance and support of Dr. Fiona Macdonald are greatly appreciated. Finally, I want to thank Susan Fox and James Yanchak for their excellent work in typesetting the book.

David R. Lide
March 2006

The 87th Edition is dedicated to the memory of Elizabeth G. Breen, 1916–2005

Note on the Ordering of Chemical Compounds: The decision on the order in which to list chemical compounds in a table is always difficult. An alphabetical list by name has the disadvantage that several different synonyms are often in common use, with the result that a reader may conclude incorrectly that a compound is not present if he looks it up under the wrong name. Listing by common formula is satisfactory for simple inorganic compounds, but is not feasible for organics. A listing by molecular formula is attractive because clear rules can be given for locating a compound, but the user may have to go to some effort to determine the molecular formula. In this book the choice is made on pragmatic grounds. The long tables, Physical Constants of Organic Compounds and Physical Constants of Inorganic Compounds, are ordered by name (generally the systematic name), but indexes to synonyms, formulas, and CAS Registry Numbers are provided. If the table is very short and includes only familiar substances, the listing is usually alphabetical by name or common formula. Tables of intermediate length are often ordered by molecular formula using a modification of the Hill convention. In this convention the molecular formula is written with C first, H second, and then all other elements in alphabetical order of their chemical symbols. For tables with organic compounds only, the sequence of entries is determined by the alphabetical order of elements in the molecular formula and the number of atoms of each element, in ascending order, e.g., C_3H_7Cl, C_3H_7N, C_3H_7NO, $C_3H_7NO_2$, etc. (For organic compounds, a quick way to determine the molecular formula is to use the Physical Constants of Organic Compounds table, which starts on Page 3-1, and its synonym index on Page 3-524.) In tables containing non-carbon compounds, those are usually listed first, followed by a separate listing of compounds that do contain carbon. This is a departure from the strict Hill convention as followed by Chemical Abstracts Service, where the molecular formulas beginning with A and B precede the formulas for carbon-containing compounds, while those beginning with D... Z follow. For tabular displays, as opposed to an index, it appears more convenient to the user if the non-carbon compounds are listed as a block, rather than being split by the longer list of carbon compounds.

CURRENT CONTRIBUTORS

Robert A. Alberty
Department of Chemistry
Massachusetts Institute of Technology
Cambridge, Massachusetts 02139

Lev I. Berger
California Institute of Electronics and
    Materials Science
2115 Flame Tree Way
Hemet, California 92545

A. K. Covington
Department of Chemistry
University of Newcastle
Newcastle upon Tyne NE1 7RU
England

K. Fischer
LTP GmbH
Oppelner Strasse 12
D-26135 Oldenburg, Germany

Jean-Claude Fontaine
ITODYS
CNRS, University of Paris VII
1 rue Guy de la Brosse
75005 Paris, France

J. R. Fuhr
Atomic Physics Division
National Institute of Standards and
    Technology
Gaithersburg, Maryland 20899

J. Gmehling
Universität Oldenburg
Fakultät V, Technische Chemie
D-26111 Oldenburg, Germany

Robert N. Goldberg
Biotechnology Division
National Institute of Standards and
    Technology
Gaithersburg, Maryland 20899

Norman E. Holden
National Nuclear Data Center
Brookhaven National Laboratory
Upton, New York 11973

Henry V. Kehiaian
ITODYS
University of Paris VII
1 rue Guy de la Brosse
75005 Paris, France

J. Alistair Kerr
School of Chemistry
University of Birmingham
Birmingham B 15 2TT England

William H. Koppenol
Dept CHAB
Lab. f. Anorg. Chemie, HCI H211
Wolfgang-Pauli-Strasse 10
ETH Hönggerberg
CH-8093 Zürich, Switzerland

J. Krafczyk
DDBST GmbH
Industriestrasse 1
D-26121 Oldenburg, Germany

Frank J. Lovas
8616 Melwood Rd.
Bethesda, Maryland 20817

Yu-Ran Luo
College of Marine Science
University of South Florida
140 7th Ave. South
St. Petersburg, Florida 33701

William C. Martin
Atomic Physics Division
National Institute of Standards and
    Technology
Gaithersburg, Maryland 20899

J. Menke
DDBST GmbH
Industriestrasse 1
D-26121 Oldenburg, Germany

Thomas M. Miller
Air Force Research Laboratory/VSBP
29 Randolph Rd.
Hanscom AFB, Massachusetts 01731-3010

Peter J. Mohr
Physics Laboratory
National Institute of Standards and
    Technology
Gaithersburg, Maryland 20899

Cedric J. Powell
Surface and Microanalysis Science
    Division
National Institute of Standards and
    Technology
Gaithersburg, Maryland 20899

Joseph Reader
Atomic Physics Division
National Institute of Standards and
    Technology
Gaithersburg, Maryland 20899

Paul A. Redhead
1958 Norway Crescent
Ottawa, Ontario K1H 5N7
Canada

Lewis E. Snyder
Astronomy Department
University of Illinois
Urbana, Illinois 61801

Barry N. Taylor
Physics Laboratory
National Institute of Standards and
    Technology
Gaithersburg, Maryland 20899

Petr Vanýsek
Department of Chemistry
Northern Illinois University
DeKalb, Illinois 60115

Wolfgang L. Wiese
Atomic Physics Division
National Institute of Standards and
    Technology
Gaithersburg, Maryland 20899

Christian Wohlfarth
Martin Luther University
Institute of Physical Chemistry
Mühlpforte 1
06108 Halle (Saale), Germany

Daniel Zwillinger
Mathematics Department
Rensselaer Polytechnic Institute
Troy, New York 12180

Piotr Zyla
Particle Data Group
Lawrence Berkeley Laboratory
Berkeley, California 94720

TABLE OF CONTENTS

SECTION 6: FLUID PROPERTIES

SECTION 7: BIOCHEMISTRY

SECTION 8: ANALYTICAL CHEMISTRY

SECTION 9: MOLECULAR STRUCTURE AND SPECTROSCOPY

SECTION 10: ATOMIC, MOLECULAR, AND OPTICAL PHYSICS

SECTION 11: NUCLEAR AND PARTICLE PHYSICS

SECTION 12: PROPERTIES OF SOLIDS

SECTION 16: HEALTH AND SAFETY INFORMATION

APPENDIX A: MATHEMATICAL TABLES

APPENDIX B: SOURCES OF PHYSICAL AND CHEMICAL DATA

INDEX

Section 1
Basic Constants, Units, and Conversion Factors

FUNDAMENTAL PHYSICAL CONSTANTS

Peter J. Mohr and Barry N. Taylor

These tables give the 2002 self-consistent set of values of the basic constants and conversion factors of physics and chemistry recommended by the Committee on Data for Science and Technology (CODATA) for international use. The 2002 set replaces the previously recommended 1998 CODATA set. The 2002 adjustment takes into account the data considered in the 1998 adjustment as well as the data that became available between 31 December 1998, the closing date of that adjustment, and 31 December 2002, the closing date of the new adjustment.

This report was prepared by the authors under the auspices of the CODATA Task Group on Fundamental Constants. The members of the Task Group are:

F. Cabiati, Istituto Elettrotecnico Nazionale "Galileo Ferraris," Italy

E. R. Cohen, Science Center, Rockwell International (retired), United States of America

K. Fujii, National Metrology Institute of Japan, Japan

S. G. Karshenboim, D. I. Mendeleyev All-Russian Research Institute for Metrology, Russian Federation

I. Lindgren, Chalmers University of Technology and Göteborg University, Sweden

B. A. Mamyrin, A. F. Ioffe Physical-Technical Institute, Russian Federation

W. Martienssen, Johann Wolfgang Goethe-Universität, Germany

P. J. Mohr, National Institute of Standards and Technology, United States of America

F. Nez, Laboratoire Kastler-Brossel, France

B. W. Petley, National Physical Laboratory, United Kingdom

T. J. Quinn, Bureau International des Poids et Mesures

B. N. Taylor, National Institute of Standards and Technology, United States of America

W. Wöger, Physikalisch-Technische Bundesanstalt, Germany

B. M. Wood, National Research Council, Canada

Z. Zhang, National Institute of Metrology, China (People's Republic of)

Reference

Mohr, P. J. and Taylor, B. N., *The 2002 CODATA Recommended Values of the Fundamental Physical Constants*, Web Version 4.0, NIST Physical Data web site <http://physics.nist.gov/cuu/constants> (December 2003); *Rev. Mod. Phys.* 76, No. 4, October 2004.

TABLE I. An Abbreviated List of the CODATA Recommended Values of the Fundamental Constants of Physics and Chemistry Based on the 2002 Adjustment

Quanity	Symbol	Numerical value	Unit	Relative std. uncert. u_r
speed of light in vacuum	c, c_0	299 792 458	m s^{-1}	(exact)
magnetic constant	μ_0	$4\pi \times 10^{-7}$	N A^{-2}	
		$= 12.566\ 370\ 614... \times 10^{-7}$	N A^{-2}	(exact)
electric constant $1/\mu_0 c^2$	ε_0	$8.854\ 187\ 817... \times 10^{-12}$	F m^{-1}	(exact)
Newtonian constant of gravitation	G	$6.6742(10) \times 10^{-11}$	m^3 kg^{-1} s^{-2}	1.5×10^{-4}
Planck constant	h	$6.626\ 0693(11) \times 10^{-34}$	J s	1.7×10^{-7}
$h/2\pi$	$\hbar$	$1.054\ 571\ 68(18) \times 10^{-34}$	J s	1.7×10^{-7}
elementary charge	e	$1.602\ 176\ 53(14) \times 10^{-19}$	C	8.5×10^{-8}
magnetic flux quantum $h/2e$	ϕ_0	$2.067\ 833\ 72(18) \times 10^{-15}$	Wb	8.5×10^{-8}
conductance quantum $2e^2/h$	G_0	$7.748\ 091\ 733(26) \times 10^{-5}$	S	3.3×10^{-9}
electron mass	m_e	$9.109\ 3826(16) \times 10^{-31}$	kg	1.7×10^{-7}
proton mass	m_p	$1.672\ 621\ 71(29) \times 10^{-27}$	kg	1.7×10^{-7}
proton-electron mass ratio	m_p/m_e	1836.152 672 61(85)		4.6×10^{-10}
fine-structure constant $e^2/4\pi\varepsilon_0 \hbar c$	α	$7.297\ 352\ 568(24) \times 10^{-3}$		3.3×10^{-9}
inverse fine-structure constant	α^{-1}	137.035 999 11(46)		3.3×10^{-9}
Rydberg constant $\alpha^2 m_e c/2h$	R_∞	10 973 731.568 525(73)	m^{-1}	6.6×10^{-12}
Avogadro constant	N_A, L	$6.022\ 1415(10) \times 10^{23}$	mol^{-1}	1.7×10^{-7}
Faraday constant $N_A e$	F	96 485.3383(83)	C mol^{-1}	8.6×10^{-8}
molar gas constant	R	8.314 472(15)	J mol^{-1} K^{-1}	1.7×10^{-6}
Boltzmann constant R/N_A	k	$1.380\ 6505(24) \times 10^{-23}$	J K^{-1}	1.8×10^{-6}
Stefan-Boltzmann constant $(\pi^2/60)k^4/\hbar^3 c^2$	σ	$5.670\ 400(40) \times 10^{-8}$	W m^{-2} K^{-4}	7.0×10^{-6}
Non-SI units accepted for use with the SI				
electron volt: (e/C) J	eV	$1.602\ 176\ 53(14) \times 10^{-19}$	J	8.5×10^{-8}
(unified) atomic mass unit	u	$1.660\ 538\ 86(28) \times 10^{-27}$	kg	1.7×10^{-7}

1 u $= m_u = (1/12)m(^{12}C) = 10^{-3}$ kg mol$^{-1}/N_A$

TABLE II.　The CODATA Recommended Values of the Fundamental Constants of Physics and Chemistry Based on the 2002 Adjustment

Quantity	Symbol	Numerical value	Unit	Relative std. uncert. u_r
Universal				
speed of light in vacuum	c, c_0	299 792 458	m s^{-1}	(exact)
magnetic constant	μ_0	$4\pi \times 10^{-7}$	N A^{-2}	
		$= 12.566\ 370\ 614... \times 10^{-7}$	N A^{-2}	(exact)
electric constant $1/\mu_0 c^2$	ε_0	$8.854\ 187\ 817... \times 10^{-12}$	F m^{-1}	(exact)
characteristic impedance of vacuum $\sqrt{\mu_0/\varepsilon_0} = \mu_0 c$	Z_0	$376.730\ 313\ 461...$	Ω	(exact)
Newtonian constant of gravitation	G	$6.6742(10) \times 10^{-11}$	m^3 kg^{-1} s^{-2}	1.5×10^{-4}
	$G/\hbar c$	$6.7087(10) \times 10^{-39}$	(GeV/c^2)$^{-2}$	1.5×10^{-4}
Planck constant	h	$6.626\ 0693(11) \times 10^{-34}$	J s	1.7×10^{-7}
in eV s		$4.135\ 667\ 43(35) \times 10^{-15}$	eV s	8.5×10^{-8}
$h/2\pi$	$\hbar$	$1.054\ 571\ 68(18) \times 10^{-34}$	J s	1.7×10^{-7}
in eV s		$6.582\ 119\ 15(56) \times 10^{-16}$	eV s	8.5×10^{-8}
$\hbar c$ in MeV fm		$197.326\ 968(17)$	MeV fm	8.5×10^{-8}
Planck mass $(\hbar c/G)^{1/2}$	m_P	$2.176\ 45(16) \times 10^{-8}$	kg	7.5×10^{-5}
Planck temperature $(\hbar c^5/G)^{1/2}/k$	T_P	$1.416\ 79(11) \times 10^{32}$	K	7.5×10^{-5}
Planck length $\hbar/m_P c = (\hbar G/c^3)^{1/2}$	l_P	$1.616\ 24(12) \times 10^{-35}$	m	7.5×10^{-5}
Planck time $l_P/c = (\hbar G/c^5)^{1/2}$	t_P	$5.391\ 21(40) \times 10^{-44}$	s	7.5×10^{-5}
Electromagnetic				
elementary charge	e	$1.602\ 176\ 53(14) \times 10^{-19}$	C	8.5×10^{-8}
	e/h	$2.417\ 989\ 40(21) \times 10^{14}$	A J^{-1}	8.5×10^{-8}
magnetic flux quantum $h/2e$	ϕ_0	$2.067\ 833\ 72(18) \times 10^{-15}$	Wb	8.5×10^{-8}
conductance quantum $2e^2/h$	G_0	$7.748\ 091\ 733(26)\times10^{-5}$	S	3.3×10^{-9}
inverse of conductance quantum	G_0^{-1}	$12\ 906.403\ 725(43)$	Ω	3.3×10^{-9}
Josephson constant[1] $2e/h$	K_J	$483\ 597.879(41)\times10^9$	Hz V^{-1}	8.5×10^{-8}
von Klitzing constant[2] $h/e^2 = \mu_0 c/2\alpha$	R_K	$25\ 812.807\ 449(86)$	Ω	3.3×10^{-9}
Bohr magneton $eh/2m_e$	μ_B	$927.400\ 949(80)\times10^{-26}$	J T^{-1}	8.6×10^{-8}
in eV T^{-1}		$5.788\ 381\ 804(39)\times10^{-5}$	eV T^{-1}	6.7×10^{-9}
	μ_B/h	$13.996\ 2458(12)\times10^9$	Hz T^{-1}	8.6×10^{-8}
	μ_B/hc	$46.686\ 4507(40)$	m^{-1} T^{-1}	8.6×10^{-8}
	μ_B/k	$0.671\ 7131(12)$	K T^{-1}	1.8×10^{-6}
nuclear magneton $eh/2m_p$	μ_N	$5.050\ 783\ 43(43)\times10^{-27}$	J T^{-1}	8.6×10^{-8}
in eV T^{-1}		$3.152\ 451\ 259(21)\times10^{-8}$	eV T^{-1}	6.7×10^{-9}
	μ_N/h	$7.622\ 593\ 71(65)$	MHz T^{-1}	8.6×10^{-8}
	μ_N/hc	$2.542\ 623\ 58(22)\times10^{-2}$	m^{-1} T^{-1}	8.6×10^{-8}
	μ_N/k	$3.658\ 2637(64)\times10^{-4}$	K T^{-1}	1.8×10^{-6}
Atomic and Nuclear				
General				
Fine-structure constant $e^2/4\pi\varepsilon_0\hbar c$	α	$7.297\ 352\ 568(24)\times10^{-3}$		3.3×10^{-9}
inverse fine-structure constant	α^{-1}	$137.035\ 999\ 11(46)$		3.3×10^{-9}
Rydberg constant $\alpha^2 m_e c/2h$	R_∞	$10\ 973\ 731.568\ 525(73)$	m^{-1}	6.6×10^{-12}
	$R_\infty c$	$3.289\ 841\ 960\ 360(22)\times10^{15}$	Hz	6.6×10^{-12}
	$R_\infty hc$	$2.179\ 872\ 09(37)\times10^{-18}$	J	1.7×10^{-7}
$R_\infty hc$ in eV		$13.605\ 6923(12)$	eV	8.5×10^{-8}
Bohr radius $\alpha/4\pi R_\infty = 4\pi\varepsilon_0\hbar^2/m_e e^2$	a_0	$0.529\ 177\ 2108(18)\times10^{-10}$	m	3.3×10^{-9}
Hartree energy $e^2/4\pi\varepsilon_0 a_0 = 2R_\infty hc = \alpha^2 m_e c^2$	E_h	$4.359\ 744\ 17(75)\times10^{-18}$	J	1.7×10^{-7}
in eV		$27.211\ 3845(23)$	eV	8.5×10^{-8}
quantum of circulation	$h/2m_e$	$3.636\ 947\ 550(24)\times10^{-4}$	m^{2}s^{-1}	6.7×10^{-9}
	h/m_e	$7.273\ 895\ 101(48)\times10^{-4}$	m^{2}s^{-1}	6.7×10^{-9}
Electroweak				
Fermi coupling constant[3]	$G_F/(\hbar c)^3$	$1.166\ 39(1)\times10^{-5}$	GeV^{-2}	8.6×10^{-6}
weak mixing angle[4] θ_W (on-shell scheme) $\sin^2\theta_W = s_W^2 \equiv 1 - (m_W/m_Z)^2$	$\sin^2\theta_W$	$0.222\ 15(76)$		3.4×10^{-3}
Electron, e^-				
electron mass	m_e	$9.109\ 3826(16)\times10^{-31}$	kg	1.7×10^{-7}
in u, $m_e = A_r(e)$ u (electron relative atomic mass times u)		$5.485\ 799\ 0945(24)\times10^{-4}$	u	4.4×10^{-10}

Quantity	Symbol	Numerical value	Unit	Relative std. uncert. u_r
energy equivalent	m_ec^2	$8.187\ 1047(14)\times10^{-14}$	J	1.7×10^{-7}
in MeV		$0.510\ 998\ 918(44)$	MeV	8.6×10^{-8}
electron-muon mass ratio	m_e/m_μ	$4.836\ 331\ 67(13)\times10^{-3}$		2.6×10^{-8}
electron-tau mass ratio	m_e/m_τ	$2.875\ 64(47)\times10^{-4}$		1.6×10^{-4}
electron-proton mass ratio	m_e/m_p	$5.446\ 170\ 2173(25)\times10^{-4}$		4.6×10^{-10}
electron-neutron mass ratio	m_e/m_n	$5.438\ 673\ 4481(38)\times10^{-4}$		7.0×10^{-10}
electron-deuteron mass ratio	m_e/m_d	$2.724\ 437\ 1095(13)\times10^{-4}$		4.8×10^{-10}
electron to alpha particle mass ratio	m_e/m_α	$1.370\ 933\ 555\ 75(61)\times10^{-4}$		4.4×10^{-10}
electron charge to mass quotient	$-e/m_e$	$-1.758\ 820\ 12(15)\times10^{11}$	C kg^{-1}	8.6×10^{-8}
electron molar mass $N_A m_e$	$M(e), M_e$	$5.485\ 799\ 0945(24)\times10^{-7}$	kg mol^{-1}	4.4×10^{-10}
Compton wavelength h/m_ec	λ_C	$2.426\ 310\ 238(16)\times10^{-12}$	m	6.7×10^{-9}
$\lambda_C/2\pi = \alpha a_0 = \alpha^2/4\pi R_\infty$	$\lambdabar_C$	$386.159\ 2678(26)\times10^{-15}$	m	6.7×10^{-9}
classical electron radius $\alpha^2 a_0$	r_e	$2.817\ 940\ 325(28)\times10^{-15}$	m	1.0×10^{-8}
Thomson cross section $(8\pi/3)r_e^2$	σ_e	$0.665\ 245\ 873(13)\times10^{-28}$	m^2	2.0×10^{-8}
electron magnetic moment	μ_e	$-928.476\ 412(80)\times10^{-26}$	J T^{-1}	8.6×10^{-8}
to Bohr magneton ratio	μ_e/μ_B	$-1.001\ 159\ 652\ 1859(38)$		3.8×10^{-12}
to nuclear magneton ratio	μ_e/μ_N	$-1838.281\ 971\ 07(85)$		4.6×10^{-10}
electron magnetic moment anomaly $|\mu_e|/\mu_B - 1$	a_e	$1.159\ 652\ 1859(38)\times10^{-3}$		3.2×10^{-9}
electron g-factor $-2(1 + a_e)$	g_e	$-2.002\ 319\ 304\ 3718(75)$		3.8×10^{-12}
electron-muon magnetic moment ratio	μ_e/μ_μ	$206.766\ 9894(54)$		2.6×10^{-8}
electron-proton magnetic moment ratio	μ_e/μ_p	$-658.210\ 6862(66)$		1.0×10^{-8}
electron to shielded proton magnetic moment ratio (H$_2$O, sphere, 25°C)	μ_e/μ_p'	$-658.227\ 5956(71)$		1.1×10^{-8}
electron-neutron magnetic moment ratio	μ_e/μ_n	$960.920\ 50(23)$		2.4×10^{-7}
electron-deuteron magnetic moment ratio	μ_e/μ_d	$-2\ 143.923\ 493(23)$		1.1×10^{-8}
electron to shielded helion[5] magnetic moment ratio (gas, sphere, 25°C)	μ_e/μ_h'	$864.058\ 255(10)$		1.2×10^{-8}
electron gyromagnetic ratio $2|\mu_e|/\hbar$	γ_e	$1.760\ 859\ 74(15)\times10^{11}$	s^{-1} T^{-1}	8.6×10^{-8}
	$\gamma_e/2\pi$	$28\ 024.9532(24)$	MHz T^{-1}	8.6×10^{-8}
Muon, μ^-				
muon mass	m_μ	$1.883\ 531\ 40(33)\times10^{-28}$	kg	1.7×10^{-7}
in u, $m_\mu = A_r(\mu)$ u (muon relative atomic mass times u)		$0.113\ 428\ 9264(30)$	u	2.6×10^{-8}
energy equivalent	$m_\mu c^2$	$1.692\ 833\ 60(29)\times10^{-11}$	J	1.7×10^{-7}
in MeV		$105.658\ 3692(94)$	MeV	8.9×10^{-8}
muon-electron mass ratio	m_μ/m_e	$206.768\ 2838(54)$		2.6×10^{-8}
muon-tau mass ratio	m_μ/m_τ	$5.945\ 92(97)\times10^{-2}$		1.6×10^{-4}
muon-proton mass ratio	m_μ/m_p	$0.112\ 609\ 5269(29)$		2.6×10^{-8}
muon-neutron mass ratio	m_μ/m_n	$0.112\ 454\ 5175(29)$		2.6×10^{-8}
muon molar mass $N_A m_\mu$	$M(\mu), M_\mu$	$0.113\ 428\ 9264(30)\times10^{-3}$	kg mol^{-1}	2.6×10^{-8}
muon Compton wavelength $h/m_\mu c$	$\lambda_{C,\mu}$	$11.734\ 441\ 05(30)\times10^{-15}$	m	2.5×10^{-8}
$\lambda_{C,\mu}/2\pi$	$\lambdabar_{C,\mu}$	$1.867\ 594\ 298(47)\times10^{-15}$	m	2.5×10^{-8}
muon magnetic moment	μ_μ	$-4.490\ 447\ 99(40)\times10^{-26}$	J T^{-1}	8.9×10^{-8}
to Bohr magneton ratio	μ_μ/μ_B	$-4.841\ 970\ 45(13)\times10^{-3}$		2.6×10^{-8}
to nuclear magneton ratio	μ_μ/μ_N	$-8.890\ 596\ 98(23)$		2.6×10^{-8}
muon magnetic moment anomaly $|\mu_\mu|/(e\hbar/2m_\mu) - 1$	a_μ	$1.165\ 919\ 81(62)\times10^{-3}$		5.3×10^{-7}
muon g-factor $-2(1 + a_\mu)$	g_μ	$-2.002\ 331\ 8396(12)$		6.2×10^{-10}
muon-proton magnetic moment ratio	μ_μ/μ_p	$-3.183\ 345\ 118(89)$		2.8×10^{-8}
Tau, τ^-				
tau mass[6]	m_τ	$3.167\ 77(52)\times10^{-27}$	kg	1.6×10^{-4}
in u, $m_\tau = A_r(\tau)$ u (tau relative atomic mass times u)		$1.907\ 68(31)$	u	1.6×10^{-4}
energy equivalent	$m_\tau c^2$	$2.847\ 05(46)\times10^{-10}$	J	1.6×10^{-4}
in MeV		$1776.99(29)$	MeV	1.6×10^{-4}
tau-electron mass ratio	m_τ/m_e	$3477.48(57)$		1.6×10^{-4}
tau-muon mass ratio	m_τ/m_μ	$16.8183(27)$		1.6×10^{-4}
tau-proton mass ratio	m_τ/m_p	$1.893\ 90(31)$		1.6×10^{-4}
tau-neutron mass ratio	m_τ/m_n	$1.891\ 29(31)$		1.6×10^{-4}
tau molar mass $N_A m_\tau$	$M(\tau), M_\tau$	$1.907\ 68(31)\times10^{-3}$	kg mol^{-1}	1.6×10^{-4}
tau Compton wavelength $h/m_\tau c$	$\lambda_{C,\tau}$	$0.697\ 72(11)\times10^{-15}$	m	1.6×10^{-4}
$\lambda_{C,\tau}/2\pi$	$\lambdabar_{C,\tau}$	$0.111\ 046(18)\times10^{-15}$	m	1.6×10^{-4}

Quantity	Symbol	Numerical value	Unit	Relative std. uncert. u_r
Proton, p				
proton mass	m_p	$1.672\ 621\ 71(29)\times10^{-27}$	kg	1.7×10^{-7}
in u, $m_p = A_r(p)$ u (proton relative atomic mass times u)		$1.007\ 276\ 466\ 88(13)$	u	1.3×10^{-10}
energy equivalent	$m_p c^2$	$1.503\ 277\ 43(26)\times10^{-10}$	J	1.7×10^{-7}
in MeV		$938.272\ 029(80)$	MeV	8.6×10^{-8}
proton-electron mass ratio	m_p/m_e	$1836.152\ 672\ 61(85)$		4.6×10^{-10}
proton-muon mass ratio	m_p/m_μ	$8.880\ 243\ 33(23)$		2.6×10^{-8}
proton-tau mass ratio	m_p/m_τ	$0.528\ 012(86)$		1.6×10^{-4}
proton-neutron mass ratio	m_p/m_n	$0.998\ 623\ 478\ 72(58)$		5.8×10^{-10}
proton charge to mass quotient	e/m_p	$9.578\ 833\ 76(82)\times10^{7}$	C kg^{-1}	8.6×10^{-8}
proton molar mass $N_A m_p$	$M(p), M_p$	$1.007\ 276\ 466\ 88(13)\times10^{-3}$	kg mol^{-1}	1.3×10^{-10}
proton Compton wavelength $h/m_p c$	$\lambda_{C,p}$	$1.321\ 409\ 8555(88)\times10^{-15}$	m	6.7×10^{-9}
$\lambda_{C,p}/2\pi$	$\lambdabar_{C,p}$	$0.210\ 308\ 9104(14)\times10^{-15}$	m	6.7×10^{-9}
proton rms charge radius	R_p	$0.8750(68)\times10^{-15}$	m	7.8×10^{-3}
proton magnetic moment	μ_p	$1.410\ 606\ 71(12)\times10^{-26}$	J T^{-1}	8.7×10^{-8}
to Bohr magneton ratio	μ_p/μ_B	$1.521\ 032\ 206(15)\times10^{-3}$		1.0×10^{-8}
to nuclear magneton ratio	μ_p/μ_N	$2.792\ 847\ 351(28)$		1.0×10^{-8}
proton g-factor $2\mu_p/\mu_N$	g_p	$5.585\ 694\ 701(56)$		1.0×10^{-8}
proton-neutron magnetic moment ratio	μ_p/μ_n	$-1.459\ 898\ 05(34)$		2.4×10^{-7}
shielded proton magnetic moment (H_2O, sphere, 25°C)	μ'_p	$1.410\ 570\ 47(12)\times10^{-26}$	J T^{-1}	8.7×10^{-8}
to Bohr magneton ratio	μ'_p/μ_B	$1.520\ 993\ 132(16)\times10^{-3}$		1.1×10^{-8}
to nuclear magneton ratio	μ'_p/μ_N	$2.792\ 775\ 604(30)$		1.1×10^{-8}
proton magnetic shielding correction $1-\mu'_p/\mu_p$ (H_2O, sphere, 25°C)	σ'_p	$25.689(15)\times10^{-6}$		5.7×10^{-4}
proton gyromagnetic ratio $2\mu_p/\hbar$	γ_p	$2.675\ 222\ 05(23)\times10^{8}$	s^{-1} T^{-1}	8.6×10^{-8}
	$\gamma_p/2\pi$	$42.577\ 4813(37)$	MHz T^{-1}	8.6×10^{-8}
shielded proton gyromagnetic ratio $2\mu'_p/\hbar$ (H_2O, sphere, 25°C)	γ'_p	$2.675\ 153\ 33(23)\times10^{8}$	s^{-1} T^{-1}	8.6×10^{-8}
	$\gamma'_p/2\pi$	$42.576\ 3875(37)$	MHz T^{-1}	8.6×10^{-8}
Neutron, n				
neutron mass	m_n	$1.674\ 927\ 28(29)\times10^{-27}$	kg	1.7×10^{-7}
in u, $m_n = A_r(n)$ u (neutron relative atomic mass times u)		$1.008\ 664\ 915\ 60(55)$	u	5.5×10^{-10}
energy equivalent	$m_n c^2$	$1.505\ 349\ 57(26)\times10^{-10}$	J	1.7×10^{-7}
in MeV		$939.565\ 360(81)$	MeV	8.6×10^{-8}
neutron-electron mass ratio	m_n/m_e	$1838.683\ 6598(13)$		7.0×10^{-10}
neutron-muon mass ratio	m_n/m_μ	$8.892\ 484\ 02(23)$		2.6×10^{-8}
neutron-tau mass ratio	m_n/m_τ	$0.528\ 740(86)$		1.6×10^{-4}
neutron-proton mass ratio	m_n/m_p	$1.001\ 378\ 418\ 70(58)$		5.8×10^{-10}
neutron molar mass $N_A m_n$	$M(n), M_n$	$1.008\ 664\ 915\ 60(55)\times10^{-3}$	kg mol^{-1}	5.5×10^{-10}
neutron Compton wavelength $h/m_n c$	$\lambda_{C,n}$	$1.319\ 590\ 9067(88)\times10^{-15}$	m	6.7×10^{-9}
$\lambda_{C,n}/2\pi$	$\lambdabar_{C,n}$	$0.210\ 019\ 4157(14)\times10^{-15}$	m	6.7×10^{-9}
neutron magnetic moment	μ_n	$-0.966\ 236\ 45(24)\times10^{-26}$	J T^{-1}	2.5×10^{-7}
to Bohr magneton ratio	μ_n/μ_B	$-1.041\ 875\ 63(25)\times10^{-3}$		2.4×10^{-7}
to nuclear magneton ratio	μ_n/μ_N	$-1.913\ 042\ 73(45)$		2.4×10^{-7}
neutron g-factor $2\mu_n/\mu_N$	g_n	$-3.826\ 085\ 46(90)$		2.4×10^{-7}
neutron-electron magnetic moment ratio	μ_n/μ_e	$1.040\ 668\ 82(25)\times10^{-3}$		2.4×10^{-7}
neutron-proton magnetic moment ratio	μ_n/μ_p	$-0.684\ 979\ 34(16)$		2.4×10^{-7}
neutron to shielded proton magnetic moment ratio (H_2O, sphere, 25°C)	μ_n/μ'_p	$-0.684\ 996\ 94(16)$		2.4×10^{-7}
neutron gyromagnetic ratio $2|\mu_n|/\hbar$	γ_n	$1.832\ 471\ 83(46)\times10^{8}$	s^{-1} T^{-1}	2.5×10^{-7}
	$\gamma_n/2\pi$	$29.164\ 6950(73)$	MHz T^{-1}	2.5×10^{-7}
Deuteron, d				
deuteron mass	m_d	$3.343\ 583\ 35(57)\times10^{-27}$	kg	1.7×10^{-7}
in u, $m_d = A_r(d)$ u (deuteron relative atomic mass times u)		$2.013\ 553\ 212\ 70(35)$	u	1.7×10^{-10}
energy equivalent	$m_d c^2$	$3.005\ 062\ 85(51)\times10^{-10}$	J	1.7×10^{-7}
in MeV		$1875.612\ 82(16)$	MeV	8.6×10^{-8}
deuteron-electron mass ratio	m_d/m_e	$3670.482\ 9652(18)$		4.8×10^{-10}
deuteron-proton mass ratio	m_d/m_p	$1.999\ 007\ 500\ 82(41)$		2.0×10^{-10}
deuteron molar mass $N_A m_d$	$M(d), M_d$	$2.013\ 553\ 212\ 70(35)\times10^{-3}$	kg mol^{-1}	1.7×10^{-10}
deuteron rms charge radius	R_d	$2.1394(28)\times10^{-15}$	m	1.3×10^{-3}

Quantity	Symbol	Numerical value	Unit	Relative std. uncert. u_r
deuteron magnetic moment	μ_d	$0.433\ 073\ 482(38)\times10^{-26}$	J T^{-1}	8.7×10^{-8}
to Bohr magneton ratio	μ_d/μ_B	$0.466\ 975\ 4567(50)\times10^{-3}$		1.1×10^{-8}
to nuclear magneton ratio	μ_d/μ_N	$0.857\ 438\ 2329(92)$		1.1×10^{-8}
deuteron-electron magnetic moment ratio	μ_d/μ_e	$-4.664\ 345\ 548(50)\times10^{-4}$		1.1×10^{-8}
deuteron-proton magnetic moment ratio	μ_d/μ_p	$0.307\ 012\ 2084(45)$		1.5×10^{-8}
deuteron-neutron magnetic moment ratio	μ_d/μ_n	$-0.448\ 206\ 52(11)$		2.4×10^{-7}

Helion, h

Quantity	Symbol	Numerical value	Unit	Relative std. uncert. u_r
helion mass[5]	m_h	$5.006\ 412\ 14(86)\times10^{-27}$	kg	1.7×10^{-7}
in u, $m_h = A_r(h)$ u (helion relative atomic mass times u)		$3.014\ 932\ 2434(58)$	u	1.9×10^{-9}
energy equivalent	$m_h c^2$	$4.499\ 538\ 84(77)\times10^{-10}$	J	1.7×10^{-7}
in MeV		$2808.391\ 42(24)$	MeV	8.6×10^{-8}
helion-electron mass ratio	m_h/m_e	$5495.885\ 269(11)$		2.0×10^{-9}
helion-proton mass ratio	m_h/m_p	$2.993\ 152\ 6671(58)$		1.9×10^{-9}
helion molar mass $N_A m_h$	$M(h), M_h$	$3.014\ 932\ 2434(58)\times10^{-3}$	kg mol^{-1}	1.9×10^{-9}
shielded helion magnetic moment (gas, sphere, 25°C)	μ_h'	$-1.074\ 553\ 024(93)\times10^{-26}$	J T^{-1}	8.7×10^{-8}
to Bohr magneton ratio	μ_h'/μ_B	$-1.158\ 671\ 474(14)\times10^{-3}$		1.2×10^{-8}
to nuclear magneton ratio	μ_h'/μ_N	$-2.127\ 497\ 723(25)$		1.2×10^{-8}
shielded helion to proton magnetic moment ratio (gas, sphere, 25°C)	μ_h'/μ_p	$-0.761\ 766\ 562(12)$		1.5×10^{-8}
shielded helion to shielded proton magnetic moment ratio (gas/H$_2$O, spheres, 25°C)	μ_h'/μ_p'	$-0.761\ 786\ 1313(33)$		4.3×10^{-9}
shielded helion gyromagnetic ratio $2\|\mu_h'\|/\hbar$ (gas, sphere, 25°C)	γ_h'	$2.037\ 894\ 70(18)\times10^8$	s^{-1} T^{-1}	8.7×10^{-8}
	$\gamma_h'/2\pi$	$32.434\ 1015(28)$	MHz T^{-1}	8.7×10^{-8}

Alpha particle, α

Quantity	Symbol	Numerical value	Unit	Relative std. uncert. u_r
alpha particle mass	m_α	$6.644\ 6565(11)\times10^{-27}$	kg	1.7×10^{-7}
in u, $m_\alpha = A_r(\alpha)$ u (alpha particle relative atomic mass times u)		$4.001\ 506\ 179\ 149(56)$	u	1.4×10^{-11}
energy equivalent	$m_\alpha c^2$	$5.971\ 9194(10)\times10^{-10}$	J	1.7×10^{-7}
in MeV		$3727.379\ 17(32)$	MeV	8.6×10^{-8}
alpha particle to electron mass ratio	m_α/m_e	$7294.299\ 5363(32)$		4.4×10^{-10}
alpha particle to proton mass ratio	m_α/m_p	$3.972\ 599\ 689\ 07(52)$		1.3×10^{-10}
alpha particle molar mass $N_A m_\alpha$	$M(\alpha), M_\alpha$	$4.001\ 506\ 179\ 149(56)\times10^{-3}$	kg mol^{-1}	1.4×10^{-11}

Physico-Chemical

Quantity	Symbol	Numerical value	Unit	Relative std. uncert. u_r
Avogadro constant	N_A, L	$6.022\ 1415(10)\times10^{23}$	mol^{-1}	1.7×10^{-7}
atomic mass constant $m_u = (1/12)m(^{12}C) = 1$ u $= 10^{-3}$ kg mol$^{-1}/N_A$	m_u	$1.660\ 538\ 86(28)\times10^{-27}$	kg	1.7×10^{-7}
energy equivalent	$m_u c^2$	$1.492\ 417\ 90(26)\times10^{-10}$	J	1.7×10^{-7}
in MeV		$931.494\ 043(80)$	MeV	8.6×10^{-8}
Faraday constant[7] $N_A e$	F	$96\ 485.3383(83)$	C mol^{-1}	8.6×10^{-8}
molar Planck constant	$N_A h$	$3.990\ 312\ 716(27)\times10^{-10}$	J s mol^{-1}	6.7×10^{-9}
	$N_A hc$	$0.119\ 626\ 565\ 72(80)$	J m mol^{-1}	6.7×10^{-9}
molar gas constant	R	$8.314\ 472(15)$	J mol^{-1} K^{-1}	1.7×10^{-6}
Boltzmann constant R/N_A	k	$1.380\ 6505(24)\times10^{-23}$	J K^{-1}	1.8×10^{-6}
in eV K^{-1}		$8.617\ 343(15)\times10^{-5}$	eV K^{-1}	1.8×10^{-6}
	k/h	$2.083\ 6644(36)\times10^{10}$	Hz K^{-1}	1.7×10^{-6}
	k/hc	$69.503\ 56(12)$	m^{-1} K^{-1}	1.7×10^{-6}
molar volume of ideal gas RT/p				
$T = 273.15$ K, $p = 101.325$ kPa	V_m	$22.413\ 996(39)\times10^{-3}$	m^3 mol^{-1}	1.7×10^{-6}
Loschmidt constant N_A/V_m	n_0	$2.686\ 7773(47)\times10^{25}$	m^{-3}	1.8×10^{-6}
$T = 273.15$ K, $p = 100$ kPa	V_m	$22.710\ 981(40)\times10^{-3}$	m^3 mol^{-1}	1.7×10^{-6}
Sackur-Tetrode constant (absolute entropy constant)[8] $5/2 + \ln[(2\pi m_u kT_1/h^2)^{3/2}kT_1/p_0]$				
$T_1 = 1$ K, $p_0 = 100$ kPa	S_0/R	$-1.151\ 7047(44)$		3.8×10^{-6}
$T_1 = 1$ K, $p_0 = 101.325$ kPa		$-1.164\ 8677(44)$		3.8×10^{-6}
Stefan-Boltzmann constant $(\pi^2/60)k^4/\hbar^3 c^2$	σ	$5.670\ 400(40)\times10^{-8}$	W m^{-2} K^{-4}	7.0×10^{-6}
first radiation constant $2\pi hc^2$	c_1	$3.741\ 771\ 38(64)\times10^{-16}$	W m^2	1.7×10^{-7}
first radiation constant for spectral radiance $2hc^2$	c_{1L}	$1.191\ 042\ 82(20)\times10^{-16}$	W m^2 sr^{-1}	1.7×10^{-7}
second radiation constant hc/k	c_2	$1.438\ 7752(25)\times10^{-2}$	m K	1.7×10^{-6}
Wien displacement law constant $b = \lambda_{max} T = c_2/4.965\ 114\ 231...$	b	$2.897\ 7685(51)\times10^{-3}$	m K	1.7×10^{-6}

[1] See the "Adopted values" table for the conventional value adopted internationally for realizing representations of the volt using the Josephson effect.
[2] See the "Adopted values" table for the conventional value adopted internationally for realizing representations of the ohm using the quantum Hall effect.
[3] Value recommended by the Particle Data Group (Hagiwara *et al.*, 2002).
[4] Based on the ratio of the masses of the W and Z bosons m_w/m_Z recommended by the Particle Data Group (Hagiwara et al., 2002). The value for $\sin^2\theta_w$ they recommend, which is based on a particular variant of the modified minimal subtraction ($\overline{\text{MS}}$) scheme, is $\sin^2\hat{\theta}_w(M_Z) = 0.231\,24(24)$.
[5] The helion, symbol h, is the nucleus of the ^{3}He atom.
[6] This and all other values involving m_τ are based on the value of $m_\tau c^2$ in MeV recommended by the Particle Data Group (Hagiwara et al., 2002), but with a standard uncertainty of 0.29 MeV rather than the quoted uncertainty of −0.26 MeV, +0.29 MeV.
[7] The numerical value of F to be used in coulometric chemical measurements is 96 485.336(16) [1.7×10^{-7}] when the relevant current is measured in terms of representations of the volt and ohm based on the Josephson and quantum Hall effects and the internationally adopted conventional values of the Josephson and von Klitzing constants K_{J-90} and R_{K-90} given in the "Adopted values" table.
[8] The entropy of an ideal monoatomic gas of relative atomic mass A_r is given by $S = S_0 + (3/2)R\ln A_r - R\ln(p/p_0) + (5/2)R\ln(T/K)$.

TABLE III. Internationally Adopted Values of Various Quantities

Quantity	Symbol	Numerical value	Unit	Relative std. uncert. u_r
molar mass of ^{12}C	$M(^{12}C)$	12×10^{-3}	kg mol^{-1}	(exact)
molar mass constant $M(^{12}C)/12$	M_u	1×10^{-3}	kg mol^{-1}	(exact)
conventional value of Josephson constant	K_{J-90}	483 597.9	GHz V^{-1}	(exact)
conventional value of von Klitzing constant	R_{K-90}	25 812.807	Ω	(exact)
standard atmosphere		101 325	Pa	(exact)
standard acceleration of gravity	g_n	9.806 65	m s^{-2}	(exact)

TABLE IV. The Values of Some Energy Equivalents Derived From the Relations $E = mc^2 = hc/\lambda = h\nu = kT$, and Based on the 2002 CODATA Adjustment of the Values of the Constants

$1\text{ eV} = (e/C)\text{ J}$, $1\text{ u} = m_u = (1/12)m(^{12}C) = 10^{-3}\text{ kg mol}^{-1}/N_A$, and $E_h = 2R_\infty hc = \alpha^2 m_e c^2$ is the Hartree Energy (hartree)

	Relevant unit		
	J	kg	m^{-1}	Hz
1 J	$(1\text{ J}) = 1\text{ J}$	$(1\text{ J})/c^2 = 1.112\,650\,056 \times 10^{-17}$ kg	$(1\text{ J})/hc = 5.034\,117\,20(86) \times 10^{24}$ m^{-1}	$(1\text{ J})/h = 1.509\,190\,37(26) \times 10^{33}$ Hz
1 kg	$(1\text{ kg})c^2 = 8.987\,551\,787 \times 10^{16}$ J	$(1\text{ kg}) = 1\text{ kg}$	$(1\text{ kg})c/h = 4.524\,438\,91(77) \times 10^{41}$ m^{-1}	$(1\text{ kg})c^2/h = 1.356\,392\,66(23) \times 10^{50}$ Hz
1 m^{-1}	$(1\text{ m}^{-1})hc = 1.986\,445\,61(34) \times 10^{-25}$ J	$(1\text{ m}^{-1})h/c = 2.210\,218\,81(38) \times 10^{-42}$ kg	$(1\text{ m}^{-1}) = 1\text{ m}^{-1}$	$(1\text{ m}^{-1})c = 299\,792\,458$ Hz
1 Hz	$(1\text{ Hz})h = 6.626\,0693(11) \times 10^{-34}$ J	$(1\text{ Hz})h/c^2 = 7.372\,4964(13) \times 10^{-51}$ kg	$(1\text{ Hz})/c = 3.335\,640\,952 \times 10^{-9}$ m^{-1}	$(1\text{ Hz}) = 1\text{ Hz}$
1 K	$(1\text{ K})k = 1.380\,6505(24) \times 10^{-23}$ J	$(1\text{ K})k/c^2 = 1.536\,1808(27) \times 10^{-40}$ kg	$(1\text{ K})k/hc = 69.503\,56(12)$ m^{-1}	$(1\text{ K})k/h = 2.083\,6644(36) \times 10^{10}$ Hz
1 eV	$(1\text{ eV}) = 1.602\,176\,53(14) \times 10^{-19}$ J	$(1\text{ eV})/c^2 = 1.782\,661\,81(15) \times 10^{-36}$ kg	$(1\text{ eV})/hc = 8.065\,544\,45(69) \times 10^5$ m^{-1}	$(1\text{ eV})/h = 2.417\,989\,40(21) \times 10^{14}$ Hz
1 u	$(1\text{ u})c^2 = 1.492\,417\,90(26) \times 10^{-10}$ J	$(1\text{ u}) = 1.660\,538\,86(28) \times 10^{-27}$ kg	$(1\text{ u})c/h = 7.513\,006\,608(50) \times 10^{14}$ m^{-1}	$(1\text{ u})c^2/h = 2.252\,342\,718(15) \times 10^{23}$ Hz
1 E_h	$(1\,E_h) = 4.359\,744\,17(75) \times 10^{-18}$ J	$(1\,E_h)/c^2 = 4.850\,869\,60(83) \times 10^{-35}$ kg	$(1\,E_h)/hc = 2.194\,746\,313\,705(15) \times 10^7$ m^{-1}	$(1\,E_h)/h = 6.579\,683\,920\,721(44) \times 10^{15}$ Hz

TABLE V. The Values of Some Energy Equivalents Derived From the Relations $E = mc^2 = hc/\lambda = h\nu = kT$, and Based on the 2002 CODATA Adjustment of the Values of the Constants

$1\text{ eV} = (e/C)\text{ J}$, $1\text{ u} = m_u = (1/12)m(^{12}C) = 10^{-3}\text{ kg mol}^{-1}/N_A$, and $E_h = 2R_\infty hc = \alpha^2 m_e c^2$ is the Hartree Energy (hartree)

	Relevant unit		
	K	eV	u	E_h
1 J	$(1\text{ J})/k = 7.242\,963(13) \times 10^{22}$ K	$(1\text{ J}) = 6.241\,509\,47(53) \times 10^{18}$ eV	$(1\text{ J})/c^2 = 6.700\,5361(11) \times 10^9$ u	$(1\text{ J}) = 2.293\,712\,57(39) \times 10^{17}\,E_h$
1 kg	$(1\text{ kg})c^2/k = 6.509\,650(11) \times 10^{39}$ K	$(1\text{ kg})c^2 = 5.609\,588\,96(48) \times 10^{35}$ eV	$(1\text{ kg}) = 6.022\,1415(10) \times 10^{26}$ u	$(1\text{ kg})c^2 = 2.061\,486\,05(35) \times 10^{34}\,E_h$
1 m^{-1}	$(1\text{ m}^{-1})hc/k = 1.438\,7752(25) \times 10^{-2}$ K	$(1\text{ m}^{-1})hc = 1.239\,841\,91(11) \times 10^{-6}$ eV	$(1\text{ m}^{-1})h/c = 1.331\,025\,0506(89) \times 10^{-15}$ u	$(1\text{ m}^{-1})hc = 4.556\,335\,252\,760(30) \times 10^{-8}\,E_h$
1 Hz	$(1\text{ Hz})h/k = 4.799\,2374(84) \times 10^{-11}$ K	$(1\text{ Hz})h = 4.135\,667\,43(35) \times 10^{-15}$ eV	$(1\text{ Hz})h/c^2 = 4.439\,821\,667(30) \times 10^{-24}$ u	$(1\text{ Hz})h = 1.519\,829\,846\,006(10) \times 10^{-16}\,E_h$
1 K	$(1\text{ K}) = 1\text{ K}$	$(1\text{ K})k = 8.617\,343(15) \times 10^{-5}$ eV	$(1\text{ K})k/c^2 = 9.251\,098(16) \times 10^{-14}$ u	$(1\text{ K})k = 3.166\,8153(55) \times 10^{-6}\,E_h$
1 eV	$(1\text{ eV})/k = 1.160\,4505(20) \times 10^4$ K	$(1\text{ eV}) = 1\text{ eV}$	$(1\text{ eV})/c^2 = 1.073\,544\,171(92) \times 10^{-9}$ u	$(1\text{ eV}) = 3.674\,932\,45(31) \times 10^{-2}\,E_h$
1 u	$(1\text{ u})c^2/k = 1.080\,9527(19) \times 10^{13}$ K	$(1\text{ u})c^2 = 931.494\,043(80) \times 10^6$ eV	$(1\text{ u}) = 1\text{ u}$	$(1\text{ u})c^2 = 3.423\,177\,686(23) \times 10^7\,E_h$
1 E_h	$(1\,E_h)/k = 3.157\,7465(55) \times 10^5$ K	$(1\,E_h) = 27.211\,3845(23)$ eV	$(1\,E_h)/c^2 = 2.921\,262\,323(19) \times 10^{-8}$ u	$(1\,E_h) = 1\,E_h$

STANDARD ATOMIC WEIGHTS (2005)

This table of atomic weights includes the changes made in 2005 by the IUPAC Commission on Isotopic Abundances and Atomic Weights. Those changes affected the following elements: Al, Au, Bi, Co, Cs, La, Mn, Na, Nd, P, Pt, Sm, Sc, Ta, Tb, and Th.

The Standard Atomic Weights apply to the elements as they exist naturally on Earth, and the uncertainties take into account the isotopic variation found in most laboratory samples. Further comments on the variability are given in the footnotes.

The number in parentheses following the atomic weight value gives the uncertainty in the last digit. An atomic weight entry in brackets indicates that the element that has no stable isotopes; the value given is the atomic mass in u (or the mass number, if the mass is not accurately known) for the isotope of longest half-life. Thorium, protactinium, and uranium have no stable isotopes, but the terrestrial isotopic composition is sufficiently uniform to permit a standard atomic weight to be specified.

References

1. *Chemistry International*, Vol. 27, No. 6, p.18, 2005.
2. Coplen, T. D., *Pure Appl. Chem.* 73, 667, 2001.
3. Coplen, T. D., *J. Phys. Chem. Ref. Data*, 30, 701, 2001.

Name	Symbol	Atomic no.	Atomic weight	Footnotes
Actinium	Ac	89	[227.0277]	a
Aluminum	Al	13	26.9815386(8)	
Americium	Am	95	[243.0614]	a
Antimony	Sb	51	121.760(1)	g
Argon	Ar	18	39.948(1)	g r
Arsenic	As	33	74.92160(2)	
Astatine	At	85	[209.9871]	a
Barium	Ba	56	137.327(7)	
Berkelium	Bk	97	[247.0703]	a
Beryllium	Be	4	9.012182(3)	
Bismuth	Bi	83	208.98040(1)	
Bohrium	Bh	107	[264.12]	a
Boron	B	5	10.811(7)	g m r
Bromine	Br	35	79.904(1)	
Cadmium	Cd	48	112.411(8)	g
Calcium	Ca	20	40.078(4)	g
Californium	Cf	98	[251.0796]	a
Carbon	C	6	12.0107(8)	g r
Cerium	Ce	58	140.116(1)	g
Cesium	Cs	55	132.9054519(2)	
Chlorine	Cl	17	35.453(2)	g m r
Chromium	Cr	24	51.9961(6)	
Cobalt	Co	27	58.933195(5)	
Copper	Cu	29	63.546(3)	r
Curium	Cm	96	[247.0704]	a
Darmstadtium	Ds	110	[271]	a
Dubnium	Db	105	[262.1141]	a
Dysprosium	Dy	66	162.500(1)	g
Einsteinium	Es	99	[252.0830]	a
Erbium	Er	68	167.259(3)	g
Europium	Eu	63	151.964(1)	g
Fermium	Fm	100	[257.0951]	a
Fluorine	F	9	18.9984032(5)	
Francium	Fr	87	[223.0197]	a
Gadolinium	Gd	64	157.25(3)	g
Gallium	Ga	31	69.723(1)	
Germanium	Ge	32	72.64(1)	
Gold	Au	79	196.966569(4)	
Hafnium	Hf	72	178.49(2)	
Hassium	Hs	108	[277]	a
Helium	He	2	4.002602(2)	g r
Holmium	Ho	67	164.93032(2)	
Hydrogen	H	1	1.00794(7)	g m r
Indium	In	49	114.818(3)	
Iodine	I	53	126.90447(3)	
Iridium	Ir	77	192.217(3)	
Iron	Fe	26	55.845(2)	
Krypton	Kr	36	83.798(2)	g m
Lanthanum	La	57	138.90547(7)	g
Lawrencium	Lr	103	[262.1097]	a
Lead	Pb	82	207.2(1)	g r
Lithium	Li	3	6.941(2)	b g m r
Lutetium	Lu	71	174.967(1)	g
Magnesium	Mg	12	24.3050(6)	
Manganese	Mn	25	54.938045(5)	
Meitnerium	Mt	109	[268.1388]	a
Mendelevium	Md	101	[258.0984]	a
Mercury	Hg	80	200.59(2)	
Molybdenum	Mo	42	95.94(2)	g
Neodymium	Nd	60	144.242(3)	g
Neon	Ne	10	20.1797(6)	g m
Neptunium	Np	93	[237.0482]	a
Nickel	Ni	28	58.6934(2)	
Niobium	Nb	41	92.90638(2)	
Nitrogen	N	7	14.0067(2)	g r
Nobelium	No	102	[259.1010]	a
Osmium	Os	76	190.23(3)	g
Oxygen	O	8	15.9994(3)	g r
Palladium	Pd	46	106.42(1)	g
Phosphorus	P	15	30.973762(2)	
Platinum	Pt	78	195.084(9)	
Plutonium	Pu	94	[244.0642]	a
Polonium	Po	84	[208.9824]	a
Potassium	K	19	39.0983(1)	g
Praseodymium	Pr	59	140.90765(2)	
Promethium	Pm	61	[144.9127]	a
Protactinium	Pa	91	231.03588(2)	
Radium	Ra	88	[226.0254]	a
Radon	Rn	86	[222.0176]	a
Rhenium	Re	75	186.207(1)	
Rhodium	Rh	45	102.90550(2)	
Roentgenium	Rg	111	[272.1535]	a
Rubidium	Rb	37	85.4678(3)	g
Ruthenium	Ru	44	101.07(2)	g
Rutherfordium	Rf	104	[261.1088]	a
Samarium	Sm	62	150.36(2)	g
Scandium	Sc	21	44.955912(6)	
Seaborgium	Sg	106	[266.1219]	a

Name	Symbol	Atomic no.	Atomic weight	Footnotes	Name	Symbol	Atomic no.	Atomic weight	Footnotes
Selenium	Se	34	78.96(3)	r	Tin	Sn	50	118.710(7)	g
Silicon	Si	14	28.0855(3)	r	Titanium	Ti	22	47.867(1)	
Silver	Ag	47	107.8682(2)	g	Tungsten	W	74	183.84(1)	
Sodium	Na	11	22.98976928(2)		Ununbium	Uub	112	[285]	a
Strontium	Sr	38	87.62(1)	g r	Ununhexium	Uuh	116	[289]	a
Sulfur	S	16	32.065(5)	g r	Ununquadium	Uuq	114	[289]	a
Tantalum	Ta	73	180.94788(2)		Uranium	U	92	238.02891(3)	g m
Technetium	Tc	43	[97.9072]	a	Vanadium	V	23	50.9415(1)	
Tellurium	Te	52	127.60(3)	g	Xenon	Xe	54	131.293(6)	g m
Terbium	Tb	65	158.92535(2)		Ytterbium	Yb	70	173.04(3)	g
Thallium	Tl	81	204.3833(2)		Yttrium	Y	39	88.90585(2)	
Thorium	Th	90	232.03806(2)	g	Zinc	Zn	30	65.409(4)	
Thulium	Tm	69	168.93421(2)		Zirconium	Zr	40	91.224(2)	g

[a] No stable isotope exists. The atomic mass in u (or the mass number, if the mass is not accurately known) is given in brackets for the isotope of longest half-life.

[b] Commercially available Li materials have atomic weights that range between 6.939 and 6.996; if a more accurate value is required, it must be determined for the specific material.

[g] Geological specimens are known in which the element has an isotopic composition outside the limits for the normal material. The difference between the atomic weight of the element in such specimens and that given in the table may exceed the stated uncertainty.

[m] Modified isotopic compositions may be found in commercially available material because it has been subject to an undisclosed or inadvertent isotopic fractionation. Substantial deviations in atomic weight of the element from that given in the table can occur.

[r] Range in isotopic composition of normal terrestrial material prevents a more precise atomic weight being given; the tabulated value should be applicable to any normal material.

ATOMIC MASSES AND ABUNDANCES

This table lists the mass (in atomic mass units, symbol u) and the natural abundance (in percent) of the stable nuclides and a few important radioactive nuclides. A complete table of all nuclides may be found in Section 11 ("Table of the Isotopes").

The atomic masses were taken from the 2003 evaluation of Audi, Wapstra, and Thibault (References 2, 3). The number in parentheses following the mass value is the uncertainty in the last digit(s) given. An asterisk * after an entry indicates the mass value was derived not purely from experimental data, but at least partly from systematic trends.

Natural abundance values were taken from the IUPAC Technical Report "Atomic Weight of the Elements: Review 2000" (Reference 4); these entries are also followed by uncertainties in the last digit(s) of the stated values. This uncertainty includes both the estimated measurement uncertainty and the reported range of variation in different terrestrial sources of the element (see Reference 4 for full details and caveats regarding elements whose abundance is variable). The absence of an entry in the Abundance column indicates a radioactive nuclide not present in nature or an element whose isotopic composition varies so widely that a meaningful natural abundance cannot be defined.

References

1. Holden, N. E., "Table of the Isotopes", in Lide, D. R., Ed., *CRC Handbook of Chemistry and Physics, 86th Ed.*, CRC Press, Boca Raton FL, 2005.
2. Audi, G., Wapstra, A. H., and Thibault, *Nucl. Phys.*, A729, 336, 2003.
3. Audi, G., and Wapstra, A. H., Atomic Mass Data Center, World Wide Web site, <http://www.nndc.bnl.gov/amdc/index.html>
4. de Laeter, J. R., Böhlke, J. K., De Bièvre, P., Hidaka, H., Peiser, H. S., Rosman, K. J. R., and Taylor, P. D. P., *Pure Appl. Chem.* 75, 683, 2003.

Z	Isotope	Mass in u	Abundance in %
1	^{1}H	1.00782503207(10)	99.9885(70)
	^{2}H	2.0141017778(4)	0.0115(70)
	^{3}H	3.0160492777(25)	
2	^{3}He	3.0160293191(26)	0.000134(3)
	^{4}He	4.00260325415(6)	99.999866(3)
3	^{6}Li	6.015122795(16)	7.59(4)
	^{7}Li	7.01600455(8)	92.41(4)
4	^{9}Be	9.0121822(4)	100
5	^{10}B	10.0129370(4)	19.9(7)
	^{11}B	11.0093054(4)	80.1(7)
6	^{11}C	11.0114336(10)	
	^{12}C	12.0000000(0)	98.93(8)
	^{13}C	13.0033548378(10)	1.07(8)
	^{14}C	14.003241989(4)	
7	^{14}N	14.0030740048(6)	99.636(7)
	^{15}N	15.0001088982(7)	0.364(7)
8	^{16}O	15.99491461956(16)	99.757(16)
	^{17}O	16.99913170(12)	0.038(1)
	^{18}O	17.9991610(7)	0.205(14)
9	^{18}F	18.0009380(6)	
	^{19}F	18.99840322(7)	100
10	^{20}Ne	19.9924401754(19)	90.48(3)
	^{21}Ne	20.99384668(4)	0.27(1)
	^{22}Ne	21.991385114(19)	9.25(3)
11	^{22}Na	21.9944364(4)	
	^{23}Na	22.9897692809(29)	100
	^{24}Na	23.99096278(8)	
12	^{24}Mg	23.985041700(14)	78.99(4)
	^{25}Mg	24.98583692(3)	10.00(1)
	^{26}Mg	25.982592929(30)	11.01(3)
13	^{27}Al	26.98153863(12)	100
14	^{28}Si	27.9769265325(19)	92.223(19)
	^{29}Si	28.976494700(22)	4.685(8)
	^{30}Si	29.97377017(3)	3.092(11)
15	^{31}P	30.97376163(20)	100
	^{32}P	31.97390727(20)	
16	^{32}S	31.97207100(15)	94.99(26)
	^{33}S	32.97145876(15)	0.75(2)
	^{34}S	33.96786690(12)	4.25(24)
	^{35}S	34.96903216(11)	
	^{36}S	35.96708076(20)	0.01(1)

Z	Isotope	Mass in u	Abundance in %
17	^{35}Cl	34.96885268(4)	75.76(10)
	^{37}Cl	36.96590259(5)	24.24(10)
18	^{36}Ar	35.967545106(29)	0.3365(30)
	^{38}Ar	37.9627324(4)	0.0632(5)
	^{40}Ar	39.9623831225(29)	99.6003(30)
19	^{39}K	38.96370668(20)	93.2581(44)
	^{40}K	39.96399848(21)	0.0117(1)
	^{41}K	40.96182576(21)	6.7302(44)
	^{42}K	41.96240281(24)	
	^{43}K	42.960716(10)	
20	^{40}Ca	39.96259098(22)	96.941(156)
	^{42}Ca	41.95861801(27)	0.647(23)
	^{43}Ca	42.9587666(3)	0.135(10)
	^{44}Ca	43.9554818(4)	2.086(110)
	^{45}Ca	44.9561866(4)	
	^{46}Ca	45.9536926(24)	0.004(3)
	^{47}Ca	46.9545460(24)	
	^{48}Ca	47.952534(4)	0.187(21)
21	^{45}Sc	44.9559119(9)	100
22	^{46}Ti	45.9526316(9)	8.25(3)
	^{47}Ti	46.9517631(9)	7.44(2)
	^{48}Ti	47.9479463(9)	73.72(3)
	^{49}Ti	48.9478700(9)	5.41(2)
	^{50}Ti	49.9447912(9)	5.18(2)
23	^{50}V	49.9471585(11)	0.250(4)
	^{51}V	50.9439595(11)	99.750(4)
24	^{50}Cr	49.9460442(11)	4.345(13)
	^{51}Cr	50.9447674(11)	
	^{52}Cr	51.9405075(8)	83.789(18)
	^{53}Cr	52.9406494(8)	9.501(17)
	^{54}Cr	53.9388804(8)	2.365(7)
25	^{54}Mn	53.9403589(14)	
	^{55}Mn	54.9380451(7)	100
26	^{52}Fe	51.948114(7)	
	^{54}Fe	53.9396105(7)	5.845(35)
	^{55}Fe	54.9382934(7)	
	^{56}Fe	55.9349375(7)	91.754(36)
	^{57}Fe	56.9353940(7)	2.119(10)
	^{58}Fe	57.9332756(8)	0.282(4)
	^{59}Fe	58.9348755(8)	
27	^{57}Co	56.9362914(8)	

Z	Isotope	Mass in u	Abundance in %	Z	Isotope	Mass in u	Abundance in %
	^{58}Co	57.9357528(13)			^{96}Zr	95.9082734(30)	2.80(9)
	^{59}Co	58.9331950(7)	100	41	^{93}Nb	92.9063781(26)	100
	^{60}Co	59.9338171(7)		42	^{92}Mo	91.906811(4)	14.77(31)
28	^{58}Ni	57.9353429(7)	68.0769(89)		^{94}Mo	93.9050883(21)	9.23(10)
	^{59}Ni	58.9343467(7)			^{95}Mo	94.9058421(21)	15.90(9)
	^{60}Ni	59.9307864(7)	26.2231(77)		^{96}Mo	95.9046795(21)	16.68(1)
	^{61}Ni	60.9310560(7)	1.1399(6)		^{97}Mo	96.9060215(21)	9.56(5)
	^{62}Ni	61.9283451(6)	3.6345(17)		^{98}Mo	97.9054082(21)	24.19(26)
	^{63}Ni	62.9296694(6)			^{99}Mo	98.9077119(21)	
	^{64}Ni	63.9279660(7)	0.9256(9)		^{100}Mo	99.907477(6)	9.67(20)
29	^{63}Cu	62.9295975(6)	69.15(3)	43	^{97}Tc	96.906365(5)	
	^{64}Cu	63.9297642(6)			^{98}Tc	97.907216(4)	
	^{65}Cu	64.9277895(7)	30.85(3)		^{99}Tc	98.9062547(21)	
30	^{64}Zn	63.9291422(7)	48.268(321)	44	^{96}Ru	95.907598(8)	5.54(14)
	^{65}Zn	64.9292410(7)			^{98}Ru	97.905287(7)	1.87(3)
	^{66}Zn	65.9260334(10)	27.975(77)		^{99}Ru	98.9059393(22)	12.76(14)
	^{67}Zn	66.9271273(10)	4.102(21)		^{100}Ru	99.9042195(22)	12.60(7)
	^{68}Zn	67.9248442(10)	19.024(123)		^{101}Ru	100.9055821(22)	17.06(2)
	^{70}Zn	69.9253193(21)	0.631(9)		^{102}Ru	101.9043493(22)	31.55(14)
31	^{67}Ga	66.9282017(14)			^{104}Ru	103.905433(3)	18.62(27)
	^{68}Ga	67.9279801(16)			^{106}Ru	105.907329(8)	
	^{69}Ga	68.9255736(13)	60.108(9)	45	^{103}Rh	102.905504(3)	100
	^{71}Ga	70.9247013(11)	39.892(9)	46	^{102}Pd	101.905609(3)	1.02(1)
32	^{68}Ge	67.928094(7)			^{104}Pd	103.904036(4)	11.14(8)
	^{70}Ge	69.9242474(11)	20.38(18)		^{105}Pd	104.905085(4)	22.33(8)
	^{72}Ge	71.9220758(18)	27.31(26)		^{106}Pd	105.903486(4)	27.33(3)
	^{73}Ge	72.9234589(18)	7.76(8)		^{108}Pd	107.903892(4)	26.46(9)
	^{74}Ge	73.9211778(18)	36.72(15)		^{110}Pd	109.905153(12)	11.72(9)
	^{76}Ge	75.9214026(18)	7.83(7)	47	^{107}Ag	106.905097(5)	51.839(8)
33	^{75}As	74.9215965(20)	100		^{109}Ag	108.904752(3)	48.161(8)
34	^{74}Se	73.9224764(18)	0.89(4)	48	^{106}Cd	105.906459(6)	1.25(6)
	^{75}Se	74.9225234(18)			^{108}Cd	107.904184(6)	0.89(3)
	^{76}Se	75.9192136(18)	9.37(29)		^{110}Cd	109.9030021(29)	12.49(18)
	^{77}Se	76.9199140(18)	7.63(16)		^{111}Cd	110.9041781(29)	12.80(12)
	^{78}Se	77.9173091(18)	23.77(28)		^{112}Cd	111.9027578(29)	24.13(21)
	^{79}Se	78.9184991(18)			^{113}Cd	112.9044017(29)	12.22(12)
	^{80}Se	79.9165213(21)	49.61(41)		^{114}Cd	113.9033585(29)	28.73(42)
	^{82}Se	81.9166994(22)	8.73(22)		^{116}Cd	115.904756(3)	7.49(18)
35	^{79}Br	78.9183371(22)	50.69(7)	49	^{111}In	110.905103(5)	
	^{81}Br	80.9162906(21)	49.31(7)		^{113}In	112.904058(3)	4.29(5)
36	^{78}Kr	77.9203648(12)	0.355(3)		^{115}In	114.903878(5)	95.71(5)
	^{80}Kr	79.9163790(16)	2.286(10)	50	^{112}Sn	111.904818(5)	0.97(1)
	^{82}Kr	81.9134836(19)	11.593(31)		^{113}Sn	112.905171(4)	
	^{83}Kr	82.914136(3)	11.500(19)		^{114}Sn	113.902779(3)	0.66(1)
	^{84}Kr	83.911507(3)	56.987(15)		^{115}Sn	114.903342(3)	0.34(1)
	^{86}Kr	85.91061073(11)	17.279(41)		^{116}Sn	115.901741(3)	14.54(9)
37	^{85}Rb	84.911789738(12)	72.17(2)		^{117}Sn	116.902952(3)	7.68(7)
	^{86}Rb	85.91116742(21)			^{118}Sn	117.901603(3)	24.22(9)
	^{87}Rb	86.909180527(13)	27.83(2)		^{119}Sn	118.903308(3)	8.59(4)
38	^{84}Sr	83.913425(3)	0.56(1)		^{120}Sn	119.9021947(27)	32.58(9)
	^{85}Sr	84.912933(3)			^{122}Sn	121.9034390(29)	4.63(3)
	^{86}Sr	85.9092602(12)	9.86(1)		^{124}Sn	123.9052739(15)	5.79(5)
	^{87}Sr	86.9088771(12)	7.00(1)	51	^{121}Sb	120.9038157(24)	57.21(5)
	^{88}Sr	87.9056121(12)	82.58(1)		^{123}Sb	122.9042140(22)	42.79(5)
	^{89}Sr	88.9074507(12)		52	^{120}Te	119.904020(10)	0.09(1)
	^{90}Sr	89.907738(3)			^{122}Te	121.9030439(16)	2.55(12)
39	^{89}Y	88.9058483(27)	100		^{123}Te	122.9042700(16)	0.89(3)
40	^{90}Zr	89.9047044(25)	51.45(40)		^{124}Te	123.9028179(16)	4.74(14)
	^{91}Zr	90.9056458(25)	11.22(5)		^{125}Te	124.9044307(16)	7.07(15)
	^{92}Zr	91.9050408(25)	17.15(8)		^{126}Te	125.9033117(16)	18.84(25)
	^{94}Zr	93.9063152(26)	17.38(28)		^{128}Te	127.9044631(19)	31.74(8)

Z	Isotope	Mass in u	Abundance in %	Z	Isotope	Mass in u	Abundance in %
	^{130}Te	129.9062244(21)	34.08(62)		^{158}Gd	157.9241039(27)	24.84(7)
53	^{123}I	122.905589(4)			^{160}Gd	159.9270541(27)	21.86(19)
	^{125}I	124.9046302(16)		65	^{159}Tb	158.9253468(27)	100
	^{127}I	126.904473(4)	100	66	^{156}Dy	155.924283(7)	0.056(3)
	^{129}I	128.904988(3)			^{158}Dy	157.924409(4)	0.095(3)
	^{131}I	130.9061246(12)			^{160}Dy	159.9251975(27)	2.329(18)
54	^{124}Xe	123.9058930(20)	0.0952(3)		^{161}Dy	160.9269334(27)	18.889(42)
	^{126}Xe	125.904274(7)	0.0890(2)		^{162}Dy	161.9267984(27)	25.475(36)
	^{128}Xe	127.9035313(15)	1.9102(8)		^{163}Dy	162.9287312(27)	24.896(42)
	^{129}Xe	128.9047794(8)	26.4006(82)		^{164}Dy	163.9291748(27)	28.260(54)
	^{130}Xe	129.9035080(8)	4.0710(13)	67	^{165}Ho	164.9303221(27)	100
	^{131}Xe	130.9050824(10)	21.2324(30)	68	^{162}Er	161.928778(4)	0.139(5)
	^{132}Xe	131.9041535(10)	26.9086(33)		^{164}Er	163.929200(3)	1.601(3)
	^{134}Xe	133.9053945(9)	10.4357(21)		^{166}Er	165.9302931(27)	33.503(36)
	^{136}Xe	135.907219(8)	8.8573(44)		^{167}Er	166.9320482(27)	22.869(9)
55	^{129}Cs	128.906064(5)			^{168}Er	167.9323702(27)	26.978(18)
	^{133}Cs	132.905451933(24)	100		^{170}Er	169.9354643(30)	14.910(36)
	^{134}Cs	133.906718475(28)		69	^{169}Tm	168.9342133(27)	100
	^{136}Cs	135.9073116(20)		70	^{168}Yb	167.933897(5)	0.13(1)
	^{137}Cs	136.9070895(5)			^{169}Yb	168.935190(5)	
56	^{130}Ba	129.9063208(30)	0.106(1)		^{170}Yb	169.9347618(26)	3.04(15)
	^{132}Ba	131.9050613(11)	0.101(1)		^{171}Yb	170.9363258(26)	14.28(57)
	^{133}Ba	132.9060075(11)			^{172}Yb	171.9363815(26)	21.83(67)
	^{134}Ba	133.9045084(4)	2.417(18)		^{173}Yb	172.9382108(26)	16.13(27)
	^{135}Ba	134.9056886(4)	6.592(12)		^{174}Yb	173.9388621(26)	31.83(92)
	^{136}Ba	135.9045759(4)	7.854(24)		^{176}Yb	175.9425717(28)	12.76(41)
	^{137}Ba	136.9058274(5)	11.232(24)	71	^{175}Lu	174.9407718(23)	97.41(2)
	^{138}Ba	137.9052472(5)	71.698(42)		^{176}Lu	175.9426863(23)	2.59(2)
	^{140}Ba	139.910605(9)		72	^{174}Hf	173.940046(3)	0.16(1)
57	^{138}La	137.907112(4)	0.090(1)		^{176}Hf	175.9414086(24)	5.26(7)
	^{139}La	138.9063533(26)	99.910(1)		^{177}Hf	176.9432207(23)	18.60(9)
58	^{136}Ce	135.907172(14)	0.185(2)		^{178}Hf	177.9436988(23)	27.28(7)
	^{138}Ce	137.905991(11)	0.251(2)		^{179}Hf	178.9458161(23)	13.62(2)
	^{140}Ce	139.9054387(26)	88.450(51)		^{180}Hf	179.9465500(23)	35.08(16)
	^{141}Ce	140.9082763(26)		73	^{180}Ta	179.9474648(24)	0.012(2)
	^{142}Ce	141.909244(3)	11.114(51)		^{181}Ta	180.9479958(19)	99.988(2)
	^{144}Ce	143.913647(4)		74	^{180}W	179.946704(4)	0.12(1)
59	^{141}Pr	140.9076528(26)	100		^{182}W	181.9482042(9)	26.50(16)
60	^{142}Nd	141.9077233(25)	27.2(5)		^{183}W	182.9502230(9)	14.31(4)
	^{143}Nd	142.9098143(25)	12.2(2)		^{184}W	183.9509312(9)	30.64(2)
	^{144}Nd	143.9100873(25)	23.8(3)		^{186}W	185.9543641(19)	28.43(19)
	^{145}Nd	144.9125736(25)	8.3(1)	75	^{185}Re	184.9529550(13)	37.40(2)
	^{146}Nd	145.9131169(25)	17.2(3)		^{187}Re	186.9557531(15)	62.60(2)
	^{148}Nd	147.916893(3)	5.7(1)	76	^{184}Os	183.9524891(14)	0.02(1)
	^{150}Nd	149.920891(3)	5.6(2)		^{186}Os	185.9538382(15)	1.59(3)
61	^{145}Pm	144.912749(3)			^{187}Os	186.9557505(15)	1.96(2)
	^{147}Pm	146.9151385(26)			^{188}Os	187.9558382(15)	13.24(8)
62	^{144}Sm	143.911999(3)	3.07(7)		^{189}Os	188.9581475(16)	16.15(5)
	^{147}Sm	146.9148979(26)	14.99(18)		^{190}Os	189.9584470(16)	26.26(2)
	^{148}Sm	147.9148227(26)	11.24(10)		^{192}Os	191.9614807(27)	40.78(19)
	^{149}Sm	148.9171847(26)	13.82(7)	77	^{191}Ir	190.9605940(18)	37.3(2)
	^{150}Sm	149.9172755(26)	7.38(1)		^{193}Ir	192.9629264(18)	62.7(2)
	^{152}Sm	151.9197324(27)	26.75(16)	78	^{190}Pt	189.959932(6)	0.014(1)
	^{154}Sm	153.9222093(27)	22.75(29)		^{192}Pt	191.9610380(27)	0.782(7)
63	^{151}Eu	150.9198502(26)	47.81(6)		^{194}Pt	193.9626803(9)	32.967(99)
	^{153}Eu	152.9212303(26)	52.19(6)		^{195}Pt	194.9647911(9)	33.832(10)
64	^{152}Gd	151.9197910(27)	0.20(1)		^{196}Pt	195.9649515(9)	25.242(41)
	^{154}Gd	153.9208656(27)	2.18(3)		^{198}Pt	197.967893(3)	7.163(55)
	^{155}Gd	154.9226220(27)	14.80(12)	79	^{197}Au	196.9665687(6)	100
	^{156}Gd	155.9221227(27)	20.47(9)		^{198}Au	197.9682423(6)	
	^{157}Gd	156.9239601(27)	15.65(2)	80	^{196}Hg	195.965833(3)	0.15(1)

Z	Isotope	Mass in u	Abundance in %	Z	Isotope	Mass in u	Abundance in %
	^{197}Hg	196.967213(3)			^{236}U	236.0455680(20)	
	^{198}Hg	197.9667690(4)	9.97(20)		^{238}U	238.0507882(20)	99.2742(10)
	^{199}Hg	198.9682799(4)	16.87(22)	93	^{237}Np	237.0481734(20)	
	^{200}Hg	199.9683260(4)	23.10(19)		^{239}Np	239.0529390(22)	
	^{201}Hg	200.9703023(6)	13.18(9)	94	^{238}Pu	238.0495599(20)	
	^{202}Hg	201.9706430(6)	29.86(26)		^{239}Pu	239.0521634(20)	
	^{203}Hg	202.9728725(18)			^{240}Pu	240.0538135(20)	
	^{204}Hg	203.9734939(4)	6.87(15)		^{241}Pu	241.0568515(20)	
81	^{201}Tl	200.970819(16)			^{242}Pu	242.0587426(20)	
	^{203}Tl	202.9723442(14)	29.52(1)		^{244}Pu	244.064204(5)	
	^{205}Tl	204.9744275(14)	70.48(1)	95	^{241}Am	241.0568291(20)	
82	^{204}Pb	203.9730436(13)	1.4(1)		^{243}Am	243.0613811(25)	
	^{206}Pb	205.9744653(13)	24.1(1)	96	^{243}Cm	243.0613891(22)	
	^{207}Pb	206.9758969(13)	22.1(1)		^{244}Cm	244.0627526(20)	
	^{208}Pb	207.9766521(13)	52.4(1)		^{245}Cm	245.0654912(22)	
	^{210}Pb	209.9841885(16)			^{246}Cm	246.0672237(22)	
83	^{207}Bi	206.9784707(26)			^{247}Cm	247.070354(5)	
	^{209}Bi	208.9803987(16)	100		^{248}Cm	248.072349(5)	
84	^{209}Po	208.9824304(20)		97	^{247}Bk	247.070307(6)	
	^{210}Po	209.9828737(13)			^{249}Bk	249.0749867(28)	
85	^{210}At	209.987148(8)		98	^{249}Cf	249.0748535(24)	
	^{211}At	210.9874963(30)			^{250}Cf	250.0764061(22)	
86	^{211}Rn	210.990601(7)			^{251}Cf	251.079587(5)	
	^{220}Rn	220.0113940(24)			^{252}Cf	252.081626(5)	
	^{222}Rn	222.0175777(25)		99	^{252}Es	252.082980(50)	
87	^{223}Fr	223.0197359(26)		100	^{257}Fm	257.095105(7)	
88	^{223}Ra	223.0185022(27)		101	^{256}Md	256.094060(60)	
	^{224}Ra	224.0202118(24)			^{258}Md	258.098431(5)	
	^{226}Ra	226.0254098(25)		102	^{259}No	259.10103(11)*	
	^{228}Ra	228.0310703(26)		103	^{262}Lr	262.10963(22)*	
89	^{227}Ac	227.0277521(26)		104	^{261}Rf	261.108770(30)*	
90	^{228}Th	228.0287411(24)		105	^{262}Db	262.11408(20)*	
	^{230}Th	230.0331338(19)		106	^{263}Sg	263.11832(13)*	
	^{232}Th	232.0380553(21)	100	107	^{264}Bh	264.12460(30)*	
91	^{231}Pa	231.0358840(24)	100	108	^{265}Hs	265.13009(15)*	
92	^{233}U	233.0396352(29)		109	^{268}Mt	268.13873(34)*	
	^{234}U	234.0409521(20)	0.0054(5)	110	^{281}Ds	281.16206(78)*	
	^{235}U	235.0439299(20)	0.7204(6)	111	^{272}Rg	273.15362(36)*	

ELECTRON CONFIGURATION AND IONIZATION ENERGY OF NEUTRAL ATOMS IN THE GROUND STATE

William C. Martin

The ground state electron configuration, ground level, and ionization energy of the elements hydrogen through rutherfordium are listed in this table. The electron configurations of elements heavier than neon are shortened by using rare-gas element symbols in brackets to represent the corresponding electrons. See the references for details of the notation for Pa, U, and Np. Ionization energies to higher states (and more precise values of the first ionization energy for certain elements) may be found in the table "Ionization Energies of Atoms and Atomic Ions" in Section 10 of this *Handbook*.

References

1. Martin, W. C., Musgrove, A., Kotochigova, S., and Sansonetti, J. E., NIST Physical Reference Data Web Site, <http://physics.nist.gov/PhysRefData/IonEnergy/ionEnergy.html>, October 2004.
2. Martin, W. C., and Wiese, W. L., "Atomic Spectroscopy", in *Atomic, Molecular, & Optical Physics Handbook*, ed. by G.W.F. Drake (AIP, Woodbury, NY, 1996) Chapter 10, pp. 135-153.

Z		Element	Ground-state configuration	Ground level	Ionization energy (eV)
1	H	Hydrogen	$1s$	$^2S_{1/2}$	13.5984
2	He	Helium	$1s^2$	$^1S_0$	24.5874
3	Li	Lithium	$1s^2\,2s$	$^2S_{1/2}$	5.3917
4	Be	Beryllium	$1s^2\,2s^2$	$^1S_0$	9.3227
5	B	Boron	$1s^2\,2s^2\,2p$	$^2P^o_{1/2}$	8.2980
6	C	Carbon	$1s^2\,2s^2\,2p^2$	$^3P_0$	11.2603
7	N	Nitrogen	$1s^2\,2s^2\,2p^3$	$^4S^o_{3/2}$	14.5341
8	O	Oxygen	$1s^2\,2s^2\,2p^4$	$^3P_2$	13.6181
9	F	Fluorine	$1s^2\,2s^2\,2p^5$	$^2P^o_{3/2}$	17.4228
10	Ne	Neon	$1s^2\,2s^2\,2p^6$	$^1S_0$	21.5645
11	Na	Sodium	[Ne] $3s$	$^2S_{1/2}$	5.1391
12	Mg	Magnesium	[Ne] $3s^2$	$^1S_0$	7.6462
13	Al	Aluminum	[Ne] $3s^2\,3p$	$^2P^o_{1/2}$	5.9858
14	Si	Silicon	[Ne] $3s^2\,3p^2$	$^3P_0$	8.1517
15	P	Phosphorus	[Ne] $3s^2\,3p^3$	$^4S^o_{3/2}$	10.4867
16	S	Sulfur	[Ne] $3s^2\,3p^4$	$^3P_2$	10.3600
17	Cl	Chlorine	[Ne] $3s^2\,3p^5$	$^2P^o_{3/2}$	12.9676
18	Ar	Argon	[Ne] $3s^2\,3p^6$	$^1S_0$	15.7596
19	K	Potassium	[Ar] $4s$	$^2S_{1/2}$	4.3407
20	Ca	Calcium	[Ar] $4s^2$	$^1S_0$	6.1132
21	Sc	Scandium	[Ar] $3d\,4s^2$	$^2D_{3/2}$	6.5615
22	Ti	Titanium	[Ar] $3d^2\,4s^2$	$^3F_2$	6.8281
23	V	Vanadium	[Ar] $3d^3\,4s^2$	$^4F_{3/2}$	6.7462
24	Cr	Chromium	[Ar] $3d^5\,4s$	$^7S_3$	6.7665
25	Mn	Manganese	[Ar] $3d^5\,4s^2$	$^6S_{5/2}$	7.4340
26	Fe	Iron	[Ar] $3d^6\,4s^2$	$^5D_4$	7.9024
27	Co	Cobalt	[Ar] $3d^7\,4s^2$	$^4F_{9/2}$	7.8810
28	Ni	Nickel	[Ar] $3d^8\,4s^2$	$^3F_4$	7.6398
29	Cu	Copper	[Ar] $3d^{10}\,4s$	$^2S_{1/2}$	7.7264
30	Zn	Zinc	[Ar] $3d^{10}\,4s^2$	$^1S_0$	9.3942
31	Ga	Gallium	[Ar] $3d^{10}\,4s^2\,4p$	$^2P^o_{1/2}$	5.9993
32	Ge	Germanium	[Ar] $3d^{10}\,4s^2\,4p^2$	$^3P_0$	7.8994
33	As	Arsenic	[Ar] $3d^{10}\,4s^2\,4p^3$	$^4S^o_{3/2}$	9.7886
34	Se	Selenium	[Ar] $3d^{10}\,4s^2\,4p^4$	$^3P_2$	9.7524
35	Br	Bromine	[Ar] $3d^{10}\,4s^2\,4p^5$	$^2P^o_{3/2}$	11.8138
36	Kr	Krypton	[Ar] $3d^{10}\,4s^2\,4p^6$	$^1S_0$	13.9996
37	Rb	Rubidium	[Kr] $5s$	$^2S_{1/2}$	4.1771
38	Sr	Strontium	[Kr] $5s^2$	$^1S_0$	5.6949
39	Y	Yttrium	[Kr] $4d\,5s^2$	$^2D_{3/2}$	6.2173
40	Zr	Zirconium	[Kr] $4d^2\,5s^2$	$^3F_2$	6.6339
41	Nb	Niobium	[Kr] $4d^4\,5s$	$^6D_{1/2}$	6.7589
42	Mo	Molybdenum	[Kr] $4d^5\,5s$	$^7S_3$	7.0924
43	Tc	Technetium	[Kr] $4d^5\,5s^2$	$^6S_{5/2}$	7.28
44	Ru	Ruthenium	[Kr] $4d^7\,5s$	$^5F_5$	7.3605

Z		Element	Ground-state configuration	Ground level	Ionization energy (eV)
45	Rh	Rhodium	[Kr] $4d^8\,5s$	$^4F_{9/2}$	7.4589
46	Pd	Palladium	[Kr] $4d^{10}$	$^1S_0$	8.3369
47	Ag	Silver	[Kr] $4d^{10}\,5s$	$^2S_{1/2}$	7.5762
48	Cd	Cadmium	[Kr] $4d^{10}\,5s^2$	$^1S_0$	8.9938
49	In	Indium	[Kr] $4d^{10}\,5s^2\,5p$	$^2P^o_{1/2}$	5.7864
50	Sn	Tin	[Kr] $4d^{10}\,5s^2\,5p^2$	$^3P_0$	7.3439
51	Sb	Antimony	[Kr] $4d^{10}\,5s^2\,5p^3$	$^4S^o_{3/2}$	8.6084
52	Te	Tellurium	[Kr] $4d^{10}\,5s^2\,5p^4$	$^3P_2$	9.0096
53	I	Iodine	[Kr] $4d^{10}\,5s^2\,5p^5$	$^2P^o_{3/2}$	10.4513
54	Xe	Xenon	[Kr] $4d^{10}\,5s^2\,5p^6$	$^1S_0$	12.1298
55	Cs	Cesium	[Xe] $6s$	$^2S_{1/2}$	3.8939
56	Ba	Barium	[Xe] $6s^2$	$^1S_0$	5.2117
57	La	Lanthanum	[Xe] $5d\,6s^2$	$^2D_{3/2}$	5.5769
58	Ce	Cerium	[Xe] $4f\,5d\,6s^2$	$^1G^o_4$	5.5387
59	Pr	Praseodymium	[Xe] $4f^3\,6s^2$	$^4I^o_{9/2}$	5.473
60	Nd	Neodymium	[Xe] $4f^4\,6s^2$	$^5I_4$	5.5250
61	Pm	Promethium	[Xe] $4f^5\,6s^2$	$^6H^o_{5/2}$	5.582
62	Sm	Samarium	[Xe] $4f^6\,6s^2$	$^7F_0$	5.6437
63	Eu	Europium	[Xe] $4f^7\,6s^2$	$^8S^o_{7/2}$	5.6704
64	Gd	Gadolinium	[Xe] $4f^7\,5d\,6s^2$	$^9D^o_2$	6.1498
65	Tb	Terbium	[Xe] $4f^9\,6s^2$	$^6H^o_{15/2}$	5.8638
66	Dy	Dysprosium	[Xe] $4f^{10}\,6s^2$	$^5I_8$	5.9389
67	Ho	Holmium	[Xe] $4f^{11}\,6s^2$	$^4I^o_{15/2}$	6.0215
68	Er	Erbium	[Xe] $4f^{12}\,6s^2$	$^3H_6$	6.1077
69	Tm	Thulium	[Xe] $4f^{13}\,6s^2$	$^2F^o_{7/2}$	6.1843
70	Yb	Ytterbium	[Xe] $4f^{14}\,6s^2$	$^1S_0$	6.2542
71	Lu	Lutetium	[Xe] $4f^{14}\,5d\,6s^2$	$^2D_{3/2}$	5.4259
72	Hf	Hafnium	[Xe] $4f^{14}\,5d^2\,6s^2$	$^3F_2$	6.8251
73	Ta	Tantalum	[Xe] $4f^{14}\,5d^3\,6s^2$	$^4F_{3/2}$	7.5496
74	W	Tungsten	[Xe] $4f^{14}\,5d^4\,6s^2$	$^5D_0$	7.8640
75	Re	Rhenium	[Xe] $4f^{14}\,5d^5\,6s^2$	$^6S_{5/2}$	7.8335
76	Os	Osmium	[Xe] $4f^{14}\,5d^6\,6s^2$	$^5D_4$	8.4382
77	Ir	Iridium	[Xe] $4f^{14}\,5d^7\,6s^2$	$^4F_{9/2}$	8.9670
78	Pt	Platinum	[Xe] $4f^{14}\,5d^9\,6s$	$^3D_3$	8.9588
79	Au	Gold	[Xe] $4f^{14}\,5d^{10}\,6s$	$^2S_{1/2}$	9.2255
80	Hg	Mercury	[Xe] $4f^{14}\,5d^{10}\,6s^2$	$^1S_0$	10.4375
81	Tl	Thallium	[Xe] $4f^{14}\,5d^{10}\,6s^2\,6p$	$^2P^o_{1/2}$	6.1082
82	Pb	Lead	[Xe] $4f^{14}\,5d^{10}\,6s^2\,6p^2$	$^3P_0$	7.4167
83	Bi	Bismuth	[Xe] $4f^{14}\,5d^{10}\,6s^2\,6p^3$	$^4S^o_{3/2}$	7.2855
84	Po	Polonium	[Xe] $4f^{14}\,5d^{10}\,6s^2\,6p^4$	$^3P_2$	8.414
85	At	Astatine	[Xe] $4f^{14}\,5d^{10}\,6s^2\,6p^5$	$^2P^o_{3/2}$	
86	Rn	Radon	[Xe] $4f^{14}\,5d^{10}\,6s^2\,6p^6$	$^1S_0$	10.7485
87	Fr	Francium	[Rn] $7s$	$^2S_{1/2}$	4.0727
88	Ra	Radium	[Rn] $7s^2$	$^1S_0$	5.2784
89	Ac	Actinium	[Rn] $6d\,7s^2$	$^2D_{3/2}$	5.17
90	Th	Thorium	[Rn] $6d^2\,7s^2$	$^3F_2$	6.3067
91	Pa	Protactinium	[Rn] $5f^2(^3H_4)\,6d\,7s^2$	$(4,3/2)_{11/2}$	5.89
92	U	Uranium	[Rn] $5f^3(^4I^o_{9/2})\,6d\,7s^2$	$(9/2,3/2)^o_6$	6.1941
93	Np	Neptunium	[Rn] $5f^4(^5I_4)\,6d\,7s^2$	$(4,3/2)_{11/2}$	6.2657
94	Pu	Plutonium	[Rn] $5f^6\,7s^2$	$^7F_0$	6.0260
95	Am	Americium	[Rn] $5f^7\,7s^2$	$^8S^o_{7/2}$	5.9738
96	Cm	Curium	[Rn] $5f^7\,6d\,7s^2$	$^9D^o_2$	5.9914
97	Bk	Berkelium	[Rn] $5f^9\,7s^2$	$^6H^o_{15/2}$	6.1979
98	Cf	Californium	[Rn] $5f^{10}\,7s^2$	$^5I_8$	6.2817
99	Es	Einsteinium	[Rn] $5f^{11}\,7s^2$	$^4I^o_{15/2}$	6.42
100	Fm	Fermium	[Rn] $5f^{12}\,7s^2$	$^3H_6$	6.50
101	Md	Mendelevium	[Rn] $5f^{13}\,7s^2$	$^2F^o_{7/2}$	6.58
102	No	Nobelium	[Rn] $5f^{14}\,7s^2$	$^1S_0$	6.65
103	Lr	Lawrencium	[Rn] $5f^{14}\,7s^2\,7p$?	$^2P^o_{1/2}$?	4.9?
104	Rf	Rutherfordium	[Rn] $5f^{14}\,6d^2\,7s^2$?	$^3F_2$?	6.0?

INTERNATIONAL TEMPERATURE SCALE OF 1990 (ITS-90)

B. W. Mangum

A new temperature scale, the International Temperature Scale of 1990 (ITS-90), was officially adopted by the Comité International des Poids et Mesures (CIPM), meeting 26—28 September 1989 at the Bureau International des Poids et Mesures (BIPM). The ITS-90 was recommended to the CIPM for its adoption following the completion of the final details of the new scale by the Comité Consultatif de Thermométrie (CCT), meeting 12—14 September 1989 at the BIPM in its 17th Session. The ITS-90 became the official international temperature scale on 1 January 1990. The ITS-90 supersedes the present scales, the International Practical Temperature Scale of 1968 (IPTS-68) and the 1976 Provisional 0.5 to 30 K Temperature Scale (EPT-76).

The ITS-90 extends upward from 0.65 K, and temperatures on this scale are in much better agreement with thermodynamic values that are those on the IPTS-68 and the EPT-76. The new scale has subranges and alternative definitions in certain ranges that greatly facilitate its use. Furthermore, its continuity, precision, and reproducibility throughout its ranges are much improved over that of the present scales. The replacement of the thermocouple with the platinum resistance thermometer at temperatures below 961.78°C resulted in the biggest improvement in reproducibility.

The ITS-90 is divided into four primary ranges:

1. Between 0.65 and 3.2 K, the ITS-90 is defined by the vapor pressure-temperature relation of ^{3}He, and between 1.25 and 2.1768 K (the λ point) and between 2.1768 and 5.0 K by the vapor pressure-temperature relations of ^{4}He. T_{90} is defined by the vapor pressure equations of the form:

$$T_{90} / K = A_0 + \sum_{i=1}^{9} A_i \left[\left(\ln(p / Pa) - B \right) / C \right]^i$$

The values of the coefficients A_i, and of the constants A_o, B, and C of the equations are given below.

2. Between 3.0 and 24.5561 K, the ITS-90 is defined in terms of a ^{3}He or ^{4}He constant volume gas thermometer (CVGT). The thermometer is calibrated at three temperatures — at the triple point of neon (24.5561 K), at the triple point of equilibrium hydrogen (13.8033 K), and at a temperature between 3.0 and 5.0 K, the value of which is determined by using either ^{3}He or ^{4}He vapor pressure thermometry.

3. Between 13.8033 K (−259.3467°C) and 1234.93 K (961.78°C), the ITS-90 is defined in terms of the specified fixed points given below, by resistance ratios of platinum resistance thermometers obtained by calibration at specified sets of the fixed points, and by reference functions and deviation functions of resistance ratios which relate to T_{90} between the fixed points.

4. Above 1234.93 K, the ITS-90 is defined in terms of Planck's radiation law, using the freezing-point temperature of either silver, gold, or copper as the reference temperature.

Full details of the calibration procedures and reference functions for various subranges are given in:

The International Temperature Scale of 1990, *Metrologia*, 27, 3, 1990; errata in *Metrologia*, 27, 107, 1990.

Defining Fixed Points of the ITS-90

Material[a]	Equilibrium state[b]	T_{90} (K)	t_{90} (°C)
He	VP	3 to 5	−270.15 to −268.15
e-H$_2$	TP	13.8033	−259.3467
e-H$_2$ (or He)	VP (or CVGT)	≈17	≈ −256.15
e-H$_2$ (or He)	VP (or CVGT)	≈20.3	≈ −252.85
Ne[c]	TP	24.5561	−248.5939
O$_2$	TP	54.3584	−218.7916
Ar	TP	83.8058	−189.3442
Hg[c]	TP	234.3156	−38.8344
H$_2$O	TP	273.16	0.01
Ga[c]	MP	302.9146	29.7646
In[c]	FP	429.7485	156.5985
Sn	FP	505.078	231.928
Zn	FP	692.677	419.527
Al[c]	FP	933.473	660.323
Ag	FP	1234.93	961.78
Au	FP	1337.33	1064.18
Cu[c]	FP	1357.77	1084.62

Values of Coefficients in the Vapor Pressure Equations for Helium

Coef. or constant	^{3}He 0.65—3.2 K	^{4}He 1.25—2.1768 K	^{4}He 2.1768—5.0 K
A_0	1.053 447	1.392 408	3.146 631
A_1	0.980 106	0.527 153	1.357 655
A_2	0.676 380	0.166 756	0.413 923
A_3	0.372 692	0.050 988	0.091 159
A_4	0.151 656	0.026 514	0.016 349
A_5	−0.002 263	0.001 975	0.001 826
A_6	0.006 596	−0.017 976	−0.004 325
A_7	0.088 966	0.005 409	−0.004 973
A_8	−0.004 770	0.013 259	0
A_9	−0.054 943	0	0
B	7.3	5.6	10.3
C	4.3	2.9	1.9

[a] e-H$_2$ indicates equilibrium hydrogen, that is, hydrogen with the equilibrium distribution of its ortho and para states. Normal hydrogen at room temperature contains 25% para hydrogen and 75% ortho hydrogen.

[b] VP indicates vapor pressure point; CVGT indicates constant volume gas thermometer point; TP indicates triple point (equilibrium temperature at which the solid, liquid, and vapor phases coexist); FP indicates freezing point, and MP indicates melting point (the equilibrium temperatures at which the solid and liquid phases coexist under a pressure of 101 325 Pa, one standard atmosphere). The isotopic composition is that naturally occurring.

[c] Previously, these were secondary fixed points.

CONVERSION OF TEMPERATURES FROM THE 1948 AND 1968 SCALES TO ITS-90

This table gives temperature corrections from older scales to the current International Temperature Scale of 1990 (see the preceding table for details on ITS-90). The first part of the table may be used for converting Celsius temperatures in the range -180 to 4000°C from IPTS-68 or IPTS-48 to ITS-90. Within the accuracy of the corrections, the temperature in the first column may be identified with either t_{68}, t_{48}, or t_{90}. The second part of the table is designed for use at lower temperatures to convert values expressed in kelvins from EPT-76 or IPTS-68 to ITS-90.

The references give analytical equations for expressing these relations. Note that Reference 1 supersedes Reference 2 with respect to corrections in the 630 to 1064°C range.

References

1. Burns, G. W. et al., in *Temperature: Its Measurement and Control in Science and Industry*, Vol. 6, Schooley, J. F., Ed., American Institute of Physics, New York, 1993.
2. Goldberg, R. N. and Weir, R. D., *Pure and Appl. Chem.*, 1545, 1992.

t/°C	$t_{90}-t_{68}$	$t_{90}-t_{48}$	t/°C	$t_{90}-t_{68}$	$t_{90}-t_{48}$	t/°C	$t_{90}-t_{68}$	$t_{90}-t_{48}$	t/°C	$t_{90}-t_{68}$	$t_{90}-t_{48}$
-180	0.008	0.020	290	-0.039	0.032	760	0.04	0.60	2400	-1.00	3.2
-170	0.010	0.017	300	-0.039	0.034	770	0.05	0.63	2500	-1.07	3.4
-160	0.012	0.007	310	-0.039	0.035	780	0.05	0.66	2600	-1.15	3.7
-150	0.013	0.000	320	-0.039	0.036	790	0.05	0.69	2700	-1.24	3.8
-140	0.014	0.001	330	-0.040	0.036	800	0.05	0.72	2800	-1.32	4.0
-130	0.014	0.008	340	-0.040	0.037	810	0.05	0.75	2900	-1.41	4.2
-120	0.014	0.017	350	-0.041	0.036	820	0.04	0.76	3000	-1.50	4.4
-110	0.013	0.026	360	-0.042	0.035	830	0.04	0.79	3100	-1.59	4.6
-100	0.013	0.035	370	-0.043	0.034	840	0.03	0.81	3200	-1.69	4.8
-90	0.012	0.041	380	-0.045	0.032	850	0.02	0.83	3300	-1.78	5.1
-80	0.012	0.045	390	-0.046	0.030	860	0.01	0.85	3400	-1.89	5.3
-70	0.011	0.045	400	-0.048	0.028	870	0.00	0.87	3500	-1.99	5.5
-60	0.010	0.042	410	-0.051	0.024	880	-0.02	0.87	3600	-2.10	5.8
-50	0.009	0.038	420	-0.053	0.022	890	-0.03	0.89	3700	-2.21	6.0
-40	0.008	0.032	430	-0.056	0.019	900	-0.05	0.90	3800	-2.32	6.3
-30	0.006	0.024	440	-0.059	0.015	910	-0.06	0.92	3900	-2.43	6.6
-20	0.004	0.016	450	-0.062	0.012	920	-0.08	0.93	4000	-2.55	6.8
-10	0.002	0.008	460	-0.065	0.009	930	-0.10	0.94			
0	0.000	0.000	470	-0.068	0.007	940	-0.11	0.96	T/K	$T_{90}-T_{76}$	$T_{90}-T_{68}$
10	-0.002	-0.006	480	-0.072	0.004	950	-0.13	0.97	5	-0.0001	
20	-0.005	-0.012	490	-0.075	0.002	960	-0.15	0.97	6	-0.0002	
30	-0.007	-0.016	500	-0.079	0.000	970	-0.16	0.99	7	-0.0003	
40	-0.010	-0.020	510	-0.083	-0.001	980	-0.18	1.00	8	-0.0004	
50	-0.013	-0.023	520	-0.087	-0.002	990	-0.19	1.02	9	-0.0005	
60	-0.016	-0.026	530	-0.090	-0.001	1000	-0.20	1.04	10	-0.0006	
70	-0.018	-0.026	540	-0.094	0.000	1010	-0.22	1.05	11	-0.0007	
80	-0.021	-0.027	550	-0.098	0.002	1020	-0.23	1.07	12	-0.0008	
90	-0.024	-0.027	560	-0.101	0.007	1030	-0.23	1.10	13	-0.0010	
100	-0.026	-0.026	570	-0.105	0.011	1040	-0.24	1.12	14	-0.0011	-0.006
110	-0.028	-0.024	580	-0.108	0.018	1050	-0.25	1.14	15	-0.0013	-0.003
120	-0.030	-0.023	590	-0.112	0.025	1060	-0.25	1.17	16	-0.0014	-0.004
130	-0.032	-0.020	600	-0.115	0.035	1070	-0.25	1.19	17	-0.0016	-0.006
140	-0.034	-0.018	610	-0.118	0.047	1080	-0.26	1.20	18	-0.0018	-0.008
150	-0.036	-0.016	620	-0.122	0.060	1090	-0.26	1.20	19	-0.0020	-0.009
160	-0.037	-0.012	630	-0.125	0.075	1100	-0.26	1.2	20	-0.0022	-0.009
170	-0.038	-0.009	640	-0.11	0.12	1200	-0.30	1.4	21	-0.0025	-0.008
180	-0.039	-0.005	650	-0.10	0.15	1300	-0.35	1.5	22	-0.0027	-0.007
190	-0.039	-0.001	660	-0.09	0.19	1400	-0.39	1.6	23	-0.0030	-0.007
200	-0.040	0.003	670	-0.07	0.24	1500	-0.44	1.8	24	-0.0032	-0.006
210	-0.040	0.007	680	-0.05	0.29	1600	-0.49	1.9	25	-0.0035	-0.005
220	-0.040	0.011	690	-0.04	0.32	1700	-0.54	2.1	26	-0.0038	-0.004
230	-0.040	0.014	700	-0.02	0.37	1800	-0.60	2.2	27	-0.0041	-0.004
240	-0.040	0.018	710	-0.01	0.41	1900	-0.66	2.3	28		-0.005
250	-0.040	0.021	720	0.00	0.45	2000	-0.72	2.5	29		-0.006
260	-0.040	0.024	730	0.02	0.49	2100	-0.79	2.7	30		-0.006
270	-0.039	0.028	740	0.03	0.53	2200	-0.85	2.9	31		-0.007
280	-0.039	0.030	750	0.03	0.56	2300	-0.93	3.1	32		-0.008

T/K	$T_{90}-T_{76}$	$T_{90}-T_{68}$	T/K	$T_{90}-T_{76}$	$T_{90}-T_{68}$	T/K	$T_{90}-T_{76}$	$T_{90}-T_{68}$	T/K	$T_{90}-T_{76}$	$T_{90}-T_{68}$
33		−0.008	57		0.000	81		0.008	150		0.014
34		−0.008	58		0.001	82		0.008	160		0.014
35		−0.007	59		0.002	83		0.008	170		0.013
36		−0.007	60		0.003	84		0.008	180		0.012
37		−0.007	61		0.003	85		0.008	190		0.012
38		−0.006	62		0.004	86		0.008	200		0.011
39		−0.006	63		0.004	87		0.008	210		0.010
40		−0.006	64		0.005	88		0.008	220		0.009
41		−0.006	65		0.005	89		0.008	230		0.008
42		−0.006	66		0.006	90		0.008	240		0.007
43		−0.006	67		0.006	91		0.008	250		0.005
44		−0.006	68		0.007	92		0.008	260		0.003
45		−0.007	69		0.007	93		0.008	270		0.001
46		−0.007	70		0.007	94		0.008	273.16		0.000
47		−0.007	71		0.007	95		0.008	300		−0.006
48		−0.006	72		0.007	96		0.008	400		−0.031
49		−0.006	73		0.007	97		0.009	500		−0.040
50		−0.006	74		0.007	98		0.009	600		−0.040
51		−0.005	75		0.008	99		0.009	700		−0.055
52		−0.005	76		0.008	100		0.009	800		−0.089
53		−0.004	77		0.008	110		0.011	900		−0.124
54		−0.003	78		0.008	120		0.013			
55		−0.002	79		0.008	130		0.014			
56		−0.001	80		0.008	140		0.014			

INTERNATIONAL SYSTEM OF UNITS (SI)

The International System of Units, abbreviated as SI (from the French name *Le Système International d'Unités*), was established in 1960 by the 11th General Conference on Weights and Measures (CGPM) as the modern metric system of measurement. The core of the SI is the seven base units for the physical quantities length, mass, time, electric current, thermodynamic temperature, amount of substance, and luminous intensity. These base units are:

Base quantity	SI base unit Name	Symbol
length	meter	m
mass	kilogram	kg
time	second	s
electric current	ampere	A
thermodynamic temperature	kelvin	K
amount of substance	mole	mol
luminous intensity	candela	cd

The SI base units are defined as follows:

ampere: The ampere is that constant current which, if maintained in two straight parallel conductors of infinite length, of negligible circular cross-section, and placed 1 meter apart in vacuum, would produce between these conductors a force equal to $2 \cdot 10^{-7}$ newton per meter of length.

candela: The candela is the luminous intensity, in a given direction, of a source that emits monochromatic radiation of frequency $540 \cdot 10^{12}$ hertz and that has a radiant intensity in that direction of 1/683 watt per steradian.

kelvin: The kelvin, unit of thermodynamic temperature, is the fraction 1/273.16 of the thermodynamic temperature of the triple point of water.

kilogram: The kilogram is the unit of mass; it is equal to the mass of the international prototype of the kilogram.

meter: The meter is the length of the path travelled by light in vacuum during a time interval of 1/299 792 458 of a second.

mole: The mole is the amount of substance of a system which contains as many elementary entities as there are atoms in 0.012 kilogram of carbon 12. When the mole is used, the elementary entities must be specified and may be atoms, molecules, ions, electrons, other particles, or specified groups of such particles.

second: The second is the duration of 9 192 631 770 periods of the radiation corresponding to the transition between the two hyperfine levels of the ground state of the cesium 133 atom.

SI derived units

Derived units are units which may be expressed in terms of base units by means of the mathematical symbols of multiplication and division (and, in the case of °C, subtraction). Certain derived units have been given special names and symbols, and these special names and symbols may themselves be used in combination with those for base and other derived units to express the units of other quantities. The next table lists some examples of derived units expressed directly in terms of base units:

Physical quantity	SI derived unit Name	Symbol
area	square meter	m^2
volume	cubic meter	m^3
speed, velocity	meter per second	m/s
acceleration	meter per second squared	m/s^2
wave number	reciprocal meter	m^{-1}
density, mass density	kilogram per cubic meter	kg/m^3
specific volume	cubic meter per kilogram	m^3/kg
current density	ampere per square meter	A/m^2
magnetic field strength	ampere per meter	A/m
concentration (of amount of substance)	mole per cubic meter	mol/m^3
luminance	candela per square meter	cd/m^2
refractive index	(the number) one	$1^{(a)}$

[a] The symbol "1" is generally omitted in combination with a numerical value.

For convenience, certain derived units, which are listed in the next table, have been given special names and symbols. These names and symbols may themselves be used to express other derived units. The special names and symbols are a compact form for the expression of units that are used frequently. The final column shows how the SI units concerned may be expressed in terms of SI base units. In this column, factors such as m^0, kg^0 ..., which are all equal to 1, are not shown explicitly.

Physical quantity	Name	Symbol	SI derived unit expressed in terms of: Other SI units	SI base units
plane angle	radian[a]	rad	$m \cdot m^{-1} = 1^{(b)}$	
solid angle	steradian[a]	sr[c]	$m^2 \cdot m^{-2} = 1^{(b)}$	
frequency	hertz	Hz	s^{-1}	
force	newton	N	$m \cdot kg \cdot s^{-2}$	
pressure, stress	pascal	Pa	N/m^2	$m^{-1} \cdot kg \cdot s^{-2}$
energy, work, quantity of heat	joule	J	$N \cdot m$	$m^2 \cdot kg \cdot s^{-2}$
power, radiant flux	watt	W	J/s	$m^2 \cdot kg \cdot s^{-3}$
electric charge, quantity of electricity	coulomb	C	$s \cdot A$	
electric potential difference, electromotive force	volt	V	W/A	$m^2 \cdot kg \cdot s^{-3} \cdot A^{-1}$
capacitance	farad	F	C/V	$m^{-2} \cdot kg^{-1} \cdot s^4 \cdot A^2$
electric resistance	ohm	Ω	V/A	$m^2 \cdot kg \cdot s^{-3} \cdot A^{-2}$
electric conductance	siemens	S	A/V	$m^{-2} \cdot kg^{-1} \cdot s^3 \cdot A^2$
magnetic flux	weber	Wb	$V \cdot s$	$m^2 \cdot kg \cdot s^{-2} \cdot A^{-1}$

magnetic flux density	tesla	T	Wb/m^2	$kg \cdot s^{-2} \cdot A^{-1}$
inductance	henry	H	Wb/A	$m^2 \cdot kg \cdot s^{-2} \cdot A^{-2}$
Celsius temperature	degree Celsius[d]	°C		K
luminous flux	lumen	lm	$cd \cdot sr$[c]	$m^2 \cdot m^{-2} \cdot cd = cd$
illuminance	lux	lx	lm/m^2	$m^2 \cdot m^{-4} \cdot cd = m^{-2} \cdot cd$
activity (of a radionuclide)	becquerel	Bq		s^{-1}
absorbed dose, specific energy (imparted), kerma	gray	Gy	J/kg	$m^2 \cdot s^{-2}$
dose equivalent, ambient dose equivalent, directional dose equivalent, personal dose equivalent, organ equivalent dose	sievert	Sv	J/kg	$m^2 \cdot s^{-2}$
catalytic activity	katal	kat		$s^{-1} \cdot mol$

[a] The radian and steradian may be used with advantage in expressions for derived units to distinguish between quantities of different nature but the same dimension. Some examples of their use in forming derived units are given in the next table.

[b] In practice, the symbols rad and sr are used where appropriate, but the derived unit "1" is generally omitted in combination with a numerical value.

[c] In photometry, the name steradian and the symbol sr are usually retained in expressions for units.

[d] It is common practice to express a thermodynamic temperature, symbol T, in terms of its difference from the reference temperature $T_0 = 273.15$ K. The numerical value of a Celsius temperature t expressed in degrees Celsius is given by $t/°C = T/K - 273.15$. The unit °C may be used in combination with SI prefixes, e.g., millidegree Celsius, m°C. Note that there should never be a space between the ° sign and the letter C, and that the symbol for kelvin is K, not °K.

The SI derived units with special names may be used in combinations to provide a convenient way to express more complex physical quantities. Examples are given in the next table:

Physical Quantity	SI derived unit		
	Name	Symbol	As SI base units
dynamic viscosity	pascal second	Pa · s	$m^{-1} \cdot kg \cdot s^{-1}$
moment of force	newton meter	N · m	$m^2 \cdot kg \cdot s^{-2}$
surface tension	newton per meter	N/m	$kg \cdot s^{-2}$
angular velocity	radian per second	rad/s	$m \cdot m^{-1} \cdot s^{-1} = s^{-1}$
angular acceleration	radian per second squared	rad/s²	$m \cdot m^{-1} \cdot s^{-2} = s^{-2}$
heat flux density, irradiance	watt per square meter	W/m²	$kg \cdot s^{-3}$
heat capacity, entropy	joule per kelvin	J/K	$m^{-3} \cdot kg \cdot s^{-2} \cdot K^{-1}$
specific heat capacity, specific entropy	joule per kilogram kelvin	J/(kg · K)	$m^2 \cdot s^{-2} \cdot K^{-1}$
specific energy	joule per kilogram	J/kg	$m^2 \cdot s^{-2}$
thermal conductivity	watt per meter kelvin	W/(m · K)	$m \cdot kg \cdot s^{-3} \cdot K^{-1}$
energy density	joule per cubic meter	J/m³	$m^{-1} \cdot kg \cdot s^{-2}$
electric field strength	volt per meter	V/m	$m \cdot kg \cdot s^{-3} \cdot A^{-1}$
electric charge density	coulomb per cubic meter	C/m³	$m^{-3} \cdot s \cdot A$
electric flux density	coulomb per square meter	C/m²	$m^{-2} \cdot s \cdot A$
permittivity	farad per meter	F/m	$m^{-3} \cdot kg^{-1} \cdot s^4 \cdot A^2$
permeability	henry per meter	H/m	$m \cdot kg \cdot s^{-2} \cdot A^{-2}$
molar energy	joule per mole	J/mol	$m^2 \cdot kg \cdot s^{-2} \cdot mol^{-1}$
molar entropy, molar heat capacity	joule per mole kelvin	J/(mol · K)	$m^2 \cdot kg \cdot s^{-2} \cdot K^{-1} \cdot mol^{-1}$
exposure (x and γ rays)	coulomb per kilogram	C/kg	$kg^{-1} \cdot s \cdot A$
absorbed dose rate	gray per second	Gy/s	$m^2 \cdot s^{-3}$
radiant intensity	watt per steradian	W/sr	$m^4 \cdot m^{-2} \cdot kg \cdot s^{-3} = m^2 \cdot kg \cdot s^{-3}$
radiance	watt per square meter steradian	W/(m² · sr)	$m^2 \cdot m^{-2} \cdot kg \cdot s^{-3} = kg \cdot s^{-3}$
catalytic (activity) concentration	katal per cubic meter	kat/m³	$m^{-3} \cdot s^{-1} \cdot mol$

In practice, with certain quantities preference is given to the use of certain special unit names, or combinations of unit names, in order to facilitate the distinction between different quantities having the same dimension. For example, the SI unit of frequency is designated the hertz, rather than the reciprocal second, and the SI unit of angular velocity is designated the radian per second rather than the reciprocal second (in this case retaining the word radian emphasizes that angular velocity is equal to 2π times the rotational frequency). Similarly the SI unit of moment of force is designated the newton meter rather than the joule.

In the field of ionizing radiation, the SI unit of activity is designated the becquerel rather than the reciprocal second, and the SI units of absorbed dose and dose equivalent the gray and sievert, respectively, rather than the joule per kilogram. In the field of catalysis, the SI unit of catalytic activity is designated the katal rather than the mole per second. The special names becquerel, gray, sievert, and katal were specifically introduced because of the dangers to human health which might arise from mistakes involving the units reciprocal second, joule per kilogram and mole per second.

Units for dimensionless quantities, quantities of dimension one

Certain quantities are defined as the ratios of two quantities of the same kind, and thus have a dimension which may be expressed by the number one. The unit of such quantities is necessarily a derived unit coherent with the other units of the SI and, since it is formed as the ratio of two identical SI units, the unit also may be expressed by the number one. Thus the SI unit of all quantities having the dimensional product one is the number one. Examples of such quantities are refractive index, relative permeability, and friction factor. Other quantities having the unit 1 include "characteristic numbers" like the Prandtl number and numbers which represent a count, such as a number of molecules, degeneracy (number of energy levels), and partition function in statistical thermodynamics. All of these quantities are described as being dimensionless, or of dimension one, and have the coherent SI unit 1. Their values are simply expressed as numbers and, in general, the unit 1 is not explicitly shown. In a few cases, however, a special name is given to this unit, mainly to avoid confusion between some compound derived units. This is the case for the radian, steradian and neper.

SI prefixes

The following prefixes have been approved by the CGPM for use with SI units. Only one prefix may be used before a unit. Thus 10^{-12} farad should be designated pF, not μμF.

Factor	Name	Symbol	Factor	Name	Symbol
10^{24}	yotta	Y	10^{-1}	deci	d
10^{21}	zetta	Z	10^{-2}	centi	c
10^{18}	exa	E	10^{-3}	milli	m
10^{15}	peta	P	10^{-6}	micro	μ
10^{12}	tera	T	10^{-9}	nano	n
10^{9}	giga	G	10^{-12}	pico	p
10^{6}	mega	M	10^{-15}	femto	f
10^{3}	kilo	k	10^{-18}	atto	a
10^{2}	hecto	h	10^{-21}	zepto	z
10^{1}	deka	da	10^{-24}	yocto	y

The kilogram

Among the base units of the International System, the unit of mass is the only one whose name, for historical reasons, contains a prefix. Names and symbols for decimal multiples and submultiples of the unit of mass are formed by attaching prefix names to the unit name "gram" and prefix symbols to the unit symbol "g".

Example : 10^{-6} kg = 1 mg (1 milligram) *but not* 1 μkg (1 microkilogram).

Units used with the SI

Many units that are not part of the SI are important and widely used in everyday life. The CGPM has adopted a classification of non-SI units: (1) units accepted for use with the SI (such as the traditional units of time and of angle); (2) units accepted for use with the SI whose values are obtained experimentally; and (3) other units currently accepted for use with the SI to satisfy the needs of special interests.

(1) Non-SI units accepted for use with the International System

Name	Symbol	Value in SI units
minute	min	1 min = 60 s
hour	h	1 h= 60 min = 3600 s
day	d	1 d = 24 h = 86 400 s
degree	°	$1° = (\pi/180)$ rad
minute	'	$1' = (1/60)° = (\pi/10\,800)$ rad
second	"	$1" = (1/60)' = (\pi/648\,000)$ rad
liter	l, L	$1 L = 1\ dm^3 = 10^{-3}\ m^3$
metric ton	t	$1\ t = 10^3$ kg
neper[a]	Np	1 Np = 1
bel[b]	B	1 B = (1/2) ln 10 Np

[a] The neper is used to express values of such logarithmic quantities as field level, power level, sound pressure level, and logarithmic decrement. Natural logarithms are used to obtain the numerical values of quantities expressed in nepers. The neper is coherent with the SI, but is not yet adopted by the CGPM as an SI unit. In using the neper, it is important to specify the quantity.

[b] The bel is used to express values of such logarithmic quantities as field level, power level, sound-pressure level, and attenuation. Logarithms to base ten are used to obtain the numerical values of quantities expressed in bels. The submultiple decibel, dB, is commonly used.

(2) Non-SI units accepted for use with the International system, whose values in SI units are obtained experimentally

Name	Symbol	Value in SI Units
electronvolt[b]	eV	1 eV = 1.602 176 53(14) $\cdot 10^{-19}$ J[a]
dalton[c]	Da	1 Da = 1.660 538 86(28) $\cdot 10^{-27}$ kg[a]
unified atomic mass unit[c]	u	1 u = 1 Da
astronomical unit[d]	ua	1 ua = 1.495 978 706 91(06) $\cdot 10^{11}$ m[a]

[a] For the electronvolt and the dalton (unified atomic mass unit), values are quoted from the 2002 CODATA set of the Fundamental Physical Constants (p. 1-1 of this Handbook). The value given for the astronomical unit is quoted from the IERS Conventions 2003 (D.D. McCarthy and G. Petit, eds., IERS Technical Note 32, Frankfurt am Main: Verlag des Bundesamts für Kartographie und Geodäsie, 200). The value of ua in meters comes from the JPL ephemerides DE403 (Standish E.M. 1995, "Report of the IAU WGAS Sub-Group on Numerical Standards", in "Highlights of Astronomy", Appenlzer ed., pp 180-184, Kluwer Academic Publishers, Dordrecht). It has been determined in "TDB" units using Barycentric Dynamical Time TDB as a time coordinate for the barycentric system.

[b] The electronvolt is the kinetic energy acquired by an electron in passing through a potential difference of 1 V in vacuum.

[c] The Dalton and unified atomic mass unit are alternative names for the same unit, equal to 1/12 of the mass of an unbound atom of the nuclide ^{12}C, at rest and in its ground state. The dalton may be combined with SI prefixes to express the masses of large molecules in kilodalton, kDa, or megadalton, MDa.

[d] The astronomical unit is a unit of length approximately equal to the mean Earth-Sun distance. It is the radius of an unperturbed circular Newtonian orbit about the Sun of a particle having infinitesimal mass, moving with a mean motion of 0.017 202 098 95 radians/day (known as the Gaussian constant).

(3) Other non-SI units currently accepted for use with the International System

Name	Symbol	Value in SI Units
nautical mile		1 nautical mile = 1852 m
knot		1 nautical mile per hour = (1852/3600) m/s
are		$1\ a = 1\ dam^2 = 10^2\ m^2$
hectare	ha	$1\ ha = 1\ hm^2 = 10^4\ m^2$
bar	bar	1 bar = 0.1 MPa = 100 kPa = 10^5 Pa
ångström	Å	$1\ Å = 0.1\ nm = 10^{-10}\ m$
barn	b	$1\ b = 100\ fm^2 = 10^{-28}\ m^2$

Other non-SI units

The SI does not encourage the use of cgs units, but these are frequently found in old scientific texts. The following table gives the relation of some common cgs units to SI units.

Name	Symbol	Value in SI units
erg	erg	1 erg = 10^{-7} J
dyne	dyn	1 dyn = 10^{-5} N
poise	P	$1P = 1 dyn \cdot s/cm^2 = 0.1\ Pa \cdot s$
stokes	St	$1\ St = 1\ cm^2/s = 10^{-4}\ m^2/s$
gauss	G	$1G \triangleq 10^{-4}$ T
oersted	Oe	1 Oe $\triangleq (1000/4\pi)$ A/m
maxwell	Mx	$1 Mx \triangleq 10^{-8}$ Wb
stilb	sb	$1\ sb = 1\ cd/cm^2 = 10^4\ cd/m^2$
phot	ph	$1\ ph = 10^4$ lx
gal	Gal	$1\ Gal = 1\ cm/s^2 = 10^{-2}\ m/s^2$

Note: The symbol $\triangleq$ should be read as "corresponds to"; these units cannot strictly be equated because of the different dimensions of the electromagnetic cgs and the SI.

Examples of other non-SI units found in the older literature and their relation to the SI are given below. Use of these units in current texts is discouraged.

Name	Symbol	Value in SI units
curie	Ci	$1\text{ Ci} = 3.7 \cdot 10^{10}\text{ Bq}$
roentgen	R	$1\text{ R} = 2.58 \cdot 10^{-4}\text{ C/kg}$
rad	rad	$1\text{ rad} = 1\text{ cGy} = 10^{-2}\text{ Gy}$
rem	rem	$1\text{ rem} = 1\text{ cSv} = 10^{-2}\text{ Sv}$
X unit		$1\text{ X unit} \approx 1.002 \cdot 10^{-4}\text{ nm}$
gamma	γ	$1\ \gamma = 1\text{ nT} = 10^{-9}\text{ T}$
jansky	Jy	$1\text{Jy} = 10^{-26}\text{ W} \cdot \text{m}^{-2} \cdot \text{Hz}^{-1}$
fermi		$1\text{ fermi} = 1\text{ fm} = 10^{-15}\text{ m}$
metric carat		$1\text{ metric carat} = 200\text{ mg} = 2 \cdot 10^{-4}\text{ kg}$
torr	Torr	$1\text{ Torr} = (101325/760)\text{ Pa}$
standard atmosphere	atm	$1\text{ atm} = 101325\text{ Pa}$
calorie[a]	cal	$1\text{ cal} = 4.184\text{ J}$
micron	μ	$1\ \mu = 1\ \mu\text{m} = 10^{-6}\text{ m}$

[a] Several types of calorie have been used; the value given here is the so-called "thermochemical calorie".

References

1. Taylor, B. N., *The International System of Units (SI)*, NIST Special Publication 330, National Institute of Standards and Technology, Gaithersburg, MD, 2001.
2. Bureau International des Poids et Mesures, *Le Système International d'Unités (SI)*, 7th French and English Edition, BIPM, Sèvres, France, 1998; 8th Edition to be published 2006.
3. Taylor, B. N., *Guide for the Use of the International System of Units (SI)*, NIST Special Publication 811, National Institute of Standards and Technology, Gaithersburg, MD, 1995.
4. NIST Physical Reference Data web site, <http://physics.nist.gov/cuu/Units/index.html>, October 2004.

UNITS FOR MAGNETIC PROPERTIES

Quantity	Symbol	Gaussian & cgs emu [a]	Conversion factor, C [b]	SI & rationalized mks [c]
Magnetic flux density, magnetic induction	B	gauss (G) [d]	10^{-4}	tesla (T), Wb/m^2
Magnetic flux	Φ	maxwell (Mx), G $\cdot$ cm^2	10^{-8}	weber (Wb), volt second (V $\cdot$ s)
Magnetic potential difference, magnetomotive force	U, F	gilbert (Gb)	$10/4\pi$	ampere (A)
Magnetic field strength, magnetizing force	H	oersted (Oe), [e] Gb/cm	$10^3/4\pi$	A/m [f]
(Volume) magnetization [g]	M	emu/cm^3 [h]	10^3	A/m
(Volume) magnetization	$4\pi M$	G	$10^3/4\pi$	A/m
Magnetic polarization, intensity of magnetization	J, I	emu/cm^3	$4\pi \times 10^{-4}$	T, Wb/m^2 [i]
(Mass) magnetization	σ, M	emu/g	1	A$\cdot$m^2/kg
			$4\pi \times 10^{-7}$	Wb$\cdot$m/kg
Magnetic moment	m	emu, erg/G	10^{-3}	A$\cdot$m^2, joule per tesla (J/T)
Magnetic dipole moment	j	emu, erg/G	$4\pi \times 10^{-10}$	Wb$\cdot$m [i]
(Volume) susceptibility	χ, κ	dimensionless, emu/cm^3	4π	dimensionless
			$(4\pi)^2 \times 10^{-7}$	henry per meter (H/m), Wb/(A$\cdot$m)
(Mass) susceptibility	χ_ρ, κ_ρ	cm^3/g, emu/g	$4\pi \times 10^{-3}$	m^3/kg
			$(4\pi)^2 \times 10^{-10}$	H$\cdot$m^2/kg
(Molar) susceptibility	χ_{mol}, κ_{mol}	cm^3/mol, emu/mol	$4\pi \times 10^{-6}$	m^3/mol
			$(4\pi)^2 \times 10^{-13}$	H$\cdot$m^2/mol
Permeability	μ	dimensionless	$4\pi \times 10^{-7}$	H/m, Wb/(A$\cdot$m)
Relative permeability [j]	μ_r	not defined		dimensionless
(Volume) energy density, energy product [k]	W	erg/cm^3	10^{-1}	J/m^3
Demagnetization factor	D, N	dimensionless	$1/4\pi$	dimensionless

[a.] Gaussian units and cgs emu are the same for magnetic properties. The defining relation is $B = H + 4\pi M$.

[b.] Multiply a number in Gaussian units by C to convert it to SI (e.g., 1 G $\times 10^{-4}$ T/G $= 10^{-4}$ T).

[c.] SI (*Système International d'Unités*) has been adopted by the National Bureau of Standards. Where two conversion factors are given, the upper one is recognized under, or consistent with, SI and is based on the definition $B = \mu_0(H + M)$, where $\mu_0 = 4\pi \times 10^{-7}$ H/m. The lower one is not recognized under SI and is based on the definition $B = \mu_0 H + J$, where the symbol I is often used in place of J.

[d.] 1 gauss $= 10^5$ gamma (γ).

[e.] Both oersted and gauss are expressed as cm$^{-1/2}\cdot$g$^{1/2}\cdot$s^{-1} in terms of base units.

[f.] A/m was often expressed as "ampere–turn per meter" when used for magnetic field strength.

[g.] Magnetic moment per unit volume.

[h.] The designation "emu" is not a unit.

[i.] Recognized under SI, even though based on the definition $B = \mu_0 H + J$. See footnote c.

[j.] $\mu_r = \mu/\mu_0 = 1+\chi$, all in SI. μ_r is equal to Gaussian μ.

[k.] $B\cdot H$ and $\mu_0 M\cdot H$ have SI units J/m^3; $M\cdot H$ and $B\cdot H/4\pi$ have Gaussian units erg/cm^3.

R. B. Goldfarb and F. R. Fickett, U.S. Department of Commerce, National Bureau of Standards, Boulder, Colorado 80303, March 1985, NBS Special Publication 696. For sale by the Superintendent of Documents, U.S. Government Printing Office, Washington, DC 20402.

CONVERSION FACTORS

The following table gives conversion factors from various units of measure to SI units. It is reproduced from NIST Special Publication 811, *Guide for the Use of the International System of Units (SI)*. The table gives the factor by which a quantity expressed in a non-SI unit should be multiplied in order to calculate its value in the SI. The SI values are expressed in terms of the base, supplementary, and derived units of SI in order to provide a coherent presentation of the conversion factors and facilitate computations (see the table "International System of Units" in this Section). If desired, powers of ten can be avoided by using SI Prefixes and shifting the decimal point if necessary.

Conversion from a non-SI unit to a different non-SI unit may be carried out by using this table in two stages, e.g.,

$$1 \; cal_{th} = 4.184 \; J$$

$$1 \; Btu_{IT} = 1.055056 \; E+03 \; J$$

Thus,

$$1 \; Btu_{IT} = (1.055056 \; E+03 \div 4.184) \; cal_{th} = 252.164 \; cal_{th}$$

Conversion factors are presented for ready adaptation to computer readout and electronic data transmission. The factors are written as a number equal to or greater than one and less than ten with six or fewer decimal places. This number is followed by the letter E (for exponent), a plus or a minus sign, and two digits which indicate the power of 10 by which the number must be multiplied to obtain the correct value. For example:

$$3.523\,907 \; E\text{-}02 \text{ is } 3.523\,907 \times 10^{-2}$$

or

$$0.035\,239\,07$$

Similarly:

$$3.386\,389 \; E+03 \text{ is } 3.386\,389 \times 10^{3}$$

or

$$3\,386.389$$

A factor in boldface is exact; i.e., all subsequent digits are zero. All other conversion factors have been rounded to the figures given in accordance with accepted practice. Where less than six digits after the decimal point are shown, more precision is not warranted.

It is often desirable to round a number obtained from a conversion of units in order to retain information on the precision of the value. The following rounding rules may be followed:

1. If the digits to be discarded begin with a digit less than 5, the digit preceding the first discarded digit is not changed.

   Example: 6.974 951 5 rounded to 3 digits is 6.97

2. If the digits to be discarded begin with a digit greater than 5, the digit preceding the first discarded digit is increased by one.

   Example: 6.974 951 5 rounded to 4 digits is 6.975

3. If the digits to be discarded begin with a 5 and at least one of the following digits is greater than 0, the digit preceding the 5 is increased by 1.

   Example: 6.974 851 rounded to 5 digits is 6.974 9

4. If the digits to be discarded begin with a 5 and all of the following digits are 0, the digit preceding the 5 is unchanged if it is even and increased by one if it is odd. (Note that this means that the final digit is always even.)

   Examples:
   6.974 951 5 rounded to 7 digits is 6.974 952
   6.974 950 5 rounded to 7 digits is 6.974 950

Reference

Taylor, B. N., *Guide for the Use of the International System of Units (SI)*, NIST Special Publication 811, 1995 Edition, Superintendent of Documents, U.S. Government Printing Office, Washington, DC 20402, 1995.

Factors in **boldface** are exact

To convert from	to	Multiply by	
abampere	ampere (A)	**1.0**	**E+01**
abcoulomb	coulomb (C)	**1.0**	**E+01**
abfarad	farad (F)	**1.0**	**E+09**
abhenry	henry (H)	**1.0**	**E−09**
abmho	siemens (S)	**1.0**	**E+09**
abohm	ohm (Ω)	**1.0**	**E−09**
abvolt	volt (V)	**1.0**	**E−08**
acceleration of free fall, standard (g_n)	meter per second squared (m/s²)	**9.806 65**	**E+00**
acre (based on U.S. survey foot)[9]	square meter (m²)	4.046 873	E+03
acre foot (based on U.S. survey foot)[9]	cubic meter (m³)	1.233 489	E+03
ampere hour (A · h)	coulomb (C)	**3.6**	**E+03**
ångström (Å)	meter (m)	**1.0**	**E−10**
ångström (Å)	nanometer (nm)	**1.0**	**E−01**
are (a)	square meter (m²)	**1.0**	**E+02**
astronomical unit (ua or AU)	meter (m)	1.495 979	E+11
atmosphere, standard (atm)	pascal (Pa)	**1.013 25**	**E+05**
atmosphere, standard (atm)	kilopascal (kPa)	**1.013 25**	**E+02**
atmosphere, technical (at)[10]	pascal (Pa)	**9.806 65**	**E+04**
atmosphere, technical (at)[10]	kilopascal (kPa)	**9.806 65**	**E+01**

[9] The U.S. survey foot equals (1200/3937) m. 1 international foot = 0.999998 survey foot.

[10] One technical atmosphere equals one kilogram-force per square centimeter (1 at = 1 kgf/cm²).

To convert from	to	Multiply by	
bar (bar)	pascal (Pa)	**1.0**	**E+05**
bar (bar)	kilopascal (kPa)	**1.0**	**E+02**
barn (b)	square meter (m^2)	**1.0**	**E−28**
barrel [for petroleum, 42 gallons (U.S.)](bbl)	cubic meter (m^3)	1.589 873	E−01
barrel [for petroleum, 42 gallons (U.S.)](bbl)	liter (L)	1.589 873	E+02
biot (Bi)	ampere (A)	**1.0**	**E+01**
British thermal unit$_{IT}$ (Btu_{IT})[11]	joule (J)	1.055 056	E+03
British thermal unit$_{th}$ (Btu_{th})[11]	joule (J)	1.054 350	E+03
British thermal unit (mean) (Btu)	joule (J)	1.055 87	E+03
British thermal unit (39 °F) (Btu)	joule (J)	1.059 67	E+03
British thermal unit (59 °F) (Btu)	joule (J)	1.054 80	E+03
British thermal unit (60 °F) (Btu)	joule (J)	1.054 68	E+03
British thermal unit$_{IT}$ foot per hour square foot degree Fahrenheit [$Btu_{IT} \cdot ft/(h \cdot ft^2 \cdot °F)$]	watt per meter kelvin [$W/(m \cdot K)$]	1.730 735	E+00
British thermal unit$_{th}$ foot per hour square foot degree Fahrenheit [$Btu_{th} \cdot ft/(h \cdot ft^2 \cdot °F)$]	watt per meter kelvin [$W/(m \cdot K)$]	1.729 577	E+00
British thermal unit$_{IT}$ inch per hour square foot degree Fahrenheit [$Btu_{IT} \cdot in/(h \cdot ft^2 \cdot °F)$]	watt per meter kelvin [$W/(m \cdot K)$]	1.442 279	E−01
British thermal unit$_{th}$ inch per hour square foot degree Fahrenheit [$Btu_{th} \cdot in/(h \cdot ft^2 \cdot °F)$]	watt per meter kelvin [$W/(m \cdot K)$]	1.441 314	E−01
British thermal unit$_{IT}$ inch per second square foot degree Fahrenheit [$Btu_{IT} \cdot in/(s \cdot ft^2 \cdot °F)$]	watt per meter kelvin [$W/(m \cdot K)$]	5.192 204	E+02
British thermal unit$_{th}$ inch per second square foot degree Fahrenheit [$Btu_{th} \cdot in/(s \cdot ft^2 \cdot °F)$]	watt per meter kelvin [$W/(m \cdot K)$]	5.188 732	E+02
British thermal unit$_{IT}$ per cubic foot (Btu_{IT}/ft^3)	joule per cubic meter (J/m^3)	3.725 895	E+04
British thermal unit$_{th}$ per cubic foot (Btu_{th}/ft^3)	joule per cubic meter (J/m^3)	3.723 403	E+04
British thermal unit$_{IT}$ per degree Fahrenheit ($Btu_{IT}/°F$)	joule per kelvin (J/k)	1.899 101	E+03
British thermal unit$_{th}$ per degree Fahrenheit ($Btu_{th}/°F$)	joule per kelvin (J/k)	1.897 830	E+03
British thermal unit$_{IT}$ per degree Rankine ($Btu_{IT}/°R$)	joule per kelvin (J/k)	1.899 101	E+03
British thermal unit$_{th}$ per degree Rankine ($Btu_{th}/°R$)	joule per kelvin (J/k)	1.897 830	E+03
British thermal unit$_{IT}$ per hour (Btu_{IT}/h)	watt (W)	2.930 711	E−01
British thermal unit$_{th}$ per hour (Btu_{th}/h)	watt (W)	2.928 751	E−01
British thermal unit$_{IT}$ per hour square foot degree Fahrenheit [$Btu_{IT}/(h \cdot ft^2 \cdot °F)$]	watt per square meter kelvin [$W/(m^2 \cdot K)$]	5.678 263	E+00
British thermal unit$_{th}$ per hour square foot degree Fahrenheit [$Btu_{th}/(h \cdot ft^2 \cdot °F)$]	watt per square meter kelvin [$W/(m^2 \cdot K)$]	5.674 466	E+00
British thermal unit$_{th}$ per minute (Btu_{th}/min)	watt (W)	1.757 250	E+01
British thermal unit$_{IT}$ per pound (Btu_{IT}/lb)	joule per kilogram (J/kg)	**2.326**	**E+03**
British thermal unit$_{th}$ per pound (Btu_{th}/lb)	joule per kilogram (J/kg)	2.324 444	E+03
British thermal unit$_{IT}$ per pound degree Fahrenheit [$Btu_{IT}/(lb \cdot °F)$]	joule per kilogram kelvin ($J/(kg \cdot K)$)	**4.1868**	**E+03**
British thermal unit$_{th}$ per pound degree Fahrenheit [$Btu_{th}/(lb \cdot °F)$]	joule per kilogram kelvin [$J/(kg \cdot K)$]	**4.184**	**E+03**
British thermal unit$_{IT}$ per pound degree Rankine [$Btu_{IT}/(lb \cdot °R)$]	joule per kilogram kelvin [$J/(kg \cdot K)$]	**4.1868**	**E+03**
British thermal unit$_{th}$ per pound degree Rankine [$Btu_{th}/(lb \cdot °R)$]	joule per kilogram kelvin [$J/(kg \cdot K)$]	**4.184**	**E+03**
British thermal unit$_{IT}$ per second (Btu_{IT}/s)	watt (W)	1.055 056	E+03
British thermal unit$_{th}$ per second (Btu_{th}/s)	watt (W)	1.054 350	E+03

---

[11] The Fifth International Conference on the Properties of Steam (London, July 1956) defined the International Table calorie as 4.1868 J. Therefore the exact conversion factor for the International Table Btu is 1.055 055 852 62 kJ. Note that the notation for International Table used in this listing is subscript "IT". Similarily, the notation for thermochemical is subscript "th." Further, the thermochemical Btu, Btu_{th}, is based on the thermochemical calorie, cal_{th}, where cal_{th} = 4.184 J exactly.

To convert from	to	Multiply by	
British thermal unit$_{IT}$ per second square foot degree Fahrenheit [Btu$_{IT}$/(s · ft^2 · °F)]	watt per square meter kelvin [W/(m^2 · K)]	2.044 175	E+04
British thermal unit$_{th}$ per second square foot degree Fahrenheit [Btu$_{th}$/(s · ft^2 · °F)]	watt per square meter kelvin [W/(m^2 · K)]	2.042 808	E+04
British thermal unit$_{IT}$ per square foot (Btu$_{IT}$/ft^2)	joule per square meter (J/m^2)	1.135 653	E+04
British thermal unit$_{th}$ per square foot (Btu$_{th}$/ft^2)	joule per square meter (J/m^2)	1.134 893	E+04
British thermal unit$_{IT}$ per square foot hour [(Btu$_{IT}$/(ft^2 · h)]	watt per square meter (W/m^2)	3.154 591	E+00
British thermal unit$_{th}$ per square foot hour [Btu$_{th}$/(ft^2 · h)]	watt per square meter (W/m^2)	3.152 481	E+00
British thermal unit$_{th}$ per square foot minute [Btu$_{th}$/(ft^2 · min)]	watt per square meter (W/m^2)	1.891 489	E+02
British thermal unit$_{IT}$ per square foot second [(Btu$_{IT}$/(ft^2 · s)]	watt per square meter (W/m^2)	1.135 653	E+04
British thermal unit$_{th}$ per square foot second [Btu$_{th}$/(ft^2 · s)]	watt per square meter (W/m^2)	1.134 893	E+04
British thermal unit$_{th}$ per square inch second [Btu$_{th}$/(in^2 · s)]	watt per square meter (W/m^2)	1.634 246	E+06
bushel (U.S.) (bu)	cubic meter (m^3)	3.523 907	E−02
bushel (U.S.) (bu)	liter (L)	3.523 907	E+01
calorie$_{IT}$ (cal$_{IT}$)[11]	joule (J)	**4.1868**	**E+00**
calorie$_{th}$ (cal$_{th}$)[11]	joule (J)	**4.184**	**E+00**
calorie (cal) (mean)	joule (J)	4.190 02	E+00
calorie (15 °C) (cal$_{15}$)	joule (J)	4.185 80	E+00
calorie (20 °C) (cal$_{20}$)	joule (J)	4.181 90	E+00
calorie$_{IT}$, kilogram (nutrition)[12]	joule (J)	**4.1868**	**E+03**
calorie$_{th}$, kilogram (nutrition)[12]	joule (J)	**4.184**	**E+03**
calorie (mean), kilogram (nutrition)[12]	joule (J)	4.190 02	E+03
calorie$_{th}$ per centimeter second degree Celsius [cal$_{th}$/(cm · s · °C)]	watt per meter kelvin [W/(m · K)]	**4.184**	**E+02**
calorie$_{IT}$ per gram (cal$_{IT}$/g)	joule per kilogram (J/kg)	**4.1868**	**E+03**
calorie$_{th}$ per gram (cal$_{th}$/g)	joule per kilogram (J/kg)	**4.184**	**E+03**
calorie$_{IT}$ per gram degree Celsius [cal$_{IT}$/(g · °C)]	joule per kilogram kelvin [J/(kg · K)]	**4.1868**	**E+03**
calorie$_{th}$ per gram degree Celsius [cal$_{th}$/(g · °C)]	joule per kilogram kelvin [J/(kg · K)]	**4.184**	**E+03**
calorie$_{IT}$ per gram kelvin [cal$_{IT}$/(g · K)]	joule per kilogram kelvin [J/(kg · K)]	**4.1868**	**E+03**
calorie$_{th}$ per gram kelvin [cal$_{th}$/(g · K)]	joule per kilogram kelvin [J/(kg · K)]	**4.184**	**E+03**
calorie$_{th}$ per minute (cal$_{th}$/min)	watt (W)	6.973 333	E−02
calorie$_{th}$ per second (cal$_{th}$/s)	watt (W)	**4.184**	**E+00**
calorie$_{th}$ per square centimeter (cal$_{th}$/cm^2)	joule per square meter (J/m^2)	**4.184**	**E+04**
calorie$_{th}$ per square centimeter minute [cal$_{th}$/(cm^2 · min)]	watt per square meter (W/m^2)	6.973 333	E+02
calorie$_{th}$ per square centimeter second [cal$_{th}$/(cm^2 · s)]	watt per square meter (W/m^2)	**4.184**	**E+04**
candela per square inch (cd/in^2)	candela per square meter (cd/m^2)	1.550 003	E+03
carat, metric	kilogram (kg)	**2.0**	**E−04**
carat, metric	gram (g)	**2.0**	**E−01**
centimeter of mercury (0 °C)[13]	pascal (Pa)	1.333 22	E+03
centimeter of mercury (0 °C)[13]	kilopascal (kPa)	1.333 22	E+00
centimeter of mercury, conventional (cmHg)[13]	pascal (Pa)	1.333 224	E+03

---

[12] The kilogram calorie or "large calorie" is an obsolete term used for the kilocalorie, which is the calorie used to express the energy content of foods. However, in practice, the prefix "kilo" is usually omitted.

[13] Conversion factors for mercury manometer pressure units are calculated using the standard value for the acceleration of gravity and the density of mercury at the stated temperature. Additional digits are not justified because the definitions of the units do not take into account the compressibility of mercury or the change in density caused by the revised practical temperature scale, ITS-90. Similar comments also apply to water manometer pressure units. Conversion factors for conventional mercury and water manometer pressure units are based on ISO 31-3.

To convert from	to	Multiply by	
centimeter of mercury, conventional (cmHg)[13]	kilopascal (kPa)	1.333 224	E+00
centimeter of water (4 °C)[13]	pascal (Pa)	9.806 38	E+01
centimeter of water, conventional (cmH$_2$O)[13]	pascal (Pa)	**9.806 65**	**E+01**
centipoise (cP)	pascal second (Pa · s)	**1.0**	**E−03**
centistokes (cSt)	meter squared per second (m^2/s)	**1.0**	**E−06**
chain (based on U.S survey foot) (ch)[9]	meter (m)	2.011 684	E+01
circular mil	square meter (m^2)	5.067 075	E−10
circular mil	square millimeter (mm^2)	5.067 075	E−04
clo	square meter kelvin per watt (m^2 · K/W)	1.55	E−01
cord (128 ft^3)	cubic meter (m^3)	3.624 556	E+00
cubic foot (ft^3)	cubic meter (m^3)	2.831 685	E−02
cubic foot per minute (ft^3/min)	cubic meter per second (m^3/s)	4.719 474	E−04
cubic foot per minute (ft^3/min)	liter per second (L/s)	4.719 474	E−01
cubic foot per second (ft^3/s)	cubic meter per second (m^3/s)	2.831 685	E−02
cubic inch (in^3)[14]	cubic meter (m^3)	1.638 706	E−05
cubic inch per minute (in^3/min)	cubic meter per second (m^3/s)	2.731 177	E−07
cubic mile (mi^3)	cubic meter (m^3)	4.168 182	E+09
cubic yard (yd^3)	cubic meter (m^3)	7.645 549	E−01
cubic yard per minute (yd^3/min)	cubic meter per second (m^3/s)	1.274 258	E−02
cup (U.S.)	cubic meter (m^3)	2.365 882	E−04
cup (U.S.)	liter (L)	2.365 882	E−01
cup (U.S.)	milliliter (mL)	2.365 882	E+02
curie (Ci)	becquerel (Bq)	**3.7**	**E+10**
darcy[15]	meter squared (m^2)	9.869 233	E−13
day (d)	second (s)	**8.64**	**E+04**
day (sidereal)	second (s)	8.616 409	E+04
debye (D)	coulomb meter (C · m)	3.335 641	E−30
degree (angle) (°)	radian (rad)	1.745 329	E−02
degree Celsius (temperature) (°C)	kelvin (K)	$T/\text{K} = t\,/\,°\text{C} + \mathbf{273.15}$	
degree Celsius (temperature interval) (°C)	kelvin (K)	**1.0**	**E+00**
degree centigrade (temperature)[16]	degree Celsius (°C)	$t\,/\,°\text{C} \approx t\,/\,\text{deg.cent.}$	
degree centigrade (temperature interval)[16]	degree Celsius (°C)	1.0	E+00
degree Fahrenheit (temperature) (°F)	degree Celsius (°C)	$t\,/\,°\text{C} = (t\,/\,°\text{F} - \mathbf{32})/\mathbf{1.8}$	
degree Fahrenheit (temperature) (°F)	kelvin (K)	$T/\text{K} = (t\,/\,°\text{F} + \mathbf{459.67})/\mathbf{1.8}$	
degree Fahrenheit (temperature interval) (°F)	degree Celsius (°C)	5.555 556	E−01
degree Fahrenheit (temperature interval) (°F)	kelvin (K)	5.555 556	E−01
degree Fahrenheit hour per British thermal unit$_{\text{IT}}$ (°F · h/Btu$_{\text{IT}}$)	kelvin per watt (K/W)	1.895 634	E+00
degree Fahrenheit hour per British thermal unit$_{\text{th}}$ (°F · h/Btu$_{\text{th}}$)	kelvin per watt (K/W)	1.896 903	E+00
degree Fahrenheit hour square foot per British thermal unit$_{\text{IT}}$ (°F · h · ft^2/Btu$_{\text{IT}}$)	square meter kelvin per watt (m^2 · K/W)	1.761 102	E−01
degree Fahrenheit hour square foot per British thermal unit$_{\text{th}}$ (°F · h · ft^2/Btu$_{\text{th}}$)	square meter kelvin per watt (m^2 · K/W)	1.762 280	E−01
degree Fahrenheit hour square foot per British thermal unit$_{\text{IT}}$ inch [°F · h · ft^2/(Btu$_{\text{IT}}$ · in)]	meter kelvin per watt (m · K/W)	6.933 472	E+00
degree Fahrenheit hour square foot per British thermal unit$_{\text{th}}$ inch [°F · h · ft^2/(Btu$_{\text{th}}$ · in)]	meter kelvin per watt (m · K/W)	6.938 112	E+00
degree Fahrenheit second per British thermal unit$_{\text{IT}}$ (°F · s /Btu$_{\text{IT}}$)	kelvin per watt (K/W)	5.265 651	E−04
degree Fahrenheit second per British thermal unit$_{\text{th}}$ (°F · s /Btu$_{\text{th}}$)	kelvin per watt (K/W)	5.269 175	E−04
degree Rankine (°R)	kelvin (K)	$T/\text{K} = (T/\,°\text{R})/\mathbf{1.8}$	
degree Rankine (temperature interval) (°R)	kelvin (K)	5.555 556	E−01
denier	kilogram per meter (kg/m)	1.111 111	E−07
denier	gram per meter (g/m)	1.111 111	E−04
dyne (dyn)	newton (N)	**1.0**	**E−05**
dyne centimeter (dyn · cm)	newton meter (N · m)	**1.0**	**E−07**
dyne per square centimeter (dyn/cm^2)	pascal (Pa)	**1.0**	**E−01**

---

[14] The exact conversion factor is 1.638 706 4 E−05.

[15] The darcy is a unit for expressing the permeability of porous solids, not area.

[16] The centigrade temperature scale is obsolete; the degree centigrade is only approximately equal to the degree Celsius.

To convert from	to	Multiply by	
electronvolt (eV)	joule (J)	1.602 177	E−19
EMU of capacitance (abfarad)	farad (F)	**1.0**	**E+09**
EMU of current (abampere)	ampere (A)	**1.0**	**E+01**
EMU of electric potential (abvolt)	volt (V)	**1.0**	**E−08**
EMU of inductance (abhenry)	henry (H)	**1.0**	**E−09**
EMU of resistance (abohm)	ohm (Ω)	**1.0**	**E−09**
erg (erg)	joule (J)	**1.0**	**E−07**
erg per second (erg/s)	watt (W)	**1.0**	**E−07**
erg per square centimeter second [1obrkt&1ru]/(cm² · s)]	watt per square meter (W/m²)	**1.0**	**E−03**
ESU of capacitance (statfarad)	farad (F)	1.112 650	E−12
ESU of current (statampere)	ampere (A)	3.335 641	E−10
ESU of electric potential (statvolt)	volt (V)	2.997 925	E+02
ESU of inductance (stathenry)	henry (H)	8.987 552	E+11
ESU of resistance (statohm)	ohm (Ω)	8.987 552	E+11
faraday (based on carbon 12)	coulomb (C)	9.648 531	E+04
fathom (based on U.S survey foot)[9]	meter (m)	1.828 804	E+00
fermi	meter (m)	**1.0**	**E−15**
fermi	femtometer (fm)	**1.0**	**E+00**
fluid ounce (U.S.) (fl oz)	cubic meter (m³)	2.957 353	E−05
fluid ounce (U.S.) (fl oz)	milliliter (mL)	2.957 353	E+01
foot (ft)	meter (m)	**3.048**	**E−01**
foot (U.S survey ft)[9]	meter (m)	3.048 006	E−01
footcandle	lux (lx)	1.076 391	E+01
footlambert	candela per square meter (cd/m²)	3.426 259	E+00
foot of mercury, conventional (ftHg)[13]	pascal (Pa)	4.063 666	E+04
foot of mercury, conventional (ftHg)[13]	kilopascal (kPa)	4.063 666	E+01
foot of water (39.2 °F)[13]	pascal (Pa)	2.988 98	E+03
foot of water (39.2 °F)[13]	kilopascal (kPa)	2.988 98	E+00
foot of water, conventional (ftH₂O)[13]	pascal (Pa)	2.989 067	E+03
foot of water, conventional (ftH₂O)[13]	kilopascal (kPa)	2.989 067	E+00
foot per hour (ft/h)	meter per second (m/s)	8.466 667	E−05
foot per minute (ft/min)	meter per second (m/s)	**5.08**	**E−03**
foot per second (ft/s)	meter per second (m/s)	**3.048**	**E−01**
foot per second squared (ft/s²)	meter per second squared (m/s²)	**3.048**	**E−01**
foot poundal	joule (J)	4.214 011	E−02
foot pound-force (ft · lbf)	joule (J)	1.355 818	E+00
foot pound-force per hour (ft · lbf/h)	watt (W)	3.766 161	E−04
foot pound-force per minute (ft · lbf/min)	watt (W)	2.259 697	E−02
foot pound-force per second (ft · lbf/s)	watt (W)	1.355 818	E+00
foot to the fourth power (ft⁴)[17]	meter to the fourth power (m⁴)	8.630 975	E−03
franklin (Fr)	coulomb (C)	3.335 641	E−10
gal (Gal)	meter per second squared (m/s²)	**1.0**	**E−02**
gallon [Canadian and U.K (Imperial)] (gal)	cubic meter (m³)	**4.546 09**	**E−03**
gallon [Canadian and U.K (Imperial)] (gal)	liter (L)	**4.546 09**	**E+00**
gallon (U.S.) (gal)	cubic meter (m³)	3.785 412	E−03
gallon (U.S.) (gal)	liter (L)	3.785 412	E+00
gallon (U.S.) per day (gal/d)	cubic meter per second (m³/s)	4.381 264	E−08
gallon (U.S.) per day (gal/d)	liter per second (L/s)	4.381 264	E−05
gallon (U.S.) per horsepower hour [gal / (hp · h)]	cubic meter per joule (m³/J)	1.410 089	E−09
gallon (U.S.) per horsepower hour [gal / (hp · h)]	liter per joule (L/ J)	1.410 089	E−06
gallon (U.S.) per minute (gpm)(gal/min)	cubic meter per second (m³/s)	6.309 020	E−05
gallon (U.S.) per minute (gpm)(gal/min)	liter per second (L/ s)	6.309 020	E−02
gamma (γ)	tesla (T)	**1.0**	**E−09**
gauss (Gs, G)	tesla (T)	**1.0**	**E−04**
gilbert (Gi)	ampere (A)	7.957 747	E−01

[17] This is a unit for the quantity second moment of area, which is sometimes called the "moment of section" or "area moment of inertia" of a plane section about a specified axis.

To convert from	to	Multiply by	
gill [Canadian and U.K (Imperial)] (gi)	cubic meter (m³)	1.420 653	E−04
gill [Canadian and U.K (Imperial)] (gi)	liter (L)	1.420 653	E−01
gill (U.S.) (gi)	cubic meter (m³)	1.182 941	E−04
gill (U.S.) (gi)	liter (L)	1.182 941	E−01
gon (also called grade) (gon)	radian (rad)	1.570 796	E−02
gon (also called grade) (gon)	degree (angle) (°)	**9.0**	**E−01**
grain (gr)	kilogram (kg)	**6.479 891**	**E−05**
grain (gr)	milligram (mg)	**6.479 891**	**E+01**
grain per gallon (U.S.) (gr/gal)	kilogram per cubic meter (kg/m³)	1.711 806	E−02
grain per gallon (U.S.) (gr/gal)	milligram per liter (mg/L)	1.711 806	E+01
gram-force per square centimeter (gf /cm²)	pascal (Pa)	**9.806 65**	**E+01**
gram per cubic centimeter (g/cm³)	kilogram per cubic meter (kg/m³)	**1.0**	**E+03**
hectare (ha)	square meter (m²)	**1.0**	**E+04**
horsepower (550 ft · lbf/s) (hp)	watt (W)	7.456 999	E+02
horsepower (boiler)	watt (W)	9.809 50	E+03
horsepower (electric)	watt (W)	**7.46**	**E+02**
horsepower (metric)	watt (W)	7.354 988	E+02
horsepower (U.K.)	watt (W)	7.4570	E+02
horsepower (water)	watt (W)	7.460 43	E+02
hour (h)	second (s)	**3.6**	**E+03**
hour (sidereal)	second (s)	3.590 170	E+03
hundredweight (long, 112 lb)	kilogram (kg)	5.080 235	E+01
hundredweight (short, 100 lb)	kilogram (kg)	4.535 924	E+01
inch (in)	meter (m)	**2.54**	**E−02**
inch (in)	centimeter (cm)	**2.54**	**E+00**
inch of mercury (32 °F)[13]	pascal (Pa)	3.386 38	E+03
inch of mercury (32 °F)[13]	kilopascal (kPa)	3.386 38	E+00
inch of mercury (60 °F)[13]	pascal (Pa)	3.376 85	E+03
inch of mercury (60 °F)[13]	kilopascal (kPa)	3.376 85	E+00
inch of mercury, conventional (inHg)[13]	pascal (Pa)	3.386 389	E+03
inch of mercury, conventional (inHg)[13]	kilopascal (kPa)	3.386 389	E+00
inch of water (39.2 °F)[13]	pascal (Pa)	2.490 82	E+02
inch of water (60 °F)[13]	pascal (Pa)	2.4884	E+02
inch of water, conventional (inH₂O)[13]	pascal (Pa)	2.490 889	E+02
inch per second (in/s)	meter per second (m/s)	**2.54**	**E−02**
inch per second squared (in/s²)	meter per second squared (m/s²)	**2.54**	**E−02**
inch to the fourth power (in⁴)[17]	meter to the fourth power (m⁴)	4.162 314	E−07
kayser (K)	reciprocal meter (m⁻¹)	**1.0**	**E+02**
kelvin (K)	degree Celsius (°C)	$t / °C = T / K -$	**273.15**
kilocalorie$_{IT}$ (kcal$_{IT}$)	joule (J)	**4.1868**	**E+03**
kilocalorie$_{th}$ (kcal$_{th}$)	joule (J)	**4.184**	**E+03**
kilocalorie (mean) (kcal)	joule (J)	4.190 02	E+03
kilocalorie$_{th}$ per minute (kcal$_{th}$ /min)	watt (W)	6.973 333	E+01
kilocalorie$_{th}$ per second (kcal$_{th}$ /s)	watt (W)	**4.184**	**E+03**
kilogram-force (kgf)	newton (N)	**9.806 65**	**E+00**
kilogram-force meter (kgf · m)	newton meter (N · m)	**9.806 65**	**E+00**
kilogram-force per square centimeter (kgf/cm²)	pascal (Pa)	**9.806 65**	**E+04**
kilogram-force per square centimeter (kgf/cm²)	kilopascal (kPa)	**9.806 65**	**E+01**
kilogram-force per square meter (kgf/m²)	pascal (Pa)	**9.806 65**	**E+00**
kilogram-force per square millimeter (kgf/mm²)	pascal (Pa)	**9.806 65**	**E+06**
kilogram-force per square millimeter (kgf/mm²)	megapascal (MPa)	**9.806 65**	**E+00**
kilogram-force second squared per meter (kgf · s²/m)	kilogram (kg)	**9.806 65**	**E+00**
kilometer per hour (km/h)	meter per second (m/s)	2.777 778	E−01
kilopond (kilogram-force) (kp)	newton (N)	**9.806 65**	**E+00**
kilowatt hour (kW · h)	joule (J)	**3.6**	**E+06**
kilowatt hour (kW · h)	megajoule (MJ)	**3.6**	**E+00**

To convert from	to	Multiply by	
kip (1 kip=1000 lbf)	newton (N)	4.448 222	E+03
kip (1 kip=1000 lbf)	kilonewton (kN)	4.448 222	E+00
kip per square inch (ksi) (kip/in²)	pascal (Pa)	6.894 757	E+06
kip per square inch (ksi) (kip/in²)	kilopascal (kPa)	6.894 757	E+03
knot (nautical mile per hour)	meter per second (m/s)	5.144 444	E−01
lambert[18]	candela per square meter (cd/m²)	3.183 099	E+03
langley (cal$_{th}$/cm²)	joule per square meter (J/m²)	**4.184**	**E+04**
light year (l.y.)[19]	meter (m)	9.460 73	E+15
liter (L)[20]	cubic meter (m³)	1.0	E−03
lumen per square foot (lm/ft²)	lux (lx)	1.076 391	E+01
maxwell (Mx)	weber (Wb)	**1.0**	**E−08**
mho	siemens (S)	**1.0**	**E+00**
microinch	meter (m)	**2.54**	**E−08**
microinch	micrometer (μm)	**2.54**	**E−02**
micron (μ)	meter (m)	**1.0**	**E−06**
micron (μ)	micrometer (μm)	**1.0**	**E+00**
mil (0.001 in)	meter (m)	**2.54**	**E−05**
mil (0.001 in)	millimeter (mm)	**2.54**	**E−02**
mil (angle)	radian (rad)	9.817 477	E−04
mil (angle)	degree (°)	**5.625**	**E−02**
mile (mi)	meter (m)	**1.609 344**	**E+03**
mile (mi)	kilometer (km)	**1.609 344**	**E+00**
mile (based on U.S survey foot) (mi)[9]	meter (m)	1.609 347	E+03
mile (based on U.S survey foot) (mi)[9]	kilometer (km)	1.609 347	E+00
mile, nautical [21]	meter (m)	**1.852**	**E+03**
mile per gallon (U.S.) (mpg) (mi/gal)	meter per cubic meter (m/m³)	4.251 437	E+05
mile per gallon (U.S.) (mpg) (mi/gal)	kilometer per liter (km/L)	4.251 437	E−01
mile per gallon (U.S.) (mpg) (mi/gal)[22]	liter per 100 kilometer (L/100 km)	divide 235.215 by number of miles per gallon	
mile per hour (mi/h)	meter per second (m/s)	**4.4704**	**E−01**
mile per hour (mi/h)	kilometer per hour (km/h)	**1.609 344**	**E+00**
mile per minute (mi/min)	meter per second (m/s)	**2.682 24**	**E+01**
mile per second (mi/s)	meter per second (m/s)	**1.609 344**	**E+03**
millibar (mbar)	pascal (Pa)	**1.0**	**E+02**
millibar (mbar)	kilopascal (kPa)	**1.0**	**E−01**
millimeter of mercury, conventional (mmHg)[13]	pascal (Pa)	1.333 224	E+02
millimeter of water, conventional (mmH$_2$O)[13]	pascal (Pa)	**9.806 65**	**E+00**
minute (angle) (')	radian (rad)	2.908 882	E−04
minute (min)	second (s)	**6.0**	**E+01**
minute (sidereal)	second (s)	5.983 617	E+01
oersted (Oe)	ampere per meter (A/m)	7.957 747	E+01
ohm centimeter (Ω · cm)	ohm meter (Ω · m)	**1.0**	**E−02**
ohm circular-mil per foot	ohm meter (Ω · m)	1.662 426	E−09
ohm circular-mil per foot	ohm square millimeter per meter (Ω · mm²/m)	1.662 426	E−03
ounce (avoirdupois) (oz)	kilogram (kg)	2.834 952	E−02
ounce (avoirdupois) (oz)	gram (g)	2.834 952	E+01
ounce (troy or apothecary) (oz)	kilogram (kg)	3.110 348	E−02
ounce (troy or apothecary) (oz)	gram (g)	3.110 348	E+01
ounce [Canadian and U.K fluid (Imperial)] (fl oz)	cubic meter (m³)	2.841 306	E−05
ounce [Canadian and U.K fluid (Imperial)] (fl oz)	milliliter (mL)	2.841 306	E+01

[18] The exact conversion factor is $10^4/\pi$.

[19] This conversion factor is based on 1 d = 86 400 s; and 1 Julian century = 36 525 d. (See *The Astronomical Almanac for the Year 1995*, page K6, U.S. Government Printing Office, Washington, DC, 1994).

[20] In 1964 the General Conference on Weights and Measures reestablished the name "liter" as a special name for the cubic decimeter. Between 1901 and 1964 the liter was slightly larger (1.000 028 dm³); when one uses high-accuracy volume data of that time, this fact must be kept in mind.

[21] The value of this unit, 1 nautical mile = 1852 m, was adopted by the First International Extraordinary Hydrographic Conference, Monaco, 1929, under the name "International nautical mile."

[22] For converting fuel economy, as used in the U.S., to fuel consumption.

To convert from	to	Multiply by	
ounce (U.S. fluid) (fl oz)	cubic meter (m^3)	2.957 353	E−05
ounce (U.S fluid) (fl oz)	milliliter (mL)	2.957 353	E+01
ounce (avoirdupois)-force (ozf)	newton (N)	2.780 139	E−01
ounce (avoirdupois)-force inch (ozf · in)	newton meter (N · m)	7.061 552	E−03
ounce (avoirdupois)-force inch (ozf · in)	millinewton meter (mN · m)	7.061 552	E+00
ounce (avoirdupois) per cubic inch (oz/ in^3)	kilogram per cubic meter (kg/m^3)	1.729 994	E+03
ounce (avoirdupois) per gallon [Canadian and U.K (Imperial)] (oz/gal)	kilogram per cubic meter (kg/m^3)	6.236 023	E+00
ounce (avoirdupois) per gallon [Canadian and U.K (Imperial)] (oz/gal)	gram per liter (g/L)	6.236 023	E+00
ounce (avoirdupois) per gallon (U.S.)(oz/gal)	kilogram per cubic meter (kg/m^3)	7.489 152	E+00
ounce (avoirdupois) per gallon (U.S.)(oz/gal)	gram per liter (g/L)	7.489 152	E+00
ounce (avoirdupois) per square foot (oz/ ft^2)	kilogram per square meter (kg/m^2)	3.051 517	E−01
ounce (avoirdupois) per square inch (oz/ in^2)	kilogram per square meter (kg/m^2)	4.394 185	E+01
ounce (avoirdupois) per square yard(oz/yd^2)	kilogram per square meter (kg/m^2)	3.390 575	E−02
parsec (pc)	meter (m)	3.085 678	E+16
peck (U.S.) (pk)	cubic meter (m^3)	8.809 768	E−03
peck (U.S.) (pk)	liter (L)	8.809 768	E+00
pennyweight (dwt)	kilogram (kg)	1.555 174	E−03
pennyweight (dwt)	gram (g)	1.555 174	E+00
perm (0 °C)	kilogram per pascal second square meter [kg/(Pa · s · m^2)]	5.721 35	E−11
perm (23 °C)	kilogram per pascal second square meter [kg/(Pa · s · m^2)]	5.745 25	E−11
perm inch (0 °C)	kilogram per pascal second meter [kg/(Pa · s · m)]	1.453 22	E−12
perm inch (23 °C)	kilogram per pascal second meter [kg/(Pa · s · m)]	1.459 29	E−12
phot (ph)	lux (lx)	**1.0**	**E+04**
pica (computer) (1/6 in)	meter (m)	4.233 333	E−03
pica (computer) (1/6 in)	millimeter (mm)	4.233 333	E+00
pica (printer's)	meter (m)	4.217 518	E−03
pica (printer's)	millimeter (mm)	4.217 518	E+00
pint (U.S dry) (dry pt)	cubic meter (m^3)	5.506 105	E−04
pint (U.S dry) (dry pt)	liter (L)	5.506 105	E−01
pint (U.S liquid) (liq pt)	cubic meter (m^3)	4.731 765	E−04
pint (U.S liquid) (liq pt)	liter (L)	4.731 765	E−01
point (computer) (1/72 in)	meter (m)	3.527 778	E−04
point (computer) (1/72 in)	millimeter (mm)	3.527 778	E−01
point (printer's)	meter (m)	3.514 598	E−04
point (printer's)	millimeter (mm)	3.514 598	E−01
poise (P)	pascal second (Pa · s)	**1.0**	**E−01**
pound (avoirdupois) (lb)[23]	kilogram (kg)	4.535 924	E−01
pound (troy or apothecary) (lb)	kilogram (kg)	3.732 417	E−01
poundal	newton (N)	1.382 550	E−01
poundal per square foot	pascal (Pa)	1.488 164	E+00
poundal second per square foot	pascal second (Pa · s)	1.488 164	E+00
pound foot squared (lb · ft^2)	kilogram meter squared (kg · m^2)	4.214 011	E−02
pound-force (lbf)[24]	newton (N)	4.448 222	E+00
pound-force foot (lbf · ft)	newton meter (N · m)	1.355 818	E+00
pound-force foot per inch (lbf · ft/in)	newton meter per meter (N · m/m)	5.337 866	E+01
pound-force inch (lbf · in)	newton meter (N · m)	1.129 848	E−01
pound-force inch per inch (lbf · in/in)	newton meter per meter (N · m/m)	4.448 222	E+00
pound-force per foot (lbf/ft)	newton per meter (N/m)	1.459 390	E+01
pound-force per inch (lbf/in)	newton per meter (N/m)	1.751 268	E+02
pound-force per pound (lbf/lb) (thrust to mass ratio)	newton per kilogram (N/kg)	**9.806 65**	**E+00**
pound-force per square foot (lbf/ft^2)	pascal (Pa)	4.788 026	E+01
pound-force per square inch (psi) (lbf/in^2)	pascal (Pa)	6.894 757	E+03

[23] The exact conversion factor is 4.535 923 7 E−01. All units that contain the pound refer to the avoirdupois pound unless otherwise specified.

[24] If the local value of the acceleration of free fall is taken as g_n=9.806 65 m/ s^2 (the standard value), the exact conversion factor is 4.448 221 615 260 5 E+00.

To convert from	to	Multiply by	
pound-force per square inch (psi) (lbf/in²)	kilopascal (kPa)	6.894 757	E+00
pound-force second per square foot (lbf · s/ft²)	pascal second (Pa · s)	4.788 026	E+01
pound-force second per square inch (lbf · s/in²)	pascal second (Pa · s)	6.894 757	E+03
pound inch squared (lb · in²)	kilogram meter squared (kg · m²)	2.926 397	E−04
pound per cubic foot (lb/ft³)	kilogram per cubic meter (kg/m³)	1.601 846	E+01
pound per cubic inch (lb/in³)	kilogram per cubic meter (kg/m³)	2.767 990	E+04
pound per cubic yard (lb/yd³)	kilogram per cubic meter (kg/m³)	5.932 764	E−01
pound per foot (lb/ft)	kilogram per meter (kg/m)	1.488 164	E+00
pound per foot hour [lb/(ft · h)]	pascal second (Pa · s)	4.133 789	E−04
pound per foot second [lb/(ft · s)]	pascal second (Pa · s)	1.488 164	E+00
pound per gallon [Canadian and U.K (Imperial)] (lb/gal)	kilogram per cubic meter (kg/m³)	9.977 637	E+01
pound per gallon [Canadian and U.K (Imperial)] (lb/gal)	kilogram per liter (kg/L)	9.977 637	E−02
pound per gallon (U.S.) (lb/gal)	kilogram per cubic meter (kg/m³)	1.198 264	E+02
pound per gallon (U.S.) (lb/gal)	kilogram per liter (kg/L)	1.198 264	E−01
pound per horsepower hour [lb/(hp · h)]	kilogram per joule (kg/J)	1.689 659	E−07
pound per hour (lb/h)	kilogram per second (kg/s)	1.259 979	E−04
pound per inch (lb/in)	kilogram per meter (kg/m)	1.785 797	E+01
pound per minute (lb/min)	kilogram per second (kg/s)	7.559 873	E−03
pound per second (lb/s)	kilogram per second (kg/s)	4.535 924	E−01
pound per square foot (lb/ft²)	kilogram per square meter (kg/m²)	4.882 428	E+00
pound per square inch (*not* pound-force) (lb/in²)	kilogram per square meter (kg/m²)	7.030 696	E+02
pound per yard (lb/yd)	kilogram per meter (kg/m)	4.960 546	E−01
psi (pound-force per square inch) (lbf/in²)	pascal (Pa)	6.894 757	E+03
psi (pound-force per square inch) (lbf/in²)	kilopascal (kPa)	6.894 757	E+00
quad (10¹⁵ Btu$_{IT}$)¹¹	joule (J)	1.055 056	E+18
quart (U.S dry) (dry qt)	cubic meter (m³)	1.101 221	E−03
quart (U.S dry) (dry qt)	liter (L)	1.101 221	E+00
quart (U.S liquid) (liq qt)	cubic meter (m³)	9.463 529	E−04
quart (U.S liquid) (liq qt)	liter (L)	9.463 529	E−01
rad (absorbed dose) (rad)	gray (Gy)	**1.0**	**E−02**
rem (rem)	sievert (Sv)	**1.0**	**E−02**
revolution (r)	radian (rad)	6.283 185	E+00
revolution per minute (rpm) (r/min)	radian per second (rad/s)	1.047 198	E−01
rhe	reciprocal pascal second [(Pa · s)⁻¹]	**1.0**	**E+01**
rod (based on U.S survey foot) (rd)⁹	meter (m)	5.029 210	E+00
roentgen (R)	coulomb per kilogram (C/kg)	**2.58**	**E−04**
rpm (revolution per minute) (r/min)	radian per second (rad/s)	1.047 198	E−01
second (angle) (")	radian (rad)	4.848 137	E−06
second (sidereal)	second (s)	9.972 696	E−01
shake	second (s)	**1.0**	**E−08**
shake	nanosecond (ns)	**1.0**	**E+01**
slug (slug)	kilogram (kg)	1.459 390	E+01
slug per cubic foot (slug/ft³)	kilogram per cubic meter (kg/m³)	5.153 788	E+02
slug per foot second [slug/(ft · s)]	pascal second (Pa · s)	4.788 026	E+01
square foot (ft²)	square meter (m²)	**9.290 304**	**E−02**
square foot per hour (ft²/h)	square meter per second (m²/s)	**2.580 64**	**E−05**
square foot per second (ft²/s)	square meter per second (m²/s)	**9.290 304**	**E−02**
square inch (in²)	square meter (m²)	**6.4516**	**E−04**
square inch (in²)	square centimeter (cm²)	**6.4516**	**E+00**
square mile (mi²)	square meter (m²)	2.589 988	E+06
square mile (mi²)	square kilometer (km²)	2.589 988	E+00
square mile (based on U.S survey foot) (mi²)⁹	square meter (m²)	2.589 998	E+06
square mile (based on U.S survey foot) (mi²)⁹	square kilometer (km²)	2.589 998	E+00

To convert from	to	Multiply by	
square yard (yd²)	square meter (m²)	8.361 274	E−01
statampere	ampere (A)	3.335 641	E−10
statcoulomb	coulomb (C)	3.335 641	E−10
statfarad	farad (F)	1.112 650	E−12
stathenry	henry (H)	8.987 552	E+11
statmho	siemens (S)	1.112 650	E−12
statohm	ohm (Ω)	8.987 552	E+11
statvolt	volt (V)	2.997 925	E+02
stere (st)	cubic meter (m³)	**1.0**	**E+00**
stilb (sb)	candela per square meter (cd/m²)	**1.0**	**E+04**
stokes (St)	meter squared per second (m²/s)	**1.0**	**E−04**
tablespoon	cubic meter (m³)	1.478 676	E−05
tablespoon	milliliter (mL)	1.478 676	E+01
teaspoon	cubic meter (m³)	4.928 922	E−06
teaspoon	milliliter (mL)	4.928 922	E+00
tex	kilogram per meter (kg/m)	**1.0**	**E−06**
therm (EC)[25]	joule (J)	**1.055 06**	**E+08**
therm (U.S.)[25]	joule (J)	**1.054 804**	**E+08**
ton, assay (AT)	kilogram (kg)	2.916 667	E−02
ton, assay (AT)	gram (g)	2.916 667	E+01
ton-force (2000 lbf)	newton (N)	8.896 443	E+03
ton-force (2000 lbf)	kilonewton (kN)	8.896 443	E+00
ton, long (2240 lb)	kilogram (kg)	1.016 047	E+03
ton, long, per cubic yard	kilogram per cubic meter (kg/m³)	1.328 939	E+03
ton, metric (t)	kilogram (kg)	**1.0**	**E+03**
tonne (called "metric ton" in U.S.) (t)	kilogram (kg)	**1.0**	**E+03**
ton of refrigeration (12 000 Btu$_{IT}$/h)	watt (W)	3.516 853	E+03
ton of TNT (energy equivalent)[26]	joule (J)	**4.184**	**E+09**
ton, register	cubic meter (m³)	2.831 685	E+00
ton, short (2000 lb)	kilogram (kg)	9.071 847	E+02
ton, short, per cubic yard	kilogram per cubic meter (kg/m³)	1.186 553	E+03
ton, short, per hour	kilogram per second (kg/s)	2.519 958	E−01
torr (Torr)	pascal (Pa)	1.333 224	E+02
unit pole	weber (Wb)	1.256 637	E−07
watt hour (W · h)	joule (J)	**3.6**	**E+03**
watt per square centimeter (W/cm²)	watt per square meter (W/m²)	**1.0**	**E+04**
watt per square inch (W/in²)	watt per square meter (W/m²)	1.550 003	E+03
watt second (W · s)	joule (J)	**1.0**	**E+00**
yard (yd)	meter (m)	**9.144**	**E−01**
year (365 days)	second (s)	**3.1536**	**E+07**
year (sidereal)	second (s)	3.155 815	E+07
year (tropical)	second (s)	3.155 693	E+07

---

[25] The therm (EC) is legally defined in the Council Directive of 20 December 1979, Council of the European Communities (now the European Union, EU). The therm (U.S.) is legally defined in the Federal Register of July 27, 1968. Although the therm (EC), which is based on the International Table Btu, is frequently used by engineers in the United States, the therm (U.S.) is the legal unit used by the U.S natural gas industry.

[26] Defined (not measured) value.

CONVERSION OF TEMPERATURES

From	To	
Celsius	Fahrenheit	$t_F/°F = (9/5)\ t/°C + 32$
	Kelvin	$T/K = t/°C + 273.15$
	Rankine	$T/°R = (9/5)\ (t/°C + 273.15)$
Fahrenheit	Celsius	$t/°C = (5/9)\ [(t_F/°F) - 32]$
	Kelvin	$T/K = (5/9)\ [(t_F/°F) - 32] + 273.15$
	Rankine	$T/°R = t_F/°F + 459.67$
Kelvin	Celsius	$t/°C = T/K - 273.15$
	Rankine	$T/°R = (9/5)\ T/K$
Rankine	Fahrenheit	$t_F/°F = T/°R - 459.67$
	Kelvin	$T/K = (5/9)\ T/°R$

Definition of symbols:
T = thermodynamic (absolute) temperature
t = Celsius temperature (the symbol θ is also used for Celsius temperature)
t_F = Fahrenheit temperature

Designation of Large Numbers

	U.S.A.	Other countries
10^6	million	million
10^9	billion	milliard
10^{12}	trillion	billion
10^{15}	quadrillion	billiard
10^{18}	quintillion	trillion
100^{100}	googol	
10^{googol}	googolplex	

CONVERSION FACTORS FOR ENERGY UNITS

If greater accuracy is required, use the *Energy Equivalents* section of the *Fundamental Physical Constants* table.

		Wavenumber $\bar{\nu}$ cm^{-1}	Frequency ν MHz	Energy E aJ	Energy E eV	Energy E E_h	Molar energy E_m kJ/mol	Molar energy E_m kcal/mol	Temperature T K
$\bar{\nu}$: 1 cm^{-1}	$\doteq$ 1		2.997925×10^4	1.986447×10^{-5}	1.239842×10^{-4}	4.556335×10^{-6}	11.96266×10^{-3}	2.85914×10^{-3}	1.438769
ν: 1 MHz	$\doteq$ 3.33564×10^{-5}		1	6.626076×10^{-10}	4.135669×10^{-9}	1.519830×10^{-10}	3.990313×10^{-7}	9.53708×10^{-8}	4.79922×10^{-5}
1 aJ	$\doteq$ 50341.1		1.509189×10^9	1	6.241506	0.2293710	602.2137	143.9325	7.24292×10^4
E: 1 eV	$\doteq$ 8065.54		2.417988×10^8	0.1602177	1	3.674931×10^{-2}	96.4853	23.0605	1.16045×10^4
E_h	$\doteq$ 219474.63		6.579684×10^9	4.359748	27.2114	1	2625.500	627.510	3.15773×10^5
E_m: 1 kJ/mol	$\doteq$ 83.5935		2.506069×10^6	1.660540×10^{-3}	1.036427×10^{-2}	3.808798×10^{-4}	1	0.239006	120.272
1 kcal/mol	$\doteq$ 349.755		1.048539×10^7	6.947700×10^{-3}	4.336411×10^{-2}	1.593601×10^{-3}	4.184	1	503.217
T: 1 K	$\doteq$ 0.695039		2.08367×10^4	1.380658×10^{-5}	8.61738×10^{-5}	3.16683×10^{-6}	8.31451×10^{-3}	1.98722×10^{-3}	1

Examples of the use of this table:

$$1 \text{ aJ} \doteq 50341 \text{ cm}^{-1}$$
$$1 \text{ eV} \doteq 96.4853 \text{ kJ mol}^{-1}$$

The symbol $\doteq$ should be read as meaning corresponds to or is equivalent to.

$E = h\nu = hc\bar{\nu} = kT$; $E_m = N_A E$; E_h is the Hartree energy

CONVERSION FACTORS FOR PRESSURE UNITS

	Pa	kPa	MPa	bar	atmos	Torr	µmHg	psi
Pa	1	0.001	0.000001	0.00001	9.8692×10^{-6}	0.0075006	7.5006	0.0001450377
kPa	1000	1	0.001	0.01	0.0098692	7.5006	7500.6	0.1450377
MPa	1000000	1000	1	10	9.8692	7500.6	7500600	145.0377
bar	100000	100	0.1	1	0.98692	750.06	750060	14.50377
atmos	101325	101.325	0.101325	1.01325	1	760	760000	14.69594
Torr	133.322	0.133322	0.000133322	0.00133322	0.00131579	1	1000	0.01933672
µmHg	0.133322	0.000133322	1.33322×10^{-7}	1.33322×10^{-6}	1.31579×10^{-6}	0.001	1	1.933672×10^{-5}
psi	6894.757	6.894757	0.006894757	0.06894757	0.068046	51.7151	51715.1	1

To convert a pressure value from a unit in the left hand column to a new unit, multiply the value by the factor appearing in the column for the new unit. For example:

$$1 \text{ kPa} = 9.8692 \times 10^{-3} \text{ atmos}$$
$$1 \text{ Torr} = 1.33322 \times 10^{-4} \text{ MPa}$$

Notes: µmHg is often referred to as "micron"

Torr is essentially identical to mmHg

psi is an abbreviation for the unit pound–force per square inch

psia (as a term for a physical quantity) implies the true (absolute) pressure

psig implies the true pressure minus the local atmospheric pressure

CONVERSION FACTORS FOR THERMAL CONDUCTIVITY UNITS

MULTIPLY ↓ by appropriate factor to OBTAIN→	Btu_{IT} $h^{-1}ft^{-1}°F^{-1}$	Btu_{IT} in. $h^{-1}ft^{-2}°F^{-1}$	Btu_{th} $h^{-1}ft^{-1}°F^{-1}$	Btu_{th} in. $h^{-1}ft^{-2}°F^{-1}$	cal_{IT} $s^{-1}cm^{-1}°C^{-1}$	cal_{th} $s^{-1}cm^{-1}°C^{-1}$	$kcal_{th}$ $h^{-1}m^{-1}°C^{-1}$	J $s^{-1}cm^{-1}K^{-1}$	W $cm^{-1}K^{-1}$	W $m^{-1}K^{-1}$	mW $cm^{-1}K^{-1}$
Btu_{IT} $h^{-1}ft^{-1}°F^{-1}$	1	12	1.00067	12.0080	$4.13379×10^{-3}$	$4.13656×10^{-3}$	1.48916	$1.73073×10^{-2}$	$1.73073×10^{-2}$	1.73073	17.3073
Btu_{IT} in. $h^{-1}ft^{-2}°F^{-1}$	$8.33333×10^{-2}$	1	$8.33891×10^{-2}$	1.00067	$3.44482×10^{-4}$	$3.44713×10^{-4}$	0.124097	$1.44228×10^{-3}$	$1.44228×10^{-3}$	0.144228	1.44228
Btu_{th} $h^{-1}ft^{-1}°F^{-1}$	0.999331	11.9920	1	12	$4.13102×10^{-3}$	$4.13379×10^{-3}$	1.48816	$1.72958×10^{-2}$	$1.72958×10^{-2}$	1.72958	17.2958
Btu_{th} in. $h^{-1}ft^{-2}°F^{-1}$	$8.32776×10^{-2}$	0.999331	$8.33333×10^{-2}$	1	$3.44252×10^{-4}$	$3.44482×10^{-4}$	0.124014	$1.44131×10^{-3}$	$1.44131×10^{-3}$	0.144131	1.44131
cal_{IT} $s^{-1}cm^{-1}°C^{-1}$	$2.41909×10^{2}$	$2.90291×10^{3}$	$2.42071×10^{2}$	$2.90485×10^{3}$	1	1.00067	$3.60241×10^{2}$	4.1868	4.1868	$4.1868×10^{2}$	$4.1868×10^{3}$
cal_{th} $s^{-1}cm^{-1}°C^{-1}$	$2.41747×10^{2}$	$2.90096×10^{3}$	$2.41909×10^{2}$	$2.90291×10^{3}$	0.999331	1	$3.6×10^{2}$	4.184	4.184	$4.184×10^{2}$	$4.184×10^{3}$
$kcal_{th}$ $h^{-1}m^{-1}°C^{-1}$	0.671520	8.05824	0.671969	8.06363	$2.77592×10^{-3}$	$2.77778×10^{-3}$	1	$1.16222×10^{-2}$	$1.16222×10^{-2}$	1.16222	11.6222
J $s^{-1}cm^{-1}K^{-1}$	57.7789	$6.93347×10^{2}$	57.8176	$6.93811×10^{2}$	0.238846	0.239006	86.0421	1	1	$1×10^{2}$	$1×10^{3}$
W $cm^{-1}K^{-1}$	57.7789	$6.93347×10^{2}$	57.8176	$6.93811×10^{2}$	0.238846	0.239006	86.0421	1	1	$1×10^{2}$	$1×10^{3}$
W $m^{-1}K^{-1}$	0.577789	6.93347	0.578176	6.93811	$2.38846×10^{-3}$	$2.39006×10^{-3}$	0.860421	$1×10^{-2}$	$1×10^{-2}$	1	10
mW $cm^{-1}K^{-1}$	$5.77789×10^{-2}$	0.693347	$5.78176×10^{-2}$	0.693811	$2.38846×10^{-4}$	$2.39006×10^{-4}$	$8.60421×10^{-2}$	$1×10^{-3}$	$1×10^{-3}$	0.1	1

CONVERSION FACTORS FOR ELECTRICAL RESISTIVITY UNITS

To convert FROM ↓ multiply by appropriate factor to OBTAIN →	abΩ cm	μΩ cm	Ω cm	StatΩ cm	Ω m	Ω cir. mil ft^{-1}	Ω in.	Ω ft
abohm centimeter	1	1×10^{-3}	10^{-9}	1.113×10^{-21}	10^{-11}	6.015×10^{-3}	3.937×10^{-10}	3.281×10^{-11}
microohm centimeter	10^3	1	10^{-6}	1.113×10^{-18}	10^{-8}	6.015	3.937×10^{-7}	3.281×10^{-6}
ohm centimeter	10^8	10^6	1	1.113×10^{-12}	1×10^{-2}	6.015×10^6	3.937×10^{-1}	3.281×10^{-2}
statohm centimeter (esu)	8.987×10^{20}	8.987×10^{17}	8.987×10^{11}	1	8.987×10^9	5.406×10^{18}	3.538×10^{11}	2.949×10^{10}
ohm meter	10^{11}	10^8	10^2	1.113×10^{-10}	1	6.015×10^8	3.937×10^1	3.281
ohm circular mil per foot	1.662×10^2	1.662×10^{-1}	1.662×10^{-7}	1.850×10^{-19}	1.662×10^{-9}	1	6.54×10^{-6}	5.45×10^{-9}
ohm inch	2.54×10^9	2.54×10^6	2.54	2.827×10^{-12}	2.54×10^{-2}	1.528×10^7	1	8.3×10^{-2}
ohm foot	3.048×10^{10}	3.048×10^7	3.048×10^{-1}	3.3924×10^{-11}	3.048×10^{-1}	1.833×10^8	12	1

CONVERSION FACTORS FOR CHEMICAL KINETICS

Equivalent Second Order Rate Constants

A \ B	$cm^3\,mol^{-1}s^{-1}$	$dm^3\,mol^{-1}s^{-1}$	$m^3\,mol^{-1}s^{-1}$	cm^3 $molecule^{-1}s^{-1}$	$(mm\,Hg)^{-1}s^{-1}$	$atm^{-1}s^{-1}$	$ppm^{-1}min^{-1}$	$m^2kN^{-1}s^{-1}$
$1\,cm^3\,mol^{-1}s^{-1} =$	1	10^{-3}	10^{-6}	1.66×10^{-24}	$1.604 \times 10^{-5}\,T^{-1}$	$1.219 \times 10^{-2}\,T^{-1}$	2.453×10^{-9}	$1.203 \times 10^{-4}\,T^{-1}$
$1\,dm^3\,mol^{-1}s^{-1} =$	10^3	1	10^{-3}	1.66×10^{-21}	$1.604 \times 10^{-2}\,T^{-1}$	$12.19\,T^{-1}$	2.453×10^{-6}	$1.203 \times 10^{-1}\,T^{-1}$
$1\,m^3\,mol^{-1}\,s^{-1} =$	10^6	10^3	1	1.66×10^{-18}	$16.04\,T^{-1}$	$1.219 \times 10^4\,T^{-1}$	2.453×10^{-3}	$120.3\,T^{-1}$
$1\,cm^3\,molecule^{-1}s^{-1} =$	6.023×10^{23}	6.023×10^{20}	6.023×10^{17}	1	$9.658 \times 10^{18}T^{-1}$	$7.34 \times 10^{21}\,T^{-1}$	1.478×10^{15}	$7.244 \times 10^{19}T^{-1}$
$1\,(mm\,Hg)^{-1}\,s^{-1} =$	$6.236 \times 10^4\,T$	$62.36\,T$	$6.236 \times 10^{-2}\,T$	$1.035 \times 10^{-19}\,T$	1	760	4.56×10^{-2}	7.500
$1\,atm^{-1}\,s^{-1} =$	$82.06\,T$	$8.206 \times 10^{-2}\,T$	$8.206 \times 10^{-5}\,T$	$1.362 \times 10^{-22}\,T$	1.316×10^{-3}	1	6×10^{-5}	9.869×10^{-3}
$1\,ppm^{-1}\,min^{-1} =$ at 298 K, 1 atm total pressure	4.077×10^8	4.077×10^5	407.7	6.76×10^{-16}	21.93	1.667×10^4	1	164.5
$1\,m^2kN^{-1}s^{-1} =$	$8314\,T$	$8.314\,T$	$8.314 \times 10^{-3}\,T$	$1.38 \times 10^{-20}\,T$	0.1333	101.325	6.079×10^{-3}	1

To convert a rate constant from one set of units A to a new set B find the conversion factor for the row A under column B and multiply the old value by it, e.g.. to convert $cm^3\,molecule^{-1}\,s^{-1}$ to $m^3\,mol^{-1}\,s^{-1}$ multiply by 6.023×10^{17}.

Table adapted from High Temperature Reaction Rate Data No. 5, The University, Leeds (1970).

Equivalent Third Order Rate Constants

A \ B	$cm^6\,mol^{-2}s^{-1}$	$dm^6\,mol^{-2}s^{-1}$	$m^6\,mol^{-2}s^{-1}$	cm^6 $molecule^{-2}s^{-1}$	$(mm\,Hg)^{-2}s^{-1}$	$atm^{-2}s^{-1}$	$ppm^{-2}min^{-1}$	$m^4kN^{-2}s^{-1}$
$1\,cm^6\,mol^{-2}s^{-1} =$	1	10^{-6}	10^{-12}	2.76×10^{-48}	$2.57 \times 10^{-10}T^{-2}$	$1.48 \times 10^{-4}\,T^{-2}$	1.003×10^{-19}	$1.477 \times 10^{-8}T^{-2}$
$1\,dm^6\,mol^{-2}s^{-1} =$	10^6	1	10^{-6}	2.76×10^{-42}	$2.57 \times 10^{-4}T^{-2}$	$148\,T^{-2}$	1.003×10^{-13}	$1.477 \times 10^{-2}T^{-2}$
$1\,m^6\,mol^{-2}\,s^{-1} =$	10^{12}	10^6	1	2.76×10^{-36}	$257\,T^{-2}$	$1.48 \times 10^8T^{-2}$	1.003×10^{-7}	$1.477 \times 10^4T^{-2}$
$1\,cm^6\,molecule^{-2}s^{-1} =$	3.628×10^{47}	3.628×10^{41}	3.628×10^{35}	1	$9.328 \times 10^{37}T^{-2}$	$5.388 \times 10^{43}T^{-2}$	3.64×10^{28}	$5.248 \times 10^{39}T^{-2}$
$1\,(mm\,Hg)^{-2}\,s^{-1} =$	$3.89 \times 10^9\,T^2$	$3.89 \times 10^3\,T^2$	$3.89 \times 10^{-3}\,T^2$	$1.07 \times 10^{-38}T^2$	1	5.776×10^5	3.46×10^{-5}	56.25
$1\,atm^{-2}\,s^{-1} =$	$6.733 \times 10^3\,T^2$	$6.733 \times 10^{-3}\,T^2$	$6.733 \times 10^{-9}\,T^2$	$1.86 \times 10^{-44}T^2$	1.73×10^{-6}	1	6×10^{-11}	9.74×10^{-5}
$1\,ppm^{-2}\,min^{-1} =$ at 298K, 1 atm total pressure	9.97×10^{18}	9.97×10^{12}	9.97×10^6	2.75×10^{-29}	2.89×10^4	1.667×10^{10}	1	1.623×10^6
$1\,m^1kN^{-2}s^{-1} =$	$6.91 \times 10^7\,T^2$	$6.91\,T^2$	$69.1 \times 10^{-5}\,T^2$	$1.904 \times 10^{-40}\,T^2$	0.0178	1.027×10^4	6.16×10^{-7}	1

From *J. Phys. Chem. Ref. Data*, 9, 470, 1980, by permission of the authors and the copyright owner, the American Institute of Physics.

CONVERSION FACTORS FOR IONIZING RADIATION

Conversion between SI and Other Units

Quantity	Symbol for quantity	Expression in SI units	Expression in symbols for SI units	Special name for SI units	Symbols using special names	Conventional units	Symbol for conventional unit	Value of conventional unit in SI units
Activity	A	1 per second	s^{-1}	becquerel	Bq	curie	Ci	3.7×10^{10} Bq
Absorbed dose	D	joule per kilogram	$J\,kg^{-1}$	gray	Gy	rad	rad	0.01 Gy
Absorbed dose rate	$\dot{D}$	joule per kilogram second	$J\,kg^{-1}\,s^{-1}$		$Gy\,s^{-1}$	rad	$rad\,s^{-1}$	0.01 $Gy\,s^{-1}$
Average energy per ion pair	W	joule	J			electronvolt	eV	1.602×10^{-19} J
Dose equivalent	H	joule per kilogram	$J\,kg^{-1}$	sievert	Sv	rem	rem	0.01 Sv
Dose equivalent rate	$\dot{H}$	joule per kilogram second	$J\,kg^{-1}\,s^{-1}$		$Sv\,s^{-1}$	rem per second	$rem\,s^{-1}$	0.01 $Sv\,s^{-1}$
Electric current	I	ampere	A			ampere	A	1.0 A
Electric potential difference	U, V	watt per ampere	$W\,A^{-1}$	volt	V	volt	V	1.0 V
Exposure	$\dot{X}$	coulomb per kilogram	$C\,kg^{-1}$			roentgen	R	2.58×10^{-4} $C\,kg^{-1}$
Exposure rate	X	coulomb per kilogram second	$C\,kg^{-1}\,s^{-1}$			roentgen	$R\,s^{-1}$	2.58×10^{-4} $C\,kg^{-1}\,s^{-1}$
Fluence	ϕ	1 per meter squared	m^{-2}			1 per centimeter squared	cm^{-2}	1.0×10^{4} m^{-2}
Fluence rate	Φ	1 per meter squared second	$m^{-2}\,s^{-1}$			1 per centimeter squared second	$cm^{-2}\,s^{-1}$	1.0×10^{4} $m^{-2}\,s^{-1}$
Kerma	K	joule per kilogram	$J\,kg^{-1}$	gray	Gy	rad	rad	0.01 Gy
Kerma rate	$\dot{K}$	joule per kilogram second	$J\,kg^{-1}\,s^{-1}$		$Gy\,s^{-1}$	rad per second	$rad\,s^{-1}$	0.01 $Gy\,s^{-1}$
Lineal energy	y	joule per meter	$J\,m^{-1}$			kiloelectron volt per micrometer	$keV\,\mu m^{-1}$	1.602×10^{-10} $J\,m^{-1}$
Linear energy transfer	L	joule per meter	$J\,m^{-1}$			kiloelectron volt per micrometer	$keV\,\mu m^{-1}$	1.602×10^{-10} $J\,m^{-1}$
Mass attenuation coefficient	μ/ρ	meter squared per kilogram	$m^2\,kg^{-1}$			centimeter squared per gram	$cm^2\,g^{-1}$	0.1 $m^2\,kg^{-1}$
Mass energy transfer coefficient	μ_{tr}/ρ	meter squared per kilogram	$m^2\,kg^{-1}$			centimeter squared per gram	$cm^2\,g^{-1}$	0.1 $m^2\,kg^{-1}$
Mass energy absorption coefficient	μ_{en}/ρ	meter squared per kilogram	$m^2\,kg^{-1}$			centimeter squared per gram	$cm^2\,g^{-1}$	0.1 $m^2\,kg^{-1}$
Mass stopping power	S/ρ	joule meter squared per kilogram	$J\,m^2\,kg^{-1}$			MeV centimeter squared per gram	$MeV\,cm^2\,g^{-1}$	1.602×10^{-14} $J\,m^2\,kg^{-1}$
Power	P	joule per second	$J\,s^{-1}$	watt	W	watt	W	1.0 W
Pressure	p	newton per meter squared	$N\,m^{-2}$	pascal	Pa	torr	torr	(101325/760)Pa
Radiation chemical yield	G	mole per joule	$mol\,J^{-1}$			molecules per 100 electron volts	molecules $(100\,eV)^{-1}$	1.04×10^{-7} $mol\,J^{-1}$
Specific energy	z	joule per kilogram	$J\,kg^{-1}$	gray	Gy	rad	rad	0.01 Gy

Conversion of Radioactivity Units from MBq to mCi and µCi

MBq	mCi	MBq	mCi	MBq	mCi	MBq	mCi	MBq	mCi
7000	189.	700	18.9	70	1.89	7	189	0.7	18.9
6000	162.	600	16.2	60	1.62	6	162	0.6	16.2
5000	135.	500	13.5	50	1.35	5	135	0.5	13.5
4000	108.	400	10.8	40	1.08	4	108	0.4	10.8
3000	81.	300	8.1	30	810	3	81	0.3	8.1
2000	54.	200	5.4	20	540	2	54	0.2	5.4
1000	27.	100	2.7	10	270	1	27	0.1	2.7
900	24.	90	2.4	9	240	0.9	24		
800	21.6	80	2.16	8	220	0.8	21.6		

Conversion of Radioactivity Units from mCi and µCi to MBq

mCi	MBq	mCi	MBq	mCi	MBq	µCi	MBq	µCi	MBq	µCi	MBq
200	7400	40	1480	5	185	1000	37.0	200	7.4	30	1.11
150	5550	30	1110	4	148	900	33.3	100	3.7	20	0.74
100	3700	20	740	3	111	800	29.6	90	3.33	10	0.37
90	3330	10	370	2	74.0	700	25.9	80	2.96	5	0.185
80	2960	9	333	1	37.0	600	22.2	70	2.59	2	0.074
70	2590	8	296			500	18.5	60	2.22	1	0.037
60	2220	7	259			400	14.8	50	1.85		
50	1850	6	222			300	11.1	40	1.48		

Conversion of Radioactivity Units

100 TBq (10^{14} Bq)	=	2.7 kCi (2.7×10^3 Ci)	100 kBq (10^5 Bq)	=	2.7 µCi (2.7×10^{-6} Ci)
10 TBq (10^{13} Bq)	=	270 Ci (2.7×10^2 Ci)	10 kBq (10^4 Bq)	=	270 nCi (2.7×10^{-7} Ci)
1 TBq (10^{12} Bq)	=	27 Ci (2.7×10^1 Ci)	1 kBq (10^3 Bq)	=	27 nCi (2.7×10^{-8} Ci)
100 GBq (10^{11} Bq)	=	2.7 Ci (2.7×10^0 Ci)	100 Bq (10^2 Bq)	=	2.7 nCi (2.7×10^{-9} Ci)
10 GBq (10^{10} Bq)	=	270 mCi (2.7×10^{-1} Ci)	10 Bq (10^1 Bq)	=	270 pCi (2.7×10^{-10} Ci)
1 GBq (10^9 Bq)	=	27 mCi (2.7×10^{-2} Ci)	1 Bq (10^0 Bq)	=	27 pCi (2.7×10^{-11} Ci)
100 MBq (10^8 Bq)	=	2.7 mCi (2.7×10^{-3} Ci)	100 mBq (10^{-1} Bq)	=	2.7 pCi (2.7×10^{-12} Ci)
10 MBq (10^7 Bq)	=	270 µCi (2.7×10^{-4} Ci)	10 mBq (10^{-2} Bq)	=	270 fCi (2.7×10^{-13} Ci)
1 MBq (10^6 Bq)	=	27 µCi (2.7×10^{-5} Ci)	1 mBq (10^{-3} Bq)	=	27 fCi (2.7×10^{-14} Ci)

Conversion of Absorbed Dose Units

SI Units		Conventional	SI Units		Conventional
100 Gy (10^2 Gy)	=	10,000 rad (10^4 rad)	100 µGy (10^{-4} Gy)	=	10 mrad (10^{-2} rad)
10 Gy (10^1 Gy)	=	1,000 rad (10^3 rad)	10 µGy (10^{-5} Gy)	=	1 mrad (10^{-3} rad)
1 Gy (10^0 Gy)	=	100 rad (10^2 rad)	1 µGy (10^{-6} Gy)	=	100 µrad (10^{-4} rad)
100 mGy (10^{-1} Gy)	=	10 rad (10^1 rad)	100 nGy (10^{-7} Gy)	=	10 µrad (10^{-5} rad)
10 mGy (10^{-2} Gy)	=	1 rad (10^0 rad)	10 nGy (10^{-8} Gy)	=	1 µrad (10^{-6} rad)
1 mGy (10^{-3} Gy)	=	100 mrad (10^{-1} rad)	1 nGy (10^{-9} Gy)	=	100 nrad (10^{-7} rad)

Conversion of Dose Equivalent Units

100 Sv (10^2 Sv)	=	10,000 rem (10^4 rem)	100 µSv (10^{-4} Sv)	=	10 mrem (10^{-2} rem)
10 Sv (10^1 Sv)	=	1,000 rem (10^3 rem)	10 µSv (10^{-5} Sv)	=	1 mrem (10^{-3} rem)
1 Sv (10^0 Sv)	=	100 rem (10^2 rem)	1 µSv (10^{-6} Sv)	=	100 µrem (10^{-4} rem)
100 mSv (10^{-1} Sv)	=	10 rem (10^1 rem)	100 nSv (10^{-7} Sv)	=	10 µrem (10^{-5} rem)
10 mSv (10^{-2} Sv)	=	1 rem (10^0 rem)	10 nSv (10^{-8} Sv)	=	1 µrem (10^{-6} rem)
1 mSv (10^{-3} Sv)	=	100 mrem (10^{-1} rem)	1 nSv (10^{-9} Sv)	=	100 nrem (10^{-7} rem)

VALUES OF THE GAS CONSTANT IN DIFFERENT UNIT SYSTEMS

In SI units the value of the gas constant, R, is:

R = 8.314472 Pa m^3 K^{-1} mol^{-1}
= 8314.472 Pa L K^{-1} mol^{-1}
= 0.08314472 bar L K^{-1} mol^{-1}

This table gives the appropriate value of R for use in the ideal gas equation, $PV = nRT$, when the variables are expressed in other units. The following conversion factors for pressure units were used in generating the table:

1 atm = 101325 Pa
1 psi = 6894.757 Pa

1 torr (mm Hg) = 133.322 Pa [at 0°C]
1 in Hg = 3386.38 Pa [at 0°C]
1 in H$_2$O = 249.082 Pa [at 4°C]
1 ft H$_2$O = 2988.98 Pa [at 4°C]

Reference

Mohr, P. J., and Taylor, B. N., "The 2002 CODATA Recommended Values of the Fundamental Physical Constants", *Rev. Mod. Phys.* 77, 1, 2005. See also <http://physics.nist.gov/constants>

Units of V, T, n			Units of P						
V	T	n	kPa	atm	psi	mmHg	in Hg	in H$_2$O	ft H$_2$O
ft^3	K	mol	0.2936228	0.00289784	0.0425864	2.20236	0.0867070	1.17881	0.0982351
		lb·mol	133.1851	1.31443	19.3168	998.973	39.3296	534.704	44.5587
	°R	mol	0.1631238	0.00160990	0.0236591	1.22353	0.0481706	0.654900	0.0545751
		lb·mol	73.99170	0.730242	10.7316	554.984	21.8498	297.058	24.7548
cm^3	K	mol	8314.472	82.0574	1205.91	62363.8	2455.27	33380.4	2781.71
		lb·mol	3771381	37220.6	546993	282878000	1113690	15141100	1261760
	°R	mol	4619.151	45.5875	669.951	34646.5	1364.03	18544.7	1545.39
		lb·mol	2095211	20678.1	303885	15715400	618717	8411730	700979
L	K	mol	8.314472	0.0820574	1.20591	62.3638	2.45527	33.3804	2.78171
		lb·mol	3771.381	37.2206	546.993	28287.8	1113.69	15141.1	1261.76
	°R	mol	4.619151	0.0455875	0.669951	34.6465	1.36403	18.5447	1.54539
		lb·mol	2095.211	20.6781	303.885	15715.4	618.717	8411.73	700.979
m^3	K	mol	0.008314472	0.0000820574	0.00120591	0.0623638	0.00245527	0.0333804	0.00278171
		lb·mol	3.771381	0.0372206	0.546993	28.2878	1.11369	15.1411	1.26176
	°R	mol	0.004619151	0.0000455875	0.000669951	0.0346465	0.00136403	0.0185447	0.00154539
		lb·mol	2.095211	0.0206781	0.303885	15.7154	0.618717	8.41173	0.700979

Section 2
Symbols, Terminology, and Nomenclature

SYMBOLS AND TERMINOLOGY FOR PHYSICAL AND CHEMICAL QUANTITIES

The International Organization for Standardization (ISO), International Union of Pure and Applied Chemistry (IUPAC), and the International Union of Pure and Applied Physics (IUPAP) have jointly developed a set of recommended symbols for physical and chemical quantities. Consistent use of these recommended symbols helps assure unambiguous scientific communication. The list below is reprinted from Reference 1 with permission from IUPAC. Full details may be found in the following references:

1. Ian Mills, Ed., *Quantities, Units, and Symbols in Physical Chemistry*, Blackwell Scientific Publications, Oxford, 1988.
2. E. R. Cohen and P. Giacomo, *Symbols, Units, Nomenclature, and Fundamental Constants in Physics*, Document IUPAP–25, 1987; also published in *Physica*, 146A 1–68, 1987.
3. *ISO Standards Handbook 2: Units of Measurement*, International Organization of Standardization, Geneva, 1982.

GENERAL RULES

The value of a physical quantity is expressed as the product of a numerical value and a unit, e.g.:

T = 300 K
V = 26.2 cm^3
C$_p$ = 45.3 J mol^{-1} K^{-1}

The symbol for a physical quantity is always given in italic (sloping) type, while symbols for units are given in roman type. Column headings in tables and axis labels on graphs may conveniently be written as the physical quantity symbol divided by the unit symbol, e.g.:

T/K
V/cm^3
$C_p/J\ mol^{-1}\ K^{-1}$

The values in the table or graph axis are then pure numbers. Subscripts to symbols for physical quantities should be italic if the subscript refers to another physical quantity or to a number, e.g.:

C_p – heat capacity at constant pressure
B_n – nth virial coefficient

Subscripts which have other meanings should be in roman type:

m_p – mass of the proton
E_k – kinetic energy

The following tables give the recommended symbols for the major classes of physical and chemical quantities. The expression in the Definition column is given as an aid in identifying the quantity but is not necessarily the complete or unique definition. The SI Unit gives one (not necessarily unique) expression for the coherent SI unit for the quantity. Other equivalent unit expressions, including those which involve SI prefixes, may be used.

Name	Symbol	Definition	SI unit
Space and Time			
cartesian space coordinates	x, y, z		m
spherical polar coordinates	r, θ, ϕ		m, 1, 1
generalized coordinate	q, q_i		(varies)
position vector	r	$r = xi + yj + zk$	m
length	l		m
special symbols:			
height	h		
breadth	b		
thickness	d, δ		
distance	d		
radius	r		
diameter	d		
path length	s		
length of arc	s		
area	A, A_s, S		m^2
volume	$V, (v)$		m^3
plane angle	$\alpha, \beta, \gamma, \theta, \phi...$	$\alpha = s/r$	rad, 1
solid angle	ω, Ω	$\omega = A/r^2$	sr, 1
time	t		s
period	T	$T = t/N$	s
frequency	ν, f	$\nu = 1/T$	Hz
circular frequency, angular frequency	ω	$\omega = 2\pi\nu$	rad s^{-1}, s^{-1}
characteristic time interval, relaxation time, time constant	τ, T	$\tau = |dt/d\ln x|$	s
angular velocity	ω	$\omega = d\phi/dt$	rad s^{-1}, s^{-1}
velocity	$v, u, w, c, \dot{r}$	$v = dr/dt$	m s^{-1}

Name	Symbol	Definition	SI unit
speed	v, u, w, c	$v = \|\boldsymbol{v}\|$	m s^{-1}
acceleration	$\boldsymbol{a}, (\boldsymbol{g})$	$\boldsymbol{a} = \mathrm{d}\boldsymbol{v}/\mathrm{d}t$	m s^{-2}
Classical Mechanics			
mass	m		kg
reduced mass	μ	$\mu = m_1 m_2/(m_1 + m_2)$	kg
density, mass density	ρ	$\rho = m/V$	kg m^{-3}
relative density	d	$d = \rho/\rho$	1
surface density	ρ_A, ρ_S	$\rho_A = m/A$	kg m^{-2}
specific volume	v	$v = V/m = 1/\rho$	m^3 kg^{-1}
momentum	$\boldsymbol{p}$	$\boldsymbol{p} = m\boldsymbol{v}$	kg m s^{-1}
angular momentum, action	$\boldsymbol{L}$	$\boldsymbol{L} = \boldsymbol{r} \times \boldsymbol{p}$	J s
moment of inertia	I, J	$I = \Sigma m_i r_i^2$	kg m^2
force	$\boldsymbol{F}$	$\boldsymbol{F} = \mathrm{d}\boldsymbol{p}/\mathrm{d}t = m\boldsymbol{a}$	N
torque, moment of a force	$\boldsymbol{T}, (\boldsymbol{M})$	$\boldsymbol{T} = \boldsymbol{r} \times \boldsymbol{F}$	N m
energy	E		J
potential energy	E_p, V, Φ	$E_p = \int \boldsymbol{F} \cdot \mathrm{d}s$	J
kinetic energy	E_k, T, K	$E_k = 1/2mv^2$	J
work	W, w	$W = \int \boldsymbol{F} \cdot \mathrm{d}s$	J
Hamilton function	H	$H(q, p) = T(q, p) + V(q)$	J
Lagrange function	L	$L(q, \dot{q}) = T(q, \dot{q}) - V(q)$	J
pressure	p, P	$p = F/A$	Pa, N m^{-2}
surface tension	γ, σ	$y = \mathrm{d}W/\mathrm{d}A$	N m^{-1}, J m^{-2}
weight	$G, (W, P)$	$G = mg$	N
gravitational constant	G	$F = Gm_1 m_2/r^2$	N m^2 kg^{-2}
normal stress	σ	$\sigma = F/A$	Pa
shear stress	τ	$\tau = F/A$	Pa
linear strain, relative elongation	ε, e	$\varepsilon = \Delta l/l$	1
modulus of elasticity, Young's modulus	E	$E = \sigma/\varepsilon$	Pa
shear strain	γ	$\gamma = \Delta x/d$	1
shear modulus	G	$G = \tau/\gamma$	Pa
volume strain, bulk strain	θ	$\theta = \Delta V/V_0$	1
bulk modulus, compression modulus	K	$K = -V_0(\mathrm{d}p/\mathrm{d}V)$	Pa
viscosity, dynamic viscosity	η, μ	$\tau_{x,z} = \eta(\mathrm{d}v_x/\mathrm{d}z)$	Pa s
fluidity	ϕ	$\phi = 1/\eta$	m kg^{-1} s
kinematic viscosity	ν	$\nu = \eta/\rho$	m^2 s^{-1}
friction coefficient	$\mu, (f)$	$F_{\mathrm{frict}} = \mu F_{\mathrm{norm}}$	1
power	P	$P = \mathrm{d}W/\mathrm{d}t$	W
sound energy flux	P, P_a	$P = \mathrm{d}E/\mathrm{d}t$	W
acoustic factors			
reflection factor	ρ	$\rho = P_r/P_0$	1
acoustic absorption factor	$\alpha_a, (\alpha)$	$\alpha_a = 1 - \rho$	1
transmission factor	τ	$\tau = P_{\mathrm{tr}}/P_0$	1
dissipation factor	δ	$\delta = \alpha_a - \tau$	1
Electricity and Magnetism			
quantity of electricity, electric charge	Q		C
charge density	ρ	$\rho = Q/V$	C m^{-3}
surface charge density	σ	$\sigma = Q/A$	C m^{-2}
electric potential	V, ϕ	$V = \mathrm{d}W/\mathrm{d}Q$	V, J C^{-1}
electric potential difference	$U, \Delta V, \Delta \phi$	$U = V_2 - V_1$	V
electromotive force	E	$E = \int (\boldsymbol{F}/Q) \cdot \mathrm{d}s$	V
electric field strength	$\boldsymbol{E}$	$\boldsymbol{E} = \boldsymbol{F}/Q = -\,\mathrm{grad}\,V$	V m^{-1}
electric flux	Ψ	$\Psi = \int \boldsymbol{D} \cdot \mathrm{d}\boldsymbol{A}$	C
electric displacement	$\boldsymbol{D}$	$\boldsymbol{D} = \varepsilon \boldsymbol{E}$	C m^{-2}
capacitance	C	$C = Q/U$	F, C V^{-1}
permittivity	ε	$\boldsymbol{D} = \varepsilon \boldsymbol{E}$	F m^{-1}
permittivity of vacuum	ε_0	$\varepsilon_0 = \mu_0^{-1} c_0^{-2}$	F m^{-1}
relative permittivity	ε_r	$\varepsilon_r = \varepsilon/\varepsilon_0$	1
dielectric polarization (dipole moment per volume)	$\boldsymbol{P}$	$\boldsymbol{P} = \boldsymbol{D} - \varepsilon_0 \boldsymbol{E}$	C m^{-2}
electric susceptibility	χ_e	$\chi_e = \varepsilon r - 1$	1
electric dipole moment	$\boldsymbol{p}, \boldsymbol{\mu}$	$\boldsymbol{p} = Q\boldsymbol{r}$	C m

Name	Symbol	Definition	SI unit
electric current	I	$I = \mathrm{d}Q/\mathrm{d}t$	A
electric current density	$\boldsymbol{j}, \boldsymbol{J}$	$I = \int \boldsymbol{j} \cdot \mathrm{d}\boldsymbol{A}$	A m^{-2}
magnetic flux density, magnetic induction	$\boldsymbol{B}$	$\boldsymbol{F} = Q\boldsymbol{v} \times \boldsymbol{B}$	T
magnetic flux	$\varPhi$	$\varPhi = \int \boldsymbol{B} \cdot \mathrm{d}\boldsymbol{A}$	A m^{-2}
magnetic field strength	$\boldsymbol{H}$	$\boldsymbol{B} = \mu \boldsymbol{H}$	A m^{-2}
permeability	μ	$\boldsymbol{B} = \mu \boldsymbol{H}$	N A^{-2}, H m^{-1}
permeability of vacuum	μ_0		H m^{-1}
relative permeability	μ_r	$\mu_r = \mu/\mu_0$	1
magnetization (magnetic dipole moment per volume)	$\boldsymbol{M}$	$\boldsymbol{M} = \boldsymbol{B}/\mu_0 - \boldsymbol{H}$	A m^{-1}
magnetic susceptibility	$\chi, \kappa, (\chi_m)$	$\chi = \mu_r - 1$	1
molar magnetic susceptibility	χ_m	$\chi_m = V_m \chi$	m^3 mol^{-1}
magnetic dipole moment	$\boldsymbol{m}, \boldsymbol{\mu}$	$E_p = -\boldsymbol{m} \cdot \boldsymbol{B}$	A m^2, J T^{-1}
electrical resistance	R	$R = U/I$	Ω
conductance	G	$G = 1/R$	S
loss angle	δ	$\delta = (\pi/2) + \phi_I - \phi_U$	1, rad
reactance	X	$X = (U/I)\sin\delta$	Ω
impedance (complex impedance)	Z	$Z = R + iX$	Ω
admittance (complex admittance)	Y	$Y = 1/Z$	S
susceptance	B	$Y = G + iB$	S
resistivity	ρ	$\rho = E/j$	Ω m
conductivity	κ, γ, σ	$\kappa = 1/\rho$	S m^{-1}
self–inductance	L	$E = -L(\mathrm{d}I/\mathrm{d}t)$	H
mutual inductance	M, L_{12}	$E_1 = L_{12}(\mathrm{d}I_2/\mathrm{d}t)$	H
magnetic vector potential	$\boldsymbol{A}$	$\boldsymbol{B} = \nabla \times \boldsymbol{A}$	Wb m^{-1}
Poynting vector	$\boldsymbol{S}$	$\boldsymbol{S} = \boldsymbol{E} \times \boldsymbol{H}$	W m^{-2}

Quantum Mechanics

Name	Symbol	Definition	SI unit
momentum operator	$\dot{\boldsymbol{p}}$	$\dot{\boldsymbol{p}} = -ih\nabla$	m^{-1} J s
kinetic energy operator	$\hat{T}$	$\hat{T} = -(h^2/2m)\nabla^2$	J
Hamiltonian operator	$\hat{H}$	$\hat{H} = \hat{T} + V$	J
wavefunction, state function	Ψ, ψ, ϕ	$\hat{H}\psi = E\psi$	(m$^{-3/2}$)
probability density	P	$P = \psi^*\psi$	(m^{-3})
charge density of electrons	ρ	$\rho = -eP$	(C m^{-3})
probability current density	$\boldsymbol{S}$	$\boldsymbol{S} = -ih(\psi^*\nabla\psi - \psi\nabla\psi^*)/2m_e$	(m^{-2} s^{-1})
electric current density of electrons	$\boldsymbol{j}$	$\boldsymbol{j} = -e\boldsymbol{S}$	(A m^{-2})
matrix element of operator $\hat{A}$	$A_{ij}, \langle i|\hat{A}|j\rangle$	$A_{ij} = \int \psi_i^* \hat{A} \psi_j \mathrm{d}\tau$	(varies)
expectation value of operator $\hat{A}$	$\langle A\rangle, \bar{A}$	$\langle A\rangle = \int \psi^* \hat{A} \Psi \mathrm{d}\tau$	(varies)
hermitian conjugate of $\hat{A}$	$\hat{A}^\dagger$	$(\hat{A}^\dagger)_{ij} = (A_{ji})^*$	(varies)
commutator of $\hat{A}$ and $\hat{B}$	$[\hat{A}, \hat{B}], [\hat{A}, \hat{B}]_-$	$[\hat{A}, \hat{B}] = \hat{A}\hat{B} - \hat{B}\hat{A}$	(varies)
anticommutator	$[\hat{A}, \hat{B}]_+$	$[\hat{A}, \hat{B}]_+ = \hat{A}\hat{B} + \hat{B}\hat{A}$	(varies)
spin wavefunction	$\alpha; \beta$		1
coulomb integral	H_{AA}	$H_{AA} = \int \psi_A^* \hat{H} \psi_A \mathrm{d}\tau$	J
resonance integral	H_{AB}	$H_{AB} = \int \psi_A^* \hat{H} \psi_B \mathrm{d}\tau$	J
overlap integral	S_{AB}	$S_{AB} = \int \psi_A^* \psi_B \mathrm{d}\tau$	1

Atoms and Molecules

Name	Symbol	Definition	SI unit
nucleon number, mass number	A		1
proton number, atomic number	Z		1
neutron number	N	$N = A - Z$	1
electron rest mass	m_e		kg
mass of atom, atomic mass	m_a, m		kg
atomic mass constant	m_u	$m_u = m_a(^{12}\mathrm{C})/12$	kg
mass excess	Δ	$\Delta = m_a - Am_u$	kg
elementary charge, proton charge	e		C
Planck constant	h		J s
Planck constant/2π	$\hbar$	$\hbar = h/2\pi$	J s
Bohr radius	a_0	$a_0 = 4\pi\varepsilon_0\hbar^2/m_e e^2$	m
Hartree energy	E_h	$E_h = \hbar^2/m_e a_0^2$	J
Rydberg constant	R_∞	$R_\infty = E_h/2hc$	m^{-1}
fine structure constant	α	$\alpha = e^2/4\pi\varepsilon_0\hbar c$	1

Name	Symbol	Definition	SI unit
ionization energy	E_i		J
electron affinity	E_{ea}		J
dissociation energy	E_d, D		J
from the ground state	D_0		J
from the potential minimum	D_e		J
principal quantum number (H atom)	n	$E = -hcR/n^2$	1
angular momentum quantum numbers	see under Spectroscopy		
magnetic dipole moment of a molecule	$\boldsymbol{m}, \boldsymbol{\mu}$	$E_p = -\boldsymbol{m} \cdot \boldsymbol{B}$	J T^{-1}
magnetizability of a molecule	ξ	$\boldsymbol{m} = \xi B$	J T^{-2}
Bohr magneton	μ_B	$\mu_B = e\hbar/2m_e$	J T^{-1}
nuclear magneton	μ_N	$\mu_N = (m_e/m_p)\mu_B$	J T^{-1}
magnetogyric ratio (gyromagnetic ratio)	γ	$\gamma = \mu/L$	C kg^{-1}
g factor	g		1
Larmor circular frequency	ω_L	$\omega_L = (e/2m)B$	s^{-1}
Larmor frequency	ν_L	$\nu_L = \omega_L/2\pi$	Hz
longitudinal relaxation time	T_1		s
transverse relaxation time	T_2		s
electric dipole moment of a molecule	$\boldsymbol{p}, \boldsymbol{\mu}$	$E_p = -\boldsymbol{p} \cdot \boldsymbol{E}$	C m
quadrupole moment of a molecule	$\boldsymbol{Q}; \boldsymbol{\Theta}$	$E_p = 1/2\boldsymbol{Q}{:}\boldsymbol{V''} = 1/3\boldsymbol{\Theta}{:}\boldsymbol{V''}$	C m^2
quadrupole moment of a nucleus	eQ	$eQ = 2 \cdot \langle \Theta_{zz} \rangle$	C m^2
electric field gradient tensor	$\boldsymbol{q}$	$q_{\alpha\beta} = -\partial^2 V/\partial\alpha\partial\beta$	V m^{-2}
quadrupole interaction energy tensor	X	$X_{\alpha\beta} = eQq_{\alpha\beta}$	J
electric polarizability of a molecule	α	p (induced) $= \alpha E$	C m^2 V^{-1}
activity (of a radioactive substance)	A	$A = -dN_B/dt$	Bq
decay (rate) constant, disintegration (rate) constant	λ	$A = \gamma N_B$	s^{-1}
half life	$t_{1/2}, T_{1/2}$		s
mean life	τ		s
level width	Γ	$\Gamma = \hbar/\tau$	J
disintegration energy	Q		J
cross section (of a nuclear reaction)	σ		m^2
Spectroscopy			
total term	T	$T = E_{tot}/hc$	m^{-1}
transition wavenumber	$\tilde{\nu}, (\nu)$	$\tilde{\nu} = T' - T''$	m^{-1}
transition frequency	ν	$\nu = (E' - E'')/h$	Hz
electronic term	T_e	$T_e = E_e/hc$	m^{-1}
vibrational term	G	$G = E_{vib}/hc$	m^{-1}
rotational term	F	$F = E_{rot}/hc$	m^{-1}
spin orbit coupling constant	A	$T_{s.o.} = A\langle \hat{\boldsymbol{L}} \cdot \hat{\boldsymbol{S}} \rangle$	m^{-1}
principal moments of inertia	$I_A; I_B; I_C$	$I_A \le I_B \le I_C$	kg m^2
rotational constants,			
in wavenumber	$\tilde{A}; \tilde{B}; \tilde{C}$	$\tilde{A} = h/8\pi^2 c I_A$	m^{-1}
in frequency	$A; B; C$	$A = h/8\pi^2 I_A$	Hz
inertial defect	Δ	$\Delta = I_C - I_A - I_B$	kg m^2
asymmetry parameter	κ	$\kappa = \dfrac{(2B - A - C)}{(A - C)}$	1
centrifugal distortion constants,			
S reduction	$D_J; D_{JK}; D_K; d_1; d_2$		m^{-1}
A reduction	$\Delta_J; \Delta_{JK}; \Delta_K; \delta_J; \delta_K$		m^{-1}
harmonic vibration wavenumber	$\omega_e; \omega_r$		m^{-1}
vibrational anharmonicity constant	$\omega_e x_e; x_{rs}; g_{u'}$		m^{-1}
vibrational quantum numbers	$v_r; l_t$		1
Coriolis zeta constant	ζ_{rs}^a		1
angular momentum quantum numbers	see additional information below		
degeneracy, statistical weight	g, d, β		1
electric dipole moment of a molecule	$\boldsymbol{p}, \boldsymbol{\mu}$	$E_p = -\boldsymbol{p} \cdot \boldsymbol{E}$	C m
transition dipole moment of a molecule	$\boldsymbol{M}, \boldsymbol{R}$	$\boldsymbol{M} = \int \psi' \boldsymbol{p} \psi'' d\tau$	C m
molecular geometry, interatomic distances,			
equilibrium distance	r_e		m
zero–point average distance	r_z		m

Name	Symbol	Definition	SI unit
ground state distance	r_0		m
substitution structure distance	r_s		m
vibrational coordinates,			
internal coordinates	R_i, r_i, θ_j, etc.		(varies)
symmetry coordinates	S_i		(varies)
normal coordinates			
mass adjusted	Q_r		$kg^{1/2}$ m
dimensionless	q_r		1
vibrational force constants,			
diatomic	$f, (k)$	$f = \partial^2 V/\partial r^2$	J m^{-2}
polyatomic,			
internal coordinates	f_{ij}	$f_{ij} = \partial^2 V/\partial r_i \partial r_j$	(varies)
symmetry coordinates	F_{ij}	$F_{ij} = \partial^2 V/\partial S_i \partial S_j$	(varies)
dimensionless normal coordinates	$\phi_{rst...}, k_{rst...}$		m^{-1}
nuclear magnetic resonance (NMR),			
magnetogyric ratio	γ	$\gamma = \mu/I\hbar$	C kg^{-1}
shielding constant	σ_A	$B_A = (1 - \sigma_A)B$	1
chemical shift, δ scale	δ	$\delta = 10^6(\nu - \nu_0)/\nu_0$	1
(indirect) spin–spin coupling constant	J_{AB}	$\hat{H}/h = J_{AB}\hat{I}_A \cdot \hat{I}_B$	Hz
direct (dipolar) coupling constant	D_{AB}		Hz
longitudinal relaxation time	T_1		s
transverse relaxation time	T_2		s
electron spin resonance, electron paramagnetic resonance (ESR, EPR),			
magnetogyric ratio	γ	$\gamma = \mu/s\hbar$	C kg^{-1}
g factor	g	$h\nu = g\mu_B B$	1
hyperfine coupling constant,			
in liquids	a, A	$\hat{H}_{hfs}/h = a\hat{S} \cdot \hat{I}$	Hz
in solids	T	$\hat{H}_{hfs}/h = \hat{S} \cdot T \cdot \hat{I}$	Hz

Angular momentum	Operator symbol	Quantum number symbol		
		Total	Z–axis	z-axis
electron orbital	$\hat{L}$	L	M_L	Λ
one electron only	$\hat{l}$	l	m_l	λ
electron spin	$\hat{S}$	S	M_S	Σ
one electron only	$\hat{s}$	s	m_s	σ
electron orbital + spin	$\hat{L} + \hat{S}$			$\Omega = \Lambda + \Sigma$
nuclear orbital (rotational)	$\hat{R}$	R		K_R, k_R
nuclear spin	$\hat{I}$	I	M_I	
internal vibrational				
spherical top	$\hat{l}$	$l(l\zeta)$		K_l
other	$\hat{j}, \hat{\pi}$			$l(l\zeta)$
sum of $R + L(+ j)$	$\hat{N}$	N		K, k
sum of $N + S$	$\hat{J}$	J	M_J	K, k
sum of $J + I$	$\hat{F}$	F	M_F	

Electromagnetic Radiation

Name	Symbol	Definition	SI unit
wavelength	λ		m
speed of light			
in vacuum	c_0		m s^{-1}
in a medium	c	$c = c_0/n$	m s^{-1}
wavenumber in vacuum	$\tilde{\nu}$	$\tilde{\nu} = \nu/c_0 = 1/n\lambda$	m^{-1}
wavenumber (in a medium)	σ	$\sigma = 1/\lambda$	m^{-1}
frequency	ν	$\nu = c/\lambda$	Hz
circular frequency, pulsatance	ω	$\omega = 2\pi\nu$	s^{-1}, rad s^{-1}
refractive index	n	$n = c_0/c$	1
Planck constant	h		J s

Name	Symbol	Definition	SI unit
Planck constant/2π	$\hbar$	$\hbar = h/2\pi$	J s
radiant energy	Q, W		J
radiant energy density	ρ, w	$\rho = Q/V$	J m^{-3}
spectral radiant energy density			
in terms of frequency	ρ_v, w_v	$\rho = d\rho/dv$	J m^{-3} Hz^{-1}
in terms of wavenumber	$\rho_{\tilde{v}}, w_{\tilde{v}}$	$\rho_{\tilde{v}} = d\rho / d\tilde{v}$	J m^{-2}
in terms of wavelength	ρ_λ, w_λ	$\rho_\lambda = d\rho/d\lambda$	J m^{-4}
Einstein transition probabilities			
spontaneous emission	A_{nm}	$dN_n/dt = -A_{nm}N_n$	s^{-1}
stimulated emission	B_{nm}	$dN_n/dt = -\rho_{\tilde{v}}(\tilde{v}_{nm}) \times B_{nm}N_n$	s kg^{-1}
stimulated absorption	B_{mn}	$dN_n/dt = -\rho_{\tilde{v}}(\tilde{v}_{nm}) B_{mn}N_m$	s kg^{-1}
radiant power, radiant energy per time	Φ, P	$\Phi = dQ/dt$	W
radiant intensity	I	$I = d\Phi/d\Omega$	W sr^{-1}
radiant exitance (emitted radiant flux)	M	$M = d\Phi/dA_{\text{source}}$	W m^{-2}
irradiance, (radiant flux received)	$E, (I)$	$E = d\Phi/dA$	W m^{-2}
emittance	ε	$\varepsilon = M/M_{bb}$	1
Stefan–Boltzmann constant	σ	$M_{bb} = \sigma T^4$	W m^{-2} K^{-4}
first radiation constant	c_1	$c_1 = 2\pi hc_0^2$	W m^2
second radiation constant	c_2	$c_2 = hc_0/k$	K m
transmittance, transmission factor	τ, T	$\tau = \Phi_{tr}/\Phi_0$	1
absorptance, absorption factor	α	$\alpha = \Phi_{abs}/\Phi_0$	1
reflectance, reflection factor	ρ	$\rho = \Phi_{refl}/\Phi_0$	1
(decadic) absorbance	A	$A = -\lg(1 - \alpha_i)$	1
napierian absorbance	B	$B = -\ln(1 - \alpha_i)$	1
absorption coefficient			
(linear) decadic	a, K	$a = A/l$	m^{-1}
(linear) napierian	α	$\alpha = B/l$	m^{-1}
molar (decadic)	ε	$\varepsilon = a/c = A/cl$	m^2 mol^{-1}
molar napierian	κ	$\kappa = \alpha/c = B/cl$	m^2 mol^{-1}
absorption index	k	$k = \alpha/4\pi\tilde{v}$	1
complex refractive index	$\hat{n}$	$\hat{n} = n + ik$	1
molar refraction	R, R_m	$R = \dfrac{(n^2 - 1)}{(n^2 + 2)} V_m$	m^3 mol^{-1}
angle of optical rotation	α		1, rad
Solid State			
lattice vector	$\boldsymbol{R}, \boldsymbol{R_0}$		m
fundamental translation vectors for the crystal lattice	$\boldsymbol{a_1}; \boldsymbol{a_2}; \boldsymbol{a_3}, \boldsymbol{a}; \boldsymbol{b}; \boldsymbol{c}$	$R = n_1\boldsymbol{a_1} + n_2\boldsymbol{a_2} + n_3\boldsymbol{a_3}$	m
(circular) reciprocal lattice vector	$\boldsymbol{G}$	$\boldsymbol{G} \cdot \boldsymbol{R} = 2\pi m$	m^{-1}
(circular) fundamental translation vectors for the reciprocal lattice	$\boldsymbol{b_1}; \boldsymbol{b_2}; \boldsymbol{b_3}, \boldsymbol{a^*}; \boldsymbol{b^*}; \boldsymbol{c^*}$	$\boldsymbol{a_i} \cdot \boldsymbol{b_k} = 2\pi\delta_{ik}$	m^{-1}
lattice plane spacing	d		m
Bragg angle	θ	$n\lambda = 2d \sin\theta$	1, rad
order of reflection	n		1
order parameters			
short range	σ		1
long range	s		1
Burgers vector	$\boldsymbol{b}$		m
particle position vector	$\boldsymbol{r}, \boldsymbol{R_j}$		m
equilibrium position vector of an ion	$\boldsymbol{R_0}$		m
displacement vector of an ion	$\boldsymbol{u}$	$\boldsymbol{u} = \boldsymbol{R} - \boldsymbol{R_0}$	m
Debye–Waller factor	B, D		1
Debye circular wavenumber	q_D		m^{-1}
Debye circular frequency	ω_D		s^{-1}
Grüneisen parameter	γ, Γ	$\gamma = \alpha V/\kappa C_v$	1
Madelung constant	$\alpha, \mathcal{M}$	$E_{\text{coul}} = \dfrac{\alpha N_A z_+ z_- e^2}{4\pi\varepsilon_0 R_0}$	1
density of states	N_E	$N_E = dN(E)/dE$	J^{-1} m^{-3}
(spectral) density of vibrational modes	N_ω, g	$N_\omega = dN(\omega)/d\omega$	s m^{-3}

Name	Symbol	Definition	SI unit
resistivity tensor	ρ_{ik}	$\boldsymbol{E} = \boldsymbol{\rho} \cdot \boldsymbol{j}$	Ω m
conductivity tensor	σ_{ik}	$\sigma = \rho^{-1}$	S m^{-1}
thermal conductivity tensor	λ_{ik}	$\boldsymbol{J}_q = -\boldsymbol{\lambda} \cdot \text{grad } T$	W m^{-1} K^{-1}
residual resistivity	ρ_R		Ω m
relaxation time	τ	$\tau = l/v_F$	s
Lorenz coefficient	L	$L = \lambda/\sigma T$	V^2 K^{-2}
Hall coefficient	$A_H, \boldsymbol{R}_H$	$\boldsymbol{E} = \boldsymbol{\rho} \cdot \boldsymbol{j} + R_H(\boldsymbol{B} \times \boldsymbol{j})$	m^3 C^{-1}
thermoelectric force	E		V
Peltier coefficient	Π		V
Thomson coefficient	$\mu, (\tau)$		V K^{-1}
work function	Φ	$\Phi = E_\infty - E_F$	J
number density, number concentration	$n, (p)$		m^{-3}
gap energy	E_g		J
donor ionization energy	E_d		J
acceptor ionization energy	E_a		J
Fermi energy	E_F, ε_F		J
circular wave vector, propagation vector	$k, \boldsymbol{q}$	$k = 2\pi/\lambda$	m^{-1}
Bloch function	$u_k(\boldsymbol{r})$	$\psi(\boldsymbol{r}) = u_k(\boldsymbol{r}) \exp(\mathrm{i}\boldsymbol{k} \cdot \boldsymbol{r})$	m$^{-3/2}$
charge density of electrons	ρ	$\rho(\boldsymbol{r}) = -e\psi^*(\boldsymbol{r})\psi(\boldsymbol{r})$	C m^{-3}
effective mass	m^*		kg
mobility	μ	$\mu = v_{\text{drift}}/E$	m^2 V^{-1} s^{-1}
mobility ratio	b	$b = \mu_n/\mu_p$	1
diffusion coefficient	D	$dN/dt = -DA(dn/dx)$	m^2 s^{-1}
diffusion length	L	$L = \sqrt{D\tau}$	m
characteristic (Weiss) temperature	θ, θ_w		K
Curie temperature	T_C		K
Néel temperature	T_N		K
Statistical Thermodynamics			
number of entities	N		1
number density of entities, number concentration	n, C	$n = N/V$	m^{-3}
Avogadro constant	L, N_A		mol^{-1}
Boltzmann constant	k, k_B		J K^{-1}
gas constant (molar)	R	$R = Lk$	J K^{-1} mol^{-1}
molecular position vector	$\boldsymbol{r}\,(x, y, z)$		m
molecular velocity vector	$\boldsymbol{c}(c_x, c_y, c_z), \boldsymbol{u}(u_x, u_y, u_z)$	$\boldsymbol{c} = d\boldsymbol{r}/dt$	m s^{-1}
molecular momentum vector	$\boldsymbol{p}(p_x, p_y, p_z)$	$\boldsymbol{p} = m\boldsymbol{c}$	kg m s^{-1}
velocity distribution function (Maxwell)	$f(c_x)$	$f(c_x) = (m/2\pi kT)^{1/2} \times \exp(-mc_x^2/2kT)$	m^{-1} s
speed distribution function (Maxwell–Boltzmann)	$F(c)$	$F(c) = (m/2\pi kT)^{3/2} \times 4\pi c^2 \exp(-mc^2/2kT)$	m^{-1} s
average speed	$\bar{c}, \bar{u}, \langle c \rangle, \langle u \rangle$	$\bar{c} = \int cF(c)dc$	m s^{-1}
generalized coordinate	q		(m)
generalized momentum	p	$p = \partial L/\partial \dot{q}$	(kg m s^{-1})
volume in phase space	Ω	$\Omega = (1/h)\int p\,dq$	1
probability	P		1
statistical weight, degeneracy	g, d, W, ω, β		1
density of states	$\rho(E)$	$\rho(E) = dN/dE$	J^{-1}
partition function, sum over states, for a single molecule	q, z	$q = \sum_i g_i \exp(-\varepsilon_i/kT)$	1
for a canonical ensemble (system, or assembly)	Q, Z		1
microcanonical ensemble	Ω		1
grand (canonical ensemble)	Ξ		1
symmetry number	σ, s		1
reciprocal temperature parameter	β	$\beta = 1/kT$	J^{-1}
characteristic temperature	Θ		K

Name	Symbol	Definition	SI unit
General Chemistry			
number of entities (e.g. molecules, atoms, ions, formula units)	N		1
amount (of substance)	n	$n_B = N_B/L$	mol
Avogadro constant	L, N_A		mol^{-1}
mass of atom, atomic mass	m_a, m		kg
mass of entity (molecule, or formula unit)	m_f, m		kg
atomic mass constant	m_u	$m_u = m_a(^{12}C)/12$	kg
molar mass	M	$M_B = m/n_B$	$kg\ mol^{-1}$
relative molecular mass (relative molar mass, molecular weight)	M_r	$M_{r,B} = m_B/m_u$	1
molar volume	V_m	$V_{m,B} = V/n_B$	$m^3\ mol^{-1}$
mass fraction	w	$w_B = m_B/\Sigma m_i$	1
volume fraction	ϕ	$\phi_B = V_B/\Sigma V_i$	1
mole fraction, amount fraction, number fraction	x, y	$x_B = n_B/\Sigma n_i$	1
(total) pressure	p, P		Pa
partial pressure	p_B	$p_B = y_B p$	Pa
mass concentration (mass density)	γ, ρ	$\gamma_B = m_B/V$	$kg\ m^{-3}$
number concentration, number density of entities	C, n	$C_B = N_B/V$	m^{-3}
amount concentration, concentration	c	$c_B = n_B/V$	$mol\ m^{-3}$
solubility	s	$s_B = c_B$ (saturated solution)	$mol\ m^{-3}$
molality (of a solute)	$m, (b)$	$m_B = n_B/m_A$	$mol\ kg^{-1}$
surface concentration	Γ	$\Gamma_B = n_B/A$	$mol\ m^{-2}$
stoichiometric number	ν		1
extent of reaction, advancement	ξ	$\Delta\xi = \Delta n_B/\nu_B$	mol
degree of dissociation	α		1
Chemical Thermodynamics			
heat	q, Q		J
work	w, W		J
internal energy	U	$\Delta U = q + w$	J
enthalpy	H	$H = U + pV$	J
thermodynamic temperature	T		K
Celsius temperature	θ, t	$\theta/°C = T/K - 273.15$	°C
entropy	S	$dS \geq dq/T$	$J\ K^{-1}$
Helmholtz energy, (Helmholtz function)	A	$A = U - TS$	J
Gibbs energy, (Gibbs function)	G	$G = H - TS$	J
Massieu function	J	$J = -A/T$	$J\ K^{-1}$
Planck function	Y	$Y = -G/T$	$J\ K^{-1}$
surface tension	γ, σ	$\gamma = (\partial G/\partial A_s)_{T,p}$	$J\ m^{-2}, N\ m^{-1}$
molar quantity X	X_m	$X_m = X/n$	(varies)
specific quantity X	x	$x = X/m$	(varies)
pressure coefficient	β	$\beta = (\partial p/\partial T)v$	$Pa\ K^{-1}$
relative pressure coefficient	α_p	$\alpha_p = (1/p)(\partial p/\partial T)_V$	K^{-1}
compressibility,			
isothermal	κ_T	$\kappa_T = -(1/V)(\partial V/\partial p)_T$	Pa^{-1}
isentropic	κ_S	$\kappa_S = -(1/V)(\partial V/\partial p)_S$	Pa^{-1}
linear expansion coefficient	α_l	$\alpha_l = (1/l)(\partial l/\partial T)$	K^{-1}
cubic expansion coefficient	α, α_V, γ	$\alpha = (1/V)(\partial V/\partial T)_p$	K^{-1}
heat capacity,			
at constant pressure	C_p	$C_p = (\partial H/\partial T)_p$	$J\ K^{-1}$
at constant volume	C_V	$C_V = (\partial U/\partial T)_V$	$J\ K^{-1}$
ratio of heat capacities	$\gamma, (\kappa)$	$\gamma = C_p/C_V$	1
Joule–Thomson coefficient	μ, μ_{JT}	$\mu = (\partial T/\partial p)_H$	$K\ Pa^{-1}$
second virial coefficient	B	$pV_m = RT(1 + B/V_m + ...)$	$m^3\ mol^{-1}$
compression factor (compressibility factor)	Z	$Z = pV_m/RT$	1
partial molar quantity X	$X_B, (X_B')$	$X_B = (\partial X/\partial n_B)_{T,p,n_j \neq B}$	(varies)
chemical potential (partial molar Gibbs energy)	μ	$\mu_B = (\partial G/\partial n_B)_{T,p,n_j \neq B}$	$J\ mol^{-1}$
absolute activity	λ	$\lambda_B = \exp(\mu_B/RT)$	1

Name	Symbol	Definition	SI unit
standard chemical potential	$\mu^{\ominus}$, μ^{o}		J mol^{-1}
standard partial molar enthalpy	$H_B^{\ominus}$	$H_B^{\ominus} = \mu_B^{\ominus} + TS_B^{\ominus}$	J mol^{-1}
standard partial molar entropy	$S_B^{\ominus}$	$S_B^{\ominus} = -(\partial\mu_B^{\ominus}/\partial T)_p$	J mol^{-1} K^{-1}
standard reaction Gibbs energy (function)	$\Delta_r G^{\ominus}$	$\Delta_r G^{\ominus} = \sum_B \nu_B \mu_B^{\ominus}$	J mol^{-1}
affinity of reaction	A, $(\mathcal{A})$	$A = -(\partial G/\partial\xi)_{p,T} = -\sum_B \nu_B \mu_B$	J mol^{-1}
standard reaction enthalpy	$\Delta_r H^{\ominus}$	$\Delta_r H^{\ominus} = \sum_B \nu_B H_B^{\ominus}$	J mol^{-1}
standard reaction entropy	$\Delta_r S^{\ominus}$	$\Delta_r S^{\ominus} = \sum_B \nu_B S_B^{\ominus}$	J mol^{-1} K^{-1}
equilibrium constant	$K^{\ominus}$, K	$K^{\ominus} = \exp(-\Delta_r G^{\ominus}/RT)$	1
equilibrium constant,			
pressure basis	K_p	$K_p = \prod_B p_B^{\nu_B}$	Pa$^{\Sigma\nu}$
concentration basis	K_c	$K_c = \prod_B c_B^{\nu_B}$	(mol m^{-3})$^{\Sigma\nu}$
molality basis	K_m	$K_m = \prod_B m_B^{\nu_B}$	(mol kg^{-1})$^{\Sigma\nu}$
fugacity	f, $\tilde{p}$	$f_B = \lambda_B \lim_{p\to 0} (p_B/\lambda_B)_T$	Pa
fugacity coefficient	ϕ	$\phi_B = f_B/p_B$	1
activity and activity coefficient referenced to Raoult's law, (relative) activity	a	$a_B = \exp\left[\dfrac{\mu_B - \mu_B^*}{RT}\right]$	1
activity coefficient	f	$f_B = a_B/x_B$	1
activities and activity coefficients referenced to Henry's law, (relative) activity,			
molality basis	a_m	$a_{m,B} = \exp\left[\dfrac{\mu_B - \mu_B^{\ominus}}{RT}\right]$	1
concentration basis	a_c	$a_{c,B} = \exp\left[\dfrac{\mu_B - \mu_B^{\ominus}}{RT}\right]$	1
mole fraction basis	a_x	$a_{x,B} = \exp\left[\dfrac{\mu_B - \mu_B^{\ominus}}{RT}\right]$	1
activity coefficient,			
molality basis	γ_m	$a_{m,B} = \gamma_{m,B} m_B/m^{\ominus}$	1
concentration basis	γ_c	$a_{c,B} = \gamma_{c,B} c_B/c^{\ominus}$	1
mole fraction basis	γ_x	$a_{x,B} = \gamma_{x,B} x_B$	1
ionic strength,			
molality basis	I_m, I	$I_m = \frac{1}{2}\sum m_B z_B^2$	mol kg^{-1}
concentration basis	I_c, I	$I_c = \frac{1}{2}\sum c_B z_B^2$	mol m^{-3}
osmotic coefficient,			
molality basis	ϕ_m	$\phi_m = (\mu_A^* - \mu_A)/(RTM_A\Sigma m_B)$	1
mole fraction basis	ϕ_x	$\phi_x = (\mu_A - \mu_A^*)/(RT\ln x_A)$	1
osmotic pressure	Π	$\Pi = c_B RT$ (ideal dilute solution)	Pa

(i) Symbols used as subscripts to denote a chemical process or reaction

These symbols should be printed in roman (upright) type, without a full stop (period).

vaporization, evaporation (liquid → gas)	vap
sublimation (solid → gas)	sub
melting, fusion (solid → liquid)	fus
transition (between two phases)	trs
mixing of fluids	mix
solution (of solute in solvent)	sol
dilution (of a solution)	dil
adsorption	ads
displacement	dpl
immersion	imm

Name	Symbol	Definition	SI unit
reaction in general	r		
atomization	at		
combustion reaction	c		
formation reaction	f		

(ii) Recommended superscripts

standard	$\ominus$, ^{o}		
pure substance	*		
infinite dilution	∞		
ideal	id		
activated complex, transition state	‡		
excess quantity	E		

Chemical Kinetics

Name	Symbol	Definition	SI unit
rate of change of quantity X	$\dot{X}$	$\dot{X} = dX/dt$	(varies)
rate of conversion	$\dot{\xi}$	$\dot{\xi} = d\xi/dt$	mol s^{-1}
rate of concentration change (due to chemical reaction)	r_B, v_B	$r_B = dc_B/dt$	mol m^{-3} s^{-1}
rate of reaction (based on amount concentration)	v	$v = \dot{\xi}/V = v_B^{-1}dc_B/dt$	mol m^{-3} s^{-1}
partial order of reaction	n_B	$v = k\Pi c_B{}^{n_B}$	1
overall order of reaction	n	$n = \Sigma n_B$	1
rate constant, rate coefficient	k	$v = k\Pi c_B{}^{n_B}$	(mol^{-1} m^3)$^{n-1}$ s^{-1}
Boltzmann constant	k, k_B		J K^{-1}
half life	$t_{1/2}$	$c(t_{1/2}) = c_0/2$	s
relaxation time	τ	$\tau = 1/(k_1 + k_{-1})$	s
energy of activation, activation energy	E_a, E	$E_a = RT^2\, d \ln k/dT$	J mol^{-1}
pre–exponential factor	A	$k = A \exp(-E_a/RT)$	(mol^{-1} m^3)$^{n-1}$ s^{-1}
volume of activation	$\Delta^{\ddagger}V$	$\Delta^{\ddagger}V = -RT \times (\partial \ln k/\partial p)_T$	m^3 mol^{-1}
collision diameter	d	$d_{AB} = r_A + r_B$	m
collision cross–section	σ	$\sigma_{AB} = \pi d_{AB}{}^2$	m^2
collision frequency	Z_A		s^{-1}
collision number	Z_{AB}, Z_{AA}		m^{-3} s^{-1}
collision frequency factor	z_{AB}, z_{AA}	$z_{AB} = Z_{AB}/Lc_Ac_B$	m^3 mol^{-1} s^{-1}
standard enthalpy of activation	$\Delta^{\ddagger}H^{\ominus}, \Delta H^{\ddagger}$		J mol^{-1}
standard entropy of activation	$\Delta^{\ddagger}S^{\ominus}, \Delta S^{\ddagger}$		J mol^{-1} K^{-1}
standard Gibbs energy of activation	$\Delta^{\ddagger}G^{\ominus}, \Delta G^{\ddagger}$		J mol^{-1}
quantum yield, photochemical yield	ϕ		1

Electrochemistry

Name	Symbol	Definition	SI unit
elementary charge (proton charge)	e		C
Faraday constant	F	$F = eL$	C mol^{-1}
charge number of an ion	z	$z_B = Q_B/e$	1
ionic strength	I_c, I	$I_c = \frac{1}{2} \Sigma c_i z_i^2$	mol m^{-3}
mean ionic activity	$a_{\pm}$	$a_{\pm} = m_{\pm}\gamma_{\pm}/m^{\ominus}$	1
mean ionic molality	$m_{\pm}$	$m_{\pm}{}^{(v_+ + v_-)} = m_+{}^{v+}m_-{}^{v-}$	mol kg^{-1}
mean ionic activity coefficient	$\gamma_{\pm}$	$\gamma_{\pm}{}^{(v_+ + v_-)} = \gamma_+{}^{v+}\gamma_-{}^{v-}$	1
charge number of electrochemical cell reaction	$n, (z)$		1
electric potential difference (of a galvanic cell)	$\Delta V, E, U$	$\Delta V = V_R - V_L$	V
emf, electromotive force	E	$E = \lim\limits_{I \to 0} \Delta V$	V
standard emf, standard potential of the electrochemical cell reaction	$E^{\ominus}$	$E^{\ominus} = -\Delta_r G^{\ominus}/nF = (RT/nF) \ln K^{\ominus}$	V
standard electrode potential	$E^{\ominus}$		V
emf of the cell, potential of the electrochemical cell reaction	E	$E = E^{\ominus} - (RT/nF) \times \Sigma v_i \ln a_i$	V
pH	pH	$pH \approx -\lg\left[\dfrac{c(\mathbf{H}^+)}{\text{mol dm}^{-3}}\right]$	1
inner electric potential	ϕ	$\nabla\phi = -\boldsymbol{E}$	V
outer electric potential	ψ	$\psi = Q/4\pi\varepsilon_0 r$	V

Name	Symbol	Definition	SI unit
surface electric potential	χ	$\chi = \phi - \psi$	V
Galvani potential difference	$\Delta\phi$	$\Delta_\alpha{}^\beta\phi = \phi^\beta - \phi^\alpha$	V
volta potential difference	$\Delta\psi$	$\Delta_\alpha{}^\beta\psi = \psi^\beta - \psi^\alpha$	V
electrochemical potential	$\bar{\mu}$	$\bar{\mu}_B{}^\alpha = (\partial G/\partial n_B{}^\alpha)$	J mol^{-1}
electric current	I	$I = dQ/dt$	A
(electric) current density	j	$j = I/A$	A m^{-2}
(surface) charge density	σ	$\sigma = Q/A$	C m^{-2}
electrode reaction rate constant	k	$k_{ox} = I_a/(nFA\prod_i c_i^{n_i})$	(varies)
mass transfer coefficient, diffusion rate constant	k_d	$k_{d,B} = \lvert \nu_B\rvert I_{l,B}/nFcA$	m s^{-1}
thickness of diffusion layer	δ	$\delta_B = D_B/k_{d,B}$	m
transfer coefficient (electrochemical)	α	$\alpha_c = \dfrac{-\lvert \nu\rvert RT\partial}{nF}\dfrac{\partial \ln\lvert I_c\rvert}{\partial E}$	1
overpotential	η	$\eta = E_I - E_{I=0} - IR_u$	V
electrokinetic potential (zeta potential)	ζ		V
conductivity	$\kappa, (\sigma)$	$\kappa = j/E$	S m^{-1}
conductivity cell constant	K_{cell}	$K_{cell} = \kappa R$	m^{-1}
molar conductivity (of an electrolyte)	Λ	$\Lambda_B = \kappa/c_B$	S m^2 mol^{-1}
ionic conductivity, molar conductivity of an ion	λ	$\lambda_B = \lvert z_B\rvert Fu_B$	S m^2 mol^{-1}
electric mobility	$u, (\mu)$	$u_B = v_B/E$	m^2 V^{-1} s^{-1}
transport number	t	$t_B = j_B/\Sigma j_i$	1
reciprocal radius of ionic atmosphere	κ	$\kappa = (2F^2I/\varepsilon RT)^{1/2}$	m^{-1}

Colloid and Surface Chemistry

Name	Symbol	Definition	SI unit
specific surface area	a, a_s, s	$a = A/m$	m^2 kg^{-1}
surface amount of B, adsorbed amount of B	$n_B{}^s, n_B{}^a$		mol
surface excess of B	$n_B{}^\sigma$		mol
surface excess concentration of B	$\Gamma_B, (\Gamma_B{}^\sigma)$	$\Gamma_B = n_B{}^\sigma/A$	mol m^{-2}
total surface excess concentration	$\Gamma, (\Gamma^\sigma)$	$\Gamma = \sum_i \Gamma_i$	mol m^{-2}
area per molecule	a, σ	$a_B = A/N_B{}^\sigma$	m^2
area per molecule in a filled monolayer	a_m, σ_m	$a_{m,B} = A/N_{m,B}$	m^2
surface coverage	θ	$\theta = N_B{}^\sigma/N_{m,B}$	1
contact angle	θ		1, rad
film thickness	t, h, δ		m
thickness of (surface or interfacial) layer	τ, δ, t		m
surface tension, interfacial tension	γ, σ	$\gamma = (\partial G/\partial A_s)_{T,p}$	N m^{-1}, J m^{-2}
film tension	Σ_f	$\Sigma_f = 2\gamma_f$	N m^{-1}
reciprocal thickness of the double layer	κ	$\kappa = [2F^2I_c/\varepsilon RT]^{1/2}$	m^{-1}
average molar masses			
number–average	M_n	$M_n = \Sigma n_i M_i/\Sigma n_i$	kg mol^{-1}
mass–average	M_m	$M_m = \Sigma n_i M_i^2/\Sigma n_i M_i$	kg mol^{-1}
Z–average	M_Z	$M_Z = \Sigma n_i M_i^3/\Sigma n_i M_i^2$	kg mol^{-1}
sedimentation coefficient	s	$s = v/a$	s
van der Waals constant	λ		J
retarded van der Waals constant	β, B		J
van der Waals–Hamaker constant	A_H		J
surface pressure	π^s, π	$\pi^s = \gamma^0 - \gamma$	N m^{-1}

Transport Properties

Name	Symbol	Definition	SI unit
flux (of a quantity X)	J_X, J	$J_X = A^{-1}\,dX/dt$	(varies)
volume flow rate	$q_V, \dot{V}$	$q_v = dV/dt$	m^3 s^{-1}
mass flow rate	$q_m, \dot{m}$	$q_m = dm/dt$	kg s^{-1}
mass transfer coefficient	k_d		m s^{-1}
heat flow rate	ϕ	$\phi = dq/dt$	W
heat flux	J_q	$J_q = \phi/A$	W m^{-2}
thermal conductance	G	$G = \phi/\Delta T$	W K^{-1}
thermal resistance	R	$R = 1/G$	K W^{-1}
thermal conductivity	λ, k	$\lambda = J_q/(dT/dl)$	W m^{-1} K^{-1}

Name	Symbol	Definition	SI unit
coefficient of heat transfer	$h, (k, K, \alpha)$	$h = J_q/\Delta T$	$\text{W m}^{-2}\text{ K}^{-1}$
thermal diffusivity	a	$a = \lambda/\rho c_p$	$\text{m}^2\text{ s}^{-1}$
diffusion coefficient	D	$D = J_n/(dc/dl)$	$\text{m}^2\text{ s}^{-1}$

The following symbols are used in the definitions of the dimensionless quantities: mass (m), time (t), volume (V), area (A), density (ρ), speed (v), length (l), viscosity (η), pressure (p), acceleration of free fall (g), cubic expansion coefficient (α), temperature (T), surface tension (γ), speed of sound (c), mean free path (λ), frequency (f), thermal diffusivity (a), coefficient of heat transfer (h), thermal conductivity (k), specific heat capacity at constant pressure (c_p), diffusion coefficient (D), mole fraction (x), mass transfer coefficient (k_d), permeability (μ), electric conductivity (κ), and magnetic flux density (B).

Name	Symbol	Definition	SI unit
Reynolds number	Re	$Re = \rho v l/\eta$	1
Euler number	Eu	$Eu = \Delta p/\rho v^2$	1
Froude number	Fr	$Fr = v/(lg)^{1/2}$	1
Grashof number	Gr	$Gr = l^3 g\alpha\Delta T\rho^2/\eta^2$	1
Weber number	We	$We = \rho v^2 l/\gamma$	1
Mach number	Ma	$Ma = v/c$	1
Knudsen number	Kn	$Kn = \lambda/l$	1
Strouhal number	Sr	$Sr = lf/v$	1
Fourier number	Fo	$Fo = at/l^2$	1
Péclet number	Pe	$Pe = vl/a$	1
Rayleigh number	Ra	$Ra = l^3 g\alpha\Delta T\rho/\eta a$	1
Nusselt number	Nu	$Nu = hl/k$	1
Stanton number	St	$St = h/\rho v c_p$	1
Fourier number for mass transfer	Fo^*	$Fo^* = Dt/l^2$	1
Péclet number for mass transfer	Pe^*	$Pe^* = vl/D$	1
Grashof number for mass transfer	Gr^*	$Gr^* = l^3 g\left(\dfrac{\partial p}{\partial x}\right)_{T,p}\left(\dfrac{\Delta xp}{\eta}\right)$	1
Nusselt number for mass transfer	Nu^*	$Nu^* = k_d l/D$	1
Stanton number for mass transfer	St^*	$St^* = k_d/v$	1
Prandtl number	Pr	$Pr = \eta/\rho a$	1
Schmidt number	Sc	$Sc = \eta/\rho D$	1
Lewis number	Le	$Le = a/D$	1
magnetic Reynolds number	Rm, Re_m	$Rm = v\mu\kappa l$	1
Alfvén number	Al	$Al = v(\rho\mu)^{1/2}/B$	1
Hartmann number	Ha	$Ha = Bl\,(\kappa/\eta)^{1/2}$	1
Cowling number	Co	$Co = B^2/\mu\rho v^2$	1

NOMENCLATURE OF CHEMICAL COMPOUNDS

The International Union of Pure and Applied Chemistry (IUPAC) maintains several commissions that deal with the naming of chemical substances. In general, the approach of IUPAC is to present rules for arriving at names in a systematic manner, rather than recommending a unique name for each compound. Thus there are often several alternative "IUPAC names", depending on which nomenclature system is used, each of which may have advantages in specific applications. However, each of these names will be unambiguous.

Organizations such as the Chemical Abstacts Service and the Beilstein Institute that prepare indexes to the chemical literature must adopt a system for selecting unique names in order to avoid excessive cross referencing. Chemical Abstracts Service uses a system which groups together compounds derived from a single parent compound. Thus most index names are inverted (e.g., Benzene, bromo rather than bromobenzene; Acetic acid, sodium salt rather than sodium acetate).

Recommended names for the most common substituent groups, ligands, ions, and organic rings are given in the two following tables, "Nomenclature for Inorganic Ions and Ligands" and "Organic Substituent Groups and Ring Systems". For the basics of macromolecular nomenclature, see "Nomenclature for Organic Polymers" in Section 13.

Some of the most useful recent guides to chemical nomenclature, prepared by IUPAC and other organizations such as the International Union of Biochemistry and Molecular Biology (IUBMB) and the American Chemical Society are listed below . These books contain citations to the more detailed nomenclature documents in each area. Two very useful web sites providing links to nomenclature documents are:

www.iupac.org/publications/index.html
www.chem.qmul.ac.uk/iupac/

Inorganic Chemistry

Nomenclature of Inorganic Chemistry - IUPAC Recommendations 2005. Connelly, N.G., Damhus, T., Hartshorn, R. M., and Hutton, A. T. The Royal Society of Chemistry, 2005.

Block, B. P., Powell, W. H., and Fernelius, W. C., *Inorganic Chemical Nomenclature, Principles and Practice,* American Chemical Society, Washington, 1990.

Organic Chemistry

International Union of Pure and Applied Chemistry, *A Guide to IUPAC Nomenclature of Organic Compounds, Recommendations 1993,* Panico, R., Powell, W. H., and Richer, J.-C., Eds., Blackwell Scientific Publications, Oxford, 1993.

International Union of Pure and Applied Chemistry, *Glossary of Class Names of Organic Compounds and Reactive Intermediates Based on Structure,* Moss, G. P., Smith, P. A. S., and Tavernier, D., Eds., *Pure & Appl. Chem,* 67, 1307, 1995.

Rhodes, P. H., *The Organic Chemist's Desk Reference,* Chapman & Hall, London, 1995.

International Union of Pure and Applied Chemistry, *Basic Terminology of Stereochemistry,* Moss, G. P., Ed., *Pure & Applied Chemistry,* 68, 2193, 1996.

Macromolecular Chemistry

International Union of Pure and Applied Chemistry, *Compendium of Macromolecular Nomenclature,* Metanomski, W. V., Ed., Blackwell Scientific Publications, Oxford, 1991.

International Union of Pure and Applied Chemistry, *Glossary of Basic Terms in Polymer Science,* Jenkins, A.D., Kratochvil, P., Stepto, R. F. T., and Suter, U. W., Eds., *Pure & Appl. Chem,* 68, 2287, 1996.

Biochemistry

International Union of Biochemistry and Molecular Biology, *Biochemical Nomenclature and Related Documents, 2nd Edition, 1992,* Portland Press, London, 1993; includes recommendations of the IUPAC-IUBMB Joint Commission on Biochemical Nomenclature.

International Union of Biochemistry and Molecular Biology, *Enzyme Nomenclature, 1992,* Academic Press, Orlando, FL, 1992.

IUPAC-IUBMB Joint Commission on Biochemical Nomenclature, *Nomenclature of Carbohydrates, Recommendations 1996,* McNaught, A. D., Ed., *Pure & Appl. Chem.,* 68, 1919, 1996.

General

Principles of Chemical Nomenclature: a Guide to IUPAC Recommendations, Leigh, G. J., Favre, H. A. and Metanomski, W. V., Blackwell Science, 1998.

Chemical Abstracts Service, *Naming and Indexing Chemical Substances for Chemical Abstracts, Appendix IV, Chemical Abstracts 1994 Index Guide.*

NOMENCLATURE FOR INORGANIC IONS AND LIGANDS

Willem H. Koppenol

The entries below were selected from Table IX of Connelly, N. G., Damhus, T., Hartshorn, R. M. and Hutton, A. T., Eds., *Nomenclature of Inorganic Chemistry. IUPAC Recommendations 2005*, The Royal Society of Chemistry, 2005. Two changes were made: in the case of the hypohalides, the oxidohalogenate names are listed, not the new halooxygenate names. Thus, for BrO⁻ the still acceptable name "oxidobromate(1−)" is listed, not the more correct, but less palatable, "bromooxygenate(1−)". Similarly, and for reasons of consistency, ClO• is not named oxygen (mono)chloride, but chlorine mono(o)oxide. The symbol '⌣' is used for dividing names when this is made necessary by a line break. When the name is reconstructed from the name given in the table, this symbol should be omitted. Thus, all *hyphens* in the table are true parts of the names. The symbols '>' and '<' placed next to an element symbol both denote two single bonds connecting the atom in question to two other atoms. For a given compound, the various systematic names, if applicable, are given in the order: stoichiometric names, substitutive names, additive names and hydrogen names. Acceptable names which are not entirely systematic (or not formed according to any of the systems mentioned above) are given at the end after a semicolon. No order of preference is implied by the order in which formulae and names are listed. Reprinted by permission of IUPAC.

Formula for uncharged atom or group	Name			
	Uncharged atoms or molecules (including zwitterions and radicals) or substituent groups[a]	*Cations (including cation radicals) or cationic substituent groups[a]*	*Anions (including anion radicals) or anionic substituent groups[b]*	*Ligands[c]*
H	hydrogen H•, hydrogen(•), monohydrogen (natural or unspecified isotopic composition) ^{1}H•, protium(•), monoprotium ^{2}H• = D•, deuterium(•), monodeuterium ^{3}H• = T•, tritium(•), monotritium	hydrogen (general) H⁺, hydrogen(1+), hydron (natural or unspecified isotopic composition) ^{1}H⁺, protium(1+), proton ^{2}H⁺ = D⁺, deuterium(1+), deuteron ^{3}H⁺ = T⁺, tritium(1+), triton	hydride (general) H⁻, hydride (natural or unspecified isotopic composition) ^{1}H⁻, protide ^{2}H⁻ = D⁻, deuteride ^{3}H⁻ = T⁻, tritide	hydrido protido deuterido tritido
H_2	H_2, dihydrogen D_2, dideuterium T_2, ditritium	H_2•⁺, dihydrogen(•1+) $^1H_2$•⁺, diprotium(•1+) D_2•⁺, dideuterium(•1+) T_2•⁺, ditritium(•1+)		
D, see H				
D_2, see H_2				
T, see H				
T_2, see H_2				
F	fluorine F•, fluorine(•), monofluorine −F, fluoro	fluorine (general) F⁺, fluorine(1+)	fluoride (general) F⁻, fluoride(1−); fluoride	fluorido (general) F⁻, fluorido(1−); fluorido
F_2	F_2, difluorine	F_2•⁺, difluorine(•1+)	F_2•⁻, difluoride(•1−)	F_2, difluorine
Cl	chlorine (general) Cl•, chlorine(•), monochlorine −Cl, chloro	chlorine (general) Cl⁺, chlorine(1+)	chloride (general) Cl⁻, chloride(1−); chloride	chlorido (general) Cl⁻, chlorido(1−); chlorido
Cl_2	Cl_2, dichlorine	Cl_2•⁺, dichlorine(•1+)	Cl_2•⁻, dichloride(•1−)	Cl_2, dichlorine Cl_2•⁻, dichlorido(•1−)
Br	bromine (general) Br•, bromine(•), monobromine −Br, bromo	bromine (general) Br⁺, bromine(1+)	bromide (general) Br⁻, bromide(1−); bromide	bromido (general) Br⁻, bromido(1−); bromido
Br_2	Br_2, dibromine	Br_2•⁺, dibromine(•1+)	Br_2•⁻, dibromide(•1−)	Br_2, dibromine
I	iodine (general) I•, iodine(•), monoiodine −I, iodo	iodine (general) I⁺, iodine(1+)	iodide (general) I⁻, iodide(1−); iodide	iodido (general) I⁻, iodido(1−); iodido
I_2	I_2, diiodine	I_2•⁺, diiodine(•1+)	I_2•⁻, diiodide(•1−)	I_2, diiodine

ClO	ClO, chlorine mon(o)oxide ClO$^{\bullet}$, oxidochlorine($\bullet$); chlorosyl –ClO, oxo-λ^3-chloranyl; chlorosyl –OCl, chlorooxy		ClO$^-$, oxidochlorate(1–); hypochlorite	ClO$^-$, oxidochlorato(1–); hypochlorito
ClO$_2$	ClO$_2$, chlorine dioxide ClO$_2^{\bullet}$, dioxidochlorine($\bullet$) ClOO$^{\bullet}$, chloridodioxygen $\circlearrowleft$ (O–O) ($\bullet$), –ClO$_2$, dioxo-λ^5-chloranyl; chloryl –OClO, oxo-λ^3-chloranyloxy	ClO$_2^+$, dioxidochlorine(1+) (*not* chloryl)	ClO$_2^-$, dioxidochlorate(1–); chlorite	ClO$_2^-$, dioxidochlorato(1–); chlorito
ClO$_3$	ClO$_3$, chlorine trioxide ClO$_3^{\bullet}$, trioxidochlorine($\bullet$) –ClO$_3$, trioxo-λ^7-chloranyl; perchloryl –OClO$_2$, dioxo-λ^5-chloranyloxy	ClO$_3^+$, trioxidochlorine(1+) (*not* perchloryl)	ClO$_3^-$, trioxidochlorate(1–); chlorate	ClO$_3^-$, trioxidochlorato(1–); chlorato
ClO$_4$	ClO$_4$, chlorine tetraoxide ClO$_4^{\bullet}$, tetraoxidochlorine($\bullet$) –OClO$_3$, trioxo-λ^7-chloranyloxy		ClO$_4^-$, tetraoxidochlorate(1–); perchlorate	ClO$_4^-$, tetraoxidochlorato(1–); perchlorato
IO	IO, iodine mon(o)oxide IO$^{\bullet}$, oxidoiodine($\bullet$); iodosyl –IO, oxo-λ^3-iodanyl; iodosyl –OI, iodooxy	IO$^+$, oxidoiodine(1+) (*not* iodosyl)	IO$^-$, oxidoiodate(1–); hypoiodite IO$^{\bullet 2-}$, oxidoiodate($\bullet$2–)	IO$^-$, oxidoiodato(1–); hypoiodito
IO$_2$	IO$_2$, iodine dioxide IO$_2^{\bullet}$, dioxidoiodine($\bullet$) –IO$_2$, dioxo-λ^5-iodanyl; iodyl –OIO, oxo-λ^3-iodanyloxy	IO$_2^+$, dioxidoiodine(1+) (*not* iodyl)	IO$_2^-$, dioxidoiodate(1–); iodite	IO$_2^-$, dioxidoiodato(1–); iodito
IO$_3$	IO$_3$, iodine trioxide IO$_3^{\bullet}$, trioxidoiodine($\bullet$) –IO^3, trioxo-λ^7-iodanyl; periodyl –OIO$_2$, dioxo-λ^5-iodanyloxy	IO$_3^+$, trioxidoiodine(1+) (*not* periodyl)	IO$_3^-$, trioxidoiodate(1–); iodate	IO$_3^-$, trioxidoiodato(1–); iodato
IO$_4$	IO$_4$, iodine tetraoxide IO$_4^{\bullet}$, tetraoxidoiodine($\bullet$) –OIO$_3$, trioxo-λ^7-iodanyloxy		IO$_4^-$, tetraoxidoiodate(1–); periodate	IO$_4^-$, tetraoxidoiodato(1–); periodato
O	oxygen (general) O, monooxygen O$^{2\bullet}$, oxidanylidene, monooxygen(2$\bullet$) >O, oxy, epoxy (in rings) =O, oxo	oxygen (general) O$^{\bullet +}$, oxygen($\bullet$1+)	oxide (general) O$^{\bullet -}$, oxidanidyl, oxide($\bullet$1–) O^{2-}, oxide(2–); oxide –O$^-$, oxido	O^{2-}, oxido
O$_2$	O$_2$, dioxygen O$_2^{2\bullet}$, dioxidanediyl, dioxygen(2$\bullet$) –OO–, dioxidanediyl; peroxy	O$_2^{\bullet +}$, dioxidanyliumyl, dioxygen($\bullet$1+) O$_2^{2+}$, dioxidanebis(ylium), dioxygen(2+)	O$_2^{\bullet -}$, dioxidanidyl, dioxide($\bullet$1–); superoxide (*not* hyperoxide) O$_2^{2-}$, dioxidanediide, dioxide(2–); peroxide	dioxido (general) O$_2$, dioxygen O$_2^{\bullet -}$, dioxido($\bullet$1–); superoxido O$_2^{2-}$, dioxidanediido, dioxido(2–); peroxido
O$_3$	O$_3$, trioxygen; ozone –OOO–, trioxidanediyl		O$_3^{\bullet -}$, trioxidanidyl, trioxide($\bullet$1–); ozonide	O$_3$, trioxygen; ozone O$_3^{\bullet -}$, trioxido($\bullet$1–); ozonido
HO	HO$^{\bullet}$, oxidanyl, hydridooxygen($\bullet$); hydroxyl –OH, oxidanyl; hydroxy	HO$^+$, oxidanylium, hydridooxygen(1+); hydroxylium	HO$^-$, oxidanide, hydroxide	HO$^-$, oxidanido; hydroxido
HO$_2$	HO$_2^{\bullet}$, dioxidanyl, hydridodioxygen($\bullet$) hydrogen dioxide –OOH, dioxidanyl; hydroperoxy	HO$_2^+$, dioxidanylium, hydridodioxygen(1+)	HO$_2^-$, dioxidanide, hydrogen(peroxide)(1–)	HO$_2^-$, dioxidanido, hydrogen(peroxido)(1–)
S	sulfur (general) S, monosulfur =S, sulfanylidene; thioxo –S–, sulfanediyl	sulfur (general) S$^+$, sulfur(1+)	sulfide (general) S$^{\bullet -}$, sulfanidyl, sulfide($\bullet$1–) S^{2-}, sulfanediide, sulfide(2–); sulfide –S$^-$, sulfido	sulfido (general) S$^{\bullet -}$, sulfanidyl, sulfido($\bullet$1–) S^{2-}, sulfanediido, sulfido(2–)

S_2	S_2, disulfur −SS−, disulfanediyl >S=S, sulfanylidene-λ^4-sulfanediyl; sulfinothioyl	$S_2^{\bullet+}$, disulfur($\bullet$1+)	$S_2^{\bullet-}$, disulfanidyl, disulfide($\bullet$1−) S_2^{2-}, disulfide(2−), disulfanediide −SS−, disulfanidyl	S_2^{2-}, disulfido(2−), disulfanediido
HS	HS$^{\bullet}$, sulfanyl, hydridosulfur($\bullet$) −SH, sulfanyl	HS$^+$, sulfanylium, hydridosulfur(1+)	HS$^-$, sulfanide, hydrogen(sulfide)(1−)	HS$^-$, sulfanido, hydrogen(sulfido)(1−)
SO	SO, sulfur mon(o)oxide [SO], oxidosulfur >SO, oxo-λ^4-sulfanediyl; sulfinyl	SO$^{\bullet+}$, oxidosulfur($\bullet$1+) (*not* sulfinyl or thionyl)	SO$^{\bullet-}$, oxidosulfate($\bullet$1−)	[SO], oxidosulfur
SO_2	SO_2, sulfur dioxide [SO_2], dioxidosulfur >SO_2, dioxo-λ^6-sulfanediyl; sulfuryl, sulfonyl		$SO_2^{\bullet-}$, dioxidosulfate($\bullet$1−) SO_2^{2-}, dioxidosulfate(2−), sulfanediolate	[SO_2], dioxidosulfur SO_2^{2-}, dioxidosulfato(2−), sulfanediolato
SO_3	SO_3, sulfur trioxide		$SO_3^{\bullet-}$, trioxidosulfate($\bullet$1−) SO_3^{2-}, trioxidosulfate(2−); sulfite −S(O)$_2$(O$^-$), oxidodioxo-λ^6-sulfanyl; sulfonato	SO_3^{2-}, trioxidosulfato(2−); sulfito
SO_4	−OS(O)$_2$O−, sulfonylbis(oxy)		$SO_4^{\bullet-}$, tetraoxidosulfate($\bullet$1−) SO_4^{2-}, tetraoxidosulfate(2−); sulfate	SO_4^{2-}, tetraoxidosulfato(2−); sulfato
S_2O_3			$S_2O_3^{\bullet-} = SO_3S^{\bullet-}$, trioxido-1$\kappa^3O$-disulfate($S$–$S$)($\bullet$1−), trioxidosulfidosulfate($\bullet$1−) $S_2O_3^{2-} = SO_3S^{2-}$, trioxido-1κ^3O-disulfate(S–S)(2−), trioxidosulfidosulfate(2−); thiosulfate, sulfurothioate	$S_2O_3^{2-} = SO_3S^{2-}$, trioxido-1κ^3O-disulfato(S–S)(2−), trioxidosulfidosulfato(2−); thiosulfato, sulfurothioato
Se	Se (general) Se, monoselenium >Se, selanediyl =Se, selanylidene; selenoxo	selenium	selenide (general) Se$^{\bullet-}$, selanidyl, selenide($\bullet$1−) Se^{2-}, selanediide, selenide(2−); selenide	selenido (general) Se$^{\bullet-}$, selanidyl, selenido($\bullet$1−) Se^{2-}, selanediido, selenido(2−)
SeO	SeO, selenium mon(o)oxide [SeO], oxidoselenium >SeO, seleninyl			[SeO], oxidoselenium
SeO_2	SeO_2, selenium dioxide [SeO_2], dioxidoselenium >SeO_2, selenonyl		SeO_2^{2-}, dioxidoselenate(2−)	[SeO_2], dioxidoselenium SeO_2^{2-}, dioxidoselenato(2−)
SeO_3	SeO_3, selenium trioxide		$SeO_3^{\bullet-}$, trioxidoselenate($\bullet$1−) SeO_3^{2-}, trioxidoselenate(2−); selenite	SeO_3^{2-}, trioxidoselenato(2−); selenito
SeO_4			SeO_4^{2-}, tetraoxidoselenate(2−); selenate	SeO_4^{2-}, tetraoxidoselenato(2−); selenato
Te	tellurium >Te, tellanediyl =Te, tellanylidene; telluroxo	tellurium	telluride (general) Te$^{\bullet-}$, tellanidyl, telluride($\bullet$1−) Te^{2-}, tellanediide, telluride(2−); telluride	tellurido (general) Te$^{\bullet-}$, tellanidyl, tellurido($\bullet$1−) Te^{2-}, tellanediido, tellurido(2−)
CrO_2	CrO_2, chromium dioxide, chromium(IV) oxide			
UO_2	UO_2, uranium dioxide	UO_2^+, dioxidouranium(1+) [*not* uranyl(1+)] UO_2^{2+}, dioxidouranium(2+) [*not* uranyl(2+)]		

NpO_2	NpO_2, neptunium dioxide	NpO_2^+, dioxidoneptunium(1+) [*not* neptunyl(1+)] NpO_2^{2+}, dioxidoneptunium(2+) [*not* neptunyl(2+)]		
PuO_2	PuO_2, plutonium dioxide	PuO_2^+, dioxidoplutonium(1+) [not plutonyl(1+)] PuO_2^{2+}, dioxidoplutonium(2+) [not plutonyl (2+)]		
N	nitrogen $N^•$, nitrogen($•$), mononitrogen $-N<$, azanetriyl; nitrilo $-N=$, azanylylidene $\equiv N$, azanylidyne	nitrogen (general) N^+, nitrogen(1+)	nitride (general) N^{3-}, nitride(3−), azanetriide; nitride $=N^-$, azanidylidene; amidylidene $-N^{2-}$, azanediidyl	N^{3-}, nitrido(3−), azanetriido
N_2	N_2, dinitrogen $=N^+=N^-$, (azanidylidene)azaniumylidene; diazo $-N=N-$, diazane-1,2-diylidene; hydrazinediylidene $=NN=$, diazene-1,2-diyl; azo	$N_2^{•+}$, dinitrogen(1+) N_2^{2+}, dinitrogen(2+) $-N^+\equiv N$, diazyn-1-ium-1-yl	N_2^{2-}, dinitride(2−) N_2^{4-}, dinitride(4−), diazanetetraide; hydrazinetetraide	N_2, dinitrogen N_2^{2-}, dinitrido(2−) N_2^{4-}, dinitrido(4−), diazanetetraido; hydrazinetetraido
N_3	$N_3^•$, trinitrogen($•$) $-N=N^+=N^-$, azido		N_3^-, trinitride(1−); azide	N_3^-, trinitrido(1−); azido
NH	$NH^{2•}$, azanylidene, hydridonitrogen(2$•$); nitrene $>NH$, azanediyl $=NH$, azanylidene; imino	NH^+, azanyliumdiyl, hydridonitrogen(1+) NH^{2+}, azanebis(ylium), hydridonitrogen(2+)	NH^-, azanidyl, hydridonitrate(1−) NH^{2-}, azanediide, hydridonitrate(2−); imide $-NH^-$, azanidyl; amidyl	NH^{2-}, azanediido, hydridonitrato(2−); imido
NH_2	$NH_2^•$, azanyl, dihydridonitrogen($•$); aminyl $-NH_2$, azanyl; amino	NH_2^+, azanylium, dihydridonitrogen(1+)	NH_2^-, azanide, dihydridonitrate(1−); amide	NH_2^-, azanido, dihydridonitrato(1−), amido
NH_3	NH_3, azane (parent hydride name), amine (parent name for certain organic derivatives), trihydridonitrogen; ammonia	$NH_3^{•+}$, azaniumyl, trihydridonitrogen($•$1+) $-NH_3^+$, azaniumyl; ammonio	$NH_3^{•-}$, azanuidyl, trihydridonitrate($•$1−)	NH_3, ammine
NH_4	$NH_4^•$, λ^5-azanyl, tetrahydridonitrogen($•$)	NH_4^+, azanium; ammonium		
H_2NO	$H_2NO^•$, aminooxidanyl, dihydridooxidonitrogen($•$); aminoxyl $HONH^•$, hydroxyazanyl, hydridohydroxidonitrogen($•$) $-NH(OH)$, hydroxyazanyl, hydroxyamino $-ONH_2$, aminooxy $-NH_2(O)$, oxo-λ^5-azanyl; azinoyl		$HONH^-$, hydroxyazanide, hydridohydroxidonitrate(1−) H_2NO^-, azanolate, aminooxidanide, dihydridooxidonitrate(1−)	$NHOH^-$, hydroxyazanido, hydridohydroxidonitrato(1−) H_2NO^-, azanolato, aminooxidanido, dihydridooxidonitrato(1−)
N_2H_2	$HN=NH$, diazene $^-N=NH_2^+$, diazen-2-ium-1-ide $H_2NN^{2•}$, diazanylidene, hydrazinylidene $=NNH_2$, diazanylidene; hydrazinylidene $^•HNNH^•$, diazane-1,2-diyl; hydrazine-1,2-diyl $-HNNH-$, diazane-1,2-diyl; hydrazine-1,2-diyl	$HNNH^{2+}$, diazynediium	$HNNH^{2-}$, diazane-1,2-diide, hydrazine-1,2-diide H_2NN^{2-}, diazane-1,1-diide, hydrazine-1,1-diide	$HN=NH$, diazene $^-N=NH_2^+$, diazen-2-ium-1-ido $HNNH^{2-}$, diazane-1,2-diido, hydrazine-1,2-diido H_2NN^{2-}, diazane-1,1-diido, hydrazine-1,1-diido

N_2H_3	$H_2NNH^•$, diazanyl, trihydrido⌒ dinitrogen(N–N)($•$); hydrazinyl $-NHNH_2$, diazanyl; hydrazinyl $^{2-}NNH_3^+$, diazan-2-ium-1,1-diide	$H_2N=NH^+$, diazenium	H_2NNH^-, diazanide, hydrazinide	$^{2-}NNH_3^+$, diazan-2-ium-1,1-diido H_2NNH^-, diazanido, hydrazinido
N_2H_4	H_2NNH_2, diazane (parent hydride name), hydrazine (parent name for organic derivatives) $^-NHNH_3^+$, diazan-2-ium-1-ide	$H_2NNH_2^{•+}$, diazaniumyl, bis(dihydridonitrogen)⌒ (N–N)($•1+$); hydraziniumyl $H_2N=NH_2^{2+}$, diazenediium		H_2NNH_2, diazane, hydrazine $^-NHNH_3^+$, diazan-2-ium-1-ido
NO	NO, nitrogen mon(o)oxide (*not* nitric oxide) $NO^•$, oxoazanyl, oxidonitrogen($•$); nitrosyl $-N=O$, oxoazanyl; nitroso $>N(O)^-$, oxo-λ^5-azanyl; azoryl $=N(O)^-$, oxo-λ^5-azanylidene; azorylidene $\equiv N(O)$, oxo-λ^5-azanylidyne; azorylidyne $-O^+=N^-$, azanidylideneoxidaniumyl	NO^+, oxidonitrogen(1+) (*not* nitrosyl) $NO^{•2+}$, oxidonitrogen(2+)	NO^-, oxidonitrate(1−) $NO^{(2•)-}$, oxidonitrate(2•1−)	NO, oxidonitrogen (general); nitrosyl = oxidonitrogen-κN (general) NO^+, oxidonitrogen(1+) NO^-, oxidonitrato(1−)
NO_2	NO_2, nitrogen dioxide $NO_2^• = ONO^•$, nitrosooxidanyl, dioxidonitrogen($•$); nitryl $-NO_2$, nitro $-ONO$, nitrosooxy	NO_2^+, dioxidonitrogen(1+) (*not* nitryl)	NO_2^-, dioxidonitrate(1−); nitrite $NO_2^{•2-}$, dioxidonitrate($•2−$)	NO_2^-, dioxidonitrato(1−); nitrito $NO_2^{•2-}$, dioxidonitrato($•2−$)
NO_3	NO_3, nitrogen trioxide $NO_3^• = O_2NO^•$, nitrooxidanyl, trioxidonitrogen($•$) $ONOO^•$, nitrosodioxidanyl, (dioxido)oxidonitrogen($•$) $-ONO_2$, nitrooxy		NO_3^-, trioxidonitrate(1−); nitrate $NO_3^{•2-}$, trioxidonitrate($•2−$) $[NO(OO)]^-$, (dioxido)oxidonitrate(1−); peroxynitrite	NO_3^-, trioxidonitrato(1−); nitrato $NO_3^{•2-}$, trioxidonitrato($•2−$) $[NO(OO)]^-$, oxidoperoxidonitrato(1−); peroxynitrito
N_2O	N_2O, dinitrogen oxide (*not* nitrous oxide) NNO, oxidodinitrogen(N—N) $-N(O)=N-$, azoxy		$N_2O^{•-}$, oxidodinitrate($•1−$)	N_2O, dinitrogen oxide (general) NNO, oxidodinitrogen(N—N) $N_2O^{•-}$, oxidodinitrato($•1−$)
N_2O_3	N_2O_3, dinitrogen trioxide O_2NNO, trioxido-1κ^2O,2κO-dinitrogen(N–N) $NO^+NO_2^-$, oxidonitrogen(1+) dioxidonitrate(1−) ONONO, dinitrosooxidane, μ-oxidobis(oxidonitrogen)		$N_2O_3^{2-} = [O_2NNO]^{2-}$, trioxido-1$\kappa_2O$,2$\kappa O$-dinitrate($N$–$N$)(2−)	
N_2O_4	N_2O_4, dinitrogen tetraoxide O_2NNO_2, bis(dioxidonitrogen)⌒ (N–N) ONOONO, 1,2-dinitrosodioxidane, 2,5-diazy-1,3,4,6-tetraoxy-[6]catena $NO^+NO_3^-$, oxidonitrogen(1+) trioxidonitrate(1−)			
N_2O_5	N_2O_5, dinitrogen pentaoxide O_2NONO_2, dinitrooxidane, $NO_2^+NO_3^-$, dioxidonitrogen(1+) trioxidonitrate(1−)			
NS	NS, nitrogen monosulfide $NS^•$, sulfidonitrogen($•$) $-N=S$, sulfanylideneazanyl; thionitroso	NS^+, sulfidonitrogen(1+) (*not* thionitrosyl)	NS^-, sulfidonitrate(1−)	NS, sulfidonitrogen, sulfidonitrato, thionitrosyl (general) NS^+, sulfidonitrogen(1+) NS^-, sulfidonitrato(1−)

P	phosphorus (general) P•, phosphorus(•), monophosphorus >P–, phosphanetriyl	phosphorus (general) P⁺, phosphorus(1+)	phosphide (general) P⁻, phosphide(1–) P³⁻, phosphide(3–), phosphanetriide; phosphide	P³⁻, phosphido, phosphanetriido
PO	PO•, oxophosphanyl, oxidophosphorus(•), phosphorus mon(o)oxide; phosphoryl >P(O)–, oxo-λ⁵-phosphanetriyl; phosphoryl =P(O)–, oxo-λ⁵-phosphanylidene; phosphorylidene ≡P(O), oxo-λ⁵-phosphanylidyne; phosphorylidyne	PO⁺, oxidophosphorus(1+) (*not* phosphoryl)	PO⁻, oxidophosphate(1–)	
PO₂	–P(O)₂, dioxo-λ⁵-phosphanyl		PO₂⁻, dioxidophosphate(1–)	PO₂⁻, dioxidophosphato(1–)
PO₃			PO₃⁻, trioxidophosphate(1–) PO₃•²⁻, trioxidophosphate(•2–) PO₃³⁻, trioxidophosphate(3–); phosphite (PO₃⁻)ₙ = ⦅P(O)₂O⦆ₙⁿ⁻, *catena*-poly[(dioxidophosphate-μ-oxido)(1–)]; metaphosphate –P(O)(O⁻)₂, dioxidooxo-λ⁵-phosphanyl; phosphonato	PO₃⁻, trioxidophosphato(1–) PO₃•²⁻, trioxidophosphato(•2–) PO₃³⁻, trioxidophosphato(3–); phosphito
PO₄			PO₄•²⁻, tetraoxidophosphate(•2–) PO₄³⁻, tetraoxidophosphate(3–); phosphate	PO₄³⁻, tetraoxidophosphato(3–); phosphato
PS	PS•, sulfidophosphorus(•); –PS, thiophosphoryl	PS⁺, sulfidophosphorus(1+) (*not* thiophosphoryl)		
AsO₃			AsO₃³⁻, trioxidoarsenate(3–); arsenite, arsorite –As(=O)(O⁻)₂, dioxidooxo-λ⁵-arsanyl; arsonato	AsO₃³⁻, trioxidoarsenato(3–); arsenito, arsorito
AsO₄			AsO₄³⁻, tetraoxidoarsenate(3–); arsenate, arsorate	AsO₄³⁻, tetraoxidoarsenato(3–); arsenato, arsorato
VO	VO, vanadium(II) oxide, vanadium mon(o)oxide	VO²⁺, oxidovanadium(2+) (*not* vanadyl)		
CO	CO, carbon mon(o)oxide >C=O, carbonyl =C=O, carbonylidene	CO•⁺, oxidocarbon(•1+) CO²⁺, oxidocarbon(2+)	CO•⁻, oxidocarbonate(•1–)	CO, oxidocarbon, oxidocarbonato (general); carbonyl = oxidocarbon-κC (general) CO•⁺, oxidocarbon(•1+) CO•⁻, oxidocarbonato(•1–)
CO₂	CO₂, carbon dioxide, dioxidocarbon		CO₂•⁻, oxidooxomethyl, dioxidocarbonate(•1–)	CO₂, dioxidocarbon CO₂•⁻, oxidooxomethyl, dioxidocarbonato(•1–)
CO₃			CO₃•⁻, trioxidocarbonate(•1–), OCOO•⁻, (dioxido)oxidocarbonate(•1–), oxidoperoxidocarbonate(•1–) CO₃²⁻, trioxidocarbonate(2–); carbonate	CO₃²⁻, trioxidocarbonato(2–); carbonato
CS	carbon monosulfide >C=S, carbonothioyl; thiocarbonyl =C=S, carbonothioylidene	CS•⁺, sulfidocarbon(•1+)	CS•⁻, sulfidocarbonate(•1–)	CS, sulfidocarbon, sulfidocarbonato, thiocarbonyl (general); CS•⁺, sulfidocarbon(•1+) CS•⁻, sulfidocarbonato(•1–)
CS₂	CS₂, disulfidocarbon, carbon disulfide		CS₂•⁻, sulfidothioxomethyl, disulfidocarbonate(•1–)	CS₂, disulfidocarbon CS₂•⁻, sulfidothioxomethyl, disulfidocarbonato(•1–)

CN	CN•, nitridocarbon(•); cyanyl –CN, cyano –NC, isocyano	CN⁺, azanylidynemethylium, nitridocarbon(1+)	CN⁻, nitridocarbonate(1−); cyanide	nitridocarbonato (general) CN⁻, nitridocarbonato(1−); cyanido = [nitridocarbonato(1−)-κ*C*]
CNO	OCN•, nitridooxidocarbon(•) –OCN, cyanato –NCO, isocyanato –ONC, λ²-methylidene ⟲ azanylylideneoxy –CNO, (oxo-λ⁵- azanylidynemethyl)		OCN⁻, nitridooxidocarbonate(1−); cyanate ONC⁻, carbidooxidonitrate(1−); fulminate OCN•²⁻, nitridooxidocarbonate(•2−)	OCN⁻, nitridooxidocarbonato(1−); cyanato ONC⁻, carbidooxidonitrato(1−); fulminato
CNS	SCN•, nitridosulfidocarbon(•) –SCN, thiocyanato –NCS, isothiocyanato –SNC, λ²-methylidene ⟲ azanylylidenesulfanediyl –CNS, (sulfanylidene-λ⁵- azanylidynemethyl)		SCN⁻, nitridosulfidocarbonate(1−); thiocyanate SNC⁻, carbidosulfidonitrate(1−)	SCN⁻, nitridosulfidocarbonato(1−); thiocyanato SNC⁻, carbidosulfidonitrato(1−)
CNSe	SeCN•, nitridoselenidocarbon(•) –SeCN, selenocyanato –NCSe, isoselenocyanato –SeNC, λ²-methylidene ⟲ azanylylideneselanediyl –CNSe, (selanylidene-λ⁵- azanylidynemethyl)		SeCN⁻, nitridoselenidocarbonate(1−); selenocyanate SeNC⁻, carbidoselenidonitrate(1−)	SeCN⁻, nitridoselenidocarbonato(1−); selenocyanato SeNC⁻, carbidoselenidonitrato(1−)

[a] Where an element symbol occurs in the first column, the unmodified element name is listed in the second and third columns. The unmodified name is generally used when the element appears as an electropositive constituent in the construction of a stoichiometric name (Sections IR-5.2 and IR-5.4). Names of homoatomic cations consisting of the element are also constructed using the element name, adding multiplicative prefixes and charge numbers as applicable (Sections IR-5.3.2.1 to IR-5.3.2.3). The sections mentioned refer to parts of Nomenclature of Inorganic Chemistry. IUPAC Recommendations 2005, see above.

[b] Where an element symbol occurs in the first column, the fourth column gives the element name appropriately modified with the ending 'ide' (hydride, nitride, etc.). The 'ide' form of the element name is generally used when the element appears as an electronegative constituent in the construction of a stoichiometric name (Sections IR-5.2 and IR-5.4). Names of homoatomic anions consisting of the element in question are also constructed using this modified form, adding multiplicative prefixes and charge numbers as applicable (Sections IR-5.3.3.1 to IR-5.3.3.3). Examples are given in the Table of names of some specific anions, e.g. chloride(1−), oxide(2−), dioxide(2−). In certain cases, a particular anion has the 'ide' form itself as an accepted short name, e.g. chloride, oxide. If specific anions are named, the 'ide' form of the element name with no further modification is given as the first entry in the fourth column, with the qualifier '(general)'. The sections mentioned refer to parts of Nomenclature of Inorganic Chemistry. IUPAC Recommendations 2005, see above.

[c] Ligand names must be placed within enclosing marks whenever necessary to avoid ambiguity, cf. Section IR-9.2.2.3. Some ligand names must always be enclosed. For example, if 'dioxido' is cited as is, it must be enclosed so as to distinguish it from two 'oxido' ligands; if combined with a multiplicative prefix it must be enclosed because it starts with a multiplicative prefix itself. A ligand name such as 'nitridocarbonato' must always be enclosed to avoid interpreting it as two separate ligand names, 'nitrido' and 'carbonato'. In this table, however, these enclosing marks are omitted for the sake of clarity. Note that the ligand names given here with a charge number can generally also be used without it if it is not desired to make any implication regarding the charge of the ligand. For example, the ligand name '[dioxido(•1−)]' may be used if one wishes explicitly to consider the ligand to be the species dioxide(•1−), whereas the ligand name '(dioxido)' can be used if no such implications are desirable. The section mentioned refer to parts of Nomenclature of Inorganic Chemistry. IUPAC Recommendations 2005, see above.

ORGANIC SUBSTITUENT GROUPS AND RING SYSTEMS

The first part of this table lists substituent groups and their line formulas. A substituent group is defined by IUPAC as a group that replaces one or more hydrogen atoms attached to a parent structure. Such groups are sometimes called radicals, but IUPAC now reserves the term radical for a free molecular species with unpaired electrons. IUPAC does not recommend some of these names, which are marked here with asterisks (e.g., amyl*), but they are included in this list because they are often encountered in the older literature. Substituent group names which are formed by systematic rules (e.g., methyl from methane, ethyl from ethane, etc.) are included here only for the first few members of a homologous series.

In the second part of the table a number of common organic ring compounds are shown, with the conventional numbering of the ring positions indicated.

The help of Warren H. Powell in preparing this table is greatly appreciated. Pertinent references may be found in the table "Nomenclature of Chemical Compounds."

Substituent Groups

acetamido (acetylamino)	CH_3CONH-	cinnamoyl	$C_6H_5CH=CHCO-$
acetoacetyl	CH_3COCH_2CO-	cinnamyl (3-phenyl-2-propenyl)	$C_6H_5CH=CHCH_2-$
acetonyl	CH_3COCH_2-	cinnamylidene	$C_6H_5CH=CHCH=$
acetyl	CH_3CO-	cresyl* (hydroxymethylphenyl)	$HO(CH_3)C_6H_4-$
acryloyl* (1-oxo-2-propenyl)	$CH_2=CHCO-$	crotonoyl	$CH_3CH=CHCO-$
alanyl (from alanine)	$CH_3CH(NH_2)CO-$	crotyl (2-butenyl)	$CH_3CH=CHCH_2-$
β-alanyl	$H_2N(CH_2)_2CO-$	cyanamido (cyanoamino)	$NCNH-$
allyl (2-propenyl)	$CH_2=CHCH_2-$	cyanato	$NCO-$
allylidene (2-propenylidene)	$CH_2=CHCH=$	cyano	$NC-$
amidino (aminoiminomethyl)	$H_2NC(=NH)-$	decanedioyl	$-OC(CH_2)_8CO-$
amino	H_2N-	decanoyl	$CH_3(CH_2)_8CO-$
amyl* (pentyl)	$CH_3(CH_2)_4-$	diazo	$N_2=$
anilino (phenylamino)	C_6H_5NH-	diazoamino	$-NHN=N-$
anisidino	$CH_3OC_6H_4NH-$	disilanyl	H_3SiSiH_2-
anthranoyl (2-aminobenzoyl)	$2-H_2NC_6H_4CO-$	disiloxanyloxy	$H_3SiOSiH_2O-$
arsino	AsH_2-	disulfinyl	$-S(O)S(O)-$
azelaoyl (from azelaic acid)	$-OC(CH_2)_7CO-$	dithio	$-SS-$
azido	N_3-	enanthoyl* (heptanoyl)	$CH_3(CH_2)_5CO-$
azino	$=N-N=$	epoxy	$-O-$
azo	$-N=N-$	ethenyl (vinyl)	$CH_2=CH-$
azoxy	$-N(O)=N-$	ethynyl	$HC≡C-$
benzal* (benzylidene)	$C_6H_5CH=$	ethoxy	C_2H_5O-
benzamido (benzoylamino)	C_6H_5CONH-	ethyl	CH_3CH_2-
benzhydryl (diphenylmethyl)	$(C_6H_5)_2CH-$	ethylene	$-CH_2CH_2-$
benzoxy* (benzoyloxy)	C_6H_5COO-	ethylidene	$CH_3CH=$
benzoyl	C_6H_5CO-	ethylthio	C_2H_5S-
benzyl	$C_6H_5CH_2-$	formamido (formylamino)	$HCONH-$
benzylidene	$C_6H_5CH=$	formyl	$HCO-$
benzylidyne	$C_6H_5C=$	furmaroyl (from fumaric acid)	$-OCCH=CHCO-$
biphenylyl	$C_6H_5C_6H_5-$	furfuryl (2-furanylmethyl)	$OC_4H_3CH_2-$
biphenylene	$-C_6H_4-C_6H_4-$	furfurylidene (2-furanylmethylene)	$OC_4H_3CH=$
butoxy	C_4H_9O-	glutamoyl (from glutamic acid)	$-OC(CH_2)_2CH(NH_2)CO-$
sec-butoxy (1-methylpropoxy)	$C_2H_5CH(CH_3)O-$	glutaryl (from glutaric acid)	$-OC(CH_2)_3CO-$
tert-butoxy (1,1-dimethylethoxy)	$(CH_3)_3CO-$	glycylamino	H_2NCH_2CONH-
butyl	$CH_3(CH_2)_3-$	glycoloyl; glycolyl (hydroxyacetyl)	$HOCH_2CO-$
sec-butyl (1-methylpropyl)	$CH_3CH_2CH(CH_3)-$	glycyl (aminoacetyl)	H_2NCH_2CO-
tert-butyl (1,1-dimethylethyl)	$(CH_3)_3C-$	glyoxyloyl; glyoxylyl (oxoacetyl)	$HCOCO-$
butyryl (1-oxobutyl)	$CH_3(CH_2)_2CO-$	guanidino	$H_2NC(=NH)NH-$
caproyl* (hexanoyl)	$CH_3(CH_2)_4CO-$	guanyl (aminoiminomethyl)	$H_2NC(=NH)-$
capryl* (decanoyl)	$CH_3(CH_2)_8CO-$	heptadecanoyl	$CH_3(CH_2)_{15}CO-$
capryloyl* (octanoyl)	$CH_3(CH_2)_6CO-$	heptanamido	$CH_3(CH_2)_5CONH-$
carbamido (carbamoylamino)	$H_2NCONH-$	heptanedioyl	$-OC(CH_2)_5CO-$
carbamoyl (aminocarbonyl)	H_2NCO-	heptanoyl	$CH_3(CH_2)_5CO-$
carbamyl (aminocarbonyl)	H_2NCO-	hexadecanoyl	$CH_3(CH_2)_{14}CO-$
carbazoyl (hydrazinocarbonyl)	$H_2NNHCO-$	hexamethylene (1,6-hexanediyl)	$-(CH_2)_6-$
carbethoxy (ethoxycarbonyl)	C_2H_5OCO-	hexanedioyl	$-OC(CH_2)_4CO-$
carbonyl	$=C=O$	hippuryl (N-benzoylglycyl)	$C_6H_5CONHCH_2CO-$
carboxy	$HOOC-$	hydrazino	H_2NNH-
cetyl* (hexadecyl)	$CH_3(CH_2)_{15}-$	hydrazo	$-HNNH-$
chloroformyl (chlorcarbonyl)	$ClCO-$	hydrocinnamoyl	$C_6H_5(CH_2)_2CO-$

Name	Formula
hydroperoxy	HOO-
hydroxyamino	HONH-
hydroxy	HO-
imino	HN=
iodoso* (iodosyl)	OI-
iodyl	O_2I-
isoamyl* (isopentyl; 3-methylbutyl)	$(CH_3)_2CH(CH_2)_2$-
isobutenyl (2-methyl-1-propenyl)	$(CH_3)_2C{=}CH$-
isobutoxy (2-methylpropoxy)	$(CH_3)_2CHCH_2O$-
isobutyl (2-methylpropyl)	$(CH_3)_2CHCH_2$-
isobutylidene (3-methylpropylidene)	$(CH_3)_2CHCH{=}$
isobutyryl (2-methyl-1-oxopropyl)	$(CH_3)_2CHCO$-
isocyanato	OCN-
isocyano	CN-
isohexyl (4-methylpentyl)	$(CH_3)_2CH(CH_2)_3$-
isoleucyl (from isoleucine)	$C_2H_5CH(CH_3)CH(NH_2)CO$-
isonitroso* (hydroxyamino)	HON=
isopentyl (3-methylbutyl)	$(CH_3)_2CH(CH_2)_2$-
isopentylidene (3-methylbutylidene)	$(CH_3)_2CHCH_2CH{=}$
isopropenyl (1-methylethenyl)	$CH_2{=}C(CH_3)$-
isopropoxy (1-methylethoxy)	$(CH_3)_2CHO$-
isopropyl (1-methylethyl)	$(CH_3)_2CH$-
isopropylidene (1-methylethylidene)	$(CH_3)_2C{=}$
isothiocyanato (isothiocyano)	SCN-
isovaleryl* (3-methyl-1-oxobutyl)	$(CH_3)_2CHCH_2CO$-
lactoyl (from lactic acid)	$CH_3CH(OH)CO$-
lauroyl (from lauric acid)	$CH_3(CH_2)_{10}CO$-
lauryl (dodecyl)	$CH_3(CH_2)_{11}$-
leucyl (from leucine)	$(CH_3)_2CHCH_2CH(NH_2)CO$-
levulinoyl (from levulinic acid)	$CH_3CO(CH_2)_2CO$-
malonyl (from malonic acid)	$-OCCH_2CO$-
mandeloyl (from mandelic acid)	$C_6H_5CH(OH)CO$-
mercapto	HS-
mesityl	$2,4,6\text{-}(CH_3)_3C_6H_2$-
methacryloyl (from methacrylic acid)	$CH_2{=}C(CH_3)CO$-
methallyl (2-methyl-2-propenyl)	$CH_2{=}C(CH_3)CH_2$-
methionyl (from methionine)	$CH_3SCH_2CH_2CH(NH_2)CO$-
methoxy	CH_3O-
methyl	H_3C-
methylene	$H_2C{=}$
methylthio	CH_3S-
myristoyl (from myristic acid)	$CH_3(CH_2)_{12}CO$-
myristyl (tetradecyl)	$CH_3(CH_2)_{13}$-
naphthyl	$(C_{10}H_7)$-
naphthylene	$-(C_{10}H_6)$-
neopentyl (2,2-dimethylpropyl)	$(CH_3)_3CCH_2$-
nitramino (nitroamino)	O_2NNH-
nitro	O_2N-
nitrosamino (nitrosoamino)	ONNH-
nitrosimino (nitrosoimino)	ONN=
nitroso	ON-
nonanoyl (from nonanoic acid)	$CH_3(CH_2)_7CO$-
oleoyl (from oleic acid)	$CH_3(CH_2)_7CH{=}CH(CH_2)_7CO$-
oxalyl (from oxalic acid)	-OCCO-
oxo	O=
palmitoyl (from palmitic acid)	$CH_3(CH_2)_{14}CO$-
pentamethylene (1,5-pentanediyl)	$-(CH_2)_5$-
pentyl	$CH_3(CH_2)_4$-
tert-pentyl	$CH_3CH_2C(CH_3)_2$-
phenacyl	$C_6H_5COCH_2$-
phenacylidene	$C_6H_5COCH{=}$
phenethyl (2-phenylethyl)	$C_6H_5CH_2CH_2$-
phenoxy	C_6H_5O-
phenyl	C_6H_5-
phenylene (benzenediyl)	$-C_6H_4$-
phosphino* (phosphanyl)	H_2P-
phosphinyl* (phosphinoyl)	$H_2P(O)$-
phospho	O_2P-
phosphono	$(HO)_2P(O)$-
phthaloyl (from phthalic acid)	$1,2\text{-}C_6H_4(CO\text{-})_2$
picryl (2,4,6-trinitrophenyl)	$2,4,6\text{-}(NO_2)_3C_6H_2$-
pimeloyl (from pimelic acid)	$-OC(CH_2)_5CO$-
piperidino (1-piperidinyl)	$C_5H_{10}N$-
pivaloyl (from pivalic acid)	$(CH_3)_3CCO$-
prenyl (3-methyl-2-butenyl)	$(CH_3)_2C{=}CHCH_2$-
propargyl (2-propynyl)	$HC'CCH_2$-
1-propenyl	$-CH{=}CHCH_3$
2-propenyl (allyl)	$CH_2{=}CHCH_2$-
propionyl* (propanyl)	CH_3CH_2CO-
propoxy	$CH_3CH_2CH_2O$-
propyl	$CH_3CH_2CH_2$-
propylidene	$CH_3CH_2CH{=}$
pyrryl (pyrrolyl)	C_4H_4N-
salicyloyl (2-hydroxybenzoyl)	$2\text{-}HOC_6H_4CO$-
selenyl* (selanyl; hydroseleno)	HSe-
seryl (from serine)	$HOCH_2CH(NH_2)CO$-
siloxy	H_3SiO-
silyl	H_3Si-
silylene	$H_2Si{=}$
sorboyl (from sorbic acid)	$CH_3CH{=}CHCH{=}CHCO$-
stearoyl (from stearic acid)	$CH_3(CH_2)_{16}CO$-
stearyl (octadecyl)	$CH_3(CH_2)_{17}$-
styryl (2-phenylethenyl)	$C_6H_5CH{=}CH$-
suberoyl (from suberic acid)	$-OC(CH_2)_6CO$-
succinyl (from succinic acid)	$-OCCH_2CH_2CO$-
sulfamino (sulfoamino)	$HOSO_2NH$-
sulfamoyl (sulfamyl)	H_2NSO_2-
sulfanilyl [(4-aminophenyl)sulfonyl]	$4\text{-}H_2NC_6H_4SO_2$-
sulfeno	HOS-
sulfhydryl (mercapto)	HS-
sulfinyl	OS=
sulfo	HO_3S-
sulfonyl (sulfuryl)	$-SO_2$-
terephthaloyl	$1,4\text{-}C_6H_4(CO\text{-})_2$
tetramethylene	$-(CH_2)_4$-
thienyl (from thiophene)	(C_4H_3S)-
thiocarbonyl (carbothionyl)	=CS
thiocarboxy	HOSC-
thiocyanato (thiocyano)	NCS-
thionyl* (sulfinyl)	-SO-
threonyl (from threonine)	$CH_3CH(OH)CH(NH_2)CO$-
toluidino [(methylphenyl)amino]	$CH_3C_6H_4NH$-
toluoyl (methylbenzoyl)	$CH_3C_6H_4CO$-
tolyl (methylphenyl)	$CH_3C_6H_4$-
α-tolyl (benzyl)	$C_6H_5CH_2$-
tolylene (methylphenylene)	$-(CH_3C_6H_3)$-
tosyl [(4-methylphenyl) sulfonyl)]	$4\text{-}CH_3C_6H_4SO_2$-
triazano	H_2NNHNH-
trimethylene (1,3-propanediyl)	$-(CH_2)_3$-
trityl (triphenylmethyl)	$(C_6H_5)_3C$-
valeryl* (pentanoyl)	$CH_3(CH_2)_3CO$-
valyl (from valine)	$(CH_3)_2CHCH(NH_2)CO$-
vinyl (ethenyl)	$CH_2{=}CH$-
vinylidene (ethenylidene)	$CH_2{=}C{=}$
xylidino [(dimethylphenyl)amino]	$(CH_3)_2C_6H_3NH$-
xylyl (dimethylphenyl)	$(CH_3)_2C_6H_3$-
xylylene [phenelenebis(methylene)]	$-CH_2C_6H_4CH_2$-

Organic Ring Compounds

Cyclopropane   Spiropentane   Cyclobutane   Cyclopentane   Furan   Thiophene   Pyrrole (Azole)   2H-Pyrrole (2H-Azole)   Pyrazole (1,2-Diazole)   3H-Pyrrole (3H-Azole)

2H-Imidazole (1,3-Diazole)   1,2,3-Triazole   1,2,4-Triazole   1,2-Dithiole   1,3-Dithiole   3H-1,2-Oxathiole   Isoxazole (1,2-Oxazole)   Thiazole (1,3-Thiazole)   Oxazole (1,3-Oxazole)

Isothiazole (1,2-Thiazole)   1,2,3-Oxadiazole   1,2,4-Oxadiazole   1,2,5-Oxadiazole (Furazan)   1,3,4-Oxadiazole   1,2,3,4-Oxatriazole   1,2,3,5-Oxatriazole

3H-1,2,3-Dioxazole   1,2,4-Dioxazole   1,3,2-Dioxazole   1,3,4-Dioxazole   5H-1,2,5-Oxathiazole   1,3-Oxathiole   Benzene   Cyclohexane

2H-Pyran   4H-Pyran   2H-Pyran-2-one (2-Pyrone)   4H-Pyran-4-one (4-Pyrone)   1,2-Dioxin   1,3-Dioxin   Pyridine   Pyridazine   Pyrimidine   Pyrazine

Piperazine   1,3,5-Triazine (s-Triazine)   1,2,4-Triazine (as-Triazine)   1,2,3-Triazine (v-Triazine)   4H-1,2-Oxazine   2H-1,3-Oxazine   6H-1,3-Oxazine   6H-1,2-Oxazine   1,4-Oxazine

2H-1,2-Oxazine   4H-1,4-Oxazine   1,2,5-Oxathiazine   1,2,6-Oxathiazine   1,2,4-Oxadiazine   1,3,5-Oxadiazine   Morpholine   Azepine

Oxepin   Thiepin   4H-1,2-Diazepine   Indene   2H-Indene (Isoindene)   Benzofuran   Isobenzofuran   Benzo[b]thiophene

Benzo[c]thiophene   Indole   3H-Indole   1H-Indole   Cyclopenta[b]pyridine   Pyrano[3,4-b]-pyrrole   Indazole   Benzisoxazole (Indoxazene)

Benzoxazole

2,1-Benzisoxazole

Naphthalene

1,2,3,4-Tetra-
hydronaphthalene
(Tetralin)

Octahydronaphthalene
(Decalin)

2H-1-Benzopyran
(2H-Chromene)

2H-1-Benzopyran-2-one
(Coumarin)

4H-1-Benzopyran-4-one
(Chromen-4-one)

1H-2-Benzopyran-1-one
(Isocoumarin)

3H-2-Benzopyran-1-one
(Isochromen-3-one)

Quinoline

Isoquinoline

Cinnoline

Quinazoline

1,8-Naphthyridine

1,7-Naphthyridine

1,5-Naphthyridine

1,6-Naphthyridine

2H-1,3-Benzoxazine

2H-1,4-Benzoxazine

1H-2,3-Benzoxazine

4H-3,1-Benzoxazine

2H-1,2-Benzoxazine

4H-1,4-Benzoxazine

Anthracene

Phenanthrene

Phenalene

Fluorene

Carbazole

Xanthene

Acridine

Norpinane
(Bicyclo[3.1.1]heptane)

7H-Purine

R   = Nearly always methyl
R′ = Usually methyl
R″ = Various groups

Steroid ring system

SCIENTIFIC ABBREVIATIONS AND SYMBOLS

This table lists some abbreviations, acronyms, and symbols encountered in the physical sciences. Most entries in italic type are symbols for physical quantities; for more details on these, see the table "Symbols and Terminology for Physical and Chemical Quantities" in this section. Additional information on units may be found in the table "International System of Units" in Section 1. Many of the terms to which these abbreviations refer are included in the tables "Definitions of Scientific Terms" in Section 2 and "Techniques for Materials Characterization" in Section 12. Useful references for further information are given below.

Publication practices vary with regard to the use of capital or lower case letters for many abbreviations. An effort has been made to follow the most common practices in this table, but much variation is found in the literature. Likewise, policies on the use of periods in an abbreviation vary considerably. Periods are generally omitted in this table unless they are necessary for clarity. Periods should never appear in SI units. The SI prefixes (m, k, M, etc.) are included here, but they should never be used alone. Selected combinations of these prefixes with SI units (e.g., mg, kV, MW) are also included.

Abbreviations are listed in alphabetical order without regard to case. Entries beginning with Greek letters fall at the end of the table.

References

1. Mills, Ian, Ed., *Quantities, Units, and Symbols in Physical Chemistry*, Blackwell Scientific Publications, Oxford, 1993.
2. Kotyk, A., *Quantities, Symbols, Units, and Abbreviations in the Life Sciences*, Humana Press, Totawa, NJ, 1999.
3. Rhodes, P. H., *The Organic Chemist's Desk Reference*, Chapman & Hall, London, 1995.
4. Minkin, V., Glossary of Terms used in Theoretical Organic Chemistry, *Pure Appl. Chem.* 71, 1919-1981, 1999.
5. Brown, R. D., Ed., Acronyms Used in Theoretical Chemistry, *Pure Appl. Chem.* 68, 387-456, 1996.
6. *Quantities and Units, ISO Standards Handbook, Third Edition*, International Organization for Standardization, Geneva, 1993.
7. Cohen, E. R., and Giacomo, P., Symbols, Units, Nomenclature, and Fundamental Constants in Physics, *Physica* 146A, 1-68, 1987.
8. *Chemical Acronyms Database*, Indiana University, < www.oscar.chem. indiana.edu/cfdocs/ libchem/acronyms/ acronymsearch.html>.
9. *Acronyms and Symbols*, <www3.interscience.wiley.com/stasa/>.

A	ampere; adenine (in genetic code)
Å	ångström
A	absorbance; area; Helmholtz energy; mass number
A_H	Hall coefficient
A_r	atomic weight (relative atomic mass)
a	atto (SI prefix for 10^{-18})
a	absorption coefficient; acceleration; activity; van der Waals constant
a_0	Bohr radius
A/D	analog to digital
AAA	acetoacetanilide
AAO	acetaldehyde oxime
AAS	atomic absorption spectroscopy
ABA	abscisic acid
Abe	abequose
ABL	α-acetylbutyrolactone
abs	absolute
Ac	acetyl; acetate
ac, AC	alternating current
Aces	2-[(2-amino-2-oxoethyl)amino]ethanesulfonic acid
ACT	activated complex theory
ACTH	adrenocorticotropic hormone
Ad	adamantyl
Ada	[(carbamoylmethyl)imino]diacetic acid
Ade	adenine
Ado	adenosine
ADP	adenosine diphosphate
ads	adsorption
ae	eon (10^9 years)
AEP	1-(2-aminoethyl)piperazine
AES	atomic emission spectroscopy; Auger electron spectroscopy
AF	audio frequency
AFM	atomic force microscopy
AI	artificial intelligence
AIBN	2,2'-azobis[isobutyronitrile]
AICA	5-amino-1*H*-imidazole-4-carboxamide
AIM	atoms in molecules (method)
AIP	aluminum isopropoxide
Al	Alfén number
Ala	alanine
alc	alcohol
ALE	atomic layer epitaxy
aliph.	aliphatic
alk.	alkaline
All	allose
Alt	altrose
AM	amplitude modulation
Am	Amyl
am	amorphous solid
AMP	adenosine monophosphate
AMPD	2-amino-2-methyl-1,3-propanediol
AMTCS	amyltrichlorosilane [trichloropentylsilane]
amu	atomic mass unit (recommended symbol is u)
AN	acetonitrile
anh, anhyd	anhydrous
ANOVA	analysis of variance
antilog	antilogarithm
ANTU	1-naphthalenylthiourea
AO	atomic orbital
AOM	angular overlap model
APAD	3-acetylpyridine adenine dinucleotide
Api	apiose
APM	atomic probe microanalysis
APS	appearance potential spectroscopy; adenosine phosphosulfate
APW	augmented plane wave
aq	aqueous
Ar	aryl
Ara	arabinose
Ara-ol	arabinitol
Arg	arginine

ARPES	angular resolved photoelectron spectroscopy
ASC	4-(acetylamino)benzenesulfonyl chloride
ASCII	American National Standard Code for Information Interchange
ASE	aromatic stabilization model
Asn	asparagine
Asp	aspartic acid
at	atomization
ATEE	N-acetyl-L-tyrosine ethyl ester
ATLC	adsorption thin layer chromatography
atm	standard atmosphere
ATP	adenosine triphosphate
ATR	attenuated total internal reflection
at.wt.	atomic weight
AU	astronomical unit (ua is also used)
av	average
avdp	avoirdupois
B	bel
B	magnetic flux density; second virial coefficient; susceptance
b	barn
b	van der Waals constant; molality
BA	benzyladenine
BAL	British anti-Lewisite [2,3-dimercapto-1-propanol]
BAP, BaP	benzo[a]pyrene
bar	bar (pressure unit)
bbl	barrel
BBP	benzyl butyl phthalate
BCB	bromocresol blue
bcc	body centered cubic
BCG	bromocresol green
BCNU	N,N'-bis(2-chloroethyl)-N-nitrosourea
BCP	bromocresol purple
BCPB	bromochlorophenol blue
BCS	Bardeen-Cooper-Schrieffer (theory)
BDE	bond dissociation energy
BDEA	butyldiethanolamime
BDMA	benzyldimethylamine
Bé	Baumé
BEBO	bond energy bond order (method)
BEI	biological exposure index
Bes	2-[bis(2-hydroxyethyl)amino]ethanesulfonic acid
BET	Brunauer-Emmett-Teller (isotherm)
BeV	billion electronvolt (GeV)
BGE	butyl glycidyl ether
BHA	$tert$-butyl-4-hydroxyanisole
BHC	benzene hexachloride [hexachlorobenzene]
Bhn	Brinell hardness number
BHT	butylated hydroxytoluene [2,6-di-$tert$-butyl-4-methylphenol]
Bi	biot
Bicine	N,N-bis(2-hydroxyethyl)glycine
Bistris	2-[bis(2-hydroxyethyl)amino]-2-(hydroxymethyl)propane-1,3-diol
Bistris-propane	1,3-bis[tris(hydroxymethyl)methylamino]propane
BLO	γ-butyrolactone
BN	bond number; benzonitrile
BNS	nuclear backscattering spectroscopy
BO	Born-Oppenheimer (approximation); bond order
BOD	biochemical oxygen demand
BON	β-hydroxynaphthoic acid
bp	boiling point

BPB	bromophenol blue
BPG	2,3-bis(phospho)-D-glycerate
bpy	2,2'-bipyridine
Bq	becquerel
Br	butyryl
BRE	bond resonance energy
BrUrd	5-bromouridine
BSSE	basis set superposition error
BTMSA	1,2-bis(trimethylsilyl)acetylene
Btu	British thermal unit
BTX	benzene, toluene, and xylene
Bu	butyl
bu	bushel
Bz	benzoyl
Bzl	benzyl
C	coulomb; cytosine (in genetic code)
°C	degree Celsius
C	capacitance; heat capacity; number concentration
c	centi (SI prefix for 10^{-2}); combustion reaction
c	amount concentration; specific heat; velocity
c_0	speed of light in vacuum
ca.	approximately
cal	calorie
calc	calculated
cAMP	adenosine cyclic 3',5'-(hydrogen phosphate)
CAN	ceric ammonium nitrate
CARS	coherent anti-Stokes Raman spectroscopy
CAS	complete active space
CASRN	Chemical Abstracts Service Registry Number
CAT	computerized axial tomography; clear air turbulence
CBE	chemical beam epitaxy
CBS	complete basis set (of orbitals)
CC	coupled cluster
cc	cubic centimeter (mL)
CCD	charge-coupled device
CD	circular dichroism
cd	candela; condensed (phase)
CDNO	complete neglect of differential overlap
CDP	cytidine 5'-diphosphate
CDT	1,5,9-cyclododecatriene
CDTA	(1,2-cyclohexylenedinitrilo)tetraacetic acid monohydrate
CDW	charge density waves
CEP	counter electrophoresis
CEPA	coupled electron-pair approximation
cf.	compare
cfm	cubic feet per minute
cgs	centimeter-gram-second system
Chaps	3-[3-(cholamidopropyl)dimethylammonio]-1-propanesulfonic acid
Ches	2-(N-cyclohexylamino)ethanesulfonic acid
CHF	coupled Hartree-Fock (method)
Chl	chlorophyll
Cho	choline
CHT	1,3,5-cycloheptatriene
Ci	curie
CI	configuration interaction; chemical ionization, color index
CIDEP	chemically induced dynamic electron polarization
CIDNP	chemically induced dynamic nuclear polarization
CIE	countercurrent immunoelectrophoresis
cir	circular
CKFF	Cotton-Kraihanzel force field

CL	cathode luminescence (spectroscopy)
cm	centimeter
c.m.	center of mass
c.m.c.	critical micelle concentration
CMO	canonical molecular orbital
CMP	cytidine 5'-monophosphate
CN	coordination number
CNDO	complete neglect of differential overlap
Co	Cowling number
COD	chemical oxygen demand; 1,4-cyclooctadiene
conc	concentrated; concentration
const	constant
COOP	crystal orbital overlap population
cos	cosine
cosh	hyperbolic cosine
COSY	correlation spectroscopy
COT	1,3,5,7-cyclooctatetraene
cot	cotangent
coth	hyperbolic cotangent
CP	chemically pure
Cp	cyclopentadienyl
cP	centipoise
cp	candle power
CPA	coherent potential approximation
CPC	centrifugal partition chromatography
cpd	contact potential difference
CPR	chlorophenol red
cps	cycles per second
CPT	charge conjugation/space inversion/time inversion (theorem)
CPU	central processing unit
cr, cryst	crystalline (phase)
CRU	constitutional repeating unit (in polymer nomenclature)
CSA	camphorsulfonic acid
csc	cosecant
CT	charge transfer
ct	carat
CTEM	conventional transmission electron microscopy
CTP	cytidine 5'-triphosphate
CTR	controlled thermonuclear reaction
cu	cubic
CV	cyclic voltammetry
CVD	chemical vapor deposition
cw	continuous wave
cwt	hundredweight (112 pounds)
Cy	cyclohexyl
Cyd	cytidine
cyl	cylinder
Cys	cysteine
Cyt	cytosine
D	debye unit
D	diffusion coefficient; dissociation energy; electric displacement
d	day; deuteron; deci (SI prefix for 10^{-1})
d	distance; density; dextrorotatory
2,4-D	2,4-dichlorophenoxyacetic acid
D/A	digital to analog
Da	dalton
DA	donor-acceptor (complex)
da	deka (SI prefix for 10^1)
DAA	diacetone alcohol
DAB	4-(dimethylamino)azobenzene
DACH	trans-1,2-diaminocyclohexane
DAP	diammonium phosphate
dB	decibel
DBA	dibenz[a,h]anthracene
DBCP	1,2-dibromo-3-chloropropane
DBMS	database management system
DBP	dibutyl phthalate
DBPC	2,6-di-tert-butyl-p-cresol
dc, DC	direct current
DCB	dicyanobenzene
DCEE	dichloroethyl ether
DCHA	dicyclohexylamine
DCM	dichloromethane
DCPD	dicyclopentadiene
DE	delocalization energy
Dec	decyl
dec	decomposes
DEET	diethyltoluamide [N,N-diethyl-3-methylbenzamide]
deg	degree
den	density
det	determinant
dev	deviation
DFT	density functional theory
dGlc	2-deoxyglucose
DHU	dihydrouridine
diam	diameter
dil	dilute; dilution
DIM	diatomics in molecules (method); digital imaging microscopy
dm	decimeter
DMA	N,N-dimethylaniline
DMAC	N,N-dimethylacetamide
DMF	N,N-dimethylformamide
DMP	dimethyl phthalate
DMS	dimethyl sulfide
DMSO	dimethyl sulfoxide
DMT	dimethyl terephthalate; dimethyl tartrate
DNA	deoxyribonucleic acid
DNase	deoxyribonuclease
DNMR	dynamic NMR spectroscopy
DNP	dinitropyrene
Dod	dodecyl
DOP	dioctyl phthalate
DOS	density of states; digital operating system
doz	dozen
d.p.	degree of polymerization
dpl	displacement
dpm	disintegrations per minute
dps	disintegrations per second
dr	dram
DRE	Dewar resonance energy
dRib	2-deoxyribose
DRIFT	diffuse reflectance infrared Fourier transform
DRP	dynamic reaction path
DRS	diffuse reflectance spectroscopy
DSC	differential scanning calorimetry
DTA	differential thermal analysis
DTBP	di-tert-butyl peroxide
DVB	divinylbenzene
dyn	dyne
DZ	double-zeta (type of basis set)
E	exa (SI prefix for 10^{18})

E	electric field strength; electromotive force; energy; modulus of elasticity; entgegen (*trans* configuration)
E_h	Hartree energy
e	electron; base of natural logarithms
e	elementary charge; linear strain
EA	electron affinity
EAN	effective atomic number
ECP	effective core potential
ECR	electron cyclotron resonance
ED	electron diffraction
EDAX	energy dispersive analysis by x-rays
EDB	ethylene dibromide [1,2-dibromoethane]
EDC	ethylene dichloride [1,2-dichloroethane]
EDS	energy-dispersive x-ray spectroscopy
EDTA	ethylenediaminetetraacetic acid
EELS	electron energy loss spectroscopy
EFF	empirical force field
EFFF	energy factored force field
EHMO, EHT	extended Hückel molecular orbital (theory)
EIS	electron impact spectroscopy; electrochemical impedance spectroscopy
ELS	energy loss spectroscopy
EM	extended molarity; electron microscopy
emf	electromotive force
EMPA, EMA	electron probe microanalysis
emu	electromagnetic unit system
en	ethylenediamine
ENDOR	electron-nuclear double resonance
EOS	equation of state
EPR	electron paramagnetic resonance
EPTC	dipropylcarbamothioic acid, *S*-ethyl ester
eq, eqn	equation
eqQ	quadrupole coupling constant
erf	error function
erg	erg (energy unit)
ESCA	electron spectroscopy for chemical analysis
ESD	electron stimulated desorption
e.s.d.	estimated standard deviation
ESR	electron spin resonance
est	estimated
esu	electrostatic unit system
ET	ephemeris time; electron transfer
Et	ethyl
Etn	ethanolamine
ETS	electron tunneling spectroscopy
Eu	Euler number
e.u.	entropy unit
eV	electronvolt
EXAFS	extended x-ray absorption fine structure (spectroscopy)
EXELFS	extended energy loss fine structure
exp	exponential function
expt	experimental
ext	external
F	farad
°F	degree Fahrenheit
F	Faraday constant; force; angular momentum
f	formation reaction; femto (SI prefix for 10^{-15})
f	activity coefficient; aperture ratio; focal length; force constant; frequency; fugacity
FAD	flavine adenine dinucleotide
fcc	face centered cubic
FEL	free electron laser
FEM	field emission microscopy
FEMO	free electron molecular orbital
FET	field effect transistor
fid	free induction decay
FIM	field ion microscopy
FIR	far infrared
fl	fluid (phase)
FM	frequency modulation
Fo	Fourier number
fp	freezing point
fpm	feet per minute
fps	feet per second; foot-pound-second system
Fr	franklin
Fr	Froude number
Fru	fructose
FSGO	floating spherical gaussian orbitals
FT	Fourier transform
ft	foot
ft-lb	foot pound
FTIR	Fourier transform infrared spectroscopy
FTMS	Fourier transform mass spectrometry
FTNMR	Fourier transform nuclear magnetic resonance
fus	fusion (melting)
FVP	flash vacuum pyrolysis
G	gauss; guanine (in genetic code); giga (SI prefix for 10^9)
G	electrical conductance; Gibbs energy; gravitational constant; sheer modulus
g	gram; gas (phase)
g	acceleration due to gravity; degeneracy; Landé g-factor; statistical weight
GABA	γ-aminobutyric acid
Gal	gal; galactose
gal	gallon
GalN	galactosamine
GC	gas chromatography
GC-MS	gas chromatography-mass spectroscopy
GDMS	glow discharge mass spectroscopy
GDP	guanosine 5'-diphosphate
gem	geminal (on the same carbon atom)
GeV	gigaelectronvolt
GIAO	gauge invariant atomic orbital
gl	glacial
GLC	gas-liquid chromatography
Glc	glucose
GlcA	gluconic acid
GlcN	glucosamine
GlcNAc	*N*-acetylglucosamine
GlcU	glucuronic acid
Gln	glutamine
Glu	glutamic acid
Gly	glycine
GMP	guanosine 5'-monophosphate
GMT	Greenwich mean time
gpm	gallons per minute
gps	gallon per second
Gr	Grashof number
gr	grain
Gra	glyceraldehyde
Gri	glyceric acid
Grn	glycerone [dihydroxyacetone]
Gro	glycerol
GTO	gaussian-type orbital

GTP	guanosine 5'-triphosphate
Gua	guanine
Gul	gulose
Guo	guanosine
GUT	grand unified theory
GVB	generalized valence bond (method)
GWS	Glashow-Weinberg-Salam (theory)
Gy	gray; gigayear
H	henry
H	enthalpy; Hamiltonian function; magnetic field
H_0	Hubble constant
h	helion; hour; hecto (SI prefix for 10^2)
h	Planck constant
Ha	Hartmann number
ha	hectare
HAM	hydrogenic atoms in molecules
hav	haversine
Hb	hemoglobin
HCA	heterocyclic amine
hcp	hexagonal closed packed
HDL	high-density lipoprotein
HEIS	high-energy ion scattering
HEP	high energy physics
Hepes	4-(2-hydroxyethyl)-1-piperazineethanesulfonic acid
Hepps	4-(2-hydroxyethyl)-1-piperazinepropanesulfonic acid
HF	high frequency; Hartree-Fock (method)
HFA	hexafluoroacetone
HFO	Hartree-Fock orbital
hfs	hyperfine structure
His	histidine
HMO	Hückel molecular orbital
HMX	cyclotetramethylenetetranitramine
HN1	2-chloro-*N*-(2-chloroethyl)-*N*-ethylethanamine
HOAc	acetic acid
HOC	halogenated organic compound(s)
HOMAS	harmonic oscillator model of aromatic stabilization
HOMO	highest occupied molecular orbital
HOSE	harmonic oscillator stabilization energy
Hp	heptyl
hp	horsepower
HPLC	high-performance liquid chromatography
HQ	*p*-hydroquinone
hr	hour
HRE	Hückel resonance energy
HREELS	high resolution electron energy loss spectroscopy
HREM	high resolution electron microscopy
HSAB	hard-soft acid-base (theory)
HSE	homodesmotic stabilization energy
Hx	hexyl
Hyp	hypoxanthine
Hz	hertz
I	inositol
I	electric current; ionic strength; moment of inertia; nuclear spin angular momentum; radiant intensity
i	square root of minus one
i	electric current
I/O	input/output
IAT	international atomic time
IC	integrated circuit
ICD	induced circular dichroism
ICP	inductive-coupled plasma
ICR	ion cyclotron resonance
ID	inside diameter
id	ideal (solution)
Ido	iodose
IdoA	iduronic acid
IDP	inosine 5'-diphosphate
IE	ionization energy
i.e.p.	isoelectric point
IEPA	independent electron pair approximation
IF	intermediate frequency
IGLO	individual gauge for localized orbitals
Ile	isoleucine
Im	imaginary part
IMFP	inelastic mean free path (of electrons)
imm	immersion
IMP	inosine 5'-monophosphate
IMPATT	impact ionization avalanche transit time
in.	inch
INDO	immediate neglect of differential overlap
Ino	inosine
INS	inelastic neutron scattering; ion neutralization spectroscopy
Ins	*myo*-inositol
int	internal
IP	ionization potential
IPA	isopropyl alcohol
IPMA	ion probe microanalysis
IPN	interpenetrating polymer network
IPR	isotope perturbation of resonance
IPTS	International Practical Temperature Scale
IQ	2-amino-3-methyl-3*H*-imidazo(4,5-f)quinoline
IR	infrared
IRAS	infrared reflection-absorption spectroscopy
IRC	intrinsic reaction coordinate
IRS	infrared spectroscopy
isc	intersystem crossing
ISE	ion-selective electrode; isodesmic stabilization energy
ISS	ion scattering spectroscopy
ITP	inosine 5'-triphosphate
ITS	International Temperature Scale (1990)
IU	international unit
J	joule
J	angular momentum; electric current density; flux; Massieu function
j	angular momentum; electric current density
JT	Jahn-Teller (effect)
K	kelvin
K	absorption coefficient; bulk modulus; equilibrium constant; kinetic energy
k	kilo (SI prefix for 10^3)
k	absorption index; Boltzmann constant; rate constant; thermal conductivity; wave vector
kat	katal (unit of catalytic activity)
kb	kilobar; kilobases (DNA or RNA)
kcal	kilocalorie
KE	kinetic energy
keV	kiloelectronvolt
kg	kilogram
kgf	kilogram force
KIE	kinetic isotope effect
kJ	kilojoule
km	kilometer
Kn	Knudsen number
kPa	kilopascal

kt	karat
kV	kilovolt
kva	kilovolt ampere
kW	kilowatt
kwh	kilowatt hour
L	liter; lambert
L	Avogadro constant; inductance; Lagrange function; angular momentum
l	liter; liquid (phase)
l	angular momentum; length; mean free path; levorotatory
Lac	lactose
LAH	lithium aluminum hydride
lat.	latitude
lb	pound
lbf	pound force
LC	liquid chromatography
lc	liquid crystal (phase)
LCAO	linear combination of atomic orbitals
LD	lethal dose
LDA	local density approximation; lithium diisopropylamide
LDL	low-density lipoprotein
LDV	laser-Doppler velocimetry
Le	Lewis function
LE	localization energy
LEC	liquid exchange chromatography
LED	light emitting diode
LEED	low-energy electron diffraction
LEIS	low-energy ion scattering
Leu	leucine
LFER	linear free energy relationships
LFL	lower flammable limit
lim	limit
LIMS	laser ionization mass spectroscopy; laboratory information management system
liq	liquid
LLCT	ligand to ligand charge transfer
lm	lumen
LMCT	ligand to metal charge transfer
LMO	localized molecular orbital
LMR	laser magnetic resonance
ln	logarithm (natural)
LNDO	local neglect of differential overlap
log	logarithm (common)
LOMO	lowest occupied molecular orbital
long.	longitude
LSFE	linear field stabilization energy
LST	local sidereal time
LT	local time
LTE	local thermodynamic equilibrium
LUMO	lowest unoccupied molecular orbital
lx	lux
ly	langley
l.y.	light year
Lys	lysine
Lyx	lyxose
M	molar (as in 0.1 M solution); mega (SI prefix for 10^6)
M	magnetization; molar mass; mutual inductance; torque; angular momentum component; median
M_r	molecular weight (relative molar mass)
m	meter; molal (as in 0.1 m solution); metastable (isotope); milli (SI prefix for 10^{-3})

m	magnetic dipole moment; mass; molality; angular momentum component; *meta* (locant on aromatic ring)
Ma	Mach number
MA	maleic anhydride
Mal	maltose
Man	mannose
MASNMR	magic angle spinning nuclear magnetic resonance
max	maximum
Mb	myoglobin
MBE	molecular beam epitaxy
MBER	molecular beam electron resonance
MBPT	many body perturbation theory
MC	Monte Carlo (method)
MCAA	monochloroacetic acid
MCD	magnetic circular dichroism
MCPA	(4-chloro-2-methylphenoxy)acetic acid
MCPF	modified coupled pair functional
MCSCF	multiconfigurational self-consistent field (approximation)
MD	molecular dynamics (method)
Me	methyl
MeCCNU	1-(2-chloroethyl)-3-(4-methylcyclohexyl)-1-nitrosourea
MeIQ	2-amino-3,4-dimethylimidazo[4,5-f]quinoline
MeIQx	2-amino-3,8-dimethylimidazo[4,5-f]quinoxaline
MEK	methyl ethyl ketone
MEP	molecular electrostatic potential
MERP	minimum energy reaction path
Mes	4-morpholineethanesulfonic acid
MESFET	metal-semiconductor field-effect transistor
Met	methionine
MeV	megaelectronvolt
meV	millielectronvolt
MF	molecular formula
mg	milligram
MHD	magnetohydrodynamics
mi	mile
min	minimum; minute
MINDO	modified INDO (method)
MIPK	methyl isopropyl ketone
MIR	mid infrared
misc	miscible
MKS	meter-kilogram-second system
MKSA	meter-kilogram-second-ampere system
mL, ml	milliliter
MM	molecular mechanics
mm	millimeter
MMDR	microwave-microwave double resonance
mmf	magnetomotive force
mmHg	millimeter of mercury
MNDO	modified neglect of diatomic overlap
MO	molecular orbital; methyl orange
MODR	microwave-optical double resonance
mol	mole
mol.wt.	molecular weight
mon	monomeric form
Mops	4-morpholinepropanesulfonic acid
MOS	metal-oxide semiconductor
MOSFET	metal-oxide semiconductor field-effect transistor
mp	melting point
MPa	megapascal
MPA	Mulliken population analysis

Mpc	megaparsec
MPTP	1,2,3,6-tetrahydro-1-methyl-4-phenylpyridine
MR	methyl red
MRD	multireference double substitution (method)
MRI	magnetic resonance imaging
mRNA	messenger RNA
MS	mass spectroscopy
ms	millisecond
MSA	methanesulfonic acid
MSDS	Material Safety Data Sheet
MSL	mean sea level
MTBE	methyl *tert*-butyl ether
Mur	muramic acid
mV	millivolt
MVK	methyl vinyl ketone
MW	megawatt; microwave; molecular weight
mW	milliwatt
Mx	maxwell
N	newton
N	angular momentum; neutron number; number density
N_A	Avogadro constant
n	neutron; nano (SI prefix for 10^{-9})
n	amount of substance; number density; principal quantum number; refractive index; normal (in chemical formulas)
NAA	nuclear activation analysis
NAAD	nicotinic acid adenine dinucleotide
NAD	nicotinamide adenine dinucleotide
NADH	reduced NAD
NADP	NAD phosphate
NANA	*N*-acetylneuraminic acid
NAO	natural atomic orbital
NBO	natural bond orbital
nbp	normal boiling point
NEDOR	nuclear electron double resonance
Neu	neuraminic acid
NEXAFS	near-edge x-ray absorption fine structure
ng	nanogram
NHO	natural hybrid orbital
NICS	nuclear independent chemical shift
NIR	near infrared; ribosylnicotinamide
nm	nanometer
NMN	β-nicotinamide mononucleotide
NMR	nuclear magnetic resonance
Nn	nonyl
NNDO	neglect of nonbonded differential overlap
NO	natural orbital
NOE	nuclear Overhauser effect
NOx	nitrogen oxides
NP	nitropyrene
NPA	natural population analysis
NQR	nuclear quadrupole resonance
NRA	nuclear reaction analysis
ns	nanosecond
NSE	neutron spin echo
NTP	normal temperature and pressure
Nu	nucleophile
Nu	Nusselt number
o	*ortho* (locant on aromatic ring)
obs, obsd	observed
Oc	octyl
OD	optical density; outside diameter
Oe	oersted

OFGF	outer valence Green's function (method)
ORD	optical rotatory dispersion
Oro	orotate; orotidine
oz	ounce
P	poise; peta (SI prefix for 10^{15})
P	power; pressure; probability; sound energy flux
p	proton; pico (SI prefix for 10^{-12})
p	dielectric polarization; electric dipole moment; momentum; pressure; bond order; *para* (as aromatic ring locant)
Pa	pascal
PA	proton affinity
PABA	*p*-aminobenzoic acid
PAH	polycyclic aromatic hydrocarbon(s)
PAM	polyacrylamide
PAN	1-(2-pyridylazo)-2-naphthol; polyacrylonitrile
PAR	4-(2'-pyridylazo)resorcinol
PAS	photoacoustic spectroscopy
PBA	poly(butyl acrylate)
PBB	polybrominated biphenyl
PBD	poly(1,3-butadiene)
PBMA	poly(butyl methacrylate)
PBT	poly(butylene terephthalate)
PC	paper chromatography
pc	parsec
PCM	polarizable continuum model
PCR	polymerase chain reaction
PD	potential difference
pdl	poundal
PDMS	poly(dimethylsiloxane)
Pe	pentyl
Pe	Péclet number
pe	probable error
PEA	poly(ethyl acrylate)
PEG	poly(ethylene glycol)
PEL	permissible exposure limit
PES	photoelectron spectroscopy; potential energy surface
PET	positron emission tomography; poly(ethylene terephthalate)
peth	petroleum ether
pf	power factor
PFOA	perfluorooctanoic acid
pg	picogram
Ph	phenyl
pH	negative log of hydrogen ion concentration
Phe	phenylalanine
PhIP	2-amino-1-methyl-6-phenylimidazo[4,5-b]pyridine
pI	isoelectric point
PIB	polyisobutylene
Pipes	1,4-piperazinediethanesulfonic acid
PIV	particle-image velocimetry
PIXE	particle induced x-ray emission
p*K*	negative log of ionization constant
PLM	principle of least motion
pm	picometer
PMA	poly(methyl acrylate)
PMMA	poly(methyl methacrylate)
PMO	perturbation MO (theory)
PNDO	partial neglect of differential overlap
PNO	pair natural orbitals
PNRA	prompt nuclear reaction analysis
POAV	π-orbital axis vector
pol	polymeric form

POx	phosphorus oxides
ppb	parts per billion
ppm	parts per million
PPO	poly(phenylene oxide)
PPP	Pariser-Parr-Pople (method)
ppt	parts per thousand; precipitate
Pr	propyl
Pr	Prandtl number
PRDDO	partial retention of diatomic differential overlap
Pro	proline
PS	photoelectron spectroscopy
ps	picosecond
PSD	photon stimulated desorption
psi	pounds per square inch
psia	pounds per square inch absolute
psig	pounds per square inch gage
PT	perturbation theory
pt	pint
PTMS	propyltrimethoxysilane
Pu	purine
PVA	poly(vinyl alcohol)
PVAc	poly(vinyl acetate)
PVC	poly(vinyl chloride)
PVDF	poly(vinylidene fluoride)
PVME	poly(methyl vinyl ether)
PVT	pressure-volume-temperature
Py	pyrimidine
Q	electric charge; heat; partition function; quadrupole moment; radiant energy; vibrational normal coordinate
q	electric field gradient; flow rate; heat; wave vector (phonons)
QCD	quantum chromodynamics
QCI	quadratic configuration interaction
QED	quantum electrodynamics
Q.E.D.	quod erat demonstrandum (which was to be proved)
QMRE	quantum mechanical resonance energy
QSAR	quantitative structure-activity relations
QSO	quasi-stellar object
qt	quart
quad	quadrillion BTU ($=1.055 \cdot 10^{18}$ joules)
Qui	quinovose
q.v.	quod vide (which you should see)
R	roentgen; alkyl radical (in chemical formulas)
°R	degree Rankine
R	electrical resistance; gas constant; molar refraction; Rydberg constant; coefficient of multiple correlation
r	reaction (as in $\Delta_r H$)
r	position vector; radius
RA	right ascension
rad	radian
RAIRS	reflection-absorption infrared spectroscopy
RAM	random access memory
RBS	Rutherford back scattering
Rbu, Rul	ribulose
RCI	ring current index
RDA	rubidium dihydrogen arsenate
Re	real part
RE	resonance energy
RED	radial electron distribution
REM	reflection electron microscopy
rem	roentgen equivalent man
REPE	resonance energy per electron
RF	radiofrequency

Rha	rhamnose
RHEED	reflection high-energy electron diffraction
RHF	restricted Hartree-Fock (theory)
RIA	radioimmunoassay
Rib	ribose
Ribulo	ribulose
rms	root-mean-square
RNA	ribonucleic acid
RNase	ribonuclease
ROHF	restricted open shell Hartree-Fock
ROM	read only memory
RPA	random phase approximation
RPH	reaction path Hamiltonian
RPLC	reversed-phase liquid chromatography
rpm	revolutions per minute
rps	revolutions per second
RRK	Rice-Ramsperger-Kassel (theory)
RRKM	Rice-Ramsperger-Kassel-Marcus (theory)
rRNA	ribosomal RNA
RRS	resonance Raman spectroscopy
RS	Raman spectroscopy
Ry	rydberg
S	siemens
S	area; entropy; probability current density; Poynting vector; symmetry coordinate; spin angular momentum
s	second; solid (phase)
s	path length; spin angular momentum; symmetry number; sedimentation coefficient; solubility; symmetrical (as stereochemical descriptor)
SALC	symmetry adapted linear combinations
SALI	surface analysis by laser ionization
SAM	scanning Auger microscopy
SAMS	self-assembled monolayers
SANS	small angle neutron scattering
Sar	sarcosine
sat, satd	saturated
SAXS	small angle x-ray scattering
Sc	Schmidt number
SC	spin-coupled (method)
SCD	state correlation diagram
SCE	saturated calomel electrode
SCF	self-consistent field (method)
SCR	silicon-controlled rectifier
SCRF	self-consistent reaction field (method)
sd	standard deviation
SDW	spin density wave
SE	strain energy
sec	secant; second
sec	secondary (in chemical name)
SECSY	spin-echo correlated spectroscopy
Sed	sedoheptulose
SEELFS	surface extended energy loss fine structure
SEM	scanning electron microscopy; standard error of the mean
sepn	separation
Ser	serine
SERS	surface-enhanced Raman spectroscopy
SET	single electron transfer
SEXAF	surface extended x-ray absorption fine structure
SFC	supercritical fluid chromatography
Sh	Sherwood number
Shy	thiohypoxanthine
SI	International System of Units

SILAR	successive ionic layer adsorption and reaction
SIMS	secondary-ion spectroscopy
sin	sine
sinh	hyperbolic sine
SIPN	semi-interpenetrating polymer network
SLAM	scanning laser acoustic microscopy
SMO	semiempirical molecular orbital
SMOW	Standard Mean Ocean Water
SNMS	sputtered neutral mass spectroscopy
Sno	thiouridine
SNU	solar neutrino unit
SOJT	second-order Jahn-Teller (effect)
sol	soluble; solution
soln, sln	solution
SOMO	singly occupied molecular orbital
Sor	sorbose
sp gr	specific gravity
SPM	scanned probe microscopy
sq	square
Sr	Strouhal number
sr	steradian
Srd	6-thioinosine
SSMS	source spark mass spectroscopy
St	stoke
St	Stanton number
std, stnd	standard (state)
STEL	short-term exposure limit
STEM	scanning transmission electron microscope
STM	scanning tunneling microscopy
STO	Slater-type orbital
STP	standard temperature and pressure
sub, subl	sublimes; sublimation
Suc, Sac	sucrose
Sur	thiouracil
Sv	sievert
T	tesla; tera (SI prefix for 10^{12})
T	kinetic energy; period; term value; temperature (thermodynamic); torque; transmittance
t	metric tonne; triton
t	Celsius temperature; thickness; time; transport number
Tal	talose
tan	tangent
tanh	hyperbolic tangent
Taps	3-{[2-hydroxy-1,1-bis(hydroxymethyl)ethyl]amino}-1-propanesulfonic acid
TBE	1,1,2,2-tetrabromoethane
TBP	tributyl phosphate
TCA	trichloroacetic acid
TCE	trichloroethylene
TCNQ	tetracyanoquinodimethane
TCP	tricresyl phosphate
TCSCF	two configuration self-consistent field
TDI	toluene diisocyanate
tDNA	transfer DNA
TE	transverse electric
TEA	triethanolamine; triethylamine
TED	transferred electron device; transmission electron diffraction
TEM	transverse electromagnetic; transmission electron microscope
temp	temperature
tert	tertiary (in chemical name)

Tes	2-{[2-hydroxy-1,1-bis(hydroxymethyl)ethyl]amino}-1-propanesulfonic acid
TFD	Thomas-Fermi-Dirac (method)
TGA	thermogravimetric analysis
Thd	ribosylthymine
theor	theoretical
thf, THF	tetrahydrofuran
Thr	threonine
Thy	thymine
TL	thermoluminescence
TLC	thin-layer chromatography
TLV	threshold limit value
TM	transverse magnetic
TMAB	tetrabutylammonium bromide
TMS	tetramethylsilane
tol	tolyl
TOPO	trioctylphosphine oxide
Torr	torr (pressure unit)
Tre	trehalose
TRE	topological resonance energy
Tricine	*N*-[2-hydroxy-1,1-bis(hydroxymethyl)ethyl]glycine
Tris	2-amino-2-(hydroxymethyl)-1,3-propanediol
tRNA	transfer RNA
Trp	tryptophan
trs	transition
TS	transition state
TSS	transition state spectroscopy
Tyr	tyrosine
U	uracil (in genetic code)
U	electric potential difference; internal energy
u	unified atomic mass unit
u	Bloch function; electric mobility; velocity
ua	astronomical unit (AU is also used)
UBFF	Urey-Bradley force field
UDMH	1,1-dimethylhydrazine
UDP	uridine 5'-diphosphate
UHF	ultrahigh frequency; unrestricted Hartree-Fock (method)
UMP	uridine 5'-monophosphate
uns, unsym	unsymmetrical (as chemical descriptor)
UPS, UPES	ultraviolet photoelectron spectroscopy
Ura	uracil
Urd	uridine
USP	United States Pharmacopeia
UT	universal time
UTP	uridine 5'-triphosphate
UV	ultraviolet
V	volt
V	electric potential; potential energy; volume
v	reaction rate; specific volume; velocity; vibrational quantum number; vicinal (as chemical descriptor)
v/v	volume per volume (volume of solute divided by volume of solution, expressed as percent)
Val	valine
vap	vaporization
VAT	vibration assisted tunneling
VB	valence band; valence bond (theory)
VCD	vibrational circular dichroism
VDW	van der Waals interaction
VHF	very high frequency
vic	vicinal (on adjacent carbon atom)
VIS	visible region of the spectrum
vit	vitreous (phase)

VOC	volatile organic compound(s)
VOFF	valence orbital force field
VPC	vapor phase chromatography
VSEPR	valence shell electron-pair repulsion (method)
VSIP	valence state ionization potential
VSLI	very large scale integrated (circuit)
VUV	vacuum ultraviolet
W	watt
W	radiant energy; statistical weight; work
w	energy density; mass fraction; velocity; work
w/v	weight per volume (mass of solute divided by volume of solution, usually expressed as g/100 mL)
w/w	weight per weight (mass of solute divided by mass of solution, expressed as percent)
WAXS	wide angle x-ray scattering
Wb	weber
We	Weber number
WKB	Wentzel-Kramers-Brillouin (approximation)
WLF	Williams-Landel-Ferry (equation)
wt	weight
X	X unit; halogen (in chemical formula)
X	reactance
x	mole fraction
XAFS	x-ray absorption fine structure
Xan	xanthine
XANES	x-ray absorption near-edge structure
Xao	xanthosine
Xlu, Xul	xylulose
XPS, XPES	x-ray photoelectron spectroscopy
XRD	x-ray diffraction
XRF	x-ray fluorescence
XRS	x-ray spectroscopy
Xyl	xylose
Y	yotta (SI prefix for 10^{24})
Y	admittance; Planck function; Young's modulus
y	yocto (SI prefix for 10^{-24})
y	mole fraction for gas (when x refers to liquid phase)
y, yr	year
yd	yard
Z	zetta (SI prefix for 10^{21})
Z	atomic number; compression factor; collision number; impedance; partition function; zusammen (*cis*-configuration)
z	zepto (SI prefix for 10^{-21})
z	charge number (of an ion); collision frequency factor
ZDO	zero differential overlap
ZINDO	Zerner's INDO method
ZPE, ZPVE	zero point vibrational energy
ZULU	Greenwich mean time
α	alpha particle
α	absorption coefficient; degree of dissociation; electric polarizability; expansion coefficient; fine structure constant
β	beta particle
β	reciprocal temperature parameter ($= 1/kT$)
γ	photon; gamma (obsolete mass unit = μg)
γ	activity coefficient; conductivity; magnetogyric ratio; mass concentration; ratio of heat capacities; surface tension
Γ	Grüneisen parameter; level width; surface concentration
δ	chemical shift; Dirac delta function; Kronecker delta; loss angle
Δ	inertial defect; mass excess
ε	emittance; Levi-Civita symbol; linear strain; molar absorption coefficient; permittivity
ζ	Coriolis coupling constant; electrokinetic potential
η	overpotential; viscosity
κ	compressibility; conductivity; magnetic susceptibility; molar absorption coefficient
λ	absolute activity; radioactive decay constant; thermal conductivity; wavelength
Λ	angular momentum; ionic conductivity
μ	muon; micro (SI prefix for 10^{-6})
μ	chemical potential; electric dipole moment; electric mobility; friction coefficient; Joule-Thompson coefficient; magnetic dipole moment; mobility; permeability
μF	microfarad
μg	microgram
μm	micrometer
μs	microsecond
ν	frequency; kinematic velocity; stoichiometric number
ν_e	neutrino
ν	wavenumber
π	pion
Π	osmotic pressure; Peltier coefficient
ρ	density; reflectance; resistivity
σ	electrical conductivity; cross section; normal stress; shielding constant (NMR); Stefan-Boltzmann constant; surface tension; standard deviation
τ	transmittance; chemical shift; shear stress; relaxation time
ϕ	electrical potential; fugacity coefficient; osmotic coefficient; quantum yield; wavefunction
Φ	magnetic flux; potential energy; radiant power; work function
χ	magnetic susceptibility
χ_e	electric susceptibility
ψ	wavefunction
ω	circular frequency; angular velocity; harmonic vibration wavenumber; statistical weight
Ω	ohm
Ω	axial angular momentum; solid angle

GREEK, RUSSIAN, AND HEBREW ALPHABETS

The following table presents the Hebrew, Greek, and Russian alphabets, their letters, the names of the letters, and the English equivalents.

Hebrew[1,3]			**Greek**[4]			**Russian**		
א	aleph	' [2]	Α α	alpha	a	А а	a
ב	beth	b, bh	Β β	beta	b	Б б	b
ג	gimel	g, gh	Γ γ	gamma	g, n	В в	v
ד	daleth	d, dh	Δ δ	delta	d	Г г	g
ה	he	h	Ε ε	epsilon	e	Д д	d
ו	waw	w	Ζ ζ	zeta	z	Е е	e
ז	zayin	z	Η η	eta	ē	Ж ж	zh
ח	heth	ḥ	Θ θ	theta	th	З з	z
ט	teth	ṭ	Ι ι	iota	i	И и Й й	i, ĭ
י	yodh	y	Κ κ	kappa	k	К к	k
כ ך	kaph	k, kh	Λ λ	lambda	l	Л л	l
ל	lamedh	l	Μ μ	mu	m	М м	m
מ ם	mem	m	Ν ν	nu	n	Н н	n
נ ן	nun	n	Ξ ξ	xi	x	О о	o
ס	samekh	s	Ο ο	omicron	o	П п	p
ע	ayin	'	Π π	pi	p	Р р	r
פ ף	pe	p, ph	Ρ ρ	rho	r, rh	С с	s
צ ץ	sadhe	ṣ	Σ σ ς	sigma	s	Т т	t
ק	qoph	q	Τ τ	tau	t	У у	u
ר	resh	r	Υ υ	upsilon	y, u	Ф ф	f
שׂ	sin	ś	Φ φ	phi	ph	Х х	kh
שׁ	shin	sh	Χ χ	chi	ch	Ц ц	ts
ת	taw	t, th	Ψ ψ	psi	ps	Ч ч	ch
			Ω ω	omega	ō	Ш ш	sh
						Щ щ	shch
						Ъ ъ[5]	"
						Ы ы	y
						Ь ь[6]	'
						Э э	e
						Ю ю	yu
						Я я	ya

[1] Where two forms of a letter are given, the second one is the form used at the end of a word.

[2] Not represented in transliteration when initial.

[3] The Hebrew letters are primarily consonants; a few of them are also used secondarily to represent certain vowels, when provided at all, by means of a system of dots or strokes adjacent to the consonated characters.

[4] The letter gamma is transliterated "n" only before velars; the letter upsilon is transliterated "u" only as the final element in diphthongs.

[5] This sign indicates that the immediately preceding consonant is not palatized even though immediately followed by a palatized vowel.

[6] This sign indicates that the immediately preceding consonant is palatized even though not immediately followed by a palatized vowel.

DEFINITIONS OF SCIENTIFIC TERMS

Brief definitions of selected terms of importance in chemistry, physics, and related fields of science are given in this section. The selection process emphasizes the following types of terms:

- ◆ Physical quantities
- ◆ Units of measure
- ◆ Classes of chemical compounds and materials
- ◆ Important theories, laws, and basic concepts.

Individual chemical compounds are not included.

Definitions have taken wherever possible from the recommendations of international or national bodies, especially the International Union of Pure and Applied Chemistry (IUPAC) and International Organization for Standardization (ISO). For physical quantities and units, the recommended symbol is also given. The source of such definitions is indicated by the reference number in brackets following the definition. In many cases these official definitions have been edited in the interest of stylistic consistency and economy of space. The user is referred to the original source for further details.

An asterisk (*) following a term indicates that further information can be found by consulting the index of this handbook under the entry for that term.

References

1. *ISO Standards Handbook 2, Units of Measurement*, International Organization for Standardization, Geneva, 1992.
2. *Quantities, Units, and Symbols in Physical Chemistry, Second Edition*, International Union of Pure and Applied Chemistry, Blackwell Scientific Publications, Oxford, 1993.
3. *Compendium of Chemical Terminology*, International Union of Pure and Applied Chemistry, Blackwell Scientific Publications, Oxford, 1987.
4. *A Guide to IUPAC Nomenclature of Organic Compounds*, International Union of Pure and Applied Chemistry, Blackwell Scientific Publications, Oxford, 1993.
5. *Glossary of Class Names of Organic Compounds and Reactive Intermediates Based on Structure, Pure and Applied Chemistry*, 67, 1307, 1995.
6. *Compendium of Analytical Nomenclature*, International Union of Pure and Applied Chemistry, Blackwell Scientific Publications, Oxford, 1987.
7. *Nomenclature of Inorganic Chemistry*, International Union of Pure and Applied Chemistry, Blackwell Scientific Publications, Oxford, 1990.
8. *Glossary of Basic Terms in Polymer Science, Pure and Applied Chemistry*, 68, 2287, 1996.
9. *The International Temperature Scale of 1990, Metrologia*, 27, 107, 1990.
10. *Compilation of ASTM Standard Definitions*, American Society of Testing and Materials, Philadelphia, 1990.
11. *ASM Metals Reference Book*, American Society for Metals, Metals Park, OH, 1983.

Ab initio method - An approach to quantum-mechanical calculations on molecules which starts with the Schrödinger equation and carries out a complete integration, without introducing empirical factors derived from experimental measurement.

Absorbance (A) - Defined as $-\log(1-\alpha) = \log(1/\tau)$, where α is the absorptance and τ the transmittance of a medium through which a light beam passes. [2]

Absorbed dose (D) - For any ionizing radiation, the mean energy imparted to an element of irradiated matter divided by the mass of that element. [1]

Absorptance (α) - Ratio of the radiant or luminous flux in a given spectral interval absorbed in a medium to that of the incident radiation. Also called absorption factor. [1]

Absorption coefficient (a) - The relative decrease in the intensity of a collimated beam of electromagnetic radiation, as a result of absorption by a medium, during traversal of an infinitesimal layer of the medium, divided by the length traversed. [1]

Absorption coefficient, molar (ε) - Absorption coefficient divided by amount-of-substance concentration of the absorbing material in the sample solution ($\varepsilon = a/c$). The SI unit is m^2/mol. Also called extinction coefficient, but usually in units of $mol^{-1}dm^3cm^{-1}$. [2]

Acceleration - Rate of change of velocity with respect to time.

Acceleration due to gravity (g)* - The standard value (9.80665 m/s^2) of the acceleration experienced by a body in the earth's gravitational field. [1]

Acenes - Polycyclic aromatic hydrocarbons consisting of fused benzene rings in a rectilinear arrangement. [5]

Acid - Historically, a substance that yields an H^+ ion when it dissociates in solution, resulting in a pH<7. In the Brönsted definition, an acid is a substance that donates a proton in any type of reaction. The most general definition, due to G.N. Lewis, classifies any chemical species capable of accepting an electron pair as an acid.

Acid dissociation constant (Ka)* - The equilibrium constant for the dissociation of an acid HA through the reaction $HA + H_2O \rightarrow A^- + H_3O^+$. The quantity $pK_a = -\log K_a$ is often used to express the acid dissociation constant.

Actinides - The elements of atomic number 89 through 103, e.g., Ac, Th, Pa, U, Np, Pu, Am, Cm, Bk, Cf, Es, Fm, Md, No, Lr. [7]

Activation energy* - In general, the energy that must be added to a system in order for a process to occur, even though the process may already be thermodynamically possible. In chemical kinetics, the activation energy is the height of the potential barrier separating the products and reactants. It determines the temperature dependence of the reaction rate.

Activity - For a mixture of substances, the absolute activity λ of substance B is defined as $\lambda_B = \exp(\mu_B/RT)$, where μ_B is the chemical potential of substance B, R the gas constant, and T the thermodynamic temperature. The relative activity a is defined as $a_B = \exp[(\mu_B - \mu_B°)/RT]$, where $\mu_B°$ designates the chemical potential in the standard state. [2]

Activity coefficient (γ)* - Ratio of the activity a_B of component B of a mixture to the concentration of that component. The value of γ depends on the method of stating the composition. For mole fraction x_B, the relation is $a_B = \gamma_B x_B$; for molarity c_B, it is $a_B = \gamma_B c_B/c°$, where $c°$ is the standard state composition (typically chosen as 1 mol/L); for molality m_B, it is $a_B = \gamma_B m_B/m°$, where $m°$ is the standard state molality (typically 1 mol/kg). [2]

Activity, of radioactive substance (A) - The average number of spontaneous nuclear transitions from a particular energy state occurring in an amount of a radionuclide in a small time interval divided by that interval. [1]

Acyl groups - Groups formed by removing the hydroxy groups from oxoacids that have the general structure RC(=O)(OH) and replacement analogues of such acyl groups. [5]

Adiabatic process - A thermodynamic process in which no heat enters or leaves the system.

Admittance (Y) - Reciprocal of impedance. $Y = G + iB$, where G is conductance and B is susceptance. [1]

Adsorption - A process in which molecules of gas, of dissolved substances in liquids, or of liquids adhere in an extremely thin layer to surfaces of solid bodies with which they are in contact. [10]

Albedo* - The ratio of the light reflected or scattered from a surface to the intensity of incident light. The term is often used in reference to specific types of terrain or to entire planets.

Alcohols - Compounds in which a hydroxy group, -OH, is attached to a saturated carbon atom. [5]

Aldehydes - Compounds RC(=O)H, in which a carbonyl group is bonded to one hydrogen atom and to one R group. [5]

Aldoses - Aldehydic parent sugars (polyhydroxyaldehydes $H[CH(OH)]_nC(=O)H$, $n>1$) and their intramolecular hemiacetals. [5]

Aldoximes - Oximes of aldehydes: RCH=NOH. [5]

Alfvén number (Al) - A dimensionless quantity used in plasma physics, defined by $Al = v(\rho\mu)^{1/2}/B$, where ρ is density, v is velocity, μ is permeability, and B is magnetic flux density. [2]

Alfvén waves - Very low frequency waves which can exist in a plasma in the presence of a uniform magnetic field. Also called magnetohydrodynamic waves.

Alicyclic compounds - Aliphatic compounds having a carbocyclic ring structure which may be saturated or unsaturated, but may not be a benzenoid or other aromatic system. [5]

Aliphatic compounds - Acyclic or cyclic, saturated or unsaturated carbon compounds, excluding aromatic compounds. [5]

Alkali metals - The elements lithium, sodium, potassium, rubidium, cesium, and francium.

Alkaline earth metals - The elements calcium, strontium, barium, and radium. [7]

Alkaloids - Basic nitrogen compounds (mostly heterocyclic) occurring mostly in the plant kingdom (but not excluding those of animal origin). Amino acids, peptides, proteins, nucleotides, nucleic acids, and amino sugars are not normally regarded as alkaloids. [5]

Alkanes - Acyclic branched or unbranched hydrocarbons having the general formula C_nH_{2n+2}, and therefore consisting entirely of hydrogen atoms and saturated carbon atoms. [5]

Alkenes - Acyclic branched or unbranched hydrocarbons having one carbon-carbon double bond and the general formula C_nH_{2n}. Acyclic branched or unbranched hydrocarbons having more than one double bond are alkadienes, alkatrienes, etc. [5]

Alkoxides - Compounds, ROM, derivatives of alcohols, ROH, in which R is saturated at the site of its attachment to oxygen and M is a metal or other cationic species. [5]

Alkyl groups - Univalent groups derived from alkanes by removal of a hydrogen atom from any carbon atom: C_nH_{2n+1}-. The groups derived by removal of a hydrogen atom from a terminal carbon atom of unbranched alkanes form a subclass of normal alkyl (*n*-alkyl) groups. The groups RCH_2-, R_2CH-, and R_3C- (R not equal to H) are primary, secondary, and tertiary alkyl groups, respectively. [5]

Alkynes - Acyclic branched or unbranched hydrocarbons having a carbon-carbon triple bond and the general formula C_nH_{2n-2}, RC≡CR′. Acyclic branched or unbranched hydrocarbons having more than one triple bond are known as alkadiynes, alkatriynes, etc. [5]

Allotropy - The occurrence of an element in two or more crystalline forms.

Allylic groups - The group CH_2=CHCH$_2$- (allyl) and derivatives formed by substitution. The term 'allylic position' or 'allylic site' refers to the saturated carbon atom. A group, such as -OH, attached at an allylic site is sometimes described as "allylic". [5]

Amagat volume unit - A non-SI unit previously used in high pressure science. It is defined as the molar volume of a real gas at one atmosphere pressure and 273.15 K. The approximate value is 22.4 L/mol.

Amides - Derivatives of oxoacids R(C=O)(OH) in which the hydroxy group has been replaced by an amino or substituted amino group. [5]

Amine oxides - Compounds derived from tertiary amines by the attachment of one oxygen atom to the nitrogen atom: R_3N^+-O^-. By extension the term includes the analogous derivatives of primary and secondary amines. [5]

Amines - Compounds formally derived from ammonia by replacing one, two, or three hydrogen atoms by hydrocarbyl groups, and having the general structures RNH_2 (primary amines), R_2NH (secondary amines), R_3N (tertiary amines). [5]

Amino acids* - Compounds containing both a carboxylic acid group (-COOH) and an amino group (-NH_2). The most important are the α-amino acids, in which the -NH_2 group in attached to the C atom adjacent to the -COOH group. In the β-amino acids, there is an intervening carbon atom. [4]

Ampere (A)* - The SI base unit of electric current. [1]

Ampere's law - The defining equation for the magnetic induction B, viz., $dF = Idl \times B$, where dF is the force produced by a current I flowing in an element of the conductor dl pointing in the direction of the current.

Ångström (Å) - A unit of length used in spectroscopy, crystallography, and molecular structure, equal to 10^{-10} m.

Angular momentum (L) - The angular momentum of a particle about a point is the vector product of the radius vector from this point to the particle and the momentum of the particle; i.e., $L = r \times p$. [1]

Angular velocity (ω) - The angle through which a body rotates per unit time.

Anilides - Compounds derived from oxoacids R(C=O)(OH) by replacing the -OH group by the -NHPh group or derivative formed by ring substitution. Also used for salts formed by replacement of a nitrogen-bound hydrogen of aniline by a metal. [5]

Anion - A negatively charged atomic or molecular particle.

Antiferroelectricity* - An effect analogous to antiferromagnetism in which electric dipoles in a crystal are ordered in two sublattices that are polarized in opposite directions, leading to zero net polarization. The effect vanishes above a critical temperature.

Antiferromagnetism* - A type of magnetism in which the magnetic moments of atoms in a solid are ordered into two antiparallel aligned sublattices. Antiferromagnets are characterized by a zero or small positive magnetic susceptibility. The susceptibility increases with temperature up to a critical value, the Néel temperature, above which the material becomes paramagnetic.

Antiparticle - A particle having the same mass as a given elementary particle and a charge equal in magnitude but opposite in sign.

Appearance potential* - The lowest energy which must be imparted to the parent molecule to cause it to produce a particular specified parent ion. This energy, usually stated in eV, may be imparted by electron impact, photon impact, or in other ways. More properly called appearance energy. [3]

Appearance potential spectroscopy (APS) - See Techniques for Materials Characterization, page 12-1.

Are (a) - A unit of area equal to 100 m^2. [1]

Arenes - Monocyclic and polycyclic aromatic hydrocarbons. See aromatic compounds. [5]

Aromatic compounds - Compounds whose structure includes a cyclic delocalized π-electron system. Historical use of the term implies a ring containing only carbon (e.g., benzene, naphthalene), but it is often generalized to include heterocyclic structures such as pyridine and thiophene. [5]

Arrhenius equation - A key equation in chemical kinetics which expresses the rate constant k as $k = A\exp(-E_a/RT)$, where E_a is the activation energy, R the molar gas constant, and T the temperature. A is called the preexponential factor and, for simple gas phase reactions, may be identified with the collision frequency.

Arsines - AsH_3 and compounds derived from it by substituting one, two or three hydrogen atoms by hydrocarbyl groups. $RAsH_2$, R_2AsH, R_3As (R not equal to H) are called primary, secondary and tertiary arsines, respectively. [5]

Aryl groups - Groups derived from arenes by removal of a hydrogen atom from a ring carbon atom. Groups similarly derived from heteroarenes are sometimes subsumed in this definition. [5]

Astronomical unit (AU)* - The mean distance of the earth from the sun, equal to $1.49597870 \times 10^{11}$ m.

Atomic absorption spectroscopy (AAS) - See Techniques for Materials Characterization, page **12**-1.

Atomic emission spectroscopy (AES) - See Techniques for Materials Characterization, page **12**-1.

Atomic force microscopy (AFM) - See Techniques for Materials Characterization, page 12-1.

Atomic mass* - The mass of a nuclide, normally expressed in unified atomic mass units (u).

Atomic mass unit (u)* - A unit of mass used in atomic, molecular, and nuclear science, defined as the mass of one atom of ^{12}C divided by 12. Its approximate value is 1.66054×10^{-27} kg. Also called the unified atomic mass unit. [1]

Atomic number (Z) - A characteristic property of an element, equal to the number of protons in the nucleus.

Atomic weight (A_r)* - The ratio of the average mass per atom of an element to 1/12 of the mass of nuclide ^{12}C. An atomic weight can be defined for a sample of any given isotopic composition. The standard atomic weight refers to a sample of normal terrestrial isotopic composition. The term relative atomic mass is synonymous with atomic weight. [2]

Attenuated total reflection (ATR) - See Techniques for Materials Characterization, page 12-1.

Auger effect - An atomic process in which an electron from a higher energy level fills a vacancy in an inner shell, transferring the released energy to another electron which is ejected.

Aurora - An atmospheric phenomenon in which streamers of light are produced when electrons from the sun are guided into the thermosphere by the earth's magnetic field. It occurs in the polar regions at altitudes of 95—300 km.

Avogadro constant (N_A)* - The number of elementary entities in one mole of a substance.

Azeotrope - A liquid mixture in a state where the variation of vapor pressure with composition at constant temperature (or, alternatively, the variation of normal boiling point with composition) shows either a maximum or a minimum. Thus when an azeotrope boils the vapor has the same composition as the liquid.

Azides - Compounds bearing the group $-N_3$, viz. $-N=N^+=N^-$; usually attached to carbon, e.g. PhN_3, phenyl azide or azidobenzene. Also used for salts of hydrazoic acid, HN_3, e.g. NaN_3, sodium azide. [5]

Azines - Condensation products, $R_2C=NN=CR_2$, of two moles of a carbonyl compound with one mole of hydrazine. [5]

Azo compounds - Derivatives of diazene (diimide), $HN=NH$, wherein both hydrogens are substituted by hydrocarbyl groups, e.g., $PhN=NPh$, azobenzene or diphenyldiazene. [5]

Balmer series - The series of lines in the spectrum of the hydrogen atom which corresponds to transitions between the state with principal quantum number $n = 2$ and successive higher states. The wavelengths are given by $1/\lambda = R_H(1/4 - 1/n^2)$, where $n = 3,4,...$ and R_H is the Rydberg constant for hydrogen. The first member of the series ($n = 2 \quad 3$), which is often called the H_α line, falls at a wavelength of 6563 Å.

Bar (bar) - A unit of pressure equal to 10^5 Pa.´

Bardeen-Cooper-Schrieffer (BCS) theory - A theory of superconductivity which is based upon the formation of electron pairs as a result of an electron-lattice interaction. The theory relates the superconducting transition temperature to the density of states and the Debye temperature.

Barn (b) - A unit used for expressing cross sections of nuclear processes, equal to 10^{-28} m^2.

Barrel - A unit of volume equal to 158.9873 L.

Baryon - Any elementary particle built up from three quarks. Examples are the proton, neutron, and various short-lived hyperons. Baryons have odd half-integer spins.

Base - Historically, a substance that yields an OH^- ion when it dissociates in solution, resulting in a pH>7. In the Brönsted definition, a base is a substance capable of accepting a proton in any type of reaction. The more general definition, due to G.N. Lewis, classifies any chemical species capable of donating an electron pair as a base.

Becquerel (Bq)* - The SI unit of radioactivity (disintegrations per unit time), equal to s^{-1}. [1]

Beer's law - An approximate expression for the change in intensity of a light beam that passes through an absorbing medium, viz., $\log(I/I_0) = -\varepsilon cl$, where I_0 is the incident intensity, I is the final intensity, ε is the molar (decadic) absorption coefficient, c is the molar concentration of the absorbing substance, and l is the path length. Also called the Beer-Lambert law

Binding energy* - A generic term for the energy required to decompose a system into two or more of its constituent parts. In nuclear physics, the binding energy is the energy difference between a nucleus and the separated nucleons of which it is composed (the energy equivalent of the mass defect). In atomic physics, it is the energy required to remove an electron from an atom.

Biot (Bi) - A name sometimes used for the unit of current in the emu system.

Birefringence - A property of certain crystals in which two refracted rays result from a single incident light ray. One, the ordinary ray, follows the normal laws of refraction, while the other, the extraordinary ray, exhibits a variable refractive index which depends on the direction in the crystal.

Black body radiation* - The radiation emitted by a perfect black body, i.e., a body which absorbs all radiation incident on it and reflects none. The wavelength dependence of the radiated energy density ρ (energy per unit volume per unit wavelength range) is given by the Planck formula

$$\rho = \frac{8\pi hc}{\lambda^5 (e^{hc/\lambda kt} - 1)}$$

where λ is the wavelength, h is Planck's constant, c is the speed of light, k is the Boltzmann constant, and T is the temperature.

Black hole - A very dense object, formed in a supernova explosion, whose gravitational field is so large that no matter or radiation can escape from the object.

Bloch wave function - A solution of the Schrödinger equation for an electron moving in a spatially periodic potential; used in the band theory of solids.

Bohr magneton (μ_B)* - The atomic unit of magnetic moment, defined as $eh/4\pi m_e$, where h is Planck's constant, m_e the electron mass, and e the elementary charge. It is the moment associated with a single electron spin.

Bohr, bohr radius (a_0)* - The radius of the lowest orbit in the Bohr model of the hydrogen atom, defined as $\varepsilon_0 h^2/\pi m_e e^2$, where ε_0 is the permittivity of a vacuum, h is Planck's constant, m_e the electron mass, and e the elementary charge. It is customarily taken as the unit of length when using atomic units.

Boiling point - The temperature at which the liquid and gas phases of a substance are in equilibrium at a specified pressure. The normal boiling point is the boiling point at normal atmospheric pressure (101.325 kPa).

Boltzmann constant (k)* - The molar gas constant R divided by Avogadro's constant.

Boltzmann distribution - An expression for the equilibrium distribution of molecules as a function of their energy, in which the number of molecules in a state of energy E is proportional to exp($-E/kT$), where k is the Boltzmann constant and T is the temperature.

Bond strength - See Dissociation energy.

Born-Haber cycle* - A thermodynamic cycle in which a crystalline solid is converted to gaseous ions and then reconverted to the solid. The cycle permits calculation of the lattice energy of the crystal.

Bose-Einstein distribution - A modification of the Boltzmann distribution which applies to a system of particles that are bosons. The number of particles of energy E is proportional to $[e^{(E-\mu)/kT}-1]^{-1}$, where μ is a normalization constant, k is the Boltzmann constant, and T is the temperature.

Boson - A particle that obeys Bose-Einstein Statistics; specifically, any particle with spin equal to zero or an integer. This includes the photon, pion, deuteron, and all nuclei of even mass number.

Boyle's law - The empirical law, exact only for an ideal gas, which states that the volume of a gas is inversely proportional to its pressure at constant temperature.

Bragg angle (θ) - Defined by the equation $n\lambda = 2d\sin\theta$, which relates the angle θ between a crystal plane and the diffracted x-ray beam, the wavelength λ of the x-rays, the crystal plane spacing d, and the diffraction order n (any integer).

Bravais lattices* - The 14 distinct crystal lattices that can exist in three dimensions. They include three in the cubic crystal system, two in the tetragonal, four in the orthorhombic, two in the monoclinic, and one each in the triclinic, hexagonal, and trigonal systems.

Breakdown voltage - The potential difference at which an insulating substance undergoes a physical or chemical change that causes it to become a conductor, thus allowing current to flow through the sample.

Bremsstrahlung - Electromagnetic radiation generated when the velocity of a charged particle is reduced (literally, "braking radiation"). An example is the x-ray continuum resulting from collisions of electrons with the target in an x-ray tube.

Brewster angle - The angle of incidence for which the maximum degree of plane polarization occurs when a beam of unpolarized light is incident on the surface of a medium of refractive index n. At this angle, the angle between the reflected and refracted beams is 90°. The value of the Brewster angle is $\tan^{-1}n$.

Brillouin scattering - The scattering of light by acoustic phonons in a solid or liquid.

Brillouin zone - A region of allowed wave vectors and energy levels in a crystalline solid, which plays a part in the propagation of waves through the lattice.

British thermal unit (Btu) - A non-SI unit of energy, equal to approximately 1055 J. Several values of the Btu, defined in slightly different ways, have been used.

Brownian motion - The random movements of small particles suspended in a fluid, which arise from collisions with the fluid molecules.

Brunauer-Emmett-Teller method (BET) - See Techniques for Materials Characterization, page 12-1.

Buffer* - A solution designed to maintain a constant pH when small amounts of a strong acid or base are added. Buffers usually consist of a fairly weak acid and its salt with a strong base. Suitable concentrations are chosen so that the pH of the solution remains close to the pK_a of the weak acid.

Calorie (cal) - A non-SI unit of energy, originally defined as the heat required to raise the temperature of 1 g of water by 1°C. Several calories of slightly different values have been used. The thermochemical calorie is now defined as 4.184 J.

Candela (cd)* - The SI base unit of luminous intensity. [1]

Capacitance (C) - Ratio of the charge acquired by a body to the change in potential. [1]

Carbamates - Salts or esters of carbamic acid, $H_2NC(=O)OH$, or of N-substituted carbamic acids: $R_2NC(=O)OR'$, (R' = hydrocarbyl or a cation). The esters are often called urethanes or urethans, a usage that is strictly correct only for the ethyl esters. [5]

Carbenes - The electrically neutral species H_2C: and its derivatives, in which the carbon is covalently bonded to two univalent groups of any kind or a divalent group and bears two non-bonding electrons, which may be spin-paired (singlet state) or spin-non-paired (triplet state). [5]

Carbinols - An obsolete term for substituted methanols, in which the name carbinol is synonymous with methanol. [5]

Carbohydrates - Originally, compounds such as aldoses and ketoses, having the stoichiometric formula $C_n(H_2O)_n$ (hence "hydrates of carbon"). The generic term carbohydrate now includes mono-, oligo-, and polysaccharides, as well as their reaction products and derivatives. [5]

Carboranes - A contraction of carbaboranes. Compounds in which a boron atom in a polyboron hydride is replaced by a carbon atom with maintenance of the skeletal structure. [5]

Carboxylic acids - Oxoacids having the structure RC(=O)OH. The term is used as a suffix in systematic name formation to denote the -C(=O)OH group including its carbon atom. [5]

Carnot cycle - A sequence of reversible changes in a heat engine using a perfect gas as the working substance, which is used to demonstrate that entropy is a state function. The Carnot cycle also provides a means to calculate the efficiency of a heat engine.

Catalyst - A substance that participates in a particular chemical reaction and thereby increases its rate but without a net change in the amount of that substance in the system. [3]

Catenanes, catena compounds - Hydrocarbons having two or more rings connected in the manner of links of a chain, without a covalent bond. More generally, the class catena compounds embraces functional derivatives and hetero analogues. [5]

Cation - A positively charged atomic or molecular particle.

Centipoise (cP) - A common non-SI unit of viscosity, equal to mPa s.

Centrifugal distortion - An effect in molecular spectroscopy in which rotational levels are lowered in energy, relative to the values of a rigid rotor, as the rotational angular momentum increases. The effect may be understood classically as a stretching of the bonds in the molecule as it rotates faster, thus increasing the moment of inertia.

Ceramic - A nonmetallic material of very high melting point.

Cerenkov radiation - Light emitted when a beam of charged particles travels through a medium at a speed greater than the speed of light in the medium. It is typically blue in color.

Cgs system of units - A system of units based upon the centimeter, gram, and second. The cgs system has been supplanted by the International System (SI).

Chalcogens - The Group VIA elements (oxygen, sulfur, selenium, tellurium, and polonium). Compounds of these elements are called chalcogenides. [7]

Chaotic system - A complex system whose behavior is governed by deterministic laws but whose evolution can vary drastically when small changes are made in the initial conditions.

Charge - See Electric charge.

Charles' law - The empirical law, exact only for an ideal gas, which states that the volume of a gas is directly proportional to its temperature at constant pressure.

Charm - A quantum number introduced in particle physics to account for certain properties of elementary particles and their reactions.

Chelate - A compound characterized by the presence of bonds from two or more bonding sites within the same ligand to a central metal atom. [3]

Chemical potential - For a mixture of substances, the chemical potential of constituent B is defined as the partial derivative of the Gibbs energy G with respect to the amount (number of moles) of B, with temperature, pressure, and amounts of all other constituents held constant. Also called partial molar Gibbs energy. [2]

Chemical shift* - A small change in the energy levels (and hence in the spectra associated with these levels) resulting from the effects of chemical binding in a molecule. The term is used in fields such as NMR, Mössbauer, and photoelectron spectroscopy, where the energy levels are determined primarily by nuclear or atomic effects.

Chiral molecule - A molecule which cannot be superimposed on its mirror image. A common example is an organic molecule containing a carbon atom to which four different atoms or groups are attached. Such molecules exhibit optical activity, i.e., they rotate the plane of a polarized light beam.

Chlorocarbons - Compounds consisting solely of chlorine and carbon. [5]

Chromatography* - A method for separation of the components of a sample in which the components are distributed between two phases, one of which is stationary while the other moves. In gas chromatography the gas moves over a liquid or solid stationary phase. In liquid chromatography the liquid mixture moves through another liquid, a solid, or a gel. The mechanism of separation of components may be adsorption, differential solubility, ion-exchange, permeation, or other mechanisms. [6]

Clapeyron equation - A relation between pressure and temperature of two phases of a pure substance that are in equilibrium, viz., $dp/dT = \Delta_{trs}S/\Delta_{trs}V$, where $\Delta_{trs}S$ is the difference in entropy between the phases and $\Delta_{trs}V$ the corresponding difference in volume.

Clathrates - Inclusion compounds in which the guest molecule is in a cage formed by the host molecule or by a lattice of host molecules. [5]

Clausius (Cl) - A non-SI unit of entropy or heat capacity defined as cal/K = 4.184 J/K. [2]

Clausius-Clapeyron equation - An approximation to the Clapeyron equation applicable to liquid-gas and solid-gas equilibrium, in which one assumes an ideal gas with volume much greater than the condensed phase volume. For the liquid-gas case, it takes the form $d(\ln p)/dT = \Delta_{vap}H/RT^2$, where R is the molar gas constant and $\Delta_{vap}H$ is the molar enthalpy of vaporization. For the solid-gas case, $\Delta_{vap}H$ is replaced by the molar enthalpy of sublimation, $\Delta_{sub}H$.

Clausius-Mosotti equation - A relation between the dielectric constant ε_r at optical frequencies and the polarizability α:

$$\frac{\varepsilon_r - 1}{\varepsilon_r + 2} = \frac{\rho N_A \alpha}{3M\varepsilon_0}$$

where ρ is density, N_A is Avogadro's number, M is molar mass, and ε_0 is the permittivity of a vacuum.

Clebsch-Gordon coefficients - A set of coefficients used to describe the vector coupling of angular momenta in atomic and nuclear physics.

Codon - A set of three bases, chosen from the four primary bases found in the DNA molecule (uracil, cytosine, adenine, and guanine), which specifies the production of a particular amino acid or carries some other genetic instruction. For example, the codon UCA specifies the amino acid serine, CAG specifies glutamine, etc. There are a total of 64 codons.

Coercive force - The magnetizing force at which the magnetic flux density is equal to zero. [10]

Coercivity* - The maximum value of coercive force that can be attained when a magnetic material is symmetrically magnetized to saturation induction. [10]

Coherent anti-Stokes Raman spectroscopy (CARS) - See Techniques for Materials Characterization, page **12**-1.

Colloid - Molecules or polymolecular particles dispersed in a medium that have, at least in one direction, a dimension roughly between 1 nm and 1 μm. [3]

Color center - A defect in a crystal that gives rise to optical absorption, thus changing the color of the material. A common type is the F-center, which results when an electron occupies the site of a negative ion.

Compressibility (κ)* - The fractional change of volume as pressure is increased, viz., $\kappa = -(1/V)(dV/dp)$. [1]

Compton wavelength (λ_C)* - In the scattering of electromagnetic radiation by a free particle (e.g., electron, proton), $\lambda_C = h/mc$ is the increase in wavelength, at a 90° scattering angle, corresponding to the transfer of energy from radiation to particle. Here h is Planck's constant, c the speed of light, and m the mass of the particle.

Conductance (G)* - For direct current, the reciprocal of resistance. More generally, the real part of admittance. [1]

Conductivity, electrical (σ)* - The reciprocal of the resistivity. [1]

Conductivity, thermal - See Thermal conductivity.

Congruent transformation - A phase transition (melting, vaporization, etc.) in which the substance preserves its exact chemical composition.

Constitutional repeating unit (CRU) - In polymer science, the smallest constitutional unit, the repetition of which constitutes a regular macromolecule, i.e., a macromolecule with all units connected identically with respect to directional sense. [8]

Copolymer - A polymer derived from more than one species of monomer. [8]

Coriolis effect - The deviation from simple trajectories when a mechanical system is described in a rotating coordinate system. It affects the motion of projectiles on the earth and in molecular spectroscopy leads to an important interaction between the rotational and vibrational motions. The effect may be described by an additional term in the equations of motion, called the Coriolis force.

Cosmic rays* - High energy nuclear particles, electrons, and photons, originating mostly outside the solar system, which continually bombard the earth's atmosphere.

Coulomb (C)* - The SI unit of electric charge, equal to A s. [1]

Coulomb's law - The statement that the force F between two electrical charges q_1 and q_2 separated by a distance r is $F = (4\pi\varepsilon_0)^{-1} q_1 q_2/r^2$, where ε_0 is the permittivity of a vacuum.

Covalent bond - A chemical bond between two atoms whose stability results from the sharing of two electrons, one from each atom.

Cowling number (Co) - A dimensionless quantity used in plasma physics, defined by $Co = B^2/\mu\rho v^2$, where ρ is density, v is velocity, μ is permeability, and B is magnetic flux density. [2]

CPT theorem - A theorem in particle physics which states that any local Lagrangian theory that is invariant under proper Lorentz transformations is also invariant under the combined operations of charge conjugation, C, space inversion, P, and time reversal, T, taken in any order.

Critical point* - In general, the point on the phase diagram of a two-phase system at which the two coexisting phases have identical properties and therefore represent a single phase. At the liquid-gas critical point of a pure substance, the distinction between liquid and gas vanishes, and the vapor pressure curve ends. The coordinates of this point are called the critical temperature and critical pressure. Above the critical temperature, it is not possible to liquefy the substance.

Cross section (σ)* - A measure of the probability of collision (or other interaction) between a beam of particles and a target which it encounters. In rough terms it is the effective area the target particles present to the incident ones; however, the precise definition depends on the nature of the interaction. A general definition of σ is the number of encounters per unit time divided by nv, where n is the concentration of incident particles and v their velocity.

Crosslink - In polymer science, a small region in a macromolecule from which at least four chains emanate, and formed by reactions involving sites or groups on existing macromolecules or by interactions between existing macromolecules. [8]

Crown compounds - Macrocyclic polydentate compounds, usually uncharged, in which three or more coordinating ring atoms (usually oxygen or nitrogen) are or may become suitably close for easy formation of chelate complexes with metal ions or other cationic species. [5]

Crust* - The outer layer of the solid earth, above the Mohorovicic discontinuity. Its thickness averages about 35 km on the continents and about 7 km below the ocean floor.

Cryoscopic constant (E_f)* - The constant that expresses the amount by which the freezing point T_f of a solvent is lowered by a non-dissociating solute, through the relation $\Delta T_f = E_f m$, where m is the molality of the solute.

Curie (Ci) - A non-SI unit of radioactivity (disintegrations per unit time), equal to 3.7×10^{10} s^{-1}.

Curie temperature (T_C)* - For a ferromagnetic material, the critical temperature above which the material becomes paramagnetic. Also applied to the temperature at which the spontaneous polarization disappears in a ferroelectric solid. [1]

Cyanohydrins - Alcohols substituted by a cyano group, most commonly, but not limited to, examples having a CN and an OH group attached to the same carbon atom. They are formally derived from aldehydes or ketones by the addition of hydrogen cyanide. [5]

Cycloalkanes - Saturated monocyclic hydrocarbons (with or without side chains). See alicyclic compounds. Unsaturated monocyclic hydrocarbons having one endocyclic double or one triple bond are called cycloalkenes and cycloalkynes, respectively. [5]

Cyclotron resonance - The resonant absorption of energy from a system in which electrons or ions that are orbiting in a uniform magnetic field are subjected to radiofrequency or microwave radiation. The resonance frequency is given by $v = eH/2\pi m^*c$, where e is the elementary charge, H is the magnetic field strength, m^* is the effective mass of the charged particle, and c is the speed of light. The effect occurs in both solids (involving electrons or holes) and in low pressure gasses (involving ions)

Dalton (Da) - A name sometimes used in biochemistry for the unified atomic mass unit (u).

De Broglie wavelength - The wavelength associated with the wave representation of a moving particle, given by h/mv, where h is Planck's constant, m the particle mass, and v the velocity.

De Haas-Van Alphen effect - An effect observed in certain metals and semiconductors at low temperatures and high magnetic fields, characterized by a periodic variation of magnetic susceptibility with field strength.

Debye equation* - The relation between the relative permittivity (dielectric constant) ε_r, polarizability α, and permanent dipole moment μ in a dielectric material whose molecules are free to rotate. It takes the form

$$\frac{\varepsilon_r - 1}{\varepsilon_r + 2} = \frac{\rho N_A}{3 M \varepsilon_0} \left(\alpha + \frac{\mu^2}{3kT} \right)$$

where ρ is density, N_A is Avogadro's number, M is molar mass, and ε_0 is the permittivity of a vacuum.

Debye length - In the Debye-Hückel theory of ionic solutions, the effective thickness of the cloud of ions of opposite charge which surrounds each given ion and shields the Coulomb potential produced by that ion.

Debye temperature (θ_D)* - In the Debye model of the heat capacity of a crystalline solid, $\theta_D = h v_D / k$, where h is Planck's constant, k is the Boltzmann constant, and v_D is the maximum vibrational frequency the crystal can support. For $T << \theta_D$, the heat capacity is proportional to T^3.

Debye unit (D) - A non-SI unit of electric dipole moment used in molecular physics, equal to 3.335641×10^{-30} C m.

Debye-Waller factor (D) - The factor by which the intensity of a diffraction line is reduced because of lattice vibrations. [1]

Defect - Any departure from the regular structure of a crystal lattice. A Frenkel defect results when an atom or ion moves to an interstitial position and leaves behind a vacancy. A Schottky defect involves either a vacancy where the atom has moved to the surface or a structure where a surface atom has moved to an interstitial position.

Degree of polymerization - The number of monomeric units in a macromolecule or an oligomer molecule. [8]

Dendrite - A tree-like crystalline pattern often observed, for example, in ice crystals and alloys in which the crystal growth branches repeatedly.

Density (ρ)* - In the most common usage, mass density or mass per unit volume. More generally, the amount of some quantity (mass, charge, energy, etc.) divided by a length, area, or volume.

Density of states (N_E, ρ) - The number of one-electron states in an infinitesimal interval of energy, divided by the range of that interval and by volume. [1]

Dew point* - The temperature at which liquid begins to condense as the temperature of a gas mixture is lowered. In meteorology, it is the temperature at which moisture begins to condense on a surface in contact with the air.

Diamagnetism - A type of magnetism characterized by a negative magnetic susceptibility, so that the material, when placed in an external magnetic field, becomes weakly magnetized in the direction opposite to the field. This magnetization is independent of temperature.

Diazo compounds - Compounds having the divalent diazo group, $=N^+=N^-$, attached to a carbon atom, e.g., $CH_2=N_2$ diazomethane. [5]

Dielectric constant (ε)* - Ratio of the electric displacement in a medium to the electric field strength. Also called permittivity. [1]

Dienes - Compounds that contain two fixed double bonds (usually assumed to be between carbon atoms). Dienes in which the two double-bond units are linked by one single bond are termed conjugated. [5]

Differential scanning calorimetry (DSC) - See Techniques for Materials Characterization, page 12-1.

Differential thermal analysis (DTA) - See Techniques for Materials Characterization, page 12-1.

Diffusion* - The migration of atoms, molecules, ions, or other particles as a result of some type of gradient (concentration, temperature, etc.).

Diopter - A unit used in optics, formally equal to m^{-1}. It is used in expressing dioptic power, which is the reciprocal of the focal length of a lens.

Dipole moment, electric (p,μ)* - For a distribution of equal positive and negative charge, the magnitude of the dipole moment vector is the positive charge multiplied by the distance between the centers of positive and negative charge distribution. The direction is given by the line from the center of negative charge to the center of positive charge.

Dipole moment, magnetic (m,μ) - Formally defined in electromagnetic theory as a vector quantity whose vector product with the magnetic flux density equals the torque. The magnetic dipole generated by a current I flowing in a small loop of area A has a magnetic moment of magnitude IA. In atomic and nuclear physics, a magnetic moment is associated with the angular momentum of a particle; e.g., an electron with orbital angular momentum l exhibits a magnetic moment of $-el/2m_e$ where e is the elementary charge and m_e the mass of the electron. [1]

Disaccharides - Compounds in which two monosaccharides are joined by a glycosidic bond. [5]

Dislocation - An extended displacement of a crystal from a regular lattice. An edge dislocation results when one portion of the crystal has partially slipped with respect to the other, resulting in an extra plane of atoms extending through part of the crystal. A screw dislocation transforms successive atomic planes into the surface of a helix.

Dispersion - Splitting of a beam of light (or other electromagnetic radiation) of mixed wavelengths into the constituent wavelengths as a result of the variation of refractive index of the medium with wavelength.

Dissociation constant* - The equilibrium constant for a chemical reaction in which a compound dissociates into its constituent parts.

Dissociation energy (D_e)* - For a diatomic molecule, the difference between the energies of the free atoms at rest and the minimum in the potential energy curve. The term bond dissociation energy (D_0), which can be applied to polyatomic molecules as well, is used for the difference between the energies of the fragments resulting when a bond is broken and the energy of the original molecule in its lowest energy state. The term bond strength implies differences in enthalpy rather than energy.

Domain - A small region of a solid in which the magnetic or electric moments of the individual units (atoms, molecules, or ions) are aligned in the same direction.

Domain wall - The transition region between adjacent ferromagnetic domains, generally a layer with a thickness of a few hundred ångström units. Also called Bloch wall.

Doppler effect - The change in the apparent frequency of a wave (sound, light, or other) when the source of the wave is moving relative to the observer.

Dose equivalent (*H*) - The product of the absorbed dose of radiation at a point of interest in tissue and various modifying factors which depend on the type of tissue and radiation. [1]

Drift velocity - The velocity of charge carriers (electrons, ions, etc.) moving under the influence of an electric field in a medium which subjects the carriers to some frictional force.

Dyne (dyn) - A non-SI (cgs) unit of force, equal to 10^{-5} N.

Ebullioscopic constant (*E*$_b$)* - The constant that expresses the amount by which the boiling point T_b of a solvent is raised by a non-dissociating solute, through the relation $\Delta T_b = E_b\, m$, where m is the molality of the solute.

Eddy currents - Circulating currents set up in conducting bulk materials or sheets by varying magnetic fields.

Effinghausen effect - The appearance of a temperature gradient in a current carrying conductor that is placed in a transverse magnetic field. The direction of the gradient is perpendicular to the current and the field.

Eigenvalue - An allowed value of the constant a in the equation $Au = au$, where A is an operator acting on a function u (which is called an eigenfunction). In quantum mechanics, the outcome of any observation is an eigenvalue of the corresponding operator. Also called characteristic value.

Einstein - A non-SI unit used in photochemistry, equal to one mole of photons.

Einstein temperature (θ$_V$) - In the Einstein theory of the heat capacity of a crystalline solid, $\theta_V = h\nu/k$, where h is Planck's constant, k is the Boltzmann constant, and ν is the vibrational frequency of the crystal.

Einstein transition probability - A constant in the Einstein relation $A_{ij} + B_{ij}\rho$ for the probability of a transition between two energy levels i and j in a radiation field of energy density ρ. The A_{ij} coefficient describes the probability of spontaneous emission, while B_{ij} and B_{ji} govern the probability of stimulated emission and absorption, respectively ($B_{ij} = B_{ji}$).

Elastic limit - The greatest stress which a material is capable of sustaining without any permanent strain remaining after complete release of the stress. [10]

Elastic modulus - See Young's modulus.

Electric charge (*Q*) - The quantity of electricity; i.e., the property that controls interactions between bodies through electrical forces.

Electric current (*I*) - The charge passing through a circuit per unit time. [1]

Electric displacement (*D*) - A vector quantity whose magnitude equals the electric field strength multiplied by the permittivity of the medium and whose direction is the same as that of the field strength.

Electric field strength (*E*) - The force exerted by an electric field on a point charge divided by the electric charge. [1]

Electric potential (*V*) - A scalar quantity whose gradient is equal to the negative of the electric field strength.

Electrical conductance - See Conductance

Electrical resistance - See Resistance

Electrical resistivity - See Resistivity.

Electrochemical series* - An arrangement of reactions which produce or consume electrons in an order based on standard electrode potentials. A common arrangement places metals in decreasing order of their tendency to give up electrons.

Electrode potential* - The electromotive force of a cell in which the electrode on the left is the standard hydrogen electrode and that on the right is the electrode in question. [2]

Electrolysis - The decomposition of a substance as a result of passing an electric current between two electrodes immersed in the sample.

Electromotive force (emf) - The energy supplied by a source divided by the charge transported through the source. [1]

Electron* - An elementary particle in the family of leptons, with negative charge and spin of 1/2.

Electron affinity* - The energy difference between the ground state of a gas-phase atom or molecule and the lowest state of the corresponding negative ion.

Electron cyclotron resonance (ECR) - See Techniques for Materials Characterization, page 12-1.

Electron energy loss spectroscopy (EELS) - See Techniques for Materials Characterization, page 12-1.

Electron nuclear double resonance (ENDOR) - See Techniques for Materials Characterization, page 12-1.

Electron paramagnetic resonance (EPR) - See Techniques for Materials Characterization, page 12-1.

Electron probe microanalysis (EPMA) - See Techniques for Materials Characterization, page 12-1.

Electron spectroscopy for chemical analysis (ESCA) - See Techniques for Materials Characterization, page 12-1.

Electron spin (*s*) - The quantum number, equal to 1/2, that specifies the intrinsic angular momentum of the electron.

Electron stimulated desorption (ESD) - See Techniques for Materials Characterization, page 12-1.

Electron volt (eV)* - A non-SI unit of energy used in atomic and nuclear physics, equal to approximately 1.602177×10^{-19} J. The electron volt is defined as the kinetic energy acquired by an electron upon acceleration through a potential difference of 1 V. [1]

Electronegativity* - A parameter originally introduced by Pauling which describes, on a relative basis, the power of an atom or group of atoms to attract electrons from the same molecular entity. [3]

Electrophoresis - The motion of macromolecules or colloidal particles in an electric field. [3]

Emissivity (ε)* - Ratio of the radiant flux emitted per unit area to that of an ideal black body at the same temperature. Also called emittance. [1]

Emu - The electromagnetic system of units, based upon the cm, g, and s plus the emu of current (sometimes called the abampere).

Enantiomers - A chiral molecule and its non-superposable mirror image. The two forms rotate the plane of polarized light by equal amounts in opposite directions. Also called optical isomers.

Energy (*E*,*U*)* - The characteristic of a system that enables it to do work.

Energy gap* - In the theory of solids, the region between two energy bands, in which no bound states can occur.

Enols, alkenols - The term refers specifically to vinylic alcohols, which have the structure $HOCR'=CR_2$. Enols are tautomeric with aldehydes (R' = H) or ketones (R' not equal to H). [5]

Enthalpy (*H*)* - A thermodynamic function, especially useful when dealing with constant-pressure processes, defined by $H = E + PV$, where E is energy, P pressure, and V volume. [1]

Enthalpy of combustion* - The enthalpy change in a combustion reaction. Its negative is the heat released in combustion.

Enthalpy of formation, standard* - The enthalpy change for the reaction in which a substance is formed from its constituent elements, each in its standard reference state (normally refers to 1 mol, sometimes to 1 g, of the substance).

Enthalpy of fusion* - The enthalpy change in the transition from solid to liquid state.

Enthalpy of sublimation - The enthalpy change in the transition from solid to gas state.

Enthalpy of vaporization* - The enthalpy change in the transition from liquid to gas state.

Entropy (*S*)* - A thermodynamic function defined such that when a small quantity of heat dQ is received by a system at temperature T, the entropy of the system is increased by dQ/T, provided that no irreversible change takes place in the system. [1]

Entropy unit (e.u.) - A non-SI unit of entropy, equal to 4.184 J/K mol.

Ephemeris time - Time measured in tropical years from January 1, 1900.

Epoxy compounds - Compounds in which an oxygen atom is directly attached to two adjacent or non-adjacent carbon atoms of a carbon chain or ring system; thus cyclic ethers. [5]

Equation of continuity - Any of a class of equations that express the fact that some quantity (mass, charge, energy, etc.) cannot be created or destroyed. Such equations typically specify that the rate of increase of the quantity in a given region of space equals the net current of the quantity flowing into the region.

Equation of state* - An equation relating the pressure, volume, and temperature of a substance or system.

Equilibrium constant (*K*)* - For a chemical reaction $aA + bB \rightleftharpoons cC + dD$, the equilibrium constant is defined by:

$$K = \frac{a_C{}^c \cdot a_D{}^d}{a_A{}^a \cdot a_B{}^b}$$

where a_i is the activity of component i. To a certain approximation, the activities can be replaced by concentrations. The equilibrium constant is related to $\Delta_r G°$, the standard Gibbs energy change in the reaction, by $RT \ln K = -\Delta_r G°$.

Equivalent conductance - See Conductivity, electrical

Erg (erg) - A non-SI (cgs) unit of energy, equal to 10^{-7} J.

Esters - Compounds formally derived from an oxoacid RC(=O)(OH) and an alcohol, phenol, heteroarenol, or enol by linking, with formal loss of water from an acidic hydroxy group of the former and a hydroxy group of the latter. [5]

Esu - The electrostatic system of units, based upon the cm, g, and s plus the esu of charge (sometimes called the statcoulomb or franklin).

Ethers - Compounds with formula ROR, where R is not equal to H. [5]

Euler number (*Eu*) - A dimensionless quantity used in fluid mechanics, defined by $Eu = \Delta p/\rho v^2$, where p is pressure, ρ is density, and v is velocity. [2]

Eutectic - The point on a two-component solid-liquid phase diagram which represents the lowest melting point of any possible mixture. A liquid having the eutectic composition will freeze at a single temperature without change of composition.

Excitance (*M*) - Radiant energy flux leaving an element of a surface divided by the area of that element. [1]

Exciton - A localized excited state consisting of a bound electron-hole pair in a molecular or ionic crystal. The exciton can propagate through the crystal.

Exosphere - The outermost part of the earth's atmosphere, beginning at about 500 to 1000 km above the surface. It is characterized by densities so low that air molecules can escape into outer space.

Expansion coefficient - See thermal expansion coefficient.

Extended electron energy loss fine structure (EXELFS) - See Techniques for Materials Characterization, page 12-1.

Extended x-ray absorption fine structure (EXAFS) - See Techniques for Materials Characterization, page 12-1.

Extinction coefficient - See Absorption coefficient, molar

F-Center - See Color center

Fahrenheit temperature (°F) - The temperature scale based on the assignment of 32°F = 0°C and a temperature interval of °F = (5/9)°C; i.e., $t/°F = (9/5)t/°C + 32$.

Farad (F)* - The SI unit of electric capacitance, equal to C/V. [1]

Faraday constant (*F*)* - The electric charge of 1 mol of singly charged positive ions; i.e., $F = N_A e$, where N_A is Avogadro's constant and e is the elementary charge. [1]

Faraday effect* - The rotation of the plane of plane-polarized light by a medium placed in a magnetic field parallel to the direction of the light beam. The effect can be observed in solids, liquids, and gasses.

Fatty acids - Aliphatic monocarboxylic acids derived from or contained in esterified form in an animal or vegetable fat, oil, or wax. Natural fatty acids commonly have a chain of 4 to 28 carbons (usually unbranched and even-numbered), which may be saturated or unsaturated. By extension, the term is sometimes used to embrace all acyclic aliphatic carboxylic acids. [5]

Fermat's principle - The law that a ray of light traversing one or more media will follow a path which minimizes the time required to pass between two given points.

Fermi (f) - Name sometimes used in nuclear physics for the femtometer.

Fermi level - The highest energy of occupied states in a solid at zero temperature. Sometimes called Fermi energy. The Fermi surface is the surface in momentum space formed by electrons occupying the Fermi level.

Fermi resonance - An effect observed in vibrational spectroscopy when an overtone of one fundamental vibration closely coincides in energy with another fundamental of the same symmetry species. It leads to a splitting of vibrational bands.

Fermi-Dirac distribution - A modification of the Boltzmann distribution which takes into account the Pauli exclusion principle. The number of particles of energy E is proportional to $[e^{(E-\mu)/kT}+1]^{-1}$, where μ is a normalization constant, k the Boltzmann constant, and T the temperature. The distribution is applicable to a system of fermions.

Fermion - A particle that obeys Fermi-Dirac statistics. Specifically, any particle with spin equal to an odd multiple of 1/2. Examples are the electron, proton, neutron, muon, etc.

Ferrimagnetism* - A type of magnetism in which the magnetic moments of atoms in a solid are ordered into two nonequivalent sublattices with unequal magnetic moments, leading to a nonzero magnetic susceptibility.

Ferrite - A ferrimagnetic material of nominal formula MFe_2O_4, where M is a divalent metal; widely used in microwave switches and other solid state devices.

Ferroelectricity* - The retention of electric polarization by certain materials after the external field that produced the polarization has been removed.

Ferromagnetism* - A type of magnetism in which the magnetic moments of atoms in a solid are aligned within domains which can in turn be aligned with each other by a weak magnetic field. Some ferromagnetic materials can retain their magnetization when the external field is removed, as long as the temperature is below a critical value, the Curie temperature. They are characterized by a large positive magnetic susceptibility.

Fick's law - The statement that the flux J of a diffusing substance is proportional to the concentration gradient, i.e., $J = -D(dc/dx)$, where D is called the diffusion coefficient.

Field - A mathematical construct which describes the interaction between particles resulting from gravity, electromagnetism, or other physical phenomena. In classical physics a field is described by equations. Quantum field theory introduces operators to represent the physical observables.

Field emission microscopy (FEM) - See Techniques for Materials Characterization, page 12-1.

Field ion microscopy (FIM) - See Techniques for Materials Characterization, page 12-1.

Fine structure - The splitting in spectral lines that results from interactions of the electron spin with the orbital angular momentum.

Fine structure constant (α)* - Defined as $e^2/2hc\varepsilon_0$, where e is the elementary charge, h Planck's constant, c the speed of light, and ε_0 the permittivity of a vacuum. It is a measure of the strength of the electromagnetic interaction between particles.

First radiation constant (c_1)* - Constant (= $2\pi hc^2$) in the equation for the radiant excitance M_λ of a black body:

$$M_\lambda = \frac{c_1 \lambda^{-5} \Delta\lambda}{e^{c_2/\lambda T} - 1}$$

where λ is the wavelength, T is the temperature, and $c_2 = hc/k$ is the second radiation constant.

Flash point - The lowest temperature at which vapors above a volatile combustible substance will ignite in air when exposed to a flame. [10]

Fluence (F) - Term used in photochemistry to specify the energy per unit area delivered in a given time interval, for example by a laser pulse. [2]

Fluorocarbons - Compounds consisting solely of fluorine and carbon. [5]

Fluxoid - The quantum of magnetic flux in superconductivity theory, equal to $hc/2e$, where h is Planck's constant, c the velocity of light, and e the elementary charge.

Force (F) - The rate of change of momentum with time. [1]

Force constants (f, k)* - In molecular vibrations, the coefficients in the expression of the potential energy in terms of atom displacements from their equilibrium positions. In a diatomic molecule, $f = d^2V/dr^2$, where $V(r)$ is the potential energy and r is the interatomic distance. [2]

Fourier number (Fo) - A dimensionless quantity used in fluid mechanics, defined by $Fo = at/l^2$, where a is thermal diffusivity, t is time, and l is length. [2]

Fourier transform infrared spectroscopy (FTIR) - A technique for obtaining an infrared spectrum by use of an interferometer in which the path length of one of the beams is varied. A Fourier transformation of the resulting interferogram yields the actual spectrum. The technique is also used for NMR and other types of spectroscopy.

Fractals - Geometrical objects that are self-similar under a change of scale; i.e., they appear similar at all levels of magnification. They can be considered to have fractional dimensionality. Examples occur in diverse fields such as geography (rivers and shorelines), biology (trees), and solid state physics (amorphous materials).

Franck-Condon principle - An important principle in molecular spectroscopy which states that the nuclei in a molecule remain essentially stationary while an electronic transition is taking place. The physical interpretation rests on the fact that the electrons move much more rapidly than the nuclei because of their much smaller mass.

Franklin (Fr) - Name sometimes given to the unit of charge in the esu system.

Fraunhofer diffraction - Diffraction of light in situations where the source and observation point are so far removed that the wave surfaces may be considered planar.

Fraunhofer lines - Sharp absorption lines in the spectrum of sunlight, caused by absorption of the solar blackbody radiation by atoms near the sun's surface.

Free radical - See Radicals. The term "free radical" is often used more broadly for molecules that have a paramagnetic ground state (e.g., O_2) and sometimes for any transient or highly reactive molecular species.

Freezing point - See Melting point.

Frequency (ν)* - Number of cycles of a periodic phenomenon divided by time. [1]

Fresnel diffraction - Diffraction of light in a situation where the source and observation point are sufficiently close together that the curvature of the wave surfaces must be taken into account.

Froude number (Fr) - A dimensionless quantity used in fluid mechanics, defined by $Fr = v/(lg)^{1/2}$, where v is velocity, l is length, and g is acceleration due to gravity. [2]

Fugacity (f_B) - For a gas mixture, the fugacity of component B is defined as the absolute activity λ_B times the limit, as the pressure p approaches zero at constant temperature, of p_B/λ_B. [2]

Fullerenes - Compounds composed solely of an even number of carbon atoms, which form a cage-like fused-ring polycyclic system with twelve five-membered rings and the rest six-membered rings. The archetypal example is [60]fullerene, where the atoms and bonds delineate a truncated icosahedron. The term has been broadened to include any closed cage structure consisting entirely of three-coordinate carbon atoms. [5]

Fulvalenes - The hydrocarbon fulvalene and its derivatives formed by substitution (and by extension, analogues formed by replacement of one or more carbon atoms of the fulvalene skeleton by a heteroatom). [5]

Fulvenes - The hydrocarbon fulvene and its derivatives formed by substitution (and by extension, analogues formed by replacement of one or more carbon atoms of the fulvene skeleton by a heteroatom). [5]

Fundamental vibrational frequencies* - In molecular spectroscopy, the characteristic vibrational frequencies obtained when the vibrational energy is expressed in normal coordinates. They determine the primary features of the infrared and Raman spectra of the molecule.

γ - Name sometimes used for microgram.

γ-rays* - Electromagnetic radiation (photons) with energy greater than about 0.1 MeV (wavelength less than about 1 pm).

g-Factor of the electron* - The proportionality factor in the equation relating the magnetic moment μ of an electron to its total angular momentum quantum number J, i.e., $\mu = -g\mu_B J$, where μ_B is the Bohr magneton. Also called Landé factor.

Gal - A non-SI unit of acceleration, equal to 0.01 m/s. Also called galileo.

Gallon (US) - A unit of volume equal to 3.785412 L.

Gallon (UK, Imperial) - A unit of volume equal to 4.546090 L.

Gauss (G) - A non-SI unit of magnetic flux density (B) equal to 10^{-4} T.

Gaussian system of units - A hybrid system used in electromagnetic theory, which combines features of both the esu and emu systems.

Gel - A colloidal system with a finite, but usually rather small, yield stress (the sheer stress at which yielding starts abruptly). [3]

Genetic code* - The set of relations between each of the 64 codons of DNA and a specific amino acid (or other genetic instruction).

Gibbs energy (G)* - An important function in chemical thermodynamics, defined by $G = H-TS$, where H is the enthalpy, S the entropy, and T the thermodynamic temperature. Sometimes called Gibbs free energy and, in older literature, simply "free energy". [2]

Gibbs phase rule - The relation $F = C - P + 2$, where C is the number of components in a mixture, P is the number of phases, and F is the degrees of freedom, i.e., the number of intensive variables that can be changed independently without affecting the number of phases.

Glass transition temperature* - The temperature at which an amorphous polymer is transformed, in a reversible way, from a viscous or rubbery condition to a hard and relatively brittle one. [10]

Glow discharge mass spectroscopy (GDMS) - See Techniques for Materials Characterization, page 12-1.

Gluon - A hypothetical particle postulated to take part in the binding of quarks, in analogy to the role of the photon in electromagnetic interactions.

Glycerides - Esters of glycerol (propane-1,2,3-triol) with fatty acids, widely distributed in nature. They are by long-established custom subdivided into triglycerides, 1,2- or 1,3-diglycerides, and 1- or 2-monoglycerides, according to the number and positions of acyl groups. [5]

Glycols - Dihydric alcohols in which two hydroxy groups are on different carbon atoms, usually but not necessarily adjacent. Also called diols. [5]

Grain (gr) - A non-SI unit of mass, equal to 64.79891 mg.

Grain boundary - The interface between two regions of different crystal orientation.

Grashof number (Gr) - A dimensionless quantity used in fluid mechanics, defined by $Gr = l^3 g\alpha\Delta T\rho^2/\eta^2$, where T is temperature, ρ is density, l is length, η is viscosity, α is cubic expansion coefficient, and g is acceleration of gravity. [2]

Gravitational constant (G)* - The universal constant in the equation for the gravitational force between two particles, $F = Gm_1m_2/r^2$, where r is the distance between the particles and m_1 and m_2 are their masses. [1]

Gray (Gy)* - The SI unit of absorbed dose of radiation, equal to J/kg. [1]

Gregorian calendar - The modification of the Julian calendar introduced in 1582 by Pope Gregory XII which specified that a year divisible by 100 is a leap year only if divisible by 400.

Grignard reagents - Organomagnesium halides, RMgX, having a carbon-magnesium bond (or their equilibrium mixtures in solution with $R_2Mg + MgX_2$). [5]

Gruneisen parameter (γ) - Defined by $\gamma = \alpha_V/\kappa\, c_V\, \rho$, where α_V is the cubic thermal expansion coefficient, κ is the isothermal compressibility, c_V is the specific heat capacity at constant volume, and ρ is the mass density. γ is independent of temperature for most crystalline solids. [1]

Gyromagnetic ratio (γ) - Ratio of the magnetic moment of a particle to its angular momentum. Also called magnetogyric ratio.

Hadron - Any elementary particle that can take part in the strong interaction. Hadrons are subdivided into baryons, with odd half integer spins, and mesons, which have zero or integral spin.

Hall effect* - The development of a transverse potential difference V in a conducting material when subjected to a magnetic field H perpendicular to the direction of the current. The potential difference is given by $V = R_H BJt$, where B is the magnetic induction, J the current density, t the thickness of the specimen in the direction of the potential difference, and R_H is called the Hall coefficient.

Halocarbon - A compound containing no elements other than carbon, hydrogen, and one or more halogens. In common practice, the term is used mainly for compounds of no more than four or five carbon atoms.

Halogens - The elements F, Cl, Br, I, and At. Compounds of these elements are called halogenides or halides. [7]

Hamiltonian (H) - An expression for the total energy of a mechanical system in terms of the momenta and positions of constituent particles. In quantum mechanics, the Hamiltonian operator appears in the eigenvalue equation $H\psi = E\psi$, where E is an energy eigenvalue and ψ the corresponding eigenfunction.

Hardness* - The resistance of a material to deformation, indentation, or scratching. Hardness is measured on various scales, such as Mohs, Brinell, Knoop, Rockwell, and Vickers. [10]

Hartmann number (Ha) - A dimensionless quantity used in plasma physics, defined by $Ha = Bl(\kappa/\eta)^{1/2}$, where B is magnetic flux density, l is length, κ is electric conductivity, and η is viscosity. [2]

Hartree (E_h)* - An energy unit used in atomic and molecular science, equal to approximately $4.3597482 \times 10^{-18}$ J.

Hartree-Fock method - A iterative procedure for solving the Schrödinger equation for an atom or molecule in which the equation is solved for each electron in an initial assumed potential from all the other electrons. The new potential that results is used to repeat the calculation and the procedure continued until convergence is reached. Also called self-consistent field (SCF) method.

Heat capacity* - Defined in general as dQ/dT, where dQ is the amount of heat that must be added to a system to increase its temperature by a small amount dT. The heat capacity at constant pressure is $C_p = (\partial H/\partial T)_p$; that at constant volume is $C_V = (\partial E/\partial T)_V$, where H is enthalpy, E is internal energy, p is pressure, V is volume, and T is temperature. An upper case C normally indicates the molar heat capacity, while a lower case c is used for the specific (per unit mass) heat capacity. [1]

Heat of formation, vaporization, etc. - See corresponding terms under Enthalpy.

Hectare (ha) - A unit of area equal to $10^4 \, m^2$. [1]

Heisenberg uncertainty principle - The statement that two observable properties of a system that are complementary, in the sense that their quantum-mechanical operators do not commute, cannot be specified simultaneously with absolute precision. An example is the position and momentum of a particle; according to this principle, the uncertainties in position Δq and momentum Δp must satisfy the relation $\Delta p \Delta q \geq h/4\pi$, where h is Planck's constant.

Heitler-London model - An early quantum-mechanical model of the hydrogen atom which introduced the concept of the exchange interaction between electrons as the primary reason for stability of the chemical bond.

Helicon - A low-frequency wave generated when a metal at low temperature is exposed to a uniform magnetic field and a circularly polarized electric field.

Helmholz energy (A) - A thermodynamic function defined by $A = E-TS$, where E is the energy, S the entropy, and T the thermodynamic temperature. [2]

Hemiacetals - Compounds having the general formula $R_2C(OH)OR'$ (R' not equal to H). [5]

Henry (H)* - The SI unit of inductance, equal to Wb/A. [1]

Henry's law * - An expression which applies to an ideal dilute solution in which one or more gasses are dissolved, viz., $p_i = Hx_i$, where p_i is the partial pressure of component i above the solution, x_i is its mole fraction in the solution, and H_i is the Henry's law constant (a characteristic of the given gas and solvent, as well as the temperature).

Hermitian operator - An operator A that satisfies the relation $\int u_m^* A u_n dx = (\int u_n^* A u_m dx)^*$, where * indicates the complex conjugate. The eigenvalues of Hermitian operators are real, and eigenfunctions belonging to different eigenvalues are orthogonal.

Hertz (Hz) - The SI unit of frequency, equal to s^{-1}. [1]

Heterocyclic compounds - Cyclic compounds having as ring members atoms of at least two different elements, e.g., quinoline, 1,2-thiazole, bicyclo[3.3.1]tetrasiloxane. [5]

Heusler alloys - Alloys of manganese, copper, aluminum, nickel, and sometimes other metals which find important uses as permanent magnets.

Holography - A technique for creating a three-dimensional image of a object by recording the interference pattern between a light beam diffracted from the object and a reference beam. The image can be reconstructed from this pattern by a suitable optical system.

Homopolymer - A polymer derived from one species of (real, implicit, or hypothetical) monomer. [8]

Hooke's law - The statement that the ratio of stress to strain is a constant in a totally elastic medium.

Horse power - A non-SI unit of energy, equal to approximately 746 W.

Hubble constant - The ratio of the recessional velocity of an extragalactic object to the distance of that object. Its value is about $2 \times 10^{-18} \, s^{-1}$.

Huckel theory - A simple approximation for calculating the energy of conjugated molecules in which only the resonance integrals between neighboring bonds are considered. Also called CNDO method (complete neglect of differential overlap).

Hume-Rothery rules - A set of empirical rules for predicting the occurrence of solid solutions in metallic systems. The rules involve size, crystal structure, and electronegativity.

Hund's rules - A series of rules for predicting the sequence of energy states in atoms and molecules. One of the important results is that when two electrons exist in different orbitals, the state with their spins parallel (triplet state) lies at lower energy than the state with antiparallel spins (singlet).

Hydrazines - Hydrazine (diazane), H_2NNH_2, and its hydrocarbyl derivatives. When one or more substituents are acyl groups, the compound is a hydrazide. [5]

Hydrocarbon - A compound containing only carbon and hydrogen. [5]

Hydrolysis - A reaction occurring in water in which a chemical bond is cleaved and a new bond formed with the oxygen atom of water.

Hyperfine structure - Splitting of energy levels and spectral lines into several closely spaced components as a result of interaction of nuclear spin angular momentum with other angular momenta in the atom or molecule.

Hysteresis* - An irreversible response of a system (parameter A) as a function of an external force (parameter F), usually symmetric with respect to the origin of the A vs. F graph after the initial application of the force. A common example is magnetic induction vs. magnetic field strength in a ferromagnet.

Ideal gas law - The equation of state $pV = RT$, which defines an ideal gas, where p is pressure, V molar volume, T temperature, and R the molar gas constant.

Ideal solution - A solution in which solvent-solvent and solvent-solute interactions are identical, so that properties such as volume and enthalpy are exactly additive. Ideal solutions follow Raoult's law, which states that the vapor pressure p_i of component i is $p_i = x_i p_i^*$, where x_i is the mole fraction of component i and p_i^* the vapor pressure of the pure substance i.

Ignition temperature* - The lowest temperature at which combustion of a material will occur spontaneously under specified conditions. Sometimes called autoignition temperature, kindling point. [10]

Imides - Diacyl derivatives of ammonia or primary amines, especially those cyclic compounds derived from diacids. Also used for salts having the anion RN_2^-. [5]

Impedence (Z) - The complex representation of potential difference divided by the complex representation of current. In terms of reactance X and resistance R, the impedance is given by $Z = R + iX$. [1]

Index of refraction (n)* - For a non-absorbing medium, the ratio of the velocity of electromagnetic radiation *in vacuo* to the phase velocity of radiation of a specified frequency in the medium. [1]

Inductance - The ratio of the electromagnetic force induced in a coil by a current to the rate of change of the current.

Inductive coupled plasma mass spectroscopy (ICPMS) - See Techniques for Materials Characterization, page 12-1.

Inertial defect - In molecular spectroscopy, the quantity I_c-I_a-I_b for a molecule whose equilibrium configuration is planar, where I_a, I_b, and I_c are the effective principal moments of inertia. The inertial defect for a rigid planar molecule would be zero, but vibration-rotation interactions in a real molecule lead to a positive inertial defect.

Insulator - A material in which the highest occupied energy band (valence band) is completely filled with electrons, while the next higher band (conduction band) is empty. Solids with an energy gap of 5 eV or more are generally considered as insulators at room temperature. Their conductivity is less than 10^{-6} S/m and increases with temperature.

Intercalation compounds - Compounds resulting from reversible inclusion, without covalent bonding, of one kind of molecule in a solid matrix of another compound, which has a laminar structure. The host compound, a solid, may be macromolecular, crystalline, or amorphous. [5]

International System of Units (SI)* - The unit system adopted by the General Conference on Weights and Measures in 1960. It consists of seven base units (meter, kilogram, second, ampere, kelvin, mole, candela), plus derived units and prefixes. [1]

International Temperature Scale (ITS-90)* - The official international temperature scale adopted in 1990. It consists of a set of fixed points and equations which enable the thermodynamic temperature to be determined from operational measurements. [9]

Ion - An atomic or molecular particle having a net electric charge. [3]

Ion exchange - A process involving the adsorption of one or several ionic species accompanied by the simultaneous desorption (displacement) of one or more other ionic species. [3]

Ion neutralization spectroscopy (INS) - See Techniques for Materials Characterization, page 12-1.

Ionic strength (I) - A measure of the total concentration of ions in a solution, defined by $I = 1/2\Sigma_i z_i^2 m_i$, where z_i is the charge of ionic species i and m_i is its molality. For a 1-1 electrolyte at molality m, $I = m$.

Ionization constant* - The equilibrium constant for a reaction in which a substance in solution dissociates into ions.

Ionization potential* - The minimum energy required to remove an electron from an isolated atom or molecule (in its vibrational ground state) in the gaseous phase. More properly called ionization energy. [3]

Irradiance (E) - The radiant energy flux incident on an element of a surface, divided by the area of that element. [1]

Isentropic process - A thermodynamic process in which the entropy of the system does not change.

Ising model - A model describing the coupling between two atoms in a ferromagnetic lattice, in which the interaction energy is proportional to the negative of the product of the spin components along a specified axis.

Isobar - A line connecting points of equal pressure on a graphical representation of a physical system.

Isochore - A line or surface of constant volume on a graphical representation of a physical system.

Isoelectric point* - The pH of a solution or dispersion at which the net charge on the macromolecules or colloidal particles is zero. In electrophoresis there is no motion of the particles in an electric field at the isoelectric point.

Isomers - In chemistry, compounds that have identical molecular formulas but differ in the nature or sequence of bonding of their atoms or in the arrangement of their atoms in space. In physics, nuclei of the same atomic number Z and mass number A but in different energy states. [3]

Isomorphs - Substances of different chemical nature but having the same crystal structure.

Isotactic macromolecule - A tactic macromolecule, essentially comprising only one species of repeating unit which has chiral or prochiral atoms in the main chain in a unique arrangement with respect to its adjacent constitutional units. [8]

Isotherm - A line connecting points of equal temperature on a graphical representation of a physical system.

Isothermal process - A thermodynamic process in which the temperature of the system does not change.

Isotones - Nuclides having the same neutron number N but different atomic number Z. [3]

Isotopes - Two or more nuclides with the same atomic number Z but different mass number A. The term is sometimes used synonymously with nuclide, but it is preferable to reserve the word nuclide for a species of specific Z and A. [3]

Jahn-Teller effect - An interaction of vibrational and electronic motions in a nonlinear molecule which removes the degeneracy of certain electronic energy levels. It can influence the spectrum, crystal structure, and magnetic properties of the substance.

Johnson noise - Electrical noise generated by random thermal motion of electrons in a conductor or semiconductor. Also called thermal noise.

Josephson effect - The tunneling of electron pairs through a thin insulating layer which separates two superconductors. When a potential difference is applied to the superconductors, an alternating current is generated whose frequency is precisely proportional to the potential difference. This effect has important applications in metrology and determination of fundamental physical constants.

Joule (J)* - The SI unit of energy, equal to N m. [1]

Joule-Thomson coefficient (μ) - A parameter which describes the temperature change when a gas expands adiabatically through a nozzle from a high pressure to a low pressure region. It is defined by $\mu = (\partial T/\partial p)_H$, where H is enthalpy.

Julian calendar - The calendar introduced by Julius Caeser in 46 B.C. which divided the year into 365 days with a leap year of 366 days every fourth year.

Julian date (JD) - The number of days elapsed since noon Greenwich Mean Time on January 1, 4713 B.C. Thus January 1, 2000, 0h (midnight) will be JD 2,451,543.5. This dating system was introduced by Joseph Scaliger in 1582.

Kaon - One of the elementary particles in the family of mesons. Kaons have a spin of zero and may be neutral or charged.

Kelvin (K)* - The SI base unit of thermodynamic temperature. [1]

Kepler's laws - The three laws of planetary motion, which established the elliptical shape of planetary orbits and the relation between orbital dimensions and the period of rotation.

Kerr effect* - An electrooptical effect in which birefringence is induced in a liquid or gas when a strong electric field is applied perpendicular to the direction of an incident light beam. The Kerr constant k is given by $n_1 - n_2 = k\lambda E^2$, where λ is the wavelength, E is the electric field strength, and n_1 and n_2 are the indices of refraction of the ordinary and extraordinary rays, respectively.

Ketenes - Compounds in which a carbonyl group is connected by a double bond to an alkylidene group: $R_2C=C=O$. [5]

Ketones - Compounds in which a carbonyl group is bonded to two carbon atoms: $R_1R_2C=O$ (neither R may be H). [5]

Kilogram (kg)* - The SI base unit of mass. [1]

Kinetic energy (E_k, T) - The energy associated with the motion of a system of particles in a specified reference frame. For a single particle of mass m moving at velocity v, $E_k = 1/2mv^2$.

Kirchhoff's laws - Basic rules for electric circuits, which state (a) the algebraic sum of the currents at a network node is zero and (b) the algebraic sum of the voltage drops around a closed path is zero.

Klein-Gordon equation - A relativistic extension of the Schrödinger equation.

Klein-Nishima formula - An expression for the scattering cross section of a photon by an unbound electron, based upon the Dirac electron theory.

Knight shift - The change in magnetic resonance frequency of a nucleus in a metal relative to the same nucleus in a diamagnetic solid. The effect is due to the polarization of the conduction electrons in the metal.

Knudsen number (Kn) - A dimensionless quantity used in fluid mechanics, defined by $Kn = \lambda/l$, where λ is mean free path and l is length. [2]

Kondo effect - A large increase in electrical resistance observed at low temperatures in certain dilute alloys of a magnetic metal in a nonmagnetic material.

Kramers-Kronig relation - A set of equations relating the real and imaginary parts of the index of refraction of a medium

Lactams - Cyclic amides of amino carboxylic acids, having a 1-azacycloalkan-2-one structure, or analogues having unsaturation or heteroatoms replacing one or more carbon atoms of the ring. [5]

Lactones - Cyclic esters of hydroxy carboxylic acids, containing a 1-oxacycloalkan-2-one structure, or analogues having unsaturation or heteroatoms replacing one or more carbon atoms of the ring. [5]

Lagrangian function (L) - A function used in classical mechanics, defined as the kinetic energy minus the potential energy for a system of particles.

Lamb shift - The small energy difference between the $^2S_{1/2}$ and $^2P_{1/2}$ levels in the hydrogen atom, which results from interactions between the electron and the radiation field.

Laminar flow - Smooth, uniform, non-turbulent flow of a gas or liquid in parallel layers, with little mixing between layers. It is characterized by small values of the Reynolds number.

Landé g-factor - See g-Factor of the electron

Langevin function - The mathematical function $L(x) = (e^x + e^{-x})/(e^x - e^{-x}) - 1/x$, which occurs in the expression for the average dipole moment of a group of rotating polar molecules in an electric field: $\mu_{av} = \mu L(\mu E/kT)$, where μ is the electric dipole moment of a single molecule, E is the electric field strength, k is the Boltzmann constant, and T is the temperature.

Lanthanides - The elements of atomic number 57 through 71, which share common chemical properties: La, Ce, Pr, Nd, Pm, Sm, Eu, Gd, Tb, Dy, Ho, Er, Tm, Yb, Lu. [7]

Larmor frequency (v_L) - The precession frequency of a magnetic dipole in an applied magnetic field. In particular, a nucleus in a magnetic field of strength B has a Larmor frequency of $\gamma B/2\pi$, where γ is the magnetogyric ratio of the nucleus.

Laser* - A device in which an optical cavity is filled with a medium where a population inversion can be produced by some means. When the resonant frequency of the cavity bears the proper relation to the separation of the inverted energy levels, stimulated emission occurs, producing a highly monochromatic, coherent beam of light.

Laser ionization mass spectroscopy (LIMS) - See Techniques for Materials Characterization, page 12-1.

Lattice constants* - Parameters specifying the dimensions of a unit cell in a crystal lattice, specifically the lengths of the cell edges and the angles between them.

Lattice energy* - The energy per ion pair required to separate completely the ions in a crystal lattice at a temperature of absolute zero.

Laue diagram - A diffraction pattern produced when an x-ray beam passes through a thin slice of a crystal and impinges on a detector behind the crystal.

Lenz's law - The statement that the current induced in a circuit by a change in magnetic flux is so directed as to oppose the change in flux

Leonard-Jones potential - A simple but useful function for approximating the interaction between two neutral atoms or molecules separated by a distance r by writing the potential energy as $U(r) = 4\varepsilon\{(r_0/r)^{12} - (r_0/r)^6\}$, where ε and r_0 are adjustable parameters. In this form the depth of the potential well is ε and the minimum occurs at $2^{1/6}r_0$. The $(1/r)^{12}$ term is often replaced by other powers of $1/r$.

Lepton - One of the class of elementary particles that do not take part in the strong interaction. Included are the electron, muon, and neutrino. All leptons have a spin of $1/2$.

Lewis number (Le) - A dimensionless quantity used in fluid mechanics, defined by $Le = a/D$, where a is thermal diffusivity and D is diffusion coefficient. [2]

Ligand field theory - A description of the structure of crystals containing a transition metal ion surrounded by nonmetallic ions (ligands). It is based on construction of molecular orbitals involving the d-orbitals of the central metal ion and combinations of atomic orbitals of the ligands.

Light year (l.y.) - A unit of distance used in astronomy, defined as the distance light travels in one year in a vacuum. Its approximate value is 9.46073×10^{15} m.

Lignins - Macromolecular constituents of wood related to lignans, composed of phenolic propylbenzene skeletal units, linked at various sites and apparently randomly. [5]

Ligroin - The petroleum fraction consisting mostly of C_7 and C_8 hydrocarbons and boiling in the range 90-140°C; commonly used as a laboratory solvent.

Lipids - A loosely defined term for substances of biological origin that are soluble in nonpolar solvents. They consist of saponifiable lipids, such as glycerides (fats and oils) and phospholipids, as well as nonsaponifiable lipids, principally steroids. [5]

Lipoproteins - Clathrate complexes consisting of a lipid enwrapped in a protein host without covalent binding, in such a way that the complex has a hydrophilic outer surface consisting of all the protein and the polar ends of any phospholipids. [5]

Liter (L)* - A synonym for cubic decimeter. [1]

Lithosphere* - The outer layer of the solid earth, extending from the base of the mantle to the surface of the crust.

Lorentz contraction - The reduction in length of a moving body in the direction of motion, given by the factor $(1-v^2/c^2)^{1/2}$, where v is the velocity of the body and c the velocity of light. Also known as the FitzGerald-Lorentz contraction.

Lorentz force - The force exerted on a point charge Q moving at velocity v in the presence of external fields E and B. It is given (in SI units) by $F = Q(E + v \times B)$.

Loss angle (δ) - For a dielectric material in an alternating electromagnetic field, δ is the phase difference between the current and the potential difference. The function tan δ is a measure of the ratio of the power dissipated in the dielectric to the power stored.

Low energy electron diffraction (LEED) - See Techniques for Materials Characterization, page 12-1.

Lumen (lm)* - The SI unit of luminous flux, equal to cd sr. [1]

Luminous flux (Φ) - The intensity of light from a source multiplied by the solid angle. The SI unit is lumen. [1]

Lux (lx)* - The SI unit of illuminance, equal to cd sr m^{-2}. [1]

Lyddane-Sachs-Teller relation - A relation between the phonon frequencies and dielectric constants of an ionic crystal which states that $(\omega_T/\omega_L)^2 = \varepsilon(\infty)/\varepsilon(0)$, where ω_T is the angular frequency of transverse optical phonons, ω_L that of longitudinal optical phonons, $\varepsilon(0)$ is the static dielectric constant, and $\varepsilon(\infty)$ the dielectric constant at optical frequencies.

Lyman series - The series of lines in the spectrum of the hydrogen atom which corresponds to transitions between the ground state (principal quantum number $n = 1$) and successive excited states. The wavelengths are given by $1/\lambda = R_H(1-1/n^2)$, where $n = 2,3,4,...$ and R_H is the Rydberg constant for hydrogen. The first member of the series ($n = 1\leftrightarrow2$), which is often called the Lyman-α line, falls at a wavelength of 1216 Å, and the series converges at 912 Å, the ionization limit of hydrogen.

Mach number (Ma) - A dimensionless quantity used in fluid mechanics, defined by $Ma = v/c$, where v is velocity and c is the speed of sound. [2]

Macromolecule - A molecule of high relative molecular mass (molecular weight), the structure of which essentially comprises the multiple repetition of units derived, actually or conceptually, from molecules of low relative molecular mass [8]

Madelung constant* - A constant characteristic of a particular crystalline material which gives a measure of the electrostatic energy binding the ions in the crystal.

Magnetic field strength (H) - An axial vector quantity, the curl of which is equal to the current density, including the displacement current. [1]

Magnetic induction (B) - An axial vector quantity such that the force exerted on an element of current is equal to the vector product of this element and the magnetic induction. [1]

Magnetic moment - See Dipole moment, magnetic.

Magnetic susceptibility (χ_m, κ)* - Defined by $\chi_m = (\mu-\mu_0)/\mu_0$, where μ is the permeability of the medium and μ_0 the permeability of a vacuum. [1]

Magnetization (M) - Defined by $M = (B/\mu_0)-H$, where B is magnetic induction, H magnetic field strength, and μ_0 the permeability of a vacuum. [1]

Magnetogyric ratio (γ) - Ratio of the magnetic moment of a particle to its angular momentum. Also called gyromagnetic ratio.

Magneton - See Bohr magneton, Nuclear magneton.

Magnetostriction* - The change in dimensions of a solid sample when it is placed in a magnetic field.

Magnon - A quantum of magnetic energy associated with a spin wave in a ferromagnetic or antiferromagnetic crystal.

Mantle - The layer of the earth between the crust and the liquid outer core, which begins about 2900 km below the earth's surface.

Maser - A device in which a microwave cavity is filled with a medium where a population inversion can be produced by some means. When the resonant frequency of the cavity bears the proper relation to the separation of the inverted energy levels, the device can serve as an amplifier or oscillator at that frequency.

Mass (m)* - Quantity of matter. Mass can also be defined as "resistance to acceleration".

Mass defect (B) - Defined by $B = Zm(^1H) + Nm_n - m_a$, where Z is the atomic number, $m(^1H)$ is the mass of the hydrogen atom, N is the neutron number, m_n is the rest mass of the neutron, and m_a is the mass of the atom in question. Thus Bc^2 can be equated to the binding energy of the nucleus if the binding energy of atomic electrons is neglected. [1]

Mass excess (Δ) - Defined by $\Delta = m_a - Am_u$, where m_a is the mass of the atom, A the number of nucleons, and m_u the unified atomic mass constant ($m_u = 1$ u). [1]

Mass fraction (w_B) - The ratio of the mass of substance B to the total mass of a mixture. [1]

Mass number (A) - A characteristic property of a specific isotope of an element, equal to the sum of the number of protons and neutrons in the nucleus.

Mass spectrometry - An analytical technique in which ions are separated according to the mass/charge ratio and detected by a suitable detector. The ions may be produced by electron impact on a gas, a chemical reaction, energetic vaporization of a solid, etc. [6]

Massieu function - A thermodynamic function defined by $J = -A/T$, where A is the Helmholz energy and T the thermodynamic temperature. [2]

Matthiessen's rule - The statement that the electrical resistivity ρ of a metal can be written as $\rho = \rho_L+\rho_i$, where ρ_L is due to scattering of conduction electrons by lattice vibrations and ρ_i to scattering by impurities and imperfections. If the impurity concentration is small, ρ_i is temperature independent.

Maxwell (Mx)* - A non-SI unit of magnetic field strength (H) equal to 10^{-8} Wb. [1]

Maxwell's equations - The fundamental equations of electromagnetism. In a form appropriate to SI units, they are:

curl $H = \partial D/\partial t + j$
div $B = 0$
curl $E = -\partial B/\partial t$
div $D = \rho$

where H is the magnetic field strength, B the magnetic induction, E the electric field strength, D the electric displacement, j the current density, ρ the charge density, and t is time.

Maxwell-Boltzmann distribution - An expression for the fraction of molecules $f(v)$ in a gas that have velocity v within a specified interval. It takes the form

$$f(v) = 4\pi(M/2\pi RT)^{3/2} v^2 e^{-Mv^2/2RT}$$

where M is the molar mass, R the molar gas constant, and T the temperature.

Mean free path* - The average distance a gas molecule travels between collisions.

Meissner effect - The complete exclusion of magnetic induction from the interior of a superconductor.

Melting point* - The temperature at which the solid and liquid phases of a substance are in equilibrium at a specified pressure (normally taken to be atmospheric unless stated otherwise).

Mercaptans - A traditional term abandoned by IUPAC, synonymous with thiols. This term is still widely used. [5]

Meson - Any elementary particle that has zero or integral spin. Mesons are responsible for the forces between protons and neutrons in the nucleus.

Mesosphere - The part of the earth's atmosphere extending from the top of the stratosphere (about 50 km above the surface) to 80-90 km. It is characterized by a decrease in temperature with increasing altitude.

Metal - A material in which the highest occupied energy band (conduction band) is only partially filled with electrons. The electrical conductivity of metals generally decreases with temperature.

Metallocenes - Organometallic coordination compounds in which one atom of a transition metal such as iron, ruthenium or osmium is bonded to and only to the face of two cyclopentadienyl ligands which lie in parallel planes. [5]

Meter (m)* - The SI base unit of length. [1]

Methine group - In organic compounds, the -C= group. [5]

Mho - An archaic name for the SI unit siemens (reciprocal ohm).

Micelle - A particle formed by the aggregation of surfactant molecules (typically, 10 to 100 molecules) in solution. For aqueous solutions, the hydrophilic end of the molecule is on the surface of the micelle, while the hydrophobic end (often a hydrocarbon chain) points toward the center. At the critical micelle concentration (cmc) the previously dissolved molecules aggregate into a micelle.

Micron (μ) - An obsolete name for micrometer.

Mie scattering - The scattering of light by spherical dielectric particles whose diameter is comparable to the wavelength of the light.

Milky way - The band of light in the night sky resulting from the stars in the galactic plane. The term is also used to denote the galaxy in which the sun is located.

Miller indices (hkl) - A set of indices used to label planes in a crystal lattice. [2]

Millimeter of mercury (mmHg) - A non-SI unit of pressure, equal to 133.322 Pa. The name is generally considered interchangeable with torr.

Mobility (μ)* - In solid state physics, the drift velocity of electrons or holes in a solid divided by the applied electric field strength. The term is used in a similar sense in other fields.

Molality (m) - A measure of concentration of a solution in which one states the amount of substance (i.e., number of moles) of solute per kilogram of solvent. Thus a 0.1 molal solution (often written as 0.1 m) has m = 0.1 mol/kg.

Molar mass - The mass of one mole of a substance. It is normally expressed in units of g/mol, in which case its numerical value is identical with the molecular weight (relative molecular mass). [1]

Molar quantity - It is often convenient to express an extensive quantity (e.g., volume, enthalpy, heat capacity, etc.) as the actual value divided by amount of substance (number of moles). The resulting quantity is called molar volume, molar enthalpy, etc

Molar refraction (R) - A property of a dielectric defined by the equation $R = V_m[(n^2-1)/(n^2+2)]$, where n is the index of refraction of the medium (at optical wavelengths) and V_m the molar volume. It is related to the polarizability α of the molecules that make up the medium by the Lorenz-Lorentz equation, $R = N_A\alpha/3\varepsilon_0$, where N_A is Avogadro's constant and ε_0 is the permittivity of a vacuum.

Molarity (c) - A measure of concentration of a solution in which one states the amount of substance (i.e., number of moles) of solute per liter of solution. Thus a 0.1 molar solution (often referred to as 0.1 M) has a concentration c = 0.1 mol/L.

Mole (mol)* - The SI base unit of amount of substance. [1]

Mole fraction (x_B) - The ratio of the amount of substance (number of moles) of substance B to the total amount of substance in a mixture. [1]

Molecular orbital - See Orbital.

Molecular weight (M_r)* - The ratio of the average mass per molecule or specified entity of a substance to 1/12 of the mass of nuclide ^{12}C. Also called relative molar (or molecular) mass. [1]

Moment of inertia (I) - The moment of inertia of a body about an axis is the sum (or integral) of the products of its elements of mass and the squares of their distances from the axis. [1]

Momentum (p) - The product of mass and velocity. [1]

Monomer - A substance consisting of molecules which can undergo polymerization, thereby contributing constitutional units to the essential structure of a macromolecule. [8]

Monosaccharides - A term which includes aldoses, ketoses, and a wide variety of derivatives. [5]

Mössbauer effect - The recoilless emission of γ-rays from nuclei bound in a crystal under conditions where the recoil energy associated with the γ emission is taken up by the crystal as a whole. This results in a very narrow line width, which can be exploited in various types of precise measurements.

Muon* - An unstable elementary particle of spin 1/2 and mass about 200 times that of the electron.

Naphtha - The petroleum fraction consisting mostly of C_6 to C_8 hydrocarbons and boiling in the range 80-120°C. Solvents derived from this fraction include ligroin and petroleum ether.

Nautical mile - A non-SI unit of length, equal to exactly 1852 m.

Navier-Stokes equations - A set of complex equations for the motion of a viscous fluid subject to external forces.

Néel temperature (T_N)* - The critical temperature above which an antiferromagnetic substance becomes paramagnetic. [1]

Nernst effect - The production of an electric field in a conductor subject to an applied magnetic field and containing a transverse temperature gradient. The electric field is perpendicular to the magnetic field and the temperature gradient.

Network - In polymer science, a highly ramified macromolecule in which essentially each constitutional unit is connected to each other constitutional unit and to the macroscopic phase boundary by many permanent paths through the macromolecule, the number of such paths increasing with the number of intervening bonds. The paths must on the average be coextensive with the macromolecule. [8]

Neutrino - A stable elementary particle in the lepton family. Neutrinos have zero (or at least near-zero) rest mass and spin 1/2.

Neutron* - An elementary particle on spin 1/2 and zero charge. The free neutron has a mean lifetime of 887 seconds. Neutrons and protons, which are collectively called nucleons, are the constituents of the nucleus.

Neutron activation analysis (NAA) - See Techniques for Materials Characterization, page 12-1.

Neutron number (N) - A characteristic property of a specific isotope of an element, equal to the number of neutrons in the nucleus.

Newton (N)* - The SI unit of force, equal to m kg s^{-2}. [1]

Nitriles - Compounds having the structure RC≡N; thus C-substituted derivatives of hydrocyanic acid, HC≡N. [5]

Nitrosamines - N-Nitroso amines: compounds of the structure R$_2$NNO. Compounds RNHNO are not ordinarily isolatable, but they, too, are nitrosamines. The name is a contraction of N-nitrosoamine and, as such, does not require the N locant. [5]

Nuclear magnetic resonance (NMR)* - A widely used technique in which the resonant absorption of radiofrequency radiation by magnetic nuclei in a magnetic field is measured. The results give important information on the local environment of each nucleus.

Nuclear magneton (μ_N)* - The unit of nuclear magnetic moment, defined as $eh/4\pi m_p$, where h is Planck's constant, m_p the proton mass, and e the elementary charge.

Nuclear quadrupole resonance (NQR) - See Techniques for Materials Characterization, page 12-1.

Nuclear reaction analysis (NRA) - See Techniques for Materials Characterization, page 12-1.

Nuclear spin (I) - The quantum number that specifies the intrinsic angular momentum of a particular nucleus. The magnitude of the angular momentum is given by $[I(I+1)]^{1/2} h/2\pi$, where h is Planck's constant.

Nucleic acids* - Macromolecules, the major organic matter of the nuclei of biological cells, made up of nucleotide units, and hydrolyzable into certain pyrimidine or purine bases (usually adenine, cytosine, guanine, thymine, uracil), D-ribose or 2-deoxy-D-ribose. [5]

Nucleon - A collective term for the proton and neutron.

Nucleosides - Ribosyl or deoxyribosyl derivatives (rarely, other glycosyl derivatives) of certain pyrimidine or purine bases. They are thus glycosylamines or N-glycosides related to nucleotides by the lack of phosphorylation. [5]

Nucleotides - Compounds formally obtained by esterification of the 3′ or 5′ hydroxy group of nucleosides with phosphoric acid. They are the monomers of nucleic acids and are formed from them by hydrolytic cleavage. [5]

Nuclide - A species of atoms in which each atom has identical atomic number Z and identical mass number A. [3]

Nusselt number (Nu) - A dimensionless quantity used in fluid mechanics, defined by $Nu = hl/k$, where h is coefficient of heat transfer, l is length, and k is thermal conductivity. [2]

Nyquist theorem - An expression for the mean square thermal noise voltage across a resistor, given by $4RkT\Delta f$ where R is the resistance, k the Boltzmann constant, T the temperature, and Δf the frequency band within which the voltage is measured.

Octanol-water partition coefficient (P)* - A measure of the way in which a compound will partition itself between the octanol and water phases in the two-phase octanol-water system, and thus an indicator of certain types of biological activity. Specifically, P is the ratio of the concentration (in moles per liter) of the compound in the octanol phase to that in the water phase at infinite dilution. The quantity normally reported is log P.

Oersted (Oe) - A non-SI unit of magnetic field (H), equal to 79.57747 A/m.

Ohm (Ω)* - The SI unit of electric resistance, equal to V/A. [1]

Ohm's law - A relation among electric current I, potential difference V, and resistance R, viz., $I = V/R$. The resistance is constant at constant temperature to high precision for many materials.

Olefins - Acyclic and cyclic hydrocarbons having one or more carbon-carbon double bonds, apart from the formal ones in aromatic compounds. The class olefins subsumes alkenes and cycloalkenes and the corresponding polyenes. [5]

Oligomer - A substance consisting of molecules of intermediate relative molecular mass (molecular weight), the structure of which essentially comprises the multiple repetition of units derived, actually or conceptually, from molecules of low relative molecular mass. In contrast to a polymer, the properties of an oligomer can vary significantly with the removal of one or a few of its units. [8]

Oligopeptides - Peptides containing from three to nine amino groups. [5]

Onsager relations - An important set of equations in the thermodynamics of irreversible processes. They express the symmetry between the transport coefficients describing reciprocal processes in systems with a linear dependence of flux on driving forces.

Optical rotary power - Angle by which the plane of polarization of a light beam is rotated by an optically active medium, divided by path length and by concentration of the active constituent. Depending on whether mass or molar concentration is used, the modifier "specific" or "molar" is attached. [2]

Orbital - A one-electron wavefunction. Atomic orbitals are classified as s-, p-, d,- or f-orbitals according to whether the angular momentum quantum number $l = 0, 1, 2$, or 3. Molecular orbitals, which are usually constructed as linear combinations of atomic orbitals, describe the distribution of electrons over the entire molecule.

Oscillator strength (f) - A measure of the intensity of a spectroscopic transition, defined by

$$f = \frac{8\pi^2 Mev}{3he^2}|\mu_{ij}|^2$$

where ν is the frequency, μ_{ij} the transition dipole moment, m_e the mass of the electron, e the elementary charge, and h Planck's constant.

Osmosis - The flow of a solvent in a system in which two solutions of different concentration are separated by a semipermeable membrane which cannot pass solute molecules. The solvent will flow from the side of lower concentration to that of higher concentration, thus tending to equalize the concentrations. The pressure that must be applied to the more concentrated side to stop the flow is called the osmotic pressure.

Osmotic coefficient (φ) - Defined by $\phi = \ln a_A/(M_A \Sigma m_B)$, where M_A is the molar mass of substance A (normally the solvent), a_A is its activity, and the m_B are molalities of the solutes. [1]

Osmotic pressure (Π) - The excess pressure necessary to maintain osmotic equilibrium between a solution and the pure solvent separated by a membrane permeable only to the solvent. In an ideal dilute solution $\Pi = c_B RT$, where c_B is the amount-of-substance concentration of the solute, R is the molar gas constant, and T the temperature. [1,2]

Ostwald dilution law - A relation for the concentration dependence of the molar conductivity Λ of an electrolyte solution, viz.,

$$\frac{1}{\Lambda} = \frac{1}{\Lambda^\circ} + \frac{\Lambda c}{K(\Lambda^\circ)^2}$$

where c is the solute concentration, K is the equilibrium constant for dissociation of the solute, and Λ° is the conductivity at $c\Lambda = 0$.

Ounce (oz) - A non-SI unit of mass. The avoirdupois ounce equals 28.34952 g, while the troy ounce equals 31.10348 g.

Overpotential (η) - In an electrochemical cell, the difference between the potential of an electrode and its zero-current value.

Oximes - Compounds of structure $R_2C=NOH$ derived from condensation of aldehydes or ketones with hydroxylamine. Oximes from aldehydes may be called aldoximes; those from ketones may be called ketoximes. [5]

Oxo compounds - Compounds containing an oxygen atom, =O, doubly bonded to carbon or another element. The term thus embraces aldehydes, carboxylic acids, ketones, sulfonic acids, amides and esters. [5]

Ozonides - The 1,2,4-trioxolanes formed by the reaction of ozone at a carbon-carbon double bond, or the analogous compounds derived from acetylenic compounds. [5]

Pair production - A process in which a photon is converted into a particle and its antiparticle (e.g., an electron and positron) in the electromagnetic field of a nucleus.

Paraffins - Obsolescent term for saturated hydrocarbons, commonly but not necessarily acyclic. Still widely used in the petrochemical industry, where the term designates acyclic saturated hydrocarbons, and stands in contradistinction to naphthenes. [5]

Paramagnetism* - A type of magnetism characterized by a positive magnetic susceptibility, so that the material becomes weakly magnetized in the direction of an external field. The magnetization disappears when the field in removed. In the simplest approximation (Curie's law) the susceptibility is inversely proportional to temperature.

Parity - The property of a quantum-mechanical wave function that describes its behavior under the symmetry operation of coordinate inversion. A parity of +1 (or even) is assigned if the wave function does not change sign when the signs of all the coordinates are changed; the parity is -1 (or odd) if the wave function changes sign under this operation.

Parsec (pc) - A unit of distance defined as the distance at which 1 astronomical unit (AU) subtends an angle of 1 second of arc. It is equal to 206264.806 AU or 3.085678×10^{16} m.

Particle induced x-ray emission (PIXE) - See Techniques for Materials Characterization, page 12-1.

Partition function (q, z) - For a single molecule, $q = \Sigma_i g_i \exp(\varepsilon_i/kT)$, where ε_i is an energy level of degeneracy g_i, k the Boltzmann constant, and T the absolute temperature; the summation extends over all energy states. For a system of N non-interacting molecules which are indistinguishable, as in an ideal gas, the canonical partition function $Q = q^N/N!$.

Pascal (Pa)* - The SI unit of pressure, equal to N/m^2. [1]

Paschen series - The series of lines in the spectrum of the hydrogen atom which corresponds to transitions between the state with principal quantum number $n = 3$ and successive higher states. The wavelengths are given by $1/\lambda = R_H(1/9-1/n^2)$, where $n = 4,5,6,...$ and R_H is the Rydberg constant. The first member of the series ($n = 3 \leftrightarrow 4$), which is often called the P_α line, falls in the infrared at a wavelength of 1.875 μm.

Paschen-Back effect - In atomic spectroscopy, the decoupling of electron spin from orbital angular momentum as the strength of an external magnetic field is increased.

Pauli exclusion principle - The statement that two electrons in an atom cannot have identical quantum numbers; thus if there are two electrons in the same orbital, their spin quantum numbers must be of opposite sign.

Pearson symbol - A code for designating crystallographic information, including the crystal system, the lattice type, and the number of atoms per unit cell.

Péclet number (Pe) - A dimensionless quantity used in fluid mechanics, defined by $Pe = vl/a$, where v is velocity, l is length, and a is thermal diffussivity. [2]

Peltier effect - The absorption or generation of heat (depending on the current direction) which occurs when an electric current is passed through a junction between two materials.

Peptides - Amides derived from two or more amino carboxylic acid molecules (the same or different) by formation of a covalent bond from the carbonyl carbon of one to the nitrogen atom of another with formal loss of water. [5]

Permeability (μ) - Magnetic induction divided by magnetic field strength; i.e. $\mu = B/H$. The relative permeability $\mu_r = \mu/\mu_0$, where μ_0 is the permeability of a vacuum. [1]

Permittivity (ε) - Ratio of the electric displacement in a medium to the electric field strength. Also called dielectric constant. [1]

Peroxides - Compounds of structure ROOR in which R may be any organic group. In inorganic chemistry, salts of the anion O_2^{-2} [5]

Peroxy acids - Acids in which an acidic -OH group has been replaced by an -OOH group; e.g. $CH_3C(=O)OOH$ peroxyacetic acid, $PhS(=O)_2OOH$ benzeneperoxysulfonic acid. [5]

Petroleum ether - The petroleum fraction consisting of C_5 and C_6 hydrocarbons and boiling in the range 35-60°C; commonly used as a laboratory solvent.

pH* - A convenient measure of the acid-base character of a solution, usually defined by pH = -log $[c(H^+)/\text{mol L}^{-1})]$, where $c(H^+)$ is the concentration of hydrogen ions. The more precise definition is in terms af activity rather than concentration. [2]

Phenols - Compounds having one or more hydroxy groups attached to a benzene or other arene ring. [5]

Phonon - A quantum of energy associated with a vibrational mode of a crystal lattice.

Phosphines - PH_3 and compounds derived from it by substituting one, two or three hydrogen atoms by hydrocarbyl groups.

RPH$_2$, R$_2$PH and R$_3$P (R not equal to H) are called primary, secondary and tertiary phosphines, respectively. [5]

Phosphonium compounds - Salts (and hydroxides) [R$_4$P]$^+$X$^-$ containing tetracoordinate phosphonium ion and the associated anion. [5]

Phosphonium ylides - Compounds having the structure R$_3$P$^+$-C$^-$R$_2$ R$_3$P=CR$_2$. Also known as Wittig reagents. [5]

Phosphoresence - The process by which a molecule is excited by light to a higher electronic state and then undergoes a radiationless transition to a state of different multiplicity from which it decays, after some delay, to the ground state. The emitted light is normally of longer wavelength than the exciting light because vibrational energy has been dissipated.

Photoelectric effect - The complete absorption of a photon by a solid with the emission of an electron.

Photon - An elementary particle of zero mass and spin 1/2. The photon is involved in electromagnetic interactions and is the quantum of electromagnetic radiation.

Photon stimulated desorption (PSD) - See Techniques for Materials Characterization, page 12-1.

Pinacols - Tetra(hydrocarbyl)ethane-1,2-diols, R$_2$C(OH)C(OH)R$_2$, of which the tetramethyl example is the simplest one and is itself commonly known as pinacol. [5]

Pion - An elementary particle in the family of mesons. Pions have zero spin and may be neutral or charged. They participate in the strong interaction which holds the nucleus together.

pK* - The negative logarithm (base 10) of an equilibrium constant K. For pK_a, see Acid dissociation constant.

Planck constant (h)* - The elementary quantum of action, which relates energy to frequency through the equation $E = h\nu$.

Planck distribution - See Black body radiation

Planck function - A thermodynamic function defined by $Y = -G/T$, where G is Gibbs energy and T thermodynamic temperature. [2]

Plasma - A highly ionized gas in which the charge of the electrons is balanced by the charge of the positive ions, so that the system as a whole is electrically neutral.

Plasmon - A quantum associated with a plasma oscillation in the electron gas of a solid.

Point group* - A group of symmetry operations (rotations, reflections, etc.) that leave a molecule invariant. Every molecular conformation can be assigned to a specific point group, which plays a major role in determining the spectrum of the molecule.

Poise (P) - A non-SI unit of viscosity, equal to 0.1 Pa s.

Poiseuille's equation - A formula for the rate of flow of a viscous fluid through a tube:

$$\frac{dV}{dt} = \frac{(p_1{}^2 - p_2{}^2)\pi r^4}{16l\eta p_0}$$

where V is the volume as measured at pressure p_0; p_1 and p_2 are the pressures at each end of the tube; r is the radius and l the length of the tube; and η is the viscosity.

Poisson ratio (μ) - The absolute value of the ratio of the transverse strain to the corresponding axial strain resulting from uniformly distributed axial stress below the proportional limit (i.e., where Hooke's law is valid). [10]

Polariton - A quantum associated with the coupled modes of photons and optical phonons in an ionic crystal.

Polarizability (α)* - The change in dipole moment of a molecule produced by an external electric field; specifically, $\alpha_{ab} = \partial p_a / \partial E_b$, where p_a is the dipole moment component on the a axis and E_b is the component of the electric field strength along the b axis. [2]

Polymer - A substance composed of molecules of high relative molecular mass (molecular weight), the structure of which essentially comprises the multiple repetition of units derived, actually or conceptually, from molecules of low relative molecular mass. A single molecule of a polymer is called a macromolecule. [8]

Polypeptides - Peptides containing 10 or more amino acid residues. See also Peptides. [5]

Polysaccharides - Compounds consisting of a large number of monosaccharides linked glycosidically. This term is commonly used only for those containing more than ten monosaccharide residues. Also called glycans. [5]

Porphyrins - Natural pigments containing a fundamental skeleton of four pyrrole nuclei united through the α-positions by four methine groups to form a macrocyclic structure (porphyrin is designated porphine in Chemical Abstracts indexes). [5]

Positron - The antiparticle of the electron. It has the same mass and spin as an electron, and an equal but opposite charge.

Positronium - The hydrogen-like "atom" formed from a positron nucleus and an electron. Its lifetime is very short because of annihilation of the positron and electron.

Potential - See Electric potential.

Potential energy (E_p, V, U) - The portion of the energy of a system that is associated with its position in a force field.

Pound (lb) - A non-SI unit of mass, equal to 0.4535924 kg.

Power (P) - Rate of energy transfer. For electrical circuits, this is equal to the product of current and potential difference, $P = IV$. [1]

Poynting vector (S) - For electromagnetic radiation, the vector product of the electric field strength and the magnetic field strength. [1]

Prandtl number (Pr) - A dimensionless quantity used in fluid mechanics, defined by $Pr = \eta/\rho a$, where η is viscosity, ρ is density, and a is thermal diffusivity. [2]

Pressure* - Force divided by area. [1]

Proteins - Naturally occurring and synthetic polypeptides having molecular weights greater than about 10,000 (the limit is not precise). See also Peptides. [5]

Proton* - A stable elementary particle of unit positive charge and spin 1/2. Protons and neutrons, which are collectively called nucleons, are the constituents of the nucleus.

Pulsar - A neutron star which rotates rapidly and emits electromagnetic radiation in regular pulses at a frequency related to the rotation period.

Purine bases* - Purine and its substitution derivatives, especially naturally occurring examples. [5]

Pyrimidine bases* - Pyrimidine and its substitution derivatives, especially naturally occurring examples. [5]

Q-switching - A technique for obtaining very high power from a laser by keeping the Q factor of the laser cavity low while the population inversion builds up, then suddenly increasing the Q to initiate the stimulated emission.

Quad - A unit of energy defined as 10^{15} Btu, equal to approximately 1.055056×10^{18} J.

Quadrupole moment - A coefficient of the third term (after monopole and dipole) in the power series expansion of the electric potential of an array of charges. A nucleus of spin greater than 1/2 has a non-vanishing nuclear quadrupole moment which can interact with the electric field gradient of the surrounding electrons. Molecular quadrupole moments have an influence on intermolecular forces.

Quality factor (Q) - The ratio of the absolute value of the reactance of an electrical system to the resistance; thus a measure of the energy stored per cycle relative to the energy dissipated.

Quantum yield - In photochemistry, the number of moles transformed in a specific process, either physically (e.g., by emission of photons) or chemically, per mole of photons absorbed by the system. [3]

Quark - An elementary entity which has not been directly observed but is considered a constituent of protons, neutrons, and other hadrons.

Quasar - An extragalactic object emitting electromagnetic radiation at a very high power level and showing a very large red shift, thus indicating that the object is receding at a speed approaching the speed of light.

Quasicrystal - A solid having conventional crystalline properties but whose lattice does not display translational periodicity.

Quaternary ammonium compounds - Derivatives of ammonium compounds, $NH_4^+ Y^-$, in which all four of the hydrogens bonded to nitrogen have been replaced with hydrocarbyl groups. Compounds having a carbon-nitrogen double bond (i.e. $R_2C=N^+R_2Y^-$) are more accurately called iminium compounds. [5]

Quinones - Compounds having a fully conjugated cyclic dione structure, such as that of benzoquinones, derived from aromatic compounds by conversion of an even number of -CH= groups into -C(=O)- groups with any necessary rearrangement of double bonds. [5]

Racemic mixture - A mixture of equal amounts of a pair of enantiomers (optical isomers); such a mixture is not optically active.

Rad - A non-SI unit of absorbed dose of radiation, equal to 0.01 Gy.

Radiance (L) - The radiant intensity in a given direction from an element of a surface, divided by the area of the orthogonal projection of this element on a plane perpendicular to the given direction. [1]

Radiant intensity (I) - The radiant energy flux leaving an element of a source within an element of solid angle, divided by that element of solid angle. [1]

Radicals - Molecular entities possessing an unpaired electron, such as $\cdot CH_3$, $\cdot SnH_3$, $\cdot Cl$. (In these formulas the dot, symbolizing the unpaired electron, should be placed so as to indicate the atom of highest spin density, if this is possible). [5]

Raman effect - The inelastic scattering of light by a molecule, in which the incident photon either gives up to, or receives energy from, one of the internal vibrational modes of the molecule. The scattered light thus has either a lower frequency (Stokes radiation) or higher frequency (anti-Stokes radiation) than the incident light. These shifts provide a measure of the normal vibrational frequencies of the molecule.

Rankine cycle - A thermodynamic cycle which can be used to calculate the ideal performance of a heat engine that uses a condensable vapor as the working fluid (e.g., a steam engine or a heat pump).

Rankine temperature - A thermodynamic temperature scale based on a temperature interval °R = (5/9) K ; i.e., $T/°R = (9/5)T/K = t/°F + 459.67$.

Raoult's law - The expression for the vapor pressure p_i of component i in an ideal solution, viz., $p_i = x_i p_{i0}$, where x_i is the mole fraction of component i and p_{i0} the vapor pressure of the pure substance i.

Rare earth elements - The elements Sc, Y, and the lanthanides (La, Ce, Pr, Nd, Pm, Sm, Eu, Gd, Tb, Dy, Ho, Er, Tm, Yb, Lu). [7]

Rayleigh number (Ra) - A dimensionless quantity used in fluid mechanics, defined by $Ra = l^3 g\alpha\Delta T\rho/\eta a$, where l is length, g is acceleration of gravity, α is cubic expansion coefficient, T is temperature, ρ is density, η is viscosity, and a is thermal diffusivity. [2]

Rayleigh scattering - The scattering of light by particles which are much smaller than the wavelength of the light. It is characterized by a scattered intensity which varies as the inverse fourth power of the wavelength.

Rayleigh wave - A guided elastic wave along the surface of a solid; also called surface acoustic wave.

Reactance (X) - The imaginary part of impedance. For an inductive reactance L and a capacitive reactance C in series, the reactance is $X = L\omega-1/(C\omega)$, where ω is 2π times the frequency of the current. [1]

Red shift - A displacement of a spectral line toward longer wavelengths. This can occur through the Doppler effect (e.g., in the light from receding galaxies) or, in the general theory of relativity, from the effects of a star's gravitational field.

Reflectance (ρ) - Ratio of the radiant or luminous flux at a given wavelength that is reflected to that of the incident radiation. Also called reflection factor. [1]

Reflection high energy electron diffraction (RHEED) - See Techniques for Materials Characterization, page 12-1.

Relative humidity* - The ratio of the partial pressure of water vapor in air to the saturation vapor pressure of water at the same temperature, expressed as a percentage. [10]

Relative molar mass - See Molecular weight.

Rem - A non-SI unit of dose equivalent, equal to 0.01 Sv.

Resistance (R) - Electric potential difference divided by current when there is no electromotive force in the conductor. This definition applies to direct current. More generally, resistance is defined as the real part of impedance. [1]

Resistivity (ρ) - Electric field strength divided by current density when there is no electromotive force in the conductor. Resistivity is an intrinsic property of a material. For a conductor of uniform cross section with area A and length L, and whose resistance is R, the resistivity is given by $\rho = RA/L$. [1]

Reynolds number (Re) - A dimensionless quantity used in fluid mechanics, defined by $Re = \rho v l/\eta$, where ρ is density, v is velocity, l is length, and η is viscosity. [2]

Rheology - The study of the flow of liquids and deformation of solids. Rheology addresses such phenomena as creep, stress relaxation, anelasticity, nonlinear stress deformation, and viscosity.

Ribonucleic acids (RNA) - Naturally occurring polyribonucleotides. See also nucleic acids, nucleosides, nucleotides, ribonucleotides. [5]

Ribonucleotides - Nucleotides in which the glycosyl group is a ribosyl group. See also nucleotides. [5]

Roentgen (R) - A unit used for expressing the charge (positive or negative) liberated by x-ray or γ radiation in air, divided by the mass of air. A roentgen is defined as 2.58×10^{-4} C/kg.

Rotational constants - In molecular spectroscopy, the constants appearing in the expression for the rotational energy levels as a function of the angular momentum quantum numbers. These constants are proportional to the reciprocals of the principal moments of inertia, averaged over the vibrational motion.

Rutherford back scattering (RBS) - See Techniques for Materials Characterization, page 12-1.

Rydberg constant (R_∞)* - The fundamental constant which appears in the equation for the energy levels of hydrogen-like atoms; i.e., $E_n = hcR_\infty Z^2\mu/n^2$, where h is Planck's constant, c the speed of light, Z the atomic number, μ the reduced mass of nucleus and electron, and n the principal quantum number ($n = 1, 2, ...$).

Rydberg series - A regular series of lines in the spectrum of an atom or molecule, with the spacing between successive lines becoming smaller as the frequency increases (wavelength decreases). The series eventually converges to a limit which usually corresponds to the complete removal of an electron from the atom or molecule.

Sackur-Tetrode equation* - An equation for the molar entropy S_m of an ideal monatomic gas: $S_m = R\ln(e^{5/2} V/N_A\Lambda^3)$, where R is the molar gas constant, V is the volume, and N_A is Avogadro's number. The constant Λ is given by $\Lambda = h/(2\pi mkT)^{1/2}$, where h is Planck's constant, m the atomic mass, k the Boltzmann constant, and T the temperature.

Salinity (S)* - A parameter used in oceanography to describe the concentration of dissolved salts in seawater. It is defined in terms of electrical conductivity relative to a standard solution of KCl. When expressed in units of parts per thousand, S may be roughly equated to the concentration of dissolved material in grams per kilogram of seawater.

Salt - An ionic compound formed by the reaction of an acid and a base.

Scanned probe microscopy (SPM) - See Techniques for Materials Characterization, page 12-1.

Scanning electron microscopy (SEM) - See Techniques for Materials Characterization, page 12-1.

Scanning laser acoustic microscopy (SLAM) - See Techniques for Materials Characterization, page 12-1.

Scanning transmission electron microscopy (STEM) - See Techniques for Materials Characterization, page 12-1.

Scanning tunneling microscopy (STM) - See Techniques for Materials Characterization, page 12-1.

Schiff bases - Imines bearing a hydrocarbyl group on the nitrogen atom: $R_2C=NR'$ (R' not equal to H). Considered by many to be synonymous with azomethines. [5]

Schmidt number (Sc) - A dimensionless quantity used in fluid mechanics, defined by $Sc = \eta/\rho D$, where η is viscosity, ρ is density, and D is diffusion coefficient. [2]

Schottky barrier - A potential barrier associated with a metal-semiconductor contact. It forms the basis for the rectifying device known as the Schottly diode.

Schrödinger equation - The basic equation of wave mechanics which, for systems not dehpendent on time, takes the form:

$$-(\hbar/2m)\nabla^2\psi + V\psi = E\psi$$

where ψ is the wavefunction, V is the potential energy expressed as a function of the spatial coordinates, E is an energy eigenvalue, ∇^2 is the Laplacian operator, $\hbar$ is Planck's constant divided by 2π, and m is the mass.

Second (s)* - The SI base unit of time. [1]

Second radiation constant (c_2)* - See First radiation constant.

Secondary ion mass spectroscopy (SIMS) - See Techniques for Materials Characterization, page 12-1.

Seebeck effect - The development of a potential difference in a circuit where two different metals or semiconductors are joined and their junctions maintained at different temperatures. It is the basis of the thermocouple.

Selenides - Compounds having the structure RSeR (R not equal to H). They are thus selenium analogues of ethers. Also used for metal salts of H_2Se. [5]

Semicarbazones - Compounds having the structure $R_2C=NNHC(=O)NH_2$, formally derived by condensation of aldehydes or ketones with semicarbazide [$NH_2NHC(=O)NH_2$]. [5]

Semiconductor - A material in which the highest occupied energy band (valence band) is completely filled with electrons at T = 0 K, and the energy gap to the next highest band (conduction band) ranges from 0 to 4 or 5 eV. With increasing temperature electrons are excited into the conduction band, leading to an increase in the electrical conductivity.

Semiquinones - Radical anions having the structure -O-Z-O· where Z is an ortho- or para-arylene group or analogous heteroarylene group; they are formally generated by the addition of an electron to a quinone. [5]

SI units* - The International System of Units adopted in 1960 and recommended for use in all scientific and technical fields. [1]

Siemens (S)* - The SI unit of electric conductance, equal to Ω^{-1}. [1]

Sievert (Sv)* - The SI unit of dose equivalent (of radiation), equal to J/kg. [1]

Silanes - Saturated silicon hydrides, analogues of the alkanes; i.e. compounds of the general formula Si_nH_{2n+2}. Silanes may be subdivided into silane, oligosilanes, and polysilanes. Hydrocarbyl derivatives are often referred to loosely as silanes. [5]

Silicones - Polymeric or oligomeric siloxanes, usually considered unbranched, of general formula $[-OSiR_2-]_n$ (R not equal to H). [5]

Siloxanes - Saturated silicon-oxygen hydrides with unbranched or branched chains of alternating silicon and oxygen atoms (each silicon atom is separated from its nearest silicon neighbors by single oxygen atoms). [5]

Skin effect - The concentration of high frequency alternating currents near the surface of a conductor.

Slater orbital - A particular mathematical expression for the radial part of the wave function of a single electron, which is used in quantum-mechanical calculations of the energy and other properties of atoms and molecules.

Small angle neutron scattering (SANS) - See Techniques for Materials Characterization, page 12-1.

Snell's law - The relation between the angle of incidence i and the angle of refraction r of a light beam which passes from a medium of refractive index n_0 to a medium of index n_1, viz., $\sin i/\sin r = n_1/n_0$.

Solar constant* - The mean radiant energy flux from the sun on a unit surface normal to the direction of the rays at the mean distance of the earth from the sun. The value is approximately 1373 W/m².

Solar wind - The stream of high velocity hydrogen and helium ions emitted by the sun which flows through the solar system and beyond.

Soliton - A spatially localized wave in a solid or liquid that can interact strongly with other solitons but will afterwards regain its original form.

Solubility* - A quantity expressing the maximum concentration of some material (the solute) that can exist in another liquid or solid material (the solvent) at thermodynamic equilibrium at specified temperature and pressure. Common measures of solubility include the mass of solute per unit mass of solution (mass fraction), mole fraction of solute, molality, molarity, and others.

Solubility product constant (K_{sp})* - The equilibrium constant for the dissolution of a sparsely soluble salt into its constituent ions.

Space group* - A group of symmetry operations (reflections, rotations, etc.) that leave a crystal invariant. A total of 230 space groups have been identified.

Spark source mass spectroscopy (SSMS) - See Techniques for Materials Characterization, page 12-1.

Specific gravity - Ratio of the mass density of a material to that of water. Since one must specify the temperature of both the sample and the water to have a precisely defined quantity, the use of this term is now discouraged.

Specific heat - Heat capacity divided by mass. See Heat capacity.

Specific quantity - It is often convenient to express an extensive quantity (e.g., volume, enthalpy, heat capacity, etc.) as the actual value divided by mass. The resulting quantity is called specific volume, specific enthalpy, etc.

Specific rotation $[\alpha]_\lambda^\theta$ - For an optically active substance, defined by $[\alpha]_\lambda^\theta = \alpha/\gamma l$, where α is the angle through which plane polarized light is rotated by a solution of mass concentration γ and path length l. Here θ is the Celsius temperature and λ the wavelength of the light at which the measurement is carried out. Also called specific optical rotatory power. [2]

Spin (s, I)* - A measure of the intrinsic angular momentum of a particle, which it possesses independent of its orbital motion. The symbol s is used for the spin quantum number of an electron, while I is generally used for nuclear spin.

Spiro compounds - Compounds having one atom (usually a quaternary carbon) as the only common member of two rings. [5]

Stacking fault - An error in the normal sequence of layer growth in a crystal.

Standard mean ocean water (SMOW) - A standard sample of pure water of accurately known isotopic composition which is maintained by the International Atomic Energy Agency. It is used for precise calibration of density and isotopic composition measurements.

Standard reduction potential ($E°$) - The zero-current potential of a cell in which the specified reduction reaction occurs at the right-hand electrode and the left-hand electrode is the standard hydrogen electrode. Also called Standard electrode potential.

Standard state - A defined state (specified temperature, pressure, concentration, etc.) for tabulating thermodynamic functions and carrying out thermodynamic calculations. The standard state pressure is usually taken as 100,000 Pa (1 bar), but various standard state temperatures are used. [2]

Stanton number (St) - A dimensionless quantity used in fluid mechanics, defined by $St = h/\rho v c_p$, where h is coefficient of heat transfer, ρ is density, v is velocity, and c_p is specific heat capacity at constant pressure. [2]

Stark effect - The splitting of an energy level of an atom or molecule, and hence a splitting of spectral lines arising from that level, as a result of the application of an external electric field.

Statistical weight (g) - The number of distinct states corresponding to the same energy level. Also called degeneracy.

Stefan-Boltzmann constant (σ)* - Constant in the equation for the radiant exitance M (radiant energy flux per unit area) from a black body at thermodynamic temperature T, viz. $M = \sigma T^4$. [1]

Stibines - SbH_3 and compounds derived from it by substituting one, two or three hydrogen atoms by hydrocarbyl groups: R_3Sb. $RSbH_2$, R_2SbH, and R_3Sb (R not equal to H) are called primary, secondary and tertiary stibines, respectively. [5]

Stochastic process - A process which involves random variables and whose outcome can thus be described only in terms of probabilities.

Stoichiometric number (ν) - The number appearing before the symbol for each compound in the equation for a chemical reaction. By convention, it is negative for reactants and positive for products. [2]

Stokes (St) - A non-SI unit of kinematic viscosity, equal to 10^{-4} m²/s.

Stokes' law - The statement, valid under certain conditions, that the viscous force F experienced by a sphere of radius a moving at velocity v in a medium of viscosity η is given by $F = -6\pi\eta a v$.

Strain - The deformation of a body that results from an applied stress.

Stratosphere - The part of the earth's atmosphere extending from the top of the troposphere (typically 10 to 15 km above the surface) to about 50 km. It is characterized by an increase in temperature with increasing altitude.

Stress - Force per unit area (pressure) applied to a body. Tensile stress tends to stretch or compress the body in the direction of the applied force. Sheer stress results from a tangential force which tends to twist the body.

Strong interaction - The short range (order of 1 fm) attractive forces between protons, neutrons, and other hadrons which are responsible for the stability of the nucleus.

Strouhal number (Sr) - A dimensionless quantity used in fluid mechanics, defined by $Sr = lf/v$, where l is length, f is frequency, and v is velocity. [2]

Structure factor - In x-ray crystallography, the sum of the scattering factors of all the atoms in a unit cell, weighted by an appropriate phase factor. The intensity of a given reflection is proportional to the square of the structure factor.

Sublimation pressure - The pressure of a gas in equilibrium with a solid at a specified temperature.

Sulfides - Compounds having the structure RSR (R not equal to H). Such compounds were once called thioethers. In an inorganic sense, salts or other derivatives of hydrogen sulfide. [5]

Sulfones - Compounds having the structure, $RS(=O)_2R$ (R not equal to H), e.g. $C_2H_5S(=O)_2CH_3$, ethyl methyl sulfone. [5]

Sulfonic acids - $HS(=O)_2OH$, sulfonic acid, and its S-hydrocarbyl derivatives. [5]

Sulfoxides - Compounds having the structure $R_2S=O$ (R not equal to H), e.g. $Ph_2S=O$, diphenyl sulfoxide. [5]

Superconductor - A material that experiences a nearly total loss of electrical resistivity below a critical temperature T_c. The effect can occur in pure metals, alloys, semiconductors, organic compounds, and certain inorganic solids.

Superfluid - A fluid with near-zero viscosity and extremely high thermal conductivity. Liquid helium exhibits these properties below 2.186 K (the λ point).

Supernova - A star in the process of exploding because of instabilities which follow the exhaustion of its nuclear fuel.

Surface analysis by laser ionization (SALI) - See Techniques for Materials Characterization, page 12-1.

Surface tension (γ,σ)* - The force per unit length in the plane of the interface between a liquid and a gas, which resists an increase in the area of that surface. It can also be equated to the surface Gibbs energy per unit area.

Surfactant - A substance which lowers the surface tension of the medium in which it is dissolved, and/or the interfacial tension with other phases, and accordingly is positively adsorbed at the liquid-vapor or other interfaces. [3]

Susceptance (B) - Imaginary part of admittance. [1]

Svedberg - A non-SI unit of time, used to express sedimentation coefficients, equal to 10^{-13} s.

Syndiotactic macromolecule - A tactic macromolecule, essentially comprising alternating enantiomeric configurational base units which have chiral or prochiral atoms in the main chain in a unique arrangement with respect to their adjacent constitutional units. In this case the repeating unit consists of two configurational base units that are enantiomeric. [8]

Tacticity - The orderliness of the succession of configurational repeating units of a macromolecule or oligomer molecule. In a tactic macromolecule essentially all the configurational repeating units are identical with respect to directional sense. See Configurational repeating unit, Isotactic, Syndiotactic. [8]

Tautomerism - Isomerism of the general form G-X-Y=Z ⇌ X=Y-Z-G, where the isomers (called tautomers) are readily interconvertible; the atoms connecting the groups X, Y, Z are typically any of C, H, O, or S, and G is a group which becomes an electrofuge (i.e., a group that does not carry away the bonding electron pair when it leaves its position in the molecule) or nucleofuge (a group that does carry away the bonding electrons when leaving) during isomerization. The commonest case, when the electrofuge is H^+, is also known as prototropy. A common example, written so as to illustrate the general pattern given above, is keto-enol tautomerism, such as

$H-O-C(CH_3)=CH-CO_2Et$ (enol) $(CH_3)C(=O)-CH_2-CO_2Et$ (keto)

In some cases the interconversion rate between tautomers is slow enough to permit isolation of the separate keto and enol forms. [5]

Tensile strength* - In tensile testing, the ratio of maximum load a body can bear before breaking to original cross-sectional area. Also called ultimate strength. [11]

Terpenes - Hydrocarbons of biological origin having carbon skeletons formally derived from isoprene $[CH_2=C(CH_3)CH=CH_2]$. [5]

Terpenoids - Natural products and related compounds formally derived from isoprene units. They contain oxygen in various functional groups. The skeleton of terpenoids may differ from strict additivity of isoprene units by the loss or shift of a methyl (or other) group. [5]

Tesla (T)* - The SI unit of magnetic flux density (B), equal to V s/m^2. [1]

Thermal conductivity* - Rate of heat flow divided by area and by temperature gradient. [1]

Thermal diffusivity - Thermal conductivity divided by density and by specific heat capacity at constant pressure. [1]

Thermal expansion coefficient (α)* - The linear expansion coefficient is defined by $\alpha_l = (1/l)(dl/dT)$; the volume expansion coefficient by $\alpha_V = (1/V)(dV/dT)$. [1]

Thermionic emission - The emission of electrons from a solid as a result of heat. The effect requires a high enough temperature to impart sufficient kinetic energy to the electrons to exceed the work function of the solid.

Thermodynamic laws - The foundation of the science of thermodynamics:

   First law: The internal energy of an isolated system is constant; if energy is supplied to the system in the form of heat dq and work dw, then the change in energy dU = dq + dw.

   Second law: No process is possible in which the only result is the transfer of heat from a reservoir and its complete conversion to work.

   Third law: The entropy of a perfect crystal approaches zero as the thermodynamic temperature approaches zero.

Thermoelectric power - For a bar of a pure material whose ends are at different temperatures, the potential difference divided by the difference in temperature of the ends. See also Seeback effect.

Thermogravimetric analysis (TGA) - See Techniques for Materials Characterization, page 12-1.

Thermosphere - The layer of the earth's atmosphere extending from the top of the mesosphere (typically 80–90 km above the surface) to about 500 km. It is characterized by a rapid increase in temperature with increasing altitude up to about 200 km, followed by a leveling off in the 300–500 km region.

Thiols - Compounds having the structure RSH (R not equal to H). Also known by the term mercaptans (abandoned by IUPAC); e.g. CH_3CH_2SH, ethanethiol. [5]

Thomson coefficient (μ, τ) - The heat power developed in the Thomson effect (whereby heat is evolved in a conductor when a current is flowing in the presence of a temperature gradient), divided by the current and the temperature difference. [1]

Tonne (t) - An alternative name for megagram (1000 kg). [1]

Torque (T) - For a force F that produces a torsional motion, $T = r \times F$, where r is a vector from some reference point to the point of application of the force.

Torr - A non-SI unit of pressure, equal to 133.322 Pa. The name is generally considered interchangeable with millimeter of mercury.

Townsend coefficient - In a radiation counter, the number of ionizing collisions by an electron per unit path length in the direction of an applied electric field.

Transducer - Any device that converts a signal from acoustical, optical, or some other form of energy into an electrical signal (or vice versa) while preserving the information content of the original signal.

Transistor - A voltage amplifier using controlled electron currents inside a semiconductor.

Transition metals - Elements characterized by a partially filled d subshell. The First Transition Series comprises Sc, Ti, V, Cr, Mn, Fe, Co, Ni, Cu. The Second and Third Transition Series include the lanthanides and actinides, respectively. [7]

Transition probability* - See Einstein transition probability.

Transmittance (τ) - Ratio of the radiant or luminous flux at a given wavelength that is transmitted to that of the incident radiation. Also called transmission factor. [1]

Tribology - The study of frictional forces between solid surfaces.

Triple point* - The point in p,T space where the solid, liquid, and gas phases of a substance are in thermodynamic equilibrium. The corresponding temperature and pressure are called the triple point temperature and triple point pressure.

Troposphere - The lowest part of the earth's atmosphere, extending to 10-15 km above the surface. It is characterized by a decrease in temperature with increasing altitude. The exact height varies with latitude and season.

Tunnel diode - A device involving a p-n junction in which both sides are so heavily doped that the Fermi level on the p-side lies in the valence band and on the n-side in the conduction band. This leads to a current-voltage curve with a maximum, so that the device exhibits a negative resistance in some regions.

Ultraviolet photoelectron spectroscopy (UPS) - See Techniques for Materials Characterization, page 12-1.

Umklapp process - A process involving the interaction of three or more waves (lattice or electron) in a solid in which the sum of the wave vectors does not equal zero.

Unified atomic mass unit (u)* - A unit of mass used in atomic, molecular, and nuclear science, defined as the mass of one atom of ^{12}C divided by 12. Its approximate value is 1.66054×10^{-27} kg. [1]

Universal time (t_U, UT) - Mean solar time counted from midnight at the Greenwich meridian. Also called Greenwich mean time (GMT). The interval of mean solar time is based on the average, over one year, of the time between successive transits of the sun across the observer's meridian.

Vacancy - A missing atom or ion in a crystal lattice.

Van Allen belts - Two toroidal regions above the earth's atmosphere containing protons and electrons. The outer belt at about 25,000 km above the surface is probably of solar origin. The inner belt at about 3000 km contains more energetic particles from outside the solar system.

Van der Waals' equation* - An equation of state for fluids which takes the form:

$$pV_m = RT\left(\frac{1}{V_m-b} - \frac{a}{V_m^2}\right)$$

where p is pressure, V_m is molar volume, T is temperature, R is the molar gas constant, and a and b are characteristic parameters of the substance which describe the effect of attractive and repulsive intermolecular forces, respectively.

Van der Waals' force - The weak attractive force between two molecules which arises from electric dipole interactions. It can lead to the formation of stable but weakly bound dimer molecules or clusters.

Van't Hoff equation - The equation expressing the temperature dependence of the equilibrium constant K of a chemical reaction:

$$\frac{d\ln K}{dT} = \frac{\Delta_r H^\circ}{RT^2}$$

where $\Delta_r H^\circ$ is the standard enthalpy of reaction, R the molar gas constant, and T the temperature. Also called van't Hoff isochore.

Vapor pressure* - The pressure of a gas in equilibrium with a liquid (or, in some usage, a solid) at a specified temperature.

Varistor - A device that utilizes the properties of certain metal oxides with small amounts of impurities, which show abrupt nonlinearities at specific voltages where the material changes from a semiconductor to an insulator.

Velocity (v) - Rate of change of distance with time.

Verdet constants (V)* - Angle of rotation of a plane polarized light beam passing through a medium in a magnetic field, divided by the field strength and by the path length.

Virial equation of state* - An equation relating the pressure p, molar volume V_m, and temperature T of a real gas in the form of an expansion in powers of the molar volume, viz., $pV_m = RT(1+BV_m^{-1}+CV_m^{-2}+ ...)$, where R is the molar gas constant. B is called the second virial coefficient, C the third virial coefficient, etc. The virial coefficients are functions of temperature.

Viscosity (η)* - The proportionality factor between sheer rate and sheer stress, defined through the equation $F = \eta A(dv/dx)$, where F is the tangential force required to move a planar surface of area A at velocity v relative to a parallel surface separated from the first by a distance x. Sometimes called dynamic or absolute viscosity. The term kinematic viscosity (symbol v) is defined as η divided by the mass density.

Volt (V)* - The SI unit of electric potential, equal to W/A. [1]

Volume fraction (ϕ_j) - Defined as $V_j/\Sigma_i V_i$, where V_j is the volume of the specified component and the V_i are the volumes of all the components of a mixture prior to mixing. [2]

Watt (W)* - The SI unit of power, equal to J/s. [1]

Wave function - A function of the coordinates of all the particles in a quantum mechanical system (and, in general, of time) which fully describes the state of the system. The product of the wave function and its complex conjugate is proportional to the probability of finding a particle at a particular point in space.

Weak interaction - The weak forces (order of 10^{-12} of the strong interaction) between elementary particles which are responsible for beta decay and other nuclear effects.

Weber (Wb)* - The SI unit of magnetic flux, equal to V s. [1]

Weber number (We) - A dimensionless quantity used in fluid mechanics, defined by $We = \rho v^2 l/\gamma$, where ρ is density, v is velocity, l is length, and γ is surface tension. [2]

Weight - That force which, when applied to a body, would give it an acceleration equal to the local acceleration of gravity. [1]

Wiedeman-Franz law - The law stating that the thermal conductivity k and electrical conductivity σ of a pure metal are related by $k = L\sigma T$, where T is the temperature and L (called the Lorenz ratio) has the approximate value 2.45×10^{-8} V²/K².

Wien displacement law - The relation, which can be derived from the Planck formula for black body radiation, that

$\lambda_{max} T = 0.0028978$ m K, where λ_{max} is the wavelength of maximum radiance at temperature T.

Wigner-Seitz method - A method of calculating electron energy levels in a solid using a model in which each electron is subject to a spherically symmetric potential.

Wittig reagents - See phosphonium ylides.

Work (W) - Force multiplied by the displacement in the direction of the force. [1]

Work function (Φ)* - The energy difference between an electron at rest at infinity and an electron at the Fermi level in the interior of a substance. It is thus the minimum energy required to remove an electron from the interior of a solid to a point just outside the surface. [1]

X unit (X) - A unit of length used in x-ray crystallography, equal to approximately 1.002×10^{-13} m.

X-ray photoelectron spectroscopy (XPS) - See Techniques for Materials Characterization, page 12-1.

Yield strength - The stress at which a material exhibits a specified deviation (often chosen as 0.2% for metals) from proportionality of stress and strain. [11]

Young's modulus (E) - In tension or compression of a body below its elastic limit, the ratio of stress to corresponding strain. Since strain is normally expressed on a fractional basis, Young's modulus has dimensions of pressure. Also called elastic modulus. [11]

Zeeman effect - The splitting of an energy level of an atom or molecule, and hence a splitting of spectral lines arising from that level, as a result of the application of an external magnetic field.

Zener diode - A control device utilizing a p-n junction with a well defined reverse-bias avalanche breakdown voltage.

Zeotrope - A liquid mixture that shows no maximum or minimum when vapor pressure is plotted against composition at constant temperature. See Azeotrope.

Zero-point energy - The energy possessed by a quantum mechanical system as a result of the uncertainty principle even when it is in its lowest energy state; e.g., the difference between the lowest energy level of a harmonic oscillator and the minimum in the potential well.

Zeta potential (ζ) - The electric potential at the surface of a colloidal particle relative to the potential in the bulk medium at a long distance. Also called electrokinetic potential.

Zwitterions - Neutral compounds having formal unit electrical charges of opposite sign. Some chemists restrict the term to compounds with the charges on non-adjacent atoms. Sometimes referred to as inner salts, dipolar ions (a misnomer). [5]

THERMODYNAMIC FUNCTIONS AND RELATIONS

p = pressure V = volume T = temperature

n_i = amount of substance i

$x_i = n_i/\Sigma_j\, n_j$ = mole fraction of substance i

Energy	U
Entropy	S
Enthalpy	$H = U + pV$
Helmholtz energy	$A = U - TS$
Gibbs energy	$G = U + pV - TS$
Isobaric heat capacity	$C_p = (\partial H/\partial T)_p$
Isochoric heat capacity	$C_V = (\partial U/\partial T)_V$
Isobaric expansivity	$\alpha = V^{-1}(\partial V/\partial T)_p$
Isothermal compressibility	$\kappa_T = -V^{-1}(\partial V/\partial p)_T$
Isentropic compressibility	$\kappa_S = -V^{-1}(\partial V/\partial p)_S$
	$\kappa_T - \kappa_S = T\alpha^2 V/C_p$
	$C_p - C_V = T\alpha^2 V/\kappa_T$
Gibbs-Helmholtz equation	$H = G - T(\partial G/\partial T)_p$
Maxwell relations	$(\partial S/\partial p)_T = -(\partial V/\partial T)_p$
	$(\partial S/\partial V)_T = -(\partial p/\partial T)_V$
Joule-Thomson expansion	$\mu_{JT} = (\partial T/\partial p)_H = -\{V - T(\partial V/\partial T)_p\}/C_p$
	$\phi_{JT} = (\partial H/\partial p)_T = V - T(\partial V/\partial T)_p$
Partial molar quantity	$X_i = (\partial X/\partial n_i)_{T,p,nj\neq i}$
Chemical potential	$\mu_i = (\partial G/\partial n_i)_{T,p,nj\neq i}$
Perfect gas [symbol ^{pg}]	$pV = (\Sigma_i\, n_i)RT$
	$\mu_i^{pg} = \mu_i^{\theta} + RT \ln(x_i p/p^{\theta})$
Fugacity	$f_i = (x_i p)\exp\{(\mu_i - \mu_i^{pg})/RT\}$
Activity coefficient	$\gamma_i = f_i/(x_i f_i^{\theta})$
Gibbs-Duhem relation	$0 = SdT - Vdp + \Sigma_i n_i d\mu_i$

[Superscript θ in above equations indicates standard state]

Notation for chemical and physical changes ($X = H, S, G$, etc.):

Chemical reaction	$\Delta_r X$
Formation from elements	$\Delta_f X$
Combustion	$\Delta_c X$
Fusion (cry→liq)	$\Delta_{fus} X$
Vaporization (liq→gas)	$\Delta_{vap} X$
Sublimation (cry→gas)	$\Delta_{sub} X$
Phase transition	$\Delta_{trs} X$
Solution	$\Delta_{sol} X$
Mixing	$\Delta_{mix} X$
Dilution	$\Delta_{dil} X$

Section 3
Physical Constants of Organic Compounds

Organic

PHYSICAL CONSTANTS OF ORGANIC COMPOUNDS

The basic physical constants and structure diagrams for about 10,900 organic compounds are presented in this table. An effort has been made to include the compounds most frequently encountered in the laboratory, the workplace, and the environment. Particular emphasis has been given to substances that are considered environmental or human health hazards. In making the selection of compounds for the table, added weight was assigned to the appearance of a compound in various lists or reference sources such as:

- Laboratory reagent lists, e.g., the ACS *Reagent Chemicals* volume (Ref. 1)
- The DIPPR list of industrially important compounds (Ref. 2) and the (much larger) TSCA Inventory of chemicals used in commerce
- The Hazardous Substance Data Bank (Ref. 3)
- The UNEP list of Persistent Organic Pollutants (Ref. 4)
- Chemicals on Reporting Rules (CORR), a database of about 7500 regulated compounds prepared by the Environmental Protection Agency (Ref. 5)
- The EPA Integrated Risk Information System (IRIS), a database of human health effects of exposure to chemicals in the environment (Ref. 6)
- Compendia of chemicals of biochemical or medical importance, such as *The Merck Index* (Ref. 10)
- Specialized tables in this *Handbook*

It should be noted that the above lists vary widely in their choice of chemical names, and even in the use of Chemical Abstracts Registry Numbers. To the extent possible, we have attempted to systematize the names and registry numbers for this table.

Clearly, criteria of this type are somewhat subjective, and compounds considered important by some users have undoubtedly been omitted. Suggestions for additional compounds or other improvements are welcomed.

The data in the table have been derived from many sources, including both the primary literature and evaluated compilations. The *Handbook of Data on Organic Compounds, Third Edition* (Reference 7) and the *Chapman & Hall/CRC Combined Chemical Dictionary* (Reference 8) were important sources. Other useful compilations of physical property data for organic compounds are listed in References 9-19. Many boiling point values (and some melting point and density values) were taken from recent physical chemistry literature dealing with fluid properties. Where conflicts were found, the value deemed most reliable was chosen.

The table is arranged alphabetically by substance name, which generally is either an IUPAC systematic name or, in the case of pesticides, pharmaceuticals, and other complex compounds, a simple trivial name. Names in ubiquitous use, such as acetic acid and formaldehyde, are adopted rather than their systematic equivalents. Synonyms are given in the column following the primary name, and structure diagrams are given on the page facing the data listing. The explanation of the data columns follows:

- **No.**: An identification number used in the indexes.
- **Name**: Primary name of the substance
- **Synonym**: A synonym in common use. When the primary name is non-systematic, a systematic name may appear here.
- **Mol. Form.**: The molecular formula written in the Hill convention.

- **CAS RN**: The Chemical Abstracts Service Registry Number for the compound.
- **Mol. Wt**: Molecular weight (relative molar mass) as calculated with the 2001 IUPAC Standard Atomic Weights.
- **Physical Form**: A notation of the physical phase, color, crystal type, or other features of the compound at ambient temperature. Abbreviations are given below.
- **mp**: Normal melting point in °C. A value is sometimes followed by "dec", indicating decomposition is observed at the stated temperature (so that it is probably not a true melting point). The notation "tp" indicates a triple point, where solid, liquid, and gas are in equilibrium.
- **bp**: Normal boiling point in °C, if it is available. This is the temperature at which the liquid phase is in equilibrium with the vapor at a pressure of 760 mmHg (101.325 kPa). A notation "sp" following the value indicates a sublimation point, where the vapor pressure of the solid phase reaches 760 mmHg. When a notation such as "dec" or "exp" (explodes) follows the value, the temperature may not be a true boiling point. A simply entry "sub" indicates the solid has a significant sublimation pressure at ambient temperatures. The boiling point at reduced pressure is listed in some cases, with or without the normal boiling point. Here the superscript indicates the pressure in mmHg.
- **den**: Density (mass per unit volume) in g/cm³. The temperature in °C is indicated by a superscript. Values refer to the liquid or solid phase, and all values are true densities, not specific gravities. The number of decimal places gives a rough estimate of the accuracy of the value.
- n_D: Refractive index, at the temperature in °C indicated by the superscript. Unless otherwise indicated, all values refer to a wavelength of 589 nm (sodium D line). Values are given only for liquids and solids.
- **Solubility**: Qualitative indication of solubility in common solvents. Abbreviations are:

i	insoluble
sl	slightly soluble
s	soluble
vs	very soluble
msc	miscible
dec	decomposes

Abbreviations for solvents are given below.

In order to facilitate the location of compounds in the table, three indexes are provided:

- **Synonym Index**: Includes common synonyms, but not the primary name by which the table is arranged.
- **Molecular Formula Index**: Lists compounds by molecular formula in the Hill Order (see Preface to this *Handbook*).
- **CAS Registry Number Index**: Lists compounds by Chemical Abstracts Service Registry Number. Note there is some redundancy in this index, because many compounds have several Registry Numbers associated with them. Thus the CAS RN in a table entry may differ from the CAS RN that points to it in the index. For example, CAS RN 1319-77-3 in the index points to all three cresol isomers, each of which has its own specific CAS RN.

The assistance of Fiona Macdonald in checking names and formulas is gratefully acknowledged, as well as the efforts of Janice Shackleton, Trupti Desai, Nazila Kamaly, Matt Griffiths, and Lawrence Braschi in preparing the structure diagrams.

List of Abbreviations

Ac	acetyl	mcl	monoclinic
Ac_2O	acetic anhydride	Me	methyl
AcOEt	ethyl acetate	MeCN	acetonitrile
ac	acid	MeOH	methanol
ace	acetone	misc, msc	miscible
al	alcohol (ethanol)	mp	melting point
alk	alkali	n	refractive index
amor	amorphous	nd	needles
anh	anhydrous	oct	octahedra, octahedral
aq	aqueous	oran	orange
bipym	bipyramidal	orth	orthorhombic
bl	blue	os	organic solvents
blk	black	pa	pale
bp	boiling point	peth	petroleum ether
br	brown	Ph	phenyl
bt	bright	PhCl	chlorobenzene
Bu	butyl	$PhNH_2$	aniline
BuOH	1-butanol	$PhNO_2$	nitrobenzene
bz	benzene	pl	plates
chl	chloroform	pow	powder
col	colorless	Pr	propyl
con, conc	concentrated	PrOH	1-propanol
cry	crystals	pr	prisms
ctc	carbon tetrachloride	purp	purple
cy, cyhex	cyclohexane	py	pyridine
dec	decomposes	pym	pyramids, pyramidal
den	density	reac	reacts
dil	dilute	rhom	rhombic
diox	dioxane	s	soluble
dk	dark	sat	saturated
DMF	dimethylformamide	sc	scales
DMSO	dimethyl sulfoxide	sl	slightly soluble
efflor	efflorescent	soln	solution
Et	ethyl	sp	sublimation point
EtOH	ethanol	stab	stable
eth	diethyl ether	sub	sublimes
exp	explodes	sulf	sulfuric acid
fl	flakes	syr	syrup
flr	fluorescent	tab	tablets
fum	fumes, fuming	tcl	triclinic
gl	glacial	tetr	tetragonal
gr	gray	tfa	trifluoroacetic acid
gran	granular	thf, THF	tetrahydrofuran
grn	green	tol	toluene
hex	hexagonal	tp	triple point
HOAc	acetic acid	trg	trigonal
hp	heptane	unstab	unstable
hx	hexane	vap	vapor
hyd	hydrate	viol	violet
hyg	hygroscopic	visc	viscous
i	insoluble	vol	volatile
i-	iso-	vs	very soluble
iso	isooctane	w	water
lf	leaves	wh	white
lig	ligroin	xyl	xylene
liq	liquid	ye	yellow
lo	long		

References

1. American Chemical Society, *Reagent Chemicals, Ninth Edition*, Oxford University Press, New York, 2000.
2. Daubert, T. E., Danner, R. P., Sibul, H. M., and Stebbins, C. C., *Physical and Thermodynamic Properties of Pure Compounds: Data Compilation*, extant 2002 (core with supplements), Taylor & Francis, Bristol, PA.
3. National Library of Medicine, *Hazardous Substances Data Bank*, <http://toxnet.nlm.nih.gov/cgi-bin/sis/htmlgen?HSDB>.
4. United Nations Environmental Program, *Persistent Organic Pollutants*, <http://www.chem.unep.ch/pops/>.

5. Environmental Protection Agency, *Chemicals on Reporting Rules*, <http://www.epa.gov/opptintr/CORR>.

6. Environmental Protection Agency, *Integrated Risk Information System*, <http://www.epa.gov/iris/index.html>.

7. Lide, D. R., and Milne, G. W. A., Editors, *Handbook of Data on Organic Compounds, Third Edition*, CRC Press, Boca Raton, FL, 1993.

8. Macdonald, F., Editor, *Chapman & Hall/CRC Combined Chemical Dictionary*, <http://www.chemnetbase.com/scripts/ccdweb.exe>.

9. Linstrom, P. J., and Mallard, W. G., Editors, *NIST Chemistry WebBook*, NIST Standard Reference Database No. 69, July 2001, National Institute of Standards and Technology, Gaithersburg, MD 20899, <http://webbook.nist.gov>.

10. Thermodynamics Research Center, National Institute of Standards and Technology, *TRC Thermodynamic Tables*, <http://trc.nist.gov>.

11. O'Neil, M. J., Editor, *The Merck Index, Thirteenth Edition*, Merck & Co., Rahway, NJ, 2001.

12. Stevenson, R. M., and Malanowski, S., *Handbook of the Thermodynamics of Organic Compounds*, Elsevier, New York, 1987.

13. Riddick, J. A., Bunger, W. B., and Sakano, T. K., *Organic Solvents*, Fourth Edition, John Wiley & Sons, New York, 1986.

14. *Physical Constants of Hydrocarbon and Non-Hydrocarbon Compounds*, ASTM Data Series DS 4B, ASTM, Philadelphia, 1988.

15. *Beilstein Database*, <http://www.mdli.com/products/xfirebeilstein.html>.

16. *Landolt-Börnstein Numerical Data and Functional Relationships in Science and Technology*, <http://www.landolt-boernstein.com>.

17. Vargaftik, N.B., Vinogradov, Y. K., and Yargin, V. S., *Handbook of Physical Properties of Liquids and Gases, Third Edition*, Begell House, New York, 1996

18. Lide, D. R., and Kehiaian, H. V., *Handbook of Thermophysical and Thermochemical Data*, CRC Press, Boca Raton, FL, 1994.

19. Lide, D. R., Editor, *Properties of Organic Compounds*, <http://www.chemnetbase.com/scripts/pocweb.exe>.

No.	Name	Synonym	Mol. Form.	CAS RN	Mol. Wt.	Physical Form	mp/°C	bp/°C	den/ g cm⁻³	n_D	Solubility
1	Abate	Temephos	$C_{16}H_{20}O_6P_2S_3$	3383-96-8	466.469		30		1.32		
2	Abietic acid		$C_{20}H_{30}O_2$	514-10-3	302.451	mcl pl (al-w)	173.5	250^9			vs ace, bz, eth, EtOH
3	Abscisic acid		$C_{15}H_{20}O_4$	21293-29-8	264.318	cry (chl-peth)	160	sub 120			vs ace, eth, chl
4	Acacetin	5,7-Dihydroxy-2-(4-methoxyphenyl)-4H-1-benzopyran-4-one	$C_{16}H_{12}O_5$	480-44-4	284.263	ye nd (95% al)	263				vs EtOH
5	Acebutolol, (±)		$C_{18}H_{28}N_2O_4$	37517-30-9	336.426	cry	121				
6	Acedapsone		$C_{16}H_{16}N_2O_4S$	77-46-3	332.374	pa ye nd (eth) lf (dil al)	290				sl H₂O
7	Acenaphthene	1,2-Dihydroacenaphthylene	$C_{12}H_{10}$	83-32-9	154.207		93.4	279	1.222^{20}	1.6048^{95}	i H₂O; sl EtOH, chl; vs bz; s HOAc
8	Acenaphthylene	Acenaphthalene	$C_{12}H_8$	208-96-8	152.192		91.8	280	0.8987^{16}		i H₂O; vs EtOH, eth, bz; sl chl
9	1,2-Acenaphthylenedione		$C_{12}H_6O_2$	82-86-0	182.175	ye nd (HOAc)	261	sub	1.4800^{20}		i H₂O; sl EtOH, bz, HOAc; s lig
10	Acenocoumarol	Nicoumalone	$C_{19}H_{15}NO_6$	152-72-7	353.325	cry (ace aq)	198				i H₂O
11	Acephate	Phosphoramidothioic acid, acetyl-, O,S-dimethyl ester	$C_4H_{10}NO_3PS$	30560-19-1	183.166		88		1.35^{20}		
12	Acepromazine		$C_{19}H_{22}N_2OS$	61-00-7	326.455	oran oil		$230^{0.5}$			
13	Acesulfame		$C_4H_5NO_4S$	33665-90-6	163.153	nd (bz)	123.2				s bz, chl
14	Acetaldehyde	Ethanal	C_2H_4O	75-07-0	44.052	vol liq or gas	-123.37	20.1	0.7834^{18}	1.3316^{20}	msc H₂O, EtOH, eth, bz; sl chl
15	Acetaldehyde phenylhydrazone		$C_8H_{10}N_2$	935-07-9	134.178		99.5	150^{40}, 135^{21}			vs EtOH
16	Acetaldoxime	Acetaldehyde oxime	C_2H_5NO	107-29-9	59.067	nd	45	115	0.9656^{20}	1.4264^{20}	s H₂O, chl; msc EtOH, eth
17	Acetamide	Ethanamide	C_2H_5NO	60-35-5	59.067	trg mcl (al-eth)	80.16	222.0	0.9986^{85}	1.4278	vs H₂O, EtOH
18	Acetanilide	N-Phenylacetamide	C_8H_9NO	103-84-4	135.163		114.3	304	1.2190^{15}		sl H₂O; vs EtOH, ace; s eth, s bz, tol
19	Acetazolamide	N-[5-(Aminosulfonyl)-1,3,4-thiadiazol-2-yl]acetamide	$C_4H_6N_4O_3S_2$	59-66-5	222.246		260.5				sl H₂O
20	Acethion		$C_8H_{17}O_4PS_2$	919-54-0	272.322	liq		$137^{1.5}$	1.18^{20}		
21	Acetic acid	Ethanoic acid	$C_2H_4O_2$	64-19-7	60.052		16.64	117.9	1.0446^{25}	1.3720^{20}	msc H₂O, EtOH, eth, ace, bz; s chl, CS₂
22	Acetic acid, 2-phenylhydrazide		$C_8H_{10}N_2O$	114-83-0	150.177	hex pr (eth)	130.0				vs H₂O, EtOH; sl eth, chl, tfa; s bz
23	Acetic anhydride		$C_4H_6O_3$	108-24-7	102.089	liq	-74.1	139.5	1.082^{20}	1.3901^{20}	vs H₂O; s EtOH, bz; msc eth; sl ctc
24	Acetoacetanilide		$C_{10}H_{11}NO_2$	102-01-2	177.200	pr or nd (bz or lig)	86				sl H₂O; s EtOH, eth, bz, chl, acid, lig
25	Acetoacetic acid		$C_4H_6O_3$	541-50-4	102.089	cry (eth)	36.5	dec 100			vs H₂O, eth, EtOH
26	2-Acetoacetoxyethyl methacrylate	2-(Methacryloyloxy)ethyl acetoacetate	$C_{10}H_{14}O_5$	21282-97-3	214.215	liq		$100^{0.8}$	1.122	1.4560^{20}	
27	Acetochlor		$C_{14}H_{20}ClNO_2$	34256-82-1	269.768	ye liq		$134^{0.4}$		1.5272^{20}	sl H₂O
28	Acetohexamide		$C_{15}H_{20}N_2O_4S$	968-81-0	324.396	cry (EtOH aq)	188				i H₂O, eth; sl EtOH, chl; s py
29	Acetohydrazide		$C_2H_6N_2O$	1068-57-1	74.081		67	137^{25}			s H₂O, EtOH; sl eth
30	Acetohydroxamic acid	N-Hydroxyacetamide	$C_2H_5NO_2$	546-88-3	75.067	hyg cry	90				
31	1-Acetonaphthone		$C_{12}H_{10}O$	941-98-0	170.206		34	297	1.1171^{21}	1.6280^{22}	i H₂O; s EtOH, eth, ace, chl
32	2-Acetonaphthone		$C_{12}H_{10}O$	93-08-3	170.206	nd (lig, dil al)	56	302			sl EtOH, ctc
33	Acetone	2-Propanone	C_3H_6O	67-64-1	58.079	liq	-94.7	56.05	0.7845^{25}	1.3588^{20}	msc H₂O, EtOH, eth, ace, bz, chl
34	Acetone cyanohydrin		C_4H_7NO	75-86-5	85.105		-19	82^{23}	0.932^{19}	1.3992^{20}	vs H₂O, EtOH, eth; s ace, bz, chl; i peth
35	Acetone (2,4-dinitrophenyl) hydrazone		$C_9H_{10}N_4O_4$	1567-89-1	238.200	ye nd or pl (al)	128				i H₂O; s EtOH, eth, bz, chl, AcOEt
36	Acetone (1-methylethylidene) hydrazone	Dimethyl ketazine	$C_6H_{12}N_2$	627-70-3	112.172	liq	-12.5	133	0.8390^{20}	1.4535^{20}	msc H₂O, EtOH, eth; s ace
37	Acetone thiosemicarbazide		$C_4H_9N_3S$	1752-30-3	131.199	ye cry	176				s ace

Abate

Abietic acid

Abscisic acid

Acacetin

Acebutolol, (±)

Acedapsone

Acenaphthene

Acenaphthylene

1,2-Acenaphthylenedione

Acenocoumarol

Acephate

Acepromazine

Acesulfame

Acetaldehyde

Acetaldehyde phenylhydrazone

Acetaldoxime

Acetamide

Acetanilide

Acetazolamide

Acethion

Acetic acid

Acetic acid, 2-phenylhydrazide

Acetic anhydride

Acetoacetanilide

Acetoacetic acid

2-Acetoacetoxyethyl methacrylate

Acetochlor

Acetohexamide

Acetohydrazide

Acetohydroxamic acid

1-Acetonaphthone

2-Acetonaphthone

Acetone

Acetone cyanohydrin

Acetone (2,4-dinitrophenyl)hydrazone

Acetone (1-methylethylidene)hydrazone

Acetone thiosemicarbazide

No.	Name	Synonym	Mol. Form.	CAS RN	Mol. Wt.	Physical Form	mp/°C	bp/°C	den/g cm⁻³	n_D	Solubility
38	Acetonitrile	Methyl cyanide	C_2H_3N	75-05-8	41.052	liq	-43.82	81.65	0.7857[20]	1.3442[30]	msc H_2O, EtOH, eth, ace, bz, ctc
39	Acetophenone	Methyl phenyl ketone	C_8H_8O	98-86-2	120.149	mcl pr or pl	20.5	202	1.0281[20]	1.5372[20]	sl H_2O; s EtOH, eth, ace, bz, con sulf, chl
40	Acetophenone azine	Methylphenyl ketazine	$C_{16}H_{16}N_2$	729-43-1	236.311		120				
41	Acetoxon	Acetophos	$C_8H_{17}O_5PS$	2425-25-4	256.257	liq		73[0.005]			
42	N-Acetylacetamide		$C_4H_7NO_2$	625-77-4	101.105	nd (eth)	79	223.5			s H_2O, EtOH, eth, chl, lig
43	N-Acetyl-L-alanine		$C_5H_9NO_3$	97-69-8	131.130		125				
44	4-(Acetylamino)benzenesulfonyl chloride	Acetylsulfanilyl chloride	$C_8H_8ClNO_3S$	121-60-8	233.673	nd (bz), pr (bz-chl)	149				vs EtOH, eth; s bz, chl
45	2-(Acetylamino)benzoic acid		$C_9H_9NO_3$	89-52-1	179.172	nd (HOAc)	187.5				sl H_2O; s EtOH; vs eth, ace, bz, HOAc
46	4-(Acetylamino)benzoic acid		$C_9H_9NO_3$	556-08-1	179.172	nd (HOAc)	256.5				i H_2O; s EtOH; sl eth, tfa
47	2-(Acetylamino)-2-deoxy-D-glucose	N-Acetyl-D-glucosamine	$C_8H_{15}NO_6$	7512-17-6	221.208		205				
48	2-(Acetylamino)-2-deoxy-D-mannose	N-Acetyl-D-mannosamine	$C_8H_{15}NO_6$	3615-17-6	221.208	cry (ace aq)	128				dec alk
49	2-(Acetylamino)fluorene		$C_{15}H_{13}NO$	53-96-3	223.270	cry (dil al)	193				i H_2O; s EtOH, eth, HOAc
50	4-(Acetylamino)fluorene		$C_{15}H_{13}NO$	28322-02-3	223.270	br cry (bz)	200				
51	6-(Acetylamino)hexanoic acid	ε-Acetamidocaproic acid	$C_8H_{15}NO_3$	57-08-9	173.210	cry (ace)	104.5				
52	4-Acetylanisole		$C_9H_{10}O_2$	100-06-1	150.174	pl (peth)	38.5	258	1.0818[41]	1.547[41]	sl H_2O; s EtOH, eth, ace, chl
53	2-Acetylbenzoic acid		$C_9H_8O_3$	577-56-0	164.158	nd (w), pr (bz)	114.5	111[2]			vs H_2O, eth, EtOH
54	3-Acetylbenzoic acid		$C_9H_8O_3$	586-42-5	164.158		172	111[2]			s H_2O; msc EtOH
55	4-Acetylbenzoic acid		$C_9H_8O_3$	586-89-0	164.158	nd (w)	208	sub			vs H_2O
56	Acetyl benzoylperoxide	Acetozone	$C_9H_8O_4$	644-31-5	180.158	wh nd (lig)	37	130[19]			vs eth
57	Acetyl bromide	Ethanoyl bromide	C_2H_3BrO	506-96-7	122.948	liq	-96	76	1.6625[16]	1.4486[20]	msc eth, bz, chl; s ace
58	Acetyl chloride	Ethanoyl chloride	C_2H_3ClO	75-36-5	78.497	liq	-112.8	50.7	1.1051[20]	1.3886[20]	msc eth, ace, bz, chl; s ctc
59	Acetylcholine bromide		$C_7H_{16}BrNO_2$	66-23-9	226.112	hyg cry	146				vs H_2O
60	Acetylcholine chloride		$C_7H_{16}ClNO_2$	60-31-1	181.661		150				s H_2O, EtOH; i eth
61	Acetylcholine iodide		$C_7H_{16}INO_2$	2260-50-6	273.112	hyg	163				
62	2-Acetylcyclohexanone		$C_8H_{12}O_2$	874-23-7	140.180		-11	112[18], 101[14]	1.0782[25]	1.5138[20]	s ctc
63	2-Acetylcyclopentanone		$C_7H_{10}O_2$	1670-46-8	126.153			73[20]	1.0431[25]	1.4906[20]	
64	N-Acetyl-L-cysteine	Acetylcysteine	$C_5H_9NO_3S$	616-91-1	163.195	cry (w)	109.5				
65	3-Acetyldihydro-2(3H)-furanone	α-Acetylbutyrolactone	$C_6H_8O_3$	517-23-7	128.126			107[5]	1.1846[20]	1.4585[20]	vs H_2O
66	1-Acetyl-2,5-dihydroxybenzene	2,5-Dihydroxyacetophenone	$C_8H_8O_3$	490-78-8	152.148	ye grn nd (dil al or w)	205.3				sl H_2O; s eth, bz; s EtOH
67	Acetylene	Ethyne	C_2H_2	74-86-2	26.037	col gas	-80.7 (triple point)	-84.7 sp	0.377[25] (p>1 atm)		sl H_2O, EtOH, CS_2; s ace, bz, chl
68	N-Acetylethanolamine		$C_4H_9NO_2$	142-26-7	103.120		63.5	168[8]	1.1079[25]	1.4674[20]	msc H_2O; s ace; sl bz, lig
69	Acetyl fluoride	Ethanoyl fluoride	C_2H_3FO	557-99-3	62.042	vol liq or gas	-84	20.8	1.032[25]		msc EtOH, eth; s bz, chl; sl CS_2
70	N-Acetylglutamic acid		$C_7H_{11}NO_5$	1188-37-0	189.166	pr (w)	199				s H_2O, EtOH
71	N-Acetylglycine	Aceturic acid	$C_4H_7NO_3$	543-24-8	117.104	lo nd (w, MeOH)	206				vs H_2O, ace, EtOH
72	trans-1-Acetyl-4-hydroxy-L-proline	Oxaceprol	$C_7H_{11}NO_4$	33996-33-7	173.167	cry (Ac)	132				vs H_2O, MeOH
73	1-Acetyl-1H-imidazole		$C_5H_6N_2O$	2466-76-4	110.114		104.5				sl H_2O; s EtOH, eth, chl, THF
74	Acetyl iodide	Ethanoyl iodide	C_2H_3IO	507-02-8	169.948			108	2.0673[20]	1.5491[20]	vs eth
75	Acetyl isothiocyanate		C_3H_3NOS	13250-46-9	101.127			132.5	1.1523[13]	1.5231[18]	s eth, CS_2
76	N6-Acetyl-L-lysine		$C_8H_{16}N_2O_3$	692-04-6	188.224		265 dec				
77	N-Acetyl-DL-methionine		$C_7H_{13}NO_3S$	1115-47-5	191.248		114.5				
78	N-Acetyl-L-methionine	Methionamine	$C_7H_{13}NO_3S$	65-82-7	191.248		105.5				
79	1-Acetyl-17-methoxyaspidospermidine	Aspidospermine	$C_{22}H_{30}N_2O_2$	466-49-9	354.485	nd or pr (al) nd (peth)	208	220[2]			sl H_2O, eth; s EtOH, bz, chl
80	N-Acetyl-N-methylacetamide		$C_5H_9NO_2$	1113-68-4	115.131	liq	-25	195; 114.5[61]	1.0663[25]	1.4502[25]	msc H_2O; i eth
81	1-Acetyl-3-methylpiperidine		$C_8H_{15}NO$	4593-16-2	141.211	liq	-13.6	239	0.9684[25]	1.4731[25]	vs H_2O
82	3-Acetyl-6-methyl-2H-pyran-2,4(3H)-dione	Dehydroacetic acid	$C_8H_8O_4$	520-45-6	168.148		109	270			vs H_2O, eth; sl EtOH, chl

Acetonitrile

Acetophenone

Acetophenone azine

Acetoxon

N-Acetylacetamide

N-Acetyl-*L*-alanine

4-(Acetylamino)benzenesulfonyl chloride

2-(Acetylamino)benzoic acid

4-(Acetylamino)benzoic acid

2-(Acetylamino)-2-deoxy-*D*-glucose

2-(Acetylamino)-2-deoxy-*D*-mannose

2-(Acetylamino)fluorene

4-(Acetylamino)fluorene

6-(Acetylamino)hexanoic acid

4-Acetylanisole

2-Acetylbenzoic acid

3-Acetylbenzoic acid

4-Acetylbenzoic acid

Acetyl benzoylperoxide

Acetyl bromide

Acetyl chloride

Acetylcholine bromide

Acetylcholine chloride

Acetylcholine iodide

2-Acetylcyclohexanone

2-Acetylcyclopentanone

N-Acetyl-*L*-cysteine

3-Acetyldihydro-2(3*H*)-furanone

1-Acetyl-2,5-dihydroxybenzene

Acetylene

N-Acetylethanolamine

Acetyl fluoride

N-Acetylglutamic acid

N-Acetylglycine

trans-1-Acetyl-4-hydroxy-*L*-proline

1-Acetyl-1*H*-imidazole

Acetyl iodide

Acetyl isothiocyanate

*N*6-Acetyl-*L*-lysine

N-Acetyl-*DL*-methionine

N-Acetyl-*L*-methionine

1-Acetyl-17-methoxyaspidospermidine

N-Acetyl-*N*-methylacetamide

1-Acetyl-3-methylpiperidine

3-Acetyl-6-methyl-2*H*-pyran-2,4(3*H*)-dione

No.	Name	Synonym	Mol. Form.	CAS RN	Mol. Wt.	Physical Form	mp/°C	bp/°C	den/ g cm⁻³	n_D	Solubility
83	4-Acetylmorpholine		$C_6H_{11}NO_2$	1696-20-4	129.157		14.5	152[50], 118[12]	1.1145[20]	1.4827[20]	msc H_2O; s EtOH, ace, ctc
84	N-Acetylneuraminic acid	Aceneuramic acid	$C_{11}H_{19}NO_9$	131-48-6	309.271		186				
85	Acetyl nitrate		$C_2H_3NO_4$	591-09-3	105.050			exp 60; 22[70]	1.24[15]		
86	2-(Acetyloxy)benzoic acid	Acetylsalicylic acid	$C_9H_8O_4$	50-78-2	180.158	nd (w), mcl tab (w)	135				s H_2O, eth, chl; vs EtOH; sl bz
87	4-(Acetyloxy)benzoic acid		$C_9H_8O_4$	2345-34-8	180.158		188.5				
88	2-(Acetyloxy)-5-bromobenzoic acid	5-Bromoacetylsalicylic acid	$C_9H_7BrO_4$	1503-53-3	259.054	nd (al)	60				i H_2O; vs EtOH, eth
89	4-(Acetyloxy)-3-methoxybenzaldehyde		$C_{10}H_{10}O_4$	881-68-5	194.184		78				sl H_2O; vs EtOH, eth
90	2-(Acetyloxy)-1-phenylethanone		$C_{10}H_{10}O_3$	2243-35-8	178.184	orth pl	49	270	1.1169[65]	1.5036[65]	i H_2O; vs EtOH, eth, chl; sl bz, lig
91	1-(Acetyloxy)-2-propanone	Acetoxyacetone	$C_5H_8O_3$	592-20-1	116.116			171; 63[11]	1.0757[20]	1.4141[20]	vs H_2O, eth, EtOH
92	(Acetyloxy)tributylstannane	Tributyltin acetate	$C_{14}H_{30}O_2Sn$	56-36-0	349.097		84.7				
93	(Acetyloxy)triphenylstannane	Triphenyltin acetate	$C_{20}H_{18}O_2Sn$	900-95-8	409.066		121.5				
94	4-Acetylphenyl acetate		$C_{10}H_{10}O_3$	13031-43-1	178.184						s ctc, CS_2
95	N-Acetyl-L-phenylalanine		$C_{11}H_{13}NO_3$	2018-61-3	207.226		173.5				s EtOH
96	N-Acetyl-L-phenylalanine, ethyl ester		$C_{13}H_{17}NO_3$	2361-96-8	235.279	cry (EtOH aq)	93				
97	N-Acetyl-L-phenylalanine, methyl ester		$C_{12}H_{15}NO_3$	3618-96-0	221.252	nd (peth) or visc oil (chl)	91				
98	Acetyl phosphate		$C_2H_5O_5P$	590-54-5	140.032	unstab in soln					
99	1-Acetylpiperidine		$C_7H_{13}NO$	618-42-8	127.184	liq	-13.4	226.5	1.011[9]	1.4790[25]	vs H_2O, EtOH
100	1-Acetyl-4-piperidinone		$C_7H_{11}NO_2$	32161-06-1	141.168			218; 124[0.2]	1.146[25]	1.5026[20]	
101	3-Acetylpyridine adenine dinucleotide	3-Acetyl NAD	$C_{22}H_{28}N_6O_{14}P_2$	86-08-8	662.436	solid					
102	4-Acetylthioanisole		$C_9H_{10}OS$	1778-09-2	166.239		81.5				
103	Acetyl thiocholine iodide		$C_7H_{16}INOS$	1866-15-5	289.177		205				
104	N-Acetyl-L-tryptophan		$C_{13}H_{14}N_2O_3$	1218-34-4	246.261	nd (dil MeOH)	189.5				s H_2O, EtOH, alk
105	N-Acetyl-L-tyrosine		$C_{11}H_{13}NO_4$	537-55-3	223.226	cry (w); pl (diox)	153				
106	N-Acetyl-L-tyrosine ethyl ester		$C_{13}H_{17}NO_4$	840-97-1	251.279		80.5				
107	N-Acetyl-L-valine		$C_7H_{13}NO_3$	96-81-1	159.183		164				
108	Acid Fuchsin	Fuchsin, acid	$C_{20}H_{17}N_3Na_2O_9S_3$	3244-88-0	585.539						sl H_2O, EtOH
109	Acifluorfen	5-[2-Chloro-4-(trifluoromethyl) phenoxy]-2-nitrobenzoic acid	$C_{14}H_7ClF_3NO_5$	50594-66-6	361.658		150				
110	Aconine		$C_{25}H_{41}NO_9$	509-20-6	499.596	amor	132				s H_2O, EtOH, chl; sl eth, lig
111	Aconitine		$C_{34}H_{47}NO_{11}$	302-27-2	645.737	orth lf	204				vs bz, EtOH, chl
112	9-Acridinamine	Aminacrine	$C_{13}H_{10}N_2$	90-45-9	194.231	ye nd (ace or al)	241				s EtOH, ace; sl DMSO; vs dil HCl
113	Acridine	Dibenzo[b,e]pyridine	$C_{13}H_9N$	260-94-6	179.217	orth nd or pr (al)	106(form a); 110(form b)	344.86	1.005[20]		i H_2O; sl ctc; vs EtOH, eth, bz
114	3,6-Acridinediamine	Proflavine	$C_{13}H_{11}N_3$	92-62-6	209.246	ye nd (al or w)	285				s H_2O; vs EtOH; sl eth, bz
115	9(10H)-Acridinone		$C_{13}H_9NO$	578-95-0	195.216	ye lf (al)	>300				i H_2O, eth, bz; sl EtOH; s HOAc, alk
116	Acrolein	2-Propenal	C_3H_4O	107-02-8	56.063	liq	-87.7	52.6	0.840[20]	1.4017[20]	vs H_2O; s EtOH, eth, ace; sl chl
117	Acrylamide	2-Propenamide	C_3H_5NO	79-06-1	71.078	lf (bz)	84.5	192.6			vs H_2O, chl; s EtOH, eth, ace
118	Acrylic acid	2-Propenoic acid	$C_3H_4O_2$	79-10-7	72.063		12.5	141	1.0511[20]	1.4224[20]	msc H_2O, EtOH, eth; s ace, bz, ctc
119	Acrylonitrile	Propenenitrile	C_3H_3N	107-13-1	53.063	liq	-83.48	77.3	0.8007[25]	1.3911[20]	s H_2O; vs ace, bz, eth, EtOH
120	Acyclovir		$C_8H_{11}N_5O_3$	59277-89-3	225.205	cry (EtOH)	225				
121	Adenine	1H-Purin-6-amine	$C_5H_5N_5$	73-24-5	135.128	orth nd (+3w)	360 dec	sub 220			s H_2O; sl EtOH; i eth, chl
122	Adenosine	β-D-Ribofuranoside, adenine-9	$C_{10}H_{13}N_5O_4$	58-61-7	267.242	n(w+3/2)	235.5				sl H_2O; i EtOH
123	Adenosine cyclic 3',5'-(hydrogen phosphate)	cAMP	$C_{10}H_{12}N_5O_6P$	60-92-4	329.206	cry	219				
124	Adenosine 3',5'-diphosphate	3'-Adenylic acid, 5'-(dihydrogen phosphate)	$C_{10}H_{15}N_5O_{10}P_2$	1053-73-2	427.202	amor pow					

4-Acetylmorpholine

N-Acetylneuraminic acid

Acetyl nitrate

2-(Acetyloxy)benzoic acid

4-(Acetyloxy)benzoic acid

2-(Acetyloxy)-5-bromobenzoic acid

4-(Acetyloxy)-3-methoxybenzaldehyde

2-(Acetyloxy)-1-phenylethanone

1-(Acetyloxy)-2-propanone

(Acetyloxy)tributylstannane

(Acetyloxy)triphenylstannane

4-Acetylphenyl acetate

N-Acetyl-*L*-phenylalanine

N-Acetyl-*L*-phenylalanine, ethyl ester

N-Acetyl-*L*-phenylalanine, methyl ester

Acetyl phosphate

1-Acetylpiperidine

1-Acetyl-4-piperidinone

3-Acetylpyridine adenine dinucleotide

4-Acetylthioanisole

Acetyl thiocholine iodide

N-Acetyl-*L*-tryptophan

N-Acetyl-*L*-tyrosine

N-Acetyl-*L*-tyrosine ethyl ester

N-Acetyl-*L*-valine

Acid Fuchsin

Acifluorfen

Aconine

Aconitine

9-Acridinamine

Acridine

3,6-Acridinediamine

9(10*H*)-Acridinone

Acrolein

Acrylamide

Acrylic acid

Acrylonitrile

Acyclovir

Adenine

Adenosine

Adenosine cyclic 3′,5′-(hydrogen phosphate)

Adenosine 3′,5′-diphosphate

No.	Name	Synonym	Mol. Form.	CAS RN	Mol. Wt.	Physical Form	mp/°C	bp/°C	den/ g cm⁻³	n_D	Solubility
125	Adenosine 5'-methylenediphosphonate	Adenosine, 5'-[hydrogen (phosphonomethyl) phosphonate]	$C_{11}H_{17}N_5O_9P_2$	3768-14-7	425.229	cry (w)	204				s H_2O
126	Adenosine 3'-phosphate	3'-Adenylic acid	$C_{10}H_{14}N_5O_7P$	84-21-9	347.222	col nd	195 dec				
127	Adenosine 5'-triphosphate	ATP	$C_{10}H_{16}N_5O_{13}P_3$	56-65-5	507.181		144 dec				
128	S-Adenosyl-L-homocysteine		$C_{14}H_{20}N_6O_5S$	979-92-0	384.411		210 dec				
129	5'-Adenylic acid	Adenosine 5'-monophosphate	$C_{10}H_{14}N_5O_7P$	61-19-8	347.222		195 dec				vs H_2O; s EtOH, 10% HCl
130	Adipamic acid		$C_6H_{11}NO_3$	334-25-8	145.156	nd (w)	161.5				
131	Adiphenine hydrochloride		$C_{20}H_{26}ClNO_2$	50-42-0	347.879	cry	113.5				vs H_2O; sl EtOH, eth
132	Adipic acid	1,6-Hexanedioic acid	$C_6H_{10}O_4$	124-04-9	146.141	mcl pr (w, ace, lig)	152.5	337.5	1.360²⁵		sl H_2O; vs EtOH; s eth; i HOAc, lig
133	Adiponitrile	Hexanedinitrile	$C_6H_8N_2$	111-69-3	108.141	nd (eth)	1	295	0.9676²⁰	1.4380²⁰	sl H_2O, eth; s chl, EtOH
134	Adrenalone		$C_9H_{11}NO_3$	99-45-6	181.188	nd	235 dec				sl H_2O, EtOH, eth
135	Affinin	N-(2-Methylpropyl)-2,6,8-decatrienamide	$C_{14}H_{23}NO$	25394-57-4	221.339	ye oil	23	162⁰·⁵		1.5134²⁵	i H_2O
136	Aflatoxin B1		$C_{17}H_{12}O_6$	1162-65-8	312.273	cry	268				
137	Aflatoxin B2		$C_{17}H_{14}O_6$	7220-81-7	314.289		287.5				
138	Aflatoxin G1		$C_{17}H_{12}O_7$	1165-39-5	328.273	cry	245				
139	Agaritine	L-Glutamic acid, 5-[2-[4-(hydroxymethyl)phenyl] hydrazide]	$C_{12}H_{17}N_3O_4$	2757-90-6	267.281	cry (dil al)	207 dec				vs H_2O
140	Ajmalan-17,21-diol, (17R,21α)	Ajmaline	$C_{20}H_{26}N_2O_2$	4360-12-7	326.432	pl (+3.5w) (aq AcOEt)	206				i H_2O; s EtOH, chl; sl eth, bz
141	Alachlor		$C_{14}H_{20}ClNO_2$	15972-60-8	269.768		40	100⁰·⁰²	1.133²⁵		
142	DL-Alanine	DL-2-Aminopropanoic acid	$C_3H_7NO_2$	302-72-7	89.094	orth pr or nd (w)	300 dec	sub 250	1.424²⁵		s H_2O; vs EtOH
143	D-Alanine	2-Aminopropanoic acid, (R)	$C_3H_7NO_2$	338-69-2	89.094	nd (w, al)	314 dec	sub			s H_2O; sl EtOH; i eth
144	L-Alanine	2-Aminopropanoic acid, (S)	$C_3H_7NO_2$	56-41-7	89.094	orth (w)	297 dec	sub 250	1.432²²		s H_2O; sl EtOH, py; i eth, ace
145	β-Alanine	3-Aminopropanoic acid	$C_3H_7NO_2$	107-95-9	89.094	nd, orth pr (al)	200 dec		1.437¹⁹		s H_2O; sl EtOH; i eth, ace
146	Alantolactone		$C_{15}H_{20}O_2$	546-43-0	232.319	nd	76	275			vs bz, eth, EtOH, chl
147	Aldicarb		$C_7H_{14}N_2O_2S$	116-06-3	190.263		99		1.195²⁵		
148	Aldosterone		$C_{21}H_{28}O_5$	52-39-1	360.444	cry (HOAc)	166.5				
149	Aldoxycarb S,S-dioxide		$C_7H_{14}N_2O_4S$	1646-88-4	222.262	cry	141				sl H_2O
150	Aldrin		$C_{12}H_8Cl_6$	309-00-2	364.910		104				i H_2O; s EtOH, eth, ace, bz
151	Alizarin	1,2-Dihydroxy-9,10-anthracenedione	$C_{14}H_8O_4$	72-48-0	240.212	oran or red tcl nd or pr (al)	289.5				sl H_2O; s EtOH, eth, ace, bz; i chl
152	Alizarin Red S	Sodium alizarinesulfonate	$C_{14}H_7NaO_7S$	130-22-3	342.257						vs H_2O; s EtOH
153	Alizarin Yellow R		$C_{13}H_9N_3O_5$	2243-76-7	287.227	oran-br nd (dil HOAc)	253 dec				vs H_2O, EtOH
154	Alizurol purple	1-Hydroxy-4-[(4-methylphenyl) amino]-9,10-anthracenedione	$C_{21}H_{15}NO_3$	81-48-1	329.349	flat viol nd					s H_2SO_4
155	Alkannin		$C_{16}H_{16}O_5$	23444-65-7	288.295	br-red pr (bz)	149	sub 140			vs EtOH
156	Allantoic acid	Bis[(aminocarbonyl)amino] acetic acid	$C_4H_8N_4O_4$	99-16-1	176.132	nd	170 dec				sl H_2O, os, dil acid
157	Allantoin		$C_4H_6N_4O_3$	97-59-6	158.116	mcl pl or	239				sl H_2O; s EtOH, NaOH; i eth, MeOH
158	Allene		C_3H_4	463-49-0	40.064	col gas	-136.6	-34.4	0.584²⁵ (p>1 atm)	1.4168	vs bz, peth
159	Allethrin		$C_{19}H_{26}O_3$	584-79-2	302.407				1.010²⁰		
160	Allicin		$C_6H_{10}OS_2$	539-86-6	162.272			dec	1.112²⁰	1.561²⁰	vs H_2O
161	Allopregnane-3β,21-diol-11,20-dione		$C_{21}H_{32}O_4$	566-02-9	348.477	cry (aq, ac +w) nd (bz, ac)	190				
162	Allopregnan-20β-ol-3-one	5α-Pregnan-20β-ol-3-one	$C_{21}H_{34}O_2$	516-58-5	318.494		185				
163	Allopurinol	1,5-Dihydro-4H-pyrazolo[3,4-d] pyrimidin-4-one	$C_5H_4N_4O$	315-30-0	136.112	cry	350				
164	D-Allose		$C_6H_{12}O_6$	2595-97-3	180.155	cry (w)	128				vs H_2O
165	Alloxanic acid		$C_4H_4N_2O_5$	470-44-0	160.085	tcl pr (eth)	162 dec				vs H_2O, EtOH
166	Alloxantin		$C_8H_6N_4O_8$	76-24-4	286.156	orth pr (w+2)	254 dec				sl H_2O, EtOH, eth

Adenosine 5'-methylenediphosphonate

Adenosine 3'-phosphate

Adenosine 5'-triphosphate

S-Adenosyl-L-homocysteine

5'-Adenylic acid

Adipamic acid

Adiphenine hydrochloride

Adipic acid

Adiponitrile

Adrenalone

Affinin

Aflatoxin B1

Aflatoxin B2

Aflatoxin G1

Agaritine

Ajmalan-17,21-diol, (17R,21α)

Alachlor

DL-Alanine

D-Alanine

L-Alanine

β-Alanine

Alantolactone

Aldicarb

Aldosterone

Aldoxycarb S,S-dioxide

Aldrin

Alizarin

Alizarin Red S

Alizarin Yellow R

Alizurol purple

Alkannin

Allantoic acid

Allantoin

Allene

Allethrin

Allicin

Allopregnane-3β,21-diol-11,20-dione

Allopregnan-20β-ol-3-one

Allopurinol

D-Allose

Alloxanic acid

Alloxantin

No.	Name	Synonym	Mol. Form.	CAS RN	Mol. Wt.	Physical Form	mp/°C	bp/°C	den/g cm⁻³	n_D	Solubility
167	Allyl acetate		$C_5H_8O_2$	591-87-7	100.117			103.5	0.9275[20]	1.4049[20]	sl H₂O; s ace; msc EtOH, eth
168	Allyl acetoacetate		$C_7H_{10}O_3$	1118-84-9	142.152	liq	-85	196; 66.5[14]	1.0366[20]	1.4398[20]	s H₂O, lig; msc EtOH, bz
169	Allyl acrylate		$C_6H_8O_2$	999-55-3	112.127			121	0.9441[20]	1.4320[20]	sl H₂O; s EtOH, eth, acid
170	Allyl alcohol	2-Propen-1-ol	C_3H_6O	107-18-6	58.079	liq	-129	97.0	0.8540[20]	1.4135[20]	msc H₂O, EtOH, eth; s chl
171	Allylamine	2-Propen-1-amine	C_3H_7N	107-11-9	57.095	liq	-88.2	53.3	0.758[20]	1.4205[20]	msc H₂O, EtOH, eth; s chl
172	N-Allylaniline	Allylphenylamine	$C_9H_{11}N$	589-09-3	133.190			219; 106[12]	0.9736[25]	1.563[20]	sl H₂O; s EtOH, ace; msc eth
173	Allylbenzene	2-Propenylbenzene	C_9H_{10}	300-57-2	118.175	liq	-40	156	0.8920[20]	1.5131[20]	i H₂O; s EtOH, eth, bz, ctc
174	α-Allylbenzenemethanol		$C_{10}H_{12}O$	936-58-3	148.201			228.5	1.004[18]	1.5289[21]	
175	Allyl benzoate		$C_{10}H_{10}O_2$	583-04-0	162.185				1.0569[15]	1.5178[20]	i H₂O; s EtOH, eth, ace, MeOH
176	Allyl butanoate		$C_7H_{12}O_2$	2051-78-7	128.169			142; 44.5[15]	0.9017[20]	1.4158[20]	i H₂O; msc EtOH, eth; sl ctc
177	Allyl carbamate		$C_4H_7NO_2$	2114-11-6	101.105						sl ctc
178	Allylchlorodimethylsilane		$C_5H_{11}ClSi$	4028-23-3	134.680			111	0.8964[20]	1.4195[20]	
179	Allyl chloroformate		$C_4H_5ClO_2$	2937-50-0	120.535	hyg liq		109.5	1.136	1.4220[20]	
180	Allyl trans-cinnamate	Allyl trans-3-phenyl-2-propenoate	$C_{12}H_{12}O_2$	1866-31-5	188.222			dec 268; 163[17]	1.048[23]	1.530[20]	i H₂O; vs EtOH; msc eth; sl ctc
181	1-Allylcyclohexanol		$C_9H_{16}O$	1123-34-8	140.222			190	0.9341[22]	1.4756[22]	
182	1-Allylcyclohexene	1-(2-Propenyl)cyclohexene	C_9H_{14}	13511-13-2	122.207	liq		156			
183	Allylcyclopentane		C_8H_{14}	3524-75-2	110.197	liq	-110.7	125	0.793[25]	1.4412[20]	s chl
184	Allyldiethoxymethylsilane		$C_8H_{18}O_2Si$	18388-45-9	174.314			155	0.8572[25]	1.4104[20]	
185	Allyldiethylamine	N,N-Diethyl-2-propen-1-amine	$C_7H_{15}N$	5666-17-1	113.201			110	0.7477[25]	1.4209[20]	
186	Allyldimethylamine	N,N-Dimethyl-2-propen-1-amine	$C_5H_{11}N$	2155-94-4	85.148			63.5	0.7094[25]	1.4010[20]	
187	Allyl ethyl ether		$C_5H_{10}O$	557-31-3	86.132			67.6	0.7651[20]	1.3881[20]	i H₂O; msc EtOH, eth; s ace
188	Allyl formate		$C_4H_6O_2$	1838-59-1	86.090			83.6	0.9460[20]		sl H₂O; s EtOH; msc eth
189	Allyl 2-furancarboxylate	Allyl 2-furanoate	$C_8H_8O_3$	4208-49-5	152.148			207.5	1.115[25]	1.4945[20]	s eth, ace; sl ctc
190	Allyl glycidyl ether		$C_6H_{10}O_2$	106-92-3	114.142			154	0.9698[20]	1.4332[20]	
191	Allyl hexanoate		$C_9H_{16}O_2$	123-68-2	156.222			186	0.8869[20]		
192	Allyl (hydroxymethyl)carbamate		$C_5H_9NO_3$	24935-97-5	131.130	cry (tol)	57				
193	Allyl isocyanate		C_4H_5NO	1476-23-9	83.089			88			
194	Allyl isothiocyanate		C_4H_5NS	57-06-7	99.155	liq	-80	152	1.0126[20]	1.5306[20]	vs bz, eth, EtOH
195	Allyl methacrylate		$C_7H_{10}O_2$	96-05-9	126.153			67[50], 55[30]	0.9335[20]	1.4360[20]	
196	4-Allyl-2-methoxyphenol	Eugenol	$C_{10}H_{12}O_2$	97-53-0	164.201	liq	-7.5	253.2	1.0652[20]	1.5405[20]	i H₂O; msc EtOH, eth; s chl, HOAc, oils
197	4-Allyl-2-methoxyphenyl acetate	1,3,4-Eugenol acetate	$C_{12}H_{14}O_3$	93-28-7	206.237	pr (al)	30.5	281; 127[6]	1.0806[20]	1.5205[20]	i H₂O; s EtOH; sl ctc
198	Allyl 3-methylbutanoate		$C_8H_{14}O_2$	2835-39-4	142.196			154			
199	Allylmethyldichlorosilane		$C_4H_8Cl_2Si$	1873-92-3	155.099			119.5	1.0758[20]	1.4419[20]	
200	2-(Allyloxy)ethanol	Ethylene glycol monoallyl ether	$C_5H_{10}O_2$	111-45-5	102.132			158.5	0.9580[20]	1.4358[20]	msc H₂O; vs EtOH; s bz, ctc, MeOH
201	2-Allylphenol		$C_9H_{10}O$	1745-81-9	134.174	liq	-6	220	1.0246[15]	1.5181[20]	vs eth
202	4-Allylphenol	Chavicol	$C_9H_{10}O$	501-92-8	134.174		15.8	238	1.0203[15]	1.5441[18]	vs eth, EtOH, chl
203	Allyl phenyl ether		$C_9H_{10}O$	1746-13-0	134.174			191.7	0.9811[20]	1.5223[20]	i H₂O; s EtOH; msc eth; sl ctc
204	Allyl propanoate	2-Propenyl propanoate	$C_6H_{10}O_2$	2408-20-0	114.142			123	0.9140[20]	1.4105[20]	s EtOH, eth, ace
205	N-Allyl-2-propen-1-amine	Diallylamine	$C_6H_{11}N$	124-02-7	97.158			111		1.4387[20]	s EtOH, eth
206	Allyl propyl disulfide		$C_6H_{12}S_2$	2179-59-1	148.289			79[13]		1.5219[20]	
207	3-(Allylsulfinyl)-L-alanine, (S)	Alliin	$C_6H_{11}NO_3S$	556-27-4	177.221	nd (dil ac)	165				vs H₂O
208	Allylthiourea	Thiosinamine	$C_4H_8N_2S$	109-57-9	116.185	mcl or orth pr (w)	78		1.217[20]	1.5936[78]	s H₂O, EtOH; sl eth; i bz
209	Allyltrichlorosilane	Trichloro-2-propenylsilane	$C_3H_5Cl_3Si$	107-37-9	175.517		35	117.5	1.2011[20]	1.4460[20]	
210	Allyltriethoxysilane		$C_9H_{20}O_3Si$	2550-04-1	204.339			100[50], 82[28]	0.9030[20]	1.4072[20]	
211	Allyltrimethylsilane		$C_6H_{14}Si$	762-72-1	114.261			85	0.7158[25]	1.4074[20]	i H₂O
212	Allylurea		$C_4H_8N_2O$	557-11-9	100.119	nd (al)	85				msc H₂O, EtOH; sl eth, chl; i peth

Allyl acetate

Allyl acetoacetate

Allyl acrylate

Allyl alcohol

Allylamine

N-Allylaniline

Allylbenzene

α-Allylbenzenemethanol

Allyl benzoate

Allyl butanoate

Allyl carbamate

Allylchlorodimethylsilane

Allyl chloroformate

Allyl *trans*-cinnamate

1-Allylcyclohexanol

1-Allylcyclohexene

Allylcyclopentane

Allyldiethoxymethylsilane

Allyldiethylamine

Allyldimethylamine

Allyl ethyl ether

Allyl formate

Allyl 2-furancarboxylate

Allyl glycidyl ether

Allyl hexanoate

Allyl (hydroxymethyl)carbamate

Allyl isocyanate

Allyl isothiocyanate

Allyl methacrylate

4-Allyl-2-methoxyphenol

4-Allyl-2-methoxyphenyl acetate

Allyl 3-methylbutanoate

Allylmethyldichlorosilane

2-(Allyloxy)ethanol

2-Allylphenol

4-Allylphenol

Allyl phenyl ether

Allyl propanoate

N-Allyl-2-propen-1-amine

Allyl propyl disulfide

3-(Allylsulfinyl)-*L*-alanine, (*S*)

Allylthiourea

Allyltrichlorosilane

Allyltriethoxysilane

Allyltrimethylsilane

Allylurea

No.	Name	Synonym	Mol. Form.	CAS RN	Mol. Wt.	Physical Form	mp/°C	bp/°C	den/ g cm^{-3}	n_D	Solubility
213	Allyl vinyl ether	3-(Ethenyloxy)-1-propene	C$_5$H$_8$O	3917-15-5	84.117			66	0.7900[20]	1.4062[20]	i H$_2$O; s eth, ace, chl
214	Aloin A		C$_{21}$H$_{22}$O$_9$	1415-73-2	418.395		149.3				s H$_2$O, EtOH, ace; sl eth, bz; i chl
215	Alphaprodine		C$_{16}$H$_{23}$NO$_2$	15867-21-7	261.360	cry	103				
216	Alstonidine		C$_{22}$H$_{24}$N$_2$O$_4$	25394-75-6	380.437	cry (eth)	189				vs ace, EtOH
217	Alstonine		C$_{21}$H$_{20}$N$_2$O$_3$	642-18-2	348.395	ye nd (ace)	207 dec				
218	D-Altrose		C$_6$H$_{12}$O$_6$	1990-29-0	180.155	pr (MeOH,al)	103.5				vs H$_2$O
219	Aluminum 2-butoxide	2-Butanol, aluminum salt	C$_{12}$H$_{27}$AlO$_3$	2269-22-9	246.322			197[20]			
220	Aluminum distearate	Hydroxyaluminum distearate	C$_{36}$H$_{71}$AlO$_5$	300-92-5	610.928	wh pow	145				i H$_2$O
221	Aluminum ethanolate	Aluminum ethoxide	C$_6$H$_{15}$AlO$_3$	555-75-9	162.163	liq/wh solid	140	200[7]			dec H$_2$O; sl xyl
222	Aluminum isopropoxide		C$_9$H$_{21}$AlO$_3$	555-31-7	204.243	hyg wh solid	119	135[10], 94[0.5]			reac H$_2$O; s EtOH, bz, peth, chl
223	Alverine	N-Ethyl-bis(3-phenylpropyl) amine	C$_{20}$H$_{27}$N	150-59-4	281.435	oil		166[0.3]			
224	α-Amanitin		C$_{39}$H$_{54}$N$_{10}$O$_{14}$S	23109-05-9	918.970	nd	254 dec				
225	Amaranth dye		C$_{20}$H$_{11}$N$_2$Na$_3$O$_{10}$S$_3$	915-67-3	604.472	dk red pow					s H$_2$O
226	Ametryn		C$_9$H$_{17}$N$_5$S	834-12-8	227.330		88				
227	Amminetrimethylboron		C$_3$H$_{12}$BN	1830-95-1	72.945		73.5				
228	19-Amino-8,11,13-abietatriene		C$_{20}$H$_{31}$N	1446-61-3	285.467	cry	44.5				
229	2-Aminoacetamide		C$_2$H$_6$N$_2$O	598-41-4	74.081	hyg nd (chl)	67.5				vs H$_2$O, EtOH; sl eth, bz; s ace, chl
230	Aminoacetonitrile		C$_2$H$_4$N$_2$	540-61-4	56.066			58[15]			vs EtOH
231	Aminoacetonitrile monohydrochloride		C$_2$H$_5$ClN$_2$	6011-14-9	92.527	hyg cry (al)	165 dec				
232	α-Aminoacetophenone hydrochloride		C$_8$H$_{10}$ClNO	5468-37-1	171.624		194 dec				
233	1-Aminoadamantane hydrochloride	Adamantanamine hydrochloride	C$_{10}$H$_{18}$ClN	665-66-7	187.710	cry (al-eth)	360 dec				vs H$_2$O, EtOH
234	2-Aminoadipic acid		C$_6$H$_{11}$NO$_4$	626-71-1	161.156	pl (w)	207.0				sl H$_2$O, EtOH, eth
235	3-Aminoalanine	2,3-Diaminopropionic acid	C$_3$H$_8$N$_2$O$_2$	515-94-6	104.108	hyg rosettes	110				vs H$_2$O
236	1-Amino-9,10-anthracenedione	1-Aminoanthraquinone	C$_{14}$H$_9$NO$_2$	82-45-1	223.227	red nd (al)	253.5	sub			vs ace, bz, EtOH, chl
237	2-Amino-9,10-anthracenedione	2-Aminoanthraquinone	C$_{14}$H$_9$NO$_2$	117-79-3	223.227	red nd (al, HOAc)	304.5	sub			i H$_2$O, eth; sl EtOH; s ace, bz, chl
238	4-Aminoazobenzene		C$_{12}$H$_{11}$N$_3$	60-09-3	197.235	oran mcl nd (al)	127	>360			sl H$_2$O, lig; s EtOH, eth, bz, chl
239	2-Aminobenzaldehyde		C$_7$H$_7$NO	529-23-7	121.137	silv lf	40.5	80[2]			sl H$_2$O; vs EtOH, eth; s bz, chl; i lig
240	3-Aminobenzaldehyde		C$_7$H$_7$NO	1709-44-0	121.137	nd (AcOEt)	29				s eth, acid
241	4-Aminobenzaldehyde		C$_7$H$_7$NO	556-18-3	121.137	pl (w)	71.5				s H$_2$O, EtOH, eth, acid
242	2-Aminobenzamide		C$_7$H$_8$N$_2$O	88-68-6	136.151		110.5 dec				s H$_2$O, EtOH; sl eth, bz; vs AcOEt
243	4-Aminobenzamide		C$_7$H$_8$N$_2$O	2835-68-9	136.151	ye cry (+1/4w)	183				sl H$_2$O; s EtOH, eth
244	α-Aminobenzeneacetic acid, (±)	α-Phenylglycine	C$_8$H$_9$NO$_2$	2835-06-5	151.163	pl	292 dec	sub 255			s alk; sl os
245	4-Aminobenzeneacetic acid	p-Aminophenylacetic acid	C$_8$H$_9$NO$_2$	1197-55-3	151.163	pl (w)	200 dec				i H$_2$O; sl EtOH, DMSO
246	5-Amino-1,3-benzenedicarboxylic acid		C$_8$H$_7$NO$_4$	99-31-0	181.147	pr(al), pl(w)	360	sub			i H$_2$O; sl EtOH
247	4-Aminobenzeneethanol		C$_8$H$_{11}$NO	104-10-9	137.179	nd (al)	108				
248	2-Aminobenzenemethanamine		C$_7$H$_{10}$N$_2$	4403-69-4	122.167		61	269			vs EtOH
249	2-Aminobenzenemethanol		C$_7$H$_9$NO	5344-90-1	123.152		83.5	273			s H$_2$O, EtOH, eth, HOAc; vs bz, chl
250	4-Aminobenzenesulfonamide	Sulfanilamide	C$_6$H$_8$N$_2$O$_2$S	63-74-1	172.205	lf (dil al)	165.5		1.08[25]		s H$_2$O, EtOH, eth, ace; sl chl, peth
251	2-Aminobenzenesulfonic acid	Orthanilic acid	C$_6$H$_7$NO$_3$S	88-21-1	173.190	pr (+ 1/2w)	>320 dec				sl H$_2$O; i EtOH, eth
252	3-Aminobenzenesulfonic acid	Metanilic acid	C$_6$H$_7$NO$_3$S	121-47-1	173.190	nd, pr (w +1)	dec				sl H$_2$O, EtOH; i eth
253	4-Aminobenzenesulfonic acid	Sulfanilic acid	C$_6$H$_7$NO$_3$S	121-57-3	173.190	orth pl or mcl (w+2)	288		1.485[25]		sl H$_2$O; i EtOH, eth
254	4-Aminobenzenesulfonyl fluoride	p-Sulfanilyl fluoride	C$_6$H$_6$FNO$_2$S	98-62-4	175.181		68.5				

Allyl vinyl ether

Aloin A

Alphaprodine

Alstonidine

Alstonine

D-Altrose

Aluminum 2-butoxide

Aluminum distearate

Aluminum ethanolate

Aluminum isopropoxide

Alverine

α-Amanitin

Amaranth dye

Ametryn

Amminetrimethylboron

19-Amino-8,11,13-abietatriene

2-Aminoacetamide

Aminoacetonitrile

Aminoacetonitrile monohydrochloride

α-Aminoacetophenone hydrochloride

1-Aminoadamantane hydrochloride

2-Aminoadipic acid

3-Aminoalanine

1-Amino-9,10-anthracenedione

2-Amino-9,10-anthracenedione

4-Aminoazobenzene

2-Aminobenzaldehyde

3-Aminobenzaldehyde

4-Aminobenzaldehyde

2-Aminobenzamide

4-Aminobenzamide

α-Aminobenzeneacetic acid, (±)

4-Aminobenzeneacetic acid

5-Amino-1,3-benzenedicarboxylic acid

4-Aminobenzeneethanol

2-Aminobenzenemethanamine

2-Aminobenzenemethanol

4-Aminobenzenesulfonamide

2-Aminobenzenesulfonic acid

3-Aminobenzenesulfonic acid

4-Aminobenzenesulfonic acid

4-Aminobenzenesulfonyl fluoride

No.	Name	Synonym	Mol. Form.	CAS RN	Mol. Wt.	Physical Form	mp/°C	bp/°C	den/ g cm⁻³	n_D	Solubility
255	2-Aminobenzenethiol		C₆H₇NS	137-07-5	125.192		26	234		1.4606²⁰	s EtOH, eth
256	4-Aminobenzenethiol		C₆H₇NS	1193-02-8	125.192		46	143¹⁷			s H₂O, EtOH
257	2-Aminobenzonitrile		C₇H₆N₂	1885-29-6	118.136	ye pr (CS₂) nd (peth)	51	263			sl H₂O; vs EtOH, eth, ace, bz; i peth
258	3-Aminobenzonitrile		C₇H₆N₂	2237-30-1	118.136	nd (dil al or CCl₄)	54.3	289			sl H₂O; vs EtOH, eth, ace, chl
259	4-Aminobenzonitrile		C₇H₆N₂	873-74-5	118.136	pr or pl (w)	87.0				sl H₂O, ctc; vs EtOH, eth, ace, bz
260	4-Aminobenzophenone		C₁₃H₁₁NO	1137-41-3	197.232	lf (dil al)	124	246¹³			sl H₂O, tfa; s EtOH, eth, HOAc
261	N-(4-Aminobenzoyl)-L-glutamic acid		C₁₂H₁₄N₂O₅	4271-30-1	266.249	cry (w)	173				
262	N-(4-Aminobenzoyl)glycine	p-Aminohippuric acid	C₉H₁₀N₂O₃	61-78-9	194.186	pr or nd (w)	198.5				vs ace, bz, EtOH
263	2-Aminobiphenyl		C₁₂H₁₁N	90-41-5	169.222	lf (dil al)	51	299			i H₂O; s EtOH, eth, bz; sl DMSO, peth
264	3-Aminobiphenyl		C₁₂H₁₁N	2243-47-2	169.222	nd	31.5				sl H₂O; s EtOH, eth, ace, bz
265	4-Aminobiphenyl	p-Biphenylamine	C₁₂H₁₁N	92-67-1	169.222	lf (dil al)	53.5	302			sl H₂O; s EtOH, eth, ace, chl
266	2-Amino-5-bromobenzoic acid	5-Bromoanthranilic acid	C₇H₆BrNO₂	5794-88-7	216.033	nd	219.5				s DMSO
267	1-Amino-4-bromo-9,10-dihydro-9,10-dioxo-2-anthracenesulfonic acid	1-Amino-4-bromoanthraquinone-2-sulfonic acid	C₁₄H₈BrNO₅S	116-81-4	382.187	red nd (w)					
268	DL-2-Aminobutanoic acid		C₄H₉NO₂	2835-81-6	103.120	lf (w)	304 dec	sub	1.2300²⁰		vs H₂O; sl EtOH; i eth, bz
269	L-2-Aminobutanoic acid		C₄H₉NO₂	1492-24-6	103.120	lf (dil al), cry (al)	292 dec				s H₂O; sl EtOH, eth; i bz
270	DL-3-Aminobutanoic acid		C₄H₉NO₂	2835-82-7	103.120	nd (al)	194.3				vs H₂O; i EtOH, eth, bz
271	4-Aminobutanoic acid	γ-Aminobutyric acid	C₄H₉NO₂	56-12-2	103.120	pr or nd (al) lf (MeOH-eth)	203 dec				vs H₂O; sl EtOH, ace; i eth, bz
272	2-Amino-1-butanol, (±)		C₄H₁₁NO	13054-87-0	89.136	liq	-1.0	178	0.9162²⁰	1.4489²⁵	msc H₂O, EtOH, eth; sl chl
273	4-Amino-1-butanol		C₄H₁₁NO	13325-10-5	89.136			205; 125³⁴	0.967¹²	1.4625²⁰	s H₂O, EtOH; i eth
274	4-Amino-N-[(butylamino)carbonyl]benzenesulfonamide	Carbutamide	C₁₁H₁₇N₃O₃S	339-43-5	271.336		144.5				
275	Aminocarb		C₁₁H₁₆N₂O₂	2032-59-9	208.257	cry	94				sl H₂O, bz; s ace
276	N-(Aminocarbonyl)acetamide		C₃H₆N₂O₂	591-07-1	102.092		218	sub 180			sl H₂O, eth; s EtOH
277	[4-[(Aminocarbonyl)amino]phenyl]arsonic acid	Carbarsone	C₇H₉AsN₂O₄	121-59-5	260.079	nd (w)	174				sl H₂O, DMSO, EtOH; i eth, chl; s alk
278	N-(Aminocarbonyl)-2-bromo-2-ethylbutanamide	Carbromal	C₇H₁₃BrN₂O₂	77-65-6	237.094	orth (dil al)	118		1.544²⁵		sl H₂O, chl; s ace, bz
279	N-(Aminocarbonyl)-2-bromo-3-methylbutanamide	Bromisovalum	C₆H₁₁BrN₂O₂	496-67-3	223.067	nd or lf (to)	154	sub	1.56¹⁵		vs ace, bz, eth, EtOH
280	[2-(Aminocarbonyl)phenoxy]acetic acid	Salicylamide O-acetic acid	C₉H₉NO₄	25395-22-6	195.172		221				s alk
281	7-Aminocephalosporanic acid		C₁₀H₁₂N₂O₅S	957-68-6	272.277	cry					
282	1-Amino-5-chloro-9,10-anthracenedione	1-Amino-5-chloroanthraquinone	C₁₄H₈ClNO₂	117-11-3	257.673		212				
283	4-Amino-6-chloro-1,3-benzenedisulfonamide	Chloraminophenamide	C₆H₈ClN₃O₄S₂	121-30-2	285.729		254.5				
284	5-Amino-2-chlorobenzenesulfonic acid	6-Chlorometanilic acid	C₆H₆ClNO₃S	88-43-7	207.635	nd (w)	280 dec				
285	2-Amino-5-chlorobenzoic acid		C₇H₆ClNO₂	635-21-2	171.582		211				
286	5-Amino-2-chlorobenzoic acid		C₇H₆ClNO₂	89-54-3	171.582		188		1.519¹⁵		vs EtOH
287	2-Amino-5-chlorobenzophenone	2-Benzoyl-4-chloroaniline	C₁₃H₁₀ClNO	719-59-5	231.677	ye nd	100.5				vs H₂O, EtOH, peth, chl
288	2-Amino-4-chloro-5-methylbenzenesulfonic acid	2-Chloro-p-toluidine-5-sulfonic acid	C₇H₈ClNO₃S	88-51-7	221.662	short nd (w)					
289	2-Amino-4-chlorophenol	2-Hydroxy-5-chloroaniline	C₆H₆ClNO	95-85-2	143.571		140				sl DMSO
290	1-Aminocyclopentanecarboxylic acid	Cycloleucine	C₆H₁₁NO₂	52-52-8	129.157	cry (al-w)	330 dec				
291	7-Aminodeacetoxycephalosporanic acid		C₈H₁₀N₂O₃S	22252-43-3	214.241		241 dec				
292	1-Amino-1-deoxy-D-glucitol	Glucamine	C₆H₁₅NO₅	488-43-7	181.187	cry (MeOH)	127				vs H₂O, EtOH
293	2-Amino-2-deoxy-D-glucose	D-Glucosamine	C₆H₁₃NO₅	3416-24-8	179.171						vs H₂O

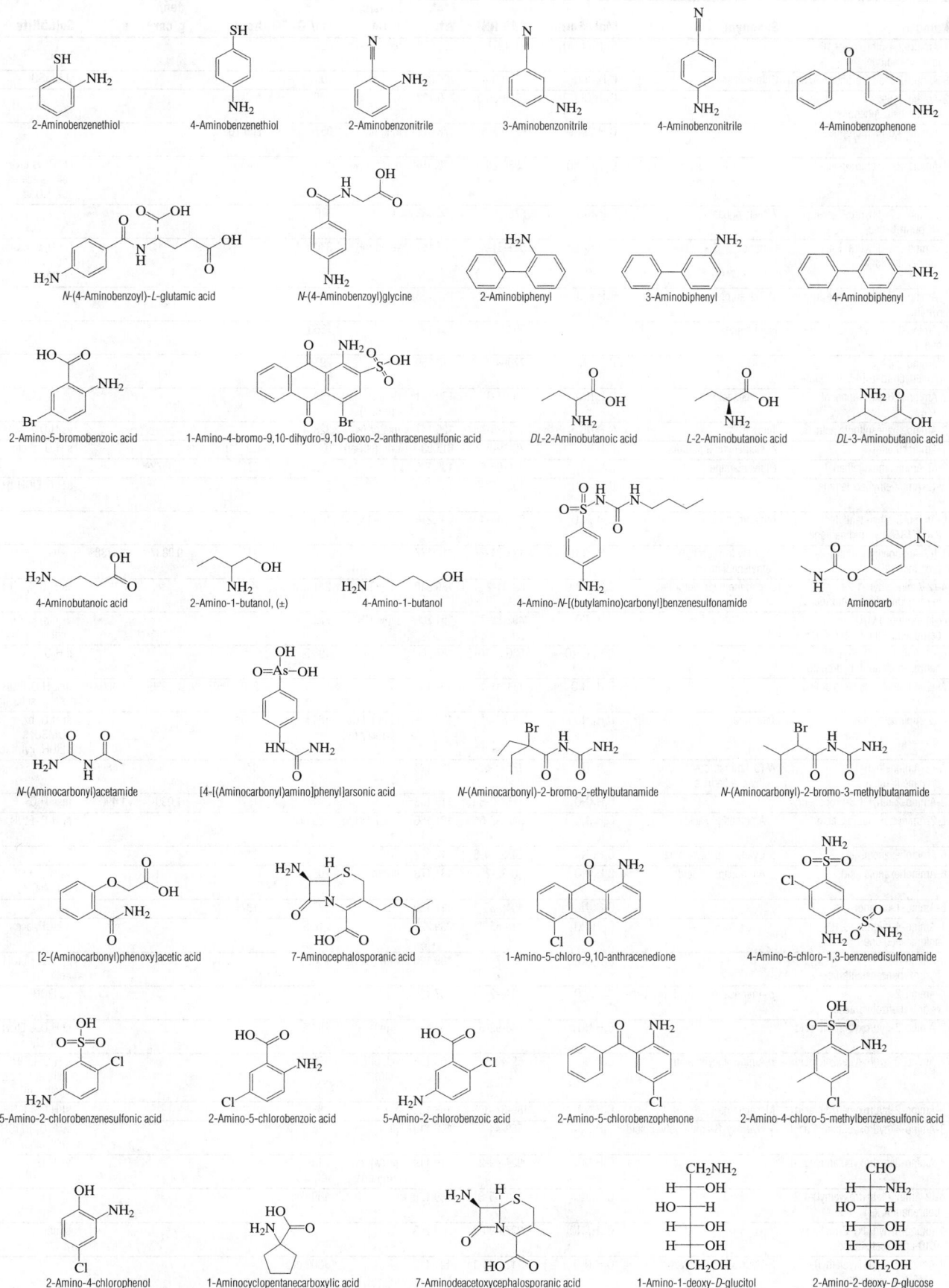

2-Aminobenzenethiol

4-Aminobenzenethiol

2-Aminobenzonitrile

3-Aminobenzonitrile

4-Aminobenzonitrile

4-Aminobenzophenone

N-(4-Aminobenzoyl)-L-glutamic acid

N-(4-Aminobenzoyl)glycine

2-Aminobiphenyl

3-Aminobiphenyl

4-Aminobiphenyl

2-Amino-5-bromobenzoic acid

1-Amino-4-bromo-9,10-dihydro-9,10-dioxo-2-anthracenesulfonic acid

DL-2-Aminobutanoic acid

L-2-Aminobutanoic acid

DL-3-Aminobutanoic acid

4-Aminobutanoic acid

2-Amino-1-butanol, (±)

4-Amino-1-butanol

4-Amino-N-[(butylamino)carbonyl]benzenesulfonamide

Aminocarb

N-(Aminocarbonyl)acetamide

[4-[(Aminocarbonyl)amino]phenyl]arsonic acid

N-(Aminocarbonyl)-2-bromo-2-ethylbutanamide

N-(Aminocarbonyl)-2-bromo-3-methylbutanamide

[2-(Aminocarbonyl)phenoxy]acetic acid

7-Aminocephalosporanic acid

1-Amino-5-chloro-9,10-anthracenedione

4-Amino-6-chloro-1,3-benzenedisulfonamide

5-Amino-2-chlorobenzenesulfonic acid

2-Amino-5-chlorobenzoic acid

5-Amino-2-chlorobenzoic acid

2-Amino-5-chlorobenzophenone

2-Amino-4-chloro-5-methylbenzenesulfonic acid

2-Amino-4-chlorophenol

1-Aminocyclopentanecarboxylic acid

7-Aminodeacetoxycephalosporanic acid

1-Amino-1-deoxy-D-glucitol

2-Amino-2-deoxy-D-glucose

No.	Name	Synonym	Mol. Form.	CAS RN	Mol. Wt.	Physical Form	mp/°C	bp/°C	den/ g cm⁻³	n_D	Solubility
294	1-Amino-2,4-dibromo-9,10-anthracenedione		$C_{14}H_7Br_2NO_2$	81-49-2	381.020	red nd (xyl)	226				
295	3-Amino-2,5-dichlorobenzoic acid	Chloramben	$C_7H_5Cl_2NO_2$	133-90-4	206.027		200				sl DMSO
296	2-Amino-2',5-dichlorobenzophenone		$C_{13}H_9Cl_2NO$	2958-36-3	266.122		≈80				
297	2-Amino-4,6-dichlorophenol		$C_6H_5Cl_2NO$	527-62-8	178.016	long nd (CS_2)	95.5	sub 70			
298	4-Amino-2,6-dichlorophenol		$C_6H_5Cl_2NO$	5930-28-9	178.016	nd or lf (w, bz)	168	sub			i H_2O; vs EtOH, eth; s ace; sl bz, HOAc
299	2-Amino-1,7-dihydro-7-methyl-6H-purin-6-one	7-Methylguanine	$C_6H_7N_5O$	578-76-7	165.153		370				
300	5-Amino-2,3-dihydro-1,4-phthalazinedione	Luminol	$C_8H_7N_3O_2$	521-31-3	177.161	ye nd (al)	330.5				i H_2O; sl EtOH, eth; vs alk; s HOAc
301	2-Amino-1,7-dihydro-6H-purine-6-thione	Thioguanine	$C_5H_5N_5S$	154-42-7	167.193		>360				
302	6-Amino-1,3-dihydro-2H-purin-2-one	Isoguanine	$C_5H_5N_5O$	3373-53-3	151.127		>360				i H_2O
303	2-Amino-3,4-dimethylimidazo[4,5-f]quinoline	Me-IQ	$C_{12}H_{12}N_4$	77094-11-2	212.250	cry	297				
304	2-Amino-4,6-dinitrophenol	Picramic acid	$C_6H_5N_3O_5$	96-91-3	199.121	dk red nd (al) pr (chl)	169				vs bz, EtOH
305	2-Aminoethanesulfonic acid	Taurine	$C_2H_7NO_3S$	107-35-7	125.147	mcl pr (w)	328				vs H_2O
306	1-Aminoethanol	Acetaldehyde ammonia	C_2H_7NO	75-39-8	61.083	orth (eth-al)	97	dec 110			s H_2O; sl eth
307	2-(2-Aminoethoxy)ethanol	Diglycolamine	$C_4H_{11}NO_2$	929-06-6	105.136		-12.5	221	1.0572²⁰		
308	N-(2-Aminoethyl)acetamide		$C_4H_{10}N_2O$	1001-53-2	102.134		51				s H_2O, EtOH, bz; i eth
309	6-Amino-3-ethyl-1-allyl-2,4(1H,3H)-pyrimidinedione	Aminometradine	$C_9H_{13}N_3O_2$	642-44-4	195.218	cry (+1w, w)	143				
310	1-[(2-Aminoethyl)amino]-2-propanol	N-(2-Hydroxypropyl) ethylenediamine	$C_5H_{14}N_2O$	123-84-2	118.177			94³	0.9837²⁵	1.4738²⁰	
311	4-(2-Aminoethyl)-1,2-benzenediol, hydrochloride	Dopamine hydrochloride	$C_8H_{12}ClNO_2$	62-31-7	189.640	nd (w)	241 dec				vs H_2O, MeOH
312	α-(1-Aminoethyl)benzenemethanol, [S-(R*,R*)]-		$C_9H_{13}NO$	492-39-7	151.205	pl(MeOH)	77.5				vs eth, EtOH, chl
313	α-(1-Aminoethyl)benzenemethanol, hydrochloride		$C_9H_{14}ClNO$	53631-70-2	187.666		198.5				s H_2O
314	N-(2-Aminoethyl)ethanolamine		$C_4H_{12}N_2O$	111-41-1	104.150			239; 105¹⁰	1.0286²⁰	1.4863²⁰	msc H_2O, EtOH; s ace; sl bz, lig
315	4-(2-Aminoethyl)phenol	Tyramine	$C_8H_{11}NO$	51-67-2	137.179	pl or nd (bz, w), cry (al)	164.5	206²⁵			sl H_2O, bz, DMSO; s EtOH, xyl; i tol
316	N-(2-Aminoethyl)-1,3-propanediamine	N-(3-Aminopropyl) ethylenediamine	$C_5H_{15}N_3$	13531-52-7	117.193			87³		1.4805²⁵	
317	2-Amino-2-ethyl-1,3-propanediol		$C_5H_{13}NO_2$	115-70-8	119.163		37.5	152¹⁰	1.099²⁰	1.490²⁰	msc H_2O
318	L-2-Aminohexanedioic acid	2-Aminoadipic acid	$C_6H_{11}NO_4$	542-32-5	161.156	cry (EtOH, w)	205 dec				sl H_2O, EtOH, eth
319	6-Aminohexanenitrile	5-Cyano-1-pentylamine	$C_5H_{12}N_2$	2432-74-8	112.172	liq		118¹⁶			
320	6-Aminohexanoic acid	ε-Aminocaproic acid	$C_6H_{13}NO_2$	60-32-2	131.173	lf (eth)	205				vs H_2O; i EtOH; sl MeOH
321	6-Amino-1-hexanol		$C_6H_{15}NO$	4048-33-3	117.189		57	137³⁰			
322	1-Amino-4-hydroxy-9,10-anthracenedione		$C_{14}H_9NO_3$	116-85-8	239.226		216.5				s EtOH, ace
323	3-Amino-4-hydroxybenzenesulfonic acid		$C_6H_7NO_4S$	98-37-3	189.190	orth (w+1)	>300				sl H_2O; i EtOH, eth
324	4-Amino-2-hydroxybenzohydrazide	p-Aminosalicylic acid hydrazide	$C_7H_9N_3O_2$	6946-29-8	167.165	nd (al)	195				vs EtOH
325	2-Amino-3-hydroxybenzoic acid		$C_7H_7NO_3$	548-93-6	153.136	lf (w)	253.5				sl H_2O; s EtOH, eth, chl
326	4-Amino-2-hydroxybenzoic acid	p-Aminosalicylic acid	$C_7H_7NO_3$	65-49-6	153.136	nd, pl (al-eth)	150 dec				s H_2O, EtOH, eth, ace; i bz, peth, chl
327	5-Amino-2-hydroxybenzoic acid	Mesalamine	$C_7H_7NO_3$	89-57-6	153.136		283				sl H_2O; i EtOH
328	3-Amino-4-hydroxybutanoic acid	γ-Hydroxy-β-aminobutyric acid	$C_4H_9NO_3$	589-44-6	119.119	pr	216				vs H_2O; sl EtOH, chl, eth, AcOEt
329	4-Amino-3-hydroxybutanoic acid, (±)		$C_4H_9NO_3$	924-49-2	119.119	pr (w), cry (dil al)	218				vs H_2O
330	4-(2-Amino-1-hydroxyethyl)-1,2-benzenediol, (±)		$C_8H_{11}NO_3$	138-65-8	169.178		189 dec				
331	1-Amino-4-hydroxy-2-methoxy-9,10-anthracenedione		$C_{15}H_{11}NO_4$	2379-90-0	269.253						sl chl
332	4-Amino-5-(hydroxymethyl)-2(1H)-pyrimidinone	5-Hydroxymethylcytosine	$C_5H_7N_3O_2$	1123-95-1	141.129		>300 dec				

1-Amino-2,4-dibromo-9,10-anthracenedione

3-Amino-2,5-dichlorobenzoic acid

2-Amino-2',5-dichlorobenzophenone

2-Amino-4,6-dichlorophenol

4-Amino-2,6-dichlorophenol

2-Amino-1,7-dihydro-7-methyl-6H-purin-6-one

5-Amino-2,3-dihydro-1,4-phthalazinedione

2-Amino-1,7-dihydro-6H-purine-6-thione

6-Amino-1,3-dihydro-2H-purin-2-one

2-Amino-3,4-dimethylimidazo[4,5-f]quinoline

2-Amino-4,6-dinitrophenol

2-Aminoethanesulfonic acid

1-Aminoethanol

2-(2-Aminoethoxy)ethanol

N-(2-Aminoethyl)acetamide

6-Amino-3-ethyl-1-allyl-2,4(1H,3H)-pyrimidinedione

1-[(2-Aminoethyl)amino]-2-propanol

4-(2-Aminoethyl)-1,2-benzenediol, hydrochloride

α-(1-Aminoethyl)benzenemethanol, [S-(R*,R*)]-

α-(1-Aminoethyl)benzenemethanol, hydrochloride

N-(2-Aminoethyl)ethanolamine

4-(2-Aminoethyl)phenol

N-(2-Aminoethyl)-1,3-propanediamine

2-Amino-2-ethyl-1,3-propanediol

L-2-Aminohexanedioic acid

6-Aminohexanenitrile

6-Aminohexanoic acid

6-Amino-1-hexanol

1-Amino-4-hydroxy-9,10-anthracenedione

3-Amino-4-hydroxybenzenesulfonic acid

4-Amino-2-hydroxybenzohydrazide

2-Amino-3-hydroxybenzoic acid

4-Amino-2-hydroxybenzoic acid

5-Amino-2-hydroxybenzoic acid

3-Amino-4-hydroxybutanoic acid

4-Amino-3-hydroxybutanoic acid, (±)

4-(2-Amino-1-hydroxyethyl)-1,2-benzenediol, (±)

1-Amino-4-hydroxy-2-methoxy-9,10-anthracenedione

4-Amino-5-(hydroxymethyl)-2(1H)-pyrimidinone

No.	Name	Synonym	Mol. Form.	CAS RN	Mol. Wt.	Physical Form	mp/°C	bp/°C	den/ g cm⁻³	n_D	Solubility
333	4-Amino-5-hydroxy-2,7-naphthalenedisulfonic acid	1-Naphthol-8-amino-3,6-disulfonic acid	$C_{10}H_9NO_7S_2$	90-20-0	319.311						sl H_2O, EtOH, eth
334	4-Amino-3-hydroxy-1-naphthalenesulfonic acid	1-Amino-2-naphthol-4-sulfonic acid	$C_{10}H_9NO_4S$	116-63-2	239.248	gray nd					i H_2O, EtOH, bz; s alk
335	2-Amino-4-hydroxypteridine		$C_6H_5N_5O$	2236-60-4	163.137	ye cry	>360				
336	5-Amino-1H-imidazole-4-carboxamide		$C_4H_6N_4O$	360-97-4	126.117	cry (EtOH)	170				
337	O-[(Aminoiminomethyl)amino]-L-homoserine	Canavanine	$C_5H_{12}N_4O_3$	543-38-4	176.174	cry (al)					vs H_2O
338	(Aminoiminomethyl)urea		$C_2H_6N_4O$	141-83-3	102.095	pr	105	dec 160			s H_2O, py; sl EtOH; i eth, bz, chl, CS_2
339	2-Amino-5-iodobenzoic acid		$C_7H_6INO_2$	5326-47-6	263.033		220 dec				sl H_2O, tfa; vs EtOH, eth, ace; s bz
340	4-Amino-1H-isoindole-1,3(2H)-dione		$C_8H_6N_2O_2$	2518-24-3	162.146		269.5				
341	4-Amino-3-isoxazolidinone, (R)	Cycloserine	$C_3H_6N_2O_2$	68-41-7	102.092		155 dec				s H_2O; sl MeOH
342	1-Amino-2-methyl-9,10-anthracenedione	1-Amino-2-methylanthraquinone	$C_{15}H_{11}NO_2$	82-28-0	237.254		205.5				i H_2O; s EtOH, bz, chl; sl eth
343	α-(Aminomethyl)benzenemethanol	Phenylethanolamine	$C_8H_{11}NO$	7568-93-6	137.179		56.5	160¹⁷			vs H_2O; s EtOH
344	β-(Aminomethyl)benzenepropanoic acid	4-Amino-3-phenylbutyric acid	$C_{10}H_{13}NO_2$	1078-21-3	179.216		252 dec				
345	2-Amino-5-methylbenzenesulfonic acid		$C_7H_9NO_3S$	88-44-8	187.216	lt ye nd	132 dec				vs H_2O
346	$trans$-4-(Aminomethyl)cyclohexanecarboxylic acid	Tranexamic acid	$C_8H_{15}NO_2$	1197-18-8	157.211		>300				vs H_2O
347	4-Amino-4-methyl-2-pentanone	Diacetonamine	$C_6H_{13}NO$	625-04-7	115.173			25⁰·¹⁴			s H_2O; msc EtOH, eth
348	2-Amino-4-methylphenol		C_7H_9NO	95-84-1	123.152	cry (w), orth (bz), lf or nd	136	sub			sl H_2O, bz; s EtOH, eth, chl; i lig
349	4-Amino-2-methylphenol		C_7H_9NO	2835-96-3	123.152	nd or lf (bz)	176.5	sub			sl H_2O, bz; s EtOH, eth
350	4-Amino-3-methylphenol		C_7H_9NO	2835-99-6	123.152	pr (dil al) cry (bz)	179				sl H_2O; vs EtOH, eth; s DMSO
351	(Aminomethyl)phosphonic acid		CH_6NO_3P	1066-51-9	111.038	cry	309				
352	2-Amino-2-methyl-1,3-propanediol		$C_4H_{11}NO_2$	115-69-5	105.136		110	151¹⁰			vs H_2O; s EtOH
353	L-3-Amino-2-methylpropanoic acid		$C_4H_9NO_2$	144-90-1	103.120	cry (w)	182				
354	2-Amino-2-methyl-1-propanol	2-Aminoisobutanol	$C_4H_{11}NO$	124-68-5	89.136		25.5	165.5	0.934²⁰	1.449²⁰	msc H_2O; s ctc
355	4-Amino-5-methyl-2(1H)-pyrimidinone	5-Methylcytosine	$C_5H_7N_3O$	554-01-8	125.129	pr (w+1/2)	270 dec				s H_2O, acid; sl EtOH; i eth
356	3-(Aminomethyl)-3,5,5-trimethylcyclohexanol	1-Hydroxy-3-aminomethyl-3,5,5-trimethylcyclohexane	$C_{10}H_{21}NO$	15647-11-7	171.280		45.5	265	0.969²⁵	1.4904²⁰	
357	3-Amino-2-naphthalenecarboxylic acid	3-Amino-2-naphthoic acid	$C_{11}H_9NO_2$	5959-52-4	187.195	ye lf (dil al)	216.5				s EtOH, eth
358	2-Amino-1,4-naphthalenedione		$C_{10}H_7NO_2$	2348-81-4	173.169		207				i H_2O, alk; s EtOH, eth, HOAc
359	7-Amino-1,3-naphthalenedisulfonic acid	Amido-G-Acid	$C_{10}H_9NO_6S_2$	86-65-7	303.311	mcl pr or nd (w+4)	274				vs H_2O, EtOH
360	2-Amino-1,5-naphthalenedisulfonic acid	2-Naphthylamine-1,5-disulfonic acid	$C_{10}H_9NO_6S_2$	117-62-4	303.311		>300				
361	4-Amino-1,6-naphthalenedisulfonic acid	1-Naphthylamine-4,7-disulfonic acid	$C_{10}H_9NO_6S_2$	85-75-6	303.311						vs H_2O
362	4-Amino-1,7-naphthalenedisulfonic acid	1-Naphthylamine-4,6-disulfonic acid	$C_{10}H_9NO_6S_2$	85-74-5	303.311						vs H_2O, EtOH
363	2-Amino-1-naphthalenesulfonic acid	2-Naphthylamine-1-sulfonic acid	$C_{10}H_9NO_3S$	81-16-3	223.248	sc(hot w)					s DMSO
364	4-Amino-1-naphthalenesulfonic acid	1-Naphthylamine-4-sulfonic acid	$C_{10}H_9NO_3S$	84-86-6	223.248	wh nd (w+1/2) red-br cry	dec		1.6703²⁵		i H_2O; sl EtOH; s MeOH, py
365	5-Amino-1-naphthalenesulfonic acid	1-Naphthylamine-5-sulfonic acid	$C_{10}H_9NO_3S$	84-89-9	223.248	wh cry					s H_2O; i eth
366	6-Amino-1-naphthalenesulfonic acid	2-Naphthylamine-5-sulfonic acid	$C_{10}H_9NO_3S$	81-05-0	223.248	nd(w)					i H_2O, EtOH, eth
367	7-Amino-1-naphthalenesulfonic acid	Badische acid	$C_{10}H_9NO_3S$	86-60-2	223.248	nd (w+1), pl (aq ace)					vs HOAc
368	8-Amino-1-naphthalenesulfonic acid	1-Naphthylamine-8-sulfonic acid	$C_{10}H_9NO_3S$	82-75-7	223.248	nd					vs gl HOAc
369	6-Amino-2-naphthalenesulfonic acid	Bronner acid	$C_{10}H_9NO_3S$	93-00-5	223.248	lf					i cold H_2O; sl hot H_2O

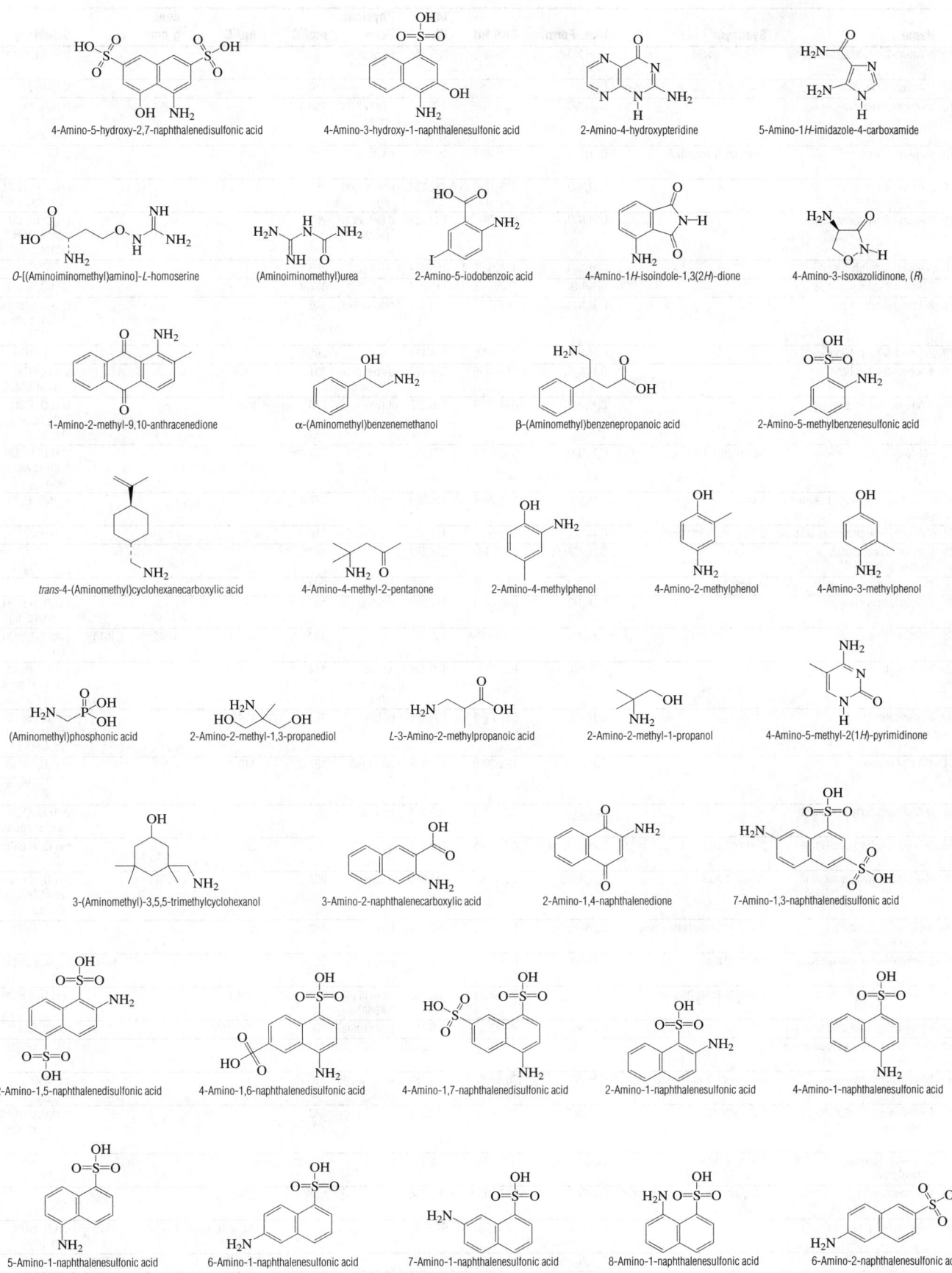

4-Amino-5-hydroxy-2,7-naphthalenedisulfonic acid

4-Amino-3-hydroxy-1-naphthalenesulfonic acid

2-Amino-4-hydroxypteridine

5-Amino-1*H*-imidazole-4-carboxamide

O-[(Aminoiminomethyl)amino]-*L*-homoserine

(Aminoiminomethyl)urea

2-Amino-5-iodobenzoic acid

4-Amino-1*H*-isoindole-1,3(2*H*)-dione

4-Amino-3-isoxazolidinone, (*R*)

1-Amino-2-methyl-9,10-anthracenedione

α-(Aminomethyl)benzenemethanol

β-(Aminomethyl)benzenepropanoic acid

2-Amino-5-methylbenzenesulfonic acid

trans-4-(Aminomethyl)cyclohexanecarboxylic acid

4-Amino-4-methyl-2-pentanone

2-Amino-4-methylphenol

4-Amino-2-methylphenol

4-Amino-3-methylphenol

(Aminomethyl)phosphonic acid

2-Amino-2-methyl-1,3-propanediol

L-3-Amino-2-methylpropanoic acid

2-Amino-2-methyl-1-propanol

4-Amino-5-methyl-2(1*H*)-pyrimidinone

3-(Aminomethyl)-3,5,5-trimethylcyclohexanol

3-Amino-2-naphthalenecarboxylic acid

2-Amino-1,4-naphthalenedione

7-Amino-1,3-naphthalenedisulfonic acid

2-Amino-1,5-naphthalenedisulfonic acid

4-Amino-1,6-naphthalenedisulfonic acid

4-Amino-1,7-naphthalenedisulfonic acid

2-Amino-1-naphthalenesulfonic acid

4-Amino-1-naphthalenesulfonic acid

5-Amino-1-naphthalenesulfonic acid

6-Amino-1-naphthalenesulfonic acid

7-Amino-1-naphthalenesulfonic acid

8-Amino-1-naphthalenesulfonic acid

6-Amino-2-naphthalenesulfonic acid

No.	Name	Synonym	Mol. Form.	CAS RN	Mol. Wt.	Physical Form	mp/°C	bp/°C	den/ g cm⁻³	n_D	Solubility
370	8-Amino-2-naphthalenesulfonic acid	1,7-Cleve's acid	C₁₀H₉NO₃S	119-28-8	223.248	nd or pr (w)					sl EtOH; s eth
371	5-Amino-1-naphthol	1-Amino-6-hydroxynaphthalene	C₁₀H₉NO	83-55-6	159.184		170				sl DMSO
372	1-Amino-2-naphthol		C₁₀H₉NO	2834-92-6	159.184	silvery lf (bz, eth)	150 dec				sl H₂O, eth; s EtOH; vs dil alk, acid
373	8-Amino-2-naphthol	8-Amino-β-naphthol	C₁₀H₉NO	118-46-7	159.184	nd (w, al)	206	sub			s H₂O, eth; vs EtOH; sl bz, lig
374	2-Amino-4-nitrobenzoic acid		C₇H₆N₂O₄	619-17-0	182.134	oran pr (dil al)	269				i H₂O; vs EtOH, eth, ace; s xyl
375	2-Amino-5-nitrobenzoic acid		C₇H₆N₂O₄	616-79-5	182.134	lf (al), ye nd (w, dil al)	269				i H₂O, bz, chl, xyl; s EtOH, eth
376	2-Amino-5-nitrobenzonitrile		C₇H₅N₃O₂	17420-30-3	163.134		203.5				sl DMSO
377	3-Amino-1-nitroguanidine		CH₅N₅O₂	18264-75-0	119.084		187.8				sl H₂O
378	2-Amino-4-nitrophenol		C₆H₆N₂O₃	99-57-0	154.123	oran pr (+w)	146				sl H₂O, ace; vs EtOH; s eth, bz, HOAc
379	2-Amino-5-nitrophenol		C₆H₆N₂O₃	121-88-0	154.123		205.8				s H₂O, EtOH, bz
380	4-Amino-2-nitrophenol		C₆H₆N₂O₃	119-34-6	154.123	dk red pl or nd (w, al)	131	110¹²			s H₂O, EtOH, eth; sl DMSO
381	2-Aminooctanoic acid, (±)		C₈H₁₇NO₂	644-90-6	159.227	lf (w)	270	sub			sl H₂O, EtOH, eth, bz; s HOAc
382	Aminooxoacetohydrazide	Semioxamazide	C₂H₅N₃O₂	515-96-8	103.080		221 dec				sl H₂O; i EtOH, eth; vs alk, acid
383	cis-4-Amino-4-oxo-2-butenoic acid	Maleamic acid	C₄H₅NO₃	557-24-4	115.088	cry (al)	172.5				vs H₂O, EtOH
384	5-Amino-4-oxopentanoic acid	5-Aminolevulinic acid	C₅H₉NO₃	106-60-5	131.130	cry (EtOH)	118				
385	(Aminooxy)acetic acid, hydrochloride (2:1)		C₄H₁₁ClN₂O₆	2921-14-4	218.592		152.5				
386	6-Aminopenicillanic acid	Penicin	C₈H₁₂N₂O₃S	551-16-6	216.257	cry (w)	208				
387	5-Aminopentanoic acid		C₅H₁₁NO₂	660-88-8	117.147	lf (dil al)	157 dec	dec			s H₂O; sl EtOH; i eth, bz, lig
388	5-Amino-1-pentanol		C₅H₁₃NO	2508-29-4	103.163		38.5	221.5	0.9488¹⁷	1.4618¹⁷	msc H₂O, EtOH, ace
389	2-Aminophenol		C₆H₇NO	95-55-6	109.126	wh orth bipym nd (bz)	174	sub 153	1.328²⁵		s H₂O, eth; vs EtOH; sl bz, tfa
390	3-Aminophenol		C₆H₇NO	591-27-5	109.126	pr (to)	123	164¹¹			s H₂O, tol; vs EtOH, eth; sl bz, DMSO
391	4-Aminophenol		C₆H₇NO	123-30-8	109.126	wh pl (w)	187.5	110⁰·³			sl H₂O, tfa; vs EtOH; i bz, chl; s alk
392	N-(3-Aminophenyl)acetamide		C₈H₁₀N₂O	102-28-3	150.177	nd or pl (bz)	88				vs H₂O, EtOH, ace; sl eth, bz
393	N-(4-Aminophenyl)acetamide	p-Aminoacetanilide	C₈H₁₀N₂O	122-80-5	150.177	nd (w)	166.5	267			s H₂O; vs EtOH, eth
394	(4-Aminophenyl)arsonic acid	Arsanilic acid	C₆H₈AsNO₃	98-50-0	217.055	mcl nd (w, al)	232		1.9571¹⁰		s H₂O, eth; sl EtOH, DMSO; i ace, bz
395	N-(4-Aminophenyl)-1,4-benzenediamine	4,4'-Diaminodiphenylamine	C₁₂H₁₃N₃	537-65-5	199.251	lf (w)	158	dec			vs eth, EtOH
396	2-Amino-1-phenylethanone	Phenacylamine	C₈H₉NO	613-89-8	135.163	ye cry	20	251		1.6160²⁰	i H₂O; s eth; sl ctc
397	1-(3-Aminophenyl)ethanone	m-Aminoacetophenone	C₈H₉NO	99-03-6	135.163	pa ye pl (al), lf (eth)	98.5	289.5			sl H₂O; s EtOH
398	1-(4-Aminophenyl)ethanone	p-Aminoacetophenone	C₈H₉NO	99-92-3	135.163	ye mcl pr (al)	106	294; 195¹⁵			vs eth, EtOH
399	1-(4-Aminophenyl)-1-pentanone		C₁₁H₁₅NO	38237-74-0	177.243	cry (bz-peth)	74.5	161³			i H₂O; s EtOH, eth
400	1-(4-Aminophenyl)-1-propanone	p-Aminopropiophenone	C₉H₁₁NO	70-69-9	149.189	pl (al, w), nd (w)	140				s DMSO
401	N-[(4-Aminophenyl)sulfonyl]acetamide	Sulfacetamide	C₈H₁₀N₂O₃S	144-80-9	214.241		183				sl H₂O; s EtOH; i eth; vs ace, alk
402	5-[(4-Aminophenyl)sulfonyl]-2-thiazolamine	Thiazolsulfone	C₉H₉N₃O₂S₂	473-30-3	255.316	nd (al)	220 dec				vs ace, eth, EtOH, diox
403	4-Aminophthalimide	5-Amino-1H-isoindole-1,3(2H)-dione	C₈H₆N₂O₂	3676-85-5	162.146			224⁰·⁵			
404	3-Amino-1,2-propanediol, (±)		C₃H₉NO₂	13552-31-3	91.109			dec 265; 145⁹	1.1752²⁰	1.4910²⁵	s H₂O, EtOH; i eth, bz
405	3-Aminopropanenitrile	3-Aminopropionitrile	C₃H₆N₂	151-18-8	70.093			185; 88²⁰	0.9584²⁰	1.4396²⁰	
406	2-Amino-1-propanol, (±)		C₃H₉NO	6168-72-5	75.109			174.5		1.4502²⁰	vs H₂O, EtOH, eth; sl chl

8-Amino-2-naphthalenesulfonic acid

5-Amino-1-naphthol

1-Amino-2-naphthol

8-Amino-2-naphthol

2-Amino-4-nitrobenzoic acid

2-Amino-5-nitrobenzoic acid

2-Amino-5-nitrobenzonitrile

3-Amino-1-nitroguanidine

2-Amino-4-nitrophenol

2-Amino-5-nitrophenol

4-Amino-2-nitrophenol

2-Aminooctanoic acid, (±)

Aminooxoacetohydrazide

cis-4-Amino-4-oxo-2-butenoic acid

5-Amino-4-oxopentanoic acid

(Aminooxy)acetic acid, hydrochloride (2:1)

6-Aminopenicillanic acid

5-Aminopentanoic acid

5-Amino-1-pentanol

2-Aminophenol

3-Aminophenol

4-Aminophenol

N-(3-Aminophenyl)acetamide

N-(4-Aminophenyl)acetamide

(4-Aminophenyl)arsonic acid

N-(4-Aminophenyl)-1,4-benzenediamine

2-Amino-1-phenylethanone

1-(3-Aminophenyl)ethanone

1-(4-Aminophenyl)ethanone

1-(4-Aminophenyl)-1-pentanone

1-(4-Aminophenyl)-1-propanone

N-[(4-Aminophenyl)sulfonyl]acetamide

5-[(4-Aminophenyl)sulfonyl]-2-thiazolamine

4-Aminophthalimide

3-Amino-1,2-propanediol, (±)

3-Aminopropanenitrile

2-Amino-1-propanol, (±)

No.	Name	Synonym	Mol. Form.	CAS RN	Mol. Wt.	Physical Form	mp/°C	bp/°C	den/ g cm⁻³	n_D	Solubility
407	3-Amino-1-propanol	Propanolamine	C_3H_9NO	156-87-6	75.109		12.4	187.5	0.9824[26]	1.4617[20]	s H_2O, EtOH, eth
408	1-Amino-2-propanol	Isopropanolamine	C_3H_9NO	1674-56-2	75.109		0.9	159.4	0.9611[20]	1.4479[20]	msc H_2O, EtOH, eth, ace, bz, ctc
409	α-(1-Aminopropyl) benzenemethanol	α-(α-Aminopropyl)benzyl alcohol	$C_{10}H_{15}NO$	5897-76-7	165.232	pl (bz-eth)	79.5				
410	N-(3-Aminopropyl)-N-methyl-1,3-propanediamine		$C_7H_{19}N_3$	105-83-9	145.246			232.5; 112[6]	0.9023[20]	1.4705[25]	
411	Aminopropylon		$C_{16}H_{22}N_4O_2$	3690-04-8	302.372	pr (bz)	181				vs H_2O
412	4-(2-Aminopropyl)phenol, (±)	Hydroxyamphetamine	$C_9H_{13}NO$	1518-86-1	151.205	cry (bz)	125.5				s H_2O, EtOH, bz, chl, AcOEt
413	N-(3-Aminopropyl)-1,3-propanediamine	Bis(3-aminopropyl)amine	$C_6H_{17}N_3$	56-18-8	131.219		-14	151[50]	0.938[25]	1.4810[20]	s chl
414	Aminopterin		$C_{19}H_{20}N_8O_5$	54-62-6	440.413	ye cry	262 dec				
415	4-Amino-N-pyrazinylbenzenesulfonamide	Sulfapyrazine	$C_{10}H_{10}N_4O_2S$	116-44-9	250.277	nd ($PhNO_2$)	251				i H_2O, EtOH, eth, bz, chl; s py; sl ace
416	3-Amino-1H-pyrazole-4-carbonitrile	3-Amino-4-cyanopyrazole	$C_4H_4N_4$	16617-46-2	108.102	cry (w)	173				
417	2-Amino-3-pyridinecarboxylic acid		$C_6H_6N_2O_2$	5345-47-1	138.124		296 dec				sl H_2O
418	6-Amino-3-pyridinecarboxylic acid	6-Aminonicotinic acid	$C_6H_6N_2O_2$	3167-49-5	138.124	cry (dil HOAc, +2w)	312				
419	4-Amino-N-2-pyridinylbenzenesulfonamide	Sulfapyridine	$C_{11}H_{11}N_3O_2S$	144-83-2	249.289	ye oran (al)	192				i H_2O, bz, ctc; s EtOH
420	5-Amino-2,4(1H,3H)-pyrimidinedione	5-Aminouracil	$C_4H_5N_3O_2$	932-52-5	127.102	nd (w)	dec				i H_2O; s alk, acid
421	6-Amino-2,4(1H,3H)-pyrimidinedione		$C_4H_5N_3O_2$	873-83-6	127.102	cry (w)	dec				vs H_2O
422	4-Amino-2(1H)-pyrimidinethione	2-Thiocytosine	$C_4H_5N_3S$	333-49-3	127.168						sl DMSO
423	5-Amino-2,4,6(1H,3H,5H)-pyrimidinetrione	Uramil	$C_4H_5N_3O_3$	118-78-5	143.101	nd or pl (w)	>400				s H_2O, chl; i eth, bz
424	4-Amino-N-2-pyrimidinylbenzenesulfonamide	Sulfadiazine	$C_{10}H_{10}N_4O_2S$	68-35-9	250.277	cry (w), wh pow	255 dec				sl H_2O, EtOH, ace, DMSO
425	Aminopyrine		$C_{13}H_{17}N_3O$	58-15-1	231.293	pr or pl (lig or AcOEt)	134.5				vs H_2O, bz, EtOH
426	4-Amino-N-2-quinoxalinylbenzenesulfonamide	Sulfaquinoxaline	$C_{14}H_{12}N_4O_2S$	59-40-5	300.336		247.5				sl H_2O, EtOH, ace; s aq alk
427	4-(Aminosulfonyl)benzoic acid	Carzenide	$C_7H_7NO_4S$	138-41-0	201.201	pr or lf (w)	291 dec				i H_2O; vs EtOH; sl eth; i bz
428	N-[4-(Aminosulfonyl)phenyl] acetamide	Acetylsulfanilamide	$C_8H_{10}N_2O_3S$	121-61-9	214.241	nd (HOAc)	219.5				s H_2O, EtOH, ace
429	5-Amino-1,3,4-thiadiazole-2(3H)-thione		$C_2H_3N_3S_2$	2349-67-9	133.195		243.0				
430	2-Amino-4(5H)-thiazolone		$C_3H_4N_2OS$	556-90-1	116.141	pr or nd (w)	256 dec				sl H_2O; i EtOH, eth
431	N-(Aminothioxomethyl)acetamide	Acetylthiourea	$C_3H_6N_2OS$	591-08-2	118.157	pr (w), orth (al)	165				sl H_2O, eth; s DMSO, EtOH
432	N-Amino-2-thioxo-4-thiazolidinone	3-Aminorhodanine	$C_3H_4N_2OS_2$	1438-16-0	148.206		101.5				s DMSO
433	1-Amino-2,2,2-trichloroethanol	Chloral ammonia	$C_2H_4Cl_3NO$	507-47-1	164.418	nd (al)	73	dec 100			vs bz, eth, EtOH
434	4-Amino-3,5,6-trichloro-2-pyridinecarboxlic acid	Picloram	$C_6H_3Cl_3N_2O_2$	1918-02-1	241.459		218.5				
435	11-Aminoundecanoic acid		$C_{11}H_{23}NO_2$	2432-99-7	201.307		189.0				
436	Amiton		$C_{10}H_{24}NO_3PS$	78-53-5	269.342	liq		76[0.01]		1.4655[27]	
437	Amitraz	N-Methylbis(2,4-xylyliminomethyl)amine	$C_{19}H_{23}N_3$	33089-61-1	293.406		86		1.128[20]		
438	Amitriptyline		$C_{20}H_{23}N$	50-48-6	277.404	cry	196 (HCl)				
439	Ammonium ferric oxalate		$C_6H_{12}FeN_3O_{12}$	14221-47-7	374.017		165 dec	1.78[17.5]			vs H_2O; i EtOH
440	Ammonium perfluorooctanoate		$C_8H_4F_{15}NO_2$	3825-26-1	431.100	solid					
441	Ammonium propanoate		$C_3H_9NO_2$	17496-08-1	91.109	hyg cry	45				s H_2O
442	Amobarbital	5-Ethyl-5-isopentyl-2,4,6(1H,3H,5H)-pyrimidinetrione	$C_{11}H_{18}N_2O_3$	57-43-2	226.272		157				vs bz, EtOH, chl
443	Amolanone	3-[2-(Diethylamino)ethyl]-3-phenyl-2(3H)-benzofuranone	$C_{20}H_{23}NO_2$	76-65-3	309.403	cry (peth)	43.4	193[2.0]		1.5614[25]	
444	Amoxicillin		$C_{16}H_{19}N_3O_5S$	26787-78-0	365.404	cry (w)					s H_2O
445	Amphecloral		$C_{11}H_{12}Cl_3N$	5581-35-1	264.579			96[0.5]		1.530	
446	Amphotericin B		$C_{47}H_{73}NO_{17}$	1397-89-3	924.080	ye pr (DMF)	170 dec				i H_2O; sl DMF; s DMSO
447	Ampicillin		$C_{16}H_{19}N_3O_4S$	69-53-4	349.405	cry	200 dec				sl H_2O
448	Ampyrone		$C_{11}H_{13}N_3O$	83-07-8	203.240	pa ye cry (bz)	109				s H_2O, EtOH, bz, chl; sl eth

3-Amino-1-propanol

1-Amino-2-propanol

α-(1-Aminopropyl)benzenemethanol

N-(3-Aminopropyl)-N-methyl-1,3-propanediamine

Aminopropylon

4-(2-Aminopropyl)phenol, (±)

N-(3-Aminopropyl)-1,3-propanediamine

Aminopterin

4-Amino-N-pyrazinylbenzenesulfonamide

3-Amino-1H-pyrazole-4-carbonitrile

2-Amino-3-pyridinecarboxylic acid

6-Amino-3-pyridinecarboxylic acid

4-Amino-N-2-pyridinylbenzenesulfonamide

5-Amino-2,4(1H,3H)-pyrimidinedione

6-Amino-2,4(1H,3H)-pyrimidinedione

4-Amino-2(1H)-pyrimidinethione

5-Amino-2,4,6(1H,3H,5H)-pyrimidinetrione

4-Amino-N-2-pyrimidinylbenzenesulfonamide

Aminopyrine

4-Amino-N-2-quinoxalinylbenzenesulfonamide

4-(Aminosulfonyl)benzoic acid

N-[4-(Aminosulfonyl)phenyl]acetamide

5-Amino-1,3,4-thiadiazole-2(3H)-thione

2-Amino-4(5H)-thiazolone

N-(Aminothioxomethyl)acetamide

N-Amino-2-thioxo-4-thiazolidinone

1-Amino-2,2,2-trichloroethanol

4-Amino-3,5,6-trichloro-2-pyridinecarboxlic acid

11-Aminoundecanoic acid

Amiton

Amitraz

Amitriptyline

Ammonium ferric oxalate

Ammonium perfluorooctanoate

Ammonium propanoate

Amobarbital

Amolanone

Amoxicillin

Amphecloral

Amphotericin B

Ampicillin

Ampyrone

No.	Name	Synonym	Mol. Form.	CAS RN	Mol. Wt.	Physical Form	mp/°C	bp/°C	den/ g cm⁻³	n_D	Solubility
449	Amygdalin		$C_{20}H_{27}NO_{11}$	29883-15-6	457.428		224.5				vs H₂O; sl EtOH; i eth, chl
450	Anacardic acid		$C_{22}H_{32}O_3$	11034-77-8	344.487	cry (ace)	35.5				vs eth, EtOH, peth
451	Anagyrine		$C_{15}H_{20}N_2O$	486-89-5	244.332	pe ye glass		265[12], 212[4]			s H₂O, eth, bz; vs EtOH, chl; i lig
452	Androstane		$C_{19}H_{32}$	24887-75-0	260.457	lf (ace-MeOH)	50	60[0.003]			vs ace, eth, EtOH, peth
453	Androstane-17-carboxylic acid, (5β,17β)	Etiocholanic acid	$C_{20}H_{32}O_2$	438-08-4	304.467	nd (gl HOAc)	228.5	sub 160			
454	Androstane-3,17-diol, (3α,5α,17β)	Epiandrostanediol	$C_{19}H_{32}O_2$	1852-53-5	292.456	nd (ace aq)	223				
455	5α-Androstane-3,17-dione		$C_{19}H_{28}O_2$	846-46-8	288.424	cry (MeOH)	135				
456	5β-Androstane-3,17-dione		$C_{19}H_{28}O_2$	1229-12-5	288.424	cry (ace-hx)	135				
457	Androst-4-ene-3,17-dione	4-Androstene-3,17-dione	$C_{19}H_{26}O_2$	63-05-8	286.408		143(form a); 173(form b)				
458	Androst-4-ene-3,11,17-trione	Adrenosterone	$C_{19}H_{24}O_3$	382-45-6	300.392	nd (al)	222	sub			sl H₂O; s EtOH, eth, ace, chl
459	Anemonin	trans-1,7-Dioxadispiro[4.0.4.2] dodeca-3,9-diene-2,8-dione	$C_{10}H_8O_4$	508-44-1	192.169	orth pl (chl) nd (al or bz)	158				vs chl
460	Anhalamine		$C_{11}H_{15}NO_3$	643-60-7	209.242	nd (al)	187.5				vs eth, EtOH
461	Anhalonidine		$C_{12}H_{17}NO_3$	17627-77-9	223.268	oct cry (bz, eth)	160.5				vs H₂O, EtOH
462	Anhalonine		$C_{12}H_{15}NO_3$	519-04-0	221.252	rhom nd	86	140[0.02]			vs EtOH, bz, chl, eth, peth
463	2,5-Anhydro-3,4-dideoxyhexitol	Tetrahydro-2,5-furandimethanol	$C_6H_{12}O_3$	104-80-3	132.157		<-50	265	1.154[20]		vs H₂O, ace, bz, EtOH
464	Anilazine	2,4-Dichloro-6-(o-chloroanilino)-s-triazine	$C_9H_5Cl_3N_4$	101-05-3	275.522		160		1.8[20]		
465	Anileridine		$C_{22}H_{28}N_2O_2$	144-14-9	352.469	cry	83				s H₂O
466	Aniline	Benzenamine	C_6H_7N	62-53-3	93.127	oily liq	-6.02	184.17	1.0217[20]	1.5863[20]	s H₂O, ctc, lig; msc EtOH, eth, ace, bz
467	Aniline-2-carboxylic acid	o-Anthranilic acid	$C_7H_7NO_2$	118-92-3	137.137	lf (al)	146.5	sub	1.412[20]		s H₂O, EtOH, eth; sl bz, tfa; vs chl, py
468	Aniline-3-carboxylic acid	m-Anthranilic acid	$C_7H_7NO_2$	99-05-8	137.137		173		1.51[25]		sl H₂O, EtOH; s eth, tfa; vs ace; i bz
469	Aniline-4-carboxylic acid	p-Anthranilic acid	$C_7H_7NO_2$	150-13-0	137.137	mcl pr (w)	188.2		1.374[20]		s H₂O, EtOH, eth; sl ace; i bz, chl
470	Aniline hydrobromide		C_6H_8BrN	542-11-0	174.039		286				
471	Aniline hydrochloride	Benzenamine hydrochloride	C_6H_8ClN	142-04-1	129.588	lf or nd	198		1.2215[4]		vs H₂O, EtOH; i eth, chl; sl DMSO
472	Aniline nitrate		$C_6H_8N_2O_3$	542-15-4	156.139	orth	190 dec		1.356[4]		vs H₂O, eth, EtOH
473	Aniline sulfate (2:1)		$C_{12}H_{16}N_2O_4S$	542-16-5	284.331				1.377[4]		s H₂O; sl EtOH, tfa; i eth
474	Anisole	Methoxybenzene	C_7H_8O	100-66-3	108.138	liq	-37.13	153.7	0.9940[20]	1.5174[20]	i H₂O; s EtOH, eth, chl; vs ace, bz
475	Anisotropine methylbromide	Octatropine methylbromide	$C_{17}H_{32}BrNO_2$	80-50-2	362.346	cry (ace)	329				
476	Antazoline		$C_{17}H_{19}N_3$	91-75-8	265.353	cry	122				
477	Anthra[9,1,2-cde]benzo[rst] pentaphene-5,10-dione		$C_{34}H_{16}O_2$	116-71-2	456.490	viol-bl or blk nd (PhNO₂)	492 dec				i EtOH, bz, HOAc; s xyl, py, sulf
478	2-Anthracenamine		$C_{14}H_{11}N$	613-13-8	193.244	ye lf (al)	238.8	sub			i H₂O; s EtOH; i con sulf
479	Anthracene		$C_{14}H_{10}$	120-12-7	178.229	tab or mcl pr (al)	215.76	339.9	1.28[25]		i H₂O; sl EtOH, eth, ace, bz, chl, ctc
480	9-Anthracenecarbonitrile		$C_{15}H_9N$	1210-12-4	203.239		177.5		1.3000[20]		
481	9-Anthracenecarboxaldehyde		$C_{15}H_{10}O$	642-31-9	206.239	oran nd (dil HOAc)	104.5				i H₂O; s bz, HOAc
482	1-Anthracenecarboxylic acid	1-Anthroic acid	$C_{15}H_{10}O_2$	607-42-1	222.239	ye nd (HOAc) ye pr (al)	251.5	sub			i H₂O; s EtOH, eth; sl bz, chl
483	2-Anthracenecarboxylic acid	2-Anthroic acid	$C_{15}H_{10}O_2$	613-08-1	222.239	ye lf (al) nd, lf (sub)	281	sub			vs HOAc
484	9-Anthracenecarboxylic acid	9-Anthroic acid	$C_{15}H_{10}O_2$	723-62-6	222.239		217 dec	sub			i H₂O; s EtOH
485	9,10-Anthracenedicarbonitrile		$C_{16}H_8N_2$	1217-45-4	228.248		337 dec				
486	9,10-Anthracenediol		$C_{14}H_{10}O_2$	4981-66-2	210.228	br or ye nd	180				vs eth, EtOH

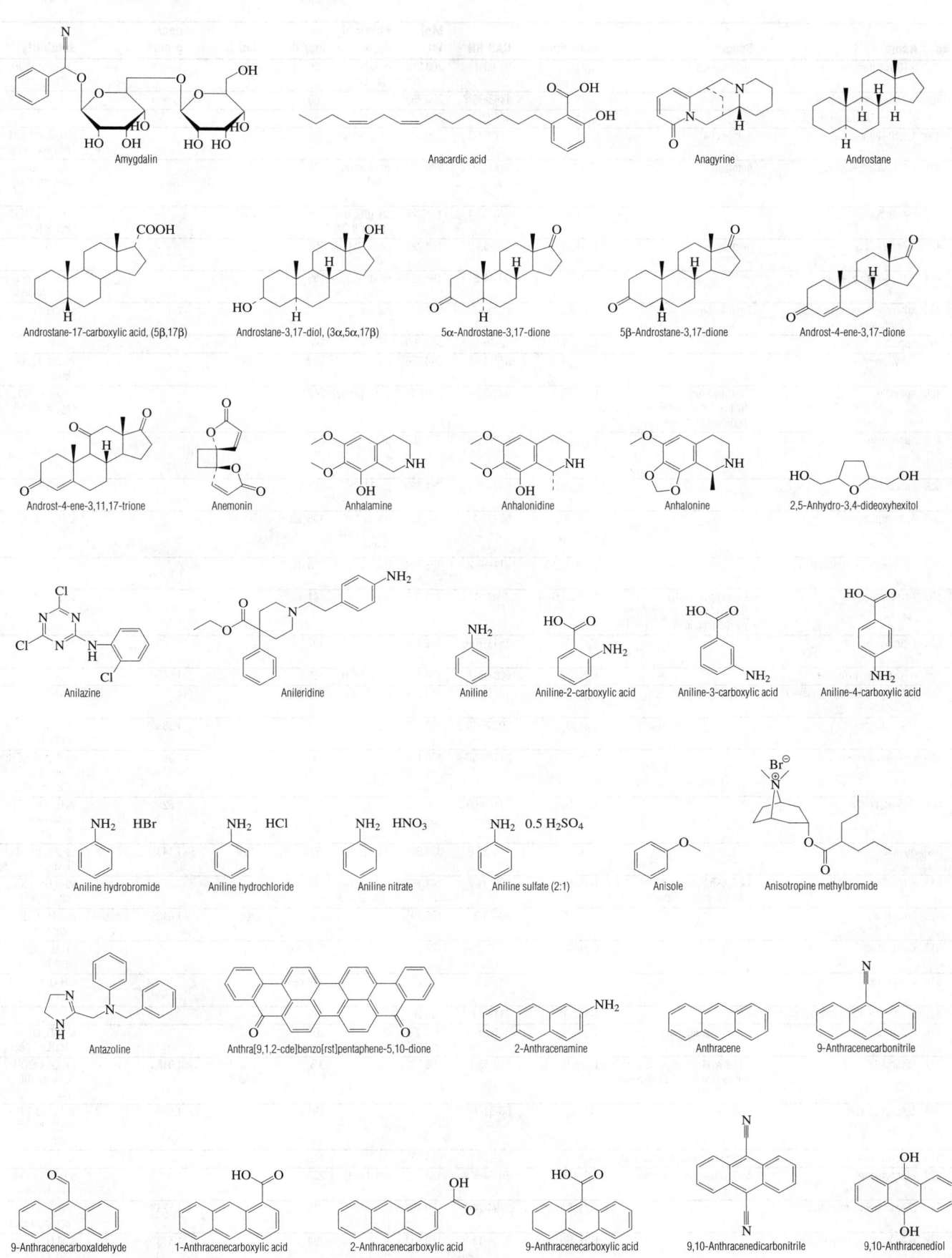

Amygdalin

Anacardic acid

Anagyrine

Androstane

Androstane-17-carboxylic acid, (5β,17β)

Androstane-3,17-diol, (3α,5α,17β)

5α-Androstane-3,17-dione

5β-Androstane-3,17-dione

Androst-4-ene-3,17-dione

Androst-4-ene-3,11,17-trione

Anemonin

Anhalamine

Anhalonidine

Anhalonine

2,5-Anhydro-3,4-dideoxyhexitol

Anilazine

Anileridine

Aniline

Aniline-2-carboxylic acid

Aniline-3-carboxylic acid

Aniline-4-carboxylic acid

Aniline hydrobromide

Aniline hydrochloride

Aniline nitrate

Aniline sulfate (2:1)

Anisole

Anisotropine methylbromide

Antazoline

Anthra[9,1,2-cde]benzo[rst]pentaphene-5,10-dione

2-Anthracenamine

Anthracene

9-Anthracenecarbonitrile

9-Anthracenecarboxaldehyde

1-Anthracenecarboxylic acid

2-Anthracenecarboxylic acid

9-Anthracenecarboxylic acid

9,10-Anthracenedicarbonitrile

9,10-Anthracenediol

No.	Name	Synonym	Mol. Form.	CAS RN	Mol. Wt.	Physical Form	mp/°C	bp/°C	den/ g cm⁻³	n_D	Solubility
487	9,10-Anthracenedione	Anthraquinone	C₁₄H₈O₂	84-65-1	208.213	ye orth nd (al, bz)	286	377	1.438²⁰		i H₂O; sl EtOH, eth, bz, chl
488	9-Anthracenemethanol		C₁₅H₁₂O	1468-95-7	208.255		160.5				
489	1,4,9,10-Anthracenetetrol		C₁₄H₁₀O₄	476-60-8	242.227		148				
490	1,2,10-Anthracenetriol	Anthrarobin	C₁₄H₁₀O₃	577-33-3	226.227	ye lf, nd (al-w)	208				sl H₂O; vs EtOH, eth, ace; s bz
491	1,8,9-Anthracenetriol	Anthralin	C₁₄H₁₀O₃	1143-38-0	226.227	ye pl or nd (lig)	179				i H₂O; s EtOH, ace, bz; sl eth; vs py
492	1-Anthracenol		C₁₄H₁₀O	610-50-4	194.228	cry (bz), br nd or lf (al)	158	234¹³			i H₂O; vs EtOH, eth; s NaOH
493	9-Anthracenol	Anthranol	C₁₄H₁₀O	529-86-2	194.228	ye red lf (dil al)	152				
494	9(10H)-Anthracenone	Anthrone	C₁₄H₁₀O	90-44-8	194.228	nd (bz-lig, HOAc)	155				s ace, bz, con sulf, dil alk
495	Antimony potassium tartrate trihydrate	Tartar emetic	C₈H₁₀K₂O₁₅Sb₂	28300-74-5	667.873	col cry			2.6		sl H₂O
496	Apholate		C₁₂H₂₄N₉P₃	52-46-0	387.300		148				
497	Aphylline		C₁₅H₂₄N₂O	577-37-7	248.364	cry	52.5	200⁴			vs ace, bz, eth, EtOH
498	Apigenin	5,7-Dihydroxy-2-(4-hydroxyphenyl)-4H-1-benzopyran-4-one	C₁₅H₁₀O₅	520-36-5	270.237	ye nd (aq py)	347.5				i H₂O; s EtOH, py; vs dil alk
499	Apoatropine		C₁₇H₂₁NO₂	500-55-0	271.355	pr (chl)	62				sl H₂O, lig; vs EtOH, eth, ace, bz
500	Apocodeine		C₁₈H₁₉NO₂	641-36-1	281.350	pr (MeOH)	123.5				sl EtOH; s eth, ace, bz, lig
501	Apomorphine		C₁₇H₁₇NO₂	58-00-4	267.323	hex pl (chl-peth) rods (eth)	195 dec				sl H₂O; s EtOH, eth, ace, bz, alk
502	Apomorphine, hydrochloride		C₁₇H₁₈ClNO₂	314-19-2	303.784	grn in air mcl pr	205 dec				
503	Aprobarbital	5-Isopropyl-5-allyl-2,4,6(1H,3H,5H)-pyrimidinetrione	C₁₀H₁₄N₂O₃	77-02-1	210.229	cry	141				vs ace, eth, EtOH, chl
504	L-Arabinitol		C₅H₁₂O₅	7643-75-6	152.146		102.5				vs H₂O; sl EtOH; i eth
505	α-D-Arabinopyranose		C₅H₁₀O₅	608-45-7	150.130	cry (MeOH)	155.5		1.585²⁵		
506	6-O-α-L-Arabinopyranosyl-D-Glucose	Vicianose	C₁₁H₂₀O₁₀	14116-69-9	312.271	nd (dil al)	210 dec				vs H₂O
507	DL-Arabinose		C₅H₁₀O₅	20235-19-2	150.130	pr, nd (al)	164.5		1.585²⁰		vs H₂O; sl EtOH; i eth, bz
508	α-D-Arabinose		C₅H₁₀O₅	31178-68-4	150.130		156		1.585²⁵		vs H₂O; sl EtOH; i eth, ace, MeOH
509	β-D-Arabinose		C₅H₁₀O₅	31178-69-5	150.130		156		1.625²⁵		vs H₂O; sl EtOH; i eth, ace, MeOH
510	Aramite		C₁₅H₂₃ClO₄S	140-57-8	334.860		-37.3	195²	1.143²⁰	1.5100²⁰	vs ace, bz, eth, EtOH
511	Arecaidine	1,2,5,6-Tetrahydro-1-methyl-3-pyridinecarboxylic acid	C₇H₁₁NO₂	499-04-7	141.168	pl (dil al) tab (dil al +1w)	232 dec				vs H₂O; i EtOH, eth, bz, chl
512	Arecoline		C₈H₁₃NO₂	63-75-2	155.195			209	1.0485²⁰	1.486⁻²⁰	msc H₂O, EtOH, eth; s chl
513	D-Arginine		C₆H₁₄N₄O₂	7200-25-1	174.201		217 dec				i H₂O, EtOH, eth, bz
514	L-Arginine		C₆H₁₄N₄O₂	74-79-3	174.201		244 dec				s H₂O; sl EtOH; i eth
515	L-Arginine, monohydrochloride		C₆H₁₅ClN₄O₂	1119-34-2	210.662		219				
516	Artemisin	8-Hydroxysantonin	C₁₅H₁₈O₄	481-05-0	262.302	cry	203	260⁰·¹			sl H₂O, chl; s AcOEt; i peth
517	Ascaridole	1-Methyl-4-isopropyl-2,3-dioxabicyclo[2.2.2]oct-5-ene	C₁₀H₁₆O₂	512-85-6	168.233	liq	3.3	exp; 115¹⁵, 39⁰·²	1.0103²⁰	1.4769²⁰	i H₂O; s EtOH, ace, bz, tol; sl chl
518	L-Ascorbic acid	Vitamin C	C₆H₈O₆	50-81-7	176.124		191 dec		1.65²⁵		vs H₂O; s EtOH; i eth, bz, chl, peth
519	Ascorbyl palmitate	6-Hexadecanoylascorbic acid	C₂₂H₃₈O₇	137-66-6	414.533		112				
520	L-Asparagine	α-Aminosuccinamic acid	C₄H₈N₂O₃	70-47-3	132.118	orth (w+1)	235		1.543¹⁵		s H₂O; i EtOH, eth, MeOH
521	D-Asparagine, monohydrate		C₄H₁₀N₂O₄	5794-24-1	150.133		215		1.523¹⁵		sl H₂O; i EtOH, eth, bz, MeOH
522	L-Asparagine, monohydrate		C₄H₁₀N₂O₄	5794-13-8	150.133		234		1.543¹⁵		sl H₂O; i EtOH, eth, bz, MeOH

9,10-Anthracenedione

9-Anthracenemethanol

1,4,9,10-Anthracenetetrol

1,2,10-Anthracenetriol

1,8,9-Anthracenetriol

1-Anthracenol

9-Anthracenol

9(10H)-Anthracenone

Antimony potassium tartrate trihydrate

Apholate

Aphylline

Apigenin

Apoatropine

Apocodeine

Apomorphine

Apomorphine, hydrochloride

Aprobarbital

L-Arabinitol

α-D-Arabinopyranose

6-O-α-L-Arabinopyranosyl-D-Glucose

DL-Arabinose

α-D-Arabinose

β-D-Arabinose

Aramite

Arecaidine

Arecoline

D-Arginine

L-Arginine

L-Arginine, monohydrochloride

Artemisin

Ascaridole

L-Ascorbic acid

Ascorbyl palmitate

L-Asparagine

D-Asparagine, monohydrate

L-Asparagine, monohydrate

No.	Name	Synonym	Mol. Form.	CAS RN	Mol. Wt.	Physical Form	mp/°C	bp/°C	den/ g cm^{-3}	n_D	Solubility
523	Aspartame	L-α-Aspartyl-L-phenylalanine, 2-methyl ester	C$_{14}$H$_{18}$N$_2$O$_5$	22839-47-0	294.303	nd (w)	246.5				
524	DL-Aspartic acid		C$_4$H$_7$NO$_4$	617-45-8	133.104	mcl pr (w)	277.5		1.6622[13]		sl H$_2$O; i EtOH, eth, bz, py
525	L-Aspartic acid	L-Aminosuccinic acid	C$_4$H$_7$NO$_4$	56-84-8	133.104	orth lf (w)	270		1.6603[13]		sl H$_2$O; i EtOH, eth, bz; s dil HCl, py
526	Aspergillic acid		C$_{12}$H$_{20}$N$_2$O$_2$	490-02-8	224.299	pa ye rods	98				vs bz, eth, EtOH
527	Astemizole		C$_{28}$H$_{31}$FN$_4$O	68844-77-9	458.570	wh cry	149.1				i H$_2$O; s os
528	Asulam	Methyl [(4-aminophenyl)sulfonyl]carbamate	C$_8$H$_{10}$N$_2$O$_4$S	3337-71-1	230.241		144				
529	Atenolol		C$_{14}$H$_{22}$N$_2$O$_3$	29122-68-7	266.336	cry (AcOEt)	147				sl H$_2$O, diox, ace; i chl; s MeOH, HOAc
530	Atisine	Anthorine	C$_{22}$H$_{33}$NO$_2$	466-43-3	343.503	orth bipym	58.5				vs eth, EtOH, chl
531	Atrazine		C$_8$H$_{14}$ClN$_5$	1912-24-9	215.684		173				
532	Atropine		C$_{17}$H$_{23}$NO$_3$	51-55-8	289.370	orth nd (dil al)	118.5	sub 95			vs H$_2$O, EtOH; i eth; sl chl
533	Auramine hydrochloride		C$_{17}$H$_{24}$ClN$_3$O	2465-27-2	321.845	ye nd (w)	267				sl H$_2$O
534	Aureothin		C$_{22}$H$_{23}$NO$_6$	2825-00-5	397.421	ye pr	158				vs ace, EtOH, chl
535	Aurin		C$_{19}$H$_{14}$O$_3$	603-45-2	290.312	dk red lf or orth	309 dec				i H$_2$O, bz; s EtOH, alk; sl eth, chl
536	Aurin tricarboxylic acid, triammonium salt	Aluminon	C$_{22}$H$_{23}$N$_3$O$_9$	569-58-4	473.433	red-br pow					s H$_2$O; sl EtOH; i peth
537	Avermectin B1a	Abamectin	C$_{48}$H$_{72}$O$_{14}$	71751-41-2	873.078		152				
538	3-Azabicyclo[3.2.2]nonane		C$_8$H$_{15}$N	283-24-9	125.212			166^{500}			
539	1-Azabicyclo[2.2.2]octane	Quinuclidine	C$_7$H$_{13}$N	100-76-5	111.185	cry (eth)	158				vs H$_2$O, ace, eth, EtOH
540	1-Azabicyclo[2.2.2]octan-3-ol	3-Quinuclidinol	C$_7$H$_{13}$NO	1619-34-7	127.184	cry (bz)	221	sub 120			s ace
541	Azacitidine	4-Amino-1-β-D-ribofuranosyl-1,3,5-triazine-2(1H)-one	C$_8$H$_{12}$N$_4$O$_5$	320-67-2	244.205	cry	229				
542	Azacyclotridecan-2-one		C$_{12}$H$_{23}$NO	947-04-6	197.317		152.5				
543	8-Azaguanine		C$_4$H$_4$N$_6$O	134-58-7	152.114		300				
544	Azaserine		C$_5$H$_7$N$_3$O$_4$	115-02-6	173.128	ye-grn orth cry	150 dec				vs H$_2$O; sl EtOH, ace, MeOH
545	Azathioprine	6-[(1-Methyl-4-nitro-1H-imidazol-5-yl)thio]-1H-purine	C$_9$H$_7$N$_7$O$_2$S	446-86-6	277.263	ye cry	243 dec				sl H$_2$O, EtOH, chl
546	6-Azauridine	2-β-D-Ribofuranosyl-1,2,4-triazine-3,5(2H,4H)-dione	C$_8$H$_{11}$N$_3$O$_6$	54-25-1	245.189		158				s H$_2$O
547	Azetidine		C$_3$H$_7$N	503-29-7	57.095	liq	-70.0	63	0.8436^{20}	1.4287^{25}	vs ace, bz, eth, EtOH
548	2-Azetidinecarboxylic acid		C$_4$H$_7$NO$_2$	2517-04-6	101.105	cry (95% MeOH)	217 dec				
549	2-Azetidinone		C$_3$H$_5$NO	930-21-2	71.078		73.5	106^{15}			vs eth, EtOH, chl
550	Azidobenzene		C$_6$H$_5$N$_3$	622-37-7	119.124	pa ye oil	-27.5	70^{11}	1.0860^{20}	1.5589^{25}	i H$_2$O; sl EtOH, eth
551	1-Azido-4-chlorobenzene		C$_6$H$_4$ClN$_3$	3296-05-7	153.569		20	96^{20}	1.2634^{25}		i H$_2$O; s eth
552	2-Azidoethanol		C$_2$H$_5$N$_3$O	1517-05-1	87.080			75^{40}	1.146^{24}		vs H$_2$O
553	1-Azido-4-methylbenzene		C$_7$H$_7$N$_3$	2101-86-2	133.151		-29.0	dec 180; 80^{10}	1.0527^{23}		vs eth, EtOH
554	(Azidomethyl)benzene		C$_7$H$_7$N$_3$	622-79-7	133.151			108^{23}, 78^{12}	1.0730^{19}	1.5341^{25}	i H$_2$O; msc EtOH, eth
555	Azinphos ethyl		C$_{12}$H$_{16}$N$_3$O$_3$PS$_2$	2642-71-9	345.377	nd	53	111$^{0.001}$	1.284^{20}		reac alk
556	Azinphos-methyl		C$_{10}$H$_{12}$N$_3$O$_3$PS$_2$	86-50-0	317.324		73		1.44^{20}		
557	1-Aziridineethanol		C$_4$H$_9$NO	1072-52-2	87.120			168	1.088^{25}	1.4560^{20}	
558	trans-Azobenzene	trans-Diphenyldiazene	C$_{12}$H$_{10}$N$_2$	17082-12-1	182.220	oran-red mcl lf (al)	67.88	293	1.203^{20}	1.6266^{78}	sl H$_2$O; s EtOH, eth, bz, chl; vs py
559	cis-Azobenzene	cis-Diphenyldiazene	C$_{12}$H$_{10}$N$_2$	1080-16-6	182.220	oran-red pl (peth)	71				sl H$_2$O; s EtOH, eth, bz, HOAc, lig
560	3,3'-Azobenzenedisulfonyl chloride		C$_{12}$H$_8$Cl$_2$N$_2$O$_4$S$_2$	104115-88-0	379.239	red nd (eth)	166.5				vs eth
561	1,1'-Azobiscyclohexanecarbonitrile		C$_{14}$H$_{20}$N$_4$	2094-98-6	244.336		100				i H$_2$O; s lig
562	2,2'-Azobis[isobutyronitrile]	2,2'-Azobis[2-methylpropionitrile]	C$_8$H$_{12}$N$_4$	78-67-1	164.208						i H$_2$O; sl EtOH, eth
563	Azobutane		C$_8$H$_{18}$N$_2$	2159-75-3	142.242			60^{18}			
564	Azopropane		C$_6$H$_{14}$N$_2$	821-67-0	114.188			114			
565	cis-Azoxybenzene	Diphenyldiazene 1-oxide, (E)	C$_{12}$H$_{10}$N$_2$O	21650-65-7	198.219		87		1.166^{20}	1.633^{20}	

Aspartame

DL-Aspartic acid

L-Aspartic acid

Aspergillic acid

Astemizole

Asulam

Atenolol

Atisine

Atrazine

Atropine

Auramine hydrochloride

Aureothin

Aurin

Aurin tricarboxylic acid, triammonium salt

Avermectin B1a

3-Azabicyclo[3.2.2]nonane

1-Azabicyclo[2.2.2]octane

1-Azabicyclo[2.2.2]octan-3-ol

Azacitidine

Azacyclotridecan-2-one

8-Azaguanine

Azaserine

Azathioprine

6-Azauridine

Azetidine

2-Azetidinecarboxylic acid

2-Azetidinone

Azidobenzene

1-Azido-4-chlorobenzene

2-Azidoethanol

1-Azido-4-methylbenzene

(Azidomethyl)benzene

Azinphos ethyl

Azinphos-methyl

1-Aziridineethanol

trans-Azobenzene

cis-Azobenzene

3,3'-Azobenzenedisulfonyl chloride

1,1'-Azobiscyclohexanecarbonitrile

2,2'-Azobis[isobutyronitrile]

Azobutane

Azopropane

cis-Azoxybenzene

No.	Name	Synonym	Mol. Form.	CAS RN	Mol. Wt.	Physical Form	mp/°C	bp/°C	den/ g cm⁻³	n_D	Solubility
566	*trans*-Azoxybenzene	Diphenyldiazene 1-oxide, (Z)	C₁₂H₁₀N₂O	20972-43-4	198.219		34.6		1.1590[26]		i H₂O; s EtOH, eth
567	Azoxyethane	Diethyldiazine 1-oxide	C₄H₁₀N₂O	16301-26-1	102.134	liq		46			
568	Azulene	Bicyclo[5.3.0]decapentaene	C₁₀H₈	275-51-4	128.171	bl or gr-blk lf (al)	99	dec 270; 125[10]			i H₂O; s EtOH, eth, ace, acid; sl chl
569	Balan	*N*-Butyl-*N*-ethyl-2,6-dinitro-4-(trifluoromethyl)aniline	C₁₃H₁₆F₃N₃O₄	1861-40-1	335.279		66	121[0.5], 148[7]			
570	Barban		C₁₁H₉Cl₂NO₂	101-27-9	258.101		75				
571	Barbital	5,5-Diethylbarbituric acid	C₈H₁₂N₂O₃	57-44-3	184.192	nd (w)	190		1.220[25]		sl H₂O; s EtOH, eth, ace, chl, lig, tfa
572	Barbituric acid		C₄H₄N₂O₃	67-52-7	128.086	orth pr (w +2)	248	dec 260			s H₂O, eth; sl EtOH
573	Bayleton	Triadimefon	C₁₄H₁₆ClN₃O₂	43121-43-3	293.749		82		1.22[20]		
574	Bebeerine		C₃₆H₃₈N₂O₆	477-60-1	594.696	cry (bz, eth, chl-MeOH)	221				s EtOH, MeOH, eth; vs ace, chl
575	Benactyzine	2-(Diethylamino)ethyl benzilate	C₂₀H₂₅NO₃	302-40-9	327.418	cry	51				
576	Benactyzine hydrochloride	2-Diethylaminoethyl benzilate hydrochloride	C₂₀H₂₆ClNO₃	57-37-4	363.878		177.5				s H₂O; i eth
577	Benalaxyl		C₂₀H₂₃NO₃	71626-11-4	325.402		79		1.27[25]		
578	Bendiocarb	1,3-Benzodioxol-4-ol, 2,2-dimethyl-, methylcarbamate	C₁₁H₁₃NO₄	22781-23-3	223.226		130		1.25[20]		
579	Bendroflumethiazide		C₁₅H₁₄F₃N₃O₄S₂	73-48-3	421.415	cry	225				i H₂O, bz, eth; s EtOH, ace
580	Benomyl		C₁₄H₁₈N₄O₃	17804-35-2	290.318		dec				
581	Bensulfuron-methyl		C₁₆H₁₈N₄O₇S	83055-99-6	410.402		187				
582	Bensulide		C₁₄H₂₄NO₄PS₃	741-58-2	397.514		34.4		1.224[20]		
583	Bentazon		C₁₀H₁₂N₂O₃S	25057-89-0	240.278		138				
584	Benz[c]acridine	12-Azabenz[a]anthracene	C₁₇H₁₁N	225-51-4	229.276	nd (dil al)	132				vs bz, eth, EtOH
585	Benzaldehyde	Benzenecarboxaldehyde	C₇H₆O	100-52-7	106.122	liq	-57.1	178.8	1.0401[25]	1.5463[20]	sl H₂O; msc EtOH, eth; vs ace, bz
586	Benzaldehyde hydrazone	Benzylidene hydrazine	C₇H₈N₂	5281-18-5	120.152	lf	16	140[14]			s EtOH
587	*cis*-Benzaldehyde oxime		C₇H₇NO	622-32-2	121.137	pr	36.5	200	1.1111[20]	1.5908[20]	vs bz, eth, EtOH
588	*trans*-Benzaldehyde oxime		C₇H₇NO	622-31-1	121.137	nd (eth)	35	119[10]	1.145[20]		s H₂O; vs EtOH, eth
589	Benzaldehyde, phenylhydrazone		C₁₃H₁₂N₂	588-64-7	196.247	nd (lig), pr	157.0				sl EtOH, eth; s ace, bz, liq NH₃
590	Benzaldehyde, (phenylmethylene) hydrazone		C₁₄H₁₂N₂	588-68-1	208.258	ye pr (al)	93				i H₂O; s EtOH, eth, ace, bz, chl; sl ctc
591	Benzamide	Benzoic acid amide	C₇H₇NO	55-21-0	121.137	mcl pr or pl (w)	127.3	290	1.0792[130]		sl H₂O, eth, bz; vs EtOH, ctc, CS₂
592	Benz[a]anthracene	1,2-Benzanthracene	C₁₈H₁₂	56-55-3	228.288	lf (al)	160.5	438			i H₂O; vs EtOH
593	Benz[a]anthracene-7,12-dione		C₁₈H₁₀O₂	2498-66-0	258.271		170.5				sl EtOH, eth, lig; s ace; vs bz, chl
594	Benzanthrone		C₁₇H₁₀O	82-05-3	230.260		170				sl bz
595	Benzene	[6]Annulene	C₆H₆	71-43-2	78.112	orth pr or liq	5.49	80.09	0.8765[20]	1.5011[20]	sl H₂O; msc EtOH, eth, ace, chl; s ctc
596	Benzeneacetaldehyde	Phenylacetaldehyde	C₈H₈O	122-78-1	120.149		33.5	195	1.0272[20]	1.5255[20]	sl H₂O; s ace; msc EtOH, eth
597	Benzeneacetamide	α-Phenylacetamide	C₈H₉NO	103-81-1	135.163		157				sl H₂O, eth, bz; s EtOH
598	Benzeneacetic acid	Phenylacetic acid	C₈H₈O₂	103-82-2	136.149	lf, pl (peth)	76.5	265.5	1.228[6]		sl H₂O, chl; vs EtOH, eth; s ace; i lig
599	Benzeneacetic acid, hydrazide		C₈H₁₀N₂O	937-39-3	150.177		115.5				
600	Benzeneacetic anhydride		C₁₆H₁₄O₃	1555-80-2	254.280	pr or nd (eth)	73.3	195[12]			vs eth, chl
601	Benzeneacetonitrile	Benzyl cyanide	C₈H₇N	140-29-4	117.149	liq	-23.8	233.5	1.0205[15]	1.5211[25]	
602	Benzeneacetyl chloride	Phenylacetyl chloride	C₈H₇ClO	103-80-0	154.594			170[250], 105[24]	1.1682[20]	1.5325[20]	vs eth
603	Benzenearsonic acid		C₆H₇AsO₃	98-05-5	202.040	cry (w)	158 dec				vs H₂O, EtOH
604	Benzeneboronic acid		C₆H₇BO₂	98-80-6	121.930		219				sl H₂O; s EtOH, eth, bz
605	Benzenebutanoic acid		C₁₀H₁₂O₂	1821-12-1	164.201	lf (w)	52	290			s H₂O, EtOH, eth
606	Benzenebutanol		C₁₀H₁₄O	3360-41-6	150.217			140[14]		1.5214[20]	
607	Benzenecarboperoxoic acid	Perbenzoic acid	C₇H₆O₃	93-59-4	138.121	mcl pl (peth)	42	100[14]			vs ace, bz, eth, EtOH
608	Benzenecarbothioamide		C₇H₇NS	2227-79-4	137.203		117				

trans-Azoxybenzene

Azoxyethane

Azulene

Balan

Barban

Barbital

Barbituric acid

Bayleton

Bebeerine

Benactyzine

Benactyzine hydrochloride

Benalaxyl

Bendiocarb

Bendroflumethiazide

Benomyl

Bensulfuron-methyl

Bensulide

Bentazon

Benz[c]acridine

Benzaldehyde

Benzaldehyde hydrazone

cis-Benzaldehyde oxime

trans-Benzaldehyde oxime

Benzaldehyde, phenylhydrazone

Benzaldehyde, (phenylmethylene)hydrazone

Benzamide

Benz[a]anthracene

Benz[a]anthracene-7,12-dione

Benzanthrone

Benzene

Benzeneacetaldehyde

Benzeneacetamide

Benzeneacetic acid

Benzeneacetic acid, hydrazide

Benzeneacetic anhydride

Benzeneacetonitrile

Benzeneacetyl chloride

Benzenearsonic acid

Benzeneboronic acid

Benzenebutanoic acid

Benzenebutanol

Benzenecarboperoxoic acid

Benzenecarbothioamide

No.	Name	Synonym	Mol. Form.	CAS RN	Mol. Wt.	Physical Form	mp/°C	bp/°C	den/ g cm⁻³	n_D	Solubility
609	Benzenecarbothioic acid		C_7H_6OS	98-91-9	138.187	ye pl (HOAc)	24	86[10]	1.28[20]	1.6040[20]	vs ace, bz, eth, EtOH
610	Benzenecarboximidamide, monohydrochloride		$C_7H_9ClN_2$	1670-14-0	156.612	orth pr (w +2)	169				s H_2O, EtOH; sl tfa
611	1,2-Benzenediamine	o-Phenylenediamine	$C_6H_8N_2$	95-54-5	108.141	brsh ye lf (w) pl (chl)	102.1	257			s H_2O, eth, bz, chl; vs EtOH
612	1,3-Benzenediamine	m-Phenylenediamine	$C_6H_8N_2$	108-45-2	108.141	orth (al)	66.0	285	1.0096[58]	1.6339[58]	vs H_2O; s EtOH, eth, bz
613	1,4-Benzenediamine	p-Phenylenediamine	$C_6H_8N_2$	106-50-3	108.141	wh pl (bz, eth)	141.1	267			sl H_2O; s EtOH, eth, bz, chl
614	1,2-Benzenediamine, dihydrochloride		$C_6H_{10}Cl_2N_2$	615-28-1	181.062		250 dec				
615	1,3-Benzenediamine, dihydrochloride		$C_6H_{10}Cl_2N_2$	541-69-5	181.062						s H_2O
616	1,4-Benzenediamine, dihydrochloride		$C_6H_{10}Cl_2N_2$	624-18-0	181.062						s H_2O
617	1,2-Benzenedicarbonyl dichloride	Phthaloyl chloride	$C_8H_4Cl_2O_2$	88-95-9	203.023		15.5	281.1	1.4089[20]	1.5684[20]	
618	1,3-Benzenedicarbonyl dichloride		$C_8H_4Cl_2O_2$	99-63-8	203.023	pr(eth)	43.5	276	1.3880[17]	1.570[47]	sl H_2O, EtOH; s eth
619	1,4-Benzenedicarbonyl dichloride		$C_8H_4Cl_2O_2$	100-20-9	203.023	nd or pl (lig)	83.5	258; 125[9]			s eth
620	1,2-Benzenedicarboxaldehyde		$C_8H_6O_2$	643-79-8	134.133	ye cry or nd (lig)	55.8	83[0.8]			vs eth, EtOH
621	1,3-Benzenedicarboxaldehyde		$C_8H_6O_2$	626-19-7	134.133	nd (dil al)	89.5	246; 136[13]			sl H_2O, eth, chl; vs EtOH; s ace, bz
622	1,4-Benzenedicarboxaldehyde		$C_8H_6O_2$	623-27-8	134.133	nd (w)	117	246			sl H_2O; vs EtOH; s eth, chl, alk
623	1,2-Benzenedicarboxamide	Phthalamide	$C_8H_8N_2O_2$	88-96-0	164.162	cry	222	dec			sl H_2O, EtOH; i eth
624	1,4-Benzenedicarboxamide		$C_8H_8N_2O_2$	3010-82-0	164.162	nd (w), pl (HOAc)	322.3				
625	1,2-Benzenedicarboxylic acid, bis(2-butoxyethyl) ester	Bis(2-butoxyethyl) phthalate	$C_{20}H_{30}O_6$	117-83-9	366.448			270			
626	1,2-Benzenedicarboxylic acid, bis(2-methoxyethyl) ester	Bis(2-methoxyethyl) phthalate	$C_{14}H_{18}O_6$	117-82-8	282.289		-60.0	230[10]	1.1596[20]		
627	1,2-Benzenedicarboxylic acid, diallyl ester	Diallyl phthalate	$C_{14}H_{14}O_4$	131-17-9	246.259			161[4]			
628	1,2-Benzenedicarboxylic acid, dipropyl ester	Dipropyl phthalate	$C_{14}H_{18}O_4$	131-16-8	250.291	liq	-31.0	304.5	1.0767[20]		i H_2O; s EtOH, eth
629	1,3-Benzenedimethanamine	m-Xylene diamine	$C_8H_{12}N_2$	1477-55-0	136.194			247	1.052[20]		vs H_2O, eth, EtOH
630	1,2-Benzenedimethanol		$C_8H_{10}O_2$	612-14-6	138.164	pl (eth, peth)	64.8	145[3]			s H_2O, EtOH; vs eth; sl bz
631	1,3-Benzenedimethanol		$C_8H_{10}O_2$	626-18-6	138.164	nd (bz)	57	156[13]	1.1610[18]		vs H_2O, eth, EtOH
632	1,4-Benzenedimethanol		$C_8H_{10}O_2$	589-29-7	138.164	nd (w)	117.5	140[1]			vs H_2O, ace, eth, EtOH
633	1,2-Benzenediol, diacetate		$C_{10}H_{10}O_4$	635-67-6	194.184	nd (al)	64.5	142[9]			i H_2O; vs EtOH, eth, chl; s peth
634	1,4-Benzenediol, diacetate		$C_{10}H_{10}O_4$	1205-91-0	194.184	pl (w, al)	123.5		0.8731[25]		s H_2O; vs EtOH, eth, chl, lig
635	1,3-Benzenediol, monobenzoate		$C_{13}H_{10}O_3$	136-36-7	214.216			134.5			
636	1,3-Benzenedisulfonic acid		$C_6H_6O_6S_2$	98-48-6	238.238	hyg cry					
637	1,3-Benzenedisulfonyl dichloride		$C_6H_4Cl_2O_4S_2$	585-47-7	275.130		61.8	195[10.5]			
638	1,2-Benzenedithiol		$C_6H_6S_2$	17534-15-5	142.242		28.5	238.5			vs EtOH, eth, bz; s AcOEt
639	1,3-Benzenedithiol		$C_6H_6S_2$	626-04-0	142.242	lf	27	245			vs bz, eth, EtOH
640	Benzeneethanamine	1-Amino-2-phenylethane	$C_8H_{11}N$	64-04-0	121.180	liq	<0	195	0.9640[25]	1.5290[25]	s H_2O, ctc; vs EtOH, eth
641	Benzeneethanamine, hydrochloride		$C_8H_{12}ClN$	156-28-5	157.641	pl or lf (al)	218.5				vs H_2O, EtOH
642	Benzeneethanol	Phenethyl alcohol	$C_8H_{10}O$	60-12-8	122.164	liq	-27	218.2	1.0202[20]	1.5325[20]	sl H_2O; msc EtOH, eth
643	Benzenehexacarboxylic acid	Mellitic acid	$C_{12}H_6O_{12}$	517-60-2	342.169	nd (al)	287 dec				vs H_2O; s EtOH, sulf
644	Benzenemethanamine, hydrochloride		$C_7H_{10}ClN$	3287-99-8	143.614		258.3				vs H_2O, EtOH
645	Benzenemethanesulfonyl chloride		$C_7H_7ClO_2S$	1939-99-7	190.648	pr (eth), nd (bz)	93				vs eth, bz
646	Benzenemethanesulfonyl fluoride		$C_7H_7FO_2S$	329-98-6	174.193		92.0				
647	Benzenemethanethiol	Thiobenzyl alcohol	C_7H_8S	100-53-8	124.204	liq	-30	194.5	1.058[20]	1.5151[20]	i H_2O; vs EtOH, eth; sl ctc; s CS_2
648	Benzenepentanoic acid	5-Phenylvaleric acid	$C_{11}H_{14}O_2$	2270-20-4	178.228	pl (w), pr (peth)	57.5	190[30]			sl H_2O; vs EtOH; s os
649	Benzenepentanol		$C_{11}H_{16}O$	10521-91-2	164.244			155[20], 150[18]	0.9725[20]	1.5156[20]	vs eth, EtOH

Benzenecarbothioic acid

Benzenecarboximidamide, monohydrochloride

1,2-Benzenediamine

1,3-Benzenediamine

1,4-Benzenediamine

1,2-Benzenediamine, dihydrochloride

1,3-Benzenediamine, dihydrochloride

1,4-Benzenediamine, dihydrochloride

1,2-Benzenedicarbonyl dichloride

1,3-Benzenedicarbonyl dichloride

1,4-Benzenedicarbonyl dichloride

1,2-Benzenedicarboxaldehyde

1,3-Benzenedicarboxaldehyde

1,4-Benzenedicarboxaldehyde

1,2-Benzenedicarboxamide

1,4-Benzenedicarboxamide

1,2-Benzenedicarboxylic acid, bis(2-butoxyethyl) ester

1,2-Benzenedicarboxylic acid, bis(2-methoxyethyl) ester

1,2-Benzenedicarboxylic acid, diallyl ester

1,2-Benzenedicarboxylic acid, dipropyl ester

1,3-Benzenedimethanamine

1,2-Benzenedimethanol

1,3-Benzenedimethanol

1,4-Benzenedimethanol

1,2-Benzenediol, diacetate

1,4-Benzenediol, diacetate

1,3-Benzenediol, monobenzoate

1,3-Benzenedisulfonic acid

1,3-Benzenedisulfonyl dichloride

1,2-Benzenedithiol

1,3-Benzenedithiol

Benzeneethanamine

Benzeneethanamine, hydrochloride

Benzeneethanol

Benzenehexacarboxylic acid

Benzenemethanamine, hydrochloride

Benzenemethanesulfonyl chloride

Benzenemethanesulfonyl fluoride

Benzenemethanethiol

Benzenepentanoic acid

Benzenepentanol

No.	Name	Synonym	Mol. Form.	CAS RN	Mol. Wt.	Physical Form	mp/°C	bp/°C	den/ g cm⁻³	n_D	Solubility
650	Benzenepropanal	Hydrocinnamic aldehyde	C₉H₁₀O	104-53-0	134.174	mcl	47	224; 117²⁸	1.0190²⁰		i H₂O; vs EtOH; msc eth
651	Benzenepropanenitrile	Hydrocinnamonitrile	C₉H₉N	645-59-0	131.174	liq	-1	261; 141²⁵	1.0016²⁰	1.5266²⁸	s EtOH, eth; sl chl
652	Benzenepropanethiol		C₉H₁₂S	24734-68-7	152.256			121²³, 109¹⁰	1.01²⁵	1.5494²⁰	
653	Benzenepropanoic acid	Hydrocinnamic acid	C₉H₁₀O₂	501-52-0	150.174	nd (w)	48	279.8	1.0712⁴⁹		s H₂O, EtOH, eth, ctc, CS₂; vs bz
654	Benzenepropanol	Hydrocinnamyl alcohol	C₉H₁₂O	122-97-4	136.190		<-18	235	0.995²⁵	1.5357²⁵	s H₂O, ctc; msc EtOH, eth
655	Benzenepropanol carbamate	Phenprobamate	C₁₀H₁₃NO₂	673-31-4	179.216		102				i H₂O; s EtOH, chl
656	Benzenepropanoyl chloride		C₉H₉ClO	645-45-4	168.619			dec 225; 105¹⁰	1.135²¹		s eth, CS₂
657	Benzeneseleninic acid	Phenylseleninic acid	C₆H₆O₂Se	6996-92-5	189.07		124.5		1.93²⁰		sl H₂O; i bz; vs alk
658	Benzeneselenol		C₆H₆Se	645-96-5	157.07			183.6; 84²⁵	1.4865¹⁵		i H₂O; s EtOH; vs eth, ctc
659	Benzenesulfinic acid		C₆H₆O₂S	618-41-7	142.176	pr (w)	84	dec			sl H₂O; s EtOH, eth, bz; i peth
660	Benzenesulfinyl chloride		C₆H₅ClOS	4972-29-6	160.621	pl (peth)	38	71¹·⁵	1.3469²⁵	1.3470²⁵	s eth, chl
661	Benzenesulfonamide		C₆H₇NO₂S	98-10-2	157.191	lf, nd (w)	156				sl H₂O, tfa; s EtOH, eth
662	Benzenesulfonic acid	Besylic acid	C₆H₆O₃S	98-11-3	158.175	nd (bz)	65				vs H₂O, EtOH; i eth; sl bz; s HOAc
663	Benzenesulfonyl chloride	Phenylsulfonyl chloride	C₆H₅ClO₂S	98-09-9	176.621		14.5	dec 251	1.3470¹⁵		i H₂O; vs EtOH; s eth, ctc
664	Benzenesulfonyl fluoride	Phenylsulfonyl fluoride	C₆H₅FO₂S	368-43-4	160.166			203.5	1.3286²⁰	1.4932¹⁸	s EtOH, eth
665	1,2,4,5-Benzenetetracarboxylic acid	Pyromellitic acid	C₁₀H₆O₈	89-05-4	254.150	tcl pr (w+2)	276				sl H₂O; s EtOH
666	Benzenethiol	Phenyl mercaptan	C₆H₆S	108-98-5	110.177	liq	-14.93	169.1	1.0775²⁰	1.5893²⁰	i H₂O; s EtOH, eth, bz; sl ctc
667	1,3,5-Benzenetricarbonyl trichloride		C₉H₃Cl₃O₃	4422-95-1	265.477		36.3	180¹⁶			s chl
668	1,2,3-Benzenetricarboxylic acid	Hemimellitic acid	C₉H₆O₆	569-51-7	210.140	pr (al)	200		1.546²⁰		vs eth, EtOH
669	1,2,4-Benzenetricarboxylic acid	Trimellitic acid	C₉H₆O₆	528-44-9	210.140	nd (w) cry (al) cry (HOAc)	219				vs H₂O, eth, EtOH
670	1,3,5-Benzenetricarboxylic acid		C₉H₆O₆	554-95-0	210.140	pr or nd (w+1)	380				sl H₂O; vs EtOH, eth
671	1,2,4-Benzenetricarboxylic acid 1,2-anhydride, 4-chloride	4-(Chloroformyl)phthalic anhydride	C₉H₃ClO₄	1204-28-0	210.571		66				
672	1,2,4-Benzenetricarboxylic acid, triallyl ester		C₁₈H₁₈O₆	2694-54-4	330.332		<-30		1.164²⁰		
673	1,2,3-Benzenetriol	Pyrogallol	C₆H₆O₃	87-66-1	126.110	lf or nd (bz)	133	309	1.453⁴	1.561¹³⁴	vs H₂O, EtOH, eth, NH₃; s ace; i bz
674	1,2,4-Benzenetriol	Hydroxyhydroquinone	C₆H₆O₃	533-73-3	126.110	pl (eth), lf or pl (w)	140.5				vs H₂O, EtOH, eth; i bz, chl
675	1,3,5-Benzenetriol	Phloroglucinol	C₆H₆O₃	108-73-6	126.110	lf or pl (w+2)	218.5	sub	1.46²⁵		sl H₂O; vs EtOH, eth, bz, py; s ace
676	1,2,4-Benzenetriol triacetate		C₁₂H₁₂O₆	613-03-6	252.219		99	300			s EtOH, chl, MeOH
677	Benzestrol		C₂₀H₂₆O₂	85-95-0	298.419	cry (al)	164				vs ace, eth, EtOH, HOAc
678	Benzethonium chloride		C₂₇H₄₂ClNO₂	121-54-0	448.081	pl (chl/eth)	165 (hyd)				vs H₂O; s ace, chl, EtOH
679	Benzidene-3,3'-dicarboxylic acid	3,3'-Dicarboxybenzidine	C₁₄H₁₂N₂O₄	2130-56-5	272.256	nd	300 dec				
680	p-Benzidine	[1,1'-Biphenyl]-4,4'-diamine	C₁₂H₁₂N₂	92-87-5	184.236	nd (w)	120	401			sl H₂O, eth, DMSO; s EtOH
681	Benzil	Diphenylethanedione	C₁₄H₁₀O₂	134-81-6	210.228	ye pr (al)	94.87	347	1.084¹⁰²		i H₂O; vs EtOH, eth; s ace; sl ctc
682	1H-Benzimidazol-2-amine		C₇H₇N₃	934-32-7	133.151	pl (w)	224				s H₂O, EtOH, ace; sl eth, bz, DMSO
683	1H-Benzimidazole	N,N'-Methenyl-o-phenylenediamine	C₇H₆N₂	51-17-2	118.136	orth bipym pl (w)	170.5	>360			sl H₂O, eth; vs EtOH; i bz; s dil alk
684	1H-Benzimidazole-2-acetonitrile		C₉H₇N₃	4414-88-4	157.172		208.4				
685	1H-Benz[de]isoquinoline-1,3(2H)-dione		C₁₂H₇NO₂	81-83-4	197.190	nd (chl-al)	300				
686	Benzo[c]chrysene		C₂₂H₁₄	194-69-4	278.346	nd (AcOH)	126.5				
687	Benzo[g]chrysene	Benzo[a]triphenylene	C₂₂H₁₄	196-78-1	278.346	nd (AcOH)	114.5				

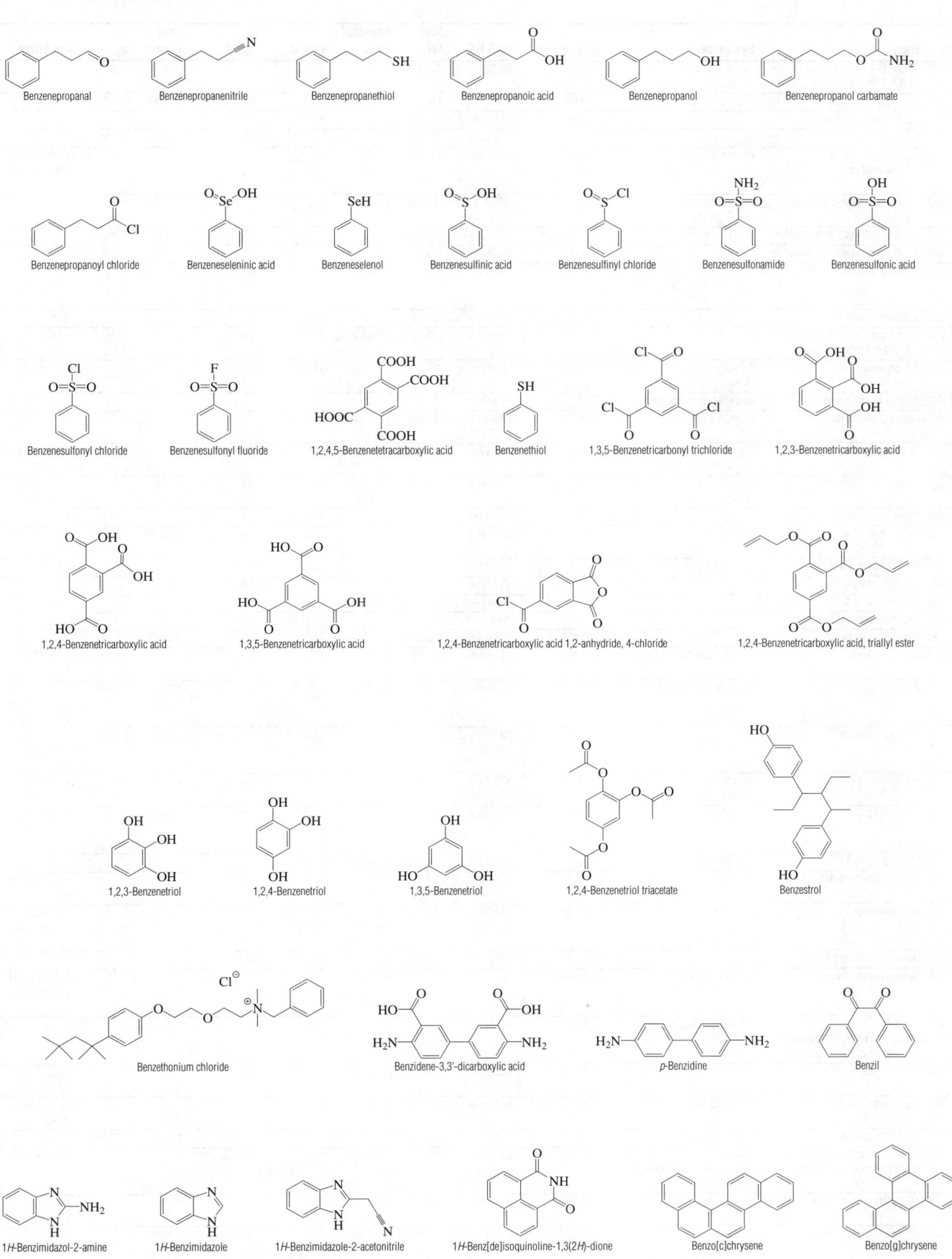

Benzenepropanal

Benzenepropanenitrile

Benzenepropanethiol

Benzenepropanoic acid

Benzenepropanol

Benzenepropanol carbamate

Benzenepropanoyl chloride

Benzeneseleninic acid

Benzeneselenol

Benzenesulfinic acid

Benzenesulfinyl chloride

Benzenesulfonamide

Benzenesulfonic acid

Benzenesulfonyl chloride

Benzenesulfonyl fluoride

1,2,4,5-Benzenetetracarboxylic acid

Benzenethiol

1,3,5-Benzenetricarbonyl trichloride

1,2,3-Benzenetricarboxylic acid

1,2,4-Benzenetricarboxylic acid

1,3,5-Benzenetricarboxylic acid

1,2,4-Benzenetricarboxylic acid 1,2-anhydride, 4-chloride

1,2,4-Benzenetricarboxylic acid, triallyl ester

1,2,3-Benzenetriol

1,2,4-Benzenetriol

1,3,5-Benzenetriol

1,2,4-Benzenetriol triacetate

Benzestrol

Benzethonium chloride

Benzidene-3,3'-dicarboxylic acid

p-Benzidine

Benzil

1H-Benzimidazol-2-amine

1H-Benzimidazole

1H-Benzimidazole-2-acetonitrile

1H-Benz[de]isoquinoline-1,3(2H)-dione

Benzo[c]chrysene

Benzo[g]chrysene

No.	Name	Synonym	Mol. Form.	CAS RN	Mol. Wt.	Physical Form	mp/°C	bp/°C	den/ g cm^{-3}	n_D	Solubility
688	1H,3H-Benzo[1,2-c:4,5-c']difuran-1,3,5,7-tetrone		$C_{10}H_2O_6$	89-32-7	218.119			285.3			
689	1,3,2-Benzodioxaborole		$C_6H_5BO_2$	274-07-7	119.914		12	88[156], 50[50]	1.2700[20]	1.5070[20]	
690	1,3-Benzodioxol-5-amine		$C_7H_7NO_2$	14268-66-7	137.137		42	144[16]			
691	1,3-Benzodioxole		$C_7H_6O_2$	274-09-9	122.122			172.5; 77[27]	1.064[25]	1.5398[20]	
692	1,3-Benzodioxole-5-carboxaldehyde	Piperonal	$C_8H_6O_3$	120-57-0	150.132		37	263			sl H_2O; vs EtOH; msc eth; s ace, chl
693	1,3-Benzodioxole-5-carboxylic acid	Piperonylic acid	$C_8H_6O_4$	94-53-1	166.132		229				
694	1,3-Benzodioxole-5-ethanamine		$C_9H_{11}NO_2$	1484-85-1	165.189			166[20], 101[1]	1.225[20]	1.5620[20]	
695	1,3-Benzodioxole-5-methanamine		$C_8H_9NO_2$	2620-50-0	151.163			139[13], 100[0.07]	1.214[25]	1.5635[20]	
696	1,3-Benzodioxole-5-methanol		$C_8H_8O_3$	495-76-1	152.148	nd (peth)	58	157[16]			sl H_2O; s EtOH, eth, bz, chl; i lig
697	1,3-Benzodioxol-5-ol		$C_7H_6O_3$	533-31-3	138.121		64.9				
698	trans,trans-5-(1,3-Benzodioxol-5-yl)-2,4-pentadienoic acid	Piperinic acid	$C_{12}H_{10}O_4$	136-72-1	218.205	nd (al), ye nd (sub)	215.8	sub			vs EtOH
699	7,8-Benzoflavone	2-Phenyl-4H-naphtho[1,2-b]pyran-4-one	$C_{19}H_{12}O_2$	604-59-1	272.297	ye pl (al)	157				sl EtOH, chl; s sulf
700	Benzo[b]fluoranthene	Benz[e]acephenanthrylene	$C_{20}H_{12}$	205-99-2	252.309	nd (bz)	168				i H_2O; msc bz
701	Benzo[j]fluoranthene	Dibenzo[a,jk]fluorene	$C_{20}H_{12}$	205-82-3	252.309	ye pl (al) nd (HOAc)	166				i H_2O; sl EtOH, HOAc
702	Benzo[k]fluoranthene	2,3,1',8'-Binaphthylene	$C_{20}H_{12}$	207-08-9	252.309	pa ye nd (bz)	217	480			i H_2O; s EtOH, bz, HOAc
703	11H-Benzo[a]fluorene		$C_{17}H_{12}$	238-84-6	216.277	pl (ace or HOAc)	189.5	405			i H_2O; sl EtOH; s eth, bz, chl
704	11H-Benzo[b]fluorene		$C_{17}H_{12}$	243-17-4	216.277		212	401			i H_2O
705	Benzofuran	Coumarone	C_8H_6O	271-89-6	118.133		<-18	174	1.0913[25]	1.5615[17]	i H_2O; s EtOH, eth
706	2-Benzofurancarboxylic acid	Coumarilic acid	$C_9H_6O_3$	496-41-3	162.142	nd (w)	192.5	312.5			vs EtOH
707	2(3H)-Benzofuranone		$C_8H_6O_2$	553-86-6	134.133		50	249	1.2236[14]		
708	3(2H)-Benzofuranone		$C_8H_6O_2$	7169-34-8	134.133	red nd (al)	102.5	152[15]			vs bz
709	1-(2-Benzofuranyl)ethanone		$C_{10}H_8O_2$	1646-26-0	160.170		76	126[11]			s H_2O
710	Benzofurazan, 1-oxide		$C_6H_4N_2O_2$	480-96-6	136.108		71.5		1.280[80]		
711	Benzohydrazide	Benzoic acid, hydrazide	$C_7H_8N_2O$	613-94-5	136.151	pl (w)	115	dec 267			s H_2O, EtOH; sl eth, ace, chl
712	Benzoic acid	Benzenecarboxylic acid	$C_7H_6O_2$	65-85-0	122.122	mcl lf or nd	122.35	249.2	1.2659[15]	1.504[132]	sl H_2O; vs EtOH, eth; s ace, bz, chl
713	Benzoic anhydride		$C_{14}H_{10}O_3$	93-97-0	226.227	pr (eth)	42.5	360	1.989[15]	1.5767[15]	i H_2O, lig; s EtOH, eth; sl chl
714	Benzoin	2-Hydroxy-1,2-diphenylethanone, (±)	$C_{14}H_{12}O_2$	579-44-2	212.244		137	344; 194[12]	1.310[20]		vs EtOH, chl
715	Benzonitrile	Phenyl cyanide	C_7H_5N	100-47-0	103.122	liq	-13.99	191.1	1.0093[15]	1.5289[20]	sl H_2O; msc EtOH; vs ace, bz; s ctc
716	Benzo[ghi]perylene	1,12-Benzperylene	$C_{22}H_{12}$	191-24-2	276.330	ye-grn lf (bz)	272.5				i H_2O
717	Benzo[c]phenanthrene	Tetrahelicene	$C_{18}H_{12}$	195-19-7	228.288		68				i H_2O; sl EtOH, lig
718	Benzophenone	Diphenyl ketone	$C_{13}H_{10}O$	119-61-9	182.217	(α) orth pr (al); (β) mcl pr	47.9 (α); 26 (β)	305.4	1.111[18]	1.6077[19]	i H_2O; vs EtOH, eth, chl, ace; s bz
719	Benzophenone hydrazone		$C_{13}H_{12}N_2$	5350-57-2	196.247		97.3	227[55]			
720	Benzophenone, oxime	Diphenyl ketoxime	$C_{13}H_{11}NO$	574-66-3	197.232	nd (al)	144				i H_2O; vs EtOH, eth, chl, ace; s bz
721	3,3',4,4'-Benzophenonetetracarboxylic acid dianhydride	4,4'-Carbonyldiphthalic anhydride	$C_{17}H_6O_7$	2421-28-5	322.226		216				
722	Benzo-2-phenylhydrazide		$C_{13}H_{12}N_2O$	532-96-7	212.246	pr (al), nd (w)	168	314			sl H_2O, eth; s EtOH, bz, chl
723	Benzopurpurine 4B	C.I. Direct Red 2, disodium salt	$C_{34}H_{26}N_6Na_2O_6S_2$	992-59-6	724.716	br pow					s H_2O, EtOH, ac, H_2SO_4
724	2H-1-Benzopyran	1,2-Chromene	C_9H_8O	254-04-6	132.159			132[102], 91[13]	1.0993[16]	1.5869[24]	i H_2O
725	[2]Benzopyrano[6,5,4-def][2]benzopyran-1,3,6,8-tetrone	1,4,5,8-Naphthalenetetracarboxylic acid anhydride	$C_{14}H_4O_6$	81-30-1	268.178	nd (al)	450	sub 320			i H_2O; s Na_2CO_3, HOAc
726	1H-2-Benzopyran-1-one	Isocoumarin	$C_9H_6O_2$	491-31-6	146.143	pl (bz)	47	286			i H_2O; vs EtOH, eth, bz, CS_2
727	2H-1-Benzopyran-2-one	Coumarin	$C_9H_6O_2$	91-64-5	146.143	orth pym (eth)	71	301.7	0.935[20]		s H_2O, EtOH, alk; vs eth, chl, py

1*H*,3*H*-Benzo[1,2-c:4,5-c']difuran-1,3,5,7-tetrone

1,3,2-Benzodioxaborole

1,3-Benzodioxol-5-amine

1,3-Benzodioxole

1,3-Benzodioxole-5-carboxaldehyde

1,3-Benzodioxole-5-carboxylic acid

1,3-Benzodioxole-5-ethanamine

1,3-Benzodioxole-5-methanamine

1,3-Benzodioxole-5-methanol

1,3-Benzodioxol-5-ol

trans,trans-5-(1,3-Benzodioxol-5-yl)-2,4-pentadienoic acid

7,8-Benzoflavone

Benzo[b]fluoranthene

Benzo[j]fluoranthene

Benzo[k]fluoranthene

11*H*-Benzo[a]fluorene

11*H*-Benzo[b]fluorene

Benzofuran

2-Benzofurancarboxylic acid

2(3*H*)-Benzofuranone

3(2*H*)-Benzofuranone

1-(2-Benzofuranyl)ethanone

Benzofurazan, 1-oxide

Benzohydrazide

Benzoic acid

Benzoic anhydride

Benzoin

Benzonitrile

Benzo[ghi]perylene

Benzo[c]phenanthrene

Benzophenone

Benzophenone hydrazone

Benzophenone, oxime

3,3',4,4'-Benzophenonetetracarboxylic acid dianhydride

Benzo-2-phenylhydrazide

Benzopurpurine 4B

2*H*-1-Benzopyran

[2]Benzopyrano[6,5,4-def][2]benzopyran-1,3,6,8-tetrone

1*H*-2-Benzopyran-1-one

2*H*-1-Benzopyran-2-one

No.	Name	Synonym	Mol. Form.	CAS RN	Mol. Wt.	Physical Form	mp/°C	bp/°C	den/ g cm⁻³	n_D	Solubility
728	4H-1-Benzopyran-4-one		C₉H₆O₂	491-38-3	146.143	nd (peth w)	59	sub	1.2900²⁰		sl H₂O; s EtOH, eth, bz, chl
729	Benzo[a]pyrene	2,3-Benzopyrene	C₂₀H₁₂	50-32-8	252.309		181.1				i H₂O; vs chl
730	Benzo[e]pyrene	1,2-Benzpyrene	C₂₀H₁₂	192-97-2	252.309	pa ye nd (bz-MeOH)	181.4	311			i H₂O
731	Benzo[f]quinoline	β-Naphthoquinoline	C₁₃H₉N	85-02-9	179.217	lf (peth or w)	94	352; 203⁸			sl H₂O; vs EtOH, bz, eth; s ace
732	Benzo[h]quinoline		C₁₃H₉N	230-27-3	179.217	lf (eth), pl (peth)	52	339; 233⁴⁷	1.2340²⁰		sl H₂O; s EtOH, eth, ace, bz, ctc
733	p-Benzoquinone	2,5-Cyclohexadiene-1,4-dione	C₆H₄O₂	106-51-4	108.095	ye mcl pr (w)	115	sub	1.318²⁰		sl H₂O, peth; s EtOH, eth, chl
734	2,1,3-Benzothiadiazole		C₆H₄N₂S	273-13-2	136.174		43	206			
735	2-Benzothiazolamine	2-Aminobenzothiazole	C₇H₆N₂S	136-95-8	150.201	pl (w), lf (w)	132				sl H₂O; s EtOH, eth, chl, con HCl
736	6-Benzothiazolamine	6-Aminobenzothiazole	C₇H₆N₂S	533-30-2	150.201	pr (w)	87				i H₂O, eth; s EtOH
737	Benzothiazole	Benzosulfonazole	C₇H₅NS	95-16-9	135.187		1.0	231	1.2460²⁰	1.6379²⁰	sl H₂O; vs EtOH, eth, CS₂; s ace
738	2(3H)-Benzothiazolethione	2-Mercaptobenzothiazole	C₇H₅NS₂	149-30-4	167.252	pa ye mcl nd(al, MeOH)	181		1.42²⁰		i H₂O; s EtOH; sl eth, bz, DMSO
739	2(3H)-Benzothiazolethione, sodium salt		C₇H₄NNaS₂	2492-26-4	189.234						sl H₂O
740	2(3H)-Benzothiazolone		C₇H₅NOS	934-34-9	151.186	pr (dil al), nd	139	360			i H₂O; vs EtOH, eth
741	2(3H)-Benzothiazolone, hydrazone		C₇H₇N₃S	615-21-4	165.216		202.8				
742	2-(2-Benzothiazolyl)phenol		C₁₃H₉NOS	3411-95-8	227.281	nd or lf (al)	131	179³			s EtOH
743	Benzo[b]thiophene	Thianaphthene	C₈H₆S	95-15-8	134.199	lf	32	221	1.1484³²	1.6374³⁷	i H₂O; vs EtOH; s eth, ace, bz; sl chl
744	Benzo[b]thiophene-2-carboxylic acid	Thionaphthene-2-carboxylic acid	C₉H₆O₂S	6314-28-9	178.208	nd (w)	240.5				vs eth
745	1H-Benzotriazole	1,2,3-Triaza-1H-indene	C₆H₅N₃	95-14-7	119.124	nd (chl or bz)	100	204¹⁵			sl H₂O; s EtOH, bz, chl, tol, DMF
746	Benzo[b]triphenylene		C₂₂H₁₄	215-58-7	278.346	nd (al, HOAc)	205				i H₂O; vs bz
747	3H-2,1-Benzoxathiol-3-one 1,1-dioxide		C₇H₄O₄S	81-08-3	184.170	nd or pr (bz)	129.5	184¹⁸			vs bz, chl
748	2H-3,1-Benzoxazine-2,4(1H)-dione		C₈H₅NO₃	118-48-9	163.131	pr (al, gl HOAc) cry (al)	243 dec				sl H₂O, EtOH, ace; i eth, bz, chl
749	Benzoxazole	1-Oxa-3-azaindene	C₇H₅NO	273-53-0	119.121	pr (dil al)	31	182.5	1.1754²⁰	1.5594²⁰	i H₂O; s EtOH, sulf
750	2(3H)-Benzoxazolethione		C₇H₅NOS	2382-96-9	151.186	nd (w)	196				sl H₂O, ace, EtOH; vs eth, HOAc
751	2(3H)-Benzoxazolone		C₇H₅NO₂	59-49-4	135.121		138	335; 230³⁰			sl H₂O; s EtOH, eth, tfa
752	2-(2-Benzoxazolyl)phenol		C₁₃H₉NO₂	835-64-3	211.216	pink nd (al, HOAc)	123.5	338			sl H₂O; vs EtOH; s eth, ace, bz
753	N-Benzoyl-DL-alanine		C₁₀H₁₁NO₃	1205-02-3	193.199	pl, pr or lf (eth)	165.5	dec			s H₂O, EtOH; sl eth, DMSO
754	4-(Benzoylamino)-2-hydroxybenzoic acid	Benzoylpas	C₁₄H₁₁NO₄	13898-58-3	257.242		260.5				
755	Benzoyl azide	Benzazide	C₇H₅N₃O	582-61-6	147.134	pl (ace)	32	exp	1.1680³⁵		vs eth, EtOH
756	2-Benzoylbenzoic acid		C₁₄H₁₀O₃	85-52-9	226.227	tcl nd (w+1)	129.0				vs EtOH, eth; s bz; sl chl
757	4-Benzoylbenzoic acid		C₁₄H₁₀O₃	611-95-0	226.227	nd (HOAc), pl (al) mcl lf (w)	199	sub			sl H₂O, tfa, bz; s EtOH, eth, HOAc
758	2-Benzoylbenzoic acid, hydrazide		C₁₄H₁₂N₂O₂	787-84-8	240.257	nd (al)	242.3				sl H₂O; i EtOH, eth, chl; s MeOH
759	4-Benzoylbiphenyl	4-Phenylbenzophenone	C₁₉H₁₄O	2128-93-0	258.313		101.5	420; 156⁰·¹			
760	Benzoyl bromide	Benzoic acid, bromide	C₇H₅BrO	618-32-6	185.018	liq	-24	218.5	1.570¹⁵	1.5868²⁵	msc eth
761	Benzoyl chloride	Benzoic acid, chloride	C₇H₅ClO	98-88-4	140.567	liq	-0.4	197.2; 71⁹	1.2120²⁰	1.5537²⁰	msc eth; s bz, ctc, CS₂
762	Benzoyl cyclohexane	Cyclohexyl phenyl ketone	C₁₃H₁₆O	712-50-5	188.265	nd (peth)	59.5	164¹⁸			
763	Benzoylecgonine		C₁₆H₁₉NO₄	519-09-5	289.327	nd (w)	195				vs bz, EtOH
764	Benzoylferrocene		C₁₇H₁₄FeO	1272-44-2	290.137		110.0				
765	Benzoyl fluoride	Benzoic acid, fluoride	C₇H₅FO	455-32-3	124.112	liq	-28	154.5	1.1400²⁰		vs EtOH, eth; s ctc

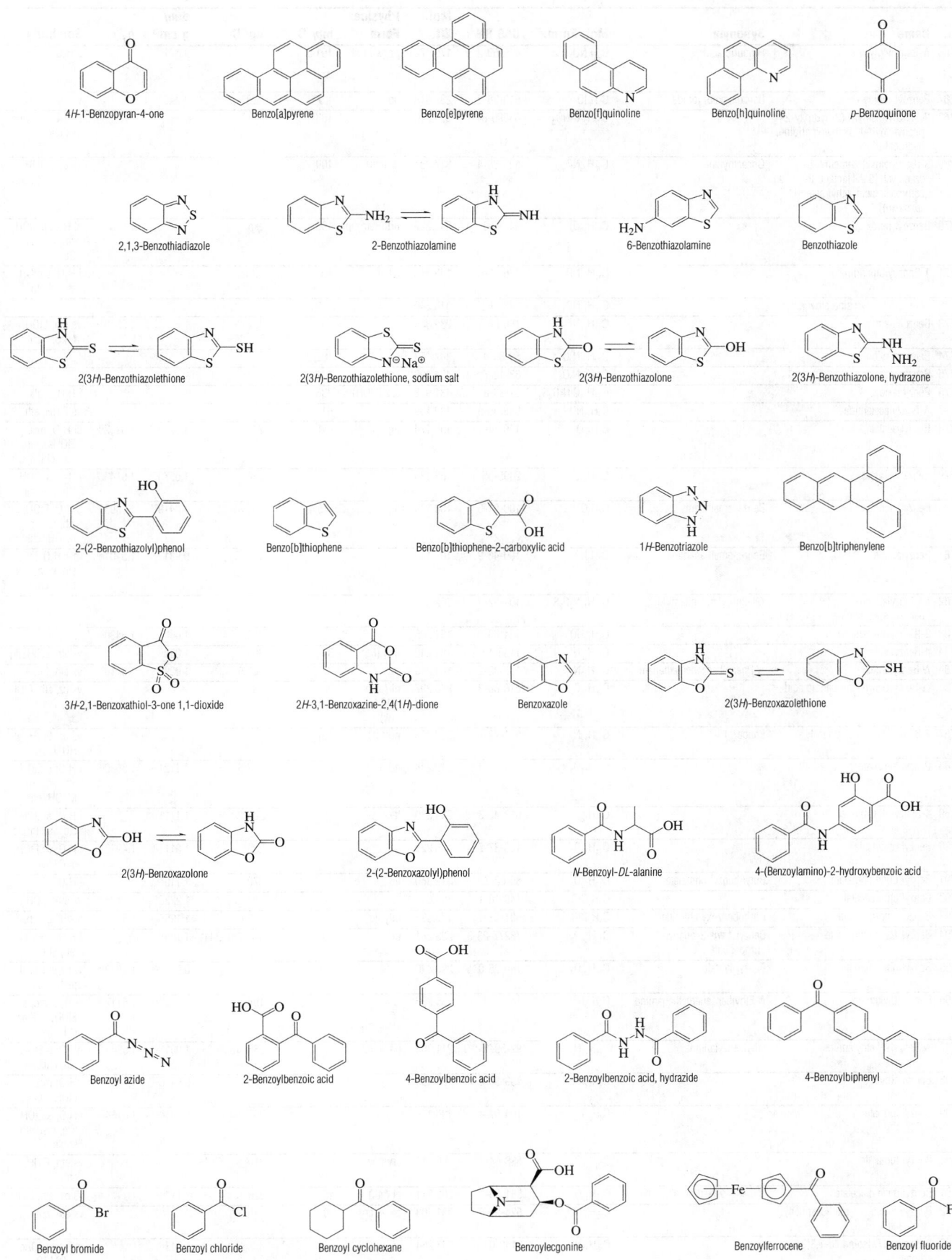

4*H*-1-Benzopyran-4-one

Benzo[a]pyrene

Benzo[e]pyrene

Benzo[f]quinoline

Benzo[h]quinoline

p-Benzoquinone

2,1,3-Benzothiadiazole

2-Benzothiazolamine

6-Benzothiazolamine

Benzothiazole

2(3*H*)-Benzothiazolethione

2(3*H*)-Benzothiazolethione, sodium salt

2(3*H*)-Benzothiazolone

2(3*H*)-Benzothiazolone, hydrazone

2-(2-Benzothiazolyl)phenol

Benzo[b]thiophene

Benzo[b]thiophene-2-carboxylic acid

1*H*-Benzotriazole

Benzo[b]triphenylene

3*H*-2,1-Benzoxathiol-3-one 1,1-dioxide

2*H*-3,1-Benzoxazine-2,4(1*H*)-dione

Benzoxazole

2(3*H*)-Benzoxazolethione

2(3*H*)-Benzoxazolone

2-(2-Benzoxazolyl)phenol

N-Benzoyl-*DL*-alanine

4-(Benzoylamino)-2-hydroxybenzoic acid

Benzoyl azide

2-Benzoylbenzoic acid

4-Benzoylbenzoic acid

2-Benzoylbenzoic acid, hydrazide

4-Benzoylbiphenyl

Benzoyl bromide

Benzoyl chloride

Benzoyl cyclohexane

Benzoylecgonine

Benzoylferrocene

Benzoyl fluoride

No.	Name	Synonym	Mol. Form.	CAS RN	Mol. Wt.	Physical Form	mp/°C	bp/°C	den/ g cm⁻³	n_D	Solubility
766	N-Benzoylglycine	Hippuric acid	$C_9H_9NO_3$	495-69-2	179.172	pr (w or al)	191.5		1.371[20]		s H₂O, EtOH; sl eth, bz, chl; i peth
767	Benzoyl iodide	Benzoic acid, iodide	C_7H_5IO	618-38-2	232.018	nd	1.5	128[20]	1.746[18]		vs eth, EtOH
768	2-Benzoylmethyl-6(2-hydroxy-2-phenylethyl)-1-methylpiperidine, hydrochloride		$C_{22}H_{28}ClNO_2$	63990-84-1	373.916		183.5				sl H₂O; s EtOH; vs chl
769	3-(Benzoyloxy)-8-methyl-8-azabicyclo[3.2.1]octane-2-carboxylic acid, ethyl ester, [1R-(exo,exo)]	Cocaethylene	$C_{18}H_{23}NO_4$	529-38-4	317.381	pr (eth)	109				vs eth, EtOH
770	Benzoyl peroxide		$C_{14}H_{10}O_4$	94-36-0	242.227	orth (eth), pr	105	exp		1.543	sl H₂O; s EtOH, eth, ace, bz, CS₂
771	1-Benzoylpiperidine		$C_{12}H_{15}NO$	776-75-0	189.253	tcl	49	320.5			i H₂O; s EtOH, eth; sl ctc
772	N-Benzoyl-L-tyrosine ethyl ester		$C_{18}H_{19}NO_4$	3483-82-7	313.349		119.5				
773	Benzphetamine		$C_{17}H_{21}N$	156-08-1	239.356			127[0.02]		1.5515[19]	vs eth, EtOH, MeOH, chl
774	Benzpiperylon		$C_{22}H_{25}N_3O$	53-89-4	347.453	cry (al)	182 dec				
775	Benzquinamide		$C_{22}H_{32}N_2O_5$	63-12-7	404.499	cry	131				
776	Benzthiazide		$C_{15}H_{14}ClN_3O_4S_3$	91-33-8	431.938	cry (EtOH)	236				i H₂O; s alk
777	N-Benzylacetamide		$C_9H_{11}NO$	588-46-5	149.189		61	157[2]			vs EtOH, eth
778	Benzyl acetate		$C_9H_{10}O_2$	140-11-4	150.174	liq	-51.3	213	1.0550[20]	1.5232[20]	sl H₂O; msc EtOH; s eth, ace, chl
779	Benzyl acrylate		$C_{10}H_{10}O_2$	2495-35-4	162.185			228	1.0573[20]	1.5143[20]	i H₂O; s EtOH, eth, ace, ctc
780	Benzyl alcohol	Benzenemethanol	C_7H_8O	100-51-6	108.138	liq	-15.4	205.31	1.0419[24]	1.5396[20]	s H₂O, EtOH, eth, ace, bz, MeOH, chl
781	Benzylamine	Benzenemethanamine	C_7H_9N	100-46-9	107.153			185; 90[12]	0.9813[20]	1.5401[20]	msc H₂O, EtOH, eth; vs ace; s bz; sl chl
782	4-(Benzylamino)benzenesulfonamide	N4-Benzylsulfanilamide	$C_{13}H_{14}N_2O_2S$	104-22-3	262.327		171				
783	2-[Benzylamino]ethanol		$C_9H_{13}NO$	104-63-2	151.205			225; 154[12]	1.065[25]	1.5430[20]	
784	4-Benzylaniline		$C_{13}H_{13}N$	1135-12-2	183.249	mcl (lig)	34.5	300	1.038[25]		vs eth, EtOH, lig
785	N-Benzylaniline	N-Phenylbenzenemethanamine	$C_{13}H_{13}N$	103-32-2	183.249	pr	37.5	306.5	1.0298[65]	1.6118[25]	vs eth, EtOH
786	α-Benzylbenzenepropanoic acid		$C_{16}H_{16}O_2$	618-68-8	240.297	pl (peth HOAc) nd (w)	90	235[18]			vs bz, eth, EtOH
787	2-Benzyl-1H-benzimidazole	Bendazol	$C_{14}H_{12}N_2$	621-72-7	208.258	nd (bz)	187				vs bz, EtOH, gl HOAc
788	Benzyl benzoate		$C_{14}H_{12}O_2$	120-51-4	212.244	nd or lf	21	323.5	1.1121[25]	1.5680[20]	i H₂O; s EtOH, eth, ace, bz, MeOH, chl
789	4-Benzyl-1,1'-biphenyl		$C_{19}H_{16}$	613-42-3	244.330	lf	85	285[110]	1.171[0]		i H₂O; s EtOH, ctc; vs eth, bz
790	Benzyl butanoate		$C_{11}H_{14}O_2$	103-37-7	178.228			239	1.0111[20]	1.4920[20]	i H₂O; vs EtOH, eth; s ctc
791	Benzyl butyl phthalate	Butyl benzyl phthalate	$C_{19}H_{20}O_4$	85-68-7	312.360	liq		370	1.119[25]		i H₂O
792	Benzyl chloroacetate		$C_9H_9ClO_2$	140-18-1	184.619			147[9], 85[0.4]	1.2223[4]	1.5426[18]	vs eth, EtOH
793	Benzyl chloroformate	Carbobenzoxy chloride	$C_8H_7ClO_2$	501-53-1	170.594	oily liq		103[20]	1.195[25]	1.5190[20]	s eth, ace, bz
794	Benzyl trans-cinnamate	Benzyl trans-3-phenyl-2-propenoate	$C_{16}H_{14}O_2$	78277-23-3	238.281	pr	39	dec 350; 244[5]	1.109[15]		i H₂O; s EtOH, eth; sl bz
795	Benzyl dodecanoate	Benzyl laurate	$C_{19}H_{30}O_2$	140-25-0	290.440		8.5	210[12]	0.9429[25]	1.4812[24]	vs bz, eth, EtOH, peth
796	Benzylethylamine	N-Ethylbenzenemethanamine	$C_9H_{13}N$	14321-27-8	135.206			194	0.9342[17]	1.5117[20]	sl H₂O, ctc; s EtOH, eth, bz, chl
797	N-Benzyl-N-ethylaniline	Ethylbenzylaniline	$C_{15}H_{17}N$	92-59-1	211.303	pa ye oil	35	288; 185[22]	1.001[55]	1.5943[23]	i H₂O; s EtOH, eth, chl
798	Benzyl ethyl ether		$C_9H_{12}O$	539-30-0	136.190			186	0.9478[20]	1.4955[20]	i H₂O; msc EtOH, eth
799	Benzyl formate		$C_8H_8O_2$	104-57-4	136.149			203; 84[10]	1.081[20]	1.5154[20]	i H₂O; s EtOH, ace; msc eth; sl ctc
800	Benzyl fumarate		$C_{18}H_{16}O_4$	538-64-7	296.318	cry pow	59	210[5]			vs eth, EtOH, chl
801	Benzylidene diacetate	Toluene-α,α-diol, diacetate	$C_{11}H_{12}O_4$	581-55-5	208.211	pl (eth)	46	220	1.11[20]		vs bz, eth, EtOH
802	Benzylimidobis(p-methoxyphenyl)methane		$C_{22}H_{21}NO_2$	524-96-9	331.408	pa ye cry	90				vs eth, chl
803	2-Benzyl-1H-isoindole-1,3(2H)-dione		$C_{15}H_{11}NO_2$	2142-01-0	237.254	ye nd (al)	116		1.343[18]		s EtOH, HOAc; sl DMSO

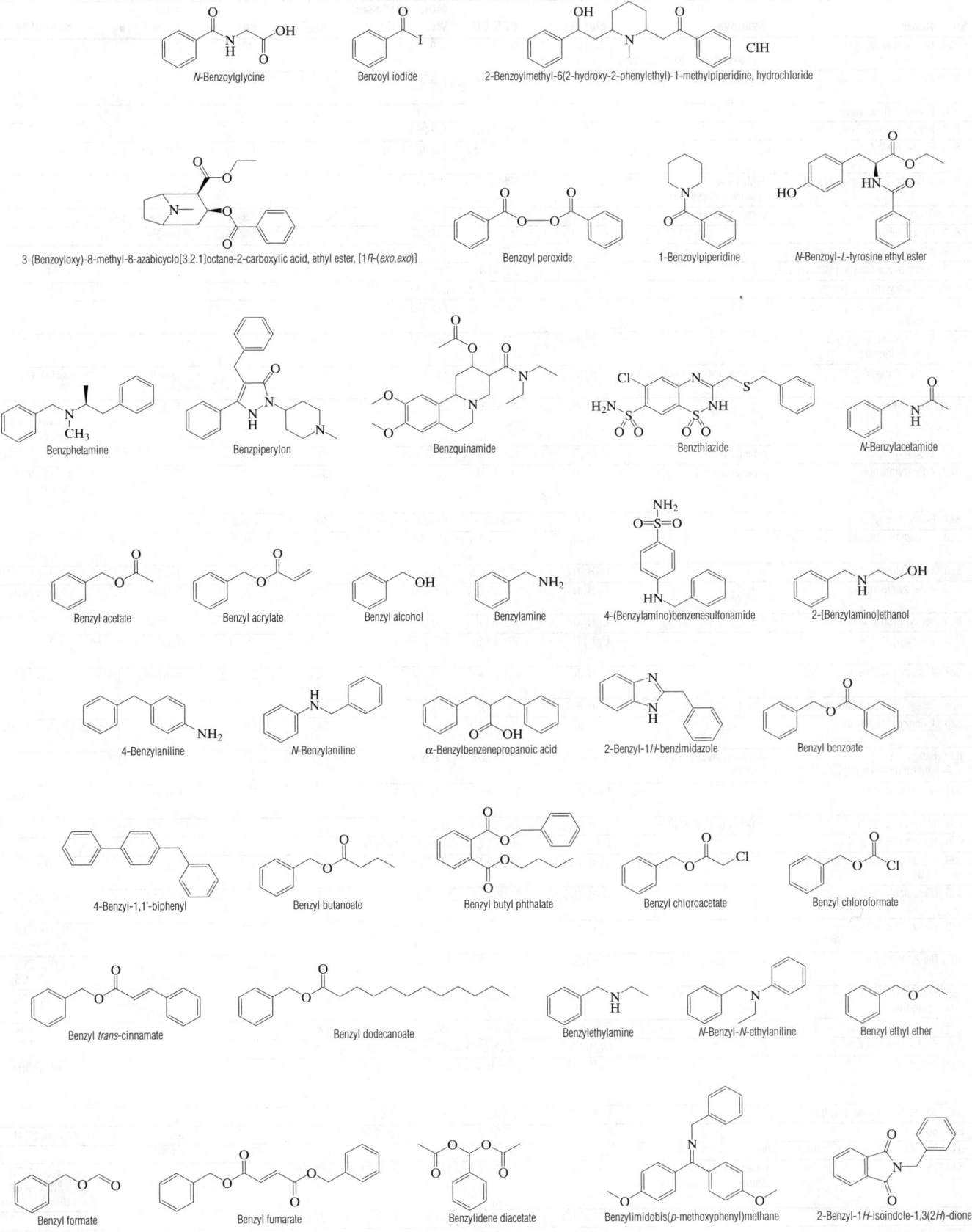

N-Benzoylglycine

Benzoyl iodide

2-Benzoylmethyl-6(2-hydroxy-2-phenylethyl)-1-methylpiperidine, hydrochloride

3-(Benzoyloxy)-8-methyl-8-azabicyclo[3.2.1]octane-2-carboxylic acid, ethyl ester, [1*R*-(*exo,exo*)]

Benzoyl peroxide

1-Benzoylpiperidine

N-Benzoyl-*L*-tyrosine ethyl ester

Benzphetamine

Benzpiperylon

Benzquinamide

Benzthiazide

N-Benzylacetamide

Benzyl acetate

Benzyl acrylate

Benzyl alcohol

Benzylamine

4-(Benzylamino)benzenesulfonamide

2-[Benzylamino]ethanol

4-Benzylaniline

N-Benzylaniline

α-Benzylbenzenepropanoic acid

2-Benzyl-1*H*-benzimidazole

Benzyl benzoate

4-Benzyl-1,1'-biphenyl

Benzyl butanoate

Benzyl butyl phthalate

Benzyl chloroacetate

Benzyl chloroformate

Benzyl *trans*-cinnamate

Benzyl dodecanoate

Benzylethylamine

N-Benzyl-N-ethylaniline

Benzyl ethyl ether

Benzyl formate

Benzyl fumarate

Benzylidene diacetate

Benzylimidobis(*p*-methoxyphenyl)methane

2-Benzyl-1*H*-isoindole-1,3(2*H*)-dione

No.	Name	Synonym	Mol. Form.	CAS RN	Mol. Wt.	Physical Form	mp/°C	bp/°C	den/ g cm⁻³	n_D	Solubility
804	Benzylisopropylamine	N-Isopropylbenzenemethanamine	C₁₀H₁₅N	102-97-6	149.233			200; 93[10]	0.892[25]	1.5025[20]	
805	Benzyl isothiocyanate	(Isothiocyanatomethyl)benzene	C₈H₇NS	622-78-6	149.214	ye oil		243	1.1246[16]	1.6049[15]	i H₂O; msc EtOH; s eth
806	Benzyl methacryalate		C₁₁H₁₂O₂	2495-37-6	176.212			144[50]			
807	Benzyl 3-methylbutanoate		C₁₂H₁₆O₂	103-38-8	192.254			245; 136[25]	0.9983[15]	1.4884[20]	
808	Benzyl methyl ether		C₈H₁₀O	538-86-3	122.164	liq	-52.6	170	0.9634[20]	1.5008[20]	i H₂O, lig; vs EtOH, eth; s bz
809	1-Benzyl-2-methylhydrazine	1-Methyl-2-phenylmethylhydrazine	C₈H₁₂N₂	10309-79-2	136.194	liq		117[20]			
810	Benzyl 2-methylpropanoate	Benzyl isobutyrate	C₁₁H₁₄O₂	103-28-6	178.228			228; 114[20]	1.0159[18]	1.4883[20]	
811	Benzyl nitrite		C₇H₇NO₂	935-05-7	137.137	oil		81[35]	1.075[25]	1.4989[25]	
812	N-Benzyloxycarbonylaspartame		C₂₂H₂₄N₂O₇	33605-72-0	428.435	cry	122				
813	Benzyloxycarbonyl-L-glutamine		C₁₃H₁₆N₂O₅	2650-64-8	280.276		134.5				s DMSO
814	Benzyloxycarbonylglycine		C₁₀H₁₁NO₄	1138-80-3	209.199		121				s ace
815	Benzyloxycarbonylglycyl-L-leucine		C₁₆H₂₂N₂O₅	1421-69-8	322.356		100				
816	Benzyloxycarbonylglycyl-L-phenylalanine		C₁₉H₂₀N₂O₅	1170-76-9	356.372		126				
817	2-(Benzyloxy)ethanol	Ethylene glycol monobenzyl ether	C₉H₁₂O₂	622-08-2	152.190	oil	<-75	256	1.0640[20]	1.5233[20]	vs H₂O, eth, EtOH
818	Benzylpenicillin sodium		C₁₆H₁₇N₂NaO₄S	69-57-8	356.372	nd (BuOH aq)	215		1.41		vs H₂O; s MeOH; i ace, eth, chl
819	2-Benzylphenol	o-Benzylphenol	C₁₃H₁₂O	28994-41-4	184.233		21	312		1.5994[20]	vs ace, bz, EtOH
820	4-Benzylphenol	p-Benzylphenol	C₁₃H₁₂O	101-53-1	184.233		84	322			s H₂O, EtOH, eth, bz, ctc, HOAc, chl
821	Benzyl phenyl ether		C₁₃H₁₂O	946-80-5	184.233	lf (al)	40	286.5			
822	1-Benzylpiperazine		C₁₁H₁₆N₂	2759-28-6	176.258			146[12]		1.5430[28]	s H₂O, EtOH, eth; sl chl
823	1-Benzylpiperidine		C₁₂H₁₇N	2905-56-8	175.270			245	0.9625[16]	1.5227[20]	
824	4-Benzylpiperidine		C₁₂H₁₇N	31252-42-3	175.270		16.8	270; 150[17]	0.9970[20]	1.5337[25]	i H₂O; s EtOH, eth
825	Benzyl propanoate		C₁₀H₁₂O₂	122-63-4	164.201			221	1.0335[20]		
826	2-Benzylpyridine		C₁₂H₁₁N	101-82-6	169.222	nd	12.5	277; 149[16]	1.067[0]	1.5785[20]	i H₂O; s EtOH, eth, chl
827	4-Benzylpyridine		C₁₂H₁₁N	2116-65-6	169.222		12.4	288; 180[31]	1.0612[20]	1.5818[20]	i H₂O; s EtOH, ctc; vs eth
828	Benzyl 3-pyridinecarboxylate	Benzyl nicotinate	C₁₃H₁₁NO₂	94-44-0	213.232			170[3]			
829	1-Benzyl-1H-pyrrole		C₁₁H₁₁N	2051-97-0	157.212		15	247	1.0183[20]	1.5655[24]	i H₂O; vs EtOH, eth
830	Benzyl 1,2-pyrrolidinedicarboxylate, (S)	N-(Benzyloxycarbonyl)-L-proline	C₁₃H₁₅NO₄	1148-11-4	249.263		78.5			1.5310[20]	sl chl
831	Benzyl salicylate		C₁₄H₁₂O₃	118-58-1	228.243			320	1.1799[20]	1.5805[20]	sl H₂O; s EtOH, eth, ctc
832	O-Benzyl-L-serine	3-(Benzyloxy)-L-alanine	C₁₀H₁₃NO₃	4726-96-9	195.215		218 dec				
833	Benzylsulfonic acid		C₇H₈O₃S	100-87-8	172.202	hyg cry					
834	4-[(Benzylsulfonyl)amino]benzoic acid	p-(Benzylsulfonamido)benzoic acid	C₁₄H₁₃NO₄S	536-95-8	291.323		229.5				vs EtOH
835	(Benzylsulfonyl)benzene		C₁₃H₁₂O₂S	3112-88-7	232.298	nd (al)	146		1.1261[153]		i H₂O; sl EtOH, eth, bz
836	(Benzylthio)benzene		C₁₃H₁₂S	831-91-4	200.299	lf (al)	43.5	197[27]			i H₂O; s EtOH, eth, con sulf
837	Benzyl thiocyanate	α-Thiocyanatotoluene	C₈H₇NS	3012-37-1	149.214	pr (al)	43	232			i H₂O; s EtOH, eth, chl, CS₂
838	Benzyltrimethylammonium chloride		C₁₀H₁₆ClN	56-93-9	185.694		243				vs H₂O; s ace
839	Benzylurea		C₈H₁₀N₂O	538-32-9	150.177	nd (al)	148	dec 200			vs ace, EtOH
840	Bephenium chloride		C₁₇H₂₂ClNO	13928-81-9	291.816	cry (ace)	135				
841	Berberine		C₂₀H₁₉NO₅	2086-83-1	353.369	red-ye nd (w+6) cry (chl)	145				vs eth, EtOH
842	Berberine chloride dihydrate		C₂₀H₂₂ClNO₆	633-65-8	407.845	ye cry					
843	Bergenin		C₁₄H₁₆O₉	477-90-7	328.272	cry (MeOH)	238				vs H₂O, EtOH
844	Beryllium 2,4-pentanedioate	Beryllium acetylacetonate	C₁₀H₁₄BeO₄	10210-64-7	207.228		108	270	1.168[20]		
845	Betaine	1-Carboxy-N,N,N-trimethylmethanaminium, inner salt	C₅H₁₁NO₂	107-43-7	117.147	pr or lf (al)	293 dec				vs H₂O, MeOH; s EtOH; sl eth, chl
846	Betaine, hydrochloride		C₅H₁₂ClNO₂	590-46-5	153.608	mcl cry (al)	227.5				vs H₂O
847	Betamethasone		C₂₂H₂₉FO₅	378-44-9	392.460	cry (AcOMe)	232 dec				
848	Bethanidine		C₁₀H₁₅N₃	55-73-2	177.246	cry (aq MeOH)	196				

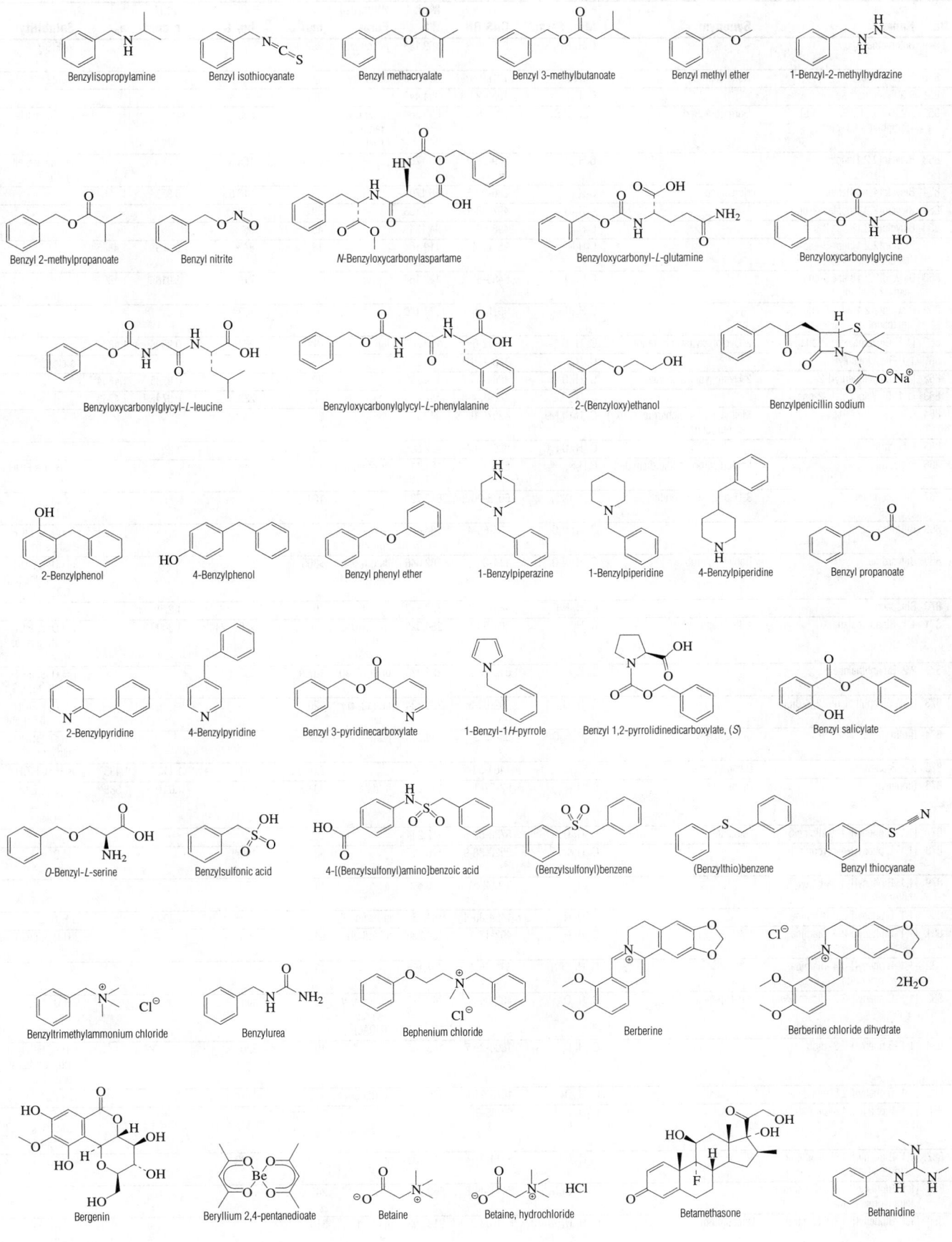

Benzylisopropylamine

Benzyl isothiocyanate

Benzyl methacryalate

Benzyl 3-methylbutanoate

Benzyl methyl ether

1-Benzyl-2-methylhydrazine

Benzyl 2-methylpropanoate

Benzyl nitrite

N-Benzyloxycarbonylaspartame

Benzyloxycarbonyl-*L*-glutamine

Benzyloxycarbonylglycine

Benzyloxycarbonylglycyl-*L*-leucine

Benzyloxycarbonylglycyl-*L*-phenylalanine

2-(Benzyloxy)ethanol

Benzylpenicillin sodium

2-Benzylphenol

4-Benzylphenol

Benzyl phenyl ether

1-Benzylpiperazine

1-Benzylpiperidine

4-Benzylpiperidine

Benzyl propanoate

2-Benzylpyridine

4-Benzylpyridine

Benzyl 3-pyridinecarboxylate

1-Benzyl-1*H*-pyrrole

Benzyl 1,2-pyrrolidinedicarboxylate, (*S*)

Benzyl salicylate

O-Benzyl-*L*-serine

Benzylsulfonic acid

4-[(Benzylsulfonyl)amino]benzoic acid

(Benzylsulfonyl)benzene

(Benzylthio)benzene

Benzyl thiocyanate

Benzyltrimethylammonium chloride

Benzylurea

Bephenium chloride

Berberine

Berberine chloride dihydrate

Bergenin

Beryllium 2,4-pentanedioate

Betaine

Betaine, hydrochloride

Betamethasone

Bethanidine

No.	Name	Synonym	Mol. Form.	CAS RN	Mol. Wt.	Physical Form	mp/°C	bp/°C	den/ g cm⁻³	n_D	Solubility
849	Betonicine		$C_7H_{13}NO_3$	515-25-3	159.183	pr (dil al, +1w)	252 dec				vs EtOH
850	Betulaprenol 9	Nonaisoprenol	$C_{45}H_{74}O$	13190-97-1	631.069	oil or cry	41				s chl
851	9,9'-Bianthracene		$C_{28}H_{18}$	1055-23-8	354.443			321.3			
852	Δ2,2'(3H,3'H)-Bibenzo[b] thiophene-3,3'-dione	Durindone Red	$C_{16}H_8O_2S_2$	522-75-8	296.364	br nd (xyl) red mcl nd (bz)	359	sub			i H_2O, EtOH; sl chl, CS_2; s bz, xyl
853	Bicyclo[2.2.1]heptane		C_7H_{12}	279-23-2	96.170		87.5	105.3			vs ace, bz, eth, EtOH
854	Bicyclo[4.1.0]heptane	Norcarane	C_7H_{12}	286-08-8	96.170			116.5	0.853²⁵	1.4564²⁰	
855	Bicyclo[2.2.1]heptan-2-one		$C_7H_{10}O$	497-38-1	110.153		89.5	170			
856	Bicyclo[2.2.1]hept-2-ene		C_7H_{10}	498-66-8	94.154		45	96			
857	Bicyclo[2.2.1]hept-5-ene-2-carbonitrile		C_8H_9N	95-11-4	119.164		13	84¹⁰	0.999²⁵	1.4885²⁰	
858	Bicyclo[2.2.1]hept-5-ene-2-carboxaldehyde		$C_8H_{10}O$	5453-80-5	122.164			71²⁰	1.018²⁵	1.4893²⁰	
859	Bicyclo[2.2.1]hept-5-ene-2-methanol		$C_8H_{12}O$	95-12-5	124.180			103²⁰			
860	[1,1'-Bicyclohexyl]-2-one	2-Cyclohexylcyclohexanone	$C_{12}H_{20}O$	90-42-6	180.286	liq	-32	264	0.9696²⁵	1.4877²⁵	
861	1,1'-Bicyclopentyl		$C_{10}H_{18}$	1636-39-1	138.250						s ctc, CS_2
862	[1,1'-Bicyclopentyl]-2-ol	2-Hydroxybicyclopentyl	$C_{10}H_{18}O$	4884-25-7	154.249		20	235.5	0.9785¹⁵	1.4884¹⁷	
863	[1,1'-Bicyclopentyl]-2-one		$C_{10}H_{16}O$	4884-24-6	152.233	liq	-13	232.5	0.9745²¹	1.4763	
864	Bifenox	Methyl 5-(2,4-dichlorophenoxy)-2-nitrobenzoate	$C_{14}H_9Cl_2NO_5$	42576-02-3	342.131		85				
865	Bifenthrin		$C_{23}H_{22}ClF_3O_2$	82657-04-3	422.868		69		1.2¹²⁵		
866	Biguanide	Imidodicarbonimidic diamide	$C_2H_7N_5$	56-03-1	101.111	pr or nd (al)	136	dec 142			vs H_2O; s EtOH; i bz, chl
867	Bikhaconitine	3-Deoxypseudaconitine	$C_{36}H_{51}NO_{11}$	6078-26-8	673.790		164				vs eth, EtOH, chl
868	Bilirubin		$C_{33}H_{36}N_4O_6$	635-65-4	584.662	red mcl pr or pl (chl)					i H_2O; sl EtOH, eth; s bz, chl
869	Biliverdine	Dehydrobilirubin	$C_{33}H_{34}N_4O_6$	114-25-0	582.646	dk grn pl or pr (MeOH)	>300				i H_2O; s EtOH, bz; sl eth, chl, CS_2
870	Binapacryl		$C_{15}H_{18}N_2O_6$	485-31-4	322.313		70		1.27²⁰		
871	1,1'-Binaphthalene	1,1'-Binaphthyl	$C_{20}H_{14}$	604-53-5	254.325	(i) pl(HOAc) (ii) orth (peth)	160	>360; 240¹²	1.3000²⁰		i H_2O; sl EtOH; s eth, ace, bz, CS_2
872	2,2'-Binaphthalene		$C_{20}H_{14}$	612-78-2	254.325	bl flr pl (al)	187.9	452			i H_2O; sl EtOH; s eth, bz, CS_2
873	[1,1'-Binaphthalene]-2,2'-diol		$C_{20}H_{14}O_2$	602-09-5	286.324	nd (al), cry (w)	220				i H_2O; s EtOH, eth, alk; sl chl
874	Biotin	Coenzyme R	$C_{10}H_{16}N_2O_3S$	58-85-5	244.310	nd (w)	232 dec				s H_2O, EtOH; sl eth, chl
875	2,2'-Bioxirane	Diepoxybutane	$C_4H_6O_2$	1464-53-5	86.090		2.0	144	1.113²⁰	1.435²⁰	vs H_2O, EtOH
876	Biphenyl	Diphenyl	$C_{12}H_{10}$	92-52-4	154.207	lf (dil al)	68.93	256.1	1.04²⁰	1.588⁷⁷	i H_2O; s EtOH, eth; vs bz, ctc, MeOH
877	[1,1'-Biphenyl]-4-acetic acid	Felbinac	$C_{14}H_{12}O_2$	5728-52-9	212.244		160.5				
878	[1,1'-Biphenyl]-4-carbonitrile		$C_{13}H_9N$	2920-38-9	179.217		88	190²⁰			i H_2O; vs EtOH, eth
879	[1,1'-Biphenyl]-4-carbonyl chloride		$C_{13}H_9ClO$	14002-51-8	216.662		111	160²			
880	[1,1'-Biphenyl]-2,2'-diamine		$C_{12}H_{12}N_2$	1454-80-4	184.236	pr or nd (al)	81	162⁴	1.3090²⁰		s H_2O, ace, bz
881	[1,1'-Biphenyl]-2,4'-diamine		$C_{12}H_{12}N_2$	492-17-1	184.236	nd (dil al)	54.5	363			i H_2O; s EtOH, eth
882	[1,1'-Biphenyl]-4,4'-diamine, dihydrochloride		$C_{12}H_{14}Cl_2N_2$	531-85-1	257.158		>300				
883	[1,1'-Biphenyl]-2,2'-dicarboxylic acid	o,o'-Diphenic acid	$C_{14}H_{10}O_4$	482-05-3	242.227	mcl pr or lf (w) cry (HOAc)	233.5	sub			i H_2O; s EtOH, eth
884	[1,1'-Biphenyl]-2,2'-diol		$C_{12}H_{10}O_2$	1806-29-7	186.206		109	320	1.3420²⁰		s H_2O, EtOH, eth, ace, bz; sl peth, chl
885	[1,1'-Biphenyl]-2,5-diol		$C_{12}H_{10}O_2$	1079-21-6	186.206	nd (dil al)	97.5				vs EtOH
886	[1,1'-Biphenyl]-4,4'-diol		$C_{12}H_{10}O_2$	92-88-6	186.206		278 dec				sl H_2O, bz, DMSO; s EtOH, eth
887	[1,1'-Biphenyl]-4,4'-disulfonic acid		$C_{12}H_{10}O_6S_2$	5314-37-4	314.333	pr	72.5	>200			vs H_2O
888	[1,1'-Biphenyl]-3,3',4,4'-tetramine, tetrahydrochloride		$C_{12}H_{18}Cl_4N_4$	7411-49-6	360.110		245 dec				
889	[1,1'-Biphenyl]-3,3',5,5'-tetrol	Diresorcinol	$C_{12}H_{10}O_4$	531-02-2	218.205	pl or nd (w+2)	310				vs H_2O, eth, EtOH

Betonicine

Betulaprenol 9

9,9'-Bianthracene

Δ2,2'(3H,3'H)-Bibenzo[b]thiophene-3,3'-dione

Bicyclo[2.2.1]heptane

Bicyclo[4.1.0]heptane

Bicyclo[2.2.1]heptan-2-one

Bicyclo[2.2.1]hept-2-ene

Bicyclo[2.2.1]hept-5-ene-2-carbonitrile

Bicyclo[2.2.1]hept-5-ene-2-carboxaldehyde

Bicyclo[2.2.1]hept-5-ene-2-methanol

[1,1'-Bicyclohexyl]-2-one

1,1'-Bicyclopentyl

[1,1'-Bicyclopentyl]-2-ol

[1,1'-Bicyclopentyl]-2-one

Bifenox

Bifenthrin

Biguanide

Bikhaconitine

Bilirubin

Biliverdine

Binapacryl

1,1'-Binaphthalene

2,2'-Binaphthalene

[1,1'-Binaphthalene]-2,2'-diol

Biotin

2,2'-Bioxirane

Biphenyl

[1,1'-Biphenyl]-4-acetic acid

[1,1'-Biphenyl]-4-carbonitrile

[1,1'-Biphenyl]-4-carbonyl chloride

[1,1'-Biphenyl]-2,2'-diamine

[1,1'-Biphenyl]-2,4'-diamine

[1,1'-Biphenyl]-4,4'-diamine, dihydrochloride

[1,1'-Biphenyl]-2,2'-dicarboxylic acid

[1,1'-Biphenyl]-2,2'-diol

[1,1'-Biphenyl]-2,5-diol

[1,1'-Biphenyl]-4,4'-diol

[1,1'-Biphenyl]-4,4'-disulfonic acid

[1,1'-Biphenyl]-3,3',4,4'-tetramine, tetrahydrochloride

[1,1'-Biphenyl]-3,3',5,5'-tetrol

No.	Name	Synonym	Mol. Form.	CAS RN	Mol. Wt.	Physical Form	mp/°C	bp/°C	den/ g cm^{-3}	n_D	Solubility
890	N-[1,1'-Biphenyl]-4-ylacetamide		C$_{14}$H$_{13}$NO	4075-79-0	211.259	cry (dil MeOH)	172.8				i H$_2$O; vs EtOH, ace, MeOH
891	1-[1,1'-Biphenyl]-4-ylethanone		C$_{14}$H$_{12}$O	92-91-1	196.244	pr (ace), cry (al)	121	326	1.2510^0		i H$_2$O; vs EtOH, ace; sl chl
892	2-[1,1'-Biphenyl]-4-yl-5-phenyl-1,3,4-oxadiazole		C$_{20}$H$_{14}$N$_2$O	852-38-0	298.337		168				
893	2,2'-Bipyridine	α,α'-Dipyridyl	C$_{10}$H$_8$N$_2$	366-18-7	156.184	pr (peth)	72	273.5			sl H$_2$O; vs EtOH, eth, bz, chl
894	2,3'-Bipyridine	2,3'-Bipyridyl	C$_{10}$H$_8$N$_2$	581-50-0	156.184			295.5	1.140^{20}	1.6223^{20}	i H$_2$O; vs EtOH, eth, bz, chl; sl peth
895	2,4'-Bipyridine	2,4'-Bipyridyl	C$_{10}$H$_8$N$_2$	581-47-5	156.184		61.5	281			sl H$_2$O; vs EtOH, eth, chl
896	3,3'-Bipyridine	3,3'-Bipyridyl	C$_{10}$H$_8$N$_2$	581-46-4	156.184		68	291.5	1.1614^{20}		vs H$_2$O, EtOH; sl eth
897	4,4'-Bipyridine	γ,γ'-Dipyridyl	C$_{10}$H$_8$N$_2$	553-26-4	156.184	nd (w+2)	114	305			sl H$_2$O; vs EtOH, bz, chl; s eth
898	2,2'-Biquinoline		C$_{18}$H$_{12}$N$_2$	119-91-5	256.301	pl or lf (al)	196				i H$_2$O; vs EtOH; s eth, ace, bz
899	4,4'-Bis(acetoacetamido)-3,3'-dimethyl-1,1'-biphenyl	N,N-Bis(acetoacetyl)-3,3'-dimethylbenzidine	C$_{22}$H$_{24}$N$_2$O$_4$	91-96-3	380.437		212				sl DMSO
900	Bisacodyl		C$_{22}$H$_{19}$NO$_4$	603-50-9	361.391		133.5				
901	Bis(4-amino-3-chlorophenyl) methane	4,4-Methylene-bis(2-chloroaniline)	C$_{13}$H$_{12}$Cl$_2$N$_2$	101-14-4	267.153						s ctc
902	Bis(4-aminocyclohexyl)methane		C$_{13}$H$_{26}$N$_2$	1761-71-3	210.358		15	320	0.92^{75}		
903	Bis(2-aminoethyl)amine	Diethylenetriamine	C$_4$H$_{13}$N$_3$	111-40-0	103.166	ye hyg liq	-39	207	0.9569^{20}	1.4810^{25}	msc H$_2$O, EtOH; i eth; s lig
904	N,N'-Bis(2-aminoethyl)-1,2-ethanediamine	Triethylenetetramine	C$_6$H$_{18}$N$_4$	112-24-3	146.234		12	266.5		1.4971^{20}	s H$_2$O, EtOH, acid
905	Bis(2-aminophenyl)disulfide		C$_{12}$H$_{12}$N$_2$S$_2$	1141-88-4	248.366		93				i H$_2$O; vs EtOH, eth
906	Bis(4-aminophenyl)disulfide		C$_{12}$H$_{12}$N$_2$S$_2$	722-27-0	248.366		85				s H$_2$O; vs EtOH, eth, chl; sl bz, lig
907	1,2-Bis(4-aminophenyl)ethane		C$_{14}$H$_{16}$N$_2$	621-95-4	212.290	pl (w)	137	sub			i H$_2$O; vs EtOH
908	Bis(4-aminophenyl) sulfone	Dapsone	C$_{12}$H$_{12}$N$_2$O$_2$S	80-08-0	248.300	cry (95% al)	175.5				s EtOH; sl DMSO
909	Bis(4-aminophenyl) sulfoxide	4,4'-Sulfinyldianiline	C$_{12}$H$_{12}$N$_2$OS	119-59-5	232.300	pr (w, al)	175 dec				s H$_2$O, EtOH
910	1,4-Bis(3-aminopropoxy)butane	1,4-Butanediol bis(3-aminopropyl) ether	C$_{10}$H$_{24}$N$_2$O$_2$	7300-34-7	204.310	liq		135^3	0.96^{20}	1.4619^{20}	
911	N,N'-Bis(3-aminopropyl)-1,4-butanediamine	Spermine	C$_{10}$H$_{26}$N$_4$	71-44-3	202.340		29	150^5			
912	N,N'-Bis(3-aminopropyl)-1,4-butanediamine, tetrahydrochloride		C$_{10}$H$_{30}$Cl$_4$N$_4$	306-67-2	348.184		301.5				s H$_2$O
913	Bis(2-bromoethyl) ether	Bromex	C$_4$H$_8$Br$_2$O	5414-19-7	231.914			115^{32}, 92^{12}	1.8452^{20}	1.5131^{27}	
914	1,2-Bis(bromomethyl)benzene		C$_8$H$_8$Br$_2$	91-13-4	263.958	orth (chl)	95	129$^{4.5}$	1.988^{25}		i H$_2$O; s EtOH, eth, ctc, chl, peth, lig
915	1,3-Bis(bromomethyl)benzene		C$_8$H$_8$Br$_2$	626-15-3	263.958	nd (chl), pr (ace)	77	137^{20}	1.959^{25}		i H$_2$O; s EtOH, eth, chl, lig
916	1,4-Bis(bromomethyl)benzene		C$_8$H$_8$Br$_2$	623-24-5	263.958	mcl pr (al), cry (chl, bz)	144.5	245	2.012^{25}		i H$_2$O; vs EtOH, chl; sl eth; s bz
917	2,2-Bis(bromomethyl)-1,3-propanediol	Pentaerythritol dibromide	C$_5$H$_{10}$Br$_2$O$_2$	3296-90-0	261.940	nd (bz)	113				
918	1,3-Bis(bromomethyl) tetramethyldisiloxane		C$_6$H$_{16}$Br$_2$OSi$_2$	2351-13-5	320.169			233; 103^{15}	1.3918^{25}	1.4719^{25}	
919	Bis(4-bromophenyl) ether		C$_{12}$H$_8$Br$_2$O	2050-47-7	327.999	lf (al)	60.5	339	1.8^{25}		i H$_2$O; s EtOH, bz; vs eth; sl chl
920	Bis(2-(2-butoxyethoxy)ethyl) adipate		C$_{22}$H$_{42}$O$_8$	141-17-3	434.563	liq			1.1^{25}		
921	1,4-Bis(α-(tert-butyldioxy)isopropyl)benzene		C$_{20}$H$_{34}$O$_4$	2781-00-2	338.482	cry	79				
922	Bis(3-tert-butyl-5-ethyl-2-hydroxyphenyl)methane		C$_{25}$H$_{36}$O$_2$	88-24-4	368.553	cry	123				
923	Bis(4-chlorobenzoyl) peroxide		C$_{14}$H$_8$Cl$_2$O$_4$	94-17-7	311.118	pr cry (bz)	141				
924	1,2-Bis(2-chloroethoxy)ethane		C$_6$H$_{12}$Cl$_2$O$_2$	112-26-5	187.064			232	1.195^{20}	1.4592^{25}	s ctc
925	Bis(2-chloroethoxy)methane		C$_5$H$_{10}$Cl$_2$O$_2$	111-91-1	173.037			215.0			
926	N,N-Bis(2-chloroethyl)aniline	Aniline mustard	C$_{10}$H$_{13}$Cl$_2$N	553-27-5	218.123	pr	45	164^{14}			sl eth; s EtOH, MeOH
927	Bis(2-chloroethyl) carbonate		C$_5$H$_8$Cl$_2$O$_3$	623-97-2	187.021		8	241	1.3506^{20}	1.461^{20}	i H$_2$O
928	Bis(2-chloroethyl) 2-chloroethylphosphonate		C$_6$H$_{12}$Cl$_3$O$_3$P	6294-34-4	269.490			170.2^5		1.488^{25}	

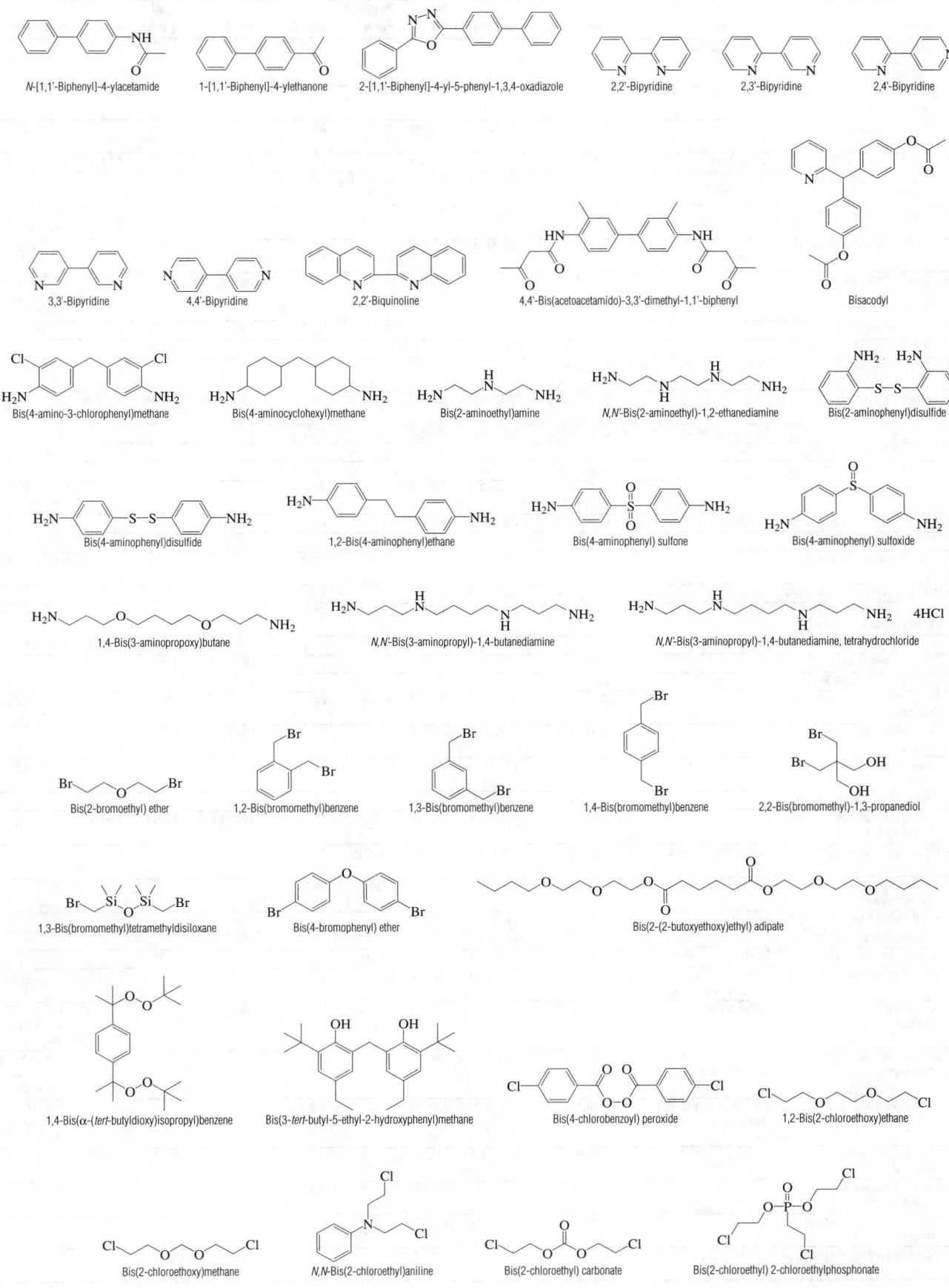

N-[1,1'-Biphenyl]-4-ylacetamide

1-[1,1'-Biphenyl]-4-ylethanone

2-[1,1'-Biphenyl]-4-yl-5-phenyl-1,3,4-oxadiazole

2,2'-Bipyridine

2,3'-Bipyridine

2,4'-Bipyridine

3,3'-Bipyridine

4,4'-Bipyridine

2,2'-Biquinoline

4,4'-Bis(acetoacetamido)-3,3'-dimethyl-1,1'-biphenyl

Bisacodyl

Bis(4-amino-3-chlorophenyl)methane

Bis(4-aminocyclohexyl)methane

Bis(2-aminoethyl)amine

N,N'-Bis(2-aminoethyl)-1,2-ethanediamine

Bis(2-aminophenyl)disulfide

Bis(4-aminophenyl)disulfide

1,2-Bis(4-aminophenyl)ethane

Bis(4-aminophenyl) sulfone

Bis(4-aminophenyl) sulfoxide

1,4-Bis(3-aminopropoxy)butane

N,N'-Bis(3-aminopropyl)-1,4-butanediamine

N,N'-Bis(3-aminopropyl)-1,4-butanediamine, tetrahydrochloride   4HCl

Bis(2-bromoethyl) ether

1,2-Bis(bromomethyl)benzene

1,3-Bis(bromomethyl)benzene

1,4-Bis(bromomethyl)benzene

2,2-Bis(bromomethyl)-1,3-propanediol

1,3-Bis(bromomethyl)tetramethyldisiloxane

Bis(4-bromophenyl) ether

Bis(2-(2-butoxyethoxy)ethyl) adipate

1,4-Bis(α-(*tert*-butyldioxy)isopropyl)benzene

Bis(3-*tert*-butyl-5-ethyl-2-hydroxyphenyl)methane

Bis(4-chlorobenzoyl) peroxide

1,2-Bis(2-chloroethoxy)ethane

Bis(2-chloroethoxy)methane

N,N-Bis(2-chloroethyl)aniline

Bis(2-chloroethyl) carbonate

Bis(2-chloroethyl) 2-chloroethylphosphonate

No.	Name	Synonym	Mol. Form.	CAS RN	Mol. Wt.	Physical Form	mp/°C	bp/°C	den/ g cm⁻³	n_D	Solubility
929	Bis(2-chloroethyl) ether	Dichloroethyl ether	$C_4H_8Cl_2O$	111-44-4	143.012	liq	-51.9	178.5	1.22[20]	1.451[20]	i H_2O; s EtOH, eth, ace; msc bz
930	Bis(2-chloroethyl)methylamine hydrochloride	Nitrogen mustard hydrochloride	$C_5H_{12}Cl_3N$	55-86-7	192.515	hyg nd	111.5				
931	N,N'-Bis(2-chloroethyl)-N-nitrosourea	Carmustine	$C_5H_9Cl_2N_3O_2$	154-93-8	214.049	lt ye pow	31				vs H_2O, EtOH
932	Bis(2-chloroethyl) sulfide	Mustard gas	$C_4H_8Cl_2S$	505-60-2	159.078	liq	13.5	216	1.2741[20]	1.5313[20]	
933	1,2-Bis(2-chloroethylsulfonyl)ethane		$C_6H_{12}Cl_2O_4S_2$	3944-87-4	283.193	cry (MeOH/HOAc)	205				
934	1,2-Bis(chloromethyl)benzene		$C_8H_8Cl_2$	612-12-4	175.056	mcl (lig)	55	239.5	1.393[25]		i H_2O; vs EtOH, eth, chl; s ctc
935	1,3-Bis(chloromethyl)benzene		$C_8H_8Cl_2$	626-16-4	175.056	cry	34.2	251.5	1.302[20]		i H_2O; vs EtOH, eth; sl chl
936	1,4-Bis(chloromethyl)benzene		$C_8H_8Cl_2$	623-25-6	175.056	pl (al)	100	dec 245; 135[16]	1.417[25]		i H_2O; vs EtOH, eth, ace, chl; sl HOAc
937	Bis(chloromethyl) ether		$C_2H_4Cl_2O$	542-88-1	114.958	liq	-41.5	106	1.323[15]	1.435[21]	msc EtOH, eth
938	3,3-Bis(chloromethyl)oxetane		$C_5H_8Cl_2O$	78-71-7	155.022	liq	18.7	101[27]	1.295[25]		
939	2,2-Bis(chloromethyl)-1,3-propanediol	Pentaerythritol dichlorohydrin	$C_5H_{10}Cl_2O_2$	2209-86-1	173.037	cry	83	159[12]			
940	1,3-Bis(chloromethyl)tetramethyldisiloxane		$C_6H_{16}Cl_2OSi_2$	2362-10-9	231.267	liq	-90	204; 92[21]	1.045[20]	1.4398[20]	
941	Bis(4-chlorophenoxy)methane	Di(4-chlorophenoxy)methane	$C_{13}H_{10}Cl_2O_2$	555-89-5	269.123	cry (peth)	70.5	191[6]			vs ace, bz
942	Bis(4-chlorophenyl) disulfide		$C_{12}H_8Cl_2S_2$	1142-19-4	287.228		72.8				s chl
943	Bis(4-chlorophenyl)ethanedione		$C_{14}H_8Cl_2O_2$	3457-46-3	279.119		197.8				
944	1,1-Bis(4-chlorophenyl)ethanol		$C_{14}H_{12}Cl_2O$	80-06-8	267.150		70				i H_2O, EtOH; s eth, bz
945	1,2-Bis(2-chlorophenyl)-hydrazine	2,2'-Dichlorohydrazobenzene	$C_{12}H_{10}Cl_2N_2$	782-74-1	253.126		87				
946	Bis(4-chlorophenyl)methane		$C_{13}H_{10}Cl_2$	101-76-8	237.124		55.5	188[18]	1.365[17]		s EtOH
947	Bis(4-chlorophenyl) sulfone		$C_{12}H_8Cl_2O_2S$	80-07-9	287.162		147.9	250[10]			sl H_2O; s EtOH, chl
948	N,N'-Bis(4-chlorophenyl)thiourea	Di(p-chlorophenyl)thiourea	$C_{13}H_{10}Cl_2N_2S$	1220-00-4	297.202	nd	176				
949	1,1-Bis(4-chlorophenyl)-2,2,2-trichloroethanol		$C_{14}H_9Cl_5O$	115-32-2	370.485	cry (petr)	77.5	180[0.1]			i H_2O, os
950	Bis(3-chloropropyl) ether	3-Chloropropyl ether	$C_6H_{12}Cl_2O$	629-36-7	171.064			216; 90.5[11]	1.136[20]	1.4158[20]	s EtOH, eth
951	Bis(2-cyanoethyl) ether		$C_6H_8N_2O$	1656-48-0	124.140			161[5], 111[0.5]	1.0504[20]	1.4405[20]	
952	Bis(2-cyanoethyl) sulfide		$C_6H_8N_2S$	111-97-7	140.206			163[1.5]		1.5047[20]	
953	Bis(η-cyclopentadienyl)titanium chloride		$C_{10}H_{10}Cl_2Ti$	1271-19-8	248.959	red cry	289	258[10]	1.60		sl H_2O, bz; s chl, EtOH, tol
954	Bis(η-cyclopentadienyl)zirconium chloride		$C_{10}H_{10}Cl_2Zr$	1291-32-3	292.316			180[0.5]			
955	1,2-Bis(dibromomethyl)benzene		$C_8H_6Br_4$	13209-15-9	421.750	mcl	116.5				sl H_2O; vs chl; i lig
956	Bis(2,4-dichlorobenzoyl) peroxide		$C_{14}H_6Cl_4O_4$	133-14-2	380.008		106				
957	1,3-Bis(dichloromethyl)tetramethyldisiloxane		$C_6H_{14}Cl_4OSi_2$	2943-70-6	300.157			149[50], 117[11]	1.2213[20]	1.4660[20]	
958	Bis(2,4-dichlorophenyl)ether	2,2',4,4'-Tetrachlorodiphenyl ether	$C_{12}H_6Cl_4O$	28076-73-5	307.987	cry (eth)	71				
959	4,4'-Bis(diethylamino) benzophenone	Michler's ethyl ketone	$C_{21}H_{28}N_2O$	90-93-7	324.459	lf (al)	95.3				
960	Bis(diethyldithiocarbamate)nickel		$C_{10}H_{20}N_2NiS_4$	14267-17-5	355.232			202[0.02]			
961	Bis(diethyldithiocarbamate)zinc		$C_{10}H_{20}N_2S_4Zn$	14324-55-1	361.948			178[0.05]			
962	Bis(difluoromethyl) ether	Difluoromethyl ether	$C_2H_2F_4O$	1691-17-4	118.030	col gas	2		1.43[20]		
963	Bis(2-dimethylaminoethyl) ether	2,2'-Oxybis[N,N-dimethylethanamine]	$C_8H_{20}N_2O$	3033-62-3	160.257	liq		80[15]			
964	Bis[4-(dimethylamino)phenyl]methane	Michler's Base	$C_{17}H_{22}N_2$	101-61-1	254.370	pl or tab (al, lig)	91.5	dec 390; 183[3]			i H_2O; sl EtOH; vs eth, bz; s acid
965	Bis[4-(dimethylamino)phenyl]methanethione	4,4'-Bis(dimethylamino)thiobenzophenone	$C_{17}H_{20}N_2S$	1226-46-6	284.419	pl	204				i H_2O, EtOH, lig; sl eth; s bz, chl, HOAc
966	Bis(4-dimethylaminophenyl)methanol	4,4'-Bis(dimethylamino)benzhydrol	$C_{17}H_{22}N_2O$	119-58-4	270.369		102.0				i H_2O; vs EtOH; s eth, bz, HOAc
967	1,3-Bis(dimethylamino)-2-propanol		$C_7H_{18}N_2O$	5966-51-8	146.230			181.5	0.8788[20]	1.4418[20]	vs H_2O
968	4,4'-Bis(dimethylamino)triphenylmethane		$C_{23}H_{26}N_2$	129-73-7	330.465	nd or lf (al, bz)	102				vs bz, eth
969	Bis(dimethyldithiocarbamate)copper		$C_6H_{12}CuN_2S_4$	137-29-1	303.978			206[0.01]			
970	Bis(dimethyldithiocarbamate)nickel		$C_6H_{12}N_2NiS_4$	15521-65-0	299.125			208[0.002]			

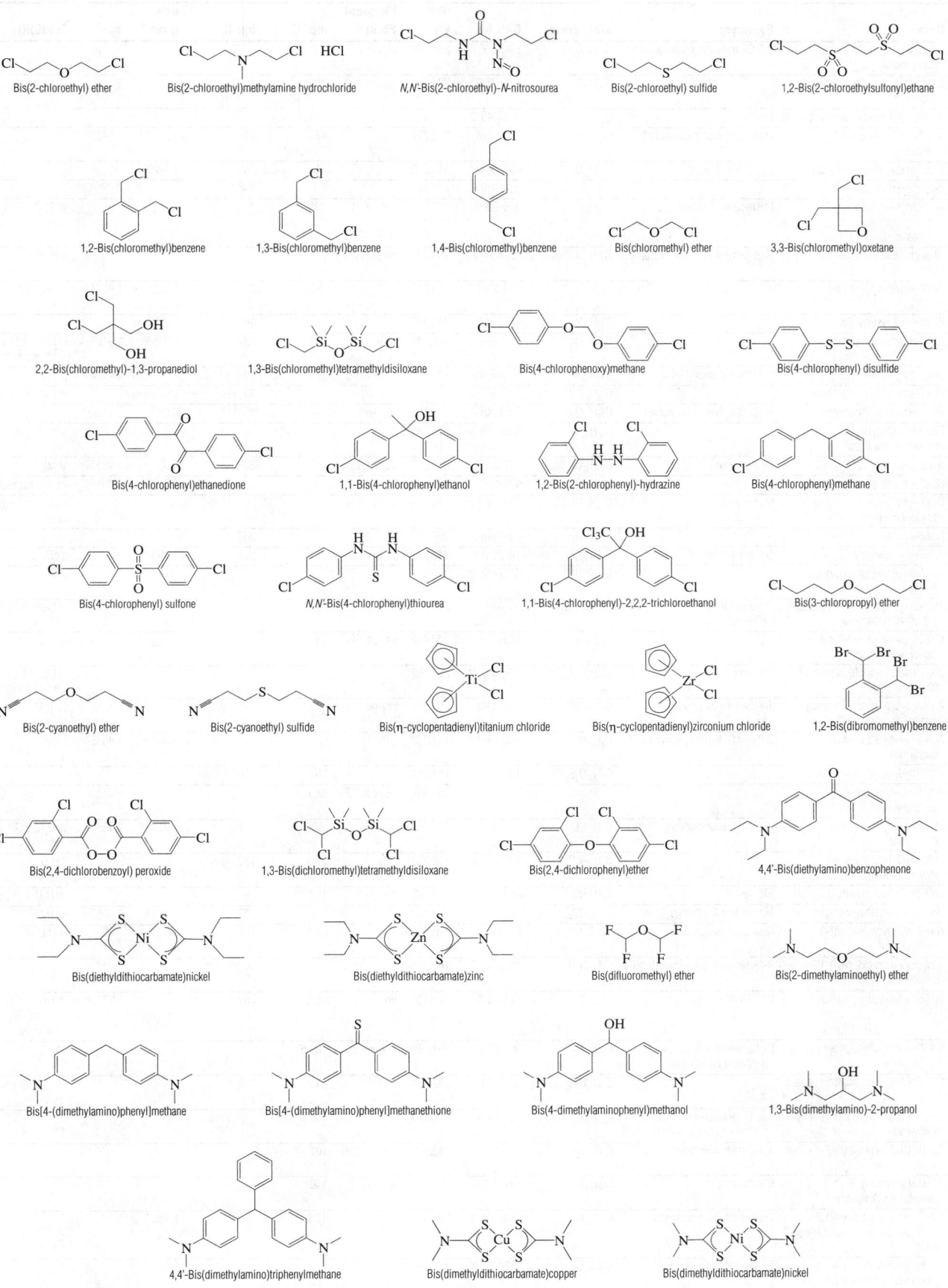

Bis(2-chloroethyl) ether

Bis(2-chloroethyl)methylamine hydrochloride

N,N'-Bis(2-chloroethyl)-N-nitrosourea

Bis(2-chloroethyl) sulfide

1,2-Bis(2-chloroethylsulfonyl)ethane

1,2-Bis(chloromethyl)benzene

1,3-Bis(chloromethyl)benzene

1,4-Bis(chloromethyl)benzene

Bis(chloromethyl) ether

3,3-Bis(chloromethyl)oxetane

2,2-Bis(chloromethyl)-1,3-propanediol

1,3-Bis(chloromethyl)tetramethyldisiloxane

Bis(4-chlorophenoxy)methane

Bis(4-chlorophenyl) disulfide

Bis(4-chlorophenyl)ethanedione

1,1-Bis(4-chlorophenyl)ethanol

1,2-Bis(2-chlorophenyl)-hydrazine

Bis(4-chlorophenyl)methane

Bis(4-chlorophenyl) sulfone

N,N'-Bis(4-chlorophenyl)thiourea

1,1-Bis(4-chlorophenyl)-2,2,2-trichloroethanol

Bis(3-chloropropyl) ether

Bis(2-cyanoethyl) ether

Bis(2-cyanoethyl) sulfide

Bis(η-cyclopentadienyl)titanium chloride

Bis(η-cyclopentadienyl)zirconium chloride

1,2-Bis(dibromomethyl)benzene

Bis(2,4-dichlorobenzoyl) peroxide

1,3-Bis(dichloromethyl)tetramethyldisiloxane

Bis(2,4-dichlorophenyl)ether

4,4'-Bis(diethylamino)benzophenone

Bis(diethyldithiocarbamate)nickel

Bis(diethyldithiocarbamate)zinc

Bis(difluoromethyl) ether

Bis(2-dimethylaminoethyl) ether

Bis[4-(dimethylamino)phenyl]methane

Bis[4-(dimethylamino)phenyl]methanethione

Bis(4-dimethylaminophenyl)methanol

1,3-Bis(dimethylamino)-2-propanol

4,4'-Bis(dimethylamino)triphenylmethane

Bis(dimethyldithiocarbamate)copper

Bis(dimethyldithiocarbamate)nickel

No.	Name	Synonym	Mol. Form.	CAS RN	Mol. Wt.	Physical Form	mp/°C	bp/°C	den/ g cm⁻³	n_D	Solubility
971	2,5-Bis(1,1-dimethylpropyl)-1,4-benzenediol	2,5-Di-*tert*-pentylhydroquinone	$C_{16}H_{26}O_2$	79-74-3	250.376		180				
972	2,4-Bis(1,1-dimethylpropyl) phenol		$C_{16}H_{26}O$	120-95-6	234.376		26.0	169[22]			
973	1,2-Bis(diphenylphosphino)ethane	Diphos	$C_{26}H_{24}P_2$	1663-45-2	398.417		143.5				
974	1,3-Bis(2,3-epoxypropoxy) benzene	Diglycidyl resorcinol ether	$C_{12}H_{14}O_4$	101-90-6	222.237		42.5	147[0.4]	1.2183[30]	1.5408[20]	
975	Bis(2-ethoxyethyl) phthalate		$C_{16}H_{22}O_6$	605-54-9	310.342		34	345	1.1229[21]		
976	Bis(ethoxymethyl) ether		$C_6H_{14}O_3$	5648-29-3	134.173			140.6			
977	N,N'-Bis(4-ethoxyphenyl) ethanimidamide monohydrochloride	Phenacaine hydrochloride	$C_{18}H_{23}ClN_2O_2$	620-99-5	334.841	cry (w+1)	191				vs H₂O, EtOH, chl
978	Bis(ethylenediamine)copper dichloride	Cupriethylenediamine dichloride	$C_4H_{16}Cl_2CuN_4$	15243-01-3	254.649	dk bl cry					s EtOH
979	Bis(2-ethylhexyl) adipate		$C_{22}H_{42}O_4$	103-23-1	370.566		-67.8	214[5]	0.922[25]	1.4474[20]	vs ace, eth, EtOH
980	Bis(2-ethylhexyl)amine		$C_{16}H_{35}N$	106-20-7	241.456			161[21]			
981	Bis(2-ethylhexyl) azelate		$C_{25}H_{48}O_4$	103-24-2	412.647		-78	237[5]	0.915[25]	1.446[25]	i H₂O; s EtOH, ace, bz; sl ctc
982	Bis(2-ethylhexyl) ether	2,2'-Diethyldihexyl ether	$C_{16}H_{34}O$	10143-60-9	242.440			269; 144[13]		1.4325[20]	sl ctc
983	Bis(2-ethylhexyl) phosphate		$C_{16}H_{35}O_4P$	298-07-7	322.420	visc liq		155[0.015]	0.975[25]		sl H₂O; s bz, hx
984	Bis(2-ethylhexyl) phosphonate	Bis(2-ethylhexyl) phosphite	$C_{16}H_{35}O_3P$	3658-48-8	306.421	liq		150[1]	0.93[25]	1.4420[20]	
985	Bis(2-ethylhexyl) phosphorodithioate		$C_{16}H_{35}O_2PS_2$	5810-88-8	354.552	cry					s bz, hp, chl
986	Bis(2-ethylhexyl) phthalate	Di-*sec*-octyl phthalate	$C_{24}H_{38}O_4$	117-81-7	390.557	liq	-55	384	0.981[25]	1.4853[20]	sl ctc
987	Bis(2-ethylhexyl) sebacate		$C_{26}H_{50}O_4$	122-62-3	426.673		-48	256[5]	0.912[25]	1.451[25]	vs ace, bz, EtOH
988	Bis(2-ethylhexyl) sodium sulfosuccinate	Docusate sodium	$C_{20}H_{37}NaO_7S$	577-11-7	444.559	waxy solid					s peth, ctc, eth, ace
989	Bis(2-ethylhexyl) terephthalate		$C_{24}H_{38}O_4$	6422-86-2	390.557			383			
990	2,2-Bis(ethylsulfonyl)butane	Sulfonethylmethane	$C_8H_{18}O_4S_2$	76-20-0	242.357	pl (w)	76	dec	1.199[85]		s chl
991	Bis[4-(hexyloxy)phenyl]diazene, 1-oxide		$C_{24}H_{34}N_2O_3$	2587-42-0	398.538						s chl
992	N,N'-Bis(2-hydroxybenzylidene)-1,2-ethanediamine	Disalicylidene-1,2-ethanediamine	$C_{16}H_{16}N_2O_2$	94-93-9	268.310		125.5				sl EtOH, eth; s bz, chl
993	Bis(2-hydroxy-3-*tert*-butyl-5-methylphenyl)methane		$C_{23}H_{32}O_2$	119-47-1	340.499	nd (peth)	131				
994	Bis(2-hydroxy-5-chlorophenyl) sulfide	Fenticlor	$C_{12}H_8Cl_2O_2S$	97-24-5	287.162		174				i H₂O; s EtOH, eth, gl HOAc
995	2-[Bis(2-hydroxyethyl)amino] ethanol hydrochloride	Triethanolamine hydrochloride	$C_6H_{16}ClNO_3$	637-39-8	185.649	cry (al)	179.5				vs H₂O
996	N,N-Bis(2-hydroxyethyl) butylamine	Butylbis(2-hydroxyethyl)amine	$C_8H_{19}NO_2$	102-79-4	161.243			275; 80[35]	0.9681[20]	1.4625[20]	s chl
997	Bis(2-hydroxyethyl) disulfide		$C_4H_{10}O_2S_2$	1892-29-1	154.251		26	160[3.5]			
998	N,N-Bis(2-hydroxyethyl) dodecanamide		$C_{16}H_{33}NO_3$	120-40-1	287.438	waxy solid	38.7				
999	N,N-Bis(2-hydroxyethyl) ethylamine	N-Ethyldiethanolamine	$C_6H_{15}NO_2$	139-87-7	133.189	ye liq	-50	247	1.0135[20]	1.4663[20]	vs H₂O, EtOH; sl eth
1000	N,N'-Bis(2-hydroxyethyl) ethylenediamine		$C_6H_{16}N_2O_2$	4439-20-7	148.203		97.5	136[1]			s H₂O
1001	N,N-Bis(2-hydroxyethyl)glycine	Bicine	$C_6H_{13}NO_4$	150-25-4	163.172	nd (al)	194 dec				vs H₂O; i EtOH
1002	Bis(2-hydroxyethyl)methylamine	Methyldiethanolamine	$C_5H_{13}NO_2$	105-59-9	119.163	liq	-21	247	1.043[25]	1.4685[20]	vs H₂O
1003	N,N-Bis(2-hydroxyethyl)-3-methylaniline	Diethanol-*m*-toluidine	$C_{11}H_{17}NO_2$	91-99-6	195.259		64.5	160[1]			sl chl
1004	N,N-Bis(2-hydroxyethyl)-1,3-propanediamine	3-(Aminopropyl)diethanolamine	$C_7H_{18}N_2O_2$	4985-85-7	162.230			160[1]			
1005	Bis(2-hydroxyethyl) sulfide	2,2'-Thiodiethanol	$C_4H_{10}O_2S$	111-48-8	122.186	liq	-10.2	282	1.1793[25]	1.5211[20]	msc H₂O, EtOH, chl, AcOEt; s eth; sl bz
1006	Bis(2-hydroxyethyl) terephthalate	Bis(2-hydroxyethyl) 1,4-benzenedicarboxylate	$C_{12}H_{14}O_6$	959-26-2	254.235	cry (w)	109.5				
1007	1,2-Bis(2-hydroxyethylthio)ethane		$C_6H_{14}O_2S_2$	5244-34-8	182.304		64.8	170[0.5]			s H₂O, EtOH, bz, peth
1008	Bis(2-hydroxy-4-methoxyphenyl) methanone	2,2'-Dihydroxy-4,4'-dimethoxybenzophenone	$C_{15}H_{14}O_5$	131-54-4	274.269		139.5				
1009	1,3-Bis(hydroxymethyl)-2-imidazolidone	1,3-Dimethylolethyleneurea	$C_5H_{10}N_2O_3$	136-84-5	146.144	cry (MeOH)	101				
1010	2,2-Bis(4-hydroxy-3-methylphenyl)propane	Bisphenol C	$C_{17}H_{20}O_2$	79-97-0	256.340	nd (xyl)	140				
1011	2,2-Bis(hydroxymethyl)-1,3-propanediol, tetra(2-propenoyl) ester	Pentaerythritol tetraacrylate	$C_{17}H_{20}O_8$	4986-89-4	352.336		17.3		1.185[25]		
1012	2,2-Bis(hydroxymethyl)-1,3-propanediol, tri(2-propenoyl) ester	Pentaerythritol triacrylate	$C_{14}H_{18}O_7$	3524-68-3	298.289				1.180[20]		

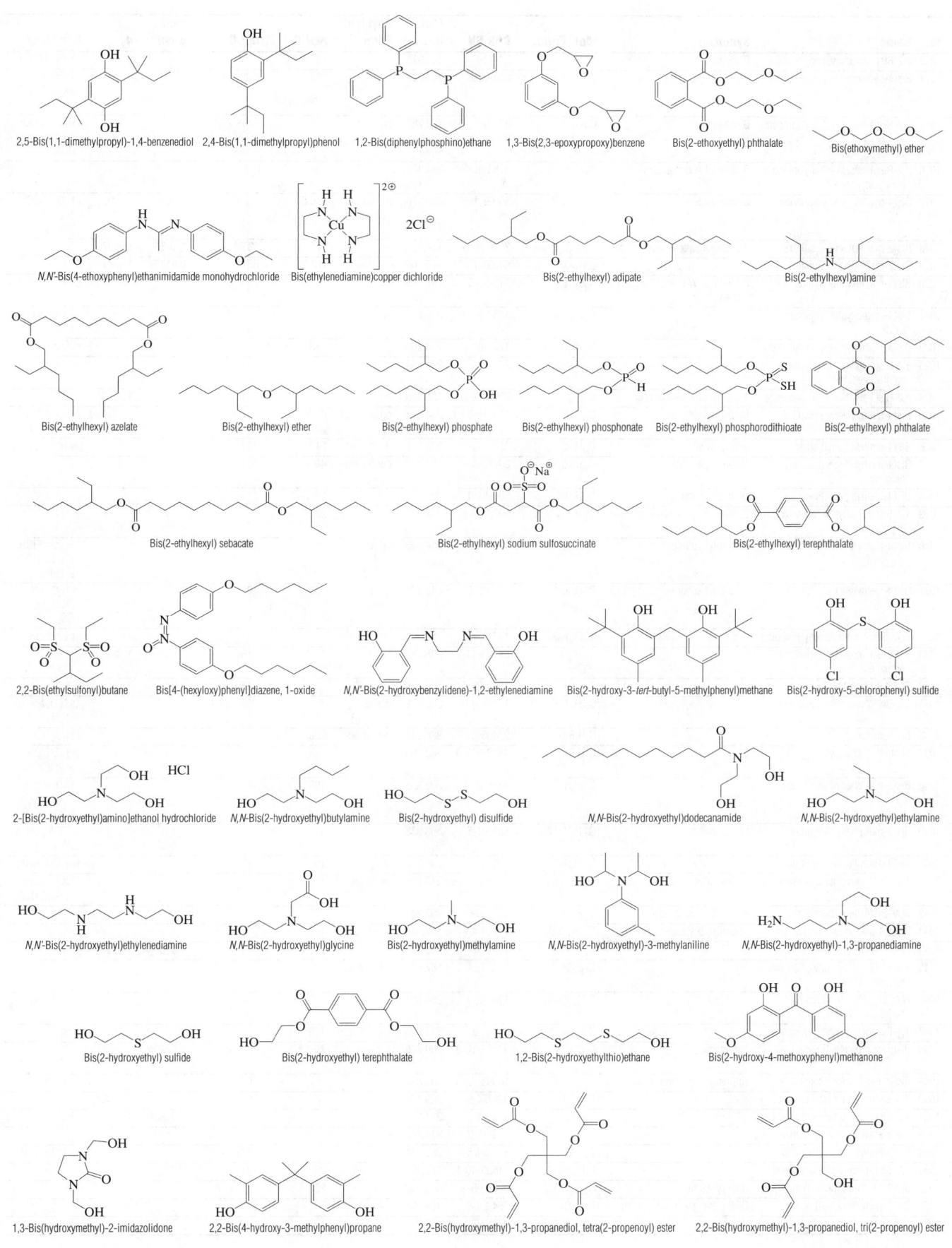

2,5-Bis(1,1-dimethylpropyl)-1,4-benzenediol

2,4-Bis(1,1-dimethylpropyl)phenol

1,2-Bis(diphenylphosphino)ethane

1,3-Bis(2,3-epoxypropoxy)benzene

Bis(2-ethoxyethyl) phthalate

Bis(ethoxymethyl) ether

N,N'-Bis(4-ethoxyphenyl)ethanimidamide monohydrochloride

Bis(ethylenediamine)copper dichloride

Bis(2-ethylhexyl) adipate

Bis(2-ethylhexyl)amine

Bis(2-ethylhexyl) azelate

Bis(2-ethylhexyl) ether

Bis(2-ethylhexyl) phosphate

Bis(2-ethylhexyl) phosphonate

Bis(2-ethylhexyl) phosphorodithioate

Bis(2-ethylhexyl) phthalate

Bis(2-ethylhexyl) sebacate

Bis(2-ethylhexyl) sodium sulfosuccinate

Bis(2-ethylhexyl) terephthalate

2,2-Bis(ethylsulfonyl)butane

Bis[4-(hexyloxy)phenyl]diazene, 1-oxide

N,N'-Bis(2-hydroxybenzylidene)-1,2-ethylenediamine

Bis(2-hydroxy-3-tert-butyl-5-methylphenyl)methane

Bis(2-hydroxy-5-chlorophenyl) sulfide

2-[Bis(2-hydroxyethyl)amino]ethanol hydrochloride

N,N-Bis(2-hydroxyethyl)butylamine

Bis(2-hydroxyethyl) disulfide

N,N-Bis(2-hydroxyethyl)dodecanamide

N,N-Bis(2-hydroxyethyl)ethylamine

N,N'-Bis(2-hydroxyethyl)ethylenediamine

N,N-Bis(2-hydroxyethyl)glycine

Bis(2-hydroxyethyl)methylamine

N,N-Bis(2-hydroxyethyl)-3-methylaniline

N,N-Bis(2-hydroxyethyl)-1,3-propanediamine

Bis(2-hydroxyethyl) sulfide

Bis(2-hydroxyethyl) terephthalate

1,2-Bis(2-hydroxyethylthio)ethane

Bis(2-hydroxy-4-methoxyphenyl)methanone

1,3-Bis(hydroxymethyl)-2-imidazolidone

2,2-Bis(4-hydroxy-3-methylphenyl)propane

2,2-Bis(hydroxymethyl)-1,3-propanediol, tetra(2-propenoyl) ester

2,2-Bis(hydroxymethyl)-1,3-propanediol, tri(2-propenoyl) ester

No.	Name	Synonym	Mol. Form.	CAS RN	Mol. Wt.	Physical Form	mp/°C	bp/°C	den/ g cm-3	n_D	Solubility
1013	2,2-Bis(4-hydroxyphenyl)butane	Bisphenol B	$C_{16}H_{18}O_2$	77-40-7	242.313		120.5				vs ace, MeOH
1014	Bis(4-hydroxyphenyl)methane	Bisphenol AD	$C_{13}H_{12}O_2$	620-92-8	200.233		162.5	sub			s EtOH, eth, chl, alk; sl DMSO; i CS_2
1015	2,2-Bis(4-hydroxyphenyl)propane	Bisphenol A	$C_{15}H_{16}O_2$	80-05-7	228.287	cry or fl	153	220[4], 222[3]			i H_2O; vs EtOH, eth, bz, alk; s HOAc
1016	2,2-Bis(4-hydroxyphenyl)propane dimethacrylate	Bisphenol A dimethacrylate	$C_{23}H_{24}O_4$	3253-39-2	364.435		73				
1017	Bis(4-hydroxyphenyl) sulfone	Bisphenol S	$C_{12}H_{10}O_4S$	80-09-1	250.270	nd (w), orth bipym	240.5		1.3663[15]		i H_2O; s EtOH, eth; sl bz, DMSO
1018	Bis(2-mercaptoethyl) sulfide	2,2'-Dimercaptodiethyl sulfide	$C_4H_{10}S_3$	3570-55-6	154.317		-11	135[18]	1.183[25]	1.5982[20]	
1019	Bis(2-methallyl) carbonate		$C_9H_{14}O_3$	64057-79-0	170.205			201.3	66[3]	0.943[25]	1.4371[20]	
1020	Bis(2-methoxyethyl)amine	2-Methoxy-N-(2-methoxyethyl)ethanamine	$C_6H_{15}NO_2$	111-95-5	133.189						s ctc
1021	Bis(4-methoxyphenyl)diazene, 1-oxide		$C_{14}H_{14}N_2O_3$	1562-94-3	258.272	ye nd (al)			1.1711[11]		s EtOH, ace, bz; sl chl
1022	Bis(4-methoxyphenyl)ethanedione		$C_{16}H_{14}O_4$	1226-42-2	270.280		133				sl EtOH, chl
1023	1,4-Bis(methylamino)-9,10-anthracenedione		$C_{16}H_{14}N_2O_2$	2475-44-7	266.294						sl chl
1024	1,3-Bis(1-methylethenyl)benzene	1,3-Diisopropenylbenzene	$C_{12}H_{14}$	3748-13-8	158.239	liq		231	0.925	1.5570[20]	
1025	Bis(4-methylphenyl) disulfide	Di-p-Tolyl disulfide	$C_{14}H_{14}S_2$	103-19-5	246.391	nd or lf (al)	47.5	212[20]	1.114[51]		i H_2O; s EtOH, ace; vs eth
1026	Bis(4-methylphenyl) ether	p-Tolyl ether	$C_{14}H_{14}O$	1579-40-4	198.260		51	285			vs bz, eth, EtOH
1027	Bis(1-methyl-1-phenylethyl) peroxide	Dicumyl peroxide	$C_{18}H_{22}O_2$	80-43-3	270.367	cry (EtOH)	40	100[0.2]			
1028	Bis(4-methylphenyl)mercury	Di-p-tolylmercury	$C_{14}H_{14}Hg$	537-64-4	382.85		245.7				
1029	1,4-Bis(4-methyl-5-phenyloxazol-2-yl)benzene	2,2'-p-Phenylenebis(4-methyl-5-phenyloxazole)	$C_{26}H_{20}N_2O_2$	3073-87-8	392.449		232				sl chl
1030	Bis(4-methylphenyl) sulfide	Di-p-tolyl sulfide	$C_{14}H_{14}S$	620-94-0	214.326	nd (al)	57.3	>300; 175[16]			i H_2O; s EtOH, ace, bz, HOAc; sl chl
1031	Bis(4-methylphenyl) sulfone	Di-p-tolyl sulfone	$C_{14}H_{14}O_2S$	599-66-6	246.325	pr(bz), nd(w,al)	159	406			sl H_2O, eth; s EtOH, bz, chl, CS_2
1032	N,N'-Bis(2-methylphenyl)thiourea		$C_{15}H_{16}N_2S$	137-97-3	256.366	nd (al, sub)					vs bz, EtOH, chl
1033	1,3-Bis(1-methyl-4-piperidyl)propane		$C_{15}H_{30}N_2$	64168-11-2	238.412		13.7	215[50]	0.8962[25]	1.4804[25]	
1034	Bis(methylthio)methane		$C_3H_8S_2$	1618-26-4	108.226			148			
1035	1,2-Bis(N-morpholino)ethane		$C_{10}H_{20}N_2O_2$	1723-94-0	200.278	wh-ye (eth,lig)	75	285; 160[25]			vs H_2O, ace, bz, EtOH
1036	Bismuth acetate		$C_6H_9BiO_6$	22306-37-2	386.111	col tablets	250				i H_2O
1037	Bismuth subsalicylate		$C_7H_5BiO_4$	14882-18-9	362.093	pr					i H_2O, EtOH; reac alk
1038	Bis(2-nitrophenyl) disulfide		$C_{12}H_8N_2O_4S_2$	1155-00-6	308.333		198.5				i H_2O, eth; sl EtOH, ace, bz, HOAc
1039	Bis(3-nitrophenyl) disulfide	Nitrophenide	$C_{12}H_8N_2O_4S_2$	537-91-7	308.333		84				sl EtOH, chl; s eth
1040	Bis(4-nitrophenyl) disulfide		$C_{12}H_8N_2O_4S_2$	100-32-3	308.333		182	255[0.1]			sl EtOH, HOAc
1041	1,2-Bis(4-nitrophenyl)ethane	4,4'-Dinitrobibenzyl	$C_{14}H_{12}N_2O_4$	736-30-1	272.256	ye nd (al,bz)	181.8				i EtOH; sl eth, bz, chl, HOAc
1042	N,N'-Bis(4-nitrophenyl)urea	4,4'-Dinitrocarbanilide	$C_{13}H_{10}N_4O_5$	587-90-6	302.242		312 dec				
1043	Bis(2,4-pentanedionato)cobalt	Cobalt(II) bis(acetylacetonate)	$C_{10}H_{14}CoO_4$	14024-48-7	257.149	bl-viol cry	167				
1044	Bis(1-phenylethyl)amine		$C_{16}H_{19}N$	10024-74-5	225.329			296.5	1.018[15]	1.573	
1045	1,2-Bis(2,4,6-tribromophenoxy)ethane		$C_{14}H_8Br_6O_2$	37853-59-1	687.637	nd (bz/EtOH)	222				
1046	N,N'-Bis(2,2,2-trichloro-1-hydroxyethyl)urea		$C_5H_6Cl_6N_2O_3$	116-52-9	354.831		196				vs ace, EtOH
1047	1,4-Bis(trichloromethyl)benzene		$C_8H_4Cl_6$	68-36-0	312.836	cry (bz, eth)	109				s chl
1048	Bis(trichloromethyl) carbonate	Triphosgene	$C_3Cl_6O_3$	32315-10-9	296.748	cry (eth, peth)	79	203	1.6290[80]		
1049	Bis(tridecyl) thiodipropanoate	Ditridecyl thiodipropionate	$C_{32}H_{62}O_4S$	10595-72-9	542.897			265[0.25]			vs EtOH
1050	3,5-Bis(trifluoromethyl)aniline		$C_8H_5F_6N$	328-74-5	229.123			85[15], 76[10]	1.487[25]	1.4335[20]	
1051	1,3-Bis(trifluoromethyl)benzene		$C_8H_4F_6$	402-31-3	214.108			116	1.3790[25]	1.3916[25]	i H_2O
1052	1,4-Bis(trifluoromethyl)benzene		$C_8H_4F_6$	433-19-2	214.108	liq		115			
1053	Bis(trifluoromethyl) disulfide		$C_2F_6S_2$	372-64-5	202.141			34.6			vs EtOH, peth
1054	1,2-Bis(trimethylsilyl)acetylene		$C_8H_{18}Si_2$	14630-40-1	170.400		26	134	0.770[20]	1.413[20]	
1055	Bis(2,4,6-trinitrophenyl) sulfide	Dipicryl sulfide	$C_{12}H_4N_6O_{12}S$	2217-06-3	456.258	ye cry	230	exp			
1056	Bis[2-(vinyloxy)ethyl] ether	Diethylene glycol divinyl ether	$C_8H_{14}O_3$	764-99-8	158.195			81[10]			
1057	Bithionol		$C_{12}H_6Cl_4O_2S$	97-18-7	356.052		188		1.73[25]		vs ace

2,2-Bis(4-hydroxyphenyl)butane

Bis(4-hydroxyphenyl)methane

2,2-Bis(4-hydroxyphenyl)propane

2,2-Bis(4-hydroxyphenyl)propane dimethacrylate

Bis(4-hydroxyphenyl) sulfone

Bis(2-mercaptoethyl) sulfide

Bis(2-methallyl) carbonate

Bis(2-methoxyethyl)amine

Bis(4-methoxyphenyl)diazene, 1-oxide

Bis(4-methoxyphenyl)ethanedione

1,4-Bis(methylamino)-9,10-anthracenedione

1,3-Bis(1-methylethenyl)benzene

Bis(4-methylphenyl) disulfide

Bis(4-methylphenyl) ether

Bis(1-methyl-1-phenylethyl)peroxide

Bis(4-methylphenyl)mercury

1,4-Bis(4-methyl-5-phenyloxazol-2-yl)benzene

Bis(4-methylphenyl) sulfide

Bis(4-methylphenyl) sulfone

N,N'-Bis(2-methylphenyl)thiourea

1,3-Bis(1-methyl-4-piperidyl)propane

Bis(methylthio)methane

1,2-Bis(N-morpholino)ethane

Bismuth acetate

Bismuth subsalicylate

Bis(2-nitrophenyl) disulfide

Bis(3-nitrophenyl) disulfide

Bis(4-nitrophenyl) disulfide

1,2-Bis(4-nitrophenyl)ethane

N,N'-Bis(4-nitrophenyl)urea

Bis(2,4-pentanedionato)cobalt

Bis(1-phenylethyl)amine

1,2-Bis(2,4,6-tribromophenoxy)ethane

N,N'-Bis(2,2,2-trichloro-1-hydroxyethyl)urea

1,4-Bis(trichloromethyl)benzene

Bis(trichloromethyl) carbonate

Bis(tridecyl) thiodipropanoate

3,5-Bis(trifluoromethyl)aniline

1,3-Bis(trifluoromethyl)benzene

1,4-Bis(trifluoromethyl)benzene

Bis(trifluoromethyl) disulfide

1,2-Bis(trimethylsilyl)acetylene

Bis(2,4,6-trinitrophenyl) sulfide

Bis[2-(vinyloxy)ethyl] ether

Bithionol

No.	Name	Synonym	Mol. Form.	CAS RN	Mol. Wt.	Physical Form	mp/°C	bp/°C	den/ g cm^{-3}	n_D	Solubility
1058	2,2'-Bithiophene		$C_8H_6S_2$	492-97-7	166.264		33	260			i H_2O; vs EtOH; s eth, ctc, HOAc
1059	Bixin		$C_{25}H_{30}O_4$	6983-79-5	394.504	viol pr (ace)	198				i H_2O; s EtOH, ace; sl eth, bz, HOAc
1060	Boldenone	Dehydrotestosterone	$C_{19}H_{26}O_2$	846-48-0	286.408		165				
1061	Boldine		$C_{19}H_{21}NO_4$	476-70-0	327.375	cry (eth)	163				vs EtOH, chl
1062	Bomyl		$C_9H_{15}O_8P$	122-10-1	282.184	ye oil		160[17]			sl H_2O; vs ace, EtOH, xyl
1063	Borane carbonyl		CH_3BO	13205-44-2	41.845	col gas	-137	-64			dec H_2O
1064	Borneol, (±)		$C_{10}H_{18}O$	6627-72-1	154.249	lf (lig)	208	sub	1.011[20]		i H_2O; vs EtOH, eth, bz
1065	l-Bornyl acetate		$C_{12}H_{20}O_2$	5655-61-8	196.286		27	223.5	0.982[25]	1.4626[20]	sl H_2O; s EtOH, eth
1066	Bornylamine		$C_{10}H_{19}N$	32511-34-5	153.265		163				vs ace, bz, eth, EtOH
1067	Bornyl chloride	2-Chloro-1,7,7-trimethylbicyclo[2.2.1]heptane, endo	$C_{10}H_{17}Cl$	464-41-5	172.695	nd	132	207.5			vs bz, eth, EtOH, peth
1068	Bornyl 3-methylbutanoate, (1R)	d-Bornyl isovalerate	$C_{15}H_{26}O_2$	53022-14-3	238.366			257.5	0.955[25]		vs eth, EtOH
1069	Boron trifluoride - dimethyl ether complex		$C_2H_6BF_3O$	353-42-4	113.874		-14	dec 127	1.2410[20]	1.302[20]	
1070	Boron trifluoride etherate		$C_4H_{10}BF_3O$	109-63-7	141.927	liq	-60.4	125.5	1.125[25]	1.348[20]	dec H_2O; vs eth, EtOH
1071	Brilliant Green		$C_{27}H_{34}N_2O_4S$	633-03-4	482.635	small gold cry					vs H_2O, EtOH
1072	Brilliant Yellow		$C_{26}H_{20}N_4Na_2O_8S_2$	3051-11-4	626.569	ye cry (w)					s H_2O, EtOH; sl ace
1073	Brodifacoum		$C_{31}H_{23}BrO_3$	56073-10-0	523.417	off-wh pow	230				i H_2O; sl EtOH, bz; s ace, chl
1074	Bromacil	5-Bromo-3-sec-butyl-6-methyluracil	$C_9H_{13}BrN_2O_2$	314-40-9	261.115		158		1.55[25]		
1075	Bromadiolone		$C_{30}H_{23}BrO_4$	28772-56-7	527.406	ye-wh pow	205				vs DMF; sl ace, chl, EtOH, eth; i hx
1076	Bromal hydrate		$C_2H_3Br_3O_2$	507-42-6	298.756	mcl pr (w+1)	53.5	dec	2.5661[40]		vs eth, EtOH
1077	Bromdian	Tetrabromobisphenol A	$C_{15}H_{12}Br_4O_2$	79-94-7	543.871		179				s EtOH, eth, bz, chl
1078	N-Bromoacetamide		C_2H_4BrNO	79-15-2	137.963	nd (chl-hx)	103.5				vs eth
1079	Bromoacetic acid		$C_2H_3BrO_2$	79-08-3	138.948	hex or orth cry	50	208	1.9335[50]	1.4804[50]	msc H_2O, EtOH, eth; s ace, bz; sl chl
1080	Bromoacetone		C_3H_5BrO	598-31-2	136.975	liq	-36.5	138; 31.5[8]	1.634[23]	1.4697[15]	sl H_2O; s EtOH, eth, ace
1081	α-Bromoacetophenone	ω-Bromoacetophenone	C_8H_7BrO	70-11-1	199.045	nd (al) orth pr (al) pl(peth)	50.5	135[18]	1.647[20]		i H_2O; s EtOH, peth; vs eth, bz, chl
1082	4-(Bromoacetyl)biphenyl	2-Bromo-4'-phenylacetophenone	$C_{14}H_{11}BrO$	135-73-9	275.140	nd (95% al)	127				
1083	Bromoacetyl bromide		$C_2H_2Br_2O$	598-21-0	201.844			148.5	2.312[22]	1.5449[20]	s ace, ctc
1084	Bromoacetylene		C_2HBr	593-61-3	104.933	col gas		4.7			vs eth
1085	5-(2-Bromoallyl)-5-sec-butylbarbituric acid	Butallylonal	$C_{11}H_{15}BrN_2O_3$	1142-70-7	303.152		131.5				vs eth, EtOH
1086	5-(2-Bromoallyl)-5-isopropylbarbituric acid	Propallylonal	$C_{10}H_{13}BrN_2O_3$	545-93-7	289.125	cry (dil HOAc, dil al)	181				sl H_2O, eth, bz; vs EtOH, ace, HOAc
1087	2-Bromoaniline		C_6H_6BrN	615-36-1	172.023		32	229	1.578[20]	1.6113[20]	i H_2O; s EtOH, eth
1088	3-Bromoaniline		C_6H_6BrN	591-19-5	172.023		18.5	251	1.5793[20]	1.6260[20]	sl H_2O; s EtOH, eth
1089	4-Bromoaniline		C_6H_6BrN	106-40-1	172.023	orth bipym nd (60% al)	66.4	dec	1.4970[100]		i H_2O; s EtOH, eth; sl chl
1090	2-Bromoanisole		C_7H_7BrO	578-57-4	187.034		1.3	216	1.5018[20]	1.5727[20]	i H_2O; s EtOH, eth
1091	3-Bromoanisole		C_7H_7BrO	2398-37-0	187.034			211; 105[16]		1.5635[20]	i H_2O; s EtOH, eth, bz, CS_2
1092	4-Bromoanisole		C_7H_7BrO	104-92-7	187.034		13.5	215	1.4564[20]	1.5642[20]	sl H_2O; vs EtOH, eth, chl; s ctc
1093	2-Bromobenzaldehyde		C_7H_5BrO	6630-33-7	185.018		21.5	230		1.5925[20]	i H_2O; vs EtOH, bz; sl ctc
1094	3-Bromobenzaldehyde		C_7H_5BrO	3132-99-8	185.018			234		1.5935[20]	i H_2O; vs EtOH, eth; sl ctc
1095	4-Bromobenzaldehyde		C_7H_5BrO	1122-91-4	185.018	lf (dil al)	58	67[2]			i H_2O; vs EtOH, bz; sl chl

2,2'-Bithiophene

Bixin

Boldenone

Boldine

Bomyl

Borane carbonyl

Borneol, (±)

l-Bornyl acetate

Bornylamine

Bornyl chloride

Bornyl 3-methylbutanoate, (1*R*)

Boron trifluoride - dimethyl ether complex

Boron trifluoride etherate

Brilliant Green

Brilliant Yellow

Brodifacoum

Bromacil

Bromadiolone

Bromal hydrate

Bromdian

N-Bromoacetamide

Bromoacetic acid

Bromoacetone

α-Bromoacetophenone

4-(Bromoacetyl)biphenyl

Bromoacetyl bromide

Bromoacetylene

5-(2-Bromoallyl)-5-*sec*-butylbarbituric acid

5-(2-Bromoallyl)-5-isopropylbarbituric acid

2-Bromoaniline

3-Bromoaniline

4-Bromoaniline

2-Bromoanisole

3-Bromoanisole

4-Bromoanisole

2-Bromobenzaldehyde

3-Bromobenzaldehyde

4-Bromobenzaldehyde

No.	Name	Synonym	Mol. Form.	CAS RN	Mol. Wt.	Physical Form	mp/°C	bp/°C	den/ g cm⁻³	n_D	Solubility
1096	Bromobenzene	Phenyl bromide	C_6H_5Br	108-86-1	157.008	liq	-30.72	156.06	1.4950[20]	1.5597[20]	i H_2O; vs EtOH, eth, bz; s ctc
1097	4-Bromobenzeneacetic acid		$C_8H_7BrO_2$	1878-68-8	215.045	nd (w)	116	sub			sl H_2O; vs EtOH, eth, CS_2
1098	4-Bromobenzeneacetonitrile		C_8H_6BrN	16532-79-9	196.045	pa ye cry (al)	48.0				vs bz, EtOH
1099	α-Bromobenzeneacetonitrile	α-Bromobenzyl cyanide	C_8H_6BrN	5798-79-8	196.045	ye cry (dil al)	29	dec 242; 133[12]	1.539[29]		i H_2O; vs EtOH, eth, ace, bz, chl
1100	2-Bromo-1,4-benzenediol		$C_6H_5BrO_2$	583-69-7	189.007	lf (lig), cry (chl)	111.5	sub			vs H_2O, EtOH, eth, bz; sl chl, lig; s HOAc
1101	4-Bromobenzenesulfonyl chloride	p-Brosyl chloride	$C_6H_4BrClO_2S$	98-58-8	255.517	tcl or mcl pl (eth)	76	153[15]			i H_2O; vs eth; s chl
1102	4-Bromobenzenethiol		C_6H_5BrS	106-53-6	189.073	lf (al)	73	230.5	1.5260[83]		sl H_2O, EtOH; vs eth, ctc, chl
1103	2-Bromobenzoic acid		$C_7H_5BrO_2$	88-65-3	201.018	mcl pr (w), nd	150	sub	1.929[25]		sl H_2O, DMSO; s EtOH, eth, ace, chl
1104	3-Bromobenzoic acid		$C_7H_5BrO_2$	585-76-2	201.018	mcl nd (dil al)	155	>280	1.845[20]		i H_2O; s EtOH, eth
1105	4-Bromobenzoic acid		$C_7H_5BrO_2$	586-76-5	201.018	nd (eth), lf (w), mcl pr	254.5		1.894[20]		sl H_2O, DMSO; s EtOH, eth
1106	2-Bromobenzonitrile		C_7H_4BrN	2042-37-7	182.018	nd (w)	55.5	252			s H_2O; vs EtOH; sl chl
1107	3-Bromobenzonitrile		C_7H_4BrN	6952-59-6	182.018		39.5	225			vs EtOH, eth; sl chl
1108	4-Bromobenzonitrile		C_7H_4BrN	623-00-7	182.018	nd (w, al)	114	236			s H_2O, EtOH, eth, chl
1109	6-Bromobenzo[a]pyrene		$C_{20}H_{11}Br$	21248-00-0	331.205	cry (ace/ MeOH)	223				
1110	2-Bromobenzoyl chloride		C_7H_4BrClO	7154-66-7	219.463	nd	11	243		1.5963[20]	sl ctc
1111	4-Bromobenzoyl chloride		C_7H_4BrClO	586-75-4	219.463	nd (peth)	42	246; 181[125]			vs EtOH, eth, bz, lig
1112	2-Bromobiphenyl		$C_{12}H_9Br$	2052-07-5	233.103		0.8	297	1.2175[26]	1.6248[25]	vs eth, EtOH
1113	3-Bromobiphenyl		$C_{12}H_9Br$	2113-57-7	233.103			300; 171[17]		1.6411[20]	i H_2O
1114	4-Bromobiphenyl		$C_{12}H_9Br$	92-66-0	233.103	pl (al)	91.5	310	0.9327[25]		i H_2O; s EtOH, eth, bz, HOAc; sl chl
1115	1-Bromo-2-(bromomethyl) benzene		$C_7H_6Br_2$	3433-80-5	249.931	cry (al, lig)	31	129[19]			vs eth, EtOH, HOAc
1116	1-Bromo-3-(bromomethyl) benzene		$C_7H_6Br_2$	823-78-9	249.931	nd or lf	42	122[12]			s chl
1117	1-Bromo-4-(bromomethyl) benzene	p-Bromobenzyl bromide	$C_7H_6Br_2$	589-15-1	249.931	nd (al)	63				sl H_2O; s EtOH, bz, chl; vs eth, CS_2
1118	2-Bromo-2-(bromomethyl) pentanedinitrile	1,2-Dibromo-2,4-dicyanobutane	$C_6H_6Br_2N_2$	35691-65-7	265.933		52				i H_2O; vs ace, bz, DMF
1119	2-Bromo-1-(4-bromophenyl) ethanone	p-Bromophenacyl bromide	$C_8H_6Br_2O$	99-73-0	277.941	nd (al)	111				i H_2O; s EtOH, eth, chl
1120	2-Bromo-1,3-butadiene		C_4H_5Br	1822-86-2	132.987			42[165]	1.397[20]	1.4988[20]	vs eth, EtOH
1121	1-Bromobutane	Butyl bromide	C_4H_9Br	109-65-9	137.018	liq	-112.6	101.6	1.2758[20]	1.4401[20]	i H_2O; msc EtOH, eth, ace; sl ctc; s chl
1122	2-Bromobutane, (±)	(±)-sec-Butyl bromide	C_4H_9Br	5787-31-5	137.018	liq	-112.65	91.3	1.2585[20]	1.4366[20]	vs ace, eth, chl
1123	Bromobutanedioic acid, (±)	Bromosuccinic acid	$C_4H_5BrO_4$	584-98-5	196.985		161		2.073[25]		s H_2O, EtOH; sl HOAc
1124	4-Bromobutanenitrile		C_4H_6BrN	5332-06-9	148.002			206	1.4967[20]	1.4818[20]	s EtOH, eth, chl
1125	2-Bromobutanoic acid, (±)	DL-α-Bromobutyric acid	$C_4H_7BrO_2$	2385-70-8	167.002		-2.0	dec 217; 127[25]	1.5641[20]		s H_2O, EtOH, eth
1126	4-Bromobutanoic acid		$C_4H_7BrO_2$	2623-87-2	167.002		33	142[25], 125[7]			
1127	3-Bromo-2-butanone		C_4H_7BrO	814-75-5	151.002			36[11]			
1128	cis-1-Bromo-1-butene		C_4H_7Br	31849-78-2	135.003			86.1	1.3265[15]	1.4536[20]	i H_2O; s eth, ace, bz, chl; sl ctc
1129	trans-1-Bromo-1-butene		C_4H_7Br	32620-08-9	135.003	liq	-100.3	94.7	1.3209[15]	1.4527[20]	i H_2O; s eth, ace, bz, chl; sl ctc
1130	2-Bromo-1-butene		C_4H_7Br	23074-36-4	135.003	liq	-133.4	88	1.3209[15]	1.4527[20]	i H_2O; s eth, ace, bz, chl; sl ctc
1131	4-Bromo-1-butene		C_4H_7Br	5162-44-7	135.003			98.5	1.3230[20]	1.4622[20]	sl H_2O; vs bz, eth, EtOH
1132	1-Bromo-2-butene		C_4H_7Br	4784-77-4	135.003			104.5	1.3371[25]	1.4822[20]	i H_2O; s EtOH, eth, ctc; vs chl, bz
1133	cis-2-Bromo-2-butene		C_4H_7Br	3017-68-3	135.003	liq	-111.5	93.9	1.3416[15]	1.4631[19]	i H_2O; s EtOH, eth, ctc; vs chl, bz

Bromobenzene

4-Bromobenzeneacetic acid

4-Bromobenzeneacetonitrile

α-Bromobenzeneacetonitrile

2-Bromo-1,4-benzenediol

4-Bromobenzenesulfonyl chloride

4-Bromobenzenethiol

2-Bromobenzoic acid

3-Bromobenzoic acid

4-Bromobenzoic acid

2-Bromobenzonitrile

3-Bromobenzonitrile

4-Bromobenzonitrile

6-Bromobenzo[a]pyrene

2-Bromobenzoyl chloride

4-Bromobenzoyl chloride

2-Bromobiphenyl

3-Bromobiphenyl

4-Bromobiphenyl

1-Bromo-2-(bromomethyl)benzene

1-Bromo-3-(bromomethyl)benzene

1-Bromo-4-(bromomethyl)benzene

2-Bromo-2-(bromomethyl)pentanedinitrile

2-Bromo-1-(4-bromophenyl)ethanone

2-Bromo-1,3-butadiene

1-Bromobutane

2-Bromobutane, (±)

Bromobutanedioic acid, (±)

4-Bromobutanenitrile

2-Bromobutanoic acid, (±)

4-Bromobutanoic acid

3-Bromo-2-butanone

cis-1-Bromo-1-butene

trans-1-Bromo-1-butene

2-Bromo-1-butene

4-Bromo-1-butene

1-Bromo-2-butene

cis-2-Bromo-2-butene

No.	Name	Synonym	Mol. Form.	CAS RN	Mol. Wt.	Physical Form	mp/°C	bp/°C	den/ g cm⁻³	n_D	Solubility
1134	*trans*-2-Bromo-2-butene		C₄H₇Br	3017-71-8	135.003	liq	-114.6	85.6	1.3323[15]	1.4602[16]	i H₂O; s EtOH, eth, ctc; vs chl, bz
1135	(4-Bromobutoxy)benzene		C₁₀H₁₃BrO	1200-03-9	229.113	cry (al)	41	154[18]			sl EtOH, ctc
1136	1-Bromo-4-*tert*-butylbenzene		C₁₀H₁₃Br	3972-65-4	213.114		19	231.5	1.2286[20]	1.5436[20]	i H₂O; s eth, bz, chl
1137	2-Bromo-3'-chloroacetophenone	3-Chlorophenacyl bromide	C₈H₆BrClO	41011-01-2	233.490	nd	40	397.5			vs EtOH
1138	1-Bromo-2-chlorobenzene		C₆H₄BrCl	694-80-4	191.453	liq	-12.3	204	1.6387[25]	1.5809[20]	i H₂O; vs bz; sl ctc
1139	1-Bromo-3-chlorobenzene		C₆H₄BrCl	108-37-2	191.453	liq	-21.5	196	1.6302[20]	1.5771[20]	i H₂O; vs EtOH, eth
1140	1-Bromo-4-chlorobenzene		C₆H₄BrCl	106-39-8	191.453	nd or pl (al, eth)	68	196	1.576[71]	1.5531[70]	i H₂O; sl EtOH; s eth, bz, ctc, chl
1141	1-Bromo-4-chlorobutane		C₄H₈BrCl	6940-78-9	171.464			175; 63[10]	1.489[20]	1.4885[20]	i H₂O; s EtOH, eth, chl; sl ctc
1142	Bromochlorodifluoromethane	Halon 1211	CBrClF₂	353-59-3	165.365	col gas	-159.5	-3.7			
1143	3-Bromo-1-chloro-5,5-dimethylhydantoin		C₅H₆BrClN₂O₂	126-06-7	241.471		162				
1144	1-Bromo-1-chloroethane		C₂H₄BrCl	593-96-4	143.410			83	1.667[10]	1.4660[20]	
1145	1-Bromo-2-chloroethane		C₂H₄BrCl	107-04-0	143.410	liq	-16.7	107	1.7392[20]	1.4908[20]	sl H₂O; s EtOH, eth, chl
1146	Bromochlorofluoromethane		CHBrClF	593-98-6	147.374	liq	-115	36	1.9771[0]	1.4144[25]	i H₂O; s eth, ace, chl
1147	Bromochloromethane	Halon 1011	CH₂BrCl	74-97-5	129.384	liq	-87.9	68.0	1.9344[20]	1.4838[20]	i H₂O; s EtOH, eth, ace, bz
1148	1-Bromo-4-(chloromethyl)benzene	*p*-Bromobenzyl chloride	C₇H₆BrCl	589-17-3	205.480	nd (al, peth)	50	236			i H₂O; vs EtOH, eth; s peth
1149	2-Bromo-1-(4-chlorophenyl)ethanone	*p*-Chlorophenacyl bromide	C₈H₆BrClO	536-38-9	233.490	nd	96.5				
1150	1-Bromo-2-chloropropane		C₃H₆BrCl	3017-96-7	157.437			118	1.531[20]	1.4745[20]	vs ace, bz, eth, EtOH
1151	1-Bromo-3-chloropropane		C₃H₆BrCl	109-70-6	157.437	liq	-58.9	143.3	1.5969[20]	1.4864[20]	i H₂O; vs EtOH, eth, chl
1152	2-Bromo-1-chloropropane		C₃H₆BrCl	3017-95-6	157.437			117	1.537[20]	1.4795[20]	i H₂O; vs EtOH, eth; s ace, bz
1153	2-Bromo-2-chloropropane		C₃H₆BrCl	2310-98-7	157.437			95	1.495[20]	1.4575[20]	vs ace, bz, eth, EtOH
1154	1-Bromo-2-chloro-1,1,2-trifluoroethane		C₂HBrClF₃	354-06-3	197.381			52.5	1.8574[25]	1.3738[20]	
1155	2-Bromo-2-chloro-1,1,1-trifluoroethane	Halothane	C₂HBrClF₃	151-67-7	197.381			50.2; 20[243]	1.8563[25]	1.3697[0]	sl H₂O; s peth
1156	Bromocresol Green	Bromcresol Green	C₂₁H₁₄Br₄O₅S	76-60-8	698.014	wh or red (+7w) ye (HOAc)	218.5				sl H₂O; vs EtOH, eth, AcOEt; s bz
1157	Bromocresol Purple	Bromcresol Purple	C₂₁H₁₆Br₂O₅S	115-40-2	540.222		241.5				
1158	Bromocycloheptane	Cycloheptyl bromide	C₇H₁₃Br	2404-35-5	177.082			101[40], 75[12]	1.3080[20]	1.4996[20]	i H₂O; vs eth, chl
1159	Bromocyclohexane	Cyclohexyl bromide	C₆H₁₁Br	108-85-0	163.055	liq	-56.5	166.2	1.3359[20]	1.4957[20]	i H₂O; msc EtOH, eth, ace, bz, lig, ctc
1160	*trans*-4-Bromocyclohexanol		C₆H₁₁BrO	32388-22-0	179.054	pl (hx)	81.5				
1161	2-Bromocyclohexanone		C₆H₉BrO	822-85-5	177.038			114[32], 90[14]	1.340[25]	1.5085[25]	
1162	3-Bromocyclohexene		C₆H₉Br	1521-51-3	161.039			81[40], 56[11]	1.3890[20]	1.5320[20]	i H₂O; s eth, bz, chl
1163	Bromocyclopentane	Cyclopentyl bromide	C₅H₉Br	137-43-9	149.029			137.5	1.3873[20]	1.4886[20]	sl ctc
1164	1-Bromodecane		C₁₀H₂₁Br	112-29-8	221.178	liq	-29.2	240.6	1.0702[20]	1.4557[20]	i H₂O; vs eth, chl; s ctc
1165	2-Bromodecanoic acid		C₁₀H₁₉BrO₂	2623-95-2	251.161		2.0	140[2]	1.1912[24]	1.4595[24]	vs eth
1166	1-Bromo-3,5-dichlorobenzene		C₆H₃BrCl₂	19752-55-7	225.898	pr (al)	83	232			i H₂O; s EtOH, eth, chl; vs bz
1167	4-Bromo-1,2-dichlorobenzene		C₆H₃BrCl₂	18282-59-2	225.898	pr	25	237			i H₂O; sl EtOH; vs eth, bz, chl
1168	Bromodichlorofluoromethane	Halon 1121	CBrCl₂F	353-58-2	181.819	liq		52.8	1.95[22]		
1169	Bromodichloromethane		CHBrCl₂	75-27-4	163.829	liq	-57	90	1.980[20]	1.4964[20]	i H₂O; vs EtOH, eth, ace, bz; sl ctc
1170	4-Bromo-2,5-dichlorophenol		C₆H₃BrCl₂O	1940-42-7	241.897	nd	71.5				
1171	2-Bromo-1,1-diethoxyethane		C₆H₁₃BrO₂	2032-35-1	197.070			170; 66[18]	1.283[20]	1.4387[20]	s EtOH, eth
1172	4-Bromo-*N,N*-diethylaniline		C₁₀H₁₄BrN	2052-06-4	228.129	nd or pr	38	270			i H₂O; vs EtOH, eth
1173	Bromodifluoromethane		CHBrF₂	1511-62-2	130.920		-145	-14.6	1.55[16]		s H₂O; vs EtOH
1174	3-Bromo-4,5-dihydro-2(3*H*)-furanone	α-Bromo-γ-butyrolactone	C₄H₅BrO₂	5061-21-2	164.986			130[20]	1.8[20]	1.5059[20]	

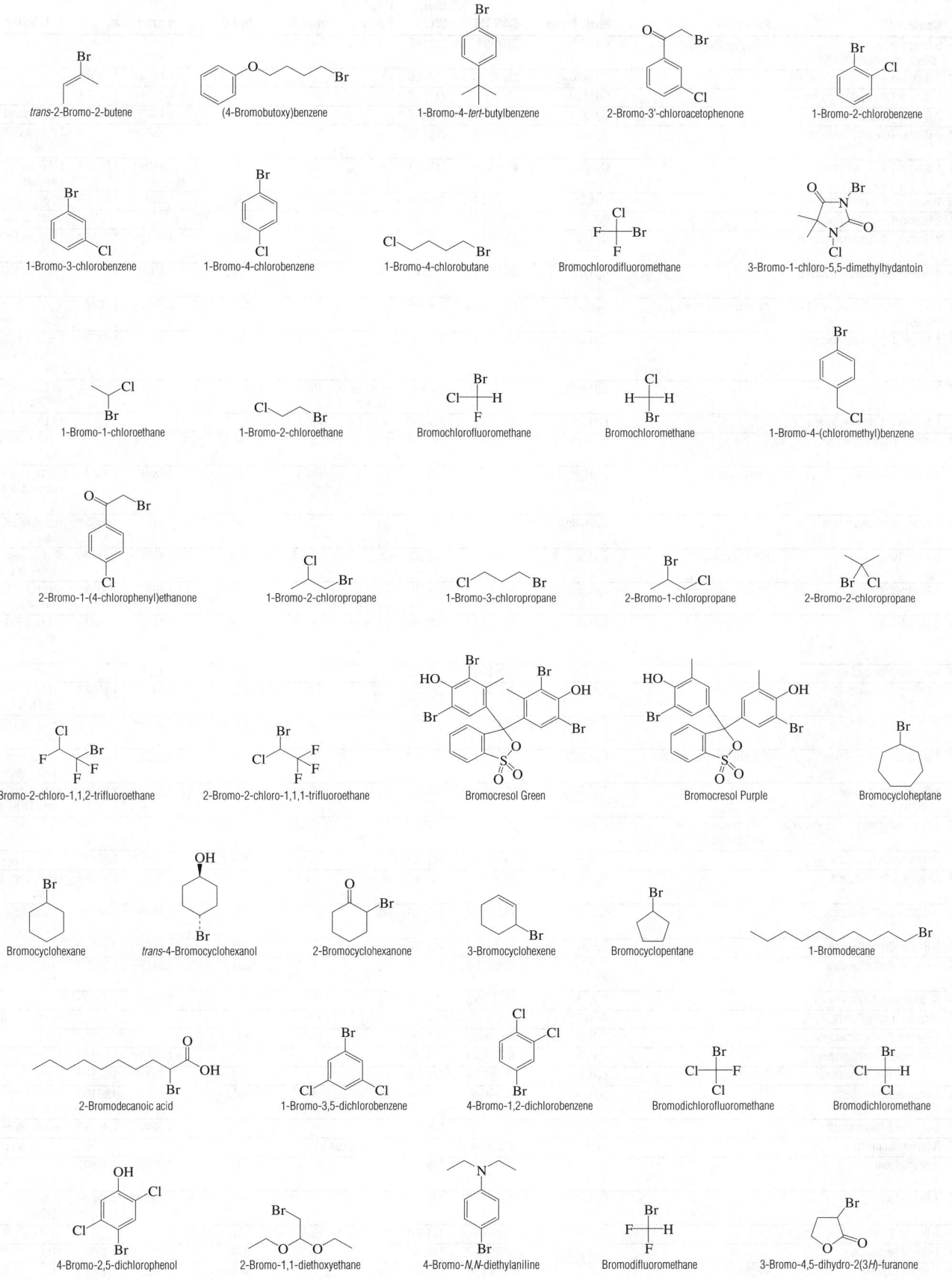

trans-2-Bromo-2-butene

(4-Bromobutoxy)benzene

1-Bromo-4-*tert*-butylbenzene

2-Bromo-3'-chloroacetophenone

1-Bromo-2-chlorobenzene

1-Bromo-3-chlorobenzene

1-Bromo-4-chlorobenzene

1-Bromo-4-chlorobutane

Bromochlorodifluoromethane

3-Bromo-1-chloro-5,5-dimethylhydantoin

1-Bromo-1-chloroethane

1-Bromo-2-chloroethane

Bromochlorofluoromethane

Bromochloromethane

1-Bromo-4-(chloromethyl)benzene

2-Bromo-1-(4-chlorophenyl)ethanone

1-Bromo-2-chloropropane

1-Bromo-3-chloropropane

2-Bromo-1-chloropropane

2-Bromo-2-chloropropane

1-Bromo-2-chloro-1,1,2-trifluoroethane

2-Bromo-2-chloro-1,1,1-trifluoroethane

Bromocresol Green

Bromocresol Purple

Bromocycloheptane

Bromocyclohexane

trans-4-Bromocyclohexanol

2-Bromocyclohexanone

3-Bromocyclohexene

Bromocyclopentane

1-Bromodecane

2-Bromodecanoic acid

1-Bromo-3,5-dichlorobenzene

4-Bromo-1,2-dichlorobenzene

Bromodichlorofluoromethane

Bromodichloromethane

4-Bromo-2,5-dichlorophenol

2-Bromo-1,1-diethoxyethane

4-Bromo-*N,N*-diethylaniline

Bromodifluoromethane

3-Bromo-4,5-dihydro-2(3*H*)-furanone

No.	Name	Synonym	Mol. Form.	CAS RN	Mol. Wt.	Physical Form	mp/°C	bp/°C	den/ g cm⁻³	n_D	Solubility
1175	5-Bromo-*N*,2-dihydroxybenzamide	5-Bromosalicylhydroxamic acid	C₇H₆BrNO₃	5798-94-7	232.032	cry (al)	232 dec				
1176	2-Bromo-1,4-dimethoxybenzene		C₈H₉BrO₂	25245-34-5	217.060	oil		262; 130¹⁰	1.445	1.5700²⁰	
1177	4-Bromo-1,2-dimethoxybenzene		C₈H₉BrO₂	2859-78-1	217.060			254.5	1.702²⁵	1.5743²⁰	
1178	2-Bromo-1,1-dimethoxyethane		C₄H₉BrO₂	7252-83-7	169.017			149	1.430²⁰	1.4450²⁰	s eth, ace, chl
1179	4-Bromo-*N*,*N*-dimethylaniline		C₈H₁₀BrN	586-77-6	200.076		55	264	1.3220¹⁰⁰		i H₂O; s EtOH; vs eth
1180	1-Bromo-2,4-dimethylbenzene		C₈H₉Br	583-70-0	185.061	liq	-17	205	1.3419²⁰	1.5501²⁰	i H₂O; vs EtOH, eth, ace
1181	1-Bromo-3,5-dimethylbenzene		C₈H₉Br	556-96-7	185.061			204	1.362²⁰	1.5462²²	vs eth; s ace, bz
1182	2-Bromo-1,3-dimethylbenzene		C₈H₉Br	576-22-7	185.061			203.5; 100²⁰		1.5552²⁰	vs eth; s ace, bz
1183	2-Bromo-1,4-dimethylbenzene		C₈H₉Br	553-94-6	185.061	lf or pl	9	199; 88¹³	1.3582¹⁸	1.5514¹⁸	i H₂O; vs EtOH; s bz
1184	4-Bromo-1,2-dimethylbenzene		C₈H₉Br	583-71-1	185.061	liq	-0.2	214.5	1.3708²⁰	1.5530²⁰	i H₂O; vs EtOH, eth
1185	*trans*-1-Bromo-3,7-dimethyl-2,6-octadiene	*trans*-Geranyl bromide	C₁₀H₁₇Br	6138-90-5	217.146			101¹²; 47⁰·⁰⁰⁵	1.0940²²	1.5027²⁰	
1186	1-Bromo-2,2-dimethylpropane		C₅H₁₁Br	630-17-1	151.045			106	1.1997²⁰	1.4370²⁰	i H₂O; s EtOH, eth, ace, bz; vs chl
1187	2-Bromo-4,6-dinitroaniline		C₆H₄BrN₃O₄	1817-73-8	262.018	ye nd (al or HOAc)	153.5	sub			vs EtOH, ace; s HOAc
1188	1-Bromo-2,4-dinitrobenzene		C₆H₃BrN₂O₄	584-48-5	247.003	ye nd (al)	75				vs EtOH
1189	α-Bromodiphenylmethane		C₁₃H₁₁Br	776-74-9	247.130		45	184²⁰; 152²			s EtOH, chl; vs bz
1190	1-Bromododecane	Lauryl bromide	C₁₂H₂₅Br	143-15-7	249.231	liq	-9.5	276	1.0399²⁰	1.4583²⁰	i H₂O; s EtOH, eth, ctc; msc ace
1191	2-Bromododecanoic acid		C₁₂H₂₃BrO₂	111-56-8	279.214	pl	32	158²	1.1474⁷⁴	1.4585²⁴	vs bz, eth, EtOH, lig
1192	Bromoethane	Ethyl bromide	C₂H₅Br	74-96-4	108.965	liq	-118.6	38.5	1.4604²⁰	1.4239²⁰	sl H₂O; msc EtOH, eth, chl
1193	2-Bromoethanol	Ethylene bromohydrin	C₂H₅BrO	540-51-2	124.964			150; 51⁴	1.7629²⁰	1.4915²⁰	msc H₂O, EtOH, eth; sl lig
1194	Bromoethene	Vinyl bromide	C₂H₃Br	593-60-2	106.949	vol liq or gas	-139.54	15.8	1.4933²⁰	1.4380²⁰	i H₂O; s EtOH, eth, ace, bz, chl
1195	1-Bromo-2-ethoxybenzene		C₈H₉BrO	583-19-7	201.060			223	1.4223²⁰		vs eth, EtOH
1196	1-Bromo-4-ethoxybenzene		C₈H₉BrO	588-96-5	201.060		2.0	231	1.4071²⁵	1.5517²⁰	i H₂O; vs EtOH, eth; s chl
1197	(2-Bromoethoxy)benzene		C₈H₉BrO	589-10-6	201.060		39	dec 240; 128²⁰	1.3555²⁰		i H₂O; vs EtOH, eth
1198	1-Bromo-2-ethoxyethane	2-Bromoethyl ethyl ether	C₄H₉BrO	592-55-2	153.017			127.5	1.385²⁰	1.4447²⁰	sl H₂O; msc EtOH, eth
1199	2-Bromoethyl acetate		C₄H₇BrO₂	927-68-4	167.002	liq	-13.8	162.5	1.514²⁰	1.457²³	vs H₂O, chl; msc EtOH, eth
1200	2-Bromoethylamine hydrobromide	2-Bromoethanamine hydrobromide	C₂H₇Br₂N	2576-47-8	204.892		174.0				
1201	(1-Bromoethyl)benzene		C₈H₉Br	585-71-7	185.061			219; 92¹¹	1.3535²⁵	1.5543²⁵	
1202	(2-Bromoethyl)benzene		C₈H₉Br	103-63-9	185.061	liq	-55.9	219; 105¹⁸	1.3643²⁰	1.5372²⁰	i H₂O; s eth, bz; sl ctc
1203	1-Bromo-2-ethylbenzene		C₈H₉Br	1973-22-4	185.061	liq	-67.9	199.3	1.3548²⁰	1.5472²⁰	vs ace, bz, eth, EtOH
1204	1-Bromo-3-ethylbenzene		C₈H₉Br	2725-82-8	185.061			202	1.3493²⁰	1.5465²⁰	
1205	1-Bromo-4-ethylbenzene		C₈H₉Br	1585-07-5	185.061	liq	-43.5	204	1.3423²⁰	1.5445²⁰	vs ace, bz, eth, EtOH
1206	(2-Bromoethyl)cyclohexane		C₈H₁₅Br	1647-26-3	191.109	liq	-57	212	1.2357²⁰	1.4899²⁰	
1207	*N*-(2-Bromoethyl)phthalimide		C₁₀H₈BrNO₂	574-98-1	254.081	nd (w)	83				vs eth; sl chl
1208	1-Bromo-4-ethynylbenzene		C₈H₅Br	766-96-1	181.030		64.5	89¹⁶			s chl
1209	1-Bromo-2-fluorobenzene		C₆H₄BrF	1072-85-1	174.998			154	1.0738²¹	1.5337²⁰	
1210	1-Bromo-3-fluorobenzene		C₆H₄BrF	1073-06-9	174.998			150	1.7081²⁰	1.5257²⁰	s ctc
1211	1-Bromo-4-fluorobenzene		C₆H₄BrF	460-00-4	174.998	liq	-17.4	151.5	1.593¹⁵	1.5310¹⁵	i H₂O; s EtOH, eth, chl
1212	1-Bromo-2-fluoroethane		C₂H₄BrF	762-49-2	126.955			71.5	1.7044²⁵	1.4236²⁰	vs eth, EtOH
1213	Bromofluoromethane		CH₂BrF	373-52-4	112.929	vol liq or gas		19			s EtOH; vs chl
1214	2-Bromofuran		C₄H₃BrO	584-12-3	146.970			103	1.6500²⁰	1.4980²⁰	sl H₂O; s EtOH, eth, ace, bz
1215	3-Bromofuran		C₄H₃BrO	22037-28-1	146.970			103	1.6606²⁰	1.4958²⁰	vs ace, bz, eth, EtOH
1216	5-Bromo-2-furancarboxaldehyde		C₅H₃BrO₂	1899-24-7	174.981	cry (50% al)	83.5	201; 112¹⁶			vs eth, EtOH
1217	1-Bromoheptadecane		C₁₇H₃₅Br	3508-00-7	319.364		29.6	349	0.9916²⁰	1.4625²⁰	i H₂O; vs chl
1218	1-Bromoheptane	Heptyl bromide	C₇H₁₅Br	629-04-9	179.098	liq	-56.1	178.9	1.1400²⁰	1.4502²⁰	i H₂O; s EtOH, eth; sl ctc; s chl

5-Bromo-N,2-dihydroxybenzamide

2-Bromo-1,4-dimethoxybenzene

4-Bromo-1,2-dimethoxybenzene

2-Bromo-1,1-dimethoxyethane

4-Bromo-N,N-dimethylaniline

1-Bromo-2,4-dimethylbenzene

1-Bromo-3,5-dimethylbenzene

2-Bromo-1,3-dimethylbenzene

2-Bromo-1,4-dimethylbenzene

4-Bromo-1,2-dimethylbenzene

trans-1-Bromo-3,7-dimethyl-2,6-octadiene

1-Bromo-2,2-dimethylpropane

2-Bromo-4,6-dinitroaniline

1-Bromo-2,4-dinitrobenzene

α-Bromodiphenylmethane

1-Bromododecane

2-Bromododecanoic acid

Bromoethane

2-Bromoethanol

Bromoethene

1-Bromo-2-ethoxybenzene

1-Bromo-4-ethoxybenzene

(2-Bromoethoxy)benzene

1-Bromo-2-ethoxyethane

2-Bromoethyl acetate

2-Bromoethylamine hydrobromide

(1-Bromoethyl)benzene

(2-Bromoethyl)benzene

1-Bromo-2-ethylbenzene

1-Bromo-3-ethylbenzene

1-Bromo-4-ethylbenzene

(2-Bromoethyl)cyclohexane

N-(2-Bromoethyl)phthalimide

1-Bromo-4-ethynylbenzene

1-Bromo-2-fluorobenzene

1-Bromo-3-fluorobenzene

1-Bromo-4-fluorobenzene

1-Bromo-2-fluoroethane

Bromofluoromethane

2-Bromofuran

3-Bromofuran

5-Bromo-2-furancarboxaldehyde

1-Bromoheptadecane

1-Bromoheptane

No.	Name	Synonym	Mol. Form.	CAS RN	Mol. Wt.	Physical Form	mp/°C	bp/°C	den/ g cm⁻³	n_D	Solubility
1219	2-Bromoheptane	2-Heptyl bromide	C₇H₁₅Br	1974-04-5	179.098		47	166	1.1277²⁰	1.4503²⁰	i H₂O; vs bz; s ctc, chl
1220	4-Bromoheptane	4-Heptyl bromide	C₇H₁₅Br	998-93-6	179.098			161; 84⁷²	1.1351²⁰	1.4495²⁰	i H₂O; s bz, ctc, chl
1221	1-Bromohexadecane		C₁₆H₃₃Br	112-82-3	305.337		18	336	0.9991²⁰	1.4618²⁵	i H₂O; s eth
1222	2-Bromohexadecanoic acid		C₁₆H₃₁BrO₂	18263-25-7	335.320		52.8				
1223	1-Bromohexane	Hexyl bromide	C₆H₁₃Br	111-25-1	165.071	liq	-83.7	155.3	1.1744²⁰	1.4478²⁰	i H₂O; msc EtOH, eth; s ace; vs chl
1224	2-Bromohexane		C₆H₁₃Br	3377-86-4	165.071			143; 78⁹⁰	1.1658²⁰	1.4832²⁵	i H₂O; vs EtOH; s eth, ace; sl ctc
1225	3-Bromohexane		C₆H₁₃Br	3377-87-5	165.071			142	1.1799²⁰	1.4472²⁰	vs ace, eth, EtOH, chl
1226	2-Bromohexanoic acid, (±)		C₆H₁₁BrO₂	2681-83-6	195.054		2.0	240; 140²³	1.2810³³		s EtOH, eth
1227	6-Bromohexanoic acid		C₆H₁₁BrO₂	4224-70-8	195.054	cry (peth)	35	167²⁰			vs peth
1228	6-Bromohexanoyl chloride		C₆H₁₀BrClO	22809-37-6	213.499			101⁶			
1229	1-Bromo-4-(hexyloxy)benzene		C₁₂H₁₇BrO	30752-19-3	257.166			156¹³	1.2306²⁰	1.5262²⁰	
1230	5-Bromo-2-hydroxybenzaldehyde		C₇H₅BrO₂	1761-61-1	201.018	nd (al), lf (eth)	105.5				i H₂O; s EtOH, eth; sl chl
1231	4-Bromo-α-hydroxybenzeneacetic acid, (±)	p-Bromomandelic acid	C₈H₇BrO₃	7021-04-7	231.044		119				vs H₂O, EtOH, eth, bz, chl
1232	5-Bromo-2-hydroxybenzenemethanol	Bromosaligenin	C₇H₇BrO₂	2316-64-5	203.034	lf (bz)	113				vs bz, eth, EtOH, chl
1233	5-Bromo-2-hydroxybenzoic acid		C₇H₅BrO₃	89-55-4	217.017	nd (w, dil al)	169.8	sub 100			sl H₂O, ace; vs EtOH, eth
1234	3-Bromo-4-hydroxy-5-methoxybenzaldehyde		C₈H₇BrO₃	2973-76-4	231.044	pl (HOAc), nd, pl (al)	167.0				i H₂O; s EtOH, DMSO; sl eth, bz
1235	1-Bromo-2-iodobenzene		C₆H₄BrI	583-55-1	282.904		9.5	257; 120¹⁵	2.2570²⁵	1.6618²⁵	i H₂O; sl EtOH, HOAc; s ace
1236	1-Bromo-3-iodobenzene		C₆H₄BrI	591-18-4	282.904	liq	-9.3	252; 120¹⁸			i H₂O; sl EtOH, HOAc
1237	1-Bromo-4-iodobenzene		C₆H₄BrI	589-87-7	282.904	pr or pl (eth-al)	92	252			i H₂O; sl EtOH, chl; s eth
1238	Bromoiodomethane		CH₂BrI	557-68-6	220.835			139.5	2.926¹⁷	1.6410²⁰	vs chl
1239	1-Bromo-4-isocyanatobenzene	p-Bromophenyl isocyanate	C₇H₄BrNO	2493-02-9	198.017	nd		226			vs eth
1240	1-Bromo-4-isopropylbenzene		C₉H₁₁Br	586-61-8	199.087	liq	-22.5	218.7	1.3145²⁰	1.5569²⁰	i H₂O; s eth, bz, chl; sl ctc
1241	4-Bromoisoquinoline		C₉H₆BrN	1532-97-4	208.055	cry (peth)	41.5	282.5			vs eth
1242	Bromomethane	Methyl bromide	CH₃Br	74-83-9	94.939	col gas	-93.68	3.5	1.6755²⁰	1.4218²⁰	sl H₂O; msc EtOH, eth, chl, CS₂
1243	1-Bromo-2-methoxyethane		C₃H₇BrO	6482-24-2	138.991			110	1.4623²⁰	1.44753²⁰	
1244	Bromomethoxymethane		C₂H₅BrO	13057-17-5	124.964			87	1.5976²⁰	1.4562²⁰	
1245	2-Bromo-4-methylaniline		C₇H₈BrN	583-68-6	186.050	lf	26	240	1.510²⁰	1.5999²⁰	i H₂O; s EtOH, eth
1246	4-Bromo-2-methylaniline		C₇H₈BrN	583-75-5	186.050	cry (al)	59.5	240			sl H₂O, chl; s EtOH; vs eth, HOAc
1247	(Bromomethyl)benzene	Benzyl bromide	C₇H₇Br	100-39-0	171.035	liq	-1.5	201	1.4380²⁵	1.5752²⁰	i H₂O; msc EtOH, eth; s ctc
1248	4-(Bromomethyl)benzoic acid		C₈H₇BrO₂	6232-88-8	215.045		226.3				
1249	3-(Bromomethyl)benzonitrile		C₈H₆BrN	28188-41-2	196.045		96.5	130⁴			
1250	4-(Bromomethyl)benzonitrile		C₈H₆BrN	17201-43-3	196.045		114				
1251	1-Bromo-2-methylbutane, DL		C₅H₁₁Br	5973-11-5	151.045			119	1.2205²⁰	1.4452²⁰	i H₂O; s EtOH, eth; vs chl
1252	1-Bromo-3-methylbutane	Isopentyl bromide	C₅H₁₁Br	107-82-4	151.045	liq	-112	120.4	1.2071²⁰	1.4420²⁰	i H₂O; s EtOH, eth; sl ctc; vs chl
1253	2-Bromo-2-methylbutane	tert-Pentyl bromide	C₅H₁₁Br	507-36-8	151.045			108	1.197¹⁸	1.4421	
1254	3-Bromo-3-methylbutanoic acid	β-Bromoisovaleric acid	C₅H₉BrO₂	5798-88-9	181.028	nd (lig)	74				vs bz, eth, EtOH
1255	1-Bromo-3-methyl-2-butene		C₅H₉Br	870-63-3	149.029			dec 131; 50⁴⁰	1.2930¹⁵	1.4930¹⁵	vs ace, bz, eth, EtOH
1256	1-(Bromomethyl)-2-chlorobenzene		C₇H₆BrCl	611-17-6	205.480			109¹⁰			
1257	(Bromomethyl)chlorodimethylsilane		C₃H₈BrClSi	16532-02-8	187.539			131	1.375²⁵	1.4630²⁵	
1258	1-Bromo-3-methylcyclohexane	3-Methylcyclohexyl bromide	C₇H₁₃Br	13905-48-1	177.082			181; 60¹¹	1.2676¹⁵	1.4979²⁰	i H₂O; vs eth; s bz
1259	(Bromomethyl)cyclohexane		C₇H₁₃Br	2550-36-9	177.082			76²⁶	1.283²⁰	1.4907³⁰	vs bz, eth, chl
1260	1-(Bromomethyl)-3-fluorobenzene		C₇H₆BrF	456-41-7	189.025			88²⁰		1.5474²⁰	
1261	3-(Bromomethyl)heptane		C₈H₁₇Br	18908-66-2	193.125			67¹⁰			

2-Bromoheptane

4-Bromoheptane

1-Bromohexadecane

2-Bromohexadecanoic acid

1-Bromohexane

2-Bromohexane

3-Bromohexane

2-Bromohexanoic acid, (±)

6-Bromohexanoic acid

6-Bromohexanoyl chloride

1-Bromo-4-(hexyloxy)benzene

5-Bromo-2-hydroxybenzaldehyde

4-Bromo-α-hydroxybenzeneacetic acid, (±)

5-Bromo-2-hydroxybenzenemethanol

5-Bromo-2-hydroxybenzoic acid

3-Bromo-4-hydroxy-5-methoxybenzaldehyde

1-Bromo-2-iodobenzene

1-Bromo-3-iodobenzene

1-Bromo-4-iodobenzene

Bromoiodomethane

1-Bromo-4-isocyanatobenzene

1-Bromo-4-isopropylbenzene

4-Bromoisoquinoline

Bromomethane

1-Bromo-2-methoxyethane

Bromomethoxymethane

2-Bromo-4-methylaniline

4-Bromo-2-methylaniline

(Bromomethyl)benzene

4-(Bromomethyl)benzoic acid

3-(Bromomethyl)benzonitrile

4-(Bromomethyl)benzonitrile

1-Bromo-2-methylbutane, *DL*

1-Bromo-3-methylbutane

2-Bromo-2-methylbutane

3-Bromo-3-methylbutanoic acid

1-Bromo-3-methyl-2-butene

1-(Bromomethyl)-2-chlorobenzene

(Bromomethyl)chlorodimethylsilane

1-Bromo-3-methylcyclohexane

(Bromomethyl)cyclohexane

1-(Bromomethyl)-3-fluorobenzene

3-(Bromomethyl)heptane

No.	Name	Synonym	Mol. Form.	CAS RN	Mol. Wt.	Physical Form	mp/°C	bp/°C	den/ g cm⁻³	n_D	Solubility
1262	1-(Bromomethyl)-2-methylbenzene		C₈H₉Br	89-92-9	185.061	pr	21	217; 108[16]	1.3811[23]	1.5730[20]	i H₂O; s EtOH, eth, ace, bz
1263	1-(Bromomethyl)-3-methylbenzene		C₈H₉Br	620-13-3	185.061			212.5	1.3711[23]	1.5660[20]	i H₂O; vs EtOH, eth
1264	1-(Bromomethyl)-4-methylbenzene		C₈H₉Br	104-81-4	185.061	nd (al)	35	220	1.324[25]		i H₂O; s EtOH; vs eth, chl
1265	1-(Bromomethyl)naphthalene		C₁₁H₉Br	3163-27-7	221.093	cry (peth, al)	56	183[18], 167[10]			vs ace, bz, eth, EtOH
1266	2-(Bromomethyl)naphthalene		C₁₁H₉Br	939-26-4	221.093	lf (al)	56	213[100], 167[14]			s EtOH, eth, chl, HOAc
1267	1-(Bromomethyl)-3-nitrobenzene		C₇H₆BrNO₂	3958-57-4	216.033	nd or pl (al)	59.3	162[13]			i H₂O; s EtOH
1268	1-(Bromomethyl)-4-nitrobenzene		C₇H₆BrNO₂	100-11-8	216.033	nd (al)	99.5				sl H₂O, chl; vs EtOH, eth; s HOAc
1269	2-(Bromomethyl)-4-nitrophenol		C₇H₆BrNO₃	772-33-8	232.032		148				
1270	(Bromomethyl)oxirane, (±)		C₃H₅BrO	82584-73-4	136.975	liq	-40	137	1.615[14]	1.4841[20]	i H₂O; s EtOH, eth, bz, chl
1271	1-Bromo-2-methylpentane	2-Methylpentyl bromide	C₆H₁₃Br	25346-33-2	165.071			141	1.1624[20]	1.4495[20]	vs eth, chl
1272	1-Bromo-4-methylpentane		C₆H₁₃Br	626-88-0	165.071			145	1.1683[20]	1.4490	vs eth, chl
1273	2-Bromo-2-methylpentane		C₆H₁₃Br	4283-80-1	165.071			142.5; 70[100]		1.442[23]	vs eth, chl
1274	3-Bromo-3-methylpentane		C₆H₁₃Br	25346-31-0	165.071			130; 76[100]	1.1835[20]	1.4525[20]	vs eth, chl
1275	2-Bromo-4-methylphenol		C₇H₇BrO	6627-55-0	187.034	nd (peth)	56.5	213.5	1.5422[25]	1.5772[20]	sl H₂O; s EtOH, bz, chl
1276	1-(Bromomethyl)-3-phenoxybenzene	3-Phenoxybenzyl bromide	C₁₃H₁₁BrO	51632-16-7	263.129	oil					
1277	2-Bromo-1-(4-methylphenyl)ethanone		C₉H₉BrO	619-41-0	213.070	nd or lf (al)	51	157[14]			vs eth, EtOH
1278	N-(Bromomethyl)phthalimide	2-(Bromomethyl)-1H-isoindole-1,3(2H)-dione	C₉H₆BrNO₂	5332-26-3	240.054	pr (chl, bz)	151.5				s ace; sl bz, chl; vs AcOEt
1279	1-Bromo-2-methylpropane	Isobutyl bromide	C₄H₉Br	78-77-3	137.018	liq	-119	91.1	1.272[15]	1.4348[20]	i H₂O; vs EtOH, eth, ace, chl, bz; s ctc
1280	2-Bromo-2-methylpropane	tert-Butyl bromide	C₄H₉Br	507-19-7	137.018	liq	-16.2	73.3	1.4278[20]	1.4278[20]	i H₂O; sl ctc
1281	2-Bromo-2-methylpropanoic acid	α-Bromoisobutyric acid	C₄H₇BrO₂	2052-01-9	167.002	cry (peth)	48.5	199; 115[24]	1.4969[60]		
1282	2-Bromo-2-methylpropanoyl bromide		C₄H₆Br₂O	20769-85-1	229.898			163	1.4067[14]		vs ace, CS₂
1283	1-Bromo-2-methylpropene		C₄H₇Br	3017-69-4	135.003			91	1.336[20]		
1284	3-Bromo-2-methylpropene		C₄H₇Br	1458-98-6	135.003			95	1.313[20]		
1285	2-(Bromomethyl)tetrahydrofuran		C₅H₉BrO	1192-30-9	165.028			170; 70[22]	1.4679[20]	1.4850[20]	s EtOH, eth
1286	(Bromomethyl)trimethylsilane		C₄H₁₁BrSi	18243-41-9	167.120			116.5	1.170[25]	1.4460[20]	
1287	1-Bromonaphthalene	1-Naphthyl bromide	C₁₀H₇Br	90-11-9	207.067	oily liq	6.1	281	1.4785[20]	1.658[20]	s H₂O, ace; msc EtOH, eth, bz; sl ctc
1288	2-Bromonaphthalene		C₁₀H₇Br	580-13-2	207.067	pl or orth lf (al)	55.9	281.5	1.605[25]	1.6382[60]	i H₂O; s EtOH, eth, bz, CS₂; sl ctc
1289	4-Bromo-1,8-naphthalenedicarboxylic anhydride		C₁₂H₅BrO₃	81-86-7	277.070		222				
1290	1-Bromo-2-naphthol	1-Bromo-β-naphthol	C₁₀H₇BrO	573-97-7	223.066	orth pr (bz-lig) nd (HOAc)	84	130			i H₂O; s EtOH, eth, bz; sl chl; vs HOAc
1291	4-Bromo-2-nitroaniline		C₆H₅BrN₂O₂	875-51-4	217.020	oran-ye nd (w)	111.5	sub			vs EtOH
1292	1-Bromo-2-nitrobenzene		C₆H₄BrNO₂	577-19-5	202.006	pa ye (al)	43	258	1.6245[80]		i H₂O; vs EtOH; s eth, ace, bz; sl chl
1293	1-Bromo-3-nitrobenzene		C₆H₄BrNO₂	585-79-5	202.006	orth	56	265	1.7036[20]	1.5979[20]	sl H₂O; s EtOH, eth, bz
1294	1-Bromo-4-nitrobenzene	p-Nitrobromobenzene	C₆H₄BrNO₂	586-78-7	202.006	orth or mcl pr (al)	127	256	1.948[25]		i H₂O; s EtOH, eth, bz; sl chl
1295	Bromonitromethane		CH₂BrNO₂	563-70-2	139.937			149; 71[40]		1.4880[20]	vs EtOH
1296	2-Bromo-2-nitro-1,3-propanediol	Bronopol	C₃H₆BrNO₄	52-51-7	199.989		131.5				
1297	1-Bromononane		C₉H₁₉Br	693-58-3	207.151	liq	-29.0	221.4; 88[4]	1.0845[25]	1.4522[25]	
1298	1-Bromooctadecane		C₁₈H₃₇Br	112-89-0	333.391	cry (al)	28.2	362; 210[10]	0.9848[20]	1.4631[20]	i H₂O; s EtOH, eth; sl ctc
1299	1-Bromooctane	Octyl bromide	C₈H₁₇Br	111-83-1	193.125	liq	-55.0	200.8	1.1072[25]	1.4503[25]	i H₂O; msc EtOH, eth; sl ctc
1300	2-Bromooctane, (±)		C₈H₁₇Br	60251-57-2	193.125			188.5	1.0878[25]	1.4442[25]	i H₂O; msc EtOH, eth
1301	8-Bromooctanoic acid		C₈H₁₅BrO₂	17696-11-6	223.108	nd (peth)	38.5	147[2]			vs bz, eth, EtOH
1302	1-Bromopentadecane		C₁₅H₃₁Br	629-72-1	291.311		19	322	1.0675[20]	1.4611[20]	i H₂O; s ace; vs chl

1-(Bromomethyl)-2-methylbenzene

1-(Bromomethyl)-3-methylbenzene

1-(Bromomethyl)-4-methylbenzene

1-(Bromomethyl)naphthalene

2-(Bromomethyl)naphthalene

1-(Bromomethyl)-3-nitrobenzene

1-(Bromomethyl)-4-nitrobenzene

2-(Bromomethyl)-4-nitrophenol

(Bromomethyl)oxirane, (±)

1-Bromo-2-methylpentane

1-Bromo-4-methylpentane

2-Bromo-2-methylpentane

3-Bromo-3-methylpentane

2-Bromo-4-methylphenol

1-(Bromomethyl)-3-phenoxybenzene

2-Bromo-1-(4-methylphenyl)ethanone

N-(Bromomethyl)phthalimide

1-Bromo-2-methylpropane

2-Bromo-2-methylpropane

2-Bromo-2-methylpropanoic acid

2-Bromo-2-methylpropanoyl bromide

1-Bromo-2-methylpropene

3-Bromo-2-methylpropene

2-(Bromomethyl)tetrahydrofuran

(Bromomethyl)trimethylsilane

1-Bromonaphthalene

2-Bromonaphthalene

4-Bromo-1,8-naphthalenedicarboxylic anhydride

1-Bromo-2-naphthol

4-Bromo-2-nitroaniline

1-Bromo-2-nitrobenzene

1-Bromo-3-nitrobenzene

1-Bromo-4-nitrobenzene

Bromonitromethane

2-Bromo-2-nitro-1,3-propanediol

1-Bromononane

1-Bromooctadecane

1-Bromooctane

2-Bromooctane, (±)

8-Bromooctanoic acid

1-Bromopentadecane

No.	Name	Synonym	Mol. Form.	CAS RN	Mol. Wt.	Physical Form	mp/°C	bp/°C	den/ g cm⁻³	n_D	Solubility
1303	Bromopentafluorobenzene		C$_6$BrF$_5$	344-04-7	246.960	liq	-31	137	1.981[25]	1.4490[20]	
1304	Bromopentafluoroethane		C$_2$BrF$_5$	354-55-2	198.917	col gas		-21	1.8098[25]		
1305	1-Bromopentane	Pentyl bromide	C$_5$H$_{11}$Br	110-53-2	151.045	liq	-88.0	129.8	1.2182[20]	1.4447[20]	i H$_2$O; s EtOH, bz, chl; sl ctc; msc eth
1306	2-Bromopentane		C$_5$H$_{11}$Br	107-81-3	151.045	liq	-95.5	117.4	1.2075[20]	1.4413[20]	vs bz, eth, EtOH, chl
1307	3-Bromopentane		C$_5$H$_{11}$Br	1809-10-5	151.045	liq	-126.2	118.6	1.214[20]	1.4441[20]	i H$_2$O; s EtOH, eth, bz, chl
1308	5-Bromopentanenitrile		C$_5$H$_8$BrN	5414-21-1	162.029			111[12], 103[10]	1.3989[20]	1.4780[20]	
1309	5-Bromopentanoic acid		C$_5$H$_9$BrO$_2$	2067-33-6	181.028		40.0	142[13]			s chl
1310	5-Bromo-1-pentene		C$_5$H$_9$Br	1119-51-3	149.029			125.5	1.2581[20]	1.4640[20]	
1311	9-Bromophenanthrene	9-Phenanthryl bromide	C$_{14}$H$_9$Br	573-17-1	257.125	pr (al)	64.5	>360	1.4093[10]		i H$_2$O; s EtOH, eth, CS$_2$; sl chl
1312	2-Bromophenol		C$_6$H$_5$BrO	95-56-7	173.007		5.6	194.5	1.4924[20]	1.589[20]	sl H$_2$O, chl; s EtOH, eth, alk
1313	3-Bromophenol		C$_6$H$_5$BrO	591-20-8	173.007		33	236.5			sl H$_2$O, ctc; vs EtOH, eth; s chl, alk
1314	4-Bromophenol		C$_6$H$_5$BrO	106-41-2	173.007		66.4	238	1.840[15]		s H$_2$O, chl; vs EtOH, eth
1315	Bromophenol Blue	Bromphenol Blue	C$_{19}$H$_{10}$Br$_4$O$_5$S	115-39-9	669.960	hex pr (HOAc-ace)	279 dec				sl H$_2$O; s EtOH, bz, HOAc
1316	1-Bromo-4-phenoxybenzene	4-Bromophenyl phenyl ether	C$_{12}$H$_9$BrO	101-55-3	249.102		18.72	126[3.5]	1.6088[20]	1.6084[20]	i H$_2$O; s eth, ctc
1317	(4-Bromophenoxy)trimethylsilane		C$_9$H$_{13}$BrOSi	17878-44-3	245.188			126[25]	1.2619[20]	1.5145[20]	
1318	N-(4-Bromophenyl)acetamide	p-Bromoacetanilide	C$_8$H$_8$BrNO	103-88-8	214.060	nd (60% al)	168		1.717[25]		i H$_2$O; s EtOH, chl; sl eth, bz
1319	1-(3-Bromophenyl)ethanone		C$_8$H$_7$BrO	2142-63-4	199.045		7.5	133[19]		1.5755[20]	i H$_2$O; s ace, bz
1320	1-(4-Bromophenyl)ethanone	p-Bromoacetophenone	C$_8$H$_7$BrO	99-90-1	199.045	lf (al)	50.5	257; 130[11]	1.647[25]	1.647	i H$_2$O; s EtOH, eth, bz, ctc, HOAc
1321	(4-Bromophenyl)hydrazine	(p-Bromophenyl)hydrazine	C$_6$H$_7$BrN$_2$	589-21-9	187.037	nd (w), lf (lig), cry (al)	108				vs eth, EtOH, lig
1322	2-(4-Bromophenyl)-1H-indene-1,3(2H)-dione	Bromindione	C$_{15}$H$_9$BrO$_2$	1146-98-1	301.135	cry (lig)	138				
1323	(4-Bromophenyl)phenylmethanone		C$_{13}$H$_9$BrO	90-90-4	261.113	lf (al)	82.5	350			i H$_2$O; sl EtOH, eth, bz, peth
1324	2-Bromo-1-phenyl-1-propanone		C$_9$H$_9$BrO	2114-00-3	213.070			247.5	1.4298[20]	1.5720[20]	i H$_2$O; s EtOH, eth, ace, bz, ctc
1325	Bromophos		C$_8$H$_8$BrCl$_2$PS	2104-96-3	317.999	ye cry	54	141[0.01]			sl H$_2$O; s eth, ctc, tol
1326	Bromophos-ethyl		C$_{10}$H$_{12}$BrCl$_2$O$_3$PS	4824-78-6	394.049	pale-ye liq		122[0.004]			
1327	1-Bromopropane	Propyl bromide	C$_3$H$_7$Br	106-94-5	122.992	liq	-110.3	71.1	1.3537[20]	1.4343[20]	sl H$_2$O; s EtOH, eth, ace, bz, chl, ctc
1328	2-Bromopropane	Isopropyl bromide	C$_3$H$_7$Br	75-26-3	122.992	liq	-89.0	59.5	1.3140[20]	1.4251[20]	sl H$_2$O; s ace, bz, chl; msc EtOH, eth
1329	3-Bromopropanenitrile		C$_3$H$_4$BrN	2417-90-5	133.975			92[25], 69[7]	1.6152[20]	1.4800[20]	vs EtOH, eth; sl ctc
1330	2-Bromopropanoic acid, (±)		C$_3$H$_5$BrO$_2$	10327-08-9	152.975	pr	25.7	203.5	1.7000[20]	1.4753[20]	vs H$_2$O, EtOH, eth; sl chl
1331	3-Bromopropanoic acid	β-Bromopropionic acid	C$_3$H$_5$BrO$_2$	590-92-1	152.975	pl (CCl$_4$)	62.5	141[45]	1.48[25]		s H$_2$O, EtOH, eth, bz, chl
1332	3-Bromo-1-propanol		C$_3$H$_7$BrO	627-18-9	138.991			105[185], 80[22]	1.5374[20]	1.4834[25]	s H$_2$O; msc EtOH, eth
1333	1-Bromo-2-propanol		C$_3$H$_7$BrO	19686-73-8	138.991			146.5	1.5585[30]	1.4801[20]	s H$_2$O; vs EtOH, eth
1334	2-Bromopropanoyl bromide		C$_3$H$_4$Br$_2$O	563-76-8	215.871			153	2.0611[16]		
1335	2-Bromopropanoyl chloride		C$_3$H$_4$BrClO	7148-74-5	171.420			132	1.697[11]	1.4780[20]	s eth, chl; sl ctc
1336	cis-1-Bromopropene		C$_3$H$_5$Br	590-13-6	120.976	liq	-113	57.8	1.4291[20]	1.4560[20]	i H$_2$O; s eth, ace, chl
1337	trans-1-Bromopropene		C$_3$H$_5$Br	590-15-8	120.976			63.2			
1338	2-Bromopropene		C$_3$H$_5$Br	557-93-7	120.976	liq	-126	48.4	1.3965[16]	1.4467[16]	i H$_2$O; s eth, ace, chl
1339	3-Bromopropene	Allyl bromide	C$_3$H$_5$Br	106-95-6	120.976	liq	-119	70.1	1.398[20]	1.4697[20]	i H$_2$O; msc EtOH, eth; s ctc, chl, CS$_2$
1340	(3-Bromo-1-propenyl)benzene		C$_9$H$_9$Br	4392-24-9	197.071	nd (al, eth)	34	130[10]	1.3428[30]	1.613[20]	vs EtOH
1341	(3-Bromopropoxy)benzene		C$_9$H$_{11}$BrO	588-63-6	215.086		10.7	127[18]	1.364[16]		vs eth
1342	3-Bromopropylamine hydrobromide	3-Bromo-1-propanamine hydrobromide	C$_3$H$_9$Br$_2$N	5003-71-4	218.918		171.5				

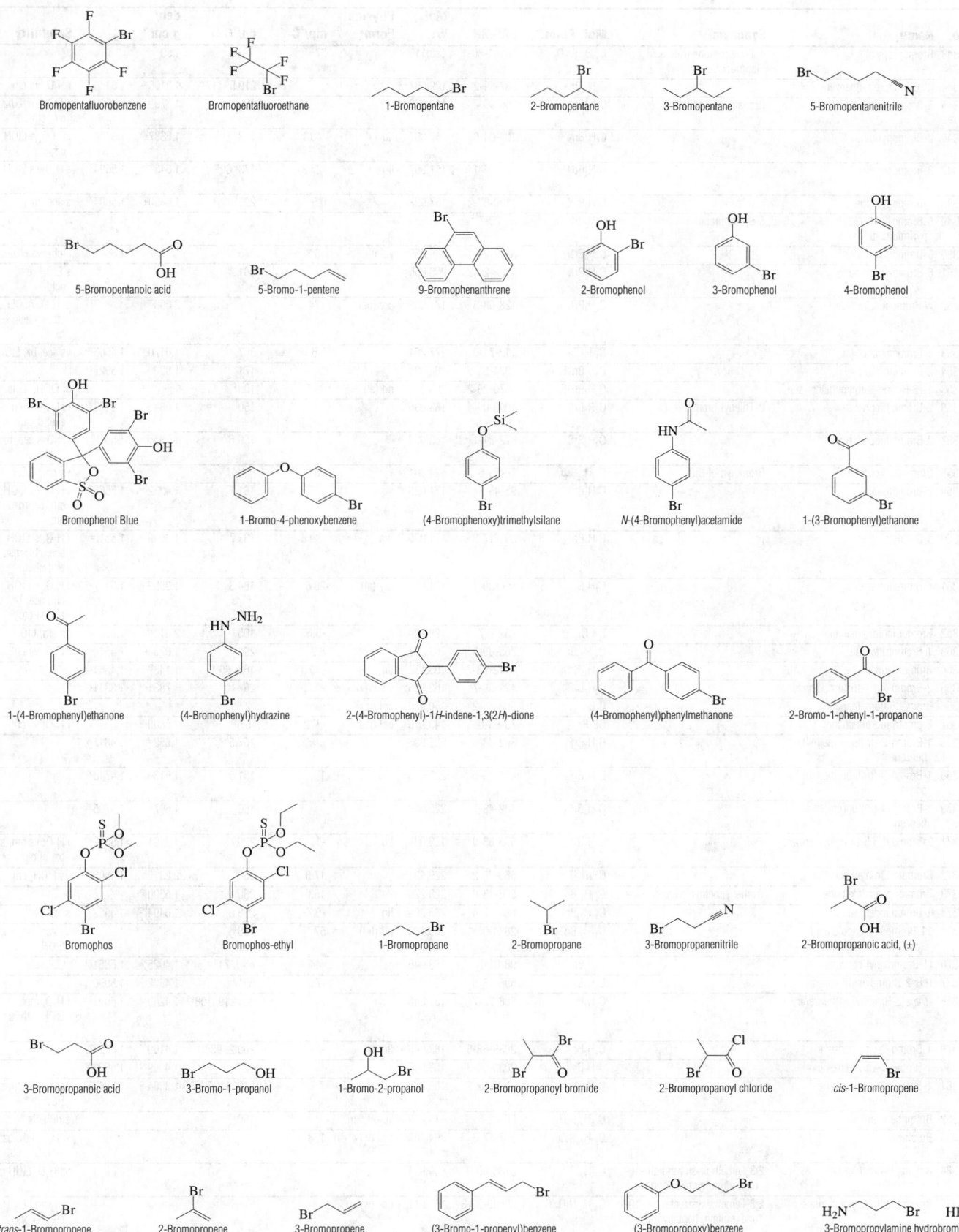

Bromopentafluorobenzene

Bromopentafluoroethane

1-Bromopentane

2-Bromopentane

3-Bromopentane

5-Bromopentanenitrile

5-Bromopentanoic acid

5-Bromo-1-pentene

9-Bromophenanthrene

2-Bromophenol

3-Bromophenol

4-Bromophenol

Bromophenol Blue

1-Bromo-4-phenoxybenzene

(4-Bromophenoxy)trimethylsilane

N-(4-Bromophenyl)acetamide

1-(3-Bromophenyl)ethanone

1-(4-Bromophenyl)ethanone

(4-Bromophenyl)hydrazine

2-(4-Bromophenyl)-1H-indene-1,3(2H)-dione

(4-Bromophenyl)phenylmethanone

2-Bromo-1-phenyl-1-propanone

Bromophos

Bromophos-ethyl

1-Bromopropane

2-Bromopropane

3-Bromopropanenitrile

2-Bromopropanoic acid, (±)

3-Bromopropanoic acid

3-Bromo-1-propanol

1-Bromo-2-propanol

2-Bromopropanoyl bromide

2-Bromopropanoyl chloride

cis-1-Bromopropene

trans-1-Bromopropene

2-Bromopropene

3-Bromopropene

(3-Bromo-1-propenyl)benzene

(3-Bromopropoxy)benzene

3-Bromopropylamine hydrobromide

No.	Name	Synonym	Mol. Form.	CAS RN	Mol. Wt.	Physical Form	mp/°C	bp/°C	den/ g cm⁻³	n_D	Solubility
1343	Bromopropylate	4,4'-Dibromobenzilic acid isopropyl ester	C₁₇H₁₆Br₂O₃	18181-80-1	428.115		77		1.59²⁰		
1344	(3-Bromopropyl)benzene		C₉H₁₁Br	637-59-2	199.087			219.5; 117²⁵	1.3106²⁵	1.5440²⁵	i H₂O; vs eth
1345	3-Bromo-1-propyne	Propargyl bromide	C₃H₃Br	106-96-7	118.960			89	1.579¹⁹	1.4922²⁰	s EtOH, eth, bz, ctc, chl
1346	2-Bromopyridine		C₅H₄BrN	109-04-6	157.997	liq	-40.1	193; 75¹³	1.6337²⁰	1.5734²⁰	sl H₂O; s EtOH, eth, ctc
1347	3-Bromopyridine		C₅H₄BrN	626-55-1	157.997	liq	-27.3	173; 69¹⁸	1.645⁰	1.5694²⁰	s H₂O; vs EtOH, eth
1348	4-Bromopyridine		C₅H₄BrN	1120-87-2	157.997		0.5	29⁰·⁴	1.6450⁰	1.5694²⁰	s ace, bz
1349	5-Bromo-2,4(1H,3H)-pyrimidinedione	5-Bromouracil	C₄H₃BrN₂O₂	51-20-7	190.983		310				
1350	3-Bromoquinoline		C₉H₆BrN	5332-24-1	208.055	ye oil	13.3	275		1.6641²⁰	s chl; vs HOAc
1351	6-Bromoquinoline		C₉H₆BrN	5332-25-2	208.055		24	281			s EtOH, eth, acid
1352	N-Bromosuccinimide		C₄H₄BrNO₂	128-08-5	177.985	cry (bz)	174		2.098²⁵		sl H₂O, AcOEt, eth; vs ace; i hx
1353	1-Bromotetradecane		C₁₄H₂₉Br	112-71-0	277.284		5.6	307	1.0170²⁰	1.4603²⁰	vs ace, bz, EtOH
1354	2-Bromothiazole		C₃H₂BrNS	3034-53-5	164.024			171	1.82²⁵	1.5927²⁰	
1355	1-(5-Bromo-2-thienyl)ethanone		C₆H₅BrOS	5370-25-2	205.072	nd (al)	94.5	103⁴			sl EtOH; s ctc
1356	2-Bromothiophene	2-Thienyl bromide	C₄H₃BrS	1003-09-4	163.036			150	1.684²⁰	1.5868²⁰	i H₂O; vs eth, ace; s ctc
1357	3-Bromothiophene		C₄H₃BrS	872-31-1	163.036			159.5	1.735²⁰	1.5919²⁰	i H₂O; s ace, bz; sl chl
1358	Bromothymol Blue	Bromthymol Blue	C₂₇H₂₈Br₂O₅S	76-59-5	624.381		201				vs eth, EtOH
1359	2-Bromotoluene		C₇H₇Br	95-46-5	171.035	liq	-27.8	181.7	1.4232²⁰	1.5565²⁰	i H₂O; vs EtOH, eth, bz; msc ctc
1360	3-Bromotoluene		C₇H₇Br	591-17-3	171.035	liq	-39.8	183.7	1.4099²⁰	1.5510²⁰	i H₂O; s EtOH, ace, chl; msc eth; sl ctc
1361	4-Bromotoluene		C₇H₇Br	106-38-7	171.035	cry (al)	28.5	184.3	1.3959³⁵	1.5477²⁰	i H₂O; s EtOH, eth, ace, bz, chl; sl ctc
1362	Bromotrichloromethane		CBrCl₃	75-62-7	198.274	liq	-5.65	105	2.012²⁵	1.5065²⁰	vs eth, EtOH
1363	1-Bromotridecane		C₁₃H₂₇Br	765-09-3	263.257		6.2	292	1.0234²⁵	1.4574²⁵	i H₂O; vs chl
1364	Bromotriethylsilane		C₆H₁₅BrSi	1112-48-7	195.173	liq	-49.3	163; 66²⁴	1.143²⁰	1.4561²⁰	
1365	2-Bromo-1,1,1-trifluoroethane		C₂H₂BrF₃	421-06-7	162.936	vol liq or gas	-93.9	26	1.7881²⁰	1.3331²⁰	
1366	Bromotrifluoroethene		C₂BrF₃	598-73-2	160.920	col gas		-2.5			
1367	Bromotrifluoromethane		CBrF₃	75-63-8	148.910	col gas	-172	-57.8	1.5800²⁰		i H₂O; vs chl
1368	1-Bromo-2-(trifluoromethyl)benzene		C₇H₄BrF₃	392-83-6	225.006			167.5	1.652²⁵	1.4817²⁰	
1369	1-Bromo-3-(trifluoromethyl)benzene		C₇H₄BrF₃	401-78-5	225.006		1	151.5	1.613²⁵	1.4716²⁰	
1370	1-Bromo-4-(trifluoromethyl)benzene		C₇H₄BrF₃	402-43-7	225.006			160	1.607²⁵	1.4705²⁵	
1371	2-Bromo-1,3,5-trimethylbenzene		C₉H₁₁Br	576-83-0	199.087	liq	-1	225	1.3191¹⁰	1.5510²⁰	i H₂O; vs eth; s bz; sl ctc
1372	Bromotrinitromethane		CBrN₃O₆	560-95-2	229.931		17.5	56¹⁰	2.0312²⁰	1.4808²⁰	vs EtOH, chl
1373	Bromotriphenylmethane	Triphenylmethyl bromide	C₁₉H₁₅Br	596-43-0	323.226		153	230¹⁵	1.5500²⁰		
1374	1-Bromoundecane		C₁₁H₂₃Br	693-67-4	235.205	liq	-9.7	258.8	1.0494²⁵	1.4552²⁵	sl ctc
1375	11-Bromoundecanoic acid		C₁₁H₂₁BrO₂	2834-05-1	265.188	nd (liq)	57	188¹⁸			vs ace, bz, eth, EtOH
1376	(1-Bromovinyl)benzene		C₈H₇Br	98-81-7	183.046		-44	86¹⁴; 71¹³	1.4025²³	1.5881²⁰	
1377	(cis-2-Bromovinyl)benzene		C₈H₇Br	588-73-8	183.046		-7	55²	1.4322¹⁰	1.5990²²	
1378	(trans-2-Bromovinyl)benzene		C₈H₇Br	588-72-7	183.046		7	dec 219; 108²⁰	1.4269¹⁶	1.6093²⁰	i H₂O; msc EtOH, eth; s chl
1379	1-Bromo-2-vinylbenzene		C₈H₇Br	2039-88-5	183.046	liq	-52.8	209.2; 98²⁰	1.4160²⁰	1.5927²⁰	
1380	1-Bromo-3-vinylbenzene		C₈H₇Br	2039-86-3	183.046			92²⁰	1.4059²⁰	1.5933²⁰	
1381	1-Bromo-4-vinylbenzene		C₈H₇Br	2039-82-9	183.046		7.7	212; 103²⁰	1.3984²⁰	1.5947²⁰	i H₂O; vs chl; s HOAc
1382	Brompheniramine		C₁₆H₁₉BrN₂	86-22-6	319.239	ye oily liq		150⁰·⁵			s dil acid
1383	Brucine		C₂₃H₂₆N₂O₄	357-57-3	394.463	mcl pr (w +4)	178				sl H₂O, eth, bz; vs EtOH, chl
1384	Brucine hydrochloride	2,3-Dimethoxystrychnidin-10-one, monohydrochloride	C₂₃H₂₇ClN₂O₄	5786-96-9	430.924	pr					vs H₂O, EtOH
1385	Brucine sulfate heptahydrate	2,3-Dimethoxystrychnidin-10-one, sulfate, heptahydrate	C₄₆H₆₈N₄O₁₉S	60583-39-3	1013.113	nd (w)					s H₂O; sl EtOH, chl, tfa; vs MeOH; i bz
1386	Bucolome	5-Butyl-1-cyclohexyl-2,4,6(1H,3H,5H)-pyrimidinetrione	C₁₄H₂₂N₂O₃	841-73-6	266.336	nd (MeOH)	84	186⁰·⁸			

Bromopropylate

(3-Bromopropyl)benzene

3-Bromo-1-propyne

2-Bromopyridine

3-Bromopyridine

4-Bromopyridine

5-Bromo-2,4(1*H*,3*H*)-pyrimidinedione

3-Bromoquinoline

6-Bromoquinoline

N-Bromosuccinimide

1-Bromotetradecane

2-Bromothiazole

1-(5-Bromo-2-thienyl)ethanone

2-Bromothiophene

3-Bromothiophene

Bromothymol Blue

2-Bromotoluene

3-Bromotoluene

4-Bromotoluene

Bromotrichloromethane

1-Bromotridecane

Bromotriethylsilane

2-Bromo-1,1,1-trifluoroethane

Bromotrifluoroethene

Bromotrifluoromethane

1-Bromo-2-(trifluoromethyl)benzene

1-Bromo-3-(trifluoromethyl)benzene

1-Bromo-4-(trifluoromethyl)benzene

2-Bromo-1,3,5-trimethylbenzene

Bromotrinitromethane

Bromotriphenylmethane

1-Bromoundecane

11-Bromoundecanoic acid

(1-Bromovinyl)benzene

(*cis*-2-Bromovinyl)benzene

(*trans*-2-Bromovinyl)benzene

1-Bromo-2-vinylbenzene

1-Bromo-3-vinylbenzene

1-Bromo-4-vinylbenzene

Brompheniramine

Brucine

Brucine hydrochloride

Brucine sulfate heptahydrate

Bucolome

No.	Name	Synonym	Mol. Form.	CAS RN	Mol. Wt.	Physical Form	mp/°C	bp/°C	den/ g cm⁻³	n_D	Solubility
1387	Bufotalin		$C_{26}H_{36}O_6$	471-95-4	444.560	cry (+1 al)	223 dec				i H_2O; s EtOH, chl
1388	Bulbocapnine		$C_{19}H_{19}NO_4$	298-45-3	325.359	pr (al)	199.5				i H_2O; s EtOH; vs chl
1389	sec-Bumeton	N-sec-Butyl-N'-ethyl-6-methoxy-1,3,5-triazine-2,4-diamine	$C_{10}H_{19}N_5O$	26259-45-0	225.291		87				
1390	BUSAN 72A	(2-Benzothiazolylthio)methyl thiocyanate	$C_9H_6N_2S_3$	21564-17-0	238.352	liq		dec			
1391	Butachlor		$C_{17}H_{26}ClNO_2$	23184-66-9	311.847		<-5	156⁵	1.070²⁵		
1392	1,2-Butadiene	Methylallene	C_4H_6	590-19-2	54.091	vol liq or gas	-136.2	10.9	0.676⁰	1.4205¹	i H_2O; msc EtOH, eth; vs bz
1393	1,3-Butadiene	Divinyl	C_4H_6	106-99-0	54.091	col gas	-108.91	-4.41	0.6149²⁵ (p>1 atm)	1.4292⁻²⁵	i H_2O; s EtOH, eth, bz; vs ace
1394	1,3-Butadien-1-ol acetate		$C_6H_8O_2$	1515-76-0	112.127			58⁴⁰	0.945²⁵	1.4690²⁰	
1395	(trans)-1,3-Butadienylbenzene		$C_{10}H_{10}$	16939-57-4	130.186		2.3	76¹¹	0.9286²⁰	1.6089²⁵	i H_2O; s EtOH, eth, ace, bz
1396	1,3-Butadiyne	Diacetylene	C_4H_2	460-12-8	50.059	vol liq or gas	-36.4	10.3	0.7364⁰	1.4189⁵	vs H_2O, eth, ace; s chl, EtOH
1397	Butalbital	5-Isobutyl-5-allyl-2,4,6(1H,3H,5H)-pyrimidinetrione	$C_{11}H_{16}N_2O_3$	77-26-9	224.256	pr	138.5				sl H_2O; s EtOH, eth, ace, chl; i lig
1398	Butanal	Butyraldehyde	C_4H_8O	123-72-8	72.106	liq	-96.86	74.8	0.8016²⁰	1.3843²⁰	s H_2O; msc EtOH; vs ace, bz; sl chl
1399	Butanal oxime		C_4H_9NO	110-69-0	87.120	liq	-29.5	154	0.923²⁰		vs H_2O, ace, bz; msc EtOH, eth; s chl
1400	Butanamide	Butyramide	C_4H_9NO	541-35-5	87.120	lf (bz)	114.8	216	0.8850¹²⁰	1.4087¹³⁰	sl H_2O, eth; i bz; s EtOH
1401	Butane		C_4H_{10}	106-97-8	58.122	col gas	-138.3	-0.5	0.573²⁵ (p>1 atm)	1.3326²⁰	i H_2O; vs EtOH, eth, chl
1402	Butanedial		$C_4H_6O_2$	638-37-9	86.090			dec 170; 58⁹	1.065²⁰	1.4262¹⁸	vs H_2O, ace, eth, EtOH
1403	1,4-Butanediamine	Putrescine	$C_4H_{12}N_2$	110-60-1	88.151	lf	21.91	158.5	0.877²⁵	1.4969²⁰	s H_2O
1404	1,4-Butanediamine dihydrochloride		$C_4H_{14}Cl_2N_2$	333-93-7	161.073	nd or lf (al, w)	280 dec	sub			vs H_2O, EtOH; i eth, bz, MeOH
1405	1,2-Butanediol, (±)		$C_4H_{10}O_2$	26171-83-5	90.121			190.5	1.0024²⁰	1.4378²⁰	s H_2O, EtOH, ace
1406	1,3-Butanediol	1,3-Butylene glycol	$C_4H_{10}O_2$	107-88-0	90.121		<-50	207.5	1.0053²⁰	1.4401²⁰	
1407	1,4-Butanediol	Tetramethylene glycol	$C_4H_{10}O_2$	110-63-4	90.121		20.4	235	1.0171²⁰	1.4460²⁰	msc H_2O; s EtOH, DMSO; sl eth
1408	2,3-Butanediol		$C_4H_{10}O_2$	6982-25-8	90.121	cry (eth)	7.6	182.5	1.0033²⁰	1.4310²⁵	msc H_2O, EtOH; s eth, ace, chl
1409	1,4-Butanediol diacetate		$C_8H_{14}O_4$	628-67-1	174.195		12	229	1.0479¹⁵	1.4251¹⁵	
1410	1,4-Butanediol diacrylate		$C_{10}H_{14}O_4$	1070-70-8	198.216			83⁰·³	1.105²⁵		
1411	1,4-Butanediol diglycidyl ether	1,4-Bis(2,3-epoxypropoxy)butane	$C_{10}H_{18}O_4$	2425-79-8	202.248			266; 155¹¹	1.1²⁵	1.4611²⁰	
1412	1,3-Butanediol dimethacrylate		$C_{12}H_{18}O_4$	1189-08-8	226.269			290		1.4495²⁵	vs ace, eth, EtOH, lig
1413	1,4-Butanediol dimethacrylate		$C_{12}H_{18}O_4$	2082-81-7	226.269	liq		133⁴, 76⁰·⁰²⁷	1.025²⁰	1.4560²⁰	sl H_2O
1414	1,4-Butanediol dimethylsulfonate	Busulfan	$C_6H_{14}O_6S_2$	55-98-1	246.301	cry	116				i H_2O; sl EtOH, ace
1415	2,3-Butanedione	Diacetyl	$C_4H_6O_2$	431-03-8	86.090	liq	-1.2	88	0.9808¹⁸	1.3951²⁰	vs H_2O; msc EtOH, eth; s bz, ctc
1416	2,3-Butanedione monooxime		$C_4H_7NO_2$	57-71-6	101.105	pr (chl), lf (w)	76.8	185.5			sl H_2O; vs EtOH, eth, chl; s alk
1417	Butanedioyl dichloride	Succinyl chloride	$C_4H_4Cl_2O_2$	543-20-4	154.980	pl or lf	20	193.3	1.3748²⁰	1.4683²⁰	s eth, ace, bz
1418	1,4-Butanedithiol	Tetramethylenedithiol	$C_4H_{10}S_2$	1191-08-8	122.252	liq	-53.9	195.5	1.0021⁰	1.5290²⁰	i H_2O; vs EtOH; sl ctc
1419	Butanenitrile	Propyl cyanide	C_4H_7N	109-74-0	69.106	liq	-111.9	117.6	0.7936²⁰	1.3842²⁰	sl H_2O, ctc; msc EtOH, eth; s bz
1420	1-Butanesulfonyl chloride		$C_4H_9ClO_2S$	2386-60-9	156.631			75¹⁰		1.4559²⁰	
1421	1,4-Butane sultone	1,2-Oxathiane 2,2-dioxide	$C_4H_8O_3S$	1633-83-6	136.170	liq	13.5	135⁴	1.331²⁰	1.4640²⁰	
1422	1,2,3,4-Butanetetracarboxylic acid		$C_8H_{10}O_8$	1703-58-8	234.160	lf (w) cry (ace)	236.5				vs H_2O, EtOH
1423	1,2,3,4-Butanetetrol	Erythritol	$C_4H_{10}O_4$	149-32-6	122.120	bipym tetr pr	121.5	330.5	1.451²⁰		s H_2O; i eth, bz
1424	1,2,3,4-Butanetetrol tetranitrate, (R*,S*)	Erythrityl tetranitrate	$C_4H_6N_4O_{12}$	7297-25-8	302.111		61				vs EtOH
1425	1-Butanethiol	Butyl mercaptan	$C_4H_{10}S$	109-79-5	90.187	liq	-115.7	98.5	0.8416²⁰	1.4440²⁰	sl H_2O, chl; vs EtOH, eth

Bufotalin

Bulbocapnine

sec-Bumeton

BUSAN 72A

Butachlor

1,2-Butadiene

1,3-Butadiene

1,3-Butadien-1-ol acetate

(trans)-1,3-Butadienylbenzene

1,3-Butadiyne

Butalbital

Butanal

Butanal oxime

Butanamide

Butane

Butanedial

1,4-Butanediamine

1,4-Butanediamine dihydrochloride

1,2-Butanediol, (±)

1,3-Butanediol

1,4-Butanediol

2,3-Butanediol

1,4-Butanediol diacetate

1,4-Butanediol diacrylate

1,4-Butanediol diglycidyl ether

1,3-Butanediol dimethacrylate

1,4-Butanediol dimethacrylate

1,4-Butanediol dimethylsulfonate

2,3-Butanedione

2,3-Butanedione monooxime

Butanedioyl dichloride

1,4-Butanedithiol

Butanenitrile

1-Butanesulfonyl chloride

1,4-Butane sultone

1,2,3,4-Butanetetracarboxylic acid

1,2,3,4-Butanetetrol

1,2,3,4-Butanetetrol tetranitrate, (R*,S*)

1-Butanethiol

No.	Name	Synonym	Mol. Form.	CAS RN	Mol. Wt.	Physical Form	mp/°C	bp/°C	den/ g cm⁻³	n_D	Solubility
1426	2-Butanethiol	*sec*-Butyl mercaptan	$C_4H_{10}S$	91840-99-2	90.187	liq	-165	85.0	0.8295²⁰	1.4366²⁰	s EtOH, eth, bz, peth; sl ctc
1427	1,2,4-Butanetriol		$C_4H_{10}O_3$	3068-00-6	106.120			190¹⁸, 172¹²	1.18²⁰	1.4688²⁰	vs H₂O, EtOH
1428	Butanilicaine	2-(Butylamino)-*N*-(2-chloro-6-methylphenyl)acetamide	$C_{13}H_{19}ClN_2O$	3785-21-5	254.755	cry	46	145⁰·⁰⁰¹			
1429	Butanoic acid	Butyric acid	$C_4H_8O_2$	107-92-6	88.106	liq	-5.1	163.75	0.9528²⁵	1.3980²⁰	msc H₂O, EtOH, eth; sl ctc
1430	Butanoic anhydride	Butyric anhydride	$C_8H_{14}O_3$	106-31-0	158.195	liq	-75	200	0.9668²⁰	1.4070²⁰	s H₂O, bz; sl ctc
1431	1-Butanol	Butyl alcohol	$C_4H_{10}O$	71-36-3	74.121	liq	-88.6	117.73	0.8095²⁰	1.3988²⁰	s H₂O, bz; msc EtOH, eth; vs ace
1432	2-Butanol	*sec*-Butyl alcohol	$C_4H_{10}O$	78-92-2	74.121	liq	-88.5	99.51	0.8063²⁰	1.3978²⁰	vs H₂O; msc EtOH, eth; s bz, ctc
1433	2-Butanone	Methyl ethyl ketone	C_4H_8O	78-93-3	72.106	liq	-86.64	79.59	0.7999²⁵	1.3788²⁰	vs H₂O; msc EtOH, eth, ace, bz; s chl
1434	2-Butanone (1-methylpropylidene) hydrazone		$C_8H_{16}N_2$	5921-54-0	140.226			171.5	0.8404²⁰	1.4511²⁰	
1435	2-Butanone oxime		C_4H_9NO	96-29-7	87.120	liq	-29.5	152.5	0.9232²⁰	1.4410²⁰	s H₂O, chl; msc EtOH, eth
1436	2-Butanone peroxide	Methyl ethyl ketone peroxide	$C_8H_{16}O_4$	1338-23-4	176.211	col liq		exp 110			sl H₂O; misc os
1437	Butanoyl chloride	*n*-Butyryl chloride	C_4H_7ClO	141-75-3	106.551	liq	-89	102	1.0277²⁰	1.4121²⁰	msc eth
1438	Butaperazine		$C_{24}H_{31}N_3OS$	653-03-2	409.587			275⁰·⁰⁵			
1439	Butazolamide	*N*-[5-(Aminosulfonyl)-1,3,4-thiadiazol-2-yl]butanamide	$C_6H_{10}N_4O_3S_2$	16790-49-1	250.298	cry	261 dec				
1440	*trans*-2-Butenal	*trans*-Crotonaldehyde	C_4H_6O	123-73-9	70.090	liq	-76	102.2	0.8516²⁰	1.4366²⁰	s H₂O, chl; vs EtOH, eth, ace; msc bz
1441	1-Butene	1-Butylene	C_4H_8	106-98-9	56.107	col gas	-185.34	-6.26	0.588²⁵ (p>1 atm)	1.3962²⁰	i H₂O; vs EtOH, eth; s bz
1442	*cis*-2-Butene		C_4H_8	590-18-1	56.107	col gas	-138.88	3.71	0.616²⁵ (p>1 atm)	1.3931⁻²⁵	i H₂O; vs EtOH, eth; s bz
1443	*trans*-2-Butene		C_4H_8	624-64-6	56.107	col gas	-105.52	0.88	0.599²⁵ (p>1 atm)	1.3848⁻²⁵	s bz
1444	*trans*-2-Butenedinitrile		$C_4H_2N_2$	764-42-1	78.072	nd (bz-peth)	96.8	186	0.9416¹¹¹	1.4349¹¹¹	s H₂O, EtOH, eth, ace, bz, chl; sl peth
1445	*cis*-2-Butene-1,4-diol		$C_4H_8O_2$	6117-80-2	88.106		2.0	235	1.0698²⁰	1.4782²⁰	s H₂O; vs EtOH
1446	*trans*-2-Butene-1,4-diol		$C_4H_8O_2$	821-11-4	88.106		25	131¹³	1.0700²⁰	1.4755²⁰	vs H₂O, EtOH
1447	*trans*-2-Butenedioyl dichloride	Fumaric acid dichloride	$C_4H_2Cl_2O_2$	627-63-4	152.964	pa ye liq		159	1.408²⁰	1.5004¹⁸	
1448	*cis*-2-Butenenitrile	Isocrotononitrile	C_4H_5N	1190-76-7	67.090	liq		107.4			
1449	*trans*-2-Butenenitrile	Crotononitrile	C_4H_5N	627-26-9	67.090	liq	-51.5	120	0.8239²⁰	1.4225²⁰	s eth, ace
1450	3-Butenenitrile	Allyl cyanide	C_4H_5N	109-75-1	67.090	liq	-87	119	0.8341²⁰	1.4060²⁰	sl H₂O; msc EtOH, eth
1451	*cis*-2-Butenoic acid	Isocrotonic acid	$C_4H_6O_2$	503-64-0	86.090	nd or pr (peth)	15	169	1.0267²⁰	1.4450²⁰	vs H₂O; s EtOH
1452	*trans*-2-Butenoic acid	Crotonic acid	$C_4H_6O_2$	107-93-7	86.090	mcl pr or nd (w, lig)	71.5	184.7	0.9604⁷⁷	1.4249⁷⁷	vs H₂O, EtOH; s eth, ace, lig
1453	3-Butenoic acid		$C_4H_6O_2$	625-38-7	86.090	liq	-35	169	1.0091²⁰	1.4239²⁰	s H₂O; msc EtOH, eth
1454	2-Butenoic anhydride	Crotonic acid anhydride	$C_8H_{10}O_3$	623-68-7	154.163			247; 129¹⁹	1.0397²⁰	1.4745²⁰	vs eth
1455	*cis*-2-Buten-1-ol	*cis*-Crotyl alcohol	C_4H_8O	4088-60-2	72.106			123	0.8662²⁰	1.4342²⁵	s H₂O
1456	*trans*-2-Buten-1-ol	*trans*-Crotyl alcohol	C_4H_8O	504-61-0	72.106		<-30	121.2	0.8521²⁰	1.4288²⁰	vs H₂O; msc EtOH, eth; s chl
1457	3-Buten-1-ol		C_4H_8O	627-27-0	72.106			113.5	0.8424²⁰	1.4224²⁰	s H₂O, ace; msc EtOH, eth; sl chl
1458	3-Buten-2-ol		C_4H_8O	598-32-3	72.106			97			
1459	3-Buten-2-one	Methyl vinyl ketone	C_4H_6O	78-94-4	70.090			81.4	0.864²⁰	1.4081²⁰	s H₂O, EtOH, bz; vs eth, ace; sl ctc
1460	2-Butenoyl chloride		C_4H_5ClO	10487-71-5	104.535			124.5	1.0905²⁰	1.460¹⁸	vs ace
1461	(*trans*-1-Butenyl)benzene		$C_{10}H_{12}$	1005-64-7	132.202	liq	-43.1	198.7	0.9019²⁰	1.5420²⁰	i H₂O; s EtOH, eth, bz, ctc
1462	2-Butenylbenzene		$C_{10}H_{12}$	1560-06-1	132.202			176	0.8831²⁰	1.5101²⁰	
1463	3-Butenylbenzene		$C_{10}H_{12}$	768-56-9	132.202	liq	-70	177	0.8831²⁰	1.5059²⁰	i H₂O; s eth, bz
1464	1-Buten-3-yne	Vinylacetylene	C_4H_4	689-97-4	52.075	col gas		5.1	0.7094⁰	1.4161¹	i H₂O; s bz
1465	Butethamine hydrochloride	2-Isobutylaminoethyl 4-aminobenzoate	$C_{13}H_{21}ClN_2O_2$	553-68-4	272.771	cry	194				s H₂O; sl EtOH, bz, chl; i eth

2-Butanethiol

1,2,4-Butanetriol

Butanilicaine

Butanoic acid

Butanoic anhydride

1-Butanol

2-Butanol

2-Butanone

2-Butanone (1-methylpropylidene)hydrazone

2-Butanone oxime

2-Butanone peroxide

Butanoyl chloride

Butaperazine

Butazolamide

trans-2-Butenal

1-Butene

cis-2-Butene

trans-2-Butene

trans-2-Butenedinitrile

cis-2-Butene-1,4-diol

trans-2-Butene-1,4-diol

trans-2-Butenedioyl dichloride

cis-2-Butenenitrile

trans-2-Butenenitrile

3-Butenenitrile

cis-2-Butenoic acid

trans-2-Butenoic acid

3-Butenoic acid

2-Butenoic anhydride

cis-2-Buten-1-ol

trans-2-Buten-1-ol

3-Buten-1-ol

3-Buten-2-ol

3-Buten-2-one

2-Butenoyl chloride

(trans-1-Butenyl)benzene

2-Butenylbenzene

3-Butenylbenzene

1-Buten-3-yne

Butethamine hydrochloride

No.	Name	Synonym	Mol. Form.	CAS RN	Mol. Wt.	Physical Form	mp/°C	bp/°C	den/ g cm⁻³	n_D	Solubility
1466	Buthalital sodium		$C_{11}H_{15}N_2NaO_2S$	510-90-7	262.304						vs H_2O; sl EtOH; i eth, bz
1467	Buthiazide		$C_{11}H_{16}ClN_3O_4S_2$	2043-38-1	353.846		221.5				
1468	Buthiobate	Denmert	$C_{21}H_{28}N_2S_2$	51308-54-4	372.590	ye oil	32		1.0865[25]	1.596[26]	i H_2O; s os
1469	Butonate		$C_6H_{14}Cl_3O_5P$	126-22-7	327.527			129[0.5]			
1470	Butoxyacetylene		$C_6H_{10}O$	3329-56-4	98.142			104	0.8200[20]	1.4067	vs eth, EtOH
1471	4-Butoxyaniline		$C_{10}H_{15}NO$	4344-55-2	165.232			132[4]			
1472	4-Butoxybenzaldehyde		$C_{11}H_{14}O_2$	5736-88-9	178.228			148[10]			
1473	2-Butoxyethanol	Ethylene glycol monobutyl ether	$C_6H_{14}O_2$	111-76-2	118.174	liq	-74.8	168.4	0.9015[20]	1.4198[20]	msc H_2O, EtOH, eth; sl ctc
1474	2-[2-(2-Butoxyethoxy)ethoxy] ethanol		$C_{10}H_{22}O_4$	143-22-6	206.280			278	0.9890[20]	1.4389[20]	vs EtOH, MeOH
1475	2-(2-Butoxyethoxy)ethyl thiocyanate	Lethane 384	$C_9H_{17}NO_2S$	112-56-1	203.302	liq		122[0.25]			i H_2O; vs os
1476	1-(2-Butoxyethoxy)-2-propanol		$C_9H_{20}O_3$	124-16-3	176.253	col liq	-90	230	0.931[20]		s H_2O
1477	2-Butoxyethyl acetate	Ethylene glycol monobutyl ether acetate	$C_8H_{16}O_3$	112-07-2	160.211	liq		192			
1478	2-Butoxyethyl (2,4-dichlorophenoxy)acetate	2,4-D 2-Butoxyethyl ester	$C_{14}H_{18}Cl_2O_4$	1929-73-3	321.197			159[1]	1.232[20]		
1479	2-Butoxyethyl (2,4,5-trichlorophenoxy)acetate	2,4,5-T Butoxyethyl ester	$C_{14}H_{17}Cl_3O_4$	2545-59-7	355.642			164[1]	1.280[20]		s ctc
1480	4-Butoxy-N-hydroxybenzeneacetamide	Bufexamac	$C_{12}H_{17}NO_3$	2438-72-4	223.268	nd (ace)	154				
1481	1-Butoxy-4-methylbenzene		$C_{11}H_{16}O$	10519-06-9	164.244			229.5	0.9205[25]	1.4970[20]	s eth
1482	4-Butoxyphenol		$C_{10}H_{14}O_2$	122-94-1	166.217		65.5	125[4]			vs ace, bz, eth, EtOH
1483	4-[3-(4-Butoxyphenoxy)propyl] morpholine	Pramoxine	$C_{17}H_{27}NO_3$	140-65-8	293.401			196[6]			
1484	1-Butoxy-2-propanol		$C_7H_{16}O_2$	5131-66-8	132.201			171.5; 71[20]	0.882[20]	1.4168[20]	s EtOH, eth, bz, ctc, MeOH
1485	Butralin	4-tert-Butyl-N-sec-butyl-2,6-dinitroaniline	$C_{14}H_{21}N_3O_4$	33629-47-9	295.335		60	135[0.5]			
1486	N-Butylacetamide		$C_6H_{13}NO$	1119-49-9	115.173			229	0.8960[25]	1.4388[25]	
1487	Butyl acetate		$C_6H_{12}O_2$	123-86-4	116.158	liq	-78	126.1	0.8825[20]	1.3941[20]	sl H_2O; msc EtOH, eth; s ace, chl
1488	sec-Butyl acetate		$C_6H_{12}O_2$	105-46-4	116.158	liq	-98.9	112	0.8748[20]	1.3888[20]	sl H_2O, ctc; s EtOH, eth
1489	tert-Butyl acetate		$C_6H_{12}O_2$	540-88-5	116.158			95.1	0.8665[20]	1.3855[20]	s EtOH, eth, chl, HOAc
1490	tert-Butylacetic acid		$C_6H_{12}O_2$	1070-83-3	116.158		6.5	190	0.9124[20]	1.4096[20]	s EtOH, eth
1491	Butyl acetoacetate		$C_8H_{14}O_3$	591-60-6	158.195		-35.6	127[50], 85[8]	0.9671[25]	1.4137[20]	sl H_2O; msc EtOH, bz, lig
1492	Butyl acrylate		$C_7H_{12}O_2$	141-32-2	128.169	liq	-64.6	145	0.8898[20]	1.4185[20]	i H_2O; s EtOH, eth, ace; sl ctc
1493	tert-Butyl acrylate		$C_7H_{12}O_2$	1663-39-4	128.169	liq		120; 62[60]	0.879[25]	1.4110[20]	
1494	Butylamine	1-Butanamine	$C_4H_{11}N$	109-73-9	73.137	liq	-49.1	77.00	0.7414[20]	1.4031[20]	msc H_2O; s EtOH, eth
1495	sec-Butylamine	2-Butanamine, (±)-	$C_4H_{11}N$	33966-50-6	73.137		<-72	62.73	0.7246[20]	1.3932[20]	s H_2O, chl; msc EtOH, eth; vs ace
1496	tert-Butylamine	2-Methyl-2-propanamine	$C_4H_{11}N$	75-64-9	73.137	liq	-66.94	44.04	0.6958[20]	1.3784[20]	msc H_2O, EtOH, eth; s chl
1497	Butylamine hydrochloride	1-Butanamine hydrochloride	$C_4H_{12}ClN$	3858-78-4	109.598		213		0.982[20]		sl H_2O, EtOH
1498	Butyl 4-aminobenzoate	Butamben	$C_{11}H_{15}NO_2$	94-25-7	193.243	cry (al or bz)	58	173[8]			i H_2O; s EtOH, eth, bz, chl
1499	2-(Butylamino)ethanol		$C_6H_{15}NO$	111-75-1	117.189			199; 91[11]	0.8907[20]	1.4437[20]	vs H_2O, EtOH, eth
1500	2-(tert-Butylamino)ethanol		$C_6H_{15}NO$	4620-70-6	117.189		44	176.5; 72[14]	0.8818[20]		
1501	N-tert-Butylaminoethyl methacrylate		$C_{10}H_{19}NO_2$	3775-90-4	185.264			102[12]			s chl
1502	2-(tert-Butylaminothio) benzothiazole	N-tert-Butyl-2-benzothiazolesulfenamide	$C_{11}H_{14}N_2S_2$	95-31-8	238.372		108				
1503	2-sec-Butylaniline		$C_{10}H_{15}N$	55751-54-7	149.233			120[16]	0.9574[20]		s EtOH, ace, bz; sl ctc
1504	4-Butylaniline		$C_{10}H_{15}N$	104-13-2	149.233	pa ye		261	0.945[20]		sl ctc
1505	4-sec-Butylaniline		$C_{10}H_{15}N$	30273-11-1	149.233			238; 118[15]	0.949[15]	1.5360[29]	vs bz, eth
1506	4-tert-Butylaniline		$C_{10}H_{15}N$	769-92-6	149.233	ye rd (peth)	17	241	0.9525[15]	1.5380[20]	sl H_2O; msc EtOH, eth; vs bz; s ctc
1507	N-Butylaniline		$C_{10}H_{15}N$	1126-78-9	149.233	liq	-14.4	243.5	0.9323[20]	1.5341[20]	vs eth, EtOH
1508	N-tert-Butylaniline		$C_{10}H_{15}N$	937-33-7	149.233			215; 95[19]		1.5270[20]	s EtOH; vs ace, bz, chl
1509	2-tert-Butyl-9,10-anthracenedione		$C_{18}H_{16}O_2$	84-47-9	264.319		99				s ctc, CS_2

Buthalital sodium

Buthiazide

Buthiobate

Butonate

Butoxyacetylene

4-Butoxyaniline

4-Butoxybenzaldehyde

2-Butoxyethanol

2-[2-(2-Butoxyethoxy)ethoxy]ethanol

2-(2-Butoxyethoxy)ethyl thiocyanate

1-(2-Butoxyethoxy)-2-propanol

2-Butoxyethyl acetate

2-Butoxyethyl (2,4-dichlorophenoxy)acetate

2-Butoxyethyl (2,4,5-trichlorophenoxy)acetate

4-Butoxy-N-hydroxybenzeneacetamide

1-Butoxy-4-methylbenzene

4-Butoxyphenol

4-[3-(4-Butoxyphenoxy)propyl]morpholine

1-Butoxy-2-propanol

Butralin

N-Butylacetamide

Butyl acetate

sec-Butyl acetate

tert-Butyl acetate

tert-Butylacetic acid

Butyl acetoacetate

Butyl acrylate

tert-Butyl acrylate

Butylamine

sec-Butylamine

tert-Butylamine

Butylamine hydrochloride

Butyl 4-aminobenzoate

2-(Butylamino)ethanol

2-(tert-Butylamino)ethanol

N-tert-Butylaminoethyl methacrylate

2-(tert-Butylaminothio)benzothiazole

2-sec-Butylaniline

4-Butylaniline

4-sec-Butylaniline

4-tert-Butylaniline

N-Butylaniline

N-tert-Butylaniline

2-tert-Butyl-9,10-anthracenedione

No.	Name	Synonym	Mol. Form.	CAS RN	Mol. Wt.	Physical Form	mp/°C	bp/°C	den/ g cm⁻³	n_D	Solubility
1510	tert-Butyl azidoformate	tert-Butyl carbonazidate	C₅H₉N₃O₂	1070-19-5	143.144	unstab >80		73⁷⁰			
1511	4-Butylbenzaldehyde		C₁₁H₁₄O	1200-14-2	162.228			123⁷		1.5265	
1512	4-tert-Butylbenzaldehyde		C₁₁H₁₄O	939-97-9	162.228	liq		107¹¹, 130²⁵	0.970	1.5270²⁰	
1513	Butylbenzene		C₁₀H₁₄	104-51-8	134.218	liq	-87.85	183.31	0.8601²⁰	1.4898²⁰	i H₂O; msc EtOH, eth, ace, bz, peth, ctc
1514	sec-Butylbenzene, (±)	2-Phenylbutane	C₁₀H₁₄	36383-15-0	134.218	liq	-82.7	173.3	0.8621²⁰	1.4902²⁰	i H₂O; msc EtOH, eth, ace, bz, peth, ctc
1515	tert-Butylbenzene		C₁₀H₁₄	98-06-6	134.218	liq	-57.8	169.1	0.8665²⁰	1.4927²⁰	i H₂O; vs EtOH, eth; msc ace, bz
1516	4-tert-Butyl-1,2-benzenediol		C₁₀H₁₄O₂	98-29-3	166.217		54.3	285; 160²²			s tfa
1517	2-tert-Butyl-1,4-benzenediol		C₁₀H₁₄O₂	1948-33-0	166.217		128				
1518	N-tert-Butylbenzenemethanamine		C₁₁H₁₇N	3378-72-1	163.260			75⁵		1.4951²⁵	
1519	4-tert-Butylbenzenemethanol		C₁₁H₁₆O	877-65-6	164.244			236; 140²⁰	0.928²⁵	1.5179²⁰	
1520	Butyl benzoate		C₁₁H₁₄O₂	136-60-7	178.228	liq	-22.4	250.3	1.000²⁰	1.4940²⁵	i H₂O; msc EtOH, eth; s ace; sl ctc
1521	2-tert-Butylbenzoic acid		C₁₁H₁₄O₂	1077-58-3	178.228	pl (dil al)	80.5				vs EtOH
1522	3-tert-Butylbenzoic acid		C₁₁H₁₄O₂	7498-54-6	178.228	nd (peth)	128.8				vs EtOH, peth
1523	4-tert-Butylbenzoic acid		C₁₁H₁₄O₂	98-73-7	178.228	nd (dil al)	164.5				i H₂O; vs EtOH, bz; s chl
1524	4-Butylbenzoyl chloride		C₁₁H₁₃ClO	28788-62-7	196.673			155²⁶	1.051²⁵	1.5351²⁰	
1525	4-tert-Butylbenzoyl chloride		C₁₁H₁₃ClO	1710-98-1	196.673			266; 135²⁰	1.007²⁵	1.5364²⁰	
1526	2-Butyl-1,1'-biphenyl		C₁₆H₁₈	54532-97-7	210.314	liq	-9.65	291.2	0.9676²⁰	1.5604²⁰	
1527	tert-Butyl bromoacetate		C₆H₁₁BrO₂	5292-43-3	195.054			73²⁵		1.4430²⁰	vs eth, EtOH
1528	Butyl butanoate		C₈H₁₆O₂	109-21-7	144.212	liq	-91.5	166	0.8700²⁰	1.4075²⁰	i H₂O; msc EtOH, eth; s ctc
1529	Butyl cis-2-butenedioate	Monobutyl maleate	C₈H₁₂O₄	925-21-3	172.179	oil			1.09²⁵		
1530	Butyl carbamate		C₅H₁₁NO₂	592-35-8	117.147	pr	53	dec 204; 108¹⁴			vs EtOH; sl chl
1531	Butyl chloroacetate		C₆H₁₁ClO₂	590-02-3	150.603			183	1.0704²⁰	1.4297²⁰	vs eth, EtOH
1532	tert-Butyl chloroacetate		C₆H₁₁ClO₂	107-59-5	150.603			150; 50¹⁰		1.4260²⁰	dec H₂O
1533	Butylchlorodimethylsilane		C₆H₁₅ClSi	1000-50-6	150.722			139	0.876²⁰	1.5145²⁰	
1534	Butyl chloroformate		C₅H₉ClO₂	592-34-7	136.577			142	1.074²⁵	1.4114²⁰	msc eth; s ace; sl ctc
1535	N-Butyl-4-chloro-2-hydroxybenzamide	Buclosamide	C₁₁H₁₄ClNO₂	575-74-6	227.688		91.5				
1536	Butyl 2-chloropropanoate		C₇H₁₃ClO₂	54819-86-2	164.630			184	1.0253²⁰	1.4263²⁰	vs eth
1537	Butyl 3-chloropropanoate		C₇H₁₃ClO₂	27387-79-7	164.630			104²², 92⁶	1.0370²⁰	1.4321²⁰	vs H₂O, eth
1538	tert-Butyl chromate		C₈H₁₈CrO₄	1189-85-1	230.223	red cry (peth)	-5				reac H₂O
1539	Butyl citrate		C₁₈H₃₂O₇	77-94-1	360.443		-20	233²²	1.043²⁰	1.4460²⁰	
1540	Butyl cyanoacetate		C₇H₁₁NO₂	5459-58-5	141.168			231; 115¹⁵	1.0010²⁰	1.4200²⁰	
1541	Butylcyclohexane		C₁₀H₂₀	1678-93-9	140.266	liq	-74.73	180.9	0.7902²⁰	1.4408²⁰	i H₂O
1542	sec-Butylcyclohexane		C₁₀H₂₀	7058-01-7	140.266			179.3	0.8131²⁰	1.4467²⁰	i H₂O; s ace
1543	tert-Butylcyclohexane		C₁₀H₂₀	3178-22-1	140.266	liq	-41.2	171.5	0.8127²⁰	1.4469²⁰	i H₂O
1544	2-tert-Butylcyclohexanol		C₁₀H₂₀O	13491-79-7	156.265		45	139⁹⁵	0.902²⁵		
1545	cis-4-tert-Butylcyclohexanol		C₁₀H₂₀O	937-05-3	156.265		83	112¹⁵			
1546	trans-4-tert-Butylcyclohexanol		C₁₀H₂₀O	21862-63-5	156.265		83	112¹⁵			
1547	4-tert-Butylcyclohexanone		C₁₀H₁₈O	98-53-3	154.249		48	90⁹			
1548	Butylcyclohexylamine	N-Butylcyclohexanamine	C₁₀H₂₁N	10108-56-2	155.281		208.3				sl H₂O, ctc; vs EtOH, eth
1549	Butyl cyclohexyl phthalate		C₁₈H₂₄O₄	84-64-0	304.382	col liq		≈205⁵	1.076²⁵		sl H₂O; misc os
1550	Butylcyclopentane		C₉H₁₈	2040-95-1	126.239	liq	-108	156.6	0.7846²⁰	1.4316²⁰	vs ace, bz, eth, EtOH
1551	Butyl dichloroacetate		C₆H₁₀Cl₂O₂	29003-73-4	185.048			193.5	1.1820²⁰	1.4420²⁰	vs eth, EtOH
1552	Butyl (2,4-dichlorophenoxy) acetate	2,4-D Butyl ester	C₁₂H₁₄Cl₂O₃	94-80-4	277.143		9	133¹			
1553	5-Butyldihydro-2(3H)-furanone		C₈H₁₄O₂	104-50-7	142.196			132²⁰	0.9796¹⁹	1.4451¹⁹	s EtOH; sl ctc
1554	Butyldimethylamine	N,N-Dimethyl-1-butanamine	C₆H₁₅N	927-62-8	101.190			95	0.7206²⁰	1.3970²⁰	msc H₂O, EtOH, eth, ace, bz
1555	1-tert-Butyl-3,5-dimethylbenzene		C₁₂H₁₈	98-19-1	162.271	liq	-18	207	0.8668²⁰		s ctc
1556	4-tert-Butyl-2,6-dimethyl-3,5-dinitroacetophenone	Musk ketone	C₁₄H₁₈N₂O₅	81-14-1	294.303	ye cry	135.5				vs chl
1557	2-tert-Butyl-4,6-dimethylphenol		C₁₂H₁₈O	1879-09-0	178.270		22.3	249	0.917⁸⁰	1.5183²⁰	i alk
1558	4-tert-Butyl-2,5-dimethylphenol		C₁₂H₁₈O	17696-37-6	178.270		71.2	264	0.939⁸⁰	1.5311²⁰	s alk
1559	4-tert-Butyl-2,6-dimethylphenol		C₁₂H₁₈O	879-97-0	178.270		82.4	248	0.916⁸⁰		s alk
1560	1-tert-Butyl-3,5-dimethyl-2,4,6-trinitrobenzene		C₁₂H₁₅N₃O₆	81-15-2	297.263	pl, nd (al)	110				i H₂O; sl EtOH; s eth, chl
1561	2-tert-Butyl-4,6-dinitrophenol		C₁₀H₁₂N₂O₅	1420-07-1	240.212	ye solid	126				

tert-Butyl azidoformate

4-Butylbenzaldehyde

4-tert-Butylbenzaldehyde

Butylbenzene

sec-Butylbenzene, (±)

tert-Butylbenzene

4-tert-Butyl-1,2-benzenediol

2-tert-Butyl-1,4-benzenediol

N-tert-Butylbenzenemethanamine

4-tert-Butylbenzenemethanol

Butyl benzoate

2-tert-Butylbenzoic acid

3-tert-Butylbenzoic acid

4-tert-Butylbenzoic acid

4-Butylbenzoyl chloride

4-tert-Butylbenzoyl chloride

2-Butyl-1,1'-biphenyl

tert-Butyl bromoacetate

Butyl butanoate

Butyl cis-2-butenedioate

Butyl carbamate

Butyl chloroacetate

tert-Butyl chloroacetate

Butylchlorodimethylsilane

Butyl chloroformate

N-Butyl-4-chloro-2-hydroxybenzamide

Butyl 2-chloropropanoate

Butyl 3-chloropropanoate

tert-Butyl chromate

Butyl citrate

Butyl cyanoacetate

Butylcyclohexane

sec-Butylcyclohexane

tert-Butylcyclohexane

2-tert-Butylcyclohexanol

cis-4-tert-Butylcyclohexanol

trans-4-tert-Butylcyclohexanol

4-tert-Butylcyclohexanone

Butylcyclohexylamine

Butyl cyclohexyl phthalate

Butylcyclopentane

Butyl dichloroacetate

Butyl (2,4-dichlorophenoxy)acetate

5-Butyldihydro-2(3H)-furanone

Butyldimethylamine

1-tert-Butyl-3,5-dimethylbenzene

4-tert-Butyl-2,6-dimethyl-3,5-dinitroacetophenone

2-tert-Butyl-4,6-dimethylphenol

4-tert-Butyl-2,5-dimethylphenol

4-tert-Butyl-2,6-dimethylphenol

1-tert-Butyl-3,5-dimethyl-2,4,6-trinitrobenzene

2-tert-Butyl-4,6-dinitrophenol

No.	Name	Synonym	Mol. Form.	CAS RN	Mol. Wt.	Physical Form	mp/°C	bp/°C	den/ g cm⁻³	n_D	Solubility
1562	5-Butyldocosane		C$_{26}$H$_{54}$	55282-16-1	366.707		208	244^{10}	0.8058^{20}	1.4503^{20}	
1563	11-Butyldocosane		C$_{26}$H$_{54}$	13475-76-8	366.707			242.5^{10}	0.8041^{20}	1.4499^{20}	
1564	Butyl dodecanoate		C$_{16}$H$_{32}$O$_2$	106-18-3	256.424			180^{18}			
1565	Butylethylamine	N-Ethyl-1-butanamine	C$_6$H$_{15}$N	13360-63-9	101.190			107.5	0.7398^{20}	1.4040^{20}	msc EtOH, eth, ace, bz
1566	1-tert-Butyl-4-ethylbenzene		C$_{12}$H$_{18}$	7364-19-4	162.271	liq	-38.4	211	0.8641^{20}		
1567	Butyl ethyl ether		C$_6$H$_{14}$O	628-81-9	102.174	liq	-124	92.3	0.7495^{20}	1.3818^{20}	i H$_2$O; msc EtOH, eth; vs ace
1568	sec-Butyl ethyl ether		C$_6$H$_{14}$O	2679-87-0	102.174			81	0.7503^{20}	1.3802^{20}	i H$_2$O; vs EtOH, eth
1569	tert-Butyl ethyl ether	Ethyl tert-butyl ether	C$_6$H$_{14}$O	637-92-3	102.174	liq	-94	72.6	0.736^{25}	1.3756^{20}	i H$_2$O; vs EtOH, eth
1570	2-tert-Butyl-4-ethylphenol		C$_{12}$H$_{18}$O	96-70-8	178.270		23	250			
1571	2-Butyl-2-ethyl-1,3-propanediol		C$_9$H$_{20}$O$_2$	115-84-4	160.254	wh cry	43.8	262	0.927^{50}	1.4587^{25}	sl H$_2$O, ace; s EtOH
1572	5-Butyl-5-ethyl-2,4,6(1H,3H,5H)-pyrimidinetrione	Butethal	C$_{10}$H$_{16}$N$_2$O$_3$	77-28-1	212.245		128.5				
1573	Butyl ethyl sulfide		C$_6$H$_{14}$S	638-46-0	118.240	liq	-95.1	144.3	0.8376^{20}	1.4492^{10}	vs EtOH; s chl
1574	tert-Butyl ethyl sulfide	2-Methyl-2-propanethiol	C$_6$H$_{14}$S	14290-92-7	118.240	liq	-88.9	120.4; 56^{109}			
1575	N-tert-Butylformamide		C$_5$H$_{11}$NO	2425-74-3	101.147	liq	16	202	0.903	1.4330^{20}	
1576	Butyl formate		C$_5$H$_{10}$O$_2$	592-84-7	102.132	liq	-91.5	106.1	0.8958^{20}	1.3887^{20}	sl H$_2$O; s ace; msc EtOH, eth
1577	sec-Butyl formate		C$_5$H$_{10}$O$_2$	589-40-2	102.132			97	0.8846^{20}	1.3865^{20}	sl H$_2$O; s ace; msc EtOH, eth
1578	tert-Butyl formate	1,1-Dimethylethyl formate	C$_5$H$_{10}$O$_2$	762-75-4	102.132	liq		82	0.872	1.3790^{20}	
1579	Butyl glycidyl ether		C$_7$H$_{14}$O$_2$	2426-08-6	130.185			169; 75^{26}	0.918^{20}		
1580	Butyl heptanoate	Butyl enanthate	C$_{11}$H$_{22}$O$_2$	5454-28-4	186.292	liq	-67.5	226.2	0.8638^{20}	1.4204^{20}	vs ace, bz, eth, EtOH
1581	Butyl hexanoate	Butyl caproate	C$_{10}$H$_{20}$O$_2$	626-82-4	172.265	liq	-64.3	208	0.8653^{20}	1.4152^{20}	i H$_2$O; s EtOH; msc eth
1582	tert-Butylhydrazine hydrochloride		C$_4$H$_{13}$ClN$_2$	7400-27-3	124.612		192.5				
1583	Butyl hydrogen succinate	Monobutyl succinate	C$_8$H$_{14}$O$_4$	5150-93-6	174.195		8.6	136.5^3	1.0732^{20}	1.4360^{20}	
1584	tert-Butyl hydroperoxide		C$_4$H$_{10}$O$_2$	75-91-2	90.121		6	dec 89; 36^{17}	0.8960^{20}	1.4015^{20}	s H$_2$O, EtOH, eth, ctc, chl
1585	tert-Butyl-4-hydroxyanisole	Butylated hydroxyanisole	C$_{11}$H$_{16}$O$_2$	25013-16-5	180.244	wax	51	268			i H$_2$O; s peth, EtOH
1586	Butyl 2-hydroxybenzoate		C$_{11}$H$_{14}$O$_3$	2052-14-4	194.227	liq	-5.9	271	1.0728^{20}	1.5115^{20}	sl ctc
1587	Butyl 4-hydroxybenzoate	Butylparaben	C$_{11}$H$_{14}$O$_3$	94-26-8	194.227		68.5				sl H$_2$O, ctc; s EtOH
1588	Butyl cis-12-hydroxy-9-octadecenoate, (R)	Butyl ricinoleate	C$_{22}$H$_{42}$O$_3$	151-13-3	354.566			275^{13}	0.9058^{22}	1.4566^{22}	vs eth
1589	tert-Butyl hypochlorite		C$_4$H$_9$ClO	507-40-4	108.566	ye liq		77.5	0.9583^{18}	1.403^{20}	i H$_2$O; vs eth, bz; s ace
1590	Butyl isobutyl ether		C$_8$H$_{18}$O	17071-47-5	130.228	liq		151	0.763^{15}	1.4077^{21}	vs ace, eth, EtOH
1591	tert-Butyl isobutyl ether		C$_8$H$_{18}$O	33021-02-2	130.228	liq		112.0			
1592	Butyl isocyanate		C$_5$H$_9$NO	111-36-4	99.131			115	0.880^{20}	1.4060^{20}	
1593	Butyl isocyanide		C$_5$H$_9$N	2769-64-4	83.132			120	0.78^{20}		vs eth, EtOH
1594	tert-Butyl isopropyl ether		C$_7$H$_{16}$O	17348-59-3	116.201	liq	-88	87.6	0.7365^{25}		s chl
1595	Butyl isothiocyanate	1-Isothiocyanatobutane	C$_5$H$_9$NS	592-82-5	115.197			168	0.9546^{20}	1.501^{20}	vs eth, EtOH
1596	sec-Butyl isothiocyanate, (±)	2-Isothiocyanatobutane, (±)	C$_5$H$_9$NS	116724-11-9	115.197			159.5	0.944^{12}		vs eth, EtOH
1597	tert-Butyl isothiocyanate	2-Isothiocyanato-2-methylpropane	C$_5$H$_9$NS	590-42-1	115.197		10.5	140	0.9187^{10}		
1598	Butyl lactate		C$_7$H$_{14}$O$_3$	34451-18-8	146.184			77^{10}	0.9744^{27}		vs eth, EtOH
1599	Butyl methacrylate		C$_8$H$_{14}$O$_2$	97-88-1	142.196			160	0.8936^{20}	1.4240^{20}	vs eth, EtOH
1600	tert-Butyl methacrylate		C$_8$H$_{14}$O$_2$	585-07-9	142.196			135.2			
1601	1-tert-Butyl-4-methoxybenzene		C$_{11}$H$_{16}$O	5396-38-3	164.244		19.0	238	0.9383^{20}	1.5039^{20}	
1602	1-tert-Butyl-2-methoxy-4-methyl-3,5-dinitrobenzene		C$_{12}$H$_{16}$N$_2$O$_5$	83-66-9	268.265	pa ye lf (al)	85	185^{16}			i H$_2$O; sl EtOH; s eth, chl
1603	2-tert-Butyl-4-methoxyphenol		C$_{11}$H$_{16}$O$_2$	121-00-6	180.244			184^{50}			
1604	3-tert-Butyl-4-methoxyphenol		C$_{11}$H$_{16}$O$_2$	88-32-4	180.244		65				
1605	Butylmethylamine	N-Methyl-1-butanamine	C$_5$H$_{13}$N	110-68-9	87.164			91	0.7637^{15}		
1606	1-tert-Butyl-2-methylbenzene	2-tert-Butyltoluene	C$_{11}$H$_{16}$	1074-92-6	148.245	liq	-50.3	200.4	0.8897^{20}	1.5076^{20}	vs ace, bz, eth, EtOH
1607	1-tert-Butyl-3-methylbenzene	3-tert-Butyltoluene	C$_{11}$H$_{16}$	1075-38-3	148.245	liq	-41.4	189.3	0.8657^{20}	1.4944^{20}	vs ace, bz, eth, EtOH
1608	1-tert-Butyl-4-methylbenzene	4-tert-Butyltoluene	C$_{11}$H$_{16}$	98-51-1	148.245	liq	-52	190	0.8612^{20}	1.4918^{20}	i H$_2$O; sl EtOH; vs eth, chl; s ace, bz
1609	Butyl 2-methylbutanoate	Butyl o-toluate	C$_9$H$_{18}$O$_2$	15706-73-7	158.238			179	0.8620^{20}	1.4135^{20}	
1610	Butyl 3-methylbutanoate	Butyl p-toluate	C$_9$H$_{18}$O$_2$	109-19-3	158.238					1.4058^{25}	

5-Butyldocosane

11-Butyldocosane

Butyl dodecanoate

Butylethylamine

1-*tert*-Butyl-4-ethylbenzene

Butyl ethyl ether

sec-Butyl ethyl ether

tert-Butyl ethyl ether

2-*tert*-Butyl-4-ethylphenol

2-Butyl-2-ethyl-1,3-propanediol

5-Butyl-5-ethyl-2,4,6(1*H*,3*H*,5*H*)-pyrimidinetrione

Butyl ethyl sulfide

tert-Butyl ethyl sulfide

N-tert-Butylformamide

Butyl formate

sec-Butyl formate

tert-Butyl formate

Butyl glycidyl ether

Butyl heptanoate

Butyl hexanoate

tert-Butylhydrazine hydrochloride

Butyl hydrogen succinate

tert-Butyl hydroperoxide

tert-Butyl-4-hydroxyanisole

Butyl 2-hydroxybenzoate

Butyl 4-hydroxybenzoate

Butyl *cis*-12-hydroxy-9-octadecenoate, (*R*)

tert-Butyl hypochlorite

Butyl isobutyl ether

tert-Butyl isobutyl ether

Butyl isocyanate

Butyl isocyanide

tert-Butyl isopropyl ether

Butyl isothiocyanate

sec-Butyl isothiocyanate, (±)

tert-Butyl isothiocyanate

Butyl lactate

Butyl methacrylate

tert-Butyl methacrylate

1-*tert*-Butyl-4-methoxybenzene

1-*tert*-Butyl-2-methoxy-4-methyl-3,5-dinitrobenzene

2-*tert*-Butyl-4-methoxyphenol

3-*tert*-Butyl-4-methoxyphenol

Butylmethylamine

1-*tert*-Butyl-2-methylbenzene

1-*tert*-Butyl-3-methylbenzene

1-*tert*-Butyl-4-methylbenzene

Butyl 2-methylbutanoate

Butyl 3-methylbutanoate

No.	Name	Synonym	Mol. Form.	CAS RN	Mol. Wt.	Physical Form	mp/°C	bp/°C	den/ g cm⁻³	n_D	Solubility
1611	Butyl methyl ether		C₅H₁₂O	628-28-4	88.148	liq	-115.7	70.16	0.7392²⁵	1.3736²⁰	i H₂O; msc EtOH, eth; s ace
1612	sec-Butyl methyl ether		C₅H₁₂O	116783-23-4	88.148			59.1	0.7415²⁰	1.3680²⁵	vs ace, eth, EtOH
1613	2-tert-Butyl-4-methylphenol		C₁₁H₁₆O	2409-55-4	164.244		51.5	237	0.9247⁷⁵	1.4969⁷⁵	sl H₂O; s ace, bz, chl
1614	2-tert-Butyl-5-methylphenol		C₁₁H₁₆O	88-60-8	164.244		46.5	127¹¹	0.922⁸⁰	1.5250²⁰	i H₂O; s EtOH, eth, ace
1615	2-tert-Butyl-6-methylphenol		C₁₁H₁₆O	2219-82-1	164.244		31	230	0.9240⁸⁰	1.5195²⁰	
1616	4-tert-Butyl-2-methylphenol		C₁₁H₁₆O	98-27-1	164.244		27.5	237; 132¹¹	0.965²⁰	1.5230²⁰	i H₂O; s eth, ace, bz
1617	Butyl methyl sulfide		C₅H₁₂S	628-29-5	104.214	liq	-97.8	123.4	0.8426²⁰	1.4477²⁰	vs EtOH, MeOH
1618	tert-Butyl methyl sulfide		C₅H₁₂S	6163-64-0	104.214	liq		98.9			
1619	4-Butylmorpholine		C₈H₁₇NO	1005-67-0	143.227	liq	-57.1	213.5	0.9068²⁰	1.4451²⁰	vs H₂O, ace, bz, EtOH
1620	1-Butylnaphthalene		C₁₄H₁₆	1634-09-9	184.277	liq	-19.8	289.3	0.9738²⁰	1.5819²⁰	i H₂O; s EtOH, eth, ace, bz
1621	2-Butylnaphthalene		C₁₄H₁₆	1134-62-9	184.277	liq	-2.5	292	0.9673²⁰	1.5777²⁰	vs ace, bz, EtOH
1622	Butyl nitrate		C₄H₉NO₃	928-45-0	119.119			133	1.0228²⁰	1.4013²³	i H₂O; s EtOH, eth; sl ctc
1623	Butyl nitrite		C₄H₉NO₂	544-16-1	103.120			78	0.9114²⁵	1.3762²⁰	msc EtOH, eth
1624	tert-Butyl nitrite		C₄H₉NO₂	540-80-7	103.120	pa ye liq		63	0.8670²⁰	1.368²⁰	sl H₂O; s EtOH, eth, chl, CS₂
1625	sec-Butyl nitrite		C₄H₉NO₂	924-43-6	103.120			68.5	0.8726²⁰	1.3710²⁰	vs eth, EtOH, chl
1626	4-(Butylnitrosoamino)-1-butanol	N-Butyl-N-(4-hydroxybutyl) nitrosamine	C₈H₁₈N₂O₂	3817-11-6	174.241			115⁰·⁰¹			
1627	5-Butylnonane		C₁₃H₂₈	17312-63-9	184.361			217.5	0.7635¹⁸	1.4273¹⁸	
1628	Butyl nonanoate	Butyl pelargonate	C₁₃H₂₆O₂	50623-57-9	214.344		-38	123²⁰	0.8520²⁵	1.4262²⁵	
1629	Butyl octanoate		C₁₂H₂₄O₂	589-75-3	200.318	liq	-42.9	240.5	0.8628²⁰	1.4232²⁵	vs ace, eth, EtOH
1630	2-Butyl-1-octanol		C₁₂H₂₆O	3913-02-8	186.333			246.5; 132¹⁵	0.891²⁰		
1631	Butyl oleate	Butyl cis-9-octadecenoate	C₂₂H₄₂O₂	142-77-8	338.567	ye cry	-26.4	227¹⁵	0.8704¹⁵	1.4480²⁵	vs EtOH
1632	tert-Butyl 3-oxobutanoate		C₈H₁₄O₃	1694-31-1	158.195			71.5¹¹	0.9756²⁰	1.4180²⁰	
1633	Butyl 4-oxopentanoate	Butyl levulinate	C₉H₁₆O₃	2052-15-5	172.221			237.5	0.9735²⁰	1.4290²⁰	sl chl
1634	Butyl palmitate	Butyl hexadecanoate	C₂₀H₄₀O₂	111-06-8	312.531	cry (dil al)	16.9			1.4312⁵⁰	i H₂O; s EtOH, eth
1635	Butyl pentanoate		C₉H₁₈O₂	591-68-4	158.238	liq	-92.8	185.8	0.8710¹⁵	1.4128²⁰	sl H₂O; s EtOH, eth
1636	sec-Butyl pentanoate		C₉H₁₈O₂	116836-32-9	158.238			174.5	0.8605²⁰	1.4070²⁰	vs bz, eth, py, EtOH
1637	4-(1-Butylpentyl)pyridine		C₁₄H₂₃N	2961-47-9	205.340			265; 181⁵⁰	0.8878²⁵	1.4846²⁵	
1638	tert-Butyl peroxybenzoate	Benzoyl tert-butyl peroxide	C₁₁H₁₄O₃	614-45-9	194.227			75⁰·²	1.021²⁵	1.4990²⁰	
1639	2-Butylphenol		C₁₀H₁₄O	3180-09-4	150.217	liq	-20	235	0.975²⁰	1.5180²⁵	i H₂O; s EtOH, eth, alk
1640	2-sec-Butylphenol		C₁₀H₁₄O	89-72-5	150.217		16	228; 116²¹	0.9804²⁵	1.5200²⁵	
1641	2-tert-Butylphenol		C₁₀H₁₄O	88-18-6	150.217	liq	-6.8	223	0.9783²⁰	1.5160²⁰	s EtOH, ctc, alk; vs eth
1642	3-Butylphenol		C₁₀H₁₄O	4074-43-5	150.217			248	0.974²⁰		vs eth, EtOH
1643	3-tert-Butylphenol		C₁₀H₁₄O	585-34-2	150.217	nd (peth)	42.3	240			s EtOH, alk; vs eth
1644	4-Butylphenol		C₁₀H₁₄O	1638-22-8	150.217		22	248	0.976²²	1.5165²⁵	i H₂O; s EtOH, eth, alk; sl ctc
1645	4-sec-Butylphenol	4-(1-Methylpropyl)phenol	C₁₀H₁₄O	99-71-8	150.217		61.5	241	0.986²⁰	1.5182²¹	i H₂O; s EtOH, alk; vs eth
1646	4-tert-Butylphenol		C₁₀H₁₄O	98-54-4	150.217	nd (lig)	98	237	0.908⁸⁰	1.4787¹¹⁴	s H₂O, EtOH, eth, chl, alk
1647	4-tert-Butylphenol, phosphate (3:1)		C₃₀H₃₉O₄P	78-33-1	494.602						i EtOH; sl eth, bz
1648	[(4-tert-Butylphenoxy)methyl] oxirane		C₁₃H₁₈O₂	3101-60-8	206.281			167¹⁴, 145⁰·⁵	1.036²⁵	1.5145²⁰	
1649	N-Butyl-N-phenylacetamide		C₁₂H₁₇NO	91-49-6	191.269		24.5	281	0.9912²⁰	1.5146²⁰	sl chl
1650	1-(4-tert-Butylphenyl)ethanone		C₁₂H₁₆O	943-27-1	176.254		17.7	263; 137²⁰	0.9635²⁰	1.518¹⁵	
1651	Butyl phenyl ether	Butoxybenzene	C₁₀H₁₄O	1126-79-0	150.217	liq	-19.4	210	0.9351²⁰	1.4969²⁰	s eth, ace
1652	N-Butylpiperidine		C₉H₁₉N	4945-48-6	141.254			176	0.8245²⁰	1.4467²⁰	
1653	Butylpropanedioic acid	Butylmalonic acid	C₇H₁₂O₄	534-59-8	160.168	pr (w)	104.5				vs H₂O; s EtOH, eth
1654	Butyl propanoate	Butyl propionate	C₇H₁₄O₂	590-01-2	130.185	liq	-89	146.8	0.8754²⁰	1.4014²⁰	sl H₂O, ctc; msc EtOH, eth
1655	sec-Butyl propanoate		C₇H₁₄O₂	591-34-4	130.185			133	0.8657²⁰	1.3952²⁰	s EtOH, eth
1656	N-tert-Butyl-2-propenamide	N-tert-Butylacrylamide	C₇H₁₃NO	107-58-4	127.184	cry (bz)	128				sl H₂O; i peth

Butyl methyl ether

sec-Butyl methyl ether

2-*tert*-Butyl-4-methylphenol

2-*tert*-Butyl-5-methylphenol

2-*tert*-Butyl-6-methylphenol

4-*tert*-Butyl-2-methylphenol

Butyl methyl sulfide

tert-Butyl methyl sulfide

4-Butylmorpholine

1-Butylnaphthalene

2-Butylnaphthalene

Butyl nitrate

Butyl nitrite

tert-Butyl nitrite

sec-Butyl nitrite

4-(Butylnitrosoamino)-1-butanol

5-Butylnonane

Butyl nonanoate

Butyl octanoate

2-Butyl-1-octanol

Butyl oleate

tert-Butyl 3-oxobutanoate

Butyl 4-oxopentanoate

Butyl palmitate

Butyl pentanoate

sec-Butyl pentanoate

4-(1-Butylpentyl)pyridine

tert-Butyl peroxybenzoate

2-Butylphenol

2-*sec*-Butylphenol

2-*tert*-Butylphenol

3-Butylphenol

3-*tert*-Butylphenol

4-Butylphenol

4-*sec*-Butylphenol

4-*tert*-Butylphenol

4-*tert*-Butylphenol, phosphate (3:1)

[(4-*tert*-Butylphenoxy)methyl]oxirane

N-Butyl-*N*-phenylacetamide

1-(4-*tert*-Butylphenyl)ethanone

Butyl phenyl ether

N-Butylpiperidine

Butylpropanedioic acid

Butyl propanoate

sec-Butyl propanoate

N-*tert*-Butyl-2-propenamide

No.	Name	Synonym	Mol. Form.	CAS RN	Mol. Wt.	Physical Form	mp/°C	bp/°C	den/ g cm⁻³	n_D	Solubility
1657	Butyl propyl ether		$C_7H_{16}O$	3073-92-5	116.201			118.1	0.7772²⁰		i H₂O; vs EtOH, eth
1658	4-tert-Butylpyridine		$C_9H_{13}N$	3978-81-2	135.206	liq	-41	196.5	0.915²⁵	1.4958²⁰	s ctc, CS₂
1659	5-Butyl-2-pyridinecarboxylic acid	Fusaric acid	$C_{10}H_{13}NO_2$	536-69-6	179.216		97				
1660	Butyl stearate		$C_{22}H_{44}O_2$	123-95-5	340.583		27	343	0.854²⁵	1.4328⁵⁰	i H₂O; s EtOH; vs ace
1661	Butyl thiocyanate	1-Thiocyanobutane	C_5H_9NS	628-83-1	115.197			186	0.9563¹⁵	1.4360²⁰	i H₂O; s EtOH, eth
1662	2-Butylthiophene		$C_8H_{12}S$	1455-20-5	140.246			181.5	0.9537²⁰	1.5090²⁰	
1663	Butyl thiophene-2-carboxylate	Butyl 2-thiophenecarboxylate	$C_9H_{12}O_2S$	56053-84-0	184.255			58⁰·¹⁵			
1664	Butyl 4-toluenesulfonate		$C_{11}H_{16}O_3S$	778-28-9	228.308			165⁶	1.1319²⁰	1.5050²⁰	i H₂O; s eth; sl ctc
1665	Butyl trichloroacetate		$C_6H_9Cl_3O_2$	3657-07-6	219.493			204	1.2778²⁰	1.4525²⁵	s ctc
1666	Butyl (2,4,5-trichlorophenoxy) acetate	2,4,5-T Butyl ester	$C_{12}H_{13}Cl_3O_3$	93-79-8	311.588		28.5	337			
1667	Butyltrichlorosilane	Trichlorobutylsilane	$C_4H_9Cl_3Si$	7521-80-4	191.559			148.5	1.1606²⁰	1.4363²⁰	s eth, bz, tol, AcOEt
1668	Butyl trifluoroacetate		$C_6H_9F_3O_2$	367-64-6	170.129			102	1.0268²²	1.353²²	s chl
1669	Butylurea		$C_5H_{12}N_2O$	592-31-4	116.161	tab (w), nd (bz)	97.0				vs H₂O, EtOH; sl chl
1670	sec-Butylurea	(1-Methylpropyl)urea	$C_5H_{12}N_2O$	689-11-2	116.161	pr (w)	169				
1671	tert-Butylurea		$C_5H_{12}N_2O$	1118-12-3	116.161		176 dec				s H₂O; vs EtOH; sl bz
1672	1-tert-Butyl-4-vinylbenzene	p-tert-Butylstyrene	$C_{12}H_{16}$	1746-23-2	160.255	liq	-36.9	99¹⁴	0.89²⁰		
1673	Butyl vinyl ether	1-(Ethenyloxy)butane	$C_6H_{12}O$	111-34-2	100.158	liq	-92	94	0.7888²⁰	1.4026²⁰	i H₂O; vs EtOH, ace; msc eth; s bz
1674	tert-Butyl vinyl ether	2-(Ethenyloxy)-2-methylpropane	$C_6H_{12}O$	926-02-3	100.158	liq	-112	75	0.7691²⁰	1.3922²⁰	
1675	1-Butyne	Ethylacetylene	C_4H_6	107-00-6	54.091	col gas	-125.7	8.08	0.6783⁰	1.3962²⁰	i H₂O; s EtOH, eth
1676	2-Butyne	Dimethylacetylene	C_4H_6	503-17-3	54.091	vol liq or gas	-32.2	26.9	0.6910²⁰	1.3921²⁰	i H₂O; s EtOH, eth, ctc
1677	2-Butynediamide	Cellocidin	$C_4H_4N_2O_2$	543-21-5	112.087	cry (dil MeOH)	217 dec				sl H₂O, chl, EtOH, eth, gl HOAc
1678	2-Butynedinitrile		C_4N_2	1071-98-3	76.056		20.5	76.5	0.9708²⁵	1.4647²⁵	
1679	2-Butynedioic acid		$C_4H_2O_4$	142-45-0	114.057		183 dec				vs H₂O, EtOH, eth
1680	2-Butyne-1,4-diol	Bis(hydroxymethyl)acetylene	$C_4H_6O_2$	110-65-6	86.090	pl (bz, AcOEt)	50	238		1.4804²⁰	vs H₂O, EtOH, ace; sl eth; i bz, peth
1681	2-Butyne-1,4-diol diacetate	1,4-Diacetoxy-2-butyne	$C_8H_{10}O_4$	1573-17-7	170.163			122¹⁰		1.4611²⁰	s ctc
1682	2-Butynoic acid		$C_4H_4O_2$	590-93-2	84.074	pl (eth, peth)	78	203	0.9641²⁰		vs H₂O, eth, EtOH, chl
1683	2-Butyn-1-ol		C_4H_6O	764-01-2	70.090	liq	-1.1	148	0.9370²⁰	1.4530²⁰	vs eth, EtOH
1684	3-Butyn-1-ol		C_4H_6O	927-74-2	70.090	liq	-63.6	129	0.9257²⁰	1.4409²⁰	vs H₂O, EtOH
1685	3-Butyn-2-ol		C_4H_6O	2028-63-9	70.090	liq	-1.5	106.5	0.8618²⁰	1.4207²⁰	vs H₂O, eth, EtOH
1686	3-Butyn-2-one	Ethynyl methyl ketone	C_4H_4O	1423-60-5	68.074			84	0.8793²⁰	1.4070²⁰	
1687	3-Butynylbenzene		$C_{10}H_{10}$	16520-62-0	130.186			190	0.9258²⁰	1.5208²⁰	
1688	γ-Butyrolactone	Oxolan-2-one	$C_4H_6O_2$	96-48-0	86.090	liq	-43.61	204	1.1296²⁰	1.4341²⁰	vs ace, bz, eth, EtOH
1689	Cacotheline		$C_{21}H_{21}N_3O_7$	561-20-6	427.408	ye cry	>300				sl H₂O
1690	γ-Cadinene		$C_{15}H_{24}$	39029-41-9	204.352			126¹²	0.9182¹⁵	1.3166²⁰	
1691	Cadmium bis(diethyldithiocarbamate)		$C_{10}H_{20}CdN_2S_4$	14239-68-0	408.950	wh cry	255				
1692	Caffeine		$C_8H_{10}N_4O_2$	58-08-2	194.191	wh nd (w+1), hex pr (sub)	238	sub 90	1.23¹⁹		sl H₂O, EtOH; i eth, ctc; s chl, py
1693	Calactin	19-Oxogomphoside	$C_{29}H_{40}O_9$	20304-47-6	532.623	small pr (ace)	271				
1694	Calcium ascorbate		$C_{12}H_{14}CaO_{12}$	5743-27-1	390.310	tricl cry (w)					s H₂O; i MeOH, EtOH
1695	Calcium citrate		$C_{12}H_{10}Ca_3O_{14}$	7693-13-2	498.433	cry (w)	≈100 dec (hyd)				sl H₂O; i EtOH
1696	Calcium cyanamide	Calcium carbimide	$CCaN_2$	156-62-7	80.102	col hex cry	≈1340	sub	2.29		dec H₂O
1697	Calcium cyclamate		$C_{12}H_{24}CaN_2O_6S_2$	139-06-0	396.535	cry					vs H₂O
1698	Calcium gluconate		$C_{12}H_{22}CaO_{14}$	299-28-5	430.373	cry					i EtOH, os
1699	Calcium iodobehenate	Iodocosanoic acid, calcium salt	$C_{44}H_{84}CaI_2O_4$	1319-91-1	971.023	wh-ye pow					i H₂O, EtOH, eth; s chl
1700	Calcium lactate		$C_6H_{10}CaO_6$	814-80-2	218.217	wh pow (w)					s H₂O; i EtOH
1701	Calcium 2,4-pentanedioate	Calcium acetylacetonate	$C_{10}H_{14}CaO_4$	19372-44-2	238.294	col cry (MeOH)	dec				

Butyl propyl ether

4-tert-Butylpyridine

5-Butyl-2-pyridinecarboxylic acid

Butyl stearate

Butyl thiocyanate

2-Butylthiophene

Butyl thiophene-2-carboxylate

Butyl 4-toluenesulfonate

Butyl trichloroacetate

Butyl (2,4,5-trichlorophenoxy)acetate

Butyltrichlorosilane

Butyl trifluoroacetate

Butylurea

sec-Butylurea

tert-Butylurea

1-tert-Butyl-4-vinylbenzene

Butyl vinyl ether

tert-Butyl vinyl ether

1-Butyne

2-Butyne

2-Butynediamide

2-Butynedinitrile

2-Butynedioic acid

2-Butyne-1,4-diol

2-Butyne-1,4-diol diacetate

2-Butynoic acid

2-Butyn-1-ol

3-Butyn-1-ol

3-Butyn-2-ol

3-Butyn-2-one

3-Butynylbenzene

γ-Butyrolactone

Cacotheline

γ-Cadinene

Cadmium bis(diethyldithiocarbamate)

Caffeine

Calactin

Calcium ascorbate

Calcium citrate

Calcium cyanamide

Calcium cyclamate

Calcium gluconate

Calcium iodobehenate

Calcium lactate

Calcium 2,4-pentanedioate

No.	Name	Synonym	Mol. Form.	CAS RN	Mol. Wt.	Physical Form	mp/°C	bp/°C	den/ g cm⁻³	n_D	Solubility
1702	Calcium thioglycollate		$C_4H_6CaO_4S_2$	814-71-1	222.297	pr (w)	220 dec				s H_2O, chl; sl EtOH; i eth, bz
1703	Calotoxin	4'β-Hydroxy-19-oxogomphoside	$C_{29}H_{40}O_{10}$	20304-49-8	548.622	cry (EtOH)	268				
1704	Calotropin		$C_{29}H_{40}O_9$	1986-70-5	532.623	pl (EtOH)	221				s H_2O, EtOH; i eth
1705	Calusterone		$C_{28}H_{48}O$	17021-26-0	400.680	cry (ace)	157.5				
1706	Camphene, (+)	2,2-Dimethyl-3-methylenebicyclo[2.2.1]heptane, (1R)-	$C_{10}H_{16}$	5794-03-6	136.234	nd	52	161	0.8950⁵⁰	1.4570²⁵	vs eth
1707	Camphene, (-)	2,2-Dimethyl-3-methylenebicyclo[2.2.1]heptane, (1S)-	$C_{10}H_{16}$	5794-04-7	136.234		52	158	0.8446⁵⁰	1.4564⁵⁴	vs eth
1708	d-Camphocarboxylic acid		$C_{11}H_{16}O_3$	18530-30-8	196.243	pr (eth, 50% al)	127.5				vs bz, eth, EtOH
1709	Camphor, (±)	1,7,7-Trimethylbicyclo[2.2.1]heptan-2-one, (±)	$C_{10}H_{16}O$	21368-68-3	152.233	wh rhom cry (EtOH)	178.3	sub			i H_2O; vs EtOH, eth; s ace, bz, ctc
1710	Camphor, (+)	1,7,7-Trimethylbicyclo[2.2.1]heptan-2-one, (1R)	$C_{10}H_{16}O$	464-49-3	152.233	pl	178.8	207.4	0.990²⁵	1.5462	i H_2O; vs EtOH, eth; s ace, bz
1711	Camphor, (-)	1,7,7-Trimethylbicyclo[2.2.1]heptan-2-one, (1S)	$C_{10}H_{16}O$	464-48-2	152.233		178.6		0.9853¹⁸		i H_2O; vs EtOH, eth, HOAc; s ace, bz
1712	Camphoric acid, (±)	1,2,2-Trimethyl-1,3-cyclopentanedicarboxylic acid, (1RS, 3SR)	$C_{10}H_{16}O_4$	5394-83-2	200.232	pr, lf	202		1.186		sl H_2O; s chl, eth, EtOH
1713	d-Camphorsulfonic acid		$C_{10}H_{16}O_4S$	3144-16-9	232.297	pr (HOAc)	195 dec				vs H_2O; i eth; sl HOAc
1714	Canadine, (±)	DL-Tetrahydroberberine	$C_{20}H_{21}NO_4$	29074-38-2	339.386	mcl nd (al)	134				vs EtOH, chl
1715	Cannabidiol		$C_{21}H_{30}O_2$	13956-29-1	314.462	rods (peth)	67	188²	1.040⁴⁰	1.5404²⁰	i H_2O; s EtOH, eth, bz, chl
1716	Cannabinol	6,6,9-Trimethyl-3-pentyl-6H-dibenzo[b,d]pyran-1-ol	$C_{21}H_{26}O_2$	521-35-7	310.430	pl, lf (peth)	77	185⁰·⁰⁵			i H_2O; s EtOH, eth, ace, bz, peth, alk
1717	Canrenone		$C_{22}H_{28}O_3$	976-71-6	340.455	cry (AcOEt)	150				
1718	Cantharidin		$C_{10}H_{12}O_4$	56-25-7	196.200	orth pl	218	sub 84			i H_2O; sl EtOH, eth, ace, bz; s HOAc
1719	Caprolactam	6-Hexanelactam	$C_6H_{11}NO$	105-60-2	113.157	lf (lig)	69.3	270			vs H_2O, bz, EtOH, chl
1720	Capsaicin		$C_{18}H_{27}NO_3$	404-86-4	305.412	mcl pl or sc (peth)	65	215⁰·⁰¹			i H_2O; vs EtOH; s eth, bz, peth; sl con HCl
1721	Capsanthin	3,3'-Dihydroxy-β,κ-caroten-6'-one, (3R,3'S,5'R)	$C_{40}H_{56}O_3$	465-42-9	584.871		176				
1722	Captafol		$C_{10}H_9Cl_4NO_2S$	2425-06-1	349.061	cry	161				
1723	Captan		$C_9H_8Cl_3NO_2S$	133-06-2	300.590	cry (CCl₄)	172.5		1.74²⁵		vs chl
1724	Captopril	1-(3-Mercapto-2-methyl-1-oxypropyl)proline	$C_9H_{15}NO_3S$	62571-86-2	217.285	cry (AcOEt)	105				s H_2O, EtOH, chl
1725	Carbachol		$C_6H_{15}ClN_2O_2$	51-83-2	182.648		210 dec				vs H_2O, MeOH; sl EtOH; i eth, chl
1726	Carbamic chloride	Carbamyl chloride	CH_2ClNO	463-72-9	79.486			dec 62			
1727	Carbamodithioic acid		CH_3NS_2	594-07-0	93.172						vs EtOH, eth
1728	Carbamoyl dihydrogen phosphate		CH_4NO_5P	590-55-6	141.021	unstab in soln					
1729	Carbaryl		$C_{12}H_{11}NO_2$	63-25-2	201.221		145		1.228²⁵		vs ace, DMF
1730	Carbazole	Dibenzopyrolle	$C_{12}H_9N$	86-74-8	167.206	pl or lf	246.3	354.69			i H_2O; sl EtOH, eth, bz, chl; s ace
1731	9H-Carbazole-9-acetic acid		$C_{14}H_{11}NO_2$	524-80-1	225.243	lf (AcOEt)	215				vs eth, EtOH, chl, HOAc
1732	Carbendazim	Carbamic acid, 1H-benzimidazol-2-yl-, methyl ester	$C_9H_9N_3O_2$	10605-21-7	191.186		300 dec		1.45		
1733	Carbetapentane	Pentoxyverine	$C_{20}H_{31}NO_3$	77-23-6	333.465			165⁰·⁰¹			
1734	N-Carbethoxyphthalimide	N-(Ethoxycarbonyl)phthalimide	$C_{11}H_9NO_4$	22509-74-6	219.194		91				
1735	Carbic anhydride		$C_9H_8O_3$	129-64-6	164.158	orth cry (peth)	164.5		1.417²⁵		vs ace, bz, EtOH, chl
1736	Carbimazole		$C_7H_{10}N_2O_2S$	22232-54-8	186.231	cry, pow	123.5				vs ace, chl
1737	Carbobenzoxyhydrazine	Benzyl carbazate	$C_8H_{10}N_2O_2$	5331-43-1	166.177		69.5				
1738	Carbofuran		$C_{12}H_{15}NO_3$	1563-66-2	221.252		151		1.18		
1739	Carboimidic difluoride		CHF_2N	2712-98-3	65.023	gas	-90	-13 dec			
1740	γ-Carboline	5H-Pyrido[4,3-b]indole	$C_{11}H_8N_2$	244-69-9	168.195	nd	225		1.352		sl H_2O, bz; vs MeOH; s EtOH

Calcium thioglycollate

Calotoxin

Calotropin

Calusterone

Camphene, (+)

Camphene, (−)

d-Camphocarboxylic acid

Camphor, (±)

Camphor, (+)

Camphor, (−)

Camphoric acid, (±)

d-Camphorsulfonic acid

Canadine, (±)

Cannabidiol

Cannabinol

Canrenone

Cantharidin

Caprolactam

Capsaicin

Capsanthin

Captafol

Captan

Captopril

Carbachol

Carbamic chloride

Carbamodithioic acid

Carbamoyl dihydrogen phosphate

Carbaryl

Carbazole

9H-Carbazole-9-acetic acid

Carbendazim

Carbetapentane

N-Carbethoxyphthalimide

Carbic anhydride

Carbimazole

Carbobenzoxyhydrazine

Carbofuran

Carboimidic difluoride

γ-Carboline

No.	Name	Synonym	Mol. Form.	CAS RN	Mol. Wt.	Physical Form	mp/°C	bp/°C	den/ g cm⁻³	n_D	Solubility
1741	Carbon dioxide	Carbonic anhydride	CO_2	124-38-9	44.010	col gas	-56.56 tp	-78.5 sp	0.720[25] (p>1 atm)		sl H_2O
1742	Carbon diselenide	Carbon selenide	CSe_2	506-80-9	169.93	ye liq	-43.7	125.5	2.6823[20]	1.8454[20]	
1743	Carbon disulfide	Carbon bisulfide	CS_2	75-15-0	76.141	col liq	-112.1	46	1.2632[20]	1.6319[20]	s H_2O, chl; msc EtOH, eth
1744	Carbonic acid		CH_2O_3	463-79-6	62.025						Aq. soln. of CO_2
1745	Carbonic dihydrazide	Carbohydrazide	CH_6N_4O	497-18-7	90.085	nd (dil al)	154		1.616[20]		vs H_2O, EtOH
1746	Carbon monoxide	Carbon oxide	CO	630-08-0	28.010	col gas	-205.02	-191.5	0.7909[-19]		sl H_2O; s bz, HOAc
1747	Carbonochloridic acid, 4-nitrophenyl ester		$C_7H_4ClNO_4$	7693-46-1	201.565		80	160[19]			
1748	Carbonochloridic acid, (4-nitrophenyl)methyl ester		$C_8H_6ClNO_4$	4457-32-3	215.592		32.8				
1749	Carbonochloridic acid, 2,2,2-trichloroethyl ester		$C_3H_2Cl_4O_2$	17341-93-4	211.859			63[11]			
1750	Carbonothioic dichloride	Thiophosgene	CCl_2S	463-71-8	114.982	red liq		73	1.508[15]	1.5442[20]	dec H_2O, EtOH; s eth
1751	Carbonothioic dihydrazide	1,3-Diamino-2-thiourea	CH_6N_4S	2231-57-4	106.151	nd, pl (w) nd, pl (w)	170 dec				vs H_2O
1752	Carbon oxyselenide	Carbonyl selenide	COSe	1603-84-5	106.97	col gas; unstab	-122	-21.5			dec H_2O
1753	Carbon oxysulfide	Carbonyl sulfide	COS	463-58-1	60.075	col gas	-138.8	-50	1.028[17]	1.24[-87]	sl H_2O; s EtOH; vs KOH
1754	Carbon suboxide	1,2-Propadiene-1,3-dione	C_3O_2	504-64-3	68.031	col gas	-107	6.8	1.114[0]	1.4538[0]	s eth, bz, CS_2
1755	Carbonyl bromide	Bromophosgene	CBr_2O	593-95-3	187.818			64.5	2.52[15]		
1756	Carbonyl chloride	Phosgene	CCl_2O	75-44-5	98.916	col gas	-127.78	8	1.3719[25] (p>1 atm)		s bz, ctc, chl, tol, HOAc
1757	Carbonyl chloride fluoride	Carbonic chloride fluoride	CClFO	353-49-1	82.461	col gas	-148	-47.2			reac H_2O
1758	Carbonyl dicyanide		C_3N_2O	1115-12-4	80.044	liq	-36	65.5	1.124[20]	1.3919[20]	s eth, ace, ctc, chl
1759	N,N'-Carbonyldiimidazole		$C_7H_6N_4O$	530-62-1	162.149	cry (bz)	119				
1760	Carbonyl fluoride		CF_2O	353-50-4	66.007	col gas	-111.2	-84.5	1.139[25]		
1761	Carbophenothion		$C_{11}H_{16}ClO_2PS_3$	786-19-6	342.866			82[0.01]	1.271[20]		
1762	Carbosulfan		$C_{20}H_{32}N_2O_3S$	55285-14-8	380.544			126	1.056[20]		
1763	Carboxin		$C_{12}H_{13}NO_2S$	5234-68-4	235.302		94				
1764	2-Carboxybenzeneacetic acid		$C_9H_8O_4$	89-51-0	180.158		184.5		1.4100[20]		s H_2O, EtOH; sl eth; i bz, chl
1765	N-(D-1-Carboxyethyl)-L-arginine	Octopine	$C_9H_{18}N_4O_4$	34522-32-2	246.264	nd (w)	281				
1766	L-γ-Carboxyglutamic acid		$C_6H_9NO_6$	53861-57-7	191.138	cry	167				
1767	S-(Carboxymethyl)-L-cysteine	Carbocysteine	$C_5H_9NO_4S$	638-23-3	179.195	nd	206				
1768	2-Carboxyphenyl 2-hydroxybenzoate	Salsalate	$C_{14}H_{10}O_5$	552-94-3	258.226		147				sl ace
1769	3-Carene, (+)		$C_{10}H_{16}$	498-15-7	136.234			171; 123[200]	0.8549[30]	1.469[3]	vs ace, bz, eth
1770	Carisoprodol		$C_{12}H_{24}N_2O_4$	78-44-4	260.330	cry	92				s os
1771	Carminic acid		$C_{22}H_{20}O_{13}$	1260-17-9	492.386	red mcl pr (aq, MeOH)	136 dec				s H_2O, EtOH; sl eth; i bz, chl
1772	Carnitine	4-Amino-3-hydroxybutanoic acid trimethylbetaine	$C_7H_{15}NO_3$	541-15-1	161.199	cry (al-ace), hyg	197 dec				vs H_2O, EtOH
1773	Carnosine	N-β-Alanyl-L-histidine	$C_9H_{14}N_4O_3$	305-84-0	226.232		260				vs H_2O
1774	α-Carotene		$C_{40}H_{56}$	7488-99-5	536.873	red pl or pr (peth, bz-MeOH)	187.5		1.00[20]		vs bz, eth, chl
1775	β-Carotene		$C_{40}H_{56}$	7235-40-7	536.873	red br hex pr (bz-MeOH)	183		1.00[20]		i H_2O; sl EtOH, chl; s eth, ace, bz
1776	β,ψ-Carotene	γ-Carotene	$C_{40}H_{56}$	472-93-5	536.873	red pr (bz-MeOH), viol pr (eth)	153				i H_2O, EtOH; sl eth, peth; s bz, chl
1777	ψ,ψ-Carotene	trans-Lycopene	$C_{40}H_{56}$	502-65-8	536.873	red pr or nd (peth)	175				sl EtOH, peth; s eth; vs bz, chl, CS_2
1778	β,β-Carotene-3,3'-diol, (3R,3'R)	Zeaxanthin	$C_{40}H_{56}O_2$	144-68-3	568.872	ye pr (MeOH) orth (chl-eth)	215.5	227[0.06]			i H_2O; sl EtOH; s eth, ace, bz, py, chl
1779	β,ε-Carotene-3,3'-diol, (3R,3'R,6'R)	Xanthophyll	$C_{40}H_{56}O_2$	127-40-2	568.872	ye or viol pr (eth-MeOH)	196				vs bz, eth, EtOH, peth
1780	β,β-Caroten-3-ol, (3R)	Cryptoxanthin	$C_{40}H_{56}O$	472-70-8	552.872	garnet red pr (bz-MeOH)	160				vs bz, chl
1781	β,ψ-Caroten-3-ol, (3R)	Rubixanthin	$C_{40}H_{56}O$	3763-55-1	552.872	dk red nd (bz-MeOH) oran-red (bz-peth)	160				sl EtOH, peth; s bz, chl

Carbon dioxide

Carbon diselenide

Carbon disulfide

Carbonic acid

Carbonic dihydrazide

Carbon monoxide

Carbonochloridic acid, 4-nitrophenyl ester

Carbonochloridic acid, (4-nitrophenyl)methyl ester

Carbonochloridic acid, 2,2,2-trichloroethyl ester

Carbonothioic dichloride

Carbonothioic dihydrazide

Carbon oxyselenide

Carbon oxysulfide

Carbon suboxide

Carbonyl bromide

Carbonyl chloride

Carbonyl chloride fluoride

Carbonyl dicyanide

N,N'-Carbonyldiimidazole

Carbonyl fluoride

Carbophenothion

Carbosulfan

Carboxin

2-Carboxybenzeneacetic acid

N-(D-1-Carboxyethyl)-L-arginine

L-γ-Carboxyglutamic acid

S-(Carboxymethyl)-L-cysteine

2-Carboxyphenyl 2-hydroxybenzoate

3-Carene, (+)

Carisoprodol

Carminic acid

Carnitine

Carnosine

α-Carotene

β-Carotene

β,ψ-Carotene

ψ,ψ-Carotene

β,β-Carotene-3,3'-diol, (3R,3'R)

β,ε-Carotene-3,3'-diol, (3R,3'R,6'R)

β,β-Caroten-3-ol, (3R)

β,ψ-Caroten-3-ol, (3R)

No.	Name	Synonym	Mol. Form.	CAS RN	Mol. Wt.	Physical Form	mp/°C	bp/°C	den/ g cm⁻³	n_D	Solubility
1782	ψ,ψ-Caroten-16-ol	Lycoxanthin	C₄₀H₅₆O	19891-74-8	552.872	red pl (bz-MeOH)	168				i H₂O; sl EtOH; s bz, CS₂
1783	Caroverine		C₂₂H₂₇N₃O₂	23465-76-1	365.468	cry	69	202⁰·⁰¹			sl i-PrOH
1784	Carpaine		C₂₈H₅₀N₂O₄	3463-92-1	478.708	mcl pr (al, ace)	121				vs ace, bz, eth, EtOH
1785	Cartap hydrochloride		C₇H₁₆ClN₃O₂S₂	22042-59-7	273.804	cry	180				s H₂O; sl EtOH, MeOH
1786	Carvenone, (S)		C₁₀H₁₆O	10395-45-6	152.233			233	0.9289²⁰	1.4805²⁰	i H₂O; s ace
1787	(R)-Carvone	p-Mentha-1,8-dien-6-one, (R)	C₁₀H₁₄O	6485-40-1	150.217		25.2	231	0.9593²⁰	1.4988²⁰	sl H₂O; vs EtOH; s eth, ctc, chl
1788	(S)-Carvone	p-Mentha-1,8-dien-6-one, (S)	C₁₀H₁₄O	2244-16-8	150.217		<15	231	0.965²⁰	1.4989²⁰	sl H₂O; vs EtOH; s eth, chl
1789	Caryophyllene		C₁₅H₂₄	87-44-5	204.352			122¹³·⁵	0.9075²⁰	1.4986²⁰	vs bz
1790	Casimiroin	6-Methoxy-9-methyl-1,3-dioxolo[4,5-h]quinolin-8(9H)-one	C₁₂H₁₁NO₄	477-89-4	233.220						sl chl
1791	Cassaine		C₂₄H₃₉NO₄	468-76-8	405.572	fl (eth)	142.5				s EtOH, ace, chl, eth, bz, MeOH
1792	Caulophylline		C₁₂H₁₆N₂O	486-86-2	204.267	cry (w+2), nd (al, bz)	137				vs H₂O, ace, bz, EtOH
1793	α-Cedrene		C₁₅H₂₄	469-61-4	204.352	oil		262.5; 125¹²			
1794	Cedrol		C₁₅H₂₆O	77-53-2	222.366		86		0.9479⁹⁰	1.4824⁹⁰	
1795	Cefazolin		C₁₄H₁₄N₈O₄S₃	25953-19-9	454.508	nd (ace aq)	200 dec				s DMF, py; sl MeOH; i chl, bz, eth
1796	β-Cellobiose		C₁₂H₂₂O₁₁	13360-52-6	342.296	cry (dil al)	225 dec				s H₂O; i EtOH, eth, ace, bz
1797	Cellotriose		C₁₈H₃₂O₁₆	33404-34-1	504.437		208				
1798	Cephalexin		C₁₆H₁₇N₃O₄S	15686-71-2	347.389	cry					
1799	Cephaloglycin	Kafocin	C₁₈H₁₉N₃O₆S	3577-01-3	405.425	cry (w)	≈220 dec				
1800	Cephaloridine		C₁₉H₁₇N₃O₄S₂	50-59-9	415.486	cry					s H₂O
1801	Cephalothin		C₁₆H₁₆N₂O₆S₂	153-61-7	396.437		160				
1802	Cephapirin		C₁₇H₁₇N₃O₆S₂	21593-23-7	423.463	cry (ace aq)	155				
1803	Cepharanthine		C₃₇H₃₈N₂O₆	481-49-2	606.707	ye amor pow	150				
1804	Cephradine		C₁₆H₁₉N₃O₄S	38821-53-3	349.405	col cry (w)	141 dec				
1805	Cerulenin	2,3-Epoxy-4-oxo-7,10-dodecadienamide, (2R,3S)-	C₁₂H₁₇NO₃	17397-89-6	223.268	wh nd	94				sl H₂O; s bz, EtOH, ace; i peth
1806	Cevadine		C₃₂H₄₉NO₉	62-59-9	591.733	flat nd (eth)	213 dec				
1807	Chavicine		C₁₇H₁₉NO₃	495-91-0	285.338						vs eth, EtOH, peth
1808	Cheirolin		C₅H₉NO₂S₂	505-34-0	179.261	cry (eth)	47.5	200³			vs EtOH, chl
1809	Chelerythrine		C₂₁H₁₉NO₅	34316-15-9	365.380	cry (chl-MeOH)		207			vs chl
1810	Chelidonine	Stylophorine	C₂₀H₁₉NO₅	476-32-4	353.369	mcl pr (al)	135.5	220⁰·⁰⁰²			i H₂O; s EtOH, eth, chl
1811	Chinomethionat		C₁₀H₆N₂OS₂	2439-01-2	234.297		170				
1812	Chloral hydrate		C₂H₃Cl₃O₂	302-17-0	165.403		57	dec 96	1.9081²⁰		vs H₂O, bz, eth, EtOH
1813	Chlorambucil		C₁₄H₁₉Cl₂NO₂	305-03-3	304.213		65				
1814	Chloramine B	N-Chlorobenzenesulfonamide sodium	C₆H₅ClNNaO₂S	127-52-6	213.618	pr (w)	190				sl EtOH; i chl, eth
1815	Chloramine T	N-Chloro-4-methylbenzenesulfonamide sodium	C₇H₇ClNNaO₂S	127-65-1	227.645	pr (hyd)	180 (hyd)				s H₂O; i bz, chl, eth
1816	Chloramphenicol		C₁₁H₁₂Cl₂N₂O₅	56-75-7	323.129	pa ye pl or nd (w)	150.5	sub			vs ace, EtOH, chl
1817	Chloramphenicol palmitate		C₂₇H₄₂Cl₂N₂O₆	530-43-8	561.537	cry (bz)	90				vs bz, eth, EtOH
1818	Chloranilic acid	2,5-Dichloro-3,6-dihydroxy-2,5-cyclohexadiene-1,4-dione	C₆H₂Cl₂O₄	87-88-7	208.984	red lf (w+2)	283.5				s H₂O
1819	Chlorbenside	1-Chloro-4[[(4-chlorophenyl)methyl]thio]benzene	C₁₃H₁₀Cl₂S	103-17-3	269.189		75		1.4210²⁰		
1820	Chlorbicyclen		C₉H₆Cl₈	2550-75-6	397.768	pow	105	174²			
1821	Chlorbromuron	N'-(4-Bromo-3-chlorophenyl)-N-methoxy-N-methylurea	C₉H₁₀BrClN₂O₂	13360-45-7	293.544		96		1.69²⁰		
1822	Chlorbufam	1-Methyl-2-propynyl(3-chlorophenyl)carbamate	C₁₁H₁₀ClNO₂	1967-16-4	223.656	cry	45.5				sl H₂O; s MeOH, EtOH, ace
1823	Chlorcyclizine		C₁₈H₂₁ClN₂	82-93-9	300.826	oil		140⁰·¹²			
1824	Chlordane		C₁₀H₆Cl₈	57-74-9	409.779		106	175¹	1.60²⁵		
1825	Chlordantoin		C₁₁H₁₇Cl₃N₂O₂S	5588-20-5	347.689						s CS₂
1826	Chlordene		C₁₀H₆Cl₆	3734-48-3	338.873	cry (EtOH)	155				
1827	Chlordimeform		C₁₀H₁₃ClN₂	6164-98-3	196.676		35	156⁰·⁴	1.105²⁵	1.5885²⁵	vs bz, eth, EtOH

ψ,ψ-Caroten-16-ol

Caroverine

Carpaine

Cartap hydrochloride

Carvenone, (S)

(R)-Carvone

(S)-Carvone

Caryophyllene

Casimiroin

Cassaine

Caulophylline

α-Cedrene

Cedrol

Cefazolin

β-Cellobiose

Cellotriose

Cephalexin

Cephaloglycin

Cephaloridine

Cephalothin

Cephapirin

Cepharanthine

Cephradine

Cerulenin

Cevadine

Chavicine

Cheirolin

Chelerythrine

Chelidonine

Chinomethionat

Chloral hydrate

Chlorambucil

Chloramine B

Chloramine T

Chloramphenicol

Chloramphenicol palmitate

Chloranilic acid

Chlorbenside

Chlorbicyclen

Chlorbromuron

Chlorbufam

Chlorcyclizine

Chlordane

Chlordantoin

Chlordene

Chlordimeform

No.	Name	Synonym	Mol. Form.	CAS RN	Mol. Wt.	Physical Form	mp/°C	bp/°C	den/g cm⁻³	n_D	Solubility
1828	Chlorendic acid	1,4,5,6,7,7-Hexachloro-5-norbornene-2,3-dicarboxylic acid	$C_9H_4Cl_6O_4$	115-28-6	388.844	cry (w)	232				
1829	Chlorendic anhydride		$C_9H_2Cl_6O_3$	115-27-5	370.828		235				
1830	Chlorfenvinphos		$C_{12}H_{14}Cl_3O_4P$	470-90-6	359.569			170$^{0.05}$			
1831	Chlorflurecol	9H-Fluorene-9-carboxylic acid, 2-chloro-9-hydroxy-	$C_{14}H_9ClO_3$	2464-37-1	260.672				1.496^{20}		
1832	Chloridazon	3(2H)-Pyridazinone, 5-amino-4-chloro-2-phenyl-	$C_{10}H_8ClN_3O$	1698-60-8	221.643		205				
1833	Chlorimuron-ethyl		$C_{15}H_{15}ClN_4O_6S$	90982-32-4	414.821		186				
1834	Chlormephos	Chloromethyl O,O-diethyl dithiophosphate	$C_5H_{12}ClO_2PS_2$	24934-91-6	234.705	oil		83$^{0.1}$		1.5244	sl H₂O; misc os
1835	Chlormequat chloride		$C_5H_{13}Cl_2N$	999-81-5	158.069		239 dec				
1836	Chlormezanone		$C_{11}H_{12}ClNO_3S$	80-77-3	273.736	cry	117				sl EtOH
1837	Chlornaphazine		$C_{14}H_{15}Cl_2N$	494-03-1	268.182	pl (peth)	55	210^5			vs ace, bz, eth, EtOH
1838	Chloroacetaldehyde		C_2H_3ClO	107-20-0	78.497	liq	-16.3	85.5	1.19		s eth
1839	2-Chloroacetamide		C_2H_4ClNO	79-07-2	93.512		121	225			s H₂O; vs EtOH; sl eth
1840	Chloroacetic acid		$C_2H_3ClO_2$	79-11-8	94.497	mcl pl	63	189.3	1.4043^{40}	1.4351^{55}	vs H₂O; s EtOH, eth, bz, chl; sl ctc
1841	Chloroacetic anhydride		$C_4H_4Cl_2O_3$	541-88-8	170.979	pr (bz)	46	203	1.5497^{20}		
1842	4-Chloroacetoacetanilide	N-Acetoacetyl-4-chloroaniline	$C_{10}H_{10}ClNO_2$	101-92-8	211.645		132				
1843	Chloroacetone		C_3H_5ClO	78-95-5	92.524	liq	-44.5	119	1.15^{20}		s H₂O, EtOH, eth, chl
1844	Chloroacetonitrile	Chloromethyl cyanide	C_2H_2ClN	107-14-2	75.497			126.5	1.1930^{20}	1.4202^{25}	vs eth, EtOH
1845	α-Chloroacetophenone	ω-Chloroacetophenone	C_8H_7ClO	532-27-4	154.594	pl(dil al), rhom, lf (peth)	56.5	247	1.324^{15}		i H₂O; vs EtOH, eth, bz; s ace, peth
1846	4-(2-Chloroacetyl)acetanilide		$C_{10}H_{10}ClNO_2$	140-49-8	211.645		218				
1847	Chloroacetyl chloride		$C_2H_2Cl_2O$	79-04-9	112.942	liq	-22	106	1.4202^{20}	1.4530^{20}	msc eth; s ace, ctc
1848	Chloroacetylene		C_2HCl	593-63-5	60.482	col gas	-126	-30			sl EtOH
1849	9-Chloroacridine		$C_{13}H_8ClN$	1207-69-8	213.663	nd (al)	121	sub			vs H₂O, EtOH
1850	2-Chloroaniline		C_6H_6ClN	95-51-2	127.572	liq	-1.9	208.8		1.5895^{20}	i H₂O; msc EtOH; s eth, ace
1851	3-Chloroaniline		C_6H_6ClN	108-42-9	127.572	liq	-10.28	230.5	1.2161^{20}	1.5941^{20}	i H₂O; msc EtOH, eth, ace, bz; s chl
1852	4-Chloroaniline		C_6H_6ClN	106-47-8	127.572	orth pr	70.5	232	1.429^{19}	1.5546^{67}	s H₂O, EtOH, eth, chl
1853	2-Chloroaniline hydrochloride		$C_6H_7Cl_2N$	137-04-2	164.033	pl (w, aq al)	235		1.505^{18}		vs H₂O
1854	3-Chloroaniline hydrochloride		$C_6H_7Cl_2N$	141-85-5	164.033	pl	222				vs H₂O, EtOH
1855	2-Chloroanisole	1-Chloro-2-methoxybenzene	C_7H_7ClO	766-51-8	142.583	liq	-26.8	198.5	1.1911^{20}	1.5480^{20}	i H₂O; s EtOH, eth; sl chl
1856	3-Chloroanisole	1-Chloro-3-methoxybenzene	C_7H_7ClO	2845-89-8	142.583			193.5	1.1759^{12}	1.5365^{20}	i H₂O; s EtOH, eth
1857	4-Chloroanisole	1-Chloro-4-methoxybenzene	C_7H_7ClO	623-12-1	142.583		<-18	197.5	1.201^{20}	1.5390^{20}	i H₂O; vs EtOH, eth, chl; s ctc
1858	1-Chloroanthracene		$C_{14}H_9Cl$	4985-70-0	212.674	lf (HOAc)	83.5		1.1707^{100}	1.6959^{100}	i H₂O; s EtOH, eth, bz, ctc
1859	1-Chloro-9,10-anthracenedione		$C_{14}H_7ClO_2$	82-44-0	242.658	ye nd (to or al)	163	sub			i H₂O; sl EtOH, ctc; msc eth; s bz
1860	2-Chloro-9,10-anthracenedione		$C_{14}H_7ClO_2$	131-09-9	242.658	pa ye nd (al, HOAc)	211	sub			i H₂O, eth; sl EtOH, bz; vs tol; s PhNO₂
1861	2-Chlorobenzaldehyde		C_7H_5ClO	89-98-5	140.567	nd	12.4	211.9	1.2483^{20}	1.5662^{20}	sl H₂O; s EtOH, eth, ace, bz, ctc
1862	3-Chlorobenzaldehyde		C_7H_5ClO	587-04-2	140.567	pr	17.5	213.5	1.2410^{20}	1.5650^{20}	sl H₂O, chl; s EtOH, eth, ace, bz
1863	4-Chlorobenzaldehyde		C_7H_5ClO	104-88-1	140.567	pl	47.5	213.5	1.196^{61}	1.555^{61}	s H₂O, ace, chl; vs EtOH, eth, bz
1864	2-Chlorobenzamide		C_7H_6ClNO	609-66-5	155.582	orth nd (w)	141.8				s H₂O, EtOH, eth
1865	Chlorobenzene	Phenyl chloride	C_6H_5Cl	108-90-7	112.557	liq	-45.31	131.72	1.1058^{20}	1.5241^{20}	i H₂O; msc EtOH, eth; vs bz, ctc
1866	2-Chlorobenzeneacetic acid		$C_8H_7ClO_2$	2444-36-2	170.594	nd (w)	96				sl H₂O; vs EtOH
1867	3-Chlorobenzeneacetic acid		$C_8H_7ClO_2$	1878-65-5	170.594	pl (dil al), nd (lig)	77.5				sl H₂O, bz, ctc, EtOH; msc eth

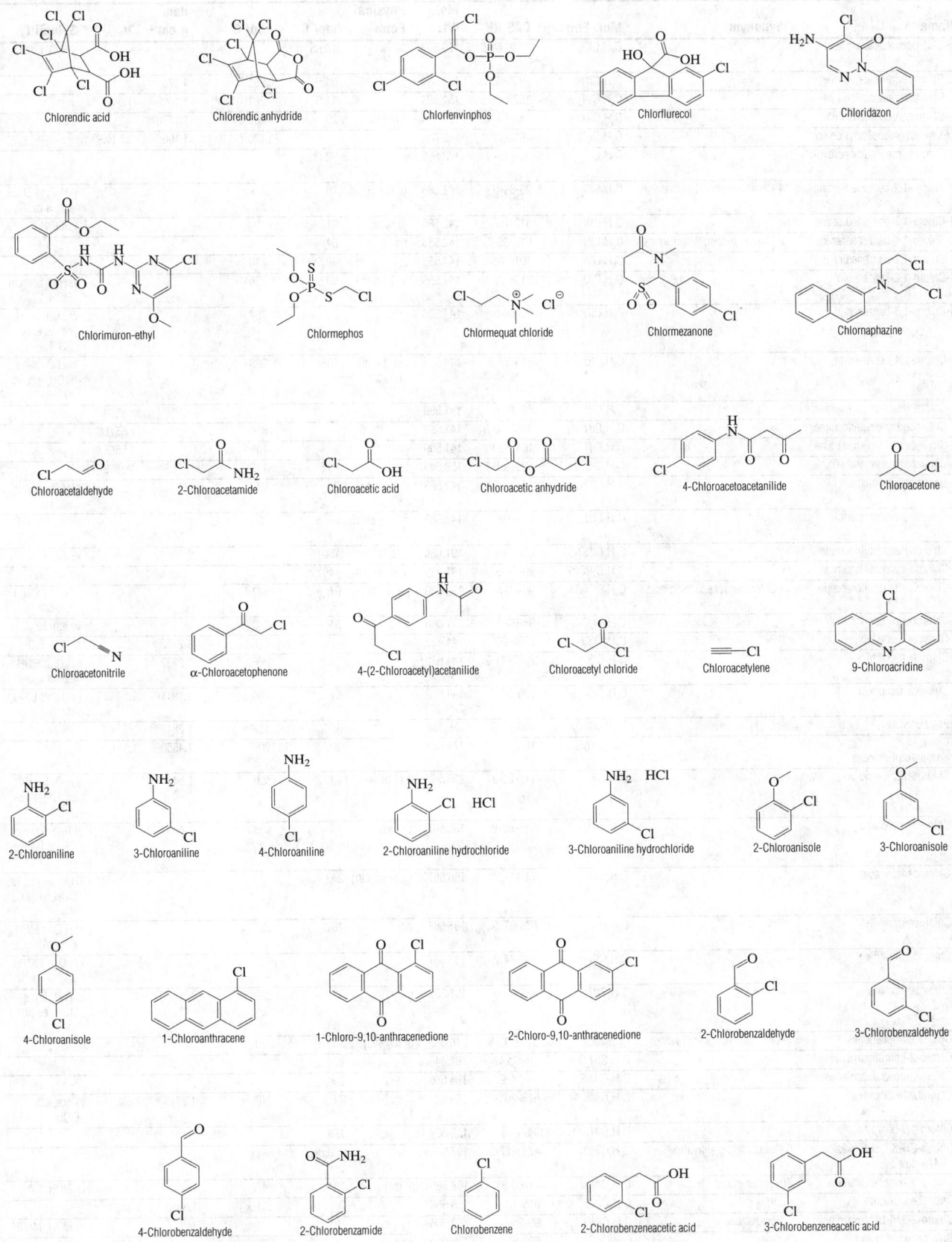

Chlorendic acid

Chlorendic anhydride

Chlorfenvinphos

Chlorflurecol

Chloridazon

Chlorimuron-ethyl

Chlormephos

Chlormequat chloride

Chlormezanone

Chlornaphazine

Chloroacetaldehyde

2-Chloroacetamide

Chloroacetic acid

Chloroacetic anhydride

4-Chloroacetoacetanilide

Chloroacetone

Chloroacetonitrile

α-Chloroacetophenone

4-(2-Chloroacetyl)acetanilide

Chloroacetyl chloride

Chloroacetylene

9-Chloroacridine

2-Chloroaniline

3-Chloroaniline

4-Chloroaniline

2-Chloroaniline hydrochloride

3-Chloroaniline hydrochloride

2-Chloroanisole

3-Chloroanisole

4-Chloroanisole

1-Chloroanthracene

1-Chloro-9,10-anthracenedione

2-Chloro-9,10-anthracenedione

2-Chlorobenzaldehyde

3-Chlorobenzaldehyde

4-Chlorobenzaldehyde

2-Chlorobenzamide

Chlorobenzene

2-Chlorobenzeneacetic acid

3-Chlorobenzeneacetic acid

No.	Name	Synonym	Mol. Form.	CAS RN	Mol. Wt.	Physical Form	mp/°C	bp/°C	den/ g cm⁻³	n_D	Solubility
1868	4-Chlorobenzeneacetic acid		C₈H₇ClO₂	1878-66-6	170.594	nd (w)	105.5				s H₂O, EtOH, eth, bz
1869	2-Chlorobenzeneacetonitrile		C₈H₆ClN	2856-63-5	151.594		24	251	1.1737[18]		
1870	3-Chlorobenzeneacetonitrile		C₈H₆ClN	1529-41-5	151.594		11.5	261; 135[10]	1.1806[30]	1.5437[20]	
1871	4-Chlorobenzeneacetonitrile		C₈H₆ClN	140-53-4	151.594		29	265.0	1.1778[30]		s ctc
1872	α-Chlorobenzeneacetyl chloride		C₈H₆Cl₂O	2912-62-1	189.039			120[23], 110[14]	1.196[25]	1.5440[20]	
1873	3-Chlorobenzenecarboperoxoic acid		C₇H₅ClO₃	937-14-4	172.566		92 dec				
1874	4-Chloro-1,2-benzenediamine	4-Chloro-o-phenylenediamine	C₆H₇ClN₂	95-83-0	142.586	pl (bz-lig) lf (w)	76				sl H₂O; vs EtOH, eth; s bz, lig
1875	4-Chloro-1,3-benzenediamine		C₆H₇ClN₂	5131-60-2	142.586	pl or nd	91				vs EtOH
1876	2-Chloro-1,4-benzenediamine	2-Chloro-p-phenylenediamine	C₆H₇ClN₂	615-66-7	142.586	nd	64				
1877	3-Chloro-1,2-benzenediol		C₆H₅ClO₂	4018-65-9	144.556	cry (lig)	48.5	110[11]			vs lig
1878	4-Chloro-1,2-benzenediol		C₆H₅ClO₂	2138-22-9	144.556	lf (bz-peth)	90.5	139[10.5]			vs H₂O, ace, eth, EtOH
1879	4-Chloro-1,3-benzenediol		C₆H₅ClO₂	95-88-5	144.556			257			vs H₂O, EtOH, eth, ace, bz, CS₂
1880	2-Chloro-1,4-benzenediol		C₆H₅ClO₂	615-67-8	144.556	red lf (chl); nd (bz)	108	263			vs H₂O, chl; s EtOH, eth; vs bz
1881	2-Chlorobenzenemethanamine		C₇H₈ClN	89-97-4	141.599			72[2]		1.5594[25]	
1882	3-Chlorobenzenemethanamine		C₇H₈ClN	4152-90-3	141.599			89[2]		1.5570[25]	
1883	4-Chlorobenzenemethanamine		C₇H₈ClN	104-86-9	141.599			109[13]		1.5566[25]	
1884	4-Chlorobenzenemethanethiol		C₇H₇ClS	6258-66-8	158.649		19.5	113[17]	1.202[25]	1.5893[20]	
1885	2-Chlorobenzenemethanol		C₇H₇ClO	17849-38-6	142.583	lf or nd (dil al)	73	230			sl H₂O; vs EtOH, eth, lig
1886	4-Chlorobenzenemethanol		C₇H₇ClO	873-76-7	142.583	nd (w), pl (bz or bz-lig)	75	235			vs bz, eth, EtOH
1887	2-Chlorobenzenesulfonamide		C₆H₆ClNO₂S	6961-82-6	191.636	lf (al)	188				vs EtOH
1888	4-Chlorobenzenesulfonamide		C₆H₆ClNO₂S	98-64-6	191.636	pr or pl (eth)	146				vs bz, eth
1889	4-Chlorobenzenesulfonic acid	p-Chlorobenzenesulfonic acid	C₆H₅ClO₃S	98-66-8	192.620	nd (w+1)	67	147[25]			s H₂O, EtOH; i eth, bz
1890	4-Chlorobenzenesulfonyl chloride		C₆H₄Cl₂O₂S	98-60-2	211.066		51	141[15]			vs eth, bz
1891	2-Chlorobenzenethiol		C₆H₅ClS	6320-03-2	144.622			205.5	1.2752[10]		sl H₂O, EtOH
1892	3-Chlorobenzenethiol		C₆H₅ClS	2037-31-2	144.622			206	1.2637[13]		i H₂O; s EtOH, eth, chl, peth
1893	4-Chlorobenzenethiol		C₆H₅ClS	106-54-7	144.622		61	206	1.1911[20]	1.5480[20]	i H₂O; vs EtOH, eth, bz; sl chl
1894	Chlorobenzilate		C₁₆H₁₄Cl₂O₃	510-15-6	325.186		37	157[0.07]	1.2816[20]		
1895	2-Chloro-1,3,2-benzodioxaphosphole		C₆H₄ClO₂P	1641-40-3	174.522		30	80[20]	1.4650[20]	1.5712[20]	
1896	2-Chlorobenzoic acid		C₇H₅ClO₂	118-91-2	156.567	mcl pr (w)	140.2	sub	1.544[20]		s H₂O, bz; vs EtOH, eth, ace; sl CS₂
1897	3-Chlorobenzoic acid		C₇H₅ClO₂	535-80-8	156.567	pr (w)	158	sub	1.496[25]		sl H₂O, bz, ctc, CS₂; s EtOH, eth
1898	4-Chlorobenzoic acid		C₇H₅ClO₂	74-11-3	156.567	tcl pr (al-eth)	243				i H₂O, bz, ctc; vs EtOH; sl eth, ace
1899	2-Chlorobenzonitrile		C₇H₄ClN	873-32-5	137.567	nd	46.3	232			sl H₂O; s EtOH, eth, chl
1900	3-Chlorobenzonitrile		C₇H₄ClN	766-84-7	137.567		41	100[15]			i H₂O; s EtOH, eth
1901	4-Chlorobenzonitrile		C₇H₄ClN	623-03-0	137.567	nd (al)	95	223; 95[5]	1.1133[17]		sl H₂O, lig; s EtOH, eth, bz, chl
1902	2-Chlorobenzophenone	2-Chlorophenyl phenyl ketone	C₁₃H₉ClO	5162-03-8	216.662	pl (chl-lig)	54	330			
1903	4-Chloro-2-benzothiazolamine		C₇H₅ClN₂S	19952-47-7	184.646		204				
1904	6-Chloro-2-benzothiazolamine		C₇H₅ClN₂S	95-24-9	184.646		200				
1905	2-Chlorobenzothiazole		C₇H₄ClNS	615-20-3	169.632		24	248	1.3715[10]	1.6338[10]	vs ace, eth, EtOH
1906	5-Chloro-1H-benzotriazole		C₆H₄ClN₃	94-97-3	153.569		158				
1907	6-Chloro-2H-3,1-benzoxazine-2,4(1H)-dione	5-Chloroisatoic anhydride	C₈H₄ClNO₃	4743-17-3	197.576		280 dec				
1908	5-Chloro-2-benzoxazolamine	Zoxazolamine	C₇H₅ClN₂O	61-80-3	168.580	pl (bz)	184.5				vs EtOH
1909	2-Chlorobenzoxazole		C₇H₄ClNO	615-18-9	153.566		7	201.5	1.3453[18]	1.5678[20]	
1910	5-Chloro-2(3H)-benzoxazolone	Chlorzoxazone	C₇H₄ClNO₂	95-25-0	169.566	cry (ace)	191.5				vs EtOH, MeOH
1911	2-Chlorobenzoyl chloride		C₇H₄Cl₂O	609-65-4	175.012	liq	-4	238		1.5726[16]	s ctc
1912	3-Chlorobenzoyl chloride		C₇H₄Cl₂O	618-46-2	175.012			225		1.5677[20]	
1913	4-Chlorobenzoyl chloride		C₇H₄Cl₂O	122-01-0	175.012		16	222	1.3770[20]	1.5756[20]	sl chl
1914	1-Chloro-4-benzylbenzene		C₁₃H₁₁Cl	831-81-2	202.679		7.5	299; 147[8]	1.1247[20]		vs ace

4-Chlorobenzeneacetic acid

2-Chlorobenzeneacetonitrile

3-Chlorobenzeneacetonitrile

4-Chlorobenzeneacetonitrile

α-Chlorobenzeneacetyl chloride

3-Chlorobenzenecarboperoxoic acid

4-Chloro-1,2-benzenediamine

4-Chloro-1,3-benzenediamine

2-Chloro-1,4-benzenediamine

3-Chloro-1,2-benzenediol

4-Chloro-1,2-benzenediol

4-Chloro-1,3-benzenediol

2-Chloro-1,4-benzenediol

2-Chlorobenzenemethanamine

3-Chlorobenzenemethanamine

4-Chlorobenzenemethanamine

4-Chlorobenzenemethanethiol

2-Chlorobenzenemethanol

4-Chlorobenzenemethanol

2-Chlorobenzenesulfonamide

4-Chlorobenzenesulfonamide

4-Chlorobenzenesulfonic acid

4-Chlorobenzenesulfonyl chloride

2-Chlorobenzenethiol

3-Chlorobenzenethiol

4-Chlorobenzenethiol

Chlorobenzilate

2-Chloro-1,3,2-benzodioxaphosphole

2-Chlorobenzoic acid

3-Chlorobenzoic acid

4-Chlorobenzoic acid

2-Chlorobenzonitrile

3-Chlorobenzonitrile

4-Chlorobenzonitrile

2-Chlorobenzophenone

4-Chloro-2-benzothiazolamine

6-Chloro-2-benzothiazolamine

2-Chlorobenzothiazole

5-Chloro-1H-benzotriazole

6-Chloro-2H-3,1-benzoxazine-2,4(1H)-dione

5-Chloro-2-benzoxazolamine

2-Chlorobenzoxazole

5-Chloro-2(3H)-benzoxazolone

2-Chlorobenzoyl chloride

3-Chlorobenzoyl chloride

4-Chlorobenzoyl chloride

1-Chloro-4-benzylbenzene

No.	Name	Synonym	Mol. Form.	CAS RN	Mol. Wt.	Physical Form	mp/°C	bp/°C	den/ g cm⁻³	n_D	Solubility
1915	o-Chlorobenzylidene malononitrile		$C_{10}H_5ClN_2$	2698-41-1	188.613	wh cry	96	312			sl H₂O; s bz, diox, EtOAc, ace
1916	2-Chlorobiphenyl		$C_{12}H_9Cl$	2051-60-7	188.652	mcl (dil al)	34	274	1.1499[32]		i H₂O; vs eth, EtOH, lig
1917	3-Chlorobiphenyl		$C_{12}H_9Cl$	2051-61-8	188.652		16	284.5	1.1579[25]	1.6181[25]	vs ace, eth, EtOH
1918	4-Chlorobiphenyl		$C_{12}H_9Cl$	2051-62-9	188.652	lf (lig or al)	78.8	292.9; 146[10]			i H₂O; s EtOH, eth, lig
1919	4'-Chloro-[1,1'-biphenyl]-4-amine	4-Amino-4'-chlorodiphenyl	$C_{12}H_{10}ClN$	135-68-2	203.667	cry (peth)	134				vs ace, bz, eth
1920	3-Chloro-[1,1'-biphenyl]-2-ol	2-Phenyl-6-chlorophenol	$C_{12}H_9ClO$	85-97-2	204.651		6	dec 317	1.24[25]	1.6237[30]	i H₂O; s EtOH, eth, ace, bz
1921	4-Chloro-1,2-butadiene		C_4H_5Cl	25790-55-0	88.536			88	0.9891[20]	1.4775[20]	vs ace, bz, eth
1922	1-Chloro-1,3-butadiene		C_4H_5Cl	627-22-5	88.536			68	0.9606[20]	1.4712[20]	vs eth, EtOH, chl
1923	2-Chloro-1,3-butadiene	Chloroprene	C_4H_5Cl	126-99-8	88.536	liq	-130	59.4	0.956[20]	1.4583[20]	sl H₂O; msc eth, ace, bz
1924	4-Chlorobutanal		C_4H_7ClO	6139-84-0	106.551			51[13]	1.106[8]	1.4466[8]	vs ace, eth, EtOH
1925	1-Chlorobutane	Butyl chloride	C_4H_9Cl	109-69-3	92.567	liq	-123.1	78.4	0.8857[20]	1.4023[20]	i H₂O; msc EtOH, eth; sl ctc
1926	2-Chlorobutane	(±)-sec-Butyl chloride	C_4H_9Cl	53178-20-4	92.567	liq	-131.3	68.2	0.8732[20]	1.3971[20]	vs bz, eth, EtOH, chl
1927	4-Chlorobutanenitrile		C_4H_6ClN	628-20-6	103.551			192	1.0934[15]	1.4413[20]	i H₂O; s EtOH, eth; sl ctc
1928	2-Chlorobutanoic acid		$C_4H_7ClO_2$	4170-24-5	122.551			189[627], 101[15]	1.1796[20]	1.441[20]	sl H₂O; vs EtOH, eth
1929	3-Chlorobutanoic acid		$C_4H_7ClO_2$	625-68-3	122.551	cry (eth)	16	116[22]	1.1898[20]	1.4221[20]	s EtOH; vs eth; sl ctc
1930	4-Chlorobutanoic acid		$C_4H_7ClO_2$	627-00-9	122.551		16	196[22], 68[0.2]	1.2236[20]	1.4642[20]	vs EtOH
1931	4-Chloro-1-butanol		C_4H_9ClO	928-51-8	108.566			84[16]	1.0883[20]	1.4518[20]	vs eth, EtOH
1932	1-Chloro-2-butanol	α-Butylene chlorohydrin	C_4H_9ClO	1873-25-2	108.566			141	1.068[25]	1.4400[20]	s EtOH, eth
1933	3-Chloro-2-butanone		C_4H_7ClO	4091-39-8	106.551			115	1.0554[25]	1.4219[20]	
1934	4-Chlorobutanoyl chloride		$C_4H_6Cl_2O$	4635-59-0	140.996			173.5	1.2581[20]	1.4616[20]	s eth
1935	2-Chloro-1-butene		C_4H_7Cl	2211-70-3	90.552			58.5	0.9107[15]	1.4165[21]	vs ace, bz, eth, EtOH
1936	3-Chloro-1-butene		C_4H_7Cl	563-52-0	90.552			64.5	0.8978[20]	1.4149[20]	vs eth, ace; s chl
1937	4-Chloro-1-butene		C_4H_7Cl	927-73-1	90.552			75	0.9211[20]	1.4233[20]	vs ace, eth, chl
1938	cis-1-Chloro-2-butene		C_4H_7Cl	4628-21-1	90.552			84.1	0.9426[20]	1.4390[20]	i H₂O; s EtOH, ace, chl
1939	trans-1-Chloro-2-butene		C_4H_7Cl	4894-61-5	90.552			85	0.9295[20]	1.4350[20]	i H₂O; s ace, chl
1940	cis-2-Chloro-2-butene		C_4H_7Cl	2211-69-0	90.552	liq	-117.3	70.6	0.9239[20]	1.4240[20]	i H₂O; msc EtOH; s ace, chl
1941	trans-2-Chloro-2-butene		C_4H_7Cl	2211-68-9	90.552	liq	-105.8	62.8	0.9138[20]	1.4190[20]	i H₂O; msc EtOH; s ace, chl
1942	1-Chloro-4-tert-butylbenzene		$C_{10}H_{13}Cl$	3972-56-3	168.663			213	1.0075[18]	1.5123[20]	
1943	Chloro-(tert-butyl)dimethylsilane		$C_6H_{15}ClSi$	18162-48-6	150.722		89.5	125			
1944	Chloro(tert-butyl)diphenylsilane		$C_{16}H_{19}ClSi$	58479-61-1	274.861			120[0.06]	1.07[20]	1.5675[20]	
1945	2-Chloro-4-tert-butylphenol		$C_{10}H_{13}ClO$	98-28-2	184.662			114[8]			
1946	3-Chloro-1-butyne		C_4H_5Cl	21020-24-6	88.536			68.5	1.4218[25]	1.4218[25]	
1947	2-Chloro-N-(2-chloroethyl) ethanamine, hydrochloride		$C_4H_{10}Cl_3N$	821-48-7	178.488		215.0				
1948	2-Chloro-N-(2-chloroethyl)-N-ethylethanamine	HN1	$C_6H_{13}Cl_2N$	538-07-8	170.080	col liq	-34	66[12]	1.0861[23]	1.4653[25]	i H₂O
1949	2-Chloro-N-(2-chloroethyl)-N-methylethanamine	Mechlorethamine	$C_5H_{11}Cl_2N$	51-75-2	156.053		-60	87[18], 64[5]			sl H₂O; msc ctc, DMF
1950	1-Chloro-2-(chloromethyl) benzene	2-Chlorobenzyl chloride	$C_7H_6Cl_2$	611-19-8	161.029	liq	-17	217	1.2699[0]	1.5530[20]	i H₂O; sl EtOH, ctc; vs eth, bz
1951	1-Chloro-3-(chloromethyl) benzene	3-Chlorobenzyl chloride	$C_7H_6Cl_2$	620-20-2	161.029			216; 110[25]	1.2695[15]	1.5554[20]	vs EtOH
1952	1-Chloro-4-(chloromethyl) benzene	4-Chlorobenzyl chloride	$C_7H_6Cl_2$	104-83-6	161.029	nd (dil al)	31	223			sl ctc
1953	Chloro(chloromethyl) dimethylsilane		$C_3H_8Cl_2Si$	1719-57-9	143.088			115.5	1.0865[20]	1.4360[20]	
1954	3-Chloro-2-(chloromethyl)-1-propene		$C_4H_6Cl_2$	1871-57-4	124.997	liq	-14	138	1.1782[20]	1.4753	vs EtOH, chl
1955	1-Chloro-4-[(chloromethyl)thio] benzene		$C_7H_6Cl_2S$	7205-90-5	193.094		21.5	128[12]	1.346[25]	1.6055[20]	
1956	2-Chloro-1-(4-chlorophenyl) ethanone		$C_8H_6Cl_2O$	937-20-2	189.039	nd (al)	101.5	270			s EtOH, bz, MeOH

o-Chlorobenzylidene malononitrile

2-Chlorobiphenyl

3-Chlorobiphenyl

4-Chlorobiphenyl

4'-Chloro-[1,1'-biphenyl]-4-amine

3-Chloro-[1,1'-biphenyl]-2-ol

4-Chloro-1,2-butadiene

1-Chloro-1,3-butadiene

2-Chloro-1,3-butadiene

4-Chlorobutanal

1-Chlorobutane

2-Chlorobutane

4-Chlorobutanenitrile

2-Chlorobutanoic acid

3-Chlorobutanoic acid

4-Chlorobutanoic acid

4-Chloro-1-butanol

1-Chloro-2-butanol

3-Chloro-2-butanone

4-Chlorobutanoyl chloride

2-Chloro-1-butene

3-Chloro-1-butene

4-Chloro-1-butene

cis-1-Chloro-2-butene

trans-1-Chloro-2-butene

cis-2-Chloro-2-butene

trans-2-Chloro-2-butene

1-Chloro-4-*tert*-butylbenzene

Chloro-(*tert*-butyl)dimethylsilane

Chloro(*tert*-butyl)diphenylsilane

2-Chloro-4-*tert*-butylphenol

3-Chloro-1-butyne

2-Chloro-*N*-(2-chloroethyl)ethanamine, hydrochloride

2-Chloro-*N*-(2-chloroethyl)-*N*-ethylethanamine

2-Chloro-*N*-(2-chloroethyl)-*N*-methylethanamine

1-Chloro-2-(chloromethyl)benzene

1-Chloro-3-(chloromethyl)benzene

1-Chloro-4-(chloromethyl)benzene

Chloro(chloromethyl)dimethylsilane

3-Chloro-2-(chloromethyl)-1-propene

1-Chloro-4-[(chloromethyl)thio]benzene

2-Chloro-1-(4-chlorophenyl)ethanone

No.	Name	Synonym	Mol. Form.	CAS RN	Mol. Wt.	Physical Form	mp/°C	bp/°C	den/ g cm⁻³	n_D	Solubility
1957	3-Chlorocholest-5-ene, (3β)		$C_{27}H_{45}Cl$	910-31-6	405.099	nd (al, ace)	96				i H_2O; s EtOH, ace, bz, chl; vs CS_2
1958	trans-o-Chlorocinnamic acid		$C_9H_7ClO_2$	939-58-2	182.604		212				vs eth, EtOH
1959	trans-m-Chlorocinnamic acid		$C_9H_7ClO_2$	14473-90-6	182.604		165				s EtOH, eth
1960	trans-p-Chlorocinnamic acid		$C_9H_7ClO_2$	940-62-5	182.604		249.5				vs ace, eth, EtOH
1961	Chlorocyclohexane	Cyclohexyl chloride	$C_6H_{11}Cl$	542-18-7	118.604	liq	-43.81	142	1.000²⁰	1.4626²⁰	i H_2O; msc EtOH, eth, ace, bz; vs chl
1962	2-Chlorocyclohexanone		C_6H_9ClO	822-87-7	132.587		23	82¹⁵	1.160²⁰	1.4825²⁰	s eth, bz, diox; sl ctc
1963	1-Chlorocyclohexene		C_6H_9Cl	930-66-5	116.588			142.5	1.0361¹⁹	1.4797²⁰	s eth, ace, ctc, chl
1964	Chlorocyclopentane	Cyclopentyl chloride	C_5H_9Cl	930-28-9	104.578			114	1.0051²⁰	1.4510²⁰	i H_2O; s eth, ace, bz, ctc
1965	2-Chlorocyclopentanone		C_5H_7ClO	694-28-0	118.562			87¹⁹, 73¹²	1.185²⁵	1.4750²⁰	
1966	3-Chlorocyclopentene		C_5H_7Cl	96-40-2	102.563			40⁴⁰, 27³⁰	1.0388²⁵	1.4708²⁶	vs eth, EtOH, chl
1967	4-Chloro-2-cyclopentylphenol	Dowicide 9	$C_{11}H_{13}ClO$	13347-42-7	196.673			183¹⁸			
1968	1-Chlorodecane		$C_{10}H_{21}Cl$	1002-69-3	176.727	liq	-31.3	225.9	0.8696²⁰	1.4380²⁰	i H_2O; vs eth, chl; s ctc
1969	10-Chloro-1-decanol		$C_{10}H_{21}ClO$	51309-10-5	192.726		12.5	187¹⁵	0.9630²⁵	1.4578²⁰	vs eth, EtOH
1970	2-Chloro-N,N-diallylacetamide	Allidochlor	$C_8H_{12}ClNO$	93-71-0	173.640	liq		116¹, 92⁰·⁷	1.088²⁵	1.4932²⁵	sl H_2O; s EtOH
1971	Chlorodiazepoxide		$C_{16}H_{14}ClN_3O$	58-25-3	299.754		236.2				
1972	Chlorodibromomethane		$CHBr_2Cl$	124-48-1	208.280	liq	-20	120	2.451²⁰	1.5482²⁰	i H_2O; s EtOH, eth, ace, bz
1973	Chloro(dichloromethyl) dimethylsilane	(Dichloromethyl) dimethylchlorosilane	$C_3H_7Cl_3Si$	18171-59-0	177.533	liq	-48	149	1.2369²⁰	1.461²⁰	
1974	5-Chloro-N-(3,4-dichlorophenyl)-2-hydroxybenzamide	3',4',5-Trichlorosalicylanilide	$C_{13}H_8Cl_3NO_2$	642-84-2	316.568		247				
1975	2-Chloro-1,1-diethoxyethane		$C_6H_{13}ClO_2$	621-62-5	152.619			157.4	1.0180²⁰	1.4170²⁰	sl H_2O, ctc; msc EtOH, eth
1976	3-Chloro-1,1-diethoxypropane		$C_7H_{15}ClO_2$	35573-93-4	166.646			84²⁵	0.9951¹⁹	1.4268²⁰	vs ace, bz
1977	2-Chloro-N,N-diethylacetamide		$C_6H_{12}ClNO$	2315-36-8	149.618			192²⁵			
1978	2-Chloro-N,N-diethylethanamine, hydrochloride		$C_6H_{15}Cl_2N$	869-24-9	172.096		200				sl H_2O
1979	Chlorodifluoroacetic acid		$C_2HClF_2O_2$	76-04-0	130.478	hyg	25	122		1.3559²⁰	s chl
1980	1-Chloro-1,1-difluoroethane	Refrigerant 142b	$C_2H_3ClF_2$	75-68-3	100.495	col gas	-130.8	-9.1	1.107²⁵		i H_2O; s bz
1981	1-Chloro-2,2-difluoroethane		$C_2H_3ClF_2$	338-65-8	100.495			35.1			
1982	1-Chloro-2,2-difluoroethene	1-Chloro-2,2-difluoroethylene	C_2HClF_2	359-10-4	98.479	col gas	-138.5	-18.5			
1983	Chlorodifluoromethane	Refrigerant 22	$CHClF_2$	75-45-6	86.469	col gas	-157.42	-40.7	1.4909⁻⁶⁹		sl H_2O; s eth, ace, chl
1984	7-Chloro-2,3-dihydro-1H-inden-4-ol	Chlorindanol	C_9H_9ClO	145-94-8	168.619	nd (peth)	92				
1985	10-Chloro-5,10-dihydrophenarsazine	Phenarsazine chloride	$C_{12}H_9AsClN$	578-94-9	277.581	ye cry	195		1.65		i H_2O; sl ctc, bz, xyl
1986	5-Chloro-2,4-dimethylaniline		$C_8H_{10}ClNO_2$	97-50-7	187.624		91				
1987	2-Chloro-1,1-dimethoxyethane		$C_4H_9ClO_2$	97-97-2	124.566			127.5	1.068²⁰	1.4150²⁰	sl EtOH, eth, bz, ctc
1988	N-(4-Chloro-2,5-dimethoxyphenyl)-3-oxobutanamide		$C_{12}H_{14}ClNO_4$	4433-79-8	271.697		107				s chl
1989	Chlorodimethylaluminum	Dimethylaluminum chloride	C_2H_6AlCl	1184-58-3	92.504	hyg liq	-21	126	0.996		reac H_2O; s hx
1990	2-Chloro-10-(3-dimethylaminopropyl)phenothiazine monohydrochloride	Aminazin hydrochloride	$C_{17}H_{20}Cl_2N_2S$	69-09-0	355.325		195 dec				s H_2O; i eth, bz; vs chl, EtOH
1991	2-Chloro-N,N-dimethylaniline		$C_8H_{10}ClN$	698-01-1	155.625			205	1.1067²⁰	1.5578²⁰	vs bz, EtOH
1992	3-Chloro-N,N-dimethylaniline		$C_8H_{10}ClN$	6848-13-1	155.625			232			sl H_2O; s EtOH, ace, bz
1993	4-Chloro-N,N-dimethylaniline		$C_8H_{10}ClN$	698-69-1	155.625	nd (al)	35.5	231	1.0480¹⁰⁰		s EtOH
1994	2-Chloro-1,4-dimethylbenzene		C_8H_9Cl	95-72-7	140.610		0.8	187	1.0589¹⁵		i H_2O; s ace, ctc; vs bz
1995	4-Chloro-1,2-dimethylbenzene		C_8H_9Cl	615-60-1	140.610	liq	-6	194	1.0682¹⁵		i H_2O; s ace, ctc; vs bz
1996	2-Chloro-N,N-dimethylethanamine, hydrochloride		$C_4H_{11}Cl_2N$	4584-46-7	144.043		201.0				sl H_2O
1997	(2-Chloro-1,1-dimethylethyl)benzene	Neophyl chloride	$C_{10}H_{13}Cl$	515-40-2	168.663			223; 105¹⁸	1.047²⁰	1.5247²⁰	vs ace, bz, eth, EtOH
1998	4-Chloro-2,5-dimethylphenol		C_8H_9ClO	1124-06-7	156.609	silv-grn nd (lig)	74.5				sl H_2O; vs bz, EtOH, peth

3-Chlorocholest-5-ene, (3β)

trans-o-Chlorocinnamic acid

trans-m-Chlorocinnamic acid

trans-p-Chlorocinnamic acid

Chlorocyclohexane

2-Chlorocyclohexanone

1-Chlorocyclohexene

Chlorocyclopentane

2-Chlorocyclopentanone

3-Chlorocyclopentene

4-Chloro-2-cyclopentylphenol

1-Chlorodecane

10-Chloro-1-decanol

2-Chloro-*N,N*-diallylacetamide

Chlorodiazepoxide

Chlorodibromomethane

Chloro(dichloromethyl)dimethylsilane

5-Chloro-*N*-(3,4-dichlorophenyl)-2-hydroxybenzamide

2-Chloro-1,1-diethoxyethane

3-Chloro-1,1-diethoxypropane

2-Chloro-*N,N*-diethylacetamide

2-Chloro-*N,N*-diethylethanamine, hydrochloride

Chlorodifluoroacetic acid

1-Chloro-1,1-difluoroethane

1-Chloro-2,2-difluoroethane

1-Chloro-2,2-difluoroethene

Chlorodifluoromethane

7-Chloro-2,3-dihydro-1*H*-inden-4-ol

10-Chloro-5,10-dihydrophenarsazine

5-Chloro-2,4-dimethoxyaniline

2-Chloro-1,1-dimethoxyethane

N-(4-Chloro-2,5-dimethoxyphenyl)-3-oxobutanamide

Chlorodimethylaluminum

2-Chloro-10-(3-dimethylaminopropyl)phenothiazine monohydrochloride

2-Chloro-*N,N*-dimethylaniline

3-Chloro-*N,N*-dimethylaniline

4-Chloro-*N,N*-dimethylaniline

2-Chloro-1,4-dimethylbenzene

4-Chloro-1,2-dimethylbenzene

2-Chloro-*N,N*-dimethylethanamine, hydrochloride

(2-Chloro-1,1-dimethylethyl)benzene

4-Chloro-2,5-dimethylphenol

No.	Name	Synonym	Mol. Form.	CAS RN	Mol. Wt.	Physical Form	mp/°C	bp/°C	den/ g cm⁻³	n_D	Solubility
1999	4-Chloro-2,6-dimethylphenol		C$_8$H$_9$ClO	1123-63-3	156.609	nd (w)	83				sl H$_2$O; vs bz, EtOH, HOAc
2000	4-Chloro-3,5-dimethylphenol	Chloroxylenol	C$_8$H$_9$ClO	88-04-0	156.609		115	246			sl H$_2$O, bz, peth; s EtOH, eth
2001	Chlorodimethylphenylsilane		C$_8$H$_{11}$ClSi	768-33-2	170.712			195; 82[16]	1.032[20]	1.5082[20]	
2002	1-Chloro-N,N-dimethyl-2-propanamine, hydrochloride		C$_5$H$_{13}$Cl$_2$N	17256-39-2	158.069						s chl
2003	1-Chloro-2,2-dimethylpropane		C$_5$H$_{11}$Cl	753-89-9	106.594	liq	-20	84.3	0.8660[20]	1.4044[20]	vs bz, eth, EtOH, chl
2004	3-Chloro-2,2-dimethylpropanoic acid		C$_5$H$_9$ClO$_2$	13511-38-1	136.577		41.5	110[10]			vs ctc
2005	Chlorodimethylsilane		C$_2$H$_7$ClSi	1066-35-9	94.616	liq	-111	34.7	0.852	1.3830[20]	
2006	2-Chloro-4,6-dinitroaniline		C$_6$H$_4$ClN$_3$O$_4$	3531-19-9	217.567	ye cry (DMF aq)	157				
2007	4-Chloro-2,6-dinitroaniline		C$_6$H$_4$ClN$_3$O$_4$	5388-62-5	217.567	oran-ye nd (al)	147				s EtOH
2008	1-Chloro-2,4-dinitrobenzene		C$_6$H$_3$ClN$_2$O$_4$	97-00-7	202.552	ye orth (eth) nd (al) ye cry	53	315	1.4982[75]	1.5857[60]	i H$_2$O; sl EtOH; s eth, bz, CS$_2$
2009	2-Chloro-1,3-dinitrobenzene		C$_6$H$_3$ClN$_2$O$_4$	606-21-3	202.552	ye nd (al, HOAc)	88	315	1.6867[16]		i H$_2$O; s EtOH, eth, tol; sl chl
2010	1-Chloro-2,4-dinitronaphthalene		C$_{10}$H$_5$ClN$_2$O$_4$	2401-85-6	252.611	ye nd (bz)	146.5				
2011	4-Chloro-2,6-dinitrophenol		C$_6$H$_3$ClN$_2$O$_5$	88-87-9	218.551	pa ye cry	81		1.74[22]		vs eth, EtOH, chl
2012	2-Chloro-3,5-dinitropyridine		C$_5$H$_2$ClN$_3$O$_4$	2578-45-2	203.541		66.5				
2013	2-Chloro-1,3-dinitro-5-(trifluoromethyl)benzene		C$_7$H$_2$ClF$_3$N$_2$O$_4$	393-75-9	270.550		57				
2014	4-Chloro-1,3-dioxolan-2-one	Chloroethylene carbonate	C$_3$H$_3$ClO$_3$	3967-54-2	122.507	liq	110	213; 122[18]	1.504	1.4540[20]	
2015	2-Chloro-1,2-diphenylethanone		C$_{14}$H$_{11}$ClO	447-31-4	230.689	nd (al)	68.5	dec			s EtOH; sl chl; i alk
2016	Chlorodiphenylmethane		C$_{13}$H$_{11}$Cl	90-99-3	202.679		16	140[3]	1.140[25]	1.5951[20]	s chl
2017	1-Chlorododecane	Lauryl chloride	C$_{12}$H$_{25}$Cl	112-52-7	204.780	liq	-9.3	263.2	0.8673[20]	1.4434[20]	i H$_2$O; vs EtOH; msc ace, ctc; s bz
2018	Chloroethane	Ethyl chloride	C$_2$H$_5$Cl	75-00-3	64.514	vol liq or gas	-138.4	12.3	0.8902[25] (p>1 atm)	1.3676[20]	sl H$_2$O, chl; vs EtOH; msc eth
2019	2-Chloroethanesulfonyl chloride		C$_2$H$_4$Cl$_2$O$_2$S	1622-32-8	163.023			201.5	1.555[20]	1.4920[20]	
2020	2-Chloroethanol	Ethylene chlorohydrin	C$_2$H$_5$ClO	107-07-3	80.513	liq	-67.5	128.6	1.2019[20]	1.4419[20]	msc H$_2$O, EtOH; sl eth; s chl
2021	2-Chloroethanol, 4-methylbenzenesulfonate		C$_9$H$_{11}$ClO$_3$S	80-41-1	234.699			210[21]			i H$_2$O; s ctc
2022	Chloroethene	Vinyl chloride	C$_2$H$_3$Cl	75-01-4	62.498	col gas	-153.84	-13.8	0.9106[20]	1.3700[20]	sl H$_2$O; s EtOH; vs eth
2023	1-Chloro-4-ethoxybenzene		C$_8$H$_9$ClO	622-61-7	156.609		21	213	1.1254[20]	1.5252[20]	s EtOH, eth, HOAc; vs bz; sl ctc
2024	(2-Chloroethoxy)benzene		C$_8$H$_9$ClO	622-86-6	156.609		28	218.5			i H$_2$O; vs EtOH, eth, ace, bz; sl ctc
2025	1-Chloro-1-ethoxyethane		C$_4$H$_9$ClO	7081-78-9	108.566			93.5	0.9655[20]	1.4053[20]	
2026	2-(2-Chloroethoxy)ethanol		C$_4$H$_9$ClO$_2$	628-89-7	124.566			180; 80[5]	1.18[25]	1.4529[20]	vs H$_2$O; msc EtOH, eth
2027	2-Chloroethyl acetate	β-Chloroethyl acetate	C$_4$H$_7$ClO$_2$	542-58-5	122.551			145	1.178[20]	1.4234[20]	i H$_2$O; msc EtOH, eth; s ctc
2028	2-Chloroethyl acetoacetate		C$_6$H$_9$ClO$_3$	54527-68-3	164.586			198; 120[19]	1.2055[21]	1.4430[20]	vs bz, eth, EtOH
2029	2-Chloroethylamine hydrochloride	2-Chloroethanamine hydrochloride	C$_2$H$_7$Cl$_2$N	870-24-6	115.990		146.3				vs H$_2$O, ace, EtOH
2030	(1-Chloroethyl)benzene		C$_8$H$_9$Cl	672-65-1	140.610			105[50]			
2031	(2-Chloroethyl)benzene		C$_8$H$_9$Cl	622-24-2	140.610			197.5; 92[20]	1.069[25]	1.5276[20]	i H$_2$O; s EtOH, eth, ace, bz, CS$_2$
2032	1-Chloro-2-ethylbenzene		C$_8$H$_9$Cl	89-96-3	140.610	liq	-82.7	178.4	1.0569[20]	1.5218[20]	i H$_2$O; s ace, bz, ctc, chl
2033	1-Chloro-3-ethylbenzene		C$_8$H$_9$Cl	620-16-6	140.610	liq	-55	183.8	1.0529[20]	1.5195[20]	vs ace, bz, eth, EtOH
2034	1-Chloro-4-ethylbenzene		C$_8$H$_9$Cl	622-98-0	140.610	liq	-62.6	184.4	1.0455[20]	1.5175[20]	i H$_2$O; msc EtOH, eth, ace, peth; s HOAc
2035	2-Chloroethyl chloroformate		C$_3$H$_4$Cl$_2$O$_2$	627-11-2	142.969			155	1.3847[20]	1.4483[20]	i H$_2$O; s EtOH, eth, ace, bz; sl ctc
2036	1-(2-Chloroethyl)-3-cyclohexyl-1-nitrosourea	Lomustine	C$_9$H$_{16}$ClN$_3$O$_2$	13010-47-4	233.695	ye pow	90				i H$_2$O; s EtOH
2037	N-(2-Chloroethyl)dibenzylamine	Dibenamine	C$_{16}$H$_{18}$ClN	51-50-3	259.774	oily liq		169[3]			

4-Chloro-2,6-dimethylphenol

4-Chloro-3,5-dimethylphenol

Chlorodimethylphenylsilane

1-Chloro-N,N-dimethyl-2-propanamine, hydrochloride

1-Chloro-2,2-dimethylpropane

3-Chloro-2,2-dimethylpropanoic acid

Chlorodimethylsilane

2-Chloro-4,6-dinitroaniline

4-Chloro-2,6-dinitroaniline

1-Chloro-2,4-dinitrobenzene

2-Chloro-1,3-dinitrobenzene

1-Chloro-2,4-dinitronaphthalene

4-Chloro-2,6-dinitrophenol

2-Chloro-3,5-dinitropyridine

2-Chloro-1,3-dinitro-5-(trifluoromethyl)benzene

4-Chloro-1,3-dioxolan-2-one

2-Chloro-1,2-diphenylethanone

Chlorodiphenylmethane

1-Chlorododecane

Chloroethane

2-Chloroethanesulfonyl chloride

2-Chloroethanol

2-Chloroethanol, 4-methylbenzenesulfonate

Chloroethene

1-Chloro-4-ethoxybenzene

(2-Chloroethoxy)benzene

1-Chloro-1-ethoxyethane

2-(2-Chloroethoxy)ethanol

2-Chloroethyl acetate

2-Chloroethyl acetoacetate

2-Chloroethylamine hydrochloride

(1-Chloroethyl)benzene

(2-Chloroethyl)benzene

1-Chloro-2-ethylbenzene

1-Chloro-3-ethylbenzene

1-Chloro-4-ethylbenzene

2-Chloroethyl chloroformate

1-(2-Chloroethyl)-3-cyclohexyl-1-nitrosourea

N-(2-Chloroethyl)dibenzylamine

No.	Name	Synonym	Mol. Form.	CAS RN	Mol. Wt.	Physical Form	mp/°C	bp/°C	den/ g cm⁻³	n_D	Solubility
2038	N-(2-Chloroethyl)dibenzylamine hydrochloride	Dibenamine hydrochloride	C₁₆H₁₉Cl₂N	55-43-6	296.235	cry	194				i H₂O; s EtOH, dil acid
2039	Chloroethyldimethylsilane		C₄H₁₁ClSi	6917-76-6	122.669			89.5	0.8675²⁰	1.4105²⁰	
2040	2-Chloroethyl ethyl ether		C₄H₉ClO	628-34-2	108.566			107.5	0.9895²⁰	1.4113²⁰	sl H₂O; msc eth; s chl
2041	2-Chloroethyl isocyanate		C₃H₄ClNO	1943-83-5	105.523			44¹⁷			
2042	1-(2-Chloroethyl)-3-(4-methylcyclohexyl)-1-nitrosourea	Semustine	C₁₀H₁₈ClN₃O₂	13909-09-6	247.722	cry	64 dec				
2043	5-(2-Chloroethyl)-4-methylthiazole	Clomethiazole	C₆H₈ClNS	533-45-9	161.653	oil		92⁷	1.233²⁵		
2044	N-(2-Chloroethyl)morpholine		C₆H₁₂ClNO	3240-94-6	149.618			42¹			
2045	4-(2-Chloroethyl)morpholine, hydrochloride		C₆H₁₃Cl₂NO	3647-69-6	186.079		185				
2046	1-Chloro-2-(ethylthio)ethane		C₄H₉ClS	693-07-2	124.632			157	1.0663²⁵		
2047	2-Chloroethyl vinyl ether		C₄H₇ClO	110-75-8	106.551	liq	-70	108	1.0495²⁰	1.4378²⁰	vs EtOH, eth; sl chl
2048	3-Chloro-4-fluoroaniline		C₆H₅ClFN	367-21-5	145.562		45.0	227.0			
2049	1-Chloro-2-fluorobenzene		C₆H₄ClF	348-51-6	130.547	liq	-43	137.6	1.2233³⁰	1.4918³⁰	i H₂O; s ace, bz
2050	1-Chloro-3-fluorobenzene		C₆H₄ClF	625-98-9	130.547			127.6	1.221²⁵	1.4911	
2051	1-Chloro-4-fluorobenzene		C₆H₄ClF	352-33-0	130.547	liq	-26.8	130	1.4990¹⁵	1.4990¹⁵	i H₂O; s EtOH, eth, bz
2052	1-Chloro-1-fluoroethane		C₂H₄ClF	1615-75-4	82.504	vol liq or gas		16.2			
2053	1-Chloro-2-fluoroethane		C₂H₄ClF	762-50-5	82.504			52.8	1.1747²⁰	1.3775²⁰	vs eth, EtOH
2054	Chlorofluoromethane		CH₂ClF	593-70-4	68.478	col gas	-135.1	-9.1			sl H₂O; vs chl
2055	1-Chloro-3-fluoro-2-methylbenzene		C₇H₆ClF	443-83-4	144.574			154	1.191²⁵	1.5026²⁰	
2056	2-Chloro-1-fluoro-4-nitrobenzene	3-Chloro-4-fluoronitrobenzene	C₆H₃ClFNO₂	350-30-1	175.545		41.5	229.5			
2057	4-Chloro-1-(4-fluorophenyl)-1-butanone		C₁₀H₁₀ClFO	3874-54-2	200.636			136⁶	1.22²⁵	1.5255²⁰	
2058	3-Chloro-2,5-furandione		C₄HClO₃	96-02-6	132.502		33	196	1.5375²⁵	1.4980²⁰	
2059	1-Chloro-1,2,2,3,3,4,4-heptafluorocyclobutane	Refrigerant C317	C₄ClF₇	377-41-3	216.485	liq or gas	-39.1	25	1.602¹⁵		
2060	1-Chloroheptane	Heptyl chloride	C₇H₁₅Cl	629-06-1	134.647	liq	-69.5	160.4	0.8762²⁰	1.4264²⁰	i H₂O; msc EtOH, eth; sl ctc; s chl
2061	2-Chloroheptane		C₇H₁₅Cl	1001-89-4	134.647			61³², 46¹⁹	0.8672²⁰	1.4221²⁰	i H₂O; vs eth; s bz, chl, HOAc
2062	3-Chloroheptane		C₇H₁₅Cl	999-52-0	134.647			144; 48²⁰	0.8690²⁰	1.4228²⁰	vs bz, eth
2063	4-Chloroheptane		C₇H₁₅Cl	998-95-8	134.647			144	0.8710²⁰	1.4237²⁰	vs bz, eth
2064	7-Chloro-1-heptanol	Heptamethylene chlorohydrin	C₇H₁₅ClO	55944-70-2	150.646	cry (peth, bz)	11	150²⁰	0.9998¹⁵	1.4537²⁵	vs EtOH, peth
2065	1-Chlorohexadecane		C₁₆H₃₃Cl	4860-03-1	260.886		17.9	326.6	0.8635²⁰	1.4503²⁰	i H₂O
2066	1-Chlorohexane	Hexyl chloride	C₆H₁₃Cl	544-10-5	120.620	liq	-94.0	135.1	0.8781²⁰	1.4200²⁰	i H₂O; s EtOH, eth, ace, bz; vs chl; sl ctc
2067	2-Chlorohexane	2-Hexyl chloride	C₆H₁₃Cl	638-28-8	120.620			122.5	0.8694²¹	1.4142²²	vs ace, bz, eth, EtOH
2068	3-Chlorohexane	3-Hexyl chloride	C₆H₁₃Cl	2346-81-8	120.620			123	0.8684²⁰	1.4163²⁰	vs ace, bz, eth, EtOH
2069	6-Chloro-1-hexanol		C₆H₁₃ClO	2009-83-8	136.619			107¹²	1.0241²⁰	1.4550²⁰	sl H₂O; vs EtOH, eth
2070	4-Chloro-17-hydroxyandrost-4-en-3-one, (17β)	Clostebol	C₁₉H₂₇ClO₂	1093-58-9	322.869		189				
2071	5-Chloro-2-hydroxybenzaldehyde		C₇H₅ClO₂	635-93-8	156.567	pl (al)	100.3	105¹²			i H₂O; vs EtOH; s eth, alk
2072	4-Chloro-α-hydroxybenzeneacetic acid		C₈H₇ClO₃	492-86-4	186.593		120.3				vs bz, EtOH
2073	3-Chloro-4-hydroxybenzoic acid		C₇H₅ClO₃	3964-58-7	172.566	nd (w)	171	sub			sl H₂O, bz, chl; vs EtOH, eth, ace
2074	5-Chloro-2-hydroxybenzoic acid		C₇H₅ClO₃	321-14-2	172.566	nd (w, al)	174.8				s H₂O, eth; vs EtOH, bz; sl ace
2075	2-Chloro-5-hydroxybenzophenone		C₁₃H₉ClO₂	85-19-8	232.662		95.3				i H₂O
2076	3-Chloro-4-hydroxy-5-methoxybenzaldehyde		C₈H₇ClO₃	19463-48-0	186.593	tetr	165				i H₂O; s EtOH, HOAc
2077	1-Chloro-2-iodobenzene		C₆H₄ClI	615-41-8	238.453		0.7	234.5	1.9515²⁵	1.6331²⁵	i H₂O; s ace; sl ctc
2078	1-Chloro-3-iodobenzene		C₆H₄ClI	625-99-0	238.453			230	1.9255²⁰		i H₂O; s ace
2079	1-Chloro-4-iodobenzene		C₆H₄ClI	637-87-6	238.453	lf (ace, al)	57	227	1.886²⁷		i H₂O; s EtOH, PhNO₂; sl chl
2080	1-Chloro-4-iodobutane		C₄H₈ClI	10297-05-9	218.464	liq		116; 89¹⁹	1.785	1.5400²⁰	
2081	Chloroiodomethane		CH₂ClI	593-71-5	176.384			109	2.422²⁰	1.5822²⁰	vs ace, bz, eth, EtOH

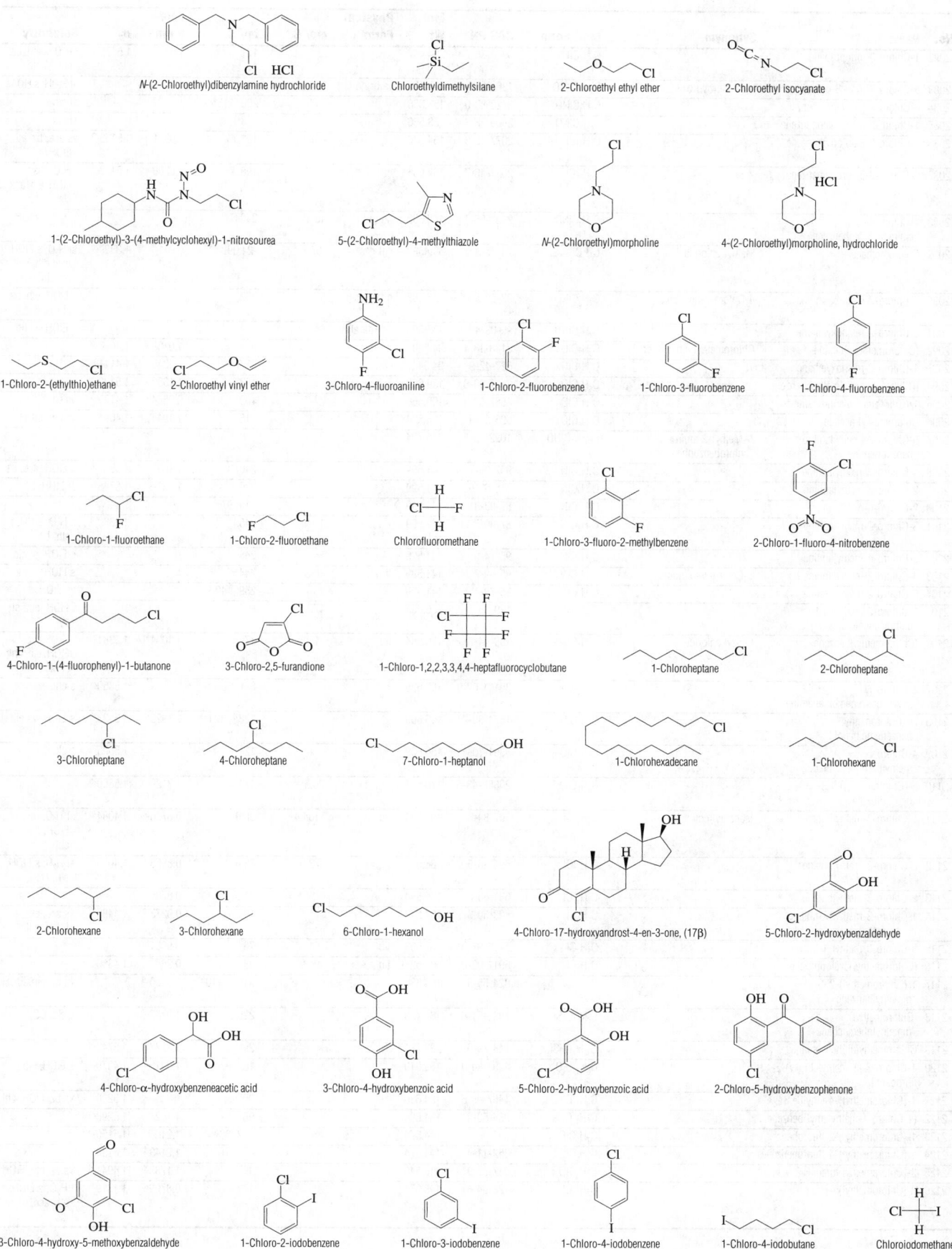

N-(2-Chloroethyl)dibenzylamine hydrochloride

Chloroethyldimethylsilane

2-Chloroethyl ethyl ether

2-Chloroethyl isocyanate

1-(2-Chloroethyl)-3-(4-methylcyclohexyl)-1-nitrosourea

5-(2-Chloroethyl)-4-methylthiazole

N-(2-Chloroethyl)morpholine

4-(2-Chloroethyl)morpholine, hydrochloride

1-Chloro-2-(ethylthio)ethane

2-Chloroethyl vinyl ether

3-Chloro-4-fluoroaniline

1-Chloro-2-fluorobenzene

1-Chloro-3-fluorobenzene

1-Chloro-4-fluorobenzene

1-Chloro-1-fluoroethane

1-Chloro-2-fluoroethane

Chlorofluoromethane

1-Chloro-3-fluoro-2-methylbenzene

2-Chloro-1-fluoro-4-nitrobenzene

4-Chloro-1-(4-fluorophenyl)-1-butanone

3-Chloro-2,5-furandione

1-Chloro-1,2,2,3,3,4,4-heptafluorocyclobutane

1-Chloroheptane

2-Chloroheptane

3-Chloroheptane

4-Chloroheptane

7-Chloro-1-heptanol

1-Chlorohexadecane

1-Chlorohexane

2-Chlorohexane

3-Chlorohexane

6-Chloro-1-hexanol

4-Chloro-17-hydroxyandrost-4-en-3-one, (17β)

5-Chloro-2-hydroxybenzaldehyde

4-Chloro-α-hydroxybenzeneacetic acid

3-Chloro-4-hydroxybenzoic acid

5-Chloro-2-hydroxybenzoic acid

2-Chloro-5-hydroxybenzophenone

3-Chloro-4-hydroxy-5-methoxybenzaldehyde

1-Chloro-2-iodobenzene

1-Chloro-3-iodobenzene

1-Chloro-4-iodobenzene

1-Chloro-4-iodobutane

Chloroiodomethane

No.	Name	Synonym	Mol. Form.	CAS RN	Mol. Wt.	Physical Form	mp/°C	bp/°C	den/ g cm⁻³	n_D	Solubility
2082	1-Chloro-3-iodopropane		C₃H₆ClI	6940-76-7	204.437			171	1.904²⁰	1.5472²⁰	i H₂O; s eth, bz, chl; sl ctc
2083	5-Chloro-7-iodo-8-quinolinol	Iodochlorhydroxyquin	C₉H₅ClINO	130-26-7	305.499	ye br nd (al)	178.5				sl EtOH; s HOAc
2084	1-Chloro-2-isocyanatobenzene		C₇H₄ClNO	3320-83-0	153.566		30.5	200; 115⁴³			sl ctc
2085	1-Chloro-3-isocyanatobenzene		C₇H₄ClNO	2909-38-8	153.566			113⁴³			sl chl
2086	1-Chloro-2-isopropylbenzene		C₉H₁₁Cl	2077-13-6	154.636	liq	-74.4	191.1	1.0341²⁰	1.5168²⁰	vs ace, bz, eth, EtOH
2087	1-Chloro-4-isopropylbenzene		C₉H₁₁Cl	2621-46-7	154.636	liq	-12.3	198.3	1.0208²⁰	1.5117²⁰	i H₂O; msc EtOH, eth, ace, ctc; vs bz
2088	1-Chloro-4-isothiocyanatobenzene		C₇H₄ClNS	2131-55-7	169.632	nd (al)	46	249.5			i H₂O; s EtOH
2089	Chloromethane	Methyl chloride	CH₃Cl	74-87-3	50.488	col gas	-97.7	-24.09	0.911²⁵ (p>1 atm)	1.3389²⁰	sl H₂O; s EtOH; msc eth, ace, bz, chl
2090	4-Chloro-2-methoxyaniline	4-Chloro-2-anisidine	C₇H₈ClNO	93-50-5	157.598	nd or pr (dil al)	52	260			s EtOH, eth, bz, chl
2091	5-Chloro-2-methoxyaniline		C₇H₈ClNO	95-03-4	157.598	nd (dil al)	84				s EtOH; sl lig
2092	(Chloromethoxy)ethane	Chloromethyl ethyl ether	C₃H₇ClO	3188-13-4	94.540			83	1.0188¹⁵	1.4040²⁰	
2093	1-Chloro-2-methoxyethane		C₃H₇ClO	627-42-9	94.540			92.5	1.0345²⁰	1.4111²⁰	vs H₂O, eth
2094	[(Chloromethoxy)methyl]benzene		C₈H₉ClO	3587-60-8	156.609			103¹³	1.1350²⁰	1.5192²⁰	
2095	1-(Chloromethoxy)propane		C₄H₉ClO	3587-57-3	108.566			109	0.9884²⁰	1.4125²⁰	vs eth, EtOH
2096	Chloromethyl acetate		C₃H₅ClO₂	625-56-9	108.524			116	1.194²⁰	1.409²⁰	vs eth, EtOH
2097	5-Chloro-2-(methylamino)benzophenone	N-Methyl-2-amino-5-chlorobenzophenone	C₁₄H₁₂ClNO	1022-13-5	245.704		92				
2098	4-Chloro-N-methylaniline		C₇H₈ClN	932-96-7	141.599			240	1.169¹¹	1.5835²⁰	s EtOH, ace, bz
2099	2-Chloro-4-methylaniline		C₇H₈ClN	615-65-6	141.599		7	220	1.151²⁰	1.5748²²	sl EtOH, bz
2100	2-Chloro-6-methylaniline		C₇H₈ClN	87-63-8	141.599			215; 97¹⁰			
2101	3-Chloro-2-methylaniline		C₇H₈ClN	87-60-5	141.599		1	245		1.5880²⁰	s H₂O, EtOH; i eth, bz
2102	3-Chloro-4-methylaniline		C₇H₈ClN	95-74-9	141.599		26	243			s EtOH; sl ctc
2103	4-Chloro-2-methylaniline	p-Chloro-o-toluidine	C₇H₈ClN	95-69-2	141.599	lf (al)	30.3	244			s EtOH; sl ctc
2104	5-Chloro-2-methylaniline		C₇H₈ClN	95-79-4	141.599		26	239; 140³⁸			vs EtOH
2105	1-Chloro-2-methyl-9,10-anthracenedione		C₁₅H₉ClO₂	129-35-1	256.684		170.5				i EtOH, eth; sl py
2106	(Chloromethyl)benzene	Benzyl chloride	C₇H₇Cl	100-44-7	126.584	liq	-45	179	1.1004²⁰	1.5391²⁰	i H₂O; msc EtOH, eth, chl; sl ctc
2107	3-Chloro-N-methylbenzenemethanamine		C₈H₁₀ClN	39191-07-6	155.625			88⁴		1.5350²⁵	s chl
2108	α-(Chloromethyl)benzenemethanol		C₈H₉ClO	1674-30-2	156.609			128¹⁷, 121¹¹	1.1926²⁰	1.5523²⁰	s EtOH; vs eth
2109	4-Chloro-α-methylbenzenemethanol		C₈H₉ClO	3391-10-4	156.609			121¹⁵		1.5505²⁰	s ctc
2110	5-(Chloromethyl)-1,3-benzodioxole		C₈H₇ClO₂	20850-43-5	170.594		20.5	134¹⁴	1.312²⁵	1.5660²⁰	
2111	1-Chloro-3-methylbutane	Isopentyl chloride	C₅H₁₁Cl	107-84-6	106.594	liq	-104.4	98.9	0.8750²⁰	1.4084²⁰	sl H₂O; msc EtOH, eth; vs chl
2112	2-Chloro-2-methylbutane		C₅H₁₁Cl	594-36-5	106.594	liq	-73.5	85.6	0.8653²⁰	1.4055²⁰	sl H₂O; s EtOH, eth, ctc
2113	2-Chloro-3-methylbutane		C₅H₁₁Cl	631-65-2	106.594			91.5	0.878²⁰		
2114	1-Chloro-3-methyl-2-butene		C₅H₉Cl	503-60-6	104.578			109	0.9273²⁰	1.4485²⁰	vs ace, eth, EtOH, chl
2115	3-Chloro-3-methyl-1-butyne		C₅H₇Cl	1111-97-3	102.563	liq	-61	76	0.9061²⁰		
2116	(Chloromethyl)cyclopropane		C₄H₇Cl	5911-08-0	90.552	liq	-90.9	88	0.98²⁵	1.4350²⁰	
2117	1-(Chloromethyl)-2,4-dimethylbenzene		C₉H₁₁Cl	824-55-5	154.636			215.5; 110²⁰	1.0580¹⁹		vs bz, eth, EtOH
2118	(Chloromethyl)dimethylphenylsilane		C₉H₁₃ClSi	1833-51-8	184.738			225	1.0240²⁵		s ctc, CS₂
2119	Chloromethyldiphenylsilane		C₁₃H₁₃ClSi	144-79-6	232.781			295	1.1277²⁰	1.5742²⁰	
2120	1-Chloro-3-(1-methylethoxy)-2-propanol		C₆H₁₃ClO₂	4288-84-0	152.619			182; 87²⁰	1.0910²⁰	1.4370²⁵	s EtOH, eth
2121	1-(Chloromethyl)-4-ethylbenzene		C₉H₁₁Cl	1467-05-6	154.636			95¹⁵		1.5290²⁰	vs bz, EtOH, chl
2122	(1-Chloro-1-methylethyl)benzene		C₉H₁₁Cl	934-53-2	154.636			98¹	1.192²⁵	1.5290²⁵	
2123	1-(Chloromethyl)-2-fluorobenzene		C₇H₆ClF	345-35-7	144.574			172; 86⁴⁰	1.216²⁵	1.5150²⁰	
2124	1-(Chloromethyl)-4-fluorobenzene		C₇H₆ClF	352-11-4	144.574			82²⁶, 76²⁰	1.2143²⁰	1.5130	
2125	2-(Chloromethyl)furan		C₅H₅ClO	617-88-9	116.546			49²⁶	1.1783²⁰	1.4941²⁰	vs bz, eth, EtOH
2126	3-(Chloromethyl)heptane		C₈H₁₇Cl	123-04-6	148.674			172	0.8769²⁰	1.4319²⁰	i H₂O; s EtOH, eth, ace, bz; sl ctc

1-Chloro-3-iodopropane

5-Chloro-7-iodo-8-quinolinol

1-Chloro-2-isocyanatobenzene

1-Chloro-3-isocyanatobenzene

1-Chloro-2-isopropylbenzene

1-Chloro-4-isopropylbenzene

1-Chloro-4-isothiocyanatobenzene

Chloromethane

4-Chloro-2-methoxyaniline

5-Chloro-2-methoxyaniline

(Chloromethoxy)ethane

1-Chloro-2-methoxyethane

[(Chloromethoxy)methyl]benzene

1-(Chloromethoxy)propane

Chloromethyl acetate

5-Chloro-2-(methylamino)benzophenone

4-Chloro-*N*-methylaniline

2-Chloro-4-methylaniline

2-Chloro-6-methylaniline

3-Chloro-2-methylaniline

3-Chloro-4-methylaniline

4-Chloro-2-methylaniline

5-Chloro-2-methylaniline

1-Chloro-2-methyl-9,10-anthracenedione

(Chloromethyl)benzene

3-Chloro-*N*-methylbenzenemethanamine

α-(Chloromethyl)benzenemethanol

4-Chloro-α-methylbenzenemethanol

5-(Chloromethyl)-1,3-benzodioxole

1-Chloro-3-methylbutane

2-Chloro-2-methylbutane

2-Chloro-3-methylbutane

1-Chloro-3-methyl-2-butene

3-Chloro-3-methyl-1-butyne

(Chloromethyl)cyclopropane

1-(Chloromethyl)-2,4-dimethylbenzene

(Chloromethyl)dimethylphenylsilane

Chloromethyldiphenylsilane

1-Chloro-3-(1-methylethoxy)-2-propanol

1-(Chloromethyl)-4-ethylbenzene

(1-Chloro-1-methylethyl)benzene

1-(Chloromethyl)-2-fluorobenzene

1-(Chloromethyl)-4-fluorobenzene

2-(Chloromethyl)furan

3-(Chloromethyl)heptane

No.	Name	Synonym	Mol. Form.	CAS RN	Mol. Wt.	Physical Form	mp/°C	bp/°C	den/ g cm⁻³	n_D	Solubility
2127	4-Chloro-5-methyl-2-isopropylphenol	Chlorothymol	C₁₀H₁₃ClO	89-68-9	184.662		63	258.5			vs H₂O; s EtOH, eth, bz, ctc, peth, alk
2128	1-(Chloromethyl)-4-methoxybenzene		C₈H₉ClO	824-94-2	156.609	nd	24.5	262.5	1.261²⁰	1.580²⁰	vs ace, bz, eth
2129	1-(Chloromethyl)-2-methylbenzene		C₈H₉Cl	552-45-4	140.610			198; 90²⁰	1.063²⁵	1.5410²⁵	vs eth, EtOH
2130	1-(Chloromethyl)-3-methylbenzene		C₈H₉Cl	620-19-9	140.610			195.5	1.064²⁰	1.5345²⁰	i H₂O; s EtOH, eth
2131	1-(Chloromethyl)-4-methylbenzene		C₈H₉Cl	104-82-5	140.610			201; 90²⁰	1.0512²⁰	1.5380	i H₂O; s EtOH; msc eth
2132	Chloromethyl methyl ether		C₂H₅ClO	107-30-2	80.513	liq	-103.5	59.5	1.063¹⁰	1.397²⁰	s EtOH, eth, ace, chl
2133	2-(Chloromethyl)-2-methyloxirane		C₄H₇ClO	598-09-4	106.551			122	1.1011²⁰	1.4310²⁰	vs H₂O, eth
2134	1-(Chloromethyl)naphthalene		C₁₁H₉Cl	86-52-2	176.642	pr	32	291.5	1.1813²⁰	1.6380²⁰	i H₂O; s EtOH, ctc, peth
2135	2-(Chloromethyl)naphthalene		C₁₁H₉Cl	2506-41-4	176.642	lf (al)	48.5	169²⁰			i H₂O; s EtOH, peth
2136	1-(Chloromethyl)-2-nitrobenzene		C₇H₆ClNO₂	612-23-7	171.582	cry (liq)	50.0	125⁴		1.5557⁶²	i H₂O; s EtOH, eth, HOAc; vs ace, bz
2137	1-(Chloromethyl)-3-nitrobenzene		C₇H₆ClNO₂	619-23-8	171.582	pa ye nd (liq)	46	173³⁴		1.5577⁶²	vs ace, bz, eth, EtOH
2138	1-(Chloromethyl)-4-nitrobenzene	4-Nitrobenzyl chloride	C₇H₆ClNO₂	100-14-1	171.582	pl or nd (al)	71			1.5647⁶²	i H₂O; s EtOH, eth; vs ace, bz, AcOEt
2139	1-Chloro-2-methyl-3-nitrobenzene		C₇H₆ClNO₂	83-42-1	171.582	nd (dil al)	37.8	238		1.5377⁶⁹	i H₂O; s EtOH
2140	1-Chloro-2-methyl-4-nitrobenzene		C₇H₆ClNO₂	13290-74-9	171.582	ye cry	42.5	249			vs eth
2141	1-Chloro-4-methyl-2-nitrobenzene	4-Chloro-3-nitrotoluene	C₇H₆ClNO₂	89-60-1	171.582		7	261; 118¹¹		1.5572⁶⁰	i H₂O; s ctc
2142	2-Chloro-1-methyl-4-nitrobenzene		C₇H₆ClNO₂	121-86-8	171.582	nd (al)	66.5	260		1.5470⁶⁹	sl H₂O, chl; s EtOH, eth, HOAc
2143	4-Chloro-1-methyl-2-nitrobenzene		C₇H₆ClNO₂	89-59-8	171.582	mcl nd	38	242; 115.5¹¹	1.2559⁸⁰		i H₂O; s EtOH, eth; sl chl
2144	2-Chloro-4-methylpentane		C₆H₁₃Cl	25346-32-1	120.620			113	0.8610²⁰	1.4113²⁰	vs eth
2145	3-(Chloromethyl)pentane		C₆H₁₃Cl	4737-41-1	120.620			126; 83²⁰²	0.8914²⁰	1.4222²⁰	vs bz, eth, chl
2146	2-Chloro-4-methylphenol	2-Chloro-p-cresol	C₇H₇ClO	6640-27-3	142.583			195.5	1.1785²⁷	1.5200²⁷	vs bz, eth, EtOH
2147	2-Chloro-5-methylphenol	6-Chloro-m-cresol	C₇H₇ClO	615-74-7	142.583	pr (peth)	45.5	196	1.215¹⁵		vs H₂O, EtOH
2148	2-Chloro-6-methylphenol	6-Chloro-o-cresol	C₇H₇ClO	87-64-9	142.583			189; 80²⁰		1.5449²⁰	sl H₂O; s eth
2149	3-Chloro-4-methylphenol	3-Chloro-p-cresol	C₇H₇ClO	615-62-3	142.583	nd (al)	55.5	228			vs bz, eth, EtOH
2150	4-Chloro-2-methylphenol	4-Chloro-o-cresol	C₇H₇ClO	1570-64-5	142.583	nd (peth)	51	223			sl H₂O; s peth
2151	4-Chloro-3-methylphenol	4-Chloro-m-cresol	C₇H₇ClO	59-50-7	142.583	nd (peth)	67	235			sl H₂O, chl; s EtOH, eth, peth
2152	(4-Chloro-2-methylphenoxy)acetic acid	MCPA	C₉H₉ClO₃	94-74-6	200.618	pl (bz, to)	120				sl H₂O; vs EtOH, eth; s bz, ctc
2153	4-(4-Chloro-2-methylphenoxy)butanoic acid		C₁₁H₁₃ClO₃	94-81-5	228.672		100				
2154	Chloromethylphenylsilane		C₇H₉ClSi	1631-82-9	156.685			113¹⁰⁰	1.043²⁰	1.5171²⁰	
2155	(Chloromethyl)phosphonic acid		CH₄ClO₃P	2565-58-4	130.468	nd (bz/MeNO₂)	90				
2156	N-Chloromethylphthalimide		C₉H₆ClNO₂	17564-64-6	195.603		135.5				
2157	2-Chloro-2-methylpropanal		C₄H₇ClO	917-93-1	106.551			90	1.053¹⁵	1.4160¹⁶	vs eth, EtOH
2158	1-Chloro-2-methylpropane	Isobutyl chloride	C₄H₉Cl	513-36-0	92.567	liq	-130.3	68.5	0.8773²⁰	1.3984²⁰	sl H₂O, ctc; s eth, ace, chl
2159	2-Chloro-2-methylpropane	tert-Butyl chloride	C₄H₉Cl	507-20-0	92.567	liq	-25.60	50.9	0.8420²⁰	1.3857²⁰	sl H₂O; msc EtOH, eth; s bz, ctc, chl
2160	1-Chloro-2-methylpropene	Dimethylvinyl chloride	C₄H₇Cl	513-37-1	90.552			68	0.9186²⁰	1.4221²⁰	sl H₂O; s chl
2161	3-Chloro-2-methylpropene		C₄H₇Cl	563-47-3	90.552			71.5	0.9165²⁰	1.4291²⁰	msc EtOH, eth; s ace; vs chl
2162	3-(Chloromethyl)pyridine, hydrochloride		C₆H₇Cl₂N	6959-48-4	164.033	hyg	143.8				
2163	Chloromethylsilane		CH₅ClSi	993-00-0	80.590	col gas	-135	7; -45⁶³			
2164	1-Chloro-4-(methylsulfonyl)benzene	4-Chlorobenzenethiol, S-methyl, S,S-dioxide	C₇H₇ClO₂S	98-57-7	190.648		98				
2165	1-Chloro-4-(methylthio)benzene		C₇H₇ClS	123-09-1	158.649			105¹⁰			
2166	1-Chloro-2-(methylthio)ethane		C₃H₇ClS	542-81-4	110.606			140; 60³⁰	1.123²⁰	1.4902²⁰	s EtOH, eth, ace
2167	Chloro(methylthio)methane		C₂H₅ClS	2373-51-5	96.579			105	1.153²⁵	1.4963²⁰	
2168	(Chloromethyl)trimethylsilane		C₄H₁₁ClSi	2344-80-1	122.669			98.5	0.879²⁵	1.4175²⁰	

4-Chloro-5-methyl-2-isopropylphenol

1-(Chloromethyl)-4-methoxybenzene

1-(Chloromethyl)-2-methylbenzene

1-(Chloromethyl)-3-methylbenzene

1-(Chloromethyl)-4-methylbenzene

Chloromethyl methyl ether

2-(Chloromethyl)-2-methyloxirane

1-(Chloromethyl)naphthalene

2-(Chloromethyl)naphthalene

1-(Chloromethyl)-2-nitrobenzene

1-(Chloromethyl)-3-nitrobenzene

1-(Chloromethyl)-4-nitrobenzene

1-Chloro-2-methyl-3-nitrobenzene

1-Chloro-2-methyl-4-nitrobenzene

1-Chloro-4-methyl-2-nitrobenzene

2-Chloro-1-methyl-4-nitrobenzene

4-Chloro-1-methyl-2-nitrobenzene

2-Chloro-4-methylpentane

3-(Chloromethyl)pentane

2-Chloro-4-methylphenol

2-Chloro-5-methylphenol

2-Chloro-6-methylphenol

3-Chloro-4-methylphenol

4-Chloro-2-methylphenol

4-Chloro-3-methylphenol

(4-Chloro-2-methylphenoxy)acetic acid

4-(4-Chloro-2-methylphenoxy)butanoic acid

Chloromethylphenylsilane

(Chloromethyl)phosphonic acid

N-Chloromethylphthalimide

2-Chloro-2-methylpropanal

1-Chloro-2-methylpropane

2-Chloro-2-methylpropane

1-Chloro-2-methylpropene

3-Chloro-2-methylpropene

3-(Chloromethyl)pyridine, hydrochloride

Chloromethylsilane

1-Chloro-4-(methylsulfonyl)benzene

1-Chloro-4-(methylthio)benzene

1-Chloro-2-(methylthio)ethane

Chloro(methylthio)methane

(Chloromethyl)trimethylsilane

No.	Name	Synonym	Mol. Form.	CAS RN	Mol. Wt.	Physical Form	mp/°C	bp/°C	den/ g cm⁻³	n_D	Solubility
2169	1-Chloronaphthalene	1-Naphthyl chloride	$C_{10}H_7Cl$	90-13-1	162.616	oily liq	-2.5	259; 106.5[5]	1.1880[25]	1.6326[20]	i H_2O; s EtOH, eth, bz, CS_2; sl ctc
2170	2-Chloronaphthalene		$C_{10}H_7Cl$	91-58-7	162.616	pl (dil al), lf	58.0	256	1.1377[71]	1.6079[13]	i H_2O; s EtOH, eth, bz, chl, CS_2
2171	4-Chloro-1-naphthol		$C_{10}H_7ClO$	604-44-4	178.615	nd (chl, aq al)	120.5				s EtOH, eth, ace, bz, chl
2172	Chloroneb	1,4-Dichloro-2,5-dimethoxybenzene	$C_8H_8Cl_2O_2$	2675-77-6	207.055		134	268			
2173	2-Chloro-4-nitroaniline		$C_6H_5ClN_2O_2$	121-87-9	172.569	ye nd (w)	108				vs eth, EtOH, HOAc
2174	2-Chloro-5-nitroaniline		$C_6H_5ClN_2O_2$	6283-25-6	172.569	ye nd (lig)	121				vs eth, EtOH, HOAc
2175	4-Chloro-2-nitroaniline		$C_6H_5ClN_2O_2$	89-63-4	172.569	dk oran-ye pr (dil al)	116.5				vs EtOH, eth, HOAc; sl ace, lig
2176	4-Chloro-3-nitroaniline		$C_6H_5ClN_2O_2$	635-22-3	172.569	ye nd or pr (w) nd (peth)	103				s H_2O, eth, chl; vs EtOH; sl lig
2177	5-Chloro-2-nitroaniline		$C_6H_5ClN_2O_2$	1635-61-6	172.569	ye nd (CS_2) ye lf (al, bz)	127.8	sub			vs eth, EtOH
2178	1-Chloro-5-nitro-9,10-anthracenedione		$C_{14}H_6ClNO_4$	129-40-8	287.656		315.3				i H_2O, EtOH, eth, lig; sl bz; s py
2179	2-Chloro-5-nitrobenzaldehyde		$C_7H_4ClNO_3$	6361-21-3	185.565	cry (al)	81.3				vs EtOH, chl
2180	4-Chloro-3-nitrobenzaldehyde		$C_7H_4ClNO_3$	16588-34-4	185.565		64.5				sl H_2O; s chl
2181	1-Chloro-2-nitrobenzene	o-Chloronitrobenzene	$C_6H_4ClNO_2$	88-73-3	157.555	mcl nd	32.1	245.5	1.368[242]		i H_2O; s EtOH, eth, bz; vs ace, tol, py
2182	1-Chloro-3-nitrobenzene	m-Chloronitrobenzene	$C_6H_4ClNO_2$	121-73-3	157.555	pa ye orth pr (al)	44.4	235.5	1.343[50]	1.5374[80]	i H_2O; s EtOH, eth, bz, chl, CS_2
2183	1-Chloro-4-nitrobenzene	p-Chloronitrobenzene	$C_6H_4ClNO_2$	100-00-5	157.555	mcl pr	82	242	1.2979[90]	1.5376[100]	i H_2O; sl EtOH; s eth, chl, CS_2
2184	5-Chloro-3-nitro-1,2-benzenediamine		$C_6H_6ClN_3O_2$	42389-30-0	187.584		167				
2185	4-Chloro-3-nitrobenzenesulfonamide		$C_6H_5ClN_2O_4S$	97-09-6	236.633	ye cry (EtOH)	175				
2186	4-Chloro-3-nitrobenzenesulfonyl chloride		$C_6H_3Cl_2NO_4S$	97-08-5	256.064		60.8				
2187	2-Chloro-4-nitrobenzoic acid		$C_7H_4ClNO_4$	99-60-5	201.565	nd (w)	141.8				s H_2O, EtOH, eth, bz
2188	2-Chloro-5-nitrobenzoic acid		$C_7H_4ClNO_4$	2516-96-3	201.565	nd or pr (w)	166.5		1.608[18]		sl H_2O, ace; s EtOH, eth, bz
2189	4-Chloro-3-nitrobenzoic acid		$C_7H_4ClNO_4$	96-99-1	201.565	nd or pl (w)	182.8		1.645[18]		i H_2O; sl EtOH, ace
2190	1-Chloro-1-nitroethane		$C_2H_4ClNO_2$	598-92-5	109.512			124.5	1.2837[20]	1.4224[20]	i H_2O; s EtOH, ctc, alk
2191	2-Chloro-4-nitrophenol		$C_6H_4ClNO_3$	619-08-9	173.554	wh nd (50% al)	111				s H_2O, EtOH, eth, chl; sl bz
2192	4-Chloro-2-nitrophenol		$C_6H_4ClNO_3$	89-64-5	173.554	ye mcl pr (al)	88.5				i H_2O; s EtOH, eth, chl; sl ace
2193	5-Chloro-2-nitrophenol		$C_6H_4ClNO_3$	611-07-4	173.554	ye pr or nd (w)	41	sub			sl H_2O; s EtOH, eth, HOAc
2194	1-Chloro-1-nitropropane		$C_3H_6ClNO_2$	600-25-9	123.539			142	1.207[20]	1.4251[20]	sl H_2O, chl; s EtOH, eth, oils
2195	2-Chloro-2-nitropropane		$C_3H_6ClNO_2$	594-71-8	123.539		-21.5	dec 134; 57[50]	1.2[20]	1.4378[19]	sl H_2O; s EtOH, eth, ctc, oils; i KOH
2196	2-Chloro-3-nitropyridine		$C_5H_3ClN_2O_2$	5470-18-8	158.543	nd (w)	104.0				
2197	1-Chloro-2-nitro-4-(trifluoromethyl)benzene		$C_7H_3ClF_3NO_2$	121-17-5	225.553	liq	-1.3	222; 95[10]	1.511[25]	1.4893[20]	
2198	1-Chloro-4-nitro-2-(trifluoromethyl)benzene		$C_7H_3ClF_3NO_2$	777-37-7	225.553		22	232	1.527[25]	1.5083[26]	
2199	1-Chlorononane		$C_9H_{19}Cl$	2473-01-0	162.700	liq	-39.4	205.2	0.8706[20]	1.4343[20]	i H_2O; s eth, chl
2200	9-Chloro-1-nonanol		$C_9H_{19}ClO$	51308-99-7	178.699		28	147[14]		1.4575[20]	vs eth, EtOH
2201	1-Chlorooctadecane		$C_{18}H_{37}Cl$	3386-33-2	288.940		28.6	352	0.8616[20]	1.4524[20]	i H_2O; sl ctc
2202	1-Chlorooctane	Octyl chloride	$C_8H_{17}Cl$	111-85-3	148.674	liq	-57.8	183.5	0.8734[20]	1.4309[20]	i H_2O; vs EtOH, eth; sl ctc
2203	2-Chlorooctane		$C_8H_{17}Cl$	628-61-5	148.674			172; 75[28]	0.8658[17]	1.4273[21]	i H_2O; vs EtOH, eth
2204	8-Chloro-1-octanol		$C_8H_{17}ClO$	23144-52-7	164.673			139[19]		1.4563[25]	vs eth, EtOH
2205	Chloropentafluoroacetone		C_3ClF_5O	79-53-8	182.476	col gas	-133	8			
2206	Chloropentafluorobenzene		C_6ClF_5	344-07-0	202.509			117.96	1.568[25]	1.4256[20]	

1-Chloronaphthalene

2-Chloronaphthalene

4-Chloro-1-naphthol

Chloroneb

2-Chloro-4-nitroaniline

2-Chloro-5-nitroaniline

4-Chloro-2-nitroaniline

4-Chloro-3-nitroaniline

5-Chloro-2-nitroaniline

1-Chloro-5-nitro-9,10-anthracenedione

2-Chloro-5-nitrobenzaldehyde

4-Chloro-3-nitrobenzaldehyde

1-Chloro-2-nitrobenzene

1-Chloro-3-nitrobenzene

1-Chloro-4-nitrobenzene

5-Chloro-3-nitro-1,2-benzenediamine

4-Chloro-3-nitrobenzenesulfonamide

4-Chloro-3-nitrobenzenesulfonyl chloride

2-Chloro-4-nitrobenzoic acid

2-Chloro-5-nitrobenzoic acid

4-Chloro-3-nitrobenzoic acid

1-Chloro-1-nitroethane

2-Chloro-4-nitrophenol

4-Chloro-2-nitrophenol

5-Chloro-2-nitrophenol

1-Chloro-1-nitropropane

2-Chloro-2-nitropropane

2-Chloro-3-nitropyridine

1-Chloro-2-nitro-4-(trifluoromethyl)benzene

1-Chloro-4-nitro-2-(trifluoromethyl)benzene

1-Chlorononane

9-Chloro-1-nonanol

1-Chlorooctadecane

1-Chlorooctane

2-Chlorooctane

8-Chloro-1-octanol

Chloropentafluoroacetone

Chloropentafluorobenzene

No.	Name	Synonym	Mol. Form.	CAS RN	Mol. Wt.	Physical Form	mp/°C	bp/°C	den/ g cm^{-3}	n_D	Solubility
2207	Chloropentafluoroethane	Refrigerant 115	C_2ClF_5	76-15-3	154.466	col gas	-99.4	-39.1	1.5678^{-42}	1.2678^{-42}	i H$_2$O; s EtOH, eth
2208	1-Chloropentane	Pentyl chloride	$C_5H_{11}Cl$	543-59-9	106.594	liq	-99.0	108.4	0.8820^{20}	1.4126^{20}	i H$_2$O; msc EtOH, eth; s bz, ctc; vs chl
2209	2-Chloropentane, (+)	sec-Pentyl chloride	$C_5H_{11}Cl$	29882-57-3	106.594	liq	-137	97.0	0.8698^{20}	1.4069^{20}	i H$_2$O; s EtOH, eth, bz; vs chl
2210	3-Chloropentane		$C_5H_{11}Cl$	616-20-6	106.594	liq	-105	97.5	0.8731^{20}	1.4082^{20}	i H$_2$O; s EtOH, eth, bz; sl ace
2211	5-Chloropentanoic acid		$C_5H_9ClO_2$	1119-46-6	136.577		18	230	1.3416^{25}	1.4555^{20}	vs eth, EtOH
2212	5-Chloro-1-pentanol		$C_5H_{11}ClO$	5259-98-3	122.593			112^{12}		1.4518^{20}	vs eth, EtOH
2213	5-Chloro-2-pentanone		C_5H_9ClO	5891-21-4	120.577			106^{110}, 76^{34}	1.0523^{20}	1.4375^{20}	s eth, ace; sl ctc
2214	1-Chloro-3-pentanone		C_5H_9ClO	32830-97-0	120.577			68^{20}		1.4361^{20}	vs eth, EtOH
2215	5-Chloropentanoyl chloride		$C_5H_8Cl_2O$	1575-61-7	155.022			83^{12}	1.210^{18}	1.4639^{20}	vs eth
2216	4-Chloro-2-pentene		C_5H_9Cl	1458-99-7	104.578			103; 47^{25}	0.8988^{20}	1.4322^{20}	vs ace, eth, chl
2217	2-Chlorophenol		C_6H_5ClO	95-57-8	128.556		9.4	174.9	1.2634^{20}	1.5524^{20}	sl H$_2$O, chl; s EtOH, eth; vs bz
2218	3-Chlorophenol		C_6H_5ClO	108-43-0	128.556		32.6	214	1.245^{45}	1.5565^{40}	sl H$_2$O, chl; s EtOH, eth; vs bz
2219	4-Chlorophenol		C_6H_5ClO	106-48-9	128.556		42.8	220	1.2651^{40}	1.5579^{40}	sl H$_2$O; vs EtOH, eth, bz; s alk
2220	Chlorophenol Red		$C_{19}H_{12}Cl_2O_5S$	4430-20-0	423.266	grn-br cry	261				sl H$_2$O; s EtOH
2221	2-Chloro-10H-phenothiazine		$C_{12}H_8ClNS$	92-39-7	233.717		198.5				
2222	2-Chlorophenoxyacetic acid		$C_8H_7ClO_3$	614-61-9	186.593	nd (w, al)	148.5				s H$_2$O, EtOH
2223	3-Chlorophenoxyacetic acid		$C_8H_7ClO_3$	588-32-9	186.593	cry (w)	110				i H$_2$O
2224	(4-Chlorophenoxy)acetic acid		$C_8H_7ClO_3$	122-88-3	186.593	pr or nd (w)	156.5				vs H$_2$O; sl chl
2225	1-Chloro-4-phenoxybenzene	4-Chlorophenyl phenyl ether	$C_{12}H_9ClO$	7005-72-3	204.651			284.5	1.2026^{15}	1.599	
2226	3-(4-Chlorophenoxy)-1,2-propanediol	Chlorphenesin	$C_9H_{11}ClO_3$	104-29-0	202.634	cry	78	214^{19}			i H$_2$O; vs EtOH, eth; s bz, con sulf
2227	2-(3-Chlorophenoxy)propanoic acid	Cloprop	$C_9H_9ClO_3$	101-10-0	200.618	cry	113	100$^{1.5}$			
2228	2-Chloro-N-phenylacetamide		C_8H_8ClNO	587-65-5	169.609	nd (dil HOAc)		sub			vs bz, eth, EtOH
2229	N-(2-Chlorophenyl)acetamide		C_8H_8ClNO	533-17-5	169.609		88.3				i H$_2$O; s EtOH, bz, chl; vs eth
2230	N-(3-Chlorophenyl)acetamide		C_8H_8ClNO	588-07-8	169.609	nd	79	333			sl H$_2$O; vs EtOH, eth, bz, CS$_2$; s chl
2231	N-(4-Chlorophenyl)acetamide		C_8H_8ClNO	539-03-7	169.609		179	333	1.385^{22}		i H$_2$O; s EtOH; vs eth; sl ctc
2232	4-Chloro-α-phenylbenzenemethanol		$C_{13}H_{11}ClO$	119-56-2	218.678		59				sl chl
2233	4-Chlorophenyl benzenesulfonate		$C_{12}H_9ClO_3S$	80-38-6	268.715	col cry	62		1.33		sl H$_2$O
2234	4-Chloro-1-phenyl-1-butanone		$C_{10}H_{11}ClO$	939-52-6	182.646		19.5	131^4	1.137^{25}	1.5459^{20}	
2235	4-Chlorophenyl 4-chlorobenzenesulfonate	Ovex	$C_{12}H_8Cl_2O_3S$	80-33-1	303.161		86.5				i H$_2$O; sl EtOH; s ace
2236	(2-Chlorophenyl)(4-chlorophenyl)methanone	2,4'-Dichlorodiphenyl ketone	$C_{13}H_8Cl_2O$	85-29-0	251.108	pr (al)	67	214^{22}	1.393^{14}		s EtOH; sl chl
2237	N'-(4-Chlorophenyl)-N,N-dimethylurea	Monuron	$C_9H_{11}ClN_2O$	150-68-5	198.648	wh pl (MeOH)	170.5				i H$_2$O; sl EtOH, ace
2238	1-(3-Chlorophenyl)ethanone	m-Chloroacetophenone	C_8H_7ClO	99-02-5	154.594			244; 129^{30}	1.2130^{40}	1.5494^{20}	s EtOH, eth, ace
2239	1-(4-Chlorophenyl)ethanone	p-Chloroacetophenone	C_8H_7ClO	99-91-2	154.594		20	232	1.1922^{20}	1.5550^{20}	i H$_2$O; msc EtOH, eth; s chl
2240	5-(4-Chlorophenyl)-6-ethyl-2,4-pyrimidinediamine	Pyrimethamine	$C_{12}H_{13}ClN_4$	58-14-0	248.711		233.5				
2241	2-(4-Chlorophenyl)-1H-indene-1,3(2H)-dione	Clorindione	$C_{15}H_9ClO_2$	1146-99-2	256.684	dk red nd (al)	145.5				vs bz, eth, EtOH
2242	4-Chlorophenyl isocyanate		C_7H_4ClNO	104-12-1	153.566		31.3	116^{45}			
2243	1-(2-Chlorophenyl)-2-methyl-2-propylamine	Clortermine	$C_{10}H_{14}ClN$	10389-73-8	183.678	liq		117^{16}			
2244	N-(2-Chlorophenyl)-3-oxobutanamide		$C_{10}H_{10}ClNO_2$	93-70-9	211.645		106.5				s EtOH; i eth, lig
2245	(4-Chlorophenyl)phenylmethanone		$C_{13}H_9ClO$	134-85-0	216.662	nd (al)	77.5	332			s EtOH, eth, ace; sl ctc
2246	3-(2-Chlorophenyl)propanoic acid		$C_9H_9ClO_2$	1643-28-3	184.619	nd or lf (w)	102				
2247	3-(3-Chlorophenyl)propanoic acid		$C_9H_9ClO_2$	21640-48-2	184.619	lf (peth)	77				
2248	3-(4-Chlorophenyl)propanoic acid		$C_9H_9ClO_2$	2019-34-3	184.619		126				
2249	3-Chloro-1-phenyl-1-propanone	2-Chloroethyl phenyl ketone	C_9H_9ClO	936-59-4	168.619	lf (eth), cry (al, peth)	49.5	113^4			

Chloropentafluoroethane

1-Chloropentane

2-Chloropentane, (+)

3-Chloropentane

5-Chloropentanoic acid

5-Chloro-1-pentanol

5-Chloro-2-pentanone

1-Chloro-3-pentanone

5-Chloropentanoyl chloride

4-Chloro-2-pentene

2-Chlorophenol

3-Chlorophenol

4-Chlorophenol

Chlorophenol Red

2-Chloro-10H-phenothiazine

2-Chlorophenoxyacetic acid

3-Chlorophenoxyacetic acid

(4-Chlorophenoxy)acetic acid

1-Chloro-4-phenoxybenzene

3-(4-Chlorophenoxy)-1,2-propanediol

2-(3-Chlorophenoxy)propanoic acid

2-Chloro-N-phenylacetamide

N-(2-Chlorophenyl)acetamide

N-(3-Chlorophenyl)acetamide

N-(4-Chlorophenyl)acetamide

4-Chloro-α-phenylbenzenemethanol

4-Chlorophenyl benzenesulfonate

4-Chloro-1-phenyl-1-butanone

4-Chlorophenyl 4-chlorobenzenesulfonate

(2-Chlorophenyl)(4-chlorophenyl)methanone

N'-(4-Chlorophenyl)-N,N-dimethylurea

1-(3-Chlorophenyl)ethanone

1-(4-Chlorophenyl)ethanone

5-(4-Chlorophenyl)-6-ethyl-2,4-pyrimidinediamine

2-(4-Chlorophenyl)-1H-indene-1,3(2H)-dione

4-Chlorophenyl isocyanate

1-(2-Chlorophenyl)-2-methyl-2-propylamine

N-(2-Chlorophenyl)-3-oxobutanamide

(4-Chlorophenyl)phenylmethanone

3-(2-Chlorophenyl)propanoic acid

3-(3-Chlorophenyl)propanoic acid

3-(4-Chlorophenyl)propanoic acid

3-Chloro-1-phenyl-1-propanone

No.	Name	Synonym	Mol. Form.	CAS RN	Mol. Wt.	Physical Form	mp/°C	bp/°C	den/ g cm^{-3}	n_D	Solubility
2250	1-(4-Chlorophenyl)-1-propanone		C$_9$H$_9$ClO	6285-05-8	168.619		37.3	135[31], 114[2]			i H$_2$O; s EtOH, CS$_2$; sl chl
2251	3-(3-Chlorophenyl)-2-propynoic acid		C$_9$H$_5$ClO$_2$	7396-28-3	180.588	cry (HOAc, bz-peth)	144.5				vs HOAc
2252	Chlorophenylsilane	Phenylchlorosilane	C$_6$H$_7$ClSi	4206-75-1	142.659			162.5	1.0683[20]	1.5340[20]	
2253	1-Chloro-4-(phenylsulfonyl) benzene	Sulphenone	C$_{12}$H$_9$ClO$_2$S	80-00-2	252.716		94				i H$_2$O; sl EtOH; s eth; vs ace, bz
2254	5-Chloro-1-phenyltetrazole		C$_7$H$_5$ClN$_4$	14210-25-4	180.595		123				
2255	(2-Chlorophenyl)thiourea		C$_7$H$_7$ClN$_2$S	5344-82-1	186.662	nd or pl	146				vs bz, EtOH
2256	α-Chlorophyll		C$_{55}$H$_{72}$MgN$_4$O$_5$	479-61-8	893.490	bl blk hex pl	152.3				i H$_2$O; vs EtOH, eth; s lig
2257	β-Chlorophyll		C$_{55}$H$_{70}$MgN$_4$O$_6$	519-62-0	907.473	bl-blk or grn pow	125				i H$_2$O; vs EtOH, eth, py; s MeOH
2258	Chloropropamide	4-Chloro-N-[(propylamino) carbonyl]benzenesulfonamide	C$_{10}$H$_{13}$ClN$_2$O$_3$S	94-20-2	276.739	cry (EtOH)	128				i H$_2$O; s EtOH; sl eth, bz
2259	2-Chloropropanal		C$_3$H$_5$ClO	683-50-1	92.524			86	1.182[15]	1.431[17]	vs bz, eth
2260	1-Chloropropane	Propyl chloride	C$_3$H$_7$Cl	540-54-5	78.541	liq	-122.9	46.5	0.8899[20]	1.3879[20]	sl H$_2$O, ctc; msc EtOH, eth; s bz, chl
2261	2-Chloropropane	Isopropyl chloride	C$_3$H$_7$Cl	75-29-6	78.541	liq	-117.18	35.7	0.8617[20]	1.3777[20]	sl H$_2$O; msc EtOH, eth; s bz, ctc, chl
2262	3-Chloro-1,2-propanediol	α-Chlorohydrin	C$_3$H$_7$ClO$_2$	96-24-2	110.540	ye liq		dec 213; 116[11]	1.325[18]	1.4809[20]	s H$_2$O, EtOH, eth
2263	2-Chloro-1,3-propanediol	Glycerol β-chlorohydrin	C$_3$H$_7$ClO$_2$	497-04-1	110.540			146[18], 124[14]	1.3219[20]	1.4831[20]	vs H$_2$O, ace, EtOH
2264	3-Chloro-1,2-propanediol dinitrate	Clonitrate	C$_3$H$_5$ClN$_2$O$_6$	2612-33-1	200.534	sl ye liq		192.5	1.5112[9]		vs ace, EtOH, chl
2265	3-Chloropropanenitrile	β-Chloropropionitrile	C$_3$H$_4$ClN	542-76-7	89.524	liq	-51	175.5	1.1573[20]	1.4360[20]	sl ctc
2266	2-Chloropropanoic acid	2-Chloropropionic acid	C$_3$H$_5$ClO$_2$	598-78-7	108.524			185	1.2585[20]	1.4380[20]	msc H$_2$O, EtOH, eth; s ace
2267	3-Chloropropanoic acid	β-Chloropropionic acid	C$_3$H$_5$ClO$_2$	107-94-8	108.524	lf (w), hyg cry (lig)	41	dec 204			s H$_2$O, EtOH, chl; msc eth
2268	2-Chloro-1-propanol	Propylene chlorohydrin	C$_3$H$_7$ClO	78-89-7	94.540			133.5	1.103[20]	1.4390[20]	vs H$_2$O, eth, EtOH
2269	3-Chloro-1-propanol		C$_3$H$_7$ClO	627-30-5	94.540			165	1.1309[20]	1.4459[20]	vs H$_2$O; s EtOH, eth; sl ctc
2270	1-Chloro-2-propanol	sec-Propylene chlorohydrin	C$_3$H$_7$ClO	127-00-4	94.540			127	1.113[20]	1.4392[20]	msc H$_2$O, EtOH, eth; sl ctc
2271	3-Chloropropanoyl chloride		C$_3$H$_4$Cl$_2$O	625-36-5	126.969			144	1.3307[13]	1.4549[20]	sl H$_2$O; vs EtOH, eth, chl
2272	cis-1-Chloropropene		C$_3$H$_5$Cl	16136-84-8	76.525	liq	-134.8	32.8	0.9347[20]	1.4055[20]	i H$_2$O; s eth, ace, bz, chl
2273	trans-1-Chloropropene		C$_3$H$_5$Cl	16136-85-9	76.525	liq	-99	37.4	0.9349[20]	1.4054[20]	i H$_2$O; s eth, ace, bz, chl
2274	2-Chloropropene	Isopropenyl chloride	C$_3$H$_5$Cl	557-98-2	76.525	vol liq or gas	-137.4	22.6	0.9017[20]	1.3973[20]	i H$_2$O; s eth, ace, bz, chl
2275	3-Chloropropene	Allyl chloride	C$_3$H$_5$Cl	107-05-1	76.525	liq	-134.5	45.1	0.9376[20]	1.4157[20]	i H$_2$O; msc EtOH, eth, ace, bz, lig; sl ctc
2276	2-Chloro-2-propenenitrile		C$_3$H$_2$ClN	920-37-6	87.508	liq	-65	88.5	1.096[25]	1.4290[20]	
2277	2-Chloropropenoic acid	2-Chloroacrylic acid	C$_3$H$_3$ClO$_2$	598-79-8	106.508		66	sub			
2278	trans-(3-Chloro-1-propenyl) benzene		C$_9$H$_9$Cl	21087-29-6	152.620		8.5	106[13]	1.0926[20]	1.5851[20]	vs ace, bz, eth, EtOH
2279	Chloropropham		C$_{10}$H$_{12}$ClNO$_2$	101-21-3	213.661		41	149[2]	1.18[30]	1.5388[20]	
2280	Chloropropylate		C$_{17}$H$_{16}$Cl$_2$O$_3$	5836-10-2	339.213	pow	73				sl H$_2$O; s os
2281	(3-Chloropropyl)benzene		C$_9$H$_{11}$Cl	104-52-9	154.636			219.5	1.056[21]	1.5160[25]	sl ctc
2282	3-Chloropropyl chloroformate		C$_4$H$_6$Cl$_2$O$_2$	628-11-5	156.996			177	1.2926[25]	1.4456[20]	i H$_2$O
2283	(3-Chloropropyl)trimethoxysilane		C$_6$H$_{15}$ClO$_3$Si	2530-87-2	198.720			91	1.077[25]	1.4183[25]	
2284	(3-Chloropropyl)trimethylsilane		C$_6$H$_{15}$ClSi	2344-83-4	150.722			151	0.8789[20]	1.4319[20]	
2285	3-Chloro-1-propyne	Propargyl chloride	C$_3$H$_3$Cl	624-65-7	74.509		-78	58	1.030[25]	1.4349[20]	i H$_2$O; msc EtOH, eth, bz; s ctc
2286	6-Chloro-1H-purine	6-Chloropurine	C$_5$H$_3$ClN$_4$	87-42-3	154.558	nd (w)	176 dec				
2287	6-Chloro-3-pyridazinamine		C$_4$H$_4$ClN$_3$	5469-69-2	129.548		220				
2288	5-Chloro-2-pyridinamine		C$_5$H$_5$ClN$_2$	1072-98-6	128.560	pl	137	127[11]			s H$_2$O, EtOH; sl DMSO; i peth, lig
2289	2-Chloropyridine		C$_5$H$_4$ClN	109-09-1	113.546	oil		170	1.205[15]	1.5320[20]	sl H$_2$O; s EtOH, eth
2290	3-Chloropyridine		C$_5$H$_4$ClN	626-60-8	113.546			148; 86[100]		1.5304[20]	sl H$_2$O
2291	4-Chloropyridine		C$_5$H$_4$ClN	626-61-9	113.546	liq	-43.5	147.5	1.2000[25]		s H$_2$O; msc EtOH

1-(4-Chlorophenyl)-1-propanone

3-(3-Chlorophenyl)-2-propynoic acid

Chlorophenylsilane

1-Chloro-4-(phenylsulfonyl)benzene

5-Chloro-1-phenyltetrazole

(2-Chlorophenyl)thiourea

α-Chlorophyll

β-Chlorophyll

Chloropropamide

2-Chloropropanal

1-Chloropropane

2-Chloropropane

3-Chloro-1,2-propanediol

2-Chloro-1,3-propanediol

3-Chloro-1,2-propanediol dinitrate

3-Chloropropanenitrile

2-Chloropropanoic acid

3-Chloropropanoic acid

2-Chloro-1-propanol

3-Chloro-1-propanol

1-Chloro-2-propanol

3-Chloropropanoyl chloride

cis-1-Chloropropene

trans-1-Chloropropene

2-Chloropropene

3-Chloropropene

2-Chloro-2-propenenitrile

2-Chloropropenoic acid

trans-(3-Chloro-1-propenyl)benzene

Chloropropham

Chloropropylate

(3-Chloropropyl)benzene

3-Chloropropyl chloroformate

(3-Chloropropyl)trimethoxysilane

(3-Chloropropyl)trimethylsilane

3-Chloro-1-propyne

6-Chloro-1H-purine

6-Chloro-3-pyridazinamine

5-Chloro-2-pyridinamine

2-Chloropyridine

3-Chloropyridine

4-Chloropyridine

No.	Name	Synonym	Mol. Form.	CAS RN	Mol. Wt.	Physical Form	mp/°C	bp/°C	den/ g cm⁻³	n_D	Solubility
2292	2-Chloro-3-pyridinecarboxylic acid		C₆H₄ClNO₂	2942-59-8	157.555		>175 dec				
2293	6-Chloro-3-pyridinecarboxylic acid		C₆H₄ClNO₂	5326-23-8	157.555		198 dec				
2294	4-Chloropyridine, hydrochloride		C₅H₅Cl₂N	7379-35-3	150.006			sub 210			
2295	Chloroquine		C₁₈H₂₆ClN₃	54-05-7	319.872		90				
2296	2-Chloroquinoline		C₉H₆ClN	612-62-4	163.604	nd (aq al)	38	266; 153²²	1.2464²⁵	1.6342²⁵	i H₂O; vs EtOH, eth; s bz, chl
2297	4-Chloroquinoline		C₉H₆ClN	611-35-8	163.604	cry	34.5	262; 130¹⁵	1.251²⁵		sl H₂O; vs EtOH, eth; s dil HCl
2298	6-Chloroquinoline		C₉H₆ClN	612-57-7	163.604	pr (eth), nd (al)	43.8	263		1.6110⁵⁶	
2299	8-Chloroquinoline		C₉H₆ClN	611-33-6	163.604	liq	-20	288.5	1.2834¹⁴	1.6408¹⁴	s H₂O; vs EtOH, eth, ace, bz, chl
2300	5-Chloro-8-quinolinol	Cloxyquin	C₉H₆ClNO	130-16-5	179.603	cry (al)	130				
2301	2-Chlorostyrene		C₈H₇Cl	2039-87-4	138.595	liq	-63.1	188.7	1.1000²⁰	1.5649²⁰	s EtOH, eth, ace, ctc, HOAc; msc peth
2302	3-Chlorostyrene		C₈H₇Cl	2039-85-2	138.595			63⁶	1.1033²⁰	1.5625²⁰	i H₂O; s EtOH, eth
2303	4-Chlorostyrene		C₈H₇Cl	1073-67-2	138.595		15.9	192	1.0868²⁰	1.5660²⁰	i H₂O; s EtOH, eth; msc ace, bz, ctc
2304	N-Chlorosuccinimide		C₄H₄ClNO₂	128-09-6	133.534	pl (CCl₄)	150		1.65²⁵		sl H₂O, EtOH, bz, lig; s ace, HOAc
2305	1-Chlorotetradecane		C₁₄H₂₉Cl	2425-54-9	232.833		4.9	296.8	0.8654²⁰	1.4474²⁰	i H₂O; s EtOH, chl; vs ace, bz; sl ctc
2306	6-Chloro-N,N,N',N'-tetraethyl-1,3,5-triazine-2,4-diamine		C₁₁H₂₀ClN₅	580-48-3	257.764	oily liq	27	155⁹	1.0956²⁰	1.5320²⁰	vs bz, chl, EtOH, lig
2307	1-Chloro-1,1,2,2-tetrafluoroethane		C₂HClF₄	354-25-6	136.476	col gas	-117	-11.7			
2308	1-Chloro-1,2,2,2-tetrafluoroethane		C₂HClF₄	2837-89-0	136.476	col gas		-12			
2309	Chlorothalonil		C₈Cl₄N₂	1897-45-6	265.911		250	350	1.7²⁵		i H₂O; sl ace, cyhex
2310	Chlorothen	Chloromethapyrilene	C₁₄H₁₈ClN₃S	148-65-2	295.831			155¹⁰, 192⁵	1.1751²⁵		
2311	Chlorothiazide		C₇H₆ClN₃O₄S₂	58-94-6	295.724		350 dec				
2312	2-Chlorothiophene	2-Thienyl chloride	C₄H₃ClS	96-43-5	118.585	liq	-71.9	128.3	1.2863²⁰	1.5487²⁰	i H₂O; msc EtOH, eth; sl chl
2313	5-Chloro-2-thiophenecarboxaldehyde		C₅H₃ClOS	7283-96-7	146.595			77.5⁵		1.6036²⁵	sl chl
2314	2-Chloro-9H-thioxanthen-9-one		C₁₃H₇ClOS	86-39-5	246.712		153.5				
2315	2-Chlorotoluene		C₇H₇Cl	95-49-8	126.584	liq	-35.8	159.0	1.0825²⁰	1.5268²⁰	i H₂O; s EtOH, bz; msc eth, ace, chl
2316	3-Chlorotoluene		C₇H₇Cl	108-41-8	126.584	liq	-47.8	161.8	1.075²⁰	1.5214¹⁹	i H₂O; s EtOH, bz, ctc, chl; msc eth
2317	4-Chlorotoluene		C₇H₇Cl	106-43-4	126.584		7.5	162.4	1.0697²⁰	1.5150²⁰	i H₂O; s EtOH, ctc, chl; msc eth
2318	6-Chloro-1,3,5-triazine-2,4-diamine		C₃H₄ClN₅	3397-62-4	145.551		>330				
2319	1-Chloro-2-(trichloromethyl) benzene		C₇H₄Cl₄	2136-89-2	229.919		29.4	264.3	1.5187²⁰	1.5836²⁰	i H₂O; s eth, ace; sl ctc
2320	1-Chloro-4-(trichloromethyl) benzene		C₇H₄Cl₄	5216-25-1	229.919			245	1.4463²⁰		vs ace, eth
2321	Chlorotriethoxysilane		C₆H₁₅ClO₃Si	4667-99-6	198.720	liq	-51	156	1.030²⁰	1.3999²⁰	vs EtOH
2322	Chlorotriethylplumbane	Lead triethyl chloride	C₆H₁₅ClPb	1067-14-7	329.8		123 dec				s H₂O
2323	Chlorotriethylsilane		C₆H₁₅ClSi	994-30-9	150.722			144.5	0.8967²⁰	1.4314²⁰	
2324	1-Chloro-1,1,2-trifluoroethane		C₂H₂ClF₃	421-04-5	118.485	vol liq or gas		12			
2325	1-Chloro-1,2,2-trifluoroethane		C₂H₂ClF₃	431-07-2	118.485	vol liq or gas		17.3			
2326	2-Chloro-1,1,1-trifluoroethane		C₂H₂ClF₃	75-88-7	118.485	col gas	-105.5	6.1	1.389⁰	1.3090⁰	
2327	Chlorotrifluoroethene	Chlorotrifluoroethylene	C₂ClF₃	79-38-9	116.469	col gas	-158.2	-27.8	1.54⁻⁶⁰	1.38⁰	s bz, chl
2328	Chlorotrifluoromethane	Refrigerant 13	CClF₃	75-72-9	104.459	col gas	-181	-81.4			i H₂O
2329	2-Chloro-5-(trifluoromethyl) aniline		C₇H₅ClF₃N	121-50-6	195.570			103²⁵	1.428²⁵	1.4975²⁰	
2330	4-Chloro-3-(trifluoromethyl) aniline		C₇H₅ClF₃N	320-51-4	195.570		36.5	132²⁷			
2331	1-Chloro-2-(trifluoromethyl) benzene	o-Chlorobenzotrifluoride	C₇H₄ClF₃	88-16-4	180.555	liq	-6	152.2	1.2540³⁰	1.4513²⁵	s chl
2332	1-Chloro-3-(trifluoromethyl) benzene	m-Chlorobenzotrifluoride	C₇H₄ClF₃	98-15-7	180.555	liq	-56	137.5	1.3311²⁵	1.4438²⁵	

2-Chloro-3-pyridinecarboxylic acid

6-Chloro-3-pyridinecarboxylic acid

4-Chloropyridine, hydrochloride

Chloroquine

2-Chloroquinoline

4-Chloroquinoline

6-Chloroquinoline

8-Chloroquinoline

5-Chloro-8-quinolinol

2-Chlorostyrene

3-Chlorostyrene

4-Chlorostyrene

N-Chlorosuccinimide

1-Chlorotetradecane

6-Chloro-N,N,N',N'-tetraethyl-1,3,5-triazine-2,4-diamine

1-Chloro-1,1,2,2-tetrafluoroethane

1-Chloro-1,2,2,2-tetrafluoroethane

Chlorothalonil

Chlorothen

Chlorothiazide

2-Chlorothiophene

5-Chloro-2-thiophenecarboxaldehyde

2-Chloro-9H-thioxanthen-9-one

2-Chlorotoluene

3-Chlorotoluene

4-Chlorotoluene

6-Chloro-1,3,5-triazine-2,4-diamine

1-Chloro-2-(trichloromethyl)benzene

1-Chloro-4-(trichloromethyl)benzene

Chlorotriethoxysilane

Chlorotriethylplumbane

Chlorotriethylsilane

1-Chloro-1,1,2-trifluoroethane

1-Chloro-1,2,2-trifluoroethane

2-Chloro-1,1,1-trifluoroethane

Chlorotrifluoroethene

Chlorotrifluoromethane

2-Chloro-5-(trifluoromethyl)aniline

4-Chloro-3-(trifluoromethyl)aniline

1-Chloro-2-(trifluoromethyl)benzene

1-Chloro-3-(trifluoromethyl)benzene

No.	Name	Synonym	Mol. Form.	CAS RN	Mol. Wt.	Physical Form	mp/°C	bp/°C	den/g cm⁻³	n_D	Solubility
2333	1-Chloro-4-(trifluoromethyl)benzene	p-Chlorobenzotrifluoride	$C_7H_4ClF_3$	98-56-6	180.555	liq	-33	138.5	1.3340[25]	1.4431[30]	
2334	3-Chloro-1,1,1-trifluoropropane		$C_3H_4ClF_3$	460-35-5	132.512	liq	-106.5	45.1	1.3253[20]	1.3350[20]	i H_2O
2335	2-Chloro-2,4,4-trimethylpentane		$C_8H_{17}Cl$	6111-88-2	148.674		-26	dec 147; 44[16]	0.8746[20]	1.4308[20]	vs EtOH
2336	Chlorotrimethylstannane		C_3H_9ClSn	1066-45-1	199.266		38.5	148			s H_2O, chl, os
2337	2-Chloro-1,3,5-trinitrobenzene	Picryl chloride	$C_6H_2ClN_3O_6$	88-88-0	247.549	wh nd or pl (chl, al-lig)	83		1.797[20]		i H_2O; s EtOH, bz; sl eth; vs ace, tol
2338	Chlorotrinitromethane		$CClN_3O_6$	1943-16-4	185.480		2.3	dec 134; 56[40]	1.6769[20]	1.4500[20]	vs eth, EtOH, chl
2339	Chlorotriphenylmethane		$C_{19}H_{15}Cl$	76-83-5	278.775	nd or pr (bz-peth)	113.5	310			i H_2O; sl EtOH; vs eth, bz, chl; s ace
2340	Chlorotriphenylsilane		$C_{18}H_{15}ClSi$	76-86-8	294.851			241[35]			s chl
2341	Chlorotriphenylstannane	Triphenyltin chloride	$C_{18}H_{15}ClSn$	639-58-7	385.475		103.5				
2342	Chlorotripropylstannane		$C_9H_{21}ClSn$	2279-76-7	283.426		-23.5	123[13]	1.2678[28]	1.4910[28]	s ctc, os
2343	Chlorovinyldimethylsilane		C_4H_9ClSi	1719-58-0	120.653			83.5	0.8744[20]	1.4141[20]	
2344	Chloroxuron	N'-[4-(4-Chlorophenoxy)phenyl]-N,N-dimethylurea	$C_{15}H_{15}ClN_2O_2$	1982-47-4	290.745		151				
2345	Chlorozotocin		$C_9H_{16}ClN_3O_7$	54749-90-5	313.692	cry	147 dec				s H_2O
2346	Chlorphenesin carbamate		$C_{10}H_{12}ClNO_4$	886-74-8	245.660	cry (bz)	90				vs ace, EtOH, diox
2347	Chlorpheniramine		$C_{16}H_{19}ClN_2$	132-22-9	274.788	oily liq		142[1]			
2348	Chlorpheniramine maleate	Chlorprophenpyridamine	$C_{20}H_{23}ClN_2O_4$	113-92-8	390.861		132.5				
2349	Chlorphentermine	2-(4-Chlorobenzyl)-2-propylamine	$C_{10}H_{14}ClN$	461-78-9	183.678	liq		231; 101[2]			
2350	Chlorpromazine	2-Chloro-N,N-dimethyl-10H-phenothiazine-10-propanamine	$C_{17}H_{19}ClN_2S$	50-53-3	318.864			202[0.8]			i H_2O; vs EtOH, eth, bz, chl; s dil HCl
2351	Chlorprothixene		$C_{18}H_{18}ClNS$	113-59-7	315.861	pale ye cry	97				i H_2O, EtOH, eth, chl
2352	Chlorpyrifos		$C_9H_{11}Cl_3NO_3PS$	2921-88-2	350.586		42				
2353	Chlorpyrifos-methyl		$C_7H_7Cl_3NO_3PS$	5598-13-0	322.534		43				
2354	Chlorsulfuron		$C_{12}H_{12}ClN_5O_4S$	64902-72-3	357.773		176				
2355	Chlortetracycline		$C_{22}H_{23}ClN_2O_8$	57-62-5	478.879	gold-ye	168.5				i H_2O, eth; sl EtOH, ace, bz; s diox
2356	Chlorthalidone		$C_{14}H_{11}ClN_2O_4S$	77-36-1	338.765	wh pow or cry	225 dec				s alk, EtOH; sl eth
2357	Chlorthion		$C_8H_9ClNO_5PS$	500-28-7	297.653	ye cry	21	125[0.1]	1.437[20]	1.5661[20]	i H_2O; vs bz, eth, EtOH
2358	Chlorthiophos		$C_{11}H_{15}Cl_2O_3PS_2$	21923-23-9	361.245			150[0.001]			
2359	Chlortoluron	N'-(3-Chloro-4-methylphenyl)-N,N-dimethylurea	$C_{10}H_{13}ClN_2O$	15545-48-9	212.675	cry	147				sl H_2O; s os
2360	Cholane		$C_{24}H_{42}$	548-98-1	330.590	pr (al)	90	190[0.001]			
2361	Cholan-24-oic acid	Cholanic acid	$C_{24}H_{40}O_2$	25312-65-6	360.574	nd (al), cry (HOAc)	163.5				s EtOH, chl, HOAc
2362	Cholesta-3,5-diene		$C_{27}H_{44}$	747-90-0	368.638	wh nd (al)	80	260[13]	0.925[100]		i H_2O; s EtOH; msc eth, bz, chl; vs lig
2363	Cholesta-5,7-dien-3-ol, (3β)	7-Dehydrocholesterol	$C_{27}H_{44}O$	434-16-2	384.637	pl (+1w), (eth-MeOH)	150.5				i H_2O; sl EtOH; s eth, ace
2364	Cholesta-8,24-dien-3-ol, (3β,5α)		$C_{27}H_{44}O$	128-33-6	384.637	pl (MeOH), nd	110	160[0.001]			s ace, chl, MeOH
2365	Cholestane, (5α)	28,29,30-Trinorlanostane	$C_{27}H_{48}$	481-21-0	372.670	sc or pl (eth-al, ace)	80	250[1]	0.9090[88]	1.4887[88]	i H_2O; sl EtOH; vs eth, bz, chl
2366	Cholestane, (5β)	Coprostane	$C_{27}H_{48}$	481-20-9	372.670	orth nd (al, ace)	72		0.9119[87]	1.4884[88]	vs eth, chl
2367	Cholestanol	Dihydrocholesterol	$C_{27}H_{48}O$	80-97-7	388.669	sc (al,+1w)	141.5				vs eth, chl
2368	Cholestan-3-ol, (3α,5α)	Epicholestanol	$C_{27}H_{48}O$	516-95-0	388.669	nd (al)	185.5				s chl
2369	Cholest-4-en-3-ol, (3β)	Allocholesterol	$C_{27}H_{46}O$	517-10-2	386.653	nd (eth-MeOH)	132				i H_2O; s EtOH; vs eth, ace, bz, chl
2370	Cholest-5-en-3-ol, (3α)	Epicholesterol	$C_{27}H_{46}O$	474-77-1	386.653	cry (al, chl-MeOH)	141.5				sl EtOH
2371	Cholest-5-en-3-ol (3β), acetate		$C_{29}H_{48}O_2$	604-35-3	428.690	wh nd (ace, al)	115.5				vs bz, eth, chl
2372	Cholest-5-en-3-ol (3β), benzoate		$C_{34}H_{50}O_2$	604-32-0	490.760	wh nd	151.3		0.9413[200]		i EtOH; s eth, chl
2373	Cholest-5-en-3-ol (3β)-, hexadecanoate		$C_{43}H_{76}O_2$	601-34-3	625.062	wh nd (eth al)	79.3				vs bz, chl
2374	Cholest-5-en-3-ol (3β)-, cis-9-octadecenoate		$C_{45}H_{78}O_2$	303-43-5	651.100		46.3				s chl

1-Chloro-4-(trifluoromethyl)benzene

3-Chloro-1,1,1-trifluoropropane

2-Chloro-2,4,4-trimethylpentane

Chlorotrimethylstannane

2-Chloro-1,3,5-trinitrobenzene

Chlorotrinitromethane

Chlorotriphenylmethane

Chlorotriphenylsilane

Chlorotriphenylstannane

Chlorotripropylstannane

Chlorovinyldimethylsilane

Chloroxuron

Chlorozotocin

Chlorphenesin carbamate

Chlorpheniramine

Chlorpheniramine maleate

Chlorphentermine

Chlorpromazine

Chlorprothixene

Chlorpyrifos

Chlorpyrifos-methyl

Chlorsulfuron

Chlortetracycline

Chlorthalidone

Chlorthion

Chlorthiophos

Chlortoluron

Cholane

Cholan-24-oic acid

Cholesta-3,5-diene

Cholesta-5,7-dien-3-ol, (3β)

Cholesta-8,24-dien-3-ol, (3β,5α)

Cholestane, (5α)

Cholestane, (5β)

Cholestanol

Cholestan-3-ol, (3α,5α)

Cholest-4-en-3-ol, (3β)

Cholest-5-en-3-ol, (3α)

Cholest-5-en-3-ol (3β), acetate

Cholest-5-en-3-ol (3β), benzoate

Cholest-5-en-3-ol (3β)-, hexadecanoate

Cholest-5-en-3-ol (3β)-, cis-9-octadecenoate

No.	Name	Synonym	Mol. Form.	CAS RN	Mol. Wt.	Physical Form	mp/°C	bp/°C	den/ g cm⁻³	n_D	Solubility
2375	Cholest-4-en-3-one		$C_{27}H_{44}O$	601-57-0	384.637	nd or pl (al)	81.5	245[0.03]			
2376	Cholesterol		$C_{27}H_{46}O$	57-88-5	386.653	orth or tcl lf (al) nd (eth)	148.5	dec 360; 233[0.5]	1.067[20]		i H_2O; sl EtOH, ace; s bz, HOAc; vs diox
2377	Cholic acid	3,7,12-Trihydroxycholan-24-oic acid, (3α,5β,7α,12α)	$C_{24}H_{40}O_5$	81-25-4	408.572		198				sl H_2O; s EtOH, ace, alk; vs eth, chl
2378	Choline chloride		$C_5H_{14}ClNO$	67-48-1	139.624	hyg cry	305 dec				vs H_2O, EtOH
2379	Choline chloride dihydrogen phosphate	Phosphorylcholine	$C_5H_{15}ClNO_4P$	107-73-3	219.605	visc liq					
2380	Chorismic acid		$C_{10}H_{10}O_6$	617-12-9	226.182	cry	148				s H_2O
2381	Chromium carbonyl		C_6CrO_6	13007-92-6	220.056	col orth cry	dec 130	sub	1.77		i H_2O, EtOH; s eth, chl
2382	Chromium(II) oxalate		C_2CrO_4	814-90-4	140.015	ye-grn pow (hyd)					i H_2O, EtOH; s dil acid
2383	Chromium(III) 2,4-pentanedioate	Chromium acetylacetonate	$C_{15}H_{21}CrO_6$	21679-31-2	349.320	red mcl cry	208	345	1.34		i H_2O; s bz
2384	Chromotrope 2B		$C_{16}H_9N_3Na_2O_{10}S_2$	548-80-1	513.366	red-br pow	300				s H_2O; i EtOH
2385	Chrysamminic acid	1,8-Dihydroxy-2,4,5,7-tetranitro-9,10-anthracenedione	$C_{14}H_4N_4O_{12}$	517-92-0	420.202	ye pl or lf	exp	dec			vs eth, EtOH
2386	6-Chrysenamine	6-Aminochrysene	$C_{18}H_{13}N$	2642-98-0	243.303	lf (al)	210.5				
2387	Chrysene	Benzo[a]phenanthrene	$C_{18}H_{12}$	218-01-9	228.288	red bl fl or orth pl (bz, HOAc)	255.5	448	1.274[20]		i H_2O; sl EtOH, eth, ace, bz, CS_2; s tol
2388	Ciafos		$C_9H_{10}NO_3PS$	2636-26-2	243.219	ye to red-ye liq	15	120[0.09] dec		1.5404[32]	sl H_2O; vs chl, EtOH, ace, MeOH
2389	Cicutoxin	8,10,12-Heptadecatriene-4,6-diyne-1,14-diol	$C_{17}H_{22}O_2$	505-75-9	258.356	pr (eth/peth)	54				s hot H_2O, EtOH, eth, chl
2390	C.I. Direct Blue 6, tetrasodium salt	Direct Blue 6	$C_{32}H_{20}N_6Na_4O_{14}S_4$	2602-46-2	932.752	dk bronze pow					
2391	Cimetidine		$C_{10}H_{16}N_6S$	51481-61-9	252.339	cry	142				
2392	Cinchonamine		$C_{19}H_{24}N_2O$	482-28-0	296.406	orth nd (al) orth pr (MeOH)	186				i H_2O; vs EtOH, eth; s bz, chl
2393	Cinchonidine		$C_{19}H_{22}N_2O$	485-71-2	294.390	or pl or pr (al)	210.5	sub			i H_2O, bz; s EtOH, chl, py; sl eth
2394	Cinchonine		$C_{19}H_{22}N_2O$	118-10-5	294.390	pr nd (al, eth)	265				
2395	Cinchotoxine		$C_{19}H_{22}N_2O$	69-24-9	294.390	nd or pr (eth)	59				i H_2O; vs EtOH, eth, ace, bz, chl
2396	trans-Cinnamaldehyde	3-Phenyl-2-propenal, (E)-	C_9H_8O	14371-10-9	132.159	ye liq	-7.5	246	1.0497[20]	1.6195[20]	sl H_2O; s EtOH, eth, chl; i lig
2397	Cinnamedrine	α-[1-[Methyl(3-phenylallyl)amino]ethyl]benzenemethanol	$C_{19}H_{23}NO$	90-86-8	281.392		75				
2398	cis-Cinnamic acid	3-Phenyl-2-propenoic acid, (Z)	$C_9H_8O_2$	102-94-3	148.159	mcl pr (w)	42				vs EtOH, HOAc, lig
2399	trans-Cinnamic acid	3-Phenyl-2-propenoic acid, (E)	$C_9H_8O_2$	140-10-3	148.159	mcl pr (dil al)	133	300	1.2475[4]		i H_2O, lig; vs EtOH; s eth, ace, bz
2400	trans-Cinnamyl anthranilate		$C_{16}H_{15}NO_2$	87-29-6	253.296	cry	64				
2401	Cinnamyl cinnamate		$C_{18}H_{16}O_2$	122-69-0	264.319	nd (al)	44		1.1565[4]		i H_2O; s EtOH, chl; vs eth
2402	Cinnamyl formate	3-Phenyl-2-propen-1-ol, formate	$C_{10}H_{10}O_2$	104-65-4	162.185		0	252	1.086[25]		
2403	Cinnoline	1,2-Benzodiazine	$C_8H_6N_2$	253-66-7	130.147	pa ye cry (lig)	38	114[0.3]			vs eth, EtOH
2404	Cinoxate	3-(4-Methoxyphenyl)-2-propenoic acid, 2-ethoxyethyl ester	$C_{14}H_{18}O_4$	104-28-9	250.291	col liq	-25	185[2]	1.102[25]	1.567[20]	i H_2O; msc EtOH
2405	Cinquasia Red	Quinacridone	$C_{20}H_{12}N_2O_2$	1047-16-1	312.321	red-viol cry	390				i H_2O, os
2406	Ciodrin		$C_{14}H_{19}O_6P$	7700-17-6	314.271			135[0.03]	1.19[25]		
2407	C.I. Pigment Red 170		$C_{26}H_{22}N_4O_4$	2786-76-7	454.478	red solid					
2408	C.I. Pigment Yellow 1		$C_{17}H_{16}N_4O_4$	2512-29-0	340.334	ye cry	256				
2409	C.I. Pigment Yellow 12		$C_{32}H_{26}Cl_2N_6O_4$	6358-85-6	629.492	ye cry	317				
2410	Cisapride		$C_{23}H_{29}ClFN_3O_4$	81098-60-4	465.945	cry (hp)	132				
2411	Citral	3,7-Dimethyl-2,6-octadienal	$C_{10}H_{16}O$	5392-40-5	152.233			228.3	0.8888[20]	1.4898[20]	i H_2O; msc EtOH, eth
2412	β-Citraurin		$C_{30}H_{40}O_2$	650-69-1	432.638	pl (bz-peth) cry (al)	147				i H_2O; vs EtOH, eth, ace, bz; sl lig
2413	Citrazinic acid	1,2-Dihydro-6-hydroxy-2-oxo-4-pyridinecarboxylic acid	$C_6H_5NO_4$	99-11-6	155.109	ye pow	>300 dec				s H_2O, alk; sl HCl

Cholest-4-en-3-one

Cholesterol

Cholic acid

Choline chloride

Choline chloride dihydrogen phosphate

Chorismic acid

Chromium carbonyl

Chromium(II) oxalate

Chromium(III) 2,4-pentanedioate

Chromotrope 2B

Chrysamminic acid

6-Chrysenamine

Chrysene

Ciafos

Cicutoxin

C.I. Direct Blue 6, tetrasodium salt

Cimetidine

Cinchonamine

Cinchonidine

Cinchonine

Cinchotoxine

trans-Cinnamaldehyde

Cinnamedrine

cis-Cinnamic acid

trans-Cinnamic acid

trans-Cinnamyl anthranilate

Cinnamyl cinnamate

Cinnamyl formate

Cinnoline

Cinoxate

Cinquasia Red

Ciodrin

C.I. Pigment Red 170

C.I. Pigment Yellow 1

C.I. Pigment Yellow 12

Cisapride

Citral

β-Citraurin

Citrazinic acid

No.	Name	Synonym	Mol. Form.	CAS RN	Mol. Wt.	Physical Form	mp/°C	bp/°C	den/ g cm⁻³	n_D	Solubility
2414	Citric acid	2-Hydroxy-1,2,3-propanetricarboxylic acid	$C_6H_8O_7$	77-92-9	192.124	orth (w+1)	153	dec	1.665^{20}		vs H_2O, EtOH; s eth, AcOEt; i bz, chl
2415	Citric acid monohydrate	2-Hydroxy-1,2,3-propanetricarboxylic acid, monohydrate	$C_6H_{10}O_8$	5949-29-1	210.138	cry (w)	135		1.542		vs H_2O; vs EtOH, eth
2416	Citrinin	Antimycin	$C_{13}H_{14}O_5$	518-75-2	250.247	ye nd (MeOH)	178 dec				i H_2O; sl EtOH, eth; s ace, bz
2417	Citrulline	N5-(Aminocarbonyl)-L-ornithine	$C_6H_{13}N_3O_3$	372-75-8	175.185	pr (aq MeOH)	222				s H_2O; i EtOH, MeOH
2418	Citrus Red 2		$C_{18}H_{16}N_2O_3$	6358-53-8	308.331	cry	156				sl H_2O; s EtOH
2419	C.I. Vat Blue 6	7,16-Dichloro-6,15-dihydro-5,9,14,18-anthrazinetetrone	$C_{28}H_{12}Cl_2N_2O_4$	130-20-1	511.312	viol-bl pow					
2420	C.I. Vat Yellow 4	Anthanthrone	$C_{24}H_{12}O_2$	128-66-5	332.351	ye cry					
2421	Clayton Yellow	Thiazol Yellow G	$C_{28}H_{19}N_5Na_2O_6S_4$	1829-00-1	695.721	ye-br pow					s H_2O, EtOH, H_2SO_4
2422	Clemastine fumarate		$C_{25}H_{30}ClNO_5$	14976-57-9	459.963		181				
2423	Clindamycin		$C_{18}H_{33}ClN_2O_5S$	18323-44-9	424.983	ye amorp solid					
2424	Cloconazole		$C_{18}H_{15}ClN_2O$	77175-51-0	310.777		73				s EtOAc
2425	Clofentezine	3,6-Bis(2-chlorophenyl)-1,2,4,5-tetrazine	$C_{14}H_8Cl_2N_4$	74115-24-5	303.147		182				
2426	Clofibrate		$C_{12}H_{15}ClO_3$	637-07-0	242.698			149^{20}			
2427	Cloforex		$C_{13}H_{18}ClNO_2$	14261-75-7	255.741	cry	52.8	$89^{0.005}$			
2428	Clomazone	2-(2-Chlorobenzyl)-4,4-dimethyl-1,2-oxazolidin-3-one	$C_{12}H_{14}ClNO_2$	81777-89-1	239.698				1.192^{20}		
2429	Clomiphene		$C_{26}H_{28}ClNO$	911-45-5	405.959		117				
2430	Clonazepam		$C_{15}H_{10}ClN_3O_3$	1622-61-3	315.711	wh cry	237.5				i H_2O, bz; sl ace, MeOH, chl
2431	Clonidine		$C_9H_9Cl_2N_3$	4205-90-7	230.093	cry	137				
2432	Clopidol		$C_7H_7Cl_2NO$	2971-90-6	192.043	pow	>320				i H_2O
2433	Clopyralid	3,6-Dichloro-2-pyridinecarboxylic acid	$C_6H_3Cl_2NO_2$	1702-17-6	192.000		151				
2434	Clorophene		$C_{13}H_{11}ClO$	120-32-1	218.678		48.5	$161^{3.5}$	1.185^{58}		s ctc, CS_2
2435	Clotrimazole		$C_{22}H_{17}ClN_2$	23593-75-1	344.836	cry	148				sl H_2O, bz; s ace, chl, AcOEt, DMF
2436	Clozapine	Clozaril	$C_{18}H_{19}ClN_4$	5786-21-0	326.824	ye cry	183.5				
2437	Cobalt carbonyl	Dicobalt octacarbonyl	$C_8Co_2O_8$	10210-68-1	341.947	oran cry	51 dec		1.78		i H_2O; s EtOH, eth, CS_2
2438	Cobalt hydrocarbonyl	Tetracarbonylhydrocobalt	C_4HCoO_4	16842-03-8	171.982	ye liq or gas	≈-30	10			s os
2439	Cobalt(III) 2,4-pentanedioate	Cobalt(III) acetylacetonate	$C_{15}H_{21}CoO_6$	21679-46-9	356.257		240				
2440	Cocaine		$C_{17}H_{21}NO_4$	50-36-2	303.354	mcl pr (al)	98	$187^{0.1}$		1.5022^{98}	sl H_2O; vs EtOH, eth, bz, py; s CS_2
2441	Coclaurine		$C_{17}H_{19}NO_3$	486-39-5	285.338	pl (al)	220.5				
2442	Codamine		$C_{20}H_{25}NO_4$	21040-59-5	343.418	pr (bz, eth)	127				vs eth, EtOH, chl
2443	Codeine		$C_{18}H_{21}NO_3$	76-57-3	299.365	orth cry (w, dil al, eth)	157.5	250^{22}, $140^{1.5}$	1.32^{25}		s H_2O, eth, bz, chl, tol; vs EtOH; i peth
2444	Codeine phosphate		$C_{18}H_{24}NO_7P$	52-28-8	397.361	lf or pr (dil al)	227 dec				vs EtOH, chl
2445	Coenzyme A		$C_{21}H_{36}N_7O_{16}P_3S$	85-61-0	767.535	pow; unstab in air					s H_2O
2446	Coenzyme I	Nicotinamide adenine dinucleotide	$C_{21}H_{27}N_7O_{14}P_2$	53-84-9	663.425	hyg pow					s H_2O
2447	Coenzyme II	Nicotinamide adenine dinucleotide phosphate	$C_{21}H_{28}N_7O_{17}P_3$	53-59-8	743.405	gray-wh pow					s H_2O
2448	Colchiceine		$C_{21}H_{23}NO_6$	477-27-0	385.411	pa ye nd (diox)	178.5		1.24^{25}		sl H_2O; vs EtOH, chl; i eth, bz
2449	Colchicine		$C_{22}H_{25}NO_6$	64-86-8	399.437	ye pl (w + 1/2) ye cry (bz)	156				vs H_2O, EtOH
2450	Colistin A		$C_{53}H_{100}N_{16}O_{13}$	7722-44-3	1169.47	amorp pow					sl H_2O, EtOH, hx; s acids, MeOH
2451	Collinomycin		$C_{27}H_{20}O_{12}$	27267-69-2	536.441	oran pr (chl-MeOH)	281				vs ace, diox, chl
2452	Columbin		$C_{20}H_{22}O_6$	546-97-4	358.385	nd (MeOH)	195.5				i H_2O; sl ace, AcOEt, MeOH; s chl
2453	Conessine		$C_{24}H_{40}N_2$	546-06-5	356.588	lf or pl (ace)	125.5	$166^{0.1}$			sl H_2O; s chl, HOAc

Citric acid

Citric acid monohydrate

Citrinin

Citrulline

Citrus Red 2

C.I. Vat Blue 6

C.I. Vat Yellow 4

Clayton Yellow

Clemastine fumarate

Clindamycin

Cloconazole

Clofentezine

Clofibrate

Cloforex

Clomazone

Clomiphene

Clonazepam

Clonidine

Clopidol

Clopyralid

Clorophene

Clotrimazole

Clozapine

Cobalt carbonyl

Cobalt hydrocarbonyl

Cobalt(III) 2,4-pentanedioate

Cocaine

Coclaurine

Codamine

Codeine

Codeine phosphate

Coenzyme A

Coenzyme I

Coenzyme II

Colchiceine

Colchicine

Collinomycin

Columbin

Conessine

No.	Name	Synonym	Mol. Form.	CAS RN	Mol. Wt.	Physical Form	mp/°C	bp/°C	den/ g cm⁻³	n_D	Solubility
2454	Congo Red		$C_{32}H_{22}N_6Na_2O_6S_2$	573-58-0	696.663	pow	>360				sl H₂O; s EtOH; i eth
2455	Conhydrine		$C_8H_{17}NO$	3238-62-8	143.227	nd (peth)	121	226			sl H₂O; vs bz, eth, EtOH
2456	Conhydrine, (+)	2-(α-Hydroxypropyl)piperidine	$C_8H_{17}NO$	495-20-5	143.227	lf (eth)	121	226			sl H₂O; vs eth, EtOH, chl
2457	Coniferin		$C_{16}H_{22}O_8$	531-29-3	342.341	nd (w+2)	186				s H₂O, py; sl EtOH; i eth
2458	Conquinamine		$C_{19}H_{24}N_2O_2$	464-86-8	312.406	ye tetr	123				sl H₂O; s EtOH, eth, chl
2459	Convallatoxin		$C_{29}H_{42}O_{10}$	508-75-8	550.637	pr (eth/ MeOH)	238				s EtOH, ace; sl chl; i eth
2460	Copaene		$C_{15}H_{24}$	3856-25-5	204.352			248.5	0.8996²⁰	1.4894²⁰	i H₂O; s eth, ace, HOAc, lig
2461	Copper(II) ethylacetoacetate	Bis(ethylacetoacetato)copper	$C_{12}H_{18}CuO_6$	14284-06-1	321.813	grn cry (EtOH)	192				s EtOH, chl
2462	Copper(II) gluconate	Cupric gluconate	$C_{12}H_{22}CuO_{14}$	527-09-3	453.841	bl-grn cry	156				sl EtOH; i os
2463	Copper(II) 2,4-pentanedioate	Copper(II) acetylacetonate	$C_{10}H_{14}CuO_4$	13395-16-9	261.762	bl pow	284 dec	sub			sl H₂O; s chl
2464	Copper phthalocyanine	Pigment Blue 15	$C_{32}H_{16}CuN_8$	147-14-8	576.069	bl-purp cry					i H₂O, EtOH; s conc H2SO4
2465	Coronene		$C_{24}H_{12}$	191-07-1	300.352	ye nd (bz)	437.4	525	1.371²⁵		i H₂O, con sulf; sl bz
2466	Corticosterone		$C_{21}H_{30}O_4$	50-22-6	346.461	nd (al, pl) (ace)	181				i H₂O; s EtOH, eth, ace
2467	Corybulbine		$C_{21}H_{25}NO_4$	518-77-4	355.429	nd (al)	237.5				i H₂O; sl EtOH, eth; s ace, bz, HCl
2468	Corycavamine		$C_{21}H_{21}NO_5$	521-85-7	367.396	pr (eth, al)	149				vs EtOH, chl
2469	Corydaline		$C_{22}H_{27}NO_4$	518-69-4	369.454	pr (al)	136				vs bz, eth, EtOH, chl
2470	Corydine		$C_{20}H_{23}NO_4$	476-69-7	341.402	tetr pr (eth)	149				vs eth, EtOH, chl
2471	Corynantheine		$C_{22}H_{26}N_2O_3$	18904-54-6	366.452		165.5				vs EtOH
2472	Cotarnine		$C_{12}H_{15}NO_4$	82-54-2	237.252	nd (bz), cry (eth)	132 dec				sl H₂O; s EtOH, eth, bz, chl, NH₄OH
2473	Coumaphos		$C_{14}H_{16}ClO_5PS$	56-72-4	362.766		93		1.474		
2474	Coumestrol	3,9-Dihydroxy-6H-benzofuro[3,2-c][1]benzopyran-6-one	$C_{15}H_8O_5$	479-13-0	268.222	cry rods	385 dec				i H₂O; sl EtOH, ace; i eth
2475	Creatine		$C_4H_9N_3O_2$	57-00-1	131.133	mcl pr (w+1)	303 dec		1.33²⁵		s H₂O; sl EtOH; i eth
2476	Creatinine		$C_4H_7N_3O$	60-27-5	113.118	orth pr (w+2) lf (w)	300 dec				s H₂O; sl EtOH; i eth, ace, chl
2477	o-Cresol	2-Methylphenol	C_7H_8O	95-48-7	108.138		31.03	191.04	1.0327³⁵	1.5386³⁵	s H₂O; vs EtOH, eth; msc ace, bz, ctc
2478	m-Cresol	3-Methylphenol	C_7H_8O	108-39-4	108.138		12.24	202.27	1.0339²⁰	1.5401²⁰	sl H₂O; msc EtOH, eth, ace, bz, ctc
2479	p-Cresol	4-Methylphenol	C_7H_8O	106-44-5	108.138	pr	34.77	201.98	1.0185⁴⁰	1.5312²⁰	sl H₂O; msc EtOH, eth, ace, bz, ctc
2480	o-Cresolphthalein		$C_{22}H_{18}O_4$	596-27-0	346.376	cry (al)	223				vs EtOH
2481	o-Cresolphthalein complexone	Metalphthalein	$C_{32}H_{32}N_2O_{12}$	2411-89-4	636.602	ye cry pow	186				i H₂O; s EtOH, ace, alk
2482	Cresol Red	o-Cresolsulfonphthalein	$C_{21}H_{18}O_5S$	1733-12-6	382.430	red-br cry pow	>300				vs H₂O, EtOH
2483	p-Cresyl diphenyl phosphate		$C_{19}H_{17}O_4P$	78-31-9	340.309	col liq	-40		1.208²⁵		i H₂O; s os
2484	Crimidine		$C_7H_{10}ClN_3$	535-89-7	171.627	br wax	87	143⁴			vs EtOH
2485	Cromolyn	Cromoglicic acid	$C_{23}H_{16}O_{11}$	16110-51-3	468.366	col cry	241 dec				
2486	Crufomate		$C_{12}H_{19}ClNO_3P$	299-86-5	291.711		60	118⁰·⁰¹			
2487	Cryptopine	Cryptocavine	$C_{21}H_{23}NO_5$	482-74-6	369.412	pr or pl (bz) nd (chl-MeOH)	223		1.315²⁰		i H₂O; sl EtOH, eth, bz; s chl, HOAc
2488	Crystal Violet	Gentian violet	$C_{25}H_{30}ClN_3$	548-62-9	407.979	grn pow	215 dec				vs H₂O, chl
2489	Cubebin		$C_{20}H_{20}O_6$	18423-69-3	356.369	nd (al, bz)	131.5				vs eth, EtOH, chl
2490	Cucurbitacin B		$C_{32}H_{46}O_8$	6199-67-3	558.702	cry (EtOH)	181				
2491	Cucurbitacin C		$C_{32}H_{48}O_8$	5988-76-1	560.718	cry (AcOEt)	207.5				
2492	Cupferron		$C_6H_9N_3O_2$	135-20-6	155.154		163.5				sl DMSO
2493	Cupreine		$C_{19}H_{22}N_2O_2$	524-63-0	310.390	pr (eth)	202				vs EtOH
2494	Curan-17-ol, (16α)	Geissoschizoline	$C_{19}H_{26}N_2O$	18397-07-4	298.421	pa ye amor pow	135 dec				i H₂O; vs EtOH, eth, chl

Congo Red

Conhydrine

Conhydrine, (+)

Coniferin

Conquinamine

Convallatoxin

Copaene

Copper(II) ethylacetoacetate

Copper(II) gluconate

Copper(II) 2,4-pentanedioate

Copper phthalocyanine

Coronene

Corticosterone

Corybulbine

Corycavamine

Corydaline

Corydine

Corynantheine

Cotarnine

Coumaphos

Coumestrol

Creatine

Creatinine

o-Cresol

m-Cresol

p-Cresol

o-Cresolphthalein

o-Cresolphthalein complexone

Cresol Red

p-Cresyl diphenyl phosphate

Crimidine

Cromolyn

Crufomate

Cryptopine

Crystal Violet

Cubebin

Cucurbitacin B

Cucurbitacin C

Cupferron

Cupreine

Curan-17-ol, (16α)

No.	Name	Synonym	Mol. Form.	CAS RN	Mol. Wt.	Physical Form	mp/°C	bp/°C	den/ g cm⁻³	n_D	Solubility
2495	Curcumin	Turmeric	$C_{21}H_{20}O_6$	458-37-7	368.380	oran ye pr, orth pr (MeOH)	183				vs EtOH, HOAc
2496	Curine		$C_{36}H_{38}N_2O_6$	436-05-5	594.696	pr, nd (chl-MeOH)	221				vs ace, bz, py
2497	Cuscohygrine		$C_{13}H_{24}N_2O$	454-14-8	224.342	oil		169[23], 122[2]	0.9733[20]	1.4832[20]	vs H_2O, bz, eth, EtOH
2498	Cusparine	2-[2-(1,3-Benzodioxol-5-yl) ethyl]-4-methoxyquinoline	$C_{19}H_{17}NO_3$	529-92-0	307.343	(α) wh or ye nd (peth); (β) amber pr	92(α form); 111(β form)				i H_2O; vs ace, bz, eth, EtOH
2499	Cyamemazine		$C_{19}H_{21}N_3S$	3546-03-0	323.455	ye pow	92	212[0.25]			i H_2O; s EtOH
2500	Cyanamide	Cyanogenamide	CH_2N_2	420-04-2	42.040	nd	45.56	140[19]	1.282[20]	1.4418[48]	vs H_2O, EtOH; s eth, ace, bz; sl CS_2
2501	Cyanazine		$C_9H_{13}ClN_6$	21725-46-2	240.692		168				
2502	Cyanic acid	Hydrogen cyanate	CHNO	420-05-3	43.025	unstab liq or gas	-86	23	1.140[20]		vs H_2O, bz, eth, chl
2503	2-Cyanoacetamide		$C_3H_4N_2O$	107-91-5	84.076	pl (w)	121.5				vs H_2O
2504	Cyanoacetic acid		$C_3H_3NO_2$	372-09-8	85.062		66	dec 160; 108[15]			s H_2O, EtOH, eth; sl chl, HOAc
2505	Cyanoacetohydrazide	Cyacetacide	$C_3H_5N_3O$	140-87-4	99.091	pr (al)	114.5				vs H_2O, EtOH
2506	Cyanoacetylene		C_3HN	1070-71-9	51.047		5	42.5	0.8167[17]	1.3868[25]	sl H_2O; s EtOH
2507	3-Cyanobenzoic acid		$C_8H_5NO_2$	1877-72-1	147.132	nd (w)	219	sub			sl H_2O; s EtOH, eth
2508	4-Cyanobenzoic acid		$C_8H_5NO_2$	619-65-8	147.132		219				s H_2O, EtOH, eth, HOAc; sl tfa
2509	4-Cyanobutanoic acid		$C_5H_7NO_2$	39201-33-7	113.116	hyg cry	45				s H_2O, EtOH, eth, bz
2510	2-Cyanoethyl acrylate		$C_6H_7NO_2$	106-71-8	125.126			108[12]	1.062[20]		
2511	Cyanofenphos		$C_{15}H_{14}NO_2PS$	13067-93-1	303.317		83			1.5839[25]	sl H_2O
2512	Cyanogen		C_2N_2	460-19-5	52.034	col gas	-27.83	-21.1	0.9537[-21]		s H_2O, EtOH, eth
2513	Cyanogen bromide	Bromine cyanide	CBrN	506-68-3	105.922	nd	52	61.5	2.015[20]		s H_2O, EtOH, eth
2514	Cyanogen chloride	Chlorine cyanide	CClN	506-77-4	61.471	col vol liq or gas	-6.5	13	1.186[20]		s H_2O, EtOH; vs eth
2515	Cyanogen fluoride	Fluorine cyanide	CFN	1495-50-7	45.016	col gas	-82	-46			
2516	Cyanogen iodide	Iodine cyanide	CIN	506-78-5	152.922	nd (al, eth)	146.7	sub	2.84[18]		vs eth, EtOH
2517	Cyanoguanidine	Dicyanodiamide	$C_2H_4N_4$	461-58-5	84.080		211		1.404[14]		s H_2O, EtOH, ace; i eth, bz, chl
2518	Cyanomethylmercury	Methylmercurynitrile	C_2H_3HgN	2597-97-9	241.64	cry (chl)	92	subl			vs H_2O, EtOH, bz; s eth
2519	(4-Cyanophenoxy)acetic acid		$C_9H_7NO_3$	1878-82-6	177.157	cry (w)	178				
2520	2-Cyano-N-phenylacetamide		$C_9H_8N_2O$	621-03-4	160.172	nd (al)	199.5				
2521	4-Cyanothiazole		$C_4H_2N_2S$	1452-15-9	110.137	nd	58				
2522	Cyanuric acid	1,3,5-Triazine-2,4,6(1H,3H,5H)-trione	$C_3H_3N_3O_3$	108-80-5	129.074	wh cry	>330	sub	1.75[25]		sl hot H_2O, ace, bz, EtOH; s conc HCl
2523	Cyanuric fluoride	2,4,6-Trifluoro-1,3,5-triazine	$C_3F_3N_3$	675-14-9	135.047			72.8			
2524	Cycasin		$C_8H_{16}N_2O_7$	14901-08-7	252.222	nd (ace aq)	154 dec				
2525	Cyclandelate		$C_{17}H_{24}O_3$	456-59-7	276.371		52	193[14]			i H_2O
2526	Cyclizine		$C_{18}H_{22}N_2$	82-92-8	266.381	cry (peth)	106				i H_2O; s chl; sl EtOH
2527	Cycloate	Carbamothioic acid, cyclohexylethyl-, S-ethyl ester	$C_{11}H_{21}NOS$	1134-23-2	215.356		11.5	145[10]	1.0156[30]		
2528	Cyclobarbital		$C_{12}H_{16}N_2O_3$	52-31-3	236.266	lf (w)	173				i H_2O; vs EtOH; s eth, dil alk; sl HOAc
2529	Cyclobutanamine	Aminocyclobutane	C_4H_9N	2516-34-9	71.121			82	0.8328[20]	1.4363[19]	
2530	Cyclobutane	Tetramethylene	C_4H_8	287-23-0	56.107	vol liq or gas	-90.7	12.6	0.7038[0]	1.375[20]	i H_2O; vs EtOH, ace; msc eth; s bz
2531	Cyclobutanecarbonitrile	Cyanocyclobutane	C_5H_7N	4426-11-3	81.117			149.6			
2532	Cyclobutanecarboxylic acid		$C_5H_8O_2$	3721-95-7	100.117	liq	-1.0	190; 74[2]	1.0599[20]	1.4400[20]	sl H_2O; msc EtOH, eth
2533	1,1-Cyclobutanedicarboxylic acid		$C_6H_8O_4$	5445-51-2	144.126	pr (w, eth)	158.0				vs H_2O; s EtOH, eth, bz; sl lig
2534	Cyclobutanol	Hydroxycyclobutane	C_4H_8O	2919-23-5	72.106			124	0.9218[15]	1.4371[20]	
2535	Cyclobutanone		C_4H_6O	1191-95-3	70.090	liq	-50.9	99	0.9547[0]	1.4215[20]	s H_2O, eth, bz, chl, tol; vs EtOH; i peth

Curcumin

Curine

Cuscohygrine

Cusparine

Cyamemazine

Cyanamide

Cyanazine

Cyanic acid

2-Cyanoacetamide

Cyanoacetic acid

Cyanoacetohydrazide

Cyanoacetylene

3-Cyanobenzoic acid

4-Cyanobenzoic acid

4-Cyanobutanoic acid

2-Cyanoethyl acrylate

Cyanofenphos

Cyanogen

Cyanogen bromide

Cyanogen chloride

Cyanogen fluoride

Cyanogen iodide

Cyanoguanidine

Cyanomethylmercury

(4-Cyanophenoxy)acetic acid

2-Cyano-N-phenylacetamide

4-Cyanothiazole

Cyanuric acid

Cyanuric fluoride

Cycasin

Cyclandelate

Cyclizine

Cycloate

Cyclobarbital

Cyclobutanamine

Cyclobutane

Cyclobutanecarbonitrile

Cyclobutanecarboxylic acid

1,1-Cyclobutanedicarboxylic acid

Cyclobutanol

Cyclobutanone

No.	Name	Synonym	Mol. Form.	CAS RN	Mol. Wt.	Physical Form	mp/°C	bp/°C	den/ g cm⁻³	n_D	Solubility
2536	Cyclobutene		C_4H_6	822-35-5	54.091	col gas		2	0.733[0]		vs ace; s bz, peth
2537	Cyclochlorotine		$C_{24}H_{31}Cl_2N_5O_7$	12663-46-6	572.439	nd (MeOH)	255 dec				
2538	Cyclodecane		$C_{10}H_{20}$	293-96-9	140.266		10	202	0.8538[25]	1.4716[20]	
2539	1,2-Cyclodecanedione	Sebacil	$C_{10}H_{16}O_2$	96-01-5	168.233		40.5	104[10]			
2540	Cyclodecanol		$C_{10}H_{20}O$	1502-05-2	156.265		40.5	125[12]	0.9606[20]	1.4926[20]	s EtOH
2541	Cyclodecanone		$C_{10}H_{18}O$	1502-06-3	154.249	amor pow	28	106[13]	0.9654[20]	1.4806[20]	vs bz, eth, chl
2542	α-Cyclodextrin	Cyclomaltohexaose	$C_{36}H_{60}O_{30}$	10016-20-3	972.843	hx pl or nd					vs cold H_2O; i hot H_2O
2543	β-Cyclodextrin	Cyclomaltoheptaose	$C_{42}H_{70}O_{35}$	7585-39-9	1134.984	mcl cry (w)	260 dec				
2544	γ-Cyclodextrin	Cyclomaltooctaose	$C_{48}H_{80}O_{40}$	17465-86-0	1297.125	sq pl or rods					
2545	Cyclododecane		$C_{12}H_{24}$	294-62-2	168.319	nd (al)	60.4	247	0.82[80]		
2546	Cyclododecanol		$C_{12}H_{24}O$	1724-39-6	184.318			286			
2547	Cyclododecanone		$C_{12}H_{22}O$	830-13-7	182.302		59	127[12]	0.9059[66]	1.4571[60]	
2548	1,5,9-Cyclododecatriene	CDT	$C_{12}H_{18}$	4904-61-4	162.271	liq	-17	240	0.84[100]		
2549	cis-Cyclododecene		$C_{12}H_{22}$	1129-89-1	166.303			133[35]; 71[2]		1.4840[20]	vs bz, chl
2550	trans-Cyclododecene		$C_{12}H_{22}$	1486-75-5	166.303			113[17]		1.4850[20]	vs bz, chl
2551	cis-9-Cycloheptadecen-1-one	Civetone	$C_{17}H_{30}O$	542-46-1	250.419		32.5	343; 159[2]			
2552	1,3-Cycloheptadiene		C_7H_{10}	4054-38-0	94.154	liq	-110.4	120.5	0.868[25]	1.4978[20]	
2553	Cycloheptanamine		$C_7H_{15}N$	5452-35-7	113.201			54[11]		1.4724[20]	
2554	Cycloheptane		C_7H_{14}	291-64-5	98.186	liq	-8.46	118.4	0.8098[20]	1.4436[20]	i H_2O; vs EtOH, eth; s bz, chl
2555	1,2-Cycloheptanedione		$C_7H_{10}O_2$	3008-39-7	126.153		-40	108[17]	1.0583[22]	1.4689[22]	s EtOH
2556	Cycloheptanol		$C_7H_{14}O$	502-41-0	114.185		7.2	185	0.9554[20]	1.40705[20]	sl H_2O; vs EtOH, eth
2557	Cycloheptanone	Suberone	$C_7H_{12}O$	502-42-1	112.169			178.5	0.9508[20]	1.4608[20]	i H_2O; vs EtOH, eth
2558	1,3,5-Cycloheptatriene	Tropilidene	C_7H_8	544-25-2	92.139	liq; cub cry (-80°C)	-79.5	117; 60.5[122]	0.8875[19]	1.5343[20]	i H_2O; s EtOH, eth; vs bz, chl
2559	2,4,6-Cycloheptatrien-1-one		C_7H_6O	539-80-0	106.122		-7	113[15]; 84[6]	1.095[22]	1.6172[20]	vs bz, chl
2560	Cycloheptene		C_7H_{12}	628-92-2	96.170	liq	-56	115	0.8228[20]	1.4552[20]	i H_2O; s EtOH, eth, bz, chl; sl ctc
2561	1,3-Cyclohexadiene		C_6H_8	592-57-4	80.128	liq	-89	80.5	0.8405[20]	1.4755[20]	i H_2O; s EtOH, bz, chl, peth; vs eth
2562	1,4-Cyclohexadiene	1,4-Dihydrobenzene	C_6H_8	628-41-1	80.128	liq	-49.2	85.5	0.8471[20]	1.4725[20]	i H_2O; msc EtOH, eth; s bz, chl, peth
2563	3,5-Cyclohexadiene-1,2-dione		$C_6H_4O_2$	583-63-1	108.095	red pl or pr	≈65 dec				s eth, ace, bz; i peth
2564	2,5-Cyclohexadiene-1,4-dione, dioxime		$C_6H_6N_2O_2$	105-11-3	138.124	pa ye nd (w)	240 dec				s H_2O
2565	Cyclohexane	Hexahydrobenzene	C_6H_{12}	110-82-7	84.159		6.59	80.73	0.7739[25]	1.4235[25]	i H_2O; msc EtOH, eth, ace, bz, lig, ctc
2566	Cyclohexaneacetic acid		$C_8H_{14}O_2$	5292-21-7	142.196	nd (HCO_2H)	33	245	1.0423[18]	1.4775[20]	sl H_2O; s eth, ace
2567	Cyclohexanecarbonitrile	Cyclohexyl cyanide	$C_7H_{11}N$	766-05-2	109.169	liq	11	184; 76[16]	0.919	1.4505[20]	
2568	Cyclohexanecarbonyl chloride		$C_7H_{11}ClO$	2719-27-9	146.614			180	1.0962[15]	1.4711[29]	
2569	Cyclohexanecarboxaldehyde		$C_7H_{12}O$	2043-61-0	112.169			159.3	0.9035[20]	1.4496[20]	s H_2O, eth
2570	Cyclohexanecarboxylic acid	Hexahydrobenzoic acid	$C_7H_{12}O_2$	98-89-5	128.169	mcl pr	31.5	232.5	1.0334[22]	1.4530[20]	sl H_2O, ctc; vs EtOH, bz, chl
2571	cis-1,2-Cyclohexanediamine	cis-1,2-Diaminocyclohexane	$C_6H_{14}N_2$	1436-59-5	114.188	liq		40[2]	0.952[20]	1.4951[20]	
2572	trans-1,2-Cyclohexanediamine	trans-1,2-Diaminocyclohexane	$C_6H_{14}N_2$	1121-22-8	114.188		14.8	80[15]; 41[2]	0.951[20]		
2573	trans-1,4-Cyclohexanedicarboxylic acid		$C_8H_{12}O_4$	619-82-9	172.179	pr (w)	312.5	sub 300			sl H_2O, eth; vs EtOH; s ace; i chl
2574	1,3-Cyclohexanedimethanamine		$C_8H_{18}N_2$	2579-20-6	142.242		<-70	220	0.945[20]		vs H_2O, eth, EtOH
2575	1,4-Cyclohexanedimethanol		$C_8H_{16}O_2$	105-08-8	144.212		43	283; 167[10]			
2576	cis-1,2-Cyclohexanediol		$C_6H_{12}O_2$	1792-81-0	116.158		100	120[15]	1.0297[101]		s EtOH, ace, bz; sl chl
2577	trans-1,4-Cyclohexanediol		$C_6H_{12}O_2$	6995-79-5	116.158	mcl pr (ace)	143		1.18[20]		s H_2O, EtOH, MeOH; i eth; sl ace
2578	1,2-Cyclohexanedione	1,2-Dioxocyclohexane	$C_6H_8O_2$	765-87-7	112.127	cry (peth)	40	194	1.1187[21]	1.4995[20]	s H_2O, EtOH, eth, bz
2579	1,3-Cyclohexanedione	Dihydroresorcinol	$C_6H_8O_2$	504-02-9	112.127	pr (bz)	105.5		1.0861[91]	1.4576[102]	s H_2O, EtOH, ace, chl; sl eth, bz
2580	1,4-Cyclohexanedione	Tetrahydroquinone	$C_6H_8O_2$	637-88-7	112.127	mcl pl (w),nd (peth)	78	132[20]	1.0861[91]		s H_2O, EtOH, eth, ace, bz, chl

Cyclobutene

Cyclochlorotine

Cyclodecane

1,2-Cyclodecanedione

Cyclodecanol

Cyclodecanone

α-Cyclodextrin

β-Cyclodextrin

γ-Cyclodextrin

Cyclododecane

Cyclododecanol

Cyclododecanone

1,5,9-Cyclododecatriene

cis-Cyclododecene

trans-Cyclododecene

cis-9-Cycloheptadecen-1-one

1,3-Cycloheptadiene

Cycloheptanamine

Cycloheptane

1,2-Cycloheptanedione

Cycloheptanol

Cycloheptanone

1,3,5-Cycloheptatriene

2,4,6-Cycloheptatrien-1-one

Cycloheptene

1,3-Cyclohexadiene

1,4-Cyclohexadiene

3,5-Cyclohexadiene-1,2-dione

2,5-Cyclohexadiene-1,4-dione, dioxime

Cyclohexane

Cyclohexaneacetic acid

Cyclohexanecarbonitrile

Cyclohexanecarbonyl chloride

Cyclohexanecarboxaldehyde

Cyclohexanecarboxylic acid

cis-1,2-Cyclohexanediamine

trans-1,2-Cyclohexanediamine

trans-1,4-Cyclohexanedicarboxylic acid

1,3-Cyclohexanedimethanamine

1,4-Cyclohexanedimethanol

cis-1,2-Cyclohexanediol

trans-1,4-Cyclohexanediol

1,2-Cyclohexanedione

1,3-Cyclohexanedione

1,4-Cyclohexanedione

No.	Name	Synonym	Mol. Form.	CAS RN	Mol. Wt.	Physical Form	mp/°C	bp/°C	den/ g cm⁻³	n_D	Solubility
2581	1,2-Cyclohexanedione dioxime	Nioxime	$C_6H_{10}N_2O_2$	492-99-9	142.155	nd (w, HOAc)	192				s H_2O, ace, chl; sl tfa
2582	Cyclohexaneethanol		$C_8H_{16}O$	4442-79-9	128.212			208	0.9229[20]	1.4641[20]	s EtOH, eth, bz
2583	Cyclohexanemethanamine		$C_7H_{15}N$	3218-02-8	113.201			160	0.87[25]	1.4630[20]	
2584	Cyclohexanemethanol	Cyclohexylcarbinol	$C_7H_{14}O$	100-49-2	114.185	liq	-43	183	0.9297[20]	1.4644[20]	vs eth, EtOH
2585	Cyclohexanepropanoic acid		$C_9H_{16}O_2$	701-97-3	156.222		16	276.5	0.912[25]	1.4638[20]	s H_2O, eth; sl ctc
2586	Cyclohexanethiol	Cyclohexyl mercaptan	$C_6H_{12}S$	1569-69-3	116.224			158.8	0.9782[20]	1.4921[20]	vs ace, bz, eth, EtOH
2587	Cyclohexanol	Cyclohexyl alcohol	$C_6H_{12}O$	108-93-0	100.158	hyg nd	25.93	160.84	0.9624[20]	1.4641[20]	s H_2O, EtOH, eth, ace; msc bz; sl chl
2588	Cyclohexanone	Pimelic ketone	$C_6H_{10}O$	108-94-1	98.142	liq	-27.9	155.43	0.9478[20]	1.4507[20]	s H_2O, EtOH, eth, ace, bz, chl, ctc
2589	Cyclohexanone oxime		$C_6H_{11}NO$	100-64-1	113.157	hex pr (lig)	90	206			s H_2O, EtOH, eth, MeOH; sl chl
2590	Cyclohexanone peroxide		$C_{12}H_{22}O_5$	78-18-2	246.300	cry or long nd	79				
2591	Cyclohexene	Tetrahydrobenzene	C_6H_{10}	110-83-8	82.143	liq	-103.5	82.98	0.8110[20]	1.4465[20]	i H_2O; msc EtOH, eth, ace, bz, lig, ctc
2592	1-Cyclohexenecarbonitrile	1-Cyanocyclohexene	C_7H_9N	1855-63-6	107.153			81[12]			
2593	1-Cyclohexene-1-carboxaldehyde		$C_7H_{10}O$	1192-88-7	110.153			69[18]	0.9694[20]	1.5005[20]	s EtOH, eth
2594	3-Cyclohexene-1-carboxaldehyde		$C_7H_{10}O$	100-50-5	110.153		1.0	105	0.9692[20]	1.4745[20]	s ace, MeOH; sl ctc
2595	1-Cyclohexene-1-carboxylic acid		$C_7H_{10}O_2$	636-82-8	126.153		38	241	1.109[20]	1.4902[20]	sl H_2O; s EtOH, ace
2596	3-Cyclohexene-1-carboxylic acid		$C_7H_{10}O_2$	4771-80-6	126.153		17	234.5	1.0820[20]	1.4814[20]	vs H_2O; s EtOH, ace
2597	4-Cyclohexene-1,2-dicarboxylic acid		$C_8H_{10}O_4$	88-98-2	170.163	pr (w)	173.0				
2598	2-Cyclohexen-1-ol		$C_6H_{10}O$	822-67-3	98.142			164	0.9923[15]	1.4790[15]	s EtOH, ace
2599	2-Cyclohexen-1-one		C_6H_8O	930-68-7	96.127	liq	-53	170	0.9620[25]	1.4883[20]	vs EtOH; s ace
2600	1-Cyclohexen-1-ylbenzene		$C_{12}H_{14}$	771-98-2	158.239	liq	-11	252	0.9939[20]	1.5718[20]	vs MeOH
2601	2-(1-Cyclohexen-1-yl)cyclohexanone		$C_{12}H_{18}O$	1502-22-3	178.270			116[3]		1.5070[20]	
2602	1-(1-Cyclohexen-1-yl)ethanone		$C_8H_{12}O$	932-66-1	124.180		73	201.5	0.9655[20]	1.4881[20]	s EtOH, eth
2603	3-Cyclohexenylmethyl 3-cyclohexenecarboxylate		$C_{14}H_{20}O_2$	2611-00-9	220.308	liq		153[7], 109[0.6]			
2604	4-(3-Cyclohexen-1-yl)pyridine		$C_{11}H_{13}N$	70644-46-1	159.228		22.1	226	1.0222[25]	1.5466[25]	
2605	Cycloheximide		$C_{15}H_{23}NO_4$	66-81-9	281.349	pl (al)	119				vs EtOH
2606	Cyclohexyl acetate		$C_8H_{14}O_2$	622-45-7	142.196			173; 96[75]	0.968[20]	1.442[20]	vs eth, EtOH
2607	Cyclohexyl acrylate		$C_9H_{14}O_2$	3066-71-5	154.206			183; 88[20]	1.0275[20]	1.4673[20]	i H_2O; msc EtOH, eth; s chl
2608	Cyclohexylamine	Cyclohexanamine	$C_6H_{13}N$	108-91-8	99.174	liq	-17.8	134	0.8191[20]	1.4625[15]	s H_2O, ctc; vs EtOH; msc eth, ace, bz
2609	Cyclohexylamine hydrochloride	Cyclohexanamine hydrochloride	$C_6H_{14}ClN$	4998-76-9	135.635	nd (w, al-eth)	206.5				vs H_2O, EtOH
2610	2-(Cyclohexylaminothio)benzothiazole		$C_{13}H_{16}N_2S_2$	95-33-0	264.409		103				
2611	N-Cyclohexylaniline		$C_{12}H_{17}N$	1821-36-9	175.270	mcl pr	16	279; 192[73]	1.0155[20]	1.5610[20]	i H_2O; s EtOH, eth, bz
2612	Cyclohexylbenzene		$C_{12}H_{16}$	827-52-1	160.255	pl	7.07	240.1	0.9427[20]	1.5329[20]	i H_2O; vs EtOH; s eth; sl ctc
2613	Cyclohexyl benzoate		$C_{13}H_{16}O_2$	2412-73-9	204.265		<-10	285	1.0429[20]	1.5200[20]	i H_2O; s EtOH, eth
2614	Cyclohexyl butanoate		$C_{10}H_{18}O_2$	1551-44-6	170.249			213	0.9572[0]		i H_2O; s EtOH; sl ctc
2615	3-Cyclohexyl-2-butenoic acid	Cicrotoic acid	$C_{10}H_{16}O_2$	25229-42-9	168.233	pr (aq-MeOH)	85.5				
2616	Cyclohexyl chloroformate		$C_7H_{11}ClO_2$	13248-54-9	162.614			87.5[27]			vs eth
2617	Cyclohexylcyclohexane		$C_{12}H_{22}$	92-51-3	166.303		4	238			sl H_2O; s EtOH, eth
2618	Cyclohexyldiethylamine	N,N-Diethylcyclohexanamine	$C_{10}H_{21}N$	91-65-6	155.281			192; 85[20]	0.8443[25]		s EtOH; sl ctc
2619	Cyclohexyldimethylamine	N,N-Dimethylcyclohexanamine	$C_8H_{17}N$	98-94-2	127.228			162			
2620	2-Cyclohexyl-4,6-dinitrophenol		$C_{12}H_{14}N_2O_5$	131-89-5	266.249	cry	104				sl H_2O; s bz, DMF
2621	(1,2-Cyclohexylenedinitrilo)tetraacetic acid monohydrate	CDTA	$C_{14}H_{24}N_2O_9$	13291-61-7	364.349	cry (w)	215				
2622	1-Cyclohexylethanone		$C_8H_{14}O$	823-76-7	126.196			180.5	0.9176[20]	1.4565[16]	i H_2O; s eth
2623	Cyclohexylethylamine	N-Ethylcyclohexanamine	$C_8H_{17}N$	5459-93-8	127.228			164	0.868[0]		sl H_2O, ctc; msc EtOH, eth

1,2-Cyclohexanedione dioxime

Cyclohexaneethanol

Cyclohexanemethanamine

Cyclohexanemethanol

Cyclohexanepropanoic acid

Cyclohexanethiol

Cyclohexanol

Cyclohexanone

Cyclohexanone oxime

Cyclohexanone peroxide

Cyclohexene

1-Cyclohexenecarbonitrile

1-Cyclohexene-1-carboxaldehyde

3-Cyclohexene-1-carboxaldehyde

1-Cyclohexene-1-carboxylic acid

3-Cyclohexene-1-carboxylic acid

4-Cyclohexene-1,2-dicarboxylic acid

2-Cyclohexen-1-ol

2-Cyclohexen-1-one

1-Cyclohexen-1-ylbenzene

2-(1-Cyclohexen-1-yl)cyclohexanone

1-(1-Cyclohexen-1-yl)ethanone

3-Cyclohexenylmethyl 3-cyclohexenecarboxylate

4-(3-Cyclohexen-1-yl)pyridine

Cycloheximide

Cyclohexyl acetate

Cyclohexyl acrylate

Cyclohexylamine

Cyclohexylamine hydrochloride

2-(Cyclohexylaminothio)benzothiazole

N-Cyclohexylaniline

Cyclohexylbenzene

Cyclohexyl benzoate

Cyclohexyl butanoate

3-Cyclohexyl-2-butenoic acid

Cyclohexyl chloroformate

Cyclohexylcyclohexane

Cyclohexyldiethylamine

Cyclohexyldimethylamine

2-Cyclohexyl-4,6-dinitrophenol

(1,2-Cyclohexylenedinitrilo)tetraacetic acid monohydrate

1-Cyclohexylethanone

Cyclohexylethylamine

No.	Name	Synonym	Mol. Form.	CAS RN	Mol. Wt.	Physical Form	mp/°C	bp/°C	den/ g cm⁻³	n_D	Solubility
2624	4-Cyclohexyl-3-ethyl-4H-1,2,4-triazole	Hexazole	$C_{10}H_{17}N_3$	4671-03-8	179.262	pr (eth)	89.5	227[10]			vs H_2O, bz, chl
2625	Cyclohexyl formate		$C_7H_{12}O_2$	4351-54-6	128.169			162	1.0057[0]	1.4430[20]	i H_2O; s EtOH, HOAc, HCOOH; vs eth
2626	Cyclohexyl hydroperoxide		$C_6H_{12}O_2$	766-07-4	116.158		-20	42[0.1]	1.019[20]	1.4645[25]	vs eth, EtOH, HOAc
2627	Cyclohexylideneacetonitrile		$C_8H_{11}N$	4435-18-1	121.180			107[22]	0.9483[15]	1.4382[25]	vs eth, EtOH
2628	2-Cyclohexylidenecyclohexanone		$C_{12}H_{18}O$	1011-12-7	178.270	cry (MeOH aq)	56.5				
2629	Cyclohexyl isocyanate	Isocyanatocyclohexane	$C_7H_{11}NO$	3173-53-3	125.168			172	0.98[25]	1.4551[20]	
2630	Cyclohexylisopropylamine	N-Isopropylcyclohexanamine	$C_9H_{19}N$	1195-42-2	141.254			62[12]	0.859[25]	1.4480[20]	
2631	Cyclohexyl isothiocyanate	Isothiocyanatocyclohexane	$C_7H_{11}NS$	1122-82-3	141.234			221	1.0339[20]	1.5375[20]	i H_2O; s EtOH, eth; sl ctc
2632	Cyclohexylmagnesium chloride		$C_6H_{11}ClMg$	931-51-1	142.909	hyg liq					s eth
2633	Cyclohexyl methacrylate		$C_{10}H_{16}O_2$	101-43-9	168.233			210	0.9626[20]	1.4578[20]	
2634	Cyclohexylmethylamine	N-Methylcyclohexanamine	$C_7H_{15}N$	100-60-7	113.201			147	0.8660[23]	1.4560[20]	sl H_2O; vs EtOH; msc eth; s chl
2635	Cyclohexyl 2-methylpropanoate		$C_{10}H_{18}O_2$	1129-47-1	170.249			204	0.9489[0]		vs eth, EtOH
2636	2-Cyclohexylphenol		$C_{12}H_{16}O$	119-42-6	176.254	nd (lig)	56.5				vs EtOH, HOAc
2637	4-Cyclohexylphenol		$C_{12}H_{16}O$	1131-60-8	176.254	nd (bz)	133	294; 133[4]			i H_2O; vs EtOH, eth; s bz; sl lig
2638	α-Cyclohexyl-α-phenyl-1-piperidinepropanol	Trihexphenidyl	$C_{20}H_{31}NO$	144-11-6	301.466		114				
2639	Cyclohexyl propanoate		$C_9H_{16}O_2$	6222-35-1	156.222			193; 93[35]	0.9359[20]	1.4403[20]	i H_2O; s EtOH, eth, ace, ctc
2640	Cyclohexylsulfamic acid	Cyclamic acid	$C_6H_{13}NO_3S$	100-88-9	179.237		169.5				vs alk
2641	Cyclononane		C_9H_{18}	293-55-0	126.239		11	178.4	0.8463[25]	1.4666[20]	
2642	Cyclononanone		$C_9H_{16}O$	3350-30-9	140.222		34	148[24], 94[12]	0.9560[20]	1.4729[20]	s EtOH
2643	1,4-Cyclooctadiene		C_8H_{12}	1073-07-0	108.181	liq	-53	145	0.8754[20]		
2644	cis,cis-1,5-Cyclooctadiene		C_8H_{12}	111-78-4	108.181	liq	-56.4	150.5	0.883[20]	1.4905[25]	vs bz
2645	Cyclooctanamine	Aminocyclooctane	$C_8H_{17}N$	5452-37-9	127.228	liq	-48	190	0.928[25]	1.4804[20]	
2646	Cyclooctane		C_8H_{16}	292-64-8	112.213		14.59	149	0.8349[20]	1.4586[20]	i H_2O; s bz, lig
2647	Cyclooctanol		$C_8H_{16}O$	696-71-9	128.212		25.1	99[16]	0.9740[20]	1.4871[20]	s EtOH
2648	Cyclooctanone		$C_8H_{14}O$	502-49-8	126.196		29	196	0.9581[20]	1.4694[20]	i H_2O; s EtOH, ace, bz; sl ctc
2649	1,3,5,7-Cyclooctatetraene	[8]Annulene	C_8H_8	629-20-9	104.150	liq	-2.4	140.5	0.9206[20]	1.5381[20]	s EtOH, eth, ace, bz
2650	1,3,5-Cyclooctatriene		C_8H_{10}	1871-52-9	106.165	liq	-83	145.5	0.8971[25]	1.5035[25]	
2651	cis-Cyclooctene		C_8H_{14}	931-87-3	110.197	liq	-12	138	0.8472[20]	1.4698[20]	s EtOH, eth, ctc
2652	trans-Cyclooctene		C_8H_{14}	931-89-5	110.197	liq	-59	143	0.8483[20]	1.4741[25]	s EtOH, chl; sl ctc
2653	Cyclooctyne		C_8H_{12}	1781-78-8	108.181			158	0.868[20]	1.4850[20]	
2654	Cyclopamine	11-Deoxojervine	$C_{27}H_{41}NO_2$	4449-51-8	411.621	nd (EtOH)	237				
2655	Cyclopentadecane		$C_{15}H_{30}$	295-48-7	210.399	nd (MeOH)	61.3		0.8364[61]	1.4592[61]	
2656	Cyclopentadecanol	Exaltol	$C_{15}H_{30}O$	4727-17-7	226.398	cry (MeOH)	80.5	177[11], 145[0.3]	0.930[20]	1.4555[98]	
2657	Cyclopentadecanone		$C_{15}H_{28}O$	502-72-7	224.382		63	120[0.3]	0.8895[25]	1.4637[60]	sl H_2O; s EtOH, ace
2658	1,3-Cyclopentadiene	Pyropentylene	C_5H_6	542-92-7	66.102	liq	-85	41	0.8021[20]	1.4440[20]	i H_2O; msc EtOH, eth, bz; s ace
2659	Cyclopentane	Pentamethylene	C_5H_{10}	287-92-3	70.133	liq	-93.4	49.3	0.7457[20]	1.4065[20]	i H_2O; msc EtOH, eth, ace, bz, peth, ctc
2660	Cyclopentaneacetic acid		$C_7H_{12}O_2$	1123-00-8	128.169	pl	13.5	228	1.0216[18]	1.4523[18]	
2661	Cyclopentanecarbonitrile	Cyanocyclopentane	C_6H_9N	4254-02-8	95.142	liq	-76	170; 67[10]	0.912	1.4410[20]	
2662	Cyclopentanecarboxaldehyde		$C_6H_{10}O$	872-53-7	98.142			133.5	0.9371[20]	1.4432[20]	vs H_2O, eth, EtOH
2663	Cyclopentanecarboxylic acid	Cyclopentanoic acid	$C_6H_{10}O_2$	3400-45-1	114.142	liq	-7	212; 104[11]	1.0527[20]	1.4532[20]	sl H_2O, ctc; s MeOH
2664	cis-1,2-Cyclopentanediol		$C_5H_{10}O_2$	5057-98-7	102.132		30	124[29], 100[10]			
2665	trans-1,2-Cyclopentanediol		$C_5H_{10}O_2$	5057-99-8	102.132		54.7	226; 136[21]			
2666	Cyclopentanemethanol		$C_6H_{12}O$	3637-61-4	100.158			163	0.9332[20]	1.4579[20]	
2667	Cyclopentanepropanoic acid		$C_8H_{14}O_2$	140-77-2	142.196			158[26], 131[12]	1.0100[17]	1.4570[20]	
2668	Cyclopentanethiol	Cyclopentyl mercaptan	$C_5H_{10}S$	1679-07-8	102.198			132.1	0.9550[20]		
2669	Cyclopentanol	Cyclopentyl alcohol	$C_5H_{10}O$	96-41-3	86.132	liq	-17.5	140.42	0.9488[20]	1.4530[20]	sl H_2O, ctc; s EtOH, eth, ace
2670	Cyclopentanone	Adipic ketone	C_5H_8O	120-92-3	84.117	liq	-51.90	130.57	0.9487[20]	1.4366[20]	i H_2O; s EtOH, ace, ctc, hx; msc eth
2671	Cyclopentanone oxime		C_5H_9NO	1192-28-5	99.131		57.8	196			vs H_2O, bz

4-Cyclohexyl-3-ethyl-4*H*-1,2,4-triazole

Cyclohexyl formate

Cyclohexyl hydroperoxide

Cyclohexylideneacetonitrile

2-Cyclohexylidenecyclohexanone

Cyclohexyl isocyanate

Cyclohexylisopropylamine

Cyclohexyl isothiocyanate

Cyclohexylmagnesium chloride

Cyclohexyl methacrylate

Cyclohexylmethylamine

Cyclohexyl 2-methylpropanoate

2-Cyclohexylphenol

4-Cyclohexylphenol

α-Cyclohexyl-α-phenyl-1-piperidinepropanol

Cyclohexyl propanoate

Cyclohexylsulfamic acid

Cyclononane

Cyclononanone

1,4-Cyclooctadiene

cis,cis-1,5-Cyclooctadiene

Cyclooctanamine

Cyclooctane

Cyclooctanol

Cyclooctanone

1,3,5,7-Cyclooctatetraene

1,3,5-Cyclooctatriene

cis-Cyclooctene

trans-Cyclooctene

Cyclooctyne

Cyclopamine

Cyclopentadecane

Cyclopentadecanol

Cyclopentadecanone

1,3-Cyclopentadiene

Cyclopentane

Cyclopentaneacetic acid

Cyclopentanecarbonitrile

Cyclopentanecarboxaldehyde

Cyclopentanecarboxylic acid

cis-1,2-Cyclopentanediol

trans-1,2-Cyclopentanediol

Cyclopentanemethanol

Cyclopentanepropanoic acid

Cyclopentanethiol

Cyclopentanol

Cyclopentanone

Cyclopentanone oxime

No.	Name	Synonym	Mol. Form.	CAS RN	Mol. Wt.	Physical Form	mp/°C	bp/°C	den/ g cm⁻³	n_D	Solubility
2672	Cyclopentene		C₅H₈	142-29-0	68.118	liq	-135.0	44.2	0.7720²⁰	1.4225²⁰	i H₂O; s EtOH, eth, bz, ctc, peth
2673	1-Cyclopentenecarbonitrile	1-Cyanocyclopentene	C₆H₇N	3047-38-9	93.127	liq		81³⁰			
2674	1-Cyclopentene-1-carboxaldehyde		C₆H₈O	6140-65-4	96.127	liq	-32	146	0.970²¹	1.4872¹⁷	
2675	2-Cyclopentene-1-tridecanoic acid, (S)	Chaulmoogric acid	C₁₈H₃₂O₂	29106-32-9	280.446	pl or lf (al, HOAc)	68.5	247²⁰			vs eth, chl
2676	2-Cyclopentene-1-undecanoic acid, (R)	Hydnocarpic acid	C₁₆H₂₈O₂	459-67-6	252.392		60.5				vs EtOH, chl, peth
2677	2-Cyclopenten-1-one		C₅H₆O	930-30-3	82.101			136; 40¹²	0.989¹⁵	1.4629¹⁵	vs eth, EtOH
2678	3-Cyclopenten-1-one		C₅H₆O	14320-37-7	82.101	liq		28¹⁷			
2679	N-(1-Cyclopenten-1-yl)pyrrolidine	1-Pyrrolidinylcyclopentene	C₉H₁₅N	7148-07-4	137.222			105¹⁵		1.5128²⁰	
2680	Cyclopenthiazide		C₁₃H₁₈ClN₃O₄S₂	742-20-1	379.883		238				
2681	Cyclopentobarbital		C₁₂H₁₄N₂O₃	76-68-6	234.250	cry (w, dil al)	139.5				sl H₂O; vs EtOH
2682	Cyclopentylamine	Cyclopentanamine	C₅H₁₁N	1003-03-8	85.148	liq	-82.7	108	0.8689²⁰	1.4728²⁵	s ace, bz, chl
2683	Cyclopentylbenzene		C₁₁H₁₄	700-88-9	146.229			219	0.9462²⁰	1.5280²⁰	vs eth
2684	2-Cyclopentylidenecyclopentanone		C₁₀H₁₄O	825-25-2	150.217			135²⁵	1.0179¹⁸	1.5215¹⁸	
2685	Cyclopentyl methyl sulfide		C₆H₁₂S	7133-36-0	116.224			156.2			
2686	Cyclophosphamide	Cyclophosphane	C₇H₁₅Cl₂N₂O₂P	50-18-0	261.086		43				vs H₂O; sl bz, chl, diox, EtOH
2687	Cycloposine		C₃₃H₅₁NO₇	23185-94-6	573.761		268				
2688	Cyclopropane	Trimethylene	C₃H₆	75-19-4	42.080	col gas	-127.58	-32.81	0.617²⁵ (p>1 atm)	1.3799⁻⁴²	s H₂O, bz, peth; vs EtOH, eth
2689	Cyclopropanecarbonitrile	Cyclopropyl cyanide	C₄H₅N	5500-21-0	67.090			135.1	0.8946²⁰	1.4229²⁰	s eth, hx; sl ctc
2690	Cyclopropanecarbonyl chloride		C₄H₅ClO	4023-34-1	104.535			119	1.1516²⁰		
2691	Cyclopropanecarboxaldehyde	Formylcyclopropane	C₄H₆O	1489-69-6	70.090	liq		100	0.938	1.4298²⁰	
2692	Cyclopropanecarboxylic acid		C₄H₆O₂	1759-53-1	86.090		18.5	183	1.0885²⁰	1.4390²⁰	s H₂O, EtOH, eth; sl ctc
2693	1,1-Cyclopropanedicarboxylic acid		C₅H₆O₄	598-10-7	130.100	pr or nd (chl) pr (w +1)	140.5				vs H₂O, eth
2694	Cyclopropanemethanol		C₄H₈O	2516-33-8	72.106			124	0.911²⁵		sl ctc
2695	Cyclopropanone		C₃H₄O	5009-27-8	56.063		stable only at low temp.				
2696	Cyclopropene		C₃H₄	2781-85-3	40.064	gas		dec -36			
2697	Cyclopropylamine	Cyclopropanamine	C₃H₇N	765-30-0	57.095	liq	-35.39	50.5	0.8240²⁰	1.4210²⁰	msc H₂O; s EtOH, eth, chl
2698	Cyclopropylbenzene		C₉H₁₀	873-49-4	118.175	liq	-31	173.6; 80³⁷	0.9317²⁰	1.5285²⁰	i H₂O; s eth, ace, chl
2699	Cyclopropyl methyl ether		C₄H₈O	540-47-6	72.106	liq	-119	44.7	0.8100²⁰	1.3802²⁰	vs H₂O, bz, eth, EtOH
2700	Cyclopropyl methyl ketone		C₅H₈O	765-43-5	84.117	liq	-68.3	111.3	0.8984²⁰	1.4251²⁰	vs H₂O, eth, EtOH
2701	Cyclotetramethylenetetranitramine	HMX	C₄H₈N₈O₈	2691-41-0	296.156	cry	286	exp			
2702	Cyclothiazide		C₁₄H₁₆ClN₃O₄S₂	2259-96-3	389.878		234				
2703	Cycluron	N'-Cyclooctyl-N,N-dimethylurea	C₁₁H₂₂N₂O	2163-69-1	198.305	cry	138				sl H₂O; s bz, ace; vs MeOH
2704	Cyfluthrin		C₂₂H₁₈Cl₂FNO₃	68359-37-5	434.287		60				
2705	Cygon		C₅H₁₂NO₃PS₂	60-51-5	229.258		52	117⁰·¹	1.277⁶⁵		
2706	Cyhalothrin	2,2-Dimethylcyclopropanecarboxylate	C₂₃H₁₉ClF₃NO₃	91465-08-6	449.850		49.2				
2707	Cyhexatin	Stannane, tricyclohexylhydroxy-	C₁₈H₃₄OSn	13121-70-5	385.172		196				
2708	Cypermethrin		C₂₂H₁₉Cl₂NO₃	52315-07-8	416.297		70		1.25²⁰		
2709	Cyprazine		C₉H₁₄ClN₅	22936-86-3	227.694		167				
2710	Cyproheptadine		C₂₁H₂₁N	129-03-3	287.399	cry (EtOH aq)	113				
2711	Cyromazine	N-Cyclopropyl-1,3,5-triazine-2,4,6-triamine	C₆H₁₀N₆	66215-27-8	166.183	cry	220				
2712	Cystamine dihydrochloride		C₄H₁₄Cl₂N₂S₂	56-17-7	225.203	nd (MeOH)	218 dec				vs H₂O, EtOH
2713	Cysteamine		C₂H₇NS	60-23-1	77.149	cry (sub)	99.5	dec			vs H₂O, EtOH
2714	L-Cysteic acid		C₃H₇NO₅S	13100-82-8	169.157	cry	260 dec				s H₂O; i EtOH
2715	L-Cysteine	Propanoic acid, 2-amino-3-mercapto-, (R)-	C₃H₇NO₂S	52-90-4	121.159	cry (w)	240 dec				vs H₂O, ace, EtOH
2716	L-Cysteine, ethyl ester, hydrochloride		C₅H₁₂ClNO₂S	868-59-7	185.673		125.8				vs H₂O
2717	L-Cysteine, hydrochloride		C₃H₈ClNO₂S	52-89-1	157.620	cry	175 dec				s H₂O
2718	L-Cystine	3,3'-Dithiobis(2-aminopropanoic acid)	C₆H₁₂N₂O₄S₂	56-89-3	240.300	hex pl or pr (w)	260 dec		1.677²⁵		sl H₂O; i EtOH, eth, bz; s acid, alk
2719	Cytarabine	Cytosine arabinoside	C₉H₁₃N₃O₅	147-94-4	243.216	pr (EtOH aq)	212				s H₂O

Cyclopentene

1-Cyclopentenecarbonitrile

1-Cyclopentene-1-carboxaldehyde

2-Cyclopentene-1-tridecanoic acid, (S)

2-Cyclopentene-1-undecanoic acid, (R)

2-Cyclopenten-1-one

3-Cyclopenten-1-one

N-(1-Cyclopenten-1-yl)pyrrolidine

Cyclopenthiazide

Cyclopentobarbital

Cyclopentylamine

Cyclopentylbenzene

2-Cyclopentylidenecyclopentanone

Cyclopentyl methyl sulfide

Cyclophosphamide

Cycloposine

Cyclopropane

Cyclopropanecarbonitrile

Cyclopropanecarbonyl chloride

Cyclopropanecarboxaldehyde

Cyclopropanecarboxylic acid

1,1-Cyclopropanedicarboxylic acid

Cyclopropanemethanol

Cyclopropanone

Cyclopropene

Cyclopropylamine

Cyclopropylbenzene

Cyclopropyl methyl ether

Cyclopropyl methyl ketone

Cyclotetramethylenetetranitramine

Cyclothiazide

Cycluron

Cyfluthrin

Cygon

Cyhalothrin

Cyhexatin

Cypermethrin

Cyprazine

Cyproheptadine

Cyromazine

Cystamine dihydrochloride

Cysteamine

L-Cysteic acid

L-Cysteine

L-Cysteine, ethyl ester, hydrochloride

L-Cysteine, hydrochloride

L-Cystine

Cytarabine

No.	Name	Synonym	Mol. Form.	CAS RN	Mol. Wt.	Physical Form	mp/°C	bp/°C	den/ g cm⁻³	n_D	Solubility
2720	Cytidine	4-Amino-1-β-D-ribofuranosyl-2(1H)-pyrimidinone	$C_9H_{13}N_3O_5$	65-46-3	243.216	nd (dil al)	230 dec				vs H_2O; sl EtOH
2721	2'-Cytidylic acid	Cytidine 2'-monophosphate	$C_9H_{14}N_3O_8P$	85-94-9	323.196		239 dec				
2722	3'-Cytidylic acid	Cytidine 3'-monophosphate	$C_9H_{14}N_3O_8P$	84-52-6	323.196		233 dec				s H_2O, EtOH
2723	5'-Cytidylic acid	Cytidine 5'-monophosphate	$C_9H_{14}N_3O_8P$	63-37-6	323.196	orth nd	233 dec				vs H_2O, EtOH
2724	Cytisine	Sophorine	$C_{11}H_{14}N_2O$	485-35-8	190.241	pr	153	218[2]			vs H_2O, EtOH, MeOH; s bz, ace
2725	Cytochalasin B		$C_{29}H_{37}NO_5$	14930-96-2	479.608	nd (ace)	219				
2726	Cytochalasin D	Zygosporin A	$C_{30}H_{37}NO_6$	22144-77-0	507.618	nd (ace/ peth)	270				
2727	Cytochalasin E		$C_{28}H_{33}NO_7$	36011-19-5	495.565		207				
2728	Cytosine		$C_4H_5N_3O$	71-30-7	111.102	mcl or tcl pl (w+1)	322 dec				s H_2O; sl EtOH, chl; i eth
2729	Dacarbazine	5-(3,3-Dimethyl-1-triazenyl)-1H-imidazole-4-carboxamide	$C_6H_{10}N_6O$	4342-03-4	182.182	cry	205				
2730	Dactinomycin		$C_{62}H_{86}N_{12}O_{16}$	50-76-0	1255.416		245 dec				
2731	Daidzein	7-Hydroxy-3-(4-hydroxyphenyl)-4H-1-benzopyran-4-one	$C_{15}H_{10}O_4$	486-66-8	254.238	pa ye pr (50% al)	323 dec	sub			s EtOH, eth
2732	Daminozide	Butanedioic acid, mono(2,2-dimethylhydrazide)	$C_6H_{12}N_2O_3$	1596-84-5	160.170		154.5				
2733	Dantrolene		$C_{14}H_{10}N_4O_5$	7261-97-4	314.253	cry (DMF aq)	280				
2734	Datiscetin		$C_{15}H_{10}O_6$	480-15-9	286.236	pa ye nd (al, aq HOAc)	277.5				vs ace, eth, EtOH
2735	Daucol		$C_{15}H_{26}O_2$	887-08-1	238.366	cry	114	128[2]			
2736	Daunorubicin		$C_{27}H_{29}NO_{10}$	20830-81-3	527.520	red nd	208				
2737	Dazomet		$C_5H_{10}N_2S_2$	533-74-4	162.276	nd (bz)	106				reac H_2O; s EtOH
2738	Decabromobiphenyl ether	Bis(pentabromophenyl) ether	$C_{12}Br_{10}O$	1163-19-5	959.167	ye pr (tol)	305				i H_2O
2739	Decachlorobiphenyl		$C_{12}Cl_{10}$	2051-24-3	498.658	cry (bz)	309				i H_2O
2740	1,3-Decadiene	1-Hexyl-1,3-butadiene	$C_{10}H_{18}$	2051-25-4	138.250			169	0.752[30]		vs bz
2741	1,9-Decadiene		$C_{10}H_{18}$	1647-16-1	138.250			167	0.75[25]	1.4325[20]	
2742	2,2',3,3',4,4',5,5',6,6'-Decafluoro-1,1'-biphenyl		$C_{12}F_{10}$	434-90-2	334.112		67.5	206	1.785[20]		
2743	cis-Decahydronaphthalene	cis-Decalin	$C_{10}H_{18}$	493-01-6	138.250	liq	-42.9	195.8	0.8965[20]	1.4810[20]	i H_2O; msc EtOH; vs eth, ace, chl
2744	trans-Decahydronaphthalene	trans-Decalin	$C_{10}H_{18}$	493-02-7	138.250	liq	-30.4	187.3	0.8659[25]	1.4695[20]	i H_2O; vs EtOH, eth, ace; msc bz; sl MeOH
2745	Decahydro-2-naphthol	Decahydro-β-naphthol	$C_{10}H_{18}O$	825-51-4	154.249			109[14]	0.996[25]	1.4992[20]	
2746	Decamethonium dibromide		$C_{16}H_{38}Br_2N_2$	541-22-0	418.294	cry (MeOH/ ace)	269 dec				i eth
2747	Decamethylcyclopentasiloxane		$C_{10}H_{30}O_5Si_5$	541-02-6	370.770	liq	-38	210	0.9593[20]	1.3982[20]	i H_2O
2748	Decamethyltetrasiloxane		$C_{10}H_{30}O_3Si_4$	141-62-8	310.685	liq	-76	194	0.8536[25]	1.3895[20]	i H_2O; sl EtOH; s bz, peth
2749	Decanal	Capraldehyde	$C_{10}H_{20}O$	112-31-2	156.265	liq	-4.0	208.5	0.830[15]	1.4287[20]	i H_2O; s EtOH, eth, ace; sl ctc
2750	Decane		$C_{10}H_{22}$	124-18-5	142.282	liq	-29.6	174.15	0.7266[25]	1.4090[25]	i H_2O; msc EtOH; s eth; sl ctc
2751	1,10-Decanediamine		$C_{10}H_{24}N_2$	646-25-3	172.311		59.73	140[12]			
2752	Decanedinitrile		$C_{10}H_{16}N_2$	1871-96-1	164.247		7.6	204[16]	0.913[20]	1.4474[20]	i H_2O; s chl
2753	1,10-Decanediol	Decamethylene glycol	$C_{10}H_{22}O_2$	112-47-0	174.281	nd (w, dil al)	74	192[20]			sl H_2O, eth; vs EtOH; s DMSO; i lig
2754	Decanedioyl dichloride		$C_{10}H_{16}Cl_2O_2$	111-19-3	239.139		-1.3	220[75], 165[11]	1.1212[20]	1.4684[18]	
2755	Decanenitrile	Caprinitrile	$C_{10}H_{19}N$	1975-78-6	153.265	liq	-17.9	243; 106[10]	0.8199[20]	1.4296[20]	vs ace, eth, EtOH, chl
2756	1-Decanethiol	Decyl mercaptan	$C_{10}H_{22}S$	143-10-2	174.347	liq	-26	240.6	0.8443[20]	1.4509[20]	i H_2O; s EtOH, eth
2757	Decanoic acid	Capric acid	$C_{10}H_{20}O_2$	334-48-5	172.265	nd	31.4	268.7	0.8858[40]	1.4288[40]	i H_2O; vs ace, bz, eth, EtOH
2758	1-Decanol	Capric alcohol	$C_{10}H_{22}O$	112-30-1	158.281		6.9	231.1	0.8297[20]	1.4372[20]	i H_2O; msc EtOH, eth, ace, bz, chl; s ctc
2759	2-Decanol		$C_{10}H_{22}O$	74742-10-2	158.281	liq	-1.2	211	0.8250[20]	1.4326[25]	s EtOH, bz; msc eth, ace; sl ctc
2760	3-Decanol		$C_{10}H_{22}O$	1565-81-7	158.281	liq	-7.5	213; 101[12]	0.827[20]	1.434[20]	
2761	4-Decanol	1-Propylheptyl alcohol	$C_{10}H_{22}O$	2051-31-2	158.281	liq	-11	210.5	0.8261[20]	1.4320[20]	i H_2O; s EtOH, ctc
2762	5-Decanol		$C_{10}H_{22}O$	5205-34-5	158.281	liq	8.7	201	0.824[20]	1.4333[20]	
2763	2-Decanone	Methyl octyl ketone	$C_{10}H_{20}O$	693-54-9	156.265	nd	14	210; 96[12]	0.8248[20]	1.4255[20]	i H_2O; s EtOH, eth; sl ctc

Cytidine

2'-Cytidylic acid

3'-Cytidylic acid

5'-Cytidylic acid

Cytisine

Cytochalasin B

Cytochalasin D

Cytochalasin E

Cytosine

Dacarbazine

Dactinomycin

Daidzein

Daminozide

Dantrolene

Datiscetin

Daucol

Daunorubicin

Dazomet

Decabromobiphenyl ether

Decachlorobiphenyl

1,3-Decadiene

1,9-Decadiene

2,2',3,3',4,4',5,5',6,6'-Decafluoro-1,1'-biphenyl

cis-Decahydronaphthalene

trans-Decahydronaphthalene

Decahydro-2-naphthol

Decamethonium dibromide

Decamethylcyclopentasiloxane

Decamethyltetrasiloxane

Decanal

Decane

1,10-Decanediamine

Decanedinitrile

1,10-Decanediol

Decanedioyl dichloride

Decanenitrile

1-Decanethiol

Decanoic acid

1-Decanol

2-Decanol

3-Decanol

4-Decanol

5-Decanol

2-Decanone

No.	Name	Synonym	Mol. Form.	CAS RN	Mol. Wt.	Physical Form	mp/°C	bp/°C	den/ g cm⁻³	n_D	Solubility
2764	3-Decanone	Ethyl heptyl ketone	$C_{10}H_{20}O$	928-80-3	156.265	liq	1.3	203	0.8251[20]	1.4252[20]	s EtOH, eth, ctc
2765	4-Decanone	Hexyl propyl ketone	$C_{10}H_{20}O$	624-16-8	156.265	liq	-9	206.5	0.824[20]	1.4240[21]	i H₂O; msc EtOH, eth
2766	Decanoyl chloride	Caprinoyl chloride	$C_{10}H_{19}ClO$	112-13-0	190.710	liq	-34.5	95	0.919[25]	1.4410[20]	s eth, ctc
2767	trans-2-Decenal		$C_{10}H_{18}O$	3913-81-3	154.249			230; 107[11]			
2768	1-Decene		$C_{10}H_{20}$	872-05-9	140.266	liq	-66.3	170.5	0.7408[20]	1.4215[20]	i H₂O; msc EtOH, eth
2769	cis-2-Decene		$C_{10}H_{20}$	20348-51-0	140.266	col liq		174.2			
2770	trans-2-Decene		$C_{10}H_{20}$	20063-97-2	140.266	col liq		173.3			
2771	cis-5-Decene		$C_{10}H_{20}$	7433-78-5	140.266	col liq	-112	171; 73[20]	0.7445[20]	1.4258[20]	i H₂O; msc EtOH, eth; sl ctc
2772	trans-5-Decene		$C_{10}H_{20}$	7433-56-9	140.266	col liq	-73	171	0.7401[20]	1.4243[20]	i H₂O; msc EtOH, eth; sl ctc
2773	9-Decenoic acid	Caproleic acid	$C_{10}H_{18}O_2$	14436-32-9	170.249		26.5	158[21], 142[4]	0.9238[15]	1.4507[15]	vs eth, EtOH
2774	9-Decen-1-ol	Decylenic alcohol	$C_{10}H_{20}O$	13019-22-2	156.265			236	0.876[25]	1.4480[20]	
2775	3-Decen-2-one	Heptylidene acetone	$C_{10}H_{18}O$	10519-33-2	154.249			102[15.3]	0.8473[20]	1.4480[20]	
2776	Declomycin	Demeclocycline	$C_{21}H_{21}ClN_2O_8$	127-33-3	464.853	cry	176 dec				
2777	Decyl acetate		$C_{12}H_{24}O_2$	112-17-4	200.318	liq	-15	244	0.8671[20]	1.4273[20]	i H₂O; s EtOH, eth, bz, ctc, HOAc
2778	Decylamine	1-Decanamine	$C_{10}H_{23}N$	2016-57-1	157.297		17	220.5	0.7936[20]	1.4369[20]	sl H₂O; msc EtOH, eth, ace, bz, chl
2779	Decylbenzene		$C_{16}H_{26}$	104-72-3	218.377	liq	-14.4	293	0.8555[20]	1.4832[20]	vs ace, bz, eth, EtOH
2780	Decylcyclohexane		$C_{16}H_{32}$	1795-16-0	224.425	liq	-0.9	299	0.8186[20]	1.4534[20]	
2781	Decylcyclopentane		$C_{15}H_{30}$	1795-21-7	210.399	liq	-22	279	0.8110[20]	1.4486[20]	vs ace, bz, eth, EtOH
2782	Decyl decanoate		$C_{20}H_{40}O_2$	1654-86-0	312.531		9.7	219[15]	0.8586[20]	1.4423[20]	vs eth
2783	Decyl formate		$C_{11}H_{22}O_2$	5451-52-5	186.292	liq		243			
2784	11-Decylheneicosane		$C_{31}H_{64}$	55320-06-4	436.840		10.0	282.0[10]	0.8116[20]	1.4540[20]	
2785	1-Decylnaphthalene		$C_{20}H_{28}$	26438-27-7	268.436		15	379	0.9322[20]	1.5435[20]	
2786	Decyloxirane		$C_{12}H_{24}O$	2855-19-8	184.318					1.4347[25]	sl ctc
2787	Decyl vinyl ether	1-(Ethenyloxy)decane	$C_{12}H_{24}O$	765-05-9	184.318		-41	101[10]	0.812[20]	1.4346[20]	
2788	1-Decyne	Octylacetylene	$C_{10}H_{18}$	764-93-2	138.250	liq	-44	174	0.7655[20]	1.4265[20]	i H₂O; s EtOH, eth
2789	5-Decyne	Dibutylacetylene	$C_{10}H_{18}$	1942-46-7	138.250	liq	-73	177; 78.8[25]	0.7690[20]	1.4331[20]	i H₂O; s EtOH, eth
2790	Dehydroabietic acid	8,11,13-Abietatrien-18-oic acid	$C_{20}H_{28}O_2$	1740-19-8	300.435	cry (EtOH aq)	172				
2791	Delphinidin		$C_{15}H_{11}ClO_7$	528-53-0	338.697		>350				vs H₂O, EtOH, MeOH; s AcOEt
2792	Delphinine		$C_{33}H_{45}NO_9$	561-07-9	599.712	orth (al)	199				i H₂O; s chl, ace, eth; vs EtOH
2793	Deltamethrin		$C_{22}H_{19}Br_2NO_3$	52918-63-5	505.199		99				
2794	Demecarium bromide		$C_{32}H_{52}Br_2N_4O_4$	56-94-0	716.588	hyg pow	165 dec				vs H₂O; sl ace; i ace, eth
2795	Demeton	Systox	$C_8H_{19}O_3PS_2$	8065-48-3	258.339	oily liq		134[2]			i H₂O; s EtOH, tol
2796	Demeton-S-methyl		$C_6H_{15}O_3PS_2$	919-86-8	230.285	ye liq		89[0.15], 118[1]	1.20[20]	1.5063[20]	i H₂O; s os
2797	2'-Deoxyadenosine		$C_{10}H_{13}N_5O_3$	958-09-8	251.242						sl H₂O
2798	2'-Deoxyadenosine 5'-triphosphate		$C_{10}H_{16}N_5O_{12}P_3$	1927-31-7	491.182	cry (EtOH aq)					
2799	6-Deoxy-L-ascorbic acid		$C_6H_8O_5$	528-81-4	160.125	pr (AcOEt)	168	sub 160			vs H₂O, ace, EtOH
2800	Deoxycholic acid	3,12-Dihydroxycholan-24-oic acid, (3α,5β,12α)	$C_{24}H_{40}O_4$	83-44-3	392.573	cry (al)	177				
2801	2'-Deoxycytidine 5'-monophosphate	2'-Deoxy-5'-cytidylic acid	$C_9H_{14}N_3O_7P$	1032-65-1	307.197	pow	183 dec				
2802	2'-Deoxy-5-fluorouridine	Floxuridine	$C_9H_{11}FN_2O_5$	50-91-9	246.191	cry	150				
2803	2-Deoxy-D-glucose		$C_6H_{12}O_5$	154-17-6	164.156		146.5				
2804	2'-Deoxyguanosine 5'-monophosphate	2'-Deoxy-5'-guanylic acid	$C_{10}H_{14}N_5O_7P$	902-04-5	347.222						s H₂O
2805	2-Deoxy-D-chiro-inositol	D-Quercitol	$C_6H_{12}O_5$	488-73-3	164.156	pr (w, dil al)	236			1.5845[13]	vs H₂O
2806	1-Deoxy-1-(methylamino)-D-glucitol	N-Methylglucamine	$C_7H_{17}NO_5$	6284-40-8	195.214	cry (MeOH)	128.5				s H₂O
2807	6-Deoxy-3-O-methylgalactose	Digitalose	$C_7H_{14}O_5$	4481-08-7	178.183	nd (AcOEt)	119				vs H₂O
2808	D-2-Deoxyribose		$C_5H_{10}O_4$	533-67-5	134.131		90				
2809	Deserpidine		$C_{32}H_{38}N_2O_8$	131-01-1	578.652	nd or pr	230.5				i H₂O; s EtOH, chl

3-Decanone

4-Decanone

Decanoyl chloride

trans-2-Decenal

1-Decene

cis-2-Decene

trans-2-Decene

cis-5-Decene

trans-5-Decene

9-Decenoic acid

9-Decen-1-ol

3-Decen-2-one

Declomycin

Decyl acetate

Decylamine

Decylbenzene

Decylcyclohexane

Decylcyclopentane

Decyl decanoate

Decyl formate

11-Decylheneicosane

1-Decylnaphthalene

Decyloxirane

Decyl vinyl ether

1-Decyne

5-Decyne

Dehydroabietic acid

Delphinidin

Delphinine

Deltamethrin

Demecarium bromide

Demeton

Demeton-*S*-methyl

2'-Deoxyadenosine

2'-Deoxyadenosine 5'-triphosphate

6-Deoxy-*L*-ascorbic acid

Deoxycholic acid

2'-Deoxycytidine 5'-monophosphate

2'-Deoxy-5-fluorouridine

2-Deoxy-*D*-glucose

2'-Deoxyguanosine 5'-monophosphate

2-Deoxy-*D*-*chiro*-inositol

1-Deoxy-1-(methylamino)-*D*-glucitol

6-Deoxy-3-*O*-methylgalactose

D-2-Deoxyribose

Deserpidine

No.	Name	Synonym	Mol. Form.	CAS RN	Mol. Wt.	Physical Form	mp/°C	bp/°C	den/ g cm⁻³	n_D	Solubility
2810	Desethyl atrazine	6-Chloro-*N*-isopropyl-1,3,5-triazine-2,4-diamine	C₆H₁₀ClN₅	6190-65-4	187.630	cry	136				
2811	Desferrioxamine	Deferoxamine	C₂₅H₄₈N₆O₈	70-51-9	560.684	cry (EtOH aq)	139				
2812	Desipramine		C₁₈H₂₂N₂	50-47-5	266.381			173[0.02]			
2813	Desmedipham		C₁₆H₁₆N₂O₄	13684-56-5	300.309		120				
2814	Desmetryne		C₈H₁₅N₅S	1014-69-3	213.304	cry	85				
2815	Desthiobiotin		C₁₀H₁₈N₂O₃	533-48-2	214.261	lo nd (H₂O)	157				s H₂O
2816	Dexamethasone		C₂₂H₂₉FO₅	50-02-2	392.460		262				
2817	Dexon	Sodium dimethylaminobenzenediazosulfonate	C₈H₁₀N₃NaO₃S	140-56-7	251.238	ye-br pow					sl H₂O; s DMF
2818	Dexpanthenol		C₉H₁₉NO₄	81-13-0	205.252	hyg oil		dec	1.20[20]	1.497[20]	vs H₂O, EtOH, MeOH; sl eth
2819	Dextroamphetamine sulfate		C₁₈H₂₈N₂O₄S	51-63-8	368.491		>300		1.15[25]		vs H₂O
2820	Dextromethorphan hydrobromide		C₁₈H₂₆BrNO	125-69-9	352.309	wh cry pow	123				s EtOH, chl; i eth
2821	Diacetone alcohol	4-Hydroxy-4-methyl-2-pentanone	C₆H₁₂O₂	123-42-2	116.158	liq	-44	167.9	0.9387[20]	1.4213[20]	msc H₂O, EtOH, eth; s chl
2822	3,3-Diacetoxy-1-propene		C₇H₁₀O₄	869-29-4	158.152	liq	-37.6	180	1.0760[20]	1.4193[20]	vs ace, bz, eth, EtOH
2823	1,3-Diacetylbenzene		C₁₀H₁₀O₂	6781-42-6	162.185		32	152[15]			sl H₂O, peth; s EtOH, bz, chl, HOAc
2824	1,4-Diacetylbenzene	4-Acetylacetophenone	C₁₀H₁₀O₂	1009-61-6	162.185		113.0	128[3]			vs EtOH; sl chl
2825	*N,N'*-Diacetyl-4,4'-diaminobiphenyl		C₁₆H₁₆N₂O₂	613-35-4	268.310	nd (HOAc)	328.3				
2826	Diacetylmorphine		C₂₁H₂₃NO₅	561-27-3	369.412	orth	173	273[12]	1.56[25]		vs bz, chl
2827	Diacetylperoxide	Acetyl peroxide	C₄H₆O₄	110-22-5	118.089	nd (eth) lf	30	63[21]			vs eth, EtOH
2828	Dialifor		C₁₄H₁₇ClNO₄PS₂	10311-84-9	393.846		68				
2829	Diallate		C₁₀H₁₇Cl₂NOS	2303-16-4	270.219			150[9]			
2830	Diallylcyanamide		C₇H₁₀N₂	538-08-9	122.167			142[90], 95[9]			s EtOH; sl eth, ctc
2831	Diallyl diethylene glycol carbonate	Diethylene glycol bis(allyl carbonate)	C₁₂H₁₈O₇	142-22-3	274.267	col liq	-4	161[2]	1.14[20]		i H₂O; s os
2832	Diallyldimethylsilane		C₈H₁₆Si	1113-12-8	140.299			137; 68[50]	0.7679[20]	1.4420[20]	
2833	Diallyl disulfide		C₆H₁₀S₂	2179-57-9	146.273			100[48], 79[16]	1.0237[15]		
2834	Diallyl ether	Allyl ether	C₆H₁₀O	557-40-4	98.142	liq	-6	94	0.8260[20]	1.4163[20]	i H₂O; msc EtOH, eth; vs ace; s chl
2835	Diallyl fumarate		C₁₀H₁₂O₄	2807-54-7	196.200			140[3]	1.0768[20]	1.4670[25]	vs ace, bz, eth, EtOH
2836	Diallyl isophthalate	Di-2-propenyl 1,3-benzenedicarboxylate	C₁₄H₁₄O₄	1087-21-4	246.259			176[5]			
2837	Diallyl maleate		C₁₀H₁₂O₄	999-21-3	196.200			129[10], 109[3]	1.075[20]	1.4699[20]	s chl
2838	Diallyl oxalate		C₈H₁₀O₄	615-99-6	170.163			217	1.1582[20]	1.4481[20]	i H₂O; s EtOH, ace, bz; sl chl
2839	*N,N*-Diallyl-2-propen-1-amine	Triallylamine	C₉H₁₅N	102-70-5	137.222		94	155.5	0.809[20]	1.4502[20]	s EtOH, eth, ace, bz, acid
2840	5,5-Diallyl-2,4,6(1*H*,3*H*,5*H*)-pyrimidinetrione	Allobarbital	C₁₀H₁₂N₂O₃	52-43-7	208.213	lf	172				sl H₂O, DMSO; s EtOH, eth, bz
2841	Diallyl sulfide		C₆H₁₀S	592-88-1	114.208	liq	-85	138.6	0.8877[27]	1.4870[25]	vs eth, EtOH
2842	Diallyl trisulfide		C₆H₁₀S₃	2050-87-5	178.338			117[16]	1.0845[15]		vs eth
2843	Diamantane	Congressane	C₁₄H₂₀	2292-79-7	188.309	cry	236				
2844	1,2-Diamino-9,10-anthracenedione		C₁₄H₁₀N₂O₂	1758-68-5	238.241	viol nd	303.5				sl EtOH, eth, chl, xyl; s py, con sulf
2845	1,4-Diamino-9,10-anthracenedione		C₁₄H₁₀N₂O₂	128-95-0	238.241	dk viol nd (py)	268				sl H₂O; s EtOH, bz, PhNO₂; vs py
2846	1,5-Diamino-9,10-anthracenedione		C₁₄H₁₀N₂O₂	129-44-2	238.241	dk red nd (al, HOAc)	319	sub			i H₂O; sl EtOH, eth, ace, bz; s PhNO₂
2847	1,8-Diamino-9,10-anthracenedione		C₁₄H₁₀N₂O₂	129-42-0	238.241	red nd (al, HOAc)	265				i H₂O; s EtOH, py; sl eth, HOAc
2848	2,6-Diamino-9,10-anthracenedione		C₁₄H₁₀N₂O₂	131-14-6	238.241	red-br pr (aq-py)	320 dec				sl H₂O; s EtOH, chl, con sulf, xyl, py
2849	4,4'-Diaminoazobenzene		C₁₂H₁₂N₄	538-41-0	212.250	ye nd (al), oran-ye pr (al)	250.5				sl H₂O, lig; s EtOH; vs bz, chl
2850	3,5-Diaminobenzoic acid		C₇H₈N₂O₂	535-87-5	152.151	nd (+1w)	228				sl H₂O, tfa; s EtOH; vs eth

Desethyl atrazine

Desferrioxamine

Desipramine

Desmedipham

Desmetryne

Desthiobiotin

Dexamethasone

Dexon

Dexpanthenol

Dextroamphetamine sulfate

Dextromethorphan hydrobromide

Diacetone alcohol

3,3-Diacetoxy-1-propene

1,3-Diacetylbenzene

1,4-Diacetylbenzene

N,N'-Diacetyl-4,4'-diaminobiphenyl

Diacetylmorphine

Diacetylperoxide

Dialifor

Diallate

Diallylcyanamide

Diallyl diethylene glycol carbonate

Diallyldimethylsilane

Diallyl disulfide

Diallyl ether

Diallyl fumarate

Diallyl isophthalate

Diallyl maleate

Diallyl oxalate

N,N-Diallyl-2-propen-1-amine

5,5-Diallyl-2,4,6(1H,3H,5H)-pyrimidinetrione

Diallyl sulfide

Diallyl trisulfide

Diamantane

1,2-Diamino-9,10-anthracenedione

1,4-Diamino-9,10-anthracenedione

1,5-Diamino-9,10-anthracenedione

1,8-Diamino-9,10-anthracenedione

2,6-Diamino-9,10-anthracenedione

4,4'-Diaminoazobenzene

3,5-Diaminobenzoic acid

No.	Name	Synonym	Mol. Form.	CAS RN	Mol. Wt.	Physical Form	mp/°C	bp/°C	den/ g cm⁻³	n_D	Solubility
2851	2,4-Diaminobutanoic acid		$C_4H_{10}N_2O_2$	305-62-4	118.134	hyg cry					s H_2O; sl EtOH, MeOH
2852	cis-2,3-Diamino-2-butenedinitrile		$C_4H_4N_4$	1187-42-4	108.102		178.5		1.41[20]		
2853	1,8-Diamino-4,5-dihydroxy-9,10-anthracenedione		$C_{14}H_{10}N_2O_4$	128-94-9	270.240	bl nd (xyl)					i H_2O; s bz, xyl, EtOH
2854	4,4'-Diaminodiphenyl ether	4,4-Oxydianiline	$C_{12}H_{12}N_2O$	101-80-4	200.235		189 dec	>300			
2855	4,4'-Diaminodiphenylmethane	4,4'-Methylenedianiline	$C_{13}H_{14}N_2$	101-77-9	198.263	pl or nd (w) pl (bz)	92.5	398; 257[18]			sl H_2O; vs EtOH, eth, bz
2856	4,4'-Diaminodiphenyl sulfide	4,4'-Thiodianiline	$C_{12}H_{12}N_2S$	139-65-1	216.301	nd (w)	108.5				sl H_2O; vs EtOH, eth, bz; s tfa
2857	3,3'-Diaminodiphenyl sulfone	3,3'-Sulfonyldianiline	$C_{12}H_{12}N_2O_2S$	599-61-1	248.300		168.5				vs H_2O, EtOH
2858	meso-2,6-Diaminoheptanedioic acid	2,6-Diaminopimelic acid	$C_7H_{14}N_2O_4$	922-54-3	190.197	nd (w)	314 dec				s H_2O
2859	1,4-Diamino-2-methoxy-9,10-anthracenedione		$C_{15}H_{12}N_2O_3$	2872-48-2	268.267		235				
2860	1,4-Diamino-5-nitro-9,10-anthracenedione		$C_{14}H_9N_3O_4$	82-33-7	283.239		278				
2861	2,4-Diaminophenol		$C_6H_8N_2O$	95-86-3	124.140	lf	79 dec				vs H_2O, ace, EtOH
2862	2,4-Diaminophenol, dihydrochloride		$C_6H_{10}Cl_2N_2O$	137-09-7	197.061	nd	235 dec				vs H_2O
2863	3,7-Diaminophenothiazin-5-ium chloride	Thionine	$C_{12}H_{10}ClN_3S$	581-64-6	263.745						sl H_2O, EtOH, eth; s bz, chl, acid
2864	4-[(2,4-Diaminophenyl)azo]benzenesulfonamide	Prontosil	$C_{12}H_{14}ClN_5O_2S$	103-12-8	327.790		249.5				sl H_2O; s EtOH, ace, oils, fats
2865	1,3-Diamino-2-propanol		$C_3H_{10}N_2O$	616-29-5	90.123	cry	42.8				i eth, bz
2866	4,4'-Diamino-2,2'-stilbenedisulfonic acid	Amsonic acid	$C_{14}H_{14}N_2O_6S_2$	81-11-8	370.400	ye nd	300				sl H_2O
2867	4,6-Diamino-1,3,5-triazin-2(1H)-one		$C_3H_5N_5O$	645-92-1	127.105	nd (aq Na_2CO_3)	dec				i H_2O, EtOH, eth, bz, HOAc; s acid, alk
2868	8,8'-Diapo-ψ,ψ-carotenedioic acid	Crocetin	$C_{20}H_{24}O_4$	27876-94-4	328.403	brick red orth	286				sl H_2O, EtOH; i eth, bz; s py; vs NaOH
2869	Diatrizoic acid	N,N'-Diacetyl-3,5-diamino-2,4,6-triiodobenzoic acid	$C_{11}H_9I_3N_2O_4$	117-96-4	613.913	cry (EtOH aq)	300				
2870	Diazenedicarboxamide	Azodicarbonamide	$C_2H_4N_4O_2$	123-77-3	116.079		212 dec				
2871	Diazinon		$C_{12}H_{21}N_2O_3PS$	333-41-5	304.345			87[0.05]	1.1088[20]	1.4922[20]	
2872	Diazomethane		CH_2N_2	334-88-3	42.040	ye gas	-145	-23			vs eth, diox
2873	Dibenz[a,h]acridine		$C_{21}H_{13}N$	226-36-8	279.335	ye cry	228				
2874	Dibenz[a,j]acridine	7-Azadibenz[a,j]anthracene	$C_{21}H_{13}N$	224-42-0	279.335		216				i H_2O
2875	Dibenz[c,h]acridine		$C_{21}H_{13}N$	224-53-3	279.335	ye cry (EtOH)	189				
2876	Dibenz[a,h]anthracene	1,2:5,6-Dibenzanthracene	$C_{22}H_{14}$	53-70-3	278.346	pl (dil ace)	269.5				i H_2O; sl EtOH; s ace, bz, CS_2
2877	Dibenz[a,j]anthracene		$C_{22}H_{14}$	224-41-9	278.346	oran lf or nd (bz)	197.5				i H_2O, HOAc; sl EtOH, eth, bz; s peth
2878	5H-Dibenz[b,f]azepine-5-carboxamide	Carbamazepine	$C_{15}H_{12}N_2O$	298-46-4	236.268		190.2				
2879	Dibenzepin		$C_{18}H_{21}N_3O$	4498-32-2	295.379		117	185[0.01]			
2880	7H-Dibenzo[c,g]carbazole		$C_{20}H_{13}N$	194-59-2	267.324	cry (EtOH)	158				
2881	13H-Dibenzo[a,i]carbazole		$C_{20}H_{13}N$	239-64-5	267.324		221.3				i H_2O
2882	Dibenzo[b,k]chrysene		$C_{26}H_{16}$	217-54-9	328.405		400				
2883	Dibenzo[b,e][1,4]dioxin	Diphenylene dioxide	$C_{12}H_8O_2$	262-12-4	184.191	nd (MeOH)	120.5				
2884	Dibenzofuran	2,2'-Biphenylene oxide	$C_{12}H_8O$	132-64-9	168.191	lf or nd (al)	86.5	287	1.0886[99]	1.6079[99]	i H_2O; s EtOH, ace, bz; vs eth, HOAc
2885	Dibenzo[a,e]pyrene	Naphtho[1,2,3,4-def]chrysene	$C_{24}H_{14}$	192-65-4	302.368	pa ye nd(xyl)	233.5				sl EtOH, ace, bz, HOAc; s tol, con sulf
2886	Dibenzo[a,h]pyrene	Dibenzo[b,def]chrysene	$C_{24}H_{14}$	189-64-0	302.368	oran pl	315				
2887	Dibenzo[a,i]pyrene	Benzo[rst]pentaphene	$C_{24}H_{14}$	189-55-9	302.368		281.5	275[0.05]			
2888	Dibenzo[a,l]pyrene	Dibenzo[def,p]chrysene	$C_{24}H_{14}$	191-30-0	302.368	ye pl (bz/EtOH)	164.5				
2889	Dibenzothiophene		$C_{12}H_8S$	132-65-0	184.257	nd (dil al, lig)	98.2	332.5			i H_2O; s chl, MeOH; vs EtOH, bz
2890	Dibenz[c,e]oxepin-5,7-dione		$C_{14}H_8O_3$	6050-13-1	224.212	nd (HOAc or bz)	217	sub			i H_2O; sl eth
2891	Dibenzoyl disulfide	Benzoyl disulfide	$C_{14}H_{10}O_2S_2$	644-32-6	274.358	pr(al), sc(chl-peth)	134.5	dec			i H_2O; sl EtOH, eth; s CS_2

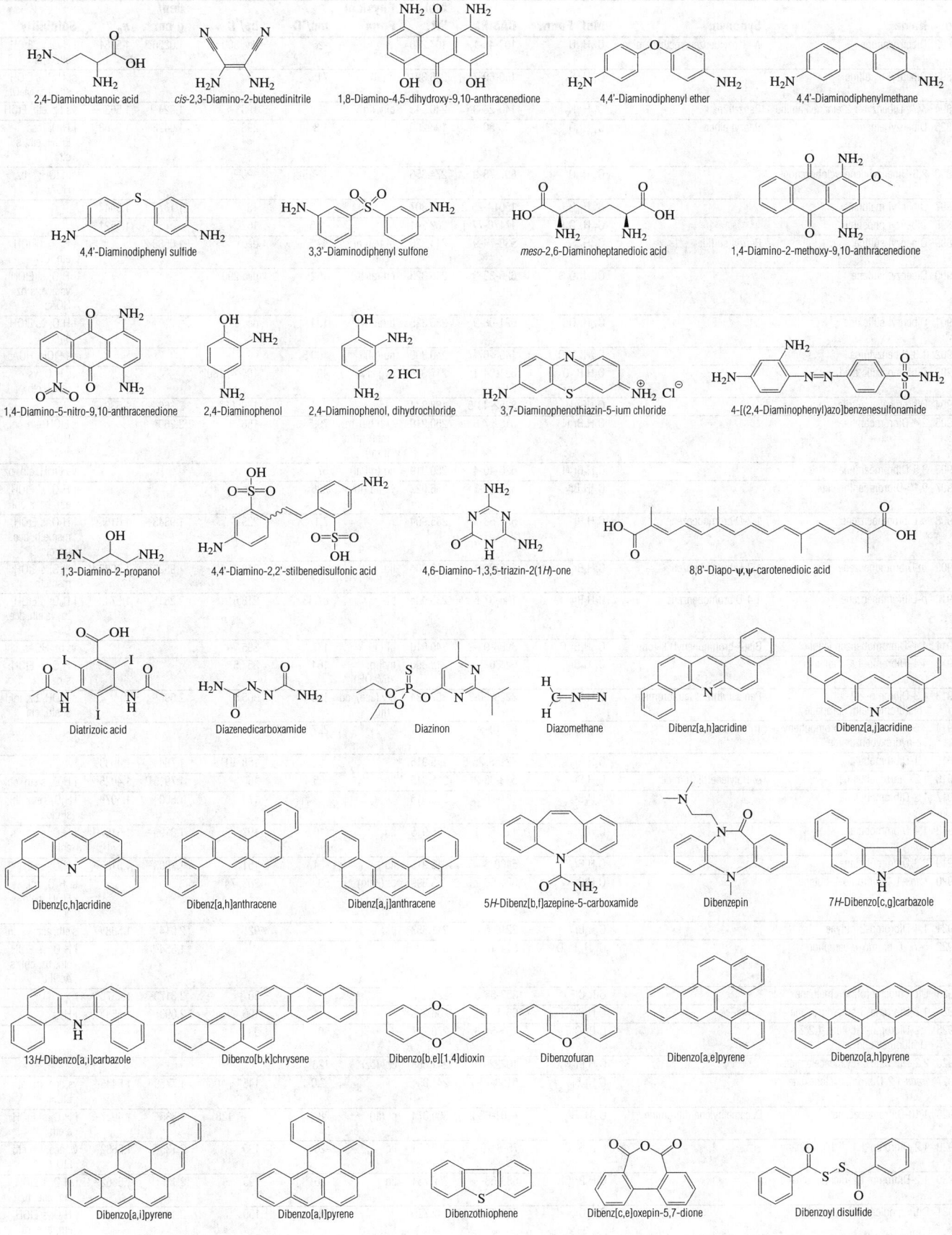

2,4-Diaminobutanoic acid

cis-2,3-Diamino-2-butenedinitrile

1,8-Diamino-4,5-dihydroxy-9,10-anthracenedione

4,4'-Diaminodiphenyl ether

4,4'-Diaminodiphenylmethane

4,4'-Diaminodiphenyl sulfide

3,3'-Diaminodiphenyl sulfone

meso-2,6-Diaminoheptanedioic acid

1,4-Diamino-2-methoxy-9,10-anthracenedione

1,4-Diamino-5-nitro-9,10-anthracenedione

2,4-Diaminophenol

2,4-Diaminophenol, dihydrochloride

3,7-Diaminophenothiazin-5-ium chloride

4-[(2,4-Diaminophenyl)azo]benzenesulfonamide

1,3-Diamino-2-propanol

4,4'-Diamino-2,2'-stilbenedisulfonic acid

4,6-Diamino-1,3,5-triazin-2(1*H*)-one

8,8'-Diapo-ψ,ψ-carotenedioic acid

Diatrizoic acid

Diazenedicarboxamide

Diazinon

Diazomethane

Dibenz[a,h]acridine

Dibenz[a,j]acridine

Dibenz[c,h]acridine

Dibenz[a,h]anthracene

Dibenz[a,j]anthracene

5*H*-Dibenz[b,f]azepine-5-carboxamide

Dibenzepin

7*H*-Dibenzo[c,g]carbazole

13*H*-Dibenzo[a,i]carbazole

Dibenzo[b,k]chrysene

Dibenzo[b,e][1,4]dioxin

Dibenzofuran

Dibenzo[a,e]pyrene

Dibenzo[a,h]pyrene

Dibenzo[a,i]pyrene

Dibenzo[a,l]pyrene

Dibenzothiophene

Dibenz[c,e]oxepin-5,7-dione

Dibenzoyl disulfide

No.	Name	Synonym	Mol. Form.	CAS RN	Mol. Wt.	Physical Form	mp/°C	bp/°C	den/ g cm⁻³	n_D	Solubility
2892	Dibenzylamine	N-Benzylbenzenemethanamine	C₁₄H₁₅N	103-49-1	197.276		-26	dec 300; 270²⁵⁰	1.0256²⁰	1.5781²⁰	i H₂O; vs EtOH, eth; s ctc
2893	Dibenzyl disulfide		C₁₄H₁₄S₂	150-60-7	246.391	lf (al)	71.5				sl H₂O; s EtOH, eth, bz, MeOH
2894	N,N'-Dibenzyl-1,2-ethanediamine	Benzathine	C₁₆H₂₀N₂	140-28-3	240.343	oily liq	26	195⁴	1.024²⁰	1.5635²⁰	vs bz, eth, EtOH
2895	Dibenzyl ether	Benzyl ether	C₁₄H₁₄O	103-50-4	198.260		1.8	298	1.0428²⁰	1.5168²⁰	i H₂O; msc EtOH, eth; s ctc
2896	2,6-Dibenzylidenecyclohexanone		C₂₀H₁₈O	897-78-9	274.356		117.5	190²⁰			sl EtOH; s bz, HOAc
2897	Dibenzyl malonate		C₁₇H₁₆O₄	15014-25-2	284.307			187²	1.137²⁵	1.5447²⁰	
2898	Dibenzyl phosphite		C₁₄H₁₅O₃P	17176-77-1	262.241		-2.5	162⁰·¹		1.5521¹⁸	
2899	Dibenzyl sulfide	Benzyl sulfide	C₁₄H₁₄S	538-74-9	214.326	pl (eth or chl)	49.5	dec	1.0583⁵⁰		i H₂O; s EtOH, eth, CS₂
2900	Dibenzyl sulfone		C₁₄H₁₄O₂S	620-32-6	246.325	nd (al-bz)	152	dec 290			i H₂O; sl EtOH; vs ace; s bz, HOAc
2901	Dibenzyl sulfoxide		C₁₄H₁₄OS	621-08-9	230.325	lf (al, w)	134	dec 210			i H₂O; vs EtOH, eth
2902	1,3-Dibenzylurea		C₁₅H₁₆N₂O	1466-67-7	240.300	nd (al)	169.5				vs EtOH, HOAc
2903	Dibromoacetic acid		C₂H₂Br₂O₂	631-64-1	217.844	hyg cry	49	195²⁵⁰, 130¹⁶			vs H₂O; vs EtOH, eth
2904	Dibromoacetonitrile		C₂HBr₂N	3252-43-5	198.844			169; 68²⁴	2.369²⁰	1.5393²⁰	
2905	2,4-Dibromoaniline		C₆H₅Br₂N	615-57-6	250.919	orth bipym (chl) nd or lf (al)	79.5	156⁷⁴	2.260²⁰		s EtOH, eth, chl, HOAc
2906	3,5-Dibromoaniline		C₆H₅Br₂N	626-40-4	250.919	nd (dil al)	57				vs EtOH, eth, bz
2907	9,10-Dibromoanthracene		C₁₄H₈Br₂	523-27-3	336.022	ye nd (to or xyl)	226	sub			i H₂O; sl EtOH, eth, bz; s chl
2908	o-Dibromobenzene	1,2-Dibromobenzene	C₆H₄Br₂	583-53-9	235.904		7.1	225	1.9843²⁰	1.6155²⁰	i H₂O; s EtOH; msc eth, ace, bz, ctc
2909	m-Dibromobenzene	1,3-Dibromobenzene	C₆H₄Br₂	108-36-1	235.904	liq	-7	218	1.9523²⁰	1.6083¹⁷	i H₂O; s EtOH; msc eth
2910	p-Dibromobenzene	1,4-Dibromobenzene	C₆H₄Br₂	106-37-6	235.904	pl	87.43	218.5	2.261¹⁷	1.5742	i H₂O; s EtOH; bz; vs eth, ace, CS₂
2911	4,4'-Dibromobenzophenone	Bis(4-bromophenyl) ketone	C₁₃H₈Br₂O	3988-03-2	340.010	pl (al)	177	395			vs bz, HOAc, chl
2912	4,4'-Dibromo-1,1'-biphenyl		C₁₂H₈Br₂	92-86-4	312.000	mcl pr (MeOH)	164	357.5			i H₂O; sl EtOH; s bz
2913	1,3-Dibromo-2,2-bis(bromomethyl)propane	Pentaerythritol tetrabromide	C₅H₈Br₄	3229-00-3	387.734	cry (ace), nd (liq)	163	305.5	2.596¹⁵		s EtOH, bz, tol; sl eth, chl
2914	3,5-Dibromo-N-(4-bromophenyl)-2-hydroxybenzamide	Tribromsalan	C₁₃H₈Br₃NO₂	87-10-5	449.921		227				
2915	1,1-Dibromobutane		C₄H₈Br₂	62168-25-6	215.915			158; 91¹⁰¹	1.784²⁵	1.4988²⁵	
2916	1,2-Dibromobutane	α-Butylene dibromide	C₄H₈Br₂	533-98-2	215.915	liq	-65.4	166.3	1.7915²⁰	1.4025²⁰	i H₂O; s eth, chl
2917	1,3-Dibromobutane		C₄H₈Br₂	107-80-2	215.915			174	1.800²⁰	1.507²⁰	i H₂O; s eth, chl; sl ctc
2918	1,4-Dibromobutane		C₄H₈Br₂	110-52-1	215.915	liq	-16.5	197	1.8199²⁵	1.5167²⁵	i H₂O; sl ctc; s chl
2919	2,3-Dibromobutane		C₄H₈Br₂	5408-86-6	215.915	liq	-24	161	1.7893²²	1.5133²⁰	i H₂O; s eth
2920	trans-1,4-Dibromo-2-butene		C₄H₆Br₂	821-06-7	213.899	pl (peth)	53.4	203; 74¹⁴			sl H₂O, chl; vs EtOH, peth; s ace
2921	1,4-Dibromo-2-butyne		C₄H₄Br₂	2219-66-1	211.883			92¹⁵	2.014¹⁸	1.588¹⁸	s eth, ace; vs chl
2922	α,α'-Dibromo-d-camphor		C₁₀H₁₄Br₂O	514-12-5	310.025		61		1.854²¹		i H₂O; vs EtOH, eth, bz, chl; s AcOEt
2923	Dibromochlorofluoromethane		CBr₂ClF	353-55-9	226.270			80.3	2.3173²²	1.4570²⁰	
2924	1,2-Dibromo-3-chloropropane		C₃H₅Br₂Cl	96-12-8	236.333			196	2.093¹⁴	1.553¹⁴	i H₂O
2925	1,2-Dibromo-1-chloro-1,2,2-trifluoroethane		C₂Br₂ClF₃	354-51-8	276.277		50	93			
2926	2,2-Dibromo-2-cyanoacetamide		C₃H₂Br₂N₂O	10222-01-2	241.868	cry (bz)	126				
2927	trans-1,2-Dibromocyclohexane, (±)		C₆H₁₀Br₂	5183-77-7	241.951		-2.0	145¹⁰⁰, 105²⁰	1.7759²⁰	1.5445¹⁹	vs ace, bz, eth, EtOH
2928	1,10-Dibromodecane	Decamethylene dibromide	C₁₀H₂₀Br₂	4101-68-2	300.074	pl (al)	28	161⁹, 128⁴	1.335³⁰	1.4927²⁵	i H₂O; sl EtOH; s eth
2929	1,2-Dibromo-1,1-dichloroethane		C₂H₂Br₂Cl₂	75-81-0	256.751	liq	-26	195	2.135²⁰	1.5662²⁰	vs ace, bz, eth, EtOH
2930	1,2-Dibromo-1,2-dichloroethane		C₂H₂Br₂Cl₂	683-68-1	256.751	liq	-26	195	2.135²⁰	1.5662²⁰	i H₂O; s EtOH, eth, ace, bz
2931	Dibromodichloromethane		CBr₂Cl₂	594-18-3	242.725		38	150.2	2.42²⁵		i H₂O; s EtOH, eth, ace, bz
2932	1,2-Dibromo-1,1-difluoroethane	Genetron 132b-B2	C₂H₂Br₂F₂	75-82-1	223.842	liq	-61.3	92.5	2.2238²⁰	1.4456²⁰	

Dibenzylamine

Dibenzyl disulfide

N,N'-Dibenzyl-1,2-ethanediamine

Dibenzyl ether

2,6-Dibenzylidenecyclohexanone

Dibenzyl malonate

Dibenzyl phosphite

Dibenzyl sulfide

Dibenzyl sulfone

Dibenzyl sulfoxide

1,3-Dibenzylurea

Dibromoacetic acid

Dibromoacetonitrile

2,4-Dibromoaniline

3,5-Dibromoaniline

9,10-Dibromoanthracene

o-Dibromobenzene

m-Dibromobenzene

p-Dibromobenzene

4,4'-Dibromobenzophenone

4,4'-Dibromo-1,1'-biphenyl

1,3-Dibromo-2,2-bis(bromomethyl)propane

3,5-Dibromo-N-(4-bromophenyl)-2-hydroxybenzamide

1,1-Dibromobutane

1,2-Dibromobutane

1,3-Dibromobutane

1,4-Dibromobutane

2,3-Dibromobutane

trans-1,4-Dibromo-2-butene

1,4-Dibromo-2-butyne

α,α'-Dibromo-d-camphor

Dibromochlorofluoromethane

1,2-Dibromo-3-chloropropane

1,2-Dibromo-1-chloro-1,2,2-trifluoroethane

2,2-Dibromo-2-cyanoacetamide

trans-1,2-Dibromocyclohexane, (±)

1,10-Dibromodecane

1,2-Dibromo-1,1-dichloroethane

1,2-Dibromo-1,2-dichloroethane

Dibromodichloromethane

1,2-Dibromo-1,1-difluoroethane

No.	Name	Synonym	Mol. Form.	CAS RN	Mol. Wt.	Physical Form	mp/°C	bp/°C	den/ g cm⁻³	n_D	Solubility
2933	Dibromodifluoromethane		CBr_2F_2	75-61-6	209.816	vol liq or gas	-110.1	22.76			s H_2O, eth, ace, bz
2934	1,3-Dibromo-5,5-dimethyl-2,4-imidazolidinedione	Dibromantine	$C_5H_6Br_2N_2O_2$	77-48-5	285.922		198 dec				
2935	1,3-Dibromo-2,2-dimethylpropane		$C_5H_{10}Br_2$	5434-27-5	229.941			184; 80[26]	1.6775[20]	1.5090	
2936	1,12-Dibromododecane		$C_{12}H_{24}Br_2$	3344-70-5	328.127	nd (al,HOAc)	41	215[15]			i H_2O; vs EtOH, chl; s eth, HOAc
2937	1,1-Dibromoethane	Ethylidene dibromide	$C_2H_4Br_2$	557-91-5	187.861	liq	-63	108.0	2.0555[20]	1.5128[20]	i H_2O; s EtOH, ace, bz; sl chl; vs eth
2938	1,2-Dibromoethane	Ethylene dibromide	$C_2H_4Br_2$	106-93-4	187.861		9.84	131.6	2.1683[25]	1.5356[25]	vs ace, bz, eth, EtOH
2939	cis-1,2-Dibromoethene	cis-1,2-Dibromoethylene	$C_2H_2Br_2$	590-11-4	185.845	liq	-53	112.5	2.2464[20]	1.5428[20]	i H_2O; vs EtOH, eth; s ace, bz, chl
2940	trans-1,2-Dibromoethene	trans-1,2-Dibromoethylene	$C_2H_2Br_2$	590-12-5	185.845	liq	-6.5	108	2.2308[20]	1.5505[18]	i H_2O; vs EtOH, eth; s ace, bz, chl
2941	1,2-Dibromo-1-ethoxyethane		$C_4H_8Br_2O$	2983-26-8	231.914			80[20]	1.7320[20]	1.5044[20]	vs EtOH, chl
2942	1,2-Dibromoethyl acetate		$C_4H_6Br_2O_2$	24442-57-7	245.898	liq		89.5[16]	1.91[20]		
2943	(1,2-Dibromoethyl)benzene		$C_8H_8Br_2$	93-52-7	263.958		75	133[19]			s EtOH, eth, bz, chl, HOAc, MeOH, lig
2944	Dibromofluoromethane		$CHBr_2F$	1868-53-7	191.825	liq	-78	64.9	2.421[20]	1.4685[20]	i H_2O; s EtOH, eth, ace, bz, chl
2945	1,2-Dibromoheptane		$C_7H_{14}Br_2$	42474-21-5	257.994			228	1.5086[20]	1.4986[20]	
2946	1,7-Dibromoheptane	Heptamethylene dibromide	$C_7H_{14}Br_2$	4549-31-9	257.994		41.7	263	1.5306[20]	1.5034[20]	i H_2O; s eth, ace, bz, ctc, chl
2947	2,3-Dibromoheptane		$C_7H_{14}Br_2$	21266-88-6	257.994			101[17]	1.5139[20]	1.4992[20]	
2948	3,4-Dibromoheptane		$C_7H_{14}Br_2$	21266-90-0	257.994			107[24]	1.5182[20]	1.5010[20]	
2949	1,2-Dibromo-1,1,2,3,3,3-hexafluoropropane		$C_3Br_2F_6$	661-95-0	309.830			72.8	2.1630[20]		i H_2O
2950	1,2-Dibromohexane		$C_6H_{12}Br_2$	624-20-4	243.967			103[36]	1.5774[20]	1.5024[20]	vs bz, eth, chl
2951	1,6-Dibromohexane		$C_6H_{12}Br_2$	629-03-8	243.967	liq	-1.2	245.5	1.6025[25]	1.5054[25]	i H_2O; s eth, ace, chl; sl ctc
2952	3,4-Dibromohexane		$C_6H_{12}Br_2$	89583-12-0	243.967			80[13]	1.6027[20]	1.5043[20]	
2953	3,5-Dibromo-2-hydroxybenzaldehyde	3,5-Dibromosalicylaldehyde	$C_7H_4Br_2O_2$	90-59-5	279.914	pa ye pr	86	sub			vs bz, eth, chl
2954	3,5-Dibromo-2-hydroxybenzoic acid	3,5-Dibromosalicylic acid	$C_7H_4Br_2O_3$	3147-55-5	295.913	nd	228				s ace
2955	3,5-Dibromo-4-hydroxybenzonitrile	Bromoxynil	$C_7H_3Br_2NO$	1689-84-5	276.913		190				
2956	Dibromomethane	Methylene bromide	CH_2Br_2	74-95-3	173.835	liq	-52.5	97	2.4969[20]	1.5420[20]	sl H_2O; msc EtOH, eth, ace; s ctc
2957	1,4-Dibromo-2-methylbenzene	2,5-Dibromotoluene	$C_7H_6Br_2$	615-59-8	249.931		5.6	236	1.8127[17]	1.5982[18]	i H_2O
2958	2,4-Dibromo-1-methylbenzene		$C_7H_6Br_2$	31543-75-6	249.931		-9.7	103[11]	1.8176[25]	1.5964[25]	
2959	(Dibromomethyl)benzene		$C_7H_6Br_2$	618-31-5	249.931		1.0	156[23]	1.8365[28]	1.6147[20]	i H_2O; msc EtOH, eth
2960	2,3-Dibromo-2-methylbutane		$C_5H_{10}Br_2$	594-51-4	229.941		7	62[17]	1.6717[20]	1.5729[25]	
2961	2,4-Dibromo-6-methylphenol		$C_7H_6Br_2O$	609-22-3	265.930	nd (peth)	58	dec 265; 105[4]			s chl
2962	1,2-Dibromo-2-methylpropane		$C_4H_8Br_2$	594-34-3	215.915		10.5	150	1.7827[20]	1.5119[20]	s EtOH, eth, chl
2963	1,4-Dibromonaphthalene		$C_{10}H_6Br_2$	83-53-4	285.963		83	310			i H_2O; s EtOH, eth; sl HOAc
2964	2,6-Dibromo-4-nitroaniline		$C_6H_4Br_2N_2O_2$	827-94-1	295.916	ye nd (al, HOAc)	207				sl H_2O; s HOAc
2965	2,6-Dibromo-4-nitrophenol		$C_6H_3Br_2NO_3$	99-28-5	296.901	pa ye pr or lf (al)	145 dec				i H_2O; vs EtOH, eth; sl ace, bz, HOAc
2966	1,9-Dibromononane		$C_9H_{18}Br_2$	4549-33-1	286.047	liq	-22.5	285; 154[10]	1.4229[20]		
2967	1,4-Dibromooctafluorobutane		$C_4Br_2F_8$	335-48-8	359.838			97			
2968	1,8-Dibromooctane	Octamethylene dibromide	$C_8H_{16}Br_2$	4549-32-0	272.021		15.5	271	1.4594[25]	1.4971[25]	i H_2O; s eth, ctc, chl
2969	1,2-Dibromopentane		$C_5H_{10}Br_2$	3234-49-9	229.941			184	1.668[18]		
2970	1,4-Dibromopentane		$C_5H_{10}Br_2$	626-87-9	229.941		-34.4	146[150], 99[14]	1.6222[20]	1.5086[20]	
2971	1,5-Dibromopentane		$C_5H_{10}Br_2$	111-24-0	229.941	liq	-39.5	222.3	1.6928[25]	1.5102[25]	i H_2O; s bz, chl; sl ctc
2972	2,4-Dibromopentane		$C_5H_{10}Br_2$	19398-53-9	229.941			75[21], 60[12]	1.6659[20]	1.4987[20]	
2973	2,4-Dibromophenol		$C_6H_4Br_2O$	615-58-7	251.903	nd (peth)	38	238.5	2.0700[20]		sl H_2O, ctc; vs EtOH, eth, bz
2974	2,6-Dibromophenol		$C_6H_4Br_2O$	608-33-3	251.903	nd (w)	56.5	255; 162[21]			s H_2O; vs EtOH, eth

Dibromodifluoromethane

1,3-Dibromo-5,5-dimethyl-2,4-imidazolidinedione

1,3-Dibromo-2,2-dimethylpropane

1,12-Dibromododecane

1,1-Dibromoethane

1,2-Dibromoethane

cis-1,2-Dibromoethene

trans-1,2-Dibromoethene

1,2-Dibromo-1-ethoxyethane

1,2-Dibromoethyl acetate

(1,2-Dibromoethyl)benzene

Dibromofluoromethane

1,2-Dibromoheptane

1,7-Dibromoheptane

2,3-Dibromoheptane

3,4-Dibromoheptane

1,2-Dibromo-1,1,2,3,3,3-hexafluoropropane

1,2-Dibromohexane

1,6-Dibromohexane

3,4-Dibromohexane

3,5-Dibromo-2-hydroxybenzaldehyde

3,5-Dibromo-2-hydroxybenzoic acid

3,5-Dibromo-4-hydroxybenzonitrile

Dibromomethane

1,4-Dibromo-2-methylbenzene

2,4-Dibromo-1-methylbenzene

(Dibromomethyl)benzene

2,3-Dibromo-2-methylbutane

2,4-Dibromo-6-methylphenol

1,2-Dibromo-2-methylpropane

1,4-Dibromonaphthalene

2,6-Dibromo-4-nitroaniline

2,6-Dibromo-4-nitrophenol

1,9-Dibromononane

1,4-Dibromooctafluorobutane

1,8-Dibromooctane

1,2-Dibromopentane

1,4-Dibromopentane

1,5-Dibromopentane

2,4-Dibromopentane

2,4-Dibromophenol

2,6-Dibromophenol

No.	Name	Synonym	Mol. Form.	CAS RN	Mol. Wt.	Physical Form	mp/°C	bp/°C	den/g cm⁻³	n_D	Solubility
2975	1,2-Dibromopropane	Propylene dibromide	C₃H₆Br₂	78-75-1	201.888	liq	-55.49	141.9	1.9324²⁰	1.5201²⁰	s EtOH, eth, chl; sl ctc
2976	1,3-Dibromopropane		C₃H₆Br₂	109-64-8	201.888	liq	-34.5	167.3	1.9701²⁵	1.5204²⁵	i H₂O; s EtOH, eth, chl; sl ctc
2977	2,2-Dibromopropane		C₃H₆Br₂	594-16-1	201.888			113	1.880²⁰		vs eth, EtOH, chl
2978	2,3-Dibromopropanoic acid		C₃H₄Br₂O₂	600-05-5	231.871		66.5	160²⁰, 138¹²			vs bz, eth, EtOH
2979	2,3-Dibromo-1-propanol		C₃H₆Br₂O	96-13-9	217.887			219	2.120²⁰		
2980	1,3-Dibromo-2-propanol		C₃H₆Br₂O	96-21-9	217.887	ye liq		dec 219; 105¹⁶	2.1364²⁰	1.5495²⁵	vs ace, eth, EtOH
2981	2,3-Dibromo-1-propanol, phosphate (3:1)	Tris(2,3-dibromopropyl) phosphate	C₉H₁₅Br₆O₄P	126-72-7	697.610						s chl
2982	1,3-Dibromo-2-propanone	1,3-Dibromoacetone	C₃H₄Br₂O	816-39-7	215.871	nd	26	97²²	2.1670¹⁸		vs eth, CS₂
2983	1,1-Dibromo-1-propene		C₃H₄Br₂	13195-80-7	199.872			125	1.9767²⁰	1.5260²⁰	sl H₂O; s bz, ctc, chl
2984	1,2-Dibromo-1-propene		C₃H₄Br₂	26391-16-2	199.872			131.5	2.0076²⁰		
2985	2,3-Dibromo-1-propene		C₃H₄Br₂	513-31-5	199.872			141; 37.7¹¹	2.0345²⁵	1.5416²⁵	i H₂O; s eth, ace, chl
2986	3,5-Dibromopyridine		C₅H₃Br₂N	625-92-3	236.893	nd (al)	112	222			sl H₂O; s EtOH, eth
2987	5,7-Dibromo-8-quinolinol	Broxyquinoline	C₉H₅Br₂NO	521-74-4	302.950	nd (al)	196	sub			i H₂O; s EtOH, ace, bz, chl, HOAc; sl eth
2988	2,6-Dibromoquinone-4-chlorimide	2,6-Dibromo-4-(chloroimino)-2,5-cyclohexadien-1-one	C₆H₂Br₂ClNO	537-45-1	299.347	ye pr (al or HOAc)	83				vs EtOH
2989	1,14-Dibromotetradecane	Tetradecamethylene dibromide	C₁₄H₂₈Br₂	37688-96-3	356.180	lf (al-eth) cry (al)	50.4	190⁸			vs eth, EtOH, chl
2990	1,2-Dibromotetrafluoroethane	Refrigerant 114B2	C₂Br₂F₄	124-73-2	259.823	liq	-110.32	47.35	2.149²⁵	1.361²⁵	i H₂O
2991	2,3-Dibromothiophene		C₄H₂Br₂S	3140-93-0	241.932	liq	-17.5	218.5; 89¹³		1.6304²²	
2992	2,5-Dibromothiophene		C₄H₂Br₂S	3141-27-3	241.932	liq	-6	210.3	2.142²³	1.6288²⁰	i H₂O; vs EtOH, eth; s ctc
2993	3,4-Dibromothiophene		C₄H₂Br₂S	3141-26-2	241.932		4.5	221.5			
2994	1,2-Dibromo-1,1,2-trifluoroethane	Halon 2302	C₂HBr₂F₃	354-04-1	241.832			76	2.274²⁷	1.4191²⁴	
2995	2,6-Dibromo-3,4,5-trihydroxybenzoic acid	Dibromogallic acid	C₇H₄Br₂O₅	602-92-6	327.912	nd, pr or lf (w+1)	150				vs H₂O, eth, EtOH
2996	3,5-Dibromo-L-tyrosine		C₉H₉Br₂NO₃	300-38-9	338.980	nd or pl	245				sl H₂O, EtOH; i eth; s alk, acid
2997	Dibucaine	Cinchocaine	C₂₀H₂₉N₃O₂	85-79-0	343.463	hyg cry	64				
2998	Dibucaine hydrochloride		C₂₀H₃₀ClN₃O₂	61-12-1	379.924		94 dec				s chl
2999	1,4-Dibutoxybenzene		C₁₄H₂₂O₂	104-36-9	222.324		45.5	158¹⁵			s ctc
3000	1,2-Dibutoxyethane	Ethylene glycol dibutyl ether	C₁₀H₂₂O₂	112-48-1	174.281	liq	-69.1	203.3	0.8319²⁵	1.4112²⁵	
3001	Dibutoxymethane	Butylal	C₉H₂₀O₂	2568-90-3	160.254	liq	-58.1	179.2	0.8339²⁰	1.4072¹⁷	
3002	Dibutyl adipate		C₁₄H₂₆O₄	105-99-7	258.354		-32.4	165¹⁰	0.9613²⁰	1.4369²⁰	i H₂O; msc EtOH, eth
3003	Dibutylamine	N-Butylbutanamine	C₈H₁₉N	111-92-2	129.244	liq	-62	159.6	0.7670²⁰	1.4177²⁰	s H₂O, ace, bz; vs EtOH, eth
3004	Di-sec-butylamine	N-sec-Butyl-2-butanamine	C₈H₁₉N	626-23-3	129.244			134	0.7534²⁰	1.4162²⁰	vs H₂O; s EtOH
3005	2-Dibutylaminoethanol		C₁₀H₂₃NO	102-81-8	173.296			114¹⁸			
3006	N,N-Dibutylaniline		C₁₄H₂₃N	613-29-6	205.340	liq	-32.2	274.8	0.9037²⁰	1.5186²⁰	i H₂O; msc EtOH, eth; vs ace, bz; s ctc
3007	1,4-Di-tert-butylbenzene		C₁₄H₂₂	1012-72-2	190.325	nd (MeOH)	79.5	238; 109¹⁵	0.9850²⁰		i H₂O; s EtOH, eth
3008	2,5-Di-tert-butyl-1,4-benzenediol		C₁₄H₂₂O₂	88-58-4	222.324	cry (aq HOAc)	213.5				
3009	Dibutylbis(dodecylthio)stannane	Dibutyltin bis(dodecyl sulfide)	C₃₂H₆₈S₂Sn	1185-81-5	635.722	col liq		122⁰·³	1.05²⁰		s tol, hp
3010	Dibutyl carbonate		C₉H₁₈O₃	542-52-9	174.237			207	0.9251²⁰	1.4117²⁰	i H₂O; s EtOH, eth
3011	Di-tert-butyl carbonate		C₉H₁₈O₃	34619-03-9	174.237	cry (al)	40	158			vs EtOH
3012	2,5-Di-tert-butyl-2,5-cyclohexadiene-1,4-dione		C₁₄H₂₀O₂	2460-77-7	220.308	ye cry (al)	152.5				i H₂O; s EtOH, eth, bz, chl, HOAc
3013	2,6-Di-tert-butyl-2,5-cyclohexadiene-1,4-dione		C₁₄H₂₀O₂	719-22-2	220.308		69	60⁰·⁰¹			
3014	2,6-Di-tert-butyl-4-(dimethylaminomethyl)phenol		C₁₇H₂₉NO	88-27-7	263.418	pl (EtOH)	94	179⁴⁰			
3015	2,2-Dibutyl-1,3,2-dioxastannepin-4,7-dione		C₁₂H₂₀O₄Sn	78-04-6	346.995	ye solid	110				
3016	Dibutyl disulfide		C₈H₁₈S₂	629-45-8	178.359	oil		226; 117²⁰	0.938²⁰	1.4923²⁰	i H₂O; msc EtOH, eth
3017	Di-tert-butyl disulfide		C₈H₁₈S₂	110-06-5	178.359		-2.5	88²¹	0.9226²⁰	1.4899²⁰	
3018	cis-1,2-Di-tert-butylethene	cis-2,2,5,5-Tetramethyl-3-hexene	C₁₀H₂₀	692-47-7	140.266	liq		144	0.744²⁰	1.4270²⁰	

1,2-Dibromopropane

1,3-Dibromopropane

2,2-Dibromopropane

2,3-Dibromopropanoic acid

2,3-Dibromo-1-propanol

1,3-Dibromo-2-propanol

2,3-Dibromo-1-propanol, phosphate (3:1)

1,3-Dibromo-2-propanone

1,1-Dibromo-1-propene

1,2-Dibromo-1-propene

2,3-Dibromo-1-propene

3,5-Dibromopyridine

5,7-Dibromo-8-quinolinol

2,6-Dibromoquinone-4-chlorimide

1,14-Dibromotetradecane

1,2-Dibromotetrafluoroethane

2,3-Dibromothiophene

2,5-Dibromothiophene

3,4-Dibromothiophene

1,2-Dibromo-1,1,2-trifluoroethane

2,6-Dibromo-3,4,5-trihydroxybenzoic acid

3,5-Dibromo-*L*-tyrosine

Dibucaine

Dibucaine hydrochloride

1,4-Dibutoxybenzene

1,2-Dibutoxyethane

Dibutoxymethane

Dibutyl adipate

Dibutylamine

Di-*sec*-butylamine

2-Dibutylaminoethanol

N,N-Dibutylaniline

1,4-Di-*tert*-butylbenzene

2,5-Di-*tert*-butyl-1,4-benzenediol

Dibutylbis(dodecylthio)stannane

Dibutyl carbonate

Di-*tert*-butyl carbonate

2,5-Di-*tert*-butyl-2,5-cyclohexadiene-1,4-dione

2,6-Di-*tert*-butyl-2,5-cyclohexadiene-1,4-dione

2,6-Di-*tert*-butyl-4-(dimethylaminomethyl)phenol

2,2-Dibutyl-1,3,2-dioxastannepin-4,7-dione

Dibutyl disulfide

Di-*tert*-butyl disulfide

cis-1,2-Di-*tert*-butylethene

No.	Name	Synonym	Mol. Form.	CAS RN	Mol. Wt.	Physical Form	mp/°C	bp/°C	den/ g cm⁻³	n_D	Solubility
3019	Dibutyl ether		$C_8H_{18}O$	142-96-1	130.228	liq	-95.2	140.28	0.7684[20]	1.3992[20]	i H_2O; msc EtOH, eth; vs ace; sl ctc
3020	Di-sec-butyl ether		$C_8H_{18}O$	6863-58-7	130.228	liq		121.1	0.756[25]		
3021	Di-tert-butyl ether		$C_8H_{18}O$	6163-66-2	130.228	liq		107.23	0.7658[20]	1.3949[20]	
3022	N,N'-Di-tert-butylethylenediamine	N,N'-Di-tert-butylethanediamine	$C_{10}H_{24}N_2$	4062-60-6	172.311	cry	53.3	189	0.69		
3023	2,6-Di-tert-butyl-4-ethylphenol		$C_{16}H_{26}O$	4130-42-1	234.376		44	272			i alk
3024	N,N-Dibutylformamide		$C_9H_{19}NO$	761-65-9	157.253						s ctc, CS_2
3025	Dibutyl fumarate		$C_{12}H_{20}O_4$	105-75-9	228.285	liq	-13.5	285; 150[4]	0.9775[20]	1.4469[20]	i H_2O; s ace, chl
3026	N,N'-Dibutyl-1,6-hexanediamine		$C_{14}H_{32}N_2$	4835-11-4	228.417			138[3.5]		1.4470[25]	
3027	3,5-Di-tert-butyl-2-hydroxybenzoic acid		$C_{15}H_{22}O_3$	19715-19-6	250.334		163.3				s chl
3028	Di-tert-butyl ketone		$C_9H_{18}O$	815-24-7	142.238	liq	-25.2	152	0.8240[18]	1.4194[20]	i H_2O; s EtOH, eth, ace, chl, HOAc
3029	Dibutyl maleate		$C_{12}H_{20}O_4$	105-76-0	228.285		<-80	280; 142[10]			
3030	Dibutyl malonate		$C_{11}H_{20}O_4$	1190-39-2	216.275	liq	-83	251.5	0.9824[20]	1.4262[20]	i H_2O; s EtOH, eth, ace, bz, HOAc, ctc
3031	Di-tert-butyl malonate		$C_{11}H_{20}O_4$	541-16-2	216.275		-6	113[31], 66[2]	1.4184[20]	1.4184[29]	s ace, chl
3032	Dibutylmercury		$C_8H_{18}Hg$	629-35-6	314.82			223; 105[10]	1.7779[20]	1.5057[20]	
3033	2,4-Di-tert-butyl-5-methylphenol	DBMC	$C_{15}H_{24}O$	497-39-2	220.351		62.1	282	0.912[80]		i H_2O; s EtOH, eth, ace, bz, ctc
3034	2,4-Di-tert-butyl-6-methylphenol		$C_{15}H_{24}O$	616-55-7	220.351		51	269	0.891[80]		i alk
3035	2,6-Di-tert-butyl-4-methylphenol		$C_{15}H_{24}O$	128-37-0	220.351		71	265	0.8937[75]	1.4859[75]	i H_2O; s EtOH, ace, bz, peth; i alk
3036	Dibutyl nonanedioate		$C_{17}H_{32}O_4$	2917-73-9	300.434			170[2]			sl chl
3037	Dibutyl oxalate		$C_{10}H_{18}O_4$	2050-60-4	202.248	liq	-30.5	241; 96[2]	0.9873[20]	1.4234[20]	i H_2O; s EtOH, eth
3038	Di-tert-butyl peroxide	DTBP	$C_8H_{18}O_2$	110-05-4	146.228	liq	-40	111	0.704[20]	1.3890[20]	i H_2O; msc ace; s ctc, lig
3039	2,6-Di-sec-butylphenol		$C_{14}H_{22}O$	5510-99-6	206.324	liq	-42	257.5		1.5080[20]	
3040	2,4-Di-tert-butylphenol		$C_{14}H_{22}O$	96-76-4	206.324		56.5	263.5		1.5080[20]	sl ctc; i alk
3041	2,6-Di-tert-butylphenol		$C_{14}H_{22}O$	128-39-2	206.324	pr (al)	39	161[50], 133[20]		1.5001[20]	sl EtOH; s ctc; i alk
3042	3,5-Di-tert-butylphenol		$C_{14}H_{22}O$	1138-52-9	206.324		88				
3043	Dibutyl phosphate		$C_8H_{19}O_4P$	107-66-4	210.208	oil		136[0.05]	1.06[20]		s ctc, BuOH
3044	Dibutyl phosphonate		$C_8H_{19}O_3P$	1809-19-4	194.209	oil		230; 131[19]	0.985[25]	1.4220[20]	
3045	Dibutyl phthalate		$C_{16}H_{22}O_4$	84-74-2	278.344	liq	-35	340	1.0465[20]	1.4911[20]	i H_2O; msc EtOH, eth, bz; s ctc
3046	2,6-Di-tert-butylpyridine		$C_{13}H_{21}N$	585-48-8	191.313			120[20]			
3047	Dibutyl sebacate		$C_{18}H_{34}O_4$	109-43-3	314.461	liq	-10	344.5	0.9405[15]	1.4433[15]	i H_2O; s eth, ctc
3048	Dibutyl succinate		$C_{12}H_{22}O_4$	141-03-7	230.301	liq	-29.2	274.5	0.9752[20]	1.4299[20]	i H_2O; s EtOH, eth, bz, ctc
3049	Di-tert-butyl succinate		$C_{12}H_{22}O_4$	926-26-1	230.301		36.5	109[9]			
3050	Dibutyl sulfate	Butyl sulfate	$C_8H_{18}O_4S$	625-22-9	210.292	liq		115[6]			
3051	Dibutyl sulfide		$C_8H_{18}S$	544-40-1	146.294	liq	-79.7	185	0.8386[20]	1.4530[20]	vs eth, EtOH, chl
3052	Di-sec-butyl sulfide		$C_8H_{18}S$	626-26-6	146.294			165	0.8348[20]	1.4506[20]	i H_2O; vs EtOH, eth
3053	Di-tert-butyl sulfide		$C_8H_{18}S$	107-47-1	146.294	liq	-9.0	149.1	0.815[25]	1.4506[20]	
3054	Dibutyl sulfite	Butyl sulfite	$C_8H_{18}O_3S$	626-85-7	194.292			230	0.9957[20]	1.4310[20]	s EtOH, eth
3055	Dibutyl sulfone		$C_8H_{18}O_2S$	598-04-9	178.293		45	291	0.9885[47]		i H_2O; s EtOH, eth
3056	Dibutyl sulfoxide		$C_8H_{18}OS$	2168-93-6	162.293	nd (dil al)	32.6	dec	0.8317[23]	1.4669[20]	i H_2O; s EtOH, eth
3057	Dibutyl tartrate		$C_{12}H_{22}O_6$	87-92-3	262.299	pr	22	320	1.0909[20]	1.4451[20]	vs H_2O, ace, EtOH
3058	N,N'-Dibutylthiourea		$C_9H_{20}N_2S$	109-46-6	188.333	nd (al)	78				
3059	Dibutyltin dichloride	Dibutyldichlorostannane	$C_8H_{18}Cl_2Sn$	683-18-1	303.845	solid	43	135[10]			s hx, eth, thf
3060	Dibutyltin dilaurate		$C_{32}H_{64}O_4Sn$	77-58-7	631.558	ye liq or cry	23				i H_2O, MeOH; s eth, bz, ctc
3061	Dicapthon		$C_8H_9ClNO_5PS$	2463-84-5	297.653	cry (MeOH)	53				i H_2O; s ace, tol, xyl, AcOEt
3062	Dicentrine		$C_{20}H_{21}NO_4$	517-66-8	339.386						s chl
3063	Dichlofenthion		$C_{10}H_{13}Cl_2O_3PS$	97-17-6	315.153						s ctc, CS_2
3064	Dichlofluanid		$C_9H_{11}Cl_2FN_2O_2S_2$	1085-98-9	333.229	wh pow	105.3				i H_2O; s ace, MeOH, xyl
3065	Dichloroacetaldehyde		$C_2H_2Cl_2O$	79-02-7	112.942			90.5		1.436[25]	sl EtOH

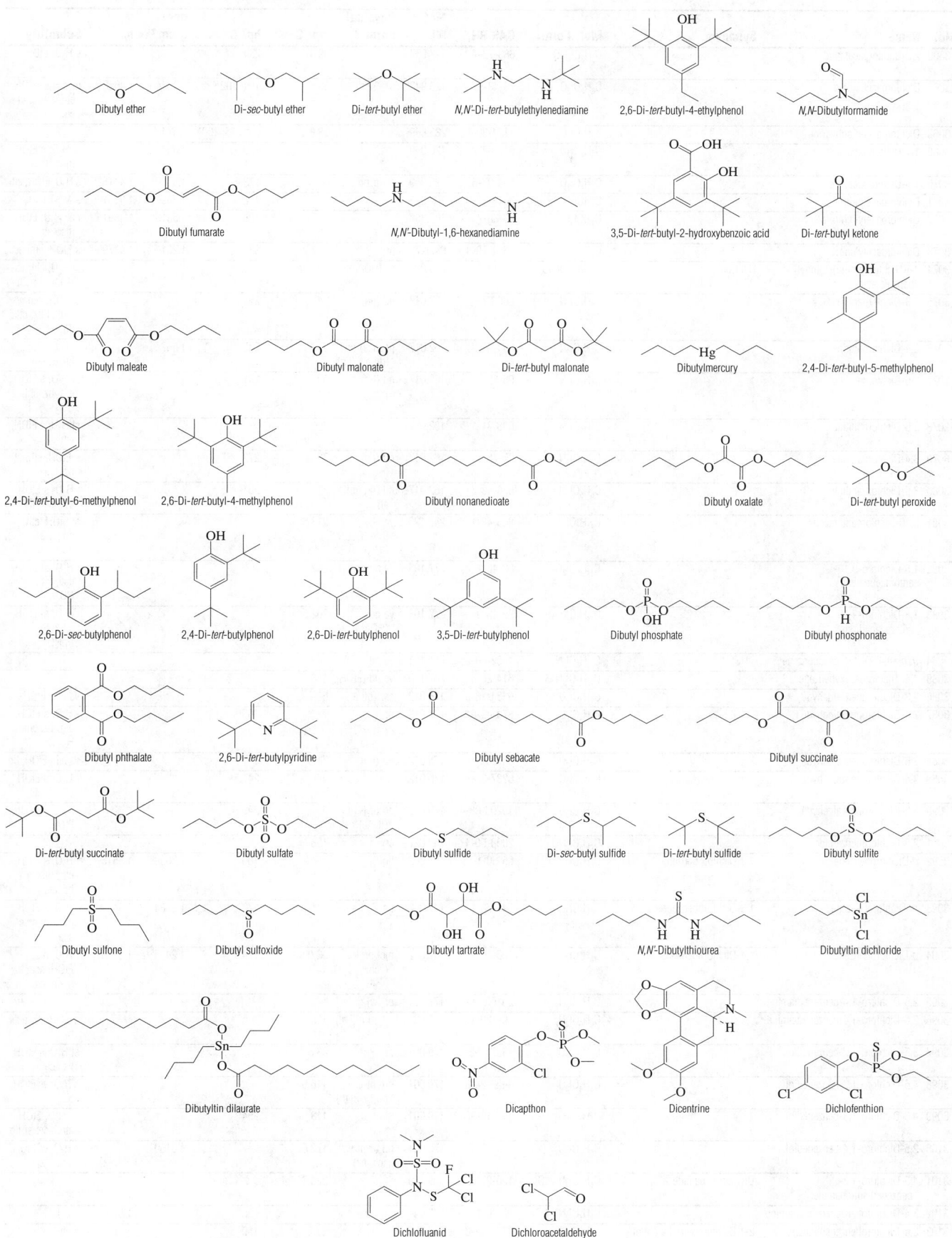

Dibutyl ether

Di-*sec*-butyl ether

Di-*tert*-butyl ether

N,N'-Di-*tert*-butylethylenediamine

2,6-Di-*tert*-butyl-4-ethylphenol

N,N-Dibutylformamide

Dibutyl fumarate

N,N'-Dibutyl-1,6-hexanediamine

3,5-Di-*tert*-butyl-2-hydroxybenzoic acid

Di-*tert*-butyl ketone

Dibutyl maleate

Dibutyl malonate

Di-*tert*-butyl malonate

Dibutylmercury

2,4-Di-*tert*-butyl-5-methylphenol

2,4-Di-*tert*-butyl-6-methylphenol

2,6-Di-*tert*-butyl-4-methylphenol

Dibutyl nonanedioate

Dibutyl oxalate

Di-*tert*-butyl peroxide

2,6-Di-*sec*-butylphenol

2,4-Di-*tert*-butylphenol

2,6-Di-*tert*-butylphenol

3,5-Di-*tert*-butylphenol

Dibutyl phosphate

Dibutyl phosphonate

Dibutyl phthalate

2,6-Di-*tert*-butylpyridine

Dibutyl sebacate

Dibutyl succinate

Di-*tert*-butyl succinate

Dibutyl sulfate

Dibutyl sulfide

Di-*sec*-butyl sulfide

Di-*tert*-butyl sulfide

Dibutyl sulfite

Dibutyl sulfone

Dibutyl sulfoxide

Dibutyl tartrate

N,N'-Dibutylthiourea

Dibutyltin dichloride

Dibutyltin dilaurate

Dicapthon

Dicentrine

Dichlofenthion

Dichlofluanid

Dichloroacetaldehyde

No.	Name	Synonym	Mol. Form.	CAS RN	Mol. Wt.	Physical Form	mp/°C	bp/°C	den/ g cm⁻³	n_D	Solubility
3066	2,2-Dichloroacetamide		$C_2H_3Cl_2NO$	683-72-7	127.957		99.4	234			s H₂O, EtOH, eth; sl ace
3067	Dichloroacetic acid		$C_2H_2Cl_2O_2$	79-43-6	128.942		13.5	194; 102[20]	1.5634[20]	1.4658[20]	msc H₂O, EtOH, eth; s ace; sl ctc
3068	Dichloroacetic anhydride		$C_4H_4Cl_4O_3$	4124-30-5	239.869		18.0	dec 215; 100[10]	1.574[24]		
3069	1,1-Dichloroacetone		$C_3H_4Cl_2O$	513-88-2	126.969			120	1.304[18]		sl H₂O; s EtOH; msc eth
3070	1,3-Dichloroacetone		$C_3H_4Cl_2O$	534-07-6	126.969	pr or nd	45	173.4	1.3826[46]	1.4716[40]	s H₂O, EtOH, eth
3071	Dichloroacetonitrile		C_2HCl_2N	3018-12-0	109.942			112.5	1.369[20]	1.4391[25]	s MeOH
3072	Dichloroacetyl chloride		C_2HCl_3O	79-36-7	147.387			108	1.5315[16]	1.4591[20]	dec H₂O, EtOH; msc eth
3073	Dichloroacetylene		C_2Cl_2	7572-29-4	94.927	liq	-66	33	1.261[20]	1.42790[20]	s EtOH, eth, ace
3074	4-[(Dichloroamino)sulfonyl] benzoic acid	Halazone	$C_7H_5Cl_2NO_4S$	80-13-7	270.091	pr (HOAc)	195 dec				sl H₂O, chl; vs HOAc; i peth
3075	2,3-Dichloroaniline		$C_6H_5Cl_2N$	608-27-5	162.017	nd (lig)	24	252			s EtOH, ace; vs eth; sl bz, ctc, lig
3076	2,4-Dichloroaniline		$C_6H_5Cl_2N$	554-00-7	162.017	pr (ace) nd (dil al) (lig)	63.5	245	1.567[20]		sl H₂O, chl; s EtOH, eth
3077	2,5-Dichloroaniline		$C_6H_5Cl_2N$	95-82-9	162.017	nd (lig)	50	251			sl H₂O; s EtOH, eth, bz, chl, CS₂
3078	2,6-Dichloroaniline		$C_6H_5Cl_2N$	608-31-1	162.017		39				sl H₂O; s EtOH, eth
3079	3,4-Dichloroaniline		$C_6H_5Cl_2N$	95-76-1	162.017	nd (lig)	72	272			s EtOH, eth; sl bz, chl
3080	3,5-Dichloroaniline		$C_6H_5Cl_2N$	626-43-7	162.017	nd (lig, dil al)	52	261			i H₂O; s EtOH, eth, ctc, lig
3081	9,10-Dichloroanthracene		$C_{14}H_8Cl_2$	605-48-1	247.120	ye nd (MeCOEt or CCl₄)	213.5				sl EtOH, eth, chl; s bz
3082	1,5-Dichloro-9,10-anthracenedione		$C_{14}H_6Cl_2O_2$	82-46-2	277.103	ye nd (to)	252				i H₂O; sl EtOH, ace; s bz, HOAc
3083	1,8-Dichloro-9,10-anthracenedione		$C_{14}H_6Cl_2O_2$	82-43-9	277.103	ye nd (HOAc)	202.5				i H₂O; sl EtOH; s bz, tol, PhNO₂
3084	trans-4,4'-Dichloroazobenzene		$C_{12}H_8Cl_2N_2$	1602-00-2	251.111	ye nd (ace)	189				
3085	4,4'-Dichloroazoxybenzene		$C_{12}H_8Cl_2N_2O$	614-26-6	267.110	ye nd (EtOH)	158				vs eth, EtOH
3086	2,3-Dichlorobenzaldehyde		$C_7H_4Cl_2O$	6334-18-5	175.012	cry (dil al)	66				vs eth, EtOH
3087	2,4-Dichlorobenzaldehyde		$C_7H_4Cl_2O$	874-42-0	175.012	pr	73.3	105[15]			i H₂O; s EtOH, eth, bz, chl, HOAc
3088	2,6-Dichlorobenzaldehyde		$C_7H_4Cl_2O$	83-38-5	175.012	nd (lig)	71.8				vs eth, EtOH, lig
3089	3,4-Dichlorobenzaldehyde		$C_7H_4Cl_2O$	6287-38-3	175.012		44	247.5			i H₂O; s EtOH, eth; sl ctc
3090	3,5-Dichlorobenzaldehyde		$C_7H_4Cl_2O$	10203-08-4	175.012	nd or lf (dil HOAc)	65	240			vs ace, bz, eth, EtOH
3091	2,6-Dichlorobenzamide		$C_7H_5Cl_2NO$	2008-58-4	190.027	cry	198				
3092	o-Dichlorobenzene	1,2-Dichlorobenzene	$C_6H_4Cl_2$	95-50-1	147.002	liq	-17.0	180	1.3059[20]	1.5515[20]	i H₂O; s EtOH, eth; msc ace, bz, ctc
3093	m-Dichlorobenzene	1,3-Dichlorobenzene	$C_6H_4Cl_2$	541-73-1	147.002	liq	-24.8	173	1.2884[20]	1.5459[20]	i H₂O; s EtOH, eth, bz; msc ace
3094	p-Dichlorobenzene	1,4-Dichlorobenzene	$C_6H_4Cl_2$	106-46-7	147.002	mcl pr, lf (ace)	53.09	174	1.2475[55]	1.5285[20]	i H₂O; msc EtOH, ace, bz; s eth, ctc
3095	2,5-Dichloro-1,4-benzenediamine		$C_6H_6Cl_2N_2$	20103-09-7	177.031	pr (w)	170				
3096	2,6-Dichloro-1,4-benzenediamine		$C_6H_6Cl_2N_2$	609-20-1	177.031	nd, pr (dil al)	125				s EtOH, eth, ace, bz
3097	3,5-Dichloro-1,2-benzenediol		$C_6H_4Cl_2O_2$	13673-92-2	179.001	pr	83.5				sl H₂O; s EtOH; vs ace
3098	4,5-Dichloro-1,2-benzenediol		$C_6H_4Cl_2O_2$	3428-24-8	179.001	pr(chl-CS₂) nd(bz-peth)	116.5				s H₂O; vs EtOH, bz
3099	4,6-Dichloro-1,3-benzenediol		$C_6H_4Cl_2O_2$	137-19-9	179.001		113	254			vs H₂O, EtOH, eth, ace; sl lig
3100	2,5-Dichloro-1,4-benzenediol		$C_6H_4Cl_2O_2$	824-69-1	179.001	nd or pr w, ace, bz)	172.5		1.8150[24]		s H₂O; vs EtOH, eth, ace
3101	4,5-Dichloro-1,3-benzenedisulfonamide	Dichlorphenamide	$C_6H_6Cl_2N_2O_4S_2$	120-97-8	305.159		228.7				
3102	2,4-Dichlorobenzenemethanamine		$C_7H_7Cl_2N$	95-00-1	176.044			125[13]		1.5762[25]	s chl
3103	2,4-Dichlorobenzenemethanol	2,4-Dichlorobenzyl alcohol	$C_7H_6Cl_2O$	1777-82-8	177.028		59.5	150[25]			s chl
3104	N,N-Dichlorobenzenesulfonamide		$C_6H_5Cl_2NO_2S$	473-29-0	226.081	ye mcl or pl	76				s EtOH; sl ctc

2,2-Dichloroacetamide   Dichloroacetic acid   Dichloroacetic anhydride   1,1-Dichloroacetone   1,3-Dichloroacetone   Dichloroacetonitrile   Dichloroacetyl chloride   Dichloroacetylene

4-[(Dichloroamino)sulfonyl]benzoic acid   2,3-Dichloroaniline   2,4-Dichloroaniline   2,5-Dichloroaniline   2,6-Dichloroaniline   3,4-Dichloroaniline   3,5-Dichloroaniline

9,10-Dichloroanthracene   1,5-Dichloro-9,10-anthracenedione   1,8-Dichloro-9,10-anthracenedione   *trans*-4,4'-Dichloroazobenzene   4,4'-Dichloroazoxybenzene

2,3-Dichlorobenzaldehyde   2,4-Dichlorobenzaldehyde   2,6-Dichlorobenzaldehyde   3,4-Dichlorobenzaldehyde   3,5-Dichlorobenzaldehyde   2,6-Dichlorobenzamide   *o*-Dichlorobenzene

m-Dichlorobenzene   *p*-Dichlorobenzene   2,5-Dichloro-1,4-benzenediamine   2,6-Dichloro-1,4-benzenediamine   3,5-Dichloro-1,2-benzenediol   4,5-Dichloro-1,2-benzenediol   4,6-Dichloro-1,3-benzenediol

2,5-Dichloro-1,4-benzenediol   4,5-Dichloro-1,3-benzenedisulfonamide   2,4-Dichlorobenzenemethanamine   2,4-Dichlorobenzenemethanol   *N,N*-Dichlorobenzenesulfonamide

No.	Name	Synonym	Mol. Form.	CAS RN	Mol. Wt.	Physical Form	mp/°C	bp/°C	den/ g cm^{-3}	n_D	Solubility
3105	2,5-Dichlorobenzenethiol		$C_6H_4Cl_2S$	5858-18-4	179.067			115^{50}			
3106	2,2'-Dichloro-p-benzidine	[1,1'-Biphenyl]-4,4'-diamine, 2,2'-dichloro-	$C_{12}H_{10}Cl_2N_2$	84-68-4	253.126	nd (w), pr (al)	165				vs eth, EtOH
3107	3,3'-Dichloro-p-benzidine	[1,1'-Biphenyl]-4,4'-diamine, 3,3'-dichloro-	$C_{12}H_{10}Cl_2N_2$	91-94-1	253.126	nd	132.5				i H_2O; s EtOH, bz, HOAc
3108	3,3'-Dichloro-p-benzidine dihydrochloride		$C_{12}H_{12}Cl_4N_2$	612-83-9	326.048						i H_2O; vs EtOH
3109	2,4-Dichlorobenzoic acid		$C_7H_4Cl_2O_2$	50-84-0	191.012	nd (w or bz)	164.2	sub			s H_2O, EtOH, eth, bz, chl; sl ace
3110	2,5-Dichlorobenzoic acid		$C_7H_4Cl_2O_2$	50-79-3	191.012	nd (w)	154.4	301			sl H_2O, DMSO; s EtOH, eth
3111	2,6-Dichlorobenzoic acid		$C_7H_4Cl_2O_2$	50-30-6	191.012	nd (al), pr (w)	144	sub			s H_2O, EtOH, eth, bz, chl
3112	3,4-Dichlorobenzoic acid		$C_7H_4Cl_2O_2$	51-44-5	191.012	nd (w, al, bz)	208.5				s H_2O, eth; vs EtOH; sl DMSO
3113	3,5-Dichlorobenzoic acid		$C_7H_4Cl_2O_2$	51-36-5	191.012	nd (al, w)	188	sub			sl H_2O, lig, DMSO; s EtOH, eth
3114	2,6-Dichlorobenzonitrile	Dichlobenil	$C_7H_3Cl_2N$	1194-65-6	172.012	cry (peth)	144.5	270			
3115	4,4'-Dichlorobenzophenone	Bis(4-chlorophenyl) ketone	$C_{13}H_8Cl_2O$	90-98-2	251.108	pl (al)	147.5	353	1.4500^{20}		i H_2O; s EtOH; vs eth, chl; sl ace
3116	3,4-Dichlorobenzotrifluoride	1,2-Dichloro-4-(trifluoromethyl) benzene	$C_7H_3Cl_2F_3$	328-84-7	215.000	liq		173.5; 64^{14}	1.4729^{25}		
3117	2,3-Dichlorobenzoyl chloride		$C_7H_3Cl_3O$	2905-60-4	209.457	liq		140^{14}			
3118	2,4-Dichlorobenzoyl chloride		$C_7H_3Cl_3O$	89-75-8	209.457		16.5	150^{34}, 111$^{7.5}$		1.5895^{20}	s ctc
3119	2,5-Dichlorobenzoyl chloride		$C_7H_3Cl_3O$	2905-61-5	209.457	liq		95.4^1			
3120	3,4-Dichlorobenzoyl chloride		$C_7H_3Cl_3O$	3024-72-4	209.457		25	242			sl ctc
3121	2,5-Dichlorobiphenyl		$C_{12}H_8Cl_2$	34883-39-1	223.098			182^{30}, 171^{15}			i H_2O
3122	2,6-Dichlorobiphenyl		$C_{12}H_8Cl_2$	33146-45-1	223.098	cry	35.5				i H_2O
3123	3,3'-Dichlorobiphenyl		$C_{12}H_8Cl_2$	2050-67-1	223.098	nd (dil al)	29	320			vs bz, eth, EtOH
3124	4,4'-Dichlorobiphenyl		$C_{12}H_8Cl_2$	2050-68-2	223.098	pr or nd (al, to-peth)	149.3	317	1.4420^0		i H_2O; sl EtOH, chl; s bz
3125	1,1-Dichloro-2,2-bis(p-chlorophenyl)ethane		$C_{14}H_{10}Cl_4$	72-54-8	320.041		109.5	193^1			sl chl
3126	2,2-Dichloro-1,1-bis(4-chlorophenyl)ethene		$C_{14}H_8Cl_4$	72-55-9	318.026		89				
3127	2,3-Dichloro-1,3-butadiene		$C_4H_4Cl_2$	1653-19-6	122.981			98	1.1829^{20}	1.4890^{20}	vs chl
3128	1,1-Dichlorobutane	Butylidene chloride	$C_4H_8Cl_2$	541-33-3	127.013			113.8	1.0863^{20}	1.4355^{20}	i H_2O; s chl
3129	1,2-Dichlorobutane		$C_4H_8Cl_2$	616-21-7	127.013			124.1	1.1116^{25}	1.4450^{20}	i H_2O; s eth, chl; sl ctc
3130	1,3-Dichlorobutane		$C_4H_8Cl_2$	1190-22-3	127.013			134	1.1158^{20}	1.4445^{20}	i H_2O; s eth, chl; sl ctc
3131	1,4-Dichlorobutane		$C_4H_8Cl_2$	110-56-5	127.013	liq	-37.3	161	1.1331^{25}	1.4522^{25}	i H_2O; vs chl
3132	2,2-Dichlorobutane		$C_4H_8Cl_2$	4279-22-5	127.013	liq	-74	104	1.1048^{25}	1.4295	i H_2O; s chl
3133	2,3-Dichlorobutane, (±)		$C_4H_8Cl_2$	2211-67-8	127.013	liq	-80	119; 53^{80}	1.105^{25}	1.4409^{25}	i H_2O
3134	1,4-Dichloro-2,3-butanediol		$C_4H_8Cl_2O_2$	2419-73-0	159.012		126.5	150^{30}			vs EtOH
3135	3,4-Dichloro-1-butene		$C_4H_6Cl_2$	760-23-6	124.997	liq	-61	116	1.1170^{20}	1.4641^{20}	i H_2O; s EtOH, eth, ctc; vs chl, bz
3136	cis-1,3-Dichloro-2-butene		$C_4H_6Cl_2$	10075-38-4	124.997			130; 34^{20}	1.1605^{20}	1.4735^{20}	vs ace, bz, eth, EtOH
3137	trans-1,3-Dichloro-2-butene		$C_4H_6Cl_2$	7415-31-8	124.997			132; 53^{50}	1.160^{20}	1.4719^{20}	vs ace, bz, eth, EtOH
3138	cis-1,4-Dichloro-2-butene		$C_4H_6Cl_2$	1476-11-5	124.997	liq	-48	152.5	1.188^{25}	1.4887^{25}	vs ace, bz, eth, EtOH
3139	trans-1,4-Dichloro-2-butene		$C_4H_6Cl_2$	110-57-6	124.997		1.0	155.4	1.183^{25}	1.4871^{25}	vs ace, bz, eth, EtOH
3140	1,4-Dichloro-2-butyne		$C_4H_4Cl_2$	821-10-3	122.981			165.5	1.258^{20}	1.5058^{20}	s eth, ace; sl ctc; vs chl
3141	2,6-Dichloro-4-(chloroimino)-2,5-cyclohexadien-1-one	Gibbs' reagent	$C_6H_2Cl_3NO$	101-38-2	210.445		66				
3142	1,2-Dichloro-4-(chloromethyl) benzene		$C_7H_5Cl_3$	102-47-6	195.474		37.5	241			i H_2O; s EtOH, ctc
3143	2,4-Dichloro-1-(chloromethyl) benzene		$C_7H_5Cl_3$	94-99-5	195.474			120^{13}			
3144	Dichloro(chloromethyl) methylsilane		$C_2H_5Cl_3Si$	1558-33-4	163.506			121.5	1.2858^{20}	1.4500^{20}	
3145	Dichloro(2-chlorovinyl)arsine		$C_2H_2AsCl_3$	541-25-3	207.318	liq	0.1	190	1.888^{20}		
3146	2,5-Dichloro-2,5-cyclohexadiene-1,4-dione		$C_6H_2Cl_2O_2$	615-93-0	176.985	pa ye mcl pr (al)	162.3				i H_2O; sl EtOH; s eth, chl

2,5-Dichlorobenzenethiol

2,2'-Dichloro-*p*-benzidine

3,3'-Dichloro-*p*-benzidine

3,3'-Dichloro-*p*-benzidine dihydrochloride

2,4-Dichlorobenzoic acid

2,5-Dichlorobenzoic acid

2,6-Dichlorobenzoic acid

3,4-Dichlorobenzoic acid

3,5-Dichlorobenzoic acid

2,6-Dichlorobenzonitrile

4,4'-Dichlorobenzophenone

3,4-Dichlorobenzotrifluoride

2,3-Dichlorobenzoyl chloride

2,4-Dichlorobenzoyl chloride

2,5-Dichlorobenzoyl chloride

3,4-Dichlorobenzoyl chloride

2,5-Dichlorobiphenyl

2,6-Dichlorobiphenyl

3,3'-Dichlorobiphenyl

4,4'-Dichlorobiphenyl

1,1-Dichloro-2,2-bis(*p*-chlorophenyl)ethane

2,2-Dichloro-1,1-bis(4-chlorophenyl)ethene

2,3-Dichloro-1,3-butadiene

1,1-Dichlorobutane

1,2-Dichlorobutane

1,3-Dichlorobutane

1,4-Dichlorobutane

2,2-Dichlorobutane

2,3-Dichlorobutane, (±)

1,4-Dichloro-2,3-butanediol

3,4-Dichloro-1-butene

cis-1,3-Dichloro-2-butene

trans-1,3-Dichloro-2-butene

cis-1,4-Dichloro-2-butene

trans-1,4-Dichloro-2-butene

1,4-Dichloro-2-butyne

2,6-Dichloro-4-(chloroimino)-2,5-cyclohexadien-1-one

1,2-Dichloro-4-(chloromethyl)benzene

2,4-Dichloro-1-(chloromethyl)benzene

Dichloro(chloromethyl)methylsilane

Dichloro(2-chlorovinyl)arsine

2,5-Dichloro-2,5-cyclohexadiene-1,4-dione

No.	Name	Synonym	Mol. Form.	CAS RN	Mol. Wt.	Physical Form	mp/°C	bp/°C	den/ g cm⁻³	n_D	Solubility
3147	2,6-Dichloro-2,5-cyclohexadiene-1,4-dione		$C_6H_2Cl_2O_2$	697-91-6	176.985	ye orth (lig, bz)	121.8				sl H_2O, EtOH; s chl
3148	1,1-Dichlorocyclohexane		$C_6H_{10}Cl_2$	2108-92-1	153.049	liq	-47	171	1.1559²⁰	1.4803²⁰	
3149	cis-1,2-Dichlorocyclohexane		$C_6H_{10}Cl_2$	10498-35-8	153.049	liq	-1.5	206.9	1.2021²⁰	1.4967²⁰	vs bz
3150	1,10-Dichlorodecane		$C_{10}H_{20}Cl_2$	2162-98-3	211.172		15.6	167²⁸	0.9945²⁵	1.4586²⁵	
3151	2,7-Dichlorodibenzo-p-dioxin		$C_{12}H_6Cl_2O_2$	33857-26-0	253.081	cry	201				
3152	1,2-Dichloro-4-(dichloromethyl)benzene		$C_7H_4Cl_4$	56961-84-3	229.919			257	1.515²²		vs bz, eth, EtOH
3153	Dichloro(dichloromethyl)methylsilane		$C_2H_4Cl_4Si$	1558-31-2	197.951			149	1.4116²⁰	1.4700²⁰	
3154	2,3-Dichloro-5,6-dicyanobenzoquinone		$C_8Cl_2N_2O_2$	84-58-2	227.004	ye-oran cry	214.5				vs bz, HOAc, diox
3155	Dichlorodiethylsilane		$C_4H_{10}Cl_2Si$	1719-53-5	157.114		-96.5	dec 129	1.0504²⁰	1.4309²⁰	
3156	1,1-Dichloro-1,2-difluoroethane		$C_2H_2Cl_2F_2$	25915-78-0	134.940	col liq		48.4			
3157	1,2-Dichloro-1,1-difluoroethane		$C_2H_2Cl_2F_2$	1649-08-7	134.940	liq	-101.2	46.2	1.4163²⁰	1.36193²⁰	sl H_2O
3158	1,2-Dichloro-1,2-difluoroethane		$C_2H_2Cl_2F_2$	431-06-1	134.940	liq	-101.2	59.6	1.4163²⁰	1.3619²⁰	
3159	1,1-Dichloro-2,2-difluoroethene	1,1-Dichloro-2,2-difluoroethylene	$C_2Cl_2F_2$	79-35-6	132.924	vol liq or gas	-116	19	1.555⁻²⁰	1.383⁻²⁰	
3160	cis-1,2-Dichloro-1,2-difluoroethene	Fluorocarbon 1112	$C_2Cl_2F_2$	311-81-9	132.924	vol liq	-119.6	21.1		1.495⁰	
3161	trans-1,2-Dichloro-1,2-difluoroethene		$C_2Cl_2F_2$	381-71-5	132.924	vol liq	-93.3	22		1.494⁰	
3162	Dichlorodifluoromethane	Refrigerant 12	CCl_2F_2	75-71-8	120.914	col gas	-158	-29.8			sl H_2O; s EtOH, eth, HOAc
3163	2,2-Dichloro-1,1-difluoro-1-methoxyethane	Methoxyflurane	$C_3H_4Cl_2F_2O$	76-38-0	164.966	col liq	-35	105	1.43²⁰	1.3861²⁰	
3164	2,2'-Dichlorodiisopropyl ether		$C_6H_{12}Cl_2O$	108-60-1	171.064			187	1.103²⁰	1.4505²⁰	i H_2O; msc EtOH, eth, ace; vs bz
3165	1,4-Dichloro-2,5-dimethylbenzene		$C_8H_8Cl_2$	1124-05-6	175.056		71	222			s chl
3166	2,5-Dichloro-2,5-dimethylhexane		$C_8H_{16}Cl_2$	6223-78-5	183.119	lf, nd	67.5		0.9543²⁰		vs bz, eth, EtOH, chl
3167	1,3-Dichloro-5,5-dimethyl hydantoin		$C_5H_6Cl_2N_2O_2$	118-52-5	197.019	pr	132		1.5²⁰		sl H_2O; s chl, ctc, bz
3168	2,4-Dichloro-3,5-dimethylphenol	Dichloroxylenol	$C_8H_8Cl_2O$	133-53-9	191.055		83				vs eth
3169	Dichlorodimethylsilane		$C_2H_6Cl_2Si$	75-78-5	129.061	liq	-16	70.3	1.064²⁵	1.4038²⁰	dec H_2O, EtOH
3170	2,3-Dichloro-1,4-dioxane		$C_4H_6Cl_2O_2$	95-59-0	156.996		30	81¹⁰	1.468²⁰	1.4928²⁰	i H_2O; vs eth, ace, bz, ctc, diox
3171	Dichlorodiphenylmethane		$C_{13}H_{10}Cl_2$	2051-90-3	237.124			dec 305; 190²¹	1.235¹⁸		s eth, bz, ctc
3172	Dichlorodiphenylsilane		$C_{12}H_{10}Cl_2Si$	80-10-4	253.199			305	1.204²⁵	1.5800²⁰	s EtOH, eth, ace, bz, ctc
3173	1,1-Dichloroethane	Ethylidene dichloride	$C_2H_4Cl_2$	75-34-3	98.959	liq	-96.9	57.3	1.1757²⁰	1.4164²⁰	sl H_2O; vs EtOH, eth; s ace, bz
3174	1,2-Dichloroethane	Ethylene dichloride	$C_2H_4Cl_2$	107-06-2	98.959	liq	-35.7	83.5	1.2454²⁵	1.4422²⁵	sl H_2O; vs EtOH; msc eth; s ace, bz, chl
3175	2,2-Dichloroethanol		$C_2H_4Cl_2O$	598-38-9	114.958			146	1.4040²⁵	1.4626²⁵	sl H_2O, ctc; s EtOH, eth
3176	1,1-Dichloroethene	Vinylidene chloride	$C_2H_2Cl_2$	75-35-4	96.943	liq	-122.56	31.6	1.213²⁰	1.4249²⁰	i H_2O; s EtOH, ace, bz; vs eth, chl
3177	cis-1,2-Dichloroethene	cis-1,2-Dichloroethylene	$C_2H_2Cl_2$	156-59-2	96.943	liq	-80.0	60.1	1.2837²⁰	1.4490²⁰	sl H_2O; msc EtOH, eth, ace; vs bz, chl
3178	trans-1,2-Dichloroethene	trans-1,2-Dichloroethylene	$C_2H_2Cl_2$	156-60-5	96.943	liq	-49.8	48.7	1.2565²⁰	1.4454²⁰	sl H_2O; msc EtOH, eth, ace; vs bz, chl
3179	1,2-Dichloro-1-ethoxyethane		$C_4H_8Cl_2O$	623-46-1	143.012			145	1.1370²⁰	1.4435²⁰	sl chl
3180	1,2-Dichloroethyl acetate		$C_4H_6Cl_2O_2$	10140-87-1	156.996	liq		79³³, 32¹⁰			
3181	Dichloroethylaluminum	Ethylaluminum chloride	$C_2H_5AlCl_2$	563-43-9	126.949	hyg solid or liq	32	115⁵⁰	1.207		reac H_2O
3182	Dichloroethylmethylsilane		$C_3H_8Cl_2Si$	4525-44-4	143.088			101	1.0047²⁰	1.4197²⁰	
3183	2',7'-Dichlorofluorescein	2',7'-Dichloro-3,6-fluorandiol	$C_{20}H_{10}Cl_2O_5$	76-54-0	401.196						sl DMSO
3184	1,1-Dichloro-1-fluoroethane		$C_2H_3Cl_2F$	1717-00-6	116.949	liq	-103.5	32.0	1.250¹⁰	1.3600¹⁰	i H_2O
3185	1,2-Dichloro-1-fluoroethane		$C_2H_3Cl_2F$	430-57-9	116.949	liq	-60	73.8	1.3814²⁰	1.4132²⁰	
3186	1,1-Dichloro-2-fluoroethene	1,1-Dichloro-2-fluoroethylene	C_2HCl_2F	359-02-4	114.933	liq	-108.8	37.5	1.3732¹⁶	1.4031¹⁶	
3187	Dichlorofluoromethane	Refrigerant 21	$CHCl_2F$	75-43-4	102.923	col gas	-135	8.9	1.405⁹	1.3724⁹	i H_2O; s EtOH, eth, ctc, chl, HOAc
3188	(Dichlorofluoromethyl)benzene		$C_7H_5Cl_2F$	498-67-9	179.019	liq	-26.8	179	1.3138¹¹	1.5180¹¹	vs EtOH
3189	1,1-Dichloro-2-fluoropropene		$C_3H_3Cl_2F$	430-95-5	128.960			78	1.3026²⁵	1.4196²⁵	
3190	1,7-Dichloroheptane		$C_7H_{14}Cl_2$	821-76-1	169.092			124³⁵	1.0408²⁵	1.4565²⁵	

2,6-Dichloro-2,5-cyclohexadiene-1,4-dione

1,1-Dichlorocyclohexane

cis-1,2-Dichlorocyclohexane

1,10-Dichlorodecane

2,7-Dichlorodibenzo-p-dioxin

1,2-Dichloro-4-(dichloromethyl)benzene

Dichloro(dichloromethyl)methylsilane

2,3-Dichloro-5,6-dicyanobenzoquinone

Dichlorodiethylsilane

1,1-Dichloro-1,2-difluoroethane

1,2-Dichloro-1,1-difluoroethane

1,2-Dichloro-1,2-difluoroethane

1,1-Dichloro-2,2-difluoroethene

cis-1,2-Dichloro-1,2-difluoroethene

trans-1,2-Dichloro-1,2-difluoroethene

Dichlorodifluoromethane

2,2-Dichloro-1,1-difluoro-1-methoxyethane

2,2'-Dichlorodiisopropyl ether

1,4-Dichloro-2,5-dimethylbenzene

2,5-Dichloro-2,5-dimethylhexane

1,3-Dichloro-5,5-dimethyl hydantoin

2,4-Dichloro-3,5-dimethylphenol

Dichlorodimethylsilane

2,3-Dichloro-1,4-dioxane

Dichlorodiphenylmethane

Dichlorodiphenylsilane

1,1-Dichloroethane

1,2-Dichloroethane

2,2-Dichloroethanol

1,1-Dichloroethene

cis-1,2-Dichloroethene

trans-1,2-Dichloroethene

1,2-Dichloro-1-ethoxyethane

1,2-Dichloroethyl acetate

Dichloroethylaluminum

Dichloroethylmethylsilane

2',7'-Dichlorofluorescein

1,1-Dichloro-1-fluoroethane

1,2-Dichloro-1-fluoroethane

1,1-Dichloro-2-fluoroethene

Dichlorofluoromethane

(Dichlorofluoromethyl)benzene

1,1-Dichloro-2-fluoropropene

1,7-Dichloroheptane

No.	Name	Synonym	Mol. Form.	CAS RN	Mol. Wt.	Physical Form	mp/°C	bp/°C	den/ g cm⁻³	n_D	Solubility
3191	1,2-Dichloro-1,2,3,3,4,4-hexafluorocyclobutane		$C_4Cl_2F_6$	356-18-3	232.939	liq	-24.2	59.5			
3192	1,2-Dichloro-3,3,4,4,5,5-hexafluorocyclopentene		$C_5Cl_2F_6$	706-79-6	244.949	liq	-105.8	90.7	1.6546²⁰	1.3676²⁰	
3193	1,2-Dichloro-1,1,2,3,3,3-hexafluoropropane		$C_3Cl_2F_6$	661-97-2	220.928			34.1			i H_2O
3194	1,3-Dichloro-1,1,2,2,3,3-hexafluoropropane	Refrigerant 216	$C_3Cl_2F_6$	662-01-1	220.928	liq	-125.4	35.7	1.573²⁰	1.3030²⁰	
3195	1,5-Dichloro-1,1,3,3,5,5-hexamethyltrisiloxane		$C_6H_{18}Cl_2O_2Si_3$	3582-71-6	277.369	liq	-53	184	1.018²⁰		dec H_2O
3196	1,2-Dichlorohexane		$C_6H_{12}Cl_2$	2162-92-7	155.065			173	1.085¹⁵		vs eth, chl
3197	1,6-Dichlorohexane		$C_6H_{12}Cl_2$	2163-00-0	155.065			204	1.0676²⁵	1.4555²⁵	i H_2O; s eth, ctc, chl
3198	3,5-Dichloro-2-hydroxybenzaldehyde		$C_7H_4Cl_2O_2$	90-60-8	191.012	ye orth (HOAc)	95				i H_2O
3199	3,5-Dichloro-2-hydroxybenzoic acid		$C_7H_4Cl_2O_3$	320-72-9	207.011	nd (dil al) orth pr	220.5	sub			sl H_2O; vs EtOH, eth
3200	2,6-Dichloroindophenol, sodium salt	Tillman's reagent	$C_{12}H_6Cl_2NNaO_2$	620-45-1	290.078	dk grn cry					s H_2O, EtOH, ace
3201	5,6-Dichloro-1,3-isobenzofurandione	4,5-Dichlorophthalic anhydride	$C_8H_2Cl_2O_3$	942-06-3	217.006	tab or pr (to)	188	313			vs eth, EtOH, tol
3202	Dichloromethane	Methylene chloride	CH_2Cl_2	75-09-2	84.933	liq	-97.2	40	1.3266²⁰	1.4242²⁰	sl H_2O; msc EtOH, eth; s ctc
3203	1,2-Dichloro-3-methoxybenzene		$C_7H_6Cl_2O$	1984-59-4	177.028		32				
3204	1,3-Dichloro-2-methoxybenzene	2,6-Dichloroanisole	$C_7H_6Cl_2O$	1984-65-2	177.028	liq	10	105²⁰	1.291	1.5430²⁰	
3205	2,4-Dichloro-1-methoxybenzene		$C_7H_6Cl_2O$	553-82-2	177.028	pr	28.5	232; 125¹⁰			sl chl
3206	3,6-Dichloro-2-methoxybenzoic acid	Dicamba	$C_8H_6Cl_2O_3$	1918-00-9	221.038	cry (pent)	115		1.57²⁵		
3207	(Dichloromethyl)benzene	Benzal chloride	$C_7H_6Cl_2$	98-87-3	161.029	liq	-17	205	1.26²⁵	1.5502²⁰	i H_2O; vs eth, EtOH
3208	N,N-Dichloro-4-methylbenzenesulfonamide	Dichloramine-T	$C_7H_7Cl_2NO_2S$	473-34-7	240.108	pr(chl-peth)	83				i H_2O; s EtOH, eth, bz, ctc, HOAc
3209	Dichloromethylborane	Methyldichloroborane	CH_3BCl_2	7318-78-7	96.752	col gas		11			
3210	2,3-Dichloro-2-methylbutane	Amylene dichloride	$C_5H_{10}Cl_2$	507-45-9	141.038			129	1.0696¹⁵	1.4450¹⁸	i H_2O; vs eth, EtOH
3211	1,1-Dichloromethyl methyl ether	Methoxydichloromethane	$C_2H_4Cl_2O$	4885-02-3	114.958			85	1.271²⁵	1.4300²⁰	
3212	2,4-Dichloro-3-methylphenol		$C_7H_6Cl_2O$	17788-00-0	177.028	pr (peth)	58	236; 77⁴			vs eth, chl
3213	2,4-Dichloro-6-methylphenol		$C_7H_6Cl_2O$	1570-65-6	177.028	nd (w, peth)	55				sl H_2O; vs EtOH, eth, chl, CS_2
3214	2,6-Dichloro-4-methylphenol		$C_7H_6Cl_2O$	2432-12-4	177.028	nd (lig)	39	231; 138²⁸			i H_2O; vs eth, EtOH, HOAc
3215	Dichloromethylphenylsilane		$C_7H_8Cl_2Si$	149-74-6	191.131			206.5	1.1866²⁰	1.5180²⁰	
3216	Dichloromethylphosphine	Methylphosphonous dichloride	CH_3Cl_2P	676-83-5	116.915			12⁵⁰	1.304²⁰	1.4940²⁰	
3217	1,2-Dichloro-2-methylpropane	1,2-Dichloroisobutane	$C_4H_8Cl_2$	594-37-6	127.013			106.5	1.093²⁰	1.4370²⁰	i H_2O; msc EtOH, eth, ace, bz, ctc
3218	2,4-Dichloro-5-methylpyrimidine		$C_5H_4Cl_2N_2$	1780-31-0	163.004	pl (al)	26	235			sl H_2O; vs EtOH, eth, bz, chl
3219	2,4-Dichloro-6-methylpyrimidine		$C_5H_4Cl_2N_2$	5424-21-5	163.004	nd (lig)	46.5	219			vs bz, eth, EtOH, chl
3220	Dichloromethylsilane		CH_4Cl_2Si	75-54-7	115.035	liq	-93	41	1.105²⁵		
3221	1,2-Dichloronaphthalene		$C_{10}H_6Cl_2$	2050-69-3	197.061	pl (al)	36	296.5	1.3147⁴⁹	1.5338⁴⁹	s EtOH, eth
3222	1,3-Dichloronaphthalene		$C_{10}H_6Cl_2$	2198-75-6	197.061	nd or pr (al)	62.3	291			s EtOH
3223	1,4-Dichloronaphthalene		$C_{10}H_6Cl_2$	1825-31-6	197.061	nd or pr (al, ace)	67.5	288; 147¹²	1.2997⁷⁶	1.6228⁷⁶	i H_2O; sl EtOH; s eth, bz, HOAc; vs ace
3224	1,5-Dichloronaphthalene		$C_{10}H_6Cl_2$	1825-30-5	197.061	nd or lf (al) pr (sub)	107	sub	1.4900²⁰		i H_2O; sl EtOH; s eth
3225	1,6-Dichloronaphthalene		$C_{10}H_6Cl_2$	2050-72-8	197.061	nd or pr (al, peth)	49	sub			
3226	1,7-Dichloronaphthalene		$C_{10}H_6Cl_2$	2050-73-9	197.061	nd or pr (al, HOAc)	63.5	285.5	1.2611¹⁰⁰	1.6092¹⁰⁰	s EtOH, eth, bz, HOAc
3227	1,8-Dichloronaphthalene		$C_{10}H_6Cl_2$	2050-74-0	197.061	orth pl (hx) nd (al, sub)	89	sub	1.2924¹⁰⁰	1.6236¹⁰⁰	s EtOH, peth
3228	2,3-Dichloronaphthalene		$C_{10}H_6Cl_2$	2050-75-1	197.061	orth lf (al)	120				i H_2O; sl EtOH; vs eth
3229	2,6-Dichloronaphthalene		$C_{10}H_6Cl_2$	2065-70-5	197.061	nd or lf (al) pl (eth, bz)	140.5	285			sl EtOH; s eth, bz, chl, HOAc
3230	2,7-Dichloronaphthalene		$C_{10}H_6Cl_2$	2198-77-8	197.061	pl or lf (al)	115.0				vs EtOH; s hx, HOAc
3231	2,3-Dichloro-1,4-naphthalenedione	Dichlone	$C_{10}H_4Cl_2O_2$	117-80-6	227.044	ye nd (al)	195				i H_2O; sl EtOH, eth, bz; s chl

1,2-Dichloro-1,2,3,3,4,4-hexafluorocyclobutane

1,2-Dichloro-3,3,4,4,5,5-hexafluorocyclopentene

1,2-Dichloro-1,1,2,3,3,3-hexafluoropropane

1,3-Dichloro-1,1,2,2,3,3-hexafluoropropane

1,5-Dichloro-1,1,3,3,5,5-hexamethyltrisiloxane

1,2-Dichlorohexane

1,6-Dichlorohexane

3,5-Dichloro-2-hydroxybenzaldehyde

3,5-Dichloro-2-hydroxybenzoic acid

2,6-Dichloroindophenol, sodium salt

5,6-Dichloro-1,3-isobenzofurandione

Dichloromethane

1,2-Dichloro-3-methoxybenzene

1,3-Dichloro-2-methoxybenzene

2,4-Dichloro-1-methoxybenzene

3,6-Dichloro-2-methoxybenzoic acid

(Dichloromethyl)benzene

N,N-Dichloro-4-methylbenzenesulfonamide

Dichloromethylborane

2,3-Dichloro-2-methylbutane

1,1-Dichloromethyl methyl ether

2,4-Dichloro-3-methylphenol

2,4-Dichloro-6-methylphenol

2,6-Dichloro-4-methylphenol

Dichloromethylphenylsilane

Dichloromethylphosphine

1,2-Dichloro-2-methylpropane

2,4-Dichloro-5-methylpyrimidine

2,4-Dichloro-6-methylpyrimidine

Dichloromethylsilane

1,2-Dichloronaphthalene

1,3-Dichloronaphthalene

1,4-Dichloronaphthalene

1,5-Dichloronaphthalene

1,6-Dichloronaphthalene

1,7-Dichloronaphthalene

1,8-Dichloronaphthalene

2,3-Dichloronaphthalene

2,6-Dichloronaphthalene

2,7-Dichloronaphthalene

2,3-Dichloro-1,4-naphthalenedione

No.	Name	Synonym	Mol. Form.	CAS RN	Mol. Wt.	Physical Form	mp/°C	bp/°C	den/ g cm⁻³	n_D	Solubility
3232	2,4-Dichloro-1-naphthol	2,4-Dichloro-α-naphthol	C$_{10}$H$_6$Cl$_2$O	2050-76-2	213.060	nd (al, bz)	107.5	180			vs bz, eth, EtOH
3233	2,6-Dichloro-4-nitroaniline		C$_6$H$_4$Cl$_2$N$_2$O$_2$	99-30-9	207.014	ye nd (al, HOAc)	191				s EtOH, acid; sl DMSO
3234	1,2-Dichloro-3-nitrobenzene		C$_6$H$_3$Cl$_2$NO$_2$	3209-22-1	192.000	mcl nd (peth, HOAc)	61.5	257.5	1.721[14]		i H$_2$O; s EtOH, eth, ace, bz, peth; sl chl
3235	1,2-Dichloro-4-nitrobenzene		C$_6$H$_3$Cl$_2$NO$_2$	99-54-7	192.000	nd (al)	43	255.5	1.4558[75]		i H$_2$O; s EtOH, eth; sl ctc
3236	1,3-Dichloro-5-nitrobenzene		C$_6$H$_3$Cl$_2$NO$_2$	618-62-2	192.000	mcl pr or lf (HOAc, al)	65.4		1.4000[100]		i H$_2$O; s EtOH, eth
3237	1,4-Dichloro-2-nitrobenzene		C$_6$H$_3$Cl$_2$NO$_2$	89-61-2	192.000	pl or pr (al) pl (AcOEt)	56	267	1.43[75]	1.4390[75]	i H$_2$O; s EtOH, eth, bz, CS$_2$; sl ctc
3238	2,4-Dichloro-1-nitrobenzene		C$_6$H$_3$Cl$_2$NO$_2$	611-06-3	192.000	nd (al)	34	258.5	1.4790[80]	1.5512[70]	i H$_2$O; s EtOH, eth; sl chl
3239	1,1-Dichloro-1-nitroethane	Ethide	C$_2$H$_3$Cl$_2$NO$_2$	594-72-9	143.957			123.5			s ctc
3240	2,6-Dichloro-4-nitrophenol		C$_6$H$_3$Cl$_2$NO$_3$	618-80-4	207.999	br nd (w)	127 exp		1.822[25]		vs eth, chl
3241	1,1-Dichloro-1-nitropropane		C$_3$H$_5$Cl$_2$NO$_2$	595-44-8	157.984			145	1.312[20]		s ctc
3242	1,9-Dichlorononane		C$_9$H$_{18}$Cl$_2$	821-99-8	197.145			260; 138[17]	1.0173[25]	1.4586[25]	
3243	1,8-Dichlorooctane		C$_8$H$_{16}$Cl$_2$	2162-99-4	183.119			241	1.0248[25]	1.4572[25]	
3244	1,3-Dichloro-1,1,2,2,3-pentafluoropropane		C$_3$HCl$_2$F$_5$	507-55-1	202.938	liq		52	1.55[25]		
3245	3,3-Dichloro-1,1,1,2,2-pentafluoropropane	Refrigerant 225ca	C$_3$HCl$_2$F$_5$	422-56-0	202.938	liq		45.5	1.54[25]		
3246	1,2-Dichloropentane		C$_5$H$_{10}$Cl$_2$	1674-33-5	141.038			148.3	1.0872[20]	1.4485[20]	i H$_2$O; s EtOH; vs chl
3247	1,5-Dichloropentane		C$_5$H$_{10}$Cl$_2$	628-76-2	141.038	liq	-72.8	179	1.0956[25]	1.4545[25]	i H$_2$O; s EtOH, eth, bz, ctc
3248	2,3-Dichloropentane		C$_5$H$_{10}$Cl$_2$	600-11-3	141.038	liq	-77.3	139	1.0789[20]	1.4464[20]	i H$_2$O
3249	Dichlorophene		C$_{13}$H$_{10}$Cl$_2$O$_2$	97-23-4	269.123	cry (bz, peth)	177.5				i H$_2$O; s EtOH, ace
3250	2,3-Dichlorophenol		C$_6$H$_4$Cl$_2$O	576-24-9	163.001	cry (lig, bz)	58				s EtOH, eth, bz, lig
3251	2,4-Dichlorophenol		C$_6$H$_4$Cl$_2$O	120-83-2	163.001	hex nd (bz)	45	210			sl H$_2$O; s EtOH, eth, bz, chl
3252	2,5-Dichlorophenol		C$_6$H$_4$Cl$_2$O	583-78-8	163.001	pr (bz, peth)	59	211			sl H$_2$O; vs EtOH, eth; s bz, peth
3253	2,6-Dichlorophenol		C$_6$H$_4$Cl$_2$O	87-65-0	163.001	nd (peth)	68.5	220; 82[4]	1.653[20]		vs EtOH, eth; s bz, peth
3254	3,4-Dichlorophenol		C$_6$H$_4$Cl$_2$O	95-77-2	163.001	nd (bz-peth)	68	253			sl H$_2$O; vs EtOH, eth; s bz, peth
3255	3,5-Dichlorophenol		C$_6$H$_4$Cl$_2$O	591-35-5	163.001	pr (peth)	68	233			sl H$_2$O; vs EtOH, eth; s peth
3256	(2,4-Dichlorophenoxy)acetic acid	2,4-D	C$_8$H$_6$Cl$_2$O$_3$	94-75-7	221.038	cry (bz)	140.5	160[0.4]			i H$_2$O; s EtOH; sl bz, DMSO
3257	4-(2,4-Dichlorophenoxy)butanoic acid	Butyrac 118	C$_{10}$H$_{10}$Cl$_2$O$_3$	94-82-6	249.090		118				
3258	2-(2,4-Dichlorophenoxy)propanoic acid	Dichlorprop	C$_9$H$_8$Cl$_2$O$_3$	120-36-5	235.064		117.5				sl H$_2$O, lig; s EtOH, eth
3259	Dichlorophenylarsine		C$_6$H$_5$AsCl$_2$	696-28-6	222.932	liq	-19	255	1.6516[20]	1.6386[15]	vs bz, eth, EtOH
3260	2,4-Dichlorophenyl benzenesulfonate	Genite	C$_{12}$H$_8$Cl$_2$O$_3$S	97-16-5	303.161		45.5				s ctc, CS$_2$
3261	2,2-Dichloro-1-phenylethanone		C$_8$H$_6$Cl$_2$O	2648-61-5	189.039	amor	20.5	249	1.340[16]	1.5686[20]	s EtOH, bz, ctc
3262	1-(2,4-Dichlorophenyl)ethanone		C$_8$H$_6$Cl$_2$O	2234-16-4	189.039		33.5			1.5640[20]	i H$_2$O
3263	1-(2,5-Dichlorophenyl)ethanone		C$_8$H$_6$Cl$_2$O	2476-37-1	189.039		12	118[12]	1.321[30]	1.5595[30]	
3264	1-(3,4-Dichlorophenyl)ethanone		C$_8$H$_6$Cl$_2$O	2642-63-9	189.039	nd (peth)	76	135[12]			i H$_2$O; s ctc, lig
3265	3,4-Dichlorophenyl isocyanate	1,2-Dichloro-5-isocyanatobenzene	C$_7$H$_3$Cl$_2$NO	102-36-3	188.011	cry	42	112[12]			
3266	3,5-Dichlorophenyl isocyanate	1,3-Dichloro-5-isocyanatobenzene	C$_7$H$_3$Cl$_2$NO	34893-92-0	188.011		33		1.380		
3267	N-(3,4-Dichlorophenyl)-2-methyl-2-propenamide	Dicryl	C$_{10}$H$_9$Cl$_2$NO	2164-09-2	230.090	cry (al-peth)	128				vs ace, EtOH
3268	3-(2,4-Dichlorophenyl)-2-propenoic acid		C$_9$H$_6$Cl$_2$O$_2$	1201-99-6	217.049		234				s DMSO
3269	Dichlorophenylsilane	Phenyldichlorosilane	C$_6$H$_6$Cl$_2$Si	1631-84-1	177.104			181	1.221[25]		dec H$_2$O
3270	1,1-Dichloropropane	Propylidene chloride	C$_3$H$_6$Cl$_2$	78-99-9	112.986			88.1	1.1321[20]	1.4289[20]	s EtOH, eth, bz, chl
3271	1,2-Dichloropropane, (±)	Propylene dichloride	C$_3$H$_6$Cl$_2$	26198-63-0	112.986	liq	-100.53	96.4	1.1560[20]	1.4394[20]	sl H$_2$O; s EtOH, eth, bz, chl
3272	1,3-Dichloropropane		C$_3$H$_6$Cl$_2$	142-28-9	112.986	liq	-99.5	120.9	1.1785[25]	1.4455[25]	sl H$_2$O; vs EtOH, eth; s bz, chl
3273	2,2-Dichloropropane		C$_3$H$_6$Cl$_2$	594-20-7	112.986	liq	-33.9	69.3	1.1136[20]	1.4148[20]	i H$_2$O; s EtOH, bz, chl; msc eth

2,4-Dichloro-1-naphthol

2,6-Dichloro-4-nitroaniline

1,2-Dichloro-3-nitrobenzene

1,2-Dichloro-4-nitrobenzene

1,3-Dichloro-5-nitrobenzene

1,4-Dichloro-2-nitrobenzene

2,4-Dichloro-1-nitrobenzene

1,1-Dichloro-1-nitroethane

2,6-Dichloro-4-nitrophenol

1,1-Dichloro-1-nitropropane

1,9-Dichlorononane

1,8-Dichlorooctane

1,3-Dichloro-1,1,2,2,3-pentafluoropropane

3,3-Dichloro-1,1,1,2,2-pentafluoropropane

1,2-Dichloropentane

1,5-Dichloropentane

2,3-Dichloropentane

Dichlorophene

2,3-Dichlorophenol

2,4-Dichlorophenol

2,5-Dichlorophenol

2,6-Dichlorophenol

3,4-Dichlorophenol

3,5-Dichlorophenol

(2,4-Dichlorophenoxy)acetic acid

4-(2,4-Dichlorophenoxy)butanoic acid

2-(2,4-Dichlorophenoxy)propanoic acid

Dichlorophenylarsine

2,4-Dichlorophenyl benzenesulfonate

2,2-Dichloro-1-phenylethanone

1-(2,4-Dichlorophenyl)ethanone

1-(2,5-Dichlorophenyl)ethanone

1-(3,4-Dichlorophenyl)ethanone

3,4-Dichlorophenyl isocyanate

3,5-Dichlorophenyl isocyanate

N-(3,4-Dichlorophenyl)-2-methyl-2-propenamide

3-(2,4-Dichlorophenyl)-2-propenoic acid

Dichlorophenylsilane

1,1-Dichloropropane

1,2-Dichloropropane, (±)

1,3-Dichloropropane

2,2-Dichloropropane

No.	Name	Synonym	Mol. Form.	CAS RN	Mol. Wt.	Physical Form	mp/°C	bp/°C	den/ g cm⁻³	n_D	Solubility
3274	2,2-Dichloropropanoic acid	2,2-Dichloropropionic acid	$C_3H_4Cl_2O_2$	75-99-0	142.969			187.5; 92[14]	1.389[12]		vs H_2O, alk, EtOH; s eth, ctc
3275	2,3-Dichloro-1-propanol		$C_3H_6Cl_2O$	616-23-9	128.985	visc		184	1.3607[20]	1.4819[20]	sl H_2O, lig; msc EtOH, eth, ace, bz
3276	1,3-Dichloro-2-propanol		$C_3H_6Cl_2O$	96-23-1	128.985			176	1.3506[17]	1.4837[20]	vs H_2O, EtOH; msc eth; s ace, chl
3277	2,3-Dichloro-1-propanol, phosphate (3:1)		$C_9H_{15}Cl_6O_4P$	78-43-3	430.904			190[0.1]	1.517[22]		
3278	2,3-Dichloropropanoyl chloride		$C_3H_3Cl_3O$	7623-13-4	161.414			53[17]	1.4757[20]	1.4764[20]	
3279	1,1-Dichloropropene		$C_3H_4Cl_2$	563-58-6	110.970			76.5	1.1864[25]	1.4430[25]	i H_2O; s eth, ace, chl
3280	cis-1,2-Dichloropropene		$C_3H_4Cl_2$	6923-20-2	110.970			93		1.4549[20]	i H_2O; s ace, bz, chl
3281	trans-1,2-Dichloropropene		$C_3H_4Cl_2$	7069-38-7	110.970			77	1.1818[20]	1.4471[20]	i H_2O; vs EtOH, ctc, MeOH
3282	cis-1,3-Dichloropropene	cis-1,3-Dichloropropylene	$C_3H_4Cl_2$	10061-01-5	110.970			104.3	1.224[20]	1.4682[20]	i H_2O; s eth, bz, chl
3283	trans-1,3-Dichloropropene	trans-1,3-Dichloropropylene	$C_3H_4Cl_2$	10061-02-6	110.970			112	1.217[20]	1.4730[20]	i H_2O; s eth, bz, chl
3284	2,3-Dichloropropene		$C_3H_4Cl_2$	78-88-6	110.970		10	94	1.211[20]	1.4603[20]	i H_2O; msc EtOH; s eth, bz, chl
3285	3,6-Dichloropyridazine		$C_4H_2Cl_2N_2$	141-30-0	148.978		68.8	89[0.2]			s chl
3286	2,6-Dichloropyridine		$C_5H_3Cl_2N$	2402-78-0	147.990		87	211			
3287	4,6-Dichloro-2-pyrimidinamine		$C_4H_3Cl_2N_3$	56-05-3	163.993		215				s DMSO
3288	2,4-Dichloropyrimidine		$C_4H_2Cl_2N_2$	3934-20-1	148.978		59	198; 101[23]			
3289	4,7-Dichloroquinoline		$C_9H_5Cl_2N$	86-98-6	198.049	cry (MeOH), nd (80% al)	93	148[10]			sl chl
3290	5,7-Dichloro-8-quinolinol	Chloroxine	$C_9H_5Cl_2NO$	773-76-2	214.048	cry (al)	179.5				sl EtOH, ace, chl, DMSO; s alk, bz, peth
3291	2,3-Dichloroquinoxaline		$C_8H_4Cl_2N_2$	2213-63-0	199.037	cry (al, bz)	152				i H_2O; vs EtOH, bz, chl, HOAc
3292	2,5-Dichlorostyrene		$C_8H_6Cl_2$	1123-84-8	173.040		8.0	93[5], 74[3]	1.246[20]	1.5798[20]	
3293	1,2-Dichloro-3,4,5,6-tetrafluorobenzene		$C_6Cl_2F_4$	1198-59-0	218.964			157.7			
3294	1,1-Dichloro-1,2,2,2-tetrafluoroethane	Refrigerant 114a	$C_2Cl_2F_4$	374-07-2	170.921	col gas	-56.6	3.4	1.455[25] (p>1 atm)	1.3092[20]	vs bz, eth, EtOH
3295	1,2-Dichloro-1,1,2,2-tetrafluoroethane	Refrigerant 114	$C_2Cl_2F_4$	76-14-2	170.921	col gas	-92.53	3.5	1.455[25] (p>1 atm)	1.3092[20]	i H_2O; vs eth, EtOH
3296	1,2-Dichloro-1,1,2,2-tetramethyldisilane		$C_4H_{12}Cl_2Si_2$	4342-61-4	187.215			148; 49[18]	1.010[20]	1.4548[20]	
3297	1,3-Dichloro-1,1,3,3-tetramethyldisiloxane		$C_4H_{12}Cl_2OSi_2$	2401-73-2	203.214	liq	-37.5	138	1.038[20]		
3298	2,5-Dichlorothiophene		$C_4H_2Cl_2S$	3172-52-9	153.030	liq	-40.5	162	1.4422[20]	1.5626[20]	i H_2O; msc EtOH, eth; s ctc
3299	2,3-Dichlorotoluene		$C_7H_6Cl_2$	32768-54-0	161.029		6	207.5	1.2458[20]	1.5511[20]	vs bz
3300	2,4-Dichlorotoluene		$C_7H_6Cl_2$	95-73-8	161.029	liq	-13.5	201	1.2476[20]	1.5511[20]	i H_2O; s ctc
3301	2,5-Dichlorotoluene		$C_7H_6Cl_2$	19398-61-9	161.029		2.5	200	1.2535[20]	1.5449[20]	i H_2O; s bz
3302	2,6-Dichlorotoluene		$C_7H_6Cl_2$	118-69-4	161.029		25.8	198	1.2686[20]	1.5507[20]	i H_2O; s chl
3303	3,4-Dichlorotoluene		$C_7H_6Cl_2$	95-75-0	161.029	liq	-15.2	208.9	1.2564[20]	1.5471[20]	i H_2O; msc EtOH, eth, ace, bz, lig, ctc
3304	1,3-Dichloro-1,3,5-triazine-2,4,6(1H,3H,5H)-trione	Dichlorocyanuric acid	$C_3HCl_2N_3O_3$	2782-57-2	197.964	cry	226.6				
3305	1,2-Dichloro-4-(trichloromethyl)benzene		$C_7H_3Cl_5$	13014-24-9	264.364		25.8	283.1	1.5913[20]	1.5886[20]	
3306	1,2-Dichloro-1,1,2-trifluoroethane	Refrigerant 123a	$C_2HCl_2F_3$	354-23-4	152.930	vol liq or gas	-78	29.5	1.50[25]		
3307	2,2-Dichloro-1,1,1-trifluoroethane		$C_2HCl_2F_3$	306-83-2	152.930	vol liq or gas	-107	27.82	1.4638[25]		sl H_2O
3308	2,2-Dichloro-1,1,2-trifluoroethane	Refrigerant 123b	$C_2HCl_2F_3$	812-04-4	152.930			30.2			
3309	2,4-Dichloro-1-(trifluoromethyl)benzene	2,4-Dichlorobenzotrifluoride	$C_7H_3Cl_2F_3$	320-60-5	215.000					1.4802[20]	
3310	4,5-Dichloro-2-(trifluoromethyl)-1H-benzimidazole	Chloroflurazole	$C_8H_3Cl_2F_3N_2$	3615-21-2	255.024		213.5				
3311	Dichlorovinylmethylsilane		$C_3H_6Cl_2Si$	124-70-9	141.072			92.5	1.0868[20]	1.4270[20]	dec H_2O
3312	Dichlorvos	Phosphoric acid, 2,2-dichloroethenyl dimethyl ester	$C_4H_7Cl_2O_4P$	62-73-7	220.976			140[20], 84[1]	1.415[25]		

2,2-Dichloropropanoic acid

2,3-Dichloro-1-propanol

1,3-Dichloro-2-propanol

2,3-Dichloro-1-propanol, phosphate (3:1)

2,3-Dichloropropanoyl chloride

1,1-Dichloropropene

cis-1,2-Dichloropropene

trans-1,2-Dichloropropene

cis-1,3-Dichloropropene

trans-1,3-Dichloropropene

2,3-Dichloropropene

3,6-Dichloropyridazine

2,6-Dichloropyridine

4,6-Dichloro-2-pyrimidinamine

2,4-Dichloropyrimidine

4,7-Dichloroquinoline

5,7-Dichloro-8-quinolinol

2,3-Dichloroquinoxaline

2,5-Dichlorostyrene

1,2-Dichloro-3,4,5,6-tetrafluorobenzene

1,1-Dichloro-1,2,2,2-tetrafluoroethane

1,2-Dichloro-1,1,2,2-tetrafluoroethane

1,2-Dichloro-1,1,2,2-tetramethyldisilane

1,3-Dichloro-1,1,3,3-tetramethyldisiloxane

2,5-Dichlorothiophene

2,3-Dichlorotoluene

2,4-Dichlorotoluene

2,5-Dichlorotoluene

2,6-Dichlorotoluene

3,4-Dichlorotoluene

1,3-Dichloro-1,3,5-triazine-2,4,6(1*H*,3*H*,5*H*)-trione

1,2-Dichloro-4-(trichloromethyl)benzene

1,2-Dichloro-1,1,2-trifluoroethane

2,2-Dichloro-1,1,1-trifluoroethane

2,2-Dichloro-1,1,2-trifluoroethane

2,4-Dichloro-1-(trifluoromethyl)benzene

4,5-Dichloro-2-(trifluoromethyl)-1*H*-benzimidazole

Dichlorovinylmethylsilane

Dichlorvos

No.	Name	Synonym	Mol. Form.	CAS RN	Mol. Wt.	Physical Form	mp/°C	bp/°C	den/ g cm⁻³	n_D	Solubility
3313	Diclofop-methyl	Methyl 2-[4-(2,4-dichlorophenoxy)phenoxy]propanoate	$C_{16}H_{14}Cl_2O_4$	51338-27-3	341.186		40	176[0.1]			
3314	Dicrotophos		$C_8H_{16}NO_5P$	141-66-2	237.191			400; 130[0.1]	1.216[15]		
3315	Dicumarol		$C_{19}H_{12}O_6$	66-76-2	336.294	nd	290				
3316	Dicyanamide	Cyanocyanamide	C_2HN_3	504-66-5	67.049	aq soln only					
3317	o-Dicyanobenzene	o-Phthalodinitrile	$C_8H_4N_2$	91-15-6	128.131	nd (w, lig)	141	150[10]	1.1250[25]		sl H₂O, lig; vs EtOH, bz; s eth, ace
3318	m-Dicyanobenzene	m-Phthalodinitrile	$C_8H_4N_2$	626-17-5	128.131	nd(al)	162	sub	0.992[40]		sl H₂O; vs EtOH; s eth, bz, chl; i peth
3319	p-Dicyanobenzene	p-Phthalodinitrile	$C_8H_4N_2$	623-26-7	128.131	nd (w, MeOH)	224	sub			i H₂O; sl EtOH, eth; s bz; vs HOAc
3320	Dicyclohexyl adipate		$C_{18}H_{30}O_4$	849-99-0	310.429		35				s chl
3321	Dicyclohexylamine	N-Cyclohexylcyclohexanamine	$C_{12}H_{23}N$	101-83-7	181.318		-0.1	dec 256; 114[9]	0.9123[20]	1.4842[20]	sl H₂O, ctc; s EtOH, eth, bz
3322	Dicyclohexylamine nitrite	N-Cyclohexylcyclohexanamine, nitrite	$C_{12}H_{24}N_2O_2$	3129-91-7	228.331	cry	182 dec				
3323	Dicyclohexylcarbodiimide		$C_{13}H_{22}N_2$	538-75-0	206.327		34.5	123[6], 99[0.5]			
3324	Dicyclohexyl disulfide		$C_{12}H_{22}S_2$	2550-40-5	230.433	liq		195[20]			
3325	Dicyclohexyl ether		$C_{12}H_{22}O$	4645-15-2	182.302	liq	-36	242.5	0.9227[20]	1.4741[20]	
3326	Dicyclohexylmethanone		$C_{13}H_{22}O$	119-60-8	194.313		57	159[20]	0.986[0]	1.4860[20]	s eth, ace, ctc
3327	Dicyclohexylphosphine		$C_{12}H_{23}P$	829-84-5	198.285			281; 129[8]	0.904[25]	1.5163[20]	
3328	Dicyclohexyl phthalate		$C_{20}H_{26}O_4$	84-61-7	330.418	pr (al)	66	225[4]	1.383[20]	1.431[20]	i H₂O; s EtOH, eth; sl chl
3329	N,N'-Dicyclohexylthiourea		$C_{13}H_{24}N_2S$	1212-29-9	240.408	cry (MeOH)	180				
3330	1,3-Dicyclohexylurea		$C_{13}H_{24}N_2O$	2387-23-7	224.342		233.8				
3331	Dicyclomine hydrochloride	Dicycloverine hydrochloride	$C_{19}H_{36}ClNO_2$	67-92-5	345.948	cry	165				
3332	Dicyclopentadiene		$C_{10}H_{12}$	1755-01-7	132.202		32	dec 170; 65[14]	0.9302[35]	1.5050[35]	vs eth, EtOH
3333	Dicyclopentyl ether	Cyclopentyl ether	$C_{10}H_{18}O$	10137-73-2	154.249	liq		80[13]			
3334	Dicyclopropyl ketone		$C_7H_{10}O$	1121-37-5	110.153			161	0.977[25]	1.4670[20]	
3335	Didecylamine	N-Decyl-1-decanamine	$C_{20}H_{43}N$	1120-49-6	297.562			359.0			
3336	Didecyl ether		$C_{20}H_{42}O$	2456-28-2	298.546		16	196[15.5]	0.8187[20]		
3337	Didecyl phthalate		$C_{28}H_{46}O_4$	84-77-5	446.663		2.5	240[3]	0.9639[20]		
3338	3',4'-Didehydro-β,ψ-caroten-16'-oic acid	Torularhodin	$C_{40}H_{52}O_2$	514-92-1	564.840	purp nd (MeOH-eth)	211				vs py, chl, CS₂
3339	2',3'-Dideoxyinosine	Didanosine	$C_{10}H_{12}N_4O_3$	69655-05-6	236.227	wh cry (EtOH aq)	162				
3340	2,6-Dideoxy-3-O-methyl-ribo-hexose	Cymarose	$C_7H_{14}O_4$	579-04-4	162.184	pr (eth-peth) nd (ace)	101				vs H₂O, ace, EtOH
3341	Didodecanoyl peroxide	Lauroyl peroxide	$C_{24}H_{46}O_4$	105-74-8	398.620	wh pl	49				i H₂O; s chl
3342	Didodecylamine	N-Dodecyl-1-dodecanamine	$C_{24}H_{51}N$	3007-31-6	353.669		53.7	263[27]			vs bz, eth, EtOH, chl
3343	Didodecyl phosphate		$C_{24}H_{51}O_4P$	7057-92-3	434.633	cry (MeOH)	59				
3344	Didodecyl phthalate	1,2-Benzenedicarboxylic acid, didodecyl ester	$C_{32}H_{54}O_4$	2432-90-8	502.769		22.0	256[1]	0.9389[20]		
3345	Dieldrin		$C_{12}H_8Cl_6O$	60-57-1	380.909		175.5		1.75[25]		i H₂O; sl EtOH; s ace, bz
3346	Dienestrol		$C_{18}H_{18}O_2$	84-17-3	266.335	cry (dil al)	227.5	sub 130			vs ace, eth, EtOH
3347	1,2:8,9-Diepoxy-p-menthane	Limonene diepoxide	$C_{10}H_{16}O_2$	96-08-2	168.233		242				
3348	Diethanolamine	Bis(2-hydroxyethyl)amine	$C_4H_{11}NO_2$	111-42-2	105.136		28	268.8	1.0966[20]	1.4776[20]	vs H₂O, EtOH; sl eth, bz
3349	Diethatyl, ethyl ester		$C_{16}H_{22}ClNO_3$	38727-55-8	311.804	cry	49.5				
3350	4,4'-Diethoxyazobenzene		$C_{16}H_{18}N_2O_2$	588-52-3	270.326	ye lf (al)	162	dec			i H₂O; sl EtOH; s eth, bz, chl; vs HOAc
3351	3,4-Diethoxybenzaldehyde		$C_{11}H_{14}O_3$	2029-94-9	194.227		22	279; 200[50]	1.0100[22]		vs EtOH
3352	1,2-Diethoxybenzene		$C_{10}H_{14}O_2$	2050-46-6	166.217	pr (peth, dil al)	44	219	1.0075[20]	1.5083[25]	s EtOH, ctc; vs eth
3353	1,4-Diethoxybenzene		$C_{10}H_{14}O_2$	122-95-2	166.217	pl (dil al)	72	246			vs EtOH; s eth, bz, ctc, chl
3354	4,4-Diethoxy-1-butanamine		$C_8H_{19}NO_2$	6346-09-4	161.243			196	0.933[25]	1.4275[20]	
3355	1,1-Diethoxy-N,N-dimethylmethanamine		$C_7H_{17}NO_2$	1188-33-6	147.216			129	0.859[25]	1.4007[20]	
3356	Diethoxydimethylsilane	Dimethyldiethoxysilane	$C_6H_{16}O_2Si$	78-62-6	148.276	liq	-87	114	0.865[25]	1.3811[20]	s ctc
3357	Diethoxydiphenylsilane		$C_{16}H_{20}O_2Si$	2553-19-7	272.415			302; 167[15]	1.0329[20]	1.5269[20]	
3358	2,2-Diethoxyethanamine		$C_6H_{15}NO_2$	645-36-3	133.189	liq	-78	163	0.9159[25]	1.4123[25]	vs H₂O, eth, EtOH, chl

Diclofop-methyl

Dicrotophos

Dicumarol

Dicyanamide

o-Dicyanobenzene

m-Dicyanobenzene

p-Dicyanobenzene

Dicyclohexyl adipate

Dicyclohexylamine

Dicyclohexylamine nitrite

HNO₂

Dicyclohexylcarbodiimide

Dicyclohexyl disulfide

Dicyclohexyl ether

Dicyclohexylmethanone

Dicyclohexylphosphine

Dicyclohexyl phthalate

N,N'-Dicyclohexylthiourea

1,3-Dicyclohexylurea

Dicyclomine hydrochloride

HCl

Dicyclopentadiene

Dicyclopentyl ether

Dicyclopropyl ketone

Didecylamine

Didecyl ether

Didecyl phthalate

3',4'-Didehydro-β,ψ-caroten-16'-oic acid

2',3'-Dideoxyinosine

2,6-Dideoxy-3-O-methyl-ribo-hexose

Didodecanoyl peroxide

Didodecylamine

Didodecyl phosphate

Didodecyl phthalate

Dieldrin

Dienestrol

1,2:8,9-Diepoxy-p-menthane

Diethanolamine

Diethatyl, ethyl ester

4,4'-Diethoxyazobenzene

3,4-Diethoxybenzaldehyde

1,2-Diethoxybenzene

1,4-Diethoxybenzene

4,4-Diethoxy-1-butanamine

1,1-Diethoxy-N,N-dimethylmethanamine

Diethoxydimethylsilane

Diethoxydiphenylsilane

2,2-Diethoxyethanamine

No.	Name	Synonym	Mol. Form.	CAS RN	Mol. Wt.	Physical Form	mp/°C	bp/°C	den/ g cm^{-3}	n_D	Solubility
3359	1,1-Diethoxyethane	Acetal	C$_6$H$_{14}$O$_2$	105-57-7	118.174	liq	-100	102.25	0.8254[20]	1.3834[20]	s H$_2$O, chl; msc EtOH, eth; vs ace
3360	1,2-Diethoxyethane	Ethylene glycol diethyl ether	C$_6$H$_{14}$O$_2$	629-14-1	118.174	liq	-74.0	121.2	0.8351[25]	1.3898[25]	vs ace, bz, eth, EtOH
3361	1,1-Diethoxyethene		C$_6$H$_{12}$O$_2$	2678-54-8	116.158			68[100]	0.7932[20]	1.3643[21]	
3362	Diethoxymethane		C$_5$H$_{12}$O$_2$	462-95-3	104.148	liq	-66.5	88	0.8319[20]	1.3748[18]	s H$_2$O; msc EtOH; vs ace, bz; sl chl
3363	2-(Diethoxymethyl)furan		C$_9$H$_{14}$O$_3$	13529-27-6	170.205			191.5	0.9976[20]	1.4451[20]	vs EtOH
3364	Diethoxymethylphenylsilane		C$_{11}$H$_{18}$O$_2$Si	775-56-4	210.346			218	0.9627[20]	1.4690[20]	
3365	Diethoxymethylsilane		C$_5$H$_{14}$O$_2$Si	2031-62-1	134.250			98	0.829[25]		
3366	1,1-Diethoxypentane		C$_9$H$_{20}$O$_2$	3658-79-5	160.254			59[12]	0.829[22]	1.4029[22]	
3367	1,1-Diethoxypropane		C$_7$H$_{16}$O$_2$	4744-08-5	132.201			123	0.825[20]	1.3924[19]	s H$_2$O, ace, bz; vs EtOH, eth
3368	2,2-Diethoxypropane		C$_7$H$_{16}$O$_2$	126-84-1	132.201			114	0.8200[21]	1.3891[20]	s EtOH, ace, bz; vs eth; sl ctc
3369	3,3-Diethoxy-1-propene	Acrolein, diethyl acetal	C$_7$H$_{14}$O$_2$	3054-95-3	130.185			123.5	0.8543[15]	1.4000[20]	sl H$_2$O; msc EtOH, eth
3370	3,3-Diethoxy-1-propyne		C$_7$H$_{12}$O$_2$	10160-87-9	128.169			139	0.8942[22]	1.4140[20]	vs ace, eth, EtOH, chl
3371	N,N-Diethylacetamide		C$_6$H$_{13}$NO	685-91-6	115.173			185.5	0.9130[17]	1.4374[17]	s H$_2$O, EtOH; msc eth, ace, bz; sl ctc
3372	Diethyl 2-acetamidomalonate		C$_9$H$_{15}$NO$_5$	1068-90-2	217.219	cry (al,bz-peth)	96.3	185[20]			sl H$_2$O, eth; s tfa, EtOH
3373	N,N-Diethylacetoacetamide		C$_8$H$_{15}$NO$_2$	2235-46-3	157.211	liq		76[13]			
3374	Diethyl acetylphosphonate		C$_6$H$_{13}$O$_4$P	919-19-7	180.138			114[20]	1.1005[20]	1.4200[26]	
3375	Diethyl 2-acetylsuccinate		C$_{10}$H$_{16}$O$_5$	1115-30-6	216.231			255; 133[17]	1.081[20]	1.4346[20]	i H$_2$O; s EtOH, eth, bz; sl chl
3376	Diethyl adipate		C$_{10}$H$_{18}$O$_4$	141-28-6	202.248	liq	-19.8	245	1.0076[20]	1.4272[20]	i H$_2$O; s EtOH, eth
3377	Diethyl 2-allylmalonate		C$_{10}$H$_{16}$O$_4$	2049-80-1	200.232			222.5; 93[6]	1.0098[20]	1.4305[20]	i H$_2$O; vs EtOH, eth; s ctc
3378	Diethylamine	N-Ethylethanamine	C$_4$H$_{11}$N	109-89-7	73.137	liq	-49.8	55.5	0.7056[20]	1.3864[20]	vs H$_2$O; msc EtOH; s eth, ctc
3379	Diethylamine hydrochloride	N-Ethylethanamine hydrochloride	C$_4$H$_{12}$ClN	660-68-4	109.598	lf (al-eth)	228.5		1.0477[22]		vs H$_2$O, EtOH
3380	(Diethylamino)acetonitrile		C$_6$H$_{12}$N$_2$	3010-02-4	112.172			170	0.8660[20]	1.4260[20]	s H$_2$O
3381	4-(Diethylamino)benzaldehyde		C$_{11}$H$_{15}$NO	120-21-8	177.243	ye nd (w)	41	172[10]			vs H$_2$O; s EtOH, eth, bz, ctc
3382	2-(Diethylamino)-N-(2,6-dimethylphenyl)acetamide	Lidocaine	C$_{14}$H$_{22}$N$_2$O	137-58-6	234.337	nd (bz, al)	68.5	181[4]			vs bz, eth, EtOH, chl
3383	2-(Diethylamino)-N-(2,6-dimethylphenyl)acetamide, monohydrochloride		C$_{14}$H$_{23}$ClN$_2$O	73-78-9	270.798		128				vs H$_2$O
3384	2-Diethylaminoethanol		C$_6$H$_{15}$NO	100-37-8	117.189	hyg		163	0.8921[20]	1.4412[20]	msc H$_2$O; s EtOH, eth, ace, bz, peth; sl ctc
3385	2-[2-(Diethylamino)ethoxy]ethanol		C$_8$H$_{19}$NO$_2$	140-82-9	161.243			221.5; 92[7]	0.9421[25]	1.4480[20]	
3386	2-(Diethylamino)ethyl acrylate		C$_9$H$_{17}$NO$_2$	2426-54-2	171.237		<-60	81[10]	0.937[20]	1.4376[25]	
3387	2-Diethylaminoethyl 4-aminobenzoate	Procaine	C$_{13}$H$_{20}$N$_2$O$_2$	59-46-1	236.310	nd (w+2) pl (lig or eth)	61				sl H$_2$O; s EtOH, eth, bz, chl
3388	2-(N,N-Diethylamino)ethyl methacrylate		C$_{10}$H$_{19}$NO$_2$	105-16-8	185.264			80[10]	0.92[30]		
3389	2-(Diethylamino)ethyl 2-phenylbutanoate	Butethamate	C$_{16}$H$_{25}$NO$_2$	14007-64-8	263.376			168[11]		1.4909[20]	
3390	4-(Diethylamino)-2-hydroxybenzaldehyde		C$_{11}$H$_{15}$NO$_2$	17754-90-4	193.243		65.0				
3391	Diethyl 2-aminomalonate		C$_7$H$_{13}$NO$_4$	6829-40-9	175.183			122[16], 116[12]	1.100[16]	1.4353[16]	vs H$_2$O, EtOH, eth; s ace, bz; i lig
3392	7-(Diethylamino)-4-methyl-2H-1-benzopyran-2-one		C$_{14}$H$_{17}$NO$_2$	91-44-1	231.291	cry (al, bz-lig)					sl H$_2$O; s EtOH, eth, ace
3393	3-(Diethylamino)phenol		C$_{10}$H$_{15}$NO	91-68-9	165.232	orth bipym (CS$_2$-lig)	78	276; 170[15]			s H$_2$O, EtOH, eth, CS$_2$; sl lig
3394	2-(Diethylamino)-1-phenyl-1-propanone	Diethylpropion	C$_{13}$H$_{19}$NO	90-84-6	205.296	liq		111[14]			
3395	3-(Diethylamino)-1-propanol		C$_7$H$_{17}$NO	622-93-5	131.216			189.5	0.8600[20]	1.4439[20]	s EtOH; s eth, ace, bz; sl chl
3396	3-(Diethylamino)-1-propyne	N,N-Diethyl-2-propargylamine	C$_7$H$_{13}$N	4079-68-9	111.185	liq		120			
3397	2,6-Diethylaniline		C$_{10}$H$_{15}$N	579-66-8	149.233		1.5	243	0.906[25]	1.5452[20]	

1,1-Diethoxyethane

1,2-Diethoxyethane

1,1-Diethoxyethene

Diethoxymethane

2-(Diethoxymethyl)furan

Diethoxymethylphenylsilane

Diethoxymethylsilane

1,1-Diethoxypentane

1,1-Diethoxypropane

2,2-Diethoxypropane

3,3-Diethoxy-1-propene

3,3-Diethoxy-1-propyne

N,N-Diethylacetamide

Diethyl 2-acetamidomalonate

N,N-Diethylacetoacetamide

Diethyl acetylphosphonate

Diethyl 2-acetylsuccinate

Diethyl adipate

Diethyl 2-allylmalonate

Diethylamine

Diethylamine hydrochloride

(Diethylamino)acetonitrile

4-(Diethylamino)benzaldehyde

2-(Diethylamino)-*N*-(2,6-dimethylphenyl)acetamide

2-(Diethylamino)-*N*-(2,6-dimethylphenyl)acetamide, monohydrochloride

2-Diethylaminoethanol

2-[2-(Diethylamino)ethoxy]ethanol

2-(Diethylamino)ethyl acrylate

2-Diethylaminoethyl 4-aminobenzoate

2-(*N,N*-Diethylamino)ethyl methacrylate

2-(Diethylamino)ethyl 2-phenylbutanoate

4-(Diethylamino)-2-hydroxybenzaldehyde

Diethyl 2-aminomalonate

7-(Diethylamino)-4-methyl-2*H*-1-benzopyran-2-one

3-(Diethylamino)phenol

2-(Diethylamino)-1-phenyl-1-propanone

3-(Diethylamino)-1-propanol

3-(Diethylamino)-1-propyne

2,6-Diethylaniline

No.	Name	Synonym	Mol. Form.	CAS RN	Mol. Wt.	Physical Form	mp/°C	bp/°C	den/ g cm⁻³	n_D	Solubility
3398	N,N-Diethylaniline		$C_{10}H_{15}N$	91-66-7	149.233	ye oil	-38.8	216.3	0.9307[20]	1.5409[20]	sl H₂O; s EtOH, ace, ctc; vs eth, chl
3399	Diethylarsine		$C_4H_{11}As$	692-42-2	134.052			105	1.1338[24]	1.4709	vs ace, bz, eth, EtOH
3400	N,N-Diethylbenzamide		$C_{11}H_{15}NO$	1696-17-9	177.243			132[5]			
3401	o-Diethylbenzene	1,2-Diethylbenzene	$C_{10}H_{14}$	135-01-3	134.218	liq	-31.2	184	0.8800[20]	1.5035[20]	i H₂O; msc EtOH, eth, ace, bz, lig, ctc
3402	m-Diethylbenzene	1,3-Diethylbenzene	$C_{10}H_{14}$	141-93-5	134.218	liq	-83.9	181.1	0.8602[20]	1.4955[20]	i H₂O; msc EtOH, eth, ace, bz, lig, ctc
3403	p-Diethylbenzene	1,4-Diethylbenzene	$C_{10}H_{14}$	105-05-5	134.218	liq	-42.83	183.7	0.8620[20]	1.4967[20]	i H₂O; msc EtOH, eth, ace, bz, lig, ctc
3404	N,N-Diethyl-1,4-benzenediamine		$C_{10}H_{16}N_2$	93-05-0	164.247			261			vs bz
3405	Diethyl benzylidenemalonate	Diethyl benzalmalonate	$C_{14}H_{16}O_4$	5292-53-5	248.275		32	216[30], 196[14]	1.1045[20]	1.5389[20]	i H₂O; s EtOH, eth, ace, bz
3406	Diethyl benzylmalonate		$C_{14}H_{18}O_4$	607-81-8	250.291			300	1.076[15]	1.4872[20]	i H₂O; sl chl
3407	Diethyl benzylphosphonate		$C_{11}H_{17}O_3P$	1080-32-6	228.225			110[2]		1.4930[20]	s ctc
3408	Diethylbromoacetamide	2-Bromo-2-ethylbutanamide	$C_6H_{12}BrNO$	511-70-6	194.069		67				sl H₂O, chl; vs EtOH, eth, bz
3409	Diethyl 2-bromomalonate	Ethyl bromomalonate	$C_7H_{11}BrO_4$	685-87-0	239.064		-54	dec 254	1.4022[25]	1.4521[20]	i H₂O; msc EtOH, eth; s ace, ctc
3410	N,N-Diethylbutanamide		$C_8H_{17}NO$	1114-76-7	143.227			206	0.8884[20]	1.4403[25]	vs H₂O, EtOH
3411	Diethyl 2-butylmalonate	Pentane-1,1-dicarboxylic acid, diethyl ester	$C_{11}H_{20}O_4$	133-08-4	216.275			238	0.9764[20]	1.4250[20]	vs EtOH, eth
3412	Diethyl 2-butynedioate		$C_8H_{10}O_4$	762-21-0	170.163		0.8	184[200]	1.0075[20]	1.4425[20]	s EtOH, eth, ctc
3413	Diethylcarbamazine citrate		$C_{16}H_{29}N_3O_8$	1642-54-2	391.416	cry	138				
3414	Diethylcarbamic chloride		$C_5H_{10}ClNO$	88-10-8	135.592			186			
3415	N,N'-Diethylcarbanilide		$C_{17}H_{20}N_2O$	85-98-3	268.353	cry (al)	79				i H₂O; vs EtOH; s chl
3416	Diethyl carbonate	Ethyl carbonate	$C_5H_{10}O_3$	105-58-8	118.131	liq	-43	126	0.9692[25]	1.3845[20]	i H₂O; s EtOH, eth, chl
3417	O,O-Diethyl chloridothionophosphate	Diethyl thiophosphoryl chloride	$C_4H_{10}ClO_2PS$	2524-04-1	188.613			45[3]			s ctc
3418	Diethylchloroaluminum		$C_4H_{10}AlCl$	96-10-6	120.557			134[70]			
3419	Diethyl chloromalonate	Ethyl chloromalonate	$C_7H_{11}ClO_4$	14064-10-9	194.613			222	1.2040[20]	1.4327[20]	i H₂O; msc EtOH, eth, chl; s CS₂
3420	Diethyl chlorophosphonate	Diethoxyphosphoryl chloride	$C_4H_{10}ClO_3P$	814-49-3	172.547			93.5	1.205[19]	1.4170[20]	
3421	Diethylcyanamide		$C_5H_{10}N_2$	617-83-4	98.146	liq	-80.6	188	0.854[20]	1.4126[25]	i H₂O; s EtOH, eth
3422	Diethyl 1,1-cyclobutanedicarboxylate		$C_{10}H_{16}O_4$	3779-29-1	200.232			224	1.0456[20]	1.4330[26]	vs EtOH; sl ctc
3423	1,1-Diethylcyclohexane		$C_{10}H_{20}$	78-01-3	140.266			179.5			
3424	Diethyl 1,1-cyclopropanedicarboxylate		$C_9H_{14}O_4$	1559-02-0	186.205			215; 100[12]	1.055[25]	1.4345[18]	vs EtOH, eth
3425	Diethyl dibutylmalonate		$C_{15}H_{28}O_4$	596-75-8	272.381			150[12]	0.9457[20]	1.4341[20]	i H₂O; s EtOH, eth, ctc
3426	Diethyl dicarbonate	Pyrocarbonic acid diethyl ester	$C_6H_{10}O_5$	1609-47-8	162.140			93[18]	1.120[20]	1.3960[20]	vs ace, EtOH, lig
3427	Diethyl [(diethanolamino)methyl] phosphonate		$C_9H_{22}NO_5P$	2781-11-5	255.249	liq		150[0.01]			
3428	5,5-Diethyldihydro-2H-1,3-oxazine-2,4(3H)-dione	Diethadione	$C_8H_{13}NO_3$	702-54-5	171.194	cry (eth)	97.5				
3429	Diethyl 1,4-dihydro-2,4,6-trimethyl-3,5-pyridinedicarboxylate	3,5-Diethoxycarbonyl-1,4-dihydrocollidine	$C_{14}H_{21}NO_4$	632-93-9	267.322	lt bl flr pl (al)	131				sl H₂O, EtOH, eth, CS₂; vs chl
3430	Diethyldimethyllead	Diethyldimethylplumbane	$C_6H_{16}Pb$	1762-27-2	295.4	col liq		51[13]	1.79[20]		
3431	Diethyl 2,6-dimethyl-3,5-pyridinedicarboxylate		$C_{13}H_{17}NO_4$	1149-24-2	251.279		71	301; 208[40]			i H₂O; s EtOH, eth, bz, chl, lig
3432	Diethyl 3,5-dimethylpyrrole-2,4-dicarboxylate		$C_{12}H_{17}NO_4$	2436-79-5	239.268	nd (dil al)	137.8				i H₂O; sl EtOH, eth; s ace, bz, HOAc
3433	Diethyl disulfide		$C_4H_{10}S_2$	110-81-6	122.252	liq	-101.5	154.0	0.9931[20]	1.5073[20]	sl H₂O; msc EtOH, eth
3434	N,N-Diethyldodecanamide		$C_{16}H_{33}NO$	3352-87-2	255.439			166[2]	0.847[25]	1.4545[20]	s chl
3435	Diethylene glycol	Diglycol	$C_4H_{10}O_3$	111-46-6	106.120	liq	-10.4	245.8	1.1197[15]	1.4472[20]	s H₂O, EtOH, eth, chl
3436	Diethylene glycol, bischloroformate	Oxydi-2,1-ethanediyl carbonochloridate	$C_6H_8Cl_2O_5$	106-75-2	231.031	liq		126[5]	1.39[20]	1.4542[20]	
3437	Diethylene glycol diacetate		$C_8H_{14}O_5$	628-68-2	190.194		18	200	1.1068[15]	1.4348[20]	vs EtOH
3438	Diethylene glycol dibenzoate		$C_{18}H_{18}O_5$	120-55-8	314.333		33.5	280[24], 250[1]	1.1690[15]		vs H₂O, EtOH

N,N-Diethylaniline

Diethylarsine

N,N-Diethylbenzamide

o-Diethylbenzene

m-Diethylbenzene

p-Diethylbenzene

N,N-Diethyl-1,4-benzenediamine

Diethyl benzylidenemalonate

Diethyl benzylmalonate

Diethyl benzylphosphonate

Diethylbromoacetamide

Diethyl 2-bromomalonate

N,N-Diethylbutanamide

Diethyl 2-butylmalonate

Diethyl 2-butynedioate

Diethylcarbamazine citrate

Diethylcarbamic chloride

N,N'-Diethylcarbanilide

Diethyl carbonate

O,O-Diethyl chloridothionophosphate

Diethylchloroaluminum

Diethyl chloromalonate

Diethyl chlorophosphonate

Diethylcyanamide

Diethyl 1,1-cyclobutanedicarboxylate

1,1-Diethylcyclohexane

Diethyl 1,1-cyclopropanedicarboxylate

Diethyl dibutylmalonate

Diethyl dicarbonate

Diethyl [(diethanolamino)methyl]phosphonate

5,5-Diethyldihydro-2H-1,3-oxazine-2,4(3H)-dione

Diethyl 1,4-dihydro-2,4,6-trimethyl-3,5-pyridinedicarboxylate

Diethyldimethyllead

Diethyl 2,6-dimethyl-3,5-pyridinedicarboxylate

Diethyl 3,5-dimethylpyrrole-2,4-dicarboxylate

Diethyl disulfide

N,N-Diethyldodecanamide

Diethylene glycol

Diethylene glycol, bischloroformate

Diethylene glycol diacetate

Diethylene glycol dibenzoate

No.	Name	Synonym	Mol. Form.	CAS RN	Mol. Wt.	Physical Form	mp/°C	bp/°C	den/ g cm⁻³	n_D	Solubility
3439	Diethylene glycol dibutyl ether	Bis(2-butoxyethyl) ether	C₁₂H₂₆O₃	112-73-2	218.332	liq	-60	256	0.885²⁵	1.4235²⁰	
3440	Diethylene glycol diethyl ether	Bis(2-ethoxyethyl) ether	C₈H₁₈O₃	112-36-7	162.227	liq	-45	188	0.9063²⁰	1.4115²⁰	vs H₂O, EtOH; s eth
3441	Diethylene glycol dimethacrylate	Oxydiethylene methacrylate	C₁₂H₁₈O₅	2358-84-1	242.268			>200; 150⁸	1.0821²⁰	1.4571²⁵	
3442	Diethylene glycol dimethyl ether	Diglyme	C₆H₁₄O₃	111-96-6	134.173	liq	-68	162	0.9434²⁰	1.4097²⁰	msc H₂O, EtOH, eth
3443	Diethylene glycol dinitrate	2,2'-Oxybisethanol, dinitrate	C₄H₈N₂O₇	693-21-0	196.116			44⁰·⁰¹			
3444	Diethylene glycol monobutyl ether		C₈H₁₈O₃	112-34-5	162.227	liq	-68	231	0.9553²⁰	1.4306²⁰	msc H₂O; vs EtOH, eth, ace; s bz
3445	Diethylene glycol monobutyl ether acetate	2-(2-Butoxyethoxy)ethyl acetate	C₁₀H₂₀O₄	124-17-4	204.264	liq	-32	245	0.985²⁰	1.4262²⁰	vs ace, eth, EtOH
3446	Diethylene glycol monododecanoate	2-(2-Hydroxyethoxy)ethyl laurate	C₁₆H₃₂O₄	141-20-8	288.423	lt ye	17.5	>270	0.96²⁵		msc EtOH, eth, ace; s bz, tol
3447	Diethylene glycol monoethyl ether	Carbitol	C₆H₁₄O₃	111-90-0	134.173	hyg liq		196	0.9885²⁰	1.4300²⁰	msc H₂O, EtOH, ace, bz; vs eth
3448	Diethylene glycol monoethyl ether acetate	Carbitol acetate	C₈H₁₆O₄	112-15-2	176.211	liq	-25	218.5	1.0096²⁰	1.4213²⁰	vs H₂O, ace, eth, EtOH
3449	Diethylene glycol monohexyl ether	2-[2-(Hexyloxy)ethoxy]ethanol	C₁₀H₂₂O₃	112-59-4	190.280	col liq	-28	258; 192¹⁰⁰			
3450	Diethylene glycol monomethyl ether	2-(2-Methoxyethoxy)ethanol	C₅H₁₂O₃	111-77-3	120.147			193	1.035²⁰	1.4264²⁰	msc H₂O, ace; vs EtOH, eth
3451	Diethylene glycol monopropyl ether		C₇H₁₆O₃	6881-94-3	148.200	liq	-53.3	213; 124⁴			
3452	N,N-Diethyl-1,2-ethanediamine	N,N-Diethylethylenediamine	C₆H₁₆N₂	100-36-7	116.204			144	0.8280²⁰	1.4340²⁰	msc H₂O; s EtOH, eth, ctc, tol
3453	N,N'-Diethyl-1,2-ethanediamine		C₆H₁₆N₂	111-74-0	116.204			146	0.8280²⁰	1.4340²⁰	vs H₂O, eth, EtOH, tol
3454	Diethyl ether	Ethyl ether	C₄H₁₀O	60-29-7	74.121	liq	-116.2	34.5	0.7138²⁰	1.3526²⁰	sl H₂O; msc EtOH, bz, eth; vs ace
3455	Diethyl (ethoxymethylene) malonate	2-Ethoxy-1,1-bis(ethoxycarbonyl)ethene	C₁₀H₁₆O₅	87-13-8	216.231			dec 280; 165¹⁹		1.4600²⁰	i H₂O; s EtOH, eth; sl chl
3456	Diethyl ethylidenemalonate		C₉H₁₄O₄	1462-12-0	186.205			116¹⁷; 86³	1.0404²⁰	1.4308¹⁷	vs eth, EtOH
3457	Diethyl ethylmalonate		C₉H₁₆O₄	133-13-1	188.221			208; 98¹²	1.006²⁰	1.4166²⁰	sl H₂O; vs EtOH, eth, ace, chl
3458	Diethyl ethylphenylmalonate		C₁₅H₂₀O₄	76-67-5	264.318			170¹⁹	1.071²⁰	1.4896²⁵	i H₂O; s EtOH, eth; sl chl
3459	Diethyl ethylphosphonate		C₆H₁₅O₃P	78-38-6	166.155			198; 90¹⁶	1.0259²⁰	1.4163²⁰	sl H₂O; s EtOH, eth
3460	N,N-Diethylformamide		C₅H₁₁NO	617-84-5	101.147			177.5	0.9080¹⁹	1.4321²⁵	msc H₂O, ace, bz; vs EtOH, eth
3461	Diethyl fumarate		C₈H₁₂O₄	623-91-6	172.179		0.8	214	1.0452²⁰	1.4412²⁰	i H₂O; s ace, chl
3462	Diethyl glutarate		C₉H₁₆O₄	818-38-2	188.221	syr liq	-24.1	236.5	1.0220²⁰	1.4241²⁰	vs eth
3463	3,4-Diethylhexane		C₁₀H₂₂	19398-77-7	142.282			163.9	0.7472²⁵	1.4190²⁰	
3464	Di-2-ethylhexyl maleate		C₂₀H₃₆O₄	142-16-5	340.498			156⁷	0.94²⁰		
3465	1,2-Diethylhydrazine		C₄H₁₂N₂	1615-80-1	88.151			85.5	0.797²⁶	1.4204²⁰	vs bz, eth, EtOH
3466	Diethyl 1,2-hydrazinedicarboxylate	Diethyl bicarbamate	C₆H₁₂N₂O₄	4114-28-7	176.170	nd (chl), pr (w)	135	dec 250	1.324⁸		vs eth, EtOH
3467	Diethyl hydrogen phosphate	Diethyl phosphate	C₄H₁₁O₄P	598-02-7	154.101	syr		dec 203; 87⁰·⁰⁰⁰¹	1.1800²⁰	1.4170²⁰	vs eth
3468	N,N-Diethyl-4-hydroxy-3-methoxybenzamide	Ethamivan	C₁₂H₁₇NO₃	304-84-7	223.268		95				s chl
3469	Diethyl iminodiacetate		C₈H₁₅NO₄	6290-05-7	189.210	orth cry	247 dec				
3470	Diethyl isobutylmalonate		C₁₁H₂₀O₄	10203-58-4	216.275				0.9804²⁰	1.4236²⁰	i H₂O; vs EtOH, eth; s chl
3471	Diethyl isophthalate		C₁₂H₁₄O₄	636-53-3	222.237		11.5	302	1.1239¹⁷	1.508¹⁸	i H₂O
3472	Diethyl isopropylidenemalonate		C₁₀H₁₆O₄	6802-75-1	200.232			176.5; 116¹⁴	1.0282¹⁸	1.4486¹⁷	vs ace, EtOH
3473	Diethyl isopropylmalonate	Ethyl isopropylmalonate	C₁₀H₁₈O₄	759-36-4	202.248			215	0.9961²⁰	1.4188²¹	sl H₂O, ctc; vs EtOH, eth; s chl
3474	Diethyl ketomalonate	Ethyl mesoxalate	C₇H₁₀O₅	609-09-6	174.151	pa ye grn oil	-30	210; 105¹⁹	1.1419¹⁶	1.4310²²	vs H₂O; s EtOH, eth, chl; i CS₂
3475	Diethyl malate	Diethyl hydroxybutanedioate	C₈H₁₄O₅	7554-12-3	190.194			253; 124¹³	1.1290²⁰		
3476	Diethyl maleate		C₈H₁₂O₄	141-05-9	172.179	liq	-8.8	223	1.0662²⁰	1.4416²⁰	i H₂O; s EtOH, eth; sl chl
3477	Diethyl malonate		C₇H₁₂O₄	105-53-3	160.168	liq	-50	200	1.0551²⁰	1.4139²⁰	sl H₂O; msc EtOH, eth; vs ace, bz
3478	Diethyl mercury		C₄H₁₀Hg	627-44-1	258.71			159; 57¹⁶	2.43²⁰		s eth; sl EtOH
3479	Diethylmethylamine	N-Ethyl-N-methylethanamine	C₅H₁₃N	616-39-7	87.164	liq	-196	66	0.703²⁵	1.3879²⁵	vs H₂O, EtOH, eth

Diethylene glycol dibutyl ether

Diethylene glycol diethyl ether

Diethylene glycol dimethacrylate

Diethylene glycol dimethyl ether

Diethylene glycol dinitrate

Diethylene glycol monobutyl ether

Diethylene glycol monobutyl ether acetate

Diethylene glycol monododecanoate

Diethylene glycol monoethyl ether

Diethylene glycol monoethyl ether acetate

Diethylene glycol monohexyl ether

Diethylene glycol monomethyl ether

Diethylene glycol monopropyl ether

N,*N*-Diethyl-1,2-ethanediamine

N,*N'*-Diethyl-1,2-ethanediamine

Diethyl ether

Diethyl (ethoxymethylene)malonate

Diethyl ethylidenemalonate

Diethyl ethylmalonate

Diethyl ethylphenylmalonate

Diethyl ethylphosphonate

N,*N*-Diethylformamide

Diethyl fumarate

Diethyl glutarate

3,4-Diethylhexane

Di-2-ethylhexyl maleate

1,2-Diethylhydrazine

Diethyl 1,2-hydrazinedicarboxylate

Diethyl hydrogen phosphate

N,*N*-Diethyl-4-hydroxy-3-methoxybenzamide

Diethyl iminodiacetate

Diethyl isobutylmalonate

Diethyl isophthalate

Diethyl isopropylidenemalonate

Diethyl isopropylmalonate

Diethyl ketomalonate

Diethyl malate

Diethyl maleate

Diethyl malonate

Diethyl mercury

Diethylmethylamine

Physical Constants of Organic Compounds

No.	Name	Synonym	Mol. Form.	CAS RN	Mol. Wt.	Physical Form	mp/°C	bp/°C	den/ g cm^{-3}	n_D	Solubility
3480	N,N-Diethyl-2-methylaniline		$C_{11}H_{17}N$	606-46-2	163.260	liq	-60	209	0.9286[20]	1.5153[20]	sl H$_2$O; msc EtOH, eth; s ctc
3481	N,N-Diethyl-4-methylaniline		$C_{11}H_{17}N$	613-48-9	163.260			229	0.9242[16]		sl H$_2$O; msc EtOH, eth
3482	N,N-Diethyl-3-methylbenzamide	DEET	$C_{12}H_{17}NO$	134-62-3	191.269			160[19], 111[1]	0.996[20]	1.5212[20]	vs H$_2$O, bz, eth, EtOH
3483	1,3-Diethyl-5-methylbenzene		$C_{11}H_{16}$	2050-24-0	148.245	liq	-74.1	205	0.8748[20]	1.5027[20]	i H$_2$O; msc EtOH, eth, ace, bz, lig, ctc
3484	N4,N4-Diethyl-2-methyl-1,4-benzenediamine, monohydrochloride	4-N,N-Diethyl-1,4-diamino-2-methylbenzene, hydrochloride	$C_{11}H_{19}ClN_2$	2051-79-8	214.735	cry	250 dec				
3485	N,N-Diethyl-3-methylbutanamide	Isovaleryl diethylamide	$C_9H_{19}NO$	533-32-4	157.253			211	0.8764[20]	1.4422[20]	vs eth, EtOH
3486	Diethyl methylenesuccinate		$C_9H_{14}O_4$	2409-52-1	186.205		58.5	228	1.0467[20]	1.4377[20]	msc EtOH; s eth, bz; vs ace
3487	Diethyl methylmalonate		$C_8H_{14}O_4$	609-08-5	174.195			201	1.0225[20]	1.4126[20]	sl H$_2$O; vs EtOH, eth, ace, chl
3488	Diethyl methylphosphonate		$C_5H_{13}O_3P$	683-08-9	152.129			194	1.0406[30]	1.4101[30]	s H$_2$O, EtOH, eth; i bz
3489	N,N-Diethyl-4-methyl-1-piperazinecarboxamide	Diethylcarbamazine	$C_{10}H_{21}N_3O$	90-89-1	199.293		48	110[3]			
3490	3,3-Diethyl-5-methyl-2,4-piperidinedione		$C_{10}H_{17}NO_2$	125-64-4	183.248		75.5				s H$_2$O, bz, chl, EtOH
3491	N,N-Diethyl-1-naphthalenamine		$C_{14}H_{17}N$	84-95-7	199.292			285	1.013[20]	1.5961[20]	s EtOH, eth, bz; sl ctc
3492	N,N-Diethyl-4-nitroaniline		$C_{10}H_{14}N_2O_2$	2216-15-1	194.230	ye nd (lig) pl (al)	77.5		1.225[25]		s EtOH; sl lig
3493	N,N-Diethyl-4-nitrosoaniline		$C_{10}H_{14}N_2O$	120-22-9	178.230	grn mcl pr (eth) grn lf (ace)	87.5		1.24[15]		sl H$_2$O; s EtOH, eth, ace, chl
3494	Diethyl nonanedioate	Diethyl azelate	$C_{13}H_{24}O_4$	624-17-9	244.328	liq	-18.5	291.5	0.9729[20]	1.4351[20]	i H$_2$O; s EtOH, eth
3495	Diethyl oxalate		$C_6H_{10}O_4$	95-92-1	146.141	liq	-40.6	185.7	1.0785[20]	1.4101[20]	sl H$_2$O; msc EtOH, eth, ace; s ctc
3496	Diethyl oxobutanedioate	Diethyl oxalacetate	$C_8H_{12}O_5$	108-56-5	188.178			131[24]	1.131[20]	1.4561[17]	i H$_2$O; msc EtOH, eth, bz; vs ace
3497	Diethyl 3-oxo-1,5-pentanedioate	Diethyl 1,3-acetonedicarboxylate	$C_9H_{14}O_5$	105-50-0	202.204			250	1.113[20]		sl H$_2$O; msc EtOH
3498	3,3-Diethylpentane	Tetraethylmethane	C_9H_{20}	1067-20-5	128.255	liq	-33.1	146.3	0.7536[20]	1.4206[20]	i H$_2$O; s eth, bz
3499	N',N-Diethyl-1,4-pentanediamine	Novoldiamine	$C_9H_{22}N_2$	140-80-7	158.284			201	0.814[20]	1.4429[20]	
3500	2,2-Diethyl-4-pentenamide	Novonal	$C_9H_{17}NO$	512-48-1	155.237	wh pow	75.5				vs eth, EtOH
3501	Diethyl 2-pentenedioate	Diethyl glutaconate	$C_9H_{14}O_4$	2049-67-4	186.205			237	1.0496[20]	1.4411[20]	vs eth, EtOH
3502	Diethylperoxide		$C_4H_{10}O_2$	628-37-5	90.121	liq	-70	65	0.8240[19]	1.3715[17]	sl H$_2$O; msc EtOH, eth
3503	N,N-Diethyl-10H-phenothiazine-10-ethanamine	Diethazine	$C_{18}H_{22}N_2S$	60-91-3	298.446	oil		167[0.5]			i H$_2$O; s dil HCl
3504	N,N-Diethyl-α-phenylbenzenemethanamine	N,N-Diethylbenzhydrylamine	$C_{17}H_{21}N$	519-72-2	239.356		58.5	170[17]			
3505	Diethyl phenylmalonate		$C_{13}H_{16}O_4$	83-13-6	236.264		16.5	dec 205; 168[12]	1.0950[20]	1.4977[20]	vs ace, EtOH
3506	Diethyl phenylphosphonite		$C_{10}H_{15}O_2P$	1638-86-4	198.199			235; 62[1]	1.032[16]		
3507	5,5-Diethyl-1-phenyl-2,4,6(1H,3H,5H)-pyrimidinetrione	Phenetharbital	$C_{14}H_{16}N_2O_3$	357-67-5	260.288		178				vs EtOH
3508	Diethylphosphine		$C_4H_{11}P$	627-49-6	90.104			85	0.786[20]		
3509	Diethyl phosphonate		$C_4H_{11}O_3P$	762-04-9	138.102			54[6]			s ctc
3510	O,O-Diethyl phosphorodithionate		$C_4H_{11}O_2PS_2$	298-06-6	186.233						s H$_2$O
3511	Diethyl phthalate		$C_{12}H_{14}O_4$	84-66-2	222.237	liq	-40.5	295	1.232[14]	1.5000[21]	i H$_2$O; msc EtOH, eth; s ace, bz, ctc
3512	3,3-Diethyl-2,4-piperidinedione	Piperidione	$C_9H_{15}NO_2$	77-03-2	169.221	nd (w)	104				vs H$_2$O, EtOH, chl, MeOH
3513	N,N-Diethylpropanamide		$C_7H_{15}NO$	1114-51-8	129.200			191	0.8972[20]	1.4425[20]	vs EtOH
3514	N,N-Diethyl-1,3-propanediamine		$C_7H_{18}N_2$	104-78-9	130.231			168.5	0.822[20]	1.443[20]	
3515	Diethylpropanedioic acid	Diethylmalonic acid	$C_7H_{12}O_4$	510-20-3	160.168	pr (w,bz)	127 dec				vs H$_2$O, EtOH, eth; sl bz, chl
3516	2,2-Diethyl-1,3-propanediol		$C_7H_{16}O_2$	115-76-4	132.201		61.5	240.5	1.050[20]	1.4574[25]	vs H$_2$O, EtOH, eth; s chl
3517	Diethyl 2-propylmalonate		$C_{10}H_{18}O_4$	2163-48-6	202.248			221; 114[22]	0.989[20]	1.4197[20]	sl H$_2$O; vs EtOH, eth
3518	N,N-Diethyl-3-pyridinecarboxamide	Nikethamide	$C_{10}H_{14}N_2O$	59-26-7	178.230	ye solid or visc liq	25	dec 280; 175[25]	1.060[25]	1.525[20]	sl DMSO

N,N-Diethyl-2-methylaniline

N,N-Diethyl-4-methylaniline

N,N-Diethyl-3-methylbenzamide

1,3-Diethyl-5-methylbenzene

N4,N4-Diethyl-2-methyl-1,4-benzenediamine, monohydrochloride

N,N-Diethyl-3-methylbutanamide

Diethyl methylenesuccinate

Diethyl methylmalonate

Diethyl methylphosphonate

N,N-Diethyl-4-methyl-1-piperazinecarboxamide

3,3-Diethyl-5-methyl-2,4-piperidinedione

N,N-Diethyl-1-naphthalenamine

N,N-Diethyl-4-nitroaniline

N,N-Diethyl-4-nitrosoaniline

Diethyl nonanedioate

Diethyl oxalate

Diethyl oxobutanedioate

Diethyl 3-oxo-1,5-pentanedioate

3,3-Diethylpentane

N',N'-Diethyl-1,4-pentanediamine

2,2-Diethyl-4-pentenamide

Diethyl 2-pentenedioate

Diethylperoxide

N,N-Diethyl-10H-phenothiazine-10-ethanamine

N,N-Diethyl-α-phenylbenzenemethanamine

Diethyl phenylmalonate

Diethyl phenylphosphonite

5,5-Diethyl-1-phenyl-2,4,6(1H,3H,5H)-pyrimidinetrione

Diethylphosphine

Diethyl phosphonate

O,O-Diethyl phosphorodithionate

Diethyl phthalate

3,3-Diethyl-2,4-piperidinedione

N,N-Diethylpropanamide

N,N-Diethyl-1,3-propanediamine

Diethylpropanedioic acid

2,2-Diethyl-1,3-propanediol

Diethyl 2-propylmalonate

N,N-Diethyl-3-pyridinecarboxamide

No.	Name	Synonym	Mol. Form.	CAS RN	Mol. Wt.	Physical Form	mp/°C	bp/°C	den/ g cm^{-3}	n_D	Solubility
3519	*N,N*-Diethyl-4-pyridinecarboxamide	Isonicotinic acid diethylamide	C$_{10}$H$_{14}$N$_2$O	530-40-5	178.230			119[1]		1.525[20]	vs H$_2$O, ace, eth, EtOH
3520	3,3-Diethyl-2,4(1*H*,3*H*)-pyridinedione	Pyrithyldione	C$_9$H$_{13}$NO$_2$	77-04-3	167.205		90.7				
3521	Diethyl sebacate		C$_{14}$H$_{26}$O$_4$	110-40-7	258.354		2.5	305; 188[19]	0.9646[20]	1.4306[20]	sl H$_2$O, ctc; s EtOH, ace; i bz
3522	Diethyl selenide		C$_4$H$_{10}$Se	627-53-2	137.08	pa ye	55	108	1.2300[20]	1.4768[20]	
3523	Diethylsilane		C$_4$H$_{12}$Si	542-91-6	88.224	liq	-134.3	57	0.6843[20]	1.3921[20]	i H$_2$O
3524	*trans*-Diethylstilbestrol		C$_{18}$H$_{20}$O$_2$	56-53-1	268.351	pl (bz)	170.5				vs eth, EtOH, chl
3525	*trans*-Diethylstilbestrol dipropanoate	Clinestrol	C$_{24}$H$_{28}$O$_4$	130-80-3	380.477	pr (MeOH)	104				vs bz, eth, EtOH
3526	*trans*-Diethylstilbestrol monomethyl ether	Mestilbol	C$_{19}$H$_{22}$O$_2$	18839-90-2	282.377	nd (bz-peth)	117.5	190[0.3]			vs ace, eth, EtOH
3527	Diethyl succinate	Ethyl succinate	C$_8$H$_{14}$O$_4$	123-25-1	174.195	liq	-21	217.7	1.0402[20]	1.4201[20]	i H$_2$O; msc EtOH, eth; s ace, chl
3528	Diethyl sulfate		C$_4$H$_{10}$O$_4$S	64-67-5	154.185	oil	-24	208	1.172[25]	1.3989[20]	i H$_2$O; msc EtOH, eth
3529	Diethyl sulfide		C$_4$H$_{10}$S	352-93-2	90.187	liq	-103.91	92.1	0.8362[20]	1.4430[20]	sl H$_2$O, ctc; s EtOH, eth
3530	Diethyl sulfite	Ethyl sulfite	C$_4$H$_{10}$O$_3$S	623-81-4	138.185			158; 51[13]	1.1[20]	1.4310[20]	s EtOH, eth
3531	Diethyl sulfone	Ethyl sulfone	C$_4$H$_{10}$O$_2$S	597-35-3	122.186	orth pl	73.5	248	1.357[20]		s H$_2$O, eth; vs bz; i peth
3532	Diethyl sulfoxide		C$_4$H$_{10}$OS	70-29-1	106.186	syr	14	104[25], 90[15]	1.0092[22]		vs H$_2$O, eth, EtOH
3533	Diethyl *DL*-tartrate		C$_8$H$_{14}$O$_6$	57968-71-5	206.193		18.7	281; 158[14]	1.2046[20]	1.4438[20]	sl H$_2$O; msc EtOH, eth; s ace, ctc
3534	Diethyl telluride		C$_4$H$_{10}$Te	627-54-3	185.72	red-ye		137.5	1.599[15]	1.5182[15]	vs EtOH
3535	Diethyl terephthalate		C$_{12}$H$_{14}$O$_4$	636-09-9	222.237	mcl pr (al, peth)	44	302	1.0989[45]		i H$_2$O; vs EtOH, eth
3536	Diethyl thiodipropionate		C$_{10}$H$_{18}$O$_4$S	673-79-0	234.313			174[15], 121[2]	1.1034[20]	1.4655[20]	
3537	*N,N*'-Diethylthiourea		C$_5$H$_{12}$N$_2$S	105-55-5	132.227		78	dec			s H$_2$O, EtOH; vs eth; sl ctc
3538	*N,N*-Diethyl-1,1,1-trimethylsilanamine	(Diethylamino)trimethylsilane	C$_7$H$_{19}$NSi	996-50-9	145.319			126.3	0.7627[20]	1.4112[20]	
3539	Diethyltrisulfide		C$_4$H$_{10}$S$_3$	3600-24-6	154.317		-72.6	85[26]	1.1082[20]	1.5689[13]	
3540	*N,N*-Diethylurea		C$_5$H$_{12}$N$_2$O	634-95-7	116.161	pl, nd (eth)	75	95[0.02]			vs H$_2$O, EtOH, bz, lig; s eth
3541	*N,N*'-Diethylurea		C$_5$H$_{12}$N$_2$O	623-76-7	116.161	tab (lig), hyg nd (al)	112.5	263	1.0415[25]	1.4616[40]	vs H$_2$O, EtOH, eth
3542	Diethyl vinylphosphonate		C$_6$H$_{13}$O$_3$P	682-30-4	164.139			110[2]	1.068[25]	1.4290[20]	
3543	Diethyl zinc	Zinc diethyl	C$_4$H$_{10}$Zn	557-20-0	123.531	col liq	-28	118; 80[200]	1.2065[20]	1.4936[20]	dec H$_2$O; msc eth, peth, bz
3544	Difenoconazole		C$_{19}$H$_{17}$Cl$_2$N$_3$O$_3$	119446-68-3	406.262		76	220[0.03]			
3545	Difenzoquat methyl sulfate	1*H*-Pyrazolium, 1,2-dimethyl-3,5-diphenyl-, methyl sulfate	C$_{18}$H$_{20}$N$_2$O$_4$S	43222-48-6	360.428		157				
3546	Diflubenzuron	*N*-[[(4-Chlorophenyl)amino]carbonyl]-2,6-difluorobenzamide	C$_{14}$H$_9$ClF$_2$N$_2$O$_2$	35367-38-5	310.683		239				
3547	Difluoroacetic acid		C$_2$H$_2$F$_2$O$_2$	381-73-7	96.033	liq	-1	133	1.526[25]	1.3470[20]	
3548	2,4-Difluoroaniline		C$_6$H$_5$F$_2$N	367-25-9	129.108	liq	-7.5	170	1.268[25]	1.5063[20]	
3549	*o*-Difluorobenzene	1,2-Difluorobenzene	C$_6$H$_4$F$_2$	367-11-3	114.093	liq	-47.1	94	1.1599[18]	1.4451[18]	i H$_2$O; s ace, bz, chl
3550	*m*-Difluorobenzene	1,3-Difluorobenzene	C$_6$H$_4$F$_2$	372-18-9	114.093	liq	-69.12	82.6	1.1572[20]	1.4374[20]	i H$_2$O; s ace, bz
3551	*p*-Difluorobenzene	1,4-Difluorobenzene	C$_6$H$_4$F$_2$	540-36-3	114.093	liq	-23.55	89	1.1701[20]	1.4422[20]	i H$_2$O; s ace, bz; sl ctc
3552	4,4'-Difluoro-1,1'-biphenyl	4,4'-Difluorodiphenyl	C$_{12}$H$_8$F$_2$	398-23-2	190.189	mcl pr (al) lf (w)	94.5	254.5			i H$_2$O; vs EtOH, bz, chl; s eth, ace
3553	1,1-Difluorocyclohexane		C$_6$H$_{10}$F$_2$	371-90-4	120.140	liq		99.5			
3554	3,3-Difluorocyclopropene		C$_3$H$_2$F$_2$	56830-75-2	76.045	liq		34			
3555	Difluorodimethylsilane		C$_2$H$_6$F$_2$Si	353-66-2	96.152	col gas	-87.5	2.5			
3556	1,5-Difluoro-2,4-dinitrobenzene		C$_6$H$_2$F$_2$N$_2$O$_4$	327-92-4	204.088		75.5	132[2]			sl EtOH
3557	Difluorodiphenylsilane		C$_{12}$H$_{10}$F$_2$Si	312-40-3	220.290			246; 157[50]	1.145[17]	1.5221[25]	
3558	1,1-Difluoroethane	Ethylidene difluoride	C$_2$H$_4$F$_2$	75-37-6	66.050	col gas	-117	-24.05	0.896[25] (p>1 atm)	1.3011[-72]	
3559	1,2-Difluoroethane	Ethylene difluoride	C$_2$H$_4$F$_2$	624-72-6	66.050	vol liq		26			vs bz, eth, chl
3560	1,1-Difluoroethene	Vinylidene fluoride	C$_2$H$_2$F$_2$	75-38-7	64.034	col gas	-144	-85.7			vs eth, EtOH
3561	*cis*-1,2-Difluoroethene	*cis*-1,2-Difluoroethylene	C$_2$H$_2$F$_2$	1630-77-9	64.034	col gas		-26			
3562	*trans*-1,2-Difluoroethene	*trans*-1,2-Difluoroethylene	C$_2$H$_2$F$_2$	1630-78-0	64.034	col gas		-53.1			
3563	Difluoromethane	Methylene fluoride	CH$_2$F$_2$	75-10-5	52.024	col gas	-136.8 tp	-51.6	1.2139[-52]		i H$_2$O; s EtOH

N,N-Diethyl-4-pyridinecarboxamide

3,3-Diethyl-2,4(1H,3H)-pyridinedione

Diethyl sebacate

Diethyl selenide

Diethylsilane

trans-Diethylstilbestrol

trans-Diethylstilbestrol dipropanoate

trans-Diethylstilbestrol monomethyl ether

Diethyl succinate

Diethyl sulfate

Diethyl sulfide

Diethyl sulfite

Diethyl sulfone

Diethyl sulfoxide

Diethyl DL-tartrate

Diethyl telluride

Diethyl terephthalate

Diethyl thiodipropionate

N,N'-Diethylthiourea

N,N-Diethyl-1,1,1-trimethylsilanamine

Diethyltrisulfide

N,N-Diethylurea

N,N'-Diethylurea

Diethyl vinylphosphonate

Diethyl zinc

Difenoconazole

Difenzoquat methyl sulfate

Diflubenzuron

Difluoroacetic acid

2,4-Difluoroaniline

o-Difluorobenzene

m-Difluorobenzene

p-Difluorobenzene

4,4'-Difluoro-1,1'-biphenyl

1,1-Difluorocyclohexane

3,3-Difluorocyclopropene

Difluorodimethylsilane

1,5-Difluoro-2,4-dinitrobenzene

Difluorodiphenylsilane

1,1-Difluoroethane

1,2-Difluoroethane

1,1-Difluoroethene

cis-1,2-Difluoroethene

trans-1,2-Difluoroethene

Difluoromethane

No.	Name	Synonym	Mol. Form.	CAS RN	Mol. Wt.	Physical Form	mp/°C	bp/°C	den/ g cm⁻³	n_D	Solubility
3564	2-(Difluoromethoxy)-1,1,1-trifluoroethane	Difluoromethyl 2,2,2-trifluoroethyl ether	C₃H₃F₅O	1885-48-9	150.047	col liq		29			
3565	Difluoromethylborane		CH₃BF₂	373-64-8	63.843	gas		-78.5[287]			reac H₂O
3566	2,4-Difluoro-1-nitrobenzene		C₆H₃F₂NO₂	446-35-5	159.091		9.8	207	1.4571[14]	1.5149[14]	sl chl
3567	2,2-Difluoropropane		C₃H₆F₂	420-45-1	80.077	col gas	-104.8	-0.4	0.9205[20] (p>1 atm)	1.2904[20]	
3568	1,3-Difluoro-2-propanol		C₃H₆F₂O	453-13-4	96.076			127; 55[34]	1.24[25]	1.3725[20]	
3569	Di-2-furanylethanedione		C₁₀H₆O₄	492-94-4	190.153	ye nd (al), cry (bz)	166.3				sl H₂O; s EtOH, eth, bz, chl
3570	Di-2-furanylethanedione dioxime	α-Furildioxime	C₁₀H₈N₂O₄	522-27-0	220.182		167				sl EtOH, eth, bz, lig
3571	1,5-Di-2-furanyl-1,4-pentadien-3-one		C₁₃H₁₀O₃	886-77-1	214.216	hyg pr (peth) ye pr (lig)	60.5	181[4]			vs eth, EtOH, chl
3572	Difurfuryl disulfide	Furfuryl disulfide	C₁₀H₁₀O₂S₂	4437-20-1	226.315		10	167[13], 112[0.5]			vs EtOH
3573	Difurfuryl ether	Furfuryl ether	C₁₀H₁₀O₃	4437-22-3	178.184			101[2]	1.1405[20]	1.5088[20]	i H₂O
3574	Digitonin		C₅₆H₉₂O₂₉	11024-24-1	1229.312		237.5				
3575	Digitoxigenin		C₂₃H₃₄O₄	143-62-4	374.514		253				s EtOH; vs MeOH
3576	Digitoxin		C₄₁H₆₄O₁₃	71-63-6	764.939	pr (dil al)	255.5				sl H₂O; vs EtOH; s eth, chl, MeOH, py
3577	Digitoxose		C₆H₁₂O₄	527-52-6	148.157	cry (MeOH+eth)	112				vs H₂O, ace; s py, AcOEt
3578	Diglycidyl ether	Bis(2,3-epoxypropyl) ether	C₆H₁₀O₃	2238-07-5	130.141			260	1.1195[20]		
3579	Diglycolic acid	2,2'-Oxydiacetic acid	C₄H₆O₅	110-99-6	134.088	mcl pr (w + 1)	148	dec			vs H₂O, eth, EtOH
3580	Digoxigenin		C₂₃H₃₄O₅	1672-46-4	390.513	pr (AcOEt)	222				vs EtOH, MeOH; sl chl
3581	Digoxin		C₄₁H₆₄O₁₄	20830-75-5	780.939	trc pl (dil al, py)	249 dec				vs EtOH
3582	Diheptylamine	N-Heptyl-1-heptanamine	C₁₄H₃₁N	2470-68-0	213.403	nd	31.5	271; 135[9]	0.7956[21]		sl H₂O; s EtOH; vs eth
3583	Diheptyl ether	Heptyl ether	C₁₄H₃₀O	629-64-1	214.387			258.5	0.8008[20]	1.4275[20]	vs eth, EtOH
3584	Diheptyl phthalate		C₂₂H₃₄O₄	3648-21-3	362.503			360			
3585	Diheptyl sulfide	Heptyl sulfide	C₁₄H₃₀S	629-65-2	230.453		70	298	0.8416[20]	1.4606[20]	i H₂O; s eth
3586	Dihexylamine	N-Hexyl-1-hexanamine	C₁₂H₂₇N	143-16-8	185.349	liq	-13.1	236; 75[1]	0.7889[20]	1.4339[20]	s EtOH, eth
3587	Dihexyl ether	Hexyl ether	C₁₂H₂₆O	112-58-3	186.333			226	0.7936[20]	1.4204[20]	i H₂O; s eth; sl ctc
3588	Dihexyl hexanedioate		C₁₈H₃₄O₄	110-33-8	314.461	liq	-9	348; 182.5[4]	0.941[20]		
3589	Dihexyl phthalate		C₂₀H₃₀O₄	84-75-3	334.450			210[5]			
3590	Dihexyl sulfide	Hexyl sulfide	C₁₂H₂₆S	6294-31-1	202.399			230; 136[20]	0.8411[20]	1.4586[20]	
3591	15,16-Dihydroaflatoxin G₁	Aflatoxin G₂	C₁₇H₁₄O₇	7241-98-7	330.289		239.3				
3592	9,10-Dihydroanthracene		C₁₄H₁₂	613-31-0	180.245	tab or pr	111	305	1.215[20]		i H₂O; s EtOH, eth, bz, chl
3593	6,15-Dihydro-5,9,14,18-anthrazinetetrone	Indanthrene	C₂₈H₁₄N₂O₄	81-77-6	442.422	bl nd	485 dec				i H₂O, EtOH, eth, ace, bz; s PhNO₂, dil alk
3594	1,2-Dihydrobenz[j]aceanthrylene	Cholanthrene	C₂₀H₁₄	479-23-2	254.325	pa ye lf (bz-al)	174				i H₂O; s EtOH, bz, HOAc, lig, tol
3595	9,10-Dihydro-9,10[1',2']-benzenoanthracene	Triptycene	C₂₀H₁₄	477-75-8	254.325	cry (cyhex)	256				
3596	1,3-Dihydro-2H-benzimidazole-2-thione	2-Benzimidazolethiol	C₇H₆N₂S	583-39-1	150.201	pl (dil al or NH₃)	298				vs EtOH
3597	1,3-Dihydro-2H-benzimidazol-2-one		C₇H₆N₂O	615-16-7	134.135	lf (w or al)	318 dec				sl H₂O, eth, bz; s ace; vs EtOH
3598	2,3-Dihydro-1,4-benzodioxin		C₈H₈O₂	493-09-4	136.149			212; 103[6]	1.180[20]	1.5485[20]	
3599	2,3-Dihydrobenzofuran	Coumaran	C₈H₈O	496-16-2	120.149	liq	-21.5	188.5	1.058[25]	1.5497[20]	vs eth, EtOH, chl
3600	3,4-Dihydro-1H-2-benzopyran	Isochroman	C₉H₁₀O	493-05-0	134.174		4	110[25], 90[12]	1.067[25]	1.5444[20]	
3601	3,4-Dihydro-2H-1-benzopyran		C₉H₁₀O	493-08-3	134.174		4.8	215; 98[18]	1.072[20]	1.5444[20]	s H₂O; msc os
3602	3,4-Dihydro-2H-1-benzopyran-2-one		C₉H₈O₂	119-84-6	148.159	lf	25	272	1.169[18]	1.5563[20]	i H₂O; sl EtOH, eth, ctc; s chl
3603	2,3-Dihydro-4H-1-benzopyran-4-one	4-Chromanone	C₉H₈O₂	491-37-2	148.159		36.5	160[50], 127[13]	1.1291[100]	1.5750	s EtOH; vs eth, ace, bz, chl; sl ctc
3604	6,7-Dihydrobenzo[b]thiophen-4(5H)-one	4,5,6,7-Tetrahydro-4-benzothiophenone	C₈H₈OS	13414-95-4	152.214						sl chl
3605	2,3-Dihydro-4H-1-benzothiopyran-4-one		C₉H₈OS	3528-17-4	164.224		29	154[12]	1.2487[14]	1.6395[20]	
3606	4,5-Dihydro-2-benzyl-1H-imidazole	Tolazoline	C₁₀H₁₂N₂	59-98-3	160.215	cry (peth)	67				
3607	7,8-Dihydrobiopterin		C₉H₁₃N₅O₃	6779-87-9	239.231	hyg nd (w)					s H₂O

2-(Difluoromethoxy)-1,1,1-trifluoroethane

Difluoromethylborane

2,4-Difluoro-1-nitrobenzene

2,2-Difluoropropane

1,3-Difluoro-2-propanol

Di-2-furanylethanedione

Di-2-furanylethanedione dioxime

1,5-Di-2-furanyl-1,4-pentadien-3-one

Difurfuryl disulfide

Difurfuryl ether

Digitonin

Digitoxigenin

Digitoxin

Digitoxose

Diglycidyl ether

Diglycolic acid

Digoxigenin

Digoxin

Diheptylamine

Diheptyl ether

Diheptyl phthalate

Diheptyl sulfide

Dihexylamine

Dihexyl ether

Dihexyl hexanedioate

Dihexyl phthalate

Dihexyl sulfide

15,16-Dihydroaflatoxin G₁

9,10-Dihydroanthracene

6,15-Dihydro-5,9,14,18-anthrazinetetrone

1,2-Dihydrobenz[j]aceanthrylene

9,10-Dihydro-9,10[1',2']-benzenoanthracene

1,3-Dihydro-2H-benzimidazole-2-thione

1,3-Dihydro-2H-benzimidazol-2-one

2,3-Dihydro-1,4-benzodioxin

2,3-Dihydrobenzofuran

3,4-Dihydro-1H-2-benzopyran

3,4-Dihydro-2H-1-benzopyran

3,4-Dihydro-2H-1-benzopyran-2-one

2,3-Dihydro-4H-1-benzopyran-4-one

6,7-Dihydrobenzo[b]thiophen-4(5H)-one

2,3-Dihydro-4H-1-benzothiopyran-4-one

4,5-Dihydro-2-benzyl-1H-imidazole

7,8-Dihydrobiopterin

No.	Name	Synonym	Mol. Form.	CAS RN	Mol. Wt.	Physical Form	mp/°C	bp/°C	den/ g cm^{-3}	n_D	Solubility
3608	Dihydrocodeine		$C_{18}H_{23}NO_3$	125-28-0	301.381	cry (aq, MeOH)	112.5	248[15]			
3609	16,17-Dihydro-15H-cyclopenta[a] phenanthrene	1,2-Cyclopentenophenanthrene	$C_{17}H_{14}$	482-66-6	218.293	nd (al, petr)	135.5				i H_2O; s EtOH, peth
3610	10,11-Dihydro-5H-dibenz[b,f] azepine		$C_{14}H_{13}N$	494-19-9	195.260						s chl
3611	10,11-Dihydro-5H-dibenzo[a,d] cyclohepten-5-one		$C_{15}H_{12}O$	1210-35-1	208.255		30	203[7]	1.1635[20]	1.6324[20]	
3612	2,5-Dihydro-2,5-dimethoxyfuran		$C_6H_{10}O_3$	332-77-4	130.141			161	1.073[25]	1.4339[20]	
3613	3,4-Dihydro-6,7-dimethoxy-1(2H)-isoquinolinone	Corydaldine	$C_{11}H_{13}NO_3$	493-49-2	207.226	mcl pr (w, al)	175				vs H_2O, bz, eth, EtOH
3614	1,2-Dihydro-1,5-dimethyl-2-phenyl-3H-pyrazol-3-one	Antipyrine	$C_{11}H_{12}N_2O$	60-80-0	188.225	lf or sc (eth, bz)	114	319			vs H_2O, EtOH
3615	2,3-Dihydro-1,4-dioxin		$C_4H_6O_2$	543-75-9	86.090			94.1	1.0836[20]	1.4372[20]	s ctc
3616	9,10-Dihydro-9,10-dioxo-2-anthracenecarboxylic acid		$C_{15}H_8O_4$	117-78-2	252.223	ye nd (HOAc)	291	sub			sl EtOH, HOAc; i eth, bz; s ace
3617	9,10-Dihydro-9,10-dioxo-1,5-anthracenedisulfonic acid		$C_{14}H_8O_8S_2$	117-14-6	368.339	ye nd (HCl +4w) pl (dil HOAc)	310 dec				vs H_2O, EtOH, HOAc
3618	9,10-Dihydro-9,10-dioxo-2,6-anthracenedisulfonic acid		$C_{14}H_8O_8S_2$	84-50-4	368.339						vs H_2O; s EtOH; i eth, bz
3619	9,10-Dihydro-9,10-dioxo-1-anthracenesulfonic acid		$C_{14}H_8O_5S$	82-49-5	288.276	lf (HOAc) ye lf (conc HCl, +3w)	216.0				vs H_2O, HOAc; s EtOH
3620	9,10-Dihydro-9,10-dioxo-2-anthracenesulfonic acid		$C_{14}H_8O_5S$	84-48-0	288.276	ye lf (+3w)					vs H_2O; s EtOH; i eth
3621	9,10-Dihydro-9,10-dioxo-1-anthracenesulfonic acid, sodium salt	Sodium anthraquinone-1-sulfonate	$C_{14}H_7NaO_5S$	128-56-3	310.258	ye lf (w)					sl H_2O
3622	9,10-Dihydro-9,10-dioxo-2-anthracenesulfonic acid, sodium salt		$C_{14}H_7NaO_5S$	131-08-8	310.258						sl DMSO
3623	7,8-Dihydrofolic acid		$C_{19}H_{21}N_7O_6$	4033-27-6	443.413	ye cry					
3624	2,3-Dihydrofuran		C_4H_6O	1191-99-7	70.090			54.5	0.927[25]	1.4239[20]	
3625	2,5-Dihydrofuran		C_4H_6O	1708-29-8	70.090					1.4311[20]	
3626	2,3-Dihydro-3-hydroxy-1-methyl-1H-indole-5,6-dione	Adrenochrome	$C_9H_9NO_3$	54-06-8	179.172		125 dec				vs H_2O, EtOH; i eth, bz
3627	2,3-Dihydro-1H-inden-5-amine		$C_9H_{11}N$	24425-40-9	133.190	nd (peth)	37.5	248; 131[15]			sl H_2O, chl; s eth, ace, bz
3628	2,3-Dihydro-1H-inden-1-ol		$C_9H_{10}O$	6351-10-6	134.174	pl (peth)	54.8	220; 128[12]			vs bz, EtOH, chl
3629	2,3-Dihydro-1H-inden-5-ol		$C_9H_{10}O$	1470-94-6	134.174		58	253			sl H_2O, peth; vs EtOH, eth; s sulf
3630	2,3-Dihydro-1H-inden-1-one		C_9H_8O	83-33-0	132.159	ta, nd (w + 3)	42	243; 129[12]	1.0943[40]	1.561[25]	sl H_2O; vs EtOH, eth, ace, chl
3631	1,3-Dihydro-2H-inden-2-one	2-Indanone	C_9H_8O	615-13-4	132.159	nd (al, eth)	59	dec 218	1.0712[69]	1.538[67]	i H_2O; vs EtOH, eth, ace, chl
3632	1a,6a-Dihydro-6H-indeno[1,2-b] oxirene		C_9H_8O	768-22-9	132.159		24.5	113[20], 98[6]	1.1255[24]		s chl
3633	2,3-Dihydro-1H-indole		C_8H_9N	496-15-1	119.164			229	1.069[20]	1.5923[20]	sl H_2O; s eth, ace, bz
3634	1,3-Dihydro-2H-indol-2-one		C_8H_7NO	59-48-3	133.148	nd (w)	128	227[23], 195[17]			s H_2O, EtOH, eth
3635	2,3-Dihydro-1H-isoindol-1-one		C_8H_7NO	480-91-1	133.148	nd (w)	151	338; 103[18]			vs eth, EtOH, chl
3636	Dihydro-α-lipoic acid	6,8-Dimercaptooctanoic acid	$C_8H_{16}O_2S_2$	462-20-4	208.342	ye liq		145[0.2]			
3637	3,4-Dihydro-6-methoxy-1(2H)-naphthalenone	6-Methoxy-α-tetralone	$C_{11}H_{12}O_2$	1078-19-9	176.212	cry (MeOH, lig)	78	171[11]			
3638	3,4-Dihydro-2-methoxy-2H-pyran		$C_6H_{10}O_2$	4454-05-1	114.142	liq		128	1.006	1.4420[20]	
3639	1,2-Dihydro-3-methylbenz[j] aceanthrylene	3-Methylcholanthrene	$C_{21}H_{16}$	56-49-5	268.352	ye nd (bz)	180	280[80]	1.28[20]		i H_2O
3640	2,3-Dihydro-2-methylbenzofuran		$C_9H_{10}O$	1746-11-8	134.174			197.5	1.061[25]	1.5308	
3641	Dihydro-3-methylene-2,5-furandione		$C_5H_4O_3$	2170-03-8	112.084	orth bipym pr (eth, chl)	69	139[30], 114[18]			sl eth; vs chl
3642	Dihydro-3-methylene-2(3H)-furanone	α-Methylene butyrolactone	$C_5H_6O_2$	547-65-9	98.101			85[10]	1.1206[20]	1.4650[20]	s H_2O, eth, ace, bz; sl ctc; vs EtOH
3643	Dihydro-3-methyl-2,5-furandione		$C_5H_6O_3$	4100-80-5	114.100		34	239	1.22[25]		
3644	Dihydro-3-methyl-2(3H)-furanone	2-Methyl-γ-butyrolactone	$C_5H_8O_2$	1679-47-6	100.117	liq		200; 79[10]	1.0570[20]	1.4325[20]	
3645	Dihydro-4-methyl-2(3H)-furanone	3-Methyl-γ-butyrolactone	$C_5H_8O_2$	1679-49-8	100.117	liq		76[11]	1.058[20]	1.4339[20]	
3646	Dihydro-5-methyl-2(3H)-furanone, (±)	(±)-γ-Valerolactone	$C_5H_8O_2$	57129-69-8	100.117	liq	-31	206	1.0551[20]	1.4328[20]	msc H_2O; s EtOH, ace; sl ctc
3647	4,5-Dihydro-2-methyl-1H-imidazole	Lysidine	$C_4H_8N_2$	534-26-9	84.120	hyg	107	196.5			vs H_2O, EtOH; i eth; s chl

Dihydrocodeine

16,17-Dihydro-15H-cyclopenta[a]phenanthrene

10,11-Dihydro-5H-dibenz[b,f]azepine

10,11-Dihydro-5H-dibenzo[a,d]cyclohepten-5-one

2,5-Dihydro-2,5-dimethoxyfuran

3,4-Dihydro-6,7-dimethoxy-1(2H)-isoquinolinone

1,2-Dihydro-1,5-dimethyl-2-phenyl-3H-pyrazol-3-one

2,3-Dihydro-1,4-dioxin

9,10-Dihydro-9,10-dioxo-2-anthracenecarboxylic acid

9,10-Dihydro-9,10-dioxo-1,5-anthracenedisulfonic acid

9,10-Dihydro-9,10-dioxo-2,6-anthracenedisulfonic acid

9,10-Dihydro-9,10-dioxo-1-anthracenesulfonic acid

9,10-Dihydro-9,10-dioxo-2-anthracenesulfonic acid

9,10-Dihydro-9,10-dioxo-1-anthracenesulfonic acid, sodium salt

9,10-Dihydro-9,10-dioxo-2-anthracenesulfonic acid, sodium salt

7,8-Dihydrofolic acid

2,3-Dihydrofuran

2,5-Dihydrofuran

2,3-Dihydro-3-hydroxy-1-methyl-1H-indole-5,6-dione

2,3-Dihydro-1H-inden-5-amine

2,3-Dihydro-1H-inden-1-ol

2,3-Dihydro-1H-inden-5-ol

2,3-Dihydro-1H-inden-1-one

1,3-Dihydro-2H-inden-2-one

1a,6a-Dihydro-6H-indeno[1,2-b]oxirene

2,3-Dihydro-1H-indole

1,3-Dihydro-2H-indol-2-one

2,3-Dihydro-1H-isoindol-1-one

Dihydro-α-lipoic acid

3,4-Dihydro-6-methoxy-1(2H)-naphthalenone

3,4-Dihydro-2-methoxy-2H-pyran

1,2-Dihydro-3-methylbenz[j]aceanthrylene

2,3-Dihydro-2-methylbenzofuran

Dihydro-3-methylene-2,5-furandione

Dihydro-3-methylene-2(3H)-furanone

Dihydro-3-methyl-2,5-furandione

Dihydro-3-methyl-2(3H)-furanone

Dihydro-4-methyl-2(3H)-furanone

Dihydro-5-methyl-2(3H)-furanone, (±)

4,5-Dihydro-2-methyl-1H-imidazole

No.	Name	Synonym	Mol. Form.	CAS RN	Mol. Wt.	Physical Form	mp/°C	bp/°C	den/ g cm⁻³	n_D	Solubility
3648	1,3-Dihydro-1-methyl-2H-imidazole-2-thione	Methimazole	C₄H₆N₂S	60-56-0	114.169	lf (al)	146	dec 280			vs H₂O; s EtOH, chl; sl eth, bz, lig
3649	2,3-Dihydro-1-methyl-1H-indene		C₁₀H₁₂	767-58-8	132.202			190.6	0.938²⁵	1.5266²⁰	i H₂O
3650	3,4-Dihydro-2-methyl-1(2H)-naphthalenone		C₁₁H₁₂O	1590-08-5	160.212		15	136¹⁶	1.057²⁵	1.5535²⁰	
3651	4-(4,5-Dihydro-3-methyl-5-oxo-1H-pyrazol-1-yl)benzenesulfonic acid		C₁₀H₁₀N₂O₄S	89-36-1	254.262	nd (w+1)	≈300 dec				
3652	1,2-Dihydro-5-methyl-2-phenyl-3H-pyrazol-3-one	5-Hydroxy-3-methyl-1-phenylpyrazole	C₁₀H₁₀N₂O	19735-89-8	174.198		128	287¹⁰⁵, 191¹⁷	1.2600²⁰	1.637	s H₂O, EtOH; sl bz; i peth
3653	2,4-Dihydro-5-methyl-2-phenyl-3H-pyrazol-3-one		C₁₀H₁₀N₂O	89-25-8	174.198	mcl pr (w)	127	287¹⁰⁵, 191¹⁷		1.637	
3654	3,6-Dihydro-4-methyl-2H-pyran		C₆H₁₀O	16302-35-5	98.142			117.5	0.912²⁵	1.4495²⁰	
3655	4,5-Dihydro-2-methylthiazole		C₄H₇NS	2346-00-1	101.171	liq	-101	145	1.067²⁵	1.5200²⁰	
3656	1,2-Dihydronaphthalene		C₁₀H₁₀	447-53-0	130.186	liq	-8	206.5	0.9974²⁰	1.5814²⁰	
3657	1,4-Dihydronaphthalene	Δ 2-Dialin	C₁₀H₁₀	612-17-9	130.186	pl	25	211.5	0.9928³³	1.5577²⁰	
3658	3,4-Dihydro-2(1H)-naphthalenone		C₁₀H₁₀O	530-93-8	146.185		18	237	1.1055²⁷	1.5598²⁰	i H₂O; s eth, bz
3659	1,2-Dihydro-5-nitroacenaphthylene		C₁₂H₉NO₂	602-87-9	199.205		103				s H₂O, EtOH, eth, lig
3660	1,6-Dihydro-6-oxo-3-pyridinecarboxylic acid		C₆H₅NO₃	5006-66-6	139.109	nd(w)	310 dec	sub			sl H₂O, tfa; i EtOH, eth, bz, chl
3661	Dihydro-5-pentyl-2(3H)-furanone	4-Hydroxynonanoic acid lactone	C₉H₁₆O₂	104-61-0	156.222	oil		134¹²			
3662	9,10-Dihydrophenanthrene		C₁₄H₁₂	776-35-2	180.245	nd (MeOH)	34.5	168¹⁵	1.0757⁴⁰	1.6415²⁰	s chl
3663	2,3-Dihydro-2-phenyl-4H-1-benzopyran-4-one		C₁₅H₁₂O₂	487-26-3	224.255	nd (lig)	76				i H₂O; s ace, bz; sl ctc
3664	4,5-Dihydro-2-(phenylmethyl)-1H-imidazole, monohydrochloride		C₁₀H₁₃ClN₂	59-97-2	196.676		174				
3665	4,5-Dihydro-5-phenyl-2-oxazolamine	Aminorex	C₉H₁₀N₂O	2207-50-3	162.187	cry (bz)	137				
3666	1,4-Dihydro-1-phenyl-5H-tetrazole-5-thione	1-Phenyl-5-mercapto-1H-tetrazole	C₇H₆N₄S	86-93-1	178.215		145				
3667	Dihydro-5-propyl-2(3H)-furanone	γ-Propyl-γ-butyrolactone	C₇H₁₂O₂	105-21-5	128.169			84⁵		1.4385²⁵	
3668	2,3-Dihydro-6-propyl-2-thioxo-4(1H)-pyrimidinone	Propylthiouracil	C₇H₁₀N₂OS	51-52-5	170.231	w pow (w)	219				sl H₂O, chl, DMSO, EtOH; i eth, bz
3669	1,7-Dihydro-6H-purine-6-thione	6-Mercaptopurine	C₅H₄N₄S	50-44-2	152.178	ye pr (w, + l w)	313 dec				i H₂O; s alk
3670	3,4-Dihydro-2H-pyran		C₅H₈O	110-87-2	84.117			86	0.921¹⁹	1.4402¹⁹	s H₂O, EtOH; sl chl
3671	3,6-Dihydro-2H-pyran		C₅H₈O	3174-74-1	84.117	liq		95	0.94¹⁹		
3672	Dihydro-2H-pyran-2,6(3H)-dione		C₅H₆O₃	108-55-4	114.100		56.3	158¹⁵	1.4110²⁰		
3673	4,5-Dihydro-1H-pyrazole	2-Pyrazoline	C₃H₆N₂	109-98-8	70.093			144	1.0200¹⁷	1.4796¹⁷	vs H₂O, eth, EtOH
3674	1,2-Dihydro-3,6-pyridazinedione	Maleic hydrazide	C₄H₄N₂O₂	123-33-1	112.087	cry (w)	307				sl H₂O, EtOH, tfa
3675	Dihydro-2,4(1H,3H)-pyrimidinedione	5,6-Dihydrouracil	C₄H₆N₂O₂	504-07-4	114.103	nd (w)	275.5				vs H₂O; s EtOH, chl, MeOH
3676	2,5-Dihydro-1H-pyrrole	3-Pyrroline	C₄H₇N	109-96-6	69.106			90.5	0.9097²⁰	1.4664²⁰	vs H₂O, ace, eth, EtOH
3677	3,4-Dihydro-2(1H)-quinolinone	Hydrocarbostyril	C₉H₉NO	553-03-7	147.173	pr (al, eth)	163.5	201⁴⁵			vs eth, EtOH
3678	1,4-Dihydro-2,3-quinoxalinedione	2,3-Quinoxalinediol	C₈H₆N₂O₂	15804-19-0	162.146	nd (w)	410				vs H₂O; sl EtOH, eth; s bz, DMSO, HOAc
3679	Dihydrotachysterol		C₂₈H₄₆O	67-96-9	398.664	cry (MeOH)	131				i H₂O; s os
3680	Dihydrothebaine		C₁₉H₂₃NO₃	561-25-1	313.391		162.5				i H₂O; s EtOH, bz, AcOEt
3681	4,5-Dihydro-2-thiazolamine		C₃H₆N₂S	1779-81-3	102.158	nd or lf (bz)	85.3	dec			vs H₂O, EtOH, bz, chl
3682	2,3-Dihydrothiophene		C₄H₆S	1120-59-8	86.156			112.1			
3683	2,5-Dihydrothiophene		C₄H₆S	1708-32-3	86.156			122.4			
3684	2,5-Dihydrothiophene 1,1-dioxide	3-Sulfolene	C₄H₆O₂S	77-79-2	118.155		64.5				s chl
3685	Dihydro-2(3H)-thiophenone		C₄H₆OS	1003-10-7	102.155			111⁵², 39¹	1.18²⁵	1.5230²⁰	
3686	Dihydro-2-thioxo-4,6(1H,5H)-pyrimidinedione	2-Thiobarbituric acid	C₄H₄N₂O₂S	504-17-6	144.152	pl (w)	235 dec				sl H₂O; s EtOH, dil alk, dil HCl
3687	2,3-Dihydro-2-thioxo-4(1H)-pyrimidinone	2-Thiouracil	C₄H₄N₂OS	141-90-2	128.152	pr (w, al)	>340 dec				sl H₂O, EtOH, DMSO; s anh HF
3688	1,2-Dihydro-3H-1,2,4-triazole-3-thione		C₂H₃N₃S	3179-31-5	101.130		222.5				s DMSO
3689	(1,3-Dihydro-1,3,3-trimethyl-2H-indol-2-ylidene)acetaldehyde		C₁₃H₁₅NO	84-83-3	201.264						s chl

1,3-Dihydro-1-methyl-2*H*-imidazole-2-thione

2,3-Dihydro-1-methyl-1*H*-indene

3,4-Dihydro-2-methyl-1(2*H*)-naphthalenone

4-(4,5-Dihydro-3-methyl-5-oxo-1*H*-pyrazol-1-yl)benzenesulfonic acid

1,2-Dihydro-5-methyl-2-phenyl-3*H*-pyrazol-3-one

2,4-Dihydro-5-methyl-2-phenyl-3*H*-pyrazol-3-one

3,6-Dihydro-4-methyl-2*H*-pyran

4,5-Dihydro-2-methylthiazole

1,2-Dihydronaphthalene

1,4-Dihydronaphthalene

3,4-Dihydro-2(1*H*)-naphthalenone

1,2-Dihydro-5-nitroacenaphthylene

1,6-Dihydro-6-oxo-3-pyridinecarboxylic acid

Dihydro-5-pentyl-2(3*H*)-furanone

9,10-Dihydrophenanthrene

2,3-Dihydro-2-phenyl-4*H*-1-benzopyran-4-one

4,5-Dihydro-2-(phenylmethyl)-1*H*-imidazole, monohydrochloride

4,5-Dihydro-5-phenyl-2-oxazolamine

1,4-Dihydro-1-phenyl-5*H*-tetrazole-5-thione

Dihydro-5-propyl-2(3*H*)-furanone

2,3-Dihydro-6-propyl-2-thioxo-4(1*H*)-pyrimidinone

1,7-Dihydro-6*H*-purine-6-thione

3,4-Dihydro-2*H*-pyran

3,6-Dihydro-2*H*-pyran

Dihydro-2*H*-pyran-2,6(3*H*)-dione

4,5-Dihydro-1*H*-pyrazole

1,2-Dihydro-3,6-pyridazinedione

Dihydro-2,4(1*H*,3*H*)-pyrimidinedione

2,5-Dihydro-1*H*-pyrrole

3,4-Dihydro-2(1*H*)-quinolinone

1,4-Dihydro-2,3-quinoxalinedione

Dihydrotachysterol

Dihydrothebaine

4,5-Dihydro-2-thiazolamine

2,3-Dihydrothiophene

2,5-Dihydrothiophene

2,5-Dihydrothiophene 1,1-dioxide

Dihydro-2(3*H*)-thiophenone

Dihydro-2-thioxo-4,6(1*H*,5*H*)-pyrimidinedione

2,3-Dihydro-2-thioxo-4(1*H*)-pyrimidinone

1,2-Dihydro-3*H*-1,2,4-triazole-3-thione

(1,3-Dihydro-1,3,3-trimethyl-2*H*-indol-2-ylidene)acetaldehyde

No.	Name	Synonym	Mol. Form.	CAS RN	Mol. Wt.	Physical Form	mp/°C	bp/°C	den/ g cm⁻³	n_D	Solubility
3690	2,3-Dihydro-1,1,3-trimethyl-3-phenyl-1H-indene		$C_{18}H_{20}$	3910-35-8	236.352	tcl pr (al)	52.5	308.5	1.0009[20]	1.5681[20]	i H_2O; s EtOH, bz, MeOH
3691	1,2-Dihydro-2,2,4-trimethylquinoline		$C_{12}H_{15}N$	147-47-7	173.254		26.5	260; 132[13]			
3692	1,4-Dihydroxy-9,10-anthracenedione	Quinizarin	$C_{14}H_8O_4$	81-64-1	240.212	ye red lf (eth) dk red nd	200				s H_2O, EtOH, eth, bz, KOH, sulf
3693	1,5-Dihydroxy-9,10-anthracenedione	Anthrarufin	$C_{14}H_8O_4$	117-12-4	240.212	pa ye pl (gl HOAc)	280	sub			i H_2O; sl EtOH, eth, ace, CS_2; s bz
3694	1,8-Dihydroxy-9,10-anthracenedione	Danthron	$C_{14}H_8O_4$	117-10-2	240.212	red or red-ye nd or lf (al)	193	sub			i H_2O; sl EtOH, eth; s ace, HOAc, alk
3695	2,6-Dihydroxy-9,10-anthracenedione		$C_{14}H_8O_4$	84-60-6	240.212	ye nd (al)	360 dec				sl H_2O, EtOH; i eth, bz, chl; s alk
3696	2,7-Dihydroxy-9,10-anthracenedione		$C_{14}H_8O_4$	572-93-0	240.212	ye nd (+1w, dil al) nd (sub)	353.8	sub			i H_2O; s EtOH; sl eth, bz, chl
3697	2,2'-Dihydroxyazobenzene		$C_{12}H_{10}N_2O_2$	2050-14-8	214.219	gold-ye lf (bz), nd (al)	173	140[0.001]			i H_2O; sl EtOH, bz; vs eth; s con alk
3698	2,3-Dihydroxybenzaldehyde		$C_7H_6O_3$	24677-78-9	138.121	ye nd	108	235; 120[16]			vs ace, EtOH, HOAc
3699	2,4-Dihydroxybenzaldehyde	β-Resorcylaldehyde	$C_7H_6O_3$	95-01-2	138.121	nd (eth-lig)	135	226[22]			s H_2O, HOAc; vs EtOH, eth, chl; sl bz
3700	2,5-Dihydroxybenzaldehyde		$C_7H_6O_3$	1194-98-5	138.121	ye nd (bz)	100.0				vs H_2O, EtOH, chl
3701	3,4-Dihydroxybenzaldehyde	Protocatechualdehyde	$C_7H_6O_3$	139-85-5	138.121	lf (w, to)	153 dec				s H_2O; vs EtOH, eth
3702	N,2-Dihydroxybenzamide	Salicylhydroxamic acid	$C_7H_7NO_3$	89-73-6	153.136	nd (HOAc)	168	sub			sl H_2O, DMSO; vs EtOH, eth; s HOAc
3703	2,5-Dihydroxybenzeneacetic acid	Homogentisic acid	$C_8H_8O_4$	451-13-8	168.148	pr (w+1), lf (al-chl)	153				vs H_2O, EtOH, eth; i bz, chl
3704	2,3-Dihydroxybenzoic acid		$C_7H_6O_4$	303-38-8	154.121	pr or nd (w+1)	205.5		1.542[20]		s H_2O, EtOH, eth; sl ace
3705	2,4-Dihydroxybenzoic acid	β-Resorcylic acid	$C_7H_6O_4$	89-86-1	154.121	cry (+w)	226 dec				s H_2O, EtOH, eth, bz; i CS_2
3706	2,5-Dihydroxybenzoic acid	Gentisic acid	$C_7H_6O_4$	490-79-9	154.121	nd or pr (w)	199.5				vs H_2O, EtOH, eth; s ace; i bz, chl, CS_2
3707	2,6-Dihydroxybenzoic acid		$C_7H_6O_4$	303-07-1	154.121	nd (+w)	167 dec				s H_2O, EtOH, eth; i chl; sl tfa
3708	3,4-Dihydroxybenzoic acid	Protocatechuic acid	$C_7H_6O_4$	99-50-3	154.121	mcl nd (w+1)	201 dec		1.524[4]		sl H_2O; vs EtOH; s eth; i bz
3709	3,5-Dihydroxybenzoic acid		$C_7H_6O_4$	99-10-5	154.121	pr or nd	239				sl H_2O, ace; vs EtOH, eth
3710	2,2'-Dihydroxybenzophenone	Bis(2-hydroxyphenyl) ketone	$C_{13}H_{10}O_3$	835-11-0	214.216		59.5	333			i H_2O; s EtOH, eth, chl
3711	4,4'-Dihydroxybenzophenone	Bis(4-hydroxyphenyl) ketone	$C_{13}H_{10}O_3$	611-99-4	214.216	nd (lig), cry (w)	210		1.133[131]		sl H_2O; s EtOH, eth, ace; i bz, CS_2
3712	6,7-Dihydroxy-2H-1-benzopyran-2-one	Esculetin	$C_9H_6O_4$	305-01-1	178.142	nd (w), pr (HOAc) lf (sub)	276	sub			sl H_2O, eth; s EtOH, ace, chl, AcOEt
3713	7,8-Dihydroxy-2H-1-benzopyran-2-one	Daphnetin	$C_9H_6O_4$	486-35-1	178.142	ye nd (dil al)	262	sub			s H_2O, EtOH; sl eth, bz, chl, CS_2
3714	2,4-Dihydroxybutanoic acid		$C_4H_8O_4$	1518-62-3	120.105	liq		96[3]			
3715	3,6-Dihydroxycholan-24-oic acid, (3α,5β,6α)	Hyodeoxycholic acid	$C_{24}H_{40}O_4$	83-49-8	392.573	cry (AcOEt)	198.5				sl H_2O, eth, ace, bz; s EtOH, HOAc
3716	3,7-Dihydroxycholan-24-oic acid, (3α,5β,7β)	Ursodiol	$C_{24}H_{40}O_4$	128-13-2	392.573	pl (al)	203				vs EtOH; sl eth
3717	3,7-Dihydroxycholan-24-oic acid, (3α,5β,7α)	Chenodiol	$C_{24}H_{40}O_4$	474-25-9	392.573	nd (EtOAc+hep)	119				i H_2O, bz; vs EtOH, ace; s eth, HOAc
3718	1,25-Dihydroxycholecalciferol	Calcitriol	$C_{27}H_{44}O_3$	32222-06-3	416.636	wh cry pow	115				sl EtOH, MeOH, thf, AcOEt
3719	2,5-Dihydroxy-2,5-cyclohexadiene-1,4-dione		$C_6H_4O_4$	615-94-1	140.094	dk ye nd	211				sl H_2O, ace, DMSO; s EtOH, HOAc; i eth

2,3-Dihydro-1,1,3-trimethyl-3-phenyl-1*H*-indene

1,2-Dihydro-2,2,4-trimethylquinoline

1,4-Dihydroxy-9,10-anthracenedione

1,5-Dihydroxy-9,10-anthracenedione

1,8-Dihydroxy-9,10-anthracenedione

2,6-Dihydroxy-9,10-anthracenedione

2,7-Dihydroxy-9,10-anthracenedione

2,2'-Dihydroxyazobenzene

2,3-Dihydroxybenzaldehyde

2,4-Dihydroxybenzaldehyde

2,5-Dihydroxybenzaldehyde

3,4-Dihydroxybenzaldehyde

N,2-Dihydroxybenzamide

2,5-Dihydroxybenzeneacetic acid

2,3-Dihydroxybenzoic acid

2,4-Dihydroxybenzoic acid

2,5-Dihydroxybenzoic acid

2,6-Dihydroxybenzoic acid

3,4-Dihydroxybenzoic acid

3,5-Dihydroxybenzoic acid

2,2'-Dihydroxybenzophenone

4,4'-Dihydroxybenzophenone

6,7-Dihydroxy-2*H*-1-benzopyran-2-one

7,8-Dihydroxy-2*H*-1-benzopyran-2-one

2,4-Dihydroxybutanoic acid

3,6-Dihydroxycholan-24-oic acid, (3α,5β,6α)

3,7-Dihydroxycholan-24-oic acid, (3α,5β,7β)

3,7-Dihydroxycholan-24-oic acid, (3α,5β,7α)

1,25-Dihydroxycholecalciferol

2,5-Dihydroxy-2,5-cyclohexadiene-1,4-dione

No.	Name	Synonym	Mol. Form.	CAS RN	Mol. Wt.	Physical Form	mp/°C	bp/°C	den/ g cm⁻³	n_D	Solubility
3720	2,3-Dihydroxy-2-cyclopenten-1-one	Reductic acid	C₅H₆O₃	80-72-8	114.100		212				s H₂O, EtOH; sl eth, ace, AcOEt; i bz
3721	2,6-Dihydroxy-2,6-dimethyl-4-heptanone	Di(2-hydroxy-2-methylpropyl) ketone	C₉H₁₈O₃	3682-91-5	174.237	pale ye cry					
3722	2,2'-Dihydroxydiphenylmethane	2,2'-Methylenebisphenol	C₁₃H₁₂O₂	2467-02-9	200.233		118.3	363	1.280²⁵		
3723	4,4'-Dihydroxydiphenyl sulfide	4,4'-Thiobisphenol	C₁₂H₁₀O₂S	2664-63-3	218.271	mcl pr or lf (al)	151				sl H₂O, EtOH, eth, CS₂
3724	1,8-Dihydroxy-3-(hydroxymethyl)-9,10-anthracenedione	Aloe-emodol	C₁₅H₁₀O₅	481-72-1	270.237	oran ye nd (to, al)	223.5	sub			vs bz, eth, EtOH
3725	2,3-Dihydroxymaleic acid	Dihydroxymaleic acid	C₄H₄O₆	526-84-1	148.071	pl (w+2)	155 dec				sl H₂O, eth, MeOH; s EtOH
3726	α,4-Dihydroxy-3-methoxybenzeneacetic acid	Vanilmandelic acid	C₉H₁₀O₅	55-10-7	198.172	sc (bz-eth)	132 dec				vs H₂O, ace, eth
3727	7,8-Dihydroxy-6-methoxy-2H-1-benzopyran-2-one	Fraxetin	C₁₀H₈O₅	574-84-5	208.168	pl (dil al)	231				vs EtOH
3728	5,7-Dihydroxy-3-(4-methoxyphenyl)-4H-1-benzopyran-4-one		C₁₆H₁₂O₅	491-80-5	284.263		214.8				
3729	(2,6-Dihydroxy-4-methoxyphenyl) phenylmethanone	Cotoin	C₁₄H₁₂O₄	479-21-0	244.243	ye pr (chl) lf or nd (w)	130.5				vs ace, bz, eth, EtOH
3730	1,7-Dihydroxy-3-methoxy-9H-xanthen-9-one	Gentisin	C₁₄H₁₀O₅	437-50-3	258.226	ye orth	266.5				i H₂O; vs EtOH; i ace; s py
3731	1,8-Dihydroxy-3-methyl-9,10-anthracenedione	Chrysophanic acid	C₁₅H₁₀O₄	481-74-3	254.238	ye hex or mcl nd (sub)	196	sub	0.92²⁵		vs bz, HOAc
3732	2,4-Dihydroxy-6-methylbenzoic acid	o-Orsellinic acid	C₈H₈O₄	480-64-8	168.148	nd (dil HOAc, +1w)	176 dec				s EtOH, eth
3733	5,7-Dihydroxy-4-methyl-2H-1-benzopyran-2-one		C₁₀H₈O₄	2107-76-8	192.169	nd (al), lf (HOAc)	283				sl H₂O, eth, bz, chl; vs EtOH, alk
3734	6,7-Dihydroxy-4-methyl-2H-1-benzopyran-2-one		C₁₀H₈O₄	529-84-0	192.169	ye nd (dil al)	275				s H₂O, EtOH, HOAc
3735	5,8-Dihydroxy-1,4-naphthalenedione		C₁₀H₆O₄	475-38-7	190.153	dk red mcl pr (bz) red-br nd (al)	232	sub			sl H₂O, EtOH, eth; s HOAc
3736	4,5-Dihydroxy-2,7-naphthalenedisulfonic acid	Chromotropic acid	C₁₀H₈O₈S₂	148-25-4	320.296	nd or lf (w+2)					s H₂O, alk; i EtOH, eth
3737	5,6-Dihydroxynaphtho[2,3-f] quinoline-7,12-dione	Alizarin Blue	C₁₇H₉NO₄	568-02-5	291.258	br-viol nd (bz)	269				vs bz, gl HOAc
3738	1,2-Dihydroxy-3-nitro-9,10-anthracenedione	Alizarin Orange	C₁₄H₇NO₆	568-93-4	285.209	oran nd or pl (HOAc)	244 dec	sub			sl H₂O; s EtOH, bz, chl, sulf, HOAc
3739	9,10-Dihydroxyoctadecanedioic acid, (R*,R*)-(±)	Phloionic acid	C₁₈H₃₄O₆	23843-52-9	346.459	cry (al)	126				
3740	9,10-Dihydroxyoctadecanoic acid	9,10-Dihydroxystearic acid	C₁₈H₃₆O₄	120-87-6	316.477		90				i H₂O; sl EtOH, eth
3741	5,7-Dihydroxy-2-phenyl-4H-1-benzopyran-4-one	Chrysin	C₁₅H₁₀O₄	480-40-0	254.238	lt ye pr (MeOH)	285.5				i H₂O; s EtOH, ace; sl eth, bz, CS₂
3742	1-(2,4-Dihydroxyphenyl)ethanone	Resacetophenone	C₈H₈O₃	89-84-9	152.148	nd or lf	146		1.18¹⁴¹		i H₂O, chl; s EtOH, py; sl eth, bz
3743	(2,4-Dihydroxyphenyl) phenylmethanone	Benzoresorcinol	C₁₃H₁₀O₃	131-56-6	214.216	nd (w)	144				i H₂O; s EtOH; vs eth; sl bz, chl
3744	3-(3,4-Dihydroxyphenyl)-2-propenoic acid	Caffeic acid	C₉H₈O₄	331-39-5	180.158	ye pr, pl (w)	225 dec				vs EtOH
3745	Dihydroxyphenylstibine oxide	Benzenestibonic acid	C₆H₇O₃Sb	535-46-6	248.878	nd (HOAc)	139				
3746	17,21-Dihydroxypregna-1,4-diene-3,11,20-trione	Prednisone	C₂₁H₂₆O₅	53-03-2	358.428		234 dec				
3747	17,21-Dihydroxypregn-4-ene-3,20-dione	11-Deoxy-17-hydrocorticosterone	C₂₁H₃₀O₄	152-58-9	346.461		215				vs ace, EtOH, chl
3748	17,21-Dihydroxypregn-4-ene-3,11,20-trione	Cortisone	C₂₁H₂₈O₅	53-06-5	360.444		222				sl H₂O, eth, bz, chl; s EtOH, ace
3749	2,3-Dihydroxypropanal, (±)		C₃H₆O₃	56-82-6	90.078	nd or pr (40% MeOH)	145	145⁰·⁸	1.453¹⁸		s H₂O; sl EtOH, eth; i bz, peth, lig
3750	2,3-Dihydroxypropanoic acid, (R)	Glyceric acid	C₃H₆O₄	6000-40-4	106.078	thick gum		dec			
3751	1,3-Dihydroxy-2-propanone	Dihydroxyacetone	C₃H₆O₃	96-26-4	90.078		90				s H₂O, EtOH, eth, ace; i lig
3752	2,3-Dihydroxypropyl decanoate	Decanoic acid glycerol monoester	C₁₃H₂₆O₄	2277-23-8	246.343	pr (peth)	53				
3753	2,3-Dihydroxypropyl octanoate	Octanoic acid glycerol monoester	C₁₁H₂₂O₄	26402-26-6	218.291	cry (peth)	40				

2,3-Dihydroxy-2-cyclopenten-1-one

2,6-Dihydroxy-2,6-dimethyl-4-heptanone

2,2'-Dihydroxydiphenylmethane

4,4'-Dihydroxydiphenyl sulfide

1,8-Dihydroxy-3-(hydroxymethyl)-9,10-anthracenedione

2,3-Dihydroxymaleic acid

α,4-Dihydroxy-3-methoxybenzeneacetic acid

7,8-Dihydroxy-6-methoxy-2H-1-benzopyran-2-one

5,7-Dihydroxy-3-(4-methoxyphenyl)-4H-1-benzopyran-4-one

(2,6-Dihydroxy-4-methoxyphenyl)phenylmethanone

1,7-Dihydroxy-3-methoxy-9H-xanthen-9-one

1,8-Dihydroxy-3-methyl-9,10-anthracenedione

2,4-Dihydroxy-6-methylbenzoic acid

5,7-Dihydroxy-4-methyl-2H-1-benzopyran-2-one

6,7-Dihydroxy-4-methyl-2H-1-benzopyran-2-one

5,8-Dihydroxy-1,4-naphthalenedione

4,5-Dihydroxy-2,7-naphthalenedisulfonic acid

5,6-Dihydroxynaphtho[2,3-f]quinoline-7,12-dione

1,2-Dihydroxy-3-nitro-9,10-anthracenedione

9,10-Dihydroxyoctadecanedioic acid, (R*,R*)-(±)

9,10-Dihydroxyoctadecanoic acid

5,7-Dihydroxy-2-phenyl-4H-1-benzopyran-4-one

1-(2,4-Dihydroxyphenyl)ethanone

(2,4-Dihydroxyphenyl)phenylmethanone

3-(3,4-Dihydroxyphenyl)-2-propenoic acid

Dihydroxyphenylstibine oxide

17,21-Dihydroxypregna-1,4-diene-3,11,20-trione

17,21-Dihydroxypregn-4-ene-3,20-dione

17,21-Dihydroxypregn-4-ene-3,11,20-trione

2,3-Dihydroxypropanal, (±)

2,3-Dihydroxypropanoic acid, (R)

1,3-Dihydroxy-2-propanone

2,3-Dihydroxypropyl decanoate

2,3-Dihydroxypropyl octanoate

No.	Name	Synonym	Mol. Form.	CAS RN	Mol. Wt.	Physical Form	mp/°C	bp/°C	den/ g cm⁻³	n_D	Solubility
3754	4,8-Dihydroxy-2-quinolinecarboxylic acid	Xanthurenic acid	C₁₀H₇NO₄	59-00-7	205.168	ye micry cry (w)	289				i H₂O; s EtOH, dil HCl; sl eth, bz
3755	Dihydroxytartaric acid		C₄H₆O₈	76-30-2	182.086		114.5				
3756	3,4-Dihydroxy-5-[(3,4,5-trihydroxybenzoyl)oxy]benzoic acid	Digallic acid	C₁₄H₁₀O₉	536-08-3	322.224	nf (dil al + 1w)	269 dec				vs ace, EtOH
3757	2-(3,6-Dihydroxy-9H-xanthen-9-yl)benzoic acid	Fluorescin	C₂₀H₁₄O₅	518-44-5	334.322	col or ye nd (eth), pl (bz)	126				i H₂O; s EtOH, eth, ace, bz, HOAc
3758	Diiodoacetylene		C₂I₂	624-74-8	277.830	orth nd (liq)	81.5	exp			vs ace, bz, eth, EtOH
3759	2,4-Diiodoaniline		C₆H₅I₂N	533-70-0	344.920	br nd or orth cry (al)	95.5		2.748²⁵		vs ace, bz, eth, EtOH
3760	o-Diiodobenzene	1,2-Diiodobenzene	C₆H₄I₂	615-42-9	329.905	pl or pr (liq)	27	287; 100³	2.54²⁰	1.7179²⁰	i H₂O; sl EtOH
3761	m-Diiodobenzene	1,3-Diiodobenzene	C₆H₄I₂	626-00-6	329.905	orth pl or pr (eth-al)	40.4	285	2.47²⁵		i H₂O; vs eth, EtOH, chl
3762	p-Diiodobenzene	1,4-Diiodobenzene	C₆H₄I₂	624-38-4	329.905	orth lf (al)	131.5	285			i H₂O; s EtOH; vs eth; sl chl
3763	1,4-Diiodobutane		C₄H₈I₂	628-21-7	309.916		5.8	125¹⁵ dec	2.3494²⁵	1.6184²⁵	i H₂O; sl ctc; s os
3764	1,2-Diiodoethane		C₂H₄I₂	624-73-7	281.862	ye mcl pr or orth (eth)	83	200	3.325²⁰	1.871²⁰	sl H₂O; s EtOH, eth, ace, chl
3765	cis-1,2-Diiodoethene	cis-1,2-Diiodoethylene	C₂H₂I₂	590-26-1	279.846		-14	72.5¹⁶	3.0625²⁰		i H₂O; s eth, chl
3766	4,4'-Diiodofluorescein		C₂₀H₁₀I₂O₅	38577-97-8	584.099	oran-red pow					sl H₂O; s alk, EtOH
3767	1,6-Diiodohexane	Hexamethylene diiodide	C₆H₁₂I₂	629-09-4	337.968	nd	9.5	163¹⁷, 141¹⁰	2.0342²⁵	1.5837²⁵	i H₂O; vs EtOH, eth
3768	Diiodomethane	Methylene iodide	CH₂I₂	75-11-6	267.836	ye nd or lf	6.1	182	3.3211²⁰	1.7411²⁰	sl H₂O, ctc; s EtOH, eth, bz, chl
3769	2,6-Diiodo-4-nitrophenol	Disophenol	C₆H₃I₂NO₃	305-85-1	390.902	lt ye cry (gl HOAc)	157				vs EtOH
3770	1,5-Diiodopentane	Pentamethylene diiodide	C₅H₁₀I₂	628-77-3	323.942		9	149²⁰, 101³	2.1692²⁵	1.5987²⁵	i H₂O; s eth, chl
3771	1,2-Diiodopropane		C₃H₆I₂	598-29-8	295.889				2.490¹⁸		vs eth, EtOH
3772	1,3-Diiodopropane	Trimethylene diiodide	C₃H₆I₂	627-31-6	295.889		-20	dec 227; 110¹⁹	2.5612²⁵	1.6391²⁰	i H₂O; s eth, ctc, chl
3773	5,7-Diiodo-8-quinolinol	Iodoquinol	C₉H₅I₂NO	83-73-8	396.951	ye nd (HOAc, xyl)	210				sl H₂O, bz, chl, eth; vs EtOH; s alk
3774	3,5-Diiodo-L-tyrosine		C₉H₉I₂NO₃	300-39-0	432.981	ye nd (w, 70% al)	213				sl H₂O; i EtOH, eth, bz
3775	Diisobutyl adipate		C₁₄H₂₆O₄	141-04-8	258.354			293; 187¹⁵	0.9543¹⁹	1.4301²⁰	
3776	Diisobutylaluminum chloride		C₈H₁₈AlCl	1779-25-5	176.664	hyg col liq	-40	152¹⁰	0.905	1.4506²⁰	s eth, hx
3777	Diisobutylaluminum hydride		C₈H₁₉Al	1191-15-7	142.219	liq		140⁴, 85⁰·⁵			s cyhex, eth, bz, tol
3778	Diisobutylamine	2-Methyl-N-(2-methylpropyl)-1-propanamine	C₈H₁₉N	110-96-3	129.244	liq	-73.5	139.6		1.4090²⁰	sl H₂O, ctc; s EtOH, eth, ace, bz
3779	Diisobutyl carbonate		C₉H₁₈O₃	539-92-4	174.237			190	0.9138²⁰	1.4072²⁰	i H₂O; msc EtOH, eth
3780	Diisobutyl ether	1,1'-Oxybis[2-methylpropane]	C₈H₁₈O	628-55-7	130.228			122.6	0.761¹⁵		i H₂O; msc EtOH, eth
3781	Diisobutyl phthalate		C₁₆H₂₂O₄	84-69-5	278.344			296.5; 159⁴	1.0490¹⁵		s ctc
3782	Diisobutyl sulfide		C₈H₁₈S	592-65-4	146.294	liq	-105.5	171	0.8363¹⁰		
3783	1,3-Diisocyanatobenzene		C₈H₄N₂O₂	123-61-5	160.130	cry	51	103⁸			
3784	1,4-Diisocyanatobenzene		C₈H₄N₂O₂	104-49-4	160.130	cry	95	117¹⁴			
3785	Diisodecyl phthalate	Bis(8-methylnonyl)phthalate	C₂₈H₄₆O₄	26761-40-0	446.663	liq	-50	253⁴	0.966²⁰		i H₂O; s os
3786	Diisononyl phthalate	Bis(7-methyloctyl)phthalate	C₂₆H₄₂O₄	28553-12-0	418.609	col liq					i H₂O; s ace, MeOH; bz, eth
3787	Diisooctyl adipate		C₂₂H₄₂O₄	1330-86-5	370.566			210⁴			
3788	Diisooctyl phthalate		C₂₄H₃₈O₄	27554-26-3	390.557			370			
3789	Diisopentylamine	3-Methyl-N-isopentyl-1-butanamine	C₁₀H₂₃N	544-00-3	157.297	liq	-44	188	0.7672²¹	1.4235²⁰	i H₂O; s EtOH; msc eth
3790	Diisopentyl ether	Diisoamyl ether	C₁₀H₂₂O	544-01-4	158.281			172.5	0.7777²⁰	1.4085²⁰	i H₂O; vs ace, EtOH, chl
3791	Diisopentyl phthalate	Diisoamyl phthalate	C₁₈H₂₆O₄	605-50-5	306.397			dec 334	1.0209¹⁶	1.4871²⁰	vs EtOH
3792	Diisopentyl sulfide		C₁₀H₂₂S	544-02-5	174.347	liq	-74.6	211	0.8323²⁰	1.4520²⁰	i H₂O; msc EtOH; vs eth
3793	Diisopropanolamine	1,1'-Iminobis-2-propanol	C₆H₁₅NO₂	110-97-4	133.189	cry	44.5	250; 151²³	0.989²⁰		s H₂O, EtOH; sl eth
3794	Diisopropyl adipate		C₁₂H₂₂O₄	6938-94-9	230.301		-0.6	120⁶·⁵	0.9569²⁰	1.4247²⁰	vs ace, eth, EtOH

4,8-Dihydroxy-2-quinolinecarboxylic acid

Dihydroxytartaric acid

3,4-Dihydroxy-5-[(3,4,5-trihydroxybenzoyl)oxy]benzoic acid

2-(3,6-Dihydroxy-9*H*-xanthen-9-yl)benzoic acid

Diiodoacetylene

2,4-Diiodoaniline

o-Diiodobenzene

m-Diiodobenzene

p-Diiodobenzene

1,4-Diiodobutane

1,2-Diiodoethane

cis-1,2-Diiodoethene

4,4'-Diiodofluorescein

1,6-Diiodohexane

Diiodomethane

2,6-Diiodo-4-nitrophenol

1,5-Diiodopentane

1,2-Diiodopropane

1,3-Diiodopropane

5,7-Diiodo-8-quinolinol

3,5-Diiodo-*L*-tyrosine

Diisobutyl adipate

Diisobutylaluminum chloride

Diisobutylaluminum hydride

Diisobutylamine

Diisobutyl carbonate

Diisobutyl ether

Diisobutyl phthalate

Diisobutyl sulfide

1,3-Diisocyanatobenzene

1,4-Diisocyanatobenzene

Diisodecyl phthalate

Diisononyl phthalate

Diisooctyl adipate

Diisooctyl phthalate

Diisopentylamine

Diisopentyl ether

Diisopentyl phthalate

Diisopentyl sulfide

Diisopropanolamine

Diisopropyl adipate

No.	Name	Synonym	Mol. Form.	CAS RN	Mol. Wt.	Physical Form	mp/°C	bp/°C	den/ g cm^{-3}	n_D	Solubility
3795	Diisopropylamine	N-Isopropyl-2-propanamine	C$_6$H$_{15}$N	108-18-9	101.190	liq	-61	83.9	0.7153[20]	1.3924[20]	vs ace, bz, eth, EtOH
3796	2,6-Diisopropylaniline		C$_{12}$H$_{19}$N	24544-04-5	177.286	liq	-45	257	0.94[25]	1.5332[20]	
3797	1,2-Diisopropylbenzene		C$_{12}$H$_{18}$	577-55-9	162.271	liq	-57	204	0.8701[20]	1.4960[20]	i H$_2$O; msc EtOH, eth, ace, bz, ctc
3798	1,3-Diisopropylbenzene		C$_{12}$H$_{18}$	99-62-7	162.271	liq	-63.1	203.2	0.8559[20]	1.4883[20]	i H$_2$O; msc EtOH, eth, ace, bz, ctc
3799	1,4-Diisopropylbenzene		C$_{12}$H$_{18}$	100-18-5	162.271	liq	-17	210.3	0.8568[20]	1.4898[20]	i H$_2$O; msc EtOH, eth, ace, bz, ctc
3800	p-Diisopropylbenzene hydroperoxide		C$_{12}$H$_{18}$O$_2$	98-49-7	194.270	waxy cry	30.1	123[1]	0.9932[20]		i H$_2$O
3801	N,N-Diisopropyl-2-benzothiazolesulfenamide		C$_{13}$H$_{18}$N$_2$S$_2$	95-29-4	266.425		59.0				
3802	N,N'-Diisopropylcarbodiimide		C$_7$H$_{14}$N$_2$	693-13-0	126.199			147	0.806[25]	1.4320[20]	
3803	Diisopropyl disulfide		C$_6$H$_{14}$S$_2$	4253-89-8	150.305	liq	-69	177	0.9435[20]	1.4916[20]	
3804	N,N-Diisopropylethanolamine	N,N-Diisopropyl-2-aminoethanol	C$_8$H$_{19}$NO	96-80-0	145.243			190	0.826[25]	1.4417[20]	
3805	Diisopropyl ether	Isopropyl ether	C$_6$H$_{14}$O	108-20-3	102.174	liq	-85.4	68.4	0.7192[25]	1.3658[25]	sl H$_2$O; msc EtOH, eth; s ace, ctc
3806	Diisopropyl methylphosphonate		C$_7$H$_{17}$O$_3$P	1445-75-6	180.182			66[3]		1.4120[16]	
3807	2,6-Diisopropylnaphthalene		C$_{16}$H$_{20}$	24157-81-1	212.330	cry (MeOH)	70				
3808	Diisopropyl oxalate		C$_8$H$_{14}$O$_4$	615-81-6	174.195			190	1.002[20]	1.4100[20]	vs eth, EtOH
3809	Diisopropyl phosphonate		C$_6$H$_{15}$O$_3$P	1809-20-7	166.155			97[40], 76[10]	0.9970[18]		
3810	O,O-Diisopropyl phosphorodithioate		C$_6$H$_{15}$O$_2$PS$_2$	107-56-2	214.286	liq		71[3]	1.09[20]		s EtOH, bz, ace, ctc, chl
3811	Diisopropyl phthalate	1,2-Benzenedicarboxylic acid, diisopropyl ester	C$_{14}$H$_{18}$O$_4$	605-45-8	250.291			130[12]	1.0615[15]	1.4900[20]	
3812	Diisopropyl sulfide		C$_6$H$_{14}$S	625-80-9	118.240	liq	-78.1	120.0	0.8142[20]	1.4438[20]	i H$_2$O; s EtOH, eth
3813	Diisopropyl tartrate, (±)		C$_{10}$H$_{18}$O$_6$	58167-01-4	234.246		34	275; 154[12]	1.1166[20]		vs ace, eth, EtOH
3814	Diisopropyl thioperoxydicarbonate	Diisopropyl dixanthogen	C$_8$H$_{14}$O$_2$S$_4$	105-65-7	270.456		52				s chl
3815	1,4-Diisothiocyanatobenzene	Bitoscanate	C$_8$H$_4$N$_2$S$_2$	4044-65-9	192.261	nd (ace, HOAc)	132				
3816	Diketene		C$_4$H$_4$O$_2$	674-82-8	84.074		-6.5	126.1	1.0877[20]	1.4379[20]	
3817	Dilactic acid	2,2'-Oxybispropanoic acid	C$_6$H$_{10}$O$_5$	19201-34-4	162.140	orth	112.5				vs H$_2$O, eth
3818	Dimefline		C$_{20}$H$_{21}$NO$_3$	1165-48-6	323.386		109.5				s chl
3819	Dimefox	Tetramethylphosphorodiamidic fluoride	C$_4$H$_{12}$FN$_2$OP	115-26-4	154.122	liq		86[15]	1.1151[20]	1.4267[20]	vs H$_2$O, bz, eth
3820	Dimemorfan	3,17-Dimethylmorphinan, (9 α,13 α,14 α)-	C$_{18}$H$_{25}$N	36309-01-0	255.399	ye oil	92	133[0.3]			
3821	2,3-Dimercaptobutanedioic acid		C$_4$H$_6$O$_4$S$_2$	2418-14-6	182.219	wh cry (MeOH)	193				
3822	1,4-Dimercapto-2,3-butanediol		C$_4$H$_{10}$O$_2$S$_2$	7634-42-6	154.251		42.5				s chl
3823	2,2'-Dimercaptodiethyl ether	2-Mercaptoethyl ether	C$_4$H$_{10}$OS$_2$	2150-02-9	138.251	liq	-80	217; 64[2]	1.114[20]		
3824	2,3-Dimercapto-1-propanol	Dimercaprol	C$_3$H$_8$OS$_2$	59-52-9	124.225			83[0.8]	1.2463[20]	1.5749[20]	s EtOH, eth, oils; sl chl
3825	Dimetan®		C$_{11}$H$_{17}$NO$_3$	122-15-6	211.258	cry	46	175[11]			s H$_2$O, cyhex; vs EtOH, eth, ace
3826	Dimethipin	2,3-Dihydro-5,6-dimethyl-1,4-dithiin, 1,1,4,4-tetraoxide	C$_6$H$_{10}$O$_4$S$_2$	55290-64-7	210.271		165				
3827	Dimethirimol	5-Butyl-2-(dimethylamino)-6-methylpyrimidin-4(1H)-one	C$_{11}$H$_{19}$N$_3$O	5221-53-4	209.288	nd	102				sl H$_2$O; vs chl, xyl; s EtOH, ace
3828	Dimethisoquin	2-[(3-Butyl-1-isoquinolinyl)oxy]-N,N-dimethylethanamine	C$_{17}$H$_{24}$N$_2$O	86-80-6	272.385		146	156[3]		1.5486[20]	s H$_2$O, EtOH
3829	Dimethoxane	2,6-Dimethyl-1,3-dioxan-4-ol acetate	C$_8$H$_{14}$O$_4$	828-00-2	174.195	liq		86[10]	1.0655[20]	1.4310[20]	msc H$_2$O; s os
3830	2',5'-Dimethoxyacetophenone		C$_{10}$H$_{12}$O$_3$	1201-38-3	180.200	cry	21	156[14]	1.139	1.5441[20]	
3831	1,2-Dimethoxy-4-allylbenzene		C$_{11}$H$_{14}$O$_2$	93-15-2	178.228	liq	-2.0	254.7	1.0396[20]	1.5340[20]	i H$_2$O; s EtOH, eth
3832	4,7-Dimethoxy-5-allyl-1,3-benzodioxole	Apiole	C$_{12}$H$_{14}$O$_4$	523-80-8	222.237	nd	29.5	294; 179[35]	1.015[20]	1.5360[20]	vs ace, bz, EtOH, lig
3833	2,4-Dimethoxyaniline		C$_8$H$_{11}$NO$_2$	2735-04-8	153.179	pl (lig)	33.5	262.0			sl H$_2$O, chl; s EtOH, eth, bz, lig
3834	2,5-Dimethoxyaniline		C$_8$H$_{11}$NO$_2$	102-56-7	153.179		82.5	270			s H$_2$O, EtOH, chl, lig
3835	3,4-Dimethoxyaniline		C$_8$H$_{11}$NO$_2$	6315-89-5	153.179	lf (eth)	87.5	159[14]			s eth, chl

Diisopropylamine

2,6-Diisopropylaniline

1,2-Diisopropylbenzene

1,3-Diisopropylbenzene

1,4-Diisopropylbenzene

p-Diisopropylbenzene hydroperoxide

N,N-Diisopropyl-2-benzothiazolesulfenamide

N,N′-Diisopropylcarbodiimide

Diisopropyl disulfide

N,N-Diisopropylethanolamine

Diisopropyl ether

Diisopropyl methylphosphonate

2,6-Diisopropylnaphthalene

Diisopropyl oxalate

Diisopropyl phosphonate

O,O-Diisopropyl phosphorodithioate

Diisopropyl phthalate

Diisopropyl sulfide

Diisopropyl tartrate, (±)

Diisopropyl thioperoxydicarbonate

1,4-Diisothiocyanatobenzene

Diketene

Dilactic acid

Dimefline

Dimefox

Dimemorfan

2,3-Dimercaptobutanedioic acid

1,4-Dimercapto-2,3-butanediol

2,2′-Dimercaptodiethyl ether

2,3-Dimercapto-1-propanol

Dimetan®

Dimethipin

Dimethirimol

Dimethisoquin

Dimethoxane

2′,5′-Dimethoxyacetophenone

1,2-Dimethoxy-4-allylbenzene

4,7-Dimethoxy-5-allyl-1,3-benzodioxole

2,4-Dimethoxyaniline

2,5-Dimethoxyaniline

3,4-Dimethoxyaniline

No.	Name	Synonym	Mol. Form.	CAS RN	Mol. Wt.	Physical Form	mp/°C	bp/°C	den/ g cm⁻³	n_D	Solubility
3836	2,4-Dimethoxybenzaldehyde		C₉H₁₀O₃	613-45-6	166.173	nd (al or liq)	72	290; 165[10]			i H₂O; s EtOH, eth, bz; sl chl
3837	2,5-Dimethoxybenzaldehyde		C₉H₁₀O₃	93-02-7	166.173		52	270; 146[10]			sl H₂O; s EtOH, eth
3838	3,4-Dimethoxybenzaldehyde	Veratraldehyde	C₉H₁₀O₃	120-14-9	166.173	nd (eth, lig, to)	43	281; 155[10]			sl H₂O, chl; vs EtOH, eth
3839	3,5-Dimethoxybenzaldehyde		C₉H₁₀O₃	7311-34-4	166.173		46.3	151[16]			sl H₂O, peth; s EtOH, bz
3840	1,2-Dimethoxybenzene	Veratrole	C₈H₁₀O₂	91-16-7	138.164		22.5	206	1.0810[25]	1.5827[21]	sl H₂O; s EtOH, eth, ctc
3841	1,3-Dimethoxybenzene		C₈H₁₀O₂	151-10-0	138.164	liq	-52	217.5	1.0521[25]	1.5231[20]	sl H₂O; s EtOH, eth, bz, ctc, sulf
3842	1,4-Dimethoxybenzene		C₈H₁₀O₂	150-78-7	138.164	lf (w)	59	212.6	1.0375[55]		sl H₂O; s EtOH, chl; vs eth, bz
3843	3,4-Dimethoxybenzeneacetic acid		C₁₀H₁₂O₄	93-40-3	196.200	cry (bz-peth) nd (w+1)	98				s H₂O, chl; vs EtOH, eth
3844	3,4-Dimethoxybenzeneethanamine		C₁₀H₁₅NO₂	120-20-7	181.232			164[14]		1.5464[20]	s ctc
3845	3,4-Dimethoxybenzenemethanamine		C₉H₁₃NO₂	5763-61-1	167.205			156[12], 120[3]	1.143[25]		s chl
3846	3,4-Dimethoxybenzenemethanol		C₉H₁₂O₃	93-03-8	168.189	visc oil		298; 172[12]	1.178[17]	1.555[17]	s H₂O, EtOH
3847	3,3'-Dimethoxybenzidine	Dianisidine	C₁₄H₁₆N₂O₂	119-90-4	244.289	lf or nd (w)	137				i H₂O; s EtOH, eth, ace, bz, chl
3848	3,3'-Dimethoxybenzidine-4,4'-diisocyanate		C₁₆H₁₂N₂O₄	91-93-0	296.277	cry	112				
3849	2,4-Dimethoxybenzoic acid		C₉H₁₀O₄	91-52-1	182.173		108.5				sl H₂O; s EtOH, eth, chl, HOAc
3850	2,6-Dimethoxybenzoic acid		C₉H₁₀O₄	1466-76-8	182.173		186 dec				
3851	3,4-Dimethoxybenzoic acid	Veratric acid	C₉H₁₀O₄	93-07-2	182.173	nd (w or HOAc) orth (sub)	181	sub			i H₂O; vs EtOH, eth; sl chl
3852	3,5-Dimethoxybenzoic acid		C₉H₁₀O₄	1132-21-4	182.173	nd (w), pr (al)	185.5	sub			vs eth, EtOH
3853	4,4'-Dimethoxybenzoin	p-Anisoin	C₁₆H₁₆O₄	119-52-8	272.296	pr (dil al)	114.0				sl H₂O, chl, EtOH, eth; s ace
3854	5,7-Dimethoxy-2H-1-benzopyran-2-one	Limettin	C₁₁H₁₀O₄	487-06-9	206.195	pr or nd (al)	149	dec 200			sl H₂O; vs EtOH, ace, chl; i eth, lig
3855	4,4'-Dimethoxy-1,1'-biphenyl		C₁₄H₁₄O₂	2132-80-1	214.260	lf (bz)	175	sub			i H₂O, peth; vs EtOH, bz, chl; sl eth
3856	Dimethoxyborane		C₂H₇BO₂	4542-61-4	73.887	vol liq or gas	-130.6	25.9			dec H₂O
3857	4,4-Dimethoxy-2-butanone		C₆H₁₂O₃	5436-21-5	132.157			50[5]			s ctc
3858	2,6-Dimethoxy-2,5-cyclohexadiene-1,4-dione	2,6-Dimethoxy-p-quinone	C₈H₈O₄	530-55-2	168.148	ye mcl pr (HOAc)	256	sub			sl H₂O, EtOH, eth; s tfa; vs alk, HOAc
3859	Dimethoxydimethylsilane		C₄H₁₂O₂Si	1112-39-6	120.223			82	0.8646[20]	1.3708[20]	dec H₂O
3860	Dimethoxydiphenylsilane		C₁₄H₁₆O₂Si	6843-66-9	244.362			286; 161[15]	1.0771[20]	1.5447[20]	
3861	1,1-Dimethoxydodecane	Lauraldehyde, dimethyl acetal	C₁₄H₃₀O₂	14620-52-1	230.387			133[5]		1.4310[25]	vs eth, EtOH
3862	2,2-Dimethoxyethanamine		C₄H₁₁NO₂	22483-09-6	105.136		-78	137[95]	0.966[25]	1.4170[20]	
3863	1,2-Dimethoxyethane	Ethylene glycol dimethyl ether	C₄H₁₀O₂	110-71-4	90.121	liq	-69.20	84.50	0.8637[25]	1.3770[25]	s H₂O, EtOH, eth, ace, bz, chl, ctc
3864	(2,2-Dimethoxyethyl)benzene		C₁₀H₁₄O₂	101-48-4	166.217			193.5			
3865	4,8-Dimethoxyfuro[2,3-b]quinoline	Fagarine	C₁₃H₁₁NO₃	524-15-2	229.231	pr (al)	142				sl H₂O, peth; s EtOH, eth, bz, chl
3866	1,1-Dimethoxyhexadecane	Palmitaldehyde, dimethyl acetal	C₁₈H₃₈O₂	2791-29-9	286.494		10	144[2]	0.8542[20]	1.4382[25]	vs ace, eth, EtOH
3867	2,4-Dimethoxy-6-hydroxyacetophenone	Xanthoxylin	C₁₀H₁₂O₄	90-24-4	196.200	cry (al)	82	185[20]			vs eth, EtOH
3868	5,6-Dimethoxy-1-indanone		C₁₁H₁₂O₃	2107-69-9	192.211		119.5				sl ctc
3869	6,7-Dimethoxy-1(3H)-isobenzofuranone	Meconin	C₁₀H₁₀O₄	569-31-3	194.184	wh nd (w)	102.5				sl H₂O; s EtOH, eth, ace, bz, HOAc, chl
3870	Dimethoxymethane	Methylal	C₃H₈O₂	109-87-5	76.095	liq	-105.1	42	0.8593[20]	1.3513[20]	s H₂O; vs ace, bz, eth, EtOH
3871	1,2-Dimethoxy-4-methylbenzene		C₉H₁₂O₂	494-99-5	152.190	pr (eth)	24	220	1.0509[25]	1.5257[25]	i H₂O; sl ctc; vs os
3872	1,3-Dimethoxy-5-methylbenzene		C₉H₁₂O₂	4179-19-5	152.190			244	1.0478[15]	1.5234[20]	vs bz, eth, EtOH
3873	1,4-Dimethoxy-2-methylbenzene		C₉H₁₂O₂	24599-58-4	152.190		21	214.0			
3874	N-(Dimethoxymethyl)dimethylamine	Dimethylformamide dimethyl acetal	C₅H₁₃NO₂	4637-24-5	119.163			104	0.897[25]	1.3972[20]	

2,4-Dimethoxybenzaldehyde

2,5-Dimethoxybenzaldehyde

3,4-Dimethoxybenzaldehyde

3,5-Dimethoxybenzaldehyde

1,2-Dimethoxybenzene

1,3-Dimethoxybenzene

1,4-Dimethoxybenzene

3,4-Dimethoxybenzeneacetic acid

3,4-Dimethoxybenzeneethanamine

3,4-Dimethoxybenzenemethanamine

3,4-Dimethoxybenzenemethanol

3,3'-Dimethoxybenzidine

3,3'-Dimethoxybenzidine-4,4'-diisocyanate

2,4-Dimethoxybenzoic acid

2,6-Dimethoxybenzoic acid

3,4-Dimethoxybenzoic acid

3,5-Dimethoxybenzoic acid

4,4'-Dimethoxybenzoin

5,7-Dimethoxy-2*H*-1-benzopyran-2-one

4,4'-Dimethoxy-1,1'-biphenyl

Dimethoxyborane

4,4-Dimethoxy-2-butanone

2,6-Dimethoxy-2,5-cyclohexadiene-1,4-dione

Dimethoxydimethylsilane

Dimethoxydiphenylsilane

1,1-Dimethoxydodecane

2,2-Dimethoxyethanamine

1,2-Dimethoxyethane

(2,2-Dimethoxyethyl)benzene

4,8-Dimethoxyfuro[2,3-b]quinoline

1,1-Dimethoxyhexadecane

2,4-Dimethoxy-6-hydroxyacetophenone

5,6-Dimethoxy-1-indanone

6,7-Dimethoxy-1(3*H*)-isobenzofuranone

Dimethoxymethane

1,2-Dimethoxy-4-methylbenzene

1,3-Dimethoxy-5-methylbenzene

1,4-Dimethoxy-2-methylbenzene

N-(Dimethoxymethyl)dimethylamine

No.	Name	Synonym	Mol. Form.	CAS RN	Mol. Wt.	Physical Form	mp/°C	bp/°C	den/ g cm^{-3}	n_D	Solubility
3875	2,2-Dimethoxy-N-methylethanamine		C$_5$H$_{13}$NO$_2$	122-07-6	119.163			140	0.928[25]	1.4115[20]	
3876	Dimethoxymethylphenylsilane		C$_9$H$_{14}$O$_2$Si	3027-21-2	182.292			129[79]		1.4795[20]	
3877	1,2-Dimethoxy-4-nitrobenzene		C$_8$H$_9$NO$_4$	709-09-1	183.162	ye nd (al-w)	98	230[15]	1.1888[133]		i H$_2$O; vs EtOH, eth; s chl; sl lig
3878	1,4-Dimethoxy-2-nitrobenzene		C$_8$H$_9$NO$_4$	89-39-4	183.162	gold-ye nd (dil al)	72.5		1.1666[132]		i H$_2$O; s EtOH, bz, chl, sulf
3879	2,6-Dimethoxyphenol		C$_8$H$_{10}$O$_3$	91-10-1	154.163	mcl pr (w)	56.5	261			vs eth, EtOH
3880	3,5-Dimethoxyphenol		C$_8$H$_{10}$O$_3$	500-99-2	154.163		37	199[35], 170[10]			s eth, bz; sl lig
3881	1-(3,4-Dimethoxyphenyl)ethanone		C$_{10}$H$_{12}$O$_3$	1131-62-0	180.200	pr (dil al)	51	287			vs H$_2$O, bz, EtOH, chl
3882	1,1-Dimethoxypropane		C$_5$H$_{12}$O$_2$	4744-10-9	104.148			86	0.8648[20]		
3883	2,2-Dimethoxypropane		C$_5$H$_{12}$O$_2$	77-76-9	104.148	liq	-47	83	0.847[25]	1.3780[20]	
3884	3,3-Dimethoxy-1-propene		C$_5$H$_{10}$O$_2$	6044-68-4	102.132			88	0.862[25]	1.3954[20]	
3885	1,2-Dimethoxy-4-(1-propenyl) benzene		C$_{11}$H$_{14}$O$_2$	93-16-3	178.228		18	270.5	1.0521[20]	1.5616[20]	
3886	4,5-Dimethoxy-6-(2-propenyl)-1,3-benzodioxole	Apiole (Dill)	C$_{12}$H$_{14}$O$_4$	484-31-1	222.237	oil	29.5	285	1.1598[15]	1.5305[17]	
3887	1,2-Dimethoxy-4-vinylbenzene		C$_{10}$H$_{12}$O$_2$	6380-23-0	164.201					1.5711[20]	s chl
3888	Dimethylacetal		C$_4$H$_{10}$O$_2$	534-15-6	90.121	liq	-113.2	64.5	0.8501[20]	1.3668[20]	s H$_2$O, EtOH, eth, ctc, chl; vs ace
3889	N,N-Dimethylacetamide	N,N-Dimethylethanamide	C$_4$H$_9$NO	127-19-5	87.120	liq	-18.59	165	0.9372[25]	1.4341[25]	msc H$_2$O, EtOH, eth, ace, bz, chl
3890	2,7-Dimethyl-3,6-acridinediamine, monohydrochloride	Acridine Yellow	C$_{15}$H$_{16}$ClN$_3$	135-49-9	273.761	red cry pow					s hot H$_2$O, EtOH
3891	Dimethyl adipate		C$_8$H$_{14}$O$_4$	627-93-0	174.195	cry	10.3	115[13]	1.0600[20]	1.4283[20]	i H$_2$O; s EtOH, eth, ctc, HOAc
3892	3,3-Dimethylallyl diphosphate	3-Methyl-2-butenyl pyrophosphate	C$_5$H$_{12}$O$_7$P$_2$	358-72-5	246.092	cry (MeOH)					
3893	Dimethylamine	N-Methylmethanamine	C$_2$H$_7$N	124-40-3	45.084	col gas	-92.18	6.88	0.6804[0]	1.350[17]	vs H$_2$O; s EtOH, eth
3894	Dimethylamine hydrochloride	N-Methylmethanamine hydrochloride	C$_2$H$_8$ClN	506-59-2	81.545	orth nd (al)	171				vs H$_2$O, EtOH, chl
3895	(Dimethylamino)acetonitrile		C$_4$H$_8$N$_2$	926-64-7	84.120			137.5	0.8649[20]	1.4095[20]	vs H$_2$O, EtOH
3896	4'-(Dimethylamino)acetophenone	4-Acetyl-N,N-dimethylaniline	C$_{10}$H$_{13}$NO	2124-31-4	163.216	nd (w, peth)	105.5				vs H$_2$O, eth, lig; sl chl
3897	10-[((Dimethylamino)acetyl]-10H-phenothiazine	Ahistan	C$_{16}$H$_{16}$N$_2$OS	518-61-6	284.375	cry	144.5				
3898	p-(Dimethylamino)azobenzene		C$_{14}$H$_{15}$N$_3$	60-11-7	225.289	ye lf (al)	117	dec			i H$_2$O; vs EtOH, py; s eth; sl chl, lig
3899	2',3-Dimethyl-4-aminoazobenzene	4-o-Tolylazo-o-toluidine	C$_{14}$H$_{15}$N$_3$	97-56-3	225.289	ye lf (al)	102				vs eth, EtOH
3900	4-(Dimethylamino)benzaldehyde	Ehrlich's reagent	C$_9$H$_{11}$NO	100-10-7	149.189	lf (w)	74.5	176[17]	1.0254[100]		sl H$_2$O, chl; s EtOH, eth, ace, bz
3901	p-(Dimethylamino) benzalrhodanine		C$_{12}$H$_{12}$N$_2$OS$_2$	536-17-4	264.365	dp red nd (xyl)	270 dec				i H$_2$O; sl EtOH, bz; vs eth, ctc; s ace
3902	2-(Dimethylamino)benzoic acid		C$_9$H$_{11}$NO$_2$	610-16-2	165.189	pr, nd (eth)	72	sub			vs H$_2$O, eth, EtOH
3903	3-(Dimethylamino)benzoic acid		C$_9$H$_{11}$NO$_2$	99-64-9	165.189	nd (w)	152.5				sl H$_2$O, chl; s EtOH, eth
3904	4-(Dimethylamino)benzoic acid		C$_9$H$_{11}$NO$_2$	619-84-1	165.189	nd (al)	242.5				s EtOH; sl eth
3905	4,4'-Dimethylaminobenzophenonimide	Brilliant Oil Yellow	C$_{17}$H$_{21}$N$_3$	492-80-8	267.369	ye or col pl (al)	136				i H$_2$O; s EtOH; sl eth
3906	(Dimethylamino)dimethylborane		C$_4$H$_{12}$BN	1113-30-0	84.956	liq	-92	65			vs eth, ace
3907	6-(Dimethylamino)-4,4-diphenyl-3-heptanone		C$_{21}$H$_{27}$NO	76-99-3	309.445		99.5				vs EtOH
3908	6-(Dimethylamino)-4,4-diphenyl-3-hexanone	Normethadone	C$_{20}$H$_{25}$NO	467-85-6	295.419	oily liq		165[3]			
3909	2-(Dimethylamino)ethyl acrylate		C$_7$H$_{13}$NO$_2$	2439-35-2	143.184		<-60	95[50]	0.938[20]		
3910	3-[2-(Dimethylamino)ethyl]-1H-indol-5-ol	Bufotenine	C$_{12}$H$_{16}$N$_2$O	487-93-4	204.267	pr (EtOAc)	146.5	320[0.1]			vs eth, EtOH
3911	2-(Dimethylamino)ethyl methacrylate		C$_8$H$_{15}$NO$_2$	2867-47-2	157.211			63[6]			
3912	4-[2-(Dimethylamino)ethyl]phenol	Hordenine	C$_{10}$H$_{15}$NO	539-15-1	165.232	orth pr (al), nd (w)	117.5	173[11]			vs eth, EtOH, chl
3913	N-[2-(Dimethylamino)ethyl]-N,N',N'-trimethyl-1,2-ethanediamine		C$_9$H$_{23}$N$_3$	3030-47-5	173.299			84[12]		1.4413[25]	
3914	5-(Dimethylamino)-1-naphthalenesulfonyl chloride	Dansyl chloride	C$_{12}$H$_{12}$ClNO$_2$S	605-65-2	269.747		70				

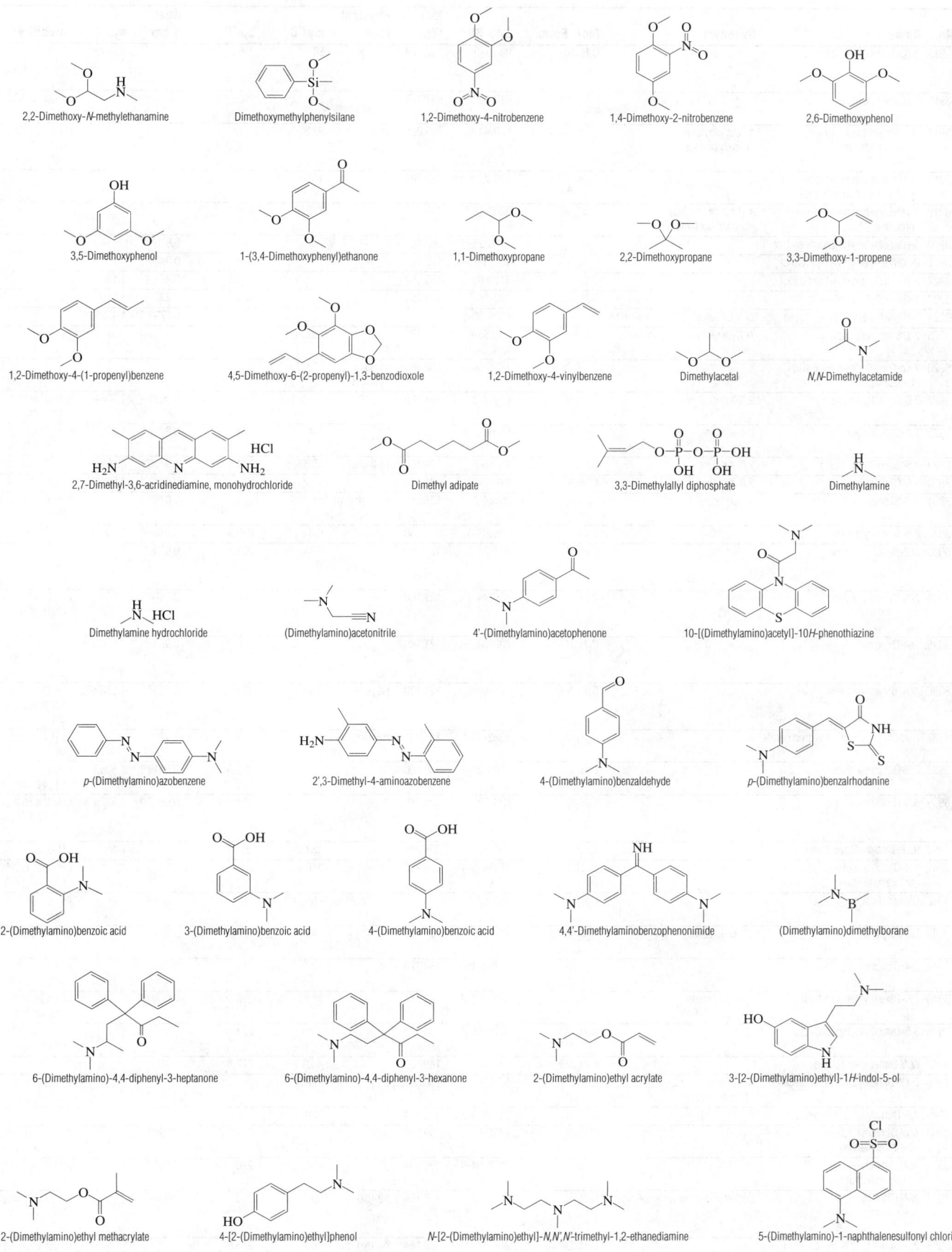

2,2-Dimethoxy-*N*-methylethanamine

Dimethoxymethylphenylsilane

1,2-Dimethoxy-4-nitrobenzene

1,4-Dimethoxy-2-nitrobenzene

2,6-Dimethoxyphenol

3,5-Dimethoxyphenol

1-(3,4-Dimethoxyphenyl)ethanone

1,1-Dimethoxypropane

2,2-Dimethoxypropane

3,3-Dimethoxy-1-propene

1,2-Dimethoxy-4-(1-propenyl)benzene

4,5-Dimethoxy-6-(2-propenyl)-1,3-benzodioxole

1,2-Dimethoxy-4-vinylbenzene

Dimethylacetal

N,N-Dimethylacetamide

2,7-Dimethyl-3,6-acridinediamine, monohydrochloride

Dimethyl adipate

3,3-Dimethylallyl diphosphate

Dimethylamine

Dimethylamine hydrochloride

(Dimethylamino)acetonitrile

4'-(Dimethylamino)acetophenone

10-[(Dimethylamino)acetyl]-10*H*-phenothiazine

p-(Dimethylamino)azobenzene

2',3-Dimethyl-4-aminoazobenzene

4-(Dimethylamino)benzaldehyde

p-(Dimethylamino)benzalrhodanine

2-(Dimethylamino)benzoic acid

3-(Dimethylamino)benzoic acid

4-(Dimethylamino)benzoic acid

4,4'-Dimethylaminobenzophenonimide

(Dimethylamino)dimethylborane

6-(Dimethylamino)-4,4-diphenyl-3-heptanone

6-(Dimethylamino)-4,4-diphenyl-3-hexanone

2-(Dimethylamino)ethyl acrylate

3-[2-(Dimethylamino)ethyl]-1*H*-indol-5-ol

2-(Dimethylamino)ethyl methacrylate

4-[2-(Dimethylamino)ethyl]phenol

N-[2-(Dimethylamino)ethyl]-*N,N',N'*-trimethyl-1,2-ethanediamine

5-(Dimethylamino)-1-naphthalenesulfonyl chloride

No.	Name	Synonym	Mol. Form.	CAS RN	Mol. Wt.	Physical Form	mp/°C	bp/°C	den/ g cm^{-3}	n_D	Solubility
3915	3-(Dimethylamino)phenol		C$_8$H$_{11}$NO	99-07-0	137.179	nd (lig)	86	266.5		1.5895[26]	i H$_2$O; s EtOH, eth, ace, bz, CS$_2$
3916	4-(Dimethylamino)phenol		C$_8$H$_{11}$NO	619-60-3	137.179		77	165[30]			sl H$_2$O; s EtOH, eth
3917	[4-(Dimethylamino)phenyl] phenylmethanone	4-(Dimethylamino) benzophenone	C$_{15}$H$_{15}$NO	530-44-9	225.286	ye lf (al) nd (peth)	92.5				i H$_2$O; sl EtOH; vs eth; s chl, peth
3918	3-(Dimethylamino)-1-phenyl-1-propanone, hydrochloride		C$_{11}$H$_{16}$ClNO	879-72-1	213.704		153.5				
3919	3-[4-(Dimethylamino)phenyl]-2-propenal	4-(Dimethylamino) cinnamaldehyde	C$_{11}$H$_{13}$NO	6203-18-5	175.227		139.5				
3920	3-(Dimethylamino)propanenitrile		C$_5$H$_{10}$N$_2$	1738-25-6	98.146			173	0.8705[20]		
3921	2-(Dimethylamino)-1-propanol		C$_5$H$_{13}$NO	15521-18-3	103.163			150.3	0.8820[25]		s H$_2$O
3922	3-(Dimethylamino)-1-propanol		C$_5$H$_{13}$NO	3179-63-3	103.163			163.5	0.872[25]	1.4360[20]	s ctc
3923	1-(Dimethylamino)-2-propanol		C$_5$H$_{13}$NO	108-16-7	103.163			124.5	0.837[25]	1.4193[20]	s ctc
3924	3-(Dimethylamino)-1-propyne	N,N-Dimethyl-2-propargylamine	C$_5$H$_9$N	7223-38-3	83.132			80	0.7792[20]	1.4195[20]	
3925	2-Dimethylaminopurine	N,N-Dimethyl-1H-purin-6-amine	C$_7$H$_9$N$_5$	938-55-6	163.180		263				
3926	2-(p-Dimethylaminostyryl) benzothiazole		C$_{17}$H$_{16}$N$_2$S	1628-58-6	280.387	ye nd (MeOH)	207 dec				
3927	2,3-Dimethylaniline	2,3-Xylidine	C$_8$H$_{11}$N	87-59-2	121.180		<-15	221.5	0.9931[20]	1.5684[20]	sl H$_2$O; vs EtOH, eth; s ctc
3928	2,4-Dimethylaniline	2,4-Xylidine	C$_8$H$_{11}$N	95-68-1	121.180	liq	-14.3	214	0.9723[20]	1.5569[20]	sl H$_2$O, ctc; s EtOH, eth, bz
3929	2,5-Dimethylaniline	2,5-Xylidine	C$_8$H$_{11}$N	95-78-3	121.180	ye lf (lig)	15.5	214	0.9790[21]	1.5591[21]	sl H$_2$O; s eth, ctc
3930	2,6-Dimethylaniline	2,6-Xylidine	C$_8$H$_{11}$N	87-62-7	121.180		11.2	215	0.9842[20]	1.5610[20]	vs eth, EtOH
3931	3,4-Dimethylaniline	3,4-Xylidine	C$_8$H$_{11}$N	95-64-7	121.180	pl or pr (lig)	51	228	1.076[18]		sl H$_2$O, chl; s eth; vs lig
3932	3,5-Dimethylaniline	3,5-Xylidine	C$_8$H$_{11}$N	108-69-0	121.180		9.8	220.5	0.9706[20]	1.5581[20]	sl H$_2$O; s eth, ctc
3933	N,2-Dimethylaniline		C$_8$H$_{11}$N	611-21-2	121.180			207.5	0.9709[20]	1.5649[20]	i H$_2$O; msc EtOH, eth; s ace
3934	N,3-Dimethylaniline		C$_8$H$_{11}$N	696-44-6	121.180			206.5	0.9660[20]	1.5557[25]	i H$_2$O; msc EtOH, eth; s ace
3935	N,4-Dimethylaniline		C$_8$H$_{11}$N	623-08-5	121.180			210	0.9348[55]	1.5568[20]	i H$_2$O; msc EtOH, eth; s ace
3936	N,N-Dimethylaniline		C$_8$H$_{11}$N	121-69-7	121.180	pa ye	2.42	194.15	0.9557[20]	1.5582[20]	sl H$_2$O; s EtOH, eth, ace, bz; vs chl
3937	N,N-Dimethylaniline hydrochloride		C$_8$H$_{12}$ClN	5882-44-0	157.641	hyg pl (w, bz)	90		1.1156[19]		vs H$_2$O, EtOH, chl
3938	2,6-Dimethylanisole		C$_9$H$_{12}$O	1004-66-6	136.190			182.5	0.9619[14]	1.5053[14]	i H$_2$O; s EtOH, eth, bz, ctc
3939	3,5-Dimethylanisole		C$_9$H$_{12}$O	874-63-5	136.190			194; 89[15]	0.9627[15]	1.5110[20]	i H$_2$O; s EtOH, eth, bz, CS$_2$; sl ctc
3940	9,10-Dimethylanthracene		C$_{16}$H$_{14}$	781-43-1	206.282		183.6	360.0			i H$_2$O
3941	1,4-Dimethyl-9,10-anthracenedione		C$_{16}$H$_{12}$O$_2$	1519-36-4	236.265	ye nd (al, sub)	140.5	sub			i H$_2$O; sl EtOH; s bz, xyl, HOAc
3942	Dimethylarsine		C$_2$H$_7$As	593-57-7	105.999	liq, ign in air	-136.1	36	1.208[29]		vs ace, bz, eth, EtOH
3943	Dimethylarsinic acid	Cacodylic acid	C$_2$H$_7$AsO$_2$	75-60-5	137.998		195	>200			vs H$_2$O; s EtOH; i eth
3944	2,4-Dimethylbenzaldehyde		C$_9$H$_{10}$O	15764-16-6	134.174	liq	-9	218			s EtOH; s eth, ace, bz; sl chl
3945	2,5-Dimethylbenzaldehyde	Isoxylaldehyde	C$_9$H$_{10}$O	5779-94-2	134.174			220	0.9500[20]		vs EtOH; s eth, ace, bz, ctc
3946	3,5-Dimethylbenzaldehyde		C$_9$H$_{10}$O	5779-95-3	134.174		9	221			vs ace, bz, eth, EtOH
3947	N,N-Dimethylbenzamide		C$_9$H$_{11}$NO	611-74-5	149.189		44.8	272.0			
3948	7,12-Dimethylbenz[a]anthracene	9,10-Dimethyl-1,2-benzanthracene	C$_{20}$H$_{16}$	57-97-6	256.341	pa ye pl (al, HOAc)	122.5				vs ace, bz
3949	4,5-Dimethyl-1,2-benzenediamine		C$_8$H$_{12}$N$_2$	3171-45-7	136.194		128				
3950	N,N-Dimethyl-1,2-benzenediamine		C$_8$H$_{12}$N$_2$	2836-03-5	136.194	oil		218; 90[22]	0.995[22]		sl H$_2$O; vs EtOH, eth, ace, bz
3951	N,N-Dimethyl-1,3-benzenediamine		C$_8$H$_{12}$N$_2$	2836-04-6	136.194		<-20	270; 138[10]	0.995[25]		sl H$_2$O; vs EtOH, eth
3952	N,N-Dimethyl-1,4-benzenediamine	Dimethyl-p-phenylenediamine	C$_8$H$_{12}$N$_2$	99-98-9	136.194	nd (bz)	53	263	1.036[20]		s H$_2$O, chl; vs EtOH, eth, bz; sl lig
3953	2,5-Dimethyl-1,3-benzenediol		C$_8$H$_{10}$O$_2$	488-87-9	138.164	nd (bz), pr (w)	163	278.5			s H$_2$O, EtOH, eth

3-(Dimethylamino)phenol

4-(Dimethylamino)phenol

[4-(Dimethylamino)phenyl]phenylmethanone

3-(Dimethylamino)-1-phenyl-1-propanone, hydrochloride

3-[4-(Dimethylamino)phenyl]-2-propenal

3-(Dimethylamino)propanenitrile

2-(Dimethylamino)-1-propanol

3-(Dimethylamino)-1-propanol

1-(Dimethylamino)-2-propanol

3-(Dimethylamino)-1-propyne

2-Dimethylaminopurine

2-(p-Dimethylaminostyryl)benzothiazole

2,3-Dimethylaniline

2,4-Dimethylaniline

2,5-Dimethylaniline

2,6-Dimethylaniline

3,4-Dimethylaniline

3,5-Dimethylaniline

N,2-Dimethylaniline

N,3-Dimethylaniline

N,4-Dimethylaniline

N,N-Dimethylaniline

N,N-Dimethylaniline hydrochloride

2,6-Dimethylanisole

3,5-Dimethylanisole

9,10-Dimethylanthracene

1,4-Dimethyl-9,10-anthracenedione

Dimethylarsine

Dimethylarsinic acid

2,4-Dimethylbenzaldehyde

2,5-Dimethylbenzaldehyde

3,5-Dimethylbenzaldehyde

N,N-Dimethylbenzamide

7,12-Dimethylbenz[a]anthracene

4,5-Dimethyl-1,2-benzenediamine

N,N-Dimethyl-1,2-benzenediamine

N,N-Dimethyl-1,3-benzenediamine

N,N-Dimethyl-1,4-benzenediamine

2,5-Dimethyl-1,3-benzenediol

No.	Name	Synonym	Mol. Form.	CAS RN	Mol. Wt.	Physical Form	mp/°C	bp/°C	den/ g cm⁻³	n_D	Solubility
3954	2,6-Dimethyl-1,4-benzenediol		$C_8H_{10}O_2$	654-42-2	138.164	nd (xyl), cry (w)	152.3				vs eth, EtOH
3955	N,β-Dimethylbenzeneethanamine	Phenylpropylmethylamine	$C_{10}H_{15}N$	93-88-9	149.233			207.5	0.915²⁵		vs bz, eth, EtOH
3956	α,α-Dimethylbenzeneethanamine	Phentermine	$C_{10}H_{15}N$	122-09-8	149.233	oily liq		205; 100²¹			
3957	α,α-Dimethylbenzenemethanamine		$C_9H_{13}N$	585-32-0	135.206			196.5	0.9423²⁰	1.5181²⁵	
3958	$\alpha,4$-Dimethylbenzenemethanol	1-(4-Methylphenyl)ethanol	$C_9H_{12}O$	536-50-5	136.190			219	0.9668²⁵	1.5246²⁰	i H_2O; vs EtOH, eth
3959	α,α-Dimethylbenzenemethanol	α-Cumyl alcohol	$C_9H_{12}O$	617-94-7	136.190	pr	36	202	0.9735²⁰	1.5325²⁰	i H_2O; s EtOH, eth, bz, HOAc
3960	α,α-Dimethylbenzenepropanol	Benzyl-*tert*-butanol	$C_{11}H_{16}O$	103-05-9	164.244	nd	24.5	121¹³	0.9626²¹	1.5077²¹	i H_2O; vs EtOH, eth, ace, bz
3961	$N,4$-Dimethylbenzenesulfonamide		$C_8H_{11}NO_2S$	640-61-9	185.244	pl (dil al)	78.5		1.340²⁵		vs eth, EtOH
3962	5,6-Dimethyl-1H-benzimidazole	Dimedazole	$C_9H_{10}N_2$	582-60-5	146.188	cry (eth)	205.5	sub			s H_2O, EtOH, eth, chl, DMSO
3963	2,4-Dimethylbenzoic acid		$C_9H_{10}O_2$	611-01-8	150.174	mcl or tcl nd (w)	90	268			sl H_2O; s EtOH, ace, bz, chl, HOAc, tol
3964	2,5-Dimethylbenzoic acid		$C_9H_{10}O_2$	610-72-0	150.174	nd (al)	132	sub	1.069²¹		i H_2O; s EtOH, eth, ace, bz
3965	2,6-Dimethylbenzoic acid		$C_9H_{10}O_2$	632-46-2	150.174	nd (lig)	116	274.5; 155¹⁷			sl H_2O, lig; s EtOH, eth
3966	3,4-Dimethylbenzoic acid		$C_9H_{10}O_2$	619-04-5	150.174	pr (al)	167.3				i H_2O; s EtOH, eth, bz
3967	3,5-Dimethylbenzoic acid	Mesitylenic acid	$C_9H_{10}O_2$	499-06-9	150.174	nd (w, al)	171.1	sub			sl H_2O; vs EtOH, eth
3968	4,4'-Dimethylbenzophenone	Bis(4-methylphenyl) ketone	$C_{15}H_{14}O$	611-97-2	210.271	orth (al)	96.5	334			vs ace, bz, eth, EtOH
3969	7,8-Dimethylbenzo[g]pteridine-2,4(1H,3H)-dione	Lumichrome	$C_{12}H_{10}N_4O_2$	1086-80-2	242.233	ye cry (chl)	300				sl H_2O, EtOH, chl
3970	2,5-Dimethylbenzoxazole		C_9H_9NO	5676-58-4	147.173			218.5	1.0880¹⁸	1.5412²⁰	s ctc
3971	N,N-Dimethylbenzylamine	Dimethylbenzylamine	$C_9H_{13}N$	103-83-3	135.206			181	0.915⁰	1.5011²⁰	sl H_2O; msc EtOH, eth
3972	N,N-Dimethyl-N'-benzyl-1,2-ethanediamine	N-Benzyl-N',N'-dimethyl-1,2-ethanediamine	$C_{11}H_{18}N_2$	103-55-9	178.274			145³⁰, 123¹¹	0.9343²⁰	1.5089²⁰	
3973	N,N-Dimethyl-N'-benzyl-N'-2-pyridinyl-1,2-ethanediamine	Tripelennamine	$C_{16}H_{21}N_3$	91-81-6	255.358	ye oil		140⁰·¹		1.576²⁵	misc H_2O
3974	6,6-Dimethylbicyclo[3.1.1]heptan-2-one, (1R)		$C_9H_{14}O$	38651-65-9	138.206	liq	-1	209	0.9807²⁰	1.4787²⁰	vs eth, EtOH
3975	2,3-Dimethylbicyclo[2.2.1]hept-2-ene	2,3-Dimethyl-2-norbornene	C_9H_{14}	529-16-8	122.207			140.5	0.8698¹⁷	1.4688¹⁷	s eth, ace, bz
3976	6,6-Dimethylbicyclo[3.1.1]hept-2-ene-2-ethanol		$C_{11}H_{18}O$	128-50-7	166.260			235; 110¹⁰	0.973²⁵	1.4930²⁰	s chl
3977	2,2'-Dimethylbiphenyl		$C_{14}H_{14}$	605-39-0	182.261	cry (al)	19.5	256	0.9906²⁰	1.5752²⁰	i H_2O; vs EtOH, eth, bz; s ace
3978	3,3'-Dimethylbiphenyl		$C_{14}H_{14}$	612-75-9	182.261		9	280	0.9995²⁰	1.5946²⁰	i H_2O; vs EtOH, eth, bz; s ace
3979	4,4'-Dimethylbiphenyl		$C_{14}H_{14}$	613-33-2	182.261	mcl pr (eth)	125	295	0.917¹²¹		i H_2O; sl EtOH; s eth, ace, bz, CS_2
3980	4,4'-Dimethyl-2,2'-bipyridine		$C_{12}H_{12}N_2$	1134-35-6	184.236		171.5				s chl
3981	2,3-Dimethyl-1,3-butadiene	Diisopropenyl	C_6H_{10}	513-81-5	82.143	liq	-76	68.8	0.7222²⁵	1.4394²⁰	s ctc
3982	N,N-Dimethylbutanamide		$C_6H_{13}NO$	760-79-2	115.173	liq	-40	186; 125¹⁰⁰	0.9064²⁵	1.4391²⁵	vs ace, bz, eth, EtOH
3983	3,3-Dimethyl-2-butanamine		$C_6H_{15}N$	3850-30-4	101.190	liq	-20	102	0.7668²⁰	1.4105²⁵	vs H_2O
3984	2,2-Dimethylbutane	Neohexane	C_6H_{14}	75-83-2	86.175	liq	-98.8	49.73	0.6444²⁵	1.3688²⁰	i H_2O; s EtOH, eth; vs ace, bz, peth, ctc
3985	2,3-Dimethylbutane		C_6H_{14}	79-29-8	86.175	liq	-128.10	57.93	0.6616²⁰	1.3750²⁰	i H_2O; s EtOH, eth; vs ace, bz, peth, ctc
3986	2,3-Dimethyl-2,3-butanediol	Pinacol	$C_6H_{14}O_2$	76-09-5	118.174	nd (al,eth)	43.32	174.4			sl H_2O, CS_2; vs EtOH, eth
3987	2,3-Dimethyl-2-butanethiol		$C_6H_{14}S$	1639-01-6	118.240	liq		126.1			
3988	2,2-Dimethylbutanoic acid		$C_6H_{12}O_2$	595-37-9	116.158	liq	-14	186	0.9276²⁰	1.4145²⁰	sl H_2O; s EtOH, eth
3989	2,2-Dimethyl-1-butanol		$C_6H_{14}O$	1185-33-7	102.174		<-15	136.5	0.8283²⁰	1.4208²⁰	sl H_2O; s EtOH, eth
3990	3,3-Dimethyl-1-butanol		$C_6H_{14}O$	624-95-3	102.174	liq	-60	143	0.844¹⁵	1.4323¹⁵	sl H_2O; s EtOH, eth, ace
3991	2,3-Dimethyl-2-butanol		$C_6H_{14}O$	594-60-5	102.174	liq	-14	118.4	0.8236²⁰	1.4176²⁰	s H_2O; msc EtOH, eth
3992	3,3-Dimethyl-2-butanol, (±)		$C_6H_{14}O$	20281-91-8	102.174		5.6	120.4	0.8122²⁵	1.4148²⁰	sl H_2O; vs EtOH, eth

2,6-Dimethyl-1,4-benzenediol

N,β-Dimethylbenzeneethanamine

α,α-Dimethylbenzeneethanamine

α,α-Dimethylbenzenemethanamine

α,4-Dimethylbenzenemethanol

α,α-Dimethylbenzenemethanol

α,α-Dimethylbenzenepropanol

N,4-Dimethylbenzenesulfonamide

5,6-Dimethyl-1H-benzimidazole

2,4-Dimethylbenzoic acid

2,5-Dimethylbenzoic acid

2,6-Dimethylbenzoic acid

3,4-Dimethylbenzoic acid

3,5-Dimethylbenzoic acid

4,4'-Dimethylbenzophenone

7,8-Dimethylbenzo[g]pteridine-2,4(1H,3H)-dione

2,5-Dimethylbenzoxazole

N,N-Dimethylbenzylamine

N,N-Dimethyl-N'-benzyl-1,2-ethanediamine

N,N-Dimethyl-N'-benzyl-N'-2-pyridinyl-1,2-ethanediamine

6,6-Dimethylbicyclo[3.1.1]heptan-2-one, (1R)

2,3-Dimethylbicyclo[2.2.1]hept-2-ene

6,6-Dimethylbicyclo[3.1.1]hept-2-ene-2-ethanol

2,2'-Dimethylbiphenyl

3,3'-Dimethylbiphenyl

4,4'-Dimethylbiphenyl

4,4'-Dimethyl-2,2'-bipyridine

2,3-Dimethyl-1,3-butadiene

N,N-Dimethylbutanamide

3,3-Dimethyl-2-butanamine

2,2-Dimethylbutane

2,3-Dimethylbutane

2,3-Dimethyl-2,3-butanediol

2,3-Dimethyl-2-butanethiol

2,2-Dimethylbutanoic acid

2,2-Dimethyl-1-butanol

3,3-Dimethyl-1-butanol

2,3-Dimethyl-2-butanol

3,3-Dimethyl-2-butanol, (±)

No.	Name	Synonym	Mol. Form.	CAS RN	Mol. Wt.	Physical Form	mp/°C	bp/°C	den/ g cm⁻³	n_D	Solubility
3993	3,3-Dimethyl-2-butanone	Pinacolone	C₆H₁₂O	75-97-8	100.158	liq	-52.5	106.1	0.7229²⁰	1.3952²⁰	sl H₂O; s EtOH, eth, ace, ctc
3994	3,3-Dimethylbutanoyl chloride		C₆H₁₁ClO	7065-46-5	134.603			130; 68¹⁰⁰	0.969²⁰	1.4210²⁰	vs eth
3995	2,3-Dimethyl-1-butene		C₆H₁₂	563-78-0	84.159	liq	-157.3	55.6	0.6803²⁰	1.3995²⁰	i H₂O; s EtOH, eth, ace, ctc, CS₂
3996	3,3-Dimethyl-1-butene		C₆H₁₂	558-37-2	84.159	liq	-115.2	41.2	0.6529²⁰	1.3763²⁰	i H₂O; s EtOH, eth, ctc, chl
3997	2,3-Dimethyl-2-butene		C₆H₁₂	563-79-1	84.159	liq	-74.19	73.3	0.7080²⁰	1.4122²⁰	i H₂O; s EtOH, eth, ace, chl
3998	N-(1,3-Dimethylbutyl)-N'-phenyl-1,4-benzenediamine		C₁₈H₂₄N₂	793-24-8	268.397		46	164¹			
3999	3,3-Dimethyl-1-butyne	tert-Butylacetylene	C₆H₁₀	917-92-0	82.143	liq	-78.2	37.7	0.6623²⁵	1.3736²⁰	
4000	Dimethyl 2-butynedioate		C₆H₆O₄	762-42-5	142.110			dec 197; 98²⁰	1.1564²⁰	1.4434²⁰	s EtOH, eth, ctc
4001	Dimethyl cadmium		C₂H₆Cd	506-82-1	142.480		-4.5	105.5 (exp 150)	1.9846¹⁸	1.5488	s peth
4002	Dimethylcarbamic chloride	Dimethylcarbamoyl chloride	C₃H₆ClNO	79-44-7	107.539	liq	-33	167	1.168²⁵	1.4540²⁰	
4003	Dimethylcarbamothioic chloride		C₃H₆ClNS	16420-13-6	123.605	pr	42.5	98¹⁰			vs eth; s chl, peth
4004	Dimethyl carbate		C₁₁H₁₄O₄	39589-98-5	210.227	cry	38	137¹²·⁵	1.164²¹	1.4852²⁰	i H₂O
4005	Dimethyl carbonate	Methyl carbonate	C₃H₆O₃	616-38-6	90.078		0.5	90.5	1.0636²⁵	1.3687²⁰	i H₂O; s EtOH, eth; sl ctc
4006	Dimethylcyanamide		C₃H₆N₂	1467-79-4	70.093			163.5		1.4089¹⁹	vs ace, eth, EtOH
4007	2,3-Dimethyl-2,5-cyclohexadiene-1,4-dione		C₈H₈O₂	526-86-3	136.149	ye nd	55	sub			sl H₂O; s EtOH, eth, chl
4008	2,5-Dimethyl-2,5-cyclohexadiene-1,4-dione		C₈H₈O₂	137-18-8	136.149	ye nd (al)	126.0				sl H₂O, EtOH; s eth, bz, chl
4009	2,6-Dimethyl-2,5-cyclohexadiene-1,4-dione		C₈H₈O₂	527-61-7	136.149	ye nd	72.5	sub	1.0479²⁸		s chl
4010	1,1-Dimethylcyclohexane		C₈H₁₆	590-66-9	112.213	liq	-33.3	119.6	0.7809²⁰	1.4290²⁰	i H₂O; s EtOH, eth, ace, bz; msc ctc
4011	cis-1,2-Dimethylcyclohexane		C₈H₁₆	2207-01-4	112.213	liq	-49.8	129.8	0.7963²⁰	1.4360²⁰	i H₂O; s EtOH, bz, ctc; msc eth, ace
4012	trans-1,2-Dimethylcyclohexane		C₈H₁₆	6876-23-9	112.213	liq	-88.15	123.5	0.7760²⁰	1.4270²⁰	i H₂O; s EtOH, eth; msc ace, bz; vs lig
4013	cis-1,3-Dimethylcyclohexane		C₈H₁₆	638-04-0	112.213	liq	-75.53	120.1	0.7660²⁰	1.4229²⁰	i H₂O; msc EtOH, eth, ace, bz, lig, ctc
4014	trans-1,3-Dimethylcyclohexane		C₈H₁₆	2207-03-6	112.213	liq	-90.07	124.5	0.79¹⁵	1.4284²⁵	
4015	cis-1,4-Dimethylcyclohexane		C₈H₁₆	624-29-3	112.213	liq	-87.39	124.4	0.7829²⁰	1.4230²⁰	i H₂O; msc EtOH, eth, ace, bz, lig, ctc
4016	trans-1,4-Dimethylcyclohexane		C₈H₁₆	2207-04-7	112.213	liq	-36.93	119.4	0.77¹⁵	1.4185²⁵	i H₂O
4017	Dimethyl trans-1,4-cyclohexanedicarboxylate		C₁₀H₁₆O₄	3399-22-2	200.232	ndl (eth)	71				s eth
4018	5,5-Dimethyl-1,3-cyclohexanedione	5,5-Dimethyldihydroresorcinol	C₈H₁₂O₂	126-81-8	140.180	nd (w)	150 dec				sl H₂O, eth; s ace, ctc; vs chl, HOAc
4019	N,α-Dimethylcyclohexaneethanamine	Propylhexedrine	C₁₀H₂₁N	101-40-6	155.281			205; 82¹⁰	0.8501²⁰	1.4600²⁰	vs EtOH
4020	3,3-Dimethylcyclohexanol		C₈H₁₆O	767-12-4	128.212		11.5	185; 99.5³⁵	0.9128¹⁴	1.4606¹⁵	
4021	2,2-Dimethylcyclohexanone		C₈H₁₄O	1193-47-1	126.196	liq	-20.5	172	0.9145²⁰	1.4486²⁰	
4022	2,6-Dimethylcyclohexanone		C₈H₁₄O	2816-57-1	126.196			175	0.925²⁵	1.4460²⁰	
4023	3,3-Dimethylcyclohexanone		C₈H₁₄O	2979-19-3	126.196			180; 72²⁵	0.909¹⁵	1.4482¹⁷	
4024	4,4-Dimethylcyclohexanone		C₈H₁₄O	4255-62-3	126.196		39	73¹⁴	0.932²⁰	1.4537²⁴	
4025	1,2-Dimethylcyclohexene		C₈H₁₄	1674-10-8	110.197	liq	-84.1	138	0.8220²⁵	1.4620²⁰	
4026	1,3-Dimethylcyclohexene		C₈H₁₄	2808-76-6	110.197			127	0.799²⁵	1.449²⁰	
4027	3,5-Dimethyl-2-cyclohexen-1-one		C₈H₁₂O	1123-09-7	124.180			208.5	0.9400²⁰	1.4812²⁰	s EtOH, eth
4028	1,1-Dimethylcyclopentane		C₇H₁₄	1638-26-2	98.186	liq	-69.8	87.5	0.7499²⁵	1.4136²⁰	
4029	cis-1,2-Dimethylcyclopentane		C₇H₁₄	1192-18-3	98.186	liq	-54	99.5	0.7680²⁵	1.4222²⁰	
4030	trans-1,2-Dimethylcyclopentane		C₇H₁₄	822-50-4	98.186	liq	-117.6	91.9	0.7468²⁵	1.4120²⁰	
4031	cis-1,3-Dimethylcyclopentane		C₇H₁₄	2532-58-3	98.186	liq	-133.7	90.8	0.7402²⁵	1.4089²⁰	
4032	trans-1,3-Dimethylcyclopentane		C₇H₁₄	1759-58-6	98.186	liq	-134	91.7	0.7443²⁵	1.4107²⁰	
4033	N,α-Dimethylcyclopentaneethanamine	Cyclopentamine	C₉H₁₉N	102-45-4	141.254			171		1.4500²⁰	
4034	1,2-Dimethylcyclopentene		C₇H₁₂	765-47-9	96.170	liq	-90.4	105.8	0.7928²⁵	1.4448²⁰	
4035	1,5-Dimethylcyclopentene		C₇H₁₂	16491-15-9	96.170	liq	-118	99	0.780²⁰	1.4331²⁰	
4036	1,1-Dimethylcyclopropane		C₅H₁₀	1630-94-0	70.133	vol liq or gas	-109	20.6	0.6604²⁰	1.3668²⁰	i H₂O; s EtOH; vs eth, sulf

3,3-Dimethyl-2-butanone

3,3-Dimethylbutanoyl chloride

2,3-Dimethyl-1-butene

3,3-Dimethyl-1-butene

2,3-Dimethyl-2-butene

N-(1,3-Dimethylbutyl)-N'-phenyl-1,4-benzenediamine

3,3-Dimethyl-1-butyne

Dimethyl 2-butynedioate

Dimethyl cadmium

Dimethylcarbamic chloride

Dimethylcarbamothioic chloride

Dimethyl carbate

Dimethyl carbonate

Dimethylcyanamide

2,3-Dimethyl-2,5-cyclohexadiene-1,4-dione

2,5-Dimethyl-2,5-cyclohexadiene-1,4-dione

2,6-Dimethyl-2,5-cyclohexadiene-1,4-dione

1,1-Dimethylcyclohexane

cis-1,2-Dimethylcyclohexane

trans-1,2-Dimethylcyclohexane

cis-1,3-Dimethylcyclohexane

trans-1,3-Dimethylcyclohexane

cis-1,4-Dimethylcyclohexane

trans-1,4-Dimethylcyclohexane

Dimethyl trans-1,4-cyclohexanedicarboxylate

5,5-Dimethyl-1,3-cyclohexanedione

N,α-Dimethylcyclohexaneethanamine

3,3-Dimethylcyclohexanol

2,2-Dimethylcyclohexanone

2,6-Dimethylcyclohexanone

3,3-Dimethylcyclohexanone

4,4-Dimethylcyclohexanone

1,2-Dimethylcyclohexene

1,3-Dimethylcyclohexene

3,5-Dimethyl-2-cyclohexen-1-one

1,1-Dimethylcyclopentane

cis-1,2-Dimethylcyclopentane

trans-1,2-Dimethylcyclopentane

cis-1,3-Dimethylcyclopentane

trans-1,3-Dimethylcyclopentane

N,α-Dimethylcyclopentaneethanamine

1,2-Dimethylcyclopentene

1,5-Dimethylcyclopentene

1,1-Dimethylcyclopropane

No.	Name	Synonym	Mol. Form.	CAS RN	Mol. Wt.	Physical Form	mp/°C	bp/°C	den/ g cm^{-3}	n_D	Solubility
4037	cis-1,2-Dimethylcyclopropane		C$_5$H$_{10}$	930-18-7	70.133	liq	-140.9	37.0	0.6889^{25}	1.3829^{20}	i H$_2$O; s EtOH; vs eth; sl ctc
4038	trans-1,2-Dimethylcyclopropane		C$_5$H$_{10}$	2402-06-4	70.133	vol liq or gas	-149.6	28.2	0.6648^{25}	1.3713^{20}	vs eth, EtOH
4039	Dimethyldecylamine	N,N-Dimethyl-1-decanamine	C$_{12}$H$_{27}$N	1120-24-7	185.349			234.5			
4040	Dimethyldiacetoxysilane	Bis(acetyloxy)dimethylsilane	C$_6$H$_{12}$O$_4$Si	2182-66-3	176.243	liq	-12.5	165	1.0540^{20}	1.4030^{20}	
4041	trans-Dimethyldiazene	Azomethane	C$_2$H$_6$N$_2$	4143-41-3	58.082	gas	-78	1.5	0.743^0	1.4199^{19}	vs ace, EtOH, eth; s ctc, hp
4042	2,2-Dimethyl-1,3-dioxane-4,6-dione	Meldrum's acid	C$_6$H$_8$O$_4$	2033-24-1	144.126		94				
4043	cis-3,6-Dimethyl-1,4-dioxane-2,5-dione		C$_6$H$_8$O$_4$	4511-42-6	144.126	orth (eth)	96.8	150^{25}			
4044	2,2-Dimethyl-1,3-dioxolane-4-methanol	Isopropylidene glycerol	C$_6$H$_{12}$O$_3$	100-79-8	132.157			82^{10}	1.064^{20}	1.4383^{20}	
4045	Dimethyldiphenoxysilane		C$_{14}$H$_{16}$O$_2$Si	3440-02-6	244.362		-23	131^5	1.0599^{25}	1.5330^{20}	
4046	2,3-Dimethyl-2,3-diphenylbutane	Dicumene	C$_{18}$H$_{22}$	1889-67-4	238.368	cry (MeOH)	119.5				
4047	3,3'-Dimethyldiphenylmethane 4,4'-diisocyanate		C$_{17}$H$_{14}$N$_2$O$_2$	139-25-3	278.305						s chl
4048	2,9-Dimethyl-4,7-diphenyl-1,10-phenanthroline		C$_{26}$H$_{20}$N$_2$	4733-39-5	360.450		280 dec				
4049	Dimethyldiphenylsilane		C$_{14}$H$_{16}$Si	778-24-5	212.363			277; 173^{45}	0.9867^{20}	1.5644^{20}	
4050	N,N'-Dimethyl-N,N'-diphenylurea		C$_{15}$H$_{16}$N$_2$O	611-92-7	240.300	pl (al)	122	350			vs H$_2$O, EtOH, ace; sl eth, bz, CS$_2$
4051	Dimethyl disulfide	Methyl disulfide	C$_2$H$_6$S$_2$	624-92-0	94.199	liq	-84.67	109.74	1.0625^{20}	1.5289^{20}	i H$_2$O; msc EtOH, eth
4052	O,O-Dimethyl dithiophosphate	O,O-Dimethyl phosphorodithionate	C$_2$H$_7$O$_2$PS$_2$	756-80-9	158.180	liq		56^4	1.29^{20}		
4053	N,N-Dimethyldodecylamine oxide		C$_{14}$H$_{31}$NO	1643-20-5	229.402	hyg nd (tol)	130.5				
4054	1,2-Dimethylenecyclohexane		C$_8$H$_{12}$	2819-48-9	108.181			127; 60^{90}	0.8361^{20}	1.4718^{25}	i H$_2$O; s EtOH, eth, bz, chl; vs ace
4055	N,N-Dimethyl-1,2-ethanediamine		C$_4$H$_{12}$N$_2$	108-00-9	88.151			104	0.803^{25}	1.4260^{20}	
4056	N,N'-Dimethyl-1,2-ethanediamine		C$_4$H$_{12}$N$_2$	110-70-3	88.151			120	0.828^{15}		s EtOH, eth, dil HCl
4057	N,N-Dimethylethanolamine	Deanol	C$_4$H$_{11}$NO	108-01-0	89.136	liq	-59	134	0.8866^{20}	1.4300^{20}	msc H$_2$O, EtOH, eth; s chl
4058	Dimethyl ether	Methyl ether	C$_2$H$_6$O	115-10-6	46.068	col gas	-141.5	-24.8			s H$_2$O, EtOH, eth, ace, chl; sl bz
4059	(1,1-Dimethylethoxy)benzene		C$_{10}$H$_{14}$O	6669-13-2	150.217	liq	-24	185.5	0.9214^{20}		
4060	[(1,1-Dimethylethoxy)methyl]oxirane		C$_7$H$_{14}$O$_2$	7665-72-7	130.185	liq	-70	152	0.898^{20}		
4061	N,N-Dimethylformamide	DMF	C$_3$H$_7$NO	68-12-2	73.094	liq	-60.48	153	0.9445^{25}	1.4305^{20}	msc H$_2$O, EtOH, eth, ace, bz; sl lig
4062	Dimethyl fumarate		C$_6$H$_8$O$_4$	624-49-7	144.126		103.5	193	1.37^{20}	1.4062^{111}	i H$_2$O; s ace, chl
4063	2,5-Dimethylfuran		C$_6$H$_8$O	625-86-5	96.127	liq	-62.8	93	0.8883^{20}	1.4363^{20}	i H$_2$O; s EtOH, eth, ace, bz, HOAc, chl
4064	3,4-Dimethyl-2,5-furandione		C$_6$H$_6$O$_3$	766-39-2	126.110	pl or lf (dil al)	96	223	1.107^{100}		sl H$_2$O; vs EtOH, eth, bz, chl
4065	Dimethyl germanium sulfide		C$_2$H$_6$GeS	16090-49-6	134.77	col cry	54.5	302			
4066	Dimethyl glutarate		C$_7$H$_{12}$O$_4$	1119-40-0	160.168	liq	-42.5	214; 109^{21}	1.0876^{20}	1.4242^{20}	vs EtOH, eth; s chl
4067	N,N-Dimethylglycine		C$_4$H$_9$NO$_2$	1118-68-9	103.120	hyg nd (PrOH)	185.5				vs H$_2$O, MeOH; s EtOH, eth, ace
4068	Dimethylglyoxime		C$_4$H$_8$N$_2$O$_2$	95-45-4	116.119	nd (to or dil al)	245.5	sub 234			i H$_2$O; vs EtOH, eth; sl bz, tol
4069	2,6-Dimethyl-1,5-heptadiene		C$_9$H$_{16}$	6709-39-3	124.223	liq	-70	143	0.7648^{25}		
4070	2,2-Dimethylheptane		C$_9$H$_{20}$	1071-26-7	128.255	liq	-113	132.7	0.7105^{20}	1.4016^{20}	i H$_2$O; s eth, ctc; vs ace, chl; msc bz
4071	2,3-Dimethylheptane		C$_9$H$_{20}$	3074-71-3	128.255	liq	-116	140.5	0.7260^{20}	1.4088^{20}	i H$_2$O; msc EtOH, eth, ace, bz, peth, chl
4072	2,4-Dimethylheptane		C$_9$H$_{20}$	2213-23-2	128.255			132.9	0.7115^{25}	1.4034^{20}	i H$_2$O; msc EtOH, eth, ace, bz, chl, peth
4073	2,5-Dimethylheptane		C$_9$H$_{20}$	2216-30-0	128.255			136	0.7198^{20}	1.4033^{20}	vs ace, bz, eth, EtOH
4074	2,6-Dimethylheptane		C$_9$H$_{20}$	1072-05-5	128.255	liq	-102.9	135.2	0.7089^{20}	1.4011^{20}	sl chl
4075	3,3-Dimethylheptane		C$_9$H$_{20}$	4032-86-4	128.255			137.3	0.7254^{20}	1.4087^{20}	i H$_2$O; msc EtOH; s eth; vs ace, bz

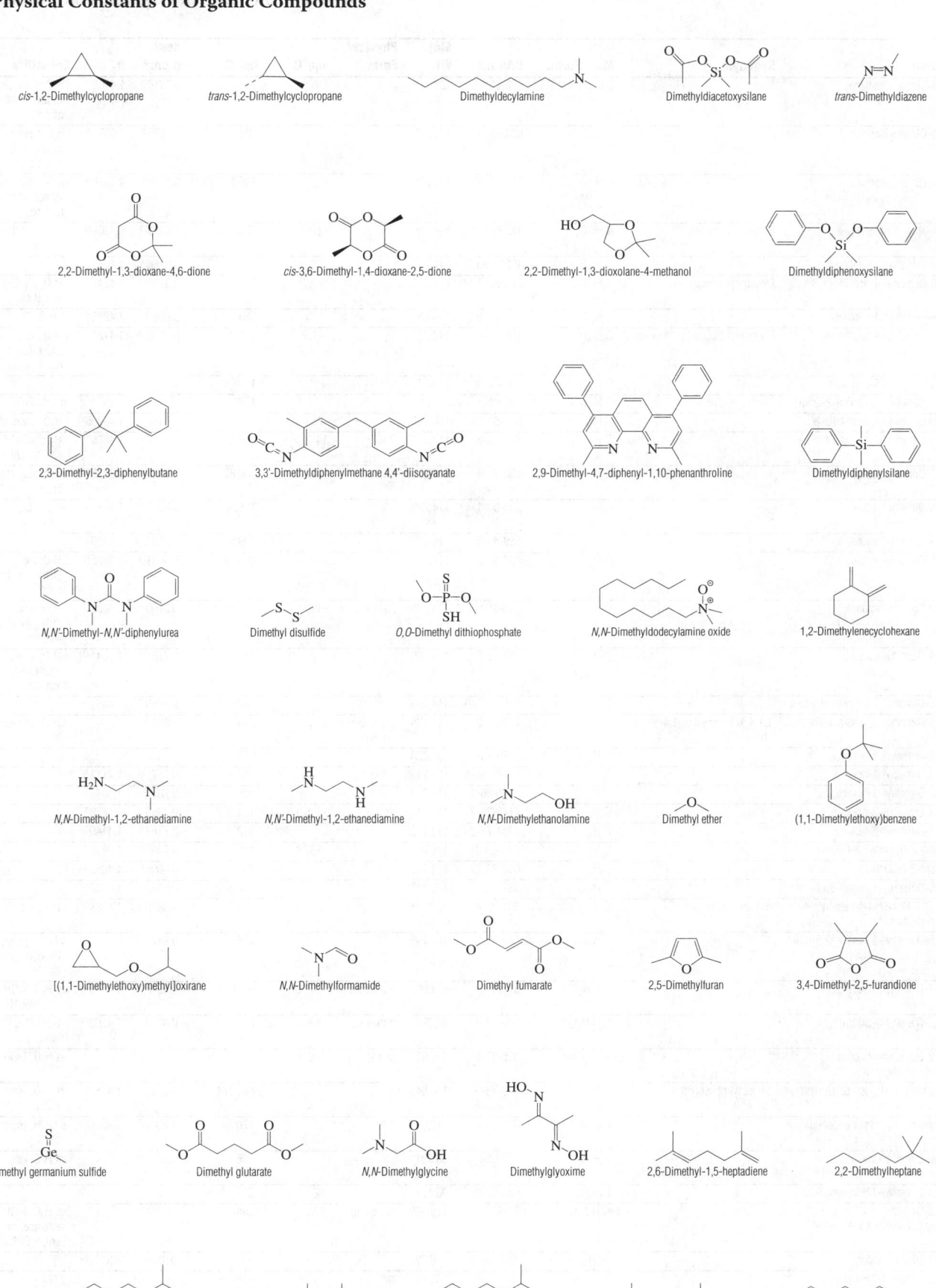

cis-1,2-Dimethylcyclopropane

trans-1,2-Dimethylcyclopropane

Dimethyldecylamine

Dimethyldiacetoxysilane

trans-Dimethyldiazene

2,2-Dimethyl-1,3-dioxane-4,6-dione

cis-3,6-Dimethyl-1,4-dioxane-2,5-dione

2,2-Dimethyl-1,3-dioxolane-4-methanol

Dimethyldiphenoxysilane

2,3-Dimethyl-2,3-diphenylbutane

3,3'-Dimethyldiphenylmethane 4,4'-diisocyanate

2,9-Dimethyl-4,7-diphenyl-1,10-phenanthroline

Dimethyldiphenylsilane

N,N'-Dimethyl-N,N'-diphenylurea

Dimethyl disulfide

O,O-Dimethyl dithiophosphate

N,N-Dimethyldodecylamine oxide

1,2-Dimethylenecyclohexane

N,N-Dimethyl-1,2-ethanediamine

N,N'-Dimethyl-1,2-ethanediamine

N,N-Dimethylethanolamine

Dimethyl ether

(1,1-Dimethylethoxy)benzene

[(1,1-Dimethylethoxy)methyl]oxirane

N,N-Dimethylformamide

Dimethyl fumarate

2,5-Dimethylfuran

3,4-Dimethyl-2,5-furandione

Dimethyl germanium sulfide

Dimethyl glutarate

N,N-Dimethylglycine

Dimethylglyoxime

2,6-Dimethyl-1,5-heptadiene

2,2-Dimethylheptane

2,3-Dimethylheptane

2,4-Dimethylheptane

2,5-Dimethylheptane

2,6-Dimethylheptane

3,3-Dimethylheptane

No.	Name	Synonym	Mol. Form.	CAS RN	Mol. Wt.	Physical Form	mp/°C	bp/°C	den/ g cm⁻³	n_D	Solubility
4076	3,4-Dimethylheptane		C₉H₂₀	922-28-1	128.255			140.6	0.7314²⁰	1.4108²⁰	i H₂O; s eth, ctc; vs ace, chl; msc bz
4077	3,5-Dimethylheptane		C₉H₂₀	926-82-9	128.255			136	0.7225²⁰	1.4083²⁰	i H₂O; s eth, ctc; vs ace, chl; msc bz
4078	4,4-Dimethylheptane		C₉H₂₀	1068-19-5	128.255			135.2	0.7221²⁰	1.4076²⁰	i H₂O; s eth, ctc; vs ace, chl; msc bz
4079	Dimethyl heptanedioate	Dimethyl pimelate	C₉H₁₆O₄	1732-08-7	188.221		-21	120¹⁰, 80¹	1.0625²⁰	1.4309²⁰	sl H₂O; s EtOH, eth, bz
4080	2,6-Dimethyl-2-heptanol		C₉H₂₀O	13254-34-7	144.254			173	0.8186²⁰	1.4242²⁰	
4081	2,6-Dimethyl-4-heptanol	Diisobutylcarbinol	C₉H₂₀O	108-82-7	144.254			174.5	0.8114²⁰	1.4242²⁰	i H₂O; s EtOH, eth; sl ctc
4082	3,5-Dimethyl-4-heptanol		C₉H₂₀O	19549-79-2	144.254			186	0.836¹⁸	1.4283²⁰	sl H₂O
4083	2,6-Dimethyl-4-heptanone	Diisobutyl ketone	C₉H₁₈O	108-83-8	142.238	liq	-41.5	169.4	0.8062²⁰	1.412²¹	i H₂O; msc EtOH, eth; s ctc
4084	2,6-Dimethyl-5-heptenal		C₉H₁₆O	106-72-9	140.222	oil		120¹⁰⁰			
4085	N,6-Dimethyl-5-hepten-2-amine	Isometheptene	C₉H₁₉N	503-01-5	141.254			177			vs eth, EtOH
4086	2,5-Dimethyl-1,5-hexadiene		C₈H₁₄	627-58-7	110.197	liq	-75.6	114.3	0.743²⁰	1.43995²¹	i H₂O; s ace, chl
4087	2,5-Dimethyl-2,4-hexadiene		C₈H₁₄	764-13-6	110.197		14	134.5	0.7577²⁵	1.4785²⁰	i H₂O; s EtOH, eth, bz, chl
4088	2,2-Dimethylhexane		C₈H₁₈	590-73-8	114.229	liq	-121.1	106.86	0.6953²⁰	1.3935²⁰	vs ace, bz, eth, EtOH
4089	2,3-Dimethylhexane		C₈H₁₈	584-94-1	114.229			115.62	0.6912²⁵	1.4011²⁰	vs ace, bz, EtOH, lig
4090	2,4-Dimethylhexane		C₈H₁₈	589-43-5	114.229			109.5	0.6962²⁵	1.3929²⁵	
4091	2,5-Dimethylhexane		C₈H₁₈	592-13-2	114.229	liq	-91	109.12	0.6901²⁵	1.3925²⁰	i H₂O; msc EtOH, ace, bz; s eth
4092	3,3-Dimethylhexane		C₈H₁₈	563-16-6	114.229	liq	-126.1	111.97	0.7100²⁰	1.4001²⁰	i H₂O; msc EtOH; vs eth, ace, bz
4093	3,4-Dimethylhexane		C₈H₁₈	583-48-2	114.229			117.73	0.7151²⁵	1.4041²⁰	i H₂O; s eth; msc EtOH, ace, bz
4094	2,5-Dimethyl-2,5-hexanediamine		C₈H₂₀N₂	23578-35-0	144.258			184; 63⁸	0.8485¹⁵	1.4459²⁰	
4095	2,5-Dimethyl-2,5-hexanediol	1,1,4,4-Tetramethyl-1,4-butanediol	C₈H₁₈O₂	110-03-2	146.228	pr (AcOEt) fl (peth)	88.50	214	0.898²⁰		s H₂O; vs EtOH, bz, chl
4096	2,2-Dimethyl-1-hexanol		C₈H₁₈O	2370-13-0	130.228			95²⁹			
4097	2,3-Dimethyl-1-hexene		C₈H₁₆	16746-86-4	112.213			110.5	0.7172²⁵	1.4113²⁰	
4098	5,5-Dimethyl-1-hexene		C₈H₁₆	7116-86-1	112.213			104	0.705²⁵	1.4049²⁰	
4099	2,3-Dimethyl-2-hexene		C₈H₁₆	7145-20-2	112.213	liq	-115.1	121.8	0.7366²⁵	1.4268²⁰	
4100	2,5-Dimethyl-2-hexene		C₈H₁₆	3404-78-2	112.213			112.2	0.7182²⁰	1.4140²⁰	
4101	cis-2,2-Dimethyl-3-hexene		C₈H₁₆	690-92-6	112.213	liq	-137.4	105.5	0.7086²⁵	1.4099²⁰	
4102	trans-2,2-Dimethyl-3-hexene		C₈H₁₆	690-93-7	112.213			100.8	0.6995²⁵	1.4063²⁰	
4103	3,5-Dimethyl-1-hexen-3-ol		C₈H₁₆O	3329-48-4	128.212			146.5	0.8382²⁰	1.4342²⁰	
4104	1-(1,5-Dimethyl-4-hexenyl)-4-methylbenzene	α-Curcumene	C₁₅H₂₂	644-30-4	202.336			140¹⁹	0.8805²⁰	1.4989²⁰	i H₂O; s bz
4105	2,5-Dimethyl-3-hexyne-2,5-diol		C₈H₁₄O₂	142-30-3	142.196		95	205	0.947²⁰		s H₂O, chl; vs EtOH, eth, ace, bz
4106	1,1-Dimethylhydrazine		C₂H₈N₂	57-14-7	60.098	liq, fumes in air	-57.20	63.9	0.791²²	1.4075²²	vs H₂O, EtOH, eth, MeOH
4107	1,2-Dimethylhydrazine		C₂H₈N₂	540-73-8	60.098	fumes (air)	-8.9	81	0.8274²⁰	1.4209²⁰	msc H₂O, EtOH, eth
4108	1,2-Dimethylhydrazine dihydrochloride		C₂H₁₀Cl₂N₂	306-37-6	133.019	pr (w)	170 dec				vs H₂O, EtOH
4109	Dimethyl hydrogen phosphate	Dimethyl phosphate	C₂H₇O₄P	813-78-5	126.048			dec 174	1.3225²⁰	1.408²⁵	vs H₂O, ace, EtOH
4110	Dimethyl hydrogen phosphite		C₂H₇O₃P	868-85-9	110.049			170.5	1.2002²⁰	1.4036²⁰	s EtOH, py; sl ctc
4111	1,2-Dimethyl-1H-imidazole		C₅H₈N₂	1739-84-0	96.131			206	1.0051¹¹		vs H₂O, eth, EtOH
4112	2,4-Dimethyl-1H-imidazole		C₅H₈N₂	930-62-1	96.131		92	267			
4113	5,5-Dimethyl-2,4-imidazolidinedione		C₅H₈N₂O₂	77-71-4	128.130	pr (dil al)	178	sub			vs H₂O, EtOH, eth, ace, bz, chl; s DMSO
4114	1,1-Dimethylindan		C₁₁H₁₄	4912-92-9	146.229			191	0.919²⁰	1.5135²⁵	
4115	1,3-Dimethyl-1H-indole		C₁₀H₁₁N	875-30-9	145.201	nd	142	258.5			s eth
4116	2,3-Dimethyl-1H-indole		C₁₀H₁₁N	91-55-4	145.201		107.5	287			
4117	N,N-Dimethyl-1H-indole-3-ethanamine	N,N-Dimethyltryptamine	C₁₂H₁₆N₂	61-50-7	188.268		46				

3,4-Dimethylheptane

3,5-Dimethylheptane

4,4-Dimethylheptane

Dimethyl heptanedioate

2,6-Dimethyl-2-heptanol

2,6-Dimethyl-4-heptanol

3,5-Dimethyl-4-heptanol

2,6-Dimethyl-4-heptanone

2,6-Dimethyl-5-heptenal

N,6-Dimethyl-5-hepten-2-amine

2,5-Dimethyl-1,5-hexadiene

2,5-Dimethyl-2,4-hexadiene

2,2-Dimethylhexane

2,3-Dimethylhexane

2,4-Dimethylhexane

2,5-Dimethylhexane

3,3-Dimethylhexane

3,4-Dimethylhexane

2,5-Dimethyl-2,5-hexanediamine

2,5-Dimethyl-2,5-hexanediol

2,2-Dimethyl-1-hexanol

2,3-Dimethyl-1-hexene

5,5-Dimethyl-1-hexene

2,3-Dimethyl-2-hexene

2,5-Dimethyl-2-hexene

cis-2,2-Dimethyl-3-hexene

trans-2,2-Dimethyl-3-hexene

3,5-Dimethyl-1-hexen-3-ol

1-(1,5-Dimethyl-4-hexenyl)-4-methylbenzene

2,5-Dimethyl-3-hexyne-2,5-diol

1,1-Dimethylhydrazine

1,2-Dimethylhydrazine

1,2-Dimethylhydrazine dihydrochloride

Dimethyl hydrogen phosphate

Dimethyl hydrogen phosphite

1,2-Dimethyl-1H-imidazole

2,4-Dimethyl-1H-imidazole

5,5-Dimethyl-2,4-imidazolidinedione

1,1-Dimethylindan

1,3-Dimethyl-1H-indole

2,3-Dimethyl-1H-indole

N,N-Dimethyl-1H-indole-3-ethanamine

No.	Name	Synonym	Mol. Form.	CAS RN	Mol. Wt.	Physical Form	mp/°C	bp/°C	den/ g cm⁻³	n_D	Solubility
4118	N,N-Dimethyl-1H-indole-3-methanamine	Gramine	$C_{11}H_{14}N_2$	87-52-5	174.242	nd or pl (ace)	138.5				i H₂O; s EtOH, eth, chl; i peth
4119	Dimethyl isophthalate		$C_{10}H_{10}O_4$	1459-93-4	194.184	nd(dil al)	67.5	282	1.194²⁰	1.5168²⁰	sl H₂O
4120	1,4-Dimethyl-7-isopropylazulene	Guaiazulene	$C_{15}H_{18}$	489-84-9	198.304	bl-viol pl (al)	31.5	167¹²	0.973²⁰		s EtOH, eth, AcOEt
4121	1,6-Dimethyl-4-isopropylnaphthalene	Cadalene	$C_{15}H_{18}$	483-78-3	198.304			294; 149¹⁰	0.9667²⁵	1.5785²⁵	vs oils
4122	2,4-Dimethyl-3-isopropylpentane		$C_{10}H_{22}$	13475-79-1	142.282	liq	-81.7	157.1	0.7545²⁵	1.4246²⁰	
4123	3,5-Dimethylisoxazole		C_5H_7NO	300-87-8	97.116			143	0.99²⁵	1.4421²⁰	
4124	Dimethylmagnesium	Magnesium dimethyl	C_2H_6Mg	2999-74-8	54.374	solid	220 dec	subl			
4125	Dimethyl maleate	Methyl cis-butenedioate	$C_6H_8O_4$	624-48-6	144.126	liq	-19	202	1.1606²⁰	1.4416²⁰	sl H₂O, lig; s eth, ctc
4126	Dimethyl malonate	Methyl malonate	$C_5H_8O_4$	108-59-8	132.116	liq	-61.9	181.4	1.528²⁰	1.4135²⁰	sl H₂O; msc EtOH; vs ace, bz; s chl
4127	Dimethylmalonic acid	Dimethylpropanedioc acid	$C_5H_8O_4$	595-46-0	132.116	pr (bz/peth)	192.5	subl			s hot H₂O
4128	Dimethyl mercury		C_2H_6Hg	593-74-8	230.66			93	3.17²⁵	1.5452²⁰	i H₂O; vs EtOH, eth
4129	Dimethyl cis-2-methyl-2-butenedioate	Dimethyl citraconate	$C_7H_{10}O_4$	617-54-9	158.152			210.5	1.1153²⁰	1.4473²⁰	vs ace, eth, EtOH
4130	Dimethyl methylenesuccinate		$C_7H_{10}O_4$	617-52-7	158.152	hyg mcl (MeOH)	38	208	1.1241¹⁸	1.4457²⁰	s EtOH, eth, MeOH; vs ace
4131	Dimethyl methylmalonate		$C_6H_{10}O_4$	609-02-9	146.141			174	1.0977²⁰	1.4128²⁰	vs ace, eth, EtOH, chl
4132	Dimethyl methylphosphonate		$C_3H_9O_3P$	756-79-6	124.075			181; 79.5²⁰	1.1684²⁰	1.4099³⁰	s H₂O, EtOH, eth
4133	trans-2,2-Dimethyl-3-(2-methyl-1-propenyl)cyclopropanecarboxylic acid		$C_{10}H_{16}O_2$	4638-92-0	168.233	pr	20.0	245			vs eth, EtOH, chl
4134	Dimethyl 2-methylsuccinate		$C_7H_{12}O_4$	1604-11-1	160.168			196	1.076²⁵	1.4200²⁰	
4135	Dimethyl p-(methylthio)phenyl phosphate		$C_9H_{13}O_4PS$	3254-63-5	248.235	liq			1.273²¹		sl H₂O; s ace, EtOH, diox, ctc, xyl
4136	2,6-Dimethylmorpholine		$C_6H_{13}NO$	141-91-3	115.173	liq	-88	146.6	0.9329²⁰	1.4460²⁰	msc H₂O, EtOH, bz, lig; s ace; sl chl
4137	Dimethyl morpholinophosphoramidate	Dimethyl 4-morpholinylphosphonate	$C_6H_{14}NO_4P$	597-25-1	195.153	liq		96¹			
4138	1,2-Dimethylnaphthalene		$C_{12}H_{12}$	573-98-8	156.223		0.8	266.5	1.0179²⁰	1.6166²⁰	i H₂O; s eth, bz
4139	1,3-Dimethylnaphthalene		$C_{12}H_{12}$	575-41-7	156.223	liq	-6	263	1.0144²⁰	1.6140²⁰	i H₂O; s eth, bz
4140	1,4-Dimethylnaphthalene		$C_{12}H_{12}$	571-58-4	156.223		7.6	268	1.0166²⁰	1.6127²⁰	i H₂O; vs EtOH; msc eth, ace, bz, ctc
4141	1,5-Dimethylnaphthalene		$C_{12}H_{12}$	571-61-9	156.223		82	265			i H₂O; vs bz, eth
4142	1,6-Dimethylnaphthalene		$C_{12}H_{12}$	575-43-9	156.223	liq	-16.9	264	1.0021²⁰	1.6166²⁰	i H₂O; s eth, bz
4143	1,7-Dimethylnaphthalene		$C_{12}H_{12}$	575-37-1	156.223	liq	-13.9	263	1.0115²⁰	1.6083²⁰	i H₂O; s eth, bz
4144	1,8-Dimethylnaphthalene		$C_{12}H_{12}$	569-41-5	156.223		65	270	1.003²⁰		i H₂O; s eth, bz
4145	2,3-Dimethylnaphthalene	Guajen	$C_{12}H_{12}$	581-40-8	156.223	lf (al)	105	268	1.003²⁰	1.5060²⁰	i H₂O; vs bz, eth
4146	2,6-Dimethylnaphthalene		$C_{12}H_{12}$	581-42-0	156.223		112	262	1.003²⁰		i H₂O
4147	2,7-Dimethylnaphthalene		$C_{12}H_{12}$	582-16-1	156.223		97	265	1.003²⁰		
4148	N,N-Dimethyl-1-naphthylamine		$C_{12}H_{13}N$	86-56-6	171.238	viol flr cry		250; 140¹³	1.0423²⁰	1.624¹⁵	i H₂O; s EtOH, eth, ctc
4149	N,N-Dimethyl-2-naphthylamine		$C_{12}H_{13}N$	2436-85-3	171.238	dk red nd	52.5	305	1.0279⁶⁰	1.6443⁵³	i H₂O; s EtOH, eth
4150	N,N-Dimethyl-2-nitroaniline		$C_8H_{10}N_2O_2$	610-17-3	166.177	ye-oran	-20	146²⁰	1.1794²⁰	1.6102²⁰	s H₂O, eth; vs EtOH, chl
4151	N,N-Dimethyl-3-nitroaniline		$C_8H_{10}N_2O_2$	619-31-8	166.177	red mcl pr (eth)	60.5	282.5	1.313¹⁷		i H₂O; s EtOH, eth
4152	N,N-Dimethyl-4-nitroaniline		$C_8H_{10}N_2O_2$	100-23-2	166.177	ye nd (al)	164.5				i H₂O; s EtOH, eth, HOAc
4153	1,2-Dimethyl-3-nitrobenzene		$C_8H_9NO_2$	83-41-0	151.163	nd (al)	15	240	1.1402²⁰	1.5441²⁰	i H₂O; s EtOH, ctc
4154	1,2-Dimethyl-4-nitrobenzene	4-Nitro-o-xylene	$C_8H_9NO_2$	99-51-4	151.163	ye pr (al)	30.5	251; 143²¹	1.112¹⁵	1.5202²⁰	i H₂O; msc EtOH
4155	1,3-Dimethyl-2-nitrobenzene		$C_8H_9NO_2$	81-20-9	151.163		15	226	1.112¹⁵	1.5202²⁰	i H₂O; vs EtOH; s ctc
4156	1,3-Dimethyl-5-nitrobenzene		$C_8H_9NO_2$	99-12-7	151.163	nd (al)	75	274			i H₂O; vs EtOH, eth
4157	1,4-Dimethyl-2-nitrobenzene		$C_8H_9NO_2$	89-58-7	151.163	pa ye liq	-25	240.5	1.132¹⁵	1.5413²⁰	i H₂O; s EtOH
4158	2,4-Dimethyl-1-nitrobenzene		$C_8H_9NO_2$	89-87-2	151.163		9	247; 122¹⁸	1.135¹⁵	1.5473²⁵	i H₂O; s eth, ace, bz, chl
4159	1,2-Dimethyl-5-nitro-1H-imidazole	Dimetridazole	$C_5H_7N_3O_2$	551-92-8	141.129	nd (w)	138.5				vs eth, EtOH
4160	N,N-Dimethyl-4-[2-(4-nitrophenyl)ethenyl]aniline		$C_{16}H_{16}N_2O_2$	4584-57-0	268.310		258.3				

N,N-Dimethyl-1*H*-indole-3-methanamine

Dimethyl isophthalate

1,4-Dimethyl-7-isopropylazulene

1,6-Dimethyl-4-isopropylnaphthalene

2,4-Dimethyl-3-isopropylpentane

3,5-Dimethylisoxazole

Dimethylmagnesium

Dimethyl maleate

Dimethyl malonate

Dimethylmalonic acid

Dimethyl mercury

Dimethyl *cis*-2-methyl-2-butenedioate

Dimethyl methylenesuccinate

Dimethyl methylmalonate

Dimethyl methylphosphonate

trans-2,2-Dimethyl-3-(2-methyl-1-propenyl)cyclopropanecarboxylic acid

Dimethyl 2-methylsuccinate

Dimethyl *p*-(methylthio)phenyl phosphate

2,6-Dimethylmorpholine

Dimethyl morpholinophosphoramidate

1,2-Dimethylnaphthalene

1,3-Dimethylnaphthalene

1,4-Dimethylnaphthalene

1,5-Dimethylnaphthalene

1,6-Dimethylnaphthalene

1,7-Dimethylnaphthalene

1,8-Dimethylnaphthalene

2,3-Dimethylnaphthalene

2,6-Dimethylnaphthalene

2,7-Dimethylnaphthalene

N,N-Dimethyl-1-naphthylamine

N,N-Dimethyl-2-naphthylamine

N,N-Dimethyl-2-nitroaniline

N,N-Dimethyl-3-nitroaniline

N,N-Dimethyl-4-nitroaniline

1,2-Dimethyl-3-nitrobenzene

1,2-Dimethyl-4-nitrobenzene

1,3-Dimethyl-2-nitrobenzene

1,3-Dimethyl-5-nitrobenzene

1,4-Dimethyl-2-nitrobenzene

2,4-Dimethyl-1-nitrobenzene

1,2-Dimethyl-5-nitro-1*H*-imidazole

N,N-Dimethyl-4-[2-(4-nitrophenyl)ethenyl]aniline

No.	Name	Synonym	Mol. Form.	CAS RN	Mol. Wt.	Physical Form	mp/°C	bp/°C	den/ g cm⁻³	n_D	Solubility
4161	N,4-Dimethyl-N-nitrosobenzenesulfonamide	p-Tolylsulfonylmethylnitrosamide	$C_8H_{10}N_2O_3S$	80-11-5	214.241	cry	60				i H_2O; vs EtOH, eth
4162	Dimethyl nonanedioate	Methyl azelate	$C_{11}H_{20}O_4$	1732-10-1	216.275		-0.8	156²⁰	1.0082²⁰	1.4367²⁰	i H_2O; s EtOH, ace, bz, ctc
4163	6,6-Dimethyl-2-norpinene-2-carboxaldehyde	Myrtenal	$C_{10}H_{14}O$	564-94-3	150.217	unstab oil		99¹⁵			
4164	cis-3,7-Dimethyl-2,6-octadienal		$C_{10}H_{16}O$	106-26-3	152.233			120²⁰	0.8869²⁰	1.4869²⁰	i H_2O; msc EtOH, eth
4165	trans-3,7-Dimethyl-2,6-octadienal		$C_{10}H_{16}O$	141-27-5	152.233			229	0.8888²⁰	1.4898²⁰	i H_2O; msc EtOH, eth
4166	3,7-Dimethyl-1,6-octadiene	Citronellene	$C_{10}H_{18}$	2436-90-0	138.250				0.7601²⁰	1.4362²⁰	
4167	3,7-Dimethyl-2,6-octadienoic acid	Geranic acid	$C_{10}H_{16}O_2$	459-80-3	168.233	oil					
4168	cis-3,7-Dimethyl-2,6-octadien-1-ol	Nerol	$C_{10}H_{18}O$	106-25-2	154.249		<-15	225; 125²⁵	0.8756²⁰	1.4746²⁰	vs EtOH
4169	cis-3,7-Dimethyl-2,6-octadien-1-ol acetate		$C_{12}H_{20}O_2$	141-12-8	196.286			134²⁵, 93³	0.905¹⁵	1.452²⁰	
4170	trans-3,7-Dimethyl-2,6-octadien-1-ol formate		$C_{11}H_{18}O_2$	105-86-2	182.260			dec 229; 113²⁵	0.9086²⁵	1.4659²⁰	i H_2O; vs EtOH; s eth, ace
4171	2,2-Dimethyloctane		$C_{10}H_{22}$	15869-87-1	142.282			155	0.7208²⁵	1.4082²⁰	
4172	2,3-Dimethyloctane		$C_{10}H_{22}$	7146-60-3	142.282			164.3	0.7377²⁰	1.4146²⁰	
4173	2,4-Dimethyloctane		$C_{10}H_{22}$	4032-94-4	142.282			156	0.7226²⁵	1.4091²⁰	
4174	2,5-Dimethyloctane		$C_{10}H_{22}$	15869-89-3	142.282			158.5	0.7264²⁵	1.4112²⁰	
4175	2,6-Dimethyloctane		$C_{10}H_{22}$	2051-30-1	142.282			160.4	0.7313²⁰	1.4097²⁰	
4176	2,7-Dimethyloctane		$C_{10}H_{22}$	1072-16-8	142.282	liq	-54.9	159.9	0.7202²⁵	1.4086²⁰	s eth, HOAc
4177	3,4-Dimethyloctane		$C_{10}H_{22}$	15869-92-8	142.282			163.4	0.7410²⁵	1.4182²⁰	
4178	3,6-Dimethyloctane		$C_{10}H_{22}$	15869-94-0	142.282			160.8	0.7324²⁵	1.4139²⁰	
4179	Dimethyl octanedioate	Dimethyl suberate	$C_{10}H_{18}O_4$	1732-09-8	202.248	liq	-1.6	268	1.0217²⁰	1.4341²⁰	i H_2O; s EtOH, eth, ace; sl ctc
4180	3,7-Dimethyl-1,7-octanediol		$C_{10}H_{22}O_2$	107-74-4	174.281			265	0.937²⁰	1.4599²⁰	sl bz, tol
4181	2,2-Dimethyloctanoic acid		$C_{10}H_{20}O_2$	29662-90-6	172.265			140¹³			
4182	2,2-Dimethyl-1-octanol		$C_{10}H_{22}O$	2370-14-1	158.281	liq		97	0.84²⁰		
4183	3,7-Dimethyl-1-octanol		$C_{10}H_{22}O$	106-21-8	158.281			212.5	0.832²⁵	1.438²⁵	s eth
4184	2,6-Dimethyl-2-octanol	Tetrahydromyrcenol	$C_{10}H_{22}O$	18479-57-7	158.281			80.5¹⁰	0.8023²⁵	1.4220²⁵	
4185	3,6-Dimethyl-3-octanol		$C_{10}H_{22}O$	151-19-9	158.281	liq	-67.5	202.2	0.8347²²	1.4370²⁰	
4186	3,7-Dimethyl-3-octanol		$C_{10}H_{22}O$	78-69-3	158.281			205.1	0.826²⁵	1.433²⁵	
4187	cis-3,7-Dimethyl-1,3,6-octatriene	cis-β-Ocimene	$C_{10}H_{16}$	3338-55-4	136.234				0.799²⁰		
4188	trans-3,7-Dimethyl-1,3,6-octatriene	trans-β-Ocimene	$C_{10}H_{16}$	3779-61-1	136.234				0.799²⁰		
4189	3,7-Dimethyl-1,3,7-octatriene	α-Ocimene	$C_{10}H_{16}$	502-99-8	136.234			dec 177	0.8000²⁰	1.4862²⁰	i H_2O; s EtOH, eth, chl, HOAc
4190	cis,cis-2,6-Dimethyl-2,4,6-octatriene	cis-allo-Ocimene	$C_{10}H_{16}$	17202-20-9	136.234	liq					
4191	trans,trans-2,6-Dimethyl-2,4,6-octatriene	trans-allo-Ocimene	$C_{10}H_{16}$	3016-19-1	136.234	liq	-35.4	188; 91²⁰	0.8118²⁰	1.5446²⁰	
4192	3,7-Dimethyl-6-octenal	Citronellal	$C_{10}H_{18}O$	106-23-0	154.249	nd or orth cry		207.5	0.853²⁰	1.4473²⁰	sl H_2O; s EtOH
4193	3,7-Dimethyl-1-octene		$C_{10}H_{20}$	4984-01-4	140.266	col liq		154	0.7396²⁰	1.4212²⁰	
4194	3,7-Dimethyl-6-octenoic acid	Citronellic acid	$C_{10}H_{18}O_2$	502-47-6	170.249			257; 157²³	0.9234²¹		
4195	3,7-Dimethyl-6-octen-1-ol, (R)	Citronellol, (+)	$C_{10}H_{20}O$	1117-61-9	156.265	oil		224; 108¹⁰	0.8550²⁰	1.4565²⁰	sl H_2O; msc EtOH, eth
4196	3,7-Dimethyl-6-octen-1-ol, (S)	Citronellol, (-)	$C_{10}H_{20}O$	7540-51-4	156.265	oil		224; 108¹⁰	0.859¹⁸	1.4576¹⁸	vs eth, EtOH
4197	3,7-Dimethyl-7-octen-1-ol, (S)	Rhodinol	$C_{10}H_{20}O$	6812-78-8	156.265			114¹²	0.8549²⁰	1.4556²⁰	vs eth, EtOH
4198	3,7-Dimethyl-6-octen-3-ol		$C_{10}H_{20}O$	18479-51-1	156.265			94¹⁴	0.8695¹⁵	1.4569¹⁵	
4199	3,7-Dimethyl-6-octen-1-ol, acetate	Citronellol acetate	$C_{12}H_{22}O_2$	150-84-5	198.302			115¹⁰			
4200	Dimethyloldihydroxyethyleneurea	4,5-Dihydroxy-1,3-bis(hydroxymethyl)-2-imidazolidinone	$C_5H_{10}N_2O_5$	1854-26-8	178.143	hyg cry					
4201	Dimethyl oxalate		$C_4H_6O_4$	553-90-2	118.089	mcl tab	54.8	163.5	1.1716⁶⁰	1.379⁸²	sl H_2O; s EtOH, eth, ace, chl
4202	5,5-Dimethyl-2,4-oxazolidinedione	Dimethadione	$C_5H_7NO_3$	695-53-4	129.115		76.5				
4203	3,3-Dimethyloxetane		$C_5H_{10}O$	6921-35-3	86.132			80.6	0.834²⁵	1.3965²⁰	
4204	3,3-Dimethyl-2-oxetanone		$C_5H_8O_2$	1955-45-9	100.117			58¹⁵			
4205	2,2-Dimethyloxirane	2-Methyl-1,2-epoxypropane	C_4H_8O	558-30-5	72.106			52	0.8112²⁰	1.3712²⁰	s EtOH, eth
4206	cis-2,3-Dimethyloxirane		C_4H_8O	1758-33-4	72.106	liq	-80	60	0.8226²⁵	1.3802²⁰	vs eth, ace, bz
4207	trans-2,3-Dimethyloxirane		C_4H_8O	6189-41-9	72.106	liq	-85	56.5	0.8010²⁵	1.3736²⁰	vs eth, ace, bz
4208	3,3-Dimethyl-2-oxobutanoic acid		$C_6H_{10}O_3$	815-17-8	130.141		90.5	189; 80¹⁵			sl H_2O; s eth, bz, chl, CS_2
4209	N-(1,1-Dimethyl-3-oxobutyl)-2-propenamide	Diacetone acrylamide	$C_9H_{15}NO_2$	2873-97-4	169.221						s chl
4210	Dimethyl 3-oxo-1,5-pentanedioate	Dimethyl 1,3-acetonedicarboxylate	$C_7H_{10}O_5$	1830-54-2	174.151			150²⁵, 77⁰·⁶	1.185²⁵	1.4434²⁰	

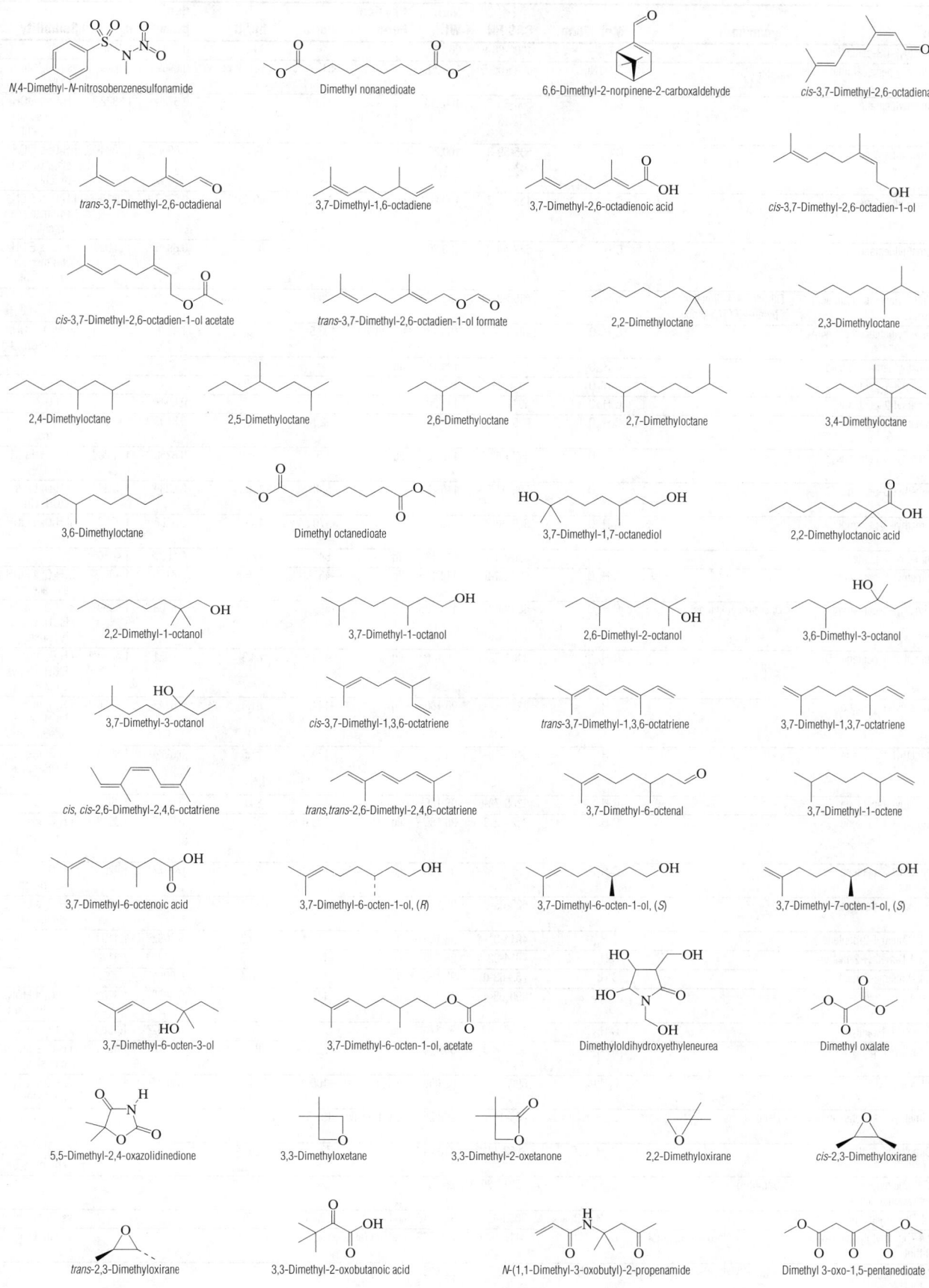

N,4-Dimethyl-N-nitrosobenzenesulfonamide

Dimethyl nonanedioate

6,6-Dimethyl-2-norpinene-2-carboxaldehyde

cis-3,7-Dimethyl-2,6-octadienal

trans-3,7-Dimethyl-2,6-octadienal

3,7-Dimethyl-1,6-octadiene

3,7-Dimethyl-2,6-octadienoic acid

cis-3,7-Dimethyl-2,6-octadien-1-ol

cis-3,7-Dimethyl-2,6-octadien-1-ol acetate

trans-3,7-Dimethyl-2,6-octadien-1-ol formate

2,2-Dimethyloctane

2,3-Dimethyloctane

2,4-Dimethyloctane

2,5-Dimethyloctane

2,6-Dimethyloctane

2,7-Dimethyloctane

3,4-Dimethyloctane

3,6-Dimethyloctane

Dimethyl octanedioate

3,7-Dimethyl-1,7-octanediol

2,2-Dimethyloctanoic acid

2,2-Dimethyl-1-octanol

3,7-Dimethyl-1-octanol

2,6-Dimethyl-2-octanol

3,6-Dimethyl-3-octanol

3,7-Dimethyl-3-octanol

cis-3,7-Dimethyl-1,3,6-octatriene

trans-3,7-Dimethyl-1,3,6-octatriene

3,7-Dimethyl-1,3,7-octatriene

cis, cis-2,6-Dimethyl-2,4,6-octatriene

trans,trans-2,6-Dimethyl-2,4,6-octatriene

3,7-Dimethyl-6-octenal

3,7-Dimethyl-1-octene

3,7-Dimethyl-6-octenoic acid

3,7-Dimethyl-6-octen-1-ol, (R)

3,7-Dimethyl-6-octen-1-ol, (S)

3,7-Dimethyl-7-octen-1-ol, (S)

3,7-Dimethyl-6-octen-3-ol

3,7-Dimethyl-6-octen-1-ol, acetate

Dimethyloldihydroxyethyleneurea

Dimethyl oxalate

5,5-Dimethyl-2,4-oxazolidinedione

3,3-Dimethyloxetane

3,3-Dimethyl-2-oxetanone

2,2-Dimethyloxirane

cis-2,3-Dimethyloxirane

trans-2,3-Dimethyloxirane

3,3-Dimethyl-2-oxobutanoic acid

N-(1,1-Dimethyl-3-oxobutyl)-2-propenamide

Dimethyl 3-oxo-1,5-pentanedioate

No.	Name	Synonym	Mol. Form.	CAS RN	Mol. Wt.	Physical Form	mp/°C	bp/°C	den/ g cm⁻³	n_D	Solubility
4211	2,4-Dimethyl-1,3-pentadiene		C₇H₁₂	1000-86-8	96.170	liq	-114	93.2	0.7343[23]	1.4390[23]	
4212	N,N-Dimethylpentanamide		C₇H₁₅NO	6225-06-5	129.200		-51	141[100]	0.8962[25]	1.4419[25]	vs H₂O, eth, EtOH
4213	2,2-Dimethylpentane		C₇H₁₆	590-35-2	100.202	liq	-123.7	79.2	0.6739[20]	1.3822[20]	i H₂O; s EtOH, eth; msc ace, bz, hp, chl
4214	2,3-Dimethylpentane		C₇H₁₆	565-59-3	100.202			89.78	0.6908[25]	1.3894[25]	i H₂O; s EtOH, eth; msc ace, bz, chl
4215	2,4-Dimethylpentane		C₇H₁₆	108-08-7	100.202	liq	-119.2	80.49	0.6727[20]	1.3815[20]	i H₂O; s EtOH, eth; msc ace, bz, chl, hp
4216	3,3-Dimethylpentane		C₇H₁₆	562-49-2	100.202	liq	-134.4	86.06	0.6936[20]	1.3909[20]	i H₂O; s EtOH, eth; msc ace, bz, hp, chl
4217	3,3-Dimethylpentanedioic acid anhydride	Dihydro-4,4-dimethyl-2H-pyran-2,6(3H)-dione	C₇H₁₀O₃	4160-82-1	142.152		125.8	181[25], 156[20]			
4218	3,3-Dimethylpentanedioic acid		C₇H₁₂O₄	4839-46-7	160.168	mcl pl, nd (bz)	103.5	126[415], 89[2]	1.4278[20]		vs H₂O, EtOH, eth; sl bz; i lig
4219	2,2-Dimethylpentanoic acid		C₇H₁₄O₂	1185-39-3	130.185	liq		98[9]	0.9189[20]		
4220	2,2-Dimethyl-1-pentanol		C₇H₁₆O	2370-12-9	116.201						s chl
4221	2,3-Dimethyl-2-pentanol		C₇H₁₆O	4911-70-0	116.201				0.804[20]		sl H₂O
4222	2,4-Dimethyl-2-pentanol		C₇H₁₆O	625-06-9	116.201		<-20	133.1	0.8103[20]	1.4172[20]	sl H₂O; s EtOH, eth, ctc
4223	2,2-Dimethyl-3-pentanol		C₇H₁₆O	3970-62-5	116.201	liq	-2.5	135	0.8253[20]	1.4223[20]	i H₂O; s EtOH, eth
4224	2,3-Dimethyl-3-pentanol		C₇H₁₆O	595-41-5	116.201		<-30	139.7	0.833[20]	1.4287[20]	sl H₂O, bz; s EtOH, eth
4225	2,4-Dimethyl-3-pentanol		C₇H₁₆O	600-36-2	116.201		<-70	138.7	0.8288[20]	1.4250[20]	sl H₂O; s EtOH, eth
4226	4,4-Dimethyl-2-pentanone		C₇H₁₄O	590-50-1	114.185	liq	-64	126	0.809[25]	1.4036[20]	
4227	2,2-Dimethyl-3-pentanone		C₇H₁₄O	564-04-5	114.185	liq	-45	125.6	0.8125[20]	1.4065[20]	sl H₂O; s EtOH, eth, ace, chl
4228	2,4-Dimethyl-3-pentanone	Diisopropyl ketone	C₇H₁₄O	565-80-0	114.185	liq	-69	125.4	0.8108[20]	1.3999[20]	sl H₂O; msc EtOH, eth; s bz; sl ctc
4229	2,3-Dimethyl-1-pentene		C₇H₁₄	3404-72-6	98.186	liq	-134.3	84.3	0.7051[20]	1.4033[20]	i H₂O; msc EtOH, eth; vs dil sulf
4230	2,4-Dimethyl-1-pentene		C₇H₁₄	2213-32-3	98.186	liq	-124.1	81.6	0.6943[20]	1.3986[20]	i H₂O; msc EtOH, eth; s bz, ctc, chl
4231	3,3-Dimethyl-1-pentene		C₇H₁₄	3404-73-7	98.186	liq	-134.3	77.5	0.6974[20]	1.3984[20]	i H₂O; msc EtOH, eth; s bz, chl
4232	3,4-Dimethyl-1-pentene		C₇H₁₄	7385-78-6	98.186			80.8	0.6934[25]	1.3992[20]	
4233	4,4-Dimethyl-1-pentene		C₇H₁₄	762-62-9	98.186	liq	-136.6	72.5	0.6827[20]	1.3818[20]	i H₂O; msc EtOH, eth; s bz, ctc, chl
4234	2,3-Dimethyl-2-pentene		C₇H₁₄	10574-37-5	98.186	liq	-118.3	97.5	0.7277[20]	1.4208[20]	i H₂O; s EtOH, eth, bz, chl
4235	2,4-Dimethyl-2-pentene		C₇H₁₄	625-65-0	98.186	liq	-127.7	83.4	0.6954[20]	1.4040[20]	i H₂O; s EtOH, eth, bz, ctc
4236	cis-3,4-Dimethyl-2-pentene		C₇H₁₄	4914-91-4	98.186	liq	-113.4	89.3	0.7092[25]	1.4104[20]	
4237	trans-3,4-Dimethyl-2-pentene		C₇H₁₄	4914-92-5	98.186	liq	-124.2	91.5	0.7124[25]	1.4128[20]	
4238	cis-4,4-Dimethyl-2-pentene		C₇H₁₄	762-63-0	98.186	liq	-135.4	80.4	0.6951[25]	1.4026[20]	
4239	trans-4,4-Dimethyl-2-pentene		C₇H₁₄	690-08-4	98.186	liq	-115.2	76.7	0.6889[20]	1.3982[20]	i H₂O; s EtOH, eth, bz, chl
4240	4,4-Dimethyl-1-pentyne		C₇H₁₂	13361-63-2	96.170	liq	-75.7	76.1	0.7142[20]	1.3983[20]	vs bz, eth, chl
4241	4,4-Dimethyl-2-pentyne		C₇H₁₂	999-78-0	96.170	liq	-82.4	83	0.7176[20]	1.4071[20]	i H₂O; s eth, bz, chl; sl ctc
4242	Dimethylperoxide		C₂H₆O₂	690-02-8	62.068	vol liq or gas	-100	14	0.8677[0]	1.3503[0]	sl EtOH, eth; s tol, HOAc
4243	2,9-Dimethyl-1,10-phenanthroline	Neocuproine	C₁₄H₁₂N₂	484-11-7	208.258	cry, 1/2w (w, lig)	159.5				
4244	3,4-Dimethylphenol phosphate (3:1)		C₂₄H₂₇O₄P	3862-11-1	410.442		72	261[7]			i H₂O; sl EtOH, chl, hx; s bz
4245	5-(2,5-Dimethylphenoxy)-2,2-dimethylpentanoic acid	Gemfibrozil	C₁₅H₂₂O₃	25812-30-0	250.334	cry	62	159[0.02]			
4246	N-(2,4-Dimethylphenyl)acetamide		C₁₀H₁₃NO	2050-43-3	163.216	nd (al)	129.3	170[10]			vs EtOH, chl
4247	1-[(2,4-Dimethylphenyl)azo]-2-naphthol	1-(2,4-Xylylazo)-2-naphthol	C₁₈H₁₆N₂O	3118-97-6	276.332	red nd (al)	166				vs eth, EtOH
4248	1-[(2,5-Dimethylphenyl)azo]-2-naphthol	1-(2,5-Xylylazo)-2-naphthol	C₁₈H₁₆N₂O	85-82-5	276.332	nd (al)	153				
4249	1-(2,4-Dimethylphenyl)ethanone	2,4-Dimethylacetophenone	C₁₀H₁₂O	89-74-7	148.201			228	1.0121[15]	1.5340[20]	vs eth, EtOH

2,4-Dimethyl-1,3-pentadiene

N,N-Dimethylpentanamide

2,2-Dimethylpentane

2,3-Dimethylpentane

2,4-Dimethylpentane

3,3-Dimethylpentane

3,3-Dimethylpentanedioic acid anhydride

3,3-Dimethylpentanedioic acid

2,2-Dimethylpentanoic acid

2,2-Dimethyl-1-pentanol

2,3-Dimethyl-2-pentanol

2,4-Dimethyl-2-pentanol

2,2-Dimethyl-3-pentanol

2,3-Dimethyl-3-pentanol

2,4-Dimethyl-3-pentanol

4,4-Dimethyl-2-pentanone

2,2-Dimethyl-3-pentanone

2,4-Dimethyl-3-pentanone

2,3-Dimethyl-1-pentene

2,4-Dimethyl-1-pentene

3,3-Dimethyl-1-pentene

3,4-Dimethyl-1-pentene

4,4-Dimethyl-1-pentene

2,3-Dimethyl-2-pentene

2,4-Dimethyl-2-pentene

cis-3,4-Dimethyl-2-pentene

trans-3,4-Dimethyl-2-pentene

cis-4,4-Dimethyl-2-pentene

trans-4,4-Dimethyl-2-pentene

4,4-Dimethyl-1-pentyne

4,4-Dimethyl-2-pentyne

Dimethylperoxide

2,9-Dimethyl-1,10-phenanthroline

3,4-Dimethylphenol phosphate (3:1)

5-(2,5-Dimethylphenoxy)-2,2-dimethylpentanoic acid

N-(2,4-Dimethylphenyl)acetamide

1-[(2,4-Dimethylphenyl)azo]-2-naphthol

1-[(2,5-Dimethylphenyl)azo]-2-naphthol

1-(2,4-Dimethylphenyl)ethanone

No.	Name	Synonym	Mol. Form.	CAS RN	Mol. Wt.	Physical Form	mp/°C	bp/°C	den/ g cm⁻³	n_D	Solubility
4250	1-(2,5-Dimethylphenyl)ethanone	2,5-Dimethylacetophenone	C₁₀H₁₂O	2142-73-6	148.201	liq	-18.1	232.5	0.9963[19]	1.5291[20]	i H₂O; vs EtOH, eth, bz, CS₂
4251	1-(3,4-Dimethylphenyl)ethanone	3,4-Dimethylacetophenone	C₁₀H₁₂O	3637-01-2	148.201	liq	-1.5	246.5	1.0090[14]	1.5413[15]	i H₂O; vs EtOH, eth, bz; s ctc, HOAc
4252	4,4-Dimethyl-1-phenyl-1-penten-3-one		C₁₃H₁₆O	538-44-3	188.265		43	154[25]	0.9508[46]	1.5523[25]	
4253	2,2-Dimethyl-1-phenyl-1-propanone		C₁₁H₁₄O	938-16-9	162.228			220	0.963[26]	1.5086[19]	s ace
4254	3,5-Dimethyl-1-phenyl-1H-pyrazole		C₁₁H₁₂N₂	1131-16-4	172.226			272; 145[12.5]	1.0566[20]	1.5738[19]	vs eth, EtOH, chl
4255	4,4-Dimethyl-1-phenyl-3-pyrazolidinone	4,4-Dimethylphenidone	C₁₁H₁₄N₂O	2654-58-2	190.241		176				
4256	N,N-Dimethyl-γ-phenyl-2-pyridinepropanamine	Pheniramine	C₁₆H₂₀N₂	86-21-5	240.343			181[13], 135[0.5]	1.0081[25]	1.5519[25]	vs bz, eth, EtOH, chl
4257	1,3-Dimethyl-3-phenyl-2,5-pyrrolidinedione	Methsuximide	C₁₂H₁₃NO₂	77-41-8	203.237		52.5	121[0.1]			
4258	Dimethylphenylsilane		C₈H₁₂Si	766-77-8	136.267			156.5	0.8891[20]	1.4995[20]	i H₂O
4259	N,N-Dimethyl-N'-phenylurea	Fenuron	C₉H₁₂N₂O	101-42-8	164.203	cry (hx)	132				
4260	Dimethylphosphine		C₂H₇P	676-59-5	62.051	vol liq or gas		25			i H₂O; s EtOH, eth
4261	Dimethylphosphinic acid		C₂H₇O₂P	3283-12-3	94.050	cry (bz)	92	377			vs H₂O, EtOH, eth; s bz
4262	O,O-Dimethyl phosphorochloridothioate	Dimethyl chlorothiophosphate	C₂H₆ClO₂PS	2524-03-0	160.560	hyg liq		68[12]	1.322	1.4820[20]	
4263	Dimethyl phthalate		C₁₀H₁₀O₄	131-11-3	194.184	pa ye	5.5	283.7	1.1905[20]	1.5138[20]	i H₂O; msc EtOH, eth; s bz; sl ctc
4264	1,4-Dimethylpiperazine		C₆H₁₄N₂	106-58-1	114.188	liq	-0.59	131	0.8600[20]	1.4474[20]	vs H₂O, EtOH, eth
4265	cis-2,5-Dimethylpiperazine		C₆H₁₄N₂	6284-84-0	114.188	orth bipym nd or pr (chl)	114	162		1.4720[20]	vs H₂O, EtOH, chl; sl eth, bz
4266	1,2-Dimethylpiperidine, (±)		C₇H₁₅N	2512-81-4	113.201			127.5	0.824[15]	1.4395[20]	vs H₂O, eth, EtOH
4267	2,6-Dimethylpiperidine		C₇H₁₅N	504-03-0	113.201			127	0.8158[25]	1.4377[20]	msc H₂O, EtOH, eth; sl ctc; s acid
4268	3,5-Dimethylpiperidine	3,5-Lupetidine	C₇H₁₅N	35794-11-7	113.201			144	0.853[25]	1.4454[20]	
4269	2,2-Dimethylpropanal	Pivaldehyde	C₅H₁₀O	630-19-3	86.132		6	77.5	0.7923[17]	1.3791[20]	s EtOH, eth
4270	2,2-Dimethylpropanamide		C₅H₁₁NO	754-10-9	101.147						s tfa
4271	N,N-Dimethylpropanamide		C₅H₁₁NO	758-96-3	101.147	liq	-45	175	0.9269[20]		
4272	N,N-Dimethyl-1-propanamine	Dimethylpropylamine	C₅H₁₃N	926-63-6	87.164			66	0.7152[20]	1.3860[20]	vs bz, eth, EtOH
4273	N,N-Dimethyl-1,3-propanediamine		C₅H₁₄N₂	109-55-7	102.178			132	0.8272[20]		
4274	2,2-Dimethyl-1,3-propanediol	Neopentyl glycol	C₅H₁₂O₂	126-30-7	104.148	nd (bz)	129.13	208			s H₂O, bz, chl; vs EtOH, eth
4275	2,2-Dimethylpropanenitrile	tert-Butyl cyanide	C₅H₉N	630-18-2	83.132		15	106.1	0.7586[25]	1.3774[20]	
4276	2,2-Dimethyl-1-propanethiol	Neopentyl mercaptan	C₅H₁₂S	1679-08-9	104.214	liq		103.7			
4277	2,2-Dimethylpropanoic acid	Trimethylacetic acid	C₅H₁₀O₂	75-98-9	102.132	nd	35	164	0.905[50]	1.3931[30]	sl H₂O; vs EtOH, eth
4278	2,2-Dimethyl-1-propanol	Neopentyl alcohol	C₅H₁₂O	75-84-3	88.148		52.5	113.5	0.812[20]		sl H₂O; vs EtOH, eth; s ctc
4279	2,2-Dimethylpropanoyl chloride	Pivalic acid chloride	C₅H₉ClO	3282-30-2	120.577			107	1.003[20]	1.4139[20]	vs eth
4280	N,N-Dimethyl-2-propenamide	N,N-Dimethylacrylamide	C₅H₉NO	2680-03-7	99.131	liq		81[20]	0.962[25]	1.4730[20]	
4281	2,2-Dimethylpropylamine	2,2-Dimethyl-1-propanamine	C₅H₁₃N	5813-64-9	87.164			82	0.7455[20]	1.4023[20]	vs eth
4282	(1,1-Dimethylpropyl)benzene		C₁₁H₁₆	2049-95-8	148.245			192.4	0.8748[20]	1.4958[20]	
4283	(2,2-Dimethylpropyl)benzene		C₁₁H₁₆	1007-26-7	148.245			185	0.8581[18]	1.4884[18]	
4284	4-(1,1-Dimethylpropyl)cyclohexanone		C₁₁H₂₀O	16587-71-6	168.276		96	125[16], 109[11]	0.920[25]	1.4677[20]	
4285	1,1-Dimethylpropyl 3-methylbutanoate	tert-Pentyl isopentanoate	C₁₀H₂₀O₂	542-37-0	172.265			173.5	0.8729[0]		vs EtOH
4286	2-(1,1-Dimethylpropyl)phenol		C₁₁H₁₆O	3279-27-4	164.244						sl ctc
4287	4-(1,1-Dimethylpropyl)phenol	p-tert-Pentylphenol	C₁₁H₁₆O	80-46-6	164.244		95	262.5			
4288	4,6-Dimethyl-2H-pyran-2-one		C₇H₈O₂	675-09-2	124.138	lf (eth)	51.5	245			vs H₂O, eth, EtOH
4289	2,6-Dimethyl-4H-pyran-4-one		C₇H₈O₂	1004-36-0	124.138	pl, nd (sub)	132	251; 140[25]	0.9953[137]		s H₂O, EtOH, eth, ace
4290	2,3-Dimethylpyrazine		C₆H₈N₂	5910-89-4	108.141			156	1.0281[0]		s H₂O, EtOH, eth
4291	2,5-Dimethylpyrazine		C₆H₈N₂	123-32-0	108.141		15	155	0.9887[20]	1.4980[20]	msc H₂O, EtOH, eth; s ace, chl
4292	2,6-Dimethylpyrazine		C₆H₈N₂	108-50-9	108.141	pr	47.5	155.6	0.9647[50]		s H₂O, EtOH, eth; sl ctc
4293	1,3-Dimethyl-1H-pyrazole		C₅H₈N₂	694-48-4	96.131			137	0.9561[17]	1.4734[15]	vs H₂O

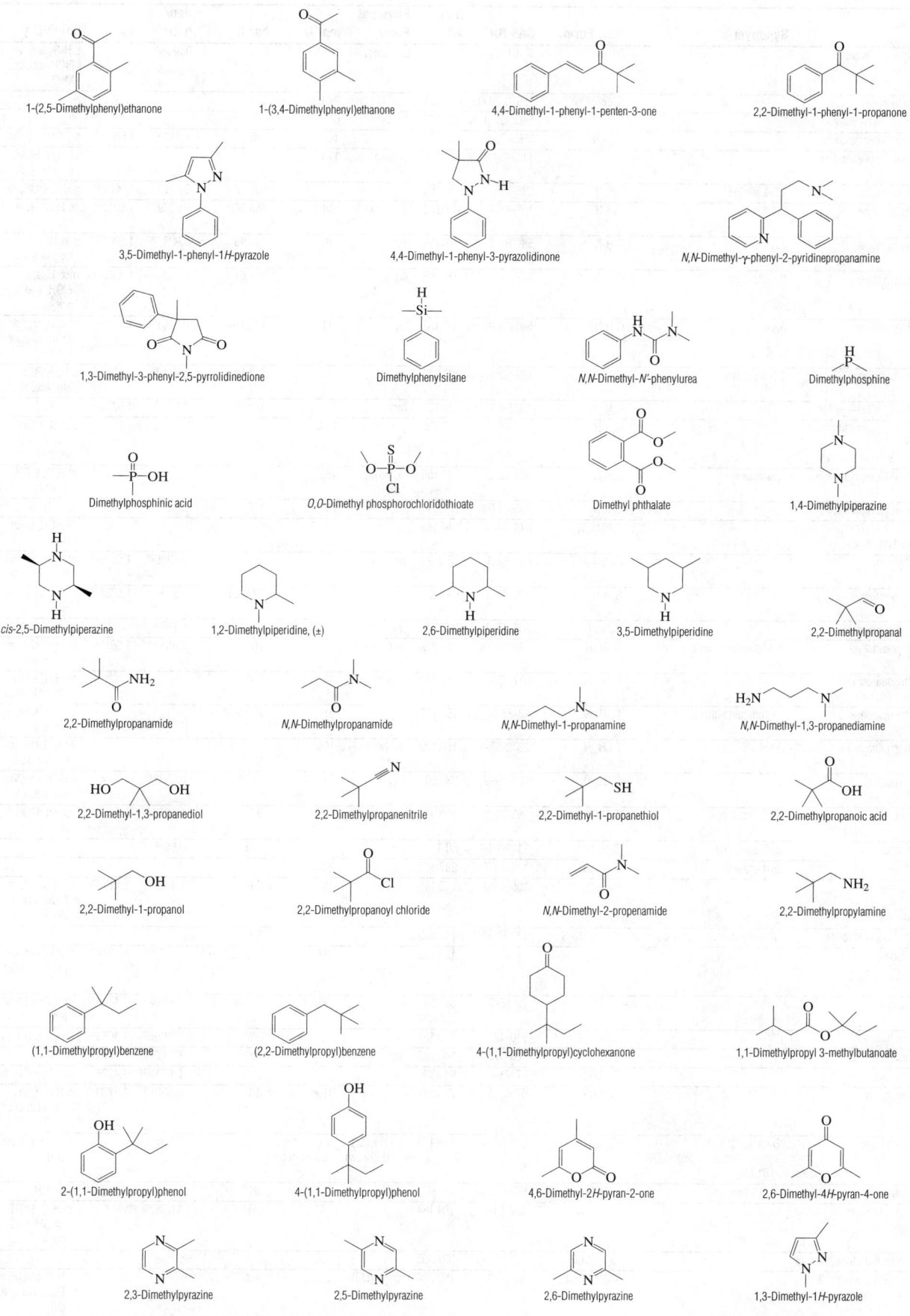

1-(2,5-Dimethylphenyl)ethanone

1-(3,4-Dimethylphenyl)ethanone

4,4-Dimethyl-1-phenyl-1-penten-3-one

2,2-Dimethyl-1-phenyl-1-propanone

3,5-Dimethyl-1-phenyl-1*H*-pyrazole

4,4-Dimethyl-1-phenyl-3-pyrazolidinone

N,N-Dimethyl-γ-phenyl-2-pyridinepropanamine

1,3-Dimethyl-3-phenyl-2,5-pyrrolidinedione

Dimethylphenylsilane

N,N-Dimethyl-*N'*-phenylurea

Dimethylphosphine

Dimethylphosphinic acid

O,O-Dimethyl phosphorochloridothioate

Dimethyl phthalate

1,4-Dimethylpiperazine

cis-2,5-Dimethylpiperazine

1,2-Dimethylpiperidine, (±)

2,6-Dimethylpiperidine

3,5-Dimethylpiperidine

2,2-Dimethylpropanal

2,2-Dimethylpropanamide

N,N-Dimethylpropanamide

N,N-Dimethyl-1-propanamine

N,N-Dimethyl-1,3-propanediamine

2,2-Dimethyl-1,3-propanediol

2,2-Dimethylpropanenitrile

2,2-Dimethyl-1-propanethiol

2,2-Dimethylpropanoic acid

2,2-Dimethyl-1-propanol

2,2-Dimethylpropanoyl chloride

N,N-Dimethyl-2-propenamide

2,2-Dimethylpropylamine

(1,1-Dimethylpropyl)benzene

(2,2-Dimethylpropyl)benzene

4-(1,1-Dimethylpropyl)cyclohexanone

1,1-Dimethylpropyl 3-methylbutanoate

2-(1,1-Dimethylpropyl)phenol

4-(1,1-Dimethylpropyl)phenol

4,6-Dimethyl-2*H*-pyran-2-one

2,6-Dimethyl-4*H*-pyran-4-one

2,3-Dimethylpyrazine

2,5-Dimethylpyrazine

2,6-Dimethylpyrazine

1,3-Dimethyl-1*H*-pyrazole

No.	Name	Synonym	Mol. Form.	CAS RN	Mol. Wt.	Physical Form	mp/°C	bp/°C	den/ g cm⁻³	n_D	Solubility
4294	3,5-Dimethyl-1*H*-pyrazole		C₅H₈N₂	67-51-6	96.131	cry (peth, al)	107.5	218	0.8839[16]		s H₂O, ace; vs EtOH, eth, bz, MeOH
4295	2,7-Dimethylpyrene		C₁₈H₁₄	15679-24-0	230.304			230			
4296	4,6-Dimethyl-2-pyridinamine		C₇H₁₀N₂	5407-87-4	122.167		61	235			
4297	*N,N*-Dimethyl-2-pyridinamine		C₇H₁₀N₂	5683-33-0	122.167		182	196	1.0149[14]	1.5663[20]	s EtOH, eth, bz
4298	*N,N*-Dimethyl-4-pyridinamine		C₇H₁₀N₂	1122-58-3	122.167	pl (eth)	114				vs H₂O, EtOH, bz, chl; s eth
4299	2,3-Dimethylpyridine	2,3-Lutidine	C₇H₉N	583-61-9	107.153			161.12	0.9319[25]	1.5057[20]	s H₂O, EtOH, eth
4300	2,4-Dimethylpyridine	2,4-Lutidine	C₇H₉N	108-47-4	107.153	liq	-64	158.38	0.9309[20]	1.5010[20]	vs H₂O, EtOH, eth; s ace
4301	2,5-Dimethylpyridine	2,5-Lutidine	C₇H₉N	589-93-5	107.153	liq	-16	156.98	0.9297[20]	1.5006[20]	sl H₂O; vs EtOH, msc eth; s ace
4302	2,6-Dimethylpyridine	2,6-Lutidine	C₇H₉N	108-48-5	107.153	liq	-6.1	144.01	0.9226[20]	1.4953[20]	msc H₂O; sl EtOH; s eth, ace, chl
4303	3,4-Dimethylpyridine	3,4-Lutidine	C₇H₉N	583-58-4	107.153	liq	-11	179.10	0.9281[20]	1.5096[20]	sl H₂O, ctc; s EtOH, eth, ace, chl
4304	3,5-Dimethylpyridine	3,5-Lutidine	C₇H₉N	591-22-0	107.153	liq	-6.6	171.84	0.9419[20]	1.5061[20]	s H₂O, EtOH, eth, ace; sl ctc
4305	2,6-Dimethylpyridine-1-oxide		C₇H₉NO	1073-23-0	123.152	hyg	35	133[22]	1.073[25]	1.5706[20]	
4306	4,6-Dimethyl-2-pyrimidinamine		C₆H₉N₃	767-15-7	123.155		153.5				s H₂O, EtOH, ace, bz; i eth; vs chl
4307	2,6-Dimethyl-4-pyrimidinamine	Kyanmethin	C₆H₉N₃	461-98-3	123.155	nd (al), pl (bz)	183	sub			sl H₂O, EtOH, bz, chl
4308	4,6-Dimethylpyrimidine		C₆H₈N₂	1558-17-4	108.141		25	159		1.4880[20]	vs H₂O
4309	1,3-Dimethyl-2,4(1*H*,3*H*)-pyrimidinedione		C₆H₈N₂O₂	874-14-6	140.140		123.5				sl EtOH; s eth, chl
4310	2,4-Dimethylpyrrole		C₆H₉N	625-82-1	95.142	pa bl flr cry		168	0.9236[20]	1.5048[20]	sl H₂O; vs EtOH, eth, bz; s chl
4311	2,5-Dimethylpyrrole		C₆H₉N	625-84-3	95.142		6.5	171; 51[8]	0.9353[20]	1.5036[20]	i H₂O; vs EtOH, eth
4312	1,2-Dimethylpyrrolidine		C₆H₁₃N	765-48-0	99.174	oil		99	0.799[20]		s H₂O
4313	2,4-Dimethylquinoline	4-Methylquinaldine	C₁₁H₁₁N	1198-37-4	157.212	orth pr (eth)		265	1.0611[15]	1.6075[20]	sl H₂O, chl; vs EtOH, eth
4314	2,6-Dimethylquinoline		C₁₁H₁₁N	877-43-0	157.212	orth pr (eth)	60	266.5			sl H₂O, EtOH, eth, chl; vs bz
4315	2,7-Dimethylquinoline	*m*-Toluquinaldine	C₁₁H₁₁N	93-37-8	157.212		61	264.5			sl H₂O; s EtOH, eth, chl
4316	2,3-Dimethylquinoxaline		C₁₀H₁₀N₂	2379-55-7	158.199	nd (w+3, ace)	106				s EtOH, eth, ace, bz, chl, acid
4317	Dimethyl sebacate		C₁₂H₂₂O₄	106-79-6	230.301	lo pr	38	175[20], 144[5]	0.9882[28]	1.4355[28]	i H₂O; s EtOH, eth, ace, ctc
4318	Dimethyl selenide	Methyl selenide	C₂H₆Se	593-79-3	109.03			57	1.4077[15]		vs eth, EtOH, chl
4319	Dimethylsilane	2-Silapropane	C₂H₈Si	1111-74-6	60.171	col gas	-150	-20	0.68[-80]		
4320	Dimethylstearylamine	Dymanthine	C₂₀H₄₃N	124-28-7	297.562		22.9				
4321	Dimethyl succinate		C₆H₁₀O₄	106-65-0	146.141		19	196.4	1.1198[20]	1.4197[20]	sl H₂O, ctc; s EtOH, ace; vs eth
4322	Dimethylsulfamoyl chloride	Dimethylaminosulfonyl chloride	C₂H₆ClNO₂S	13360-57-1	143.593			80[16]			
4323	Dimethyl sulfate		C₂H₆O₄S	77-78-1	126.132		-27	dec 188; 76[15]	1.3322[20]	1.3874[20]	s H₂O, eth, bz, ctc; msc EtOH; i CS₂
4324	Dimethyl sulfide		C₂H₆S	75-18-3	62.134	liq	-98.24	37.33	0.8483[20]	1.4438[20]	sl H₂O; s EtOH, eth
4325	Dimethyl sulfite		C₂H₆O₃S	616-42-2	110.132			126	1.2129[20]	1.4083[20]	s H₂O, EtOH, eth
4326	2,4-Dimethylsulfolane		C₆H₁₂O₂S	1003-78-7	148.223	liq	-1.5	281	1.1362[20]	1.4732[20]	vs lig
4327	Dimethyl sulfone		C₂H₆O₂S	67-71-0	94.133	pr	108.9	238	1.1700[110]	1.4226	s H₂O, EtOH, bz
4328	Dimethyl sulfoxide	DMSO	C₂H₆OS	67-68-5	78.133		17.89	189	1.1010[25]	1.4793[20]	s H₂O, EtOH, eth, ace, ctc, AcOEt
4329	Dimethyl *L*-tartrate	Dimethyl 2,3-dihydroxybutanedioate, [*R*-(*R**,*R**)]-	C₆H₁₀O₆	608-68-4	178.139	(i) cry (bz) (ii) cry (w)	50(form a); 61(form b)	280	1.306[45]		vs H₂O, ace, eth, EtOH
4330	Dimethyl telluride		C₂H₆Te	593-80-6	157.67	pa ye		94			vs EtOH
4331	Dimethyl terephthalate		C₁₀H₁₀O₄	120-61-6	194.184		141	288	1.075[141]		sl H₂O, EtOH, MeOH; s eth, chl
4332	Dimethyl tetrachloroterephthalate		C₁₀H₆Cl₄O₄	1861-32-1	331.965		155				
4333	2,7-Dimethylthiachromine-8-ethanol		C₁₂H₁₄N₄OS	92-35-3	262.330	ye pr (chl)	228.8	sub			s H₂O, MeOH; sl EtOH, eth, ace, chl

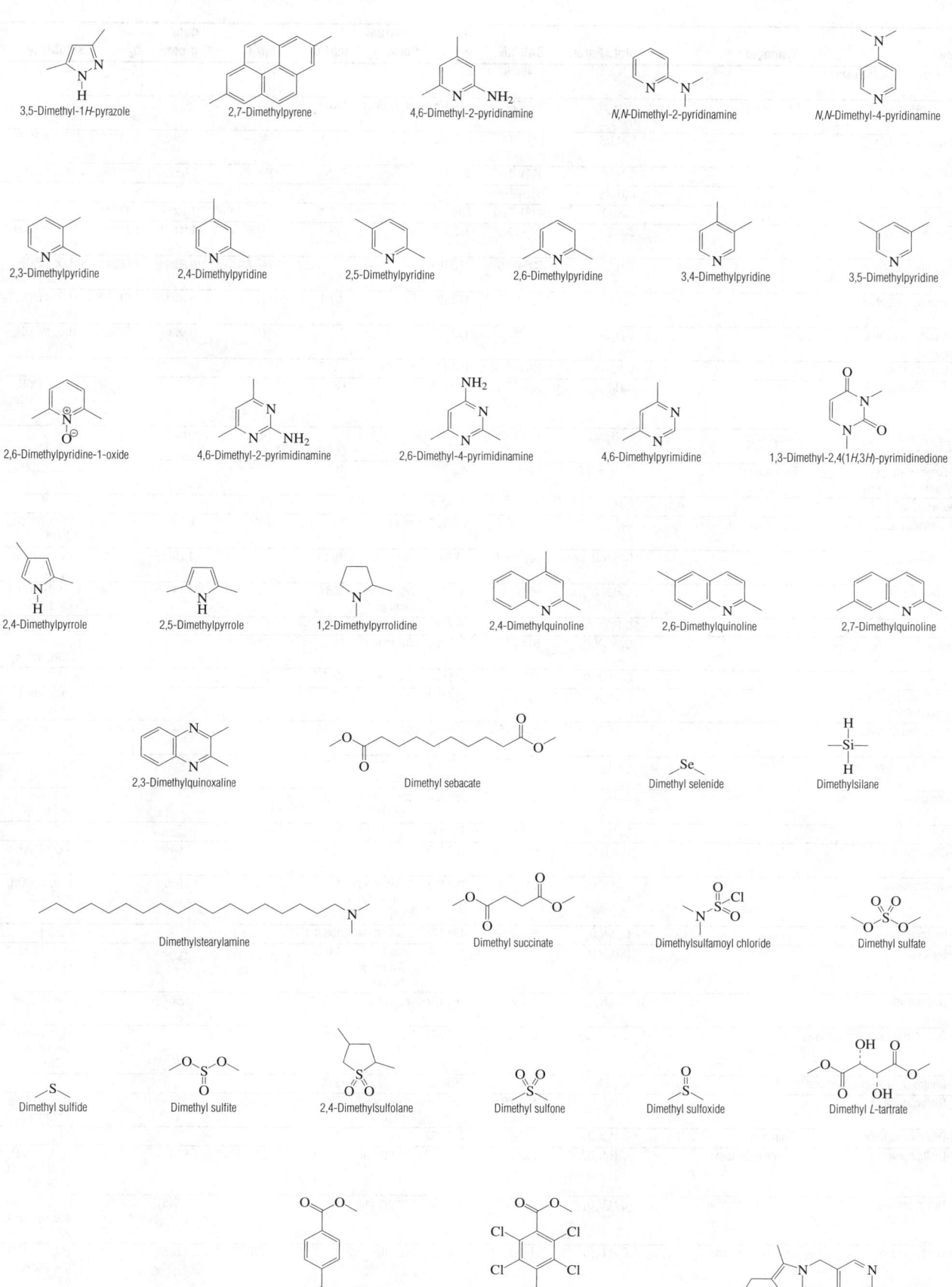

3,5-Dimethyl-1*H*-pyrazole

2,7-Dimethylpyrene

4,6-Dimethyl-2-pyridinamine

N,N-Dimethyl-2-pyridinamine

N,N-Dimethyl-4-pyridinamine

2,3-Dimethylpyridine

2,4-Dimethylpyridine

2,5-Dimethylpyridine

2,6-Dimethylpyridine

3,4-Dimethylpyridine

3,5-Dimethylpyridine

2,6-Dimethylpyridine-1-oxide

4,6-Dimethyl-2-pyrimidinamine

2,6-Dimethyl-4-pyrimidinamine

4,6-Dimethylpyrimidine

1,3-Dimethyl-2,4(1*H*,3*H*)-pyrimidinedione

2,4-Dimethylpyrrole

2,5-Dimethylpyrrole

1,2-Dimethylpyrrolidine

2,4-Dimethylquinoline

2,6-Dimethylquinoline

2,7-Dimethylquinoline

2,3-Dimethylquinoxaline

Dimethyl sebacate

Dimethyl selenide

Dimethylsilane

Dimethylstearylamine

Dimethyl succinate

Dimethylsulfamoyl chloride

Dimethyl sulfate

Dimethyl sulfide

Dimethyl sulfite

2,4-Dimethylsulfolane

Dimethyl sulfone

Dimethyl sulfoxide

Dimethyl *L*-tartrate

Dimethyl telluride

Dimethyl terephthalate

Dimethyl tetrachloroterephthalate

2,7-Dimethylthiachromine-8-ethanol

No.	Name	Synonym	Mol. Form.	CAS RN	Mol. Wt.	Physical Form	mp/°C	bp/°C	den/ g cm⁻³	n_D	Solubility
4334	2,5-Dimethyl-1,3,4-thiadiazole		$C_4H_6N_2S$	27464-82-0	114.169		65	202.5			sl H_2O, EtOH, eth
4335	2,7-Dimethylthianthrene	Mesulphen	$C_{14}H_{12}S_2$	135-58-0	244.375	nd (HOAc,al)	123	184[3]			vs ace, eth, peth, chl
4336	2,4-Dimethylthiazole		C_5H_7NS	541-58-2	113.182			146; 71[50]	1.0562[15]	1.5091[20]	sl H_2O; s EtOH, eth, chl
4337	4,5-Dimethylthiazole		C_5H_7NS	3581-91-7	113.182		83.5	158	1.0699[20]		vs eth, EtOH
4338	N,N-Dimethylthioacetamide		C_4H_9NS	631-67-4	103.186		74.5				
4339	Dimethyl thiodipropionate		$C_8H_{14}O_4S$	4131-74-2	206.260			162[18], 148[18]	1.1559[20]	1.4740[20]	
4340	2,3-Dimethylthiophene		C_6H_8S	632-16-6	112.193	liq	-49	141.6	1.0021[20]	1.5192[20]	i H_2O; vs EtOH, eth; s bz
4341	2,4-Dimethylthiophene		C_6H_8S	638-00-6	112.193			140.7	0.9938[20]	1.5104[20]	i H_2O; s EtOH, eth, bz
4342	2,5-Dimethylthiophene		C_6H_8S	638-02-8	112.193	liq	-62.6	136.5	0.9850[20]	1.5129[20]	i H_2O; s EtOH, eth, bz
4343	3,4-Dimethylthiophene		C_6H_8S	632-15-5	112.193			145	0.993[25]	1.5206[20]	i H_2O; s EtOH; vs eth
4344	N,N-Dimethylthiourea		$C_3H_8N_2S$	6972-05-0	104.174	cry (w)	161.5				
4345	N,N'-Dimethylthiourea		$C_3H_8N_2S$	534-13-4	104.174	hyg pl	62				vs H_2O, EtOH, ace; sl eth, bz; i CS_2
4346	2,6-Dimethyl-4-tridecylmorpholine	Tridemorph	$C_{19}H_{39}NO$	24602-86-6	297.519			141[1.3]	0.86		
4347	N,N-Dimethyl-N'-[3-(trifluoromethyl)phenyl]urea	Fluometuron	$C_{10}H_{11}F_3N_2O$	2164-17-2	232.201		164				vs ace, EtOH
4348	Dimethyl trisulfide		$C_2H_6S_3$	3658-80-8	126.264			41[6]			
4349	6,10-Dimethyl-3,5,9-undecatrien-2-one	Pseudoionone	$C_{13}H_{20}O$	141-10-6	192.297	pa ye oil		144[12]	0.8984[20]	1.5335[20]	s EtOH, eth, chl, MeOH
4350	N,N-Dimethylurea		$C_3H_8N_2O$	598-94-7	88.108	mcl pr (al, chl)	182.1		1.2555[25]		s H_2O; sl EtOH, tfa; i eth
4351	N,N'-Dimethylurea		$C_3H_8N_2O$	96-31-1	88.108	orth bipym (chl-eth)	106.6	269	1.142[25]		vs H_2O, EtOH; i eth; sl chl
4352	Dimethyl zinc		C_2H_6Zn	544-97-8	95.478	liq, ign in air	-43.0	46	1.386[10]		s eth; msc peth
4353	Dimetilan		$C_{10}H_{16}N_4O_3$	644-64-4	240.259	col solid	69	205[13]			s H_2O, chl, EtOH, ace, xyl
4354	Dimorpholamine		$C_{20}H_{38}N_4O_4$	119-48-2	398.541	cry (peth)	41.5	229[0.4]			vs H_2O
4355	N,N'-Di-2-naphthyl-1,4-benzenediamine		$C_{26}H_{20}N_2$	93-46-9	360.450		235				i EtOH, eth, bz
4356	Di-2-naphthyl disulfide		$C_{20}H_{14}S_2$	5586-15-2	318.455	nd	139.5		1.144[145]	1.4555[20]	i H_2O; vs EtOH, eth; i lig
4357	N,N'-Di-1-naphthylurea		$C_{21}H_{16}N_2O$	607-56-7	312.364	nd (py, HOAc)	296	sub			vs py
4358	Diniconazole		$C_{15}H_{17}Cl_2N_3O$	83657-24-3	326.221	cry	149				s H_2O, ace, MeOH, xyl
4359	Dinitramine		$C_{11}H_{13}F_3N_4O_4$	29091-05-2	322.241		98				
4360	2,3-Dinitroaniline		$C_6H_5N_3O_4$	602-03-9	183.122		128		1.646[50]		i H_2O; s EtOH; sl eth
4361	2,4-Dinitroaniline		$C_6H_5N_3O_4$	97-02-9	183.122	ye nd (ace) grn ye tab (al)	180.0		1.615[14]		i H_2O; sl EtOH, ace, HCl
4362	2,5-Dinitroaniline		$C_6H_5N_3O_4$	619-18-1	183.122	oran nd (al)	138.0				vs EtOH
4363	2,6-Dinitroaniline		$C_6H_5N_3O_4$	606-22-4	183.122	gold lf (HOAc) ye nd (al)	141.5				i H_2O, lig; sl EtOH; s eth, bz
4364	3,5-Dinitroaniline		$C_6H_5N_3O_4$	618-87-1	183.122	ye nd (dil al)	163		1.601[50]		i H_2O; s EtOH, eth; sl ace, bz
4365	1,5-Dinitro-9,10-anthracenedione		$C_{14}H_6N_2O_6$	82-35-9	298.207	pa ye nd (xyl)	385	sub			i H_2O; sl EtOH, eth, bz; vs $PhNO_2$
4366	1,8-Dinitro-9,10-anthracenedione		$C_{14}H_6N_2O_6$	129-39-5	298.207		312				
4367	2,4-Dinitrobenzaldehyde		$C_7H_4N_2O_5$	528-75-6	196.117	pa ye pr (al), pl (bz)	72	200[15]			sl H_2O, chl, lig; s EtOH, eth, bz
4368	3,5-Dinitrobenzamide	Nitromide	$C_7H_5N_3O_5$	121-81-3	211.132	lf (w)	184				vs H_2O
4369	1,2-Dinitrobenzene	o-Dinitrobenzene	$C_6H_4N_2O_4$	528-29-0	168.107	nd (bz), pl (al)	116.5	318; 194[30]	1.3119[120]	1.565[17]	i H_2O; s EtOH, bz, chl, AcOEt; sl DMSO
4370	1,3-Dinitrobenzene	m-Dinitrobenzene	$C_6H_4N_2O_4$	99-65-0	168.107	orth pl (al)	90.3	291; 167[14]	1.5751[18]		sl H_2O; vs EtOH, ace, py; s eth, tol
4371	1,4-Dinitrobenzene	p-Dinitrobenzene	$C_6H_4N_2O_4$	100-25-4	168.107	nd (al)	173.5	297; 183[34]	1.625[18]		i H_2O; sl EtOH, chl; s ace, bz, tol
4372	2,4-Dinitro-1,3-benzenediol	2,4-Dinitroresorcinol	$C_6H_4N_2O_6$	519-44-8	200.105	ye lf (al)	147.5				sl H_2O, EtOH
4373	2,4-Dinitrobenzenesulfenyl chloride		$C_6H_3ClN_2O_4S$	528-76-7	234.617	ye pr (bz-peth)	99				vs bz, chl, HOAc; sl peth

2,5-Dimethyl-1,3,4-thiadiazole

2,7-Dimethylthianthrene

2,4-Dimethylthiazole

4,5-Dimethylthiazole

N,N-Dimethylthioacetamide

Dimethyl thiodipropionate

2,3-Dimethylthiophene

2,4-Dimethylthiophene

2,5-Dimethylthiophene

3,4-Dimethylthiophene

N,N-Dimethylthiourea

N,N'-Dimethylthiourea

2,6-Dimethyl-4-tridecylmorpholine

N,N-Dimethyl-N'-[3-(trifluoromethyl)phenyl]urea

Dimethyl trisulfide

6,10-Dimethyl-3,5,9-undecatrien-2-one

N,N-Dimethylurea

N,N'-Dimethylurea

Dimethyl zinc

Dimetilan

Dimorpholamine

N,N'-Di-2-naphthyl-1,4-benzenediamine

Di-2-naphthyl disulfide

N,N'-Di-1-naphthylurea

Diniconazole

Dinitramine

2,3-Dinitroaniline

2,4-Dinitroaniline

2,5-Dinitroaniline

2,6-Dinitroaniline

3,5-Dinitroaniline

1,5-Dinitro-9,10-anthracenedione

1,8-Dinitro-9,10-anthracenedione

2,4-Dinitrobenzaldehyde

3,5-Dinitrobenzamide

1,2-Dinitrobenzene

1,3-Dinitrobenzene

1,4-Dinitrobenzene

2,4-Dinitro-1,3-benzenediol

2,4-Dinitrobenzenesulfenyl chloride

No.	Name	Synonym	Mol. Form.	CAS RN	Mol. Wt.	Physical Form	mp/°C	bp/°C	den/ g cm⁻³	n_D	Solubility
4374	2,4-Dinitrobenzenesulfonic acid		$C_6H_4N_2O_7S$	89-02-1	248.170	nd (w+3)	108				vs H₂O, EtOH; sl eth; i bz, peth
4375	2,4-Dinitrobenzoic acid		$C_7H_4N_2O_6$	610-30-0	212.116	nd (w)	183		1.672²⁰		sl H₂O, EtOH, bz
4376	3,4-Dinitrobenzoic acid		$C_7H_4N_2O_6$	528-45-0	212.116	cry (dil al)	166				sl H₂O; vs EtOH, eth
4377	3,5-Dinitrobenzoic acid		$C_7H_4N_2O_6$	99-34-3	212.116	mcl pr (al)	205				sl H₂O; vs EtOH, HOAc
4378	3,5-Dinitrobenzoyl chloride		$C_7H_3ClN_2O_5$	99-33-2	230.562	ye nd (bz)	74	196¹²			s eth, chl
4379	2,2'-Dinitro-1,1'-biphenyl		$C_{12}H_8N_2O_4$	2436-96-6	244.203	ye mcl pr or nd (al)	126	305	1.45²⁵		i H₂O; vs EtOH; s eth, bz; sl ace, lig
4380	4,4'-Dinitro-1,1'-biphenyl		$C_{12}H_8N_2O_4$	1528-74-1	244.203	nd (al)	242.3				i H₂O; sl EtOH; s bz, HOAc
4381	1,4-Dinitrobutane		$C_4H_8N_2O_4$	4286-49-1	148.118	pl (al)	33.5	176¹³			i H₂O; sl EtOH; s eth, bz, MeOH
4382	4,4'-Dinitrodiphenylamine	4-Nitro-N-(4-nitrophenyl)aniline	$C_{12}H_9N_3O_4$	1821-27-8	259.217	ye nd(al)	217.5				i H₂O, tol; sl EtOH, bz; s ace, HOAc
4383	4,4'-Dinitrodiphenyl ether	Bis(4-nitrophenyl) ether	$C_{12}H_8N_2O_5$	101-63-3	260.202		146.0				i H₂O; sl EtOH, eth; s bz, HOAc
4384	4,4'-Dinitrodiphenyl sulfide	Bis(4-nitrophenyl) sulfide	$C_{12}H_8N_2O_4S$	1223-31-0	276.268	oran pl (HOAc)	160.5				i H₂O; sl EtOH; s con sulf
4385	1,1-Dinitroethane		$C_2H_4N_2O_4$	600-40-8	120.064	ye mcl (bz, MeOH)		185.5	1.349²⁴		sl H₂O; s EtOH, eth
4386	1,2-Dinitroethane		$C_2H_4N_2O_4$	7570-26-5	120.064		39.5	95⁵	1.4597²⁰	1.4468²⁰	vs eth, EtOH
4387	Dinitromethane		$CH_2N_2O_4$	625-76-3	106.038	ye nd	<-15	exp 100			i H₂O; s EtOH, eth
4388	1,3-Dinitronaphthalene		$C_{10}H_6N_2O_4$	606-37-1	218.166	ye nd (bz, py-w)	148	sub			i H₂O; s EtOH, ace
4389	1,5-Dinitronaphthalene		$C_{10}H_6N_2O_4$	605-71-0	218.166	hex nd (ace, HOAc)	219	sub	1.5860²⁰		i H₂O; sl EtOH, ace; s bz, py; vs eth
4390	1,8-Dinitronaphthalene		$C_{10}H_6N_2O_4$	602-38-0	218.166	ye orth pl (chl)	173	dec 445			i H₂O; sl EtOH, bz; s ace, chl, py
4391	2,4-Dinitro-1-naphthol		$C_{10}H_6N_2O_5$	605-69-6	234.165	ye nd (al, chl)	138.8				
4392	2,3-Dinitrophenol		$C_6H_4N_2O_5$	66-56-8	184.106	ye nd (w)	144.5		1.681²⁰		sl H₂O, DMSO; vs EtOH, eth; s bz
4393	2,4-Dinitrophenol		$C_6H_4N_2O_5$	51-28-5	184.106	pa ye pl or lf (w)	114.8	sub	1.683²⁴		sl H₂O; s EtOH, eth, ace, bz, tol, chl, py
4394	2,5-Dinitrophenol		$C_6H_4N_2O_5$	329-71-5	184.106	ye mcl pr or nd (w,lig)	108				vs bz, eth
4395	2,6-Dinitrophenol		$C_6H_4N_2O_5$	573-56-8	184.106	pa ye orth nd or lf (dil al)	63.5				i H₂O; vs EtOH, eth; s bz, chl; sl ctc
4396	3,4-Dinitrophenol		$C_6H_4N_2O_5$	577-71-9	184.106	tcl nd (w)	134		1.672²⁵		vs bz, eth, EtOH
4397	2,4-Dinitrophenol, acetate		$C_8H_6N_2O_6$	4232-27-3	226.143	cry (MeOH)	72.5				
4398	4-[(2,4-Dinitrophenyl)amino] phenol		$C_{12}H_9N_3O_5$	119-15-3	275.216	red lf	195.5				s alk
4399	2,4-Dinitro-N-phenylaniline		$C_{12}H_9N_3O_4$	961-68-2	259.217	ye red nd (al)	157.8				i H₂O; s EtOH, ace; sl eth, bz, DMSO
4400	2,4-Dinitrophenyl dimethylcarbamodithioate		$C_9H_9N_3O_4S_2$	89-37-2	287.315		152.5		1.54²⁰		i H₂O; s EtOH, ace, bz
4401	(2,4-Dinitrophenyl)hydrazine		$C_6H_6N_4O_4$	119-26-6	198.137	blsh-red (al)	194				i H₂O; s EtOH; sl eth, bz, chl, DMSO
4402	1,1-Dinitropropane		$C_3H_6N_2O_4$	601-76-3	134.091	liq	-42	184	1.2610²⁵	1.4339²⁰	s alk
4403	1,3-Dinitropropane		$C_3H_6N_2O_4$	6125-21-9	134.091		-21.4	103¹	1.353²⁶	1.4654²⁰	i H₂O; s eth
4404	2,2-Dinitropropane		$C_3H_6N_2O_4$	595-49-3	134.091		53	185.5	1.30²⁵		sl H₂O
4405	2,2-Dinitro-1,3-propanediol		$C_3H_6N_2O_6$	2736-80-3	166.089	wh pl (bz)	142				
4406	1,6-Dinitropyrene		$C_{16}H_8N_2O_4$	42397-64-8	292.246		>300				
4407	1,8-Dinitropyrene		$C_{16}H_8N_2O_4$	42397-65-9	292.246		300				
4408	Dinitrosopentamethylenetetramine		$C_5H_{10}N_6O_2$	101-25-7	186.172	cry (MeOH)	207				
4409	1,4-Dinitrosopiperazine		$C_4H_8N_4O_2$	140-79-4	144.133	pa ye pl (w)	159.0				vs EtOH
4410	4,4'-Dinitro-2,2'-stilbenedisulfonic acid		$C_{14}H_{10}N_2O_{10}S_2$	128-42-7	430.366	cry (AcOH)	266				
4411	Dinobuton	Dessin	$C_{14}H_{18}N_2O_7$	973-21-7	326.302	ye cry (EtOH)	60				
4412	Dinocap		$C_{18}H_{24}N_2O_6$	6119-92-2	364.393			136^{0.01}			
4413	Dinonyl adipate		$C_{24}H_{46}O_4$	151-32-6	398.620			205¹			

2,4-Dinitrobenzenesulfonic acid

2,4-Dinitrobenzoic acid

3,4-Dinitrobenzoic acid

3,5-Dinitrobenzoic acid

3,5-Dinitrobenzoyl chloride

2,2'-Dinitro-1,1'-biphenyl

4,4'-Dinitro-1,1'-biphenyl

1,4-Dinitrobutane

4,4'-Dinitrodiphenylamine

4,4'-Dinitrodiphenyl ether

4,4'-Dinitrodiphenyl sulfide

1,1-Dinitroethane

1,2-Dinitroethane

Dinitromethane

1,3-Dinitronaphthalene

1,5-Dinitronaphthalene

1,8-Dinitronaphthalene

2,4-Dinitro-1-naphthol

2,3-Dinitrophenol

2,4-Dinitrophenol

2,5-Dinitrophenol

2,6-Dinitrophenol

3,4-Dinitrophenol

2,4-Dinitrophenol, acetate

4-[(2,4-Dinitrophenyl)amino]phenol

2,4-Dinitro-N-phenylaniline

2,4-Dinitrophenyl dimethylcarbamodithioate

(2,4-Dinitrophenyl)hydrazine

1,1-Dinitropropane

1,3-Dinitropropane

2,2-Dinitropropane

2,2-Dinitro-1,3-propanediol

1,6-Dinitropyrene

1,8-Dinitropyrene

Dinitrosopentamethylenetetramine

1,4-Dinitrosopiperazine

4,4'-Dinitro-2,2'-stilbenedisulfonic acid

Dinobuton

R = NO$_2$, R' = C$_8$H$_{17}$
and
R = C$_8$H$_{17}$, R' = NO$_2$

Dinocap

Dinonyl adipate

No.	Name	Synonym	Mol. Form.	CAS RN	Mol. Wt.	Physical Form	mp/°C	bp/°C	den/g cm⁻³	n_D	Solubility
4414	Dinonyl ether		$C_{18}H_{38}O$	2456-27-1	270.494	liq		318	0.81	1.4356[20]	
4415	Dinonyl phthalate		$C_{26}H_{42}O_4$	84-76-4	418.609			413			
4416	Dinoseb	Phenol, 2-(1-methylpropyl)-4,6-dinitro-	$C_{10}H_{12}N_2O_5$	88-85-7	240.212		40		1.265[45]		
4417	Dioctadecylamine	Distearylamine	$C_{36}H_{75}N$	112-99-2	521.988		72.9	268[2]			vs chl
4418	Dioctylamine	N-Octyl-1-octanamine	$C_{16}H_{35}N$	1120-48-5	241.456	nd	35.5	297.5	0.7963[26]	1.4415[26]	vs eth, EtOH
4419	Dioctyl ether		$C_{16}H_{34}O$	629-82-3	242.440	liq	-7.6	283	0.8063[20]	1.4327[20]	sl H_2O; s EtOH, eth, ctc
4420	Dioctyl hexanedioate		$C_{22}H_{42}O_4$	123-79-5	370.566		9.6	191[2]	0.922[25]		
4421	Dioctyl maleate		$C_{20}H_{36}O_4$	2915-53-9	340.498	liq		242[0.002]	0.94[20]	1.4539[20]	
4422	Dioctyl phthalate		$C_{24}H_{38}O_4$	117-84-0	390.557		25	220[4]			
4423	Dioctyl sebacate	Dioctyl decanedioate	$C_{26}H_{50}O_4$	2432-87-3	426.673		18	218[0.5]	0.9074[25]		s ctc
4424	Dioctyl sulfide	Octyl sulfide	$C_{16}H_{34}S$	2690-08-6	258.506			202[29], 180[10]	0.842[20]	1.4610[20]	
4425	Dioctyl terephthalate		$C_{24}H_{38}O_4$	4654-26-6	390.557	cry		425	1.21[62]		
4426	Dioscorine		$C_{13}H_{19}NO_2$	3329-91-7	221.296	grn-ye pr (eth)	34				s H_2O, ace, chl, EtOH; sl eth, bz
4427	1,3-Dioxane	1,3-Dioxacyclohexane	$C_4H_8O_2$	505-22-6	88.106	liq	-45	106.1	1.0286[25]	1.4165[20]	msc H_2O, EtOH, eth, ace, bz
4428	1,4-Dioxane	1,4-Dioxacyclohexane	$C_4H_8O_2$	123-91-1	88.106		11.85	101.5	1.0337[20]	1.4224[20]	msc H_2O, EtOH, eth, ace, bz; s ctc
4429	1,4-Dioxane-2,5-dione		$C_4H_4O_4$	502-97-6	116.073	lf (al, al-chl)	85.4				vs ace
4430	1,4-Dioxane-2,6-dione	Diglycollic anhydride	$C_4H_4O_4$	4480-83-5	116.073	cry (bz)	92.5	240.5; 120[12]			
4431	Dioxathion		$C_{12}H_{26}O_6P_2S_4$	78-34-2	456.538		-20		1.257[26]		
4432	1,3-Dioxepane		$C_5H_{10}O_2$	505-65-7	102.132						s chl
4433	1,3-Dioxolane	1,3-Dioxacyclopentane	$C_3H_6O_2$	646-06-0	74.079	liq	-97.22	78	1.060[20]	1.3974[20]	msc H_2O; s EtOH, eth, ace
4434	1,3-Dioxol-2-one		$C_3H_2O_3$	872-36-6	86.046	liq	22	162; 73[32]	1.35[25]		
4435	Dioxybenzone	(2-Hydroxy-4-methoxyphenyl)(2-hydroxyphenyl)methanone	$C_{14}H_{12}O_4$	131-53-3	244.243			172[1]			
4436	Dioxypyramidon		$C_{13}H_{17}N_3O_3$	519-65-3	263.292	pr	105.5	197[2]			s H_2O, EtOH
4437	Dipentaerythritol		$C_{10}H_{22}O_7$	126-58-9	254.278	cry (w)	221		1.366[15]		s hot H_2O
4438	Dipentene	p-Menthadiene	$C_{10}H_{16}$	7705-14-8	136.234	liq	-95.5	178	0.8402[21]	1.4727[20]	
4439	Dipentylamine	Diamylamine	$C_{10}H_{23}N$	2050-92-2	157.297			202.5	0.7771[20]	1.4272[20]	sl H_2O; vs EtOH; msc eth; s ace
4440	Dipentyl cis-2-butenedioate	Dipentyl maleate	$C_{14}H_{24}O_4$	10099-71-5	256.339	liq		161[10]	0.974[20]		
4441	Dipentyl ether	Amyl ether	$C_{10}H_{22}O$	693-65-2	158.281	liq	-69	190	0.7833[20]	1.4119[20]	i H_2O; msc EtOH, eth; s chl
4442	2,6-Di-tert-pentyl-4-methylphenol	2,6-Bis(1,1-dimethylpropyl)-4-methylphenol	$C_{17}H_{28}O$	56103-67-4	248.403			283	0.931[25]	1.4950[20]	
4443	Di-tert-pentyl peroxide		$C_{10}H_{22}O_2$	10508-09-5	174.281		-55	58[14], 38[9]	0.808[20]	1.4095[20]	
4444	Dipentyl phthalate		$C_{18}H_{26}O_4$	131-18-0	306.397			205[11]			s ctc, CS_2
4445	Dipentyl sulfide		$C_{10}H_{22}S$	872-10-6	174.347		-51.3	86[3.7]	0.8407[20]	1.4561[20]	i H_2O; s eth
4446	Dipentyl sulfoxide		$C_{10}H_{22}OS$	1986-90-9	190.346		58	120[1]			
4447	Diphenamid	Benzeneacetamide, N,N-dimethyl-α-phenyl-	$C_{16}H_{17}NO$	957-51-7	239.312		135		1.17[23.3]		
4448	Diphenidol	1,1-Diphenyl-4-piperidinyl-1-butanol	$C_{21}H_{27}NO$	972-02-1	309.445	nd (peth)	104.5				
4449	Diphenolic acid		$C_{17}H_{18}O_4$	126-00-1	286.323	cry (w)	171.5				vs H_2O, ace, EtOH
4450	1,2-Diphenoxyethane	Ethylene glycol diphenyl ether	$C_{14}H_{14}O_2$	104-66-5	214.260	lf (al)	98	182[12]			i H_2O; sl EtOH; s eth, chl
4451	N,N-Diphenylacetamide		$C_{14}H_{13}NO$	519-87-9	211.259	wh cry pow	103	sub			sl H_2O, eth, chl; s EtOH
4452	Diphenylacetylene		$C_{14}H_{10}$	501-65-5	178.229	mcl pr or pl (al)	62.5	300	0.9657[100]		i H_2O; sl EtOH, chl; vs eth
4453	2-(Diphenylacetyl)-1H-indene-1,3(2H)-dione	Diphenadione	$C_{23}H_{16}O_3$	82-66-6	340.371	pa ye mcl (al)	146.5			1.670	vs ace, HOAc
4454	Diphenylamine	N-Phenylbenzenamine	$C_{12}H_{11}N$	122-39-4	169.222	mcl lf(dil al)	53.2	302	1.158[22]		i H_2O; vs EtOH, ace; s eth; sl chl
4455	Diphenylamine-2,2'-dicarboxylic acid		$C_{14}H_{11}NO_4$	579-92-0	257.242	ye cry (al)	296 dec				
4456	Diphenylamine-4-sulfonic acid, sodium salt	Sodium diphenylamine-4-sulfonate	$C_{12}H_{10}NNaO_3S$	6152-67-6	271.267	ye cry					
4457	9,10-Diphenylanthracene		$C_{26}H_{18}$	1499-10-1	330.421		246.5				
4458	Diphenylarsinous chloride	Chlorodiphenylarsine	$C_{12}H_{10}AsCl$	712-48-1	264.582	orth pl (peth)	44	337	1.4820[16]	1.6332[56]	vs ace, bz, eth, EtOH
4459	N,N'-Diphenyl-1,4-benzenediamine	N,N-Diphenyl-p-phenylenediamine	$C_{18}H_{16}N_2$	74-31-7	260.333		150	222[0.5]			sl EtOH, eth, bz, chl; i acid

Dinonyl ether

Dinonyl phthalate

Dinoseb

Dioctadecylamine

Dioctylamine

Dioctyl ether

Dioctyl hexanedioate

Dioctyl maleate

Dioctyl phthalate

Dioctyl sebacate

Dioctyl sulfide

Dioctyl terephthalate

Dioscorine

1,3-Dioxane

1,4-Dioxane

1,4-Dioxane-2,5-dione

1,4-Dioxane-2,6-dione

Dioxathion

1,3-Dioxepane

1,3-Dioxolane

1,3-Dioxol-2-one

Dioxybenzone

Dioxypyramidon

Dipentaerythritol

Dipentene

Dipentylamine

Dipentyl cis-2-butenedioate

Dipentyl ether

2,6-Di-tert-pentyl-4-methylphenol

Di-tert-pentyl peroxide

Dipentyl phthalate

Dipentyl sulfide

Dipentyl sulfoxide

Diphenamid

Diphenidol

Diphenolic acid

1,2-Diphenoxyethane

N,N-Diphenylacetamide

Diphenylacetylene

2-(Diphenylacetyl)-1H-indene-1,3(2H)-dione

Diphenylamine

Diphenylamine-2,2'-dicarboxylic acid

Diphenylamine-4-sulfonic acid, sodium salt

9,10-Diphenylanthracene

Diphenylarsinous chloride

N,N'-Diphenyl-1,4-benzenediamine

No.	Name	Synonym	Mol. Form.	CAS RN	Mol. Wt.	Physical Form	mp/°C	bp/°C	den/ g cm⁻³	n_D	Solubility
4460	α,α-Diphenylbenzeneethanol		$C_{20}H_{18}O$	4428-13-1	274.356	nd(bz-lig) pr (peth)	89.5	222[11]			i H₂O; vs EtOH; sl eth, chl, peth
4461	α,α-Diphenylbenzenemethanethiol	Triphenylmethyl mercaptan	$C_{19}H_{16}S$	3695-77-0	276.395		105.8				
4462	N,N'-Diphenyl-[1,1'-biphenyl]-4,4'-diamine	N,N'-Diphenylbenzidine	$C_{24}H_{20}N_2$	531-91-9	336.429	lf or pl	247				i H₂O; sl EtOH, eth, bz; vs tol, HOAc
4463	trans,trans-1,4-Diphenyl-1,3-butadiene		$C_{16}H_{14}$	538-81-8	206.282	lf (al, HOAc)	154.3	352			vs bz, eth, EtOH, peth
4464	1,4-Diphenyl-1,3-butadiyne	Diphenyldiacetylene	$C_{16}H_{10}$	886-66-8	202.250		86.5				
4465	1,1-Diphenylbutane		$C_{16}H_{18}$	719-79-9	210.314		27	287	0.9928[20]	1.5664[20]	i H₂O; s EtOH, eth, bz, chl
4466	1,2-Diphenylbutane		$C_{16}H_{18}$	5223-59-6	210.314			291; 152[11]	0.9673[20]	1.5554[20]	i H₂O; s EtOH, eth, bz, chl
4467	1,4-Diphenylbutane		$C_{16}H_{18}$	1083-56-3	210.314		52.5	317	0.9880[20]		i H₂O; s EtOH, eth, chl
4468	1,3-Diphenyl-1-butene		$C_{16}H_{16}$	7614-93-9	208.298		47.5	311	0.9996[20]	1.590[15]	
4469	trans-1,4-Diphenyl-2-butene-1,4-dione		$C_{16}H_{12}O_2$	959-28-4	236.265	ye nd (al, bz)	111				sl EtOH; s bz, HOAc; vs chl; i lig
4470	1,3-Diphenyl-2-buten-1-one	Dypnone	$C_{16}H_{14}O$	495-45-4	222.281			342.5	1.1080[15]	1.6343[20]	vs eth, EtOH
4471	Diphenylcarbamic chloride		$C_{13}H_{10}ClNO$	83-01-2	231.677	lf (al)	84.5				
4472	Diphenylcarbazone		$C_{13}H_{12}N_4O$	538-62-5	240.260	oran oran nd (bz) pr (al)	157 dec				i H₂O; vs EtOH, bz, chl
4473	N,N'-Diphenylcarbodiimide		$C_{13}H_{10}N_2$	622-16-2	194.231		169	331; 175[20]			sl H₂O, EtOH, eth; s bz
4474	Diphenyl carbonate	Phenyl carbonate	$C_{13}H_{10}O_3$	102-09-0	214.216	nd (al, bz)	83	306	1.1215[87]		i H₂O; s EtOH, eth, ctc, HOAc
4475	2,2'-Diphenylcarbonic dihydrazide	sym-Diphenylcarbazide	$C_{13}H_{14}N_4O$	140-22-7	242.276	cry (al + 1) cry (HOAc)	170	dec			sl H₂O, eth; s EtOH, ace, bz
4476	Diphenyl chlorophosphonate		$C_{12}H_{10}ClO_3P$	2524-64-3	268.632			314[272]	1.296[25]	1.5500[20]	s tfa
4477	Diphenyl diselenide	Phenyl diselenide	$C_{12}H_{10}Se_2$	1666-13-3	312.13	ye nd	63.5	202[11]	1.557[80]	1.743[20]	s EtOH, eth, xyl, MeOH
4478	Diphenyl disulfide	Phenyl disulfide	$C_{12}H_{10}S_2$	882-33-7	218.337	nd(al) or orth	62	310	1.353[20]		i H₂O; s EtOH, eth, bz, CS₂
4479	1,1-Diphenylethane		$C_{14}H_{14}$	612-00-0	182.261	liq	-17.9	272.6	0.9997[20]	1.5756[20]	i H₂O; msc EtOH, eth; s bz
4480	1,2-Diphenylethane	Dibenzyl	$C_{14}H_{14}$	103-29-7	182.261	mcl pr (MeOH)	52.5	284	0.9780[25]	1.5476[60]	i H₂O; s EtOH, eth, CS₂
4481	N,N'-Diphenylethanediamide		$C_{14}H_{12}N_2O_2$	620-81-5	240.257	lf (bz)	254	>360			vs bz
4482	N,N'-Diphenyl-1,2-ethanediamine	1,2-Dianilinoethane	$C_{14}H_{16}N_2$	150-61-8	212.290	cry (dil al)	74	229[12], 178[2]			i H₂O; s EtOH, eth; sl tfa
4483	1,2-Diphenyl-1,2-ethanediol, (R*,R*)-(±)		$C_{14}H_{14}O_2$	655-48-1	214.260	nd (w,al),tab (eth)	122.5	>300			i H₂O, lig; vs EtOH, eth; s ace
4484	1,1-Diphenylethene		$C_{14}H_{12}$	530-48-3	180.245		8.2	277	1.0232[20]	1.6085[20]	i H₂O; s eth, chl
4485	Diphenyl ether		$C_{12}H_{10}O$	101-84-8	170.206		26.87	258.0	1.0661[30]	1.5787[25]	i H₂O; s EtOH, eth, bz, HOAc; sl chl
4486	Diphenyl 2-ethylhexyl phosphate		$C_{20}H_{27}O_4P$	1241-94-7	362.399			232[5]	1.090[25]	1.510[25]	
4487	N,N-Diphenylformamide		$C_{13}H_{11}NO$	607-00-1	197.232	orth (dil al)	73.5	337.5; 189[13]			i H₂O; s EtOH, eth, bz; sl ctc
4488	2,5-Diphenylfuran		$C_{16}H_{12}O$	955-83-9	220.265	nd or lf (dil al)	91	344			i H₂O; vs EtOH, eth; s ace, bz
4489	N,N'-Diphenylguanidine	1,3-Diphenylguanidine	$C_{13}H_{13}N_3$	102-06-7	211.262	mcl nd (al, to)	150	dec 170	1.13[20]		sl H₂O; s EtOH, ctc chl, tol; vs eth
4490	1,6-Diphenyl-1,3,5-hexatriene		$C_{18}H_{16}$	1720-32-7	232.320	lf (ace)	202.3				i H₂O, EtOH, eth, HOAc; s ace; sl bz, chl
4491	1,1-Diphenylhydrazine		$C_{12}H_{12}N_2$	530-50-7	184.236	tab (lig)	50.5	220[40]	1.190[16]		vs bz, eth, EtOH, chl
4492	1,2-Diphenylhydrazine	Hydrazobenzene	$C_{12}H_{12}N_2$	122-66-7	184.236	tab (al-eth)	131		1.158[16]		vs EtOH; sl bz, DMSO; i HOAc
4493	5,5-Diphenyl-4-imidazolidinone	Doxenitoin	$C_{15}H_{14}N_2O$	3254-93-1	238.284	pl (MeOH)	183				
4494	Diphenyl isophthalate		$C_{20}H_{14}O_4$	744-45-6	318.323		138				s chl
4495	Diphenylketene	Diphenylethenone	$C_{14}H_{10}O$	525-06-4	194.228	red-ye liq		267.5	1.1107[13]	1.615[14]	
4496	Diphenyl maleate		$C_{16}H_{12}O_4$	7242-17-3	268.264	pl (lig)	73	226[15]			vs ace, bz, eth, EtOH
4497	Diphenylmercury	Mercuriodibenzene	$C_{12}H_{10}Hg$	587-85-9	354.80			204[10]	2.318[25]		i H₂O; sl EtOH, eth; s bz, chl
4498	Diphenylmethane		$C_{13}H_{12}$	101-81-5	168.234	pr nd	25.4	265.0	1.001[26]	1.5753[20]	i H₂O; s EtOH, eth, chl

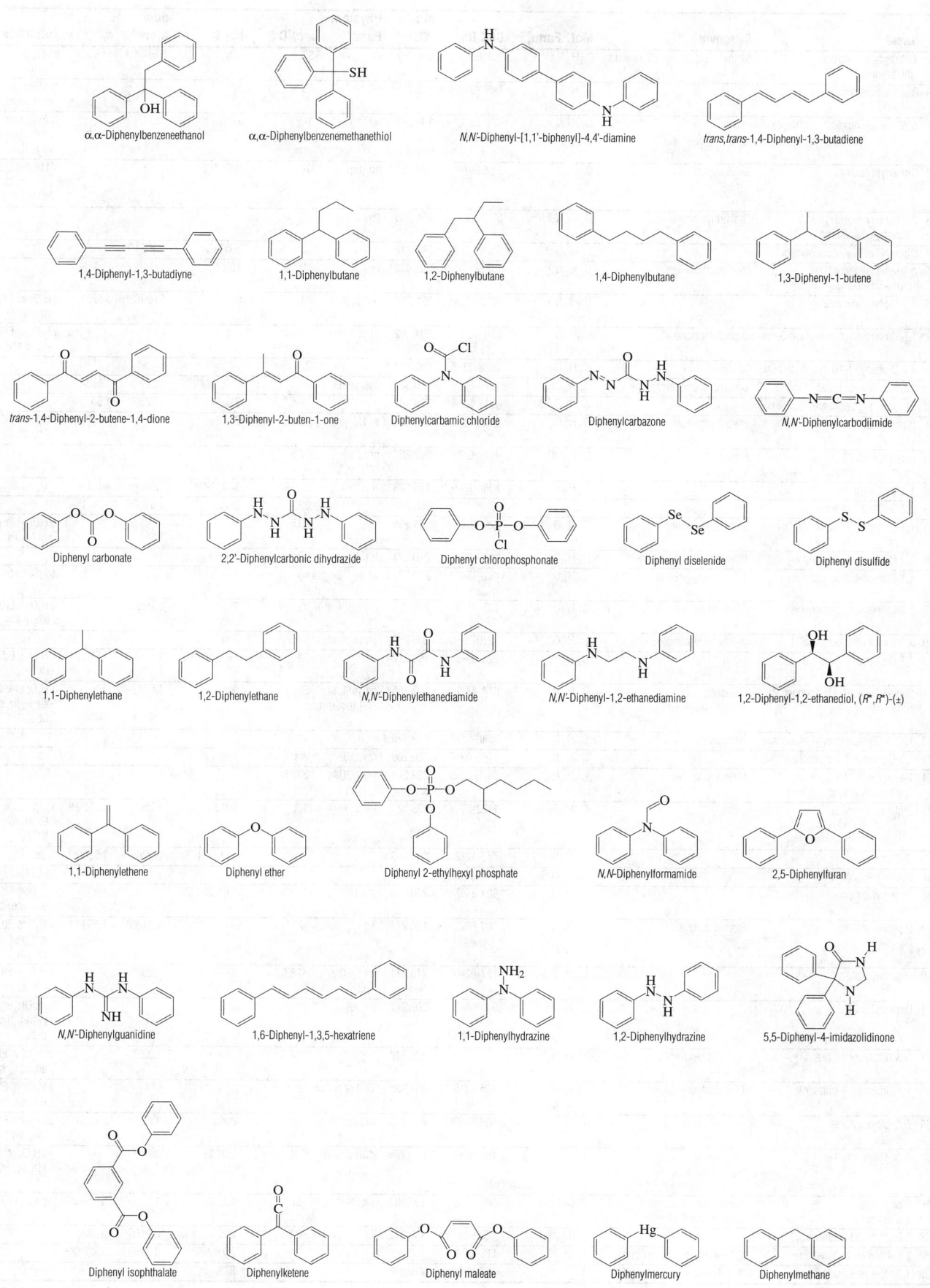

α,α-Diphenylbenzeneethanol

α,α-Diphenylbenzenemethanethiol

N,N'-Diphenyl-[1,1'-biphenyl]-4,4'-diamine

trans,trans-1,4-Diphenyl-1,3-butadiene

1,4-Diphenyl-1,3-butadiyne

1,1-Diphenylbutane

1,2-Diphenylbutane

1,4-Diphenylbutane

1,3-Diphenyl-1-butene

trans-1,4-Diphenyl-2-butene-1,4-dione

1,3-Diphenyl-2-buten-1-one

Diphenylcarbamic chloride

Diphenylcarbazone

N,N'-Diphenylcarbodiimide

Diphenyl carbonate

2,2'-Diphenylcarbonic dihydrazide

Diphenyl chlorophosphonate

Diphenyl diselenide

Diphenyl disulfide

1,1-Diphenylethane

1,2-Diphenylethane

N,N'-Diphenylethanediamide

N,N'-Diphenyl-1,2-ethanediamine

1,2-Diphenyl-1,2-ethanediol, (*R**,*R**)-(±)

1,1-Diphenylethene

Diphenyl ether

Diphenyl 2-ethylhexyl phosphate

N,N-Diphenylformamide

2,5-Diphenylfuran

N,N'-Diphenylguanidine

1,6-Diphenyl-1,3,5-hexatriene

1,1-Diphenylhydrazine

1,2-Diphenylhydrazine

5,5-Diphenyl-4-imidazolidinone

Diphenyl isophthalate

Diphenylketene

Diphenyl maleate

Diphenylmercury

Diphenylmethane

No.	Name	Synonym	Mol. Form.	CAS RN	Mol. Wt.	Physical Form	mp/°C	bp/°C	den/ g cm⁻³	n_D	Solubility
4499	4,4'-Diphenylmethane diisocyanate	Methylene diphenyl diisocyanate	$C_{15}H_{10}N_2O_2$	101-68-8	250.252		37	196[5]	1.197[70]	1.5906[50]	s ace, bz, PhNO₂
4500	Diphenylmethanethione		$C_{13}H_{10}S$	1450-31-3	198.283		53.5	174[14]			sl EtOH, eth, peth; vs bz, chl
4501	N,N'-Diphenylmethanimidamide		$C_{13}H_{12}N_2$	622-15-1	196.247	nd (al)	142	>250			sl H₂O, peth; s EtOH, ace, bz; vs eth
4502	Diphenylmethanol	Benzohydrol	$C_{13}H_{12}O$	91-01-0	184.233	nd (lig)	69	298; 180[20]			sl H₂O; vs EtOH, eth, ctc, chl; s HOAc
4503	2-(Diphenylmethoxy)-N,N-dimethylethanamine	Diphenhydramine	$C_{17}H_{21}NO$	58-73-1	255.355	oil		165[3]			
4504	Diphenyl methylphosphonate		$C_{13}H_{13}O_3P$	7526-26-3	248.214		35	205[13]	1.2051[20]		i H₂O
4505	2-(Diphenylmethyl)-1-piperidineethanol	Diphemethoxidine	$C_{20}H_{25}NO$	13862-07-2	295.419		106.5	180[0.1]			
4506	2,5-Diphenyloxazole		$C_{15}H_{11}NO$	92-71-7	221.254	nd (lig)	74	360	1.0940[100]	1.6231[100]	i H₂O; vs EtOH, eth; sl chl
4507	1,5-Diphenyl-1,4-pentadien-3-one	Dibenzalacetone	$C_{17}H_{14}O$	538-58-9	234.292	pl or lf (ace, AcOEt)	113 dec	dec			i H₂O; sl EtOH, eth; s ace, chl
4508	4,7-Diphenyl-1,10-phenanthroline		$C_{24}H_{16}N_2$	1662-01-7	332.397		220 dec				
4509	Diphenylphosphinous chloride	Chlorodiphenylphosphine	$C_{12}H_{10}ClP$	1079-66-9	220.634	hyg ye liq		320; 174[5]	1.229	1.6360[20]	
4510	Diphenyl phosphonate		$C_{12}H_{11}O_3P$	4712-55-4	234.187		12	218[26]	1.223[25]	1.5575[20]	
4511	Diphenyl phthalate	Phenyl phthalate	$C_{20}H_{14}O_4$	84-62-8	318.323	pr (al, lig)	73	253[14]			i H₂O; sl EtOH, eth, ctc
4512	α,α-Diphenyl-2-piperidinemethanol	Pipradrol	$C_{18}H_{21}NO$	467-60-7	267.366	cry (hx)	97.5				
4513	1,3-Diphenylpropane		$C_{15}H_{16}$	1081-75-0	196.288	liq	6	300; 123[1.7]	1.007[20]	1.5760[20]	
4514	2,2-Diphenylpropane		$C_{15}H_{16}$	778-22-3	196.288		29	282.5	0.9980[20]		
4515	1,3-Diphenyl-1,3-propanedione	Dibenzoylmethane	$C_{15}H_{12}O_2$	120-46-7	224.255		70.5				s EtOH, eth, chl, dil NaOH
4516	1,3-Diphenyl-1-propanone	Phenethyl phenyl ketone	$C_{15}H_{14}O$	1083-30-3	210.271	lf (EtOH)	72.5	360			
4517	1,1-Diphenyl-2-propanone	1,1-Diphenylacetone	$C_{15}H_{14}O$	781-35-1	210.271		46	307; 174[10]		1.5361[16]	s EtOH, eth, bz, chl, lig
4518	1,3-Diphenyl-2-propanone	Dibenzyl ketone	$C_{15}H_{14}O$	102-04-5	210.271	cry (al, peth)	35	331	1.195[0]		i H₂O; s EtOH, eth, peth
4519	3,3-Diphenyl-2-propenal	β-Phenylcinnamaldehyde	$C_{15}H_{12}O$	1210-39-5	208.255	pa ye pr (lig)	44.8	205[14]			
4520	1,1-Diphenyl-1-propene		$C_{15}H_{14}$	778-66-5	194.272		52	280; 149[11]	1.0250[20]	1.5880[20]	i H₂O; s EtOH, bz
4521	trans-1,3-Diphenyl-2-propen-1-one	Chalcone	$C_{15}H_{12}O$	614-47-1	208.255	pa ye lf, pr, nd (peth)	59	dec 346	1.0712[62]		i H₂O; sl EtOH; s eth, bz, chl, CS₂
4522	1-(3,3-Diphenylpropyl)piperidine	Fenpiprane	$C_{20}H_{25}N$	3540-95-2	279.420		41.5	215[8]			
4523	3,5-Diphenyl-1H-pyrazole		$C_{15}H_{12}N_2$	1145-01-3	220.269	cry (al)	200				
4524	1,4-Diphenyl-3,5-pyrazolidinedione	Phenopyrazone	$C_{15}H_{12}N_2O_2$	3426-01-5	252.268	cry (EtOAc, Diox)	233.5				
4525	Diphenyl selenide		$C_{12}H_{10}Se$	1132-39-4	233.17	ye nd (bz)	1.3	301.5	1.351[20]	1.5500[20]	i H₂O; msc EtOH, eth; s bz, xyl
4526	Diphenylsilane		$C_{12}H_{12}Si$	775-12-2	184.309			134[16], 96[13]	0.9969[20]	1.5800[20]	s ctc, CS₂
4527	Diphenylsilanediol		$C_{12}H_{12}O_2Si$	947-42-2	216.308						sl DMSO
4528	Diphenyl succinate		$C_{16}H_{14}O_4$	621-14-7	270.280	lf (al)	121	330; 222.5[15]			i H₂O; s EtOH, eth, ace, bz
4529	Diphenyl sulfide	Phenyl sulfide	$C_{12}H_{10}S$	139-66-2	186.272	liq	-25.9	296	1.1136[20]	1.6334[20]	i H₂O; s EtOH, ctc; msc eth, bz, CS₂
4530	Diphenyl sulfone		$C_{12}H_{10}O_2S$	127-63-9	218.271	mcl pr(bz) pl(al)	128.5	379	1.252[20]		i H₂O; s EtOH, eth, bz
4531	Diphenyl sulfoxide		$C_{12}H_{10}OS$	945-51-7	202.271	pr(lig)	71.2	340[16]			vs EtOH, eth, bz, HOAc; sl chl; i peth
4532	N,N'-Diphenylthiourea	sym-Diphenylthiourea	$C_{13}H_{12}N_2S$	102-08-9	228.312		154.5		1.32[25]		sl H₂O; vs EtOH, eth, chl, oils
4533	1,3-Diphenyl-1-triazene	Diazoaminobenzene	$C_{12}H_{11}N_3$	136-35-6	197.235	ye lf or pr (al)	98				i H₂O; vs EtOH, eth, bz, py
4534	N,N-Diphenylurea		$C_{13}H_{12}N_2O$	603-54-3	212.246	tab (al)	189	dec	1.276[25]		sl H₂O; s EtOH, eth, chl
4535	N,N'-Diphenylurea	Carbanilide	$C_{13}H_{12}N_2O$	102-07-8	212.246	orth pr (al)	239	260 dec	1.239[25]		sl H₂O, EtOH; s eth, py, HOAc; i bz
4536	Diphosgene	Carbonochloridic acid, trichloromethyl ester	$C_2Cl_4O_2$	503-38-8	197.832	liq	-57	128	1.6525[14]	1.4566[22]	vs eth, EtOH
4537	1,2-Dipiperidinoethane		$C_{12}H_{24}N_2$	1932-04-3	196.332	liq	-0.5	265	0.9160[25]	1.4853[25]	
4538	1,1'-Dipiperidinomethane	1,1'-Methylenedipiperidine	$C_{11}H_{22}N_2$	880-09-1	182.306			230; 122[15]	0.9269[20]	1.4820[20]	
4539	1,3-Di-4-piperidylpropane	4,4'-Trimethylenedipiperidine	$C_{13}H_{26}N_2$	16898-52-5	210.358		67.1	329			vs H₂O

4,4'-Diphenylmethane diisocyanate

Diphenylmethanethione

N,N'-Diphenylmethanimidamide

Diphenylmethanol

2-(Diphenylmethoxy)-*N,N*-dimethylethanamine

Diphenyl methylphosphonate

2-(Diphenylmethyl)-1-piperidineethanol

2,5-Diphenyloxazole

1,5-Diphenyl-1,4-pentadien-3-one

4,7-Diphenyl-1,10-phenanthroline

Diphenylphosphinous chloride

Diphenyl phosphonate

Diphenyl phthalate

α,α-Diphenyl-2-piperidinemethanol

1,3-Diphenylpropane

2,2-Diphenylpropane

1,3-Diphenyl-1,3-propanedione

1,3-Diphenyl-1-propanone

1,1-Diphenyl-2-propanone

1,3-Diphenyl-2-propanone

3,3-Diphenyl-2-propenal

1,1-Diphenyl-1-propene

trans-1,3-Diphenyl-2-propen-1-one

1-(3,3-Diphenylpropyl)piperidine

3,5-Diphenyl-1*H*-pyrazole

1,4-Diphenyl-3,5-pyrazolidinedione

Diphenyl selenide

Diphenylsilane

Diphenylsilanediol

Diphenyl succinate

Diphenyl sulfide

Diphenyl sulfone

Diphenyl sulfoxide

N,N'-Diphenylthiourea

1,3-Diphenyl-1-triazene

N,N-Diphenylurea

N,N'-Diphenylurea

Diphosgene

1,2-Dipiperidinoethane

1,1'-Dipiperidinomethane

1,3-Di-4-piperidylpropane

No.	Name	Synonym	Mol. Form.	CAS RN	Mol. Wt.	Physical Form	mp/°C	bp/°C	den/ g cm⁻³	n_D	Solubility
4540	Diploicin		$C_{16}H_{10}Cl_4O_5$	527-93-5	424.059		232				
4541	Di-2-propenoyldiethyleneglycol		$C_{10}H_{14}O_5$	4074-88-8	214.215			200	1.1110[25]	1.4595[25]	
4542	Di-2-propenoyl-2,2-dimethyl-1,3-propanediol	2-Propenoic acid, 2,2-dimethyl-1,3-propanediyl ester	$C_{11}H_{16}O_4$	2223-82-7	212.243					1.4542[25]	
4543	Di-2-propenoyl-1,6-hexanediol	2-Propenoic acid, 1,6-hexanediyl ester	$C_{12}H_{18}O_4$	13048-33-4	226.269				1.010[25]		
4544	Dipropetryn	6-(Ethylthio)-N,N'-diisopropyl-1,3,5-triazine-2,4-diamine	$C_{11}H_{21}N_5S$	4147-51-7	255.384		105				
4545	1,2-Dipropoxyethane		$C_8H_{18}O_2$	18854-56-3	146.228	liq		163.2 dec	0.8312[25]	1.4013[25]	
4546	Dipropoxymethane	Formaldehyde, dipropyl acetal	$C_7H_{16}O_2$	505-84-0	132.201	liq	-97.3	140.5	0.8345[20]	1.3939[19]	vs ace, bz, eth, EtOH
4547	N,N-Dipropylacetamide		$C_8H_{17}NO$	1116-24-1	143.227			209.5	0.8992[17]	1.4419[17]	vs EtOH
4548	Dipropyl adipate		$C_{12}H_{22}O_4$	106-19-4	230.301		-15.7	151[11]	0.9790[20]	1.4314[20]	vs eth, EtOH, chl
4549	Dipropylamine	N-Propyl-1-propanamine	$C_6H_{15}N$	142-84-7	101.190	liq	-63	109.3	0.7400[20]	1.4050[20]	s H₂O, EtOH; msc eth; vs ace, bz
4550	4-[(Dipropylamino)sulfonyl]benzoic acid	Probenecid	$C_{13}H_{19}NO_4S$	57-66-9	285.360		195				
4551	N,N-Dipropylaniline		$C_{12}H_{19}N$	2217-07-4	177.286	ye lf		242	0.9104[20]	1.5271[20]	i H₂O; s EtOH, eth, ace, bz; sl ctc
4552	Dipropylcarbamothioic acid, S-ethyl ester	EPTC	$C_9H_{19}NOS$	759-94-4	189.318			127[20]	0.9546[30]		
4553	Dipropyl carbonate		$C_7H_{14}O_3$	623-96-1	146.184			168	0.9435[20]	1.4008[20]	sl H₂O; msc EtOH, eth
4554	Dipropyl disulfide		$C_6H_{14}S_2$	629-19-6	150.305	liq	-85.6	195.8	0.9599[20]	1.4981[20]	
4555	Dipropylene glycol		$C_6H_{14}O_3$	25265-71-8	134.173			230.5	1.0206[20]		msc H₂O; s EtOH
4556	Dipropylene glycol dibenzoate		$C_{20}H_{22}O_5$	27138-31-4	342.386			197[1]			
4557	Dipropylene glycol monomethyl ether	1-(2-Methoxyisopropoxy)-2-propanol	$C_7H_{16}O_3$	34590-94-8	148.200	liq	-80	188.3	0.95	1.4190[20]	
4558	Dipropyl ether	Propyl ether	$C_6H_{14}O$	111-43-3	102.174	liq	-114.8	90.08	0.7466[20]	1.3809[20]	sl H₂O; vs eth, EtOH
4559	Dipropyl fumarate		$C_{10}H_{16}O_4$	14595-35-8	200.232			110[5]	1.0129[20]	1.4435[20]	s EtOH, eth
4560	Dipropyl maleate		$C_{10}H_{16}O_4$	2432-63-5	200.232			126[12]	1.0245[20]	1.4434[20]	i H₂O; s EtOH, eth, ace, bz
4561	Dipropyl oxalate		$C_8H_{14}O_4$	615-98-5	174.195	liq	-44.3	211	1.0188[20]	1.4158[20]	sl H₂O; msc EtOH; s eth
4562	5,5-Dipropyl-2,4-oxazolidinedione		$C_9H_{15}NO_3$	512-12-9	185.220		42.5	149[3]			
4563	Dipropyl succinate		$C_{10}H_{18}O_4$	925-15-5	202.248	liq	-5.9	250.8	1.0020[20]	1.4250[20]	vs ace, bz, eth
4564	Dipropyl sulfate		$C_6H_{14}O_4S$	598-05-0	182.238			120[20]	1.1064[20]	1.4135[20]	vs peth
4565	Dipropyl sulfide		$C_6H_{14}S$	111-47-7	118.240	liq	-102.5	142.9	0.814[17]	1.4487[20]	i H₂O; s EtOH, eth
4566	Dipropyl sulfone		$C_6H_{14}O_2S$	598-03-8	150.239	cry	29.5		1.0278[50]	1.4456[30]	sl H₂O; s EtOH, eth
4567	Dipropyl sulfoxide		$C_6H_{14}OS$	4253-91-2	134.239	nd	22.5	80[2]	0.9654[20]	1.4663[20]	vs eth, EtOH
4568	Dipyridamole		$C_{24}H_{40}N_8O_4$	58-32-2	504.627		163				
4569	Di-2-pyridinyl disulfide, N,N'-dioxide	Dipyrithione	$C_{10}H_8N_2O_2S_2$	3696-28-4	252.313	cry (MeOH)	205				
4570	2,2'-Dipyrrolylmethane		$C_9H_{10}N_2$	21211-65-4	146.188	lf or nd (al)	73	164[12]			vs bz, eth, EtOH
4571	Diquat		$C_{12}H_{12}N_2$	2764-72-9	184.236	Cation					
4572	Diquat dibromide		$C_{12}H_{12}Br_2N_2$	85-00-7	344.044		337		1.24[20]		
4573	Disodium calcium EDTA	Edetate calcium disodium	$C_{10}H_{12}CaN_2Na_2O_8$	62-33-9	374.268	pow					s H₂O
4574	Disodium hydrogen citrate	Sodium acid citrate	$C_6H_6Na_2O_7$	144-33-2	236.088	wh pow (w)	149 dec				vs H₂O
4575	Disperse Blue No. 1	1,4,5,8-Tetraamino-9,10-anthracenedione	$C_{14}H_{12}N_4O_2$	2475-45-8	268.271	red-br nd	331				
4576	Distearyl thiodipropionate	Dioctadecyl thiobispropanoate	$C_{42}H_{82}O_4S$	693-36-7	683.163	cry	61				
4577	Disulfiram		$C_{10}H_{20}N_2S_4$	97-77-8	296.539		71.5	117[17]			i H₂O; s EtOH; sl eth; vs chl
4578	Disulfoton		$C_8H_{19}O_2PS_3$	298-04-4	274.405		-25	108[0.01], 128[1]	1.144[20]		
4579	1,2-Dithiane		$C_4H_8S_2$	505-20-4	120.237	nd	32.5	80[14], 60[5]		1.5981[25]	s eth, bz, chl
4580	1,3-Dithiane		$C_4H_8S_2$	505-23-7	120.237		54	89[14]		1.5981[25]	vs bz, eth, chl
4581	1,4-Dithiane		$C_4H_8S_2$	505-29-3	120.237	mcl pr	112.3	199.5			sl H₂O; s EtOH, eth, ctc, CS₂, HOAc
4582	Dithianone		$C_{14}H_4N_2O_2S_2$	3347-22-6	296.324	nd (ace)	220				
4583	Dithiazanine iodide		$C_{23}H_{23}IN_2S_2$	514-73-8	518.476	grn nd (MeOH)	248 dec				i H₂O
4584	2,2'-Dithiobisbenzoic acid	Diphenyl disulfide-2,2'-dicarboxylic acid	$C_{14}H_{10}O_4S_2$	119-80-2	306.357		289.5				i H₂O; s EtOH, eth
4585	3,3'-Dithiobispropanoic acid		$C_6H_{10}O_4S_2$	1119-62-6	210.271		158				
4586	3,3'-Dithiobis-D-valine		$C_{10}H_{20}N_2O_4S_2$	20902-45-8	296.407		204.5				

Diploicin

Di-2-propenoyldiethyleneglycol

Di-2-propenoyl-2,2-dimethyl-1,3-propanediol

Di-2-propenoyl-1,6-hexanediol

Dipropetryn

1,2-Dipropoxyethane

Dipropoxymethane

N,N-Dipropylacetamide

Dipropyl adipate

Dipropylamine

4-[(Dipropylamino)sulfonyl]benzoic acid

N,N-Dipropylaniline

Dipropylcarbamothioic acid, *S*-ethyl ester

Dipropyl carbonate

Dipropyl disulfide

Dipropylene glycol

Dipropylene glycol dibenzoate

Dipropylene glycol monomethyl ether

Dipropyl ether

Dipropyl fumarate

Dipropyl maleate

Dipropyl oxalate

5,5-Dipropyl-2,4-oxazolidinedione

Dipropyl succinate

Dipropyl sulfate

Dipropyl sulfide

Dipropyl sulfone

Dipropyl sulfoxide

Dipyridamole

Di-2-pyridinyl disulfide, *N,N'*-dioxide

2,2'-Dipyrrolylmethane

Diquat

Diquat dibromide

Disodium calcium EDTA

Disodium hydrogen citrate

Disperse Blue No. 1

Distearyl thiodipropionate

Disulfiram

Disulfoton

1,2-Dithiane

1,3-Dithiane

1,4-Dithiane

Dithianone

Dithiazanine iodide

2,2'-Dithiobisbenzoic acid

3,3'-Dithiobispropanoic acid

3,3'-Dithiobis-*D*-valine

No.	Name	Synonym	Mol. Form.	CAS RN	Mol. Wt.	Physical Form	mp/°C	bp/°C	den/ g cm⁻³	n_D	Solubility
4587	2,5-Dithiobiurea	1,2-Hydrazinedicarbothioamide	$C_2H_6N_4S_2$	142-46-1	150.226	nd (w)	214				
4588	4,4'-Dithiodimorpholine		$C_8H_{16}N_2O_2S_2$	103-34-4	236.355		124.5				s chl
4589	1,2-Dithiolane		$C_3H_6S_2$	557-22-2	106.210		77	90²⁷			
4590	1,3-Dithiolane	1,3-Dithiacyclopentane	$C_3H_6S_2$	4829-04-3	106.210	liq	-50	175	1.259¹⁷	1.5975¹⁵	s EtOH, eth, xyl
4591	1,3-Dithiolane-2-thione		$C_3H_4S_3$	822-38-8	136.259		35	307			
4592	Dithiopyr		$C_{15}H_{16}F_5NO_2S_2$	97886-45-8	401.416		65				
4593	Dithizone		$C_{13}H_{12}N_4S$	60-10-6	256.326	bl-blk (chl-al)	167 dec				i H_2O; sl EtOH, eth; s chl, alk
4594	Di(p-tolyl)carbodiimide		$C_{15}H_{14}N_2$	726-42-1	222.285		58.5	221²⁰	1.1500²⁰		
4595	1,2-Di(p-tolyl)ethane	1,2-Bis(p-tolyl)ethane	$C_{16}H_{18}$	538-39-6	210.314	lf (al)	85	178¹⁸			i H_2O; sl EtOH; s bz, peth
4596	N,N'-Di(o-tolyl)guanidine		$C_{15}H_{17}N_3$	97-39-2	239.316	cry (dil al)	179		1.10²⁰		sl H_2O, tfa, EtOH; vs eth; s chl
4597	Ditridecyl phthalate		$C_{34}H_{58}O_4$	119-06-2	530.823	liq		285³·⁵	0.952²⁵		
4598	Diundecyl phthalate		$C_{30}H_{50}O_4$	3648-20-2	474.716	cry (EtOH)	35.5				
4599	Diuron		$C_9H_{10}Cl_2N_2O$	330-54-1	233.093		158				
4600	o-Divinylbenzene	1,2-Divinylbenzene	$C_{10}H_{10}$	91-14-5	130.186			82¹⁴	0.9325²²	1.5767²⁰	s ace, bz
4601	m-Divinylbenzene	1,3-Divinylbenzene	$C_{10}H_{10}$	108-57-6	130.186		-52.3	121⁷⁶, 52³	0.9294²⁰	1.5760²⁰	s ace, bz
4602	p-Divinylbenzene	1,4-Divinylbenzene	$C_{10}H_{10}$	105-06-6	130.186		31	95¹⁸, 34⁰·²	0.913⁴⁰	1.5835²⁵	s ace, bz
4603	cis-1,2-Divinylcyclobutane		C_8H_{12}	16177-46-1	108.181			38³⁸	0.8010²⁰	1.4563²⁰	
4604	trans-1,2-Divinylcyclobutane		C_8H_{12}	6553-48-6	108.181			112.5	0.7817²⁰	1.4451²⁰	
4605	Divinyl ether		C_4H_6O	109-93-3	70.090	vol liq or gas	-100.6	28.3	0.773²⁰	1.3989²⁰	i H_2O; msc EtOH, eth, ace, chl
4606	Divinyl sulfide	Vinyl sulfide	C_4H_6S	627-51-0	86.156		20	84	0.9174¹⁵		sl H_2O; s ace; msc EtOH, eth
4607	Divinyl sulfone	Vinyl sulfone	$C_4H_6O_2S$	77-77-0	118.155	liq	-26	234.5	1.177²⁵	1.4765²⁰	
4608	1,3-Divinyl-1,1,3,3-tetramethyldisiloxane		$C_8H_{18}OSi_2$	2627-95-4	186.399	liq	-99.7	39	0.811²⁰	1.4123²⁰	
4609	Djenkolic acid		$C_7H_{14}N_2O_4S_2$	498-59-9	254.327	nd(w)	≈325 dec				
4610	DMPA		$C_{10}H_{14}Cl_2NO_2PS$	299-85-4	314.169	solid	51.4	150²			sl H_2O; vs bz, ctc, ace
4611	Docosane		$C_{22}H_{46}$	629-97-0	310.600	pl(to), cry (eth)	43.6	368.6	0.7944²⁰	1.4455²⁰	i H_2O; s EtOH, chl; vs eth
4612	Docosanoic acid	Behenic acid	$C_{22}H_{44}O_2$	112-85-6	340.583	nd	81.5	306⁶⁰	0.8223⁹⁰	1.4270¹⁰⁰	sl H_2O, EtOH, eth
4613	1-Docosanol		$C_{22}H_{46}O$	661-19-8	326.599	cry (ace, chl)	72.5	180⁰·²²			sl H_2O, eth; vs EtOH, MeOH; s chl
4614	13-Docosenamide	Erucamide	$C_{22}H_{43}NO$	112-84-5	337.582	cry	94				
4615	1-Docosene		$C_{22}H_{44}$	1599-67-3	308.584		38	367	0.794²⁵		
4616	cis-13-Docosenoic acid	Erucic acid	$C_{22}H_{42}O_2$	112-86-7	338.567	nd (al)	34.7	265¹⁵	0.860⁵⁵	1.4758²⁰	i H_2O; s EtOH, ctc; vs eth, MeOH
4617	trans-13-Docosenoic acid	Brassidic acid	$C_{22}H_{42}O_2$	506-33-2	338.567	pl (al)	61.9	282³⁰, 256¹⁰	0.8585⁵⁷	1.4347¹⁰⁰	
4618	5,7-Dodecadiyne	Dibutylbutadiyne	$C_{12}H_{18}$	1120-29-2	162.271			103⁸			
4619	Dodecamethylcyclohexasiloxane		$C_{12}H_{36}O_6Si_6$	540-97-6	444.923	liq	-1.5	245	0.9672²⁵	1.4015²⁰	i H_2O
4620	Dodecamethylpentasiloxane		$C_{12}H_{36}O_4Si_5$	141-63-9	384.840	liq	-80	232; 105¹²	0.8755²⁰	1.3925²⁰	s ctc, CS_2
4621	Dodecanal	Lauraldehyde	$C_{12}H_{24}O$	112-54-9	184.318	lf	44.5	185¹⁰⁰, 100²·⁵	0.8352¹⁵	1.435²²	i H_2O; sl EtOH; s eth
4622	Dodecanamide		$C_{12}H_{25}NO$	1120-16-7	199.333	nd	110	199¹²		1.4287¹¹⁰	i H_2O; s EtOH, ace, ctc; sl eth, bz
4623	Dodecane		$C_{12}H_{26}$	112-40-3	170.334	liq	-9.57	216.32	0.7495²⁰	1.4210²⁰	i H_2O; vs EtOH, eth, ace, ctc, chl
4624	1,12-Dodecanediamine		$C_{12}H_{28}N_2$	2783-17-7	200.363		67.38	135³			
4625	Dodecanedioic acid		$C_{12}H_{22}O_4$	693-23-2	230.301		128	222²⁵	1.15²⁵		s tfa
4626	1,12-Dodecanediol		$C_{12}H_{26}O_2$	5675-51-4	202.333	cry (bz, dil al)	81.3	189¹²			s tfa
4627	Dodecanenitrile	Lauronitrile	$C_{12}H_{23}N$	2437-25-4	181.318		4	277; 198¹⁰⁰	0.8240²⁰	1.4361²⁰	i H_2O; msc EtOH, eth, ace, bz, chl
4628	1-Dodecanethiol		$C_{12}H_{26}S$	112-55-0	202.399	liq	-6.7	277; 143¹⁵	0.844²⁰	1.4589²⁰	i H_2O; s EtOH, eth, chl
4629	Dodecanoic acid	Lauric acid	$C_{12}H_{24}O_2$	143-07-7	200.318	nd (al)	43.8	225¹⁰⁰	0.8679⁵⁰	1.4183⁸²	i H_2O; vs EtOH, eth; s ace; msc bz
4630	Dodecanoic anhydride		$C_{24}H_{46}O_3$	645-66-9	382.620	lf (al, eth)	41.8		0.8533⁷⁰	1.4292⁷⁰	vs EtOH
4631	1-Dodecanol	Lauryl alcohol	$C_{12}H_{26}O$	112-53-8	186.333	lf (dil al)	23.9	260	0.8309²⁴		i H_2O; s EtOH, eth; sl bz
4632	2-Dodecanol		$C_{12}H_{26}O$	10203-28-8	186.333		19	252	0.8286²⁰	1.4400²⁰	

2,5-Dithiobiurea

4,4'-Dithiodimorpholine

1,2-Dithiolane

1,3-Dithiolane

1,3-Dithiolane-2-thione

Dithiopyr

Dithizone

Di(p-tolyl)carbodiimide

1,2-Di(p-tolyl)ethane

N,N'-Di(o-tolyl)guanidine

Ditridecyl phthalate

Diundecyl phthalate

Diuron

o-Divinylbenzene

m-Divinylbenzene

p-Divinylbenzene

cis-1,2-Divinylcyclobutane

trans-1,2-Divinylcyclobutane

Divinyl ether

Divinyl sulfide

Divinyl sulfone

1,3-Divinyl-1,1,3,3-tetramethyldisiloxane

Djenkolic acid

DMPA

Docosane

Docosanoic acid

1-Docosanol

13-Docosenamide

1-Docosene

cis-13-Docosenoic acid

trans-13-Docosenoic acid

5,7-Dodecadiyne

Dodecamethylcyclohexasiloxane

Dodecamethylpentasiloxane

Dodecanal

Dodecanamide

Dodecane

1,12-Dodecanediamine

Dodecanedioic acid

1,12-Dodecanediol

Dodecanenitrile

1-Dodecanethiol

Dodecanoic acid

Dodecanoic anhydride

1-Dodecanol

2-Dodecanol

No.	Name	Synonym	Mol. Form.	CAS RN	Mol. Wt.	Physical Form	mp/°C	bp/°C	den/ g cm⁻³	n_D	Solubility
4633	2-Dodecanone	Decyl methyl ketone	$C_{12}H_{24}O$	6175-49-1	184.318		21	246.5	0.8198²⁰	1.4330²⁰	i H₂O; s EtOH, eth, ace; sl ctc
4634	Dodecanoyl chloride		$C_{12}H_{23}ClO$	112-16-3	218.763		-17	145¹⁸	0.9169²⁵	1.4458²⁰	vs eth
4635	1-Dodecene		$C_{12}H_{24}$	112-41-4	168.319	liq	-35.2	213.8	0.7584²⁰	1.4300²⁰	i H₂O; s EtOH, eth, ace, ctc, peth
4636	trans-2-Dodecenedioic acid	Traumatic acid	$C_{12}H_{20}O_4$	6402-36-4	228.285	cry (al,ace)	165.5				vs eth, EtOH, chl
4637	2-Dodecenylsuccinic anhydride		$C_{16}H_{26}O_3$	19780-11-1	266.375	hyg cry	42	181⁵			
4638	Dodecyl acetate		$C_{14}H_{28}O_2$	112-66-3	228.371		0.7	265; 180⁴⁰	0.8652²²	1.4439²⁰	
4639	Dodecyl acrylate	Lauryl 2-propenoate	$C_{15}H_{28}O_2$	2156-97-0	240.382		4	120⁰·⁸	0.8727²⁰		
4640	Dodecylamine	1-Dodecanamine	$C_{12}H_{27}N$	124-22-1	185.349		28.3	259	0.8015²⁰	1.4421²⁰	sl H₂O; msc EtOH, eth, bz, chl
4641	Dodecylamine, acetate	1-Dodecanamine, acetate	$C_{14}H_{31}NO_2$	2016-56-0	245.402		69.5				vs H₂O, EtOH
4642	Dodecylamine hydrochloride	Lauryl amine hydrochloride	$C_{12}H_{28}ClN$	929-73-7	221.810		186				vs H₂O, EtOH
4643	4-Dodecylaniline		$C_{18}H_{31}N$	104-42-7	261.446		41.5	211¹⁰			
4644	Dodecylbenzene	Laurylbenzene	$C_{18}H_{30}$	123-01-3	246.431		3	328	0.8551²⁰	1.4824²⁰	i H₂O
4645	4-Dodecylbenzenesulfonic acid		$C_{18}H_{30}O_3S$	121-65-3	326.494			>205			
4646	Dodecylcyclohexane		$C_{18}H_{36}$	1795-17-1	252.479		12.5	331	0.8223²⁰	1.4559²⁰	
4647	Dodecyl mercaptoacetate		$C_{14}H_{28}O_2S$	3746-39-2	260.436		1.5	171³			
4648	Dodecyl methacrylate		$C_{16}H_{30}O_2$	142-90-5	254.408			142⁴	0.866²⁰		
4649	Dodecyloxirane	1,2-Epoxytetradecane	$C_{14}H_{28}O$	3234-28-4	212.371	oil		95⁰·⁴	0.845	1.4408²⁰	
4650	2-(Dodecyloxy)ethanol		$C_{14}H_{30}O_2$	4536-30-5	230.387			143⁰·⁸			
4651	4-Dodecyloxy-2-hydroxybenzophenone		$C_{25}H_{34}O_3$	2985-59-3	382.536		43.5				
4652	4-Dodecylphenol		$C_{18}H_{30}O$	104-43-8	262.430	nd (bz)	66	175²			
4653	1-Dodecylpiperidine		$C_{17}H_{35}N$	5917-47-5	253.467	pa ye		161⁵, 115⁰·⁶	0.8378²⁰	1.4588²⁰	
4654	Dodecyl sulfate	Lauryl sulfate	$C_{12}H_{26}O_4S$	151-41-7	266.397	cry					s H₂O
4655	Dodecyltetraethylene glycol monoether	3,6,9,12-Tetraoxatetracosan-1-ol	$C_{20}H_{42}O_5$	5274-68-0	362.544			247¹⁰			
4656	Dodecyl 3,4,5-trihydroxybenzoate		$C_{19}H_{30}O_5$	1166-52-5	338.438		96.5				s ace
4657	Dodecyltrimethylammonium chloride		$C_{15}H_{34}ClN$	112-00-5	263.891		246 dec				vs H₂O, ace, EtOH, chl
4658	1-Dodecyne	Decylacetylene	$C_{12}H_{22}$	765-03-7	166.303	liq	-19	215	0.7788²⁰	1.4340²⁰	
4659	6-Dodecyne		$C_{12}H_{22}$	6975-99-1	166.303			210; 100¹⁴	0.785²⁰	1.4442²⁰	vs ace, eth, EtOH
4660	Dodine	Dodecylguanidine, monoacetate	$C_{15}H_{33}N_3O_2$	2439-10-3	287.442		136				
4661	Dopamine	4(2-Aminoethyl)-1,2-benzenediol	$C_8H_{11}NO_2$	51-61-6	153.179	pr					
4662	Dothiepin		$C_{19}H_{21}NS$	113-53-1	295.442		56	172⁰·⁰⁵			
4663	Dotriacontane	Bicetyl	$C_{32}H_{66}$	544-85-4	450.866	pl (bz,chl,HO Ac,eth)	69.4	467	0.8124²⁰	1.4550²⁰	i H₂O; sl EtOH, chl; s eth, ctc; vs bz
4664	Doxepin		$C_{19}H_{21}NO$	1668-19-5	279.376	oily liq		155⁰·⁰³, 265⁰·²			
4665	Doxorubicin	Adriamycin	$C_{27}H_{29}NO_{11}$	23214-92-8	543.519	cry	230				
4666	Doxorubicin hydrochloride	Adriamycin hydrochloride	$C_{27}H_{30}ClNO_{11}$	25316-40-9	579.980	oran-red nd	204 dec				s H₂O, MeOH; i ace, bz, chl, eth, peth
4667	Doxylamine		$C_{17}H_{22}N_2O$	469-21-6	270.369	liq		139⁰·⁵			
4668	Drimenin		$C_{15}H_{22}O_2$	2326-89-8	234.335	cry	133	110⁰·¹			i H₂O
4669	Dromostanolone propanoate	2-Methyl-17-(1-oxopropoxy)androstan-3-one, (2α,5α,17β)	$C_{23}H_{36}O_3$	521-12-0	360.530		128				
4670	Droperidol	Dehydrobenzperidol	$C_{22}H_{22}FN_3O_2$	548-73-2	379.427	cry (w)	146 (hyd)				i H₂O; sl EtOH, eth, bz; s chl, DMF
4671	Dydrogesterone		$C_{21}H_{28}O_2$	152-62-5	312.446	cry (ace/hx)	170				
4672	Dyphylline		$C_{10}H_{14}N_4O_4$	479-18-5	254.243		161.5				
4673	Ecgonidine		$C_9H_{13}NO_2$	484-93-5	167.205	cry (MeOH) (MeOH-eth)	228 dec				vs H₂O
4674	Ecgonine		$C_9H_{15}NO_3$	481-37-8	185.220	mcl pr	205				vs H₂O, EtOH
4675	Echimidine		$C_{20}H_{31}NO_7$	520-68-3	397.463	glass					
4676	Echinochrome A	2-Ethyl-3,5,6,7,8-pentahydroxy-1,4-naphthalenedione	$C_{12}H_{10}O_7$	517-82-8	266.203	red nd (Diox-w)	220 dec	sub 120			sl H₂O; s EtOH, ace; vs eth, bz
4677	Echitamine		$C_{22}H_{30}N_2O_5$	6871-44-9	402.483		206				s H₂O, EtOH, eth, chl, con sulf; i peth
4678	Edrophonium chloride		$C_{10}H_{16}ClNO$	116-38-1	201.693	cry	162				vs H₂O; s EtOH; i eth, chl
4679	Efloxate	Ethyl [(4-oxo-2-phenyl-4H-1-benzopyran-7-yl)oxy]acetate	$C_{19}H_{16}O_5$	119-41-5	324.327		123.7				s chl
4680	Eicosamethylnonasiloxane		$C_{20}H_{60}O_8Si_9$	2652-13-3	681.455			307.5; 198¹⁶	0.9173²⁰	1.3980²⁰	vs bz

2-Dodecanone

Dodecanoyl chloride

1-Dodecene

trans-2-Dodecenedioic acid

2-Dodecenylsuccinic anhydride

Dodecyl acetate

Dodecyl acrylate

Dodecylamine

Dodecylamine, acetate

Dodecylamine hydrochloride

4-Dodecylaniline

Dodecylbenzene

4-Dodecylbenzenesulfonic acid

Dodecylcyclohexane

Dodecyl mercaptoacetate

Dodecyl methacrylate

Dodecyloxirane

2-(Dodecyloxy)ethanol

4-Dodecyloxy-2-hydroxybenzophenone

4-Dodecylphenol

1-Dodecylpiperidine

Dodecyl sulfate

Dodecyltetraethylene glycol monoether

Dodecyl 3,4,5-trihydroxybenzoate

Dodecyltrimethylammonium chloride

1-Dodecyne

6-Dodecyne

Dodine

Dopamine

Dothiepin

Dotriacontane

Doxepin

Doxorubicin

Doxorubicin hydrochloride

Doxylamine

Drimenin

Dromostanolone propanoate

Droperidol

Dydrogesterone

Dyphylline

Ecgonidine

Ecgonine

Echimidine

Echinochrome A

Echitamine

Edrophonium chloride

Efloxate

Eicosamethylnonasiloxane

No.	Name	Synonym	Mol. Form.	CAS RN	Mol. Wt.	Physical Form	mp/°C	bp/°C	den/ g cm^{-3}	n_D	Solubility
4681	Eicosane	Icosane	C$_{20}$H$_{42}$	112-95-8	282.547	lf (al)	36.6	343	0.7886[20]	1.4425[20]	i H$_2$O; s eth, peth, bz; sl chl; vs ace
4682	Eicosanedioic acid	1,18-Octadecanedicarboxylic acid	C$_{20}$H$_{38}$O$_4$	2424-92-2	342.514	cry (bz,al)	125.5	233[2]			s eth
4683	Eicosanoic acid	Arachidic acid	C$_{20}$H$_{40}$O$_2$	506-30-9	312.531	pl (al)	76.5	dec 328; 204[1]	0.8240[100]	1.425[100]	i H$_2$O; sl EtOH; vs eth; s bz, chl
4684	1-Eicosanol	Arachic alcohol	C$_{20}$H$_{42}$O	629-96-9	298.546	wax (al), cry (chl)	65.4	356; 222[3]	0.8405[20]	1.4350[20]	i H$_2$O; sl EtOH, chl; vs ace; s bz, peth
4685	5,8,11,14-Eicosatetraenoic acid, (all-*cis*)	Arachidonic acid	C$_{20}$H$_{32}$O$_2$	506-32-1	304.467		-49.5	163[1]	0.9082[20]	1.4824[20]	i H$_2$O; vs ace, eth, EtOH, peth
4686	1-Eicosene		C$_{20}$H$_{40}$	3452-07-1	280.532		28.5	341; 151[2]	0.7882[30]	1.4440[30]	i H$_2$O; s bz, peth
4687	*cis*-9-Eicosenoic acid		C$_{20}$H$_{38}$O$_2$	29204-02-2	310.515		24.5	220[6]	0.8882[25]		
4688	*trans*-9-Eicosenoic acid		C$_{20}$H$_{38}$O$_2$	506-31-0	310.515		54				
4689	11-Eicosenoic acid		C$_{20}$H$_{38}$O$_2$	2462-94-4	310.515		24	267[15]	0.8826[25]		vs EtOH, MeOH
4690	Elaidic acid	*trans*-9-Octadecenoic acid	C$_{18}$H$_{34}$O$_2$	112-79-8	282.462	pl (al)	45	288[100]; 234[15]	0.8734[45]	1.4499[45]	i H$_2$O; s EtOH, eth, bz, chl
4691	Elaiomycin		C$_{13}$H$_{26}$N$_2$O$_3$	23315-05-1	258.356	ye oil				1.4798[25]	sl H$_2$O; s os
4692	1,3-Elemadien-11-ol	Elemol	C$_{15}$H$_{26}$O	639-99-6	222.366	cry (al)	52.5	142[12]	0.9345[18]	1.4980[18]	
4693	β-Elemene		C$_{15}$H$_{24}$	33880-83-0	204.352			120[16]; 104[11]	0.8749[20]	1.4935[20]	
4694	Embelin	2,5-Dihydroxy-3-undecyl-2,5-cyclohexadiene-1,4-dione	C$_{17}$H$_{26}$O$_4$	550-24-3	294.386	oran pl (al)	142.5				vs bz, eth, EtOH
4695	Emetine	6',7',10,11-Tetramethoxyemetan	C$_{29}$H$_{40}$N$_2$O$_4$	483-18-1	480.639	amor pow	74				i H$_2$O; s EtOH, eth, ace; sl bz, chl
4696	Emylcamate	3-Methyl-3-pentanol, carbamate	C$_7$H$_{15}$NO$_2$	78-28-4	145.200	nd	57	35[1]			sl H$_2$O; vs bz, eth, EtOH
4697	Enallylpropymal		C$_{11}$H$_{16}$N$_2$O$_3$	1861-21-8	224.256	cry (w, dil al)	56.5	177[12]			vs bz, eth, EtOH, chl
4698	Endosulfan		C$_9$H$_6$Cl$_6$O$_3$S	115-29-7	406.925		106	106[0.7]	1.745[20]		
4699	Endosulfan sulfate		C$_9$H$_6$Cl$_6$O$_4$S	1031-07-8	422.925	cry (cyhex)	181				
4700	Endothall disodium		C$_8$H$_{10}$Na$_2$O$_5$	145-73-3	232.142		144		1.431[20]		
4701	Endrin		C$_{12}$H$_8$Cl$_6$O	72-20-8	380.909	cry	dec 245				vs ace, bz, xyl; s ctc, hx
4702	Enflurane		C$_3$H$_2$ClF$_5$O	13838-16-9	184.492	liq		56.5	1.5121[25]	1.3025[20]	vs os
4703	Ephedrine, (±)	α-[1-(Methylamino)ethyl] benzenemethanol, (R*,S*)-(±)-	C$_{10}$H$_{15}$NO	90-81-3	165.232	nd (eth, peth)	76.5	135[12]	1.1220[20]		s H$_2$O, EtOH, eth, bz, chl
4704	*d*-Ephedrine	α-[1-(Methylamino)ethyl] benzenemethanol, [S-(R*,S*)]-	C$_{10}$H$_{15}$NO	321-98-2	165.232	pl (w)	40	225			s H$_2$O, EtOH, eth, bz, chl
4705	*l*-Ephedrine	α-[1-(Methylamino)ethyl] benzenemethanol, [R-(R*,S*)]-	C$_{10}$H$_{15}$NO	299-42-3	165.232	pl (w + 1)	40	225	1.0085[22]		s H$_2$O, EtOH, eth, bz, chl
4706	Ephedrine hydrochloride	2-(Methylamino)-1-phenyl-1-propanol, hydrochloride	C$_{10}$H$_{16}$ClNO	50-98-6	201.693	orth nd	219		1.0208[20]		
4707	Epichlorohydrin	(Chloromethyl)oxirane	C$_3$H$_5$ClO	13403-37-7	92.524	liq	-26	118; 62[100]	1.1812[20]	1.4358[25]	sl H$_2$O; msc EtOH, eth; s bz, ctc
4708	Epinephrine	*D*-Adrenaline	C$_9$H$_{13}$NO$_3$	51-43-4	183.204	br (in air)	211.5				sl H$_2$O; i EtOH; s HOAc, acid
4709	Epiquinidine		C$_{20}$H$_{24}$N$_2$O$_2$	572-59-8	324.417	cry (AcOEt) lf (eth)	113				vs EtOH; s eth
4710	1,2-Epoxybutane	Ethyloxirane	C$_4$H$_8$O	106-88-7	72.106	liq	-150	63.4	0.8297[20]	1.3851[20]	vs EtOH, ace; msc eth
4711	1,2-Epoxy-4-(epoxyethyl) cyclohexane	4-Vinyl-1-cyclohexene dioxide	C$_8$H$_{12}$O$_2$	106-87-6	140.180		<-55	227	1.0966[20]	1.4787[20]	vs H$_2$O
4712	1,2-Epoxyhexadecane	Tetradecyloxirane	C$_{16}$H$_{32}$O	7320-37-8	240.424	hyg cry or liq	24.1	178[12]	0.846	1.2240	
4713	1,2-Epoxyoctadecane	Hexadecyloxirane	C$_{18}$H$_{36}$O	7390-81-0	268.478	hyg cry	26.1	137[0.5]			
4714	2,3-Epoxy-α-pinane		C$_{10}$H$_{16}$O	1686-14-2	152.233			85[24]			
4715	2,3-Epoxypropyl acrylate	Glycidyl acrylate	C$_6$H$_8$O$_3$	106-90-1	128.126			53[10]	1.1109[20]	1.4490[20]	vs bz
4716	2,3-Epoxypropyl methacrylate	Glycidol methacrylate	C$_7$H$_{10}$O$_3$	106-91-2	142.152			189; 75[10]	1.042[20]	1.448[25]	vs bz, eth, EtOH
4717	Equol		C$_{14}$H$_{14}$O$_3$	531-95-3	230.259	cry (aq, al)	189.5				
4718	Ergocornine		C$_{31}$H$_{39}$N$_5$O$_5$	564-36-3	561.673	cry (MeOH)	183 dec				i H$_2$O; s EtOH, ace, bz, chl, AcOEt
4719	Ergocorninine		C$_{31}$H$_{39}$N$_5$O$_5$	564-37-4	561.673	lo pr (al)	228 dec				vs ace, bz, EtOH, chl
4720	Ergocristine		C$_{35}$H$_{39}$N$_5$O$_5$	511-08-0	609.716	orth (bz)	175 dec				i H$_2$O; s EtOH, ace, chl
4721	Ergocristinine		C$_{35}$H$_{39}$N$_5$O$_5$	511-07-9	609.716	pr (al)	237 dec				i H$_2$O; sl EtOH, ace, chl
4722	Ergocryptine		C$_{32}$H$_{41}$N$_5$O$_5$	511-09-1	575.699	pr (al)	213 dec				i H$_2$O; s EtOH, chl
4723	Ergocryptinine		C$_{33}$H$_{41}$N$_5$O$_5$	511-10-4	587.710	lo pr (al)	245 dec				vs ace, chl

Eicosane

Eicosanedioic acid

Eicosanoic acid

1-Eicosanol

5,8,11,14-Eicosatetraenoic acid, (all-*cis*)

1-Eicosene

cis-9-Eicosenoic acid

trans-9-Eicosenoic acid

11-Eicosenoic acid

Elaidic acid

Elaiomycin

1,3-Elemadien-11-ol

β-Elemene

Embelin

Emetine

Emylcamate

Enallylpropymal

Endosulfan

Endosulfan sulfate

Endothall disodium

Endrin

Enflurane

Ephedrine, (±)

d-Ephedrine

l-Ephedrine

Ephedrine hydrochloride

Epichlorohydrin

Epinephrine

Epiquinidine

1,2-Epoxybutane

1,2-Epoxy-4-(epoxyethyl)cyclohexane

1,2-Epoxyhexadecane

1,2-Epoxyoctadecane

2,3-Epoxy-α-pinane

2,3-Epoxypropyl acrylate

2,3-Epoxypropyl methacrylate

Equol

Ergocornine

Ergocorninine

Ergocristine

Ergocristinine

Ergocryptine

Ergocryptinine

No.	Name	Synonym	Mol. Form.	CAS RN	Mol. Wt.	Physical Form	mp/°C	bp/°C	den/ g cm⁻³	n_D	Solubility
4724	Ergometrinine		C₁₉H₂₃N₃O₂	479-00-5	325.405	pr (ace)	196 dec				vs chl
4725	Ergonovine	Ergometrine	C₁₉H₂₃N₃O₂	60-79-7	325.405	pl or nd	162 dec				s H₂O, ace; vs EtOH; sl chl
4726	Ergosine		C₃₀H₃₇N₅O₅	561-94-4	547.646	pr (MeOH, AcOEt)	228 dec				s ace, chl; sl MeOH
4727	Ergostane, (5α)		C₂₈H₅₀	511-20-6	386.697	lf or pl (ace, eth- MeOH)	85				vs ace, eth, chl
4728	Ergostane, (5β)	Coproergostane	C₂₈H₅₀	511-21-7	386.697	nd (ace)	64				vs eth, chl
4729	Ergostan-3-ol, (3β,5α)	Ergostanol	C₂₈H₅₀O	6538-02-9	402.696	nd (MeOH-eth)	144.5				i H₂O; s eth, chl
4730	Ergosta-5,7,9(11),22-tetraen-3-ol, (3β,22E)	Dehydroergosterol	C₂₈H₄₂O	516-85-8	394.632	lf (al) nd (eth),pl (al)	146	230^0.5			vs ace, bz, eth, EtOH
4731	Ergosta-5,7,22-trien-3-ol, (3β,22E)	Ergosterol	C₂₈H₄₄O	57-87-4	396.648	pl (+w, al) nd (eth)	170	250^0.01			i H₂O; sl EtOH, eth, peth; s bz, chl
4732	Ergosta-5,7,22-trien-3-ol, (3β,10α,22E)	Pyrocalciferol	C₂₈H₄₄O	128-27-8	396.648	nd (MeOH)	94				i H₂O; s EtOH, chl, MeOH
4733	Ergosta-5,7,22-trien-3-ol, (3β,9β,10α,22E)	Lumisterol	C₂₈H₄₄O	474-69-1	396.648	nd (ace-MeOH)	118				i H₂O; s EtOH, HOAc; vs eth, ace, chl
4734	Ergost-5-en-3-ol, (3β,24R)	Campesterol	C₂₈H₄₈O	474-62-4	400.680	cry (ace)	157.5				
4735	Ergost-7-en-3-ol, (3β,5α)	γ-Ergostenol	C₂₈H₄₈O	516-78-9	400.680	nd (MeOH) cry (PrOH	146				s eth
4736	Ergost-8(14)-en-3-ol, (3β,5α)	α-Ergostenol	C₂₈H₄₈O	632-32-6	400.680	lf or nd (MeOH)	131				sl EtOH; s eth, bz, chl
4737	Ergotamine		C₃₃H₃₅N₅O₅	113-15-5	601.821	nd (al), pr (bz) pl (ace)	213 dec				vs bz, eth, chl
4738	Ergotamine tartrate (2:1)	Gynergen	C₃₅H₃₈N₅O₈	379-79-3	656.706		192 dec				
4739	Ergotaminine		C₃₃H₃₅N₅O₅	639-81-6	581.662	orth pl (MeOH) pl (al)	252 dec				i H₂O; sl EtOH, ace, bz; s chl; vs py
4740	Ergothioneine		C₉H₁₅N₃O₂S	497-30-3	229.299	nd or lf (dil EtOH)	290 dec				vs H₂O; sl EtOH, ace; i eth, bz, chl
4741	Eriochrome Black T		C₂₀H₁₂N₃NaO₇S	1787-61-7	461.380	br-blk pow					s H₂O, EtOH, MeOH
4742	Eriodictyol	3',4',5,7-Tetrahydroxyflavanone, (S)	C₁₅H₁₂O₆	552-58-9	288.252	pl or nd (EtOH)	267 dec				vs EtOH, HOAc
4743	Erythorbic acid	Isoascorbic acid	C₆H₈O₆	89-65-6	176.124	gran cry	168				s H₂O, py; sl ace
4744	β-Erythroidine		C₁₆H₁₉NO₃	466-81-9	273.327	cry (al)	99.5				s H₂O, eth, chl; vs EtOH, bz
4745	Erythromycin	Propiocine	C₃₇H₆₇NO₁₃	114-07-8	733.927	cry (w)	191				vs ace, eth, EtOH, chl
4746	Erythromycin ethyl succinate		C₄₃H₇₅NO₁₆	1264-62-6	862.053	cry (ace aq)	222				
4747	Erythromycin stearate		C₅₅H₁₀₃NO₁₅	643-22-1	1018.405	cry	92				i H₂O; sl EtOH, eth, chl
4748	Erythrophleine	Norcassamidine	C₂₄H₃₉NO₅	36150-73-9	421.571	glass	115				s H₂O, EtOH
4749	D-Erythrose		C₄H₈O₄	583-50-6	120.105	syr					s H₂O; vs EtOH
4750	L-Erythrose		C₄H₈O₄	533-49-3	120.105	syr					vs H₂O, EtOH
4751	D-Erythrose 4-phosphate	2,3-Dihydroxy-4-(phosphonooxy)butanal	C₄H₉O₇P	585-18-2	200.084	stab in aq soln only					s H₂O
4752	Erythrosine		C₂₀H₈I₄O₅	15905-32-5	835.893	br pow (Na salt)					s H₂; vs eth, EtOH
4753	L-Erythrulose		C₄H₈O₄	533-50-6	120.105	syr	dec	dec			vs H₂O, EtOH
4754	Esaprazole	N-Cyclohexyl-1-piperazineacetamide	C₁₂H₂₃N₃O	64204-55-3	225.330			112 190^0.5			
4755	Esculin	6-(β-D-Glucopyranosyloxy)-7-hydroxy-2H-1-benzopyran-2-one	C₁₅H₁₆O₉	531-75-9	340.283	pr (w+2)	205 (pentahydrate)				sl H₂O, EtOH, eth; s chl, py, HOAc
4756	Eserine sulfate	Physostigmine sulfate	C₃₀H₄₄N₆O₆S	64-47-1	648.770	hyg cry (ace-eth)	141				vs ace, EtOH
4757	Estra-1,3,5(10)-triene-3,17-diol, (17α)	α-Estradiol	C₁₈H₂₄O₂	57-91-0	272.383	nd (+1/2 w) (80% al)	221.5				i H₂O; s EtOH, ace; sl eth, bz
4758	Estra-1,3,5(10)-triene-3,17-diol, (17β)	β-Estradiol	C₁₈H₂₄O₂	50-28-2	272.383	pr (80% al),	178.5				vs ace, EtOH, Diox
4759	Estra-1,3,5(10)-triene-3,17-diol, (8α,17β)	Isoestradiol	C₁₈H₂₄O₂	517-04-4	272.383	cry (dil MeOH-chl)	181				s EtOH, diox
4760	Estra-1,3,5(10)-triene-3,17-diol 3-benzoate, (17β)	Estradiol benzoate	C₂₅H₂₈O₃	50-50-0	376.488		196				
4761	Estra-1,3,5(10)-triene-3,16,17-triol, (16α,17β)	Estriol	C₁₈H₂₄O₃	50-27-1	288.382	lf (al), mcl (dil al)	288 dec		1.27^25		s EtOH; sl eth, bz, tfa; vs py
4762	Estra-1,3,5(10)-triene-3,16,17-triol, (16β,17β)	16-Epiestriol	C₁₈H₂₄O₃	547-81-9	288.382	cry (MeOH-bz)	290				

Ergometrinine

Ergonovine

Ergosine

Ergostane, (5α)

Ergostane, (5β)

Ergostan-3-ol, (3β,5α)

Ergosta-5,7,9(11),22-tetraen-3-ol, (3β,22E)

Ergosta-5,7,22-trien-3-ol, (3β,22E)

Ergosta-5,7,22-trien-3-ol, (3β,10α,22E)

Ergosta-5,7,22-trien-3-ol, (3β,9β,10α,22E)

Ergost-5-en-3-ol, (3β,24R)

Ergost-7-en-3-ol, (3β,5α)

Ergost-8(14)-en-3-ol, (3β,5α)

Ergotamine

Ergotamine tartrate (2:1)

Ergotaminine

Ergothioneine

Eriochrome Black T

Eriodictyol

Erythorbic acid

β-Erythroidine

Erythromycin

Erythromycin ethyl succinate

Erythromycin stearate

Erythrophleine

D-Erythrose

L-Erythrose

D-Erythrose 4-phosphate

Erythrosine

L-Erythrulose

Esaprazole

Esculin

Eserine sulfate

Estra-1,3,5(10)-triene-3,17-diol, (17α)

Estra-1,3,5(10)-triene-3,17-diol (17β)

Estra-1,3,5(10)-triene-3,17-diol, (8α,17β)

Estra-1,3,5(10)-triene-3,17-diol 3-benzoate, (17β)

Estra-1,3,5(10)-triene-3,16,17-triol, (16α,17β)

Estra-1,3,5(10)-triene-3,16,17-triol, (16β,17β)

No.	Name	Synonym	Mol. Form.	CAS RN	Mol. Wt.	Physical Form	mp/°C	bp/°C	den/ g cm⁻³	n_D	Solubility
4763	Estrone		$C_{18}H_{22}O_2$	53-16-7	270.367	mcl, orth (al)	260.2		1.236²⁵		i H₂O; sl EtOH, eth, bz; s ace, diox
4764	Ethacrynic acid		$C_{13}H_{12}Cl_2O_4$	58-54-8	303.138		122.5				
4765	Ethalfluralin		$C_{13}H_{14}F_3N_3O_4$	55283-68-6	333.263		57	dec 256			
4766	Ethambutol		$C_{10}H_{24}N_2O_2$	74-55-5	204.310	cry	89				sl H₂O; s bz, chl
4767	Ethane		C_2H_6	74-84-0	30.069	col gas	-182.79	-88.6	0.5446⁻⁸⁹		i H₂O; vs bz
4768	Ethanearsonic acid		$C_2H_7AsO_3$	507-32-4	153.997	nd (al), orth nd (w)	99.5	210¹²			vs H₂O, EtOH
4769	Ethanedial dioxime		$C_2H_4N_2O_2$	557-30-2	88.065	orth pl (w)	178 dec	sub			vs H₂O, EtOH, eth
4770	1,2-Ethanediamine	Ethylenediamine	$C_2H_8N_2$	107-15-3	60.098		11.14	117	0.8979²⁰	1.4565²⁰	vs H₂O; msc EtOH; i eth, bz; s ctc
4771	1,2-Ethanediamine, dihydrochloride	Ethylenediamine dihydrochloride	$C_2H_{10}Cl_2N_2$	333-18-6	133.019					1.633	vs H₂O
4772	1,2-Ethanediol	Ethylene glycol	$C_2H_6O_2$	107-21-1	62.068	liq	-12.69	197.3	1.1135²⁰	1.4318²⁰	msc H₂O, EtOH, ace; s eth, chl; sl bz
4773	1,2-Ethanediol, bis(4-methylbenzenesulfonate)		$C_{16}H_{18}O_6S_2$	6315-52-2	370.440	cry (bz)	128				
4774	1,1-Ethanediol, diacetate	Ethylidene diacetate	$C_6H_{10}O_4$	542-10-9	146.141		18.9	169	1.070²⁵	1.3985²⁵	vs eth, EtOH
4775	1,2-Ethanediol, diacetate	Ethylene glycol diacetate	$C_6H_{10}O_4$	111-55-7	146.141	liq	-31	190	1.1043²⁰	1.4159²⁰	vs H₂O; msc EtOH, eth, ace, bz, CS₂
4776	1,2-Ethanediol, diacrylate	Ethylene glycol diacrylate	$C_8H_{10}O_4$	2274-11-5	170.163	liq		55⁰·⁶	1.0935²⁶		
4777	1,2-Ethanediol, dibenzoate	Ethylene glycol dibenzoate	$C_{16}H_{14}O_4$	94-49-5	270.280	orth pr (eth)	73.5	dec 360			i H₂O; s eth, chl
4778	1,2-Ethanediol, didodecanoate	Ethylene glycol didodecanoate	$C_{26}H_{50}O_4$	624-04-4	426.673	pl (al)	56.6	188²⁰			vs eth, EtOH
4779	1,2-Ethanediol, diformate	Ethylene glycol diformate	$C_4H_6O_4$	629-15-2	118.089			174	1.193⁰	1.3580	sl H₂O; s EtOH, eth
4780	1,2-Ethanediol, dihexadecanoate	Ethylene glycol dipalmitate	$C_{34}H_{66}O_4$	624-03-3	538.886	lf or nd (al-mcl)	72		0.8594⁷⁸		i H₂O, EtOH; s eth; vs ace
4781	1,2-Ethanediol, dimethacrylate	Ethylene glycol dimethacrylate	$C_{10}H_{14}O_4$	97-90-5	198.216	liq	-40	260	1.053²⁰	1.4532²⁵	vs bz, EtOH, lig
4782	1,2-Ethanediol, dinitrate	Ethylene glycol dinitrate	$C_2H_4N_2O_6$	628-96-6	152.062	ye liq	-22.3	198.5	1.4918²⁰		vs eth, EtOH
4783	1,2-Ethanediol, distearate	Ethylene glycol distearate	$C_{38}H_{74}O_4$	627-83-8	594.993	lf	79	241²⁰	0.8581⁷⁸		i H₂O, EtOH; vs eth, ace
4784	1,2-Ethanediol, ditetradecanoate	Ethylene glycol ditetradecanoate	$C_{30}H_{58}O_4$	627-84-9	482.780	cry (eth, ace)	65	208²⁰	0.8600⁸⁰		i H₂O, EtOH; s eth; vs ace, bz, ctc
4785	1,2-Ethanediol, dithiocyanate	Ethylene glycol dithiocyanate	$C_4H_4N_2S_2$	629-17-4	144.218	orth pl or nd (w)	90	dec	1.4200⁰		sl H₂O, bz; s EtOH, eth; vs ace
4786	1,2-Ethanediol, monoacetate	Ethylene glycol monoacetate	$C_4H_8O_3$	542-59-6	104.105			188	1.108¹⁵		msc H₂O, EtOH, eth
4787	1,2-Ethanediol, monobenzoate	Ethylene glycol monobenzoate	$C_9H_{10}O_3$	94-33-7	166.173		45	150¹⁰	1.1101³⁰		vs EtOH
4788	1,2-Ethanediol, monostearate	Ethylene glycol monostearate	$C_{20}H_{40}O_3$	111-60-4	328.530	cry (peth)	60.5	190³	0.8780⁶⁰	1.4310⁶⁰	sl EtOH; s eth
4789	1,2-Ethanediol, monosulfite	Ethylene glycol monosulfite	$C_2H_4O_3S$	3741-38-6	108.116	liq	-11	173	1.4402²⁰	1.4463²⁰	vs H₂O, EtOH, eth, ace, bz, AcOEt; sl chl
4790	1,2-Ethanediphosphonic acid	1,2-Diphosphonoethane	$C_2H_8O_6P_2$	6145-31-9	190.029	nd (EtOH/eth)	223				
4791	1,2-Ethanedisulfonic acid	Ethylene disulfonic acid	$C_2H_6O_6S_2$	110-04-3	190.195		173				vs diox
4792	Ethanedithioamide	Rubeanic acid	$C_2H_4N_2S_2$	79-40-3	120.196	red cry	170 dec				sl H₂O, EtOH; s con sulf
4793	1,2-Ethanedithiol	Ethylene dimercaptan	$C_2H_6S_2$	540-63-6	94.199	liq	-41.2	146.1	1.234²⁰	1.5590²⁰	i H₂O; s EtOH, eth, ace, bz; vs alk
4794	1,2-Ethanediyl mercaptoacetate		$C_6H_{10}O_4S_2$	123-81-9	210.271			138¹·⁵			
4795	Ethanesulfonic acid	Ethylsulfonic acid	$C_2H_6O_3S$	594-45-6	110.132	hyg	-17	123¹	1.3341²⁵	1.4335²⁰	vs H₂O, EtOH
4796	Ethanesulfonyl chloride		$C_2H_5ClO_2S$	594-44-5	128.578	pa ye		174	1.357²²	1.4531²⁰	vs eth; s CS₂
4797	Ethanethiol	Ethyl mercaptan	C_2H_6S	75-08-1	62.134	liq	-147.88	35.0	0.8315²⁵	1.4310²⁰	sl H₂O; s EtOH, eth, ace, dil alk
4798	Ethanimidamide		$C_2H_6N_2$	143-37-3	58.082		-35				sl H₂O; s EtOH, acid
4799	Ethanimidamide monohydrochloride	Acetamidine hydrochloride	$C_2H_7ClN_2$	124-42-5	94.543	nd or pr (al) hyg lo pr (al)	177.5				vs H₂O, EtOH
4800	Ethanol	Ethyl alcohol	C_2H_6O	64-17-5	46.068	liq	-114.14	78.29	0.7893²⁰	1.3611²⁰	msc H₂O, EtOH, eth, ace, chl; s bz
4801	Ethanolamine	Glycinol	C_2H_7NO	141-43-5	61.083		10.5	171	1.0180²⁰	1.4541²⁰	msc H₂O, EtOH; sl eth, lig, bz; s chl
4802	Ethanolamine hydrochloride	2-Aminoethanol hydrochloride	C_2H_8ClNO	2002-24-6	97.544	hyg cry (EtOH)	85				

Estrone

Ethacrynic acid

Ethalfluralin

Ethambutol

Ethane

Ethanearsonic acid

Ethanedial dioxime

1,2-Ethanediamine

1,2-Ethanediamine, dihydrochloride

1,2-Ethanediol

1,2-Ethanediol, bis(4-methylbenzenesulfonate)

1,1-Ethanediol, diacetate

1,2-Ethanediol, diacetate

1,2-Ethanediol, diacrylate

1,2-Ethanediol, dibenzoate

1,2-Ethanediol, didodecanoate

1,2-Ethanediol, diformate

1,2-Ethanediol, dihexadecanoate

1,2-Ethanediol, dimethacrylate

1,2-Ethanediol, dinitrate

1,2-Ethanediol, distearate

1,2-Ethanediol, ditetradecanoate

1,2-Ethanediol, dithiocyanate

1,2-Ethanediol, monoacetate

1,2-Ethanediol, monobenzoate

1,2-Ethanediol, monostearate

1,2-Ethanediol, monosulfite

1,2-Ethanediphosphonic acid

1,2-Ethanedisulfonic acid

Ethanedithioamide

1,2-Ethanedithiol

1,2-Ethanediyl mercaptoacetate

Ethanesulfonic acid

Ethanesulfonyl chloride

Ethanethiol

Ethanimidamide

Ethanimidamide monohydrochloride

Ethanol

Ethanolamine

Ethanolamine hydrochloride

No.	Name	Synonym	Mol. Form.	CAS RN	Mol. Wt.	Physical Form	mp/°C	bp/°C	den/ g cm⁻³	n_D	Solubility
4803	Ethanolamine *O*-sulfate	2-Aminoethyl sulfate	C₂H₇NO₄S	926-39-6	141.147		230 dec				s H₂O; i EtOH
4804	Ethaverine	1-[(3,4-Diethoxyphenyl)methyl]-6,7-diethoxyisoquinoline	C₂₄H₂₉NO₄	486-47-5	395.492		100				i H₂O; s EtOH; sl eth, chl
4805	Ethchlorvynol	1-Chloro-3-ethyl-1-penten-4-yn-ol	C₇H₉ClO	113-18-8	144.598	liq		181; 30⁰·¹	1.07²⁵	1.474²⁵	i H₂O; s os
4806	Ethephon	Phosphonic acid, (2-chloroethyl)	C₂H₆ClO₃P	16672-87-0	144.494		74		1.2		
4807	Ethinylestradiol	19-Norpregna-1,3,5(10)-trien-20-yne-3,17-diol, (17α)-	C₂₀H₂₄O₂	57-63-6	296.404						sl chl
4808	Ethion		C₉H₂₂O₄P₂S₄	563-12-2	384.476		-13	165⁰·³	1.22²⁰		
4809	*D*-Ethionine	3-Ethylhomocysteine, (*R*)	C₆H₁₃NO₂S	535-32-0	163.238	cry (H₂O)	278 dec				
4810	*L*-Ethionine	3-Ethylhomocysteine, (*S*)	C₆H₁₃NO₂S	13073-35-3	163.238	cry (H₂O)	273 dec				
4811	Ethirimol	4(1*H*)-Pyrimidinone, 5-butyl-2-(ethylamino)-6-methyl-	C₁₁H₁₉N₃O	23947-60-6	209.288		160		1.21²⁵		
4812	Ethisterone		C₂₁H₂₈O₂	434-03-7	312.446		272				
4813	Ethoate-methyl		C₆H₁₄NO₃PS₂	116-01-8	243.284	cry (tol/hp)	67				
4814	Ethofumesate		C₁₃H₁₈O₅S	26225-79-6	286.344		71		1.14		
4815	Ethoheptazine	4-Carbethoxymethyl-4-phenylazacycloheptane	C₁₆H₂₃NO₂	77-15-6	261.360	liq		134¹	1.038²⁶	1.5210²⁶	
4816	Ethoprop	Phosphorodithioic acid, *O*-ethyl *S,S*-dipropyl ester	C₈H₁₉O₂PS₂	13194-48-4	242.340			88⁰·²	1.094²⁰		
4817	Ethotoin		C₁₁H₁₂N₂O₂	86-35-1	204.225	pr (w)	94				s hot H₂O; vs EtOH, bz, eth
4818	Ethoxyacetic acid		C₄H₈O₃	627-03-2	104.105			206.5	1.1021²⁰	1.4194²⁰	vs H₂O, EtOH, eth; s chl
4819	4'-Ethoxyacetophenone		C₁₀H₁₂O₂	1676-63-7	164.201	pl (eth)	39	268			
4820	Ethoxyacetylene		C₄H₆O	927-80-0	70.090			50	0.8000²⁰	1.3796²⁰	vs eth, EtOH
4821	7-Ethoxy-3,9-acridinediamine	Ethacridine	C₁₅H₁₅N₃O	442-16-0	253.299	ye nd	226				
4822	2-Ethoxyaniline	*o*-Phenetidine	C₈H₁₁NO	94-70-2	137.179		<-21	232.5		1.5560²⁰	sl H₂O, ctc; s EtOH, eth
4823	3-Ethoxyaniline	*m*-Phenetidine	C₈H₁₁NO	621-33-0	137.179			248			vs eth, EtOH
4824	4-Ethoxyaniline	*p*-Phenetidine	C₈H₁₁NO	156-43-4	137.179		1.2	254	1.0652¹⁶	1.5528²⁰	sl H₂O; s EtOH, eth, chl
4825	2-Ethoxybenzaldehyde		C₉H₁₀O₂	613-69-4	150.174		21	248			msc EtOH, eth; sl chl
4826	4-Ethoxybenzaldehyde		C₉H₁₀O₂	10031-82-0	150.174		13.5	249	1.08²¹		vs EtOH, eth, bz
4827	2-Ethoxybenzamide	Ethenzamide	C₉H₁₁NO₂	938-73-8	165.189	nd (w, al)	133				sl H₂O, chl; vs EtOH, eth
4828	Ethoxybenzene	Phenetole	C₈H₁₀O	103-73-1	122.164	liq	-29.43	169.81	0.9651²⁰	1.5076²⁰	i H₂O; s EtOH, eth, ctc
4829	4-Ethoxy-1,2-benzenediamine		C₈H₁₂N₂O	1197-37-1	152.193		71.5	295			vs H₂O; s EtOH, eth, chl
4830	2-Ethoxybenzoic acid		C₉H₁₀O₃	134-11-2	166.173		20.7	211³⁹			sl H₂O, EtOH, ctc
4831	4-Ethoxybenzoic acid		C₉H₁₀O₃	619-86-3	166.173	nd (w)	198.5				sl H₂O, tfa; s EtOH, eth, bz
4832	6-Ethoxy-2-benzothiazolesulfonamide	Ethoxzolamide	C₉H₁₀N₂O₃S₂	452-35-7	258.316		189				
4833	3-Ethoxy-*N,N*-diethylaniline		C₁₂H₁₉NO	1864-92-2	193.285			286; 97⁰·⁶		1.5325²⁵	s EtOH, bz, HOAc
4834	2-Ethoxy-3,4-dihydro-2*H*-pyran		C₇H₁₂O₂	103-75-3	128.169			132; 42¹⁶	0.9658²⁵	1.4394²⁰	
4835	6-Ethoxy-1,2-dihydro-2,2,4-trimethylquinoline	Ethoxyquin	C₁₄H₁₉NO	91-53-2	217.307			124²	1.026²⁵	1.569²⁵	
4836	Ethoxydimethylsilane	Dimethylethoxysilane	C₄H₁₂OSi	14857-34-2	104.223	liq		54	0.76²⁰		
4837	2-Ethoxy-1,2-diphenylethanone		C₁₆H₁₆O₂	574-09-4	240.297	nd (lig)	62	194²⁰	1.1016¹⁷	1.5727¹⁷	vs bz, eth, EtOH, lig
4838	2-Ethoxyethanamine		C₄H₁₁NO	110-76-9	89.136			107	0.8512²⁰	1.4101²⁰	msc H₂O, EtOH, eth; s ace, bz; sl chl
4839	2-Ethoxyethanol	Ethylene glycol monoethyl ether	C₄H₁₀O₂	110-80-5	90.121	liq	-70	135	0.9253²⁵	1.4054²⁵	vs H₂O, ace, eth, EtOH
4840	2-(2-Ethoxyethoxy)ethyl 2-propenoate	Diethylene glycol ethyl ether acrylate	C₉H₁₆O₄	7328-17-8	188.221				1.13²⁵		
4841	2-Ethoxyethyl acetate	Ethylene glycol monoethyl ether acetate	C₆H₁₂O₃	111-15-9	132.157	liq	-61.7	156.4	0.9740²⁰	1.4054²⁰	vs H₂O, ace, eth, EtOH
4842	2-Ethoxyethyl acrylate	Ethylene glycol monoethyl ether acrylate	C₇H₁₂O₃	106-74-1	144.168	liq	-47	174	0.983²⁰	1.4274²⁰	
4843	3-Ethoxy-2-hydroxybenzaldehyde		C₉H₁₀O₃	492-88-6	166.173		65.3	264			
4844	3-Ethoxy-4-hydroxybenzaldehyde	Ethyl vanillin	C₉H₁₀O₃	121-32-4	166.173		77.5	285			sl H₂O; s EtOH, eth, bz, chl
4845	4-Ethoxy-3-methoxybenzaldehyde		C₁₀H₁₂O₃	120-25-2	180.200	mcl pr	64.5	168¹³			sl H₂O; s EtOH, eth, bz, chl, HOAc
4846	1-Ethoxy-2-methoxyethane		C₅H₁₂O₂	5137-45-1	104.148	liq		103.5	0.8460²⁵	1.3843²⁵	

Ethanolamine *O*-sulfate

Ethaverine

Ethchlorvynol

Ethephon

Ethinylestradiol

Ethion

D-Ethionine

L-Ethionine

Ethirimol

Ethisterone

Ethoate-methyl

Ethofumesate

Ethoheptazine

Ethoprop

Ethotoin

Ethoxyacetic acid

4'-Ethoxyacetophenone

Ethoxyacetylene

7-Ethoxy-3,9-acridinediamine

2-Ethoxyaniline

3-Ethoxyaniline

4-Ethoxyaniline

2-Ethoxybenzaldehyde

4-Ethoxybenzaldehyde

2-Ethoxybenzamide

Ethoxybenzene

4-Ethoxy-1,2-benzenediamine

2-Ethoxybenzoic acid

4-Ethoxybenzoic acid

6-Ethoxy-2-benzothiazolesulfonamide

3-Ethoxy-*N*,*N*-diethylaniline

2-Ethoxy-3,4-dihydro-2*H*-pyran

6-Ethoxy-1,2-dihydro-2,2,4-trimethylquinoline

Ethoxydimethylsilane

2-Ethoxy-1,2-diphenylethanone

2-Ethoxyethanamine

2-Ethoxyethanol

2-(2-Ethoxyethoxy)ethyl 2-propenoate

2-Ethoxyethyl acetate

2-Ethoxyethyl acrylate

3-Ethoxy-2-hydroxybenzaldehyde

3-Ethoxy-4-hydroxybenzaldehyde

4-Ethoxy-3-methoxybenzaldehyde

1-Ethoxy-2-methoxyethane

No.	Name	Synonym	Mol. Form.	CAS RN	Mol. Wt.	Physical Form	mp/°C	bp/°C	den/g cm⁻³	n_D	Solubility
4847	1-Ethoxy-3-methylbenzene		$C_9H_{12}O$	621-32-9	136.190			192	0.949²⁰	1.513²⁰	i H₂O; s EtOH, eth
4848	1-Ethoxy-4-methylbenzene		$C_9H_{12}O$	622-60-6	136.190			188.5	0.9509¹⁸	1.5058¹⁸	i H₂O; s EtOH, eth; sl ctc
4849	2-Ethoxy-2-methylbutane	Ethyl tert-pentyl ether	$C_7H_{16}O$	919-94-8	116.201			101.4	0.7606²⁵	1.3886²⁵	vs eth, EtOH
4850	(Ethoxymethylene)propanedinitrile		$C_6H_6N_2O$	123-06-8	122.124		66	160¹²			s EtOH, eth; sl chl
4851	(Ethoxymethyl)oxirane	2,3-Epoxypropyl ethyl ether	$C_5H_{10}O_2$	4016-11-9	102.132			128	0.9700²⁰	1.4320²⁰	s H₂O, EtOH, eth; sl ctc
4852	1-Ethoxynaphthalene		$C_{12}H_{12}O$	5328-01-8	172.222	nd	5.5	280.5	1.060²⁰	1.5953²⁵	i H₂O; vs EtOH, eth
4853	2-Ethoxynaphthalene		$C_{12}H_{12}O$	93-18-5	172.222	pl (al)	37.5	282	1.0640²⁰	1.5975³⁶	s H₂O; s EtOH, eth, tol, lig, CS₂
4854	2-Ethoxy-5-nitroaniline	5-Nitro-o-phenetidine	$C_8H_{10}N_2O_3$	136-79-8	182.176	ye nd (dil al)	96.5	205¹⁴			vs eth, EtOH
4855	1-Ethoxy-2-nitrobenzene		$C_8H_9NO_3$	610-67-3	167.162	br ye	1.1	267	1.1903¹⁵	1.5425²⁰	vs eth, EtOH
4856	1-Ethoxy-4-nitrobenzene		$C_8H_9NO_3$	100-29-8	167.162	pr (dil al, eth)	60	283	1.1176¹⁰⁰		sl H₂O, EtOH; vs eth; msc ace, bz; s peth
4857	N-(4-Ethoxy-3-nitrophenyl)acetamide		$C_{10}H_{12}N_2O_4$	1777-84-0	224.213	nd (dil al)	124.0				vs ace, bz, EtOH
4858	2-Ethoxyphenol	Catechol monoethyl ether	$C_8H_{10}O_2$	94-71-3	138.164		29	217	1.0903²⁵		sl H₂O, ctc; msc EtOH, eth
4859	3-Ethoxyphenol	Resorcinol monoethyl ether	$C_8H_{10}O_2$	621-34-1	138.164			246; 131¹⁰	1.105¹⁵		i H₂O; s EtOH, eth, bz; sl chl
4860	4-Ethoxyphenol	Hydroquinone monoethyl ether	$C_8H_{10}O_2$	622-62-8	138.164	pr or lf (w)	66.5	246.5			sl H₂O; vs EtOH, eth; s chl
4861	N-(2-Ethoxyphenyl)acetamide		$C_{10}H_{13}NO_2$	581-08-8	179.216	lf(dil al)	79	>240			i H₂O; s EtOH, eth, chl
4862	N-(4-Ethoxyphenyl)acetamide	Phenacetin	$C_{10}H_{13}NO_2$	62-44-2	179.216	mcl pr	137.5			1.571	sl H₂O, eth, bz; s EtOH, ace; vs py
4863	N-(4-Ethoxyphenyl)-2-hydroxypropanamide	p-Lactophenetide	$C_{11}H_{15}NO_3$	539-08-2	209.242		118				s H₂O; vs EtOH; sl eth, bz, chl, peth
4864	(4-Ethoxyphenyl)urea	Dulcin	$C_9H_{12}N_2O_2$	150-69-6	180.203	lf (dil al), pl (w)	173.5	dec			sl H₂O; s EtOH; vs AcOEt
4865	3-Ethoxypropanal		$C_5H_{10}O_2$	2806-85-1	102.132			135.2	0.9165²⁰		
4866	3-Ethoxypropanenitrile		C_5H_9NO	2141-62-0	99.131			171	0.9285¹⁵	1.4068²⁰	vs eth, EtOH
4867	8-Ethoxy-5-quinolinesulfonic acid	Actinoquinol	$C_{11}H_{11}NO_4S$	15301-40-3	253.275	br nd (w)	286 dec				s alk
4868	Ethoxytrimethylsilane		$C_5H_{14}OSi$	1825-62-3	118.250			76	0.7573²⁰	1.3741²⁰	i H₂O; s EtOH, eth, ace
4869	Ethoxytriphenylsilane		$C_{20}H_{20}OSi$	1516-80-9	304.458		65	344			s chl
4870	N-Ethylacetamide		C_4H_9NO	625-50-3	87.120			205; 104¹⁸	0.942⁴	1.4338²⁰	msc H₂O, EtOH; s chl, HOAc
4871	Ethyl acetate		$C_4H_8O_2$	141-78-6	88.106	liq	-83.8	77.11	0.9003²⁰	1.3723²⁰	s H₂O; msc EtOH, eth; vs ace, bz
4872	Ethyl acetoacetate		$C_6H_{10}O_3$	141-97-9	130.141	liq	-45	180.8	1.0368¹⁰	1.4171²⁰	s H₂O; msc EtOH, eth; s bz, chl
4873	4'-Ethylacetophenone		$C_{10}H_{12}O$	937-30-4	148.201			114¹¹			
4874	Ethyl 2-acetylhexanoate		$C_{10}H_{18}O_3$	1540-29-0	186.248			221.5	0.9523²⁰	1.4301²⁰	vs ace, eth
4875	Ethyl 2-acetyl-3-methylbutanoate		$C_9H_{16}O_3$	1522-46-9	172.221			201; 97²⁰	0.9648¹⁸	1.4256¹⁸	i H₂O; msc EtOH, eth
4876	Ethyl 2-acetylpentanoate		$C_9H_{16}O_3$	1540-28-9	172.221			224; 90¹⁵	0.9661²⁰	1.4255²⁰	vs eth, EtOH
4877	Ethyl 2-acetyl-4-pentenoate	Ethyl 2-allylacetoacetate	$C_9H_{14}O_3$	610-89-9	170.205			208	0.9898²⁰	1.4388¹⁸	msc EtOH, eth, bz
4878	Ethyl acrylate	Ethyl propenoate	$C_5H_8O_2$	140-88-5	100.117	liq	-71.2	99.4	0.9234²⁰	1.4068²⁰	sl H₂O, DMSO; msc EtOH, eth; s chl
4879	Ethylamine	Ethanamine	C_2H_7N	75-04-7	45.084	vol liq or gas	-80.5	16.5	0.677²⁵ (p>1 atm)	1.3663²⁰	msc H₂O, EtOH, eth
4880	Ethylamine hydrochloride	Ethanamine hydrochloride	C_2H_8ClN	557-66-4	81.545	mcl pl (al)	109.5		1.2160²⁰		vs H₂O, EtOH
4881	Ethyl 2-aminoacetate	Glycine, ethyl ester	$C_4H_9NO_2$	459-73-4	103.120			149; 58¹⁸	1.0275¹⁰	1.4242¹⁰	msc H₂O, EtOH, eth, ace, bz; vs lig
4882	Ethyl 2-aminobenzoate		$C_9H_{11}NO_2$	87-25-2	165.189		13	268	1.1174²⁰	1.5646²⁰	vs eth, EtOH
4883	Ethyl 3-aminobenzoate		$C_9H_{11}NO_2$	582-33-2	165.189			294; 160⁵	1.171²⁰	1.5600²²	sl H₂O; vs EtOH, eth; s ctc
4884	Ethyl 4-aminobenzoate	Ethyl aminobenzoate	$C_9H_{11}NO_2$	94-09-7	165.189	nd (w), orth (eth)	92	310			i H₂O; vs EtOH, eth; s chl, acid
4885	Ethyl (aminocarbonyl)carbamate		$C_4H_8N_2O_3$	626-36-8	132.118	nd (w, bz)	196.5	dec			i H₂O, eth; sl EtOH, bz, tfa

Physical Constants of Organic Compounds

1-Ethoxy-3-methylbenzene

1-Ethoxy-4-methylbenzene

2-Ethoxy-2-methylbutane

(Ethoxymethylene)propanedinitrile

(Ethoxymethyl)oxirane

1-Ethoxynaphthalene

2-Ethoxynaphthalene

2-Ethoxy-5-nitroaniline

1-Ethoxy-2-nitrobenzene

1-Ethoxy-4-nitrobenzene

N-(4-Ethoxy-3-nitrophenyl)acetamide

2-Ethoxyphenol

3-Ethoxyphenol

4-Ethoxyphenol

N-(2-Ethoxyphenyl)acetamide

N-(4-Ethoxyphenyl)acetamide

N-(4-Ethoxyphenyl)-2-hydroxypropanamide

(4-Ethoxyphenyl)urea

3-Ethoxypropanal

3-Ethoxypropanenitrile

8-Ethoxy-5-quinolinesulfonic acid

Ethoxytrimethylsilane

Ethoxytriphenylsilane

N-Ethylacetamide

Ethyl acetate

Ethyl acetoacetate

4'-Ethylacetophenone

Ethyl 2-acetylhexanoate

Ethyl 2-acetyl-3-methylbutanoate

Ethyl 2-acetylpentanoate

Ethyl 2-acetyl-4-pentenoate

Ethyl acrylate

Ethylamine

Ethylamine hydrochloride

Ethyl 2-aminoacetate

Ethyl 2-aminobenzoate

Ethyl 3-aminobenzoate

Ethyl 4-aminobenzoate

Ethyl (aminocarbonyl)carbamate

No.	Name	Synonym	Mol. Form.	CAS RN	Mol. Wt.	Physical Form	mp/°C	bp/°C	den/g cm⁻³	n_D	Solubility
4886	2-(Ethylamino)ethanol		C₄H₁₁NO	110-73-6	89.136			169.5	0.914[20]	1.444[20]	vs H₂O, EtOH, eth; s chl
4887	2-Ethylaniline		C₈H₁₁N	578-54-1	121.180	liq	-43	209.5	0.983[22]	1.5584[22]	sl H₂O, chl; vs EtOH, eth
4888	3-Ethylaniline		C₈H₁₁N	587-02-0	121.180	liq	-64	214; 94[6]	0.9896[25]		vs eth, EtOH
4889	4-Ethylaniline		C₈H₁₁N	589-16-2	121.180	liq	-2.4	217.5	0.9679[20]	1.5554[20]	sl H₂O, ctc; vs EtOH, eth
4890	N-Ethylaniline		C₈H₁₁N	103-69-5	121.180	liq	-63.5	203.0	0.9625[20]	1.5559[20]	i H₂O; msc EtOH, eth; vs ace, bz; s ctc
4891	2-Ethyl-9,10-anthracenedione		C₁₆H₁₂O₂	84-51-5	236.265		108.8				
4892	4-Ethylbenzaldehyde		C₉H₁₀O	4748-78-1	134.174			221	0.9790[20]		
4893	N-Ethylbenzamide		C₉H₁₁NO	614-17-5	149.189	nd (w)	70.5				
4894	Ethylbenzene	Phenylethane	C₈H₁₀	100-41-4	106.165	liq	-94.96	136.19	0.8626[25]	1.4959[20]	i H₂O; msc EtOH, eth; sl chl
4895	α-Ethylbenzeneacetamide	α-Phenylbutyramide	C₁₀H₁₃NO	90-26-6	163.216	cry	86	185[16]			s H₂O, ctc; sl ace
4896	α-Ethylbenzeneacetic acid		C₁₀H₁₂O₂	90-27-7	164.201	pl (eth)	47.5	271			s eth, bz, ctc
4897	α-Ethylbenzeneacetonitrile		C₁₀H₁₁N	769-68-6	145.201			241	0.977[14]		i H₂O; s EtOH, eth, bz
4898	4-Ethyl-1,3-benzenediol		C₈H₁₀O₂	2896-60-8	138.164	pr (chl, bz)	98.5	160[24], 131[15]			sl H₂O, EtOH, eth
4899	α-Ethylbenzenemethanol	α-Ethylbenzyl alcohol	C₉H₁₂O	93-54-9	136.190			219	0.9915[25]	1.5169[23]	vs bz, eth, EtOH, MeOH
4900	Ethyl benzenesulfonate		C₈H₁₀O₃S	515-46-8	186.228			156[15]	1.2167[20]	1.5081[20]	sl H₂O; s EtOH; vs eth, chl
4901	4-Ethylbenzenesulfonic acid		C₈H₁₀O₃S	98-69-1	186.228				1.23		
4902	2-Ethyl-1H-benzimidazole		C₉H₁₀N₂	1848-84-6	146.188		176.5				sl chl
4903	Ethyl benzoate		C₉H₁₀O₂	93-89-0	150.174	liq	-34	212	1.0415[25]	1.5007[20]	i H₂O; s EtOH, ace, bz; msc eth; sl ctc
4904	Ethyl 1,3-benzodioxole-5-carboxylate		C₁₀H₁₀O₄	6951-08-2	194.184	pr	18.5	285.5; 135[6]			vs eth, EtOH, peth
4905	Ethyl benzoylacetate		C₁₁H₁₂O₃	94-02-0	192.211		<0	dec 267; 167[20]	1.1202[15]	1.5317[15]	sl H₂O; s EtOH, eth
4906	Ethyl N-benzylglycinate		C₁₁H₁₅NO₂	6436-90-4	193.243			177[50]		1.5041[20]	vs EtOH, eth, bz
4907	Ethyl 2-benzylideneacetoacetate		C₁₃H₁₄O₃	620-80-4	218.248	orth pl (dil al)	60.5	296; 180[17]			i H₂O; sl EtOH, eth, bz; vs chl
4908	Ethyl bromoacetate		C₄H₇BrO₂	105-36-2	167.002			168.5	1.5032[20]	1.4489[20]	i H₂O; msc EtOH, eth; s ace; sl ctc
4909	Ethyl 4-bromoacetoacetate		C₆H₉BrO₃	13176-46-0	209.037			115[14], 110[10]	1.5278[18]	1.5281[20]	vs eth, EtOH
4910	Ethyl 4-bromobenzoate		C₉H₉BrO₂	5798-75-4	229.070	liq	-18	263; 125[15]	1.4332[17]	1.5438[17]	sl H₂O; s EtOH, eth, ace, bz
4911	Ethyl 2-bromobutanoate		C₆H₁₁BrO₂	533-68-6	195.054			177; 43[5]	1.3273[20]	1.4475[20]	i H₂O; msc EtOH, eth; s chl
4912	Ethyl 4-bromobutanoate		C₆H₁₁BrO₂	2969-81-5	195.054			192; 82[10]	1.3540[20]	1.4559[20]	
4913	Ethyl trans-4-bromo-2-butenoate		C₆H₉BrO₂	37746-78-4	193.038			100[14]	1.402[16]	1.4925[20]	vs EtOH
4914	Ethyl 6-bromohexanoate	Ethyl 6-bromocaproate	C₈H₁₅BrO₂	25542-62-5	223.108	cry (peth)	33	126[21]	1.238[23]	1.4566[21]	
4915	Ethyl 2-bromo-3-methylbutanoate		C₇H₁₃BrO₂	609-12-1	209.081			186	1.2760[20]	1.4496[20]	vs eth, EtOH
4916	Ethyl 2-bromo-2-methylpropanoate		C₆H₁₁BrO₂	600-00-0	195.054			163	1.3263[20]	1.4446[20]	i H₂O; s EtOH; msc eth
4917	Ethyl 3-bromo-2-oxopropanoate	Ethyl 3-bromopyruvate	C₅H₇BrO₃	70-23-5	195.012			87[9]			
4918	Ethyl 2-bromopentanoate		C₇H₁₃BrO₂	615-83-8	209.081			191	1.226[18]	1.4496[20]	i H₂O; s EtOH, eth
4919	Ethyl 5-bromopentanoate		C₇H₁₃BrO₂	14660-52-7	209.081			129[35], 107[20]	1.3085[20]	1.4543[20]	sl ctc
4920	Ethyl 2-bromopropanoate	Ethyl α-bromopropionate	C₅H₉BrO₂	535-11-5	181.028			dec 160; 71[26]	1.4135[20]	1.4490[20]	i H₂O; msc EtOH, eth; s chl
4921	Ethyl 3-bromopropanoate		C₅H₉BrO₂	539-74-2	181.028			179; 65[15]	1.4123[18]	1.4516[20]	s EtOH, eth, ace; sl ctc
4922	2-Ethylbutanal	Diethylacetaldehyde	C₆H₁₂O	97-96-1	100.158			118[160]	0.8110[20]	1.4025[20]	sl H₂O, ctc; msc EtOH, eth
4923	Ethyl butanoate		C₆H₁₂O₂	105-54-4	116.158	liq	-98	121.3	0.8735[25]	1.3898[25]	sl H₂O, ctc; s EtOH, eth
4924	2-Ethylbutanoic acid	Diethylacetic acid	C₆H₁₂O₂	88-09-5	116.158	liq	-31.8	194	0.9239[20]	1.4132[20]	sl H₂O, ctc; msc EtOH, eth
4925	2-Ethylbutanoic acid, triethyleneglycol diester		C₁₈H₃₄O₆	95-08-9	346.459			181[3.5]			
4926	2-Ethyl-1-butanol		C₆H₁₄O	97-95-0	102.174		<-15	147	0.8326[20]	1.4220[20]	sl H₂O; s EtOH, eth, chl
4927	2-Ethylbutanoyl chloride		C₆H₁₁ClO	2736-40-5	134.603			140	0.9825[20]	1.4234[20]	vs eth

Physical Constants of Organic Compounds

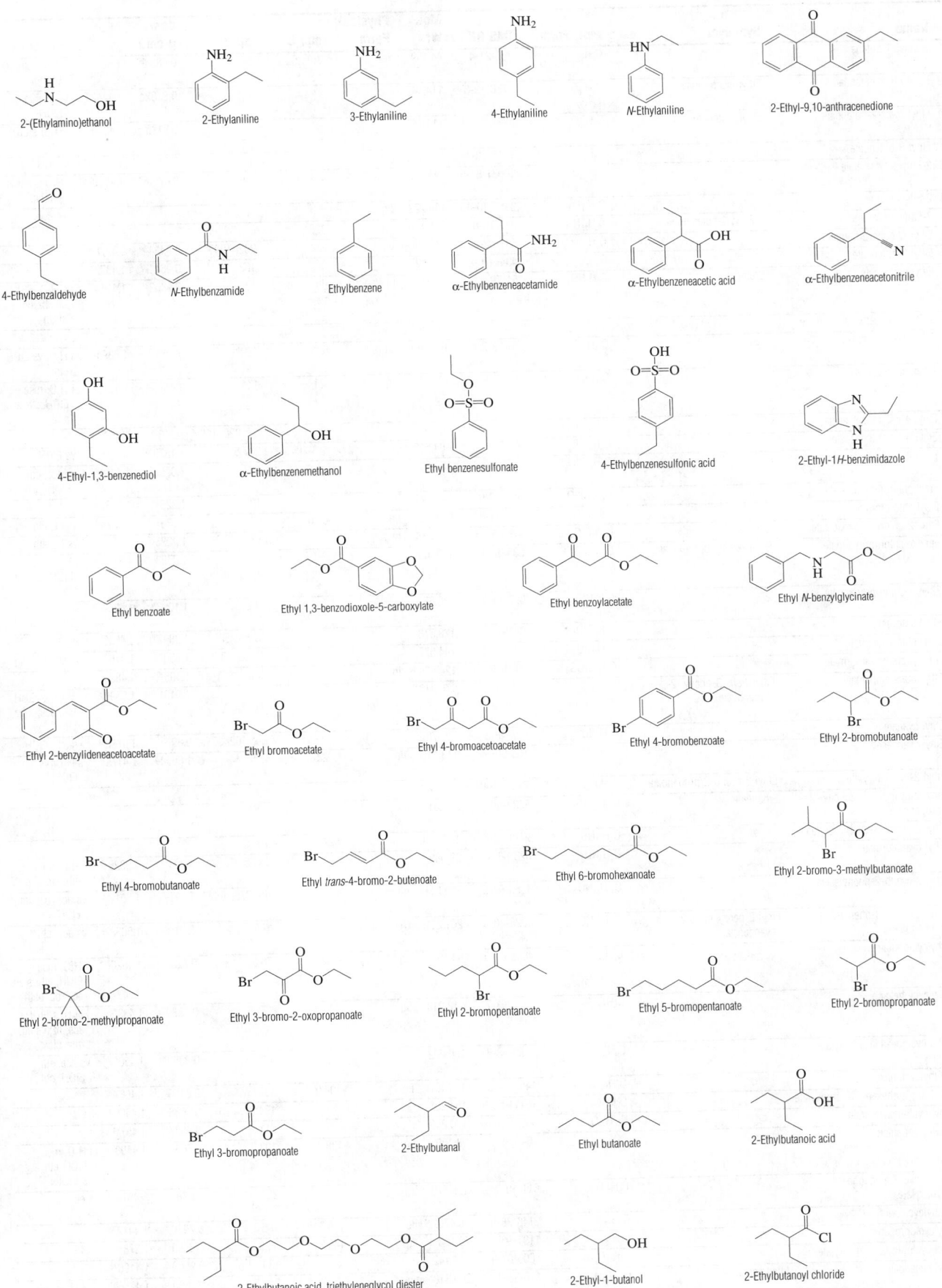

2-(Ethylamino)ethanol

2-Ethylaniline

3-Ethylaniline

4-Ethylaniline

N-Ethylaniline

2-Ethyl-9,10-anthracenedione

4-Ethylbenzaldehyde

N-Ethylbenzamide

Ethylbenzene

α-Ethylbenzeneacetamide

α-Ethylbenzeneacetic acid

α-Ethylbenzeneacetonitrile

4-Ethyl-1,3-benzenediol

α-Ethylbenzenemethanol

Ethyl benzenesulfonate

4-Ethylbenzenesulfonic acid

2-Ethyl-1H-benzimidazole

Ethyl benzoate

Ethyl 1,3-benzodioxole-5-carboxylate

Ethyl benzoylacetate

Ethyl N-benzylglycinate

Ethyl 2-benzylideneacetoacetate

Ethyl bromoacetate

Ethyl 4-bromoacetoacetate

Ethyl 4-bromobenzoate

Ethyl 2-bromobutanoate

Ethyl 4-bromobutanoate

Ethyl trans-4-bromo-2-butenoate

Ethyl 6-bromohexanoate

Ethyl 2-bromo-3-methylbutanoate

Ethyl 2-bromo-2-methylpropanoate

Ethyl 3-bromo-2-oxopropanoate

Ethyl 2-bromopentanoate

Ethyl 5-bromopentanoate

Ethyl 2-bromopropanoate

Ethyl 3-bromopropanoate

2-Ethylbutanal

Ethyl butanoate

2-Ethylbutanoic acid

2-Ethylbutanoic acid, triethyleneglycol diester

2-Ethyl-1-butanol

2-Ethylbutanoyl chloride

No.	Name	Synonym	Mol. Form.	CAS RN	Mol. Wt.	Physical Form	mp/°C	bp/°C	den/ g cm^{-3}	n_D	Solubility
4928	2-Ethyl-1-butene		C_6H_{12}	760-21-4	84.159	liq	-131.5	64.7	0.6894[20]	1.3969[20]	i H_2O; s eth, ace, bz, chl
4929	Ethyl cis-2-butenoate	Ethyl isocrotonate	$C_6H_{10}O_2$	6776-19-8	114.142			136	0.9182[20]	1.4242[20]	vs ace, eth, EtOH
4930	Ethyl trans-2-butenoate	Ethyl crotonate	$C_6H_{10}O_2$	623-70-1	114.142			138	0.9175[20]	1.4243[20]	i H_2O; s EtOH, eth
4931	Ethyl 3-butenoate		$C_6H_{10}O_2$	1617-18-1	114.142			119	0.9122[20]	1.4105[20]	s EtOH
4932	2-Ethylbutyl acetate		$C_8H_{16}O_2$	10031-87-5	144.212		<-100	162.5	0.8790[20]	1.4109[20]	i H_2O; s EtOH, eth, ctc
4933	2-Ethylbutyl acrylate		$C_9H_{16}O_2$	3953-10-4	156.222	liq		80[20]			
4934	2-Ethylbutylamine	2-Ethyl-1-butanamine	$C_6H_{15}N$	617-79-8	101.190	liq	125				
4935	Ethyl N-butylcarbamate		$C_7H_{15}NO_2$	591-62-8	145.200	liq	-22	202; 100[15]	0.9434[26]	1.4278[26]	
4936	Ethyl 2-butynoate		$C_6H_8O_2$	4341-76-8	112.127			163	0.9641[20]	1.4372[20]	
4937	Ethyl carbamate	Urethane	$C_3H_7NO_2$	51-79-6	89.094	pr (bz, to)	49	185	0.9862[21]	1.4144[51]	vs H_2O, EtOH, eth, bz, chl, py; sl lig
4938	9-Ethyl-9H-carbazol-3-amine		$C_{14}H_{14}N_2$	132-32-1	210.274		99				
4939	9-Ethyl-9H-carbazole		$C_{14}H_{13}N$	86-28-2	195.260	nd (al)	68	190[10]	1.059[80]	1.6394[80]	i H_2O; vs EtOH, eth
4940	Ethyl chloroacetate		$C_4H_7ClO_2$	105-39-5	122.551	liq	-21	144.3	1.1585[20]	1.4215[20]	i H_2O; msc EtOH, eth, ace; s bz
4941	Ethyl 4-chloroacetoacetate		$C_6H_9ClO_3$	638-07-3	164.586		-8	dec 220; 115[14]	1.218[25]	1.4520[20]	
4942	Ethyl 4-chlorobenzoate		$C_9H_9ClO_2$	7335-27-5	184.619			237.5	1.1873[14]		vs EtOH
4943	Ethyl 4-chlorobutanoate		$C_6H_{11}ClO_2$	3153-36-4	150.603			184	1.0756[20]	1.4311[20]	vs ace, eth, EtOH
4944	Ethyl chlorofluoroacetate		$C_4H_6ClFO_2$	401-56-9	140.541			129	1.225[20]	1.3927[20]	
4945	Ethyl chloroformate		$C_3H_5ClO_2$	541-41-3	108.524	liq	-80.6	95	1.1352[20]	1.3974[20]	vs bz, eth, chl
4946	Ethyl 2-chloro-2-oxoacetate	Ethyl oxalyl chloride	$C_4H_5ClO_3$	4755-77-5	136.534	hyg		137	1.2226[20]		vs bz, eth
4947	Ethyl 2-chloropropanoate	Ethyl α-chloropropionate	$C_5H_9ClO_2$	535-13-7	136.577			147	1.0793[20]	1.4178[20]	i H_2O; msc EtOH, eth; sl ctc
4948	Ethyl 3-chloropropanoate		$C_5H_9ClO_2$	623-71-2	136.577			162	1.1086[20]	1.4254[20]	sl H_2O; msc EtOH, eth
4949	Ethyl chlorosulfinate		$C_2H_5ClO_2S$	6378-11-6	128.578			52.5[44], 32[16]	1.2837[20]	1.4550[25]	vs eth
4950	Ethyl chlorosulfonate		$C_2H_5ClO_3S$	625-01-4	144.577			152.5; 93[100]	1.3502[25]	1.416[20]	vs eth, chl, lig
4951	S-Ethyl chlorothioformate		C_3H_5ClOS	2941-64-2	124.589	liq		136	1.195[20]	1.4820[20]	
4952	Ethyl trans-cinnamate	Ethyl trans-3-phenyl-2-propenoate	$C_{11}H_{12}O_2$	4192-77-2	176.212		10	271.5	1.0491[20]	1.5598[20]	i H_2O; vs EtOH, eth, ace; s bz, ctc
4953	Ethyl cyanate		C_3H_5NO	627-48-5	71.078			dec 162; 30[12]	0.89[20]	1.3788[25]	vs eth, EtOH
4954	Ethyl cyanoacetate		$C_5H_7NO_2$	105-56-6	113.116	liq	-22.5	205	1.0654[20]	1.4175[20]	s H_2O; vs eth, EtOH
4955	Ethyl 2-cyanoacrylate	Ethyl 2-cyano-2-propenoate	$C_6H_7NO_2$	7085-85-0	125.126	liq		55[3]			
4956	Ethyl 2-cyano-3,3-diphenyl-2-propenoate	Etocrilene	$C_{18}H_{15}NO_2$	5232-99-5	277.318		110.5	195[3]			
4957	Ethyl 2-cyano-3-ethoxyacrylate		$C_9H_{11}NO_3$	94-05-3	169.178		52	190.5			
4958	Ethyl cyanoformate		$C_4H_5NO_2$	623-49-4	99.089			115.5	1.003[25]	1.3820[20]	i H_2O; s EtOH, eth, ctc
4959	Ethyl 2-cyano-2-phenylacetate		$C_{11}H_{11}NO_2$	4553-07-5	189.211	oil		dec 275; 165[20]	1.091[20]	1.5012[25]	vs ace, bz, eth, EtOH
4960	Ethyl 2-cyano-3-phenyl-2-propenoate	Ethyl 2-benzylidene-2-cyanoacetate	$C_{12}H_{11}NO_2$	2025-40-3	201.221	(i) nd (al) (ii) oil	51	188[15]	1.1076[25]	1.5033	vs ace, chl
4961	Ethylcyclobutane		C_6H_{12}	4806-61-5	84.159	liq	-142.9	70.8	0.7284[20]	1.4020[20]	i H_2O; msc EtOH, eth; s ace, bz, peth
4962	Ethylcyclohexane		C_8H_{16}	1678-91-7	112.213	liq	-111.3	131.9	0.7880[20]	1.4330[20]	i H_2O; s EtOH, ace, bz; vs lig; msc ctc
4963	Ethyl cyclohexanecarboxylate		$C_9H_{16}O_2$	3289-28-9	156.222			196	0.9362[20]	1.4501[15]	vs ace, eth, EtOH, chl
4964	1-Ethylcyclohexene		C_8H_{14}	1453-24-3	110.197	liq	-109.9	137	0.8176[25]	1.4567[20]	
4965	Ethyl 3-cyclohexene-1-carboxylate		$C_9H_{14}O_2$	15111-56-5	154.206			194.5	0.9688[20]	1.4578[20]	
4966	Ethyl cyclohexylacetate		$C_{10}H_{18}O_2$	5452-75-5	170.249			211	0.9537[14]	1.451[14]	
4967	Ethylcyclopentane		C_7H_{14}	1640-89-7	98.186	liq	-138.4	103.5	0.7665[20]	1.4198[20]	i H_2O; msc EtOH, eth, ace; s bz, tol
4968	Ethyl 2-cyclopentanone-1-carboxylate		$C_8H_{12}O_3$	611-10-9	156.179			221; 110[16]	1.0781[21]	1.4519[20]	s eth, bz
4969	1-Ethylcyclopentene		C_7H_{12}	2146-38-5	96.170	liq	-118.5	106.3	0.7936[25]	1.4412[20]	
4970	Ethylcyclopropane		C_5H_{10}	1191-96-4	70.133	liq	-149.2	35.9	0.6790[25]	1.3786[20]	
4971	Ethyl cyclopropanecarboxylate		$C_6H_{10}O_2$	4606-07-9	114.142			134	0.9608[15]	1.4190[20]	
4972	Ethyl decanoate	Ethyl caprate	$C_{12}H_{24}O_2$	110-38-3	200.318	liq	-20	241.5	0.8650[20]	1.4256[20]	i H_2O; vs eth, EtOH, chl

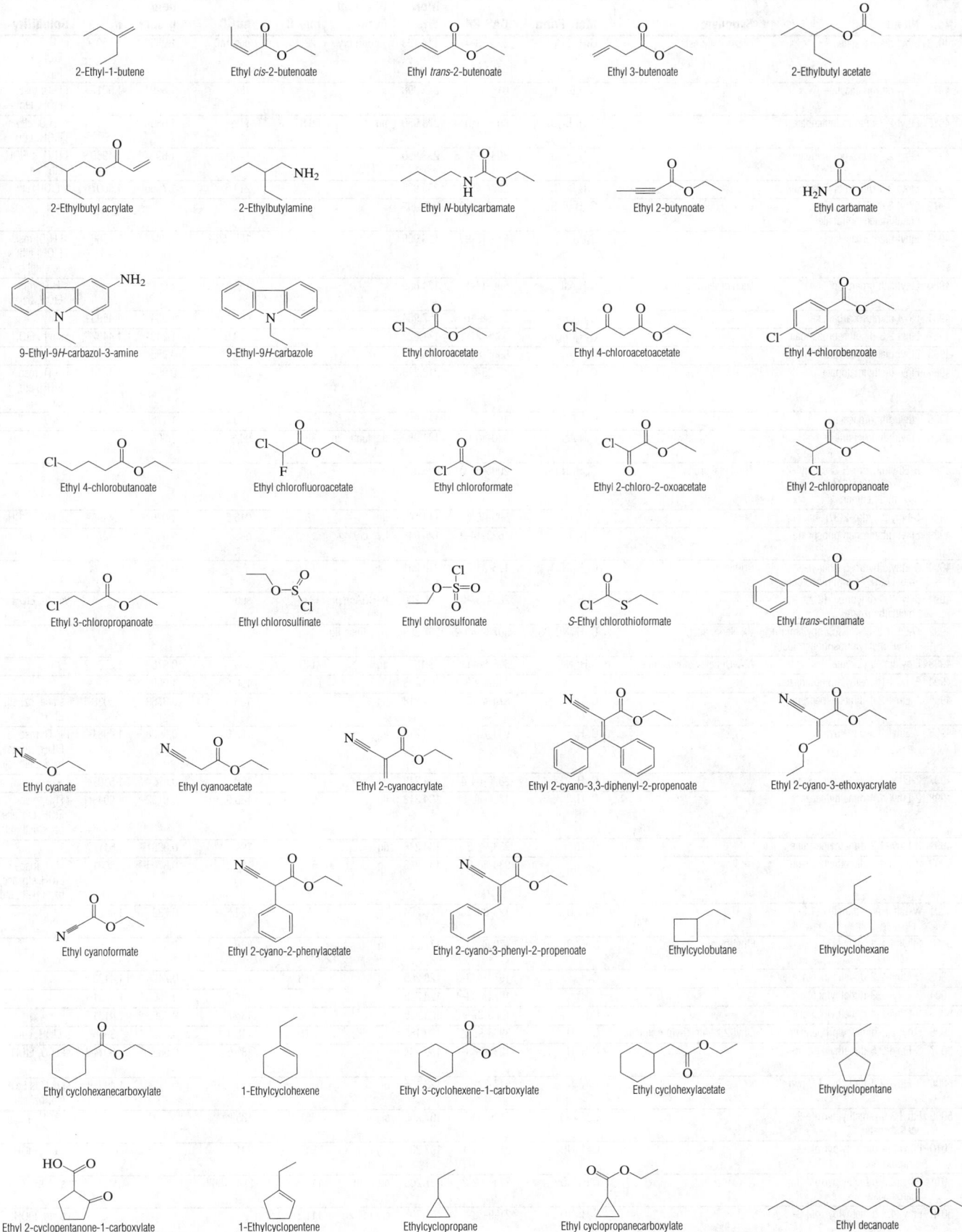

2-Ethyl-1-butene

Ethyl *cis*-2-butenoate

Ethyl *trans*-2-butenoate

Ethyl 3-butenoate

2-Ethylbutyl acetate

2-Ethylbutyl acrylate

2-Ethylbutylamine

Ethyl *N*-butylcarbamate

Ethyl 2-butynoate

Ethyl carbamate

9-Ethyl-9*H*-carbazol-3-amine

9-Ethyl-9*H*-carbazole

Ethyl chloroacetate

Ethyl 4-chloroacetoacetate

Ethyl 4-chlorobenzoate

Ethyl 4-chlorobutanoate

Ethyl chlorofluoroacetate

Ethyl chloroformate

Ethyl 2-chloro-2-oxoacetate

Ethyl 2-chloropropanoate

Ethyl 3-chloropropanoate

Ethyl chlorosulfinate

Ethyl chlorosulfonate

S-Ethyl chlorothioformate

Ethyl *trans*-cinnamate

Ethyl cyanate

Ethyl cyanoacetate

Ethyl 2-cyanoacrylate

Ethyl 2-cyano-3,3-diphenyl-2-propenoate

Ethyl 2-cyano-3-ethoxyacrylate

Ethyl cyanoformate

Ethyl 2-cyano-2-phenylacetate

Ethyl 2-cyano-3-phenyl-2-propenoate

Ethylcyclobutane

Ethylcyclohexane

Ethyl cyclohexanecarboxylate

1-Ethylcyclohexene

Ethyl 3-cyclohexene-1-carboxylate

Ethyl cyclohexylacetate

Ethylcyclopentane

Ethyl 2-cyclopentanone-1-carboxylate

1-Ethylcyclopentene

Ethylcyclopropane

Ethyl cyclopropanecarboxylate

Ethyl decanoate

No.	Name	Synonym	Mol. Form.	CAS RN	Mol. Wt.	Physical Form	mp/°C	bp/°C	den/ g cm⁻³	n_D	Solubility
4973	Ethyl diazoacetate	Diazoacetic ester	$C_4H_6N_2O_2$	623-73-4	114.103	ye orth cry	-22	dec 140	1.0852[18]	1.4605[20]	sl H_2O; msc EtOH, eth, bz, lig
4974	Ethyl dibromoacetate		$C_4H_6Br_2O_2$	617-33-4	245.898			194	1.8991[20]	1.5017[13]	i H_2O; msc EtOH, eth
4975	Ethyl 2,3-dibromobutanoate		$C_6H_{10}Br_2O_2$	609-11-0	273.950	nd	58.5	113[30]	1.6800[20]	sl H_2O, ctc; s EtOH, eth
4976	Ethyl 2,4-dibromobutanoate		$C_6H_{10}Br_2O_2$	36847-51-5	273.950			149[52]	1.6987[20]	1.4960[20]	i H_2O; s EtOH, eth
4977	Ethyl 2,3-dibromopropanoate		$C_5H_8Br_2O_2$	3674-13-3	259.925			214.5	1.7966[20]	1.5007[20]	s EtOH, eth
4978	Ethyl 3,6-di(tert-butyl)-1-naphthalenesulfonate	Ethyl dibunate	$C_{20}H_{28}O_3S$	5560-69-0	348.499						s chl
4979	Ethyl dichloroacetate		$C_4H_6Cl_2O_2$	535-15-9	156.996			155; 56[10]	1.2827[20]	1.4386[20]	sl H_2O; msc EtOH, eth; s ace, chl
4980	Ethyldichloroarsine	Dichloroethylarsine	$C_2H_5AsCl_2$	598-14-1	174.889			155.3; 74[50]	1.66[20]		s H_2O; misc EtOH, bz
4981	Ethyl dichlorocarbamate		$C_3H_5Cl_2NO_2$	13698-16-3	157.984			66[18], 55[15]	1.304[30]	1.4595[20]	
4982	Ethyl 2,3-dichloropropanoate		$C_5H_8Cl_2O_2$	6628-21-3	171.022			183.5	1.2401[20]	1.4482[20]	vs eth, EtOH
4983	Ethyl diethoxyacetate		$C_8H_{16}O_4$	6065-82-3	176.211			199	0.985[25]	1.4100[20]	
4984	Ethyl diethylmalonate		$C_{11}H_{20}O_4$	77-25-8	216.275			230	0.9643[30]	1.4240[20]	i H_2O; msc EtOH, eth; s ctc
4985	Ethyl difluoroacetate		$C_4H_6F_2O_2$	454-31-9	124.087			100	1.1765[20]		i H_2O
4986	Ethyldifluoroarsine		$C_2H_5AsF_2$	430-40-0	141.980	liq, fumes in air	-38.7	94.3	1.708[17]		
4987	5-Ethyldihydro-5-sec-butyl-2-thioxo-4,6(1H,5H)-pyrimidinedione	Thiobutabarbital	$C_{10}H_{16}N_2O_2S$	2095-57-0	228.311		169				
4988	5-Ethyldihydro-2(3H)-furanone		$C_6H_{10}O_2$	695-06-7	114.142	liq	-18	215.5	1.0261[20]	1.4495[20]	vs H_2O, EtOH
4989	Ethyl dihydrogen phosphate		$C_2H_7O_4P$	1623-14-9	126.048	hyg cry		dec	1.430[25]	1.427	vs H_2O, ace, eth, EtOH
4990	5-Ethyldihydro-5-phenyl-4,6(1H,5H)-pyrimidinedione	Primidone	$C_{12}H_{14}N_2O_2$	125-33-7	218.251		281.5				
4991	Ethyl 2,4-dihydroxy-6-methylbenzoate		$C_{10}H_{12}O_4$	2524-37-0	196.200	lf (HOAc), pr (al)	132	sub			vs eth, EtOH
4992	O-Ethyl S-[2-(diisopropylamino)ethyl] methylphosphonothioate	VX Nerve agent	$C_{11}H_{26}NO_2PS$	50782-69-9	267.369	very toxic liq					
4993	Ethyldimethylamine	N,N-Dimethylethanamine	$C_4H_{11}N$	598-56-1	73.137	liq	-140	36.5	0.675[20]	1.3705[25]	
4994	Ethyl 4-(dimethylamino)benzoate		$C_{11}H_{15}NO_2$	10287-53-3	193.243		66.5	190[14]	1.0099[100]		
4995	1-Ethyl-2,4-dimethylbenzene		$C_{10}H_{14}$	874-41-9	134.218	liq	-62.9	188.4	0.8763[20]	1.5038[20]	vs ace, bz, eth, EtOH
4996	1-Ethyl-3,5-dimethylbenzene		$C_{10}H_{14}$	934-74-7	134.218	liq	-84.3	183.6	0.8608[25]	1.4981[20]	i H_2O; msc EtOH, eth, ace, bz; s peth, ctc
4997	2-Ethyl-1,3-dimethylbenzene		$C_{10}H_{14}$	2870-04-4	134.218	liq	-16.2	190	0.8864[25]	1.5107[20]	
4998	2-Ethyl-1,4-dimethylbenzene		$C_{10}H_{14}$	1758-88-9	134.218	liq	-53.7	186.9	0.8732[25]	1.5043[20]	i H_2O; msc EtOH, eth, ace, bz; s peth, ctc
4999	3-Ethyl-1,2-dimethylbenzene		$C_{10}H_{14}$	933-98-2	134.218	liq	-49.5	194	0.8881[25]	1.5117[20]	
5000	4-Ethyl-1,2-dimethylbenzene		$C_{10}H_{14}$	934-80-5	134.218	liq	-66.9	189.5	0.8706[25]	1.5031[20]	i H_2O; msc EtOH, eth, ace, bz; s peth, ctc
5001	N'-Ethyl-N,N-dimethyl-1,2-ethanediamine		$C_6H_{16}N_2$	123-83-1	116.204			134.5	0.738[25]	1.4222[20]	
5002	Ethyl 4,4-dimethyl-3-oxopentanoate	Ethyl pivaloylacetate	$C_9H_{16}O_3$	17094-34-7	172.221	liq		83[17]	0.97[18]		
5003	3-Ethyl-2,2-dimethylpentane		C_9H_{20}	16747-32-3	128.255	liq	-99.3	133.8	0.7438[20]	1.4123[20]	
5004	3-Ethyl-2,3-dimethylpentane		C_9H_{20}	16747-33-4	128.255			144.7	0.7508[25]	1.4221[20]	
5005	3-Ethyl-2,4-dimethylpentane		C_9H_{20}	1068-87-7	128.255	liq	-122.4	136.7	0.7365[20]	1.4131[20]	
5006	Ethyl 2,2-dimethylpropanoate	Ethyl 2,2-dimethylpropionate	$C_7H_{14}O_2$	3938-95-2	130.185	liq	-89.5	118.4	0.856[20]	1.3906[20]	s EtOH, eth
5007	3-Ethyl-2,5-dimethylpyrazine		$C_8H_{12}N_2$	13360-65-1	136.194			180.5	0.9657[24]	1.5014[24]	sl H_2O, EtOH, eth
5008	3-Ethyl-2,4-dimethyl-1H-pyrrole		$C_8H_{13}N$	517-22-6	123.196	pr	0	199; 96[16]	0.913[20]	1.4961[20]	sl H_2O; s EtOH, eth, bz, chl
5009	Ethyl 3,5-dimethylpyrrole-2-carboxylate		$C_9H_{13}NO_2$	2199-44-2	167.205	cry (al)	125	135[10.5]			s EtOH, ace
5010	Ethyl 2,4-dimethylpyrrole-3-carboxylate		$C_9H_{13}NO_2$	2199-51-1	167.205	cry (eth-lig, peth)	78.5	291			vs eth, EtOH
5011	Ethyl 2,5-dimethylpyrrole-3-carboxylate		$C_9H_{13}NO_2$	2199-52-2	167.205	orth (al)	117.5	291; 130[15]			vs EtOH
5012	Ethyl 4,5-dimethylpyrrole-3-carboxylate		$C_9H_{13}NO_2$	2199-53-3	167.205	cry (dil al)	111.3				vs eth, EtOH, chl
5013	Ethyl 2,4-dioxopentanoate		$C_7H_{10}O_4$	615-79-2	158.152		18	214	1.1251[20]	1.4757[17]	vs eth, EtOH

Ethyl diazoacetate

Ethyl dibromoacetate

Ethyl 2,3-dibromobutanoate

Ethyl 2,4-dibromobutanoate

Ethyl 2,3-dibromopropanoate

Ethyl 3,6-di(*tert*-butyl)-1-naphthalenesulfonate

Ethyl dichloroacetate

Ethyldichloroarsine

Ethyl dichlorocarbamate

Ethyl 2,3-dichloropropanoate

Ethyl diethoxyacetate

Ethyl diethylmalonate

Ethyl difluoroacetate

Ethyldifluoroarsine

5-Ethyldihydro-5-*sec*-butyl-2-thioxo-4,6(1*H*,5*H*)-pyrimidinedione

5-Ethyldihydro-2(3*H*)-furanone

Ethyl dihydrogen phosphate

5-Ethyldihydro-5-phenyl-4,6(1*H*,5*H*)-pyrimidinedione

Ethyl 2,4-dihydroxy-6-methylbenzoate

O-Ethyl *S*-[2-(diisopropylamino)ethyl] methylphosphonothioate

Ethyldimethylamine

Ethyl 4-(dimethylamino)benzoate

1-Ethyl-2,4-dimethylbenzene

1-Ethyl-3,5-dimethylbenzene

2-Ethyl-1,3-dimethylbenzene

2-Ethyl-1,4-dimethylbenzene

3-Ethyl-1,2-dimethylbenzene

4-Ethyl-1,2-dimethylbenzene

N'-Ethyl-*N*,*N*-dimethyl-1,2-ethanediamine

Ethyl 4,4-dimethyl-3-oxopentanoate

3-Ethyl-2,2-dimethylpentane

3-Ethyl-2,3-dimethylpentane

3-Ethyl-2,4-dimethylpentane

Ethyl 2,2-dimethylpropanoate

3-Ethyl-2,5-dimethylpyrazine

3-Ethyl-2,4-dimethyl-1*H*-pyrrole

Ethyl 3,5-dimethylpyrrole-2-carboxylate

Ethyl 2,4-dimethylpyrrole-3-carboxylate

Ethyl 2,5-dimethylpyrrole-3-carboxylate

Ethyl 4,5-dimethylpyrrole-3-carboxylate

Ethyl 2,4-dioxopentanoate

No.	Name	Synonym	Mol. Form.	CAS RN	Mol. Wt.	Physical Form	mp/°C	bp/°C	den/g cm^{-3}	n_D	Solubility
5014	O-Ethyl dithiocarbonate	Xanthogenic acid	$C_3H_6OS_2$	151-01-9	122.209	unstab liq	-53	25			
5015	Ethylene	Ethene	C_2H_4	74-85-1	28.053	col gas	-169.15	-103.77	0.5678^{-104}	1.363^{-100}	i H_2O; sl EtOH, bz, ace; s eth
5016	Ethylenebisdithiocarbamic acid		$C_4H_8N_2S_4$	111-54-6	212.380	unstab liq					
5017	Ethylene carbonate	Vinylene carbonate	$C_3H_4O_3$	96-49-1	88.062	mcl pl (al)	36.4	248	1.3214^{39}	1.4148^{50}	msc H_2O, EtOH, eth, bz, chl, AcOEt
5018	Ethylenediaminetetraacetic acid	EDTA	$C_{10}H_{16}N_2O_8$	60-00-4	292.242	cry (w)	245 dec				
5019	Ethylenediaminetetraacetic acid, disodium salt, dihydrate	EDTA disodium	$C_{10}H_{18}N_2Na_2O_{10}$	6381-92-6	372.237		242 dec				
5020	N,N'-Ethylene distearylamide	N,N'-Dioctadecanoylethanediamine	$C_{38}H_{76}N_2O_2$	110-30-5	593.022	cry (EtOH)	149				
5021	Ethyleneimine	Aziridine	C_2H_5N	151-56-4	43.068	liq	-77.9	56	0.832^{25}		msc H_2O; s EtOH; vs eth; sl chl
5022	Ethylestrenol		$C_{20}H_{32}O$	965-90-2	288.467	cry	77				
5023	N-Ethyl-1,2-ethanediamine		$C_4H_{12}N_2$	110-72-5	88.151			129	0.837^{25}	1.4385^{20}	
5024	Ethyl ethoxyacetate		$C_6H_{12}O_3$	817-95-8	132.157			158	0.9702^{20}	1.4039^{20}	s EtOH, eth, ace
5025	Ethyl 3-ethoxypropanoate		$C_7H_{14}O_3$	763-69-9	146.184			166; 48^5	0.9490^{20}	1.4065^{20}	s chl
5026	Ethyl 2-ethoxy-1(2H)-quinolinecarboxylate	EEDQ	$C_{14}H_{17}NO_3$	16357-59-8	247.290		56.5	126$^{0.1}$			
5027	Ethyl 2-ethylacetoacetate		$C_8H_{14}O_3$	607-97-6	158.195			198.0; 80^{10}	0.9847^{16}	1.4214^{25}	msc EtOH, eth
5028	Ethyl ethylcarbamate		$C_5H_{11}NO_2$	623-78-9	117.147			176	0.9813^{20}	1.4215^{20}	vs H_2O, eth, EtOH
5029	Ethyl 2-ethylhexanoate	Ethyl 2-ethylcaproate	$C_{10}H_{20}O_2$	2983-37-1	172.265			90^{28}	0.8586^{25}	1.4123^{25}	
5030	2-Ethyl-N-(2-ethylphenyl)aniline		$C_{16}H_{19}N$	64653-59-4	225.329		29	336^{603}, 173^{10}		1.5550^{25}	i H_2O; vs EtOH, eth; sl chl; s acid
5031	O-Ethyl ethylthiophosphonyl chloride		$C_4H_{10}ClOPS$	1497-68-3	172.613	liq		35$^{0.7}$	1.15^{20}		
5032	Ethyl fluoroacetate		$C_4H_7FO_2$	459-72-3	106.096			120	1.0912^{20}	1.3755^{20}	vs H_2O
5033	Ethyl 4-fluorobenzoate		$C_9H_9FO_2$	451-46-7	168.164	mcl pr (w)	26	210	1.146^{25}	1.4864^{20}	vs eth, EtOH
5034	N-Ethylformamide		C_3H_7NO	627-45-2	73.094			198	0.9552^{20}	1.4320^{20}	msc H_2O, EtOH, eth
5035	Ethyl formate		$C_3H_6O_2$	109-94-4	74.079	liq	-79.6	54.4	0.9208^{20}	1.3609^{20}	s H_2O; msc EtOH, eth; vs ace; sl ctc
5036	2-Ethylfuran		C_6H_8O	3208-16-0	96.127			92.5	0.9018^{20}	1.4403^{20}	s EtOH, eth, ace, bz
5037	Ethyl 2-furancarboxylate	Ethyl 2-furanoate	$C_7H_8O_3$	614-99-3	140.137	lf or pr	34.5	196.8	1.1174^{21}	1.4797^{21}	i H_2O; msc EtOH, eth, ace; s bz
5038	γ-Ethyl L-glutamate		$C_7H_{13}NO_4$	1119-33-1	175.183		191				sl H_2O
5039	Ethyl heptafluorobutanoate		$C_6H_5F_7O_2$	356-27-4	242.092			95	1.394^{20}	1.3011^{20}	sl H_2O; s eth, ace
5040	3-Ethylheptane		C_9H_{20}	15869-80-4	128.255	liq	-114.9	143.0	0.7225^{25}	1.4093^{20}	
5041	4-Ethylheptane		C_9H_{20}	2216-32-2	128.255			141.2	0.7241^{25}	1.4096^{20}	i H_2O; s eth; msc EtOH, ace, bz
5042	Ethyl heptanoate	Ethyl oenanthate	$C_9H_{18}O_2$	106-30-9	158.238	liq	-66.1	187	0.8817^{20}	1.4100^{20}	sl H_2O, ctc; s EtOH, eth
5043	2-Ethylheptanoic acid		$C_9H_{18}O_2$	3274-29-1	158.238	liq		153^{31}			
5044	4-Ethyl-4-heptanol		$C_9H_{20}O$	597-90-0	144.254			182	0.8350^{20}	1.4255^{27}	vs eth, EtOH
5045	Ethyl trans,trans-2,4-hexadienoate	Ethyl sorbate	$C_8H_{12}O_2$	2396-84-1	140.180			195.5	0.9506^{20}	1.4951^{20}	vs eth, EtOH, chl
5046	2-Ethylhexanal		$C_8H_{16}O$	123-05-7	128.212		<-100	163	0.8540^{20}	1.4142^{20}	i H_2O; s EtOH, eth; sl ctc
5047	3-Ethylhexane		C_8H_{18}	619-99-8	114.229			118.6	0.7136^{20}	1.4018^{20}	i H_2O; msc EtOH, eth, ace, bz, chl; s ctc
5048	2-Ethyl-1,3-hexanediol	Ethohexadiol	$C_8H_{18}O_2$	94-96-2	146.228	liq	-40	244	0.9325^{22}	1.4497^{20}	sl H_2O; s EtOH, eth
5049	Ethyl hexanoate		$C_8H_{16}O_2$	123-66-0	144.212	liq	-67	167	0.873^{20}	1.4073^{20}	sl H_2O; vs eth, EtOH
5050	2-Ethylhexanoic acid		$C_8H_{16}O_2$	149-57-5	144.212			228; 120^{13}	0.9031^{25}	1.4241^{20}	s H_2O, eth, ctc; sl EtOH
5051	2-Ethyl-1-hexanol		$C_8H_{18}O$	104-76-7	130.228	liq	-70	184.6	0.8319^{25}	1.4300^{20}	i H_2O; s EtOH, eth, ace, bz, chl
5052	2-Ethylhexanoyl chloride		$C_8H_{15}ClO$	760-67-8	162.657			101^{40}, 67^{11}	0.939^{25}	1.4335^{20}	
5053	2-Ethyl-2-hexenal		$C_8H_{14}O$	645-62-5	126.196			175	0.8554^{20}		
5054	Ethyl 3-hexenoate	Ethyl hydrosorbate	$C_8H_{14}O_2$	2396-83-0	142.196			166.5	0.8957^{20}	1.4255^{20}	
5055	2-Ethylhexyl acetate		$C_{10}H_{20}O_2$	103-09-3	172.265	liq	-80	199	0.8718^{20}	1.4204^{20}	i H_2O; s EtOH, eth

O-Ethyl dithiocarbonate

Ethylene

Ethylenebisdithiocarbamic acid

Ethylene carbonate

Ethylenediaminetetraacetic acid

Ethylenediaminetetraacetic acid, disodium salt, dihydrate

$2H_2O$

N,N'-Ethylene distearylamide

Ethyleneimine

Ethylestrenol

N-Ethyl-1,2-ethanediamine

Ethyl ethoxyacetate

Ethyl 3-ethoxypropanoate

Ethyl 2-ethoxy-1(2H)-quinolinecarboxylate

Ethyl 2-ethylacetoacetate

Ethyl ethylcarbamate

Ethyl 2-ethylhexanoate

2-Ethyl-N-(2-ethylphenyl)aniline

O-Ethyl ethylthiophosphonyl chloride

Ethyl fluoroacetate

Ethyl 4-fluorobenzoate

N-Ethylformamide

Ethyl formate

2-Ethylfuran

Ethyl 2-furancarboxylate

γ-Ethyl L-glutamate

Ethyl heptafluorobutanoate

3-Ethylheptane

4-Ethylheptane

Ethyl heptanoate

2-Ethylheptanoic acid

4-Ethyl-4-heptanol

Ethyl trans,trans-2,4-hexadienoate

2-Ethylhexanal

3-Ethylhexane

2-Ethyl-1,3-hexanediol

Ethyl hexanoate

2-Ethylhexanoic acid

2-Ethyl-1-hexanol

2-Ethylhexanoyl chloride

2-Ethyl-2-hexenal

Ethyl 3-hexenoate

2-Ethylhexyl acetate

No.	Name	Synonym	Mol. Form.	CAS RN	Mol. Wt.	Physical Form	mp/°C	bp/°C	den/ g cm⁻³	n_D	Solubility
5056	2-Ethylhexyl acrylate		$C_{11}H_{20}O_2$	103-11-7	184.276		-90	125[60]	0.880[25]	1.4332[25]	
5057	2-Ethylhexylamine	2-Ethyl-1-hexanamine	$C_8H_{19}N$	104-75-6	129.244			169.2			sl H₂O
5058	2-Ethylhexyl butyl phthalate	Butyl 2-ethylhexyl phthalate	$C_{20}H_{30}O_4$	85-69-8	334.450	col liq					sl H₂O
5059	2-Ethylhexyl dihydrogen phosphate	Mono(2-ethylhexyl) phosphate	$C_8H_{19}O_4P$	1070-03-7	210.208	liq					s H₂O, bz
5060	2-Ethylhexyl diphenyl phosphite	Forstab	$C_{20}H_{27}O_3P$	15647-08-2	346.400			152[0.15]	1.054[20]	1.5207[27]	
5061	Ethyl hexyl ether	1-Ethoxyhexane	$C_8H_{18}O$	5756-43-4	130.228			143	0.7722[20]	1.4008[20]	vs eth, EtOH
5062	2-Ethylhexyl 2-hydroxybenzoate	Octisalate	$C_{15}H_{22}O_3$	118-60-5	250.334	liq		190[21]	1.01		
5063	2-Ethylhexyl methacrylate		$C_{12}H_{22}O_2$	688-84-6	198.302			120[18], 110[14]	0.880[25]	1.436[25]	
5064	2-[(2-Ethylhexyl)oxy]ethanol	Ethylene glycol mono(2-ethylhexyl) ether	$C_{10}H_{22}O_2$	1559-35-9	174.281			227.7			
5065	Ethylhydrazine		$C_2H_8N_2$	624-80-6	60.098			101			vs H₂O, ace, eth, EtOH
5066	Ethyl hydrazinecarboxylate	Ethyl carbazate	$C_3H_8N_2O_2$	4114-31-2	104.108	cry	46	dec 198; 93[9]			s EtOH, eth; sl chl
5067	Ethyl hydrogen adipate		$C_8H_{14}O_4$	626-86-8	174.195	hyg cry (eth, peth)	29	285	0.9796[20]	1.4311[20]	s EtOH, eth, peth
5068	Ethyl hydrogen fumarate		$C_6H_8O_4$	2459-05-4	144.126		70	147[16]	1.1109[87]		s EtOH, ace; sl chl
5069	Ethyl hydrogen succinate	Butanedioic acid, monoethyl ester	$C_6H_{10}O_4$	1070-34-4	146.141	pr or nd	8	172[42], 119[3]	1.1466[20]	1.4327[20]	vs H₂O, eth, EtOH
5070	Ethyl hydroperoxide	Ethyl hydrogen peroxide	$C_2H_6O_2$	3031-74-1	62.068	liq	-100	95	0.9332[20]	1.3800[20]	vs H₂O, bz, eth, EtOH
5071	Ethyl hydroxyacetate		$C_4H_8O_3$	623-50-7	104.105			160	1.0826[23]	1.4180[20]	vs eth, EtOH
5072	Ethyl 3-hydroxybenzoate		$C_9H_{10}O_3$	7781-98-8	166.173	pl (bz)	74		1.0680[131]		sl H₂O, chl; s EtOH, eth
5073	Ethyl 4-hydroxybenzoate	Ethylparaben	$C_9H_{10}O_3$	120-47-8	166.173	cry (dil al)	117	297.5			sl H₂O, chl, tfa; vs EtOH, eth; i CS₂
5074	Ethyl 3-hydroxybutanoate, (±)		$C_6H_{12}O_3$	35608-64-1	132.157			185; 76[15]	1.017[20]	1.4182[20]	s H₂O, EtOH; sl ctc
5075	Ethyl 2-hydroxy-3-butenoate		$C_6H_{10}O_3$	91890-87-8	130.141			dec 173; 68[15]	1.0470[15]	1.436[13]	vs H₂O, eth, EtOH
5076	α-Ethyl-1-hydroxycyclohexaneacetic acid	Cyclobutyrol	$C_{10}H_{18}O_3$	512-16-3	186.248	cry (eth-peth)	81.5	164[24]	1.0010[18]	1.4680[18]	vs ace, eth, EtOH, chl
5077	N-Ethyl-N-hydroxyethanamine	N,N-Diethylhydroxylamine	$C_4H_{11}NO$	3710-84-7	89.136		10	133	0.8669[20]	1.4195[20]	
5078	2-Ethyl-3-hydroxyhexanal		$C_8H_{16}O_2$	496-03-7	144.212			138[50], 101[9]			
5079	Ethyl 4-hydroxy-3-methoxybenzoate		$C_{10}H_{12}O_4$	617-05-0	196.200	nd (dil al)	44	292			i H₂O; vs EtOH, eth; s chl
5080	Ethyl cis-12-hydroxy-9-octadecenoate, (R)	Ethyl ricinoleate	$C_{20}H_{38}O_3$	55066-53-0	326.514			258[13]	0.9180[20]	1.4618[22]	
5081	Ethylidenecyclohexane		C_8H_{14}	1003-64-1	110.197			136	0.822[25]	1.4618[20]	
5082	5-Ethylidene-2-norbornene	5-Ethylidenebicyclo[2.2.1]hept-2-ene	C_9H_{12}	16219-75-3	120.191	liq		146	0.893	1.4900[20]	
5083	1-Ethyl-1H-imidazole		$C_5H_8N_2$	7098-07-9	96.131			208	0.999[25]		msc H₂O
5084	Ethyl iodoacetate		$C_4H_7IO_2$	623-48-3	214.002	oil		179	1.8173[13]	1.5079[13]	s EtOH, eth
5085	Ethyl isobutylcarbamate	Isobutyl urethane	$C_7H_{15}NO_2$	539-89-9	145.200		<-65	110[30]	0.9432[20]	1.4288[20]	vs eth, EtOH
5086	Ethyl isocyanate		C_3H_5NO	109-90-0	71.078			60	0.9031[20]	1.3808[20]	i H₂O; msc EtOH, eth
5087	Ethyl isocyanide		C_3H_5N	624-79-3	55.079		<-66	79	0.7402[20]	1.3622[20]	vs H₂O; msc EtOH, eth; s ace
5088	N-Ethyl-1H-isoindole-1,3(2H)-dione		$C_{10}H_9NO_2$	5022-29-7	175.184	nd (al)	79	285.5			s EtOH, eth
5089	Ethyl isopentyl ether		$C_7H_{16}O$	628-04-6	116.201			112.5	0.7688[21]		vs eth, EtOH
5090	Ethylisopropylamine	N-Ethyl-2-propanamine	$C_5H_{13}N$	19961-27-4	87.164			69.6		1.3872[25]	
5091	1-Ethyl-2-isopropylbenzene		$C_{11}H_{16}$	18970-44-0	148.245			193	0.888[20]	1.508[20]	vs ace, bz, eth, EtOH
5092	Ethyl isopropyl ether		$C_5H_{12}O$	625-54-7	88.148			54.1	0.720[25]	1.3698[25]	s H₂O, ace, chl; msc EtOH, eth
5093	N-Ethyl-N-isopropyl-2-propanamine		$C_8H_{19}N$	7087-68-5	129.244			126.5	0.742[25]	1.4138[20]	s ctc
5094	Ethyl isopropyl sulfide		$C_5H_{12}S$	5145-99-3	104.214	liq	-122.2	107.5	0.8246[20]		
5095	Ethyl isothiocyanate		C_3H_5NS	542-85-8	87.144	liq	-5.9	131.5	0.9990[20]	1.5130[20]	i H₂O; msc EtOH, eth
5096	Ethyl lactate	Ethyl 2-hydroxypropionate	$C_5H_{10}O_3$	2676-33-7	118.131	liq	-26	154.5	1.0328[20]	1.4124[20]	vs H₂O, eth, EtOH
5097	Ethyl laurate		$C_{14}H_{28}O_2$	106-33-2	228.371	liq	-10	271; 154[15]	0.8618[20]	1.4311[20]	i H₂O; vs EtOH; msc eth; sl ctc
5098	Ethyl levulinate		$C_7H_{12}O_3$	539-88-8	144.168			205.8	1.0111[20]	1.4229[20]	vs H₂O, EtOH
5099	Ethyl mercaptoacetate		$C_4H_8O_2S$	623-51-8	120.171			157	1.0964[15]	1.4582[20]	s EtOH, eth; sl ctc
5100	Ethyl methacrylate		$C_6H_{10}O_2$	97-63-2	114.142			117	0.9135[20]	1.4147[20]	sl H₂O, chl; msc EtOH, eth

2-Ethylhexyl acrylate

2-Ethylhexylamine

2-Ethylhexyl butyl phthalate

2-Ethylhexyl dihydrogen phosphate

2-Ethylhexyl diphenyl phosphite

Ethyl hexyl ether

2-Ethylhexyl 2-hydroxybenzoate

2-Ethylhexyl methacrylate

2-[(2-Ethylhexyl)oxy]ethanol

Ethylhydrazine

Ethyl hydrazinecarboxylate

Ethyl hydrogen adipate

Ethyl hydrogen fumarate

Ethyl hydrogen succinate

Ethyl hydroperoxide

Ethyl hydroxyacetate

Ethyl 3-hydroxybenzoate

Ethyl 4-hydroxybenzoate

Ethyl 3-hydroxybutanoate, (±)

Ethyl 2-hydroxy-3-butenoate

α-Ethyl-1-hydroxycyclohexaneacetic acid

N-Ethyl-N-hydroxyethanamine

2-Ethyl-3-hydroxyhexanal

Ethyl 4-hydroxy-3-methoxybenzoate

Ethyl cis-12-hydroxy-9-octadecenoate, (R)

Ethylidenecyclohexane

5-Ethylidene-2-norbornene

1-Ethyl-1H-imidazole

Ethyl iodoacetate

Ethyl isobutylcarbamate

Ethyl isocyanate

Ethyl isocyanide

N-Ethyl-1H-isoindole-1,3(2H)-dione

Ethyl isopentyl ether

Ethylisopropylamine

1-Ethyl-2-isopropylbenzene

Ethyl isopropyl ether

N-Ethyl-N-isopropyl-2-propanamine

Ethyl isopropyl sulfide

Ethyl isothiocyanate

Ethyl lactate

Ethyl laurate

Ethyl levulinate

Ethyl mercaptoacetate

Ethyl methacrylate

No.	Name	Synonym	Mol. Form.	CAS RN	Mol. Wt.	Physical Form	mp/°C	bp/°C	den/ g cm⁻³	n_D	Solubility
5101	Ethyl methanesulfonate		$C_3H_8O_3S$	62-50-0	124.159			86[10]			
5102	1-Ethyl-4-methoxybenzene		$C_9H_{12}O$	1515-95-3	136.190			198	0.9624[15]	1.5120[20]	vs bz, eth
5103	α-Ethyl-4-methoxybenzenemethanol		$C_{10}H_{14}O_2$	5349-60-0	166.217			143[20]		1.5277[20]	s ctc
5104	Ethyl 2-methoxybenzoate		$C_{10}H_{12}O_3$	7335-26-4	180.200			261	1.1124[20]	1.5224[20]	vs eth, EtOH
5105	Ethyl 4-methoxybenzoate		$C_{10}H_{12}O_3$	94-30-4	180.200		7.5	269.5	1.1038[20]	1.5254[20]	i H₂O; s EtOH, eth
5106	4-Ethyl-2-methoxyphenol		$C_9H_{12}O_2$	2785-89-9	152.190	liq	-7	236.5	1.0931[18]		
5107	Ethyl (4-methoxyphenyl)acetate		$C_{11}H_{14}O_3$	14062-18-1	194.227			139[70]	1.097[25]	1.5075[20]	
5108	Ethyl 2-methylacetoacetate		$C_7H_{12}O_3$	609-14-3	144.168			187	0.9941[20]	1.4185[20]	sl H₂O; s EtOH, eth; vs ace
5109	N-Ethyl-2-methylallylamine	N-Ethyl-2-methyl-2-propen-1-amine	$C_6H_{13}N$	18328-90-0	99.174	liq		104.7	0.753	1.4221[20]	msc H₂O
5110	5-Ethyl-5-(2-methylallyl)-2-thiobarbituric acid	Methallatal	$C_{10}H_{14}N_2O_2S$	115-56-0	226.295		160.5				
5111	Ethylmethylamine	N-Methylethanamine	C_3H_9N	624-78-2	59.110			36.7			vs H₂O, ace, eth, EtOH
5112	Ethylmethylamine hydrochloride	N-Methylethanamine hydrochloride	$C_3H_{10}ClN$	624-60-2	95.571	pl (al-eth)	128		1.0874[20]		vs H₂O, EtOH; i eth; s chl
5113	2-Ethyl-6-methylaniline		$C_9H_{13}N$	24549-06-2	135.206	liq	-33	231	0.968[25]	1.5525[20]	
5114	N-Ethyl-2-methylaniline		$C_9H_{13}N$	94-68-8	135.206		<-15	216	0.948[25]	1.5456[20]	s EtOH, eth
5115	N-Ethyl-3-methylaniline		$C_9H_{13}N$	102-27-2	135.206			221	0.9263[15]	1.5451[20]	s EtOH, eth
5116	N-Ethyl-4-methylaniline	N-Ethyl-4-toluidine	$C_9H_{13}N$	622-57-1	135.206			217	0.9391[16]		s EtOH, eth
5117	N-Ethyl-N-methylaniline		$C_9H_{13}N$	613-97-8	135.206			204	0.92[55]		i H₂O; msc EtOH, eth; s ctc
5118	N-Ethyl-α-methylbenzeneethanamine	N-Ethylamphetamine	$C_{11}H_{17}N$	457-87-4	163.260			105[14]		1.4986[25]	
5119	N-Ethyl-4-methylbenzenesulfonamide		$C_9H_{13}NO_2S$	80-39-7	199.270		64				s EtOH
5120	1-Ethyl-2-methyl-1H-benzimidazole		$C_{10}H_{12}N_2$	5805-76-5	160.215		51	296	1.073[25]		
5121	Ethyl 2-methylbenzoate		$C_{10}H_{12}O_2$	87-24-1	164.201		<-10	227; 113[18]	1.0325[21]	1.507[22]	i H₂O; msc EtOH, eth
5122	Ethyl 4-methylbenzoate		$C_{10}H_{12}O_2$	94-08-6	164.201			232	1.0269[18]	1.5089[18]	i H₂O; msc EtOH, eth
5123	Ethyl 3-methylbutanoate	Ethyl isovalerate	$C_7H_{14}O_2$	108-64-5	130.185	liq	-99.3	135.0	0.8656[20]	1.3962[20]	sl H₂O; vs EtOH, eth
5124	2-Ethyl-2-methylbutanoic acid		$C_7H_{14}O_2$	19889-37-3	130.185		<-20	207		1.4250[20]	vs EtOH
5125	2-Ethyl-3-methyl-1-butene		C_7H_{14}	7357-93-9	98.186			89	0.7150[20]	1.410[20]	i H₂O; s eth, ace, bz, chl
5126	Ethyl trans-2-methyl-2-butenoate		$C_7H_{12}O_2$	5837-78-5	128.169			156	0.9200[20]	1.4340[20]	
5127	Ethyl 3-methyl-2-butenoate		$C_7H_{12}O_2$	638-10-8	128.169			153.5	0.9199[21]	1.4345[20]	
5128	5-Ethyl-5-(1-methylbutyl)-2,4,6(1H,3H,5H)-pyrimidinetrione		$C_{11}H_{18}N_2O_3$	76-74-4	226.272		130				sl H₂O; s EtOH, eth
5129	Ethyl N-methylcarbamate		$C_4H_9NO_2$	105-40-8	103.120			170	1.0115[20]	1.4183[20]	vs H₂O, EtOH
5130	Ethyl methyl carbonate		$C_4H_8O_3$	623-53-0	104.105	liq	-14	107.5	1.012[20]	1.3778[20]	vs eth, EtOH
5131	trans-1-Ethyl-4-methylcyclohexane		C_9H_{18}	6236-88-0	126.239	liq	-80.8	149	0.7798[20]	1.4304[20]	
5132	1-Ethyl-1-methylcyclopentane		C_8H_{16}	16747-50-5	112.213	liq	-143.8	121.6	0.7767[25]	1.4272[20]	vs ace, bz, eth, EtOH
5133	cis-1-Ethyl-2-methylcyclopentane		C_8H_{16}	930-89-2	112.213	liq	-106	128	0.7852[20]	1.4293[20]	
5134	trans-1-Ethyl-2-methylcyclopentane		C_8H_{16}	930-90-5	112.213	liq	-105.9	121.2	0.7649[25]	1.4219[20]	
5135	cis-1-Ethyl-3-methylcyclopentane		C_8H_{16}	2613-66-3	112.213			121	0.7724[20]	1.4203[20]	vs ace, bz, eth, EtOH
5136	trans-1-Ethyl-3-methylcyclopentane		C_8H_{16}	2613-65-2	112.213	liq	-108	121	0.7619[20]	1.4186[20]	
5137	1-Ethyl-1-methylcyclopropane		C_6H_{12}	53778-43-1	84.159	liq	-130.2	56.8	0.6968[25]	1.3887[20]	
5138	2-Ethyl-2-methyl-1,3-dioxolane		$C_6H_{12}O_2$	126-39-6	116.158			118	0.9360[20]		
5139	Ethyl methyl ether		C_3H_8O	540-67-0	60.095	col gas	-113	7.4	0.7251[0]	1.3420[4]	s H₂O, ace, chl; msc EtOH, eth
5140	3-Ethyl-2-methylhexane		C_9H_{20}	16789-46-1	128.255			138	0.7310[20]	1.4106[20]	
5141	3-Ethyl-3-methylhexane		C_9H_{20}	3074-76-8	128.255			140.6	0.7371[25]	1.4140[20]	
5142	3-Ethyl-4-methylhexane	2,3-Diethylpentane	C_9H_{20}	3074-77-9	128.255			140	0.7420[20]	1.4134[20]	
5143	4-Ethyl-2-methylhexane		C_9H_{20}	3074-75-7	128.255			133.8	0.7195[25]	1.4063[20]	
5144	Ethyl 4-methylhexanoate	Ethyl 4-methylcaproate	$C_9H_{18}O_2$	1561-10-0	158.238			180	0.8708[20]	1.4051[20]	
5145	Ethyl 4-methyl-3-oxopentanoate		$C_8H_{14}O_3$	7152-15-0	158.195	liq	-9	173	0.98[25]	1.250[20]	
5146	3-Ethyl-2-methylpentane		C_8H_{18}	609-26-7	114.229	liq	-114.9	115.66	0.7193[20]	1.4040[20]	i H₂O; s eth; msc EtOH, ace, bz

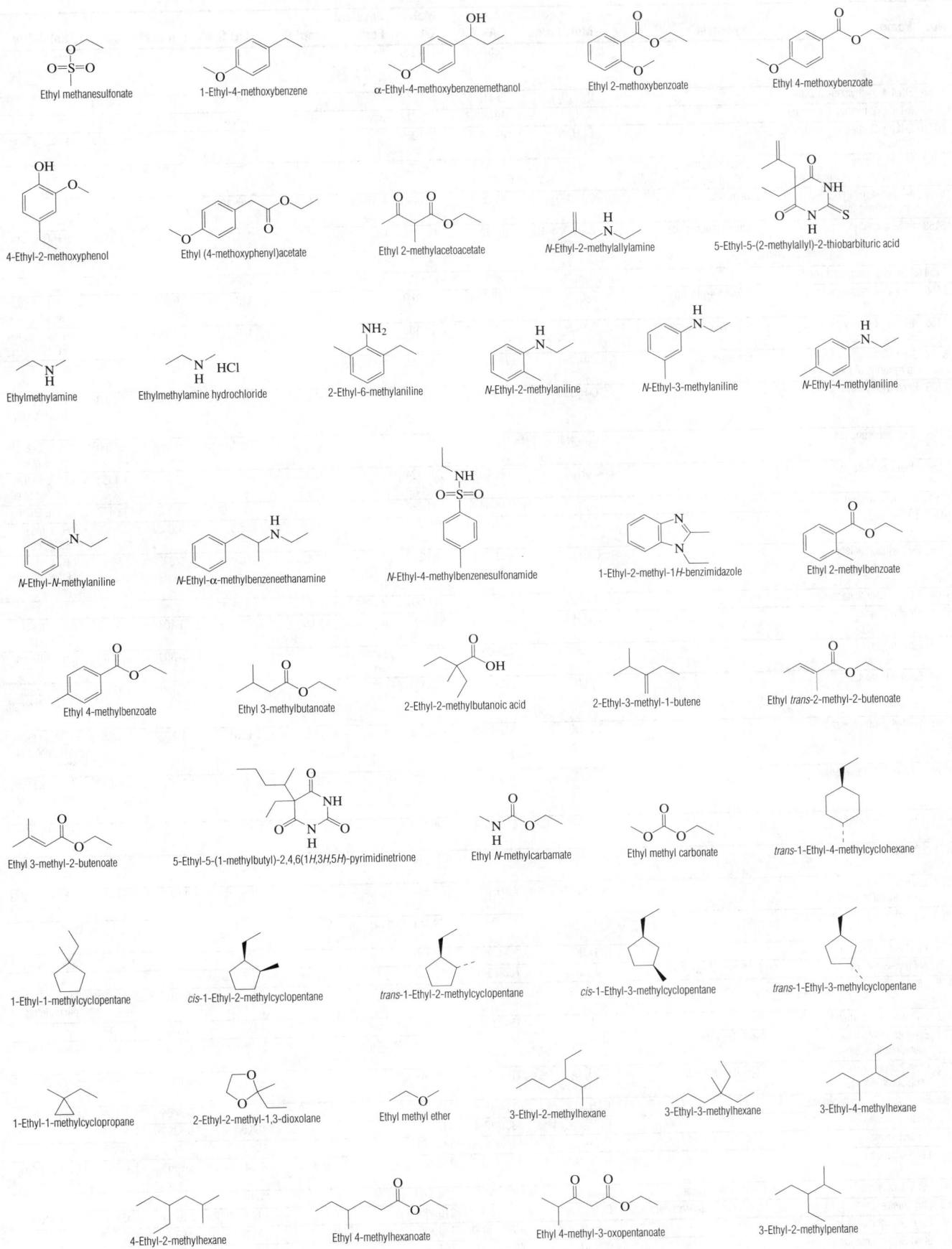

Ethyl methanesulfonate

1-Ethyl-4-methoxybenzene

α-Ethyl-4-methoxybenzenemethanol

Ethyl 2-methoxybenzoate

Ethyl 4-methoxybenzoate

4-Ethyl-2-methoxyphenol

Ethyl (4-methoxyphenyl)acetate

Ethyl 2-methylacetoacetate

N-Ethyl-2-methylallylamine

5-Ethyl-5-(2-methylallyl)-2-thiobarbituric acid

Ethylmethylamine

Ethylmethylamine hydrochloride

2-Ethyl-6-methylaniline

N-Ethyl-2-methylaniline

N-Ethyl-3-methylaniline

N-Ethyl-4-methylaniline

N-Ethyl-N-methylaniline

N-Ethyl-α-methylbenzeneethanamine

N-Ethyl-4-methylbenzenesulfonamide

1-Ethyl-2-methyl-1H-benzimidazole

Ethyl 2-methylbenzoate

Ethyl 4-methylbenzoate

Ethyl 3-methylbutanoate

2-Ethyl-2-methylbutanoic acid

2-Ethyl-3-methyl-1-butene

Ethyl trans-2-methyl-2-butenoate

Ethyl 3-methyl-2-butenoate

5-Ethyl-5-(1-methylbutyl)-2,4,6(1H,3H,5H)-pyrimidinetrione

Ethyl N-methylcarbamate

Ethyl methyl carbonate

trans-1-Ethyl-4-methylcyclohexane

1-Ethyl-1-methylcyclopentane

cis-1-Ethyl-2-methylcyclopentane

trans-1-Ethyl-2-methylcyclopentane

cis-1-Ethyl-3-methylcyclopentane

trans-1-Ethyl-3-methylcyclopentane

1-Ethyl-1-methylcyclopropane

2-Ethyl-2-methyl-1,3-dioxolane

Ethyl methyl ether

3-Ethyl-2-methylhexane

3-Ethyl-3-methylhexane

3-Ethyl-4-methylhexane

4-Ethyl-2-methylhexane

Ethyl 4-methylhexanoate

Ethyl 4-methyl-3-oxopentanoate

3-Ethyl-2-methylpentane

No.	Name	Synonym	Mol. Form.	CAS RN	Mol. Wt.	Physical Form	mp/°C	bp/°C	den/g cm⁻³	n_D	Solubility
5147	3-Ethyl-3-methylpentane		C₈H₁₈	1067-08-9	114.229	liq	-90.9	118.27	0.7274²⁰	1.4078²⁰	i H₂O; s eth; msc EtOH, ace, bz
5148	Ethyl 4-methylpentanoate		C₈H₁₆O₂	25415-67-2	144.212			163; 52¹⁰	0.8705²⁰	1.4050²⁰	
5149	3-Ethyl-2-methyl-1-pentene		C₈H₁₆	19780-66-6	112.213	liq	-112.9	109.5	0.7262²⁰	1.4140²⁰	
5150	2-[Ethyl(3-methylphenyl)amino]ethanol		C₁₁H₁₇NO	91-88-3	179.259			118¹·⁵		1.5540²⁰	s ctc
5151	Ethyl 3-methyl-3-phenyloxiranecarboxylate	Ethyl 3-methyl-3-phenylglycidate	C₁₂H₁₄O₃	77-83-8	206.237			273.5	1.044²⁰	1.5182²⁰	
5152	4-Ethyl-4-methyl-2,6-piperidinedione	Bemegride	C₈H₁₃NO₂	64-65-3	155.195	pl (w, ace-eth)	126.5	sub 100			s chl
5153	Ethyl 2-methylpropanoate	Ethyl isobutanoate	C₆H₁₂O₂	97-62-1	116.158	liq	-88.2	110.1	0.868²⁰	1.3869¹⁸	sl H₂O, ctc; msc EtOH, eth; s ace
5154	2-Ethyl-5-methylpyrazine		C₇H₁₀N₂	13360-64-0	122.167	liq		79⁵⁶			
5155	3-Ethyl-4-methylpyridine	3-Ethyl-4-picoline	C₈H₁₁N	529-21-5	121.180			198	0.9286¹⁷		sl H₂O; s EtOH, eth, chl; vs ace
5156	4-Ethyl-2-methylpyridine	4-Ethyl-2-picoline	C₈H₁₁N	536-88-9	121.180			179	0.9130²⁵		vs ace, bz, eth, EtOH
5157	3-Ethyl-3-methyl-2,5-pyrrolidinedione	Ethosuximide	C₇H₁₁NO₂	77-67-8	141.168	cry (ace-eth)	64.5				vs H₂O
5158	Ethyl methyl sulfide		C₃H₈S	624-89-5	76.161	liq	-105.93	66.7	0.8422²⁰	1.4404²⁰	i H₂O; msc EtOH; s eth; chl
5159	N-Ethylmorpholine		C₆H₁₃NO	100-74-3	115.173			138.5	0.8996²⁰	1.4400²⁰	msc H₂O, EtOH, eth; s ace, bz
5160	Ethyl myristate		C₁₆H₃₂O₂	124-06-1	256.424		12.3	295	0.8573²⁵	1.4362²⁰	i H₂O; s EtOH, ctc, lig; sl eth
5161	N-Ethyl-1-naphthalenamine		C₁₂H₁₃N	118-44-5	171.238			305; 191¹⁶	1.0652¹⁵	1.6477¹⁵	vs eth, EtOH
5162	1-Ethylnaphthalene		C₁₂H₁₂	1127-76-0	156.223	liq	-13.9	258.6	1.0082²⁰	1.6062²⁰	i H₂O; msc EtOH, eth
5163	2-Ethylnaphthalene		C₁₂H₁₂	939-27-5	156.223	liq	-7.4	258	0.9922²⁰	1.5999²⁰	i H₂O; msc EtOH, eth; sl chl
5164	Ethyl 1-naphthylacetate		C₁₄H₁₄O₂	2122-70-5	214.260	oil	88.5	222²⁰, 118¹³			s EtOH, eth
5165	Ethyl nitrate		C₂H₅NO₃	625-58-1	91.066	liq	-94.6	87.2	1.1084²⁰	1.3852²⁰	s H₂O; msc EtOH, eth
5166	Ethyl nitrite		C₂H₅NO₂	109-95-5	75.067	ye vol liq or gas		18	0.899¹⁵	1.3418¹⁰	msc EtOH, eth
5167	Ethyl nitroacetate		C₄H₇NO₄	626-35-7	133.104			106²⁵, 83⁶	1.1953²⁰	1.4250²⁰	sl H₂O; msc EtOH; vs eth; s dil alk
5168	1-Ethyl-2-nitrobenzene		C₈H₉NO₂	612-22-6	151.163	liq	-12.3	232.5	1.1207²⁰	1.5356²⁰	i H₂O; vs EtOH, eth; s ace; sl ctc
5169	1-Ethyl-4-nitrobenzene		C₈H₉NO₂	100-12-9	151.163	liq	-12.3	245.5	1.1192²⁰	1.5455²⁰	i H₂O; vs EtOH, eth; s ace; sl ctc
5170	Ethyl 3-nitrobenzoate		C₉H₉NO₄	618-98-4	195.172		47	297			i H₂O; vs EtOH, eth
5171	Ethyl 4-nitrobenzoate		C₉H₉NO₄	99-77-4	195.172		57	186.3; 153⁸			i H₂O; s EtOH, eth
5172	Ethyl p-nitrophenyl benzenethiophosphate		C₁₄H₁₄NO₄PS	2104-64-5	323.304		36		1.27²⁵	1.5978³⁰	vs bz, eth, EtOH
5173	2-Ethyl-2-nitro-1,3-propanediol		C₅H₁₁NO₄	597-09-1	149.146	nd (w)	57.5	dec			vs H₂O, eth, EtOH
5174	Ethyl 2-nitropropanoate		C₅H₉NO₄	2531-80-8	147.130			190.5		1.4210²⁰	vs bz, eth, EtOH
5175	N-Ethyl-N-nitrosourea	N-Nitroso-N-ethylurea	C₃H₇N₃O₂	759-73-9	117.107		100 dec				s chl
5176	Ethyl nonanoate		C₁₁H₂₂O₂	123-29-5	186.292	liq	-36.7	227.0	0.8657²⁰	1.4220²⁰	i H₂O; s EtOH, eth, ace, ctc
5177	5-Ethyl-2-norbornene		C₉H₁₄	15403-89-1	122.207	liq		143.6	0.86	1.4630²⁰	
5178	Ethyl cis,cis-9,12-octadecadienoate	Ethyl linoleate	C₂₀H₃₆O₂	544-35-4	308.499	ye or col		272¹⁸⁰, 212¹²	0.8865²⁰		vs eth, EtOH
5179	Ethyl cis,cis,cis-9,12,15-octadecatrienoate	Ethyl linolenate	C₂₀H₃₄O₂	1191-41-9	306.483			218¹⁵	0.8919²⁰	1.4694²⁰	vs eth, EtOH
5180	Ethyl trans-9-octadecenoate		C₂₀H₃₈O₂	6114-18-7	310.515		5.8	218¹⁵	0.8664²⁵	1.4480²⁵	vs eth, EtOH
5181	3-Ethyloctane		C₁₀H₂₂	5881-17-4	142.282			166.5	0.7359²⁵	1.4156²⁰	
5182	4-Ethyloctane		C₁₀H₂₂	15869-86-0	142.282			163.7	0.7343²⁵	1.4151²⁰	
5183	Ethyl octanoate		C₁₀H₂₀O₂	106-32-1	172.265	liq	-43.1	208.5	0.866¹⁸	1.4178²⁰	i H₂O; vs EtOH, eth; sl ctc
5184	Ethyl 1-octyl sulfide	1-(Ethylthio)octane	C₁₀H₂₂S	3698-94-0	174.347	liq		109¹⁴			
5185	Ethyl oleate	Ethyl cis-9-octadecenoate	C₂₀H₃₈O₂	111-62-6	310.515			216¹⁵, 207¹³	0.8720²⁰	1.4515²⁰	vs eth, EtOH
5186	Ethyl 5-oxohexanoate		C₈H₁₄O₃	13984-57-1	158.195			221.5	0.989²⁵	1.4277²⁰	
5187	Ethyl 3-oxopentanoate		C₇H₁₂O₃	4949-44-4	144.168			191	1.0120²⁰	1.4230²⁰	vs bz, eth, EtOH

3-Ethyl-3-methylpentane

Ethyl 4-methylpentanoate

3-Ethyl-2-methyl-1-pentene

2-[Ethyl(3-methylphenyl)amino]ethanol

Ethyl 3-methyl-3-phenyloxiranecarboxylate

4-Ethyl-4-methyl-2,6-piperidinedione

Ethyl 2-methylpropanoate

2-Ethyl-5-methylpyrazine

3-Ethyl-4-methylpyridine

4-Ethyl-2-methylpyridine

3-Ethyl-3-methyl-2,5-pyrrolidinedione

Ethyl methyl sulfide

N-Ethylmorpholine

Ethyl myristate

N-Ethyl-1-naphthalenamine

1-Ethylnaphthalene

2-Ethylnaphthalene

Ethyl 1-naphthylacetate

Ethyl nitrate

Ethyl nitrite

Ethyl nitroacetate

1-Ethyl-2-nitrobenzene

1-Ethyl-4-nitrobenzene

Ethyl 3-nitrobenzoate

Ethyl 4-nitrobenzoate

Ethyl p-nitrophenyl benzenethiophosphate

2-Ethyl-2-nitro-1,3-propanediol

Ethyl 2-nitropropanoate

N-Ethyl-N-nitrosourea

Ethyl nonanoate

5-Ethyl-2-norbornene

Ethyl cis,cis-9,12-octadecadienoate

Ethyl cis,cis,cis-9,12,15-octadecatrienoate

Ethyl trans-9-octadecenoate

3-Ethyloctane

4-Ethyloctane

Ethyl octanoate

Ethyl 1-octyl sulfide

Ethyl oleate

Ethyl 5-oxohexanoate

Ethyl 3-oxopentanoate

No.	Name	Synonym	Mol. Form.	CAS RN	Mol. Wt.	Physical Form	mp/°C	bp/°C	den/ g cm^{-3}	n_D	Solubility
5188	Ethyl 2-oxo-2-phenylacetate	Ethyl phenylglyoxylate	C$_{10}$H$_{10}$O$_3$	1603-79-8	178.184			256.5	1.1222[25]	1.5190[25]	
5189	Ethyl 2-oxopropanoate	Ethyl pyruvate	C$_5$H$_8$O$_3$	617-35-6	116.116	liq	-50	155	1.0596[15]	1.4052[20]	sl H$_2$O; s ace; msc EtOH, eth
5190	Ethyl palmitate		C$_{18}$H$_{36}$O$_2$	628-97-7	284.478	nd	24	191[10]	0.8577[25]	1.4347[34]	i H$_2$O; s EtOH, eth, ace, bz, chl
5191	3-Ethylpentane		C$_7$H$_{16}$	617-78-7	100.202	liq	-118.55	93.5	0.6982[20]	1.3934[20]	i H$_2$O; s EtOH, eth; msc ace, bz, hp, chl
5192	3-Ethyl-2,4-pentanedione		C$_7$H$_{12}$O$_2$	1540-34-7	128.169			178.5	0.9531[19]	1.4408[19]	vs eth, EtOH, chl
5193	Ethyl pentanoate	Ethyl valerate	C$_7$H$_{14}$O$_2$	539-82-2	130.185	liq	-91.2	146.1	0.8770[20]	1.4120[20]	i H$_2$O; msc EtOH, eth; sl ctc
5194	3-Ethyl-3-pentanol		C$_7$H$_{16}$O	597-49-9	116.201	liq	-12.5	142	0.8407[22]	1.4294[20]	sl H$_2$O; s EtOH, eth
5195	2-Ethyl-1-pentene		C$_7$H$_{14}$	3404-71-5	98.186			94	0.7079[20]	1.405[20]	vs bz, eth, EtOH
5196	3-Ethyl-1-pentene		C$_7$H$_{14}$	4038-04-4	98.186	liq	-127.5	84.1	0.6917[25]	1.3982[20]	
5197	3-Ethyl-2-pentene		C$_7$H$_{14}$	816-79-5	98.186			96	0.7204[20]	1.4148[20]	i H$_2$O; s EtOH, eth, bz, chl
5198	Ethyl pentyl ether		C$_7$H$_{16}$O	17952-11-3	116.201			117.6	0.7622[20]	1.3927[20]	vs eth, EtOH
5199	Ethyl 2-pentynoate		C$_7$H$_{10}$O$_2$	55314-57-3	126.153			67[18]	0.962[25]		
5200	2-Ethylphenol		C$_8$H$_{10}$O	90-00-6	122.164		18	204.5	1.0146[25]	1.5367[20]	vs ace, bz, eth, EtOH
5201	3-Ethylphenol		C$_8$H$_{10}$O	620-17-7	122.164	liq	-4	218.4	1.0283[20]		sl H$_2$O, chl; vs EtOH, eth
5202	4-Ethylphenol		C$_8$H$_{10}$O	123-07-9	122.164	nd	45.0	217.9		1.5239[25]	sl H$_2$O, chl; vs EtOH, eth, bz; s ace
5203	Ethyl phenoxyacetate		C$_{10}$H$_{12}$O$_3$	2555-49-9	180.200			247; 110[3]	1.0958[30]	1.5080[20]	
5204	N-Ethyl-N-phenylacetamide		C$_{10}$H$_{13}$NO	529-65-7	163.216		55	260	0.9938[60]		s H$_2$O, eth, ctc
5205	Ethyl phenylacetate	Benzeneacetic acid, ethyl ester	C$_{10}$H$_{12}$O$_2$	101-97-3	164.201	liq	-29.4	227	1.0333[20]	1.4980[20]	vs eth, EtOH
5206	2-(Ethylphenylamino)ethanol		C$_{10}$H$_{15}$NO	92-50-2	165.232						s chl
5207	Ethyl phenylcarbamate	Phenylurethane	C$_9$H$_{11}$NO$_2$	101-99-5	165.189	wh nd (w) pl (dil al)	53	dec 237	1.1064[30]	1.5376[30]	i H$_2$O; vs EtOH, eth; s bz; sl ctc
5208	Ethyl N-phenylformimidate		C$_9$H$_{11}$NO	6780-49-0	149.189			214; 87[10]	1.0051[20]	1.5279[20]	s eth, bz
5209	Ethyl N-phenylglycinate		C$_{10}$H$_{13}$NO$_2$	2216-92-4	179.216	lf (dil al)	58	273.5			vs eth, EtOH
5210	1-(4-Ethylphenyl)-2-phenylethane		C$_{16}$H$_{18}$	7439-15-8	210.314	cry		294	1.028[50]		
5211	Ethyl 3-phenylpropanoate		C$_{11}$H$_{14}$O$_2$	2021-28-5	178.228			247.2	1.0147[20]	1.4954[20]	vs eth, EtOH
5212	Ethyl 3-phenylpropynoate	Ethyl phenylacetylenecarboxylate	C$_{11}$H$_{10}$O$_2$	2216-94-6	174.196			265; 128[1.6]	1.055[25]	1.5520[20]	s eth
5213	Ethyl phenyl sulfone		C$_8$H$_{10}$O$_2$S	599-70-2	170.229	lf (dil al)	42	160[12]	1.1410[20]		vs bz, eth, EtOH, chl
5214	Ethylphosphonic acid		C$_2$H$_7$O$_3$P	6779-09-5	110.049	hyg pl or nd	61.5	335[8]			vs H$_2$O, eth, EtOH
5215	Ethyl phosphorodichloridate	Ethylphosphoric acid dichloride	C$_2$H$_5$Cl$_2$O$_2$P	1498-51-7	162.940			62[10]		1.4338[20]	
5216	5-Ethyl-2-picoline		C$_8$H$_{11}$N	104-90-5	121.180			178.3	0.9202[20]	1.4971[20]	sl H$_2$O; s EtOH, eth, bz; vs ace
5217	Ethyl 1-piperazinecarboxylate	1-Carbethoxypiperazine	C$_7$H$_{14}$N$_2$O$_2$	120-43-4	158.198			237		1.4760[25]	vs H$_2$O, eth, EtOH
5218	1-Ethylpiperidine		C$_7$H$_{15}$N	766-09-6	113.201			130.8	0.8237[20]	1.4480[20]	
5219	Ethyl 4-piperidinecarboxylate		C$_8$H$_{15}$NO$_2$	1126-09-6	157.211	col oil		100[10]		1.4591[20]	vs H$_2$O, bz, eth, EtOH
5220	Ethyl 1-piperidinepropanoate		C$_{10}$H$_{19}$NO$_2$	19653-33-9	185.264			217; 139[50]	0.9627[25]	1.4525[25]	vs H$_2$O
5221	1-Ethyl-3-piperidinol		C$_7$H$_{15}$NO	13444-24-1	129.200			94[15]		1.4777[14]	
5222	N-Ethyl-1-propanamine		C$_5$H$_{13}$N	20193-20-8	87.164			81	0.7204[17]	1.3858[25]	sl H$_2$O; vs ace, EtOH
5223	Ethylpropanedioic acid		C$_5$H$_8$O$_4$	601-75-2	132.116	pr (w+1)	114	180[0.05]			vs H$_2$O; s EtOH, eth, bz; i ace; sl tfa
5224	Ethyl propanoate	Ethyl propionate	C$_5$H$_{10}$O$_2$	105-37-3	102.132	liq	-73.9	99.1	0.8843[25]	1.3839[20]	sl H$_2$O, ctc; msc EtOH, eth; s ace
5225	Ethyl propyl ether		C$_5$H$_{12}$O	628-32-0	88.148	liq	-127.5	63.21	0.7386[20]	1.3695[20]	vs eth, EtOH, HOAc
5226	2-(1-Ethylpropyl)pyridine		C$_{10}$H$_{15}$N	7399-50-0	149.233			195.4	0.8981[20]	1.4850[25]	
5227	4-(1-Ethylpropyl)pyridine		C$_{10}$H$_{15}$N	35182-51-5	149.233		125.5	217; 80[12]	0.9085[25]	1.40905[25]	
5228	Ethyl propyl sulfide		C$_5$H$_{12}$S	4110-50-3	104.214	liq	-117	118.6	0.8370[20]	1.4462[20]	s EtOH
5229	Ethyl 2-propynoate	(Ethoxycarbonyl)acetylene	C$_5$H$_6$O$_2$	623-47-2	98.101			120	0.9645[16]	1.4105[20]	i H$_2$O; vs EtOH, eth, chl
5230	2-Ethylpyrazine		C$_6$H$_8$N$_2$	13925-00-3	108.141			112[200]			
5231	2-Ethylpyridine		C$_7$H$_9$N	100-71-0	107.153	liq	-63.1	148.6	0.9502[25]	1.4964[20]	s H$_2$O; msc EtOH; vs eth, ace; sl ctc

Ethyl 2-oxo-2-phenylacetate

Ethyl 2-oxopropanoate

Ethyl palmitate

3-Ethylpentane

3-Ethyl-2,4-pentanedione

Ethyl pentanoate

3-Ethyl-3-pentanol

2-Ethyl-1-pentene

3-Ethyl-1-pentene

3-Ethyl-2-pentene

Ethyl pentyl ether

Ethyl 2-pentynoate

2-Ethylphenol

3-Ethylphenol

4-Ethylphenol

Ethyl phenoxyacetate

N-Ethyl-N-phenylacetamide

Ethyl phenylacetate

2-(Ethylphenylamino)ethanol

Ethyl phenylcarbamate

Ethyl N-phenylformimidate

Ethyl N-phenylglycinate

1-(4-Ethylphenyl)-2-phenylethane

Ethyl 3-phenylpropanoate

Ethyl 3-phenylpropynoate

Ethyl phenyl sulfone

Ethylphosphonic acid

Ethyl phosphorodichloridate

5-Ethyl-2-picoline

Ethyl 1-piperazinecarboxylate

1-Ethylpiperidine

Ethyl 4-piperidinecarboxylate

Ethyl 1-piperidinepropanoate

1-Ethyl-3-piperidinol

N-Ethyl-1-propanamine

Ethylpropanedioic acid

Ethyl propanoate

Ethyl propyl ether

2-(1-Ethylpropyl)pyridine

4-(1-Ethylpropyl)pyridine

Ethyl propyl sulfide

Ethyl 2-propynoate

2-Ethylpyrazine

2-Ethylpyridine

No.	Name	Synonym	Mol. Form.	CAS RN	Mol. Wt.	Physical Form	mp/°C	bp/°C	den/ g cm⁻³	n_D	Solubility
5232	3-Ethylpyridine		C$_7$H$_9$N	536-78-7	107.153	liq	-76.9	165	0.9539[25]	1.5021[20]	s H$_2$O, EtOH, eth; vs ace; sl ctc
5233	4-Ethylpyridine		C$_7$H$_9$N	536-75-4	107.153	liq	-90.5	168.3	0.9417[20]	1.5009[20]	s H$_2$O, EtOH, eth; vs ace; sl ctc
5234	2-Ethyl-4-pyridinecarbothioamide	Ethionamide	C$_8$H$_{10}$N$_2$S	536-33-4	166.243		163				
5235	Ethyl 2-pyridinecarboxylate	Ethyl 2-picolinate	C$_8$H$_9$NO$_2$	2524-52-9	151.163	ye cry in air	1	243	1.1194[20]	1.5104[20]	vs H$_2$O, eth, EtOH
5236	Ethyl 3-pyridinecarboxylate	Ethyl nicotinate	C$_8$H$_9$NO$_2$	614-18-6	151.163		8.5	224	1.1070[20]	1.5024[20]	vs H$_2$O, EtOH, eth, bz; sl ctc
5237	Ethyl 4-pyridinecarboxylate		C$_8$H$_9$NO$_2$	1570-45-2	151.163		23	219.5	1.0091[15]	1.5017[20]	sl H$_2$O; s EtOH, bz; vs eth, chl
5238	N-Ethylpyridinium bromide		C$_7$H$_{10}$BrN	1906-79-2	188.065	cry (al)	111.5				s H$_2$O, EtOH; i eth
5239	1-Ethyl-1H-pyrrole		C$_6$H$_9$N	617-92-5	95.142			129.5	0.9009[20]	1.4841[20]	vs EtOH
5240	1-Ethyl-1H-pyrrole-2,5-dione	N-Ethylmaleimide	C$_6$H$_7$NO$_2$	128-53-0	125.126	cry (bz)	45.5				sl H$_2$O; vs EtOH, eth; s chl
5241	1-Ethyl-2-pyrrolidinemethanamine		C$_7$H$_{16}$N$_2$	26116-12-1	128.215			59[16], 40[10]	0.887[25]	1.4665[20]	
5242	Ethyl Red	2-(4-Diethylaminophenylazo) benzoic acid	C$_{17}$H$_{19}$N$_3$O$_2$	76058-33-8	297.352		135				
5243	Ethyl salicylate		C$_9$H$_{10}$O$_3$	118-61-6	166.173		45	150[10]	1.1326[20]	1.5296[20]	i H$_2$O; msc EtOH; vs eth; s ctc
5244	Ethyl silicate		C$_8$H$_{20}$O$_4$Si	78-10-4	208.329	liq	-82.5	168.8	0.9320[20]	1.3928[20]	dec H$_2$O
5245	Ethyl stearate	Ethyl octadecanoate	C$_{20}$H$_{40}$O$_2$	111-61-5	312.531		33	199[10]	1.057[20]	1.4349[40]	i H$_2$O; s EtOH, eth, chl; vs ace
5246	2-Ethylstyrene		C$_{10}$H$_{12}$	7564-63-8	132.202	liq	-75.5	187.3; 68[12]	0.9017[25]	1.5380[20]	
5247	3-Ethylstyrene		C$_{10}$H$_{12}$	7525-62-4	132.202	liq	-101	190.0	0.8945[20]	1.5351[20]	
5248	4-Ethylstyrene		C$_{10}$H$_{12}$	3454-07-7	132.202	liq	-49.7	192.3; 86[20]	0.8884[25]	1.5376[20]	
5249	Ethyl sulfate		C$_2$H$_6$O$_4$S	540-82-9	126.132			dec 280	1.3657[20]	1.4105[20]	vs H$_2$O
5250	2-(Ethylsulfonyl)ethanol	Ethylsulfonylethyl alcohol	C$_4$H$_{10}$O$_3$S	513-12-2	138.185						sl chl
5251	2-Ethyl-5-(3-sulfophenyl) isoxazolium hydroxide, inner salt	Woodward's Reagent K	C$_{11}$H$_{11}$NO$_4$S	4156-16-5	253.275		dec 207				
5252	Ethyl tartrate	Ethyl tartrate, acid	C$_6$H$_{10}$O$_6$	608-89-9	178.139		90				vs H$_2$O, EtOH
5253	2-Ethyltetrahydrofuran		C$_6$H$_{12}$O	1003-30-1	100.158			109	0.8570[19]	1.4147[19]	vs ace, bz, eth, EtOH
5254	5-Ethyl-1,3,4-thiadiazol-2-amine		C$_4$H$_7$N$_3$S	14068-53-2	129.184		200.8				
5255	S-Ethyl thioacetate		C$_4$H$_8$OS	625-60-5	104.171			116.4	0.9792[20]	1.4583[21]	i H$_2$O; vs EtOH, eth
5256	(Ethylthio)acetic acid		C$_4$H$_8$O$_2$S	627-04-3	120.171		-8.5	164[83], 109[5]	1.1497[20]		vs H$_2$O, EtOH, eth
5257	(Ethylthio)benzene	Thiophenetole	C$_8$H$_{10}$S	622-38-8	138.230			205	1.0211[20]	1.5670[20]	s EtOH
5258	Ethyl thiocyanate		C$_3$H$_5$NS	542-90-5	87.144	liq	-85.5	146	1.007[23]	1.4684[15]	i H$_2$O; msc EtOH, eth; s chl
5259	2-(Ethylthio)ethanol		C$_4$H$_{10}$OS	110-77-0	106.186	liq	-100	184	1.0166[20]	1.4867[20]	sl H$_2$O; s EtOH; vs ace
5260	1-(Ethylthio)-4-methylbenzene		C$_9$H$_{12}$S	622-63-9	152.256			220	0.9996[20]	1.555[20]	
5261	2-Ethylthiophene		C$_6$H$_8$S	872-55-9	112.193			134	0.9930[20]	1.5122[20]	i H$_2$O; vs EtOH, eth
5262	Ethyl thiophene-2-carboxylate		C$_7$H$_8$O$_2$S	2810-04-0	156.203			218	1.1623[16]	1.5248[20]	s EtOH, ace; sl ctc
5263	3-Ethyl-2-thioxo-4-thiazolidinone	3-Ethylrhodanine	C$_5$H$_7$NOS$_2$	7648-01-3	161.246		35.5				
5264	2-Ethyltoluene		C$_9$H$_{12}$	611-14-3	120.191	liq	-79.83	165.2	0.8807[20]	1.5046[20]	i H$_2$O; msc EtOH, eth, ace, bz, peth, ctc
5265	3-Ethyltoluene		C$_9$H$_{12}$	620-14-4	120.191	liq	-95.6	161.3	0.8645[20]	1.4966[20]	i H$_2$O; vs EtOH, eth; msc ace, bz
5266	4-Ethyltoluene		C$_9$H$_{12}$	622-96-8	120.191	liq	-62.35	162	0.8614[20]	1.4959[20]	i H$_2$O; vs EtOH, eth; msc ace, bz
5267	Ethyl p-toluenesulfonate		C$_9$H$_{12}$O$_3$S	80-40-0	200.254		34.5	173[15]	1.166[48]		i H$_2$O; s EtOH, eth, AcOEt; sl ctc
5268	Ethyl trichloroacetate		C$_4$H$_5$Cl$_3$O$_2$	515-84-4	191.441			167.5	1.3836[20]	1.4505[20]	i H$_2$O; s EtOH, eth, bz; sl chl
5269	Ethyl trifluoroacetate		C$_4$H$_5$F$_3$O$_2$	383-63-1	142.077			61	1.194[20]	1.308[20]	
5270	Ethyl 4,4,4-trifluoroacetoacetate		C$_6$H$_7$F$_3$O$_3$	372-31-6	184.113	liq	-39.1	132	1.2586[15]	1.3783[15]	s EtOH, eth
5271	Ethyl trifluoromethanesulfonate		C$_3$H$_5$F$_3$O$_3$S	425-75-2	178.130			115	1.3740[0]		s eth
5272	Ethyl 3,4,5-trihydroxybenzoate		C$_9$H$_{10}$O$_5$	831-61-8	198.172	mcl pr (w+2 1/2) nd (chl)	163.0				sl H$_2$O, chl; s EtOH, eth, AcOEt

3-Ethylpyridine

4-Ethylpyridine

2-Ethyl-4-pyridinecarbothioamide

Ethyl 2-pyridinecarboxylate

Ethyl 3-pyridinecarboxylate

Ethyl 4-pyridinecarboxylate

N-Ethylpyridinium bromide

1-Ethyl-1H-pyrrole

1-Ethyl-1H-pyrrole-2,5-dione

1-Ethyl-2-pyrrolidinemethanamine

Ethyl Red

Ethyl salicylate

Ethyl silicate

Ethyl stearate

2-Ethylstyrene

3-Ethylstyrene

4-Ethylstyrene

Ethyl sulfate

2-(Ethylsulfonyl)ethanol

2-Ethyl-5-(3-sulfophenyl)isoxazolium hydroxide, inner salt

Ethyl tartrate

2-Ethyltetrahydrofuran

5-Ethyl-1,3,4-thiadiazol-2-amine

S-Ethyl thioacetate

(Ethylthio)acetic acid

(Ethylthio)benzene

Ethyl thiocyanate

2-(Ethylthio)ethanol

1-(Ethylthio)-4-methylbenzene

2-Ethylthiophene

Ethyl thiophene-2-carboxylate

3-Ethyl-2-thioxo-4-thiazolidinone

2-Ethyltoluene

3-Ethyltoluene

4-Ethyltoluene

Ethyl p-toluenesulfonate

Ethyl trichloroacetate

Ethyl trifluoroacetate

Ethyl 4,4,4-trifluoroacetoacetate

Ethyl trifluoromethanesulfonate

Ethyl 3,4,5-trihydroxybenzoate

No.	Name	Synonym	Mol. Form.	CAS RN	Mol. Wt.	Physical Form	mp/°C	bp/°C	den/ g cm^{-3}	n_D	Solubility
5273	Ethyltrimethoxysilane		$C_5H_{14}O_3Si$	5314-55-6	150.249			124.3	0.9488^{20}	1.3838^{20}	vs EtOH
5274	1-Ethyl-2,4,5-trimethylbenzene		$C_{11}H_{16}$	17851-27-3	148.245	liq	-13.5	213	0.883^{20}	1.5075^{20}	vs ace, bz, eth, EtOH
5275	2-Ethyl-1,3,5-trimethylbenzene		$C_{11}H_{16}$	3982-67-0	148.245	liq	-15.5	212.4	0.883^{20}	1.5074^{20}	vs ace, bz, eth, EtOH
5276	Ethyltrimethyllead	Ethyltrimethylplumbane	$C_5H_{14}Pb$	1762-26-1	281.4	col liq		27$^{10.5}$	1.88^{20}		
5277	3-Ethyl-2,4,5-trimethylpyrrole		$C_9H_{15}N$	520-69-4	137.222	lf (eth)	66.5	214; 110^{35}			
5278	4-Ethyl-2,6,7-trioxa-1-phosphabicyclo[2.2.2]octane	Trimethylolpropane phosphite	$C_6H_{11}O_3P$	824-11-3	162.123		53.7				s chl
5279	Ethyl undecanoate	Ethyl undecylate	$C_{13}H_{26}O_2$	627-90-7	214.344		-15	131^{14}	0.8633^{20}	1.4285^{20}	i H_2O; s EtOH, eth, ace, bz
5280	Ethyl 10-undecenoate		$C_{13}H_{24}O_2$	692-86-4	212.329	liq	-38	264.5	0.8827^{15}	1.4449^{25}	i H_2O; s EtOH, eth, HOAc; sl ctc
5281	N-Ethylurea		$C_3H_8N_2O$	625-52-5	88.108	nd (bz, al-eth)	92.5	dec	1.2130^{18}		vs H_2O, EtOH, bz; s eth; i CS_2
5282	Ethyl vinyl ether		C_4H_8O	109-92-2	72.106	liq	-115.8	35.5	0.7589^{20}	1.3767^{20}	sl H_2O, ctc; s EtOH; msc eth
5283	Ethyl Violet		$C_{31}H_{42}ClN_3$	2390-59-2	492.138	gray-viol cry					s H_2O, EtOH
5284	α-Ethynylbenzenemethanol	1-Phenylpropargyl alcohol	C_9H_8O	4187-87-5	132.159	pr	22	114^{12}	1.0655^{20}	1.5508^{20}	
5285	α-Ethynylbenzenemethanol carbamate	Carfimate	$C_{10}H_9NO_2$	3567-38-2	175.184	cry (al)	86.5				
5286	1-Ethynylcyclohexanamine		$C_8H_{13}N$	30389-18-5	123.196			65^{20}	0.913^{25}	1.4817^{20}	
5287	1-Ethynylcyclohexanol		$C_8H_{12}O$	78-27-3	124.180	cry (peth)	31.5	174	0.9873^{20}	1.4822^{20}	i H_2O; s EtOH, bz, peth; sl chl
5288	1-Ethynylcyclohexanol, carbamate	Ethinamate	$C_9H_{13}NO_2$	126-52-3	167.205	nd	97	120^3		1.4441^{21}	sl H_2O; vs EtOH; s hx
5289	1-Ethynylcyclopentanol		$C_7H_{10}O$	17356-19-3	110.153		27	157.5	0.962^{25}	1.4751^{20}	
5290	α-Ethynyl-α-methylbenzenemethanol		$C_{10}H_{10}O$	127-66-2	146.185		52.3	217.5; 102^{12}	1.0314^{20}		
5291	Ethynylsilane	Silylacetylene	C_2H_4Si	1066-27-9	56.139	col gas		-22.5			
5292	Etioporphyrin		$C_{32}H_{38}N_4$	448-71-5	478.671		362				
5293	Etofylline		$C_9H_{12}N_4O_3$	519-37-9	224.216		158				vs H_2O; s EtOH; sl eth, bz
5294	Etoglucid	Oxirane, 2,2'-(2,5,8,11-tetraoxadodecane-1,12-diyl) bis-	$C_{12}H_{22}O_6$	1954-28-5	262.299	col liq	-13	196^2	1.1312^{20}	1.4622^{20}	
5295	Etoposide		$C_{29}H_{32}O_{13}$	33419-42-0	588.556	cry (MeOH)	≈243				s MeOH
5296	Etrimfos		$C_{10}H_{17}N_2O_4PS$	38260-54-7	292.291		-1.7		1.195^{20}		
5297	Eucalyptol	Cineole	$C_{10}H_{18}O$	470-82-6	154.249		0.8	176.4	0.9267^{20}	1.4586^{20}	i H_2O; s EtOH, eth, chl; sl ctc
5298	Euparin	1-[6-Hydroxy-2-(1-methylvinyl)-5-benzofuranyl]ethanone	$C_{13}H_{12}O_3$	532-48-9	216.232		121.5				s eth, bz, chl; sl NaOH
5299	Evan's Blue		$C_{34}H_{24}N_6Na_4O_{14}S_4$	314-13-6	960.806						s H_2O, EtOH, acid
5300	Evodiamine		$C_{19}H_{17}N_3O$	518-17-2	303.357	ye lf (al)	28				
5301	Famotidine		$C_8H_{15}N_7O_2S_3$	76824-35-6	337.446	cry	163				i EtOH, chl; vs DMF; s HOAc; sl MeOH
5302	Famphur		$C_{10}H_{16}NO_5PS_2$	52-85-7	325.342		53				
5303	α-Farnesene		$C_{15}H_{24}$	502-61-4	204.352			130^{12}	0.8410^{20}	1.4836^{20}	i H_2O; s eth, ace; msc peth, lig
5304	β-Farnesene		$C_{15}H_{24}$	18794-84-8	204.352			121^9	0.8363^{20}	1.4899^{20}	vs ace, eth, chl
5305	Farnesic acid		$C_{15}H_{24}O_2$	7548-13-2	236.351	oil		204^{16}			
5306	2-cis,6-trans-Farnesol		$C_{15}H_{26}O$	3790-71-4	222.366	oil		156^{12}, 120$^{0.3}$	0.8908^{20}	1.4877^{20}	vs ace, eth, EtOH
5307	2-trans,6-trans-Farnesol		$C_{15}H_{26}O$	106-28-5	222.366	oil		160^{10}, 137^3	0.888^{20}	1.4877^{20}	i H_2O; vs EtOH; s eth, ace
5308	Farnesol acetate		$C_{17}H_{28}O_2$	29548-30-9	264.403			168^{10}			
5309	Fenadiazole	2-(1,2,4-Oxadiazol-2-yl)phenol	$C_8H_6N_2O_4$	1008-65-7	194.145	cry	112	180$^{0.1}$			
5310	Fenamiphos		$C_{13}H_{22}NO_3PS$	22224-92-6	303.358		49				
5311	Fenarimol		$C_{17}H_{12}Cl_2N_2O$	60168-88-9	331.195		118		1.15^{20}		
5312	Fenbuconazole		$C_{20}H_{19}ClN_4$	114369-43-6	350.845		125				
5313	Fenbutatin oxide	Distannoxane, hexakis(2-methyl-2-phenylpropyl)-	$C_{60}H_{78}OSn_2$	13356-08-6	1052.680		138				
5314	α-Fenchol, (±)	1,3,3-Trimethylbicyclo[2.2.1]heptan-2-ol, endo-(±)	$C_{10}H_{18}O$	36386-49-9	154.249		39	199.5	0.9420^{40}		vs eth, EtOH
5315	(±)-Fenchone		$C_{10}H_{16}O$	18492-37-0	152.233	oily liq	6.1	193.5	0.9492^{15}	1.4702^{20}	i H_2O; vs EtOH; s eth, ace
5316	Fenfluramine		$C_{12}H_{16}F_3N$	458-24-2	231.257	cry (AcOEt)		110^{12}			
5317	Fenitrothion		$C_9H_{12}NO_5PS$	122-14-5	277.234			118$^{0.05}$, 164^1	1.3227^{25}		
5318	Fenoxaprop-ethyl		$C_{18}H_{16}ClNO_5$	82110-72-3	361.777		85	200$^{0.001}$			sl H_2O, hx; s eth; vs ace, tol

Ethyltrimethoxysilane

1-Ethyl-2,4,5-trimethylbenzene

2-Ethyl-1,3,5-trimethylbenzene

Ethyltrimethyllead

3-Ethyl-2,4,5-trimethylpyrrole

4-Ethyl-2,6,7-trioxa-1-phosphabicyclo[2.2.2]octane

Ethyl undecanoate

Ethyl 10-undecenoate

N-Ethylurea

Ethyl vinyl ether

Ethyl Violet

α-Ethynylbenzenemethanol

α-Ethynylbenzenemethanol carbamate

1-Ethynylcyclohexanamine

1-Ethynylcyclohexanol

1-Ethynylcyclohexanol, carbamate

1-Ethynylcyclopentanol

α-Ethynyl-α-methylbenzenemethanol

Ethynylsilane

Etioporphyrin

Etofylline

Etoglucid

Etoposide

Etrimfos

Eucalyptol

Euparin

Evan's Blue

Evodiamine

Famotidine

Famphur

α-Farnesene

β-Farnesene

Farnesic acid

2-cis,6-trans-Farnesol

2-trans,6-trans-Farnesol

Farnesol acetate

Fenadiazole

Fenamiphos

Fenarimol

Fenbuconazole

Fenbutatin oxide

α-Fenchol, (±)

(±)-Fenchone

Fenfluramine

Fenitrothion

Fenoxaprop-ethyl

No.	Name	Synonym	Mol. Form.	CAS RN	Mol. Wt.	Physical Form	mp/°C	bp/°C	den/ g cm^{-3}	n_D	Solubility
5319	Fenoxycarb	Ethyl 2-(4-phenoxyphenoxy) ethylcarbamate	C$_{17}$H$_{19}$NO$_4$	79127-80-3	301.338		53				
5320	Fenpropathrin		C$_{22}$H$_{23}$NO$_3$	64257-84-7	349.423		47		1.15^{25}		
5321	Fensulfothion		C$_{11}$H$_{17}$O$_4$PS$_2$	115-90-2	308.354			140$^{0.01}$	1.202^{20}		
5322	Fentanyl		C$_{22}$H$_{28}$N$_2$O	437-38-7	336.469		87.5				
5323	Fenthion		C$_{10}$H$_{15}$O$_3$PS$_2$	55-38-9	278.328		7.5	87$^{0.01}$	1.246^{20}		
5324	Fenvalerate		C$_{25}$H$_{22}$ClNO$_3$	51630-58-1	419.901			dec	1.15^{25}		
5325	Ferbam	Iron, tris(dimethylcarbamodithioato-S,S')-, (OC-6-11)-	C$_9$H$_{18}$FeN$_3$S$_6$	14484-64-1	416.494		180 dec				
5326	Ferrocene	Dicyclopentadienyl iron	C$_{10}$H$_{10}$Fe	102-54-5	186.031		172.5	249			i H$_2$O
5327	Ferrous gluconate		C$_{12}$H$_{22}$FeO$_{14}$	299-29-6	446.140	ye-gray pow (w)					s H$_2$O; i EtOH
5328	Ferrous lactate		C$_6$H$_{10}$FeO$_6$	5905-52-2	233.984	grn-wh pow (hyd)					s H$_2$O; i EtOH
5329	Fichtelite	18-Norabietane	C$_{19}$H$_{34}$	2221-95-6	262.473	cry	46	236^{43}	0.9380^{22}	1.5052^{20}	
5330	Finasteride	Proscar	C$_{23}$H$_{36}$N$_2$O$_2$	98319-26-7	372.544	wh cry	252				sl H$_2$O; s chl, EtOH, MeOH, DMSO
5331	Fisetin		C$_{15}$H$_{10}$O$_6$	528-48-3	286.236	lt ye nd (dil al, + 1 w)	330				i H$_2$O; s EtOH, ace; sl eth, bz, peth
5332	Flavine adenine dinucleotide	FAD	C$_{27}$H$_{33}$N$_9$O$_{15}$P$_2$	146-14-5	785.550	ye cry (w)					
5333	Florantyrone		C$_{20}$H$_{14}$O$_3$	519-95-9	302.323	ye cry (HOAc)	208				s EtOH, MeOH
5334	Fluazipop-butyl		C$_{19}$H$_{20}$F$_3$NO$_4$	79241-46-6	383.362	pale ye liq	5				
5335	Flubenzimine		C$_{17}$H$_{10}$F$_6$N$_4$S	37893-02-0	416.343	ye cry	119				sl H$_2$O
5336	Fluchloralin		C$_{12}$H$_{13}$ClF$_3$N$_3$O$_4$	33245-39-5	355.697		42				
5337	Flucythrinate	Cythrin	C$_{26}$H$_{23}$F$_2$NO$_4$	70124-77-5	451.463			108$^{0.35}$	1.189^{22}		
5338	Fludrocortisone		C$_{21}$H$_{29}$FO$_5$	127-31-1	380.450	cry (EtOH)	261 dec				
5339	Flumethiazide	Trifluoromethylthiazide	C$_8$H$_6$F$_3$N$_3$O$_4$S$_2$	148-56-1	329.277	cry	306				sl H$_2$O; i bz, tol; s MeOH, EtOH, DMF
5340	Fluocinolone acetonide		C$_{24}$H$_{30}$F$_2$O$_6$	67-73-2	452.488	cry (ace/hx)	266 dec				
5341	Fluoranthene	1,2-(1,8-Naphthylene)benzene	C$_{16}$H$_{10}$	206-44-0	202.250	pa ye nd or pl (al)	110.19	384	1.252^0		i H$_2$O; s EtOH, eth, bz, chl, CS$_2$
5342	9H-Fluoren-2-amine		C$_{13}$H$_{11}$N	153-78-6	181.233	pl or nd (dil al)	130.3				i H$_2$O; s EtOH, eth, ctc, CS$_2$
5343	9H-Fluorene	2,2'-Methylenebiphenyl	C$_{13}$H$_{10}$	86-73-7	166.218	lf (al)	114.77	295	1.203^0		i H$_2$O; sl EtOH; s eth, ace, bz, CS$_2$
5344	9H-Fluorene-9-carboxylic acid		C$_{14}$H$_{10}$O$_2$	1989-33-9	210.228		226				
5345	9H-Fluorene-2,7-diamine	2,7-Diaminofluorene	C$_{13}$H$_{12}$N$_2$	525-64-4	196.247	nd (w), pr (bz), pl (eth)	166				i H$_2$O; s EtOH, chl
5346	9H-Fluorene-9-methanol		C$_{14}$H$_{12}$O	24324-17-2	196.244		105.0				
5347	9H-Fluoren-9-ol		C$_{13}$H$_{10}$O	1689-64-1	182.217	hex nd (w, peth)	156.0				sl H$_2$O, peth, EtOH; s eth, ace; vs bz
5348	9H-Fluoren-9-one		C$_{13}$H$_8$O	486-25-9	180.202	ye orth bipym (al, bz-peth)	84	341.5	1.1300^{99}	1.6309^{99}	i H$_2$O; s EtOH, ace, bz; vs tol; sl ctc
5349	Fluorescein		C$_{20}$H$_{12}$O$_5$	2321-07-5	332.306	red orth pr	315 dec				sl H$_2$O, EtOH, eth; vs ace; s py, MeOH
5350	Fluorescein sodium	CI Acid Yellow 73	C$_{20}$H$_{10}$Na$_2$O$_5$	518-47-8	376.270	ye pow					s H$_2$O, EtOH, glycerol, dil acid
5351	2-Fluoroacetamide	Fluoroacetic acid amide	C$_2$H$_4$FNO	640-19-7	77.057		108	sub			s H$_2$O, ace; sl chl
5352	Fluoroacetic acid	Fluoroethanoic acid	C$_2$H$_3$FO$_2$	144-49-0	78.042	nd	35.2	168	1.3693^{36}		s H$_2$O, EtOH
5353	Fluoroacetyl chloride		C$_2$H$_2$ClFO	359-06-8	96.487	liq		72; 23^{105}			
5354	Fluoroacetylene	Fluoroethyne	C$_2$HF	2713-09-9	44.027	gas	-196	-105 exp			
5355	2-Fluoroaniline		C$_6$H$_6$FN	348-54-9	111.117	pa ye liq	-34.6	175; 55^{12}	1.1513^{21}	1.5421^{20}	i H$_2$O; s EtOH, eth; sl ctc
5356	3-Fluoroaniline		C$_6$H$_6$FN	372-19-0	111.117			188	1.1561^{19}	1.5436^{20}	sl H$_2$O, chl; s EtOH, eth
5357	4-Fluoroaniline		C$_6$H$_6$FN	371-40-4	111.117	pa ye liq	-0.8	182; 85^{19}	1.1725^{20}	1.5195^{20}	sl H$_2$O, ctc; s EtOH, eth
5358	2-Fluorobenzaldehyde		C$_7$H$_5$FO	446-52-6	124.112	liq	-44.5	175	1.178^{25}	1.5234^{20}	
5359	3-Fluorobenzaldehyde		C$_7$H$_5$FO	456-48-4	124.112			173	1.17^{25}	1.5206^{20}	
5360	4-Fluorobenzaldehyde		C$_7$H$_5$FO	459-57-4	124.112	liq	-10	181.5	1.1810^{19}		

Fenoxycarb

Fenpropathrin

Fensulfothion

Fentanyl

Fenthion

Fenvalerate

Ferbam

Ferrocene

Ferrous gluconate

Ferrous lactate

Fichtelite

Finasteride

Fisetin

Flavine adenine dinucleotide

Florantyrone

Fluazipop-butyl

Flubenzimine

Fluchloralin

Flucythrinate

Fludrocortisone

Flumethiazide

Fluocinolone acetonide

Fluoranthene

9H-Fluoren-2-amine

9H-Fluorene

9H-Fluorene-9-carboxylic acid

9H-Fluorene-2,7-diamine

9H-Fluorene-9-methanol

9H-Fluoren-9-ol

9H-Fluoren-9-one

Fluorescein

Fluorescein sodium

2-Fluoroacetamide

Fluoroacetic acid

Fluoroacetyl chloride

Fluoroacetylene

2-Fluoroaniline

3-Fluoroaniline

4-Fluoroaniline

2-Fluorobenzaldehyde

3-Fluorobenzaldehyde

4-Fluorobenzaldehyde

No.	Name	Synonym	Mol. Form.	CAS RN	Mol. Wt.	Physical Form	mp/°C	bp/°C	den/g cm⁻³	n_D	Solubility
5361	Fluorobenzene		C_6H_5F	462-06-6	96.102	liq	-42.18	84.73	1.0225[20]	1.4684[30]	sl H_2O; vs bz, eth, EtOH, lig
5362	4-Fluorobenzeneacetic acid		$C_8H_7FO_2$	405-50-5	154.139	cry (chl)	86	164[2]			
5363	2-Fluorobenzeneacetonitrile		C_8H_6FN	326-62-5	135.139			232; 102[10]	1.059[25]	1.5009[20]	
5364	4-Fluorobenzeneacetonitrile		C_8H_6FN	459-22-3	135.139		86.0	228; 119[18]	1.1390[20]	1.5002[20]	
5365	4-Fluorobenzenemethanamine		C_7H_8FN	140-75-0	125.144			183		1.5139[20]	
5366	4-Fluorobenzenemethanol		C_7H_7FO	459-56-3	126.128		23	210		1.5080[20]	
5367	4-Fluorobenzenesulfonyl chloride		$C_6H_4ClFO_2S$	349-88-2	194.611	pl or nd	30	106[9]			vs bz, eth, chl
5368	2-Fluorobenzoic acid		$C_7H_5FO_2$	445-29-4	140.112	nd (a)	126.5		1.460[25]		sl H_2O; vs EtOH, eth; i bz; s chl
5369	3-Fluorobenzoic acid		$C_7H_5FO_2$	455-38-9	140.112	lf (w)	124		1.474[25]		sl H_2O; s eth
5370	4-Fluorobenzoic acid		$C_7H_5FO_2$	456-22-4	140.112	pr (w), mcl pr (w)	185		1.479[25]		sl H_2O, ace; s EtOH, eth
5371	2-Fluorobenzonitrile		C_7H_4FN	394-47-8	121.112			93[22]			
5372	4-Fluorobenzonitrile		C_7H_4FN	1194-02-1	121.112	nd (peth)	34.8	188.8	1.1070[55]	1.4925[55]	sl chl; s peth
5373	2-Fluorobenzoyl chloride		C_7H_4ClFO	393-52-2	158.557		2.0	91[15]	1.328[25]	1.5365[20]	
5374	3-Fluorobenzoyl chloride		C_7H_4ClFO	1711-07-5	158.557	liq	-30	189	1.304[25]	1.5285[20]	
5375	4-Fluorobenzoyl chloride		C_7H_4ClFO	403-43-0	158.557		9	82[20]	1.342[25]	1.5296[20]	
5376	2-Fluoro-1,1'-biphenyl		$C_{12}H_9F$	321-60-8	172.197		73.5	248	1.2452[25]		s EtOH, eth, chl, peth; sl lig
5377	4-Fluoro-1,1'-biphenyl		$C_{12}H_9F$	324-74-3	172.197	pr	74.2	253	1.247[25]		sl EtOH; s eth, gl HOAc
5378	1-Fluorobutane	Butyl fluoride	C_4H_9F	2366-52-1	76.112	liq	-134	32.5	0.7789[20]	1.3396[20]	vs EtOH
5379	2-Fluorobutane	sec-Butyl fluoride	C_4H_9F	359-01-3	76.112	vol liq or gas	-121.4	25.1	0.7559[25]		
5380	Fluorocyclohexane	Cyclohexyl fluoride	$C_6H_{11}F$	372-46-3	102.149		13	101	0.9279[20]	1.4146[20]	i H_2O; s py
5381	1-Fluorocyclohexene		C_6H_9F	694-51-9	100.133			96.5		1.4441[25]	
5382	5-Fluorocytosine	4-Amino-5-fluoro-2-hydroxypyrimidine	$C_4H_4FN_3O$	2022-85-7	129.092	wh cry	296 dec				
5383	1-Fluorodecane	Decyl fluoride	$C_{10}H_{21}F$	334-56-5	160.272	liq	-35	186.2	0.8194[20]	1.4085	vs eth
5384	Fluorodifen	2-Nitro-1-(4-nitrophenoxy)-4-(trifluoromethyl)benzene	$C_{13}H_7F_3N_2O_5$	15457-05-3	328.200		94				
5385	1-Fluoro-2,4-dinitrobenzene	2,4-Dinitrophenyl fluoride	$C_6H_3FN_2O_4$	70-34-8	186.097		25.8	296	1.4718[54]	1.5690[20]	s EtOH; sl chl
5386	Fluoroethane	Ethyl fluoride	C_2H_5F	353-36-6	48.059	col gas	-143.2	-37.7	0.7182[20] (p>1 atm)	1.2656[20]	sl H_2O; vs EtOH, eth
5387	2-Fluoroethanol	Ethylene fluorohydrin	C_2H_5FO	371-62-0	64.058	liq	-26.4	103.5	1.1040[20]	1.3647[18]	msc H_2O, EtOH, eth; vs ace; sl chl
5388	Fluoroethene	Vinyl fluoride	C_2H_3F	75-02-5	46.043	col gas	-160.5	-72			i H_2O; s EtOH, ace
5389	1-Fluoroheptane		$C_7H_{15}F$	661-11-0	118.192	liq	-73	117.9	0.8062[20]	1.3854[20]	i H_2O; s eth, ace, bz; vs peth
5390	1-Fluorohexane	Hexyl fluoride	$C_6H_{13}F$	373-14-8	104.165	liq	-103	91.5	0.7995[20]	1.3738[20]	s eth, bz
5391	1-Fluoro-2-iodobenzene		C_6H_4FI	348-52-7	221.998	liq	-41.5	188.6		1.5910[20]	s ace, bz, chl
5392	1-Fluoro-4-iodobenzene		C_6H_4FI	352-34-1	221.998	liq	-27	183	1.9523[15]	1.5270[22]	i H_2O; s EtOH, eth, ace
5393	1-Fluoro-3-isothiocyanatobenzene		C_7H_4FNS	404-72-8	153.177			227	1.27[25]	1.6186[20]	
5394	1-Fluoro-4-isothiocyanatobenzene		C_7H_4FNS	1544-68-9	153.177		27	228			
5395	Fluoromethane	Methyl fluoride	CH_3F	593-53-3	34.033	col gas	-141.8	-78.4	0.557[25] (p>1 atm)	1.1674[25]	sl H_2O, bz, chl; vs EtOH, eth
5396	1-Fluoro-2-methoxybenzene		C_7H_7FO	321-28-8	126.128	liq	-39	154.5	1.5489[17]	1.4969[17]	i H_2O; s eth, ctc
5397	1-Fluoro-3-methoxybenzene		C_7H_7FO	456-49-5	126.128	liq	-35	159; 51[14]	1.104[25]	1.4876[20]	
5398	1-Fluoro-4-methoxybenzene		C_7H_7FO	459-60-9	126.128	liq	-45	157	1.1781[18]	1.4886[18]	s eth
5399	4-Fluoro-2-methylaniline		C_7H_8FN	452-71-1	125.144		14.2	94[16]	1.1263[18]	1.5363[18]	s eth, ace, bz, ctc
5400	(Fluoromethyl)benzene		C_7H_7F	350-50-5	110.129	liq	-35	140; 40[14]	1.0228[25]	1.4892[25]	s ctc
5401	2-Fluoro-4-methyl-1-nitrobenzene	3-Fluoro-4-nitrotoluene	$C_7H_6FNO_2$	446-34-4	155.127	nd (al)	53.2	97[3]	1.4380[25]		
5402	2-Fluoro-2-methylpropane	tert-Butyl fluoride	C_4H_9F	353-61-7	76.112	col gas		12.1			
5403	1-Fluoronaphthalene		$C_{10}H_7F$	321-38-0	146.161	liq	-9	215; 80[11]	1.1322[20]	1.5939[20]	i H_2O; s EtOH, eth, bz, chl, HOAc
5404	2-Fluoronaphthalene		$C_{10}H_7F$	323-09-1	146.161	nd (al)	61	212; 90[16]			i H_2O; s EtOH, eth, bz, chl, HOAc
5405	1-Fluoro-2-nitrobenzene	o-Fluoronitrobenzene	$C_6H_4FNO_2$	1493-27-2	141.100	ye liq	-6	dec 215	1.3285[18]	1.5489[17]	vs eth, EtOH
5406	1-Fluoro-3-nitrobenzene	m-Fluoronitrobenzene	$C_6H_4FNO_2$	402-67-5	141.100	ye cry	41	199; 86[19]	1.3254[19]	1.5262[15]	i H_2O; s EtOH, eth; sl bz
5407	1-Fluoro-4-nitrobenzene	p-Fluoronitrobenzene	$C_6H_4FNO_2$	350-46-9	141.100	ye nd	21	205	1.3300[20]	1.5316[20]	i H_2O; s EtOH, eth; sl ctc
5408	1-Fluorooctane	Octyl fluoride	$C_8H_{17}F$	463-11-6	132.219	liq	-64	142.3	0.8116[20]	1.3946[20]	
5409	1-Fluoropentane	Pentyl fluoride	$C_5H_{11}F$	592-50-7	90.139	liq	-120	62.8	0.7907[20]	1.3591[2-]	vs eth, EtOH
5410	2-Fluorophenol		C_6H_5FO	367-12-4	112.101		16.1	151.5	1.120[25]	1.5144[20]	s H_2O

Fluorobenzene

4-Fluorobenzeneacetic acid

2-Fluorobenzeneacetonitrile

4-Fluorobenzeneacetonitrile

4-Fluorobenzenemethanamine

4-Fluorobenzenemethanol

4-Fluorobenzenesulfonyl chloride

2-Fluorobenzoic acid

3-Fluorobenzoic acid

4-Fluorobenzoic acid

2-Fluorobenzonitrile

4-Fluorobenzonitrile

2-Fluorobenzoyl chloride

3-Fluorobenzoyl chloride

4-Fluorobenzoyl chloride

2-Fluoro-1,1'-biphenyl

4-Fluoro-1,1'-biphenyl

1-Fluorobutane

2-Fluorobutane

Fluorocyclohexane

1-Fluorocyclohexene

5-Fluorocytosine

1-Fluorodecane

Fluorodifen

1-Fluoro-2,4-dinitrobenzene

Fluoroethane

2-Fluoroethanol

Fluoroethene

1-Fluoroheptane

1-Fluorohexane

1-Fluoro-2-iodobenzene

1-Fluoro-4-iodobenzene

1-Fluoro-3-isothiocyanatobenzene

1-Fluoro-4-isothiocyanatobenzene

Fluoromethane

1-Fluoro-2-methoxybenzene

1-Fluoro-3-methoxybenzene

1-Fluoro-4-methoxybenzene

4-Fluoro-2-methylaniline

(Fluoromethyl)benzene

2-Fluoro-4-methyl-1-nitrobenzene

2-Fluoro-2-methylpropane

1-Fluoronaphthalene

2-Fluoronaphthalene

1-Fluoro-2-nitrobenzene

1-Fluoro-3-nitrobenzene

1-Fluoro-4-nitrobenzene

1-Fluorooctane

1-Fluoropentane

2-Fluorophenol

No.	Name	Synonym	Mol. Form.	CAS RN	Mol. Wt.	Physical Form	mp/°C	bp/°C	den/ g cm⁻³	n_D	Solubility
5411	3-Fluorophenol		C₆H₅FO	372-20-3	112.101		13.7	178	1.238²⁵	1.5140²⁰	
5412	4-Fluorophenol		C₆H₅FO	371-41-5	112.101		48	185.5	1.1889⁵⁶		sl H₂O; s ace, peth
5413	2-Fluoro-1-phenylethanone		C₈H₇FO	450-95-3	138.139	pl	29	90¹²	1.152²⁰	1.5200²⁰	
5414	1-(4-Fluorophenyl)ethanone		C₈H₇FO	403-42-9	138.139	liq	-45	196	1.1382²⁵	1.5081²⁵	i H₂O; s bz, chl
5415	1-Fluoropropane	Propyl fluoride	C₃H₇F	460-13-9	62.086	col gas	-159	-2.5	0.7596²⁰ (p>1 atm)	1.3115²⁰	sl H₂O; vs EtOH, eth
5416	2-Fluoropropane	Isopropyl fluoride	C₃H₇F	420-26-8	62.086	gas		-9.4			sl H₂O
5417	1-Fluoro-2-propanone	Fluoroacetone	C₃H₅FO	430-51-3	76.069			77	1.0288²⁰	1.3700²⁰	
5418	cis-1-Fluoropropene		C₃H₅F	19184-10-2	60.070	col gas		≈-20			
5419	trans-1-Fluoropropene		C₃H₅F	20327-65-5	60.070	col gas	≈-20				
5420	2-Fluoropropene		C₃H₅F	1184-60-7	60.070	col gas		-24			
5421	3-Fluoropropene		C₃H₅F	818-92-8	60.070	col gas		-3			sl H₂O; vs EtOH, eth; s chl
5422	2-Fluoropyridine		C₅H₄FN	372-48-5	97.091			125	1.1280²⁰	1.4574²⁰	
5423	3-Fluoropyridine		C₅H₄FN	372-47-4	97.091	liq		107	1.130	1.4720²⁰	
5424	2-Fluorotoluene		C₇H₇F	95-52-3	110.129	liq	-62	115	1.0041¹³	1.4704²⁰	i H₂O; vs EtOH, eth
5425	3-Fluorotoluene		C₇H₇F	352-70-5	110.129	liq	-87	115	0.9974²⁰	1.4691²⁰	i H₂O; vs EtOH, eth
5426	4-Fluorotoluene		C₇H₇F	352-32-9	110.129	liq	-56	116.6	0.9975²⁰	1.4699²⁰	i H₂O; vs EtOH, eth
5427	1-Fluoro-2-(trichloromethyl) benzene		C₇H₄Cl₃F	488-98-2	213.464			95¹², 75⁵	1.453²⁵	1.5432²⁰	
5428	1-Fluoro-2-(trifluoromethyl) benzene		C₇H₄F₄	392-85-8	164.101			114.5	1.293²⁵	1.4040²⁵	
5429	1-Fluoro-3-(trifluoromethyl) benzene		C₇H₄F₄	401-80-9	164.101	liq	-81.5	101.5	1.3021¹⁷		
5430	1-Fluoro-4-(trifluoromethyl) benzene		C₇H₄F₄	402-44-8	164.101	liq	-41.7	103.5	1.293²⁵	1.4025²⁰	
5431	Fluorotrimethylsilane		C₃H₉FSi	420-56-4	92.187	vol liq or gas		16.4			
5432	5-Fluorouracil	5-Fluoro-2,4(1H,3H)-Pyrimidinedione	C₄H₃FN₂O₂	51-21-8	130.077	cry (w, MeOH-eth)	283	sub 190			
5433	Fluoxetine		C₁₇H₁₈F₃NO	54910-89-3	309.326	oil					
5434	Fluoxymesterone		C₂₀H₂₉FO₃	76-43-7	336.440		270				
5435	Fluphenazine		C₂₂H₂₆F₃N₃OS	69-23-8	437.520			251⁰·³			
5436	Fluprednisolone		C₂₁H₂₇FO₅	53-34-9	378.434		210				
5437	Flurandrenolide	Fludroxycortide	C₂₄H₃₃FO₆	1524-88-5	436.513	cry (ace/hx)	251				
5438	Flurazepam		C₂₁H₂₃ClFN₃O	17617-23-1	387.878	wh rods (eth/ peth)	80				
5439	Fluridone		C₁₉H₁₄F₃NO	59756-60-4	329.315		155				
5440	Fluroxypyr	[(4-Amino-3,5-dichloro-6-fluoro-2-pyridyl)oxy]acetic acid	C₇H₅Cl₂FN₂O₃	69377-81-7	255.030		232				
5441	Fluvalinate		C₂₆H₂₂ClF₃N₂O₃	102851-06-9	502.912			>450	1.29²⁵		
5442	Folic acid	Vitamin Bc	C₁₉H₁₉N₇O₆	59-30-3	441.397	ye-oran nd (w)	250 dec				vs py, EtOH, HOAc
5443	Folinic acid	5-Formyl-5,6,7,8-tetrahydrofolic acid	C₂₀H₂₃N₇O₇	58-05-9	473.440	cry (w + 3)	245 dec				sl H₂O
5444	Folpet	1H-Isoindole-1,3(2H)-dione, 2-[(trichloromethyl)thio]-	C₉H₄Cl₃NO₂S	133-07-3	296.558		177				
5445	Fomesafen		C₁₅H₁₀ClF₃N₂O₆S	72178-02-0	438.762		220		1.28²⁰		
5446	Fomocaine	4-[3-[4-(Phenoxymethyl)phenyl] propyl]morpholine	C₂₀H₂₅NO₂	17692-39-6	311.419	col cry	53	239¹·¹			
5447	Fonofos	Phosphonodithioic acid, ethyl-, O-ethyl S-phenyl ester	C₁₀H₁₅OPS₂	944-22-9	246.329			130⁰·¹	1.16²⁵		
5448	Formaldehyde	Methanal	CH₂O	50-00-0	30.026	col gas	-92	-19.1	0.815⁻²⁰		s H₂O, EtOH, chl; msc eth, ace, bz
5449	Formaldehyde oxime		CH₃NO	75-17-2	45.041		1.3	109¹⁵	1.133²⁵		s H₂O; vs EtOH, eth
5450	Formamide	Methanamide	CH₃NO	75-12-7	45.041		2.49	220	1.1334²⁰	1.4472²⁰	msc H₂O, EtOH; sl eth; s ace; i bz, chl
5451	Formamidinesulfinic acid	Aminoiminomethanesulfinic acid	CH₄N₂O₂S	1758-73-2	108.120	nd (al)	144 dec				vs H₂O; i eth, bz
5452	Formetanate hydrochloride		C₁₁H₁₆ClN₃O₂	23422-53-9	257.717	pow	201 dec				vs H₂O; s MeOH; sl ace, hx, chl
5453	Formic acid	Methanoic acid	CH₂O₂	64-18-6	46.026		8.3	101	1.220²⁰	1.3714²⁰	msc H₂O, EtOH, eth; vs ace; s bz, tol
5454	N-Formimidoyl-L-glutamic acid	N-(Iminomethyl)-L-glutamic acid	C₆H₁₀N₂O₄	816-90-0	174.154		90				
5455	Formononetin	7-Hydroxy-3-(4-methoxyphenyl)-4H-1-benzopyran-4-one	C₁₆H₁₂O₄	485-72-3	268.264		256.5				

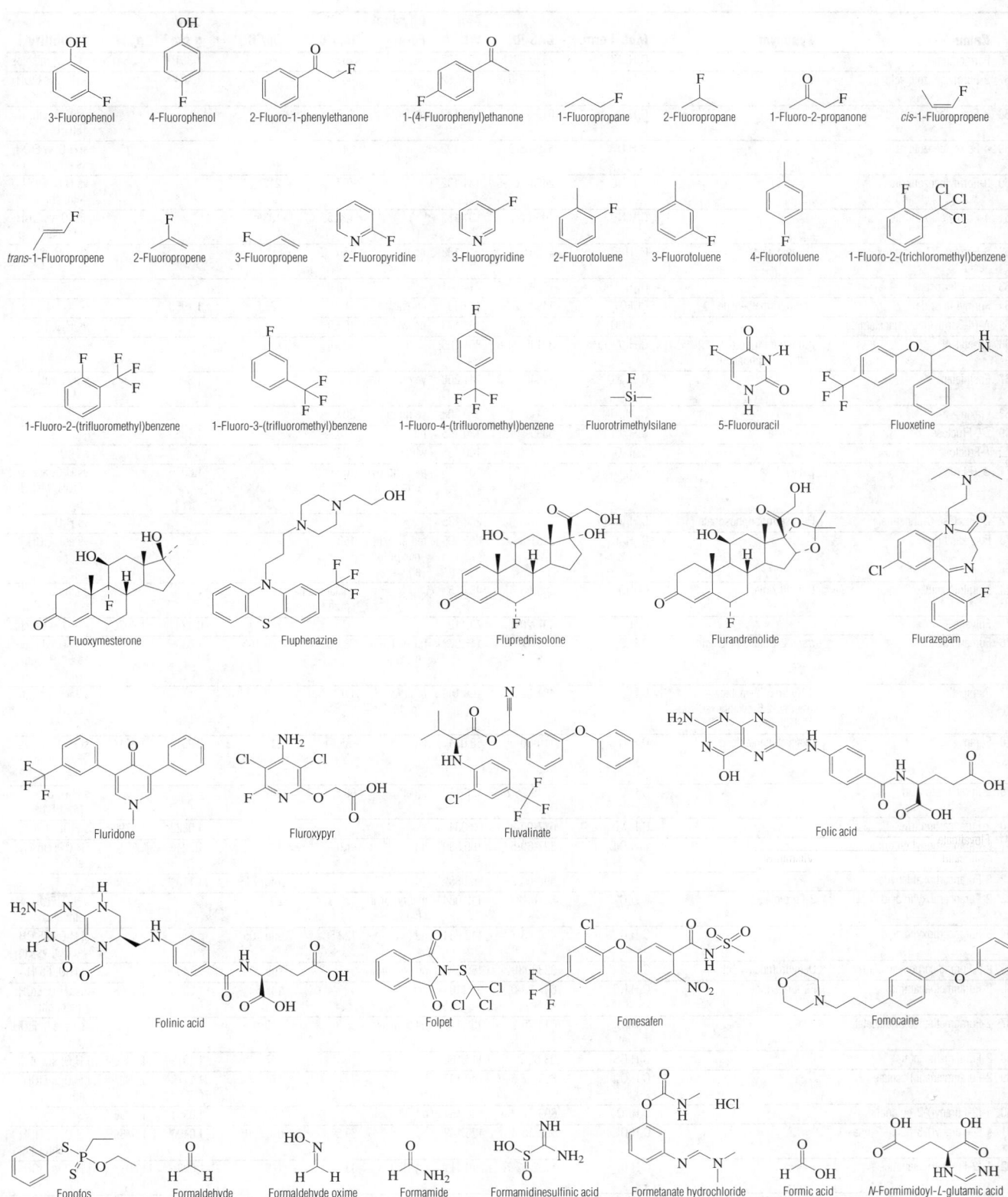

3-Fluorophenol

4-Fluorophenol

2-Fluoro-1-phenylethanone

1-(4-Fluorophenyl)ethanone

1-Fluoropropane

2-Fluoropropane

1-Fluoro-2-propanone

cis-1-Fluoropropene

trans-1-Fluoropropene

2-Fluoropropene

3-Fluoropropene

2-Fluoropyridine

3-Fluoropyridine

2-Fluorotoluene

3-Fluorotoluene

4-Fluorotoluene

1-Fluoro-2-(trichloromethyl)benzene

1-Fluoro-2-(trifluoromethyl)benzene

1-Fluoro-3-(trifluoromethyl)benzene

1-Fluoro-4-(trifluoromethyl)benzene

Fluorotrimethylsilane

5-Fluorouracil

Fluoxetine

Fluoxymesterone

Fluphenazine

Fluprednisolone

Flurandrenolide

Flurazepam

Fluridone

Fluroxypyr

Fluvalinate

Folic acid

Folinic acid

Folpet

Fomesafen

Fomocaine

Fonofos

Formaldehyde

Formaldehyde oxime

Formamide

Formamidinesulfinic acid

Formetanate hydrochloride

Formic acid

N-Formimidoyl-*L*-glutamic acid

Formononetin

No.	Name	Synonym	Mol. Form.	CAS RN	Mol. Wt.	Physical Form	mp/°C	bp/°C	den/ g cm-3	n_D	Solubility
5456	Formothion		$C_6H_{12}NO_4PS_2$	2540-82-1	257.267	visc ye oil	25.5	dec	1.361[20]	1.5541[20]	sl H_2O; misc os
5457	2-Formylbenzoic acid		$C_8H_6O_3$	119-67-5	150.132		98		1.404[25]		s H_2O; vs EtOH, eth
5458	3-Formylbenzoic acid		$C_8H_6O_3$	619-21-6	150.132	nd (w)	175				vs H_2O, eth, EtOH
5459	4-Formylbenzoic acid		$C_8H_6O_3$	619-66-9	150.132		247				sl H_2O; vs EtOH; s eth, chl
5460	3-Formylbenzonitrile		C_8H_5NO	24964-64-5	131.132		76.5	210			vs H_2O, EtOH, eth, chl
5461	4-Formylbenzonitrile		C_8H_5NO	105-07-7	131.132		100.5	133[12]			s H_2O; vs EtOH, eth, chl
5462	6-Formyl-2,3-dimethoxybenzoic acid	Opianic acid	$C_{10}H_{10}O_5$	519-05-1	210.183	nd (w)	150				s EtOH, eth
5463	Formylferrocene		$C_{11}H_{10}FeO$	12093-10-6	214.041		118.5	70[0.1]			
5464	Formyl fluoride	Fluoroformaldehyde	CHFO	1493-02-3	48.016	col gas	-142.2	-26.5	1.1950[-30]		
5465	N-(4-Formylphenyl)acetamide		$C_9H_9NO_2$	122-85-0	163.173	pr (w)	158.0				vs H_2O, bz
5466	Fosetyl-Al	Aluminum tris(O-ethylphosphonate)	$C_6H_{18}AlO_9P_3$	39148-24-8	354.105		>300				
5467	Fosthietan		$C_6H_{12}NO_3PS_2$	21548-32-3	241.268	ye oil			1.3[25]	1.5348[25]	s ace, chl, MeOH, tol
5468	Fraxin		$C_{16}H_{18}O_{10}$	524-30-1	370.308	ye nd (al)	205				
5469	DL-Fructose	α-Acrose	$C_6H_{12}O_6$	6035-50-3	180.155	nd	130		1.665[16]		
5470	L-Fructose		$C_6H_{12}O_6$	7776-48-9	180.155	wh cry	102				s H_2O
5471	β-D-Fructose	β-Levulose	$C_6H_{12}O_6$	53188-23-1	180.155	pr or nd (w) orth pr (al)	103 dec		1.60[20]		vs H_2O, ace; s EtOH, MeOH, py
5472	D-Fructose 6-phosphate	Hexose monophosphate	$C_6H_{13}O_9P$	643-13-0	260.135						vs H_2O
5473	Fucoxanthin		$C_{42}H_{58}O_6$	3351-86-8	658.905	red pl (eth) hex pl (dil al)	168				vs eth, EtOH
5474	Fulminic acid	Carbyloxime	CHNO	506-85-4	43.025		unstable in pure form				s eth
5475	Fulvene		C_6H_6	497-20-1	78.112			7[56]	0.8241[20]	1.4920[20]	i H_2O; s bz, chl
5476	Fumaric acid	trans-2-Butenedioic acid	$C_4H_4O_4$	110-17-8	116.073	nd, mcl pr or lf (w)	287 dec	sub 165	1.635[20]		sl H_2O, eth, ace; s EtOH, con sulf
5477	Fumigatin	3-Hydroxy-2-methoxy-5-methyl-2,5-cyclohexadiene-1,4-dione	$C_8H_8O_4$	484-89-9	168.148	br nd or pl (peth)	116				vs ace, bz, eth, EtOH
5478	Furan	Oxacyclopentadiene	C_4H_4O	110-00-9	68.074	liq	-85.61	31.5	0.9514[20]	1.4214[20]	sl H_2O, chl; vs EtOH, eth; s ace, bz
5479	2-Furanacetic acid		$C_6H_6O_3$	2745-26-8	126.110	lf(peth)	68.5	102[0.4]			s H_2O, bz, MeOH, peth
5480	2-Furancarbonitrile		C_5H_3NO	617-90-3	93.084			147	1.0822[20]	1.4798[20]	s EtOH, eth
5481	2-Furancarbonyl chloride		$C_5H_3ClO_2$	527-69-5	130.530	liq	-1.0	173	1.324[25]	1.5310[20]	i H_2O; s eth, chl; sl ctc
5482	3-Furancarboxaldehyde		$C_5H_4O_2$	498-60-2	96.085			145; 71[43]	1.110[20]	1.4945[20]	
5483	2-Furancarboxylic acid	2-Furoic acid	$C_5H_4O_3$	88-14-2	112.084	mcl nd or lf (w)	133.5	231			s H_2O, EtOH; vs eth; sl ace
5484	3-Furancarboxylic acid		$C_5H_4O_3$	488-93-7	112.084	nd (w)	122.5	sub 105			sl H_2O; s EtOH, AcOEt; vs eth
5485	2,5-Furandicarboxylic acid	Dehydromucic acid	$C_6H_4O_5$	3238-40-2	156.093	nd (w), lf (al)	342	sub	1.7400[20]		sl H_2O, EtOH
5486	2-Furanmethanamine	Furfurylamine	C_5H_7NO	617-89-0	97.116			145.5	1.0995[20]	1.4908[20]	msc H_2O, EtOH; s eth, chl
5487	2-Furanmethanediol diacetate		$C_9H_{10}O_5$	613-75-2	198.172	nd or pl (eth-peth)	53.3	220			vs bz, eth, EtOH
5488	2-Furanmethanethiol		C_5H_6OS	98-02-2	114.166			157	1.1319[20]	1.5329[20]	i H_2O; sl chl
5489	2-Furanmethanol acetate		$C_7H_8O_3$	623-17-6	140.137			179	1.1175[20]	1.4327[20]	i H_2O; s EtOH, eth
5490	4-(2-Furanyl)-2-butanone		$C_8H_{10}O_2$	699-17-2	138.164	oil		203	1.0361[19]	1.4696[17]	
5491	4-(2-Furanyl)-3-buten-2-one		$C_8H_8O_2$	623-15-4	136.149		39.5	dec 229; 113[10]	1.0496[57]	1.5788[45]	i H_2O; vs EtOH, eth, chl; s peth
5492	1-(2-Furanyl)ethanone		$C_6H_6O_2$	1192-62-7	110.111	cry (lig)	33	175	1.098[20]	1.5017[20]	i H_2O; s EtOH, eth
5493	2-Furanylmethyl pentanoate	Furfuryl valerate	$C_{10}H_{14}O_3$	36701-01-6	182.216			228; 82[1]	1.0284[20]		vs eth, EtOH
5494	3-(2-Furanyl)-1-phenyl-2-propen-1-one		$C_{13}H_{10}O_2$	717-21-5	198.217		47	317	1.1140[20]		s EtOH, eth
5495	1-(2-Furanyl)-1-propanone		$C_7H_8O_2$	3194-15-8	124.138	cry	28	88[14]	1.0626[28]	1.4922[25]	s eth; sl ctc
5496	1-(2-Furanyl)-2-propanone	2-Furfuryl methyl ketone	$C_7H_8O_2$	6975-60-6	124.138		29	179.5	1.104[20]	1.5035[20]	
5497	3-(2-Furanyl)-2-propenal		$C_7H_6O_2$	623-30-3	122.122		54	135[14]			i H_2O; msc EtOH; s eth; sl chl
5498	3-(2-Furanyl)-2-propenenitrile	2-Furanacrylonitrile	C_7H_5NO	7187-01-1	119.121		38	96		1.5824[25]	vs tol

Formothion

2-Formylbenzoic acid

3-Formylbenzoic acid

4-Formylbenzoic acid

3-Formylbenzonitrile

4-Formylbenzonitrile

6-Formyl-2,3-dimethoxybenzoic acid

Formylferrocene

Formyl fluoride

N-(4-Formylphenyl)acetamide

Fosetyl-Al

Fosthietan

Fraxin

DL-Fructose

L-Fructose

β-D-Fructose

D-Fructose 6-phosphate

Fucoxanthin

Fulminic acid

Fulvene

Fumaric acid

Fumigatin

Furan

2-Furanacetic acid

2-Furancarbonitrile

2-Furancarbonyl chloride

3-Furancarboxaldehyde

2-Furancarboxylic acid

3-Furancarboxylic acid

2,5-Furandicarboxylic acid

2-Furanmethanamine

2-Furanmethanediol diacetate

2-Furanmethanethiol

2-Furanmethanol acetate

4-(2-Furanyl)-2-butanone

4-(2-Furanyl)-3-buten-2-one

1-(2-Furanyl)ethanone

2-Furanylmethyl pentanoate

3-(2-Furanyl)-1-phenyl-2-propen-1-one

1-(2-Furanyl)-1-propanone

1-(2-Furanyl)-2-propanone

3-(2-Furanyl)-2-propenal

3-(2-Furanyl)-2-propenenitrile

No.	Name	Synonym	Mol. Form.	CAS RN	Mol. Wt.	Physical Form	mp/°C	bp/°C	den/ g cm^{-3}	n_D	Solubility
5499	3-(2-Furanyl)-2-propenoic acid	2-Furanacrylic acid	$C_7H_6O_3$	539-47-9	138.121	nd (w)	141	286			vs eth, EtOH
5500	Furazolidone	3-[[(5-Nitro-2-furanyl) methylene]amino]-2-oxazolidinone	$C_8H_7N_3O_5$	67-45-8	225.159		255				
5501	Furethidine		$C_{21}H_{31}NO_4$	2385-81-1	361.476		28	210[0.5]		1.5219[20]	
5502	Furfural	2-Furaldehyde	$C_5H_4O_2$	98-01-1	96.085	liq	-38.1	161.7	1.1594[20]	1.5261[20]	s H_2O, bz, chl; vs EtOH, ace; msc eth
5503	Furfuryl alcohol	2-Furanmethanol	$C_5H_6O_2$	98-00-0	98.101	col-ye liq	-14.6	171	1.1296[20]	1.4869[20]	msc H_2O; vs EtOH, eth; s chl
5504	Furfuryl propanoate	2-Furanmethanol, propanoate	$C_8H_{10}O_3$	623-19-8	154.163			195	1.1085[20]		sl H_2O; s EtOH, ace; msc eth
5505	Furoin	1,2-Di-2-furanyl-2-hydroxyethanone	$C_{10}H_8O_4$	552-86-3	192.169	nd (al)	138.5				sl H_2O, EtOH, chl; s eth, MeOH
5506	Furonazide		$C_{12}H_{11}N_3O_2$	3460-67-1	229.234		202.3				
5507	Furosemide		$C_{12}H_{11}ClN_2O_5S$	54-31-9	330.743		204 dec				
5508	Fursultiamine		$C_{17}H_{26}N_4O_3S_2$	804-30-8	398.543	col pr	132 dec		1.29		sl H_2O
5509	Furylfuramide, (E)	2-(2-Furanyl)-3-(5-nitro-2-furanyl)-2-propenamide	$C_{11}H_8N_2O_5$	18819-45-9	248.192	cry	154				
5510	Fusarenon X		$C_{17}H_{22}O_8$	23255-69-8	354.352	cry	182				
5511	Galactaric acid	Mucic acid	$C_6H_{10}O_8$	526-99-8	210.138	pr (w)	255 dec				
5512	Galactitol	Dulcose	$C_6H_{14}O_6$	608-66-2	182.171	cry (dil MeOH)	189.5	277[1]	1.47[20]		s H_2O; sl EtOH, py; i eth, bz
5513	D-Galactonic acid, γ-lactone		$C_6H_{10}O_6$	2782-07-2	178.139	nd (w+1), nd (al)	112				vs H_2O
5514	α-D-Galactopyranose		$C_6H_{12}O_6$	3646-73-9	180.155		167				
5515	4-O-β-D-Galactopyranosyl-D-gluconic acid	Lactobionic acid	$C_{12}H_{22}O_{12}$	96-82-2	358.296	syr					vs H_2O; sl EtOH, MeOH, HOAc; i eth
5516	D-Galactose		$C_6H_{12}O_6$	59-23-4	180.155	pl or pr (al)pr or nd (w+1)	170				vs H_2O; sl EtOH; i eth, bz; s py
5517	D-Galacturonic acid		$C_6H_{10}O_7$	685-73-4	194.139	nd (w)	166 (β)				s H_2O, EtOH; i eth
5518	Galanthamine	Lycoremine	$C_{17}H_{21}NO_3$	357-70-0	287.354	cry (bz)	126.5				vs ace, EtOH, chl
5519	Galipine	2-[2-(3,4-Dimethoxyphenyl) ethyl]-4-methoxyquinoline	$C_{20}H_{21}NO_3$	525-68-8	323.386	pr (al, eth) nd (peth)	115.5				vs ace, bz, eth, EtOH
5520	Gallamine triethiodide		$C_{30}H_{60}I_3N_3O_3$	65-29-2	891.528		147.5				vs H_2O, EtOH; sl eth, ace, bz, chl
5521	Gallein		$C_{20}H_{12}O_5$	2103-64-2	332.306	br-red pow (+1.5w) red (anh)	>300				vs ace, EtOH
5522	Ganciclovir		$C_9H_{13}N_5O_4$	82410-32-0	255.231	cry (MeOH)	250 dec				
5523	Gardol		$C_{15}H_{28}NNaO_3$	137-16-6	293.378						sl H_2O
5524	Gelsemine		$C_{20}H_{22}N_2O_2$	509-15-9	322.401	cry (ace)	178				vs ace, bz, eth, EtOH
5525	Gelsemine, monohydrochloride		$C_{20}H_{23}ClN_2O_2$	35306-33-3	358.862		326				s H_2O; sl EtOH
5526	Genistein	5,7-Dihydroxy-3-(4-hydroxyphenyl)-4H-1-benzopyran-4-one	$C_{15}H_{10}O_5$	446-72-0	270.237	nd(eth), pr(dil al)	301 dec				
5527	β-Gentiobiose	6-O-β-D-Glucopyranosyl-D-glucose	$C_{12}H_{22}O_{11}$	554-91-6	342.296	cry (EtOH)	192				s hot H_2O, hot MeOH
5528	trans-Geraniol		$C_{10}H_{18}O$	106-24-1	154.249		<-15	230	0.8894[20]	1.4766[20]	i H_2O; s EtOH, eth, ace, chl
5529	Geranyl 2-methylpropanoate		$C_{14}H_{24}O_2$	2345-26-8	224.340			136[13]	0.8997[15]	1.4576[20]	
5530	Geranyl acetate		$C_{12}H_{20}O_2$	16409-44-2	196.286			115[12]	0.9163[15]	1.4624[20]	
5531	Germine		$C_{27}H_{43}NO_8$	508-65-6	509.632	pr or cry (MeOH)	220				s bz, MeOH, alk, acid
5532	Gibberellic acid		$C_{19}H_{22}O_6$	77-06-5	346.374	cry (EtOAc)	234				vs ace, EtOH, MeOH
5533	Gitoxigenin		$C_{23}H_{34}O_5$	545-26-6	390.513	pr (AcOEt) pr (+w, dil al)	234				i H_2O; sl eth; s chl
5534	Gitoxin		$C_{41}H_{64}O_{14}$	4562-36-1	780.939	pr (chl-MeOH)	285 dec				
5535	d-Glaucine		$C_{21}H_{25}NO_4$	475-81-0	355.429	pl, pr (eth, AcOEt)	120				vs ace, EtOH, chl
5536	D-Glucaric acid	D-Tetrahydroxyadipic acid	$C_6H_{10}O_8$	87-73-0	210.138	nd (45% al)	125.5				vs H_2O, EtOH; sl eth, chl
5537	D-Glucitol	Sorbitol	$C_6H_{14}O_6$	50-70-4	182.171	nd (w)	111	295[3.5]	1.489[20]	1.3330[20]	vs H_2O, ace
5538	D-Glucitol, hexaacetate	Sorbitol hexaacetate	$C_{18}H_{26}O_{12}$	7208-47-1	434.392	pr (w)	100.8		1.30[20]		sl H_2O, eth; vs EtOH; s chl, AcOEt

3-(2-Furanyl)-2-propenoic acid

Furazolidone

Furethidine

Furfural

Furfuryl alcohol

Furfuryl propanoate

Furoin

Furonazide

Furosemide

Fursultiamine

Furylfuramide, (E)

Fusarenon X

Galactaric acid

Galactitol

D-Galactonic acid, γ-lactone

α-D-Galactopyranose

4-O-β-D-Galactopyranosyl-D-gluconic acid

D-Galactose

D-Galacturonic acid

Galanthamine

Galipine

Gallamine triethiodide

Gallein

Ganciclovir

Gardol

Gelsemine

Gelsemine, monohydrochloride

Genistein

β-Gentiobiose

trans-Geraniol

Geranyl 2-methylpropanoate

Geranyl acetate

Germine

Gibberellic acid

Gitoxigenin

Gitoxin

d-Glaucine

D-Glucaric acid

D-Glucitol

D-Glucitol, hexaacetate

No.	Name	Synonym	Mol. Form.	CAS RN	Mol. Wt.	Physical Form	mp/°C	bp/°C	den/ g cm⁻³	n_D	Solubility
5539	*D*-Gluconic acid		$C_6H_{12}O_7$	526-95-4	196.155	nd (al-eth)	131				s H_2O; sl EtOH; i eth, bz
5540	β-*D*-Glucopyranose		$C_6H_{12}O_6$	492-61-5	180.155	cry (hot EtOH)	149				
5541	6-*O*-α-*D*-Glucopyranosyl-*D*-fructose	Palatinose	$C_{12}H_{22}O_{11}$	13718-94-0	342.296						s H_2O
5542	2-(β-*D*-Glucopyranosyloxy) benzaldehyde	Helicin	$C_{13}H_{16}O_7$	618-65-5	284.262	nd (w)	175				vs H_2O, EtOH
5543	7-(β-*D*-Glucopyranosyloxy)-2*H*-1-benzopyran-2-one	Skimmin	$C_{15}H_{16}O_8$	93-39-0	324.283	cry (w + 1)	220				s H_2O, EtOH; i eth, chl
5544	2-(β-*D*-Glucopyranosyloxy)-2-methylpropanenitrile	Linamarin	$C_{10}H_{17}NO_6$	554-35-8	247.245	nd (w, al)	145				vs ace
5545	1-[4-(β-*D*-Glucopyranosyloxy) phenyl]ethanone	Picein	$C_{14}H_{18}O_7$	530-14-3	298.289	nd (w+1), nd (MeOH)	195.5				sl H_2O; s EtOH, eth, HOAc; i chl
5546	α-*D*-Glucose		$C_6H_{12}O_6$	26655-34-5	180.155		146 dec		1.5620[18]		vs H_2O; sl EtOH; i ace, AcOEt; s py
5547	α-*D*-Glucose pentaacetate		$C_{16}H_{22}O_{11}$	604-68-2	390.339	pl or nd (al)	113.3	sub			sl H_2O, EtOH, CS_2; s eth, chl, HOAc
5548	β-*D*-Glucose pentaacetate		$C_{16}H_{22}O_{11}$	604-69-3	390.339	nd (al)	134	sub	1.2740[20]		i H_2O; sl EtOH, peth, eth; s bz; msc chl
5549	α-*D*-Glucose 1-phosphate		$C_6H_{13}O_9P$	59-56-3	260.135						vs H_2O
5550	*D*-Glucuronic acid		$C_6H_{10}O_7$	6556-12-3	194.139	nd (al)	165				vs H_2O, EtOH
5551	*D*-Glucuronic acid γ-lactone	*D*-Glucuronolactone	$C_6H_8O_6$	32449-92-6	176.124	mcl pl (w) cry (al)	177.5		1.76[20]		s H_2O; sl EtOH, DMSO, MeOH; i bz
5552	*DL*-Glutamic acid		$C_5H_9NO_4$	617-65-2	147.130	orth (al,w)	199 dec		1.4601[20]		sl H_2O, eth; i EtOH, CS_2, lig
5553	*D*-Glutamic acid		$C_5H_9NO_4$	6893-26-1	147.130	lf (w)	213 dec		1.538[20]		sl H_2O; i EtOH, eth, ace, bz, HOAc, MeOH
5554	*L*-Glutamic acid	(*S*)-2-Aminopentanedioic acid	$C_5H_9NO_4$	56-86-0	147.130	orth (dil al)	160 dec	sub 175	1.538[20]		sl H_2O
5555	*L*-Glutamic acid, hydrochloride		$C_5H_{10}ClNO_4$	138-15-8	183.591	orth pl (w)	214 dec				vs H_2O, EtOH
5556	*L*-Glutamine	2-Aminoglutaramic acid	$C_5H_{10}N_2O_3$	56-85-9	146.144	nd (w, dil al)	185 dec				s H_2O; i EtOH, eth, bz, MeOH
5557	Glutaric acid	Pentanedioic acid	$C_5H_8O_4$	110-94-1	132.116	nd (bz)	97.8	dec 303	1.429[15]	1.4188[106]	vs H_2O, EtOH, eth; i bz; s chl, lig
5558	Glutathione	*L*-γ-Glutamyl-*L*-cysteinylglycine	$C_{10}H_{17}N_3O_6S$	70-18-8	307.323	cry (50% al)	195				vs H_2O; i EtOH, eth; s DMF
5559	Glutathione disulfide	*L*-γ-Glutamyl-*L*-cysteinylglycine disulfide	$C_{20}H_{32}N_6O_{12}S_2$	27025-41-8	612.631	cry (EtOH aq)	179				
5560	Glutethimide		$C_{13}H_{15}NO_2$	77-21-4	217.264	cry (eth)	84				i H_2O; s EtOH; vs eth, ace
5561	Glycerol	1,2,3-Propanetriol	$C_3H_8O_3$	56-81-5	92.094	syr, orth pl	18.1	290	1.2613[20]	1.4746[20]	msc H_2O, EtOH; sl eth; i bz, ctc, chl
5562	Glycerol 1-acetate, (*DL*)	1,2,3-Propanetriol 1-acetate, (±)	$C_5H_{10}O_4$	93713-40-7	134.131			158[165], 130[3]	1.2060[20]	1.4157[20]	s H_2O, EtOH; sl eth; i bz
5563	Glycerol 1-butanoate		$C_7H_{14}O_4$	557-25-5	162.184			280; 117[10]	1.129[18]	1.4531[20]	vs H_2O, EtOH
5564	Glycerol 1,3-dinitrate	1,2,3-Propanetriol, 1,3-dinitrate	$C_3H_6N_2O_7$	623-87-0	182.089	pr (w), cry (eth)	26	148[15], 116[0.6]	1.523[20]	1.4715[20]	vs H_2O, eth, EtOH
5565	Glycerol 1,3-di-9-octadecenoate, *cis,cis*		$C_{39}H_{72}O_5$	2465-32-9	620.986	cry (eth/ EtOH)	50.1				
5566	Glycerol 1-oleate	1-Monoolein	$C_{21}H_{40}O_4$	111-03-5	356.541	pl (al)	35	239[3]	0.9420[20]	1.4626[20]	i H_2O; s EtOH, eth, chl
5567	*L*-Glycerol 1-phosphate	α-Glycerophosphoric acid	$C_3H_9O_6P$	5746-57-6	172.073	syr		dec			dec H_2O
5568	Glycerol tridecanoate	Decanoic acid glycerol triester	$C_{33}H_{62}O_6$	621-71-6	554.841	cry (peth)	32				
5569	Glycerol trielaidate	Trielaidin	$C_{57}H_{104}O_6$	537-39-3	885.432						vs bz, eth, chl
5570	Glycerol trilaurate	Trilaurin	$C_{39}H_{74}O_6$	538-24-9	639.001	nd (al)			0.8986[55]	1.4404[60]	i H_2O; s EtOH, eth, peth; vs ace, bz
5571	Glycerol tri-3-methylbutanoate	Triisovalerin	$C_{18}H_{32}O_6$	620-63-3	344.443			332.5	0.9984[20]	1.4354[20]	vs eth, EtOH
5572	Glycerol trioleate	Triolein	$C_{57}H_{104}O_6$	122-32-7	885.432	col-ye oil	-4	237[18]	0.915[15]	1.4676[15]	i H_2O; sl EtOH; vs eth; s chl, peth
5573	Glycerol tripalmitate	Tripalmitin	$C_{51}H_{98}O_6$	555-44-2	807.320	nd (eth)	66.5	315	0.8752[70]	1.4381[80]	i H_2O; sl EtOH; vs eth; s bz, chl
5574	Glycerol tristearate	Tristearin	$C_{57}H_{110}O_6$	555-43-1	891.479				0.8559[90]	1.4395[80]	i H_2O, EtOH; sl bz, ctc; s ace, chl

D-Gluconic acid

β-D-Glucopyranose

6-O-α-D-Glucopyranosyl-D-fructose

2-(β-D-Glucopyranosyloxy)benzaldehyde

7-(β-D-Glucopyranosyloxy)-2H-1-benzopyran-2-one

2-(β-D-Glucopyranosyloxy)-2-methylpropanenitrile

1-[4-(β-D-Glucopyranosyloxy)phenyl]ethanone

α-D-Glucose

α-D-Glucose pentaacetate

β-D-Glucose pentaacetate

α-D-Glucose 1-phosphate

D-Glucuronic acid

D-Glucuronic acid γ-lactone

DL-Glutamic acid

D-Glutamic acid

L-Glutamic acid

L-Glutamic acid, hydrochloride

L-Glutamine

Glutaric acid

Glutathione

Glutathione disulfide

Glutethimide

Glycerol

Glycerol 1-acetate, (DL)

Glycerol 1-butanoate

Glycerol 1,3-dinitrate

Glycerol 1,3-di-9-octadecenoate, cis,cis

Glycerol 1-oleate

L-Glycerol 1-phosphate

Glycerol tridecanoate

Glycerol trielaidate

Glycerol trilaurate

Glycerol tri-3-methylbutanoate

Glycerol trioleate

Glycerol tripalmitate

Glycerol tristearate

No.	Name	Synonym	Mol. Form.	CAS RN	Mol. Wt.	Physical Form	mp/°C	bp/°C	den/ g cm⁻³	n_D	Solubility
5575	Glycerol tritetradecanoate	Trimyristin	C$_{45}$H$_{86}$O$_6$	555-45-3	723.161	wh-ye solid	58.5	311	0.8848[60]	1.4428[60]	i H$_2$O; sl EtOH, liq; s eth, ace, bz
5576	Glycerone phosphate	1-Hydroxy-3-(phosphonooxy)-2-propanone	C$_3$H$_7$O$_6$P	57-04-5	170.058						dec H$_2$O
5577	Glycine	Aminoacetic acid	C$_2$H$_5$NO$_2$	56-40-6	75.067	mcl or trg pr (dil al)	290 dec		1.161[20]		vs H$_2$O; i EtOH, eth; sl ace, py
5578	Glycine, ethyl ester, hydrochloride	Ethyl aminoacetate hydrochloride	C$_4$H$_{10}$ClNO$_2$	623-33-6	139.581		144				vs H$_2$O, EtOH
5579	Glycine, hydrochloride		C$_2$H$_6$ClNO$_2$	6000-43-7	111.528	hyg orth nd (w)	200.5				vs H$_2$O
5580	Glycocholic acid		C$_{26}$H$_{43}$NO$_6$	475-31-0	465.622	nd (w)	166.5				sl H$_2$O, eth; vs EtOH
5581	Glycocyamine		C$_3$H$_7$N$_3$O$_2$	352-97-6	117.107	pl or nd (w)	282				sl H$_2$O, EtOH, eth
5582	Glycogen		(C$_6$H$_{10}$O$_5$)$_x$	9005-79-2	162.140	wh pow					vs H$_2$O; i EtOH, eth
5583	Glycolaldehyde		C$_2$H$_4$O$_2$	141-46-8	60.052	pl	97		1.366[100]	1.4772[19]	s chl
5584	Glycolic acid		C$_2$H$_4$O$_3$	79-14-1	76.051	orth nd (w) lf (eth)	79.5	100			s H$_2$O, EtOH, eth
5585	N-Glycolylneuraminic acid	N-(Hydroxyacetyl)neuraminic acid	C$_{11}$H$_{19}$NO$_{10}$	1113-83-3	325.270		186				
5586	Glycopyrrolate		C$_{19}$H$_{28}$BrNO$_3$	596-51-0	398.334		192.5				
5587	Glycylalanine	N-Alanylglycine	C$_5$H$_{10}$N$_2$O$_3$	1188-01-8	146.144		237 dec				s H$_2$O; i EtOH, eth
5588	L-Glycylasparagine		C$_6$H$_{11}$N$_3$O$_4$	1999-33-3	189.169	nd (EtOH aq)	216				s H$_2$O; sl EtOH
5589	N-Glycylglycine	2-(Aminoacetamido)acetic acid	C$_4$H$_8$N$_2$O$_3$	556-50-3	132.118		263 dec				s H$_2$O
5590	N-(N-Glycylglycyl)glycine		C$_6$H$_{11}$N$_3$O$_4$	556-33-2	189.169	nd (dil al)	246 dec				s H$_2$O; i EtOH, eth
5591	N-Glycyl-L-leucine		C$_8$H$_{16}$N$_2$O$_3$	869-19-2	188.224	pl (dil al) pl (dil al)	256 dec				vs H$_2$O; i EtOH
5592	N-Glycyl-L-phenylalanine		C$_{11}$H$_{14}$N$_2$O$_3$	3321-03-7	222.240		266				s H$_2$O
5593	N-Glycylserine, (DL)		C$_5$H$_{10}$N$_2$O$_4$	687-38-7	162.144		198 dec				
5594	Glycyrrhizic acid		C$_{42}$H$_{62}$O$_{16}$	1405-86-3	822.931	pl or pr (HOAc)	220 dec				vs H$_2$O, EtOH; i eth
5595	Glyodin	1H-Imidazole, 2-heptadecyl-4,5-dihydro-, monoacetate	C$_{22}$H$_{44}$N$_2$O$_2$	556-22-9	368.596				1.035[20]		
5596	Glyoxal		C$_2$H$_2$O$_2$	107-22-2	58.036	ye pr	15	50.4	1.14[20]	1.3826[20]	vs H$_2$O; s EtOH, eth
5597	Glyoxal bis(2-hydroxyanil)	2,2'-Benzoxazoline	C$_{14}$H$_{12}$N$_2$O$_2$	1149-16-2	240.257		202				s DMSO
5598	Glyoxylic acid		C$_2$H$_2$O$_3$	298-12-4	74.035	orth pr (w+1/2)	98				vs H$_2$O; sl EtOH, eth, bz
5599	Glyphosate	Glycine, N-(phosphonomethyl)-	C$_3$H$_8$NO$_5$P	1071-83-6	169.074		230 dec				vs H$_2$O
5600	Glyphosate isopropylamine salt		C$_6$H$_{17}$N$_2$O$_5$P	38641-94-0	228.183	cry					
5601	Glyphosine	Glycine, N,N-bis(phosphonomethyl)-	C$_4$H$_{11}$NO$_8$P$_2$	2439-99-8	263.080	wh cry					s H$_2$O
5602	Grayanotoxin I		C$_{22}$H$_{36}$O$_7$	4720-09-6	412.517	cry (AcOEt/C$_5$H$_{12}$)	268				
5603	Griseofulvin, (+)		C$_{17}$H$_{17}$ClO$_6$	126-07-8	352.766	oct or orth cry (bz)	220				i H$_2$O; sl EtOH, eth, ace, bz, AcOEt, chl
5604	Guaiol		C$_{15}$H$_{26}$O	489-86-1	222.366	trg pr (al)	91	dec 288; 165[17]	0.9074[20]	1.4716[100]	i H$_2$O; s EtOH, eth
5605	Guanabenz		C$_8$H$_8$Cl$_2$N$_4$	5051-62-7	231.083	wh solid	228 dec				
5606	Guanadrel sulfate (2:1)		C$_{20}$H$_{40}$N$_6$O$_8$S	22195-34-2	524.632	cry (MeOH/EtOH)	214				
5607	Guanethidine		C$_{10}$H$_{22}$N$_4$	55-65-2	198.309	wh cry (MeOH)	226				
5608	Guanidine	Aminomethanamidine	CH$_5$N$_3$	113-00-8	59.071	cry	50				vs H$_2$O, EtOH
5609	Guanidine monohydrochloride		CH$_6$ClN$_3$	50-01-1	95.532	orth bipym (al)	182.3		1.354[20]		vs H$_2$O, EtOH
5610	Guanidine mononitrate		CH$_6$N$_4$O$_3$	506-93-4	122.084	lf (w)	217	dec			vs H$_2$O, EtOH
5611	Guanidine, sulfate (2:1)		C$_2$H$_{12}$N$_6$O$_4$S	594-14-9	216.219		292 dec				
5612	2-Guanidinoethanesulfonic acid	Taurocyamine	C$_3$H$_9$N$_3$O$_3$S	543-18-0	167.186	cry (EtOH, ace)	227				
5613	3-Guanidinopropanoic acid	N-Amidino-β-alanine	C$_4$H$_9$N$_3$O$_2$	353-09-3	131.133	cry (EtOH)	210				
5614	Guanine		C$_5$H$_5$N$_5$O	73-40-5	151.127	nd or pl (aq NH$_3$)	360 dec	sub			i H$_2$O, HOAc; sl EtOH, eth; s alk, acid
5615	Guanosine	2-Amino-1,9-dihydro-9-β-D-ribofuranosyl-6H-purin-6-one	C$_{10}$H$_{13}$N$_5$O$_5$	118-00-3	283.241	nd (w)	239 dec				sl H$_2$O; i EtOH, eth; vs HOAc
5616	Guanosine 5'-diphosphate	Guanosine 5'-(trihydrogen diphosphate)	C$_{10}$H$_{15}$N$_5$O$_{11}$P$_2$	146-91-8	443.201	amorp solid					
5617	Guanosine 5'-monophosphate	5'-Guanylic acid	C$_{10}$H$_{14}$N$_5$O$_8$P	85-32-5	363.221	hyg cry	190 dec				sl H$_2$O

Glycerol tritetradecanoate

Glycerone phosphate

Glycine

Glycine, ethyl ester, hydrochloride

Glycine, hydrochloride

Glycocholic acid

Glycocyamine

Glycogen

Glycolaldehyde

Glycolic acid

N-Glycolylneuraminic acid

Glycopyrrolate

Glycylalanine

L-Glycylasparagine

N-Glycylglycine

N-(N-Glycylglycyl)glycine

N-Glycyl-L-leucine

N-Glycyl-L-phenylalanine

N-Glycylserine, (DL)

Glycyrrhizic acid

Glyodin

Glyoxal

Glyoxal bis(2-hydroxyanil)

Glyoxylic acid

Glyphosate

Glyphosate isopropylamine salt

Glyphosine

Grayanotoxin I

Griseofulvin, (+)

Guaiol

Guanabenz

Guanadrel sulfate (2:1)

Guanethidine

Guanidine

Guanidine monohydrochloride

Guanidine mononitrate

Guanidine, sulfate (2:1)

2-Guanidinoethanesulfonic acid

3-Guanidinopropanoic acid

Guanine

Guanosine

Guanosine 5'-diphosphate

Guanosine 5'-monophosphate

No.	Name	Synonym	Mol. Form.	CAS RN	Mol. Wt.	Physical Form	mp/°C	bp/°C	den/ g cm⁻³	n_D	Solubility
5618	Guanosine 5'-monophosphate, disodium salt	5'-Guanylic acid, disodium salt	$C_{10}H_{12}N_5Na_2O_8P$	5550-12-9	407.185		195 dec				sl H_2O
5619	Guinea Green B	C.I. Acid Green 3	$C_{37}H_{35}N_2NaO_6S_2$	4680-78-8	690.803	dk grn pow					s H_2O; sl EtOH
5620	D-Gulose		$C_6H_{12}O_6$	4205-23-6	180.155	syr		dec			vs H_2O
5621	L-Gulose		$C_6H_{12}O_6$	6027-89-0	180.155	syr		dec			vs H_2O
5622	Haloperidol		$C_{21}H_{23}ClFNO_2$	52-86-8	375.865		151.5				
5623	Harmaline	4,9-Dihydro-7-methoxy-1-methyl-3H-pyrido[3,4-b]indole	$C_{13}H_{14}N_2O$	304-21-2	214.262	tab (MeOH) orth pr (al)	230				sl H_2O, EtOH, eth; s chl, py
5624	Harman	1-Methyl-9H-pyrido[3,4-b]indole	$C_{12}H_{10}N_2$	486-84-0	182.220	bl flr orth cry (hp)	236.5				
5625	Harmine	7-Methoxy-1-methyl-9H-pyrido[3,4-b]indole	$C_{13}H_{12}N_2O$	442-51-3	212.246	orth (al), pr (MeOH)	273	sub			sl H_2O, chl, EtOH, eth; s py
5626	HC Blue No. 1		$C_{11}H_{17}N_3O_4$	2784-94-3	255.271	blk cry	100				
5627	HC Blue No. 2		$C_{12}H_{19}N_3O_5$	33229-34-4	285.296	dk bl-blk cry	110				
5628	Hectane		$C_{100}H_{202}$	6703-98-6	1404.67		117				
5629	Hederagenin		$C_{30}H_{48}O_4$	465-99-6	472.700	pr (al)	333				
5630	Helenalin		$C_{15}H_{18}O_4$	6754-13-8	262.302	cry (EtOH)	226				sl H_2O; s EtOH, chl
5631	Helminthosporal		$C_{15}H_{22}O_2$	723-61-5	234.335		58	117[0.015]			
5632	Helvolic acid		$C_{33}H_{44}O_8$	29400-42-8	568.697	nd (dil HOAc)	212 dec				sl H_2O, EtOH; s eth, ace, bz, diox
5633	Hematein		$C_{16}H_{12}O_6$	475-25-2	300.262	red-br cry	250 dec				i H_2O, eth, bz, chl; sl EtOH, HOAc
5634	Hematin		$C_{34}H_{33}FeN_4O_5$	15489-90-4	633.495	br pow (py)	>200				i H_2O, eth; s EtOH, alk; sl py, HOAc
5635	Hematoporphyrin		$C_{34}H_{38}N_4O_6$	14459-29-1	598.689	deep red cry	172.5				i H_2O; s EtOH; sl eth, chl
5636	Hematoxylin		$C_{16}H_{14}O_6$	517-28-2	302.278	ye cry	140				sl H_2O, eth; s alk, EtOH
5637	Hemin		$C_{34}H_{32}ClFeN_4O_4$	16009-13-5	651.941	long blades (gl HOAc)	>300				
5638	Heneicosane		$C_{21}H_{44}$	629-94-7	296.574	cry (w)	40.01	356.5	0.7919[20]	1.4441[20]	i H_2O; sl EtOH; s peth
5639	Hentriacontane	Untriacontane	$C_{31}H_{64}$	630-04-6	436.840	lf (AcOEt)	67.9	458	0.781[68]	1.4278[90]	sl EtOH, eth, bz, chl; s peth
5640	Heptachlor		$C_{10}H_5Cl_7$	76-44-8	373.318	wh cry	95.5		1.57[9]		vs bz, eth, EtOH, lig
5641	Heptachlor epoxide		$C_{10}H_5Cl_7O$	1024-57-3	389.317		160				
5642	2,2',3,3',4,4',6-Heptachlorobiphenyl		$C_{12}H_3Cl_7$	52663-71-5	395.323	cry	117.5				i H_2O
5643	1,1,1,2,3,3,3-Heptachloropropane		C_3HCl_7	3849-33-0	285.211		11	249	1.7921[34]	1.5427[21]	vs chl
5644	Heptacontane		$C_{70}H_{142}$	7719-93-9	983.876		107	647			
5645	Heptacosane		$C_{27}H_{56}$	593-49-7	380.734	cry (al, bz) lf (AcOEt)	59.23	442	0.7796[60]	1.4345[65]	i H_2O, EtOH; sl eth
5646	Heptadecanal	Margaric aldehyde	$C_{17}H_{34}O$	629-90-3	254.451	nd (peth), cry (al)	36	204[26]			vs bz, eth
5647	1-Heptadecanamine		$C_{17}H_{37}N$	4200-95-7	255.483		49	336	0.8510[20]	1.4510[20]	i H_2O; s EtOH, eth
5648	Heptadecane		$C_{17}H_{36}$	629-78-7	240.468	hex lf	22.0	302.0	0.7780[20]	1.4369[20]	i H_2O; sl EtOH, ctc; s eth
5649	Heptadecanenitrile		$C_{17}H_{33}N$	5399-02-0	251.451	cry (al)	34	349	0.8315[20]	1.4467[20]	i H_2O; sl EtOH, chl; vs eth
5650	Heptadecanoic acid	Margaric acid	$C_{17}H_{34}O_2$	506-12-7	270.451	pl (peth)	61.3	227[100]	0.8532[60]	1.4342[60]	i H_2O; sl EtOH; s eth, ace, bz, chl
5651	1-Heptadecanol	Margaryl alcohol	$C_{17}H_{36}O$	1454-85-9	256.467	lf (al), cry (ace)	53.9	324	0.8475[20]		i H_2O; s EtOH, eth
5652	2-Heptadecanone	Pentadecyl methyl ketone	$C_{17}H_{34}O$	2922-51-2	254.451	pl (dil al)	48	320	0.8049[48]		i H_2O; sl EtOH; s ace, peth; vs bz, eth
5653	9-Heptadecanone		$C_{17}H_{34}O$	540-08-9	254.451	pl (MeOH)	53	251.5; 142[1.5]	0.8140[48]		sl EtOH; s MeOH
5654	1-Heptadecene	Hexahydroaplotaxene	$C_{17}H_{34}$	6765-39-5	238.452		11.5	300	0.7852[20]	1.4432[20]	i H_2O; vs eth; s bz; msc lig
5655	Heptadecylbenzene	1-Phenylheptadecane	$C_{23}H_{40}$	14752-75-1	316.564		32	397	0.8546[20]	1.4810[20]	
5656	trans,trans-2,4-Heptadienal		$C_7H_{10}O$	4313-03-5	110.153			84.5	0.881[25]	1.5315[20]	
5657	1,6-Heptadiene		C_7H_{12}	3070-53-9	96.170	liq		90			
5658	1,6-Heptadiyne		C_7H_8	2396-63-6	92.139	liq	-85	112	0.8164[17]	1.451[17]	i H_2O; s bz, HOAc
5659	Heptafluorobutanoic acid		$C_4HF_7O_2$	375-22-4	214.039	liq	-17.5	121	1.651[20]	1.295[25]	s H_2O, eth, tol; i peth

Guanosine 5'-monophosphate, disodium salt

Guinea Green B

D-Gulose

L-Gulose

Haloperidol

Harmaline

Harman

Harmine

HC Blue No. 1

HC Blue No. 2

$H_3C(CH_2)_{98}CH_3$ Hectane

Hederagenin

Helenalin

Helminthosporal

Helvolic acid

Hematein

Hematin

Hematoporphyrin

Hematoxylin

Hemin

Heneicosane

Hentriacontane

Heptachlor

Heptachlor epoxide

2,2',3,3',4,4',6-Heptachlorobiphenyl

1,1,1,2,3,3,3-Heptachloropropane

$H_3C(CH_2)_{68}CH_3$ Heptacontane

Heptacosane

Heptadecanal

1-Heptadecanamine

Heptadecane

Heptadecanenitrile

Heptadecanoic acid

1-Heptadecanol

2-Heptadecanone

9-Heptadecanone

1-Heptadecene

Heptadecylbenzene

trans,trans-2,4-Heptadienal

1,6-Heptadiene

1,6-Heptadiyne

Heptafluorobutanoic acid

No.	Name	Synonym	Mol. Form.	CAS RN	Mol. Wt.	Physical Form	mp/°C	bp/°C	den/g cm^{-3}	n_D	Solubility
5660	Heptafluorobutanoic anhydride		$C_8F_{14}O_3$	336-59-4	410.062	liq	-43	106.5	1.665[20]	1.285[20]	
5661	2,2,3,3,4,4,4-Heptafluoro-1-butanol		$C_4H_3F_7O$	375-01-9	200.055			95	1.600[20]	1.294[20]	s EtOH, ace
5662	Heptafluorobutanoyl chloride		C_4ClF_7O	375-16-6	232.484			38.5	1.55[20]	1.288[20]	
5663	6,6,7,7,8,8,8-Heptafluoro-2,2-dimethyl-3,5-octanedione		$C_{10}H_{11}F_7O_2$	17587-22-3	296.182		38	46[5]	1.273[25]	1.3766[20]	
5664	Heptafluoro-2-iodopropane	Perfluoroisopropyl iodide	C_3F_7I	677-69-0	295.925			38	1.3298[20]		
5665	1,1,1,2,3,3,3-Heptafluoropropane	Refrigerant 227ea	C_3HF_7	431-89-0	170.029	col gas	-131	-16.4			
5666	2,2,4,4,6,8,8-Heptamethylnonane		$C_{16}H_{34}$	4390-04-9	226.441			246.3			
5667	1,1,1,3,5,5,5-Heptamethyltrisiloxane		$C_7H_{22}O_2Si_3$	1873-88-7	222.506			142	0.8194[20]	1.3818[20]	
5668	Heptanal	Heptaldehyde	$C_7H_{14}O$	111-71-7	114.185	liq	-43.4	152.8	0.8132[25]	1.4113[20]	sl H$_2$O, ctc; msc EtOH, eth
5669	Heptanal oxime	Enanthaldoxime	$C_7H_{15}NO$	629-31-2	129.200	pl (al)	57.5	195	0.8583[55]	1.4210[20]	sl H$_2$O; s EtOH, eth
5670	2-Heptanamine	Tuaminoheptane	$C_7H_{17}N$	123-82-0	115.217			142	0.7665[19]	1.4199[19]	sl H$_2$O, chl; s EtOH, eth, peth
5671	4-Heptanamine		$C_7H_{17}N$	16751-59-0	115.217			139.5	0.767[20]	1.4172[20]	
5672	Heptane		C_7H_{16}	142-82-5	100.202	liq	-90.55	98.4	0.6795[25]	1.3855[25]	i H$_2$O; vs EtOH; msc eth, bz, chl; s ctc
5673	1,7-Heptanediamine		$C_7H_{18}N_2$	646-19-5	130.231		25.32	224			s EtOH, eth, ace
5674	Heptanedinitrile		$C_7H_{10}N_2$	646-20-8	122.167		-31.4	155[14]	0.949[18]	1.4472[20]	i H$_2$O; msc EtOH, eth, chl
5675	Heptanedioic acid	Pimelic acid	$C_7H_{12}O_4$	111-16-0	160.168	pr (w)	106	272[100], 212[10]	1.329[15]		s H$_2$O, EtOH, eth; i bz
5676	1,7-Heptanediol		$C_7H_{16}O_2$	629-30-1	132.201		22.5	262	0.9569[25]	1.4520[25]	vs eth, EtOH
5677	2,3-Heptanedione	Acetyl valeryl	$C_7H_{12}O_2$	96-04-8	128.169			144; 46[13]	0.919[18]	1.4150[18]	
5678	3,5-Heptanedione	Dipropionylmethane	$C_7H_{12}O_2$	7424-54-6	128.169			175; 79[30]	0.945[20]		
5679	Heptanedioyl dichloride		$C_7H_{10}Cl_2O_2$	142-79-0	197.059			137[15]			
5680	Heptanenitrile		$C_7H_{13}N$	629-08-3	111.185	liq	-64	183; 71[10]	0.8106[20]	1.4104[30]	i H$_2$O; s eth, ace, bz, HOAc
5681	1-Heptanethiol	Heptyl mercaptan	$C_7H_{16}S$	1639-09-4	132.267	liq	-43	176.9	0.8427[20]	1.4521[20]	i H$_2$O; msc EtOH, eth; s chl
5682	2,4,6-Heptanetrione		$C_7H_{10}O_3$	626-53-9	142.152	lf	49	121[10]	1.0599[40]	1.4930[20]	vs H$_2$O, eth, EtOH
5683	Heptanoic acid	Enanthic acid	$C_7H_{14}O_2$	111-14-8	130.185	liq	-7.17	222.2	0.9124[25]	1.4170[20]	sl H$_2$O, ctc; s EtOH, eth, ace
5684	Heptanoic anhydride		$C_{14}H_{26}O_3$	626-27-7	242.354	liq	-12.4	269.5	0.9321[20]	1.4335[15]	i H$_2$O; s EtOH, eth
5685	1-Heptanol	Heptyl alcohol	$C_7H_{16}O$	111-70-6	116.201	liq	-33.2	176.45	0.8219[20]	1.4249[20]	sl H$_2$O, ctc; msc EtOH, eth
5686	2-Heptanol, (±)		$C_7H_{16}O$	52390-72-4	116.201			159	0.8167[20]	1.4210[20]	sl H$_2$O, ctc; s EtOH, eth
5687	3-Heptanol, (S)		$C_7H_{16}O$	26549-25-7	116.201	liq	-70	157; 66[18]	0.8227[20]	1.4201[20]	sl H$_2$O, ctc; s EtOH, eth
5688	4-Heptanol	Dipropylcarbinol	$C_7H_{16}O$	589-55-9	116.201	liq	-41.2	156	0.8183[20]	1.4205[20]	sl H$_2$O; s EtOH, eth
5689	2-Heptanone	Methyl pentyl ketone	$C_7H_{14}O$	110-43-0	114.185	liq	-35	151.05	0.8111[20]	1.4088[20]	vs H$_2$O; s EtOH, eth
5690	3-Heptanone	Ethyl butyl ketone	$C_7H_{14}O$	106-35-4	114.185	liq	-39	147	0.8183[20]	1.4057[20]	sl H$_2$O, ctc; msc EtOH, eth
5691	4-Heptanone	Dipropyl ketone	$C_7H_{14}O$	123-19-3	114.185	liq	-33	144	0.8174[20]	1.4069[20]	i H$_2$O; msc EtOH, eth; s ctc
5692	Heptanoyl chloride		$C_7H_{13}ClO$	2528-61-2	148.630	liq	-83.8	125.2	0.9590[20]	1.4345[18]	s eth; sl ctc; vs lig
5693	2-Heptenal	Butylacrolein	$C_7H_{12}O$	2463-63-0	112.169			166	0.864[17]	1.4468[17]	
5694	1-Heptene		C_7H_{14}	592-76-7	98.186	liq	-118.9	93.64	0.6970[20]	1.3998[20]	i H$_2$O; s EtOH, eth; sl ctc
5695	cis-2-Heptene		C_7H_{14}	6443-92-1	98.186			98.4	0.708[20]	1.406[20]	i H$_2$O; s EtOH, eth, ace, bz, chl; sl ctc
5696	trans-2-Heptene		C_7H_{14}	14686-13-6	98.186	liq	-109.5	98	0.7012[20]	1.4045[20]	i H$_2$O; s EtOH, eth, ace, bz, peth, chl
5697	cis-3-Heptene		C_7H_{14}	7642-10-6	98.186	liq	-136.6	95.8	0.7030[20]	1.4059[20]	i H$_2$O; s EtOH, eth, ace, bz, peth, chl
5698	trans-3-Heptene		C_7H_{14}	14686-14-7	98.186	liq	-136.6	95.7	0.6981[20]	1.4043[20]	i H$_2$O; s EtOH, eth, ace, bz, chl; sl ctc
5699	6-Heptenoic acid		$C_7H_{12}O_2$	1119-60-4	128.169	liq	-6.5	226	0.9515[14]	1.4404[14]	
5700	1-Hepten-4-ol		$C_7H_{14}O$	3521-91-3	114.185			152.1	0.8384[22]	1.4347[20]	

Heptafluorobutanoic anhydride

2,2,3,3,4,4,4-Heptafluoro-1-butanol

Heptafluorobutanoyl chloride

6,6,7,7,8,8,8-Heptafluoro-2,2-dimethyl-3,5-octanedione

Heptafluoro-2-iodopropane

1,1,1,2,3,3,3-Heptafluoropropane

2,2,4,6,8,8-Heptamethylnonane

1,1,1,3,5,5,5-Heptamethyltrisiloxane

Heptanal

Heptanal oxime

2-Heptanamine

4-Heptanamine

Heptane

1,7-Heptanediamine

Heptanedinitrile

Heptanedioic acid

1,7-Heptanediol

2,3-Heptanedione

3,5-Heptanedione

Heptanedioyl dichloride

Heptanenitrile

1-Heptanethiol

2,4,6-Heptanetrione

Heptanoic acid

Heptanoic anhydride

1-Heptanol

2-Heptanol, (±)

3-Heptanol, (S)

4-Heptanol

2-Heptanone

3-Heptanone

4-Heptanone

Heptanoyl chloride

2-Heptenal

1-Heptene

cis-2-Heptene

trans-2-Heptene

cis-3-Heptene

trans-3-Heptene

6-Heptenoic acid

1-Hepten-4-ol

No.	Name	Synonym	Mol. Form.	CAS RN	Mol. Wt.	Physical Form	mp/°C	bp/°C	den/ g cm⁻³	n_D	Solubility
5701	*trans*-2-Hepten-1-ol		$C_7H_{14}O$	33467-76-4	114.185			178; 75[10]	0.8516[20]	1.4460[20]	s EtOH, ace
5702	Heptyl acetate		$C_9H_{18}O_2$	112-06-1	158.238	liq	-50.2	193	0.8750[15]	1.4150[20]	i H₂O; s EtOH, eth, ctc
5703	Heptylamine	1-Heptanamine	$C_7H_{17}N$	111-68-2	115.217	liq	-18	156	0.7754[20]	1.4251[20]	sl H₂O, chl; msc EtOH, eth
5704	Heptylbenzene		$C_{13}H_{20}$	1078-71-3	176.298	liq	-48	240; 109[10]	0.8567[20]	1.4865[20]	i H₂O; s bz, chl
5705	Heptyl butanoate		$C_{11}H_{22}O_2$	5870-93-9	186.292	liq	-57.5	225.8	0.8637[20]	1.4231[20]	vs EtOH
5706	Heptylcyclohexane		$C_{13}H_{26}$	5617-41-4	182.345	liq	-30	244	0.8109[20]	1.4484[20]	
5707	Heptylcyclopentane		$C_{12}H_{24}$	5617-42-5	168.319	liq	-53	224	0.8010[20]	1.4421[20]	vs ace, bz, eth, EtOH
5708	5-Heptyldihydro-2(3*H*)-furanone	4-Hydroxyundecanoic acid lactone	$C_{11}H_{20}O_2$	104-67-6	184.276			286	0.9494[20]	1.4512[20]	vs EtOH
5709	Heptyl formate		$C_8H_{16}O_2$	112-23-2	144.212			178.1	0.8784[20]	1.4140[20]	i H₂O; msc EtOH, eth
5710	Heptyl pentanoate		$C_{12}H_{24}O_2$	5451-80-9	200.318	liq	-46.4	245.2	0.8623[20]	1.4254[15]	vs ace, eth, EtOH
5711	6-Heptyltetrahydro-2*H*-pyran-2-one	5-Dodecanolide	$C_{12}H_{22}O_2$	713-95-1	198.302	liq	-12	101[0.03]			
5712	1-Heptyne		C_7H_{12}	628-71-7	96.170	liq	-81	99.7	0.7328[20]	1.4087[20]	sl H₂O; msc EtOH, eth; s bz, chl, peth
5713	2-Heptyne	1-Methyl-2-butylacetylene	C_7H_{12}	1119-65-9	96.170			112	0.744[25]	1.4230[20]	i H₂O; msc EtOH, eth; s bz, chl, peth
5714	3-Heptyne	1-Ethyl-2-propylacetylene	C_7H_{12}	2586-89-2	96.170	liq	-130.5	107.2	0.7336[25]	1.4189[20]	i H₂O; msc EtOH, eth; s bz, chl, peth
5715	Hesperetin		$C_{16}H_{14}O_6$	520-33-2	302.278	pl (dil al + 1/2 w)	227.5	sub 205			vs eth, EtOH
5716	Hesperidin		$C_{28}H_{34}O_{15}$	520-26-3	610.561	wh nd (dil MeOH, HOAc)	262				vs py, EtOH, HOAc
5717	Hexabromobenzene		C_6Br_6	87-82-1	551.488	mcl nd (bz)	327				i H₂O; sl EtOH, eth; s bz, chl
5718	2,2',4,4',5,5'-Hexabromobiphenyl		$C_{12}H_4Br_6$	59080-40-9	627.584	cry (ctc)	160				
5719	1,2,5,6,9,10-Hexabromocyclododecane		$C_{12}H_{18}Br_6$	3194-55-6	641.695	cry	167				
5720	Hexabromoethane		C_2Br_6	594-73-0	503.445	orth pr (bz)		dec 200	3.823[20]	1.863	sl EtOH, eth, CS₂
5721	Hexabutyldistannoxane	Bis(tributyltin) oxide	$C_{24}H_{54}OSn_2$	56-35-9	596.105		45	180[2]			
5722	Hexacene		$C_{26}H_{16}$	258-31-1	328.405	dk bl-grn cry (sub)	380	sub			i H₂O, EtOH
5723	Hexachloroacetone		C_3Cl_6O	116-16-5	264.749	liq	-1.0	203	1.7434[12]	1.5112[20]	sl H₂O; s ace
5724	Hexachlorobenzene	Perchlorobenzene	C_6Cl_6	118-74-1	284.782	nd (sub)	228.83	325	2.044[23]	1.5691[23]	i H₂O; sl EtOH; s eth, chl; vs bz
5725	2,2',3,3',4,4'-Hexachlorobiphenyl		$C_{12}H_4Cl_6$	38380-07-3	360.878	cry	151				i H₂O
5726	2,2',4,4',6,6'-Hexachlorobiphenyl		$C_{12}H_4Cl_6$	33979-03-2	360.878	cry	112.5				i H₂O
5727	2,2',3,3',6,6'-Hexachlorobiphenyl		$C_{12}H_4Cl_6$	38411-22-2	360.878	cry (hx)	114.2				i H₂O
5728	2,2',4,4',5,5'-Hexachlorobiphenyl		$C_{12}H_4Cl_6$	35065-27-1	360.878	cry	103.5				
5729	Hexachloro-1,3-butadiene		C_4Cl_6	87-68-3	260.761	liq	-21	215	1.556[25]	1.5542[20]	i H₂O; s EtOH, eth
5730	1,2,3,4,5,6-Hexachlorocyclohexane, (1α,2α,3β,4α,5α,6β)	Lindane	$C_6H_6Cl_6$	58-89-9	290.830	nd (al)	112.5	323.4			vs ace, bz
5731	1,2,3,4,5,6-Hexachlorocyclohexane, (1α,2α,3β,4α,5β,6β)	α-Hexachlorocyclohexane	$C_6H_6Cl_6$	319-84-6	290.830	cry	158				
5732	1,2,3,4,5,6-Hexachlorocyclohexane, (1α,2β,3α,4β,5α,6β)	β-Hexachlorocyclohexane	$C_6H_6Cl_6$	319-85-7	290.830	cry (bz, al, xyl)		60[0.50]	1.89[19]		i H₂O; sl EtOH, bz, chl, HOAc
5733	1,2,3,4,5,6-Hexachlorocyclohexane, (1α,2α,3α,4β,5α,6β)	δ-Lindane	$C_6H_6Cl_6$	319-86-8	290.830	pl	141.5	60[0.36]			
5734	Hexachloro-1,3-cyclopentadiene	Perchlorocyclopentadiene	C_5Cl_6	77-47-4	272.772	ye grn liq	-9	239; 48[0.3]	1.7019[25]	1.5658[20]	
5735	1,2,3,6,7,8-Hexachlorodibenzo-*p*-dioxin		$C_{12}H_2Cl_6O_2$	57653-85-7	390.861	cry	285				
5736	1,2,3,7,8,9-Hexachlorodibenzo-*p*-dioxin		$C_{12}H_2Cl_6O_2$	19408-74-3	390.861	cry	243				
5737	Hexachloroethane	Perchloroethane	C_2Cl_6	67-72-1	236.739	orth (al-eth)	186.8 tp	184.7 sp	2.091[20]		i H₂O; vs EtOH, eth; s bz; sl liq HF
5738	Hexachlorophene		$C_{13}H_6Cl_6O_2$	70-30-4	406.904	nd (bz)	166.5				i H₂O; s EtOH, eth, ace, chl, dil alk

trans-2-Hepten-1-ol

Heptyl acetate

Heptylamine

Heptylbenzene

Heptyl butanoate

Heptylcyclohexane

Heptylcyclopentane

5-Heptyldihydro-2(3H)-furanone

Heptyl formate

Heptyl pentanoate

6-Heptyltetrahydro-2H-pyran-2-one

1-Heptyne

2-Heptyne

3-Heptyne

Hesperetin

Hesperidin

Hexabromobenzene

2,2',4,4',5,5'-Hexabromobiphenyl

1,2,5,6,9,10-Hexabromocyclododecane

Hexabromoethane

Hexabutyldistannoxane

Hexacene

Hexachloroacetone

Hexachlorobenzene

2,2',3,3',4,4'-Hexachlorobiphenyl

2,2',4,4',6,6'-Hexachlorobiphenyl

2,2',3,3',6,6'-Hexachlorobiphenyl

2,2',4,4',5,5'-Hexachlorobiphenyl

Hexachloro-1,3-butadiene

1,2,3,4,5,6-Hexachlorocyclohexane, (1α,2α,3β,4α,5α,6β)

1,2,3,4,5,6-Hexachlorocyclohexane, (1α,2α,3β,4α,5β,6β)

1,2,3,4,5,6-Hexachlorocyclohexane, (1α,2β,3α,4β,5α,6β)

1,2,3,4,5,6-Hexachlorocyclohexane, (1α,2α,3α,4β,5α,6β)

Hexachloro-1,3-cyclopentadiene

1,2,3,6,7,8-Hexachlorodibenzo-p-dioxin

1,2,3,7,8,9-Hexachlorodibenzo-p-dioxin

Hexachloroethane

Hexachlorophene

No.	Name	Synonym	Mol. Form.	CAS RN	Mol. Wt.	Physical Form	mp/°C	bp/°C	den/ g cm⁻³	n_D	Solubility
5739	Hexachloropropene		C_3Cl_6	1888-71-7	248.750	liq	-72.9	209.5	1.7632[20]	1.5091[20]	i H_2O; s ctc, chl
5740	Hexacontane		$C_{60}H_{122}$	7667-80-3	843.611		99.3				
5741	Hexacosane		$C_{26}H_{54}$	630-01-3	366.707	mcl, tcl or orth (bz) cry (eth)	56.1	412.2	0.7783[60]	1.4357[60]	vs bz, lig, chl
5742	Hexacosanoic acid	Cerotic acid	$C_{26}H_{52}O_2$	506-46-7	396.690		88.5		0.8198[100]	1.4301[100]	i H_2O; vs EtOH, eth
5743	1-Hexacosanol		$C_{26}H_{54}O$	506-52-5	382.706	orth pl (dil al)	80	305[20] dec			i H_2O; s EtOH, eth
5744	Hexadecamethylheptasiloxane		$C_{16}H_{48}O_6Si_7$	541-01-5	533.147	liq	-78	270	0.9012[20]	1.3965[20]	vs bz, lig
5745	Hexadecanal		$C_{16}H_{32}O$	629-80-1	240.424	pl (eth), nd (peth)	35	200[29]			i H_2O; s EtOH, eth, ace, bz
5746	Hexadecanamide		$C_{16}H_{33}NO$	629-54-9	255.439	lf	107	236[12]	1.0000[20]		i H_2O; sl EtOH, bz, ace, eth
5747	Hexadecane	Cetane	$C_{16}H_{34}$	544-76-3	226.441	lf (HOAc)	18.12	286.86	0.7701[25]	1.4329[25]	i H_2O; sl EtOH; msc eth; s ctc
5748	Hexadecanedioic acid		$C_{16}H_{30}O_4$	505-54-4	286.407	pl (al)	126.6				vs ace, EtOH
5749	Hexadecanenitrile		$C_{16}H_{31}N$	629-79-8	237.424	hex	31	333	0.8303[20]	1.4450[20]	i H_2O; vs EtOH, eth, ace, bz, chl
5750	1-Hexadecanethiol	Cetyl mercaptan	$C_{16}H_{34}S$	2917-26-2	258.506	cry (lig)	19	125[0.5]			i H_2O; sl EtOH, ctc; s eth
5751	Hexadecanoic acid	Palmitic acid	$C_{16}H_{32}O_2$	57-10-3	256.424	nd (al)	62.5	351.5	0.8527[62]	1.43345[60]	i H_2O; s EtOH, ace, bz; msc eth; vs chl
5752	Hexadecanoic anhydride		$C_{32}H_{62}O_3$	623-65-4	494.832	lf (peth)	64		0.8388[83]	1.4364[68]	vs eth
5753	1-Hexadecanol	Cetyl alcohol	$C_{16}H_{34}O$	36653-82-4	242.440	fl (AcOEt)	49.2	312	0.8187[50]	1.4283[79]	i H_2O; sl EtOH; vs eth, bz, chl; s ace
5754	3-Hexadecanone		$C_{16}H_{32}O$	18787-64-9	240.424	lf (peth)	43	184[17], 140[2]			s chl
5755	Hexadecanoyl chloride		$C_{16}H_{31}ClO$	112-67-4	274.869		12	199[20]	0.9016[25]	1.4514[20]	vs eth
5756	1-Hexadecene	1-Cetene	$C_{16}H_{32}$	629-73-2	224.425	lf	2.1	284.9	0.7811[20]	1.4412[20]	i H_2O; s EtOH, eth, ctc, peth
5757	cis-9-Hexadecenoic acid	Palmitoleic acid	$C_{16}H_{30}O_2$	373-49-9	254.408		0.5	182[1]			
5758	Hexadecyl acetate		$C_{18}H_{36}O_2$	629-70-9	284.478		-18.5	222[205]	0.8574[25]	1.4438[20]	i H_2O; sl EtOH; s ctc
5759	Hexadecylamine	1-Hexadecanamine	$C_{16}H_{35}N$	143-27-1	241.456	lf	46.8	322.5	0.8129[20]	1.4496[20]	i H_2O; vs EtOH, eth, bz; s ace
5760	Hexadecylbenzene		$C_{22}H_{38}$	1459-09-2	302.537		27	385	0.8547[20]	1.4813[20]	i H_2O; sl EtOH; vs eth, bz, CS_2
5761	Hexadecyldimethylamine	N,N-Dimethyl-1-hexadecanamine	$C_{18}H_{39}N$	112-69-6	269.510			330.0			
5762	Hexadecyl hexadecanoate	Cetyl palmitate	$C_{32}H_{64}O_2$	540-10-3	480.849	mcl lf	54		0.989[20]	1.4398[70]	vs eth, EtOH
5763	Hexadecyl 3-hydroxy-2-naphthalenecarboxylate	Hexadecyl 3-hydroxy-2-naphthoate	$C_{27}H_{40}O_3$	531-84-0	412.605	grn-wh fl	72.5				vs bz, HOAc
5764	Hexadecyl 2-hydroxypropanoate	Cetyl lactate	$C_{19}H_{38}O_3$	35274-05-6	314.503	wax	41	219[10], 170[1]		1.4410[40]	
5765	Hexadecyl 2-methyl-2-propenoate		$C_{20}H_{38}O_2$	2495-27-4	310.515		24	183[2]	0.87[20]		
5766	3-(Hexadecyloxy)-1,2-propanediol, (S)	Chimyl alcohol	$C_{19}H_{40}O_3$	506-03-6	316.519	lf (hx)	64	120[0.005]			vs ace, peth, chl
5767	1-Hexadecylpyridinium bromide		$C_{21}H_{38}BrN$	140-72-7	384.438		61				
5768	1-Hexadecylpyridinium chloride	Cetylpyridinium chloride	$C_{21}H_{38}ClN$	123-03-5	339.987	wh pow	80				vs H_2O, chl
5769	Hexadecyl stearate	Cetyl stearate	$C_{34}H_{68}O_2$	1190-63-2	508.903	lf or pl (eth, HOAc)	57			1.4410[70]	vs ace, eth, chl
5770	Hexadecyltrichlorosilane		$C_{16}H_{33}Cl_3Si$	5894-60-0	359.878			269			
5771	Hexadecyl vinyl ether	1-(Ethenyloxy)hexadecane	$C_{18}H_{36}O$	822-28-6	268.478		16	160[2]	0.821[27]	1.4444[25]	
5772	1-Hexadecyne		$C_{16}H_{30}$	629-74-3	222.409		15	284	0.7965[20]	1.4440[20]	vs bz
5773	trans,trans-2,4-Hexadienal	Sorbinaldehyde	C_6H_8O	142-83-6	96.127	liq	-16.5	174; 76[30]	0.898[20]	1.5384[20]	
5774	1,2-Hexadiene	Propylallene	C_6H_{10}	592-44-9	82.143			76	0.7149[20]	1.4282[20]	vs eth, chl
5775	cis-1,3-Hexadiene		C_6H_{10}	14596-92-0	82.143			73.1	0.7033[25]	1.4379[20]	
5776	trans-1,3-Hexadiene		C_6H_{10}	20237-34-7	82.143	liq	-102.4	73.2	0.6995[25]	1.4406[20]	
5777	cis-1,4-Hexadiene		C_6H_{10}	7318-67-4	82.143			66.3	0.695[25]	1.4049[20]	vs eth
5778	trans-1,4-Hexadiene		C_6H_{10}	7319-00-8	82.143	liq	-138.7	65.0	0.695[25]	1.4104[20]	
5779	1,5-Hexadiene	Biallyl	C_6H_{10}	592-42-7	82.143	liq	-140.7	59.4	0.6878[25]	1.4042[20]	i H_2O; s EtOH, eth, bz, chl; sl ctc
5780	cis,cis-2,4-Hexadiene		C_6H_{10}	6108-61-8	82.143	liq		85	0.7298[25]	1.4606[20]	i H_2O; s EtOH, eth, chl
5781	trans,cis-2,4-Hexadiene		C_6H_{10}	5194-50-3	82.143	liq	-96.1	83.5	0.7185[25]	1.4560[20]	i H_2O; s EtOH, eth, chl
5782	trans,trans-2,4-Hexadiene		C_6H_{10}	5194-51-4	82.143	liq	-44.9	82.2	0.7101[25]	1.4510[20]	i H_2O; s EtOH, eth, chl
5783	2,4-Hexadienoic acid	Sorbic acid	$C_6H_8O_2$	110-44-1	112.127	nd (dil al) nd (w)	134.5	dec 228; 153[50]	1.204[19]		s H_2O, EtOH, chl; vs eth

Hexachloropropene

Hexacontane

Hexacosane

Hexacosanoic acid

1-Hexacosanol

Hexadecamethylheptasiloxane

Hexadecanal

Hexadecanamide

Hexadecane

Hexadecanedioic acid

Hexadecanenitrile

1-Hexadecanethiol

Hexadecanoic acid

Hexadecanoic anhydride

1-Hexadecanol

3-Hexadecanone

Hexadecanoyl chloride

1-Hexadecene

cis-9-Hexadecenoic acid

Hexadecyl acetate

Hexadecylamine

Hexadecylbenzene

Hexadecyldimethylamine

Hexadecyl hexadecanoate

Hexadecyl 3-hydroxy-2-naphthalenecarboxylate

Hexadecyl 2-hydroxypropanoate

Hexadecyl 2-methyl-2-propenoate

3-(Hexadecyloxy)-1,2-propanediol, (S)

1-Hexadecylpyridinium bromide

1-Hexadecylpyridinium chloride

Hexadecyl stearate

Hexadecyltrichlorosilane

Hexadecyl vinyl ether

1-Hexadecyne

trans,trans-2,4-Hexadienal

1,2-Hexadiene

cis-1,3-Hexadiene

trans-1,3-Hexadiene

cis-1,4-Hexadiene

trans-1,4-Hexadiene

1,5-Hexadiene

cis,cis-2,4-Hexadiene

trans,cis-2,4-Hexadiene

trans,trans-2,4-Hexadiene

2,4-Hexadienoic acid

No.	Name	Synonym	Mol. Form.	CAS RN	Mol. Wt.	Physical Form	mp/°C	bp/°C	den/ g cm^{-3}	n_D	Solubility
5784	2,4-Hexadien-1-ol	Sorbic alcohol	C$_6$H$_{10}$O	111-28-4	98.142	nd	30.5	76[12]	0.8967[23]	1.4981[20]	i H$_2$O; s EtOH, eth
5785	*trans,trans*-2,4-Hexadienoyl chloride		C$_6$H$_7$ClO	2614-88-2	130.572			82[22]	1.0666[19]	1.5545[20]	vs ace
5786	1,5-Hexadien-3-yne	Divinylacetylene	C$_6$H$_6$	821-08-9	78.112	liq	-88	85	0.7851[20]	1.5035[20]	i H$_2$O; s bz
5787	1,5-Hexadiyne	Bipropargyl	C$_6$H$_6$	628-16-0	78.112	liq	-6	86	0.8049[20]	1.4380[23]	i H$_2$O; s EtOH, eth, ace, bz
5788	2,4-Hexadiyne	Dimethyldiacetylene	C$_6$H$_6$	2809-69-0	78.112	pr (sub)	67.8	129.5			vs EtOH, eth
5789	Hexaethylbenzene		C$_{18}$H$_{30}$	604-88-6	246.431	mcl pr (al or bz)	129	298	0.8305[130]	1.4736[130]	i H$_2$O; s EtOH, sulf; vs eth, bz
5790	Hexaethyldisiloxane		C$_{12}$H$_{30}$OSi$_2$	994-49-0	246.536			233; 129[30]	0.8457[20]	1.4340[20]	
5791	Hexaethyl tetraphosphate	Ethyl tetraphosphate	C$_{12}$H$_{30}$O$_{13}$P$_4$	757-58-4	506.253	hyg	-40	dec 150	1.2917[27]	1.4273[27]	vs ace, bz, EtOH
5792	Hexafluorenium bromide		C$_{36}$H$_{42}$Br$_2$N$_2$	317-52-2	662.539	cry (PrOH)	188				
5793	Hexafluoroacetylacetone		C$_5$H$_2$F$_6$O$_2$	1522-22-1	208.059			54.15	1.485[20]	1.3333[20]	
5794	Hexafluorobenzene	Perfluorobenzene	C$_6$F$_6$	392-56-3	186.054		5.03	80.32	1.6175[20]	1.3777[20]	
5795	1,1,2,3,4,4-Hexafluoro-1,3-butadiene		C$_4$F$_6$	685-63-2	162.033	col gas	-132	6	1.553[-20]	1.378[-20]	
5796	1,1,1,4,4,4-Hexafluoro-2-butyne		C$_4$F$_6$	692-50-2	162.033	col gas	-117.4	-24.6			s EtOH, eth, ace, ctc, HOAc
5797	Hexafluorocyclobutene		C$_4$F$_6$	697-11-0	162.033	col gas	-60	5.5	1.602[-20]	1.298[-20]	
5798	Hexafluoroethane	Perfluoroethane	C$_2$F$_6$	76-16-4	138.011	col gas	-100.05	-78.1	1.590[-78]		i H$_2$O; sl EtOH, eth
5799	1,1,1,2,3,3-Hexafluoropropane	Refrigerant 236ea	C$_3$H$_2$F$_6$	431-63-0	152.038	col gas		6.1	1.5026[0]		
5800	1,1,1,3,3,3-Hexafluoropropane	Refrigerant 236fa	C$_3$H$_2$F$_6$	690-39-1	152.038	col gas	-93.6	-1.0	1.4343[0]		
5801	1,1,1,3,3,3-Hexafluoro-2-propanol		C$_3$H$_2$F$_6$O	920-66-1	168.037	liq	-2.0	59	1.4600[21]		
5802	Hexahydro-1*H*-azepine	Hexamethylenimine	C$_6$H$_{13}$N	111-49-9	99.174			138	0.8643[22]	1.4631[20]	s H$_2$O; vs EtOH, eth
5803	Hexahydro-1*H*-1,4-diazepine		C$_5$H$_{12}$N$_2$	505-66-8	100.162	hyg	40.5	169			
5804	1,5a,6,9,9a,9b-Hexahydro-4a(4*H*)-dibenzofurancarboxaldehyde		C$_{13}$H$_{16}$O$_2$	126-15-8	204.265	liq	-80	307	1.10[20]	1.5254[20]	i H$_2$O
5805	*cis*-1,2,3,5,6,8a-Hexahydro-4,7-dimethyl-1-isopropylnaphthalene, (1*S*)		C$_{15}$H$_{24}$	483-76-1	204.352			125[12]	0.9160[15]	1.5089[15]	
5806	1,2,4a,5,8,8a-Hexahydro-4,7-dimethyl-1-isopropylnaphthalene, [1*S*-(1α,4aβ,8aα)]		C$_{15}$H$_{24}$	523-47-7	204.352			274; 136[11]	0.9230[20]	1.5059[20]	vs eth, lig
5807	Hexahydro-1,3-isobenzofurandione	Hexahydrophthalic anhydride	C$_8$H$_{10}$O$_3$	85-42-7	154.163		32	145[18]			
5808	Hexahydro-1-methyl-1*H*-1,4-diazepine		C$_6$H$_{14}$N$_2$	4318-37-0	114.188			154	0.9111[20]	1.4769[20]	
5809	2,3,4,6,7,8-Hexahydropyrrolo[1,2-a]pyrimidine		C$_7$H$_{12}$N$_2$	3001-72-7	124.183			96[7], 81[2]	1.005[25]	1.5190[20]	
5810	Hexahydro-1,3,5-trinitro-1,3,5-triazine	Cyclonite	C$_3$H$_6$N$_6$O$_6$	121-82-4	222.116	orth cry (ace)	205.5		1.82[20]		i H$_2$O, EtOH, bz; sl eth, MeOH; s ace, HOAc
5811	Hexahydro-1,3,5-triphenyl-1,3,5-triazine		C$_{21}$H$_{21}$N$_3$	91-78-1	315.412		144	185; 60[29]			i H$_2$O; sl EtOH; s eth, ace, bz, tol
5812	1,2,3,5,6,7-Hexahydroxy-9,10-anthracenedione	Rufigallol	C$_{14}$H$_8$O$_8$	82-12-2	304.209	red rhom, red-ye nd (sub)		sub			i H$_2$O; sl EtOH, eth; s ace, alk
5813	Hexamethylbenzene	Mellitene	C$_{12}$H$_{18}$	87-85-4	162.271	orth pr or nd (al)	165.5	263.4	1.0630[25]		i H$_2$O; s EtOH, eth, ace, bz, HOAc, chl
5814	2,2,4,4,6,6-Hexamethylcyclotrisilazane		C$_6$H$_{21}$N$_3$Si$_3$	1009-93-4	219.508	liq	-10	188	0.9196[20]	1.448[20]	
5815	Hexamethylcyclotrisiloxane	Dimethylsiloxane cyclic trimer	C$_6$H$_{18}$O$_3$Si$_3$	541-05-9	222.462		64.5	134	1.1200[20]		i H$_2$O
5816	Hexamethyldisilane		C$_6$H$_{18}$Si$_2$	1450-14-2	146.378		13.5	113.5	0.7247[22]	1.4229[20]	i H$_2$O; s eth, ace, bz; dec alk
5817	Hexamethyldisilathiane		C$_6$H$_{18}$SSi$_2$	3385-94-2	178.443			162.5	0.851[20]		
5818	Hexamethyldisilazane		C$_6$H$_{19}$NSi$_2$	999-97-3	161.393			125	0.7741[25]	1.4090[20]	
5819	Hexamethyldisiloxane		C$_6$H$_{18}$OSi$_2$	107-46-0	162.377	liq	-66	99	0.7638[20]	1.3774[20]	i H$_2$O
5820	Hexamethylenediamine carbamate	(6-Aminohexyl)carbamic acid	C$_7$H$_{16}$N$_2$O$_2$	143-06-6	160.214	cry	150				
5821	Hexamethylene diisocyanate		C$_8$H$_{12}$N$_2$O$_2$	822-06-0	168.193			122[10], 94[1]	1.0528[20]	1.4585[20]	
5822	Hexamethylenetetramine	Methenamine	C$_6$H$_{12}$N$_4$	100-97-0	140.186	orth (al)	>250	sub	1.331[-5]		vs H$_2$O; s EtOH, ace, chl; sl eth, bz
5823	Hexamethylolmelamine		C$_9$H$_{18}$N$_6$O$_6$	531-18-0	306.275		137				vs H$_2$O
5824	Hexamethylphosphoric triamide	Tris(dimethylamino)phosphine oxide	C$_6$H$_{18}$N$_3$OP	680-31-9	179.200			232.5	1.03[20]	1.4579[20]	s EtOH, eth
5825	Hexamethylphosphorous triamide	Tris(dimethylamino)phosphine	C$_6$H$_{18}$N$_3$P	1608-26-0	163.201						s chl

2,4-Hexadien-1-ol

trans,trans-2,4-Hexadienoyl chloride

1,5-Hexadien-3-yne

1,5-Hexadiyne

2,4-Hexadiyne

Hexaethylbenzene

Hexaethyldisiloxane

Hexaethyl tetraphosphate

Hexafluorenium bromide

Hexafluoroacetylacetone

Hexafluorobenzene

1,1,2,3,4,4-Hexafluoro-1,3-butadiene

1,1,1,4,4,4-Hexafluoro-2-butyne

Hexafluorocyclobutene

Hexafluoroethane

1,1,1,2,3,3-Hexafluoropropane

1,1,1,3,3,3-Hexafluoropropane

1,1,1,3,3,3-Hexafluoro-2-propanol

Hexahydro-1*H*-azepine

Hexahydro-1*H*-1,4-diazepine

1,5a,6,9,9a,9b-Hexahydro-4a(4*H*)-dibenzofurancarboxaldehyde

cis-1,2,3,5,6,8a-Hexahydro-4,7-dimethyl-1-isopropylnaphthalene, (1*S*)

1,2,4a,5,8,8a-Hexahydro-4,7-dimethyl-1-isopropylnaphthalene, [1*S*-(1α,4aβ,8aα)]

Hexahydro-1,3-isobenzofurandione

Hexahydro-1-methyl-1*H*-1,4-diazepine

2,3,4,6,7,8-Hexahydropyrrolo[1,2-a]pyrimidine

Hexahydro-1,3,5-trinitro-1,3,5-triazine

Hexahydro-1,3,5-triphenyl-1,3,5-triazine

1,2,3,5,6,7-Hexahydroxy-9,10-anthracenedione

Hexamethylbenzene

2,2,4,4,6,6-Hexamethylcyclotrisilazane

Hexamethylcyclotrisiloxane

Hexamethyldisilane

Hexamethyldisilathiane

Hexamethyldisilazane

Hexamethyldisiloxane

Hexamethylenediamine carbamate

Hexamethylene diisocyanate

Hexamethylenetetramine

Hexamethylolmelamine

Hexamethylphosphoric triamide

Hexamethylphosphorous triamide

No.	Name	Synonym	Mol. Form.	CAS RN	Mol. Wt.	Physical Form	mp/°C	bp/°C	den/ g cm⁻³	n_D	Solubility
5826	2,6,10,15,19,23-Hexamethyltetracosane	Squalane	$C_{30}H_{62}$	111-01-3	422.813	liq	-38	350	0.8115^{15}	1.4530^{15}	i H_2O; sl EtOH, ace; s eth, chl; msc bz
5827	Hexanal	Caproaldehyde	$C_6H_{12}O$	66-25-1	100.158	liq	-56	131	0.8335^{20}	1.4039^{20}	sl H_2O; vs EtOH, eth; s ace, bz
5828	Hexanamide		$C_6H_{13}NO$	628-02-4	115.173	cry (ace)	101	255	0.999^{20}	1.4200^{110}	vs bz, eth, EtOH, chl
5829	Hexane		C_6H_{14}	110-54-3	86.175	liq	-95.35	68.73	0.6606^{25}	1.3727^{25}	i H_2O; vs EtOH; s eth, chl
5830	Hexanedial		$C_6H_{10}O_2$	1072-21-5	114.142		-8	93^9	1.003^{19}	1.4350^{20}	vs bz, eth, EtOH
5831	Hexanediamide		$C_6H_{12}N_2O_2$	628-94-4	144.171	pl	220				vs EtOH
5832	1,6-Hexanediamine	Hexamethylenediamine	$C_6H_{16}N_2$	124-09-4	116.204	orth bipym pl	39.13	205			vs H_2O; s EtOH, bz
5833	Hexanedioic acid, dihydrazide		$C_6H_{14}N_4O_2$	1071-93-8	174.201		181.8				
5834	1,2-Hexanediol		$C_6H_{14}O_2$	6920-22-5	118.174		45	224; $87^{1.5}$		1.4431^{20}	
5835	1,6-Hexanediol	Hexamethylene glycol	$C_6H_{14}O_2$	629-11-8	118.174		41.5	208		1.4579^{25}	s H_2O, EtOH, ace; sl eth; i bz
5836	2,5-Hexanediol	Diisopropanol	$C_6H_{14}O_2$	2935-44-6	118.174	cry (eth)	43	218; 86^1	0.9610^{20}	1.4475^{20}	s H_2O, EtOH, eth; sl ctc
5837	1,6-Hexanediol dimethacrylate	Hexamethylene methacrylate	$C_{14}H_{22}O_4$	6606-59-3	254.323				0.998^{25}		
5838	2,3-Hexanedione	Acetylbutyryl	$C_6H_{10}O_2$	3848-24-6	114.142			128	0.934^{19}		
5839	2,4-Hexanedione	Propionylacetone	$C_6H_{10}O_2$	3002-24-2	114.142	oil		160	0.959^{20}	1.4516^{20}	
5840	2,5-Hexanedione	Acetonylacetone	$C_6H_{10}O_2$	110-13-4	114.142	liq	-5.5	194	0.7370^{20}	1.4232^{20}	vs H_2O, bz, eth, EtOH
5841	3,4-Hexanedione	Bipropionyl	$C_6H_{10}O_2$	4437-51-8	114.142	liq	-10	130	0.941^{21}	1.4130^{21}	
5842	Hexanedioyl dichloride		$C_6H_8Cl_2O_2$	111-50-2	183.033			126^{12}			sl chl
5843	1,6-Hexanedithiol		$C_6H_{14}S_2$	1191-43-1	150.305	liq	-21	237; 118^{15}	0.9886^{25}	1.5110^{20}	
5844	Hexanenitrile	Capronitrile	$C_6H_{11}N$	628-73-9	97.158	liq	-80.3	163.65	0.8051^{20}	1.4068^{20}	i H_2O; s EtOH, eth; sl chl
5845	1-Hexanethiol	Hexyl mercaptan	$C_6H_{14}S$	111-31-9	118.240	liq	-81	152.7	0.8424^{20}	1.4496^{20}	i H_2O; vs EtOH, eth
5846	2-Hexanethiol		$C_6H_{14}S$	1679-06-7	118.240	liq	-147	142	0.8345^{20}	1.4451^{20}	i H_2O; s EtOH, eth, bz
5847	1,2,6-Hexanetriol	1,2,6-Trihydroxyhexane	$C_6H_{14}O_3$	106-69-4	134.173			170^3; 161^1	1.1049^{20}	1.58^{20}	
5848	Hexanoic acid	Caproic acid	$C_6H_{12}O_2$	142-62-1	116.158	liq	-3	205.2	0.9212^{25}	1.4163^{20}	sl H_2O; s EtOH, eth, chl
5849	Hexanoic anhydride		$C_{12}H_{22}O_3$	2051-49-2	214.301		-41	dec 255	0.9240^{15}	1.4297^{20}	vs eth, EtOH
5850	1-Hexanol	Caproyl alcohol	$C_6H_{14}O$	111-27-3	102.174	liq	-47.4	157.6	0.8136^{20}	1.4178^{20}	sl H_2O; s EtOH, ace, chl; msc eth, bz
5851	2-Hexanol		$C_6H_{14}O$	20281-86-1	102.174			140	0.8159^{20}	1.4144^{20}	sl H_2O, ctc; s EtOH, eth
5852	3-Hexanol		$C_6H_{14}O$	17015-11-1	102.174			135	0.8182^{20}	1.4167^{20}	sl H_2O; s EtOH, ace; msc eth
5853	2-Hexanone	Butyl methyl ketone	$C_6H_{12}O$	591-78-6	100.158	liq	-55.5	127.6	0.8113^{20}	1.4007^{20}	sl H_2O; s ace; msc EtOH, eth
5854	3-Hexanone	Ethyl propyl ketone	$C_6H_{12}O$	589-38-8	100.158	liq	-55.4	123.5	0.8118^{20}	1.4004^{20}	sl H_2O; s ace; msc EtOH, eth
5855	Hexanoyl chloride	Caproyl chloride	$C_6H_{11}ClO$	142-61-0	134.603	liq	-87	153	0.9784^{20}	1.4264^{20}	s eth, ace
5856	Hexatriacontane		$C_{36}H_{74}$	630-06-8	506.973		75.8	298.4^3	0.7803^{80}	1.4397^{80}	
5857	cis-1,3,5-Hexatriene		C_6H_8	2612-46-6	80.128	liq	-12	78	0.7175^{20}	1.4577^{20}	i H_2O; s EtOH, ace, chl, peth
5858	trans-1,3,5-Hexatriene		C_6H_8	821-07-8	80.128	liq	-12	78.5	0.7369^{15}	1.5135^{20}	i H_2O; s EtOH, ace, chl, peth
5859	Hexazinone		$C_{12}H_{20}N_4O_2$	51235-04-2	252.313		99	dec	1.25		
5860	trans-2-Hexenal		$C_6H_{10}O$	6728-26-3	98.142			146.5; 50^{20}	0.8491^{20}	1.4480^{20}	
5861	cis-3-Hexenal		$C_6H_{10}O$	6789-80-6	98.142			121	0.8533^{22}	1.4300^{21}	
5862	1-Hexene		C_6H_{12}	592-41-6	84.159	liq	-139.76	63.48	0.6685^{25}	1.3852^{25}	i H_2O; vs bz, eth, EtOH, peth
5863	cis-2-Hexene		C_6H_{12}	7688-21-3	84.159	liq	-141.11	68.8	0.6824^{25}	1.3979^{20}	i H_2O; s EtOH, eth, bz, chl, lig
5864	trans-2-Hexene		C_6H_{12}	4050-45-7	84.159	liq	-133	67.9	0.6733^{25}	1.3936^{20}	i H_2O; s EtOH, eth, bz, chl, lig
5865	cis-3-Hexene		C_6H_{12}	7642-09-3	84.159	liq	-137.8	66.4	0.6778^{20}	1.3947^{20}	i H_2O; s EtOH, eth, bz, chl, lig
5866	trans-3-Hexene		C_6H_{12}	13269-52-8	84.159	liq	-115.4	67.1	0.6772^{20}	1.3943^{20}	i H_2O; s EtOH, eth, bz, chl, lig
5867	trans-3-Hexenedinitrile	trans-1,4-Dicyano-2-butene	$C_6H_6N_2$	1119-85-3	106.125	cry	76				
5868	2-Hexenoic acid		$C_6H_{10}O_2$	1191-04-4	114.142	nd (w, al)	36.5	216.5	0.965^{20}	1.4460^{40}	vs eth
5869	3-Hexenoic acid	Hydrosorbic acid	$C_6H_{10}O_2$	4219-24-3	114.142		12	208	0.9640^{23}	1.4935^{20}	
5870	5-Hexenoic acid		$C_6H_{10}O_2$	1577-22-6	114.142	liq	-37	203	0.9610^{20}	1.4343^{20}	vs eth, EtOH
5871	1-Hexen-3-ol		$C_6H_{12}O$	4798-44-1	100.158			134	0.834^{22}	1.4297^{18}	sl H_2O; vs ace, eth, EtOH

2,6,10,15,19,23-Hexamethyltetracosane

Hexanal

Hexanamide

Hexane

Hexanedial

Hexanediamide

1,6-Hexanediamine

Hexanedioic acid, dihydrazide

1,2-Hexanediol

1,6-Hexanediol

2,5-Hexanediol

1,6-Hexanediol dimethacrylate

2,3-Hexanedione

2,4-Hexanedione

2,5-Hexanedione

3,4-Hexanedione

Hexanedioyl dichloride

1,6-Hexanedithiol

Hexanenitrile

1-Hexanethiol

2-Hexanethiol

1,2,6-Hexanetriol

Hexanoic acid

Hexanoic anhydride

1-Hexanol

2-Hexanol

3-Hexanol

2-Hexanone

3-Hexanone

Hexanoyl chloride

Hexatriacontane

cis-1,3,5-Hexatriene

trans-1,3,5-Hexatriene

Hexazinone

trans-2-Hexenal

cis-3-Hexenal

1-Hexene

cis-2-Hexene

trans-2-Hexene

cis-3-Hexene

trans-3-Hexene

trans-3-Hexenedinitrile

2-Hexenoic acid

3-Hexenoic acid

5-Hexenoic acid

1-Hexen-3-ol

No.	Name	Synonym	Mol. Form.	CAS RN	Mol. Wt.	Physical Form	mp/°C	bp/°C	den/ g cm^{-3}	n_D	Solubility
5872	cis-2-Hexen-1-ol		$C_6H_{12}O$	928-94-9	100.158			157	0.8472[20]	1.4397[20]	s H$_2$O; vs EtOH; s eth, ace; sl ctc
5873	trans-2-Hexen-1-ol		$C_6H_{12}O$	928-95-0	100.158			157	0.8490[16]	1.4340[20]	
5874	cis-3-Hexen-1-ol		$C_6H_{12}O$	928-96-1	100.158			156.5	0.8478[22]	1.4380[20]	s H$_2$O; vs EtOH, eth
5875	trans-3-Hexen-1-ol		$C_6H_{12}O$	928-97-2	100.158			154.5		1.4374[20]	
5876	trans-4-Hexen-1-ol		$C_6H_{12}O$	928-92-7	100.158			159	0.8513[20]	1.4402[20]	
5877	4-Hexen-2-ol		$C_6H_{12}O$	52387-50-5	100.158			137.5	0.8405[18]	1.4392[20]	sl H$_2$O
5878	5-Hexen-2-ol		$C_6H_{12}O$	626-94-8	100.158			139	0.842[16]		sl H$_2$O
5879	cis-3-Hexen-1-ol, acetate		$C_8H_{14}O_2$	3681-71-8	142.196	liq		66[12]			
5880	trans-2-Hexen-1-ol, acetate		$C_8H_{14}O_2$	2497-18-9	142.196	liq		166; 68[15]	0.898	1.4270[20]	
5881	5-Hexen-2-one		$C_6H_{10}O$	109-49-9	98.142			129.5	0.833[27]	1.4178[27]	
5882	4-Hexen-3-one		$C_6H_{10}O$	2497-21-4	98.142			138.5	0.8559[20]	1.4388[20]	s EtOH, eth; vs ace
5883	Hexestrol		$C_{18}H_{22}O_2$	84-16-2	270.367	nd (bz)	186.5				vs ace, eth, EtOH
5884	Hexobarbital		$C_{12}H_{16}N_2O_3$	56-29-1	236.266		146.5				
5885	Hexocyclium methyl sulfate		$C_{21}H_{36}N_2O_5S$	115-63-9	428.586	cry	205				sl chl; i eth
5886	Hexyl acetate		$C_8H_{16}O_2$	142-92-7	144.212	liq	-80.9	171.5	0.8779[15]	1.4092[20]	i H$_2$O; vs eth, EtOH
5887	sec-Hexyl acetate		$C_8H_{16}O_2$	108-84-9	144.212			147.5	0.8805[25]	1.3980[20]	sl H$_2$O; vs eth, EtOH
5888	Hexyl acrylate		$C_9H_{16}O_2$	2499-95-8	156.222		-45	40[1]	0.878[20]		
5889	Hexylamine	1-Hexanamine	$C_6H_{15}N$	111-26-2	101.190	liq	-22.9	132.8	0.7660[20]	1.4180[20]	sl H$_2$O; msc EtOH, eth; s chl
5890	Hexylbenzene		$C_{12}H_{18}$	1077-16-3	162.271	liq	-61	226.1	0.8575[20]	1.4864[20]	i H$_2$O; msc eth; s bz, peth
5891	4-Hexyl-1,3-benzenediol	4-Hexylresorcinol	$C_{12}H_{18}O_2$	136-77-6	194.270	nd (bz)	68	334			vs ace, eth, EtOH, chl
5892	Hexyl benzoate		$C_{13}H_{18}O_2$	6789-88-4	206.281			272; 139[8]	0.9793[20]		i H$_2$O; s EtOH, ace
5893	Hexyl butanoate		$C_{10}H_{20}O_2$	2639-63-6	172.265	liq	-78	208	0.8652[20]	1.4160[15]	i H$_2$O; s EtOH; sl chl
5894	Hexylcyclohexane		$C_{12}H_{24}$	4292-75-5	168.319	liq	-43	224	0.8076[20]	1.4462[20]	
5895	Hexylcyclopentane		$C_{11}H_{22}$	4457-00-5	154.293	liq	-73	203	0.7965[20]	1.4392[20]	vs ace, bz, eth, EtOH
5896	2-Hexyldecanoic acid		$C_{16}H_{32}O_2$	25354-97-6	256.424	visc oil		145[0.02]		1.4432[24]	
5897	Hexyl formate		$C_7H_{14}O_2$	629-33-4	130.185	liq	-62.6	155.5	0.8813[20]	1.4071[20]	i H$_2$O; msc EtOH, eth
5898	Hexyl hexanoate	Hexyl caproate	$C_{12}H_{24}O_2$	6378-65-0	200.318	liq	-55	246	0.865[18]	1.4264[15]	vs ace, bz, eth, EtOH
5899	Hexyl isocyanate		$C_7H_{13}NO$	2525-62-4	127.184			44[7]			
5900	Hexyl methacrylate		$C_{10}H_{18}O_2$	142-09-6	170.249			162; 86[17]	0.880[25]	1.429[25]	vs ace, bz, eth, EtOH
5901	Hexyl methyl ether		$C_7H_{16}O$	4747-07-3	116.201			126.1			
5902	1-Hexylnaphthalene		$C_{16}H_{20}$	2876-53-1	212.330	liq	-18	322	0.9566[20]	1.5647[20]	
5903	Hexyl octanoate		$C_{14}H_{28}O_2$	1117-55-1	228.371	liq	-30.6	277.4	0.8603[20]	1.4323[25]	i H$_2$O; s EtOH, eth, ace
5904	4-(Hexyloxy)benzoic acid		$C_{13}H_{18}O_3$	1142-39-8	222.280	cry	106				
5905	2-(Hexyloxy)ethanol	Ethylene glycol monohexyl ether	$C_8H_{18}O_2$	112-25-4	146.228	liq	-45.1	208	0.8878[20]	1.4291[20]	sl H$_2$O; vs EtOH, eth
5906	Hexyl pentanoate		$C_{11}H_{22}O_2$	1117-59-5	186.292	liq	-63.1	226.3	0.8635[20]	1.4228[15]	vs ace, eth, EtOH
5907	4-Hexylphenol		$C_{12}H_{18}O$	2446-69-7	178.270			148[9]			
5908	1-Hexyl propanoate		$C_9H_{18}O_2$	2445-76-3	158.238	liq	-57.5	190	0.8698[20]	1.4162[15]	i H$_2$O; s EtOH, eth, ace, AcOEt
5909	1-Hexyl-1,2,3,4-tetrahydronaphthalene		$C_{16}H_{24}$	66325-11-9	216.362	liq		305	0.9176[25]	1.5127[25]	
5910	1-Hexyne	Butylacetylene	C_6H_{10}	693-02-7	82.143	liq	-131.9	71.3	0.7155[25]	1.3989[20]	i H$_2$O; s EtOH, eth, bz, chl; sl ctc
5911	2-Hexyne	1-Methyl-2-propylacetylene	C_6H_{10}	764-35-2	82.143	liq	-89.6	84.5	0.7315[20]	1.4138[20]	i H$_2$O; msc EtOH, eth; s bz, chl, peth
5912	3-Hexyne	Diethylacetylene	C_6H_{10}	928-49-4	82.143	liq	-103	81	0.7231[20]	1.4115[20]	i H$_2$O; s EtOH, eth, bz, chl, peth
5913	3-Hexyne-2,5-diol		$C_6H_{10}O_2$	3031-66-1	114.142			121[15]	1.0180[20]	1.4691[20]	
5914	3-Hexyn-1-ol	3-Hexynol	$C_6H_{10}O$	1002-28-4	98.142			162; 65[12]	0.8982[20]	1.4530[20]	
5915	1-Hexyn-3-ol		$C_6H_{10}O$	105-31-7	98.142	liq	-80	142	0.8704[20]	1.4340[25]	s ctc
5916	5-Hexyn-2-one		C_6H_8O	2550-28-9	96.127			149	0.9065[20]	1.4366[20]	

cis-2-Hexen-1-ol

trans-2-Hexen-1-ol

cis-3-Hexen-1-ol

trans-3-Hexen-1-ol

trans-4-Hexen-1-ol

4-Hexen-2-ol

5-Hexen-2-ol

cis-3-Hexen-1-ol, acetate

trans-2-Hexen-1-ol, acetate

5-Hexen-2-one

4-Hexen-3-one

Hexestrol

Hexobarbital

Hexocyclium methyl sulfate

Hexyl acetate

sec-Hexyl acetate

Hexyl acrylate

Hexylamine

Hexylbenzene

4-Hexyl-1,3-benzenediol

Hexyl benzoate

Hexyl butanoate

Hexylcyclohexane

Hexylcyclopentane

2-Hexyldecanoic acid

Hexyl formate

Hexyl hexanoate

Hexyl isocyanate

Hexyl methacrylate

Hexyl methyl ether

1-Hexylnaphthalene

Hexyl octanoate

4-(Hexyloxy)benzoic acid

2-(Hexyloxy)ethanol

Hexyl pentanoate

4-Hexylphenol

1-Hexyl propanoate

1-Hexyl-1,2,3,4-tetrahydronaphthalene

1-Hexyne

2-Hexyne

3-Hexyne

3-Hexyne-2,5-diol

3-Hexyn-1-ol

1-Hexyn-3-ol

5-Hexyn-2-one

No.	Name	Synonym	Mol. Form.	CAS RN	Mol. Wt.	Physical Form	mp/°C	bp/°C	den/ g cm⁻³	n_D	Solubility
5917	Histamine		$C_5H_9N_3$	51-45-6	111.145	wh nd (chl)	83	209[18]			s H_2O, EtOH, chl; sl eth
5918	L-Histidine	Glyoxaline-5-alanine	$C_6H_9N_3O_2$	71-00-1	155.154	nd or pl (dil al)	287 dec				s H_2O; sl EtOH; i eth, ace, bz, chl
5919	L-Histidine, monohydrochloride		$C_6H_{10}ClN_3O_2$	645-35-2	191.615		245 dec				s H_2O
5920	Homatropine		$C_{16}H_{21}NO_3$	87-00-3	275.343	pr (al, eth)	99.5				sl H_2O, bz; s EtOH, eth, ace, chl
5921	Homatropine hydrobromide	Tropanol mandelate	$C_{16}H_{22}BrNO_3$	51-56-9	356.255	orth pym or pl (w)	217 dec				vs H_2O, EtOH
5922	Homochlorocyclizine		$C_{19}H_{23}ClN_2$	848-53-3	314.852	oil		177[0.8]			
5923	DL-Homocysteine	DL-2-Amino-4-mercaptobutanoic acid	$C_4H_9NO_2S$	454-29-5	135.185		272 dec				s H_2O; i eth, bz
5924	L-Homocysteine	L-2-Amino-4-mercaptobutanoic acid	$C_4H_9NO_2S$	6027-13-0	135.185	platelets	232				
5925	Homocystine		$C_8H_{16}N_2O_4S_2$	870-93-9	268.354		264				sl H_2O; i eth, bz
5926	L-Homoserine	2-Amino-4-hydroxybutanoic acid, (S)	$C_4H_9NO_3$	672-15-1	119.119	pr (90% al)	203 dec				vs H_2O; sl EtOH; i eth, bz
5927	Humulene		$C_{15}H_{24}$	6753-98-6	204.352			123[10]	0.8905[20]	1.5038[20]	
5928	Humulon		$C_{21}H_{30}O_5$	26472-41-3	362.460	ye cry (eth)	66.5				sl H_2O; s EtOH, eth, ace, bz, alk
5929	Hydralazine	1-Hydrazinophthalazine	$C_8H_8N_4$	86-54-4	160.177	ye cry (MeOH)	172				s acid
5930	Hydramethylnon		$C_{25}H_{24}F_6N_4$	67485-29-4	494.476		190				
5931	Hydrastine		$C_{21}H_{21}NO_6$	118-08-1	383.395	ye pr (al)	132				i H_2O; s ace, bz
5932	Hydrastinine		$C_{11}H_{13}NO_3$	6592-85-4	207.226	nd (lig), cry (eth)	116.5				s H_2O; vs EtOH, eth, chl
5933	Hydrazinecarbothioamide	Thiosemicarbazide	CH_5N_3S	79-19-6	91.136	lo nd (w)	183				vs H_2O, EtOH
5934	Hydrazinecarboxaldehyde		CH_4N_2O	624-84-0	60.055	ye lf or nd (al)	54				vs bz, eth, EtOH, chl
5935	Hydrazinecarboxamide		CH_5N_3O	57-56-7	75.070	pr (al)	96		1.484[8]		vs H_2O; s EtOH; i eth, bz, chl
5936	Hydrazinecarboximidamide	Aminoguanidine	CH_6N_4	79-17-4	74.086	cry	dec				vs H_2O, EtOH
5937	1,2-Hydrazinedicarboxaldehyde		$C_2H_4N_2O_2$	628-36-4	88.065	pr (al)	161.0				vs H_2O; sl EtOH, DMSO; i eth
5938	1,2-Hydrazinedicarboxamide		$C_2H_6N_4O_2$	110-21-4	118.095	pl (w)	258		1.604[17]		
5939	4-Hydrazinobenzenesulfonic acid	Phenylhydrazine-4-sulfonic acid	$C_6H_8N_2O_3S$	98-71-5	188.204	nd, lf (w)	286				sl H_2O, EtOH
5940	4-Hydrazinobenzoic acid		$C_7H_8N_2O_2$	619-67-0	152.151	ye nd or pl (w)	221 dec				sl H_2O; i eth
5941	2-Hydrazinoethanol		$C_2H_8N_2O$	109-84-2	76.097	liq	-70	219; 120[17.5]	1.119[25]		vs H_2O, EtOH, MeOH
5942	Hydrindantin		$C_{18}H_{10}O_6$	5103-42-4	322.268	pr (ace)	250 dec				
5943	Hydrochlorothiazide		$C_7H_8ClN_3O_4S_2$	58-93-5	297.740		274				
5944	Hydrocinchonidine		$C_{19}H_{24}N_2O$	485-64-3	296.406	lf (al)	229				vs EtOH
5945	Hydrocinchonine		$C_{19}H_{24}N_2O$	485-65-4	296.406	pr	268.5				s H_2O; sl EtOH; i eth
5946	Hydrocodone		$C_{18}H_{21}NO_3$	125-29-1	299.365		198				i H_2O; s EtOH
5947	Hydrocortisone		$C_{21}H_{30}O_5$	50-23-7	362.460	pl (al or i-PrOH)	220				sl H_2O; s EtOH, diox, HOAc
5948	Hydrocortisone 21-acetate	Cortisol acetate	$C_{23}H_{32}O_6$	50-03-3	404.496		223 dec		1.289[20]		
5949	Hydrocotarnine		$C_{12}H_{15}NO_3$	550-10-7	221.252		56				i H_2O; s EtOH, eth, ace, bz, chl
5950	Hydroflumethiazide		$C_8H_8F_3N_3O_4S_2$	135-09-1	331.293		270.5				
5951	Hydrofuramide		$C_{15}H_{12}N_2O_3$	494-47-3	268.267	nd (al)	117				i H_2O; vs EtOH, eth
5952	Hydrogen cyanide	Hydrocyanic acid	CHN	74-90-8	27.026	vol liq or gas	-13.29	26	0.6876[20]	1.2614[20]	msc H_2O, EtOH, eth
5953	Hydrohydrastinine		$C_{11}H_{13}NO_2$	494-55-3	191.227	nd (lig), cry (peth)	66	303			vs ace, bz, eth, EtOH
5954	Hydromorphone	7,8-Dihydromorphin-6-one	$C_{17}H_{19}NO_3$	466-99-9	285.338	cry (EtOH)	266.5				
5955	Hydroprene		$C_{17}H_{30}O_2$	41096-46-2	266.419			174[19]	0.8955[20]		
5956	Hydroquinidine		$C_{20}H_{26}N_2O_2$	1435-55-8	326.432	nd (al)	168.5				s EtOH, eth, ace, chl
5957	Hydroquinine		$C_{20}H_{26}N_2O_2$	522-66-7	326.432	nd (eth, chl)	172.5				vs ace, eth, EtOH, chl
5958	p-Hydroquinone	1,4-Benzenediol	$C_6H_6O_2$	123-31-9	110.111	mcl pr (sub) nd(w) pr (MeOH)	172.4	285	1.330[20]	1.632[25]	s H_2O, eth; vs EtOH, ace; i bz
5959	Hydroxocobalamin	Vitamin B-12a	$C_{62}H_{89}CoN_{13}O_{15}P$	13422-51-0	1346.355	red cry (ace aq)	200 dec				s H_2O, EtOH; i ace, eth, bz

Histamine

L-Histidine

L-Histidine, monohydrochloride

Homatropine

Homatropine hydrobromide

Homochlorocyclizine

DL-Homocysteine

L-Homocysteine

Homocystine

L-Homoserine

Humulene

Humulon

Hydralazine

Hydramethylnon

Hydrastine

Hydrastinine

Hydrazinecarbothioamide

Hydrazinecarboxaldehyde

Hydrazinecarboxamide

Hydrazinecarboximidamide

1,2-Hydrazinedicarboxaldehyde

1,2-Hydrazinedicarboxamide

4-Hydrazinobenzenesulfonic acid

4-Hydrazinobenzoic acid

2-Hydrazinoethanol

Hydrindantin

Hydrochlorothiazide

Hydrocinchonidine

Hydrocinchonine

Hydrocodone

Hydrocortisone

Hydrocortisone 21-acetate

Hydrocotarnine

Hydroflumethiazide

Hydrofuramide

Hydrogen cyanide

Hydrohydrastinine

Hydromorphone

Hydroprene

Hydroquinidine

Hydroquinine

p-Hydroquinone

Hydroxocobalamin

No.	Name	Synonym	Mol. Form.	CAS RN	Mol. Wt.	Physical Form	mp/°C	bp/°C	den/ g cm^{-3}	n_D	Solubility
5960	Hydroxyacetonitrile	Glyconitrile	C$_2$H$_3$NO	107-16-4	57.051		<-72	dec 183; 119[24]		1.4117[19]	vs H$_2$O, EtOH, eth; i bz, chl
5961	(Hydroxyacetyl)benzene		C$_8$H$_8$O$_2$	582-24-1	136.149	hex pl (al), pl (w or dil al)	90	125[12], 56[1]	1.0963[99]		s H$_2$O, EtOH, eth, chl; sl lig
5962	17-Hydroxyandrostan-3-one, (5α,17β)	Stanolone	C$_{19}$H$_{30}$O$_2$	521-18-6	290.440		181	sub 135			
5963	3-Hydroxyandrostan-17-one, (3α,5α)	Androsterone	C$_{19}$H$_{30}$O$_2$	53-41-8	290.440	lf or nd (al, ace)	185				sl H$_2$O, chl; s EtOH, eth, ace, bz
5964	3-Hydroxyandrostan-17-one, (3β,5α)	Epiandrosterone	C$_{19}$H$_{30}$O$_2$	481-29-8	290.440	cry (bz-peth, ace)	178				
5965	17-Hydroxyandrost-4-en-3-one, (17β)	Testosterone	C$_{19}$H$_{28}$O$_2$	58-22-0	288.424	nd (dil ace)	155				i H$_2$O; s EtOH, eth, ace
5966	1-Hydroxy-9,10-anthracenedione		C$_{14}$H$_8$O$_3$	129-43-1	224.212	red-oran nd (al)	193.8	sub			i H$_2$O; s EtOH, eth, bz; sl liq NH$_3$
5967	2-Hydroxy-9,10-anthracenedione		C$_{14}$H$_8$O$_3$	605-32-3	224.212	ye pl or nd (al or HOAc)	306	sub			i H$_2$O; s EtOH, eth, aq NH$_3$, KOH
5968	3-Hydroxybenzaldehyde	3-Formylphenol	C$_7$H$_6$O$_2$	100-83-4	122.122	nd (w)	108	240	1.1179[130]		sl H$_2$O; s EtOH, eth, ace, bz; i lig
5969	4-Hydroxybenzaldehyde	4-Formylphenol	C$_7$H$_6$O$_2$	123-08-0	122.122	nd (w)	117		1.129[130]	1.5705[130]	sl H$_2$O, ace; vs EtOH, eth; s bz
5970	2-Hydroxybenzaldehyde, [(2-hydroxyphenyl)methylene]hydrazone		C$_{14}$H$_{12}$N$_2$O$_2$	959-36-4	240.257		214				i H$_2$O; s EtOH, chl; vs bz, alk
5971	2-Hydroxybenzamide	Salicylamide	C$_7$H$_7$NO$_2$	65-45-2	137.137		142	181.5[14]	1.175[140]		sl H$_2$O, eth, DMSO; s EtOH
5972	N-Hydroxybenzamide		C$_7$H$_7$NO$_2$	495-18-1	137.137	orth ta, lf (eth)	131 exp				s H$_2$O, EtOH; sl eth, bz
5973	α-Hydroxybenzeneacetic acid, (±)	DL-Mandelic acid	C$_8$H$_8$O$_3$	611-72-3	152.148	orth pl	119		1.2890[20]		s H$_2$O, eth, EtOH, i-PrOH
5974	2-Hydroxybenzeneacetic acid		C$_8$H$_8$O$_3$	614-75-5	152.148		148	240			sl H$_2$O, chl; s eth
5975	3-Hydroxybenzeneacetic acid		C$_8$H$_8$O$_3$	621-37-4	152.148	nd (bz-lig)	132	190[11]			vs H$_2$O, EtOH, eth; s bz; sl lig
5976	4-Hydroxybenzeneacetic acid		C$_8$H$_8$O$_3$	156-38-7	152.148	nd (w)	152	sub			sl H$_2$O; vs EtOH, eth
5977	α-Hydroxybenzeneacetonitrile	Mandelonitrile	C$_8$H$_7$NO	532-28-5	133.148	ye oily liq	-10		1.12		i H$_2$O; vs chl, eth, EtOH
5978	2-Hydroxybenzenecarbodithioic acid	Dithiosalicylic acid	C$_7$H$_6$OS$_2$	527-89-9	170.252	oran-ye nd	49				vs bz, eth, EtOH
5979	4-Hydroxy-1,3-benzenedicarboxylic acid	4-Hydroxyisophthalic acid	C$_8$H$_6$O$_5$	636-46-4	182.131	nd(w), lf (dil al)	310				i H$_2$O, chl; vs EtOH, eth; s HOAc
5980	5-Hydroxy-1,3-benzenedicarboxylic acid		C$_8$H$_6$O$_5$	618-83-7	182.131	nd(w+2) cr(aq-al)		sub			vs bz, eth, EtOH
5981	4-Hydroxy-1,3-benzenedisulfonic acid	Phenoldisulfonic acid	C$_6$H$_6$O$_7$S$_2$	96-77-5	254.238	nd (w)	>100 dec				vs H$_2$O, EtOH
5982	4-Hydroxybenzeneethanol		C$_8$H$_{10}$O$_2$	501-94-0	138.164		91.8	310.0			
5983	2-Hydroxybenzenemethanol	Salicyl alcohol	C$_7$H$_8$O$_2$	90-01-7	124.138	lf (bz), nd or pl (w, eth)	87	sub	1.1613[25]		s H$_2$O, EtOH, eth, bz; vs chl
5984	3-Hydroxybenzenemethanol	3-Hydroxybenzyl alcohol	C$_7$H$_8$O$_2$	620-24-6	124.138	nd (bz), cry (CCl$_4$)	73	dec 300	1.161[25]		vs H$_2$O, EtOH, eth; sl chl
5985	4-Hydroxybenzenemethanol	4-Hydroxybenzyl alcohol	C$_7$H$_8$O$_2$	623-05-2	124.138	pr or nd (w)	124.5	252			vs H$_2$O, EtOH, bz, chl; s eth; sl DMSO
5986	4-Hydroxybenzenepropanoic acid	p-Hydroxyhydrocinnamic acid	C$_9$H$_{10}$O$_3$	501-97-3	166.173		130.8	209[14]			s H$_2$O, EtOH, eth, bz; i CS$_2$
5987	α-Hydroxybenzenepropanoic acid, (±)	(±)-3-Phenyllactic acid	C$_9$H$_{10}$O$_3$	828-01-3	166.173	cry (chl, bz), pr (w)	98	149[15]			vs H$_2$O, ace, eth, EtOH
5988	3-Hydroxybenzenesulfonic acid	m-Phenolsulfonic acid	C$_6$H$_6$O$_4$S	585-38-6	174.175	nd (w+2)					
5989	4-Hydroxybenzenesulfonic acid	p-Phenolsulfonic acid	C$_6$H$_6$O$_4$S	98-67-9	174.175	nd					vs H$_2$O, EtOH
5990	2-Hydroxybenzoic acid	Salicylic acid	C$_7$H$_6$O$_3$	69-72-7	138.121	nd (w), mcl pr (al)	159.0	211[20]	1.443[20]	1.565	sl H$_2$O, bz, chl, ctc; vs EtOH, eth, ace
5991	3-Hydroxybenzoic acid		C$_7$H$_6$O$_3$	99-06-9	138.121	nd (w) pl, pr (al)	202.5		1.485[25]		sl H$_2$O; s EtOH, eth, ace; i bz
5992	4-Hydroxybenzoic acid		C$_7$H$_6$O$_3$	99-96-7	138.121	pr or pl (w, al) cry (ace)	214.5		1.46[25]		sl H$_2$O, bz; vs EtOH; s eth, ace
5993	2-Hydroxybenzoic acid, hydrazide		C$_7$H$_8$N$_2$O$_2$	936-02-7	152.151		148				vs bz, EtOH
5994	2-Hydroxybenzonitrile		C$_7$H$_5$NO	611-20-1	119.121		98	149[14]	1.1052[100]	1.5372[100]	sl H$_2$O; vs EtOH, eth, bz, chl

Hydroxyacetonitrile

(Hydroxyacetyl)benzene

17-Hydroxyandrostan-3-one, (5α,17β)

3-Hydroxyandrostan-17-one, (3α,5α)

3-Hydroxyandrostan-17-one, (3β,5α)

17-Hydroxyandrost-4-en-3-one, (17β)

1-Hydroxy-9,10-anthracenedione

2-Hydroxy-9,10-anthracenedione

3-Hydroxybenzaldehyde

4-Hydroxybenzaldehyde

2-Hydroxybenzaldehyde, [(2-hydroxyphenyl)methylene]hydrazone

2-Hydroxybenzamide

N-Hydroxybenzamide

α-Hydroxybenzeneacetic acid, (±)

2-Hydroxybenzeneacetic acid

3-Hydroxybenzeneacetic acid

4-Hydroxybenzeneacetic acid

α-Hydroxybenzeneacetonitrile

2-Hydroxybenzenecarbodithioic acid

4-Hydroxy-1,3-benzenedicarboxylic acid

5-Hydroxy-1,3-benzenedicarboxylic acid

4-Hydroxy-1,3-benzenedisulfonic acid

4-Hydroxybenzeneethanol

2-Hydroxybenzenemethanol

3-Hydroxybenzenemethanol

4-Hydroxybenzenemethanol

4-Hydroxybenzenepropanoic acid

α-Hydroxybenzenepropanoic acid, (±)

3-Hydroxybenzenesulfonic acid

4-Hydroxybenzenesulfonic acid

2-Hydroxybenzoic acid

3-Hydroxybenzoic acid

4-Hydroxybenzoic acid

2-Hydroxybenzoic acid, hydrazide

2-Hydroxybenzonitrile

No.	Name	Synonym	Mol. Form.	CAS RN	Mol. Wt.	Physical Form	mp/°C	bp/°C	den/ g cm⁻³	n_D	Solubility
5995	3-Hydroxybenzonitrile		C_7H_5NO	873-62-1	119.121	pr (al, eth) lf (w)	82.8				vs H_2O, EtOH, eth, bz, chl
5996	4-Hydroxybenzonitrile		C_7H_5NO	767-00-0	119.121	lf (w)	113	148[1]			sl H_2O, DMSO; vs EtOH, eth, chl
5997	4-Hydroxybenzophenone	4-Hydroxyphenyl phenyl ketone	$C_{13}H_{10}O_2$	1137-42-4	198.217	nd (al), pr (dil al)	135		1.133[172]		sl H_2O; vs EtOH, eth, HOAc
5998	4-Hydroxy-2H-1-benzopyran-2-one		$C_9H_6O_3$	1076-38-6	162.142	nd (w)	213.5				s H_2O, EtOH, eth; sl DMSO
5999	7-Hydroxy-2H-1-benzopyran-2-one	Umbelliferone	$C_9H_6O_3$	93-35-6	162.142	nd (w)	230.5	sub			vs EtOH, HOAc, chl
6000	1-Hydroxy-1H-benzotriazole		$C_6H_5N_3O$	2592-95-2	135.123		157.8				
6001	2-Hydroxybenzoyl chloride		$C_7H_5ClO_2$	1441-87-8	156.567		19	92[15]	1.3112[20]	1.5812[20]	vs eth
6002	4-(2-Hydroxybenzoyl)morpholine	4-Salicyloylmorpholine	$C_{11}H_{10}NO_3$	3202-84-4	204.202						s DMSO
6003	2-Hydroxybiphenyl	[1,1'-Biphenyl]-2-ol	$C_{12}H_{10}O$	90-43-7	170.206		57.5	286	1.213[25]		i H_2O; s EtOH, ace, bz; vs eth, py
6004	3-Hydroxybiphenyl	[1,1'-Biphenyl]-3-ol	$C_{12}H_{10}O$	580-51-8	170.206		78	>300			sl H_2O; vs EtOH, eth, bz, py; s chl
6005	4-Hydroxybiphenyl	[1,1'-Biphenyl]-4-ol	$C_{12}H_{10}O$	92-69-3	170.206		166	305			sl H_2O, DMSO; vs EtOH, eth, chl, py
6006	3-Hydroxybutanal	Aldol	$C_4H_8O_2$	107-89-1	88.106			83[20]	1.103[20]	1.4238[20]	msc H_2O, EtOH; s eth; vs ace
6007	2-Hydroxybutanoic acid, (±)		$C_4H_8O_3$	600-15-7	104.105		44.2	dec 260; 140[14]	1.125[20]		s H_2O, EtOH, eth
6008	3-Hydroxybutanoic acid, (±)		$C_4H_8O_3$	625-71-8	104.105		49	130[12], 94[0.1]		1.4424[20]	vs H_2O, EtOH, eth; i bz
6009	4-Hydroxybutanoic acid		$C_4H_8O_3$	591-81-1	104.105		<-17	dec 180			
6010	1-Hydroxy-2-butanone		$C_4H_8O_2$	5077-67-8	88.106			160; 78[60]	1.0272[20]	1.4189[20]	vs H_2O, EtOH, eth
6011	3-Hydroxy-2-butanone, (±)	Acetoin	$C_4H_8O_2$	52217-02-4	88.106		15	148	1.0044[20]	1.4171[20]	msc H_2O; sl EtOH, eth; s ace, chl; i lig
6012	4-Hydroxy-2-butanone		$C_4H_8O_2$	590-90-9	88.106			182; 90[11]	1.0233[20]	1.4585[14]	msc H_2O, EtOH, eth; vs ace
6013	2-Hydroxy-3-butenenitrile		C_4H_5NO	5809-59-6	83.089	liq		94[17]			
6014	4-Hydroxybutyramide		$C_4H_9NO_2$	927-60-6	103.120		52				
6015	3-Hydroxycamphor	3-Hydroxy-1,7,7-trimethylbicyclo[2.2.1]heptan-2-one	$C_{10}H_{16}O_2$	10373-81-6	168.233	nd (bz-peth)	205.5				vs eth, EtOH, chl
6016	3-Hydroxycholan-24-oic acid, (3α,5β)	Lithocholic acid	$C_{24}H_{40}O_3$	434-13-9	376.573	hex lf (al) pr (dil al)	186				i H_2O, lig; s EtOH, chl, HOAc; sl eth
6017	Hydroxycodeinone		$C_{18}H_{19}NO_4$	508-54-3	313.349		275 dec				
6018	2-Hydroxycyclodecanone	Sebacoin	$C_{10}H_{18}O_2$	96-00-4	170.249	cry (peth)	38.5	136[14]			
6019	2-Hydroxy-2,4,6-cycloheptatrien-1-one		$C_7H_6O_2$	533-75-5	122.122	nd	50.8	sub 40			s H_2O, eth, ace
6020	1-Hydroxycyclohexanecarbonitrile		$C_7H_{11}NO$	931-97-5	125.168		35	132[20]	1.0172[20]	1.4693[20]	vs H_2O, eth
6021	2-Hydroxycyclohexanone		$C_6H_{10}O_2$	533-60-8	114.142	nd (al)				1.4785[21]	vs H_2O, EtOH; i eth, bz, peth
6022	1-(1-Hydroxycyclohexyl)ethanone		$C_8H_{14}O_2$	1123-27-9	142.196			125.5; 91[11]	1.0248[25]	1.4670[25]	vs eth, EtOH
6023	4-Hydroxydecanoic acid γ-lactone	5-Hexyldihydro-2(3H)-furanone	$C_{10}H_{18}O_2$	706-14-9	170.249	liq		281			
6024	2-Hydroxy-3,5-diiodobenzoic acid	3,5-Diiodosalicylic acid	$C_7H_4I_2O_3$	133-91-5	389.914	nd (al)	235.5				sl H_2O; vs EtOH, eth; i bz, chl
6025	4-Hydroxy-3,5-diiodobenzoic acid		$C_7H_4I_2O_3$	618-76-8	389.914		237	dec 260			i H_2O; vs EtOH, eth; sl bz, chl, lig
6026	4-Hydroxy-3,5-diiodobenzonitrile		$C_7H_3I_2NO$	1689-83-4	370.914		201 dec				
6027	4-Hydroxy-3,5-diiodo-α-phenylbenzenepropanoic acid	Iodoalphionic acid	$C_{15}H_{12}I_2O_3$	577-91-3	494.063		164				i H_2O; s EtOH, eth; sl bz, chl
6028	2-Hydroxy-4,6-dimethoxybenzaldehyde		$C_9H_{10}O_4$	708-76-9	182.173		70	193[25], 165[10]			i H_2O; vs EtOH, eth, bz, chl, HOAc
6029	4-Hydroxy-3,5-dimethoxybenzaldehyde	Syringaldehyde	$C_9H_{10}O_4$	134-96-3	182.173	br nd (lig)	113	192[14]			sl H_2O, lig; vs EtOH, eth, bz, chl
6030	4-Hydroxy-3,5-dimethoxybenzoic acid		$C_9H_{10}O_5$	530-57-4	198.172	nd (w)	204.5				sl H_2O; vs EtOH
6031	7-Hydroxy-3,7-dimethyloctanal		$C_{10}H_{20}O_2$	107-75-5	172.265			103[3]	0.9220[20]	1.4494[20]	sl H_2O; s EtOH, ace
6032	3-Hydroxy-2,2-dimethylpropanal	Hydroxypivaldehyde	$C_5H_{10}O_2$	597-31-9	102.132	nd (w)	89.5	173; 68[14]			
6033	2-Hydroxy-3,5-dinitrobenzoic acid		$C_7H_4N_2O_7$	609-99-4	228.116	ye nd or pl (+1w)	182				s H_2O, EtOH, eth, bz

3-Hydroxybenzonitrile

4-Hydroxybenzonitrile

4-Hydroxybenzophenone

4-Hydroxy-2*H*-1-benzopyran-2-one

7-Hydroxy-2*H*-1-benzopyran-2-one

1-Hydroxy-1*H*-benzotriazole

2-Hydroxybenzoyl chloride

4-(2-Hydroxybenzoyl)morpholine

2-Hydroxybiphenyl

3-Hydroxybiphenyl

4-Hydroxybiphenyl

3-Hydroxybutanal

2-Hydroxybutanoic acid, (±)

3-Hydroxybutanoic acid, (±)

4-Hydroxybutanoic acid

1-Hydroxy-2-butanone

3-Hydroxy-2-butanone, (±)

4-Hydroxy-2-butanone

2-Hydroxy-3-butenenitrile

4-Hydroxybutyramide

3-Hydroxycamphor

3-Hydroxycholan-24-oic acid, (3α,5β)

Hydroxycodeinone

2-Hydroxycyclodecanone

2-Hydroxy-2,4,6-cycloheptatrien-1-one

1-Hydroxycyclohexanecarbonitrile

2-Hydroxycyclohexanone

1-(1-Hydroxycyclohexyl)ethanone

4-Hydroxydecanoic acid γ-lactone

2-Hydroxy-3,5-diiodobenzoic acid

4-Hydroxy-3,5-diiodobenzoic acid

4-Hydroxy-3,5-diiodobenzonitrile

4-Hydroxy-3,5-diiodo-α-phenylbenzenepropanoic acid

2-Hydroxy-4,6-dimethoxybenzaldehyde

4-Hydroxy-3,5-dimethoxybenzaldehyde

4-Hydroxy-3,5-dimethoxybenzoic acid

7-Hydroxy-3,7-dimethyloctanal

3-Hydroxy-2,2-dimethylpropanal

2-Hydroxy-3,5-dinitrobenzoic acid

No.	Name	Synonym	Mol. Form.	CAS RN	Mol. Wt.	Physical Form	mp/°C	bp/°C	den/ g cm⁻³	n_D	Solubility
6034	11-Hydroxy-9,15-dioxoprosta-5,13-dien-1-oic acid, (5Z,11α,13E)	15-Oxo-prostaglandin E2	$C_{20}H_{30}O_5$	26441-05-4	350.449	cry					
6035	1-Hydroxy-1,1-diphosphonoethane	Etidronic acid	$C_2H_8O_7P_2$	2809-21-4	206.028	cry (w)	105				s H_2O, EtOH, MeOH
6036	3-Hydroxyestra-1,3,5,7,9-pentaen-17-one	Equilenin	$C_{18}H_{18}O_2$	517-09-9	266.335		258.5	sub 170			sl EtOH, ace, chl
6037	3-Hydroxyestra-1,3,5(10),7-tetraen-17-one	Equilin	$C_{18}H_{20}O_2$	474-86-2	268.351	pl (AcOEt)	239	sub 170			sl H_2O; s EtOH, ace, diox, AcOEt
6038	2-Hydroxyethyl acrylate	2-Hydroxyethyl 2-propenoate	$C_5H_8O_3$	818-61-1	116.116	liq		191; 91[12]	1.011[23]		
6039	N-(2-Hydroxyethyl)dodecanamide		$C_{14}H_{29}NO_2$	142-78-9	243.386		88.5				
6040	N-(2-Hydroxyethyl)ethylenediaminetriacetic acid		$C_{10}H_{18}N_2O_7$	150-39-0	278.259	cry	165 dec				
6041	2-Hydroxyethyl 2-hydroxybenzoate	Glycol salicylate	$C_9H_{10}O_4$	87-28-5	182.173		37	173[15]	1.2526[15]		sl H_2O; vs EtOH, eth, bz, chl
6042	2-Hydroxyethyl methacrylate	Ethylene glycol monomethacrylate	$C_6H_{10}O_3$	868-77-9	130.141			103[13], 67[3]	1.079[20]	1.4515[20]	vs H_2O, EtOH, chl
6043	N-(2-Hydroxyethyl)phthalimide		$C_{10}H_9NO_3$	3891-07-4	191.183	nd (al), lf (w)	130.3				sl H_2O
6044	1-(2-Hydroxyethyl)-2-pyrrolidinone		$C_6H_{11}NO_2$	3445-11-2	129.157		20	295	1.1435[20]		
6045	4-Hydroxy-4H-furo[3,2-c]pyran-2(6H)-one	Patulin	$C_7H_6O_4$	149-29-1	154.121	pl or pr (eth, chl)	111				s H_2O, EtOH, eth, ace, bz; i peth
6046	16-Hydroxyhexadecanoic acid	16-Hydroxypalmitic acid	$C_{16}H_{32}O_3$	506-13-8	272.423		96.5				i H_2O; s EtOH, ace; sl eth, bz
6047	2-Hydroxyhexanoic acid		$C_6H_{12}O_3$	6064-63-7	132.157	pr (eth)	60				vs H_2O
6048	6-Hydroxyhexanoic acid		$C_6H_{12}O_3$	1191-25-9	132.157	liq					
6049	3-Hydroxy-2-(hydroxymethyl)-2-methylpropanoic acid	Dimethylolpropionic acid	$C_5H_{10}O_4$	4767-03-7	134.131		190				
6050	5-Hydroxy-2-(hydroxymethyl)-4H-pyran-4-one	Kojic acid	$C_6H_6O_4$	501-30-4	142.110	pr nd (ace)	153.5				sl H_2O, bz; s EtOH, eth, ace, DMSO
6051	8-Hydroxy-7-iodo-5-quinolinesulfonic acid	Ferron	$C_9H_6INO_4S$	547-91-1	351.118	ye pr, lf (al)	260 dec				sl H_2O, EtOH; i eth, bz, chl; s con sulf
6052	2-Hydroxy-1H-isoindole-1,3(2H)-dione		$C_8H_5NO_3$	524-38-9	163.131		232				s DMSO
6053	2-Hydroxy-4-isopropyl-2,4,6-cycloheptatrien-1-one		$C_{10}H_{12}O_2$	499-44-5	164.201	pa ye (peth)	51.5	137[10]	1.0606[65]		sl H_2O, bz, lig; s ctc
6054	Hydroxylupanine		$C_{15}H_{24}N_2O_2$	15358-48-2	264.364	cry (ace)	169.5				vs H_2O, EtOH, chl
6055	N-Hydroxymethanamine	N-Methylhydroxylamine	CH_5NO	593-77-1	47.057	hyg nd	87.5	62.5[15]	1.0003[20]	1.4164[20]	vs H_2O, EtOH
6056	2-Hydroxy-3-methoxybenzaldehyde		$C_8H_8O_3$	148-53-8	152.148	lt ye lf, grn nd (w, lig)	44.5	265.5			sl H_2O, lig; vs EtOH, eth, ctc
6057	2-Hydroxy-4-methoxybenzaldehyde		$C_8H_8O_3$	673-22-3	152.148	nd (w), cry (al)	42.0				s EtOH, eth, bz, lig
6058	2-Hydroxy-5-methoxybenzaldehyde		$C_8H_8O_3$	672-13-9	152.148	ye liq (w)	4	247.5			vs eth, EtOH
6059	3-Hydroxy-4-methoxybenzaldehyde		$C_8H_8O_3$	621-59-0	152.148		114	179[15]	1.196[25]		sl H_2O; s EtOH, eth, bz, HOAc; vs chl
6060	4-Hydroxy-3-methoxybenzaldehyde	Vanillin	$C_8H_8O_3$	121-33-5	152.148	tetr (w, lig)	81.5	285	1.056[25]		sl H_2O; vs EtOH, eth, ace; s bz, lig
6061	4-Hydroxy-3-methoxybenzeneacetic acid	Homovanillic acid	$C_9H_{10}O_4$	306-08-1	182.173		143.5				
6062	4-Hydroxy-3-methoxybenzenemethanol		$C_8H_{10}O_3$	498-00-0	154.163	pr (w), nd (bz)	115	dec			s H_2O, EtOH, eth, bz
6063	4-Hydroxy-3-methoxybenzenepropanol		$C_{10}H_{14}O_3$	2305-13-7	182.216		65	197[15]		1.5545[25]	vs eth, EtOH
6064	2-Hydroxy-5-methoxybenzoic acid		$C_8H_8O_4$	2612-02-4	168.148		142				
6065	4-Hydroxy-3-methoxybenzoic acid	Vanillic acid	$C_8H_8O_4$	121-34-6	168.148	wh nd	211.5	sub			sl H_2O; vs EtOH; s eth, DMSO
6066	7-Hydroxy-6-methoxy-2H-1-benzopyran-2-one	Scopoletin	$C_{10}H_8O_4$	92-61-5	192.169	nd or pr (al)	204				sl H_2O, EtOH; s chl; i bz, CS_2
6067	4-(4-Hydroxy-3-methoxyphenyl)-2-butanone	Zingerone	$C_{11}H_{14}O_3$	122-48-5	194.227	cry (ace, eth)	40.5	187[14]			vs eth
6068	1-(2-Hydroxy-4-methoxyphenyl)ethanone		$C_9H_{10}O_3$	552-41-0	166.173	nd (al)	52.5	158[20]	1.3102[81]	1.5452[81]	vs bz, eth, EtOH, chl
6069	1-(4-Hydroxy-3-methoxyphenyl)ethanone	Apocynin	$C_9H_{10}O_3$	498-02-2	166.173	pr (w)	115	297; 234[15]			sl H_2O; s EtOH, ace, bz; vs eth, chl
6070	(2-Hydroxy-4-methoxyphenyl)phenylmethanone	Oxybenzone	$C_{14}H_{12}O_3$	131-57-7	228.243		65.5				s ctc

11-Hydroxy-9,15-dioxoprosta-5,13-dien-1-oic acid, (5Z,11α,13E)

1-Hydroxy-1,1-diphosphonoethane

3-Hydroxyestra-1,3,5,7,9-pentaen-17-one

3-Hydroxyestra-1,3,5(10),7-tetraen-17-one

2-Hydroxyethyl acrylate

N-(2-Hydroxyethyl)dodecanamide

N-(2-Hydroxyethyl)ethylenediaminetriacetic acid

2-Hydroxyethyl 2-hydroxybenzoate

2-Hydroxyethyl methacrylate

N-(2-Hydroxyethyl)phthalimide

1-(2-Hydroxyethyl)-2-pyrrolidinone

4-Hydroxy-4H-furo[3,2-c]pyran-2(6H)-one

16-Hydroxyhexadecanoic acid

2-Hydroxyhexanoic acid

6-Hydroxyhexanoic acid

3-Hydroxy-2-(hydroxymethyl)-2-methylpropanoic acid

5-Hydroxy-2-(hydroxymethyl)-4H-pyran-4-one

8-Hydroxy-7-iodo-5-quinolinesulfonic acid

2-Hydroxy-1H-isoindole-1,3(2H)-dione

2-Hydroxy-4-isopropyl-2,4,6-cycloheptatrien-1-one

Hydroxylupanine

N-Hydroxymethanamine

2-Hydroxy-3-methoxybenzaldehyde

2-Hydroxy-4-methoxybenzaldehyde

2-Hydroxy-5-methoxybenzaldehyde

3-Hydroxy-4-methoxybenzaldehyde

4-Hydroxy-3-methoxybenzaldehyde

4-Hydroxy-3-methoxybenzeneacetic acid

4-Hydroxy-3-methoxybenzenemethanol

4-Hydroxy-3-methoxybenzenepropanol

2-Hydroxy-5-methoxybenzoic acid

4-Hydroxy-3-methoxybenzoic acid

7-Hydroxy-6-methoxy-2H-1-benzopyran-2-one

4-(4-Hydroxy-3-methoxyphenyl)-2-butanone

1-(2-Hydroxy-4-methoxyphenyl)ethanone

1-(4-Hydroxy-3-methoxyphenyl)ethanone

(2-Hydroxy-4-methoxyphenyl)phenylmethanone

No.	Name	Synonym	Mol. Form.	CAS RN	Mol. Wt.	Physical Form	mp/°C	bp/°C	den/ g cm^{-3}	n_D	Solubility
6071	3-(4-Hydroxy-3-methoxyphenyl)-2-propenal		$C_{10}H_{10}O_3$	458-36-6	178.184	cry (bz)	84		1.1562[102]		vs bz, eth, EtOH
6072	N-Hydroxymethylamine hydrochloride	N-Methylhydroxylamine hydrochloride	CH_6ClNO	4229-44-1	83.518		83.5				
6073	4-Hydroxy-α-[(methylamino)methyl]benzenemethanol	Synephrine	$C_9H_{13}NO_2$	94-07-5	167.205		184.5				
6074	17-Hydroxy-17-methylandrostan-3-one, (5α,17β)-	Mestanolone	$C_{20}H_{32}O_2$	521-11-9	304.467		192.5				sl AcOEt
6075	N-Hydroxy-4-methylaniline		C_7H_9NO	623-10-9	123.152	lf (bz)	96	dec 117			vs eth, EtOH, chl
6076	2-Hydroxy-5-methylbenzaldehyde		$C_8H_8O_2$	613-84-3	136.149	pl (aq, al)	56	217.5	1.0913[59]	1.547[59]	vs eth, EtOH, chl
6077	α-(Hydroxymethyl)benzeneacetic acid, (±)	Tropic acid	$C_9H_{10}O_3$	552-63-6	166.173	nd, pl (al, bz, w)	118	dec			vs H_2O, eth, EtOH
6078	α-Hydroxy-α-methylbenzeneacetic acid, (±)	Atrolactic acid	$C_9H_{10}O_3$	4607-38-9	166.173	nd, pl (lig)	94				vs ace, bz
6079	2-Hydroxy-5-methyl-1,3-benzenedimethanol		$C_9H_{12}O_3$	91-04-3	168.189		130.5				
6080	2-(Hydroxymethyl)-1,4-benzenediol	Gentisyl alcohol	$C_7H_8O_3$	495-08-9	140.137	nd (chl)	100	sub 75			vs H_2O, EtOH, chl
6081	2-Hydroxy-5-methylbenzoic acid	p-Cresotic acid	$C_8H_8O_3$	89-56-5	152.148		151				sl H_2O; s EtOH, eth, bz, chl; i CS_2
6082	2-Hydroxy-3-methylbenzoic acid	o-Cresotic acid	$C_8H_8O_3$	83-40-9	152.148		165.5				sl H_2O; s EtOH, eth, bz, chl
6083	2-Hydroxy-4-methylbenzoic acid	m-Cresotic acid	$C_8H_8O_3$	50-85-1	152.148	cry, lf	177				sl H_2O; s EtOH, bz, chl; vs eth
6084	7-Hydroxy-4-methyl-2H-1-benzopyran-2-one	Hymecromone	$C_{10}H_8O_3$	90-33-5	176.169	nd (al)	194.5				sl H_2O, eth, chl; s EtOH, alk, HOAc
6085	3-Hydroxy-3-methylbutanoic acid		$C_5H_{10}O_3$	625-08-1	118.131		<-32	162[12]	0.9384[20]	1.5081[20]	vs H_2O, eth, EtOH
6086	3-Hydroxy-3-methyl-2-butanone		$C_5H_{10}O_2$	115-22-0	102.132			140	0.9526[20]		s chl
6087	2-Hydroxy-3-methyl-2-cyclopenten-1-one		$C_6H_8O_2$	80-71-7	112.127		104.8				
6088	5-(Hydroxymethyl)-2-furancarboxaldehyde	5-(Hydroxymethyl)-2-furaldehyde	$C_6H_6O_3$	67-47-0	126.110	nd (eth-peth)	31.5	115[1]	1.2062[25]	1.5627[18]	s H_2O, EtOH, bz, chl; sl eth, ctc
6089	2-Hydroxy-6-methyl-3-isopropylbenzoic acid	o-Thymotic acid	$C_{11}H_{14}O_3$	548-51-6	194.227	nd (w, bz, lig)	127	sub			vs bz, eth, EtOH
6090	2-Hydroxy-3-methyl-6-isopropyl-2-cyclohexen-1-one	Diosphenol	$C_{10}H_{16}O_2$	490-03-9	168.233		83	109[10]			
6091	2-(Hydroxymethyl)-2-methyl-1,3-propanediol		$C_5H_{12}O_3$	77-85-0	120.147	wh pow or nd (al)	204	136[15]			msc H_2O, EtOH; i eth, bz; vs HOAc
6092	2-Hydroxy-3-methyl-1,4-naphthalenedione	Phthiocol	$C_{11}H_8O_3$	483-55-6	188.180	ye pr (eth-peth)	173.5	sub			vs ace, eth
6093	5-Hydroxy-2-methyl-1,4-naphthalenedione	Plumbagin	$C_{11}H_8O_3$	481-42-5	188.180	gold pr or oran-ye nd (dil al)	78.5	sub			vs ace, bz, eth, EtOH
6094	2-(Hydroxymethyl)-2-nitro-1,3-propanediol	Tris(hydroxymethyl)nitromethane	$C_4H_9NO_5$	126-11-4	151.118	nd or pr	165	dec			vs H_2O, eth, EtOH
6095	2-Hydroxy-4-methylpentanoic acid, (S)	L-Leucic acid	$C_6H_{12}O_3$	13748-90-8	132.157	orth (eth)	81.5				vs H_2O, eth, EtOH
6096	1-(2-Hydroxy-4-methylphenyl)ethanone		$C_9H_{10}O_2$	6921-64-8	150.174		21	245	1.1012[10]	1.5527[13]	
6097	1-(2-Hydroxy-5-methylphenyl)ethanone		$C_9H_{10}O_2$	1450-72-2	150.174	pr (lig)	50	210; 120[20]	1.0797[53]		vs bz, eth, EtOH, chl
6098	2-(Hydroxymethyl)phenyl-β-D-glucopyranoside	Salicin	$C_{13}H_{18}O_7$	138-52-3	286.278	orth nd or lf (w)	207	dec 240	1.434[20]		vs H_2O, EtOH, HOAc
6099	1-(2-Hydroxy-5-methylphenyl)-1-propanone		$C_{10}H_{12}O_2$	938-45-4	164.201		1.0	129[16.5]	1.0841[14]	1.549[13]	s chl
6100	N-(Hydroxymethyl)phthalimide		$C_9H_7NO_3$	118-29-6	177.157	lf, pr (to)	141.5				i H_2O, eth, ctc; sl EtOH, bz; s tol
6101	3-Hydroxy-2-methylpropanal		$C_4H_8O_2$	38433-80-6	88.106	oil					
6102	2-Hydroxy-2-methylpropanoic acid		$C_4H_8O_3$	594-61-6	104.105	hyg pr (eth) nd (bz)	82.5	212			vs H_2O, EtOH, eth; sl bz
6103	3-Hydroxy-2-methylpropanoic acid		$C_4H_8O_3$	2068-83-9	104.105	oil					
6104	N-(Hydroxymethyl)-2-propenamide	N-(Hydroxymethyl)acrylamide	$C_4H_7NO_2$	924-42-5	101.105	cry	76				
6105	4-Hydroxy-6-methyl-2H-pyran-2-one	Triacetic acid lactone	$C_6H_6O_3$	675-10-5	126.110		189 dec				
6106	3-Hydroxy-2-methyl-4H-pyran-4-one	Maltol	$C_6H_6O_3$	118-71-8	126.110	mcl pr (chl)	161.5	sub 93			sl H_2O, eth, bz; vs chl; s alk; peth

3-(4-Hydroxy-3-methoxyphenyl)-2-propenal

N-Hydroxymethylamine hydrochloride

4-Hydroxy-α-[(methylamino)methyl]benzenemethanol

17-Hydroxy-17-methylandrostan-3-one, (5α,17β)

N-Hydroxy-4-methylaniline

2-Hydroxy-5-methylbenzaldehyde

α-(Hydroxymethyl)benzeneacetic acid, (±)

α-Hydroxy-α-methylbenzeneacetic acid, (±)

2-Hydroxy-5-methyl-1,3-benzenedimethanol

2-(Hydroxymethyl)-1,4-benzenediol

2-Hydroxy-5-methylbenzoic acid

2-Hydroxy-3-methylbenzoic acid

2-Hydroxy-4-methylbenzoic acid

7-Hydroxy-4-methyl-2H-1-benzopyran-2-one

3-Hydroxy-3-methylbutanoic acid

3-Hydroxy-3-methyl-2-butanone

2-Hydroxy-3-methyl-2-cyclopenten-1-one

5-(Hydroxymethyl)-2-furancarboxaldehyde

2-Hydroxy-6-methyl-3-isopropylbenzoic acid

2-Hydroxy-3-methyl-6-isopropyl-2-cyclohexen-1-one

2-(Hydroxymethyl)-2-methyl-1,3-propanediol

2-Hydroxy-3-methyl-1,4-naphthalenedione

5-Hydroxy-2-methyl-1,4-naphthalenedione

2-(Hydroxymethyl)-2-nitro-1,3-propanediol

2-Hydroxy-4-methylpentanoic acid, (S)

1-(2-Hydroxy-4-methylphenyl)ethanone

1-(2-Hydroxy-5-methylphenyl)ethanone

2-(Hydroxymethyl)phenyl-β-D-glucopyranoside

1-(2-Hydroxy-5-methylphenyl)-1-propanone

N-(Hydroxymethyl)phthalimide

3-Hydroxy-2-methylpropanal

2-Hydroxy-2-methylpropanoic acid

3-Hydroxy-2-methylpropanoic acid

N-(Hydroxymethyl)-2-propenamide

4-Hydroxy-6-methyl-2H-pyran-2-one

3-Hydroxy-2-methyl-4H-pyran-4-one

No.	Name	Synonym	Mol. Form.	CAS RN	Mol. Wt.	Physical Form	mp/°C	bp/°C	den/ g cm⁻³	n_D	Solubility
6107	5-Hydroxy-6-methyl-3,4-pyridinedimethanol	Pyridoxin	$C_8H_{11}NO_3$	65-23-6	169.178	nd (HOAc)	160	140$^{0.0001}$			
6108	4-Hydroxy-1-methyl-2-quinolinone	4-Hydroxy-*N*-methylcarbostyril	$C_{10}H_9NO_2$	1677-46-9	175.184		265				sl DMSO
6109	2-Hydroxy-4-(methylthio)butanoic acid	Methionine hydroxy analog	$C_5H_{10}O_3S$	583-91-5	150.196	oil					
6110	3-Hydroxy-α-methyl-*L*-tyrosine	Methyldopa	$C_{10}H_{13}NO_4$	555-30-6	211.215	cry (MeOH)	300 dec				
6111	(Hydroxymethyl)urea		$C_2H_6N_2O_2$	1000-82-4	90.081	pr (al)	111				vs H_2O; s EtOH, MeOH, HOAc; i eth
6112	2-Hydroxy-1-naphthalenecarboxaldehyde		$C_{11}H_8O_2$	708-06-5	172.181	pr (al), nd (AcOEt)	83	192^{27}			i H_2O; s EtOH, eth, aq alk, sulf, peth
6113	2-Hydroxy-1-naphthalenecarboxylic acid	2-Hydroxy-1-naphthoic acid	$C_{11}H_8O_3$	2283-08-1	188.180		157.3				sl H_2O; vs EtOH; s eth, ace, bz, lig, chl
6114	1-Hydroxy-2-naphthalenecarboxylic acid	1-Hydroxy-2-naphthoic acid	$C_{11}H_8O_3$	86-48-6	188.180	cry (al) nd (al, eth, bz)	195				sl H_2O; vs EtOH, eth; s bz
6115	3-Hydroxy-2-naphthalenecarboxylic acid	3-Hydroxy-2-naphthoic acid	$C_{11}H_8O_3$	92-70-6	188.180	nd (dil al) ye lf (dil al)	222.5				sl H_2O; vs EtOH, eth; s bz, chl, tol
6116	2-Hydroxy-1,4-naphthalenedione	Lawsone	$C_{10}H_6O_3$	83-72-7	174.153	ye pr (HOAc)	195 dec				vs EtOH; i eth, bz, chl; s HOAc
6117	5-Hydroxy-1,4-naphthalenedione	Juglone	$C_{10}H_6O_3$	481-39-0	174.153	ye nd (bz) peth)	155	sub			i H_2O; s EtOH, eth, bz; vs chl; sl lig
6118	7-Hydroxy-1,3-naphthalenedisulfonic acid	2-Naphthol-6,8-disulfonic acid	$C_{10}H_8O_7S_2$	118-32-1	304.297						s H_2O
6119	3-Hydroxy-2,7-naphthalenedisulfonic acid	2-Naphthol-3,6-disulfonic acid	$C_{10}H_8O_7S_2$	148-75-4	304.297	hyg nd	dec				vs H_2O, EtOH
6120	6-Hydroxy-2-naphthalenepropanoic acid	Allenolic acid	$C_{13}H_{12}O_3$	553-39-9	216.232	cry (dil MeOH)	180.5				vs py, EtOH, MeOH
6121	4-Hydroxy-1-naphthalenesulfonic acid	1-Naphthol-4-sulfonic acid	$C_{10}H_8O_4S$	84-87-7	224.234	tab or pl (w)	170 dec				vs H_2O; i eth
6122	7-Hydroxy-1-naphthalenesulfonic acid	Croceic acid	$C_{10}H_8O_4S$	132-57-0	224.234						s H_2O
6123	1-Hydroxy-2-naphthalenesulfonic acid	1-Naphthol-2-sulfonic acid	$C_{10}H_8O_4S$	567-18-0	224.234	pl (w)	>250				sl H_2O, dil HCl; s EtOH; i eth
6124	6-Hydroxy-2-naphthalenesulfonic acid	2-Naphthol-6-sulfonic acid	$C_{10}H_8O_4S$	93-01-6	224.234	lf, cry (w+1)	125				vs H_2O, EtOH; i eth; s HOAc
6125	Hydroxynaphthol blue, trisodium salt		$C_{20}H_{14}N_2Na_3O_{11}S_3$	63451-35-4	623.495	dk red cry					
6126	*N*-(2-Hydroxy-1-naphthyl)acetamide		$C_{12}H_{11}NO_2$	117-93-1	201.221	lf (w, dil al)	235 dec	sub			vs ace, bz, eth, EtOH
6127	1-(1-Hydroxy-2-naphthyl)ethanone		$C_{12}H_{10}O_2$	711-79-5	186.206	pr (bz, lig) grn-ye nd (al)	101	dec 325			vs bz, HOAc
6128	2-Hydroxy-3-nitrobenzaldehyde		$C_7H_5NO_4$	5274-70-4	167.120	nd (HOAc)	109.5				vs bz, EtOH
6129	2-Hydroxy-5-nitrobenzaldehyde		$C_7H_5NO_4$	97-51-8	167.120	cry (dil HOAc)	127.0				s ace
6130	2-Hydroxy-3-nitrobenzoic acid	3-Nitrosalicylic acid	$C_7H_5NO_5$	85-38-1	183.119	ye nd (HOAc, w+1)	148				sl H_2O; vs EtOH, eth; s ace, bz, chl
6131	2-Hydroxy-5-nitrobenzoic acid	5-Nitrosalicylic acid	$C_7H_5NO_5$	96-97-9	183.119	nd (w)	229.5		1.650^{20}		sl H_2O; vs EtOH, eth, ace, bz; s chl
6132	2-Hydroxy-1,2,3-nonadecanetricarboxylic acid	Agaricic acid	$C_{22}H_{40}O_7$	666-99-9	416.549	cry pow	142 dec				s H_2O; sl EtOH, eth; i bz, chl
6133	12-Hydroxyoctadecanoic acid	12-Hydroxysteric acid	$C_{18}H_{36}O_3$	106-14-9	300.477	cry (al)	82				i H_2O; s EtOH, eth, chl
6134	*cis*-12-Hydroxy-9-octadecenoic acid, (*R*)	Ricinoleic acid	$C_{18}H_{34}O_3$	141-22-0	298.461	visc liq	5.5	227^{10}	0.9450^{21}	1.4716^{21}	i H_2O; vs eth, EtOH
6135	2-Hydroxyoctanoic acid		$C_9H_{16}O_3$	617-73-2	160.211	pl	70	162^{10}			sl H_2O, chl; vs EtOH, eth
6136	5-Hydroxy-4-octanone	Butyroin	$C_9H_{16}O_2$	496-77-5	144.212	liq	-10	185	0.9107^{16}	1.4345^{16}	
6137	[2-Hydroxy-4-(octyloxy)phenyl]phenylmethanone	Octabenzone	$C_{21}H_{26}O_3$	1843-05-6	326.429		48.5				
6138	3-Hydroxy-2-oxopropanoic acid	Hydroxypyruvic acid	$C_3H_4O_4$	1113-60-6	104.062		81 dec				
6139	3-Hydroxy-6-oxo-4*H*-pyran-2,6-dicarboxylic acid	Meconic acid	$C_7H_4O_7$	497-59-6	200.103	orth pl (w, dil HCl) (+3w)	120 dec				sl H_2O, MeOH, ace, eth; s EtOH, bz
6140	2-Hydroxypentanoic acid		$C_5H_{10}O_3$	617-31-2	118.131	hyg pl	34	sub			s H_2O, EtOH, eth
6141	5-Hydroxy-2-pentanone		$C_5H_{10}O_2$	1071-73-4	102.132			209; 117^{33}	1.0071^{20}	1.4390^{20}	msc H_2O; s EtOH, eth

5-Hydroxy-6-methyl-3,4-pyridinedimethanol

4-Hydroxy-1-methyl-2-quinolinone

2-Hydroxy-4-(methylthio)butanoic acid

3-Hydroxy-α-methyl-L-tyrosine

(Hydroxymethyl)urea

2-Hydroxy-1-naphthalenecarboxaldehyde

2-Hydroxy-1-naphthalenecarboxylic acid

1-Hydroxy-2-naphthalenecarboxylic acid

3-Hydroxy-2-naphthalenecarboxylic acid

2-Hydroxy-1,4-naphthalenedione

5-Hydroxy-1,4-naphthalenedione

7-Hydroxy-1,3-naphthalenedisulfonic acid

3-Hydroxy-2,7-naphthalenedisulfonic acid

6-Hydroxy-2-naphthalenepropanoic acid

4-Hydroxy-1-naphthalenesulfonic acid

7-Hydroxy-1-naphthalenesulfonic acid

1-Hydroxy-2-naphthalenesulfonic acid

6-Hydroxy-2-naphthalenesulfonic acid

Hydroxynaphthol blue, trisodium salt

N-(2-Hydroxy-1-naphthyl)acetamide

1-(1-Hydroxy-2-naphthyl)ethanone

2-Hydroxy-3-nitrobenzaldehyde

2-Hydroxy-5-nitrobenzaldehyde

2-Hydroxy-3-nitrobenzoic acid

2-Hydroxy-5-nitrobenzoic acid

2-Hydroxy-1,2,3-nonadecanetricarboxylic acid

12-Hydroxyoctadecanoic acid

cis-12-Hydroxy-9-octadecenoic acid, (R)

2-Hydroxyoctanoic acid

5-Hydroxy-4-octanone

[2-Hydroxy-4-(octyloxy)phenyl]phenylmethanone

3-Hydroxy-2-oxopropanoic acid

3-Hydroxy-4-oxo-4H-pyran-2,6-dicarboxylic acid

2-Hydroxypentanoic acid

5-Hydroxy-2-pentanone

No.	Name	Synonym	Mol. Form.	CAS RN	Mol. Wt.	Physical Form	mp/°C	bp/°C	den/ g cm⁻³	n_D	Solubility
6142	7-Hydroxy-3*H*-phenoxazin-3-one	Resorufine	C₁₂H₇NO₃	635-78-9	213.189	br nd (PhNO₂) pr (HCl)					i H₂O; sl EtOH; i eth; vs alk
6143	*N*-(2-Hydroxyphenyl)acetamide		C₈H₉NO₂	614-80-2	151.163	pl (dil al)	209				sl H₂O; vs EtOH, eth, bz; s DMSO
6144	*N*-(3-Hydroxyphenyl)acetamide		C₈H₉NO₂	621-42-1	151.163	nd (w)	148.5				vs H₂O, EtOH; sl eth, bz, chl, DMSO
6145	*N*-(4-Hydroxyphenyl)acetamide	Acetaminophen	C₈H₉NO₂	103-90-2	151.163	mcl pr (w)	170		1.293²¹		i H₂O; vs EtOH
6146	2-[(4-Hydroxyphenyl)azo]benzoic acid		C₁₃H₁₀N₂O₃	1634-82-8	242.229		206				sl DMSO
6147	2-Hydroxy-*N*-phenylbenzamide	Salicylanilide	C₁₃H₁₁NO₂	87-17-2	213.232	pr (w, al)	136.5				s H₂O; sl EtOH, eth, bz, chl
6148	*N*-Hydroxy-*N*-phenylbenzamide		C₁₃H₁₁NO₂	304-88-1	213.232		120.3				
6149	α-Hydroxy-α-phenylbenzeneacetic acid	Benzilic acid	C₁₄H₁₂O₃	76-93-7	228.243	mcl nd (w)	150	dec 180			sl H₂O, ace; vs EtOH, eth; s con sulf
6150	3-Hydroxy-2-phenyl-4*H*-1-benzopyran-4-one		C₁₅H₁₀O₃	577-85-5	238.238	pa ye nd (al)	169.5				s EtOH
6151	*N*-(4-Hydroxyphenyl)butanamide	4'-Hydroxybutyranilide	C₁₀H₁₃NO₂	101-91-7	179.216	nd (w)	139.5				vs H₂O, EtOH
6152	4-(4-Hydroxyphenyl)-2-butanone		C₁₀H₁₂O₂	5471-51-2	164.201		82.5				
6153	1-(2-Hydroxyphenyl)ethanone		C₈H₈O₂	118-93-4	136.149		2.5	218	1.1307²⁰	1.5584²⁰	vs eth, EtOH, HOAc
6154	1-(3-Hydroxyphenyl)ethanone		C₈H₈O₂	121-71-1	136.149	nd or lf	96	296; 153⁵	1.0992¹⁰⁹	1.5348¹⁰⁹	sl H₂O; vs EtOH, eth, bz, chl; i lig
6155	1-(4-Hydroxyphenyl)ethanone		C₈H₈O₂	99-93-4	136.149	nd (eth, dil al)	109.5	147³	1.1090¹⁰⁹	1.5577¹⁰⁹	sl H₂O, DMSO; vs EtOH, eth
6156	4-Hydroxyphenyl-β-*D*-glucopyranoside	Arbutin	C₁₂H₁₆O₇	497-76-7	272.251	nd (w+1)	199.5				vs H₂O; s EtOH; sl eth; i bz, chl, CS₂
6157	2-(4-Hydroxyphenyl)-*D*-glycine	Oxfenicine	C₈H₉NO₃	22818-40-2	167.162	cry	240 dec				
6158	*N*-(4-Hydroxyphenyl)glycine		C₈H₉NO₃	122-87-2	167.162	lf (w) pl (w)	246 dec				sl H₂O, EtOH; i eth; s AcOEt, chl
6159	2(2-Hydroxyphenyl)-2(4-hydroxyphenyl)propane	2,4'-Isopropylidenediphenol	C₁₅H₁₆O₂	837-08-1	228.287	cry (bz)	111				
6160	2-[[(2-Hydroxyphenyl)imino]methyl]phenol	*N*-Salicylidene-*o*-aminophenol	C₁₃H₁₁NO₂	1761-56-4	213.232		185				
6161	*N*-Hydroxy-*N*-(phenylmethyl)benzenemethanamine		C₁₄H₁₅NO	621-07-8	213.275		122.5				s chl
6162	*N*-(4-Hydroxyphenyl)octadecanamide		C₂₄H₄₁NO₂	103-99-1	375.589		133.8	239.5¹⁰			i H₂O; sl eth, bz, chl; s ace
6163	3-(4-Hydroxyphenyl)-2-oxopropanoic acid	4-Hydroxy-α-oxobenzenepropanoic acid	C₉H₈O₄	156-39-8	180.158	cry (w)	220 dec				s H₂O; dec alk
6164	(2-Hydroxyphenyl)phenylmethanone		C₁₃H₁₀O₂	117-99-7	198.217	pl (dil al)	40	250⁵⁶⁰			i H₂O; vs EtOH, eth, bz; sl chl, peth
6165	1-(2-Hydroxyphenyl)-3-phenyl-2-propen-1-one	2'-Hydroxychalcone	C₁₅H₁₂O₂	1214-47-7	224.255		90				
6166	2-Hydroxy-1-phenyl-1-propanone		C₉H₁₀O₂	5650-40-8	150.174	ye oil		251	1.1085¹⁸	1.536²³	
6167	1-(2-Hydroxyphenyl)-1-propanone		C₉H₁₀O₂	610-99-1	150.174		150⁸⁰; 115¹⁵		1.5501²⁰		sl H₂O; s EtOH, eth, ctc, alk
6168	1-(4-Hydroxyphenyl)-1-propanone	Paroxypropione	C₉H₁₀O₂	70-70-2	150.174	wh nd or pl (w)	149				sl H₂O, ace; s EtOH, eth, alk
6169	3-(4-Hydroxyphenyl)-2-propenoic acid	*p*-Coumaric acid	C₉H₈O₃	7400-08-0	164.158	nd	211.5				vs eth, EtOH
6170	3-Hydroxy-2-phenyl-4-quinolinecarboxylic acid	Oxycinchophen	C₁₆H₁₁NO₃	485-89-2	265.263	ye pr (al)	206 dec				vs bz, EtOH, HOAc
6171	*N*-Hydroxypiperidine	1-Piperidinol	C₅H₁₁NO	4801-58-5	101.147	hyg	39.3	110⁵⁵			
6172	3-Hydroxypregnan-20-one, (3α,5α)	Allopregnan-3α-ol-20-one	C₂₁H₃₄O₂	516-54-1	318.494	cry (al)	177				
6173	3-Hydroxypregnan-20-one, (3β,5α)	Allopregnan-3β-ol-20-one	C₂₁H₃₄O₂	516-55-2	318.494		189.5				
6174	17-Hydroxypregn-4-ene-3,20-dione	17α-Hydroxyprogesterone	C₂₁H₃₀O₃	68-96-2	330.461						sl chl
6175	21-Hydroxypregn-4-ene-3,20-dione	Deoxycorticosterone	C₂₁H₃₀O₃	64-85-7	330.461	pl (eth)	141.5				sl H₂O, eth; vs EtOH, ace; s chl
6176	21-Hydroxypregn-4-ene-3,11,20-trione	11-Dehydrocorticosterone	C₂₁H₂₈O₄	72-23-1	344.445	pr (ace-w, al, ace-eth)	183.5				i H₂O; s EtOH, ace, bz
6177	*cis*-4-Hydroxy-*L*-proline		C₅H₉NO₃	618-27-9	131.130	nd (w+1)	239.5				vs H₂O
6178	*trans*-4-Hydroxy-*L*-proline		C₅H₉NO₃	51-35-4	131.130	lf (dil al) pr (w)	274				vs H₂O; sl EtOH

7-Hydroxy-3H-phenoxazin-3-one

N-(2-Hydroxyphenyl)acetamide

N-(3-Hydroxyphenyl)acetamide

N-(4-Hydroxyphenyl)acetamide

2-[(4-Hydroxyphenyl)azo]benzoic acid

2-Hydroxy-N-phenylbenzamide

N-Hydroxy-N-phenylbenzamide

α-Hydroxy-α-phenylbenzeneacetic acid

3-Hydroxy-2-phenyl-4H-1-benzopyran-4-one

N-(4-Hydroxyphenyl)butanamide

4-(4-Hydroxyphenyl)-2-butanone

1-(2-Hydroxyphenyl)ethanone

1-(3-Hydroxyphenyl)ethanone

1-(4-Hydroxyphenyl)ethanone

4-Hydroxyphenyl-β-D-glucopyranoside

2-(4-Hydroxyphenyl)-D-glycine

N-(4-Hydroxyphenyl)glycine

2(2-Hydroxyphenyl)-2(4-hydroxyphenyl)propane

2-[[(2-Hydroxyphenyl)imino]methyl]phenol

N-Hydroxy-N-(phenylmethyl)benzenemethanamine

N-(4-Hydroxyphenyl)octadecanamide

3-(4-Hydroxyphenyl)-2-oxopropanoic acid

(2-Hydroxyphenyl)phenylmethanone

1-(2-Hydroxyphenyl)-3-phenyl-2-propen-1-one

2-Hydroxy-1-phenyl-1-propanone

1-(2-Hydroxyphenyl)-1-propanone

1-(4-Hydroxyphenyl)-1-propanone

3-(4-Hydroxyphenyl)-2-propenoic acid

3-Hydroxy-2-phenyl-4-quinolinecarboxylic acid

N-Hydroxypiperidine

3-Hydroxypregnan-20-one, (3α,5α)

3-Hydroxypregnan-20-one, (3β,5α)

17-Hydroxypregn-4-ene-3,20-dione

21-Hydroxypregn-4-ene-3,20-dione

21-Hydroxypregn-4-ene-3,11,20-trione

cis-4-Hydroxy-L-proline

trans-4-Hydroxy-L-proline

No.	Name	Synonym	Mol. Form.	CAS RN	Mol. Wt.	Physical Form	mp/°C	bp/°C	den/ g cm⁻³	n_D	Solubility
6179	3-Hydroxypropanal	Hydracrolein	$C_3H_6O_2$	2134-29-4	74.079			90^{18}, $38^{0.2}$			vs ace, eth, EtOH
6180	Hydroxypropanedioic acid	Tartronic acid	$C_3H_4O_5$	80-69-3	120.061	pr (w+1)	157	sub			s H_2O, EtOH; sl eth
6181	2-Hydroxypropanenitrile	Acetaldehyde cyanohydrin	C_3H_5NO	78-97-7	71.078	liq	-40	183	0.9877^{20}	1.4058^{18}	msc H_2O, EtOH; s eth, chl; i CS_2, peth
6182	3-Hydroxypropanenitrile	Hydracrylonitrile	C_3H_5NO	109-78-4	71.078	liq	-46	221	1.0404^{25}	1.4248^{20}	msc H_2O, EtOH; sl eth; s chl; i CS_2
6183	3-Hydroxypropanoic acid	Hydracrylic acid	$C_3H_6O_3$	503-66-2	90.078	syr		dec		1.4489^{20}	vs H_2O; s EtOH; msc eth
6184	1-Hydroxy-2-propanone	Acetone alcohol	$C_3H_6O_2$	116-09-6	74.079	hyg liq	-17	145.5	1.0805^{20}	1.4295^{20}	vs H_2O, EtOH, eth
6185	4-(3-Hydroxy-1-propenyl)-2-methoxyphenol	Coniferyl alcohol	$C_{10}H_{12}O_3$	458-35-5	180.200	pr (eth-lig)	74	164^3			i H_2O; s EtOH, alk; vs eth
6186	2-Hydroxypropyl acrylate		$C_6H_{10}O_3$	999-61-1	130.141	liq		70^2			
6187	(2-Hydroxypropyl) trimethylammonium chloride		$C_6H_{16}ClNO$	2382-43-6	153.650	pr (Bu OH)	165	dec			vs H_2O, EtOH
6188	3-Hydroxy-1H-pyridin-2-one		$C_5H_5NO_2$	16867-04-2	111.100			245 dec			
6189	1-Hydroxy-2,5-pyrrolidinedione	N-Hydroxysuccinimide	$C_4H_5NO_3$	6066-82-6	115.088	hyg	96.3				sl DMSO
6190	4-Hydroxy-2-quinolinecarboxylic acid	Kynurenic acid	$C_{10}H_7NO_3$	492-27-3	189.168	ye nd (+w, dil al)	282.5				sl H_2O; s EtOH; i eth; vs alk
6191	8-Hydroxy-5-quinolinesulfonic acid		$C_9H_7NO_4S$	84-88-8	225.222	ye lf, nd (+1w) (dil HCl)	322.5				sl H_2O
6192	4-Hydroxy-2-quinolinone	2,4-Quinolinediol	$C_9H_7NO_2$	86-95-3	161.158		360 dec				sl EtOH, $PhNO_2$, gl HOAc
6193	3-Hydroxyspirostan-12-one, (3β,5α,25R)	Hecogenin	$C_{27}H_{42}O_4$	467-55-0	430.620	pl (eth)	266.5				vs ace, eth, EtOH
6194	4-Hydroxystyrene	4-Vinylphenol	C_8H_8O	2628-17-3	120.149		73.5				
6195	2-Hydroxy-5-sulfobenzoic acid	5-Sulfosalicylic acid	$C_7H_6O_6S$	97-05-2	218.184	hyg nd	120				vs H_2O; vs EtOH, eth
6196	2-Hydroxy-5-sulfobenzoic acid dihydrate	5-Sulfosalicylic acid dihydrate	$C_7H_{10}O_8S$	5965-83-3	254.214	wh cry (w)					vs H_2O; vs EtOH, eth
6197	4-Hydroxy-2,2,6,6-tetramethylpiperidine	2,2,6,6-Tetramethyl-4-piperidinol	$C_9H_{19}NO$	2403-88-5	157.253		130	213.5			
6198	5-Hydroxytryptamine	3-(2-Aminoethyl)indol-5-ol	$C_{10}H_{12}N_2O$	50-67-9	176.214						s H_2O
6199	5-Hydroxy-DL-tryptophan		$C_{11}H_{12}N_2O_3$	114-03-4	220.224	rod or nd (al)	300 dec				
6200	Hydroxyurea		$CH_4N_2O_2$	127-07-1	76.055	nd (al)	141	dec			vs H_2O
6201	Hydroxyzine		$C_{21}H_{27}ClN_2O_2$	68-88-2	374.904	oil		$220^{0.5}$			
6202	Hymecromone O,O-diethyl phosphorothioate		$C_{14}H_{17}O_5PS$	299-45-6	328.321	nd	38	$210^{1.0}$ dec	1.260^{38}	1.5685^{37}	vs H_2O; sl peth
6203	Hymenoxone		$C_{15}H_{22}O_5$	57377-32-9	282.333	cry					
6204	Hyoscyamine		$C_{17}H_{23}NO_3$	101-31-5	289.370	tetr nd (dil al)	108.5				sl H_2O, eth, bz; vs EtOH, chl
6205	Hypoglycin A		$C_7H_{11}NO_2$	156-56-9	141.168	ye pl (Me aq)	282				
6206	Hypoxanthine		$C_5H_4N_4O$	68-94-0	136.112	oct nd (w)	150 dec				sl H_2O; s alk, dil acid
6207	Ibuprofen	2-(4-Isobutylphenyl)propanoic acid	$C_{13}H_{18}O_2$	15687-27-1	206.281	col cry	76				sl H_2O; s os
6208	Icosylamine	1-Eicosanamine	$C_{20}H_{43}N$	10525-37-8	297.562			372.4			
6209	D-Idose		$C_6H_{12}O_6$	5978-95-0	180.155	syr					vs H_2O
6210	L-Idose		$C_6H_{12}O_6$	5934-56-5	180.155	syr					vs H_2O
6211	Imazalil		$C_{14}H_{14}Cl_2N_2O$	35554-44-0	297.179		50	dec	1.243^{23}		
6212	Imazapyr		$C_{13}H_{15}N_3O_3$	81334-34-1	261.276		171				
6213	Imazaquin		$C_{17}H_{17}N_3O_3$	81335-37-7	311.335		221				
6214	Imazethapyr		$C_{15}H_{19}N_3O_3$	81335-77-5	289.330		173				
6215	Imidazole	1,3-Diazole	$C_3H_4N_2$	288-32-4	68.077	mcl pr (bz)	89.5	257	1.0303^{101}	1.4801^{101}	vs H_2O, EtOH; s eth, ace, py; sl bz
6216	1H-Imidazole-4,5-dicarboxylic acid		$C_5H_4N_2O_4$	570-22-9	156.097	pr	290 dec		1.749^{25}		sl H_2O, py; i EtOH, eth, bz
6217	1H-Imidazole-4-ethanamine, dihydrochloride		$C_5H_{11}Cl_2N_3$	56-92-8	184.066	pl (eth-HOAc), pr (w)	251.3		1.43^{20}		vs H_2O, MeOH
6218	2,4-Imidazolidinedione	Hydantoin	$C_3H_4N_2O_2$	461-72-3	100.076	nd (MeOH), lf (w)	220				s H_2O, EtOH, alk; sl eth; i peth
6219	2-Imidazolidinethione	Ethylene thiourea	$C_3H_6N_2S$	96-45-7	102.158	nd (al), pr (al)	203				vs H_2O; s EtOH; i eth, bz, chl; sl DMSO
6220	Imidazolidinetrione	Parabanic acid	$C_3H_2N_2O_3$	120-89-8	114.059	mcl nd (w)	244 dec	sub 100			s H_2O; vs EtOH

3-Hydroxypropanal

Hydroxypropanedioic acid

2-Hydroxypropanenitrile

3-Hydroxypropanenitrile

3-Hydroxypropanoic acid

1-Hydroxy-2-propanone

4-(3-Hydroxy-1-propenyl)-2-methoxyphenol

2-Hydroxypropyl acrylate

(2-Hydroxypropyl)trimethylammonium chloride

3-Hydroxy-1*H*-pyridin-2-one

1-Hydroxy-2,5-pyrrolidinedione

4-Hydroxy-2-quinolinecarboxylic acid

8-Hydroxy-5-quinolinesulfonic acid

4-Hydroxy-2-quinolinone

3-Hydroxyspirostan-12-one, (3β,5α,25*R*)

4-Hydroxystyrene

2-Hydroxy-5-sulfobenzoic acid

2-Hydroxy-5-sulfobenzoic acid dihydrate

4-Hydroxy-2,2,6,6-tetramethylpiperidine

5-Hydroxytryptamine

5-Hydroxy-*DL*-tryptophan

Hydroxyurea

Hydroxyzine

Hymecromone O,O-diethyl phosphorothioate

Hymenoxone

Hyoscyamine

Hypoglycin A

Hypoxanthine

Ibuprofen

Icosylamine

D-Idose

L-Idose

Imazalil

Imazapyr

Imazaquin

Imazethapyr

Imidazole

1*H*-Imidazole-4,5-dicarboxylic acid

1*H*-Imidazole-4-ethanamine, dihydrochloride

2,4-Imidazolidinedione

2-Imidazolidinethione

Imidazolidinetrione

No.	Name	Synonym	Mol. Form.	CAS RN	Mol. Wt.	Physical Form	mp/°C	bp/°C	den/ g cm⁻³	n_D	Solubility
6221	2-Imidazolidinone	Ethylene urea	$C_3H_6N_2O$	120-93-4	86.092		131				vs H_2O, EtOH; sl eth, chl
6222	Imidodicarbonic diamide	Biuret	$C_2H_5N_3O_2$	108-19-0	103.080	pl (al), nd (w+1)	190 dec				sl H_2O; vs EtOH; i eth
6223	3,3'-Iminobispropanenitrile	Bis(2-cyanoethyl)amine	$C_6H_9N_3$	111-94-4	123.155		-6	162[5]	1.0165[20]		
6224	Iminodiacetic acid		$C_4H_7NO_4$	142-73-4	133.104	orth pr	247.5				sl H_2O; i EtOH, eth
6225	Iminodiacetic acid, dinitrile	2,2'-Iminobisacetonitrile	$C_4H_5N_3$	628-87-5	95.103		78				s H_2O, EtOH; sl eth, bz, chl
6226	Imipramine		$C_{19}H_{24}N_2$	50-49-7	280.407			160[0.1]			
6227	Imipramine hydrochloride	Tofranil	$C_{19}H_{25}ClN_2$	113-52-0	316.868		174.5				vs H_2O; s EtOH; sl ace
6228	Imperatorin		$C_{16}H_{14}O_4$	482-44-0	270.280	cry (al)	102				sl H_2O; s EtOH, eth, bz, peth; vs chl
6229	Indaconitine		$C_{34}H_{47}NO_{10}$	4491-19-4	629.738	cry	202 dec				vs eth, EtOH, chl
6230	Indalone	Butopyronoxyl	$C_{12}H_{18}O_4$	532-34-3	226.269	ye-red liq		263	1.057[20]	1.475[25]	i H_2O; vs EtOH, eth, chl
6231	Indan		C_9H_{10}	496-11-7	118.175	liq	-51.38	177.97	0.9639[20]	1.5378[20]	i H_2O; msc EtOH, eth; sl chl
6232	1-Indanamine	1-Aminoindane	$C_9H_{11}N$	34698-41-4	133.190			221; 96[8]	1.038[15]	1.5613[20]	sl H_2O; s eth, ace, bz
6233	1H-Indazole	1H-Benzopyrazole	$C_7H_6N_2$	271-44-3	118.136	nd (al, w)	148	269			s H_2O, EtOH, eth
6234	1H-Indazol-3-ol	1,2-Dihydro-3H-indazol-3-one	$C_7H_6N_2O$	7364-25-2	134.135	nd or lf (MeOH) pl or nd (al)	252.5				sl H_2O, eth; s MeOH, EtOH
6235	Indene	Indonaphthene	C_9H_8	95-13-6	116.160	liq	-1.5	182	0.9960[25]	1.5768[20]	i H_2O; msc EtOH, eth; s ace, bz, py; sl chl
6236	1H-Indene-1,3(2H)-dione		$C_9H_6O_2$	606-23-5	146.143	nd (eth, lig)	131 dec		1.37[21]		sl H_2O, ctc; vs EtOH; s eth, bz, alk
6237	1H-Indene-1,2,3-trione monohydrate	Ninhydrin	$C_9H_6O_4$	485-47-2	178.142	pa ye pr (w, al)	242 dec				vs H_2O; s EtOH, alk; sl eth
6238	Indeno[1,2,3-cd]pyrene	1,10-(1,2-Phenylene)pyrene	$C_{22}H_{12}$	193-39-5	276.330	ye cry (cy)	162				
6239	Indigo		$C_{16}H_{10}N_2O_2$	482-89-3	262.262	dk bl pow	390 dec	sub 300			
6240	5,5'-Indigodisulfonic acid, disodium salt	Indigo Carmine	$C_{16}H_8N_2Na_2O_8S_2$	860-22-0	466.353	dk-bl pow					sl H_2O, EtOH; i os
6241	Indocyanine green		$C_{43}H_{47}N_2NaO_6S_2$	3599-32-4	774.962	grn pow	244 dec				
6242	1H-Indol-5-amine		$C_8H_8N_2$	5192-03-0	132.163		132				
6243	1H-Indole	2,3-Benzopyrrole	C_8H_7N	120-72-9	117.149	lf (w, peth) cry (eth)	52.5	253.6	1.22[25]		s H_2O, bz; vs EtOH, eth, tol; sl ctc
6244	1H-Indole-3-acetic acid	Indoleacetic acid	$C_{10}H_9NO_2$	87-51-4	175.184	lf (bz), pl (chl)	168.5				i H_2O; vs EtOH; s eth, ace, bz; sl chl
6245	1H-Indole-3-acetonitrile		$C_{10}H_8N_2$	771-51-7	156.184		36	160[0.2]			
6246	1H-Indole-3-butanoic acid	Indolebutyric acid	$C_{12}H_{13}NO_2$	133-32-4	203.237		124.5				vs bz; s DMSO; i peth
6247	1H-Indole-3-carboxaldehyde		C_9H_7NO	487-89-8	145.158		197.8				
6248	1H-Indole-2,3-dione	Isatin	$C_8H_5NO_2$	91-56-5	147.132	oran mcl pr	203 dec				s H_2O, ace, bz; vs EtOH; sl eth
6249	1H-Indole-2,3-dione, 3-thiosemicarbazone	Isatin, 3-thiosemicarbazone	$C_9H_8N_4OS$	487-16-1	220.251		283				
6250	1H-Indole-3-ethanamine, monohydrochloride	Tryptamine hydrochloride	$C_{10}H_{13}ClN_2$	343-94-2	196.676	nd (al-bz or lig)	255				vs ace, EtOH
6251	1H-Indole-3-ethanol	Tryptophol	$C_{10}H_{11}NO$	526-55-6	161.200	pr (bz-peth)	59	174[2]			vs ace, eth, EtOH, chl
6252	1H-Indole-3-lactic acid, (S)	α-Hydroxy-1H-indole-3-propanoic acid	$C_{11}H_{11}NO_3$	7417-65-4	205.210	cry (peth)	100				
6253	1H-Indole-3-propanoic acid		$C_{11}H_{11}NO_2$	830-96-6	189.211		134.5				sl H_2O, DMSO; vs EtOH, eth, ace, bz
6254	Indolizine		C_8H_7N	274-40-8	117.149	pl	75	205			i H_2O; s EtOH
6255	1H-Indol-3-ol, acetate		$C_{10}H_9NO_2$	608-08-2	175.184		129				
6256	1-(1H-Indol-3-yl)ethanone		$C_{10}H_9NO$	703-80-0	159.184	nd (bz)	192.3	144[10]			vs EtOH
6257	1-(1H-Indol-3-yl)-2-propanone	3-Indolylacetone	$C_{11}H_{11}NO$	1201-26-9	173.211	br orth (bz), nd (aq MeOH)	116				
6258	3-(1H-Indol-3-yl)-2-propenoic acid	3-Indolylacrylic acid	$C_{11}H_9NO_2$	1204-06-4	187.195		185 dec				

2-Imidazolidinone

Imidodicarbonic diamide

3,3'-Iminobispropanenitrile

Iminodiacetic acid

Iminodiacetic acid, dinitrile

Imipramine

Imipramine hydrochloride

Imperatorin

Indaconitine

Indalone

Indan

1-Indanamine

1H-Indazole

1H-Indazol-3-ol

Indene

1H-Indene-1,3(2H)-dione

1H-Indene-1,2,3-trione monohydrate

Indeno[1,2,3-cd]pyrene

Indigo

5,5'-Indigodisulfonic acid, disodium salt

Indocyanine green

1H-Indol-5-amine

1H-Indole

1H-Indole-3-acetic acid

1H-Indole-3-acetonitrile

1H-Indole-3-butanoic acid

1H-Indole-3-carboxaldehyde

1H-Indole-2,3-dione

1H-Indole-2,3-dione, 3-thiosemicarbazone

1H-Indole-3-ethanamine, monohydrochloride

1H-Indole-3-ethanol

1H-Indole-3-lactic acid, (S)

1H-Indole-3-propanoic acid

Indolizine

1H-Indol-3-ol, acetate

1-(1H-Indol-3-yl)ethanone

1-(1H-Indol-3-yl)-2-propanone

3-(1H-Indol-3-yl)-2-propenoic acid

No.	Name	Synonym	Mol. Form.	CAS RN	Mol. Wt.	Physical Form	mp/°C	bp/°C	den/ g cm⁻³	n_D	Solubility
6259	Indomethacin		$C_{19}H_{16}ClNO_4$	53-86-1	357.788		155 (form a); 162 (form b				
6260	Inosine	Hypoxanthine riboside	$C_{10}H_{12}N_4O_5$	58-63-9	268.226	pl (w + 2), nd (80% al)	218 dec				sl H_2O; vs EtOH
6261	Inosine 5'-monophosphate	5'-Inosinic acid	$C_{10}H_{13}N_4O_8P$	131-99-7	348.206	visc liq or glass					vs H_2O; sl EtOH, eth
6262	*myo*-Inositol	(1α,2α,3α,4β,5α,6β)-Cyclohexanehexol	$C_6H_{12}O_6$	87-89-8	180.155	cry (w)	225		1.752		s H_2O
6263	Iocetamic acid		$C_{12}H_{13}I_3N_2O_3$	16034-77-8	613.955	wh-ye pow	225				i H_2O; sl EtOH, bz, eth, ace
6264	Iodipamide		$C_{20}H_{14}I_6N_2O_6$	606-17-7	1139.761		307 dec				i H_2O, bz; sl EtOH, eth, ace
6265	2-Iodoacetamide		C_2H_4INO	144-48-9	184.963		93.0				s H_2O; sl tfa
6266	Iodoacetic acid		$C_2H_3IO_2$	64-69-7	185.948		82.5	dec			s H_2O, EtOH, peth; sl eth, chl
6267	Iodoacetone		C_3H_5IO	3019-04-3	183.975			62¹²	2.17¹⁵		s EtOH
6268	Iodoacetonitrile		C_2H_2IN	624-75-9	166.948			185	2.307²⁵	1.5744²⁰	
6269	Iodoacetylene		C_2HI	14545-08-5	151.933			32			
6270	2-Iodoaniline		C_6H_6IN	615-43-0	219.023	nd (dil al)	60.5				sl H_2O; vs EtOH, eth, ace
6271	3-Iodoaniline		C_6H_6IN	626-01-7	219.023	lf	33	145¹⁵		1.6811²⁰	i H_2O; s EtOH, chl
6272	4-Iodoaniline		C_6H_6IN	540-37-4	219.023	nd (w)	67.5				sl H_2O, peth; s EtOH, eth
6273	2-Iodobenzaldehyde		C_7H_5IO	26260-02-6	232.018		37	129¹⁴			sl H_2O; s ace
6274	4-Iodobenzaldehyde		C_7H_5IO	15164-44-0	232.018		77.5	265			sl H_2O; s EtOH, bz
6275	Iodobenzene		C_6H_5I	591-50-4	204.008	liq	-31.3	188.4	1.8308²⁰	1.6200²⁰	i H_2O; s EtOH; msc eth, ace, bz, ctc
6276	2-Iodobenzenemethanol		C_7H_7IO	5159-41-1	234.034		92	148³²		1.6349²⁰	
6277	4-Iodobenzenesulfonyl chloride	Pipsyl chloride	$C_6H_4ClIO_2S$	98-61-3	302.517		85				
6278	2-Iodobenzoic acid		$C_7H_5IO_2$	88-67-5	248.018	nd (w)	163	exp 233	2.25²⁵		sl H_2O, ace; vs EtOH, eth
6279	3-Iodobenzoic acid		$C_7H_5IO_2$	618-51-9	248.018	mcl pr (ace)	188.3	sub			sl H_2O, eth; vs EtOH
6280	4-Iodobenzoic acid		$C_7H_5IO_2$	619-58-9	248.018	mcl pr (dil al) lf (sub)	270	sub	2.184²⁰		i H_2O; sl EtOH; s eth, DMSO
6281	4-Iodobenzonitrile		C_7H_4IN	3058-39-7	229.018		127.5				
6282	2-Iodobenzoyl chloride		C_7H_4ClIO	609-67-6	266.463		38.3	159²⁷, 135¹⁹			
6283	4-Iodobenzoyl chloride		C_7H_4ClIO	1711-02-0	266.463		65.5	164³²			
6284	2-Iodo-1,1'-biphenyl		$C_{12}H_9I$	2113-51-1	280.103			190³⁶, 169¹⁷	1.5511²⁵	1.6620²⁰	i H_2O; s EtOH, eth, bz, HOAc
6285	3-Iodo-1,1'-biphenyl		$C_{12}H_9I$	20442-79-9	280.103		26.5	188¹⁶	1.5967²⁵		
6286	4-Iodo-1,1'-biphenyl		$C_{12}H_9I$	1591-31-7	280.103	nd (al, HOAc)	113.5	320; 183¹¹			i H_2O; s EtOH, eth, bz, HOAc
6287	1-Iodobutane	Butyl iodide	C_4H_9I	542-69-8	184.018	liq	-103	130.5	1.6154²⁰	1.5001²⁰	i H_2O; msc EtOH, eth; vs chl
6288	2-Iodobutane, (±)	(±)-*sec*-Butyl iodide	C_4H_9I	52152-71-3	184.018	liq	-104.2	120.1	1.5920²⁰	1.4991²⁰	i H_2O; msc EtOH, eth; vs chl
6289	Iodocyclohexane	Cyclohexyl iodide	$C_6H_{11}I$	626-62-0	210.055			dec 180; 81²⁰	1.6244²⁰	1.5477²⁰	i H_2O; s EtOH, eth, ace, bz
6290	Iodocyclopentane	Cyclopentyl iodide	C_5H_9I	1556-18-9	196.029			166.5	1.7096²⁰	1.5447²⁰	i H_2O; s eth, bz; sl ctc
6291	1-Iododecane		$C_{10}H_{21}I$	2050-77-3	268.178	liq	-16.3	263.7; 132¹⁵	1.2546²⁰	1.4858²⁰	i H_2O; s EtOH, eth, ctc
6292	1-Iodo-2,4-dimethylbenzene		C_8H_9I	4214-28-2	232.061			dec 231; 111¹⁴	1.6282¹⁶	1.6008¹⁶	i H_2O; s ace, bz
6293	2-Iodo-1,3-dimethylbenzene		C_8H_9I	608-28-6	232.061	oil	11.2	229.5	1.6158²⁰	1.6035²⁰	i H_2O; s ace, bz
6294	1-Iodo-1,4-dimethylbenzene		C_8H_9I	1122-42-5	232.061			dec 227	1.6168¹⁷	1.5992¹⁷	i H_2O; s ace, bz
6295	1-Iodo-2,2-dimethylpropane		$C_5H_{11}I$	15501-33-4	198.045			dec 128	1.4940²⁰	1.4890²⁰	i H_2O; s EtOH, eth
6296	1-Iododododecane	Lauryl iodide	$C_{12}H_{25}I$	4292-19-7	296.231		0.3	298.2	1.1999²⁰	1.4840²⁰	i H_2O; s EtOH, MeOH; msc eth, ace, ctc
6297	Iodoethane	Ethyl iodide	C_2H_5I	75-03-6	155.965	liq	-111.1	72.3	1.9357²⁰	1.5133²⁰	sl H_2O; msc EtOH; s eth, chl
6298	2-Iodoethanol		C_2H_5IO	624-76-0	171.964			dec 176	2.1967²⁰	1.5713²⁰	vs H_2O, eth, EtOH
6299	Iodoethene	Vinyl iodide	C_2H_3I	593-66-8	153.949			56	2.037²⁰	1.5385²⁰	vs eth, EtOH
6300	(2-Iodoethyl)benzene		C_8H_9I	17376-04-4	232.061	liq		122¹³	1.603	1.6010²⁰	

Indomethacin

Inosine

Inosine 5'-monophosphate

myo-Inositol

Iocetamic acid

Iodipamide

2-Iodoacetamide

Iodoacetic acid

Iodoacetone

Iodoacetonitrile

Iodoacetylene

2-Iodoaniline

3-Iodoaniline

4-Iodoaniline

2-Iodobenzaldehyde

4-Iodobenzaldehyde

Iodobenzene

2-Iodobenzenemethanol

4-Iodobenzenesulfonyl chloride

2-Iodobenzoic acid

3-Iodobenzoic acid

4-Iodobenzoic acid

4-Iodobenzonitrile

2-Iodobenzoyl chloride

4-Iodobenzoyl chloride

2-Iodo-1,1'-biphenyl

3-Iodo-1,1'-biphenyl

4-Iodo-1,1'-biphenyl

1-Iodobutane

2-Iodobutane, (±)

Iodocyclohexane

Iodocyclopentane

1-Iododecane

1-Iodo-2,4-dimethylbenzene

2-Iodo-1,3-dimethylbenzene

2-Iodo-1,4-dimethylbenzene

1-Iodo-2,2-dimethylpropane

1-Iodododecane

Iodoethane

2-Iodoethanol

Iodoethene

(2-Iodoethyl)benzene

No.	Name	Synonym	Mol. Form.	CAS RN	Mol. Wt.	Physical Form	mp/°C	bp/°C	den/ g cm^{-3}	n_D	Solubility
6301	2-(1-Iodoethyl)-1,3-dioxolane-4-methanol	Iodinated glycerol	C$_6$H$_{11}$IO$_3$	5634-39-9	258.053	pale ye liq			1.797	1.547	s eth, chl, thf, AcOEt
6302	Iodofenphos		C$_8$H$_8$Cl$_2$IO$_3$PS	18181-70-9	412.997	wh cry	76				i H$_2$O; s ace, xyl; sl EtOH
6303	1-Iodoheptane		C$_7$H$_{15}$I	4282-40-0	226.098	liq	-48.2	204.0	1.3791[20]	1.4904[20]	i H$_2$O; s EtOH, eth, ace, chl; sl ctc
6304	3-Iodoheptane		C$_7$H$_{15}$I	31294-92-5	226.098			89[30]	1.3676[20]		
6305	1-Iodohexadecane		C$_{16}$H$_{33}$I	544-77-4	352.337	pa ye liq	24.7	357; 212[15]	1.1213[25]	1.4797[20]	i H$_2$O; sl EtOH; s eth, ace; msc bz; vs chl
6306	1-Iodohexane	Hexyl iodide	C$_6$H$_{13}$I	638-45-9	212.071	liq	-74.2	181.3	1.4397[20]	1.4928[20]	i H$_2$O
6307	Iodomethane	Methyl iodide	CH$_3$I	74-88-4	141.939	liq	-66.4	42.43	2.2789[20]	1.5308[20]	sl H$_2$O; s ace, bz, chl; msc EtOH, eth
6308	1-Iodo-2-methoxybenzene	o-Iodoanisole	C$_7$H$_7$IO	529-28-2	234.034			241; 91[2]	1.8[20]		vs EtOH, eth, ace, bz, chl, lig
6309	1-Iodo-3-methoxybenzene	m-Iodoanisole	C$_7$H$_7$IO	766-85-8	234.034			244.5	1.9650[20]		vs EtOH, eth
6310	1-Iodo-4-methoxybenzene	p-Iodoanisole	C$_7$H$_7$IO	696-62-8	234.034	lf (al), nd (MeOH)	53	238; 138[25]			s EtOH, eth, chl
6311	1-Iodo-2-methylbenzene		C$_7$H$_7$I	615-37-2	218.035			211.5	1.713[20]	1.6079[20]	i H$_2$O; msc EtOH, eth
6312	1-Iodo-3-methylbenzene		C$_7$H$_7$I	625-95-6	218.035	liq	-27.2	213	1.705[20]	1.6053[20]	i H$_2$O; msc EtOH, eth
6313	(Iodomethyl)benzene		C$_7$H$_7$I	620-05-3	218.035	col or ye nd (MeOH)	24.5	93[10]	1.7335[25]	1.6334[25]	vs bz, eth, EtOH
6314	1-Iodo-3-methylbutane	Isopentyl iodide	C$_5$H$_{11}$I	541-28-6	198.045			147	1.5118[20]	1.4939[20]	sl H$_2$O, ctc; msc EtOH, eth
6315	2-Iodo-2-methylbutane		C$_5$H$_{11}$I	594-38-7	198.045			124.5	1.4937[20]	1.4981[20]	i H$_2$O; msc EtOH, eth
6316	1-Iodo-2-methylpropane	Isobutyl iodide	C$_4$H$_9$I	513-38-2	184.018			121.1	1.6035[20]	1.4959[20]	
6317	2-Iodo-2-methylpropane	tert-Butyl iodide	C$_4$H$_9$I	558-17-8	184.018	liq	-38.2	100.1	1.571[25]	1.4918[20]	msc EtOH, eth
6318	Iodomethylsilane		CH$_5$ISi	18089-64-0	172.041	col liq	-109.5	71.8			
6319	1-Iodonaphthalene		C$_{10}$H$_7$I	90-14-2	254.067		2.1	302	1.7399[20]	1.7026[20]	i H$_2$O; msc EtOH, eth, bz, CS$_2$
6320	2-Iodonaphthalene		C$_{10}$H$_7$I	612-55-5	254.067	lf (dil al)	54.5	308	1.6319[99]	1.6662[99]	i H$_2$O; vs EtOH, eth, HOAc
6321	1-Iodo-2-nitrobenzene		C$_6$H$_4$INO$_2$	609-73-4	249.006	ye orth nd (al)	54	290; 162[18]	1.9186[75]		i H$_2$O; s EtOH, eth
6322	1-Iodo-3-nitrobenzene		C$_6$H$_4$INO$_2$	645-00-1	249.006	mcl pr	38.5	280	1.9477[50]		i H$_2$O; s EtOH, eth
6323	1-Iodo-4-nitrobenzene		C$_6$H$_4$INO$_2$	636-98-6	249.006	ye nd (al)	174.7	288	1.8090[155]		i H$_2$O; s EtOH, HOAc; sl DMSO
6324	1-Iodononane		C$_9$H$_{19}$I	4282-42-2	254.151	col liq	-20	245.0	1.2836[25]	1.4848[25]	
6325	1-Iodooctadecane		C$_{18}$H$_{37}$I	629-93-6	380.391	lf (lig), nd (ace, al-ace)	34.0	383	1.0994[20]	1.4810[20]	i H$_2$O; sl EtOH, eth
6326	1-Iodooctane		C$_8$H$_{17}$I	629-27-6	240.125	liq	-45.7	225.1	1.3298[20]	1.4885[20]	s EtOH, eth
6327	2-Iodooctane, (±)	2-Octyl iodide, (±)	C$_8$H$_{17}$I	36049-78-2	240.125			210; 95[16]	1.3251[20]	1.4896[20]	i H$_2$O; s EtOH, eth, lig
6328	1-Iodopentane	Pentyl iodide	C$_5$H$_{11}$I	628-17-1	198.045	liq	-85.6	157.0	1.5161[20]	1.4959[20]	s chl
6329	3-Iodopentane		C$_5$H$_{11}$I	1809-05-8	198.045			145.5	1.5176[20]	1.4974[20]	vs ace, bz, eth
6330	2-Iodophenol		C$_6$H$_5$IO	533-58-4	220.007	nd	43	186[160]; 91[2]	1.8757[80]		s H$_2$O; vs EtOH, eth, CS$_2$
6331	3-Iodophenol		C$_6$H$_5$IO	626-02-8	220.007	nd (lig)	118	186[100]			sl H$_2$O; s EtOH, eth
6332	4-Iodophenol		C$_6$H$_5$IO	540-38-5	220.007	nd (w or sub)	93.5	139[5] dec	1.8573[112]		sl H$_2$O; vs EtOH, eth
6333	1-(3-Iodophenyl)ethanone	3-Iodoacetophenone	C$_8$H$_7$IO	14452-30-3	246.045			129[8]; 117[4]		1.622[20]	s bz
6334	1-(4-Iodophenyl)ethanone	4-Iodoacetophenone	C$_8$H$_7$IO	13329-40-3	246.045		86	153[18]			s EtOH, bz, CS$_2$, HOAc; sl lig, eth
6335	1-Iodopropane	Propyl iodide	C$_3$H$_7$I	107-08-4	169.992	liq	-101.3	102.5	1.7489[20]	1.5058[20]	sl H$_2$O, ctc; msc EtOH, eth
6336	2-Iodopropane	Isopropyl iodide	C$_3$H$_7$I	75-30-9	169.992	liq	-90	89.5	1.7042[20]	1.5028[20]	sl H$_2$O; msc EtOH, eth, bz, chl
6337	3-Iodopropanoic acid		C$_3$H$_5$IO$_2$	141-76-4	199.975	lf (w)	85				sl H$_2$O, chl; vs EtOH; s eth, ace
6338	3-Iodo-1-propanol		C$_3$H$_7$IO	627-32-7	185.991	visc oil		226; 115[38]	1.9976[20]	1.5585[20]	
6339	3-Iodopropene	Allyl iodide	C$_3$H$_5$I	556-56-9	167.976	ye liq	-99.3	103	1.848[12]	1.5540[21]	i H$_2$O; s EtOH, eth, chl

2-(1-Iodoethyl)-1,3-dioxolane-4-methanol

Iodofenphos

1-Iodoheptane

3-Iodoheptane

1-Iodohexadecane

1-Iodohexane

Iodomethane

1-Iodo-2-methoxybenzene

1-Iodo-3-methoxybenzene

1-Iodo-4-methoxybenzene

1-Iodo-2-methylbenzene

1-Iodo-3-methylbenzene

(Iodomethyl)benzene

1-Iodo-3-methylbutane

2-Iodo-2-methylbutane

1-Iodo-2-methylpropane

2-Iodo-2-methylpropane

Iodomethylsilane

1-Iodonaphthalene

2-Iodonaphthalene

1-Iodo-2-nitrobenzene

1-Iodo-3-nitrobenzene

1-Iodo-4-nitrobenzene

1-Iodononane

1-Iodooctadecane

1-Iodooctane

2-Iodooctane, (±)

1-Iodopentane

3-Iodopentane

2-Iodophenol

3-Iodophenol

4-Iodophenol

1-(3-Iodophenyl)ethanone

1-(4-Iodophenyl)ethanone

1-Iodopropane

2-Iodopropane

3-Iodopropanoic acid

3-Iodo-1-propanol

3-Iodopropene

No.	Name	Synonym	Mol. Form.	CAS RN	Mol. Wt.	Physical Form	mp/°C	bp/°C	den/ g cm⁻³	n_D	Solubility
6340	2-Iodopyridine		C_5H_4IN	5029-67-4	204.997			$100^{15}, 93^{13}$	1.928^{25}	1.6366^{20}	s EtOH, eth, ace, bz
6341	5-Iodo-2,4(1H,3H)-pyrimidinedione	5-Iodouracil	$C_4H_3IN_2O_2$	696-07-1	237.983		275 dec				
6342	1-Iodo-2,5-pyrrolidinedione	N-Iodosuccinimide	$C_4H_4INO_2$	516-12-1	224.985	cry (ace)	200.5		2.245^{25}		vs H_2O; s EtOH, ace; sl eth, DMSO
6343	Iodosylbenzene		C_6H_5IO	536-80-1	220.007	ye pow	210 exp				s H_2O, EtOH; i eth, ace, bz, peth
6344	2-Iodothiophene		C_4H_3IS	3437-95-4	210.036	liq	-40	181	2.0595^{25}	1.6465^{20}	vs EtOH, eth; sl chl
6345	4-Iodotoluene		C_7H_7I	624-31-7	218.035	lf (al)	36.5	211	1.678^{20}		i H_2O; s EtOH, eth, CS_2; sl chl
6346	L-3-Iodotyrosine		$C_9H_{10}INO_3$	70-78-0	307.084	cry (w)	205 dec				
6347	trans-α-Ionone, (±)		$C_{13}H_{20}O$	30685-95-1	192.297			146^{28}	0.9298^{21}	1.5041^{20}	vs ace, eth, EtOH
6348	trans-β-Ionone		$C_{13}H_{20}O$	79-77-6	192.297			$124^{10}, 73^{0.1}$	0.945^{20}	1.5198^{20}	sl H_2O; msc EtOH, eth; s chl
6349	Iopanoic acid		$C_{11}H_{12}I_3NO_2$	96-83-3	570.932	wh solid	156				i H_2O; s dil alk, EtOH
6350	Iophendylate	Ethyl 10-(4-iodophenyl) undecanoate	$C_{19}H_{29}IO_2$	99-79-6	416.336	visc liq		197^1	1.25^{20}	1.525^{25}	sl H_2O; s EtOH, bz, chl
6351	Iopodic acid	Ipodate	$C_{12}H_{13}I_3N_2O_2$	5587-89-3	597.956	cry	168				i H_2O; vs EtOH, MeOH, chl, ace
6352	Iprodione		$C_{13}H_{13}Cl_2N_3O_3$	36734-19-7	330.166		136				
6353	Iridomyrmecin	Hexahydro-4,7-dimethylcyclopenta[c]pyran-3(1H)-one	$C_{10}H_{16}O_2$	485-43-8	168.233	pr	61	$106^{1.5}$		1.4607^{65}	sl H_2O; s eth
6354	α-Irone	4-(2,5,6,6-Tetramethyl-2-cyclohexen-1-yl)-3-buten-2-one	$C_{14}H_{22}O$	79-69-6	206.324			$90^{0.4}$	0.9362^{20}	1.5002^{20}	
6355	β-Irone	4-(2,5,6,6-Tetramethyl-1-cyclohexen-1-yl)-3-buten-2-one	$C_{14}H_{22}O$	79-70-9	206.324			125^{11}	0.9434^{21}	1.5162^{25}	sl H_2O; vs EtOH, eth, bz, chl
6356	Iron hydrocarbonyl	Hydrogen tetracarbonylferrate(II)	$C_4H_2FeO_4$	17440-90-3	169.902	col liq; unstab	-70	dec			s alk
6357	Iron nonacarbonyl	Diiron nonacarbonyl	$C_9Fe_2O_9$	15321-51-4	363.781	oran-ye cry	100 dec		2.85		
6358	Iron(III) NTA	Nitrilotriacetatoiron(III)	$C_6H_6FeNO_6$	16448-54-7	243.960	solid					s H_2O
6359	Iron pentacarbonyl		C_5FeO_5	13463-40-6	195.896	col to ye oily liq	-20	103	1.5^{20}	1.453^{22}	i H_2O; sl EtOH; s bz, ace, ctc
6360	Iron(III) 2,4-pentanedioate	Ferric acetylacetonate	$C_{15}H_{21}FeO_6$	14024-18-1	353.169		179		5.24		
6361	Isanic acid	17-Octadecene-9,11-diynoic acid	$C_{18}H_{26}O_2$	506-25-2	274.398	cry	39.5		0.9309^{45}	1.49148^{50}	s ace, EtOH, i-PrOH; sl peth
6362	Isatidine	Retrorsine N-oxide	$C_{18}H_{25}NO_7$	15503-86-3	367.395	cry	145				
6363	Isaxonine	N-Isopropyl-2-pyrimidineamine	$C_7H_{11}N_3$	4214-72-6	137.182		28	93^{12}			
6364	Isazophos		$C_9H_{17}ClN_3O_3PS$	67329-04-8	313.741			$170; 100^{0.001}$	1.22^{20}		
6365	Isobenzan		$C_9H_4Cl_6O$	297-78-9	411.751	cry (hp)	121				s eth, bz, xyl, tol
6366	1(3H)-Isobenzofuranone		$C_8H_6O_2$	87-41-2	134.133	nd or pl (w)	75	290	1.1636^{99}	1.536^{99}	s H_2O; vs EtOH, eth; sl chl
6367	Isoborneol	1,7,7-Trimethylbicyclo[2.2.1]heptan-2-ol, exo-(±)	$C_{10}H_{18}O$	24393-70-2	154.249	tab (peth)	212	sub	1.10^{20}		i H_2O; vs EtOH, eth, chl; sl bz
6368	Isobornyl thiocyanoacetate		$C_{13}H_{19}NO_2S$	115-31-1	253.361	ye oily liq		$95^{0.06}$	1.1465^{25}	1.512^{25}	i H_2O; vs EtOH, bz, chl, peth
6369	6-Isobornyl-3,4-xylenol	Xibornol	$C_{18}H_{26}O$	13741-18-9	258.398	cry	95	167^3	1.0240^{20}	1.5382^{20}	
6370	Isobutanal	2-Methyl-1-propanal	C_4H_8O	78-84-2	72.106	liq	-65.9	64.5	0.7891^{20}	1.3730^{20}	s H_2O, eth, ace, chl; sl ctc
6371	Isobutane	2-Methylpropane	C_4H_{10}	75-28-5	58.122	col gas	-159.4	-11.73	0.5510^{20} (p>1 atm)	1.3518^{-25}	sl H_2O; s EtOH, eth, chl
6372	Isobutene		C_4H_8	115-11-7	56.107	col gas	-140.7	-6.9	0.589^{25} (p>1 atm)	1.3926^{-25}	i H_2O; vs EtOH, eth; s bz, sulf
6373	Isobutyl acetate		$C_6H_{12}O_2$	110-19-0	116.158	liq	-98.8	116.5	0.8712^{20}	1.3902^{20}	sl H_2O, ctc; msc EtOH, eth; s ace
6374	Isobutyl acrylate		$C_7H_{12}O_2$	106-63-8	128.169	liq	-61	132	0.8896^{20}	1.4150^{20}	sl H_2O; s EtOH, eth, MeOH
6375	5-Isobutyl-3-allyl-2-thioxo-4-imidazolidinone	Albutoin	$C_{10}H_{16}N_2OS$	830-89-7	212.311		210.5				
6376	Isobutylamine	2-Methyl-1-propanamine	$C_4H_{11}N$	78-81-9	73.137	liq	-86.7	67.75	0.724^{25}	1.3988^{19}	
6377	Isobutyl 4-aminobenzoate	Isobutyl p-aminobenzoate	$C_{11}H_{15}NO_2$	94-14-4	193.243		64.5				

2-Iodopyridine

5-Iodo-2,4(1*H*,3*H*)-pyrimidinedione

1-Iodo-2,5-pyrrolidinedione

Iodosylbenzene

2-Iodothiophene

4-Iodotoluene

L-3-Iodotyrosine

trans-α-Ionone, (±)

trans-β-Ionone

Iopanoic acid

Iophendylate

Iopodic acid

Iprodione

Iridomyrmecin

α-Irone

β-Irone

H₂Fe(CO)₄
Iron hydrocarbonyl

Iron nonacarbonyl

Iron(III) NTA

Iron pentacarbonyl

Iron(III) 2,4-pentanedioate

Isanic acid

Isatidine

Isaxonine

Isazophos

Isobenzan

1(3*H*)-Isobenzofuranone

Isoborneol

Isobornyl thiocyanoacetate

6-Isobornyl-3,4-xylenol

Isobutanal

Isobutane

Isobutene

Isobutyl acetate

Isobutyl acrylate

5-Isobutyl-3-allyl-2-thioxo-4-imidazolidinone

Isobutylamine

Isobutyl 4-aminobenzoate

No.	Name	Synonym	Mol. Form.	CAS RN	Mol. Wt.	Physical Form	mp/°C	bp/°C	den/ g cm^{-3}	n_D	Solubility
6378	Isobutylbenzene		$C_{10}H_{14}$	538-93-2	134.218	liq	-51.4	172.79	0.8532[20]	1.4866[20]	i H_2O; msc EtOH, eth, ace, bz, peth, ctc
6379	Isobutyl benzoate		$C_{11}H_{14}O_2$	120-50-3	178.228			242	0.9990[20]		i H_2O; msc EtOH, eth; s ace, chl
6380	Isobutyl butanoate		$C_8H_{16}O_2$	539-90-2	144.212			156.9	0.8364[18]	1.4032[20]	sl H_2O; msc EtOH, eth
6381	Isobutyl carbamate		$C_5H_{11}NO_2$	543-28-2	117.147	lf	67	207		1.4098[76]	vs eth, EtOH
6382	Isobutyl chlorocarbonate		$C_5H_9ClO_2$	543-27-1	136.577			128.8	1.0426[18]	1.4071[18]	s EtOH, bz, chl; msc eth
6383	Isobutyl 2-chloropropanoate		$C_7H_{13}ClO_2$	114489-96-2	164.630			176	1.0312[20]	1.4247[20]	
6384	Isobutyl 3-chloropropanoate		$C_7H_{13}ClO_2$	62108-68-3	164.630			191.3	1.0323[20]	1.4295[20]	vs eth, EtOH
6385	Isobutylcyclohexane		$C_{10}H_{20}$	1678-98-4	140.266	liq	-95	171.3	0.7952[20]	1.4386[20]	i H_2O; s EtOH, ace, chl; vs eth, bz
6386	Isobutylcyclopentane		C_9H_{18}	3788-32-7	126.239	liq	-115.2	148	0.7769[25]	1.4298[20]	
6387	Isobutyldimethylamine	N,N,2-Trimethyl-1-propanamine	$C_6H_{15}N$	7239-24-9	101.190			80.5	0.7097[20]	1.3907[20]	vs H_2O
6388	Isobutyl formate		$C_5H_{10}O_2$	542-55-2	102.132	liq	-95.8	98.2	0.8776[20]	1.3857[20]	sl H_2O, chl; msc EtOH, eth; vs ace
6389	Isobutyl heptanoate	Isobutyl enanthate	$C_{11}H_{22}O_2$	7779-80-8	186.292			208	0.8593[20]		vs ace, bz, eth, EtOH
6390	Isobutyl 2-hydroxybenzoate	Isobutyl salicylate	$C_{11}H_{14}O_3$	87-19-4	194.227		5.9	261	1.0639[20]	1.5087[20]	i H_2O; s EtOH, eth, ctc
6391	Isobutyl isobutanoate		$C_8H_{16}O_2$	97-85-8	144.212	liq	-80.7	148.6	0.8542[20]	1.3999[20]	sl H_2O, ctc; s EtOH, ace; msc eth
6392	Isobutyl isocyanate		C_5H_9NO	1873-29-6	99.131			106			
6393	Isobutyl isothiocyanate	1-Isothiocyanato-2-methylpropane	C_5H_9NS	591-82-2	115.197			160	0.9631[14]	1.5005[14]	
6394	Isobutyl methacrylate		$C_8H_{14}O_2$	97-86-9	142.196			155	0.8858[20]	1.4199[20]	i H_2O; msc EtOH, eth
6395	Isobutyl 3-methylbutanoate	Isobutyl isovalerate	$C_9H_{18}O_2$	589-59-3	158.238			168.5	0.853[20]	1.4057[20]	i H_2O; msc EtOH, eth; vs ace; s chl
6396	Isobutyl methyl ether		$C_5H_{12}O$	625-44-5	88.148			58.6	0.7311[20]		vs eth, EtOH
6397	Isobutyl nitrate		$C_4H_9NO_3$	543-29-3	119.119			123.4	1.0152[20]	1.4028[20]	
6398	Isobutyl nitrite		$C_4H_9NO_2$	542-56-3	103.120	col liq		67	0.8699[22]	1.3715[22]	sl H_2O; s EtOH, eth
6399	Isobutyl pentanoate		$C_9H_{18}O_2$	10588-10-0	158.238			179	0.8625[25]	1.4046[20]	i H_2O; msc EtOH; s eth, ace
6400	Isobutyl phenylacetate		$C_{12}H_{16}O_2$	102-13-6	192.254			247	0.999[18]		i H_2O; s EtOH, eth
6401	Isobutyl propanoate	Isobutyl propionate	$C_7H_{14}O_2$	540-42-1	130.185	liq	-71.4	137	0.888[0]	1.3973[20]	sl H_2O; vs EtOH, eth; s ace, bz, chl, ctc
6402	Isobutyl stearate		$C_{22}H_{44}O_2$	646-13-9	340.583	wax	28.9	223[15]	0.8498[20]		vs eth
6403	Isobutyl thiocyanate		C_5H_9NS	591-84-4	115.197	liq	-59	175.4			vs eth, EtOH
6404	Isobutyl trichloroacetate		$C_6H_9Cl_3O_2$	33560-15-5	219.493			188	1.2636[20]	1.4483[20]	vs bz, eth, EtOH
6405	Isobutyl vinyl ether		$C_6H_{12}O$	109-53-5	100.158	liq	-112	83	0.7645[20]	1.3966[20]	sl H_2O; vs EtOH, ace, bz; msc eth
6406	Isocitric acid		$C_6H_8O_7$	320-77-4	192.124	ye syr	105				
6407	Isocorybulbine		$C_{21}H_{25}NO_4$	22672-74-8	355.429	lf (al)	187.5		1.045[20]		i H_2O; s EtOH, chl, acid
6408	Isocorydine		$C_{20}H_{23}NO_4$	475-67-2	341.402	pl	185				vs chl
6409	2-Isocyanato-1,3-dimethylbenzene	2,6-Dimethylphenyl isocyanate	C_9H_9NO	28556-81-2	147.173	liq		100[13]			
6410	1-Isocyanato-2-methoxybenzene		$C_8H_7NO_2$	700-87-8	149.148			94[17]			
6411	1-Isocyanato-3-methoxybenzene		$C_8H_7NO_2$	18908-07-1	149.148			102[15]			
6412	1-Isocyanato-2-methylbenzene	2-Tolyl isocyanate	C_8H_7NO	614-68-6	133.148			185		1.5282[20]	i H_2O; s eth
6413	1-Isocyanato-3-methylbenzene	3-Tolyl isocyanate	C_8H_7NO	621-29-4	133.148			196.5	1.0330[20]		vs bz, eth
6414	1-Isocyanato-4-methylbenzene	4-Tolyl isocyanate	C_8H_7NO	622-58-2	133.148			187			vs bz, eth
6415	2-Isocyanato-2-methylpropane	tert-Butyl isocyanate	C_5H_9NO	1609-86-5	99.131		85.5		0.8670[7]	1.4061[20]	
6416	1-Isocyanatonaphthalene	1-Naphthyl isocyanate	$C_{11}H_7NO$	86-84-0	169.180			269	1.1774[20]		s eth, bz
6417	1-Isocyanato-2-nitrobenzene	2-Nitrophenyl isocyanate	$C_7H_4N_2O_3$	3320-86-3	164.118	wh nd (peth)	41	137[18]			vs bz, eth, chl
6418	1-Isocyanato-3-nitrobenzene	3-Nitrophenyl isocyanate	$C_7H_4N_2O_3$	3320-87-4	164.118	wh lf (lig)	51	130[11]			vs bz, eth, chl
6419	1-Isocyanato-4-nitrobenzene	4-Nitrophenyl isocyanate	$C_7H_4N_2O_3$	100-28-7	164.118	pa ye nd	57	162[20], 137[11]			vs bz, eth, chl
6420	2-Isocyanatopropane	Isopropyl isocyanate	C_4H_7NO	1795-48-8	85.105			74.5	0.866[25]	1.3825[20]	
6421	1-Isocyanato-3-(trifluoromethyl)benzene	3-(Trifluoromethyl)phenyl isocyanate	$C_8H_4F_3NO$	329-01-1	187.119			54[11]	1.3455[20]	1.4690[20]	
6422	Isocyanobenzene	Phenyl isocyanide	C_7H_5N	931-54-4	103.122	unstab liq		80[40]	0.98[15]		

Isobutylbenzene

Isobutyl benzoate

Isobutyl butanoate

Isobutyl carbamate

Isobutyl chlorocarbonate

Isobutyl 2-chloropropanoate

Isobutyl 3-chloropropanoate

Isobutylcyclohexane

Isobutylcyclopentane

Isobutyldimethylamine

Isobutyl formate

Isobutyl heptanoate

Isobutyl 2-hydroxybenzoate

Isobutyl isobutanoate

Isobutyl isocyanate

Isobutyl isothiocyanate

Isobutyl methacrylate

Isobutyl 3-methylbutanoate

Isobutyl methyl ether

Isobutyl nitrate

Isobutyl nitrite

Isobutyl pentanoate

Isobutyl phenylacetate

Isobutyl propanoate

Isobutyl stearate

Isobutyl thiocyanate

Isobutyl trichloroacetate

Isobutyl vinyl ether

Isocitric acid

Isocorybulbine

Isocorydine

2-Isocyanato-1,3-dimethylbenzene

1-Isocyanato-2-methoxybenzene

1-Isocyanato-3-methoxybenzene

1-Isocyanato-2-methylbenzene

1-Isocyanato-3-methylbenzene

1-Isocyanato-4-methylbenzene

2-Isocyanato-2-methylpropane

1-Isocyanatonaphthalene

1-Isocyanato-2-nitrobenzene

1-Isocyanato-3-nitrobenzene

1-Isocyanato-4-nitrobenzene

2-Isocyanatopropane

1-Isocyanato-3-(trifluoromethyl)benzene

Isocyanobenzene

No.	Name	Synonym	Mol. Form.	CAS RN	Mol. Wt.	Physical Form	mp/°C	bp/°C	den/g cm⁻³	n_D	Solubility
6423	Isocyanomethane	Methyl isocyanide	C_2H_3N	593-75-9	41.052		-45	exp 59.6	0.756[4]		
6424	(Isocyanomethyl)benzene	Benzyl isocyanide	C_8H_7N	10340-91-7	117.149			dec 199; 93[55]	0.972[15]	1.5193[20]	
6425	2-Isocyanopropane	Isopropyl isocyanide	C_4H_7N	598-45-8	69.106			87	0.7596[25]		i H_2O; msc EtOH, eth
6426	Isodecyl acrylate		$C_{13}H_{24}O_2$	1330-61-6	212.329		-100	158[50]	0.885[20]	1.4416[20]	
6427	Isodecyl diphenyl phosphate		$C_{22}H_{31}O_4P$	29761-21-5	390.452			249[10] dec			
6428	Isodecyl methacrylate		$C_{14}H_{26}O_2$	29964-84-9	226.355			126[10]	0.876[20]		
6429	8-Isoestrone		$C_{18}H_{22}O_2$	517-06-6	270.367	pr (MeOH)	254				vs eth, Diox
6430	Isoeugenol		$C_{10}H_{12}O_2$	97-54-1	164.201			266	1.080[25]	1.5739[19]	vs eth, EtOH
6431	Isofenphos		$C_{15}H_{24}NO_4PS$	25311-71-1	345.395		<-12	120[0.01]	1.134[20]		
6432	Isoflurophate		$C_6H_{14}FO_3P$	55-91-4	184.145			62[9]	1.055[25]	1.3830[25]	sl H_2O; lig; s eth; vs oils
6433	1H-Isoindole-1,3(2H)-dione	Phthalimide	$C_8H_5NO_2$	85-41-6	147.132	nd (w), pr (HOAc) lf (sub)	238				vs bz
6434	Isolan		$C_{10}H_{17}N_3O_2$	119-38-0	211.261	col liq		118[2.5]	1.07[20]		msc H_2O; s EtOH, xyl
6435	DL-Isoleucine		$C_6H_{13}NO_2$	443-79-8	131.173		292 dec				
6436	L-Isoleucine	2-Amino-3-methylpentanoic acid	$C_6H_{13}NO_2$	73-32-5	131.173		284 dec				s H_2O; i EtOH
6437	Isolongifolene		$C_{15}H_{24}$	1135-66-6	204.352	liq		82[0.4]			
6438	Isolysergic acid		$C_{16}H_{16}N_2O_2$	478-95-5	268.310	cry (w+2)	218 dec				sl H_2O, EtOH; s py
6439	α-Isomaltose	6-O-α-D-Glucopyranosyl-D-glucose	$C_{12}H_{22}O_{11}$	499-40-1	342.296		120				
6440	Isoniazid	4-Pyridinecarboxylic acid hydrazide	$C_6H_7N_3O$	54-85-3	137.139	cry (al)	171.4				vs H_2O, EtOH
6441	Isopentane	2-Methylbutane	C_5H_{12}	78-78-4	72.149	vol liq or gas	-159.77	27.88	0.6201[20]	1.3537[20]	i H_2O; msc EtOH, eth
6442	Isopentyl acetate		$C_7H_{14}O_2$	123-92-2	130.185	liq	-78.5	142.5	0.876[15]	1.4000[20]	sl H_2O; msc EtOH, eth; s ace, chl
6443	Isopentylbenzene		$C_{11}H_{16}$	2049-94-7	148.245			195	0.856[20]	1.4867[10]	i H_2O; s EtOH, eth; vs bz
6444	Isopentyl butanoate		$C_9H_{18}O_2$	106-27-4	158.238			179	0.865[19]	1.4110[20]	i H_2O; vs EtOH, eth
6445	Isopentyl formate		$C_6H_{12}O_2$	110-45-2	116.158	liq	-93.5	123.5	0.877[20]	1.3967[20]	sl H_2O, ctc; s EtOH; msc eth
6446	Isopentyl hexanoate	Isopentyl caproate	$C_{11}H_{22}O_2$	2198-61-0	186.292			225.5	0.861[20]		i H_2O; s EtOH, eth
6447	Isopentyl α-hydroxybenzeneacetate	Isopentyl mandelate	$C_{13}H_{18}O_3$	5421-04-5	222.280	oily liq		172[11]			
6448	Isopentyl isopentanoate	Isoamyl isovalerate	$C_{10}H_{20}O_2$	659-70-1	172.265			190.4	0.8583[19]	1.4130[19]	
6449	Isopentyl lactate		$C_8H_{16}O_3$	19329-89-6	160.211			202.4	0.9589[25]	1.4240[25]	vs eth, EtOH
6450	Isopentyl 2-methylpropanoate	Isopentyl isobutyrate	$C_9H_{18}O_2$	2050-01-3	158.238			168.5	0.8627[20]		sl H_2O; s EtOH, eth, ace
6451	Isopentyl nitrite	Isoamyl nitrite	$C_5H_{11}NO_2$	110-46-3	117.147			99.2	0.8828[20]	1.3918[20]	sl H_2O; msc EtOH, eth
6452	Isopentyl pentanoate		$C_{10}H_{20}O_2$	2050-09-1	172.265			193			
6453	Isopentyl propanoate		$C_8H_{16}O_2$	105-68-0	144.212			160.2	0.8697[20]	1.4069[20]	vs eth, EtOH
6454	Isopentyl salicylate		$C_{12}H_{16}O_3$	87-20-7	208.253			278; 151[15]	1.0535[20]	1.5080[20]	i H_2O; vs EtOH; s eth, chl; sl ctc
6455	Isopentyl trichloroacetate		$C_7H_{11}Cl_3O_2$	57392-55-9	233.520			217	1.2314[20]	1.4521[20]	vs eth, EtOH
6456	Isophorone	3,5,5-Trimethyl-2-cyclohexen-1-one	$C_9H_{14}O$	78-59-1	138.206	liq	-8.1	215.2	0.9255[20]	1.4766[18]	
6457	Isophorone diisocyanate		$C_{12}H_{18}N_2O_2$	4098-71-9	222.283		60	217[100]	1.062[20]		
6458	Isophthalic acid	1,3-Benzenedicarboxylic acid	$C_8H_6O_4$	121-91-5	166.132	nd (w, al)	347	sub			sl H_2O; s EtOH, HOAc; i eth, bz, lig
6459	Isopilosine		$C_{16}H_{18}N_2O_3$	491-88-3	286.325	pl (al), pr (w, dil al)	187				vs EtOH
6460	Isopropalin	Benzenamine, 4-(1-methylethyl)-2,6-dinitro-N,N-dipropyl-	$C_{15}H_{23}N_3O_4$	33820-53-0	309.362	red-oran liq					i H_2O; s os
6461	Isopropamide iodide		$C_{23}H_{33}IN_2O$	71-81-8	480.424	cry or pow	190				s H_2O, EtOH, MeOH; i chl
6462	Isopropenyl acetate		$C_5H_8O_2$	108-22-5	100.117	liq	-92.9	94	0.9090[20]	1.4033[20]	sl H_2O; s EtOH, chl, ace; vs eth
6463	Isopropenylbenzene	α-Methyl styrene	C_9H_{10}	98-83-9	118.175	liq	-23.2	165.4	0.9106[20]	1.5386[20]	i H_2O; s EtOH, eth; msc ace, bz, ctc
6464	p-Isopropenylisopropylbenzene		$C_{12}H_{16}$	2388-14-9	160.255	liq	-30.6	220.8	0.8936[20]	1.5238[20]	vs ace, bz, eth, EtOH
6465	p-Isopropenylstyrene		$C_{11}H_{12}$	16262-48-9	144.213	liq		242	0.93	1.5684[20]	

Isocyanomethane

(Isocyanomethyl)benzene

2-Isocyanopropane

Isodecyl acrylate

Isodecyl diphenyl phosphate

Isodecyl methacrylate

8-Isoestrone

Isoeugenol

Isofenphos

Isoflurophate

1*H*-Isoindole-1,3(2*H*)-dione

Isolan

DL-Isoleucine

L-Isoleucine

Isolongifolene

Isolysergic acid

α-Isomaltose

Isoniazid

Isopentane

Isopentyl acetate

Isopentylbenzene

Isopentyl butanoate

Isopentyl formate

Isopentyl hexanoate

Isopentyl α-hydroxybenzeneacetate

Isopentyl isopentanoate

Isopentyl lactate

Isopentyl 2-methylpropanoate

Isopentyl nitrite

Isopentyl pentanoate

Isopentyl propanoate

Isopentyl salicylate

Isopentyl trichloroacetate

Isophorone

Isophorone diisocyanate

Isophthalic acid

Isopilosine

Isopropalin

Isopropamide iodide

Isopropenyl acetate

Isopropenylbenzene

p-Isopropenylisopropylbenzene

p-Isopropenylstyrene

No.	Name	Synonym	Mol. Form.	CAS RN	Mol. Wt.	Physical Form	mp/°C	bp/°C	den/ g cm^{-3}	n_D	Solubility
6466	4-Isopropoxydiphenylamine	4-Isopropoxy-*N*-phenylaniline	C$_{15}$H$_{17}$NO	101-73-5	227.302		83				
6467	2-Isopropoxyethanol		C$_5$H$_{12}$O$_2$	109-59-1	104.148			145	0.9030[20]	1.4095[20]	msc H$_2$O, EtOH, eth; s ace
6468	3-Isopropoxypropanenitrile	1-Cyano-2-isopropoxyethane	C$_6$H$_{11}$NO	110-47-4	113.157			65[10]			s chl
6469	Isopropyl acetate		C$_5$H$_{10}$O$_2$	108-21-4	102.132	liq	-73.4	88.6	0.8718[20]	1.3773[20]	s H$_2$O, EtOH, ace, chl; msc eth
6470	Isopropyl acrylate	Isopropyl 2-propenoate	C$_6$H$_{10}$O$_2$	689-12-3	114.142	liq		51[103]			
6471	Isopropylamine	2-Propanamine	C$_3$H$_9$N	75-31-0	59.110	liq	-95.13	31.76	0.6891[20]	1.3742[20]	msc H$_2$O, EtOH, eth; vs ace; s bz, chl
6472	Isopropylamine hydrochloride	2-Propanamine hydrochloride	C$_3$H$_{10}$ClN	15572-56-2	95.571		164				s DMSO
6473	2-(Isopropylamino)ethanol		C$_5$H$_{13}$NO	109-56-8	103.163		128.5	173	0.8970[20]	1.4395[20]	msc H$_2$O, EtOH, eth
6474	2-Isopropylaniline		C$_9$H$_{13}$N	643-28-7	135.206			221; 95[13]	0.9760[12]		i H$_2$O; s eth, bz, ctc
6475	4-Isopropylaniline	Cumidine	C$_9$H$_{13}$N	99-88-7	135.206			225	0.953[20]		
6476	*N*-Isopropylaniline		C$_9$H$_{13}$N	768-52-5	135.206			203	0.9526[25]	1.5380[20]	s EtOH, eth, ace, bz
6477	4-Isopropylbenzaldehyde	Cuminaldehyde	C$_{10}$H$_{12}$O	122-03-2	148.201			235.5	0.9755[20]	1.5301[20]	i H$_2$O; s EtOH, eth; sl ctc
6478	Isopropylbenzene	Cumene	C$_9$H$_{12}$	98-82-8	120.191	liq	-96.02	152.41	0.8640[25]	1.4915[20]	i H$_2$O; msc EtOH, eth, ace, bz, peth, ctc
6479	Isopropylbenzene hydroperoxide	Cumene hydroperoxide	C$_9$H$_{12}$O$_2$	80-15-9	152.190	liq		153; 84[8]	1.03[20]		
6480	4-Isopropylbenzenemethanol	Cumic alcohol	C$_{10}$H$_{14}$O	536-60-7	150.217		28	249	0.9818[20]	1.5210[20]	i H$_2$O; msc EtOH, eth; vs bz
6481	α-Isopropylbenzenemethanol	1-Phenyl-2-methylpropyl alcohol	C$_{10}$H$_{14}$O	611-69-8	150.217			223	0.9869[14]	1.5193[14]	i H$_2$O; s EtOH, ace
6482	Isopropyl benzoate		C$_{10}$H$_{12}$O$_2$	939-48-0	164.201			216	1.0163[15]	1.4890[20]	i H$_2$O; s EtOH, eth, ace
6483	4-Isopropylbenzoic acid	Cumic acid	C$_{10}$H$_{12}$O$_2$	536-66-3	164.201	tcl pl (al)	117.5	sub	1.162[4]		sl H$_2$O; vs EtOH, eth; s peth
6484	Isopropyl butanoate		C$_7$H$_{14}$O$_2$	638-11-9	130.185			130.5	0.8588[20]	1.3936[20]	i H$_2$O; s EtOH
6485	Isopropyl carbamate		C$_4$H$_9$NO$_2$	1746-77-6	103.120	nd	93	183	0.9951[66]		
6486	Isopropyl chloroacetate		C$_5$H$_9$ClO$_2$	105-48-6	136.577			150.5	1.0888[20]	1.4382[20]	vs eth
6487	Isopropyl chloroformate		C$_4$H$_7$ClO$_2$	108-23-6	122.551			105		1.4013[20]	vs eth
6488	Isopropyl 2-chloropropanoate		C$_6$H$_{11}$ClO$_2$	40058-87-5	150.603			151.5	1.0315[20]	1.4149[20]	i H$_2$O; s EtOH, eth
6489	Isopropylcyclohexane		C$_9$H$_{18}$	696-29-7	126.239	liq	-89.4	154.8	0.8023[20]	1.4410[20]	i H$_2$O; vs EtOH, eth; msc ace, bz
6490	4-Isopropylcyclohexanone		C$_9$H$_{16}$O	5432-85-9	140.222			214; 139[100]	0.9099[30]	1.4552[25]	
6491	Isopropylcyclopentane		C$_8$H$_{16}$	3875-51-2	112.213	liq	-111.4	126.5	0.7765[20]	1.4258[20]	i H$_2$O; msc EtOH, ace, ctc; s eth, bz
6492	Isopropylcyclopropane		C$_6$H$_{12}$	3638-35-5	84.159	liq	-112.9	58.3	0.6936[25]	1.3865[20]	
6493	Isopropyl (2,4-dichlorophenoxy) acetate		C$_{11}$H$_{12}$Cl$_2$O$_3$	94-11-1	263.117		5	140[1]	1.26[25]	1.5209[25]	
6494	*N*-Isopropyl-4,4-diphenylcyclohexanamine	Pramiverin	C$_{21}$H$_{27}$N	14334-40-8	293.446		70	165[0.05]			
6495	Isopropyl dodecanoate	Isopropyl laurate	C$_{15}$H$_{30}$O$_2$	10233-13-3	242.398			196[60], 105[0.8]	0.8536[20]	1.4280[25]	vs eth, EtOH
6496	Isopropyl formate		C$_4$H$_8$O$_2$	625-55-8	88.106			68.2	0.8728[20]	1.3678[20]	sl H$_2$O; msc EtOH, eth; vs ace; s chl
6497	Isopropyl 2-furancarboxylate	Isopropyl 2-furanoate	C$_8$H$_{10}$O$_3$	6270-34-4	154.163			198.5	1.0655[24]	1.4682[24]	i H$_2$O; s EtOH, eth, ace, bz
6498	Isopropyl glycidyl ether	(1-Methylethoxy)methyloxirane	C$_6$H$_{12}$O$_2$	4016-14-2	116.158			137	0.9186[20]		s H$_2$O, ace, EtOH
6499	4-Isopropylheptane		C$_{10}$H$_{22}$	52896-87-4	142.282			158.9	0.7354[25]	1.4153[20]	
6500	Isopropylhydrazine		C$_3$H$_{10}$N$_2$	2257-52-5	74.124	liq		107			s H$_2$O, bz, EtOH; sl eth
6501	Isopropyl 2-hydroxybenzoate	Isopropyl salicylate	C$_{10}$H$_{12}$O$_3$	607-85-2	180.200			238	1.0729[20]	1.5065[20]	i H$_2$O; msc EtOH, eth
6502	Isopropyl isobutanoate	Isopropyl isobutyrate	C$_7$H$_{14}$O$_2$	617-50-5	130.185			120.7	0.8471[21]		i H$_2$O; s EtOH, eth, ace
6503	Isopropyl lactate		C$_6$H$_{12}$O$_3$	617-51-6	132.157			167	0.9980[20]	1.4082[25]	vs H$_2$O, bz, eth, EtOH
6504	Isopropyl methacrylate	Isopropyl 2-methyl-2-propenoate	C$_7$H$_{12}$O$_2$	4655-34-9	128.169			125	0.8847[20]	1.4122[20]	vs ace, bz, eth, EtOH
6505	Isopropyl methanesulfonate		C$_4$H$_{10}$O$_3$S	926-06-7	138.185			82[6]			
6506	Isopropylmethylamine	Methylisopropylamine	C$_4$H$_{11}$N	4747-21-1	73.137			50.4			

4-Isopropoxydiphenylamine

2-Isopropoxyethanol

3-Isopropoxypropanenitrile

Isopropyl acetate

Isopropyl acrylate

Isopropylamine

Isopropylamine hydrochloride

2-(Isopropylamino)ethanol

2-Isopropylaniline

4-Isopropylaniline

N-Isopropylaniline

4-Isopropylbenzaldehyde

Isopropylbenzene

Isopropylbenzene hydroperoxide

4-Isopropylbenzenemethanol

α-Isopropylbenzenemethanol

Isopropyl benzoate

4-Isopropylbenzoic acid

Isopropyl butanoate

Isopropyl carbamate

Isopropyl chloroacetate

Isopropyl chloroformate

Isopropyl 2-chloropropanoate

Isopropylcyclohexane

4-Isopropylcyclohexanone

Isopropylcyclopentane

Isopropylcyclopropane

Isopropyl (2,4-dichlorophenoxy)acetate

N-Isopropyl-4,4-diphenylcyclohexanamine

Isopropyl dodecanoate

Isopropyl formate

Isopropyl 2-furancarboxylate

Isopropyl glycidyl ether

4-Isopropylheptane

Isopropylhydrazine

Isopropyl 2-hydroxybenzoate

Isopropyl isobutanoate

Isopropyl lactate

Isopropyl methacrylate

Isopropyl methanesulfonate

Isopropylmethylamine

No.	Name	Synonym	Mol. Form.	CAS RN	Mol. Wt.	Physical Form	mp/°C	bp/°C	den/ g cm^{-3}	n_D	Solubility
6507	1-Isopropyl-2-methylbenzene	o-Cymene	C$_{10}$H$_{14}$	527-84-4	134.218	liq	-71.5	178.1	0.8766[20]	1.5006[20]	i H$_2$O; msc EtOH, eth, ace, bz, peth, ctc
6508	1-Isopropyl-3-methylbenzene	m-Cymene	C$_{10}$H$_{14}$	535-77-3	134.218	liq	-63.7	175.1	0.8610[20]	1.4930[20]	i H$_2$O; msc EtOH, eth, ace, bz, peth, ctc
6509	1-Isopropyl-4-methylbenzene	p-Cymene	C$_{10}$H$_{14}$	99-87-6	134.218	liq	-67.94	177.1	0.8573[20]	1.4909[20]	i H$_2$O; msc EtOH, eth, ace, bz, peth, ctc
6510	Isopropyl 3-methylbutanoate		C$_8$H$_{16}$O$_2$	32665-23-9	144.212			142; 70[55]	0.8538[17]	1.3960[20]	vs ace, eth, EtOH
6511	5-Isopropyl-2-methyl-1,3-cyclohexadiene, (R)		C$_{10}$H$_{16}$	4221-98-1	136.234			173	0.8421[20]	1.4772[19]	
6512	5-Isopropyl-3-methyl-2-cyclohexen-1-one, (±)	Homocamfin	C$_{10}$H$_{16}$O	535-86-4	152.233	pa ye		244; 121[15]	0.9340[21]	1.4865[21]	vs ace, EtOH
6513	6-Isopropyl-3-methyl-2-cyclohexen-1-one, (±)	(±)-Piperitone	C$_{10}$H$_{16}$O	6091-52-7	152.233	liq	-19	232.5	0.9331[20]	1.4845[20]	vs ace, EtOH
6514	Isopropyl methyl ether	2-Methoxypropane	C$_4$H$_{10}$O	598-53-8	74.121			30.77	0.7237[15]	1.3576[20]	sl H$_2$O; msc EtOH, eth
6515	5-Isopropyl-2-methylphenol	Carvacrol	C$_{10}$H$_{14}$O	499-75-2	150.217	nd	1	237.7	0.9772[20]	1.5230[20]	sl H$_2$O; s EtOH, eth, ctc; vs ace
6516	2-Isopropyl-6-methyl-4-pyrimidinol		C$_8$H$_{12}$N$_2$O	2814-20-2	152.193	cry	173				
6517	Isopropyl methyl sulfide		C$_4$H$_{10}$S	1551-21-9	90.187	liq	-101.5	84.8	0.8291[20]	1.4932[20]	s EtOH, eth, ace
6518	1-Isopropylnaphthalene		C$_{13}$H$_{14}$	6158-45-8	170.250	liq	-16	268	0.9956[20]	1.5952[20]	
6519	2-Isopropylnaphthalene		C$_{13}$H$_{14}$	2027-17-0	170.250		14.5	268.2	0.9753[20]	1.5848[20]	i H$_2$O; vs EtOH, eth; s bz
6520	Isopropyl nitrate		C$_3$H$_7$NO$_3$	1712-64-7	105.093			100	1.034[19]	1.3912[16]	s EtOH, eth
6521	Isopropyl nitrite		C$_3$H$_7$NO$_2$	541-42-4	89.094	pa ye oil		40	0.8684[15]		i H$_2$O; s EtOH, eth
6522	1-Isopropyl-4-nitrobenzene		C$_9$H$_{11}$NO$_2$	1817-47-6	165.189	pa ye oil		122[9]	1.084[20]	1.5367[20]	i H$_2$O; s ace, bz, lig
6523	N-Isopropyl-N-nitroso-2-propanamine		C$_6$H$_{14}$N$_2$O	601-77-4	130.187	cry (eth,w)	48	194.5	0.9422[20]		sl H$_2$O; s EtOH, eth, bz
6524	Isopropyl 3-oxobutanoate	Isopropyl acetoacetate	C$_7$H$_{12}$O$_3$	542-08-5	144.168	liq	-27.3	186	0.9835[20]	1.4173[20]	vs eth, EtOH, lig
6525	Isopropyl palmitate	Isopropyl hexadecanoate	C$_{19}$H$_{38}$O$_2$	142-91-6	298.504		13.5	160[2]	0.8404[38]	1.4364[25]	vs ace, bz, eth, EtOH
6526	Isopropyl pentanoate		C$_8$H$_{16}$O$_2$	18362-97-5	144.212				0.8579[20]	1.4061[20]	i H$_2$O; s EtOH, eth, ace
6527	2-Isopropylphenol		C$_9$H$_{12}$O	88-69-7	136.190		15.5	213.5	1.012[20]	1.5315[20]	sl H$_2$O; s EtOH, eth, bz, ctc
6528	3-Isopropylphenol		C$_9$H$_{12}$O	618-45-1	136.190		26	228		1.5261[20]	vs eth
6529	4-Isopropylphenol		C$_9$H$_{12}$O	99-89-8	136.190	nd (peth)	62.3	230; 110[10]	0.990[20]	1.5228[20]	sl H$_2$O; s EtOH, chl
6530	N-Isopropyl-N'-phenyl-1,4-benzenediamine		C$_{15}$H$_{18}$N$_2$	101-72-4	226.317		72.5	148[2]			
6531	Isopropyl phenylcarbamate	Propham	C$_{10}$H$_{13}$NO$_2$	122-42-9	179.216	wh nd (al)	90		1.09[20]	1.4989[91]	vs bz, EtOH
6532	1-(4-Isopropylphenyl)ethanone		C$_{11}$H$_{14}$O	645-13-6	162.228			254	0.9753[15]	1.5235[20]	
6533	Isopropyl propanoate		C$_6$H$_{12}$O$_2$	637-78-5	116.158			109.5	0.8660[20]	1.3872[20]	sl H$_2$O; msc EtOH, eth
6534	N-Isopropyl-2-propenamide		C$_6$H$_{11}$NO	2210-25-5	113.157		64.5	110[15]			
6535	Isopropylpropylamine	N-Propyl-2-propanamine	C$_6$H$_{15}$N	21968-17-2	101.190			96.9			
6536	Isopropyl propyl sulfide		C$_6$H$_{14}$S	5008-73-1	118.240			132.1	0.8269[20]		
6537	4-Isopropylpyridine		C$_8$H$_{11}$N	696-30-0	121.180	liq	-54.9	178	0.9382[25]	1.4962[20]	sl H$_2$O; msc EtOH, eth; vs ace
6538	Isopropyl 3-pyridinecarboxylate	Isopropyl nicotinate	C$_9$H$_{11}$NO$_2$	553-60-6	165.189			126[30], 92.5[5]	1.0624[20]	1.4926[20]	
6539	Isopropyl silicate	Tetra(isopropoxy)silane	C$_{12}$H$_{28}$O$_4$Si	1992-48-9	264.434			184	0.8770[20]		s ctc, CS$_2$
6540	Isopropyl stearate		C$_{21}$H$_{42}$O$_2$	112-10-7	326.557		28	207[6]	0.8403[38]		vs ace, eth, EtOH, chl
6541	4-Isopropylstyrene		C$_{11}$H$_{14}$	2055-40-5	146.229	liq	-44.7	204.1	0.8850[20]	1.5289[20]	vs ace, bz, eth, EtOH
6542	Isopropyl tetradecanoate	Isopropyl myristate	C$_{17}$H$_{34}$O$_2$	110-27-0	270.451			193[20], 140[2]	0.8532[20]	1.4325[25]	i H$_2$O; s EtOH, eth, chl; vs ace, bz
6543	(Isopropylthio)benzene		C$_9$H$_{12}$S	3019-20-3	152.256			208	0.9852[20]	1.5464[20]	
6544	Isopropyl trichloroacetate		C$_5$H$_7$Cl$_3$O$_2$	3974-99-0	205.468			175; 66[15]	1.2911[25]	1.4428[20]	vs bz, eth, EtOH
6545	Isopropylurea		C$_4$H$_{10}$N$_2$O	691-60-1	102.134	nd		103[0.1]			s H$_2$O, EtOH, chl, ace; sl eth
6546	Isopropyl vinyl ether	2-(Ethenyloxy)propane	C$_5$H$_{10}$O	926-65-8	86.132	liq	-140	55.5	0.7534[20]	1.3840[20]	vs ace, bz, eth, EtOH
6547	Isoproterenol	4-[1-Hydroxy-2-[isopropylamino]ethyl]-1,2-benzenediol	C$_{11}$H$_{17}$NO$_3$	7683-59-2	211.258		170.5				

1-Isopropyl-2-methylbenzene

1-Isopropyl-3-methylbenzene

1-Isopropyl-4-methylbenzene

Isopropyl 3-methylbutanoate

5-Isopropyl-2-methyl-1,3-cyclohexadiene, (R)

5-Isopropyl-3-methyl-2-cyclohexen-1-one, (±)

6-Isopropyl-3-methyl-2-cyclohexen-1-one, (±)

Isopropyl methyl ether

5-Isopropyl-2-methylphenol

2-Isopropyl-6-methyl-4-pyrimidinol

Isopropyl methyl sulfide

1-Isopropylnaphthalene

2-Isopropylnaphthalene

Isopropyl nitrate

Isopropyl nitrite

1-Isopropyl-4-nitrobenzene

N-Isopropyl-N-nitroso-2-propanamine

Isopropyl 3-oxobutanoate

Isopropyl palmitate

Isopropyl pentanoate

2-Isopropylphenol

3-Isopropylphenol

4-Isopropylphenol

N-Isopropyl-N'-phenyl-1,4-benzenediamine

Isopropyl phenylcarbamate

1-(4-Isopropylphenyl)ethanone

Isopropyl propanoate

N-Isopropyl-2-propenamide

Isopropylpropylamine

Isopropyl propyl sulfide

4-Isopropylpyridine

Isopropyl 3-pyridinecarboxylate

Isopropyl silicate

Isopropyl stearate

4-Isopropylstyrene

Isopropyl tetradecanoate

(Isopropylthio)benzene

Isopropyl trichloroacetate

Isopropylurea

Isopropyl vinyl ether

Isoproterenol

No.	Name	Synonym	Mol. Form.	CAS RN	Mol. Wt.	Physical Form	mp/°C	bp/°C	den/ g cm⁻³	n_D	Solubility
6548	Isopsoralen		$C_{11}H_6O_3$	523-50-2	186.164		139				
6549	1-Isoquinolinamine		$C_9H_8N_2$	1532-84-9	144.173	pl(w)	123	164[8]			sl H₂O, eth; vs EtOH
6550	3-Isoquinolinamine		$C_9H_8N_2$	25475-67-6	144.173		178.5				
6551	Isoquinoline	Benzo[c]pyridine	C_9H_7N	119-65-3	129.159	hyg pl	26.47	243.22	1.0910[30]	1.6148[20]	i H₂O; vs EtOH, chl; msc eth, bz
6552	7-Isoquinolinol		C_9H_7NO	7651-83-4	145.158		230				sl H₂O, eth; s EtOH
6553	Isosorbide		$C_6H_{10}O_4$	652-67-5	146.141		63	170[2]			
6554	Isosorbide dinitrate	1,4:3,6-Dianhydroglucitol	$C_6H_8N_2O_8$	87-33-2	236.136	col cry	52				vs EtOH, eth, ace
6555	Isosystox	Demeton-S	$C_8H_{19}O_3PS_2$	126-75-0	258.339	liq		133[2]	1.132[21]		s H₂O
6556	Isothebaine		$C_{19}H_{21}NO_3$	568-21-8	311.375	orth cry (al)	203.5				i H₂O; msc EtOH, chl; sl eth; s MeOH
6557	Isothiocyanic acid		CHNS	3129-90-6	59.091	unstab gas					
6558	L-Isovaline	2-Amino-2-methylbutyric acid	$C_5H_{11}NO_2$	595-40-4	117.147	nd (w)	≈300				s EtOH; sl eth
6559	Isoxaben		$C_{18}H_{24}N_2O_4$	82558-50-7	332.395	wh cry	173				s EtOAc, MeCN, MeOH
6560	Isoxazole	1-Oxa-2-azacyclopentadiene	C_3H_3NO	288-14-2	69.062			95	1.078[20]	1.4298[17]	s H₂O
6561	Isoxsuprine		$C_{18}H_{23}NO_3$	395-28-8	301.381	cry	103.0				
6562	Jacobine		$C_{18}H_{25}NO_6$	6870-67-3	351.395	pl (EtOH)	228				
6563	Javanicin		$C_{15}H_{14}O_6$	476-45-9	290.268	red cry (al)	208 dec				s alk
6564	Jervine		$C_{27}H_{39}NO_3$	469-59-0	425.604		243 dec				i H₂O; s EtOH, ace, chl; sl eth
6565	Kaempferol		$C_{15}H_{10}O_6$	520-18-3	286.236	ye nd (al, +1 w)	277				sl H₂O, chl; vs EtOH, eth, ace; i bz
6566	Kainic acid		$C_{10}H_{15}NO_4$	487-79-6	213.231	cry (EtOH aq)	253 dec				s H₂O; i EtOH
6567	Kanamycin A		$C_{18}H_{36}N_4O_{11}$	59-01-8	484.499	cry (EtOH)					
6568	Kepone	Chlordecone	$C_{10}Cl_{10}O$	143-50-0	490.636		350 dec		1.61[25]		
6569	Ketamine	2-(2-Chlorophenyl)-2-(methylamino)cyclohexanone, (±)	$C_{13}H_{16}ClNO$	6740-88-1	237.725	cry (eth-pentane)	92.5				
6570	Ketene		C_2H_2O	463-51-4	42.036	col gas	-151	-49.8			sl eth, ace
6571	Khellin	4,9-Dimethoxy-7-methyl-5H-furo[3,2-g][1]benzopyran-5-one	$C_{14}H_{12}O_5$	82-02-0	260.242	eth, al	154 dec	190[0.05]			i H₂O, chl; s EtOH, ace; sl eth, chl
6572	L-Kynurenine	Benzenebutanoic acid, α,2-diamino-γ-oxo-	$C_{10}H_{12}N_2O_3$	343-65-7	208.213	lf (+l/2w)	191 dec				sl H₂O
6573	Labetalol		$C_{19}H_{24}N_2O_3$	36894-69-6	328.405	cry (MeOH)	164				
6574	DL-Lactic acid	2-Hydroxypropanoic acid, (±)	$C_3H_6O_3$	598-82-3	90.078	ye cry	16.8	122[15]	1.2060[21]	1.4392[20]	vs H₂O, EtOH; sl eth
6575	D-Lactic acid	D-2-Hydroxypropanoic acid	$C_3H_6O_3$	10326-41-7	90.078	pl (chl)	53	103[2]			vs H₂O, EtOH
6576	L-Lactic acid	L-2-Hydroxypropanoic acid	$C_3H_6O_3$	79-33-4	90.078	hyg pr (eth)	25.5				vs H₂O, EtOH
6577	Lactofen		$C_{19}H_{15}ClF_3NO_7$	77501-63-4	461.773	ye pow (bz)	93				
6578	δ-Lactone-D-gluconic acid	δ-D-Gluconolactone	$C_6H_{10}O_6$	90-80-2	178.139	nd (al)					
6579	α-Lactose		$C_{12}H_{22}O_{11}$	14641-93-1	342.296	wh pow	222.8				vs H₂O; sl EtOH; i eth, chl
6580	β-D-Lactose		$C_{12}H_{22}O_{11}$	5965-66-2	342.296		254		1.59[20]		vs H₂O; sl EtOH; i eth, chl
6581	α-Lactose monohydrate		$C_{12}H_{24}O_{12}$	5989-81-1	360.312	mcl (w)	201 dec		1.547[20]		vs H₂O; i EtOH, eth, chl, MeOH
6582	Lactulose	4-O-β-D-Galactopyranosyl-D-fructose	$C_{12}H_{22}O_{11}$	4618-18-2	342.296	hx pl (MeOH)	169				vs H₂O
6583	Laminaribiose	3-O-β-D-Glucopyranosyl-D-glucose	$C_{12}H_{22}O_{11}$	34980-39-7	342.296		205				
6584	Lanosta-8,24-dien-3-ol, (3β)	Lanosterol	$C_{30}H_{50}O$	79-63-0	426.717	nd (eth), cry (MeOH-ace)	140.5				vs eth, EtOH, chl
6585	Lantadene A	Rehmannic acid	$C_{35}H_{52}O_5$	467-81-2	552.785	cry (MeOH)	297				
6586	Lantadene B		$C_{35}H_{52}O_5$	467-82-3	552.785	cry (EtOH)	302				
6587	L-Lanthionine	L-Cysteine, S-(2-amino-2-carboxyethyl)-, (R)-	$C_6H_{12}N_2O_4S$	922-55-4	208.235	hex pl	294 dec				sl H₂O
6588	Lapachol	2-Hydroxy-3-(3-methyl-2-butenyl)-1,4-naphthalenedione	$C_{15}H_{14}O_3$	84-79-7	242.270	ye pr (eth, bz) pl (al)	139.5				i H₂O; s EtOH, eth, bz, chl; vs HOAc
6589	Lappaconitine		$C_{32}H_{44}N_2O_8$	32854-75-4	584.699	hex pl (al)	217.5				i H₂O; sl EtOH, eth; s bz, chl
6590	Lasiocarpine		$C_{21}H_{33}NO_7$	303-34-4	411.490	col pl (peth)	95.5				sl H₂O; s EtOH, bz, eth
6591	Laudanidine		$C_{20}H_{25}NO_4$	301-21-3	343.418	hex pr (al)	184.5				vs H₂O, bz

Isopsoralen

1-Isoquinolinamine

3-Isoquinolinamine

Isoquinoline

7-Isoquinolinol

Isosorbide

Isosorbide dinitrate

Isosystox

Isothebaine

Isothiocyanic acid

L-Isovaline

Isoxaben

Isoxazole

Isoxsuprine

Jacobine

Javanicin

Jervine

Kaempferol

Kainic acid

Kanamycin A

Kepone

Ketamine

Ketene

Khellin

L-Kynurenine

Labetalol

DL-Lactic acid

D-Lactic acid

L-Lactic acid

Lactofen

δ-Lactone-D-gluconic acid

α-Lactose

β-D-Lactose

α-Lactose monohydrate

Lactulose

Laminaribiose

Lanosta-8,24-dien-3-ol, (3β)

Lantadene A

Lantadene B

L-Lanthionine

Lapachol

Lappaconitine

Lasiocarpine

Laudanidine

No.	Name	Synonym	Mol. Form.	CAS RN	Mol. Wt.	Physical Form	mp/°C	bp/°C	den/ g cm⁻³	n_D	Solubility
6592	Laudanine		$C_{20}H_{25}NO_4$	85-64-3	343.418	ye wh pr (dil al, al-chl)	167		1.26^{20}		sl H₂O, EtOH, eth; s bz, chl
6593	Laudanosine		$C_{21}H_{27}NO_4$	2688-77-9	357.444	nd (peth), pr (al)	89				vs ace, eth, EtOH, chl
6594	Laureline		$C_{19}H_{19}NO_3$	81-38-9	309.359	tab (al) cubes (peth)	114				i H₂O; s EtOH, eth, dil acid, con sulf
6595	Laurocapram	1-Dodecylhexahydro-2H-azepin-2-one	$C_{18}H_{35}NO$	59227-89-3	281.477	col liq	-7	160^{50}	0.91	1.4701	i H₂O
6596	Lead bis(dimethyldithiocarbamate)		$C_6H_{12}N_2PbS_4$	19010-66-3	447.6	pale ye nd	258				
6597	Ledol		$C_{15}H_{26}O$	577-27-5	222.366	nd (al)	105	292	0.9078^{100}	1.4667^{110}	vs ace, eth, EtOH
6598	Lenacil		$C_{13}H_{18}N_2O_2$	2164-08-1	234.294		290		1.32^{25}		vs py
6599	Leptophos		$C_{13}H_{10}BrCl_2O_2PS$	21609-90-5	412.066	tan waxy solid	71		1.53^{25}		i H₂O; vs bz; s ace, 2-PrOH, xyl
6600	DL-Leucine		$C_6H_{13}NO_2$	328-39-2	131.173	lf (w)	293	sub	1.293^{18}		s H₂O; sl EtOH; i eth
6601	D-Leucine		$C_6H_{13}NO_2$	328-38-1	131.173	pl (al)	293	sub			sl H₂O
6602	L-Leucine	2-Amino-4-methylpentanoic acid	$C_6H_{13}NO_2$	61-90-5	131.173	hex pl (dil al)	293	sub	1.293^{18}		sl H₂O; i EtOH, eth
6603	N-Leucylglycine		$C_8H_{16}N_2O_3$	686-50-0	188.224		248 dec				s H₂O; sl EtOH, eth; i ace, bz, chl
6604	Leuprolide		$C_{59}H_{84}N_{16}O_{12}$	53714-56-0	1209.398	fluffy solid					
6605	Leurosine		$C_{46}H_{56}N_4O_9$	23360-92-1	808.959	cry	203				
6606	Levallorphan	17-Allylmorphinan-3-ol	$C_{19}H_{25}NO$	152-02-3	283.408	cry (EtOH aq)	181				
6607	Levodopa	L-3,4-Dihydroxyphenylalanine	$C_9H_{11}NO_4$	59-92-7	197.188	pl (dil al) pr or nd (w+SO₂)	277 dec				s H₂O; i EtOH, eth, ace, bz; s alk, MeOH
6608	Levopimaric acid		$C_{20}H_{30}O_2$	79-54-9	302.451	orth cry	150				
6609	Levorphanol	17-Methylmorphinan-3-ol	$C_{17}H_{23}NO$	77-07-6	257.371	cry	198				
6610	d-Limonene	p-Mentha-1,8-diene, (R)	$C_{10}H_{16}$	5989-27-5	136.234	oil	-74.0	178	0.8411^{20}	1.4730^{20}	i H₂O; msc EtOH, eth; s ctc
6611	l-Limonene	p-Mentha-1,8-diene, (S)	$C_{10}H_{16}$	5989-54-8	136.234	oil		$178; 64.4^{15}$	0.843^{20}	1.4746^{20}	i H₂O; vs eth, EtOH
6612	Linalol	3,7-Dimethyl-1,6-octadien-3-ol, (±)-	$C_{10}H_{18}O$	22564-99-4	154.249			$198; 86^{13}$	0.870^{15}	1.4627	
6613	Linalyl acetate	3,7-Dimethyl-1,6-octadien-3-yl acetate	$C_{12}H_{20}O_2$	115-95-7	196.286	liq		$220; 44^{0.2}$	0.895^{20}	1.4460^{20}	i H₂O; msc EtOH, eth
6614	Lincomycin		$C_{18}H_{34}N_2O_6S$	154-21-2	406.537	amor solid					sl H₂O; s EtOH, ace, chl
6615	Linoleic acid	cis,cis-9,12-Octadecadienoic acid	$C_{18}H_{32}O_2$	60-33-3	280.446		-7	229^{16}	0.9022^{20}	1.4699^{20}	vs ace, bz, eth, EtOH
6616	Linolenic acid	cis,cis,cis-9,12,15-Octadecatrienoic acid	$C_{18}H_{30}O_2$	463-40-1	278.430		-11	$231^{17}, 129^{0.05}$	0.9164^{20}	1.4800^{20}	i H₂O; s EtOH, eth; sl bz
6617	Linuron	N-(3,4-Dichlorophenyl)-N'-methoxy-N'-methylurea	$C_9H_{10}Cl_2N_2O_2$	330-55-2	249.093		93				
6618	Liothyronine		$C_{15}H_{12}I_3NO_4$	6893-02-3	650.974	cry	236 dec				i H₂O, EtOH; s dil alk
6619	Lipoamide	1,2-Dithiolane-3-pentanamide	$C_8H_{15}NOS_2$	940-69-2	205.341	cry	128				
6620	α-Lipoic acid	1,2-Dithiolane-3-pentanoic acid	$C_8H_{14}O_2S_2$	1077-28-7	206.326	ye pl (cy)	60	87^{25}			i H₂O
6621	Lisinopril		$C_{21}H_{35}N_3O_7$	83915-83-7	441.519	wh cry pow	159				i EtOH, chl, ace; sl MeOH
6622	Lithium oxalate		$C_2Li_2O_4$	30903-87-8	101.901		dec		2.121^{17}		s H₂O; i EtOH, eth
6623	Lobelanidine		$C_{22}H_{29}NO_2$	552-72-7	339.471	sc (al, eth)	150				i H₂O; s EtOH; sl eth; vs ace, bz, py
6624	Lobelanine		$C_{22}H_{25}NO_2$	579-21-5	335.440	nd (eth, peth)	99				vs ace, bz, EtOH, chl
6625	Lobeline		$C_{22}H_{27}NO_2$	90-69-7	337.455	nd (al, bz)	130.5				sl H₂O; s EtOH, eth, bz, chl; vs ace
6626	Loflucarban		$C_{13}H_9Cl_2FN_2S$	790-69-2	315.192		163.5				
6627	Longifolene	Kuromatsuene	$C_{15}H_{24}$	475-20-7	204.352			$258; 126^{15}$	0.9319^{18}	1.5040^{20}	i H₂O; s bz
6628	Loratadine	Claritin	$C_{22}H_{23}ClN_2O_2$	79794-75-5	382.883	cry (MeCN)	132				
6629	Lovastatin	Mevacor	$C_{24}H_{36}O_5$	75330-75-5	404.540	wh cry (ace aq)	174				i H₂O; vs chl; s DMF; sl ace, EtOH
6630	Lovozal		$C_{15}H_7Cl_2F_3N_2O_2$	14255-88-0	375.130	ye cry	103				s ace, diox
6631	Loxapine		$C_{18}H_{18}ClN_3O$	1977-10-2	327.808	ye cry (peth)	109.5				

Laudanine

Laudanosine

Laureline

Laurocapram

Lead bis(dimethyldithiocarbamate)

Ledol

Lenacil

Leptophos

DL-Leucine

D-Leucine

L-Leucine

N-Leucylglycine

H-5-oxoPro-His-Trp-Ser-Tyr-D-Leu-Leu-Arg-Pro-NHEt
Leuprolide

Leurosine

Levallorphan

Levodopa

Levopimaric acid

Levorphanol

d-Limonene

l-Limonene

Linalol

Linalyl acetate

Lincomycin

Linoleic acid

Linolenic acid

Linuron

Liothyronine

Lipoamide

α-Lipoic acid

Lisinopril

Lithium oxalate

Lobelanidine

Lobelanine

Lobeline

Loflucarban

Longifolene

Loratadine

Lovastatin

Lovozal

Loxapine

No.	Name	Synonym	Mol. Form.	CAS RN	Mol. Wt.	Physical Form	mp/°C	bp/°C	den/ g cm⁻³	n_D	Solubility
6632	Loxoprofen	α-Methyl-4-[(2-oxocyclopentyl) methyl]benzeneacetic acid	$C_{15}H_{18}O_3$	68767-14-6	246.302	col oil	110	192[0.3]			
6633	Luciculine	Napelline	$C_{22}H_{35}NO_3$	5008-52-6	361.518	cry (+1w, ace)	149	165[0.02]			vs EtOH
6634	Lunacrine		$C_{16}H_{19}NO_3$	82-40-6	273.327						s chl
6635	Lup-20(29)-ene-3,28-diol, (3β)	Betulin	$C_{30}H_{50}O_2$	473-98-3	442.717	nd (al +1)	250	sub 240			i H_2O; sl EtOH, bz; s eth, AcOEt, lig
6636	Lup-20(29)-en-3-ol, (3β)	Lupeol	$C_{30}H_{50}O$	545-47-1	426.717	nd (al, ace)	216		0.9457[218]	1.4910[218]	i H_2O; vs EtOH, eth, ace, bz, chl
6637	Lupulon		$C_{26}H_{38}O_4$	468-28-0	414.578	pr (MeOH)	93				i H_2O; s EtOH, peth, hx
6638	Luteolin		$C_{15}H_{10}O_6$	491-70-3	286.236	ye nd (dil al, + 1 w)	329 dec				sl H_2O; s EtOH, eth, alk, con sulf
6639	Luteoskyrin	8,8'-Dihydroxyrugulosin	$C_{30}H_{22}O_{12}$	21884-44-6	574.489	ye nd (EtOH)	278 dec				
6640	Lycodine		$C_{16}H_{22}N_2$	20316-18-1	242.359	orth pr	99	190[1.0]			s H_2O, chl, eth, EtOH; i peth
6641	Lycomarasmine		$C_9H_{15}N_3O_7$	7611-43-0	277.231			228 dec			
6642	Lycorine		$C_{16}H_{17}NO_4$	476-28-8	287.311	pr (al, py)	280	sub			i H_2O; sl EtOH, eth, chl
6643	Lysergamide		$C_{16}H_{17}N_3O$	478-94-4	267.325	cry (MeOH), pr (aq, ace)	137.5				sl EtOH, ace, os
6644	Lysergic acid		$C_{16}H_{16}N_2O_2$	82-58-6	268.310	lf or hex sc (w)	240 dec				sl H_2O, eth, bz; s EtOH, py
6645	Lysergide		$C_{20}H_{25}N_3O$	50-37-3	323.432		82				
6646	DL-Lysine	2,6-Diaminohexanoic acid, (±)	$C_6H_{14}N_2O_2$	70-54-2	146.187		224				sl H_2O
6647	D-Lysine	2,6-Diaminohexanoic acid, (D)	$C_6H_{14}N_2O_2$	923-27-3	146.187		218 dec				s H_2O
6648	L-Lysine	2,6-Diaminohexanoic acid, (L)	$C_6H_{14}N_2O_2$	56-87-1	146.187	nd (w, dil al)	224 dec				s H_2O; i EtOH, eth, ace, bz
6649	L-Lysine, hydrochloride		$C_6H_{15}ClN_2O_2$	10098-89-2	182.648		263 dec				
6650	D-Lyxose		$C_5H_{10}O_5$	1114-34-7	150.130		108		1.545[20]		
6651	L-Lyxose		$C_5H_{10}O_5$	1949-78-6	150.130		110				
6652	Maclurin	(3,4-Dihydroxyphenyl)(2,4,6-trihydroxyphenyl)methanone	$C_{13}H_{10}O_6$	519-34-6	262.214	ye nd (al)	222.5				vs eth, EtOH
6653	Magenta base	Rosaniline	$C_{20}H_{19}N_3$	3248-93-9	301.385	br-red cry	186 dec				
6654	Magenta I	Rosaniline hydrochloride	$C_{20}H_{20}ClN_3$	632-99-5	337.846	grn cry	200 dec				sl H_2O, EtOH; i eth
6655	Magnesium stearate	Magnesium octadecanoate	$C_{36}H_{70}MgO_4$	557-04-0	591.244	wh pow	132				i H_2O; reac acid
6656	Malachite Green		$C_{23}H_{25}ClN_2$	569-64-2	364.911	grn cry					vs H_2O, EtOH, MeOH
6657	Malaoxon	(Dimethoxyphosphinylthio) butanedioic acid	$C_{10}H_{19}O_7PS$	1634-78-2	314.293	liq		132[0.1]			
6658	Malathion		$C_{10}H_{19}O_6PS_2$	121-75-5	330.358	ye-br liq	1.4	156[0.7] dec	1.2076[20]	1.4960[20]	sl H_2O; s EtOH, eth, bz
6659	Maleic acid	cis-2-Butenedioic acid	$C_4H_4O_4$	110-16-7	116.073	mcl pr (w)	139		1.590[20]		vs H_2O, EtOH, ace; s eth; i bz, chl
6660	Maleic anhydride		$C_4H_2O_3$	108-31-6	98.057	nd (chl, eth)	52.56	202	1.314[60]		s H_2O; s eth, ace, chl; sl lig
6661	Maleonitrile	cis-Butenedinitrile	$C_4H_2N_2$	928-53-0	78.072	pr (EtOH)	31.5	111[20]			
6662	Malic acid	Hydroxybutanedioic acid	$C_4H_6O_5$	617-48-1	134.088		132		1.601[20]		s H_2O; vs eth, EtOH, MeOH
6663	Malonaldehyde	1,3-Propanedial	$C_3H_4O_2$	542-78-9	72.063	hyg nd	73				
6664	Malonic acid		$C_3H_4O_4$	141-82-2	104.062	tcl (al)	135 dec	sub	1.619[10]		vs H_2O, py; s EtOH, eth; i bz
6665	Malononitrile		$C_3H_2N_2$	109-77-3	66.061		32	218.5	1.1910[20]	1.4146[34]	s H_2O, ace, bz, chl; vs EtOH, eth
6666	Maltopentaose		$C_{30}H_{52}O_{26}$	34620-76-3	828.718	cry (w)	78 (hyd)				
6667	α-Maltose		$C_{12}H_{22}O_{11}$	4482-75-1	342.296	nd (al)	162.5		1.546[20]		vs H_2O
6668	6-O-α-Maltosyl-β-cyclodextrin		$C_{54}H_{90}O_{45}$	104723-60-6	1459.266	cry (MeOH)					
6669	Maltotetraose		$C_{24}H_{42}O_{21}$	34612-38-9	666.577	amorp solid	170 dec				
6670	Malvidin chloride		$C_{17}H_{15}ClO_7$	643-84-5	366.750		>300				sl H_2O; s EtOH, MeOH
6671	Mandelic acid	α-Hydroxybenzeneacetic acid	$C_8H_8O_3$	90-64-2	152.148	orth pl	119		1.2890[20]		s H_2O, eth, EtOH, i-PrOH
6672	Mandelonitrile glucoside		$C_{14}H_{17}NO_6$	138-53-4	295.288	wh nd or pl (al)	122				vs H_2O, EtOH
6673	Maneb	Manganese, [[1,2-ethanediylbis[carbamodithioato]](2-)]-	$C_4H_6MnN_2S_4$	12427-38-2	265.302		dec 200				

Loxoprofen

Luciculine

Lunacrine

Lup-20(29)-ene-3,28-diol, (3β)

Lup-20(29)-en-3-ol, (3β)

Lupulon

Luteolin

Luteoskyrin

Lycodine

Lycomarasmine

Lycorine

Lysergamide

Lysergic acid

Lysergide

DL-Lysine

D-Lysine

L-Lysine

L-Lysine, hydrochloride

D-Lyxose

L-Lyxose

Maclurin

Magenta base

Magenta I

Magnesium stearate

Malachite Green

Malaoxon

Malathion

Maleic acid

Maleic anhydride

Maleonitrile

Malic acid

Malonaldehyde

Malonic acid

Malononitrile

Maltopentaose

α-Maltose

6-O-α-Maltosyl-β-cyclodextrin

Maltotetraose

Malvidin chloride

Mandelic acid

Mandelonitrile glucoside

Maneb

No.	Name	Synonym	Mol. Form.	CAS RN	Mol. Wt.	Physical Form	mp/°C	bp/°C	den/ g cm⁻³	n_D	Solubility
6674	Manganese(II) acetate		$C_4H_6O_4Mn$	638-38-0	173.027	red cry (w)	210				s H_2O, MeOH, HOAc; i ace
6675	Manganese carbonyl	Dimanganese decacarbonyl	$C_{10}Mn_2O_{10}$	10170-69-1	389.977	ye mcl cry	154		1.75		i H_2O; s os
6676	Manganese cyclopentadienyl tricarbonyl		$C_8H_5MnO_3$	12079-65-1	204.062	pale ye cry	77.0	subl			s os
6677	Manganese 2-methylcyclopentadienyl tricarbonyl		$C_9H_7MnO_3$	12108-13-3	218.088	ye liq	1.5	233; 102¹⁰	1.388²⁰		i H_2O; misc bz
6678	D-Mannitol	Cordycepic acid	$C_6H_{14}O_6$	69-65-8	182.171	orth nd or pr (w)	168	295³·⁵	1.489²⁰	1.3330	vs H_2O; sl EtOH, py; i eth
6679	D-Mannitol hexanitrate		$C_6H_8N_6O_{18}$	15825-70-4	452.157	nd (al)	107	exp	1.8²⁰		vs bz, eth, EtOH
6680	D-Mannose	Seminose	$C_6H_{12}O_6$	3458-28-4	180.155	nd or orth pr (al)	132 dec		1.539²⁰		vs H_2O; sl EtOH, MeOH; i eth, bz
6681	L-Mannose		$C_6H_{12}O_6$	10030-80-5	180.155	cry (al)	132				vs H_2O
6682	Matridin-15-one	Matrine	$C_{15}H_{24}N_2O$	519-02-8	248.364	α-nd or pl; β-orth pr,		223⁶		1.5286²⁵	s H_2O, eth, ace; vs EtOH, bz; sl peth
6683	Mazindol		$C_{16}H_{13}ClN_2O$	22232-71-9	284.739	cry (ace/hx)	198				i H_2O; s EtOH
6684	Mebendazole		$C_{16}H_{13}N_3O_3$	31431-39-7	295.292	cry (HOAc/ MeOH)	288.5				i H_2O, EtOH, eth, chl
6685	Mebhydroline		$C_{19}H_{20}N_2$	524-81-2	276.375	cry	95	211¹			i H_2O; sl eth; vs EtOH, ace, MeOH
6686	Mecarbam		$C_{10}H_{20}NO_5PS_2$	2595-54-2	329.374	ye oil		144⁰·⁰²	1.223²⁰		sl H_2O
6687	Meclizine		$C_{25}H_{27}ClN_2$	569-65-3	390.948			230			s CS_2
6688	Medroxyprogesterone		$C_{22}H_{32}O_3$	520-85-4	344.487		214.5				vs chl
6689	Mefenamic acid	2-[(2,3-Dimethylphenyl)amino]benzoic acid	$C_{15}H_{15}NO_2$	61-68-7	241.286	hyg cry	230 dec				s alk; sl eth, chl
6690	Mefloquine		$C_{17}H_{16}F_6N_2O$	53230-10-7	378.311	cry (MeOH aq)	178.2				
6691	Mefluidide		$C_{11}H_{13}F_3N_2O_3S$	53780-34-0	310.292		184				
6692	Melezitose		$C_{18}H_{32}O_{16}$	597-12-6	504.437	cry (w+2)	153		1.5565²⁵		vs H_2O
6693	α-D-Melibiose	6-O-α-D-Galactopyranosyl-D-glucose	$C_{12}H_{22}O_{11}$	585-99-9	342.296						vs H_2O; sl EtOH; dec acid
6694	Melinamide	N-(1-Phenylethyl)-9,12-octadecadieneamide, (Z,Z)-	$C_{26}H_{41}NO$	14417-88-0	383.610	oil	<4	202⁰·⁰⁷		1.5050²³	
6695	Melphalan	L-Phenylalanine, 4-[bis(2-chloroethyl)amino]-	$C_{13}H_{18}Cl_2N_2O_2$	148-82-3	305.200	nd	183 dec				i H_2O; s EtOH
6696	Menaquinone 7	Vitamin K_2(35)	$C_{46}H_{64}O_2$	2124-57-4	648.999	cry	54				
6697	Menazon		$C_6H_{12}N_5O_2PS_2$	78-57-9	281.296	cry (MeOH)	160				sl H_2O; s thf
6698	p-Menthane hydroperoxide	1-Methyl-1-(4-methylcyclohexyl)ethyl hydroperoxide	$C_{10}H_{20}O_2$	80-47-7	172.265			259	0.92		
6699	p-Menth-8-en-2-one	2-Methyl-5-(1-methylethenyl)cyclohexanone	$C_{10}H_{16}O$	7764-50-3	152.233			223.0			
6700	Menthol 3-methylbutanoate	Menthol, isovalerate	$C_{15}H_{28}O_2$	16409-46-4	240.382			129⁹	0.908¹⁵	1.4486²⁰	i H_2O; s EtOH, ace
6701	Meperidine	Pethidine	$C_{15}H_{21}NO_2$	57-42-1	247.334		30	155⁵			
6702	Mephenytoin		$C_{12}H_{14}N_2O_2$	50-12-4	218.251		136				
6703	Mephobarbital		$C_{13}H_{14}N_2O_3$	115-38-8	246.261	wh cry (w)	176				sl H_2O, eth, chl; vs EtOH
6704	Mephosfolan		$C_8H_{16}NO_3PS_2$	950-10-7	269.322	ye liq		120⁰·⁰⁰¹		1.5354²⁶	s ace, EtOH, bz
6705	Mepiquat chloride	Piperidinium, 1,1-dimethyl-, chloride	$C_7H_{16}ClN$	24307-26-4	149.662		223				
6706	Mepivacaine	N-(2,6-Dimethylphenyl)-1-methyl-2-piperidinecarboxamide	$C_{15}H_{22}N_2O$	96-88-8	246.348	cry (eth)	150.5				s CS_2
6707	Mepivacaine monohydrochloride	Carbocaine hydrochloride	$C_{15}H_{23}ClN_2O$	1722-62-9	282.809	cry	263				s H_2O
6708	Mercaptoacetic acid, 2-ethylhexyl ester		$C_{10}H_{20}O_2S$	7659-86-1	204.330			133.5	0.97²⁰		
6709	2-Mercaptobenzoic acid	o-Thiosalicylic acid	$C_7H_6O_2S$	147-93-3	154.187	lf or nd (al, w, HOAc)	168.5	sub			s H_2O, EtOH, eth; sl DMSO, lig
6710	Mercaptobenzthiazyl ether	2,2'-Dithiobis[benzothiazole]	$C_{14}H_8N_2S_4$	120-78-5	332.487	ye nd	180		1.50		i H_2O; sl EtOH, bz, ctc, ace
6711	2-Mercaptoethanol		C_2H_6OS	60-24-2	78.133			158; 55¹³	1.1143²⁰	1.4996²⁰	s H_2O, EtOH, eth, bz
6712	2-Mercapto-2-methylpropanoic acid		$C_4H_8O_2S$	4695-31-2	120.171		47	101¹⁵			vs H_2O
6713	2-Mercapto-N-2-naphthylacetamide	Thionalide	$C_{12}H_{11}NOS$	93-42-5	217.286		111.5				i H_2O; vs EtOH, os
6714	2-Mercaptophenol		C_6H_6OS	1121-24-0	126.176	oil	5.5	217; 89⁸	1.2371⁰		vs bz, eth, EtOH

Manganese(II) acetate

Manganese carbonyl

Manganese cyclopentadienyl tricarbonyl

Manganese 2-methylcyclopentadienyl tricarbonyl

D-Mannitol

D-Mannitol hexanitrate

D-Mannose

L-Mannose

Matridin-15-one

Mazindol

Mebendazole

Mebhydroline

Mecarbam

Meclizine

Medroxyprogesterone

Mefenamic acid

Mefloquine

Mefluidide

Melezitose

α-D-Melibiose

Melinamide

Melphalan

Menaquinone 7

Menazon

p-Menthane hydroperoxide

p-Menth-8-en-2-one

Menthol 3-methylbutanoate

Meperidine

Mephenytoin

Mephobarbital

Mephosfolan

Mepiquat chloride

Mepivacaine

Mepivacaine monohydrochloride

Mercaptoacetic acid, 2-ethylhexyl ester

2-Mercaptobenzoic acid

Mercaptobenzthiazyl ether

2-Mercaptoethanol

2-Mercapto-2-methylpropanoic acid

2-Mercapto-N-2-naphthylacetamide

2-Mercaptophenol

No.	Name	Synonym	Mol. Form.	CAS RN	Mol. Wt.	Physical Form	mp/°C	bp/°C	den/ g cm⁻³	n_D	Solubility
6715	4-Mercaptophenol		C₆H₆OS	637-89-8	126.176	cry	29.5	167[45], 135[11]	1.1285[25]	1.5101[25]	s H₂O, EtOH, alk, con sulf
6716	3-Mercapto-1,2-propanediol	Thioglycerol	C₃H₈O₂S	96-27-5	108.160	visc		100[1]	1.2455[20]	1.5268[20]	sl H₂O, eth, bz, chl; msc EtOH; vs ace
6717	3-Mercaptopropanoic acid		C₃H₆O₂S	107-96-0	106.144	amor	18	111[15], 86[3]	1.218[21]	1.494[20]	s H₂O, EtOH, eth, ctc
6718	3-Mercapto-D-valine	Penicillamine	C₅H₁₁NO₂S	52-67-5	149.212		198.5				
6719	Mercury(II) benzoate	Mercuric benzoate	C₁₄H₁₀HgO₄	583-15-3	442.81	cry pow (w)	≈125				i EtOH
6720	Mercury(II) oleate	Mercuric oleate	C₃₆H₆₆HgO₄	1191-80-6	763.35	ye-br solid					i H₂O; sl EtOH, eth
6721	Mercury(II) phenyl acetate	Phenylmercuric acetate	C₈H₈HgO₂	62-38-4	336.74		153				i H₂O; s chl
6722	Merphos	Phosphorotrithious acid, S,S,S-tributyl ester	C₁₂H₂₇PS₃	150-50-5	298.511		100	137[0.7], 176[15]	1.02[20]		
6723	Mesityl oxide	Isobutenyl methyl ketone	C₆H₁₀O	141-79-7	98.142	liq	-59	130	0.8653[20]	1.4440[20]	s H₂O, ace; msc EtOH, eth
6724	Mesoridazine		C₂₁H₂₆N₂OS₂	5588-33-0	386.573	oil					
6725	Mestranol		C₂₁H₂₆O₂	72-33-3	310.430	cry	151				i H₂O; s diox, eth, EtOH, chl
6726	[2.2]Metacyclophane	Tricyclo[9.3.1.1]hexadeca-1(15),4,6,8(16),11,13-hexaene	C₁₆H₁₆	2319-97-3	208.298	orth pr	132.5	290			sl EtOH; s bz, eth
6727	Metalaxyl		C₁₅H₂₁NO₄	57837-19-1	279.333		71				
6728	Metaldehyde	Metacetaldehyde (polymer)	(C₂H₄O)ₓ	37273-91-9		tetr nd or pr (al)	246	sub 115			i H₂O, ace; sl EtOH, eth, bz, chl
6729	Metanil Yellow		C₁₈H₁₄N₃NaO₃S	587-98-4	375.377	br-ye pow					vs H₂O, EtOH; s bz, eth; sl ace
6730	Metaraminol	2-Amino-1-(3-hydroxyphenyl)-1-propanol, (1R,2S)	C₉H₁₃NO₂	54-49-9	167.205	hyg cry (HCl)					s H₂O
6731	Metaxalone		C₁₂H₁₅NO₃	1665-48-1	221.252	cry (AcOEt)	122	223[1.5]			
6732	Methacholine chloride		C₈H₁₈ClNO₂	62-51-1	195.688	hyg cry	172				vs H₂O, EtOH, chl
6733	Methacrylic acid	2-Methylpropenoic acid	C₄H₆O₂	79-41-4	86.090	pr	16	162.5	1.0153[20]	1.4314[20]	s H₂O, chl; msc EtOH, eth
6734	Methacycline		C₂₂H₂₂N₂O₈	914-00-1	442.418	cry	205 dec				
6735	Methadone hydrochloride	6-(Dimethylamino)-4,4-diphenyl-3-heptanone hydrochloride	C₂₁H₂₈ClNO	1095-90-5	345.906	pl (al-eth)	235				vs H₂O, EtOH
6736	Methallenestril		C₁₈H₂₂O₃	517-18-0	286.366	cry (MeOH aq)	139				s eth
6737	Methamidophos	Phosphoramidothioic acid, O,S-dimethyl ester	C₂H₈NO₂PS	10265-92-6	141.130		46		1.31[20]		
6738	Methamphetamine		C₁₀H₁₅N	537-46-2	149.233			212			
6739	Methamphetamine hydrochloride	N,α-Dimethylbenzeneethanamine, hydrochloride, (S)-	C₁₀H₁₆ClN	51-57-0	185.694		173.8				vs H₂O, EtOH, chl
6740	Methandrostenolone		C₂₀H₂₈O₂	72-63-9	300.435		166				
6741	Methane		CH₄	74-82-8	16.043	col gas	-182.47	-161.48	0.4228[-162]		sl H₂O, ace; s EtOH, eth, bz, tol, MeOH
6742	Methanearsonic acid		CH₅AsO₃	124-58-3	139.971		160.5				s H₂O, EtOH
6743	Methanedisulfonic acid	Methionic acid	CH₄O₆S₂	503-40-2	176.169		98				i H₂O; s HNO₃
6744	Methanesulfonic acid	Methylsulfonic acid	CH₄O₃S	75-75-2	96.106		20	167[10]	1.4812[18]	1.4317[18]	s H₂O
6745	Methanesulfonyl chloride		CH₃ClO₂S	124-63-0	114.552			162; 55[11]	1.4805[18]	1.4573[20]	i H₂O; s EtOH, eth
6746	Methanesulfonyl fluoride		CH₃FO₂S	558-25-8	98.097			123.5			
6747	Methanethiol	Methyl mercaptan	CH₄S	74-93-1	48.108	col gas	-123	5.9	0.8665[20]		sl H₂O, chl; vs EtOH, eth
6748	Methanimidamide	Formamidine	CH₄N₂	463-52-5	44.056	pr	81	dec			vs H₂O, EtOH
6749	Methanimidamide, monoacetate	Formamidine acetate	C₃H₈N₂O₂	3473-63-0	104.108		161.5				vs H₂O
6750	Methanol	Methyl alcohol	CH₄O	67-56-1	32.042	liq	-97.53	64.6	0.7914[20]	1.3288[20]	msc H₂O, EtOH, eth, ace; vs bz; s chl
6751	Methantheline bromide		C₂₁H₂₆BrNO₃	53-46-3	420.340	cry (i-PrOH)	174.5				s H₂O, EtOH, chl; i eth
6752	Methapyrilene		C₁₄H₁₉N₃S	91-80-5	261.386			174[3]		1.5915[20]	
6753	Metharbital	5,5-Diethyl-1-methyl-2,4,6(1H,3H,5H)-pyrimidinetrione	C₉H₁₄N₂O₃	50-11-3	198.218	nd	150.5				s H₂O; sl chl
6754	Methazolamide		C₅H₈N₄O₃S₂	554-57-4	236.273	cry (w)	213 dec				
6755	Methazole		C₉H₆Cl₂N₂O₃	20354-26-1	261.061		123		1.24[25]		
6756	Methenamine allyl iodide	Allylhexamethylenetetramine iodide	C₉H₁₇IN₄	36895-62-2	308.162	cry	148 dec				vs H₂O; i chl, eth

4-Mercaptophenol

3-Mercapto-1,2-propanediol

3-Mercaptopropanoic acid

3-Mercapto-D-valine

Mercury(II) benzoate

Mercury(II) oleate

Mercury(II) phenyl acetate

Merphos

Mesityl oxide

Mesoridazine

Mestranol

[2.2]Metacyclophane

Metalaxyl

Metaldehyde

Metanil Yellow

Metaraminol

Metaxalone

Methacholine chloride

Methacrylic acid

Methacycline

Methadone hydrochloride

Methallenestril

Methamidophos

Methamphetamine

Methamphetamine hydrochloride

Methandrostenolone

Methane

Methanearsonic acid

Methanedisulfonic acid

Methanesulfonic acid

Methanesulfonyl chloride

Methanesulfonyl fluoride

Methanethiol

Methanimidamide

Methanimidamide, monoacetate

Methanol

Methantheline bromide

Methapyrilene

Metharbital

Methazolamide

Methazole

Methenamine allyl iodide

No.	Name	Synonym	Mol. Form.	CAS RN	Mol. Wt.	Physical Form	mp/°C	bp/°C	den/ g cm^{-3}	n_D	Solubility
6757	Methestrol		$C_{20}H_{26}O_2$	130-73-4	298.419	cry (dil HOAc)	145				
6758	Methidathion		$C_6H_{11}N_2O_4PS_3$	950-37-8	302.330		39				
6759	Methiocarb	Phenol, 3,5-dimethyl-4-(methylthio)-, methylcarbamate	$C_{11}H_{15}NO_2S$	2032-65-7	225.308		120				
6760	L-Methionine		$C_5H_{11}NO_2S$	63-68-3	149.212	hex pl (dil al)	281 dec				s H_2O; i EtOH, eth, ace, bz, peth; sl HOAc
6761	Methocarbamol	Guaifenesin-1-carbamate	$C_{11}H_{15}NO_5$	532-03-6	241.241	cry (bz)	93				s EtOH
6762	Methomyl		$C_5H_{10}N_2O_2S$	16752-77-5	162.210		78		1.2946[24]		
6763	Methoprene		$C_{19}H_{34}O_3$	40596-69-8	310.471			100[0.05]	0.926[20]		
6764	Methoprotryne		$C_{11}H_{21}N_5OS$	841-06-5	271.383	cry	69				sl H_2O; s os
6765	Methotrexate		$C_{20}H_{22}N_8O_5$	59-05-2	454.440	ye cry (w)	190 dec				
6766	Methoxamine hydrochloride		$C_{11}H_{18}ClNO_3$	61-16-5	247.719	cry	214				vs H_2O; i eth, bz, chl
6767	Methoxsalen	9-Methoxy-7H-furo[3,2-g][1]benzopyran-7-one	$C_{12}H_8O_4$	298-81-7	216.190	pr (dil al) nd (peth)	148				sl H_2O, eth, ace, peth; vs EtOH
6768	Methoxyacetaldehyde		$C_3H_6O_2$	10312-83-1	74.079			92	1.005[25]	1.3950[20]	vs H_2O, ace, eth, EtOH
6769	Methoxyacetic acid		$C_3H_6O_3$	625-45-6	90.078	hyg		203.5	1.1768[20]	1.4168[20]	s H_2O, EtOH, eth
6770	Methoxyacetonitrile		C_3H_5NO	1738-36-9	71.078			119	0.9492[20]	1.3831[20]	sl H_2O; s EtOH, eth, ace, chl, alk, acid
6771	Methoxyacetyl chloride		$C_3H_5ClO_2$	38870-89-2	108.524			112.5	1.1871[20]	1.4199[20]	s eth, ace, ctc; vs chl
6772	2-Methoxyaniline	o-Anisidine	C_7H_9NO	90-04-0	123.152		6.2	224	1.0923[20]	1.5715[10]	sl H_2O; s EtOH, eth, ace, bz
6773	3-Methoxyaniline	m-Anisidine	C_7H_9NO	536-90-3	123.152	liq	-1	251	1.096[20]	1.5794[20]	sl H_2O, ctc; s EtOH, eth, ace, bz
6774	4-Methoxyaniline	p-Anisidine	C_7H_9NO	104-94-9	123.152	orth pl	57.2	243	1.071[57]	1.5559[60]	s H_2O, ace, bz; vs EtOH, eth
6775	2-Methoxyaniline hydrochloride	o-Anisidine hydrochloride	$C_7H_{10}ClNO$	134-29-2	159.613	nd	225				
6776	1-Methoxy-9,10-anthracenedione		$C_{15}H_{10}O_3$	82-39-3	238.238		170.3				sl EtOH; vs bz, chl
6777	2-Methoxybenzaldehyde		$C_8H_8O_2$	135-02-4	136.149	pr	37.5	243.5	1.1326[20]	1.5600[20]	i H_2O; s EtOH, bz, ctc; vs eth, ace, chl
6778	3-Methoxybenzaldehyde		$C_8H_8O_2$	591-31-1	136.149			231	1.1187[20]	1.5530[20]	i H_2O; s EtOH, bz; vs eth, ace, chl
6779	4-Methoxybenzaldehyde	p-Anisaldehyde	$C_8H_8O_2$	123-11-5	136.149		0	248; 134[12]	1.119[15]	1.5730[20]	i H_2O; msc EtOH, eth; vs ace, chl; s bz
6780	4-Methoxybenzamide		$C_8H_9NO_2$	3424-93-9	151.163	nd or tab (w)	166.5	295			vs H_2O, EtOH
6781	4-Methoxybenzeneacetaldehyde		$C_9H_{10}O_2$	5703-26-4	150.174			255.5	1.096[20]	1.5359[20]	
6782	2-Methoxybenzeneacetic acid		$C_9H_{10}O_3$	93-25-4	166.173	nd (w)	124	100[2]			s H_2O; vs EtOH, eth, ace, bz, chl
6783	4-Methoxybenzeneacetic acid		$C_9H_{10}O_3$	104-01-8	166.173	pl (w)	87	138[2]			i H_2O; vs EtOH; s eth, bz; sl chl, lig
6784	4-Methoxybenzeneacetonitrile		C_9H_9NO	104-47-2	147.173			286.5	1.0845[20]	1.5309[20]	s EtOH, eth, chl
6785	4-Methoxy-1,2-benzenediamine	4-Methoxy-o-phenylenediamine	$C_7H_{10}N_2O$	102-51-2	138.166	grn pl	51	200[21], 168[11]			vs eth
6786	4-Methoxy-1,3-benzenediamine	4-Methoxy-m-phenylenediamine	$C_7H_{10}N_2O$	615-05-4	138.166	nd (eth)	67.5				s EtOH, eth; sl DMSO
6787	2-Methoxy-1,4-benzenediamine	2,5-Diaminoanisole	$C_7H_{10}N_2O$	5307-02-8	138.166	cry	107				
6788	3-Methoxy-1,2-benzenediol		$C_7H_8O_3$	934-00-9	140.137	nd	42.8	163[48], 129[10]			s chl
6789	4-Methoxybenzeneethanamine		$C_9H_{13}NO$	55-81-2	151.205			139[20]		1.5379[20]	
6790	4-Methoxybenzeneethanol		$C_9H_{12}O_2$	702-23-8	152.190		29	335			
6791	2-Methoxybenzenemethanamine		$C_8H_{11}NO$	6850-57-3	137.179			228	1.051[25]	1.5475[20]	
6792	4-Methoxybenzenemethanamine		$C_8H_{11}NO$	2393-23-9	137.179			236.5	1.050[15]	1.5462[20]	sl H_2O, EtOH, eth
6793	2-Methoxybenzenemethanol		$C_8H_{10}O_2$	612-16-8	138.164			249	1.0386[25]	1.5455[20]	i H_2O; s EtOH; msc eth
6794	3-Methoxybenzenemethanol		$C_8H_{10}O_2$	6971-51-3	138.164		30	252	1.112[25]	1.5440[20]	
6795	4-Methoxybenzenemethanol	Anise alcohol	$C_8H_{10}O_2$	105-13-5	138.164	nd	25	259.1	1.109[26]	1.5420[25]	s H_2O, ctc; vs EtOH, eth
6796	4-Methoxybenzenesulfonyl chloride		$C_7H_7ClO_3S$	98-68-0	206.647	nd or pr (bz)	42.5	103[0.25]			s EtOH, eth, bz
6797	3-Methoxybenzenethiol		C_7H_8OS	15570-12-4	140.203			224.5; 114[20]		1.5874[20]	s chl
6798	4-Methoxybenzenethiol		C_7H_8OS	696-63-9	140.203			228	1.1313[25]	1.5801[25]	s EtOH, eth, bz; sl chl

Methestrol

Methidathion

Methiocarb

L-Methionine

Methocarbamol

Methomyl

Methoprene

Methoprotryne

Methotrexate

Methoxamine hydrochloride

Methoxsalen

Methoxyacetaldehyde

Methoxyacetic acid

Methoxyacetonitrile

Methoxyacetyl chloride

2-Methoxyaniline

3-Methoxyaniline

4-Methoxyaniline

2-Methoxyaniline hydrochloride

1-Methoxy-9,10-anthracenedione

2-Methoxybenzaldehyde

3-Methoxybenzaldehyde

4-Methoxybenzaldehyde

4-Methoxybenzamide

4-Methoxybenzeneacetaldehyde

2-Methoxybenzeneacetic acid

4-Methoxybenzeneacetic acid

4-Methoxybenzeneacetonitrile

4-Methoxy-1,2-benzenediamine

4-Methoxy-1,3-benzenediamine

2-Methoxy-1,4-benzenediamine

3-Methoxy-1,2-benzenediol

4-Methoxybenzeneethanamine

4-Methoxybenzeneethanol

2-Methoxybenzenemethanamine

4-Methoxybenzenemethanamine

2-Methoxybenzenemethanol

3-Methoxybenzenemethanol

4-Methoxybenzenemethanol

4-Methoxybenzenesulfonyl chloride

3-Methoxybenzenethiol

4-Methoxybenzenethiol

No.	Name	Synonym	Mol. Form.	CAS RN	Mol. Wt.	Physical Form	mp/°C	bp/°C	den/ g cm⁻³	n_D	Solubility
6799	2-Methoxybenzoic acid		C₈H₈O₃	579-75-9	152.148	pl (w)	101	200			sl H₂O; vs EtOH, eth, chl; s bz, ctc
6800	3-Methoxybenzoic acid		C₈H₈O₃	586-38-9	152.148	nd (w)	107	170[10]			sl H₂O, ctc; s EtOH, eth, bz; vs chl
6801	4-Methoxybenzoic acid	p-Anisic acid	C₈H₈O₃	100-09-4	152.148		185	276.5			i H₂O; vs EtOH, MeOH, eth; s chl
6802	2-Methoxybenzonitrile		C₈H₇NO	6609-56-9	133.148		24.5	255.5	1.1063[20]		s EtOH; vs eth
6803	3-Methoxybenzonitrile		C₈H₇NO	1527-89-5	133.148			140[34], 111[13]	1.089[25]	1.5402[20]	
6804	4-Methoxybenzonitrile		C₈H₇NO	874-90-8	133.148	nd (w) lf (al)	61.5	256.5			i H₂O; vs EtOH, eth; s bz
6805	7-Methoxy-2H-1-benzopyran-2-one		C₁₀H₈O₃	531-59-9	176.169	lf (w, MeOH)	118.3				sl H₂O; s EtOH, eth, con sulf, alk
6806	6-Methoxy-2-benzothiazolamine		C₈H₈N₂OS	1747-60-0	180.227		166				
6807	2-(4-Methoxybenzoyl)benzoic acid	o-(p-Anisoyl)benzoic acid	C₁₅H₁₂O₄	1151-15-1	256.254	lf (w), cry (al, to)	146				vs eth, EtOH, tol
6808	2-Methoxybenzoyl chloride		C₈H₇ClO₂	21615-34-9	170.594			254			
6809	4-Methoxybenzoyl chloride	p-Anisoyl chloride	C₈H₇ClO₂	100-07-2	170.594	nd	24.5	262.5	1.261[20]	1.580[20]	s eth, ace; vs bz; sl ctc
6810	4-Methoxybenzyl acetate		C₁₀H₁₂O₃	104-21-2	180.200		84	270; 150[23]	1.105[25]		s ctc
6811	2-Methoxy-1,1'-biphenyl		C₁₃H₁₂O	86-26-0	184.233	pr (peth)	29	274	1.0233[99]	1.5641[99]	i H₂O; s EtOH, peth; sl ctc
6812	4-Methoxy-1,1'-biphenyl		C₁₃H₁₂O	613-37-6	184.233	pl (al)	90	157[10]	1.0278[100]	1.5744[100]	i H₂O; s EtOH, eth
6813	1-Methoxy-1,3-butadiene		C₅H₈O	3036-66-6	84.117			91.5	0.8296[20]	1.4594[20]	s H₂O, EtOH
6814	2-Methoxy-1,3-butadiene		C₅H₈O	3588-30-5	84.117			75	0.8272[20]	1.4442[20]	vs ace, bz, eth, EtOH
6815	3-Methoxy-1-butanol		C₅H₁₂O₂	2517-43-3	104.148			157	0.923[23]	1.4148[25]	vs EtOH, ace; s eth; sl chl
6816	1-Methoxy-1-buten-3-yne		C₅H₆O	2798-73-4	82.101			dec 123; 39[23]	0.906[20]	1.4818[20]	i H₂O; s chl
6817	Methoxychlor		C₁₆H₁₅Cl₃O₂	72-43-5	345.648	cry (dil al)	87		1.41[25]		i H₂O; s EtOH, ctc; vs eth, bz
6818	Methoxycyclohexane		C₇H₁₄O	931-56-6	114.185	liq	-74.4	133	0.8756[20]	1.4355[20]	vs eth, EtOH
6819	1-Methoxy-2,4-dinitrobenzene		C₇H₆N₂O₅	119-27-7	198.133	nd (al or w)	94.5	206[12]	1.3364[131]	1.546[15]	sl H₂O; s EtOH, eth, ace, bz; vs py
6820	1-Methoxy-3,5-dinitrobenzene	3,5-Dinitroanisole	C₇H₆N₂O₅	5327-44-6	198.133	nd (al)	105.3		1.558[12]		vs ace, bz, MeOH
6821	2-Methoxy-1,2-diphenylethanone		C₁₅H₁₄O₂	3524-62-7	226.271	nd (lig)	49.5	188[15]	1.1278[14]		vs bz, eth, EtOH
6822	2-Methoxyethanol	Ethylene glycol monomethyl ether	C₃H₈O₂	109-86-4	76.095	liq	-85.1	124.1	0.9647[20]	1.4024[20]	msc H₂O, eth, bz; vs EtOH; s ace; sl chl
6823	(2-Methoxyethoxy)ethene		C₅H₁₀O₂	1663-35-0	102.132			107			
6824	2-[2-(2-Methoxyethoxy)ethoxy]ethanol	Triethyleneglycol monomethyl ether	C₇H₁₆O₄	112-35-6	164.200			246			
6825	2-Methoxyethyl acetate	Ethylene glycol monomethyl ether acetate	C₅H₁₀O₃	110-49-6	118.131	liq	-70	143	1.0074[19]	1.4002[20]	s H₂O, EtOH, eth; sl ctc
6826	2-Methoxyethyl acrylate	2-Methoxyethyl 2-propenoate	C₆H₁₀O₃	3121-61-7	130.141			67[16], 56[12]	1.012[20]		
6827	2-Methoxyethylamine	1-Amino-2-methoxyethane	C₃H₉NO	109-85-3	75.109			95			vs H₂O, EtOH; sl chl
6828	Methoxyethylmercuric acetate		C₅H₁₀HgO₃	151-38-2	318.72	nd (peth)	42				
6829	2-(2-Methoxyethyl)pyridine	Metyridine	C₈H₁₁NO	114-91-0	137.179			203; 96[17]	0.988[20]	1.4975[20]	vs H₂O, EtOH
6830	2-Methoxyfuran		C₅H₆O₂	25414-22-6	98.101			110.5	1.0646[25]	1.4468[25]	
6831	4-Methoxyfuro[2,3-b]quinoline	Dictamnine	C₁₂H₉NO₂	484-29-7	199.205	pr (al)	133.5				sl H₂O; vs EtOH; s eth, chl, AcOEt
6832	12-Methoxyibogamine	Ibogaine	C₂₀H₂₆N₂O	83-74-9	310.432		148				s chl
6833	5-Methoxy-1H-indole-3-ethanamine	5-Methoxytryptamine	C₁₁H₁₄N₂O	608-07-1	190.241	cry (al)	121.5				
6834	N-[2-(5-Methoxy-1H-indol-3-yl)ethyl]acetamide	Melatonin	C₁₃H₁₆N₂O₂	73-31-4	232.278	pa ye lf (bz)	117				
6835	3-Methoxyisopropylamine	1-Methoxy-2-propanamine	C₄H₁₁NO	37143-54-7	89.136			97		1.4031[25]	
6836	4-Methoxy-N-(4-methoxyphenyl)aniline	4,4'-Dimethoxydiphenylamine	C₁₄H₁₅NO₂	101-70-2	229.275	lf (EtOH)	103				
6837	N-Methoxymethylamine	N-Methoxymethanamine	C₂H₇NO	1117-97-1	61.083	liq		42.4			
6838	2-Methoxy-5-methylaniline	5-Methyl-o-anisidine	C₈H₁₁NO	120-71-8	137.179		53	235			sl H₂O, chl; s EtOH, eth, bz, peth
6839	4-Methoxy-2-methylaniline		C₈H₁₁NO	102-50-1	137.179	cry (lig)	29.5	248.5	1.065[25]	1.5647[20]	vs EtOH
6840	4-Methoxy-α-methylbenzenemethanol		C₉H₁₂O₂	3319-15-1	152.190			dec 310; 140[17]	1.0794[20]	1.5310[25]	s ctc

2-Methoxybenzoic acid

3-Methoxybenzoic acid

4-Methoxybenzoic acid

2-Methoxybenzonitrile

3-Methoxybenzonitrile

4-Methoxybenzonitrile

7-Methoxy-2*H*-1-benzopyran-2-one

6-Methoxy-2-benzothiazolamine

2-(4-Methoxybenzoyl)benzoic acid

2-Methoxybenzoyl chloride

4-Methoxybenzoyl chloride

4-Methoxybenzyl acetate

2-Methoxy-1,1'-biphenyl

4-Methoxy-1,1'-biphenyl

1-Methoxy-1,3-butadiene

2-Methoxy-1,3-butadiene

3-Methoxy-1-butanol

1-Methoxy-1-buten-3-yne

Methoxychlor

Methoxycyclohexane

1-Methoxy-2,4-dinitrobenzene

1-Methoxy-3,5-dinitrobenzene

2-Methoxy-1,2-diphenylethanone

2-Methoxyethanol

(2-Methoxyethoxy)ethene

2-[2-(2-Methoxyethoxy)ethoxy]ethanol

2-Methoxyethyl acetate

2-Methoxyethyl acrylate

2-Methoxyethylamine

Methoxyethylmercuric acetate

2-(2-Methoxyethyl)pyridine

2-Methoxyfuran

4-Methoxyfuro[2,3-b]quinoline

12-Methoxyibogamine

5-Methoxy-1*H*-indole-3-ethanamine

N-[2-(5-Methoxy-1*H*-indol-3-yl)ethyl]acetamide

3-Methoxyisopropylamine

4-Methoxy-*N*-(4-methoxyphenyl)aniline

N-Methoxymethylamine

2-Methoxy-5-methylaniline

4-Methoxy-2-methylaniline

4-Methoxy-α-methylbenzenemethanol

No.	Name	Synonym	Mol. Form.	CAS RN	Mol. Wt.	Physical Form	mp/°C	bp/°C	den/ g cm⁻³	n_D	Solubility
6841	2-Methoxy-2-methylbutane	Methyl *tert*-pentyl ether	C₆H₁₄O	994-05-8	102.174			86.1	0.7660²⁵	1.3862²⁵	sl H₂O; vs eth, EtOH
6842	2-(Methoxymethyl)furan		C₆H₈O₂	13679-46-4	112.127			132	1.0163²⁰	1.4570²⁰	i H₂O; s EtOH; vs eth
6843	2-(Methoxymethyl)-5-nitrofuran		C₆H₇NO₄	586-84-5	157.125			104³	1.281²⁰	1.5325²⁰	vs EtOH
6844	(Methoxymethyl)oxirane		C₄H₈O₂	930-37-0	88.106			113	0.9890²⁰	1.4320²⁰	vs H₂O, ace, eth, EtOH
6845	3-Methoxy-5-methyl-4-oxo-2,5-hexadienoic acid	Penicillic acid	C₈H₁₀O₄	90-65-3	170.163	orth or hex pl (+ 1w)	83				s H₂O, ace; vs EtOH, eth, bz; sl peth
6846	4-Methoxy-4-methyl-2-pentanone	Pentoxone	C₇H₁₄O₂	107-70-0	130.185			160	0.8980²⁵	1.418²⁰	
6847	2-Methoxy-4-methylphenol	Creosol	C₈H₁₀O₂	93-51-6	138.164	pr	5.5	221	1.098²⁰	1.5353²⁵	vs eth, EtOH
6848	1-Methoxynaphthalene		C₁₁H₁₀O	2216-69-5	158.196		<-10	269	1.0963¹⁴	1.6940²⁵	i H₂O; s EtOH, eth, bz, chl; vs CS₂
6849	2-Methoxynaphthalene		C₁₁H₁₀O	93-04-9	158.196	lf (eth), pl (peth)	73.5	274			vs bz, eth, chl
6850	2-Methoxy-1,4-naphthalenedione		C₁₁H₈O₃	2348-82-5	188.180		183.0				
6851	4-Methoxy-1-naphthol		C₁₁H₁₀O₂	84-85-5	174.196		129.8				
6852	2-Methoxy-4-nitroaniline		C₇H₈N₂O₃	97-52-9	168.150		141.0				s DMSO
6853	2-Methoxy-5-nitroaniline	5-Nitro-*o*-anisidine	C₇H₈N₂O₃	99-59-2	168.150		118		1.2068¹⁵		s H₂O, eth; vs EtOH, ace, bz; sl lig
6854	4-Methoxy-2-nitroaniline		C₇H₈N₂O₃	96-96-8	168.150	dk red pr (w or al)	129				vs H₂O, ace, eth, EtOH
6855	2-Methoxyphenol	Guaiacol	C₇H₈O₂	90-05-1	124.138	hex pr	32	205	1.1287²¹	1.5429²⁰	sl H₂O; s EtOH, eth, ctc, chl
6856	3-Methoxyphenol		C₇H₈O₂	150-19-6	124.138		<-17	114⁵	1.131²⁵	1.5510²⁰	sl H₂O, chl; msc EtOH, eth
6857	4-Methoxyphenol		C₇H₈O₂	150-76-5	124.138	pl	57	243			s H₂O, bz, ctc; vs EtOH, eth
6858	2-Methoxyphenol benzoate	Guaiacol benzoate	C₁₄H₁₂O₃	531-37-3	228.243		57.5				vs eth, chl
6859	2-Methoxyphenol carbonate (2:1)	Guaiacol carbonate	C₁₅H₁₄O₅	553-17-3	274.269	cry (al)	89				i H₂O; sl EtOH; s eth; vs chl
6860	2-Methoxyphenol phosphate (3:1)	Guaiacol phosphate	C₂₁H₂₁O₇P	563-03-1	416.362		91	277³			vs ace, tol, chl
6861	5-[(2-Methoxyphenoxy)methyl]-2-oxazolidinone	Mephenoxalone	C₁₁H₁₃NO₄	70-07-5	223.226		144				
6862	3-(2-Methoxyphenoxy)-1,2-propanediol	Guaifenesin	C₁₀H₁₄O₄	93-14-1	198.216	orth pr (eth, eth-peth)	78.5	215¹⁹, 127⁰·²			s H₂O, bz, chl; vs EtOH; i peth
6863	N-(2-Methoxyphenyl)acetamide	*o*-Acetanisidine	C₉H₁₁NO₂	93-26-5	165.189	nd (w)	87.5	304			vs H₂O, EtOH; s eth, ace, HOAc
6864	N-(3-Methoxyphenyl)acetamide	*m*-Acetanisidine	C₉H₁₁NO₂	588-16-9	165.189	nd or pl (w)	81				vs H₂O, EtOH; s eth, ace
6865	N-(4-Methoxyphenyl)acetamide	*p*-Acetanisidine	C₉H₁₁NO₂	51-66-1	165.189	pl (w)	131				vs ace, EtOH, chl
6866	2-Methoxyphenyl acetate	2-Acetoxyanisole	C₉H₁₀O₃	613-70-7	166.173		31.5	123¹³	1.1285²⁵	1.5101²⁵	i H₂O; s EtOH, eth
6867	4-(4-Methoxyphenyl)-3-buten-2-one		C₁₁H₁₂O₂	943-88-4	176.212	lf (al, eth, HOAc)	74.0	187.5¹⁹			i H₂O; vs EtOH, eth; s bz, HOAc, sulf
6868	2-Methoxy-1-phenylethanone		C₉H₁₀O₂	4079-52-1	150.174	ye liq	8	245; 125¹⁹	1.0897²⁰	1.5393²⁰	sl H₂O; s EtOH, ace
6869	1-(3-Methoxyphenyl)ethanone		C₉H₁₀O₂	586-37-8	150.174		95.5	240	1.0343¹⁹	1.5410²⁰	s H₂O, EtOH, ace, ctc
6870	2-(4-Methoxyphenyl)-1H-indene-1,3(2H)-dione	Anisindione	C₁₆H₁₂O₃	117-37-3	252.264	pa ye cry (HOAc, al)	156.5				
6871	4-Methoxyphenyl isocyanate		C₈H₇NO₂	5416-93-3	149.148			110¹⁰			
6872	2-Methoxyphenyl isothiocyanate	1-Isothiocyanato-2-methoxybenzene	C₈H₇NOS	3288-04-8	165.213			264; 131¹¹	1.1878²⁰	1.6458²⁰	
6873	N-(4-Methoxyphenyl)-3-oxobutanamide		C₁₁H₁₃NO₃	5437-98-9	207.226		117.3				s EtOH, chl; sl eth
6874	2-Methoxyphenyl pentanoate	Guaiacol valerate	C₁₂H₁₆O₃	531-39-5	208.253			265	1.05²⁵		vs bz, eth, EtOH
6875	(4-Methoxyphenyl)phenyldiazene		C₁₃H₁₂N₂O	2396-60-3	212.246	oran-red pl, lf (al, peth)	56	340	1.12⁷⁵		i H₂O; s EtOH, eth, ace
6876	N-(*p*-Methoxyphenyl)-*p*-phenylenediamine	N-(4-Methoxyphenyl)-1,4-benzenediamine	C₁₃H₁₄N₂O	101-64-4	214.262	nd	102	238¹²			sl H₂O, peth; vs bz, eth, EtOH
6877	N-(4-Methoxyphenyl)-*p*-phenylenediamine hydrochloride		C₁₃H₁₅ClN₂O	3566-44-7	250.723	cry	245 dec				
6878	(4-Methoxyphenyl)phenylmethanone		C₁₄H₁₂O₂	611-94-9	212.244	pr (eth)	61.5	355; 168¹²			i H₂O; vs EtOH, eth; s ace, bz, HOAc
6879	3-(4-Methoxyphenyl)-1-phenyl-2-propen-1-one		C₁₆H₁₄O₂	959-33-1	238.281	ye nd (al)	79	187¹⁹			i H₂O; vs EtOH; s eth, ctc, chl, HOAc

2-Methoxy-2-methylbutane

2-(Methoxymethyl)furan

2-(Methoxymethyl)-5-nitrofuran

(Methoxymethyl)oxirane

3-Methoxy-5-methyl-4-oxo-2,5-hexadienoic acid

4-Methoxy-4-methyl-2-pentanone

2-Methoxy-4-methylphenol

1-Methoxynaphthalene

2-Methoxynaphthalene

2-Methoxy-1,4-naphthalenedione

4-Methoxy-1-naphthol

2-Methoxy-4-nitroaniline

2-Methoxy-5-nitroaniline

4-Methoxy-2-nitroaniline

2-Methoxyphenol

3-Methoxyphenol

4-Methoxyphenol

2-Methoxyphenol benzoate

2-Methoxyphenol carbonate (2:1)

2-Methoxyphenol phosphate (3:1)

5-[(2-Methoxyphenoxy)methyl]-2-oxazolidinone

3-(2-Methoxyphenoxy)-1,2-propanediol

N-(2-Methoxyphenyl)acetamide

N-(3-Methoxyphenyl)acetamide

N-(4-Methoxyphenyl)acetamide

2-Methoxyphenyl acetate

4-(4-Methoxyphenyl)-3-buten-2-one

2-Methoxy-1-phenylethanone

1-(3-Methoxyphenyl)ethanone

2-(4-Methoxyphenyl)-1H-indene-1,3(2H)-dione

4-Methoxyphenyl isocyanate

2-Methoxyphenyl isothiocyanate

N-(4-Methoxyphenyl)-3-oxobutanamide

2-Methoxyphenyl pentanoate

(4-Methoxyphenyl)phenyldiazene

N-(p-Methoxyphenyl)-p-phenylenediamine

N-(4-Methoxyphenyl)-p-phenylenediamine hydrochloride

(4-Methoxyphenyl)phenylmethanone

3-(4-Methoxyphenyl)-1-phenyl-2-propen-1-one

No.	Name	Synonym	Mol. Form.	CAS RN	Mol. Wt.	Physical Form	mp/°C	bp/°C	den/ g cm⁻³	n_D	Solubility
6880	1-(4-Methoxyphenyl)-1-propanone	Ethyl 4-methoxyphenyl ketone	$C_{10}H_{12}O_2$	121-97-1	164.201		25.5	266	1.0798[16]		s ctc
6881	1-(4-Methoxyphenyl)-2-propanone	Anisyl methyl ketone	$C_{10}H_{12}O_2$	122-84-9	164.201		<-15	268	1.0694[17]	1.5253[20]	vs eth, EtOH
6882	trans-3-(4-Methoxyphenyl)-2-propenoic acid	trans-4-Methoxycinnamic acid	$C_{10}H_{10}O_3$	943-89-5	178.184		173.5				sl H_2O, EtOH, bz, DMSO; s ctc, HOAc
6883	trans-1-Methoxy-4-(2-phenylvinyl)benzene		$C_{15}H_{14}O$	1694-19-5	210.271		136.5	142.5[15]			i H_2O; vs EtOH, eth, ace, bz; s peth
6884	1-Methoxy-1,2-propadiene	Methoxyallene	C_4H_6O	13169-00-1	70.090	oil		51.5			
6885	3-Methoxy-1-propanamine		$C_4H_{11}NO$	5332-73-0	89.136			117.5	0.8727[20]	1.4391[20]	s H_2O, ace, bz, ctc, chl, MeOH
6886	3-Methoxy-1,2-propanediol	Glycerol 3-methyl ether	$C_4H_{10}O_3$	623-39-2	106.120	hyg liq		220	1.114[20]	1.442[25]	vs H_2O, EtOH, ace; s eth
6887	3-Methoxypropanenitrile		C_4H_7NO	110-67-8	85.105			163	0.9379[20]	1.4043[20]	s EtOH, eth, chl
6888	2-Methoxy-1-propanol		$C_4H_{10}O_2$	1589-47-5	90.121			130	0.938[20]	1.4070[20]	
6889	1-Methoxy-2-propanone	Methoxyacetone	$C_4H_8O_2$	5878-19-3	88.106			116	0.957[25]	1.3970[20]	
6890	2-Methoxy-1-propene		C_4H_8O	116-11-0	72.106			38	0.7372[20]		
6891	3-Methoxy-1-propene		C_4H_8O	627-40-7	72.106			44	0.77[11]	1.3778[20]	i H_2O; msc EtOH, eth; s ace
6892	trans-1-Methoxy-4-(1-propenyl)benzene	Anethole	$C_{10}H_{12}O$	4180-23-8	148.201	col oily liq	22.5	235; 81[2.3]	0.9882[20]	1.5615[20]	sl H_2O; msc EtOH, eth; s ace; vs bz
6893	1-Methoxy-4-(2-propenyl)benzene	Estragole	$C_{10}H_{12}O$	140-67-0	148.201			215.5	0.965[25]	1.5195[20]	vs EtOH, chl
6894	cis-2-Methoxy-4-(1-propenyl)phenol		$C_{10}H_{12}O_2$	5912-86-7	164.201			134[13]	1.0837[20]	1.5726[20]	sl H_2O; s EtOH, eth
6895	trans-2-Methoxy-4-(1-propenyl)phenol		$C_{10}H_{12}O_2$	5932-68-3	164.201		33.5	141[13]	1.0852[20]	1.5784[20]	sl H_2O; s EtOH, eth, chl
6896	1-Methoxy-4-propylbenzene		$C_{10}H_{14}O$	104-45-0	150.217			211.5	0.9472[20]	1.5045[20]	sl H_2O; s EtOH, ace, bz, chl; vs eth
6897	2-Methoxy-4-propylphenol		$C_{10}H_{14}O_2$	2785-87-7	166.217			121[10]			
6898	3-Methoxy-1-propyne		C_4H_6O	627-41-8	70.090			63	0.83[12]	1.5035[20]	vs eth, EtOH
6899	5-Methoxypsoralen	Bergaptene	$C_{12}H_8O_4$	484-20-8	216.190	nd (EtOH)	188				i H_2O; sl EtOH, bz, chl
6900	6-Methoxy-3-pyridinamine		$C_6H_8N_2O$	6628-77-9	124.140		30	125[10], 87[1]		1.5745[20]	
6901	2-Methoxypyridine		C_6H_7NO	1628-89-3	109.126			142.5	1.0457[20]	1.5042[20]	
6902	3-Methoxypyridine		C_6H_7NO	7295-76-3	109.126	liq		178.5; 65[15]	1.083	1.5180[20]	
6903	4-Methoxypyridine		C_6H_7NO	620-08-6	109.126			192; 95[45]			msc H_2O
6904	6-Methoxyquinoline		$C_{10}H_9NO$	5263-87-6	159.184	hyg lf	26.5	306; 153[12]	1.152[20]		s EtOH, eth, chl, dil HCl
6905	6-Methoxy-4-quinolinecarboxylic acid	Quininic acid	$C_{11}H_9NO_3$	86-68-0	203.194	pa ye pr (dil al)	285 dec	sub			sl H_2O, eth, bz, tfa; i chl; s EtOH
6906	2-Methoxy-1,3,5-trinitrobenzene	Methyl picrate	$C_7H_5N_3O_7$	606-35-9	243.131	nd (dil MeOH)	69			1.4947[80]	i H_2O; vs EtOH, chl, bz; s eth
6907	(2-Methoxyvinyl)benzene		$C_9H_{10}O$	4747-15-3	134.174			211.5	0.9894[23]	1.5620[24]	
6908	Methscopolamine bromide	Scopolamine methobromide	$C_{18}H_{24}BrNO_4$	155-41-9	398.293	cry (EtOH)	215 dec				s H_2O; sl EtOH
6909	Methyl abietate		$C_{21}H_{32}O_2$	127-25-3	316.478	pa ye lf (liq)		225[16]	1.049[20]	1.5344	i H_2O; s EtOH, HOAc
6910	N-Methylacetamide		C_3H_7NO	79-16-3	73.094		28	205	0.9371[25]	1.4301[20]	vs ace, bz, eth, EtOH
6911	4-Methylacetanilide		$C_9H_{11}NO$	103-89-9	149.189	mcl cry or nd (dil al)	152	307	1.2120[15]		vs eth, EtOH
6912	Methyl acetate		$C_3H_6O_2$	79-20-9	74.079	liq	-98.25	56.87	0.9342[20]	1.3614[20]	vs H_2O, eth, EtOH
6913	Methyl acetoacetate		$C_5H_8O_3$	105-45-3	116.116		27.5	171.7	1.0762[20]	1.4184[20]	vs H_2O; msc EtOH, eth; s ctc
6914	4-Methylacetophenone		$C_9H_{10}O$	122-00-9	134.174	nd	28	226; 93.5[7]	1.0051[20]	1.5335[20]	vs bz, eth, EtOH, chl
6915	Methyl 2-(acetyloxy)benzoate	Methyl o-acetylsalicylate	$C_{10}H_{10}O_4$	580-02-9	194.184	pl (peth)	51.5	135[9]			vs eth, EtOH, chl
6916	Methyl acrylate	Methyl propenoate	$C_4H_6O_2$	96-33-3	86.090	liq	<-75	80.7	0.9535[20]	1.4040[20]	sl H_2O; s EtOH, eth, ace, bz, chl
6917	2-Methylacrylonitrile	2-Methylpropenenitrile	C_4H_5N	126-98-7	67.090	liq	-35.8	90.3	0.8001[20]	1.4003[20]	sl H_2O, chl; msc EtOH, eth, ace, tol
6918	2-Methylalanine	α-Aminoisobutyric acid	$C_4H_9NO_2$	62-57-7	103.120	mcl pr	335	sub 280			vs H_2O; sl EtOH; i eth
6919	5-Methyl-3-allyl-2,4-oxazolidinedione	Aloxidone	$C_7H_9NO_3$	526-35-2	155.151			138[35], 86[0.5]		1.4688[25]	

1-(4-Methoxyphenyl)-1-propanone

1-(4-Methoxyphenyl)-2-propanone

trans-3-(4-Methoxyphenyl)-2-propenoic acid

trans-1-Methoxy-4-(2-phenylvinyl)benzene

1-Methoxy-1,2-propadiene

3-Methoxy-1-propanamine

3-Methoxy-1,2-propanediol

3-Methoxypropanenitrile

2-Methoxy-1-propanol

1-Methoxy-2-propanone

2-Methoxy-1-propene

3-Methoxy-1-propene

trans-1-Methoxy-4-(1-propenyl)benzene

1-Methoxy-4-(2-propenyl)benzene

cis-2-Methoxy-4-(1-propenyl)phenol

trans-2-Methoxy-4-(1-propenyl)phenol

1-Methoxy-4-propylbenzene

2-Methoxy-4-propylphenol

3-Methoxy-1-propyne

5-Methoxypsoralen

6-Methoxy-3-pyridinamine

2-Methoxypyridine

3-Methoxypyridine

4-Methoxypyridine

6-Methoxyquinoline

6-Methoxy-4-quinolinecarboxylic acid

2-Methoxy-1,3,5-trinitrobenzene

(2-Methoxyvinyl)benzene

Methscopolamine bromide

Methyl abietate

N-Methylacetamide

4-Methylacetanilide

Methyl acetate

Methyl acetoacetate

4-Methylacetophenone

Methyl 2-(acetyloxy)benzoate

Methyl acrylate

2-Methylacrylonitrile

2-Methylalanine

5-Methyl-3-allyl-2,4-oxazolidinedione

No.	Name	Synonym	Mol. Form.	CAS RN	Mol. Wt.	Physical Form	mp/°C	bp/°C	den/ g cm⁻³	n_D	Solubility
6920	Methylamine	Methanamine	CH_5N	74-89-5	31.058	col gas	-93.5	-6.32	0.656[25] (p>1 atm)		vs H_2O; s EtOH, ace, bz; msc eth
6921	Methylamine hydrochloride	Methanamine hydrochloride	CH_6ClN	593-51-1	67.519	hyg tetr tab (al)	227.5	227[15]			s H_2O, EtOH; i chl, ace
6922	1-(Methylamino)-9,10-anthracenedione		$C_{15}H_{11}NO_2$	82-38-2	237.254	ye-red nd	171.0				s EtOH, bz, chl, HOAc
6923	Methyl 2-aminobenzoate	Methyl anthranilate	$C_8H_9NO_2$	134-20-3	151.163		24.5	256	1.1682[10]	1.5810	sl H_2O; vs EtOH, eth
6924	Methyl 3-aminobenzoate		$C_8H_9NO_2$	4518-10-9	151.163		39	152[11]	1.232[20]		vs EtOH, eth, bz, chl; s lig; sl peth
6925	Methyl 4-aminobenzoate		$C_8H_9NO_2$	619-45-4	151.163	lf or nd (aq MeOH)	113.0				s chl
6926	2-(Methylamino)benzoic acid		$C_8H_9NO_2$	119-68-6	151.163	pl (al or lig)	180.5	80[0.01]			sl H_2O; vs EtOH, eth, bz, chl
6927	3-(Methylamino)benzoic acid		$C_8H_9NO_2$	51524-84-6	151.163	pl (peth)	127				vs ace, bz, EtOH, chl
6928	4-(Methylamino)benzoic acid		$C_8H_9NO_2$	10541-83-0	151.163	nd (bz, w, dil al)	168				s H_2O, bz, AcOEt; vs EtOH, eth; sl tfa
6929	Methyl 3-amino-2-butenoate		$C_5H_9NO_2$	14205-39-1	115.131						s chl
6930	N-[(Methylamino)carbonyl] acetamide		$C_4H_8N_2O_2$	623-59-6	116.119	tcl (w, al), pr (w)	180.5	dec			s H_2O, chl; sl EtOH, eth
6931	2-(Methylamino)-2-deoxy-α-L-glucopyranose	N-Methyl-α-L-glucosamine	$C_7H_{15}NO_5$	42852-95-9	193.198	glass					s MeOH
6932	2-(Methylamino)ethanesulfonic acid	N-Methyltaurine	$C_3H_9NO_3S$	107-68-6	139.173		241.5				vs H_2O; i EtOH, eth
6933	4-[2-(Methylamino)ethyl]-1,2-benzenediol	Deoxyepinephrine	$C_9H_{13}NO_2$	501-15-5	167.205		188.5				
6934	Methyl 3-amino-4-hydroxybenzoate	Orthocaine	$C_8H_9NO_3$	536-25-4	167.162	nd (bz or HOAc)	143				i H_2O; vs EtOH; s eth, alk; sl bz
6935	4-(Methylamino)phenol sulfate		$C_{14}H_{20}N_2O_6S$	1936-57-8	344.383	cry	260 dec				sl EtOH; i eth
6936	3-(Methylamino)propanenitrile		$C_4H_8N_2$	693-05-0	84.120			102[49], 74[16]	0.8992[20]	1.4320[20]	s H_2O, ace, bz, chl, MeOH
6937	4-[2-(Methylamino)propyl]phenol	Pholedrine	$C_{10}H_{15}NO$	370-14-9	165.232	cry (MeOH)	161				vs eth, EtOH
6938	2-Methylaniline	o-Toluidine	C_7H_9N	95-53-4	107.153	liq	-14.41	200.3	0.9984[20]	1.5725[20]	sl H_2O; msc EtOH, eth, ctc
6939	3-Methylaniline	m-Toluidine	C_7H_9N	108-44-1	107.153	liq	-31.3	203.3	0.9889[20]	1.5681[20]	vs ace, bz, eth, EtOH
6940	4-Methylaniline	p-Toluidine	C_7H_9N	106-49-0	107.153	lf (w+1)	43.6	200.4	0.9619[20]	1.5534[45]	sl H_2O; vs EtOH, py; s eth, ace, ctc
6941	N-Methylaniline		C_7H_9N	100-61-8	107.153	liq	-57	196.2	0.9891[20]	1.5684[20]	i H_2O; s EtOH, eth, ctc, chl
6942	2-Methylaniline, hydrochloride	o-Toluidine, hydrochloride	$C_7H_{10}ClN$	636-21-5	143.614	mcl pr (w)	215				vs H_2O, EtOH
6943	4-Methylaniline, hydrochloride		$C_7H_{10}ClN$	540-23-8	143.614	mcl nd (eth-HOAc)	244.5	258	1.1930[18]		vs H_2O, EtOH, HOAc
6944	2-Methylanisole		$C_8H_{10}O$	578-58-5	122.164	liq	-34.1	171	0.985[25]	1.5161[20]	i H_2O; s EtOH, eth, ace, ctc
6945	3-Methylanisole		$C_8H_{10}O$	100-84-5	122.164	liq	-47	175.5	0.969[25]	1.5130[20]	i H_2O; s EtOH, eth, ace, bz; sl ctc
6946	4-Methylanisole		$C_8H_{10}O$	104-93-8	122.164	liq	-32	175.5	0.969[25]	1.5112[20]	i H_2O; s EtOH, eth, chl
6947	1-Methylanthracene		$C_{15}H_{12}$	610-48-0	192.256	bl nd (MeOH) lf (al)	85.5	199.5	1.0471[99]	1.6802[99]	i H_2O; s EtOH, eth, bz, chl, sulf
6948	2-Methylanthracene		$C_{15}H_{12}$	613-12-7	192.256	grn bl flr lf (sub)	209	sub	1.80[0]		i H_2O, ace; sl EtOH, eth; s bz, CS_2
6949	9-Methylanthracene		$C_{15}H_{12}$	779-02-2	192.256	ye nd (dil al) pr (bz, al)	81.5	196[12]	1.065[99]	1.6959[99]	s EtOH, eth, ace, bz, chl
6950	2-Methyl-9,10-anthracenedione	2-Methylanthraquinone	$C_{15}H_{10}O_2$	84-54-8	222.239	ye nd (al, HOAc)	177	sub			vs bz, EtOH, HOAc
6951	Methylarsine		CH_5As	593-52-2	91.973	col gas	-143	2			vs ace, eth, EtOH
6952	9-Methyl-9-azabicyclo[3.3.1] nonan-3-one	Pseudopelletierine	$C_9H_{15}NO$	552-70-5	153.221	orth pr (peth)	54	246	1.001[100]	1.4760[100]	vs H_2O, eth, EtOH
6953	8-Methyl-8-azabicyclo[3.2.1] octane	Tropane	$C_8H_{15}N$	529-17-9	125.212			166	0.9251[15]		
6954	8-Methyl-8-azabicyclo[3.2.1] octan-3-one		$C_8H_{13}NO$	532-24-1	139.195		43	227; 113[25]	1.9872[100]	1.4598[100]	s EtOH, eth, ace, bz, peth; sl chl
6955	Methyl azide		CH_3N_3	624-90-8	57.055			exp 20.5	0.869[15]		
6956	Methylazoxymethanol acetate		$C_4H_8N_2O_3$	592-62-1	132.118			191; 49[0.45]			

Methylamine

Methylamine hydrochloride

1-(Methylamino)-9,10-anthracenedione

Methyl 2-aminobenzoate

Methyl 3-aminobenzoate

Methyl 4-aminobenzoate

2-(Methylamino)benzoic acid

3-(Methylamino)benzoic acid

4-(Methylamino)benzoic acid

Methyl 3-amino-2-butenoate

N-[(Methylamino)carbonyl]acetamide

2-(Methylamino)-2-deoxy-α-L-glucopyranose

2-(Methylamino)ethanesulfonic acid

4-[2-(Methylamino)ethyl]-1,2-benzenediol

Methyl 3-amino-4-hydroxybenzoate

4-(Methylamino)phenol sulfate

3-(Methylamino)propanenitrile

4-[2-(Methylamino)propyl]phenol

2-Methylaniline

3-Methylaniline

4-Methylaniline

N-Methylaniline

2-Methylaniline, hydrochloride

4-Methylaniline, hydrochloride

2-Methylanisole

3-Methylanisole

4-Methylanisole

1-Methylanthracene

2-Methylanthracene

9-Methylanthracene

2-Methyl-9,10-anthracenedione

Methylarsine

9-Methyl-9-azabicyclo[3.3.1]nonan-3-one

8-Methyl-8-azabicyclo[3.2.1]octane

8-Methyl-8-azabicyclo[3.2.1]octan-3-one

Methyl azide

Methylazoxymethanol acetate

No.	Name	Synonym	Mol. Form.	CAS RN	Mol. Wt.	Physical Form	mp/°C	bp/°C	den/ g cm⁻³	n_D	Solubility
6957	2-Methylbenzaldehyde	o-Tolualdehyde	C₈H₈O	529-20-4	120.149			200; 94[10]	1.0328[20]	1.5462[20]	sl H₂O, ctc; s EtOH, eth, bz; vs ace
6958	3-Methylbenzaldehyde	m-Tolualdehyde	C₈H₈O	620-23-5	120.149			199	1.0189[21]	1.5413[21]	sl H₂O; msc EtOH, eth; vs ace; s bz, chl
6959	4-Methylbenzaldehyde	p-Tolualdehyde	C₈H₈O	104-87-0	120.149			204; 106[10]	1.0194[17]	1.5454[20]	sl H₂O; msc EtOH, eth, ace; vs chl
6960	2-Methylbenzamide	o-Toluamide	C₈H₉NO	527-85-5	135.163		147				sl H₂O, eth, tfa, bz; vs EtOH
6961	4-Methylbenzamide	p-Toluamide	C₈H₉NO	619-55-6	135.163		162.5				sl H₂O, bz, chl; vs EtOH, eth; s tfa
6962	N-Methylbenzamide		C₈H₉NO	613-93-4	135.163		82	291; 167[12]			s EtOH, ace
6963	7-Methylbenz[a]anthracene		C₁₉H₁₄	2541-69-7	242.314	ye pl (al)	141				i H₂O; s EtOH, eth, ace, ctc, HOAc, CS₂
6964	8-Methylbenz[a]anthracene		C₁₉H₁₄	2381-31-9	242.314	pl (bz-al), nd (bz-lig)	156.5	272[3], 160[0.1]	1.2310[0]		i H₂O; s EtOH, eth, bz, xyl
6965	9-Methylbenz[a]anthracene		C₁₉H₁₄	2381-16-0	242.314	nd (al)	152.5				i H₂O; s EtOH, eth, ctc, chl, CS₂, xyl
6966	10-Methylbenz[a]anthracene		C₁₉H₁₄	2381-15-9	242.314		184				i H₂O; s EtOH, HOAc
6967	12-Methylbenz[a]anthracene		C₁₉H₁₄	2422-79-9	242.314	pl (al)	150.5				i H₂O; s EtOH, CS₂, HOAc
6968	2-Methylbenzeneacetaldehyde		C₉H₁₀O	10166-08-2	134.174			221; 92[10]	1.0241[10]		vs eth, EtOH, chl
6969	4-Methylbenzeneacetaldehyde		C₉H₁₀O	104-09-6	134.174		40	221.5	1.0052[20]	1.5255[20]	vs eth, EtOH, chl
6970	α-Methylbenzeneacetaldehyde		C₉H₁₀O	93-53-8	134.174			203.5	1.0089[20]	1.5176[20]	vs EtOH
6971	2-Methylbenzeneacetic acid		C₉H₁₀O₂	644-36-0	150.174	nd (w)	89				s H₂O, chl
6972	3-Methylbenzeneacetic acid		C₉H₁₀O₂	621-36-3	150.174	nd (w)	62	121[26]			s H₂O, chl
6973	4-Methylbenzeneacetic acid		C₉H₁₀O₂	622-47-9	150.174	nd or pl (al, w)	93	265			vs bz, eth, EtOH
6974	α-Methylbenzeneacetic acid, (±)		C₉H₁₀O₂	2328-24-7	150.174		<-20	263	1.1[0]	1.5237[20]	
6975	4-Methylbenzeneacetonitrile		C₉H₉N	2947-61-7	131.174		18	242.5	0.992[25]	1.5190[20]	i H₂O; s EtOH, eth, bz, ctc
6976	α-Methylbenzeneacetonitrile		C₉H₉N	1823-91-2	131.174			231	0.9854[20]	1.5095[25]	vs eth, EtOH
6977	3-Methyl-1,2-benzenediamine	Toluene-2,3-diamine	C₇H₁₀N₂	2687-25-4	122.167		63.5	255			vs ace, bz, EtOH
6978	4-Methyl-1,2-benzenediamine	Toluene-3,4-diamine	C₇H₁₀N₂	496-72-0	122.167	pl (lig)	89.5	265			vs H₂O; s lig
6979	2-Methyl-1,3-benzenediamine	Toluene-2,6-diamine	C₇H₁₀N₂	823-40-5	122.167	pr (bz, w)	106				s H₂O, EtOH, bz
6980	2-Methyl-1,4-benzenediamine	Toluene-2,5-diamine	C₇H₁₀N₂	95-70-5	122.167	pl (bz)	64	273.5			s H₂O, EtOH, eth; sl bz, HOAc
6981	3-Methyl-1,2-benzenediol		C₇H₈O₂	488-17-5	124.138	lf (bz)	68	248			s H₂O, EtOH, bz, chl
6982	4-Methyl-1,2-benzenediol		C₇H₈O₂	452-86-8	124.138	lf (bz-lig), pr (bz)	65	258	1.1287[74]	1.5425[74]	s H₂O, EtOH, eth, ace, chl; sl lig
6983	2-Methyl-1,3-benzenediol		C₇H₈O₂	608-25-3	124.138	pr (bz)	120	265			vs H₂O, bz, eth, EtOH
6984	4-Methyl-1,3-benzenediol		C₇H₈O₂	496-73-1	124.138	cry (bz-peth)	105	270			s H₂O, EtOH, eth; sl bz, peth
6985	5-Methyl-1,3-benzenediol	Orcinol	C₇H₈O₂	504-15-4	124.138	pr(w+1), lf(chl)	107	287	1.290[4]		s H₂O, EtOH, eth, bz; sl lig, peth
6986	2-Methyl-1,4-benzenediol		C₇H₈O₂	95-71-6	124.138		125	283; 163[11]			vs H₂O, EtOH, eth; s ace; sl bz, lig
6987	4-Methyl-1,2-benzenedithiol	Toluene-3,4-dithiol	C₇H₈S₂	496-74-2	156.269		29				s chl
6988	β-Methylbenzeneethanamine		C₉H₁₃N	582-22-9	135.206			210	0.9433[4]	1.5255[20]	vs bz, eth, EtOH
6989	N-Methylbenzeneethanamine		C₉H₁₃N	589-08-2	135.206			206	0.93[25]	1.5162[20]	
6990	2-Methylbenzeneethanol		C₉H₁₂O	19819-98-8	136.190		1.0	243.5	1.016[25]	1.5355[20]	
6991	4-Methylbenzeneethanol		C₉H₁₂O	699-02-5	136.190			244.5; 94[6]	1.0028[20]	1.5267[20]	
6992	2-Methylbenzenemethanamine		C₈H₁₁N	89-93-0	121.180	liq	-30	206; 81[15]	0.9766[19]	1.5436[19]	
6993	3-Methylbenzenemethanamine		C₈H₁₁N	100-81-2	121.180			203.5	0.966[25]	1.5360[20]	
6994	4-Methylbenzenemethanamine		C₈H₁₁N	104-84-7	121.180		12.5	195	0.952[20]	1.5340[20]	
6995	N-Methylbenzenemethanamine		C₈H₁₁N	103-67-3	121.180			180.5	0.9442[18]		vs H₂O
6996	α-Methylbenzenemethanol	1-Phenylethanol	C₈H₁₀O	98-85-1	122.164		20	205	1.013[25]	1.5265[20]	i H₂O; vs EtOH, eth
6997	2-Methylbenzenemethanol	o-Tolyl alcohol	C₈H₁₀O	89-95-2	122.164	nd (peth-eth)	38	224; 118[20]	1.023[40]		vs eth, EtOH, chl

2-Methylbenzaldehyde

3-Methylbenzaldehyde

4-Methylbenzaldehyde

2-Methylbenzamide

4-Methylbenzamide

N-Methylbenzamide

7-Methylbenz[a]anthracene

8-Methylbenz[a]anthracene

9-Methylbenz[a]anthracene

10-Methylbenz[a]anthracene

12-Methylbenz[a]anthracene

2-Methylbenzeneacetaldehyde

4-Methylbenzeneacetaldehyde

α-Methylbenzeneacetaldehyde

2-Methylbenzeneacetic acid

3-Methylbenzeneacetic acid

4-Methylbenzeneacetic acid

α-Methylbenzeneacetic acid, (±)

4-Methylbenzeneacetonitrile

α-Methylbenzeneacetonitrile

3-Methyl-1,2-benzenediamine

4-Methyl-1,2-benzenediamine

2-Methyl-1,3-benzenediamine

2-Methyl-1,4-benzenediamine

3-Methyl-1,2-benzenediol

4-Methyl-1,2-benzenediol

2-Methyl-1,3-benzenediol

4-Methyl-1,3-benzenediol

5-Methyl-1,3-benzenediol

2-Methyl-1,4-benzenediol

4-Methyl-1,2-benzenedithiol

β-Methylbenzeneethanamine

N-Methylbenzeneethanamine

2-Methylbenzeneethanol

4-Methylbenzeneethanol

2-Methylbenzenemethanamine

3-Methylbenzenemethanamine

4-Methylbenzenemethanamine

N-Methylbenzenemethanamine

α-Methylbenzenemethanol

2-Methylbenzenemethanol

No.	Name	Synonym	Mol. Form.	CAS RN	Mol. Wt.	Physical Form	mp/°C	bp/°C	den/ g cm⁻³	n_D	Solubility
6998	3-Methylbenzenemethanol	m-Tolyl alcohol	C₈H₁₀O	587-03-1	122.164		<-20	215.5	0.9157[17]		sl H₂O; vs EtOH, eth; s chl
6999	4-Methylbenzenemethanol	p-Tolyl alcohol	C₈H₁₀O	589-18-4	122.164	nd (liq)	61.5	217	0.978[22]		vs eth, EtOH
7000	α-Methylbenzenemethanol, acetate		C₁₀H₁₂O₂	93-92-5	164.201	oil		109[18]			
7001	4-Methylbenzenepropanal		C₁₀H₁₂O	5406-12-2	148.201			223	0.999[14]	1.525[14]	
7002	α-Methylbenzenepropanamine	1-Methyl-3-phenylpropylamine	C₁₀H₁₅N	22374-89-6	149.233		143	223; 101[14]	0.9289[15]	1.5152[20]	vs EtOH
7003	β-Methylbenzenepropanoic acid, (±)		C₁₀H₁₂O₂	772-17-8	164.201		46.5	168[14]	1.0701[20]	1.5155[20]	sl H₂O; s peth
7004	α-Methylbenzenepropanol		C₁₀H₁₄O	2344-70-9	150.217			239; 123[15]	0.9899[16]	1.517[16]	
7005	4-Methylbenzenesulfinic acid	p-Toluenesulfinic acid	C₇H₈O₂S	536-57-2	156.203	orth pl or nd (w)	86.5				s H₂O; vs EtOH, eth; sl bz
7006	4-Methylbenzenesulfinyl chloride		C₇H₇ClOS	10439-23-3	174.648	nd	57	113[3.5]			vs chl
7007	2-Methylbenzenesulfonamide		C₇H₉NO₂S	88-19-7	171.217	oct cry (al), pr (w)	158.7	214[10]			sl H₂O, eth, DMSO; s EtOH
7008	4-Methylbenzenesulfonamide	p-Toluenesulfonamide	C₇H₉NO₂S	70-55-3	171.217	mcl pl (w+2)	138	214[10]			sl H₂O, eth; s EtOH
7009	Methyl benzenesulfonate		C₇H₈O₃S	80-18-2	172.202		4.5	150[15]	1.2730[17]	1.5151[20]	sl H₂O; vs EtOH, eth, chl
7010	2-Methylbenzenesulfonic acid		C₇H₈O₃S	88-20-0	172.202	hyg pl (w+2)	67.5	128.8[25]			vs H₂O; s EtOH; i eth
7011	2-Methylbenzenesulfonyl chloride	o-Toluenesulfonyl chloride	C₇H₇ClO₂S	133-59-5	190.648		10.2	154[36]	1.3383[20]	1.5565[20]	i H₂O; s EtOH, eth, bz, ctc
7012	2-Methylbenzenethiol		C₇H₈S	137-06-4	124.204		15	195	1.041[20]	1.570[20]	i H₂O; s EtOH; vs eth
7013	3-Methylbenzenethiol		C₇H₈S	108-40-7	124.204	liq	-20	195	1.044[20]	1.572[20]	i H₂O; s EtOH; msc eth
7014	4-Methylbenzenethiol		C₇H₈S	106-45-6	124.204		43	195	1.0220[51]		i H₂O; s EtOH, chl; vs eth
7015	1-Methyl-1H-benzimidazole		C₈H₈N₂	1632-83-3	132.163	nd (peth), pl (al)	66	286	1.1254[20]	1.6013[7]	s peth
7016	2-Methyl-1H-benzimidazole		C₈H₈N₂	615-15-6	132.163	pr or nd (w)	177.8				s H₂O; sl EtOH, eth; i bz
7017	Methyl benzoate		C₈H₈O₂	93-58-3	136.149	liq	-12.4	199	1.0837[25]	1.5164[20]	i H₂O; s EtOH, ctc, MeOH; msc eth
7018	Methyl 1,3-benzodioxole-5-carboxylate		C₉H₈O₄	326-56-7	180.158	nd or lf (peth)	53	dec 273			vs eth, EtOH
7019	2-Methylbenzofuran		C₉H₈O	4265-25-2	132.159			197.5	1.0540[20]	1.5495[22]	vs eth, EtOH
7020	2-Methylbenzonitrile	o-Tolunitrile	C₈H₇N	529-19-1	117.149	liq	-13.5	205	0.9955[20]	1.5279[20]	i H₂O; msc EtOH, eth; sl ctc
7021	3-Methylbenzonitrile	m-Tolunitrile	C₈H₇N	620-22-4	117.149	liq	-23	213	1.0316[20]	1.5252[20]	i H₂O; msc EtOH, eth; sl ctc
7022	4-Methylbenzonitrile	p-Tolunitrile	C₈H₇N	104-85-8	117.149		29.5	217.0	0.9762[30]		i H₂O; vs EtOH, eth; sl ctc
7023	6-Methyl-2H-1-benzopyran-2-one		C₁₀H₈O₂	92-48-8	160.170		76.5	304; 174[14]			vs EtOH, eth, bz; sl chl, peth
7024	7-Methyl-2H-1-benzopyran-2-one	7-Methylcoumarin	C₁₀H₈O₂	2445-83-2	160.170	nd, (pl) (aq al)	128	171.5[11]			sl H₂O; vs EtOH, HOAc; s eth
7025	3-Methyl-4H-1-benzopyran-4-one	Tricromyl	C₁₀H₈O₂	85-90-5	160.170						s chl
7026	6-Methyl-2-benzothiazolamine		C₈H₈N₂S	2536-91-6	164.228	nd (w) pr (dil al)	142				sl H₂O; s EtOH
7027	2-Methylbenzothiazole		C₈H₇NS	120-75-2	149.214		14	238	1.1763[19]	1.6092[19]	i H₂O; s EtOH, chl
7028	3-Methyl-2(3H)-benzothiazolethione		C₈H₇NS₂	2254-94-6	181.279	nd (al), pr (HOAc)	90	335			i H₂O; sl EtOH, eth; vs bz, chl
7029	4-(6-Methyl-2-benzothiazolyl)aniline		C₁₄H₁₂N₂S	92-36-4	240.323		194.8	434			sl EtOH, eth, bz, HOAc
7030	1-Methyl-1H-benzotriazole		C₇H₇N₃	13351-73-0	133.151	pl (bz-lig)	64.5	270.5			vs bz, EtOH, HOAc
7031	1-Methyl-2H-3,1-benzoxazine-2,4(1H)-dione		C₉H₇NO₃	10328-92-4	177.157		180				
7032	2-Methylbenzoxazole		C₈H₇NO	95-21-6	133.148		9.5	200.5	1.1211[20]	1.5497[20]	i H₂O; vs EtOH; msc eth
7033	Methyl benzoylacetate		C₁₀H₁₀O₃	614-27-7	178.184	pa ye		dec 265; 151[12]	1.158[29]	1.537[20]	vs ace, eth, EtOH
7034	Methyl 2-benzoylbenzoate		C₁₅H₁₂O₃	606-28-0	240.254	pl or mcl pr (dil al)	52	351	1.1903[19]	1.591[20]	i H₂O; vs EtOH, eth; s sulf
7035	2-(4-Methylbenzoyl)benzoic acid	2-(p-Toluoyl)benzoic acid	C₁₅H₁₂O₃	85-55-2	240.254		146				sl H₂O, DMSO; vs EtOH, eth, ace, bz
7036	2-Methylbenzoyl chloride		C₈H₇ClO	933-88-0	154.594			213.5		1.5549[20]	vs eth, EtOH
7037	3-Methylbenzoyl chloride		C₈H₇ClO	1711-06-4	154.594	liq	-23	219.5	1.0265[21]	1.505[22]	vs eth, EtOH

3-Methylbenzenemethanol

4-Methylbenzenemethanol

α-Methylbenzenemethanol, acetate

4-Methylbenzenepropanal

α-Methylbenzenepropanamine

β-Methylbenzenepropanoic acid, (±)

α-Methylbenzenepropanol

4-Methylbenzenesulfinic acid

4-Methylbenzenesulfinyl chloride

2-Methylbenzenesulfonamide

4-Methylbenzenesulfonamide

Methyl benzenesulfonate

2-Methylbenzenesulfonic acid

2-Methylbenzenesulfonyl chloride

2-Methylbenzenethiol

3-Methylbenzenethiol

4-Methylbenzenethiol

1-Methyl-1H-benzimidazole

2-Methyl-1H-benzimidazole

Methyl benzoate

Methyl 1,3-benzodioxole-5-carboxylate

2-Methylbenzofuran

2-Methylbenzonitrile

3-Methylbenzonitrile

4-Methylbenzonitrile

6-Methyl-2H-1-benzopyran-2-one

7-Methyl-2H-1-benzopyran-2-one

3-Methyl-4H-1-benzopyran-4-one

6-Methyl-2-benzothiazolamine

2-Methylbenzothiazole

3-Methyl-2(3H)-benzothiazolethione

4-(6-Methyl-2-benzothiazolyl)aniline

1-Methyl-1H-benzotriazole

1-Methyl-2H-3,1-benzoxazine-2,4(1H)-dione

2-Methylbenzoxazole

Methyl benzoylacetate

Methyl 2-benzoylbenzoate

2-(4-Methylbenzoyl)benzoic acid

2-Methylbenzoyl chloride

3-Methylbenzoyl chloride

No.	Name	Synonym	Mol. Form.	CAS RN	Mol. Wt.	Physical Form	mp/°C	bp/°C	den/ g cm⁻³	n_D	Solubility
7038	4-Methylbenzoyl chloride		C_8H_7ClO	874-60-2	154.594	liq	-1.5	226	1.1686[20]	1.5547[20]	s ctc
7039	Methyl benzoylsalicylate	2-(Benzoyloxy)benzoic acid, methyl ester	$C_{15}H_{12}O_4$	610-60-6	256.254	cry	85	385			i H_2O; s bz, chl, eth, EtOH
7040	α-Methylbenzylamine, (±)	1-Amino-1-phenylethane	$C_8H_{11}N$	618-36-0	121.180		32	187	0.9395[15]	1.5238[25]	s H_2O, chl; msc EtOH, eth
7041	1-Methyl-2-benzylbenzene		$C_{14}H_{14}$	713-36-0	182.261		6.6	280.5	1.0020[20]	1.5763[20]	
7042	1-Methyl-4-benzylbenzene		$C_{14}H_{14}$	620-83-7	182.261	liq	-30	286	0.9976[20]	1.5712[20]	vs eth, bz, EtOH, chl
7043	α-Methylbenzyl formate		$C_9H_{10}O_2$	7775-38-4	150.174	liq					
7044	1-Methyl-2-benzyl-4(1H)-quinazolinone	Glycosine	$C_{16}H_{14}N_2O$	6873-15-0	250.294		161.5				
7045	1-Methylbicyclo[3.1.0]hexane		C_7H_{12}	4625-24-5	96.170			93.1			
7046	2-Methylbiphenyl		$C_{13}H_{12}$	643-58-3	168.234	liq	-0.2	255.3	1.0113[20]	1.5914[20]	i H_2O; s EtOH, eth
7047	3-Methylbiphenyl		$C_{13}H_{12}$	643-93-6	168.234		2.3	272.7	1.0182[17]	1.5972[20]	i H_2O; s EtOH, eth, ctc
7048	4-Methylbiphenyl		$C_{13}H_{12}$	644-08-6	168.234	pl (lig, MeOH)	49.5	267.5	1.015[27]		i H_2O; s EtOH, eth; sl ctc
7049	4-Methyl-N,N-bis(4-methylphenyl)aniline		$C_{21}H_{21}N$	1159-53-1	287.399	cry (HOAc)	117				vs ace, bz, eth, chl
7050	Methyl bromoacetate		$C_3H_5BrO_2$	96-32-2	152.975			132	1.6350[20]	1.4520[20]	i H_2O; s EtOH, eth, ace, bz
7051	Methyl 2-bromobenzoate		$C_8H_7BrO_2$	610-94-6	215.045			244			i H_2O; s EtOH
7052	Methyl 3-bromobenzoate		$C_8H_7BrO_2$	618-89-3	215.045	pl	32	125[15]			sl H_2O; s EtOH, eth
7053	Methyl 4-bromobenzoate		$C_8H_7BrO_2$	619-42-1	215.045	lf (dil al), nd (eth)	81		1.689[25]		s EtOH, eth, ace, peth; vs bz, chl
7054	Methyl 2-bromobutanoate		$C_5H_9BrO_2$	3196-15-4	181.028			168	1.4528[20]	1.4029[25]	vs EtOH
7055	Methyl 4-bromobutanoate		$C_5H_9BrO_2$	4897-84-1	181.028			186.5	1.4[25]	1.4567[25]	vs EtOH
7056	Methyl 4-bromo-2-butenoate		$C_5H_7BrO_2$	1117-71-1	179.013			84[12]	1.490[19]	1.498[19]	
7057	Methyl 5-bromopentanoate		$C_6H_{11}BrO_2$	5454-83-1	195.054	liq		101[14]	1.363	1.4630[20]	
7058	Methyl 3-bromopropanoate		$C_4H_7BrO_2$	3395-91-3	167.002			105[60], 62[12]	1.4123[18]	1.4542[20]	s EtOH, eth, ace
7059	3-Methyl-1,2-butadiene		C_5H_8	598-25-4	68.118	liq	-113.6	40.83	0.6806[25]	1.4203[20]	vs ace, bz, eth, EtOH
7060	2-Methyl-1,3-butadiene	Isoprene	C_5H_8	78-79-5	68.118	liq	-145.9	34.0	0.679[20]	1.4219[20]	i H_2O; msc EtOH, eth, ace, bz
7061	3-Methylbutanal	Isovaleraldehyde	$C_5H_{10}O$	590-86-3	86.132	liq	-51	92.5	0.7977[20]	1.3902[20]	sl H_2O; s EtOH, eth
7062	3-Methylbutanamide	Isovaleramide	$C_5H_{11}NO$	541-46-8	101.147	mcl lf (al)	137	226			s H_2O, EtOH, eth; vs peth
7063	3-Methyl-1-butanamine	Isopentylamine	$C_5H_{13}N$	107-85-7	87.164			96	0.7505[20]	1.4083[20]	msc H_2O, EtOH, eth; s ace, chl
7064	2-Methyl-2-butanamine		$C_5H_{13}N$	594-39-8	87.164	liq	-105	77	0.731[25]	1.3954[25]	vs H_2O, ace, eth, EtOH
7065	3-Methyl-2-butanamine		$C_5H_{13}N$	598-74-3	87.164	liq	-50	85.5	0.7574[19]	1.4096[18]	vs H_2O; s EtOH
7066	3-Methyl-1,3-butanediol		$C_5H_{12}O_2$	2568-33-4	104.148			202.5	0.9448[20]	1.4452[20]	s H_2O, EtOH
7067	2-Methylbutanenitrile		C_5H_9N	18936-17-9	83.132			125	0.7913[15]	1.3933[20]	vs eth, EtOH
7068	3-Methylbutanenitrile	Isobutyl cyanide	C_5H_9N	625-28-5	83.132	liq	-101	127.5	0.7914[20]	1.3927[20]	sl H_2O; msc EtOH, eth; vs ace
7069	2-Methyl-1-butanethiol, (+)		$C_5H_{12}S$	20089-07-0	104.214	liq		119.1	0.8420[20]	1.4440[20]	
7070	3-Methyl-1-butanethiol	Isopentyl mercaptan	$C_5H_{12}S$	541-31-1	104.214	liq		116	0.8350[20]	1.4412[20]	i H_2O; msc EtOH, eth; s ctc
7071	2-Methyl-2-butanethiol		$C_5H_{12}S$	1679-09-0	104.214	liq		99.1	0.8120[20]	1.4385[20]	
7072	3-Methyl-2-butanethiol		$C_5H_{12}S$	2084-18-6	104.214	liq	-127.1	109.8			
7073	Methyl butanoate		$C_5H_{10}O_2$	623-42-7	102.132	liq	-85.8	102.8	0.8984[20]	1.3878[20]	sl H_2O, ctc; msc EtOH, eth
7074	2-Methylbutanoic acid	(±)-2-Methylbutyric acid	$C_5H_{10}O_2$	600-07-7	102.132		<-80	177	0.934[20]	1.4051[20]	sl H_2O; msc EtOH, eth; s chl
7075	3-Methylbutanoic acid	Isovaleric acid	$C_5H_{10}O_2$	503-74-2	102.132	liq	-29.3	176.5	0.931[20]	1.4033[20]	s H_2O; msc EtOH, eth, chl
7076	3-Methylbutanoic anhydride		$C_{10}H_{18}O_3$	1468-39-9	186.248			215	0.9327[20]	1.4043[20]	vs eth
7077	2-Methyl-1-butanol, (±)		$C_5H_{12}O$	34713-94-5	88.148			127.5	0.8152[25]	1.4092[20]	sl H_2O; msc EtOH, eth; vs ace
7078	3-Methyl-1-butanol	Isopentyl alcohol	$C_5H_{12}O$	123-51-3	88.148	liq	-117.2	131.1	0.8104[20]	1.4053[20]	sl H_2O; vs ace, eth, EtOH
7079	2-Methyl-2-butanol	tert-Pentyl alcohol	$C_5H_{12}O$	75-85-4	88.148	liq	-9.1	102.4	0.8096[20]	1.4052[20]	s H_2O, bz, chl; msc EtOH, eth; vs ace

4-Methylbenzoyl chloride

Methyl benzoylsalicylate

α-Methylbenzylamine, (±)

1-Methyl-2-benzylbenzene

1-Methyl-4-benzylbenzene

α-Methylbenzyl formate

1-Methyl-2-benzyl-4(1*H*)-quinazolinone

1-Methylbicyclo[3.1.0]hexane

2-Methylbiphenyl

3-Methylbiphenyl

4-Methylbiphenyl

4-Methyl-*N*,*N*-bis(4-methylphenyl)aniline

Methyl bromoacetate

Methyl 2-bromobenzoate

Methyl 3-bromobenzoate

Methyl 4-bromobenzoate

Methyl 2-bromobutanoate

Methyl 4-bromobutanoate

Methyl 4-bromo-2-butenoate

Methyl 5-bromopentanoate

Methyl 3-bromopropanoate

3-Methyl-1,2-butadiene

2-Methyl-1,3-butadiene

3-Methylbutanal

3-Methylbutanamide

3-Methyl-1-butanamine

2-Methyl-2-butanamine

3-Methyl-2-butanamine

3-Methyl-1,3-butanediol

2-Methylbutanenitrile

3-Methylbutanenitrile

2-Methyl-1-butanethiol, (+)

3-Methyl-1-butanethiol

2-Methyl-2-butanethiol

3-Methyl-2-butanethiol

Methyl butanoate

2-Methylbutanoic acid

3-Methylbutanoic acid

3-Methylbutanoic anhydride

2-Methyl-1-butanol, (±)

3-Methyl-1-butanol

2-Methyl-2-butanol

No.	Name	Synonym	Mol. Form.	CAS RN	Mol. Wt.	Physical Form	mp/°C	bp/°C	den/ g cm⁻³	n_D	Solubility
7080	3-Methyl-2-butanol, (±)		$C_5H_{12}O$	70116-68-6	88.148			112.9	0.8180[20]	1.4089[20]	sl H_2O; msc EtOH, eth; vs ace; s bz, ctc
7081	2-Methyl-1-butanol acetate		$C_7H_{14}O_2$	624-41-9	130.185			140	0.8740[20]	1.4040[20]	vs ace, eth, EtOH
7082	3-Methyl-2-butanone	Methyl isopropyl ketone	$C_5H_{10}O$	563-80-4	86.132	liq	-93.1	94.33	0.8051[20]	1.3880[20]	sl H_2O; msc EtOH, eth; vs ace; s ctc
7083	2-Methylbutanoyl chloride, (±)		C_5H_9ClO	57526-28-0	120.577			116	0.9917[20]	1.4170[20]	
7084	3-Methylbutanoyl chloride	Isovaleryl chloride	C_5H_9ClO	108-12-3	120.577			114	0.9844[20]	1.4149[20]	s eth
7085	trans-2-Methyl-2-butenal	Tiglic aldehyde	C_5H_8O	497-03-0	84.117	liq		117; 64[119]	0.8710[20]	1.4475[20]	sl H_2O; vs EtOH
7086	3-Methyl-2-butenal	Senecialdehyde	C_5H_8O	107-86-8	84.117			134	0.8722[20]	1.4528[20]	s H_2O, EtOH, eth
7087	2-Methyl-1-butene		C_5H_{10}	563-46-2	70.133	liq	-137.53	31.2	0.6504[20]	1.3778[20]	i H_2O; s EtOH, eth, bz, ctc
7088	3-Methyl-1-butene		C_5H_{10}	563-45-1	70.133	vol liq or gas	-168.43	20.1	0.6213[25]	1.3643[20]	i H_2O; msc EtOH, eth; s bz
7089	2-Methyl-2-butene		C_5H_{10}	513-35-9	70.133	liq	-133.72	38.56	0.6623[20]	1.3874[20]	i H_2O; s EtOH, eth, bz, ctc; vs lig
7090	cis-2-Methyl-2-butenedioic acid	Citraconic acid	$C_5H_6O_4$	498-23-7	130.100	nd (eth-lig) tcl pr (eth-bz)	93.5		1.617[25]		vs H_2O; sl eth, chl; i bz, CS_2
7091	3-Methyl-2-butenenitrile		C_5H_7N	4786-24-7	81.117	liq		141			
7092	Methyl cis-2-butenoate	Methyl isocrotonate	$C_5H_8O_2$	4358-59-2	100.117			118		1.4175[20]	
7093	Methyl trans-2-butenoate	Methyl crotonate	$C_5H_8O_2$	623-43-8	100.117	liq	-42	121	0.9444[20]	1.4242[20]	i H_2O; vs EtOH, eth
7094	cis-2-Methyl-2-butenoic acid	Angelic acid	$C_5H_8O_2$	565-63-9	100.117	mcl pr or nd	45.5	185	0.9834[49]	1.4434[47]	sl H_2O; s EtOH; vs eth
7095	trans-2-Methyl-2-butenoic acid	Tiglic acid	$C_5H_8O_2$	80-59-1	100.117	tab (w)	64.5	198.5	0.9641[76]	1.4330[76]	s H_2O; vs EtOH, eth
7096	3-Methyl-2-butenoic acid		$C_5H_8O_2$	541-47-9	100.117		69.5	197	1.0062[24]		
7097	3-Methyl-2-buten-1-ol		$C_5H_{10}O$	556-82-1	86.132			140	0.848[25]	1.4412[20]	
7098	3-Methyl-3-buten-1-ol		$C_5H_{10}O$	763-32-6	86.132			129.9			
7099	3-Methyl-2-buten-2-ol		$C_5H_{10}O$	115-18-4	86.132	liq	-28	97	0.82[20]		
7100	3-Methyl-3-buten-2-ol		$C_5H_{10}O$	10473-14-0	86.132			114	0.8531[17]	1.4288[17]	
7101	3-Methyl-3-buten-2-one	Isopropenyl methyl ketone	C_5H_8O	814-78-8	84.117	liq	-54	98	0.8527[20]	1.4220[20]	vs EtOH
7102	3-Methyl-2-butenoyl chloride		C_5H_7ClO	3350-78-5	118.562			146	1.065[25]	1.4770[20]	
7103	(3-Methyl-2-butenyl)guanidine	Galegine	$C_6H_{13}N_3$	543-83-9	127.187	hyg	62.5	dec			vs H_2O, EtOH
7104	2-Methyl-1-buten-3-yne	Isopropenylacetylene	C_5H_6	78-80-8	66.102	liq	-113	32	0.6801[11]	1.4140[20]	s chl
7105	[(3-Methylbutoxy)methyl]benzene		$C_{12}H_{18}O$	122-73-6	178.270			236; 118[19]	0.909[20]	1.4792[20]	vs eth, EtOH
7106	1-[2-(3-Methylbutoxy)-2-phenylethyl]pyrrolidine	Amixetrine	$C_{17}H_{27}NO$	24622-72-8	261.402			121[2]		1.4978[22]	
7107	2-Methylbutyl acrylate		$C_8H_{14}O_2$	44914-03-6	142.196			160; 45[10]	0.8936[20]	1.4240[20]	vs eth, EtOH
7108	3-Methylbutyl benzoate	Isopentyl benzoate	$C_{12}H_{16}O_2$	94-46-2	192.254			261	0.993[15]		vs EtOH
7109	3-Methylbutyl 2-chloropropanoate		$C_8H_{15}ClO_2$	62108-69-4	178.657			208	1.0050[20]	1.4289[20]	
7110	3-Methylbutyl 3-chloropropanoate		$C_8H_{15}ClO_2$	62108-70-7	178.657			208; 87[12]	1.0171[20]	1.4343[20]	vs eth, EtOH
7111	Methyl tert-butyl ether	tert-Butyl methyl ether	$C_5H_{12}O$	1634-04-4	88.148	liq	-108.6	55.0	0.7353[25]	1.3664[20]	s H_2O; vs EtOH, eth
7112	3-Methylbutyl nitrate	Isopentyl nitrate	$C_5H_{11}NO_3$	543-87-3	133.146			148	0.996[22]	1.4122[21]	
7113	2-Methyl-3-butyn-2-amine		C_5H_9N	2978-58-7	83.132		18	79.5	0.79[25]	1.4235[20]	
7114	3-Methyl-1-butyne		C_5H_8	598-23-2	68.118	vol liq or gas	-89.7	26.3	0.6660[20]	1.3723[20]	i H_2O; msc EtOH, eth
7115	2-Methyl-3-butyn-2-ol	1,1-Dimethylpropargyl alcohol	C_5H_8O	115-19-5	84.117		1.5	104	0.8618[20]	1.4207[20]	
7116	Methyl carbamate		$C_2H_5NO_2$	598-55-0	75.067	nd	54	177	1.1361[56]	1.4125[56]	vs H_2O; vs EtOH, eth
7117	3-Methyl-9H-carbazole		$C_{13}H_{11}N$	4630-20-0	181.233	pl (HOAc)	208.5	365			vs bz, eth
7118	9-Methyl-9H-carbazole		$C_{13}H_{11}N$	1484-12-4	181.233	nd, lf (al)	89.34	343.64; 195[12]			vs eth
7119	Methyl chloroacetate		$C_3H_5ClO_2$	96-34-4	108.524	liq	-32.1	129.5	1.236[20]	1.4218[20]	vs ace, bz, eth, EtOH
7120	Methyl 2-chloroacrylate		$C_4H_5ClO_2$	80-63-7	120.535			52[51]	1.189[20]	1.4420[20]	vs eth
7121	Methyl 2-chlorobenzoate		$C_8H_7ClO_2$	610-96-8	170.594			234			s EtOH
7122	Methyl 3-chlorobenzoate		$C_8H_7ClO_2$	2905-65-9	170.594		21	229			
7123	Methyl 4-chlorobenzoate		$C_8H_7ClO_2$	1126-46-1	170.594	nd or mcl pr	43.5		1.382[20]		vs EtOH
7124	Methyl 4-chlorobutanoate		$C_5H_9ClO_2$	3153-37-5	136.577			174; 55[4]	1.1293[20]	1.4321[20]	i H_2O; vs EtOH, eth; s ace
7125	Methyl chlorocarbonate		$C_2H_3ClO_2$	79-22-1	94.497			70.5	1.2231[20]	1.3868[20]	msc EtOH, eth; s bz, ctc, chl
7126	Methyl 5-chloro-2-hydroxybenzoate		$C_8H_7ClO_3$	4068-78-4	186.593	nd (al)	50	dec 249; 120[12]			vs EtOH
7127	Methyl 5-chloro-2-nitrobenzoate		$C_8H_6ClNO_4$	51282-49-6	215.592	pl (MeOH)	48.5		1.453[18]		vs MeOH
7128	Methyl chlorooxoacetate		$C_3H_3ClO_3$	5781-53-3	122.507			119	1.3316[20]	1.4189[20]	
7129	Methyl 2-chloropropanoate		$C_4H_7ClO_2$	17639-93-9	122.551			132.5	1.0750[25]		

3-Methyl-2-butanol, (±)

2-Methyl-1-butanol acetate

3-Methyl-2-butanone

2-Methylbutanoyl chloride, (±)

3-Methylbutanoyl chloride

trans-2-Methyl-2-butenal

3-Methyl-2-butenal

2-Methyl-1-butene

3-Methyl-1-butene

2-Methyl-2-butene

cis-2-Methyl-2-butenedioic acid

3-Methyl-2-butenenitrile

Methyl cis-2-butenoate

Methyl trans-2-butenoate

cis-2-Methyl-2-butenoic acid

trans-2-Methyl-2-butenoic acid

3-Methyl-2-butenoic acid

3-Methyl-2-buten-1-ol

3-Methyl-3-buten-1-ol

2-Methyl-3-buten-2-ol

3-Methyl-3-buten-2-ol

3-Methyl-3-buten-2-one

3-Methyl-2-butenoyl chloride

(3-Methyl-2-butenyl)guanidine

2-Methyl-1-buten-3-yne

[(3-Methylbutoxy)methyl]benzene

1-[2-(3-Methylbutoxy)-2-phenylethyl]pyrrolidine

2-Methylbutyl acrylate

3-Methylbutyl benzoate

3-Methylbutyl 2-chloropropanoate

3-Methylbutyl 3-chloropropanoate

Methyl tert-butyl ether

3-Methylbutyl nitrate

2-Methyl-3-butyn-2-amine

3-Methyl-1-butyne

2-Methyl-3-butyn-2-ol

Methyl carbamate

3-Methyl-9H-carbazole

9-Methyl-9H-carbazole

Methyl chloroacetate

Methyl 2-chloroacrylate

Methyl 2-chlorobenzoate

Methyl 3-chlorobenzoate

Methyl 4-chlorobenzoate

Methyl 4-chlorobutanoate

Methyl chlorocarbonate

Methyl 5-chloro-2-hydroxybenzoate

Methyl 5-chloro-2-nitrobenzoate

Methyl chlorooxoacetate

Methyl 2-chloropropanoate

No.	Name	Synonym	Mol. Form.	CAS RN	Mol. Wt.	Physical Form	mp/°C	bp/°C	den/ g cm⁻³	n_D	Solubility
7130	3-Methylchrysene		C₁₉H₁₄	3351-31-3	242.314	lf (bz-peth)	173.3				vs EtOH
7131	5-Methylchrysene		C₁₉H₁₄	3697-24-3	242.314		118.3				i H₂O
7132	6-Methylchrysene		C₁₉H₁₄	1705-85-7	242.314		161				
7133	Methyl *trans*-cinnamate	Methyl *trans*-3-phenyl-2-propenoate	C₁₀H₁₀O₂	1754-62-7	162.185	cry (peth, dil al)	36.5	261.9	1.042³⁶	1.5766²²	i H₂O; vs EtOH, eth; s ace, bz; sl chl
7134	*trans-o*-Methylcinnamic acid		C₁₀H₁₀O₂	2373-76-4	162.185	cry (EtOH)	175				
7135	*trans-m*-Methylcinnamic acid		C₁₀H₁₀O₂	3029-79-6	162.185	cry (w)	115				
7136	*trans-p*-Methylcinnamic acid		C₁₀H₁₀O₂	1866-39-3	162.185		198.5				
7137	Methyclothiazide		C₉H₁₁Cl₂N₃O₄S₂	135-07-9	360.237	cry (EtOH aq)	225				i H₂O, bz, chl; sl MeOH; vs ace, py
7138	Methyl cyanate		C₂H₃NO	1768-34-9	57.051	unstab gas	-30	exp			
7139	Methyl cyanoacetate		C₄H₅NO₂	105-34-0	99.089	liq	-22.5	200.5	1.1225²⁵	1.4176²⁰	vs eth, EtOH
7140	Methyl 2-cyanoacrylate	Mecrylate	C₅H₅NO₂	137-05-3	111.100			47²	1.1012²⁰	1.4430	
7141	Methylcyclobutane		C₅H₁₀	598-61-8	70.133	liq	-161.5	36.3	0.6884²⁰	1.3866²⁰	i H₂O; msc EtOH, eth; s ace, bz, peth
7142	Methyl cyclobutanecarboxylate		C₆H₁₀O₂	765-85-5	114.142			135.5			
7143	2-Methyl-1,3-cyclohexadiene	4,5-Dihydrotoluene	C₇H₁₀	1489-57-2	94.154			107.5	0.8260¹⁸	1.4662¹⁸	
7144	2-Methyl-2,5-cyclohexadiene-1,4-dione		C₇H₆O₂	553-97-9	122.122	ye pl or nd	69	sub	1.08⁷⁵		sl H₂O; s EtOH, eth
7145	Methylcyclohexane		C₇H₁₄	108-87-2	98.186	liq	-126.6	100.93	0.7694²⁰	1.4231²⁰	i H₂O; s EtOH, eth; msc ace, bz, lig
7146	Methyl cyclohexanecarboxylate		C₈H₁₄O₂	4630-82-4	142.196			183	0.9954¹⁵	1.4433²⁰	i H₂O; s EtOH, eth, ace, chl
7147	α-Methylcyclohexanemethanol		C₈H₁₆O	1193-81-3	128.212			189	0.928²⁵	1.4656²⁰	vs EtOH, eth; sl ctc
7148	4-Methylcyclohexanemethanol		C₈H₁₆O	34885-03-5	128.212			75²·⁵	0.9074²⁰	1.4617²⁰	
7149	1-Methylcyclohexanol		C₇H₁₄O	590-67-0	114.185		25	155; 70²⁵	0.9194²⁰	1.4595²⁰	i H₂O; s EtOH, bz, chl
7150	*cis*-2-Methylcyclohexanol		C₇H₁₄O	615-38-3	114.185		7	165	0.9360²⁰	1.4640²⁰	vs EtOH
7151	*trans*-2-Methylcyclohexanol, (±)		C₇H₁₄O	615-39-4	114.185	liq	-2.0	167.5	0.9247²⁰	1.4616²⁰	vs eth, EtOH
7152	*cis*-3-Methylcyclohexanol, (±)		C₇H₁₄O	5454-79-5	114.185	liq	-5.5	168; 94¹²	0.9155²⁰	1.4752²⁰	vs eth, EtOH
7153	*trans*-3-Methylcyclohexanol, (±)		C₇H₁₄O	7443-55-2	114.185	liq	-0.5	167; 84¹³	0.9214³⁰	1.4580²⁰	vs eth, EtOH
7154	*cis*-4-Methylcyclohexanol		C₇H₁₄O	7731-28-4	114.185	liq	-9.2	173	0.9170²⁰	1.4614²⁰	vs EtOH
7155	*trans*-4-Methylcyclohexanol		C₇H₁₄O	7731-29-5	114.185			174	0.9118²¹	1.4561²⁰	sl H₂O; msc EtOH; s eth
7156	2-Methylcyclohexanone, (±)		C₇H₁₂O	24965-84-2	112.169	liq	-13.9	165	0.9250²⁰	1.4483²⁵	i H₂O; s EtOH, eth
7157	3-Methylcyclohexanone, (±)		C₇H₁₂O	625-96-7	112.169	liq	-73.5	169; 65¹⁵	0.9136²⁰	1.4456²⁰	i H₂O; s EtOH, eth
7158	4-Methylcyclohexanone		C₇H₁₂O	589-92-4	112.169	liq	-40.6	170	0.9138²⁰	1.4451²⁰	i H₂O; s EtOH, eth; sl ctc
7159	1-Methylcyclohexene		C₇H₁₂	591-49-1	96.170	liq	-120.4	110.3	0.8102²⁰	1.4503²⁰	i H₂O; s eth, bz, ctc
7160	3-Methylcyclohexene, (±)		C₇H₁₂	56688-75-6	96.170	liq	-115.5	104	0.7990²⁰	1.4414²⁰	vs bz, eth, chl, peth
7161	4-Methylcyclohexene		C₇H₁₂	591-47-9	96.170	liq	-115.5	102.7	0.7991²⁰	1.4414²⁰	i H₂O; s EtOH, eth
7162	Methyl 3-cyclohexene-1-carboxylate		C₈H₁₂O₂	6493-77-2	140.180			182; 80²⁰	1.0130²⁰	1.4610²⁰	
7163	2-Methyl-2-cyclohexen-1-one		C₇H₁₀O	1121-18-2	110.153			178.5	0.966²⁰	1.4833²⁰	s bz
7164	3-Methyl-2-cyclohexen-1-one		C₇H₁₀O	1193-18-6	110.153	liq	-21	201	0.9693²⁰	1.49475²⁰	msc H₂O; s bz
7165	3-Methylcyclopentadecanone	Muscone	C₁₆H₃₀O	541-91-3	238.408	oily liq		329; 130⁰·⁵	0.9221¹⁷	1.4802¹⁷	vs ace, eth, EtOH
7166	1-Methyl-1,3-cyclopentadiene		C₆H₈	96-39-9	80.128	liq		73	0.81²⁰	1.4512²⁰	
7167	Methylcyclopentane		C₆H₁₂	96-37-7	84.159	liq	-142.42	71.8	0.7486²⁰	1.4097²⁰	i H₂O; msc EtOH, eth, ace, bz, lig, ctc
7168	1-Methylcyclopentanol		C₆H₁₂O	1462-03-9	100.158	nd	36	136; 53³⁰	0.9044²³	1.4429²³	
7169	*cis*-2-Methylcyclopentanol		C₆H₁₂O	25144-05-2	100.158			148.5	0.9379¹⁶	1.4504¹⁶	
7170	2-Methylcyclopentanone		C₆H₁₀O	1120-72-5	98.142	liq	-75	139.5	0.9139²⁰	1.4364²⁰	s H₂O; vs EtOH, eth, ace
7171	3-Methylcyclopentanone, (±)		C₆H₁₀O	6195-92-2	98.142	liq	-58.4	144	0.913²²	1.4329²⁰	s H₂O; vs EtOH, eth, ace, HOAc
7172	1-Methylcyclopentene		C₆H₁₀	693-89-0	82.143	liq	-126.5	75.5	0.7748²⁵	1.4322²⁰	
7173	3-Methylcyclopentene		C₆H₁₀	1120-62-3	82.143			64.9	0.7572²⁵	1.4216²⁰	
7174	4-Methylcyclopentene		C₆H₁₀	1759-81-5	82.143	liq	-160.8	65.7	0.7634²⁵	1.4209²⁰	
7175	2-Methyl-2-cyclopenten-1-one		C₆H₈O	1120-73-6	96.127			157	0.9808¹⁶	1.4762¹⁵	
7176	3-Methyl-2-cyclopenten-1-one		C₆H₈O	2758-18-1	96.127			157.5	0.9712²⁰	1.4714²⁰	

3-Methylchrysene

5-Methylchrysene

6-Methylchrysene

Methyl *trans*-cinnamate

trans-o-Methylcinnamic acid

trans-m-Methylcinnamic acid

trans-p-Methylcinnamic acid

Methyclothiazide

Methyl cyanate

Methyl cyanoacetate

Methyl 2-cyanoacrylate

Methylcyclobutane

Methyl cyclobutanecarboxylate

2-Methyl-1,3-cyclohexadiene

2-Methyl-2,5-cyclohexadiene-1,4-dione

Methylcyclohexane

Methyl cyclohexanecarboxylate

α-Methylcyclohexanemethanol

4-Methylcyclohexanemethanol

1-Methylcyclohexanol

cis-2-Methylcyclohexanol

trans-2-Methylcyclohexanol, (±)

cis-3-Methylcyclohexanol, (±)

trans-3-Methylcyclohexanol, (±)

cis-4-Methylcyclohexanol

trans-4-Methylcyclohexanol

2-Methylcyclohexanone, (±)

3-Methylcyclohexanone, (±)

4-Methylcyclohexanone

1-Methylcyclohexene

3-Methylcyclohexene, (±)

4-Methylcyclohexene

Methyl 3-cyclohexene-1-carboxylate

2-Methyl-2-cyclohexen-1-one

3-Methyl-2-cyclohexen-1-one

3-Methylcyclopentadecanone

1-Methyl-1,3-cyclopentadiene

Methylcyclopentane

1-Methylcyclopentanol

cis-2-Methylcyclopentanol

2-Methylcyclopentanone

3-Methylcyclopentanone, (±)

1-Methylcyclopentene

3-Methylcyclopentene

4-Methylcyclopentene

2-Methyl-2-cyclopenten-1-one

3-Methyl-2-cyclopenten-1-one

No.	Name	Synonym	Mol. Form.	CAS RN	Mol. Wt.	Physical Form	mp/°C	bp/°C	den/ g cm⁻³	n_D	Solubility
7177	Methylcyclopropane		C$_4$H$_8$	594-11-6	56.107	col gas	-177.6	0.7	0.6912^{20}		vs eth, EtOH
7178	Methyl cyclopropanecarboxylate		C$_5$H$_8$O$_2$	2868-37-3	100.117			114.9	0.9848^{20}	1.4144^{19}	s ace, chl
7179	α-Methylcyclopropanemethanol		C$_5$H$_{10}$O	765-42-4	86.132	liq	-32.1	123.5	0.8805^{20}	1.4316^{20}	
7180	Methyl L-cysteine hydrochloride		C$_4$H$_{10}$ClNO$_2$S	18598-63-5	171.646	cry (MeOH)	140.5				
7181	Methyl trans-2,cis-4-decadienoate		C$_{11}$H$_{18}$O$_2$	4493-42-9	182.260			71$^{0.15}$	0.9128^{22}	1.4874^{22}	
7182	Methyl trans-2,trans-4-decadienoate		C$_{11}$H$_{18}$O$_2$	7328-33-8	182.260			87^{13}, 70$^{0.2}$	0.9082^{22}	1.4918^{22}	
7183	2-Methyldecane		C$_{11}$H$_{24}$	6975-98-0	156.309	liq	-48.9	189.3	0.7368^{20}	1.4154^{20}	
7184	3-Methyldecane		C$_{11}$H$_{24}$	13151-34-3	156.309	liq	-92.9	188.1	0.7422^{20}	1.4177^{20}	
7185	4-Methyldecane		C$_{11}$H$_{24}$	2847-72-5	156.309	liq	-77.5	187		1.4352^{20}	
7186	Methyl decanoate		C$_{11}$H$_{22}$O$_2$	110-42-9	186.292	liq	-18	224	0.8730^{20}	1.4259^{20}	i H$_2$O; vs EtOH, eth; sl ctc; msc chl
7187	Methyl demeton		C$_6$H$_{15}$O$_3$PS$_2$	8022-00-2	230.285	ye liq		89$^{0.15}$, 118^1	1.20^{20}	1.5063^{20}	i H$_2$O; s os
7188	Methyldiborane(6)		CH$_8$B$_2$	23777-55-1	41.697	unstab gas					s eth
7189	Methyl 2,3-dibromopropanoate		C$_4$H$_6$Br$_2$O$_2$	1729-67-5	245.898			206	1.9333^{20}	1.5127^{20}	s EtOH
7190	Methyl dichloroacetate		C$_3$H$_4$Cl$_2$O$_2$	116-54-1	142.969	liq	-51.9	142.9	1.3774^{20}	1.4429^{20}	i H$_2$O; s EtOH, ctc
7191	Methyl 2,5-dichlorobenzoate		C$_8$H$_6$Cl$_2$O$_2$	2905-69-3	205.039	cry	38				
7192	Methyl (2,4-dichlorophenoxy) acetate	2,4-D methyl ester	C$_9$H$_8$Cl$_2$O$_3$	1928-38-7	235.064		119	141^{18}			
7193	Methyl (3,4-dichlorophenyl) carbamate	Swep	C$_8$H$_7$Cl$_2$NO$_2$	1918-18-9	220.054	nd	114				
7194	Methyl 2,3-dichloropropanoate		C$_4$H$_6$Cl$_2$O$_2$	3674-09-7	156.996			92^{50}, 63^{10}	1.3282^{20}		vs ace, eth, EtOH
7195	Methyldifluoroarsine		CH$_3$AsF$_2$	420-24-6	127.954	liq, fumes in air	-29.7	76.5	1.924^{18}		
7196	Methyldifluorophosphine	(Difluoro)methylphosphine	CH$_3$F$_2$P	753-59-3	84.006	gas	-110	-28			
7197	Methyl 2,4-dihydroxybenzoate		C$_8$H$_8$O$_4$	2150-47-2	168.148		116.5				sl EtOH, ace
7198	Methyl 3,5-dihydroxybenzoate		C$_8$H$_8$O$_4$	2150-44-9	168.148		165				
7199	Methyl 3,4-dimethoxybenzoate		C$_{10}$H$_{12}$O$_4$	2150-38-1	196.200	nd (dil al)	60.8	283			vs bz, eth, EtOH
7200	Methyldimethoxysilane		C$_3$H$_{10}$O$_2$Si	16881-77-9	106.196			61			
7201	3-Methyl-4'-(dimethylamino) azobenzene		C$_{15}$H$_{17}$N$_3$	55-80-1	239.316	oran cry	122				
7202	2-Methyl-N,N-dimethylaniline	N,N-Dimethyl-o-toluidine	C$_9$H$_{13}$N	609-72-3	135.206	liq	-60	194.1	0.9286^{20}	1.5152^{20}	vs eth, EtOH
7203	3-Methyl-N,N-dimethylaniline	N,N-Dimethyl-m-toluidine	C$_9$H$_{13}$N	121-72-2	135.206			212	0.9410^{20}	1.5492^{20}	msc EtOH, eth
7204	4-Methyl-N,N-dimethylaniline	N,N-Dimethyl-p-toluidine	C$_9$H$_{13}$N	99-97-8	135.206			211	0.9366^{20}	1.5366^{20}	i H$_2$O; msc EtOH, eth; s ctc
7205	Methyl 2,2-dimethylpropanoate	Methyl 2,2-dimethylpropionate	C$_6$H$_{12}$O$_2$	598-98-1	116.158			101.1	0.891^0	1.3905^{20}	vs eth, EtOH
7206	Methyl dimethylthioborane	Dimethyl(methylthio)borane	C$_3$H$_9$BS	19163-05-4	87.979	liq	-84	71			vs ace, eth
7207	2-Methyl-3,5-dinitrobenzamide	Dinitolmide	C$_8$H$_7$N$_3$O$_5$	148-01-6	225.159	cry	181				
7208	1-Methyl-2,3-dinitrobenzene	2,3-Dinitrotoluene	C$_7$H$_6$N$_2$O$_4$	602-01-7	182.134		63				i H$_2$O; s EtOH, eth; sl chl
7209	1-Methyl-2,4-dinitrobenzene	2,4-Dinitrotoluene	C$_7$H$_6$N$_2$O$_4$	121-14-2	182.134	ye nd or mcl pr (CS$_2$)	70.5	dec 300	1.3208^{71}	1.442	i H$_2$O; s EtOH, eth, chl, bz; vs ace, py
7210	1-Methyl-3,5-dinitrobenzene	3,5-Dinitrotoluene	C$_7$H$_6$N$_2$O$_4$	618-85-9	182.134	ye orth nd (HOAc)	93	sub	1.2772^{111}		sl H$_2$O; s EtOH, eth, bz, chl, CS$_2$
7211	2-Methyl-1,3-dinitrobenzene	2,6-Dinitrotoluene	C$_7$H$_6$N$_2$O$_4$	606-20-2	182.134	orth nd (al)	66.0	285	1.2833^{111}	1.479	s EtOH, bz; vs CS$_2$
7212	2-Methyl-1,4-dinitrobenzene	2,5-Dinitrotoluene	C$_7$H$_6$N$_2$O$_4$	619-15-8	182.134	nd (al)	52.5		1.282^{111}		s EtOH, bz; vs CS$_2$
7213	4-Methyl-1,2-dinitrobenzene	3,4-Dinitrotoluene	C$_7$H$_6$N$_2$O$_4$	610-39-9	182.134	ye nd (CS$_2$)	59.0		1.2594^{111}		i H$_2$O; s EtOH, CS$_2$; sl chl
7214	2-Methyl-4,6-dinitrophenol	4,6-Dinitro-o-cresol	C$_7$H$_6$N$_2$O$_5$	534-52-1	198.133	ye pr or nd (al)	86.5				sl H$_2$O, peth; s EtOH, eth, ace, chl
7215	4-Methyl-2,6-dinitrophenol	2,6-Dinitro-p-cresol	C$_7$H$_6$N$_2$O$_5$	609-93-8	198.133	ye nd (eth, peth)	85				i H$_2$O; s EtOH, eth, bz
7216	Methyldioctylamine	N-Methyl-N-octyl-1-octanamine	C$_{17}$H$_{37}$N	4455-26-9	255.483		-30.1	158^{10}		1.4424^{20}	
7217	4-Methyl-1,3-dioxane		C$_5$H$_{10}$O$_2$	1120-97-4	102.132	liq	-44.5	114	0.9758^{20}	1.4159^{20}	sl H$_2$O; vs os
7218	2-Methyl-1,3-dioxolane		C$_4$H$_8$O$_2$	497-26-7	88.106			81.5	0.9811^{20}	1.4035^{17}	vs H$_2$O; msc EtOH, eth
7219	4-Methyl-1,3-dioxolane		C$_4$H$_8$O$_2$	1072-47-5	88.106	liq		85	0.99^{20}	1.3980^{20}	
7220	Methyldiphenylamine	N-Methyl-N-phenylbenzenamine	C$_{13}$H$_{13}$N	552-82-9	183.249	liq	-7.5	293.5	1.0476^{20}	1.6193^{20}	i H$_2$O; sl EtOH, MeOH; s ctc
7221	4-Methyl-2,4-diphenyl-1-pentene		C$_{18}$H$_{20}$	6362-80-7	236.352	liq		172^8, 102$^{0.2}$	0.99^{25}		
7222	Methyldiphenylsilane		C$_{13}$H$_{14}$Si	776-76-1	198.336			93.5^1	0.996^{20}	1.5694^{20}	s ctc
7223	Methyldiphenylsilanol		C$_{13}$H$_{14}$OSi	778-25-6	214.335		167	184^{24}, 148^3	1.0840^{25}		s ctc, CS$_2$
7224	2-Methyl-1,2-di-3-pyridinyl-1-propanone	Metyrapone	C$_{14}$H$_{14}$N$_2$O	54-36-4	226.273		50.5				
7225	Methyl docosanoate	Methyl behenate	C$_{23}$H$_{46}$O$_2$	929-77-1	354.610	nd (ace)	54			1.4339^{60}	vs eth, EtOH

Methylcyclopropane

Methyl cyclopropanecarboxylate

α-Methylcyclopropanemethanol

Methyl L-cysteine hydrochloride

Methyl trans-2,cis-4-decadienoate

Methyl trans-2,trans-4-decadienoate

2-Methyldecane

3-Methyldecane

4-Methyldecane

Methyl decanoate

Methyl demeton

Methyldiborane(6)

Methyl 2,3-dibromopropanoate

Methyl dichloroacetate

Methyl 2,5-dichlorobenzoate

Methyl (2,4-dichlorophenoxy)acetate

Methyl (3,4-dichlorophenyl)carbamate

Methyl 2,3-dichloropropanoate

Methyldifluoroarsine

Methyldifluorophosphine

Methyl 2,4-dihydroxybenzoate

Methyl 3,5-dihydroxybenzoate

Methyl 3,4-dimethoxybenzoate

Methyldimethoxysilane

3-Methyl-4'-(dimethylamino)azobenzene

2-Methyl-N,N-dimethylaniline

3-Methyl-N,N-dimethylaniline

4-Methyl-N,N-dimethylaniline

Methyl 2,2-dimethylpropanoate

Methyl dimethylthioborane

2-Methyl-3,5-dinitrobenzamide

1-Methyl-2,3-dinitrobenzene

1-Methyl-2,4-dinitrobenzene

1-Methyl-3,5-dinitrobenzene

2-Methyl-1,3-dinitrobenzene

2-Methyl-1,4-dinitrobenzene

4-Methyl-1,2-dinitrobenzene

2-Methyl-4,6-dinitrophenol

4-Methyl-2,6-dinitrophenol

Methyldioctylamine

4-Methyl-1,3-dioxane

2-Methyl-1,3-dioxolane

4-Methyl-1,3-dioxolane

Methyldiphenylamine

4-Methyl-2,4-diphenyl-1-pentene

Methyldiphenylsilane

Methyldiphenylsilanol

2-Methyl-1,2-di-3-pyridinyl-1-propanone

Methyl docosanoate

No.	Name	Synonym	Mol. Form.	CAS RN	Mol. Wt.	Physical Form	mp/°C	bp/°C	den/ g cm⁻³	n_D	Solubility
7226	Methyl cis-13-docosenoate		C₂₃H₄₄O₂	1120-34-9	352.594		-1.2	220[5]			
7227	Methyl dodecanoate	Methyl laurate	C₁₃H₂₆O₂	111-82-0	214.344		5.2	267	0.8702[20]	1.4319[20]	i H₂O; msc EtOH, eth, ace, bz; s chl, ctc
7228	2-Methyldodecanoic acid		C₁₃H₂₆O₂	2874-74-0	214.344	pl	22	153[1]	0.890[18]		
7229	Methyl eicosanoate	Methyl arachidate	C₂₁H₄₂O₂	1120-28-1	326.557	lf (MeOH)	54.5	215[10]		1.4317[60]	vs bz, eth, EtOH, chl
7230	(Methyleneamino)acetonitrile		C₃H₄N₂	109-82-0	68.077		129				
7231	α-Methylenebenzeneacetic acid	Atropic acid	C₉H₈O₂	492-38-6	148.159	lf (al), nd (w)	106.5	dec 267			sl H₂O; s EtOH, eth, bz, chl, CS₂
7232	Methylenebis(4-cyclohexylisocyanate)		C₁₅H₂₂N₂O₂	5124-30-1	262.348	liq			1.066	1.4970[20]	
7233	4,4'-Methylenebis[2,6-di-tert-butylphenol]	Bis(3,5-di-tert-butyl-4-hydroxyphenyl)methane	C₂₉H₄₄O₂	118-82-1	424.658		154	289[40], 250[10]			
7234	4,4'-Methylenebis(N-methylaniline)	N,N'-Dimethyl-4,4'-diaminodiphenylmethane	C₁₅H₁₈N₂	1807-55-2	226.317						s ctc, CS₂
7235	Methylene blue		C₁₆H₁₈ClN₃S	61-73-4	319.852	dk grn cry or pow (chl-eth)					s H₂O, EtOH, chl; i eth; sl py
7236	Methylenecyclobutane		C₅H₈	1120-56-5	68.118	liq	-134.7	42.2	0.7401[20]	1.4210[20]	
7237	Methylenecyclohexane		C₇H₁₂	1192-37-6	96.170	liq	-106.7	102.5	0.8074[20]	1.4523[20]	i H₂O; s eth, bz, chl
7238	2-Methylenecyclohexanol		C₇H₁₂O	4065-80-9	112.169			83[13]	0.955[20]	1.4843[20]	
7239	Methylenecyclopentane		C₆H₁₀	1528-30-9	82.143			75.5	0.7787[20]	1.4355[20]	s bz, chl
7240	Methylenecyclopropene		C₄H₄	4095-06-1	52.075	solid stab at -196					
7241	2,4'-Methylenedianiline	2,4'-Diaminodiphenylmethane	C₁₃H₁₄N₂	1208-52-2	198.263	lf (bz)	88.5	222[9]			
7242	5,5'-Methylenedisalicylic acid		C₁₅H₁₂O₆	122-25-8	288.252	nd (bz)	243.5				vs ace, eth, EtOH
7243	5-Methylene-2(5H)-furanone	Protoanemonin	C₅H₄O₂	108-28-1	96.085	pa ye oil		73[11]			sl H₂O; s chl
7244	3-Methyleneheptane		C₈H₁₆	1632-16-2	112.213			120	0.7270[20]	1.4157[20]	i H₂O; vs eth, bz, peth
7245	4-Methylene-1-isopropylbicyclo[3.1.0]hexan-3-ol, [1S-(1α,3β,5α)]	4(10)-Thujene-3-ol	C₁₀H₁₆O	471-16-9	152.233			208	0.9488[19]	1.4871[25]	s eth
7246	4-Methylene-1-isopropylcyclohexene		C₁₀H₁₆	99-84-3	136.234			173.5	0.838[22]	1.4754[22]	
7247	2-Methylenepentanedinitrile	2,4-Dicyano-1-butene	C₆H₆N₂	1572-52-7	106.125			103[5]		1.4561[20]	s chl
7248	Methylene thiocyanate	Dithiocyanatomethane	C₃H₂N₂S₂	6317-18-6	130.191	solid	102				
7249	2-Methylene-1,3,3-trimethylindoline	Fischer's base	C₁₂H₁₅N	118-12-7	173.254			244			sl H₂O; s EtOH, eth, bz, chl
7250	N-Methylephedrine, [R-(R*,S*)]	(1R,2S)-N-Methylephedrine	C₁₁H₁₇NO	552-79-4	179.259	nd or pl (al, eth)	87.5				i H₂O; s EtOH, eth, MeOH
7251	Methylergonovine	Methylergometrine	C₂₀H₂₅N₃O₂	113-42-8	339.432	pr (MeOH,ace)	172				i H₂O; s EtOH, ace
7252	N-Methyl-1,2-ethanediamine		C₃H₁₀N₂	109-81-9	74.124			115	0.841[25]	1.4395[20]	
7253	N-Methyl-2-ethanolamine		C₃H₉NO	109-83-1	75.109			158	0.937[20]	1.4385[20]	msc H₂O, EtOH, eth
7254	1-(1-Methylethoxy)butane	Butyl isopropyl ether	C₇H₁₆O	1860-27-1	116.201			108	0.7594[15]	1.3870[15]	i H₂O; s EtOH, eth, ace, con sulf
7255	2-[2-(1-Methylethoxy)ethyl]pyridine		C₁₀H₁₅NO	70715-19-4	165.232			133[50]	0.9502[25]	1.4820[25]	vs H₂O
7256	1-(1-Methylethoxy)propane		C₆H₁₄O	627-08-7	102.174			83	0.7370[20]	1.376[21]	sl H₂O; vs EtOH; s eth, ace
7257	1-(1-Methylethoxy)-2-propanol	1-Isopropoxy-2-propanol	C₆H₁₄O₂	3944-36-3	118.174			137.5	0.879[20]	1.4070[20]	
7258	Methyl 2-ethylacetoacetate		C₇H₁₂O₃	51756-08-2	144.168			182	0.995[14]		vs ace, eth, EtOH
7259	5-(1-Methylethylidene)-1,3-cyclopentadiene		C₈H₁₀	2175-91-9	106.165		1.4	155; 49[11]	0.881[20]	1.5474[20]	
7260	1-Methyl-9H-fluorene		C₁₄H₁₂	1730-37-6	180.245		87				
7261	9-Methyl-9H-fluorene		C₁₄H₁₂	2523-37-7	180.245	pr	46.5	155[15]	1.0263[66]	1.610[66]	i H₂O; s EtOH, eth, ace, bz, chl
7262	Methyl fluorosulfonate		CH₃FO₃S	421-20-5	114.096	col liq	-95	93	1.412	1.3326[20]	
7263	N-Methylformamide		C₂H₅NO	123-39-7	59.067	liq	-3.8	199.51	1.011[19]	1.4319[20]	vs H₂O, ace, EtOH
7264	Methyl formate		C₂H₄O₂	107-31-3	60.052	liq	-99	31.7	0.9713[20]	1.3419[20]	vs H₂O; msc EtOH, eth, chl, MeOH
7265	Methyl 4-formylbenzoate		C₉H₈O₃	1571-08-0	164.158	nd (w)	63	265			
7266	2-Methylfuran		C₅H₆O	534-22-5	82.101	liq	-91.3	64.7	0.9132[20]	1.4342[20]	sl H₂O, ctc; s EtOH, eth

Methyl *cis*-13-docosenoate

Methyl dodecanoate

2-Methyldodecanoic acid

Methyl eicosanoate

(Methyleneamino)acetonitrile

α-Methylenebenzeneacetic acid

Methylenebis(4-cyclohexylisocyanate)

4,4'-Methylenebis[2,6-di-*tert*-butylphenol]

4,4'-Methylenebis(*N*-methylaniline)

Methylene blue

Methylenecyclobutane

Methylenecyclohexane

2-Methylenecyclohexanol

Methylenecyclopentane

Methylenecyclopropene

2,4'-Methylenedianiline

5,5'-Methylenedisalicylic acid

5-Methylene-2(5*H*)-furanone

3-Methyleneheptane

4-Methylene-1-isopropylbicyclo[3.1.0]hexan-3-ol, [1*S*-(1α,3β,5α)]

4-Methylene-1-isopropylcyclohexene

2-Methylenepentanedinitrile

Methylene thiocyanate

2-Methylene-1,3,3-trimethylindoline

N-Methylephedrine, [*R*-(*R**,*S**)]

Methylergonovine

N-Methyl-1,2-ethanediamine

N-Methyl-2-ethanolamine

1-(1-Methylethoxy)butane

2-[2-(1-Methylethoxy)ethyl]pyridine

1-(1-Methylethoxy)propane

1-(1-Methylethoxy)-2-propanol

Methyl 2-ethylacetoacetate

5-(1-Methylethylidene)-1,3-cyclopentadiene

1-Methyl-9*H*-fluorene

9-Methyl-9*H*-fluorene

Methyl fluorosulfonate

N-Methylformamide

Methyl formate

Methyl 4-formylbenzoate

2-Methylfuran

No.	Name	Synonym	Mol. Form.	CAS RN	Mol. Wt.	Physical Form	mp/°C	bp/°C	den/g cm⁻³	n_D	Solubility
7267	3-Methylfuran		C_5H_6O	930-27-8	82.101			65.5	0.923[18]	1.4330[19]	i H_2O; s EtOH, eth
7268	5-Methyl-2-furancarboxaldehyde		$C_6H_6O_2$	620-02-0	110.111			187; 89[26]	1.1072[18]	1.5264[20]	s H_2O; vs EtOH; msc eth; sl ctc
7269	Methyl 2-furancarboxylate	Methyl 2-furanoate	$C_6H_6O_3$	611-13-2	126.110			181.3	1.1786[21]	1.4860[20]	i H_2O; s EtOH, eth, bz, chl
7270	3-Methyl-2,5-furandione		$C_5H_4O_3$	616-02-4	112.084		7.5	213.5	1.2469[16]	1.4710[21]	vs ace, eth, EtOH
7271	N-Methyl-2-furanmethanamine		C_6H_9NO	4753-75-7	111.141			149	0.989[25]	1.4729[20]	
7272	5-Methyl-2-furanmethanol		$C_6H_8O_2$	3857-25-8	112.127			dec 195; 81[23]	1.0769[20]	1.4853[20]	vs eth, EtOH
7273	α-Methyl-2-furanmethanol		$C_6H_8O_2$	4208-64-4	112.127			162.5	1.0739[25]	1.4827[15]	
7274	5-Methyl-2(3H)-furanone		$C_5H_6O_2$	591-12-8	98.101	nd	18	56[12]	1.084[20]	1.4476[20]	s H_2O, EtOH, eth, CS_2; sl ctc
7275	5-Methyl-2(5H)-furanone		$C_5H_6O_2$	591-11-7	98.101		<-17	209; 98[15]	1.0810[20]	1.4454[20]	msc H_2O; s EtOH, eth
7276	Methylgermane		CH_6Ge	1449-65-6	90.70	col gas	-158	-23			
7277	Methyl β-D-glucopyranoside		$C_7H_{14}O_6$	709-50-2	194.182		109				s H_2O
7278	Methyl α-D-glucopyranoside	α-Methylglucoside	$C_7H_{14}O_6$	97-30-3	194.182	orth nd (al)	168	200[0.2]	1.46[30]		vs H_2O
7279	3-Methylglutaric acid	3-Methylpentanedioic acid	$C_6H_{10}O_4$	626-51-7	146.141		87	166[0.5]			s H_2O, EtOH, eth; sl bz, chl; i lig
7280	Methyl Green		$C_{27}H_{35}BrClN_3$	14855-76-6	516.944	grn pow (al)					vs H_2O
7281	Methyl heptadecanoate		$C_{18}H_{36}O_2$	1731-92-6	284.478	pl (al)	30	185[9], 152[0.05]			i H_2O; s EtOH, ace, ctc; vs eth, bz
7282	Methyl heptafluorobutanoate		$C_5H_3F_7O_2$	356-24-1	228.066	liq	-86	80	1.483[20]	1.295[20]	sl H_2O; s eth, ace
7283	6-Methyl-2-heptanamine, (±)	Octodrine	$C_8H_{19}N$	5984-58-7	129.244	visc liq		155	0.767[25]	1.4209[20]	
7284	N-Methyl-2-heptanamine		$C_8H_{19}N$	540-43-2	129.244			155			
7285	2-Methylheptane		C_8H_{18}	592-27-8	114.229	liq	-109.02	117.66	0.6980[20]	1.3949[20]	i H_2O; msc EtOH, ace, bz; s eth, ctc
7286	3-Methylheptane		C_8H_{18}	589-81-1	114.229	col liq	-120.48	118.9	0.7017[25]	1.3961[25]	i H_2O; s EtOH; eth; msc ace, bz, chl
7287	4-Methylheptane		C_8H_{18}	589-53-7	114.229	liq	-121.0	117.72	0.7046[20]	1.3979[20]	i H_2O; s eth; msc EtOH, ace, bz
7288	Methyl heptanoate		$C_8H_{16}O_2$	106-73-0	144.212	liq	-56	174	0.8815[20]	1.4152[20]	sl H_2O, ctc, ace; s EtOH, eth
7289	2-Methyl-1-heptanol, (±)		$C_8H_{18}O$	111675-77-5	130.228	col liq	-112	175.6	0.8022[20]	1.424[20]	
7290	3-Methyl-1-heptanol		$C_8H_{18}O$	1070-32-2	130.228	liq	-90	186; 101[20]	0.824[24]	1.4295[25]	
7291	4-Methyl-1-heptanol		$C_8H_{18}O$	817-91-4	130.228			188	0.8065[25]	1.4253[25]	vs EtOH
7292	5-Methyl-1-heptanol, (±)		$C_8H_{18}O$	111767-95-4	130.228	col liq	-104	186.6	0.8153[25]	1.4272[25]	
7293	6-Methyl-1-heptanol	Isooctyl alcohol	$C_8H_{18}O$	1653-40-3	130.228	liq	-106	188; 95.8[20]	0.8176[25]	1.4251[25]	i H_2O; s EtOH, eth
7294	2-Methyl-2-heptanol		$C_8H_{18}O$	625-25-2	130.228	liq	-50.4	156	0.8142[20]	1.4250[20]	i H_2O; s EtOH, eth
7295	3-Methyl-2-heptanol		$C_8H_{18}O$	31367-46-1	130.228			166.1	0.8177[25]	1.4199[25]	i H_2O; s EtOH, eth, ctc
7296	4-Methyl-2-heptanol		$C_8H_{18}O$	56298-90-9	130.228	col liq	-102	171.6	0.8027[20]	1.424[20]	
7297	5-Methyl-2-heptanol		$C_8H_{18}O$	54630-50-1	130.228	liq	-61	170	0.8174[21]		
7298	6-Methyl-2-heptanol		$C_8H_{18}O$	4730-22-7	130.228	liq	-105	174	0.8218[20]	1.4238[10]	
7299	2-Methyl-3-heptanol, (±)		$C_8H_{18}O$	100296-26-2	130.228	liq	-85	167.5	0.8235[20]	1.4265[20]	sl H_2O; s EtOH, eth, ctc
7300	3-Methyl-3-heptanol	2-Ethyl-2-hexanol	$C_8H_{18}O$	5582-82-1	130.228	liq	-83	163	0.8282[20]	1.4279[20]	i H_2O; s EtOH, eth, ctc
7301	4-Methyl-3-heptanol		$C_8H_{18}O$	14979-39-6	130.228	liq	-123	170	0.827[25]	1.4300[20]	
7302	5-Methyl-3-heptanol		$C_8H_{18}O$	18720-65-5	130.228	liq	-91.2	172	0.8425[25]	1.433[24]	
7303	6-Methyl-3-heptanol, (±)		$C_8H_{18}O$	100295-85-0	130.228	col liq	-61	169	0.8220[20]	1.4254[20]	
7304	2-Methyl-4-heptanol		$C_8H_{18}O$	21570-35-4	130.228	liq	-81	164	0.8207[20]	1.4203	vs eth, EtOH
7305	3-Methyl-4-heptanol		$C_8H_{18}O$	1838-73-9	130.228	liq		164.7	0.8329[25]	1.4211[25]	sl H_2O; s EtOH, eth, ctc
7306	4-Methyl-4-heptanol		$C_8H_{18}O$	598-01-6	130.228	liq	-82	161	0.8248[20]	1.4258[20]	i H_2O; s EtOH, eth, ctc
7307	6-Methyl-2-heptanol acetate		$C_{10}H_{20}O_2$	67952-57-2	172.265			187	0.8474[20]	1.413[20]	vs EtOH
7308	6-Methyl-2-heptanone		$C_8H_{16}O$	928-68-7	128.212			167	0.8151[20]	1.4162[20]	sl H_2O; vs EtOH, eth; msc ace, bz, chl
7309	5-Methyl-3-heptanone		$C_8H_{16}O$	541-85-5	128.212	liq		161			
7310	6-Methyl-3-heptanone		$C_8H_{16}O$	624-42-0	128.212			164	0.8304[20]	1.4209[20]	i H_2O; s EtOH, eth, bz, ctc
7311	2-Methyl-4-heptanone	Isobutyl propyl ketone	$C_8H_{16}O$	626-33-5	128.212			154	0.813[22]		i H_2O; s EtOH, eth

3-Methylfuran

5-Methyl-2-furancarboxaldehyde

Methyl 2-furancarboxylate

3-Methyl-2,5-furandione

N-Methyl-2-furanmethanamine

5-Methyl-2-furanmethanol

α-Methyl-2-furanmethanol

5-Methyl-2(3H)-furanone

5-Methyl-2(5H)-furanone

Methylgermane

Methyl β-D-glucopyranoside

Methyl α-D-glucopyranoside

3-Methylglutaric acid

Methyl Green

Methyl heptadecanoate

Methyl heptafluorobutanoate

6-Methyl-2-heptanamine, (±)

N-Methyl-2-heptanamine

2-Methylheptane

3-Methylheptane

4-Methylheptane

Methyl heptanoate

2-Methyl-1-heptanol, (±)

3-Methyl-1-heptanol

4-Methyl-1-heptanol

5-Methyl-1-heptanol, (±)

6-Methyl-1-heptanol

2-Methyl-2-heptanol

3-Methyl-2-heptanol

4-Methyl-2-heptanol

5-Methyl-2-heptanol

6-Methyl-2-heptanol

2-Methyl-3-heptanol, (±)

3-Methyl-3-heptanol

4-Methyl-3-heptanol

5-Methyl-3-heptanol

6-Methyl-3-heptanol, (±)

2-Methyl-4-heptanol

3-Methyl-4-heptanol

4-Methyl-4-heptanol

6-Methyl-2-heptanol acetate

6-Methyl-2-heptanone

5-Methyl-3-heptanone

6-Methyl-3-heptanone

2-Methyl-4-heptanone

No.	Name	Synonym	Mol. Form.	CAS RN	Mol. Wt.	Physical Form	mp/°C	bp/°C	den/ g cm⁻³	n_D	Solubility
7312	2-Methyl-1-heptene		C₈H₁₆	15870-10-7	112.213	liq	-90	119.3	0.7104²⁵	1.4123²⁰	
7313	6-Methyl-1-heptene		C₈H₁₆	5026-76-6	112.213			113.2	0.7079²⁵	1.4070²⁰	
7314	2-Methyl-2-heptene		C₈H₁₆	627-97-4	112.213			122.6	0.7200²⁵	1.4170²⁰	i H₂O; s eth, bz, ctc, chl
7315	cis-3-Methyl-2-heptene		C₈H₁₆	22768-19-0	112.213			122	0.725²⁵	1.419²⁰	
7316	6-Methyl-5-hepten-2-ol		C₈H₁₆O	1569-60-4	128.212			175	0.8545²⁰	1.4505²⁰	
7317	3-Methyl-5-hepten-2-one		C₈H₁₄O	38552-72-6	126.196			63²⁰	0.8463¹⁸	1.4345¹⁸	
7318	6-Methyl-5-hepten-2-one		C₈H₁₄O	110-93-0	126.196		173.5		0.8546¹⁶	1.4445²⁰	vs eth, EtOH
7319	2-Methylheptyl acetate, (±)		C₁₀H₂₀O₂	74112-36-0	172.265			195	0.8626¹⁴	1.4146²⁰	vs eth, EtOH
7320	2-Methyl-1,5-hexadiene		C₇H₁₂	4049-81-4	96.170	liq	-128.8	88.1	0.7153²⁵	1.4183²⁰	
7321	Methyl trans,trans-2,4-hexadienoate	Methyl sorbate	C₇H₁₀O₂	689-89-4	126.153	lf	15	180; 70²⁰	0.9777²⁰	1.5025²²	i H₂O; s EtOH, eth
7322	2-Methylhexanal		C₇H₁₄O	925-54-2	114.185	liq		141; 132⁶⁰			
7323	3-Methylhexanal	3-Methylcaproaldehyde	C₇H₁₄O	19269-28-4	114.185			143	0.8203²⁰	1.4122²⁰	i H₂O; s EtOH, eth
7324	3-Methyl-1-hexanamine		C₇H₁₇N	65530-93-0	115.217			149; 67⁴⁵	0.772²⁶	1.4249²⁵	
7325	4-Methyl-2-hexanamine		C₇H₁₇N	105-41-9	115.217			132.5	0.7655²⁰	1.4150²⁵	sl H₂O; vs EtOH, eth, chl, dil acid
7326	2-Methylhexane		C₇H₁₆	591-76-4	100.202	liq	-118.2	90.04	0.6787²⁰	1.3848²⁰	i H₂O; s EtOH; msc eth, ace, bz, lig, chl
7327	3-Methylhexane		C₇H₁₆	78918-91-9	100.202	liq	-119.4	92	0.687²¹	1.3854²⁵	i H₂O; s EtOH; msc eth, ace, bz, lig, chl
7328	5-Methyl-2,3-hexanedione	2-Methylhexa-4,5-dione	C₇H₁₂O₂	13706-86-0	128.169			138	0.908²²	1.4119²⁰	
7329	Methyl hexanoate	Methyl caproate	C₇H₁₄O₂	106-70-7	130.185	liq	-71	149.5	0.8846²⁰	1.4049²⁰	i H₂O; vs EtOH, eth; s ace, bz, ctc
7330	2-Methylhexanoic acid		C₇H₁₄O₂	4536-23-6	130.185			215.5	0.918²⁰	1.4193²⁰	vs ace, bz, eth, EtOH
7331	2-Methyl-1-hexanol, (±)		C₇H₁₆O	111768-04-8	116.201			164; 71¹⁵	0.826²⁰	1.4226²⁰	vs eth, EtOH
7332	5-Methyl-1-hexanol		C₇H₁₆O	627-98-5	116.201			169; 54¹⁵	0.8192²⁴	1.4175²⁰	vs eth, EtOH
7333	2-Methyl-2-hexanol		C₇H₁₆O	625-23-0	116.201			143	0.8119²⁰	1.4175²⁰	sl H₂O; msc EtOH, eth
7334	3-Methyl-2-hexanol		C₇H₁₆O	2313-65-7	116.201			151; 80⁵²	0.8220²⁵	1.4198¹⁸	i H₂O; vs EtOH, eth; s ace
7335	5-Methyl-2-hexanol		C₇H₁₆O	627-59-8	116.201			151; 78²⁸	0.814²⁰	1.4180²⁰	sl H₂O; s EtOH, eth
7336	3-Methyl-3-hexanol		C₇H₁₆O	597-96-6	116.201			143	0.8233²⁰	1.4231²⁰	sl H₂O; s EtOH, eth, ctc
7337	5-Methyl-2-hexanone	Methyl isopentyl ketone	C₇H₁₄O	110-12-3	114.185			144	0.888²⁰	1.4062²⁰	sl H₂O; msc EtOH; vs ace, bz; s ctc
7338	2-Methyl-3-hexanone	Propyl isopropyl ketone	C₇H₁₄O	7379-12-6	114.185			135	0.8091²⁰	1.4042²⁰	s EtOH, eth, chl; vs ace
7339	5-Methyl-2-hexanone oxime		C₇H₁₅NO	624-44-2	129.200			195.5	0.8881²⁰	1.4448²⁰	sl chl
7340	2-Methyl-1-hexene		C₇H₁₄	6094-02-6	98.186	liq	-102.8	92	0.7000²⁰	1.4035²⁰	
7341	3-Methyl-1-hexene		C₇H₁₄	3404-61-3	98.186			83.9	0.6871²⁵	1.3965²⁰	
7342	4-Methyl-1-hexene		C₇H₁₄	3769-23-1	98.186	liq	-141.5	86.7	0.6942²⁵	1.4000²⁰	
7343	5-Methyl-1-hexene		C₇H₁₄	3524-73-0	98.186			85.3	0.6877²⁵	1.3967²⁰	
7344	2-Methyl-2-hexene		C₇H₁₄	2738-19-4	98.186	liq	-130.4	95.4	0.7038²⁵	1.4106²⁰	
7345	cis-3-Methyl-2-hexene		C₇H₁₄	10574-36-4	98.186	liq	-118.5	95.6	0.712²⁰	1.4126²⁰	
7346	cis-4-Methyl-2-hexene		C₇H₁₄	3683-19-0	98.186			86.3	0.6952²⁵	1.4026²⁰	
7347	trans-4-Methyl-2-hexene		C₇H₁₄	3683-22-5	98.186	liq	-125.7	87.6	0.6925²⁵	1.4025²⁰	
7348	cis-5-Methyl-2-hexene		C₇H₁₄	13151-17-2	98.186			89.5	0.697²⁵	1.404²⁰	
7349	trans-5-Methyl-2-hexene		C₇H₁₄	7385-82-2	98.186	liq	-124.3	88.1	0.6883²⁵	1.4006²⁰	
7350	cis-2-Methyl-3-hexene		C₇H₁₄	15840-60-5	98.186			86	0.690²⁵	1.401²⁰	
7351	trans-2-Methyl-3-hexene		C₇H₁₄	692-24-0	98.186	liq	-141.6	85.9	0.6853²⁵	1.4001²⁰	
7352	cis-3-Methyl-3-hexene		C₇H₁₄	4914-89-0	98.186			95.4	0.7079²⁵	1.4126²⁰	
7353	trans-3-Methyl-3-hexene		C₇H₁₄	3899-36-3	98.186			93.5	0.7050²⁵	1.4109²⁰	
7354	Methyl 3-hexenoate		C₇H₁₂O₂	2396-78-3	128.169			67³⁴	0.9132²⁵	1.4240²³	
7355	5-Methyl-3-hexen-2-one	2-Oxo-5-methylhex-3-ene	C₇H₁₂O	5166-53-0	112.169			77⁵⁰, 65¹³	0.8549²⁸	1.4395²²	
7356	5-Methyl-5-hexen-2-one		C₇H₁₂O	3240-09-3	112.169			150	0.8460²⁰	1.4348²⁰	vs ace, eth, EtOH
7357	5-Methyl-1-hexyne		C₇H₁₂	2203-80-7	96.170	liq	-125	92	0.7274²⁰	1.4059⁻²⁰	i H₂O; s EtOH, eth, bz, chl, peth
7358	5-Methyl-2-hexyne		C₇H₁₂	53566-37-3	96.170	liq	-92.9	102.5	0.7378²⁰	1.4176²⁰	i H₂O; s eth, ace, bz, chl, peth
7359	2-Methyl-3-hexyne		C₇H₁₂	36566-80-0	96.170	liq	-116.7	95.2	0.7263²⁰	1.4120²⁰	vs bz, eth, chl, peth
7360	Methyl 2-hexynoate		C₇H₁₀O₂	18937-79-6	126.153			80²³	0.9648²⁵		

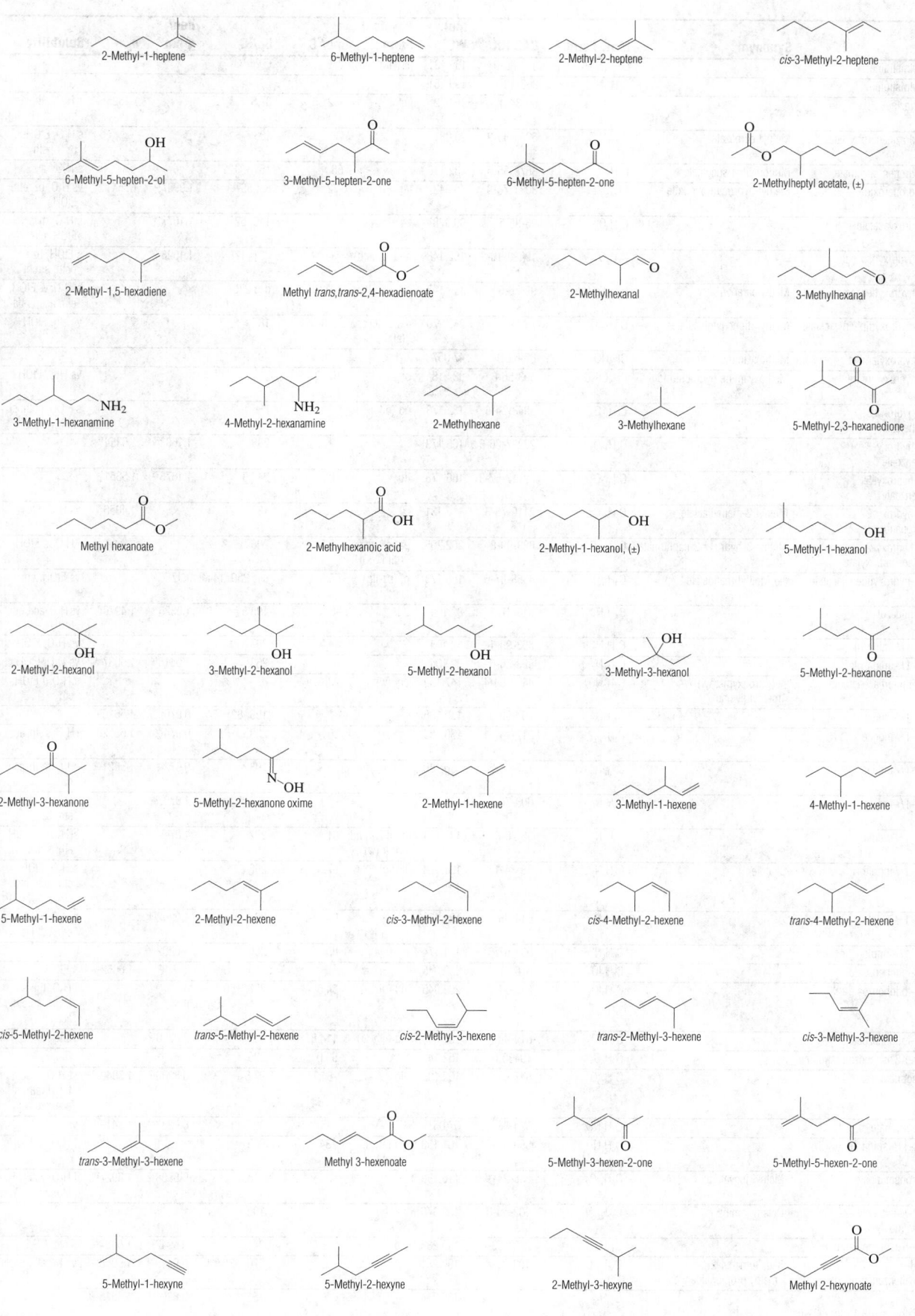

2-Methyl-1-heptene

6-Methyl-1-heptene

2-Methyl-2-heptene

cis-3-Methyl-2-heptene

6-Methyl-5-hepten-2-ol

3-Methyl-5-hepten-2-one

6-Methyl-5-hepten-2-one

2-Methylheptyl acetate, (±)

2-Methyl-1,5-hexadiene

Methyl trans,trans-2,4-hexadienoate

2-Methylhexanal

3-Methylhexanal

3-Methyl-1-hexanamine

4-Methyl-2-hexanamine

2-Methylhexane

3-Methylhexane

5-Methyl-2,3-hexanedione

Methyl hexanoate

2-Methylhexanoic acid

2-Methyl-1-hexanol, (±)

5-Methyl-1-hexanol

2-Methyl-2-hexanol

3-Methyl-2-hexanol

5-Methyl-2-hexanol

3-Methyl-3-hexanol

5-Methyl-2-hexanone

2-Methyl-3-hexanone

5-Methyl-2-hexanone oxime

2-Methyl-1-hexene

3-Methyl-1-hexene

4-Methyl-1-hexene

5-Methyl-1-hexene

2-Methyl-2-hexene

cis-3-Methyl-2-hexene

cis-4-Methyl-2-hexene

trans-4-Methyl-2-hexene

cis-5-Methyl-2-hexene

trans-5-Methyl-2-hexene

cis-2-Methyl-3-hexene

trans-2-Methyl-3-hexene

cis-3-Methyl-3-hexene

trans-3-Methyl-3-hexene

Methyl 3-hexenoate

5-Methyl-3-hexen-2-one

5-Methyl-5-hexen-2-one

5-Methyl-1-hexyne

5-Methyl-2-hexyne

2-Methyl-3-hexyne

Methyl 2-hexynoate

No.	Name	Synonym	Mol. Form.	CAS RN	Mol. Wt.	Physical Form	mp/°C	bp/°C	den/ g cm⁻³	n_D	Solubility
7361	*L*-1-Methylhistidine		C₇H₁₁N₃O₂	332-80-9	169.181	pl (DMF aq)	249				
7362	*L*-3-Methylhistidine		C₇H₁₁N₃O₂	368-16-1	169.181		250				
7363	Methylhydrazine		CH₆N₂	60-34-4	46.072	liq	-52.36	87.5		1.4325²⁰	s H₂O, eth, ctc; msc EtOH; i lig
7364	Methyl hydrazinecarboxylate	Methyl carbazate	C₂H₆N₂O₂	6294-89-9	90.081		73	108¹²			s H₂O, EtOH; sl bz; i peth
7365	Methyl hydrogen succinate	Monomethyl succinate	C₅H₈O₄	3878-55-5	132.116		58	151²⁰, 122⁴			s H₂O
7366	Methyl hydroperoxide	Methyl hydrogen peroxide	CH₄O₂	3031-73-0	48.042	liq	-72	86; 39⁶⁵	1.9967¹⁵	1.3641¹⁵	vs H₂O, bz, eth, EtOH
7367	Methyl hydroxyacetate		C₃H₆O₃	96-35-5	90.078			149; 52¹⁷	1.1677¹⁸		s H₂O; msc EtOH, eth
7368	Methyl 3-hydroxybenzoate		C₈H₈O₃	19438-10-9	152.148	nd (bz-peth)	73	281; 178¹⁷	1.1528¹⁰⁰		s EtOH, bz, peth; sl chl
7369	Methyl 4-hydroxybenzoate	Methylparaben	C₈H₈O₃	99-76-3	152.148	nd (dil al)	131	dec 275			sl H₂O; vs EtOH, eth, ace; s tfa
7370	Methyl α-hydroxydiphenylacetate	Methyl diphenylglycolate	C₁₅H₁₄O₃	76-89-1	242.270	mcl or tcl cry (al)	75.8	187¹³			vs eth, EtOH
7371	*O*-Methylhydroxylamine	Methoxyamine	CH₅NO	67-62-9	47.057			49			
7372	*O*-Methylhydroxylamine hydrochloride	Methoxyamine hydrochloride	CH₆ClNO	593-56-6	83.518	pr	150.0				vs H₂O, EtOH
7373	Methyl 4-hydroxy-3-methoxybenzoate		C₉H₁₀O₄	3943-74-6	182.173	nd (dil al)	64	286			s EtOH, peth; sl chl
7374	Methyl 2-hydroxy-3-methylbenzoate		C₉H₁₀O₃	23287-26-5	166.173		29	235	1.1683²⁵	1.5354¹⁶	
7375	Methyl 2-hydroxy-5-methylbenzoate		C₉H₁₀O₃	22717-57-3	166.173	liq	-1	244.5	1.1673²⁵	1.5351¹⁵	
7376	Methyl 2-hydroxy-2-methylpropanoate	Methyl 2-methyllactate	C₅H₁₀O₃	2110-78-3	118.131			137		1.4056²⁰	vs H₂O, EtOH
7377	Methyl 3-hydroxy-2-naphthalenecarboxylate	Methyl 3-hydroxy-2-naphthoate	C₁₂H₁₀O₃	883-99-8	202.205	pa ye orth nd (dil MeOH)	75.5	206			i H₂O; s EtOH
7378	Methyl α-hydroxyphenylacetate, (±)	(±)-Methyl mandelate	C₉H₁₀O₃	4358-87-6	166.173	pl (bz-lig)	58	dec 250; 144²⁰	1.1756²⁰		vs EtOH, chl
7379	1-Methylimidazol		C₄H₆N₂	616-47-7	82.104	liq	-6	195.5	1.0325²⁰	1.4970²⁰	vs H₂O, ace, eth, EtOH
7380	2-Methyl-1*H*-imidazole		C₄H₆N₂	693-98-1	82.104		144	267			vs H₂O, EtOH
7381	4-Methyl-1*H*-imidazole		C₄H₆N₂	822-36-6	82.104		56	263	1.0416¹⁴	1.5037¹⁴	vs H₂O, EtOH
7382	*N*-Methyliminodiacetic acid	*N*-(Carboxymethyl)-*N*-methylglycine	C₅H₉NO₄	4408-64-4	147.130	cry (w)	226				s H₂O; i EtOH, eth
7383	1-Methyl-1*H*-indene		C₁₀H₁₀	767-59-9	130.186			199; 82¹⁵	0.970²⁵	1.5616²⁰	
7384	2-Methyl-1*H*-indene		C₁₀H₁₀	2177-47-1	130.186		80	208	0.974²⁵	1.5652²⁰	i H₂O; s eth, ace, bz
7385	3-Methyl-1*H*-indene		C₁₀H₁₀	767-60-2	130.186			198	0.972²⁵	1.5621²⁰	i H₂O; s eth, ace, bz
7386	1-Methyl-1*H*-indole		C₉H₉N	603-76-9	131.174			237	1.0707²⁵		i H₂O; s EtOH, eth, bz
7387	2-Methyl-1*H*-indole		C₉H₉N	95-20-5	131.174	pl (dil al) nd or lf (w)	61	272	1.07²⁰		sl H₂O; vs EtOH, eth; s ace, bz
7388	3-Methyl-1*H*-indole	Skatole	C₉H₉N	83-34-1	131.174	lf (lig)	97.5	266			s H₂O, EtOH, eth, ace, bz, chl
7389	5-Methyl-1*H*-indole		C₉H₉N	614-96-0	131.174		60	267	1.0202⁷⁸		s H₂O, EtOH, eth, bz, lig
7390	7-Methyl-1*H*-indole		C₉H₉N	933-67-5	131.174		85	266	1.0202¹⁰⁰		
7391	Methyl 2-iodobenzoate		C₈H₇IO₂	610-97-9	262.045			280; 146¹⁶		1.6052²⁰	s EtOH
7392	Methyl 3-iodobenzoate		C₈H₇IO₂	618-91-7	262.045	nd (dil al)	54.5	277; 150¹⁸			i H₂O, lig; s EtOH; vs eth, ace
7393	Methyl 4-iodobenzoate		C₈H₇IO₂	619-44-3	262.045	nd (eth-al)	114.8	sub	2.0200¹⁰		s EtOH, eth
7394	5-Methyl-1,3-isobenzofurandione		C₉H₆O₃	19438-61-0	162.142		93.0	295			
7395	Methyl isobutanoate		C₅H₁₀O₂	547-63-7	102.132	liq	-84.7	92.5	0.8906²⁰	1.3840²⁰	sl H₂O; msc EtOH, eth; s ace, ctc
7396	Methyl isocyanate		C₂H₃NO	624-83-9	57.051	liq	-45	39.5	0.9230²⁷	1.3419¹⁸	vs H₂O
7397	2-Methyl-1*H*-isoindole-1,3(2*H*)-dione		C₉H₇NO₂	550-44-7	161.158	nd (al), lf (sub)	134	286			i H₂O; sl EtOH
7398	Methyl isopentanoate	Methyl isovalerate	C₆H₁₂O₂	556-24-1	116.158			116.5	0.8808²⁰	1.3927²⁰	i H₂O; vs EtOH, eth, ace
7399	6-Methyl-*N*-isopentyl-2-heptanamine	Octamylamine	C₁₃H₂₉N	502-59-0	199.376			100⁷			
7400	2-Methyl-5-isopropylaniline		C₁₀H₁₅N	2051-53-8	149.233	liq	-16	241	0.9942²⁰	1.5387²⁰	s ctc, CS₂
7401	α-Methyl-4-isopropylbenzenepropanal	3-*p*-Cumenyl-2-methylpropionaldehyde	C₁₃H₁₈O	103-95-7	190.281			270; 135⁹⁹	0.9459²⁰	1.5068²⁰	vs bz, eth, EtOH
7402	2-Methyl-5-isopropylbicyclo[3.1.0]hex-2-ene		C₁₀H₁₆	2867-05-2	136.234			151	0.8301²⁰	1.4515²⁰	

L-1-Methylhistidine

L-3-Methylhistidine

Methylhydrazine

Methyl hydrazinecarboxylate

Methyl hydrogen succinate

Methyl hydroperoxide

Methyl hydroxyacetate

Methyl 3-hydroxybenzoate

Methyl 4-hydroxybenzoate

Methyl α-hydroxydiphenylacetate

O-Methylhydroxylamine

O-Methylhydroxylamine hydrochloride

Methyl 4-hydroxy-3-methoxybenzoate

Methyl 2-hydroxy-3-methylbenzoate

Methyl 2-hydroxy-5-methylbenzoate

Methyl 2-hydroxy-2-methylpropanoate

Methyl 3-hydroxy-2-naphthalenecarboxylate

Methyl α-hydroxyphenylacetate, (±)

1-Methylimidazol

2-Methyl-1H-imidazole

4-Methyl-1H-imidazole

N-Methyliminodiacetic acid

1-Methyl-1H-indene

2-Methyl-1H-indene

3-Methyl-1H-indene

1-Methyl-1H-indole

2-Methyl-1H-indole

3-Methyl-1H-indole

5-Methyl-1H-indole

7-Methyl-1H-indole

Methyl 2-iodobenzoate

Methyl 3-iodobenzoate

Methyl 4-iodobenzoate

5-Methyl-1,3-isobenzofurandione

Methyl isobutanoate

Methyl isocyanate

2-Methyl-1H-isoindole-1,3(2H)-dione

Methyl isopentanoate

6-Methyl-N-isopentyl-2-heptanamine

2-Methyl-5-isopropylaniline

α-Methyl-4-isopropylbenzenepropanal

2-Methyl-5-isopropylbicyclo[3.1.0]hex-2-ene

No.	Name	Synonym	Mol. Form.	CAS RN	Mol. Wt.	Physical Form	mp/°C	bp/°C	den/ g cm⁻³	n_D	Solubility
7403	2-Methyl-5-isopropyl-2,5-cyclohexadiene-1,4-dione		$C_{10}H_{12}O_2$	490-91-5	164.201		45.5	232			s chl
7404	cis-1-Methyl-4-isopropylcyclohexane		$C_{10}H_{20}$	6069-98-3	140.266	liq	-89.9	172	0.8039²⁰	1.4431²⁰	i H₂O; vs EtOH, eth; s bz, peth
7405	trans-1-Methyl-4-isopropylcyclohexane	trans-p-Menthane	$C_{10}H_{20}$	1678-82-6	140.266	oil	-86.3	170.6	0.7928²⁰	1.4366²⁰	vs bz, eth, EtOH, lig
7406	1-Methyl-4-isopropylcyclohexanol		$C_{10}H_{20}O$	21129-27-1	156.265			208.5	0.90²⁰	1.4619²⁰	
7407	5-Methyl-2-isopropylcyclohexanol, [1S-(1α,2β,5α)]-	(+)-Menthol	$C_{10}H_{20}O$	15356-60-2	156.265		39	103⁹			vs ace, bz, eth, EtOH
7408	5-Methyl-2-isopropylcyclohexanol, [1R-(1α,2β,5α)]-	(-)-Menthol	$C_{10}H_{20}O$	2216-51-5	156.265	nd (MeOH)	43	216	0.903¹⁵	1.460²²	sl H₂O; vs EtOH, eth, ace, bz; s peth
7409	5-Methyl-2-isopropylcyclohexanol, [1S-(1α,2α,5β)]-	(+)-Neomenthol	$C_{10}H_{20}O$	2216-52-6	156.265	oil	-22	211.7	0.897²²	1.4600²⁰	vs ace, EtOH
7410	5-Methyl-2-isopropylcyclohexanol, [1S-(1α,2β,5β)]-	(+)-Isomenthol	$C_{10}H_{20}O$	23283-97-8	156.265	nd(dil al)	82.5	218			vs eth, EtOH
7411	5-Methyl-2-isopropylcyclohexanol acetate, [1R-(1α,2α,5β)]		$C_{12}H_{22}O_2$	2623-23-6	198.302			222; 109¹⁰	0.9244²⁰	1.4469²⁰	
7412	cis-5-Methyl-2-isopropylcyclohexanone	Menthone	$C_{10}H_{18}O$	491-07-6	154.249			205; 89¹⁵	0.8995²⁰	1.4527²⁰	
7413	trans-5-Methyl-2-isopropylcyclohexanone, (2S)	l-Menthone	$C_{10}H_{18}O$	14073-97-3	154.249	liq	-6	207	0.8954²⁰	1.4505²⁰	sl H₂O; msc EtOH, eth, bz, CS₂; s ace
7414	1-Methyl-4-isopropylcyclohexene		$C_{10}H_{18}$	5502-88-5	138.250			174.5	0.8457¹⁵	1.4735²⁰	
7415	3-Methyl-6-isopropyl-2-cyclohexen-1-ol		$C_{10}H_{18}O$	491-04-3	154.249			97¹⁵·⁵	0.9119²⁵	1.4729²⁵	
7416	4-Methyl-1-isopropyl-3-cyclohexen-1-ol		$C_{10}H_{18}O$	562-74-3	154.249			209	0.926²⁰	1.4785¹⁹	
7417	5-Methyl-2-isopropylcyclohexyl ethoxyacetate, (1α,2β,5α)		$C_{14}H_{26}O_3$	579-94-2	242.354			155²⁰, 144¹⁴	0.9545²⁰		vs eth, EtOH, chl
7418	1-Methyl-4-isopropyl-2-nitrobenzene		$C_{10}H_{13}NO_2$	943-15-7	179.216			126¹⁰	1.0744²⁰	1.5301²⁰	vs eth, EtOH
7419	1-Methyl-4-isopropyl-7-oxabicyclo[2.2.1]heptane		$C_{10}H_{18}O$	470-67-7	154.249		1	173.5	0.8997²⁰	1.4562²⁰	sl H₂O; msc EtOH, eth; s bz, lig
7420	1-Methyl-7-isopropylphenanthrene	Retene	$C_{18}H_{18}$	483-65-8	234.336		101	390	1.035²⁵		i H₂O; s EtOH, eth, bz, CS₂, HOAc
7421	4-Methyl-2-isopropylphenol		$C_{10}H_{14}O$	4427-56-9	150.217	nd (HOAc)	36.5	228.5	0.9910²⁰	1.5275²⁰	sl H₂O; s EtOH, bz, chl
7422	5-Methyl-2-isopropylphenyl acetate	Thymol, acetate	$C_{12}H_{16}O_2$	528-79-0	192.254			245	1.009⁹		vs bz, eth, EtOH, chl
7423	1-Methylisoquinoline	Isoquinaldine	$C_{10}H_9N$	1721-93-3	143.185		10	248	1.0777²⁰	1.6095²⁰	sl H₂O; s eth, ace, bz
7424	3-Methylisoquinoline		$C_{10}H_9N$	1125-80-0	143.185	cry (eth)	68	249			sl H₂O, chl; s eth, ace
7425	Methyl isothiocyanate		C_2H_3NS	556-61-6	73.117		36	119	1.0691³⁷	1.5258	sl H₂O; msc EtOH; vs eth
7426	5-Methyl-3-isoxazolamine		$C_4H_6N_2O$	1072-67-9	98.103		62				
7427	4-Methylisoxazole		C_4H_5NO	6454-84-8	83.089	liq		127			
7428	5-Methylisoxazole		C_4H_5NO	5765-44-6	83.089			122	1.023²⁰	1.4386²⁰	s DMSO
7429	Methyl lactate, (±)	Methyl 2-hydroxypropanoate, (±)	$C_4H_8O_3$	2155-30-8	104.105	oil		144.8	1.0928²⁰	1.4141²⁰	vs H₂O, eth, EtOH
7430	Methyl linoleate		$C_{19}H_{34}O_2$	112-63-0	294.472		-35	215²⁰	0.8886¹⁰	1.4638²⁰	vs eth, EtOH
7431	Methyl linolenate		$C_{19}H_{32}O_2$	301-00-8	292.456		-45.5	207¹⁴, 182³	0.895²⁵	1.4709²⁰	
7432	Methyl magnesium bromide	Bromomethylmagnesium	CH_3BrMg	75-16-1	119.244						s eth, thf; i hx, bz
7433	Methylmagnesium chloride	Chloromethylmagnesium	CH_3ClMg	676-58-4	74.793	stab in thf soln					i peth, bz
7434	Methylmalonic acid		$C_4H_6O_4$	516-05-2	118.089	nd (bz-AcOEt) pr (eth-bz)	135 dec		1.455²⁰		vs H₂O, EtOH, eth; sl bz, tfa; s AcOEt
7435	Methyl mercaptoacetate		$C_3H_6O_2S$	2365-48-2	106.144			42¹⁰		1.4657²⁰	vs eth, EtOH
7436	Methyl 3-mercaptopropanoate		$C_4H_8O_2S$	2935-90-2	120.171			54¹⁴	1.085²⁵	1.4640²⁰	
7437	Methylmercuric dicyanamide	1-Cyano-3-(methylmercurio) guanidine	$C_3H_6HgN_4$	502-39-6	298.70		157				
7438	Methyl methacrylate		$C_5H_8O_2$	80-62-6	100.117	liq	-47.55	100.5	0.9377²⁵	1.4142²⁰	sl H₂O; msc EtOH, eth, ace; s chl
7439	Methyl methanesulfonate		$C_2H_6O_3S$	66-27-3	110.132		20	202.5	1.2943²⁰	1.4138²⁰	
7440	Methyl methoxyacetate		$C_4H_8O_3$	6290-49-9	104.105			131	1.0511²⁰	1.3962²⁰	sl H₂O; vs EtOH, eth, ace

2-Methyl-5-isopropyl-2,5-cyclohexadiene-1,4-dione

cis-1-Methyl-4-isopropylcyclohexane

trans-1-Methyl-4-isopropylcyclohexane

1-Methyl-4-isopropylcyclohexanol

5-Methyl-2-isopropylcyclohexanol, [1S-(1α,2β,5α)]-

5-Methyl-2-isopropylcyclohexanol, [1R-(1α,2β,5α)]-

5-Methyl-2-isopropylcyclohexanol, [1S-(1α,2α,5β)]-

5-Methyl-2-isopropylcyclohexanol, [1S-(1α,2β,5β)]-

5-Methyl-2-isopropylcyclohexanol acetate, [1R-(1α,2α,5β)]

cis-5-Methyl-2-isopropylcyclohexanone

trans-5-Methyl-2-isopropylcyclohexanone, (2S)

1-Methyl-4-isopropylcyclohexene

3-Methyl-6-isopropyl-2-cyclohexen-1-ol

4-Methyl-1-isopropyl-3-cyclohexen-1-ol

5-Methyl-2-isopropylcyclohexyl ethoxyacetate, (1α,2β,5α)

1-Methyl-4-isopropyl-2-nitrobenzene

1-Methyl-4-isopropyl-7-oxabicyclo[2.2.1]heptane

1-Methyl-7-isopropylphenanthrene

4-Methyl-2-isopropylphenol

5-Methyl-2-isopropylphenyl acetate

1-Methylisoquinoline

3-Methylisoquinoline

Methyl isothiocyanate

5-Methyl-3-isoxazolamine

4-Methylisoxazole

5-Methylisoxazole

Methyl lactate, (±)

Methyl linoleate

Methyl linolenate

Methyl magnesium bromide

Methylmagnesium chloride

Methylmalonic acid

Methyl mercaptoacetate

Methyl 3-mercaptopropanoate

Methylmercuric dicyanamide

Methyl methacrylate

Methyl methanesulfonate

Methyl methoxyacetate

No.	Name	Synonym	Mol. Form.	CAS RN	Mol. Wt.	Physical Form	mp/°C	bp/°C	den/ g cm⁻³	n_D	Solubility
7441	Methyl 2-methoxybenzoate		$C_9H_{10}O_3$	606-45-1	166.173			246.5	1.1571[19]	1.534[19]	i H_2O; s EtOH
7442	Methyl 3-methoxybenzoate		$C_9H_{10}O_3$	5368-81-0	166.173			248	1.1310[20]	1.5224[20]	i H_2O; s EtOH
7443	Methyl 4-methoxybenzoate		$C_9H_{10}O_3$	121-98-2	166.173	lf (al or eth)	49	244			i H_2O; s EtOH, eth, chl
7444	Methyl 3-methoxy-2-(methylamino)benzoate	Damascenine	$C_{10}H_{13}NO_3$	483-64-7	195.215	pr (al)	28	271; 147[10]			vs bz, eth, EtOH, lig
7445	Methyl 3-methoxypropanoate		$C_5H_{10}O_3$	3852-09-3	118.131			142.8	1.0139[15]	1.4030[20]	
7446	Methyl 2-methylacetoacetate		$C_6H_{10}O_3$	17094-21-2	130.141			177.4	1.0217[25]	1.416[24]	vs eth, EtOH
7447	Methyl 2-(methylamino)benzoate		$C_9H_{11}NO_2$	85-91-6	165.189	cry (peth)	19	255	1.120[15]	1.5839[15]	i H_2O; s EtOH, eth
7448	Methyl 2-methylbenzoate		$C_9H_{10}O_2$	89-71-4	150.174		<-50	215	1.068[20]		i H_2O; msc EtOH, eth
7449	Methyl 3-methylbenzoate		$C_9H_{10}O_2$	99-36-5	150.174			221	1.061[20]		i H_2O; s EtOH; sl ctc
7450	Methyl 4-methylbenzoate		$C_9H_{10}O_2$	99-75-2	150.174	cry (aq MeOH, peth)	33.2	220			i H_2O; vs EtOH, eth
7451	Methyl 2-methyl-2-butenoate, (E)		$C_6H_{10}O_2$	6622-76-0	114.142			139	0.9349[12]	1.4370[20]	
7452	Methyl 3-methyl-2-butenoate		$C_6H_{10}O_2$	924-50-5	114.142		114	136.5	0.9337[20]	1.432[20]	
7453	3-Methyl-4-methylenehexane		C_8H_{16}	3404-67-9	112.213			112.5	0.725[25]	1.4142[20]	
7454	2-Methyl-5-(1-methylethenyl) cyclohexanone, (2R-trans)		$C_{10}H_{16}O$	5524-05-0	152.233			221.5	0.928[19]	1.4724	vs ace, eth
7455	5-Methyl-2-(1-methylethylidene) cyclohexanone		$C_{10}H_{16}O$	15932-80-6	152.233			93[10]	0.9367[20]	1.4869[20]	
7456	3-Methyl-6-(1-methylethylidene)-2-cyclohexen-1-one	Piperitenone	$C_{10}H_{14}O$	491-09-8	150.217			120[14]	0.9774[20]	1.5294[20]	vs EtOH, eth
7457	1-Methyl-4-(5-methyl-1-methylene-4-hexenyl) cyclohexene, (S)		$C_{15}H_{24}$	495-61-4	204.352			129[10]	0.8673[20]	1.4880[20]	
7458	N-Methyl-N-(2-methylphenyl) acetamide		$C_{10}H_{13}NO$	573-26-2	163.216		55.5	260			s EtOH, chl
7459	4-Methyl-N-(4-methylphenyl) aniline		$C_{14}H_{15}N$	620-93-9	197.276	nd (peth)	79.8	330.5			vs eth, peth
7460	2-Methyl-3-(2-methylphenyl)-4(3H)-quinazolinone	Methaqualone	$C_{16}H_{14}N_2O$	72-44-6	250.294		120				vs eth, EtOH, chl
7461	Methyl 3-(methylthio)propanoate	2-Methoxycarbonylethyl methyl sulfide	$C_5H_{10}O_2S$	13532-18-8	134.197			75[13], 69[11]	1.077[25]	1.4650[20]	
7462	1-Methyl-4-(1-methylvinyl) benzene		$C_{10}H_{12}$	1195-32-0	132.202	liq	-20	185.3	0.8936[23]	1.5283[23]	
7463	1-Methyl-4-(1-methylvinyl) cyclohexanol	β-Terpineol	$C_{10}H_{18}O$	138-87-4	154.249	nd	32.5	210; 90[10]	0.917[20]	1.4747[20]	
7464	5-Methyl-2-(1-methylvinyl) cyclohexanol, [1R-(1α,2β,5α)]		$C_{10}H_{18}O$	89-79-2	154.249		78	93[14]	0.911[20]	1.4723[20]	sl H_2O; s EtOH, eth
7465	5-Methyl-2-(1-methylvinyl) cyclohexanol acetate, [1R-(1α,2β,5α)]		$C_{12}H_{20}O_2$	57576-09-7	196.286		85	113[8]	0.925[25]	1.4566[20]	
7466	trans-5-Methyl-2-(1-methylvinyl) cyclohexanone		$C_{10}H_{16}O$	29606-79-9	152.233			100[18]	0.9198[20]	1.4675[20]	
7467	2-Methyl-5-(1-methylvinyl)-2-cyclohexen-1-ol		$C_{10}H_{16}O$	99-48-9	152.233			228	0.9484[25]	1.4942[25]	
7468	4-Methylmorpholine		$C_5H_{11}NO$	109-02-4	101.147	liq	-64.40	116	0.9051[20]	1.4332[20]	s H_2O, EtOH, eth
7469	α-Methyl-4-morpholineethanol		$C_7H_{15}NO_2$	2109-66-2	145.200			121[18], 93[13]	1.0174[20]	1.4638[20]	vs H_2O, ace, bz, EtOH
7470	1-Methylnaphthalene		$C_{11}H_{10}$	90-12-0	142.197	liq	-30.43	244.7	1.0202[20]	1.6170[20]	i H_2O; vs EtOH, eth; s bz
7471	2-Methylnaphthalene		$C_{11}H_{10}$	91-57-6	142.197	mcl (al)	34.6	241.1	1.0058[20]	1.6015[40]	i H_2O; vs EtOH, eth; s bz, chl
7472	Methyl 1-naphthalenecarboxylate	Methyl 1-naphthoate	$C_{12}H_{10}O_2$	2459-24-7	186.206		59.5	168[20], 101[0.04]	1.1290[20]	1.6086[20]	vs bz, EtOH
7473	Methyl 2-naphthalenecarboxylate	Methyl 2-naphthoate	$C_{12}H_{10}O_2$	2459-25-8	186.206	lf (MeOH)	77	290			vs bz, eth, EtOH, chl
7474	2-Methyl-1,4-naphthalenediol diacetate	Menadiol diacetate	$C_{15}H_{14}O_4$	573-20-6	258.270	pr (al)	113				vs EtOH
7475	2-Methyl-1,4-naphthalenedione	Menadione	$C_{11}H_8O_2$	58-27-5	172.181	ye nd (al, peth)	107				i H_2O; sl EtOH, HOAc; s eth, bz, chl
7476	Methyl-1-naphthylamine	N-Methyl-1-naphthalenamine	$C_{11}H_{11}N$	2216-68-4	157.212	oil	174	294.5		1.6722[20]	vs eth, EtOH
7477	Methyl nitrate		CH_3NO_3	598-58-3	77.040	exp gas	-83.0	exp 64.6	1.2075[20]	1.3748[20]	sl H_2O; s EtOH, eth
7478	Methyl nitrite		CH_3NO_2	624-91-9	61.041	ye gas	-16	-12	0.991[15]		s EtOH, eth
7479	Methyl nitroacetate		$C_3H_5NO_4$	2483-57-0	119.077			107[28]	1.320[0]		
7480	2-Methyl-3-nitroaniline		$C_7H_8N_2O_2$	603-83-8	152.151	ye orth nd (w), ye lf (al)	92	305	1.3780[15]		sl H_2O; s EtOH, eth, bz, chl

Methyl 2-methoxybenzoate

Methyl 3-methoxybenzoate

Methyl 4-methoxybenzoate

Methyl 3-methoxy-2-(methylamino)benzoate

Methyl 3-methoxypropanoate

Methyl 2-methylacetoacetate

Methyl 2-(methylamino)benzoate

Methyl 2-methylbenzoate

Methyl 3-methylbenzoate

Methyl 4-methylbenzoate

Methyl 2-methyl-2-butenoate, (E)

Methyl 3-methyl-2-butenoate

3-Methyl-4-methylenehexane

2-Methyl-5-(1-methylethenyl)cyclohexanone, (2R-trans)

5-Methyl-2-(1-methylethylidene)cyclohexanone

3-Methyl-6-(1-methylethylidene)-2-cyclohexen-1-one

1-Methyl-4-(5-methyl-1-methylene-4-hexenyl)cyclohexene, (S)

N-Methyl-N-(2-methylphenyl)acetamide

4-Methyl-N-(4-methylphenyl)aniline

2-Methyl-3-(2-methylphenyl)-4(3H)-quinazolinone

Methyl 3-(methylthio)propanoate

1-Methyl-4-(1-methylvinyl)benzene

1-Methyl-4-(1-methylvinyl)cyclohexanol

5-Methyl-2-(1-methylvinyl)cyclohexanol, [1R-(1α,2β,5α)]

5-Methyl-2-(1-methylvinyl)cyclohexanol acetate, [1R-(1α,2β,5α)]

trans-5-Methyl-2-(1-methylvinyl)cyclohexanone

2-Methyl-5-(1-methylvinyl)-2-cyclohexen-1-ol

4-Methylmorpholine

α-Methyl-4-morpholineethanol

1-Methylnaphthalene

2-Methylnaphthalene

Methyl 1-naphthalenecarboxylate

Methyl 2-naphthalenecarboxylate

2-Methyl-1,4-naphthalenediol diacetate

2-Methyl-1,4-naphthalenedione

Methyl-1-naphthylamine

Methyl nitrate

Methyl nitrite

Methyl nitroacetate

2-Methyl-3-nitroaniline

No.	Name	Synonym	Mol. Form.	CAS RN	Mol. Wt.	Physical Form	mp/°C	bp/°C	den/ g cm⁻³	n_D	Solubility
7481	2-Methyl-4-nitroaniline		$C_7H_8N_2O_2$	99-52-5	152.151		133.5		1.1586[140]		sl H_2O, DMSO; s EtOH, bz, HOAc
7482	2-Methyl-5-nitroaniline		$C_7H_8N_2O_2$	99-55-8	152.151		105.5				sl H_2O; s EtOH, eth, ace, bz, chl
7483	2-Methyl-6-nitroaniline		$C_7H_8N_2O_2$	570-24-1	152.151		96		1.1900[100]		sl H_2O; s EtOH, eth, bz, chl
7484	4-Methyl-2-nitroaniline		$C_7H_8N_2O_2$	89-62-3	152.151		116.3		1.16[121]		sl H_2O; s EtOH, chl
7485	4-Methyl-3-nitroaniline		$C_7H_8N_2O_2$	119-32-4	152.151		79.8				sl H_2O, CS_2; s EtOH, eth, bz
7486	N-Methyl-2-nitroaniline		$C_7H_8N_2O_2$	612-28-2	152.151	red or oran nd (peth)	38	158[18]			sl H_2O, lig; s EtOH, eth, ace, bz
7487	N-Methyl-4-nitroaniline		$C_7H_8N_2O_2$	100-15-2	152.151	br-ye pr (al) cry (eth)	152	dec	1.201[155]		i H_2O; s EtOH, bz, chl; sl eth, lig
7488	2-Methyl-1-nitro-9,10-anthracenedione		$C_{15}H_9NO_4$	129-15-7	267.237	pa ye nd (HOAc)	273.0				i H_2O, EtOH; sl eth, bz, chl; s $PhNO_2$
7489	2-Methyl-5-nitrobenzenesulfonic acid		$C_7H_7NO_5S$	121-03-9	217.200		135.8				vs H_2O, EtOH, eth, chl
7490	Methyl 2-nitrobenzoate		$C_8H_7NO_4$	606-27-9	181.147	liq	-13	275	1.2855[20]		i H_2O; s EtOH, eth, bz, chl; i lig
7491	Methyl 3-nitrobenzoate		$C_8H_7NO_4$	618-95-1	181.147		78	279[60]			i H_2O; sl EtOH, eth, MeOH
7492	Methyl 4-nitrobenzoate		$C_8H_7NO_4$	619-50-1	181.147		96				i H_2O; s EtOH, eth, chl
7493	2-Methyl-4-nitro-1H-imidazole		$C_4H_5N_3O_2$	696-23-1	127.102		253				
7494	N-Methyl-N-nitromethanamine		$C_2H_6N_2O_2$	4164-28-7	90.081	nd(eth)	58	187	1.1090[72]	1.4462[72]	vs H_2O, ace, eth, EtOH
7495	2-Methyl-1-nitronaphthalene		$C_{11}H_9NO_2$	881-03-8	187.195	ye pr or nd (al)	81.5	188[20]			i H_2O; s EtOH; vs ace
7496	N-Methyl-N'-nitro-N-nitrosoguanidine		$C_2H_5N_5O_3$	70-25-7	147.093						s DMSO
7497	3-Methyl-4-nitrophenol		$C_7H_7NO_3$	2581-34-2	153.136	nd or pr (w)	129				sl H_2O; s EtOH, eth, bz, chl
7498	4-Methyl-2-nitrophenol		$C_7H_7NO_3$	119-33-5	153.136	ye nd (al, w)	36.5	125[22]	1.2399[20]	1.5744[40]	vs ace, bz, eth, EtOH
7499	1-Methyl-2-(4-nitrophenoxy)benzene	2-Methylphenyl 4-nitrophenyl ether	$C_{13}H_{11}NO_3$	2444-29-3	229.231	ye cry (peth)		220[27]			vs bz, eth, EtOH
7500	2-Methyl-2-nitro-1,3-propanediol		$C_4H_9NO_4$	77-49-6	135.119	mcl	150.1	dec			vs H_2O, EtOH; sl DMSO
7501	2-Methyl-2-nitro-1-propanol		$C_4H_9NO_3$	76-39-1	119.119	nd or pl (MeOH)	89.5	94[10]			sl H_2O; vs EtOH, eth; s chl
7502	3-Methyl-4-nitroquinoline-N-oxide		$C_{10}H_8N_2O_3$	14073-00-8	204.182	cry (MeOH)	179				
7503	N-Methyl-N-nitrosoaniline		$C_7H_8N_2O$	614-00-6	136.151	ye cry	14.7	dec 225; 121[13]	1.1240[20]	1.5769[20]	i H_2O; s EtOH, eth
7504	N-Methyl-N-nitrosourea	N-Nitroso-N-methylurea	$C_2H_5N_3O_2$	684-93-5	103.080	col or ye pl (eth)	123 dec				sl H_2O, EtOH, eth
7505	Methyl nonadecanoate		$C_{20}H_{40}O_2$	1731-94-8	312.531		41.3	190[4]			
7506	2-Methylnonane		$C_{10}H_{22}$	871-83-0	142.282	liq	-74.6	167.1	0.7281[20]	1.4099[20]	i H_2O; s eth, bz, chl
7507	3-Methylnonane		$C_{10}H_{22}$	5911-04-6	142.282	liq	-84.8	167.9	0.7354[20]	1.4125[20]	vs bz, eth, chl
7508	4-Methylnonane		$C_{10}H_{22}$	17301-94-9	142.282	liq	-99	165.7	0.7323[20]	1.4123[20]	vs bz, eth, chl
7509	5-Methylnonane		$C_{10}H_{22}$	15869-85-9	142.282	liq	-87.7	165.1	0.7326[20]	1.4116[20]	i H_2O; s eth, bz, chl
7510	Methyl nonanoate		$C_{10}H_{20}O_2$	1731-84-6	172.265			213.5	0.8799[15]	1.4214[20]	i H_2O; s EtOH, eth; sl ctc
7511	8-Methyl-1-nonanol		$C_{10}H_{22}O$	55505-26-5	158.281			108[10]			
7512	2-Methyl-1-nonene		$C_{10}H_{20}$	2980-71-4	140.266	liq	-64.2	168.4	0.7412[25]	1.4241[20]	
7513	2-Methyl-2-norbornene	2-Methylbicyclo[2.2.1]hept-2-ene	C_8H_{12}	694-92-8	108.181	liq		122			
7514	Methyl trans-9-octadecenoate		$C_{19}H_{36}O_2$	1937-62-8	296.488	liq	13.5	218[24]	0.8730[20]	1.4513[20]	vs eth, EtOH
7515	2-Methyloctane		C_9H_{20}	3221-61-2	128.255	liq	-80.3	143.2	0.7095[25]	1.4031[20]	i H_2O; s EtOH, eth; sl ctc; vs peth
7516	3-Methyloctane		C_9H_{20}	2216-33-3	128.255	liq	-107.6	144.2	0.717[25]	1.4040[25]	
7517	4-Methyloctane		C_9H_{20}	2216-34-4	128.255	liq	-113.3	142.4	0.716[25]	1.4039[25]	i H_2O
7518	Methyl octanoate	Methyl caprylate	$C_9H_{18}O_2$	111-11-5	158.238	liq	-40	192.9	0.8775[20]	1.4170[20]	i H_2O; vs EtOH, eth; sl ctc
7519	2-Methyloctanoic acid		$C_9H_{18}O_2$	3004-93-1	158.238			138[14], 88[4]		1.4281[25]	

2-Methyl-4-nitroaniline

2-Methyl-5-nitroaniline

2-Methyl-6-nitroaniline

4-Methyl-2-nitroaniline

4-Methyl-3-nitroaniline

N-Methyl-2-nitroaniline

N-Methyl-4-nitroaniline

2-Methyl-1-nitro-9,10-anthracenedione

2-Methyl-5-nitrobenzenesulfonic acid

Methyl 2-nitrobenzoate

Methyl 3-nitrobenzoate

Methyl 4-nitrobenzoate

2-Methyl-4-nitro-1H-imidazole

N-Methyl-N-nitromethanamine

2-Methyl-1-nitronaphthalene

N-Methyl-N'-nitro-N-nitrosoguanidine

3-Methyl-4-nitrophenol

4-Methyl-2-nitrophenol

1-Methyl-2-(4-nitrophenoxy)benzene

2-Methyl-2-nitro-1,3-propanediol

2-Methyl-2-nitro-1-propanol

3-Methyl-4-nitroquinoline-N-oxide

N-Methyl-N-nitrosoaniline

N-Methyl-N-nitrosourea

Methyl nonadecanoate

2-Methylnonane

3-Methylnonane

4-Methylnonane

5-Methylnonane

Methyl nonanoate

8-Methyl-1-nonanol

2-Methyl-1-nonene

2-Methyl-2-norbornene

Methyl trans-9-octadecenoate

2-Methyloctane

3-Methyloctane

4-Methyloctane

Methyl octanoate

2-Methyloctanoic acid

No.	Name	Synonym	Mol. Form.	CAS RN	Mol. Wt.	Physical Form	mp/°C	bp/°C	den/ g cm⁻³	n_D	Solubility
7520	2-Methyl-2-octanol		C₉H₂₀O	628-44-4	144.254			178	0.8210²⁰	1.4280²⁰	i H₂O; s EtOH, eth
7521	3-Methyl-3-octanol		C₉H₂₀O	5340-36-3	144.254			83¹⁸, 36³	0.8108²⁵	1.4257²⁵	
7522	5-Methyl-2-octanone		C₉H₁₈O	58654-67-4	142.238			101⁵⁰			
7523	2-Methyl-1-octene		C₉H₁₈	4588-18-5	126.239	liq	-77.8	144.8	0.7343²⁰	1.4184²⁰	
7524	7-Methyl-1-octene		C₉H₁₈	13151-09-9	126.239	liq		138.9			
7525	Methyloctylamine	N-Methyl-1-octanamine	C₉H₂₁N	2439-54-5	143.270			68⁸			
7526	Methyl 2-octynoate		C₉H₁₄O₂	111-12-6	154.206			217; 107²⁰	0.926²⁰	1.4464²⁰	
7527	3-Methyl-1-octyn-3-ol		C₉H₁₆O	23580-51-0	140.222			174; 75¹⁰	0.8547²⁰	1.443¹⁰	
7528	Methyl oleate		C₁₉H₃₆O₂	112-62-9	296.488		-19.9	218.5²⁰	0.8739²⁰	1.4522²⁰	i H₂O; msc EtOH, eth; s chl
7529	Methyl Orange	Sodium p-dimethylaminoazobenzenesulfonate	C₁₄H₁₄N₃NaO₃S	547-58-0	327.334	oran, ye pl or sc (w)	dec				sl H₂O, EtOH, py; i eth
7530	2-Methyloxazole		C₄H₅NO	23012-10-4	83.089	liq		87.5			
7531	4-Methyloxazole		C₄H₅NO	693-93-6	83.089			88	1.015²⁵	1.4317²⁰	
7532	5-Methyloxazole		C₄H₅NO	66333-88-8	83.089	liq		88			
7533	2-Methyl-2-oxazoline		C₄H₇NO	1120-64-5	85.105			111	1.005²⁵	1.4340²⁰	
7534	2-Methyloxetane		C₄H₈O	2167-39-7	72.106	hyg		59	0.841²⁵	1.3885²⁰	
7535	4-Methyl-2-oxetanone	3-Hydroxybutyric acid lactone	C₄H₆O₂	3068-88-0	86.090			86⁵⁰, 57⁹	1.0555²⁰		
7536	Methyloxirane	1,2-Propylene oxide	C₃H₆O	16033-71-9	58.079	liq	-111.9	35	0.859⁰	1.3660²⁰	vs H₂O, EtOH, eth; s chl
7537	3-Methyl-2-oxobutanoic acid		C₅H₈O₃	759-05-7	116.116		31.5	170.5	0.9968²⁰	1.3850¹⁶	s H₂O, EtOH, eth
7538	N-Methyl-N-(1-oxododecyl) glycine	N-Dodecanoylsarcosine	C₁₅H₂₉NO₃	97-78-9	271.396		44.5				s chl
7539	Methyl 4-oxopentanoate	Methyl levulinate	C₆H₁₀O₃	624-45-3	130.141			196	1.0511²⁰	1.4233²⁰	sl H₂O; s EtOH, ace, bz, ctc; msc eth
7540	4-Methyl-2-oxopentanoic acid		C₆H₁₀O₃	816-66-0	130.141	liq	10	84¹⁵			
7541	Methyl 2-oxopropanoate	Methyl pyruvate	C₄H₆O₃	600-22-6	102.089			135.5	1.154⁰	1.4046²⁵	sl H₂O; s ace; msc EtOH, eth
7542	Methyl palmitate	Methyl hexadecanoate	C₁₇H₃₄O₂	112-39-0	270.451		30	417; 148²	0.8247⁷⁵		i H₂O; vs EtOH, ace, bz; s eth
7543	Methyl parathion		C₈H₁₀NO₅PS	298-00-0	263.208	cry	38		1.358²⁰	1.5367²⁵	i H₂O; s os
7544	Methyl pentachlorophenyl sulfide	S-Methyl pentachlorobenzenethiol	C₇H₃Cl₅S	1825-19-0	296.429	cry (EtOH)	95.5				
7545	Methyl pentadecanoate		C₁₆H₃₂O₂	7132-64-1	256.424	nd (dil al)	18.5	153.5	0.8618²⁵	1.4390²⁵	s EtOH, eth
7546	cis-2-Methyl-1,3-pentadiene		C₆H₁₀	1501-60-6	82.143	liq	-117.6	75.8	0.714²⁵	1.446²⁰	
7547	3-Methyl-1,3-pentadiene		C₆H₁₀	4549-74-0	82.143			77	0.730²⁵	1.452²⁰	
7548	4-Methyl-1,3-pentadiene	1,1-Dimethyl-1,3-butadiene	C₆H₁₀	926-56-7	82.143			76.5	0.7181²⁰	1.4532²⁰	
7549	Methyl pentafluoroethyl ether	1-Methoxyperfluoroethane	C₃H₃F₅O	22410-44-2	150.047	col gas		5.59			
7550	Methyl pentafluoropropanoate		C₄H₃F₅O₂	378-75-6	178.058			59.5	1.390²⁵	1.2869²⁵	
7551	2-Methylpentanal	2-Methylvaleraldehyde	C₆H₁₂O	123-15-9	100.158			117			s H₂O; s eth, ace; sl ctc
7552	2-Methylpentane	Isohexane	C₆H₁₄	107-83-5	86.175	liq	-153.6	60.26	0.650²⁵	1.3715²⁰	i H₂O; s EtOH, eth; msc ace, bz, chl
7553	3-Methylpentane		C₆H₁₄	96-14-0	86.175	liq	-162.90	63.27	0.6598²⁵	1.3765²⁰	i H₂O; s EtOH, ctc; msc eth, ace, bz, hp
7554	2-Methylpentanedinitrile	2-Methylglutaronitrile	C₆H₈N₂	4553-62-2	108.141	liq	-45	270; 134¹³	0.950	1.4340²⁰	s H₂O
7555	2-Methyl-2,4-pentanediol	Hexylene glycol	C₆H₁₄O₂	107-41-5	118.174	liq	-50	197.1	0.923¹⁵	1.4276²⁰	s H₂O, EtOH, eth; sl ctc
7556	4-Methylpentanenitrile	Isopentyl cyanide	C₆H₁₁N	542-54-1	97.158	liq	-51	156.5	0.8030²⁰	1.4059²⁰	i H₂O; s EtOH; msc eth; sl ctc
7557	2-Methyl-2-pentanethiol		C₆H₁₄S	1633-97-2	118.240	liq		125.0; 36³⁰			
7558	Methyl pentanoate	Methyl valerate	C₆H₁₂O₂	624-24-8	116.158			127.4	0.8947²⁰	1.4003²⁰	sl H₂O, ctc; msc EtOH, eth; s ace
7559	2-Methylpentanoic acid, (±)		C₆H₁₂O₂	22160-39-0	116.158			195.6	0.9230²⁰	1.413²⁰	s H₂O, EtOH, eth; sl ctc
7560	3-Methylpentanoic acid, (±)		C₆H₁₂O₂	22160-40-3	116.158	liq	-41.6	197.5	0.9262²⁰	1.4159²⁰	vs eth, EtOH
7561	4-Methylpentanoic acid		C₆H₁₂O₂	646-07-1	116.158	liq	-33	200.5	0.9225²⁰	1.4144²⁰	sl H₂O; s EtOH, eth, chl
7562	2-Methyl-1-pentanol		C₆H₁₄O	105-30-6	102.174			149	0.8263²⁰	1.4182²⁰	sl H₂O; s EtOH, eth, ace, ctc
7563	3-Methyl-1-pentanol, (±)		C₆H₁₄O	20281-83-8	102.174			153	0.8242²⁰	1.4112²³	i H₂O; s EtOH, eth
7564	4-Methyl-1-pentanol	Isohexyl alcohol	C₆H₁₄O	626-89-1	102.174			151.9	0.8131²⁰	1.4134²⁵	i H₂O; s EtOH, eth
7565	2-Methyl-2-pentanol		C₆H₁₄O	590-36-3	102.174	liq	-103	121.1	0.8350¹⁶	1.4100²⁰	sl H₂O; s EtOH, eth

2-Methyl-2-octanol

3-Methyl-3-octanol

5-Methyl-2-octanone

2-Methyl-1-octene

7-Methyl-1-octene

Methyloctylamine

Methyl 2-octynoate

3-Methyl-1-octyn-3-ol

Methyl oleate

Methyl Orange

2-Methyloxazole

4-Methyloxazole

5-Methyloxazole

2-Methyl-2-oxazoline

2-Methyloxetane

4-Methyl-2-oxetanone

Methyloxirane

3-Methyl-2-oxobutanoic acid

N-Methyl-N-(1-oxododecyl)glycine

Methyl 4-oxopentanoate

4-Methyl-2-oxopentanoic acid

Methyl 2-oxopropanoate

Methyl palmitate

Methyl parathion

Methyl pentachlorophenyl sulfide

Methyl pentadecanoate

cis-2-Methyl-1,3-pentadiene

3-Methyl-1,3-pentadiene

4-Methyl-1,3-pentadiene

Methyl pentafluoroethyl ether

Methyl pentafluoropropanoate

2-Methylpentanal

2-Methylpentane

3-Methylpentane

2-Methylpentanedinitrile

2-Methyl-2,4-pentanediol

4-Methylpentanenitrile

2-Methyl-2-pentanethiol

Methyl pentanoate

2-Methylpentanoic acid, (±)

3-Methylpentanoic acid, (±)

4-Methylpentanoic acid

2-Methyl-1-pentanol

3-Methyl-1-pentanol, (±)

4-Methyl-1-pentanol

2-Methyl-2-pentanol

No.	Name	Synonym	Mol. Form.	CAS RN	Mol. Wt.	Physical Form	mp/°C	bp/°C	den/ g cm⁻³	n_D	Solubility
7566	3-Methyl-2-pentanol		$C_6H_{14}O$	565-60-6	102.174			134.3	0.8307²⁰	1.4182²⁰	sl H_2O; s EtOH, eth
7567	4-Methyl-2-pentanol		$C_6H_{14}O$	108-11-2	102.174	liq	-90	131.6	0.8075²⁰	1.4100²⁰	sl H_2O, ctc; s EtOH, eth
7568	2-Methyl-3-pentanol		$C_6H_{14}O$	565-67-3	102.174			126.5	0.8243²⁰	1.4175²⁰	sl H_2O; msc EtOH, eth
7569	3-Methyl-3-pentanol		$C_6H_{14}O$	77-74-7	102.174	liq	-23.6	122.4	0.8286²⁰	1.4186²⁰	sl H_2O, ctc; msc EtOH, eth
7570	2-Methyl-1-pentanol acetate		$C_8H_{16}O_2$	7789-99-3	144.212			163	0.870²⁵		vs eth, EtOH
7571	3-Methyl-2-pentanone, (±)	(±)-sec-Butyl methyl ketone	$C_6H_{12}O$	55156-16-6	100.158			117.5	0.8130²⁰	1.4002²⁰	sl H_2O; msc EtOH, eth; s chl
7572	4-Methyl-2-pentanone	Isobutyl methyl ketone	$C_6H_{12}O$	108-10-1	100.158	liq	-84	116.5	0.7965²⁵	1.3962²⁰	sl H_2O; msc EtOH, eth, ace, bz; s chl
7573	2-Methyl-3-pentanone	Ethyl isopropyl ketone	$C_6H_{12}O$	565-69-5	100.158			113.5	0.814¹⁸	1.3975²⁰	sl H_2O; vs EtOH, bz; msc eth, ace; s chl
7574	4-Methylpentanoyl chloride		$C_6H_{11}ClO$	38136-29-7	134.603			143	0.9725²⁰		
7575	2-Methyl-2-pentenal		$C_6H_{10}O$	623-36-9	98.142			136.5	0.8581²⁰	1.4488²⁰	i H_2O; s EtOH, eth, bz, MeOH
7576	2-Methyl-1-pentene		C_6H_{12}	763-29-1	84.159	liq	-135.7	62.1	0.6799²⁰	1.3920²⁰	i H_2O; s EtOH, bz, chl; sl ctc
7577	3-Methyl-1-pentene		C_6H_{12}	760-20-3	84.159	liq	-153	54.2	0.6675²⁰	1.3841²⁰	i H_2O; s EtOH, bz, chl, peth
7578	4-Methyl-1-pentene		C_6H_{12}	691-37-2	84.159	liq	-153.6	53.9	0.6642²⁰	1.3828²⁰	i H_2O; s EtOH, bz, chl, peth
7579	2-Methyl-2-pentene		C_6H_{12}	625-27-4	84.159	liq	-135	67.3	0.6863²⁰	1.4004²⁰	i H_2O; s EtOH, bz, ctc, chl
7580	3-Methyl-cis-2-pentene		C_6H_{12}	922-62-3	84.159	liq	-134.8	67.7	0.6886²⁵	1.4016²⁰	i H_2O; s EtOH, bz, chl, peth
7581	3-Methyl-trans-2-pentene		C_6H_{12}	616-12-6	84.159	liq	-138.5	70.4	0.6930²⁵	1.4045²⁰	i H_2O; s EtOH, bz, ctc, chl, peth
7582	4-Methyl-cis-2-pentene		C_6H_{12}	691-38-3	84.159	liq	-134.8	56.3	0.6690²⁰	1.3800²⁰	i H_2O; s EtOH, bz, chl, peth
7583	4-Methyl-trans-2-pentene		C_6H_{12}	674-76-0	84.159	liq	-140.8	58.6	0.6686²⁰	1.3889²⁰	i H_2O; s EtOH, bz, chl; sl ctc
7584	trans-2-Methyl-2-pentenoic acid		$C_6H_{10}O_2$	16957-70-3	114.142	pr	24.4	214; 112¹²	0.9751²⁰	1.4513²⁰	sl H_2O; s eth, chl, CS_2
7585	4-Methyl-2-pentenoic acid	4,4-Dimethyl-2-butenoic acid	$C_6H_{10}O_2$	10321-71-8	114.142		35	217	0.9529²¹	1.4489²¹	vs ace, eth, EtOH
7586	2-Methyl-3-pentenoic acid		$C_6H_{10}O_2$	37674-63-8	114.142			199	0.966¹⁵	1.4402²⁵	
7587	4-Methyl-3-penten-2-ol		$C_6H_{12}O$	4325-82-0	100.158			134	0.840¹⁵	1.9377¹⁵	
7588	3-Methyl-2-penten-4-one		$C_6H_{10}O$	565-62-8	98.142			138		1.4508²⁰	
7589	4-Methyl-4-penten-2-one		$C_6H_{10}O$	3744-02-3	98.142	liq	-72.6	124.2	0.8411²⁰		
7590	cis-3-Methyl-2-(2-pentenyl)-2-cyclopenten-1-one	Jasmone	$C_{11}H_{16}O$	488-10-8	164.244	ye oil		258; 134¹²	0.9437²²	1.4979²²	sl H_2O; s EtOH, eth, ctc, lig
7591	3-(4-Methyl-3-pentenyl)furan		$C_{10}H_{14}O$	539-52-6	150.217			185.5	0.9017²⁰	1.4705²¹	
7592	3-Methyl-3-penten-1-yne		C_6H_8	1574-33-0	80.128			66.5	0.739²⁰	1.4332²⁰	s eth, bz
7593	3-Methyl-2-pentyl-2-cyclopenten-1-one		$C_{11}H_{18}O$	1128-08-1	166.260			143²², 116¹²	0.9165¹⁸	1.4767²⁰	
7594	Methyl pentyl ether		$C_6H_{14}O$	628-80-8	102.174			99	0.759²²	1.3862²²	vs ace, eth, EtOH
7595	5-Methyl-2-pentylphenol	6-Pentyl-m-cresol	$C_{12}H_{18}O$	1300-94-3	178.270		24	138¹⁵			vs ace, eth, EtOH
7596	Methyl pentyl sulfide		$C_6H_{14}S$	1741-83-9	118.240	liq	-94	145.1	0.8431²⁰	1.4506²⁰	s EtOH, eth, ace, bz, chl
7597	Methyl tert-pentyl sulfide	2-Methyl-2-(methylthio)butane	$C_6H_{14}S$	13286-92-5	118.240	liq		150	0.84	1.4570²⁰	
7598	4-Methyl-1-pentyne		C_6H_{10}	7154-75-8	82.143	liq	-104.6	61.2	0.7000²⁵	1.3936²⁰	i H_2O; s bz, chl
7599	4-Methyl-2-pentyne		C_6H_{10}	21020-27-9	82.143	liq	-110.3	73.1	0.7112²⁵	1.4057²⁰	vs bz, chl
7600	3-Methyl-1-pentyn-3-ol	Meparfynol	$C_6H_{10}O$	77-75-8	98.142		30.5	120.5	0.8688²⁰	1.4310²⁰	
7601	Methyl perfluorooctanoate		$C_9H_3F_{15}O_2$	376-27-2	428.095			158	1.684²⁰	1.304²⁷	
7602	1-Methylphenanthrene		$C_{15}H_{12}$	832-69-9	192.256	lf, pl (dil al)	123	354			i H_2O; s EtOH
7603	3-Methylphenanthrene		$C_{15}H_{12}$	832-71-3	192.256	pr or nd (al)	65	350; 145⁶			i H_2O; s EtOH, ace; sl chl
7604	4-Methylphenanthrene		$C_{15}H_{12}$	832-64-4	192.256	pl (90% al)	53.5	177¹⁰			i H_2O; s EtOH, ctc
7605	Methylphenidate		$C_{14}H_{19}NO_2$	113-45-1	233.307			136⁰·⁶			i H_2O, peth; s chl, EtOH, eth, AcOEt
7606	10-Methyl-10H-phenothiazine		$C_{13}H_{11}NS$	1207-72-3	213.298		101				
7607	10-Methyl-10H-phenothiazine-2-acetic acid	Metiazinic acid	$C_{15}H_{13}NO_2S$	13993-65-2	271.335		144				s chl

3-Methyl-2-pentanol

4-Methyl-2-pentanol

2-Methyl-3-pentanol

3-Methyl-3-pentanol

2-Methyl-1-pentanol acetate

3-Methyl-2-pentanone, (±)

4-Methyl-2-pentanone

2-Methyl-3-pentanone

4-Methylpentanoyl chloride

2-Methyl-2-pentenal

2-Methyl-1-pentene

3-Methyl-1-pentene

4-Methyl-1-pentene

2-Methyl-2-pentene

3-Methyl-cis-2-pentene

3-Methyl-trans-2-pentene

4-Methyl-cis-2-pentene

4-Methyl-trans-2-pentene

trans-2-Methyl-2-pentenoic acid

4-Methyl-2-pentenoic acid

2-Methyl-3-pentenoic acid

4-Methyl-3-penten-2-ol

3-Methyl-2-penten-4-one

4-Methyl-4-penten-2-one

cis-3-Methyl-2-(2-pentenyl)-2-cyclopenten-1-one

3-(4-Methyl-3-pentenyl)furan

3-Methyl-3-penten-1-yne

3-Methyl-2-pentyl-2-cyclopenten-1-one

Methyl pentyl ether

5-Methyl-2-pentylphenol

Methyl pentyl sulfide

Methyl tert-pentyl sulfide

4-Methyl-1-pentyne

4-Methyl-2-pentyne

3-Methyl-1-pentyn-3-ol

Methyl perfluorooctanoate

1-Methylphenanthrene

3-Methylphenanthrene

4-Methylphenanthrene

Methylphenidate

10-Methyl-10H-phenothiazine

10-Methyl-10H-phenothiazine-2-acetic acid

No.	Name	Synonym	Mol. Form.	CAS RN	Mol. Wt.	Physical Form	mp/°C	bp/°C	den/ g cm^{-3}	n_D	Solubility
7608	Methyl phenoxyacetate		$C_9H_{10}O_3$	2065-23-8	166.173			245	1.1493^{20}	1.5155^{20}	vs eth, EtOH
7609	1-Methyl-3-phenoxybenzene		$C_{13}H_{12}O$	3586-14-9	184.233			272	1.051^{25}	1.5727^{20}	
7610	[(2-Methylphenoxy)methyl]oxirane		$C_{10}H_{12}O_2$	2210-79-9	164.201			123^2	1.0884^{20}		
7611	3-(2-Methylphenoxy)-1,2-propanediol	Mephenesin	$C_{10}H_{14}O_3$	59-47-2	182.216		70 dec				sl H_2O, eth; s EtOH
7612	N-(2-Methylphenyl)acetamide		$C_9H_{11}NO$	120-66-1	149.189	nc (al)	110	296	1.168^{15}		sl H_2O, bz; s EtOH, eth, ace, HOAc
7613	N-(3-Methylphenyl)acetamide		$C_9H_{11}NO$	537-92-8	149.189	nd (w)	65.5	303	1.141^{15}		sl H_2O; vs EtOH, eth; s chl
7614	N-Methyl-N-phenylacetamide	N-Methylacetanilide	$C_9H_{11}NO$	579-10-2	149.189	nd (eth), pr (al)	103	256	1.0036^{105}	1.576	s H_2O, EtOH, eth, chl, lig
7615	2-Methylphenyl acetate	o-Cresyl acetate	$C_9H_{10}O_2$	533-18-6	150.174			208	1.0533^{15}	1.5002^{20}	vs eth, EtOH
7616	3-Methylphenyl acetate	m-Cresyl acetate	$C_9H_{10}O_2$	122-46-3	150.174		12	212	1.043^{20}	1.4978^{20}	vs bz, eth, EtOH
7617	4-Methylphenyl acetate	p-Cresyl acetate	$C_9H_{10}O_2$	140-39-6	150.174			212.5	1.0512^{17}	1.5163^{22}	sl H_2O, ctc; s EtOH, eth, chl
7618	Methyl 2-phenylacetate		$C_9H_{10}O_2$	101-41-7	150.174			216.5	1.0622^{16}	1.5075^{20}	i H_2O; msc EtOH, eth; s ace, ctc
7619	2-(Methylphenylamino)ethanol		$C_9H_{13}NO$	93-90-3	151.205			$218^{110}, 150^{14}$	1.0143^0		s H_2O; vs EtOH, eth, ace, bz
7620	2-[(2-Methylphenyl)amino]ethanol		$C_9H_{13}NO$	136-80-1	151.205			285.5	1.0794^{20}	1.5675^{20}	vs eth, EtOH
7621	3-Methyl-N-phenylaniline		$C_{13}H_{13}N$	1205-64-7	183.249		30	$316; 183^{17}$		1.6350^{20}	vs bz, eth, EtOH
7622	N-(4-Methylphenyl)benzamide		$C_{14}H_{13}NO$	582-78-5	211.259	orth nd (al)	158		1.202^{15}		vs eth, EtOH
7623	N-Methyl-N-phenylbenzenemethanamine		$C_{14}H_{15}N$	614-30-2	197.276						s ctc
7624	4-Methyl-α-phenylbenzenemethanol		$C_{14}H_{14}O$	1517-63-1	198.260		52				
7625	α-Methyl-α-phenylbenzenemethanol		$C_{14}H_{14}O$	599-67-7	198.260			$285; 190^{12}$	1.1059^{15}		
7626	4-Methyl-N-phenylbenzenesulfonamide		$C_{13}H_{13}NO_2S$	68-34-8	247.313	(α) tcl, (β) mcl pr (al, bz)	103.5				i H_2O; vs EtOH; s bz, HOAc
7627	4-Methylphenyl benzoate		$C_{14}H_{12}O_2$	614-34-6	212.244	pl (eth-al)	71.5	316			vs eth, EtOH
7628	1-Methyl-N-phenyl-N-benzyl-4-piperidinamine	Bamipine	$C_{19}H_{24}N_2$	4945-47-5	280.407	cry (MeOH)	115				
7629	Methyl 2-phenylbutanoate		$C_{11}H_{14}O_2$	2294-71-5	178.228	nd (dil al)	77.5	228			vs eth, EtOH
7630	3-Methyl-1-phenyl-1-butanone		$C_{11}H_{14}O$	582-62-7	162.228			236.5	0.9701^{16}	1.5139^{15}	i H_2O; msc EtOH, eth; vs ace
7631	3-Methyl-4-phenyl-3-butenamide	β-Benzalbutyramide	$C_{11}H_{13}NO$	7236-47-7	175.227		133				
7632	Methylphenylcarbamic chloride		C_8H_8ClNO	4285-42-1	169.609	pl (al)	88.5	280			vs eth, EtOH
7633	1-(2-Methylphenyl)ethanone		$C_9H_{10}O$	577-16-2	134.174			214	1.026^{20}	1.5276^{20}	
7634	1-(3-Methylphenyl)ethanone		$C_9H_{10}O$	585-74-0	134.174			220	1.0165^0	1.533^{15}	s EtOH, eth, ace; sl ctc
7635	4-(1-Methyl-1-phenylethyl)phenol		$C_{15}H_{16}O$	599-64-4	212.287	pr (peth)	74.5	335			
7636	N-Methyl-N-phenylformamide		C_8H_9NO	93-61-8	135.163		14.5	243	1.0948^{20}	1.5589^{20}	sl H_2O, ctc; s EtOH, ace
7637	N-(2-Methylphenyl)formamide		C_8H_9NO	94-69-9	135.163	lf (al)	62	288	1.086^{55}		s H_2O; vs EtOH
7638	5-Methyl-1-phenyl-1-hexen-3-one		$C_{13}H_{16}O$	2892-18-4	188.265	cry	43	154^{25}	0.9509^{46}	1.5523^{25}	sl H_2O; s EtOH, bz, chl
7639	1-Methyl-1-phenylhydrazine		$C_7H_{10}N_2$	618-40-6	122.167			$228; 131^{35}$	1.0404^{20}	1.5691^{20}	sl H_2O; msc EtOH, eth, bz, chl
7640	3-Methyl-5-phenyl-2,4-imidazolidinedione	3-Methyl-5-phenylhydantoin	$C_{10}H_{10}N_2O_2$	6846-11-3	190.198		164.5				s chl
7641	1-Methyl-6-phenylimidazo[4,5-b]pyridin-2-amine	PhIP	$C_{13}H_{12}N_4$	105650-23-5	224.261	solid	327				
7642	2-[Methyl(phenylmethyl)amino]ethanol		$C_{10}H_{15}NO$	101-98-4	165.232			134^{14}			
7643	4-Methyl-N-(phenylmethylene)aniline		$C_{14}H_{13}N$	2272-45-9	195.260	ye cry	35	$318; 178^{11}$			vs ace
7644	3-Methyl-2-phenylmorpholine	Phenmetrazine	$C_{11}H_{15}NO$	134-49-6	177.243			$139^{12}, 104^1$			
7645	2-Methyl-2-phenyloxirane		$C_9H_{10}O$	2085-88-3	134.174			84^{17}	1.0228^{20}	1.5232^{20}	
7646	N-(2-Methylphenyl)-3-oxo-butanamide		$C_{11}H_{13}NO_2$	93-68-5	191.227	pr (AcOEt)	107.5				vs bz, EtOH
7647	N-(4-Methylphenyl)-3-oxobutanamide		$C_{11}H_{13}NO_2$	2415-85-2	191.227	pr (AcOEt)	95				sl H_2O, lig; s EtOH, bz
7648	(2-Methylphenyl)phenylmethanone		$C_{14}H_{12}O$	131-58-8	196.244		<-18	$308; 128^{12}$	1.1098^{20}		i H_2O; vs EtOH
7649	(3-Methylphenyl)phenylmethanone		$C_{14}H_{12}O$	643-65-2	196.244	oil	2	$317; 170^9$	1.095^{20}		i H_2O; s EtOH, eth, bz, chl, HOAc

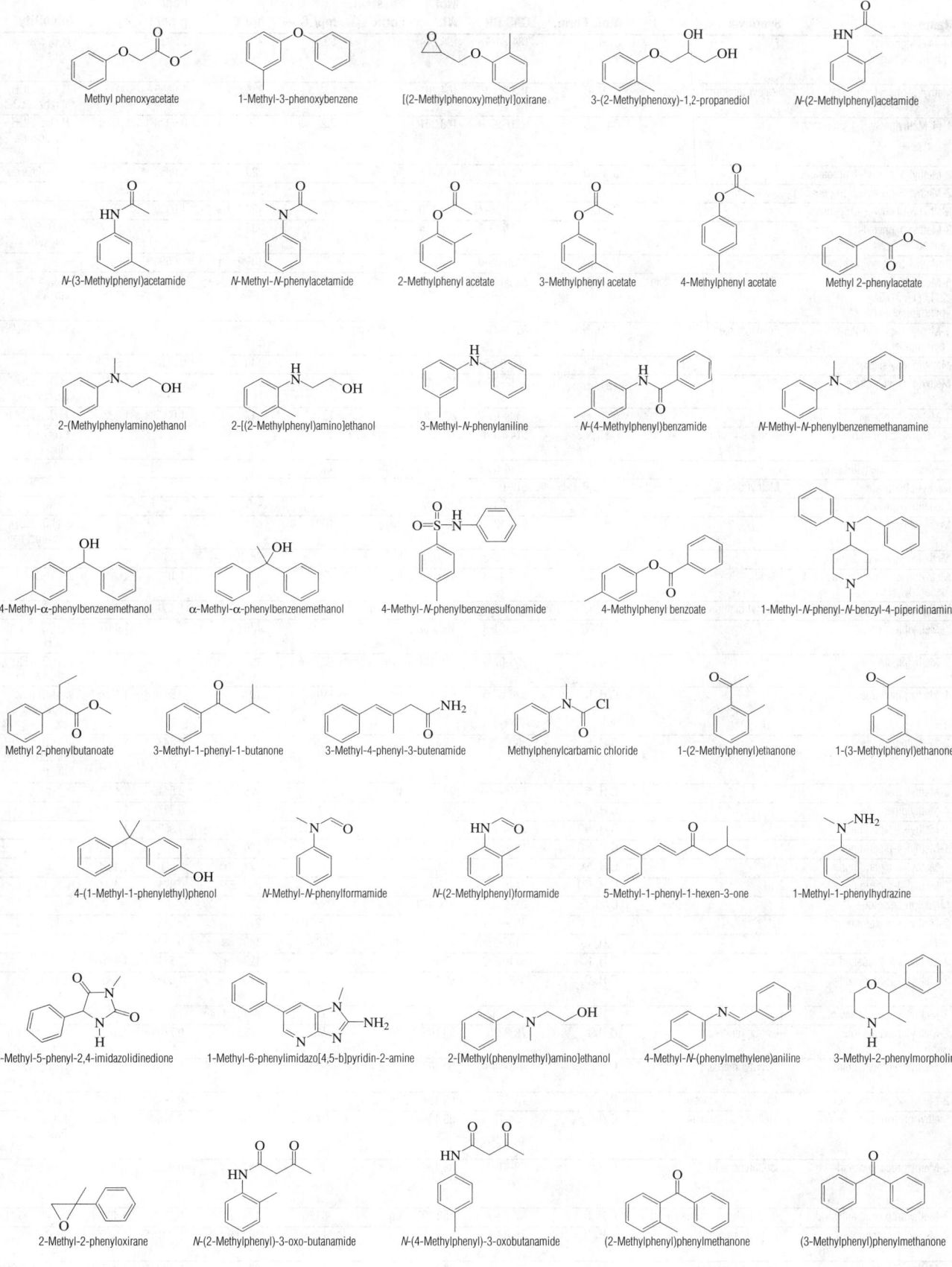

Methyl phenoxyacetate

1-Methyl-3-phenoxybenzene

[(2-Methylphenoxy)methyl]oxirane

3-(2-Methylphenoxy)-1,2-propanediol

N-(2-Methylphenyl)acetamide

N-(3-Methylphenyl)acetamide

N-Methyl-*N*-phenylacetamide

2-Methylphenyl acetate

3-Methylphenyl acetate

4-Methylphenyl acetate

Methyl 2-phenylacetate

2-(Methylphenylamino)ethanol

2-[(2-Methylphenyl)amino]ethanol

3-Methyl-*N*-phenylaniline

N-(4-Methylphenyl)benzamide

N-Methyl-*N*-phenylbenzenemethanamine

4-Methyl-α-phenylbenzenemethanol

α-Methyl-α-phenylbenzenemethanol

4-Methyl-*N*-phenylbenzenesulfonamide

4-Methylphenyl benzoate

1-Methyl-*N*-phenyl-*N*-benzyl-4-piperidinamine

Methyl 2-phenylbutanoate

3-Methyl-1-phenyl-1-butanone

3-Methyl-4-phenyl-3-butenamide

Methylphenylcarbamic chloride

1-(2-Methylphenyl)ethanone

1-(3-Methylphenyl)ethanone

4-(1-Methyl-1-phenylethyl)phenol

N-Methyl-*N*-phenylformamide

N-(2-Methylphenyl)formamide

5-Methyl-1-phenyl-1-hexen-3-one

1-Methyl-1-phenylhydrazine

3-Methyl-5-phenyl-2,4-imidazolidinedione

1-Methyl-6-phenylimidazo[4,5-b]pyridin-2-amine

2-[Methyl(phenylmethyl)amino]ethanol

4-Methyl-*N*-(phenylmethylene)aniline

3-Methyl-2-phenylmorpholine

2-Methyl-2-phenyloxirane

N-(2-Methylphenyl)-3-oxo-butanamide

N-(4-Methylphenyl)-3-oxobutanamide

(2-Methylphenyl)phenylmethanone

(3-Methylphenyl)phenylmethanone

No.	Name	Synonym	Mol. Form.	CAS RN	Mol. Wt.	Physical Form	mp/°C	bp/°C	den/ g cm^{-3}	n_D	Solubility
7650	(4-Methylphenyl) phenylmethanone		$C_{14}H_{12}O$	134-84-9	196.244	mcl pr	59.5	228[70]	0.9926[0]		i H_2O; sl EtOH, liq; s eth, bz, chl
7651	Methyl 3-phenylpropanoate	Methyl dihydrocinnamate	$C_{10}H_{12}O_2$	103-25-3	164.201			238.5; 91[4]	1.0455[25]		i H_2O; s EtOH, eth, bz, AcOEt
7652	1-(4-Methylphenyl)-1-propanone		$C_{10}H_{12}O$	5337-93-9	148.201		7.2	236	0.9926[20]	1.5278[20]	i H_2O; s EtOH, eth, ace, bz, CS_2
7653	2-Methyl-1-phenyl-1-propanone		$C_{10}H_{12}O$	611-70-1	148.201	liq	-0.7	220	0.9863[11]	1.5172[20]	vs eth, EtOH
7654	2-Methyl-3-phenyl-2-propenal		$C_{10}H_{10}O$	101-39-3	146.185			248; 150[100]	1.0407[17]	1.6057[17]	
7655	Methyl 3-phenyl-2-propynoate		$C_{10}H_8O_2$	4891-38-7	160.170		26	158[48], 132[16]	1.0830[25]	1.5618[25]	
7656	3-Methyl-1-phenyl-1H-pyrazol-5-amine		$C_{10}H_{11}N_3$	1131-18-6	173.214		116	333			s H_2O, EtOH, chl; sl bz
7657	2-Methyl-5-phenylpyridine		$C_{12}H_{11}N$	3256-88-0	169.222			189[50]	1.0590[25]	1.6055[25]	
7658	5-Methyl-5-phenyl-2,4,6(1H,3H,5H)-pyrimidinetrione	Phenylmethylbarbituric acid	$C_{11}H_{10}N_2O_3$	76-94-8	218.208	cry	220				i H_2O; s EtOH, eth, alk
7659	1-Methyl-3-phenyl-2,5-pyrrolidinedione	Phensuximide	$C_{11}H_{11}NO_2$	86-34-0	189.211	cry (hot al)	72				vs EtOH, MeOH
7660	Methylphenylsilane		$C_7H_{10}Si$	766-08-5	122.240			140	0.8895[20]	1.5058[20]	
7661	Methyl phenyl sulfone		$C_7H_8O_2S$	3112-85-4	156.203		88				i H_2O; s EtOH, bz, chl; sl ctc
7662	1-Methyl-4-(phenylthio)benzene		$C_{13}H_{12}S$	3699-01-2	200.299		15.7	317	1.0986[25]	1.6225[25]	i H_2O; s ace, bz
7663	(2-Methylphenyl)thiourea	o-Tolylthiourea	$C_8H_{10}N_2S$	614-78-8	166.243	nd (dil al, w)	162				vs H_2O, EtOH; sl eth
7664	N-Methyl-N'-phenylthiourea		$C_8H_{10}N_2S$	2724-69-8	166.243	ta, pl	112.5				vs EtOH
7665	Methyl phosphate	Methyl dihydrogen phosphate	CH_5O_4P	812-00-0	112.022	oil					
7666	Methylphosphine		CH_5P	593-54-4	48.025	col gas		-16			vs eth
7667	Methylphosphonic acid		CH_5O_3P	993-13-5	96.023	hyg pl	108.5	dec			vs H_2O, EtOH, eth; i bz, peth
7668	Methylphosphonic difluoride		CH_3F_2OP	676-99-3	100.005	liq		98; 22[27]	1.3314[20]		
7669	Methylphosphonofluoridic acid, isopropyl ester	Sarin	$C_4H_{10}FO_2P$	107-44-8	140.093	liq	-57	147	1.10[20]		dec H_2O
7670	Methyl phosphorodichloridite	Methyl dichlorophosphite	CH_3Cl_2OP	3279-26-3	132.914	hyg liq	-91	93	1.406	1.4740[20]	
7671	1-Methylpiperazine		$C_5H_{12}N_2$	109-01-3	100.162			138		1.4378[20]	vs H_2O, eth, EtOH
7672	2-Methylpiperazine		$C_5H_{12}N_2$	109-07-9	100.162	hyg lf (al)	62	153			vs H_2O; s EtOH, eth, bz, chl
7673	1-Methylpiperidine		$C_6H_{13}N$	626-67-5	99.174	liq	-102.7	107	0.8159[20]	1.4355[20]	vs H_2O; msc EtOH, eth; s ctc
7674	2-Methylpiperidine, (±)		$C_6H_{13}N$	3000-79-1	99.174	liq	-2.5	118	0.8436[24]	1.4459[20]	vs H_2O; s EtOH, eth; sl chl; i dil KOH
7675	3-Methylpiperidine, (±)		$C_6H_{13}N$	53152-98-0	99.174	liq	-24	125.5	0.8446[26]	1.4470[20]	vs H_2O; sl chl
7676	4-Methylpiperidine		$C_6H_{13}N$	626-58-4	99.174			130	0.8674[25]	1.4458[20]	vs H_2O; sl chl
7677	1-Methyl-3-piperidinol		$C_6H_{13}NO$	3554-74-3	115.173			93[26], 77[11]	0.9635[16]	1.4735[20]	
7678	1-Methyl-4-piperidinol		$C_6H_{13}NO$	106-52-5	115.173		29	200		1.4775[20]	
7679	1-Methyl-2-piperidinone		$C_6H_{11}NO$	931-20-4	113.157			221; 105[12]	1.0263[25]	1.4820[20]	
7680	1-Methyl-4-piperidinone		$C_6H_{11}NO$	1445-73-4	113.157			85[45], 57[11]	0.971[25]	1.4580[25]	
7681	Methylprednisolone		$C_{22}H_{30}O_5$	83-43-2	374.470	cry	232				
7682	2-Methylpropanamide		C_4H_9NO	563-83-7	87.120		129.0	217	1.013[20]		s chl
7683	N-Methylpropanamide		C_4H_9NO	1187-58-2	87.120	liq	-30.9	148	0.9305[25]	1.4345[25]	
7684	2-Methyl-1,2-propanediamine		$C_4H_{12}N_2$	811-93-8	88.151			123	0.841[25]	1.4410[20]	s ctc
7685	2-Methyl-1,2-propanediol		$C_4H_{10}O_2$	558-43-0	90.121			176	1.0024[20]	1.4350[20]	vs H_2O, eth, EtOH
7686	2-Methyl-1,3-propanediol		$C_4H_{10}O_2$	2163-42-0	90.121	liq	-91	211.6; 124[20]	1.015[20]	1.4450[20]	
7687	2-Methylpropanenitrile	Isobutyronitrile	C_4H_7N	78-82-0	69.106	liq	-71.5	103.9	0.7704[20]	1.3720[20]	sl H_2O; vs EtOH, eth, ace, chl
7688	2-Methyl-1-propanethiol	Isobutyl mercaptan	$C_4H_{10}S$	513-44-0	90.187		<-70	88.5	0.8357[20]	1.4387[20]	sl H_2O; vs EtOH, ace; s ctc
7689	2-Methyl-2-propanethiol	tert-Butyl mercaptan	$C_4H_{10}S$	75-66-1	90.187	liq	-0.5	64.2	0.7943[25]	1.4232[20]	i H_2O; s ctc, hp
7690	Methyl propanoate	Methyl propionate	$C_4H_8O_2$	554-12-1	88.106	liq	-87.5	79.8	0.9150[20]	1.3775[20]	sl H_2O; msc EtOH, eth; s ace, ctc
7691	2-Methylpropanoic acid	Isobutyric acid	$C_4H_8O_2$	79-31-2	88.106	liq	-46	154.45	0.9681[20]	1.3930[20]	vs H_2O; msc EtOH, eth; sl ctc
7692	2-Methylpropanoic anhydride	Isobutyric anhydride	$C_8H_{14}O_3$	97-72-3	158.195	liq	-53.5	183; 89[32]	0.9535[20]	1.4061[19]	msc eth; s chl
7693	2-Methyl-1-propanol	Isobutyl alcohol	$C_4H_{10}O$	78-83-1	74.121	liq	-101.9	107.89	0.8018[20]	1.3955[20]	s H_2O, EtOH, eth, ace, ctc
7694	2-Methyl-2-propanol	tert-Butyl alcohol	$C_4H_{10}O$	75-65-0	74.121		25.69	82.4	0.7887[20]	1.3878[20]	msc H_2O, EtOH, eth; s chl

(4-Methylphenyl)phenylmethanone

Methyl 3-phenylpropanoate

1-(4-Methylphenyl)-1-propanone

2-Methyl-1-phenyl-1-propanone

2-Methyl-3-phenyl-2-propenal

Methyl 3-phenyl-2-propynoate

3-Methyl-1-phenyl-1*H*-pyrazol-5-amine

2-Methyl-5-phenylpyridine

5-Methyl-5-phenyl-2,4,6(1*H*,3*H*,5*H*)-pyrimidinetrione

1-Methyl-3-phenyl-2,5-pyrrolidinedione

Methylphenylsilane

Methyl phenyl sulfone

1-Methyl-4-(phenylthio)benzene

(2-Methylphenyl)thiourea

N-Methyl-*N*'-phenylthiourea

Methyl phosphate

Methylphosphine

Methylphosphonic acid

Methylphosphonic difluoride

Methylphosphonofluoridic acid, isopropyl ester

Methyl phosphorodichloridite

1-Methylpiperazine

2-Methylpiperazine

1-Methylpiperidine

2-Methylpiperidine, (±)

3-Methylpiperidine, (±)

4-Methylpiperidine

1-Methyl-3-piperidinol

1-Methyl-4-piperidinol

1-Methyl-2-piperidinone

1-Methyl-4-piperidinone

Methylprednisolone

2-Methylpropanamide

N-Methylpropanamide

2-Methyl-1,2-propanediamine

2-Methyl-1,2-propanediol

2-Methyl-1,3-propanediol

2-Methylpropanenitrile

2-Methyl-1-propanethiol

2-Methyl-2-propanethiol

Methyl propanoate

2-Methylpropanoic acid

2-Methylpropanoic anhydride

2-Methyl-1-propanol

2-Methyl-2-propanol

No.	Name	Synonym	Mol. Form.	CAS RN	Mol. Wt.	Physical Form	mp/°C	bp/°C	den/ g cm^{-3}	n_D	Solubility
7695	2-Methylpropanoyl chloride	Isobutyric acid chloride	C$_4$H$_7$ClO	79-30-1	106.551	liq	-90	92		1.4079[20]	s eth
7696	2-Methylpropenal	Methacrolein	C$_4$H$_6$O	78-85-3	70.090			68.4	0.840[25]	1.4144[20]	msc H$_2$O, EtOH, eth
7697	2-Methyl-2-propenamide		C$_4$H$_7$NO	79-39-0	85.105	cry (bz)	110.5				sl eth, chl; s EtOH, CH$_2$Cl$_2$
7698	N-Methyl-2-propen-1-amine		C$_4$H$_9$N	627-37-2	71.121			64		1.4065[20]	vs H$_2$O, ace, eth, EtOH
7699	2-Methyl-2-propene-1,1-diol diacetate	Methacrolein diacetate	C$_8$H$_{12}$O$_4$	10476-95-6	172.179			191		1.4241[20]	
7700	2-Methyl-1-propene, tetramer		C$_{16}$H$_{32}$	15220-85-6	224.425	liq	-98	244; 109[15]	0.7944[20]	1.4482[20]	
7701	2-Methyl-2-propenoic anhydride	Methacrylic acid anhydride	C$_8$H$_{10}$O$_3$	760-93-0	154.163			89[5]		1.4540[20]	msc EtOH, eth
7702	2-Methyl-2-propenol	Methallyl alcohol	C$_4$H$_8$O	513-42-8	72.106			114.5	0.8515[20]	1.4255[20]	vs H$_2$O; msc EtOH, eth
7703	2-Methyl-2-propenoyl chloride	Methacrylic acid chloride	C$_4$H$_5$ClO	920-46-7	104.535	liq	-60	96	1.0871[20]	1.4435[20]	s eth, ace, chl
7704	cis-(1-Methyl-1-propenyl)benzene		C$_{10}$H$_{12}$	767-99-7	132.202			194.7; 177[500]	0.9191[25]	1.5402[25]	i H$_2$O; s bz, chl
7705	trans-(1-Methyl-1-propenyl) benzene		C$_{10}$H$_{12}$	768-00-3	132.202	liq	-23.5	194.7	0.9138[25]	1.5425[20]	i H$_2$O; s bz, chl
7706	(2-Methyl-1-propenyl)benzene		C$_{10}$H$_{12}$	768-49-0	132.202	liq	-48.0	183; 99[43]	0.900[20]	1.5388[20]	
7707	4-(2-Methylpropenyl)morpholine	1-Morpholinoisobutene	C$_8$H$_{15}$NO	2403-55-6	141.211		120	89[20]		1.4663[20]	
7708	2-(2-Methylpropoxy)ethanol		C$_6$H$_{14}$O$_2$	4439-24-1	118.174			160	0.8900[20]	1.4143[20]	
7709	Methylpropylamine	N-Methyl-1-propanamine	C$_4$H$_{11}$N	627-35-0	73.137			63	0.7204[17]		
7710	1-Methyl-2-propylbenzene		C$_{10}$H$_{14}$	1074-17-5	134.218	liq	-60.3	185	0.8697[25]	1.4996[20]	
7711	1-Methyl-3-propylbenzene		C$_{10}$H$_{14}$	1074-43-7	134.218	liq	-82.5	182	0.8569[25]	1.4935[20]	
7712	1-Methyl-4-propylbenzene		C$_{10}$H$_{14}$	1074-55-1	134.218	liq	-63.6	183.4	0.8544[25]	1.4922[20]	i H$_2$O; s EtOH, eth
7713	cis-1-Methyl-2-propylcyclopentane		C$_9$H$_{18}$	932-43-4	126.239	liq	-104	152.6	0.7881[25]	1.4343[20]	
7714	trans-1-Methyl-2-propylcyclopentane		C$_9$H$_{18}$	932-44-5	126.239	liq	-123	146.4	0.7735[25]	1.4274[20]	
7715	Methyl propyl disulfide		C$_4$H$_{10}$S$_2$	2179-60-4	122.252	liq		70[43]	0.980	1.5080[20]	
7716	Methyl propyl ether	1-Methoxypropane	C$_4$H$_{10}$O	557-17-5	74.121			39.1	0.7356[13]	1.3579[25]	s H$_2$O, ace; msc EtOH, eth
7717	1-Methyl-2-propylpiperidine, (S)	Methylconiine	C$_9$H$_{19}$N	35305-13-6	141.254			174	0.8326[22]	1.4538[12]	vs ace, EtOH
7718	2-Methyl-2-propyl-1,3-propanediol		C$_7$H$_{16}$O$_2$	78-26-2	132.201	cry (hx)	62.5	234; 121[10]			s H$_2$O, hx; sl chl
7719	2-Methyl-2-propyl-1,3-propanediol dicarbamate	Meprobamate	C$_9$H$_{18}$N$_2$O$_4$	57-53-4	218.250	cry (w)	105				vs bz, eth, EtOH
7720	Methyl propyl sulfide		C$_4$H$_{10}$S	3877-15-4	90.187	liq	-113	95.6	0.8424[20]	1.4442[20]	s H$_2$O, EtOH, eth, ace
7721	N-Methyl-2-propyn-1-amine		C$_4$H$_7$N	35161-71-8	69.106			83	0.819[25]	1.4332[20]	
7722	N-Methyl-N-2-propynylbenzenemethanamine	Pargyline	C$_{11}$H$_{13}$N	555-57-7	159.228			96[11]	0.944[25]	1.5213[20]	
7723	3-Methylpyrazinamine	2-Amino-3-methylpyrazine	C$_5$H$_7$N$_3$	19838-08-5	109.130	nd (hx/ AcOEt)	174				
7724	2-Methylpyrazine		C$_5$H$_6$N$_2$	109-08-0	94.115	liq	-29	137	1.03[20]	1.5042[20]	msc H$_2$O, EtOH, eth; s ace; sl ctc
7725	1-Methyl-1H-pyrazole		C$_4$H$_6$N$_2$	930-36-9	82.104			127	0.9929[13]	1.4787[13]	
7726	3-Methyl-1H-pyrazole		C$_4$H$_6$N$_2$	1453-58-3	82.104		36.5	204; 108[25]	1.0203[16]	1.4915[20]	msc H$_2$O, EtOH, eth
7727	4-Methyl-1H-pyrazole	Fomepizole	C$_4$H$_6$N$_2$	7554-65-6	82.104			206; 95[13]	1.015[20]		
7728	3-Methyl-2-pyrazolin-5-one		C$_4$H$_6$N$_2$O	108-26-9	98.103		215				vs H$_2$O; sl EtOH
7729	1-Methylpyrene		C$_{17}$H$_{12}$	2381-21-7	216.277		71.3	410			
7730	2-Methylpyrene		C$_{17}$H$_{12}$	3442-78-2	216.277	fl (EtOH)	143	409.8			
7731	3-Methylpyridazine	3-Methyl-1,2-diazine	C$_5$H$_6$N$_2$	1632-76-4	94.115		184	214	1.0450[26]	1.5145[20]	
7732	3-Methyl-2-pyridinamine	2-Amino-3-picoline	C$_6$H$_8$N$_2$	1603-40-3	108.141	hyg	33.5	222; 95[8]			vs H$_2$O; s EtOH, eth, ace, bz, ctc; sl lig
7733	4-Methyl-2-pyridinamine	2-Amino-4-picoline	C$_6$H$_8$N$_2$	695-34-1	108.141	lf or pl (lig)	100	116[11]			vs H$_2$O; s EtOH, eth, ace, bz; i lig; sl chl
7734	5-Methyl-2-pyridinamine		C$_6$H$_8$N$_2$	1603-41-4	108.141		76.5	227			
7735	6-Methyl-2-pyridinamine	2-Amino-6-picoline	C$_6$H$_8$N$_2$	1824-81-3	108.141	hyg (lig)	41	208.5			vs H$_2$O; s EtOH, eth, ace, bz, lig
7736	N-Methyl-2-pyridinamine		C$_6$H$_8$N$_2$	4597-87-9	108.141		15	200.5	1.048[29]		s H$_2$O, bz; vs EtOH, eth, HOAc
7737	N-Methyl-4-pyridinamine		C$_6$H$_8$N$_2$	1121-58-0	108.141	pl (eth)	118.8				vs H$_2$O, ace, eth, EtOH
7738	2-Methylpyridine	2-Picoline	C$_6$H$_7$N	109-06-8	93.127	liq	-66.68	129.38	0.9443[20]	1.4957[20]	vs H$_2$O, ace; msc EtOH, eth; s ctc

2-Methylpropanoyl chloride

2-Methylpropenal

2-Methyl-2-propenamide

N-Methyl-2-propen-1-amine

2-Methyl-2-propene-1,1-diol diacetate

2-Methyl-1-propene, tetramer

2-Methyl-2-propenoic anhydride

2-Methyl-2-propenol

2-Methyl-2-propenoyl chloride

cis-(1-Methyl-1-propenyl)benzene

trans-(1-Methyl-1-propenyl)benzene

(2-Methyl-1-propenyl)benzene

4-(2-Methylpropenyl)morpholine

2-(2-Methylpropoxy)ethanol

Methylpropylamine

1-Methyl-2-propylbenzene

1-Methyl-3-propylbenzene

1-Methyl-4-propylbenzene

cis-1-Methyl-2-propylcyclopentane

trans-1-Methyl-2-propylcyclopentane

Methyl propyl disulfide

Methyl propyl ether

1-Methyl-2-propylpiperidine, (S)

2-Methyl-2-propyl-1,3-propanediol

2-Methyl-2-propyl-1,3-propanediol dicarbamate

Methyl propyl sulfide

N-Methyl-2-propyn-1-amine

N-Methyl-N-2-propynylbenzenemethanamine

3-Methylpyrazinamine

2-Methylpyrazine

1-Methyl-1H-pyrazole

3-Methyl-1H-pyrazole

4-Methyl-1H-pyrazole

3-Methyl-2-pyrazolin-5-one

1-Methylpyrene

2-Methylpyrene

3-Methylpyridazine

3-Methyl-2-pyridinamine

4-Methyl-2-pyridinamine

5-Methyl-2-pyridinamine

6-Methyl-2-pyridinamine

N-Methyl-2-pyridinamine

N-Methyl-4-pyridinamine

2-Methylpyridine

No.	Name	Synonym	Mol. Form.	CAS RN	Mol. Wt.	Physical Form	mp/°C	bp/°C	den/ g cm⁻³	n_D	Solubility
7739	3-Methylpyridine	3-Picoline	C₆H₇N	108-99-6	93.127	liq	-18.14	144.14	0.9566²⁰	1.5040²⁰	msc H₂O, EtOH, eth; vs ace; s ctc
7740	4-Methylpyridine	4-Picoline	C₆H₇N	108-89-4	93.127		3.67	145.36	0.9548²⁰	1.5037²⁰	msc H₂O, EtOH, eth; s ace, ctc
7741	6-Methyl-2-pyridinecarboxaldehyde		C₇H₇NO	1122-72-1	121.137		32	77¹²			
7742	Methyl 3-pyridinecarboxylate	Methyl nicotinate	C₇H₇NO₂	93-60-7	137.137	cry	42.5	204			s H₂O, EtOH, bz
7743	Methyl 4-pyridinecarboxylate	Methyl isonicotinate	C₇H₇NO₂	2459-09-8	137.137		16.1	208	1.1599²⁰	1.5135²⁰	sl H₂O, ctc; s EtOH, eth, bz
7744	2-Methylpyridine-1-oxide		C₆H₇NO	931-19-1	109.126		49	260			
7745	3-Methylpyridine-1-oxide		C₆H₇NO	1003-73-2	109.126		39	148¹⁵			s chl
7746	4-Methylpyridine-1-oxide		C₆H₇NO	1003-67-4	109.126		185.8				
7747	1-Methyl-2(1H)-pyridinone		C₆H₇NO	694-85-9	109.126	nd	31	250	1.1120²⁰		msc H₂O; sl peth, lig
7748	1-(6-Methyl-3-pyridinyl)ethanone		C₈H₉NO	36357-38-7	135.163		17.6	144⁵⁰	1.0168²⁵	1.5302²⁵	vs H₂O
7749	4-Methyl-2-pyrimidinamine		C₅H₇N₃	108-52-1	109.130	pl (w), nd (sub)	160.3	sub			s H₂O, EtOH; sl chl
7750	2-Methylpyrimidine	2-Methyl-1,3-diazine	C₅H₆N₂	5053-43-0	94.115	liq	-4	138			msc H₂O
7751	4-Methylpyrimidine	4-Methyl-1,3-diazine	C₅H₆N₂	3438-46-8	94.115		32	142	1.030¹⁶	1.500²⁰	msc H₂O
7752	5-Methylpyrimidine	5-Methyl-1,3-diazine	C₅H₆N₂	2036-41-1	94.115		30.5	153			vs H₂O
7753	6-Methyl-2,4(1H,3H)-pyrimidinedione	6-Methyluracil	C₅H₆N₂O₂	626-48-2	126.114	oct pr or nd (w, al)	275 dec				s H₂O, EtOH; sl eth, tfa; vs NH₃
7754	1-Methylpyrrole		C₅H₇N	96-54-8	81.117	liq	-56.32	112.81	0.9145¹⁵	1.4875¹⁶	i H₂O; msc EtOH, eth
7755	2-Methylpyrrole		C₅H₇N	636-41-9	81.117	liq	-35.6	147.6	0.9446¹⁵	1.5035¹⁶	i H₂O; msc EtOH, eth
7756	3-Methylpyrrole		C₅H₇N	616-43-3	81.117	liq	-48.4	142.9; 45¹¹		1.4970²⁰	msc EtOH, eth
7757	N-Methylpyrrolidine		C₅H₁₁N	120-94-5	85.148			81	0.8188²⁰	1.4247²⁰	vs H₂O, eth
7758	1-Methyl-2,5-pyrrolidinedione		C₅H₇NO₂	1121-07-9	113.116	nd (eth-peth, al, ace)	71	234			s H₂O, EtOH; vs eth
7759	N-Methyl-2-pyrrolidinethione		C₅H₉NS	10441-57-3	115.197	oil		100⁰·⁰⁸			
7760	5-Methyl-2-pyrrolidinone		C₅H₉NO	108-27-0	99.131		43	248	1.0458²⁰		
7761	1-(1-Methyl-2-pyrrolidinyl)-2-propanone, (R)	Hygrine	C₈H₁₅NO	496-49-1	141.211			76.5¹¹		1.4555²⁰	vs EtOH, chl
7762	3-(1-Methyl-2-pyrrolidinyl)pyridine, (±)		C₁₀H₁₄N₂	22083-74-5	162.231			244	1.0082²⁰	1.5289²⁰	msc H₂O; vs EtOH, eth, chl; s lig
7763	N-Methyl-2-pyrrolidinone		C₅H₉NO	872-50-4	99.131	liq	-23.09	202	1.0230²⁵	1.4684²⁰	vs H₂O; s eth, ace, chl
7764	1-(1-Methyl-1H-pyrrol-2-yl)ethanone		C₇H₉NO	932-16-1	123.152			201²⁵²; 93²²	1.0445¹⁵	1.5403¹⁵	s EtOH, bz, chl
7765	6-Methyl-8-quinolinamine	8-Amino-6-methylquinoline	C₁₀H₁₀N₂	68420-93-9	158.199	nd	73	sub			vs ace, bz, eth, EtOH
7766	2-Methylquinoline	Quinaldine	C₁₀H₉N	91-63-4	143.185	col oily liq	-0.8	246.5	1.06²⁵	1.6116²⁰	sl H₂O; s EtOH, eth, ace, ctc, chl
7767	3-Methylquinoline		C₁₀H₉N	612-58-8	143.185	pr	16.5	259.8	1.0673²⁰	1.6171²⁰	vs ace, eth, EtOH
7768	4-Methylquinoline	Lepidine	C₁₀H₉N	491-35-0	143.185	col oily liq	9.5	262	1.083²⁰	1.6200²⁰	sl H₂O; s EtOH, eth, ace; i alk
7769	5-Methylquinoline		C₁₀H₉N	7661-55-4	143.185	col cry	19	262.7	1.0832²⁰	1.6219²⁰	sl H₂O; s ace; msc EtOH, eth
7770	6-Methylquinoline		C₁₀H₉N	91-62-3	143.185	col oily liq	-22	258.6	1.0654²⁰	1.6157²⁰	sl H₂O; s EtOH, eth, ace
7771	7-Methylquinoline	m-Toluquinoline	C₁₀H₉N	612-60-2	143.185	ye cry	39	257.6	1.0609²⁰	1.6150²⁰	sl H₂O; s EtOH, eth, ace
7772	8-Methylquinoline		C₁₀H₉N	611-32-5	143.185	col liq	-80	247.5	1.0719²⁰	1.6164²⁰	sl H₂O; s ace; msc EtOH, eth
7773	2-Methyl-8-quinolinol		C₁₀H₉NO	826-81-3	159.184		73.8	267			i H₂O; s EtOH, eth, bz, ctc
7774	1-Methyl-2(1H)-quinolinone		C₁₀H₉NO	606-43-9	159.184	nd (lig)	74	325			sl H₂O, lig; s EtOH, eth, ace; vs bz
7775	1-Methyl-4(1H)-quinolinone	Echinopsine	C₁₀H₉NO	83-54-5	159.184	α-nd (bz); β-cry (al)					s H₂O; vs EtOH, bz, chl; sl eth
7776	2-Methylquinoxaline		C₉H₈N₂	7251-61-8	144.173	ye cry	180.5	244			msc H₂O, eth, ace, bz; vs EtOH; s ctc
7777	Methyl Red	Benzoic acid, 2-[[4-(dimethylamino)phenyl]azo]-	C₁₅H₁₅N₃O₂	493-52-7	269.299	viol or red pr (to, bz)	183				sl H₂O, lig; s EtOH, eth; vs ace, bz, chl
7778	Methyl β-D-ribofuranoside		C₆H₁₂O₅	7473-45-2	164.156		80				

3-Methylpyridine

4-Methylpyridine

6-Methyl-2-pyridinecarboxaldehyde

Methyl 3-pyridinecarboxylate

Methyl 4-pyridinecarboxylate

2-Methylpyridine-1-oxide

3-Methylpyridine-1-oxide

4-Methylpyridine-1-oxide

1-Methyl-2(1*H*)-pyridinone

1-(6-Methyl-3-pyridinyl)ethanone

4-Methyl-2-pyrimidinamine

2-Methylpyrimidine

4-Methylpyrimidine

5-Methylpyrimidine

6-Methyl-2,4(1*H*,3*H*)-pyrimidinedione

1-Methylpyrrole

2-Methylpyrrole

3-Methylpyrrole

N-Methylpyrrolidine

1-Methyl-2,5-pyrrolidinedione

N-Methyl-2-pyrrolidinethione

5-Methyl-2-pyrrolidinone

1-(1-Methyl-2-pyrrolidinyl)-2-propanone, (*R*)

3-(1-Methyl-2-pyrrolidinyl)pyridine, (±)

N-Methyl-2-pyrrolidinone

1-(1-Methyl-1*H*-pyrrol-2-yl)ethanone

6-Methyl-8-quinolinamine

2-Methylquinoline

3-Methylquinoline

4-Methylquinoline

5-Methylquinoline

6-Methylquinoline

7-Methylquinoline

8-Methylquinoline

2-Methyl-8-quinolinol

1-Methyl-2(1*H*)-quinolinone

1-Methyl-4(1*H*)-quinolinone

2-Methylquinoxaline

Methyl Red

Methyl β-*D*-ribofuranoside

No.	Name	Synonym	Mol. Form.	CAS RN	Mol. Wt.	Physical Form	mp/°C	bp/°C	den/ g cm^{-3}	n_D	Solubility
7779	Methyl salicylate	Methyl 2-hydroxybenzoate	C$_8$H$_8$O$_3$	119-36-8	152.148	liq	-8	222.9	1.181[25]	1.535[20]	sl H$_2$O; vs eth, EtOH, chl
7780	Methylsilane		CH$_6$Si	992-94-9	46.145	col gas	-156.5	-57.5			
7781	Methyl silyl ether		CH$_6$OSi	2171-96-2	62.144	col gas	-98.5	-21; -87[10]			
7782	Methylstannane		CH$_6$Sn	1631-78-3	136.769	col gas		0			dec H$_2$O
7783	Methyl stearate		C$_{19}$H$_{38}$O$_2$	112-61-8	298.504		39.1	443; 215[15]	0.8498[40]	1.4367[40]	vs eth, chl
7784	2-Methylstyrene		C$_9$H$_{10}$	611-15-4	118.175	liq	-68.5	169.8	0.9077[25]	1.5437[20]	i H$_2$O; s bz, chl
7785	3-Methylstyrene		C$_9$H$_{10}$	100-80-1	118.175	liq	-86.3	164	0.9076[25]	1.5411[20]	i H$_2$O; s EtOH, eth, bz
7786	4-Methylstyrene		C$_9$H$_{10}$	622-97-9	118.175	liq	-34.1	172.8	0.9173[25]	1.5420	i H$_2$O; s bz
7787	Methylsuccinic acid		C$_5$H$_8$O$_4$	636-60-2	132.116	pr	115	dec	1.4200[0]	1.4303	vs H$_2$O, EtOH, MeOH; s eth; sl chl
7788	Methyl sulfate		CH$_4$O$_4$S	75-93-4	112.106		<-30	dec 135			vs H$_2$O, eth, EtOH
7789	(Methylsulfinyl)benzene		C$_7$H$_8$OS	1193-82-4	140.203		32.0	263.5; 140[13]		1.5885[20]	
7790	1-(Methylsulfinyl)decane	Decyl methyl sulfoxide	C$_{11}$H$_{24}$OS	3079-28-5	204.373	cry	52.5				
7791	3-Methyl sulfolane		C$_5$H$_{10}$O$_2$S	872-93-5	134.197		1	276	1.188[25]	1.4772[20]	
7792	(Methylsulfonyl)ethene		C$_3$H$_6$O$_2$S	3680-02-2	106.144			122[24]	1.2117[20]	1.4636[20]	s eth, ace
7793	Methyl terephthalate	Methyl 1,4-benzenedicarboxylate	C$_9$H$_8$O$_4$	1679-64-7	180.158	nd (w)	222	subl ≈ 230			
7794	17-Methyltestosterone	17-Hydroxy-17-methylandrost-4-en-3-one, (17β)	C$_{20}$H$_{30}$O$_2$	58-18-4	302.451		163.5				vs eth, EtOH
7795	Methyl tetradecanoate		C$_{15}$H$_{30}$O$_2$	124-10-7	242.398		19	295; 155[7]	0.8671[20]	1.425[45]	i H$_2$O; msc EtOH, eth, ace, bz, chl, ctc
7796	5-N-Methyl-5,6,7,8-tetrahydrofolic acid		C$_{20}$H$_{25}$N$_7$O$_6$	134-35-0	459.456	cry (w)					
7797	2-Methyltetrahydrofuran		C$_5$H$_{10}$O	96-47-9	86.132			78	0.8552[20]	1.4059[21]	s H$_2$O; vs EtOH, eth, ace, bz; sl ctc
7798	N-Methyl-N,2,4,6-tetranitroaniline	Tetryl	C$_7$H$_5$N$_5$O$_8$	479-45-8	287.144	ye pr (al)	131.5	exp 180	1.57[10]		i H$_2$O; sl EtOH, eth, chl; s ace, bz, py
7799	4-Methyl-2-thiazolamine	2-Amino-4-methylthiazole	C$_4$H$_6$N$_2$S	1603-91-4	114.169		45.5	125[20], 70[0.4]			vs H$_2$O, EtOH, eth
7800	2-Methylthiazole		C$_4$H$_5$NS	3581-87-1	99.155			128		1.510	msc H$_2$O; s EtOH, ace
7801	4-Methylthiazole		C$_4$H$_5$NS	693-95-8	99.155			133.3	1.112[25]		s H$_2$O, EtOH, eth
7802	4-Methyl-5-thiazoleethanol		C$_6$H$_9$NOS	137-00-8	143.206	col to pa ye		135[7]	1.196[24]		vs H$_2$O; s EtOH, eth, bz, chl
7803	4-Methyl-2(3H)-thiazolethione		C$_4$H$_5$NS$_2$	5685-06-3	131.220	ye cry (dil al)	89.3	188[3]			vs EtOH
7804	Methylthiirane		C$_3$H$_6$S	1072-43-1	74.145	liq	-91	72.5	0.941[20]	1.472[20]	s chl
7805	(Methylthio)acetic acid		C$_3$H$_6$O$_2$S	2444-37-3	106.144		13.0	130[27]	1.221[20]	1.495[20]	
7806	2-(Methylthio)aniline		C$_7$H$_9$NS	2987-53-3	139.218			234	1.111[25]	1.6239[20]	
7807	4-(Methylthio)aniline		C$_7$H$_9$NS	104-96-1	139.218			272.5	1.1379[20]	1.6395[20]	s EtOH, eth, ace, bz
7808	(Methylthio)benzene	Methyl phenyl sulfide	C$_7$H$_8$S	100-68-5	124.204			193	1.0579[20]	1.5868[20]	i H$_2$O; s EtOH; vs ace
7809	2-(Methylthio)benzothiazole		C$_8$H$_7$NS$_2$	615-22-5	181.279	pr (dil al)	52	174[22]			s EtOH, chl
7810	Methyl thiocyanate		C$_2$H$_3$NS	556-64-9	73.117	col liq	-2.5	132.9	1.0678[25]	1.4669[25]	sl H$_2$O; msc EtOH, eth; s ctc
7811	2-(Methylthio)ethanol		C$_3$H$_8$OS	5271-38-5	92.160			70[20]	1.063[20]	1.4861[30]	vs H$_2$O, eth, EtOH
7812	(Methylthio)ethene		C$_3$H$_6$S	1822-74-8	74.145			69.5	0.9026[20]	1.4837[20]	s eth, ace, chl
7813	[(Methylthio)methyl]benzene		C$_8$H$_{10}$S	766-92-7	138.230	liq	-30	210; 120[48]	1.0274[20]	1.5620[20]	
7814	4-(Methylthio)-2-oxobutanoic acid		C$_5$H$_8$O$_3$S	583-92-6	148.181	oil					
7815	2-Methylthiophene		C$_5$H$_6$S	554-14-3	98.167	liq	-63.4	112.6	1.0193[20]	1.5203[20]	i H$_2$O; msc EtOH, eth, ace, bz, hp, ctc
7816	3-Methylthiophene		C$_5$H$_6$S	616-44-4	98.167	liq	-69	115.5	1.0218[20]	1.5204[20]	i H$_2$O; msc EtOH, eth, ace, bz; vs chl
7817	5-Methyl-2-thiophenecarboxaldehyde		C$_6$H$_6$OS	13679-70-4	126.176			114[25]		1.5825[20]	s chl
7818	4-(Methylthio)phenol		C$_7$H$_8$OS	1073-72-9	140.203		84	154[20], 113[6]			
7819	3-(Methylthio)propanal		C$_4$H$_8$OS	3268-49-3	104.171			62[11]			
7820	3-(Methylthio)propanoic acid	S-Methylpropiothetin	C$_4$H$_8$O$_2$S	646-01-5	120.171	ye oil or fl (hx)	21	132[13]			
7821	3-(Methylthio)-1-propene		C$_4$H$_8$S	10152-76-8	88.172			92	0.8767[20]	1.4714[20]	
7822	N-Methylthiosemicarbazide	N-Methylhydrazinecarbothioamide	C$_2$H$_7$N$_3$S	6610-29-3	105.162		136.5				s H$_2$O, EtOH, DMSO; i eth, bz, lig

Methyl salicylate

Methylsilane

Methyl silyl ether

Methylstannane

Methyl stearate

2-Methylstyrene

3-Methylstyrene

4-Methylstyrene

Methylsuccinic acid

Methyl sulfate

(Methylsulfinyl)benzene

1-(Methylsulfinyl)decane

3-Methyl sulfolane

(Methylsulfonyl)ethene

Methyl terephthalate

17-Methyltestosterone

Methyl tetradecanoate

5-N-Methyl-5,6,7,8-tetrahydrofolic acid

2-Methyltetrahydrofuran

N-Methyl-N,2,4,6-tetranitroaniline

4-Methyl-2-thiazolamine

2-Methylthiazole

4-Methylthiazole

4-Methyl-5-thiazoleethanol

4-Methyl-2(3H)-thiazolethione

Methylthiirane

(Methylthio)acetic acid

2-(Methylthio)aniline

4-(Methylthio)aniline

(Methylthio)benzene

2-(Methylthio)benzothiazole

Methyl thiocyanate

2-(Methylthio)ethanol

(Methylthio)ethene

[(Methylthio)methyl]benzene

4-(Methylthio)-2-oxobutanoic acid

2-Methylthiophene

3-Methylthiophene

5-Methyl-2-thiophenecarboxaldehyde

4-(Methylthio)phenol

3-(Methylthio)propanal

3-(Methylthio)propanoic acid

3-(Methylthio)-1-propene

N-Methylthiosemicarbazide

No.	Name	Synonym	Mol. Form.	CAS RN	Mol. Wt.	Physical Form	mp/°C	bp/°C	den/ g cm^{-3}	n_D	Solubility
7823	Methylthiouracil		C$_5$H$_6$N$_2$OS	56-04-2	142.179		330 dec	sub			i H$_2$O; sl EtOH, eth, MeOH, bz
7824	Methylthiourea		C$_2$H$_6$N$_2$S	598-52-7	90.147	pr (EtOH)	121				vs H$_2$O, EtOH; sl eth; s ace
7825	1-Methylthymine	1,5-Dimethyl-2,4(1H,3H)-pyrimidinedione	C$_6$H$_8$N$_2$O$_2$	4160-72-9	140.140	nd (w)	295				s H$_2$O
7826	Methylthymol blue, sodium salt		C$_{37}$H$_{40}$N$_2$O$_{13}$Na$_4$S	1945-77-3	844.743	bl-viol cry					s H$_2$O
7827	Methyl 4-toluenesulfonate		C$_8$H$_{10}$O$_3$S	80-48-8	186.228		28.5	292; 186[22]	1.2087[40]		i H$_2$O; vs EtOH, bz; s eth, ctc; sl lig
7828	Methyltriacetoxysilane	Methylsilanetriol, triacetate	C$_7$H$_{12}$O$_6$Si	4253-34-3	220.252		40.5	111[17]	1.1750[20]	1.4083[20]	
7829	6-Methyl-1,2,4-triazine-3,5(2H,4H)-dione	6-Azathymine	C$_4$H$_5$N$_3$O$_2$	932-53-6	127.102	cry (w)	211				s H$_2$O, EtOH, ace
7830	5-Methyl-[1,2,4]triazolo[1,5-a]pyrimidin-7-ol		C$_6$H$_6$N$_4$O	2503-56-2	150.138		>245				
7831	Methyl trichloroacetate		C$_3$H$_3$Cl$_3$O$_2$	598-99-2	177.414	liq	-17.5	153.8	1.4874[20]	1.4572[20]	i H$_2$O; vs EtOH, eth; s ctc
7832	Methyltrichlorosilane		CH$_3$Cl$_3$Si	75-79-6	149.480	liq	-90	65.6	1.273[20]	1.4106[20]	dec H$_2$O, EtOH
7833	Methyl tridecanoate		C$_{14}$H$_{28}$O$_2$	1731-88-0	228.371		6.5	92[1]		1.4405[20]	msc EtOH; s ctc
7834	Methyltriethyllead	Triethylmethylplumbane	C$_7$H$_{18}$Pb	1762-28-3	309.4	col liq		70[16]	1.71[20]		
7835	Methyl trifluoroacetate		C$_3$H$_3$F$_3$O$_2$	431-47-0	128.050			43.0	1.28[20]		
7836	Methyl trifluoromethyl ether		C$_2$H$_3$F$_3$O	421-14-7	100.039	col gas	-149	-23.66			
7837	Methyl 3,4,5-trihydroxybenzoate		C$_8$H$_8$O$_5$	99-24-1	184.147	mcl pr (MeOH)	202				sl H$_2$O; vs EtOH, MeOH
7838	Methyl 3,4,5-trimethoxybenzoate		C$_{11}$H$_{14}$O$_5$	1916-07-0	226.226		83	274.5			
7839	Methyltriphenoxysilane		C$_{19}$H$_{18}$O$_3$Si	3439-97-2	322.430			269[100], 179[2]	1.135[20]	1.5599[20]	
7840	Methyl trithion		C$_9$H$_{12}$ClO$_2$PS$_3$	953-17-3	314.812	ye liq	-18				sl H$_2$O; misc os
7841	N-Methyl-L-tryptophan	L-Abrine	C$_{12}$H$_{14}$N$_2$O$_2$	526-31-8	218.251	pr (w)	295 dec				sl H$_2$O, MeOH; i eth; s alk
7842	N-Methyl-L-tyrosine	Surinamine	C$_{10}$H$_{13}$NO$_3$	537-49-5	195.215	nd	293				
7843	α-Methyl-DL-tyrosine, methyl ester, hydrochloride		C$_{11}$H$_{16}$ClNO$_3$	7361-31-1	245.703		190 dec				s H$_2$O
7844	2-Methylundecanal		C$_{12}$H$_{24}$O	110-41-8	184.318			119[16], 114[10]	0.832[15]	1.4321[20]	sl H$_2$O; s EtOH, eth
7845	2-Methylundecane		C$_{12}$H$_{26}$	7045-71-8	170.334	liq	-45.6	210.2		1.4191[20]	
7846	3-Methylundecane		C$_{12}$H$_{26}$	1002-43-3	170.334	col liq	-58.0	211.2	0.7485[25]	1.4208[25]	
7847	Methyl undecanoate		C$_{12}$H$_{24}$O$_2$	1731-86-8	200.318			123[10]			
7848	2-Methyl-1-undecanol		C$_{12}$H$_{26}$O	10522-26-6	186.333			129[12]	0.8300[15]	1.4382[20]	vs eth, EtOH
7849	Methyl 10-undecenoate		C$_{12}$H$_{22}$O$_2$	111-81-9	198.302	liq	-27.5	248	0.889[15]	1.4393[20]	i H$_2$O; s EtOH, eth, HOAc; sl ctc
7850	N-Methylurea		C$_2$H$_6$N$_2$O	598-50-5	74.081	orth pr (w, al)	104.9	dec	1.2040[0]		vs H$_2$O, EtOH; i eth, bz; s CS$_2$, lig
7851	5-Methyluridine	Thymine riboside	C$_{10}$H$_{14}$N$_2$O$_6$	1463-10-1	258.227	cry (EtOH)	184				
7852	3-Methyl-L-valine	L-$tert$-Leucine	C$_6$H$_{13}$NO$_2$	20859-02-3	131.173		248 dec				
7853	2-(1-Methylvinyl)aniline		C$_9$H$_{11}$N	52562-19-3	133.190			115[20], 95[13]	0.977[25]	1.5722[20]	
7854	1-Methyl-4-vinylcyclohexene		C$_9$H$_{14}$	17699-86-4	122.207	liq		152	0.85	1.4701[20]	
7855	4-(1-Methylvinyl)-1-cyclohexene-1-carboxaldehyde, (R)	d-Perillaldehyde	C$_{10}$H$_{14}$O	5503-12-8	150.217	oil		238; 99[7]	0.953[20]	1.5058[20]	s ctc
7856	4-(1-Methylvinyl)-1-cyclohexene-1-carboxaldehyde, (S)	l-Perillaldehyde	C$_{10}$H$_{14}$O	18031-40-8	150.217	oil		104[10]	0.9645[20]	1.5072[20]	
7857	4-(1-Methylvinyl)-1-cyclohexene-1-methanol		C$_{10}$H$_{16}$O	536-59-4	152.233			244; 12.5[12]	0.9690[20]	1.5005[20]	
7858	(1-Methylvinyl)cyclopropane		C$_6$H$_{10}$	4663-22-3	82.143	liq	-102.3	70	0.751[20]	1.4252[20]	
7859	Methyl vinyl ether		C$_3$H$_6$O	107-25-5	58.079	col gas	-122	5.5	0.7725[0]	1.3730[0]	sl H$_2$O; vs EtOH, eth, ace, bz
7860	Methyl Violet	C.I. Basic Violet 1	C$_{24}$H$_{28}$ClN$_3$	8004-87-3	393.952	bl-viol pow	137 dec				s H$_2$O, EtOH
7861	Methysergide		C$_{21}$H$_{27}$N$_3$O$_2$	361-37-5	353.458	cry	195				
7862	Methysticin		C$_{15}$H$_{14}$O$_5$	495-85-2	274.269	nd (MeOH), pr (ace)	137				
7863	Metobromuron	3-(p-Bromophenyl)-1-methoxy-1-methylurea	C$_9$H$_{11}$BrN$_2$O$_2$	3060-89-7	259.099		95		1.60[20]		
7864	Metolachlor		C$_{15}$H$_{22}$ClNO$_2$	51218-45-2	283.795			100[0.001]	1.12[20]		
7865	Metolazone		C$_{16}$H$_{16}$ClN$_3$O$_3$S	17560-51-9	365.834	cry (EtOH)	254				
7866	Metoprolol tartrate		C$_{34}$H$_{56}$N$_2$O$_{12}$	56392-17-7	684.815	cry	121				
7867	Metribuzin		C$_8$H$_{14}$N$_4$OS	21087-64-9	214.288		126		1.31[20]		
7868	Metronidazole	2-Methyl-5-nitro-1H-imidazole-1-ethanol	C$_6$H$_9$N$_3$O$_3$	443-48-1	171.153		160.5				
7869	Metsulfuron-methyl		C$_{14}$H$_{15}$N$_5$O$_6$S	74223-64-6	381.364	wh cry	163				sl H$_2$O
7870	Mevinphos		C$_7$H$_{13}$O$_6$P	7786-34-7	224.148		21 (E), 6.9 (Z)	101[0.3]			
7871	Mexacarbate	4-(Dimethylamino)-3,5-xylyl methylcarbamate	C$_{12}$H$_{18}$N$_2$O$_2$	315-18-4	222.283	cry	85				vs EtOH, bz, ace

Methylthiouracil

Methylthiourea

1-Methylthymine

Methylthymol blue, sodium salt

Methyl 4-toluenesulfonate

Methyltriacetoxysilane

6-Methyl-1,2,4-triazine-3,5(2H,4H)-dione

5-Methyl-[1,2,4]triazolo[1,5-a]pyrimidin-7-ol

Methyl trichloroacetate

Methyltrichlorosilane

Methyl tridecanoate

Methyltriethyllead

Methyl trifluoroacetate

Methyl trifluoromethyl ether

Methyl 3,4,5-trihydroxybenzoate

Methyl 3,4,5-trimethoxybenzoate

Methyltriphenoxysilane

Methyl trithion

N-Methyl-L-tryptophan

N-Methyl-L-tyrosine

α-Methyl-DL-tyrosine, methyl ester, hydrochloride

2-Methylundecanal

2-Methylundecane

3-Methylundecane

Methyl undecanoate

2-Methyl-1-undecanol

Methyl 10-undecenoate

N-Methylurea

5-Methyluridine

3-Methyl-L-valine

2-(1-Methylvinyl)aniline

1-Methyl-4-vinylcyclohexene

4-(1-Methylvinyl)-1-cyclohexene-1-carboxaldehyde, (R)

4-(1-Methylvinyl)-1-cyclohexene-1-carboxaldehyde, (S)

4-(1-Methylvinyl)-1-cyclohexene-1-methanol

(1-Methylvinyl)cyclopropane

Methyl vinyl ether

Methyl Violet

Methysergide

Methysticin

Metobromuron

Metolachlor

Metolazone

Metoprolol tartrate

Metribuzin

Metronidazole

Metsulfuron-methyl

Mevinphos

Mexacarbate

No.	Name	Synonym	Mol. Form.	CAS RN	Mol. Wt.	Physical Form	mp/°C	bp/°C	den/ g cm⁻³	n_D	Solubility
7872	MGK 264		$C_{17}H_{25}NO_2$	113-48-4	275.387		<-20	157	1.04		
7873	Mifepristone	RU-486	$C_{29}H_{35}NO_2$	84371-65-3	429.594	cry	150				
7874	Mimosine		$C_8H_{10}N_2O_4$	500-44-7	198.176	tab (w)	228 dec				sl H_2O; i EtOH, eth, ace, bz; s dil alk
7875	Minocycline		$C_{23}H_{27}N_3O_7$	10118-90-8	457.476	ye-oran amorp solid					
7876	Minoxidil		$C_9H_{15}N_5O$	38304-91-5	209.248	cry	248				i ace, bz, chl, sl; EtOH, MeOH
7877	Mipafox	Bis(isopropylamido) fluorophosphate	$C_6H_{16}FN_2OP$	371-86-8	182.175	cry (peth)	65	125^2			sl H_2O
7878	Mirex	Hexachloropentadiene dimer	$C_{10}Cl_{12}$	2385-85-5	545.543	cry (bz)	485 dec				vs bz, diox
7879	Misoprostol		$C_{22}H_{38}O_5$	59122-46-2	382.534	ye oil					s H_2O
7880	Mithramycin	Plicamycin	$C_{52}H_{76}O_{24}$	18378-89-7	1085.145	ye cry (ace)	182				s H_2O, EtOH, AcOEt; sl bz, eth
7881	Mitomycin A		$C_{16}H_{19}N_3O_6$	4055-39-4	349.338	purp nd	160 dec				
7882	Mitomycin B		$C_{16}H_{19}N_3O_6$	4055-40-7	349.338	purp-bl nd	dec				
7883	Mitomycin C		$C_{15}H_{18}N_4O_5$	50-07-7	334.328	bl-viol cry	360				s H_2O, MeOH, ace
7884	Mitotane		$C_{14}H_{10}Cl_4$	53-19-0	320.041		77				
7885	Mitragynine	9-Methoxycorynantheidine	$C_{23}H_{30}N_2O_4$	4098-40-2	398.495	wh amor pow	104	235^5			s EtOH, chl, HOAc
7886	Molinate	Ethyl 1-hexamethyleneiminecarbothiolate	$C_9H_{17}NOS$	2212-67-1	187.302			202^{10}	1.063^{20}		
7887	Molindone		$C_{16}H_{24}N_2O_2$	7416-34-4	276.374	cry	180				
7888	Molybdenum hexacarbonyl		C_6MoO_6	13939-06-5	264.00		dec 150				s os
7889	Monobutyl phthalate	1,2-Benzenedicarboxylic acid, monobutyl ester	$C_{12}H_{14}O_4$	131-70-4	222.237	pl (ace, al)	73.5				vs EtOH, chl
7890	Monobutyltin trichloride		$C_4H_9Cl_3Sn$	1118-46-3	282.183	hyg liq	-63	93^{10}	0.85^{20}		s bz, CH_2Cl_2
7891	Monocrotaline		$C_{16}H_{23}NO_6$	315-22-0	325.357	wh pr (EtOH)	198 dec				
7892	Monocrotophos		$C_7H_{14}NO_5P$	6923-22-4	223.164		55	$125^{0.0005}$	1.33^{20}		
7893	Monolinuron	N'-(4-Chlorophenyl)-N-methoxy-N-methylurea	$C_9H_{11}ClN_2O_2$	1746-81-2	214.648	solid	77				
7894	Monomethyl adipate		$C_7H_{12}O_4$	627-91-8	160.168	lf (Me₃N-MeOH)	9	158^{10}	1.0623^{20}	1.4283^{20}	s EtOH
7895	Monomethyl glutarate		$C_6H_{10}O_4$	1501-27-5	146.141			158^{27}, 150^{10}	1.169^{25}	1.4381^{20}	
7896	Monosodium L-glutamate		$C_5H_8NNaO_4$	142-47-2	169.113						s H_2O
7897	Moquizone		$C_{20}H_{21}N_3O_3$	19395-58-5	351.399		136				s chl
7898	Morin		$C_{15}H_{10}O_7$	480-16-0	302.236	pa ye nd (+1 w, dil al)	303.5				sl H_2O, eth; vs EtOH; s bz, alk; i CS_2
7899	Morphine		$C_{17}H_{19}NO_3$	57-27-2	285.338	pr	255	sub 190			i H_2O, eth, ace; s MeOH, py; sl EtOH
7900	4-Morpholinamine		$C_4H_{10}N_2O$	4319-49-7	102.134			166	1.059^{25}	1.4772^{20}	
7901	Morpholine	Tetrahydro-1,4-oxazine	C_4H_9NO	110-91-8	87.120	hyg liq	-4.8	128	1.0005^{20}	1.4548^{20}	msc H_2O; s EtOH, eth, ace, bz; sl chl
7902	4-Morpholinecarboxaldehyde		$C_5H_9NO_2$	4394-85-8	115.131		21	239	1.1520^{20}	1.4845^{20}	
7903	4-Morpholineethanamine		$C_6H_{14}N_2O$	2038-03-1	130.187		25.6	205	0.9897^{20}	1.4715^{20}	msc H_2O, EtOH, bz, lig; s ace
7904	4-Morpholineethanol		$C_6H_{13}NO_2$	622-40-2	131.173	liq	-0.8	227	1.0710^{20}	1.4763^{20}	s H_2O, EtOH; sl ctc
7905	4-Morpholinepropanamine	4-(3-Aminopropyl)morpholine	$C_7H_{16}N_2O$	123-00-2	144.214	liq	-15	220; 134^{50}	0.9854^{20}	1.4762^{20}	msc H_2O, EtOH, bz, lig; s ace; sl ctc
7906	2-(4-Morpholinothio) benzothiazole	4-(2-Benzothiazolylthio) morpholine	$C_{11}H_{12}N_2OS_2$	102-77-2	252.355	cry (EtOH)	85				
7907	4-(4-Morpholinyl)aniline		$C_{10}H_{14}N_2O$	2524-67-6	178.230		131.6				
7908	2-(4-Morpholinyldithio) benzothiazole		$C_{11}H_{12}N_2OS_3$	95-32-9	284.420		135				
7909	Muldamine		$C_{29}H_{47}NO_3$	36069-45-1	457.688		210				
7910	Murexide	5,5'-Nitrilobarbituric acid, ammonium salt	$C_8H_{10}N_6O_7$	3051-09-0	302.201						sl H_2O; i EtOH, eth; s alk
7911	Muscimol	5-(Aminomethyl)-3(2H)-isoxazolone	$C_4H_6N_2O_2$	2763-96-4	114.103	cry (EtOH)	175 dec				
7912	Myclobutanil		$C_{15}H_{17}ClN_4$	88671-89-0	288.776	ye cry	65	$205^{1.0}$			i H_2O, peth; s EtOH
7913	Mycophenolic acid		$C_{17}H_{20}O_6$	24280-93-1	320.337	nd (w)	141				i H_2O; vs EtOH, eth, chl; sl bz, tol

MGK 264

Mifepristone

Mimosine

Minocycline

Minoxidil

Mipafox

Mirex

Misoprostol

Mithramycin

Mitomycin A

Mitomycin B

Mitomycin C

Mitotane

Mitragynine

Molinate

Molindone

Molybdenum hexacarbonyl

Monobutyl phthalate

Monobutyltin trichloride

Monocrotaline

Monocrotophos

Monolinuron

Monomethyl adipate

Monomethyl glutarate

Monosodium *L*-glutamate

Moquizone

Morin

Morphine

4-Morpholinamine

Morpholine

4-Morpholinecarboxaldehyde

4-Morpholineethanamine

4-Morpholineethanol

4-Morpholinepropanamine

2-(4-Morpholinothio)benzothiazole

4-(4-Morpholinyl)aniline

2-(4-Morpholinyldithio)benzothiazole

Muldamine

Murexide

Muscimol

Myclobutanil

Mycophenolic acid

No.	Name	Synonym	Mol. Form.	CAS RN	Mol. Wt.	Physical Form	mp/°C	bp/°C	den/ g cm⁻³	n_D	Solubility
7914	β-Myrcene	7-Methyl-3-methylene-1,6-octadiene	C₁₀H₁₆	123-35-3	136.234			167	0.8013[15]	1.4722[20]	i H₂O; s EtOH, eth, bz, chl, HOAc
7915	Myristicin		C₁₁H₁₂O₃	607-91-0	192.211		<-20	276.5	1.1416[20]	1.5403[20]	i H₂O; sl EtOH; s eth, bz
7916	Nabam	Sodium ethylenebisdithiocarbamic acid	C₄H₆N₂Na₂S₄	142-59-6	256.344	cry (w)					s H₂O
7917	Nadolol		C₁₇H₂₇NO₄	42200-33-9	309.401	cry (bz)	≈130				s EtOH; sl chl; i ace, eth, hx
7918	Naled	1,2-Dibromo-2,2-dichloroethylphosphoric acid, dimethyl ester	C₄H₇Br₂Cl₂O₄P	300-76-5	380.784		27	110[0.5]	1.96[20]		
7919	Nalidixic acid		C₁₂H₁₂N₂O₃	389-08-2	232.234		229.5				sl EtOH, eth; s chl
7920	Nalmefene		C₂₁H₂₅NO₃	55096-26-9	339.429	cry (AcOEt)	189				
7921	Nalorphine	Acetorphin	C₁₉H₂₁NO₃	62-67-9	311.375	cry (eth)	208				sl H₂O; s alk, ace, EtOH
7922	Naloxone		C₁₉H₂₁NO₄	465-65-6	327.375	cry (AcOEt)	178				i peth; s chl
7923	Naltrexone		C₂₀H₂₃NO₄	16590-41-3	341.402	cry (ace)	169				
7924	Nandrolone	17-Hydroxyestr-4-en-3-one	C₁₈H₂₆O₂	434-22-0	274.398	cry	112				s EtOH, eth, chl
7925	Naphazoline hydrochloride		C₁₄H₁₄ClN₂	550-99-2	245.727						sl H₂O
7926	Naphthacene	2,3-Benzanthracene	C₁₈H₁₂	92-24-0	228.288	oran-ye lf (bz, xyl)	357	sub			i H₂O; sl bz; s con sulf
7927	5,12-Naphthacenedione		C₁₈H₁₀O₂	1090-13-7	258.271		285 dec				sl ace, bz, gl HOAc
7928	Naphthalene		C₁₀H₈	91-20-3	128.171	mcl pl (al)	80.26	217.9	1.0253[20]	1.5898[25]	i H₂O; s EtOH; vs eth, ace, bz, CS₂
7929	1-Naphthaleneacetamide		C₁₂H₁₁NO	86-86-2	185.221	nd(w, al)		sub 180			i H₂O; s eth, bz, CS₂, HOAc
7930	1-Naphthaleneacetic acid	1-Naphthylacetic acid	C₁₂H₁₀O₂	86-87-3	186.206	nd (w)	133	dec			sl H₂O, EtOH; vs eth, ace, chl; s bz
7931	2-Naphthaleneacetic acid	2-Naphthylacetic acid	C₁₂H₁₀O₂	581-96-4	186.206	lf(w) cry (bz)	143				vs eth, lig, chl
7932	1-Naphthaleneacetonitrile		C₁₂H₉N	132-75-2	167.206		32.5	192[18], 163[12]		1.6192[20]	s EtOH
7933	1-Naphthalenecarbonitrile		C₁₁H₇N	86-53-3	153.181	nd (lig)	37.5	299	1.1080[25]	1.6298[18]	i H₂O; vs EtOH, eth; s lig
7934	2-Naphthalenecarbonitrile		C₁₁H₇N	613-46-7	153.181	lf (lig)	66	306.5	1.0755[60]		sl H₂O, chl; s EtOH, eth, lig
7935	1-Naphthalenecarbonyl chloride		C₁₁H₇ClO	879-18-5	190.626		20	297.5			
7936	2-Naphthalenecarbonyl chloride		C₁₁H₇ClO	2243-83-6	190.626	cry (peth)	51	305			vs bz, eth, chl
7937	1-Naphthalenecarboxaldehyde		C₁₁H₈O	66-77-3	156.181	pa ye	33.5	292	1.1503[20]	1.6507[20]	i H₂O; s EtOH, eth, ace, bz, sulf
7938	2-Naphthalenecarboxaldehyde		C₁₁H₈O	66-99-9	156.181	lf (w)	62	160[19]	1.0775[94]	1.6211[99]	sl H₂O; vs EtOH, eth; s ace
7939	1-Naphthalenecarboxylic acid	1-Naphthoic acid	C₁₁H₈O₂	86-55-5	172.181	nd (HOAc-w, w, al)	161	>300	1.398[25]	1.46	i H₂O; vs eth, EtOH, chl
7940	2-Naphthalenecarboxylic acid	2-Naphthoic acid	C₁₁H₈O₂	93-09-4	172.181	nd (lig, chl, sub) pl (ace)	185.5	>300	1.077[100]		sl H₂O, DMSO, lig; s EtOH, eth, chl
7941	1,5-Naphthalenediamine	1,5-Diaminonaphthalene	C₁₀H₁₀N₂	2243-62-1	158.199	pr (eth, al, w)	190	sub	1.4[25]		s H₂O, EtOH, eth; vs chl
7942	1,8-Naphthalenediamine	1,8-Diaminonaphthalene	C₁₀H₁₀N₂	479-27-6	158.199		66.5	205[12]	1.1265[90]	1.6828[99]	vs eth, EtOH
7943	2,3-Naphthalenediamine	2,3-Diaminonaphthalene	C₁₀H₁₀N₂	771-97-1	158.199	lf (eth, w)	199		1.0968[26]	1.6392[26]	sl H₂O, DMSO; vs EtOH; s eth
7944	1,8-Naphthalenedicarboxylic acid	Naphthalic acid	C₁₂H₈O₄	518-05-8	216.190		260				i H₂O; sl EtOH, eth
7945	2,3-Naphthalenedicarboxylic acid		C₁₂H₈O₄	2169-87-1	216.190	pr (HOAc, w, sub)	244.5				i H₂O, bz, chl; sl EtOH, eth, DMSO
7946	2,6-Naphthalenedicarboxylic acid		C₁₂H₈O₄	1141-38-4	216.190	nd (al or sub)	>300 dec				vs EtOH
7947	2,6-Naphthalenedicarboxylic acid, dimethyl ester		C₁₄H₁₂O₄	840-65-3	244.243		190.0				
7948	1,5-Naphthalene diisocyanate	1,5-Diisocyanatonaphthalene	C₁₂H₆N₂O₂	3173-72-6	210.188	cry	127	183[10]			
7949	1,3-Naphthalenediol	Naphthoresorcinol	C₁₀H₈O₂	132-86-5	160.170	lf (w)	123.5				s H₂O, EtOH, eth; sl ace, bz, lig
7950	1,4-Naphthalenediol		C₁₀H₈O₂	571-60-8	160.170	mcl nd (bz, w)	192				s H₂O, EtOH, eth; sl ace; i bz
7951	1,5-Naphthalenediol		C₁₀H₈O₂	83-56-7	160.170	pr (w), nd (sub)	262 dec	sub			sl H₂O, EtOH; vs eth, ace; i bz; s HOAc

β-Myrcene

Myristicin

Nabam

Nadolol

Naled

Nalidixic acid

Nalmefene

Nalorphine

Naloxone

Naltrexone

Nandrolone

Naphazoline hydrochloride

Naphthacene

5,12-Naphthacenedione

Naphthalene

1-Naphthaleneacetamide

1-Naphthaleneacetic acid

2-Naphthaleneacetic acid

1-Naphthaleneacetonitrile

1-Naphthalenecarbonitrile

2-Naphthalenecarbonitrile

1-Naphthalenecarbonyl chloride

2-Naphthalenecarbonyl chloride

1-Naphthalenecarboxaldehyde

2-Naphthalenecarboxaldehyde

1-Naphthalenecarboxylic acid

2-Naphthalenecarboxylic acid

1,5-Naphthalenediamine

1,8-Naphthalenediamine

2,3-Naphthalenediamine

1,8-Naphthalenedicarboxylic acid

2,3-Naphthalenedicarboxylic acid

2,6-Naphthalenedicarboxylic acid

2,6-Naphthalenedicarboxylic acid, dimethyl ester

1,5-Naphthalene diisocyanate

1,3-Naphthalenediol

1,4-Naphthalenediol

1,5-Naphthalenediol

No.	Name	Synonym	Mol. Form.	CAS RN	Mol. Wt.	Physical Form	mp/°C	bp/°C	den/ g cm⁻³	n_D	Solubility
7952	1,6-Naphthalenediol		$C_{10}H_8O_2$	575-44-0	160.170	pr (bz)	138	sub			sl H_2O, EtOH; s eth, ace, bz, DMSO
7953	1,7-Naphthalenediol		$C_{10}H_8O_2$	575-38-2	160.170	nd (bz or sub)	180.5	sub			sl H_2O; vs EtOH, eth; s bz, HOAc
7954	2,3-Naphthalenediol		$C_{10}H_8O_2$	92-44-4	160.170	lf (w)	163.5				s H_2O, EtOH, eth, ace, bz, lig, HOAc
7955	2,6-Naphthalenediol		$C_{10}H_8O_2$	581-43-1	160.170	orth pl (w)	220	sub			sl H_2O, bz; s EtOH, eth, ace; i lig
7956	2,7-Naphthalenediol		$C_{10}H_8O_2$	582-17-2	160.170	nd, (w, dil al), pl (dil al)	193	sub			s H_2O, EtOH, eth, bz, chl; sl ace; i lig
7957	1,2-Naphthalenedione	1,2-Naphthoquinone	$C_{10}H_6O_2$	524-42-5	158.154	ye-red nd (eth) oran lf (bz)	146		1.450[25]		s H_2O, EtOH, eth, sulf; sl lig
7958	1,4-Naphthalenedione	1,4-Naphthoquinone	$C_{10}H_6O_2$	130-15-4	158.154	bt ye nd (al, peth) ye (sub)	128.5	sub			sl H_2O; vs EtOH; s eth, bz, chl, CS_2
7959	1,5-Naphthalenedisulfonic acid	Armstrong's acid	$C_{10}H_8O_6S_2$	81-04-9	288.297	pl (+4w, dil HOAc)	242 dec		1.493[25]		vs H_2O; s EtOH; i eth
7960	1,6-Naphthalenedisulfonic acid	Naphthalene-1,6-disulfonic acid	$C_{10}H_8O_6S_2$	525-37-1	288.297	oran pr (+4w, HOAc or w)	125 dec				vs H_2O; s EtOH; i eth
7961	2,7-Naphthalenedisulfonic acid	Naphthalene-2,7-disulfonic acid	$C_{10}H_8O_6S_2$	92-41-1	288.297	hyg nd (conc HCl)	199				s H_2O; sl con HCl
7962	1-Naphthalenemethanamine		$C_{11}H_{11}N$	118-31-0	157.212			292	1.0958[20]		s EtOH, eth, sulf, CS_2
7963	1-Naphthalenemethanol		$C_{11}H_{10}O$	4780-79-4	158.196	nd (w, al), cry (bz-lig)	64	304; 163[12]	1.1039[80]		sl H_2O; vs EtOH, eth
7964	2-Naphthalenemethanol		$C_{11}H_{10}O$	1592-38-7	158.196	lf	81.3	178[12]			sl H_2O; s EtOH, eth
7965	1-Naphthalenesulfonic acid	α-Naphthylsulfonic acid	$C_{10}H_8O_3S$	85-47-2	208.234	pr (+2 w, dil HCl)	140				s H_2O, EtOH; sl eth
7966	2-Naphthalenesulfonic acid	β-Naphthylsulfonic acid	$C_{10}H_8O_3S$	120-18-3	208.234	hyg pl (+1w), cry (+3w, HCl)	91	dec	1.441[25]		vs H_2O, EtOH; s eth; sl bz
7967	1-Naphthalenesulfonyl chloride		$C_{10}H_7ClO_2S$	85-46-1	226.680	lf (eth)	68	209[20], 147[0.9]			vs bz, eth, EtOH
7968	2-Naphthalenesulfonyl chloride		$C_{10}H_7ClO_2S$	93-11-8	226.680	pow or lf (bz-peth)	81	201[13], 148[0.5]			i H_2O; s EtOH, bz, chl; sl peth; vs eth
7969	1,4,5,8-Naphthalenetetracarboxylic acid		$C_{14}H_8O_8$	128-97-2	304.209	lf or nd (w, dil HCl)	320				sl H_2O, bz, chl, EtOH; vs ace
7970	1-Naphthalenethiol	1-Naphthyl mercaptan	$C_{10}H_8S$	529-36-2	160.236			dec 285; 161[20]	1.1607[20]	1.6802[20]	sl H_2O, dil alk; vs EtOH, eth
7971	2-Naphthalenethiol	2-Naphthyl mercaptan	$C_{10}H_8S$	91-60-1	160.236	pl (al)	81	288	1.550[25]		sl H_2O; vs EtOH, eth, lig
7972	N-(1-Naphthalenyl)-1,2-ethanediamine, dihydrochloride		$C_{12}H_{16}Cl_2N_2$	1465-25-4	259.174	hex pr	189				vs H_2O, EtOH
7973	1-Naphthalenylthiourea	ANTU	$C_{11}H_{10}N_2S$	86-88-4	202.275	pr (al)	198				i H_2O; sl EtOH, eth, ace
7974	Naphtho[2,3-c]furan-1,3-dione	2,3-Naphthalenedicarboxylic acid anhydride	$C_{12}H_6O_3$	716-39-2	198.174		246				sl EtOH, chl; s eth, bz
7975	1-Naphthol	1-Naphthalenol	$C_{10}H_8O$	90-15-3	144.170	ye nd (w)	95.0	288; 184[40]	1.0989[99]	1.6224[99]	i H_2O; vs EtOH, eth; s ace, bz; sl ctc
7976	2-Naphthol	2-Naphthalenol	$C_{10}H_8O$	135-19-3	144.170	mcl lf (w)	121.5	285	1.28[20]		i H_2O; vs EtOH, eth; s bz, chl; sl lig
7977	1-Naphthol, acetate	1-Naphthyl acetate	$C_{12}H_{10}O_2$	830-81-9	186.206	nd or pl (al)	49	114[1]			i H_2O; s EtOH, eth
7978	2-Naphthol, acetate	2-Naphthyl acetate	$C_{12}H_{10}O_2$	1523-11-1	186.206	nd (al)	71.0	132[2]			i H_2O; s EtOH, eth, chl
7979	p-Naphtholbenzein		$C_{27}H_{18}O_2$	145-50-6	374.431		123				
7980	1H,3H-Naphtho[1,8-cd]pyran-1,3-dione		$C_{12}H_6O_3$	81-84-5	198.174		275.0				i H_2O, eth, bz; sl EtOH; s HOAc
7981	1-Naphthylamine	α-Naphthylamine	$C_{10}H_9N$	134-32-7	143.185		49.2	300.7	1.0228[20]	1.6140[20]	s chl
7982	2-Naphthylamine	β-Naphthylamine	$C_{10}H_9N$	91-59-8	143.185		113	306.2	1.6414[98]	1.6493[98]	s H_2O, EtOH, eth
7983	2-[(1-Naphthylamino)carbonyl]benzoic acid	Naptalam	$C_{18}H_{13}NO_3$	132-66-1	291.301		185		1.4[20]		i H_2O; sl EtOH, ace, bz, tfa
7984	2-Naphthyl benzoate	2-Naphthalenol benzoate	$C_{17}H_{12}O_2$	93-44-7	248.276	nd or pr (al)	107				i H_2O; s EtOH; sl eth, HOAc
7985	N-1-Naphthalenylacetamide		$C_{12}H_{11}NO$	575-36-0	185.221		160				s H_2O, EtOH; sl eth
7986	N-1-Naphthyl-1,2-ethanediamine	N-(1-Naphthyl)ethylenediamine	$C_{12}H_{14}N_2$	551-09-7	186.252	visc liq		204[9]	1.114[25]	1.6648[25]	

Physical Constants of Organic Compounds

1,6-Naphthalenediol

1,7-Naphthalenediol

2,3-Naphthalenediol

2,6-Naphthalenediol

2,7-Naphthalenediol

1,2-Naphthalenedione

1,4-Naphthalenedione

1,5-Naphthalenedisulfonic acid

1,6-Naphthalenedisulfonic acid

2,7-Naphthalenedisulfonic acid

1-Naphthalenemethanamine

1-Naphthalenemethanol

2-Naphthalenemethanol

1-Naphthalenesulfonic acid

2-Naphthalenesulfonic acid

1-Naphthalenesulfonyl chloride

2-Naphthalenesulfonyl chloride

1,4,5,8-Naphthalenetetracarboxylic acid

1-Naphthalenethiol

2-Naphthalenethiol

N-(1-Naphthalenyl)-1,2-ethanediamine, dihydrochloride

1-Naphthalenylthiourea

Naphtho[2,3-c]furan-1,3-dione

1-Naphthol

2-Naphthol

1-Naphthol, acetate

2-Naphthol, acetate

p-Naphtholbenzein

1H,3H-Naphtho[1,8-cd]pyran-1,3-dione

1-Naphthylamine

2-Naphthylamine

2-[(1-Naphthylamino)carbonyl]benzoic acid

2-Naphthyl benzoate

N-1-Naphthalenylacetamide

N-1-Naphthyl-1,2-ethanediamine

No.	Name	Synonym	Mol. Form.	CAS RN	Mol. Wt.	Physical Form	mp/°C	bp/°C	den/ g cm^-3	n_D	Solubility
7987	1-Naphthyl 2-hydroxybenzoate	1-Naphthyl salicylate	$C_{17}H_{12}O_3$	550-97-0	264.275		83				vs eth
7988	1-Naphthylhydroxylamine	N-Hydroxyl-1-naphthalenamine	$C_{10}H_9NO$	607-30-7	159.184		79				
7989	1-Naphthyl isothiocyanate	1-Isothiocyanatonaphthalene	$C_{11}H_7NS$	551-06-4	185.246	wh nd (al)	58				vs bz, eth, EtOH, chl
7990	N-2-Naphthyl-2-naphthalenamine	β,β'-Dinaphthylamine	$C_{20}H_{15}N$	532-18-3	269.340	lf(bz)	172.2	471			i H_2O; sl EtOH, bz, DMSO; s eth, HOAc
7991	(2-Naphthyloxy)acetic acid	2-Naphthoxyacetic acid	$C_{12}H_{10}O_3$	120-23-0	202.205	pr(w)	156				s H_2O, EtOH, eth; sl DMSO
7992	1-Naphthyl phosphate	1-Naphthalenol, dihydrogen phosphate	$C_{10}H_9O_4P$	1136-89-6	224.149	cry	160				
7993	2-Naphthyl salicylate	2-Naphthyl 2-hydroxybenzoate	$C_{17}H_{12}O_3$	613-78-5	264.275	cry (al)	95.5		1.11[116]		i H_2O; sl EtOH; s eth, bz
7994	1,5-Naphthyridine	1,5-Diazanaphthalene	$C_8H_6N_2$	254-79-5	130.147	ye nd (peth)	75	112[12]	1.2100[20]		
7995	1,6-Naphthyridine		$C_8H_6N_2$	253-72-5	130.147		29.5				
7996	Napropamide	Propanamide, N,N-diethyl-2-(1-naphthalenyloxy)-	$C_{17}H_{21}NO_2$	15299-99-7	271.355		75				
7997	Naproxen	6-Methoxy-α-methyl-2-naphthaleneacetic acid	$C_{14}H_{14}O_3$	22204-53-1	230.259	cry (ace/hx)	155				i H_2O; sl eth; s MeOH, chl
7998	Narceine		$C_{23}H_{27}NO_8$	131-28-2	445.462		138				i H_2O
7999	Narcobarbital		$C_{11}H_{15}BrN_2O_3$	125-55-3	303.152		115				sl H_2O; s EtOH, py
8000	Naringenin		$C_{15}H_{12}O_5$	480-41-1	272.253	nd (dil al)	251				vs bz, eth, EtOH
8001	Naringin		$C_{27}H_{32}O_{14}$	10236-47-2	580.535	nd (w+8)					sl H_2O, EtOH; i eth, bz, chl; s HOAc
8002	Nealbarbital		$C_{12}H_{18}N_2O_3$	561-83-1	238.282		156				vs ace, eth, EtOH
8003	Nellite	Diamidafos	$C_8H_{13}N_2O_2P$	1754-58-1	200.175	cry (ctc)	103.5				sl AcOEt, bz
8004	Neoabietic acid	8(14),13(15)-Abietadien-18-oic acid	$C_{20}H_{30}O_2$	471-77-2	302.451	cry (EtOH aq)	173				
8005	Neobornylamine		$C_{10}H_{19}N$	2223-67-8	153.265	pow	184				vs ace, eth
8006	Neopentane	2,2-Dimethylpropane	C_5H_{12}	463-82-1	72.149	col gas	-16.4	9.48	0.5852[25] (p>1 atm	1.3476[6]	i H_2O; s EtOH, eth, ctc
8007	Neopine		$C_{18}H_{21}NO_3$	467-14-1	299.365	nd (peth)	127.5				s H_2O, EtOH, eth, bz; vs chl; sl lig
8008	Neostigmine bromide		$C_{12}H_{19}BrN_2O_2$	114-80-7	303.195	cry (al-eth)	167 dec				vs H_2O; s EtOH
8009	Nepetalactone		$C_{10}H_{14}O_2$	490-10-8	166.217			71[0.05]	1.0663[25]	1.4859[25]	
8010	cis-Nerolidol		$C_{15}H_{26}O$	142-50-7	222.366			276; 70[0.1]	0.8778[20]	1.4898[20]	vs EtOH; s eth, ace, HOAc
8011	Neurine		$C_5H_{13}NO$	463-88-7	103.163	syr					vs H_2O, eth, EtOH
8012	Neutral Red		$C_{15}H_{17}ClN_4$	553-24-2	288.776	grn pow					s H_2O, ethylene glycol, EtOH; i xyl
8013	Nialamide		$C_{16}H_{18}N_4O_2$	51-12-7	298.340		151.6				
8014	Nickel(II) acetate		$C_4H_6NiO_4$	373-02-4	176.782						vs H_2O; s EtOH
8015	Nickel bis(dibutyldithiocarbamate)		$C_{18}H_{36}N_2NiS_4$	13927-77-0	467.445	grn cry (bz/ EtOH)	91				s bz, ace
8016	Nickel bis(2,4-pentanedioate)	Nickel acetylacetonate	$C_{10}H_{14}NiO_4$	3264-82-2	256.909	grn orth cry	230	227[11]			s H_2O, bz, chl, EtOH; i eth
8017	Nickel carbonyl	Nickel tetracarbonyl	C_4NiO_4	13463-39-3	170.734	col liq	-19.3	43 (exp 60)	1.31[25]		
8018	Nickelocene	Bis(η5-2,4-cyclopentadien-1-yl)nickel	$C_{10}H_{10}Ni$	1271-28-9	188.879		172				
8019	Niclosamide		$C_{13}H_8Cl_2N_2O_4$	50-65-7	327.120		227				
8020	Nicofibrate		$C_{16}H_{16}ClNO_3$	31980-29-7	305.756		49	180[0.4]			
8021	Nicosulfuron		$C_{15}H_{18}N_6O_6S$	111991-09-4	410.405		172				
8022	Nicotelline	3,2':4',3''-Terpyridine	$C_{15}H_{11}N_3$	494-04-2	233.268	prismatic nd	148	>300			sl H_2O, eth; s bz, chl, EtOH
8023	Nicotinamide hypoxanthine dinucleotide	Nicotinic acid adenine dinucleotide	$C_{21}H_{26}N_6O_{15}P_2$	1851-07-6	664.410	pow					
8024	β-Nicotinamide mononucleotide	NMN	$C_{11}H_{15}N_2O_8P$	1094-61-7	334.219	amor pow					vs H_2O; i ace
8025	L-Nicotine	3-(1-Methyl-2-pyrrolidinyl)pyridine, (S)-	$C_{10}H_{14}N_2$	54-11-5	162.231	hyg liq	-79	247; 125[18]	1.0097[20]	1.5282[20]	msc H_2O; vs EtOH, eth, chl; s lig
8026	Nifurthiazole		$C_8H_6N_4O_4S$	3570-75-0	254.224	cry	215 dec				
8027	Nitralin	4-(Methylsulfonyl)-2,6-dinitro-N,N-dipropylaniline	$C_{13}H_{19}N_3O_6S$	4726-14-1	345.371		150				
8028	Nitranilic acid	2,5-Dihydroxy-3,6-dinitro-2,5-cyclohexadiene-1,4-dione	$C_6H_2N_2O_8$	479-22-1	230.088	gold-ye pl (+w, dil HNO_3	170 dec				vs H_2O, EtOH; i eth
8029	Nitrapyrin	Pyridine, 2-chloro-6-(trichloromethyl)-	$C_6H_3Cl_4N$	1929-82-4	230.907		63	136[11]			

1-Naphthyl 2-hydroxybenzoate

1-Naphthylhydroxylamine

1-Naphthyl isothiocyanate

N-2-Naphthyl-2-naphthalenamine

(2-Naphthyloxy)acetic acid

1-Naphthyl phosphate

2-Naphthyl salicylate

1,5-Naphthyridine

1,6-Naphthyridine

Napropamide

Naproxen

Narceine

Narcobarbital

Naringenin

Naringin

Nealbarbital

Nellite

Neoabietic acid

Neobornylamine

Neopentane

Neopine

Neostigmine bromide

Nepetalactone

cis-Nerolidol

Neurine

Neutral Red

Nialamide

Nickel(II) acetate

Nickel bis(dibutyldithiocarbamate)

Nickel bis(2,4-pentanedioate)

Nickel carbonyl

Nickelocene

Niclosamide

Nicofibrate

Nicosulfuron

Nicotelline

Nicotinamide hypoxanthine dinucleotide

β-Nicotinamide mononucleotide

L-Nicotine

Nifurthiazole

Nitralin

Nitranilic acid

Nitrapyrin

No.	Name	Synonym	Mol. Form.	CAS RN	Mol. Wt.	Physical Form	mp/°C	bp/°C	den/ g cm⁻³	n_D	Solubility
8030	Nitrilotriacetic acid	N,N-Bis(carboxymethyl)glycine	$C_6H_9NO_6$	139-13-9	191.138	pr cry (w)	242 dec				sl H_2O, DMSO; s EtOH
8031	2,2',2''-Nitrilotriacetonitrile	Tricyanotrimethylamine	$C_6H_6N_4$	7327-60-8	134.139	nd (EtOH)	125.5				
8032	Nitroacetic acid		$C_2H_3NO_4$	625-75-2	105.050	nd (chl)	92 dec				vs bz, eth, EtOH, chl
8033	Nitroacetone		$C_3H_5NO_3$	10230-68-9	103.077	pl, nd (eth, bz)	50.3	103[24]			vs bz, eth, EtOH
8034	2-Nitroaniline		$C_6H_6N_2O_2$	88-74-4	138.124		71.0	284	0.9015[25]		sl H_2O; s EtOH; vs eth, ace, bz, chl
8035	3-Nitroaniline		$C_6H_6N_2O_2$	99-09-2	138.124		113.4	dec 306	0.9011[25]		sl H_2O, bz; s EtOH, eth, ace; vs MeOH
8036	4-Nitroaniline		$C_6H_6N_2O_2$	100-01-6	138.124	pa ye mcl nd (w)	147.5	332	1.424[20]		i H_2O; s EtOH, eth, ace; sl bz, DMSO
8037	2-Nitroanisole	1-Methoxy-2-nitrobenzene	$C_7H_7NO_3$	91-23-6	153.136		10.5	277; 144[4]	1.2540[20]	1.5161[20]	i H_2O; msc EtOH, eth; s ctc
8038	3-Nitroanisole	1-Methoxy-3-nitrobenzene	$C_7H_7NO_3$	555-03-3	153.136	nd (al), pl (bz-lig)	38.5	258	1.373[18]		i H_2O; s EtOH; vs eth
8039	4-Nitroanisole	1-Methoxy-4-nitrobenzene	$C_7H_7NO_3$	100-17-4	153.136	pr (al), nd (dil al)	54	274	1.2192[60]	1.5070[60]	i H_2O; vs EtOH, eth; s ctc; sl peth
8040	9-Nitroanthracene		$C_{14}H_9NO_2$	602-60-8	223.227	ye nd (al) pr (HOAc or xyl)	146	275[17]			i H_2O; sl EtOH, chl; vs ace, CS_2
8041	1-Nitro-9,10-anthracenedione		$C_{14}H_7NO_4$	82-34-8	253.211	nd (HOAc) ye pr (ace)	231.5	270[7]			i H_2O; sl EtOH, eth; s ace, bz
8042	2-Nitrobenzaldehyde		$C_7H_5NO_3$	552-89-6	151.120	ye nd (w)	43.5	153[23]	1.2844[20]		sl H_2O, chl; vs EtOH, eth, ace, bz
8043	3-Nitrobenzaldehyde		$C_7H_5NO_3$	99-61-6	151.120	lt ye nd (w)	58.5	164[23]	1.2792[20]		sl H_2O; s EtOH, eth, chl; vs ace, bz
8044	4-Nitrobenzaldehyde		$C_7H_5NO_3$	555-16-8	151.120	lf, pr (w)	107	sub	1.496[25]		sl H_2O, lig; vs EtOH, eth; s bz, chl, HOAc
8045	3-Nitrobenzamide		$C_7H_6N_2O_3$	645-09-0	166.134		142.7	312.5			s H_2O, EtOH, eth
8046	4-Nitrobenzamide		$C_7H_6N_2O_3$	619-80-7	166.134	nd (w)	200.7				i H_2O; s EtOH, eth
8047	Nitrobenzene		$C_6H_5NO_2$	98-95-3	123.110		5.7	210.8	1.2037[20]	1.5562[20]	sl H_2O, ctc; vs EtOH, eth, ace, bz
8048	2-Nitrobenzeneacetic acid	o-Nitrophenylacetic acid	$C_8H_7NO_4$	3740-52-1	181.147	nd (w, pl (dil al)	141.5				s H_2O, EtOH
8049	3-Nitrobenzeneacetic acid	m-Nitrophenylacetic acid	$C_8H_7NO_4$	1877-73-2	181.147	nd (w)	122				vs EtOH
8050	4-Nitrobenzeneacetic acid	p-Nitrophenylacetic acid	$C_8H_7NO_4$	104-03-0	181.147	pa ye nd (w)	154				sl H_2O; s EtOH, eth, bz
8051	2-Nitrobenzeneacetonitrile	2-Nitrobenzyl cyanide	$C_8H_6N_2O_2$	610-66-2	162.146	nd (dil al), pr (HOAc, al)	84	178[12], 138[1]			vs ace, bz, eth, EtOH
8052	4-Nitrobenzeneacetonitrile	4-Nitrobenzyl cyanide	$C_8H_6N_2O_2$	555-21-5	162.146	pr (al)	117	196[12]			sl H_2O; s EtOH, eth, bz, chl
8053	4-Nitro-1,2-benzenediamine	4-Nitro-o-phenylenediamine	$C_6H_7N_3O_2$	99-56-9	153.139	dk red nd (dil al)	199.5				s acid
8054	4-Nitro-1,3-benzenediamine		$C_6H_7N_3O_2$	5131-58-8	153.139	oran pr (w)	161				
8055	5-Nitro-1,3-benzenediamine		$C_6H_7N_3O_2$	5042-55-7	153.139	red cry (w)	143				
8056	2-Nitro-1,4-benzenediamine		$C_6H_7N_3O_2$	5307-14-2	153.139		140.0				
8057	3-Nitro-1,2-benzenedicarboxylic acid		$C_8H_5NO_6$	603-11-2	211.129	pa ye pr (w)	218				sl H_2O, ace; s EtOH; i bz, peth, chl
8058	4-Nitro-1,2-benzenedicarboxylic acid		$C_8H_5NO_6$	610-27-5	211.129	pa ye nd (w, eth)	164.8				s H_2O, EtOH; i bz, chl, CS_2, peth
8059	2-Nitrobenzeneethanol		$C_8H_9NO_3$	15121-84-3	167.162		1.0	267	1.19[25]	1.5637[20]	
8060	4-Nitrobenzeneethanol		$C_8H_9NO_3$	100-27-6	167.162		63	148[2]			
8061	2-Nitrobenzenemethanol	2-Nitrobenzyl alcohol	$C_7H_7NO_3$	612-25-9	153.136	nd (w)	74	270; 168[20]			sl H_2O; s EtOH, eth
8062	3-Nitrobenzenemethanol	3-Nitrobenzyl alcohol	$C_7H_7NO_3$	619-25-0	153.136	orth nd (w)	30.5	177[3]	1.296[19]		s H_2O, EtOH, eth; sl chl
8063	4-Nitrobenzenemethanol	4-Nitrobenzyl alcohol	$C_7H_7NO_3$	619-73-8	153.136	nd (w)	96.5	dec 255; 185[12]			sl H_2O, ace; s EtOH, eth
8064	2-Nitrobenzenesulfenyl chloride		$C_6H_4ClNO_2S$	7669-54-7	189.620	ye nd (bz)	75				vs eth, bz, chl
8065	4-Nitrobenzenesulfenyl chloride		$C_6H_4ClNO_2S$	937-32-6	189.620	ye lf (peth)	52	125[0.1]			vs bz
8066	4-Nitrobenzenesulfonamide		$C_6H_6N_2O_4S$	6325-93-5	202.188		180 dec				

Nitrilotriacetic acid

2,2',2''-Nitrilotriacetonitrile

Nitroacetic acid

Nitroacetone

2-Nitroaniline

3-Nitroaniline

4-Nitroaniline

2-Nitroanisole

3-Nitroanisole

4-Nitroanisole

9-Nitroanthracene

1-Nitro-9,10-anthracenedione

2-Nitrobenzaldehyde

3-Nitrobenzaldehyde

4-Nitrobenzaldehyde

3-Nitrobenzamide

4-Nitrobenzamide

Nitrobenzene

2-Nitrobenzeneacetic acid

3-Nitrobenzeneacetic acid

4-Nitrobenzeneacetic acid

2-Nitrobenzeneacetonitrile

4-Nitrobenzeneacetonitrile

4-Nitro-1,2-benzenediamine

4-Nitro-1,3-benzenediamine

5-Nitro-1,3-benzenediamine

2-Nitro-1,4-benzenediamine

3-Nitro-1,2-benzenedicarboxylic acid

4-Nitro-1,2-benzenedicarboxylic acid

2-Nitrobenzeneethanol

4-Nitrobenzeneethanol

2-Nitrobenzenemethanol

3-Nitrobenzenemethanol

4-Nitrobenzenemethanol

2-Nitrobenzenesulfenyl chloride

4-Nitrobenzenesulfenyl chloride

4-Nitrobenzenesulfonamide

No.	Name	Synonym	Mol. Form.	CAS RN	Mol. Wt.	Physical Form	mp/°C	bp/°C	den/ g cm⁻³	n_D	Solubility
8067	3-Nitrobenzenesulfonic acid		C₆H₅NO₅S	98-47-5	203.173	pl	48				vs H₂O; s EtOH; i eth, bz
8068	4-Nitrobenzenesulfonic acid		C₆H₅NO₅S	138-42-1	203.173		95				vs H₂O
8069	2-Nitrobenzenesulfonyl chloride		C₆H₄ClNO₄S	1694-92-4	221.619	pr (lig, eth-peth)	68.5				s eth; sl peth
8070	3-Nitrobenzenesulfonyl chloride		C₆H₄ClNO₄S	121-51-7	221.619	mcl pr (eth) nd (lig)	64				i H₂O; s EtOH
8071	4-Nitrobenzenesulfonyl chloride		C₆H₄ClNO₄S	98-74-8	221.619	mcl pr (peth)	79.5	143¹·⁵			s peth
8072	5-Nitro-1H-benzimidazole		C₇H₅N₃O₂	94-52-0	163.134	nd (w)	207.8				i H₂O, eth, bz, chl; s acid; vs EtOH
8073	2-Nitrobenzoic acid		C₇H₅NO₄	552-16-9	167.120	tcl nd (w)	147.5		1.575²⁰		s H₂O, eth; vs EtOH, ace; sl bz, lig
8074	3-Nitrobenzoic acid		C₇H₅NO₄	121-92-6	167.120	mcl pr (w)	141.1		1.494²⁰		sl H₂O, bz; vs EtOH, eth, ace; s chl
8075	4-Nitrobenzoic acid		C₇H₅NO₄	62-23-7	167.120	mcl lf (w)	242	sub	1.610²⁰		vs ace, eth, EtOH, chl, MeOH
8076	3-Nitrobenzoic acid, hydrazide		C₇H₇N₃O₃	618-94-0	181.149		153.5				sl H₂O, EtOH; i eth, bz, chl
8077	4-Nitrobenzoic acid, hydrazide		C₇H₇N₃O₃	636-97-5	181.149		215.5				sl H₂O, EtOH; i eth, bz, chl
8078	3-Nitrobenzonitrile		C₇H₄N₂O₂	619-24-9	148.119		118	165¹⁶			s H₂O, EtOH, bz; vs eth, ace; i peth
8079	4-Nitrobenzonitrile		C₇H₄N₂O₂	619-72-7	148.119		150.0				sl H₂O, EtOH, eth; s chl, HOAc
8080	5-Nitro-1H-benzotriazole		C₆H₄N₄O₂	2338-12-7	164.122		217				
8081	2-Nitrobenzoyl chloride		C₇H₄ClNO₃	610-14-0	185.565	pl	20				vs eth; sl ctc
8082	3-Nitrobenzoyl chloride		C₇H₄ClNO₃	121-90-4	185.565		36	276.5			vs eth
8083	4-Nitrobenzoyl chloride		C₇H₄ClNO₃	122-04-3	185.565	ye nd (lig)	75	203¹⁰⁵, 151¹⁵			s eth
8084	2-Nitrobiphenyl	2-Nitro-1,1'-biphenyl	C₁₂H₉NO₂	86-00-0	199.205	pl (al, MeOH)	37.2	320	1.44²⁵		i H₂O; s EtOH, eth, chl
8085	3-Nitrobiphenyl	3-Nitro-1,1'-biphenyl	C₁₂H₉NO₂	2113-58-8	199.205	ye pl or nd (dil al)	62	227³⁵, 143⁹			i H₂O; s EtOH, eth, HOAc, lig
8086	4-Nitrobiphenyl	4-Nitro-1,1'-biphenyl	C₁₂H₉NO₂	92-93-3	199.205	ye nd (al)	114	340			i H₂O; sl EtOH; s eth, bz, chl, HOAc
8087	2-Nitro-1,1-bis(p-chlorophenyl)propane		C₁₅H₁₃Cl₂NO₂	117-27-1	310.176	cry	81	180⁰·¹⁶			
8088	1-Nitrobutane		C₄H₉NO₂	627-05-4	103.120			153	0.970²⁵	1.4303²⁰	sl H₂O; msc EtOH, eth; s alk
8089	2-Nitro-1-butanol		C₄H₉NO₃	609-31-4	119.119		-47	105¹⁰	1.1332²⁵	1.4390²⁰	s H₂O, ace; msc EtOH, eth; sl ctc
8090	3-Nitro-2-butanol		C₄H₉NO₃	6270-16-2	119.119			91⁹, 55⁰·⁵	1.1260²⁰	1.4414²⁰	
8091	6-Nitrochrysene		C₁₈H₁₁NO₂	7496-02-8	273.286	ye nd (bz)	≈215 dec				
8092	Nitrocyclohexane		C₆H₁₁NO₂	1122-60-7	129.157	liq	-34	205; 95²²	1.0610²⁰	1.4612¹⁹	i H₂O; s EtOH, lig
8093	1-Nitrodecane		C₁₀H₂₁NO₂	4609-87-4	187.280			86¹		1.4337²⁰	
8094	N-Nitrodiethylamine	N-Ethyl-N-nitroethanamine	C₄H₁₀N₂O₂	7119-92-8	118.134			206.5	1.057¹⁵		vs eth, EtOH
8095	Nitroethane		C₂H₅NO₂	79-24-3	75.067	liq	-89.5	114.0	1.0448²⁵	1.3917²⁰	sl H₂O; msc EtOH, eth; s ace, chl
8096	2-Nitroethanol		C₂H₅NO₃	625-48-9	91.066	liq	-80	194; 102¹⁰	1.270¹⁵	1.4438¹⁹	msc H₂O, EtOH, eth; i bz
8097	Nitroethene		C₂H₃NO₂	3638-64-0	73.051	liq	-55.5	98.5	1.2212¹⁴	1.4282²⁰	vs EtOH, eth, ace, bz, chl
8098	(2-Nitroethyl)benzene		C₈H₉NO₂	6125-24-2	151.163	liq	-23	250; 137¹⁶	1.126²⁴	1.5407¹⁹	
8099	Nitrofen	2,4-Dichloro-1-(4-nitrophenoxy)benzene	C₁₂H₇Cl₂NO₃	1836-75-5	284.095		70				
8100	2-Nitro-9H-fluorene		C₁₃H₉NO₂	607-57-8	211.216	nd (50% HOAc ace)	159.3				i H₂O; s ace, bz
8101	2-Nitro-9H-fluoren-9-one		C₁₃H₇NO₃	3096-52-4	225.200	ye nd or lf (HOAc)	224.3	sub			sl EtOH; s ace, sulf, HOAc
8102	5-Nitro-2-furaldehyde diacetate		C₉H₉NO₇	92-55-7	243.170		92.0				s chl
8103	2-Nitrofuran		C₄H₃NO₃	609-39-2	113.072	ye mcl cry (peth)	30	134¹²³, 84¹³			s H₂O, EtOH, eth
8104	5-Nitro-2-furancarboxaldehyde		C₅H₃NO₄	698-63-5	141.083	pa ye (peth)	35.5	130¹⁰			sl H₂O; s peth

3-Nitrobenzenesulfonic acid

4-Nitrobenzenesulfonic acid

2-Nitrobenzenesulfonyl chloride

3-Nitrobenzenesulfonyl chloride

4-Nitrobenzenesulfonyl chloride

5-Nitro-1*H*-benzimidazole

2-Nitrobenzoic acid

3-Nitrobenzoic acid

4-Nitrobenzoic acid

3-Nitrobenzoic acid, hydrazide

4-Nitrobenzoic acid, hydrazide

3-Nitrobenzonitrile

4-Nitrobenzonitrile

5-Nitro-1*H*-benzotriazole

2-Nitrobenzoyl chloride

3-Nitrobenzoyl chloride

4-Nitrobenzoyl chloride

2-Nitrobiphenyl

3-Nitrobiphenyl

4-Nitrobiphenyl

2-Nitro-1,1-bis(*p*-chlorophenyl)propane

1-Nitrobutane

2-Nitro-1-butanol

3-Nitro-2-butanol

6-Nitrochrysene

Nitrocyclohexane

1-Nitrodecane

N-Nitrodiethylamine

Nitroethane

2-Nitroethanol

Nitroethene

(2-Nitroethyl)benzene

Nitrofen

2-Nitro-9*H*-fluorene

2-Nitro-9*H*-fluoren-9-one

5-Nitro-2-furaldehyde diacetate

2-Nitrofuran

5-Nitro-2-furancarboxaldehyde

No.	Name	Synonym	Mol. Form.	CAS RN	Mol. Wt.	Physical Form	mp/°C	bp/°C	den/ g cm⁻³	n_D	Solubility
8105	5-Nitro-2-furancarboxylic acid		C₅H₃NO₅	645-12-5	157.082	pa ye pl (w)	186	sub			s H₂O, EtOH, eth; sl ace, bz; i chl
8106	Nitrofurantoin		C₈H₆N₄O₅	67-20-9	238.158			263			
8107	Nitrofurazone	2-[(5-Nitro-2-furanyl) methylene] hydrazinecarboxamide	C₆H₆N₄O₄	59-87-0	198.137	pa ye nd	238 dec				i H₂O, eth; sl EtOH, DMSO; s alk
8108	Nitrogen mustard N-oxide hydrochloride	Mechlorethamine oxide hydrochloride	C₅H₁₂Cl₃NO	302-70-5	208.514	pr (ace)	110				s H₂O
8109	Nitroguanidine		CH₄N₄O₂	556-88-7	104.069	nd or pr (w)	239 dec				sl H₂O, EtOH; i eth; vs alk
8110	1-Nitrohexane		C₆H₁₃NO₂	646-14-0	131.173			193; 84²¹	0.9396²⁰	1.4270²⁰	i H₂O; s EtOH, eth, ace, bz, alk
8111	3-Nitro-4-hydroxyphenylarsonic acid	Roxarsone	C₆H₆AsNO₆	121-19-7	263.037	ye nd or pl (w)	300				sl hot H₂O; i eth, EtOAc; vs MeOH, EtOH
8112	2-Nitro-1H-imidazole	Azomycin	C₃H₃N₃O₂	527-73-1	113.075	cry (MeOH)	287 dec				
8113	4-Nitro-1H-imidazole		C₃H₃N₃O₂	3034-38-6	113.075		303 dec				
8114	5-Nitro-1H-indazole		C₇H₅N₃O₂	5401-94-5	163.134	ye nd or col nd (al)	208				s EtOH, eth, bz; vs ace, HOAc; i lig
8115	6-Nitro-1H-indazole		C₇H₅N₃O₂	7597-18-4	163.134	nd (w, al, ace)	181 dec				s H₂O, EtOH, eth, bz; vs ace; i lig
8116	4-Nitro-1,3-isobenzofurandione		C₈H₃NO₅	641-70-3	193.114	nd (ace, al)	164				i H₂O; s EtOH, ace, HOAc; sl bz
8117	5-Nitro-1,3-isobenzofurandione		C₈H₃NO₅	5466-84-2	193.114		120.3	196⁸			i H₂O, peth; s EtOH, ace; sl eth
8118	2-Nitroisobutane		C₄H₉NO₂	594-70-7	103.120		26.23	127.16	0.9501²⁸	1.4015²⁰	msc EtOH, eth, ace, bz; vs chl; i alk
8119	5-Nitro-1H-isoindole-1,3(2H)-dione		C₈H₄N₂O₄	89-40-7	192.129	col nd (w), ye lf (al-ace)	202				vs ace
8120	Nitromersol		C₇H₅HgNO₃	133-58-4	351.71						i H₂O; sl ace, EtOH; s alk
8121	N-Nitromethanamine		CH₄N₂O₂	598-57-2	76.055		38	82¹⁰	1.2433⁴⁹	1.4616⁴⁹	vs H₂O, EtOH, bz, chl; s eth; sl peth
8122	Nitromethane		CH₃NO₂	75-52-5	61.041	liq	-28.38	101.19	1.1371²⁰	1.3817²⁰	s H₂O, EtOH, eth, ace, ctc, alk
8123	(Nitromethyl)benzene		C₇H₇NO₂	622-42-4	137.137	ye liq		226; 135²⁵	1.1596²⁰	1.5323²⁰	vs ace, eth
8124	Nitron		C₂₀H₁₆N₄	2218-94-2	312.368	ye lf (al), nd (chl)	189 dec				vs ace, bz, EtOH, chl
8125	1-Nitronaphthalene		C₁₀H₇NO₂	86-57-7	173.169	ye nd (al)	61	180¹⁴	1.332²⁰		i H₂O; vs EtOH, eth, bz, chl, py
8126	2-Nitronaphthalene		C₁₀H₇NO₂	581-89-5	173.169	ye orth nd or pl (al)	79	314; 165¹⁵			i H₂O; vs EtOH, eth
8127	1-Nitro-2-naphthol		C₁₀H₇NO₃	550-60-7	189.168	ye nd, lf or pr (al)	104	115⁰·⁰⁵			s H₂O, EtOH; vs eth; sl chl
8128	1-Nitrooctane		C₈H₁₇NO₂	629-37-8	159.227		15	208.5	0.9346²⁰	1.4322²⁰	
8129	1-Nitropentane		C₅H₁₁NO₂	628-05-7	117.147			172.5	0.9525²⁰	1.4175²⁰	s EtOH, eth, bz
8130	3-Nitropentane		C₅H₁₁NO₂	551-88-2	117.147			154	0.957⁰		vs ace, eth, EtOH
8131	5-Nitro-1,10-phenanthroline		C₁₂H₇N₃O₂	4199-88-6	225.203		202.3				
8132	2-Nitrophenol		C₆H₅NO₃	88-75-5	139.109	ye nd or pr (eth, al)	44.8	216	1.2942⁴⁰	1.5723⁵⁰	sl H₂O; vs EtOH, eth, ace, bz, py
8133	3-Nitrophenol		C₆H₅NO₃	554-84-7	139.109	ye mcl (eth, aq Hcl)	96.8	194⁷⁰	1.2797¹⁰⁰		sl H₂O, DMSO; vs EtOH, eth, ace, bz
8134	4-Nitrophenol		C₆H₅NO₃	100-02-7	139.109	ye mcl pr (to)	113.6		1.479²⁰		sl H₂O; vs EtOH, eth, ace; s tol, py
8135	1-Nitro-2-phenoxybenzene		C₁₂H₉NO₃	2216-12-8	215.204	ye liq	<-20	235⁶⁰, 184⁸	1.2539²²	1.575²⁰	vs bz, eth, EtOH, chl
8136	1-Nitro-4-phenoxybenzene		C₁₂H₉NO₃	620-88-2	215.204	pl (peth), MeOH)	61	320; 225³⁰			i H₂O; sl EtOH, ctc; s eth, bz
8137	N-(2-Nitrophenyl)acetamide		C₈H₈N₂O₃	552-32-9	180.161		94	100⁰·¹	1.419¹⁵		s H₂O, EtOH, bz, chl, lig; vs eth
8138	N-(3-Nitrophenyl)acetamide		C₈H₈N₂O₃	122-28-1	180.161	wh lf (al)	155	100⁰·⁰⁰⁷⁴			s H₂O, EtOH, chl; i eth; sl tfa

5-Nitro-2-furancarboxylic acid

Nitrofurantoin

Nitrofurazone

Nitrogen mustard N-oxide hydrochloride

Nitroguanidine

1-Nitrohexane

3-Nitro-4-hydroxyphenylarsonic acid

2-Nitro-1H-imidazole

4-Nitro-1H-imidazole

5-Nitro-1H-indazole

6-Nitro-1H-indazole

4-Nitro-1,3-isobenzofurandione

5-Nitro-1,3-isobenzofurandione

2-Nitroisobutane

5-Nitro-1H-isoindole-1,3(2H)-dione

Nitromersol

N-Nitromethanamine

Nitromethane

(Nitromethyl)benzene

Nitron

1-Nitronaphthalene

2-Nitronaphthalene

1-Nitro-2-naphthol

1-Nitrooctane

1-Nitropentane

3-Nitropentane

5-Nitro-1,10-phenanthroline

2-Nitrophenol

3-Nitrophenol

4-Nitrophenol

1-Nitro-2-phenoxybenzene

1-Nitro-4-phenoxybenzene

N-(2-Nitrophenyl)acetamide

N-(3-Nitrophenyl)acetamide

No.	Name	Synonym	Mol. Form.	CAS RN	Mol. Wt.	Physical Form	mp/°C	bp/°C	den/ g cm⁻³	n_D	Solubility
8139	*N*-(4-Nitrophenyl)acetamide		C₈H₈N₂O₃	104-04-1	180.161	ye pr (w)	216	100[0.008]			sl H₂O, eth, chl; s EtOH, tfa, alk
8140	2-Nitrophenyl acetate		C₈H₇NO₄	610-69-5	181.147	nd or pr (lig)	40.5	dec 253; 141[11]			s H₂O; vs EtOH, eth, ace, bz; sl lig
8141	4-Nitrophenyl acetate		C₈H₇NO₄	830-03-5	181.147	lf (dil al)	82.3				vs H₂O; bz; s EtOH, chl, lig
8142	2-Nitro-*N*-phenylaniline		C₁₂H₁₀N₂O₂	119-75-5	214.219		75.5	215[15]	1.3660[20]		i H₂O; s EtOH; sl ctc
8143	4-Nitro-*N*-phenylaniline		C₁₂H₁₀N₂O₂	836-30-6	214.219		135.3	211[30]			i H₂O; vs EtOH; sl ace; s con sulf
8144	(4-Nitrophenyl)arsonic acid	Nitarsone	C₆H₆AsNO₅	98-72-6	247.038	lf or nd (w)	>310 dec				sl H₂O, EtOH, DMSO
8145	4-[(4-Nitrophenyl)azo]-1,3-benzenediol	Magneson	C₁₂H₉N₃O₄	74-39-5	259.217	red pow (al or MeOH)	200				i H₂O; sl EtOH, bz, HOAc, tol
8146	1-[(4-Nitrophenyl)azo]-2-naphthol		C₁₆H₁₁N₃O₃	6410-10-2	293.276	br-oran pl (to or bz)	257				vs bz, EtOH
8147	(3-Nitrophenyl)boronic acid		C₆H₆BNO₄	13331-27-6	166.928		274.5				
8148	1-(2-Nitrophenyl)ethanone	2-Nitroacetophenone	C₈H₇NO₃	577-59-3	165.147		28.5	178[32], 158[16]	1.2370[25]	1.5468[20]	i H₂O; vs EtOH, eth, chl
8149	1-(3-Nitrophenyl)ethanone	3-Nitroacetophenone	C₈H₇NO₃	121-89-1	165.147	nd (al)	81	202; 167[18]			vs H₂O; eth; sl EtOH, chl
8150	1-(4-Nitrophenyl)ethanone	4-Nitroacetophenone	C₈H₇NO₃	100-19-6	165.147	ye pr (al)	81.8	165[5]			vs eth, EtOH
8151	2-Nitro-1-phenylethanone		C₈H₇NO₃	614-21-1	165.147		106	158[16], 142[10]		1.5468[30]	vs eth, EtOH
8152	(4-Nitrophenyl)hydrazine		C₆H₇N₃O₂	100-16-3	153.139	oran-red lf or nd (al)	158 dec				sl H₂O; s EtOH, eth, bz, chl, AcOEt
8153	(4-Nitrophenyl)phenylmethanone		C₁₃H₉NO₃	1144-74-7	227.215	nd or lf (al)	138		1.406[9]		vs bz
8154	3-(4-Nitrophenyl)-1-phenyl-2-propen-1-one	Nitrochalcone	C₁₅H₁₁NO₃	1222-98-6	253.253	pa ye nd (al) pl (bz)	164				s EtOH, chl; i eth, lig
8155	4-Nitrophenyl phosphate	4-Nitrophenyl dihydrogen phosphate	C₆H₆NO₆P	330-13-2	219.089	ye-wh nd	155				i cold H₂O; s EtOH, chl, bz
8156	3-(2-Nitrophenyl)propanoic acid	2-Nitrobeazenepropanoic acid	C₉H₉NO₄	2001-32-3	195.172	ye cry	115				
8157	3-(4-Nitrophenyl)propanoic acid	4-Nitrobenzenepropanoic acid	C₉H₉NO₄	16642-79-8	195.172	nd (w)	163				
8158	3-(4-Nitrophenyl)-2-propenal	4-Nitrocinnamaldehyde	C₉H₇NO₃	1734-79-8	177.157	nd (w, al)	141.5				s H₂O, eth, ace, bz; vs EtOH
8159	3-(2-Nitrophenyl)-2-propynoic acid	*o*-Nitrophenylpropiolic acid	C₉H₅NO₄	530-85-8	191.141		≈157 dec; may explode				sl H₂O; vs EtOH, eth; i CS₂O
8160	1-Nitro-4-(phenylthio)benzene		C₁₂H₉NO₂S	952-97-6	231.270	pa ye mcl pr (lig)	56	288[100], 240[25]			vs eth, EtOH
8161	(4-Nitrophenyl)urea	*p*-Nitrophenylurea	C₇H₇N₃O₃	556-10-5	181.149	pr (al), nd (dil al)	238				vs H₂O, EtOH
8162	*N*-Nitropiperidine		C₅H₁₀N₂O₂	7119-94-0	130.145	liq	-5.5	245; 121[20]	1.1519[26]	1.4954[26]	
8163	1-Nitropropane		C₃H₇NO₂	108-03-2	89.094	liq	-108	131.1	0.9961[25]	1.4018[20]	sl H₂O; msc EtOH, eth; s chl
8164	2-Nitropropane		C₃H₇NO₂	79-46-9	89.094	liq	-91.3	120.2	0.9821[25]	1.3944[20]	sl H₂O; s chl
8165	3-Nitropropanoic acid		C₃H₅NO₄	504-88-1	119.077		62		1.59[20]		vs H₂O, EtOH, eth; s chl; i lig
8166	2-Nitro-1-propanol		C₃H₇NO₃	2902-96-7	105.093			120[32], 100[12]	1.1841[25]	1.4379[20]	s H₂O, EtOH, eth; sl chl
8167	1-Nitro-1-propene		C₃H₅NO₂	3156-70-5	87.078			60[34], 37[10]	1.0661[20]	1.4527[20]	s eth, ace, chl
8168	2-Nitro-1-propene		C₃H₅NO₂	4749-28-4	87.078	ye-grn liq		52[80], 32[30]	1.0559[25]	1.4358[20]	s eth, ace, chl
8169	5-Nitro-2-propoxyaniline		C₉H₁₂N₂O₃	553-79-7	196.202	oran (PrOH-peth)	49				vs EtOH
8170	*N*-(5-Nitro-2-propoxyphenyl)acetamide	5'-Nitro-2'-propoxyacetanilide	C₁₁H₁₄N₂O₄	553-20-8	238.240	cry (PrOH)	102.5				
8171	1-Nitropyrene		C₁₆H₉NO₂	5522-43-0	247.248	ye nd (MeCN)	152				
8172	5-Nitro-2-pyridinamine		C₅H₅N₃O₂	4214-76-0	139.113	ye lf (dil al)	188				sl H₂O, eth, bz, lig; s EtOH
8173	4-Nitropyridine		C₅H₄N₂O₂	1122-61-8	124.098	pl (aq al)	50				
8174	4-Nitropyridine 1-oxide		C₅H₄N₂O₃	1124-33-0	140.097		160.5				
8175	5-Nitropyrimidinamine		C₄H₄N₄O₂	3073-77-6	140.101	nd (al)	236.5				sl H₂O, DMSO; s EtOH, ace; i eth, bz
8176	5-Nitro-2,4(1*H*,3*H*)-pyrimidinedione	5-Nitrouracil	C₄H₃N₃O₄	611-08-5	157.085	gold nd (al)	>300 exp				sl H₂O; s EtOH
8177	5-Nitro-2,4,6(1*H*,3*H*,5*H*)-pyrimidinetrione	5-Nitrobarbituric acid	C₄H₃N₃O₅	480-68-2	173.084	pr, lf (w+3)	180.5				s H₂O, EtOH; i eth
8178	5-Nitroquinoline		C₉H₆N₂O₂	607-34-1	174.156	pl (w, al) nd (+w)	74	sub			sl H₂O, chl; s EtOH, bz

N-(4-Nitrophenyl)acetamide

2-Nitrophenyl acetate

4-Nitrophenyl acetate

2-Nitro-N-phenylaniline

4-Nitro-N-phenylaniline

(4-Nitrophenyl)arsonic acid

4-[(4-Nitrophenyl)azo]-1,3-benzenediol

1-[(4-Nitrophenyl)azo]-2-naphthol

(3-Nitrophenyl)boronic acid

1-(2-Nitrophenyl)ethanone

1-(3-Nitrophenyl)ethanone

1-(4-Nitrophenyl)ethanone

2-Nitro-1-phenylethanone

(4-Nitrophenyl)hydrazine

(4-Nitrophenyl)phenylmethanone

3-(4-Nitrophenyl)-1-phenyl-2-propen-1-one

4-Nitrophenyl phosphate

3-(2-Nitrophenyl)propanoic acid

3-(4-Nitrophenyl)propanoic acid

3-(4-Nitrophenyl)-2-propenal

3-(2-Nitrophenyl)-2-propynoic acid

1-Nitro-4-(phenylthio)benzene

(4-Nitrophenyl)urea

N-Nitropiperidine

1-Nitropropane

2-Nitropropane

3-Nitropropanoic acid

2-Nitro-1-propanol

1-Nitro-1-propene

2-Nitro-1-propene

5-Nitro-2-propoxyaniline

N-(5-Nitro-2-propoxyphenyl)acetamide

1-Nitropyrene

5-Nitro-2-pyridinamine

4-Nitropyridine

4-Nitropyridine 1-oxide

5-Nitropyrimidinamine

5-Nitro-2,4(1H,3H)-pyrimidinedione

5-Nitro-2,4,6(1H,3H,5H)-pyrimidinetrione

5-Nitroquinoline

No.	Name	Synonym	Mol. Form.	CAS RN	Mol. Wt.	Physical Form	mp/°C	bp/°C	den/ g cm⁻³	n_D	Solubility
8179	6-Nitroquinoline		$C_9H_6N_2O_2$	613-50-3	174.156	ye pl (HCl-HOAc)	153.5	170[0.2]			s H_2O, EtOH; sl eth, chl; vs bz
8180	8-Nitroquinoline		$C_9H_6N_2O_2$	607-35-2	174.156	mcl pr (al)	91.5				sl H_2O, chl; s EtOH, eth, bz, acid
8181	4-Nitroquinoline 1-oxide		$C_9H_6N_2O_3$	56-57-5	190.155	ye nd, pl (ace)	154				
8182	5-Nitro-8-quinolinol	Nitroxoline	$C_9H_6N_2O_3$	4008-48-4	190.155		180				
8183	Nitrosobenzene		C_6H_5NO	586-96-9	107.110	orth or mcl (al-eth)	67	58[18]			i H_2O; s EtOH, eth, bz, lig
8184	N-Nitrosodibutylamine	Dibutylnitrosamine	$C_8H_{18}N_2O$	924-16-3	158.241			105[8]			
8185	N-Nitrosodiethanolamine	2,2'-(Nitrosoimino)ethanol	$C_4H_{10}N_2O_3$	1116-54-7	134.133	wh-ye oil		125[0.01]		1.4849[20]	
8186	N-Nitrosodiethylamine	Diethylnitrosamine	$C_4H_{10}N_2O$	55-18-5	102.134	ye oil		176.9	0.9422[20]	1.4386[20]	s H_2O, EtOH, eth; sl chl
8187	N-Nitrosodimethylamine	Dimethylnitrosamine	$C_2H_6N_2O$	62-75-9	74.081	ye liq		152	1.0048[20]	1.4368[20]	vs H_2O, EtOH, eth; s chl
8188	p-Nitroso-N,N-dimethylaniline		$C_8H_{10}N_2O$	138-89-6	150.177	grn pl (eth)	92.5		1.145[20]		sl H_2O; s EtOH, eth, chl, HCONH₂
8189	N-Nitrosodiphenylamine	N,N-Diphenylnitrosamine	$C_{12}H_{10}N_2O$	86-30-6	198.219	ye pl(lig)	66.5				i H_2O; sl EtOH, chl; s bz
8190	4-(N-Nitrosomethylamino)-1-(3-pyridyl)-1-butanone	Ketone, 3-pyridyl-3-(N-methyl-N-nitrosamino)propyl	$C_{10}H_{13}N_3O_2$	64091-91-4	207.229		63				sl H_2O
8191	N-Nitrosomethylethylamine		$C_3H_8N_2O$	10595-95-6	88.108	ye liq		67[40]			
8192	N-Nitroso-N-methylvinylamine	N-Methyl-N-nitrosoethenamine	$C_3H_6N_2O$	4549-40-0	86.092	ye liq		47			sl H_2O
8193	4-Nitrosomorpholine	N-Nitrosomorpholine	$C_4H_8N_2O_2$	59-89-2	116.119		29	225; 140[25]			s H_2O
8194	2-Nitroso-1-naphthol		$C_{10}H_7NO_2$	132-53-6	173.169		157 dec				sl H_2O, eth, bz, chl; s EtOH, ace, HOAc
8195	1-Nitroso-2-naphthol	1-Nitroso-β-naphthol	$C_{10}H_7NO_2$	131-91-9	173.169	ye-br nd (peth)	109.5				vs bz, eth
8196	N-Nitrosonornicotine	N'-Nitroso-3-(2-pyrrolidinyl)pyridine	$C_9H_{11}N_3O$	16543-55-8	177.202			155[0.2]			
8197	4-Nitrosophenol		$C_6H_5NO_2$	104-91-6	123.110	pa ye orth nd (ace, bz)	144 dec				sl H_2O; s EtOH, eth, ace, bz, dil alk
8198	4-Nitroso-N-phenylaniline	p-Nitrosodiphenylamine	$C_{12}H_{10}N_2O$	156-10-5	198.219		143				sl H_2O, lig; vs EtOH, eth, bz
8199	N-Nitrosopiperidine	1-Nitrosopiperidine	$C_5H_{10}N_2O$	100-75-4	114.145	pa ye		219; 109[20]	1.0631[18]	1.4933[18]	s H_2O, HCl
8200	N-Nitroso-N-propyl-1-propanamine	N-Nitrosodipropylamine	$C_6H_{14}N_2O$	621-64-7	130.187	gold		206; 113[40]	0.9163[20]	1.4437[20]	sl H_2O; msc EtOH, eth
8201	N-Nitrosopyrrolidine		$C_4H_8N_2O$	930-55-2	100.119			214	1.085[25]	1.4880[25]	
8202	5-Nitro-2-thiazolamine	2-Amino-5-nitrothiazole	$C_3H_3N_3O_2S$	121-66-4	145.140	oran-ye pow	202 dec				
8203	N-(5-Nitro-2-thiazolyl)acetamide	Aminitrozole	$C_5H_5N_3O_3S$	140-40-9	187.177	nd (al), pl (HOAc)	264.5				s alk
8204	4-Nitrothioanisole		$C_7H_7NO_2S$	701-57-5	169.202		72	137[2]	1.2391[80]	1.6401[20]	i H_2O; s ace, bz
8205	2-Nitrothiophene		$C_4H_3NO_2S$	609-40-5	129.138	lt ye mcl nd (peth)	46.5	224.5	1.3644[43]		i H_2O; vs EtOH; s alk; sl peth
8206	2-Nitrotoluene		$C_7H_7NO_2$	88-72-2	137.137	liq	-10.4	222	1.1611[19]	1.5450[20]	i H_2O; msc EtOH, eth; s ctc
8207	3-Nitrotoluene		$C_7H_7NO_2$	99-08-1	137.137	pa ye	15.5	232	1.1581[20]	1.5466[20]	i H_2O; s EtOH, bz, ctc; msc eth
8208	4-Nitrotoluene		$C_7H_7NO_2$	99-99-0	137.137	orth cry (al, eth)	51.63	238.3	1.1038[75]		i H_2O; s EtOH; vs eth, ace, bz, chl
8209	1-Nitro-2-(trifluoromethyl)benzene		$C_7H_4F_3NO_2$	384-22-5	191.108	cry (al)	32.5	217; 105[20]			i H_2O; vs EtOH, HOAc, bz; sl ctc
8210	1-Nitro-3-(trifluoromethyl)benzene		$C_7H_4F_3NO_2$	98-46-4	191.108	liq	-2.4	202.8; 103[40]	1.4357[15]	1.4719[20]	i H_2O; s EtOH, eth; sl ctc
8211	Nitrourea		$CH_3N_3O_3$	556-89-8	105.053	pl (al-peth)	158 dec				vs ace, EtOH
8212	trans-(2-Nitrovinyl)benzene		$C_8H_7NO_2$	5153-67-3	149.148	ye pr (peth, al)	60	255			i H_2O; s EtOH, ace; vs eth, chl, CS₂
8213	Nivalenol		$C_{15}H_{20}O_7$	23282-20-4	312.316	cry (MeOH)	224 dec				sl H_2O; s EtOH, MeOH
8214	Nizatidine		$C_{12}H_{21}N_5O_2S_2$	76963-41-2	331.458	cry (EtOH/AcOEt)	131				sl H_2O; s MeOH; vs chl; i bz, eth
8215	2,2',3,3',4,5,5',6,6'-Nonachlorobiphenyl		$C_{12}HCl_9$	52663-77-1	464.213	cry	180.5				i H_2O
8216	Nonacontane		$C_{90}H_{182}$	7667-51-8	1264.408			612[200]			

6-Nitroquinoline

8-Nitroquinoline

4-Nitroquinoline 1-oxide

5-Nitro-8-quinolinol

Nitrosobenzene

N-Nitrosodibutylamine

N-Nitrosodiethanolamine

N-Nitrosodiethylamine

N-Nitrosodimethylamine

p-Nitroso-N,N-dimethylaniline

N-Nitrosodiphenylamine

4-(N-Nitrosomethylamino)-1-(3-pyridyl)-1-butanone

N-Nitrosomethylethylamine

N-Nitroso-N-methylvinylamine

4-Nitrosomorpholine

2-Nitroso-1-naphthol

1-Nitroso-2-naphthol

N-Nitrosonornicotine

4-Nitrosophenol

4-Nitroso-N-phenylaniline

N-Nitrosopiperidine

N-Nitroso-N-propyl-1-propanamine

N-Nitrosopyrrolidine

5-Nitro-2-thiazolamine

N-(5-Nitro-2-thiazolyl)acetamide

4-Nitrothioanisole

2-Nitrothiophene

2-Nitrotoluene

3-Nitrotoluene

4-Nitrotoluene

1-Nitro-2-(trifluoromethyl)benzene

1-Nitro-3-(trifluoromethyl)benzene

Nitrourea

trans-(2-Nitrovinyl)benzene

Nivalenol

Nizatidine

2,2',3,3',4,5,5',6,6'-Nonachlorobiphenyl

$H_3C(CH_2)_{88}CH_3$
Nonacontane

No.	Name	Synonym	Mol. Form.	CAS RN	Mol. Wt.	Physical Form	mp/°C	bp/°C	den/ g cm⁻³	n_D	Solubility
8217	Nonacosane		$C_{29}H_{60}$	630-03-5	408.786	orth cry (peth)	63.7	440.8	0.8083[20]	1.4529[20]	i H₂O; vs EtOH, eth, ace; s bz; sl chl
8218	Nonadecafluorodecanoic acid		$C_{10}HF_{19}O_2$	335-76-2	514.084			219			
8219	Nonadecane		$C_{19}H_{40}$	629-92-5	268.521	wax	32.0	329.9	0.7855[20]	1.4409[20]	i H₂O; sl EtOH; s eth, ace, ctc
8220	Nonadecanoic acid		$C_{19}H_{38}O_2$	646-30-0	298.504	lf (al)	69.4	297[100], 228[10]	0.8468[70]		i H₂O; vs EtOH, eth, bz, chl, lig
8221	1-Nonadecanol		$C_{19}H_{40}O$	1454-84-8	284.520	cry (ace)	61.7	345; 166[0.3]		1.4328[75]	s eth, ace
8222	2-Nonadecanone		$C_{19}H_{38}O$	629-66-3	282.504	pr (al)	57	266[110], 165[2]	0.8108[56]		i H₂O; sl EtOH; s eth, bz; vs eth, ctc
8223	10-Nonadecanone		$C_{19}H_{38}O$	504-57-4	282.504	lf(al)	65.5	>350; 156[1.1]			i H₂O; sl EtOH; s eth, ace, lig; vs bz
8224	1-Nonadecene		$C_{19}H_{38}$	18435-45-5	266.505		23.4	329.0	0.7886[25]	1.4445[25]	
8225	Nonadecylbenzene		$C_{25}H_{44}$	29136-19-4	344.617		40	419	0.8545[20]	1.4807[20]	
8226	trans,trans-2,4-Nonadienal		$C_9H_{14}O$	5910-87-2	138.206			98[10]	0.862[25]	1.5207[20]	
8227	1,8-Nonadiene		C_9H_{16}	4900-30-5	124.223			142.5	0.7511[20]	1.4302[20]	
8228	2,6-Nonadien-1-ol		$C_9H_{16}O$	7786-44-9	140.222			108[24], 98[11]	0.8604[25]	1.4598[25]	
8229	1,8-Nonadiyne		C_9H_{12}	2396-65-8	120.191	liq	-27.3	162	0.8158[20]	1.4490[20]	i H₂O; s eth, ace
8230	Nonanal	Nonaldehyde	$C_9H_{18}O$	124-19-6	142.238		-19.3	191	0.8264[22]	1.4273[20]	s eth, chl
8231	Nonane		C_9H_{20}	111-84-2	128.255	liq	-53.46	150.82	0.7192[20]	1.4058[20]	i H₂O; vs EtOH, eth; msc ace, bz, hp
8232	Nonanedioic acid	Azelaic acid	$C_9H_{16}O_4$	123-99-9	188.221	lf or nd	106.5	287[100], 225[10]	1.225[25]	1.4303[111]	sl H₂O, eth, bz, DMSO; s EtOH
8233	1,9-Nonanediol		$C_9H_{20}O_2$	3937-56-2	160.254	cry (bz)	45.8	173[20], 150[3]			sl H₂O; vs EtOH, eth; s bz; i lig
8234	Nonanedioyl dichloride		$C_9H_{14}Cl_2O_2$	123-98-8	225.112			166[18]	1.143	1.4680[20]	s eth; vs bz
8235	Nonanenitrile		$C_9H_{17}N$	2243-27-8	139.238	liq	-34.2	224.4	0.8178[20]	1.4255[20]	i H₂O; s EtOH, eth; sl ctc
8236	1-Nonanethiol	Nonyl mercaptan	$C_9H_{20}S$	1455-21-6	160.320	liq	-20.1	220	0.842[25]	1.4548[20]	
8237	Nonanoic acid	Pelargonic acid	$C_9H_{18}O_2$	112-05-0	158.238		12.4	254.5	0.9052[20]	1.4343[19]	i H₂O; s EtOH, eth, chl
8238	1-Nonanol	Nonyl alcohol	$C_9H_{20}O$	143-08-8	144.254	liq	-5	213.37	0.8280[20]	1.4333[20]	i H₂O; s EtOH, eth; sl ctc
8239	2-Nonanol, (±)		$C_9H_{20}O$	74683-66-2	144.254	liq	-35	193.5	0.8471[20]	1.4353[20]	i H₂O; vs eth, EtOH
8240	3-Nonanol, (±)		$C_9H_{20}O$	74742-08-8	144.254		22	195; 93[18]	0.8250[20]	1.4289[20]	i H₂O; s EtOH, eth
8241	4-Nonanol		$C_9H_{20}O$	52708-03-9	144.254			192.5; 94[18]	0.8282[20]	1.4197[20]	i H₂O; s EtOH, eth
8242	5-Nonanol	Dibutylcarbinol	$C_9H_{20}O$	623-93-8	144.254		5.6	193; 97[20]	0.8220[20]	1.4289[20]	i H₂O; s EtOH
8243	2-Nonanone	Heptyl methyl ketone	$C_9H_{18}O$	821-55-6	142.238	liq	-7.5	195.3	0.8208[20]	1.4210[20]	i H₂O; s EtOH, eth, bz; vs ace, chl
8244	3-Nonanone	Ethyl hexyl ketone	$C_9H_{18}O$	925-78-0	142.238	liq	-8	190; 86[20]	0.8241[20]	1.4208[20]	i H₂O; s EtOH, eth, bz, chl; vs ace
8245	4-Nonanone	Pentyl propyl ketone	$C_9H_{18}O$	4485-09-0	142.238			187.5	0.8190[25]	1.4189[20]	i H₂O; s EtOH, eth, chl; vs ace
8246	5-Nonanone	Dibutyl ketone	$C_9H_{18}O$	502-56-7	142.238	liq	-3.8	188.45	0.8217[20]	1.4195[20]	i H₂O; s EtOH; vs eth, chl
8247	Nonanoyl chloride		$C_9H_{17}ClO$	764-85-2	176.683	liq	-60.5	215.3	0.9463[15]		s eth, ace
8248	trans-2-Nonenal		$C_9H_{16}O$	18829-56-6	140.222			101[16], 89[12]	0.846	1.4531[20]	
8249	1-Nonene		C_9H_{18}	124-11-8	126.239	liq	-81.3	146.9	0.7253[25]	1.4257[20]	
8250	2-Nonenoic acid		$C_9H_{16}O_2$	3760-11-0	156.222			173[20], 136[5]			
8251	3-Nonenoic acid		$C_9H_{16}O_2$	4124-88-3	156.222		-4.4	156[18], 106[1]	0.9254[20]	1.4454[25]	
8252	1-Nonen-3-ol	1-Vinylheptanol	$C_9H_{18}O$	21964-44-3	142.238			193.5	0.824[21]	1.4382[15]	
8253	Nonyl acetate		$C_{11}H_{22}O_2$	143-13-5	186.292	liq	-26	210	0.8785[15]	1.426[20]	
8254	Nonylamine	1-Nonanamine	$C_9H_{21}N$	112-20-9	143.270	liq	-1	202.2	0.7886[20]	1.4336[20]	sl H₂O, chl; s EtOH, eth
8255	Nonylbenzene		$C_{15}H_{24}$	1081-77-2	204.352	liq	-24	280.5	0.8584[20]	1.4816[20]	
8256	Nonylcyclohexane		$C_{15}H_{30}$	2883-02-5	210.399	liq	-10	282	0.8163[20]	1.4519[20]	
8257	Nonylcyclopentane		$C_{14}H_{28}$	2882-98-6	196.372	liq	-29	262	0.8081[20]	1.4467[20]	vs ace, bz, eth, EtOH
8258	Nonyl formate		$C_{10}H_{20}O_2$	5451-92-3	172.265	liq	-33	214	0.86	1.4216[20]	
8259	1-Nonylnaphthalene		$C_{19}H_{26}$	26438-26-6	254.409		8	366	0.9371[20]	1.5477[20]	
8260	4-Nonylphenol		$C_{15}H_{24}O$	104-40-5	220.351	visc ye liq	42	≈295; 180[10]	0.950[20]	1.513[20]	i H₂O; s bz, ctc, hp
8261	1-Nonyne	Heptylacetylene	C_9H_{16}	3452-09-3	124.223	liq	-50	150.8	0.7658[20]	1.4217[20]	i H₂O; s eth, bz, ctc

Nonacosane

Nonadecafluorodecanoic acid

Nonadecane

Nonadecanoic acid

1-Nonadecanol

2-Nonadecanone

10-Nonadecanone

1-Nonadecene

Nonadecylbenzene

trans,trans-2,4-Nonadienal

1,8-Nonadiene

2,6-Nonadien-1-ol

1,8-Nonadiyne

Nonanal

Nonane

Nonanedioic acid

1,9-Nonanediol

Nonanedioyl dichloride

Nonanenitrile

1-Nonanethiol

Nonanoic acid

1-Nonanol

2-Nonanol, (±)

3-Nonanol, (±)

4-Nonanol

5-Nonanol

2-Nonanone

3-Nonanone

4-Nonanone

5-Nonanone

Nonanoyl chloride

trans-2-Nonenal

1-Nonene

2-Nonenoic acid

3-Nonenoic acid

1-Nonen-3-ol

Nonyl acetate

Nonylamine

Nonylbenzene

Nonylcyclohexane

Nonylcyclopentane

Nonyl formate

1-Nonylnaphthalene

4-Nonylphenol

1-Nonyne

No.	Name	Synonym	Mol. Form.	CAS RN	Mol. Wt.	Physical Form	mp/°C	bp/°C	den/g cm⁻³	n_D	Solubility
8262	Norbormide		$C_{33}H_{25}N_3O_3$	991-42-4	511.570	cry (eth)	194				
8263	2,5-Norbornadiene	Bicyclo[2.2.1]hepta-2,5-diene	C_7H_8	121-46-0	92.139	liq	-19.1	89.5	0.9064[20]	1.4702[20]	i H_2O; s EtOH, eth, ace, bz; msc tol
8264	5-Norbornene-2,3-dicarboxylic acid anhydride		$C_9H_8O_3$	826-62-0	164.158		166				
8265	5-Norbornene-2-methylolacrylate		$C_{11}H_{14}O_2$	95-39-6	178.228	col liq		104	1.029[25]		s os
8266	24-Norcholan-23-oic acid, (5β)	Norcholanic acid	$C_{23}H_{38}O_2$	511-18-2	346.547	nd(HOAc)	177				
8267	Nordazepam	7-Chloro-1,3-dihydro-5-phenyl-2H-1,4-benzodiazepin-2-one	$C_{15}H_{11}ClN_2O$	1088-11-5	270.713		216.5				
8268	Nordihydroguaiaretic acid		$C_{18}H_{22}O_4$	500-38-9	302.366	nd(w, al, HOAc)	185.5				sl H_2O; s EtOH, eth, ace, alk; i bz
8269	Norea		$C_{13}H_{22}N_2O$	18530-56-8	222.326		177				
8270	Norepinephrine	Noradrenaline	$C_8H_{11}NO_3$	51-41-2	169.178		217 dec				sl H_2O, EtOH, eth; vs alk, dil HCl
8271	Norethisterone	19-Norpregn-4-en-20-yn-3-one, 17-hydroxy-, (17 α)-	$C_{20}H_{26}O_2$	68-22-4	298.419	cry	204				
8272	Norethynodrel		$C_{20}H_{26}O_2$	68-23-5	298.419	cry (MeOH)	170				
8273	Norflurazon		$C_{12}H_9ClF_3N_3O$	27314-13-2	303.666		184				
8274	Norhyoscyamine		$C_{16}H_{21}NO_3$	537-29-1	275.343	nd	140.5				vs EtOH, chl
8275	DL-Norleucine	2-Aminohexanoic acid, (DL)	$C_6H_{13}NO_2$	616-06-8	131.173	lf(w)	327 dec		1.172[25]		s H_2O; sl EtOH; i eth
8276	L-Norleucine	2-Aminohexanoic acid, (L)	$C_6H_{13}NO_2$	327-57-1	131.173		301 dec				sl H_2O
8277	Normorphine		$C_{16}H_{17}NO_3$	466-97-7	271.311		273				
8278	Norplant	Norgestrel, (-)	$C_{21}H_{28}O_2$	797-63-7	312.446	cry (MeOH)	206				
8279	19-Nortestosterone phenylpropionate	Nandrolone phenpropionate	$C_{27}H_{34}O_3$	62-90-8	406.557	cry	95				
8280	Nortriptyline hydrochloride		$C_{19}H_{22}ClN$	894-71-3	299.838	cry (eth)	214				s H_2O, EtOH; i bz, eth, ace
8281	DL-Norvaline	2-Aminopentanoic acid, (±)	$C_5H_{11}NO_2$	760-78-1	117.147	lf(al, w)	303	sub			s H_2O; i EtOH, eth, chl, AcOEt, lig
8282	L-Norvaline	2-Aminopentanoic acid, (S)	$C_5H_{11}NO_2$	6600-40-4	117.147	cry (dil al)	307				s H_2O
8283	Noscapine		$C_{22}H_{23}NO_7$	128-62-1	413.421	pr or nd (al)	176				i H_2O; s EtOH, bz, chl; sl eth; vs ace
8284	Novobiocin	Streptonivicin	$C_{31}H_{36}N_2O_{11}$	303-81-1	612.624	wh-ye orth cry	154		1.3448		i H_2O; s EtOH, EtOAc, ace, py
8285	Nuarimol		$C_{17}H_{12}ClFN_2O$	63284-71-9	314.740		126				
8286	Nylidrin	Buphenine	$C_{19}H_{25}NO_2$	447-41-6	299.408	cry (MeOH)	111				
8287	Ochratoxin A		$C_{20}H_{18}ClNO_6$	303-47-9	403.813	cry (xyl)	169				
8288	Ochratoxin B		$C_{20}H_{19}NO_6$	4825-86-9	369.368	cry (MeOH)	221				
8289	Ochratoxin C		$C_{22}H_{22}ClNO_6$	4865-85-4	431.866	amorp solid					
8290	Octacaine	3-(Diethylamino)-N-phenylbutanamide	$C_{14}H_{22}N_2O$	13912-77-1	234.337	cry	47	200[1]			vs EtOH, bz, eth
8291	2,2',3,3',5,5',6,6'-Octachlorobiphenyl		$C_{12}H_2Cl_8$	2136-99-4	429.768	cry	161				i H_2O
8292	Octachlorocyclopentene	Perchlorocyclopentene	C_5Cl_8	706-78-5	343.678	nd	40	283	1.8200[50]	1.5660[50]	i H_2O; vs EtOH
8293	Octachlorodibenzo-p-dioxin		$C_{12}Cl_8O_2$	3268-87-9	459.751	nd	331				
8294	Octachloronaphthalene	Perchloronaphthalene	$C_{10}Cl_8$	2234-13-1	403.731	nd (bz-CCl₄)	197.5	441[7], 248[0.5]			sl EtOH; vs bz, chl, lig
8295	Octachlorostyrene	Perchlorostyrene	C_8Cl_8	29082-74-4	379.710	cry (ace/EtOH)	99				
8296	Octacontane		$C_{80}H_{162}$	7667-88-1	1124.142		112	672			
8297	Octacosane		$C_{28}H_{58}$	630-02-4	394.761	mcl or orth (bz-al)	61.1	431.6	0.8067[20]	1.4330[70]	i H_2O; msc ace; s bz, chl
8298	Octacosanoic acid	Montanic acid	$C_{28}H_{56}O_2$	506-48-9	424.744		90.9		0.8191[100]	1.4313[100]	vs bz, chl
8299	1-Octacosanol	Montanyl alcohol	$C_{28}H_{58}O$	557-61-9	410.760	cry (ace, peth)	83.4	200[1]			i H_2O; s CS_2
8300	trans, trans-9,12-Octadecadienoic acid	Linolelaidic acid	$C_{18}H_{32}O_2$	506-21-8	280.446	cry (MeOH)	28.5	181[0.8]			sl H_2O; s ace, hx
8301	Octadecahydrochrysene		$C_{18}H_{30}$	2090-14-4	246.431		115	353			vs EtOH
8302	Octadecamethyloctasiloxane		$C_{18}H_{54}O_7Si_8$	556-69-4	607.302		-63	186[20], 153[5]	0.913[25]	1.3970[20]	vs bz, peth, lig
8303	Octadecanamide		$C_{18}H_{37}NO$	124-26-5	283.493	lf (al)	109	250[12]			vs eth, chl
8304	Octadecane		$C_{18}H_{38}$	593-45-3	254.495	nd (al, eth-MeOH)	28.2	316.3	0.7768[28]	1.4390[20]	i H_2O; sl EtOH; s eth, ace, chl, lig
8305	Octadecanenitrile		$C_{18}H_{35}N$	638-65-3	265.478		41	362	0.8325[20]	1.4389[45]	i H_2O; s EtOH; vs eth, ace, chl
8306	1-Octadecanethiol	Stearyl mercaptan	$C_{18}H_{38}S$	2885-00-9	286.560		30	207[11]	0.8475[20]	1.4645[20]	vs eth

Norbormide

2,5-Norbornadiene

5-Norbornene-2,3-dicarboxylic acid anhydride

5-Norbornene-2-methylolacrylate

24-Norcholan-23-oic acid, (5β)

Nordazepam

Nordihydroguaiaretic acid

Norea

Norepinephrine

Norethisterone

Norethynodrel

Norflurazon

Norhyoscyamine

DL-Norleucine

L-Norleucine

Normorphine

Norplant

19-Nortestosterone phenylpropionate

Nortriptyline hydrochloride

DL-Norvaline

L-Norvaline

Noscapine

Novobiocin

Nuarimol

Nylidrin

Ochratoxin A

Ochratoxin B

Ochratoxin C

Octacaine

2,2',3,3',5,5',6,6'-Octachlorobiphenyl

Octachlorocyclopentene

Octachlorodibenzo-p-dioxin

Octachloronaphthalene

Octachlorostyrene

Octacontane

Octacosane

Octacosanoic acid

1-Octacosanol

trans, trans-9,12-Octadecadienoic acid

Octadecahydrochrysene

Octadecamethyloctasiloxane

Octadecanamide

Octadecane

Octadecanenitrile

1-Octadecanethiol

No.	Name	Synonym	Mol. Form.	CAS RN	Mol. Wt.	Physical Form	mp/°C	bp/°C	den/g cm⁻³	n_D	Solubility
8307	1-Octadecanol	Stearyl alcohol	$C_{18}H_{38}O$	112-92-5	270.494	lf (al)	57.9	335; 210.5[15]	0.8124[59]		i H_2O; s EtOH, eth; sl ace, bz
8308	Octadecanoyl chloride		$C_{18}H_{35}ClO$	112-76-5	302.923		23	215[15]	0.8969[0]	1.4523[24]	sl EtOH
8309	trans,cis,trans-9,11,13-Octadecatrienoic acid	cis-Eleostearic acid	$C_{18}H_{30}O_2$	506-23-0	278.430	nd (al)	49	235[12] dec, 170[1]	0.9028[50]	1.5112[50]	vs eth, EtOH
8310	trans,trans,trans-9,11,13-Octadecatrienoic acid	trans-Eleostearic acid	$C_{18}H_{30}O_2$	544-73-0	278.430	lf (al)	71.5	188[1]	0.8839[80]	1.5000[80]	vs EtOH
8311	cis-9-Octadecenamide		$C_{18}H_{35}NO$	301-02-0	281.477		76				vs eth
8312	1-Octadecene		$C_{18}H_{36}$	112-88-9	252.479		17.5	179[15]; 145[8]	0.7891[20]	1.4448[20]	i H_2O; s ace, ctc
8313	cis-9-Octadecenenitrile		$C_{18}H_{33}N$	112-91-4	263.462		-1	dec 332	0.847[17]	1.4566[20]	vs EtOH
8314	cis-6-Octadecenoic acid	Petroselinic acid	$C_{18}H_{34}O_2$	593-39-5	282.462	lf	29.8	238[18]	0.8700[40]	1.4533[40]	s eth; sl hp, MeOH
8315	trans-11-Octadecenoic acid	Vaccenic acid	$C_{18}H_{34}O_2$	693-72-1	282.462		44			1.4499[60]	s ace
8316	cis-9-Octadecen-1-ol	Oleyl alcohol	$C_{18}H_{36}O$	143-28-2	268.478		6.5	207[15]	0.8489[20]	1.4606[20]	i H_2O; s EtOH, eth; sl ctc
8317	cis-9-Octadecenylamine	Oleylamine	$C_{18}H_{37}N$	112-90-3	267.494	oil	25	147[2]			
8318	Octadecyl acetate		$C_{20}H_{40}O_2$	822-23-1	312.531		34.5	208[9]	0.8510[30]		vs EtOH
8319	Octadecyl acrylate	Stearyl 2-propenoate	$C_{21}H_{40}O_2$	4813-57-4	324.542						s ctc, CS_2
8320	Octadecylamine	1-Octadecanamine	$C_{18}H_{39}N$	124-30-1	269.510	cry (w)	52.9	346.8	0.8618[20]	1.4522[20]	i H_2O; s EtOH, eth, bz; sl ace
8321	Octadecylbenzene		$C_{24}H_{42}$	4445-07-2	330.590		36	400	0.85[36]	1.479[36]	
8322	Octadecylcyclohexane		$C_{24}H_{48}$	4445-06-1	336.638		41.6	409; 175[1]	0.8300[20]	1.4610[20]	
8323	Octadecyl 3-(3,5-di-tert-butyl-4-hydroxyphenyl)propanoate	Irganox 1076	$C_{35}H_{62}O_3$	2082-79-3	530.865	cry (MeOH/AcOEt)	50				
8324	Octadecyl isocyanate	1-Isocyanatooctadecane	$C_{19}H_{37}NO$	112-96-9	295.503		15.5	172[5]			
8325	Octadecyl methacrylate	Stearyl methacrylate	$C_{22}H_{42}O_2$	32360-05-7	338.567			195[6]	0.880[25]	1.429[25]	
8326	Octadecyl octadecanoate	Octadecyl stearate	$C_{36}H_{72}O_2$	2778-96-3	536.956	cry (EtOH)	60				
8327	3-(Octadecyloxy)-1,2-propanediol	Batyl alcohol	$C_{21}H_{44}O_3$	544-62-7	344.572		70.5	217[2]			vs eth
8328	Octadecyl vinyl ether	1-(Ethenyloxy)octadecane	$C_{20}H_{40}O$	930-02-9	296.531		30	182[3]	0.8138[40]		sl chl
8329	1,7-Octadiene		C_8H_{14}	3710-30-3	110.197			115.5	0.734[20]	1.4245[20]	
8330	1,7-Octadiyne		C_8H_{10}	871-84-1	106.165			135.5; 59[35]	0.8169[21]	1.4521[18]	s eth
8331	2,2,3,3,4,4,5,5-Octafluoro-1-pentanol		$C_5H_4F_8O$	355-80-6	232.072			140.5	1.6647[20]	1.3178[20]	
8332	1,2,3,4,5,6,7,8-Octahydroanthracene		$C_{14}H_{18}$	1079-71-6	186.293	pl (al)	78	294	0.9703[80]	1.5372[80]	i H_2O; s EtOH, HOAc; vs bz; sl ctc
8333	Octahydroazocine		$C_7H_{15}N$	1121-92-2	113.201		29	52[15]	0.896[25]	1.4720[20]	
8334	Octahydroindene		C_9H_{16}	496-10-6	124.223	liq	-53	167	0.876[25]	1.4702[20]	
8335	Octahydroindolizine		$C_8H_{15}N$	13618-93-4	125.212			75[43]	0.9074[10]	1.4748	vs eth, EtOH
8336	trans-Octahydro-1(2H)-naphthalenone		$C_{10}H_{16}O$	21370-71-8	152.233		33	122[20]	0.986[20]	1.4849[21]	
8337	1,2,3,4,5,6,7,8-Octahydrophenanthrene		$C_{14}H_{18}$	5325-97-3	186.293		16.7	295	1.026[20]	1.5569[17]	i H_2O; s ace, bz, CS_2, HOAc
8338	trans-Octahydro-2H-quinolizine-1-methanol, (1R)	Lupinine	$C_{10}H_{19}NO$	486-70-4	169.264	orth (peth)	70	270			s H_2O, EtOH, eth, bz, chl; sl peth
8339	2,2,4,4,6,6,8,8-Octamethylcyclotetrasilazane		$C_8H_{28}N_4Si_4$	1020-84-4	292.677		97				
8340	Octamethylcyclotetrasiloxane		$C_8H_{24}O_4Si_4$	556-67-2	296.617		17.5	175.8	0.9561[20]	1.3968[20]	i H_2O; s ctc
8341	1,1,1,3,5,7,7,7-Octamethyltetrasiloxane		$C_8H_{26}O_3Si_4$	16066-09-4	282.632			170	0.8559[20]	1.3854[20]	
8342	Octamethyltrisiloxane		$C_8H_{24}O_2Si_3$	107-51-7	236.533	liq	-80	153; 51[17]	0.8200[20]	1.3840[20]	sl EtOH; s bz, peth
8343	Octanal	Caprylic aldehyde	$C_8H_{16}O$	124-13-0	128.212			171	0.8211[20]	1.4217[20]	vs ace, bz, eth, EtOH
8344	Octanamide		$C_8H_{17}NO$	629-01-6	143.227	lf, pl	108	239	0.8450[110]		sl H_2O, bz, chl; vs EtOH; s eth, ace
8345	2-Octanamine, (±)		$C_8H_{19}N$	44855-57-4	129.244		97	164	0.7744[20]	1.4232[20]	vs eth, EtOH
8346	Octane		C_8H_{18}	111-65-9	114.229	liq	-56.82	125.67	0.6986[25]	1.3944[25]	i H_2O; s eth; msc EtOH, ace, bz
8347	1,8-Octanediamine		$C_8H_{20}N_2$	373-44-4	144.258	pl	51.64	225			vs H_2O, eth, EtOH
8348	Octanedinitrile	Suberonitrile	$C_8H_{12}N_2$	629-40-3	136.194		-1.8	185[15]	0.954[25]	1.4436[20]	
8349	Octanedioic acid	Suberic acid	$C_8H_{14}O_4$	505-48-6	174.195	lo nd or pl (w)	144	219[20]			i H_2O; msc eth, bz; sl DMSO
8350	1,2-Octanediol		$C_8H_{18}O_2$	1117-86-8	146.228		30	131[10], 104[0.2]			
8351	1,8-Octanediol		$C_8H_{18}O_2$	629-41-4	146.228	nd (bz-lig), pr	63	172[20]			sl H_2O, eth, chl, lig; vs EtOH; s bz
8352	Octanenitrile	Caprylnitrile	$C_8H_{15}N$	124-12-9	125.212	liq	-45.6	205.25	0.8136[20]	1.4203[20]	vs eth
8353	1-Octanethiol	Octyl mercaptan	$C_8H_{18}S$	111-88-6	146.294	liq	-49.2	199.1	0.8433[20]	1.4540[20]	s EtOH; sl ctc

1-Octadecanol

Octadecanoyl chloride

trans,cis,trans-9,11,13-Octadecatrienoic acid

trans,trans,trans-9,11,13-Octadecatrienoic acid

cis-9-Octadecenamide

1-Octadecene

cis-9-Octadecenenitrile

cis-6-Octadecenoic acid

trans-11-Octadecenoic acid

cis-9-Octadecen-1-ol

cis-9-Octadecenylamine

Octadecyl acetate

Octadecyl acrylate

Octadecylamine

Octadecylbenzene

Octadecylcyclohexane

Octadecyl 3-(3,5-di-*tert*-butyl-4-hydroxyphenyl)propanoate

Octadecyl isocyanate

Octadecyl methacrylate

Octadecyl octadecanoate

3-(Octadecyloxy)-1,2-propanediol

Octadecyl vinyl ether

1,7-Octadiene

1,7-Octadiyne

2,2,3,3,4,4,5,5-Octafluoro-1-pentanol

1,2,3,4,5,6,7,8-Octahydroanthracene

Octahydroazocine

Octahydroindene

Octahydroindolizine

trans-Octahydro-1(2H)-naphthalenone

1,2,3,4,5,6,7,8-Octahydrophenanthrene

trans-Octahydro-2H-quinolizine-1-methanol, (1R)

2,2,4,4,6,6,8,8-Octamethylcyclotetrasilazane

Octamethylcyclotetrasiloxane

1,1,1,3,5,7,7,7-Octamethyltetrasiloxane

Octamethyltrisiloxane

Octanal

Octanamide

2-Octanamine, (±)

Octane

1,8-Octanediamine

Octanedinitrile

Octanedioic acid

1,2-Octanediol

1,8-Octanediol

Octanenitrile

1-Octanethiol

No.	Name	Synonym	Mol. Form.	CAS RN	Mol. Wt.	Physical Form	mp/°C	bp/°C	den/ g cm^{-3}	n_D	Solubility
8354	Octanoic acid	Caprylic acid	$C_8H_{16}O_2$	124-07-2	144.212		16.5	239	0.9073[25]	1.4285[20]	sl H_2O; msc EtOH, chl, CH_3CN
8355	Octanoic anhydride		$C_{16}H_{30}O_3$	623-66-5	270.407	liq	-1	282.5	0.9065[18]	1.4358[18]	vs ace, eth, EtOH
8356	1-Octanol	Capryl alcohol	$C_8H_{18}O$	111-87-5	130.228	liq	-14.8	195.16	0.8262[25]	1.4295[20]	i H_2O; msc EtOH, eth; s ctc
8357	2-Octanol	(±)-sec-Caprylic alcohol	$C_8H_{18}O$	4128-31-8	130.228	liq	-31.6	179.3	0.8193[20]	1.4203[20]	sl H_2O; s EtOH, eth, ace
8358	3-Octanol		$C_8H_{18}O$	589-98-0	130.228	liq	-45	171	0.8258[20]		
8359	4-Octanol		$C_8H_{18}O$	74778-22-6	130.228	liq	-40.7	176.3	0.8186[20]	1.4248[20]	sl H_2O, ctc; s EtOH
8360	2-Octanone	Hexyl methyl ketone	$C_8H_{16}O$	111-13-7	128.212	liq	-16	172.5	0.820[20]	1.4151[20]	sl H_2O; msc EtOH, eth
8361	3-Octanone	Ethyl pentyl ketone	$C_8H_{16}O$	106-68-3	128.212			167.5	0.822[25]	1.4150[20]	i H_2O; msc EtOH, eth
8362	4-Octanone	Butyl propyl ketone	$C_8H_{16}O$	589-63-9	128.212			163	0.8146[25]	1.4173[14]	i H_2O; msc EtOH, eth; s ctc
8363	Octanoyl chloride		$C_8H_{15}ClO$	111-64-8	162.657	liq	-63	195.6	0.9535[15]	1.4335[20]	s eth
8364	Octaphenylcyclotetrasiloxane		$C_{48}H_{40}O_4Si_4$	546-56-5	793.172	nd (bz-al, HOAc)	200.5	330[1]			i H_2O; sl EtOH; s bz, chl, HOAc
8365	1,3,5,7-Octatetraene		C_8H_{10}	1482-91-3	106.165	cry (bz)	50	sub			s peth, HOAc
8366	trans-2-Octenal		$C_8H_{14}O$	2548-87-0	126.196	liq		85[19]	0.846	1.4500[20]	
8367	1-Octene	Caprylene	C_8H_{16}	111-66-0	112.213	liq	-101.7	121.29	0.7149[20]	1.4087[20]	i H_2O; msc EtOH; s eth, ace; sl ctc
8368	cis-2-Octene		C_8H_{16}	7642-04-8	112.213	liq	-100.2	125.6	0.7243[20]	1.4150[20]	i H_2O; s EtOH, eth, ace, bz, chl
8369	trans-2-Octene		C_8H_{16}	13389-42-9	112.213	liq	-87.7	125	0.7199[20]	1.4132[20]	i H_2O; s EtOH, eth, ace, bz; vs chl
8370	cis-3-Octene		C_8H_{16}	14850-22-7	112.213	liq	-126	122.9	0.7159[20]	1.4135[20]	vs ace, bz, eth, EtOH
8371	trans-3-Octene		C_8H_{16}	14919-01-8	112.213	liq	-110	123.3	0.7152[20]	1.4126[20]	i H_2O; s EtOH, eth, ace, bz, lig, ctc
8372	cis-4-Octene		C_8H_{16}	7642-15-1	112.213	liq	-118.7	122.5	0.7212[20]	1.4148[20]	vs ace, bz, eth, EtOH
8373	trans-4-Octene		C_8H_{16}	14850-23-8	112.213	liq	-93.8	122.3	0.7141[20]	1.4114[20]	i H_2O; s EtOH, eth, ace, bz, lig; sl ctc
8374	1-Octen-3-ol		$C_8H_{16}O$	3391-86-4	128.212			174; 69[12]	0.8395[13]	1.4391[12]	
8375	2-Octen-1-ol		$C_8H_{16}O$	22104-78-5	128.212			88[11]	0.850[20]	1.4470[20]	
8376	1-Octen-3-yne		C_8H_{12}	17679-92-4	108.181			134; 62[60]	0.7749[20]	1.4592[20]	vs eth
8377	Octhilinone	2-Octyl-3(2H)-isothiazolone	$C_{11}H_{19}NOS$	26530-20-1	213.340			120[0.01]			
8378	Octyl acetate		$C_{10}H_{20}O_2$	112-14-1	172.265	liq	-38.5	210	0.8705[20]	1.4150[20]	i H_2O; s EtOH, eth; sl ctc
8379	Octyl acrylate	Octyl 2-propenoate	$C_{11}H_{20}O_2$	2499-59-4	184.276			229; 57[0.05]	0.8810[20]		
8380	Octylamine	1-Octanamine	$C_8H_{19}N$	111-86-4	129.244		0	179.6	0.7826[20]	1.4292[20]	sl H_2O; vs EtOH, eth; s ctc
8381	Octylamine hydrochloride	1-Octanamine hydrochloride	$C_8H_{20}ClN$	142-95-0	165.705		196.5				s H_2O
8382	4-Octylaniline		$C_{14}H_{23}N$	16245-79-7	205.340		20	310; 138[5]	0.9128[20]		vs eth
8383	Octylbenzene		$C_{14}H_{22}$	2189-60-8	190.325	liq	-36	264	0.8562[20]	1.4845[20]	i H_2O; msc eth, bz
8384	Octyl butanoate		$C_{12}H_{24}O_2$	110-39-4	200.318	liq	-55.6	244.1	0.8629[20]	1.4267[15]	vs EtOH
8385	Octylcyclohexane		$C_{14}H_{28}$	1795-15-9	196.372	liq	-20	264	0.8138[20]	1.4503[20]	
8386	Octylcyclopentane		$C_{13}H_{26}$	1795-20-6	182.345	liq	-44	243	0.8048[20]	1.4446[20]	
8387	2-Octyldecanoic acid		$C_{18}H_{36}O_2$	619-39-6	284.478	nd or lf (al)	38.5	215[13]	0.8447[70]		vs eth, EtOH
8388	Octyldimethylamine	N,N-Dimethyl-1-octanamine	$C_{10}H_{23}N$	7378-99-6	157.297			194			
8389	Octyl diphenyl phosphate		$C_{20}H_{27}O_4P$	115-88-8	362.399				1.09[25]		
8390	Octyl formate		$C_9H_{18}O_2$	112-32-3	158.238	liq	-39.1	198.8	0.8744[20]	1.4208[15]	i H_2O; s EtOH; msc eth; sl ctc
8391	Octyl isocyanate		$C_9H_{17}NO$	3158-26-7	155.237			78[6]			
8392	Octyl methacrylate		$C_{12}H_{22}O_2$	2157-01-9	198.302			239.5			
8393	Octyl nitrate		$C_8H_{17}NO_3$	629-39-0	175.226			110[20]	0.975[0]		sl H_2O; s EtOH, eth
8394	Octyl nitrite		$C_8H_{17}NO_2$	629-46-9	159.227			174.5	0.862[17]	1.4127[20]	sl H_2O; vs EtOH, eth
8395	Octyl octanoate		$C_{16}H_{32}O_2$	2306-88-9	256.424	liq	-18.1	306.8	0.8554[20]	1.4352[20]	vs ace, eth, EtOH

Octanoic acid

Octanoic anhydride

1-Octanol

2-Octanol

3-Octanol

4-Octanol

2-Octanone

3-Octanone

4-Octanone

Octanoyl chloride

Octaphenylcyclotetrasiloxane

1,3,5,7-Octatetraene

trans-2-Octenal

1-Octene

cis-2-Octene

trans-2-Octene

cis-3-Octene

trans-3-Octene

cis-4-Octene

trans-4-Octene

1-Octen-3-ol

2-Octen-1-ol

1-Octen-3-yne

Octhilinone

Octyl acetate

Octyl acrylate

Octylamine

Octylamine hydrochloride

4-Octylaniline

Octylbenzene

Octyl butanoate

Octylcyclohexane

Octylcyclopentane

2-Octyldecanoic acid

Octyldimethylamine

Octyl diphenyl phosphate

Octyl formate

Octyl isocyanate

Octyl methacrylate

Octyl nitrate

Octyl nitrite

Octyl octanoate

No.	Name	Synonym	Mol. Form.	CAS RN	Mol. Wt.	Physical Form	mp/°C	bp/°C	den/ g cm⁻³	n_D	Solubility
8396	Octyloxirane		$C_{10}H_{20}O$	2404-44-6	156.265	liq		128[95], 97[30]			
8397	4-(Octyloxy)benzaldehyde		$C_{15}H_{22}O_2$	24083-13-4	234.335			131[0.5]			
8398	4-Octylphenol		$C_{14}H_{22}O$	1806-26-4	206.324		43.0	169[10], 150[4]			
8399	Octyl phenyl ether	(Octyloxy)benzene	$C_{14}H_{22}O$	1818-07-1	206.324		8	285	0.9131[15]	1.4875[20]	i H₂O; s EtOH, eth
8400	4-Octylphenyl salicylate	2-Hydroxybenzoic acid, 4-octylphenyl ester	$C_{21}H_{26}O_3$	2512-56-3	326.429	wh cry	73				
8401	Octyl propanoate		$C_{11}H_{22}O_2$	142-60-9	186.292	liq	-42.6	228	0.8663[20]	1.4221[15]	i H₂O; s EtOH, eth, bz; sl ctc
8402	1-Octyne	Hexylacetylene	C_8H_{14}	629-05-0	110.197	liq	-79.3	126.3	0.7461[20]	1.4159[20]	i H₂O; s EtOH, eth
8403	2-Octyne	Methylpentylacetylene	C_8H_{14}	2809-67-8	110.197	liq	-61.6	137.6	0.7596[20]	1.4278[20]	i H₂O; s EtOH, eth
8404	3-Octyne		C_8H_{14}	15232-76-5	110.197	liq	-103.9	133.1	0.7529[20]	1.4250[20]	i H₂O; s EtOH, eth
8405	4-Octyne	Dipropylacetylene	C_8H_{14}	1942-45-6	110.197	liq	-101	131.6	0.7509[20]	1.4248[20]	i H₂O; s EtOH, eth
8406	2-Octyn-1-ol	2-Octynol	$C_8H_{14}O$	20739-58-6	126.196		-18	98[15]	0.8805[20]	1.4556[20]	vs eth
8407	Oleandrin		$C_{32}H_{48}O_9$	465-16-7	576.718	cry (EtOH)	250 dec				i H₂O; s EtOH, chl
8408	Olean-12-en-3-ol, (3β)	β-Amyrin	$C_{30}H_{50}O$	559-70-6	426.717	nd (lig or al)	197	260[05]			i H₂O; sl EtOH, chl, lig; s eth, bz
8409	Oleanolic acid		$C_{30}H_{48}O_3$	508-02-1	456.700	nd or pr (al)	310 dec	sub 280			i H₂O; sl EtOH, eth, ace; vs py, HOAc
8410	Oleic acid	cis-9-Octadecenoic acid	$C_{18}H_{34}O_2$	112-80-1	282.462		13.4	360; 286[100]	0.8935[20]	1.4582[20]	i H₂O; msc EtOH, eth, ace, bz, chl, ctc
8411	Omeprazole		$C_{17}H_{19}N_3O_3S$	73590-58-6	345.416	cry (MeCN)	156				
8412	Omethoate		$C_5H_{12}NO_4PS$	1113-02-6	213.192	oil	≈135 dec		1.32[20]	1.4987[20]	msc H₂O; i hx
8413	Orange I	1-Naphthol Orange	$C_{16}H_{11}N_2NaO_4S$	523-44-4	350.324	red-br pow					s H₂O; sl EtOH; i bz
8414	Orange IV	Tropaeolin OO	$C_{18}H_{14}N_3NaO_3S$	554-73-4	375.377	ye pow					s H₂O
8415	Orcein			1400-62-0		br-red pow					
8416	L-Ornithine	2,5-Diaminopentanoic acid, (S)	$C_5H_{12}N_2O_2$	70-26-8	132.161	micro cry (al-eth)	140				vs H₂O, EtOH
8417	L-Ornithine, monohydrochloride		$C_5H_{13}ClN_2O_2$	3184-13-2	168.622	nd	215				vs H₂O
8418	Orotic acid	1,2,3,6-Tetrahydro-2,6-dioxo-4-pyrimidinecarboxylic acid	$C_5H_4N_2O_4$	65-86-1	156.097	cry (w)	345.5				sl H₂O; i os
8419	Oroxylin A	5,7-Dihydroxy-6-methoxy-2-phenyl-4H-1-benzopyran-4-one	$C_{16}H_{12}O_5$	480-11-5	284.263	ye nd (al)	231.5				vs ace, eth, EtOH
8420	Orphenadrine		$C_{18}H_{23}NO$	83-98-7	269.382			195[12]			
8421	Oryzalin	Benzenesulfonamide, 4-(dipropylamino)-3,5-dinitro-	$C_{12}H_{18}N_4O_6S$	19044-88-3	346.359		141				
8422	Ouabain		$C_{29}H_{44}O_{12}$	630-60-4	584.652	hyg pl (+9w)	200				sl H₂O; vs EtOH
8423	7-Oxabicyclo[4.1.0]heptane		$C_6H_{10}O$	286-20-4	98.142		<-10	131.5	0.9663[20]	1.4519[20]	i H₂O; vs EtOH, eth, ace, bz; s chl; sl ctc
8424	6-Oxabicyclo[3.1.0]hexane		C_5H_8O	285-67-6	84.117			102	0.964[25]	1.4336[20]	
8425	Oxacyclohexadecan-2-one	Exaltolide	$C_{15}H_{28}O_2$	106-02-5	240.382	thick oil		176[15]	0.9549[20]	1.4708[20]	
8426	1,3,4-Oxadiazole	1-Oxa-3,4-diazacyclopentadiene	$C_2H_2N_2O$	288-99-3	70.049			150		1.4300[25]	
8427	Oxadiazon		$C_{15}H_{18}Cl_2N_2O_3$	19666-30-9	345.221		90				
8428	Oxadixyl		$C_{14}H_{18}N_2O_4$	77732-09-3	278.304		104				
8429	Oxalic acid		$C_2H_2O_4$	144-62-7	90.035	orth pym or oct	189.5 dec	sub 157	1.900[17]		s H₂O; vs EtOH; sl eth; i bz, chl, peth
8430	Oxalic acid dihydrate		$C_2H_6O_6$	6153-56-6	126.065	mcl tab or pr	101.5		1.653[18]		s H₂O, EtOH; sl eth
8431	Oxaloacetic acid	Oxalacetic acid	$C_4H_4O_5$	328-42-7	132.072		161 dec				
8432	Oxalyl chloride	Oxalyl dichloride	$C_2Cl_2O_2$	79-37-8	126.926	liq	-16	63.5	1.4785[20]	1.4316[20]	s eth
8433	Oxalyl dihydrazide		$C_2H_6N_4O_2$	996-98-5	118.095	nd (w)	244.0		1.458[22]		s H₂O; sl EtOH, eth, bz, chl
8434	Oxamic acid		$C_2H_3NO_3$	471-47-6	89.050	cry (w)	210 dec				sl H₂O; i EtOH, eth
8435	Oxamide		$C_2H_4N_2O_2$	471-46-5	88.065	nd (w)	350 dec		1.667[20]		sl H₂O, EtOH; i eth
8436	Oxamniquine		$C_{14}H_{21}N_3O_3$	21738-42-1	279.335	ye-oran cry	149				s ace, chl, MeOH
8437	Oxamyl		$C_7H_{13}N_3O_3S$	23135-22-0	219.261		109	dec	0.97[25]		
8438	Oxandrolone		$C_{19}H_{30}O_3$	53-39-4	306.439		236				
8439	1,4-Oxathiane		C_4H_8OS	15980-15-1	104.171	liq	-17	147	1.1174[20]		sl H₂O

Octyloxirane

4-(Octyloxy)benzaldehyde

4-Octylphenol

Octyl phenyl ether

4-Octylphenyl salicylate

Octyl propanoate

1-Octyne

2-Octyne

3-Octyne

4-Octyne

2-Octyn-1-ol

Oleandrin

Olean-12-en-3-ol, (3β)

Oleanolic acid

Oleic acid

Omeprazole

Omethoate

Orange I

Orange IV

L-Ornithine

L-Ornithine, monohydrochloride

Orotic acid

Oroxylin A

Orphenadrine

Oryzalin

Ouabain

7-Oxabicyclo[4.1.0]heptane

6-Oxabicyclo[3.1.0]hexane

Oxacyclohexadecan-2-one

1,3,4-Oxadiazole

Oxadiazon

Oxadixyl

Oxalic acid

Oxalic acid dihydrate

Oxaloacetic acid

Oxalyl chloride

Oxalyl dihydrazide

Oxamic acid

Oxamide

Oxamniquine

Oxamyl

Oxandrolone

1,4-Oxathiane

No.	Name	Synonym	Mol. Form.	CAS RN	Mol. Wt.	Physical Form	mp/°C	bp/°C	den/ g cm⁻³	n_D	Solubility
8440	Oxazepam		$C_{15}H_{11}ClN_2O_2$	604-75-1	286.713	cry (EtOH)	205.5				i H_2O; s EtOH, chl, diox
8441	Oxazole		C_3H_3NO	288-42-6	69.062			69.5		1.4285[17]	
8442	Oxepane		$C_6H_{12}O$	592-90-5	100.158			119	0.89[25]	1.4400[20]	
8443	2-Oxepanone	Caprolactone	$C_6H_{10}O_2$	502-44-3	114.142	liq	-1.0	215	1.0761[20]	1.4611[20]	s EtOH, eth, ace
8444	Oxetane	Trimethylene oxide	C_3H_6O	503-30-0	58.079	liq	-97	47.6	0.8930[25]	1.3961[20]	msc H_2O, EtOH; s eth; vs ace
8445	2-Oxetanone	β-Propiolactone	$C_3H_4O_2$	57-57-8	72.063	liq	-33.4	162	1.1460[20]	1.4105[20]	msc eth; s chl
8446	3-Oxetanone		$C_3H_4O_2$	6704-31-0	72.063	unstab liq		106	1.137		
8447	Oxirane	Ethylene oxide	C_2H_4O	75-21-8	44.052	vol liq or gas	-112.5	10.6	0.8821[10]	1.3597[7]	s H_2O, EtOH, eth, ace, bz
8448	Oxiranecarboxaldehyde	Glycidaldehyde	$C_3H_4O_2$	765-34-4	72.063	liq	-62	112.5	1.1403[20]	1.4265[20]	vs H_2O; s EtOH, eth, ace, bz
8449	Oxiranemethanol, (±)	Glycidol	$C_3H_6O_2$	61915-27-3	74.079		-45	dec 167; 66[2.5]	1.1143[25]	1.4287[20]	vs H_2O, ace, eth, EtOH; s bz, chl
8450	α-Oxobenzeneacetaldehyde aldoxime	Isonitrosoacetophenone	$C_8H_7NO_2$	532-54-7	149.148		129				sl H_2O; s chl
8451	α-Oxobenzeneacetic acid		$C_8H_6O_3$	611-73-4	150.132	pr (CCl_4)	66	163[15]			vs H_2O; s EtOH, eth; sl ctc; i CS_2
8452	α-Oxobenzeneacetic acid, methyl ester		$C_9H_8O_3$	15206-55-0	164.158			247		1.5268[20]	
8453	α-Oxobenzeneacetonitrile		C_8H_5NO	613-90-1	131.132		32.5	206			i H_2O; vs EtOH, eth; sl chl
8454	γ-Oxobenzenebutanoic acid		$C_{10}H_{10}O_3$	2051-95-8	178.184	lf (dil al)	116.5				s H_2O, EtOH, eth, bz, chl, CS_2
8455	β-Oxobenzenepropanenitrile	Benzoylacetonitrile	C_9H_7NO	614-16-4	145.158		80.5	160[10]			sl H_2O; s EtOH, eth, bz, chl, alk, aq KCN
8456	α-Oxobenzenepropanoic acid	3-Phenylpyruvic acid	$C_9H_8O_3$	156-06-9	164.158	lf (bz, chl)	157.5				sl H_2O; vs EtOH, eth; s bz, chl; i lig
8457	2-Oxo-2H-1-benzopyran-3-carboxylic acid	Coumarin-3-carboxylic acid	$C_{10}H_6O_4$	531-81-7	162.142	nd (w, bz)	190 dec				vs EtOH
8458	Oxobis(2,4-pentanedione) vanadium	Vanadyl acetylacetonate	$C_{10}H_{14}O_5V$	3153-26-2	265.157	bl cry	258	174[0.2]			i H_2O; s EtOH, MeOH, bz, chl
8459	2-Oxobutanoic acid		$C_4H_6O_3$	600-18-0	102.089		33	81[16]	1.200[17]	1.3972[20]	vs H_2O, EtOH; sl eth
8460	4-Oxobutanoic acid		$C_4H_6O_3$	692-29-5	102.089	oil		135[14]			s H_2O, EtOH, eth, bz
8461	2-Oxoglutaric acid	α-Ketoglutaric acid	$C_5H_6O_5$	328-50-7	146.099	cry (ace-bz)	115.5				vs H_2O, EtOH, eth; s ace
8462	6-Oxoheptanoic acid		$C_7H_{12}O_3$	3128-07-2	144.168		40.2	251[280], 135[1]		1.4306[25]	vs H_2O, ace, eth, EtOH
8463	5-Oxohexanoic acid		$C_6H_{10}O_3$	3128-06-1	130.141		13.5	274.5	1.09[25]	1.4451[20]	s H_2O, EtOH, eth; sl ctc
8464	α-Oxo-1H-indole-3-propanoic acid	Indole-3-pyruvic acid	$C_{11}H_9NO_3$	392-12-1	203.194	gray cry	211				
8465	Oxolinic acid		$C_{13}H_{11}NO_5$	14698-29-4	261.230	cry (DMF)	313 dec				
8466	4-Oxopentanal		$C_5H_8O_2$	626-96-0	100.117		<-21	dec 187	1.0134[21]	1.4257[22]	vs H_2O, ace, eth, EtOH
8467	3-Oxopentanedioic acid	Acetonedicarboxylic acid	$C_5H_6O_5$	542-05-2	146.099	nd (AcOEt)	138 dec				s H_2O, EtOH; sl eth; i bz, chl, lig
8468	2-Oxopentanoic acid		$C_5H_8O_3$	1821-02-9	116.116		6.5	179	1.0970[14]		sl H_2O; s eth, bz, chl, lig, CS_2
8469	4-Oxopentanoic acid	Levulinic acid	$C_5H_8O_3$	123-76-2	116.116	lf or pl	33	dec 245	1.1335[20]	1.4396[20]	vs H_2O, EtOH, eth; s chl
8470	4-Oxo-4-(phenylamino)butanoic acid	Succinanilic acid	$C_{10}H_{11}NO_3$	102-14-7	193.199	nd (w)	148.5				sl H_2O; s EtOH; vs eth
8471	cis-4-Oxo-4-(phenylamino)-2-butenoic acid	Maleanilic acid	$C_{10}H_9NO_3$	555-59-9	191.183	mcl ye cry	192 dec		1.418[30]		
8472	Oxophenylarsine	Phenylarsine oxide	C_6H_5AsO	637-03-6	168.025	cry (bz-eth) or (chl-eth)	145				i H_2O, eth; sl EtOH; vs bz, chl
8473	4-Oxo-4-phenyl-2-butenoic acid		$C_{10}H_8O_3$	583-06-2	176.169	nd or pr (tol)	99				sl H_2O, chl, lig; s EtOH, eth, tol
8474	2-Oxopropanal oxime	Isonitrosoacetone	$C_3H_5NO_2$	306-44-5	87.078	nd(CCl_4) lf (eth-peth)	69	sub	1.0744[67]		s H_2O, eth; sl bz, ctc, chl
8475	2-Oxopropanenitrile		C_3H_3NO	631-57-2	69.062			92.3	0.9745[20]	1.3764[20]	s eth, ace, CH_3CN
8476	17-(1-Oxopropoxy)-androst-4-en-3-one, (17β)	Testosterone-17-propionate	$C_{22}H_{32}O_3$	57-85-2	344.487		120				vs eth, py, EtOH

Oxazepam

Oxazole

Oxepane

2-Oxepanone

Oxetane

2-Oxetanone

3-Oxetanone

Oxirane

Oxiranecarboxaldehyde

Oxiranemethanol, (±)

α-Oxobenzeneacetaldehyde aldoxime

α-Oxobenzeneacetic acid

α-Oxobenzeneacetic acid, methyl ester

α-Oxobenzeneacetonitrile

γ-Oxobenzenebutanoic acid

β-Oxobenzenepropanenitrile

α-Oxobenzenepropanoic acid

2-Oxo-2H-1-benzopyran-3-carboxylic acid

Oxobis(2,4-pentanedione)vanadium

2-Oxobutanoic acid

4-Oxobutanoic acid

2-Oxoglutaric acid

6-Oxoheptanoic acid

5-Oxohexanoic acid

α-Oxo-1H-indole-3-propanoic acid

Oxolinic acid

4-Oxopentanal

3-Oxopentanedioic acid

2-Oxopentanoic acid

4-Oxopentanoic acid

4-Oxo-4-(phenylamino)butanoic acid

cis-4-Oxo-4-(phenylamino)-2-butenoic acid

Oxophenylarsine

4-Oxo-4-phenyl-2-butenoic acid

2-Oxopropanal oxime

2-Oxopropanenitrile

17-(1-Oxopropoxy)-androst-4-en-3-one, (17β)

No.	Name	Synonym	Mol. Form.	CAS RN	Mol. Wt.	Physical Form	mp/°C	bp/°C	den/ g cm⁻³	n_D	Solubility
8477	2-Oxo-2H-pyran-5-carboxylic acid	Coumalic acid	$C_6H_4O_4$	500-05-0	140.094	pr (MeOH)	207 dec	218[120]			sl H_2O, eth, ace; i bz, chl; s EtOH, HOAc
8478	4-Oxo-4H-pyran-2,6-dicarboxylic acid	Chelidonic acid	$C_7H_4O_6$	99-32-1	184.103	rose mcl nd (al-w,+1w)	262				sl H_2O, EtOH
8479	17-Oxosparteine		$C_{15}H_{24}N_2O$	489-72-5	248.364	ye to col hyg nd (peth)	84	209[12]			vs H_2O, EtOH, eth; s chl
8480	4,4'-Oxybis(benzenesulfonyl chloride)	Diphenyl ether 4,4'-disulfonyl chloride	$C_{12}H_8Cl_2O_5S_2$	121-63-1	367.225	cry (peth)	128				
8481	4,4'-Oxybis(benzenesulfonyl hydrazide)		$C_{12}H_{14}N_4O_5S_2$	80-51-3	358.393	cry (H_2O)	164 dec				
8482	Oxybutynin		$C_{22}H_{31}NO_3$	5633-20-5	357.486	cry	114				
8483	Oxycarboxin	Carboxin S,S-dioxide	$C_{12}H_{13}NO_4S$	5259-88-1	267.301	pr (EtOH)	129				sl H_2O; s bz, EtOH; vs ace
8484	Oxychlordane		$C_{10}H_4Cl_6O$	27304-13-8	423.762	cry (pentane)	100				
8485	Oxycodone	Dihydro-14-hydroxycodeinone	$C_{18}H_{21}NO_4$	76-42-6	315.365	rods (EtOH)	219				i H_2O, eth; s EtOH, chl
8486	Oxydemeton-methyl		$C_6H_{15}O_4PS_2$	301-12-2	246.284		<-20	106[0.01]	1.289[20]		
8487	10,10'-Oxydiphenoxarsine	10,10'-Oxybis[10H-phenoxarsine]	$C_{24}H_{16}As_2O_3$	58-36-6	502.225	col mcl cry	185		1.41		i H_2O; s EtOH, chl; i CH_2Cl_2
8488	Oxyfluorfen		$C_{15}H_{11}ClF_3NO_4$	42874-03-3	361.701		84	dec 358	1.35[73]		
8489	Oxymetazoline		$C_{16}H_{24}N_2O$	1491-59-4	260.374	cry (bz)	182				i eth, chl
8490	Oxymetholone		$C_{21}H_{32}O_3$	434-07-1	332.477	cry	179				
8491	Oxymethurea		$C_3H_8N_2O_3$	140-95-4	120.107	pr(al)	126	149[25]			s H_2O, EtOH, MeOH; i eth; sl DMSO
8492	Oxyphenbutazone		$C_{19}H_{20}N_2O_3$	129-20-4	324.373	cry (eth/peth)	124				s EtOH, MeOH, chl, bz, eth
8493	Oxyphenonium bromide		$C_{21}H_{34}BrNO_3$	50-10-2	428.404		191.5				vs H_2O; sl EtOH
8494	Oxytetracycline		$C_{22}H_{24}N_2O_9$	79-57-2	460.434		184.5		1.634[20]		
8495	Oxytocin		$C_{43}H_{66}N_{12}O_{12}S_2$	50-56-6	1007.187	wh pow					s H_2O, BuOH
8496	Paclobutrazol		$C_{15}H_{20}ClN_3O$	76738-62-0	293.792	wh cry	166		1.22		i H_2O; vs ace, MeOH; s xyl, hx
8497	Palustric acid		$C_{20}H_{30}O_2$	1945-53-5	302.451	cry (MeOH)	164.5				
8498	Pamoic acid		$C_{23}H_{16}O_6$	130-85-8	388.369		315				
8499	Pancuronium dibromide		$C_{35}H_{60}Br_2N_2O_4$	15500-66-0	732.670	cry	215				sl chl
8500	Panose	4-α-Isomaltosylglucose	$C_{18}H_{32}O_{16}$	33401-87-5	504.437		223 dec				
8501	Panthesin		$C_{18}H_{32}N_2O_5S$	135-44-4	388.522	pa ye pow (al)	158				vs H_2O, EtOH
8502	Pantolactone		$C_6H_{10}O_3$	599-04-2	130.141		92				
8503	Pantothenic acid		$C_9H_{17}NO_5$	79-83-4	219.235	ye visc oil					vs H_2O, bz, eth
8504	Papaveraldine		$C_{20}H_{19}NO_5$	522-57-6	353.369	nd (al),cry (bz, peth)	210.5				i H_2O; sl EtOH, eth; s bz, chl
8505	Papaverine		$C_{20}H_{21}NO_4$	58-74-2	339.386	wh pr (al-eth), nd (chl-peth)	147.5	sub 135	1.337[20]	1.625	sl H_2O; vs EtOH, chl; s ace, bz, py
8506	Papaverine hydrochloride	Cerespan	$C_{20}H_{22}ClNO_4$	61-25-6	375.847	wh mcl pr (w)	224.5				vs H_2O, EtOH
8507	Paraformaldehyde		$(CH_2O)_x$	30525-89-4	30.026		164 dec				
8508	Paraldehyde	2,4,6-Trimethyl-1,3,5-trioxane	$C_6H_{12}O_3$	123-63-7	132.157		12.6	124.3	0.9943[20]	1.4049[20]	sl H_2O; msc EtOH, eth, chl
8509	Paramethadione		$C_7H_{11}NO_3$	115-67-3	157.167	liq			1.121[25]	1.449[25]	sl H_2O; s EtOH, chl, bz, eth
8510	Paraoxon	O,O-Diethyl O-(4-nitrophenyl) phosphate	$C_{10}H_{14}NO_6P$	311-45-5	275.195	oily liq		161[0.5]	1.2683[25]	1.5096	s eth
8511	Paraquat		$C_{12}H_{14}N_2$	4685-14-7	186.252	cation					
8512	Pararosaniline hydrochloride	Basic fuchsin	$C_{19}H_{18}ClN_3$	569-61-9	323.819	pale viol pow	269 dec				
8513	Parasorbic acid		$C_6H_8O_2$	10048-32-5	112.127	oily liq		100[15]	1.079[18]	1.4736[20]	vs H_2O, eth, EtOH
8514	Parathion		$C_{10}H_{14}NO_5PS$	56-38-2	291.261	ye liq	6.1	375	1.2681[20]	1.5370[25]	i H_2O; s eth, ace; sl ctc; vs EtOH, AcOEt
8515	Patchouli alcohol		$C_{15}H_{26}O$	5986-55-0	222.366		56		0.9906[65]	1.5029[65]	i H_2O; s EtOH, eth
8516	Pebulate		$C_{10}H_{21}NOS$	1114-71-2	203.345			142[20]	0.9458[20]	1.4752[20]	vs ace, bz, MeOH
8517	Pelargonidin chloride		$C_{15}H_{11}ClO_5$	134-04-3	306.698	red br hyg (anh) pr or pl	>350				s H_2O; vs EtOH; sl chl, MeOH
8518	Pellotine		$C_{13}H_{19}NO_3$	83-14-7	237.295	pl (al, peth)	111.5				vs ace, eth, EtOH, peth
8519	Pemoline	2-Amino-5-phenyl-4(5H)-oxazolone	$C_9H_8N_2O_2$	2152-34-3	176.172	cry	256 dec				i H_2O, eth, ace; sl hot EtOH

2-Oxo-2*H*-pyran-5-carboxylic acid

4-Oxo-4*H*-pyran-2,6-dicarboxylic acid

17-Oxosparteine

4,4'-Oxybis(benzenesulfonyl chloride)

4,4'-Oxybis(benzenesulfonyl hydrazide)

Oxybutynin

Oxycarboxin

Oxychlordane

Oxycodone

Oxydemeton-methyl

10,10'-Oxydiphenoxarsine

Oxyfluorfen

Oxymetazoline

Oxymetholone

Oxymethurea

Oxyphenbutazone

Oxyphenonium bromide

Oxytetracycline

Cys-Tyr-Ile-Gln-Asn-Cys-Pro-Leu-Gly(NH₂)

Oxytocin

Paclobutrazol

Palustric acid

Pamoic acid

Pancuronium dibromide

Panose

Panthesin

Pantolactone

Pantothenic acid

Papaveraldine

Papaverine

Papaverine hydrochloride

Paraformaldehyde

Paraldehyde

Paramethadione

Paraoxon

Paraquat

Pararosaniline hydrochloride

Parasorbic acid

Parathion

Patchouli alcohol

Pebulate

Pelargonidin chloride

Pellotine

Pemoline

No.	Name	Synonym	Mol. Form.	CAS RN	Mol. Wt.	Physical Form	mp/°C	bp/°C	den/ g cm^{-3}	n_D	Solubility
8520	Pendimethalin	N-(1-Ethylpropyl)-3,4-dimethyl-2,6-dinitroaniline	$C_{13}H_{19}N_3O_4$	40487-42-1	281.308		56	dec	1.19[25]		
8521	Penicillamine cysteine disulfide		$C_8H_{16}N_2O_4S_2$	18840-45-4	268.354		195				
8522	Penicillin G	Benzylpenicillinic acid	$C_{16}H_{18}N_2O_4S$	61-33-6	334.390	amor wh pow					sl H_2O; s MeOH, EtOH, eth, chl, bz, ace
8523	Penicillin G procaine		$C_{29}H_{38}N_4O_6S$	54-35-3	570.700		108 dec		1.2555[25]		s H_2O, EtOH, chl
8524	Penicillin V	Phenoxymethylpenicillin	$C_{16}H_{18}N_2O_5S$	87-08-1	350.389	cry	124 dec				sl H_2O; s os
8525	1,2,3,4,5-Pentabromo-6-chlorocyclohexane		$C_6H_6Br_5Cl$	87-84-3	513.085	cry	204				
8526	Pentabromomethylbenzene		$C_7H_3Br_5$	87-83-2	486.619		288		2.97[17]		i H_2O; sl EtOH, HOAc; s bz
8527	Pentabromophenol		C_6HBr_5O	608-71-9	488.591	mcl pr (HOAc) nd (al)	229.5	sub			i H_2O; s EtOH, bz, HOAc; sl eth
8528	1,1,1,3,3-Pentabromo-2-propanone	Pentabromoacetone	C_3HBr_5O	79-49-2	452.559	nd (w, al) pr (eth)	79.5	sub			i H_2O; vs EtOH, eth, ace, chl
8529	Pentac	Dienochlor	$C_{10}Cl_{10}$	2227-17-0	474.637	tan cry (peth)	122				
8530	Pentacene	Benzo[b]naphthacene	$C_{22}H_{14}$	135-48-8	278.346	ye grn nd or lf (xyl)	>300 dec				i H_2O; sl bz; s $PhNO_2$
8531	2,3,4,5,6-Pentachloroaniline		$C_6H_2Cl_5N$	527-20-8	265.352	nd (al)	233.0				vs eth, EtOH, lig
8532	2,3,4,5,6-Pentachloroanisole	Methyl pentachlorophenyl ether	$C_7H_3Cl_5O$	1825-21-4	280.363	nd MeOH	108.5				
8533	Pentachlorobenzene		C_6HCl_5	608-93-5	250.337	nd (al)	86	277	1.8342[16]		i H_2O, EtOH; sl eth, bz, chl, CS_2
8534	Pentachlorobenzenethiol	Pentachlorophenyl mercaptan	C_6HCl_5S	133-49-3	282.402		231.5				
8535	2,3,4,5,6-Pentachlorobiphenyl		$C_{12}H_5Cl_5$	18259-05-7	326.433	nd (peth)	123.5				i H_2O
8536	2,2',4,5,5'-Pentachlorobiphenyl		$C_{12}H_5Cl_5$	37680-73-2	326.433	cry (EtOH)	78.5				i H_2O
8537	1,2,3,4,7-Pentachlorodibenzo-p-dioxin		$C_{12}H_3Cl_5O_2$	39227-61-7	356.416	cry (bz/MeOH)	195				
8538	Pentachloroethane	Refrigerant 120	C_2HCl_5	76-01-7	202.294	liq	-28.78	162.0	1.6796[20]	1.5025[20]	i H_2O; msc EtOH, eth
8539	Pentachlorofluoroethane		C_2Cl_5F	354-56-3	220.284	col liq	101.3	138	1.74[25]		i H_2O; s EtOH, eth
8540	Pentachloronitrobenzene	Quintozene	$C_6Cl_5NO_2$	82-68-8	295.335	cry (al)	144	dec 328	1.718[25]		i H_2O; sl EtOH; s bz, chl
8541	Pentachlorophenol		C_6HCl_5O	87-86-5	266.336	mcl pr (al + 1w) nd (bz)	174	dec 310	1.978[22]		i H_2O; sl lig; vs EtOH, eth; s bz
8542	1,1,2,2,3-Pentachloropropane		$C_3H_3Cl_5$	16714-68-4	216.321			181[500]	1.633[25]	1.5098[25]	
8543	1,1,2,3,3-Pentachloro-1-propene		C_3HCl_5	1600-37-9	214.305			185	1.6317[34]	1.5313[20]	vs eth
8544	Pentachloropyridine		C_5Cl_5N	2176-62-7	251.326		125.5	280			vs bz, EtOH, lig
8545	2,3,4,5,6-Pentachlorotoluene		$C_7H_3Cl_5$	877-11-2	264.364	nd (bz, peth)	224.8	301			sl EtOH, eth, CS_2; s bz, tol, peth
8546	Pentacontane		$C_{50}H_{102}$	6596-40-3	703.345		92.1	575.0			
8547	Pentacosane		$C_{25}H_{52}$	629-99-2	352.681		53.93	401.9; 282[40]	0.8012[20]	1.4491[20]	s bz, chl
8548	1H-Pentadecafluoroheptane		C_7HF_{15}	375-83-7	370.059			96.0	1.725[25]	1.2690[25]	
8549	Pentadecafluorooctanoic acid		$C_8HF_{15}O_2$	335-67-1	414.069		54.3	188			
8550	2,2,3,3,4,4,5,5,6,6,7,7,8,8,8-Pentadecafluoro-1-octanol	1,1-Dihydroperfluorooctanol	$C_8H_3F_{15}O$	307-30-2	400.085	waxy solid	47	164; 68[8]			
8551	Pentadecanal		$C_{15}H_{30}O$	2765-11-9	226.398	nd	24.5	185[25]			vs ace, eth, EtOH
8552	Pentadecane		$C_{15}H_{32}$	629-62-9	212.415		9.95	270.6	0.7685[20]	1.4315[20]	i H_2O; vs EtOH, eth
8553	Pentadecanoic acid	Pentadecylic acid	$C_{15}H_{30}O_2$	1002-84-2	242.398	pl (dil al, HOAc) cry (peth)	52.3	257[100], 158[1]	0.8423[80]	1.4254[80]	i H_2O; vs EtOH, ace; s eth; sl tfa
8554	1-Pentadecanol		$C_{15}H_{32}O$	629-76-5	228.414		43.9	300	0.8347[25]		i H_2O
8555	2-Pentadecanone		$C_{15}H_{30}O$	2345-28-0	226.398		39.5	294	0.8182[39]		
8556	8-Pentadecanone		$C_{15}H_{30}O$	818-23-5	226.398	cry (al)	43	291	0.8180[39]		s EtOH, eth, bz, ctc, chl
8557	1-Pentadecene		$C_{15}H_{30}$	13360-61-7	210.399	liq	-1.4	268.2	0.7764[20]	1.4389[20]	i H_2O; s ace
8558	Pentadecylamine	Pentadecanamine	$C_{15}H_{33}N$	2570-26-5	227.430		37.3	307.6	0.8104[20]	1.4480[20]	vs eth, EtOH
8559	Pentadecylbenzene		$C_{21}H_{36}$	2131-18-2	288.511		22	373	0.8548[20]	1.4815[20]	
8560	3-Pentadecyl-1,2-benzenediol	3-Pentadecylcatechol	$C_{21}H_{36}O_2$	492-89-7	320.510	nd (to, peth)	59.5				vs bz, eth, EtOH
8561	Pentadecylcyclohexane		$C_{21}H_{42}$	6006-95-7	294.558		29	373	0.8267[20]	1.4588[20]	
8562	3-Pentadecylphenol		$C_{21}H_{36}O$	501-24-6	304.510	nd (peth)	53.5	230[8], 197[1.5]			vs ace, bz, EtOH
8563	1-Pentadecyne		$C_{15}H_{28}$	765-13-9	208.383		10	268	0.7928[20]	1.4419[20]	vs ace
8564	1,2-Pentadiene	Ethylallene	C_5H_8	591-95-7	68.118	liq	-137.3	44.9	0.6926[20]	1.4209[20]	msc EtOH, eth, ace, bz, ctc, hp
8565	cis-1,3-Pentadiene	cis-Piperylene	C_5H_8	1574-41-0	68.118	liq	-140.8	44.1	0.6910[20]	1.4363[20]	msc EtOH, eth, ace, bz, ctc, hp
8566	trans-1,3-Pentadiene	trans-Piperylene	C_5H_8	2004-70-8	68.118	liq	-87.4	42	0.6710[25]	1.4301[20]	

Pendimethalin

Penicillamine cysteine disulfide

Penicillin G

Penicillin G procaine

Penicillin V

1,2,3,4,5-Pentabromo-6-chlorocyclohexane

Pentabromomethylbenzene

Pentabromophenol

1,1,1,3,3-Pentabromo-2-propanone

Pentac

Pentacene

2,3,4,5,6-Pentachloroaniline

2,3,4,5,6-Pentachloroanisole

Pentachlorobenzene

Pentachlorobenzenethiol

2,3,4,5,6-Pentachlorobiphenyl

2,2',4,5,5'-Pentachlorobiphenyl

1,2,3,4,7-Pentachlorodibenzo-p-dioxin

Pentachloroethane

Pentachlorofluoroethane

Pentachloronitrobenzene

Pentachlorophenol

1,1,2,2,3-Pentachloropropane

1,1,2,3,3-Pentachloro-1-propene

Pentachloropyridine

2,3,4,5,6-Pentachlorotoluene

$H_3C(CH_2)_{48}CH_3$
Pentacontane

Pentacosane

1H-Pentadecafluoroheptane

Pentadecafluorooctanoic acid

2,2,3,3,4,4,5,5,6,6,7,7,8,8,8-Pentadecafluoro-1-octanol

Pentadecanal

Pentadecane

Pentadecanoic acid

1-Pentadecanol

2-Pentadecanone

8-Pentadecanone

1-Pentadecene

Pentadecylamine

Pentadecylbenzene

3-Pentadecyl-1,2-benzenediol

Pentadecylcyclohexane

3-Pentadecylphenol

1-Pentadecyne

1,2-Pentadiene

cis-1,3-Pentadiene

trans-1,3-Pentadiene

No.	Name	Synonym	Mol. Form.	CAS RN	Mol. Wt.	Physical Form	mp/°C	bp/°C	den/ g cm⁻³	n_D	Solubility
8567	1,4-Pentadiene		C_5H_8	591-93-5	68.118	vol liq or gas	-148.2	26	0.6608²⁰	1.3888²⁰	i H₂O; vs EtOH, eth, ace, bz
8568	2,3-Pentadiene	1,3-Dimethylallene	C_5H_8	591-96-8	68.118	liq	-125.6	48.2	0.6950²⁰	1.4284²⁰	i H₂O; msc EtOH, eth, ace, bz, hp, ctc
8569	1,4-Pentadien-3-ol		C_5H_8O	922-65-6	84.117			115.5	0.860²³	1.4400¹⁷	
8570	1,3-Pentadiyne	Methyldiacetylene	C_5H_4	4911-55-1	64.086	liq	-38.5	55	0.7909²⁰	1.4431²¹	i H₂O; s eth, bz, chl
8571	Pentaerythritol		$C_5H_{12}O_4$	115-77-5	136.147	cry (dil HCl)	258	sub		1.548	s H₂O; i eth, bz
8572	Pentaerythritol tetraacetate	2,2-Bis[(acetyloxy)methyl]-1,3-propanediol diacetate	$C_{13}H_{20}O_8$	597-71-7	304.293	tetr nd (w, bz)	83.5		1.273¹⁸		s H₂O; vs EtOH, eth
8573	Pentaerythritol tetrakis(2-mercaptoacetate)		$C_{13}H_{20}O_8S_4$	10193-99-4	432.553	liq		250¹	1.385²⁵	1.5470²⁰	
8574	Pentaerythritol tetramethacrylate	Tetramethylolmethane tetramethacrylate	$C_{21}H_{28}O_8$	3253-41-6	408.442		53.5				
8575	Pentaerythritol tetranitrate		$C_5H_8N_4O_{12}$	78-11-5	316.138	tetr (ace) pr (ace-al)	140.5		1.773²⁰		sl H₂O, EtOH, eth; vs ace; s bz, py
8576	Pentaethylbenzene		$C_{16}H_{26}$	605-01-6	218.377		<-20	277	0.8971¹⁹	1.5127²⁰	
8577	Pentaethyl tantalate	Ethanol, tantalum(5+) salt	$C_{10}H_{25}O_5Ta$	6074-84-6	406.251			151¹			
8578	2,3,4,5,6-Pentafluoroaniline		$C_6H_2F_5N$	771-60-8	183.079		34	153.5			
8579	Pentafluorobenzaldehyde		C_7HF_5O	653-37-2	196.074		20	167		1.4506²⁰	
8580	Pentafluorobenzene		C_6HF_5	363-72-4	168.064	liq	-47.4	85.74	1.514²⁵	1.3905²⁰	
8581	Pentafluorobenzenethiol		C_6HF_5S	771-62-0	200.129	liq	-24	143	1.501²⁵	1.4645²⁰	
8582	Pentafluorobenzoic acid		$C_7HF_5O_2$	602-94-8	212.074		101	220			
8583	Pentafluorobenzonitrile		C_7F_5N	773-82-0	193.074		1.2	162	1.563²⁰	1.4402²⁵	
8584	Pentafluoroethane		C_2HF_5	354-33-6	120.021	col gas	-103	-48.1			
8585	Pentafluoroiodobenzene		C_6F_5I	827-15-6	293.960	liq	-29	166	2.212²⁰	1.4950²⁵	
8586	Pentafluoromethoxybenzene	Methyl pentafluorophenyl ether	$C_7H_3F_5O$	389-40-2	198.090	liq	-37	138.5	1.493²⁰	1.4087²⁰	
8587	Pentafluorophenol		C_6HF_5O	771-61-9	184.063		37.5	145.6		1.4263²⁰	
8588	1,1,1,2,2-Pentafluoropropane	Refrigerant 245cb	$C_3H_3F_5$	1814-88-6	134.048	col gas		-17.4			
8589	2,2,3,3,3-Pentafluoro-1-propanol		$C_3H_3F_5O$	422-05-9	150.047			26⁵⁰			
8590	2,3,4,5,6-Pentafluorotoluene		$C_7H_3F_5$	771-56-2	182.091	liq	-29.78	117.5	1.440²⁰	1.4016²⁵	
8591	1,1,2,4,4-Pentafluoro-3-(trifluoromethyl)-1,3-butadiene		C_5F_8	384-04-3	212.041			39	1.527⁰	1.3000⁰	vs ace, bz, eth
8592	Pentagastrin		$C_{37}H_{49}N_7O_9S$	5534-95-2	767.892	col nd	230 dec				i H₂O, bz, EtOH, eth
8593	trans-3,3',4',5,7-Pentahydroxyflavanone, (±)	Taxifolin	$C_{15}H_{12}O_7$	480-18-2	304.252		227 dec				s chl
8594	Pentamethonium bromide		$C_{11}H_{28}Br_2N_2$	541-20-8	348.161		301				sl H₂O
8595	Pentamethylbenzene		$C_{11}H_{16}$	700-12-9	148.245	pr (al)	54.5	232	0.917²⁰	1.527²⁰	i H₂O; vs EtOH, bz; s chl
8596	2,4,6,8,10-Pentamethylcyclopentasiloxane		$C_5H_{20}O_5Si_5$	6166-86-5	300.638	liq	-108	169	0.9985²⁰	1.3912²⁰	
8597	2,2,4,6,6-Pentamethylheptane		$C_{12}H_{26}$	13475-82-6	170.334	liq	-67	177.8	0.7463²⁰	1.4440²⁰	
8598	2,2,4,6,6-Pentamethyl-3-heptene		$C_{12}H_{24}$	123-48-8	168.319	liq		180.5			
8599	2,2,3,3,4-Pentamethylpentane		$C_{10}H_{22}$	16747-44-7	142.282	liq	-36.4	166.1	0.7767²⁵	1.4361²⁰	
8600	2,2,3,4,4-Pentamethylpentane		$C_{10}H_{22}$	16747-45-8	142.282	liq	-38.7	159.3	0.7636²⁵	1.4307²⁰	
8601	Pentamethylphenol		$C_{11}H_{16}O$	2819-86-5	164.244	nd (al, peth, ace)	128	267			i H₂O; s EtOH
8602	1,2,2,6,6-Pentamethylpiperidine	Pempidine	$C_{10}H_{21}N$	79-55-0	155.281			147	0.8580⁰	1.4550²¹	
8603	Pentamethylsilanamine		$C_5H_{15}NSi$	2083-91-2	117.266			86	0.7400²⁰	1.4379²⁴	
8604	Pentanal	Valeraldehyde	$C_5H_{10}O$	110-62-3	86.132	liq	-91.5	103	0.8095²⁰	1.3944²⁰	sl H₂O; s EtOH, eth
8605	Pentanamide		$C_5H_{11}NO$	626-97-1	101.147	mcl pl (peth, al)	106	225	0.8735¹¹⁰	1.4183¹¹⁰	vs H₂O, EtOH, eth; sl chl
8606	3-Pentanamine		$C_5H_{13}N$	616-24-0	87.164			89	0.7487²⁰	1.4063²⁰	s EtOH; sl chl
8607	Pentane		C_5H_{12}	109-66-0	72.149	liq	-129.67	36.06	0.6262²⁰	1.3575²⁰	sl H₂O; msc EtOH, eth, ace, bz, chl; s ctc
8608	Pentanedial	Glutaraldehyde	$C_5H_8O_2$	111-30-8	100.117			dec 188			msc H₂O, EtOH; s bz
8609	1,5-Pentanediamine	Cadaverine	$C_5H_{14}N_2$	462-94-2	102.178		11.83	179	0.873²⁵	1.463²⁰	s H₂O, EtOH; sl eth
8610	Pentanedinitrile	Glutaronitrile	$C_5H_6N_2$	544-13-8	94.115	liq	-29	286	0.9911¹⁵	1.4295²⁰	vs EtOH, chl
8611	1,2-Pentanediol, (±)		$C_5H_{12}O_2$	91049-43-3	104.148			209	0.9723²⁰	1.4397¹⁹	
8612	1,4-Pentanediol		$C_5H_{12}O_2$	626-95-9	104.148			202; 125¹⁰	0.9883²⁰	1.4452²³	vs H₂O, EtOH, chl
8613	1,5-Pentanediol	Pentamethylene glycol	$C_5H_{12}O_2$	111-29-5	104.148	liq	-18	239	0.9914²⁰	1.4494²⁰	s H₂O, EtOH; sl eth, bz
8614	2,3-Pentanediol		$C_5H_{12}O_2$	42027-23-6	104.148			187.5; 100¹⁷	0.9798¹⁹	1.4412²⁵	s H₂O, EtOH; sl eth
8615	2,4-Pentanediol	2,4-Amylene glycol	$C_5H_{12}O_2$	625-69-4	104.148			199; 97¹³	0.9635²⁰	1.4349²⁰	vs H₂O, EtOH

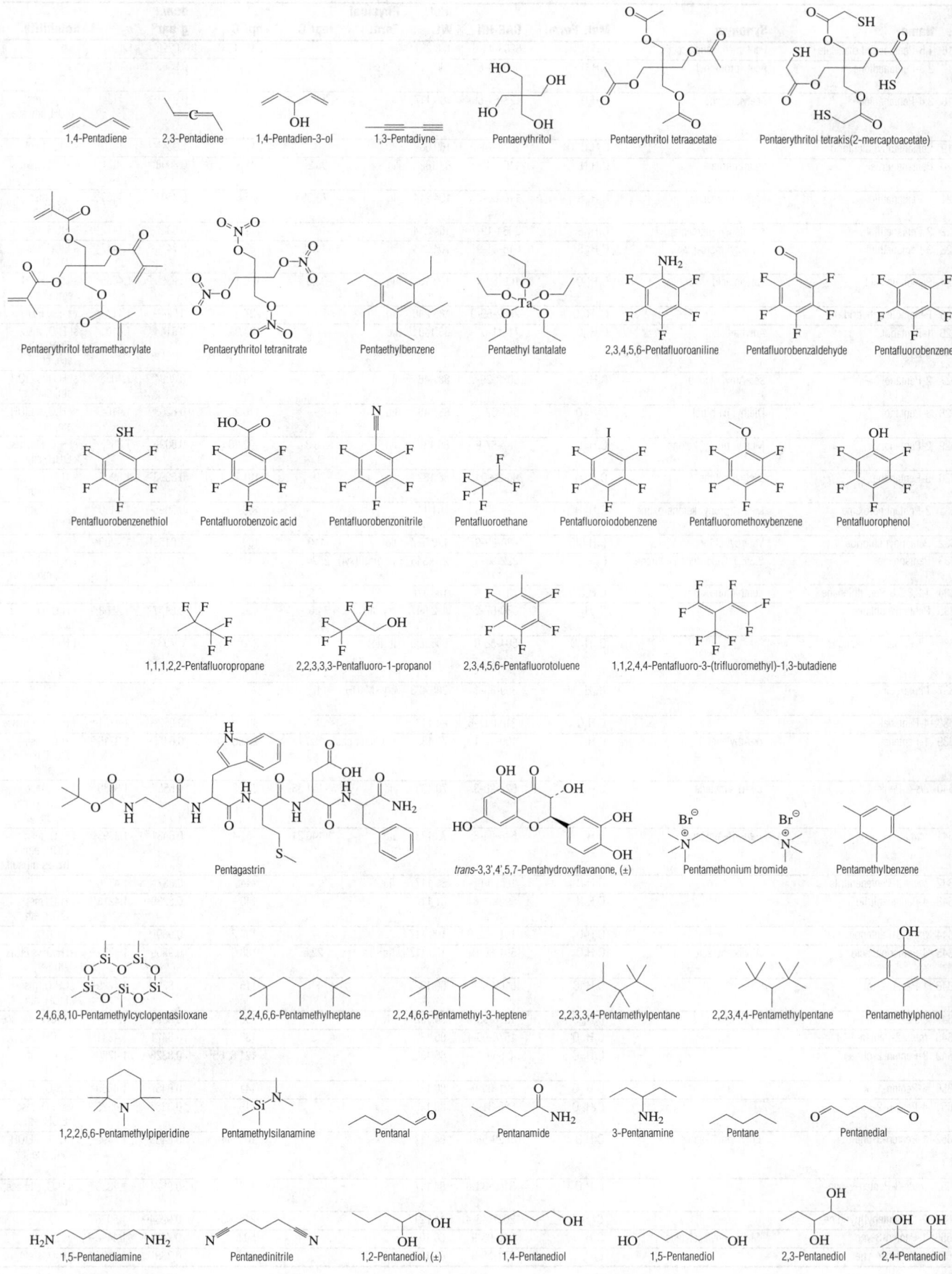

1,4-Pentadiene

2,3-Pentadiene

1,4-Pentadien-3-ol

1,3-Pentadiyne

Pentaerythritol

Pentaerythritol tetraacetate

Pentaerythritol tetrakis(2-mercaptoacetate)

Pentaerythritol tetramethacrylate

Pentaerythritol tetranitrate

Pentaethylbenzene

Pentaethyl tantalate

2,3,4,5,6-Pentafluoroaniline

Pentafluorobenzaldehyde

Pentafluorobenzene

Pentafluorobenzenethiol

Pentafluorobenzoic acid

Pentafluorobenzonitrile

Pentafluoroethane

Pentafluoroiodobenzene

Pentafluoromethoxybenzene

Pentafluorophenol

1,1,1,2,2-Pentafluoropropane

2,2,3,3,3-Pentafluoro-1-propanol

2,3,4,5,6-Pentafluorotoluene

1,1,2,4,4-Pentafluoro-3-(trifluoromethyl)-1,3-butadiene

Pentagastrin

trans-3,3',4',5,7-Pentahydroxyflavanone, (±)

Pentamethonium bromide

Pentamethylbenzene

2,4,6,8,10-Pentamethylcyclopentasiloxane

2,2,4,6,6-Pentamethylheptane

2,2,4,6,6-Pentamethyl-3-heptene

2,2,3,3,4-Pentamethylpentane

2,2,3,4,4-Pentamethylpentane

Pentamethylphenol

1,2,2,6,6-Pentamethylpiperidine

Pentamethylsilanamine

Pentanal

Pentanamide

3-Pentanamine

Pentane

Pentanedial

1,5-Pentanediamine

Pentanedinitrile

1,2-Pentanediol, (±)

1,4-Pentanediol

1,5-Pentanediol

2,3-Pentanediol

2,4-Pentanediol

No.	Name	Synonym	Mol. Form.	CAS RN	Mol. Wt.	Physical Form	mp/°C	bp/°C	den/ g cm⁻³	n_D	Solubility
8616	1,5-Pentanediol diacetate	Pentamethylene acetate	$C_9H_{16}O_4$	6963-44-6	188.221		2	241; 123³	1.0296²⁰	1.4261¹⁹	
8617	2,3-Pentanedione	Acetylpropionyl	$C_5H_8O_2$	600-14-6	100.117	ye liq		108	0.9565¹⁹	1.4014¹⁹	s H_2O; msc EtOH, eth, ace
8618	2,4-Pentanedione	Acetylacetone	$C_5H_8O_2$	123-54-6	100.117	liq	-23	138	0.9721²⁵	1.4494²⁰	vs H_2O; msc EtOH, eth, ace, chl
8619	Pentanedioyl dichloride		$C_5H_6Cl_2O_2$	2873-74-7	169.006			217	1.324²⁰	1.4728²⁰	s eth; sl chl
8620	Pentanenitrile	Valeronitrile	C_5H_9N	110-59-8	83.132	liq	-96.2	141.3	0.8008²⁰	1.3971²⁰	s eth, ace, bz; sl ctc
8621	1-Pentanethiol	Pentyl mercaptan	$C_5H_{12}S$	110-66-7	104.214	liq	-75.65	126.6	0.850²⁰	1.4469²⁰	i H_2O; msc EtOH, eth
8622	2-Pentanethiol	sec-Pentyl mercaptan	$C_5H_{12}S$	2084-19-7	104.214	liq	-169	112.9	0.8327²⁰	1.4412²⁰	s EtOH, lig
8623	3-Pentanethiol	3-Pentyl mercaptan	$C_5H_{12}S$	616-31-9	104.214	liq	-110.8	105	0.8410²⁰	1.4447²⁰	s EtOH; sl DMSO
8624	Pentanoic acid	Valeric acid	$C_5H_{10}O_2$	109-52-4	102.132	liq	-33.6	186.1	0.9339²⁵	1.4085²⁰	s H_2O, EtOH, eth; sl ctc
8625	Pentanoic anhydride		$C_{10}H_{18}O_3$	2082-59-9	186.248	liq	-56.1	227	0.924²⁰	1.4171²⁶	vs eth, EtOH
8626	1-Pentanol	Amyl alcohol	$C_5H_{12}O$	71-41-0	88.148	liq	-77.6	137.98	0.8144²⁰	1.4101²⁰	sl H_2O; msc EtOH, eth; s ace, chl
8627	2-Pentanol	sec-Amyl alcohol	$C_5H_{12}O$	6032-29-7	88.148	liq	-73	119.3	0.8094²⁰	1.4053²⁰	sl H_2O; s EtOH, eth, ctc, chl
8628	3-Pentanol	Diethyl carbinol	$C_5H_{12}O$	584-02-1	88.148	liq	-69	116.25	0.8203²⁰	1.4104²⁰	sl H_2O; s EtOH, eth, ace, ctc
8629	2-Pentanone	Methyl propyl ketone	$C_5H_{10}O$	107-87-9	86.132	liq	-76.8	102.26	0.809²⁰	1.3895²⁰	sl H_2O, ctc; msc EtOH, eth
8630	3-Pentanone	Diethyl ketone	$C_5H_{10}O$	96-22-0	86.132	liq	-39	101.7	0.8098²⁵	1.3905²⁵	s H_2O, ctc; msc EtOH, eth
8631	2-Pentanone oxime	Methyl propyl ketone oxime	$C_5H_{11}NO$	623-40-5	101.147			168	0.9095²⁰	1.4450²⁰	vs H_2O, eth, EtOH
8632	Pentanoyl chloride	Valeroyl chloride	C_5H_9ClO	638-29-9	120.577	liq	-110	109	1.0155¹⁵	1.4200²⁰	
8633	Pentaphene	2,3:6,7-Dibenzphenanthrene	$C_{22}H_{14}$	222-93-5	278.346	ye grn lf(xyl)	257				i H_2O; sl EtOH, xyl, eth; s bz
8634	1,2,3,5,6-Pentathiepane	Lenthionine	$C_2H_4S_5$	292-46-6	188.378		60.5				
8635	Pentatriacontane		$C_{35}H_{72}$	630-07-9	492.947	cry (al)	74.6	490	0.8157²⁰	1.4568²⁰	i H_2O; sl eth; s ace
8636	18-Pentatriacontanone		$C_{35}H_{70}O$	504-53-0	506.930	lf (lig)	89.0	270⁰·¹	0.793⁹⁵		i H_2O; sl EtOH, eth, ace, bz, lig, chl
8637	Pentazocine		$C_{19}H_{27}NO$	359-83-1	285.423	cry (MeOH aq)	147				
8638	4-Pentenal		C_5H_8O	2100-17-6	84.117			99	0.852²⁰	1.4191²⁰	i H_2O; s eth, ace
8639	1-Pentene	α-Amylene	C_5H_{10}	109-67-1	70.133	vol liq or gas	-165.12	29.96	0.6405²⁰	1.3715²⁰	i H_2O; msc EtOH, eth; s bz; sl ctc
8640	cis-2-Pentene	cis-β-Amylene	C_5H_{10}	627-20-3	70.133	liq	-151.36	36.93	0.6556²⁰	1.3830²⁰	i H_2O; msc EtOH, eth; s bz, dil sulf
8641	trans-2-Pentene	trans-β-Amylene	C_5H_{10}	646-04-8	70.133	liq	-140.21	36.34	0.6431²⁵	1.3793²⁰	i H_2O; msc EtOH, eth; s bz; vs dil sulf
8642	trans-3-Pentenenitrile		C_5H_7N	16529-66-1	81.117	liq		144	0.837	1.4220²⁰	
8643	4-Pentenenitrile		C_5H_7N	592-51-8	81.117			140	0.8239²⁴	1.4213¹⁴	i H_2O; msc EtOH, eth
8644	trans-3-Pentenoic acid		$C_5H_8O_2$	1617-32-9	100.117			193.2	0.989¹⁹		
8645	4-Pentenoic acid	Allylacetic acid	$C_5H_8O_2$	591-80-0	100.117	liq	-22.5	188.5	0.9809²⁰	1.4281²⁰	sl H_2O; vs EtOH, eth
8646	1-Penten-3-ol		$C_5H_{10}O$	616-25-1	86.132			115	0.839²⁰	1.4239²⁰	sl H_2O; msc EtOH, eth
8647	cis-2-Penten-1-ol		$C_5H_{10}O$	1576-95-0	86.132			138	0.8529²⁰	1.4354²⁰	s EtOH, eth, ace
8648	trans-2-Penten-1-ol		$C_5H_{10}O$	1576-96-1	86.132			138	0.8471²⁰	1.4341²⁰	s EtOH, eth, ace
8649	3-Penten-2-ol, (±)		$C_5H_{10}O$	42569-16-4	86.132			121.6; 65⁷⁰	0.8328²⁵	1.4280²⁰	vs ace, eth, EtOH
8650	4-Penten-1-ol		$C_5H_{10}O$	821-09-0	86.132			141	0.8457²⁰	1.4309²⁰	sl H_2O, ctc; s eth
8651	4-Penten-2-ol		$C_5H_{10}O$	625-31-0	86.132			116	0.8367²⁰	1.4225²⁰	vs H_2O; msc EtOH, eth
8652	1-Penten-3-one	Ethyl vinyl ketone	C_5H_8O	1629-58-9	84.117			103; 44⁹⁰	0.8468²⁰	1.4195²⁰	i H_2O; s EtOH, eth, ace, bz, chl
8653	trans-3-Penten-2-one		C_5H_8O	3102-33-8	84.117			122	0.8624²⁰	1.4350²⁰	s H_2O, eth, ace, ctc
8654	2-(3-Pentenyl)pyridine		$C_{10}H_{13}N$	2057-43-4	147.217			216; 93¹²	0.9234²⁵	1.5076²⁵	
8655	1-Penten-3-yne	Methylvinylacetylene	C_5H_6	646-05-9	66.102			59.5	0.7401²⁰	1.4496²⁰	vs bz, eth
8656	1-Penten-4-yne		C_5H_6	871-28-3	66.102			42.5	0.738¹⁶	1.4125¹⁶	i H_2O; s eth, bz

1,5-Pentanediol diacetate

2,3-Pentanedione

2,4-Pentanedione

Pentanedioyl dichloride

Pentanenitrile

1-Pentanethiol

2-Pentanethiol

3-Pentanethiol

Pentanoic acid

Pentanoic anhydride

1-Pentanol

2-Pentanol

3-Pentanol

2-Pentanone

3-Pentanone

2-Pentanone oxime

Pentanoyl chloride

Pentaphene

1,2,3,5,6-Pentathiepane

Pentatriacontane

18-Pentatriacontanone

Pentazocine

4-Pentenal

1-Pentene

cis-2-Pentene

trans-2-Pentene

trans-3-Pentenenitrile

4-Pentenenitrile

trans-3-Pentenoic acid

4-Pentenoic acid

1-Penten-3-ol

cis-2-Penten-1-ol

trans-2-Penten-1-ol

3-Penten-2-ol, (±)

4-Penten-1-ol

4-Penten-2-ol

1-Penten-3-one

trans-3-Penten-2-one

2-(3-Pentenyl)pyridine

1-Penten-3-yne

1-Penten-4-yne

No.	Name	Synonym	Mol. Form.	CAS RN	Mol. Wt.	Physical Form	mp/°C	bp/°C	den/ g cm^{-3}	n_D	Solubility
8657	cis-3-Penten-1-yne		C_5H_6	1574-40-9	66.102			44.6			
8658	trans-3-Penten-1-yne		C_5H_6	2004-69-5	66.102			52.2			
8659	Pentetic acid	Diethylenetriaminepentaacetic acid	$C_{14}H_{23}N_3O_{10}$	67-43-6	393.347	cry (w)	219				s H_2O, alk
8660	Pentostatin		$C_{11}H_{16}N_4O_4$	53910-25-1	268.270	wh cry (MeOH aq)	222				
8661	Pentryl	2-(N,2,4,6-Tetranitroanilino)ethanol	$C_8H_6N_6O_{11}$	4481-55-4	362.167	wh-ye cry	129		1.82		i H_2O, ctc; s chl; vs eth, bz
8662	Pentyl acetate	Amyl acetate	$C_7H_{14}O_2$	628-63-7	130.185	liq	-70.8	149.2	0.8756[20]	1.4023[20]	sl H_2O; msc EtOH, eth; s ctc
8663	sec-Pentyl acetate (R)	sec-Amyl acetate (R)	$C_7H_{14}O_2$	54638-10-7	130.185			142	0.8803[18]	1.4012[20]	vs eth, EtOH
8664	Pentylamine	Amylamine	$C_5H_{13}N$	110-58-7	87.164	liq	-55	104.3	0.7544[20]	1.448[20]	msc H_2O, EtOH, eth; vs ace, bz; sl chl
8665	4-tert-Pentylaniline		$C_{11}H_{17}N$	2049-92-5	163.260			260.5			
8666	Pentylbenzene	Amylbenzene	$C_{11}H_{16}$	538-68-1	148.245	liq	-75	205.4	0.8585[20]	1.4878[20]	i H_2O; msc EtOH, eth, ace, bz, peth, ctc
8667	Pentyl benzoate		$C_{12}H_{16}O_2$	2049-96-9	192.254			137[15]			
8668	4-Pentylbenzoyl chloride		$C_{12}H_{15}ClO$	49763-65-7	210.699			144[10], 121[8]	1.036[25]	1.5300[20]	
8669	Pentyl butanoate	Amyl butyrate	$C_9H_{18}O_2$	540-18-1	158.238	liq	-73.2	186.4	0.8713[15]	1.4123[20]	i H_2O; vs EtOH, eth
8670	tert-Pentyl carbamate	tert-Amyl carbamate	$C_6H_{13}NO_2$	590-60-3	131.173	nd (dil al)	86				vs ace, bz
8671	Pentyl chloroformate		$C_6H_{11}ClO_2$	638-41-5	150.603			61[15]		1.4181[18]	s eth
8672	Pentylcyclohexane		$C_{11}H_{22}$	4292-92-6	154.293	liq	-57.5	203.7	0.8037[20]	1.4437[20]	vs ace, bz, eth, EtOH
8673	Pentylcyclopentane		$C_{10}H_{20}$	3741-00-2	140.266	liq	-83	180	0.7912[20]	1.4356[20]	i H_2O; vs ace, bz, eth, EtOH
8674	Pentyl formate	Amyl formate	$C_6H_{12}O_2$	638-49-3	116.158	liq	-73.5	130.4	0.8853[20]	1.3992[20]	sl H_2O; msc EtOH, eth
8675	Pentyl heptanoate	Amyl enanthate	$C_{12}H_{24}O_2$	7493-82-5	200.318	liq	-50	245.4	0.8623[20]	1.4263[15]	vs ace, bz, eth, EtOH
8676	Pentyl hexanoate	Amyl caproate	$C_{11}H_{22}O_2$	540-07-8	186.292	liq	-47	226	0.8612[25]	1.4202[25]	s EtOH, eth, ace; sl ctc
8677	1-Pentylnaphthalene		$C_{15}H_{18}$	86-89-5	198.304	liq	-22	307	0.9656[20]	1.5725[20]	
8678	Pentyl nitrite	Amyl nitrite	$C_5H_{11}NO_2$	463-04-7	117.147			104.5	0.8817[20]	1.3851[20]	sl H_2O; msc EtOH, eth
8679	Pentyl nonanoate	Pentyl pelargonate	$C_{14}H_{28}O_2$	61531-45-1	228.371		-27	131[20]	0.8506[25]	1.4318[20]	
8680	Pentyl octanoate	Amyl octanoate	$C_{13}H_{26}O_2$	638-25-5	214.344	liq	-34.8	260.2	0.8613[20]	1.4262[25]	i H_2O; s EtOH, eth, ace
8681	4-(Pentyloxy)benzoyl chloride		$C_{12}H_{15}ClO_2$	36823-84-4	226.699			198[30], 182[25]	1.087[25]	1.5434[20]	
8682	Pentyl pentanoate		$C_{10}H_{20}O_2$	2173-56-0	172.265	liq	-78.8	203.7	0.8638[20]	1.4164[20]	sl H_2O; msc EtOH, eth
8683	4-Pentylphenol		$C_{11}H_{16}O$	14938-35-3	164.244		23	250.5	0.960[20]	1.5272[25]	vs eth, EtOH
8684	Pentyl propanoate		$C_8H_{16}O_2$	624-54-4	144.212	liq	-73.1	168.6	0.8761[25]	1.4096[15]	i H_2O; msc EtOH, eth; s bz; sl ctc
8685	Pentyl salicylate		$C_{12}H_{16}O_3$	2050-08-0	208.253			270	1.064[15]	1.506[20]	sl H_2O; msc EtOH, eth
8686	Pentyl stearate		$C_{23}H_{46}O_2$	6382-13-4	354.610	pl	30			1.4342[50]	vs eth, EtOH
8687	1-Pentyne	Propylacetylene	C_5H_8	627-19-0	68.118	liq	-90	40.1	0.6901[20]	1.3852[20]	i H_2O; vs EtOH; msc eth; s bz, chl; sl ctc
8688	2-Pentyne		C_5H_8	627-21-4	68.118	liq	-109.3	56.1	0.7058[20]	1.4039[20]	i H_2O; vs EtOH; msc eth; s bz, chl
8689	4-Pentynoic acid	Propargylacetic acid	$C_5H_6O_2$	6089-09-4	98.101		57.7	110[30], 102[17]			vs eth, EtOH
8690	2-Pentyn-1-ol		C_5H_8O	6261-22-9	84.117	liq	-49.7	154; 61[15]	0.909[20]	1.4518[17]	
8691	3-Pentyn-1-ol		C_5H_8O	10229-10-4	84.117			154	0.9002[20]	1.4454[20]	
8692	4-Pentyn-1-ol		C_5H_8O	5390-04-5	84.117			154	0.913[20]	1.4414[20]	
8693	Perazine	10-[3-(4-Methyl-1-piperazinyl)propyl]-10H-phenothiazine	$C_{20}H_{25}N_3S$	84-97-9	339.498	cry	52	165[0.001]			
8694	Perfluidone		$C_{14}H_{12}F_3NO_4S_2$	37924-13-3	379.375		143				
8695	Perfluoroacetone	Hexafluoroacetone	C_3F_6O	684-16-2	166.021	col gas	-125.45	-27.4			
8696	Perfluorobutane	Decafluorobutane	C_4F_{10}	355-25-9	238.027	col gas	-129.1	-1.9	1.6484[25]		s bz, chl
8697	Perfluoro-2-butene		C_4F_8	360-89-4	200.030	col gas	-129	1.5	1.5297[25]		
8698	Perfluoro-2-butyltetrahydrofuran		$C_8F_{16}O$	335-36-4	416.059			102.6			
8699	Perfluorocyclobutane	Octafluorocyclobutane	C_4F_8	115-25-3	200.030	col gas	-40.19	-5.9	1.500[25] (p>1 atm)		i H_2O; s eth
8700	Perfluorocyclohexane		C_6F_{12}	355-68-0	300.045		62.5 (triple point)	52.8 sp			
8701	Perfluorocyclohexene		C_6F_{10}	355-75-9	262.048			52.0	1.6650[25]	1.293[20]	

cis-3-Penten-1-yne

trans-3-Penten-1-yne

Pentetic acid

Pentostatin

Pentryl

Pentyl acetate

sec-Pentyl acetate (*R*)

Pentylamine

4-*tert*-Pentylaniline

Pentylbenzene

Pentyl benzoate

4-Pentylbenzoyl chloride

Pentyl butanoate

tert-Pentyl carbamate

Pentyl chloroformate

Pentylcyclohexane

Pentylcyclopentane

Pentyl formate

Pentyl heptanoate

Pentyl hexanoate

1-Pentylnaphthalene

Pentyl nitrite

Pentyl nonanoate

Pentyl octanoate

4-(Pentyloxy)benzoyl chloride

Pentyl pentanoate

4-Pentylphenol

Pentyl propanoate

Pentyl salicylate

Pentyl stearate

1-Pentyne

2-Pentyne

4-Pentynoic acid

2-Pentyn-1-ol

3-Pentyn-1-ol

4-Pentyn-1-ol

Perazine

Perfluidone

Perfluoroacetone

Perfluorobutane

Perfluoro-2-butene

Perfluoro-2-butyltetrahydrofuran

Perfluorocyclobutane

Perfluorocyclohexane

Perfluorocyclohexene

No.	Name	Synonym	Mol. Form.	CAS RN	Mol. Wt.	Physical Form	mp/°C	bp/°C	den/ g cm^{-3}	n_D	Solubility
8702	Perfluorodecalin		$C_{10}F_{18}$	306-94-5	462.078	liq	-10	142.02	1.9305[25]		
8703	Perfluorodecane		$C_{10}F_{22}$	307-45-9	538.072			144.2			i H_2O
8704	Perfluorodimethoxymethane		$C_3F_8O_2$	53772-78-4	220.018	col gas	-161	-10			
8705	Perfluoro-2,3-dimethylbutane		C_6F_{14}	354-96-1	338.042	liq	-15	59.8			
8706	Perfluoroethyl ethyl ether		$C_4H_5F_5O$	22052-81-9	164.074	vol liq or gas		28.11			
8707	Perfluoroethyl 2,2,2-trifluoroethyl ether		$C_4H_2F_8O$	156053-88-2	218.045	vol liq or gas		27.89			
8708	Perfluoroheptane		C_7F_{16}	335-57-9	388.049	liq	-51.2	82.5	1.7333[20]	1.2618[20]	i H_2O; vs ace, eth, EtOH, chl
8709	Perfluoro-1-heptene		C_7F_{14}	355-63-5	350.053			81.0			
8710	Perfluorohexane		C_6F_{14}	355-42-0	338.042	liq	-88.2	57.14	1.6910[20]	1.2515[20]	i H_2O; s eth, bz, chl
8711	Perfluoro-1-hexene		C_6F_{12}	755-25-9	300.045			57.0			vs chl
8712	Perfluoroisobutane		C_4F_{10}	354-92-7	238.027	col gas		0			
8713	Perfluoroisobutene	Perfluoroisobutylene	C_4F_8	382-21-8	200.030	col gas	-130	7	1.5922[0]		
8714	Perfluoroisopropyl methyl ether		$C_4H_3F_7O$	22052-84-2	200.055	vol liq or gas		29.34	1.4205[20]		
8715	Perfluoromethylcyclohexane		C_7F_{14}	355-02-2	350.053	liq	-44.7	76.3	1.7878[25]	1.285[17]	s ace, bz, ctc, tol, AcOEt
8716	Perfluoro-2-methylpentane		C_6F_{14}	355-04-4	338.042			57.6	1.7326[20]	1.2564[22]	i H_2O; s bz
8717	Perfluoro-3-methylpentane		C_6F_{14}	865-71-4	338.042	liq	-115	58.4			s bz
8718	Perfluoronaphthalene		$C_{10}F_8$	313-72-4	272.094		87.5	209			
8719	Perfluorononane		C_9F_{20}	375-96-2	488.064			117.61	1.8001[25]		
8720	Perfluorooctane		C_8F_{18}	307-34-6	438.057			105.9	1.73[20]	1.282[20]	i H_2O
8721	Perfluorooctylsulfonyl fluoride		$C_8F_{18}O_2S$	307-35-7	502.121	liq		154			
8722	Perfluorooxetane		C_3F_6O	425-82-1	166.021	col gas	-117	-28.4			
8723	Perfluoropentane		C_5F_{12}	678-26-2	288.035	vol liq or gas	-10	29.2			i H_2O
8724	Perfluoropropane		C_3F_8	76-19-7	188.019	col gas	-147.70	-36.6			i H_2O
8725	Perfluoropropene		C_3F_6	116-15-4	150.022	col gas	-156.5	-29.6		1.583[-40]	i H_2O
8726	Perfluoropropyl methyl ether		$C_4H_3F_7O$	375-03-1	200.055			34.23	1.4092[20]		
8727	Perfluoropyridine	Pentafluoropyridine	C_5F_5N	700-16-3	169.053			83.7			
8728	Perfluorotoluene		C_7F_8	434-64-0	236.062	liq	-65.49	103.55	1.6616[25]	1.3670[20]	
8729	Perfluorotripropylamine		$C_9F_{21}N$	338-83-0	521.069			130	1.822[4]	1.279[25]	
8730	1H-Perimidine		$C_{11}H_8N_2$	204-02-4	168.195	grn cry (dil al)	223.0				i H_2O; s EtOH, eth, ace, bz; sl DMSO
8731	Permethrin		$C_{21}H_{20}Cl_2O_3$	52645-53-1	391.288	cry or ye liq	34	200[0.01]	1.23[20]		i H_2O; s os
8732	Peroxyacetic acid	Ethaneperoxoic acid	$C_2H_4O_3$	79-21-0	76.051	liq	-0.2	110	1.226[15]	1.3974[20]	vs H_2O, eth, sulf; s EtOH
8733	Peroxypropanoic acid	Propaneperoxoic acid	$C_3H_6O_3$	4212-43-5	90.078			exp 119.7		1.4148[15]	
8734	Perphenazine		$C_{21}H_{26}ClN_3OS$	58-39-9	403.968		97				
8735	Perthane	Ethane, 1,1-dichloro-2,2-bis(p-ethylphenyl)-	$C_{18}H_{20}Cl_2$	72-56-0	307.258		56				
8736	Perylene	Dibenz[de,kl]anthracene	$C_{20}H_{12}$	198-55-0	252.309	gold-br, ye pl (bz, HOAc)	277.76		1.35[25]		i H_2O; sl EtOH, eth; vs ace, chl; s bz
8737	Peucedanin	3-Methoxy-2-isopropyl-7H-furo[3,2-g][1]benzopyran-7-one	$C_{15}H_{14}O_4$	133-26-6	258.270	pr or pl (bz-peth)	85	278[17]			sl H_2O, bz; s EtOH, eth; vs chl, CS_2
8738	Phalloidin		$C_{35}H_{48}N_8O_{11}S$	17466-45-4	788.868	nd (w)	281 (hyd)				s EtOH, MeOH, py
8739	Phalloin		$C_{35}H_{48}N_8O_{10}S$	28227-92-1	772.869	cry (w)	250 dec				
8740	α-Phellandrene	2-Methyl-5-(1-methylethyl)-1,3-cyclohexadiene	$C_{10}H_{16}$	99-83-2	136.234		238	174.9	0.8410[20]	1.471[25]	i H_2O; s eth
8741	β-Phellandrene	p-Mentha-1(7),2-diene	$C_{10}H_{16}$	555-10-2	136.234			171.5	0.8520[20]	1.4788[20]	i H_2O, EtOH; s eth
8742	9-Phenanthrenamine		$C_{14}H_{11}N$	947-73-9	193.244	lt ye cry (al)	138.3	sub			sl eth, bz, chl
8743	Phenanthrene		$C_{14}H_{10}$	85-01-8	178.229	mcl pl (al), lf (sub)	99.24	340	0.9800[4]	1.5943	i H_2O; s EtOH, eth, ace, bz, CS_2
8744	9,10-Phenanthrenedione	Phenanthrenequinone	$C_{14}H_8O_2$	84-11-7	208.213	oran nd (to) oran-red pl (sub)	209		1.405[22]		i H_2O; sl EtOH, bz; s eth
8745	Phenanthridine		$C_{13}H_9N$	229-87-8	179.217	nd (dil al)	107.4	348.9			sl H_2O; vs EtOH, eth, bz, CS_2; s ace
8746	1,7-Phenanthroline		$C_{12}H_8N_2$	230-46-6	180.205	pl (anh), nd (w+2)	78	360			s H_2O; vs EtOH; i eth, bz, lig
8747	1,10-Phenanthroline	o-Phenanthroline	$C_{12}H_8N_2$	66-71-7	180.205	wh nd (bz) cry (w+1)	117	>300			vs H_2O; s EtOH, ace, bz; i peth
8748	4,7-Phenanthroline		$C_{12}H_8N_2$	230-07-9	180.205	nd (w)	177	sub 100			s H_2O, lig; vs EtOH; sl eth, bz, CS_2

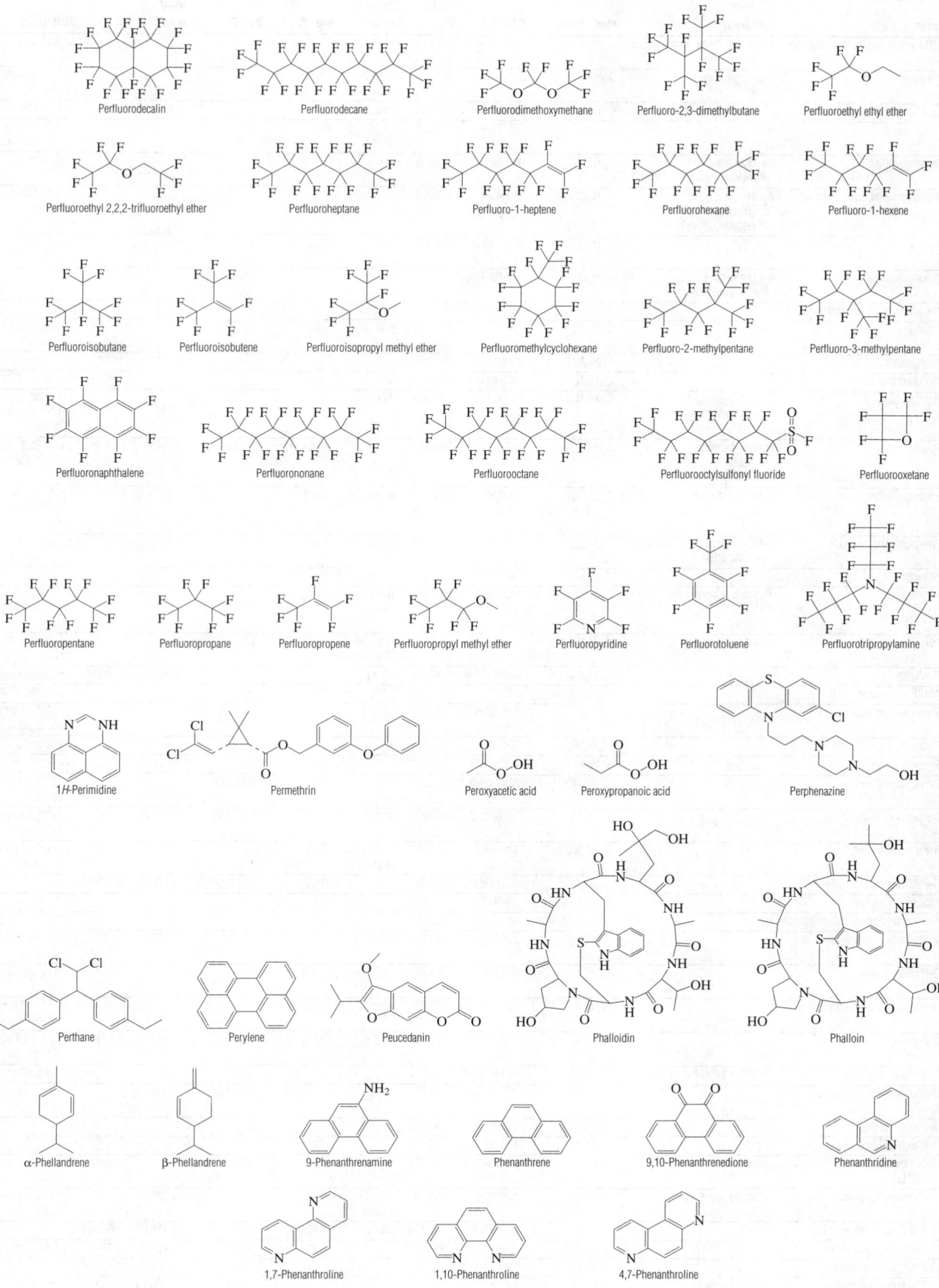

Perfluorodecalin

Perfluorodecane

Perfluorodimethoxymethane

Perfluoro-2,3-dimethylbutane

Perfluoroethyl ethyl ether

Perfluoroethyl 2,2,2-trifluoroethyl ether

Perfluoroheptane

Perfluoro-1-heptene

Perfluorohexane

Perfluoro-1-hexene

Perfluoroisobutane

Perfluoroisobutene

Perfluoroisopropyl methyl ether

Perfluoromethylcyclohexane

Perfluoro-2-methylpentane

Perfluoro-3-methylpentane

Perfluoronaphthalene

Perfluorononane

Perfluorooctane

Perfluorooctylsulfonyl fluoride

Perfluorooxetane

Perfluoropentane

Perfluoropropane

Perfluoropropene

Perfluoropropyl methyl ether

Perfluoropyridine

Perfluorotoluene

Perfluorotripropylamine

1H-Perimidine

Permethrin

Peroxyacetic acid

Peroxypropanoic acid

Perphenazine

Perthane

Perylene

Peucedanin

Phalloidin

Phalloin

α-Phellandrene

β-Phellandrene

9-Phenanthrenamine

Phenanthrene

9,10-Phenanthrenedione

Phenanthridine

1,7-Phenanthroline

1,10-Phenanthroline

4,7-Phenanthroline

No.	Name	Synonym	Mol. Form.	CAS RN	Mol. Wt.	Physical Form	mp/°C	bp/°C	den/ g cm⁻³	n_D	Solubility
8749	1,10-Phenanthroline monohydrate	o-Phenanthroline monohydrate	$C_{12}H_{10}N_2O$	5144-89-8	198.219	wh cry pow	93				s EtOH, ace; sl bz
8750	Phenazine	Dibenzopyrazine	$C_{12}H_8N_2$	92-82-0	180.205	ye-red nd (HOAc)	176.5				sl H_2O, eth; s bz, EtOH
8751	2,3-Phenazinediamine	2,3-Diaminophenazine	$C_{12}H_{10}N_4$	655-86-7	210.234	ye nd	264	sub			vs bz, EtOH
8752	1-Phenazinol	Hemipyocyanine	$C_{12}H_8N_2O$	528-71-2	196.204	ye nd (bz, dil MeOH)	158	sub			sl H_2O, EtOH; s bz, py, dil alk
8753	Phenazopyridine	2,6-Diamino-3-phenylazopyridine	$C_{11}H_{11}N_5$	94-78-0	213.239	red cry	139				
8754	Phenazopyridine hydrochloride	3-(Phenylazo)-2,6-pyridinediamine, monohydrochloride	$C_{11}H_{12}ClN_5$	136-40-3	249.700	ye-red cry					sl H_2O, EtOH; i bz, ace; s HOAc
8755	Phencarbamide		$C_{19}H_{24}N_2OS$	3735-90-8	328.471		48.5	121[0.01]			vs eth, chl, MeOH, peth
8756	Phendimetrazine	3,4-Dimethyl-2-phenylmorpholine	$C_{12}H_{17}NO$	634-03-7	191.269			134[12], 78[0.35]			
8757	Phenethicillin potassium		$C_{17}H_{19}KN_2O_5S$	132-93-4	402.506	cry (ace)	235				s H_2O
8758	Phenicin		$C_{14}H_{10}O_6$	128-68-7	274.225	ye-br (al)	230.5				sl H_2O; vs EtOH, chl, HOAc
8759	Phenindamine		$C_{19}H_{19}N$	82-88-2	261.361	cry	91		1.17		
8760	Phenmedipham		$C_{16}H_{16}N_2O_4$	13684-63-4	300.309		143				
8761	Phenobarbital	5-Ethyl-5-phenyl-2,4,6(1H,3H,5H)-pyrimidinetrione	$C_{12}H_{12}N_2O_3$	50-06-6	232.234	pl (w)	174				i H_2O, bz; s EtOH, eth; sl DMSO
8762	Phenol	Hydroxybenzene	C_6H_6O	108-95-2	94.111		40.89	181.87	1.0545[45]	1.5408[41]	s H_2O, EtOH; vs eth; msc ace, bz
8763	Phenolphthalein	3,3-Bis(4-hydroxyphenyl)-1(3H)-isobenzofuranone	$C_{20}H_{14}O_4$	77-09-8	318.323	wh orth nd	262.5		1.277[32]		i H_2O, bz; vs EtOH, ace; s eth, chl
8764	Phenolphthalin	2-[Bis(4-hydroxyphenyl)methyl]benzoic acid	$C_{20}H_{16}O_4$	81-90-3	320.339	nd (w)	230.5				vs eth, EtOH
8765	Phenolphthalol		$C_{20}H_{18}O_3$	81-92-5	306.355	cry (dil al)	201.5				
8766	Phenol Red	Phenolsulfonphthalein	$C_{19}H_{14}O_5S$	143-74-8	354.376	dk red nd or pl	>300				sl H_2O, EtOH, ace, bz; i eth, chl
8767	10H-Phenothiazine	Thiodiphenylamine	$C_{12}H_9NS$	92-84-2	199.271	ye pr (al) ye lf or pl (tol)	187.5	371			vs ace, bz, eth, EtOH
8768	Phenothrin		$C_{23}H_{26}O_3$	26002-80-2	350.450	col liq			1.061[25]	1.5483[25]	i H_2O; s ace, xyl
8769	10H-Phenoxazine		$C_{12}H_9NO$	135-67-1	183.205	lf (dil al, bz)	156	dec			vs bz, eth, EtOH
8770	Phenoxyacetic acid		$C_8H_8O_3$	122-59-8	152.148	nd or pl (w)	98.5	dec 285			s H_2O; vs EtOH, eth, bz, CS_2
8771	Phenoxyacetyl chloride		$C_8H_7ClO_2$	701-99-5	170.594			225.5			s eth
8772	Phenoxyacetylene		C_8H_6O	4279-76-9	118.133		-36	61[25]	1.0614[20]	1.5125[20]	vs eth, EtOH
8773	2-Phenoxyaniline		$C_{12}H_{11}NO$	2688-84-8	185.221	cry (lig)	45.8	308; 172[14]			s EtOH; s eth, ace, bz
8774	3-Phenoxyaniline		$C_{12}H_{11}NO$	3586-12-7	185.221	pr (lig)	37	315; 180[10]	1.1583[25]		s EtOH, eth, ace, bz; sl lig
8775	4-Phenoxyaniline		$C_{12}H_{11}NO$	139-59-3	185.221	nd (w), cry (dil al)	85.5				s H_2O; vs EtOH, eth; sl lig
8776	3-Phenoxybenzaldehyde		$C_{13}H_{10}O_2$	39515-51-0	198.217		14.0	169[11], 140[0.1]	1.147[25]	1.5954[20]	
8777	Phenoxybenzamine		$C_{18}H_{22}ClNO$	59-96-1	303.827		39				s bz
8778	Phenoxybenzamine hydrochloride		$C_{18}H_{23}Cl_2NO$	63-92-3	340.288		139				sl H_2O; s EtOH
8779	2-Phenoxybenzoic acid		$C_{13}H_{10}O_3$	2243-42-7	214.216	lf (dil al)	113	355	1.1553[50]		i H_2O; vs EtOH, eth; s chl
8780	3-Phenoxybenzoic acid		$C_{13}H_{10}O_3$	3739-38-6	214.216	nd (aq al)	145.8				i H_2O; s EtOH, eth
8781	4-Phenoxybenzoic acid		$C_{13}H_{10}O_3$	2215-77-2	214.216	pr (chl)	161				sl H_2O; s EtOH, eth, chl
8782	2-Phenoxyethanol		$C_8H_{10}O_2$	122-99-6	138.164	oil	14	245	1.102[22]	1.534[20]	i H_2O; s EtOH, eth, chl, alk
8783	2-Phenoxyethyl acrylate	Phenyl Cellosolve acrylate	$C_{11}H_{12}O_3$	48145-04-6	192.211			110[2]	1.090[25]		vs ace, eth, chl
8784	2-Phenoxyethyl butanoate		$C_{12}H_{16}O_3$	23511-70-8	208.253			251; 88[2]	1.0388[21]		vs ace, eth, EtOH
8785	3-Phenoxyphenol		$C_{12}H_{10}O_2$	713-68-8	186.206			175[7]			
8786	4-Phenoxyphenol		$C_{12}H_{10}O_2$	831-82-3	186.206		84.0				
8787	2-(3-Phenoxyphenyl)propanoic acid, (±)	Fenoprofen	$C_{15}H_{14}O_3$	31879-05-7	242.270	visc oil		170[0.11]		1.5742[25]	
8788	3-Phenoxy-1,2-propanediol	Phenylglyceryl ether	$C_9H_{12}O_3$	538-43-2	168.189	nd (eth, peth)	67.5	200[22]	1.225[20]		vs H_2O, bz, eth, EtOH
8789	2-Phenoxypropanoic acid		$C_9H_{10}O_3$	940-31-8	166.173	nd (w)	115.5	266; 105[5]	1.1865[20]	1.5184[20]	
8790	2-Phenoxy-1-propanol		$C_9H_{12}O_2$	4169-04-4	152.190			244	0.9801[25]	1.4760[25]	s EtOH, eth
8791	1-Phenoxy-2-propanol		$C_9H_{12}O_2$	770-35-4	152.190			233; 134[20]	1.0622[20]	1.5232[20]	
8792	1-Phenoxy-2-propanone	Phenoxyacetone	$C_9H_{10}O_2$	621-87-4	150.174			229.5	1.0903[20]	1.5228[20]	s eth, ace

1,10-Phenanthroline monohydrate

Phenazine

2,3-Phenazinediamine

1-Phenazinol

Phenazopyridine

Phenazopyridine hydrochloride

Phencarbamide

Phendimetrazine

Phenethicillin potassium

Phenicin

Phenindamine

Phenmedipham

Phenobarbital

Phenol

Phenolphthalein

Phenolphthalin

Phenolphthalol

Phenol Red

10H-Phenothiazine

Phenothrin

10H-Phenoxazine

Phenoxyacetic acid

Phenoxyacetyl chloride

Phenoxyacetylene

2-Phenoxyaniline

3-Phenoxyaniline

4-Phenoxyaniline

3-Phenoxybenzaldehyde

Phenoxybenzamine

Phenoxybenzamine hydrochloride

2-Phenoxybenzoic acid

3-Phenoxybenzoic acid

4-Phenoxybenzoic acid

2-Phenoxyethanol

2-Phenoxyethyl acrylate

2-Phenoxyethyl butanoate

3-Phenoxyphenol

4-Phenoxyphenol

2-(3-Phenoxyphenyl)propanoic acid, (±)

3-Phenoxy-1,2-propanediol

2-Phenoxypropanoic acid

2-Phenoxy-1-propanol

1-Phenoxy-2-propanol

1-Phenoxy-2-propanone

No.	Name	Synonym	Mol. Form.	CAS RN	Mol. Wt.	Physical Form	mp/°C	bp/°C	den/ g cm⁻³	n_D	Solubility
8793	2-Phenoxypropanoyl chloride		C₉H₉ClO₂	122-35-0	184.619			147; 116[10]	1.1865[20]	1.5178[20]	s eth
8794	Phenprocoumon	3-(α-Ethylbenzyl)-4-hydroxycoumarin	C₁₈H₁₆O₃	435-97-2	280.318	pr (MeOH aq)	179				
8795	Phenthoate		C₁₂H₁₇O₄PS₂	2597-03-7	320.364	ye oil		123[0.01]			sl H₂O; s hx
8796	Phentolamine		C₁₇H₁₉N₃O	50-60-2	281.352		175				
8797	Phenyl acetate		C₈H₈O₂	122-79-2	136.149			196; 75[8]	1.0780[20]	1.5035[20]	sl H₂O; msc EtOH, eth, chl; s ctc
8798	2-Phenylacetophenone		C₁₄H₁₂O	451-40-1	196.244	pl (al)	60	320	1.201[0]		sl H₂O; s EtOH, eth, ctc, chl
8799	N-(Phenylacetyl)-7-aminodeacetoxycephalosporanic acid	7-Phenylacetamidodeacetoxycephalosporanic acid	C₁₆H₁₆N₂O₄S	27255-72-7	332.374	cry (2-PrOH/ peth)	200				
8800	Phenylacetylene	Ethynylbenzene	C₈H₆	536-74-3	102.134	liq	-44.8	143	0.9300[20]	1.5470[20]	i H₂O; msc EtOH, eth; s ace; sl chl
8801	(N-Phenylacetyl)glycine	Phenaceturic acid	C₁₀H₁₁NO₃	500-98-1	193.199	lf (EtOH)	143				
8802	Phenyl 2-(acetyloxy)benzoate	Phenyl acetylsalicylate	C₁₅H₁₂O₄	134-55-4	256.254		96				
8803	(Phenylacetyl)urea	Phenacemide	C₉H₁₀N₂O₂	63-98-9	178.187	cry (al)	215				vs bz, eth, EtOH
8804	9-Phenylacridine		C₁₉H₁₃N	602-56-2	255.313	ye nd, lf (al)	184	404			i H₂O; sl EtOH; s eth; vs bz
8805	L-Phenylalaninamide	α-Aminobenzenepropanamide, (S)-	C₉H₁₂N₂O	5241-58-7	164.203		82				
8806	L-Phenylalanine	α-Aminobenzenepropanoic acid, (S)	C₉H₁₁NO₂	63-91-2	165.189	pr (w)	283 dec				sl H₂O; i EtOH, eth, bz, acid
8807	L-Phenylalanine, ethyl ester	Ethyl 2-amino-3-phenylpropionate	C₁₁H₁₅NO₂	3081-24-1	193.243		136	148[13]	1.065[15]		sl H₂O
8808	L-Phenylalanylglycine		C₁₁H₁₄N₂O₃	721-90-4	222.240		262 dec				s H₂O
8809	3-Phenylallyl acetate		C₁₁H₁₂O₂	103-54-8	176.212			123[5]			
8810	5-Phenyl-5-allyl-2,4,6(1H,3H,5H)-pyrimidinetrione	Phenallymal	C₁₃H₁₂N₂O₃	115-43-5	244.245		156.5				sl H₂O, bz, DMSO; vs EtOH, eth; i lig
8811	4-(Phenylamino)benzenesulfonic acid	N-Phenylsulfanilic acid	C₁₂H₁₁NO₃S	101-57-5	249.285	pl (al-eth)	206				vs H₂O, EtOH
8812	2-(Phenylamino)benzoic acid	N-Phenylanthranilic acid	C₁₃H₁₁NO₂	91-40-7	213.232	lf (al)	183.5				i H₂O; vs EtOH; sl eth, bz
8813	Phenyl 4-amino-3-hydroxybenzoate	Phenyl p-aminosalicylate	C₁₃H₁₁NO₃	133-11-9	229.231		153				
8814	3-(Phenylamino)phenol		C₁₂H₁₁NO	101-18-8	185.221	lf (w)	81.5	340			sl H₂O; vs EtOH, eth, ace; s bz, acid
8815	4-(Phenylamino)phenol		C₁₂H₁₁NO	122-37-2	185.221	lf (w)	73	330			sl H₂O; vs EtOH, eth, bz, chl; s acid
8816	9-Phenylanthracene		C₂₀H₁₄	602-55-1	254.325	bl lf (al) (HOAc)	156	417			i H₂O; s EtOH, eth, bz, chl, CS₂
8817	Phenylarsonous diiodide		C₆H₅AsI₂	6380-34-3	405.835		15	205[14], 185[10]	1.6264[15]		
8818	4-(Phenylazo)-1,3-benzenediamine monohydrochloride	Chrysoidine hydrochloride	C₁₂H₁₃ClN₄	532-82-1	248.711	red-br cry pow	118.5				vs ace
8819	4-(Phenylazo)-1,3-benzenediol		C₁₂H₁₀N₂O₂	2051-85-6	214.219	dk red nd (dil al)	170				i H₂O; vs EtOH, eth, bz, HOAc
8820	4-Phenylazodiphenylamine	N-Phenyl-4-(phenylazo)benzenamine	C₁₈H₁₅N₃	101-75-7	273.332	ye pl or pr	84.0				i H₂O; vs EtOH, eth, lig
8821	4-(Phenylazo)-1-naphthalenamine	α-Naphthyl Red	C₁₆H₁₃N₃	131-22-6	247.294	red-viol cry (EtOH)	123				s EtOH, dil HCl, bz
8822	1-(Phenylazo)-2-naphthalenamine	Yellow AB	C₁₆H₁₃N₃	85-84-7	247.294	red pl (al)	103				vs EtOH, HOAc
8823	1-(Phenylazo)-2-naphthol	Sudan I	C₁₆H₁₂N₂O	842-07-9	248.278	ye cry	132				
8824	4-(Phenylazo)phenol		C₁₂H₁₀N₂O	1689-82-3	198.219	ye lf (bz) oran pr (al)	155	225[20] dec			i H₂O; vs EtOH, eth; s bz, con sulf
8825	1-[[4-(Phenylazo)phenyl]azo]-2-naphthol	Sudan III	C₂₂H₁₆N₄O	85-86-9	352.388	br lf (grn lustre) (HOAc)	195				i H₂O; s EtOH, eth, ace, bz, xyl, chl
8826	N-Phenylbenzamide	Benzanilide	C₁₃H₁₁NO	93-98-1	197.232	lf (al)	163	sub 117	1.315[25]		i H₂O; sl EtOH, eth, HOAc
8827	α-Phenylbenzeneacetaldehyde		C₁₄H₁₂O	947-91-1	196.244			dec 315; 157[7]	1.1061[21]	1.5920[21]	i H₂O; vs EtOH, eth, bz
8828	α-Phenylbenzeneacetic acid	Diphenylacetic acid	C₁₄H₁₂O₂	117-34-0	212.244	nd (w), lf (al)	147.29	194[25]	1.257[15]		sl H₂O; vs EtOH; s eth, chl
8829	α-Phenylbenzeneacetonitrile		C₁₄H₁₁N	86-29-3	193.244	pr (eth), lf (dil al)	74.3	184[16]			s EtOH, chl; vs eth; sl lig
8830	α-Phenylbenzeneacetyl chloride		C₁₄H₁₁ClO	1871-76-7	230.689		56.5	170[16]			s lig

2-Phenoxypropanoyl chloride

Phenprocoumon

Phenthoate

Phentolamine

Phenyl acetate

2-Phenylacetophenone

N-(Phenylacetyl)-7-aminodeacetoxycephalosporanic acid

Phenylacetylene

(*N*-Phenylacetyl)glycine

Phenyl 2-(acetyloxy)benzoate

(Phenylacetyl)urea

9-Phenylacridine

L-Phenylalaninamide

L-Phenylalanine

L-Phenylalanine, ethyl ester

L-Phenylalanylglycine

3-Phenylallyl acetate

5-Phenyl-5-allyl-2,4,6(1*H*,3*H*,5*H*)-pyrimidinetrione

4-(Phenylamino)benzenesulfonic acid

2-(Phenylamino)benzoic acid

Phenyl 4-amino-3-hydroxybenzoate

3-(Phenylamino)phenol

4-(Phenylamino)phenol

9-Phenylanthracene

Phenylarsonous diiodide

4-(Phenylazo)-1,3-benzenediamine monohydrochloride

4-(Phenylazo)-1,3-benzenediol

4-Phenylazodiphenylamine

4-(Phenylazo)-1-naphthalenamine

1-(Phenylazo)-2-naphthalenamine

1-(Phenylazo)-2-naphthol

4-(Phenylazo)phenol

1-[[4-(Phenylazo)phenyl]azo]-2-naphthol

N-Phenylbenzamide

α-Phenylbenzeneacetaldehyde

α-Phenylbenzeneacetic acid

α-Phenylbenzeneacetonitrile

α-Phenylbenzeneacetyl chloride

No.	Name	Synonym	Mol. Form.	CAS RN	Mol. Wt.	Physical Form	mp/°C	bp/°C	den/ g cm⁻³	n_D	Solubility
8831	N-Phenylbenzenecarbothioamide		C₁₃H₁₁NS	636-04-4	213.298	ye pl or pr (al)	102	dec			i H₂O; vs EtOH; s eth, bz, chl; sl lig
8832	N-Phenyl-1,2-benzenediamine		C₁₂H₁₂N₂	534-85-0	184.236	nd(w)	79.5	313			sl H₂O, lig; s ace, bz, chl
8833	N-Phenyl-1,4-benzenediamine	p-Aminodiphenylamine	C₁₂H₁₂N₂	101-54-2	184.236	nd(al)	66	354			sl H₂O, chl; vs EtOH; s eth, lig
8834	α-Phenylbenzeneethanamine		C₁₄H₁₅N	25611-78-3	197.276			311; 175[15]	1.031[15]		vs eth, EtOH
8835	α-Phenylbenzeneethanol		C₁₄H₁₄O	614-29-9	198.260	nd (peth-bz)	67	177[15]	1.0360[70]		
8836	α-Phenylbenzenemethanamine	Benzhydrylamine	C₁₃H₁₃N	91-00-9	183.249	hex pl	34	304; 176[23]	1.0633[20]	1.5963	sl H₂O; s ace
8837	α-Phenylbenzenemethanimine		C₁₃H₁₁N	1013-88-3	181.233			282	1.0847[19]	1.6191[19]	vs eth
8838	β-Phenylbenzenepropanoic acid		C₁₅H₁₄O₂	606-83-7	226.271	nd (dil al)	156.0				sl H₂O; vs EtOH; s eth, ace
8839	2-Phenylbenzimidazole	Phenzidole	C₁₃H₁₀N₂	716-79-0	194.231	pl (HOAc) (al-w) nd (bz, w)	293				sl H₂O, bz; s EtOH, chl, HOAc
8840	Phenyl benzoate		C₁₃H₁₀O₂	93-99-2	198.217	mcl pr (eth-al)	71	314	1.235[20]		i H₂O; s EtOH, eth, chl
8841	2-Phenylbenzoic acid		C₁₃H₁₀O₂	947-84-2	198.217	lf (dil al)	114.3	343.5			i H₂O; vs EtOH, bz, HOAc
8842	4-Phenylbenzoic acid		C₁₃H₁₀O₂	92-92-2	198.217	nd (bz, al)	228	sub			i H₂O; s EtOH, eth, bz
8843	2-Phenyl-4H-1-benzopyran-4-one	Flavone	C₁₅H₁₀O₂	525-82-6	222.239	nd (lig), cry (30% al)	100				i H₂O; s EtOH, eth, ace, bz
8844	3-Phenyl-4H-1-benzopyran-4-one	Isoflavone	C₁₅H₁₀O₂	574-12-9	222.239		148				
8845	2-Phenylbenzothiazole		C₁₃H₉NS	883-93-2	211.282	nd (dil al)	115	371			i H₂O; s EtOH, eth, CS₂
8846	N-Phenyl-N-benzylbenzenemethanamine		C₂₀H₁₉N	91-73-6	273.372		69	226[10]	1.0444[80]	1.6065[80]	i H₂O; sl EtOH, HOAc; s eth, bz
8847	Phenyl biguanide	N-Phenylimidodicarbonimidic diamide	C₈H₁₁N₅	102-02-3	177.207		143				
8848	2-Phenyl-1,3-butadiene		C₁₀H₁₀	2288-18-8	130.186			60[17]	0.925[20]	1.5489[20]	i H₂O; s eth, bz, chl
8849	N-Phenylbutanamide		C₁₀H₁₃NO	1129-50-6	163.216	mcl pr (al, bz, eth)	97	189[15]	1.134[25]		i H₂O; vs EtOH, eth; sl chl
8850	Phenylbutanedioic acid, (±)		C₁₀H₁₀O₄	10424-29-0	194.184	lf or nd (w)	168	dec			sl H₂O, chl; vs EtOH, eth, ace; i bz
8851	1-Phenyl-1,3-butanedione		C₁₀H₁₀O₂	93-91-4	162.185	pr	56	261.5	1.0599[74]	1.5678[78]	i H₂O; s eth; sl chl
8852	Phenyl butanoate	Phenyl butyrate	C₁₀H₁₂O₂	4346-18-3	164.201			225	1.0382[15]		i H₂O; s EtOH, eth
8853	1-Phenyl-1-butanone		C₁₀H₁₂O	495-40-9	148.201		12	228.5	0.988[20]	1.5203[20]	i H₂O; msc EtOH, eth; ace; s ctc
8854	1-Phenyl-2-butanone		C₁₀H₁₂O	1007-32-5	148.201			228; 111[16]	0.9877[20]		i H₂O; s EtOH, ctc; msc eth; vs ace
8855	4-Phenyl-2-butanone		C₁₀H₁₂O	2550-26-7	148.201	liq	-13	233.5	0.9849[22]	1.511[22]	i H₂O; s EtOH, eth, ctc; vs ace
8856	Phenylbutazone		C₁₉H₂₀N₂O₂	50-33-9	308.374		105				
8857	2-Phenyl-1-butene	α-Ethylstyrene	C₁₀H₁₂	2039-93-2	132.202			182	0.887[25]	1.5288[20]	
8858	1-Phenyl-2-buten-1-one		C₁₀H₁₀O	495-41-0	146.185		20.5	111[9]	1.025[15]	1.5626[18]	
8859	trans-4-Phenyl-3-buten-2-one	Benzilideneacetone	C₁₀H₁₀O	1896-62-4	146.185	pl	41.5	261	1.0097[45]	1.5836[45]	i H₂O; vs EtOH; s eth, ace, bz; sl peth
8860	4-Phenyl-3-butyn-2-one		C₁₀H₈O	1817-57-8	144.170		4.5	79[2]	1.0215[20]	1.5762[20]	
8861	Phenyl chloroacetate		C₈H₇ClO₂	620-73-5	170.594	nd or pl (al)	44.5	232.5	1.2202[44]	1.5146[44]	i H₂O; vs EtOH, eth
8862	Phenyl chloroformate		C₇H₅ClO₂	1885-14-9	156.567			71[9]			
8863	4-Phenyl-2-chlorophenol	3-Chloro-(1,1'-biphenyl)-4-ol	C₁₂H₉ClO	92-04-6	204.651	wh-ye cry	77	161[7]			
8864	2-Phenyl-2,5-cyclohexadiene-1,4-dione		C₁₂H₈O₂	363-03-1	184.191	ye lf (peth, al)	114				sl H₂O; s EtOH, bz, peth; vs chl
8865	4-Phenylcyclohexanone		C₁₂H₁₄O	4894-75-1	174.238	cry (peth)	79	158[12]			
8866	1-(1-Phenylcyclohexyl)piperidine	Phencyclidine	C₁₇H₂₅N	77-10-1	243.388		46.5	136[1.0]			
8867	3-Phenyl-2-cyclopenten-1-one		C₁₁H₁₀O	3810-26-2	158.196	liq	-23	234.2	0.9711[20]	1.5440[20]	s EtOH, ace, chl; sl eth
8868	N-Phenyl-N,N-diethanolamine		C₁₀H₁₅NO₂	120-07-0	181.232		57	200[10]	1.201[60]		vs ace, bz, eth, EtOH
8869	2-Phenyl-1,3-dioxane		C₁₀H₁₂O₂	772-01-0	164.201	nd (peth)	41	253	1.6053[60]		vs EtOH, eth
8870	4-Phenyl-1,3-dioxane		C₁₀H₁₂O₂	772-00-9	164.201			247	1.1038[20]	1.5306[18]	i H₂O; s os
8871	1-Phenyl-1-dodecanone		C₁₈H₂₈O	1674-38-0	260.414		47	201[9], 181[5]	0.8794[18]	1.4700[18]	i H₂O; s ace; sl ctc

N-Phenylbenzenecarbothioamide

N-Phenyl-1,2-benzenediamine

N-Phenyl-1,4-benzenediamine

α-Phenylbenzeneethanamine

α-Phenylbenzeneethanol

α-Phenylbenzenemethanamine

α-Phenylbenzenemethanimine

β-Phenylbenzenepropanoic acid

2-Phenylbenzimidazole

Phenyl benzoate

2-Phenylbenzoic acid

4-Phenylbenzoic acid

2-Phenyl-4H-1-benzopyran-4-one

3-Phenyl-4H-1-benzopyran-4-one

2-Phenylbenzothiazole

N-Phenyl-N-benzylbenzenemethanamine

Phenyl biguanide

2-Phenyl-1,3-butadiene

N-Phenylbutanamide

Phenylbutanedioic acid, (±)

1-Phenyl-1,3-butanedione

Phenyl butanoate

1-Phenyl-1-butanone

1-Phenyl-2-butanone

4-Phenyl-2-butanone

Phenylbutazone

2-Phenyl-1-butene

1-Phenyl-2-buten-1-one

trans-4-Phenyl-3-buten-2-one

4-Phenyl-3-butyn-2-one

Phenyl chloroacetate

Phenyl chloroformate

4-Phenyl-2-chlorophenol

2-Phenyl-2,5-cyclohexadiene-1,4-dione

4-Phenylcyclohexanone

1-(1-Phenylcyclohexyl)piperidine

3-Phenyl-2-cyclopenten-1-one

N-Phenyl-N,N-diethanolamine

2-Phenyl-1,3-dioxane

4-Phenyl-1,3-dioxane

1-Phenyl-1-dodecanone

No.	Name	Synonym	Mol. Form.	CAS RN	Mol. Wt.	Physical Form	mp/°C	bp/°C	den/ g cm⁻³	n_D	Solubility
8872	1-Phenyl-1,2-ethanediol	Styrene glycol	C₈H₁₀O₂	93-56-1	138.164	nd (liq)	67.5	273			vs H₂O, eth, bz, EtOH; sl lig
8873	N-Phenylethanolamine		C₈H₁₁NO	122-98-5	137.179			279.5; 150¹⁰	1.0945²⁰	1.5760²⁰	sl H₂O; vs EtOH, eth, chl
8874	1-Phenylethanone oxime		C₈H₉NO	613-91-2	135.163	nd (w)	60	245	1.0515⁷⁸		sl H₂O; vs EtOH, eth, ace, bz; s ctc
8875	2-Phenylethyl acetate		C₁₀H₁₂O₂	103-45-7	164.201	liq	-31.1	232.6	1.0883²⁰	1.5171²⁰	vs eth, EtOH
8876	1-Phenylethyl hydroperoxide		C₈H₁₀O₂	3071-32-7	138.164	liq		50⁰·⁰¹			
8877	N-(2-Phenylethyl) imidodicarbonimidic diamide, monohydrochloride	Phenformin hydrochloride	C₁₀H₁₆ClN₅	834-28-6	241.721	cry	177.3				s H₂O
8878	2-Phenylethyl 2-methylpropanoate	Benzylcarbinol isobutyrate	C₁₂H₁₆O₂	103-48-0	192.254			250; 123¹⁵	0.9950¹⁵	1.4871²⁰	
8879	2-Phenylethyl phenylacetate		C₁₆H₁₆O₂	102-20-5	240.297		26.5	177⁴·⁵	1.077²⁵		vs EtOH
8880	2-Phenylethyl propanoate	Phenethyl propionate	C₁₁H₁₄O₂	122-70-3	178.228	liq		244	1.02²⁵	1.4950²⁰	
8881	2-(2-Phenylethyl)pyridine		C₁₃H₁₃N	2116-62-3	183.249	liq	-1.5	289	1.0465⁰		
8882	N-Phenylformamide	Formanilide	C₇H₇NO	103-70-8	121.137	mcl pr (lig-xyl)	46	271	1.1186⁵⁰		s H₂O, eth, bz; vs EtOH
8883	Phenyl formate		C₇H₆O₂	1864-94-4	122.122	liq		178; 82¹⁵			
8884	2-Phenylfuran		C₁₀H₈O	17113-33-6	144.170			108¹⁸, 82⁵	1.083²⁰	1.5920²⁰	vs ace, bz
8885	Phenyl α-D-glucopyranoside		C₁₂H₁₆O₆	4630-62-0	256.251		174				
8886	Phenyl glycidyl ether		C₉H₁₀O₂	122-60-1	150.174			243	1.1109²¹	1.5307²¹	
8887	N-Phenylglycine	Phenylaminoacetic acid	C₈H₉NO₂	103-01-5	151.163		127.5				vs H₂O, EtOH
8888	1-Phenyl-1-heptanone		C₁₃H₁₈O	1671-75-6	190.281	lf	16.4	283.3	0.9516²⁰	1.5060²⁰	vs ace, eth, EtOH
8889	1-Phenyl-1-hexanone		C₁₂H₁₆O	942-92-7	176.254	fl	27	265	0.9576²⁰	1.5027²⁵	sl H₂O, ctc; s EtOH, eth, ace
8890	Phenylhydrazine		C₆H₈N₂	100-63-0	108.141	mcl pr or pl	20.6	243.5	1.0986²⁰	1.6084¹⁰	s H₂O; msc EtOH, eth, bz; vs ace
8891	2-Phenylhydrazinecarboxamide	Phenicarbazide	C₇H₉N₃O	103-03-7	151.165		172				sl H₂O, eth, bz, lig; s EtOH, ace
8892	N-Phenylhydrazinecarboxamide	4-Phenylsemicarbazide	C₇H₉N₃O	537-47-3	151.165	nd (bz), pl (w)	128				sl H₂O; vs EtOH, chl; i eth
8893	Phenylhydrazine monohydrochloride		C₆H₉ClN₂	59-88-1	144.601	lf (al)	244 dec	sub			vs H₂O, EtOH
8894	Phenylhydroxylamine	N-Hydroxybenzenamine	C₆H₇NO	100-65-2	109.126	nd (w, bz, peth)	83.5				vs bz, eth, EtOH, chl
8895	Phenyl 1-hydroxy-2-naphthalenecarboxylate		C₁₇H₁₂O₃	132-54-7	264.275		96				vs bz, EtOH
8896	1-Phenyl-1H-imidazole		C₉H₈N₂	7164-98-9	144.173		13	276	1.1397¹⁵	1.6025²⁵	i H₂O; vs eth, ace, chl
8897	2-Phenyl-1H-imidazole		C₉H₈N₂	670-96-2	144.173	lf (bz)	149.3	340			vs EtOH
8898	5-Phenyl-2,4-imidazolidinedione	5-Phenylhydantoin	C₉H₈N₂O₂	89-24-7	176.172		184.5				
8899	Phenylimidocarbonyl chloride		C₇H₅Cl₂N	622-44-6	174.028	liq		210; 105³⁰	1.28¹⁵		
8900	2-[(Phenylimino)methyl]phenol		C₁₃H₁₁NO	779-84-0	197.232		49.5		1.087²⁵		i H₂O; s EtOH
8901	4-[(Phenylimino)methyl]phenol	N-(4-Hydroxybenzilidene)aniline	C₁₃H₁₁NO	1689-73-2	197.232		196.0				i H₂O; s EtOH, eth; sl bz, chl
8902	1-Phenyl-1H-indene		C₁₅H₁₂	1961-96-2	192.256	oil		158⁷			
8903	2-Phenyl-1H-indene-1,3(2H)-dione	Phenindione	C₁₅H₁₀O₂	83-12-5	222.239	lf (al, bz)	150				i H₂O; s EtOH, eth, ace, bz, MeOH, chl
8904	2-Phenyl-1H-indole		C₁₄H₁₁N	948-65-2	193.244		190.5	250¹⁰			sl H₂O; s eth, bz, chl, HOAc, CS₂
8905	Phenyliodine diacetate	Iodobenzene diacetate	C₁₀H₁₁IO₄	3240-34-4	322.096	cry	161				
8906	Phenyl isocyanate		C₇H₅NO	103-71-9	119.121			163; 55¹³	1.0956²⁰	1.5368²⁰	vs eth; sl chl
8907	2-Phenyl-1H-isoindole-1,3(2H)-dione		C₁₄H₉NO₂	520-03-6	223.227	wh nd (al)	210	sub			i H₂O; sl EtOH; msc chl
8908	Phenyl isopropyl ether	Isopropoxybenzene	C₉H₁₂O	2741-16-4	136.190	liq	-33	176.8	0.9408²⁵	1.4975²⁰	s H₂O, EtOH, ace, bz
8909	Phenyl isothiocyanate		C₇H₅NS	103-72-0	135.187	liq	-21	221	1.1303²⁰	1.6492²³	i H₂O; s EtOH, eth, ctc
8910	3-Phenyl-2-isoxazolin-5-one		C₉H₇NO₂	1076-59-1	161.158		151				sl chl
8911	Phenyl laurate	Phenyl dodecanoate	C₁₈H₂₈O₂	4228-00-6	276.414	lf (al)	24.5	210¹⁵	0.9354³⁰		vs ace, eth, EtOH
8912	Phenylmagnesium chloride	Chlorophenylmagnesium	C₆H₅ClMg	100-59-4	136.862	cry					reac H₂O; s thf, eth
8913	Phenylmercuric chloride	Chlorophenylmercury	C₆H₅ClHg	100-56-1	313.15	pl (bz)	251				i H₂O; sl EtOH, bz
8914	Phenylmercuric nitrate		C₆H₅HgNO₃	55-68-5	339.70		≈181				
8915	4-(Phenylmethoxy)benzaldehyde		C₁₄H₁₂O₂	4397-53-9	212.244		73	217¹³			

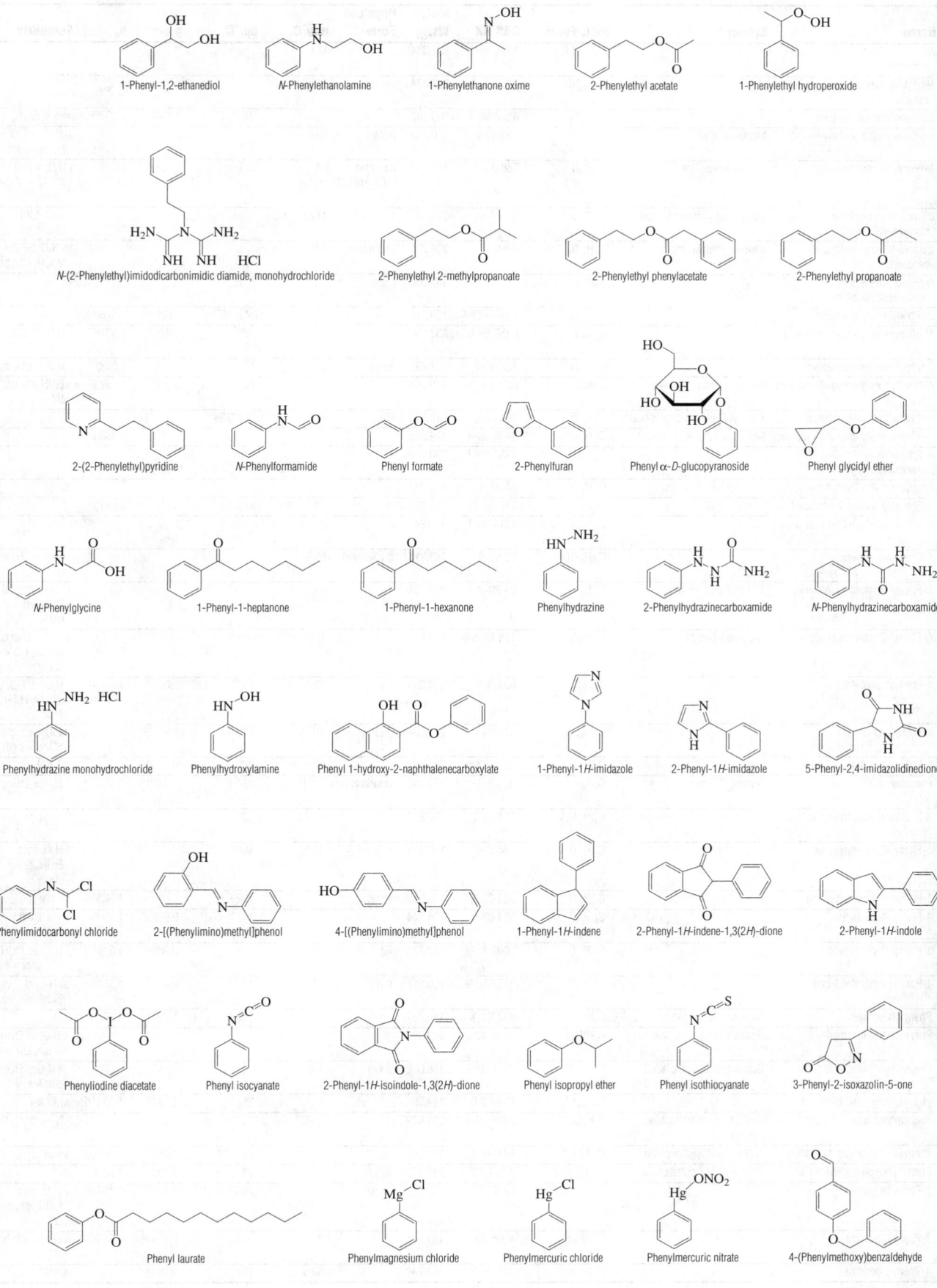

1-Phenyl-1,2-ethanediol

N-Phenylethanolamine

1-Phenylethanone oxime

2-Phenylethyl acetate

1-Phenylethyl hydroperoxide

N-(2-Phenylethyl)imidodicarbonimidic diamide, monohydrochloride

2-Phenylethyl 2-methylpropanoate

2-Phenylethyl phenylacetate

2-Phenylethyl propanoate

2-(2-Phenylethyl)pyridine

N-Phenylformamide

Phenyl formate

2-Phenylfuran

Phenyl α-D-glucopyranoside

Phenyl glycidyl ether

N-Phenylglycine

1-Phenyl-1-heptanone

1-Phenyl-1-hexanone

Phenylhydrazine

2-Phenylhydrazinecarboxamide

N-Phenylhydrazinecarboxamide

Phenylhydrazine monohydrochloride

Phenylhydroxylamine

Phenyl 1-hydroxy-2-naphthalenecarboxylate

1-Phenyl-1H-imidazole

2-Phenyl-1H-imidazole

5-Phenyl-2,4-imidazolidinedione

Phenylimidocarbonyl chloride

2-[(Phenylimino)methyl]phenol

4-[(Phenylimino)methyl]phenol

1-Phenyl-1H-indene

2-Phenyl-1H-indene-1,3(2H)-dione

2-Phenyl-1H-indole

Phenyliodine diacetate

Phenyl isocyanate

2-Phenyl-1H-isoindole-1,3(2H)-dione

Phenyl isopropyl ether

Phenyl isothiocyanate

3-Phenyl-2-isoxazolin-5-one

Phenyl laurate

Phenylmagnesium chloride

Phenylmercuric chloride

Phenylmercuric nitrate

4-(Phenylmethoxy)benzaldehyde

No.	Name	Synonym	Mol. Form.	CAS RN	Mol. Wt.	Physical Form	mp/°C	bp/°C	den/ g cm⁻³	n_D	Solubility
8916	N2-[(Phenylmethoxy)carbonyl]-L-arginine		$C_{14}H_{20}N_4O_4$	1234-35-1	308.334		174				
8917	N-[(Phenylmethoxy)carbonyl]-L-aspartic acid		$C_{12}H_{13}NO_6$	1152-61-0	267.234		117.0				
8918	2-(Phenylmethoxy)phenol		$C_{13}H_{12}O_2$	6272-38-4	200.233			205²⁰, 173¹³	1.154²²	1.5906¹⁸	vs eth, EtOH
8919	4-(Phenylmethoxy)phenol	Monobenzone	$C_{13}H_{12}O_2$	103-16-2	200.233	pl (w)	122				sl H₂O; vs EtOH, eth; s ace
8920	N-(Phenylmethylene)aniline	Benzylideneaniline	$C_{13}H_{11}N$	538-51-2	181.233	pa ye nd (CS₂) pl (dil al)	54	310	1.038⁵⁵	1.600¹⁰⁰	i H₂O; s EtOH, eth, NH₃; sl chl
8921	cis-α-(Phenylmethylene)benzeneacetic acid	cis-α-Phenylcinnamic acid	$C_{15}H_{12}O_2$	91-47-4	224.255	silky needles	174				s H₂O, EtOH, MeOH, eth, bz
8922	trans-α-(Phenylmethylene)benzeneacetic acid	trans-α-Phenylcinnamic acid	$C_{15}H_{12}O_2$	91-48-5	224.255	prisms	138				vs H₂O; s EtOH, MeOH, eth, bz
8923	N-(Phenylmethylene)benzenemethanamine		$C_{14}H_{13}N$	780-25-6	195.260			205²⁰			
8924	2-(Phenylmethylene)butanal		$C_{11}H_{12}O$	28467-92-7	160.212		18	243; 157⁵	1.0201²²	1.578²⁰	
8925	N-(Phenylmethylene)ethanamine		$C_9H_{11}N$	6852-54-6	133.190			195	0.937²⁰	1.5378¹⁵	i H₂O; s EtOH, eth
8926	2-(Phenylmethylene)heptanal		$C_{14}H_{18}O$	122-40-7	202.292	ye oil	80	174²⁰	0.9711²⁰	1.5381²⁰	i H₂O; s ace, ctc
8927	N-(Phenylmethylene)methanamine	Benzylidenemethylamine	C_8H_9N	622-29-7	119.164			185; 92³⁴	0.9671¹⁴	1.5526²⁰	s EtOH, eth, ace, chl
8928	2-(Phenylmethylene)octanal	2-Hexyl-3-phenyl-2-propenal	$C_{15}H_{20}O$	101-86-0	216.319	liq	4	252; 169²⁰			
8929	3-(Phenylmethylene)-2-pentanone	Methyl α-ethylstyryl ketone	$C_{12}H_{14}O$	3437-89-6	174.238			137¹²	1.0005²²	1.5650²²	
8930	N-(Phenylmethyl)-1,2-ethanediamine		$C_9H_{14}N_2$	4152-09-4	150.220			130¹¹			
8931	Phenylmethyl 4-hydroxybenzoate		$C_{14}H_{12}O_3$	94-18-8	228.243						sl chl
8932	1-Phenyl-2-methyl-2-propanol		$C_{10}H_{14}O$	100-86-7	150.217	nd	24	215	0.9787¹⁶	1.5173¹⁶	
8933	N-(Phenylmethyl)-1H-purin-6-amine		$C_{12}H_{11}N_5$	1214-39-7	225.249		232.8				
8934	4-Phenylmorpholine		$C_{10}H_{13}NO$	92-53-5	163.216	cry (al-eth)	58.3				i H₂O, EtOH; vs eth
8935	N-Phenyl-1-naphthalenamine	1-Naphthylphenylamine	$C_{16}H_{13}N$	90-30-2	219.281		61				sl H₂O, ctc; s EtOH, eth, bz, HOAc
8936	N-Phenyl-2-naphthalenamine	N-Phenyl-β-naphthylamine	$C_{16}H_{13}N$	135-88-6	219.281		108	395.5			i H₂O; s EtOH, eth, bz, HOAc; sl chl
8937	1-Phenylnaphthalene		$C_{16}H_{12}$	605-02-7	204.266	cry	45	334	1.096²⁰	1.6664²⁰	i H₂O; vs EtOH, eth, bz, HOAc; s ctc
8938	2-Phenylnaphthalene		$C_{16}H_{12}$	612-94-2	204.266	lf (al)	103.5	345.5	1.2180²⁰		s EtOH, bz, chl, HOAc; vs eth
8939	1-Phenyl-1-octanone		$C_{14}H_{20}O$	1674-37-9	204.308		22.8	285; 164¹⁵	0.9360³⁰		s EtOH, eth
8940	Phenyloxirane	Styrene-7,8-oxide	C_8H_8O	96-09-3	120.149	colorless liq	-35.6	194.1	1.0490²⁵	1.5342²⁰	i H₂O; s EtOH, eth, chl
8941	3-Phenyloxiranecarboxylic acid, ethyl ester		$C_{11}H_{12}O_3$	121-39-1	192.211			136⁵			
8942	5-Phenyl-2,4-pentadienal		$C_{11}H_{10}O$	13466-40-5	158.196		42.5	160³, 133¹·⁰			i H₂O; msc EtOH, bz; vs eth
8943	1-Phenyl-1,4-pentanedione		$C_{11}H_{12}O_2$	583-05-1	176.212	ye oil		162¹²		1.5250³⁰	vs ace
8944	1-Phenyl-1-pentanol		$C_{11}H_{16}O$	583-03-9	164.244			141²⁵, 102³	0.9655²⁰	1.4086²⁵	vs ace, eth, EtOH
8945	1-Phenyl-1-pentanone		$C_{11}H_{14}O$	1009-14-9	162.228	liq	-9.4	245	0.986²⁰	1.5158²⁰	i H₂O; vs EtOH, eth; sl ctc
8946	1-Phenyl-1-penten-3-one		$C_{11}H_{12}O$	3152-68-9	160.212	lf (lig)	38.5	142¹²	0.8697²⁰	1.5684²⁰	sl H₂O, chl; vs EtOH, eth, bz
8947	Phenylphosphine	Monophenylphosphine	C_6H_7P	638-21-1	110.094			160.5	1.001¹⁵	1.5796²⁰	
8948	Phenylphosphinic acid	Benzenephosphinic acid	$C_6H_7O_2P$	1779-48-2	142.093		83.8				s H₂O; vs EtOH; sl eth, chl
8949	Phenylphosphonic acid	Benzenephosphonic acid	$C_6H_7O_3P$	1571-33-1	158.092	lf (w)	160				vs H₂O; s EtOH, eth, ace; i bz
8950	Phenylphosphonic dichloride		$C_6H_5Cl_2OP$	824-72-6	194.983		1	258	1.197²⁵	1.5581²⁵	sl DMSO
8951	Phenylphosphonothioic dichloride	Dichlorophenylphosphine sulfide	$C_6H_5Cl_2PS$	3497-00-5	211.049			205¹³⁰	1.376¹³		
8952	Phenylphosphonous dichloride	Dichlorophenylphosphine	$C_6H_5Cl_2P$	644-97-3	178.984	liq	-51	225; 142⁵⁷	1.356²⁰	1.6030²⁰	vs bz
8953	Phenyl phosphorodichloridate	Phenyl dichlorophosphate	$C_6H_5Cl_2O_2P$	770-12-7	210.983	hyg liq		242; 100⁵	1.412²⁰	1.5230²⁰	
8954	1-Phenylpiperazine		$C_{10}H_{14}N_2$	92-54-6	162.231	pa ye oil		286.5; 161¹⁵	1.0621²⁰	1.5875²⁰	i H₂O; msc EtOH, eth; s chl
8955	1-Phenylpiperidine		$C_{11}H_{15}N$	4096-20-2	161.244		4.7	258	0.9944²⁵	1.5598²⁵	vs EtOH, eth, bz, chl
8956	4-Phenylpiperidine		$C_{11}H_{15}N$	771-99-3	161.244		60.5	257	0.9996¹⁶		s chl

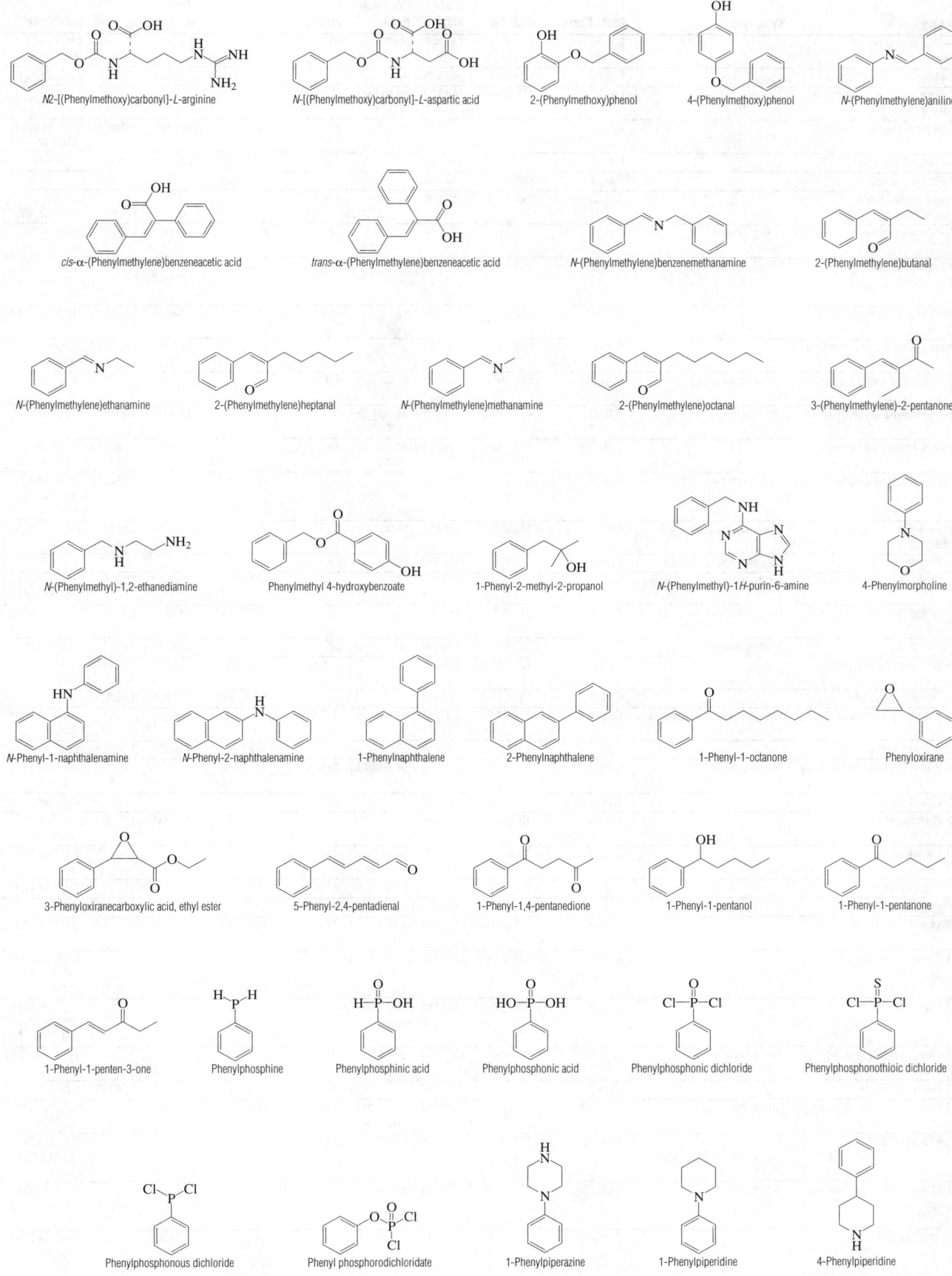

N2-[(Phenylmethoxy)carbonyl]-L-arginine

N-[(Phenylmethoxy)carbonyl]-L-aspartic acid

2-(Phenylmethoxy)phenol

4-(Phenylmethoxy)phenol

N-(Phenylmethylene)aniline

cis-α-(Phenylmethylene)benzeneacetic acid

trans-α-(Phenylmethylene)benzeneacetic acid

N-(Phenylmethylene)benzenemethanamine

2-(Phenylmethylene)butanal

N-(Phenylmethylene)ethanamine

2-(Phenylmethylene)heptanal

N-(Phenylmethylene)methanamine

2-(Phenylmethylene)octanal

3-(Phenylmethylene)-2-pentanone

N-(Phenylmethyl)-1,2-ethanediamine

Phenylmethyl 4-hydroxybenzoate

1-Phenyl-2-methyl-2-propanol

N-(Phenylmethyl)-1H-purin-6-amine

4-Phenylmorpholine

N-Phenyl-1-naphthalenamine

N-Phenyl-2-naphthalenamine

1-Phenylnaphthalene

2-Phenylnaphthalene

1-Phenyl-1-octanone

Phenyloxirane

3-Phenyloxiranecarboxylic acid, ethyl ester

5-Phenyl-2,4-pentadienal

1-Phenyl-1,4-pentanedione

1-Phenyl-1-pentanol

1-Phenyl-1-pentanone

1-Phenyl-1-penten-3-one

Phenylphosphine

Phenylphosphinic acid

Phenylphosphonic acid

Phenylphosphonic dichloride

Phenylphosphonothioic dichloride

Phenylphosphonous dichloride

Phenyl phosphorodichloridate

1-Phenylpiperazine

1-Phenylpiperidine

4-Phenylpiperidine

No.	Name	Synonym	Mol. Form.	CAS RN	Mol. Wt.	Physical Form	mp/°C	bp/°C	den/g cm⁻³	n_D	Solubility
8957	N-Phenylpropanamide		$C_9H_{11}NO$	620-71-3	149.189	pl (eth, al, bz)	105.5	222.2	1.175^{25}		sl H_2O; vs EtOH, eth
8958	1-Phenyl-1,2-propanedione		$C_9H_8O_2$	579-07-7	148.159	ye oil	<20	222; 102^{12}	1.1006^{20}	1.537^{10}	s H_2O, EtOH, eth
8959	1-Phenyl-1,2-propanedione, 2-oxime		$C_9H_9NO_2$	119-51-7	163.173	wh nd (w)	115				
8960	Phenyl propanoate		$C_9H_{10}O_2$	637-27-4	150.174	pr	20	211	1.0436^{25}	1.4980^{20}	i H_2O; vs EtOH, eth; s bz
8961	2-Phenyl-1-propanol		$C_9H_{12}O$	1123-85-9	136.190			121^{26}, 105^{11}	0.975^{25}	1.5582^{2}	i H_2O; s EtOH
8962	1-Phenyl-2-propanol		$C_9H_{12}O$	698-87-3	136.190			125^{25}, 120^{20}	0.991^{20}	1.5190^{20}	
8963	Phenylpropanolamine hydrochloride		$C_9H_{14}ClNO$	154-41-6	187.666		194				vs H_2O; s EtOH; i eth, bz, chl
8964	1-Phenyl-1-propanone	Propiophenone	$C_9H_{10}O$	93-55-0	134.174		18.6	217.5	1.0096^{20}	1.5269^{20}	i H_2O; s EtOH, eth, chl
8965	1-Phenyl-2-propanone	Phenylacetone	$C_9H_{10}O$	103-79-7	134.174	liq	-15	216.5	1.0157^{20}	1.5168^{20}	i H_2O; vs EtOH, eth; msc bz, xyl; s chl
8966	cis-3-Phenyl-2-propenenitrile		C_9H_7N	24840-05-9	129.159	liq	-4.4	249; 139^{30}	1.0289^{20}	1.5843^{20}	i H_2O; s EtOH; vs bz
8967	trans-3-Phenyl-2-propenenitrile		C_9H_7N	1885-38-7	129.159		22	263.8	1.0304^{20}	1.6013^{20}	i H_2O; s EtOH, ace, ctc
8968	3-Phenyl-2-propenoic anhydride	Cinnamic anhydride	$C_{18}H_{14}O_3$	538-56-7	278.302	nd (bz or al) pt (al)	136				vs bz
8969	cis-3-Phenyl-2-propen-1-ol		$C_9H_{10}O$	4510-34-3	134.174	wh nd (eth-peth)	34	257.5	1.0440^{20}	1.5819^{20}	vs eth, EtOH
8970	trans-3-Phenyl-2-propen-1-ol		$C_9H_{10}O$	4407-36-7	134.174	wh nd (eth-peth)	34	257.5	1.0440^{20}	1.5819^{20}	sl H_2O, chl; vs EtOH, eth
8971	trans-3-Phenyl-2-propen-1-ol acetate	trans-Cinnamyl acetate	$C_{11}H_{12}O_2$	21040-45-9	176.212			265; 145^{15}	1.0567^{20}	1.5425^{20}	i H_2O; s EtOH, eth, ace, bz, chl
8972	trans-3-Phenyl-2-propenoyl chloride	Cinnamoyl chloride	C_9H_7ClO	17082-09-6	166.604	ye cry	37.5	257.5	1.1617^{45}	1.614^{42}	i H_2O; s EtOH, ctc, lig
8973	3-Phenylpropyl acetate	Benzenepropanol, acetate	$C_{11}H_{14}O_2$	122-72-5	178.228	col liq	-40	69^1			
8974	1-Phenyl-2-propylamine, (±)	Amphetamine	$C_9H_{13}N$	300-62-9	135.206	oil		203	0.9306^{25}	1.518^{26}	sl H_2O, eth; s chl, EtOH
8975	1-Phenyl-2-propylamine, (S)	Dexamphetamine	$C_9H_{13}N$	51-64-9	135.206	oil	27.5	203.5; 80^{12}	0.949^{15}	1.4704^{20}	sl H_2O; s EtOH, eth
8976	Phenyl propyl ether	Propoxybenzene	$C_9H_{12}O$	622-85-5	136.190	liq	-27	189.9	0.9474^{20}	1.5014^{20}	s EtOH, eth
8977	4-(3-Phenylpropyl)pyridine		$C_{14}H_{15}N$	2057-49-0	197.276			322; 150^5	1.024^{25}	1.5616^{25}	vs bz, eth, py, EtOH
8978	3-Phenyl-2-propynal		C_9H_6O	2579-22-8	130.143			127^{28}, 104^{11}	1.0622^{20}	1.6079^{12}	
8979	3-Phenyl-2-propynoic acid	Phenylacetylenecarboxylic acid	$C_9H_6O_2$	637-44-5	146.143	nd (w)	137.5		1.28^{20}		sl H_2O; vs EtOH, eth
8980	3-Phenyl-2-propyn-1-ol		C_9H_8O	1504-58-1	132.159			137^{15}	1.078^{20}	1.5873^{28}	s eth, ace, bz
8981	6-Phenyl-2,4,7-pteridinetriamine	Triamterene	$C_{12}H_{11}N_7$	396-01-0	253.262	ye pl (BuOH)	316				i eth; sl EtOH, chl
8982	1-Phenyl-3-pyrazolidinone		$C_9H_{10}N_2O$	92-43-3	162.187		126				i eth, lig
8983	2-Phenylpyridine		$C_{11}H_9N$	1008-89-5	155.196			271	1.0833^{25}	1.6210^{20}	sl H_2O; msc EtOH, eth
8984	3-Phenylpyridine		$C_{11}H_9N$	1008-88-4	155.196	pa ye oil	164	272		1.6123^{25}	sl H_2O; s EtOH, eth
8985	4-Phenylpyridine		$C_{11}H_9N$	939-23-1	155.196	pl (w)	77.5	281			s H_2O, EtOH, eth
8986	Phenyl-2-pyridinylmethanone		$C_{12}H_9NO$	91-02-1	183.205		42	317	1.1556^{20}		s chl
8987	Phenyl-4-pyridinylmethanone		$C_{12}H_9NO$	14548-46-0	183.205	nd (peth), pl (w)	72	315; 170^{10}			sl H_2O; s EtOH, eth, bz
8988	1-Phenyl-1H-pyrrole		$C_{10}H_9N$	635-90-5	143.185	pl (sub), red in air	62	234			i H_2O; s EtOH, eth, ace, bz; vs peth
8989	2-Phenyl-1H-pyrrole		$C_{10}H_9N$	3042-22-6	143.185	pl (al, sub)	129	272			i H_2O; vs EtOH, eth, bz, chl; sl lig
8990	1-Phenyl-1H-pyrrole-2,5-dione	N-Phenylmaleimide	$C_{10}H_7NO_2$	941-69-5	173.169	ye nd (bz-lig)	90.5	162^{12}			vs bz, eth, EtOH
8991	1-Phenylpyrrolidine		$C_{10}H_{13}N$	4096-21-3	147.217		11	119^{12}, 102^5	1.018^{20}	1.5813^{20}	s eth
8992	1-Phenyl-2,5-pyrrolidinedione	Succinanil	$C_{10}H_9NO_2$	83-25-0	175.184	mcl pr or nd (w, al)	156	400	1.356^{25}		i H_2O; s EtOH, eth
8993	2-Phenylquinoline		$C_{15}H_{11}N$	612-96-4	205.255	nd (dil al)	86	363; 194^6			sl H_2O, peth; vs EtOH, eth, ace, bz
8994	2-Phenyl-4-quinolinecarboxylic acid	Cinchophen	$C_{16}H_{11}NO_2$	132-60-5	249.264	nd	214.5				i H_2O; s EtOH, eth, alk; sl ace, bz
8995	Phenyl salicylate		$C_{13}H_{10}O_3$	118-55-8	214.216		43	173^{12}	1.2614^{30}		i H_2O; vs EtOH, ace, bz; s eth, HOAc
8996	Phenylsilane		C_6H_8Si	694-53-1	108.214			119	0.8681^{20}	1.5125^{20}	i H_2O

N-Phenylpropanamide

1-Phenyl-1,2-propanedione

1-Phenyl-1,2-propanedione, 2-oxime

Phenyl propanoate

2-Phenyl-1-propanol

1-Phenyl-2-propanol

Phenylpropanolamine hydrochloride

1-Phenyl-1-propanone

1-Phenyl-2-propanone

cis-3-Phenyl-2-propenenitrile

trans-3-Phenyl-2-propenenitrile

3-Phenyl-2-propenoic anhydride

cis-3-Phenyl-2-propen-1-ol

trans-3-Phenyl-2-propen-1-ol

trans-3-Phenyl-2-propen-1-ol acetate

trans-3-Phenyl-2-propenoyl chloride

3-Phenylpropyl acetate

1-Phenyl-2-propylamine, (±)

1-Phenyl-2-propylamine, (S)

Phenyl propyl ether

4-(3-Phenylpropyl)pyridine

3-Phenyl-2-propynal

3-Phenyl-2-propynoic acid

3-Phenyl-2-propyn-1-ol

6-Phenyl-2,4,7-pteridinetriamine

1-Phenyl-3-pyrazolidinone

2-Phenylpyridine

3-Phenylpyridine

4-Phenylpyridine

Phenyl-2-pyridinylmethanone

Phenyl-4-pyridinylmethanone

1-Phenyl-1H-pyrrole

2-Phenyl-1H-pyrrole

1-Phenyl-1H-pyrrole-2,5-dione

1-Phenylpyrrolidine

1-Phenyl-2,5-pyrrolidinedione

2-Phenylquinoline

2-Phenyl-4-quinolinecarboxylic acid

Phenyl salicylate

Phenylsilane

No.	Name	Synonym	Mol. Form.	CAS RN	Mol. Wt.	Physical Form	mp/°C	bp/°C	den/ g cm⁻³	n_D	Solubility
8997	1-Phenylsilatrane		$C_{12}H_{17}NO_3Si$	2097-19-0	251.354	pr or nd (ace)	209				
8998	Phenyl stearate		$C_{24}H_{40}O_2$	637-55-8	360.574		52	267[15]			i H_2O; s EtOH, eth
8999	5'-Phenyl-1,1':3',1''-terphenyl		$C_{24}H_{18}$	612-71-5	306.400	orth nd (al or HOAc)	176	462	1.199[30]		i H_2O; s EtOH, eth, HOAc; vs bz; sl chl
9000	5-Phenyl-2,4-thiazolediamine	Amiphenazole	$C_9H_9N_3S$	490-55-1	191.252	fl (dil al) br in air	163 dec				
9001	Phenyl-2-thienylmethanone		$C_{11}H_8OS$	135-00-2	188.246	nd (dil al)	56.5	300	1.1890[54]	1.6181[54]	i H_2O; s EtOH, eth
9002	N-Phenylthioacetamide	Thioacetanilide	C_8H_9NS	637-53-6	151.229	nd (w)	75.5	dec			
9003	Phenyl thiocyanate		C_7H_5NS	5285-87-0	135.187			232.5	1.153[18]		i H_2O; s EtOH, eth
9004	2-Phenylthiosemicarbazide	2-Phenylhydrazinecarbothioamide	$C_7H_9N_3S$	645-48-7	167.231	pr (al)	200 dec				
9005	4-Phenyl-3-thiosemicarbazide	N-Phenylhydrazinecarbothioamide	$C_7H_9N_3S$	5351-69-9	167.231	pl (al)	140 dec				i EtOH, lig; sl bz
9006	Phenylthiourea		$C_7H_8N_2S$	103-85-5	152.217	nd (w), pr (al)	154				sl H_2O; s EtOH, NaOH
9007	3-Phenyl-2-thioxo-4-thiazolidinone	3-Phenylrhodanine	$C_9H_7NOS_2$	1457-46-1	209.288	ye pr (HOAc) nd or pr (al)	194.5				i H_2O; sl EtOH, eth; s ace, chl, HOAc
9008	6-Phenyl-1,3,5-triazine-2,4-diamine	Benzoguanamine	$C_9H_9N_5$	91-76-9	187.201	nd, pl (al)	226.5				s EtOH, eth; sl tfa
9009	N-Phenyl-1,3,5-triazine-2,4-diamine	Amanozine	$C_9H_9N_5$	537-17-7	187.201	cry (diox, 50% al)	235.5				
9010	4-Phenyl-1,2,4-triazolidine-3,5-dione		$C_8H_7N_3O_2$	15988-11-1	177.161		205.5				
9011	Phenyltrimethylammonium iodide		$C_9H_{14}IN$	98-04-4	263.118	lf (al)	224				vs H_2O; s EtOH, HOAc; sl ace; i chl
9012	Phenyl(triphenylmethyl)diazene		$C_{25}H_{20}N_2$	981-18-0	348.440		111 dec				
9013	Phenylurea		$C_7H_8N_2O$	64-10-8	136.151	mcl pr (w, al)	147	238	1.302[25]		sl H_2O, eth, DMSO; s EtOH, AcOEt
9014	trans-5-(2-Phenylvinyl)-1,3-benzenediol	Pinosylvin	$C_{14}H_{12}O_2$	22139-77-1	212.244	nd (HOAc)	156				vs ace, bz, chl, HOAc
9015	Phenyl vinyl ether		C_8H_8O	766-94-9	120.149			155.5	0.9770[20]	1.5224[20]	i H_2O; vs eth
9016	Phenytoin	5,5-Diphenyl-2,4-imidazolidinedione	$C_{15}H_{12}N_2O_2$	57-41-0	252.268	nd (al)	286				i H_2O; s EtOH, ace; sl eth, bz
9017	Phloretin		$C_{15}H_{14}O_5$	60-82-2	274.269	nd (dil al), cry (ace)	263 dec				sl H_2O, chl; msc EtOH, bz; i eth; s ace
9018	Phorate		$C_7H_{17}O_2PS_3$	298-02-2	260.378		<-15	119[0.8]	1.16[25]		
9019	Phorbol		$C_{20}H_{28}O_6$	17673-25-5	364.432	cry (EtOH)	250 dec				s H_2O, ace
9020	Phorone		$C_9H_{14}O$	504-20-1	138.206	ye-grn pr	28	197.5	0.8850[20]	1.4998[20]	sl H_2O; s EtOH, eth, ace, ctc
9021	Phosalone		$C_{12}H_{15}ClNO_4PS_2$	2310-17-0	367.808		46				
9022	Phosfolan		$C_7H_{14}NO_3PS_2$	947-02-4	255.295		36.5	117[0.001]			vs H_2O, bz, ace; sl eth; s hx
9023	Phosmet		$C_{11}H_{12}NO_4PS_2$	732-11-6	317.321		72	dec			
9024	Phosphamidon		$C_{10}H_{19}ClNO_5P$	13171-21-6	299.689	oil	-45	162[1.5]	1.2132[25]	1.4718[25]	msc H_2O; s hx
9025	N-Phospho-L-arginine		$C_6H_{15}N_4O_5P$	1189-11-3	254.181	cry (ace aq)	177				
9026	O-Phosphorylethanolamine	Ethanolamine O-phosphate	$C_2H_8NO_4P$	1071-23-4	141.063	cry (EtOH aq)	242				
9027	O-Phosphoserine		$C_3H_8NO_6P$	407-41-0	185.073	cry	166 dec				
9028	Phthalazine	2,3-Benzodiazine	$C_8H_6N_2$	253-52-1	130.147		90.5	316			s H_2O, EtOH, bz; sl eth; i lig
9029	Phthalic acid	1,2-Benzenedicarboxylic acid	$C_8H_6O_4$	88-99-3	166.132	pl (w)	230 dec	dec	2.18[191]		sl H_2O, eth; i chl; s EtOH
9030	Phthalic anhydride		$C_8H_4O_3$	85-44-9	148.116	wh nd (al, bz)	130.8	295	1.527[4]		sl H_2O, eth; s EtOH, ace, bz
9031	29H,31H-Phthalocyanine		$C_{32}H_{18}N_8$	574-93-6	514.539	grsh-bl mcl (quinoline)		sub 550			i H_2O, EtOH, eth; s $PhNH_2$
9032	Phthalylsulphathiazole		$C_{17}H_{13}N_3O_5S_2$	85-73-4	403.432		273				i H_2O, eth, chl; sl EtOH; s acid, alk
9033	Physostigmine		$C_{15}H_{21}N_3O_2$	57-47-6	275.347	orth pr (eth, bz)	105.5				sl H_2O; s EtOH, eth, bz, chl
9034	Phytol	3,7,11,15-Tetramethyl-2-hexadecen-1-ol, [R-[R*,R*-(E)]]	$C_{20}H_{40}O$	150-86-7	296.531	oily liq		203[10]	0.8497[25]	1.4595[25]	

1-Phenylsilatrane

Phenyl stearate

5'-Phenyl-1,1':3',1"-terphenyl

5-Phenyl-2,4-thiazolediamine

Phenyl-2-thienylmethanone

N-Phenylthioacetamide

Phenyl thiocyanate

2-Phenylthiosemicarbazide

4-Phenyl-3-thiosemicarbazide

Phenylthiourea

3-Phenyl-2-thioxo-4-thiazolidinone

6-Phenyl-1,3,5-triazine-2,4-diamine

N-Phenyl-1,3,5-triazine-2,4-diamine

4-Phenyl-1,2,4-triazolidine-3,5-dione

Phenyltrimethylammonium iodide

Phenyl(triphenylmethyl)diazene

Phenylurea

trans-5-(2-Phenylvinyl)-1,3-benzenediol

Phenyl vinyl ether

Phenytoin

Phloretin

Phorate

Phorbol

Phorone

Phosalone

Phosfolan

Phosmet

Phosphamidon

N-Phospho-L-arginine

O-Phosphorylethanolamine

O-Phosphoserine

Phthalazine

Phthalic acid

Phthalic anhydride

29H,31H-Phthalocyanine

Phthalylsulphathiazole

Physostigmine

Phytol

No.	Name	Synonym	Mol. Form.	CAS RN	Mol. Wt.	Physical Form	mp/°C	bp/°C	den/ g cm⁻³	n_D	Solubility
9035	Picene	Benzo[a]chrysene	$C_{22}H_{14}$	213-46-7	278.346	lf, pl (xyl, py, sub)	368	519			i H_2O; sl EtOH, bz, chl; s con sulf
9036	Picrolonic acid		$C_{10}H_8N_4O_5$	550-74-3	264.195	ye nd (al)	116	dec			sl H_2O; s EtOH, eth, MeOH
9037	Picropodophyllin		$C_{22}H_{22}O_8$	477-47-4	414.405	col nd (al, bz)	228				vs ace, bz, eth, EtOH
9038	Picrotoxin		$C_{30}H_{34}O_{13}$	124-87-8	602.583	orth lf	203.5				vs py, EtOH
9039	Pilocarpine		$C_{11}H_{16}N_2O_2$	92-13-7	208.257	nd	34	260[5]			s H_2O, EtOH; sl eth, bz; vs chl; i peth
9040	Pilocarpine, monohydrochloride		$C_{11}H_{17}ClN_2O_2$	54-71-7	244.718	hyg cry	204.5				vs H_2O, EtOH
9041	Pilocarpine, mononitrate		$C_{11}H_{17}N_3O_5$	148-72-1	271.270	wh pow or cry (al)	178				vs H_2O
9042	Pilosine		$C_{16}H_{18}N_2O_3$	13640-28-3	286.325	nd (al)	179				
9043	Pimaric acid	Dextropimaric acid	$C_{20}H_{30}O_2$	127-27-5	302.451	orth (ace) pr (al)	218.5	282[18]			vs eth, py, EtOH
9044	Pinane	2,6,6-Trimethylbicyclo[3.1.1] heptane	$C_{10}H_{18}$	473-55-2	138.250	oil	-53	169	0.8467[21]	1.4605[21]	
9045	trans-2-Pinanol	Pinene hydrate	$C_{10}H_{18}O$	35408-04-9	154.249		60	81[10]			
9046	Pindolol		$C_{14}H_{20}N_2O_2$	13523-86-9	248.321	cry (EtOH)	172				
9047	α-Pinene	2-Pinene	$C_{10}H_{16}$	80-56-8	136.234	liq	-64	156.2	0.8539[25]	1.4632[25]	i H_2O; msc EtOH, eth, chl
9048	β-Pinene	Nopinene	$C_{10}H_{16}$	127-91-3	136.234	liq	-61.5	166	0.860[25]	1.4768[25]	i H_2O; s bz, EtOH, eth, chl
9049	Piperazine	Diethylenediamine	$C_4H_{10}N_2$	110-85-0	86.135	hyg pl or lf (al)	106	146		1.446[113]	vs H_2O; s EtOH, chl; i eth
9050	1-Piperazinecarboxaldehyde		$C_5H_{10}N_2O$	7755-92-2	114.145			95[0.5]		1.5094[20]	
9051	1,4-Piperazinediethanol		$C_8H_{18}N_2O_2$	122-96-3	174.241		135	217[30]			
9052	Piperazine dihydrochloride	Diethylenediamine dihydrochloride	$C_4H_{12}Cl_2N_2$	142-64-3	159.057						sl H_2O; i EtOH
9053	2,5-Piperazinedione		$C_4H_6N_2O_2$	106-57-0	114.103	tab or pl (w)	312 dec	sub 260			sl H_2O, EtOH; s HCl
9054	1,4-Piperazinedipropanamine	1,4-Bis(3-aminopropyl) piperazine	$C_{10}H_{24}N_4$	7209-38-3	200.325		15	151[2]	0.973[25]	1.5015[20]	
9055	1-Piperazineethanamine	1-(2-Aminoethyl)piperazine	$C_6H_{15}N_3$	140-31-8	129.203			220	0.985[25]	1.4983[20]	
9056	1-Piperazineethanol		$C_6H_{14}N_2O$	103-76-4	130.187			246	1.061[25]	1.5065[20]	
9057	1-Piperidinamine		$C_5H_{12}N_2$	2213-43-6	100.162			147	0.928[25]	1.4750[20]	
9058	Piperidine	Azacyclohexane	$C_5H_{11}N$	110-89-4	85.148	liq	-11.02	106.22	0.8606[20]	1.4530[20]	msc H_2O, EtOH; s eth, ace, bz, chl
9059	1-Piperidinecarboxaldehyde		$C_6H_{11}NO$	2591-86-8	113.157	liq	-30.8	222.5	1.0158[25]	1.4805[25]	msc H_2O, EtOH, eth, bz, chl, lig
9060	4-Piperidinecarboxamide		$C_6H_{12}N_2O$	39546-32-2	128.171		138.5				
9061	2-Piperidinecarboxylic acid, (S)	L-Pipecolic acid	$C_6H_{11}NO_2$	3105-95-1	129.157	nd (MeOH/ eth)	260				
9062	3-Piperidinecarboxylic acid	Nipecotic acid	$C_6H_{11}NO_2$	498-95-3	129.157		261 dec				vs H_2O
9063	4-Piperidinecarboxylic acid	Isonipecotic acid	$C_6H_{11}NO_2$	498-94-2	129.157	nd	336				
9064	1-Piperidineethanol		$C_7H_{15}NO$	3040-44-6	129.200		17.9	202; 90[12]	0.9703[25]	1.4749[20]	msc H_2O; vs EtOH
9065	2-Piperidineethanol	2-(2-Hydroxyethyl)piperidine	$C_7H_{15}NO$	1484-84-0	129.200		69	202; 145[36]	1.01[27]		vs H_2O
9066	4-Piperidineethanol	4-(2-Hydroxyethyl)piperidine	$C_7H_{15}NO$	622-26-4	129.200	syr	132.5	227.5	1.0059[15]	1.4907[20]	vs H_2O, eth, EtOH
9067	Piperidine, hydrochloride	Piperidinium chloride	$C_5H_{12}ClN$	6091-44-7	121.609		142 dec				vs H_2O, chl
9068	4-Piperidinemethanamine	4-(Aminomethyl)piperidine	$C_6H_{14}N_2$	7144-05-0	114.188		25	200; 31[10]		1.4900[20]	
9069	2-Piperidinemethanol		$C_6H_{13}NO$	3433-37-2	115.173		69	104[10], 80[1]			sl chl
9070	3-Piperidinemethanol		$C_6H_{13}NO$	4606-65-9	115.173		61	106[3.5]	1.0263[20]	1.4964[20]	sl chl
9071	1-Piperidinepropanenitrile		$C_8H_{14}N_2$	3088-41-3	138.210		-6.8	145[50]	0.9403[25]	1.4676[25]	
9072	2-Piperidinone		C_5H_9NO	675-20-7	99.131	hyg	39.5	256			vs H_2O, EtOH, eth; s dil acid; i con alk
9073	2-(1-Piperidinylmethyl) cyclohexanone	Pimeclone	$C_{12}H_{21}NO$	534-84-9	195.301			119[14]			
9074	1-(2-Piperidinyl)-2-propanone, (±)		$C_8H_{15}NO$	539-00-4	141.211	oil		91[14]	0.9624[20]	1.4683[20]	vs EtOH, chl
9075	3-(2-Piperidinyl)pyridine, (S)	Anabasine	$C_{10}H_{14}N_2$	494-52-0	162.231	liq	9	276; 146[14]	1.0455[20]	1.5430[20]	msc H_2O; s EtOH, eth, bz
9076	Piperine		$C_{17}H_{19}NO_3$	94-62-2	285.338	pr (AcOEt) pl or mcl pr (al), cry	131.5				i H_2O; s EtOH, bz, py; sl eth; vs chl
9077	Piperonyl butoxide		$C_{19}H_{30}O_5$	51-03-6	338.438			180[1]	1.05[25]		
9078	Piperonyl sulfoxide	Isosafrole octyl sulfoxide	$C_{18}H_{28}O_3S$	120-62-7	324.478	ye-br liq		dec		1.530[25]	sl H_2O; misc os
9079	Pipobroman		$C_{10}H_{16}Br_2N_2O_2$	54-91-1	356.054		106				
9080	Piprotal	Tropital	$C_{24}H_{40}O_8$	5281-13-0	456.570	liq		215[0.04]			

Picene

Picrolonic acid

Picropodophyllin

Picrotoxin

Pilocarpine

Pilocarpine, monohydrochloride

Pilocarpine, mononitrate

Pilosine

Pimaric acid

Pinane

trans-2-Pinanol

Pindolol

α-Pinene

β-Pinene

Piperazine

1-Piperazinecarboxaldehyde

1,4-Piperazinediethanol

Piperazine dihydrochloride

2,5-Piperazinedione

1,4-Piperazinedipropanamine

1-Piperazineethanamine

1-Piperazineethanol

1-Piperidinamine

Piperidine

1-Piperidinecarboxaldehyde

4-Piperidinecarboxamide

2-Piperidinecarboxylic acid, (S)

3-Piperidinecarboxylic acid

4-Piperidinecarboxylic acid

1-Piperidineethanol

2-Piperidineethanol

4-Piperidineethanol

Piperidine, hydrochloride

4-Piperidinemethanamine

2-Piperidinemethanol

3-Piperidinemethanol

1-Piperidinepropanenitrile

2-Piperidinone

2-(1-Piperidinylmethyl)cyclohexanone

1-(2-Piperidinyl)-2-propanone, (±)

3-(2-Piperidinyl)pyridine, (S)

Piperine

Piperonyl butoxide

Piperonyl sulfoxide

Pipobroman

Piprotal

No.	Name	Synonym	Mol. Form.	CAS RN	Mol. Wt.	Physical Form	mp/°C	bp/°C	den/ g cm⁻³	n_D	Solubility
9081	Pirimicarb		$C_{11}H_{18}N_4O_2$	23103-98-2	238.287		90.5				
9082	Pirimiphos-ethyl		$C_{13}H_{24}N_3O_3PS$	23505-41-1	333.387			dec >130	1.14[20]		
9083	Pirimiphos-methyl		$C_{11}H_{20}N_3O_3PS$	29232-93-7	305.334		15	dec	1.17[20]		
9084	Pithecolobine		$C_{22}H_{46}N_4O_2$	22368-82-7	398.626	cry	68	230[0.007]			s H₂O, chl, eth, EtOH, peth
9085	2-Pivaloyl-1,3-indandione	Pindone	$C_{14}H_{14}O_3$	83-26-1	230.259	ye cry	109				
9086	Plasmocid		$C_{17}H_{25}N_3O$	551-01-9	287.400			182[1.0]	1.0569[24]	1.5855[24]	
9087	Plumericin		$C_{15}H_{14}O_6$	77-16-7	290.268						s chl
9088	Podophyllotoxin		$C_{22}H_{22}O_8$	518-28-5	414.405		183				sl H₂O; vs EtOH; i eth; s ace, bz, HOAc
9089	Polythiazide		$C_{11}H_{13}ClF_3N_3O_4S_3$	346-18-9	439.882		214				
9090	Ponceau 3R	C.I. Food Red 6	$C_{19}H_{16}N_2Na_2O_7S_2$	3564-09-8	494.449	dk red pow					s H₂O; sl EtOH
9091	Populin		$C_{20}H_{22}O_8$	99-17-2	390.384	nd (w+2), pr (al)	180				
9092	21H,23H-Porphine		$C_{20}H_{14}N_4$	101-60-0	310.352	red or oran lf (chl-MeOH)	360	sub 300	1.336[25]		i H₂O, eth, ace, bz; sl EtOH; s diox
9093	Potassium benzoate		$C_7H_5KO_2$	582-25-2	160.212	hyg cry					
9094	Potassium dichloroisocyanurate	Troclosene potassium	$C_3Cl_2KN_3O_3$	2244-21-5	236.054	hyg cry	250 dec				
9095	Potassium D-gluconate		$C_6H_{11}KO_7$	299-27-4	234.245	ye-wh cry	183 dec				vs H₂O; i EtOH, eth, bz, chl
9096	Potassium trans,trans-2,4-hexadienoate	Potassium sorbate	$C_6H_7KO_2$	24634-61-5	150.217		>270 dec		1.361[25]		vs H₂O; s EtOH
9097	Potassium hydrogen phthalate	Potassium biphthalate	$C_8H_5KO_4$	877-24-7	204.222				1.636[25]		s H₂O; sl EtOH
9098	Potassium cis-9-octadecenoate	Potassium oleate	$C_{18}H_{33}KO_2$	143-18-0	320.552	ye-br solid					s H₂O, EtOH
9099	Prazosin		$C_{19}H_{21}N_5O_4$	19216-56-9	383.402	cry	279				
9100	Prednisolone		$C_{21}H_{28}O_5$	50-24-8	360.444		235				
9101	5α-Pregnane	Allopregnane	$C_{21}H_{36}$	641-85-0	288.511		84.5				
9102	5β-Pregnane	17β-Ethyletiocholane	$C_{21}H_{36}$	481-26-5	288.511	mcl sc or pl (MeOH)	83.5		1.032[15]		i H₂O; s chl, MeOH
9103	5α-Pregnane-3α,20α-diol	Allopregnane-3α,20α-diol	$C_{21}H_{36}O_2$	566-58-5	320.510	cry (MeOH)	244				
9104	5β-Pregnane-3α,20S-diol	Pregnanediol	$C_{21}H_{36}O_2$	80-92-2	320.510	pl (ace)	243.5		1.15[25]		sl EtOH, eth; s ace
9105	5α-Pregnane-3,20-dione	3,20-Allopregnanedione	$C_{21}H_{32}O_2$	566-65-4	316.478	cry	200				
9106	5β-Pregnane-3,20-dione		$C_{21}H_{32}O_2$	128-23-4	316.478	nd (dil al) cry (dil ace)	123				i H₂O; vs EtOH; s eth, ace
9107	5-Pregnane-3,17,21-triol-20-one	3,17,21-Trihydroxypregnan-20-one, (3α,5β)	$C_{21}H_{34}O_4$	68-60-0	350.493	cry (EtOAc)	226				
9108	Pregnan-3α-ol-20-one		$C_{21}H_{34}O_2$	128-20-1	318.494	nd (bz), cry (dil al)	149.5				vs EtOH
9109	Pregnenolone		$C_{21}H_{32}O_2$	145-13-1	316.478	nd (dil al)	192				
9110	Prenoxdiazine hydrochloride		$C_{23}H_{28}ClN_3O$	982-43-4	397.940		186.5				
9111	Prephenic acid		$C_{10}H_{10}O_6$	126-49-8	226.182	free acid unstab					
9112	Pridinol	1,1-Diphenyl-3-(1-piperidinyl)-1-propanol	$C_{20}H_{25}NO$	511-45-5	295.419	cry	120				s ace
9113	Prilocaine	N-(2-Methylphenyl)-2-(propylamino)propanamide	$C_{13}H_{20}N_2O$	721-50-6	220.310	nd	38	160[1]		1.5299[20]	
9114	Procainamide	4-Amino-N-[2-(diethylamino)ethyl]benzamide	$C_{13}H_{21}N_3O$	51-06-9	235.325		47	212[2]			
9115	Procainamide hydrochloride		$C_{13}H_{22}ClN_3O$	614-39-1	271.786		166				vs H₂O; s EtOH; i eth, bz; sl chl
9116	Procarbazine hydrochloride		$C_{12}H_{20}ClN_3O$	366-70-1	257.759	cry (MeOH)	225				
9117	Prochlorperazine		$C_{20}H_{24}ClN_3S$	58-38-8	373.943		228				
9118	Procymidone		$C_{13}H_{11}Cl_2NO_2$	32809-16-8	284.138		166		1.452[25]		
9119	Prodiamine		$C_{13}H_{17}F_3N_4O_4$	29091-21-2	350.294		124		1.47[25]		
9120	Profenofos		$C_{11}H_{15}BrClO_3PS$	41198-08-7	373.631			110[0.001]	1.455[20]		
9121	Profluralin		$C_{14}H_{16}F_3N_3O_4$	26399-36-0	347.290		34				
9122	Progesterone	Pregn-4-ene-3,20-dione	$C_{21}H_{30}O_2$	57-83-0	314.462	pr	129		1.166[23]		i H₂O; s EtOH, diox, ace
9123	DL-Proline		$C_5H_9NO_2$	609-36-9	115.131	hyg nd (al-eth) cry (+w)	205 dec				vs H₂O, EtOH
9124	L-Proline	2-Pyrrolidinecarboxylic acid	$C_5H_9NO_2$	147-85-3	115.131	nd (al-eth) pr (w)	221 dec				vs H₂O; sl EtOH, ace, bz; i eth, PrOH
9125	Promazine		$C_{17}H_{20}N_2S$	58-40-2	284.419			206[0.3]			
9126	Promecarb	Phenol, 3-methyl-5-(1-methylethyl)-, methylcarbamate	$C_{12}H_{17}NO_2$	2631-37-0	207.269		87	117[0.01]			
9127	Promethazine	N,N,α-Trimethyl-10H-phenothiazine-10-ethanamine	$C_{17}H_{20}N_2S$	60-87-7	284.419		60	191[0.5]			i H₂O; vs dil HCl

Pirimicarb

Pirimiphos-ethyl

Pirimiphos-methyl

Pithecolobine

2-Pivaloyl-1,3-indandione

Plasmocid

Plumericin

Podophyllotoxin

Polythiazide

Ponceau 3R

Populin

21H,23H-Porphine

Potassium benzoate

Potassium dichloroisocyanurate

Potassium D-gluconate

Potassium trans,trans-2,4-hexadienoate

Potassium hydrogen phthalate

Potassium cis-9-octadecenoate

Prazosin

Prednisolone

5α-Pregnane

5β-Pregnane

5α-Pregnane-3α,20α-diol

5β-Pregnane-3α,20S-diol

5α-Pregnane-3,20-dione

5β-Pregnane-3,20-dione

5-Pregnane-3,17,21-triol-20-one

Pregnan-3α-ol-20-one

Pregnenolone

Prenoxdiazine hydrochloride

Prephenic acid

Pridinol

Prilocaine

Procainamide

Procainamide hydrochloride

Procarbazine hydrochloride

Prochlorperazine

Procymidone

Prodiamine

Profenofos

Profluralin

Progesterone

DL-Proline

L-Proline

Promazine

Promecarb

Promethazine

No.	Name	Synonym	Mol. Form.	CAS RN	Mol. Wt.	Physical Form	mp/°C	bp/°C	den/ g cm⁻³	n_D	Solubility
9128	Promethazine hydrochloride	Diprazin	$C_{17}H_{21}ClN_2S$	58-33-3	320.880		231				vs H_2O, EtOH, chl
9129	Prometone		$C_{10}H_{19}N_5O$	1610-18-0	225.291	solid	91.5				
9130	Prometryn	N,N'-Diisopropyl-6-(methylthio) -1,3,5-triazine-2,4-diamine	$C_{10}H_{19}N_5S$	7287-19-6	241.357		119		1.157[20]		
9131	Propachlor	Acetamide, 2-chloro-N-(1-methylethyl)-N-phenyl-	$C_{11}H_{14}ClNO$	1918-16-7	211.688		77	110[0.03]	1.242[25]		
9132	Propanal	Propionaldehyde	C_3H_6O	123-38-6	58.079	liq	-80	48	0.8657[25]	1.3636[20]	s H_2O; msc EtOH, eth
9133	Propanal oxime		C_3H_7NO	627-39-4	73.094		40	131.5	0.9258[20]	1.4287[20]	
9134	Propanamide	Propionamide	C_3H_7NO	79-05-0	73.094	rhom, pl (bz)	81.3	213	0.9262[110]	1.4180[110]	vs H_2O, EtOH, eth, chl
9135	Propane		C_3H_8	74-98-6	44.096	col gas	-187.63	-42.1	0.493[25] (p>1 atm)		s H_2O, EtOH; vs eth, bz; sl ace
9136	Propanediamide		$C_3H_6N_2O_2$	108-13-4	102.092	mcl pr(w)	170.8				s H_2O; i EtOH, eth, bz; sl DMSO
9137	1,2-Propanediamine, (±)	Propylenediamine	$C_3H_{10}N_2$	10424-38-1	74.124	hyg		119.5	0.878[15]	1.4460[20]	vs H_2O; i eth; vs chl
9138	1,3-Propanediamine	1,3-Diaminopropane	$C_3H_{10}N_2$	109-76-2	74.124	liq	-10.8	139.8	0.884[25]	1.4600[20]	s H_2O; msc EtOH, eth
9139	1,2-Propanediol diacetate		$C_7H_{12}O_4$	623-84-7	160.168			190.5	1.059[20]	1.4173[20]	vs H_2O; s EtOH, eth
9140	1,3-Propanediol diacetate		$C_7H_{12}O_4$	628-66-0	160.168			209.5	1.070[14]	1.4192	vs H_2O; s EtOH
9141	1,2-Propanediol 1-methacrylate	2-Hydroxypropyl methacrylate	$C_7H_{12}O_3$	923-26-2	144.168			90[9], 57[0.5]	1.066[25]	1.4458[20]	
9142	1,2-Propanedione	Pyruvaldehyde	$C_3H_4O_2$	78-98-8	72.063	ye hyg liq		72	1.0455[20]	1.4002[18]	s EtOH, eth, bz
9143	Propanedioyl dichloride		$C_3H_2Cl_2O_2$	1663-67-8	140.953			57[28]	1.4509[20]	1.4639[20]	s eth, AcOEt
9144	1,2-Propanedithiol		$C_3H_8S_2$	814-67-5	108.226			152	1.08[20]	1.532[20]	s chl
9145	1,3-Propanedithiol	Trimethylene dimercaptan	$C_3H_8S_2$	109-80-8	108.226	liq	-79	172.9	1.0772[20]	1.5392[20]	sl H_2O, ctc; msc EtOH, eth, bz
9146	2,2'-[1,3-Propanediylbis(nitrilomethylidyne)]bisphenol	Disalicylidene-1,3-propanediamine	$C_{17}H_{18}N_2O_2$	120-70-7	282.337		54.3				
9147	Propanenitrile	Ethyl cyanide	C_3H_5N	107-12-0	55.079	liq	-92.78	97.14	0.7818[20]	1.3655[20]	vs H_2O; s EtOH, eth, ace, bz, ctc
9148	1-Propanesulfonic acid		$C_3H_8O_3S$	5284-66-2	124.159		8	136[1]	1.2516[25]		
9149	1-Propanesulfonyl chloride		$C_3H_7ClO_2S$	10147-36-1	142.605			dec 180; 77[12]	1.267[20]	1.452[20]	
9150	1,3-Propane sultone	1,2-Oxathiolane, 2,2-dioxide	$C_3H_6O_3S$	1120-71-4	122.143						s chl
9151	1-Propanethiol	Propyl mercaptan	C_3H_8S	107-03-9	76.161	liq	-113.13	67.8	0.8411[20]	1.4380[20]	sl H_2O; s EtOH, eth, ace, bz
9152	2-Propanethiol	Isopropyl mercaptan	C_3H_8S	75-33-2	76.161	liq	-130.5	52.6	0.8143[20]	1.4255[20]	sl H_2O; msc EtOH, eth; vs ace; s chl
9153	1,2,3-Propanetriamine	1,2,3-Triaminopropane	$C_3H_{11}N_3$	21291-99-6	89.139	visc oil		190; 92[9]			s H_2O
9154	1,2,3-Propanetricarboxylic acid	Tricarballylic acid	$C_6H_8O_6$	99-14-9	176.124	orth (eth)	166				vs H_2O, EtOH; sl eth
9155	1,2,3-Propanetriol-1-acetate		$C_5H_{10}O_4$	106-61-6	134.131			158[165], 129[3]	1.2060[20]	1.4157[20]	vs H_2O, EtOH
9156	1,2,3-Propanetriol 1-(4-aminobenzoate)	Glyceryl p-aminobenzoate	$C_{10}H_{13}NO_4$	136-44-7	211.215						i H_2O; s EtOH
9157	1,2,3-Propanetriol-1,3-diacetate	1,3-Diacetin	$C_7H_{12}O_5$	105-70-4	176.167	hyg liq		260; 149[12]	1.179[15]	1.4395[20]	vs H_2O, EtOH; sl eth; i CS_2
9158	1,2,3-Propanetriol tribenzoate		$C_{24}H_{20}O_6$	614-33-5	404.412	nd (MeOH)	76		1.228[12]		i H_2O; s EtOH; vs eth, ace, bz, chl
9159	1,2,3-Propanetriol tripropanoate		$C_{12}H_{20}O_6$	139-45-7	260.283			175[20], 157[13]	1.108[15]	1.4318[19]	i H_2O; s EtOH, chl; vs eth
9160	1,2,3-Propanetriyl hexanoate		$C_{21}H_{38}O_6$	621-70-5	386.523		-60	>200	0.9867[20]	1.4427[20]	i H_2O; msc EtOH, eth, bz; vs ace
9161	1,2,3-Propanetriyl octanoate		$C_{27}H_{50}O_6$	538-23-8	470.682		10	233	0.9540[20]	1.4482[20]	i H_2O; msc EtOH; vs eth, bz, chl, lig
9162	Propanidid		$C_{18}H_{27}NO_5$	1421-14-3	337.411			211[0.7]			i H_2O; s EtOH, chl
9163	Propanil	Propanamide, N-(3,4-dichlorophenyl)-	$C_9H_9Cl_2NO$	709-98-8	218.079		92		1.25[25]		
9164	Propanoic acid	Propionic acid	$C_3H_6O_2$	79-09-4	74.079	liq	-20.5	141.15	0.9882[25]	1.3809[20]	msc H_2O, EtOH; s eth; sl chl
9165	Propanoic anhydride	Propionic anhydride	$C_6H_{10}O_3$	123-62-6	130.141	liq	-45	170; 67.5[18]	1.0110[20]	1.4038[20]	msc eth; sl ctc
9166	1-Propanol	Propyl alcohol	C_3H_8O	71-23-8	60.095	liq	-124.39	97.2	0.7997[25]	1.3850[20]	msc H_2O, EtOH, eth; s ace, chl; vs bz

Promethazine hydrochloride

Prometone

Prometryn

Propachlor

Propanal

Propanal oxime

Propanamide

Propane

Propanediamide

1,2-Propanediamine, (±)

1,3-Propanediamine

1,2-Propanediol diacetate

1,3-Propanediol diacetate

1,2-Propanediol 1-methacrylate

1,2-Propanedione

Propanedioyl dichloride

1,2-Propanedithiol

1,3-Propanedithiol

2,2'-[1,3-Propanediylbis(nitrilomethylidyne)]bisphenol

Propanenitrile

1-Propanesulfonic acid

1-Propanesulfonyl chloride

1,3-Propane sultone

1-Propanethiol

2-Propanethiol

1,2,3-Propanetriamine

1,2,3-Propanetricarboxylic acid

1,2,3-Propanetriol-1-acetate

1,2,3-Propanetriol 1-(4-aminobenzoate)

1,2,3-Propanetriol-1,3-diacetate

1,2,3-Propanetriol tribenzoate

1,2,3-Propanetriol tripropanoate

1,2,3-Propanetriyl hexanoate

1,2,3-Propanetriyl octanoate

Propanidid

Propanil

Propanoic acid

Propanoic anhydride

1-Propanol

No.	Name	Synonym	Mol. Form.	CAS RN	Mol. Wt.	Physical Form	mp/°C	bp/°C	den/ g cm⁻³	n_D	Solubility
9167	2-Propanol	Isopropyl alcohol	C₃H₈O	67-63-0	60.095	liq	-87.9	82.3	0.7809²⁵	1.3776²⁰	msc H₂O, EtOH, eth; s ace, chl; vs bz
9168	2-Propanone oxime	Acetoxime	C₃H₇NO	127-06-0	73.094	pr (al)	61	136; 61²⁰	0.9113⁶²	1.4156²⁰	s H₂O, EtOH, eth, chl, lig
9169	2-Propanone phenylhydrazone	Acetone, phenylhydrazone	C₉H₁₂N₂	103-02-6	148.204	orth	42	163⁵⁰			s EtOH, eth, dil acid
9170	Propanoyl chloride	Propionyl chloride	C₃H₅ClO	79-03-8	92.524	liq	-94	80	1.0646²⁰	1.4032²⁰	s eth
9171	Propanoyl fluoride	Propionyl fluoride	C₃H₅FO	430-71-7	76.069			44	0.972¹⁵	1.329¹³	
9172	Propantheline bromide		C₂₃H₃₀BrNO₃	50-34-0	448.393	cry	160				vs H₂O, EtOH, chl; i eth, bz
9173	Propargite		C₁₉H₂₆O₄S	2312-35-8	350.472				1.10²⁵		
9174	Propargyl acetate		C₅H₆O₂	627-09-8	98.101			121.5	0.9982²⁰	1.4187²⁰	sl H₂O; s EtOH, eth
9175	Propargyl alcohol	3-Hydroxy-1-propyne	C₃H₄O	107-19-7	56.063	liq	-51.8	113.6	0.9478²⁰	1.4322²⁰	s H₂O, chl; msc EtOH, eth
9176	Propatyl nitrate	2-Ethyl-2-[(nitrooxy)methyl]-1,3-propanediol, dinitrate	C₆H₁₁N₃O₉	2921-92-8	269.166	wh pow	52		1.49		i H₂O; s EtOH, ace
9177	Propazine	6-Chloro-N,N'-diisopropyl-1,3,5-triazine-2,4-diamine	C₉H₁₆ClN₅	139-40-2	229.710		213		1.162²⁰		
9178	Propene	Propylene	C₃H₆	115-07-1	42.080	col gas	-185.24	-47.69	0.505²⁵ (p>1 atm)	1.3567⁻⁷⁰	sl H₂O; vs EtOH, HOAc
9179	trans-1-Propene-1,2-dicarboxylic acid	Mesaconic acid	C₅H₆O₄	498-24-8	130.100	orth nd or mcl pr (eth)	204.5	sub	1.466²⁰		sl H₂O, bz, CS₂; vs EtOH; s eth, tfa
9180	1-Propene-2,3-dicarboxylic acid	Itaconic acid	C₅H₆O₄	97-65-4	130.100	rhom (bz)	175	dec	1.632²⁵		s H₂O, EtOH, ace; sl eth, bz, peth
9181	2-Propene-1-thiol		C₃H₆S	870-23-5	74.145			65	0.925²³	1.4832²⁰	i H₂O; msc EtOH, eth; s chl
9182	cis-1-Propene-1,2,3-tricarboxylic acid	cis-Aconitic acid	C₆H₆O₆	585-84-2	174.108	nd (w)	125				s H₂O; sl eth
9183	trans-1-Propene-1,2,3-tricarboxylic acid	trans-Aconitic acid	C₆H₆O₆	4023-65-8	174.108	lf (w) nd (w, eth)	196 dec				vs H₂O, EtOH
9184	1-Propen-1-one	Methylketene	C₃H₄O	6004-44-0	56.063	col gas	-80	-23			vs eth
9185	2-Propenoyl chloride	Acrylic acid chloride	C₃H₃ClO	814-68-6	90.508			75.5	1.1136²⁰	1.4343²⁰	vs chl
9186	cis-1-Propenylbenzene		C₉H₁₀	766-90-5	118.175	liq	-61.6	167.5	0.9088²⁰	1.5420²⁰	i H₂O; msc EtOH, eth, ace, bz, peth, ctc
9187	trans-1-Propenylbenzene		C₉H₁₀	873-66-5	118.175	liq	-29.3	178.3	0.9023²⁵	1.5506²⁰	i H₂O; msc EtOH, eth, ace, bz
9188	trans-5-(1-Propenyl)-1,3-benzodioxole		C₁₀H₁₀O₂	4043-71-4	162.185		6.8	253	1.1224²⁰	1.5782²⁰	i H₂O; msc EtOH, eth; vs ace; s chl
9189	4-(1-Propenyl)phenol	p-Anol	C₉H₁₀O	539-12-8	134.174	lf	94	dec 250			sl H₂O; vs DMF
9190	2-(1-Propenyl)piperidine	β-Coniceine	C₈H₁₅N	538-90-9	125.212		8	168	0.8716¹⁵		
9191	Propetamphos		C₁₀H₂₀NO₄PS	31218-83-4	281.309			88⁰·⁰⁰⁵	1.1294²⁰		
9192	Propiconazole		C₁₅H₁₇Cl₂N₃O₂	60207-90-1	342.221			180⁰·¹	1.27²⁰		
9193	Propiomazine		C₂₀H₂₄N₂OS	362-29-8	340.482			240⁰·⁵			
9194	Propionyl-L-carnitine	Carnitine, O-propanoyl	C₁₀H₁₉NO₄	20064-19-1	217.263	hyg pr (2-PrOH)	147 dec				
9195	Propofol		C₁₂H₁₈O	2078-54-8	178.270		19	256; 136³⁰	0.955²⁰	1.5140²⁰	
9196	Propoxur	Phenol, 2-(1-methylethoxy)-, methylcarbamate	C₁₁H₁₅NO₃	114-26-1	209.242		87	dec	1.12²⁰		
9197	2-Propoxyethanol	Ethylene glycol monopropyl ether	C₅H₁₂O₂	2807-30-9	104.148			149.8	0.9112²⁰	1.4133²⁰	s H₂O; vs EtOH, eth
9198	D-Propoxyphene	Dextropropoxyphene	C₂₂H₂₉NO₂	469-62-5	339.471	cry (peth)	75.5				
9199	L-Propoxyphene	Levopropoxyphene	C₂₂H₂₉NO₂	2338-37-6	339.471	cry (peth)	75.5				
9200	1-Propoxy-2-propanol	1,2-Propylene glycol 1-propyl ether	C₆H₁₄O₂	1569-01-3	118.174			150	0.8886²⁰	1.4130²⁰	
9201	3-Propoxy-1-propene		C₆H₁₂O	1471-03-0	100.158			91	0.7764²⁰	1.3919²⁰	vs ace, eth, EtOH
9202	Propranolol		C₁₆H₂₁NO₂	525-66-6	259.344	cry (cyhex)	96				
9203	Propyl acetate		C₅H₁₀O₂	109-60-4	102.132	liq	-93	101.54	0.8878²⁰	1.3842²⁰	sl H₂O; msc EtOH, eth; s ctc
9204	Propyl acrylate	2-Propenoic acid, propyl ester	C₆H₁₀O₂	925-60-0	114.142			122; 63¹⁰⁰			
9205	Propylamine	1-Propanamine	C₃H₉N	107-10-8	59.110	liq	-84.75	47.22	0.7173²⁰	1.3870²⁰	msc H₂O; vs EtOH, ace; s bz, chl; sl ctc
9206	Propylamine hydrochloride	1-Propanamine hydrochloride	C₃H₁₀ClN	556-53-6	95.571		163.5				s DMSO

2-Propanol

2-Propanone oxime

2-Propanone phenylhydrazone

Propanoyl chloride

Propanoyl fluoride

Propantheline bromide

Propargite

Propargyl acetate

Propargyl alcohol

Propatyl nitrate

Propazine

Propene

trans-1-Propene-1,2-dicarboxylic acid

1-Propene-2,3-dicarboxylic acid

2-Propene-1-thiol

cis-1-Propene-1,2,3-tricarboxylic acid

trans-1-Propene-1,2,3-tricarboxylic acid

1-Propen-1-one

2-Propenoyl chloride

cis-1-Propenylbenzene

trans-1-Propenylbenzene

trans-5-(1-Propenyl)-1,3-benzodioxole

4-(1-Propenyl)phenol

2-(1-Propenyl)piperidine

Propetamphos

Propiconazole

Propiomazine

Propionyl-*L*-carnitine

Propofol

Propoxur

2-Propoxyethanol

D-Propoxyphene

L-Propoxyphene

1-Propoxy-2-propanol

3-Propoxy-1-propene

Propranolol

Propyl acetate

Propyl acrylate

Propylamine

Propylamine hydrochloride

No.	Name	Synonym	Mol. Form.	CAS RN	Mol. Wt.	Physical Form	mp/°C	bp/°C	den/ g cm⁻³	n_D	Solubility
9207	Propyl 4-aminobenzoate	Risocaine	C₁₀H₁₃NO₂	94-12-2	179.216	pr	75				vs bz, eth, EtOH, chl
9208	2-(Propylamino)ethanol		C₅H₁₃NO	16369-21-4	103.163			182	0.9005[20]	1.4428[20]	
9209	4-Propylaniline		C₉H₁₃N	2696-84-6	135.206			227			
9210	N-Propylaniline		C₉H₁₃N	622-80-0	135.206			222; 98[11]	0.9443[20]	1.5428[20]	vs eth, EtOH
9211	Propylarsonic acid	1-Propanearsonic acid	C₃H₉AsO₃	107-34-6	168.023	nd (al), pl (w)	134.5				vs H₂O, EtOH; i eth
9212	Propylbenzene	Isocumene	C₉H₁₂	103-65-1	120.191	liq	-99.6	159.24	0.8593[25]	1.4895[25]	i H₂O; msc EtOH, eth, ace, bz, peth, ctc
9213	α-Propylbenzenemethanol, (R)		C₁₀H₁₄O	22144-60-1	150.217		16	232	0.9740[20]	1.5139[20]	vs eth, EtOH
9214	Propyl benzenesulfonate		C₉H₁₂O₃S	80-42-2	200.254			162[15]	1.1804[17]	1.5035[25]	sl H₂O; s EtOH; vs eth, chl
9215	Propyl benzoate		C₁₀H₁₂O₂	2315-68-6	164.201	liq	-51.6	211	1.0230[20]	1.5000[20]	i H₂O; msc EtOH, eth
9216	5-Propyl-1,3-benzodioxole	Dihydrosafrole	C₁₀H₁₂O₂	94-58-6	164.201			228			s ctc
9217	Propyl butanoate		C₇H₁₄O₂	105-66-8	130.185	liq	-95.2	143.0	0.8730[20]	1.4001[20]	sl H₂O; msc EtOH, eth
9218	Propyl carbamate		C₄H₉NO₂	627-12-3	103.120	pr	60	196			vs ace, eth, EtOH
9219	Propyl chloroacetate		C₅H₉ClO₂	5396-24-7	136.577			161	1.104[20]	1.4261[20]	vs eth
9220	Propyl 2-chlorobutanoate		C₇H₁₃ClO₂	62108-71-8	164.630			183	1.0252[20]		
9221	Propyl chlorocarbonate		C₄H₇ClO₂	109-61-5	122.551			115.2	1.0901[20]	1.4035[20]	msc EtOH, eth
9222	Propyl 3-chloropropanoate		C₆H₁₁ClO₂	62108-66-1	150.603			180	1.0656[20]	1.4290[20]	vs eth, EtOH
9223	S-Propyl chlorothioformate	S-Propyl carbonochloridothioate	C₄H₇ClOS	13889-92-4	138.616	liq		59[26]			
9224	Propyl trans-cinnamate	Propyl trans-3-phenyl-2-propenoate	C₁₂H₁₄O₂	74513-58-9	190.238			285	1.0433[0]		i H₂O
9225	Propylcyclohexane		C₉H₁₈	1678-92-8	126.239	liq	-94.9	156.7	0.7936[20]	1.4370[20]	i H₂O; msc EtOH, ace, ctc; s eth, bz
9226	2-Propylcyclohexanone		C₉H₁₆O	94-65-5	140.222			197	0.927[20]	1.4538[20]	i H₂O; s EtOH, ace; vs eth, bz
9227	Propylcyclopentane		C₈H₁₆	2040-96-2	112.213	liq	-117.3	131	0.7763[20]	1.4266[20]	i H₂O; msc EtOH, eth, ace; s bz; vs ctc
9228	1-Propylcyclopentanol		C₈H₁₆O	1604-02-0	128.212	liq	-37.5	173.5	0.9040[25]	1.4502[25]	
9229	Propylene carbonate	4-Methyl-1,3-dioxolan-2-one	C₄H₆O₃	108-32-7	102.089	liq	-48.8	242	1.2047[20]	1.4189[20]	vs H₂O, EtOH, eth, ace, bz
9230	1,2-Propylene glycol	1,2-Propanediol	C₃H₈O₂	57-55-6	76.095	liq	-60	187.6	1.0361[20]	1.4324[20]	msc H₂O, EtOH; s eth, bz, chl
9231	1,3-Propylene glycol	Trimethylene glycol	C₃H₈O₂	504-63-2	76.095	liq	-27.7	214.4	1.0538[20]	1.4398[20]	msc H₂O, EtOH; vs eth; sl bz
9232	1,2-Propylene glycol 2-tert-butyl ether	2-(1,1-Dimethylethoxy)-1-propanol	C₇H₁₆O₂	94023-15-1	132.201	liq		152	0.87		
9233	1,2-Propylene glycol dinitrate		C₃H₆N₂O₆	6423-43-4	166.089	liq	exp	92[10]			
9234	1,2-Propylene glycol monomethyl ether	1-Methoxy-2-propanol	C₄H₁₀O₂	107-98-2	90.121			119	0.9620[20]	1.4034[20]	
9235	1,2-Propylene glycol monomethyl ether acetate	2-Acetoxy-1-methoxypropane	C₆H₁₂O₃	108-65-6	132.157	liq		147			
9236	Propyleneimine	2-Methylaziridine	C₃H₇N	75-55-8	57.095			67	0.812[16]		
9237	Propyl formate		C₄H₈O₂	110-74-7	88.106	liq	-92.9	80.9	0.9073[20]	1.377[20]	sl H₂O; ctc; msc EtOH, eth
9238	Propyl 3-(2-furyl)acrylate		C₁₀H₁₂O₃	623-22-3	180.200			113[16], 92[3]	1.0744[20]	1.5392[24]	vs bz, eth, EtOH
9239	4-Propylheptane		C₁₀H₂₂	3178-29-8	142.282			157.5	0.7321[25]	1.4135[20]	
9240	Propyl hexanoate		C₉H₁₈O₂	626-77-7	158.238	liq	-68.7	187	0.8672[20]	1.4170[20]	vs eth, EtOH
9241	Propyl 2-hydroxybenzoate		C₁₀H₁₂O₃	607-90-9	180.200		97	239	1.0979[20]	1.5161[20]	s ctc, CS₂
9242	Propyl 4-hydroxybenzoate	Propylparaben	C₁₀H₁₂O₃	94-13-3	180.200	pr (eth)	97		1.0630[102]	1.5050[102]	i H₂O; s EtOH, eth; sl chl
9243	Propyliodone		C₁₀H₁₁I₂NO₃	587-61-1	447.008		186				
9244	Propyl isobutanoate		C₇H₁₄O₂	644-49-5	130.185			135.4	0.8843[0]	1.3955[20]	sl H₂O; s EtOH, ace; vs eth
9245	Propyl isocyanate	1-Isocyanatopropane	C₄H₇NO	110-78-1	85.105			83.5	0.908[25]	1.3970[20]	
9246	Propyl isothiocyanate	1-Isothiocyanatopropane	C₄H₇NS	628-30-8	101.171			153	0.9781[16]	1.5085[16]	sl H₂O; msc EtOH, eth
9247	Propyl methacrylate		C₇H₁₂O₂	2210-28-8	128.169			141	0.9022[20]	1.4190[20]	i H₂O; msc EtOH, eth
9248	Propyl 3-methylbutanoate	Propyl isopentanoate	C₈H₁₆O₂	557-00-6	144.212			155.9	0.8617[20]	1.4031[20]	vs eth, EtOH
9249	1-Propylnaphthalene		C₁₃H₁₄	2765-18-6	170.250	liq	-8.6	274.5	0.9897[20]	1.5923[20]	
9250	Propyl nitrate		C₃H₇NO₃	627-13-4	105.093			110	1.0538[20]	1.3973[20]	sl H₂O; s EtOH, eth, ctc
9251	Propyl nitrite		C₃H₇NO₂	543-67-9	89.094	liq		48	0.886[20]	1.3604[20]	sl H₂O; s EtOH, eth

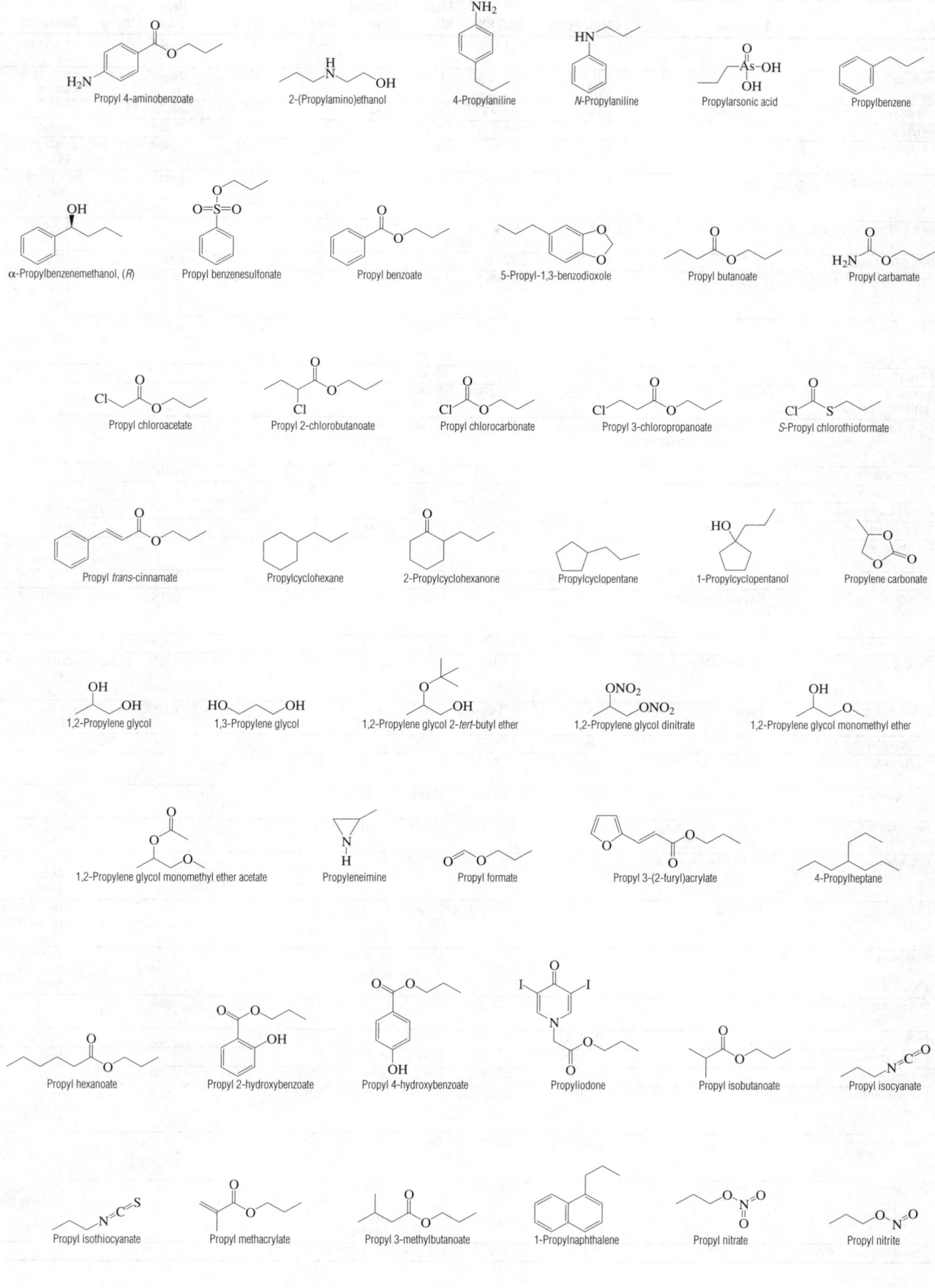

Propyl 4-aminobenzoate

2-(Propylamino)ethanol

4-Propylaniline

N-Propylaniline

Propylarsonic acid

Propylbenzene

α-Propylbenzenemethanol, (*R*)

Propyl benzenesulfonate

Propyl benzoate

5-Propyl-1,3-benzodioxole

Propyl butanoate

Propyl carbamate

Propyl chloroacetate

Propyl 2-chlorobutanoate

Propyl chlorocarbonate

Propyl 3-chloropropanoate

S-Propyl chlorothioformate

Propyl *trans*-cinnamate

Propylcyclohexane

2-Propylcyclohexanone

Propylcyclopentane

1-Propylcyclopentanol

Propylene carbonate

1,2-Propylene glycol

1,3-Propylene glycol

1,2-Propylene glycol 2-*tert*-butyl ether

1,2-Propylene glycol dinitrate

1,2-Propylene glycol monomethyl ether

1,2-Propylene glycol monomethyl ether acetate

Propyleneimine

Propyl formate

Propyl 3-(2-furyl)acrylate

4-Propylheptane

Propyl hexanoate

Propyl 2-hydroxybenzoate

Propyl 4-hydroxybenzoate

Propyliodone

Propyl isobutanoate

Propyl isocyanate

Propyl isothiocyanate

Propyl methacrylate

Propyl 3-methylbutanoate

1-Propylnaphthalene

Propyl nitrate

Propyl nitrite

No.	Name	Synonym	Mol. Form.	CAS RN	Mol. Wt.	Physical Form	mp/°C	bp/°C	den/ g cm⁻³	n_D	Solubility
9252	Propyl octanoate		$C_{11}H_{22}O_2$	624-13-5	186.292	liq	-46.2	226.4	0.8659[20]	1.4191[25]	vs ace, eth, EtOH
9253	Propyl pentanoate		$C_8H_{16}O_2$	141-06-0	144.212	liq	-70.7	167.5	0.8699[20]	1.4065[20]	i H₂O; s EtOH, eth, chl
9254	2-Propylpentanoic acid	Valproic acid	$C_8H_{16}O_2$	99-66-1	144.212	col liq		221; 120[14]	0.904[25]	1.425[25]	sl H₂O
9255	2-Propylphenol		$C_9H_{12}O$	644-35-9	136.190		7	220	1.015[20]		vs eth, EtOH
9256	4-Propylphenol		$C_9H_{12}O$	645-56-7	136.190		22	232.6	1.009[20]	1.5379[25]	sl H₂O, ctc; s EtOH
9257	2-Propylpiperidine, (S)	Coniine	$C_8H_{17}N$	458-88-8	127.228	liq	-1.0	166.5	0.8440[20]	1.4512[22]	sl H₂O, chl; msc EtOH; vs eth; s bz
9258	trans-6-Propyl-3-piperidinol, (3S)	Pseudoconhydrine	$C_8H_{17}NO$	140-55-6	143.227	hyg nd (eth)	106	236			vs H₂O, eth, EtOH
9259	N-Propylpropanamide		$C_6H_{13}NO$	3217-86-5	115.173		154	215; 108[9]	0.8985[25]		sl H₂O, eth
9260	Propyl propanoate	Propyl propionate	$C_6H_{12}O_2$	106-36-5	116.158	liq	-75.9	122.5	0.8809[20]	1.3935[20]	sl H₂O, ctc; msc EtOH, eth; s ace
9261	2-Propylpyridine		$C_8H_{11}N$	622-39-9	121.180		1.0	167	0.9119[20]	1.4925[20]	sl H₂O; msc EtOH, eth; vs ace
9262	4-Propylpyridine		$C_8H_{11}N$	1122-81-2	121.180			185	0.9381[15]	1.4966[20]	vs eth, EtOH
9263	2-Propyl-4-pyridinecarbothioamide	Protionamide	$C_9H_{12}N_2S$	14222-60-7	180.269		136.7				
9264	Propyl Red	Benzoic acid, 2-[[4-(dipropylamino)phenyl]azo]-	$C_{19}H_{23}N_3O_2$	2641-01-2	325.405	viol-bl or purp red cry (al)					s EtOH, KOH
9265	(Propylthio)benzene		$C_9H_{12}S$	874-79-3	152.256	liq	-45	220	0.9995[20]	1.5571[20]	
9266	Propyl 4-toluenesulfonate		$C_{10}H_{14}O_3S$	599-91-7	214.281		<-20	189[9]	1.144[20]	1.4998[20]	
9267	Propyl trichloroacetate		$C_5H_7Cl_3O_2$	13391-92-1	205.468			187	1.3221[20]	1.4501[20]	vs eth, EtOH
9268	Propyl 3,4,5-trihydroxybenzoate	Propyl gallate	$C_{10}H_{12}O_5$	121-79-9	212.199	nd (w)	130				sl H₂O
9269	Propylurea		$C_4H_{10}N_2O$	627-06-5	102.134	pr (al)	108.5				sl H₂O, DMSO; s EtOH
9270	Propyl vinyl ether	1-(Ethenyloxy)propane	$C_5H_{10}O$	764-47-6	86.132			65	0.7674[20]	1.3908[20]	
9271	2-Propynal	Propargyl aldehyde	C_3H_2O	624-67-9	54.047			60	0.9152[20]	1.4033[25]	msc H₂O; s EtOH, eth, ace, bz, tol
9272	2-Propyn-1-amine		C_3H_5N	2450-71-7	55.079			83	0.803[25]	1.4480[20]	
9273	Propyne	Methylacetylene	C_3H_4	74-99-7	40.064	col gas	-102.7	-23.2	0.607[25] (p>1 atm)	1.3863[-40]	sl H₂O; vs EtOH; s bz, chl
9274	2-Propynoic acid	Propiolic acid	$C_3H_2O_2$	471-25-0	70.047	cry (CS₂)	9	dec 144; 72[50]	1.1380[20]	1.4306[20]	vs H₂O, eth, EtOH, chl
9275	1-Propynylbenzene		C_9H_8	673-32-5	116.160			183	0.942[15]	1.563[15]	
9276	Propyzamide	N-(1,1-Dimethyl-2-propynyl)-3,5-dichlorobenzamide	$C_{12}H_{11}Cl_2NO$	23950-58-5	256.127		155				
9277	Prostaglandin E₁	11,15-Dihydroxy-9-oxo-13-prostenoic acid	$C_{20}H_{34}O_5$	745-65-3	354.481	cry (EtOAc)	115				s H₂O
9278	Prostaglandin E₂	11,15-Dihydroxy-9-oxo-5,13-prostadienoic acid	$C_{20}H_{32}O_5$	363-24-6	352.465	col cry	67				s H₂O, thf
9279	Prostaglandin F₂α	9,11,15-Trihydroxyprosta-5,13-dienoic acid	$C_{20}H_{34}O_5$	551-11-1	354.481	oil or solid	≈30				sl H₂O; s EtOH, MeOH, chl, AcOEt
9280	Protopine	Fumarine	$C_{20}H_{19}NO_5$	130-86-9	353.369	mcl pr (al-chl)	208				i H₂O; sl EtOH, eth, bz, peth; s chl
9281	Protoverine		$C_{27}H_{43}NO_9$	76-45-9	525.632	nd (MeOH)	221				i H₂O; s EtOH, bz, aq acid, MeOH
9282	Protriptyline hydrochloride	Triptil	$C_{19}H_{22}ClN$	1225-55-4	299.838	cry (2-PrOH/ eth)	170				
9283	Prunetin		$C_{16}H_{12}O_5$	552-59-0	284.263		239.5				
9284	Pseudoaconitine		$C_{36}H_{51}NO_{12}$	127-29-7	689.790	tcl (MeOH)	214				vs eth, EtOH
9285	Pseudocodeine		$C_{18}H_{21}NO_3$	466-96-6	299.365	wh nd	181.5		1.290[80]	1.574	
9286	Pseudojervine		$C_{33}H_{49}NO_8$	36069-05-3	587.744	wh nd or hex cry	304 dec				i H₂O, eth, bz, chl, tol, peth; s EtOH
9287	Pseudomorphine		$C_{34}H_{36}N_2O_6$	125-24-6	568.659	cry (aq NH₃, + 3 w)	282.5				i H₂O, EtOH, eth, chl, sulf; s py, NH₃
9288	Pseudotropine	8-Methyl-8-azabicyclo[3.2.1] octan-3-ol, exo	$C_8H_{15}NO$	135-97-7	141.211	orth cry (eth), orth bipym (peth-bz)	109	241			vs H₂O, EtOH; sl eth; s bz, chl
9289	Psoralen		$C_{11}H_6O_3$	66-97-7	186.164	nd (w, EtOH)	171				

Propyl octanoate

Propyl pentanoate

2-Propylpentanoic acid

2-Propylphenol

4-Propylphenol

2-Propylpiperidine, (S)

trans-6-Propyl-3-piperidinol, (3S)

N-Propylpropanamide

Propyl propanoate

2-Propylpyridine

4-Propylpyridine

2-Propyl-4-pyridinecarbothioamide

Propyl Red

(Propylthio)benzene

Propyl 4-toluenesulfonate

Propyl trichloroacetate

Propyl 3,4,5-trihydroxybenzoate

Propylurea

Propyl vinyl ether

2-Propynal

2-Propyn-1-amine

Propyne

2-Propynoic acid

1-Propynylbenzene

Propyzamide

Prostaglandin E$_1$

Prostaglandin E$_2$

Prostaglandin F$_{2\alpha}$

Protopine

Protoverine

Protriptyline hydrochloride

Prunetin

Pseudoaconitine

Pseudocodeine

Pseudojervine

Pseudomorphine

Pseudotropine

Psoralen

No.	Name	Synonym	Mol. Form.	CAS RN	Mol. Wt.	Physical Form	mp/°C	bp/°C	den/ g cm⁻³	n_D	Solubility
9290	Pteridine	Pyrazino[2,3-d]pyrimidine	C₆H₄N₄	91-18-9	132.123	ye pl (bz, sub)	139.5	sub 125			vs H₂O; s EtOH; sl eth, bz
9291	2,4(1H,3H)-Pteridinedione	Lumazine	C₆H₄N₄O₂	487-21-8	164.122	ye-oran nd (w)	348.5				vs HOAc
9292	Pulegone		C₁₀H₁₆O	89-82-7	152.233			224	0.9346⁴⁵	1.4894²⁰	i H₂O; msc EtOH, eth, chl; s ctc
9293	1H-Purine	6H-Imidazo[4,5-d]pyrimidine	C₅H₄N₄	120-73-0	120.113		216.5				vs H₂O, EtOH; sl eth, chl; s ace
9294	1H-Purine-2,6-diamine	2,6-Diaminopurine	C₅H₆N₆	1904-98-9	150.142	cry (dil al)	302				
9295	Pyocyanine		C₁₃H₁₀N₂O	85-66-5	210.230	dk bl nd (w + 1) (chl- peth)	133 dec				sl H₂O, bz; s EtOH, ace; i eth; vs chl
9296	4H-Pyran	1,4-Pyran	C₅H₆O	289-65-6	82.101	unstab oil		80		1.4559²⁰	s EtOH, eth, bz
9297	2H-Pyran-2-one		C₅H₄O₂	504-31-4	96.085		8.5	207.5	1.200²⁰	1.5270²⁵	msc H₂O; vs ace
9298	4H-Pyran-4-one		C₅H₄O₂	108-97-4	96.085		32.5	212.5	1.190²⁵	1.5238	vs H₂O, chl, eth; s EtOH, bz; sl CS₂
9299	Pyrantel		C₁₁H₁₄N₂S	15686-83-6	206.307	cry (MeOH)	178				
9300	4H-Pyran-4-thione		C₅H₄OS	1120-93-0	112.150		49				s H₂O
9301	8,16-Pyranthrenedione		C₃₀H₁₄O₂	128-70-1	406.431	red-ye or red-br nd (PhNO₂)	dec	sub			
9302	Pyrazine	1,4-Diazine	C₄H₄N₂	290-37-9	80.088	pr (w)	51.0	115	1.0311⁶¹	1.4953⁶¹	s H₂O, EtOH, eth, ace; sl ctc
9303	Pyrazinecarboxamide	Pyrazinamide	C₅H₅N₃O	98-96-4	123.113	wh nd (w, al)	192	sub			s H₂O, EtOH
9304	Pyrazinecarboxylic acid	Pyrazinoic acid	C₅H₄N₂O₂	98-97-5	124.098	wh nd (w)	225 dec	sub			
9305	2,3-Pyrazinedicarboxylic acid	2,3-Dicarboxypyrazine	C₆H₄N₂O₄	89-01-0	168.107	pr (w+2)	193 dec				vs H₂O; sl EtOH, eth, bz; s ace, MeOH
9306	1H-Pyrazole	1,2-Diazole	C₃H₄N₂	288-13-1	68.077	nd or pr (lig)	70.7	187		1.4203	s H₂O, EtOH, eth, bz; sl chl
9307	1-Pyrenamine		C₁₆H₁₁N	1606-67-3	217.265	ye nd (hx) lf (dil al)	117.5				s EtOH, ace, hx, acid; sl chl
9308	Pyrene	Benzo[def]phenanthrene	C₁₆H₁₀	129-00-0	202.250	pa ye pl (to, sub)	150.62	404	1.271²³		i H₂O; s EtOH, eth, bz, tol; sl ctc
9309	Pyrethrin I		C₂₁H₂₈O₃	121-21-1	328.445	visc liq		170⁰·¹ dec	1.5192¹⁸	1.5192¹⁸	i H₂O; s EtOH, eth, ctc, peth
9310	Pyrethrin II		C₂₂H₂₈O₅	121-29-9	372.454	visc liq		200⁰·¹ dec		1.5258²⁰	i H₂O; s EtOH, eth, ctc, peth
9311	Pyridate		C₁₉H₂₃ClN₂O₂S	55512-33-9	378.916	br oil	27	220⁰·¹	1.555²⁰	1.568²⁰	i H₂O
9312	Pyridazine	1,2-Diazabenzene	C₄H₄N₂	289-80-5	80.088	liq	-8	208	1.1035²³	1.5218²⁰	msc H₂O, EtOH; vs eth, ace, bz; i peth
9313	2-Pyridinamine	2-Aminopyridine	C₅H₆N₂	504-29-0	94.115	lf (lig)	57.5	105²⁰			s EtOH, eth, ace, bz; sl chl
9314	3-Pyridinamine	3-Aminopyridine	C₅H₆N₂	462-08-8	94.115	lf (bz-lig)	64.5	252			s H₂O, EtOH, eth; sl lig
9315	4-Pyridinamine	4-Aminopyridine	C₅H₆N₂	504-24-5	94.115	nd (bz)	158.5	273			s H₂O, eth, bz; vs EtOH; sl lig
9316	Pyridine	Azine	C₅H₅N	110-86-1	79.101	liq	-41.70	115.23	0.9819²⁰	1.5095²⁰	msc H₂O, EtOH, eth, ace, bz, chl
9317	2-Pyridinecarbonitrile		C₆H₄N₂	100-70-9	104.109	nd or pr (eth)	29	224.5	1.0810²⁵	1.5242²⁵	s H₂O, chl; vs EtOH, eth, bz; sl ctc
9318	3-Pyridinecarbonitrile		C₆H₄N₂	100-54-9	104.109	nd (lig), peth-eth)	51	206.9; 170³⁰⁰	1.1590²⁵		vs H₂O, EtOH, eth, bz; s chl; sl lig
9319	4-Pyridinecarbonitrile		C₆H₄N₂	100-48-1	104.109	nd(lig-eth)	83	186			s H₂O, EtOH, eth, bz, chl; sl lig
9320	3-Pyridinecarbothioamide		C₆H₆N₂S	4621-66-3	138.190		192				
9321	4-Pyridinecarbothioamide		C₆H₆N₂S	2196-13-6	138.190		198 dec				
9322	2-Pyridinecarboxaldehyde		C₆H₅NO	1121-60-4	107.110			180; 62¹³	1.1181²⁵	1.5389¹⁸	s H₂O, EtOH, eth, AcOEt; sl ctc
9323	3-Pyridinecarboxaldehyde	Nicotinaldehyde	C₆H₅NO	500-22-1	107.110			92²³	1.1394²⁵		s H₂O, EtOH, ace, chl; sl eth, peth
9324	4-Pyridinecarboxaldehyde		C₆H₅NO	872-85-5	107.110			77¹²		1.5423²⁰	s H₂O, eth, ctc
9325	2-Pyridinecarboxaldehyde oxime		C₆H₆N₂O	873-69-8	122.124		112.5				
9326	2-Pyridinecarboxamide		C₆H₆N₂O	1452-77-3	122.124	mcl pr (w)	108.3				sl H₂O, chl; s EtOH, bz

Pteridine

2,4(1H,3H)-Pteridinedione

Pulegone

1H-Purine

1H-Purine-2,6-diamine

Pyocyanine

4H-Pyran

2H-Pyran-2-one

4H-Pyran-4-one

Pyrantel

4H-Pyran-4-thione

8,16-Pyranthrenedione

Pyrazine

Pyrazinecarboxamide

Pyrazinecarboxylic acid

2,3-Pyrazinedicarboxylic acid

1H-Pyrazole

1-Pyrenamine

Pyrene

Pyrethrin I

Pyrethrin II

Pyridate

Pyridazine

2-Pyridinamine

3-Pyridinamine

4-Pyridinamine

Pyridine

2-Pyridinecarbonitrile

3-Pyridinecarbonitrile

4-Pyridinecarbonitrile

3-Pyridinecarbothioamide

4-Pyridinecarbothioamide

2-Pyridinecarboxaldehyde

3-Pyridinecarboxaldehyde

4-Pyridinecarboxaldehyde

2-Pyridinecarboxaldehyde oxime

2-Pyridinecarboxamide

No.	Name	Synonym	Mol. Form.	CAS RN	Mol. Wt.	Physical Form	mp/°C	bp/°C	den/g cm⁻³	n_D	Solubility
9327	3-Pyridinecarboxamide	Niacinamide	$C_6H_6N_2O$	98-92-0	122.124	wh pw, nd (bz)	130	157[0.0005]	1.400[25]	1.466	vs H_2O, EtOH, glycerol; sl chl
9328	4-Pyridinecarboxamide		$C_6H_6N_2O$	1453-82-3	122.124		157.5				
9329	2-Pyridinecarboxylic acid	Picolinic acid	$C_6H_5NO_2$	98-98-6	123.110	nd (w, al, bz)	136.5	sub			sl H_2O, bz; s EtOH; i eth, chl, CS_2
9330	3-Pyridinecarboxylic acid	Nicotinic acid	$C_6H_5NO_2$	59-67-6	123.110	nd (al, w)	236.6	sub	1.473[25]		sl H_2O, EtOH, eth
9331	4-Pyridinecarboxylic acid	Isonicotinic acid	$C_6H_5NO_2$	55-22-1	123.110	nd(w)	315	sub 260			sl H_2O, EtOH, eth, bz
9332	3-Pyridinecarboxylic acid 1-oxide	Oxiniacid acid	$C_6H_5NO_3$	2398-81-4	139.109	nd	254 dec				vs H_2O, MeOH
9333	4-Pyridinecarboxylic acid 1-oxide		$C_6H_5NO_3$	13602-12-5	139.109		273 dec				
9334	2,3-Pyridinediamine		$C_5H_7N_3$	452-58-4	109.130	lf or pl (dil al)	120.8	149[5]			s H_2O, EtOH, bz
9335	2,5-Pyridinediamine	2,5-Diaminopyridine	$C_5H_7N_3$	4318-76-7	109.130	nd	110.3	182[12]			vs H_2O, EtOH
9336	2,6-Pyridinediamine		$C_5H_7N_3$	141-86-6	109.130		121.5	285; 148[5]			sl H_2O, ace
9337	3,4-Pyridinediamine		$C_5H_7N_3$	54-96-6	109.130	nd or lf	219.3				
9338	2,3-Pyridinedicarboxylic acid	Quinolinic acid	$C_7H_5NO_4$	89-00-9	167.120	mcl pr (w)	228.5				sl H_2O, tfa; i EtOH, eth, bz
9339	2,4-Pyridinedicarboxylic acid	Lutidinic acid	$C_7H_5NO_4$	499-80-9	167.120	lf (w+1)	249		0.942[25]		sl H_2O; s EtOH; i eth, bz, CS_2
9340	2,5-Pyridinedicarboxylic acid	Isocinchomeronic acid	$C_7H_5NO_4$	100-26-5	167.120	lf or pr (dil HCl)	254				s H_2O, HCl; sl EtOH; i eth, bz
9341	2,6-Pyridinedicarboxylic acid	Dipicolinic acid	$C_7H_5NO_4$	499-83-2	167.120	nd (w+3/2)	252				sl H_2O, EtOH, HOAc
9342	3,4-Pyridinedicarboxylic acid	Cinchomeronic acid	$C_7H_5NO_4$	490-11-9	167.120	cry (w)	256	sub			sl H_2O, EtOH, bz; i eth, i chl
9343	3,5-Pyridinedicarboxylic acid	Dinicotinic acid	$C_7H_5NO_4$	499-81-0	167.120	cry (w)	324	sub			i H_2O; sl eth, HOAc; s DMSO, HCl
9344	2,3-Pyridinedicarboxylic acid anhydride	Furo[3,4-b]pyridine-5,7-dione	$C_7H_3NO_3$	699-98-9	149.104		138				
9345	2-Pyridineethanamine		$C_7H_{10}N_2$	2706-56-1	122.167			213; 131[50]	1.0220[25]	1.5335[25]	
9346	4-Pyridineethanamine		$C_7H_{10}N_2$	13258-63-4	122.167			121[10]	1.0302[25]	1.5381[25]	vs H_2O
9347	2-Pyridineethanol		C_7H_9NO	103-74-2	123.152		-7.8	190[200], 170[100]	1.091[25]	1.5366[20]	vs H_2O, EtOH, chl; sl eth
9348	Pyridine hydrochloride		C_5H_6ClN	628-13-7	115.562	hyg pl or sc (al)	146	222			vs H_2O, EtOH, chl
9349	2-Pyridinemethanamine		$C_6H_8N_2$	3731-51-9	108.141			203; 91[17]	1.0525[25]	1.5431[25]	vs H_2O
9350	3-Pyridinemethanamine		$C_6H_8N_2$	3731-52-0	108.141	liq	-21.1	226	1.064[20]	1.552[20]	vs H_2O, eth, EtOH
9351	4-Pyridinemethanamine		$C_6H_8N_2$	3731-53-1	108.141	liq	-7.6	230; 103[11]	1.072[20]	1.5495[25]	vs H_2O
9352	2-Pyridinemethanol		C_6H_7NO	586-98-1	109.126			112[16], 102.5[8]	1.1317[20]	1.5444[20]	msc H_2O; vs EtOH, eth, ace, bz
9353	3-Pyridinemethanol	Nicotinyl alcohol	C_6H_7NO	100-55-0	109.126	liq	-6.5	266	1.131[20]	1.5455[20]	vs H_2O, eth
9354	4-Pyridinemethanol	4-Picolyl alcohol	C_6H_7NO	586-95-8	109.126		53	141[12]			s chl
9355	Pyridine-1-oxide	Pyridine N-oxide	C_5H_5NO	694-59-7	95.100		65.5	146[13]			
9356	2-Pyridinepropanol		$C_8H_{11}NO$	2859-68-9	137.179		34	260.2; 116[4]	1.060[25]	1.5298[20]	vs H_2O
9357	3-Pyridinepropanol		$C_8H_{11}NO$	2859-67-8	137.179			284; 130[3]	1.063[25]	1.5313[20]	vs H_2O
9358	3-Pyridinesulfonic acid	3-Pyridylsulfonic acid	$C_5H_5NO_3S$	636-73-7	159.164	orth	357 dec		1.713[25]		vs H_2O; sl EtOH; i eth
9359	2-Pyridinethiol, 1-oxide		C_5H_5NOS	1121-31-9	127.165		70.5				
9360	2(1H)-Pyridinethione		C_5H_5NS	2637-34-5	111.166		130.0				s H_2O, EtOH, bz, chl
9361	2-Pyridinol		C_5H_5NO	72762-00-6	95.100	nd (bz)	107.8		1.3910[20]		vs H_2O, bz, EtOH
9362	3-Pyridinol		C_5H_5NO	109-00-2	95.100	nd (bz)	129				s H_2O, EtOH; sl eth, chl
9363	4-Pyridinol		C_5H_5NO	626-64-2	95.100	pr or nd (w+1)	149.8	>350; 257[10]			s H_2O, EtOH; i eth, bz
9364	2(1H)-Pyridinone		C_5H_5NO	142-08-5	95.100	nd (bz)	107.8	280	1.3910[20]		s H_2O, EtOH, bz, chl; sl eth, DMSO
9365	2(1H)-Pyridinone hydrazone	2-Pyridinylhydrazine	$C_5H_7N_3$	4930-98-7	109.130		46.6	185[140], 90[1]			s chl
9366	α-[(2-Pyridinylamino)methyl] benzenemethanol	Phenyramidol	$C_{13}H_{14}N_2O$	553-69-5	214.262	cry (dil MeOH)	83.5				
9367	1-(2-Pyridinyl)ethanone		C_7H_7NO	1122-62-9	121.137	ye in air		192	1.077[25]	1.5203[20]	s EtOH, eth, HOAc; sl ctc
9368	1-(3-Pyridinyl)ethanone	Methyl pyridyl ketone	C_7H_7NO	350-03-8	121.137		13.5	220		1.5341[20]	s H_2O, EtOH, eth, acid
9369	1-(4-Pyridinyl)ethanone		C_7H_7NO	1122-54-9	121.137		16	212	1.097[25]	1.5282[25]	sl EtOH, eth, acid

3-Pyridinecarboxamide

4-Pyridinecarboxamide

2-Pyridinecarboxylic acid

3-Pyridinecarboxylic acid

4-Pyridinecarboxylic acid

3-Pyridinecarboxylic acid 1-oxide

4-Pyridinecarboxylic acid 1-oxide

2,3-Pyridinediamine

2,5-Pyridinediamine

2,6-Pyridinediamine

3,4-Pyridinediamine

2,3-Pyridinedicarboxylic acid

2,4-Pyridinedicarboxylic acid

2,5-Pyridinedicarboxylic acid

2,6-Pyridinedicarboxylic acid

3,4-Pyridinedicarboxylic acid

3,5-Pyridinedicarboxylic acid

2,3-Pyridinedicarboxylic acid anhydride

2-Pyridineethanamine

4-Pyridineethanamine

2-Pyridineethanol

Pyridine hydrochloride

2-Pyridinemethanamine

3-Pyridinemethanamine

4-Pyridinemethanamine

2-Pyridinemethanol

3-Pyridinemethanol

4-Pyridinemethanol

Pyridine-1-oxide

2-Pyridinepropanol

3-Pyridinepropanol

3-Pyridinesulfonic acid

2-Pyridinethiol, 1-oxide

2(1H)-Pyridinethione

2-Pyridinol

3-Pyridinol

4-Pyridinol

2(1H)-Pyridinone

2(1H)-Pyridinone hydrazone

α-[(2-Pyridinylamino)methyl]benzenemethanol

1-(2-Pyridinyl)ethanone

1-(3-Pyridinyl)ethanone

1-(4-Pyridinyl)ethanone

No.	Name	Synonym	Mol. Form.	CAS RN	Mol. Wt.	Physical Form	mp/°C	bp/°C	den/ g cm^{-3}	n_D	Solubility
9370	N-(2-Pyridinylmethyl)-2-pyridinemethanamine		C$_{12}$H$_{13}$N$_3$	1539-42-0	199.251			200[10], 139[1]	1.1074[25]	1.5757[25]	
9371	N-2-Pyridinyl-2-pyridinamine		C$_{10}$H$_9$N$_3$	1202-34-2	171.198		90.5	307.5			sl H$_2$O, chl; vs EtOH, eth, ace, bz
9372	Pyridoxal hydrochloride	Vitamin B6	C$_8$H$_{10}$ClNO$_3$	65-22-5	203.623	orth	165 dec				vs H$_2$O; sl EtOH
9373	Pyridoxal 5-phosphate	Pyridoxal 5-(dihydrogen phosphate)	C$_8$H$_{10}$NO$_6$P	54-47-7	247.142	wh-ye pow or cry	141				
9374	Pyridoxamine	4-(Aminomethyl)-5-hydroxy-6-methyl-3-pyridinemethanol	C$_8$H$_{12}$N$_2$O$_2$	85-87-0	168.193	cry	198				s EtOH, acid
9375	Pyridoxamine dihydrochloride		C$_8$H$_{14}$Cl$_2$N$_2$O$_2$	524-36-7	241.115	pl (al)	226 dec				vs H$_2$O; sl EtOH
9376	Pyridoxine hydrochloride	5-Hydroxy-6-methyl-3,4-pyridinedimethanol hydrochloride	C$_8$H$_{12}$ClNO$_3$	58-56-0	205.639	pl (al, ace)	207	sub			vs H$_2$O
9377	1-(2-Pyridylazo)-2-naphthol	PAN	C$_{15}$H$_{11}$N$_3$O	85-85-8	249.267	red-br cry	130				i H$_2$O; s EtOH, eth, chl
9378	4-(2'-Pyridylazo)resorcinol	PAR	C$_{11}$H$_9$N$_3$O$_2$	1141-59-9	215.208	red-br cry	187 dec				
9379	Pyrilamine		C$_{17}$H$_{23}$N$_3$O	91-84-9	285.384			201[5]			
9380	2-Pyrimidinamine		C$_4$H$_5$N$_3$	109-12-6	95.103	nd (AcOEt)	127.5	sub			s H$_2$O; sl chl
9381	4-Pyrimidinamine		C$_4$H$_5$N$_3$	591-54-8	95.103	pl (AcOEt)	151.5				vs H$_2$O, EtOH
9382	Pyrimidine	1,3-Diazine	C$_4$H$_4$N$_2$	289-95-2	80.088		22	123.8		1.4998[20]	msc H$_2$O; s EtOH
9383	2,4,5,6(1H,3H)-Pyrimidinetetrone	Alloxan	C$_4$H$_2$N$_2$O$_4$	50-71-5	142.070		256 dec	sub			vs H$_2$O; s EtOH, ace, bz, HOAc
9384	2,4,5,6(1H,3H)-Pyrimidinetetrone 5-oxime	Violuric acid	C$_4$H$_3$N$_3$O$_4$	87-39-8	157.085	pa ye orth	203 dec				sl H$_2$O; s EtOH
9385	2,4,6-Pyrimidinetriamine		C$_4$H$_7$N$_5$	1004-38-2	125.133		248 dec				
9386	Pyriminil		C$_{13}$H$_{12}$N$_4$O$_3$	53558-25-1	272.259	solid	224 dec				
9387	Pyrithione zinc		C$_{10}$H$_8$N$_2$O$_2$S$_2$Zn	13463-41-7	317.722	wh solid	262				s chl, DMSO, DMF
9388	Pyrocatechol	1,2-Benzenediol	C$_6$H$_6$O$_2$	120-80-9	110.111	cry	104.6	245	1.344[20]	1.604[25]	vs H$_2$O, bz, eth, EtOH
9389	L-Pyroglutamic acid	5-Oxo-L-proline	C$_5$H$_7$NO$_3$	98-79-3	129.115		162				s DMSO
9390	Pyrolan		C$_{13}$H$_{15}$N$_3$O$_2$	87-47-8	245.277		50	161[0.2]			s ctc, CS$_2$
9391	Pyrrobutamine	1-[4-(4-Chlorophenyl)-3-phenyl-2-butenyl]pyrrolidine	C$_{20}$H$_{22}$ClN	91-82-7	311.849	cry	49	192[0.3]			
9392	Pyrrole	Imidole	C$_4$H$_5$N	109-97-7	67.090	liq	-23.39	129.79	0.9698[20]	1.5085[20]	sl H$_2$O; s EtOH, eth, ace, bz, chl
9393	1H-Pyrrole-2-carboxaldehyde		C$_5$H$_5$NO	1003-29-8	95.100	orth pr (peth)	46.5	218		1.5939[16]	sl chl, lig
9394	1H-Pyrrole-2-carboxylic acid		C$_5$H$_5$NO$_2$	634-97-9	111.100	lf (w)	208 dec				s H$_2$O, EtOH, eth
9395	1H-Pyrrole-3-carboxylic acid	3-Pyrrolecarboxylic acid	C$_5$H$_5$NO$_2$	931-03-3	111.100	nd (lig)	161.5				
9396	1H-Pyrrole-2,5-dione		C$_4$H$_3$NO$_2$	541-59-3	97.073	pl (bz)	94	sub	1.2493[106]		s H$_2$O, EtOH, eth
9397	Pyrrolidine	Azacyclopentane	C$_4$H$_9$N	123-75-1	71.121	col liq	-57.79	86.56	0.8586[20]	1.4431[20]	msc H$_2$O; s EtOH, eth; sl bz, chl
9398	1-Pyrrolidineethanamine		C$_6$H$_{14}$N$_2$	7154-73-6	114.188			166; 68[23]	0.901[25]	1.4687[20]	
9399	1-Pyrrolidineethanol		C$_6$H$_{13}$NO	2955-88-6	115.173			187; 80[13]	0.9750[20]	1.4713[20]	
9400	1-[4-(1-Pyrrolidinyl)-2-butynyl]-2-pyrrolidinone	Oxotremorine	C$_{12}$H$_{18}$N$_2$O	70-22-4	206.283	pa ye liq		124[0.1]	0.991[25]	1.5160[20]	
9401	3-(2-Pyrrolidinyl)pyridine, (S)	Nornicotine	C$_9$H$_{12}$N$_2$	494-97-3	148.204	hyg		270	1.0737[19]	1.5378[18]	vs H$_2$O, ace, eth, EtOH
9402	2-Pyrrolidone	γ-Butyrolactam	C$_4$H$_7$NO	616-45-5	85.105	cry (peth)	25	251; 133[12]	1.120[20]	1.4806[30]	vs H$_2$O, EtOH, eth, bz, chl, CS$_2$
9403	1-(1H-Pyrrol-2-yl)ethanone		C$_6$H$_7$NO	1072-83-9	109.126	mcl nd (w)	90	220			s H$_2$O, EtOH, eth
9404	Pyruvic acid		C$_3$H$_4$O$_3$	127-17-3	88.062		13.8	dec 165; 54[10]	1.2272[20]	1.4280[20]	msc H$_2$O, EtOH, eth; s ace
9405	Pyrvinium chloride		C$_{26}$H$_{28}$ClN$_3$	548-84-5	417.973	red pow (w)	250 dec				
9406	1,1':4',1":4",1'''-Quaterphenyl		C$_{24}$H$_{18}$	135-70-6	306.400		320	428[18]			i H$_2$O, EtOH, eth, chl; s bz, PhNO$_2$, HOAc
9407	Quercetin		C$_{15}$H$_{10}$O$_7$	117-39-5	302.236	ye nd (dil al, + 2 w)	316.5	sub			sl H$_2$O, eth, MeOH; s EtOH, ace, py
9408	Quercitrin	Quercetin-3-L-rhamnoside	C$_{21}$H$_{20}$O$_{11}$	522-12-3	448.377	pa ye nd or pl (+2w, dil al)	170				i H$_2$O, eth; s EtOH, HOAc, MeOH, alk
9409	Quillaic acid		C$_{30}$H$_{46}$O$_5$	631-01-6	486.683	nd (dil al)	294				vs ace, eth, py, EtOH
9410	Quinacrine	Mepacrine	C$_{23}$H$_{30}$ClN$_3$O	83-89-6	399.956	ye oil	87				
9411	Quinaldine Red		C$_{21}$H$_{23}$IN$_2$	117-92-0	430.325	dk red pow					s H$_2$O; vs EtOH
9412	Quinamine		C$_{19}$H$_{24}$N$_2$O$_2$	464-85-7	312.406	pr (bz), nd (80% al)	185.5				i H$_2$O; vs EtOH, bz; s eth, ace

N-(2-Pyridinylmethyl)-2-pyridinemethanamine

N-2-Pyridinyl-2-pyridinamine

Pyridoxal hydrochloride

Pyridoxal 5-phosphate

Pyridoxamine

Pyridoxamine dihydrochloride

Pyridoxine hydrochloride

1-(2-Pyridylazo)-2-naphthol

4-(2'-Pyridylazo)resorcinol

Pyrilamine

2-Pyrimidinamine

4-Pyrimidinamine

Pyrimidine

2,4,5,6(1H,3H)-Pyrimidinetetrone

2,4,5,6(1H,3H)-Pyrimidinetetrone 5-oxime

2,4,6-Pyrimidinetriamine

Pyriminil

Pyrithione zinc

Pyrocatechol

L-Pyroglutamic acid

Pyrolan

Pyrrobutamine

Pyrrole

1H-Pyrrole-2-carboxaldehyde

1H-Pyrrole-2-carboxylic acid

1H-Pyrrole-3-carboxylic acid

1H-Pyrrole-2,5-dione

Pyrrolidine

1-Pyrrolidineethanamine

1-Pyrrolidineethanol

1-[4-(1-Pyrrolidinyl)-2-butynyl]-2-pyrrolidinone

3-(2-Pyrrolidinyl)pyridine, (S)

2-Pyrrolidone

1-(1H-Pyrrol-2-yl)ethanone

Pyruvic acid

Pyrvinium chloride

1,1':4',1":4",1'''-Quaterphenyl

Quercetin

Quercitrin

Quillaic acid

Quinacrine

Quinaldine Red

Quinamine

No.	Name	Synonym	Mol. Form.	CAS RN	Mol. Wt.	Physical Form	mp/°C	bp/°C	den/ g cm⁻³	n_D	Solubility
9413	Quinazoline	1,3-Benzodiazine	$C_8H_6N_2$	253-82-7	130.147	ye pl (peth)	48	241			vs H_2O; s EtOH, eth, ace, bz; sl chl
9414	Quinclorac	3,7-Dichloroquinoline-8-carboxylic acid	$C_{10}H_5Cl_2NO_2$	84087-01-4	242.059		274		1.75		
9415	Quinethazone		$C_{10}H_{12}ClN_3O_3S$	73-49-4	289.738						s tfa
9416	Quinic acid		$C_7H_{12}O_6$	77-95-2	192.166		162.5		1.64[25]		vs H_2O, EtOH, HOAc
9417	Quinidine		$C_{20}H_{24}N_2O_2$	56-54-2	324.417	cry (+2.5w, dil al)	174				sl H_2O, eth; s EtOH, bz; vs chl; i peth
9418	Quinine	6'-Methoxycinchonan-9-ol, (8α,9R)	$C_{20}H_{24}N_2O_2$	130-95-0	324.417		57			1.625[15]	sl H_2O, ace; vs EtOH, py; s eth, chl
9419	Quinine hydrochloride	6'-Methoxycinchonan-9-ol monohydrochloride, (8α,9R)	$C_{20}H_{25}ClN_2O_2$	130-89-2	360.878	silky efflor nd (w)	159				vs H_2O, EtOH, chl
9420	Quinine sulfate		$C_{40}H_{50}N_4O_8S$	804-63-7	746.912	silky nd (w)	235.2				vs EtOH
9421	Quininone		$C_{20}H_{22}N_2O_2$	84-31-1	322.401	nd, lf (eth)	108				vs bz, eth, EtOH
9422	2-Quinolinamine	2-Aminoquinoline	$C_9H_8N_2$	580-22-3	144.173	lf (w)	131.5	sub			vs H_2O; s EtOH, eth, ace, chl; sl bz
9423	3-Quinolinamine	3-Aminoquinoline	$C_9H_8N_2$	580-17-6	144.173	orth (w, dil al)	94				vs eth, EtOH, chl
9424	4-Quinolinamine	4-Aminoquinoline	$C_9H_8N_2$	578-68-7	144.173	nd (bz, dil al)	154.8	180[12]			s H_2O, bz, chl; vs EtOH, eth
9425	5-Quinolinamine	5-Aminoquinoline	$C_9H_8N_2$	611-34-7	144.173	ye nd (al) lf (eth)	110	310; 184[10]			sl H_2O; vs EtOH, eth; s bz; i lig
9426	6-Quinolinamine	6-Aminoquinoline	$C_9H_8N_2$	580-15-4	144.173	cry (w+2), pr (eth)	114	187[12]			sl H_2O, eth; s NH_3, EtOH
9427	8-Quinolinamine	8-Aminoquinoline	$C_9H_8N_2$	578-66-5	144.173	pa ye nd (sub) cry (al, lig)	70	157[19]			vs H_2O, EtOH
9428	Quinoline	1-Azanaphthalene	C_9H_7N	91-22-5	129.159	liq	-14.78	237.16	1.0977[15]	1.6268[20]	sl H_2O; msc EtOH, eth, ace, bz, CS_2; s ctc
9429	4-Quinolinecarboxaldehyde	Cinchoninaldehyde	$C_{10}H_7NO$	4363-93-3	157.169	nd (to-peth)	51	122[4]			vs eth, tol
9430	2-Quinolinecarboxylic acid	Quinaldic acid	$C_{10}H_7NO_2$	93-10-7	173.169		156				s H_2O; vs bz
9431	8-Quinolinecarboxylic acid	8-Carboxyquinoline	$C_{10}H_7NO_2$	86-59-9	173.169	nd (w)	187	sub			vs EtOH
9432	2(1H)-Quinolinethione		C_9H_7NS	2637-37-8	161.224		187				i H_2O; vs EtOH, eth, bz; sl DMSO
9433	2-Quinolinol	2-Hydroxyquinoline	C_9H_7NO	59-31-4	145.158	pr (MeOH)	199.5	sub			sl H_2O, DMSO; vs EtOH, eth; s dil HCl
9434	3-Quinolinol	3-Hydroxyquinoline	C_9H_7NO	580-18-7	145.158	cry (bz, dil al)	201.3				i H_2O; s EtOH; sl eth, chl; vs bz
9435	4-Quinolinol	4-Hydroxyquinoline	C_9H_7NO	611-36-9	145.158	nd (w+3)	210				vs H_2O, EtOH; sl eth, bz, peth
9436	5-Quinolinol	5-Hydroxyquinoline	C_9H_7NO	578-67-6	145.158	nd (al), pl	226 dec	sub			s H_2O, bz, chl; sl EtOH; vs MeOH; i lig
9437	6-Quinolinol	6-Hydroxyquinoline	C_9H_7NO	580-16-5	145.158	pr (al, eth)	195	360			i H_2O, bz, chl; sl EtOH, eth; s alk
9438	7-Quinolinol	7-Hydroxyquinoline	C_9H_7NO	580-20-1	145.158	pr (al), nd (dil al-eth)	239	sub			vs EtOH
9439	8-Quinolinol	8-Hydroxyquinoline	C_9H_7NO	148-24-3	145.158	nd (dil al)	75.5	267	1.034[20]		i H_2O, eth; vs EtOH, bz, chl; s ace
9440	8-Quinolinol benzoate	Benzoxiquine	$C_{16}H_{11}NO_2$	86-75-9	249.264						sl chl
9441	8-Quinolinol sulfate (2:1)	8-Hydroxyquinoline sulfate	$C_{18}H_{16}N_2O_6S$	134-31-6	388.934		177.5				vs H_2O; s EtOH; i eth
9442	Quinovic acid		$C_{30}H_{46}O_5$	465-74-7	486.683	pl or nd	298 dec				
9443	Quinovose		$C_6H_{12}O_5$	7658-08-4	164.156	cry (AcOEt)	139.5				vs H_2O, EtOH
9444	Quinoxaline	1,4-Benzodiazine	$C_8H_6N_2$	91-19-0	130.147	cry (peth)	28	229.5	1.1334[48]	1.6231[48]	s H_2O; msc EtOH, eth, ace, bz; sl chl
9445	2(1H)-Quinoxalinone		$C_8H_6N_2O$	1196-57-2	146.146	lf (al)	271	sub 200			
9446	Quizalofop-Ethyl		$C_{19}H_{17}ClN_2O_4$	76578-14-8	372.802	wh cry	93	220[0.2]			i H_2O; s bz, EtOH, ace, xyl
9447	Radicinin		$C_{12}H_{12}O_5$	10088-95-6	236.220		221.5				sl chl
9448	Raffinose		$C_{18}H_{32}O_{16}$	512-69-6	504.437		80		1.465[25]		s H_2O, py; vs MeOH; sl EtOH; i eth
9449	Ranitidine		$C_{13}H_{22}N_4O_3S$	66357-35-5	314.404	solid	69.5				

Quinazoline

Quinclorac

Quinethazone

Quinic acid

Quinidine

Quinine

Quinine hydrochloride

Quinine sulfate

Quininone

2-Quinolinamine

3-Quinolinamine

4-Quinolinamine

5-Quinolinamine

6-Quinolinamine

8-Quinolinamine

Quinoline

4-Quinolinecarboxaldehyde

2-Quinolinecarboxylic acid

8-Quinolinecarboxylic acid

2(1H)-Quinolinethione

2-Quinolinol

3-Quinolinol

4-Quinolinol

5-Quinolinol

6-Quinolinol

7-Quinolinol

8-Quinolinol

8-Quinolinol benzoate

8-Quinolinol sulfate (2:1)

Quinovic acid

Quinovose

Quinoxaline

2(1H)-Quinoxalinone

Quizalofop-Ethyl

Radicinin

Raffinose

Ranitidine

No.	Name	Synonym	Mol. Form.	CAS RN	Mol. Wt.	Physical Form	mp/°C	bp/°C	den/ g cm⁻³	n_D	Solubility
9450	Raubasine		$C_{21}H_{24}N_2O_3$	483-04-5	352.427		258 dec				i H_2O; s MeOH
9451	Raunescine		$C_{31}H_{36}N_2O_8$	117-73-7	564.626		165				i H_2O; s EtOH, chl, HOAc
9452	Reinecke salt		$C_4H_{12}CrN_7OS_4$	13573-16-5	354.440	red cry (w)	270 dec				s H_2O, EtOH, ace; i bz
9453	Resazurin	7-Hydroxy-3H-phenoxazin-3-one, 10-oxide	$C_{12}H_7NO_4$	550-82-3	229.189	dk red to gr pr or pl (HOAc)		sub			i H_2O, eth; sl EtOH, HOAc; s alk
9454	Rescinnamine		$C_{35}H_{42}N_2O_9$	24815-24-5	634.716	nd (bz)	238.5				i H_2O; sl EtOH; s ace, chl, AcOEt
9455	Reserpic acid		$C_{22}H_{28}N_2O_5$	83-60-3	400.467	cry (MeOH)	242				
9456	Reserpine		$C_{33}H_{40}N_2O_9$	50-55-5	608.679	lo pr (dil ace)	264.5				sl H_2O, eth, ace; s EtOH, bz, AcOEt
9457	cis-Resmethrin, (-)		$C_{22}H_{26}O_3$	10453-86-8	338.439		75				
9458	Resorcinol	1,3-Benzenediol	$C_6H_6O_2$	108-46-3	110.111	nd (bz), pl (w)	109.4	276.5; 178[16]	1.278[20]	1.578[25]	vs H_2O, ctc; s EtOH, eth; sl bz, chl
9459	11-cis-Retinal	Vitamin A_1 aldehyde	$C_{20}H_{28}O$	564-87-4	284.435	cry					
9460	Retinal (all trans)		$C_{20}H_{28}O$	116-31-4	284.435	oran cry	64				i H_2O; s EtOH, chl, cy, peth
9461	13-cis-Retinoic acid	Accutane	$C_{20}H_{28}O_2$	4759-48-2	300.435	cry (EtOH)	189				
9462	13-trans-Retinoic acid		$C_{20}H_{28}O_2$	302-79-4	300.435	cry (MeOH)	181.5				
9463	Retinol	Vitamin A	$C_{20}H_{30}O$	68-26-8	286.451	ye pr (peth)	63.5	137[0.000001]			i H_2O; s EtOH, eth, ace, bz
9464	Retinyl palmitate	Retinol, hexadecanoate	$C_{36}H_{60}O_2$	79-81-2	524.860		28				
9465	Retronecine, (+)		$C_8H_{13}NO_2$	480-85-3	155.195	cry (ace)	121				s H_2O, EtOH; sl eth
9466	Retrorsine		$C_{18}H_{25}NO_6$	480-54-6	351.395	cry (AcOEt)	212				sl H_2O, ace; s EtOH, chl; i eth
9467	Rhamnetin		$C_{16}H_{12}O_7$	90-19-7	316.262	ye nd (al)	295				sl H_2O; s EtOH, ace, PhOH; vs dil alk
9468	DL-α-Rhamnose		$C_6H_{12}O_5$	116908-82-8	164.156	cry (w)	151				vs H_2O, EtOH
9469	D-Rhamnose	6-Deoxy-D-mannose	$C_6H_{12}O_5$	634-74-2	164.156						s H_2O
9470	Rheadine		$C_{21}H_{21}NO_6$	2718-25-4	383.395	nd (chl, eth, al)	257	sub			
9471	Rhein		$C_{15}H_8O_6$	478-43-3	284.221	ye or oran nd (MeOH, py)	321	sub			sl H_2O, EtOH, eth, ace, bz; vs py
9472	Rhenium carbonyl	Dirhenium decacarbonyl	$C_{10}O_{10}Re_2$	14285-68-8	652.515	ye-wh cry	170 dec		2.87		s os
9473	Rhizopterin		$C_{15}H_{12}N_6O_4$	119-20-0	340.294	lt ye pl (w)	>300				i H_2O, EtOH, eth; s aq alk, aq NH_3, py
9474	Rhodamine B		$C_{28}H_{32}ClN_2O_3$	81-88-9	480.018		165				s H_2O, EtOH, eth, bz, xyl
9475	Rhodium carbonyl chloride	Dirhodium tetracarbonyl dichloride	$C_4Cl_2O_4Rh_2$	14523-22-9	388.758	red-oran cry	124				s os
9476	Ribavirin	Tribavirin	$C_8H_{12}N_4O_5$	36791-04-5	244.205	col cry (EtOH)	175				s H_2O
9477	Ribitol	Adonitol	$C_5H_{12}O_5$	488-81-3	152.146	pr (w), nd (al)	104				s H_2O, EtOH; i eth, lig
9478	Riboflavin		$C_{17}H_{20}N_4O_6$	83-88-5	376.364	ye or oran-ye nd (w)	280 dec				i H_2O, eth, ace, chl; sl EtOH
9479	Riboflavin-5'-phosphate		$C_{17}H_{20}N_4O_9P$	146-17-8	455.336	ye cry (w)					
9480	D-Ribose		$C_5H_{10}O_5$	50-69-1	150.130	pl (al)	88				s H_2O; sl EtOH
9481	L-Ribose		$C_5H_{10}O_5$	24259-59-4	150.130		81				
9482	D-Ribulose	erythro-2-Pentulose	$C_5H_{10}O_5$	488-84-6	150.130	syrup					vs H_2O
9483	Ricinine	1,2-Dihydro-4-methoxy-1-methyl-2-oxo-3-pyridinecarbonitrile	$C_8H_8N_2O_2$	524-40-3	164.162	pr or lf (w, al)	201.5	sub 170			s H_2O, chl; sl EtOH, bz; vs py; i peth
9484	Rifabutin		$C_{46}H_{62}N_4O_{11}$	72559-06-9	847.004	viol-red cry					i H_2O; vs chl; s MeOH; sl EtOH
9485	Rifampin		$C_{43}H_{58}N_4O_{12}$	13292-46-1	822.941	red-oran pl (ace)	185 dec				
9486	Rinderine	Echinatine-3'-epimer	$C_{15}H_{25}NO_5$	6029-84-1	299.364	cry (ace)	100.5				
9487	Ronnel		$C_8H_8Cl_3O_3PS$	299-84-3	321.546		41	152[0.4]	1.44[32]	1.5335[35]	
9488	Rotenone		$C_{23}H_{22}O_6$	83-79-4	394.417	nd or lf (al, aq-ace)	176	215[0.5]			i H_2O; s EtOH, ace, bz; sl eth; vs chl
9489	Rubijervine		$C_{27}H_{43}NO_2$	79-58-3	413.636	nd (+1w, dil al)	242				vs bz, EtOH, chl
9490	Rubratoxin B		$C_{26}H_{30}O_{11}$	21794-01-4	518.509	cry (MeCN)	169 dec				

Raubasine

Raunescine

Reinecke salt

$NH_4^{\oplus} \cdot H_2O$

Resazurin

Rescinnamine

Reserpic acid

Reserpine

cis-Resmethrin, (-)

Resorcinol

11-cis-Retinal

Retinal (all trans)

13-cis-Retinoic acid

13-trans-Retinoic acid

Retinol

Retinyl palmitate

Retronecine, (+)

Retrorsine

Rhamnetin

DL-α-Rhamnose

D-Rhamnose

Rheadine

Rhein

Rhenium carbonyl

Rhizopterin

Rhodamine B

Rhodium carbonyl chloride

Ribavirin

Ribitol

Riboflavin

Riboflavin-5'-phosphate

D-Ribose

L-Ribose

D-Ribulose

Ricinine

Rifabutin

Rifampin

Rinderine

Ronnel

Rotenone

Rubijervine

Rubratoxin B

No.	Name	Synonym	Mol. Form.	CAS RN	Mol. Wt.	Physical Form	mp/°C	bp/°C	den/ g cm^{-3}	n_D	Solubility
9491	Rutecarpine		$C_{18}H_{13}N_3O$	84-26-4	287.315	ye nd (al, AcOEt)	259.5				sl EtOH, ace, bz
9492	Ruthenium dodecacarbonyl	Triruthenium dodecacarbonyl	$C_{12}O_{12}Ru_3$	15243-33-1	639.33	oran cry	dec 150				
9493	Ruthenium(III) 2,4-pentanedioate	Ruthenium(III) acetylacetonate	$C_{15}H_{21}O_6Ru$	14284-93-6	398.39		230				
9494	Rutinose		$C_{12}H_{22}O_{10}$	90-74-4	326.297	hyg pow (al, eth)	190 dec				vs H_2O, EtOH
9495	Sabadine		$C_{29}H_{47}NO_8$	124-80-1	537.685	nd (eth)	258				vs ace, EtOH
9496	Saccharin		$C_7H_5NO_3S$	81-07-2	183.185	nd (ace) pr (al), lf (w)	228 dec	sub	0.828^{25}		sl H_2O, bz, eth, chl; s ace, EtOH
9497	Saccharin sodium	1,2-Benzisothiazolin-3-one, 1,1-dioxide, sodium salt	$C_7H_4NNaO_3S$	128-44-9	205.168	wh cry	229				s H_2O
9498	Safranal	2,6,6-Trimethyl-1,3-cyclohexadiene-1-carboxaldehyde	$C_{10}H_{14}O$	116-26-7	150.217			70^1	0.9734^{19}	1.5281^{19}	vs EtOH, peth
9499	Safrole	5-(2-Propenyl)-1,3-benzodioxole	$C_{10}H_{10}O_2$	94-59-7	162.185	mcl	11.2	234.5	1.1000^{20}	1.5381^{20}	i H_2O; vs EtOH; msc eth, chl
9500	Salcomine	N,N'-Bis(salicylidene)ethylenediaminocobalt(II)	$C_{16}H_{14}CoN_2O_2$	14167-18-1	325.227	red cry (DMF)					s bz, chl, py
9501	Salicylaldehyde	2-Hydroxybenzaldehyde	$C_7H_6O_2$	90-02-8	122.122	liq	-7	197	1.1674^{20}	1.5740^{20}	sl H_2O, chl; msc EtOH; vs ace, bz
9502	Salicylaldoxime		$C_7H_7NO_2$	94-67-7	137.137		57				sl H_2O; vs EtOH, eth, bz; s chl; i lig
9503	Salsoline		$C_{11}H_{15}NO_2$	89-31-6	193.243	pow or cry (al)	221.5				sl H_2O, EtOH; i eth, peth; s chl, alk
9504	Salvarsan dihydrochloride	Arsphenamine	$C_{12}H_{14}As_2Cl_2N_2O_2$	139-93-5	439.001	ye hyg pow	190 dec				vs H_2O
9505	Sanguinarine		$C_{20}H_{15}NO_5$	2447-54-3	349.337	cry (eth, al)	266				vs ace, bz, eth, EtOH
9506	α-Santalol		$C_{15}H_{24}O$	115-71-9	220.351			301.5	0.9679^{20}	1.5023^{20}	i H_2O; s EtOH
9507	β-Santalol		$C_{15}H_{24}O$	77-42-9	220.351			167^{10}	0.9750^{20}	1.5115^{20}	
9508	Santonic acid		$C_{15}H_{20}O_4$	510-35-0	264.318	cry	171	285^{15}			sl H_2O; s chl, eth, HOAc, EtOH
9509	α-Santonin		$C_{15}H_{18}O_3$	481-06-1	246.302	orth (w, eth)	175		1.590^{25}		sl H_2O, EtOH, eth; s bz, chl; i peth
9510	Sarcosine	N-Methylglycine	$C_3H_7NO_2$	107-97-1	89.094	cry (al)	212 dec				s H_2O
9511	Sarmentogenin		$C_{23}H_{34}O_5$	76-28-8	390.513	pr (95% al, MeOH-eth)	280				i H_2O, eth, bz; s EtOH; sl ace, chl
9512	Sarpagan-17-al	Vellosimine	$C_{19}H_{20}N_2O$	6874-98-2	292.374	cry (MeOH)	305.5	sub 180			
9513	Sarpagan-10,17-diol	Sarpagine	$C_{19}H_{22}N_2O_2$	482-68-8	310.390	nd	320				i H_2O; s EtOH
9514	Saxitoxin dihydrochloride		$C_{10}H_{19}Cl_2N_7O_4$	35554-08-6	372.209	hyg wh solid					vs H_2O, MeOH, EtOH
9515	Scarlet red		$C_{24}H_{20}N_4O$	85-83-6	380.442	dk br pow or nd	185; dec 260				i H_2O; sl ace, bz; vs chl, peth
9516	Schradan		$C_8H_{24}N_4O_3P_2$	152-16-9	286.250		17	154$^{2.0}$	1.09^{25}	1.462^{25}	vs H_2O, EtOH, chl
9517	Scilliroside		$C_{32}H_{44}O_{12}$	507-60-8	620.684	lo pr (dil MeOH)	169	dec			sl H_2O, ace, chl; vs EtOH, diox; i eth
9518	Scopolamine		$C_{17}H_{21}NO_4$	51-34-3	303.354	visc liq					vs hot H_2O, EtOH, ace; sl bz
9519	Scopoline		$C_8H_{13}NO_2$	487-27-4	155.195	hyg nd (lig, eth, chl, peth)	108.5	248	1.0891^{134}		s H_2O
9520	Sebacic acid		$C_{10}H_{18}O_4$	111-20-6	202.248	lf	130.9	295^{100}, 232^{10}	1.2705^{20}	1.422^{133}	sl H_2O; s EtOH, eth; i bz
9521	Selenium methionine	Selenomethionine	$C_5H_{11}NO_2Se$	1464-42-2	196.11	hex pl (MeOH aq)	265 dec				
9522	Selenoformaldehyde		CH_2Se	6596-50-5	92.99	unstab gas					
9523	Selenourea	Carbamimidoselenoic acid	CH_4N_2Se	630-10-4	123.02	pr or nd (w)	123.02	dec 200			vs H_2O
9524	Semicarbazide hydrochloride		CH_6ClN_3O	563-41-7	111.531	pr (dil al)	176 dec				vs H_2O
9525	Senecionine		$C_{18}H_{25}NO_5$	130-01-8	335.396	pl	232				i H_2O; sl EtOH, eth; s chl
9526	Seneciphylline		$C_{18}H_{23}NO_5$	480-81-9	333.380	pl (AcOEt)	217 dec				s chl; sl EtOH, ace; i eth
9527	Senkirkin		$C_{19}H_{27}NO_6$	2318-18-5	365.420	pl (ace)	197				
9528	L-Sepiapterin	6-Lactoyl-7,8-dihydropterin	$C_9H_{11}N_5O_3$	17094-01-8	237.215	ye pow or cry					
9529	DL-Serine		$C_3H_7NO_3$	302-84-1	105.093	mcl pr or lf (w)	246 dec		1.603^{22}		s H_2O; i EtOH, eth, bz, HOAc

Rutecarpine

Ruthenium dodecacarbonyl

Ruthenium(III) 2,4-pentanedioate

Rutinose

Sabadine

Saccharin

Saccharin sodium

Safranal

Safrole

Salcomine

Salicylaldehyde

Salicylaldoxime

Salsoline

Salvarsan dihydrochloride 2HCl

Sanguinarine

α-Santalol

β-Santalol

Santonic acid

α-Santonin

Sarcosine

Sarmentogenin

Sarpagan-17-al

Sarpagan-10,17-diol

Saxitoxin dihydrochloride 2HCl

Scarlet red

Schradan

Scilliroside

Scopolamine

Scopoline

Sebacic acid

Selenium methionine

Selenoformaldehyde

Selenourea

Semicarbazide hydrochloride HCl

Senecionine

Seneciphylline

Senkirkin

L-Sepiapterin

DL-Serine

No.	Name	Synonym	Mol. Form.	CAS RN	Mol. Wt.	Physical Form	mp/°C	bp/°C	den/ g cm⁻³	n_D	Solubility
9530	D-Serine		$C_3H_7NO_3$	312-84-5	105.093	nd or hex pr (w)	229 dec	dec			vs H_2O; i EtOH, eth, bz, HOAc
9531	L-Serine	2-Amino-3-hydroxypropanoic acid, (S)	$C_3H_7NO_3$	56-45-1	105.093	hex pl or pr (w)	228 dec	sub 150	1.6^{22}		s H_2O; i EtOH, eth, bz, HOAc
9532	Serpentine alkaloid		$C_{21}H_{20}N_2O_3$	18786-24-8	348.395		175				i H_2O; s EtOH, eth, ace
9533	Sesin	2,4-Dichlorophenoxyethyl benzoate	$C_{15}H_{12}Cl_2O_3$	94-83-7	311.160	cry	66	$185^{1.5}$			
9534	Sesone	Sodium 2-(2,4-dichlorophenoxy)ethyl sulfate	$C_8H_7Cl_2NaO_5S$	136-78-7	309.100		245 dec				
9535	Sethoxydim		$C_{17}H_{29}NO_3S$	74051-80-2	327.482			$>90^{0.00003}$	1.043^{25}		
9536	Shikimic acid		$C_7H_{10}O_5$	138-59-0	174.151	nd	184	subl			sl EtOH; i eth, bz, chl
9537	Siduron		$C_{14}H_{20}N_2O$	1982-49-6	232.321	cry solid	135				s EtOH, DMF, CH_2Cl_2
9538	Silvex	Propanoic acid, 2-(2,4,5-trichlorophenoxy)-	$C_9H_7Cl_3O_3$	93-72-1	269.509		181.6				
9539	Simazine	1,3,5-Triazine-2,4-diamine, 6-chloro-N,N'-diethyl-	$C_7H_{12}ClN_5$	122-34-9	201.657		226		1.302^{20}		
9540	Simfibrate		$C_{23}H_{26}Cl_2O_6$	14929-11-4	469.354	cry	52	$225^{0.15}$			
9541	Sinapinic acid	3-(4-Hydroxy-3,5-dimethoxyphenyl)-2-propenoic acid	$C_{11}H_{12}O_5$	530-59-6	224.210	wh pow					i H_2O; s MeOH, ace
9542	Sinomenine		$C_{19}H_{23}NO_4$	115-53-7	329.391	nd (bz)	162				sl H_2O, eth, bz; s EtOH, ace, dil alk
9543	α₁-Sitosterol	4-Methylstigmasta-7,24(28)-dien-3-ol, (3β,4α,5α,24Z)	$C_{30}H_{50}O$	474-40-8	426.717	nd (al)	166				vs EtOH, chl
9544	Sodium arsanilate	Sodium (4-aminophenyl) arsonate	$C_6H_7AsNNaO_3$	127-85-5	239.037	wh cry					s H_2O
9545	Sodium ascorbate		$C_6H_7NaO_6$	134-03-2	198.106	cry	218 dec				
9546	Sodium benzenesulfinate		$C_6H_5NaO_2S$	873-55-2	164.158	cry	300				
9547	Sodium benzenesulfonate	Monosodium benzenesulfonate	$C_6H_5NaO_3S$	515-42-4	180.157		>300				s H_2O; sl EtOH
9548	Sodium benzoate		$C_7H_5NaO_2$	532-32-1	144.104		>300				s H_2O
9549	Sodium cacodylate	Sodium dimethylarsonate	$C_2H_6AsNaO_2$	124-65-2	159.980	gran cry	60 (hyd)				vs H_2O; s EtOH
9550	Sodium 2,2-dichloropropanoate		$C_3H_3Cl_2NaO_2$	127-20-8	164.951	hyg pow	166 dec				
9551	Sodium diethyldithiocarbamate	Dithiocarb sodium	$C_5H_{10}NNaS_2$	148-18-5	171.260	cry (EtOH)	95				s H_2O, EtOH, MeOH, ace; i eth, bz
9552	Sodium diethyldithiocarbamate trihydrate	Diethyldithiocarbamate sodium salt trihydrate	$C_5H_{16}NNaO_3S_2$	20624-25-3	225.306	orth cry (ace)	95				vs H_2O; s EtOH, ace; i bz, eth
9553	Sodium 4,5-dihydroxy-2,7-naphthalenedisulfonic acid	Chromotropic acid disodium salt	$C_{10}H_6Na_2O_8S_2$	129-96-4	364.260	wh nd or lf (w)					vs H_2O
9554	Sodium dimethyldithiocarbamate		$C_3H_7NNaS_2$	128-04-1	144.215	col cry (w)	121 (hyd)				
9555	Sodium 4-dodecylbenzenesulfonate		$C_{18}H_{29}NaO_3S$	2211-98-5	348.476	cry	144				
9556	Sodium dodecyl sulfate	Sodium lauryl sulfate	$C_{12}H_{25}NaO_4S$	151-21-3	288.379	wh pow	205				
9557	Sodium ethanolate	Sodium ethoxide	C_2H_5NaO	141-52-6	68.050	hyg wh pow	260 dec				reac H_2O; s EtOH
9558	Sodium fluoroacetate		$C_2H_2FNaO_2$	62-74-8	100.024	wh mcl cry	200				i ace, chl; sl EtOH, MeOH
9559	Sodium formaldehyde bisulfite	Sodium hydroxymethanesulfonate	CH_3NaO_4S	870-72-4	134.088	cry (EtOH aq)					
9560	Sodium formaldehydesulfoxylate	Sodium hydroxymethanesulfinate	CH_3NaO_3S	149-44-0	118.088	cry (w)	63 (hyd)				s H_2O; i EtOH, bz, eth
9561	Sodium gluconate		$C_6H_{11}NaO_7$	527-07-1	218.137						s H_2O
9562	Sodium 2-hydroxyethanesulfonate	Monosodium 2-hydroxyethanesulfonate	$C_2H_5NaO_4S$	1562-00-1	148.114						s H_2O
9563	Sodium 2-hydroxy-2-propanesulfonate	Monosodium 2-hydroxy-2-propanesulfonate	$C_3H_7NaO_4S$	540-92-1	162.141	cry					s H_2O; sl EtOH
9564	Sodium iodomethanesulfonate	Methiodal sodium	CH_2INaO_3S	126-31-8	243.984	cry					sl EtOH, ace, bz
9565	Sodium O-isopropyl xanthate		$C_4H_7NaOS_2$	140-93-2	158.218	hyg wh-ye pow	150 dec				
9566	Sodium methanolate	Sodium methoxide	CH_3NaO	124-41-4	54.024	wh hyg tetr cry	300				reac H_2O; s MeOH, EtOH
9567	Sodium methylarsonate		CH_4AsNaO_3	2163-80-6	161.953	cry (w)	115				vs H_2O; s MeOH; i os
9568	Sodium methyldithiocarbamate	Metham sodium	$C_2H_4NNaS_2$	137-42-8	129.180	cry (w)					vs H_2O
9569	Sodium β-naphthoquinone-4-sulfonate	Sodium 3,4-dihydro-3,4-dioxo-1-naphthalenesulfonate	$C_{10}H_5NaO_5S$	521-24-4	260.199		287 dec				
9570	Sodium 2-oxopropanoate		$C_3H_3NaO_3$	113-24-6	110.044						s H_2O; sl abs EtOH
9571	Sodium phenolate	Sodium phenoxide	C_6H_5NaO	139-02-6	116.093	hyg cry	384				vs H_2O; s EtOH, thf

D-Serine

L-Serine

Serpentine alkaloid

Sesin

Sesone

Sethoxydim

Shikimic acid

Siduron

Silvex

Simazine

Simfibrate

Sinapinic acid

Sinomenine

α₁-Sitosterol

Sodium arsanilate

Sodium ascorbate

Sodium benzenesulfinate

Sodium benzenesulfonate

Sodium benzoate

Sodium cacodylate

Sodium 2,2-dichloropropanoate

Sodium diethyldithiocarbamate

Sodium diethyldithiocarbamate trihydrate

Sodium 4,5-dihydroxy-2,7-naphthalenedisulfonic acid

Sodium dimethyldithiocarbamate

Sodium 4-dodecylbenzenesulfonate

Sodium dodecyl sulfate

Sodium ethanolate

Sodium fluoroacetate

Sodium formaldehyde bisulfite

Sodium formaldehydesulfoxylate

Sodium gluconate

Sodium 2-hydroxyethanesulfonate

Sodium 2-hydroxy-2-propanesulfonate

Sodium iodomethanesulfonate

Sodium O-isopropyl xanthate

Sodium methanolate

Sodium methylarsonate

Sodium methyldithiocarbamate

Sodium β-naphthoquinone-4-sulfonate

Sodium 2-oxopropanoate

Sodium phenolate

No.	Name	Synonym	Mol. Form.	CAS RN	Mol. Wt.	Physical Form	mp/°C	bp/°C	den/ g cm⁻³	n_D	Solubility
9572	Sodium propanoate		$C_3H_5NaO_2$	137-40-6	96.061						sl H_2O
9573	Sodium sulfobromophthalein	Sulfobromophthalein sodium	$C_{20}H_8Br_4Na_2O_{10}S_2$	71-67-0	837.998	hyg cry					s H_2O; i EtOH, ace
9574	Sodium tartrate		$C_4H_4Na_2O_6$	868-18-8	194.051						s H_2O
9575	Sodium tartrate dihydrate		$C_4H_8Na_2O_8$	6106-24-7	230.082				1.545[25]		s H_2O; i EtOH
9576	Sodium tetraphenylborate		$C_{24}H_{20}BNa$	143-66-8	342.217	nd	300				s H_2O, EtOH, ace; sl eth, chl; i peth
9577	Sodium trichloroacetate		$C_2Cl_3NaO_2$	650-51-1	185.369	ye-wh pow	300				s H_2O, EtOH
9578	Sodium trifluoroacetate		$C_2F_3NaO_2$	2923-18-4	136.005	cry	207 dec				
9579	Solanid-5-ene-3,18-diol, (3β)	Isorubijervine	$C_{27}H_{43}NO_2$	468-45-1	413.636	pr(al)	242.5				vs bz, chl
9580	Solanine		$C_{45}H_{73}NO_{15}$	20562-02-1	868.060	nd (EtOH aq)	286 dec				i H_2O, eth, chl; s hot EtOH
9581	Solanone		$C_{13}H_{22}O$	1937-54-8	194.313			60[1]	0.870[20]	1.4755[20]	
9582	Soman		$C_7H_{16}FO_2P$	96-64-0	182.173	liq					
9583	Sophoricoside		$C_{21}H_{20}O_{10}$	152-95-4	432.378		274				
9584	Sorbitan oleate		$C_{24}H_{44}O_6$	1338-43-8	428.602	ye oil			0.986	1.4800[20]	i H_2O; s EtOH
9585	L-Sorbose	L-Sorbinose	$C_6H_{12}O_6$	87-79-6	180.155	orth (al)	165		1.612[17]		s H_2O; sl EtOH, eth, MeOH
9586	Sparteine		$C_{15}H_{26}N_2$	90-39-1	234.380		30.5	325; 173[8]	1.0196[20]	1.5312[20]	vs eth, EtOH, chl
9587	Spinulosin		$C_8H_8O_5$	85-23-4	184.147	red-bl	202.5	sub 120			sl H_2O; s alk
9588	Spironolactone		$C_{24}H_{32}O_4S$	52-01-7	416.574			134			
9589	Spiro[2.2]pentane		C_5H_8	157-40-4	68.118	liq	-107.0	39	0.7266[20]	1.4120[20]	
9590	Spirosolan-3-ol, (3β,5α,22β,25S)	Tomatidine	$C_{27}H_{45}NO_2$	77-59-8	415.652	pl	210.5				s EtOH, eth
9591	Spirosol-5-en-3-ol, (3β,22α,25R)	Solasodine	$C_{27}H_{43}NO_2$	126-17-0	413.636	hex pl (sub)	202				s EtOH, ace, bz, diox, py; sl eth; vs chl
9592	Spirostan-2,3-diol, (2α,3β,5α,25R)	Gitogenin	$C_{27}H_{44}O_4$	511-96-6	432.636	lf (bz), nd (eth)	271.5				i H_2O; s EtOH, chl; sl eth
9593	Spirostan-3-ol, (3β,5α,25R)	Tigogenin	$C_{27}H_{44}O_3$	77-60-1	416.636	lf (al +1w) pr (ace)	205.5				s EtOH, eth, ace, ctc, MeOH, peth
9594	Spirostan-3-ol, (3β,5β,25R)	Smilagenin	$C_{27}H_{44}O_3$	126-18-1	416.636	nd (ace)	185				vs ace, bz, EtOH
9595	Spirostan-3-ol, (3β,5β,25S)	Sarsasapogenin	$C_{27}H_{44}O_3$	126-19-2	416.636	lo pr, nd (ace)	200.5				s EtOH, ace, bz, chl
9596	Spirostan-2,3,15-triol, (2α,3β,5α,15β,25R)	Digitogenin	$C_{27}H_{44}O_5$	511-34-2	448.635	nd (al)	281.5				vs chl
9597	Spirost-5-en-3-ol, (3β,25R)	Diosgenin	$C_{27}H_{42}O_3$	512-04-9	414.620	cry (ace)	205.5				vs EtOH
9598	Spiro[5.5]undecane		$C_{11}H_{20}$	180-43-8	152.277			208	0.8783[20]	1.4731	
9599	S-Propyl thioacetate		$C_5H_{10}OS$	2307-10-0	118.197			137.9	0.9535[25]		
9600	Squalene		$C_{30}H_{50}$	111-02-4	410.718	oil	-4.8	421.3; 280[17]	0.8584[20]	1.4990[20]	i H_2O; sl EtOH; s eth, ace, ctc
9601	Stachydrine		$C_7H_{13}NO_2$	471-87-4	143.184	cry (w+1)	235				vs H_2O, EtOH
9602	Stanozolol		$C_{21}H_{32}N_2O$	10418-03-8	328.491	cry (EtOH)	≈236				
9603	Stearaldehyde		$C_{18}H_{36}O$	638-66-4	268.478	nd (peth)		261			
9604	Stearic acid	Octadecanoic acid	$C_{18}H_{36}O_2$	57-11-4	284.478	mcl lf (al)	69.3	dec 350; 232[15]	0.9408[20]	1.4299[80]	i H_2O; sl EtOH, bz; s ace, chl, CS_2
9605	Stearic acid anhydride	Octadecanoic anhydride	$C_{36}H_{70}O_3$	638-08-4	550.939		72		0.8365[82]	1.4362[80]	i H_2O, EtOH; sl eth, bz
9606	Sterigmatocystin		$C_{18}H_{12}O_6$	10048-13-2	324.284	ye nd	246 dec				
9607	Stigmasta-5,7-dien-3-ol, (3β)	7-Dehydrositosterol	$C_{29}H_{48}O$	521-04-0	412.690		144.5				vs bz, eth, EtOH
9608	Stigmasta-5,22-dien-3-ol, (3β,22E)	Stigmasterol	$C_{29}H_{48}O$	83-48-7	412.690		170				vs bz, eth, EtOH
9609	Stigmastan-3-ol, (3β,5α)		$C_{29}H_{52}O$	83-45-4	416.722		144				
9610	Stigmast-5-en-3-ol, (3β,24R)	β-Sitosterol	$C_{29}H_{50}O$	83-46-5	414.706	pl (al)	137				s EtOH, eth, HOAc
9611	Stigmast-5-en-3-ol, (3β,24S)	γ-Sitosterol	$C_{29}H_{50}O$	83-47-6	414.706	cry (EtOH)	148				s EtOH
9612	cis-Stilbene	cis-1,2-Diphenylethene	$C_{14}H_{12}$	645-49-8	180.245		-5	141[12]	1.0143[20]	1.6130[20]	i H_2O; s EtOH, eth, ace, bz, peth, chl
9613	trans-Stilbene	trans-1,2-Diphenylethene	$C_{14}H_{12}$	103-30-0	180.245	cry (al)	124.2	307; 166[12]	0.9707[20]	1.6264[17]	i H_2O; sl EtOH, chl; vs eth, bz
9614	Streptomycin	N-Methyl-L-glucosamidinostreptosidostreptidine	$C_{21}H_{39}N_7O_{12}$	57-92-1	581.575	hyg pow					s H_2O
9615	Streptomycin sulfate		$C_{42}H_{84}N_{14}O_{36}S_3$	3810-74-0	1457.383	pow	≈230 dec				
9616	Streptozotocin		$C_8H_{15}N_3O_7$	18883-66-4	265.221	pl	115 dec				s H_2O, EtOH
9617	Strophanthidin		$C_{23}H_{32}O_6$	66-28-4	404.496	orth tab (MeOH-w) lf (w+2)	173 dec				i H_2O, eth; s EtOH, ace, bz, HOAc, chl

Sodium propanoate

Sodium sulfobromophthalein

Sodium tartrate

Sodium tartrate dihydrate

2H₂O

Sodium tetraphenylborate

Sodium trichloroacetate

Sodium trifluoroacetate

Solanid-5-ene-3,18-diol, (3β)

Solanine

Solanone

Soman

Sophoricoside

Sorbitan oleate

L-Sorbose

Sparteine

Spinulosin

Spironolactone

Spiro[2.2]pentane

Spirosolan-3-ol, (3β,5α,22β,25S)

Spirosol-5-en-3-ol, (3β,22α,25R)

Spirostan-2,3-diol, (2α,3β,5α,25R)

Spirostan-3-ol, (3β,5α,25R)

Spirostan-3-ol, (3β,5β,25R)

Spirostan-3-ol, (3β,5β,25S)

Spirostan-2,3,15-triol, (2α,3β,5α,15β,25R)

Spirost-5-en-3-ol, (3β,25R)

Spiro[5.5]undecane

S-Propyl thioacetate

Squalene

Stachydrine

Stanozolol

Stearaldehyde

Stearic acid

Stearic acid anhydride

Sterigmatocystin

Stigmasta-5,7-dien-3-ol, (3β)

Stigmasta-5,22-dien-3-ol, (3β,22E)

Stigmastan-3-ol, (3β,5α)

Stigmast-5-en-3-ol, (3β,24R)

Stigmast-5-en-3-ol, (3β,24S)

cis-Stilbene

trans-Stilbene

Streptomycin

3H₂SO₄

Streptomycin sulfate

Streptozotocin

Strophanthidin

No.	Name	Synonym	Mol. Form.	CAS RN	Mol. Wt.	Physical Form	mp/°C	bp/°C	den/ g cm⁻³	n_D	Solubility
9618	Strychnidin-10-one mononitrate	Strychnine nitrate	$C_{21}H_{23}N_3O_5$	66-32-0	397.425	nd (w)	295		1.627^{25}		vs H_2O, MeOH; sl bz; s chl, EtOH
9619	Strychnidin-10-one sulfate (2:1)	Strychnine sulfate	$C_{42}H_{46}N_4O_8S$	60-41-3	766.901		200 dec				s H_2O, EtOH, MeOH; i eth; sl chl
9620	Strychnine		$C_{21}H_{22}N_2O_2$	57-24-9	334.412	orth pr (al)	287	270^5	1.36^{20}		sl H_2O, EtOH, ace, bz; i eth; s chl
9621	Styrene	Vinylbenzene	C_8H_8	100-42-5	104.150	liq	-30.65	145	0.9016^{25}	1.5440^{25}	i H_2O; s EtOH, eth, ace; msc bz; sl ctc
9622	Succimer	2,3-Dimercaptobutanedioic acid, (R*,S*)	$C_4H_6O_4S_2$	304-55-2	182.219	wh cry (MeOH)	193				
9623	Succinamide		$C_4H_8N_2O_2$	110-14-5	116.119	orth nd (w)	268 dec	sub 125			s H_2O
9624	Succinic acid		$C_4H_6O_4$	110-15-6	118.089	tcl or mcl pr	187.9	dec 235	1.572^{25}	1.450	sl H_2O, DMSO; s EtOH, eth, ace; i bz
9625	Succinic anhydride		$C_4H_4O_3$	108-30-5	100.073	nd (al), orth pym (chl)	119	261	1.2^{20}		i H_2O; s EtOH, chl; sl eth
9626	Succinimide		$C_4H_5NO_2$	123-56-8	99.089	pl (+1w, al) orth (ace)	126.5	dec 287	1.418^{25}		s H_2O; sl EtOH, eth, ace
9627	Succinonitrile	Butanedinitrile	$C_4H_4N_2$	110-61-2	80.088		57.98	266	0.9867^{60}	1.4173^{60}	vs H_2O; s EtOH, ace, bz, chl; sl eth
9628	Succinylcholine chloride	Suxamethonium chloride	$C_{14}H_{30}Cl_2N_2O_4$	71-27-2	361.305	cry (w)	190				sl EtOH, bz, chl; i eth
9629	Succinylsulphathiazole		$C_{13}H_{13}N_3O_5S_2$	116-43-8	355.389	cry	193.5				i H_2O, eth, chl; sl EtOH, ace; s alk
9630	Sucralfate		$C_{12}H_{54}Al_{16}O_{75}S_8$	54182-58-0	2086.737	wh amorp pow					i H_2O, EtOH, chl; s dil HCl; alk
9631	Sucrose		$C_{12}H_{22}O_{11}$	57-50-1	342.296	mcl	185.5		1.5805^{17}	1.5376	s H_2O, py; sl EtOH; i eth
9632	Sucrose monohexadecanoate	Sucrose palmitate	$C_{28}H_{52}O_{12}$	26446-38-8	580.706	cry	61				s H_2O
9633	Sucrose octaacetate		$C_{28}H_{38}O_{19}$	126-14-7	678.591	nd (al)	86.5	250^1	1.27^{16}	1.4660	sl H_2O; s EtOH, eth, ace, bz, chl
9634	Sufentanil		$C_{22}H_{30}N_2O_2S$	56030-54-7	386.550	cry (peth)	96.6				
9635	Sulfabenzamide	N-[(4-Aminophenyl)sulfonyl] benzamide	$C_{13}H_{12}N_2O_3S$	127-71-9	276.310	hex pr (60% al)	181.5				
9636	Sulfachlorpyridazine		$C_{10}H_9ClN_4O_2S$	80-32-0	284.722		187·				
9637	Sulfacytine		$C_{12}H_{14}N_4O_3S$	17784-12-2	294.329	cry (MeOH/BuOH)	167				i H_2O; s alk
9638	Sulfadimethoxine		$C_{12}H_{14}N_4O_4S$	122-11-2	310.329		203.5				
9639	Sulfaguanidine		$C_7H_{10}N_4O_2S$	57-67-0	214.245	nd (w)	191.5				
9640	Sulfallate	Carbamodithioic acid, diethyl-, 2-chloro-2-propenyl ester	$C_8H_{14}ClNS_2$	95-06-7	223.787			129^1	1.088		
9641	Sulfamerazine		$C_{11}H_{12}N_4O_2S$	127-79-7	264.304	cry	236				sl H_2O, EtOH, ace, DMSO; i eth, chl
9642	Sulfamethazine		$C_{12}H_{14}N_4O_2S$	57-68-1	278.330	pa ye (w+1/2) cry (diox-w)	198.5				s H_2O, acid, alk; sl DMSO
9643	Sulfamethiazole		$C_9H_{10}N_4O_2S_2$	144-82-1	270.331	cry (w)	210				sl hot H_2O
9644	Sulfamethoxazole		$C_{10}H_{11}N_3O_3S$	723-46-6	253.277	ye-wh pow	171				i eth
9645	Sulfamethoxypyridazine		$C_{11}H_{12}N_4O_3S$	80-35-3	280.303		182.5				
9646	Sulfamethylthiazole		$C_{10}H_{11}N_3O_2S_2$	515-59-3	269.343		237				vs EtOH
9647	N⁴-Sulfanilylsulfanilamide	4-Amino-N-[4-(aminosulfonyl)phenyl]benzenesulfonamide	$C_{12}H_{13}N_3O_4S_2$	547-52-4	327.379		137				sl H_2O; s EtOH, eth, ace; i chl, peth
9648	Sulfanilylurea		$C_7H_9N_3O_3S$	547-44-4	215.229	cry (w)	147 dec				
9649	Sulfaphenazole		$C_{15}H_{14}N_4O_2S$	526-08-9	314.363	cry (EtOH)	181				sl EtOH, MeOH, gl HOAc
9650	Sulfasalazine		$C_{18}H_{14}N_4O_5S$	599-79-1	398.393		220 dec				
9651	Sulfathiazole	4-Amino-N-2-thiazolylbenzenesulfonamide	$C_9H_9N_3O_2S_2$	72-14-0	255.316	br pl, rods or pow (45% al)	175(form a); 202(form b)				sl H_2O, EtOH, DMSO
9652	Sulfathiourea		$C_7H_9N_3O_2S_2$	515-49-1	231.295		182				i H_2O; sl EtOH
9653	Sulfinpyrazone		$C_{23}H_{20}N_2O_3S$	57-96-5	404.481		137				
9654	N-Sulfinylaniline		C_6H_5NOS	1122-83-4	139.175			200	1.236^{25}	1.6270^{20}	
9655	Sulfisoxazole		$C_{11}H_{13}N_3O_3S$	127-69-5	267.304		191				
9656	Sulfoacetic acid		$C_2H_4O_5S$	123-43-3	140.115	hyg tab (w+1)	85	dec 245			vs H_2O, ace, EtOH

Strychnidin-10-one mononitrate

Strychnidin-10-one sulfate (2:1)

Strychnine

Styrene

Succimer

Succinamide

Succinic acid

Succinic anhydride

Succinimide

Succinonitrile

Succinylcholine chloride

Succinylsulphathiazole

Sucralfate
R = SO₃[Al₂(OH)₅]

Sucrose

Sucrose octaacetate

Sufentanil

Sulfabenzamide

Sulfachlorpyridazine

Sulfacytine

Sulfadimethoxine

Sulfaguanidine

Sulfallate

Sulfamerazine

Sulfamethazine

Sulfamethiazole

Sulfamethoxazole

Sulfamethoxypyridazine

Sulfamethylthiazole

N-Sulfanilylsulfanilamide

Sulfanilylurea

Sulfaphenazole

Sulfasalazine

Sulfathiazole

Sulfathiourea

Sulfinpyrazone

N-Sulfinylaniline

Sulfisoxazole

Sulfoacetic acid

No.	Name	Synonym	Mol. Form.	CAS RN	Mol. Wt.	Physical Form	mp/°C	bp/°C	den/ g cm⁻³	n_D	Solubility
9657	2-Sulfobenzoic acid		C₇H₆O₅S	632-25-7	202.185	nd (w+3)	141				vs H₂O, EtOH
9658	Sulfolane	Tetrahydrothiophene, 1-1-dioxide	C₄H₈O₂S	126-33-0	120.171		27.6	287.3	1.2723¹⁸	1.4833¹⁸	s chl
9659	Sulfometuron methyl		C₁₅H₁₆N₄O₅S	74222-97-2	364.377	wh solid	202				
9660	Sulfonmethane	2,2-Bis(ethylsulfonyl)propane	C₇H₁₆O₄S₂	115-24-2	228.330	mcl (w), pr (al)	125.8	dec 300			vs bz, EtOH, chl
9661	Sulfonyldiacetic acid		C₄H₆O₆S	123-45-5	182.152		187				vs H₂O, EtOH; s eth, sulf
9662	4-Sulfophthalic acid	4-Sulfo-1,2-benzenedicarboxylic acid	C₈H₆O₇S	89-08-7	246.195	cry	139				
9663	Sulfotep		C₈H₂₀O₅P₂S₂	3689-24-5	322.320			137²	1.196²⁵	1.4753²⁵	i H₂O; s EtOH
9664	Sulfuryl chloride isocyanate		CClNO₃S	1189-71-5	141.534	liq	-44	107	1.626²⁵	1.4467²⁰	
9665	Sulphan Blue		C₂₇H₃₁N₂NaO₆S₂	129-17-9	566.664	viol pow					s EtOH
9666	Sulprofos		C₁₂H₁₉O₂PS₃	35400-43-2	322.447			156⁰·¹	1.20²⁰	1.5859	sl H₂O
9667	Sunset Yellow FCF	C.I. Food Yellow 3	C₁₆H₁₀N₂Na₂O₇S₂	2783-94-0	452.369	cry	>300				s H₂O; sl EtOH
9668	Suprasterol II		C₂₈H₄₄O	562-71-0	396.648	pr	110	190⁰·⁰⁰⁵			s MeOH
9669	Sutan	Carbamothioic acid, bis(2-methylpropyl)-, S-ethyl ester	C₁₁H₂₃NOS	2008-41-5	217.372			138²¹	0.9402²⁵		
9670	Symclosene	1,3,5-Trichloro-1,3,5-triazine-2,4,6(1H,3H,5H)-trione	C₃Cl₃N₃O₃	87-90-1	232.409		246.7 dec				
9671	Syringin		C₁₇H₂₄O₉	118-34-3	372.368	cry (w), nd (al)	192				vs EtOH
9672	Tabun	Dimethylphosphoroamidocyanidic acid, ethyl ester	C₅H₁₁N₂O₂P	77-81-6	162.127	liq	-50	240	1.077	1.4250²⁰	msc H₂O
9673	Tachysterol	9,10-Secoergosta-5(10),6,8,22-tetraen-3-ol, (3β,6E,22E)-	C₂₈H₄₄O	115-61-7	396.648						i H₂O, MeOH; s EtOH, eth, ace, bz
9674	D-Tagatose		C₆H₁₂O₆	87-81-0	180.155	cry (dil al)	134.5				vs H₂O
9675	Talbutal		C₁₁H₁₆N₂O₃	115-44-6	224.256	cry	109				i H₂O, peth; s EtOH, ace, eth, chl
9676	Tamoxifen		C₂₆H₂₉NO	10540-29-1	371.514	cry (peth)	97				
9677	Tannic acid	Tannin	C₇₆H₅₂O₄₆	1401-55-4	1701.198	ye-br amorp pow	≈210 dec				vs EtOH, ace; i bz, chl, eth, ctc
9678	DL-Tartaric acid	2,3-Dihydroxybutanedioic acid, (R*, R*)-(±)-	C₄H₆O₆	133-37-9	150.087	mcl pr (w, al +1w)	206		1.788²⁵		s H₂O, EtOH; sl eth; i bz
9679	meso-Tartaric acid		C₄H₆O₆	147-73-9	150.087	tcl pl (w)	147		1.666²⁰		vs H₂O, EtOH
9680	D-Tartaric acid	2,3-Dihydroxybutanedioic acid, [S-(R*,R*)]-	C₄H₆O₆	147-71-7	150.087	mcl, orth pr (w+1)	172.5		1.7598²⁰	1.4955²⁰	sl DMSO
9681	L-Tartaric acid	2,3-Dihydroxybutanedioic acid, [R-(R*,R*)]-	C₄H₆O₆	87-69-4	150.087		169				
9682	Taurocholic acid	Cholaic acid	C₂₆H₄₅NO₇S	81-24-3	515.703	pr (al-eth)	125 dec				vs H₂O, EtOH; sl eth, AcOEt
9683	Taxine A		C₃₅H₄₇NO₁₀	1361-49-5	641.749	cry (ace)	205				i H₂O; s EtOH, eth, chl
9684	Taxol	Paclitaxel	C₄₇H₅₁NO₁₄	33069-62-4	853.907	nd (MeOH aq)	214 dec				
9685	Tebuconazole		C₁₆H₂₂ClN₃O	107534-96-3	308.826		102.4				
9686	Tebuthiuron		C₉H₁₆N₄OS	34014-18-1	228.314		163 dec				
9687	Teniposide		C₃₂H₃₂O₁₃S	29767-20-2	656.653	cry (EtOH)	244				
9688	Tephrosin		C₂₃H₂₂O₇	76-80-2	410.417	pr (chl-MeOH)	198				vs ace, eth, chl
9689	Terbacil	5-Chloro-3-tert-butyl-6-methyl-2,4(1H,3H)-pyrimidinedione	C₉H₁₃ClN₂O₂	5902-51-2	216.664		176	sub 175	1.34²⁵		
9690	Terbufos		C₉H₂₁O₂PS₃	13071-79-9	288.431		-29.2	69⁰·⁰¹	1.105²⁴		
9691	Terbuthylazine	6-Chloro-N-tert-butyl-N'-ethyl-1,3,5-triazine-2,4-diamine	C₉H₁₆ClN₅	5915-41-3	229.710		178		1.188²⁰		
9692	Terbutryn		C₁₀H₁₉N₅S	886-50-0	241.357		104	157⁰·⁰⁶	1.115²⁰		
9693	Terebic acid	Tetrahydro-2,2-dimethyl-5-oxo-3-furancarboxylic acid	C₇H₁₀O₄	79-91-4	158.152	cry	175		0.815		sl H₂O; s EtOH
9694	Terephthalic acid	1,4-Benzenedicarboxylic acid	C₈H₆O₄	100-21-0	166.132	nd (sub)		sub 300			i H₂O, EtOH, eth, chl, HOAc; sl ctc
9695	Terfenadine	Seldane	C₃₂H₄₁NO₂	50679-08-8	471.674		147				i H₂O; s EtOH; sl hx
9696	o-Terphenyl		C₁₈H₁₄	84-15-1	230.304	mcl pr (MeOH)	56.20	332			i H₂O; s ace, bz, chl, MeOH
9697	m-Terphenyl		C₁₈H₁₄	92-06-8	230.304	ye nd (al)	87	363	1.199²⁰		i H₂O; s EtOH, eth, bz, HOAc; sl chl
9698	p-Terphenyl		C₁₈H₁₄	92-94-4	230.304		213.9	376			i H₂O; sl EtOH; s eth, bz, CS₂
9699	α-Terpinene	4-Isopropyl-1-methyl-1,3-cyclohexadiene	C₁₀H₁₆	99-86-5	136.234			174	0.8375¹⁹	1.477¹⁹	i H₂O; msc EtOH, eth

2-Sulfobenzoic acid

Sulfolane

Sulfometuron methyl

Sulfonmethane

Sulfonyldiacetic acid

4-Sulfophthalic acid

Sulfotep

Sulfuryl chloride isocyanate

Sulphan Blue

Sulprofos

Sunset Yellow FCF

Suprasterol II

Sutan

Symclosene

Syringin

Tabun

Tachysterol

D-Tagatose

Talbutal

Tamoxifen

R =

Tannic acid

DL-Tartaric acid

meso-Tartaric acid

D-Tartaric acid

L-Tartaric acid

Taurocholic acid

Taxine A

Taxol

Tebuconazole

Tebuthiuron

Teniposide

Tephrosin

Terbacil

Terbufos

Terbuthylazine

Terbutryn

Terebic acid

Terephthalic acid

Terfenadine

o-Terphenyl

m-Terphenyl

p-Terphenyl

α-Terpinene

No.	Name	Synonym	Mol. Form.	CAS RN	Mol. Wt.	Physical Form	mp/°C	bp/°C	den/ g cm⁻³	n_D	Solubility
9700	γ-Terpinene		$C_{10}H_{16}$	99-85-4	136.234			183	0.849²⁰	1.4765¹⁴	
9701	α-Terpineol		$C_{10}H_{18}O$	2438-12-2	154.249	cry (peth)	40.5	220	0.9337²⁰	1.4831²⁰	sl H₂O; vs ace, bz, eth, EtOH
9702	α-Terpineol acetate		$C_{12}H_{20}O_2$	80-26-2	196.286			140⁴⁰, 105¹¹	0.9659²¹	1.4689²¹	i H₂O; s EtOH, eth, bz
9703	Terpinolene	p-Mentha-1,4(8)-diene	$C_{10}H_{16}$	586-62-9	136.234			186	0.8632¹⁵	1.4883²⁰	i H₂O; msc EtOH, eth; s bz, ctc
9704	2,2':6',2"-Terpyridine		$C_{15}H_{11}N_3$	1148-79-4	233.268		88.0	370			
9705	Terrazole	1,2,4-Thiadiazole, 5-ethoxy-3-(trichloromethyl)-	$C_5H_5Cl_3N_2OS$	2593-15-9	247.530		19.9	95¹	1.503²⁵		
9706	2,2':5',2"-Terthiophene	α-Terthienyl	$C_{12}H_8S_3$	1081-34-1	248.387	ye-oran pl (MeOH)	93				i H₂O; sl EtOH; s bz, eth, ace, peth
9707	Testolactone		$C_{19}H_{24}O_3$	968-93-4	300.392	cry (ace)	218				
9708	3,6,9,12-Tetraazatetradecane-1,14-diamine	Pentaethylenehexamine	$C_{10}H_{28}N_6$	4067-16-7	232.369	liq			0.950	1.5096²⁰	
9709	Tetrabenazine		$C_{19}H_{27}NO_3$	58-46-8	317.422		128				s chl
9710	1,2,4,5-Tetrabromobenzene		$C_6H_2Br_4$	636-28-2	393.696	mcl pr (CS₂)	182	3.1²⁰	3.072²⁰		i H₂O; vs eth
9711	1,1,2,2-Tetrabromoethane	Acetylene tetrabromide	$C_2H_2Br_4$	79-27-6	345.653	ye visc liq	0	243.5; 151⁵⁴	2.9655²⁰	1.6353²⁰	i H₂O; msc EtOH, eth; s ace, bz; sl ctc
9712	Tetrabromoethene	Tetrabromoethylene	C_2Br_4	79-28-7	343.637	pl (dil al), nd (al)	56.5	226			i H₂O; s EtOH, eth, ace; vs chl
9713	2',4',5',7'-Tetrabromofluorescein, disodium salt	Eosine YS	$C_{20}H_6Br_4Na_2O_5$	17372-87-1	691.855	ye-red cry	295.5				vs EtOH
9714	4,5,6,7-Tetrabromo-1,3-isobenzofurandione		$C_8Br_4O_3$	632-79-1	463.700	nd (xyl, HOAc)	280				i H₂O, EtOH; sl bz; s PhNO₂
9715	Tetrabromomethane	Carbon tetrabromide	CBr_4	558-13-4	331.627	mcl tab (dil al)	92.3	189.5	2.9608¹⁰⁰	1.5942¹⁰⁰	i H₂O; s EtOH, eth, chl; vs CS₂
9716	2,3,4,5-Tetrabromo-6-methylphenol	3,4,5,6-Tetrabromo-o-cresol	$C_7H_4Br_4O$	576-55-6	423.722	ye nd (chl, HOAc)	208	dec			i H₂O; s EtOH, eth, bz, chl; sl lig, HOAc
9717	3',3",5',5"-Tetrabromophenolphthalein		$C_{20}H_{10}Br_4O_4$	76-62-0	633.907	nd (al, eth)	296				i H₂O; sl EtOH; vs eth; s alk, HOAc
9718	3',3",5',5"-Tetrabromophenolphthalein ethyl ester		$C_{22}H_{14}Br_4O_4$	1176-74-5	661.960	ye cry (bz)	210				
9719	3',3",5',5"-Tetrabromophenolphthalein ethyl ester, potassium salt		$C_{22}H_{13}Br_4KO_4$	62637-91-6	700.050		210				
9720	Tetrabutylammonium bromide	TMAB	$C_{16}H_{36}BrN$	1643-19-2	322.368		99				s chl
9721	Tetrabutylammonium chloride		$C_{16}H_{36}ClN$	1112-67-0	277.917	cry	74				
9722	Tetrabutylammonium fluoride		$C_{16}H_{36}FN$	429-41-4	261.462	cry (w)	37				
9723	Tetrabutylammonium hydroxide		$C_{16}H_{37}NO$	2052-49-5	259.471	stab in soln					s H₂O, MeOH
9724	Tetrabutylammonium iodide		$C_{16}H_{36}IN$	311-28-4	369.368	lf (w, bz)	148				sl H₂O, chl; vs EtOH
9725	Tetrabutylammonium sulfate		$C_{32}H_{72}N_2O_4S$	32503-27-8	580.990		170				sl chl
9726	Tetrabutylphosphonium bromide		$C_{16}H_{36}BrP$	3115-68-2	339.335	cry (ace/eth)	102				
9727	Tetrabutyl silicate	Silicic acid, tetrabutyl ester	$C_{16}H_{36}O_4Si$	4766-57-8	320.541			256; 120³	0.8990²⁰	1.4128²⁰	
9728	Tetrabutylstannane		$C_{16}H_{36}Sn$	1461-25-2	347.167	liq	-97	145¹⁰, 95⁰·²⁸	1.06²⁰		
9729	N,N,N',N'-Tetrabutylthioperoxydicarbonic diamide	Bis(dibutylthiocarbamoyl) disulfide	$C_{18}H_{36}N_2S_4$	1634-02-2	408.752		39.5		1.03²⁰		i H₂O; sl EtOH; s eth
9730	Tetrabutyl titanate	Titanium(IV) butoxide	$C_{16}H_{36}O_4Ti$	5593-70-4	340.322			292.4			
9731	Tetracaine hydrochloride		$C_{15}H_{25}ClN_2O_2$	136-47-0	300.825		147				
9732	1,2,3,4-Tetrachlorobenzene		$C_6H_2Cl_4$	634-66-2	215.892	nd (al)	47.5	254			i H₂O; sl EtOH; vs eth, CS₂
9733	1,2,3,5-Tetrachlorobenzene		$C_6H_2Cl_4$	634-90-2	215.892	nd (al)	54.5	246			i H₂O
9734	1,2,4,5-Tetrachlorobenzene		$C_6H_2Cl_4$	95-94-3	215.892	nd, mcl pr (eth, al or bz)	139.5	244.5	1.858²²		i H₂O; sl EtOH; s eth, bz, chl, CS₂
9735	3,4,5,6-Tetrachloro-1,2-benzenediol		$C_6H_2Cl_4O_2$	1198-55-6	247.891	cry (dil al, bz)	194				sl H₂O
9736	2,3,5,6-Tetrachloro-1,4-benzenediol		$C_6H_2Cl_4O_2$	87-87-6	247.891	nd (HOAc)		sub			i H₂O, bz, ctc; vs EtOH, eth; sl HOAc
9737	2,2',4',5-Tetrachlorobiphenyl		$C_{12}H_6Cl_4$	41464-40-8	291.988	cry (MeOH)	66.5				i H₂O
9738	2,3,4,5-Tetrachlorobiphenyl		$C_{12}H_6Cl_4$	33284-53-6	291.988	cry	92.2				i H₂O
9739	3,3',4,4'-Tetrachlorobiphenyl		$C_{12}H_6Cl_4$	32598-13-3	291.988	cry (EtOH)	180				
9740	2,2',6,6'-Tetrachlorobisphenol A		$C_{15}H_{12}Cl_4O_2$	79-95-8	366.067	cry (HOAc)	136				

γ-Terpinene

α-Terpineol

α-Terpineol acetate

Terpinolene

2,2':6',2"-Terpyridine

Terrazole

2,2':5',2"-Terthiophene

Testolactone

3,6,9,12-Tetraazatetradecane-1,14-diamine

Tetrabenazine

1,2,4,5-Tetrabromobenzene

1,1,2,2-Tetrabromoethane

Tetrabromoethene

2',4',5',7'-Tetrabromofluorescein, disodium salt

4,5,6,7-Tetrabromo-1,3-isobenzofurandione

Tetrabromomethane

2,3,4,5-Tetrabromo-6-methylphenol

3',3",5',5"-Tetrabromophenolphthalein

3',3",5',5"-Tetrabromophenolphthalein ethyl ester

3',3",5',5"-Tetrabromophenolphthalein ethyl ester, potassium salt

Tetrabutylammonium bromide

Tetrabutylammonium chloride

Tetrabutylammonium fluoride

Tetrabutylammonium hydroxide

Tetrabutylammonium iodide

Tetrabutylammonium sulfate

Tetrabutylphosphonium bromide

Tetrabutyl silicate

Tetrabutylstannane

N,N,N',N'-Tetrabutylthioperoxydicarbonic diamide

Tetrabutyl titanate

Tetracaine hydrochloride

1,2,3,4-Tetrachlorobenzene

1,2,3,5-Tetrachlorobenzene

1,2,4,5-Tetrachlorobenzene

3,4,5,6-Tetrachloro-1,2-benzenediol

2,3,5,6-Tetrachloro-1,4-benzenediol

2,2',4',5-Tetrachlorobiphenyl

2,3,4,5-Tetrachlorobiphenyl

3,3',4,4'-Tetrachlorobiphenyl

2,2',6,6'-Tetrachlorobisphenol A

No.	Name	Synonym	Mol. Form.	CAS RN	Mol. Wt.	Physical Form	mp/°C	bp/°C	den/ g cm⁻³	n_D	Solubility
9741	2,3,5,6-Tetrachloro-2,5-cyclohexadiene-1,4-dione	Chloranil	$C_6Cl_4O_2$	118-75-2	245.875	ye mcl, pr (bz) ye lf (HOAc)	290	sub			i H_2O, liq; sl EtOH, chl; s eth
9742	3,4,5,6-Tetrachloro-3,5-cyclohexadiene-1,2-dione		$C_6Cl_4O_2$	2435-53-2	245.875		130.5				
9743	2,3,7,8-Tetrachlorodibenzo-p-dioxin	Dioxin	$C_{12}H_4Cl_4O_2$	1746-01-6	321.971	nd	295				
9744	2,3,7,8-Tetrachlorodibenzofuran		$C_{12}H_4Cl_4O$	51207-31-9	305.971	cry	227				
9745	1,1,1,2-Tetrachloro-2,2-difluoroethane		$C_2Cl_4F_2$	76-11-9	203.830		41.0	92.8	1.649²⁵		i H_2O; s EtOH, eth, chl
9746	1,1,2,2-Tetrachloro-1,2-difluoroethane		$C_2Cl_4F_2$	76-12-0	203.830		24.8	92.8	1.5951⁵⁰	1.4130²⁵	i H_2O; s EtOH, eth, chl
9747	1,2,3,4-Tetrachloro-5,5-dimethoxy-1,3-cyclopentadiene		$C_7H_6Cl_4O_2$	2207-27-4	263.934			109	1.501²⁵	1.5282²⁰	
9748	1,2,3,4-Tetrachloro-5,6-dimethylbenzene		$C_8H_6Cl_4$	877-08-7	243.946		228				i H_2O; s EtOH, eth, bz
9749	1,2,3,5-Tetrachloro-4,6-dimethylbenzene		$C_8H_6Cl_4$	877-09-8	243.946		223		1.703²⁵		i H_2O, EtOH, eth, bz, chl
9750	1,1,2,2-Tetrachloro-1,2-dimethyldisilane		$C_2H_6Cl_4Si_2$	4518-98-3	228.052			154			
9751	1,1,1,2-Tetrachloroethane		$C_2H_2Cl_4$	630-20-6	167.849	liq	-70.2	130.2	1.5406²⁰	1.4821²⁰	sl H_2O; s ace, bz, chl; msc EtOH, eth
9752	1,1,2,2-Tetrachloroethane	Acetylene tetrachloride	$C_2H_2Cl_4$	79-34-5	167.849	liq	-42.4	145.2	1.5953²⁰	1.4940²⁰	sl H_2O; s ace, bz, chl; msc EtOH, eth
9753	Tetrachloroethene	Perchloroethylene	C_2Cl_4	127-18-4	165.833	liq	-22.3	121.3	1.6230²⁰	1.5059²⁰	i H_2O; msc EtOH, eth, bz
9754	1,1,1,2-Tetrachloro-2-fluoroethane		C_2HCl_4F	354-11-0	185.839	liq	-95.3	117.1			
9755	1,1,2,2-Tetrachloro-1-fluoroethane		C_2HCl_4F	354-14-3	185.839	liq	-82.6	116.7	1.5497¹⁷	1.4390²⁰	
9756	Tetrachloromethane	Carbon tetrachloride	CCl_4	56-23-5	153.823	liq	-22.62	76.8	1.5940²⁰	1.4601²⁰	i H_2O; s EtOH, ace; msc eth, bz, chl
9757	2,3,5,6-Tetrachloro-4-methoxyphenol	Drosophilin A	$C_7H_4Cl_4O_2$	484-67-3	261.918		116				
9758	2,3,4,6-Tetrachloro-5-methylphenol		$C_7H_4Cl_4O$	10460-33-0	245.918	nd (peth)	189.5				i H_2O; s EtOH, eth, ace, bz, KOH
9759	1,2,3,4-Tetrachloronaphthalene		$C_{10}H_4Cl_4$	20020-02-4	265.951		199				
9760	1,2,3,4-Tetrachloro-5-nitrobenzene		$C_6HCl_4NO_2$	879-39-0	260.890		66				
9761	1,2,4,5-Tetrachloro-3-nitrobenzene		$C_6HCl_4NO_2$	117-18-0	260.890		99.5	304	1.744²⁵		i H_2O; s EtOH, bz, chl
9762	2,3,4,5-Tetrachlorophenol		$C_6H_2Cl_4O$	4901-51-3	231.891	nd (peth, sub)	116.5	sub			vs EtOH
9763	2,3,4,6-Tetrachlorophenol		$C_6H_2Cl_4O$	58-90-2	231.891	nd (lig)	70	150¹⁵			i H_2O; s EtOH, bz, chl, HOAc; vs NaOH
9764	2,3,5,6-Tetrachlorophenol		$C_6H_2Cl_4O$	935-95-5	231.891	lf (lig)	115				sl H_2O; vs bz; s lig
9765	Tetrachlorophthalic anhydride		$C_8Cl_4O_3$	117-08-8	285.896		254.5	sub	1.49²⁷⁵		sl eth
9766	1,1,1,2-Tetrachloropropane		$C_3H_4Cl_4$	812-03-3	181.876	liq	-64	152.5	1.473²⁰	1.4867²⁰	i H_2O; vs EtOH; s eth, chl
9767	1,1,1,3-Tetrachloropropane		$C_3H_4Cl_4$	1070-78-6	181.876			157	1.4509²⁰	1.4825²⁰	i H_2O; vs EtOH, eth, bz, chl
9768	1,1,2,3-Tetrachloropropane		$C_3H_4Cl_4$	18495-30-2	181.876			179.5	1.513¹⁷	1.5037¹⁷	i H_2O; s EtOH, chl; vs eth
9769	1,2,2,3-Tetrachloropropane		$C_3H_4Cl_4$	13116-53-5	181.876			165	1.500¹⁸	1.4940¹⁸	i H_2O; vs EtOH, eth; s chl
9770	1,1,2,3-Tetrachloropropene		$C_3H_2Cl_4$	10436-39-2	179.860	liq		167.2; 59¹⁷	1.55²⁰		
9771	2,3,5,6-Tetrachloropyridine		C_5HCl_4N	2402-79-1	216.881	cry (aq al)	90.5	250.5			vs eth, EtOH, peth
9772	Tetrachloropyrimidine		$C_4Cl_4N_2$	1780-40-1	217.868		69.0				
9773	3,3',4',5-Tetrachlorosalicylanilide	3,5-Dichloro-N-(3,4-dichlorophenyl)-2-hydroxybenzamide	$C_{13}H_7Cl_4NO_2$	1154-59-2	351.013		161				
9774	2,3,5,6-Tetrachloroterphthaloyl dichloride		$C_8Cl_6O_2$	719-32-4	340.803	cry (ctc)	146.5				
9775	Tetrachlorothiophene		C_4Cl_4S	6012-97-1	221.920	nd (dil al)	30.5	233.4	1.7036³⁰	1.5915³⁰	i H_2O; vs EtOH; msc eth
9776	Tetrachlorovinphos		$C_{10}H_9Cl_4O_4P$	961-11-5	365.961		97				
9777	Tetracontane		$C_{40}H_{82}$	4181-95-7	563.079		81.5	522; 400⁵⁰	0.8171²⁵	1.4572²⁵	
9778	Tetracosamethylundecasiloxane	Tetracosamethylhendecasiloxane	$C_{24}H_{72}O_{10}Si_{11}$	107-53-9	829.764			322.8; 202⁴⁷	0.9247²⁵	1.3994²⁰	vs bz
9779	Tetracosane		$C_{24}H_{50}$	646-31-1	338.654	cry (eth)	50.4	391.3	0.7991²⁰	1.4283⁷⁰	i H_2O; sl EtOH; vs eth

2,3,5,6-Tetrachloro-2,5-cyclohexadiene-1,4-dione

3,4,5,6-Tetrachloro-3,5-cyclohexadiene-1,2-dione

2,3,7,8-Tetrachlorodibenzo-*p*-dioxin

2,3,7,8-Tetrachlorodibenzofuran

1,1,1,2-Tetrachloro-2,2-difluoroethane

1,1,2,2-Tetrachloro-1,2-difluoroethane

1,2,3,4-Tetrachloro-5,5-dimethoxy-1,3-cyclopentadiene

1,2,3,4-Tetrachloro-5,6-dimethylbenzene

1,2,3,5-Tetrachloro-4,6-dimethylbenzene

1,1,2,2-Tetrachloro-1,2-dimethyldisilane

1,1,1,2-Tetrachloroethane

1,1,2,2-Tetrachloroethane

Tetrachloroethene

1,1,1,2-Tetrachloro-2-fluoroethane

1,1,2,2-Tetrachloro-1-fluoroethane

Tetrachloromethane

2,3,5,6-Tetrachloro-4-methoxyphenol

2,3,4,6-Tetrachloro-5-methylphenol

1,2,3,4-Tetrachloronaphthalene

1,2,3,4-Tetrachloro-5-nitrobenzene

1,2,4,5-Tetrachloro-3-nitrobenzene

2,3,4,5-Tetrachlorophenol

2,3,4,6-Tetrachlorophenol

2,3,5,6-Tetrachlorophenol

Tetrachlorophthalic anhydride

1,1,1,2-Tetrachloropropane

1,1,1,3-Tetrachloropropane

1,1,2,3-Tetrachloropropane

1,2,2,3-Tetrachloropropane

1,1,2,3-Tetrachloropropene

2,3,5,6-Tetrachloropyridine

Tetrachloropyrimidine

3,3',4',5-Tetrachlorosalicylanilide

2,3,5,6-Tetrachloroterphthaloyl dichloride

Tetrachlorothiophene

Tetrachlorovinphos

H₃C(CH₂)₃₈CH₃
Tetracontane

Tetracosamethylundecasiloxane

Tetracosane

No.	Name	Synonym	Mol. Form.	CAS RN	Mol. Wt.	Physical Form	mp/°C	bp/°C	den/g cm⁻³	n_D	Solubility
9780	Tetracosanoic acid	Lignoceric acid	$C_{24}H_{48}O_2$	557-59-5	368.637		87.5	272[10]	0.8207[100]	1.4287[100]	vs bz, eth
9781	1-Tetracosanol		$C_{24}H_{50}O$	506-51-4	354.653		77	210[0.4]			
9782	cis-15-Tetracosenoic acid	Nervonic acid	$C_{24}H_{46}O_2$	506-37-6	366.621		43				
9783	Tetracyanoethene	Tetracyanoethylene	C_6N_4	670-54-2	128.091		199	223	1.348[25]	1.560[25]	sl eth, bz, ctc, chl; s ace
9784	Tetracycline		$C_{22}H_{24}N_2O_8$	60-54-8	444.434	cry (+3w)	172 dec				
9785	Tetracycline hydrochloride		$C_{22}H_{25}ClN_2O_8$	64-75-5	480.895		214				
9786	Tetradecahydrophenanthrene		$C_{14}H_{24}$	5743-97-5	192.341	liq	-3	270; 87[2]	0.944[20]	1.5011[20]	i H_2O; s eth, ace, bz
9787	Tetradecamethylhexasiloxane		$C_{14}H_{42}O_5Si_6$	107-52-8	458.993	liq	-59	245.5	0.8910[20]	1.3948[20]	vs bz
9788	Tetradecanal		$C_{14}H_{28}O$	124-25-4	212.371	lf	30				i H_2O; s EtOH, eth, ace
9789	Tetradecanamide		$C_{14}H_{29}NO$	638-58-4	227.386	lf (ace)	104	217[12]			vs EtOH
9790	Tetradecane		$C_{14}H_{30}$	629-59-4	198.388		5.82	253.58	0.7596[20]	1.4290[20]	i H_2O; vs EtOH, eth; s ctc
9791	Tetradecanedioic acid		$C_{14}H_{26}O_4$	821-38-5	258.354		125.5				
9792	1,14-Tetradecanediol		$C_{14}H_{30}O_2$	19812-64-7	230.387	nd (bz)	85.8	200[9]			vs eth, EtOH
9793	Tetradecanenitrile	Myristonitrile	$C_{14}H_{27}N$	629-63-0	209.371		19	226[100], 119[1]	0.8281[19]	1.4392[23]	i H_2O; msc EtOH, eth, ace, bz; sl ctc
9794	1-Tetradecanethiol		$C_{14}H_{30}S$	2079-95-0	230.453		7	310; 178[22]	0.8641[20]	1.4597[20]	i H_2O; s EtOH, eth, ctc
9795	Tetradecanoic acid	Myristic acid	$C_{14}H_{28}O_2$	544-63-8	228.371	lf (eth)	54.2	250[100]	0.8622[54]	1.4723[70]	i H_2O; s EtOH, ace, chl; sl eth; vs bz
9796	Tetradecanoic anhydride		$C_{28}H_{54}O_3$	626-29-9	438.727	lf (peth)	53.4		0.8502[70]	1.4335[70]	vs eth, EtOH
9797	1-Tetradecanol	Tetradecyl alcohol	$C_{14}H_{30}O$	112-72-1	214.387	lf	38.2	287	0.8236[38]		i H_2O; vs EtOH, eth, ace, bz, chl
9798	2-Tetradecanone	Dodecyl methyl ketone	$C_{14}H_{28}O$	2345-27-9	212.371	cry (dil al)	33.5	205[100], 134[13]			i H_2O; s EtOH, ace
9799	Tetradecanoyl chloride	Myristoyl chloride	$C_{14}H_{27}ClO$	112-64-1	246.816		-1	171[16]	0.9078[25]		s eth
9800	12-O-Tetradecanoylphorbol-13-acetate	Cocarcinogen A1	$C_{36}H_{56}O_8$	16561-29-8	616.825	oil					
9801	1-Tetradecene		$C_{14}H_{28}$	1120-36-1	196.372	liq	-12	233	0.7745[25]	1.4351[20]	i H_2O; vs EtOH, eth; s bz; sl ctc
9802	Tetradecyl acetate	1-Tetradecanol, acetate	$C_{16}H_{32}O_2$	638-59-5	256.424			173[10]			
9803	Tetradecylamine	1-Tetradecanamine	$C_{14}H_{31}N$	2016-42-4	213.403		83.1	291.2	0.8079[20]	1.4463[20]	i H_2O; vs EtOH, eth, bz, chl; s ace
9804	Tetradecylbenzene		$C_{20}H_{34}$	1459-10-5	274.484		16	359	0.8549[20]	1.4818[20]	
9805	Tetradecylcyclohexane		$C_{20}H_{40}$	1795-18-2	280.532		24	360	0.8254[20]	1.4579[20]	
9806	Tetradifon	1,2,4-Trichloro-5-[(4-chlorophenyl)sulfonyl]benzene	$C_{12}H_6Cl_4O_2S$	116-29-0	356.052		146		1.151[20]		
9807	Tetraethoxygermane	Ethanol, germanium(4+) salt	$C_8H_{20}GeO_4$	14165-55-0	252.88			139[200]			
9808	Tetraethoxymethane	Tetraethyl orthocarbonate	$C_9H_{20}O_4$	78-09-1	192.253			159.5	0.9186[20]	1.3905[25]	msc EtOH, eth; s ctc
9809	Tetraethylammonium bromide		$C_8H_{20}BrN$	71-91-0	210.156	hyg (al)	286 dec		1.3970[20]		vs H_2O, EtOH, chl, MeOH
9810	Tetraethylammonium chloride		$C_8H_{20}ClN$	56-34-8	165.705	hyg cry					vs H_2O, EtOH, ace, chl
9811	Tetraethylammonium iodide		$C_8H_{20}IN$	68-05-3	257.156	cry (w)	300 dec				s H_2O
9812	1,2,3,5-Tetraethylbenzene		$C_{14}H_{22}$	38842-05-6	190.325			249.0			
9813	N,N,N',N'-Tetraethyl-1,2-benzenedicarboxamide	N,N,N',N'-Tetraethylphthalamide	$C_{16}H_{24}N_2O_2$	83-81-8	276.374		36	204[16]			
9814	Tetra(2-ethylbutyl) silicate	Silicic acid, tetrakis(2-ethylbutyl) ester	$C_{24}H_{52}O_4Si$	78-13-7	432.754	liq			0.8920[20]	1.4307[20]	i H_2O; sl EtOH, ctc; s eth, bz
9815	Tetraethylene glycol	3,6,9-Trioxaundecane-1,11-diol	$C_8H_{18}O_5$	112-60-7	194.226	liq	-6.2	328	1.1285[15]	1.4577[20]	vs H_2O; s EtOH, eth, ctc, diox
9816	Tetraethylene glycol diacrylate		$C_{14}H_{22}O_7$	17831-71-9	302.321				1.125[25]		
9817	Tetraethylene glycol dimethacrylate		$C_{16}H_{26}O_7$	109-17-1	330.373			220[1]		1.4610[25]	
9818	Tetraethylene glycol dimethyl ether		$C_{10}H_{22}O_5$	143-24-8	222.279			275.3	1.0114[20]		msc H_2O; s EtOH, eth, ctc
9819	Tetraethylene glycol monostearate		$C_{26}H_{52}O_6$	106-07-0	460.687		40	328	1.1285[15]	1.4593[20]	
9820	Tetraethylenepentamine		$C_8H_{23}N_5$	112-57-2	189.303			341.5		1.5042[20]	s H_2O
9821	N,N,N',N'-Tetraethyl-1,2-ethanediamine		$C_{10}H_{24}N_2$	150-77-6	172.311			192	0.808[25]	1.4343[20]	
9822	Tetraethylgermane		$C_8H_{20}Ge$	597-63-7	188.89			164.5	1.199		
9823	Tetraethyl lead		$C_8H_{20}Pb$	78-00-2	323.4			dec 200	1.653[20]	1.5198[20]	i H_2O; s bz; sl EtOH
9824	N,N,N',N'-Tetraethylmethanediamine		$C_9H_{22}N_2$	102-53-4	158.284			165.8	0.8000[20]	1.4420[25]	

Tetracosanoic acid

1-Tetracosanol

cis-15-Tetracosenoic acid

Tetracyanoethene

Tetracycline

Tetracycline hydrochloride

HCl

Tetradecahydrophenanthrene

Tetradecamethylhexasiloxane

Tetradecanal

Tetradecanamide

Tetradecane

Tetradecanedioic acid

1,14-Tetradecanediol

Tetradecanenitrile

1-Tetradecanethiol

Tetradecanoic acid

Tetradecanoic anhydride

1-Tetradecanol

2-Tetradecanone

Tetradecanoyl chloride

12-*O*-Tetradecanoylphorbol-13-acetate

1-Tetradecene

Tetradecyl acetate

Tetradecylamine

Tetradecylbenzene

Tetradecylcyclohexane

Tetradifon

Tetraethoxygermane

Tetraethoxymethane

Tetraethylammonium bromide

Tetraethylammonium chloride

Tetraethylammonium iodide

1,2,3,5-Tetraethylbenzene

N,N,N′,N′-Tetraethyl-1,2-benzenedicarboxamide

Tetra(2-ethylbutyl) silicate

Tetraethylene glycol

Tetraethylene glycol diacrylate

Tetraethylene glycol dimethacrylate

Tetraethylene glycol dimethyl ether

Tetraethylene glycol monostearate

Tetraethylenepentamine

N,N,N′,N′-Tetraethyl-1,2-ethanediamine

Tetraethylgermane

Tetraethyl lead

N,N,N′,N′-Tetraethylmethanediamine

No.	Name	Synonym	Mol. Form.	CAS RN	Mol. Wt.	Physical Form	mp/°C	bp/°C	den/ g cm⁻³	n_D	Solubility
9825	Tetraethyl pyrophosphate		$C_8H_{20}O_7P_2$	107-49-3	290.188		170 dec	155³	1.1847²⁰	1.4180²⁰	msc H₂O, EtOH, eth, ace, xyl, chl; sl ctc
9826	Tetraethylsilane		$C_8H_{20}Si$	631-36-7	144.331			154.7	0.7658²⁰	1.4268²⁰	i H₂O
9827	Tetraethylstannane	Tin tetraethyl	$C_8H_{20}Sn$	597-64-8	234.955	liq	-112	181; 64¹²	1.187²⁵	1.4730²⁰	
9828	Tetraethylthiodicarbonic diamide	Sulfiram	$C_{10}H_{20}N_2S_3$	95-05-6	264.474			232³	1.12²⁰		s chl
9829	Tetraethylurea		$C_9H_{20}N_2O$	1187-03-7	172.267			209	0.919²⁰	1.4474²⁰	i H₂O, alk, acid
9830	1,2,3,4-Tetrafluorobenzene		$C_6H_2F_4$	551-62-2	150.074			94.3		1.4054²⁰	
9831	1,2,3,5-Tetrafluorobenzene		$C_6H_2F_4$	2367-82-0	150.074	liq	-46.25	84.4	1.319²⁵	1.4035²⁰	
9832	1,2,4,5-Tetrafluorobenzene		$C_6H_2F_4$	327-54-8	150.074		3.88	90.2	1.4255²⁰	1.4075²⁰	
9833	3,3,4,4-Tetrafluorodihydro-2,5-furandione		$C_4F_4O_3$	699-30-9	172.035			54.5	1.6209²⁰	1.3240²⁰	
9834	1,1,2,2-Tetrafluoro-1,2-dinitroethane		$C_2F_4N_2O_4$	356-16-1	192.026	liq	-41.5	58.5	1.6024²⁵	1.3265²⁵	i H₂O; s ace
9835	1,1,1,2-Tetrafluoroethane		$C_2H_2F_4$	811-97-2	102.031	col gas	-103.3	-26.08	1.2072²⁵		i H₂O; s eth
9836	1,1,2,2-Tetrafluoroethane		$C_2H_2F_4$	359-35-3	102.031	col gas	-89	-19.9			
9837	Tetrafluoroethene	Tetrafluoroethylene	C_2F_4	116-14-3	100.015	col gas	-131.15	-75.9	1.519⁻⁷⁶		i H₂O
9838	1,2,2,2-Tetrafluoroethyl difluoromethyl ether	Refrigerant 236me	$C_3H_2F_6O$	57041-67-5	168.037	vol liq or gas		23.35	1.4540²³		
9839	Tetrafluoromethane	Carbon tetrafluoride	CF_4	75-73-0	88.005	col gas	-183.60	-128.0	3.034²⁵		i H₂O; s bz, chl
9840	2,2,3,3-Tetrafluoro-1-propanol		$C_3H_4F_4O$	76-37-9	132.057	liq	-15	109.5	1.4853²⁰	1.3197²⁰	s EtOH, ace, chl
9841	6,7,8,9-Tetrahydro-5H-benzocyclohepten-5-one		$C_{11}H_{12}O$	826-73-3	160.212			175⁴⁰, 124⁷	1.080²⁰	1.5698²⁰	s EtOH
9842	2,3,6,7-Tetrahydro-1H,5H-benzo[ij]quinolizine	Julolidine	$C_{12}H_{15}N$	479-59-4	173.254		40	dec 280; 155¹⁷	1.003²⁰	1.568²⁵	
9843	1,2,3,6-Tetrahydro-2,3'-bipyridine, (S)	Anatabine	$C_{10}H_{12}N_2$	581-49-7	160.215			145¹⁰	1.091¹⁹	1.5676²⁰	msc H₂O; s EtOH, eth, bz
9844	2,3,4,9-Tetrahydro-1H-carbazole		$C_{12}H_{13}N$	942-01-8	171.238	lf (dil al)	120	327.5			i H₂O; s EtOH; vs eth, bz, MeOH
9845	Tetrahydrocortisone		$C_{21}H_{32}O_5$	53-05-4	364.476	cry (EtOAc)	190				
9846	1,2,3,4-Tetrahydro-6,7-dimethoxy-1,2-dimethylisoquinoline, (±)	Carnegine	$C_{13}H_{19}NO_2$	490-53-9	221.296	pa br syr		170¹			vs H₂O, eth, EtOH
9847	Tetrahydro-2,5-dimethoxyfuran		$C_6H_{12}O_3$	696-59-3	132.157			145.7	1.02²⁵	1.4180²⁰	
9848	4,5,6,7-Tetrahydro-3,6-dimethylbenzofuran		$C_{10}H_{14}O$	494-90-6	150.217		86	80¹⁸	0.972¹⁵		
9849	1,2,3,4-Tetrahydro-1,5-dimethylnaphthalene		$C_{12}H_{16}$	21564-91-0	160.255			239	0.941²⁰	1.526²⁰	
9850	Tetrahydro-2,2-dimethyl-5-oxo-3-furanacetic acid	Terpenylic acid	$C_8H_{12}O_4$	26754-48-3	172.179	lf or pr (w+1)	90				vs H₂O
9851	cis-Tetrahydro-2,5-dimethylthiophene		$C_6H_{12}S$	5161-13-7	116.224	liq	-89	142.3	0.9222²⁰	1.4799²⁰	vs ace, bz, eth, EtOH
9852	1,2,3,4-Tetrahydro-9H-fluoren-9-one	Phentydrone	$C_{13}H_{12}O$	634-19-5	184.233	lt ye nd or pr (pentane)	81.5	139⁰·⁰⁵			
9853	5,6,7,8-Tetrahydrofolic acid		$C_{19}H_{23}N_7O_6$	135-16-0	445.429	pow					s H₂O
9854	Tetrahydrofuran	Tetramethylene oxide	C_4H_8O	109-99-9	72.106	liq	-108.44	65	0.8833²⁵	1.4050²⁰	s H₂O, chl; vs EtOH, eth, ace, bz
9855	Tetrahydro-2-furanmethanamine	Tetrahydrofurfurylamine	$C_5H_{11}NO$	4795-29-3	101.147			153	0.9752²⁰	1.4551²⁰	vs H₂O, eth, EtOH
9856	Tetrahydro-2-furanmethanol propanoate		$C_8H_{14}O_3$	637-65-0	158.195			205.5	1.044²⁰		vs eth, EtOH, chl
9857	Tetrahydro-3-furanol		$C_4H_8O_2$	453-20-3	88.106			181	1.09²⁵	1.4500²⁰	
9858	Tetrahydrofurfuryl acetate		$C_7H_{12}O_3$	637-64-9	144.168			193; 89¹⁸	1.0624²⁰	1.4350²⁵	vs H₂O, eth, EtOH, chl
9859	Tetrahydrofurfuryl acrylate		$C_8H_{12}O_3$	2399-48-6	156.179		<-60	96⁶	1.061²⁰		
9860	Tetrahydrofurfuryl alcohol	Tetrahydro-2-furancarbinol	$C_5H_{10}O_2$	97-99-4	102.132		<-80	178	1.0524²⁰	1.4520²⁰	vs ace, eth
9861	Tetrahydrofurfuryl methacrylate		$C_9H_{14}O_3$	2455-24-5	170.205			265; 81⁴	1.040²⁵	1.4554²⁵	
9862	Tetrahydroimidazo[4,5-d]imidazole-2,5(1H,3H)-dione	Acetyleneurea	$C_4H_6N_4O_2$	496-46-8	142.117	nd or pr (w)	300 dec				sl H₂O; i EtOH, HOAc; s eth, HCl, alk
9863	cis-3a,4,7,7a-Tetrahydro-1,3-isobenzofurandione	4-Cyclohexene-1,2-dicarboxylic acid, anhydride	$C_8H_8O_3$	935-79-5	152.148	cry (peth)	103.5				s EtOH, ace, chl, bz; sl peth
9864	4,5,6,7-Tetrahydro-1,3-isobenzofurandione	1-Cyclohexene-1,2-dicarboxylic acid, anhydride	$C_8H_8O_3$	2426-02-0	152.148	pl (EtOH)	74		1.2¹⁰⁵		s EtOH, ace, chl; vs eth
9865	1,2,3,4-Tetrahydroisoquinoline		$C_9H_{11}N$	91-21-4	133.190		<-15	232.5	1.0642²⁴	1.5668²⁰	i H₂O; s EtOH, bz, acid, xyl
9866	3,4,5,6-Tetrahydro-7-methoxy-2H-azepine		$C_7H_{13}NO$	2525-16-8	127.184	liq		49¹⁶, 66²⁴	0.887	1.4630²⁰	
9867	1,2,3,4-Tetrahydro-6-methoxyquinoline		$C_{10}H_{13}NO$	120-15-0	163.216	pr (peth, al) orth pym (w)	42.5	284; 128¹		1.5718²⁰	s chl
9868	1,2,3,4-Tetrahydro-1-methylnaphthalene		$C_{11}H_{14}$	1559-81-5	146.229			220.6	0.9583²⁰	1.5353²⁰	

Tetraethyl pyrophosphate

Tetraethylsilane

Tetraethylstannane

Tetraethylthiodicarbonic diamide

Tetraethylurea

1,2,3,4-Tetrafluorobenzene

1,2,3,5-Tetrafluorobenzene

1,2,4,5-Tetrafluorobenzene

3,3,4,4-Tetrafluorodihydro-2,5-furandione

1,1,2,2-Tetrafluoro-1,2-dinitroethane

1,1,1,2-Tetrafluoroethane

1,1,2,2-Tetrafluoroethane

Tetrafluoroethene

1,2,2,2-Tetrafluoroethyl difluoromethyl ether

Tetrafluoromethane

2,2,3,3-Tetrafluoro-1-propanol

6,7,8,9-Tetrahydro-5H-benzocyclohepten-5-one

2,3,6,7-Tetrahydro-1H,5H-benzo[ij]quinolizine

1,2,3,6-Tetrahydro-2,3'-bipyridine, (S)

2,3,4,9-Tetrahydro-1H-carbazole

Tetrahydrocortisone

1,2,3,4-Tetrahydro-6,7-dimethoxy-1,2-dimethylisoquinoline, (±)

Tetrahydro-2,5-dimethoxyfuran

4,5,6,7-Tetrahydro-3,6-dimethylbenzofuran

1,2,3,4-Tetrahydro-1,5-dimethylnaphthalene

Tetrahydro-2,2-dimethyl-5-oxo-3-furanacetic acid

cis-Tetrahydro-2,5-dimethylthiophene

1,2,3,4-Tetrahydro-9H-fluoren-9-one

5,6,7,8-Tetrahydrofolic acid

Tetrahydrofuran

Tetrahydro-2-furanmethanamine

Tetrahydro-2-furanmethanol propanoate

Tetrahydro-3-furanol

Tetrahydrofurfuryl acetate

Tetrahydrofurfuryl acrylate

Tetrahydrofurfuryl alcohol

Tetrahydrofurfuryl methacrylate

Tetrahydroimidazo[4,5-d]imidazole-2,5(1H,3H)-dione

cis-3a,4,7,7a-Tetrahydro-1,3-isobenzofurandione

4,5,6,7-Tetrahydro-1,3-isobenzofurandione

1,2,3,4-Tetrahydroisoquinoline

3,4,5,6-Tetrahydro-7-methoxy-2H-azepine

1,2,3,4-Tetrahydro-6-methoxyquinoline

1,2,3,4-Tetrahydro-1-methylnaphthalene

No.	Name	Synonym	Mol. Form.	CAS RN	Mol. Wt.	Physical Form	mp/°C	bp/°C	den/ g cm⁻³	n_D	Solubility
9869	1,2,3,4-Tetrahydro-5-methylnaphthalene		C₁₁H₁₄	2809-64-5	146.229	liq	-23	234	0.9720²⁰	1.5439²⁰	
9870	1,2,3,4-Tetrahydro-6-methylnaphthalene		C₁₁H₁₄	1680-51-9	146.229	liq	-40	229	0.9537²⁰	1.5357²⁰	
9871	1,2,3,6-Tetrahydro-1-methyl-4-phenylpyridine	MPTP	C₁₂H₁₅N	28289-54-5	173.254	cry	41	87⁰·⁸			
9872	Tetrahydro-3-methyl-2H-thiopyran		C₆H₁₂S	5258-50-4	116.224	liq	-60	158	0.9473²⁰	1.4922²⁰	
9873	5,6,7,8-Tetrahydro-1-naphthalenamine		C₁₀H₁₃N	2217-41-6	147.217		38	279	1.0625¹⁶	1.5900²⁰	sl H₂O; s EtOH, eth, acid
9874	1,2,3,4-Tetrahydronaphthalene	Tetralin	C₁₀H₁₂	119-64-2	132.202	liq	-35.7	207.6	0.9645²⁵	1.5413²⁰	i H₂O; vs EtOH, eth; s chl, PhNH₂
9875	1,2,3,4-Tetrahydro-1-naphthol	1,2,3,4-Tetrahydro-α-naphthol	C₁₀H₁₂O	529-33-9	148.201		34.5	255; 103²	1.0996²⁰	1.5638²⁰	
9876	5,6,7,8-Tetrahydro-1-naphthol	5,6,7,8-Tetrahydro-α-naphthol	C₁₀H₁₂O	529-35-1	148.201		70	266; 143¹¹	1.0556⁷⁵		
9877	1,2,3,4-Tetrahydro-2-naphthol	Tetralol	C₁₀H₁₂O	530-91-6	148.201		15.5	140¹²			
9878	5,6,7,8-Tetrahydro-2-naphthol	5,6,7,8-Tetrahydro-β-naphthol	C₁₀H₁₂O	1125-78-6	148.201		57	275.5	1.0552⁶⁵		
9879	Tetrahydro-6-pentyl-2H-pyran-2-one	5-Hydroxydecanoic acid lactone	C₁₀H₁₈O₂	705-86-2	170.249	liq	-27	121³			
9880	1,2,3,4-Tetrahydrophenanthrene		C₁₄H₁₄	1013-08-7	182.261	lf (MeOH)	33.5	173¹¹	1.0601⁴⁰		i H₂O; s EtOH, eth, ace, bz, HOAc, chl, lig
9881	1,2,3,6-Tetrahydrophthalimide		C₈H₉NO₂	85-40-5	151.163	cry (EtOH)	137				
9882	Tetrahydro-6-propyl-2H-pyran-2-one	5-Hydroxyoctanoic acid lactone	C₈H₁₄O₂	698-76-0	142.196	liq	-13	126¹⁵			
9883	2,3,4,5-Tetrahydro-6-propylpyridine	γ-Coniceine	C₈H₁₅N	1604-01-9	125.212			174	0.8753¹⁵	1.4661¹⁶	
9884	Tetrahydropyran	Oxane	C₅H₁₀O	142-68-7	86.132	liq	-49.1	88	0.8814²⁰	1.4200²⁰	s EtOH, eth, bz, ctc
9885	Tetrahydro-2H-pyran-2-methanol		C₆H₁₂O₂	100-72-1	116.158			185	1.027²⁵	1.458²⁰	
9886	Tetrahydro-2H-pyran-2-one		C₅H₈O₂	542-28-9	100.117	liq	-12.5	219	1.1082²⁰	1.4503²⁰	s H₂O; msc EtOH, eth; sl ctc
9887	Tetrahydro-4H-pyran-4-one		C₅H₈O₂	29943-42-8	100.117			166.5	1.084²⁵	1.4520²⁰	
9888	1,2,5,6-Tetrahydropyridine	Δ³-Piperidine	C₅H₉N	694-05-3	83.132	liq	-48	108	0.911²⁵	1.4800²⁰	s chl
9889	1,2,5,6-Tetrahydro-3-pyridinecarboxylic acid	Guvacine	C₆H₉NO₂	498-96-4	127.141	pr (w), rods (+1w dil al)	295 dec				vs H₂O
9890	3,4,5,6-Tetrahydro-2(1H)-pyrimidinethione	Hexahydropyrimidine-2-thione	C₄H₈N₂S	2055-46-1	116.185		211		1.33²⁰		
9891	1,2,3,4-Tetrahydroquinoline		C₉H₁₁N	635-46-1	133.190	nd	20	251	1.0588²⁰	1.6062¹⁹	s H₂O, chl; msc EtOH, eth
9892	5,6,7,8-Tetrahydroquinoline	2,3-Cyclohexenopyridine	C₉H₁₁N	10500-57-9	133.190			222	1.0304¹³	1.5435²⁰	sl H₂O; s EtOH, eth, ace, bz
9893	1,2,3,4-Tetrahydroquinoxaline		C₈H₁₀N₂	3476-89-9	134.178	lf (w, eth, peth)	99	289			s H₂O, chl; vs EtOH, eth, bz; sl peth
9894	6,7,8,9-Tetrahydro-5H-tetrazolo[1,5-a]azepine	Pentylenetetrazole	C₆H₁₀N₄	54-95-5	138.170	cry (bz-lig)	59.5	194¹²			vs H₂O, EtOH, ace; s eth, bz; sl chl
9895	Tetrahydrothiophene	Thiacyclopentane	C₄H₈S	110-01-0	88.172	liq	-96.2	121.1	0.9987²⁰	1.4871¹⁸	i H₂O; msc EtOH, eth, ace, bz; s chl
9896	1,2,3,4-Tetrahydro-1,1,6-trimethylnaphthalene		C₁₃H₁₈	475-03-6	174.282			240; 90⁴	0.9303²⁰	1.5257²⁰	s EtOH, eth, bz, chl
9897	1,2,5,8-Tetrahydroxy-9,10-anthracenedione	Quinalizarin	C₁₄H₈O₆	81-61-8	272.210	oran nd	>275				sl H₂O, ace, bz, EtOH, eth
9898	2,3,4,6-Tetrahydroxy-5H-benzocyclohepten-5-one	Purpurogallin	C₁₁H₈O₅	569-77-7	220.179	red nd (gl HOAc)	274 dec				
9899	2,2',4,4'-Tetrahydroxybenzophenone		C₁₃H₁₀O₅	131-55-5	246.215	ye nd (w+1)	197				vs H₂O, ace, eth, EtOH
9900	2,3,5,6-Tetrahydroxy-2,5-cyclohexadiene-1,4-dione	Tetroquinone	C₆H₄O₆	319-89-1	172.092	bl-blk cry					sl H₂O, eth, ctc; vs EtOH
9901	11,17,20,21-Tetrahydroxypregn-4-en-3-one, (11β,20R)	4-Pregnene-11β,17α,20β,21-tetrol-3-on	C₂₁H₃₂O₅	116-58-5	364.476	cry (aq ace)	125 dec				vs ace, EtOH
9902	N,N,N',N'-Tetra(2-hydroxypropyl)ethylenediamine	ENTPROL	C₁₄H₃₂N₂O₄	102-60-3	292.415				1.030²⁵	1.478²⁵	sl chl
9903	Tetraiodoethene	Tetraiodoethylene	C₂I₄	513-92-8	531.639	ye lf, pr (eth)	187	sub	2.983²⁰		vs bz, chl
9904	4,5,6,7-Tetraiodo-1,3-isobenzofurandione		C₈I₄O₃	632-80-4	651.702	ye pr, nd (HOAc) nd (sub)	327.5	sub			i H₂O, EtOH, bz; sl HOAc
9905	Tetraiodomethane	Carbon tetraiodide	CI₄	507-25-5	519.629	red lf (bz, chl)	171	135¹·⁵	4.23²⁰		vs py, chl
9906	2,3,4,5-Tetraiodo-1H-pyrrole	Iodopyrrole	C₄HI₄N	87-58-1	570.676	ye nd (al)	150 dec				vs ace, eth, chl
9907	Tetraisobutyl titanate	2-Methyl-1-propanol, titanium(4+) salt	C₁₆H₃₆O₄Ti	7425-80-1	340.322			256⁵⁰⁰	0.960⁵⁰		dec H₂O

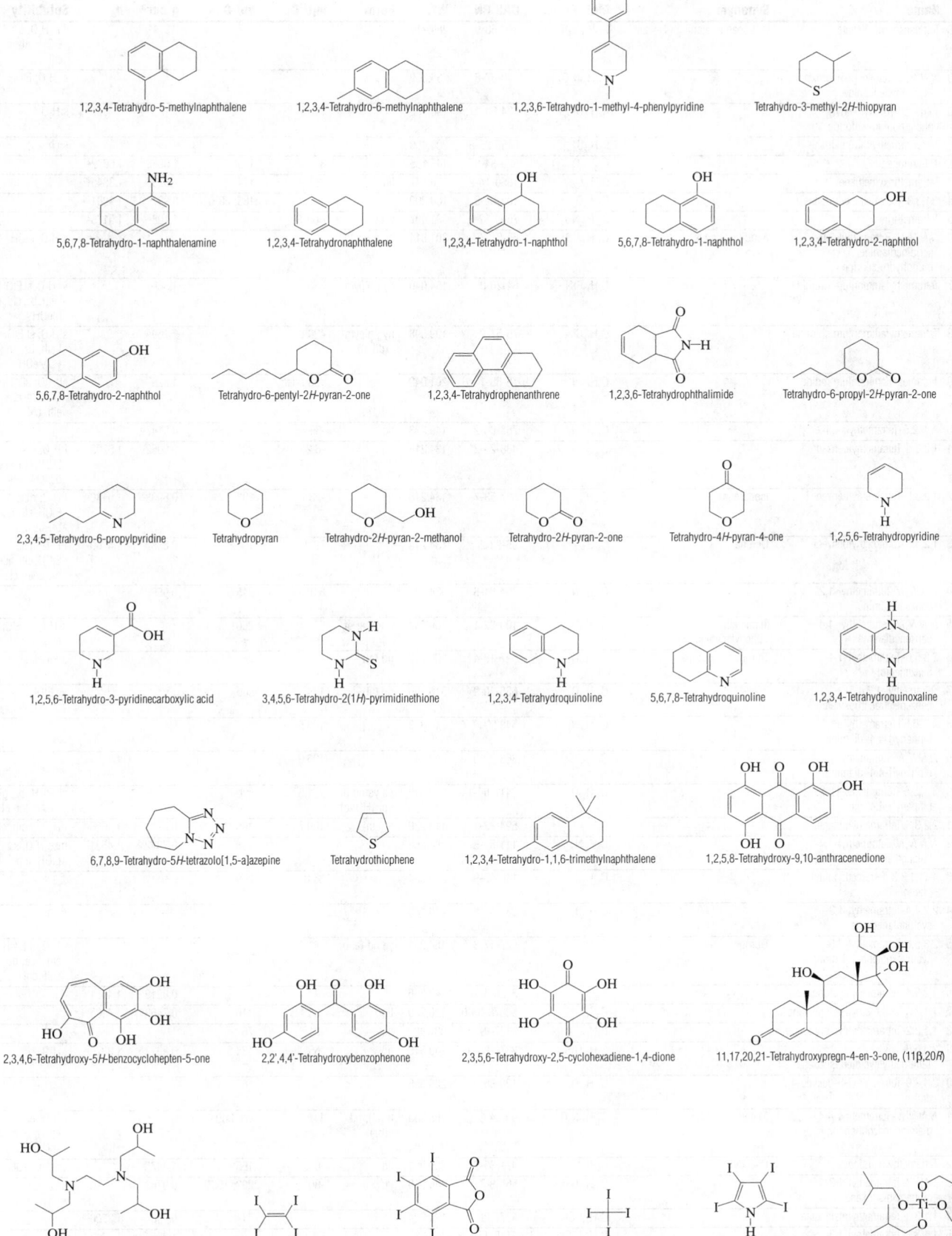

1,2,3,4-Tetrahydro-5-methylnaphthalene

1,2,3,4-Tetrahydro-6-methylnaphthalene

1,2,3,6-Tetrahydro-1-methyl-4-phenylpyridine

Tetrahydro-3-methyl-2H-thiopyran

5,6,7,8-Tetrahydro-1-naphthalenamine

1,2,3,4-Tetrahydronaphthalene

1,2,3,4-Tetrahydro-1-naphthol

5,6,7,8-Tetrahydro-1-naphthol

1,2,3,4-Tetrahydro-2-naphthol

5,6,7,8-Tetrahydro-2-naphthol

Tetrahydro-6-pentyl-2H-pyran-2-one

1,2,3,4-Tetrahydrophenanthrene

1,2,3,6-Tetrahydrophthalimide

Tetrahydro-6-propyl-2H-pyran-2-one

2,3,4,5-Tetrahydro-6-propylpyridine

Tetrahydropyran

Tetrahydro-2H-pyran-2-methanol

Tetrahydro-2H-pyran-2-one

Tetrahydro-4H-pyran-4-one

1,2,5,6-Tetrahydropyridine

1,2,5,6-Tetrahydro-3-pyridinecarboxylic acid

3,4,5,6-Tetrahydro-2(1H)-pyrimidinethione

1,2,3,4-Tetrahydroquinoline

5,6,7,8-Tetrahydroquinoline

1,2,3,4-Tetrahydroquinoxaline

6,7,8,9-Tetrahydro-5H-tetrazolo[1,5-a]azepine

Tetrahydrothiophene

1,2,3,4-Tetrahydro-1,1,6-trimethylnaphthalene

1,2,5,8-Tetrahydroxy-9,10-anthracenedione

2,3,4,6-Tetrahydroxy-5H-benzocyclohepten-5-one

2,2',4,4'-Tetrahydroxybenzophenone

2,3,5,6-Tetrahydroxy-2,5-cyclohexadiene-1,4-dione

11,17,20,21-Tetrahydroxypregn-4-en-3-one, (11β,20R)

N,N,N',N'-Tetra(2-hydroxypropyl)ethylenediamine

Tetraiodoethene

4,5,6,7-Tetraiodo-1,3-isobenzofurandione

Tetraiodomethane

2,3,4,5-Tetraiodo-1H-pyrrole

Tetraisobutyl titanate

No.	Name	Synonym	Mol. Form.	CAS RN	Mol. Wt.	Physical Form	mp/°C	bp/°C	den/ g cm⁻³	n_D	Solubility
9908	Tetraisopropyl titanate	2-Propanol, titanium(4+) salt	$C_{12}H_{28}O_4Ti$	546-68-9	284.215			227.5	0.9711[20]		dec H_2O; s EtOH, eth, bz, chl
9909	N,N,N',N'-Tetrakis(2-hydroxyethyl)-1,2-ethanediamine		$C_{10}H_{24}N_2O_4$	140-07-8	236.309						sl H_2O, EtOH
9910	Tetrakis(hydroxymethyl) phosphonium chloride		$C_4H_{12}ClO_4P$	124-64-1	190.562		152.5				s H_2O
9911	Tetrakis(methylthio)methane		$C_5H_{12}S_4$	6156-25-8	200.409						s chl
9912	1-Tetralone		$C_{10}H_{10}O$	529-34-0	146.185		8	115[6]	1.0988[16]	1.5672[20]	
9913	Tetramethoxymethane		$C_5H_{12}O_4$	1850-14-2	136.147	liq	-2.5	114	1.023[25]	1.3845[20]	
9914	1,1,3,3-Tetramethoxypropane		$C_7H_{16}O_4$	102-52-3	164.200			183; 66[12]	0.997[25]	1.4081[20]	
9915	Tetramethrin		$C_{19}H_{25}NO_4$	7696-12-0	331.407	wh cry	≈65-80		1.108[20]	1.5175[21]	
9916	N,N,N',N'-Tetramethyl-3,6-acridinediamine, monohydrochloride	Acridine Orange	$C_{17}H_{20}ClN_3$	65-61-2	301.814	oran-ye soln					s H_2O, EtOH
9917	Tetramethylammonium bromide		$C_4H_{12}BrN$	64-20-0	154.049	hyg bipym	230 dec		1.56[25]		vs H_2O; sl EtOH; i eth, bz, chl; s MeOH
9918	Tetramethylammonium chloride		$C_4H_{12}ClN$	75-57-0	109.598	hyg bipym (dil al)	420 dec		1.169[20]		s H_2O; sl EtOH; i eth, bz, chl; vs MeOH
9919	Tetramethylammonium iodide		$C_4H_{12}IN$	75-58-1	201.049		>230 dec		1.829[25]		sl H_2O, alk, EtOH, ace; i eth, chl
9920	N,N,2,6-Tetramethylaniline		$C_{10}H_{15}N$	769-06-2	149.233	liq	-36	196; 88[20]	0.9147[20]		
9921	1,2,3,4-Tetramethylbenzene		$C_{10}H_{14}$	488-23-3	134.218	liq	-6.2	205	0.9052[20]	1.5203[20]	i H_2O; msc EtOH, eth, ace, bz, peth, ctc
9922	1,2,3,5-Tetramethylbenzene	Isodurene	$C_{10}H_{14}$	527-53-7	134.218	liq	-23.7	198	0.8903[20]	1.5130[20]	i H_2O; msc EtOH, eth, ace, bz, peth, ctc
9923	1,2,4,5-Tetramethylbenzene	Durene	$C_{10}H_{14}$	95-93-2	134.218		79.3	196.8	0.8380[81]	1.4790[81]	i H_2O; msc EtOH, eth, ace, bz, peth, ctc
9924	N,N,N',N'-Tetramethyl-1,2-benzenediamine		$C_{10}H_{16}N_2$	704-01-8	164.247		8.9	215.5	0.9560[20]		
9925	N,N,N',N'-Tetramethyl-1,4-benzenediamine	Tetramethyl-p-phenylenediamine	$C_{10}H_{16}N_2$	100-22-1	164.247	lf (dil al or lig)	51	260			sl H_2O; vs EtOH, eth, bz, chl
9926	2,3,5,6-Tetramethyl-1,4-benzenediol	Durohydroquinone	$C_{10}H_{14}O_2$	527-18-4	166.217	nd (al)	233				s EtOH; sl eth
9927	Tetramethyl 1,2,4,5-benzenetetracarboxylate		$C_{14}H_{14}O_8$	635-10-9	310.256	nd (al)	144	sub			vs EtOH
9928	3,3',5,5'-Tetramethyl-[1,1'-biphenyl]-4,4'-diamine		$C_{16}H_{20}N_2$	54827-17-7	240.343		168.5				
9929	N,N,N',N'-Tetramethyl-[1,1'-biphenyl]-4,4'-diamine		$C_{16}H_{20}N_2$	366-29-0	240.343		196.0				
9930	3,3',5,5'-Tetramethyl-[1,1'-biphenyl]-4,4'-diol		$C_{16}H_{18}O_2$	2417-04-1	242.313	pa ye nd or pr (HOAc)	221.8	sub			sl EtOH, bz, gl HOAc, tol; i lig
9931	2,2,3,3-Tetramethylbutane		C_8H_{18}	594-82-1	114.229	lf (eth)	100.7	106.45	0.8242[20]	1.4695[20]	i H_2O; s eth, chl
9932	N,N,N',N'-Tetramethyl-1,4-butanediamine		$C_8H_{20}N_2$	111-51-3	144.258			168	0.7942[15]	1.4621[25]	msc H_2O; s EtOH, eth
9933	4-(1,1,3,3-Tetramethylbutyl) phenol		$C_{14}H_{22}O$	140-66-9	206.324		85.8	279			
9934	2,2,4,4-Tetramethyl-1,3-cyclobutanedione		$C_8H_{12}O_2$	933-52-8	140.180						s chl
9935	2,3,5,6-Tetramethyl-2,5-cyclohexadiene-1,4-dione	Duroquinone	$C_{10}H_{12}O_2$	527-17-3	164.201	ye nd (al or lig)	111.5				i H_2O; s EtOH, eth, ace, bz, sulf, chl
9936	1,2,3,4-Tetramethylcyclohexane		$C_{10}H_{20}$	3726-45-2	140.266				0.8219[20]	1.4531[20]	
9937	1,1,3,3-Tetramethylcyclopentane		C_9H_{18}	50876-33-0	126.239	liq	-88.4	118	0.7469[25]	1.4125[20]	
9938	1,1,2,2-Tetramethylcyclopropane		C_7H_{14}	4127-47-3	98.186	liq	-81	76			
9939	2,4,6,8-Tetramethylcyclotetrasiloxane		$C_4H_{16}O_4Si_4$	2370-88-9	240.510	liq	-65	134.5	0.9912[20]	1.3870[20]	i H_2O
9940	2,4,7,9-Tetramethyl-5-decyne-4,7-diol		$C_{14}H_{26}O_2$	126-86-3	226.355		47	165[40]			
9941	N,N,N',N'-Tetramethyl-4,4'-diaminobenzophenone	Michler's ketone	$C_{17}H_{20}N_2O$	90-94-8	268.353	lf (al), nd (bz)	179	dec 360			i H_2O, eth; sl EtOH; vs bz; s chl
9942	Tetramethyldiarsine	Cacodyl	$C_4H_{12}As_2$	471-35-2	209.981	liq	-6	165	1.447[15]		vs eth, EtOH
9943	1,1,3,3-Tetramethyl-1,3-diphenyldisiloxane		$C_{16}H_{22}OSi_2$	56-33-7	286.516	liq	-80	292; 156[13]	0.9763[20]	1.5176[20]	s ctc
9944	1,1,3,3-Tetramethyldisiloxane		$C_4H_{14}OSi_2$	3277-26-7	134.324			71	0.756[20]	1.3700[20]	
9945	1,1,3,3-Tetramethyl-1,3-disiloxanediol		$C_4H_{14}O_3Si_2$	1118-15-6	166.323		66		1.095[25]		

Tetraisopropyl titanate

N,N,N',N'-Tetrakis(2-hydroxyethyl)-1,2-ethanediamine

Tetrakis(hydroxymethyl)phosphonium chloride

Tetrakis(methylthio)methane

1-Tetralone

Tetramethoxymethane

1,1,3,3-Tetramethoxypropane

Tetramethrin

N,N,N',N'-Tetramethyl-3,6-acridinediamine, monohydrochloride

Tetramethylammonium bromide

Tetramethylammonium chloride

Tetramethylammonium iodide

N,N,2,6-Tetramethylaniline

1,2,3,4-Tetramethylbenzene

1,2,3,5-Tetramethylbenzene

1,2,4,5-Tetramethylbenzene

N,N,N',N'-Tetramethyl-1,2-benzenediamine

N,N,N',N'-Tetramethyl-1,4-benzenediamine

2,3,5,6-Tetramethyl-1,4-benzenediol

Tetramethyl 1,2,4,5-benzenetetracarboxylate

3,3',5,5'-Tetramethyl-[1,1'-biphenyl]-4,4'-diamine

N,N,N',N'-Tetramethyl-[1,1'-biphenyl]-4,4'-diamine

3,3',5,5'-Tetramethyl-[1,1'-biphenyl]-4,4'-diol

2,2,3,3-Tetramethylbutane

N,N,N',N'-Tetramethyl-1,4-butanediamine

4-(1,1,3,3-Tetramethylbutyl)phenol

2,2,4,4-Tetramethyl-1,3-cyclobutanedione

2,3,5,6-Tetramethyl-2,5-cyclohexadiene-1,4-dione

1,2,3,4-Tetramethylcyclohexane

1,1,3,3-Tetramethylcyclopentane

1,1,2,2-Tetramethylcyclopropane

2,4,6,8-Tetramethylcyclotetrasiloxane

2,4,7,9-Tetramethyl-5-decyne-4,7-diol

N,N,N',N'-Tetramethyl-4,4'-diaminobenzophenone

Tetramethyldiarsine

1,1,3,3-Tetramethyl-1,3-diphenyldisiloxane

1,1,3,3-Tetramethyldisiloxane

1,1,3,3-Tetramethyl-1,3-disiloxanediol

No.	Name	Synonym	Mol. Form.	CAS RN	Mol. Wt.	Physical Form	mp/°C	bp/°C	den/ g cm⁻³	n_D	Solubility
9946	N,N,N',N'-Tetramethyl-1,2-ethanediamine	1,2-Dimethylaminoethane	C₆H₁₆N₂	110-18-9	116.204	liq	-55	121	0.77²⁵	1.4179²⁰	
9947	Tetramethylgermane	Germanium tetramethyl	C₄H₁₂Ge	865-52-1	132.78			32⁵⁰⁰	1.006		
9948	1,1,3,3-Tetramethylguanidine		C₅H₁₃N₃	80-70-6	115.177						s ctc
9949	2,2,6,6-Tetramethyl-3,5-heptanedione	Dipivaloylmethane	C₁₁H₂₀O₂	1118-71-4	184.276			93³⁵, 72⁶	0.883²⁵	1.4589²⁰	sl ctc
9950	3,7,11,15-Tetramethylhexadecanoic acid	Phytanic acid	C₂₀H₄₀O₂	14721-66-5	312.531		-65				
9951	3,7,11,15-Tetramethyl-1-hexadecen-3-ol	Isophytol	C₂₀H₄₀O	505-32-8	296.531	oil		108⁰·⁰¹	0.8519²⁰	1.4571²⁰	vs bz, eth, EtOH
9952	2,2,3,3-Tetramethylhexane		C₁₀H₂₂	13475-81-5	142.282	liq	-54	160.3	0.7609²⁵	1.4282²⁰	
9953	2,2,5,5-Tetramethylhexane		C₁₀H₂₂	1071-81-4	142.282	liq	-12.6	137.4	0.7148²⁵	1.4055²⁰	
9954	3,3,4,4-Tetramethylhexane		C₁₀H₂₂	5171-84-6	142.282			170.0	0.7789²⁵	1.4368²⁰	
9955	N,N,N',N'-Tetramethyl-1,6-hexanediamine		C₁₀H₂₄N₂	111-18-2	172.311			209.5	0.806²⁵	1.4359²⁰	
9956	Tetramethyl lead		C₄H₁₂Pb	75-74-1	267.3	liq	-30.2	110	1.995²⁰		
9957	N,N,N',N'-Tetramethylmethanediamine		C₅H₁₄N₂	51-80-9	102.178			83	0.7491¹⁸		s H₂O
9958	Tetramethyloxirane		C₆H₁₂O	5076-20-0	100.158			90.4	0.8156¹⁶	1.3984¹⁶	s H₂O
9959	2,6,10,14-Tetramethylpentadecane	Pristane	C₁₉H₄₀	1921-70-6	268.521			296	0.7791²⁵	1.4370²⁵	vs bz, eth, chl, peth
9960	2,2,3,3-Tetramethylpentane		C₉H₂₀	7154-79-2	128.255	liq	-9.75	140.2	0.7530²⁵	1.4236²⁰	
9961	2,2,3,4-Tetramethylpentane		C₉H₂₀	1186-53-4	128.255	liq	-121.0	133.0	0.7389²⁰	1.4147²⁰	
9962	2,2,4,4-Tetramethylpentane	Di-tert-butylmethane	C₉H₂₀	1070-87-7	128.255	liq	-66.54	122.29	0.7195²⁰	1.4069²⁰	i H₂O; vs EtOH, bz
9963	2,3,3,4-Tetramethylpentane		C₉H₂₀	16747-38-9	128.255	liq	-102.1	141.5	0.7547²⁰	1.4222²⁰	
9964	2,2,4,4-Tetramethyl-3-pentanol		C₉H₂₀O	14609-79-1	144.254		52	165.5			
9965	2,3,4,5-Tetramethylphenol	Prehnitenol	C₁₀H₁₄O	488-70-0	150.217	nd (lig, aq al)	85.3	266			sl H₂O, lig; vs EtOH, eth
9966	2,3,4,6-Tetramethylphenol		C₁₀H₁₄O	3238-38-8	150.217	cry (peth)	80.5	240			s EtOH
9967	2,3,5,6-Tetramethylphenol		C₁₀H₁₄O	527-35-5	150.217	nd (lig), pr (al)	118.5	247			s chl, peth, HOAc
9968	2,2,6,6-Tetramethyl-4-piperidinamine		C₉H₂₀N₂	36768-62-4	156.268		17	188.5	0.912²⁵	1.4706²⁰	
9969	2,2,6,6-Tetramethylpiperidine	Norpempidine	C₉H₁₉N	768-66-1	141.254		28	156	0.8367¹⁶	1.4455²⁰	vs eth
9970	2,2,6,6-Tetramethyl-4-piperidinone		C₉H₁₇NO	826-36-8	155.237	orth pl (eth-w) nd (eth)	36	205			s H₂O, EtOH, eth; sl chl
9971	N,N,N',N'-Tetramethyl-1,3-propanediamine		C₇H₁₈N₂	110-95-2	130.231			144	0.7837¹⁸		msc H₂O, EtOH, eth
9972	Tetramethylpyrazine		C₈H₁₂N₂	1124-11-4	136.194	cry (w)	86	190			
9973	Tetramethylsilane	TMS	C₄H₁₂Si	75-76-3	88.224	vol liq or gas	-99.06	26.6	0.648¹⁹	1.3587²⁰	i H₂O; vs EtOH, eth; i sulf
9974	Tetramethyl silicate	Methyl silicate	C₄H₁₂O₄Si	681-84-5	152.222	liq	-1.0	121	1.0232²⁰	1.3683²⁰	vs EtOH
9975	Tetramethylstannane		C₄H₁₂Sn	594-27-4	178.848	liq	-55.1	78	1.314²⁵	1.4386	i H₂O; s ctc, CS₂
9976	Tetramethylsuccinonitrile	Tetramethylbutanedinitrile	C₈H₁₂N₂	3333-52-6	136.194	mcl pl, lf, pr (dil al)	170.5		1.070²⁵		s EtOH
9977	2,4,6,8-Tetramethyl-2,4,6,8-tetraphenylcyclotetrasiloxane		C₂₈H₃₂O₄Si₄	77-63-4	544.894	cry (HOAc)	99	237¹¹⁵	1.1183²⁰	1.5461²⁰	i H₂O; msc ace, hp
9978	Tetramethylthiodicarbonic diamide		C₆H₁₂N₂S₃	97-74-5	208.367		109.5		1.37²⁵		i H₂O; s EtOH, ace, bz, chl; sl eth
9979	Tetramethylthiourea		C₅H₁₂N₂S	2782-91-4	132.227		79.3	245			s H₂O, EtOH, chl; sl eth
9980	Tetramethylurea		C₅H₁₂N₂O	632-22-4	116.161	liq	-0.6	176.5	0.9687²⁰	1.4496²³	sl EtOH, eth, ctc
9981	Tetranitromethane		CN₄O₈	509-14-8	196.033		13.8	126.1	1.6380²⁰	1.4384²⁰	i H₂O; s EtOH, eth
9982	2,4,8,10-Tetraoxaspiro[5.5]undecane		C₇H₁₂O₄	126-54-5	160.168		48.3	147⁵³, 68¹			vs H₂O, ace, eth, EtOH
9983	2,5,8,11-Tetraoxatridecan-13-ol		C₉H₂₀O₅	23783-42-8	208.252			164¹¹	0.987²⁵	1.4453²⁰	
9984	Tetraphenoxysilane		C₂₄H₂₀O₄Si	1174-72-7	400.500		49	417; 236¹	1.1412⁶⁰		
9985	1,1,4,4-Tetraphenyl-1,3-butadiene		C₂₈H₂₂	1450-63-1	358.475		203.5				s EtOH, bz, chl, HOAc
9986	2,3,4,5-Tetraphenyl-2,4-cyclopentadien-1-one		C₂₉H₂₀O	479-33-4	384.468	blk-viol lf (HOAc, xyl)	222.3				s EtOH, bz, xyl, HOAc
9987	1,1,2,2-Tetraphenylethane		C₂₆H₂₂	632-50-8	334.453	cry (bz), orth nd (chl)	214.5	360			sl EtOH; s bz, HOAc
9988	1,1,2,2-Tetraphenyl-1,2-ethanediol	Benzopinacol	C₂₆H₂₂O₂	464-72-2	366.452	pr (bz), cry (ace)	182				i H₂O, peth; sl EtOH; s eth, ace, CS₂
9989	1,1,2,2-Tetraphenylethene		C₂₆H₂₀	632-51-9	332.437	mcl or orth (bz-eth or chl-al)	225	420	1.155⁰		i H₂O; sl EtOH, chl, eth; vs bz
9990	Tetraphenylgermane	Germanium tetraphenyl	C₂₄H₂₀Ge	1048-05-1	381.06			229.0			

N,N,N',N'-Tetramethyl-1,2-ethanediamine

Tetramethylgermane

1,1,3,3-Tetramethylguanidine

2,2,6,6-Tetramethyl-3,5-heptanedione

3,7,11,15-Tetramethylhexadecanoic acid

3,7,11,15-Tetramethyl-1-hexadecen-3-ol

2,2,3,3-Tetramethylhexane

2,2,5,5-Tetramethylhexane

3,3,4,4-Tetramethylhexane

N,N,N',N'-Tetramethyl-1,6-hexanediamine

Tetramethyl lead

N,N,N',N'-Tetramethylmethanediamine

Tetramethyloxirane

2,6,10,14-Tetramethylpentadecane

2,2,3,3-Tetramethylpentane

2,2,3,4-Tetramethylpentane

2,2,4,4-Tetramethylpentane

2,3,3,4-Tetramethylpentane

2,2,4,4-Tetramethyl-3-pentanol

2,3,4,5-Tetramethylphenol

2,3,4,6-Tetramethylphenol

2,3,5,6-Tetramethylphenol

2,2,6,6-Tetramethyl-4-piperidinamine

2,2,6,6-Tetramethylpiperidine

2,2,6,6-Tetramethyl-4-piperidinone

N,N,N',N'-Tetramethyl-1,3-propanediamine

Tetramethylpyrazine

Tetramethylsilane

Tetramethyl silicate

Tetramethylstannane

Tetramethylsuccinonitrile

2,4,6,8-Tetramethyl-2,4,6,8-tetraphenylcyclotetrasiloxane

Tetramethylthiodicarbonic diamide

Tetramethylthiourea

Tetramethylurea

Tetranitromethane

2,4,8,10-Tetraoxaspiro[5.5]undecane

2,5,8,11-Tetraoxatridecan-13-ol

Tetraphenoxysilane

1,1,4,4-Tetraphenyl-1,3-butadiene

2,3,4,5-Tetraphenyl-2,4-cyclopentadien-1-one

1,1,2,2-Tetraphenylethane

1,1,2,2-Tetraphenyl-1,2-ethanediol

1,1,2,2-Tetraphenylethene

Tetraphenylgermane

No.	Name	Synonym	Mol. Form.	CAS RN	Mol. Wt.	Physical Form	mp/°C	bp/°C	den/ g cm⁻³	n_D	Solubility
9991	Tetraphenylmethane		$C_{25}H_{20}$	630-76-2	320.427	orth nd (bz, sub)	282	431			i H_2O, EtOH, eth, lig, HOAc; s bz, tol
9992	5,6,11,12-Tetraphenylnaphthacene	Rubrene	$C_{42}H_{28}$	517-51-1	532.671	oran-red (bz-lig)	332.5				i H_2O; sl EtOH, eth, ace, py; s bz
9993	Tetraphenylplumbane		$C_{24}H_{20}Pb$	595-89-1	515.6		228.3	126¹³	1.5298²⁰		s chl
9994	Tetraphenylsilane		$C_{24}H_{20}Si$	1048-08-4	336.502		236.5	228³	1.078²⁰		s ctc, CS_2
9995	Tetraphenylstannane		$C_{24}H_{20}Sn$	595-90-4	427.126		228	420			sl chl
9996	Tetrapropoxysilane		$C_{12}H_{28}O_4Si$	682-01-9	264.434			226	0.9158²⁰	1.4012²⁰	s ctc, CS_2
9997	Tetrapropylammonium bromide	N,N,N-Tripropyl-1-propanaminium bromide	$C_{12}H_{28}BrN$	1941-30-6	266.261		252				vs H_2O, chl
9998	Tetrapropylammonium iodide		$C_{12}H_{28}IN$	631-40-3	313.261	orth bipym	280 dec		1.3138²⁵		vs H_2O, chl; s EtOH, HOAc; sl eth
9999	Tetrapropylstannane		$C_{12}H_{28}Sn$	2176-98-9	291.060	liq	-109.1	228	1.1065²⁰	1.4745²⁰	
10000	Tetrapropyl thiodiphosphate	Aspon	$C_{12}H_{28}O_5P_2S_2$	3244-90-4	378.425	amber liq		104⁰·¹	1.12²⁵	1.4710²¹	sl H_2O, peth
10001	Tetrapropyl titanate	1-Propanol, titanium(4+) salt	$C_{12}H_{28}O_4Ti$	3087-37-4	284.215			206¹⁰⁰			
10002	Tetrasodium EDTA	Edetate sodium	$C_{10}H_{12}N_2Na_4O_8$	64-02-8	380.169	amorp pow	300 (dihydrate)				sl EtOH
10003	Tetratetracontane		$C_{44}H_{90}$	7098-22-8	619.186		85.6				
10004	Tetratriacontane		$C_{34}H_{70}$	14167-59-0	478.920	pl (eth)	72.5	285.4³	0.7728⁹⁰	1.4296⁹⁰	
10005	Tetravinylsilane		$C_8H_{12}Si$	1112-55-6	136.267			130.2	0.7999²⁰	1.4625²⁰	
10006	2,4,6,8-Tetravinyl-2,4,6,8-tetramethylcyclotetrasiloxane		$C_{12}H_{24}O_4Si_4$	2554-06-5	344.659	liq	-43.5	224; 111¹²	0.9875²⁰		s ctc, CS_2
10007	1,2,4,5-Tetrazine	sym-Tetrazine	$C_2H_2N_4$	290-96-0	82.064	dk red pr	99	sub			s H_2O, EtOH, eth, sulf
10008	1H-Tetrazol-5-amine		CH_3N_5	4418-61-5	85.069		204 dec				
10009	1H-Tetrazole		CH_2N_4	288-94-8	70.054	pl (al)	157.3	sub	1.4060²⁰		sl H_2O
10010	Tetrodotoxin		$C_{11}H_{17}N_3O_8$	4368-28-9	319.268	cry	225 dec				sl H_2O,eth, EtOH; s dil HOAc
10011	Thalidomide	2-(2,6-Dioxo-3-piperidinyl)-1H-isoindole-1,3(2H)-dione	$C_{13}H_{10}N_2O_4$	50-35-1	258.229	nd	270				vs py, diox
10012	Thallium(I) ethanolate	Thallous ethoxide	C_2H_5OTl	20398-06-5	249.443	cloudy liq	-3	dec 130	3.49		dec H_2O
10013	Thebaine		$C_{19}H_{21}NO_3$	115-37-7	311.375	pl (eth), pr (dil al)	193	sub 91	1.305²⁰		i H_2O; vs EtOH, chl; sl eth; s bz
10014	Thebainone		$C_{18}H_{21}NO_3$	467-98-1	299.365	nd or pr (al)	151.5				sl H_2O, EtOH, eth; s ace, bz, AcOEt
10015	Thenaldine	1-Methyl-N-phenyl-N-(2-thienylmethyl)-4-piperidinamine	$C_{17}H_{22}N_2S$	86-12-4	286.435		96	159⁰·⁰²			
10016	Thenyldiamine		$C_{14}H_{19}N_3S$	91-79-2	261.386			170¹		1.5915²⁰	
10017	Theobromine		$C_7H_8N_4O_2$	83-67-0	180.165	orth or mcl nd (w)	357	sub 290			sl H_2O, EtOH; i eth, bz, ctc, lig, chl
10018	Theophylline	3,7-Dihydro-1,3-dimethyl-1H-purine-2,6-dione	$C_7H_8N_4O_2$	58-55-9	180.165	nd or pl (w+1)	273				s H_2O; sl EtOH, eth, chl
10019	Thiabendazole	1H-Benzimidazole, 2-(4-thiazolyl)-	$C_{10}H_7N_3S$	148-79-8	201.248			sub 305			
10020	Thiacetazone		$C_{10}H_{12}N_4OS$	104-06-3	236.293		225 dec				i H_2O, os, CS_2
10021	Thiacyclohexane		$C_5H_{10}S$	1613-51-0	102.198		19	141.8	0.9861²⁰	1.5067²⁰	i H_2O; s EtOH, eth, ace, bz
10022	1,2,5-Thiadiazole	Piazthiole	$C_2H_2N_2S$	288-39-1	86.115	liq	-50.1	94	1.268²⁵	1.5150²⁵	
10023	1,3,4-Thiadiazole		$C_2H_2N_2S$	289-06-5	86.115	cry (sub)	42.5	204			
10024	1,3,4-Thiadiazolidine-2,5-dithione		$C_2H_2N_2S_3$	1072-71-5	150.245	ye cry (MeOH)	168				s H_2O
10025	Thiamine chloride		$C_{12}H_{17}ClN_4OS$	59-43-8	300.807	cry	164				s H_2O
10026	Thiamine hydrochloride		$C_{12}H_{18}Cl_2N_4OS$	67-03-8	337.268	mcl pl	248 dec				vs H_2O; sl EtOH; i eth, bz, chl
10027	Thiamine O-phosphate, chloride		$C_{12}H_{18}ClN_4O_4PS$	532-40-1	380.787		200				
10028	Thianthrene		$C_{12}H_8S_2$	92-85-3	216.322	mcl pr or pl (al)	159.3	365	1.4420²⁰		i H_2O; sl EtOH; s eth, bz, CS_2
10029	2-Thiazolamine	2-Aminothiazole	$C_3H_4N_2S$	96-50-4	100.142	ye pl (al)	93	140¹¹			sl H_2O, EtOH, eth, chl; vs dil HCl
10030	Thiazole		C_3H_3NS	288-47-1	85.128		-33.62	118	1.1998¹⁷	1.5969²⁰	sl H_2O; s EtOH, eth, ace
10031	Thiazolidine		C_3H_7NS	504-78-9	89.160			164.5	1.131²⁵	1.551²⁰	msc H_2O; s EtOH, ctc; vs eth, ace
10032	4-Thiazolidinecarboxylic acid	Timonacic	$C_4H_7NO_2S$	444-27-9	133.170	cry (w)	196.5				vs H_2O
10033	2,4-Thiazolidinedione		$C_3H_3NO_2S$	2295-31-0	117.127	pl (w), pr (al)	128	179¹⁹			vs eth

Tetraphenylmethane

5,6,11,12-Tetraphenylnaphthacene

Tetraphenylplumbane

Tetraphenylsilane

Tetraphenylstannane

Tetrapropoxysilane

Tetrapropylammonium bromide

Tetrapropylammonium iodide

Tetrapropylstannane

Tetrapropyl thiodiphosphate

Tetrapropyl titanate

Tetrasodium EDTA

Tetratetracontane

Tetratriacontane

Tetravinylsilane

2,4,6,8-Tetravinyl-2,4,6,8-tetramethylcyclotetrasiloxane

1,2,4,5-Tetrazine

1*H*-Tetrazol-5-amine

1*H*-Tetrazole

Tetrodotoxin

Thalidomide

Thallium(I) ethanolate

Thebaine

Thebainone

Thenaldine

Thenyldiamine

Theobromine

Theophylline

Thiabendazole

Thiacetazone

Thiacyclohexane

1,2,5-Thiadiazole

1,3,4-Thiadiazole

1,3,4-Thiadiazolidine-2,5-dithione

Thiamine chloride

Thiamine hydrochloride

Thiamine *O*-phosphate, chloride

Thianthrene

2-Thiazolamine

Thiazole

Thiazolidine

4-Thiazolidinecarboxylic acid

2,4-Thiazolidinedione

No.	Name	Synonym	Mol. Form.	CAS RN	Mol. Wt.	Physical Form	mp/°C	bp/°C	den/ g cm⁻³	n_D	Solubility
10034	2-Thiazolidinethione		$C_3H_5NS_2$	96-53-7	119.209	nd (w, MeOH)	107.3				s H_2O, bz, chl; sl EtOH; i eth, CS_2
10035	Thidiazuron	N-Phenyl-N'-1,2,3-thiadiazol-5-yl-urea	$C_9H_8N_4OS$	51707-55-2	220.251			211 dec			
10036	1-(2-Thienyl)ethanone		C_6H_6OS	88-15-3	126.176		10.5	213.5	1.1679[20]	1.5667[20]	sl H_2O; msc EtOH, eth; s ctc
10037	Thiepane	Hexamethylene sulfide	$C_6H_{12}S$	4753-80-4	116.224	liq	0.5	173.5	0.991[20]	1.5044[18]	i H_2O; s eth, ace, chl
10038	Thietane	Trimethylene sulfide	C_3H_6S	287-27-4	74.145	liq	-73.24	95.0	1.0200[20]	1.5102[20]	i H_2O; vs EtOH, bz; s ace
10039	Thietane 1,1-dioxide	Trimethylene sulfone	$C_3H_6O_2S$	5687-92-3	106.144		75.5	91.2[14]		1.5156[20]	s H_2O, EtOH; sl eth, peth
10040	Thiethylperazine		$C_{22}H_{29}N_3S_2$	1420-55-9	399.615	cry	63	227[0.01]			sl ace
10041	Thiirane	Ethylene sulfide	C_2H_4S	420-12-2	60.118		-109	dec 57	1.0130[20]	1.4935[20]	sl EtOH, eth; s ace, chl
10042	Thioacetaldehyde trimer	2,4,6-Trimethyl-1,3,5-trithiane	$C_6H_{12}S_3$	2765-04-0	180.354	α-mcl pl; β-nd (ace)	101	246.5			i H_2O; s EtOH, eth, ace; vs bz, chl
10043	Thioacetamide	Ethanethioamide	C_2H_5NS	62-55-5	75.133		115.5				vs H_2O, EtOH; sl eth, bz; s DMSO
10044	Thioacetic acid		C_2H_4OS	507-09-5	76.117	ye fuming liq	<-17	93; 26[35]	1.064[20]	1.4648[20]	s H_2O, chl; vs EtOH, ace; msc eth
10045	Thiobencarb		$C_{12}H_{16}ClNOS$	28249-77-6	257.779		1.7	127[0.008]	1.16[20]		
10046	4,4'-Thiobis(6-tert-butyl-m-cresol)	Bis(5-tert-butyl-4-hydroxy-2-methylphenyl) sulfide	$C_{22}H_{30}O_2S$	96-69-5	358.537	cry	163				
10047	2,2'-Thiobisethanamine	Bis(2-aminoethyl) sulfide	$C_4H_{12}N_2S$	871-76-1	120.216	ye cry		232; 119[17]			
10048	3,3'-Thiobispropanoic acid, didodecyl ester	Didodecyl thiobispropanoate	$C_{30}H_{58}O_4S$	123-28-4	514.845		39				
10049	Thioctic acid	1,2-Dithiolane-3-pentanoic acid	$C_8H_{14}O_2S_2$	62-46-4	206.326	ye nd	61	162			i H_2O
10050	Thiocyanic acid		CHNS	463-56-9	59.091		dec 0				vs H_2O; s os
10051	Thiodicarb		$C_{10}H_{18}N_4O_4S_3$	59669-26-0	354.470		173		1.4[20]		
10052	Thiodiglycolic acid	Thiodiacetic acid	$C_4H_6O_4S$	123-93-3	150.154	cry (w)	129				sl H_2O; vs EtOH; s bz
10053	3,3'-Thiodipropionic acid		$C_6H_{10}O_4S$	111-17-1	178.206	cry wh pow	129				vs H_2O, EtOH
10054	Thiofanox		$C_9H_{18}N_2O_2S$	39196-18-4	218.316		57				
10055	Thioformaldehyde	Methanethial	CH_2S	865-36-1	46.092	unstab gas					
10056	Thioglycolic acid		$C_2H_4O_2S$	68-11-1	92.117		-16.5	120[20]	1.3253[20]	1.5080[20]	msc H_2O, EtOH, eth; sl chl
10057	Thioimidodicarbonic diamide	2,4-Dithiobiuret	$C_2H_5N_3S_2$	541-53-7	135.211	mcl cry	181 dec				vs ace
10058	Thiolactic acid		$C_3H_6O_2S$	71563-86-5	106.144		12	106[15]	1.1938[20]	1.4810[20]	s H_2O, EtOH, eth; sl chl
10059	Thiometon		$C_6H_{15}O_2PS_3$	640-15-3	246.351	oil		110[0.1], 77[0.01]	1.209[20]		sl H_2O; s os
10060	Thiomorpholine	Thiamorpholine	C_4H_9NS	123-90-0	103.186			175; 110[100]	1.0882[20]	1.5386[20]	vs H_2O, ace, eth, EtOH
10061	Thionazin	Phosphorothioic acid, O,O-diethyl O-pyrazinyl ester	$C_8H_{13}N_2O_3PS$	297-97-2	248.239	liq	-0.9	80			
10062	Thiophanate-methyl		$C_{12}H_{14}N_4O_4S_2$	23564-05-8	342.394			172 dec			
10063	Thiophene	Thiofuran	C_4H_4S	110-02-1	84.140	liq	-38.21	84.0	1.0649[20]	1.5289[20]	msc EtOH, eth, ace, bz, ctc, diox, py; sl chl
10064	2-Thiopheneacetic acid		$C_6H_6O_2S$	1918-77-0	142.176	cry (w)	76				vs H_2O, eth, EtOH
10065	2-Thiopheneacetonitrile		C_6H_5NS	20893-30-5	123.176			120[23]	1.155[25]	1.5425[20]	
10066	2-Thiophenecarbonitrile	2-Cyanothiophene	C_5H_3NS	1003-31-2	109.150			192	1.172[25]	1.5629[20]	s chl
10067	3-Thiophenecarbonitrile	3-Cyanothiophene	C_5H_3NS	1641-09-4	109.150	oil		204; 85[15]			
10068	2-Thiophenecarbonyl chloride		C_5H_3ClOS	5271-67-0	146.595			280			
10069	2-Thiophenecarboxaldehyde		C_5H_4OS	98-03-3	112.150	pa ye liq		197; 85[16]	1.2127[21]	1.5920[20]	i H_2O; vs EtOH; s eth; sl chl
10070	3-Thiophenecarboxaldehyde	3-Formylthiophene	C_5H_4OS	498-62-4	112.150			86.7[20]		1.5855[20]	i H_2O; vs EtOH, eth
10071	2-Thiophenecarboxylic acid	2-Carboxythiophene	$C_5H_4O_2S$	527-72-0	128.150	nd (w)	129.5	dec 260			vs H_2O, EtOH, eth; s chl; sl peth
10072	3-Thiophenecarboxylic acid	3-Thenoic acid	$C_5H_4O_2S$	88-13-1	128.150		138				s H_2O
10073	2,5-Thiophenedicarboxylic acid	2,5-Dicarboxythiophene	$C_6H_4O_4S$	4282-31-9	172.159		359	sub 150			sl H_2O; s EtOH, eth
10074	2-Thiophenemethanol		C_5H_6OS	636-72-6	114.166			207; 86[10]	1.2053[16]	1.5280[20]	s EtOH, ace
10075	2-Thiophenesulfonyl chloride		$C_4H_3ClO_2S_2$	16629-19-9	182.649		28	100[6]			s eth
10076	Thiopropazate		$C_{23}H_{28}ClN_3O_2S$	84-06-0	446.005			216[0.1]			
10077	4H-Thiopyran-4-thione		$C_5H_4S_2$	1120-94-1	128.216		47				

2-Thiazolidinethione

Thidiazuron

1-(2-Thienyl)ethanone

Thiepane

Thietane

Thietane 1,1-dioxide

Thiethylperazine

Thiirane

Thioacetaldehyde trimer

Thioacetamide

Thioacetic acid

Thiobencarb

4,4'-Thiobis(6-tert-butyl-m-cresol)

2,2'-Thiobisethanamine

3,3'-Thiobispropanoic acid, didodecyl ester

Thioctic acid

Thiocyanic acid

Thiodicarb

Thiodiglycolic acid

3,3'-Thiodipropionic acid

Thiofanox

Thioformaldehyde

Thioglycolic acid

Thioimidodicarbonic diamide

Thiolactic acid

Thiometon

Thiomorpholine

Thionazin

Thiophanate-methyl

Thiophene

2-Thiopheneacetic acid

2-Thiopheneacetonitrile

2-Thiophenecarbonitrile

3-Thiophenecarbonitrile

2-Thiophenecarbonyl chloride

2-Thiophenecarboxaldehyde

3-Thiophenecarboxaldehyde

2-Thiophenecarboxylic acid

3-Thiophenecarboxylic acid

2,5-Thiophenedicarboxylic acid

2-Thiophenemethanol

2-Thiophenesulfonyl chloride

Thiopropazate

4H-Thiopyran-4-thione

No.	Name	Synonym	Mol. Form.	CAS RN	Mol. Wt.	Physical Form	mp/°C	bp/°C	den/ g cm⁻³	n_D	Solubility
10078	Thioquinox		$C_9H_4N_2S_3$	93-75-4	236.336	br-ye pow	180				i H_2O; sl ace, EtOH, peth
10079	Thioridazine		$C_{12}H_{26}N_2S_2$	50-52-2	262.477	cry	73	230[0.02]			sl ace
10080	cis-Thiothixene		$C_{23}H_{29}N_3O_2S_2$	3313-26-6	443.625	cry	148				
10081	Thiourea	Thiocarbamide	CH_4N_2S	62-56-6	76.121	orth (al)	178		1.405[25]		s H_2O, EtOH; i eth
10082	9H-Thioxanthene	Dibenzothiapyran	$C_{13}H_{10}S$	261-31-4	198.283	nd (al-chl)	128.5	341			s chl
10083	9H-Thioxanthen-9-one	Thioxanthone	$C_{13}H_8OS$	492-22-8	212.267	ye nd (chl)	209	373			i H_2O, peth; sl EtOH; s bz, chl, CS_2
10084	2-Thioxo-4-imidazolidinone	2-Thiohydantoin	$C_3H_4N_2OS$	503-87-7	116.141	wh nd (w)	230 dec				vs H_2O, EtOH; s eth, alk
10085	2-Thioxo-4-thiazolidinone	Rhodanine	$C_3H_3NOS_2$	141-84-4	133.192	lt ye pr (al, w)	170		0.868[25]		sl H_2O, DMSO; vs EtOH, eth
10086	Thiram		$C_6H_{12}N_2S_4$	137-26-8	240.432	wh or ye mcl (chl-al)	155.6	129[20]			vs chl
10087	L-Threonine	2-Amino-3-hydroxybutanoic acid, [R-(R*,S*)]	$C_4H_9NO_3$	72-19-5	119.119		256 dec				s H_2O; i EtOH, eth, chl
10088	D-Threose		$C_4H_8O_4$	95-43-2	120.105	hyg-syr or nd (w)	129				
10089	L-Threose		$C_4H_8O_4$	95-44-3	120.105						vs H_2O
10090	Thujic acid	5,5-Dimethyl-1,3,6-cycloheptatriene-1-carboxylic acid	$C_{10}H_{12}O_2$	499-89-8	164.201	cry (peth)	88.5				
10091	α-Thujone	4-Methyl-1-(1-methylethyl)-bicyclo[3.1.0]hexan-3-one, (l)	$C_{10}H_{16}O$	546-80-5	152.233			201.2	0.9109[25]	1.4490[15]	i H_2O; s EtOH
10092	3-Thujopsene	Widdrene	$C_{15}H_{24}$	470-40-6	204.352	liq		122[12]	0.932[24]	1.5031[25]	
10093	Thymidine	Thymine 2-desoxyriboside	$C_{10}H_{14}N_2O_5$	50-89-5	242.228	nd (AcOEt)	186.5				s H_2O, EtOH, ace, py, HOAc; sl chl
10094	Thymine		$C_5H_6N_2O_2$	65-71-4	126.114		316				sl H_2O, EtOH, eth, DMSO
10095	Thymol	2-Isopropyl-5-methylphenol	$C_{10}H_{14}O$	89-83-8	150.217		49.5	232.5	0.970[25]	1.5227[20]	i H_2O; vs EtOH, eth, chl, AcOEt
10096	Thymol Blue		$C_{27}H_{30}O_5S$	76-61-9	466.589	grn-red (al, eth)	222 dec				sl H_2O, ace, bz; s EtOH, HOAc, PhNH₂
10097	Thymol iodide		$C_{20}H_{24}I_2O_2$	552-22-7	550.213	amorp					i H_2O; s eth; vs EtOH
10098	Thymolphthalein		$C_{28}H_{30}O_4$	125-20-2	430.536	pr or nd (al)	253				i H_2O; s EtOH, eth, ace; sl DMSO
10099	L-Thyroxine		$C_{15}H_{11}I_4NO_4$	51-48-9	776.871	nd	235				sl H_2O; i EtOH, bz
10100	Timolol		$C_{13}H_{24}N_4O_3S$	26839-75-8	316.420	oil					
10101	Tiocarlide		$C_{23}H_{32}N_2O_2S$	910-86-1	400.577		146				
10102	Tipepidine	3-(Di-2-thienylmethylene)-1-methylpiperidine	$C_{15}H_{17}NS_2$	5169-78-8	275.433	ye cry	65	181[4.5]			
10103	Tobramycin		$C_{18}H_{37}N_5O_9$	32986-56-4	467.516	cry					s H_2O
10104	β-Tocopherol	5,8-Dimethyltocol	$C_{28}H_{48}O_2$	148-03-8	416.680	pa ye visc oil		205[0.1]			vs ace, eth, EtOH, chl
10105	γ-Tocopherol	7,8-Dimethyltocol	$C_{28}H_{48}O_2$	7616-22-0	416.680	pa ye visc oil	-1.5	205[0.1]			i H_2O; msc EtOH, eth, ace, chl
10106	δ-Tocopherol	8-Methyltocol	$C_{27}H_{46}O_2$	119-13-1	402.653	pa ye visc oil		150[0.001]			i H_2O; vs EtOH, eth, ace, chl
10107	Tolazamide		$C_{14}H_{21}N_3O_3S$	1156-19-0	311.400	cry	172				
10108	Tolbutamide	N-[(Butylamino)carbonyl]-4-methylbenzenesulfonamide	$C_{12}H_{18}N_2O_3S$	64-77-7	270.347	orth cry	128.5		1.245[25]		sl H_2O; s EtOH, eth, chl
10109	o-Tolidine	3,3'-Dimethylbenzidine	$C_{14}H_{16}N_2$	119-93-7	212.290	wh-red lf (EtOH aq)	131				sl H_2O, chl; vs EtOH, eth
10110	Tolmetin		$C_{15}H_{15}NO_3$	26171-23-3	257.285	cry (MeCN)	156 dec				
10111	Toluene	Methylbenzene	C_7H_8	108-88-3	92.139	liq	-94.95	110.63	0.8668[20]	1.4961[20]	i H_2O; msc EtOH, eth; s ace, CS_2
10112	Toluene-2,4-diamine	4-Methyl-1,3-benzenediamine	$C_7H_{10}N_2$	95-80-7	122.167	nd (w), cry (al)	99	292			vs H_2O, EtOH, eth, bz; s chl
10113	Toluene-3,5-diamine	5-Methyl-1,3-benzenediamine	$C_7H_{10}N_2$	108-71-4	122.167	oil		284			
10114	Toluene-2,4-diisocyanate		$C_9H_6N_2O_2$	584-84-9	174.156		20.5	251	1.2244[20]		vs ace, bz, eth
10115	Toluene-2,6-diisocyanate		$C_9H_6N_2O_2$	91-08-7	174.156		18.3				dec H_2O; s ace, bz
10116	p-Toluenesulfonic acid		$C_7H_8O_3S$	104-15-4	172.202	hyg pl (w+1) mcl lf or pl	104.5	140[20]			vs H_2O; s EtOH, eth
10117	p-Toluenesulfonic acid monohydrate	4-Methylbenzenesulfonic acid, monohydrate	$C_7H_{10}O_4S$	6192-52-5	190.217		105.3				s H_2O

Thioquinox

Thioridazine

cis-Thiothixene

Thiourea

9H-Thioxanthene

9H-Thioxanthen-9-one

2-Thioxo-4-imidazolidinone

2-Thioxo-4-thiazolidinone

Thiram

L-Threonine

D-Threose

L-Threose

Thujic acid

α-Thujone

3-Thujopsene

Thymidine

Thymine

Thymol

Thymol Blue

Thymol iodide

Thymolphthalein

L-Thyroxine

Timolol

Tiocarlide

Tipepidine

Tobramycin

β-Tocopherol

γ-Tocopherol

δ-Tocopherol

Tolazamide

Tolbutamide

o-Tolidine

Tolmetin

Toluene

Toluene-2,4-diamine

Toluene-3,5-diamine

Toluene-2,4-diisocyanate

Toluene-2,6-diisocyanate

p-Toluenesulfonic acid

p-Toluenesulfonic acid monohydrate

No.	Name	Synonym	Mol. Form.	CAS RN	Mol. Wt.	Physical Form	mp/°C	bp/°C	den/ g cm⁻³	n_D	Solubility
10118	p-Toluenesulfonyl chloride		C₇H₇ClO₂S	98-59-9	190.648	tcl (eth, peth)	71	145[15]			i H₂O; s EtOH, eth, chl; vs bz
10119	o-Toluic acid		C₈H₈O₂	118-90-1	136.149	pr or nd (w)	103.5	259	1.062[115]	1.512[115]	i H₂O; vs EtOH, eth; s chl
10120	m-Toluic acid		C₈H₈O₂	99-04-7	136.149		109.9		1.054[112]	1.509	sl H₂O, chl; vs EtOH, eth
10121	p-Toluic acid		C₈H₈O₂	99-94-5	136.149		179.6				i H₂O; vs EtOH, eth, MeOH; sl tfa
10122	N-o-Tolylbiguanide	N-(2-Methylphenyl) imidodicarbonimidic diamide	C₉H₁₃N₅	93-69-6	191.233	nd or pl (w+1)	145.0				sl H₂O; vs EtOH, ace; i bz, chl, eth
10123	Tomatine		C₅₀H₈₃NO₂₁	17406-45-0	1034.188	nd (MeOH)	270				vs EtOH, diox
10124	Tralomethrin		C₂₂H₁₉Br₄NO₃	66841-25-6	665.007	oran-ye solid					
10125	Tranylcypromine	2-Phenylcyclopropylamine	C₉H₁₁N	155-09-9	133.190	cry	44	127[32]			
10126	Trehalose		C₁₂H₂₂O₁₁	99-20-7	342.296	orth cry	203		1.58[24]		vs H₂O; s EtOH; i eth, bz
10127	Triacetamide		C₆H₉NO₃	641-06-5	143.140	nd (eth)	79				vs eth
10128	Triacetin	Glycerol triacetate	C₉H₁₄O₆	102-76-1	218.203	col oily liq	-78	259	1.1583[20]	1.4301[20]	sl H₂O; msc EtOH, eth, bz; vs ace
10129	Triacontane		C₃₀H₆₂	638-68-6	422.813	orth (eth, bz)	65.1	449.7	0.8097[20]	1.4352[70]	i H₂O; sl EtOH; s eth; vs bz
10130	Triacontanoic acid		C₃₀H₆₀O₂	506-50-3	452.796	sc, nd (al, ace)	93.6			1.4323[100]	vs bz, CS₂, chl
10131	1-Triacontanol	Myricyl alcohol	C₃₀H₆₂O	593-50-0	438.812	nd (eth),pl (bz)	88		0.777[95]		vs bz, eth, EtOH
10132	Triadimenol	Mercury, chloro(2-methoxyethyl)-	C₃H₇ClHgO	123-88-6	295.13	cry	115				i H₂O; s EtOH, ace
10133	Triallate		C₁₀H₁₆Cl₃NOS	2303-17-5	304.664		29	117[0.0003]	1.273[25]		
10134	Triallyl phosphate		C₉H₁₅O₄P	1623-19-4	218.186		-50	108[7]	1.0815[20]		sl chl
10135	1,3,5-Triallyl-1,3,5-triazine-2,4,6(1H,3H,5H)-trione		C₁₂H₁₅N₃O₃	1025-15-6	249.265		20.5	149[4], 105[0.5]	1.1590[20]		
10136	Triamcinolone	Fluoxiprednisolone	C₂₁H₂₇FO₆	124-94-7	394.433	cry	270				
10137	Triamiphos		C₁₂H₁₉N₆OP	1031-47-6	294.292	cry (EtOH aq)	167				sl H₂O; s os
10138	Triasulfuron		C₁₄H₁₆ClN₅O₅S	82097-50-5	401.826		186				
10139	1,2,4-Triazine		C₃H₃N₃	290-38-0	81.076	pa ye oil	16.5	157		1.5149[25]	
10140	1,3,5-Triazine		C₃H₃N₃	290-87-9	81.076		80.3	114	1.38[25]		s EtOH, eth
10141	1,2,4-Triazine-3,5(2H,4H)-dione		C₃H₃N₃O₂	461-89-2	113.075		276.8				
10142	1,3,5-Triazine-2,4,6-triamine	Melamine	C₃H₆N₆	108-78-1	126.120	mcl pr (w)	345 dec	sub	1.573[16]	1.872[20]	sl H₂O, EtOH; i eth
10143	1,3,5-Triazine-2,4,6(1H,3H,5H)-trithione	Trithiocyanuric acid	C₃H₃N₃S₃	638-16-4	177.271	ye pr	>300	100[22]			
10144	Triazofos		C₁₂H₁₆N₃O₃PS	24017-47-8	313.312	ye-br oil	5		1.2514[20]		i H2O; s os
10145	Triazolam		C₁₇H₁₂Cl₂N₄	28911-01-5	343.210	tan cry (2-PrOH)	234				
10146	1H-1,2,4-Triazol-3-amine	Amitrole	C₂H₄N₄	61-82-5	84.080	cry (w, al)	159				vs H₂O, EtOH; i eth, ace; s chl; sl AcOEt
10147	1H-1,2,3-Triazole		C₂H₃N₃	288-36-8	69.065	hyg cry	23	204	1.1861[25]	1.4854[25]	s H₂O; s eth, ace; i lig
10148	1H-1,2,4-Triazole	Pyrrodiazole	C₂H₃N₃	288-88-0	69.065	nd (bz/EtOH)	120.5	260 dec			s H₂O, EtOH
10149	1H-1,2,4-Triazole-3,5-diamine		C₂H₅N₅	1455-77-2	99.095		211.5				s H₂O, EtOH; i eth, bz
10150	Tribenuron-methyl		C₁₅H₁₇N₅O₆S	101200-48-0	395.391	solid	141				
10151	Tribenzylamine	N,N-Bis(phenylmethyl) benzenemethanamine	C₂₁H₂₁N	620-40-6	287.399	pl (eth), mcl (al)	91.5	385	0.9912[95]		sl H₂O, EtOH; s eth, ctc
10152	Tribromoacetaldehyde	Bromal	C₂HBr₃O	115-17-3	280.740			174	2.6649[25]	1.5939[20]	vs ace, eth, EtOH
10153	Tribromoacetic acid		C₂HBr₃O₂	75-96-7	296.740	mcl	132	dec 245			s H₂O, EtOH, eth
10154	2,4,6-Tribromoaniline		C₆H₄Br₃N	147-82-0	329.815	nd (al, bz)	122	300	2.35[20]		i H₂O; sl EtOH; s eth, chl
10155	1,2,4-Tribromobenzene		C₆H₃Br₃	615-54-3	314.800		44.5	275			i H₂O; s EtOH; vs eth, ace; sl bz
10156	1,3,5-Tribromobenzene		C₆H₃Br₃	626-39-1	314.800	nd or pr (al)	122.8	271			i H₂O; sl EtOH; s eth, bz, chl
10157	1,1,2-Tribromobutane		C₄H₇Br₃	3675-68-1	294.811			216.2	2.1835[20]	1.5626[17]	vs eth, EtOH, chl
10158	1,2,2-Tribromobutane		C₄H₇Br₃	3675-69-2	294.811			213.8	2.1692[20]	1.568[20]	vs eth, EtOH, chl

p-Toluenesulfonyl chloride

o-Toluic acid

m-Toluic acid

p-Toluic acid

N-o-Tolylbiguanide

Tomatine

Tralomethrin

Tranylcypromine

Trehalose

Triacetamide

Triacetin

Triacontane

Triacontanoic acid

1-Triacontanol

Triadimenol

Triallate

Triallyl phosphate

1,3,5-Triallyl-1,3,5-triazine-2,4,6(1H,3H,5H)-trione

Triamcinolone

Triamiphos

Triasulfuron

1,2,4-Triazine

1,3,5-Triazine

1,2,4-Triazine-3,5(2H,4H)-dione

1,3,5-Triazine-2,4,6-triamine

1,3,5-Triazine-2,4,6(1H,3H,5H)-trithione

Triazofos

Triazolam

1H-1,2,4-Triazol-3-amine

1H-1,2,3-Triazole

1H-1,2,4-Triazole

1H-1,2,4-Triazole-3,5-diamine

Tribenuron-methyl

Tribenzylamine

Tribromoacetaldehyde

Tribromoacetic acid

2,4,6-Tribromoaniline

1,2,4-Tribromobenzene

1,3,5-Tribromobenzene

1,1,2-Tribromobutane

1,2,2-Tribromobutane

No.	Name	Synonym	Mol. Form.	CAS RN	Mol. Wt.	Physical Form	mp/°C	bp/°C	den/ g cm⁻³	n_D	Solubility
10159	1,2,3-Tribromobutane		C$_4$H$_7$Br$_3$	632-05-3	294.811	liq	-19	220	2.1907^{20}	1.5680^{20}	vs eth, EtOH, chl
10160	1,2,4-Tribromobutane		C$_4$H$_7$Br$_3$	38300-67-3	294.811	liq	-18	215	2.170^{20}	1.5608^{20}	vs eth, EtOH, chl
10161	2,2,3-Tribromobutane		C$_4$H$_7$Br$_3$	62127-47-3	294.811		0.9	206	2.1723^{20}	1.5602^{20}	i H$_2$O; s EtOH, eth, chl; sl ctc
10162	Tribromochloromethane		CBr$_3$Cl	594-15-0	287.176	lf (eth)	55	158.5	2.71^{15}		vs eth
10163	1,1,2-Tribromoethane		C$_2$H$_3$Br$_3$	78-74-0	266.757	liq	-29.3	188.93	2.6210^{20}	1.5933^{20}	i H$_2$O; s EtOH, eth, chl; ctc
10164	2,2,2-Tribromoethanol		C$_2$H$_3$Br$_3$O	75-80-9	282.756	nd or pr (peth)	81	92^{10}			vs bz, eth, EtOH
10165	Tribromoethene		C$_2$HBr$_3$	598-16-3	264.741			164	2.708^{20}	1.6045^{16}	sl H$_2$O; vs EtOH; s eth, ace, chl
10166	Tribromofluoromethane		CBr$_3$F	353-54-8	270.721	liq	-73.6	108			i H$_2$O; s EtOH
10167	Tribromomethane	Bromoform	CHBr$_3$	75-25-2	252.731		8.69	149.1	2.8788^{25}	1.5948^{25}	sl H$_2$O; msc EtOH, eth; s bz, lig, chl
10168	1,3,5-Tribromo-2-methoxybenzene		C$_7$H$_5$Br$_3$O	607-99-8	344.826	nd (al)	88	298	2.491^{25}		sl H$_2$O, EtOH; vs ace, bz; s ctc
10169	2,4,6-Tribromo-3-methylphenol	2,4,6-Tribromo-*m*-cresol	C$_7$H$_5$Br$_3$O	4619-74-3	344.826		84				s EtOH, eth, bz, HOAc; sl chl, peth
10170	1,1,1-Tribromo-2-methyl-2-propanol	1,1,1-Tribromo-*tert*-butyl alcohol	C$_4$H$_7$Br$_3$O	76-08-4	310.810	nd (lig) cry (dil al)	169	sub			sl H$_2$O, chl; s EtOH, eth
10171	Tribromonitromethane		CBr$_3$NO$_2$	464-10-8	297.729	pr	10	127^{18}	2.811^{12}	1.5790^{20}	i H$_2$O; s EtOH, eth; vs ace, bz
10172	2,4,6-Tribromophenol		C$_6$H$_3$Br$_3$O	118-79-6	330.799	nd (al), pr (bz)	95.5	286	2.55^{20}		i H$_2$O; vs EtOH; s eth, bz, HOAc, chl
10173	1,1,2-Tribromopropane		C$_3$H$_5$Br$_3$	14602-62-1	280.784			200.5	2.3547^{20}	1.5790^{20}	i H$_2$O; s EtOH, chl, HOAc; vs eth
10174	1,2,2-Tribromopropane		C$_3$H$_5$Br$_3$	14476-30-3	280.784			190.5	2.2984^{20}	1.5670^{20}	vs eth, EtOH, chl
10175	1,2,3-Tribromopropane		C$_3$H$_5$Br$_3$	96-11-7	280.784		16.9	222.1	2.4208^{20}	1.5862^{20}	i H$_2$O; vs EtOH, eth; sl ctc
10176	2,3,5-Tribromothiophene		C$_4$HBr$_3$S	3141-24-0	320.828	nd (al)	29	260			s chl
10177	Tribromotrimethyldialuminum	Methyl aluminum sesquibromide	C$_3$H$_9$Al$_2$Br$_3$	12263-85-3	338.778	hyg col liq		110^{50}			
10178	Tributyl 2-(acetyloxy)-1,2,3-propanetricarboxylate		C$_{20}$H$_{34}$O$_8$	77-90-7	402.479			173^1			sl chl
10179	Tributyl aluminate	1-Butanol, aluminum salt	C$_{12}$H$_{27}$AlO$_3$	3085-30-1	246.322			260^5			
10180	Tributylaluminum		C$_{12}$H$_{27}$Al	1116-70-7	198.324			102^2			
10181	Tributylamine	*N,N*-Dibutyl-1-butanamine	C$_{12}$H$_{27}$N	102-82-9	185.349	liq	-70	216.5	0.7770^{20}	1.4299^{20}	sl H$_2$O, ctc; vs EtOH, eth; s ace, bz
10182	Tributyl borate		C$_{12}$H$_{27}$BO$_3$	688-74-4	230.151	oil	<-70	234	0.8567^{20}	1.4106^{18}	s EtOH, bz; vs eth, MeOH
10183	Tributylfluorostannane	Tributyltin fluoride	C$_{12}$H$_{27}$FSn	1983-10-4	309.050	nd	≈260	sub >200			
10184	2,4,6-Tri-*tert*-butylphenol		C$_{18}$H$_{30}$O	732-26-3	262.430	cry (al, peth)	131	278	0.864^{27}		i H$_2$O, alk; s EtOH, ace, ctc
10185	Tributyl phosphate		C$_{12}$H$_{27}$O$_4$P	126-73-8	266.313			289	0.9727^{25}	1.4224^{25}	s H$_2$O, eth, bz, CS$_2$; msc EtOH
10186	Tributylphosphine		C$_{12}$H$_{27}$P	998-40-3	202.316			240; 150^{50}	0.812^{25}	1.4619^{20}	
10187	Tributyl phosphite	Tributoxyphosphine	C$_{12}$H$_{27}$O$_3$P	102-85-2	250.314			137^{26}, 122^{12}	0.9259^{20}	1.4321^{19}	s EtOH; sl ctc; vs eth
10188	*S,S,S*-Tributyl phosphorotrithioate	*S,S,S*-Tributyl trithiophosphate	C$_{12}$H$_{27}$OPS$_3$	78-48-8	314.510		<-25	150$^{0.3}$	1.057^{20}		
10189	Tributylsilane		C$_{12}$H$_{28}$Si	998-41-4	200.436			221	0.7794^{20}	1.4380^{20}	
10190	Tributylstannane	Tributyltin hydride	C$_{12}$H$_{28}$Sn	688-73-3	291.060	liq		113^8, 76$^{0.7}$	1.103^{20}		
10191	Tributyrin	Butanoic acid, 1,2,3-propanetriyl ester	C$_{15}$H$_{26}$O$_6$	60-01-5	302.363	liq	-75	307.5	1.0350^{20}	1.4359^{20}	i H$_2$O; s EtOH, ace, bz; sl ctc; vs eth
10192	Tricalcium citrate	Calcium citrate	C$_{12}$H$_{10}$Ca$_3$O$_{14}$	813-94-5	498.433	cry (w)	≈100 dec (hyd)				sl H$_2$O; i EtOH
10193	Trichlorfon		C$_4$H$_8$Cl$_3$O$_4$P	52-68-6	257.437		77	100$^{0.1}$	1.73^{20}		
10194	Trichloroacetaldehyde	Chloral	C$_2$HCl$_3$O	75-87-6	147.387	liq	-57.5	97.8	1.512^{20}	1.4580^{20}	vs H$_2$O; s EtOH, eth
10195	2,2,2-Trichloroacetamide		C$_2$H$_2$Cl$_3$NO	594-65-0	162.402		142	240			sl H$_2$O; vs EtOH, eth
10196	Trichloroacetic acid		C$_2$HCl$_3$O$_2$	76-03-9	163.387	hyg cry	59.2	196.5	1.6126^{64}	1.4603^{61}	vs H$_2$O; s EtOH, eth; sl ctc
10197	Trichloroacetic anhydride		C$_4$Cl$_6$O$_3$	4124-31-6	308.759			dec 223; 139^{60}	1.6908^{20}		vs eth, HOAc
10198	Trichloroacetonitrile		C$_2$Cl$_3$N	545-06-2	144.387	liq	-42	85.7	1.4403^{25}	1.4409^{20}	i H$_2$O
10199	Trichloroacetyl chloride		C$_2$Cl$_4$O	76-02-8	181.832			117.9	1.6202^{20}	1.4695^{20}	msc eth
10200	2,3,4-Trichloroaniline		C$_6$H$_4$Cl$_3$N	634-67-3	196.462	nd (lig)	73	292			vs EtOH

1,2,3-Tribromobutane

1,2,4-Tribromobutane

2,2,3-Tribromobutane

Tribromochloromethane

1,1,2-Tribromoethane

2,2,2-Tribromoethanol

Tribromoethene

Tribromofluoromethane

Tribromomethane

1,3,5-Tribromo-2-methoxybenzene

2,4,6-Tribromo-3-methylphenol

1,1,1-Tribromo-2-methyl-2-propanol

Tribromonitromethane

2,4,6-Tribromophenol

1,1,2-Tribromopropane

1,2,2-Tribromopropane

1,2,3-Tribromopropane

2,3,5-Tribromothiophene

Tribromotrimethyldialuminum

Tributyl 2-(acetyloxy)-1,2,3-propanetricarboxylate

Tributyl aluminate

Tributylaluminum

Tributylamine

Tributyl borate

Tributylfluorostannane

2,4,6-Tri-tert-butylphenol

Tributyl phosphate

Tributylphosphine

Tributyl phosphite

S,S,S-Tributyl phosphorotrithioate

Tributylsilane

Tributylstannane

Tributyrin

Tricalcium citrate

Trichlorfon

Trichloroacetaldehyde

2,2,2-Trichloroacetamide

Trichloroacetic acid

Trichloroacetic anhydride

Trichloroacetonitrile

Trichloroacetyl chloride

2,3,4-Trichloroaniline

No.	Name	Synonym	Mol. Form.	CAS RN	Mol. Wt.	Physical Form	mp/°C	bp/°C	den/ g cm⁻³	n_D	Solubility
10201	2,4,5-Trichloroaniline		$C_6H_4Cl_3N$	636-30-6	196.462	nd (liq)	96.5	270			s EtOH, eth; vs CS_2; sl lig
10202	2,4,6-Trichloroaniline		$C_6H_4Cl_3N$	634-93-5	196.462	cry (al), nd (lig or peth)	78.5	262			i H_2O; s EtOH, eth, chl; vs CS_2
10203	2,3,6-Trichlorobenzaldehyde		$C_7H_3Cl_3O$	4659-47-6	209.457	nd (liq)	87.3				vs ace, bz, eth
10204	1,2,3-Trichlorobenzene		$C_6H_3Cl_3$	87-61-6	181.447	pl (al)	51.3	218.5	1.4533²⁵		i H_2O; sl EtOH, chl; vs eth, bz
10205	1,2,4-Trichlorobenzene		$C_6H_3Cl_3$	120-82-1	181.447	orth	16.92	213.5	1.459²⁵	1.5717²⁰	i H_2O; sl EtOH, chl; vs eth
10206	1,3,5-Trichlorobenzene		$C_6H_3Cl_3$	108-70-3	181.447	nd	62.8	208			i H_2O; sl EtOH; vs eth, bz; s chl
10207	2,3,6-Trichlorobenzeneacetic acid	Chlorfenac	$C_8H_5Cl_3O_2$	85-34-7	239.484			161			
10208	3,4,5-Trichloro-1,2-benzenediol		$C_6H_3Cl_3O_2$	56961-20-7	213.446	(i) pr (HOAc) (ii) pr (bz)	115(form a); 134(form b)				sl H_2O; vs eth, EtOH, HOAc
10209	2,3,6-Trichlorobenzoic acid		$C_7H_3Cl_3O_2$	50-31-7	225.457		124.5				sl H_2O; s eth
10210	2,4,5-Trichlorobiphenyl		$C_{12}H_7Cl_3$	15862-07-4	257.543	cry	78.5				i H_2O
10211	2,4,6-Trichlorobiphenyl		$C_{12}H_7Cl_3$	35693-92-6	257.543	cry (EtOH aq)	62.5	172¹⁵			i H_2O
10212	1,1,1-Trichloro-2,2-bis(4-chlorophenyl)ethane	Dichlorodiphenyltrichloroethane (DDT)	$C_{14}H_9Cl_5$	50-29-3	354.486	nd (al)	108.5	260; 186⁰·⁰⁵			i H_2O; sl EtOH; vs eth, ace, bz, py
10213	2,2,3-Trichlorobutanal	2,2,3-Trichlorobutyraldehyde	$C_4H_5Cl_3O$	76-36-8	175.441			164	1.3956²⁰	1.4755²⁰	vs H_2O, eth, EtOH
10214	2,3,4-Trichloro-1-butene		$C_4H_5Cl_3$	2431-50-7	159.442			60²⁰, 40¹⁰	1.3430²⁰	1.4944²⁰	vs ace, chl
10215	3,4,4'-Trichlorocarbanilide	Triclocarban	$C_{13}H_9Cl_3N_2O$	101-20-2	315.581	fine pl	256				
10216	1,2,4-Trichloro-5-(chloromethyl)benzene		$C_7H_4Cl_4$	3955-26-8	229.919			273	1.547²⁰		vs ace, eth, EtOH
10217	Trichloro(chloromethyl)silane	(Chloromethyl)trichlorosilane	CH_2Cl_4Si	1558-25-4	183.925			118	1.4650²⁰	1.4555²⁰	
10218	Trichloro(4-chlorophenyl)silane		$C_6H_4Cl_4Si$	825-94-5	245.994			233; 116²⁰	1.4062²⁰	1.5418²⁰	
10219	Trichloro(3-chloropropyl)silane		$C_3H_6Cl_4Si$	2550-06-3	211.978			181.5	1.3590²⁰	1.4668²⁰	
10220	Trichloro(dichloromethyl)silane	(Dichloromethyl)trichlorosilane	$CHCl_5Si$	1558-24-3	218.370			145	1.5518²⁰	1.4714²⁰	
10221	1,1,1-Trichloro-2,2-difluoroethane		$C_2HCl_3F_2$	354-12-1	169.385			73			
10222	1,2,2-Trichloro-1,1-difluoroethane		$C_2HCl_3F_2$	354-21-2	169.385		-140	71.9	1.5447²⁰	1.3889²⁰	
10223	1,2,2-Trichloro-1,2-difluoroethane		$C_2HCl_3F_2$	354-15-4	169.385		-174	72.5			
10224	2,4,6-Trichloro-3,5-dimethylphenol		$C_8H_7Cl_3O$	6972-47-0	225.500	ye nd (peth)	175				i H_2O; s chl; vs peth
10225	1,1,1-Trichloro-2,2-diphenylethane		$C_{14}H_{11}Cl_3$	2971-22-4	285.596		65				s EtOH; sl chl
10226	Trichlorododecylsilane	Dodecyltrichlorosilane	$C_{12}H_{25}Cl_3Si$	4484-72-4	303.772			155¹⁰		1.4581²⁰	
10227	1,1,1-Trichloro-3,4-epoxybutane	(2,2,2-Trichloroethyl)oxirane	$C_4H_5Cl_3O$	3083-25-8	175.441	liq		110¹⁰⁰			
10228	1,1,1-Trichloroethane	Methyl chloroform	$C_2H_3Cl_3$	71-55-6	133.404	liq	-30.01	74.09	1.3390²⁰	1.4379²⁰	sl H_2O; s EtOH, chl; msc eth
10229	1,1,2-Trichloroethane	Vinyl trichloride	$C_2H_3Cl_3$	79-00-5	133.404	liq	-36.3	113.8	1.4397²⁰	1.4714²⁰	i H_2O; s EtOH, eth, chl
10230	2,2,2-Trichloroethanol		$C_2H_3Cl_3O$	115-20-8	149.403	hyg orth tab or pl	19	152; 52¹¹		1.4861²⁰	sl H_2O, ctc; msc EtOH, eth; s alk
10231	Trichloroethene	Trichloroethylene	C_2HCl_3	79-01-6	131.388	liq	-84.7	87.21	1.4642²⁰	1.4773²⁰	sl H_2O, ctc; msc EtOH, eth; s ace
10232	2,2,2-Trichloro-1-ethoxyethanol	Chloral alcoholate	$C_4H_7Cl_3O_2$	515-83-3	193.457		56.5	115.5	1.143⁴⁰		s H_2O, EtOH, eth
10233	Trichloroethoxysilane		$C_2H_5Cl_3OSi$	1825-82-7	179.505	liq	-135	101.9	1.2274²⁰	1.4045²⁰	vs EtOH
10234	2,2,2-Trichloroethyl-β-D-glucopyranosiduronic acid	Urochloralic acid	$C_8H_{11}Cl_3O_7$	97-25-6	325.528	nd	142				vs H_2O, EtOH
10235	Trichloroethylsilane	Ethyltrichlorosilane	$C_2H_5Cl_3Si$	115-21-9	163.506	liq	-105.6	100.5	1.2373²⁰	1.4256²⁰	s ctc
10236	1,1,1-Trichloro-2-fluoroethane	Refrigerant 131b	$C_2H_2Cl_3F$	2366-36-1	151.394	liq		86.5			
10237	1,1,2-Trichloro-1-fluoroethane	Refrigerant 131a	$C_2H_2Cl_3F$	811-95-0	151.394		-104.7	88.0	1.492²⁰		
10238	1,1,2-Trichloro-2-fluoroethane		$C_2H_2Cl_3F$	359-28-4	151.394			102.4	1.5393²⁰	1.4390²⁰	i H_2O
10239	Trichlorofluoromethane	Refrigerant 11	CCl_3F	75-69-4	137.368	vol liq or gas	-110.44	23.7			i H_2O
10240	2,2,3-Trichloro-1,1,1,3,4,4,4-heptafluorobutane		$C_4Cl_3F_7$	335-44-4	287.391		2.0	98	1.7484²⁰	1.3530²⁰	
10241	Trichlorohexylsilane	Hexyltrichlorosilane	$C_6H_{13}Cl_3Si$	928-65-4	219.612			190	1.1100²⁰		dec H_2O
10242	N-(2,2,2-Trichloro-1-hydroxyethyl)formamide	Chloral formamide	$C_3H_4Cl_3NO_2$	515-82-2	192.429	cry	120				vs ace, eth, EtOH
10243	3,3,3-Trichloro-2-hydroxypropanenitrile	Chlorocyanohydrin	$C_3H_2Cl_3NO$	513-96-2	174.413	pl (w)	61	dec 217			vs H_2O, eth, EtOH
10244	Trichloroisobutylsilane		$C_4H_9Cl_3Si$	18169-57-8	191.559			143.3	1.154²⁰		dec H_2O
10245	Trichloromethane	Chloroform	$CHCl_3$	67-66-3	119.378	liq	-63.41	61.17	1.4788²⁵	1.4459²⁰	sl H_2O; msc EtOH, eth, bz; s ace, ctc
10246	Trichloromethanesulfenyl chloride	Perchloromethyl mercaptan	CCl_4S	594-42-3	185.888	ye oil		149	1.6947²⁰	1.5484²⁰	s eth

2,4,5-Trichloroaniline

2,4,6-Trichloroaniline

2,3,6-Trichlorobenzaldehyde

1,2,3-Trichlorobenzene

1,2,4-Trichlorobenzene

1,3,5-Trichlorobenzene

2,3,6-Trichlorobenzeneacetic acid

3,4,5-Trichloro-1,2-benzenediol

2,3,6-Trichlorobenzoic acid

2,4,5-Trichlorobiphenyl

2,4,6-Trichlorobiphenyl

1,1,1-Trichloro-2,2-bis(4-chlorophenyl)ethane

2,2,3-Trichlorobutanal

2,3,4-Trichloro-1-butene

3,4,4'-Trichlorocarbanilide

1,2,4-Trichloro-5-(chloromethyl)benzene

Trichloro(chloromethyl)silane

Trichloro(4-chlorophenyl)silane

Trichloro(3-chloropropyl)silane

Trichloro(dichloromethyl)silane

1,1,1-Trichloro-2,2-difluoroethane

1,2,2-Trichloro-1,1-difluoroethane

1,2,2-Trichloro-1,2-difluoroethane

2,4,6-Trichloro-3,5-dimethylphenol

1,1,1-Trichloro-2,2-diphenylethane

Trichlorododecylsilane

1,1,1-Trichloro-3,4-epoxybutane

1,1,1-Trichloroethane

1,1,2-Trichloroethane

2,2,2-Trichloroethanol

Trichloroethene

2,2,2-Trichloro-1-ethoxyethanol

Trichloroethoxysilane

2,2,2-Trichloroethyl-β-D-glucopyranosiduronic acid

Trichloroethylsilane

1,1,1-Trichloro-2-fluoroethane

1,1,2-Trichloro-1-fluoroethane

1,1,2-Trichloro-2-fluoroethane

Trichlorofluoromethane

2,2,3-Trichloro-1,1,1,3,4,4,4-heptafluorobutane

Trichlorohexylsilane

N-(2,2,2-Trichloro-1-hydroxyethyl)formamide

3,3,3-Trichloro-2-hydroxypropanenitrile

Trichloroisobutylsilane

Trichloromethane

Trichloromethanesulfenyl chloride

No.	Name	Synonym	Mol. Form.	CAS RN	Mol. Wt.	Physical Form	mp/°C	bp/°C	den/ g cm⁻³	n_D	Solubility
10247	Trichloromethanesulfonyl chloride		CCl₄O₂S	2547-61-7	217.887	cry (al-w)	140.5	170			i H₂O; s EtOH, eth, CS₂
10248	Trichloromethanethiol	Trichloromethyl mercaptan	CHCl₃S	75-70-7	151.443	oran oil		125¹⁵			
10249	Trichloromethiazide		C₈H₈Cl₃N₃O₄S₂	133-67-5	380.657		270 dec				sl H₂O; s EtOH
10250	1,2,4-Trichloro-5-methoxybenzene		C₇H₅Cl₃O	6130-75-2	211.473	nd (dil al)	77.5	254			vs EtOH, ace
10251	1,3,5-Trichloro-2-methoxybenzene	2,4,6-Trichloroanisole	C₇H₅Cl₃O	87-40-1	211.473	mcl nd (al)	61.5	241	1.640²⁵		s EtOH, bz, chl; vs ace
10252	1,2,4-Trichloro-5-methylbenzene	2,4,5-Trichlorotoluene	C₇H₅Cl₃	6639-30-1	195.474	nd or lf (al)	82.4	231			i H₂O; s EtOH, ace
10253	(Trichloromethyl)benzene	Benzotrichloride	C₇H₅Cl₃	98-07-7	195.474	liq	-4.42	221	1.3723²⁰	1.5580²⁰	i H₂O; s EtOH, eth, bz
10254	(Trichloromethyl)oxirane		C₃H₃Cl₃O	3083-23-6	161.414			149; 44¹³	1.495²⁰	1.4737²⁵	vs eth; s chl
10255	2,3,4-Trichloro-6-methylphenol	4,5,6-Trichloro-o-cresol	C₇H₅Cl₃O	551-78-0	211.473	nd (peth)	77				
10256	2,3,6-Trichloro-4-methylphenol	2,3,6-Trichloro-p-cresol	C₇H₅Cl₃O	551-77-9	211.473	nd (HOAc, peth)	66.5				vs EtOH
10257	2,4,6-Trichloro-3-methylphenol	2,4,6-Trichloro-m-cresol	C₇H₅Cl₃O	551-76-8	211.473	nd or pl (w, peth)	46	265			i H₂O; vs EtOH, MeOH, chl
10258	1,1,1-Trichloro-2-methyl-2-propanol	1,1,1-Trichloro-tert-butyl alcohol	C₄H₇Cl₃O	57-15-8	177.457	hyg nd (w + 1)	97	167			i H₂O; s EtOH, eth, ace, bz, lig, chl
10259	Trichloronate		C₁₀H₁₂Cl₃O₂PS	327-98-0	333.599	ye liq		108⁰·⁰¹	1.365²⁰		
10260	1,2,4-Trichloro-5-nitrobenzene		C₆H₂Cl₃NO₂	89-69-0	226.445	pr (al), nd (al)	57.5	288	1.790²³		i H₂O; sl EtOH; s eth, bz, chl, CS₂
10261	Trichloronitromethane	Chloropicrin	CCl₃NO₂	76-06-2	164.376	liq	-64	112	1.6558²⁰	1.4611²⁰	s H₂O; msc EtOH, ace, bz, MeOH, HOAc
10262	3,4,6-Trichloro-2-nitrophenol		C₆H₂Cl₃NO₃	82-62-2	242.444	pa ye cry (peth)	92.5				
10263	Trichlorooctadecylsilane	Octadecyltrichlorosilane	C₁₈H₃₇Cl₃Si	112-04-9	387.932			223¹⁰	0.984²⁵	1.4602²⁰	
10264	Trichlorooctylsilane	Octyltrichlorosilane	C₈H₁₇Cl₃Si	5283-66-9	247.666			232		1.4480²⁰	dec H₂O, EtOH; s ctc
10265	1,2,3-Trichloro-1,1,2,3,3-pentafluoropropane		C₃Cl₃F₅	76-17-5	237.383	liq	-72	73.7	1.6631²⁰	1.3512²⁰	
10266	Trichloropentylsilane	Amyltrichlorosilane	C₅H₁₁Cl₃Si	107-72-2	205.586			172; 60.5¹⁵	1.1330²⁰	1.4503²⁰	
10267	2,3,4-Trichlorophenol		C₆H₃Cl₃O	15950-66-0	197.446	nd (bz, lig, sub)	83.5	sub			s EtOH, eth, bz, alk, HOAc
10268	2,3,5-Trichlorophenol		C₆H₃Cl₃O	933-78-8	197.446	nd (al)	62	248²⁵⁰			vs eth, EtOH
10269	2,3,6-Trichlorophenol		C₆H₃Cl₃O	933-75-5	197.446	nd (dil al, lig)	58				sl H₂O; vs EtOH, eth, bz; s HOAc
10270	2,4,5-Trichlorophenol		C₆H₃Cl₃O	95-95-4	197.446	nd (al, peth)	69	247			sl H₂O; vs EtOH, eth, bz; s HOAc
10271	2,4,6-Trichlorophenol		C₆H₃Cl₃O	88-06-2	197.446	orth nd (HOAc)	69	246	1.4901⁷⁵		sl H₂O; s EtOH, eth, HOAc
10272	3,4,5-Trichlorophenol		C₆H₃Cl₃O	609-19-8	197.446	nd (lig)	101	275			sl H₂O, lig; s eth
10273	2,4,5-Trichlorophenoxyacetic acid	2,4,5-T	C₈H₅Cl₃O₃	93-76-5	255.483	cry (bz)	153	dec			i H₂O; s EtOH; vs bz
10274	2-(2,4,5-Trichlorophenoxy)ethyl 2,2-dichloropropanoate	Pentanate	C₁₁H₉Cl₅O₃	136-25-4	366.452		49	162⁰·⁵	1.55⁵⁰		i H₂O; s EtOH, ace, xyl
10275	Trichloro(2-phenylethyl)silane		C₈H₉Cl₃Si	940-41-0	239.602			242; 98⁵	1.2397²⁰	1.5185²⁰	
10276	(2,4,6-Trichlorophenyl)hydrazine		C₆H₅Cl₃N₂	5329-12-4	211.476	cry (bz)	143				s H₂O, bz
10277	Trichlorophenylsilane		C₆H₅Cl₃Si	98-13-5	211.549			201	1.321²⁰	1.5230²⁰	s ctc, chl, CS₂
10278	1,1,2-Trichloropropane		C₃H₅Cl₃	598-77-6	147.431			132.0; 117⁵⁰⁰	1.372¹⁵		i H₂O; s EtOH, chl; vs eth; sl ctc
10279	1,1,3-Trichloropropane		C₃H₅Cl₃	20395-25-9	147.431	liq	-59	145.5	1.3557²⁰	1.4718²⁰	vs eth, EtOH, chl
10280	1,2,2-Trichloropropane		C₃H₅Cl₃	3175-23-3	147.431			124	1.318²⁵	1.4609²⁰	i H₂O; s EtOH, eth; vs chl
10281	1,2,3-Trichloropropane		C₃H₅Cl₃	96-18-4	147.431	liq	-14.7	157	1.3889²⁰	1.4852²⁰	sl H₂O, ctc; s EtOH, eth; vs chl
10282	1,1,1-Trichloro-2-propanol		C₃H₅Cl₃O	76-00-6	163.430		50.5	163; 54¹²			vs ace, bz, eth, EtOH
10283	1,1,1-Trichloro-2-propanone	1,1,1-Trichloroacetone	C₃H₃Cl₃O	918-00-3	161.414			149; 28¹⁰	1.435²⁰	1.4635¹⁷	i H₂O; vs EtOH, eth
10284	1,2,3-Trichloro-1-propene		C₃H₃Cl₃	96-19-5	145.415			142	1.412²⁰	1.5030²⁰	i H₂O; vs EtOH, eth; s bz, chl
10285	3,3,3-Trichloro-1-propene		C₃H₃Cl₃	2233-00-3	145.415	liq	-30	114.5	1.367²⁰	1.4827²⁰	i H₂O; s EtOH, eth, bz, chl
10286	2,3,3-Trichloro-2-propenoyl chloride		C₃Cl₄O	815-58-7	193.843			158		1.5271¹⁸	vs bz
10287	Trichloropropylsilane	Propyltrichlorosilane	C₃H₇Cl₃Si	141-57-1	177.533			123.5	1.195²⁰	1.4310²⁰	

Trichloromethanesulfonyl chloride

Trichloromethanethiol

Trichloromethiazide

1,2,4-Trichloro-5-methoxybenzene

1,3,5-Trichloro-2-methoxybenzene

1,2,4-Trichloro-5-methylbenzene

(Trichloromethyl)benzene

(Trichloromethyl)oxirane

2,3,4-Trichloro-6-methylphenol

2,3,6-Trichloro-4-methylphenol

2,4,6-Trichloro-3-methylphenol

1,1,1-Trichloro-2-methyl-2-propanol

Trichloronate

1,2,4-Trichloro-5-nitrobenzene

Trichloronitromethane

3,4,6-Trichloro-2-nitrophenol

Trichlorooctadecylsilane

Trichlorooctylsilane

1,2,3-Trichloro-1,1,2,3,3-pentafluoropropane

Trichloropentylsilane

2,3,4-Trichlorophenol

2,3,5-Trichlorophenol

2,3,6-Trichlorophenol

2,4,5-Trichlorophenol

2,4,6-Trichlorophenol

3,4,5-Trichlorophenol

2,4,5-Trichlorophenoxyacetic acid

2-(2,4,5-Trichlorophenoxy)ethyl 2,2-dichloropropanoate

Trichloro(2-phenylethyl)silane

(2,4,6-Trichlorophenyl)hydrazine

Trichlorophenylsilane

1,1,2-Trichloropropane

1,1,3-Trichloropropane

1,2,2-Trichloropropane

1,2,3-Trichloropropane

1,1,1-Trichloro-2-propanol

1,1,1-Trichloro-2-propanone

1,2,3-Trichloro-1-propene

3,3,3-Trichloro-1-propene

2,3,3-Trichloro-2-propenoyl chloride

Trichloropropylsilane

No.	Name	Synonym	Mol. Form.	CAS RN	Mol. Wt.	Physical Form	mp/°C	bp/°C	den/g cm^{-3}	n_D	Solubility
10288	2,4,6-Trichloropyrimidine		$C_4HCl_3N_2$	3764-01-0	183.423		22.5	212.5		1.5700^{20}	
10289	3-(Trichlorosilyl)propanenitrile		$C_3H_4Cl_3NSi$	1071-22-3	188.516			109^{30}			
10290	2,4,6-Trichloro-1,3,5-triazine	Cyanuric acic trichloride	$C_3Cl_3N_3$	108-77-0	184.411	cry (eth, bz)	154	192			vs EtOH
10291	2,2',2"-Trichlorotriethylamine		$C_6H_{12}Cl_3N$	555-77-1	204.525	pa ye	-2.0	143^{15}			vs bz, eth, EtOH
10292	Trichlorotriethyldialuminum	Ethylaluminum sesquichloride	$C_6H_{15}Al_2Cl_3$	12075-68-2	247.505	ye liq		115.5^{50}, $36.2^{0.2}$			
10293	1,3,5-Trichloro-2,4,6-trifluorobenzene		$C_6Cl_3F_3$	319-88-0	235.418			198.4			
10294	1,1,1-Trichloro-2,2,2-trifluoroethane		$C_2Cl_3F_3$	354-58-5	187.375		14.37	45.5	1.5790^{20}	1.3610^{35}	i H_2O; s EtOH, eth, chl
10295	1,1,2-Trichloro-1,2,2-trifluoroethane		$C_2Cl_3F_3$	76-13-1	187.375	liq	-36.22	47.7	1.5635^{25}	1.3557^{25}	i H_2O; s EtOH; msc eth, bz
10296	Trichlorovinylsilane	Vinyltrichlorosilane	$C_2H_3Cl_3Si$	75-94-5	161.490	liq	-95	91.5	1.2426^{20}	1.4295^{20}	vs chl
10297	Trichodermin	12,13-Epoxytrichothec-9-en-4-ol acetate	$C_{17}H_{24}O_4$	4682-50-2	292.371	cry	59	$111^{0.05}$			sl H_2O; s EtOH, chl
10298	Triclofos	2,2,2-Trichloroethanol dihydrogen phosphate	$C_2H_4Cl_3O_4P$	306-52-5	229.383	cry (bz)	120.5				
10299	Triclopyr	Acetic acid, [(3,5,6-trichloro-2-pyridinyl)oxy]-	$C_7H_4Cl_3NO_3$	55335-06-3	256.471		149	dec 290			
10300	Tricosane		$C_{23}H_{48}$	638-67-5	324.627	lf (eth-al)	47.76	380	0.7785^{48}	1.4468^{20}	i H_2O; sl EtOH; s eth, ctc
10301	12-Tricosanone	Diundecyl ketone	$C_{23}H_{46}O$	540-09-0	338.610	lf (al)	70.2		0.8086^{69}	1.4283^{60}	vs bz, eth, chl
10302	Tri-o-cresyl phosphate	Tri-o-tolyl phosphate	$C_{21}H_{21}O_4P$	78-30-8	368.363	col or pa ye	11	410	1.1955^{20}	1.5575^{20}	i H_2O; vs EtOH, eth, ctc, tol; s HOAc
10303	Tri-m-cresyl phosphate	Tri-m-tolyl phosphate	$C_{21}H_{21}O_4P$	563-04-2	368.363	wax	25.5	260^{15}	1.150^{25}	1.5575^{20}	i H_2O; sl EtOH; s eth; vs ctc, tol
10304	Tri-p-cresyl phosphate	Tri-p-tolyl phosphate	$C_{21}H_{21}O_4P$	78-32-0	368.363	nd (al), tab (eth)	77.5	224^{35}	1.247^{25}		s EtOH, eth, bz, chl, HOAc
10305	1,3,6-Tricyanohexane		$C_9H_{11}N_3$	1772-25-4	161.203	br liq		257^2	1.040	1.4660^{20}	
10306	Tricyclazole	1,2,4-Triazolo[3,4-b]benzothiazole, 5-methyl-	$C_9H_7N_3S$	41814-78-2	189.237		187				
10307	Tricyclene	1,7,7-Trimethyltricyclo[2.2.1.0^{2,6}]heptane	$C_{10}H_{16}$	508-32-7	136.234	cry (al)	67.5	152.5	0.8668^{80}	1.4296^{80}	
10308	Tricyclo[3.3.1.1^{3,7}]decan-1-amine	Amantadine	$C_{10}H_{17}N$	768-94-5	151.249		180				sl H_2O
10309	Tricyclo[3.3.1.1^{3,7}]decane	Adamantane	$C_{10}H_{16}$	281-23-2	136.234	nd (sub)	268	sub	1.07^{25}	1.568	s bz, ctc
10310	Tridecanal		$C_{13}H_{26}O$	10486-19-8	198.344		14	156^{13}	0.8356^{18}	1.4384^{18}	i H_2O; s EtOH
10311	Tridecane		$C_{13}H_{28}$	629-50-5	184.361	liq	-5.4	235.47	0.7564^{20}	1.4256^{20}	i H_2O; vs EtOH, eth; s ctc
10312	Tridecanedioic acid		$C_{13}H_{24}O_4$	505-52-2	244.328		114				sl H_2O, bz, tfa; s EtOH, eth, chl
10313	Tridecanenitrile		$C_{13}H_{25}N$	629-60-7	195.345		9.7	293	0.8257^{20}	1.4378^{20}	vs EtOH, eth
10314	Tridecanoic acid	Tridecylic acid	$C_{13}H_{26}O_2$	638-53-9	214.344	cry (peth ace)	41.5	236^{100}, 140^1	0.8458^{80}	1.4286^{60}	i H_2O; vs EtOH, eth, HOAc; s ace
10315	1-Tridecanol	Tridecyl alcohol	$C_{13}H_{28}O$	112-70-9	200.360	cry (al)	31.7	274; 152^{14}	0.8223^{31}		i H_2O; s EtOH, eth
10316	2-Tridecanone	Methyl undecyl ketone	$C_{13}H_{26}O$	593-08-8	198.344		30.5	263	0.8217^{30}	1.4318^{20}	i H_2O; vs EtOH, eth, ace, bz, chl
10317	7-Tridecanone	Dihexyl ketone	$C_{13}H_{26}O$	462-18-0	198.344	lf (al)	33	261	0.825^{30}		s EtOH, chl, lig; vs eth
10318	1-Tridecene		$C_{13}H_{26}$	2437-56-1	182.345	liq	-13	232.8	0.7658^{20}	1.4340^{20}	i H_2O; vs EtOH, eth; s bz
10319	Tridecyl acrylate		$C_{16}H_{30}O_2$	3076-04-8	254.408	liq		150^{10}	0.88^{20}		
10320	Tridecylaluminum		$C_{30}H_{63}Al$	1726-66-5	450.803	hyg visc liq	-38				
10321	Tridecylamine	N,N-Didecyl-1-decanamine	$C_{30}H_{63}N$	1070-01-5	437.828			406			
10322	(Tridecyl)amine	1-Tridecanamine	$C_{13}H_{29}N$	2869-34-3	199.376		27.4	275.8	0.8049^{20}	1.4443^{20}	sl H_2O; s EtOH, eth
10323	Tridecylbenzene	1-Phenyltridecane	$C_{19}H_{32}$	123-02-4	260.457		10	346	0.8550^{20}	1.4821^{20}	
10324	Tridecylcyclohexane		$C_{19}H_{38}$	6006-33-3	266.505		18.5	346	0.8239^{20}	1.4570^{20}	
10325	Tridecyl methacrylate		$C_{17}H_{32}O_2$	2495-25-2	268.435			118^1	0.881^{20}	1.448^{25}	
10326	Tri(decyl) phosphite		$C_{30}H_{63}O_3P$	2929-86-4	502.793	liq		255^3; $180^{0.1}$			
10327	1-Tridecyne		$C_{13}H_{24}$	26186-02-7	180.330		2.5	234; 94^{25}	0.7842^{20}	1.4309^{20}	vs bz, eth
10328	Tridiphane	2-(3,5-Dichlorophenyl)-2-(2,2,2-trichloroethyl)oxirane, (±)	$C_{10}H_7Cl_5O$	58138-08-2	320.427		42.8				
10329	Tridodecylamine	N,N-Didodecyl-1-dodecanamine	$C_{36}H_{75}N$	102-87-4	521.988		16.4	$220^{0.03}$			
10330	Triethanolamine	Tris(2-hydroxyethyl)amine	$C_6H_{15}NO_3$	102-71-6	149.188	hyg cry	20.5	335.4	1.1242^{20}	1.4852^{20}	msc H_2O, EtOH; sl eth, bz; s chl
10331	1,3,5-Triethoxybenzene		$C_{12}H_{18}O_3$	2437-88-9	210.269	cry (al, dil al)	43.5	170^{24}			vs eth, EtOH
10332	Triethoxy(3-chloropropyl)silane	(3-Chloropropyl)triethoxysilane	$C_9H_{21}ClO_3Si$	5089-70-3	240.800	col gas		-149			

2,4,6-Trichloropyrimidine

3-(Trichlorosilyl)propanenitrile

2,4,6-Trichloro-1,3,5-triazine

2,2',2''-Trichlorotriethylamine

Trichlorotriethyldialuminum

1,3,5-Trichloro-2,4,6-trifluorobenzene

1,1,1-Trichloro-2,2,2-trifluoroethane

1,1,2-Trichloro-1,2,2-trifluoroethane

Trichlorovinylsilane

Trichodermin

Triclofos

Triclopyr

Tricosane

12-Tricosanone

Tri-o-cresyl phosphate

Tri-m-cresyl phosphate

Tri-p-cresyl phosphate

1,3,6-Tricyanohexane

Tricyclazole

Tricyclene

Tricyclo[3.3.1.1³,⁷]decan-1-amine

Tricyclo[3.3.1.1³,⁷]decane

Tridecanal

Tridecane

Tridecanedioic acid

Tridecanenitrile

Tridecanoic acid

1-Tridecanol

2-Tridecanone

7-Tridecanone

1-Tridecene

Tridecyl acrylate

Tridecylaluminum

Tridecylamine

(Tridecyl)amine

Tridecylbenzene

Tridecylcyclohexane

Tridecyl methacrylate

Tri(decyl) phosphite

1-Tridecyne

Tridiphane

Tridodecylamine

Triethanolamine

1,3,5-Triethoxybenzene

Triethoxy(3-chloropropyl)silane

No.	Name	Synonym	Mol. Form.	CAS RN	Mol. Wt.	Physical Form	mp/°C	bp/°C	den/ g cm⁻³	n_D	Solubility
10333	1,1,1-Triethoxyethane		$C_8H_{18}O_3$	78-39-7	162.227			145	0.8847^{25}	1.3980^{20}	i H_2O; msc EtOH, eth, ctc, chl
10334	Triethoxyethylsilane		$C_8H_{20}O_3Si$	78-07-9	192.329			158.5	0.8963^{20}	1.3955^{20}	i H_2O; msc EtOH, eth; s chl
10335	Triethoxymethane		$C_7H_{16}O_3$	122-51-0	148.200			143; 60^{20}	0.8909^{20}	1.3922^{20}	s EtOH, eth
10336	Triethoxymethylsilane		$C_7H_{18}O_3Si$	2031-67-6	178.302			142	0.8948^{25}	1.3832^{20}	
10337	Triethoxypentylsilane		$C_{11}H_{26}O_3Si$	2761-24-2	234.408			$100^{30}, 95^{13}$	0.8862^{20}	1.4059^{20}	
10338	Triethoxyphenylsilane		$C_{12}H_{20}O_3Si$	780-69-8	240.371			232; 113^{10}	0.996^{25}	1.4604^{20}	
10339	1,1,1-Triethoxypropane		$C_9H_{20}O_3$	115-80-0	176.253			171		1.4000^{25}	vs eth, EtOH
10340	Triethoxysilane		$C_6H_{16}O_3Si$	998-30-1	164.275			133.5	0.8745^{20}		
10341	3-(Triethoxysilyl)-1-propanamine		$C_9H_{23}NO_3Si$	919-30-2	221.370			119^{29}	0.9506^{20}	1.4225^{20}	
10342	3-(Triethoxysilyl)propanenitrile		$C_9H_{19}NO_3Si$	919-31-3	217.338	liq		109^{10}	0.974^{20}		
10343	Triethyl 2-acetoxy-1,2,3-propanetricarboxylate	Triethyl acetylcitrate	$C_{14}H_{22}O_8$	77-89-4	318.320			214^{40}	1.135^{25}	1.4380	
10344	Triethylaluminum	Hexaethyldialuminum	$C_6H_{15}Al$	97-93-8	114.165	col hyg liq	-46	194; 100^{13}	0.832^{25}		
10345	Triethylamine	N,N-Diethylethanamine	$C_6H_{15}N$	121-44-8	101.190	liq	-114.7	89	0.7275^{20}	1.4010^{20}	s H_2O, EtOH, eth, ctc; vs ace, bz, chl
10346	Triethylamine hydrochloride	N,N-Diethylethanamine hydrochloride	$C_6H_{16}ClN$	554-68-7	137.651	hex (al)	260 dec	sub 245	1.0689^{21}		vs H_2O, EtOH, chl
10347	Triethylarsine		$C_6H_{15}As$	617-75-4	162.105			138.5	1.150^{20}	1.467^{20}	vs ace, eth, EtOH
10348	1,2,3-Triethylbenzene		$C_{12}H_{18}$	42205-08-3	162.271	col liq	-26	172			
10349	1,2,4-Triethylbenzene		$C_{12}H_{18}$	877-44-1	162.271			218; 99^{15}	0.8738^{20}	1.5024^{20}	i H_2O; s EtOH, eth
10350	1,3,5-Triethylbenzene		$C_{12}H_{18}$	102-25-0	162.271	liq	-66.5	215.9	0.8631^{20}	1.4969^{20}	i H_2O; vs EtOH, eth
10351	Triethylborane		$C_6H_{15}B$	97-94-9	97.994	liq	-93	95	0.70^{23}	1.3971	s EtOH, eth
10352	Triethyl borate	Boric acid, triethyl ester	$C_6H_{15}BO_3$	150-46-9	145.992	liq	-84.8	120	0.8546^{20}	1.3749^{20}	msc EtOH, eth
10353	Triethyl citrate		$C_{12}H_{20}O_7$	77-93-0	276.283			294	1.1369^{20}	1.4455^{20}	i H_2O; s EtOH, eth; sl ctc
10354	Triethylenediamine		$C_6H_{12}N_2$	280-57-9	112.172		159				s chl
10355	Triethylene glycol	Triglycol	$C_6H_{14}O_4$	112-27-6	150.173	hyg liq	-7	285	1.1274^{15}	1.4531^{20}	msc H_2O, EtOH, bz; sl eth, chl; i peth
10356	Triethylene glycol bis(2-ethylhexanoate)		$C_{22}H_{42}O_6$	94-28-0	402.564						s chl
10357	Triethylene glycol diacetate		$C_{10}H_{18}O_6$	111-21-7	234.246	liq	-50	286	1.1153^{20}		vs H_2O, eth, EtOH
10358	Triethylene glycol dimethacrylate		$C_{14}H_{22}O_6$	109-16-0	286.321			170^5	1.092^{20}	1.4595^{25}	vs ace, eth, EtOH, peth
10359	Triethylene glycol dimethyl ether	Triglyme	$C_8H_{18}O_4$	112-49-2	178.227	liq	-45	216	0.986^{20}	1.4224^{20}	vs H_2O, bz
10360	Triethylene glycol dinitrate	Ethanol, 2,2'-[1,2-ethanediylbis(oxy)]bis-, dinitrate	$C_6H_{12}N_2O_8$	111-22-8	240.167			$82^{0.03}$			
10361	Triethylene glycol monoethyl ether	2-[2-(2-Ethoxyethoxy)ethoxy]ethanol	$C_8H_{18}O_4$	112-50-5	178.227			256	1.0209^{20}		
10362	Triethylenephosphoramide	Tris(1-aziridinyl)phosphine, oxide	$C_6H_{12}N_3OP$	545-55-1	173.152	cry	41	91^{23}			vs H_2O, EtOH, eth, ace
10363	Triethylenethiophosphoramide	Thiotepa	$C_6H_{12}N_3PS$	52-24-4	189.218	cry	51.5				vs H_2O; s bz, chl, eth, EtOH
10364	1,3,5-Triethylhexahydro-1,3,5-triazine		$C_9H_{21}N_3$	7779-27-3	171.283			78^6		1.4580^{25}	
10365	Triethyl phosphate	Ethyl phosphate	$C_6H_{15}O_4P$	78-40-0	182.154	liq	-56.4	215.5	1.0695^{20}	1.4053^{20}	s H_2O, eth, bz; vs EtOH, sl chl
10366	Triethylphosphine		$C_6H_{15}P$	554-70-1	118.157	liq	-88	129	0.8006^{19}	1.458^{15}	i H_2O; msc EtOH, eth
10367	Triethylphosphine oxide		$C_6H_{15}OP$	597-50-2	134.156	wh hyg nd	48	243			vs H_2O, eth, EtOH
10368	Triethylphosphine sulfide		$C_6H_{15}PS$	597-51-3	150.222	cry (al)	94				s H_2O; sl ctc
10369	Triethyl phosphite	Triethoxyphosphine	$C_6H_{15}O_3P$	122-52-1	166.155			157.9	0.9629^{20}	1.4127^{20}	i H_2O; vs EtOH, eth
10370	O,O,O-Triethyl phosphorothioate	O,O,O-Triethyl thiophosphate	$C_6H_{15}O_3PS$	126-68-1	198.220			217; 100^{16}	1.0768^{20}	1.4480^{20}	
10371	Triethylsilane		$C_6H_{16}Si$	617-86-7	116.277	liq	-156.9	109	0.7302^{20}	1.447^{20}	i H_2O, sulf
10372	Triethylsilanol		$C_6H_{16}OSi$	597-52-4	132.276			154	0.8647^{20}	1.4329^{20}	i H_2O; msc EtOH, eth
10373	Triethylstibine		$C_6H_{15}Sb$	617-85-6	208.943	liq	-98	161.4	1.3224^{15}		i H_2O; s EtOH, eth
10374	Trifenmorph	4-(Triphenylmethyl)morpholine	$C_{23}H_{23}NO$	1420-06-0	329.435	cry (EtOH)	176				i H_2O; s chl, ctc
10375	Triflumizole		$C_{15}H_{15}ClF_3N_3O$	68694-11-1	345.747		63.5				
10376	Trifluoperazine		$C_{21}H_{24}F_3N_3S$	117-89-5	407.496	cry		$206^{0.7}$			
10377	Trifluoperazine dihydrochloride	Stelazine	$C_{21}H_{26}Cl_2F_3N_3S$	440-17-5	480.417		241.5				

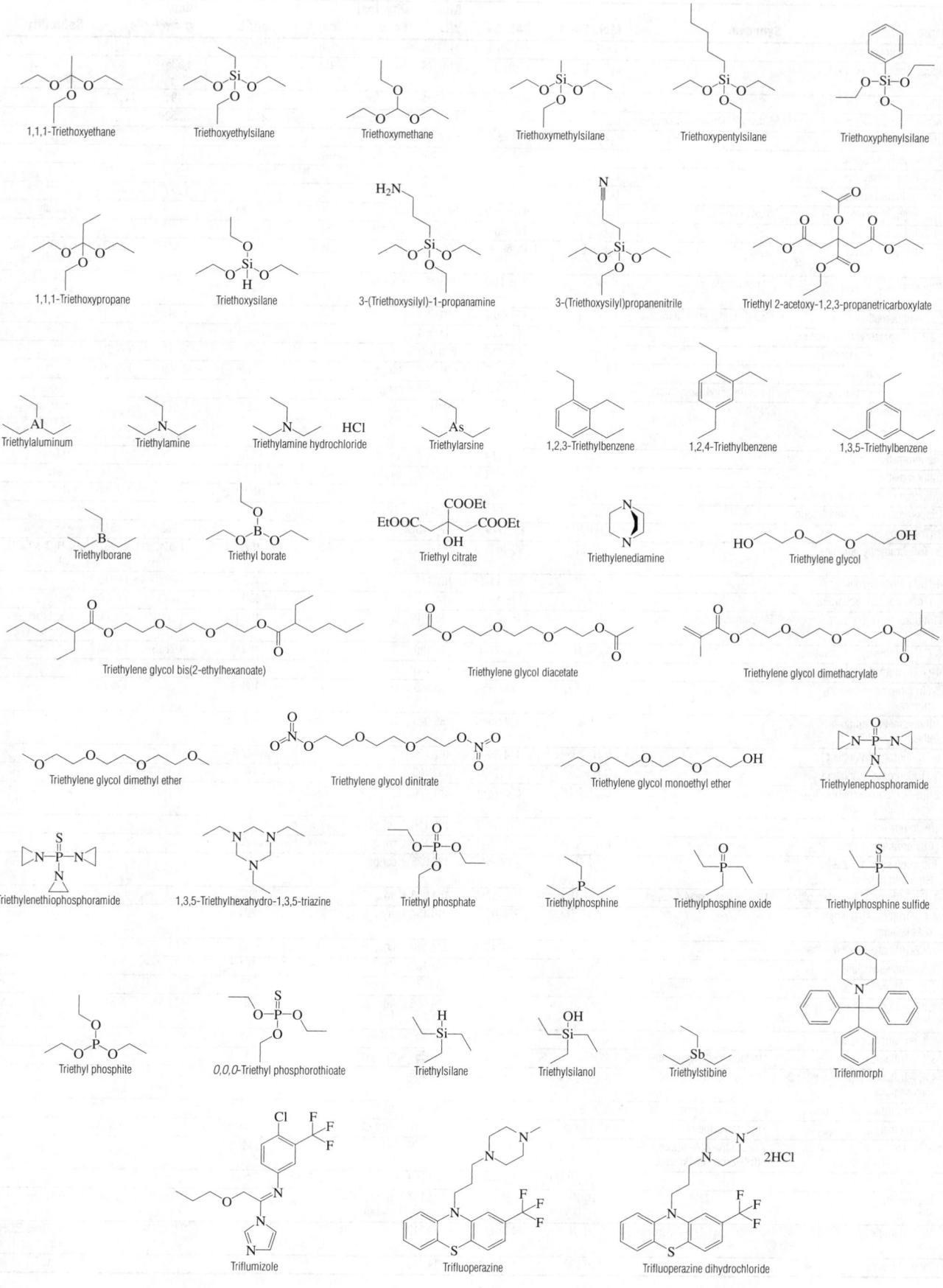

1,1,1-Triethoxyethane

Triethoxyethylsilane

Triethoxymethane

Triethoxymethylsilane

Triethoxypentylsilane

Triethoxyphenylsilane

1,1,1-Triethoxypropane

Triethoxysilane

3-(Triethoxysilyl)-1-propanamine

3-(Triethoxysilyl)propanenitrile

Triethyl 2-acetoxy-1,2,3-propanetricarboxylate

Triethylaluminum

Triethylamine

Triethylamine hydrochloride

Triethylarsine

1,2,3-Triethylbenzene

1,2,4-Triethylbenzene

1,3,5-Triethylbenzene

Triethylborane

Triethyl borate

Triethyl citrate

Triethylenediamine

Triethylene glycol

Triethylene glycol bis(2-ethylhexanoate)

Triethylene glycol diacetate

Triethylene glycol dimethacrylate

Triethylene glycol dimethyl ether

Triethylene glycol dinitrate

Triethylene glycol monoethyl ether

Triethylenephosphoramide

Triethylenethiophosphoramide

1,3,5-Triethylhexahydro-1,3,5-triazine

Triethyl phosphate

Triethylphosphine

Triethylphosphine oxide

Triethylphosphine sulfide

Triethyl phosphite

O,O,O-Triethyl phosphorothioate

Triethylsilane

Triethylsilanol

Triethylstibine

Trifenmorph

Triflumizole

Trifluoperazine

Trifluoperazine dihydrochloride

No.	Name	Synonym	Mol. Form.	CAS RN	Mol. Wt.	Physical Form	mp/°C	bp/°C	den/ g cm⁻³	n_D	Solubility
10378	2,2,2-Trifluoroacetamide		$C_2H_2F_3NO$	354-38-1	113.038		73.8	162.5			
10379	Trifluoroacetic acid		$C_2HF_3O_2$	76-05-1	114.023	liq	-15.2	73	1.5351[25]		s H_2O, EtOH, eth, ace
10380	Trifluoroacetic acid anhydride		$C_4F_6O_3$	407-25-0	210.031	liq	-65	39.5	1.490[25]	1.269[25]	
10381	1,1,1-Trifluoroacetone	Methyl trifluoromethyl ketone	$C_3H_3F_3O$	421-50-1	112.050	vol liq or gas		21.5	1.252[25]		
10382	Trifluoroacetonitrile		C_2F_3N	353-85-5	95.023	col gas		-68.8			
10383	Trifluoroacetyl chloride		C_2ClF_3O	354-32-5	132.468	col gas	-146	-18			
10384	1,2,4-Trifluorobenzene		$C_6H_3F_3$	367-23-7	132.083			90	1.264[25]	1.4171[20]	
10385	1,3,5-Trifluorobenzene		$C_6H_3F_3$	372-38-3	132.083	liq	-5.5	75.5	1.277[25]	1.4140[20]	
10386	1,1,1-Trifluoroethane	Methyl fluoroform	$C_2H_3F_3$	420-46-2	84.040	col gas	-111.3	-47.25			s eth, chl
10387	1,1,2-Trifluoroethane		$C_2H_3F_3$	430-66-0	84.040	col gas	-84	3.7			
10388	2,2,2-Trifluoroethanol		$C_2H_3F_3O$	75-89-8	100.039	liq	-43.5	74	1.3842[20]	1.2907[22]	vs EtOH; s eth, ace, bz, chl
10389	Trifluoroethene	Trifluoroethylene	C_2HF_3	359-11-5	82.024	col gas		-51	1.26[-70]		i H_2O; sl EtOH; s eth
10390	2,2,2-Trifluoroethylamine	2,2,2-Trifluoroethanamine	$C_2H_4F_3N$	753-90-2	99.055			36	1.245[25]		
10391	2,2,2-Trifluoroethyl methyl ether		$C_3H_5F_3O$	460-43-5	114.066			31.62			
10392	1,1,1-Trifluoro-2-iodoethane		$C_2H_2F_3I$	353-83-3	209.936			54.5	2.13[25]	1.4009[20]	
10393	Trifluoroiodomethane		CF_3I	2314-97-8	195.910	col gas		-22.5	2.3607[-32]	1.3790[-32]	
10394	Trifluoroisocyanomethane	Trifluoromethyl isocyanide	C_2F_3N	19480-01-4	95.023	col gas		-80			
10395	Trifluoromethane	Fluoroform	CHF_3	75-46-7	70.014	col gas	-155.2	-82.1	0.673[25] (p>1 atm)		s H_2O, ace, bz; vs EtOH; sl chl
10396	Trifluoromethanesulfenyl chloride		$CClF_3S$	421-17-0	136.524	col gas		-0.7			i H_2O
10397	Trifluoromethanesulfonic acid		CHF_3O_3S	1493-13-6	150.077	hyg liq	45	162			vs eth
10398	Trifluoromethanesulfonyl chloride		$CClF_3O_2S$	421-83-0	168.523			162; 62[18]		1.3344[20]	i H_2O
10399	Trifluoromethanesulfonyl fluoride		CF_4O_2S	335-05-7	152.069	col gas		-21.7			
10400	2-(Trifluoromethyl)aniline		$C_7H_6F_3N$	88-17-5	161.125		35.5	68[15]	1.282[25]	1.4810[20]	
10401	3-(Trifluoromethyl)aniline		$C_7H_6F_3N$	98-16-8	161.125		5.5	187; 74[10]	1.3047[12]	1.4787[20]	sl H_2O; s EtOH, eth
10402	4-(Trifluoromethyl)aniline		$C_7H_6F_3N$	455-14-1	161.125		38	117.5[60]	1.283[27]	1.4815[25]	
10403	4-(Trifluoromethyl)benzaldehyde		$C_8H_5F_3O$	455-19-6	174.120			80[25]		1.4630[20]	
10404	(Trifluoromethyl)benzene	Benzotrifluoride	$C_7H_5F_3$	98-08-8	146.110	liq	-28.95	102.1	1.1884[20]	1.4146[20]	msc EtOH, eth, ace, bz, ctc
10405	3-(Trifluoromethyl)benzonitrile		$C_8H_4F_3N$	368-77-4	171.120		14.5	189	1.2813[20]	1.4508[20]	
10406	4-(Trifluoromethyl)benzonitrile		$C_8H_4F_3N$	455-18-5	171.120		37.5				
10407	3-(Trifluoromethyl)benzoyl chloride		$C_8H_4ClF_3O$	2251-65-2	208.565	oil		186; 80[16]	1.383	1.4770[20]	
10408	Trifluoromethyl difluoromethyl ether		C_2HF_5O	3822-68-2	136.020	col gas	-157	-38			
10409	2-(Trifluoromethyl)phenol		$C_7H_5F_3O$	444-30-4	162.109		45.5	147.5			
10410	3-(Trifluoromethyl)phenol		$C_7H_5F_3O$	98-17-9	162.109	liq	-0.9	178	1.3418[25]		
10411	2-[[3-(Trifluoromethyl)phenyl] amino]benzoic acid	Flufenamic acid	$C_{14}H_{10}F_3NO_2$	530-78-9	281.230		133.5				s DMSO
10412	Trifluoromethylsilane		CH_3F_3Si	373-74-0	100.116	col gas	-73	-30			
10413	(Trifluoromethyl)silane		CH_3F_3Si	10112-11-5	100.116	col gas	-124	-38.3			
10414	Trifluoromethyl 1,1,2,2-tetrafluoroethyl ether		C_3HF_7O	2356-61-8	186.028	col gas	-141	-3			
10415	1,1,1-Trifluoro-2,4-pentanedione	1,1,1-Trifluoroacetylacetone	$C_5H_5F_3O_2$	367-57-7	154.088	liq		107			s os
10416	4,4,4-Trifluoro-1-phenyl-1,3-butanedione		$C_{10}H_7F_3O_2$	326-06-7	216.157	cry	39	224			i H_2O; s EtOH, ace
10417	2,2,2-Trifluoro-1-phenylethanone		$C_8H_5F_3O$	434-45-7	174.120	liq	-40	153	1.279[20]	1.4583[20]	
10418	Trifluorophenylsilane		$C_6H_5F_3Si$	368-47-8	162.185	liq	-18	101.5	1.2169[20]	1.4110[20]	vs bz, EtOH
10419	1,1,1-Trifluoropropane		$C_3H_5F_3$	421-07-8	98.067	col gas		-13			
10420	1,1,1-Trifluoro-2-propanol, (±)		$C_3H_5F_3O$	17556-48-8	114.066	liq	-52	78	1.2632[25]	1.3130[25]	vs EtOH, eth; s ace, bz; sl ctc
10421	3,3,3-Trifluoropropene		$C_3H_3F_3$	677-21-4	96.051	col gas		-17			
10422	3,3,3-Trifluoro-1-propyne	(Trifluoromethyl)acetylene	C_3HF_3	661-54-1	94.035	col gas		-48.3			
10423	4,4,4-Trifluoro-1-(2-thienyl)-1,3-butanedione	Thenoyltrifluoroacetone	$C_8H_5F_3O_2S$	326-91-0	222.185		42.8	97[8]			
10424	Trifluoro(trifluoromethyl)oxirane	Perfluoropropylene oxide	C_3F_6O	428-59-1	166.021	gas		-27.4			
10425	Triflupromazine	Fluopromazine	$C_{18}H_{19}F_3N_2S$	146-54-3	352.417	visc oil		176[0.7]		1.5780[23]	
10426	Trifluralin	2,6-Dinitro-N,N-dipropyl-4-(trifluoromethyl)aniline	$C_{13}H_{16}F_3N_3O_4$	1582-09-8	335.279		49	140[4.2]			
10427	Triforine		$C_{10}H_{14}Cl_6N_4O_2$	26644-46-2	434.962		155 dec				
10428	Trigonelline		$C_7H_7NO_2$	535-83-1	137.137	pr (aq, al, +1w)					vs H_2O
10429	Trihexylamine	N,N-Dihexyl-1-hexanamine	$C_{18}H_{39}N$	102-86-3	269.510			261.7	0.7976[21]		i H_2O; vs EtOH, eth; s acid
10430	Trihexyl borate		$C_{18}H_{39}BO_3$	5337-36-0	314.312			143[2]			sl ctc

2,2,2-Trifluoroacetamide

Trifluoroacetic acid

Trifluoroacetic acid anhydride

1,1,1-Trifluoroacetone

Trifluoroacetonitrile

Trifluoroacetyl chloride

1,2,4-Trifluorobenzene

1,3,5-Trifluorobenzene

1,1,1-Trifluoroethane

1,1,2-Trifluoroethane

2,2,2-Trifluoroethanol

Trifluoroethene

2,2,2-Trifluoroethylamine

2,2,2-Trifluoroethyl methyl ether

1,1,1-Trifluoro-2-iodoethane

Trifluoroiodomethane

Trifluoroisocyanomethane

Trifluoromethane

Trifluoromethanesulfenyl chloride

Trifluoromethanesulfonic acid

Trifluoromethanesulfonyl chloride

Trifluoromethanesulfonyl fluoride

2-(Trifluoromethyl)aniline

3-(Trifluoromethyl)aniline

4-(Trifluoromethyl)aniline

4-(Trifluoromethyl)benzaldehyde

(Trifluoromethyl)benzene

3-(Trifluoromethyl)benzonitrile

4-(Trifluoromethyl)benzonitrile

3-(Trifluoromethyl)benzoyl chloride

Trifluoromethyl difluoromethyl ether

2-(Trifluoromethyl)phenol

3-(Trifluoromethyl)phenol

2-[[3-(Trifluoromethyl)phenyl]amino]benzoic acid

Trifluoromethylsilane

(Trifluoromethyl)silane

Trifluoromethyl 1,1,2,2-tetrafluoroethyl ether

1,1,1-Trifluoro-2,4-pentanedione

4,4,4-Trifluoro-1-phenyl-1,3-butanedione

2,2,2-Trifluoro-1-phenylethanone

Trifluorophenylsilane

1,1,1-Trifluoropropane

1,1,1-Trifluoro-2-propanol, (±)

3,3,3-Trifluoropropene

3,3,3-Trifluoro-1-propyne

4,4,4-Trifluoro-1-(2-thienyl)-1,3-butanedione

Trifluoro(trifluoromethyl)oxirane

Triflupromazine

Trifluralin

Triforine

Trigonelline

Trihexylamine

Trihexyl borate

No.	Name	Synonym	Mol. Form.	CAS RN	Mol. Wt.	Physical Form	mp/°C	bp/°C	den/ g cm⁻³	n_D	Solubility
10431	Trihexyphenidyl hydrochloride	α-Cyclohexyl-α-phenyl-1-piperidinepropanol hydrochloride	$C_{20}H_{32}ClNO$	52-49-3	337.927		258.5				
10432	Trihydro(pyridine)boron	Borane pyridine	C_5H_8BN	110-51-0	92.936		10.5		0.920²⁰	1.5280²⁵	i H_2O; dec acid
10433	1,2,3-Trihydroxy-9,10-anthracenedione	Anthragallol	$C_{14}H_8O_5$	602-64-2	256.211	ye nd (dil al)	313	sub 290			sl H_2O; s EtOH, eth, HOAc, CS_2
10434	1,2,4-Trihydroxy-9,10-anthracenedione	Purpurin	$C_{14}H_8O_5$	81-54-9	256.211	oran red or oran-ye nd (al)	259	sub			sl H_2O; vs EtOH, bz, HOAc; s eth
10435	2,3,4-Trihydroxybenzoic acid		$C_7H_6O_5$	610-02-6	170.120	nd (+w)	221	sub			sl H_2O; s EtOH, eth, ace; i bz, CS_2
10436	2,4,6-Trihydroxybenzoic acid		$C_7H_6O_5$	83-30-7	170.120	cry (w+1)	100 dec				sl H_2O; s EtOH; vs eth; i bz
10437	3,4,5-Trihydroxybenzoic acid	Gallic acid	$C_7H_6O_5$	149-91-7	170.120	pr (w+1)	253 dec		1.694⁶		sl H_2O, eth; vs EtOH; s ace; i bz, chl
10438	2,3,4-Trihydroxybenzophenone	Alizarin Yellow A	$C_{13}H_{10}O_4$	1143-72-2	230.216	ye nd (dil al)	140.5				sl H_2O, bz; s EtOH, eth, ace, HOAc
10439	2′,4,4′-Trihydroxychalcone	Isoliquiritigenin	$C_{15}H_{12}O_4$	961-29-5	256.254	ye nd (EtOH-w)	200				
10440	9,10,16-Trihydroxyhexadecanoic acid	Aleuritic acid	$C_{16}H_{32}O_5$	6949-98-0	304.422	lf (dil al), nd (w)	102				sl H_2O
10441	1,3,8-Trihydroxy-6-methyl-9,10-anthracenedione	Emodin	$C_{15}H_{10}O_5$	518-82-1	270.237	oran-red mcl nd (HOAc)	257	sub			vs eth, EtOH
10442	9,10,18-Trihydroxyoctadecanoic acid, (R*,R*)	Phloionolic acid	$C_{18}H_{36}O_5$	583-86-8	332.476	cry (dil al)	101.5				
10443	5,6,7-Trihydroxy-2-phenyl-4H-1-benzopyran-4-one	Baicalein	$C_{15}H_{10}O_5$	491-67-8	270.237	ye pr (al)	264 dec				sl H_2O, bz; s EtOH, eth, ace, HOAc
10444	1-(2,4,5-Trihydroxyphenyl)-1-butanone		$C_{10}H_{12}O_4$	1421-63-2	196.200		153.8				
10445	1-(2,3,4-Trihydroxyphenyl)ethanone	Gallacetophenone	$C_8H_8O_4$	528-21-2	168.148		173				s H_2O, eth; vs EtOH, ace; sl bz, chl
10446	1-(2,4,6-Trihydroxyphenyl)ethanone	2′,4′,6′-Trihydroxyacetophenone	$C_8H_8O_4$	480-66-0	168.148		221.0				sl H_2O, chl, bz; vs EtOH, eth, ace
10447	1-(2,4,6-Trihydroxyphenyl)-1-propanone	Flopropione	$C_9H_{10}O_4$	2295-58-1	182.173	nd (w, +1w)	175.5				vs eth, EtOH
10448	2,6,7-Trihydroxy-9-phenyl-3H-xanthen-3-one	Phenylfluorone	$C_{19}H_{12}O_5$	975-17-7	320.295	oran red (al-HCl)	>300				
10449	2,3,5-Triiodobenzoic acid		$C_7H_3I_3O_2$	88-82-4	499.811	pr (al)	225				i H_2O; vs EtOH, eth; sl bz
10450	Triiodomethane	Iodoform	CHI_3	75-47-8	393.732	ye cry	121.2	218	4.008²⁵		i H_2O, bz; s EtOH, eth, ace; sl DMSO
10451	2,4,6-Triiodophenol		$C_6H_3I_3O$	609-23-4	471.800	nd (dil al)	159.8	sub			i H_2O; sl EtOH; s eth, ace
10452	3,3′,5-Triiodothyropropanoic acid		$C_{15}H_{11}I_3O_4$	51-26-3	635.959	cry (EtOH)	200				sl EtOH
10453	Triisobutyl aluminate	2-Methyl-1-propanol, aluminum salt	$C_{12}H_{27}AlO_3$	3453-79-0	246.322			275⁵⁰			
10454	Triisobutylaluminum		$C_{12}H_{27}Al$	100-99-2	198.324	liq	6	86¹⁰			
10455	Triisobutylamine	2-Methyl-N,N-bis(2-methylpropyl)-1-propanamine	$C_{12}H_{27}N$	1116-40-1	185.349	liq	-21.8	191.5	0.7684²⁰	1.4252¹⁷	vs eth, EtOH
10456	Triisobutylborane		$C_{12}H_{27}B$	1116-39-8	182.153			188; 86²⁰	0.7380²⁵	1.4188²³	vs bz, eth, EtOH
10457	Triisobutyl phosphate		$C_{12}H_{27}O_4P$	126-71-6	266.313			264	0.9681²⁰	1.4193²⁰	vs H_2O, bz, eth, EtOH
10458	Triisopentylamine	3-Methyl-N,N-bis(3-methylbutyl)-1-butanamine	$C_{15}H_{33}N$	645-41-0	227.430			235	0.7848²⁰	1.4331²⁰	i H_2O; vs EtOH; msc eth, bz, ctc
10459	Triisopropanolamine		$C_9H_{21}NO_3$	122-20-3	191.268		45	175¹⁰	1.0²⁰		s H_2O, EtOH; sl chl
10460	Triisopropoxymethane	Isopropyl orthoformate	$C_{10}H_{22}O_3$	4447-60-3	190.280			167	0.8621²⁰	1.4000²⁰	vs eth, EtOH
10461	Triisopropoxyvinylsilane		$C_{11}H_{24}O_3Si$	18023-33-1	232.393			179.5; 77²⁰	0.8627²⁵	1.3981²⁰	s ctc
10462	1,2,4-Triisopropylbenzene		$C_{15}H_{24}$	948-32-3	204.352			244	0.8574²⁵	1.4896²⁵	
10463	1,3,5-Triisopropylbenzene		$C_{15}H_{24}$	717-74-8	204.352	liq	-7.4	238	0.8545²⁰	1.4882²⁰	s ace, bz, chl
10464	Triisopropyl borate		$C_9H_{21}BO_3$	5419-55-6	188.072			140; 75⁷⁶	0.8251²⁰	1.3777²⁰	vs EtOH, eth, bz, PrOH
10465	Triisopropyl phosphate		$C_9H_{21}O_4P$	513-02-0	224.234			219	0.9867²⁰	1.4057²⁰	vs EtOH
10466	Triisopropyl phosphite		$C_9H_{21}O_3P$	116-17-6	208.235			74²⁰, 60¹⁰	0.9063²⁰	1.4085²⁵	s EtOH, eth, chl
10467	Triisopropyl vanadate	Vanadium, oxotris(2-propanolato)-, (T-4)-	$C_9H_{21}O_4V$	5588-84-1	244.203			104¹⁰			

Trihexyphenidyl hydrochloride

Trihydro(pyridine)boron

1,2,3-Trihydroxy-9,10-anthracenedione

1,2,4-Trihydroxy-9,10-anthracenedione

2,3,4-Trihydroxybenzoic acid

2,4,6-Trihydroxybenzoic acid

3,4,5-Trihydroxybenzoic acid

2,3,4-Trihydroxybenzophenone

2',4,4'-Trihydroxychalcone

9,10,16-Trihydroxyhexadecanoic acid

1,3,8-Trihydroxy-6-methyl-9,10-anthracenedione

9,10,18-Hydroxyoctadecanoic acid, (R*,R*)

5,6,7-Trihydroxy-2-phenyl-4H-1-benzopyran-4-one

1-(2,4,5-Trihydroxyphenyl)-1-butanone

1-(2,3,4-Trihydroxyphenyl)ethanone

1-(2,4,6-Trihydroxyphenyl)ethanone

1-(2,4,6-Trihydroxyphenyl)-1-propanone

2,6,7-Trihydroxy-9-phenyl-3H-xanthen-3-one

2,3,5-Triiodobenzoic acid

Triiodomethane

2,4,6-Triiodophenol

3,3',5-Triiodothyropropanoic acid

Triisobutyl aluminate

Triisobutylaluminum

Triisobutylamine

Triisobutylborane

Triisobutyl phosphate

Triisopentylamine

Triisopropanolamine

Triisopropoxymethane

Triisopropoxyvinylsilane

1,2,4-Triisopropylbenzene

1,3,5-Triisopropylbenzene

Triisopropyl borate

Triisopropyl phosphate

Triisopropyl phosphite

Triisopropyl vanadate

No.	Name	Synonym	Mol. Form.	CAS RN	Mol. Wt.	Physical Form	mp/°C	bp/°C	den/g cm⁻³	n_D	Solubility
10468	Trimecaine	2-Diethylamino-2',4',6'-trimethylacetanilide	$C_{15}H_{24}N_2O$	616-68-2	248.364	cry	44	187^6			
10469	Trimellitic anhydride		$C_9H_4O_5$	552-30-7	192.125		162	241^{14}			
10470	Trimeprazine	N,N,β-Trimethyl-10H-phenothiazine-10-propanamine	$C_{18}H_{22}N_2S$	84-96-8	298.446	cry	68	$162^{0.3}$			
10471	Trimethoate		$C_9H_{20}NO_3PS_2$	2275-18-5	285.364	solid	28.5	$135^{0.1}$			sl H_2O
10472	Trimethobenzamide hydrochloride		$C_{21}H_{29}ClN_2O_5$	554-92-7	424.918	cry	188				vs H_2O
10473	Trimethoprim		$C_{14}H_{18}N_4O_3$	738-70-5	290.318	ye cry	201				sl chl, MeOH; i eth, bz
10474	3,4,5-Trimethoxyaniline		$C_9H_{13}NO_3$	24313-88-0	183.204		112.8				
10475	2,3,4-Trimethoxybenzaldehyde		$C_{10}H_{12}O_4$	2103-57-3	196.200			$122^{0.5}$		1.5547^{20}	
10476	2,4,5-Trimethoxybenzaldehyde		$C_{10}H_{12}O_4$	4460-86-0	196.200		114				s H_2O, eth, chl, lig
10477	3,4,5-Trimethoxybenzaldehyde		$C_{10}H_{12}O_4$	86-81-7	196.200		72.5	148^5			s chl
10478	1,2,3-Trimethoxybenzene		$C_9H_{12}O_3$	634-36-6	168.189	orth nd (al)	48.5	235	1.1009^{45}		i H_2O; s EtOH, eth, bz
10479	1,3,5-Trimethoxybenzene		$C_9H_{12}O_3$	621-23-8	168.189	pr (al), lf (peth)	54.5	255.5			i H_2O; s EtOH, eth, bz
10480	3,4,5-Trimethoxybenzeneethanamine	Mescaline	$C_{11}H_{17}NO_3$	54-04-6	211.258	cry	35.5	180^{12}			s H_2O, EtOH, bz, chl; i eth, peth
10481	3,4,5-Trimethoxybenzenemethanol	3,4,5-Trimethoxybenzyl alcohol	$C_{10}H_{14}O_4$	3840-31-1	198.216		3	228^{25}	1.1427^{20}	1.5439^{20}	
10482	2,4,5-Trimethoxybenzoic acid		$C_{10}H_{12}O_5$	490-64-2	212.199	nd (al or bz-peth)	145	300			vs H_2O, bz, EtOH, peth
10483	3,4,5-Trimethoxybenzoic acid		$C_{10}H_{12}O_5$	118-41-2	212.199	mcl nd (w)	172.3	226^{10}			sl H_2O; vs EtOH, eth, chl
10484	3,4,5-Trimethoxybenzoyl chloride		$C_{10}H_{11}ClO_4$	4521-61-3	230.645		82	185^{18}			
10485	Trimethoxyboroxin		$C_3H_9B_3O_6$	102-24-9	173.532					1.40^{25}	
10486	6,6',7-Trimethoxy-2,2'-dimethylberbaman-12-ol	Berbamine	$C_{37}H_{40}N_2O_6$	478-61-5	608.723	lf (+2w, al) cry (peth)	198.5				sl H_2O; s EtOH, eth, chl, peth
10487	6,6',7-Trimethoxy-2,2'-dimethyloxyacanthan-12'-ol	Oxyacanthine	$C_{37}H_{40}N_2O_6$	548-40-3	608.723	nd (al, eth)	216.5				i H_2O; s EtOH, eth, bz, chl; i lig
10488	7',10,11-Trimethoxyemetan-6'-ol	Cephaeline	$C_{28}H_{38}N_2O_4$	483-17-0	466.613	nd (eth)	115.5				vs ace, EtOH, MeOH, chl
10489	1,1,1-Trimethoxyethane		$C_5H_{12}O_3$	1445-45-0	120.147			108	0.9438^{25}	1.3859^{25}	vs eth, EtOH
10490	4,7,8-Trimethoxyfuro[2,3-b]quinoline	Skimmianine	$C_{14}H_{13}NO_4$	83-95-4	259.258	pym (al)	177				i H_2O, peth; s EtOH, chl; sl eth, CS_2
10491	Trimethoxymethane		$C_4H_{10}O_3$	149-73-5	106.120		15	104	0.9676^{20}	1.3793^{20}	s EtOH, eth
10492	Trimethoxymethylsilane		$C_4H_{12}O_3Si$	1185-55-3	136.222			102.5	0.9548^{20}	1.3696^{20}	s chl
10493	Trimethoxyphenylsilane		$C_9H_{14}O_3Si$	2996-92-1	198.291			130^{45}, 110^{20}	1.064^{20}	1.4734^{20}	s ctc, CS_2
10494	Trimethoxysilane		$C_3H_{10}O_3Si$	2487-90-3	122.195			32^{100}			
10495	3-(Trimethoxysilyl)-1-propanethiol	(3-Mercaptopropyl)trimethoxysilane	$C_6H_{16}O_3SSi$	4420-74-0	196.340			128^{50}, 93^{10}	1.015^{25}	1.4420^{25}	
10496	N-[3-(Trimethoxysilyl)propyl]-1,2-ethanediamine		$C_8H_{22}N_2O_3Si$	1760-24-3	222.358			140.5^{15}	1.01^{25}	1.4416^{25}	
10497	3-(Trimethoxysilyl)propyl methacrylate		$C_{10}H_{20}O_5Si$	2530-85-0	248.349	liq		107^5, 95^1			
10498	Trimethyl aluminum		C_3H_9Al	75-24-1	72.085		15.4	130; 20^8	0.752^{20}		
10499	Trimethylamine	N,N-Dimethylmethanamine	C_3H_9N	75-50-3	59.110	col gas	-117.1	2.87	0.627^{25} (p>1 atm)	1.3631^0	vs H_2O, chl, tol; s EtOH, eth, bz
10500	Trimethylamine borane	N,N-Dimethylmethanamine borane	$C_3H_{12}BN$	75-22-9	72.945		94	172	0.792^{25}		vs eth, EtOH
10501	Trimethylamine hydrochloride	N,N-Dimethylmethanamine hydrochloride	$C_3H_{10}ClN$	593-81-7	95.571	mcl hyg nd (al)	277.5	sub 200			vs H_2O, EtOH, chl
10502	Trimethylamine oxide	N,N-Dimethylmethanamine oxide	C_3H_9NO	1184-78-7	75.109	hyg nd (w+2)	256				vs H_2O, EtOH
10503	2,4,5-Trimethylaniline		$C_9H_{13}N$	137-17-7	135.206	nd (w)	68	234.5	0.957^{25}		vs EtOH
10504	2,4,6-Trimethylaniline	Mesitylamine	$C_9H_{13}N$	88-05-1	135.206	liq	-2.5	232.5	0.9633^{25}	1.5495^{20}	sl ctc
10505	Trimethylarsine		C_3H_9As	593-88-4	120.025	liq	-87.3	52	1.144^{15}		vs bz, eth, EtOH
10506	2,4,6-Trimethylbenzaldehyde		$C_{10}H_{12}O$	487-68-3	148.201		14	238.5	1.0154^{25}		i H_2O; s EtOH, eth, ace, bz
10507	1,2,3-Trimethylbenzene	Hemimellitene	C_9H_{12}	526-73-8	120.191	liq	-25.4	176.12	0.8944^{20}	1.5139^{20}	i H_2O; msc EtOH, eth, ace, bz, peth, ctc
10508	1,2,4-Trimethylbenzene	Pseudocumene	C_9H_{12}	95-63-6	120.191	liq	-43.77	169.38	0.8758^{20}	1.5048^{20}	i H_2O; msc EtOH, eth, ace, bz, peth, ctc
10509	1,3,5-Trimethylbenzene	Mesitylene	C_9H_{12}	108-67-8	120.191	liq	-44.72	164.74	0.8615^{25}	1.4994^{20}	i H_2O; msc EtOH, eth, ace, bz, peth, ctc
10510	2,3,5-Trimethyl-1,4-benzenediol		$C_9H_{12}O_2$	700-13-0	152.190	nd (w)	169 dec				sl H_2O; vs EtOH, eth, bz

Trimecaine

Trimellitic anhydride

Trimeprazine

Trimethoate

Trimethobenzamide hydrochloride

Trimethoprim

3,4,5-Trimethoxyaniline

2,3,4-Trimethoxybenzaldehyde

2,4,5-Trimethoxybenzaldehyde

3,4,5-Trimethoxybenzaldehyde

1,2,3-Trimethoxybenzene

1,3,5-Trimethoxybenzene

3,4,5-Trimethoxybenzeneethanamine

3,4,5-Trimethoxybenzenemethanol

2,4,5-Trimethoxybenzoic acid

3,4,5-Trimethoxybenzoic acid

3,4,5-Trimethoxybenzoyl chloride

Trimethoxyboroxin

6,6',7-Trimethoxy-2,2'-dimethylberbaman-12-ol

6,6',7-Trimethoxy-2,2'-dimethyloxyacanthan-12'-ol

7',10,11-Trimethoxyemetan-6'-ol

1,1,1-Trimethoxyethane

4,7,8-Trimethoxyfuro[2,3-b]quinoline

Trimethoxymethane

Trimethoxymethylsilane

Trimethoxyphenylsilane

Trimethoxysilane

3-(Trimethoxysilyl)-1-propanethiol

N-[3-(Trimethoxysilyl)propyl]-1,2-ethanediamine

3-(Trimethoxysilyl)propyl methacrylate

Trimethyl aluminum

Trimethylamine

Trimethylamine borane

Trimethylamine hydrochloride

Trimethylamine oxide

2,4,5-Trimethylaniline

2,4,6-Trimethylaniline

Trimethylarsine

2,4,6-Trimethylbenzaldehyde

1,2,3-Trimethylbenzene

1,2,4-Trimethylbenzene

1,3,5-Trimethylbenzene

2,3,5-Trimethyl-1,4-benzenediol

No.	Name	Synonym	Mol. Form.	CAS RN	Mol. Wt.	Physical Form	mp/°C	bp/°C	den/ g cm⁻³	n_D	Solubility
10511	N,α,α-Trimethylbenzeneethanamine	Mephentermine	$C_{11}H_{17}N$	100-92-5	163.260	liq		95[9]			i H_2O; s eth; vs EtOH
10512	Trimethyl 1,2,4-benzenetricarboxylate	Trimethyl trimellitate	$C_{12}H_{12}O_6$	2459-10-1	252.219	visc oil	-13	194[12]	1.261	1.5230[20]	
10513	2,4,6-Trimethylbenzoic acid		$C_{10}H_{12}O_2$	480-63-7	164.201	pr (lig)	156.5				sl H_2O; s EtOH, eth, ace, chl
10514	Trimethylbenzylsilane		$C_{10}H_{16}Si$	770-09-2	164.320			190.5	0.8933[20]	1.4941[20]	
10515	1,7,7-Trimethylbicyclo[2.2.1]heptane		$C_{10}H_{18}$	464-15-3	138.250	hex pl(al), pr(MeOH)		161			i H_2O; s EtOH, eth, AcOEt, MeOH
10516	1,3,3-Trimethylbicyclo[2.2.1]heptan-2-ol, (1S-endo)	α-Fenchyl alcohol, (l)	$C_{10}H_{18}O$	512-13-0	154.249	pr	48	94[20]	0.9034[84]		
10517	1,7,7-Trimethylbicyclo[2.2.1]heptan-2-ol acetate, endo	Bornyl acetate	$C_{12}H_{20}O_2$	76-49-3	196.286		29	221			
10518	1,7,7-Trimethylbicyclo[2.2.1]hept-2-ene		$C_{10}H_{16}$	464-17-5	136.234	cry (al)	113	146			vs bz, eth, EtOH
10519	4,6,6-Trimethylbicyclo[3.1.1]hept-3-en-2-ol, (1α,2α,5α)		$C_{10}H_{16}O$	1820-09-3	152.233		24	92[10]	0.9657[25]	1.4908[25]	
10520	4,6,6-Trimethylbicyclo[3.1.1]hept-3-en-2-ol, (1α,2β,5α)		$C_{10}H_{16}O$	1845-30-3	152.233		15.5	90[10]	0.9684[25]	1.4912[25]	
10521	2,7,7-Trimethylbicyclo[3.1.1]hept-2-en-6-one	Chrysanthenone	$C_{10}H_{14}O$	473-06-3	150.217			88[12]		1.4720[22]	vs EtOH
10522	Trimethylborane		C_3H_9B	593-90-8	55.914	col gas	-161.5	-20.2			
10523	Trimethyl borate		$C_3H_9BO_3$	121-43-7	103.912	liq	-29.3	67.5	0.915[25]	1.3568[20]	vs eth, MeOH
10524	2,2,3-Trimethylbutane	Triptane	C_7H_{16}	464-06-2	100.202	liq	-24.6	80.86	0.6901[20]	1.3864[20]	i H_2O; s EtOH, eth; vs ace, bz, peth, ctc
10525	2,3,3-Trimethyl-2-butanol		$C_7H_{16}O$	594-83-2	116.201	cry (dil al +1/2w)	17	131	0.8380[25]	1.4233[22]	sl H_2O; vs ace, eth, EtOH
10526	2,3,3-Trimethyl-1-butene		C_7H_{14}	594-56-9	98.186	liq	-109.9	77.9	0.7050[20]	1.4025[20]	i H_2O; s eth, bz, chl, MeOH
10527	Trimethylchlorosilane		C_3H_9ClSi	75-77-4	108.642	liq	-40	60	0.856[25]	1.3870[20]	
10528	Trimethyl citrate		$C_9H_{14}O_7$	1587-20-8	234.203	tcl	79.3	285; 176[16]			vs eth, EtOH
10529	2,6,6-Trimethyl-2,4-cycloheptadien-1-one	Eucarvone	$C_{10}H_{14}O$	503-93-5	150.217			210; 105[22]	0.9490[20]	1.5087[20]	s eth, ace
10530	1,1,2-Trimethylcyclohexane		C_9H_{18}	7094-26-0	126.239	liq	-29	145.2	0.7963[25]	1.4382[20]	
10531	1,1,3-Trimethylcyclohexane		C_9H_{18}	3073-66-3	126.239	liq	-65.7	136.6	0.7749[25]	1.4295[20]	i H_2O
10532	1α,2β,4β-1,2,4-Trimethylcyclohexane		C_9H_{18}	7667-60-9	126.239	liq	-83.5	142.9	0.7870[25]	1.4341[20]	
10533	1α,3α,5β-1,3,5-Trimethylcyclohexane	trans-1,3,5-Trimethylcyclohexane	C_9H_{18}	1795-26-2	126.239	liq	-107.4	140.5	0.7794[20]	1.4307[20]	vs bz, eth, lig
10534	cis-3,3,5-Trimethylcyclohexanol		$C_9H_{18}O$	933-48-2	142.238		37.3	202; 92[12]	0.9006[16]	1.4550[16]	i H_2O; s EtOH, eth, chl
10535	trans-3,3,5-Trimethylcyclohexanol		$C_9H_{18}O$	767-54-4	142.238	cry (eth)	55.8	189.2	0.8631[60]		i H_2O; s EtOH, eth, chl
10536	2,2,6-Trimethylcyclohexanone		$C_9H_{16}O$	2408-37-9	140.222	liq	-31.8	178.5	0.9043[18]	1.4470[20]	
10537	2,4,4-Trimethylcyclohexanone		$C_9H_{16}O$	2230-70-8	140.222			191	0.902[20]	1.4493[20]	
10538	3,3,5-Trimethylcyclohexanone	Dihydroisophorone	$C_9H_{16}O$	873-94-9	140.222	ye oil		189	0.8919[19]	1.4454[15]	
10539	2,6,6-Trimethyl-1-cyclohexene-1-carboxaldehyde	β-Cyclocitral	$C_{10}H_{16}O$	432-25-7	152.233			112[29], 97[15]	0.959[15]	1.4971[15]	
10540	3,5,5-Trimethyl-2-cyclohexen-1-ol	Isophorol	$C_9H_{16}N$	470-99-5	140.222			69[5]	0.914[20]	1.4717[20]	
10541	4-(2,6,6-Trimethyl-1-cyclohexen-1-yl)-3-buten-2-ol	β-Ionol	$C_{13}H_{22}O$	22029-76-1	194.313			130[14]	0.9243[20]	1.4969[20]	s EtOH, eth, ace
10542	4-(2,6,6-Trimethyl-2-cyclohexen-1-yl)-3-buten-2-ol	α-Ionol	$C_{13}H_{22}O$	25312-34-9	194.313	oil		127[14]	0.9189[20]	1.4735[20]	
10543	1,1,2-Trimethylcyclopentane		C_8H_{16}	4259-00-1	112.213	liq	-21.6	114; 53[100]	0.7660[20]	1.4199[20]	
10544	1,1,3-Trimethylcyclopentane		C_8H_{16}	4516-69-2	112.213	liq	-142.4	104.9	0.7439[25]	1.4112[20]	i H_2O
10545	1α,2α,4β-1,2,4-Trimethylcyclopentane		C_8H_{16}	4850-28-6	112.213	liq	-132.6	116.7	0.7592[25]	1.4186[20]	
10546	1α,2β,4α-1,2,4-Trimethylcyclopentane		C_8H_{16}	16883-48-0	112.213	liq	-130.8	109.3	0.7430[25]	1.4106[20]	
10547	1,2,2-Trimethyl-1,3-cyclopentanedicarboxylic acid, (1R, 3S)	(+)-Camphoric acid	$C_{10}H_{16}O_4$	124-83-4	200.232	pr, lf (w)	187		1.186[20]		sl H_2O; vs EtOH, eth; s ace; i bz, chl
10548	2,2,4-Trimethylcyclopentanone		$C_8H_{14}O$	28056-54-4	126.196	liq	-40.6	158	0.877[25]	1.4300[20]	
10549	2,4,4-Trimethylcyclopentanone		$C_8H_{14}O$	4694-12-6	126.196	liq	-25.6	160.5	0.8785[18]	1.433[18]	
10550	1,1,2-Trimethylcyclopropane		C_6H_{12}	4127-45-1	84.159	liq	-138.2	54	0.6897[25]	1.3864[20]	
10551	3,7,11-Trimethyl-2,6,10-dodecatrienal		$C_{15}H_{24}O$	19317-11-4	220.351			172[14]	0.893[18]	1.4995	
10552	Trimethylgallium		C_3H_9Ga	1445-79-0	114.826			55.7			dec H_2O (exp)
10553	2,2,6-Trimethylheptane		$C_{10}H_{22}$	1190-83-6	142.282	liq	-105	148.9	0.7200[25]	1.4078[20]	
10554	2,5,5-Trimethylheptane		$C_{10}H_{22}$	1189-99-7	142.282			152.8	0.7362[25]	1.4149[20]	
10555	3,3,5-Trimethylheptane		$C_{10}H_{22}$	7154-80-5	142.282			155.7	0.7248[20]	1.4170[20]	i H_2O; s bz, ctc, chl

N,α,α-Trimethylbenzeneethanamine

Trimethyl 1,2,4-benzenetricarboxylate

2,4,6-Trimethylbenzoic acid

Trimethylbenzylsilane

1,7,7-Trimethylbicyclo[2.2.1]heptane

1,3,3-Trimethylbicyclo[2.2.1]heptan-2-ol, (1S-endo)

1,7,7-Trimethylbicyclo[2.2.1]heptan-2-ol acetate, endo

1,7,7-Trimethylbicyclo[2.2.1]hept-2-ene

4,6,6-Trimethylbicyclo[3.1.1]hept-3-en-2-ol, (1α,2α,5α)

4,6,6-Trimethylbicyclo[3.1.1]hept-3-en-2-ol, (1α,2β,5α)

2,7,7-Trimethylbicyclo[3.1.1]hept-2-en-6-one

Trimethylborane

Trimethyl borate

2,2,3-Trimethylbutane

2,3,3-Trimethyl-2-butanol

2,3,3-Trimethyl-1-butene

Trimethylchlorosilane

Trimethyl citrate

2,6,6-Trimethyl-2,4-cycloheptadien-1-one

1,1,2-Trimethylcyclohexane

1,1,3-Trimethylcyclohexane

1α,2β,4β-1,2,4-Trimethylcyclohexane

1α,3α,5β-1,3,5-Trimethylcyclohexane

cis-3,3,5-Trimethylcyclohexanol

trans-3,3,5-Trimethylcyclohexanol

2,2,6-Trimethylcyclohexanone

2,4,4-Trimethylcyclohexanone

3,3,5-Trimethylcyclohexanone

2,6,6-Trimethyl-1-cyclohexene-1-carboxaldehyde

3,5,5-Trimethyl-2-cyclohexen-1-ol

4-(2,6,6-Trimethyl-1-cyclohexen-1-yl)-3-buten-2-ol

4-(2,6,6-Trimethyl-2-cyclohexen-1-yl)-3-buten-2-ol

1,1,2-Trimethylcyclopentane

1,1,3-Trimethylcyclopentane

1α,2α,4β-1,2,4-Trimethylcyclopentane

1α,2β,4α-1,2,4-Trimethylcyclopentane

1,2,2-Trimethyl-1,3-cyclopentanedicarboxylic acid, (1R, 3S)

2,2,4-Trimethylcyclopentanone

2,4,4-Trimethylcyclopentanone

1,1,2-Trimethylcyclopropane

3,7,11-Trimethyl-2,6,10-dodecatrienal

Trimethylgallium

2,2,6-Trimethylheptane

2,5,5-Trimethylheptane

3,3,5-Trimethylheptane

No.	Name	Synonym	Mol. Form.	CAS RN	Mol. Wt.	Physical Form	mp/°C	bp/°C	den/ g cm⁻³	n_D	Solubility
10556	3,4,5-Trimethylheptane		C₁₀H₂₂	20278-89-1	142.282			162.5	0.7519²⁵	1.4229²⁰	
10557	2,2,3-Trimethylhexane		C₉H₂₀	16747-25-4	128.255			133.6	0.7257²⁵	1.4106²⁰	
10558	2,2,4-Trimethylhexane		C₉H₂₀	16747-26-5	128.255	liq	-120	126.5	0.711²⁰	1.4033²⁰	
10559	2,2,5-Trimethylhexane		C₉H₂₀	3522-94-9	128.255	liq	-105.7	124.09	0.7072²⁰	1.3997²⁰	i H₂O; vs EtOH, eth, ace, bz; s ctc
10560	2,3,3-Trimethylhexane		C₉H₂₀	16747-28-7	128.255	liq	-116.8	137.7	0.7345²⁵	1.4141²⁰	
10561	2,3,4-Trimethylhexane		C₉H₂₀	921-47-1	128.255			139.1	0.7354²⁵	1.4144²⁰	
10562	2,3,5-Trimethylhexane		C₉H₂₀	1069-53-0	128.255	liq	-127.9	131.4	0.7218²⁰	1.4051²⁰	
10563	2,4,4-Trimethylhexane		C₉H₂₀	16747-30-1	128.255	liq	-113.4	130.7	0.7201²⁵	1.4074²⁰	
10564	3,3,4-Trimethylhexane		C₉H₂₀	16747-31-2	128.255	liq	-101.2	140.5	0.7414²⁵	1.4178²⁰	
10565	3,5,5-Trimethylhexanoic acid	Isononanoic acid	C₉H₁₈O₂	3302-10-1	158.238	liq		121¹⁰, 85⁴			
10566	3,5,5-Trimethyl-1-hexanol		C₉H₂₀O	3452-97-9	144.254			194	0.8236²⁵	1.4300²⁵	
10567	1,2,3-Trimethylindene		C₁₂H₁₄	4773-83-5	158.239	liq		100.5¹⁰	0.9714²⁰	1.5521²⁰	
10568	Trimethylindium	Indium trimethyl	C₃H₉In	3385-78-2	159.921			135.7	1.568¹⁹		
10569	2,3,3-Trimethyl-3H-indole		C₁₁H₁₃N	1640-39-7	159.228			107¹¹			
10570	Trimethyl(4-methylphenyl)silane		C₁₀H₁₆Si	3728-43-6	164.320			192; 73¹⁰	0.8666²⁰	1.4900²⁰	
10571	1,4,5-Trimethylnaphthalene		C₁₃H₁₄	2131-41-1	170.250	lf (MeOH)	63	145¹²			i H₂O
10572	1,3,5-Trimethyl-2-nitrobenzene		C₉H₁₁NO₂	603-71-4	165.189	orth pr (al)	44	255	1.51²⁵		vs EtOH
10573	2,6,8-Trimethyl-4-nonanol		C₁₂H₂₆O	123-17-1	186.333			225.4	0.8178²⁰		sl ctc
10574	2,4,7-Trimethyloctane		C₁₁H₂₄	62016-38-0	156.309			168.1			
10575	Trimethylolpropane		C₆H₁₄O₃	77-99-6	134.173	wh pow or pl	58	160⁵			vs H₂O, EtOH
10576	3,5,5-Trimethyl-2,4-oxazolidinedione	Trimethadione	C₆H₉NO₃	127-48-0	143.140		46	79⁵			s H₂O; vs EtOH, eth, ace, bz; i peth
10577	Trimethyloxonium fluoborate		C₃H₉BF₄O	420-37-1	147.907	hyg nd	148 dec				vs ace, chl
10578	2,4,4-Trimethyl-2-pentanamine		C₈H₁₉N	107-45-9	129.244						s chl
10579	2,2,3-Trimethylpentane		C₈H₁₈	564-02-3	114.229	liq	-112.2	110	0.7161²⁰	1.4030²⁰	i H₂O; msc EtOH, eth, ace, hp; s bz
10580	2,2,4-Trimethylpentane	Isooctane	C₈H₁₈	540-84-1	114.229	liq	-107.3	99.22	0.6878²⁵	1.3884²⁵	i H₂O; msc EtOH, ace, hp; s eth, ctc
10581	2,3,3-Trimethylpentane		C₈H₁₈	560-21-4	114.229	liq	-100.9	114.8	0.7262²⁰	1.4075²⁰	i H₂O; vs EtOH; msc eth, ace, bz, hp
10582	2,3,4-Trimethylpentane		C₈H₁₈	565-75-3	114.229	liq	-109.2	113.5	0.7191²⁰	1.4042²⁰	i H₂O; vs EtOH; msc eth, ace, bz; sl ctc
10583	2,2,4-Trimethyl-1,3-pentanediol		C₈H₁₈O₂	144-19-4	146.228	pl (bz)	51.5	235; 81¹	0.936¹⁵	1.4513¹⁵	sl H₂O; vs EtOH, eth; s bz, chl
10584	2,4,4-Trimethyl-2-pentanethiol		C₈H₁₈S	141-59-3	146.294	liq		76⁵⁰			
10585	2,4,4-Trimethyl-2-pentanol		C₈H₁₈O	690-37-9	130.228	liq	-20	147.5	0.8225²⁰	1.4284²⁰	i H₂O; sl EtOH; s eth
10586	2,2,4-Trimethyl-3-pentanol		C₈H₁₈O	5162-48-1	130.228	liq	-13	150.5	0.8297²⁰	1.4288²⁰	
10587	2,2,4-Trimethyl-3-pentanone	tert-Butyl isopropyl ketone	C₈H₁₆O	5857-36-3	128.212			135.1	0.8065²⁰	1.4060	i H₂O; s eth, ace
10588	2,3,3-Trimethyl-1-pentene		C₈H₁₆	560-23-6	112.213	liq	-69	108.3	0.7308²⁵	1.4174²⁰	
10589	2,4,4-Trimethyl-1-pentene		C₈H₁₆	107-39-1	112.213	liq	-93.5	101.4	0.7150²⁰	1.4086²⁰	i H₂O; s eth, bz, ctc, chl, lig
10590	2,3,4-Trimethyl-2-pentene		C₈H₁₆	565-77-5	112.213	liq	-113.4	116.5	0.7434²⁰	1.4274²⁰	
10591	2,4,4-Trimethyl-2-pentene		C₈H₁₆	107-40-4	112.213	liq	-106.3	104.9	0.7218²⁰	1.4160²⁰	i H₂O; s eth, bz, ctc, chl; vs lig
10592	2,3,4-Trimethylphenol		C₉H₁₂O	526-85-2	136.190	nd (peth)	81	236			vs bz, eth, EtOH
10593	2,3,5-Trimethylphenol		C₉H₁₂O	697-82-5	136.190		94.5	233			
10594	2,3,6-Trimethylphenol		C₉H₁₂O	2416-94-6	136.190		63				
10595	2,4,5-Trimethylphenol		C₉H₁₂O	496-78-6	136.190	nd (lig)	72	232			i H₂O; vs EtOH, eth
10596	2,4,6-Trimethylphenol		C₉H₁₂O	527-60-6	136.190	nd (peth, MeOH)	73	220			vs eth, EtOH
10597	3,4,5-Trimethylphenol		C₉H₁₂O	527-54-8	136.190	nd (peth)	108	248.5			
10598	Trimethylphenoxysilane		C₉H₁₄OSi	1529-17-5	166.292	liq	-55	119	0.8681²⁰	1.5125²⁰	
10599	Trimethylphenylammonium chloride	Phenyltrimethylammonium chloride	C₉H₁₄ClN	138-24-9	171.667						vs H₂O, EtOH
10600	1-(2,4,6-Trimethylphenyl)ethanone		C₁₁H₁₄O	1667-01-2	162.228			241; 120¹²	0.9754²⁰	1.5175²⁰	i H₂O; s EtOH, eth, ace, bz, ctc
10601	1,1,1-Trimethyl-N-phenylsilanamine	Phenyl(trimethylsilyl)amine	C₉H₁₅NSi	3768-55-6	165.308			206	0.940²⁰		
10602	Trimethylphenylsilane		C₉H₁₄Si	768-32-1	150.293			169.5	0.8722²⁰	1.4907²⁰	s ctc, CS₂
10603	Trimethyl phosphate	Methyl phosphate	C₃H₉O₄P	512-56-1	140.074	liq	-46	197.2	1.2144²⁰	1.3967²⁰	vs H₂O; sl EtOH; s eth
10604	Trimethylphosphine		C₃H₉P	594-09-2	76.077	liq	-85	37.5			i H₂O; s eth

3,4,5-Trimethylheptane

2,2,3-Trimethylhexane

2,2,4-Trimethylhexane

2,2,5-Trimethylhexane

2,3,3-Trimethylhexane

2,3,4-Trimethylhexane

2,3,5-Trimethylhexane

2,4,4-Trimethylhexane

3,3,4-Trimethylhexane

3,5,5-Trimethylhexanoic acid

3,5,5-Trimethyl-1-hexanol

1,2,3-Trimethylindene

Trimethylindium

2,3,3-Trimethyl-3H-indole

Trimethyl(4-methylphenyl)silane

1,4,5-Trimethylnaphthalene

1,3,5-Trimethyl-2-nitrobenzene

2,6,8-Trimethyl-4-nonanol

2,4,7-Trimethyloctane

Trimethylolpropane

3,5,5-Trimethyl-2,4-oxazolidinedione

Trimethyloxonium fluoborate

2,4,4-Trimethyl-2-pentanamine

2,2,3-Trimethylpentane

2,2,4-Trimethylpentane

2,3,3-Trimethylpentane

2,3,4-Trimethylpentane

2,2,4-Trimethyl-1,3-pentanediol

2,4,4-Trimethyl-2-pentanethiol

2,4,4-Trimethyl-2-pentanol

2,2,4-Trimethyl-3-pentanol

2,2,4-Trimethyl-3-pentanone

2,3,3-Trimethyl-1-pentene

2,4,4-Trimethyl-1-pentene

2,3,4-Trimethyl-2-pentene

2,4,4-Trimethyl-2-pentene

2,3,4-Trimethylphenol

2,3,5-Trimethylphenol

2,3,6-Trimethylphenol

2,4,5-Trimethylphenol

2,4,6-Trimethylphenol

3,4,5-Trimethylphenol

Trimethylphenoxysilane

Trimethylphenylammonium chloride

1-(2,4,6-Trimethylphenyl)ethanone

1,1,1-Trimethyl-N-phenylsilanamine

Trimethylphenylsilane

Trimethyl phosphate

Trimethylphosphine

No.	Name	Synonym	Mol. Form.	CAS RN	Mol. Wt.	Physical Form	mp/°C	bp/°C	den/ g cm⁻³	n_D	Solubility
10605	Trimethyl phosphite		$C_3H_9O_3P$	121-45-9	124.075			111.5	1.0518[20]	1.4095[20]	vs EtOH, eth; sl ctc
10606	1,2,4-Trimethylpiperazine		$C_7H_{16}N_2$	120-85-4	128.215			149.5		1.4433[20]	s ctc
10607	2,2,4-Trimethylpiperidine		$C_8H_{17}N$	101257-71-0	127.228			148	0.832[15]	1.4458[20]	vs eth, EtOH
10608	Trimethylpyrazine		$C_7H_{10}N_2$	14667-55-1	122.167			87[35]			
10609	1,3,5-Trimethyl-1H-pyrazole		$C_6H_{10}N_2$	1072-91-9	110.156		37	170	0.9269[40]	1.4589[57]	
10610	2,3,6-Trimethylpyridine	2,3,6-Collidine	$C_8H_{11}N$	1462-84-6	121.180			171.6	0.9220[25]	1.5053[20]	s H₂O, EtOH, eth, ace, bz
10611	2,4,6-Trimethylpyridine	2,4,6-Collidine	$C_8H_{11}N$	108-75-8	121.180	liq	-46	170.6	0.9166[22]	1.4959[25]	s H₂O, EtOH, eth, ace, ctc
10612	1,2,5-Trimethyl-1H-pyrrole		$C_7H_{11}N$	930-87-0	109.169			171	0.807[25]	1.4969[20]	
10613	N,N,2-Trimethyl-6-quinolinamine		$C_{12}H_{14}N_2$	92-99-9	186.252	ye pr (HOAc, AcOEt)	101	319			s ctc, CS₂
10614	Trimethylsilane		$C_3H_{10}Si$	993-07-7	74.197	col gas	-135.9	6.7			
10615	1-(Trimethylsilyl)-1H-imidazole		$C_6H_{12}N_2Si$	18156-74-6	140.258						s chl
10616	3-(Trimethylsilyl)-1-propanol		$C_6H_{16}OSi$	2917-47-7	132.276			141; 82[24]	0.822[25]	1.4298[20]	
10617	Trimethylstibine		C_3H_9Sb	594-10-5	166.863	liq	-62	80.6	1.523[15]	1.42[15]	i H₂O; s EtOH, eth, CS₂
10618	Trimethylsulfonium iodide		C_3H_9IS	2181-42-2	204.072	cry (eth)	211 dec				
10619	Trimethylthiourea		$C_4H_{10}N_2S$	2489-77-2	118.200	pr (bz-lig)	87.5				vs bz, EtOH, chl
10620	2,4,6-Trimethyl-2,4,6-triphenylcyclotrisiloxane		$C_{21}H_{24}O_3Si_3$	546-45-2	408.671		100	190[1.5]	1.1062[20]	1.5397[20]	
10621	Trimethylurea		$C_4H_{10}N_2O$	632-14-4	102.134	pr (eth)	75.5	232.5	1.1900[20]		s H₂O, EtOH; sl eth, bz
10622	Trinitroacetonitrile		$C_2N_4O_6$	630-72-8	176.044	wax	41.5	exp 220			vs eth
10623	2,4,6-Trinitroaniline		$C_6H_4N_4O_6$	489-98-5	228.119	dk ye pr (HOAc)	193.5	exp	1.762[10]		i H₂O; sl EtOH, eth; s ace, bz, AcOEt
10624	1,3,5-Trinitrobenzene	sym-Trinitrobenzene	$C_6H_3N_3O_6$	99-35-4	213.104	orth pl (bz) lf (w)	122.9	315	1.4775[152]		sl H₂O, EtOH, eth; vs ace; s bz, py
10625	2,4,6-Trinitro-1,3-benzenediol	Styphnic acid	$C_6H_3N_3O_8$	82-71-3	245.103	hex ye cry (dil al)	175.5	sub			vs eth, EtOH
10626	2,4,6-Trinitrobenzoic acid		$C_7H_3N_3O_8$	129-66-8	257.114	orth (w)	228 dec				sl H₂O, bz; vs EtOH; s eth, ace
10627	2,4,7-Trinitro-9H-fluoren-9-one		$C_{13}H_5N_3O_7$	129-79-3	315.195	pa ye nd (bz, HOAc)	175.8				sl H₂O; vs ace, bz, chl
10628	Trinitrofluoromethane	Fluorotrinitromethane	CFN_3O_6	1840-42-2	169.025			86.3	1.59[20]		
10629	Trinitroglycerol	Nitroglycerin	$C_3H_5N_3O_9$	55-63-0	227.087	pa ye tcl or orth	13.5	exp 218; 93[0.31]	1.5931[20]	1.4786[12]	sl H₂O; s EtOH, bz; msc eth; vs ace, chl
10630	Trinitromethane		CHN_3O_6	517-25-9	151.035		15	exp	1.479[20]	1.4451[24]	vs ace, EtOH
10631	2,4,6-Trinitrophenol	Picric acid	$C_6H_3N_3O_7$	88-89-1	229.104	ye lf (w), pr (eth) pl (al)	122.5	exp 300		1.763	sl H₂O; s EtOH, eth, bz, chl; vs ace
10632	2,4,6-Trinitrophenol, sodium salt	Sodium picrate	$C_6H_2N_3NaO_7$	3324-58-1	251.086	nd (w)	270.4				
10633	2,4,6-Trinitrotoluene	2-Methyl-1,3,5-trinitrobenzene	$C_7H_5N_3O_6$	118-96-7	227.131	orth (al)	80.5	exp 240	1.654[25]		i H₂O; sl EtOH; s eth; vs ace, bz
10634	2,4,6-Trinitro-N-(2,4,6-trinitrophenyl)aniline	Dipicrylamine	$C_{12}H_5N_7O_{12}$	131-73-7	439.208	pa ye pr(HOAc)	244 dec				i H₂O, EtOH, bz, ctc; sl eth, ace; vs py
10635	Trioctylaluminum		$C_{24}H_{51}Al$	1070-00-4	366.644	hyg visc liq	-62		0.701		
10636	Trioctylamine	N,N-Dioctyl-1-octanamine	$C_{24}H_{51}N$	1116-76-3	353.669	liq	-34.6	366	0.8110[20]	1.4510[19]	
10637	Trioctylphosphine oxide	TOPO	$C_{24}H_{51}OP$	78-50-2	386.635		52	201[2]			
10638	1,3,5-Trioxane	Formaldehyde, trimer	$C_3H_6O_3$	110-88-3	90.078	orth nd (eth)	60.29	114.5	1.17[65]		vs H₂O; s EtOH, eth, bz, CS₂; i peth
10639	1,3,5-Trioxane-2,4,6-triimine	Cyamelide	$C_3H_3N_3O_3$	462-02-2	129.074	amor pow	dec	dec	1.127[15]		vs eth, EtOH
10640	4,7,10-Trioxatridecane-1,13-diamine	Diethyleneglycol diaminopropyl ether	$C_{10}H_{24}N_2O_3$	4246-51-9	220.309	liq		147[4]	1.005	1.4640[20]	
10641	3,7,12-Trioxocholan-24-oic acid, (5β)	Dehydrocholic acid	$C_{24}H_{34}O_5$	81-23-2	402.524		237				i H₂O, eth; sl EtOH, bz; s ace, AcOEt
10642	Tripentylamine	N,N-Dipentyl-1-pentanamine	$C_{15}H_{33}N$	621-77-2	227.430			242.5	0.7907[20]	1.4366[20]	i H₂O; s EtOH, eth, acid
10643	Triphenylamine	N,N-Diphenylbenzenamine	$C_{18}H_{15}N$	603-34-9	245.319	mcl (MeOH, bz)	126.5	365	0.774[0]	1.353[16]	i H₂O; sl EtOH; s eth, bz, MeOH
10644	Triphenylarsine		$C_{18}H_{15}As$	603-32-7	306.234		61	360	1.2634[18]	1.6888[21]	i H₂O; sl EtOH; vs eth, bz; s chl
10645	Triphenylarsine oxide		$C_{18}H_{15}AsO$	1153-05-5	322.233		192	324.0			

Trimethyl phosphite

1,2,4-Trimethylpiperazine

2,2,4-Trimethylpiperidine

Trimethylpyrazine

1,3,5-Trimethyl-1*H*-pyrazole

2,3,6-Trimethylpyridine

2,4,6-Trimethylpyridine

1,2,5-Trimethyl-1*H*-pyrrole

N,N,2-Trimethyl-6-quinolinamine

Trimethylsilane

1-(Trimethylsilyl)-1*H*-imidazole

3-(Trimethylsilyl)-1-propanol

Trimethylstibine

Trimethylsulfonium iodide

Trimethylthiourea

2,4,6-Trimethyl-2,4,6-triphenylcyclotrisiloxane

Trimethylurea

Trinitroacetonitrile

2,4,6-Trinitroaniline

1,3,5-Trinitrobenzene

2,4,6-Trinitro-1,3-benzenediol

2,4,6-Trinitrobenzoic acid

2,4,7-Trinitro-9*H*-fluoren-9-one

Trinitrofluoromethane

Trinitroglycerol

Trinitromethane

2,4,6-Trinitrophenol

2,4,6-Trinitrophenol, sodium salt

2,4,6-Trinitrotoluene

2,4,6-Trinitro-*N*-(2,4,6-trinitrophenyl)aniline

Trioctylaluminum

Trioctylamine

Trioctylphosphine oxide

1,3,5-Trioxane

1,3,5-Trioxane-2,4,6-triimine

4,7,10-Trioxatridecane-1,13-diamine

3,7,12-Trioxocholan-24-oic acid, (5β)

Tripentylamine

Triphenylamine

Triphenylarsine

Triphenylarsine oxide

No.	Name	Synonym	Mol. Form.	CAS RN	Mol. Wt.	Physical Form	mp/°C	bp/°C	den/ g cm⁻³	n_D	Solubility
10646	Triphenylbismuthine		C₁₈H₁₅Bi	603-33-8	440.292		77.6	242[14]	1.715[75]	1.7040[75]	sl EtOH, chl; s eth, ace, bz, CS₂
10647	Triphenylborane		C₁₈H₁₅B	960-71-4	242.123	wh cry	142				i H₂O; sl eth; s bz, lig
10648	Triphenylene	Benzo[1]phenanthrene	C₁₈H₁₂	217-59-4	228.288	nd (al, chl, bz)	197.8	425			i H₂O; s EtOH, HOAc; vs bz, chl
10649	1,1,2-Triphenylethane		C₂₀H₁₈	1520-42-9	258.357	mcl lf (dil al), nd (al)	57				i H₂O; vs EtOH, eth, bz; sl MeOH
10650	1,1,2-Triphenylethene		C₂₀H₁₆	58-72-0	256.341	lf (al)	72.5	220[14]	1.0373[78]	1.6292[78]	i H₂O; s EtOH, chl, MeOH; vs eth
10651	N,N',N''-Triphenylguanidine		C₁₉H₁₇N₃	101-01-9	287.358	nd or pr (al)	146.5	dec	1.163[20]		sl H₂O; s EtOH
10652	2,4,5-Triphenyl-1H-imidazole		C₂₁H₁₆N₂	484-47-9	296.365	nd (al)	275	sub			i H₂O; s EtOH, eth
10653	Triphenylmethane		C₁₉H₁₆	519-73-3	244.330	orth (al)	93.4	359; 200[10]	1.014[99]	1.5839[99]	i H₂O; sl EtOH; vs eth, py, chl; s bz
10654	Triphenylmethanol		C₁₉H₁₆O	76-84-6	260.329	pl (al), trg (bz)	164.2	380	1.199[0]		i H₂O, peth; vs EtOH, eth; s ace, bz
10655	Triphenyl phosphate		C₁₈H₁₅O₄P	115-86-6	326.283	cry (lig), pr (al) nd (eth)	50.5	245[11]	1.2055[50]		i H₂O; s EtOH; vs eth, bz, ctc, chl
10656	Triphenylphosphine		C₁₈H₁₅P	603-35-0	262.286		80	188[1]	1.0749[80]	1.6358[80]	i H₂O; s EtOH, bz, chl; vs eth
10657	Triphenylphosphine oxide		C₁₈H₁₅OP	791-28-6	278.285	pr	156.5	>360	1.2124[23]		sl H₂O, eth, chl; vs EtOH, bz
10658	Triphenyl phosphite		C₁₈H₁₅O₃P	101-02-0	310.284		25	360	1.1842[20]	1.5900[20]	i H₂O; vs EtOH
10659	Triphenylsilane		C₁₈H₁₆Si	789-25-3	260.406						s ctc, CS₂
10660	Triphenylsilanol		C₁₈H₁₆OSi	791-31-1	276.405		154.8		1.1777[20]		s ctc, CS₂
10661	Triphenylstibine		C₁₈H₁₅Sb	603-36-1	353.072	pr (peth)	53.5	>360	1.4343[25]	1.6948[42]	i H₂O; s EtOH; vs eth, ace, bz, chl
10662	Triphenyltetrazolium chloride		C₁₉H₁₅ClN₄	298-96-4	334.802	nd (al,chl)	243 dec				s H₂O, EtOH, ace, chl; i eth
10663	Triphenyltin hydroxide	Stannane, hydroxytriphenyl-	C₁₈H₁₆OSn	76-87-9	367.029		119		1.54[20]		
10664	2,4,6-Triphenyl-1,3,5-triazine		C₂₁H₁₅N₃	493-77-6	309.364		257				
10665	Tripotassium citrate	Potassium citrate	C₆H₅K₃O₇	866-84-2	306.395	wh cry (w)	275 dec				vs H₂O; i EtOH
10666	Triprolidine		C₁₉H₂₂N₂	486-12-4	278.391	cry (peth)	60				
10667	Tri-2-propenoyl-2-ethyl-2-(hydroxymethyl)-1,3-propanediol	Trimethylolpropane triacrylate	C₁₅H₂₀O₆	15625-89-5	296.316			>200[1]		1.4735[20]	
10668	Tripropylamine	N,N-Dipropyl-1-propanamine	C₉H₂₁N	102-69-2	143.270	liq	-93.5	156	0.7558[20]	1.4181[20]	vs eth, EtOH
10669	Tripropylborane		C₉H₂₁B	1116-61-6	140.074	liq	-56	159	0.7204[25]	1.4135[22]	
10670	Tripropyl borate	Boric acid, tripropyl ester	C₉H₂₁BO₃	688-71-1	188.072			179.5	0.8576[20]	1.3948[20]	vs EtOH; msc eth; s PrOH
10671	Tripropylene glycol	[(1-Methyl-1,2-ethanediyl) bis(oxy)]bispropanol	C₉H₂₀O₄	24800-44-0	192.253	liq		268; 115[2]	1.02[20]	1.4440[20]	
10672	Tripropylene glycol diacrylate		C₁₅H₂₄O₆	42978-66-5	300.348			>120[1]			
10673	Tripropylene glycol monomethyl ether	1-[2-(2-Methoxy-1-methylethoxy)-1-methylethoxy]-2-propanol	C₁₀H₂₂O₄	20324-33-8	206.280			241.3			
10674	Tripropyl phosphate		C₉H₂₁O₄P	513-08-6	224.234			252	1.0121[20]	1.4165[20]	sl H₂O, chl; s EtOH, eth, tol, CS₂
10675	Tripropyl phosphite	Tripropoxyphosphine	C₉H₂₁O₃P	923-99-9	208.235			206.5	0.9417[20]	1.4282[20]	vs eth, EtOH
10676	Tripropylsilane		C₉H₂₂Si	998-29-8	158.357			172	0.7723[0]	1.4280[20]	i H₂O
10677	Tris(4-aminophenyl)methanol	C.I. Basic Red 9	C₁₉H₁₉N₃O	467-62-9	305.373	purp cry	205				
10678	2,4,6-Tris(1-aziridinyl)-1,3,5-triazine	Triethylenemelamine	C₉H₁₂N₆	51-18-3	204.231	cry pow	139 dec				s H₂O
10679	Tris(2-butoxyethyl) phosphate		C₁₈H₃₉O₇P	78-51-3	398.473	liq		255[10]	1.02[25]		i H₂O
10680	Tris(2-chloroethyl) phosphate		C₆H₁₂Cl₃O₄P	115-96-8	285.489			330; 194[10]	1.39[25]	1.4721[20]	s ctc
10681	Tris(2-chloroethyl) phosphite		C₆H₁₂Cl₃O₃P	140-08-9	269.490			120[3]	1.3443[26]	1.4868[20]	
10682	Tris(1,3-dichloro-2-propyl) phosphate	Fyrol FR-2	C₉H₁₅Cl₆O₄P	13674-87-8	430.904	visc liq		236[5]		1.5022[20]	i H₂O
10683	Tris(4-dimethylaminophenyl) methane	Paraleucaniline	C₂₅H₃₁N₃	603-48-5	373.534	lf (al), nd (bz)	176.5				vs bz, eth, chl
10684	Tris(2,4-dimethylphenyl) phosphate	2,4-Xylenol, phosphate (3:1)	C₂₄H₂₇O₄P	3862-12-2	410.442			233.5	1.142[38]	1.5550[20]	i H₂O; s bz, chl, hx
10685	Tris(2,5-dimethylphenyl) phosphate	2,5-Xylenol, phosphate (3:1)	C₂₄H₂₇O₄P	19074-59-0	410.442		79.8	262[8]	1.197[25]		i H₂O; sl EtOH, hx; s eth, bz, ctc

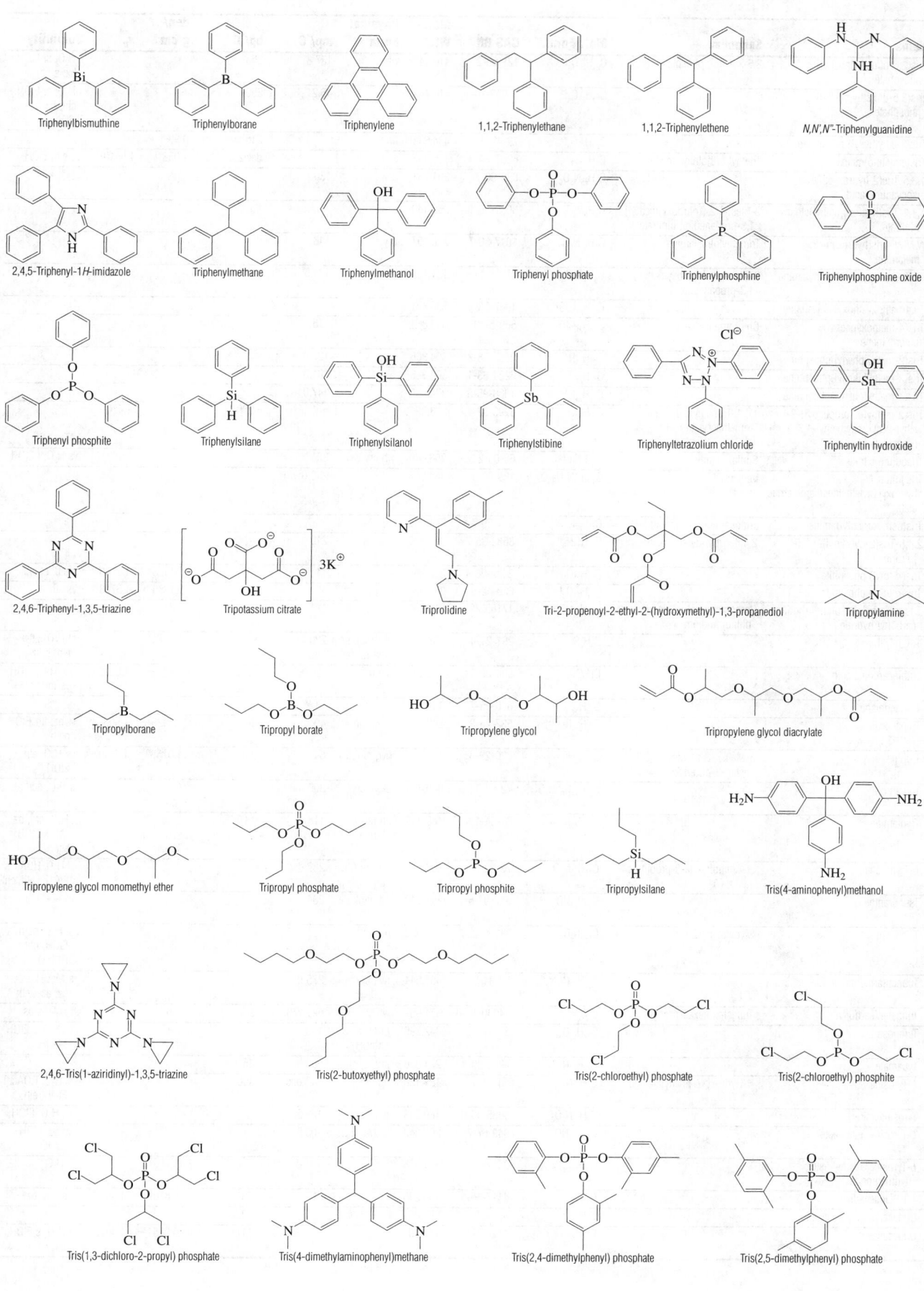

Triphenylbismuthine

Triphenylborane

Triphenylene

1,1,2-Triphenylethane

1,1,2-Triphenylethene

N,N',N''-Triphenylguanidine

2,4,5-Triphenyl-1H-imidazole

Triphenylmethane

Triphenylmethanol

Triphenyl phosphate

Triphenylphosphine

Triphenylphosphine oxide

Triphenyl phosphite

Triphenylsilane

Triphenylsilanol

Triphenylstibine

Triphenyltetrazolium chloride

Triphenyltin hydroxide

2,4,6-Triphenyl-1,3,5-triazine

Tripotassium citrate

Triprolidine

Tri-2-propenoyl-2-ethyl-2-(hydroxymethyl)-1,3-propanediol

Tripropylamine

Tripropylborane

Tripropyl borate

Tripropylene glycol

Tripropylene glycol diacrylate

Tripropylene glycol monomethyl ether

Tripropyl phosphate

Tripropyl phosphite

Tripropylsilane

Tris(4-aminophenyl)methanol

2,4,6-Tris(1-aziridinyl)-1,3,5-triazine

Tris(2-butoxyethyl) phosphate

Tris(2-chloroethyl) phosphate

Tris(2-chloroethyl) phosphite

Tris(1,3-dichloro-2-propyl) phosphate

Tris(4-dimethylaminophenyl)methane

Tris(2,4-dimethylphenyl) phosphate

Tris(2,5-dimethylphenyl) phosphate

No.	Name	Synonym	Mol. Form.	CAS RN	Mol. Wt.	Physical Form	mp/°C	bp/°C	den/ g cm^{-3}	n_D	Solubility
10686	Tris(2,6-dimethylphenyl) phosphate	2,6-Xylenol, phosphate (3:1)	$C_{24}H_{27}O_4P$	121-06-2	410.442	wax	137.8	263[6]			i H_2O; sl EtOH, hx; s bz
10687	Tris(3,5-dimethylphenyl) phosphate		$C_{24}H_{27}O_4P$	25653-16-1	410.442		46.2	290[10]			i H_2O; sl EtOH, chl, hx; s HOAc
10688	Tris(2-ethylhexyl) phosphate		$C_{24}H_{51}O_4P$	78-42-2	434.633	liq		215[5]	0.99[20]		
10689	Tris(ethylthio)methane	Triethyl orthothioformate	$C_7H_{16}S_3$	6267-24-9	196.397			dec 235; 127[12]	1.053[20]	1.5410[15]	vs eth, EtOH
10690	1,3,5-Tris(2-hydroxyethyl) isocyanuric acid		$C_9H_{15}N_3O_6$	839-90-7	261.231	cry	136				
10691	1,1,1-Tris(hydroxymethyl)ethane trinitrate	2-Methyl-2-[(nitrooxy)methyl]-1,3-propanediol, dinitrate	$C_5H_9N_3O_9$	3032-55-1	255.140			83[0.05]			
10692	*N,N',N''*-Tris(hydroxymethyl) melamine	Trimethylolmelamine	$C_6H_{12}N_6O_3$	1017-56-7	216.197	cry	148				
10693	Tris(hydroxymethyl)methylamine	2-Amino-2-(hydroxymethyl)-1,3-propanediol	$C_4H_{11}NO_3$	77-86-1	121.135		171.5	219[10]			vs H_2O; s MeOH
10694	Tris(methoxyethoxy)vinylsilane		$C_{11}H_{24}O_6Si$	1067-53-4	280.391						s ctc
10695	Tris(4-methoxyphenyl) chloroethene	Chlorotrianisene	$C_{23}H_{21}ClO_3$	569-57-3	380.864		115				
10696	Tris(2-methylphenyl)phosphine		$C_{21}H_{21}P$	6163-58-2	304.366		127.0				
10697	Tris(3-methylphenyl)phosphine		$C_{21}H_{21}P$	6224-63-1	304.366		101.0				
10698	Tris(4-methylphenyl)phosphine		$C_{21}H_{21}P$	1038-95-5	304.366		147.0				
10699	Tris(2-methyl-2-propenoyl)-2-ethyl-2-hydroxymethyl-1,3-propanediol	1,1,1-Trimethylolpropane trimethacrylate	$C_{18}H_{26}O_6$	3290-92-4	338.395			>200[1]		1.470[25]	
10700	Trisodium citrate	Sodium citrate	$C_6H_5Na_3O_7$	68-04-2	258.069	wh cry (w)	300				vs H_2O; i EtOH
10701	Trisodium *N*-hydroxyethylethylenediaminetriac etate	Versen-Ol	$C_{10}H_{15}N_2Na_3O_7$	139-89-9	344.204		288 (hyd)				
10702	Tris(perfluorobutyl)amine	Trinonafluorobutylamine	$C_{12}F_{27}N$	311-89-7	671.092			178	1.884[25]	1.291[25]	s ace
10703	2,4,6-Tris(2-pyridinyl)-1,3,5-triazine	2,4,6-Tripyridyl-*s*-triazine	$C_{18}H_{12}N_6$	3682-35-7	312.328		210				
10704	Tris(*o*-tolyl) phosphite		$C_{21}H_{21}O_3P$	2622-08-4	352.364		11	238[11], 197[2]	1.1423[20]	1.5740[28]	s eth; sl chl
10705	Tris(*p*-tolyl) phosphite		$C_{21}H_{21}O_3P$	620-42-8	352.364	pa ye	52	252[10]	1.1280[25]	1.5703[28]	vs eth
10706	Tris(triphenylphosphine) rhodium carbonyl hydride	Carbonylhydrotris(triphenylphosphine)rhodium	$C_{55}H_{46}OP_3Rh$	17185-29-4	918.781	ye cry	121		1.33		sl bz, chl
10707	1,3,5-Trithiane		$C_3H_6S_3$	291-21-4	138.275	hex (bz), pr (w) nd (al)	220	sub	1.6374[24]		sl H_2O, EtOH, eth; s bz
10708	Trithiocarbonic acid		CH_2S_3	594-08-1	110.222	red oil	-26.9	57.8	1.476[25]	1.8225[20]	dec H_2O, EtOH; vs tol, chl
10709	Tritriacontane		$C_{33}H_{68}$	630-05-7	464.893		71.2				
10710	Tropacocaine		$C_{15}H_{19}NO_2$	537-26-8	245.318	pl or tab	49	dec	1.0426[100]	1.5080[100]	vs bz, eth, EtOH, peth
10711	Tropine	8-Methyl-8-azabicyclo[3.2.1] octan-3-ol, *endo*	$C_8H_{15}NO$	120-29-6	141.211	hyg pl (eth)	64	233	1.016[100]	1.4811[100]	vs H_2O, eth, EtOH
10712	Trypan blue		$C_{34}H_{24}N_6Na_4O_{14}S_4$	72-57-1	960.805	dk bl cry	300				s H_2O, acid; i EtOH
10713	Tryptamine		$C_{10}H_{12}N_2$	61-54-1	160.215	nd (al-bz, lig)	118	137[0.15]			i H_2O, eth, bz, chl; s EtOH, ace
10714	*L*-Tryptophan	α-Aminoindole-3-propionic acid, (*l*)	$C_{11}H_{12}N_2O_2$	73-22-3	204.225	lf or pl (dil al)	289 dec				sl H_2O, HOAc; s EtOH; i eth, chl
10715	Tsuduranine		$C_{18}H_{19}NO_3$	517-97-5	297.349	nd (eth)	204				vs ace, eth, EtOH
10716	T-2 Toxin	Mycotoxin T2	$C_{24}H_{34}O_9$	21259-20-1	466.522	nd	151				sl H_2O, peth; s EtOH, chl, DMSO
10717	Tubocurarine dichloride		$C_{37}H_{42}Cl_2N_2O_6$	57-94-3	681.644	hyg cry	275 dec				s MeOH; i py, bz, ace, eth
10718	Tungsten carbonyl	Tungsten hexacarbonyl	C_6O_6W	14040-11-0	351.90	wh cry	dec 170	sub	2.65		i H_2O; s os
10719	Turanose		$C_{12}H_{22}O_{11}$	547-25-1	342.296	pr (w-al, MeOH)	168				vs H_2O; s EtOH, MeOH
10720	Tybamate		$C_{13}H_{26}N_2O_4$	4268-36-4	274.356	cry	50	151[0.06]			
10721	*L*-Tyrosine	4-Hydroxy-*L*-phenylalanine	$C_9H_{11}NO_3$	60-18-4	181.188	nd (w)	343 dec	sub			sl H_2O, HOAc; i EtOH, eth
10722	Tyrosineamide		$C_9H_{12}N_2O_2$	4985-46-0	180.203	pl or pl (al)	153.5				vs H_2O, EtOH
10723	*L*-Tyrosine, ethyl ester		$C_{11}H_{15}NO_3$	949-67-7	209.242	pr (AcOEt)	108.5				vs bz, EtOH, AcOEt
10724	*L*-Tyrosine, methyl ester, hydrochloride		$C_{10}H_{14}ClNO_3$	3417-91-2	231.676		191.0				s H_2O
10725	1,10-Undecadiyne		$C_{11}H_{16}$	4117-15-1	148.245		-17	83[12]	0.8182[21]	1.453[21]	vs ace, bz
10726	Undecafluorocyclohexane		C_6HF_{11}	308-24-7	282.054			62.0			
10727	Undecanal		$C_{11}H_{22}O$	112-44-7	170.292		-2.0	117[18]	0.8251[23]	1.4520[20]	i H_2O; s EtOH, eth

Tris(2,6-dimethylphenyl) phosphate

Tris(3,5-dimethylphenyl) phosphate

Tris(2-ethylhexyl) phosphate

Tris(ethylthio)methane

1,3,5-Tris(2-hydroxyethyl) isocyanuric acid

1,1,1-Tris(hydroxymethyl)ethane trinitrate

N,N',N''-Tris(hydroxymethyl)melamine

Tris(hydroxymethyl)methylamine

Tris(methoxyethoxy)vinylsilane

Tris(4-methoxyphenyl)chloroethene

Tris(2-methylphenyl)phosphine

Tris(3-methylphenyl)phosphine

Tris(4-methylphenyl)phosphine

Tris(2-methyl-2-propenoyl)-2-ethyl-2-hydroxymethyl-1,3-propanediol

Trisodium citrate

Trisodium N-hydroxyethylethylenediaminetriacetate

Tris(perfluorobutyl)amine

2,4,6-Tris(2-pyridinyl)-1,3,5-triazine

Tris(o-tolyl) phosphite

Tris(p-tolyl) phosphite

Tris(triphenylphosphine) rhodium carbonyl hydride

1,3,5-Trithiane

Trithiocarbonic acid

Tritriacontane

Tropacocaine

Tropine

Trypan blue

Tryptamine

L-Tryptophan

Tsuduranine

T-2 Toxin

Tubocurarine dichloride

Tungsten carbonyl

Turanose

Tybamate

L-Tyrosine

Tyrosineamide

L-Tyrosine, ethyl ester

L-Tyrosine, methyl ester, hydrochloride

1,10-Undecadiyne

Undecafluorocyclohexane

Undecanal

No.	Name	Synonym	Mol. Form.	CAS RN	Mol. Wt.	Physical Form	mp/°C	bp/°C	den/ g cm⁻³	n_D	Solubility
10728	Undecane	Hendecane	$C_{11}H_{24}$	1120-21-4	156.309	liq	-25.5	195.9	0.7402²⁰	1.4164²⁰	i H₂O; msc EtOH, eth
10729	Undecanenitrile	Decyl cyanide	$C_{11}H_{21}N$	2244-07-7	167.292			253	0.8254³⁰	1.4293³⁰	i H₂O; s EtOH, eth, ctc
10730	1-Undecanethiol	Undecyl mercaptan	$C_{11}H_{24}S$	5332-52-5	188.374	liq	-1.5	257.4	0.8448²⁰	1.4585²⁰	
10731	Undecanoic acid		$C_{11}H_{22}O_2$	112-37-8	186.292	cry (ace)	28.6	280	0.8907²⁰	1.4294⁴⁵	i H₂O; vs EtOH, ace; s eth; msc bz
10732	1-Undecanol	Undecyl alcohol	$C_{11}H_{24}O$	112-42-5	172.308		15.9	245	0.8298²⁰	1.4392²⁰	i H₂O; s EtOH; vs eth
10733	2-Undecanol	sec-Undecyl alcohol	$C_{11}H_{24}O$	1653-30-1	172.308	col liq	0	229.7	0.8234²⁵	1.4352²⁵	
10734	2-Undecanone	Methyl nonyl ketone	$C_{11}H_{22}O$	112-12-9	170.292		15	231.5	0.8250²⁰	1.4291²⁰	i H₂O; s EtOH, eth, ace, bz, chl
10735	6-Undecanone	Butyl hexyl ketone	$C_{11}H_{22}O$	927-49-1	170.292		14.5	228	0.8308²⁰	1.4270²⁰	i H₂O; vs EtOH, eth
10736	Undecanoyl chloride		$C_{11}H_{21}ClO$	17746-05-3	204.737						sl ctc
10737	10-Undecenal		$C_{11}H_{20}O$	112-45-8	168.276						sl ctc
10738	1-Undecene		$C_{11}H_{22}$	821-95-4	154.293	liq	-49.2	192.7	0.7503²⁰	1.4261²⁰	i H₂O; s eth, chl, lig
10739	cis-2-Undecene		$C_{11}H_{22}$	821-96-5	154.293	liq	-66.5	196.1	0.7576²⁰		
10740	trans-2-Undecene		$C_{11}H_{22}$	693-61-8	154.293	liq	-48.3	192.5	0.7528²⁰	1.4292²⁰	
10741	cis-4-Undecene		$C_{11}H_{22}$	821-98-7	154.293	liq	-97	192.6	0.7541²⁰	1.4302²⁰	
10742	trans-4-Undecene		$C_{11}H_{22}$	693-62-9	154.293	liq	-63.7	193	0.7508²⁰	1.4285²⁰	
10743	cis-5-Undecene		$C_{11}H_{22}$	764-96-5	154.293	liq	-106.5	192.3	0.7537²⁰	1.4302²⁰	
10744	trans-5-Undecene		$C_{11}H_{22}$	764-97-6	154.293	liq	-61.1	192	0.7497²⁰	1.4285²⁰	vs eth, chl, lig
10745	10-Undecenoic acid	Undecylenic acid	$C_{11}H_{20}O_2$	112-38-9	184.276	cry	24.5	275	0.9072²⁴	1.4486²⁴	i H₂O; s EtOH, eth; sl ctc
10746	10-Undecen-1-ol		$C_{11}H_{22}O$	112-43-6	170.292	liq	-1.0	250	0.8495¹⁵	1.4500²⁰	i H₂O; s EtOH, eth; sl ctc
10747	10-Undecenoyl chloride		$C_{11}H_{19}ClO$	38460-95-6	202.721			127¹³	0.944²⁰	1.4542⁰	
10748	Undecylamine	1-Undecanamine	$C_{11}H_{25}N$	7307-55-3	171.324	cry (eth, al)	17	242	0.7979²⁰	1.4398²⁰	s H₂O, EtOH; i eth; sl ctc
10749	Undecylbenzene		$C_{17}H_{28}$	6742-54-7	232.404	liq	-5	316	0.8553²⁰	1.4828²⁰	
10750	1-Undecyne		$C_{11}H_{20}$	2243-98-3	152.277	liq	-25	196	0.7728²⁰	1.4306²⁰	vs ace, bz, eth, EtOH
10751	2-Undecyne		$C_{11}H_{20}$	60212-29-5	152.277	liq	-30.1	204.2	0.7827²⁰	1.4391²⁰	
10752	Uracil		$C_4H_4N_2O_2$	66-22-8	112.087	nd (w)	338				sl H₂O; vs EtOH, eth; s dil NH₃
10753	Uracil mustard		$C_8H_{11}Cl_2N_3O_2$	66-75-1	252.098		206 dec				sl H₂O
10754	Uranyl acetate dihydrate		$C_4H_{10}O_8U$	6159-44-0	424.146	ye cry (HOAc)	80 dec		2.89		sl EtOH
10755	Urazole		$C_2H_3N_3O_2$	3232-84-6	101.064	lf (w)	249 dec				
10756	Urea	Carbamide	CH_4N_2O	57-13-6	60.055	tetr pr (al)	133.3	dec	1.3230²⁰	1.484	vs H₂O, EtOH; i eth, bz; s HOAc, py
10757	Urea hydrochloride		CH_5ClN_2O	506-89-8	96.516		145 dec				s H₂O
10758	Urea nitrate		$CH_5N_3O_4$	124-47-0	123.069	mcl lf (w)	152 dec		1.690²⁰		vs EtOH
10759	Uric acid		$C_5H_4N_4O_3$	69-93-2	168.111	orth pr or pl	dec	dec	1.89²⁵		i H₂O, EtOH, eth; s alk, glycerol; sl acid
10760	Uridine	1-β-D-Ribofuranosyluracil	$C_9H_{12}N_2O_6$	58-96-8	244.200	nd (aq al)	165				s H₂O, EtOH, py
10761	5'-Uridylic acid	Uridine 5'-phosphoric acid	$C_9H_{13}N_2O_9P$	58-97-9	324.180	pr (MeOH)	202 dec				vs H₂O; s MeOH
10762	Urocanic acid	Imidazole-4-acrylic acid	$C_6H_6N_2O_2$	104-98-3	138.124		227				s H₂O, ace; i EtOH, eth
10763	Urs-12-en-3-ol, (3β)	α-Amyrin	$C_{30}H_{50}O$	638-95-9	426.717	nd (al)	186	243⁰·⁵			s EtOH, eth, bz, chl, HOAc; sl peth
10764	Ursolic acid		$C_{30}H_{48}O_3$	77-52-1	456.700	pl (al)	284				vs ace, eth, chl
10765	Uzarin		$C_{35}H_{54}O_{14}$	20231-81-6	698.796	pr	269				
10766	Vacciniin	D-Glucose, 6-benzoate	$C_{13}H_{16}O_7$	14200-76-1	284.262	amor (aq ace, +1w)	122				vs H₂O, ace, EtOH, eth
10767	Validamycin A		$C_{20}H_{35}NO_{13}$	37248-47-8	497.491	amorp pow	95 dec				
10768	L-Valine	2-Aminoisovaleric acid	$C_5H_{11}NO_2$	72-18-4	117.147	lf (w-al)	315	sub	1.23²⁵		s H₂O
10769	Valinomycin		$C_{54}H_{90}N_6O_{18}$	2001-95-8	1111.322	cry	187				
10770	Valium		$C_{16}H_{13}ClN_2O$	439-14-5	284.739		132				
10771	Vamidothion		$C_8H_{18}NO_4PS_2$	2275-23-2	287.337	oil					i peth; s os
10772	Vanadium carbonyl	Vanadium hexacarbonyl	C_6O_6V	14024-00-1	219.002	bl-grn cry	dec 60	sub			
10773	Vanadium(III) 2,4-pentanedioate	Vanadium(III) acetylacetonate	$C_{15}H_{21}O_6V$	13476-99-8	348.266	brn cry	≈185	sub	≈1.0		s MeOH, ace, bz, chl

Undecane

Undecanenitrile

1-Undecanethiol

Undecanoic acid

1-Undecanol

2-Undecanol

2-Undecanone

6-Undecanone

Undecanoyl chloride

10-Undecenal

1-Undecene

cis-2-Undecene

trans-2-Undecene

cis-4-Undecene

trans-4-Undecene

cis-5-Undecene

trans-5-Undecene

10-Undecenoic acid

10-Undecen-1-ol

10-Undecenoyl chloride

Undecylamine

Undecylbenzene

1-Undecyne

2-Undecyne

Uracil

Uracil mustard

Uranyl acetate dihydrate 2H₂O

Urazole

Urea

Urea hydrochloride HCl

Urea nitrate HNO₃

Uric acid

Uridine

5'-Uridylic acid

Urocanic acid

Urs-12-en-3-ol, (3β)

Ursolic acid

Uzarin

Vacciniin

Validamycin A

L-Valine

Valinomycin

Valium

Vamidothion

Vanadium carbonyl

Vanadium(III) 2,4-pentanedioate

No.	Name	Synonym	Mol. Form.	CAS RN	Mol. Wt.	Physical Form	mp/°C	bp/°C	den/ g cm⁻³	n_D	Solubility
10774	*DL*-Vasicine	*DL*-Peganine	C₁₁H₁₂N₂O	6159-56-4	188.225	nd (al)	210.8				sl H₂O, eth, bz; s EtOH, ace, chl
10775	*L*-Vasicine	*L*-Peganine	C₁₁H₁₂N₂O	6159-55-3	188.225	nd (al)	211.5				sl H₂O, eth, bz; s EtOH, ace, chl
10776	Verapamil		C₂₇H₃₈N₂O₄	52-53-9	454.602	ye oil		245⁰·⁰¹		1.5448²⁵	i H₂O; vs EtOH, ace; sl bz, hx
10777	Veratramine		C₂₇H₃₉NO₂	60-70-8	409.605	nd	206				s EtOH, bz, chl, dil acid; i dil alk
10778	Veratramine, 3-glucoside		C₃₃H₄₉NO₇	475-00-3	571.745	nd (aq. MeOH)	242 dec				
10779	Veratridine		C₃₆H₅₁NO₁₁	71-62-5	673.790	ye amorp pow	180				i H₂O; sl eth
10780	*d*-Verbenone		C₁₀H₁₄O	18309-32-5	150.217		9.8	227.5	0.9978²⁰	1.4993¹⁸	s H₂O, EtOH, ace, bz
10781	Vernolate	Carbamothioic acid, dipropyl-, *S*-propyl ester	C₁₀H₂₁NOS	1929-77-7	203.345			150³⁰	0.952²⁰		
10782	Versalide		C₁₈H₂₆O	88-29-9	258.398	cry	46.5	130²			s EtOH
10783	α-Vetivone	Isonootkatone	C₁₅H₂₂O	15764-04-2	218.335	cry (peth)	51.5	144²	1.0035²⁰	1.5370²⁰	vs ace
10784	β-Vetivone		C₁₅H₂₂O	18444-79-6	218.335	cry (peth)	44.5	141²	1.0001²⁰	1.5309²⁰	s ace
10785	Vicine	2,6-Diamino-5-(β-*D*-glucopyranosyloxy)-4(1*H*)-pyrimidinone	C₁₀H₁₆N₄O₇	152-93-2	304.257	nd (w, dil al, +1 w)	240 dec				sl H₂O, EtOH; vs acid, alk
10786	Vidarabine	β-*D*-9-Arabinofuranosyladenine	C₁₀H₁₅N₅O₅	5536-17-4	285.257	nd (w)	257				
10787	Vinblastine		C₄₆H₅₈N₄O₉	865-21-4	810.975	nd (MeOH)	216				i H₂O; s EtOH, ace, chl, AcOEt
10788	Vincamine		C₂₁H₂₆N₂O₃	1617-90-9	354.442		231.5				
10789	Vinclozolin		C₁₂H₉Cl₂NO₃	50471-44-8	286.110		108	131⁰·⁰⁵	1.51		
10790	Vincristine		C₄₆H₅₆N₄O₁₀	57-22-7	824.958		219				
10791	Vinyl acetate		C₄H₆O₂	108-05-4	86.090	liq	-93.2	72.8	0.9256²⁵	1.3926²⁵	sl H₂O; msc EtOH; s eth, ace, bz, chl
10792	4-Vinylaniline		C₈H₉N	1520-21-4	119.164		23.5	116⁹	1.010²⁰	1.6250²²	s ace, bz
10793	α-Vinylbenzenemethanol	1-Phenylallyl alcohol	C₉H₁₀O	4393-06-0	134.174			1.0249²¹		1.5406²⁰	sl H₂O; s EtOH, eth, bz, chl
10794	Vinyl butanoate		C₆H₁₀O₂	123-20-6	114.142			116.7; 64¹³⁰	0.9006²⁰		
10795	Vinyl *trans*-2-butenoate	Vinyl crotonate	C₆H₈O₂	3234-54-6	112.127						s ctc
10796	9-Vinyl-9*H*-carbazole		C₁₄H₁₁N	1484-13-5	193.244	cry (al)	66				i H₂O; sl EtOH; vs eth
10797	Vinylcyclohexane		C₈H₁₄	695-12-5	110.197			128	0.8166¹⁹	1.455¹⁹	
10798	1-Vinylcyclohexene		C₈H₁₂	2622-21-1	108.181			145	0.8623¹⁵	1.4915²⁰	i H₂O; s eth, bz; vs MeOH
10799	4-Vinylcyclohexene		C₈H₁₂	100-40-3	108.181	liq	-108.9	128	0.8299²⁰	1.4639²⁰	i H₂O; s eth, bz, peth
10800	Vinylcyclopentane		C₇H₁₂	3742-34-5	96.170	liq	-126.5	97	0.7834²⁰	1.4360²⁰	
10801	Vinyldiethoxymethylsilane		C₇H₁₆O₂Si	5507-44-8	160.287			133	0.8620²⁰	1.4001²⁰	
10802	Vinylethoxydimethylsilane		C₆H₁₄OSi	5356-83-2	130.260			99	0.790²⁰	1.3983²⁰	
10803	1-Vinyl-4-fluorobenzene		C₈H₇F	405-99-2	122.140		-34.5	67.4⁵⁰, 30⁴	1.0220²⁰	1.5150²⁰	i H₂O; s EtOH, eth, bz
10804	Vinyl formate		C₃H₄O₂	692-45-5	72.063	visc liq	-78	46	0.965²⁰	1.3842²⁰	
10805	2-Vinylfuran		C₆H₆O	1487-18-9	94.111	liq	-94	99.5	0.9445¹⁹	1.4992¹⁹	
10806	1-Vinyl-2-methoxybenzene		C₉H₁₀O	612-15-7	134.174	nd	29	197; 83¹²	1.0049¹⁷	1.5388²⁰	vs ace, bz, eth, EtOH
10807	1-Vinyl-3-methoxybenzene		C₉H₁₀O	626-20-0	134.174			91¹⁵, 70⁵	0.9919²⁰	1.5586²³	i H₂O; s EtOH, eth, bz
10808	1-Vinyl-4-methoxybenzene		C₉H₁₀O	637-69-4	134.174		2.0	205; 91¹³	1.0001¹³	1.5642¹³	i H₂O; s EtOH, eth, bz; sl ctc
10809	6-Vinyl-6-methyl-1-isopropyl-3-(1-methylethylidene)cyclohexene, (*S*)		C₁₅H₂₄	5951-67-7	204.352			125⁸	0.8782²⁰	1.5130²⁶	vs ace, bz
10810	1-Vinylnaphthalene		C₁₂H₁₀	826-74-4	154.207			124¹⁵	1.0656²⁰	1.644²⁰	
10811	2-Vinylnaphthalene		C₁₂H₁₀	827-54-3	154.207		66	135¹⁸, 95²			i H₂O; s EtOH, ace, bz
10812	1-Vinyl-3-nitrobenzene		C₈H₇NO₂	586-39-0	149.148		-10	120¹¹	1.1552³²	1.5836²⁰	i H₂O; s EtOH, eth, bz, chl, lig, HOAc
10813	1-Vinyl-4-nitrobenzene		C₈H₇NO₂	100-13-0	149.148	pr (lig)	29	dec			vs EtOH, eth; s chl, HOAc, lig
10814	2-Vinyl-5-norbornene	5-Vinylbicyclo[2.2.1]hept-2-ene	C₉H₁₂	3048-64-4	120.191	liq	-80	139	0.841	1.4810²⁰	
10815	Vinyl octadecanoate	Vinyl stearate	C₂₀H₃₈O₂	111-63-7	310.515		29	167²	0.8517²⁰		sl chl
10816	3-Vinyl-7-oxabicyclo[4.1.0]heptane		C₈H₁₂O	106-86-5	124.180		<-100	169; 70²⁰	0.9581²⁰	1.4700²⁰	

DL-Vasicine

L-Vasicine

Verapamil

Veratramine

Veratramine, 3-glucoside

Veratridine

d-Verbenone

Vernolate

Versalide

α-Vetivone

β-Vetivone

Vicine

Vidarabine

Vinblastine

Vincamine

Vinclozolin

Vincristine

Vinyl acetate

4-Vinylaniline

α-Vinylbenzenemethanol

Vinyl butanoate

Vinyl *trans*-2-butenoate

9-Vinyl-9*H*-carbazole

Vinylcyclohexane

1-Vinylcyclohexene

4-Vinylcyclohexene

Vinylcyclopentane

Vinyldiethoxymethylsilane

Vinylethoxydimethylsilane

1-Vinyl-4-fluorobenzene

Vinyl formate

2-Vinylfuran

1-Vinyl-2-methoxybenzene

1-Vinyl-3-methoxybenzene

1-Vinyl-4-methoxybenzene

6-Vinyl-6-methyl-1-isopropyl-3-(1-methylethylidene)cyclohexene, (*S*)

1-Vinylnaphthalene

2-Vinylnaphthalene

1-Vinyl-3-nitrobenzene

1-Vinyl-4-nitrobenzene

2-Vinyl-5-norbornene

Vinyl octadecanoate

3-Vinyl-7-oxabicyclo[4.1.0]heptane

No.	Name	Synonym	Mol. Form.	CAS RN	Mol. Wt.	Physical Form	mp/°C	bp/°C	den/ g cm⁻³	n_D	Solubility
10817	Vinyloxirane		C_4H_6O	930-22-3	70.090			68	0.9006^{25}	1.4168^{20}	s EtOH, eth, bz
10818	2-(Vinyloxy)ethanol	Ethylene glycol monovinyl ether	$C_4H_8O_2$	764-48-7	88.106			141.6	0.9821^{20}	1.4564^{17}	s H_2O, EtOH, eth, bz; i lig
10819	Vinyl propanoate	Vinyl propionate	$C_5H_8O_2$	105-38-4	100.117			91.2			
10820	2-Vinylpyridine		C_7H_7N	100-69-6	105.138			159.5	0.9983^{20}	1.5495^{20}	sl H_2O; vs EtOH, eth, ace, chl
10821	3-Vinylpyridine		C_7H_7N	1121-55-7	105.138			162	0.9879^{20}	1.5530^{20}	sl H_2O; s EtOH, eth
10822	4-Vinylpyridine		C_7H_7N	100-43-6	105.138	red to dk-br		121^{150}, 79^{33}	0.9879^{20}	1.5449^{20}	s H_2O, EtOH, chl; sl eth
10823	1-Vinyl-2-pyrrolidinone		C_6H_9NO	88-12-0	111.141		13.5	193^{400}, 93^{11}	1.04^{20}		
10824	Vinylsilane		C_2H_6Si	7291-09-0	58.155	col gas	-171.6	-22.8			
10825	Vinyl sulfoxide		C_4H_6OS	1115-15-7	102.155	liq		86^{18}			
10826	Vinyltriacetoxysilane	Vinylsilanetriol, triacetate	$C_8H_{12}O_6Si$	4130-08-9	232.263			115^{10}	1.169^{20}	1.4226^{20}	
10827	Vinyltriethoxysilane		$C_8H_{18}O_3Si$	78-08-0	190.313			160; 62^{20}	0.901^{20}	1.3960^{25}	s chl
10828	Vinyltrimethylsilane		$C_5H_{12}Si$	754-05-2	100.235			55	0.65^{20}	1.3914^{20}	i H_2O
10829	Violaxanthin		$C_{40}H_{56}O_4$	126-29-4	600.871	red pr (MeOH, al-eth)	208				s EtOH, eth, CS_2; i peth
10830	Viquidil		$C_{20}H_{24}N_2O_2$	84-55-9	324.417	red ye amor	60				vs eth, EtOH, chl
10831	Visnadine		$C_{21}H_{24}O_7$	477-32-7	388.412	nd	85.5				i H_2O; s EtOH, eth
10832	Visnagin	4-Methoxy-7-methyl-5H-furo[3,2-g][1]benzopyran-5-one	$C_{13}H_{10}O_4$	82-57-5	230.216	nd (w, MeOH)	144.5				sl H_2O, EtOH; vs chl
10833	Vitamin B12	Cyanocobalamin	$C_{63}H_{88}CoN_{14}O_{14}P$	68-19-9	1355.365		>300				
10834	Vitamin D2		$C_{28}H_{44}O$	50-14-6	396.648	pr (ace)	116.5	sub			i H_2O; s EtOH, eth, ace, chl
10835	Vitamin D3	9,10-Secocholesta-5,7,10(19)-trien-3-ol, (3β,5Z,7E)-	$C_{27}H_{44}O$	67-97-0	384.637		84.5				i H_2O; s os
10836	Vitamin E	α-Tocopherol	$C_{29}H_{50}O_2$	59-02-9	430.706	pale ye oil	3.0	$210^{0.1}$	0.950^{25}	1.5045^{25}	i H_2O; s EtOH, eth, ace, chl
10837	Vitamin E acetate		$C_{31}H_{52}O_3$	58-95-7	472.743		-27.5	$184^{0.01}$	0.9533^{21}	1.497^{20}	i H_2O; sl EtOH; s eth, ace, chl
10838	Vitamin K1		$C_{31}H_{46}O_2$	84-80-0	450.696		-20	$142^{0.001}$	0.964^{25}	1.5250^{25}	i H_2O; s EtOH, eth, ace, bz, peth, chl
10839	Vomicine	4-Hydroxy-19-methyl-16,19-secostrychnidine-10,16-dione	$C_{22}H_{24}N_2O_4$	125-15-5	380.437	nd (80% al) pr (ace)	282				sl EtOH, eth, ace; vs chl; s AcOEt
10840	Warfarin	Coumadin	$C_{19}H_{16}O_4$	81-81-2	308.328	cry (al)	161				i H_2O; s EtOH, ace, diox
10841	9H-Xanthene	10H-9-Oxaanthracene	$C_{13}H_{10}O$	92-83-1	182.217	ye lf (al)	100.5	311			i H_2O; sl EtOH, ctc; s eth, bz, chl
10842	9H-Xanthen-9-ol		$C_{13}H_{10}O_2$	90-46-0	198.217	nd (aq al)	125				sl H_2O; s EtOH, eth, chl
10843	Xanthine		$C_5H_4N_4O_2$	69-89-6	152.112	ye pl (w)	dec	sub			i H_2O
10844	Xanthone		$C_{13}H_8O_2$	90-47-1	196.202	nd (al)	174	351; 146^3			i H_2O; s EtOH, eth, bz, chl; sl peth
10845	Xanthopterin		$C_6H_5N_5O_2$	119-44-8	179.137	hyg ye amor or oran pow (HOAc)	>410 dec	99^{18}	1.559^{25}		i H_2O; sl EtOH, eth; vs acid, alk
10846	Xanthosine		$C_{10}H_{12}N_4O_6$	146-80-5	284.225	pr cry (w)					sl cold H_2O; vs hot H_2O; dec acid
10847	Xanthoxyletin		$C_{15}H_{14}O_4$	84-99-1	258.270	pr (MeOH, peth)	133				i H_2O; s EtOH, ace; sl eth; vs bz, alk
10848	Xanthyletin	8,8-Dimethyl-2H,8H-benzo[1,2-b:5,4-b']dipyran-2-one	$C_{14}H_{12}O_3$	553-19-5	228.243	pr (MeOH)	131.5	$142^{0.1}$			s EtOH, peth
10849	p-Xenylcarbimide	4-Isocyanato-1,1'-biphenyl	$C_{13}H_9NO$	92-95-5	195.216	nd	56	dec 283			vs eth
10850	Xibenolol		$C_{15}H_{25}NO_2$	81584-06-7	251.366	cry	57	$135^{0.7}$			s EtOH
10851	o-Xylene	1,2-Dimethylbenzene	C_8H_{10}	95-47-6	106.165	liq	-25.2	144.5	0.8802^{10}	1.5055^{20}	i H_2O; msc EtOH, eth, ace, bz, peth, ctc
10852	m-Xylene	1,3-Dimethylbenzene	C_8H_{10}	108-38-3	106.165	liq	-47.8	139.12	0.8596^{25}	1.4972^{10}	i H_2O; msc EtOH, eth, ace, bz; s chl
10853	p-Xylene	1,4-Dimethylbenzene	C_8H_{10}	106-42-3	106.165	mcl pr (al)	13.25	138.37	0.8566^{25}	1.4958^{20}	i H_2O; msc EtOH, eth, ace, bz; s chl

Vinyloxirane

2-(Vinyloxy)ethanol

Vinyl propanoate

2-Vinylpyridine

3-Vinylpyridine

4-Vinylpyridine

1-Vinyl-2-pyrrolidinone

Vinylsilane

Vinyl sulfoxide

Vinyltriacetoxysilane

Vinyltriethoxysilane

Vinyltrimethylsilane

Violaxanthin

R = CN

Viquidil

Visnadine

Visnagin

Vitamin B12

Vitamin D2

Vitamin D3

Vitamin E

Vitamin E acetate

Vitamin K1

Vomicine

Warfarin

9H-Xanthene

9H-Xanthen-9-ol

Xanthine

Xanthone

Xanthopterin

Xanthosine

Xanthoxyletin

Xanthyletin

p-Xenylcarbimide

Xibenolol

o-Xylene

m-Xylene

p-Xylene

No.	Name	Synonym	Mol. Form.	CAS RN	Mol. Wt.	Physical Form	mp/°C	bp/°C	den/ g cm⁻³	n_D	Solubility
10854	2,3-Xylenol	2,3-Dimethylphenol	$C_8H_{10}O$	526-75-0	122.164	nd (w, dil al)	72.5	216.9		1.5420[20]	sl H_2O; s EtOH, eth
10855	2,4-Xylenol	2,4-Dimethylphenol	$C_8H_{10}O$	105-67-9	122.164	nd (w)	24.5	210.98	0.9650[20]	1.5420[14]	sl H_2O; msc EtOH, eth; s ctc
10856	2,5-Xylenol	2,5-Dimethylphenol	$C_8H_{10}O$	95-87-4	122.164	nd (w), pr (al-eth)	74.8	211.1			s H_2O, EtOH; vs eth; sl chl
10857	2,6-Xylenol	2,6-Dimethylphenol	$C_8H_{10}O$	576-26-1	122.164	lf or nd (al)	45.8	201.07			s H_2O, EtOH, eth, ctc
10858	3,4-Xylenol	3,4-Dimethylphenol	$C_8H_{10}O$	95-65-8	122.164		65.1	227	0.9830[20]		sl H_2O; s EtOH, ctc; msc eth
10859	3,5-Xylenol	3,5-Dimethylphenol	$C_8H_{10}O$	108-68-9	122.164	nd (w, peth)	63.4	221.74	0.9680[20]		s H_2O, EtOH, ctc
10860	Xylenol orange		$C_{31}H_{32}N_2O_{13}S$	1611-35-4	672.656	dk red cry	286 dec				s H_2O
10861	Xylitol	Xylite	$C_5H_{12}O_5$	87-99-0	152.146	mcl (al)	93.5	216			vs H_2O, py, EtOH
10862	6-O-β-D-Xylopyranosyl-D-glucose	Primeverose	$C_{11}H_{20}O_{10}$	26531-85-1	312.271	cry (MeOH)	210				vs H_2O, MeOH
10863	D-Xylose		$C_5H_{10}O_5$	58-86-6	150.130	mcl nd	90.5		1.525[20]		vs H_2O; s EtOH; sl eth
10864	D-Xylulose	D-threo-2-Pentulose	$C_5H_{10}O_5$	551-84-8	150.130	visc liq					s H_2O
10865	L-Xylulose	L-threo-2-Pentulose	$C_5H_{10}O_5$	527-50-4	150.130	syrup					vs H_2O
10866	3,5-Xylyl methylcarbamate	3,5-Dimethylphenyl methylcarbamate	$C_{10}H_{13}NO_2$	2655-14-3	179.216	cry	99				sl H_2O; s os
10867	Yohimbine		$C_{21}H_{26}N_2O_3$	146-48-5	354.442	nd (dil al)	241	sub 160			sl H_2O, bz; s EtOH, eth, chl
10868	Yohimbine hydrochloride	Tosanpin	$C_{21}H_{27}ClN_2O_3$	65-19-0	390.903	orth nd or pl (w, dil HCl)	302				vs H_2O
10869	Zearalenone		$C_{18}H_{22}O_5$	17924-92-4	318.365	cry	164				i H_2O; s alk, bz, EtOH, eth
10870	Zidovudine	3'-Azido-3'-deoxythymidine	$C_{10}H_{13}N_5O_4$	30516-87-1	267.242	cry (w)	121				
10871	Zinc benzoate		$C_{14}H_{10}O_4Zn$	553-72-0	307.636						sl H_2O
10872	Zinc bis(dibutyldithiocarbamate)		$C_{18}H_{36}N_2S_4Zn$	136-23-2	474.161	cry	138				
10873	Zinc N,N'-ethylenebisdithiocarbamate	Zineb	$C_4H_6N_2S_4Zn$	12122-67-7	275.773		157 dec				
10874	Zinc gluconate		$C_{12}H_{22}O_{14}Zn$	4468-02-4	455.704	pow					
10875	Zinc 2,4-pentanedioate	Zinc acetylacetonate	$C_{10}H_{14}O_4Zn$	14024-63-6	263.625						s DMSO
10876	Zinc propanoate		$C_6H_{10}O_4Zn$	557-28-8	211.550	hyg pl or nd					sl EtOH
10877	Ziram	Zinc, bis(dimethylcarbamodithioato-S,S)-, (T-4)-	$C_6H_{12}N_2S_4Zn$	137-30-4	305.841	cry	250		1.66[25]		i H_2O; sl bz; s chl

2,3-Xylenol   2,4-Xylenol   2,5-Xylenol   2,6-Xylenol   3,4-Xylenol   3,5-Xylenol   Xylenol orange   Xylitol

6-*O*-β-*D*-Xylopyranosyl-*D*-glucose   *D*-Xylose   *D*-Xylulose   *L*-Xylulose   3,5-Xylyl methylcarbamate   Yohimbine

Yohimbine hydrochloride   Zearalenone   Zidovudine   Zinc benzoate   Zinc bis(dibutyldithiocarbamate)

Zinc *N,N'*-ethylenebisdithiocarbamate   Zinc gluconate   Zinc 2,4-pentanedioate   Zinc propanoate   Ziram

SYNONYM INDEX OF ORGANIC COMPOUNDS

1-Cyanocyclohexene	: 2592
Cyanocyclopentane	: 2661
1-Cyanocyclopentene	: 2673
Cyanogenamide	: 2500
1-Cyano-2-isopropoxyethane	: 6468
1-Cyano-3-(methylmercurio) guanidine	: 7437
5-Cyano-1-pentylamine	: 319
2-Cyanothiophene	: 10066
3-Cyanothiophene	: 10067
Cyanuric acic trichloride	: 10290
Cyclamic acid	: 2640
Cyclobutyrol	: 5076
β-Cyclocitral	: 10539
Cycloheptyl bromide	: 1158
2,5-Cyclohexadiene-1,4-dione	: 733
Cyclohexanamine	: 2608
Cyclohexanamine hydrochloride	: 2609
(1α,2α,3α,4β,5α,6β)- Cyclohexanehexol	: 6262
1-Cyclohexene-1,2-dicarboxylic acid, anhydride	: 9864
4-Cyclohexene-1,2-dicarboxylic acid, anhydride	: 9863
2,3-Cyclohexenopyridine	: 9892
Cyclohexyl alcohol	: 2587
Cyclohexyl bromide	: 1159
Cyclohexylcarbinol	: 2584
Cyclohexyl chloride	: 1961
Cyclohexyl cyanide	: 2567
N-Cyclohexylcyclohexanamine	: 3321
N-Cyclohexylcyclohexanamine, nitrite	: 3322
2-Cyclohexylcyclohexanone	: 860
Cyclohexyl fluoride	: 5380
Cyclohexyl iodide	: 6289
Cyclohexyl mercaptan	: 2586
Cyclohexyl phenyl ketone	: 762
α-Cyclohexyl-α-phenyl-1- piperidinepropanol hydrochloride	: 10431
N-Cyclohexyl-1-piperazineacetamide	: 4754
Cycloleucine	: 290
Cyclomaltoheptaose	: 2543
Cyclomaltohexaose	: 2542
Cyclomaltooctaose	: 2544
Cyclonite	: 5810
N'-Cyclooctyl-N,N-dimethylurea	: 2703
Cyclopentamine	: 4033
Cyclopentanamine	: 2682
Cyclopentanoic acid	: 2663
1,2-Cyclopentenophenanthrene	: 3609
Cyclopentyl alcohol	: 2669
Cyclopentyl bromide	: 1163
Cyclopentyl chloride	: 1964
Cyclopentyl ether	: 3333
Cyclopentyl iodide	: 6290
Cyclopentyl mercaptan	: 2668
Cyclophosphane	: 2686
Cyclopropanamine	: 2697
Cyclopropyl cyanide	: 2689
N-Cyclopropyl-1,3,5-triazine-2,4,6- triamine	: 2711
Cycloserine	: 341

Cymarose	: 3340
o-Cymene	: 6507
m-Cymene	: 6508
p-Cymene	: 6509
L-Cysteine, S-(2-amino-2- carboxyethyl)-, (R)-	: 6587
Cythrin	: 5337
Cytidine 2'-monophosphate	: 2721
Cytidine 3'-monophosphate	: 2722
Cytidine 5'-monophosphate	: 2723
Cytosine arabinoside	: 2719
2,4-D	: 3256
2,4-D 2-Butoxyethyl ester	: 1478
2,4-D Butyl ester	: 1552
2,4-D Methyl ester	: 7192
Damascenine	: 7444
Dansyl chloride	: 3914
Danthron	: 3694
Daphnetin	: 3713
Dapsone	: 908
DBMC	: 3033
Deanol	: 4057
Decafluorobutane	: 8696
Decahydro-β-naphthol	: 2745
cis-Decalin	: 2743
trans-Decalin	: 2744
Decamethylene dibromide	: 2928
Decamethylene glycol	: 2753
1-Decanamine	: 2778
Decanoic acid glycerol monoester	: 3752
Decanoic acid glycerol triester	: 5568
Decylacetylene	: 4658
Decyl cyanide	: 10729
N-Decyl-1-decanamine	: 3335
Decylenic alcohol	: 2774
Decyl fluoride	: 5383
Decyl mercaptan	: 2756
Decyl methyl ketone	: 4633
Decyl methyl sulfoxide	: 7790
DEET	: 3482
Deferoxamine	: 2811
Dehydroacetic acid	: 82
Dehydrobenzperidol	: 4670
Dehydrobilirubin	: 869
7-Dehydrocholesterol	: 2363
Dehydrocholic acid	: 10641
11-Dehydrocorticosterone	: 6176
Dehydroergosterol	: 4730
Dehydromucic acid	: 5485
7-Dehydrositosterol	: 9607
Dehydrotestosterone	: 1060
Demeclocycline	: 2776
Demeton-S	: 6555
Denmert	: 1468
11-Deoxojervine	: 2654
Deoxycorticosterone	: 6175
2'-Deoxy-5'-cytidylic acid	: 2801
Deoxyepinephrine	: 6933
2'-Deoxy-5'-guanylic acid	: 2804
11-Deoxy-17-hydrocorticosterone	: 3747
6-Deoxy-D-mannose	: 9469
3-Deoxypseudaconitine	: 867

Dessin	: 4411
Dexamphetamine	: 8975
Dextropimaric acid	: 9043
Dextropropoxyphene	: 9198
1,3-Diacetin	: 9157
Diacetonamine	: 347
Diacetone acrylamide	: 4209
1,4-Diacetoxy-2-butyne	: 1681
Diacetyl	: 1415
N,N'-Diacetyl-3,5-diamino-2,4,6- triiodobenzoic acid	: 2869
Diacetylene	: 1396
Δ 2-Dialin	: 3657
Diallylamine	: 205
Diallyl phthalate	: 627
Diamidafos	: 8003
2,5-Diaminoanisole	: 6787
cis-1,2-Diaminocyclohexane	: 2571
trans-1,2-Diaminocyclohexane	: 2572
4,4'-Diaminodiphenylamine	: 395
2,4'-Diaminodiphenylmethane	: 7241
2,7-Diaminofluorene	: 5345
2,6-Diamino-5-(β-D- glucopyranosyloxy)-4(1H)- pyrimidinone	: 10785
2,6-Diaminohexanoic acid, (±)	: 6646
2,6-Diaminohexanoic acid, (D)	: 6647
2,6-Diaminohexanoic acid, (L)	: 6648
1,5-Diaminonaphthalene	: 7941
1,8-Diaminonaphthalene	: 7942
2,3-Diaminonaphthalene	: 7943
2,5-Diaminopentanoic acid, (S)	: 8416
2,3-Diaminophenazine	: 8751
2,6-Diamino-3-phenylazopyridine	: 8753
2,6-Diaminopimelic acid	: 2858
1,3-Diaminopropane	: 9138
2,3-Diaminopropionic acid	: 235
2,6-Diaminopurine	: 9294
2,5-Diaminopyridine	: 9335
1,3-Diamino-2-thiourea	: 1751
Diamylamine	: 4439
1,4:3,6-Dianhydroglucitol	: 6554
1,2-Dianilinoethane	: 4482
Dianisidine	: 3847
1,2-Diazabenzene	: 9312
1,5-Diazanaphthalene	: 7994
1,3-Diazine	: 9382
1,4-Diazine	: 9302
Diazoacetic ester	: 4973
Diazoaminobenzene	: 4533
1,2-Diazole	: 9306
1,3-Diazole	: 6215
Dibenamine	: 2037
Dibenamine hydrochloride	: 2038
Dibenzalacetone	: 4507
1,2:5,6-Dibenzanthracene	: 2876
Dibenz[de,kl]anthracene	: 8736
Dibenzo[b,def]chrysene	: 2886
Dibenzo[def,p]chrysene	: 2888
Dibenzo[a,jk]fluorene	: 701
Dibenzopyrazine	: 8750
Dibenzo[b,e]pyridine	: 113

Magnesium dimethyl	: 4124	
Magnesium octadecanoate	: 6655	
Magneson	: 8145	
Maleamic acid	: 383	
Maleanilic acid	: 8471	
Maleic hydrazide	: 3674	
Maltol	: 6106	
DL-Mandelic acid	: 5973	
Mandelonitrile	: 5977	
Manganese, [[1,2-ethanediylbis[carbamodithioato]](2-)]-	: 6673	
Margaric acid	: 5650	
Margaric aldehyde	: 5646	
Margaryl alcohol	: 5651	
Matrine	: 6682	
MCPA	: 2152	
Mechlorethamine	: 1949	
Mechlorethamine oxide hydrochloride	: 8108	
Meconic acid	: 6139	
Meconin	: 3869	
Mecrylate	: 7140	
Me-IQ	: 303	
Melamine	: 10142	
Melatonin	: 6834	
Meldrum's acid	: 4042	
Mellitene	: 5813	
Mellitic acid	: 643	
Menadiol diacetate	: 7474	
Menadione	: 7475	
p-Menthadiene	: 4438	
p-Mentha-1,8-diene, (R)	: 6610	
p-Mentha-1,8-diene, (S)	: 6611	
p-Mentha-1(7),2-diene	: 8741	
p-Mentha-1,4(8)-diene	: 9703	
p-Mentha-1,8-dien-6-one, (R)	: 1787	
p-Mentha-1,8-dien-6-one, (S)	: 1788	
trans-p-Menthane	: 7405	
(+)-Menthol	: 7407	
(-)-Menthol	: 7408	
Menthol, isovalerate	: 6700	
Menthone	: 7412	
l-Menthone	: 7413	
Mepacrine	: 9410	
Meparfynol	: 7600	
Mephenesin	: 7611	
Mephenoxalone	: 6861	
Mephentermine	: 10511	
Meprobamate	: 7719	
2-Mercaptobenzothiazole	: 738	
2-Mercaptoethyl ether	: 3823	
1-(3-Mercapto-2-methyl-1-oxypropyl)proline	: 1724	
(3-Mercaptopropyl)trimethoxysilane	: 10495	
6-Mercaptopurine	: 3669	
Mercuric benzoate	: 6719	
Mercuric oleate	: 6720	
Mercuriodibenzene	: 4497	
Mercury, chloro(2-methoxyethyl)-	: 10132	
Mesaconic acid	: 9179	
Mesalamine	: 327	
Mescaline	: 10480	
Mesitylamine	: 10504	
Mesitylene	: 10509	
Mesitylenic acid	: 3967	
Mestanolone	: 6074	
Mestilbol	: 3526	
Mesulphen	: 4335	
Metacetaldehyde (polymer)	: 6728	
Metalphthalein	: 2481	
Metanilic acid	: 252	
Methacrolein	: 7696	
Methacrolein diacetate	: 7699	
Methacrylic acid anhydride	: 7701	
Methacrylic acid chloride	: 7703	
2-(Methacryloyloxy)ethyl acetoacetate	: 26	
Methallatal	: 5110	
Methallyl alcohol	: 7702	
Metham sodium	: 9568	
Methanal	: 5448	
Methanamide	: 5450	
Methanamine	: 6920	
Methanamine hydrochloride	: 6921	
Methanethial	: 10055	
Methanoic acid	: 5453	
Methaqualone	: 7460	
Methenamine	: 5822	
N,N'-Methenyl-o-phenylenediamine	: 683	
Methimazole	: 3648	
Methiodal sodium	: 9564	
Methionamine	: 78	
Methionic acid	: 6743	
Methionine hydroxy analog	: 6109	
Methoxyacetone	: 6889	
Methoxyallene	: 6884	
Methoxyamine	: 7371	
Methoxyamine hydrochloride	: 7372	
Methoxybenzene	: 474	
2-Methoxycarbonylethyl methyl sulfide	: 7461	
6'-Methoxycinchonan-9-ol, (8α,9R)	: 9418	
6'-Methoxycinchonan-9-ol monohydrochloride, (8α,9R)	: 9419	
trans-4-Methoxycinnamic acid	: 6882	
9-Methoxycorynantheidine	: 7885	
Methoxydichloromethane	: 3211	
2-(2-Methoxyethoxy)ethanol	: 3450	
2-Methoxyethyl 2-propenoate	: 6826	
Methoxyflurane	: 3163	
9-Methoxy-7H-furo[3,2-g][1]benzopyran-7-one	: 6767	
1-(2-Methoxyisopropoxy)-2-propanol	: 4557	
3-Methoxy-2-isopropyl-7H-furo[3,2-g][1]benzopyran-7-one	: 8737	
N-Methoxymethanamine	: 6837	
2-Methoxy-N-(2-methoxyethyl)ethanamine	: 1020	
6-Methoxy-9-methyl-1,3-dioxolo[4,5-h]quinolin-8(9H)-one	: 1790	
1-[2-(2-Methoxy-1-methylethoxy)-1-methylethoxy]-2-propanol	: 10673	
4-Methoxy-7-methyl-5H-furo[3,2-g][1]benzopyran-5-one	: 10832	
6-Methoxy-α-methyl-2-naphthaleneacetic acid	: 7997	
7-Methoxy-1-methyl-9H-pyrido[3,4-b]indole	: 5625	
1-Methoxy-2-nitrobenzene	: 8037	
1-Methoxy-3-nitrobenzene	: 8038	
1-Methoxy-4-nitrobenzene	: 8039	
1-Methoxyperfluoroethane	: 7549	
N-(4-Methoxyphenyl)-1,4-benzenediamine	: 6876	
4-Methoxy-o-phenylenediamine	: 6785	
4-Methoxy-m-phenylenediamine	: 6786	
3-(4-Methoxyphenyl)-2-propenoic acid, 2-ethoxyethyl ester	: 2404	
1-Methoxy-2-propanamine	: 6835	
2-Methoxypropane	: 6514	
1-Methoxypropane	: 7716	
1-Methoxy-2-propanol	: 9234	
6-Methoxy-α-tetralone	: 3637	
5-Methoxytryptamine	: 6833	
Methsuximide	: 4257	
N-Methylacetanilide	: 7614	
Methylacetylene	: 9273	
Methyl o-acetylsalicylate	: 6915	
Methylal	: 3870	
Methyl alcohol	: 6750	
Methylallene	: 1392	
Methyl aluminum sesquibromide	: 10177	
N-Methyl-2-amino-5-chlorobenzophenone	: 2097	
α-[1-(Methylamino)ethyl]benzenemethanol, (R*,S*)-(±)-	: 4703	
α-[1-(Methylamino)ethyl]benzenemethanol, [S-(R*,S*)]-	: 4704	
α-[1-(Methylamino)ethyl]benzenemethanol, [R-(R*,S*)]-	: 4705	
2-(Methylamino)-1-phenyl-1-propanol, hydrochloride	: 4706	
Methyl [(4-aminophenyl)sulfonyl]carbamate	: 528	
5-Methyl-o-anisidine	: 6838	
Methyl anthranilate	: 6923	
2-Methylanthraquinone	: 6950	
Methyl arachidate	: 7229	
8-Methyl-8-azabicyclo[3.2.1]octan-3-ol, endo	: 10711	
8-Methyl-8-azabicyclo[3.2.1]octan-3-ol, exo	: 9288	
Methyl azelate	: 4162	
2-Methylaziridine	: 9236	
Methyl behenate	: 7225	
Methylbenzene	: 10111	
4-Methyl-1,3-benzenediamine	: 10112	
5-Methyl-1,3-benzenediamine	: 10113	
Methyl 1,4-benzenedicarboxylate	: 7793	
4-Methylbenzenesulfonic acid, monohydrate	: 10117	
2-Methylbicyclo[2.2.1]hept-2-ene	: 7513	
3-Methyl-N,N-bis(3-methylbutyl)-1-butanamine	: 10458	
2-Methyl-N,N-bis(2-methylpropyl)-1-propanamine	: 10455	

Phosphoramidothioic acid, O,S-dimethyl ester	: 6737
Phosphoric acid, 2,2-dichloroethenyl dimethyl ester	: 3312
Phosphorodithioic acid, O-ethyl S,S-dipropyl ester	: 4816
Phosphorothioic acid, O,O-diethyl O-pyrazinyl ester	: 10061
Phosphorotrithious acid, S,S,S-tributyl ester	: 6722
Phosphorylcholine	: 2379
Phthalamide	: 623
Phthalimide	: 6433
o-Phthalodinitrile	: 3317
m-Phthalodinitrile	: 3318
p-Phthalodinitrile	: 3319
Phthaloyl chloride	: 617
Phthiocol	: 6092
Physostigmine sulfate	: 4756
Phytanic acid	: 9950
Piazthiole	: 10022
Picein	: 5545
Picloram	: 434
2-Picoline	: 7738
3-Picoline	: 7739
4-Picoline	: 7740
Picolinic acid	: 9329
4-Picolyl alcohol	: 9354
Picramic acid	: 304
Picric acid	: 10631
Picryl chloride	: 2337
Pigment Blue 15	: 2464
Pimeclone	: 9073
Pimelic acid	: 5675
Pimelic ketone	: 2588
Pinacol	: 3986
Pinacolone	: 3993
Pindone	: 9085
2-Pinene	: 9047
Pinene hydrate	: 9045
Pinosylvin	: 9014
L-Pipecolic acid	: 9061
Δ³-Piperidine	: 9888
Piperidinium chloride	: 9067
Piperidinium, 1,1-dimethyl-, chloride	: 6705
1-Piperidinol	: 6171
Piperidione	: 3512
Piperinic acid	: 698
Piperitenone	: 7456
(±)-Piperitone	: 6513
Piperonal	: 692
Piperonylic acid	: 693
cis-Piperylene	: 8565
trans-Piperylene	: 8566
Pipradrol	: 4512
Pipsyl chloride	: 6277
Pivaldehyde	: 4269
Pivalic acid chloride	: 4279
Plicamycin	: 7880
Plumbagin	: 6093
Potassium biphthalate	: 9097
Potassium citrate	: 10665

Potassium oleate	: 9098
Potassium sorbate	: 9096
Pramiverin	: 6494
Pramoxine	: 1483
Prednisone	: 3746
Pregnanediol	: 9104
5α-Pregnan-20β-ol-3-one	: 162
Pregn-4-ene-3,20-dione	: 9122
4-Pregnene-11β,17α,20β,21-tetrol-3-on	: 9901
Prehnitenol	: 9965
Primeverose	: 10862
Primidone	: 4990
Pristane	: 9959
Probenecid	: 4550
Procaine	: 3387
Proflavine	: 114
Prontosil	: 2864
1,2-Propadiene-1,3-dione	: 1754
Propallylonal	: 1086
Propanamide, N-(3,4-dichlorophenyl)-	: 9163
Propanamide, N,N-diethyl-2-(1-naphthalenyloxy)-	: 7996
1-Propanamine	: 9205
2-Propanamine	: 6471
1-Propanamine hydrochloride	: 9206
2-Propanamine hydrochloride	: 6472
1-Propanearsonic acid	: 9211
1,3-Propanedial	: 6663
1,2-Propanediol	: 9230
Propaneperoxoic acid	: 8733
1,2,3-Propanetriol	: 5561
1,2,3-Propanetriol, 1,3-dinitrate	: 5564
Propanoic acid, 2-amino-3-mercapto-, (R)-	: 2715
Propanoic acid, 2-(2,4,5-trichlorophenoxy)-	: 9538
Propanolamine	: 407
1-Propanol, titanium(4+) salt	: 10001
2-Propanol, titanium(4+) salt	: 9908
2-Propanone	: 33
Propargylacetic acid	: 8689
Propargyl aldehyde	: 9271
Propargyl bromide	: 1345
Propargyl chloride	: 2285
2-Propenal	: 116
2-Propenamide	: 117
2-Propen-1-amine	: 171
Propenenitrile	: 119
2-Propenoic acid	: 118
2-Propenoic acid, 2,2-dimethyl-1,3-propanediyl ester	: 4542
2-Propenoic acid, 1,6-hexanediyl ester	: 4543
2-Propenoic acid, propyl ester	: 9204
2-Propen-1-ol	: 170
2-Propenylbenzene	: 173
5-(2-Propenyl)-1,3-benzodioxole	: 9499
1-(2-Propenyl)cyclohexene	: 182
2-Propenyl propanoate	: 204
Propham	: 6531
Propiocine	: 4745

β-Propiolactone	: 8445
Propiolic acid	: 9274
Propionaldehyde	: 9132
Propionamide	: 9134
Propionic acid	: 9164
Propionic anhydride	: 9165
Propionylacetone	: 5839
Propionyl chloride	: 9170
Propionyl fluoride	: 9171
Propiophenone	: 8964
Propoxybenzene	: 8976
Propylacetylene	: 8687
Propyl alcohol	: 9166
Propylallene	: 5774
Propyl bromide	: 1327
gamma-Propyl-gamma-butyrolactone	: 3667
S-Propyl carbonochloridothioate	: 9223
Propyl chloride	: 2260
Propyl cyanide	: 1419
Propylene	: 9178
Propylene chlorohydrin	: 2268
sec-Propylene chlorohydrin	: 2270
Propylenediamine	: 9137
Propylene dibromide	: 2975
Propylene dichloride	: 3271
1,2-Propylene glycol 1-propyl ether	: 9200
1,2-Propylene oxide	: 7536
Propyl ether	: 4558
Propyl fluoride	: 5415
Propyl gallate	: 9268
1-Propylheptyl alcohol	: 2761
Propylhexedrine	: 4019
Propylidene chloride	: 3270
Propyl iodide	: 6335
Propyl isopentanoate	: 9248
Propyl isopropyl ketone	: 7338
Propyl mercaptan	: 9151
Propylparaben	: 9242
Propyl trans-3-phenyl-2-propenoate	: 9224
N-Propyl-1-propanamine	: 4549
N-Propyl-2-propanamine	: 6535
Propyl propionate	: 9260
Propylthiouracil	: 3668
Propyltrichlorosilane	: 10287
Proscar	: 5330
Protionamide	: 9263
Protoanemonin	: 7243
Protocatechualdehyde	: 3701
Protocatechuic acid	: 3708
Pseudoconhydrine	: 9258
Pseudocumene	: 10508
Pseudoionone	: 4349
Pseudopelletierine	: 6952
1H-Purin-6-amine	: 121
1H-Purine, 6-[(1-methyl-4-nitro-1H-imidazol-5-yl)thio]-	: 545
Purpurin	: 10434
Purpurogallin	: 9898
Putrescine	: 1403
1,4-Pyran	: 9296
Pyrazinamide	: 9303
Pyrazinoic acid	: 9304

MOLECULAR FORMULA INDEX OF ORGANIC COMPOUNDS

$CBrClF_2$
1142 - Bromochlorodifluoromethane
$CBrCl_2F$
1168 - Bromodichlorofluoromethane
$CBrCl_3$
1362 - Bromotrichloromethane
$CBrF_3$
1367 - Bromotrifluoromethane
$CBrN$
2513 - Cyanogen bromide
$CBrN_3O_6$
1372 - Bromotrinitromethane
CBr_2ClF
2923 - Dibromochlorofluoromethane
CBr_2Cl_2
2931 - Dibromodichloromethane
CBr_2F_2
2933 - Dibromodifluoromethane
CBr_2O
1755 - Carbonyl bromide
CBr_3Cl
10162 - Tribromochloromethane
CBr_3F
10166 - Tribromofluoromethane
CBr_3NO_2
10171 - Tribromonitromethane
CBr_4
9715 - Tetrabromomethane
$CCaN_2$
1696 - Calcium cyanamide
$CClFO$
1757 - Carbonyl chloride fluoride
$CClF_3$
2328 - Chlorotrifluoromethane
$CClF_3O_2S$
10398 - Trifluoromethanesulfonyl chloride
$CClF_3S$
10396 - Trifluoromethanesulfenyl chloride
$CClN$
2514 - Cyanogen chloride
$CClNO_3S$
9664 - Sulfuryl chloride isocyanate
$CClN_3O_6$
2338 - Chlorotrinitromethane
CCl_2F_2
3162 - Dichlorodifluoromethane
CCl_2O
1756 - Carbonyl chloride
CCl_2S
1750 - Carbonothioic dichloride
CCl_3F
10239 - Trichlorofluoromethane
CCl_3NO_2
10261 - Trichloronitromethane
CCl_4
9756 - Tetrachloromethane
CCl_4O_2S
10247 - Trichloromethanesulfonyl chloride
CCl_4S
10246 - Trichloromethanesulfenyl chloride
CFN
2515 - Cyanogen fluoride

CFN_3O_6
10628 - Trinitrofluoromethane
CF_2O
1760 - Carbonyl fluoride
CF_3I
10393 - Trifluoroiodomethane
CF_4
9839 - Tetrafluoromethane
CF_4O_2S
10399 - Trifluoromethanesulfonyl fluoride
$CHBrClF$
1146 - Bromochlorofluoromethane
$CHBrCl_2$
1169 - Bromodichloromethane
$CHBrF_2$
1173 - Bromodifluoromethane
$CHBr_2Cl$
1972 - Chlorodibromomethane
$CHBr_2F$
2944 - Dibromofluoromethane
$CHBr_3$
10167 - Tribromomethane
$CHClF_2$
1983 - Chlorodifluoromethane
$CHCl_2F$
3187 - Dichlorofluoromethane
$CHCl_3$
10245 - Trichloromethane
$CHCl_3S$
10248 - Trichloromethanethiol
$CHCl_5Si$
10220 - Trichloro(dichloromethyl)silane
$CHFO$
5464 - Formyl fluoride
CHF_2N
1739 - Carboimidic difluoride
CHF_3
10395 - Trifluoromethane
CHF_3O_3S
10397 - Trifluoromethanesulfonic acid
CHI_3
10450 - Triiodomethane
CHN
5952 - Hydrogen cyanide
$CHNO$
2502 - Cyanic acid
5474 - Fulminic acid
$CHNS$
6557 - Isothiocyanic acid
10050 - Thiocyanic acid
CHN_3O_6
10630 - Trinitromethane
CH_2BrCl
1147 - Bromochloromethane
CH_2BrF
1213 - Bromofluoromethane
CH_2BrI
1238 - Bromoiodomethane
CH_2BrNO_2
1295 - Bromonitromethane
CH_2Br_2
2956 - Dibromomethane

CH_2ClF
2054 - Chlorofluoromethane
CH_2ClI
2081 - Chloroiodomethane
CH_2ClNO
1726 - Carbamic chloride
CH_2Cl_2
3202 - Dichloromethane
CH_2Cl_4Si
10217 - Trichloro(chloromethyl)silane
CH_2F_2
3563 - Difluoromethane
CH_2INaO_3S
9564 - Sodium iodomethanesulfonate
CH_2I_2
3768 - Diiodomethane
CH_2N_2
2500 - Cyanamide
2872 - Diazomethane
$CH_2N_2O_4$
4387 - Dinitromethane
CH_2N_4
10009 - $1H$-Tetrazole
CH_2O
5448 - Formaldehyde
$(CH_2O)_x$
8507 - Paraformaldehyde
CH_2O_2
5453 - Formic acid
CH_2O_3
1744 - Carbonic acid
CH_2S
10055 - Thioformaldehyde
CH_2S_3
10708 - Trithiocarbonic acid
CH_2Se
9522 - Selenoformaldehyde
CH_3AsF_2
7195 - Methyldifluoroarsine
CH_3BCl_2
3209 - Dichloromethylborane
CH_3BF_2
3565 - Difluoromethylborane
CH_3BO
1063 - Borane carbonyl
CH_3Br
1242 - Bromomethane
CH_3BrMg
7432 - Methyl magnesium bromide
CH_3Cl
2089 - Chloromethane
CH_3ClMg
7433 - Methylmagnesium chloride
CH_3ClO_2S
6745 - Methanesulfonyl chloride
CH_3Cl_2OP
7670 - Methyl phosphorodichloridite
CH_3Cl_2P
3216 - Dichloromethylphosphine
CH_3Cl_3Si
7832 - Methyltrichlorosilane

CH₃F
  5395 - Fluoromethane
CH₃FO₂S
  6746 - Methanesulfonyl fluoride
CH₃FO₃S
  7262 - Methyl fluorosulfonate
CH₃F₂OP
  7668 - Methylphosphonic difluoride
CH₃F₂P
  7196 - Methyldifluorophosphine
CH₃F₃Si
  10412 - Trifluoromethylsilane
  10413 - (Trifluoromethyl)silane
CH₃I
  6307 - Iodomethane
CH₃NO
  5449 - Formaldehyde oxime
  5450 - Formamide
CH₃NO₂
  7478 - Methyl nitrite
  8122 - Nitromethane
CH₃NO₃
  7477 - Methyl nitrate
CH₃NS₂
  1727 - Carbamodithioic acid
CH₃N₃
  6955 - Methyl azide
CH₃N₃O₃
  8211 - Nitrourea
CH₃N₅
  10008 - 1*H*-Tetrazol-5-amine
CH₃NaO
  9566 - Sodium methanolate
CH₃NaO₃S
  9560 - Sodium formaldehydesulfoxylate
CH₃NaO₄S
  9559 - Sodium formaldehyde bisulfite
CH₄
  6741 - Methane
CH₄AsNaO₃
  9567 - Sodium methylarsonate
CH₄ClO₃P
  2155 - (Chloromethyl)phosphonic acid
CH₄Cl₂Si
  3220 - Dichloromethylsilane
CH₄NO₅P
  1728 - Carbamoyl dihydrogen phosphate
CH₄N₂
  6748 - Methanimidamide
CH₄N₂O
  5934 - Hydrazinecarboxaldehyde
  10756 - Urea
CH₄N₂O₂
  6200 - Hydroxyurea
  8121 - *N*-Nitromethanamine
CH₄N₂O₂S
  5451 - Formamidinesulfinic acid
CH₄N₂S
  10081 - Thiourea
CH₄N₂Se
  9523 - Selenourea
CH₄N₄O₂
  8109 - Nitroguanidine
CH₄O
  6750 - Methanol

CH₄O₂
  7366 - Methyl hydroperoxide
CH₄O₃S
  6744 - Methanesulfonic acid
CH₄O₄S
  7788 - Methyl sulfate
CH₄O₆S₂
  6743 - Methanedisulfonic acid
CH₄S
  6747 - Methanethiol
CH₅As
  6951 - Methylarsine
CH₅AsO₃
  6742 - Methanearsonic acid
CH₅ClN₂O
  10757 - Urea hydrochloride
CH₅ClSi
  2163 - Chloromethylsilane
CH₅ISi
  6318 - Iodomethylsilane
CH₅N
  6920 - Methylamine
CH₅NO
  6055 - *N*-Hydroxymethanamine
  7371 - *O*-Methylhydroxylamine
CH₅N₃
  5608 - Guanidine
CH₅N₃O
  5935 - Hydrazinecarboxamide
CH₅N₃O₄
  10758 - Urea nitrate
CH₅N₃S
  5933 - Hydrazinecarbothioamide
CH₅N₅O₂
  377 - 3-Amino-1-nitroguanidine
CH₅O₃P
  7667 - Methylphosphonic acid
CH₅O₄P
  7665 - Methyl phosphate
CH₅P
  7666 - Methylphosphine
CH₆ClN
  6921 - Methylamine hydrochloride
CH₆ClNO
  6072 - *N*-Hydroxymethylamine hydrochloride
  7372 - *O*-Methylhydroxylamine hydrochloride
CH₆ClN₃
  5609 - Guanidine monohydrochloride
CH₆ClN₃O
  9524 - Semicarbazide hydrochloride
CH₆Ge
  7276 - Methylgermane
CH₆NO₃P
  351 - (Aminomethyl)phosphonic acid
CH₆N₂
  7363 - Methylhydrazine
CH₆N₄
  5936 - Hydrazinecarboximidamide
CH₆N₄O
  1745 - Carbonic dihydrazide
CH₆N₄O₃
  5610 - Guanidine mononitrate
CH₆N₄S
  1751 - Carbonothioic dihydrazide
CH₆OSi
  7781 - Methyl silyl ether

CH₆Si
  7780 - Methylsilane
CH₆Sn
  7782 - Methylstannane
CH₈B₂
  7188 - Methyldiborane(6)
CIN
  2516 - Cyanogen iodide
CI₄
  9905 - Tetraiodomethane
CN₄O₈
  9981 - Tetranitromethane
CO
  1746 - Carbon monoxide
COS
  1753 - Carbon oxysulfide
COSe
  1752 - Carbon oxyselenide
CO₂
  1741 - Carbon dioxide
CS₂
  1743 - Carbon disulfide
CSe₂
  1742 - Carbon diselenide
C₂BrF₃
  1366 - Bromotrifluoroethene
C₂BrF₅
  1304 - Bromopentafluoroethane
C₂Br₂ClF₃
  2925 - 1,2-Dibromo-1-chloro-1,2,2-trifluoro-
    ethane
C₂Br₂F₄
  2990 - 1,2-Dibromotetrafluoroethane
C₂Br₄
  9712 - Tetrabromoethene
C₂Br₆
  5720 - Hexabromoethane
C₂ClF₃
  2327 - Chlorotrifluoroethene
C₂ClF₃O
  10383 - Trifluoroacetyl chloride
C₂ClF₅
  2207 - Chloropentafluoroethane
C₂Cl₂
  3073 - Dichloroacetylene
C₂Cl₂F₂
  3159 - 1,1-Dichloro-2,2-difluoroethene
  3160 - *cis*-1,2-Dichloro-1,2-difluoroethene
  3161 - *trans*-1,2-Dichloro-1,2-difluoroethene
C₂Cl₂F₄
  3294 - 1,1-Dichloro-1,2,2,2-tetrafluoroethane
  3295 - 1,2-Dichloro-1,1,2,2-tetrafluoroethane
C₂Cl₂O₂
  8432 - Oxalyl chloride
C₂Cl₃F₃
  10294 - 1,1,1-Trichloro-2,2,2-trifluoroethane
  10295 - 1,1,2-Trichloro-1,2,2-trifluoroethane
C₂Cl₃N
  10198 - Trichloroacetonitrile
C₂Cl₃NaO₂
  9577 - Sodium trichloroacetate
C₂Cl₄
  9753 - Tetrachloroethene
C₂Cl₄F₂
  9745 - 1,1,1,2-Tetrachloro-2,2-difluoroethane
  9746 - 1,1,2,2-Tetrachloro-1,2-difluoroethane

C_2Cl_4O
10199 - Trichloroacetyl chloride
$C_2Cl_4O_2$
4536 - Diphosgene
C_2Cl_5F
8539 - Pentachlorofluoroethane
C_2Cl_6
5737 - Hexachloroethane
C_2CrO_4
2382 - Chromium(II) oxalate
C_2F_3N
10382 - Trifluoroacetonitrile
10394 - Trifluoroisocyanomethane
$C_2F_3NaO_2$
9578 - Sodium trifluoroacetate
C_2F_4
9837 - Tetrafluoroethene
$C_2F_4N_2O_4$
9834 - 1,1,2,2-Tetrafluoro-1,2-dinitroethane
C_2F_6
5798 - Hexafluoroethane
$C_2F_6S_2$
1053 - Bis(trifluoromethyl) disulfide
C_2HBr
1084 - Bromoacetylene
$C_2HBrClF_3$
1154 - 1-Bromo-2-chloro-1,1,2-trifluoroethane
1155 - 2-Bromo-2-chloro-1,1,1-trifluoroethane
$C_2HBr_2F_3$
2994 - 1,2-Dibromo-1,1,2-trifluoroethane
C_2HBr_2N
2904 - Dibromoacetonitrile
C_2HBr_3
10165 - Tribromoethene
C_2HBr_3O
10152 - Tribromoacetaldehyde
$C_2HBr_3O_2$
10153 - Tribromoacetic acid
C_2HCl
1848 - Chloroacetylene
C_2HClF_2
1982 - 1-Chloro-2,2-difluoroethene
$C_2HClF_2O_2$
1979 - Chlorodifluoroacetic acid
C_2HClF_4
2307 - 1-Chloro-1,1,2,2-tetrafluoroethane
2308 - 1-Chloro-1,2,2,2-tetrafluoroethane
C_2HCl_2F
3186 - 1,1-Dichloro-2-fluoroethene
$C_2HCl_2F_3$
3306 - 1,2-Dichloro-1,1,2-trifluoroethane
3307 - 2,2-Dichloro-1,1,1-trifluoroethane
3308 - 2,2-Dichloro-1,1,2-trifluoroethane
C_2HCl_2N
3071 - Dichloroacetonitrile
C_2HCl_3
10231 - Trichloroethene
$C_2HCl_3F_2$
10221 - 1,1,1-Trichloro-2,2-difluoroethane
10222 - 1,2,2-Trichloro-1,1-difluoroethane
10223 - 1,2,2-Trichloro-1,2-difluoroethane
C_2HCl_3O
3072 - Dichloroacetyl chloride
10194 - Trichloroacetaldehyde

$C_2HCl_3O_2$
10196 - Trichloroacetic acid
C_2HCl_4F
9754 - 1,1,1,2-Tetrachloro-2-fluoroethane
9755 - 1,1,2,2-Tetrachloro-1-fluoroethane
C_2HCl_5
8538 - Pentachloroethane
C_2HF
5354 - Fluoroacetylene
C_2HF_3
10389 - Trifluoroethene
$C_2HF_3O_2$
10379 - Trifluoroacetic acid
C_2HF_5
8584 - Pentafluoroethane
C_2HF_5O
10408 - Trifluoromethyl difluoromethyl ether
C_2HI
6269 - Iodoacetylene
C_2HN_3
3316 - Dicyanamide
C_2H_2
67 - Acetylene
$C_2H_2AsCl_3$
3145 - Dichloro(2-chlorovinyl)arsine
$C_2H_2BrF_3$
1365 - 2-Bromo-1,1,1-trifluoroethane
$C_2H_2Br_2$
2939 - cis-1,2-Dibromoethene
2940 - trans-1,2-Dibromoethene
$C_2H_2Br_2Cl_2$
2929 - 1,2-Dibromo-1,1-dichloroethane
2930 - 1,2-Dibromo-1,2-dichloroethane
$C_2H_2Br_2F_2$
2932 - 1,2-Dibromo-1,1-difluoroethane
$C_2H_2Br_2O$
1083 - Bromoacetyl bromide
$C_2H_2Br_2O_2$
2903 - Dibromoacetic acid
$C_2H_2Br_4$
9711 - 1,1,2,2-Tetrabromoethane
C_2H_2ClFO
5353 - Fluoroacetyl chloride
$C_2H_2ClF_3$
2324 - 1-Chloro-1,1,2-trifluoroethane
2325 - 1-Chloro-1,2,2-trifluoroethane
2326 - 2-Chloro-1,1,1-trifluoroethane
C_2H_2ClN
1844 - Chloroacetonitrile
$C_2H_2Cl_2$
3176 - 1,1-Dichloroethene
3177 - cis-1,2-Dichloroethene
3178 - trans-1,2-Dichloroethene
$C_2H_2Cl_2F_2$
3156 - 1,1-Dichloro-1,2-difluoroethane
3157 - 1,2-Dichloro-1,1-difluoroethane
3158 - 1,2-Dichloro-1,2-difluoroethane
$C_2H_2Cl_2O$
1847 - Chloroacetyl chloride
3065 - Dichloroacetaldehyde
$C_2H_2Cl_2O_2$
3067 - Dichloroacetic acid
$C_2H_2Cl_3F$
10236 - 1,1,1-Trichloro-2-fluoroethane
10237 - 1,1,2-Trichloro-1-fluoroethane
10238 - 1,1,2-Trichloro-2-fluoroethane

$C_2H_2Cl_3NO$
10195 - 2,2,2-Trichloroacetamide
$C_2H_2Cl_4$
9751 - 1,1,1,2-Tetrachloroethane
9752 - 1,1,2,2-Tetrachloroethane
$C_2H_2FNaO_2$
9558 - Sodium fluoroacetate
$C_2H_2F_2$
3560 - 1,1-Difluoroethene
3561 - cis-1,2-Difluoroethene
3562 - trans-1,2-Difluoroethene
$C_2H_2F_2O_2$
3547 - Difluoroacetic acid
$C_2H_2F_3I$
10392 - 1,1,1-Trifluoro-2-iodoethane
$C_2H_2F_3NO$
10378 - 2,2,2-Trifluoroacetamide
$C_2H_2F_4$
9835 - 1,1,1,2-Tetrafluoroethane
9836 - 1,1,2,2-Tetrafluoroethane
$C_2H_2F_4O$
962 - Bis(difluoromethyl) ether
C_2H_2IN
6268 - Iodoacetonitrile
$C_2H_2I_2$
3765 - cis-1,2-Diiodoethene
$C_2H_2N_2O$
8426 - 1,3,4-Oxadiazole
$C_2H_2N_2S$
10022 - 1,2,5-Thiadiazole
10023 - 1,3,4-Thiadiazole
$C_2H_2N_2S_3$
10024 - 1,3,4-Thiadiazolidine-2,5-dithione
$C_2H_2N_4$
10007 - 1,2,4,5-Tetrazine
C_2H_2O
6570 - Ketene
$C_2H_2O_2$
5596 - Glyoxal
$C_2H_2O_3$
5598 - Glyoxylic acid
$C_2H_2O_4$
8429 - Oxalic acid
C_2H_3Br
1194 - Bromoethene
C_2H_3BrO
57 - Acetyl bromide
$C_2H_3BrO_2$
1079 - Bromoacetic acid
$C_2H_3Br_3$
10163 - 1,1,2-Tribromoethane
$C_2H_3Br_3O$
10164 - 2,2,2-Tribromoethanol
$C_2H_3Br_3O_2$
1076 - Bromal hydrate
C_2H_3Cl
2022 - Chloroethene
$C_2H_3ClF_2$
1980 - 1-Chloro-1,1-difluoroethane
1981 - 1-Chloro-2,2-difluoroethane
C_2H_3ClO
58 - Acetyl chloride
1838 - Chloroacetaldehyde
$C_2H_3ClO_2$
1840 - Chloroacetic acid
7125 - Methyl chlorocarbonate

$C_2H_3Cl_2F$
3184 - 1,1-Dichloro-1-fluoroethane
3185 - 1,2-Dichloro-1-fluoroethane

$C_2H_3Cl_2NO$
3066 - 2,2-Dichloroacetamide

$C_2H_3Cl_2NO_2$
3239 - 1,1-Dichloro-1-nitroethane

$C_2H_3Cl_3$
10228 - 1,1,1-Trichloroethane
10229 - 1,1,2-Trichloroethane

$C_2H_3Cl_3O$
10230 - 2,2,2-Trichloroethanol

$C_2H_3Cl_3O_2$
1812 - Chloral hydrate

$C_2H_3Cl_3Si$
10296 - Trichlorovinylsilane

C_2H_3F
5388 - Fluoroethene

C_2H_3FO
69 - Acetyl fluoride

$C_2H_3FO_2$
5352 - Fluoroacetic acid

$C_2H_3F_3$
10386 - 1,1,1-Trifluoroethane
10387 - 1,1,2-Trifluoroethane

$C_2H_3F_3O$
7836 - Methyl trifluoromethyl ether
10388 - 2,2,2-Trifluoroethanol

C_2H_3HgN
2518 - Cyanomethylmercury

C_2H_3I
6299 - Iodoethene

C_2H_3IO
74 - Acetyl iodide

$C_2H_3IO_2$
6266 - Iodoacetic acid

C_2H_3N
38 - Acetonitrile
6423 - Isocyanomethane

C_2H_3NO
5960 - Hydroxyacetonitrile
7138 - Methyl cyanate
7396 - Methyl isocyanate

$C_2H_3NO_2$
8097 - Nitroethene

$C_2H_3NO_3$
8434 - Oxamic acid

$C_2H_3NO_4$
85 - Acetyl nitrate
8032 - Nitroacetic acid

C_2H_3NS
7425 - Methyl isothiocyanate
7810 - Methyl thiocyanate

$C_2H_3N_3$
10147 - 1H-1,2,3-Triazole
10148 - 1H-1,2,4-Triazole

$C_2H_3N_3O_2$
10755 - Urazole

$C_2H_3N_3S$
3688 - 1,2-Dihydro-3H-1,2,4-triazole-3-thione

$C_2H_3N_3S_2$
429 - 5-Amino-1,3,4-thiadiazole-2(3H)-thione

C_2H_4
5015 - Ethylene

C_2H_4BrCl
1144 - 1-Bromo-1-chloroethane

1145 - 1-Bromo-2-chloroethane

C_2H_4BrF
1212 - 1-Bromo-2-fluoroethane

C_2H_4BrNO
1078 - N-Bromoacetamide

$C_2H_4Br_2$
2937 - 1,1-Dibromoethane
2938 - 1,2-Dibromoethane

C_2H_4ClF
2052 - 1-Chloro-1-fluoroethane
2053 - 1-Chloro-2-fluoroethane

C_2H_4ClNO
1839 - 2-Chloroacetamide

$C_2H_4ClNO_2$
2190 - 1-Chloro-1-nitroethane

$C_2H_4Cl_2$
3173 - 1,1-Dichloroethane
3174 - 1,2-Dichloroethane

$C_2H_4Cl_2O$
937 - Bis(chloromethyl) ether
3175 - 2,2-Dichloroethanol
3211 - 1,1-Dichloromethyl methyl ether

$C_2H_4Cl_2O_2S$
2019 - 2-Chloroethanesulfonyl chloride

$C_2H_4Cl_3NO$
433 - 1-Amino-2,2,2-trichloroethanol

$C_2H_4Cl_3O_4P$
10298 - Triclofos

$C_2H_4Cl_4Si$
3153 - Dichloro(dichloromethyl)methylsilane

C_2H_4FNO
5351 - 2-Fluoroacetamide

$C_2H_4F_2$
3558 - 1,1-Difluoroethane
3559 - 1,2-Difluoroethane

$C_2H_4F_3N$
10390 - 2,2,2-Trifluoroethylamine

C_2H_4INO
6265 - 2-Iodoacetamide

$C_2H_4I_2$
3764 - 1,2-Diiodoethane

$C_2H_4NNaS_2$
9568 - Sodium methyldithiocarbamate

$C_2H_4N_2$
230 - Aminoacetonitrile

$C_2H_4N_2O_2$
4769 - Ethanedial dioxime
5937 - 1,2-Hydrazinedicarboxaldehyde
8435 - Oxamide

$C_2H_4N_2O_4$
4385 - 1,1-Dinitroethane
4386 - 1,2-Dinitroethane

$C_2H_4N_2O_6$
4782 - 1,2-Ethanediol, dinitrate

$C_2H_4N_2S_2$
4792 - Ethanedithioamide

$C_2H_4N_4$
2517 - Cyanoguanidine
10146 - 1H-1,2,4-Triazol-3-amine

$C_2H_4N_4O_2$
2870 - Diazenedicarboxamide

C_2H_4O
14 - Acetaldehyde
8447 - Oxirane

$(C_2H_4O)_x$
6728 - Metaldehyde

C_2H_4OS
10044 - Thioacetic acid

$C_2H_4O_2$
21 - Acetic acid
5583 - Glycolaldehyde
7264 - Methyl formate

$C_2H_4O_2S$
10056 - Thioglycolic acid

$C_2H_4O_3$
5584 - Glycolic acid
8732 - Peroxyacetic acid

$C_2H_4O_3S$
4789 - 1,2-Ethanediol, monosulfite

$C_2H_4O_5S$
9656 - Sulfoacetic acid

C_2H_4S
10041 - Thiirane

$C_2H_4S_5$
8634 - 1,2,3,5,6-Pentathiepane

C_2H_4Si
5291 - Ethynylsilane

$C_2H_5AlCl_2$
3181 - Dichloroethylaluminum

$C_2H_5AsCl_2$
4980 - Ethyldichloroarsine

$C_2H_5AsF_2$
4986 - Ethyldifluoroarsine

C_2H_5Br
1192 - Bromoethane

C_2H_5BrO
1193 - 2-Bromoethanol
1244 - Bromomethoxymethane

C_2H_5Cl
2018 - Chloroethane

$C_2H_5ClN_2$
231 - Aminoacetonitrile monohydrochloride

C_2H_5ClO
2020 - 2-Chloroethanol
2132 - Chloromethyl methyl ether

$C_2H_5ClO_2S$
4796 - Ethanesulfonyl chloride
4949 - Ethyl chlorosulfinate

$C_2H_5ClO_3S$
4950 - Ethyl chlorosulfonate

C_2H_5ClS
2167 - Chloro(methylthio)methane

$C_2H_5Cl_2O_2P$
5215 - Ethyl phosphorodichloridate

$C_2H_5Cl_3OSi$
10233 - Trichloroethoxysilane

$C_2H_5Cl_3Si$
3144 - Dichloro(chloromethyl)methylsilane
10235 - Trichloroethylsilane

C_2H_5F
5386 - Fluoroethane

C_2H_5FO
5387 - 2-Fluoroethanol

C_2H_5I
6297 - Iodoethane

C_2H_5IO
6298 - 2-Iodoethanol

C_2H_5N
5021 - Ethyleneimine

C_2H_5NO
16 - Acetaldoxime
17 - Acetamide

C₂I₂
  3758 - Diiodoacetylene
C₂I₄
  9903 - Tetraiodoethene
C₂Li₂O₄
  6622 - Lithium oxalate
C₂N₂
  2512 - Cyanogen
C₂N₄O₆
  10622 - Trinitroacetonitrile
C₃Br₂F₆
  2949 - 1,2-Dibromo-1,1,2,3,3,3-hexafluoropro-
    pane
C₃ClF₅O
  2205 - Chloropentafluoroacetone
C₃Cl₂F₆
  3193 - 1,2-Dichloro-1,1,2,3,3,3-hexafluoropro-
    pane
  3194 - 1,3-Dichloro-1,1,2,2,3,3-hexafluoropro-
    pane
C₃Cl₂KN₃O₃
  9094 - Potassium dichloroisocyanurate
C₃Cl₃F₅
  10265 - 1,2,3-Trichloro-1,1,2,3,3-pentafluoro-
    propane
C₃Cl₃N₃
  10290 - 2,4,6-Trichloro-1,3,5-triazine
C₃Cl₃N₃O₃
  9670 - Symclosene
C₃Cl₄O
  10286 - 2,3,3-Trichloro-2-propenoyl chloride
C₃Cl₆
  5739 - Hexachloropropene
C₃Cl₆O
  5723 - Hexachloroacetone
C₃Cl₆O₃
  1048 - Bis(trichloromethyl) carbonate
C₃F₃N₃
  2523 - Cyanuric fluoride
C₃F₆
  8725 - Perfluoropropene
C₃F₆O
  8695 - Perfluoroacetone
  8722 - Perfluorooxetane
  10424 - Trifluoro(trifluoromethyl)oxirane
C₃F₇I
  5664 - Heptafluoro-2-iodopropane
C₃F₈
  8724 - Perfluoropropane
C₃F₈O₂
  8704 - Perfluorodimethoxymethane
C₃HBr₅O
  8528 - 1,1,1,3,3-Pentabromo-2-propanone
C₃HCl₂F₅
  3244 - 1,3-Dichloro-1,1,2,2,3-pentafluoropro-
    pane
  3245 - 3,3-Dichloro-1,1,1,2,2-pentafluoropro-
    pane
C₃HCl₂N₃O₃
  3304 - 1,3-Dichloro-1,3,5-triazine-
    2,4,6(1H,3H,5H)-trione
C₃HCl₅
  8543 - 1,1,2,3,3-Pentachloro-1-propene
C₃HCl₇
  5643 - 1,1,1,2,3,3,3-Heptachloropropane

C₃HF₃
  10422 - 3,3,3-Trifluoro-1-propyne
C₃HF₇
  5665 - 1,1,1,2,3,3,3-Heptafluoropropane
C₃HF₇O
  10414 - Trifluoromethyl 1,1,2,2-tetrafluoroet-
    hyl ether
C₃HN
  2506 - Cyanoacetylene
C₃H₂BrNS
  1354 - 2-Bromothiazole
C₃H₂Br₂N₂O
  2926 - 2,2-Dibromo-2-cyanoacetamide
C₃H₂ClF₅O
  4702 - Enflurane
C₃H₂ClN
  2276 - 2-Chloro-2-propenenitrile
C₃H₂Cl₂O₂
  9143 - Propanedioyl dichloride
C₃H₂Cl₃NO
  10243 - 3,3,3-Trichloro-2-hydroxypropaneni-
    trile
C₃H₂Cl₄
  9770 - 1,1,2,3-Tetrachloropropene
C₃H₂Cl₄O₂
  1749 - Carbonochloridic acid, 2,2,2-trichloro-
    ethyl ester
C₃H₂F₂
  3554 - 3,3-Difluorocyclopropene
C₃H₂F₆
  5799 - 1,1,1,2,3,3-Hexafluoropropane
  5800 - 1,1,1,3,3,3-Hexafluoropropane
C₃H₂F₆O
  5801 - 1,1,1,3,3,3-Hexafluoro-2-propanol
  9838 - 1,2,2,2-Tetrafluoroethyl difluoromethyl
    ether
C₃H₂N₂
  6665 - Malononitrile
C₃H₂N₂O₃
  6220 - Imidazolidinetrione
C₃H₂N₂S₂
  7248 - Methylene thiocyanate
C₃H₂O
  9271 - 2-Propynal
C₃H₂O₂
  9274 - 2-Propynoic acid
C₃H₂O₃
  4434 - 1,3-Dioxol-2-one
C₃H₃Br
  1345 - 3-Bromo-1-propyne
C₃H₃Cl
  2285 - 3-Chloro-1-propyne
C₃H₃ClO
  9185 - 2-Propenoyl chloride
C₃H₃ClO₂
  2277 - 2-Chloropropenoic acid
C₃H₃ClO₃
  2014 - 4-Chloro-1,3-dioxolan-2-one
  7128 - Methyl chlorooxoacetate
C₃H₃Cl₂F
  3189 - 1,1-Dichloro-2-fluoropropene
C₃H₃Cl₂NaO₂
  9550 - Sodium 2,2-dichloropropanoate
C₃H₃Cl₃
  10284 - 1,2,3-Trichloro-1-propene
  10285 - 3,3,3-Trichloro-1-propene

C₃H₃Cl₃O
  3278 - 2,3-Dichloropropanoyl chloride
  10254 - (Trichloromethyl)oxirane
  10283 - 1,1,1-Trichloro-2-propanone
C₃H₃Cl₃O₂
  7831 - Methyl trichloroacetate
C₃H₃Cl₅
  8542 - 1,1,2,2,3-Pentachloropropane
C₃H₃F₃
  10421 - 3,3,3-Trifluoropropene
C₃H₃F₃O
  10381 - 1,1,1-Trifluoroacetone
C₃H₃F₃O₂
  7835 - Methyl trifluoroacetate
C₃H₃F₅
  8588 - 1,1,1,2,2-Pentafluoropropane
C₃H₃F₅O
  3564 - 2-(Difluoromethoxy)-1,1,1-trifluoroet-
    hane
  7549 - Methyl pentafluoroethyl ether
  8589 - 2,2,3,3,3-Pentafluoro-1-propanol
C₃H₃N
  119 - Acrylonitrile
C₃H₃NO
  6560 - Isoxazole
  8441 - Oxazole
  8475 - 2-Oxopropanenitrile
C₃H₃NOS
  75 - Acetyl isothiocyanate
C₃H₃NOS₂
  10085 - 2-Thioxo-4-thiazolidinone
C₃H₃NO₂
  2504 - Cyanoacetic acid
C₃H₃NO₂S
  10033 - 2,4-Thiazolidinedione
C₃H₃NS
  10030 - Thiazole
C₃H₃N₃
  10139 - 1,2,4-Triazine
  10140 - 1,3,5-Triazine
C₃H₃N₃O₂
  8112 - 2-Nitro-1H-imidazole
  8113 - 4-Nitro-1H-imidazole
  10141 - 1,2,4-Triazine-3,5(2H,4H)-dione
C₃H₃N₃O₂S
  8202 - 5-Nitro-2-thiazolamine
C₃H₃N₃O₃
  2522 - Cyanuric acid
  10639 - 1,3,5-Trioxane-2,4,6-triimine
C₃H₃N₃S₃
  10143 - 1,3,5-Triazine-2,4,6(1H,3H,5H)-
    trithione
C₃H₃NaO₃
  9570 - Sodium 2-oxopropanoate
C₃H₄
  158 - Allene
  2696 - Cyclopropene
  9273 - Propyne
C₃H₄BrClO
  1335 - 2-Bromopropanoyl chloride
C₃H₄BrN
  1329 - 3-Bromopropanenitrile
C₃H₄Br₂
  2983 - 1,1-Dibromo-1-propene
  2984 - 1,2-Dibromo-1-propene
  2985 - 2,3-Dibromo-1-propene

$C_3H_4Br_2O$
1334 - 2-Bromopropanoyl bromide
2982 - 1,3-Dibromo-2-propanone

$C_3H_4Br_2O_2$
2978 - 2,3-Dibromopropanoic acid

$C_3H_4ClF_3$
2334 - 3-Chloro-1,1,1-trifluoropropane

C_3H_4ClN
2265 - 3-Chloropropanenitrile

C_3H_4ClNO
2041 - 2-Chloroethyl isocyanate

$C_3H_4ClN_5$
2318 - 6-Chloro-1,3,5-triazine-2,4-diamine

$C_3H_4Cl_2$
3279 - 1,1-Dichloropropene
3280 - *cis*-1,2-Dichloropropene
3281 - *trans*-1,2-Dichloropropene
3282 - *cis*-1,3-Dichloropropene
3283 - *trans*-1,3-Dichloropropene
3284 - 2,3-Dichloropropene

$C_3H_4Cl_2F_2O$
3163 - 2,2-Dichloro-1,1-difluoro-1-methoxy-
ethane

$C_3H_4Cl_2O$
2271 - 3-Chloropropanoyl chloride
3069 - 1,1-Dichloroacetone
3070 - 1,3-Dichloroacetone

$C_3H_4Cl_2O_2$
2035 - 2-Chloroethyl chloroformate
3274 - 2,2-Dichloropropanoic acid
7190 - Methyl dichloroacetate

$C_3H_4Cl_3NO_2$
10242 - *N*-(2,2,2-Trichloro-1-hydroxyethyl)
formamide

$C_3H_4Cl_3NSi$
10289 - 3-(Trichlorosilyl)propanenitrile

$C_3H_4Cl_4$
9766 - 1,1,1,2-Tetrachloropropane
9767 - 1,1,1,3-Tetrachloropropane
9768 - 1,1,2,3-Tetrachloropropane
9769 - 1,2,2,3-Tetrachloropropane

$C_3H_4F_4O$
9840 - 2,2,3,3-Tetrafluoro-1-propanol

$C_3H_4N_2$
6215 - Imidazole
7230 - (Methyleneamino)acetonitrile
9306 - 1*H*-Pyrazole

$C_3H_4N_2O$
2503 - 2-Cyanoacetamide

$C_3H_4N_2OS$
430 - 2-Amino-4(5*H*)-thiazolone
10084 - 2-Thioxo-4-imidazolidinone

$C_3H_4N_2OS_2$
432 - *N*-Amino-2-thioxo-4-thiazolidinone

$C_3H_4N_2O_2$
6218 - 2,4-Imidazolidinedione

$C_3H_4N_2S$
10029 - 2-Thiazolamine

C_3H_4O
116 - Acrolein
2695 - Cyclopropanone
9175 - Propargyl alcohol
9184 - 1-Propen-1-one

$C_3H_4O_2$
118 - Acrylic acid
6663 - Malonaldehyde

8445 - 2-Oxetanone
8446 - 3-Oxetanone
8448 - Oxiranecarboxaldehyde
9142 - 1,2-Propanedione
10804 - Vinyl formate

$C_3H_4O_3$
5017 - Ethylene carbonate
9404 - Pyruvic acid

$C_3H_4O_4$
6138 - 3-Hydroxy-2-oxopropanoic acid
6664 - Malonic acid

$C_3H_4O_5$
6180 - Hydroxypropanedioic acid

$C_3H_4S_3$
4591 - 1,3-Dithiolane-2-thione

C_3H_5Br
1336 - *cis*-1-Bromopropene
1337 - *trans*-1-Bromopropene
1338 - 2-Bromopropene
1339 - 3-Bromopropene

C_3H_5BrO
1080 - Bromoacetone
1270 - (Bromomethyl)oxirane, (±)

$C_3H_5BrO_2$
1330 - 2-Bromopropanoic acid, (±)
1331 - 3-Bromopropanoic acid
7050 - Methyl bromoacetate

$C_3H_5Br_2Cl$
2924 - 1,2-Dibromo-3-chloropropane

$C_3H_5Br_3$
10173 - 1,1,2-Tribromopropane
10174 - 1,2,2-Tribromopropane
10175 - 1,2,3-Tribromopropane

C_3H_5Cl
2272 - *cis*-1-Chloropropene
2273 - *trans*-1-Chloropropene
2274 - 2-Chloropropene
2275 - 3-Chloropropene

$C_3H_5ClN_2O_6$
2264 - 3-Chloro-1,2-propanediol dinitrate

C_3H_5ClO
1843 - Chloroacetone
2259 - 2-Chloropropanal
4707 - Epichlorohydrin
9170 - Propanoyl chloride

C_3H_5ClOS
4951 - *S*-Ethyl chlorothioformate

$C_3H_5ClO_2$
2096 - Chloromethyl acetate
2266 - 2-Chloropropanoic acid
2267 - 3-Chloropropanoic acid
4945 - Ethyl chloroformate
6771 - Methoxyacetyl chloride
7119 - Methyl chloroacetate

$C_3H_5Cl_2NO_2$
3241 - 1,1-Dichloro-1-nitropropane
4981 - Ethyl dichlorocarbamate

$C_3H_5Cl_3$
10278 - 1,1,2-Trichloropropane
10279 - 1,1,3-Trichloropropane
10280 - 1,2,2-Trichloropropane
10281 - 1,2,3-Trichloropropane

$C_3H_5Cl_3O$
10282 - 1,1,1-Trichloro-2-propanol

$C_3H_5Cl_3Si$
209 - Allyltrichlorosilane

C_3H_5F
5418 - *cis*-1-Fluoropropene
5419 - *trans*-1-Fluoropropene
5420 - 2-Fluoropropene
5421 - 3-Fluoropropene

C_3H_5FO
5417 - 1-Fluoro-2-propanone
9171 - Propanoyl fluoride

$C_3H_5F_3$
10419 - 1,1,1-Trifluoropropane

$C_3H_5F_3O$
10391 - 2,2,2-Trifluoroethyl methyl ether
10420 - 1,1,1-Trifluoro-2-propanol, (±)

$C_3H_5F_3O_3S$
5271 - Ethyl trifluoromethanesulfonate

C_3H_5I
6339 - 3-Iodopropene

C_3H_5IO
6267 - Iodoacetone

$C_3H_5IO_2$
6337 - 3-Iodopropanoic acid

C_3H_5N
5087 - Ethyl isocyanide
9147 - Propanenitrile
9272 - 2-Propyn-1-amine

C_3H_5NO
117 - Acrylamide
549 - 2-Azetidinone
4953 - Ethyl cyanate
5086 - Ethyl isocyanate
6181 - 2-Hydroxypropanenitrile
6182 - 3-Hydroxypropanenitrile
6770 - Methoxyacetonitrile

$C_3H_5NO_2$
8167 - 1-Nitro-1-propene
8168 - 2-Nitro-1-propene
8474 - 2-Oxopropanal oxime

$C_3H_5NO_3$
8033 - Nitroacetone

$C_3H_5NO_4$
7479 - Methyl nitroacetate
8165 - 3-Nitropropanoic acid

C_3H_5NS
5095 - Ethyl isothiocyanate
5258 - Ethyl thiocyanate

$C_3H_5NS_2$
10034 - 2-Thiazolidinethione

$C_3H_5N_3O$
2505 - Cyanoacetohydrazide

$C_3H_5N_3O_9$
10629 - Trinitroglycerol

$C_3H_5N_5O$
2867 - 4,6-Diamino-1,3,5-triazin-2(1*H*)-one

$C_3H_5NaO_2$
9572 - Sodium propanoate

C_3H_6
2688 - Cyclopropane
9178 - Propene

C_3H_6BrCl
1150 - 1-Bromo-2-chloropropane
1151 - 1-Bromo-3-chloropropane
1152 - 2-Bromo-1-chloropropane
1153 - 2-Bromo-2-chloropropane

$C_3H_6BrNO_4$
1296 - 2-Bromo-2-nitro-1,3-propanediol

$C_3H_6Br_2$
2975 - 1,2-Dibromopropane
2976 - 1,3-Dibromopropane
2977 - 2,2-Dibromopropane

$C_3H_6Br_2O$
2979 - 2,3-Dibromo-1-propanol
2980 - 1,3-Dibromo-2-propanol

C_3H_6ClI
2082 - 1-Chloro-3-iodopropane

C_3H_6ClNO
4002 - Dimethylcarbamic chloride

$C_3H_6ClNO_2$
2194 - 1-Chloro-1-nitropropane
2195 - 2-Chloro-2-nitropropane

C_3H_6ClNS
4003 - Dimethylcarbamothioic chloride

$C_3H_6Cl_2$
3270 - 1,1-Dichloropropane
3271 - 1,2-Dichloropropane, (±)
3272 - 1,3-Dichloropropane
3273 - 2,2-Dichloropropane

$C_3H_6Cl_2O$
3275 - 2,3-Dichloro-1-propanol
3276 - 1,3-Dichloro-2-propanol

$C_3H_6Cl_2Si$
3311 - Dichlorovinylmethylsilane

$C_3H_6Cl_4Si$
10219 - Trichloro(3-chloropropyl)silane

$C_3H_6F_2$
3567 - 2,2-Difluoropropane

$C_3H_6F_2O$
3568 - 1,3-Difluoro-2-propanol

$C_3H_6HgN_4$
7437 - Methylmercuric dicyanamide

$C_3H_6I_2$
3771 - 1,2-Diiodopropane
3772 - 1,3-Diiodopropane

$C_3H_6N_2$
405 - 3-Aminopropanenitrile
3673 - 4,5-Dihydro-1*H*-pyrazole
4006 - Dimethylcyanamide

$C_3H_6N_2O$
6221 - 2-Imidazolidinone
8192 - *N*-Nitroso-*N*-methylvinylamine

$C_3H_6N_2OS$
431 - *N*-(Aminothioxomethyl)acetamide

$C_3H_6N_2O_2$
276 - *N*-(Aminocarbonyl)acetamide
341 - 4-Amino-3-isoxazolidinone, (*R*)
9136 - Propanediamide

$C_3H_6N_2O_4$
4402 - 1,1-Dinitropropane
4403 - 1,3-Dinitropropane
4404 - 2,2-Dinitropropane

$C_3H_6N_2O_6$
4405 - 2,2-Dinitro-1,3-propanediol
9233 - 1,2-Propylene glycol dinitrate

$C_3H_6N_2O_7$
5564 - Glycerol 1,3-dinitrate

$C_3H_6N_2S$
3681 - 4,5-Dihydro-2-thiazolamine
6219 - 2-Imidazolidinethione

$C_3H_6N_6$
10142 - 1,3,5-Triazine-2,4,6-triamine

$C_3H_6N_6O_6$
5810 - Hexahydro-1,3,5-trinitro-1,3,5-triazine

C_3H_6O
33 - Acetone
170 - Allyl alcohol
7536 - Methyloxirane
7859 - Methyl vinyl ether
8444 - Oxetane
9132 - Propanal

$C_3H_6OS_2$
5014 - *O*-Ethyl dithiocarbonate

$C_3H_6O_2$
4433 - 1,3-Dioxolane
5035 - Ethyl formate
6179 - 3-Hydroxypropanal
6184 - 1-Hydroxy-2-propanone
6768 - Methoxyacetaldehyde
6912 - Methyl acetate
8449 - Oxiranemethanol, (±)
9164 - Propanoic acid

$C_3H_6O_2S$
6717 - 3-Mercaptopropanoic acid
7435 - Methyl mercaptoacetate
7792 - (Methylsulfonyl)ethene
7805 - (Methylthio)acetic acid
10039 - Thietane 1,1-dioxide
10058 - Thiolactic acid

$C_3H_6O_3$
3749 - 2,3-Dihydroxypropanal, (±)
3751 - 1,3-Dihydroxy-2-propanone
4005 - Dimethyl carbonate
6183 - 3-Hydroxypropanoic acid
6574 - *DL*-Lactic acid
6575 - *D*-Lactic acid
6576 - *L*-Lactic acid
6769 - Methoxyacetic acid
7367 - Methyl hydroxyacetate
8733 - Peroxypropanoic acid
10638 - 1,3,5-Trioxane

$C_3H_6O_3S$
9150 - 1,3-Propane sultone

$C_3H_6O_4$
3750 - 2,3-Dihydroxypropanoic acid, (*R*)

C_3H_6S
7804 - Methylthiirane
7812 - (Methylthio)ethene
9181 - 2-Propene-1-thiol
10038 - Thietane

$C_3H_6S_2$
4589 - 1,2-Dithiolane
4590 - 1,3-Dithiolane

$C_3H_6S_3$
10707 - 1,3,5-Trithiane

C_3H_7Br
1327 - 1-Bromopropane
1328 - 2-Bromopropane

C_3H_7BrO
1243 - 1-Bromo-2-methoxyethane
1332 - 3-Bromo-1-propanol
1333 - 1-Bromo-2-propanol

C_3H_7Cl
2260 - 1-Chloropropane
2261 - 2-Chloropropane

C_3H_7ClHgO
10132 - Triadimenol

C_3H_7ClO
2092 - (Chloromethoxy)ethane
2093 - 1-Chloro-2-methoxyethane
2268 - 2-Chloro-1-propanol

2269 - 3-Chloro-1-propanol
2270 - 1-Chloro-2-propanol

$C_3H_7ClO_2$
2262 - 3-Chloro-1,2-propanediol
2263 - 2-Chloro-1,3-propanediol

$C_3H_7ClO_2S$
9149 - 1-Propanesulfonyl chloride

C_3H_7ClS
2166 - 1-Chloro-2-(methylthio)ethane

$C_3H_7Cl_3Si$
1973 - Chloro(dichloromethyl)dimethylsilane
10287 - Trichloropropylsilane

C_3H_7F
5415 - 1-Fluoropropane
5416 - 2-Fluoropropane

C_3H_7I
6335 - 1-Iodopropane
6336 - 2-Iodopropane

C_3H_7IO
6338 - 3-Iodo-1-propanol

C_3H_7N
171 - Allylamine
547 - Azetidine
2697 - Cyclopropylamine
9236 - Propyleneimine

$C_3H_7NNaS_2$
9554 - Sodium dimethyldithiocarbamate

C_3H_7NO
4061 - *N,N*-Dimethylformamide
5034 - *N*-Ethylformamide
6910 - *N*-Methylacetamide
9133 - Propanal oxime
9134 - Propanamide
9168 - 2-Propanone oxime

$C_3H_7NO_2$
142 - *DL*-Alanine
143 - *D*-Alanine
144 - *L*-Alanine
145 - β-Alanine
4937 - Ethyl carbamate
6521 - Isopropyl nitrite
8163 - 1-Nitropropane
8164 - 2-Nitropropane
9251 - Propyl nitrite
9510 - Sarcosine

$C_3H_7NO_2S$
2715 - *L*-Cysteine

$C_3H_7NO_3$
6520 - Isopropyl nitrate
8166 - 2-Nitro-1-propanol
9250 - Propyl nitrate
9529 - *DL*-Serine
9530 - *D*-Serine
9531 - *L*-Serine

$C_3H_7NO_5S$
2714 - *L*-Cysteic acid

C_3H_7NS
10031 - Thiazolidine

$C_3H_7N_3O_2$
5175 - *N*-Ethyl-*N*-nitrosourea
5581 - Glycocyamine

$C_3H_7NaO_4S$
9563 - Sodium 2-hydroxy-2-propanesulfonate

$C_3H_7O_6P$
5576 - Glycerone phosphate

C_3H_8
9135 - Propane

C₃H₈BrClSi
1257 - (Bromomethyl)chlorodimethylsilane

C₃H₈ClNO₂S
2717 - L-Cysteine, hydrochloride

C₃H₈Cl₂Si
1953 - Chloro(chloromethyl)dimethylsilane
3182 - Dichloroethylmethylsilane

C₃H₈NO₅P
5599 - Glyphosate

C₃H₈NO₆P
9027 - O-Phosphoserine

C₃H₈N₂O
4350 - N,N-Dimethylurea
4351 - N,N'-Dimethylurea
5281 - N-Ethylurea
8191 - N-Nitrosomethylethylamine

C₃H₈N₂O₂
235 - 3-Aminoalanine
5066 - Ethyl hydrazinecarboxylate
6749 - Methanimidamide, monoacetate

C₃H₈N₂O₃
8491 - Oxymethurea

C₃H₈N₂S
4344 - N,N-Dimethylthiourea
4345 - N,N'-Dimethylthiourea

C₃H₈O
5139 - Ethyl methyl ether
9166 - 1-Propanol
9167 - 2-Propanol

C₃H₈OS
7811 - 2-(Methylthio)ethanol

C₃H₈OS₂
3824 - 2,3-Dimercapto-1-propanol

C₃H₈O₂
3870 - Dimethoxymethane
6822 - 2-Methoxyethanol
9230 - 1,2-Propylene glycol
9231 - 1,3-Propylene glycol

C₃H₈O₂S
6716 - 3-Mercapto-1,2-propanediol

C₃H₈O₃
5561 - Glycerol

C₃H₈O₃S
5101 - Ethyl methanesulfonate
9148 - 1-Propanesulfonic acid

C₃H₈S
5158 - Ethyl methyl sulfide
9151 - 1-Propanethiol
9152 - 2-Propanethiol

C₃H₈S₂
1034 - Bis(methylthio)methane
9144 - 1,2-Propanedithiol
9145 - 1,3-Propanedithiol

C₃H₉Al
10498 - Trimethyl aluminum

C₃H₉Al₂Br₃
10177 - Tribromotrimethyldialuminum

C₃H₉As
10505 - Trimethylarsine

C₃H₉AsO₃
9211 - Propylarsonic acid

C₃H₉B
10522 - Trimethylborane

C₃H₉BF₄O
10577 - Trimethyloxonium fluoborate

C₃H₉BO₃
10523 - Trimethyl borate

C₃H₉BS
7206 - Methyl dimethylthioborane

C₃H₉B₃O₆
10485 - Trimethoxyboroxin

C₃H₉Br₂N
1342 - 3-Bromopropylamine hydrobromide

C₃H₉ClSi
10527 - Trimethylchlorosilane

C₃H₉ClSn
2336 - Chlorotrimethylstannane

C₃H₉FSi
5431 - Fluorotrimethylsilane

C₃H₉Ga
10552 - Trimethylgallium

C₃H₉IS
10618 - Trimethylsulfonium iodide

C₃H₉In
10568 - Trimethylindium

C₃H₉N
5111 - Ethylmethylamine
6471 - Isopropylamine
9205 - Propylamine
10499 - Trimethylamine

C₃H₉NO
406 - 2-Amino-1-propanol, (±)
407 - 3-Amino-1-propanol
408 - 1-Amino-2-propanol
6827 - 2-Methoxyethylamine
7253 - N-Methyl-2-ethanolamine
10502 - Trimethylamine oxide

C₃H₉NO₂
404 - 3-Amino-1,2-propanediol, (±)
441 - Ammonium propanoate

C₃H₉NO₃S
6932 - 2-(Methylamino)ethanesulfonic acid

C₃H₉N₃O₃S
5612 - 2-Guanidinoethanesulfonic acid

C₃H₉O₃P
4132 - Dimethyl methylphosphonate
10605 - Trimethyl phosphite

C₃H₉O₄P
10603 - Trimethyl phosphate

C₃H₉O₆P
5567 - L-Glycerol 1-phosphate

C₃H₉P
10604 - Trimethylphosphine

C₃H₉Sb
10617 - Trimethylstibine

C₃H₁₀ClN
5112 - Ethylmethylamine hydrochloride
6472 - Isopropylamine hydrochloride
9206 - Propylamine hydrochloride
10501 - Trimethylamine hydrochloride

C₃H₁₀N₂
6500 - Isopropylhydrazine
7252 - N-Methyl-1,2-ethanediamine
9137 - 1,2-Propanediamine, (±)
9138 - 1,3-Propanediamine

C₃H₁₀N₂O
2865 - 1,3-Diamino-2-propanol

C₃H₁₀O₂Si
7200 - Methyldimethoxysilane

C₃H₁₀O₃Si
10494 - Trimethoxysilane

C₃H₁₀Si
10614 - Trimethylsilane

C₃H₁₁N₃
9153 - 1,2,3-Propanetriamine

C₃H₁₂BN
227 - Amminetrimethylboron
10500 - Trimethylamine borane

C₃N₂O
1758 - Carbonyl dicyanide

C₃O₂
1754 - Carbon suboxide

C₄Br₂F₈
2967 - 1,4-Dibromooctafluorobutane

C₄ClF₇
2059 - 1-Chloro-1,2,2,3,3,4,4-heptafluorocyclobutane

C₄ClF₇O
5662 - Heptafluorobutanoyl chloride

C₄Cl₂F₆
3191 - 1,2-Dichloro-1,2,3,3,4,4-hexafluorocyclobutane

C₄Cl₂O₄Rh₂
9475 - Rhodium carbonyl chloride

C₄Cl₃F₇
10240 - 2,2,3-Trichloro-1,1,1,3,4,4,4-heptafluorobutane

C₄Cl₄N₂
9772 - Tetrachloropyrimidine

C₄Cl₄S
9775 - Tetrachlorothiophene

C₄Cl₆
5729 - Hexachloro-1,3-butadiene

C₄Cl₆O₃
10197 - Trichloroacetic anhydride

C₄F₄O₃
9833 - 3,3,4,4-Tetrafluorodihydro-2,5-furandione

C₄F₆
5795 - 1,1,2,3,4,4-Hexafluoro-1,3-butadiene
5796 - 1,1,1,4,4,4-Hexafluoro-2-butyne
5797 - Hexafluorocyclobutene

C₄F₆O₃
10380 - Trifluoroacetic acid anhydride

C₄F₈
8697 - Perfluoro-2-butene
8699 - Perfluorocyclobutane
8713 - Perfluoroisobutene

C₄F₁₀
8696 - Perfluorobutane
8712 - Perfluoroisobutane

C₄HBr₃S
10176 - 2,3,5-Tribromothiophene

C₄HClO₃
2058 - 3-Chloro-2,5-furandione

C₄HCl₃N₂
10288 - 2,4,6-Trichloropyrimidine

C₄HCoO₄
2438 - Cobalt hydrocarbonyl

C₄HF₇O₂
5659 - Heptafluorobutanoic acid

C₄HI₄N
9906 - 2,3,4,5-Tetraiodo-1H-pyrrole

C₄H₂
1396 - 1,3-Butadiyne

C₄H₂Br₂S
2991 - 2,3-Dibromothiophene

2992 - 2,5-Dibromothiophene
2993 - 3,4-Dibromothiophene

$C_4H_2Cl_2N_2$
3285 - 3,6-Dichloropyridazine
3288 - 2,4-Dichloropyrimidine

$C_4H_2Cl_2O_2$
1447 - *trans*-2-Butenedioyl dichloride

$C_4H_2Cl_2S$
3298 - 2,5-Dichlorothiophene

$C_4H_2Cl_4O_3$
3068 - Dichloroacetic anhydride

$C_4H_2F_8O$
8707 - Perfluoroethyl 2,2,2-trifluoroethyl ether

$C_4H_2FeO_4$
6356 - Iron hydrocarbonyl

$C_4H_2N_2$
1444 - *trans*-2-Butenedinitrile
6661 - Maleonitrile

$C_4H_2N_2O_4$
9383 - 2,4,5,6(1H,3H)-Pyrimidinetetrone

$C_4H_2N_2S$
2521 - 4-Cyanothiazole

$C_4H_2O_3$
6660 - Maleic anhydride

$C_4H_2O_4$
1679 - 2-Butynedioic acid

$C_4H_3BrN_2O_2$
1349 - 5-Bromo-2,4(1H,3H)-pyrimidinedione

C_4H_3BrO
1214 - 2-Bromofuran
1215 - 3-Bromofuran

C_4H_3BrS
1356 - 2-Bromothiophene
1357 - 3-Bromothiophene

$C_4H_3ClO_2S_2$
10075 - 2-Thiophenesulfonyl chloride

C_4H_3ClS
2312 - 2-Chlorothiophene

$C_4H_3Cl_2N_3$
3287 - 4,6-Dichloro-2-pyrimidinamine

$C_4H_3FN_2O_2$
5432 - 5-Fluorouracil

$C_4H_3F_5O_2$
7550 - Methyl pentafluoropropanoate

$C_4H_3F_7O$
5661 - 2,2,3,3,4,4,4-Heptafluoro-1-butanol
8714 - Perfluoroisopropyl methyl ether
8726 - Perfluoropropyl methyl ether

$C_4H_3IN_2O_2$
6341 - 5-Iodo-2,4(1H,3H)-pyrimidinedione

C_4H_3IS
6344 - 2-Iodothiophene

$C_4H_3NO_2$
9396 - 1H-Pyrrole-2,5-dione

$C_4H_3NO_2S$
8205 - 2-Nitrothiophene

$C_4H_3NO_3$
8103 - 2-Nitrofuran

$C_4H_3N_3O_4$
8176 - 5-Nitro-2,4(1H,3H)-pyrimidinedione
9384 - 2,4,5,6(1H,3H)-Pyrimidinetetrone 5-oxime

$C_4H_3N_3O_5$
8177 - 5-Nitro-2,4,6(1H,3H,5H)-pyrimidinetrione

C_4H_4
1464 - 1-Buten-3-yne
7240 - Methylenecyclopropene

$C_4H_4BrNO_2$
1352 - *N*-Bromosuccinimide

$C_4H_4Br_2$
2921 - 1,4-Dibromo-2-butyne

$C_4H_4ClNO_2$
2304 - *N*-Chlorosuccinimide

$C_4H_4ClN_3$
2287 - 6-Chloro-3-pyridazinamine

$C_4H_4Cl_2$
3127 - 2,3-Dichloro-1,3-butadiene
3140 - 1,4-Dichloro-2-butyne

$C_4H_4Cl_2O_2$
1417 - Butanedioyl dichloride

$C_4H_4Cl_2O_3$
1841 - Chloroacetic anhydride

$C_4H_4FN_3O$
5382 - 5-Fluorocytosine

$C_4H_4INO_2$
6342 - 1-Iodo-2,5-pyrrolidinedione

$C_4H_4N_2$
9302 - Pyrazine
9312 - Pyridazine
9382 - Pyrimidine
9627 - Succinonitrile

$C_4H_4N_2OS$
3687 - 2,3-Dihydro-2-thioxo-4(1H)-pyrimidinone

$C_4H_4N_2O_2$
1677 - 2-Butynediamide
3674 - 1,2-Dihydro-3,6-pyridazinedione
10752 - Uracil

$C_4H_4N_2O_2S$
3686 - Dihydro-2-thioxo-4,6(1H,5H)-pyrimidinedione

$C_4H_4N_2O_3$
572 - Barbituric acid

$C_4H_4N_2O_5$
165 - Alloxanic acid

$C_4H_4N_2S_2$
4785 - 1,2-Ethanediol, dithiocyanate

$C_4H_4N_4$
416 - 3-Amino-1H-pyrazole-4-carbonitrile
2852 - *cis*-2,3-Diamino-2-butenedinitrile

$C_4H_4N_4O_2$
8175 - 5-Nitropyrimidinamine

$C_4H_4N_6O$
543 - 8-Azaguanine

$C_4H_4Na_2O_6$
9574 - Sodium tartrate

C_4H_4O
1686 - 3-Butyn-2-one
5478 - Furan

$C_4H_4O_2$
1682 - 2-Butynoic acid
3816 - Diketene

$C_4H_4O_3$
9625 - Succinic anhydride

$C_4H_4O_4$
4429 - 1,4-Dioxane-2,5-dione
4430 - 1,4-Dioxane-2,6-dione
5476 - Fumaric acid
6659 - Maleic acid

$C_4H_4O_5$
8431 - Oxaloacetic acid

$C_4H_4O_6$
3725 - 2,3-Dihydroxymaleic acid

C_4H_4S
10063 - Thiophene

C_4H_5Br
1120 - 2-Bromo-1,3-butadiene

$C_4H_5BrO_2$
1174 - 3-Bromo-4,5-dihydro-2(3H)-furanone

$C_4H_5BrO_4$
1123 - Bromobutanedioic acid, (±)

C_4H_5Cl
1921 - 4-Chloro-1,2-butadiene
1922 - 1-Chloro-1,3-butadiene
1923 - 2-Chloro-1,3-butadiene
1946 - 3-Chloro-1-butyne

C_4H_5ClO
1460 - 2-Butenoyl chloride
2690 - Cyclopropanecarbonyl chloride
7703 - 2-Methyl-2-propenoyl chloride

$C_4H_5ClO_2$
179 - Allyl chloroformate
7120 - Methyl 2-chloroacrylate

$C_4H_5ClO_3$
4946 - Ethyl 2-chloro-2-oxoacetate

$C_4H_5Cl_3$
10214 - 2,3,4-Trichloro-1-butene

$C_4H_5Cl_3O$
10213 - 2,2,3-Trichlorobutanal
10227 - 1,1,1-Trichloro-3,4-epoxybutane

$C_4H_5Cl_3O_2$
5268 - Ethyl trichloroacetate

$C_4H_5F_3O_2$
5269 - Ethyl trifluoroacetate

$C_4H_5F_5O$
8706 - Perfluoroethyl ethyl ether

C_4H_5N
1448 - *cis*-2-Butenenitrile
1449 - *trans*-2-Butenenitrile
1450 - 3-Butenenitrile
2689 - Cyclopropanecarbonitrile
6917 - 2-Methylacrylonitrile
9392 - Pyrrole

C_4H_5NO
193 - Allyl isocyanate
6013 - 2-Hydroxy-3-butenenitrile
7427 - 4-Methylisoxazole
7428 - 5-Methylisoxazole
7530 - 2-Methyloxazole
7531 - 4-Methyloxazole
7532 - 5-Methyloxazole

$C_4H_5NO_2$
4958 - Ethyl cyanoformate
7139 - Methyl cyanoacetate
9626 - Succinimide

$C_4H_5NO_3$
383 - *cis*-4-Amino-4-oxo-2-butenoic acid
6189 - 1-Hydroxy-2,5-pyrrolidinedione

$C_4H_5NO_4S$
13 - Acesulfame

C_4H_5NS
194 - Allyl isothiocyanate
7800 - 2-Methylthiazole
7801 - 4-Methylthiazole

$C_4H_5NS_2$
7803 - 4-Methyl-2(3H)-thiazolethione

$C_4H_5N_3$
6225 - Iminodiacetic acid, dinitrile
9380 - 2-Pyrimidinamine
9381 - 4-Pyrimidinamine

$C_4H_5N_3O$
2728 - Cytosine

$C_4H_5N_3O_2$
420 - 5-Amino-2,4(1H,3H)-pyrimidinedione
421 - 6-Amino-2,4(1H,3H)-pyrimidinedione
7493 - 2-Methyl-4-nitro-1H-imidazole
7829 - 6-Methyl-1,2,4-triazine-3,5(2H,4H)-dione

$C_4H_5N_3O_3$
423 - 5-Amino-2,4,6(1H,3H,5H)-pyrimidine-trione

$C_4H_5N_3S$
422 - 4-Amino-2(1H)-pyrimidinethione

C_4H_6
1392 - 1,2-Butadiene
1393 - 1,3-Butadiene
1675 - 1-Butyne
1676 - 2-Butyne
2536 - Cyclobutene

C_4H_6BrN
1124 - 4-Bromobutanenitrile

$C_4H_6Br_2$
2920 - trans-1,4-Dibromo-2-butene

$C_4H_6Br_2O$
1282 - 2-Bromo-2-methylpropanoyl bromide

$C_4H_6Br_2O_2$
2942 - 1,2-Dibromoethyl acetate
4974 - Ethyl dibromoacetate
7189 - Methyl 2,3-dibromopropanoate

$C_4H_6CaO_4S_2$
1702 - Calcium thioglycollate

$C_4H_6ClFO_2$
4944 - Ethyl chlorofluoroacetate

C_4H_6ClN
1927 - 4-Chlorobutanenitrile

$C_4H_6Cl_2$
1954 - 3-Chloro-2-(chloromethyl)-1-propene
3135 - 3,4-Dichloro-1-butene
3136 - cis-1,3-Dichloro-2-butene
3137 - trans-1,3-Dichloro-2-butene
3138 - cis-1,4-Dichloro-2-butene
3139 - trans-1,4-Dichloro-2-butene

$C_4H_6Cl_2O$
1934 - 4-Chlorobutanoyl chloride

$C_4H_6Cl_2O_2$
2282 - 3-Chloropropyl chloroformate
3170 - 2,3-Dichloro-1,4-dioxane
3180 - 1,2-Dichloroethyl acetate
4979 - Ethyl dichloroacetate
7194 - Methyl 2,3-dichloropropanoate

$C_4H_6F_2O_2$
4985 - Ethyl difluoroacetate

$C_4H_6MnN_2S_4$
6673 - Maneb

$C_4H_6N_2$
7379 - 1-Methylimidazol
7380 - 2-Methyl-1H-imidazole
7381 - 4-Methyl-1H-imidazole
7725 - 1-Methyl-1H-pyrazole
7726 - 3-Methyl-1H-pyrazole
7727 - 4-Methyl-1H-pyrazole

$C_4H_6N_2Na_2S_4$
7916 - Nabam

$C_4H_6N_2O$
7426 - 5-Methyl-3-isoxazolamine
7728 - 3-Methyl-2-pyrazolin-5-one

$C_4H_6N_2O_2$
3675 - Dihydro-2,4(1H,3H)-pyrimidinedione
4973 - Ethyl diazoacetate
7911 - Muscimol
9053 - 2,5-Piperazinedione

$C_4H_6N_2S$
3648 - 1,3-Dihydro-1-methyl-2H-imidazole-2-thione
4334 - 2,5-Dimethyl-1,3,4-thiadiazole
7799 - 4-Methyl-2-thiazolamine

$C_4H_6N_2S_4Zn$
10873 - Zinc N,N'-ethylenebisdithiocarbamate

$C_4H_6N_4O$
336 - 5-Amino-1H-imidazole-4-carboxamide

$C_4H_6N_4O_2$
9862 - Tetrahydroimidazo[4,5-d]imidazole-2,5(1H,3H)-dione

$C_4H_6N_4O_3$
157 - Allantoin

$C_4H_6N_4O_3S_2$
19 - Acetazolamide

$C_4H_6N_4O_{12}$
1424 - 1,2,3,4-Butanetetrol tetranitrate, (R^*,S^*)

$C_4H_6NiO_4$
8014 - Nickel(II) acetate

C_4H_6O
1440 - trans-2-Butenal
1459 - 3-Buten-2-one
1683 - 2-Butyn-1-ol
1684 - 3-Butyn-1-ol
1685 - 3-Butyn-2-ol
2535 - Cyclobutanone
2691 - Cyclopropanecarboxaldehyde
3624 - 2,3-Dihydrofuran
3625 - 2,5-Dihydrofuran
4605 - Divinyl ether
4820 - Ethoxyacetylene
6884 - 1-Methoxy-1,2-propadiene
6898 - 3-Methoxy-1-propyne
7696 - 2-Methylpropenal
10817 - Vinyloxirane

C_4H_6OS
3685 - Dihydro-2(3H)-thiophenone
10825 - Vinyl sulfoxide

$C_4H_6O_2$
188 - Allyl formate
875 - 2,2'-Bioxirane
1402 - Butanedial
1415 - 2,3-Butanedione
1451 - cis-2-Butenoic acid
1452 - trans-2-Butenoic acid
1453 - 3-Butenoic acid
1680 - 2-Butyne-1,4-diol
1688 - γ-Butyrolactone
2692 - Cyclopropanecarboxylic acid
3615 - 2,3-Dihydro-1,4-dioxin
6733 - Methacrylic acid
6916 - Methyl acrylate
7535 - 4-Methyl-2-oxetanone
10791 - Vinyl acetate

$C_4H_6O_2S$
3684 - 2,5-Dihydrothiophene 1,1-dioxide

4607 - Divinyl sulfone

$C_4H_6O_3$
23 - Acetic anhydride
25 - Acetoacetic acid
7541 - Methyl 2-oxopropanoate
8459 - 2-Oxobutanoic acid
8460 - 4-Oxobutanoic acid
9229 - Propylene carbonate

$C_4H_6O_4$
2827 - Diacetylperoxide
4201 - Dimethyl oxalate
4779 - 1,2-Ethanediol, diformate
7434 - Methylmalonic acid
9624 - Succinic acid

$C_4H_6O_4Mn$
6674 - Manganese(II) acetate

$C_4H_6O_4S$
10052 - Thiodiglycolic acid

$C_4H_6O_4S_2$
9622 - Succimer

$C_4H_6O_5$
3579 - Diglycolic acid
6662 - Malic acid

$C_4H_6O_6$
9678 - DL-Tartaric acid
9679 - meso-Tartaric acid
9680 - D-Tartaric acid
9681 - L-Tartaric acid

$C_4H_6O_6S$
9661 - Sulfonyldiacetic acid

$C_4H_6O_8$
3755 - Dihydroxytartaric acid

C_4H_6S
3682 - 2,3-Dihydrothiophene
3683 - 2,5-Dihydrothiophene
4606 - Divinyl sulfide

C_4H_7Br
1128 - cis-1-Bromo-1-butene
1129 - trans-1-Bromo-1-butene
1130 - 2-Bromo-1-butene
1131 - 4-Bromo-1-butene
1132 - 1-Bromo-2-butene
1133 - cis-2-Bromo-2-butene
1134 - trans-2-Bromo-2-butene
1283 - 1-Bromo-2-methylpropene
1284 - 3-Bromo-2-methylpropene

C_4H_7BrO
1127 - 3-Bromo-2-butanone

$C_4H_7BrO_2$
1125 - 2-Bromobutanoic acid, (±)
1126 - 4-Bromobutanoic acid
1199 - 2-Bromoethyl acetate
1281 - 2-Bromo-2-methylpropanoic acid
4908 - Ethyl bromoacetate
7058 - Methyl 3-bromopropanoate

$C_4H_7Br_2Cl_2O_4P$
7918 - Naled

$C_4H_7Br_3$
10157 - 1,1,2-Tribromobutane
10158 - 1,2,2-Tribromobutane
10159 - 1,2,3-Tribromobutane
10160 - 1,2,4-Tribromobutane
10161 - 2,2,3-Tribromobutane

$C_4H_7Br_3O$
10170 - 1,1,1-Tribromo-2-methyl-2-propanol

C_4H_7Cl
1935 - 2-Chloro-1-butene

1936 - 3-Chloro-1-butene
1937 - 4-Chloro-1-butene
1938 - cis-1-Chloro-2-butene
1939 - trans-1-Chloro-2-butene
1940 - cis-2-Chloro-2-butene
1941 - trans-2-Chloro-2-butene
2116 - (Chloromethyl)cyclopropane
2160 - 1-Chloro-2-methylpropene
2161 - 3-Chloro-2-methylpropene

C₄H₇ClO

1437 - Butanoyl chloride
1924 - 4-Chlorobutanal
1933 - 3-Chloro-2-butanone
2047 - 2-Chloroethyl vinyl ether
2133 - 2-(Chloromethyl)-2-methyloxirane
2157 - 2-Chloro-2-methylpropanal
7695 - 2-Methylpropanoyl chloride

C₄H₇ClOS

9223 - S-Propyl chlorothioformate

C₄H₇ClO₂

1928 - 2-Chlorobutanoic acid
1929 - 3-Chlorobutanoic acid
1930 - 4-Chlorobutanoic acid
2027 - 2-Chloroethyl acetate
4940 - Ethyl chloroacetate
6487 - Isopropyl chloroformate
7129 - Methyl 2-chloropropanoate
9221 - Propyl chlorocarbonate

C₄H₇Cl₂O₄P

3312 - Dichlorvos

C₄H₇Cl₃O

10258 - 1,1,1-Trichloro-2-methyl-2-propanol

C₄H₇Cl₃O₂

10232 - 2,2,2-Trichloro-1-ethoxyethanol

C₄H₇FO₂

5032 - Ethyl fluoroacetate

C₄H₇IO₂

5084 - Ethyl iodoacetate

C₄H₇N

1419 - Butanenitrile
3676 - 2,5-Dihydro-1H-pyrrole
6425 - 2-Isocyanopropane
7687 - 2-Methylpropanenitrile
7721 - N-Methyl-2-propyn-1-amine

C₄H₇NO

34 - Acetone cyanohydrin
6420 - 2-Isocyanatopropane
6887 - 3-Methoxypropanenitrile
7533 - 2-Methyl-2-oxazoline
7697 - 2-Methyl-2-propenamide
9245 - Propyl isocyanate
9402 - 2-Pyrrolidone

C₄H₇NO₂

42 - N-Acetylacetamide
177 - Allyl carbamate
548 - 2-Azetidinecarboxylic acid
1416 - 2,3-Butanedione monooxime
6104 - N-(Hydroxymethyl)-2-propenamide

C₄H₇NO₂S

10032 - 4-Thiazolidinecarboxylic acid

C₄H₇NO₃

71 - N-Acetylglycine

C₄H₇NO₄

524 - DL-Aspartic acid
525 - L-Aspartic acid
5167 - Ethyl nitroacetate
6224 - Iminodiacetic acid

C₄H₇NS

3655 - 4,5-Dihydro-2-methylthiazole
9246 - Propyl isothiocyanate

C₄H₇N₃O

2476 - Creatinine

C₄H₇N₃S

5254 - 5-Ethyl-1,3,4-thiadiazol-2-amine

C₄H₇N₅

9385 - 2,4,6-Pyrimidinetriamine

C₄H₇NaOS₂

9565 - Sodium O-isopropyl xanthate

C₄H₈

1441 - 1-Butene
1442 - cis-2-Butene
1443 - trans-2-Butene
2530 - Cyclobutane
6372 - Isobutene
7177 - Methylcyclopropane

C₄H₈BrCl

1141 - 1-Bromo-4-chlorobutane

C₄H₈Br₂

2915 - 1,1-Dibromobutane
2916 - 1,2-Dibromobutane
2917 - 1,3-Dibromobutane
2918 - 1,4-Dibromobutane
2919 - 2,3-Dibromobutane
2962 - 1,2-Dibromo-2-methylpropane

C₄H₈Br₂O

913 - Bis(2-bromoethyl) ether
2941 - 1,2-Dibromo-1-ethoxyethane

C₄H₈ClI

2080 - 1-Chloro-4-iodobutane

C₄H₈Cl₂

3128 - 1,1-Dichlorobutane
3129 - 1,2-Dichlorobutane
3130 - 1,3-Dichlorobutane
3131 - 1,4-Dichlorobutane
3132 - 2,2-Dichlorobutane
3133 - 2,3-Dichlorobutane, (±)
3217 - 1,2-Dichloro-2-methylpropane

C₄H₈Cl₂O

929 - Bis(2-chloroethyl) ether
3179 - 1,2-Dichloro-1-ethoxyethane

C₄H₈Cl₂O₂

3134 - 1,4-Dichloro-2,3-butanediol

C₄H₈Cl₂S

932 - Bis(2-chloroethyl) sulfide

C₄H₈Cl₂Si

199 - Allylmethyldichlorosilane

C₄H₈Cl₃O₄P

10193 - Trichlorfon

C₄H₈I₂

3763 - 1,4-Diiodobutane

C₄H₈N₂

3647 - 4,5-Dihydro-2-methyl-1H-imidazole
3895 - (Dimethylamino)acetonitrile
6936 - 3-(Methylamino)propanenitrile

C₄H₈N₂O

212 - Allylurea
8201 - N-Nitrosopyrrolidine

C₄H₈N₂O₂

4068 - Dimethylglyoxime
6930 - N-[(Methylamino)carbonyl]acetamide
8193 - 4-Nitrosomorpholine
9623 - Succinamide

C₄H₈N₂O₃

520 - L-Asparagine

4885 - Ethyl (aminocarbonyl)carbamate
5589 - N-Glycylglycine
6956 - Methylazoxymethanol acetate

C₄H₈N₂O₄

4381 - 1,4-Dinitrobutane

C₄H₈N₂O₇

3443 - Diethylene glycol dinitrate

C₄H₈N₂S

208 - Allylthiourea
9890 - 3,4,5,6-Tetrahydro-2(1H)-pyrimidineth-ione

C₄H₈N₂S₄

5016 - Ethylenebisdithiocarbamic acid

C₄H₈N₄O₂

4409 - 1,4-Dinitrosopiperazine

C₄H₈N₄O₄

156 - Allantoic acid

C₄H₈N₈O₈

2701 - Cyclotetramethylenetetranitramine

C₄H₈Na₂O₈

9575 - Sodium tartrate dihydrate

C₄H₈O

1398 - Butanal
1433 - 2-Butanone
1455 - cis-2-Buten-1-ol
1456 - trans-2-Buten-1-ol
1457 - 3-Buten-1-ol
1458 - 3-Buten-2-ol
2534 - Cyclobutanol
2694 - Cyclopropanemethanol
2699 - Cyclopropyl methyl ether
4205 - 2,2-Dimethyloxirane
4206 - cis-2,3-Dimethyloxirane
4207 - trans-2,3-Dimethyloxirane
4710 - 1,2-Epoxybutane
5282 - Ethyl vinyl ether
6370 - Isobutanal
6890 - 2-Methoxy-1-propene
6891 - 3-Methoxy-1-propene
7534 - 2-Methyloxetane
7702 - 2-Methyl-2-propenol
9854 - Tetrahydrofuran

C₄H₈OS

5255 - S-Ethyl thioacetate
7819 - 3-(Methylthio)propanal
8439 - 1,4-Oxathiane

C₄H₈O₂

1429 - Butanoic acid
1445 - cis-2-Butene-1,4-diol
1446 - trans-2-Butene-1,4-diol
4427 - 1,3-Dioxane
4428 - 1,4-Dioxane
4871 - Ethyl acetate
6006 - 3-Hydroxybutanal
6010 - 1-Hydroxy-2-butanone
6011 - 3-Hydroxy-2-butanone, (±)
6012 - 4-Hydroxy-2-butanone
6101 - 3-Hydroxy-2-methylpropanal
6496 - Isopropyl formate
6844 - (Methoxymethyl)oxirane
6889 - 1-Methoxy-2-propanone
7218 - 2-Methyl-1,3-dioxolane
7219 - 4-Methyl-1,3-dioxolane
7690 - Methyl propanoate
7691 - 2-Methylpropanoic acid
9237 - Propyl formate
9857 - Tetrahydro-3-furanol

10818 - 2-(Vinyloxy)ethanol

C₄H₈O₂S

5099 - Ethyl mercaptoacetate
5256 - (Ethylthio)acetic acid
6712 - 2-Mercapto-2-methylpropanoic acid
7436 - Methyl 3-mercaptopropanoate
7820 - 3-(Methylthio)propanoic acid
9658 - Sulfolane

C₄H₈O₃

4786 - 1,2-Ethanediol, monoacetate
4818 - Ethoxyacetic acid
5071 - Ethyl hydroxyacetate
5130 - Ethyl methyl carbonate
6007 - 2-Hydroxybutanoic acid, (±)
6008 - 3-Hydroxybutanoic acid, (±)
6009 - 4-Hydroxybutanoic acid
6102 - 2-Hydroxy-2-methylpropanoic acid
6103 - 3-Hydroxy-2-methylpropanoic acid
7429 - Methyl lactate, (±)
7440 - Methyl methoxyacetate

C₄H₈O₃S

1421 - 1,4-Butane sultone

C₄H₈O₄

3714 - 2,4-Dihydroxybutanoic acid
4749 - *D*-Erythrose
4750 - *L*-Erythrose
4753 - *L*-Erythrulose
10088 - *D*-Threose
10089 - *L*-Threose

C₄H₈S

7821 - 3-(Methylthio)-1-propene
9895 - Tetrahydrothiophene

C₄H₈S₂

4579 - 1,2-Dithiane
4580 - 1,3-Dithiane
4581 - 1,4-Dithiane

C₄H₉Br

1121 - 1-Bromobutane
1122 - 2-Bromobutane, (±)
1279 - 1-Bromo-2-methylpropane
1280 - 2-Bromo-2-methylpropane

C₄H₉BrO

1198 - 1-Bromo-2-ethoxyethane

C₄H₉BrO₂

1178 - 2-Bromo-1,1-dimethoxyethane

C₄H₉Cl

1925 - 1-Chlorobutane
1926 - 2-Chlorobutane
2158 - 1-Chloro-2-methylpropane
2159 - 2-Chloro-2-methylpropane

C₄H₉ClO

1589 - *tert*-Butyl hypochlorite
1931 - 4-Chloro-1-butanol
1932 - 1-Chloro-2-butanol
2025 - 1-Chloro-1-ethoxyethane
2040 - 2-Chloroethyl ethyl ether
2095 - 1-(Chloromethoxy)propane

C₄H₉ClO₂

1987 - 2-Chloro-1,1-dimethoxyethane
2026 - 2-(2-Chloroethoxy)ethanol

C₄H₉ClO₂S

1420 - 1-Butanesulfonyl chloride

C₄H₉ClS

2046 - 1-Chloro-2-(ethylthio)ethane

C₄H₉ClSi

2343 - Chlorovinyldimethylsilane

C₄H₉Cl₃Si

1667 - Butyltrichlorosilane
10244 - Trichloroisobutylsilane

C₄H₉Cl₃Sn

7890 - Monobutyltin trichloride

C₄H₉F

5378 - 1-Fluorobutane
5379 - 2-Fluorobutane
5402 - 2-Fluoro-2-methylpropane

C₄H₉I

6287 - 1-Iodobutane
6288 - 2-Iodobutane, (±)
6316 - 1-Iodo-2-methylpropane
6317 - 2-Iodo-2-methylpropane

C₄H₉N

2529 - Cyclobutanamine
7698 - *N*-Methyl-2-propen-1-amine
9397 - Pyrrolidine

C₄H₉NO

557 - 1-Aziridineethanol
1399 - Butanal oxime
1400 - Butanamide
1435 - 2-Butanone oxime
3889 - *N,N*-Dimethylacetamide
4870 - *N*-Ethylacetamide
7682 - 2-Methylpropanamide
7683 - *N*-Methylpropanamide
7901 - Morpholine

C₄H₉NO₂

68 - *N*-Acetylethanolamine
268 - *DL*-2-Aminobutanoic acid
269 - *L*-2-Aminobutanoic acid
270 - *DL*-3-Aminobutanoic acid
271 - 4-Aminobutanoic acid
353 - *L*-3-Amino-2-methylpropanoic acid
1623 - Butyl nitrite
1624 - *tert*-Butyl nitrite
1625 - *sec*-Butyl nitrite
4067 - *N,N*-Dimethylglycine
4881 - Ethyl 2-aminoacetate
5129 - Ethyl *N*-methylcarbamate
6014 - 4-Hydroxybutyramide
6398 - Isobutyl nitrite
6485 - Isopropyl carbamate
6918 - 2-Methylalanine
8088 - 1-Nitrobutane
8118 - 2-Nitroisobutane
9218 - Propyl carbamate

C₄H₉NO₂S

5923 - *DL*-Homocysteine
5924 - *L*-Homocysteine

C₄H₉NO₃

328 - 3-Amino-4-hydroxybutanoic acid
329 - 4-Amino-3-hydroxybutanoic acid, (±)
1622 - Butyl nitrate
5926 - *L*-Homoserine
6397 - Isobutyl nitrate
7501 - 2-Methyl-2-nitro-1-propanol
8089 - 2-Nitro-1-butanol
8090 - 3-Nitro-2-butanol
10087 - *L*-Threonine

C₄H₉NO₄

7500 - 2-Methyl-2-nitro-1,3-propanediol

C₄H₉NO₅

6094 - 2-(Hydroxymethyl)-2-nitro-1,3-propanediol

C₄H₉NS

4338 - *N,N*-Dimethylthioacetamide
10060 - Thiomorpholine

C₄H₉N₃O₂

2475 - Creatine
5613 - 3-Guanidinopropanoic acid

C₄H₉N₃S

37 - Acetone thiosemicarbazide

C₄H₉O₇P

4751 - *D*-Erythrose 4-phosphate

C₄H₁₀

1401 - Butane
6371 - Isobutane

C₄H₁₀AlCl

3418 - Diethylchloroaluminum

C₄H₁₀BF₃O

1070 - Boron trifluoride etherate

C₄H₁₀ClNO₂

5578 - Glycine, ethyl ester, hydrochloride

C₄H₁₀ClNO₂S

7180 - Methyl *L*-cysteine hydrochloride

C₄H₁₀ClOPS

5031 - *O*-Ethyl ethylthiophosphonyl chloride

C₄H₁₀ClO₂PS

3417 - *O,O*-Diethyl chloridothionophosphate

C₄H₁₀ClO₃P

3420 - Diethyl chlorophosphonate

C₄H₁₀Cl₂Si

3155 - Dichlorodiethylsilane

C₄H₁₀Cl₃N

1947 - 2-Chloro-*N*-(2-chloroethyl)ethanamine, hydrochloride

C₄H₁₀FO₂P

7669 - Methylphosphonofluoridic acid, isopropyl ester

C₄H₁₀Hg

3478 - Diethyl mercury

C₄H₁₀NO₃PS

11 - Acephate

C₄H₁₀N₂

9049 - Piperazine

C₄H₁₀N₂O

308 - *N*-(2-Aminoethyl)acetamide
567 - Azoxyethane
6545 - Isopropylurea
7900 - 4-Morpholinamine
8186 - *N*-Nitrosodiethylamine
9269 - Propylurea
10621 - Trimethylurea

C₄H₁₀N₂O₂

2851 - 2,4-Diaminobutanoic acid
8094 - *N*-Nitrodiethylamine

C₄H₁₀N₂O₃

8185 - *N*-Nitrosodiethanolamine

C₄H₁₀N₂O₄

521 - *D*-Asparagine, monohydrate
522 - *L*-Asparagine, monohydrate

C₄H₁₀N₂S

10619 - Trimethylthiourea

C₄H₁₀O

1431 - 1-Butanol
1432 - 2-Butanol
3454 - Diethyl ether
6514 - Isopropyl methyl ether
7693 - 2-Methyl-1-propanol
7694 - 2-Methyl-2-propanol

7716 - Methyl propyl ether

$C_4H_{10}OS$

3532 - Diethyl sulfoxide

5259 - 2-(Ethylthio)ethanol

$C_4H_{10}OS_2$

3823 - 2,2'-Dimercaptodiethyl ether

$C_4H_{10}O_2$

1405 - 1,2-Butanediol, (±)

1406 - 1,3-Butanediol

1407 - 1,4-Butanediol

1408 - 2,3-Butanediol

1584 - tert-Butyl hydroperoxide

3502 - Diethylperoxide

3863 - 1,2-Dimethoxyethane

3888 - Dimethylacetal

4839 - 2-Ethoxyethanol

6888 - 2-Methoxy-1-propanol

7685 - 2-Methyl-1,2-propanediol

7686 - 2-Methyl-1,3-propanediol

9234 - 1,2-Propylene glycol monomethyl ether

$C_4H_{10}O_2S$

1005 - Bis(2-hydroxyethyl) sulfide

3531 - Diethyl sulfone

$C_4H_{10}O_2S_2$

997 - Bis(2-hydroxyethyl) disulfide

3822 - 1,4-Dimercapto-2,3-butanediol

$C_4H_{10}O_3$

1427 - 1,2,4-Butanetriol

3435 - Diethylene glycol

6886 - 3-Methoxy-1,2-propanediol

10491 - Trimethoxymethane

$C_4H_{10}O_3S$

3530 - Diethyl sulfite

5250 - 2-(Ethylsulfonyl)ethanol

6505 - Isopropyl methanesulfonate

$C_4H_{10}O_4$

1423 - 1,2,3,4-Butanetetrol

$C_4H_{10}O_4S$

3528 - Diethyl sulfate

$C_4H_{10}O_8U$

10754 - Uranyl acetate dihydrate

$C_4H_{10}S$

1425 - 1-Butanethiol

1426 - 2-Butanethiol

3529 - Diethyl sulfide

6517 - Isopropyl methyl sulfide

7688 - 2-Methyl-1-propanethiol

7689 - 2-Methyl-2-propanethiol

7720 - Methyl propyl sulfide

$C_4H_{10}S_2$

1418 - 1,4-Butanedithiol

3433 - Diethyl disulfide

7715 - Methyl propyl disulfide

$C_4H_{10}S_3$

1018 - Bis(2-mercaptoethyl) sulfide

3539 - Diethyltrisulfide

$C_4H_{10}Se$

3522 - Diethyl selenide

$C_4H_{10}Te$

3534 - Diethyl telluride

$C_4H_{10}Zn$

3543 - Diethyl zinc

$C_4H_{11}As$

3399 - Diethylarsine

$C_4H_{11}BrSi$

1286 - (Bromomethyl)trimethylsilane

$C_4H_{11}ClN_2O_6$

385 - (Aminooxy)acetic acid, hydrochloride (2:1)

$C_4H_{11}ClSi$

2039 - Chloroethyldimethylsilane

2168 - (Chloromethyl)trimethylsilane

$C_4H_{11}Cl_2N$

1996 - 2-Chloro-N,N-dimethylethanamine, hydrochloride

$C_4H_{11}N$

1494 - Butylamine

1495 - sec-Butylamine

1496 - tert-Butylamine

3378 - Diethylamine

4993 - Ethyldimethylamine

6376 - Isobutylamine

6506 - Isopropylmethylamine

7709 - Methylpropylamine

$C_4H_{11}NO$

272 - 2-Amino-1-butanol, (±)

273 - 4-Amino-1-butanol

354 - 2-Amino-2-methyl-1-propanol

4057 - N,N-Dimethylethanolamine

4838 - 2-Ethoxyethanamine

4886 - 2-(Ethylamino)ethanol

5077 - N-Ethyl-N-hydroxyethanamine

6835 - 3-Methoxyisopropylamine

6885 - 3-Methoxy-1-propanamine

$C_4H_{11}NO_2$

307 - 2-(2-Aminoethoxy)ethanol

352 - 2-Amino-2-methyl-1,3-propanediol

3348 - Diethanolamine

3862 - 2,2-Dimethoxyethanamine

$C_4H_{11}NO_3$

10693 - Tris(hydroxymethyl)methylamine

$C_4H_{11}NO_8P_2$

5601 - Glyphosine

$C_4H_{11}O_2PS_2$

3510 - O,O'-Diethyl phosphorodithionate

$C_4H_{11}O_3P$

3509 - Diethyl phosphonate

$C_4H_{11}O_4P$

3467 - Diethyl hydrogen phosphate

$C_4H_{11}P$

3508 - Diethylphosphine

$C_4H_{12}As_2$

9942 - Tetramethyldiarsine

$C_4H_{12}BN$

3906 - (Dimethylamino)dimethylborane

$C_4H_{12}BrN$

9917 - Tetramethylammonium bromide

$C_4H_{12}ClN$

1497 - Butylamine hydrochloride

3379 - Diethylamine hydrochloride

9918 - Tetramethylammonium chloride

$C_4H_{12}ClO_4P$

9910 - Tetrakis(hydroxymethyl)phosphonium chloride

$C_4H_{12}Cl_2N_2$

9052 - Piperazine dihydrochloride

$C_4H_{12}Cl_2OSi_2$

3297 - 1,3-Dichloro-1,1,3,3-tetramethyldisiloxane

$C_4H_{12}Cl_2Si_2$

3296 - 1,2-Dichloro-1,1,2,2-tetramethyldisilane

$C_4H_{12}CrN_7OS_4$

9452 - Reinecke salt

$C_4H_{12}FN_2OP$

3819 - Dimefox

$C_4H_{12}Ge$

9947 - Tetramethylgermane

$C_4H_{12}IN$

9919 - Tetramethylammonium iodide

$C_4H_{12}N_2$

1403 - 1,4-Butanediamine

3465 - 1,2-Diethylhydrazine

4055 - N,N-Dimethyl-1,2-ethanediamine

4056 - N,N'-Dimethyl-1,2-ethanediamine

5023 - N-Ethyl-1,2-ethanediamine

7684 - 2-Methyl-1,2-propanediamine

$C_4H_{12}N_2O$

314 - N-(2-Aminoethyl)ethanolamine

$C_4H_{12}N_2S$

10047 - 2,2'-Thiobisethanamine

$C_4H_{12}OSi$

4836 - Ethoxydimethylsilane

$C_4H_{12}O_2Si$

3859 - Dimethoxydimethylsilane

$C_4H_{12}O_3Si$

10492 - Trimethoxymethylsilane

$C_4H_{12}O_4Si$

9974 - Tetramethyl silicate

$C_4H_{12}Pb$

9956 - Tetramethyl lead

$C_4H_{12}Si$

3523 - Diethylsilane

9973 - Tetramethylsilane

$C_4H_{12}Sn$

9975 - Tetramethylstannane

$C_4H_{13}ClN_2$

1582 - tert-Butylhydrazine hydrochloride

$C_4H_{13}N_3$

903 - Bis(2-aminoethyl)amine

$C_4H_{14}Cl_2N_2$

1404 - 1,4-Butanediamine dihydrochloride

$C_4H_{14}Cl_2N_2S_2$

2712 - Cystamine dihydrochloride

$C_4H_{14}OSi_2$

9944 - 1,1,3,3-Tetramethyldisiloxane

$C_4H_{14}O_3Si_2$

9945 - 1,1,3,3-Tetramethyl-1,3-disiloxanediol

$C_4H_{16}Cl_2CuN_4$

978 - Bis(ethylenediamine)copper dichloride

$C_4H_{16}O_4Si_4$

9939 - 2,4,6,8-Tetramethylcyclotetrasiloxane

C_4N_2

1678 - 2-Butynedinitrile

C_4NiO_4

8017 - Nickel carbonyl

$C_5Cl_2F_6$

3192 - 1,2-Dichloro-3,3,4,4,5,5-hexafluorocyclopentene

C_5Cl_5N

8544 - Pentachloropyridine

C_5Cl_6

5734 - Hexachloro-1,3-cyclopentadiene

C_5Cl_8

8292 - Octachlorocyclopentene

C_5F_5N

8727 - Perfluoropyridine

C_5F_8
8591 - 1,1,2,4,4-Pentafluoro-3-(trifluoromethyl)-1,3-butadiene

C_5F_{12}
8723 - Perfluoropentane

C_5FeO_5
6359 - Iron pentacarbonyl

C_5HCl_4N
9771 - 2,3,5,6-Tetrachloropyridine

$C_5H_2ClN_3O_4$
2012 - 2-Chloro-3,5-dinitropyridine

$C_5H_2F_6O_2$
5793 - Hexafluoroacetylacetone

$C_5H_3BrO_2$
1216 - 5-Bromo-2-furancarboxaldehyde

$C_5H_3Br_2N$
2986 - 3,5-Dibromopyridine

$C_5H_3ClN_2O_2$
2196 - 2-Chloro-3-nitropyridine

$C_5H_3ClN_4$
2286 - 6-Chloro-1H-purine

C_5H_3ClOS
2313 - 5-Chloro-2-thiophenecarboxaldehyde
10068 - 2-Thiophenecarbonyl chloride

$C_5H_3ClO_2$
5481 - 2-Furancarbonyl chloride

$C_5H_3Cl_2N$
3286 - 2,6-Dichloropyridine

$C_5H_3F_7O_2$
7282 - Methyl heptafluorobutanoate

C_5H_3NO
5480 - 2-Furancarbonitrile

$C_5H_3NO_4$
8104 - 5-Nitro-2-furancarboxaldehyde

$C_5H_3NO_5$
8105 - 5-Nitro-2-furancarboxylic acid

C_5H_3NS
10066 - 2-Thiophenecarbonitrile
10067 - 3-Thiophenecarbonitrile

C_5H_4
8570 - 1,3-Pentadiyne

C_5H_4BrN
1346 - 2-Bromopyridine
1347 - 3-Bromopyridine
1348 - 4-Bromopyridine

C_5H_4ClN
2289 - 2-Chloropyridine
2290 - 3-Chloropyridine
2291 - 4-Chloropyridine

$C_5H_4Cl_2N_2$
3218 - 2,4-Dichloro-5-methylpyrimidine
3219 - 2,4-Dichloro-6-methylpyrimidine

C_5H_4FN
5422 - 2-Fluoropyridine
5423 - 3-Fluoropyridine

$C_5H_4F_8O$
8331 - 2,2,3,3,4,4,5,5-Octafluoro-1-pentanol

C_5H_4IN
6340 - 2-Iodopyridine

$C_5H_4N_2O_2$
8173 - 4-Nitropyridine
9304 - Pyrazinecarboxylic acid

$C_5H_4N_2O_3$
8174 - 4-Nitropyridine 1-oxide

$C_5H_4N_2O_4$
6216 - 1H-Imidazole-4,5-dicarboxylic acid

8418 - Orotic acid

$C_5H_4N_4$
9293 - 1H-Purine

$C_5H_4N_4O$
163 - Allopurinol
6206 - Hypoxanthine

$C_5H_4N_4O_2$
10843 - Xanthine

$C_5H_4N_4O_3$
10759 - Uric acid

$C_5H_4N_4S$
3669 - 1,7-Dihydro-6H-purine-6-thione

C_5H_4OS
9300 - 4H-Pyran-4-thione
10069 - 2-Thiophenecarboxaldehyde
10070 - 3-Thiophenecarboxaldehyde

$C_5H_4O_2$
5482 - 3-Furancarboxaldehyde
5502 - Furfural
7243 - 5-Methylene-2(5H)-furanone
9297 - 2H-Pyran-2-one
9298 - 4H-Pyran-4-one

$C_5H_4O_2S$
10071 - 2-Thiophenecarboxylic acid
10072 - 3-Thiophenecarboxylic acid

$C_5H_4O_3$
3641 - Dihydro-3-methylene-2,5-furandione
5483 - 2-Furancarboxylic acid
5484 - 3-Furancarboxylic acid
7270 - 3-Methyl-2,5-furandione

$C_5H_4S_2$
10077 - 4H-Thiopyran-4-thione

$C_5H_5ClN_2$
2288 - 5-Chloro-2-pyridinamine

C_5H_5ClO
2125 - 2-(Chloromethyl)furan

$C_5H_5Cl_2N$
2294 - 4-Chloropyridine, hydrochloride

$C_5H_5Cl_3N_2OS$
9705 - Terrazole

$C_5H_5F_3O_2$
10415 - 1,1,1-Trifluoro-2,4-pentanedione

C_5H_5N
9316 - Pyridine

C_5H_5NO
9355 - Pyridine-1-oxide
9361 - 2-Pyridinol
9362 - 3-Pyridinol
9363 - 4-Pyridinol
9364 - 2(1H)-Pyridinone
9393 - 1H-Pyrrole-2-carboxaldehyde

C_5H_5NOS
9359 - 2-Pyridinethiol, 1-oxide

$C_5H_5NO_2$
6188 - 3-Hydroxy-1H-pyridin-2-one
7140 - Methyl 2-cyanoacrylate
9394 - 1H-Pyrrole-2-carboxylic acid
9395 - 1H-Pyrrole-3-carboxylic acid

$C_5H_5NO_3S$
9358 - 3-Pyridinesulfonic acid

C_5H_5NS
9360 - 2(1H)-Pyridinethione

$C_5H_5N_3O$
9303 - Pyrazinecarboxamide

$C_5H_5N_3O_2$
8172 - 5-Nitro-2-pyridinamine

$C_5H_5N_3O_3S$
8203 - N-(5-Nitro-2-thiazolyl)acetamide

$C_5H_5N_5$
121 - Adenine

$C_5H_5N_5O$
302 - 6-Amino-1,3-dihydro-2H-purin-2-one
5614 - Guanine

$C_5H_5N_5S$
301 - 2-Amino-1,7-dihydro-6H-purine-6-
thione

C_5H_6
2658 - 1,3-Cyclopentadiene
7104 - 2-Methyl-1-buten-3-yne
8655 - 1-Penten-3-yne
8656 - 1-Penten-4-yne
8657 - cis-3-Penten-1-yne
8658 - trans-3-Penten-1-yne

$C_5H_6BrClN_2O_2$
1143 - 3-Bromo-1-chloro-5,5-dimethylhydan-
toin

$C_5H_6Br_2N_2O_2$
2934 - 1,3-Dibromo-5,5-dimethyl-2,4-imida-
zolidinedione

C_5H_6ClN
9348 - Pyridine hydrochloride

$C_5H_6Cl_2N_2O_2$
3167 - 1,3-Dichloro-5,5-dimethyl hydantoin

$C_5H_6Cl_2O_2$
8619 - Pentanedioyl dichloride

$C_5H_6Cl_6N_2O_3$
1046 - N,N'-Bis(2,2,2-trichloro-1-hydroxy-
ethyl)urea

$C_5H_6N_2$
7724 - 2-Methylpyrazine
7731 - 3-Methylpyridazine
7750 - 2-Methylpyrimidine
7751 - 4-Methylpyrimidine
7752 - 5-Methylpyrimidine
8610 - Pentanedinitrile
9313 - 2-Pyridinamine
9314 - 3-Pyridinamine
9315 - 4-Pyridinamine

$C_5H_6N_2O$
73 - 1-Acetyl-1H-imidazole

$C_5H_6N_2OS$
7823 - Methylthiouracil

$C_5H_6N_2O_2$
7753 - 6-Methyl-2,4(1H,3H)-pyrimidinedione
10094 - Thymine

$C_5H_6N_6$
9294 - 1H-Purine-2,6-diamine

C_5H_6O
2677 - 2-Cyclopenten-1-one
2678 - 3-Cyclopenten-1-one
6816 - 1-Methoxy-1-buten-3-yne
7266 - 2-Methylfuran
7267 - 3-Methylfuran
9296 - 4H-Pyran

C_5H_6OS
5488 - 2-Furanmethanethiol
10074 - 2-Thiophenemethanol

$C_5H_6O_2$
3642 - Dihydro-3-methylene-2(3H)-furanone
5229 - Ethyl 2-propynoate
5503 - Furfuryl alcohol
6830 - 2-Methoxyfuran
7274 - 5-Methyl-2(3H)-furanone

C$_5$H$_9$ClO$_2$
1534 - Butyl chloroformate
2004 - 3-Chloro-2,2-dimethylpropanoic acid
2211 - 5-Chloropentanoic acid
4947 - Ethyl 2-chloropropanoate
4948 - Ethyl 3-chloropropanoate
6382 - Isobutyl chlorocarbonate
6486 - Isopropyl chloroacetate
7124 - Methyl 4-chlorobutanoate
9219 - Propyl chloroacetate

C$_5$H$_9$Cl$_2$N$_3$O$_2$
931 - N,N'-Bis(2-chloroethyl)-N-nitrosourea

C$_5$H$_9$I
6290 - Iodocyclopentane

C$_5$H$_9$N
1593 - Butyl isocyanide
3924 - 3-(Dimethylamino)-1-propyne
4275 - 2,2-Dimethylpropanenitrile
7067 - 2-Methylbutanenitrile
7068 - 3-Methylbutanenitrile
7113 - 2-Methyl-3-butyn-2-amine
8620 - Pentanenitrile
9888 - 1,2,5,6-Tetrahydropyridine

C$_5$H$_9$NO
1592 - Butyl isocyanate
2671 - Cyclopentanone oxime
4280 - N,N-Dimethyl-2-propenamide
4866 - 3-Ethoxypropanenitrile
6392 - Isobutyl isocyanate
6415 - 2-Isocyanato-2-methylpropane
7760 - 5-Methyl-2-pyrrolidinone
7763 - N-Methyl-2-pyrrolidone
9072 - 2-Piperidinone

C$_5$H$_9$NO$_2$
80 - N-Acetyl-N-methylacetamide
6929 - Methyl 3-amino-2-butenoate
7902 - 4-Morpholinecarboxaldehyde
9123 - DL-Proline
9124 - L-Proline

C$_5$H$_9$NO$_2$S$_2$
1808 - Cheirolin

C$_5$H$_9$NO$_3$
43 - N-Acetyl-L-alanine
192 - Allyl (hydroxymethyl)carbamate
384 - 5-Amino-4-oxopentanoic acid
6177 - cis-4-Hydroxy-L-proline
6178 - trans-4-Hydroxy-L-proline

C$_5$H$_9$NO$_3$S
64 - N-Acetyl-L-cysteine

C$_5$H$_9$NO$_4$
5174 - Ethyl 2-nitropropanoate
5552 - DL-Glutamic acid
5553 - D-Glutamic acid
5554 - L-Glutamic acid
7382 - N-Methyliminodiacetic acid

C$_5$H$_9$NO$_4$S
1767 - S-(Carboxymethyl)-L-cysteine

C$_5$H$_9$NS
1595 - Butyl isothiocyanate
1596 - sec-Butyl isothiocyanate, (±)
1597 - tert-Butyl isothiocyanate
1661 - Butyl thiocyanate
6393 - Isobutyl isothiocyanate
6403 - Isobutyl thiocyanate
7759 - N-Methyl-2-pyrrolidinethione

C$_5$H$_9$N$_3$
5917 - Histamine

C$_5$H$_9$N$_3$O$_2$
1510 - tert-Butyl azidoformate

C$_5$H$_9$N$_3$O$_9$
10691 - 1,1,1-Tris(hydroxymethyl)ethane trinitrate

C$_5$H$_{10}$
2659 - Cyclopentane
4036 - 1,1-Dimethylcyclopropane
4037 - cis-1,2-Dimethylcyclopropane
4038 - trans-1,2-Dimethylcyclopropane
4970 - Ethylcyclopropane
7087 - 2-Methyl-1-butene
7088 - 3-Methyl-1-butene
7089 - 2-Methyl-2-butene
7141 - Methylcyclobutane
8639 - 1-Pentene
8640 - cis-2-Pentene
8641 - trans-2-Pentene

C$_5$H$_{10}$Br$_2$
2935 - 1,3-Dibromo-2,2-dimethylpropane
2960 - 2,3-Dibromo-2-methylbutane
2969 - 1,2-Dibromopentane
2970 - 1,4-Dibromopentane
2971 - 1,5-Dibromopentane
2972 - 2,4-Dibromopentane

C$_5$H$_{10}$Br$_2$O$_2$
917 - 2,2-Bis(bromomethyl)-1,3-propanediol

C$_5$H$_{10}$ClNO
3414 - Diethylcarbamic chloride

C$_5$H$_{10}$ClNO$_4$
5555 - L-Glutamic acid, hydrochloride

C$_5$H$_{10}$Cl$_2$
3210 - 2,3-Dichloro-2-methylbutane
3246 - 1,2-Dichloropentane
3247 - 1,5-Dichloropentane
3248 - 2,3-Dichloropentane

C$_5$H$_{10}$Cl$_2$O$_2$
925 - Bis(2-chloroethoxy)methane
939 - 2,2-Bis(chloromethyl)-1,3-propanediol

C$_5$H$_{10}$HgO$_3$
6828 - Methoxyethylmercuric acetate

C$_5$H$_{10}$I$_2$
3770 - 1,5-Diiodopentane

C$_5$H$_{10}$NNaS$_2$
9551 - Sodium diethyldithiocarbamate

C$_5$H$_{10}$N$_2$
3421 - Diethylcyanamide
3920 - 3-(Dimethylamino)propanenitrile

C$_5$H$_{10}$N$_2$O
8199 - N-Nitrosopiperidine
9050 - 1-Piperazinecarboxaldehyde

C$_5$H$_{10}$N$_2$O$_2$
8162 - N-Nitropiperidine

C$_5$H$_{10}$N$_2$O$_2$S
6762 - Methomyl

C$_5$H$_{10}$N$_2$O$_3$
1009 - 1,3-Bis(hydroxymethyl)-2-imidazolidone
5556 - L-Glutamine
5587 - Glycylalanine

C$_5$H$_{10}$N$_2$O$_4$
5593 - N-Glycylserine, (DL)

C$_5$H$_{10}$N$_2$O$_5$
4200 - Dimethyloldihydroxyethyleneurea

C$_5$H$_{10}$N$_2$S$_2$
2737 - Dazomet

C$_5$H$_{10}$N$_6$O$_2$
4408 - Dinitrosopentamethylenetetramine

C$_5$H$_{10}$O
187 - Allyl ethyl ether
2669 - Cyclopentanol
4203 - 3,3-Dimethyloxetane
4269 - 2,2-Dimethylpropanal
6546 - Isopropyl vinyl ether
7061 - 3-Methylbutanal
7082 - 3-Methyl-2-butanone
7097 - 3-Methyl-2-buten-1-ol
7098 - 3-Methyl-3-buten-1-ol
7099 - 2-Methyl-3-buten-2-ol
7100 - 3-Methyl-3-buten-2-ol
7179 - α-Methylcyclopropanemethanol
7797 - 2-Methyltetrahydrofuran
8604 - Pentanal
8629 - 2-Pentanone
8630 - 3-Pentanone
8646 - 1-Penten-3-ol
8647 - cis-2-Penten-1-ol
8648 - trans-2-Penten-1-ol
8649 - 3-Penten-2-ol, (±)
8650 - 4-Penten-1-ol
8651 - 4-Penten-2-ol
9270 - Propyl vinyl ether
9884 - Tetrahydropyran

C$_5$H$_{10}$OS
9599 - S-Propyl thioacetate

C$_5$H$_{10}$O$_2$
200 - 2-(Allyloxy)ethanol
1576 - Butyl formate
1577 - sec-Butyl formate
1578 - tert-Butyl formate
2664 - cis-1,2-Cyclopentanediol
2665 - trans-1,2-Cyclopentanediol
3884 - 3,3-Dimethoxy-1-propene
4277 - 2,2-Dimethylpropanoic acid
4432 - 1,3-Dioxepane
4851 - (Ethoxymethyl)oxirane
4865 - 3-Ethoxypropanal
5224 - Ethyl propanoate
6032 - 3-Hydroxy-2,2-dimethylpropanal
6086 - 3-Hydroxy-3-methyl-2-butanone
6141 - 5-Hydroxy-2-pentanone
6388 - Isobutyl formate
6469 - Isopropyl acetate
6823 - (2-Methoxyethoxy)ethene
7073 - Methyl butanoate
7074 - 2-Methylbutanoic acid
7075 - 3-Methylbutanoic acid
7217 - 4-Methyl-1,3-dioxane
7395 - Methyl isobutanoate
8624 - Pentanoic acid
9203 - Propyl acetate
9860 - Tetrahydrofurfuryl alcohol

C$_5$H$_{10}$O$_2$S
7461 - Methyl 3-(methylthio)propanoate
7791 - 3-Methyl sulfolane

C$_5$H$_{10}$O$_3$
3416 - Diethyl carbonate
5096 - Ethyl lactate
6085 - 3-Hydroxy-3-methylbutanoic acid
6140 - 2-Hydroxypentanoic acid
6825 - 2-Methoxyethyl acetate
7376 - Methyl 2-hydroxy-2-methylpropanoate
7445 - Methyl 3-methoxypropanoate

$C_5H_{10}O_3S$
6109 - 2-Hydroxy-4-(methylthio)butanoic acid

$C_5H_{10}O_4$
2808 - *D*-2-Deoxyribose
6049 - 3-Hydroxy-2-(hydroxymethyl)-2-meth-ylpropanoic acid
9155 - 1,2,3-Propanetriol-1-acetate

$C_5H_{10}O_5$
505 - α-*D*-Arabinopyranose
507 - *DL*-Arabinose
508 - α-*D*-Arabinose
509 - β-*D*-Arabinose
6650 - *D*-Lyxose
6651 - *L*-Lyxose
9480 - *D*-Ribose
9481 - *L*-Ribose
9482 - *D*-Ribulose
10863 - *D*-Xylose
10864 - *D*-Xylulose
10865 - *L*-Xylulose

$C_5H_{10}S$
2668 - Cyclopentanethiol
10021 - Thiacyclohexane

$C_5H_{11}Br$
1186 - 1-Bromo-2,2-dimethylpropane
1251 - 1-Bromo-2-methylbutane, *DL*
1252 - 1-Bromo-3-methylbutane
1253 - 2-Bromo-2-methylbutane
1305 - 1-Bromopentane
1306 - 2-Bromopentane
1307 - 3-Bromopentane

$C_5H_{11}Cl$
2003 - 1-Chloro-2,2-dimethylpropane
2111 - 1-Chloro-3-methylbutane
2112 - 2-Chloro-2-methylbutane
2113 - 2-Chloro-3-methylbutane
2208 - 1-Chloropentane
2209 - 2-Chloropentane, (+)
2210 - 3-Chloropentane

$C_5H_{11}ClO$
2212 - 5-Chloro-1-pentanol

$C_5H_{11}ClSi$
178 - Allylchlorodimethylsilane

$C_5H_{11}Cl_2N$
1949 - 2-Chloro-*N*-(2-chloroethyl)-*N*-methyle-thanamine

$C_5H_{11}Cl_2N_3$
6217 - 1*H*-Imidazole-4-ethanamine, dihydro-chloride

$C_5H_{11}Cl_3Si$
10266 - Trichloropentylsilane

$C_5H_{11}F$
5409 - 1-Fluoropentane

$C_5H_{11}I$
6295 - 1-Iodo-2,2-dimethylpropane
6314 - 1-Iodo-3-methylbutane
6315 - 2-Iodo-2-methylbutane
6328 - 1-Iodopentane
6329 - 3-Iodopentane

$C_5H_{11}N$
186 - Allyldimethylamine
2682 - Cyclopentylamine
7757 - *N*-Methylpyrrolidine
9058 - Piperidine

$C_5H_{11}NO$
1575 - *N-tert*-Butylformamide
3460 - *N,N*-Diethylformamide

4270 - 2,2-Dimethylpropanamide
4271 - *N,N*-Dimethylpropanamide
6171 - *N*-Hydroxypiperidine
7062 - 3-Methylbutanamide
7468 - 4-Methylmorpholine
8605 - Pentanamide
8631 - 2-Pentanone oxime
9855 - Tetrahydro-2-furanmethanamine

$C_5H_{11}NO_2$
387 - 5-Aminopentanoic acid
845 - Betaine
1530 - Butyl carbamate
5028 - Ethyl ethylcarbamate
6381 - Isobutyl carbamate
6451 - Isopentyl nitrite
6558 - *L*-Isovaline
8129 - 1-Nitropentane
8130 - 3-Nitropentane
8281 - *DL*-Norvaline
8282 - *L*-Norvaline
8678 - Pentyl nitrite
10768 - *L*-Valine

$C_5H_{11}NO_2S$
6718 - 3-Mercapto-*D*-valine
6760 - *L*-Methionine

$C_5H_{11}NO_2Se$
9521 - Selenium methionine

$C_5H_{11}NO_3$
7112 - 3-Methylbutyl nitrate

$C_5H_{11}NO_4$
5173 - 2-Ethyl-2-nitro-1,3-propanediol

$C_5H_{11}N_2O_2P$
9672 - Tabun

C_5H_{12}
6441 - Isopentane
8006 - Neopentane
8607 - Pentane

$C_5H_{12}ClN$
9067 - Piperidine, hydrochloride

$C_5H_{12}ClNO_2$
846 - Betaine, hydrochloride

$C_5H_{12}ClNO_2S$
2716 - *L*-Cysteine, ethyl ester, hydrochloride

$C_5H_{12}ClO_2PS_2$
1834 - Chlormephos

$C_5H_{12}Cl_3N$
930 - Bis(2-chloroethyl)methylamine hydro-chloride

$C_5H_{12}Cl_3NO$
8108 - Nitrogen mustard *N*-oxide hydrochlo-ride

$C_5H_{12}NO_3PS_2$
2705 - Cygon

$C_5H_{12}NO_4PS$
8412 - Omethoate

$C_5H_{12}N_2$
5803 - Hexahydro-1*H*-1,4-diazepine
7671 - 1-Methylpiperazine
7672 - 2-Methylpiperazine
9057 - 1-Piperidinamine

$C_5H_{12}N_2O$
1669 - Butylurea
1670 - *sec*-Butylurea
1671 - *tert*-Butylurea
3540 - *N,N*-Diethylurea
3541 - *N,N'*-Diethylurea
9980 - Tetramethylurea

$C_5H_{12}N_2O_2$
8416 - *L*-Ornithine

$C_5H_{12}N_2S$
3537 - *N,N'*-Diethylthiourea
9979 - Tetramethylthiourea

$C_5H_{12}N_4O_3$
337 - *O*-[(Aminoiminomethyl)amino]-*L*-homoserine

$C_5H_{12}O$
1611 - Butyl methyl ether
1612 - *sec*-Butyl methyl ether
4278 - 2,2-Dimethyl-1-propanol
5092 - Ethyl isopropyl ether
5225 - Ethyl propyl ether
6396 - Isobutyl methyl ether
7077 - 2-Methyl-1-butanol, (±)
7078 - 3-Methyl-1-butanol
7079 - 2-Methyl-2-butanol
7080 - 3-Methyl-2-butanol, (±)
7111 - Methyl *tert*-butyl ether
8626 - 1-Pentanol
8627 - 2-Pentanol
8628 - 3-Pentanol

$C_5H_{12}O_2$
3362 - Diethoxymethane
3882 - 1,1-Dimethoxypropane
3883 - 2,2-Dimethoxypropane
4274 - 2,2-Dimethyl-1,3-propanediol
4846 - 1-Ethoxy-2-methoxyethane
6467 - 2-Isopropoxyethanol
6815 - 3-Methoxy-1-butanol
7066 - 3-Methyl-1,3-butanediol
8611 - 1,2-Pentanediol, (±)
8612 - 1,4-Pentanediol
8613 - 1,5-Pentanediol
8614 - 2,3-Pentanediol
8615 - 2,4-Pentanediol
9197 - 2-Propoxyethanol

$C_5H_{12}O_3$
3450 - Diethylene glycol monomethyl ether
6091 - 2-(Hydroxymethyl)-2-methyl-1,3-pro-panediol
10489 - 1,1,1-Trimethoxyethane

$C_5H_{12}O_4$
8571 - Pentaerythritol
9913 - Tetramethoxymethane

$C_5H_{12}O_5$
504 - *L*-Arabinitol
9477 - Ribitol
10861 - Xylitol

$C_5H_{12}O_7P_2$
3892 - 3,3-Dimethylallyl diphosphate

$C_5H_{12}S$
1617 - Butyl methyl sulfide
1618 - *tert*-Butyl methyl sulfide
4276 - 2,2-Dimethyl-1-propanethiol
5094 - Ethyl isopropyl sulfide
5228 - Ethyl propyl sulfide
7069 - 2-Methyl-1-butanethiol, (+)
7070 - 3-Methyl-1-butanethiol
7071 - 2-Methyl-2-butanethiol
7072 - 3-Methyl-2-butanethiol
8621 - 1-Pentanethiol
8622 - 2-Pentanethiol
8623 - 3-Pentanethiol

$C_5H_{12}S_4$
9911 - Tetrakis(methylthio)methane

$C_5H_{12}Si$
10828 - Vinyltrimethylsilane
$C_5H_{13}ClN_2O_2$
8417 - L-Ornithine, monohydrochloride
$C_5H_{13}Cl_2N$
1835 - Chlormequat chloride
2002 - 1-Chloro-N,N-dimethyl-2-propan-amine, hydrochloride
$C_5H_{13}N$
1605 - Butylmethylamine
3479 - Diethylmethylamine
4272 - N,N-Dimethyl-1-propanamine
4281 - 2,2-Dimethylpropylamine
5090 - Ethylisopropylamine
5222 - N-Ethyl-1-propanamine
7063 - 3-Methyl-1-butanamine
7064 - 2-Methyl-2-butanamine
7065 - 3-Methyl-2-butanamine
8606 - 3-Pentanamine
8664 - Pentylamine
$C_5H_{13}NO$
388 - 5-Amino-1-pentanol
3921 - 2-(Dimethylamino)-1-propanol
3922 - 3-(Dimethylamino)-1-propanol
3923 - 1-(Dimethylamino)-2-propanol
6473 - 2-(Isopropylamino)ethanol
8011 - Neurine
9208 - 2-(Propylamino)ethanol
$C_5H_{13}NO_2$
317 - 2-Amino-2-ethyl-1,3-propanediol
1002 - Bis(2-hydroxyethyl)methylamine
3874 - N-(Dimethoxymethyl)dimethylamine
3875 - 2,2-Dimethoxy-N-methylethanamine
$C_5H_{13}N_3$
9948 - 1,1,3,3-Tetramethylguanidine
$C_5H_{13}O_3P$
3488 - Diethyl methylphosphonate
$C_5H_{14}ClNO$
2378 - Choline chloride
$C_5H_{14}N_2$
4273 - N,N-Dimethyl-1,3-propanediamine
8609 - 1,5-Pentanediamine
9957 - N,N,N',N'-Tetramethylmethanediamine
$C_5H_{14}N_2O$
310 - 1-[(2-Aminoethyl)amino]-2-propanol
$C_5H_{14}OSi$
4868 - Ethoxytrimethylsilane
$C_5H_{14}O_2Si$
3365 - Diethoxymethylsilane
$C_5H_{14}O_3Si$
5273 - Ethyltrimethoxysilane
$C_5H_{14}Pb$
5276 - Ethyltrimethyllead
$C_5H_{15}ClNO_4P$
2379 - Choline chloride dihydrogen phosphate
$C_5H_{15}NSi$
8603 - Pentamethylsilanamine
$C_5H_{15}N_3$
316 - N-(2-Aminoethyl)-1,3-propanediamine
$C_5H_{16}NNaO_3S_2$
9552 - Sodium diethyldithiocarbamate trihydrate
$C_5H_{20}O_5Si_5$
8596 - 2,4,6,8,10-Pentamethylcyclopentasiloxane

C_6BrF_5
1303 - Bromopentafluorobenzene
C_6Br_6
5717 - Hexabromobenzene
C_6ClF_5
2206 - Chloropentafluorobenzene
$C_6Cl_2F_4$
3293 - 1,2-Dichloro-3,4,5,6-tetrafluorobenzene
$C_6Cl_3F_3$
10293 - 1,3,5-Trichloro-2,4,6-trifluorobenzene
$C_6Cl_4O_2$
9741 - 2,3,5,6-Tetrachloro-2,5-cyclohexadiene-1,4-dione
9742 - 3,4,5,6-Tetrachloro-3,5-cyclohexadiene-1,2-dione
$C_6Cl_5NO_2$
8540 - Pentachloronitrobenzene
C_6Cl_6
5724 - Hexachlorobenzene
C_6CrO_6
2381 - Chromium carbonyl
C_6F_5I
8585 - Pentafluoroiodobenzene
C_6F_6
5794 - Hexafluorobenzene
C_6F_{10}
8701 - Perfluorocyclohexene
C_6F_{12}
8700 - Perfluorocyclohexane
8711 - Perfluoro-1-hexene
C_6F_{14}
8705 - Perfluoro-2,3-dimethylbutane
8710 - Perfluorohexane
8716 - Perfluoro-2-methylpentane
8717 - Perfluoro-3-methylpentane
C_6HBr_5O
8527 - Pentabromophenol
$C_6HCl_4NO_2$
9760 - 1,2,3,4-Tetrachloro-5-nitrobenzene
9761 - 1,2,4,5-Tetrachloro-3-nitrobenzene
C_6HCl_5
8533 - Pentachlorobenzene
C_6HCl_5O
8541 - Pentachlorophenol
C_6HCl_5S
8534 - Pentachlorobenzenethiol
C_6HF_5
8580 - Pentafluorobenzene
C_6HF_5O
8587 - Pentafluorophenol
C_6HF_5S
8581 - Pentafluorobenzenethiol
C_6HF_{11}
10726 - Undecafluorocyclohexane
$C_6H_2Br_2ClNO$
2988 - 2,6-Dibromoquinone-4-chlorimide
$C_6H_2Br_4$
9710 - 1,2,4,5-Tetrabromobenzene
$C_6H_2ClN_3O_6$
2337 - 2-Chloro-1,3,5-trinitrobenzene
$C_6H_2Cl_2O_2$
3146 - 2,5-Dichloro-2,5-cyclohexadiene-1,4-dione
3147 - 2,6-Dichloro-2,5-cyclohexadiene-1,4-dione

$C_6H_2Cl_2O_4$
1818 - Chloranilic acid
$C_6H_2Cl_3NO$
3141 - 2,6-Dichloro-4-(chloroimino)-2,5-cyclohexadien-1-one
$C_6H_2Cl_3NO_2$
10260 - 1,2,4-Trichloro-5-nitrobenzene
$C_6H_2Cl_3NO_3$
10262 - 3,4,6-Trichloro-2-nitrophenol
$C_6H_2Cl_4$
9732 - 1,2,3,4-Tetrachlorobenzene
9733 - 1,2,3,5-Tetrachlorobenzene
9734 - 1,2,4,5-Tetrachlorobenzene
$C_6H_2Cl_4O$
9762 - 2,3,4,5-Tetrachlorophenol
9763 - 2,3,4,6-Tetrachlorophenol
9764 - 2,3,5,6-Tetrachlorophenol
$C_6H_2Cl_4O_2$
9735 - 3,4,5,6-Tetrachloro-1,2-benzenediol
9736 - 2,3,5,6-Tetrachloro-1,4-benzenediol
$C_6H_2Cl_5N$
8531 - 2,3,4,5,6-Pentachloroaniline
$C_6H_2F_2N_2O_4$
3556 - 1,5-Difluoro-2,4-dinitrobenzene
$C_6H_2F_4$
9830 - 1,2,3,4-Tetrafluorobenzene
9831 - 1,2,3,5-Tetrafluorobenzene
9832 - 1,2,4,5-Tetrafluorobenzene
$C_6H_2F_5N$
8578 - 2,3,4,5,6-Pentafluoroaniline
$C_6H_2N_2O_8$
8028 - Nitranilic acid
$C_6H_2N_3NaO_7$
10632 - 2,4,6-Trinitrophenol, sodium salt
$C_6H_3BrCl_2$
1166 - 1-Bromo-3,5-dichlorobenzene
1167 - 4-Bromo-1,2-dichlorobenzene
$C_6H_3BrCl_2O$
1170 - 4-Bromo-2,5-dichlorophenol
$C_6H_3BrN_2O_4$
1188 - 1-Bromo-2,4-dinitrobenzene
$C_6H_3Br_2NO_3$
2965 - 2,6-Dibromo-4-nitrophenol
$C_6H_3Br_3$
10155 - 1,2,4-Tribromobenzene
10156 - 1,3,5-Tribromobenzene
$C_6H_3Br_3O$
10172 - 2,4,6-Tribromophenol
$C_6H_3ClFNO_2$
2056 - 2-Chloro-1-fluoro-4-nitrobenzene
$C_6H_3ClN_2O_4$
2008 - 1-Chloro-2,4-dinitrobenzene
2009 - 2-Chloro-1,3-dinitrobenzene
$C_6H_3ClN_2O_4S$
4373 - 2,4-Dinitrobenzenesulfenyl chloride
$C_6H_3ClN_2O_5$
2011 - 4-Chloro-2,6-dinitrophenol
$C_6H_3Cl_2NO_2$
2433 - Clopyralid
3234 - 1,2-Dichloro-3-nitrobenzene
3235 - 1,2-Dichloro-4-nitrobenzene
3236 - 1,3-Dichloro-5-nitrobenzene
3237 - 1,4-Dichloro-2-nitrobenzene
3238 - 2,4-Dichloro-1-nitrobenzene
$C_6H_3Cl_2NO_3$
3240 - 2,6-Dichloro-4-nitrophenol

$C_6H_5AsCl_2$
3259 - Dichlorophenylarsine

$C_6H_5AsI_2$
8817 - Phenylarsonous diiodide

C_6H_5AsO
8472 - Oxophenylarsine

$C_6H_5BO_2$
689 - 1,3,2-Benzodioxaborole

C_6H_5Br
1096 - Bromobenzene

$C_6H_5BrN_2O_2$
1291 - 4-Bromo-2-nitroaniline

C_6H_5BrO
1312 - 2-Bromophenol
1313 - 3-Bromophenol
1314 - 4-Bromophenol

C_6H_5BrOS
1355 - 1-(5-Bromo-2-thienyl)ethanone

$C_6H_5BrO_2$
1100 - 2-Bromo-1,4-benzenediol

C_6H_5BrS
1102 - 4-Bromobenzenethiol

$C_6H_5Br_2N$
2905 - 2,4-Dibromoaniline
2906 - 3,5-Dibromoaniline

C_6H_5Cl
1865 - Chlorobenzene

C_6H_5ClFN
2048 - 3-Chloro-4-fluoroaniline

C_6H_5ClHg
8913 - Phenylmercuric chloride

C_6H_5ClMg
8912 - Phenylmagnesium chloride

$C_6H_5ClNNaO_2S$
1814 - Chloramine B

$C_6H_5ClN_2O_2$
2173 - 2-Chloro-4-nitroaniline
2174 - 2-Chloro-5-nitroaniline
2175 - 4-Chloro-2-nitroaniline
2176 - 4-Chloro-3-nitroaniline
2177 - 5-Chloro-2-nitroaniline

$C_6H_5ClN_2O_4S$
2185 - 4-Chloro-3-nitrobenzenesulfonamide

C_6H_5ClO
2217 - 2-Chlorophenol
2218 - 3-Chlorophenol
2219 - 4-Chlorophenol

C_6H_5ClOS
660 - Benzenesulfinyl chloride

$C_6H_5ClO_2$
1877 - 3-Chloro-1,2-benzenediol
1878 - 4-Chloro-1,2-benzenediol
1879 - 4-Chloro-1,3-benzenediol
1880 - 2-Chloro-1,4-benzenediol

$C_6H_5ClO_2S$
663 - Benzenesulfonyl chloride

$C_6H_5ClO_3S$
1889 - 4-Chlorobenzenesulfonic acid

C_6H_5ClS
1891 - 2-Chlorobenzenethiol
1892 - 3-Chlorobenzenethiol
1893 - 4-Chlorobenzenethiol

$C_6H_5Cl_2N$
3075 - 2,3-Dichloroaniline
3076 - 2,4-Dichloroaniline
3077 - 2,5-Dichloroaniline

3078 - 2,6-Dichloroaniline
3079 - 3,4-Dichloroaniline
3080 - 3,5-Dichloroaniline

$C_6H_5Cl_2NO$
297 - 2-Amino-4,6-dichlorophenol
298 - 4-Amino-2,6-dichlorophenol

$C_6H_5Cl_2NO_2S$
3104 - N,N-Dichlorobenzenesulfonamide

$C_6H_5Cl_2OP$
8950 - Phenylphosphonic dichloride

$C_6H_5Cl_2O_2P$
8953 - Phenyl phosphorodichloridate

$C_6H_5Cl_2P$
8952 - Phenylphosphonous dichloride

$C_6H_5Cl_2PS$
8951 - Phenylphosphonothioic dichloride

$C_6H_5Cl_3N_2$
10276 - (2,4,6-Trichlorophenyl)hydrazine

$C_6H_5Cl_3Si$
10277 - Trichlorophenylsilane

C_6H_5F
5361 - Fluorobenzene

C_6H_5FO
5410 - 2-Fluorophenol
5411 - 3-Fluorophenol
5412 - 4-Fluorophenol

$C_6H_5FO_2S$
664 - Benzenesulfonyl fluoride

$C_6H_5F_2N$
3548 - 2,4-Difluoroaniline

$C_6H_5F_3Si$
10418 - Trifluorophenylsilane

$C_6H_5F_7O_2$
5039 - Ethyl heptafluorobutanoate

$C_6H_5HgNO_3$
8914 - Phenylmercuric nitrate

C_6H_5I
6275 - Iodobenzene

C_6H_5IO
6330 - 2-Iodophenol
6331 - 3-Iodophenol
6332 - 4-Iodophenol
6343 - Iodosylbenzene

$C_6H_5I_2N$
3759 - 2,4-Diiodoaniline

$C_6H_5K_3O_7$
10665 - Tripotassium citrate

C_6H_5NO
8183 - Nitrosobenzene
9322 - 2-Pyridinecarboxaldehyde
9323 - 3-Pyridinecarboxaldehyde
9324 - 4-Pyridinecarboxaldehyde

C_6H_5NOS
9654 - N-Sulfinylaniline

$C_6H_5NO_2$
8047 - Nitrobenzene
8197 - 4-Nitrosophenol
9329 - 2-Pyridinecarboxylic acid
9330 - 3-Pyridinecarboxylic acid
9331 - 4-Pyridinecarboxylic acid

$C_6H_5NO_3$
3660 - 1,6-Dihydro-6-oxo-3-pyridinecarboxylic acid
8132 - 2-Nitrophenol
8133 - 3-Nitrophenol
8134 - 4-Nitrophenol

9332 - 3-Pyridinecarboxylic acid 1-oxide
9333 - 4-Pyridinecarboxylic acid 1-oxide

$C_6H_5NO_4$
2413 - Citrazinic acid

$C_6H_5NO_5S$
8067 - 3-Nitrobenzenesulfonic acid
8068 - 4-Nitrobenzenesulfonic acid

C_6H_5NS
10065 - 2-Thiopheneacetonitrile

$C_6H_5N_3$
550 - Azidobenzene
745 - 1H-Benzotriazole

$C_6H_5N_3O$
6000 - 1-Hydroxy-1H-benzotriazole

$C_6H_5N_3O_4$
4360 - 2,3-Dinitroaniline
4361 - 2,4-Dinitroaniline
4362 - 2,5-Dinitroaniline
4363 - 2,6-Dinitroaniline
4364 - 3,5-Dinitroaniline

$C_6H_5N_3O_5$
304 - 2-Amino-4,6-dinitrophenol

$C_6H_5N_5O$
335 - 2-Amino-4-hydroxypteridine

$C_6H_5N_5O_2$
10845 - Xanthopterin

C_6H_5NaO
9571 - Sodium phenolate

$C_6H_5NaO_2S$
9546 - Sodium benzenesulfinate

$C_6H_5NaO_3S$
9547 - Sodium benzenesulfonate

$C_6H_5Na_3O_7$
10700 - Trisodium citrate

C_6H_6
595 - Benzene
5475 - Fulvene
5786 - 1,5-Hexadien-3-yne
5787 - 1,5-Hexadiyne
5788 - 2,4-Hexadiyne

$C_6H_6AsNO_5$
8144 - (4-Nitrophenyl)arsonic acid

$C_6H_6AsNO_6$
8111 - 3-Nitro-4-hydroxyphenylarsonic acid

$C_6H_6BNO_4$
8147 - (3-Nitrophenyl)boronic acid

C_6H_6BrN
1087 - 2-Bromoaniline
1088 - 3-Bromoaniline
1089 - 4-Bromoaniline

$C_6H_6Br_2N_2$
1118 - 2-Bromo-2-(bromomethyl)pentanedinitrile

$C_6H_6Br_5Cl$
8525 - 1,2,3,4,5-Pentabromo-6-chlorocyclohexane

C_6H_6ClN
1850 - 2-Chloroaniline
1851 - 3-Chloroaniline
1852 - 4-Chloroaniline

C_6H_6ClNO
289 - 2-Amino-4-chlorophenol

$C_6H_6ClNO_2S$
1887 - 2-Chlorobenzenesulfonamide
1888 - 4-Chlorobenzenesulfonamide

$C_6H_6ClNO_3S$
284 - 5-Amino-2-chlorobenzenesulfonic acid

$C_6H_6ClN_3O_2$
2184 - 5-Chloro-3-nitro-1,2-benzenediamine

$C_6H_6Cl_2N_2$
3095 - 2,5-Dichloro-1,4-benzenediamine
3096 - 2,6-Dichloro-1,4-benzenediamine

$C_6H_6Cl_2N_2O_4S_2$
3101 - 4,5-Dichloro-1,3-benzenedisulfonamide

$C_6H_6Cl_2Si$
3269 - Dichlorophenylsilane

$C_6H_6Cl_6$
5730 - 1,2,3,4,5,6-Hexachlorocyclohexane, (1α,2α,3β,4α,5α,6β)
5731 - 1,2,3,4,5,6-Hexachlorocyclohexane, (1α,2α,3β,4α,5β,6β)
5732 - 1,2,3,4,5,6-Hexachlorocyclohexane, (1α,2β,3α,4β,5α,6β)
5733 - 1,2,3,4,5,6-Hexachlorocyclohexane, (1α,2α,3α,4β,5α,6β)

C_6H_6FN
5355 - 2-Fluoroaniline
5356 - 3-Fluoroaniline
5357 - 4-Fluoroaniline

$C_6H_6FNO_2S$
254 - 4-Aminobenzenesulfonyl fluoride

$C_6H_6FeNO_6$
6358 - Iron(III)NTA

C_6H_6IN
6270 - 2-Iodoaniline
6271 - 3-Iodoaniline
6272 - 4-Iodoaniline

$C_6H_6NO_6P$
8155 - 4-Nitrophenyl phosphate

$C_6H_6N_2$
5867 - trans-3-Hexenedinitrile
7247 - 2-Methylenepentanedinitrile

$C_6H_6N_2O$
4850 - (Ethoxymethylene)propanedinitrile
9325 - 2-Pyridinecarboxaldehyde oxime
9326 - 2-Pyridinecarboxamide
9327 - 3-Pyridinecarboxamide
9328 - 4-Pyridinecarboxamide

$C_6H_6N_2O_2$
417 - 2-Amino-3-pyridinecarboxylic acid
418 - 6-Amino-3-pyridinecarboxylic acid
2564 - 2,5-Cyclohexadiene-1,4-dione, dioxime
8034 - 2-Nitroaniline
8035 - 3-Nitroaniline
8036 - 4-Nitroaniline
10762 - Urocanic acid

$C_6H_6N_2O_3$
378 - 2-Amino-4-nitrophenol
379 - 2-Amino-5-nitrophenol
380 - 4-Amino-2-nitrophenol

$C_6H_6N_2O_4S$
8066 - 4-Nitrobenzenesulfonamide

$C_6H_6N_2S$
9320 - 3-Pyridinecarbothioamide
9321 - 4-Pyridinecarbothioamide

$C_6H_6N_4$
8031 - 2,2',2"-Nitrilotriacetonitrile

$C_6H_6N_4O$
7830 - 5-Methyl-[1,2,4]triazolo[1,5-a]pyrimidin-7-ol

$C_6H_6N_4O_4$
4401 - (2,4-Dinitrophenyl)hydrazine

8107 - Nitrofurazone

$C_6H_6Na_2O_7$
4574 - Disodium hydrogen citrate

C_6H_6O
8762 - Phenol
10805 - 2-Vinylfuran

C_6H_6OS
6714 - 2-Mercaptophenol
6715 - 4-Mercaptophenol
7817 - 5-Methyl-2-thiophenecarboxaldehyde
10036 - 1-(2-Thienyl)ethanone

$C_6H_6O_2$
5492 - 1-(2-Furanyl)ethanone
5958 - p-Hydroquinone
7268 - 5-Methyl-2-furancarboxaldehyde
9388 - Pyrocatechol
9458 - Resorcinol

$C_6H_6O_2S$
659 - Benzenesulfinic acid
10064 - 2-Thiopheneacetic acid

$C_6H_6O_2Se$
657 - Benzeneseleninic acid

$C_6H_6O_3$
673 - 1,2,3-Benzenetriol
674 - 1,2,4-Benzenetriol
675 - 1,3,5-Benzenetriol
4064 - 3,4-Dimethyl-2,5-furandione
5479 - 2-Furanacetic acid
6088 - 5-(Hydroxymethyl)-2-furancarboxaldehyde
6105 - 4-Hydroxy-6-methyl-2H-pyran-2-one
6106 - 3-Hydroxy-2-methyl-4H-pyran-4-one
7269 - Methyl 2-furancarboxylate

$C_6H_6O_3S$
662 - Benzenesulfonic acid

$C_6H_6O_4$
4000 - Dimethyl 2-butynedioate
6050 - 5-Hydroxy-2-(hydroxymethyl)-4H-pyran-4-one

$C_6H_6O_4S$
5988 - 3-Hydroxybenzenesulfonic acid
5989 - 4-Hydroxybenzenesulfonic acid

$C_6H_6O_6$
9182 - cis-1-Propene-1,2,3-tricarboxylic acid
9183 - trans-1-Propene-1,2,3-tricarboxylic acid

$C_6H_6O_6S_2$
636 - 1,3-Benzenedisulfonic acid

$C_6H_6O_7S_2$
5981 - 4-Hydroxy-1,3-benzenedisulfonic acid

C_6H_6S
666 - Benzenethiol

$C_6H_6S_2$
638 - 1,2-Benzenedithiol
639 - 1,3-Benzenedithiol

C_6H_6Se
658 - Benzeneselenol

$C_6H_7AsNNaO_3$
9544 - Sodium arsanilate

$C_6H_7AsO_3$
603 - Benzenearsonic acid

$C_6H_7BO_2$
604 - Benzeneboronic acid

$C_6H_7BrN_2$
1321 - (4-Bromophenyl)hydrazine

$C_6H_7ClN_2$
1874 - 4-Chloro-1,2-benzenediamine

1875 - 4-Chloro-1,3-benzenediamine
1876 - 2-Chloro-1,4-benzenediamine

C_6H_7ClO
5785 - trans,trans-2,4-Hexadienoyl chloride

C_6H_7ClSi
2252 - Chlorophenylsilane

$C_6H_7Cl_2N$
1853 - 2-Chloroaniline hydrochloride
1854 - 3-Chloroaniline hydrochloride
2162 - 3-(Chloromethyl)pyridine, hydrochloride

$C_6H_7F_3O_3$
5270 - Ethyl 4,4,4-trifluoroacetoacetate

$C_6H_7KO_2$
9096 - Potassium trans,trans-2,4-hexadienoate

C_6H_7N
466 - Aniline
2673 - 1-Cyclopentenecarbonitrile
7738 - 2-Methylpyridine
7739 - 3-Methylpyridine
7740 - 4-Methylpyridine

C_6H_7NO
389 - 2-Aminophenol
390 - 3-Aminophenol
391 - 4-Aminophenol
6901 - 2-Methoxypyridine
6902 - 3-Methoxypyridine
6903 - 4-Methoxypyridine
7744 - 2-Methylpyridine-1-oxide
7745 - 3-Methylpyridine-1-oxide
7746 - 4-Methylpyridine-1-oxide
7747 - 1-Methyl-2(1H)-pyridinone
8894 - Phenylhydroxylamine
9352 - 2-Pyridinemethanol
9353 - 3-Pyridinemethanol
9354 - 4-Pyridinemethanol
9403 - 1-(1H-Pyrrol-2-yl)ethanone

$C_6H_7NO_2$
2510 - 2-Cyanoethyl acrylate
4955 - Ethyl 2-cyanoacrylate
5240 - 1-Ethyl-1H-pyrrole-2,5-dione

$C_6H_7NO_2S$
661 - Benzenesulfonamide

$C_6H_7NO_3S$
251 - 2-Aminobenzenesulfonic acid
252 - 3-Aminobenzenesulfonic acid
253 - 4-Aminobenzenesulfonic acid

$C_6H_7NO_4$
6843 - 2-(Methoxymethyl)-5-nitrofuran

$C_6H_7NO_4S$
323 - 3-Amino-4-hydroxybenzenesulfonic acid

C_6H_7NS
255 - 2-Aminobenzenethiol
256 - 4-Aminobenzenethiol

$C_6H_7N_3O$
6440 - Isoniazid

$C_6H_7N_3O_2$
8053 - 4-Nitro-1,2-benzenediamine
8054 - 4-Nitro-1,3-benzenediamine
8055 - 5-Nitro-1,3-benzenediamine
8056 - 2-Nitro-1,4-benzenediamine
8152 - (4-Nitrophenyl)hydrazine

$C_6H_7N_5O$
299 - 2-Amino-1,7-dihydro-7-methyl-6H-purin-6-one

$C_6H_7NaO_6$
9545 - Sodium ascorbate

$C_6H_7O_2P$
8948 - Phenylphosphinic acid
$C_6H_7O_3P$
8949 - Phenylphosphonic acid
$C_6H_7O_3Sb$
3745 - Dihydroxyphenylstibine oxide
C_6H_7P
8947 - Phenylphosphine
C_6H_8
2561 - 1,3-Cyclohexadiene
2562 - 1,4-Cyclohexadiene
5857 - *cis*-1,3,5-Hexatriene
5858 - *trans*-1,3,5-Hexatriene
7166 - 1-Methyl-1,3-cyclopentadiene
7592 - 3-Methyl-3-penten-1-yne
$C_6H_8AsNO_3$
394 - (4-Aminophenyl)arsonic acid
C_6H_8BrN
470 - Aniline hydrobromide
C_6H_8ClN
471 - Aniline hydrochloride
C_6H_8ClNS
2043 - 5-(2-Chloroethyl)-4-methylthiazole
$C_6H_8ClN_3O_4S_2$
283 - 4-Amino-6-chloro-1,3-benzenedisulfonamide
$C_6H_8Cl_2O_2$
5842 - Hexanedioyl dichloride
$C_6H_8Cl_2O_5$
3436 - Diethylene glycol, bischloroformate
$C_6H_8N_2$
133 - Adiponitrile
611 - 1,2-Benzenediamine
612 - 1,3-Benzenediamine
613 - 1,4-Benzenediamine
4290 - 2,3-Dimethylpyrazine
4291 - 2,5-Dimethylpyrazine
4292 - 2,6-Dimethylpyrazine
4308 - 4,6-Dimethylpyrimidine
5230 - 2-Ethylpyrazine
7554 - 2-Methylpentanedinitrile
7732 - 3-Methyl-2-pyridinamine
7733 - 4-Methyl-2-pyridinamine
7734 - 5-Methyl-2-pyridinamine
7735 - 6-Methyl-2-pyridinamine
7736 - *N*-Methyl-2-pyridinamine
7737 - *N*-Methyl-4-pyridinamine
8890 - Phenylhydrazine
9349 - 2-Pyridinemethanamine
9350 - 3-Pyridinemethanamine
9351 - 4-Pyridinemethanamine
$C_6H_8N_2O$
951 - Bis(2-cyanoethyl) ether
2861 - 2,4-Diaminophenol
6900 - 6-Methoxy-3-pyridinamine
$C_6H_8N_2O_2$
4309 - 1,3-Dimethyl-2,4(1*H*,3*H*)-pyrimidinedione
7825 - 1-Methylthymine
$C_6H_8N_2O_2S$
250 - 4-Aminobenzenesulfonamide
$C_6H_8N_2O_3$
472 - Aniline nitrate
$C_6H_8N_2O_3S$
5939 - 4-Hydrazinobenzenesulfonic acid
$C_6H_8N_2O_8$
6554 - Isosorbide dinitrate

$C_6H_8N_2S$
952 - Bis(2-cyanoethyl) sulfide
$C_6H_8N_6O_{18}$
6679 - *D*-Mannitol hexanitrate
C_6H_8O
2599 - 2-Cyclohexen-1-one
2674 - 1-Cyclopentene-1-carboxaldehyde
4063 - 2,5-Dimethylfuran
5036 - 2-Ethylfuran
5773 - *trans,trans*-2,4-Hexadienal
5916 - 5-Hexyn-2-one
7175 - 2-Methyl-2-cyclopenten-1-one
7176 - 3-Methyl-2-cyclopenten-1-one
$C_6H_8O_2$
169 - Allyl acrylate
1394 - 1,3-Butadien-1-ol acetate
2578 - 1,2-Cyclohexanedione
2579 - 1,3-Cyclohexanedione
2580 - 1,4-Cyclohexanedione
4936 - Ethyl 2-butynoate
5783 - 2,4-Hexadienoic acid
6087 - 2-Hydroxy-3-methyl-2-cyclopenten-1-one
6842 - 2-(Methoxymethyl)furan
7272 - 5-Methyl-2-furanmethanol
7273 - α-Methyl-2-furanmethanol
8513 - Parasorbic acid
10795 - Vinyl *trans*-2-butenoate
$C_6H_8O_3$
65 - 3-Acetyldihydro-2(3*H*)-furanone
4715 - 2,3-Epoxypropyl acrylate
$C_6H_8O_4$
2533 - 1,1-Cyclobutanedicarboxylic acid
4042 - 2,2-Dimethyl-1,3-dioxane-4,6-dione
4043 - *cis*-3,6-Dimethyl-1,4-dioxane-2,5-dione
4062 - Dimethyl fumarate
4125 - Dimethyl maleate
5068 - Ethyl hydrogen fumarate
$C_6H_8O_5$
2799 - 6-Deoxy-*L*-ascorbic acid
$C_6H_8O_6$
518 - *L*-Ascorbic acid
4743 - Erythorbic acid
5551 - *D*-Glucuronic acid γ-lactone
9154 - 1,2,3-Propanetricarboxylic acid
$C_6H_8O_7$
2414 - Citric acid
6406 - Isocitric acid
C_6H_8S
4340 - 2,3-Dimethylthiophene
4341 - 2,4-Dimethylthiophene
4342 - 2,5-Dimethylthiophene
4343 - 3,4-Dimethylthiophene
5261 - 2-Ethylthiophene
C_6H_8Si
8996 - Phenylsilane
$C_6H_9BiO_6$
1036 - Bismuth acetate
C_6H_9Br
1162 - 3-Bromocyclohexene
C_6H_9BrO
1161 - 2-Bromocyclohexanone
$C_6H_9BrO_2$
4913 - Ethyl *trans*-4-bromo-2-butenoate
$C_6H_9BrO_3$
4909 - Ethyl 4-bromoacetoacetate

C_6H_9Cl
1963 - 1-Chlorocyclohexene
$C_6H_9ClN_2$
8893 - Phenylhydrazine monohydrochloride
C_6H_9ClO
1962 - 2-Chlorocyclohexanone
$C_6H_9ClO_3$
2028 - 2-Chloroethyl acetoacetate
4941 - Ethyl 4-chloroacetoacetate
$C_6H_9Cl_3O_2$
1665 - Butyl trichloroacetate
6404 - Isobutyl trichloroacetate
C_6H_9F
5381 - 1-Fluorocyclohexene
$C_6H_9F_3O_2$
1668 - Butyl trifluoroacetate
C_6H_9N
2661 - Cyclopentanecarbonitrile
4310 - 2,4-Dimethylpyrrole
4311 - 2,5-Dimethylpyrrole
5239 - 1-Ethyl-1*H*-pyrrole
C_6H_9NO
7271 - *N*-Methyl-2-furanmethanamine
10823 - 1-Vinyl-2-pyrrolidinone
C_6H_9NOS
7802 - 4-Methyl-5-thiazoleethanol
$C_6H_9NO_2$
9889 - 1,2,5,6-Tetrahydro-3-pyridinecarboxylic acid
$C_6H_9NO_3$
10127 - Triacetamide
10576 - 3,5,5-Trimethyl-2-oxazolidinedione
$C_6H_9NO_6$
1766 - *L*-γ-Carboxyglutamic acid
8030 - Nitrilotriacetic acid
$C_6H_9N_3$
4306 - 4,6-Dimethyl-2-pyrimidinamine
4307 - 2,6-Dimethyl-4-pyrimidinamine
6223 - 3,3'-Iminobispropanenitrile
$C_6H_9N_3O_2$
2492 - Cupferron
5918 - *L*-Histidine
$C_6H_9N_3O_3$
7868 - Metronidazole
C_6H_{10}
2591 - Cyclohexene
3981 - 2,3-Dimethyl-1,3-butadiene
3999 - 3,3-Dimethyl-1-butyne
5774 - 1,2-Hexadiene
5775 - *cis*-1,3-Hexadiene
5776 - *trans*-1,3-Hexadiene
5777 - *cis*-1,4-Hexadiene
5778 - *trans*-1,4-Hexadiene
5779 - 1,5-Hexadiene
5780 - *cis,cis*-2,4-Hexadiene
5781 - *trans,cis*-2,4-Hexadiene
5782 - *trans,trans*-2,4-Hexadiene
5910 - 1-Hexyne
5911 - 2-Hexyne
5912 - 3-Hexyne
7172 - 1-Methylcyclopentene
7173 - 3-Methylcyclopentene
7174 - 4-Methylcyclopentene
7239 - Methylenecyclopentane
7546 - *cis*-2-Methyl-1,3-pentadiene
7547 - 3-Methyl-1,3-pentadiene
7548 - 4-Methyl-1,3-pentadiene

7598 - 4-Methyl-1-pentyne
7599 - 4-Methyl-2-pentyne
7858 - (1-Methylvinyl)cyclopropane

$C_6H_{10}BrClO$
1228 - 6-Bromohexanoyl chloride

$C_6H_{10}Br_2$
2927 - trans-1,2-Dibromocyclohexane, (±)

$C_6H_{10}Br_2O_2$
4975 - Ethyl 2,3-dibromobutanoate
4976 - Ethyl 2,4-dibromobutanoate

$C_6H_{10}CaO_6$
1700 - Calcium lactate

$C_6H_{10}ClN_3O_2$
5919 - L-Histidine, monohydrochloride

$C_6H_{10}ClN_5$
2810 - Desethyl atrazine

$C_6H_{10}Cl_2$
3148 - 1,1-Dichlorocyclohexane
3149 - cis-1,2-Dichlorocyclohexane

$C_6H_{10}Cl_2N_2$
614 - 1,2-Benzenediamine, dihydrochloride
615 - 1,3-Benzenediamine, dihydrochloride
616 - 1,4-Benzenediamine, dihydrochloride

$C_6H_{10}Cl_2N_2O$
2862 - 2,4-Diaminophenol, dihydrochloride

$C_6H_{10}Cl_2O_2$
1551 - Butyl dichloroacetate

$C_6H_{10}F_2$
3553 - 1,1-Difluorocyclohexane

$C_6H_{10}FeO_6$
5328 - Ferrous lactate

$C_6H_{10}N_2$
10609 - 1,3,5-Trimethyl-1H-pyrazole

$C_6H_{10}N_2O_2$
2581 - 1,2-Cyclohexanedione dioxime

$C_6H_{10}N_2O_4$
5454 - N-Formimidoyl-L-glutamic acid

$C_6H_{10}N_4$
9894 - 6,7,8,9-Tetrahydro-5H-tetrazolo[1,5-a]
    azepine

$C_6H_{10}N_4O_3S_2$
1439 - Butazolamide

$C_6H_{10}N_6$
2711 - Cyromazine

$C_6H_{10}N_6O$
2729 - Dacarbazine

$C_6H_{10}O$
1470 - Butoxyacetylene
2588 - Cyclohexanone
2598 - 2-Cyclohexen-1-ol
2662 - Cyclopentanecarboxaldehyde
2834 - Diallyl ether
3654 - 3,6-Dihydro-4-methyl-2H-pyran
5784 - 2,4-Hexadien-1-ol
5860 - trans-2-Hexenal
5861 - cis-3-Hexenal
5881 - 5-Hexen-2-one
5882 - 4-Hexen-3-one
5914 - 3-Hexyn-1-ol
5915 - 1-Hexyn-3-ol
6723 - Mesityl oxide
7170 - 2-Methylcyclopentanone
7171 - 3-Methylcyclopentanone, (±)
7575 - 2-Methyl-2-pentenal
7588 - 3-Methyl-2-penten-4-one
7589 - 4-Methyl-4-penten-2-one
7600 - 3-Methyl-1-pentyn-3-ol

8423 - 7-Oxabicyclo[4.1.0]heptane

$C_6H_{10}OS_2$
160 - Allicin

$C_6H_{10}O_2$
190 - Allyl glycidyl ether
204 - Allyl propanoate
2663 - Cyclopentanecarboxylic acid
3638 - 3,4-Dihydro-2-methoxy-2H-pyran
4929 - Ethyl cis-2-butenoate
4930 - Ethyl trans-2-butenoate
4931 - Ethyl 3-butenoate
4971 - Ethyl cyclopropanecarboxylate
4988 - 5-Ethyldihydro-2(3H)-furanone
5100 - Ethyl methacrylate
5830 - Hexanedial
5838 - 2,3-Hexanedione
5839 - 2,4-Hexanedione
5840 - 2,5-Hexanedione
5841 - 3,4-Hexanedione
5868 - 2-Hexenoic acid
5869 - 3-Hexenoic acid
5870 - 5-Hexenoic acid
5913 - 3-Hexyne-2,5-diol
6021 - 2-Hydroxycyclohexanone
6470 - Isopropyl acrylate
7142 - Methyl cyclobutanecarboxylate
7451 - Methyl 2-methyl-2-butenoate, (E)
7452 - Methyl 3-methyl-2-butenoate
7584 - trans-2-Methyl-2-pentenoic acid
7585 - 4-Methyl-2-pentenoic acid
7586 - 2-Methyl-3-pentenoic acid
8443 - 2-Oxepanone
9204 - Propyl acrylate
10794 - Vinyl butanoate

$C_6H_{10}O_3$
3578 - Diglycidyl ether
3612 - 2,5-Dihydro-2,5-dimethoxyfuran
4208 - 3,3-Dimethyl-2-oxobutanoic acid
4872 - Ethyl acetoacetate
5075 - Ethyl 2-hydroxy-3-butenoate
6042 - 2-Hydroxyethyl methacrylate
6186 - 2-Hydroxypropyl acrylate
6826 - 2-Methoxyethyl acrylate
7446 - Methyl 2-methylacetoacetate
7539 - Methyl 4-oxopentanoate
7540 - 4-Methyl-2-oxopentanoic acid
8463 - 5-Oxohexanoic acid
8502 - Pantolactone
9165 - Propanoic anhydride

$C_6H_{10}O_4$
132 - Adipic acid
3495 - Diethyl oxalate
4131 - Dimethyl methylmalonate
4321 - Dimethyl succinate
4774 - 1,1-Ethanediol, diacetate
4775 - 1,2-Ethanediol, diacetate
5069 - Ethyl hydrogen succinate
6553 - Isosorbide
7279 - 3-Methylglutaric acid
7895 - Monomethyl glutarate

$C_6H_{10}O_4S$
10053 - 3,3'-Thiodipropionic acid

$C_6H_{10}O_4S_2$
3826 - Dimethipin
4585 - 3,3'-Dithiobispropanoic acid
4794 - 1,2-Ethanediyl mercaptoacetate

$C_6H_{10}O_4Zn$
10876 - Zinc propanoate

$C_6H_{10}O_5$
3426 - Diethyl dicarbonate
3817 - Dilactic acid

$(C_6H_{10}O_5)_x$
5582 - Glycogen

$C_6H_{10}O_6$
4329 - Dimethyl L-tartrate
5252 - Ethyl tartrate
5513 - D-Galactonic acid, γ-lactone
6578 - δ-Lactone-D-gluconic acid

$C_6H_{10}O_7$
5517 - D-Galacturonic acid
5550 - D-Glucuronic acid

$C_6H_{10}O_8$
2415 - Citric acid monohydrate
5511 - Galactaric acid
5536 - D-Glucaric acid

$C_6H_{10}S$
2841 - Diallyl sulfide

$C_6H_{10}S_2$
2833 - Diallyl disulfide

$C_6H_{10}S_3$
2842 - Diallyl trisulfide

$C_6H_{11}Br$
1159 - Bromocyclohexane

$C_6H_{11}BrN_2O_2$
279 - N-(Aminocarbonyl)-2-bromo-3-meth-
    ylbutanamide

$C_6H_{11}BrO$
1160 - trans-4-Bromocyclohexanol

$C_6H_{11}BrO_2$
1226 - 2-Bromohexanoic acid, (±)
1227 - 6-Bromohexanoic acid
1527 - tert-Butyl bromoacetate
4911 - Ethyl 2-bromobutanoate
4912 - Ethyl 4-bromobutanoate
4916 - Ethyl 2-bromo-2-methylpropanoate
7057 - Methyl 5-bromopentanoate

$C_6H_{11}Cl$
1961 - Chlorocyclohexane

$C_6H_{11}ClMg$
2632 - Cyclohexylmagnesium chloride

$C_6H_{11}ClO$
3994 - 3,3-Dimethylbutanoyl chloride
4927 - 2-Ethylbutanoyl chloride
5855 - Hexanoyl chloride
7574 - 4-Methylpentanoyl chloride

$C_6H_{11}ClO_2$
1531 - Butyl chloroacetate
1532 - tert-Butyl chloroacetate
4943 - Ethyl 4-chlorobutanoate
6488 - Isopropyl 2-chloropropanoate
8671 - Pentyl chloroformate
9222 - Propyl 3-chloropropanoate

$C_6H_{11}F$
5380 - Fluorocyclohexane

$C_6H_{11}I$
6289 - Iodocyclohexane

$C_6H_{11}IO_3$
6301 - 2-(1-Iodoethyl)-1,3-dioxolane-4-metha-
    nol

$C_6H_{11}KO_7$
9095 - Potassium D-gluconate

$C_6H_{11}N$
205 - N-Allyl-2-propen-1-amine
5844 - Hexanenitrile
7556 - 4-Methylpentanenitrile

C$_6$H$_{11}$NO
 1719 - Caprolactam
 2589 - Cyclohexanone oxime
 6468 - 3-Isopropoxypropanenitrile
 6534 - N-Isopropyl-2-propenamide
 7679 - 1-Methyl-2-piperidinone
 7680 - 1-Methyl-4-piperidinone
 9059 - 1-Piperidinecarboxaldehyde
C$_6$H$_{11}$NO$_2$
 83 - 4-Acetylmorpholine
 290 - 1-Aminocyclopentanecarboxylic acid
 6044 - 1-(2-Hydroxyethyl)-2-pyrrolidinone
 8092 - Nitrocyclohexane
 9061 - 2-Piperidinecarboxylic acid, (S)
 9062 - 3-Piperidinecarboxylic acid
 9063 - 4-Piperidinecarboxylic acid
C$_6$H$_{11}$NO$_3$
 130 - Adipamic acid
C$_6$H$_{11}$NO$_3$S
 207 - 3-(Allylsulfinyl)-L-alanine, (S)
C$_6$H$_{11}$NO$_4$
 234 - 2-Aminoadipic acid
 318 - L-2-Aminohexanedioic acid
C$_6$H$_{11}$N$_2$O$_4$PS$_3$
 6758 - Methidathion
C$_6$H$_{11}$N$_3$O$_4$
 5588 - L-Glycylasparagine
 5590 - N-(N-Glycylglycyl)glycine
C$_6$H$_{11}$N$_3$O$_9$
 9176 - Propatyl nitrate
C$_6$H$_{11}$NaO$_7$
 9561 - Sodium gluconate
C$_6$H$_{11}$O$_3$P
 5278 - 4-Ethyl-2,6,7-trioxa-1-phosphabicy-
    clo[2.2.2]octane
C$_6$H$_{12}$
 2565 - Cyclohexane
 3995 - 2,3-Dimethyl-1-butene
 3996 - 3,3-Dimethyl-1-butene
 3997 - 2,3-Dimethyl-2-butene
 4928 - 2-Ethyl-1-butene
 4961 - Ethylcyclobutane
 5137 - 1-Ethyl-1-methylcyclopropane
 5862 - 1-Hexene
 5863 - cis-2-Hexene
 5864 - trans-2-Hexene
 5865 - cis-3-Hexene
 5866 - trans-3-Hexene
 6492 - Isopropylcyclopropane
 7167 - Methylcyclopentane
 7576 - 2-Methyl-1-pentene
 7577 - 3-Methyl-1-pentene
 7578 - 4-Methyl-1-pentene
 7579 - 2-Methyl-2-pentene
 7580 - 3-Methyl-cis-2-pentene
 7581 - 3-Methyl-trans-2-pentene
 7582 - 4-Methyl-cis-2-pentene
 7583 - 4-Methyl-trans-2-pentene
 10550 - 1,1,2-Trimethylcyclopropane
C$_6$H$_{12}$BrNO
 3408 - Diethylbromoacetamide
C$_6$H$_{12}$Br$_2$
 2950 - 1,2-Dibromohexane
 2951 - 1,6-Dibromohexane
 2952 - 3,4-Dibromohexane
C$_6$H$_{12}$ClNO
 1977 - 2-Chloro-N,N-diethylacetamide

 2044 - N-(2-Chloroethyl)morpholine
C$_6$H$_{12}$Cl$_2$
 3196 - 1,2-Dichlorohexane
 3197 - 1,6-Dichlorohexane
C$_6$H$_{12}$Cl$_2$O
 950 - Bis(3-chloropropyl) ether
 3164 - 2,2'-Dichlorodiisopropyl ether
C$_6$H$_{12}$Cl$_2$O$_2$
 924 - 1,2-Bis(2-chloroethoxy)ethane
C$_6$H$_{12}$Cl$_2$O$_4$S$_2$
 933 - 1,2-Bis(2-chloroethylsulfonyl)ethane
C$_6$H$_{12}$Cl$_3$N
 10291 - 2,2',2''-Trichlorotriethylamine
C$_6$H$_{12}$Cl$_3$O$_3$P
 928 - Bis(2-chloroethyl) 2-chloroethylphos-
    phonate
 10681 - Tris(2-chloroethyl) phosphite
C$_6$H$_{12}$Cl$_3$O$_4$P
 10680 - Tris(2-chloroethyl) phosphate
C$_6$H$_{12}$CuN$_2$S$_4$
 969 - Bis(dimethyldithiocarbamate)copper
C$_6$H$_{12}$FeN$_3$O$_{12}$
 439 - Ammonium ferric oxalate
C$_6$H$_{12}$I$_2$
 3767 - 1,6-Diiodohexane
C$_6$H$_{12}$NO$_3$PS$_2$
 5467 - Fosthietan
C$_6$H$_{12}$NO$_4$PS$_2$
 5456 - Formothion
C$_6$H$_{12}$N$_2$
 36 - Acetone (1-methylethylidene)hydrazone
 319 - 6-Aminohexanenitrile
 3380 - (Diethylamino)acetonitrile
 10354 - Triethylenediamine
C$_6$H$_{12}$N$_2$NiS$_4$
 970 - Bis(dimethyldithiocarbamate)nickel
C$_6$H$_{12}$N$_2$O
 9060 - 4-Piperidinecarboxamide
C$_6$H$_{12}$N$_2$O$_2$
 5831 - Hexanediamide
C$_6$H$_{12}$N$_2$O$_3$
 2732 - Daminozide
C$_6$H$_{12}$N$_2$O$_4$
 3466 - Diethyl 1,2-hydrazinedicarboxylate
C$_6$H$_{12}$N$_2$O$_4$S
 6587 - L-Lanthionine
C$_6$H$_{12}$N$_2$O$_4$S$_2$
 2718 - L-Cystine
C$_6$H$_{12}$N$_2$O$_8$
 10360 - Triethylene glycol dinitrate
C$_6$H$_{12}$N$_2$PbS$_4$
 6596 - Lead bis(dimethyldithiocarbamate)
C$_6$H$_{12}$N$_2$S$_3$
 9978 - Tetramethylthiodicarbonic diamide
C$_6$H$_{12}$N$_2$S$_4$
 10086 - Thiram
C$_6$H$_{12}$N$_2$S$_4$Zn
 10877 - Ziram
C$_6$H$_{12}$N$_2$Si
 10615 - 1-(Trimethylsilyl)-1H-imidazole
C$_6$H$_{12}$N$_3$OP
 10362 - Triethylenephosphoramide
C$_6$H$_{12}$N$_3$PS
 10363 - Triethylenethiophosphoramide
C$_6$H$_{12}$N$_4$
 5822 - Hexamethylenetetramine

C$_6$H$_{12}$N$_5$O$_2$PS$_2$
 6697 - Menazon
C$_6$H$_{12}$N$_6$O$_3$
 10692 - N,N',N''-Tris(hydroxymethyl)melamin
C$_6$H$_{12}$O
 1673 - Butyl vinyl ether
 1674 - tert-Butyl vinyl ether
 2587 - Cyclohexanol
 2666 - Cyclopentanemethanol
 3993 - 3,3-Dimethyl-2-butanone
 4922 - 2-Ethylbutanal
 5253 - 2-Ethyltetrahydrofuran
 5827 - Hexanal
 5853 - 2-Hexanone
 5854 - 3-Hexanone
 5871 - 1-Hexen-3-ol
 5872 - cis-2-Hexen-1-ol
 5873 - trans-2-Hexen-1-ol
 5874 - cis-3-Hexen-1-ol
 5875 - trans-3-Hexen-1-ol
 5876 - trans-4-Hexen-1-ol
 5877 - 4-Hexen-2-ol
 5878 - 5-Hexen-2-ol
 6405 - Isobutyl vinyl ether
 7168 - 1-Methylcyclopentanol
 7169 - cis-2-Methylcyclopentanol
 7551 - 2-Methylpentanal
 7571 - 3-Methyl-2-pentanone, (±)
 7572 - 4-Methyl-2-pentanone
 7573 - 2-Methyl-3-pentanone
 7587 - 4-Methyl-3-penten-2-ol
 8442 - Oxepane
 9201 - 3-Propoxy-1-propene
 9958 - Tetramethyloxirane
C$_6$H$_{12}$O$_2$
 1487 - Butyl acetate
 1488 - sec-Butyl acetate
 1489 - tert-Butyl acetate
 1490 - tert-Butylacetic acid
 2576 - cis-1,2-Cyclohexanediol
 2577 - trans-1,4-Cyclohexanediol
 2626 - Cyclohexyl hydroperoxide
 2821 - Diacetone alcohol
 3361 - 1,1-Diethoxyethene
 3988 - 2,2-Dimethylbutanoic acid
 4923 - Ethyl butanoate
 4924 - 2-Ethylbutanoic acid
 5138 - 2-Ethyl-2-methyl-1,3-dioxolane
 5153 - Ethyl 2-methylpropanoate
 5848 - Hexanoic acid
 6373 - Isobutyl acetate
 6445 - Isopentyl formate
 6498 - Isopropyl glycidyl ether
 6533 - Isopropyl propanoate
 7205 - Methyl 2,2-dimethylpropanoate
 7398 - Methyl isopentanoate
 7558 - Methyl pentanoate
 7559 - 2-Methylpentanoic acid, (±)
 7560 - 3-Methylpentanoic acid, (±)
 7561 - 4-Methylpentanoic acid
 8674 - Pentyl formate
 9260 - Propyl propanoate
 9885 - Tetrahydro-2H-pyran-2-methanol
C$_6$H$_{12}$O$_2$S
 4326 - 2,4-Dimethylsulfolane
C$_6$H$_{12}$O$_3$
 463 - 2,5-Anhydro-3,4-dideoxyhexitol
 3857 - 4,4-Dimethoxy-2-butanone

4044 - 2,2-Dimethyl-1,3-dioxolane-4-metha-
  nol
4841 - 2-Ethoxyethyl acetate
5024 - Ethyl ethoxyacetate
5074 - Ethyl 3-hydroxybutanoate, (±)
6047 - 2-Hydroxyhexanoic acid
6048 - 6-Hydroxyhexanoic acid
6095 - 2-Hydroxy-4-methylpentanoic acid, (S)
6503 - Isopropyl lactate
8508 - Paraldehyde
9235 - 1,2-Propylene glycol monomethyl ether
  acetate
9847 - Tetrahydro-2,5-dimethoxyfuran

$C_6H_{12}O_4$
3577 - Digitoxose

$C_6H_{12}O_4Si$
4040 - Dimethyldiacetoxysilane

$C_6H_{12}O_5$
2803 - 2-Deoxy-D-glucose
2805 - 2-Deoxy-D-chiro-inositol
7778 - Methyl β-D-ribofuranoside
9443 - Quinovose
9468 - DL-α-Rhamnose
9469 - D-Rhamnose

$C_6H_{12}O_6$
164 - D-Allose
218 - D-Altrose
5469 - DL-Fructose
5470 - L-Fructose
5471 - β-D-Fructose
5514 - α-D-Galactopyranose
5516 - D-Galactose
5540 - β-D-Glucopyranose
5546 - α-D-Glucose
5620 - D-Gulose
5621 - L-Gulose
6209 - D-Idose
6210 - L-Idose
6262 - myo-Inositol
6680 - D-Mannose
6681 - L-Mannose
9585 - L-Sorbose
9674 - D-Tagatose

$C_6H_{12}O_7$
5539 - D-Gluconic acid

$C_6H_{12}S$
2586 - Cyclohexanethiol
2685 - Cyclopentyl methyl sulfide
9851 - cis-Tetrahydro-2,5-dimethylthiophene
9872 - Tetrahydro-3-methyl-2H-thiopyran
10037 - Thiepane

$C_6H_{12}S_2$
206 - Allyl propyl disulfide

$C_6H_{12}S_3$
10042 - Thioacetaldehyde trimer

$C_6H_{13}Br$
1223 - 1-Bromohexane
1224 - 2-Bromohexane
1225 - 3-Bromohexane
1271 - 1-Bromo-2-methylpentane
1272 - 1-Bromo-4-methylpentane
1273 - 2-Bromo-2-methylpentane
1274 - 3-Bromo-3-methylpentane

$C_6H_{13}BrO_2$
1171 - 2-Bromo-1,1-diethoxyethane

$C_6H_{13}Cl$
2066 - 1-Chlorohexane

2067 - 2-Chlorohexane
2068 - 3-Chlorohexane
2144 - 2-Chloro-4-methylpentane
2145 - 3-(Chloromethyl)pentane

$C_6H_{13}ClO$
2069 - 6-Chloro-1-hexanol

$C_6H_{13}ClO_2$
1975 - 2-Chloro-1,1-diethoxyethane
2120 - 1-Chloro-3-(1-methylethoxy)-2-pro-
  panol

$C_6H_{13}Cl_2N$
1948 - 2-Chloro-N-(2-chloroethyl)-N-ethyle-
  thanamine

$C_6H_{13}Cl_2NO$
2045 - 4-(2-Chloroethyl)morpholine, hydrochloride

$C_6H_{13}Cl_3Si$
10241 - Trichlorohexylsilane

$C_6H_{13}F$
5390 - 1-Fluorohexane

$C_6H_{13}I$
6306 - 1-Iodohexane

$C_6H_{13}N$
2608 - Cyclohexylamine
4312 - 1,2-Dimethylpyrrolidine
5109 - N-Ethyl-2-methylallylamine
5802 - Hexahydro-1H-azepine
7673 - 1-Methylpiperidine
7674 - 2-Methylpiperidine, (±)
7675 - 3-Methylpiperidine, (±)
7676 - 4-Methylpiperidine

$C_6H_{13}NO$
347 - 4-Amino-4-methyl-2-pentanone
1486 - N-Butylacetamide
3371 - N,N-Diethylacetamide
3982 - N,N-Dimethylbutanamide
4136 - 2,6-Dimethylmorpholine
5159 - N-Ethylmorpholine
5828 - Hexanamide
7677 - 1-Methyl-3-piperidinol
7678 - 1-Methyl-4-piperidinol
9069 - 2-Piperidinemethanol
9070 - 3-Piperidinemethanol
9259 - N-Propylpropanamide
9399 - 1-Pyrrolidineethanol

$C_6H_{13}NO_2$
320 - 6-Aminohexanoic acid
6435 - DL-Isoleucine
6436 - L-Isoleucine
6600 - DL-Leucine
6601 - D-Leucine
6602 - L-Leucine
7852 - 3-Methyl-L-valine
7904 - 4-Morpholineethanol
8110 - 1-Nitrohexane
8275 - DL-Norleucine
8276 - L-Norleucine
8670 - tert-Pentyl carbamate

$C_6H_{13}NO_2S$
4809 - D-Ethionine
4810 - L-Ethionine

$C_6H_{13}NO_3S$
2640 - Cyclohexylsulfamic acid

$C_6H_{13}NO_4$
1001 - N,N-Bis(2-hydroxyethyl)glycine

$C_6H_{13}NO_5$
293 - 2-Amino-2-deoxy-D-glucose

$C_6H_{13}N_3$
7103 - (3-Methyl-2-butenyl)guanidine

$C_6H_{13}N_3O_3$
2417 - Citrulline

$C_6H_{13}O_3P$
3542 - Diethyl vinylphosphonate

$C_6H_{13}O_4P$
3374 - Diethyl acetylphosphonate

$C_6H_{13}O_9P$
5472 - D-Fructose 6-phosphate
5549 - α-D-Glucose 1-phosphate

C_6H_{14}
3984 - 2,2-Dimethylbutane
3985 - 2,3-Dimethylbutane
5829 - Hexane
7552 - 2-Methylpentane
7553 - 3-Methylpentane

$C_6H_{14}ClN$
2609 - Cyclohexylamine hydrochloride

$C_6H_{14}Cl_4OSi_2$
957 - 1,3-Bis(dichloromethyl)tetramethyldisi-
  loxane

$C_6H_{14}FO_3P$
6432 - Isoflurophate

$C_6H_{14}NO_3PS_2$
4813 - Ethoate-methyl

$C_6H_{14}NO_4P$
4137 - Dimethyl morpholinophosphoramidate

$C_6H_{14}N_2$
564 - Azopropane
2571 - cis-1,2-Cyclohexanediamine
2572 - trans-1,2-Cyclohexanediamine
4264 - 1,4-Dimethylpiperazine
4265 - cis-2,5-Dimethylpiperazine
5808 - Hexahydro-1-methyl-1H-1,4-diazepine
9068 - 4-Piperidinemethanamine
9398 - 1-Pyrrolidineethanamine

$C_6H_{14}N_2O$
6523 - N-Isopropyl-N-nitroso-2-propanamine
7903 - 4-Morpholineethanamine
8200 - N-Nitroso-N-propyl-1-propanamine
9056 - 1-Piperazineethanol

$C_6H_{14}N_2O_2$
6646 - DL-Lysine
6647 - D-Lysine
6648 - L-Lysine

$C_6H_{14}N_4O_2$
513 - D-Arginine
514 - L-Arginine
5833 - Hexanedioic acid, dihydrazide

$C_6H_{14}O$
1567 - Butyl ethyl ether
1568 - sec-Butyl ethyl ether
1569 - tert-Butyl ethyl ether
3805 - Diisopropyl ether
3989 - 2,2-Dimethyl-1-butanol
3990 - 3,3-Dimethyl-1-butanol
3991 - 2,3-Dimethyl-2-butanol
3992 - 3,3-Dimethyl-2-butanol, (±)
4558 - Dipropyl ether
4926 - 2-Ethyl-1-butanol
5850 - 1-Hexanol
5851 - 2-Hexanol
5852 - 3-Hexanol
6841 - 2-Methoxy-2-methylbutane
7256 - 1-(1-Methylethoxy)propane
7562 - 2-Methyl-1-pentanol

C₆H₁₈N₃P

$C_6H_{18}N_3P$

5825 - Hexamethylphosphorous triamide

$C_6H_{18}N_4$

904 - N,N'-Bis(2-aminoethyl)-1,2-ethanediamine

$C_6H_{18}OSi_2$

5819 - Hexamethyldisiloxane

$C_6H_{18}O_3Si_3$

5815 - Hexamethylcyclotrisiloxane

$C_6H_{18}SSi_2$

5817 - Hexamethyldisilathiane

$C_6H_{18}Si_2$

5816 - Hexamethyldisilane

$C_6H_{19}NSi_2$

5818 - Hexamethyldisilazane

$C_6H_{21}N_3Si_3$

5814 - 2,2,4,4,6,6-Hexamethylcyclotrisilazane

C_6MoO_6

7888 - Molybdenum hexacarbonyl

C_6N_4

9783 - Tetracyanoethene

C_6O_6V

10772 - Vanadium carbonyl

C_6O_6W

10718 - Tungsten carbonyl

C_7F_5N

8583 - Pentafluorobenzonitrile

C_7F_8

8728 - Perfluorotoluene

C_7F_{14}

8709 - Perfluoro-1-heptene

8715 - Perfluoromethylcyclohexane

C_7F_{16}

8708 - Perfluoroheptane

C_7HF_5O

8579 - Pentafluorobenzaldehyde

$C_7HF_5O_2$

8582 - Pentafluorobenzoic acid

C_7HF_{15}

8548 - 1H-Pentadecafluoroheptane

$C_7H_2ClF_3N_2O_4$

2013 - 2-Chloro-1,3-dinitro-5-(trifluoromethyl)benzene

$C_7H_3Br_2NO$

2955 - 3,5-Dibromo-4-hydroxybenzonitrile

$C_7H_3Br_5$

8526 - Pentabromomethylbenzene

$C_7H_3ClF_3NO_2$

2197 - 1-Chloro-2-nitro-4-(trifluoromethyl)benzene

2198 - 1-Chloro-4-nitro-2-(trifluoromethyl)benzene

$C_7H_3ClN_2O_5$

4378 - 3,5-Dinitrobenzoyl chloride

$C_7H_3Cl_2F_3$

3116 - 3,4-Dichlorobenzotrifluoride

3309 - 2,4-Dichloro-1-(trifluoromethyl)benzene

$C_7H_3Cl_2N$

3114 - 2,6-Dichlorobenzonitrile

$C_7H_3Cl_2NO$

3265 - 3,4-Dichlorophenyl isocyanate

3266 - 3,5-Dichlorophenyl isocyanate

$C_7H_3Cl_3O$

3117 - 2,3-Dichlorobenzoyl chloride

3118 - 2,4-Dichlorobenzoyl chloride

3119 - 2,5-Dichlorobenzoyl chloride

3120 - 3,4-Dichlorobenzoyl chloride

10203 - 2,3,6-Trichlorobenzaldehyde

$C_7H_3Cl_3O_2$

10209 - 2,3,6-Trichlorobenzoic acid

$C_7H_3Cl_5$

3305 - 1,2-Dichloro-4-(trichloromethyl)benzene

8545 - 2,3,4,5,6-Pentachlorotoluene

$C_7H_3Cl_5O$

8532 - 2,3,4,5,6-Pentachloroanisole

$C_7H_3Cl_5S$

7544 - Methyl pentachlorophenyl sulfide

$C_7H_3F_5$

8590 - 2,3,4,5,6-Pentafluorotoluene

$C_7H_3F_5O$

8586 - Pentafluoromethoxybenzene

$C_7H_3I_2NO$

6026 - 4-Hydroxy-3,5-diiodobenzonitrile

$C_7H_3I_3O_2$

10449 - 2,3,5-Triiodobenzoic acid

$C_7H_3NO_3$

9344 - 2,3-Pyridinedicarboxylic acid anhydride

$C_7H_3N_3O_8$

10626 - 2,4,6-Trinitrobenzoic acid

C_7H_4BrClO

1110 - 2-Bromobenzoyl chloride

1111 - 4-Bromobenzoyl chloride

$C_7H_4BrF_3$

1368 - 1-Bromo-2-(trifluoromethyl)benzene

1369 - 1-Bromo-3-(trifluoromethyl)benzene

1370 - 1-Bromo-4-(trifluoromethyl)benzene

C_7H_4BrN

1106 - 2-Bromobenzonitrile

1107 - 3-Bromobenzonitrile

1108 - 4-Bromobenzonitrile

C_7H_4BrNO

1239 - 1-Bromo-4-isocyanatobenzene

$C_7H_4Br_2O_2$

2953 - 3,5-Dibromo-2-hydroxybenzaldehyde

$C_7H_4Br_2O_3$

2954 - 3,5-Dibromo-2-hydroxybenzoic acid

$C_7H_4Br_2O_5$

2995 - 2,6-Dibromo-3,4,5-trihydroxybenzoic acid

$C_7H_4Br_4O$

9716 - 2,3,4,5-Tetrabromo-6-methylphenol

C_7H_4ClFO

5373 - 2-Fluorobenzoyl chloride

5374 - 3-Fluorobenzoyl chloride

5375 - 4-Fluorobenzoyl chloride

$C_7H_4ClF_3$

2331 - 1-Chloro-2-(trifluoromethyl)benzene

2332 - 1-Chloro-3-(trifluoromethyl)benzene

2333 - 1-Chloro-4-(trifluoromethyl)benzene

C_7H_4ClIO

6282 - 2-Iodobenzoyl chloride

6283 - 4-Iodobenzoyl chloride

C_7H_4ClN

1899 - 2-Chlorobenzonitrile

1900 - 3-Chlorobenzonitrile

1901 - 4-Chlorobenzonitrile

C_7H_4ClNO

1909 - 2-Chlorobenzoxazole

2084 - 1-Chloro-2-isocyanatobenzene

2085 - 1-Chloro-3-isocyanatobenzene

2242 - 4-Chlorophenyl isocyanate

$C_7H_4ClNO_2$

1910 - 5-Chloro-2(3H)-benzoxazolone

$C_7H_4ClNO_3$

2179 - 2-Chloro-5-nitrobenzaldehyde

2180 - 4-Chloro-3-nitrobenzaldehyde

8081 - 2-Nitrobenzoyl chloride

8082 - 3-Nitrobenzoyl chloride

8083 - 4-Nitrobenzoyl chloride

$C_7H_4ClNO_4$

1747 - Carbonochloridic acid, 4-nitrophenyl ester

2187 - 2-Chloro-4-nitrobenzoic acid

2188 - 2-Chloro-5-nitrobenzoic acid

2189 - 4-Chloro-3-nitrobenzoic acid

C_7H_4ClNS

1905 - 2-Chlorobenzothiazole

2088 - 1-Chloro-4-isothiocyanatobenzene

$C_7H_4Cl_2O$

1911 - 2-Chlorobenzoyl chloride

1912 - 3-Chlorobenzoyl chloride

1913 - 4-Chlorobenzoyl chloride

3086 - 2,3-Dichlorobenzaldehyde

3087 - 2,4-Dichlorobenzaldehyde

3088 - 2,6-Dichlorobenzaldehyde

3089 - 3,4-Dichlorobenzaldehyde

3090 - 3,5-Dichlorobenzaldehyde

$C_7H_4Cl_2O_2$

3109 - 2,4-Dichlorobenzoic acid

3110 - 2,5-Dichlorobenzoic acid

3111 - 2,6-Dichlorobenzoic acid

3112 - 3,4-Dichlorobenzoic acid

3113 - 3,5-Dichlorobenzoic acid

3198 - 3,5-Dichloro-2-hydroxybenzaldehyde

$C_7H_4Cl_2O_3$

3199 - 3,5-Dichloro-2-hydroxybenzoic acid

$C_7H_4Cl_3F$

5427 - 1-Fluoro-2-(trichloromethyl)benzene

$C_7H_4Cl_3NO_3$

10299 - Triclopyr

$C_7H_4Cl_4$

2319 - 1-Chloro-2-(trichloromethyl)benzene

2320 - 1-Chloro-4-(trichloromethyl)benzene

3152 - 1,2-Dichloro-4-(dichloromethyl)benzene

10216 - 1,2,4-Trichloro-5-(chloromethyl)benzene

$C_7H_4Cl_4O$

9758 - 2,3,4,6-Tetrachloro-5-methylphenol

$C_7H_4Cl_4O_2$

9757 - 2,3,5,6-Tetrachloro-4-methoxyphenol

C_7H_4FN

5371 - 2-Fluorobenzonitrile

5372 - 4-Fluorobenzonitrile

C_7H_4FNS

5393 - 1-Fluoro-3-isothiocyanatobenzene

5394 - 1-Fluoro-4-isothiocyanatobenzene

$C_7H_4F_3NO_2$

8209 - 1-Nitro-2-(trifluoromethyl)benzene

8210 - 1-Nitro-3-(trifluoromethyl)benzene

$C_7H_4F_4$

5428 - 1-Fluoro-2-(trifluoromethyl)benzene

5429 - 1-Fluoro-3-(trifluoromethyl)benzene

5430 - 1-Fluoro-4-(trifluoromethyl)benzene

C_7H_4IN

6281 - 4-Iodobenzonitrile

C₇H₄I₂O₃
6024 - 2-Hydroxy-3,5-diiodobenzoic acid
6025 - 4-Hydroxy-3,5-diiodobenzoic acid

C₇H₄NNaS₂
739 - 2(3*H*)-Benzothiazolethione, sodium salt

C₇H₄N₂O₂
8078 - 3-Nitrobenzonitrile
8079 - 4-Nitrobenzonitrile

C₇H₄N₂O₃
6417 - 1-Isocyanato-2-nitrobenzene
6418 - 1-Isocyanato-3-nitrobenzene
6419 - 1-Isocyanato-4-nitrobenzene

C₇H₄N₂O₅
4367 - 2,4-Dinitrobenzaldehyde

C₇H₄N₂O₆
4375 - 2,4-Dinitrobenzoic acid
4376 - 3,4-Dinitrobenzoic acid
4377 - 3,5-Dinitrobenzoic acid

C₇H₄N₂O₇
6033 - 2-Hydroxy-3,5-dinitrobenzoic acid

C₇H₄O₄S
747 - 3*H*-2,1-Benzoxathiol-3-one 1,1-dioxide

C₇H₄O₆
8478 - 4-Oxo-4*H*-pyran-2,6-dicarboxylic acid

C₇H₄O₇
6139 - 3-Hydroxy-4-oxo-4*H*-pyran-2,6-dicar-
boxylic acid

C₇H₅BiO₄
1037 - Bismuth subsalicylate

C₇H₅BrO
760 - Benzoyl bromide
1093 - 2-Bromobenzaldehyde
1094 - 3-Bromobenzaldehyde
1095 - 4-Bromobenzaldehyde

C₇H₅BrO₂
1103 - 2-Bromobenzoic acid
1104 - 3-Bromobenzoic acid
1105 - 4-Bromobenzoic acid
1230 - 5-Bromo-2-hydroxybenzaldehyde

C₇H₅BrO₃
1233 - 5-Bromo-2-hydroxybenzoic acid

C₇H₅Br₃O
10168 - 1,3,5-Tribromo-2-methoxybenzene
10169 - 2,4,6-Tribromo-3-methylphenol

C₇H₅ClF₃N
2329 - 2-Chloro-5-(trifluoromethyl)aniline
2330 - 4-Chloro-3-(trifluoromethyl)aniline

C₇H₅ClN₂O
1908 - 5-Chloro-2-benzoxazolamine

C₇H₅ClN₂S
1903 - 4-Chloro-2-benzothiazolamine
1904 - 6-Chloro-2-benzothiazolamine

C₇H₅ClN₄
2254 - 5-Chloro-1-phenyltetrazole

C₇H₅ClO
761 - Benzoyl chloride
1861 - 2-Chlorobenzaldehyde
1862 - 3-Chlorobenzaldehyde
1863 - 4-Chlorobenzaldehyde

C₇H₅ClO₂
1896 - 2-Chlorobenzoic acid
1897 - 3-Chlorobenzoic acid
1898 - 4-Chlorobenzoic acid
2071 - 5-Chloro-2-hydroxybenzaldehyde
6001 - 2-Hydroxybenzoyl chloride
8862 - Phenyl chloroformate

C₇H₅ClO₃
1873 - 3-Chlorobenzenecarboperoxoic acid
2073 - 3-Chloro-4-hydroxybenzoic acid
2074 - 5-Chloro-2-hydroxybenzoic acid

C₇H₅Cl₂F
3188 - (Dichlorofluoromethyl)benzene

C₇H₅Cl₂FN₂O₃
5440 - Fluroxypyr

C₇H₅Cl₂N
8899 - Phenylimidocarbonyl chloride

C₇H₅Cl₂NO
3091 - 2,6-Dichlorobenzamide

C₇H₅Cl₂NO₂
295 - 3-Amino-2,5-dichlorobenzoic acid

C₇H₅Cl₂NO₄S
3074 - 4-[(Dichloroamino)sulfonyl]benzoic
acid

C₇H₅Cl₃
3142 - 1,2-Dichloro-4-(chloromethyl)benzene
3143 - 2,4-Dichloro-1-(chloromethyl)benzene
10252 - 1,2,4-Trichloro-5-methylbenzene
10253 - (Trichloromethyl)benzene

C₇H₅Cl₃O
10250 - 1,2,4-Trichloro-5-methoxybenzene
10251 - 1,3,5-Trichloro-2-methoxybenzene
10255 - 2,3,4-Trichloro-6-methylphenol
10256 - 2,3,6-Trichloro-4-methylphenol
10257 - 2,4,6-Trichloro-3-methylphenol

C₇H₅FO
765 - Benzoyl fluoride
5358 - 2-Fluorobenzaldehyde
5359 - 3-Fluorobenzaldehyde
5360 - 4-Fluorobenzaldehyde

C₇H₅FO₂
5368 - 2-Fluorobenzoic acid
5369 - 3-Fluorobenzoic acid
5370 - 4-Fluorobenzoic acid

C₇H₅F₃
10404 - (Trifluoromethyl)benzene

C₇H₅F₃O
10409 - 2-(Trifluoromethyl)phenol
10410 - 3-(Trifluoromethyl)phenol

C₇H₅HgNO₃
8120 - Nitromersol

C₇H₅IO
767 - Benzoyl iodide
6273 - 2-Iodobenzaldehyde
6274 - 4-Iodobenzaldehyde

C₇H₅IO₂
6278 - 2-Iodobenzoic acid
6279 - 3-Iodobenzoic acid
6280 - 4-Iodobenzoic acid

C₇H₅KO₂
9093 - Potassium benzoate

C₇H₅N
715 - Benzonitrile
6422 - Isocyanobenzene

C₇H₅NO
749 - Benzoxazole
5498 - 3-(2-Furanyl)-2-propenenitrile
5994 - 2-Hydroxybenzonitrile
5995 - 3-Hydroxybenzonitrile
5996 - 4-Hydroxybenzonitrile
8906 - Phenyl isocyanate

C₇H₅NOS
740 - 2(3*H*)-Benzothiazolone
750 - 2(3*H*)-Benzoxazolethione

C₇H₅NO₂
751 - 2(3*H*)-Benzoxazolone

C₇H₅NO₃
8042 - 2-Nitrobenzaldehyde
8043 - 3-Nitrobenzaldehyde
8044 - 4-Nitrobenzaldehyde

C₇H₅NO₃S
9496 - Saccharin

C₇H₅NO₄
6128 - 2-Hydroxy-3-nitrobenzaldehyde
6129 - 2-Hydroxy-5-nitrobenzaldehyde
8073 - 2-Nitrobenzoic acid
8074 - 3-Nitrobenzoic acid
8075 - 4-Nitrobenzoic acid
9338 - 2,3-Pyridinedicarboxylic acid
9339 - 2,4-Pyridinedicarboxylic acid
9340 - 2,5-Pyridinedicarboxylic acid
9341 - 2,6-Pyridinedicarboxylic acid
9342 - 3,4-Pyridinedicarboxylic acid
9343 - 3,5-Pyridinedicarboxylic acid

C₇H₅NO₅
6130 - 2-Hydroxy-3-nitrobenzoic acid
6131 - 2-Hydroxy-5-nitrobenzoic acid

C₇H₅NS
737 - Benzothiazole
8909 - Phenyl isothiocyanate
9003 - Phenyl thiocyanate

C₇H₅NS₂
738 - 2(3*H*)-Benzothiazolethione

C₇H₅N₃O
755 - Benzoyl azide

C₇H₅N₃O₂
376 - 2-Amino-5-nitrobenzonitrile
8072 - 5-Nitro-1*H*-benzimidazole
8114 - 5-Nitro-1*H*-indazole
8115 - 6-Nitro-1*H*-indazole

C₇H₅N₃O₅
4368 - 3,5-Dinitrobenzamide

C₇H₅N₃O₆
10633 - 2,4,6-Trinitrotoluene

C₇H₅N₃O₇
6906 - 2-Methoxy-1,3,5-trinitrobenzene

C₇H₅N₅O₈
7798 - *N*-Methyl-*N*,2,4,6-tetranitroaniline

C₇H₅NaO₂
9548 - Sodium benzoate

C₇H₆BrCl
1148 - 1-Bromo-4-(chloromethyl)benzene
1256 - 1-(Bromomethyl)-2-chlorobenzene

C₇H₆BrF
1260 - 1-(Bromomethyl)-3-fluorobenzene

C₇H₆BrNO₂
266 - 2-Amino-5-bromobenzoic acid
1267 - 1-(Bromomethyl)-3-nitrobenzene
1268 - 1-(Bromomethyl)-4-nitrobenzene

C₇H₆BrNO₃
1175 - 5-Bromo-*N*,2-dihydroxybenzamide
1269 - 2-(Bromomethyl)-4-nitrophenol

C₇H₆Br₂
1115 - 1-Bromo-2-(bromomethyl)benzene
1116 - 1-Bromo-3-(bromomethyl)benzene
1117 - 1-Bromo-4-(bromomethyl)benzene
2957 - 1,4-Dibromo-2-methylbenzene
2958 - 2,4-Dibromo-1-methylbenzene
2959 - (Dibromomethyl)benzene

C₇H₆Br₂O
2961 - 2,4-Dibromo-6-methylphenol

C₇H₆ClF

C_7H_6ClF

2055 - 1-Chloro-3-fluoro-2-methylbenzene
2123 - 1-(Chloromethyl)-2-fluorobenzene
2124 - 1-(Chloromethyl)-4-fluorobenzene

C_7H_6ClNO

1864 - 2-Chlorobenzamide

$C_7H_6ClNO_2$

285 - 2-Amino-5-chlorobenzoic acid
286 - 5-Amino-2-chlorobenzoic acid
2136 - 1-(Chloromethyl)-2-nitrobenzene
2137 - 1-(Chloromethyl)-3-nitrobenzene
2138 - 1-(Chloromethyl)-4-nitrobenzene
2139 - 1-Chloro-2-methyl-3-nitrobenzene
2140 - 1-Chloro-2-methyl-4-nitrobenzene
2141 - 1-Chloro-4-methyl-2-nitrobenzene
2142 - 2-Chloro-1-methyl-4-nitrobenzene
2143 - 4-Chloro-1-methyl-2-nitrobenzene

$C_7H_6ClN_3O_4S_2$

2311 - Chlorothiazide

$C_7H_6Cl_2$

1950 - 1-Chloro-2-(chloromethyl)benzene
1951 - 1-Chloro-3-(chloromethyl)benzene
1952 - 1-Chloro-4-(chloromethyl)benzene
3207 - (Dichloromethyl)benzene
3299 - 2,3-Dichlorotoluene
3300 - 2,4-Dichlorotoluene
3301 - 2,5-Dichlorotoluene
3302 - 2,6-Dichlorotoluene
3303 - 3,4-Dichlorotoluene

$C_7H_6Cl_2O$

3103 - 2,4-Dichlorobenzenemethanol
3203 - 1,2-Dichloro-3-methoxybenzene
3204 - 1,3-Dichloro-2-methoxybenzene
3205 - 2,4-Dichloro-1-methoxybenzene
3212 - 2,4-Dichloro-3-methylphenol
3213 - 2,4-Dichloro-6-methylphenol
3214 - 2,6-Dichloro-4-methylphenol

$C_7H_6Cl_2S$

1955 - 1-Chloro-4-[(chloromethyl)thio]benzene

$C_7H_6Cl_4O_2$

9747 - 1,2,3,4-Tetrachloro-5,5-dimethoxy-1,3-cyclopentadiene

$C_7H_6FNO_2$

5401 - 2-Fluoro-4-methyl-1-nitrobenzene

$C_7H_6F_3N$

10400 - 2-(Trifluoromethyl)aniline
10401 - 3-(Trifluoromethyl)aniline
10402 - 4-(Trifluoromethyl)aniline

$C_7H_6INO_2$

339 - 2-Amino-5-iodobenzoic acid

$C_7H_6N_2$

257 - 2-Aminobenzonitrile
258 - 3-Aminobenzonitrile
259 - 4-Aminobenzonitrile
683 - 1H-Benzimidazole
6233 - 1H-Indazole

$C_7H_6N_2O$

3597 - 1,3-Dihydro-2H-benzimidazol-2-one
6234 - 1H-Indazol-3-ol

$C_7H_6N_2O_3$

8045 - 3-Nitrobenzamide
8046 - 4-Nitrobenzamide

$C_7H_6N_2O_4$

374 - 2-Amino-4-nitrobenzoic acid
375 - 2-Amino-5-nitrobenzoic acid
7208 - 1-Methyl-2,3-dinitrobenzene

7209 - 1-Methyl-2,4-dinitrobenzene
7210 - 1-Methyl-3,5-dinitrobenzene
7211 - 2-Methyl-1,3-dinitrobenzene
7212 - 2-Methyl-1,4-dinitrobenzene
7213 - 4-Methyl-1,2-dinitrobenzene

$C_7H_6N_2O_5$

6819 - 1-Methoxy-2,4-dinitrobenzene
6820 - 1-Methoxy-3,5-dinitrobenzene
7214 - 2-Methyl-4,6-dinitrophenol
7215 - 4-Methyl-2,6-dinitrophenol

$C_7H_6N_2S$

735 - 2-Benzothiazolamine
736 - 6-Benzothiazolamine
3596 - 1,3-Dihydro-2H-benzimidazole-2-thione

$C_7H_6N_4O$

1759 - N,N'-Carbonyldiimidazole

$C_7H_6N_4S$

3666 - 1,4-Dihydro-1-phenyl-5H-tetrazole-5-thione

C_7H_6O

585 - Benzaldehyde
2559 - 2,4,6-Cycloheptatrien-1-one

C_7H_6OS

609 - Benzenecarbothioic acid

$C_7H_6OS_2$

5978 - 2-Hydroxybenzenecarbodithioic acid

$C_7H_6O_2$

691 - 1,3-Benzodioxole
712 - Benzoic acid
5497 - 3-(2-Furanyl)-2-propenal
5968 - 3-Hydroxybenzaldehyde
5969 - 4-Hydroxybenzaldehyde
6019 - 2-Hydroxy-2,4-cycloheptatrien-1-one
7144 - 2-Methyl-2,5-cyclohexadiene-1,4-dione
8883 - Phenyl formate
9501 - Salicylaldehyde

$C_7H_6O_2S$

6709 - 2-Mercaptobenzoic acid

$C_7H_6O_3$

607 - Benzenecarboperoxoic acid
697 - 1,3-Benzodioxol-5-ol
3698 - 2,3-Dihydroxybenzaldehyde
3699 - 2,4-Dihydroxybenzaldehyde
3700 - 2,5-Dihydroxybenzaldehyde
3701 - 3,4-Dihydroxybenzaldehyde
5499 - 3-(2-Furanyl)-2-propenoic acid
5990 - 2-Hydroxybenzoic acid
5991 - 3-Hydroxybenzoic acid
5992 - 4-Hydroxybenzoic acid

$C_7H_6O_4$

3704 - 2,3-Dihydroxybenzoic acid
3705 - 2,4-Dihydroxybenzoic acid
3706 - 2,5-Dihydroxybenzoic acid
3707 - 2,6-Dihydroxybenzoic acid
3708 - 3,4-Dihydroxybenzoic acid
3709 - 3,5-Dihydroxybenzoic acid
6045 - 4-Hydroxy-4H-furo[3,2-c]pyran-2(6H)-one

$C_7H_6O_5$

10435 - 2,3,4-Trihydroxybenzoic acid
10436 - 2,4,6-Trihydroxybenzoic acid
10437 - 3,4,5-Trihydroxybenzoic acid

$C_7H_6O_5S$

9657 - 2-Sulfobenzoic acid

$C_7H_6O_6S$

6195 - 2-Hydroxy-5-sulfobenzoic acid

C_7H_7Br

1247 - (Bromomethyl)benzene
1359 - 2-Bromotoluene
1360 - 3-Bromotoluene
1361 - 4-Bromotoluene

C_7H_7BrO

1090 - 2-Bromoanisole
1091 - 3-Bromoanisole
1092 - 4-Bromoanisole
1275 - 2-Bromo-4-methylphenol

$C_7H_7BrO_2$

1232 - 5-Bromo-2-hydroxybenzenemethanol

C_7H_7Cl

2106 - (Chloromethyl)benzene
2315 - 2-Chlorotoluene
2316 - 3-Chlorotoluene
2317 - 4-Chlorotoluene

$C_7H_7ClNNaO_2S$

1815 - Chloramine T

$C_7H_7ClN_2S$

2255 - (2-Chlorophenyl)thiourea

C_7H_7ClO

1855 - 2-Chloroanisole
1856 - 3-Chloroanisole
1857 - 4-Chloroanisole
1885 - 2-Chlorobenzenemethanol
1886 - 4-Chlorobenzenemethanol
2146 - 2-Chloro-4-methylphenol
2147 - 2-Chloro-5-methylphenol
2148 - 2-Chloro-6-methylphenol
2149 - 3-Chloro-4-methylphenol
2150 - 4-Chloro-2-methylphenol
2151 - 4-Chloro-3-methylphenol

C_7H_7ClOS

7006 - 4-Methylbenzenesulfinyl chloride

$C_7H_7ClO_2S$

645 - Benzenemethanesulfonyl chloride
2164 - 1-Chloro-4-(methylsulfonyl)benzene
7011 - 2-Methylbenzenesulfonyl chloride
10118 - p-Toluenesulfonyl chloride

$C_7H_7ClO_3S$

6796 - 4-Methoxybenzenesulfonyl chloride

C_7H_7ClS

1884 - 4-Chlorobenzenemethanethiol
2165 - 1-Chloro-4-(methylthio)benzene

$C_7H_7Cl_2N$

3102 - 2,4-Dichlorobenzenemethanamine

$C_7H_7Cl_2NO$

2432 - Clopidol

$C_7H_7Cl_2NO_2S$

3208 - N,N-Dichloro-4-methylbenzenesulfonamide

$C_7H_7Cl_3NO_3PS$

2353 - Chlorpyrifos-methyl

C_7H_7F

5400 - (Fluoromethyl)benzene
5424 - 2-Fluorotoluene
5425 - 3-Fluorotoluene
5426 - 4-Fluorotoluene

C_7H_7FO

5366 - 4-Fluorobenzenemethanol
5396 - 1-Fluoro-2-methoxybenzene
5397 - 1-Fluoro-3-methoxybenzene
5398 - 1-Fluoro-4-methoxybenzene

$C_7H_7FO_2S$

646 - Benzenemethanesulfonyl fluoride

C_7H_7I
6311 - 1-Iodo-2-methylbenzene
6312 - 1-Iodo-3-methylbenzene
6313 - (Iodomethyl)benzene
6345 - 4-Iodotoluene

C_7H_7IO
6276 - 2-Iodobenzenemethanol
6308 - 1-Iodo-2-methoxybenzene
6309 - 1-Iodo-3-methoxybenzene
6310 - 1-Iodo-4-methoxybenzene

C_7H_7N
10820 - 2-Vinylpyridine
10821 - 3-Vinylpyridine
10822 - 4-Vinylpyridine

C_7H_7NO
239 - 2-Aminobenzaldehyde
240 - 3-Aminobenzaldehyde
241 - 4-Aminobenzaldehyde
587 - cis-Benzaldehyde oxime
588 - trans-Benzaldehyde oxime
591 - Benzamide
7741 - 6-Methyl-2-pyridinecarboxaldehyde
8882 - N-Phenylformamide
9367 - 1-(2-Pyridinyl)ethanone
9368 - 1-(3-Pyridinyl)ethanone
9369 - 1-(4-Pyridinyl)ethanone

$C_7H_7NO_2$
467 - Aniline-2-carboxylic acid
468 - Aniline-3-carboxylic acid
469 - Aniline-4-carboxylic acid
690 - 1,3-Benzodioxol-5-amine
811 - Benzyl nitrite
5971 - 2-Hydroxybenzamide
5972 - N-Hydroxybenzamide
7742 - Methyl 3-pyridinecarboxylate
7743 - Methyl 4-pyridinecarboxylate
8123 - (Nitromethyl)benzene
8206 - 2-Nitrotoluene
8207 - 3-Nitrotoluene
8208 - 4-Nitrotoluene
9502 - Salicylaldoxime
10428 - Trigonelline

$C_7H_7NO_2S$
8204 - 4-Nitrothioanisole

$C_7H_7NO_3$
325 - 2-Amino-3-hydroxybenzoic acid
326 - 4-Amino-2-hydroxybenzoic acid
327 - 5-Amino-2-hydroxybenzoic acid
3702 - N,2-Dihydroxybenzamide
7497 - 3-Methyl-4-nitrophenol
7498 - 4-Methyl-2-nitrophenol
8037 - 2-Nitroanisole
8038 - 3-Nitroanisole
8039 - 4-Nitroanisole
8061 - 2-Nitrobenzenemethanol
8062 - 3-Nitrobenzenemethanol
8063 - 4-Nitrobenzenemethanol

$C_7H_7NO_4S$
427 - 4-(Aminosulfonyl)benzoic acid

$C_7H_7NO_5S$
7489 - 2-Methyl-5-nitrobenzenesulfonic acid

C_7H_7NS
608 - Benzenecarbothioamide

$C_7H_7N_3$
553 - 1-Azido-4-methylbenzene
554 - (Azidomethyl)benzene
682 - 1H-Benzimidazol-2-amine

7030 - 1-Methyl-1H-benzotriazole

$C_7H_7N_3O_3$
8076 - 3-Nitrobenzoic acid, hydrazide
8077 - 4-Nitrobenzoic acid, hydrazide
8161 - (4-Nitrophenyl)urea

$C_7H_7N_3S$
741 - 2(3H)-Benzothiazolone, hydrazone

C_7H_8
2558 - 1,3,5-Cycloheptatriene
5658 - 1,6-Heptadiyne
8263 - 2,5-Norbornadiene
10111 - Toluene

C_7H_8BrN
1245 - 2-Bromo-4-methylaniline
1246 - 4-Bromo-2-methylaniline

C_7H_8ClN
1881 - 2-Chlorobenzenemethanamine
1882 - 3-Chlorobenzenemethanamine
1883 - 4-Chlorobenzenemethanamine
2098 - 4-Chloro-N-methylaniline
2099 - 2-Chloro-4-methylaniline
2100 - 2-Chloro-6-methylaniline
2101 - 3-Chloro-2-methylaniline
2102 - 3-Chloro-4-methylaniline
2103 - 4-Chloro-2-methylaniline
2104 - 5-Chloro-2-methylaniline

C_7H_8ClNO
2090 - 4-Chloro-2-methoxyaniline
2091 - 5-Chloro-2-methoxyaniline

$C_7H_8ClNO_3S$
288 - 2-Amino-4-chloro-5-methylbenzene-
sulfonic acid

$C_7H_8ClN_3O_4S_2$
5943 - Hydrochlorothiazide

$C_7H_8Cl_2Si$
3215 - Dichloromethylphenylsilane

C_7H_8FN
5365 - 4-Fluorobenzenemethanamine
5399 - 4-Fluoro-2-methylaniline

$C_7H_4NNaO_3S$
9497 - Saccharin sodium

$C_7H_8N_2$
586 - Benzaldehyde hydrazone

$C_7H_8N_2O$
242 - 2-Aminobenzamide
243 - 4-Aminobenzamide
711 - Benzohydrazide
7503 - N-Methyl-N-nitrosoaniline
9013 - Phenylurea

$C_7H_8N_2O_2$
2850 - 3,5-Diaminobenzoic acid
5940 - 4-Hydrazinobenzoic acid
5993 - 2-Hydroxybenzoic acid, hydrazide
7480 - 2-Methyl-3-nitroaniline
7481 - 2-Methyl-4-nitroaniline
7482 - 2-Methyl-5-nitroaniline
7483 - 2-Methyl-6-nitroaniline
7484 - 4-Methyl-2-nitroaniline
7485 - 4-Methyl-3-nitroaniline
7486 - N-Methyl-2-nitroaniline
7487 - N-Methyl-4-nitroaniline

$C_7H_8N_2O_3$
6852 - 2-Methoxy-4-nitroaniline
6853 - 2-Methoxy-5-nitroaniline
6854 - 4-Methoxy-2-nitroaniline

$C_7H_8N_2S$
9006 - Phenylthiourea

$C_7H_8N_4O_2$
10017 - Theobromine
10018 - Theophylline

C_7H_8O
474 - Anisole
780 - Benzyl alcohol
2477 - o-Cresol
2478 - m-Cresol
2479 - p-Cresol

C_7H_8OS
6797 - 3-Methoxybenzenethiol
6798 - 4-Methoxybenzenethiol
7789 - (Methylsulfinyl)benzene
7818 - 4-(Methylthio)phenol

$C_7H_8O_2$
4288 - 4,6-Dimethyl-2H-pyran-2-one
4289 - 2,6-Dimethyl-4H-pyran-4-one
5495 - 1-(2-Furanyl)-1-propanone
5496 - 1-(2-Furanyl)-2-propanone
5983 - 2-Hydroxybenzenemethanol
5984 - 3-Hydroxybenzenemethanol
5985 - 4-Hydroxybenzenemethanol
6855 - 2-Methoxyphenol
6856 - 3-Methoxyphenol
6857 - 4-Methoxyphenol
6981 - 3-Methyl-1,2-benzenediol
6982 - 4-Methyl-1,2-benzenediol
6983 - 2-Methyl-1,3-benzenediol
6984 - 4-Methyl-1,3-benzenediol
6985 - 5-Methyl-1,3-benzenediol
6986 - 2-Methyl-1,4-benzenediol

$C_7H_8O_2S$
5262 - Ethyl thiophene-2-carboxylate
7005 - 4-Methylbenzenesulfinic acid
7661 - Methyl phenyl sulfone

$C_7H_8O_3$
5037 - Ethyl 2-furancarboxylate
5489 - 2-Furanmethanol acetate
6080 - 2-(Hydroxymethyl)-1,4-benzenediol
6788 - 3-Methoxy-1,2-benzenediol

$C_7H_8O_3S$
833 - Benzylsulfonic acid
7009 - Methyl benzenesulfonate
7010 - 2-Methylbenzenesulfonic acid
10116 - p-Toluenesulfonic acid

C_7H_8S
647 - Benzenemethanethiol
7012 - 2-Methylbenzenethiol
7013 - 3-Methylbenzenethiol
7014 - 4-Methylbenzenethiol
7808 - (Methylthio)benzene

$C_7H_8S_2$
6987 - 4-Methyl-1,2-benzenedithiol

$C_7H_9AsN_2O_4$
277 - [4-[(Aminocarbonyl)amino]phenyl]
arsonic acid

$C_7H_9ClN_2$
610 - Benzenecarboximidamide, monohydrochloride

C_7H_9ClO
4805 - Ethchlorvynol

C_7H_9ClSi
2154 - Chloromethylphenylsilane

C_7H_9N
781 - Benzylamine
2592 - 1-Cyclohexenecarbonitrile
4299 - 2,3-Dimethylpyridine

5006 - Ethyl 2,2-dimethylpropanoate
5123 - Ethyl 3-methylbutanoate
5124 - 2-Ethyl-2-methylbutanoic acid
5193 - Ethyl pentanoate
5683 - Heptanoic acid
5897 - Hexyl formate
6401 - Isobutyl propanoate
6442 - Isopentyl acetate
6484 - Isopropyl butanoate
6502 - Isopropyl isobutanoate
6846 - 4-Methoxy-4-methyl-2-pentanone
7081 - 2-Methyl-1-butanol acetate
7329 - Methyl hexanoate
7330 - 2-Methylhexanoic acid
8662 - Pentyl acetate
8663 - sec-Pentyl acetate (R)
9217 - Propyl butanoate
9244 - Propyl isobutanoate

C₇H₁₄O₃
1598 - Butyl lactate
4553 - Dipropyl carbonate
5025 - Ethyl 3-ethoxypropanoate

C₇H₁₄O₄
3340 - 2,6-Dideoxy-3-O-methyl-ribo-hexose
5563 - Glycerol 1-butanoate

C₇H₁₄O₅
2807 - 6-Deoxy-3-O-methylgalactose

C₇H₁₄O₆
7277 - Methyl β-D-glucopyranoside
7278 - Methyl α-D-glucopyranoside

C₇H₁₅Br
1218 - 1-Bromoheptane
1219 - 2-Bromoheptane
1220 - 4-Bromoheptane

C₇H₁₅Cl
2060 - 1-Chloroheptane
2061 - 2-Chloroheptane
2062 - 3-Chloroheptane
2063 - 4-Chloroheptane

C₇H₁₅ClO
2064 - 7-Chloro-1-heptanol

C₇H₁₅ClO₂
1976 - 3-Chloro-1,1-diethoxypropane

C₇H₁₅Cl₂N₂O₂P
2686 - Cyclophosphamide

C₇H₁₅F
5389 - 1-Fluoroheptane

C₇H₁₅I
6303 - 1-Iodoheptane
6304 - 3-Iodoheptane

C₇H₁₅N
185 - Allyldiethylamine
2553 - Cycloheptanamine
2583 - Cyclohexanemethanamine
2634 - Cyclohexylmethylamine
4266 - 1,2-Dimethylpiperidine, (±)
4267 - 2,6-Dimethylpiperidine
4268 - 3,5-Dimethylpiperidine
5218 - 1-Ethylpiperidine
8333 - Octahydroazocine

C₇H₁₅NO
3513 - N,N-Diethylpropanamide
4212 - N,N-Dimethylpentanamide
5221 - 1-Ethyl-3-piperidinol
5669 - Heptanal oxime
7339 - 5-Methyl-2-hexanone oxime
9064 - 1-Piperidineethanol

9065 - 2-Piperidineethanol
9066 - 4-Piperidineethanol

C₇H₁₅NO₂
4696 - Emylcamate
4935 - Ethyl N-butylcarbamate
5085 - Ethyl isobutylcarbamate
7469 - α-Methyl-4-morpholineethanol

C₇H₁₅NO₃
1772 - Carnitine

C₇H₁₅NO₅
6931 - 2-(Methylamino)-2-deoxy-α-L-glu-
     copyranose

C₇H₁₆
4213 - 2,2-Dimethylpentane
4214 - 2,3-Dimethylpentane
4215 - 2,4-Dimethylpentane
4216 - 3,3-Dimethylpentane
5191 - 3-Ethylpentane
5672 - Heptane
7326 - 2-Methylhexane
7327 - 3-Methylhexane
10524 - 2,2,3-Trimethylbutane

C₇H₁₆BrNO₂
59 - Acetylcholine bromide

C₇H₁₆ClN
6705 - Mepiquat chloride

C₇H₁₆ClNO₂
60 - Acetylcholine chloride

C₇H₁₆ClN₃O₂S₂
1785 - Cartap hydrochloride

C₇H₁₆FO₂P
9582 - Soman

C₇H₁₆INOS
103 - Acetyl thiocholine iodide

C₇H₁₆INO₂
61 - Acetylcholine iodide

C₇H₁₆N₂
5241 - 1-Ethyl-2-pyrrolidinemethanamine
10606 - 1,2,4-Trimethylpiperazine

C₇H₁₆N₂O
7905 - 4-Morpholinepropanamine

C₇H₁₆N₂O₂
5820 - Hexamethylenediamine carbamate

C₇H₁₆O
1594 - tert-Butyl isopropyl ether
1657 - Butyl propyl ether
4220 - 2,2-Dimethyl-1-pentanol
4221 - 2,3-Dimethyl-2-pentanol
4222 - 2,4-Dimethyl-2-pentanol
4223 - 2,2-Dimethyl-3-pentanol
4224 - 2,3-Dimethyl-3-pentanol
4225 - 2,4-Dimethyl-3-pentanol
4849 - 2-Ethoxy-2-methylbutane
5089 - Ethyl isopentyl ether
5194 - 3-Ethyl-3-pentanol
5198 - Ethyl pentyl ether
5685 - 1-Heptanol
5686 - 2-Heptanol, (±)
5687 - 3-Heptanol, (S)
5688 - 4-Heptanol
5901 - Hexyl methyl ether
7254 - 1-(1-Methylethoxy)butane
7331 - 2-Methyl-1-hexanol, (±)
7332 - 5-Methyl-1-hexanol
7333 - 2-Methyl-2-hexanol
7334 - 3-Methyl-2-hexanol
7335 - 5-Methyl-2-hexanol

7336 - 3-Methyl-3-hexanol
10525 - 2,3,3-Trimethyl-2-butanol

C₇H₁₆O₂
1484 - 1-Butoxy-2-propanol
3367 - 1,1-Diethoxypropane
3368 - 2,2-Diethoxypropane
3516 - 2,2-Diethyl-1,3-propanediol
4546 - Dipropoxymethane
5676 - 1,7-Heptanediol
7718 - 2-Methyl-2-propyl-1,3-propanediol
9232 - 1,2-Propylene glycol 2-tert-butyl ether

C₇H₁₆O₂Si
10801 - Vinyldiethoxymethylsilane

C₇H₁₆O₃
3451 - Diethylene glycol monopropyl ether
4557 - Dipropylene glycol monomethyl ether
10335 - Triethoxymethane

C₇H₁₆O₄
6824 - 2-[2-(2-Methoxyethoxy)ethoxy]ethanol
9914 - 1,1,3,3-Tetramethoxypropane

C₇H₁₆O₄S₂
9660 - Sulfonmethane

C₇H₁₆S
5681 - 1-Heptanethiol

C₇H₁₆S₃
10689 - Tris(ethylthio)methane

C₇H₁₇N
5670 - 2-Heptanamine
5671 - 4-Heptanamine
5703 - Heptylamine
7324 - 3-Methyl-1-hexanamine
7325 - 4-Methyl-2-hexanamine

C₇H₁₇NO
3395 - 3-(Diethylamino)-1-propanol

C₇H₁₇NO₂
3355 - 1,1-Diethoxy-N,N-dimethylmetha-
     namine

C₇H₁₇NO₅
2806 - 1-Deoxy-1-(methylamino)-D-glucitol

C₇H₁₇O₂PS₃
9018 - Phorate

C₇H₁₇O₃P
3806 - Diisopropyl methylphosphonate

C₇H₁₈N₂
3514 - N,N-Diethyl-1,3-propanediamine
5673 - 1,7-Heptanediamine
9971 - N,N,N',N'-Tetramethyl-1,3-propanedi-
     amine

C₇H₁₈N₂O
967 - 1,3-Bis(dimethylamino)-2-propanol

C₇H₁₈N₂O₂
1004 - N,N-Bis(2-hydroxyethyl)-1,3-pro-
     panediamine

C₇H₁₈O₃Si
10336 - Triethoxymethylsilane

C₇H₁₈Pb
7834 - Methyltriethyllead

C₇H₁₉NSi
3538 - N,N-Diethyl-1,1,1-trimethylsilanamine

C₇H₁₉N₃
410 - N-(3-Aminopropyl)-N-methyl-1,3-pro-
     panediamine

C₇H₂₂O₂Si₃
5667 - 1,1,1,3,5,5,5-Heptamethyltrisiloxane

$C_8Br_4O_3$

  9714 - 4,5,6,7-Tetrabromo-1,3-isobenzofuran-dione

$C_8Cl_2N_2O_2$

  3154 - 2,3-Dichloro-5,6-dicyanobenzoquinone

$C_8Cl_4N_2$

  2309 - Chlorothalonil

$C_8Cl_4O_3$

  9765 - Tetrachlorophthalic anhydride

$C_8Cl_6O_2$

  9774 - 2,3,5,6-Tetrachloroterphthaloyl dichloride

C_8Cl_8

  8295 - Octachlorostyrene

$C_8Co_2O_8$

  2437 - Cobalt carbonyl

$C_8F_{14}O_3$

  5660 - Heptafluorobutanoic anhydride

$C_8F_{16}O$

  8698 - Perfluoro-2-butyltetrahydrofuran

C_8F_{18}

  8720 - Perfluorooctane

$C_8F_{18}O_2S$

  8721 - Perfluorooctylsulfonyl fluoride

$C_8HF_{15}O_2$

  8549 - Pentadecafluorooctanoic acid

$C_8H_2Cl_2O_3$

  3201 - 5,6-Dichloro-1,3-isobenzofurandione

$C_8H_3Cl_2F_3N_2$

  3310 - 4,5-Dichloro-2-(trifluoromethyl)-1H-benzimidazole

$C_8H_3F_{15}O$

  8550 - 2,2,3,3,4,4,5,5,6,6,7,7,8,8,8-Pentadecafluoro-1-octanol

$C_8H_3NO_5$

  8116 - 4-Nitro-1,3-isobenzofurandione

  8117 - 5-Nitro-1,3-isobenzofurandione

$C_8H_4ClF_3O$

  10407 - 3-(Trifluoromethyl)benzoyl chloride

$C_8H_4ClNO_3$

  1907 - 6-Chloro-2H-3,1-benzoxazine-2,4(1H)-dione

$C_8H_4Cl_2N_2$

  3291 - 2,3-Dichloroquinoxaline

$C_8H_4Cl_2O_2$

  617 - 1,2-Benzenedicarbonyl dichloride

  618 - 1,3-Benzenedicarbonyl dichloride

  619 - 1,4-Benzenedicarbonyl dichloride

$C_8H_4Cl_6$

  1047 - 1,4-Bis(trichloromethyl)benzene

$C_8H_4F_3N$

  10405 - 3-(Trifluoromethyl)benzonitrile

  10406 - 4-(Trifluoromethyl)benzonitrile

$C_8H_4F_3NO$

  6421 - 1-Isocyanato-3-(trifluoromethyl)benzene

$C_8H_4F_6$

  1051 - 1,3-Bis(trifluoromethyl)benzene

  1052 - 1,4-Bis(trifluoromethyl)benzene

$C_8H_4F_{15}NO_2$

  440 - Ammonium perfluorooctanoate

$C_8H_4N_2$

  3317 - o-Dicyanobenzene

  3318 - m-Dicyanobenzene

  3319 - p-Dicyanobenzene

$C_8H_4N_2O_2$

  3783 - 1,3-Diisocyanatobenzene

  3784 - 1,4-Diisocyanatobenzene

$C_8H_4N_2O_4$

  8119 - 5-Nitro-1H-isoindole-1,3(2H)-dione

$C_8H_4N_2S_2$

  3815 - 1,4-Diisothiocyanatobenzene

$C_8H_4O_3$

  9030 - Phthalic anhydride

C_8H_5Br

  1208 - 1-Bromo-4-ethynylbenzene

$C_8H_5Cl_3O_2$

  10207 - 2,3,6-Trichlorobenzeneacetic acid

$C_8H_5Cl_3O_3$

  10273 - 2,4,5-Trichlorophenoxyacetic acid

$C_8H_5F_3O$

  10403 - 4-(Trifluoromethyl)benzaldehyde

  10417 - 2,2,2-Trifluoro-1-phenylethanone

$C_8H_5F_3O_2S$

  10423 - 4,4,4-Trifluoro-1-(2-thienyl)-1,3-butanedione

$C_8H_5F_6N$

  1050 - 3,5-Bis(trifluoromethyl)aniline

$C_8H_5KO_4$

  9097 - Potassium hydrogen phthalate

$C_8H_5MnO_3$

  6676 - Manganese cyclopentadienyl tricarbonyl

C_8H_5NO

  5460 - 3-Formylbenzonitrile

  5461 - 4-Formylbenzonitrile

  8453 - α-Oxobenzeneacetonitrile

$C_8H_5NO_2$

  2507 - 3-Cyanobenzoic acid

  2508 - 4-Cyanobenzoic acid

  6248 - 1H-Indole-2,3-dione

  6433 - 1H-Isoindole-1,3(2H)-dione

$C_8H_5NO_3$

  748 - 2H-3,1-Benzoxazine-2,4(1H)-dione

  6052 - 2-Hydroxy-1H-isoindole-1,3(2H)-dione

$C_8H_5NO_6$

  8057 - 3-Nitro-1,2-benzenedicarboxylic acid

  8058 - 4-Nitro-1,2-benzenedicarboxylic acid

C_8H_6

  8800 - Phenylacetylene

C_8H_6BrClO

  1137 - 2-Bromo-3'-chloroacetophenone

  1149 - 2-Bromo-1-(4-chlorophenyl)ethanone

C_8H_6BrN

  1098 - 4-Bromobenzeneacetonitrile

  1099 - α-Bromobenzeneacetonitrile

  1249 - 3-(Bromomethyl)benzonitrile

  1250 - 4-(Bromomethyl)benzonitrile

$C_8H_6Br_2O$

  1119 - 2-Bromo-1-(4-bromophenyl)ethanone

$C_8H_6Br_4$

  955 - 1,2-Bis(dibromomethyl)benzene

C_8H_6ClN

  1869 - 2-Chlorobenzeneacetonitrile

  1870 - 3-Chlorobenzeneacetonitrile

  1871 - 4-Chlorobenzeneacetonitrile

$C_8H_6ClNO_4$

  1748 - Carbonochloridic acid, (4-nitrophenyl) methyl ester

  7127 - Methyl 5-chloro-2-nitrobenzoate

$C_8H_6Cl_2$

  3292 - 2,5-Dichlorostyrene

$C_8H_6Cl_2O$

  1872 - α-Chlorobenzeneacetyl chloride

  1956 - 2-Chloro-1-(4-chlorophenyl)ethanone

  3261 - 2,2-Dichloro-1-phenylethanone

  3262 - 1-(2,4-Dichlorophenyl)ethanone

  3263 - 1-(2,5-Dichlorophenyl)ethanone

  3264 - 1-(3,4-Dichlorophenyl)ethanone

$C_8H_6Cl_2O_2$

  7191 - Methyl 2,5-dichlorobenzoate

$C_8H_6Cl_2O_3$

  3206 - 3,6-Dichloro-2-methoxybenzoic acid

  3256 - (2,4-Dichlorophenoxy)acetic acid

$C_8H_6Cl_4$

  9748 - 1,2,3,4-Tetrachloro-5,6-dimethylbenzene

  9749 - 1,2,3,5-Tetrachloro-4,6-dimethylbenzene

C_8H_6FN

  5363 - 2-Fluorobenzeneacetonitrile

  5364 - 4-Fluorobenzeneacetonitrile

$C_8H_6F_3N_3O_4S_2$

  5339 - Flumethiazide

$C_8H_6N_2$

  2403 - Cinnoline

  7994 - 1,5-Naphthyridine

  7995 - 1,6-Naphthyridine

  9028 - Phthalazine

  9413 - Quinazoline

  9444 - Quinoxaline

$C_8H_6N_2O$

  9445 - 2(1H)-Quinoxalinone

$C_8H_6N_2O_2$

  340 - 4-Amino-1H-isoindole-1,3(2H)-dione

  403 - 4-Aminophthalimide

  3678 - 1,4-Dihydro-2,3-quinoxalinedione

  8051 - 2-Nitrobenzeneacetonitrile

  8052 - 4-Nitrobenzeneacetonitrile

$C_8H_6N_2O_4$

  5309 - Fenadiazole

$C_8H_6N_2O_6$

  4397 - 2,4-Dinitrophenol, acetate

$C_8H_6N_4O_4S$

  8026 - Nifurthiazole

$C_8H_6N_4O_5$

  8106 - Nitrofurantoin

$C_8H_6N_4O_8$

  166 - Alloxantin

$C_8H_6N_6O_{11}$

  8661 - Pentryl

C_8H_6O

  705 - Benzofuran

  8772 - Phenoxyacetylene

$C_8H_6O_2$

  620 - 1,2-Benzenedicarboxaldehyde

  621 - 1,3-Benzenedicarboxaldehyde

  622 - 1,4-Benzenedicarboxaldehyde

  707 - 2(3H)-Benzofuranone

  708 - 3(2H)-Benzofuranone

  6366 - 1(3H)-Isobenzofuranone

$C_8H_6O_3$

  692 - 1,3-Benzodioxole-5-carboxaldehyde

  5457 - 2-Formylbenzoic acid

  5458 - 3-Formylbenzoic acid

  5459 - 4-Formylbenzoic acid

  8451 - α-Oxobenzeneacetic acid

C₈H₈N₂O₃

8137 - N-(2-Nitrophenyl)acetamide
8138 - N-(3-Nitrophenyl)acetamide
8139 - N-(4-Nitrophenyl)acetamide

C₈H₈N₂S

7026 - 6-Methyl-2-benzothiazolamine

C₈H₈N₄

5929 - Hydralazine

C₈H₈O

39 - Acetophenone
596 - Benzeneacetaldehyde
3599 - 2,3-Dihydrobenzofuran
6194 - 4-Hydroxystyrene
6957 - 2-Methylbenzaldehyde
6958 - 3-Methylbenzaldehyde
6959 - 4-Methylbenzaldehyde
8940 - Phenyloxirane
9015 - Phenyl vinyl ether

C₈H₈OS

3604 - 6,7-Dihydrobenzo[b]thiophen-4(5H)-
one

C₈H₈O₂

598 - Benzeneacetic acid
799 - Benzyl formate
3598 - 2,3-Dihydro-1,4-benzodioxin
4007 - 2,3-Dimethyl-2,5-cyclohexadiene-1,4-
dione
4008 - 2,5-Dimethyl-2,5-cyclohexadiene-1,4-
dione
4009 - 2,6-Dimethyl-2,5-cyclohexadiene-1,4-
dione
5491 - 4-(2-Furanyl)-3-buten-2-one
5961 - (Hydroxyacetyl)benzene
6076 - 2-Hydroxy-5-methylbenzaldehyde
6153 - 1-(2-Hydroxyphenyl)ethanone
6154 - 1-(3-Hydroxyphenyl)ethanone
6155 - 1-(4-Hydroxyphenyl)ethanone
6777 - 2-Methoxybenzaldehyde
6778 - 3-Methoxybenzaldehyde
6779 - 4-Methoxybenzaldehyde
7017 - Methyl benzoate
8797 - Phenyl acetate
10119 - o-Toluic acid
10120 - m-Toluic acid
10121 - p-Toluic acid

C₈H₈O₃

66 - 1-Acetyl-2,5-dihydroxybenzene
189 - Allyl 2-furancarboxylate
696 - 1,3-Benzodioxole-5-methanol
3742 - 1-(2,4-Dihydroxyphenyl)ethanone
5973 - α-Hydroxybenzeneacetic acid, (±)
5974 - 2-Hydroxybenzeneacetic acid
5975 - 3-Hydroxybenzeneacetic acid
5976 - 4-Hydroxybenzeneacetic acid
6056 - 2-Hydroxy-3-methoxybenzaldehyde
6057 - 2-Hydroxy-4-methoxybenzaldehyde
6058 - 2-Hydroxy-5-methoxybenzaldehyde
6059 - 3-Hydroxy-4-methoxybenzaldehyde
6060 - 4-Hydroxy-3-methoxybenzaldehyde
6081 - 2-Hydroxy-5-methylbenzoic acid
6082 - 2-Hydroxy-3-methylbenzoic acid
6083 - 2-Hydroxy-4-methylbenzoic acid
6799 - 2-Methoxybenzoic acid
6800 - 3-Methoxybenzoic acid
6801 - 4-Methoxybenzoic acid
7368 - Methyl 3-hydroxybenzoate
7369 - Methyl 4-hydroxybenzoate

7779 - Methyl salicylate
8770 - Phenoxyacetic acid
9863 - cis-3a,4,7,7a-Tetrahydro-1,3-isobenzo-
furandione
9864 - 4,5,6,7-Tetrahydro-1,3-isobenzofuran-
dione

C₈H₈O₄

82 - 3-Acetyl-6-methyl-2H-pyran-2,4(3H)-
dione
3703 - 2,5-Dihydroxybenzeneacetic acid
3732 - 2,4-Dihydroxy-6-methylbenzoic acid
3858 - 2,6-Dimethoxy-2,5-cyclohexadiene-1,4-
dione
5477 - Fumigatin
6064 - 2-Hydroxy-5-methoxybenzoic acid
6065 - 4-Hydroxy-3-methoxybenzoic acid
7197 - Methyl 2,4-dihydroxybenzoate
7198 - Methyl 3,5-dihydroxybenzoate
10445 - 1-(2,3,4-Trihydroxyphenyl)ethanone
10446 - 1-(2,4,6-Trihydroxyphenyl)ethanone

C₈H₈O₅

7837 - Methyl 3,4,5-trihydroxybenzoate
9587 - Spinulosin

C₈H₉Br

1180 - 1-Bromo-2,4-dimethylbenzene
1181 - 1-Bromo-3,5-dimethylbenzene
1182 - 2-Bromo-1,3-dimethylbenzene
1183 - 2-Bromo-1,4-dimethylbenzene
1184 - 4-Bromo-1,2-dimethylbenzene
1201 - (1-Bromoethyl)benzene
1202 - (2-Bromoethyl)benzene
1203 - 1-Bromo-2-ethylbenzene
1204 - 1-Bromo-3-ethylbenzene
1205 - 1-Bromo-4-ethylbenzene
1262 - 1-(Bromomethyl)-2-methylbenzene
1263 - 1-(Bromomethyl)-3-methylbenzene
1264 - 1-(Bromomethyl)-4-methylbenzene

C₈H₉BrO

1195 - 1-Bromo-2-ethoxybenzene
1196 - 1-Bromo-4-ethoxybenzene
1197 - (2-Bromoethoxy)benzene

C₈H₉BrO₂

1176 - 2-Bromo-1,4-dimethoxybenzene
1177 - 4-Bromo-1,2-dimethoxybenzene

C₈H₉Cl

1994 - 2-Chloro-1,4-dimethylbenzene
1995 - 4-Chloro-1,2-dimethylbenzene
2030 - (1-Chloroethyl)benzene
2031 - (2-Chloroethyl)benzene
2032 - 1-Chloro-2-ethylbenzene
2033 - 1-Chloro-3-ethylbenzene
2034 - 1-Chloro-4-ethylbenzene
2129 - 1-(Chloromethyl)-2-methylbenzene
2130 - 1-(Chloromethyl)-3-methylbenzene
2131 - 1-(Chloromethyl)-4-methylbenzene

C₈H₉ClNO₅PS

2357 - Chlorthion
3061 - Dicapthon

C₈H₉ClO

1998 - 4-Chloro-2,5-dimethylphenol
1999 - 4-Chloro-2,6-dimethylphenol
2000 - 4-Chloro-3,5-dimethylphenol
2023 - 1-Chloro-4-ethoxybenzene
2024 - (2-Chloroethoxy)benzene
2094 - [(Chloromethoxy)methyl]benzene
2108 - α-(Chloromethyl)benzenemethanol
2109 - 4-Chloro-α-methylbenzenemethanol

2128 - 1-(Chloromethyl)-4-methoxybenzene

C₈H₉Cl₃Si

10275 - Trichloro(2-phenylethyl)silane

C₈H₉I

6292 - 1-Iodo-2,4-dimethylbenzene
6293 - 2-Iodo-1,3-dimethylbenzene
6294 - 2-Iodo-1,4-dimethylbenzene
6300 - (2-Iodoethyl)benzene

C₈H₉N

857 - Bicyclo[2.2.1]hept-5-ene-2-carbonitrile
3633 - 2,3-Dihydro-1H-indole
8927 - N-(Phenylmethylene)methanamine
10792 - 4-Vinylaniline

C₈H₉NO

18 - Acetanilide
396 - 2-Amino-1-phenylethanone
397 - 1-(3-Aminophenyl)ethanone
398 - 1-(4-Aminophenyl)ethanone
597 - Benzeneacetamide
6960 - 2-Methylbenzamide
6961 - 4-Methylbenzamide
6962 - N-Methylbenzamide
7636 - N-Methyl-N-phenylformamide
7637 - N-(2-Methylphenyl)formamide
7748 - 1-(6-Methyl-3-pyridinyl)ethanone
8874 - 1-Phenylethanone oxime

C₈H₉NO₂

244 - α-Aminobenzeneacetic acid, (±)
245 - 4-Aminobenzeneacetic acid
695 - 1,3-Benzodioxole-5-methanamine
4153 - 1,2-Dimethyl-3-nitrobenzene
4154 - 1,2-Dimethyl-4-nitrobenzene
4155 - 1,3-Dimethyl-2-nitrobenzene
4156 - 1,3-Dimethyl-5-nitrobenzene
4157 - 1,4-Dimethyl-2-nitrobenzene
4158 - 2,4-Dimethyl-1-nitrobenzene
5168 - 1-Ethyl-2-nitrobenzene
5169 - 1-Ethyl-4-nitrobenzene
5235 - Ethyl 2-pyridinecarboxylate
5236 - Ethyl 3-pyridinecarboxylate
5237 - Ethyl 4-pyridinecarboxylate
6143 - N-(2-Hydroxyphenyl)acetamide
6144 - N-(3-Hydroxyphenyl)acetamide
6145 - N-(4-Hydroxyphenyl)acetamide
6780 - 4-Methoxybenzamide
6923 - Methyl 2-aminobenzoate
6924 - Methyl 3-aminobenzoate
6925 - Methyl 4-aminobenzoate
6926 - 2-(Methylamino)benzoic acid
6927 - 3-(Methylamino)benzoic acid
6928 - 4-(Methylamino)benzoic acid
8098 - (2-Nitroethyl)benzene
8887 - N-Phenylglycine
9881 - 1,2,3,6-Tetrahydrophthalimide

C₈H₉NO₃

4855 - 1-Ethoxy-2-nitrobenzene
4856 - 1-Ethoxy-4-nitrobenzene
6157 - 2-(4-Hydroxyphenyl)-D-glycine
6158 - N-(4-Hydroxyphenyl)glycine
6934 - Methyl 3-amino-4-hydroxybenzoate
8059 - 2-Nitrobenzeneethanol
8060 - 4-Nitrobenzeneethanol

C₈H₉NO₄

3877 - 1,2-Dimethoxy-4-nitrobenzene
3878 - 1,4-Dimethoxy-2-nitrobenzene

C₈H₉NS

9002 - N-Phenylthioacetamide

$C_8H_{11}NO_2S$
3961 - *N*,4-Dimethylbenzenesulfonamide
$C_8H_{11}NO_3$
330 - 4-(2-Amino-1-hydroxyethyl)-1,2-ben-
zenediol, (±)
4957 - Ethyl 2-cyano-3-ethoxyacrylate
6107 - 5-Hydroxy-6-methyl-3,4-
pyridinedimethanol
8270 - Norepinephrine
$C_8H_{11}N_3O_6$
546 - 6-Azauridine
$C_8H_{11}N_5$
8847 - Phenyl biguanide
$C_8H_{11}N_5O_3$
120 - Acyclovir
C_8H_{12}
2643 - 1,4-Cyclooctadiene
2644 - *cis*,*cis*-1,5-Cyclooctadiene
2653 - Cyclooctyne
4054 - 1,2-Dimethylenecyclohexane
4603 - *cis*-1,2-Divinylcyclobutane
4604 - *trans*-1,2-Divinylcyclobutane
7513 - 2-Methyl-2-norbornene
8376 - 1-Octen-3-yne
10798 - 1-Vinylcyclohexene
10799 - 4-Vinylcyclohexene
$C_8H_{12}ClN$
641 - Benzeneethanamine, hydrochloride
3937 - *N*,*N*-Dimethylaniline hydrochloride
$C_8H_{12}ClNO$
1970 - 2-Chloro-*N*,*N*-diallylacetamide
$C_8H_{12}ClNO_2$
311 - 4-(2-Aminoethyl)-1,2-benzenediol,
hydrochloride
$C_8H_{12}ClNO_3$
9376 - Pyridoxine hydrochloride
$C_8H_{12}N_2$
629 - 1,3-Benzenedimethanamine
809 - 1-Benzyl-2-methylhydrazine
3949 - 4,5-Dimethyl-1,2-benzenediamine
3950 - *N*,*N*-Dimethyl-1,2-benzenediamine
3951 - *N*,*N*-Dimethyl-1,3-benzenediamine
3952 - *N*,*N*-Dimethyl-1,4-benzenediamine
5007 - 3-Ethyl-2,5-dimethylpyrazine
8348 - Octanedinitrile
9972 - Tetramethylpyrazine
9976 - Tetramethylsuccinonitrile
$C_8H_{12}N_2O$
4829 - 4-Ethoxy-1,2-benzenediamine
6516 - 2-Isopropyl-6-methyl-4-pyrimidinol
$C_8H_{12}N_2O_2$
5821 - Hexamethylene diisocyanate
9374 - Pyridoxamine
$C_8H_{12}N_2O_3$
571 - Barbital
$C_8H_{12}N_2O_3S$
386 - 6-Aminopenicillanic acid
$C_8H_{12}N_4$
562 - 2,2'-Azobis[isobutyronitrile]
$C_8H_{12}N_4O_5$
541 - Azacitidine
9476 - Ribavirin
$C_8H_{12}O$
859 - Bicyclo[2.2.1]hept-5-ene-2-methanol
2602 - 1-(1-Cyclohexen-1-yl)ethanone
4027 - 3,5-Dimethyl-2-cyclohexen-1-one
5287 - 1-Ethynylcyclohexanol

10816 - 3-Vinyl-7-oxabicyclo[4.1.0]heptane
$C_8H_{12}O_2$
62 - 2-Acetylcyclohexanone
4018 - 5,5-Dimethyl-1,3-cyclohexanedione
4711 - 1,2-Epoxy-4-(epoxyethyl)cyclohexane
5045 - Ethyl *trans*,*trans*-2,4-hexadienoate
7162 - Methyl 3-cyclohexene-1-carboxylate
9934 - 2,2,4,4-Tetramethyl-1,3-cyclobutanedi-
one
$C_8H_{12}O_3$
4968 - Ethyl 2-cyclopentanone-1-carboxylate
9859 - Tetrahydrofurfuryl acrylate
$C_8H_{12}O_4$
1529 - Butyl *cis*-2-butenedioate
2573 - *trans*-1,4-Cyclohexanedicarboxylic acid
3461 - Diethyl fumarate
3476 - Diethyl maleate
7699 - 2-Methyl-2-propene-1,1-diol diacetate
9850 - Tetrahydro-2,2-dimethyl-5-oxo-3-
furanacetic acid
$C_8H_{12}O_5$
3496 - Diethyl oxobutanedioate
$C_8H_{12}O_6Si$
10826 - Vinyltriacetoxysilane
$C_8H_{12}S$
1662 - 2-Butylthiophene
$C_8H_{12}Si$
4258 - Dimethylphenylsilane
10005 - Tetravinylsilane
$C_8H_{13}N$
5008 - 3-Ethyl-2,4-dimethyl-1*H*-pyrrole
5286 - 1-Ethynylcyclohexanamine
$C_8H_{13}NO$
6954 - 8-Methyl-8-azabicyclo[3.2.1]octan-3-
one
$C_8H_{13}NO_2$
512 - Arecoline
5152 - 4-Ethyl-4-methyl-2,6-piperidinedione
9465 - Retronecine, (+)
9519 - Scopoline
$C_8H_{13}NO_3$
3428 - 5,5-Diethyldihydro-2*H*-1,3-oxazine-
2,4(3*H*)-dione
$C_8H_{13}N_2O_2P$
8003 - Nellite
$C_8H_{13}N_2O_3PS$
10061 - Thionazin
C_8H_{14}
183 - Allylcyclopentane
2651 - *cis*-Cyclooctene
2652 - *trans*-Cyclooctene
4025 - 1,2-Dimethylcyclohexene
4026 - 1,3-Dimethylcyclohexene
4086 - 2,5-Dimethyl-1,5-hexadiene
4087 - 2,5-Dimethyl-2,4-hexadiene
4964 - 1-Ethylcyclohexene
5081 - Ethylidenecyclohexane
8329 - 1,7-Octadiene
8402 - 1-Octyne
8403 - 2-Octyne
8404 - 3-Octyne
8405 - 4-Octyne
10797 - Vinylcyclohexane
$C_8H_{14}ClNS_2$
9640 - Sulfallate
$C_8H_{14}ClN_5$
531 - Atrazine

$C_8H_{14}Cl_2N_2O_2$
9375 - Pyridoxamine dihydrochloride
$C_8H_{14}Cl_3O_5P$
1469 - Butonate
$C_8H_{14}N_2$
9071 - 1-Piperidinepropanenitrile
$C_8H_{14}N_4OS$
7867 - Metribuzin
$C_8H_{14}O$
2622 - 1-Cyclohexylethanone
2648 - Cyclooctanone
4021 - 2,2-Dimethylcyclohexanone
4022 - 2,6-Dimethylcyclohexanone
4023 - 3,3-Dimethylcyclohexanone
4024 - 4,4-Dimethylcyclohexanone
5053 - 2-Ethyl-2-hexenal
7317 - 3-Methyl-5-hepten-2-one
7318 - 6-Methyl-5-hepten-2-one
8366 - *trans*-2-Octenal
8406 - 2-Octyn-1-ol
10548 - 2,2,4-Trimethylcyclopentanone
10549 - 2,4,4-Trimethylcyclopentanone
$C_8H_{14}O_2$
198 - Allyl 3-methylbutanoate
1553 - 5-Butyldihydro-2(3*H*)-furanone
1599 - Butyl methacrylate
1600 - *tert*-Butyl methacrylate
2566 - Cyclohexaneacetic acid
2606 - Cyclohexyl acetate
2667 - Cyclopentanepropanoic acid
4105 - 2,5-Dimethyl-3-hexyne-2,5-diol
5054 - Ethyl 3-hexenoate
5879 - *cis*-3-Hexen-1-ol, acetate
5880 - *trans*-2-Hexen-1-ol, acetate
6022 - 1-(1-Hydroxycyclohexyl)ethanone
6394 - Isobutyl methacrylate
7107 - 2-Methylbutyl acrylate
7146 - Methyl cyclohexanecarboxylate
9882 - Tetrahydro-6-propyl-2*H*-pyran-2-one
$C_8H_{14}O_2S_2$
10049 - Thioctic acid
$C_8H_{14}O_2S_4$
3814 - Diisopropyl thioperoxydicarbonate
$C_8H_{14}O_3$
1056 - Bis[2-(vinyloxy)ethyl] ether
1430 - Butanoic anhydride
1491 - Butyl acetoacetate
1632 - *tert*-Butyl 3-oxobutanoate
5027 - Ethyl 2-ethylacetoacetate
5145 - Ethyl 4-methyl-3-oxopentanoate
5186 - Ethyl 5-oxohexanoate
7692 - 2-Methylpropanoic anhydride
9856 - Tetrahydro-2-furanmethanol propanoate
$C_8H_{14}O_4$
1409 - 1,4-Butanediol diacetate
1583 - Butyl hydrogen succinate
3487 - Diethyl methylmalonate
3527 - Diethyl succinate
3808 - Diisopropyl oxalate
3829 - Dimethoxane
3891 - Dimethyl adipate
4561 - Dipropyl oxalate
5067 - Ethyl hydrogen adipate
8349 - Octanedioic acid
$C_8H_{14}O_4S$
4339 - Dimethyl thiodipropionate

$C_8H_{14}O_5$
3437 - Diethylene glycol diacetate
3475 - Diethyl malate

$C_8H_{14}O_6$
3533 - Diethyl *DL*-tartrate

$C_8H_{15}Br$
1206 - (2-Bromoethyl)cyclohexane

$C_8H_{15}BrO_2$
1301 - 8-Bromooctanoic acid
4914 - Ethyl 6-bromohexanoate

$C_8H_{15}ClO$
5052 - 2-Ethylhexanoyl chloride
8363 - Octanoyl chloride

$C_8H_{15}ClO_2$
7109 - 3-Methylbutyl 2-chloropropanoate
7110 - 3-Methylbutyl 3-chloropropanoate

$C_8H_{15}N$
538 - 3-Azabicyclo[3.2.2]nonane
6953 - 8-Methyl-8-azabicyclo[3.2.1]octane
8335 - Octahydroindolizine
8352 - Octanenitrile
9190 - 2-(1-Propenyl)piperidine
9883 - 2,3,4,5-Tetrahydro-6-propylpyridine

$C_8H_{15}NO$
81 - 1-Acetyl-3-methylpiperidine
7707 - 4-(2-Methylpropenyl)morpholine
7761 - 1-(1-Methyl-2-pyrrolidinyl)-2-pro-
   panone, (*R*)
9074 - 1-(2-Piperidinyl)-2-propanone, (±)
9288 - Pseudotropine
10711 - Tropine

$C_8H_{15}NOS_2$
6619 - Lipoamide

$C_8H_{15}NO_2$
346 - *trans*-4-(Aminomethyl)cyclohexanecar-
   boxylic acid
3373 - *N,N*-Diethylacetoacetamide
3911 - 2-(Dimethylamino)ethyl methacrylate
5219 - Ethyl 4-piperidinecarboxylate

$C_8H_{15}NO_3$
51 - 6-(Acetylamino)hexanoic acid

$C_8H_{15}NO_4$
3469 - Diethyl iminodiacetate

$C_8H_{15}NO_6$
47 - 2-(Acetylamino)-2-deoxy-*D*-glucose
48 - 2-(Acetylamino)-2-deoxy-*D*-mannose

$C_8H_{15}N_3O_7$
9616 - Streptozotocin

$C_8H_{15}N_5S$
2814 - Desmetryne

$C_8H_{15}N_7O_2S_3$
5301 - Famotidine

C_8H_{16}
2646 - Cyclooctane
4010 - 1,1-Dimethylcyclohexane
4011 - *cis*-1,2-Dimethylcyclohexane
4012 - *trans*-1,2-Dimethylcyclohexane
4013 - *cis*-1,3-Dimethylcyclohexane
4014 - *trans*-1,3-Dimethylcyclohexane
4015 - *cis*-1,4-Dimethylcyclohexane
4016 - *trans*-1,4-Dimethylcyclohexane
4097 - 2,3-Dimethyl-1-hexene
4098 - 5,5-Dimethyl-1-hexene
4099 - 2,3-Dimethyl-2-hexene
4100 - 2,5-Dimethyl-2-hexene
4101 - *cis*-2,2-Dimethyl-3-hexene
4102 - *trans*-2,2-Dimethyl-3-hexene

4962 - Ethylcyclohexane
5132 - 1-Ethyl-1-methylcyclopentane
5133 - *cis*-1-Ethyl-2-methylcyclopentane
5134 - *trans*-1-Ethyl-2-methylcyclopentane
5135 - *cis*-1-Ethyl-3-methylcyclopentane
5136 - *trans*-1-Ethyl-3-methylcyclopentane
5149 - 3-Ethyl-2-methyl-1-pentene
6491 - Isopropylcyclopentane
7244 - 3-Methyleneheptane
7312 - 2-Methyl-1-heptene
7313 - 6-Methyl-1-heptene
7314 - 2-Methyl-2-heptene
7315 - *cis*-3-Methyl-2-heptene
7453 - 3-Methyl-4-methylenehexane
8367 - 1-Octene
8368 - *cis*-2-Octene
8369 - *trans*-2-Octene
8370 - *cis*-3-Octene
8371 - *trans*-3-Octene
8372 - *cis*-4-Octene
8373 - *trans*-4-Octene
9227 - Propylcyclopentane
10543 - 1,1,2-Trimethylcyclopentane
10544 - 1,1,3-Trimethylcyclopentane
10545 - 1α,2α,4β-1,2,4-Trimethylcyclopentane
10546 - 1α,2β,4α-1,2,4-Trimethylcyclopentane
10588 - 2,3,3-Trimethyl-1-pentene
10589 - 2,4,4-Trimethyl-1-pentene
10590 - 2,3,4-Trimethyl-2-pentene
10591 - 2,4,4-Trimethyl-2-pentene

$C_8H_{16}Br_2$
2968 - 1,8-Dibromooctane

$C_8H_{16}Cl_2$
3166 - 2,5-Dichloro-2,5-dimethylhexane
3243 - 1,8-Dichlorooctane

$C_8H_{16}NO_3PS_2$
6704 - Mephosfolan

$C_8H_{16}NO_5P$
3314 - Dicrotophos

$C_8H_{16}N_2$
1434 - 2-Butanone (1-methylpropylidene)
   hydrazone

$C_8H_{16}N_2O_2S_2$
4588 - 4,4'-Dithiodimorpholine

$C_8H_{16}N_2O_3$
76 - *N*6-Acetyl-*L*-lysine
5591 - *N*-Glycyl-*L*-leucine
6603 - *N*-Leucylglycine

$C_8H_{16}N_2O_4S_2$
5925 - Homocystine
8521 - Penicillamine cysteine disulfide

$C_8H_{16}N_2O_7$
2524 - Cycasin

$C_8H_{16}O$
2582 - Cyclohexaneethanol
2647 - Cyclooctanol
4020 - 3,3-Dimethylcyclohexanol
4103 - 3,5-Dimethyl-1-hexen-3-ol
5046 - 2-Ethylhexanal
7147 - α-Methylcyclohexanemethanol
7148 - 4-Methylcyclohexanemethanol
7308 - 6-Methyl-2-heptanone
7309 - 5-Methyl-3-heptanone
7310 - 6-Methyl-3-heptanone
7311 - 2-Methyl-4-heptanone
7316 - 6-Methyl-5-hepten-2-ol
8343 - Octanal

8360 - 2-Octanone
8361 - 3-Octanone
8362 - 4-Octanone
8374 - 1-Octen-3-ol
8375 - 2-Octen-1-ol
9228 - 1-Propylcyclopentanol
10587 - 2,2,4-Trimethyl-3-pentanone

$C_8H_{16}O_2$
1528 - Butyl butanoate
2575 - 1,4-Cyclohexanedimethanol
4932 - 2-Ethylbutyl acetate
5049 - Ethyl hexanoate
5050 - 2-Ethylhexanoic acid
5078 - 2-Ethyl-3-hydroxyhexanal
5148 - Ethyl 4-methylpentanoate
5709 - Heptyl formate
5886 - Hexyl acetate
5887 - *sec*-Hexyl acetate
6136 - 5-Hydroxy-4-octanone
6380 - Isobutyl butanoate
6391 - Isobutyl isobutanoate
6453 - Isopentyl propanoate
6510 - Isopropyl 3-methylbutanoate
6526 - Isopropyl pentanoate
7288 - Methyl heptanoate
7570 - 2-Methyl-1-pentanol acetate
8354 - Octanoic acid
8684 - Pentyl propanoate
9248 - Propyl 3-methylbutanoate
9253 - Propyl pentanoate
9254 - 2-Propylpentanoic acid

$C_8H_{16}O_2S_2$
3636 - Dihydro-α-lipoic acid

$C_8H_{16}O_3$
1477 - 2-Butoxyethyl acetate
6135 - 2-Hydroxyoctanoic acid
6449 - Isopentyl lactate

$C_8H_{16}O_4$
1436 - 2-Butanone peroxide
3448 - Diethylene glycol monoethyl ether acetate
4983 - Ethyl diethoxyacetate

$C_8H_{16}Si$
2832 - Diallyldimethylsilane

$C_8H_{17}Br$
1261 - 3-(Bromomethyl)heptane
1299 - 1-Bromooctane
1300 - 2-Bromooctane, (±)

$C_8H_{17}Cl$
2126 - 3-(Chloromethyl)heptane
2202 - 1-Chlorooctane
2203 - 2-Chlorooctane
2335 - 2-Chloro-2,4,4-trimethylpentane

$C_8H_{17}ClO$
2204 - 8-Chloro-1-octanol

$C_8H_{17}Cl_3Si$
10264 - Trichlorooctylsilane

$C_8H_{17}F$
5408 - 1-Fluorooctane

$C_8H_{17}I$
6326 - 1-Iodooctane
6327 - 2-Iodooctane, (±)

$C_8H_{17}N$
2619 - Cyclohexyldimethylamine
2623 - Cyclohexylethylamine
2645 - Cyclooctanamine
9257 - 2-Propylpiperidine, (*S*)

$C_8H_{20}Sn$
9827 - Tetraethylstannane

$C_8H_{22}N_2O_3Si$
10496 - N-[3-(Trimethoxysilyl)propyl]-1,2-ethanediamine

$C_8H_{23}N_5$
9820 - Tetraethylenepentamine

$C_8H_{24}N_4O_3P_2$
9516 - Schradan

$C_8H_{24}O_2Si_3$
8342 - Octamethyltrisiloxane

$C_8H_{24}O_4Si_4$
8340 - Octamethylcyclotetrasiloxane

$C_8H_{26}O_3Si_4$
8341 - 1,1,1,3,5,7,7,7-Octamethyltetrasiloxane

$C_8H_{28}N_4Si_4$
8339 - 2,2,4,4,6,6,8,8-Octamethylcyclotetrasilazane

$C_8I_4O_3$
9904 - 4,5,6,7-Tetraiodo-1,3-isobenzofurandione

C_9F_{20}
8719 - Perfluorononane

$C_9F_{21}N$
8729 - Perfluorotripropylamine

$C_9Fe_2O_9$
6357 - Iron nonacarbonyl

$C_9H_2Cl_6O_3$
1829 - Chlorendic anhydride

$C_9H_3ClO_4$
671 - 1,2,4-Benzenetricarboxylic acid 1,2-anhydride, 4-chloride

$C_9H_3Cl_3O_3$
667 - 1,3,5-Benzenetricarbonyl trichloride

$C_9H_3F_{15}O_2$
7601 - Methyl perfluorooctanoate

$C_9H_4Cl_3NO_2S$
5444 - Folpet

$C_9H_4Cl_6O_4$
1828 - Chlorendic acid

$C_9H_4Cl_8O$
6365 - Isobenzan

$C_9H_4N_2S_3$
10078 - Thioquinox

$C_9H_4O_5$
10469 - Trimellitic anhydride

$C_9H_5Br_2NO$
2987 - 5,7-Dibromo-8-quinolinol

C_9H_5ClINO
2083 - 5-Chloro-7-iodo-8-quinolinol

$C_9H_5ClO_2$
2251 - 3-(3-Chlorophenyl)-2-propynoic acid

$C_9H_5Cl_2N$
3289 - 4,7-Dichloroquinoline

$C_9H_5Cl_2NO$
3290 - 5,7-Dichloro-8-quinolinol

$C_9H_5Cl_3N_4$
464 - Anilazine

$C_9H_5I_2NO$
3773 - 5,7-Diiodo-8-quinolinol

$C_9H_5NO_4$
8159 - 3-(2-Nitrophenyl)-2-propynoic acid

C_9H_6BrN
1241 - 4-Bromoisoquinoline
1350 - 3-Bromoquinoline
1351 - 6-Bromoquinoline

$C_9H_6BrNO_2$
1278 - N-(Bromomethyl)phthalimide

C_9H_6ClN
2296 - 2-Chloroquinoline
2297 - 4-Chloroquinoline
2298 - 6-Chloroquinoline
2299 - 8-Chloroquinoline

C_9H_6ClNO
2300 - 5-Chloro-8-quinolinol

$C_9H_6ClNO_2$
2156 - N-Chloromethylphthalimide

$C_9H_6Cl_2N_2O_3$
6755 - Methazole

$C_9H_6Cl_2O_2$
3268 - 3-(2,4-Dichlorophenyl)-2-propenoic acid

$C_9H_6Cl_6O_3S$
4698 - Endosulfan

$C_9H_6Cl_6O_4S$
4699 - Endosulfan sulfate

$C_9H_6Cl_8$
1820 - Chlorbicyclen

$C_9H_6INO_4S$
6051 - 8-Hydroxy-7-iodo-5-quinolinesulfonic acid

$C_9H_6N_2O_2$
8178 - 5-Nitroquinoline
8179 - 6-Nitroquinoline
8180 - 8-Nitroquinoline
10114 - Toluene-2,4-diisocyanate
10115 - Toluene-2,6-diisocyanate

$C_9H_6N_2O_3$
8181 - 4-Nitroquinoline 1-oxide
8182 - 5-Nitro-8-quinolinol

$C_9H_6N_2S_3$
1390 - BUSAN 72A

C_9H_6O
8978 - 3-Phenyl-2-propynal

$C_9H_6O_2$
726 - 1H-2-Benzopyran-1-one
727 - 2H-1-Benzopyran-2-one
728 - 4H-1-Benzopyran-4-one
6236 - 1H-Indene-1,3(2H)-dione
8979 - 3-Phenyl-2-propynoic acid

$C_9H_6O_2S$
744 - Benzo[b]thiophene-2-carboxylic acid

$C_9H_6O_3$
706 - 2-Benzofurancarboxylic acid
5998 - 4-Hydroxy-2H-1-benzopyran-2-one
5999 - 7-Hydroxy-2H-1-benzopyran-2-one
7394 - 5-Methyl-1,3-isobenzofurandione
8457 - 2-Oxo-2H-1-benzopyran-3-carboxylic acid

$C_9H_6O_4$
3712 - 6,7-Dihydroxy-2H-1-benzopyran-2-one
3713 - 7,8-Dihydroxy-2H-1-benzopyran-2-one
6237 - 1H-Indene-1,2,3-trione monohydrate

$C_9H_6O_6$
668 - 1,2,3-Benzenetricarboxylic acid
669 - 1,2,4-Benzenetricarboxylic acid
670 - 1,3,5-Benzenetricarboxylic acid

$C_9H_7BrO_4$
88 - 2-(Acetyloxy)-5-bromobenzoic acid

C_9H_7ClO
8972 - trans-3-Phenyl-2-propenoyl chloride

$C_9H_7ClO_2$
1958 - trans-o-Chlorocinnamic acid
1959 - trans-m-Chlorocinnamic acid
1960 - trans-p-Chlorocinnamic acid

$C_9H_7Cl_3O_3$
9538 - Silvex

$C_9H_7MnO_3$
6677 - Manganese 2-methylcyclopentadienyl tricarbonyl

C_9H_7N
6551 - Isoquinoline
8966 - cis-3-Phenyl-2-propenenitrile
8967 - trans-3-Phenyl-2-propenenitrile
9428 - Quinoline

C_9H_7NO
6247 - 1H-Indole-3-carboxaldehyde
6552 - 7-Isoquinolinol
8455 - β-Oxobenzenepropanenitrile
9433 - 2-Quinolinol
9434 - 3-Quinolinol
9435 - 4-Quinolinol
9436 - 5-Quinolinol
9437 - 6-Quinolinol
9438 - 7-Quinolinol
9439 - 8-Quinolinol

$C_9H_7NOS_2$
9007 - 3-Phenyl-2-thioxo-4-thiazolidinone

$C_9H_7NO_2$
6192 - 4-Hydroxy-2-quinolinone
7397 - 2-Methyl-1H-isoindole-1,3(2H)-dione
8910 - 3-Phenyl-2-isoxazolin-5-one

$C_9H_7NO_3$
2519 - (4-Cyanophenoxy)acetic acid
6100 - N-(Hydroxymethyl)phthalimide
7031 - 1-Methyl-2H-3,1-benzoxazine-2,4(1H)-dione
8158 - 3-(4-Nitrophenyl)-2-propenal

$C_9H_7NO_4S$
6191 - 8-Hydroxy-5-quinolinesulfonic acid

C_9H_7NS
9432 - 2(1H)-Quinolinethione

$C_9H_7N_3$
684 - 1H-Benzimidazole-2-acetonitrile

$C_9H_7N_3S$
10306 - Tricyclazole

$C_9H_7N_7O_2S$
545 - Azathioprine

C_9H_8
6235 - Indene
9275 - 1-Propynylbenzene

$C_9H_8Cl_2O_3$
3258 - 2-(2,4-Dichlorophenoxy)propanoic acid
7192 - Methyl (2,4-dichlorophenoxy)acetate

$C_9H_8Cl_3NO_2S$
1723 - Captan

$C_9H_8N_2$
6549 - 1-Isoquinolinamine
6550 - 3-Isoquinolinamine
7776 - 2-Methylquinoxaline
8896 - 1-Phenyl-1H-imidazole
8897 - 2-Phenyl-1H-imidazole
9422 - 2-Quinolinamine
9423 - 3-Quinolinamine
9424 - 4-Quinolinamine
9425 - 5-Quinolinamine
9426 - 6-Quinolinamine
9427 - 8-Quinolinamine

$C_9H_8N_2O$
2520 - 2-Cyano-*N*-phenylacetamide
$C_9H_8N_2O_2$
8519 - Pemoline
8898 - 5-Phenyl-2,4-imidazolidinedione
$C_9H_8N_4OS$
6249 - 1*H*-Indole-2,3-dione, 3-thiosemicarbazone
10035 - Thidiazuron
C_9H_8O
724 - 2*H*-1-Benzopyran
2396 - *trans*-Cinnamaldehyde
3630 - 2,3-Dihydro-1*H*-inden-1-one
3631 - 1,3-Dihydro-2*H*-inden-2-one
3632 - 1a,6a-Dihydro-6*H*-indeno[1,2-b]oxirene
5284 - α-Ethynylbenzenemethanol
7019 - 2-Methylbenzofuran
8980 - 3-Phenyl-2-propyn-1-ol
C_9H_8OS
3605 - 2,3-Dihydro-4*H*-1-benzothiopyran-4-one
$C_9H_8O_2$
2398 - *cis*-Cinnamic acid
2399 - *trans*-Cinnamic acid
3602 - 3,4-Dihydro-2*H*-1-benzopyran-2-one
3603 - 2,3-Dihydro-4*H*-1-benzopyran-4-one
7231 - α-Methylenebenzeneacetic acid
8958 - 1-Phenyl-1,2-propanedione
$C_9H_8O_3$
53 - 2-Acetylbenzoic acid
54 - 3-Acetylbenzoic acid
55 - 4-Acetylbenzoic acid
1735 - Carbic anhydride
6169 - 3-(4-Hydroxyphenyl)-2-propenoic acid
7265 - Methyl 4-formylbenzoate
8264 - 5-Norbornene-2,3-dicarboxylic acid anhydride
8452 - α-Oxobenzeneacetic acid, methyl ester
8456 - α-Oxobenzenepropanoic acid
$C_9H_8O_4$
56 - Acetyl benzoylperoxide
86 - 2-(Acetyloxy)benzoic acid
87 - 4-(Acetyloxy)benzoic acid
1764 - 2-Carboxybenzeneacetic acid
3744 - 3-(3,4-Dihydroxyphenyl)-2-propenoic acid
6163 - 3-(4-Hydroxyphenyl)-2-oxopropanoic acid
7018 - Methyl 1,3-benzodioxole-5-carboxylate
7793 - Methyl terephthalate
C_9H_9Br
1340 - (3-Bromo-1-propenyl)benzene
C_9H_9BrO
1277 - 2-Bromo-1-(4-methylphenyl)ethanone
1324 - 2-Bromo-1-phenyl-1-propanone
$C_9H_9BrO_2$
4910 - Ethyl 4-bromobenzoate
$C_9H_9Br_2NO_3$
2996 - 3,5-Dibromo-*L*-tyrosine
C_9H_9Cl
2278 - *trans*-(3-Chloro-1-propenyl)benzene
C_9H_9ClO
656 - Benzenepropanoyl chloride
1984 - 7-Chloro-2,3-dihydro-1*H*-inden-4-ol
2249 - 3-Chloro-1-phenyl-1-propanone
2250 - 1-(4-Chlorophenyl)-1-propanone

$C_9H_9ClO_2$
792 - Benzyl chloroacetate
2246 - 3-(2-Chlorophenyl)propanoic acid
2247 - 3-(3-Chlorophenyl)propanoic acid
2248 - 3-(4-Chlorophenyl)propanoic acid
4942 - Ethyl 4-chlorobenzoate
8793 - 2-Phenoxypropanoyl chloride
$C_9H_9ClO_3$
2152 - (4-Chloro-2-methylphenoxy)acetic acid
2227 - 2-(3-Chlorophenoxy)propanoic acid
$C_9H_9Cl_2NO$
9163 - Propanil
$C_9H_9Cl_2N_3$
2431 - Clonidine
$C_9H_9FO_2$
5033 - Ethyl 4-fluorobenzoate
$C_9H_9I_2NO_3$
3774 - 3,5-Diiodo-*L*-tyrosine
C_9H_9N
651 - Benzenepropanenitrile
6975 - 4-Methylbenzeneacetonitrile
6976 - α-Methylbenzeneacetonitrile
7386 - 1-Methyl-1*H*-indole
7387 - 2-Methyl-1*H*-indole
7388 - 3-Methyl-1*H*-indole
7389 - 5-Methyl-1*H*-indole
7390 - 7-Methyl-1*H*-indole
C_9H_9NO
3677 - 3,4-Dihydro-2(1*H*)-quinolinone
3970 - 2,5-Dimethylbenzoxazole
6409 - 2-Isocyanato-1,3-dimethylbenzene
6784 - 4-Methoxybenzeneacetonitrile
$C_9H_9NO_2$
5465 - *N*-(4-Formylphenyl)acetamide
8959 - 1-Phenyl-1,2-propanedione, 2-oxime
$C_9H_9NO_3$
45 - 2-(Acetylamino)benzoic acid
46 - 4-(Acetylamino)benzoic acid
766 - *N*-Benzoylglycine
3626 - 2,3-Dihydro-3-hydroxy-1-methyl-1*H*-indole-5,6-dione
$C_9H_9NO_4$
280 - [2-(Aminocarbonyl)phenoxy]acetic acid
5170 - Ethyl 3-nitrobenzoate
5171 - Ethyl 4-nitrobenzoate
8156 - 3-(2-Nitrophenyl)propanoic acid
8157 - 3-(4-Nitrophenyl)propanoic acid
$C_9H_9NO_7$
8102 - 5-Nitro-2-furaldehyde diacetate
$C_9H_9N_3O_2$
1732 - Carbendazim
$C_9H_9N_3O_2S_2$
402 - 5-[(4-Aminophenyl)sulfonyl]-2-thiazolamine
9651 - Sulfathiazole
$C_9H_9N_3O_4S_2$
4400 - 2,4-Dinitrophenyl dimethylcarbamodithioate
$C_9H_9N_3S$
9000 - 5-Phenyl-2,4-thiazolediamine
$C_9H_9N_5$
9008 - 6-Phenyl-1,3,5-triazine-2,4-diamine
9009 - *N*-Phenyl-1,3,5-triazine-2,4-diamine
C_9H_{10}
173 - Allylbenzene
2698 - Cyclopropylbenzene
6231 - Indan

6463 - Isopropenylbenzene
7784 - 2-Methylstyrene
7785 - 3-Methylstyrene
7786 - 4-Methylstyrene
9186 - *cis*-1-Propenylbenzene
9187 - *trans*-1-Propenylbenzene
$C_9H_{10}BrClN_2O_2$
1821 - Chlorbromuron
$C_9H_{10}Cl_2N_2O$
4599 - Diuron
$C_9H_{10}Cl_2N_2O_2$
6617 - Linuron
$C_9H_{10}INO_3$
6346 - *L*-3-Iodotyrosine
$C_9H_{10}NO_3PS$
2388 - Ciafos
$C_9H_{10}N_2$
3962 - 5,6-Dimethyl-1*H*-benzimidazole
4570 - 2,2'-Dipyrrolylmethane
4902 - 2-Ethyl-1*H*-benzimidazole
$C_9H_{10}N_2O$
3665 - 4,5-Dihydro-5-phenyl-2-oxazolamine
8982 - 1-Phenyl-3-pyrazolidinone
$C_9H_{10}N_2O_2$
8803 - (Phenylacetyl)urea
$C_9H_{10}N_2O_3$
262 - *N*-(4-Aminobenzoyl)glycine
$C_9H_{10}N_2O_3S_2$
4832 - 6-Ethoxy-2-benzothiazolesulfonamide
$C_9H_{10}N_4O_2S_2$
9643 - Sulfamethiazole
$C_9H_{10}N_4O_4$
35 - Acetone (2,4-dinitrophenyl)hydrazone
$C_9H_{10}O$
201 - 2-Allylphenol
202 - 4-Allylphenol
203 - Allyl phenyl ether
650 - Benzenepropanal
3600 - 3,4-Dihydro-1*H*-2-benzopyran
3601 - 3,4-Dihydro-2*H*-1-benzopyran
3628 - 2,3-Dihydro-1*H*-inden-1-ol
3629 - 2,3-Dihydro-1*H*-inden-5-ol
3640 - 2,3-Dihydro-2-methylbenzofuran
3944 - 2,4-Dimethylbenzaldehyde
3945 - 2,5-Dimethylbenzaldehyde
3946 - 3,5-Dimethylbenzaldehyde
4892 - 4-Ethylbenzaldehyde
6907 - (2-Methoxyvinyl)benzene
6914 - 4-Methylacetophenone
6968 - 2-Methylbenzeneacetaldehyde
6969 - 4-Methylbenzeneacetaldehyde
6970 - α-Methylbenzeneacetaldehyde
7633 - 1-(2-Methylphenyl)ethanone
7634 - 1-(3-Methylphenyl)ethanone
7645 - 2-Methyl-2-phenyloxirane
8964 - 1-Phenyl-1-propanone
8965 - 1-Phenyl-2-propanone
8969 - *cis*-3-Phenyl-2-propen-1-ol
8970 - *trans*-3-Phenyl-2-propen-1-ol
9189 - 4-(1-Propenyl)phenol
10793 - α-Vinylbenzenemethanol
10806 - 1-Vinyl-2-methoxybenzene
10807 - 1-Vinyl-3-methoxybenzene
10808 - 1-Vinyl-4-methoxybenzene
$C_9H_{10}OS$
102 - 4-Acetylthioanisole

$C_9H_{10}O_2$
52 - 4-Acetylanisole
653 - Benzenepropanoic acid
778 - Benzyl acetate
3963 - 2,4-Dimethylbenzoic acid
3964 - 2,5-Dimethylbenzoic acid
3965 - 2,6-Dimethylbenzoic acid
3966 - 3,4-Dimethylbenzoic acid
3967 - 3,5-Dimethylbenzoic acid
4825 - 2-Ethoxybenzaldehyde
4826 - 4-Ethoxybenzaldehyde
4903 - Ethyl benzoate
6096 - 1-(2-Hydroxy-4-methylphenyl)etha-
　　none
6097 - 1-(2-Hydroxy-5-methylphenyl)etha-
　　none
6166 - 2-Hydroxy-1-phenyl-1-propanone
6167 - 1-(2-Hydroxyphenyl)-1-propanone
6168 - 1-(4-Hydroxyphenyl)-1-propanone
6781 - 4-Methoxybenzeneacetaldehyde
6868 - 2-Methoxy-1-phenylethanone
6869 - 1-(3-Methoxyphenyl)ethanone
6971 - 2-Methylbenzeneacetic acid
6972 - 3-Methylbenzeneacetic acid
6973 - 4-Methylbenzeneacetic acid
6974 - α-Methylbenzeneacetic acid, (±)
7043 - α-Methylbenzyl formate
7448 - Methyl 2-methylbenzoate
7449 - Methyl 3-methylbenzoate
7450 - Methyl 4-methylbenzoate
7615 - 2-Methylphenyl acetate
7616 - 3-Methylphenyl acetate
7617 - 4-Methylphenyl acetate
7618 - Methyl 2-phenylacetate
8792 - 1-Phenoxy-2-propanone
8886 - Phenyl glycidyl ether
8960 - Phenyl propanoate
$C_9H_{10}O_3$
3836 - 2,4-Dimethoxybenzaldehyde
3837 - 2,5-Dimethoxybenzaldehyde
3838 - 3,4-Dimethoxybenzaldehyde
3839 - 3,5-Dimethoxybenzaldehyde
4787 - 1,2-Ethanediol, monobenzoate
4830 - 2-Ethoxybenzoic acid
4831 - 4-Ethoxybenzoic acid
4843 - 3-Ethoxy-2-hydroxybenzaldehyde
4844 - 3-Ethoxy-4-hydroxybenzaldehyde
5072 - Ethyl 3-hydroxybenzoate
5073 - Ethyl 4-hydroxybenzoate
5243 - Ethyl salicylate
5986 - 4-Hydroxybenzenepropanoic acid
5987 - α-Hydroxybenzenepropanoic acid, (±)
6068 - 1-(2-Hydroxy-4-methoxyphenyl)etha-
　　none
6069 - 1-(4-Hydroxy-3-methoxyphenyl)etha-
　　none
6077 - α-(Hydroxymethyl)benzeneacetic acid,
　　(±)
6078 - α-Hydroxy-α-methylbenzeneacetic
　　acid, (±)
6782 - 2-Methoxybenzeneacetic acid
6783 - 4-Methoxybenzeneacetic acid
6866 - 2-Methoxyphenyl acetate
7374 - Methyl 2-hydroxy-3-methylbenzoate
7375 - Methyl 2-hydroxy-5-methylbenzoate
7378 - Methyl α-hydroxyphenylacetate, (±)
7441 - Methyl 2-methoxybenzoate
7442 - Methyl 3-methoxybenzoate

7443 - Methyl 4-methoxybenzoate
7608 - Methyl phenoxyacetate
8789 - 2-Phenoxypropanoic acid
$C_9H_{10}O_4$
3849 - 2,4-Dimethoxybenzoic acid
3850 - 2,6-Dimethoxybenzoic acid
3851 - 3,4-Dimethoxybenzoic acid
3852 - 3,5-Dimethoxybenzoic acid
6028 - 2-Hydroxy-4,6-dimethoxybenzaldehyde
6029 - 4-Hydroxy-3,5-dimethoxybenzaldehyde
6041 - 2-Hydroxyethyl 2-hydroxybenzoate
6061 - 4-Hydroxy-3-methoxybenzeneacetic
　　acid
7373 - Methyl 4-hydroxy-3-methoxybenzoate
10447 - 1-(2,4,6-Trihydroxyphenyl)-1-pro-
　　panone
$C_9H_{10}O_5$
3726 - α,4-Dihydroxy-3-methoxybenzeneace-
　　tic acid
5272 - Ethyl 3,4,5-trihydroxybenzoate
5487 - 2-Furanmethanediol diacetate
6030 - 4-Hydroxy-3,5-dimethoxybenzoic acid
$C_9H_{11}Br$
1240 - 1-Bromo-4-isopropylbenzene
1344 - (3-Bromopropyl)benzene
1371 - 2-Bromo-1,3,5-trimethylbenzene
$C_9H_{11}BrN_2O_2$
7863 - Metobromuron
$C_9H_{11}BrO$
1341 - (3-Bromopropoxy)benzene
$C_9H_{11}Cl$
2086 - 1-Chloro-2-isopropylbenzene
2087 - 1-Chloro-4-isopropylbenzene
2117 - 1-(Chloromethyl)-2,4-dimethylbenzene
2121 - 1-(Chloromethyl)-4-ethylbenzene
2122 - (1-Chloro-1-methylethyl)benzene
2281 - (3-Chloropropyl)benzene
$C_9H_{11}ClN_2O$
2237 - N'-(4-Chlorophenyl)-N,N-dimethylurea
$C_9H_{11}ClN_2O_2$
7893 - Monolinuron
$C_9H_{11}ClO_3$
2226 - 3-(4-Chlorophenoxy)-1,2-propanediol
$C_9H_{11}ClO_3S$
2021 - 2-Chloroethanol, 4-methylbenzene-
　　sulfonate
$C_9H_{11}Cl_2FN_2O_2S_2$
3064 - Dichlofluanid
$C_9H_{11}Cl_2N_3O_4S_2$
7137 - Methyclothiazide
$C_9H_{11}Cl_3NO_3PS$
2352 - Chlorpyrifos
$C_9H_{11}FN_2O_5$
2802 - 2'-Deoxy-5-fluorouridine
$C_9H_{11}N$
172 - N-Allylaniline
3627 - 2,3-Dihydro-1H-inden-5-amine
6232 - 1-Indanamine
7853 - 2-(1-Methylvinyl)aniline
8925 - N-(Phenylmethylene)ethanamine
9865 - 1,2,3,4-Tetrahydroisoquinoline
9891 - 1,2,3,4-Tetrahydroquinoline
9892 - 5,6,7,8-Tetrahydroquinoline
10125 - Tranylcypromine
$C_9H_{11}NO$
400 - 1-(4-Aminophenyl)-1-propanone
777 - N-Benzylacetamide

3900 - 4-(Dimethylamino)benzaldehyde
3947 - N,N-Dimethylbenzamide
4893 - N-Ethylbenzamide
5208 - Ethyl N-phenylformimidate
6911 - 4-Methylacetanilide
7612 - N-(2-Methylphenyl)acetamide
7613 - N-(3-Methylphenyl)acetamide
7614 - N-Methyl-N-phenylacetamide
8957 - N-Phenylpropanamide
$C_9H_{11}NO_2$
694 - 1,3-Benzodioxole-5-ethanamine
3902 - 2-(Dimethylamino)benzoic acid
3903 - 3-(Dimethylamino)benzoic acid
3904 - 4-(Dimethylamino)benzoic acid
4827 - 2-Ethoxybenzamide
4882 - Ethyl 2-aminobenzoate
4883 - Ethyl 3-aminobenzoate
4884 - Ethyl 4-aminobenzoate
5207 - Ethyl phenylcarbamate
6522 - 1-Isopropyl-4-nitrobenzene
6538 - Isopropyl 3-pyridinecarboxylate
6863 - N-(2-Methoxyphenyl)acetamide
6864 - N-(3-Methoxyphenyl)acetamide
6865 - N-(4-Methoxyphenyl)acetamide
7447 - Methyl 2-(methylamino)benzoate
8806 - L-Phenylalanine
10572 - 1,3,5-Trimethyl-2-nitrobenzene
$C_9H_{11}NO_3$
134 - Adrenalone
10721 - L-Tyrosine
$C_9H_{11}NO_4$
6607 - Levodopa
$C_9H_{11}N_3$
10305 - 1,3,6-Tricyanohexane
$C_9H_{11}N_3O$
8196 - N-Nitrosonornicotine
$C_9H_{11}N_5O_3$
9528 - L-Sepiapterin
C_9H_{12}
5082 - 5-Ethylidene-2-norbornene
5264 - 2-Ethyltoluene
5265 - 3-Ethyltoluene
5266 - 4-Ethyltoluene
6478 - Isopropylbenzene
8229 - 1,8-Nonadiyne
9212 - Propylbenzene
10507 - 1,2,3-Trimethylbenzene
10508 - 1,2,4-Trimethylbenzene
10509 - 1,3,5-Trimethylbenzene
10814 - 2-Vinyl-5-norbornene
$C_9H_{12}ClO_2PS_3$
7840 - Methyl trithion
$C_9H_{12}NO_5PS$
5317 - Fenitrothion
$C_9H_{12}N_2$
9169 - 2-Propanone phenylhydrazone
9401 - 3-(2-Pyrrolidinyl)pyridine, (S)
$C_9H_{12}N_2O$
4259 - N,N-Dimethyl-N'-phenylurea
8805 - L-Phenylalaninamide
$C_9H_{12}N_2O_2$
4864 - (4-Ethoxyphenyl)urea
10722 - Tyrosineamide
$C_9H_{12}N_2O_3$
8169 - 5-Nitro-2-propoxyaniline
$C_9H_{12}N_2O_6$
10760 - Uridine

C₉H₁₂N₂S
9263 - 2-Propyl-4-pyridinecarbothioamide

C₉H₁₂N₄O₃
5293 - Etofylline

C₉H₁₂N₆
10678 - 2,4,6-Tris(1-aziridinyl)-1,3,5-triazine

C₉H₁₂O
654 - Benzenepropanol
798 - Benzyl ethyl ether
3938 - 2,6-Dimethylanisole
3939 - 3,5-Dimethylanisole
3958 - α,4-Dimethylbenzenemethanol
3959 - α,α-Dimethylbenzenemethanol
4847 - 1-Ethoxy-3-methylbenzene
4848 - 1-Ethoxy-4-methylbenzene
4899 - α-Ethylbenzenemethanol
5102 - 1-Ethyl-4-methoxybenzene
6527 - 2-Isopropylphenol
6528 - 3-Isopropylphenol
6529 - 4-Isopropylphenol
6990 - 2-Methylbenzeneethanol
6991 - 4-Methylbenzeneethanol
8908 - Phenyl isopropyl ether
8961 - 2-Phenyl-1-propanol
8962 - 1-Phenyl-2-propanol
8976 - Phenyl propyl ether
9255 - 2-Propylphenol
9256 - 4-Propylphenol
10592 - 2,3,4-Trimethylphenol
10593 - 2,3,5-Trimethylphenol
10594 - 2,3,6-Trimethylphenol
10595 - 2,4,5-Trimethylphenol
10596 - 2,4,6-Trimethylphenol
10597 - 3,4,5-Trimethylphenol

C₉H₁₂O₂
817 - 2-(Benzyloxy)ethanol
3871 - 1,2-Dimethoxy-4-methylbenzene
3872 - 1,3-Dimethoxy-5-methylbenzene
3873 - 1,4-Dimethoxy-2-methylbenzene
5106 - 4-Ethyl-2-methoxyphenol
6479 - Isopropylbenzene hydroperoxide
6790 - 4-Methoxybenzeneethanol
6840 - 4-Methoxy-α-methylbenzenemethanol
8790 - 2-Phenoxy-1-propanol
8791 - 1-Phenoxy-2-propanol
10510 - 2,3,5-Trimethyl-1,4-benzenediol

C₉H₁₂O₂S
1663 - Butyl thiophene-2-carboxylate

C₉H₁₂O₃
3846 - 3,4-Dimethoxybenzenemethanol
6079 - 2-Hydroxy-5-methyl-1,3-benzene-
dimethanol
8788 - 3-Phenoxy-1,2-propanediol
10478 - 1,2,3-Trimethoxybenzene
10479 - 1,3,5-Trimethoxybenzene

C₉H₁₂O₃S
5267 - Ethyl p-toluenesulfonate
9214 - Propyl benzenesulfonate

C₉H₁₂S
652 - Benzenepropanethiol
5260 - 1-(Ethylthio)-4-methylbenzene
6543 - (Isopropylthio)benzene
9265 - (Propylthio)benzene

C₉H₁₃BrN₂O₂
1074 - Bromacil

C₉H₁₃BrOSi
1317 - (4-Bromophenoxy)trimethylsilane

C₉H₁₃ClN₂O₂
9689 - Terbacil

C₉H₁₃ClN₆
2501 - Cyanazine

C₉H₁₃ClSi
2118 - (Chloromethyl)dimethylphenylsilane

C₉H₁₃N
796 - Benzylethylamine
1658 - 4-tert-Butylpyridine
3957 - α,α-Dimethylbenzenemethanamine
3971 - N,N-Dimethylbenzylamine
5113 - 2-Ethyl-6-methylaniline
5114 - N-Ethyl-2-methylaniline
5115 - N-Ethyl-3-methylaniline
5116 - N-Ethyl-4-methylaniline
5117 - N-Ethyl-N-methylaniline
6474 - 2-Isopropylaniline
6475 - 4-Isopropylaniline
6476 - N-Isopropylaniline
6988 - β-Methylbenzeneethanamine
6989 - N-Methylbenzeneethanamine
7202 - 2-Methyl-N,N-dimethylaniline
7203 - 3-Methyl-N,N-dimethylaniline
7204 - 4-Methyl-N,N-dimethylaniline
8974 - 1-Phenyl-2-propylamine, (±)
8975 - 1-Phenyl-2-propylamine, (S)
9209 - 4-Propylaniline
9210 - N-Propylaniline
10503 - 2,4,5-Trimethylaniline
10504 - 2,4,6-Trimethylaniline

C₉H₁₃NO
312 - α-(1-Aminoethyl)benzenemethanol, [S-
(R*,R*)]-
412 - 4-(2-Aminopropyl)phenol, (±)
783 - 2-[Benzylamino]ethanol
6789 - 4-Methoxybenzeneethanamine
7619 - 2-(Methylphenylamino)ethanol
7620 - 2-[(2-Methylphenyl)amino]ethanol

C₉H₁₃NO₂
3520 - 3,3-Diethyl-2,4(1H,3H)-pyridinedione
3845 - 3,4-Dimethoxybenzenemethanamine
4673 - Ecgonidine
5009 - Ethyl 3,5-dimethylpyrrole-2-carboxylate
5010 - Ethyl 2,4-dimethylpyrrole-3-carboxylate
5011 - Ethyl 2,5-dimethylpyrrole-3-carboxylate
5012 - Ethyl 4,5-dimethylpyrrole-3-carboxylate
5288 - 1-Ethynylcyclohexanol, carbamate
6073 - 4-Hydroxy-α-[(methylamino)methyl]
benzenemethanol
6730 - Metaraminol
6933 - 4-[2-(Methylamino)ethyl]-1,2-benzene-
diol

C₉H₁₃NO₂S
5119 - N-Ethyl-4-methylbenzenesulfonamide

C₉H₁₃NO₃
4708 - Epinephrine
10474 - 3,4,5-Trimethoxyaniline

C₉H₁₃N₂O₉P
10761 - 5'-Uridylic acid

C₉H₁₃N₃O₂
309 - 6-Amino-3-ethyl-1-allyl-2,4(1H,3H)-
pyrimidinedione

C₉H₁₃N₃O₅
2719 - Cytarabine
2720 - Cytidine

C₉H₁₃N₅
10122 - N-o-Tolylbiguanide

C₉H₁₃N₅O₃
3607 - 7,8-Dihydrobiopterin

C₉H₁₃N₅O₄
5522 - Ganciclovir

C₉H₁₃O₄PS
4135 - Dimethyl p-(methylthio)phenyl phosphate

C₉H₁₄
182 - 1-Allylcyclohexene
3975 - 2,3-Dimethylbicyclo[2.2.1]hept-2-ene
5177 - 5-Ethyl-2-norbornene
7854 - 1-Methyl-4-vinylcyclohexene

C₉H₁₄ClN
10599 - Trimethylphenylammonium chloride

C₉H₁₄ClNO
313 - α-(1-Aminoethyl)benzenemethanol,
hydrochloride
8963 - Phenylpropanolamine hydrochloride

C₉H₁₄ClN₅
2709 - Cyprazine

C₉H₁₄Cl₂O₂
8234 - Nonanedioyl dichloride

C₉H₁₄IN
9011 - Phenyltrimethylammonium iodide

C₉H₁₄N₂
8930 - N-(Phenylmethyl)-1,2-ethanediamine

C₉H₁₄N₂O₃
6753 - Metharbital

C₉H₁₄N₃O₇P
2801 - 2'-Deoxycytidine 5'-monophosphate

C₉H₁₄N₃O₈P
2721 - 2'-Cytidylic acid
2722 - 3'-Cytidylic acid
2723 - 5'-Cytidylic acid

C₉H₁₄N₄O₃
1773 - Carnosine

C₉H₁₄O
3974 - 6,6-Dimethylbicyclo[3.1.1]heptan-2-
one, (1R)
6456 - Isophorone
8226 - trans,trans-2,4-Nonadienal
9020 - Phorone

C₉H₁₄OSi
10598 - Trimethylphenoxysilane

C₉H₁₄O₂
2607 - Cyclohexyl acrylate
4965 - Ethyl 3-cyclohexene-1-carboxylate
7526 - Methyl 2-octynoate

C₉H₁₄O₂Si
3876 - Dimethoxymethylphenylsilane

C₉H₁₄O₃
1019 - Bis(2-methallyl) carbonate
3363 - 2-(Diethoxymethyl)furan
4877 - Ethyl 2-acetyl-4-pentenoate
9861 - Tetrahydrofurfuryl methacrylate

C₉H₁₄O₃Si
10493 - Trimethoxyphenylsilane

C₉H₁₄O₄
3424 - Diethyl 1,1-cyclopropanedicarboxylate
3456 - Diethyl ethylidenemalonate
3486 - Diethyl methylenesuccinate
3501 - Diethyl 2-pentenedioate

C₉H₁₄O₅
3497 - Diethyl 3-oxo-1,5-pentanedioate

C₉H₁₄O₆
10128 - Triacetin

7717 - 1-Methyl-2-propylpiperidine, (S)
9969 - 2,2,6,6-Tetramethylpiperidine

$C_9H_{19}NO$
3024 - N,N-Dibutylformamide
3485 - N,N-Diethyl-3-methylbutanamide
6197 - 4-Hydroxy-2,2,6,6-tetramethylpiperidine

$C_9H_{19}NOS$
4552 - Dipropylcarbamothioic acid, S-ethyl ester

$C_9H_{19}NO_3Si$
10342 - 3-(Triethoxysilyl)propanenitrile

$C_9H_{19}NO_4$
2818 - Dexpanthenol

C_9H_{20}
3498 - 3,3-Diethylpentane
4070 - 2,2-Dimethylheptane
4071 - 2,3-Dimethylheptane
4072 - 2,4-Dimethylheptane
4073 - 2,5-Dimethylheptane
4074 - 2,6-Dimethylheptane
4075 - 3,3-Dimethylheptane
4076 - 3,4-Dimethylheptane
4077 - 3,5-Dimethylheptane
4078 - 4,4-Dimethylheptane
5003 - 3-Ethyl-2,2-dimethylpentane
5004 - 3-Ethyl-2,3-dimethylpentane
5005 - 3-Ethyl-2,4-dimethylpentane
5040 - 3-Ethylheptane
5041 - 4-Ethylheptane
5140 - 3-Ethyl-2-methylhexane
5141 - 3-Ethyl-3-methylhexane
5142 - 3-Ethyl-4-methylhexane
5143 - 4-Ethyl-2-methylhexane
7515 - 2-Methyloctane
7516 - 3-Methyloctane
7517 - 4-Methyloctane
8231 - Nonane
9960 - 2,2,3,3-Tetramethylpentane
9961 - 2,2,3,4-Tetramethylpentane
9962 - 2,2,4,4-Tetramethylpentane
9963 - 2,3,3,4-Tetramethylpentane
10557 - 2,2,3-Trimethylhexane
10558 - 2,2,4-Trimethylhexane
10559 - 2,2,5-Trimethylhexane
10560 - 2,3,3-Trimethylhexane
10561 - 2,3,4-Trimethylhexane
10562 - 2,3,5-Trimethylhexane
10563 - 2,4,4-Trimethylhexane
10564 - 3,3,4-Trimethylhexane

$C_9H_{20}NO_3PS_2$
10471 - Trimethoate

$C_9H_{20}N_2$
9968 - 2,2,6,6-Tetramethyl-4-piperidinamine

$C_9H_{20}N_2O$
9829 - Tetraethylurea

$C_9H_{20}N_2S$
3058 - N,N'-Dibutylthiourea

$C_9H_{20}O$
4080 - 2,6-Dimethyl-2-heptanol
4081 - 2,6-Dimethyl-4-heptanol
4082 - 3,5-Dimethyl-4-heptanol
5044 - 4-Ethyl-4-heptanol
7520 - 2-Methyl-2-octanol
7521 - 3-Methyl-3-octanol
8238 - 1-Nonanol
8239 - 2-Nonanol, (±)

8240 - 3-Nonanol, (±)
8241 - 4-Nonanol
8242 - 5-Nonanol
9964 - 2,2,4,4-Tetramethyl-3-pentanol
10566 - 3,5,5-Trimethyl-1-hexanol

$C_9H_{20}O_2$
1571 - 2-Butyl-2-ethyl-1,3-propanediol
3001 - Dibutoxymethane
3366 - 1,1-Diethoxypentane
8233 - 1,9-Nonanediol

$C_9H_{20}O_3$
1476 - 1-(2-Butoxyethoxy)-2-propanol
10339 - 1,1,1-Triethoxypropane

$C_9H_{20}O_3Si$
210 - Allyltriethoxysilane

$C_9H_{20}O_4$
9808 - Tetraethoxymethane
10671 - Tripropylene glycol

$C_9H_{20}O_5$
9983 - 2,5,8,11-Tetraoxatridecan-13-ol

$C_9H_{20}S$
8236 - 1-Nonanethiol

$C_9H_{21}AlO_3$
222 - Aluminum isopropoxide

$C_9H_{21}B$
10669 - Tripropylborane

$C_9H_{21}BO_3$
10464 - Triisopropyl borate
10670 - Tripropyl borate

$C_9H_{21}ClO_3Si$
10332 - Triethoxy(3-chloropropyl)silane

$C_9H_{21}ClSn$
2342 - Chlorotripropylstannane

$C_9H_{21}N$
7525 - Methyloctylamine
8254 - Nonylamine
10668 - Tripropylamine

$C_9H_{21}NO_3$
10459 - Triisopropanolamine

$C_9H_{21}N_3$
10364 - 1,3,5-Triethylhexahydro-1,3,5-triazine

$C_9H_{21}O_2PS_3$
9690 - Terbufos

$C_9H_{21}O_3P$
10466 - Triisopropyl phosphite
10675 - Tripropyl phosphite

$C_9H_{21}O_4P$
10465 - Triisopropyl phosphate
10674 - Tripropyl phosphate

$C_9H_{21}O_4V$
10467 - Triisopropyl vanadate

$C_9H_{22}NO_5P$
3427 - Diethyl [(diethanolamino)methyl]phosphonate

$C_9H_{22}N_2$
3499 - N^1,N^1-Diethyl-1,4-pentanediamine
9824 - N,N,N',N'-Tetraethylmethanediamine

$C_9H_{22}O_4P_2S_4$
4808 - Ethion

$C_9H_{22}Si$
10676 - Tripropylsilane

$C_9H_{23}NO_3Si$
10341 - 3-(Triethoxysilyl)-1-propanamine

$C_9H_{23}N_3$
3913 - N-[2-(Dimethylamino)ethyl]-N,N',N'-trimethyl-1,2-ethanediamine

$C_{10}Cl_8$
8294 - Octachloronaphthalene

$C_{10}Cl_{10}$
8529 - Pentac

$C_{10}Cl_{10}O$
6568 - Kepone

$C_{10}Cl_{12}$
7878 - Mirex

$C_{10}F_8$
8718 - Perfluoronaphthalene

$C_{10}F_{18}$
8702 - Perfluorodecalin

$C_{10}F_{22}$
8703 - Perfluorodecane

$C_{10}HF_{19}O_2$
8218 - Nonadecafluorodecanoic acid

$C_{10}H_2O_6$
688 - 1H,3H-Benzo[1,2-c:4,5-c']difuran-1,3,5,7-tetrone

$C_{10}H_4Cl_2O_2$
3231 - 2,3-Dichloro-1,4-naphthalenedione

$C_{10}H_4Cl_4$
9759 - 1,2,3,4-Tetrachloronaphthalene

$C_{10}H_4Cl_8O$
8484 - Oxychlordane

$C_{10}H_5ClN_2$
1915 - o-Chlorobenzylidene malononitrile

$C_{10}H_5ClN_2O_4$
2010 - 1-Chloro-2,4-dinitronaphthalene

$C_{10}H_5Cl_2NO_2$
9414 - Quinclorac

$C_{10}H_5Cl_7$
5640 - Heptachlor

$C_{10}H_5Cl_7O$
5641 - Heptachlor epoxide

$C_{10}H_5NaO_5S$
9569 - Sodium β-naphthoquinone-4-sulfonate

$C_{10}H_6Br_2$
2963 - 1,4-Dibromonaphthalene

$C_{10}H_6Cl_2$
3221 - 1,2-Dichloronaphthalene
3222 - 1,3-Dichloronaphthalene
3223 - 1,4-Dichloronaphthalene
3224 - 1,5-Dichloronaphthalene
3225 - 1,6-Dichloronaphthalene
3226 - 1,7-Dichloronaphthalene
3227 - 1,8-Dichloronaphthalene
3228 - 2,3-Dichloronaphthalene
3229 - 2,6-Dichloronaphthalene
3230 - 2,7-Dichloronaphthalene

$C_{10}H_6Cl_2O$
3232 - 2,4-Dichloro-1-naphthol

$C_{10}H_6Cl_4O_4$
4332 - Dimethyl tetrachloroterephthalate

$C_{10}H_6Cl_6$
1826 - Chlordene

$C_{10}H_6Cl_8$
1824 - Chlordane

$C_{10}H_6N_2OS_2$
1811 - Chinomethionat

$C_{10}H_6N_2O_4$
4388 - 1,3-Dinitronaphthalene
4389 - 1,5-Dinitronaphthalene
4390 - 1,8-Dinitronaphthalene

$C_{10}H_6N_2O_5$
4391 - 2,4-Dinitro-1-naphthol

$C_{10}H_6Na_2O_8S_2$

9553 - Sodium 4,5-dihydroxy-2,7-naphtha-
lenedisulfonic acid

$C_{10}H_6O_2$

7957 - 1,2-Naphthalenedione
7958 - 1,4-Naphthalenedione

$C_{10}H_6O_3$

6116 - 2-Hydroxy-1,4-naphthalenedione
6117 - 5-Hydroxy-1,4-naphthalenedione

$C_{10}H_6O_4$

3569 - Di-2-furanylethanedione
3735 - 5,8-Dihydroxy-1,4-naphthalenedione

$C_{10}H_6O_8$

665 - 1,2,4,5-Benzenetetracarboxylic acid

$C_{10}H_7Br$

1287 - 1-Bromonaphthalene
1288 - 2-Bromonaphthalene

$C_{10}H_7BrO$

1290 - 1-Bromo-2-naphthol

$C_{10}H_7Cl$

2169 - 1-Chloronaphthalene
2170 - 2-Chloronaphthalene

$C_{10}H_7ClO$

2171 - 4-Chloro-1-naphthol

$C_{10}H_7ClO_2S$

7967 - 1-Naphthalenesulfonyl chloride
7968 - 2-Naphthalenesulfonyl chloride

$C_{10}H_7Cl_5O$

10328 - Tridiphane

$C_{10}H_7F$

5403 - 1-Fluoronaphthalene
5404 - 2-Fluoronaphthalene

$C_{10}H_7F_3O_2$

10416 - 4,4,4-Trifluoro-1-phenyl-1,3-butane-
dione

$C_{10}H_7I$

6319 - 1-Iodonaphthalene
6320 - 2-Iodonaphthalene

$C_{10}H_7NO$

9429 - 4-Quinolinecarboxaldehyde

$C_{10}H_7NO_2$

358 - 2-Amino-1,4-naphthalenedione
8125 - 1-Nitronaphthalene
8126 - 2-Nitronaphthalene
8194 - 2-Nitroso-1-naphthol
8195 - 1-Nitroso-2-naphthol
8990 - 1-Phenyl-1H-pyrrole-2,5-dione
9430 - 2-Quinolinecarboxylic acid
9431 - 8-Quinolinecarboxylic acid

$C_{10}H_7NO_3$

6190 - 4-Hydroxy-2-quinolinecarboxylic acid
8127 - 1-Nitro-2-naphthol

$C_{10}H_7NO_4$

3754 - 4,8-Dihydroxy-2-quinolinecarboxylic
acid

$C_{10}H_7N_3S$

10019 - Thiabendazole

$C_{10}H_8$

568 - Azulene
7928 - Naphthalene

$C_{10}H_8BrNO_2$

1207 - N-(2-Bromoethyl)phthalimide

$C_{10}H_8ClN_3O$

1832 - Chloridazon

$C_{10}H_8N_2$

893 - 2,2'-Bipyridine

894 - 2,3'-Bipyridine
895 - 2,4'-Bipyridine
896 - 3,3'-Bipyridine
897 - 4,4'-Bipyridine
6245 - 1H-Indole-3-acetonitrile

$C_{10}H_8N_2O_2S_2$

4569 - Di-2-pyridinyl disulfide, N,N'-dioxide

$C_{10}H_8N_2O_2S_2Zn$

9387 - Pyrithione zinc

$C_{10}H_8N_2O_3$

7502 - 3-Methyl-4-nitroquinoline-N-oxide

$C_{10}H_8N_2O_4$

3570 - Di-2-furanylethanedione dioxime

$C_{10}H_8N_4O_5$

9036 - Picrolonic acid

$C_{10}H_8O$

7975 - 1-Naphthol
7976 - 2-Naphthol
8860 - 4-Phenyl-3-butyn-2-one
8884 - 2-Phenylfuran

$C_{10}H_8O_2$

709 - 1-(2-Benzofuranyl)ethanone
7023 - 6-Methyl-2H-1-benzopyran-2-one
7024 - 7-Methyl-2H-1-benzopyran-2-one
7025 - 3-Methyl-4H-1-benzopyran-4-one
7655 - Methyl 3-phenyl-2-propynoate
7949 - 1,3-Naphthalenediol
7950 - 1,4-Naphthalenediol
7951 - 1,5-Naphthalenediol
7952 - 1,6-Naphthalenediol
7953 - 1,7-Naphthalenediol
7954 - 2,3-Naphthalenediol
7955 - 2,6-Naphthalenediol
7956 - 2,7-Naphthalenediol

$C_{10}H_8O_3$

6084 - 7-Hydroxy-4-methyl-2H-1-benzopy-
ran-2-one
6805 - 7-Methoxy-2H-1-benzopyran-2-one
8473 - 4-Oxo-4-phenyl-2-butenoic acid

$C_{10}H_8O_3S$

7965 - 1-Naphthalenesulfonic acid
7966 - 2-Naphthalenesulfonic acid

$C_{10}H_8O_4$

459 - Anemonin
3733 - 5,7-Dihydroxy-4-methyl-2H-1-ben-
zopyran-2-one
3734 - 6,7-Dihydroxy-4-methyl-2H-1-ben-
zopyran-2-one
5505 - Furoin
6066 - 7-Hydroxy-6-methoxy-2H-1-benzopy-
ran-2-one

$C_{10}H_8O_4S$

6121 - 4-Hydroxy-1-naphthalenesulfonic acid
6122 - 7-Hydroxy-1-naphthalenesulfonic acid
6123 - 1-Hydroxy-2-naphthalenesulfonic acid
6124 - 6-Hydroxy-2-naphthalenesulfonic acid

$C_{10}H_8O_5$

3727 - 7,8-Dihydroxy-6-methoxy-2H-1-ben-
zopyran-2-one

$C_{10}H_8O_6S_2$

7959 - 1,5-Naphthalenedisulfonic acid
7960 - 1,6-Naphthalenedisulfonic acid
7961 - 2,7-Naphthalenedisulfonic acid

$C_{10}H_8O_7S_2$

6118 - 7-Hydroxy-1,3-naphthalenedisulfonic
acid

6119 - 3-Hydroxy-2,7-naphthalenedisulfonic
acid

$C_{10}H_8O_8S_2$

3736 - 4,5-Dihydroxy-2,7-naphthalenedisul-
fonic acid

$C_{10}H_8S$

7970 - 1-Naphthalenethiol
7971 - 2-Naphthalenethiol

$C_{10}H_9ClN_4O_2S$

9636 - Sulfachlorpyridazine

$C_{10}H_9Cl_2NO$

3267 - N-(3,4-Dichlorophenyl)-2-methyl-2-
propenamide

$C_{10}H_9Cl_4NO_2S$

1722 - Captafol

$C_{10}H_9Cl_4O_4P$

9776 - Tetrachlorovinphos

$C_{10}H_9N$

7423 - 1-Methylisoquinoline
7424 - 3-Methylisoquinoline
7766 - 2-Methylquinoline
7767 - 3-Methylquinoline
7768 - 4-Methylquinoline
7769 - 5-Methylquinoline
7770 - 6-Methylquinoline
7771 - 7-Methylquinoline
7772 - 8-Methylquinoline
7981 - 1-Naphthylamine
7982 - 2-Naphthylamine
8988 - 1-Phenyl-1H-pyrrole
8989 - 2-Phenyl-1H-pyrrole

$C_{10}H_9NO$

371 - 5-Amino-1-naphthol
372 - 1-Amino-2-naphthol
373 - 8-Amino-2-naphthol
6256 - 1-(1H-Indol-3-yl)ethanone
6904 - 6-Methoxyquinoline
7773 - 2-Methyl-8-quinolinol
7774 - 1-Methyl-2(1H)-quinolinone
7775 - 1-Methyl-4(1H)-quinolinone
7988 - 1-Naphthylhydroxylamine

$C_{10}H_9NO_2$

5088 - N-Ethyl-1H-isoindole-1,3(2H)-dione
5285 - α-Ethynylbenzenemethanol carbamate
6108 - 4-Hydroxy-1-methyl-2-quinolinone
6244 - 1H-Indole-3-acetic acid
6255 - 1H-Indol-3-ol, acetate
8992 - 1-Phenyl-2,5-pyrrolidinedione

$C_{10}H_9NO_3$

6043 - N-(2-Hydroxyethyl)phthalimide
8471 - cis-4-Oxo-4-(phenylamino)-2-butenoic
acid

$C_{10}H_9NO_3S$

363 - 2-Amino-1-naphthalenesulfonic acid
364 - 4-Amino-1-naphthalenesulfonic acid
365 - 5-Amino-1-naphthalenesulfonic acid
366 - 6-Amino-1-naphthalenesulfonic acid
367 - 7-Amino-1-naphthalenesulfonic acid
368 - 8-Amino-1-naphthalenesulfonic acid
369 - 6-Amino-2-naphthalenesulfonic acid
370 - 8-Amino-2-naphthalenesulfonic acid

$C_{10}H_9NO_4S$

334 - 4-Amino-3-hydroxy-1-naphthalene-
sulfonic acid

$C_{10}H_9NO_6S_2$

359 - 7-Amino-1,3-naphthalenedisulfonic acid
360 - 2-Amino-1,5-naphthalenedisulfonic acid

361 - 4-Amino-1,6-naphthalenedisulfonic acid
362 - 4-Amino-1,7-naphthalenedisulfonic acid

$C_{10}H_9NO_7S_2$
333 - 4-Amino-5-hydroxy-2,7-naphthalenedisulfonic acid

$C_{10}H_9N_3$
9371 - N-2-Pyridinyl-2-pyridinamine

$C_{10}H_9O_4P$
7992 - 1-Naphthyl phosphate

$C_{10}H_{10}$
1395 - (trans-1,3-Butadienyl)benzene
1687 - 3-Butynylbenzene
3656 - 1,2-Dihydronaphthalene
3657 - 1,4-Dihydronaphthalene
4600 - o-Divinylbenzene
4601 - m-Divinylbenzene
4602 - p-Divinylbenzene
7383 - 1-Methyl-1H-indene
7384 - 2-Methyl-1H-indene
7385 - 3-Methyl-1H-indene
8848 - 2-Phenyl-1,3-butadiene

$C_{10}H_{10}ClFO$
2057 - 4-Chloro-1-(4-fluorophenyl)-1-butanone

$C_{10}H_{10}ClNO_2$
1842 - 4-Chloroacetoacetanilide
1846 - 4-(2-Chloroacetyl)acetanilide
2244 - N-(2-Chlorophenyl)-3-oxo-butanamide

$C_{10}H_{10}Cl_2O_3$
3257 - 4-(2,4-Dichlorophenoxy)butanoic acid

$C_{10}H_{10}Cl_2Ti$
953 - Bis(η-cyclopentadienyl)titanium chloride

$C_{10}H_{10}Cl_2Zr$
954 - Bis(η-cyclopentadienyl)zirconium chloride

$C_{10}H_{10}Fe$
5326 - Ferrocene

$C_{10}H_{10}N_2$
4316 - 2,3-Dimethylquinoxaline
7765 - 6-Methyl-8-quinolinamine
7941 - 1,5-Naphthalenediamine
7942 - 1,8-Naphthalenediamine
7943 - 2,3-Naphthalenediamine

$C_{10}H_{10}N_2O$
3652 - 1,2-Dihydro-5-methyl-2-phenyl-3H-pyrazol-3-one
3653 - 2,4-Dihydro-5-methyl-2-phenyl-3H-pyrazol-3-one

$C_{10}H_{10}N_2O_2$
7640 - 3-Methyl-5-phenyl-2,4-imidazolidinedione

$C_{10}H_{10}N_2O_4S$
3651 - 4-(4,5-Dihydro-3-methyl-5-oxo-1H-pyrazol-1-yl)benzenesulfonic acid

$C_{10}H_{10}N_4O_2S$
415 - 4-Amino-N-pyrazinylbenzenesulfonamide
424 - 4-Amino-N-2-pyrimidinylbenzenesulfonamide

$C_{10}H_{10}Ni$
8018 - Nickelocene

$C_{10}H_{10}O$
3658 - 3,4-Dihydro-2(1H)-naphthalenone
5290 - α-Ethynyl-α-methylbenzenemethanol
7654 - 2-Methyl-3-phenyl-2-propenal
8858 - 1-Phenyl-2-buten-1-one
8859 - trans-4-Phenyl-3-buten-2-one

9912 - 1-Tetralone

$C_{10}H_{10}O_2$
175 - Allyl benzoate
779 - Benzyl acrylate
2402 - Cinnamyl formate
2823 - 1,3-Diacetylbenzene
2824 - 1,4-Diacetylbenzene
7133 - Methyl trans-cinnamate
7134 - trans-o-Methylcinnamic acid
7135 - trans-m-Methylcinnamic acid
7136 - trans-p-Methylcinnamic acid
8851 - 1-Phenyl-1,3-butanedione
9188 - trans-5-(1-Propenyl)-1,3-benzodioxole
9499 - Safrole

$C_{10}H_{10}O_2S_2$
3572 - Difurfuryl disulfide

$C_{10}H_{10}O_3$
90 - 2-(Acetyloxy)-1-phenylethanone
94 - 4-Acetylphenyl acetate
3573 - Difurfuryl ether
5188 - Ethyl 2-oxo-2-phenylacetate
6071 - 3-(4-Hydroxy-3-methoxyphenyl)-2-propenal
6882 - trans-3-(4-Methoxyphenyl)-2-propenoic acid
7033 - Methyl benzoylacetate
8454 - γ-Oxobenzenebutanoic acid

$C_{10}H_{10}O_4$
89 - 4-(Acetyloxy)-3-methoxybenzaldehyde
633 - 1,2-Benzenediol, diacetate
634 - 1,4-Benzenediol, diacetate
3869 - 6,7-Dimethoxy-1(3H)-isobenzofuranone
4119 - Dimethyl isophthalate
4263 - Dimethyl phthalate
4331 - Dimethyl terephthalate
4904 - Ethyl 1,3-benzodioxole-5-carboxylate
6915 - Methyl 2-(acetyloxy)benzoate
8850 - Phenylbutanedioic acid, (±)

$C_{10}H_{10}O_5$
5462 - 6-Formyl-2,3-dimethoxybenzoic acid

$C_{10}H_{10}O_6$
2380 - Chorismic acid
9111 - Prephenic acid

$C_{10}H_{11}ClO$
2234 - 4-Chloro-1-phenyl-1-butanone

$C_{10}H_{11}ClO_4$
10484 - 3,4,5-Trimethoxybenzoyl chloride

$C_{10}H_{11}F_3N_2O$
4347 - N,N-Dimethyl-N'-[3-(trifluoromethyl)phenyl]urea

$C_{10}H_{11}F_7O_2$
5663 - 6,6,7,7,8,8,8-Heptafluoro-2,2-dimethyl-3,5-octanedione

$C_{10}H_{11}IO_4$
8905 - Phenyliodine diacetate

$C_{10}H_{11}I_2NO_3$
9243 - Propyliodone

$C_{10}H_{11}N$
4115 - 1,3-Dimethyl-1H-indole
4116 - 2,3-Dimethyl-1H-indole
4897 - α-Ethylbenzeneacetonitrile

$C_{10}H_{11}NO$
6251 - 1H-Indole-3-ethanol

$C_{10}H_{11}NO_2$
24 - Acetoacetanilide

$C_{10}H_{11}NO_3$
753 - N-Benzoyl-DL-alanine
8470 - 4-Oxo-4-(phenylamino)butanoic acid
8801 - (N-Phenylacetyl)glycine

$C_{10}H_{11}NO_4$
814 - Benzyloxycarbonylglycine

$C_{10}H_{11}N_3$
7656 - 3-Methyl-1-phenyl-1H-pyrazol-5-amine

$C_{10}H_{11}N_3O_2S_2$
9646 - Sulfamethylthiazole

$C_{10}H_{11}N_3O_3S$
9644 - Sulfamethoxazole

$C_{10}H_{12}$
1461 - trans-1-Butenylbenzene
1462 - 2-Butenylbenzene
1463 - 3-Butenylbenzene
3332 - Dicyclopentadiene
3649 - 2,3-Dihydro-1-methyl-1H-indene
5246 - 2-Ethylstyrene
5247 - 3-Ethylstyrene
5248 - 4-Ethylstyrene
7462 - 1-Methyl-4-(1-methylvinyl)benzene
7704 - cis-(1-Methyl-1-propenyl)benzene
7705 - trans-(1-Methyl-1-propenyl)benzene
7706 - (2-Methyl-1-propenyl)benzene
8857 - 2-Phenyl-1-butene
9874 - 1,2,3,4-Tetrahydronaphthalene

$C_{10}H_{12}BrCl_2O_3PS$
1326 - Bromophos-ethyl

$C_{10}H_{12}CaN_2Na_2O_8$
4573 - Disodium calcium EDTA

$C_{10}H_{12}ClNO_2$
2279 - Chloropropham

$C_{10}H_{12}ClNO_4$
2346 - Chlorphenesin carbamate

$C_{10}H_{12}ClN_3O_3S$
9415 - Quinethazone

$C_{10}H_{12}Cl_3O_2PS$
10259 - Trichloronate

$C_{10}H_{12}N_2$
3606 - 4,5-Dihydro-2-benzyl-1H-imidazole
5120 - 1-Ethyl-2-methyl-1H-benzimidazole
9843 - 1,2,3,6-Tetrahydro-2,3'-bipyridine, (S)
10713 - Tryptamine

$C_{10}H_{12}N_2Na_4O_8$
10002 - Tetrasodium EDTA

$C_{10}H_{12}N_2O$
6198 - 5-Hydroxytryptamine

$C_{10}H_{12}N_2O_3$
2840 - 5,5-Diallyl-2,4,6(1H,3H,5H)-pyrimidinetrione
6572 - L-Kynurenine

$C_{10}H_{12}N_2O_3S$
583 - Bentazon

$C_{10}H_{12}N_2O_4$
4857 - N-(4-Ethoxy-3-nitrophenyl)acetamide

$C_{10}H_{12}N_2O_5$
1561 - 2-tert-Butyl-4,6-dinitrophenol
4416 - Dinoseb

$C_{10}H_{12}N_2O_5S$
281 - 7-Aminocephalosporanic acid

$C_{10}H_{12}N_3O_3PS_2$
556 - Azinphos-methyl

$C_{10}H_{12}N_4OS$
10020 - Thiacetazone

$C_{10}H_{14}BeO_4$
844 - Beryllium 2,4-pentanedioate

$C_{10}H_{14}BrN$
1172 - 4-Bromo-N,N-diethylaniline

$C_{10}H_{14}Br_2O$
2922 - α,α'-Dibromo-d-camphor

$C_{10}H_{14}CaO_4$
1701 - Calcium 2,4-pentanedioate

$C_{10}H_{14}ClN$
2243 - 1-(2-Chlorophenyl)-2-methyl-2-propylamine
2349 - Chlorphentermine

$C_{10}H_{14}ClNO_3$
10724 - L-Tyrosine, methyl ester, hydrochloride

$C_{10}H_{14}Cl_2NO_2PS$
4610 - DMPA

$C_{10}H_{14}Cl_6N_4O_2$
10427 - Triforine

$C_{10}H_{14}CoO_4$
1043 - Bis(2,4-pentanedionato)cobalt

$C_{10}H_{14}CuO_4$
2463 - Copper(II) 2,4-pentanedioate

$C_{10}H_{14}NO_5PS$
8514 - Parathion

$C_{10}H_{14}NO_6P$
8510 - Paraoxon

$C_{10}H_{14}N_2$
7762 - 3-(1-Methyl-2-pyrrolidinyl)pyridine, (±)
8025 - L-Nicotine
8954 - 1-Phenylpiperazine
9075 - 3-(2-Piperidinyl)pyridine, (S)

$C_{10}H_{14}N_2O$
3493 - N,N-Diethyl-4-nitrosoaniline
3518 - N,N-Diethyl-3-pyridinecarboxamide
3519 - N,N-Diethyl-4-pyridinecarboxamide
7907 - 4-(4-Morpholinyl)aniline

$C_{10}H_{14}N_2O_2$
3492 - N,N-Diethyl-4-nitroaniline

$C_{10}H_{14}N_2O_2S$
5110 - 5-Ethyl-5-(2-methylallyl)-2-thiobarbituric acid

$C_{10}H_{14}N_2O_3$
503 - Aprobarbital

$C_{10}H_{14}N_2O_5$
10093 - Thymidine

$C_{10}H_{14}N_2O_6$
7851 - 5-Methyluridine

$C_{10}H_{14}N_4O_4$
4672 - Dyphylline

$C_{10}H_{14}N_5O_7P$
126 - Adenosine 3'-phosphate
129 - 5'-Adenylic acid
2804 - 2'-Deoxyguanosine 5'-monophosphate

$C_{10}H_{14}N_5O_8P$
5617 - Guanosine 5'-monophosphate

$C_{10}H_{14}NiO_4$
8016 - Nickel bis(2,4-pentanedioate)

$C_{10}H_{14}O$
606 - Benzenebutanol
1639 - 2-Butylphenol
1640 - 2-sec-Butylphenol
1641 - 2-tert-Butylphenol
1642 - 3-Butylphenol
1643 - 3-tert-Butylphenol
1644 - 4-Butylphenol
1645 - 4-sec-Butylphenol
1646 - 4-tert-Butylphenol
1651 - Butyl phenyl ether
1787 - (R)-Carvone
1788 - (S)-Carvone
2684 - 2-Cyclopentylidenecyclopentanone
4059 - (1,1-Dimethylethoxy)benzene
4163 - 6,6-Dimethyl-2-norpinene-2-carboxaldehyde
6480 - 4-Isopropylbenzenemethanol
6481 - α-Isopropylbenzenemethanol
6515 - 5-Isopropyl-2-methylphenol
6896 - 1-Methoxy-4-propylbenzene
7004 - α-Methylbenzenepropanol
7421 - 4-Methyl-2-isopropylphenol
7456 - 3-Methyl-6-(1-methylethylidene)-2-cyclohexen-1-one
7591 - 3-(4-Methyl-3-pentenyl)furan
7855 - 4-(1-Methylvinyl)-1-cyclohexene-1-carboxaldehyde, (R)
7856 - 4-(1-Methylvinyl)-1-cyclohexene-1-carboxaldehyde, (S)
8932 - 1-Phenyl-2-methyl-2-propanol
9213 - α-Propylbenzenemethanol, (R)
9498 - Safranal
9848 - 4,5,6,7-Tetrahydro-3,6-dimethylbenzofuran
9965 - 2,3,4,5-Tetramethylphenol
9966 - 2,3,4,6-Tetramethylphenol
9967 - 2,3,5,6-Tetramethylphenol
10095 - Thymol
10521 - 2,7,7-Trimethylbicyclo[3.1.1]hept-2-en-6-one
10529 - 2,6,6-Trimethyl-2,4-cycloheptadien-1-one
10780 - d-Verbenone

$C_{10}H_{14}O_2$
1482 - 4-Butoxyphenol
1516 - 4-tert-Butyl-1,2-benzenediol
1517 - 2-tert-Butyl-1,4-benzenediol
3352 - 1,2-Diethoxybenzene
3353 - 1,4-Diethoxybenzene
3864 - (2,2-Dimethoxyethyl)benzene
5103 - α-Ethyl-4-methoxybenzenemethanol
6897 - 2-Methoxy-4-propylphenol
8009 - Nepetalactone
9926 - 2,3,5,6-Tetramethyl-1,4-benzenediol

$C_{10}H_{14}O_3$
5493 - 2-Furanylmethyl pentanoate
6063 - 4-Hydroxy-3-methoxybenzenepropanol
7611 - 3-(2-Methylphenoxy)-1,2-propanediol

$C_{10}H_{14}O_3S$
9266 - Propyl 4-toluenesulfonate

$C_{10}H_{14}O_4$
1410 - 1,4-Butanediol diacrylate
4781 - 1,2-Ethanediol, dimethacrylate
6862 - 3-(2-Methoxyphenoxy)-1,2-propanediol
10481 - 3,4,5-Trimethoxybenzenemethanol

$C_{10}H_{14}O_4Zn$
10875 - Zinc 2,4-pentanedioate

$C_{10}H_{14}O_5$
26 - 2-Acetoacetoxyethyl methacrylate
4541 - Di-2-propenoyldiethyleneglycol

$C_{10}H_{14}O_5V$
8458 - Oxobis(2,4-pentanedione)vanadium

$C_{10}H_{15}N$
804 - Benzylisopropylamine
1503 - 2-sec-Butylaniline
1504 - 4-Butylaniline
1505 - 4-sec-Butylaniline
1506 - 4-tert-Butylaniline
1507 - N-Butylaniline
1508 - N-tert-Butylaniline
3397 - 2,6-Diethylaniline
3398 - N,N-Diethylaniline
3955 - N,β-Dimethylbenzeneethanamine
3956 - α,α-Dimethylbenzeneethanamine
5226 - 2-(1-Ethylpropyl)pyridine
5227 - 4-(1-Ethylpropyl)pyridine
6738 - Methamphetamine
7002 - α-Methylbenzenepropanamine
7400 - 2-Methyl-5-isopropylaniline
9920 - N,N,2,6-Tetramethylaniline

$C_{10}H_{15}NO$
409 - α-(1-Aminopropyl)benzenemethanol
1471 - 4-Butoxyaniline
3393 - 3-(Diethylamino)phenol
3912 - 4-[2-(Dimethylamino)ethyl]phenol
4703 - Ephedrine, (±)
4704 - d-Ephedrine
4705 - l-Ephedrine
5206 - 2-(Ethylphenylamino)ethanol
6937 - 4-[2-(Methylamino)propyl]phenol
7255 - 2-[2-(1-Methylethoxy)ethyl]pyridine
7642 - 2-[Methyl(phenylmethyl)amino]ethanol

$C_{10}H_{15}NO_2$
3844 - 3,4-Dimethoxybenzeneethanamine
8868 - N-Phenyl-N,N-diethanolamine

$C_{10}H_{15}NO_4$
6566 - Kainic acid

$C_{10}H_{15}N_2Na_3O_7$
10701 - Trisodium N-hydroxyethylethylenediaminetriacetate

$C_{10}H_{15}N_3$
848 - Bethanidine

$C_{10}H_{15}N_5O_5$
10786 - Vidarabine

$C_{10}H_{15}N_5O_{10}P_2$
124 - Adenosine 3',5'-diphosphate

$C_{10}H_{15}N_5O_{11}P_2$
5616 - Guanosine 5'-diphosphate

$C_{10}H_{15}OPS_2$
5447 - Fonofos

$C_{10}H_{15}O_2P$
3506 - Diethyl phenylphosphonite

$C_{10}H_{15}O_3PS_2$
5323 - Fenthion

$C_{10}H_{16}$
1706 - Camphene, (+)
1707 - Camphene, (-)
1769 - 3-Carene, (+)
4187 - cis-3,7-Dimethyl-1,3,6-octatriene
4188 - trans-3,7-Dimethyl-1,3,6-octatriene
4189 - 3,7-Dimethyl-1,3,7-octatriene
4190 - cis, cis-2,6-Dimethyl-2,4,6-octatriene
4191 - trans,trans-2,6-Dimethyl-2,4,6-octatriene
4438 - Dipentene
6511 - 5-Isopropyl-2-methyl-1,3-cyclohexadiene, (R)
6610 - d-Limonene
6611 - l-Limonene
7246 - 4-Methylene-1-isopropylcyclohexene

7402 - 2-Methyl-5-isopropylbicyclo[3.1.0]hex-2-ene

7914 - β-Myrcene

8740 - α-Phellandrene

8741 - β-Phellandrene

9047 - α-Pinene

9048 - β-Pinene

9699 - α-Terpinene

9700 - γ-Terpinene

9703 - Terpinolene

10307 - Tricyclene

10309 - Tricyclo[3.3.1.13,7]decane

10518 - 1,7,7-Trimethylbicyclo[2.2.1]hept-2-ene

$C_{10}H_{16}Br_2N_2O_2$

9079 - Pipobroman

$C_{10}H_{16}ClN$

838 - Benzyltrimethylammonium chloride

6739 - Methamphetamine hydrochloride

$C_{10}H_{16}ClNO$

4678 - Edrophonium chloride

4706 - Ephedrine hydrochloride

$C_{10}H_{16}ClN_5$

8877 - N-(2-Phenylethyl)imidodicarbonimidic diamide, monohydrochloride

$C_{10}H_{16}Cl_2O_2$

2754 - Decanedioyl dichloride

$C_{10}H_{16}Cl_3NOS$

10133 - Triallate

$C_{10}H_{16}NO_5PS_2$

5302 - Famphur

$C_{10}H_{16}N_2$

2752 - Decanedinitrile

3404 - N,N-Diethyl-1,4-benzenediamine

9924 - N,N,N',N'-Tetramethyl-1,2-benzenediamine

9925 - N,N,N',N'-Tetramethyl-1,4-benzenediamine

$C_{10}H_{16}N_2OS$

6375 - 5-Isobutyl-3-allyl-2-thioxo-4-imidazolidinone

$C_{10}H_{16}N_2O_2S$

4987 - 5-Ethyldihydro-5-sec-butyl-2-thioxo-4,6(1H,5H)-pyrimidinedione

$C_{10}H_{16}N_2O_3$

1572 - 5-Butyl-5-ethyl-2,4,6(1H,3H,5H)-pyrimidinetrione

$C_{10}H_{16}N_2O_3S$

874 - Biotin

$C_{10}H_{16}N_2O_8$

5018 - Ethylenediaminetetraacetic acid

$C_{10}H_{16}N_4O_3$

4353 - Dimetilan

$C_{10}H_{16}N_4O_7$

10785 - Vicine

$C_{10}H_{16}N_5O_{12}P_3$

2798 - 2'-Deoxyadenosine 5'-triphosphate

$C_{10}H_{16}N_5O_{13}P_3$

127 - Adenosine 5'-triphosphate

$C_{10}H_{16}N_6S$

2391 - Cimetidine

$C_{10}H_{16}O$

863 - [1,1'-Bicyclopentyl]-2-one

1709 - Camphor, (±)

1710 - Camphor, (+)

1711 - Camphor, (-)

1786 - Carvenone, (S)

4164 - cis-3,7-Dimethyl-2,6-octadienal

4165 - trans-3,7-Dimethyl-2,6-octadienal

4714 - 2,3-Epoxy-α-pinane

5315 - (±)-Fenchone

6512 - 5-Isopropyl-3-methyl-2-cyclohexen-1-one, (±)

6513 - 6-Isopropyl-3-methyl-2-cyclohexen-1-one, (±)

6699 - p-Menth-8-en-2-one

7245 - 4-Methylene-1-isopropylbicyclo[3.1.0]hexan-3-ol, [1S-(1α,3β,5α)]

7454 - 2-Methyl-5-(1-methylethenyl)cyclohexanone, (2R-trans)

7455 - 5-Methyl-2-(1-methylethylidene)cyclohexanone

7466 - trans-5-Methyl-2-(1-methylvinyl)cyclohexanone

7467 - 2-Methyl-5-(1-methylvinyl)-2-cyclohexen-1-ol

7857 - 4-(1-Methylvinyl)-1-cyclohexene-1-methanol

8336 - trans-Octahydro-1(2H)-naphthalenone

9292 - Pulegone

10091 - α-Thujone

10519 - 4,6,6-Trimethylbicyclo[3.1.1]hept-3-en-2-ol, (1α,2α,5α)

10520 - 4,6,6-Trimethylbicyclo[3.1.1]hept-3-en-2-ol, (1α,2β,5α)

10539 - 2,6,6-Trimethyl-1-cyclohexene-1-carboxaldehyde

$C_{10}H_{16}O_2$

517 - Ascaridole

2539 - 1,2-Cyclodecanedione

2615 - 3-Cyclohexyl-2-butenoic acid

2633 - Cyclohexyl methacrylate

3347 - 1,2:8,9-Diepoxy-p-menthane

4133 - trans-2,2-Dimethyl-3-(2-methyl-1-propenyl)cyclopropanecarboxylic acid

4167 - 3,7-Dimethyl-2,6-octadienoic acid

6015 - 3-Hydroxycamphor

6090 - 2-Hydroxy-3-methyl-6-isopropyl-2-cyclohexen-1-one

6353 - Iridomyrmecin

$C_{10}H_{16}O_4$

1712 - (±)-Camphoric acid

3377 - Diethyl 2-allylmalonate

3422 - Diethyl 1,1-cyclobutanedicarboxylate

3472 - Diethyl isopropylidenemalonate

4017 - Dimethyl trans-1,4-cyclohexanedicarboxylate

4559 - Dipropyl fumarate

4560 - Dipropyl maleate

10547 - 1,2,2-Trimethyl-1,3-cyclopentanedicarboxylic acid, (1R,3S)

$C_{10}H_{16}O_4S$

1713 - d-Camphorsulfonic acid

$C_{10}H_{16}O_5$

3375 - Diethyl 2-acetylsuccinate

3455 - Diethyl (ethoxymethylene)malonate

$C_{10}H_{16}Si$

10514 - Trimethylbenzylsilane

10570 - Trimethyl(4-methylphenyl)silane

$C_{10}H_{17}Br$

1185 - trans-1-Bromo-3,7-dimethyl-2,6-octadiene

$C_{10}H_{17}Cl$

1067 - Bornyl chloride

$C_{10}H_{17}Cl_2NOS$

2829 - Diallate

$C_{10}H_{17}N$

10308 - Tricyclo[3.3.1.13,7]decan-1-amine

$C_{10}H_{17}NO_2$

3490 - 3,3-Diethyl-5-methyl-2,4-piperidinedione

$C_{10}H_{17}NO_6$

5544 - 2-(β-D-Glucopyranosyloxy)-2-methylpropanenitrile

$C_{10}H_{17}N_2O_4PS$

5296 - Etrimfos

$C_{10}H_{17}N_3$

2624 - 4-Cyclohexyl-3-ethyl-4H-1,2,4-triazole

$C_{10}H_{17}N_3O_2$

6434 - Isolan

$C_{10}H_{17}N_3O_6S$

5558 - Glutathione

$C_{10}H_{18}$

861 - 1,1'-Bicyclopentyl

2740 - 1,3-Decadiene

2741 - 1,9-Decadiene

2743 - cis-Decahydronaphthalene

2744 - trans-Decahydronaphthalene

2788 - 1-Decyne

2789 - 5-Decyne

4166 - 3,7-Dimethyl-1,6-octadiene

7414 - 1-Methyl-4-isopropylcyclohexene

9044 - Pinane

10515 - 1,7,7-Trimethylbicyclo[2.2.1]heptane

$C_{10}H_{18}ClN$

233 - 1-Aminoadamantane hydrochloride

$C_{10}H_{18}ClN_3O_2$

2042 - 1-(2-Chloroethyl)-3-(4-methylcyclohexyl)-1-nitrosourea

$C_{10}H_{18}N_2Na_2O_{10}$

5019 - Ethylenediaminetetraacetic acid, disodium salt, dihydrate

$C_{10}H_{18}N_2O_3$

2815 - Desthiobiotin

$C_{10}H_{18}N_2O_7$

6040 - N-(2-Hydroxyethyl)ethylenediaminetriacetic acid

$C_{10}H_{18}N_4O_4S_3$

10051 - Thiodicarb

$C_{10}H_{18}O$

862 - [1,1'-Bicyclopentyl]-2-ol

1064 - Borneol, (±)

1547 - 4-tert-Butylcyclohexanone

2541 - Cyclodecanone

2745 - Decahydro-2-naphthol

2767 - trans-2-Decenal

2775 - 3-Decen-2-one

3333 - Dicyclopentyl ether

4168 - cis-3,7-Dimethyl-2,6-octadien-1-ol

4192 - 3,7-Dimethyl-6-octenal

5297 - Eucalyptol

5314 - α-Fenchol, (±)

5528 - trans-Geraniol

6367 - Isoborneol

6612 - Linalol

7412 - cis-5-Methyl-2-isopropylcyclohexanone

7413 - trans-5-Methyl-2-isopropylcyclohexanone, (2S)

7415 - 3-Methyl-6-isopropyl-2-cyclohexen-1-ol

7416 - 4-Methyl-1-isopropyl-3-cyclohexen-1-ol

7419 - 1-Methyl-4-isopropyl-7-oxabicyclo[2.2.1]heptane

7463 - 1-Methyl-4-(1-methylvinyl)cyclohexanol

7464 - 5-Methyl-2-(1-methylvinyl)cyclohexanol, [1R-(1α,2β,5α)]

9045 - trans-2-Pinanol

9701 - α-Terpineol

10516 - 1,3,3-Trimethylbicyclo[2.2.1]heptan-2-ol, (1S-endo)

$C_{10}H_{18}O_2$

2614 - Cyclohexyl butanoate

2635 - Cyclohexyl 2-methylpropanoate

2773 - 9-Decenoic acid

4194 - 3,7-Dimethyl-6-octenoic acid

4966 - Ethyl cyclohexylacetate

5900 - Hexyl methacrylate

6018 - 2-Hydroxycyclodecanone

6023 - 4-Hydroxydecanoic acid γ-lactone

9879 - Tetrahydro-6-pentyl-2H-pyran-2-one

$C_{10}H_{18}O_3$

4874 - Ethyl 2-acetylhexanoate

5076 - α-Ethyl-1-hydroxycyclohexaneacetic acid

7076 - 3-Methylbutanoic anhydride

8625 - Pentanoic anhydride

$C_{10}H_{18}O_4$

1411 - 1,4-Butanediol diglycidyl ether

3037 - Dibutyl oxalate

3376 - Diethyl adipate

3473 - Diethyl isopropylmalonate

3517 - Diethyl 2-propylmalonate

4179 - Dimethyl octanedioate

4563 - Dipropyl succinate

9520 - Sebacic acid

$C_{10}H_{18}O_4S$

3536 - Diethyl thiodipropionate

$C_{10}H_{18}O_6$

3813 - Diisopropyl tartrate, (±)

10357 - Triethylene glycol diacetate

$C_{10}H_{19}BrO_2$

1165 - 2-Bromodecanoic acid

$C_{10}H_{19}ClNO_5P$

9024 - Phosphamidon

$C_{10}H_{19}ClO$

2766 - Decanoyl chloride

$C_{10}H_{19}Cl_2N_7O_4$

9514 - Saxitoxin dihydrochloride

$C_{10}H_{19}N$

1066 - Bornylamine

2755 - Decanenitrile

8005 - Neobornylamine

$C_{10}H_{19}NO$

8338 - trans-Octahydro-2H-quinolizine-1-methanol, (1R)

$C_{10}H_{19}NO_2$

1501 - N-tert-Butylaminoethyl methacrylate

3388 - 2-(N,N-Diethylamino)ethyl methacrylate

5220 - Ethyl 1-piperidinepropanoate

$C_{10}H_{19}NO_4$

9194 - Propionyl-L-carnitine

$C_{10}H_{19}N_5O$

1389 - sec-Bumeton

9129 - Prometone

$C_{10}H_{19}N_5S$

9130 - Prometryn

9692 - Terbutryn

$C_{10}H_{19}O_6PS_2$

6658 - Malathion

$C_{10}H_{19}O_7PS$

6657 - Malaoxon

$C_{10}H_{20}$

1541 - Butylcyclohexane

1542 - sec-Butylcyclohexane

1543 - tert-Butylcyclohexane

2538 - Cyclodecane

2768 - 1-Decene

2769 - cis-2-Decene

2770 - trans-2-Decene

2771 - cis-5-Decene

2772 - trans-5-Decene

3018 - cis-1,2-Di-tert-butylethene

3423 - 1,1-Diethylcyclohexane

4193 - 3,7-Dimethyl-1-octene

6385 - Isobutylcyclohexane

7404 - cis-1-Methyl-4-isopropylcyclohexane

7405 - trans-1-Methyl-4-isopropylcyclohexane

7512 - 2-Methyl-1-nonene

8673 - Pentylcyclopentane

9936 - 1,2,3,4-Tetramethylcyclohexane

$C_{10}H_{20}Br_2$

2928 - 1,10-Dibromodecane

$C_{10}H_{20}CdN_2S_4$

1691 - Cadmium bis(diethyldithiocarbamate)

$C_{10}H_{20}Cl_2$

3150 - 1,10-Dichlorodecane

$C_{10}H_{20}NO_4PS$

9191 - Propetamphos

$C_{10}H_{20}NO_5PS_2$

6686 - Mecarbam

$C_{10}H_{20}N_2NiS_4$

960 - Bis(diethyldithiocarbamate)nickel

$C_{10}H_{20}N_2O_2$

1035 - 1,2-Bis(N-morpholino)ethane

$C_{10}H_{20}N_2O_4S_2$

4586 - 3,3'-Dithiobis-D-valine

$C_{10}H_{20}N_2S_3$

9828 - Tetraethylthiodicarbonic diamide

$C_{10}H_{20}N_2S_4$

4577 - Disulfiram

$C_{10}H_{20}N_2S_4Zn$

961 - Bis(diethyldithiocarbamate)zinc

$C_{10}H_{20}O$

1544 - 2-tert-Butylcyclohexanol

1545 - cis-4-tert-Butylcyclohexanol

1546 - trans-4-tert-Butylcyclohexanol

2540 - Cyclodecanol

2749 - Decanal

2763 - 2-Decanone

2764 - 3-Decanone

2765 - 4-Decanone

2774 - 9-Decen-1-ol

4195 - 3,7-Dimethyl-6-octen-1-ol, (R)

4196 - 3,7-Dimethyl-6-octen-1-ol, (S)

4197 - 3,7-Dimethyl-7-octen-1-ol, (S)

4198 - 3,7-Dimethyl-6-octen-3-ol

7406 - 1-Methyl-4-isopropylcyclohexanol

7407 - 5-Methyl-2-isopropylcyclohexanol, [1S-(1α,2β,5α)]-

7408 - 5-Methyl-2-isopropylcyclohexanol, [1R-(1α,2β,5α)]-

7409 - 5-Methyl-2-isopropylcyclohexanol, [1S-(1α,2α,5β)]-

7410 - 5-Methyl-2-isopropylcyclohexanol, [1S-(1α,2β,5β)]-

8396 - Octyloxirane

$C_{10}H_{20}O_2$

1581 - Butyl hexanoate

2757 - Decanoic acid

4181 - 2,2-Dimethyloctanoic acid

4285 - 1,1-Dimethylpropyl 3-methylbutanoate

5029 - Ethyl 2-ethylhexanoate

5055 - 2-Ethylhexyl acetate

5183 - Ethyl octanoate

5893 - Hexyl butanoate

6031 - 7-Hydroxy-3,7-dimethyloctanal

6448 - Isopentyl isopentanoate

6452 - Isopentyl pentanoate

6698 - p-Menthane hydroperoxide

7307 - 6-Methyl-2-heptanol acetate

7319 - 2-Methylheptyl acetate, (±)

7510 - Methyl nonanoate

8258 - Nonyl formate

8378 - Octyl acetate

8682 - Pentyl pentanoate

$C_{10}H_{20}O_2S$

6708 - Mercaptoacetic acid, 2-ethylhexyl ester

$C_{10}H_{20}O_4$

3445 - Diethylene glycol monobutyl ether acetate

$C_{10}H_{20}O_5Si$

10497 - 3-(Trimethoxysilyl)propyl methacrylate

$C_{10}H_{21}Br$

1164 - 1-Bromodecane

$C_{10}H_{21}Cl$

1968 - 1-Chlorodecane

$C_{10}H_{21}ClO$

1969 - 10-Chloro-1-decanol

$C_{10}H_{21}F$

5383 - 1-Fluorodecane

$C_{10}H_{21}I$

6291 - 1-Iododecane

$C_{10}H_{21}N$

1548 - Butylcyclohexylamine

2618 - Cyclohexyldiethylamine

4019 - N,α-Dimethylcyclohexaneethanamine

8602 - 1,2,2,6,6-Pentamethylpiperidine

$C_{10}H_{21}NO$

356 - 3-(Aminomethyl)-3,5,5-trimethylcyclohexanol

$C_{10}H_{21}NOS$

8516 - Pebulate

10781 - Vernolate

$C_{10}H_{21}NO_2$

8093 - 1-Nitrodecane

$C_{10}H_{21}N_3O$

3489 - N,N-Diethyl-4-methyl-1-piperazinecarboxamide

$C_{10}H_{22}$

2750 - Decane

3463 - 3,4-Diethylhexane

4122 - 2,4-Dimethyl-3-isopropylpentane

4171 - 2,2-Dimethyloctane

4172 - 2,3-Dimethyloctane

4173 - 2,4-Dimethyloctane

4174 - 2,5-Dimethyloctane

4175 - 2,6-Dimethyloctane

C$_{11}$H$_{10}$O$_4$
3854 - 5,7-Dimethoxy-2*H*-1-benzopyran-2-one

C$_{11}$H$_{11}$N
829 - 1-Benzyl-1*H*-pyrrole
4313 - 2,4-Dimethylquinoline
4314 - 2,6-Dimethylquinoline
4315 - 2,7-Dimethylquinoline
7476 - Methyl-1-naphthylamine
7962 - 1-Naphthalenemethanamine

C$_{11}$H$_{11}$NO
6257 - 1-(1*H*-Indol-3-yl)-2-propanone

C$_{11}$H$_{11}$NO$_2$
4959 - Ethyl 2-cyano-2-phenylacetate
6253 - 1*H*-Indole-3-propanoic acid
7659 - 1-Methyl-3-phenyl-2,5-pyrrolidinedione

C$_{11}$H$_{11}$NO$_3$
6252 - 1*H*-Indole-3-lactic acid, (*S*)

C$_{11}$H$_{11}$NO$_4$S
4867 - 8-Ethoxy-5-quinolinesulfonic acid
5251 - 2-Ethyl-5-(3-sulfophenyl)isoxazolium hydroxide, inner salt

C$_{11}$H$_{11}$N$_3$O$_2$S
419 - 4-Amino-*N*-2-pyridinylbenzenesulfonamide

C$_{11}$H$_{11}$N$_5$
8753 - Phenazopyridine

C$_{11}$H$_{12}$
6465 - *p*-Isopropenylstyrene

C$_{11}$H$_{12}$ClNO$_3$S
1836 - Chlormezanone

C$_{11}$H$_{12}$ClN$_5$
8754 - Phenazopyridine hydrochloride

C$_{11}$H$_{12}$Cl$_2$N$_2$O$_5$
1816 - Chloramphenicol

C$_{11}$H$_{12}$Cl$_2$O$_3$
6493 - Isopropyl (2,4-dichlorophenoxy)acetate

C$_{11}$H$_{12}$Cl$_3$N
445 - Amphecloral

C$_{11}$H$_{12}$I$_3$NO$_2$
6349 - Iopanoic acid

C$_{11}$H$_{12}$NO$_4$PS$_2$
9023 - Phosmet

C$_{11}$H$_{12}$N$_2$
4254 - 3,5-Dimethyl-1-phenyl-1*H*-pyrazole

C$_{11}$H$_{12}$N$_2$O
3614 - 1,2-Dihydro-1,5-dimethyl-2-phenyl-3*H*-pyrazol-3-one
10774 - *DL*-Vascicine
10775 - Vasicine

C$_{11}$H$_{12}$N$_2$OS$_2$
7906 - 2-(4-Morpholinothio)benzothiazole

C$_{11}$H$_{12}$N$_2$OS$_3$
7908 - 2-(4-Morpholinyldithio)benzothiazole

C$_{11}$H$_{12}$N$_2$O$_2$
4817 - Ethotoin
10714 - *L*-Tryptophan

C$_{11}$H$_{12}$N$_2$O$_3$
6199 - 5-Hydroxy-*DL*-tryptophan

C$_{11}$H$_{12}$N$_4$O$_2$S
9641 - Sulfamerazine

C$_{11}$H$_{12}$N$_4$O$_3$S
9645 - Sulfamethoxypyridazine

C$_{11}$H$_{12}$O
3650 - 3,4-Dihydro-2-methyl-1(2*H*)-naphthalenone
8924 - 2-(Phenylmethylene)butanal
8946 - 1-Phenyl-1-penten-3-one
9841 - 6,7,8,9-Tetrahydro-5*H*-benzocyclohepten-5-one

C$_{11}$H$_{12}$O$_2$
806 - Benzyl methacryalate
3637 - 3,4-Dihydro-6-methoxy-1(2*H*)-naphthalenone
4952 - Ethyl *trans*-cinnamate
6867 - 4-(4-Methoxyphenyl)-3-buten-2-one
8943 - 1-Phenyl-1,4-pentanedione
8971 - *trans*-3-Phenyl-2-propen-1-ol acetate

C$_{11}$H$_{12}$O$_3$
3868 - 5,6-Dimethoxy-1-indanone
4905 - Ethyl benzoylacetate
7915 - Myristicin
8783 - 2-Phenoxyethyl acrylate
8941 - 3-Phenyloxiranecarboxylic acid, ethyl ester

C$_{11}$H$_{12}$O$_4$
801 - Benzylidene diacetate

C$_{11}$H$_{12}$O$_5$
9541 - Sinapinic acid

C$_{11}$H$_{13}$ClF$_3$N$_3$O$_4$S$_3$
9089 - Polythiazide

C$_{11}$H$_{13}$ClO
1524 - 4-Butylbenzoyl chloride
1525 - 4-*tert*-Butylbenzoyl chloride
1967 - 4-Chloro-2-cyclopentylphenol

C$_{11}$H$_{13}$ClO$_3$
2153 - 4-(4-Chloro-2-methylphenoxy)butanoic acid

C$_{11}$H$_{13}$F$_3$N$_2$O$_3$S
6691 - Mefluidide

C$_{11}$H$_{13}$F$_3$N$_4$O$_4$
4359 - Dinitramine

C$_{11}$H$_{13}$N
2604 - 4-(3-Cyclohexen-1-yl)pyridine
7722 - *N*-Methyl-*N*-2-propynylbenzenemethanamine
10569 - 2,3,3-Trimethyl-3*H*-indole

C$_{11}$H$_{13}$NO
3919 - 3-[4-(Dimethylamino)phenyl]-2-propenal
7631 - 3-Methyl-4-phenyl-3-butenamide

C$_{11}$H$_{13}$NO$_2$
5953 - Hydrohydrastinine
7646 - *N*-(2-Methylphenyl)-3-oxo-butanamide
7647 - *N*-(4-Methylphenyl)-3-oxobutanamide

C$_{11}$H$_{13}$NO$_3$
95 - *N*-Acetyl-*L*-phenylalanine
3613 - 3,4-Dihydro-6,7-dimethoxy-1(2*H*)-isoquinolinone
5932 - Hydrastinine
6873 - *N*-(4-Methoxyphenyl)-3-oxobutanamide

C$_{11}$H$_{13}$NO$_4$
105 - *N*-Acetyl-*L*-tyrosine
578 - Bendiocarb
6861 - 5-[(2-Methoxyphenoxy)methyl]-2-oxazolidinone

C$_{11}$H$_{13}$N$_3$O
448 - Ampyrone

C$_{11}$H$_{13}$N$_3$O$_3$S
9655 - Sulfisoxazole

C$_{11}$H$_{14}$
2683 - Cyclopentylbenzene
4114 - 1,1-Dimethylindan
6541 - 4-Isopropylstyrene
9868 - 1,2,3,4-Tetrahydro-1-methylnaphthalene
9869 - 1,2,3,4-Tetrahydro-5-methylnaphthalene
9870 - 1,2,3,4-Tetrahydro-6-methylnaphthalene

C$_{11}$H$_{14}$ClNO
9131 - Propachlor

C$_{11}$H$_{14}$ClNO$_2$
1535 - *N*-Butyl-4-chloro-2-hydroxybenzamide

C$_{11}$H$_{14}$N$_2$
4118 - *N,N*-Dimethyl-1*H*-indole-3-methanamine

C$_{11}$H$_{14}$N$_2$O
2724 - Cytisine
4255 - 4,4-Dimethyl-1-phenyl-3-pyrazolidinone
6833 - 5-Methoxy-1*H*-indole-3-ethanamine

C$_{11}$H$_{14}$N$_2$O$_3$
5592 - *N*-Glycyl-*L*-phenylalanine
8808 - *L*-Phenylalanylglycine

C$_{11}$H$_{14}$N$_2$O$_4$
8170 - *N*-(5-Nitro-2-propoxyphenyl)acetamide

C$_{11}$H$_{14}$N$_2$S
9299 - Pyrantel

C$_{11}$H$_{14}$N$_2$S$_2$
1502 - 2-(*tert*-Butylaminothio)benzothiazole

C$_{11}$H$_{14}$O
1511 - 4-Butylbenzaldehyde
1512 - 4-*tert*-Butylbenzaldehyde
4253 - 2,2-Dimethyl-1-phenyl-1-propanone
6532 - 1-(4-Isopropylphenyl)ethanone
7630 - 3-Methyl-1-phenyl-1-butanone
8945 - 1-Phenyl-1-pentanone
10600 - 1-(2,4,6-Trimethylphenyl)ethanone

C$_{11}$H$_{14}$O$_2$
648 - Benzenepentanoic acid
790 - Benzyl butanoate
810 - Benzyl 2-methylpropanoate
1472 - 4-Butoxybenzaldehyde
1520 - Butyl benzoate
1521 - 2-*tert*-Butylbenzoic acid
1522 - 3-*tert*-Butylbenzoic acid
1523 - 4-*tert*-Butylbenzoic acid
3831 - 1,2-Dimethoxy-4-allylbenzene
3885 - 1,2-Dimethoxy-4-(1-propenyl)benzene
5211 - Ethyl 3-phenylpropanoate
6379 - Isobutyl benzoate
7629 - Methyl 2-phenylbutanoate
8265 - 5-Norbornene-2-methylolacrylate
8880 - 2-Phenylethyl propanoate
8973 - 3-Phenylpropyl acetate

C$_{11}$H$_{14}$O$_3$
1586 - Butyl 2-hydroxybenzoate
1587 - Butyl 4-hydroxybenzoate
1638 - *tert*-Butyl peroxybenzoate
3351 - 3,4-Diethoxybenzaldehyde
5107 - Ethyl (4-methoxyphenyl)acetate
6067 - 4-(4-Hydroxy-3-methoxyphenyl)-2-butanone

6089 - 2-Hydroxy-6-methyl-3-isopropylbenzoic acid

6390 - Isobutyl 2-hydroxybenzoate

$C_{11}H_{14}O_4$

4004 - Dimethyl carbate

$C_{11}H_{14}O_5$

7838 - Methyl 3,4,5-trimethoxybenzoate

$C_{11}H_{15}BrClO_3PS$

9120 - Profenofos

$C_{11}H_{15}BrN_2O_3$

1085 - 5-(2-Bromoallyl)-5-sec-butylbarbituric acid

7999 - Narcobarbital

$C_{11}H_{15}Cl_2O_3PS_2$

2358 - Chlorthiophos

$C_{11}H_{15}N$

8955 - 1-Phenylpiperidine

8956 - 4-Phenylpiperidine

$C_{11}H_{15}NO$

399 - 1-(4-Aminophenyl)-1-pentanone

3381 - 4-(Diethylamino)benzaldehyde

3400 - N,N-Diethylbenzamide

7644 - 3-Methyl-2-phenylmorpholine

$C_{11}H_{15}NO_2$

1498 - Butyl 4-aminobenzoate

3390 - 4-(Diethylamino)-2-hydroxybenzaldehyde

4906 - Ethyl N-benzylglycinate

4994 - Ethyl 4-(dimethylamino)benzoate

6377 - Isobutyl 4-aminobenzoate

8807 - L-Phenylalanine, ethyl ester

9503 - Salsoline

$C_{11}H_{15}NO_2S$

6759 - Methiocarb

$C_{11}H_{15}NO_3$

460 - Anhalamine

4863 - N-(4-Ethoxyphenyl)-2-hydroxypropanamide

9196 - Propoxur

10723 - L-Tyrosine, ethyl ester

$C_{11}H_{15}NO_5$

6761 - Methocarbamol

$C_{11}H_{15}N_2NaO_2S$

1466 - Buthalital sodium

$C_{11}H_{15}N_2O_8P$

8024 - β-Nicotinamide mononucleotide

$C_{11}H_{16}$

1606 - 1-tert-Butyl-2-methylbenzene

1607 - 1-tert-Butyl-3-methylbenzene

1608 - 1-tert-Butyl-4-methylbenzene

3483 - 1,3-Diethyl-5-methylbenzene

4282 - (1,1-Dimethylpropyl)benzene

4283 - (2,2-Dimethylpropyl)benzene

5091 - 1-Ethyl-2-isopropylbenzene

5274 - 1-Ethyl-2,4,5-trimethylbenzene

5275 - 2-Ethyl-1,3,5-trimethylbenzene

6443 - Isopentylbenzene

8595 - Pentamethylbenzene

8666 - Pentylbenzene

10725 - 1,10-Undecadiyne

$C_{11}H_{16}ClNO$

3918 - 3-(Dimethylamino)-1-phenyl-1-propanone, hydrochloride

$C_{11}H_{16}ClNO_3$

7843 - α-Methyl-DL-tyrosine, methyl ester, hydrochloride

$C_{11}H_{16}ClN_3O_2$

5452 - Formetanate hydrochloride

$C_{11}H_{16}ClN_3O_4S_2$

1467 - Buthiazide

$C_{11}H_{16}ClO_2PS_3$

1761 - Carbophenothion

$C_{11}H_{16}N_2$

822 - 1-Benzylpiperazine

$C_{11}H_{16}N_2O_2$

275 - Aminocarb

9039 - Pilocarpine

$C_{11}H_{16}N_2O_3$

1397 - Butalbital

4697 - Enallylpropymal

9675 - Talbutal

$C_{11}H_{16}N_4O_4$

8660 - Pentostatin

$C_{11}H_{16}O$

649 - Benzenepentanol

1481 - 1-Butoxy-4-methylbenzene

1519 - 4-tert-Butylbenzenemethanol

1601 - 1-tert-Butyl-4-methoxybenzene

1613 - 2-tert-Butyl-4-methylphenol

1614 - 2-tert-Butyl-5-methylphenol

1615 - 2-tert-Butyl-6-methylphenol

1616 - 4-tert-Butyl-2-methylphenol

3960 - α,α-Dimethylbenzenepropanol

4286 - 2-(1,1-Dimethylpropyl)phenol

4287 - 4-(1,1-Dimethylpropyl)phenol

7590 - cis-3-Methyl-2-(2-pentenyl)-2-cyclopenten-1-one

8601 - Pentamethylphenol

8683 - 4-Pentylphenol

8944 - 1-Phenyl-1-pentanol

$C_{11}H_{16}O_2$

1585 - tert-Butyl-4-hydroxyanisole

1603 - 2-tert-Butyl-4-methoxyphenol

1604 - 3-tert-Butyl-4-methoxyphenol

$C_{11}H_{16}O_3$

1708 - d-Camphocarboxylic acid

$C_{11}H_{16}O_3S$

1664 - Butyl 4-toluenesulfonate

$C_{11}H_{16}O_4$

4542 - Di-2-propenoyl-2,2-dimethyl-1,3-propanediol

$C_{11}H_{17}ClN_2O_2$

9040 - Pilocarpine, monohydrochloride

$C_{11}H_{17}Cl_3N_2O_2S$

1825 - Chlordantoin

$C_{11}H_{17}N$

1518 - N-tert-Butylbenzenemethanamine

3480 - N,N-Diethyl-2-methylaniline

3481 - N,N-Diethyl-4-methylaniline

5118 - N-Ethyl-α-methylbenzeneethanamine

8665 - 4-tert-Pentylaniline

10511 - N,α,α-Trimethylbenzeneethanamine

$C_{11}H_{17}NO$

5150 - 2-[Ethyl(3-methylphenyl)amino]ethanol

7250 - N-Methylephedrine, [R-(R*,S*)]

$C_{11}H_{17}NO_2$

1003 - N,N-Bis(2-hydroxyethyl)-3-methylaniline

$C_{11}H_{17}NO_3$

3825 - Dimetan (R)

6547 - Isoproterenol

10480 - 3,4,5-Trimethoxybenzeneethanamine

$C_{11}H_{17}N_3O_3S$

274 - 4-Amino-N-[(butylamino)carbonyl]benzenesulfonamide

$C_{11}H_{17}N_3O_4$

5626 - HC Blue No. 1

$C_{11}H_{17}N_3O_5$

9041 - Pilocarpine, mononitrate

$C_{11}H_{17}N_3O_8$

10010 - Tetrodotoxin

$C_{11}H_{17}N_5O_9P_2$

125 - Adenosine 5'-methylenediphosphonate

$C_{11}H_{17}O_3P$

3407 - Diethyl benzylphosphonate

$C_{11}H_{17}O_4PS_2$

5321 - Fensulfothion

$C_{11}H_{18}ClNO_3$

6766 - Methoxamine hydrochloride

$C_{11}H_{18}N_2$

3972 - N,N-Dimethyl-N'-benzyl-1,2-ethanediamine

$C_{11}H_{18}N_2O_3$

442 - Amobarbital

5128 - 5-Ethyl-5-(1-methylbutyl)-2,4,6(1H,3H,5H)-pyrimidinetrione

$C_{11}H_{18}N_4O_2$

9081 - Pirimicarb

$C_{11}H_{18}O$

3976 - 6,6-Dimethylbicyclo[3.1.1]hept-2-ene-2-ethanol

7593 - 3-Methyl-2-pentyl-2-cyclopenten-1-one

$C_{11}H_{18}O_2$

4170 - trans-3,7-Dimethyl-2,6-octadien-1-ol formate

7181 - Methyl trans-2,cis-4-decadienoate

7182 - Methyl trans-2,trans-4-decadienoate

$C_{11}H_{18}O_2Si$

3364 - Diethoxymethylphenylsilane

$C_{11}H_{19}ClN_2$

3484 - N4,N4-Diethyl-2-methyl-1,4-benzenediamine, monohydrochloride

$C_{11}H_{19}ClO$

10747 - 10-Undecenoyl chloride

$C_{11}H_{19}NOS$

8377 - Octhilinone

$C_{11}H_{19}NO_9$

84 - N-Acetylneuraminic acid

$C_{11}H_{19}NO_{10}$

5585 - N-Glycolylneuraminic acid

$C_{11}H_{19}N_3O$

3827 - Dimethirimol

4811 - Ethirimol

$C_{11}H_{20}$

9598 - Spiro[5.5]undecane

10750 - 1-Undecyne

10751 - 2-Undecyne

$C_{11}H_{20}ClN_5$

2306 - 6-Chloro-N,N,N',N'-tetraethyl-1,3,5-triazine-2,4-diamine

$C_{11}H_{20}N_3O_3PS$

9083 - Pirimiphos-methyl

$C_{11}H_{20}O$

4284 - 4-(1,1-Dimethylpropyl)cyclohexanone

10737 - 10-Undecenal

$C_{11}H_{20}O_2$

5056 - 2-Ethylhexyl acrylate

5708 - 5-Heptyldihydro-2(3*H*)-furanone
8379 - Octyl acrylate
9949 - 2,2,6,6-Tetramethyl-3,5-heptanedione
10745 - 10-Undecenoic acid

$C_{11}H_{20}O_4$
3030 - Dibutyl malonate
3031 - Di-*tert*-butyl malonate
3411 - Diethyl 2-butylmalonate
3470 - Diethyl isobutylmalonate
4162 - Dimethyl nonanedioate
4984 - Ethyl diethylmalonate

$C_{11}H_{20}O_{10}$
506 - 6-*O*-α-*L*-Arabinopyranosyl-*D*-Glucose
10862 - 6-*O*-β-*D*-Xylopyranosyl-*D*-glucose

$C_{11}H_{21}BrO_2$
1375 - 11-Bromoundecanoic acid

$C_{11}H_{21}ClO$
10736 - Undecanoyl chloride

$C_{11}H_{21}N$
10729 - Undecanenitrile

$C_{11}H_{21}NOS$
2527 - Cycloate

$C_{11}H_{21}N_5OS$
6764 - Methoprotryne

$C_{11}H_{21}N_5S$
4544 - Dipropetryn

$C_{11}H_{22}$
5895 - Hexylcyclopentane
8672 - Pentylcyclohexane
10738 - 1-Undecene
10739 - *cis*-2-Undecene
10740 - *trans*-2-Undecene
10741 - *cis*-4-Undecene
10742 - *trans*-4-Undecene
10743 - *cis*-5-Undecene
10744 - *trans*-5-Undecene

$C_{11}H_{22}N_2$
4538 - 1,1'-Dipiperidinomethane

$C_{11}H_{22}N_2O$
2703 - Cycluron

$C_{11}H_{22}O$
10727 - Undecanal
10734 - 2-Undecanone
10735 - 6-Undecanone
10746 - 10-Undecen-1-ol

$C_{11}H_{22}O_2$
1580 - Butyl heptanoate
2783 - Decyl formate
5176 - Ethyl nonanoate
5705 - Heptyl butanoate
5906 - Hexyl pentanoate
6389 - Isobutyl heptanoate
6446 - Isopentyl hexanoate
7186 - Methyl decanoate
8253 - Nonyl acetate
8401 - Octyl propanoate
8676 - Pentyl hexanoate
9252 - Propyl octanoate
10731 - Undecanoic acid

$C_{11}H_{22}O_4$
3753 - 2,3-Dihydroxypropyl octanoate

$C_{11}H_{23}Br$
1374 - 1-Bromoundecane

$C_{11}H_{23}NOS$
9669 - Sutan

$C_{11}H_{23}NO_2$
435 - 11-Aminoundecanoic acid

$C_{11}H_{24}$
7183 - 2-Methyldecane
7184 - 3-Methyldecane
7185 - 4-Methyldecane
10574 - 2,4,7-Trimethyloctane
10728 - Undecane

$C_{11}H_{24}O$
10732 - 1-Undecanol
10733 - 2-Undecanol

$C_{11}H_{24}OS$
7790 - 1-(Methylsulfinyl)decane

$C_{11}H_{24}O_3Si$
10461 - Triisopropoxyvinylsilane

$C_{11}H_{24}O_6Si$
10694 - Tris(methoxyethoxy)vinylsilane

$C_{11}H_{24}S$
10730 - 1-Undecanethiol

$C_{11}H_{25}N$
10748 - Undecylamine

$C_{11}H_{26}NO_2PS$
4992 - *O*-Ethyl *S*-[2-(diisopropylamino)ethyl]
methylphosphonothioate

$C_{11}H_{26}O_3Si$
10337 - Triethoxypentylsilane

$C_{11}H_{28}Br_2N_2$
8594 - Pentamethonium bromide

$C_{12}Br_{10}O$
2738 - Decabromobiphenyl ether

$C_{12}Cl_8O_2$
8293 - Octachlorodibenzo-*p*-dioxin

$C_{12}Cl_{10}$
2739 - Decachlorobiphenyl

$C_{12}F_{10}$
2742 - 2,2',3,3',4,4',5,5',6,6'-Decafluoro-1,1'-
biphenyl

$C_{12}F_{27}N$
10702 - Tris(perfluorobutyl)amine

$C_{12}HCl_9$
8215 - 2,2',3,3',4,5,5',6,6'-Nonachlorobiphenyl

$C_{12}H_2Cl_6O_2$
5735 - 1,2,3,6,7,8-Hexachlorodibenzo-*p*-dioxin
5736 - 1,2,3,7,8,9-Hexachlorodibenzo-*p*-dioxin

$C_{12}H_2Cl_8$
8291 - 2,2',3,3',5,5',6,6'-Octachlorobiphenyl

$C_{12}H_3Cl_5O_2$
8537 - 1,2,3,4,7-Pentachlorodibenzo-*p*-dioxin

$C_{12}H_3Cl_7$
5642 - 2,2',3,3',4,4',6-Heptachlorobiphenyl

$C_{12}H_4Br_6$
5718 - 2,2',4,4',5,5'-Hexabromobiphenyl

$C_{12}H_4Cl_4O$
9744 - 2,3,7,8-Tetrachlorodibenzofuran

$C_{12}H_4Cl_4O_2$
9743 - 2,3,7,8-Tetrachlorodibenzo-*p*-dioxin

$C_{12}H_4Cl_6$
5725 - 2,2',3,3',4,4'-Hexachlorobiphenyl
5726 - 2,2',4,4',6,6'-Hexachlorobiphenyl
5727 - 2,2',3,3',6,6'-Hexachlorobiphenyl
5728 - 2,2',4,4',5,5'-Hexachlorobiphenyl

$C_{12}H_4N_6O_{12}S$
1055 - Bis(2,4,6-trinitrophenyl) sulfide

$C_{12}H_5BrO_3$
1289 - 4-Bromo-1,8-naphthalenedicarboxylic
anhydride

$C_{12}H_5Cl_5$
8535 - 2,3,4,5,6-Pentachlorobiphenyl

8536 - 2,2',4,5,5'-Pentachlorobiphenyl

$C_{12}H_5N_7O_{12}$
10634 - 2,4,6-Trinitro-*N*-(2,4,6-trinitrophenyl)
aniline

$C_{12}H_6Cl_2NNaO_2$
3200 - 2,6-Dichloroindophenol, sodium salt

$C_{12}H_6Cl_2O_2$
3151 - 2,7-Dichlorodibenzo-*p*-dioxin

$C_{12}H_6Cl_4$
9737 - 2,2',4',5-Tetrachlorobiphenyl
9738 - 2,3,4,5-Tetrachlorobiphenyl
9739 - 3,3',4',4'-Tetrachlorobiphenyl

$C_{12}H_6Cl_4O$
958 - Bis(2,4-dichlorophenyl)ether

$C_{12}H_6Cl_4O_2S$
1057 - Bithionol
9806 - Tetradifon

$C_{12}H_6N_2O_2$
7948 - 1,5-Naphthalene diisocyanate

$C_{12}H_6O_2$
9 - 1,2-Acenaphthylenedione

$C_{12}H_6O_3$
7974 - Naphtho[2,3-c]furan-1,3-dione
7980 - 1*H*,3*H*-Naphtho[1,8-cd]pyran-1,3-
dione

$C_{12}H_6O_{12}$
643 - Benzenehexacarboxylic acid

$C_{12}H_7Cl_2NO_3$
8099 - Nitrofen

$C_{12}H_7Cl_3$
10210 - 2,4,5-Trichlorobiphenyl
10211 - 2,4,6-Trichlorobiphenyl

$C_{12}H_7NO_2$
685 - 1*H*-Benz[de]isoquinoline-1,3(2*H*)-dione

$C_{12}H_7NO_3$
6142 - 7-Hydroxy-3*H*-phenoxazin-3-one

$C_{12}H_7NO_4$
9453 - Resazurin

$C_{12}H_7N_3O_2$
8131 - 5-Nitro-1,10-phenanthroline

$C_{12}H_8$
8 - Acenaphthylene

$C_{12}H_8Br_2$
2912 - 4,4'-Dibromo-1,1'-biphenyl

$C_{12}H_8Br_2O$
919 - Bis(4-bromophenyl) ether

$C_{12}H_8ClNS$
2221 - 2-Chloro-10*H*-phenothiazine

$C_{12}H_8Cl_2$
3121 - 2,5-Dichlorobiphenyl
3122 - 2,6-Dichlorobiphenyl
3123 - 3,3'-Dichlorobiphenyl
3124 - 4,4'-Dichlorobiphenyl

$C_{12}H_8Cl_2N_2$
3084 - *trans*-4,4'-Dichloroazobenzene

$C_{12}H_8Cl_2N_2O$
3085 - 4,4'-Dichloroazoxybenzene

$C_{12}H_8Cl_2N_2O_4S_2$
560 - 3,3'-Azobenzenedisulfonyl chloride

$C_{12}H_8Cl_2O_2S$
947 - Bis(4-chlorophenyl) sulfone
994 - Bis(2-hydroxy-5-chlorophenyl) sulfide

$C_{12}H_8Cl_2O_3S$
2235 - 4-Chlorophenyl 4-chlorobenzene-
sulfonate
3260 - 2,4-Dichlorophenyl benzenesulfonate

$C_{12}H_8Cl_2O_5S_2$
8480 - 4,4'-Oxybis(benzenesulfonyl chloride)
$C_{12}H_8Cl_2S_2$
942 - Bis(4-chlorophenyl) disulfide
$C_{12}H_8Cl_6$
150 - Aldrin
$C_{12}H_8Cl_6O$
3345 - Dieldrin
4701 - Endrin
$C_{12}H_8F_2$
3552 - 4,4'-Difluoro-1,1'-biphenyl
$C_{12}H_8N_2$
8746 - 1,7-Phenanthroline
8747 - 1,10-Phenanthroline
8748 - 4,7-Phenanthroline
8750 - Phenazine
$C_{12}H_8N_2O$
8752 - 1-Phenazinol
$C_{12}H_8N_2O_4$
4379 - 2,2'-Dinitro-1,1'-biphenyl
4380 - 4,4'-Dinitro-1,1'-biphenyl
$C_{12}H_8N_2O_4S$
4384 - 4,4'-Dinitrodiphenyl sulfide
$C_{12}H_8N_2O_4S_2$
1038 - Bis(2-nitrophenyl) disulfide
1039 - Bis(3-nitrophenyl) disulfide
1040 - Bis(4-nitrophenyl) disulfide
$C_{12}H_8N_2O_5$
4383 - 4,4'-Dinitrodiphenyl ether
$C_{12}H_8O$
2884 - Dibenzofuran
$C_{12}H_8O_2$
2883 - Dibenzo[b,e][1,4]dioxin
8864 - 2-Phenyl-2,5-cyclohexadiene-1,4-dione
$C_{12}H_8O_4$
6767 - Methoxsalen
6899 - 5-Methoxypsoralen
7944 - 1,8-Naphthalenedicarboxylic acid
7945 - 2,3-Naphthalenedicarboxylic acid
7946 - 2,6-Naphthalenedicarboxylic acid
$C_{12}H_8S$
2889 - Dibenzothiophene
$C_{12}H_8S_2$
10028 - Thianthrene
$C_{12}H_8S_3$
9706 - 2,2':5',2''-Terthiophene
$C_{12}H_9AsClN$
1985 - 10-Chloro-5,10-dihydrophenarsazine
$C_{12}H_9Br$
1112 - 2-Bromobiphenyl
1113 - 3-Bromobiphenyl
1114 - 4-Bromobiphenyl
$C_{12}H_9BrO$
1316 - 1-Bromo-4-phenoxybenzene
$C_{12}H_9Cl$
1916 - 2-Chlorobiphenyl
1917 - 3-Chlorobiphenyl
1918 - 4-Chlorobiphenyl
$C_{12}H_9ClF_3N_3O$
8273 - Norflurazon
$C_{12}H_9ClO$
1920 - 3-Chloro-[1,1'-biphenyl]-2-ol
2225 - 1-Chloro-4-phenoxybenzene
8863 - 4-Phenyl-2-chlorophenol
$C_{12}H_9ClO_2S$
2253 - 1-Chloro-4-(phenylsulfonyl)benzene

$C_{12}H_9ClO_3S$
2233 - 4-Chlorophenyl benzenesulfonate
$C_{12}H_9Cl_2NO_3$
10789 - Vinclozolin
$C_{12}H_9F$
5376 - 2-Fluoro-1,1'-biphenyl
5377 - 4-Fluoro-1,1'-biphenyl
$C_{12}H_9I$
6284 - 2-Iodo-1,1'-biphenyl
6285 - 3-Iodo-1,1'-biphenyl
6286 - 4-Iodo-1,1'-biphenyl
$C_{12}H_9N$
1730 - Carbazole
7932 - 1-Naphthaleneacetonitrile
$C_{12}H_9NO$
8769 - 10H-Phenoxazine
8986 - Phenyl-2-pyridinylmethanone
8987 - Phenyl-4-pyridinylmethanone
$C_{12}H_9NO_2$
3659 - 1,2-Dihydro-5-nitroacenaphthylene
6831 - 4-Methoxyfuro[2,3-b]quinoline
8084 - 2-Nitrobiphenyl
8085 - 3-Nitrobiphenyl
8086 - 4-Nitrobiphenyl
$C_{12}H_9NO_2S$
8160 - 1-Nitro-4-(phenylthio)benzene
$C_{12}H_9NO_3$
8135 - 1-Nitro-2-phenoxybenzene
8136 - 1-Nitro-4-phenoxybenzene
$C_{12}H_9NS$
8767 - 10H-Phenothiazine
$C_{12}H_9N_3O_4$
4382 - 4,4'-Dinitrodiphenylamine
4399 - 2,4-Dinitro-N-phenylaniline
8145 - 4-[(4-Nitrophenyl)azo]-1,3-benzenediol
$C_{12}H_9N_3O_5$
4398 - 4-[(2,4-Dinitrophenyl)amino]phenol
$C_{12}H_{10}$
7 - Acenaphthene
876 - Biphenyl
10810 - 1-Vinylnaphthalene
10811 - 2-Vinylnaphthalene
$C_{12}H_{10}AsCl$
4458 - Diphenylarsinous chloride
$C_{12}H_{10}Ca_3O_{14}$
10192 - Tricalcium citrate
$C_{12}H_{10}ClN$
1919 - 4'-Chloro-[1,1'-biphenyl]-4-amine
$C_{12}H_{10}ClN_3S$
2863 - 3,7-Diaminophenothiazin-5-ium chloride
$C_{12}H_{10}ClO_3P$
4476 - Diphenyl chlorophosphonate
$C_{12}H_{10}ClP$
4509 - Diphenylphosphinous chloride
$C_{12}H_{10}Cl_2N_2$
945 - 1,2-Bis(2-chlorophenyl)-hydrazine
3106 - 2,2'-Dichloro-p-benzidine
3107 - 3,3'-Dichloro-p-benzidine
$C_{12}H_{10}Cl_2Si$
3172 - Dichlorodiphenylsilane
$C_{12}H_{10}F_2Si$
3557 - Difluorodiphenylsilane
$C_{12}H_{10}Hg$
4497 - Diphenylmercury

$C_{12}H_{10}NNaO_3S$
4456 - Diphenylamine-4-sulfonic acid, sodium salt
$C_{12}H_{10}N_2$
558 - trans-Azobenzene
559 - cis-Azobenzene
5624 - Harman
$C_{12}H_{10}N_2O$
565 - cis-Azoxybenzene
566 - trans-Azoxybenzene
8189 - N-Nitrosodiphenylamine
8198 - 4-Nitroso-N-phenylaniline
8749 - 1,10-Phenanthroline monohydrate
8824 - 4-(Phenylazo)phenol
$C_{12}H_{10}N_2O_2$
3697 - 2,2'-Dihydroxyazobenzene
8142 - 2-Nitro-N-phenylaniline
8143 - 4-Nitro-N-phenylaniline
8819 - 4-(Phenylazo)-1,3-benzenediol
$C_{12}H_{10}N_4$
8751 - 2,3-Phenazinediamine
$C_{12}H_{10}N_4O_2$
3969 - 7,8-Dimethylbenzo[g]pteridine-2,4(1H,3H)-dione
$C_{12}H_{10}O$
31 - 1-Acetonaphthone
32 - 2-Acetonaphthone
4485 - Diphenyl ether
6003 - 2-Hydroxybiphenyl
6004 - 3-Hydroxybiphenyl
6005 - 4-Hydroxybiphenyl
$C_{12}H_{10}OS$
4531 - Diphenyl sulfoxide
$C_{12}H_{10}O_2$
884 - [1,1'-Biphenyl]-2,2'-diol
885 - [1,1'-Biphenyl]-2,5-diol
886 - [1,1'-Biphenyl]-4,4'-diol
6127 - 1-(1-Hydroxy-2-naphthyl)ethanone
7472 - Methyl 1-naphthalenecarboxylate
7473 - Methyl 2-naphthalenecarboxylate
7930 - 1-Naphthaleneacetic acid
7931 - 2-Naphthaleneacetic acid
7977 - 1-Naphthol, acetate
7978 - 2-Naphthol, acetate
8785 - 3-Phenoxyphenol
8786 - 4-Phenoxyphenol
$C_{12}H_{10}O_2S$
3723 - 4,4'-Dihydroxydiphenyl sulfide
4530 - Diphenyl sulfone
$C_{12}H_{10}O_3$
7377 - Methyl 3-hydroxy-2-naphthalenecarboxylate
7991 - (2-Naphthyloxy)acetic acid
$C_{12}H_{10}O_4$
698 - trans,trans-5-(1,3-Benzodioxol-5-yl)-2,4-pentadienoic acid
889 - [1,1'-Biphenyl]-3,3',5,5'-tetrol
$C_{12}H_{10}O_4S$
1017 - Bis(4-hydroxyphenyl) sulfone
$C_{12}H_{10}O_6S_2$
887 - [1,1'-Biphenyl]-4,4'-disulfonic acid
$C_{12}H_{10}O_7$
4676 - Echinochrome A
$C_{12}H_{10}S$
4529 - Diphenyl sulfide
$C_{12}H_{10}S_2$
4478 - Diphenyl disulfide

$C_{12}H_{10}Se$
4525 - Diphenyl selenide
$C_{12}H_{10}Se_2$
4477 - Diphenyl diselenide
$C_{12}H_{11}ClN_2O_5S$
5507 - Furosemide
$C_{12}H_{11}Cl_2NO$
9276 - Propyzamide
$C_{12}H_{11}N$
263 - 2-Aminobiphenyl
264 - 3-Aminobiphenyl
265 - 4-Aminobiphenyl
826 - 2-Benzylpyridine
827 - 4-Benzylpyridine
4454 - Diphenylamine
7657 - 2-Methyl-5-phenylpyridine
$C_{12}H_{11}NO$
7929 - 1-Naphthaleneacetamide
7985 - N-1-Naphthalenylacetamide
8773 - 2-Phenoxyaniline
8774 - 3-Phenoxyaniline
8775 - 4-Phenoxyaniline
8814 - 3-(Phenylamino)phenol
8815 - 4-(Phenylamino)phenol
$C_{12}H_{11}NOS$
6713 - 2-Mercapto-N-2-naphthylacetamide
$C_{12}H_{11}NO_2$
1729 - Carbaryl
4960 - Ethyl 2-cyano-3-phenyl-2-propenoate
6126 - N-(2-Hydroxy-1-naphthyl)acetamide
$C_{12}H_{11}NO_3S$
8811 - 4-(Phenylamino)benzenesulfonic acid
$C_{12}H_{11}NO_4$
1790 - Casimiroin
$C_{12}H_{11}N_3$
238 - 4-Aminoazobenzene
4533 - 1,3-Diphenyl-1-triazene
$C_{12}H_{11}N_3O_2$
5506 - Furonazide
$C_{12}H_{11}N_5$
8933 - N-(Phenylmethyl)-1H-purin-6-amine
$C_{12}H_{11}N_7$
8981 - 6-Phenyl-2,4,7-pteridinetriamine
$C_{12}H_{11}O_3P$
4510 - Diphenyl phosphonate
$C_{12}H_{12}$
4138 - 1,2-Dimethylnaphthalene
4139 - 1,3-Dimethylnaphthalene
4140 - 1,4-Dimethylnaphthalene
4141 - 1,5-Dimethylnaphthalene
4142 - 1,6-Dimethylnaphthalene
4143 - 1,7-Dimethylnaphthalene
4144 - 1,8-Dimethylnaphthalene
4145 - 2,3-Dimethylnaphthalene
4146 - 2,6-Dimethylnaphthalene
4147 - 2,7-Dimethylnaphthalene
5162 - 1-Ethylnaphthalene
5163 - 2-Ethylnaphthalene
$C_{12}H_{12}Br_2N_2$
4572 - Diquat dibromide
$C_{12}H_{12}ClNO_2S$
3914 - 5-(Dimethylamino)-1-naphthalene-
sulfonyl chloride
$C_{12}H_{12}ClN_5O_4S$
2354 - Chlorsulfuron

$C_{12}H_{12}Cl_4N_2$
3108 - 3,3'-Dichloro-p-benzidine dihydrochloride
$C_{12}H_{12}N_2$
680 - p-Benzidine
880 - [1,1'-Biphenyl]-2,2'-diamine
881 - [1,1'-Biphenyl]-2,4'-diamine
3980 - 4,4'-Dimethyl-2,2'-bipyridine
4491 - 1,1-Diphenylhydrazine
4492 - 1,2-Diphenylhydrazine
4571 - Diquat
8832 - N-Phenyl-1,2-benzenediamine
8833 - N-Phenyl-1,4-benzenediamine
$C_{12}H_{12}N_2O$
2854 - 4,4'-Diaminodiphenyl ether
$C_{12}H_{12}N_2OS$
909 - Bis(4-aminophenyl) sulfoxide
$C_{12}H_{12}N_2OS_2$
3901 - p-(Dimethylamino)benzalrhodanine
$C_{12}H_{12}N_2O_2S$
908 - Bis(4-aminophenyl) sulfone
2857 - 3,3'-Diaminodiphenyl sulfone
$C_{12}H_{12}N_2O_3$
7919 - Nalidixic acid
8761 - Phenobarbital
$C_{12}H_{12}N_2S$
2856 - 4,4'-Diaminodiphenyl sulfide
$C_{12}H_{12}N_2S_2$
905 - Bis(2-aminophenyl)disulfide
906 - Bis(4-aminophenyl)disulfide
$C_{12}H_{12}N_4$
303 - 2-Amino-3,4-dimethylimidazo[4,5-f]
quinoline
2849 - 4,4'-Diaminoazobenzene
$C_{12}H_{12}O$
4852 - 1-Ethoxynaphthalene
4853 - 2-Ethoxynaphthalene
$C_{12}H_{12}O_2$
180 - Allyl $trans$-cinnamate
$C_{12}H_{12}O_2Si$
4527 - Diphenylsilanediol
$C_{12}H_{12}O_5$
9447 - Radicinin
$C_{12}H_{12}O_6$
676 - 1,2,4-Benzenetriol triacetate
10512 - Trimethyl 1,2,4-benzenetricarboxylate
$C_{12}H_{12}Si$
4526 - Diphenylsilane
$C_{12}H_{13}ClF_3N_3O_4$
5336 - Fluchloralin
$C_{12}H_{13}ClN_4$
2240 - 5-(4-Chlorophenyl)-6-ethyl-2,4-pyrim-
idinediamine
8818 - 4-(Phenylazo)-1,3-benzenediamine
monohydrochloride
$C_{12}H_{13}Cl_3O_3$
1666 - Butyl (2,4,5-trichlorophenoxy)acetate
$C_{12}H_{13}I_3N_2O_2$
6351 - Iopodic acid
$C_{12}H_{13}I_3N_2O_3$
6263 - Iocetamic acid
$C_{12}H_{13}N$
4148 - N,N-Dimethyl-1-naphthylamine
4149 - N,N-Dimethyl-2-naphthylamine
5161 - N-Ethyl-1-naphthalenamine
9844 - 2,3,4,9-Tetrahydro-1H-carbazole

$C_{12}H_{13}NO_2$
4257 - 1,3-Dimethyl-3-phenyl-2,5-pyrro-
lidinedione
6246 - 1H-Indole-3-butanoic acid
$C_{12}H_{13}NO_2S$
1763 - Carboxin
$C_{12}H_{13}NO_4S$
8483 - Oxycarboxin
$C_{12}H_{13}NO_6$
8917 - N-[(Phenylmethoxy)carbonyl]-L-aspar-
tic acid
$C_{12}H_{13}N_3$
395 - N-(4-Aminophenyl)-1,4-benzenediamine
9370 - N-(2-Pyridinylmethyl)-2-
pyridinemethanamine
$C_{12}H_{13}N_3O_4S_2$
9647 - N^4-Sulfanilylsulfanilamide
$C_{12}H_{14}$
1024 - 1,3-Bis(1-methylethenyl)benzene
2600 - 1-Cyclohexen-1-ylbenzene
10567 - 1,2,3-Trimethylindene
$C_{12}H_{14}As_2Cl_2N_2O_2$
9504 - Salvarsan dihydrochloride
$C_{12}H_{14}CaO_{12}$
1694 - Calcium ascorbate
$C_{12}H_{14}ClNO_2$
2428 - Clomazone
$C_{12}H_{14}ClNO_4$
1988 - N-(4-Chloro-2,5-dimethoxyphenyl)-3-
oxobutanamide
$C_{12}H_{14}ClN_5O_2S$
2864 - 4-[(2,4-Diaminophenyl)azo]benzene-
sulfonamide
$C_{12}H_{14}Cl_2N_2$
882 - [1,1'-Biphenyl]-4,4'-diamine, dihydrochloride
$C_{12}H_{14}Cl_2O_3$
1552 - Butyl (2,4-dichlorophenoxy)acetate
$C_{12}H_{14}Cl_3O_4P$
1830 - Chlorfenvinphos
$C_{12}H_{14}N_2$
7986 - N-1-Naphthyl-1,2-ethanediamine
8511 - Paraquat
10613 - $N,N,2$-Trimethyl-6-quinolinamine
$C_{12}H_{14}N_2O_2$
4990 - 5-Ethyldihydro-5-phenyl-4,6(1H,5H)-
pyrimidinedione
6702 - Mephenytoin
7841 - N-Methyl-L-tryptophan
$C_{12}H_{14}N_2O_3$
2681 - Cyclopentobarbital
$C_{12}H_{14}N_2O_5$
261 - N-(4-Aminobenzoyl)-L-glutamic acid
2620 - 2-Cyclohexyl-4,6-dinitrophenol
$C_{12}H_{14}N_4OS$
4333 - 2,7-Dimethylthiachromine-8-ethanol
$C_{12}H_{14}N_4O_2S$
9642 - Sulfamethazine
$C_{12}H_{14}N_4O_3S$
9637 - Sulfacytine
$C_{12}H_{14}N_4O_4S$
9638 - Sulfadimethoxine
$C_{12}H_{14}N_4O_4S_2$
10062 - Thiophanate-methyl
$C_{12}H_{14}N_4O_5S_2$
8481 - 4,4'-Oxybis(benzenesulfonyl hydrazide)

C₁₂H₁₄O

8865 - 4-Phenylcyclohexanone
8929 - 3-(Phenylmethylene)-2-pentanone

C₁₂H₁₄O₂

9224 - Propyl *trans*-cinnamate

C₁₂H₁₄O₃

197 - 4-Allyl-2-methoxyphenyl acetate
5151 - Ethyl 3-methyl-3-phenyloxiranecarboxylate

C₁₂H₁₄O₄

974 - 1,3-Bis(2,3-epoxypropoxy)benzene
3471 - Diethyl isophthalate
3511 - Diethyl phthalate
3535 - Diethyl terephthalate
3832 - 4,7-Dimethoxy-5-allyl-1,3-benzodioxole
3886 - 4,5-Dimethoxy-6-(2-propenyl)-1,3-benzodioxole
7889 - Monobutyl phthalate

C₁₂H₁₄O₆

1006 - Bis(2-hydroxyethyl) terephthalate

C₁₂H₁₅ClNO₄PS₂

9021 - Phosalone

C₁₂H₁₅ClO

8668 - 4-Pentylbenzoyl chloride

C₁₂H₁₅ClO₂

8681 - 4-(Pentyloxy)benzoyl chloride

C₁₂H₁₅ClO₃

2426 - Clofibrate

C₁₂H₁₅N

3691 - 1,2-Dihydro-2,2,4-trimethylquinoline
7249 - 2-Methylene-1,3,3-trimethylindoline
9842 - 2,3,6,7-Tetrahydro-1H,5H-benzo[ij]quinolizine
9871 - 1,2,3,6-Tetrahydro-1-methyl-4-phenylpyridine

C₁₂H₁₅NO

771 - 1-Benzoylpiperidine

C₁₂H₁₅NO₃

97 - N-Acetyl-L-phenylalanine, methyl ester
462 - Anhalonine
1738 - Carbofuran
5949 - Hydrocotarnine
6731 - Metaxalone

C₁₂H₁₅NO₄

2472 - Cotarnine

C₁₂H₁₅N₃O₃

10135 - 1,3,5-Triallyl-1,3,5-triazine-2,4,6(1H,3H,5H)-trione

C₁₂H₁₅N₃O₆

1560 - 1-*tert*-Butyl-3,5-dimethyl-2,4,6-trinitrobenzene

C₁₂H₁₆

1672 - 1-*tert*-Butyl-4-vinylbenzene
2612 - Cyclohexylbenzene
6464 - *p*-Isopropenylisopropylbenzene
9849 - 1,2,3,4-Tetrahydro-1,5-dimethylnaphthalene

C₁₂H₁₆ClNOS

10045 - Thiobencarb

C₁₂H₁₆Cl₂N₂

7972 - N-(1-Naphthalenyl)-1,2-ethanediamine, dihydrochloride

C₁₂H₁₆F₃N

5316 - Fenfluramine

C₁₂H₁₆N₂

4117 - N,N-Dimethyl-1H-indole-3-ethanamine

C₁₂H₁₆N₂O

1792 - Caulophylline
3910 - 3-[2-(Dimethylamino)ethyl]-1H-indol-5-ol

C₁₂H₁₆N₂O₃

2528 - Cyclobarbital
5884 - Hexobarbital

C₁₂H₁₆N₂O₄S

473 - Aniline sulfate (2:1)

C₁₂H₁₆N₂O₅

1602 - 1-*tert*-Butyl-2-methoxy-4-methyl-3,5-dinitrobenzene

C₁₂H₁₆N₃O₃PS

10144 - Triazofos

C₁₂H₁₆N₃O₃PS₂

555 - Azinphos ethyl

C₁₂H₁₆O

1650 - 1-(4-*tert*-Butylphenyl)ethanone
2636 - 2-Cyclohexylphenol
2637 - 4-Cyclohexylphenol
8889 - 1-Phenyl-1-hexanone

C₁₂H₁₆O₂

807 - Benzyl 3-methylbutanoate
6400 - Isobutyl phenylacetate
7108 - 3-Methylbutyl benzoate
7422 - 5-Methyl-2-isopropylphenyl acetate
8667 - Pentyl benzoate
8878 - 2-Phenylethyl 2-methylpropanoate

C₁₂H₁₆O₃

6454 - Isopentyl salicylate
6874 - 2-Methoxyphenyl pentanoate
8685 - Pentyl salicylate
8784 - 2-Phenoxyethyl butanoate

C₁₂H₁₆O₆

8885 - Phenyl α-D-glucopyranoside

C₁₂H₁₆O₇

6156 - 4-Hydroxyphenyl-β-D-glucopyranoside

C₁₂H₁₇BrO

1229 - 1-Bromo-4-(hexyloxy)benzene

C₁₂H₁₇ClN₄OS

10025 - Thiamine chloride

C₁₂H₁₇N

823 - 1-Benzylpiperidine
824 - 4-Benzylpiperidine
2611 - N-Cyclohexylaniline

C₁₂H₁₇NO

1649 - N-Butyl-N-phenylacetamide
3482 - N,N-Diethyl-3-methylbenzamide
8756 - Phendimetrazine

C₁₂H₁₇NO₂

9126 - Promecarb

C₁₂H₁₇NO₃

461 - Anhalonidine
1480 - 4-Butoxy-N-hydroxybenzeneacetamide
1805 - Cerulenin
3468 - N,N-Diethyl-4-hydroxy-3-methoxybenzamide

C₁₂H₁₇NO₃Si

8997 - 1-Phenylsilatrane

C₁₂H₁₇NO₄

3432 - Diethyl 3,5-dimethylpyrrole-2,4-dicarboxylate

C₁₂H₁₇N₃O₄

139 - Agaritine

C₁₂H₁₇O₄PS₂

8795 - Phenthoate

C₁₂H₁₈

1555 - 1-*tert*-Butyl-3,5-dimethylbenzene
1566 - 1-*tert*-Butyl-4-ethylbenzene
2548 - 1,5,9-Cyclododecatriene
3797 - 1,2-Diisopropylbenzene
3798 - 1,3-Diisopropylbenzene
3799 - 1,4-Diisopropylbenzene
4618 - 5,7-Dodecadiyne
5813 - Hexamethylbenzene
5890 - Hexylbenzene
10348 - 1,2,3-Triethylbenzene
10349 - 1,2,4-Triethylbenzene
10350 - 1,3,5-Triethylbenzene

C₁₂H₁₈Br₆

5719 - 1,2,5,6,9,10-Hexabromocyclododecane

C₁₂H₁₈ClN₄O₄PS

10027 - Thiamine O-phosphate, chloride

C₁₂H₁₈Cl₂N₄OS

10026 - Thiamine hydrochloride

C₁₂H₁₈Cl₄N₄

888 - [1,1'-Biphenyl]-3,3',4,4'-tetramine, tetrahydrochloride

C₁₂H₁₈CuO₆

2461 - Copper(II) ethylacetoacetate

C₁₂H₁₈N₂O

9400 - 1-[4-(1-Pyrrolidinyl)-2-butynyl]-2-pyrrolidinone

C₁₂H₁₈N₂O₂

6457 - Isophorone diisocyanate
7871 - Mexacarbate

C₁₂H₁₈N₂O₃

8002 - Nealbarbital

C₁₂H₁₈N₂O₃S

10108 - Tolbutamide

C₁₂H₁₈N₄O₆S

8421 - Oryzalin

C₁₂H₁₈O

1557 - 2-*tert*-Butyl-4,6-dimethylphenol
1558 - 4-*tert*-Butyl-2,5-dimethylphenol
1559 - 4-*tert*-Butyl-2,6-dimethylphenol
1570 - 2-*tert*-Butyl-4-ethylphenol
2601 - 2-(1-Cyclohexen-1-yl)cyclohexanone
2628 - 2-Cyclohexylidenecyclohexanone
5907 - 4-Hexylphenol
7105 - [(3-Methylbutoxy)methyl]]benzene
7595 - 5-Methyl-2-pentylphenol
9195 - Propofol

C₁₂H₁₈O₂

3800 - *p*-Diisopropylbenzene hydroperoxide
5891 - 4-Hexyl-1,3-benzenediol

C₁₂H₁₈O₃

10331 - 1,3,5-Triethoxybenzene

C₁₂H₁₈O₄

1412 - 1,3-Butanediol dimethacrylate
1413 - 1,4-Butanediol dimethacrylate
4543 - Di-2-propenoyl-1,6-hexanediol
6230 - Indalone

C₁₂H₁₈O₅

3441 - Diethylene glycol dimethacrylate

C₁₂H₁₈O₇

2831 - Diallyl diethylene glycol carbonate

C₁₂H₁₉BrN₂O₂

8008 - Neostigmine bromide

C₁₂H₁₉ClNO₃P

2486 - Crufomate

C₁₂H₁₉N

3796 - 2,6-Diisopropylaniline

$C_{12}H_{27}Al$
 10180 - Tributylaluminum
 10454 - Triisobutylaluminum

$C_{12}H_{27}AlO_3$
 219 - Aluminum 2-butoxide
 10179 - Tributyl aluminate
 10453 - Triisobutyl aluminate

$C_{12}H_{27}B$
 10456 - Triisobutylborane

$C_{12}H_{27}BO_3$
 10182 - Tributyl borate

$C_{12}H_{27}FSn$
 10183 - Tributylfluorostannane

$C_{12}H_{27}N$
 3586 - Dihexylamine
 4039 - Dimethyldecylamine
 4640 - Dodecylamine
 10181 - Tributylamine
 10455 - Triisobutylamine

$C_{12}H_{27}OPS_3$
 10188 - S,S,S-Tributyl phosphorotrothioate

$C_{12}H_{27}O_3P$
 10187 - Tributyl phosphite

$C_{12}H_{27}O_4P$
 10185 - Tributyl phosphate
 10457 - Triisobutyl phosphate

$C_{12}H_{27}P$
 10186 - Tributylphosphine

$C_{12}H_{27}PS_3$
 6722 - Merphos

$C_{12}H_{28}BrN$
 9997 - Tetrapropylammonium bromide

$C_{12}H_{28}ClN$
 4642 - Dodecylamine hydrochloride

$C_{12}H_{28}IN$
 9998 - Tetrapropylammonium iodide

$C_{12}H_{28}N_2$
 4624 - 1,12-Dodecanediamine

$C_{12}H_{28}O_4Si$
 6539 - Isopropyl silicate
 9996 - Tetrapropoxysilane

$C_{12}H_{28}O_4Ti$
 9908 - Tetraisopropyl titanate
 10001 - Tetrapropyl titanate

$C_{12}H_{28}O_5P_2S_2$
 10000 - Tetrapropyl thiodiphosphate

$C_{12}H_{28}Si$
 10189 - Tributylsilane

$C_{12}H_{28}Sn$
 9999 - Tetrapropylstannane
 10190 - Tributylstannane

$C_{12}H_{30}OSi_2$
 5790 - Hexaethyldisiloxane

$C_{12}H_{30}O_{13}P_4$
 5791 - Hexaethyl tetraphosphate

$C_{12}H_{36}O_4Si_5$
 4620 - Dodecamethylpentasiloxane

$C_{12}H_{36}O_6Si_6$
 4619 - Dodecamethylcyclohexasiloxane

$C_{12}H_{54}Al_{16}O_{75}S_8$
 9630 - Sucralfate

$C_{12}O_{12}Ru_3$
 9492 - Ruthenium dodecacarbonyl

$C_{13}H_5N_3O_7$
 10627 - 2,4,7-Trinitro-9H-fluoren-9-one

$C_{13}H_6Cl_6O_2$
 5738 - Hexachlorophene

$C_{13}H_7ClOS$
 2314 - 2-Chloro-9H-thioxanthen-9-one

$C_{13}H_7Cl_4NO_2$
 9773 - 3,3',4',5-Tetrachlorosalicylanilide

$C_{13}H_7F_3N_2O_5$
 5384 - Fluorodifen

$C_{13}H_7NO_3$
 8101 - 2-Nitro-9H-fluoren-9-one

$C_{13}H_8Br_2O$
 2911 - 4,4'-Dibromobenzophenone

$C_{13}H_8Br_3NO_2$
 2914 - 3,5-Dibromo-N-(4-bromophenyl)-2-
   hydroxybenzamide

$C_{13}H_8ClN$
 1849 - 9-Chloroacridine

$C_{13}H_8Cl_2N_2O_4$
 8019 - Niclosamide

$C_{13}H_8Cl_2O$
 2236 - (2-Chlorophenyl)(4-chlorophenyl)
   methanone
 3115 - 4,4'-Dichlorobenzophenone

$C_{13}H_8Cl_3NO_2$
 1974 - 5-Chloro-N-(3,4-dichlorophenyl)-2-
   hydroxybenzamide

$C_{13}H_8O$
 5348 - 9H-Fluoren-9-one

$C_{13}H_8OS$
 10083 - 9H-Thioxanthen-9-one

$C_{13}H_8O_2$
 10844 - Xanthone

$C_{13}H_9BrO$
 1323 - (4-Bromophenyl)phenylmethanone

$C_{13}H_9ClO$
 879 - [1,1'-Biphenyl]-4-carbonyl chloride
 1902 - 2-Chlorobenzophenone
 2245 - (4-Chlorophenyl)phenylmethanone

$C_{13}H_9ClO_2$
 2075 - 2-Chloro-5-hydroxybenzophenone

$C_{13}H_9Cl_2FN_2S$
 6626 - Loflucarban

$C_{13}H_9Cl_2NO$
 296 - 2-Amino-2',5-dichlorobenzophenone

$C_{13}H_9Cl_3N_2O$
 10215 - 3,4,4'-Trichlorocarbanilide

$C_{13}H_9N$
 113 - Acridine
 731 - Benzo[f]quinoline
 732 - Benzo[h]quinoline
 878 - [1,1'-Biphenyl]-4-carbonitrile
 8745 - Phenanthridine

$C_{13}H_9NO$
 115 - 9(10H)-Acridinone
 10849 - p-Xenylcarbimide

$C_{13}H_9NOS$
 742 - 2-(2-Benzothiazolyl)phenol

$C_{13}H_9NO_2$
 752 - 2-(2-Benzoxazolyl)phenol
 8100 - 2-Nitro-9H-fluorene

$C_{13}H_9NO_3$
 8153 - (4-Nitrophenyl)phenylmethanone

$C_{13}H_9NS$
 8845 - 2-Phenylbenzothiazole

$C_{13}H_9N_3O_5$
 153 - Alizarin Yellow R

$C_{13}H_{10}$
 5343 - 9H-Fluorene

$C_{13}H_{10}BrCl_2O_2PS$
 6599 - Leptophos

$C_{13}H_{10}ClNO$
 287 - 2-Amino-5-chlorobenzophenone
 4471 - Diphenylcarbamic chloride

$C_{13}H_{10}Cl_2$
 946 - Bis(4-chlorophenyl)methane
 3171 - Dichlorodiphenylmethane

$C_{13}H_{10}Cl_2N_2S$
 948 - N,N'-Bis(4-chlorophenyl)thiourea

$C_{13}H_{10}Cl_2O_2$
 941 - Bis(4-chlorophenoxy)methane
 3249 - Dichlorophene

$C_{13}H_{10}Cl_2S$
 1819 - Chlorbenside

$C_{13}H_{10}N_2$
 112 - 9-Acridinamine
 4473 - N,N'-Diphenylcarbodiimide
 8839 - 2-Phenylbenzimidazole

$C_{13}H_{10}N_2O$
 9295 - Pyocyanine

$C_{13}H_{10}N_2O_3$
 6146 - 2-[(4-Hydroxyphenyl)azo]benzoic acid

$C_{13}H_{10}N_2O_4$
 10011 - Thalidomide

$C_{13}H_{10}N_4O_5$
 1042 - N,N'-Bis(4-nitrophenyl)urea

$C_{13}H_{10}O$
 718 - Benzophenone
 5347 - 9H-Fluoren-9-ol
 10841 - 9H-Xanthene

$C_{13}H_{10}O_2$
 5494 - 3-(2-Furanyl)-1-phenyl-2-propen-1-one
 5997 - 4-Hydroxybenzophenone
 6164 - (2-Hydroxyphenyl)phenylmethanone
 8776 - 3-Phenoxybenzaldehyde
 8840 - Phenyl benzoate
 8841 - 2-Phenylbenzoic acid
 8842 - 4-Phenylbenzoic acid
 10842 - 9H-Xanthen-9-ol

$C_{13}H_{10}O_3$
 635 - 1,3-Benzenediol, monobenzoate
 3571 - 1,5-Di-2-furanyl-1,4-pentadien-3-one
 3710 - 2,2'-Dihydroxybenzophenone
 3711 - 4,4'-Dihydroxybenzophenone
 3743 - (2,4-Dihydroxyphenyl)phenylmetha-
   none
 4474 - Diphenyl carbonate
 8779 - 2-Phenoxybenzoic acid
 8780 - 3-Phenoxybenzoic acid
 8781 - 4-Phenoxybenzoic acid
 8995 - Phenyl salicylate

$C_{13}H_{10}O_4$
 10438 - 2,3,4-Trihydroxybenzophenone
 10832 - Visnagin

$C_{13}H_{10}O_5$
 9899 - 2,2',4,4'-Tetrahydroxybenzophenone

$C_{13}H_{10}O_6$
 6652 - Maclurin

$C_{13}H_{10}S$
 4500 - Diphenylmethanethione
 10082 - 9H-Thioxanthene

$C_{13}H_{11}Br$
 1189 - α-Bromodiphenylmethane

788 - Benzyl benzoate
877 - [1,1'-Biphenyl]-4-acetic acid
6878 - (4-Methoxyphenyl)phenylmethanone
7627 - 4-Methylphenyl benzoate
8828 - α-Phenylbenzeneacetic acid
8915 - 4-(Phenylmethoxy)benzaldehyde
9014 - *trans*-5-(2-Phenylvinyl)-1,3-benzene-
diol

$C_{14}H_{12}O_3$
831 - Benzyl salicylate
6070 - (2-Hydroxy-4-methoxyphenyl)phenyl-
methanone
6149 - α-Hydroxy-α-phenylbenzeneacetic acid
6858 - 2-Methoxyphenol benzoate
8931 - Phenylmethyl 4-hydroxybenzoate
10848 - Xanthyletin

$C_{14}H_{12}O_4$
3729 - (2,6-Dihydroxy-4-methoxyphenyl)phe-
nylmethanone
4435 - Dioxybenzone
7947 - 2,6-Naphthalenedicarboxylic acid, dimethyl ester

$C_{14}H_{12}O_5$
6571 - Khellin

$C_{14}H_{12}S_2$
4335 - 2,7-Dimethylthianthrene

$C_{14}H_{13}N$
3610 - 10,11-Dihydro-5*H*-dibenz[b,f]azepine
4939 - 9-Ethyl-9*H*-carbazole
7643 - 4-Methyl-*N*-(phenylmethylene)aniline
8923 - *N*-(Phenylmethylene)benzenemetha-
namine

$C_{14}H_{13}NO$
890 - *N*-[1,1'-Biphenyl]-4-ylacetamide
4451 - *N,N*-Diphenylacetamide
7622 - *N*-(4-Methylphenyl)benzamide

$C_{14}H_{13}NO_4$
10490 - 4,7,8-Trimethoxyfuro[2,3-b]quinoline

$C_{14}H_{13}NO_4S$
834 - 4-[(Benzylsulfonyl)amino]benzoic acid

$C_{14}H_{14}$
3977 - 2,2'-Dimethylbiphenyl
3978 - 3,3'-Dimethylbiphenyl
3979 - 4,4'-Dimethylbiphenyl
4479 - 1,1-Diphenylethane
4480 - 1,2-Diphenylethane
7041 - 1-Methyl-2-benzylbenzene
7042 - 1-Methyl-4-benzylbenzene
9880 - 1,2,3,4-Tetrahydrophenanthrene

$C_{14}H_{14}ClN_2$
7925 - Naphazoline hydrochloride

$C_{14}H_{14}Cl_2N_2O$
6211 - Imazalil

$C_{14}H_{14}Hg$
1028 - Bis(4-methylphenyl)mercury

$C_{14}H_{14}NO_4PS$
5172 - Ethyl *p*-nitrophenyl benzenethiophosphate

$C_{14}H_{14}N_2$
4938 - 9-Ethyl-9*H*-carbazol-3-amine

$C_{14}H_{14}N_2O$
7224 - 2-Methyl-1,2-di-3-pyridinyl-1-pro-
panone

$C_{14}H_{14}N_2O_3$
1021 - Bis(4-methoxyphenyl)diazene, 1-oxide

$C_{14}H_{14}N_2O_6S_2$
2866 - 4,4'-Diamino-2,2'-stilbenedisulfonic
acid

$C_{14}H_{14}N_3NaO_3S$
7529 - Methyl Orange

$C_{14}H_{14}N_8O_4S_3$
1795 - Cefazolin

$C_{14}H_{14}O$
1026 - Bis(4-methylphenyl) ether
2895 - Dibenzyl ether
7624 - 4-Methyl-α-phenylbenzenemethanol
7625 - α-Methyl-α-phenylbenzenemethanol
8835 - α-Phenylbenzeneethanol

$C_{14}H_{14}OS$
2901 - Dibenzyl sulfoxide

$C_{14}H_{14}O_2$
3855 - 4,4'-Dimethoxy-1,1'-biphenyl
4450 - 1,2-Diphenoxyethane
4483 - 1,2-Diphenyl-1,2-ethanediol, (R*,R*)-(±)
5164 - Ethyl 1-naphthylacetate

$C_{14}H_{14}O_2S$
1031 - Bis(4-methylphenyl) sulfone
2900 - Dibenzyl sulfone

$C_{14}H_{14}O_3$
4717 - Equol
7997 - Naproxen
9085 - 2-Pivaloyl-1,3-indandione

$C_{14}H_{14}O_4$
627 - 1,2-Benzenedicarboxylic acid, diallyl
ester
2836 - Diallyl isophthalate

$C_{14}H_{14}O_8$
9927 - Tetramethyl 1,2,4,5-benzenetetracar-
boxylate

$C_{14}H_{14}S$
1030 - Bis(4-methylphenyl) sulfide
2899 - Dibenzyl sulfide

$C_{14}H_{14}S_2$
1025 - Bis(4-methylphenyl) disulfide
2893 - Dibenzyl disulfide

$C_{14}H_{15}Cl_2N$
1837 - Chlornaphazine

$C_{14}H_{15}N$
2892 - Dibenzylamine
7459 - 4-Methyl-*N*-(4-methylphenyl)aniline
7623 - *N*-Methyl-*N*-phenylbenzenemetha-
namine
8834 - α-Phenylbenzeneethanamine
8977 - 4-(3-Phenylpropyl)pyridine

$C_{14}H_{15}NO$
6161 - *N*-Hydroxy-*N*-(phenylmethyl)benzen-
emethanamine

$C_{14}H_{15}NO_2$
6836 - 4-Methoxy-*N*-(4-methoxyphenyl)
aniline

$C_{14}H_{15}N_3$
3898 - *p*-(Dimethylamino)azobenzene
3899 - 2',3-Dimethyl-4-aminoazobenzene

$C_{14}H_{15}N_5O_6S$
7869 - Metsulfuron-methyl

$C_{14}H_{15}O_3P$
2898 - Dibenzyl phosphite

$C_{14}H_{16}$
1620 - 1-Butylnaphthalene
1621 - 2-Butylnaphthalene

$C_{14}H_{16}ClN_3O_2$
573 - Bayleton

$C_{14}H_{16}ClN_3O_4S_2$
2702 - Cyclothiazide

$C_{14}H_{16}ClN_5O_5S$
10138 - Triasulfuron

$C_{14}H_{16}ClO_5PS$
2473 - Coumaphos

$C_{14}H_{16}F_3N_3O_4$
9121 - Profluralin

$C_{14}H_{16}N_2$
907 - 1,2-Bis(4-aminophenyl)ethane
4482 - *N,N'*-Diphenyl-1,2-ethanediamine
10109 - *o*-Tolidine

$C_{14}H_{16}N_2O_2$
3847 - 3,3'-Dimethoxybenzidine

$C_{14}H_{16}N_2O_3$
3507 - 5,5-Diethyl-1-phenyl-2,4,6(1*H*,3*H*,5*H*)-
pyrimidinetrione

$C_{14}H_{16}O_2Si$
3860 - Dimethoxydiphenylsilane
4045 - Dimethyldiphenoxysilane

$C_{14}H_{16}O_4$
3405 - Diethyl benzylidenemalonate

$C_{14}H_{16}O_9$
843 - Bergenin

$C_{14}H_{16}Si$
4049 - Dimethyldiphenylsilane

$C_{14}H_{17}ClNO_4PS_2$
2828 - Dialifor

$C_{14}H_{17}Cl_3O_4$
1479 - 2-Butoxyethyl (2,4,5-trichlorophenoxy)
acetate

$C_{14}H_{17}N$
3491 - *N,N*-Diethyl-1-naphthalenamine

$C_{14}H_{17}NO_2$
3392 - 7-(Diethylamino)-4-methyl-2*H*-1-ben-
zopyran-2-one

$C_{14}H_{17}NO_3$
5026 - Ethyl 2-ethoxy-1(2*H*)-quinolinecarbox-
ylate

$C_{14}H_{17}NO_6$
6672 - Mandelonitrile glucoside

$C_{14}H_{17}O_5PS$
6202 - Hymecromone O,O-diethyl phosphorothioate

$C_{14}H_{18}$
8332 - 1,2,3,4,5,6,7,8-Octahydroanthracene
8337 - 1,2,3,4,5,6,7,8-Octahydrophenanthrene

$C_{14}H_{18}ClN_3S$
2310 - Chlorothen

$C_{14}H_{18}Cl_2O_4$
1478 - 2-Butoxyethyl (2,4-dichlorophenoxy)
acetate

$C_{14}H_{18}N_2O_4$
8428 - Oxadixyl

$C_{14}H_{18}N_2O_5$
523 - Aspartame
1556 - 4-*tert*-Butyl-2,6-dimethyl-3,5-dinitroac-
etophenone

$C_{14}H_{18}N_2O_7$
4411 - Dinobuton

$C_{14}H_{18}N_4O_3$
580 - Benomyl
10473 - Trimethoprim

$C_{14}H_{18}O$
8926 - 2-(Phenylmethylene)heptanal

$C_{14}H_{18}O_4$
  628 - 1,2-Benzenedicarboxylic acid, dipropyl ester
  2404 - Cinoxate
  3406 - Diethyl benzylmalonate
  3811 - Diisopropyl phthalate

$C_{14}H_{18}O_6$
  626 - 1,2-Benzenedicarboxylic acid, bis(2-methoxyethyl) ester

$C_{14}H_{18}O_7$
  1012 - 2,2-Bis(hydroxymethyl)-1,3-propanediol, tri(2-propenoyl) ester
  5545 - 1-[4-(β-D-Glucopyranosyloxy)phenyl]ethanone

$C_{14}H_{19}Cl_2NO_2$
  1813 - Chlorambucil

$C_{14}H_{19}NO$
  4835 - 6-Ethoxy-1,2-dihydro-2,2,4-trimethylquinoline

$C_{14}H_{19}NO_2$
  7605 - Methylphenidate

$C_{14}H_{19}N_3S$
  6752 - Methapyrilene
  10016 - Thenyldiamine

$C_{14}H_{19}O_6P$
  2406 - Ciodrin

$C_{14}H_{20}$
  2843 - Diamantane

$C_{14}H_{20}ClNO_2$
  27 - Acetochlor
  141 - Alachlor

$C_{14}H_{20}N_2O$
  9537 - Siduron

$C_{14}H_{20}N_2O_2$
  9046 - Pindolol

$C_{14}H_{20}N_2O_6S$
  6935 - 4-(Methylamino)phenol sulfate

$C_{14}H_{20}N_4$
  561 - 1,1'-Azobiscyclohexanecarbonitrile

$C_{14}H_{20}N_4O_4$
  8916 - N2-[(Phenylmethoxy)carbonyl]-L-arginine

$C_{14}H_{20}N_6O_5S$
  128 - S-Adenosyl-L-homocysteine

$C_{14}H_{20}O$
  8939 - 1-Phenyl-1-octanone

$C_{14}H_{20}O_2$
  2603 - 3-Cyclohexenylmethyl 3-cyclohexenecarboxylate
  3012 - 2,5-Di-tert-butyl-2,5-cyclohexadiene-1,4-dione
  3013 - 2,6-Di-tert-butyl-2,5-cyclohexadiene-1,4-dione

$C_{14}H_{21}NO_4$
  3429 - Diethyl 1,4-dihydro-2,4,6-trimethyl-3,5-pyridinedicarboxylate

$C_{14}H_{21}N_3O_3$
  8436 - Oxamniquine

$C_{14}H_{21}N_3O_3S$
  10107 - Tolazamide

$C_{14}H_{21}N_3O_4$
  1485 - Butralin

$C_{14}H_{22}$
  3007 - 1,4-Di-tert-butylbenzene
  8383 - Octylbenzene
  9812 - 1,2,3,5-Tetraethylbenzene

$C_{14}H_{22}N_2O$
  3382 - 2-(Diethylamino)-N-(2,6-dimethylphenyl)acetamide
  8290 - Octacaine

$C_{14}H_{22}N_2O_3$
  529 - Atenolol
  1386 - Bucolome

$C_{14}H_{22}O$
  3039 - 2,6-Di-sec-butylphenol
  3040 - 2,4-Di-tert-butylphenol
  3041 - 2,6-Di-tert-butylphenol
  3042 - 3,5-Di-tert-butylphenol
  6354 - α-Irone
  6355 - β-Irone
  8398 - 4-Octylphenol
  8399 - Octyl phenyl ether
  9933 - 4-(1,1,3,3-Tetramethylbutyl)phenol

$C_{14}H_{22}O_2$
  2999 - 1,4-Dibutoxybenzene
  3008 - 2,5-Di-tert-butyl-1,4-benzenediol

$C_{14}H_{22}O_4$
  5837 - 1,6-Hexanediol dimethacrylate

$C_{14}H_{22}O_6$
  10358 - Triethylene glycol dimethacrylate

$C_{14}H_{22}O_7$
  9816 - Tetraethylene glycol diacrylate

$C_{14}H_{22}O_8$
  10343 - Triethyl 2-acetoxy-1,2,3-propanetricarboxylate

$C_{14}H_{23}ClN_2O$
  3383 - 2-(Diethylamino)-N-(2,6-dimethylphenyl)acetamide, monohydrochloride

$C_{14}H_{23}N$
  1637 - 4-(1-Butylpentyl)pyridine
  3006 - N,N-Dibutylaniline
  8382 - 4-Octylaniline

$C_{14}H_{23}NO$
  135 - Affinin

$C_{14}H_{23}N_3O_{10}$
  8659 - Pentetic acid

$C_{14}H_{24}$
  9786 - Tetradecahydrophenanthrene

$C_{14}H_{24}NO_4PS_3$
  582 - Bensulide

$C_{14}H_{24}N_2O_9$
  2621 - (1,2-Cyclohexylenedinitrilo)tetraacetic acid monohydrate

$C_{14}H_{24}O_2$
  5529 - Geranyl 2-methylpropanoate

$C_{14}H_{24}O_4$
  4440 - Dipentyl cis-2-butenedioate

$C_{14}H_{26}O_2$
  6428 - Isodecyl methacrylate
  9940 - 2,4,7,9-Tetramethyl-5-decyne-4,7-diol

$C_{14}H_{26}O_3$
  5684 - Heptanoic anhydride
  7417 - 5-Methyl-2-isopropylcyclohexyl ethoxyacetate, (1α,2β,5α)

$C_{14}H_{26}O_4$
  3002 - Dibutyl adipate
  3521 - Diethyl sebacate
  3775 - Diisobutyl adipate
  9791 - Tetradecanedioic acid

$C_{14}H_{27}ClO$
  9799 - Tetradecanoyl chloride

$C_{14}H_{27}N$
  9793 - Tetradecanenitrile

$C_{14}H_{28}$
  8257 - Nonylcyclopentane
  8385 - Octylcyclohexane
  9801 - 1-Tetradecene

$C_{14}H_{28}Br_2$
  2989 - 1,14-Dibromotetradecane

$C_{14}H_{28}O$
  4649 - Dodecyloxirane
  9788 - Tetradecanal
  9798 - 2-Tetradecanone

$C_{14}H_{28}O_2$
  4638 - Dodecyl acetate
  5097 - Ethyl laurate
  5903 - Hexyl octanoate
  7833 - Methyl tridecanoate
  8679 - Pentyl nonanoate
  9795 - Tetradecanoic acid

$C_{14}H_{28}O_2S$
  4647 - Dodecyl mercaptoacetate

$C_{14}H_{29}Br$
  1353 - 1-Bromotetradecane

$C_{14}H_{29}Cl$
  2305 - 1-Chlorotetradecane

$C_{14}H_{29}NO$
  9789 - Tetradecanamide

$C_{14}H_{29}NO_2$
  6039 - N-(2-Hydroxyethyl)dodecanamide

$C_{14}H_{30}$
  9790 - Tetradecane

$C_{14}H_{30}Cl_2N_2O_4$
  9628 - Succinylcholine chloride

$C_{14}H_{30}O$
  3583 - Diheptyl ether
  9797 - 1-Tetradecanol

$C_{14}H_{30}O_2$
  3861 - 1,1-Dimethoxydodecane
  4650 - 2-(Dodecyloxy)ethanol
  9792 - 1,14-Tetradecanediol

$C_{14}H_{30}O_2Sn$
  92 - (Acetyloxy)tributylstannane

$C_{14}H_{30}S$
  3585 - Diheptyl sulfide
  9794 - 1-Tetradecanethiol

$C_{14}H_{31}N$
  3582 - Diheptylamine
  9803 - Tetradecylamine

$C_{14}H_{31}NO$
  4053 - N,N-Dimethyldodecylamine oxide

$C_{14}H_{31}NO_2$
  4641 - Dodecylamine, acetate

$C_{14}H_{32}N_2$
  3026 - N,N'-Dibutyl-1,6-hexanediamine

$C_{14}H_{32}N_2O_4$
  9902 - N,N,N',N'-Tetra(2-hydroxypropyl)ethylenediamine

$C_{14}H_{42}O_5Si_6$
  9787 - Tetradecamethylhexasiloxane

$C_{15}H_7Cl_2F_3N_2O_2$
  6630 - Lovozal

$C_{15}H_8O_4$
  3616 - 9,10-Dihydro-9,10-dioxo-2-anthracenecarboxylic acid

$C_{15}H_8O_5$
  2474 - Coumestrol

7862 - Methysticin
9017 - Phloretin

$C_{15}H_{14}O_6$
6563 - Javanicin
9087 - Plumericin

$C_{15}H_{15}ClF_3N_3O$
10375 - Triflumizole

$C_{15}H_{15}ClN_2O_2$
2344 - Chloroxuron

$C_{15}H_{15}ClN_4O_6S$
1833 - Chlorimuron-ethyl

$C_{15}H_{15}NO$
3917 - [4-(Dimethylamino)phenyl]phenyl-
methanone

$C_{15}H_{15}NO_2$
6689 - Mefenamic acid

$C_{15}H_{15}NO_3$
10110 - Tolmetin

$C_{15}H_{15}N_3O$
4821 - 7-Ethoxy-3,9-acridinediamine

$C_{15}H_{15}N_3O_2$
7777 - Methyl Red

$C_{15}H_{16}$
4513 - 1,3-Diphenylpropane
4514 - 2,2-Diphenylpropane

$C_{15}H_{16}ClN_3$
3890 - 2,7-Dimethyl-3,6-acridinediamine,
monohydrochloride

$C_{15}H_{16}F_5NO_2S_2$
4592 - Dithiopyr

$C_{15}H_{16}N_2O$
2902 - 1,3-Dibenzylurea
4050 - N,N'-Dimethyl-N,N'-diphenylurea

$C_{15}H_{16}N_2S$
1032 - N,N'-Bis(2-methylphenyl)thiourea

$C_{15}H_{16}N_4O_5S$
9659 - Sulfometuron methyl

$C_{15}H_{16}O$
7635 - 4-(1-Methyl-1-phenylethyl)phenol

$C_{15}H_{16}O_2$
1015 - 2,2-Bis(4-hydroxyphenyl)propane
6159 - 2(2-Hydroxyphenyl)-2-(4-hydroxyphe-
nyl)propane

$C_{15}H_{16}O_8$
5543 - 7-(β-D-Glucopyranosyloxy)-2H-1-ben-
zopyran-2-one

$C_{15}H_{16}O_9$
4755 - Esculin

$C_{15}H_{17}ClN_4$
7912 - Myclobutanil
8012 - Neutral Red

$C_{15}H_{17}Cl_2N_3O$
4358 - Diniconazole

$C_{15}H_{17}Cl_2N_3O_2$
9192 - Propiconazole

$C_{15}H_{17}N$
797 - N-Benzyl-N-ethylaniline

$C_{15}H_{17}NO$
6466 - 4-Isopropoxydiphenylamine

$C_{15}H_{17}NS_2$
10102 - Tipepidine

$C_{15}H_{17}N_3$
4596 - N,N'-Di(o-tolyl)guanidine
7201 - 3-Methyl-4'-(dimethylamino)azoben-
zene

$C_{15}H_{17}N_5O_6S$
10150 - Tribenuron-methyl

$C_{15}H_{18}$
4120 - 1,4-Dimethyl-7-isopropylazulene
4121 - 1,6-Dimethyl-4-isopropylnaphthalene
8677 - 1-Pentylnaphthalene

$C_{15}H_{18}Cl_2N_2O_3$
8427 - Oxadiazon

$C_{15}H_{18}N_2$
6530 - N-Isopropyl-N'-phenyl-1,4-benzenedi-
amine
7234 - 4,4'-Methylenebis(N-methylaniline)

$C_{15}H_{18}N_2O_6$
870 - Binapacryl

$C_{15}H_{18}N_4O_5$
7883 - Mitomycin C

$C_{15}H_{18}N_6O_6S$
8021 - Nicosulfuron

$C_{15}H_{18}O_3$
6632 - Loxoprofen
9509 - α-Santonin

$C_{15}H_{18}O_4$
516 - Artemisin
5630 - Helenalin

$C_{15}H_{19}NO_2$
10710 - Tropacocaine

$C_{15}H_{19}N_3O_3$
6214 - Imazethapyr

$C_{15}H_{20}ClN_3O$
8496 - Paclobutrazol

$C_{15}H_{20}N_2O$
451 - Anagyrine

$C_{15}H_{20}N_2O_4S$
28 - Acetohexamide

$C_{15}H_{20}O$
8928 - 2-(Phenylmethylene)octanal

$C_{15}H_{20}O_2$
146 - Alantolactone

$C_{15}H_{20}O_4$
3 - Abscisic acid
3458 - Diethyl ethylphenylmalonate
9508 - Santonic acid

$C_{15}H_{20}O_6$
10667 - Tri-2-propenoyl-2-ethyl-2-
(hydroxymethyl)-1,3-propanediol

$C_{15}H_{20}O_7$
8213 - Nivalenol

$C_{15}H_{21}CoO_6$
2439 - Cobalt(III) 2,4-pentanedioate

$C_{15}H_{21}CrO_6$
2383 - Chromium(III) 2,4-pentanedioate

$C_{15}H_{21}FeO_6$
6360 - Iron(III) 2,4-pentanedioate

$C_{15}H_{21}NO_2$
6701 - Meperidine

$C_{15}H_{21}NO_4$
6727 - Metalaxyl

$C_{15}H_{21}N_3O_2$
9033 - Physostigmine

$C_{15}H_{21}O_6Ru$
9493 - Ruthenium(III) 2,4-pentanedioate

$C_{15}H_{21}O_6V$
10773 - Vanadium(III) 2,4-pentanedioate

$C_{15}H_{22}$
4104 - 1-(1,5-Dimethyl-4-hexenyl)-4-methyl-
benzene

$C_{15}H_{22}ClNO_2$
7864 - Metolachlor

$C_{15}H_{22}N_2O$
6706 - Mepivacaine

$C_{15}H_{22}N_2O_2$
7232 - Methylenebis(4-cyclohexylisocyanate)

$C_{15}H_{22}O$
10783 - α-Vetivone
10784 - β-Vetivone

$C_{15}H_{22}O_2$
4668 - Drimenin
5631 - Helminthosporal
8397 - 4-(Octyloxy)benzaldehyde

$C_{15}H_{22}O_3$
3027 - 3,5-Di-$tert$-butyl-2-hydroxybenzoic acid
4245 - 5-(2,5-Dimethylphenoxy)-2,2-dimethyl-
pentanoic acid
5062 - 2-Ethylhexyl 2-hydroxybenzoate

$C_{15}H_{22}O_5$
6203 - Hymenoxone

$C_{15}H_{23}ClN_2O$
6707 - Mepivacaine monohydrochloride

$C_{15}H_{23}ClO_4S$
510 - Aramite

$C_{15}H_{23}NO_4$
2605 - Cycloheximide

$C_{15}H_{23}N_3O_4$
6460 - Isopropalin

$C_{15}H_{24}$
1690 - γ-Cadinene
1789 - Caryophyllene
1793 - α-Cedrene
2460 - Copaene
4693 - β-Elemene
5303 - α-Farnesene
5304 - β-Farnesene
5805 - cis-1,2,3,5,6,8a-Hexahydro-4,7-dime-
thyl-1-isopropylnaphthalene, (1S)
5806 - 1,2,4a,5,8,8a-Hexahydro-4,7-dimethyl-
1-isopropylnaphthalene, [1S-(1α,4aβ,8aα)]
5927 - Humulene
6437 - Isolongifolene
6627 - Longifolene
7457 - 1-Methyl-4-(5-methyl-1-methylene-4-
hexenyl)cyclohexene, (S)
8255 - Nonylbenzene
10092 - 3-Thujopsene
10462 - 1,2,4-Triisopropylbenzene
10463 - 1,3,5-Triisopropylbenzene
10809 - 6-Vinyl-6-methyl-1-isopropyl-3-(1-
methylethylidene)cyclohexene, (S)

$C_{15}H_{24}NO_4PS$
6431 - Isofenphos

$C_{15}H_{24}N_2O$
497 - Aphylline
6682 - Matridin-15-one
8479 - 17-Oxosparteine
10468 - Trimecaine

$C_{15}H_{24}N_2O_2$
6054 - Hydroxylupanine

$C_{15}H_{24}O$
3033 - 2,4-Di-$tert$-butyl-5-methylphenol
3034 - 2,4-Di-$tert$-butyl-6-methylphenol
3035 - 2,6-Di-$tert$-butyl-4-methylphenol
8260 - 4-Nonylphenol
9506 - α-Santalol
9507 - β-Santalol

$C_{16}H_{16}N_2OS$

3897 - 10-[(Dimethylamino)acetyl]-10H-phe-
nothiazine

$C_{16}H_{16}N_2O_2$

992 - N,N'-Bis(2-hydroxybenzylidene)-1,2-eth-
ylenediamine

2825 - N,N'-Diacetyl-4,4'-diaminobiphenyl

4160 - N,N-Dimethyl-4-[2-(4-nitrophenyl)
ethenyl]aniline

6438 - Isolysergic acid

6644 - Lysergic acid

$C_{16}H_{16}N_2O_4$

2813 - Desmedipham

8760 - Phenmedipham

$C_{16}H_{16}N_2O_4S$

6 - Acedapsone

8799 - N-(Phenylacetyl)-7-aminodeacetox-
ycephalosporanic acid

$C_{16}H_{16}N_2O_6S_2$

1801 - Cephalothin

$C_{16}H_{16}O_2$

786 - α-Benzylbenzenepropanoic acid

4837 - 2-Ethoxy-1,2-diphenylethanone

8879 - 2-Phenylethyl phenylacetate

$C_{16}H_{16}O_4$

3853 - 4,4'-Dimethoxybenzoin

$C_{16}H_{16}O_5$

155 - Alkannin

$C_{16}H_{17}NO$

4447 - Diphenamid

$C_{16}H_{17}NO_3$

8277 - Normorphine

$C_{16}H_{17}NO_4$

6642 - Lycorine

$C_{16}H_{17}N_2NaO_4S$

818 - Benzylpenicillin sodium

$C_{16}H_{17}N_3O$

6643 - Lysergamide

$C_{16}H_{17}N_3O_4S$

1798 - Cephalexin

$C_{16}H_{18}$

1526 - 2-Butyl-1,1'-biphenyl

4465 - 1,1-Diphenylbutane

4466 - 1,2-Diphenylbutane

4467 - 1,4-Diphenylbutane

4595 - 1,2-Di(p-tolyl)ethane

5210 - 1-(4-Ethylphenyl)-2-phenylethane

$C_{16}H_{18}ClN$

2037 - N-(2-Chloroethyl)dibenzylamine

$C_{16}H_{18}ClN_3S$

7235 - Methylene blue

$C_{16}H_{18}N_2O_2$

3350 - 4,4'-Diethoxyazobenzene

$C_{16}H_{18}N_2O_3$

6459 - Isopilosine

9042 - Pilosine

$C_{16}H_{18}N_2O_4S$

8522 - Penicillin G

$C_{16}H_{18}N_2O_5S$

8524 - Penicillin V

$C_{16}H_{18}N_4O_2$

8013 - Nialamide

$C_{16}H_{18}N_4O_7S$

581 - Bensulfuron-methyl

$C_{16}H_{18}O_2$

1013 - 2,2-Bis(4-hydroxyphenyl)butane

9930 - 3,3',5,5'-Tetramethyl-[1,1'-biphenyl]-
4,4'-diol

$C_{16}H_{18}O_6S_2$

4773 - 1,2-Ethanediol, bis(4-methylbenzene-
sulfonate)

$C_{16}H_{18}O_{10}$

5468 - Fraxin

$C_{16}H_{19}BrN_2$

1382 - Brompheniramine

$C_{16}H_{19}ClN_2$

2347 - Chlorpheniramine

$C_{16}H_{19}ClSi$

1944 - Chloro(tert-butyl)diphenylsilane

$C_{16}H_{19}Cl_2N$

2038 - N-(2-Chloroethyl)dibenzylamine
hydrochloride

$C_{16}H_{19}N$

1044 - Bis(1-phenylethyl)amine

5030 - 2-Ethyl-N-(2-ethylphenyl)aniline

$C_{16}H_{19}NO_3$

4744 - β-Erythroidine

6634 - Lunacrine

$C_{16}H_{19}NO_4$

763 - Benzoylecgonine

$C_{16}H_{19}N_3O_4S$

447 - Ampicillin

1804 - Cephradine

$C_{16}H_{19}N_3O_5S$

444 - Amoxicillin

$C_{16}H_{19}N_3O_6$

7881 - Mitomycin A

7882 - Mitomycin B

$C_{16}H_{20}$

3807 - 2,6-Diisopropylnaphthalene

5902 - 1-Hexylnaphthalene

$C_{16}H_{20}N_2$

2894 - N,N'-Dibenzyl-1,2-ethanediamine

4256 - N,N-Dimethyl-γ-phenyl-2-pyridinepro-
panamine

9928 - 3,3',5,5'-Tetramethyl-[1,1'-biphenyl]-
4,4'-diamine

9929 - N,N,N',N'-Tetramethyl-[1,1'-biphenyl]-
4,4'-diamine

$C_{16}H_{20}O_2Si$

3357 - Diethoxydiphenylsilane

$C_{16}H_{20}O_6P_2S_3$

1 - Abate

$C_{16}H_{21}NO_2$

9202 - Propranolol

$C_{16}H_{21}NO_3$

5920 - Homatropine

8274 - Norhyoscyamine

$C_{16}H_{21}N_3$

3973 - N,N-Dimethyl-N'-benzyl-N'-2-pyridi-
nyl-1,2-ethanediamine

$C_{16}H_{22}BrNO_3$

5921 - Homatropine hydrobromide

$C_{16}H_{22}ClNO_3$

3349 - Diethatyl, ethyl ester

$C_{16}H_{22}N_2$

6640 - Lycodine

$C_{16}H_{22}N_2O_5$

815 - Benzyloxycarbonylglycyl-L-leucine

$C_{16}H_{22}N_4O_2$

411 - Aminopropylon

$C_{16}H_{22}OSi_2$

9943 - 1,1,3,3-Tetramethyl-1,3-diphenyldisi-
loxane

$C_{16}H_{22}O_4$

3045 - Dibutyl phthalate

3781 - Diisobutyl phthalate

$C_{16}H_{22}O_6$

975 - Bis(2-ethoxyethyl) phthalate

$C_{16}H_{22}O_8$

2457 - Coniferin

$C_{16}H_{22}O_{11}$

5547 - α-D-Glucose pentaacetate

5548 - β-D-Glucose pentaacetate

$C_{16}H_{23}ClN_3O$

9685 - Tebuconazole

$C_{16}H_{23}NO_2$

215 - Alphaprodine

4815 - Ethoheptazine

$C_{16}H_{23}NO_6$

7891 - Monocrotaline

$C_{16}H_{24}$

5909 - 1-Hexyl-1,2,3,4-tetrahydronaphthalene

$C_{16}H_{24}N_2O$

8489 - Oxymetazoline

$C_{16}H_{24}N_2O_2$

7887 - Molindone

9813 - N,N,N',N'-Tetraethyl-1,2-benzenedicar-
boxamide

$C_{16}H_{25}NO_2$

3389 - 2-(Diethylamino)ethyl 2-phenylbu-
tanoate

$C_{16}H_{26}$

2779 - Decylbenzene

8576 - Pentaethylbenzene

$C_{16}H_{26}O$

972 - 2,4-Bis(1,1-dimethylpropyl)phenol

3023 - 2,6-Di-tert-butyl-4-ethylphenol

$C_{16}H_{26}O_2$

971 - 2,5-Bis(1,1-dimethylpropyl)-1,4-ben-
zenediol

$C_{16}H_{26}O_3$

4637 - 2-Dodecenylsuccinic anhydride

$C_{16}H_{26}O_7$

9817 - Tetraethylene glycol dimethacrylate

$C_{16}H_{28}O_2$

2676 - 2-Cyclopentene-1-undecanoic acid, (R)

$C_{16}H_{29}N_3O_8$

3413 - Diethylcarbamazine citrate

$C_{16}H_{30}$

5772 - 1-Hexadecyne

$C_{16}H_{30}O$

7165 - 3-Methylcyclopentadecanone

$C_{16}H_{30}O_2$

4648 - Dodecyl methacrylate

5757 - cis-9-Hexadecenoic acid

10319 - Tridecyl acrylate

$C_{16}H_{30}O_3$

8355 - Octanoic anhydride

$C_{16}H_{30}O_4$

5748 - Hexadecanedioic acid

$C_{16}H_{31}BrO_2$

1222 - 2-Bromohexadecanoic acid

$C_{16}H_{31}ClO$

5755 - Hexadecanoyl chloride

$C_{16}H_{31}N$

5749 - Hexadecanenitrile

$C_{16}H_{32}$
2780 - Decylcyclohexane
5756 - 1-Hexadecene
7700 - 2-Methyl-1-propene, tetramer
$C_{16}H_{32}O$
4712 - 1,2-Epoxyhexadecane
5745 - Hexadecanal
5754 - 3-Hexadecanone
$C_{16}H_{32}O_2$
1564 - Butyl dodecanoate
5160 - Ethyl myristate
5751 - Hexadecanoic acid
5896 - 2-Hexyldecanoic acid
7545 - Methyl pentadecanoate
8395 - Octyl octanoate
9802 - Tetradecyl acetate
$C_{16}H_{32}O_3$
6046 - 16-Hydroxyhexadecanoic acid
$C_{16}H_{32}O_4$
3446 - Diethylene glycol monododecanoate
$C_{16}H_{32}O_5$
10440 - 9,10,16-Trihydroxyhexadecanoic acid
$C_{16}H_{33}Br$
1221 - 1-Bromohexadecane
$C_{16}H_{33}Cl$
2065 - 1-Chlorohexadecane
$C_{16}H_{33}Cl_3Si$
5770 - Hexadecyltrichlorosilane
$C_{16}H_{33}I$
6305 - 1-Iodohexadecane
$C_{16}H_{33}NO$
3434 - N,N-Diethyldodecanamide
5746 - Hexadecanamide
$C_{16}H_{33}NO_3$
998 - N,N-Bis(2-hydroxyethyl)dodecanamide
$C_{16}H_{34}$
5666 - 2,2,4,4,6,8,8-Heptamethylnonane
5747 - Hexadecane
$C_{16}H_{34}O$
982 - Bis(2-ethylhexyl) ether
4419 - Dioctyl ether
5753 - 1-Hexadecanol
$C_{16}H_{34}S$
4424 - Dioctyl sulfide
5750 - 1-Hexadecanethiol
$C_{16}H_{35}N$
980 - Bis(2-ethylhexyl)amine
4418 - Dioctylamine
5759 - Hexadecylamine
$C_{16}H_{35}O_2PS_2$
985 - Bis(2-ethylhexyl) phosphorodithioate
$C_{16}H_{35}O_3P$
984 - Bis(2-ethylhexyl) phosphonate
$C_{16}H_{35}O_4P$
983 - Bis(2-ethylhexyl) phosphate
$C_{16}H_{36}BrN$
9720 - Tetrabutylammonium bromide
$C_{16}H_{36}BrP$
9726 - Tetrabutylphosphonium bromide
$C_{16}H_{36}ClN$
9721 - Tetrabutylammonium chloride
$C_{16}H_{36}FN$
9722 - Tetrabutylammonium fluoride
$C_{16}H_{36}IN$
9724 - Tetrabutylammonium iodide

$C_{16}H_{36}O_4Si$
9727 - Tetrabutyl silicate
$C_{16}H_{36}O_4Ti$
9730 - Tetrabutyl titanate
9907 - Tetraisobutyl titanate
$C_{16}H_{36}Sn$
9728 - Tetrabutylstannane
$C_{16}H_{37}NO$
9723 - Tetrabutylammonium hydroxide
$C_{16}H_{38}Br_2N_2$
2746 - Decamethonium dibromide
$C_{16}H_{48}O_6Si_7$
5744 - Hexadecamethylheptasiloxane
$C_{17}H_6O_7$
721 - 3,3',4,4'-Benzophenonetetracarboxylic
     acid dianhydride
$C_{17}H_9NO_4$
3737 - 5,6-Dihydroxynaphtho[2,3-f]quinoline-
     7,12-dione
$C_{17}H_{10}F_6N_4S$
5335 - Flubenzimine
$C_{17}H_{10}O$
594 - Benzanthrone
$C_{17}H_{11}N$
584 - Benz[c]acridine
$C_{17}H_{12}$
703 - 11H-Benzo[a]fluorene
704 - 11H-Benzo[b]fluorene
7729 - 1-Methylpyrene
7730 - 2-Methylpyrene
$C_{17}H_{12}ClFN_2O$
8285 - Nuarimol
$C_{17}H_{12}Cl_2N_2O$
5311 - Fenarimol
$C_{17}H_{12}Cl_2N_4$
10145 - Triazolam
$C_{17}H_{12}O_2$
7984 - 2-Naphthyl benzoate
$C_{17}H_{12}O_3$
7987 - 1-Naphthyl 2-hydroxybenzoate
8895 - Phenyl 1-hydroxy-2-naphthalenecar-
     boxylate
$C_{17}H_{12}O_6$
136 - Aflatoxin B1
$C_{17}H_{12}O_7$
138 - Aflatoxin G1
$C_{17}H_{13}N_3O_5S_2$
9032 - Phthalylsulphathiazole
$C_{17}H_{14}$
3609 - 16,17-Dihydro-15H-cyclopenta[a]
     phenanthrene
$C_{17}H_{14}FeO$
764 - Benzoylferrocene
$C_{17}H_{14}N_2O_2$
4047 - 3,3'-Dimethyldiphenylmethane 4,4'-
     diisocyanate
$C_{17}H_{14}O$
4507 - 1,5-Diphenyl-1,4-pentadien-3-one
$C_{17}H_{14}O_6$
137 - Aflatoxin B2
$C_{17}H_{14}O_7$
3591 - 15,16-Dihydroaflatoxin G_1
$C_{17}H_{15}ClO_7$
6670 - Malvidin chloride
$C_{17}H_{16}Br_2O_3$
1343 - Bromopropylate

$C_{17}H_{16}Cl_2O_3$
2280 - Chloropropylate
$C_{17}H_{16}F_6N_2O$
6690 - Mefloquine
$C_{17}H_{16}N_2S$
3926 - 2-(p-Dimethylaminostyryl)benzothiaz-
     ole
$C_{17}H_{16}N_4O_4$
2408 - C.I. Pigment Yellow 1
$C_{17}H_{16}O_4$
2897 - Dibenzyl malonate
$C_{17}H_{17}ClO_6$
5603 - Griseofulvin, (+)
$C_{17}H_{17}NO_2$
501 - Apomorphine
$C_{17}H_{17}N_3O_3$
6213 - Imazaquin
$C_{17}H_{17}N_3O_6S_2$
1802 - Cephapirin
$C_{17}H_{18}ClNO_2$
502 - Apomorphine, hydrochloride
$C_{17}H_{18}F_3NO$
5433 - Fluoxetine
$C_{17}H_{18}N_2O_2$
9146 - 2,2'-[1,3-Propanediylbis(nitrilomethyli-
     dyne)]bisphenol
$C_{17}H_{18}O_4$
4449 - Diphenolic acid
$C_{17}H_{19}ClN_2S$
2350 - Chlorpromazine
$C_{17}H_{19}KN_2O_5S$
8757 - Phenethicillin potassium
$C_{17}H_{19}NO_3$
1807 - Chavicine
2441 - Coclaurine
5954 - Hydromorphone
7899 - Morphine
9076 - Piperine
$C_{17}H_{19}NO_4$
5319 - Fenoxycarb
$C_{17}H_{19}N_3$
476 - Antazoline
$C_{17}H_{19}N_3O$
8796 - Phentolamine
$C_{17}H_{19}N_3O_2$
5242 - Ethyl Red
$C_{17}H_{19}N_3O_3S$
8411 - Omeprazole
$C_{17}H_{20}ClN_3$
9916 - N,N,N',N'-Tetramethyl-3,6-acridinedi-
     amine, monohydrochloride
$C_{17}H_{20}Cl_2N_2S$
1990 - 2-Chloro-10-(3-dimethylaminopropyl)
     phenothiazine monohydrochloride
$C_{17}H_{20}N_2O$
3415 - N,N-Diethylcarbanilide
9941 - N,N,N',N'-Tetramethyl-4,4'-diami-
     nobenzophenone
$C_{17}H_{20}N_2S$
965 - Bis[4-(dimethylamino)phenyl]meth-
     anethione
9125 - Promazine
9127 - Promethazine
$C_{17}H_{20}N_4O_6$
9478 - Riboflavin

$C_{17}H_{20}N_4O_9P$
9479 - Riboflavin-5'-phosphate
$C_{17}H_{20}O_2$
1010 - 2,2-Bis(4-hydroxy-3-methylphenyl)propane
$C_{17}H_{20}O_6$
7913 - Mycophenolic acid
$C_{17}H_{20}O_8$
1011 - 2,2-Bis(hydroxymethyl)-1,3-propanediol, tetra(2-propenoyl) ester
$C_{17}H_{21}ClN_2S$
9128 - Promethazine hydrochloride
$C_{17}H_{21}N$
773 - Benzphetamine
3504 - N,N-Diethyl-α-phenylbenzenemethanamine
$C_{17}H_{21}NO$
4503 - 2-(Diphenylmethoxy)-N,N-dimethylethanamine
$C_{17}H_{21}NO_2$
499 - Apoatropine
7996 - Napropamide
$C_{17}H_{21}NO_3$
5518 - Galanthamine
$C_{17}H_{21}NO_4$
2440 - Cocaine
9518 - Scopolamine
$C_{17}H_{21}N_3$
3905 - 4,4'-Dimethylaminobenzophenonimide
$C_{17}H_{22}ClNO$
840 - Bephenium chloride
$C_{17}H_{22}N_2$
964 - Bis[4-(dimethylamino)phenyl]methane
$C_{17}H_{22}N_2O$
966 - Bis(4-dimethylaminophenyl)methanol
4667 - Doxylamine
$C_{17}H_{22}N_2S$
10015 - Thenaldine
$C_{17}H_{22}O_2$
2389 - Cicutoxin
$C_{17}H_{22}O_8$
5510 - Fusarenon X
$C_{17}H_{23}NO$
6609 - Levorphanol
$C_{17}H_{23}NO_3$
532 - Atropine
6204 - Hyoscyamine
$C_{17}H_{23}N_3O$
9379 - Pyrilamine
$C_{17}H_{24}ClN_3O$
533 - Auramine hydrochloride
$C_{17}H_{24}N_2O$
3828 - Dimethisoquin
$C_{17}H_{24}O_3$
2525 - Cyclandelate
$C_{17}H_{24}O_4$
10297 - Trichodermin
$C_{17}H_{24}O_9$
9671 - Syringin
$C_{17}H_{25}N$
8866 - 1-(1-Phenylcyclohexyl)piperidine
$C_{17}H_{25}NO_2$
7872 - MGK 264
$C_{17}H_{25}N_3O$
9086 - Plasmocid

$C_{17}H_{26}ClNO_2$
1391 - Butachlor
$C_{17}H_{26}N_4O_3S_2$
5508 - Fursultiamine
$C_{17}H_{26}O_4$
4694 - Embelin
$C_{17}H_{27}NO$
7106 - 1-[2-(3-Methylbutoxy)-2-phenylethyl]pyrrolidine
$C_{17}H_{27}NO_3$
1483 - 4-[3-(4-Butoxyphenoxy)propyl]morpholine
$C_{17}H_{27}NO_4$
7917 - Nadolol
$C_{17}H_{28}$
10749 - Undecylbenzene
$C_{17}H_{28}O$
4442 - 2,6-Di-tert-pentyl-4-methylphenol
$C_{17}H_{28}O_2$
5308 - Farnesol acetate
$C_{17}H_{29}NO$
3014 - 2,6-Di-tert-butyl-4-(dimethylaminomethyl)phenol
$C_{17}H_{29}NO_3S$
9535 - Sethoxydim
$C_{17}H_{30}O$
2551 - cis-9-Cycloheptadecen-1-one
$C_{17}H_{30}O_2$
5955 - Hydroprene
$C_{17}H_{32}BrNO_2$
475 - Anisotropine methylbromide
$C_{17}H_{32}O_2$
10325 - Tridecyl methacrylate
$C_{17}H_{32}O_4$
3036 - Dibutyl nonanedioate
$C_{17}H_{33}N$
5649 - Heptadecanenitrile
$C_{17}H_{34}$
5654 - 1-Heptadecene
$C_{17}H_{34}O$
5646 - Heptadecanal
5652 - 2-Heptadecanone
5653 - 9-Heptadecanone
$C_{17}H_{34}O_2$
5650 - Heptadecanoic acid
6542 - Isopropyl tetradecanoate
7542 - Methyl palmitate
$C_{17}H_{35}Br$
1217 - 1-Bromoheptadecane
$C_{17}H_{35}N$
4653 - 1-Dodecylpiperidine
$C_{17}H_{36}$
5648 - Heptadecane
$C_{17}H_{36}O$
5651 - 1-Heptadecanol
$C_{17}H_{37}N$
5647 - 1-Heptadecanamine
7216 - Methyldioctylamine
$C_{18}H_{10}O_2$
593 - Benz[a]anthracene-7,12-dione
7927 - 5,12-Naphthacenedione
$C_{18}H_{10}O_6$
5942 - Hydrindantin
$C_{18}H_{11}NO_2$
8091 - 6-Nitrochrysene

$C_{18}H_{12}$
592 - Benz[a]anthracene
717 - Benzo[c]phenanthrene
2387 - Chrysene
7926 - Naphthacene
10648 - Triphenylene
$C_{18}H_{12}N_2$
898 - 2,2'-Biquinoline
$C_{18}H_{12}N_6$
10703 - 2,4,6-Tris(2-pyridinyl)-1,3,5-triazine
$C_{18}H_{12}O_6$
9606 - Sterigmatocystin
$C_{18}H_{13}N$
2386 - 6-Chrysenamine
$C_{18}H_{13}NO_3$
7983 - 2-[(1-Naphthylamino)carbonyl]benzoic acid
$C_{18}H_{13}N_3O$
9491 - Rutecarpine
$C_{18}H_{14}$
4295 - 2,7-Dimethylpyrene
9696 - o-Terphenyl
9697 - m-Terphenyl
9698 - p-Terphenyl
$C_{18}H_{14}N_3NaO_3S$
6729 - Metanil Yellow
8414 - Orange IV
$C_{18}H_{14}N_4O_5S$
9650 - Sulfasalazine
$C_{18}H_{14}O_3$
8968 - 3-Phenyl-2-propenoic anhydride
$C_{18}H_{15}As$
10644 - Triphenylarsine
$C_{18}H_{15}AsO$
10645 - Triphenylarsine oxide
$C_{18}H_{15}B$
10647 - Triphenylborane
$C_{18}H_{15}Bi$
10646 - Triphenylbismuthine
$C_{18}H_{15}ClN_2O$
2424 - Cloconazole
$C_{18}H_{15}ClSi$
2340 - Chlorotriphenylsilane
$C_{18}H_{15}ClSn$
2341 - Chlorotriphenylstannane
$C_{18}H_{15}N$
10643 - Triphenylamine
$C_{18}H_{15}NO_2$
4956 - Ethyl 2-cyano-3,3-diphenyl-2-propenoate
$C_{18}H_{15}N_3$
8820 - 4-Phenylazodiphenylamine
$C_{18}H_{15}OP$
10657 - Triphenylphosphine oxide
$C_{18}H_{15}O_3P$
10658 - Triphenyl phosphite
$C_{18}H_{15}O_4P$
10655 - Triphenyl phosphate
$C_{18}H_{15}P$
10656 - Triphenylphosphine
$C_{18}H_{15}Sb$
10661 - Triphenylstibine
$C_{18}H_{16}$
4490 - 1,6-Diphenyl-1,3,5-hexatriene
$C_{18}H_{16}ClNO_5$
5318 - Fenoxaprop-ethyl

$C_{18}H_{30}O_4$
  3320 - Dicyclohexyl adipate
$C_{18}H_{31}N$
  4643 - 4-Dodecylaniline
$C_{18}H_{32}N_2O_5S$
  8501 - Panthesin
$C_{18}H_{32}O_2$
  2675 - 2-Cyclopentene-1-tridecanoic acid, (S)
  6615 - Linoleic acid
  8300 - trans,trans-9,12-Octadecadienoic acid
$C_{18}H_{32}O_6$
  5571 - Glycerol tri-3-methylbutanoate
$C_{18}H_{32}O_7$
  1539 - Butyl citrate
$C_{18}H_{32}O_{16}$
  1797 - Cellotriose
  6692 - Melezitose
  8500 - Panose
  9448 - Raffinose
$C_{18}H_{33}ClN_2O_5S$
  2423 - Clindamycin
$C_{18}H_{33}KO_2$
  9098 - Potassium cis-9-octadecenoate
$C_{18}H_{33}N$
  8313 - cis-9-Octadecenenitrile
$C_{18}H_{34}N_2O_6S$
  6614 - Lincomycin
$C_{18}H_{34}OSn$
  2707 - Cyhexatin
$C_{18}H_{34}O_2$
  4690 - Elaidic acid
  8314 - cis-6-Octadecenoic acid
  8315 - trans-11-Octadecenoic acid
  8410 - Oleic acid
$C_{18}H_{34}O_3$
  6134 - cis-12-Hydroxy-9-octadecenoic acid, (R)
$C_{18}H_{34}O_4$
  3047 - Dibutyl sebacate
  3588 - Dihexyl hexanedioate
$C_{18}H_{34}O_6$
  3739 - 9,10-Dihydroxyoctadecanedioic acid, (R*,R*)-(±)
  4925 - 2-Ethylbutanoic acid, triethyleneglycol diester
$C_{18}H_{35}ClO$
  8308 - Octadecanoyl chloride
$C_{18}H_{35}N$
  8305 - Octadecanenitrile
$C_{18}H_{35}NO$
  6595 - Laurocapram
  8311 - cis-9-Octadecenamide
$C_{18}H_{36}$
  4646 - Dodecylcyclohexane
  8312 - 1-Octadecene
$C_{18}H_{36}N_2NiS_4$
  8015 - Nickel bis(dibutyldithiocarbamate)
$C_{18}H_{36}N_2S_4$
  9729 - N,N,N′,N′-Tetrabutylthioperoxydicarbonic diamide
$C_{18}H_{36}N_2S_4Zn$
  10872 - Zinc bis(dibutyldithiocarbamate)
$C_{18}H_{36}N_4O_{11}$
  6567 - Kanamycin A
$C_{18}H_{36}O$
  4713 - 1,2-Epoxyoctadecane

  5771 - Hexadecyl vinyl ether
  8316 - cis-9-Octadecen-1-ol
  9603 - Stearaldehyde
$C_{18}H_{36}O_2$
  5190 - Ethyl palmitate
  5758 - Hexadecyl acetate
  7281 - Methyl heptadecanoate
  8387 - 2-Octyldecanoic acid
  9604 - Stearic acid
$C_{18}H_{36}O_3$
  6133 - 12-Hydroxyoctadecanoic acid
$C_{18}H_{36}O_4$
  3740 - 9,10-Dihydroxyoctadecanoic acid
$C_{18}H_{36}O_5$
  10442 - 9,10,18-Trihydroxyoctadecanoic acid, (R*,R*)
$C_{18}H_{37}Br$
  1298 - 1-Bromooctadecane
$C_{18}H_{37}Cl$
  2201 - 1-Chlorooctadecane
$C_{18}H_{37}Cl_3Si$
  10263 - Trichlorooctadecylsilane
$C_{18}H_{37}I$
  6325 - 1-Iodooctadecane
$C_{18}H_{37}N$
  8317 - cis-9-Octadecenylamine
$C_{18}H_{37}NO$
  8303 - Octadecanamide
$C_{18}H_{37}N_5O_9$
  10103 - Tobramycin
$C_{18}H_{38}$
  8304 - Octadecane
$C_{18}H_{38}O$
  4414 - Dinonyl ether
  8307 - 1-Octadecanol
$C_{18}H_{38}O_2$
  3866 - 1,1-Dimethoxyhexadecane
$C_{18}H_{38}S$
  8306 - 1-Octadecanethiol
$C_{18}H_{39}BO_3$
  10430 - Trihexyl borate
$C_{18}H_{39}N$
  5761 - Hexadecyldimethylamine
  8320 - Octadecylamine
  10429 - Trihexylamine
$C_{18}H_{39}O_7P$
  10679 - Tris(2-butoxyethyl) phosphate
$C_{18}H_{54}O_7Si_8$
  8302 - Octadecamethyloctasiloxane
$C_{19}H_{10}Br_4O_5S$
  1315 - Bromophenol Blue
$C_{19}H_{12}Cl_2O_5S$
  2220 - Chlorophenol Red
$C_{19}H_{12}O_2$
  699 - 7,8-Benzoflavone
$C_{19}H_{12}O_5$
  10448 - 2,6,7-Trihydroxy-9-phenyl-3H-xanthen-3-one
$C_{19}H_{12}O_6$
  3315 - Dicumarol
$C_{19}H_{13}N$
  8804 - 9-Phenylacridine
$C_{19}H_{14}$
  6963 - 7-Methylbenz[a]anthracene
  6964 - 8-Methylbenz[a]anthracene
  6965 - 9-Methylbenz[a]anthracene

  6966 - 10-Methylbenz[a]anthracene
  6967 - 12-Methylbenz[a]anthracene
  7130 - 3-Methylchrysene
  7131 - 5-Methylchrysene
  7132 - 6-Methylchrysene
$C_{19}H_{14}F_3NO$
  5439 - Fluridone
$C_{19}H_{14}O$
  759 - 4-Benzoylbiphenyl
$C_{19}H_{14}O_3$
  535 - Aurin
$C_{19}H_{14}O_5S$
  8766 - Phenol Red
$C_{19}H_{15}Br$
  1373 - Bromotriphenylmethane
$C_{19}H_{15}Cl$
  2339 - Chlorotriphenylmethane
$C_{19}H_{15}ClF_3NO_7$
  6577 - Lactofen
$C_{19}H_{15}ClN_4$
  10662 - Triphenyltetrazolium chloride
$C_{19}H_{15}NO_6$
  10 - Acenocoumarol
$C_{19}H_{16}$
  789 - 4-Benzyl-1,1′-biphenyl
  10653 - Triphenylmethane
$C_{19}H_{16}ClNO_4$
  6259 - Indomethacin
$C_{19}H_{16}N_2Na_2O_7S_2$
  9090 - Ponceau 3R
$C_{19}H_{16}O$
  10654 - Triphenylmethanol
$C_{19}H_{16}O_4$
  10840 - Warfarin
$C_{19}H_{16}O_5$
  4679 - Efloxate
$C_{19}H_{16}S$
  4461 - α,α-Diphenylbenzenemethanethiol
$C_{19}H_{17}ClN_2O_4$
  9446 - Quizalofop-Ethyl
$C_{19}H_{17}Cl_2N_3O_3$
  3544 - Difenoconazole
$C_{19}H_{17}NO_3$
  2498 - Cusparine
$C_{19}H_{17}N_3$
  10651 - N,N′,N″-Triphenylguanidine
$C_{19}H_{17}N_3O$
  5300 - Evodiamine
$C_{19}H_{17}N_3O_4S_2$
  1800 - Cephaloridine
$C_{19}H_{17}O_4P$
  2483 - p-Cresyl diphenyl phosphate
$C_{19}H_{18}ClN_3$
  8512 - Pararosaniline hydrochloride
$C_{19}H_{18}O_3Si$
  7839 - Methyltriphenoxysilane
$C_{19}H_{19}N$
  8759 - Phenindamine
$C_{19}H_{19}NO_3$
  6594 - Laureline
$C_{19}H_{19}NO_4$
  1388 - Bulbocapnine
$C_{19}H_{19}N_3O$
  10677 - Tris(4-aminophenyl)methanol
$C_{19}H_{19}N_7O_6$
  5442 - Folic acid

$C_{19}H_{20}F_3NO_4$
5334 - Fluazipop-butyl

$C_{19}H_{20}N_2$
6685 - Mebhydroline

$C_{19}H_{20}N_2O$
9512 - Sarpagan-17-al

$C_{19}H_{20}N_2O_2$
8856 - Phenylbutazone

$C_{19}H_{20}N_2O_3$
8492 - Oxyphenbutazone

$C_{19}H_{20}N_2O_5$
816 - Benzyloxycarbonylglycyl-L-phenylalanine

$C_{19}H_{20}N_8O_5$
414 - Aminopterin

$C_{19}H_{20}O_4$
791 - Benzyl butyl phthalate

$C_{19}H_{21}NO$
4664 - Doxepin

$C_{19}H_{21}NO_3$
6556 - Isothebaine
7921 - Nalorphine
10013 - Thebaine

$C_{19}H_{21}NO_4$
1061 - Boldine
7922 - Naloxone

$C_{19}H_{21}NS$
4662 - Dothiepin

$C_{19}H_{21}N_3S$
2499 - Cyamemazine

$C_{19}H_{21}N_5O_4$
9099 - Prazosin

$C_{19}H_{21}N_7O_6$
3623 - 7,8-Dihydrofolic acid

$C_{19}H_{22}ClN$
8280 - Nortriptyline hydrochloride
9282 - Protriptyline hydrochloride

$C_{19}H_{22}N_2$
10666 - Triprolidine

$C_{19}H_{22}N_2O$
2393 - Cinchonidine
2394 - Cinchonine
2395 - Cinchotoxine

$C_{19}H_{22}N_2OS$
12 - Acepromazine

$C_{19}H_{22}N_2O_2$
2493 - Cupreine
9513 - Sarpagan-10,17-diol

$C_{19}H_{22}O_2$
3526 - trans-Diethylstilbestrol monomethyl ether

$C_{19}H_{22}O_6$
5532 - Gibberellic acid

$C_{19}H_{23}ClN_2$
5922 - Homochlorocyclizine

$C_{19}H_{23}ClN_2O_2S$
9311 - Pyridate

$C_{19}H_{23}NO$
2397 - Cinnamedrine

$C_{19}H_{23}NO_3$
3680 - Dihydrothebaine

$C_{19}H_{23}NO_4$
9542 - Sinomenine

$C_{19}H_{23}N_3$
437 - Amitraz

$C_{19}H_{23}N_3O_2$
4724 - Ergometrinine
4725 - Ergonovine
9264 - Propyl Red

$C_{19}H_{23}N_7O_6$
9853 - 5,6,7,8-Tetrahydrofolic acid

$C_{19}H_{24}N_2$
6226 - Imipramine
7628 - 1-Methyl-N-phenyl-N-benzyl-4-piperidinamine

$C_{19}H_{24}N_2O$
2392 - Cinchonamine
5944 - Hydrocinchonidine
5945 - Hydrocinchonine

$C_{19}H_{24}N_2OS$
8755 - Phencarbamide

$C_{19}H_{24}N_2O_2$
2458 - Conquinamine
9412 - Quinamine

$C_{19}H_{24}N_2O_3$
6573 - Labetalol

$C_{19}H_{24}O_3$
458 - Androst-4-ene-3,11,17-trione
9707 - Testolactone

$C_{19}H_{25}ClN_2$
6227 - Imipramine hydrochloride

$C_{19}H_{25}NO$
6606 - Levallorphan

$C_{19}H_{25}NO_2$
8286 - Nylidrin

$C_{19}H_{25}NO_4$
9915 - Tetramethrin

$C_{19}H_{26}$
8259 - 1-Nonylnaphthalene

$C_{19}H_{26}N_2O$
2494 - Curan-17-ol, (16α)

$C_{19}H_{26}O_2$
457 - Androst-4-ene-3,17-dione
1060 - Boldenone

$C_{19}H_{26}O_3$
159 - Allethrin

$C_{19}H_{26}O_4S$
9173 - Propargite

$C_{19}H_{27}ClO_2$
2070 - 4-Chloro-17-hydroxyandrost-4-en-3-one, (17β)

$C_{19}H_{27}NO$
8637 - Pentazocine

$C_{19}H_{27}NO_3$
9709 - Tetrabenazine

$C_{19}H_{27}NO_6$
9527 - Senkirkin

$C_{19}H_{28}BrNO_3$
5586 - Glycopyrrolate

$C_{19}H_{28}O_2$
455 - 5α-Androstane-3,17-dione
456 - 5β-Androstane-3,17-dione
5965 - 17-Hydroxyandrost-4-en-3-one, (17β)

$C_{19}H_{29}IO_2$
6350 - Iophendylate

$C_{19}H_{30}O_2$
795 - Benzyl dodecanoate
5962 - 17-Hydroxyandrostan-3-one, (5α,17β)
5963 - 3-Hydroxyandrostan-17-one, (3α,5α)
5964 - 3-Hydroxyandrostan-17-one, (3β,5α)

$C_{19}H_{30}O_3$
8438 - Oxandrolone

$C_{19}H_{30}O_5$
4656 - Dodecyl 3,4,5-trihydroxybenzoate
9077 - Piperonyl butoxide

$C_{19}H_{32}$
452 - Androstane
10323 - Tridecylbenzene

$C_{19}H_{32}O_2$
454 - Androstane-3,17-diol, (3α,5α,17β)
7431 - Methyl linolenate

$C_{19}H_{34}$
5329 - Fichtelite

$C_{19}H_{34}O_2$
7430 - Methyl linoleate

$C_{19}H_{34}O_3$
6763 - Methoprene

$C_{19}H_{36}ClNO_2$
3331 - Dicyclomine hydrochloride

$C_{19}H_{36}O_2$
7514 - Methyl trans-9-octadecenoate
7528 - Methyl oleate

$C_{19}H_{37}NO$
8324 - Octadecyl isocyanate

$C_{19}H_{38}$
8224 - 1-Nonadecene
10324 - Tridecylcyclohexane

$C_{19}H_{38}O$
8222 - 2-Nonadecanone
8223 - 10-Nonadecanone

$C_{19}H_{38}O_2$
6525 - Isopropyl palmitate
7783 - Methyl stearate
8220 - Nonadecanoic acid

$C_{19}H_{38}O_3$
5764 - Hexadecyl 2-hydroxypropanoate

$C_{19}H_{39}NO$
4346 - 2,6-Dimethyl-4-tridecylmorpholine

$C_{19}H_{40}$
8219 - Nonadecane
9959 - 2,6,10,14-Tetramethylpentadecane

$C_{19}H_{40}O$
8221 - 1-Nonadecanol

$C_{19}H_{40}O_3$
5766 - 3-(Hexadecyloxy)-1,2-propanediol, (S)

$C_{20}H_6Br_4Na_2O_5$
9713 - 2',4',5',7'-Tetrabromofluorescein, disodium salt

$C_{20}H_8Br_4Na_2O_{10}S_2$
9573 - Sodium sulfobromophthalein

$C_{20}H_8I_4O_5$
4752 - Erythrosine

$C_{20}H_{10}Br_4O_4$
9717 - 3',3'',5',5''-Tetrabromophenolphthalein

$C_{20}H_{10}Cl_2O_5$
3183 - 2',7'-Dichlorofluorescein

$C_{20}H_{10}I_2O_5$
3766 - 4,4'-Diiodofluorescein

$C_{20}H_{10}Na_2O_5$
5350 - Fluorescein sodium

$C_{20}H_{11}Br$
1109 - 6-Bromobenzo[a]pyrene

$C_{20}H_{11}N_2Na_3O_{10}S_3$
225 - Amaranth dye

$C_{20}H_{12}$
700 - Benzo[b]fluoranthene

701 - Benzo[j]fluoranthene
702 - Benzo[k]fluoranthene
729 - Benzo[a]pyrene
730 - Benzo[e]pyrene
8736 - Perylene

$C_{20}H_{12}N_2O_2$
2405 - Cinquasia Red

$C_{20}H_{12}N_3NaO_7S$
4741 - Eriochrome Black T

$C_{20}H_{12}O_5$
5349 - Fluorescein
5521 - Gallein

$C_{20}H_{13}N$
2880 - 7H-Dibenzo[c,g]carbazole
2881 - 13H-Dibenzo[a,i]carbazole

$C_{20}H_{14}$
871 - 1,1'-Binaphthalene
872 - 2,2'-Binaphthalene
3594 - 1,2-Dihydrobenz[j]aceanthrylene
3595 - 9,10-Dihydro-9,10[1',2']-benzenoan-
thracene
8816 - 9-Phenylanthracene

$C_{20}H_{14}I_6N_2O_6$
6264 - Iodipamide

$C_{20}H_{14}N_2Na_3O_{11}S_3$
6125 - Hydroxynaphthol blue, trisodium salt

$C_{20}H_{14}N_2O$
892 - 2-[1,1'-Biphenyl]-4-yl-5-phenyl-1,3,4-
oxadiazole

$C_{20}H_{14}N_4$
9092 - 21H,23H-Porphine

$C_{20}H_{14}O_2$
873 - [1,1'-Binaphthalene]-2,2'-diol

$C_{20}H_{14}O_3$
5333 - Florantyrone

$C_{20}H_{14}O_4$
4494 - Diphenyl isophthalate
4511 - Diphenyl phthalate
8763 - Phenolphthalein

$C_{20}H_{14}O_5$
3757 - 2-(3,6-Dihydroxy-9H-xanthen-9-yl)
benzoic acid

$C_{20}H_{14}S_2$
4356 - Di-2-naphthyl disulfide

$C_{20}H_{15}N$
7990 - N-2-Naphthyl-2-naphthalenamine

$C_{20}H_{15}NO_5$
9505 - Sanguinarine

$C_{20}H_{16}$
3948 - 7,12-Dimethylbenz[a]anthracene
10650 - 1,1,2-Triphenylethene

$C_{20}H_{16}N_4$
8124 - Nitron

$C_{20}H_{16}O_4$
8764 - Phenolphthalin

$C_{20}H_{17}N_3Na_2O_9S_3$
108 - Acid Fuchsin

$C_{20}H_{18}$
10649 - 1,1,2-Triphenylethane

$C_{20}H_{18}ClNO_6$
8287 - Ochratoxin A

$C_{20}H_{18}O$
2896 - 2,6-Dibenzylidenecyclohexanone
4460 - α,α-Diphenylbenzeneethanol

$C_{20}H_{18}O_2Sn$
93 - (Acetyloxy)triphenylstannane

$C_{20}H_{18}O_3$
8765 - Phenolphthalol

$C_{20}H_{19}ClN_4$
5312 - Fenbuconazole

$C_{20}H_{19}N$
8846 - N-Phenyl-N-benzylbenzenemethanamine

$C_{20}H_{19}NO_5$
841 - Berberine
1810 - Chelidonine
8504 - Papaveraldine
9280 - Protopine

$C_{20}H_{19}NO_6$
8288 - Ochratoxin B

$C_{20}H_{19}N_3$
6653 - Magenta base

$C_{20}H_{20}ClN_3$
6654 - Magenta I

$C_{20}H_{20}OSi$
4869 - Ethoxytriphenylsilane

$C_{20}H_{20}O_6$
2489 - Cubebin

$C_{20}H_{21}NO_3$
3818 - Dimefline
5519 - Galipine

$C_{20}H_{21}NO_4$
1714 - Canadine, (±)
3062 - Dicentrine
8505 - Papaverine

$C_{20}H_{21}N_3O_3$
7897 - Moquizone

$C_{20}H_{22}ClN$
9391 - Pyrrobutamine

$C_{20}H_{22}ClNO_4$
8506 - Papaverine hydrochloride

$C_{20}H_{22}ClNO_6$
842 - Berberine chloride dihydrate

$C_{20}H_{22}N_2O_2$
5524 - Gelsemine
9421 - Quininone

$C_{20}H_{22}N_8O_5$
6765 - Methotrexate

$C_{20}H_{22}O_5$
4556 - Dipropylene glycol dibenzoate

$C_{20}H_{22}O_6$
2452 - Columbin

$C_{20}H_{22}O_8$
9091 - Populin

$C_{20}H_{23}ClN_2O_2$
5525 - Gelsemine, monohydrochloride

$C_{20}H_{23}ClN_2O_4$
2348 - Chlorpheniramine maleate

$C_{20}H_{23}N$
438 - Amitriptyline

$C_{20}H_{23}NO_2$
443 - Amolanone

$C_{20}H_{23}NO_3$
577 - Benalaxyl

$C_{20}H_{23}NO_4$
2470 - Corydine
6408 - Isocorydine
7923 - Naltrexone

$C_{20}H_{23}N_7O_7$
5443 - Folinic acid

$C_{20}H_{24}ClN_3S$
9117 - Prochlorperazine

$C_{20}H_{24}I_2O_2$
10097 - Thymol iodide

$C_{20}H_{24}N_2OS$
9193 - Propiomazine

$C_{20}H_{24}N_2O_2$
4709 - Epiquinidine
9417 - Quinidine
9418 - Quinine
10830 - Viquidil

$C_{20}H_{24}O_2$
4807 - Ethinylestradiol

$C_{20}H_{24}O_4$
2868 - 8,8'-Diapo-ψ,ψ-carotenedioic acid

$C_{20}H_{25}ClN_2O_2$
9419 - Quinine hydrochloride

$C_{20}H_{25}N$
4522 - 1-(3,3-Diphenylpropyl)piperidine

$C_{20}H_{25}NO$
3908 - 6-(Dimethylamino)-4,4-diphenyl-3-
hexanone
4505 - 2-(Diphenylmethyl)-1-piperidineetha-
nol
9112 - Pridinol

$C_{20}H_{25}NO_2$
5446 - Fomocaine

$C_{20}H_{25}NO_3$
575 - Benactyzine

$C_{20}H_{25}NO_4$
2442 - Codamine
6591 - Laudanidine
6592 - Laudanine

$C_{20}H_{25}N_3O$
6645 - Lysergide

$C_{20}H_{25}N_3O_2$
7251 - Methylergonovine

$C_{20}H_{25}N_3S$
8693 - Perazine

$C_{20}H_{25}N_7O_6$
7796 - 5-N-Methyl-5,6,7,8-tetrahydrofolic acid

$C_{20}H_{26}ClNO_2$
131 - Adiphenine hydrochloride

$C_{20}H_{26}ClNO_3$
576 - Benactyzine hydrochloride

$C_{20}H_{26}N_2O$
6832 - 12-Methoxyibogamine

$C_{20}H_{26}N_2O_2$
140 - Ajmalan-17,21-diol, (17R,21α)
5956 - Hydroquinidine
5957 - Hydroquinine

$C_{20}H_{26}O_2$
677 - Benzestrol
6757 - Methestrol
8271 - Norethisterone
8272 - Norethynodrel

$C_{20}H_{26}O_4$
3328 - Dicyclohexyl phthalate

$C_{20}H_{27}N$
223 - Alverine

$C_{20}H_{27}NO_{11}$
449 - Amygdalin

$C_{20}H_{27}O_3P$
5060 - 2-Ethylhexyl diphenyl phosphite

$C_{20}H_{27}O_4P$
4486 - Diphenyl 2-ethylhexyl phosphate
8389 - Octyl diphenyl phosphate

$C_{20}H_{28}$
2785 - 1-Decylnaphthalene

$C_{20}H_{28}O$
9459 - 11-*cis*-Retinal
9460 - Retinal (all *trans*)

$C_{20}H_{28}O_2$
2790 - Dehydroabietic acid
6740 - Methandrostenolone
9461 - 13-*cis*-Retinoic acid
9462 - 13-*trans*-Retinoic acid

$C_{20}H_{28}O_3S$
4978 - Ethyl 3,6-di(*tert*-butyl)-1-naphthalene-
　　　sulfonate

$C_{20}H_{28}O_6$
9019 - Phorbol

$C_{20}H_{29}FO_3$
5434 - Fluoxymesterone

$C_{20}H_{29}N_3O_2$
2997 - Dibucaine

$C_{20}H_{30}ClN_3O_2$
2998 - Dibucaine hydrochloride

$C_{20}H_{30}O$
9463 - Retinol

$C_{20}H_{30}O_2$
2 - Abietic acid
6608 - Levopimaric acid
7794 - 17-Methyltestosterone
8004 - Neoabietic acid
8497 - Palustric acid
9043 - Pimaric acid

$C_{20}H_{30}O_4$
3589 - Dihexyl phthalate
5058 - 2-Ethylhexyl butyl phthalate

$C_{20}H_{30}O_5$
6034 - 11-Hydroxy-9,15-dioxoprosta-5,13-
　　　dien-1-oic acid, (5Z,11α,13E)

$C_{20}H_{30}O_6$
625 - 1,2-Benzenedicarboxylic acid, bis(2-
　　　butoxyethyl) ester

$C_{20}H_{31}N$
228 - 19-Amino-8,11,13-abietatriene

$C_{20}H_{31}NO$
2638 - α-Cyclohexyl-α-phenyl-1-pipe-
　　　ridinepropanol

$C_{20}H_{31}NO_3$
1733 - Carbetapentane

$C_{20}H_{31}NO_7$
4675 - Echimidine

$C_{20}H_{32}ClNO$
10431 - Trihexyphenidyl hydrochloride

$C_{20}H_{32}N_2O_3S$
1762 - Carbosulfan

$C_{20}H_{32}N_6O_{12}S_2$
5559 - Glutathione disulfide

$C_{20}H_{32}O$
5022 - Ethylestrenol

$C_{20}H_{32}O_2$
453 - Androstane-17-carboxylic acid, (5β,17β)
4685 - 5,8,11,14-Eicosatetraenoic acid, (all-*cis*)
6074 - 17-Hydroxy-17-methylandrostan-3-
　　　one, (5α,17β)

$C_{20}H_{32}O_5$
9278 - Prostaglandin E₂

$C_{20}H_{34}$
9804 - Tetradecylbenzene

$C_{20}H_{34}O_2$
5179 - Ethyl *cis,cis,cis*-9,12,15-octadecatri-
　　　enoate

$C_{20}H_{34}O_4$
921 - 1,4-Bis(α-(*tert*-butyldioxy)isopropyl)
　　　benzene

$C_{20}H_{34}O_5$
9277 - Prostaglandin E₁
9279 - Prostaglandin F2α

$C_{20}H_{34}O_8$
10178 - Tributyl 2-(acetyloxy)-1,2,3-propanet-
　　　ricarboxylate

$C_{20}H_{35}NO_{13}$
10767 - Validamycin A

$C_{20}H_{36}O_2$
5178 - Ethyl *cis,cis*-9,12-octadecadienoate

$C_{20}H_{36}O_4$
3464 - Di-2-ethylhexyl maleate
4421 - Dioctyl maleate

$C_{20}H_{37}NaO_7S$
988 - Bis(2-ethylhexyl) sodium sulfosuccinate

$C_{20}H_{38}N_4O_4$
4354 - Dimorpholamine

$C_{20}H_{38}O_2$
4687 - *cis*-9-Eicosenoic acid
4688 - *trans*-9-Eicosenoic acid
4689 - 11-Eicosenoic acid
5180 - Ethyl *trans*-9-octadecenoate
5185 - Ethyl oleate
5765 - Hexadecyl 2-methyl-2-propenoate
10815 - Vinyl octadecanoate

$C_{20}H_{38}O_3$
5080 - Ethyl *cis*-12-hydroxy-9-octadecenoate,
　　　(R)

$C_{20}H_{38}O_4$
4682 - Eicosanedioic acid

$C_{20}H_{40}$
4686 - 1-Eicosene
9805 - Tetradecylcyclohexane

$C_{20}H_{40}N_6O_8S$
5606 - Guanadrel sulfate (2:1)

$C_{20}H_{40}O$
8328 - Octadecyl vinyl ether
9034 - Phytol
9951 - 3,7,11,15-Tetramethyl-1-hexadecen-3-
　　　ol

$C_{20}H_{40}O_2$
1634 - Butyl palmitate
2782 - Decyl decanoate
4683 - Eicosanoic acid
5245 - Ethyl stearate
7505 - Methyl nonadecanoate
8318 - Octadecyl acetate
9950 - 3,7,11,15-Tetramethylhexadecanoic
　　　acid

$C_{20}H_{40}O_3$
4788 - 1,2-Ethanediol, monostearate

$C_{20}H_{42}$
4681 - Eicosane

$C_{20}H_{42}O$
3336 - Didecyl ether
4684 - 1-Eicosanol

$C_{20}H_{42}O_5$
4655 - Dodecyltetraethylene glycol monoether

$C_{20}H_{43}N$
3335 - Didecylamine

4320 - Dimethylstearylamine
6208 - Icosylamine

$C_{20}H_{60}O_8Si_9$
4680 - Eicosamethylnonasiloxane

$C_{21}H_{13}N$
2873 - Dibenz[a,h]acridine
2874 - Dibenz[a,j]acridine
2875 - Dibenz[c,h]acridine

$C_{21}H_{14}Br_4O_5S$
1156 - Bromocresol Green

$C_{21}H_{15}NO_3$
154 - Alizurol purple

$C_{21}H_{15}N_3$
10664 - 2,4,6-Triphenyl-1,3,5-triazine

$C_{21}H_{16}$
3639 - 1,2-Dihydro-3-methylbenz[j]acean-
　　　thrylene

$C_{21}H_{16}Br_2O_5S$
1157 - Bromocresol Purple

$C_{21}H_{16}N_2$
10652 - 2,4,5-Triphenyl-1*H*-imidazole

$C_{21}H_{16}N_2O$
4357 - *N,N'*-Di-1-naphthylurea

$C_{21}H_{18}O_5S$
2482 - Cresol Red

$C_{21}H_{19}NO_5$
1809 - Chelerythrine

$C_{21}H_{20}Cl_2O_3$
8731 - Permethrin

$C_{21}H_{20}N_2O_3$
217 - Alstonine
9532 - Serpentine alkaloid

$C_{21}H_{20}O_6$
2495 - Curcumin

$C_{21}H_{20}O_{10}$
9583 - Sophoricoside

$C_{21}H_{20}O_{11}$
9408 - Quercitrin

$C_{21}H_{21}ClN_2O_8$
2776 - Declomycin

$C_{21}H_{21}N$
2710 - Cyproheptadine
7049 - 4-Methyl-*N,N*-bis(4-methylphenyl)
　　　aniline
10151 - Tribenzylamine

$C_{21}H_{21}NO_5$
2468 - Corycavamine

$C_{21}H_{21}NO_6$
5931 - Hydrastine
9470 - Rheadine

$C_{21}H_{21}N_3$
5811 - Hexahydro-1,3,5-triphenyl-1,3,5-triaz-
　　　ine

$C_{21}H_{21}N_3O_7$
1689 - Cacotheline

$C_{21}H_{21}O_3P$
10704 - Tris(*o*-tolyl) phosphite
10705 - Tris(*p*-tolyl) phosphite

$C_{21}H_{21}O_4P$
10302 - Tri-*o*-cresyl phosphate
10303 - Tri-*m*-cresyl phosphate
10304 - Tri-*p*-cresyl phosphate

$C_{21}H_{21}O_7P$
6860 - 2-Methoxyphenol phosphate (3:1)

$C_{21}H_{21}P$
10696 - Tris(2-methylphenyl)phosphine

10697 - Tris(3-methylphenyl)phosphine
10698 - Tris(4-methylphenyl)phosphine

$C_{21}H_{22}N_2O_2$
9620 - Strychnine

$C_{21}H_{22}O_9$
214 - Aloin A

$C_{21}H_{23}ClFNO_2$
5622 - Haloperidol

$C_{21}H_{23}ClFN_3O$
5438 - Flurazepam

$C_{21}H_{23}IN_2$
9411 - Quinaldine Red

$C_{21}H_{23}NO_5$
2487 - Cryptopine
2826 - Diacetylmorphine

$C_{21}H_{23}NO_6$
2448 - Colchiceine

$C_{21}H_{23}N_3O_5$
9618 - Strychnidin-10-one mononitrate

$C_{21}H_{24}F_3N_3S$
10376 - Trifluoperazine

$C_{21}H_{24}N_2O_3$
9450 - Raubasine

$C_{21}H_{24}O_3Si_3$
10620 - 2,4,6-Trimethyl-2,4,6-triphenylcyclotrisiloxane

$C_{21}H_{24}O_7$
10831 - Visnadine

$C_{21}H_{25}NO_3$
7920 - Nalmefene

$C_{21}H_{25}NO_4$
2467 - Corybulbine
5535 - d-Glaucine
6407 - Isocorybulbine

$C_{21}H_{26}BrNO_3$
6751 - Methantheline bromide

$C_{21}H_{26}ClN_3OS$
8734 - Perphenazine

$C_{21}H_{26}Cl_2F_3N_3S$
10377 - Trifluoperazine dihydrochloride

$C_{21}H_{26}N_2OS_2$
6724 - Mesoridazine

$C_{21}H_{26}N_2O_3$
10788 - Vincamine
10867 - Yohimbine

$C_{21}H_{26}N_6O_{15}P_2$
8023 - Nicotinamide hypoxanthine dinucleotide

$C_{21}H_{26}O_2$
1716 - Cannabinol
6725 - Mestranol

$C_{21}H_{26}O_3$
6137 - [2-Hydroxy-4-(octyloxy)phenyl]phenylmethanone
8400 - 4-Octylphenyl salicylate

$C_{21}H_{26}O_5$
3746 - 17,21-Dihydroxypregna-1,4-diene-3,11,20-trione

$C_{21}H_{27}ClN_2O_2$
6201 - Hydroxyzine

$C_{21}H_{27}ClN_2O_3$
10868 - Yohimbine hydrochloride

$C_{21}H_{27}FO_5$
5436 - Fluprednisolone

$C_{21}H_{27}FO_6$
10136 - Triamcinolone

$C_{21}H_{27}N$
6494 - N-Isopropyl-4,4-diphenylcyclohexanamine

$C_{21}H_{27}NO$
3907 - 6-(Dimethylamino)-4,4-diphenyl-3-heptanone
4448 - Diphenidol

$C_{21}H_{27}NO_4$
6593 - Laudanosine

$C_{21}H_{27}N_3O_2$
7861 - Methysergide

$C_{21}H_{27}N_7O_{14}P_2$
2446 - Coenzyme I

$C_{21}H_{28}ClNO$
6735 - Methadone hydrochloride

$C_{21}H_{28}N_2O$
959 - 4,4'-Bis(diethylamino)benzophenone

$C_{21}H_{28}N_2S_2$
1468 - Buthiobate

$C_{21}H_{28}N_7O_{17}P_3$
2447 - Coenzyme II

$C_{21}H_{28}O_2$
4671 - Dydrogesterone
4812 - Ethisterone
8278 - Norplant

$C_{21}H_{28}O_3$
9309 - Pyrethrin I

$C_{21}H_{28}O_4$
6176 - 21-Hydroxypregn-4-ene-3,11,20-trione

$C_{21}H_{28}O_5$
148 - Aldosterone
3748 - 17,21-Dihydroxypregn-4-ene-3,11,20-trione
9100 - Prednisolone

$C_{21}H_{28}O_8$
8574 - Pentaerythritol tetramethacrylate

$C_{21}H_{29}ClN_2O_5$
10472 - Trimethobenzamide hydrochloride

$C_{21}H_{29}FO_5$
5338 - Fludrocortisone

$C_{21}H_{30}O_2$
1715 - Cannabidiol
9122 - Progesterone

$C_{21}H_{30}O_3$
6174 - 17-Hydroxypregn-4-ene-3,20-dione
6175 - 21-Hydroxypregn-4-ene-3,20-dione

$C_{21}H_{30}O_4$
2466 - Corticosterone
3747 - 17,21-Dihydroxypregn-4-ene-3,20-dione

$C_{21}H_{30}O_5$
5928 - Humulon
5947 - Hydrocortisone

$C_{21}H_{31}NO_4$
5501 - Furethidine

$C_{21}H_{32}N_2O$
9602 - Stanozolol

$C_{21}H_{32}O_2$
6909 - Methyl abietate
9105 - 5α-Pregnane-3,20-dione
9106 - 5β-Pregnane-3,20-dione
9109 - Pregnenolone

$C_{21}H_{32}O_3$
8490 - Oxymetholone

$C_{21}H_{32}O_4$
161 - Allopregnane-3β,21-diol-11,20-dione

$C_{21}H_{32}O_5$
9845 - Tetrahydrocortisone
9901 - 11,17,20,21-Tetrahydroxypregn-4-en-3-one, (11β,20R)

$C_{21}H_{33}NO_7$
6590 - Lasiocarpine

$C_{21}H_{34}BrNO_3$
8493 - Oxyphenonium bromide

$C_{21}H_{34}O_2$
162 - Allopregnan-20β-ol-3-one
6172 - 3-Hydroxypregnan-20-one, (3α,5α)
6173 - 3-Hydroxypregnan-20-one, (3β,5α)
9108 - Pregnan-3α-ol-20-one

$C_{21}H_{34}O_4$
9107 - 5-Pregnane-3,17,21-triol-20-one

$C_{21}H_{35}N_3O_7$
6621 - Lisinopril

$C_{21}H_{36}$
8559 - Pentadecylbenzene
9101 - 5α-Pregnane
9102 - 5β-Pregnane

$C_{21}H_{36}N_2O_5S$
5885 - Hexocyclium methyl sulfate

$C_{21}H_{36}N_7O_{16}P_3S$
2445 - Coenzyme A

$C_{21}H_{36}O$
8562 - 3-Pentadecylphenol

$C_{21}H_{36}O_2$
8560 - 3-Pentadecyl-1,2-benzenediol
9103 - 5α-Pregnane-3α,20α-diol
9104 - 5β-Pregnane-3α,20S-diol

$C_{21}H_{38}BrN$
5767 - 1-Hexadecylpyridinium bromide

$C_{21}H_{38}ClN$
5768 - 1-Hexadecylpyridinium chloride

$C_{21}H_{38}O_6$
9160 - 1,2,3-Propanetriyl hexanoate

$C_{21}H_{39}N_7O_{12}$
9614 - Streptomycin

$C_{21}H_{40}O_2$
8319 - Octadecyl acrylate

$C_{21}H_{40}O_4$
5566 - Glycerol 1-oleate

$C_{21}H_{42}$
8561 - Pentadecylcyclohexane

$C_{21}H_{42}O_2$
6540 - Isopropyl stearate
7229 - Methyl eicosanoate

$C_{21}H_{44}$
5638 - Heneicosane

$C_{21}H_{44}O_3$
8327 - 3-(Octadecyloxy)-1,2-propanediol

$C_{22}H_{12}$
716 - Benzo[ghi]perylene
6238 - Indeno[1,2,3-cd]pyrene

$C_{22}H_{13}Br_4KO_4$
9719 - 3',3'',5',5''-Tetrabromophenolphthalein ethyl ester, potassium salt

$C_{22}H_{14}$
686 - Benzo[c]chrysene
687 - Benzo[g]chrysene
746 - Benzo[b]triphenylene
2876 - Dibenz[a,h]anthracene
2877 - Dibenz[a,j]anthracene
8530 - Pentacene
8633 - Pentaphene

9035 - Picene

$C_{22}H_{14}Br_4O_4$

9718 - 3',3",5',5"-Tetrabromophenolphthalein
   ethyl ester

$C_{22}H_{16}N_4O$

8825 - 1-[[4-(Phenylazo)phenyl]azo]-2-naph-
   thol

$C_{22}H_{17}ClN_2$

2435 - Clotrimazole

$C_{22}H_{18}Cl_2FNO_3$

2704 - Cyfluthrin

$C_{22}H_{18}O_4$

2480 - *o*-Cresolphthalein

$C_{22}H_{19}Br_2NO_3$

2793 - Deltamethrin

$C_{22}H_{19}Br_4NO_3$

10124 - Tralomethrin

$C_{22}H_{19}Cl_2NO_3$

2708 - Cypermethrin

$C_{22}H_{19}NO_4$

900 - Bisacodyl

$C_{22}H_{20}O_{13}$

1771 - Carminic acid

$C_{22}H_{21}NO_2$

802 - Benzylimidobis(*p*-methoxyphenyl)meth-
   ane

$C_{22}H_{22}ClNO_6$

8289 - Ochratoxin C

$C_{22}H_{22}FN_3O_2$

4670 - Droperidol

$C_{22}H_{22}N_4O_8$

6734 - Methacycline

$C_{22}H_{22}O_8$

9037 - Picropodophyllin

9088 - Podophyllotoxin

$C_{22}H_{23}ClN_2O_2$

6628 - Loratadine

$C_{22}H_{23}ClN_2O_8$

2355 - Chlortetracycline

$C_{22}H_{23}NO_3$

5320 - Fenpropathrin

$C_{22}H_{23}NO_6$

534 - Aureothin

$C_{22}H_{23}NO_7$

8283 - Noscapine

$C_{22}H_{23}N_3O_9$

536 - Aurin tricarboxylic acid, triammonium
   salt

$C_{22}H_{24}N_2O_4$

216 - Alstonidine

899 - 4,4'-Bis(acetoacetamido)-3,3'-dimethyl-
   1,1'-biphenyl

10839 - Vomicine

$C_{22}H_{24}N_2O_7$

812 - *N*-Benzyloxycarbonylaspartame

$C_{22}H_{24}N_2O_8$

9784 - Tetracycline

$C_{22}H_{24}N_2O_9$

8494 - Oxytetracycline

$C_{22}H_{25}ClN_2O_8$

9785 - Tetracycline hydrochloride

$C_{22}H_{25}NO_2$

6624 - Lobelanine

$C_{22}H_{25}NO_6$

2449 - Colchicine

$C_{22}H_{25}N_3O$

774 - Benzpiperylon

$C_{22}H_{26}F_3N_3OS$

5435 - Fluphenazine

$C_{22}H_{26}N_2O_3$

2471 - Corynantheine

$C_{22}H_{26}O_3$

9457 - *cis*-Resmethrin, (-)

$C_{22}H_{27}NO_2$

6625 - Lobeline

$C_{22}H_{27}NO_4$

2469 - Corydaline

$C_{22}H_{27}N_3O_2$

1783 - Caroverine

$C_{22}H_{28}ClNO_2$

768 - 2-Benzoylmethyl-6(2-hydroxy-2-phenyl-
   ethyl)-1-methylpiperidine, hydrochloride

$C_{22}H_{28}N_2O$

5322 - Fentanyl

$C_{22}H_{28}N_2O_2$

465 - Anileridine

$C_{22}H_{28}N_2O_5$

9455 - Reserpic acid

$C_{22}H_{28}N_6O_{14}P_2$

101 - 3-Acetylpyridine adenine dinucleotide

$C_{22}H_{28}O_3$

1717 - Canrenone

$C_{22}H_{28}O_5$

9310 - Pyrethrin II

$C_{22}H_{29}FO_5$

847 - Betamethasone

2816 - Dexamethasone

$C_{22}H_{29}NO_2$

6623 - Lobelanidine

9198 - *D*-Propoxyphene

9199 - *L*-Propoxyphene

$C_{22}H_{29}N_3S_2$

10040 - Thiethylperazine

$C_{22}H_{30}N_2O_2$

79 - 1-Acetyl-17-methoxyaspidospermidine

$C_{22}H_{30}N_2O_2S$

9634 - Sufentanil

$C_{22}H_{30}N_2O_5$

4677 - Echitamine

$C_{22}H_{30}O_2S$

10046 - 4,4'-Thiobis(6-*tert*-butyl-*m*-cresol)

$C_{22}H_{30}O_5$

7681 - Methylprednisolone

$C_{22}H_{31}NO_3$

8482 - Oxybutynin

$C_{22}H_{31}O_4P$

6427 - Isodecyl diphenyl phosphate

$C_{22}H_{32}N_2O_5$

775 - Benzquinamide

$C_{22}H_{32}O_3$

450 - Anacardic acid

6688 - Medroxyprogesterone

8476 - 17-(1-Oxopropoxy)-androst-4-en-3-
   one, (17β)

$C_{22}H_{33}NO_2$

530 - Atisine

$C_{22}H_{34}O_4$

3584 - Diheptyl phthalate

$C_{22}H_{35}NO_3$

6633 - Luciculine

$C_{22}H_{36}O_7$

5602 - Grayanotoxin I

$C_{22}H_{38}$

5760 - Hexadecylbenzene

$C_{22}H_{38}O_5$

7879 - Misoprostol

$C_{22}H_{38}O_7$

519 - Ascorbyl palmitate

$C_{22}H_{40}O_7$

6132 - 2-Hydroxy-1,2,3-nonadecanetricarbox-
   ylic acid

$C_{22}H_{42}O_2$

1631 - Butyl oleate

4616 - *cis*-13-Docosenoic acid

4617 - *trans*-13-Docosenoic acid

8325 - Octadecyl methacrylate

$C_{22}H_{42}O_3$

1588 - Butyl *cis*-12-hydroxy-9-octadecenoate,
   (*R*)

$C_{22}H_{42}O_4$

979 - Bis(2-ethylhexyl) adipate

3787 - Diisooctyl adipate

4420 - Dioctyl hexanedioate

$C_{22}H_{42}O_6$

10356 - Triethylene glycol bis(2-ethylhex-
   anoate)

$C_{22}H_{42}O_8$

920 - Bis(2-(2-butoxyethoxy)ethyl) adipate

$C_{22}H_{43}NO$

4614 - 13-Docosenamide

$C_{22}H_{44}$

4615 - 1-Docosene

$C_{22}H_{44}N_2O_2$

5595 - Glyodin

$C_{22}H_{44}O_2$

1660 - Butyl stearate

4612 - Docosanoic acid

6402 - Isobutyl stearate

$C_{22}H_{46}$

4611 - Docosane

$C_{22}H_{46}N_4O_2$

9084 - Pithecolobine

$C_{22}H_{46}O$

4613 - 1-Docosanol

$C_{23}H_{16}O_3$

4453 - 2-(Diphenylacetyl)-1*H*-indene-1,3(2*H*)-
   dione

$C_{23}H_{16}O_6$

8498 - Pamoic acid

$C_{23}H_{16}O_{11}$

2485 - Cromolyn

$C_{23}H_{19}ClF_3NO_3$

2706 - Cyhalothrin

$C_{23}H_{20}N_2O_3S$

9653 - Sulfinpyrazone

$C_{23}H_{21}ClO_3$

10695 - Tris(4-methoxyphenyl)chloroethene

$C_{23}H_{22}ClF_3O_2$

865 - Bifenthrin

$C_{23}H_{22}O_6$

9488 - Rotenone

$C_{23}H_{22}O_7$

9688 - Tephrosin

$C_{23}H_{23}IN_2S_2$

4583 - Dithiazanine iodide

C$_{23}$H$_{23}$NO
  10374 - Trifenmorph
C$_{23}$H$_{24}$O$_4$
  1016 - 2,2-Bis(4-hydroxyphenyl)propane
    dimethacrylate
C$_{23}$H$_{25}$ClN$_2$
  6656 - Malachite Green
C$_{23}$H$_{26}$Cl$_2$O$_6$
  9540 - Simfibrate
C$_{23}$H$_{26}$N$_2$
  968 - 4,4'-Bis(dimethylamino)triphenyl-
    methane
C$_{23}$H$_{26}$N$_2$O$_4$
  1383 - Brucine
C$_{23}$H$_{26}$O$_3$
  8768 - Phenothrin
C$_{23}$H$_{27}$ClN$_2$O$_4$
  1384 - Brucine hydrochloride
C$_{23}$H$_{27}$NO$_8$
  7998 - Narceine
C$_{23}$H$_{27}$N$_3$O$_7$
  7875 - Minocycline
C$_{23}$H$_{28}$ClN$_3$O
  9110 - Prenoxdiazine hydrochloride
C$_{23}$H$_{28}$ClN$_3$O$_2$S
  10076 - Thiopropazate
C$_{23}$H$_{29}$ClFN$_3$O$_4$
  2410 - Cisapride
C$_{23}$H$_{29}$N$_3$O$_2$S$_2$
  10080 - cis-Thiothixene
C$_{23}$H$_{30}$BrNO$_3$
  9172 - Propantheline bromide
C$_{23}$H$_{30}$ClN$_3$O
  9410 - Quinacrine
C$_{23}$H$_{30}$N$_2$O$_4$
  7885 - Mitragynine
C$_{23}$H$_{32}$N$_2$O$_2$S
  10101 - Tiocarlide
C$_{23}$H$_{32}$O$_2$
  993 - Bis(2-hydroxy-3-tert-butyl-5-methylphe-
    nyl)methane
C$_{23}$H$_{32}$O$_6$
  5948 - Hydrocortisone 21-acetate
  9617 - Strophanthidin
C$_{23}$H$_{33}$IN$_2$O
  6461 - Isopropamide iodide
C$_{23}$H$_{34}$O$_4$
  3575 - Digitoxigenin
C$_{23}$H$_{34}$O$_5$
  3580 - Digoxigenin
  5533 - Gitoxigenin
  9511 - Sarmentogenin
C$_{23}$H$_{36}$N$_2$O$_2$
  5330 - Finasteride
C$_{23}$H$_{36}$O$_3$
  4669 - Dromostanolone propanoate
C$_{23}$H$_{38}$O$_2$
  8266 - 24-Norcholan-23-oic acid, (5β)
C$_{23}$H$_{40}$
  5655 - Heptadecylbenzene
C$_{23}$H$_{44}$O$_2$
  7226 - Methyl cis-13-docosenoate
C$_{23}$H$_{46}$O
  10301 - 12-Tricosanone
C$_{23}$H$_{46}$O$_2$
  7225 - Methyl docosanoate

  8686 - Pentyl stearate
C$_{23}$H$_{48}$
  10300 - Tricosane
C$_{24}$H$_{12}$
  2465 - Coronene
C$_{24}$H$_{12}$O$_2$
  2420 - C.I. Vat Yellow 4
C$_{24}$H$_{14}$
  2885 - Dibenzo[a,e]pyrene
  2886 - Dibenzo[a,h]pyrene
  2887 - Dibenzo[a,i]pyrene
  2888 - Dibenzo[a,l]pyrene
C$_{24}$H$_{16}$As$_2$O$_3$
  8487 - 10,10'-Oxydiphenoxarsine
C$_{24}$H$_{16}$N$_2$
  4508 - 4,7-Diphenyl-1,10-phenanthroline
C$_{24}$H$_{18}$
  8999 - 5'-Phenyl-1,1':3',1''-terphenyl
  9406 - 1,1':4',1'':4'',1'''-Quaterphenyl
C$_{24}$H$_{20}$BNa
  9576 - Sodium tetraphenylborate
C$_{24}$H$_{20}$Ge
  9990 - Tetraphenylgermane
C$_{24}$H$_{20}$N$_2$
  4462 - N,N'-Diphenyl-[1,1'-biphenyl]-4,4'-
    diamine
C$_{24}$H$_{20}$N$_4$O
  9515 - Scarlet red
C$_{24}$H$_{20}$O$_4$Si
  9984 - Tetraphenoxysilane
C$_{24}$H$_{20}$O$_6$
  9158 - 1,2,3-Propanetriol tribenzoate
C$_{24}$H$_{20}$Pb
  9993 - Tetraphenylplumbane
C$_{24}$H$_{20}$Si
  9994 - Tetraphenylsilane
C$_{24}$H$_{20}$Sn
  9995 - Tetraphenylstannane
C$_{24}$H$_{27}$O$_4$P
  4244 - 3,4-Dimethylphenol phosphate (3:1)
  10684 - Tris(2,4-dimethylphenyl) phosphate
  10685 - Tris(2,5-dimethylphenyl) phosphate
  10686 - Tris(2,6-dimethylphenyl) phosphate
  10687 - Tris(3,5-dimethylphenyl) phosphate
C$_{24}$H$_{28}$ClN$_3$
  7860 - Methyl Violet
C$_{24}$H$_{28}$O$_4$
  3525 - trans-Diethylstilbestrol dipropanoate
C$_{24}$H$_{29}$NO$_4$
  4804 - Ethaverine
C$_{24}$H$_{30}$F$_2$O$_6$
  5340 - Fluocinolone acetonide
C$_{24}$H$_{31}$Cl$_2$N$_5$O$_7$
  2537 - Cyclochlorotine
C$_{24}$H$_{31}$N$_3$OS
  1438 - Butaperazine
C$_{24}$H$_{32}$O$_4$S
  9588 - Spironolactone
C$_{24}$H$_{33}$FO$_6$
  5437 - Flurandrenolide
C$_{24}$H$_{34}$N$_2$O$_3$
  991 - Bis[4-(hexyloxy)phenyl]diazene, 1-oxide
C$_{24}$H$_{34}$O$_5$
  10641 - 3,7,12-Trioxocholan-24-oic acid, (5β)
C$_{24}$H$_{34}$O$_9$
  10716 - T-2 Toxin

C$_{24}$H$_{36}$O$_5$
  6629 - Lovastatin
C$_{24}$H$_{38}$O$_4$
  986 - Bis(2-ethylhexyl) phthalate
  989 - Bis(2-ethylhexyl) terephthalate
  3788 - Diisooctyl phthalate
  4422 - Dioctyl phthalate
  4425 - Dioctyl terephthalate
C$_{24}$H$_{39}$NO$_4$
  1791 - Cassaine
C$_{24}$H$_{39}$NO$_5$
  4748 - Erythrophleine
C$_{24}$H$_{40}$N$_2$
  2453 - Conessine
C$_{24}$H$_{40}$N$_8$O$_4$
  4568 - Dipyridamole
C$_{24}$H$_{40}$O$_2$
  2361 - Cholan-24-oic acid
  8998 - Phenyl stearate
C$_{24}$H$_{40}$O$_3$
  6016 - 3-Hydroxycholan-24-oic acid, (3α,5β)
C$_{24}$H$_{40}$O$_4$
  2800 - Deoxycholic acid
  3715 - 3,6-Dihydroxycholan-24-oic acid,
    (3α,5β,6α)
  3716 - 3,7-Dihydroxycholan-24-oic acid,
    (3α,5β,7β)
  3717 - 3,7-Dihydroxycholan-24-oic acid,
    (3α,5β,7α)
C$_{24}$H$_{40}$O$_5$
  2377 - Cholic acid
C$_{24}$H$_{40}$O$_8$
  9080 - Piprotal
C$_{24}$H$_{41}$NO$_2$
  6162 - N-(4-Hydroxyphenyl)octadecanamide
C$_{24}$H$_{42}$
  2360 - Cholane
  8321 - Octadecylbenzene
C$_{24}$H$_{42}$O$_{21}$
  6669 - Maltotetraose
C$_{24}$H$_{44}$O$_6$
  9584 - Sorbitan oleate
C$_{24}$H$_{46}$O$_2$
  9782 - cis-15-Tetracosenoic acid
C$_{24}$H$_{46}$O$_3$
  4630 - Dodecanoic anhydride
C$_{24}$H$_{46}$O$_4$
  3341 - Didodecanoyl peroxide
  4413 - Dinonyl adipate
C$_{24}$H$_{48}$
  8322 - Octadecylcyclohexane
C$_{24}$H$_{48}$O$_2$
  9780 - Tetracosanoic acid
C$_{24}$H$_{50}$
  9779 - Tetracosane
C$_{24}$H$_{50}$O
  9781 - 1-Tetracosanol
C$_{24}$H$_{51}$Al
  10635 - Trioctylaluminum
C$_{24}$H$_{51}$N
  3342 - Didodecylamine
  10636 - Trioctylamine
C$_{24}$H$_{51}$OP
  10637 - Trioctylphosphine oxide
C$_{24}$H$_{51}$O$_4$P
  3343 - Didodecyl phosphate

10688 - Tris(2-ethylhexyl) phosphate

$C_{24}H_{52}O_4Si$

9814 - Tetra(2-ethylbutyl) silicate

$C_{24}H_{54}OSn_2$

5721 - Hexabutyldistannoxane

$C_{24}H_{72}O_{10}Si_{11}$

9778 - Tetracosamethylundecasiloxane

$C_{25}H_{20}$

9991 - Tetraphenylmethane

$C_{25}H_{20}N_2$

9012 - Phenyl(triphenylmethyl)diazene

$C_{25}H_{22}ClNO_3$

5324 - Fenvalerate

$C_{25}H_{24}F_6N_4$

5930 - Hydramethylnon

$C_{25}H_{27}ClN_2$

6687 - Meclizine

$C_{25}H_{28}O_3$

4760 - Estra-1,3,5(10)-triene-3,17-diol 3-benzoate, (17β)

$C_{25}H_{30}ClNO_5$

2422 - Clemastine fumarate

$C_{25}H_{30}ClN_3$

2488 - Crystal Violet

$C_{25}H_{30}O_4$

1059 - Bixin

$C_{25}H_{31}N_3$

10683 - Tris(4-dimethylaminophenyl)methane

$C_{25}H_{34}O_3$

4651 - 4-Dodecyloxy-2-hydroxybenzophenone

$C_{25}H_{36}O_2$

922 - Bis(3-*tert*-butyl-5-ethyl-2-hydroxyphenyl)methane

$C_{25}H_{41}NO_9$

110 - Aconine

$C_{25}H_{44}$

8225 - Nonadecylbenzene

$C_{25}H_{48}N_6O_8$

2811 - Desferrioxamine

$C_{25}H_{48}O_4$

981 - Bis(2-ethylhexyl) azelate

$C_{25}H_{52}$

8547 - Pentacosane

$C_{26}H_{16}$

2882 - Dibenzo[b,k]chrysene

5722 - Hexacene

$C_{26}H_{18}$

4457 - 9,10-Diphenylanthracene

$C_{26}H_{20}$

9989 - 1,1,2,2-Tetraphenylethene

$C_{26}H_{20}N_2$

4048 - 2,9-Dimethyl-4,7-diphenyl-1,10-phenanthroline

4355 - *N,N*'-Di-2-naphthyl-1,4-benzenediamine

$C_{26}H_{20}N_2O_2$

1029 - 1,4-Bis(4-methyl-5-phenyloxazol-2-yl)benzene

$C_{26}H_{20}N_4Na_2O_8S_2$

1072 - Brilliant Yellow

$C_{26}H_{22}$

9987 - 1,1,2,2-Tetraphenylethane

$C_{26}H_{22}ClF_3N_2O_3$

5441 - Fluvalinate

$C_{26}H_{22}N_4O_4$

2407 - C.I. Pigment Red 170

$C_{26}H_{22}O_2$

9988 - 1,1,2,2-Tetraphenyl-1,2-ethanediol

$C_{26}H_{23}F_2NO_4$

5337 - Flucythrinate

$C_{26}H_{24}P_2$

973 - 1,2-Bis(diphenylphosphino)ethane

$C_{26}H_{28}ClNO$

2429 - Clomiphene

$C_{26}H_{28}ClN_3$

9405 - Pyrvinium chloride

$C_{26}H_{29}NO$

9676 - Tamoxifen

$C_{26}H_{30}O_{11}$

9490 - Rubratoxin B

$C_{26}H_{36}O_6$

1387 - Bufotalin

$C_{26}H_{38}O_4$

6637 - Lupulon

$C_{26}H_{41}NO$

6694 - Melinamide

$C_{26}H_{42}O_4$

3786 - Diisononyl phthalate

4415 - Dinonyl phthalate

$C_{26}H_{43}NO_6$

5580 - Glycocholic acid

$C_{26}H_{45}NO_7S$

9682 - Taurocholic acid

$C_{26}H_{50}O_4$

987 - Bis(2-ethylhexyl) sebacate

4423 - Dioctyl sebacate

4778 - 1,2-Ethanediol, didodecanoate

$C_{26}H_{52}O_2$

5742 - Hexacosanoic acid

$C_{26}H_{52}O_6$

9819 - Tetraethylene glycol monostearate

$C_{26}H_{54}$

1562 - 5-Butyldocosane

1563 - 11-Butyldocosane

5741 - Hexacosane

$C_{26}H_{54}O$

5743 - 1-Hexacosanol

$C_{27}H_{18}O_2$

7979 - *p*-Naphtholbenzein

$C_{27}H_{20}O_{12}$

2451 - Collinomycin

$C_{27}H_{28}Br_2O_5S$

1358 - Bromothymol Blue

$C_{27}H_{29}NO_{10}$

2736 - Daunorubicin

$C_{27}H_{29}NO_{11}$

4665 - Doxorubicin

$C_{27}H_{30}ClNO_{11}$

4666 - Doxorubicin hydrochloride

$C_{27}H_{30}O_5S$

10096 - Thymol Blue

$C_{27}H_{31}N_2NaO_6S_2$

9665 - Sulphan Blue

$C_{27}H_{32}O_{14}$

8001 - Naringin

$C_{27}H_{33}N_9O_{15}P_2$

5332 - Flavine adenine dinucleotide

$C_{27}H_{34}N_2O_4S$

1071 - Brilliant Green

$C_{27}H_{34}O_3$

8279 - 19-Nortestosterone phenylpropionate

$C_{27}H_{35}BrClN_3$

7280 - Methyl Green

$C_{27}H_{38}N_2O_4$

10776 - Verapamil

$C_{27}H_{39}NO_2$

10777 - Veratramine

$C_{27}H_{39}NO_3$

6564 - Jervine

$C_{27}H_{40}O_3$

5763 - Hexadecyl 3-hydroxy-2-naphthalenecarboxylate

$C_{27}H_{41}NO_2$

2654 - Cyclopamine

$C_{27}H_{42}ClNO_2$

678 - Benzethonium chloride

$C_{27}H_{42}Cl_2N_2O_6$

1817 - Chloramphenicol palmitate

$C_{27}H_{42}O_3$

9597 - Spirost-5-en-3-ol, (3β,25R)

$C_{27}H_{42}O_4$

6193 - 3-Hydroxyspirostan-12-one, (3β,5α,25R)

$C_{27}H_{43}NO_2$

9489 - Rubijervine

9579 - Solanid-5-ene-3,18-diol, (3β)

9591 - Spirosol-5-en-3-ol, (3β,22α,25R)

$C_{27}H_{43}NO_8$

5531 - Germine

$C_{27}H_{43}NO_9$

9281 - Protoverine

$C_{27}H_{44}$

2362 - Cholesta-3,5-diene

$C_{27}H_{44}O$

2363 - Cholesta-5,7-dien-3-ol, (3β)

2364 - Cholesta-8,24-dien-3-ol, (3β,5α)

2375 - Cholest-4-en-3-one

10835 - Vitamin D3

$C_{27}H_{44}O_3$

3718 - 1,25-Dihydroxycholecalciferol

9593 - Spirostan-3-ol, (3β,5α,25R)

9594 - Spirostan-3-ol, (3β,5β,25R)

9595 - Spirostan-3-ol, (3β,5β,25S)

$C_{27}H_{44}O_4$

9592 - Spirostan-2,3-diol, (2α,3β,5α,25R)

$C_{27}H_{44}O_5$

9596 - Spirostan-2,3,15-triol, (2α,3β,5α,15β,25R)

$C_{27}H_{45}Cl$

1957 - 3-Chlorocholest-5-ene, (3β)

$C_{27}H_{45}NO_2$

9590 - Spirosolan-3-ol, (3β,5α,22β,25S)

$C_{27}H_{46}O$

2369 - Cholest-4-en-3-ol, (3β)

2370 - Cholest-5-en-3-ol, (3α)

2376 - Cholesterol

$C_{27}H_{46}O_2$

10106 - δ-Tocopherol

$C_{27}H_{48}$

2365 - Cholestane, (5α)

2366 - Cholestane, (5β)

$C_{27}H_{48}O$

2367 - Cholestanol

2368 - Cholestan-3-ol, (3α,5α)

$C_{27}H_{50}O_6$

9161 - 1,2,3-Propanetriyl octanoate

$C_{27}H_{56}$
5645 - Heptacosane

$C_{28}H_{12}Cl_2N_2O_4$
2419 - C.I. Vat Blue 6

$C_{28}H_{14}N_2O_4$
3593 - 6,15-Dihydro-5,9,14,18-anthrazine-tetrone

$C_{28}H_{18}$
851 - 9,9'-Bianthracene

$C_{28}H_{19}N_5Na_2O_6S_4$
2421 - Clayton Yellow

$C_{28}H_{22}$
9985 - 1,1,4,4-Tetraphenyl-1,3-butadiene

$C_{28}H_{30}O_4$
10098 - Thymolphthalein

$C_{28}H_{31}FN_4O$
527 - Astemizole

$C_{28}H_{32}ClN_2O_3$
9474 - Rhodamine B

$C_{28}H_{32}O_4Si_4$
9977 - 2,4,6,8-Tetramethyl-2,4,6,8-tetraphenyl-cyclotetrasiloxane

$C_{28}H_{33}NO_7$
2727 - Cytochalasin E

$C_{28}H_{34}O_{15}$
5716 - Hesperidin

$C_{28}H_{38}N_2O_4$
10488 - 7',10,11-Trimethoxyemetan-6'-ol

$C_{28}H_{38}O_{19}$
9633 - Sucrose octaacetate

$C_{28}H_{42}O$
4730 - Ergosta-5,7,9(11),22-tetraen-3-ol, (3β,22E)

$C_{28}H_{44}O$
4731 - Ergosta-5,7,22-trien-3-ol, (3β,22E)
4732 - Ergosta-5,7,22-trien-3-ol, (3β,10α,22E)
4733 - Ergosta-5,7,22-trien-3-ol, (3β,9β,10α,22E)
9668 - Suprasterol II
9673 - Tachysterol
10834 - Vitamin D2

$C_{28}H_{46}O$
3679 - Dihydrotachysterol

$C_{28}H_{46}O_4$
3337 - Didecyl phthalate
3785 - Diisodecyl phthalate

$C_{28}H_{48}O$
1705 - Calusterone
4734 - Ergost-5-en-3-ol, (3β,24R)
4735 - Ergost-7-en-3-ol, (3β,5α)
4736 - Ergost-8(14)-en-3-ol, (3β,5α)

$C_{28}H_{48}O_2$
10104 - β-Tocopherol
10105 - γ-Tocopherol

$C_{28}H_{50}$
4727 - Ergostane, (5α)
4728 - Ergostane, (5β)

$C_{28}H_{50}N_2O_4$
1784 - Carpaine

$C_{28}H_{50}O$
4729 - Ergostan-3-ol, (3β,5α)

$C_{28}H_{52}O_{12}$
9632 - Sucrose monohexadecanoate

$C_{28}H_{54}O_3$
9796 - Tetradecanoic anhydride

$C_{28}H_{56}O_2$
8298 - Octacosanoic acid

$C_{28}H_{58}$
8297 - Octacosane

$C_{28}H_{58}O$
8299 - 1-Octacosanol

$C_{29}H_{20}O$
9986 - 2,3,4,5-Tetraphenyl-2,4-cyclopentadien-1-one

$C_{29}H_{32}O_{13}$
5295 - Etoposide

$C_{29}H_{35}NO_2$
7873 - Mifepristone

$C_{29}H_{37}NO_5$
2725 - Cytochalasin B

$C_{29}H_{38}N_4O_6S$
8523 - Penicillin G procaine

$C_{29}H_{40}N_2O_4$
4695 - Emetine

$C_{29}H_{40}O_9$
1693 - Calactin
1704 - Calotropin

$C_{29}H_{40}O_{10}$
1703 - Calotoxin

$C_{29}H_{42}O_{10}$
2459 - Convallatoxin

$C_{29}H_{44}O_2$
7233 - 4,4'-Methylenebis[2,6-di-*tert*-butylphenol]

$C_{29}H_{44}O_{12}$
8422 - Ouabain

$C_{29}H_{47}NO_3$
7909 - Muldamine

$C_{29}H_{47}NO_8$
9495 - Sabadine

$C_{29}H_{48}O$
9607 - Stigmasta-5,7-dien-3-ol, (3β)
9608 - Stigmasta-5,22-dien-3-ol, (3β,22E)

$C_{29}H_{48}O_2$
2371 - Cholest-5-en-3-ol (3β), acetate

$C_{29}H_{50}O$
9610 - Stigmast-5-en-3-ol, (3β,24R)
9611 - Stigmast-5-en-3-ol, (3β,24S)

$C_{29}H_{50}O_2$
10836 - Vitamin E

$C_{29}H_{52}O$
9609 - Stigmastan-3-ol, (3β,5α)

$C_{29}H_{60}$
8217 - Nonacosane

$C_{30}H_{14}O_2$
9301 - 8,16-Pyranthrenedione

$C_{30}H_{22}O_{12}$
6639 - Luteoskyrin

$C_{30}H_{23}BrO_4$
1075 - Bromadiolone

$C_{30}H_{34}O_{13}$
9038 - Picrotoxin

$C_{30}H_{37}NO_6$
2726 - Cytochalasin D

$C_{30}H_{37}N_5O_5$
4726 - Ergosine

$C_{30}H_{39}O_4P$
1647 - 4-*tert*-Butylphenol, phosphate (3:1)

$C_{30}H_{40}O_2$
2412 - β-Citraurin

$C_{30}H_{44}N_6O_8S$
4756 - Eserine sulfate

$C_{30}H_{46}O_5$
9409 - Quillaic acid
9442 - Quinovic acid

$C_{30}H_{48}O_3$
8409 - Oleanolic acid
10764 - Ursolic acid

$C_{30}H_{48}O_4$
5629 - Hederagenin

$C_{30}H_{50}$
9600 - Squalene

$C_{30}H_{50}O$
6584 - Lanosta-8,24-dien-3-ol, (3β)
6636 - Lup-20(29)-en-3-ol, (3β)
8408 - Olean-12-en-3-ol, (3β)
9543 - α₁-Sitosterol
10763 - Urs-12-en-3-ol, (3β)

$C_{30}H_{50}O_2$
6635 - Lup-20(29)-ene-3,28-diol, (3β)

$C_{30}H_{50}O_4$
4598 - Diundecyl phthalate

$C_{30}H_{52}O_{26}$
6666 - Maltopentaose

$C_{30}H_{58}O_4$
4784 - 1,2-Ethanediol, ditetradecanoate

$C_{30}H_{58}O_4S$
10048 - 3,3'-Thiobispropanoic acid, didodecyl ester

$C_{30}H_{60}I_3N_3O_3$
5520 - Gallamine triethiodide

$C_{30}H_{60}O_2$
10130 - Triacontanoic acid

$C_{30}H_{62}$
5826 - 2,6,10,15,19,23-Hexamethyltetracosane
10129 - Triacontane

$C_{30}H_{62}O$
10131 - 1-Triacontanol

$C_{30}H_{63}Al$
10320 - Tridecylaluminum

$C_{30}H_{63}N$
10321 - Tridecylamine

$C_{30}H_{63}O_3P$
10326 - Tri(decyl) phosphite

$C_{31}H_{23}BrO_3$
1073 - Brodifacoum

$C_{31}H_{32}N_2O_{13}S$
10860 - Xylenol orange

$C_{31}H_{36}N_2O_8$
9451 - Raunescine

$C_{31}H_{36}N_2O_{11}$
8284 - Novobiocin

$C_{31}H_{39}N_5O_5$
4718 - Ergocornine
4719 - Ergocorninine

$C_{31}H_{42}ClN_3$
5283 - Ethyl Violet

$C_{31}H_{46}O_2$
10838 - Vitamin K1

$C_{31}H_{52}O_3$
10837 - Vitamin E acetate

$C_{31}H_{64}$
2784 - 11-Decylheneicosane
5639 - Hentriacontane

$C_{32}H_{16}CuN_8$
2464 - Copper phthalocyanine

$C_{32}H_{18}N_8$
9031 - 29H,31H-Phthalocyanine

$C_{32}H_{20}N_6Na_4O_{14}S_4$
2390 - C.I. Direct Blue 6, tetrasodium salt

$C_{32}H_{22}N_6Na_2O_6S_2$
2454 - Congo Red

$C_{32}H_{26}Cl_2N_6O_4$
2409 - C.I. Pigment Yellow 12

$C_{32}H_{32}N_2O_{12}$
2481 - o-Cresolphthalein complexone

$C_{32}H_{32}O_{13}S$
9687 - Teniposide

$C_{32}H_{38}N_2O_8$
2809 - Deserpidine

$C_{32}H_{38}N_4$
5292 - Etioporphyrin

$C_{32}H_{41}NO_2$
9695 - Terfenadine

$C_{32}H_{41}N_5O_5$
4722 - Ergocryptine

$C_{32}H_{44}N_2O_8$
6589 - Lappaconitine

$C_{32}H_{44}O_{12}$
9517 - Scilliroside

$C_{32}H_{46}O_8$
2490 - Cucurbitacin B

$C_{32}H_{48}O_8$
2491 - Cucurbitacin C

$C_{32}H_{48}O_9$
8407 - Oleandrin

$C_{32}H_{49}NO_9$
1806 - Cevadine

$C_{32}H_{52}Br_2N_4O_4$
2794 - Demecarium bromide

$C_{32}H_{54}O_4$
3344 - Didodecyl phthalate

$C_{32}H_{62}O_3$
5752 - Hexadecanoic anhydride

$C_{32}H_{62}O_4S$
1049 - Bis(tridecyl) thiodipropanoate

$C_{32}H_{64}O_2$
5762 - Hexadecyl hexadecanoate

$C_{32}H_{64}O_4Sn$
3060 - Dibutyltin dilaurate

$C_{32}H_{66}$
4663 - Dotriacontane

$C_{32}H_{68}S_2Sn$
3009 - Dibutylbis(dodecylthio)stannane

$C_{32}H_{72}N_2O_4S$
9725 - Tetrabutylammonium sulfate

$C_{33}H_{25}N_3O_3$
8262 - Norbormide

$C_{33}H_{34}N_4O_6$
869 - Biliverdine

$C_{33}H_{35}N_5O_5$
4739 - Ergotaminine

$C_{33}H_{36}N_4O_6$
868 - Bilirubin

$C_{33}H_{40}N_2O_9$
9456 - Reserpine

$C_{33}H_{41}N_5O_5$
4723 - Ergocryptinine

$C_{33}H_{44}O_8$
5632 - Helvolic acid

$C_{33}H_{45}NO_9$
2792 - Delphinine

$C_{33}H_{49}NO_7$
10778 - Veratramine, 3-glucoside

$C_{33}H_{49}NO_8$
9286 - Pseudojervine

$C_{33}H_{51}NO_7$
2687 - Cycloposine

$C_{33}H_{55}N_5O_5$
4737 - Ergotamine

$C_{33}H_{62}O_6$
5568 - Glycerol tridecanoate

$C_{33}H_{68}$
10709 - Tritriacontane

$C_{34}H_{16}O_2$
477 - Anthra[9,1,2-cde]benzo[rst]pentaphene-5,10-dione

$C_{34}H_{24}N_6Na_4O_{14}S_4$
5299 - Evan's Blue
10712 - Trypan blue

$C_{34}H_{26}N_6Na_2O_6S_2$
723 - Benzopurpurine 4B

$C_{34}H_{32}ClFeN_4O_4$
5637 - Hemin

$C_{34}H_{33}FeN_4O_5$
5634 - Hematin

$C_{34}H_{36}N_2O_6$
9287 - Pseudomorphine

$C_{34}H_{38}N_4O_6$
5635 - Hematoporphyrin

$C_{34}H_{47}NO_{10}$
6229 - Indaconitine

$C_{34}H_{47}NO_{11}$
111 - Aconitine

$C_{34}H_{50}O_2$
2372 - Cholest-5-en-3-ol (3β), benzoate

$C_{34}H_{56}N_2O_{12}$
7866 - Metoprolol tartrate

$C_{34}H_{58}O_4$
4597 - Ditridecyl phthalate

$C_{34}H_{66}O_4$
4780 - 1,2-Ethanediol, dihexadecanoate

$C_{34}H_{68}O_2$
5769 - Hexadecyl stearate

$C_{34}H_{70}$
10004 - Tetratriacontane

$C_{35}H_{38}N_5O_8$
4738 - Ergotamine tartrate (2:1)

$C_{35}H_{39}N_5O_5$
4720 - Ergocristine
4721 - Ergocristinine

$C_{35}H_{42}N_2O_9$
9454 - Rescinnamine

$C_{35}H_{47}NO_{10}$
9683 - Taxine A

$C_{35}H_{48}N_8O_{10}S$
8739 - Phalloin

$C_{35}H_{48}N_8O_{11}S$
8738 - Phalloidin

$C_{35}H_{52}O_5$
6585 - Lantadene A
6586 - Lantadene B

$C_{35}H_{54}O_{14}$
10765 - Uzarin

$C_{35}H_{60}Br_2N_2O_4$
8499 - Pancuronium dibromide

$C_{35}H_{62}O_3$
8323 - Octadecyl 3-(3,5-di-$tert$-butyl-4-hydroxyphenyl)propanoate

$C_{35}H_{70}O$
8636 - 18-Pentatriacontanone

$C_{35}H_{72}$
8635 - Pentatriacontane

$C_{36}H_{38}N_2O_6$
574 - Bebeerine
2496 - Curine

$C_{36}H_{42}Br_2N_2$
5792 - Hexafluorenium bromide

$C_{36}H_{51}NO_{11}$
867 - Bikhaconitine
10779 - Veratridine

$C_{36}H_{51}NO_{12}$
9284 - Pseudoaconitine

$C_{36}H_{56}O_8$
9800 - 12-O-Tetradecanoylphorbol-13-acetate

$C_{36}H_{60}O_2$
9464 - Retinyl palmitate

$C_{36}H_{60}O_{30}$
2542 - α-Cyclodextrin

$C_{36}H_{66}HgO_4$
6720 - Mercury(II) oleate

$C_{36}H_{70}MgO_4$
6655 - Magnesium stearate

$C_{36}H_{70}O_3$
9605 - Stearic acid anhydride

$C_{36}H_{71}AlO_5$
220 - Aluminum distearate

$C_{36}H_{72}O_2$
8326 - Octadecyl octadecanoate

$C_{36}H_{74}$
5856 - Hexatriacontane

$C_{36}H_{75}N$
4417 - Dioctadecylamine
10329 - Tridodecylamine

$C_{37}H_{35}N_2NaO_6S_2$
5619 - Guinea Green B

$C_{37}H_{38}N_2O_6$
1803 - Cepharanthine

$C_{37}H_{40}N_2O_6$
10486 - 6,6',7-Trimethoxy-2,2'-dimethylberbaman-12-ol
10487 - 6,6',7-Trimethoxy-2,2'-dimethyloxyacanthan-12'-ol

$C_{37}H_{40}N_2O_{13}Na_4S$
7826 - Methylthymol blue, sodium salt

$C_{37}H_{42}Cl_2N_2O_6$
10717 - Tubocurarine dichloride

$C_{37}H_{49}N_7O_9S$
8592 - Pentagastrin

$C_{37}H_{67}NO_{13}$
4745 - Erythromycin

$C_{38}H_{74}O_4$
4783 - 1,2-Ethanediol, distearate

$C_{38}H_{76}N_2O_2$
5020 - N,N'-Ethylene distearylamide

$C_{39}H_{54}N_{10}O_{14}S$
224 - α-Amanitin

$C_{39}H_{72}O_5$
5565 - Glycerol 1,3-di-9-octadecenoate, cis,cis

$C_{39}H_{74}O_6$
5570 - Glycerol trilaurate

$C_{40}H_{50}N_4O_8S$
9420 - Quinine sulfate

$C_{40}H_{52}O_2$
3338 - 3',4'-Didehydro-β,ψ-caroten-16'-oic acid

$C_{40}H_{56}$
1774 - α-Carotene
1775 - β-Carotene
1776 - β,ψ-Carotene
1777 - ψ,ψ-Carotene

$C_{40}H_{56}O$
1780 - β,β-Caroten-3-ol, (3R)
1781 - β,ψ-Caroten-3-ol, (3R)
1782 - ψ,ψ-Caroten-16-ol

$C_{40}H_{56}O_2$
1778 - β,β-Carotene-3,3'-diol, (3R,3'R)
1779 - β,ε-Carotene-3,3'-diol, (3R,3'R,6'R)

$C_{40}H_{56}O_3$
1721 - Capsanthin

$C_{40}H_{56}O_4$
10829 - Violaxanthin

$C_{40}H_{82}$
9777 - Tetracontane

$C_{41}H_{64}O_{13}$
3576 - Digitoxin

$C_{41}H_{64}O_{14}$
3581 - Digoxin
5534 - Gitoxin

$C_{42}H_{28}$
9992 - 5,6,11,12-Tetraphenylnaphthacene

$C_{42}H_{46}N_4O_8S$
9619 - Strychnidin-10-one sulfate (2:1)

$C_{42}H_{58}O_6$
5473 - Fucoxanthin

$C_{42}H_{62}O_{16}$
5594 - Glycyrrhizic acid

$C_{42}H_{70}O_{35}$
2543 - β-Cyclodextrin

$C_{42}H_{82}O_4S$
4576 - Distearyl thiodipropionate

$C_{42}H_{84}N_{14}O_{36}S_3$
9615 - Streptomycin sulfate

$C_{43}H_{47}N_2NaO_6S_2$
6241 - Indocyanine green

$C_{43}H_{58}N_4O_{12}$
9485 - Rifampin

$C_{43}H_{66}N_{12}O_{12}S_2$
8495 - Oxytocin

$C_{43}H_{75}NO_{16}$
4746 - Erythromycin ethyl succinate

$C_{43}H_{76}O_2$
2373 - Cholest-5-en-3-ol (3β)-, hexadecanoate

$C_{44}H_{84}CaI_2O_4$
1699 - Calcium iodobehenate

$C_{44}H_{90}$
10003 - Tetratetracontane

$C_{45}H_{73}NO_{15}$
9580 - Solanine

$C_{45}H_{74}O$
850 - Betulaprenol 9

$C_{45}H_{78}O_2$
2374 - Cholest-5-en-3-ol (3β)-, cis-9-octadecenoate

$C_{45}H_{86}O_6$
5575 - Glycerol tritetradecanoate

$C_{46}H_{56}N_4O_9$
6605 - Leurosine

$C_{46}H_{56}N_4O_{10}$
10790 - Vincristine

$C_{46}H_{58}N_4O_9$
10787 - Vinblastine

$C_{46}H_{62}N_4O_{11}$
9484 - Rifabutin

$C_{46}H_{64}O_2$
6696 - Menaquinone 7

$C_{46}H_{68}N_4O_{19}S$
1385 - Brucine sulfate heptahydrate

$C_{47}H_{51}NO_{14}$
9684 - Taxol

$C_{47}H_{73}NO_{17}$
446 - Amphotericin B

$C_{48}H_{40}O_4Si_4$
8364 - Octaphenylcyclotetrasiloxane

$C_{48}H_{72}O_{14}$
537 - Avermectin B1a

$C_{48}H_{80}O_{40}$
2544 - γ-Cyclodextrin

$C_{50}H_{83}NO_{21}$
10123 - Tomatine

$C_{50}H_{102}$
8546 - Pentacontane

$C_{51}H_{98}O_6$
5573 - Glycerol tripalmitate

$C_{52}H_{76}O_{24}$
7880 - Mithramycin

$C_{53}H_{100}N_{16}O_{13}$
2450 - Colistin A

$C_{54}H_{90}N_6O_{18}$
10769 - Valinomycin

$C_{54}H_{90}O_{45}$
6668 - 6-O-α-Maltosyl-β-cyclodextrin

$C_{55}H_{46}OP_3Rh$
10706 - Tris(triphenylphosphine) rhodium carbonyl hydride

$C_{55}H_{70}MgN_4O_6$
2257 - β-Chlorophyll

$C_{55}H_{72}MgN_4O_5$
2256 - α-Chlorophyll

$C_{55}H_{103}NO_{15}$
4747 - Erythromycin stearate

$C_{56}H_{92}O_{29}$
3574 - Digitonin

$C_{57}H_{104}O_6$
5569 - Glycerol trielaidate
5572 - Glycerol trioleate

$C_{57}H_{110}O_6$
5574 - Glycerol tristearate

$C_{59}H_{84}N_{16}O_{12}$
6604 - Leuprolide

$C_{60}H_{78}OSn_2$
5313 - Fenbutatin oxide

$C_{60}H_{122}$
5740 - Hexacontane

$C_{62}H_{86}N_{12}O_{16}$
2730 - Dactinomycin

$C_{62}H_{89}CoN_{13}O_{15}P$
5959 - Hydroxocobalamin

$C_{63}H_{88}CoN_{14}O_{14}P$
10833 - Vitamin B12

$C_{70}H_{142}$
5644 - Heptacontane

$C_{76}H_{52}O_{46}$
9677 - Tannic acid

$C_{80}H_{162}$
8296 - Octacontane

$C_{90}H_{182}$
8216 - Nonacontane

$C_{100}H_{202}$
5628 - Hectane

CAS	No.	CAS	No.	CAS	No.	CAS	No.	CAS	No.
50-00-0	: 5448	51-20-7	: 1349	54-04-6	: 10480	56-84-8	: 525	58-25-3	: 1971
50-01-1	: 5609	51-21-8	: 5432	54-05-7	: 2295	56-85-9	: 5556	58-27-5	: 7475
50-02-2	: 2816	51-26-3	: 10452	54-06-8	: 3626	56-86-0	: 5554	58-32-2	: 4568
50-03-3	: 5948	51-28-5	: 4393	54-11-5	: 8025	56-87-1	: 6648	58-33-3	: 9128
50-06-6	: 8761	51-34-3	: 9518	54-25-1	: 546	56-89-3	: 2718	58-36-6	: 8487
50-07-7	: 7883	51-35-4	: 6178	54-31-9	: 5507	56-92-8	: 6217	58-38-8	: 9117
50-10-2	: 8493	51-36-5	: 3113	54-35-3	: 8523	56-93-9	: 838	58-39-9	: 8734
50-11-3	: 6753	51-41-2	: 8270	54-36-4	: 7224	56-94-0	: 2794	58-40-2	: 9125
50-12-4	: 6702	51-43-4	: 4708	54-47-7	: 9373	57-00-1	: 2475	58-46-8	: 9709
50-14-6	: 10834	51-44-5	: 3112	54-49-9	: 6730	57-04-5	: 5576	58-54-8	: 4764
50-18-0	: 2686	51-45-6	: 5917	54-62-6	: 414	57-06-7	: 194	58-55-9	: 10018
50-21-5	: 6575	51-48-9	: 10099	54-71-7	: 9040	57-08-9	: 51	58-56-0	: 9376
50-22-6	: 2466	51-50-3	: 2037	54-85-3	: 6440	57-10-3	: 5751	58-61-7	: 122
50-23-7	: 5947	51-52-5	: 3668	54-91-1	: 9079	57-11-4	: 9604	58-63-9	: 6260
50-24-8	: 9100	51-55-8	: 532	54-95-5	: 9894	57-12-5	: 4809	58-72-0	: 10650
50-27-1	: 4761	51-56-9	: 5921	54-96-6	: 9337	57-12-5	: 4810	58-73-1	: 4503
50-28-2	: 4758	51-57-0	: 6739	55-10-7	: 3726	57-13-6	: 10756	58-74-2	: 8505
50-29-3	: 10212	51-61-6	: 4661	55-18-5	: 8186	57-14-7	: 4106	58-85-5	: 874
50-30-6	: 3111	51-63-8	: 2819	55-21-0	: 591	57-15-8	: 10258	58-86-6	: 10863
50-31-7	: 10209	51-64-9	: 8975	55-22-1	: 9331	57-22-7	: 10790	58-89-9	: 5730
50-32-8	: 729	51-66-1	: 6865	55-38-9	: 5323	57-24-9	: 9620	58-90-2	: 9763
50-33-9	: 8856	51-67-2	: 315	55-43-6	: 2038	57-27-2	: 7899	58-93-5	: 5943
50-34-0	: 9172	51-75-2	: 1949	55-63-0	: 10629	57-37-4	: 576	58-94-6	: 2311
50-35-1	: 10011	51-79-6	: 4937	55-65-2	: 5607	57-41-0	: 9016	58-95-7	: 10837
50-36-2	: 2440	51-80-9	: 9957	55-68-5	: 8914	57-42-1	: 6701	58-96-8	: 10760
50-37-3	: 6645	51-83-2	: 1725	55-73-2	: 848	57-43-2	: 442	58-97-9	: 10761
50-42-0	: 131	52-01-7	: 9588	55-80-1	: 7201	57-44-3	: 571	59-00-7	: 3754
50-44-2	: 3669	52-24-4	: 10363	55-81-2	: 6789	57-47-6	: 9033	59-01-8	: 6567
50-47-5	: 2812	52-28-8	: 2444	55-86-7	: 930	57-48-7	: 5471	59-02-9	: 10836
50-48-6	: 438	52-31-3	: 2528	55-91-4	: 6432	57-50-1	: 9631	59-05-2	: 6765
50-49-7	: 6226	52-39-1	: 148	55-98-1	: 1414	57-53-4	: 7719	59-23-4	: 5516
50-50-0	: 4760	52-43-7	: 2840	56-03-1	: 866	57-55-6	: 9230	59-26-7	: 3518
50-52-2	: 10079	52-46-0	: 496	56-04-2	: 7823	57-56-7	: 5935	59-30-3	: 5442
50-53-3	: 2350	52-49-3	: 10431	56-05-3	: 3287	57-57-8	: 8445	59-31-4	: 9433
50-55-5	: 9456	52-51-7	: 1296	56-12-2	: 271	57-62-5	: 2355	59-40-5	: 426
50-56-6	: 8495	52-52-8	: 290	56-17-7	: 2712	57-63-6	: 4807	59-43-8	: 10025
50-59-9	: 1800	52-53-9	: 10776	56-18-8	: 413	57-66-9	: 4550	59-43-8	: 10026
50-60-2	: 8796	52-67-5	: 6718	56-23-5	: 9756	57-67-0	: 9639	59-46-1	: 3387
50-65-7	: 8019	52-68-6	: 10193	56-25-7	: 1718	57-68-1	: 9642	59-47-2	: 7611
50-67-9	: 6198	52-85-7	: 5302	56-29-1	: 5884	57-71-6	: 1416	59-48-3	: 3634
50-69-1	: 9480	52-86-8	: 5622	56-33-7	: 9943	57-74-9	: 1824	59-49-4	: 751
50-70-4	: 5537	52-89-1	: 2717	56-34-8	: 9810	57-83-0	: 9122	59-50-7	: 2151
50-71-5	: 9383	52-90-4	: 2715	56-35-9	: 5721	57-85-2	: 8476	59-52-9	: 3824
50-76-0	: 2730	53-03-2	: 3746	56-36-0	: 92	57-87-4	: 4731	59-56-3	: 5549
50-78-2	: 86	53-05-4	: 9845	56-38-2	: 8514	57-88-5	: 2376	59-66-5	: 19
50-79-3	: 3110	53-06-5	: 3748	56-40-6	: 5577	57-91-0	: 4757	59-67-6	: 9330
50-81-7	: 518	53-16-7	: 4763	56-41-7	: 144	57-92-1	: 9614	59-87-0	: 8107
50-84-0	: 3109	53-19-0	: 7884	56-45-1	: 9531	57-94-3	: 10717	59-88-1	: 8893
50-85-1	: 6083	53-34-9	: 5436	56-49-5	: 3639	57-95-4	: 10717	59-89-2	: 8193
50-89-5	: 10093	53-39-4	: 8438	56-53-1	: 3524	57-96-5	: 9653	59-92-7	: 6607
50-91-9	: 2802	53-41-8	: 5963	56-54-2	: 9417	57-97-6	: 3948	59-96-1	: 8777
50-98-6	: 4706	53-46-3	: 6751	56-55-3	: 592	58-00-4	: 501	59-97-2	: 3664
50-99-7	: 5546	53-59-8	: 2447	56-57-5	: 8181	58-05-9	: 5443	59-98-3	: 3606
51-03-6	: 9077	53-70-3	: 2876	56-65-5	: 127	58-08-2	: 1692	59-99-4	: 8008
51-06-9	: 9114	53-84-9	: 2446	56-72-4	: 2473	58-14-0	: 2240	60-00-4	: 5018
51-12-7	: 8013	53-86-1	: 6259	56-75-7	: 1816	58-15-1	: 425	60-01-5	: 10191
51-17-2	: 683	53-89-4	: 774	56-81-5	: 5561	58-18-4	: 7794	60-09-3	: 238
51-18-3	: 10678	53-96-3	: 49	56-82-6	: 3749	58-22-0	: 5965	60-10-6	: 4593

75-94-5	: 10296	77-09-8	: 8763	78-31-9	: 2483	79-19-6	: 5933	80-43-3	: 1027
75-96-7	: 10153	77-10-1	: 8866	78-32-0	: 10304	79-20-9	: 6912	80-46-6	: 4287
75-97-8	: 3993	77-15-6	: 4815	78-33-1	: 1647	79-21-0	: 8732	80-47-7	: 6698
75-98-9	: 4277	77-16-7	: 9087	78-34-2	: 4431	79-22-1	: 7125	80-48-8	: 7827
75-99-0	: 3274	77-19-0	: 3331	78-38-6	: 3459	79-24-3	: 8095	80-50-2	: 475
76-00-6	: 10282	77-20-3	: 215	78-39-7	: 10333	79-27-6	: 9711	80-51-3	: 8481
76-01-7	: 8538	77-21-4	: 5560	78-40-0	: 10365	79-28-7	: 9712	80-56-8	: 9047
76-02-8	: 10199	77-23-6	: 1733	78-42-2	: 10688	79-29-8	: 3985	80-58-0	: 1125
76-03-9	: 10196	77-25-8	: 4984	78-43-3	: 3277	79-30-1	: 7695	80-59-1	: 7095
76-04-0	: 1979	77-26-9	: 1397	78-44-4	: 1770	79-31-2	: 7691	80-62-6	: 7438
76-05-1	: 10379	77-28-1	: 1572	78-48-8	: 10188	79-33-4	: 6576	80-63-7	: 7120
76-06-2	: 10261	77-36-1	: 2356	78-50-2	: 10637	79-34-5	: 9752	80-69-3	: 6180
76-08-4	: 10170	77-40-7	: 1013	78-51-3	: 10679	79-35-6	: 3159	80-70-6	: 9948
76-09-5	: 3986	77-41-8	: 4257	78-53-5	: 436	79-36-7	: 3072	80-71-7	: 6087
76-11-9	: 9745	77-42-9	: 9507	78-57-9	: 6697	79-37-8	: 8432	80-72-8	: 3720
76-12-0	: 9746	77-46-3	: 6	78-59-1	: 6456	79-38-9	: 2327	80-77-3	: 1836
76-13-1	: 10295	77-47-4	: 5734	78-62-6	: 3356	79-39-0	: 7697	80-92-2	: 9104
76-14-2	: 3295	77-48-5	: 2934	78-67-1	: 562	79-40-3	: 4792	80-97-7	: 2367
76-15-3	: 2207	77-49-6	: 7500	78-69-3	: 4186	79-41-4	: 6733	81-04-9	: 7959
76-16-4	: 5798	77-50-9	: 9198	78-70-6	: 6612	79-42-5	: 10058	81-05-0	: 366
76-17-5	: 10265	77-50-9	: 9199	78-71-7	: 938	79-43-6	: 3067	81-07-2	: 9496
76-19-7	: 8724	77-52-1	: 10764	78-74-0	: 10163	79-44-7	: 4002	81-08-3	: 747
76-20-0	: 990	77-53-2	: 1794	78-75-1	: 2975	79-46-9	: 8164	81-11-8	: 2866
76-22-2	: 1709	77-58-7	: 3060	78-76-2	: 1122	79-49-2	: 8528	81-13-0	: 2818
76-22-2	: 1710	77-59-8	: 9590	78-77-3	: 1279	79-53-8	: 2205	81-14-1	: 1556
76-22-2	: 1711	77-60-1	: 9593	78-78-4	: 6441	79-54-9	: 6608	81-15-2	: 1560
76-24-4	: 166	77-63-4	: 9977	78-79-5	: 7060	79-55-0	: 8602	81-16-3	: 363
76-28-8	: 9511	77-65-6	: 278	78-80-8	: 7104	79-57-2	: 8494	81-20-9	: 4155
76-30-2	: 3755	77-67-8	: 5157	78-81-9	: 6376	79-58-3	: 9489	81-23-2	: 10641
76-36-8	: 10213	77-71-4	: 4113	78-82-0	: 7687	79-63-0	: 6584	81-24-3	: 9682
76-37-9	: 9840	77-73-6	: 3332	78-83-1	: 7693	79-69-6	: 6354	81-25-4	: 2377
76-38-0	: 3163	77-74-7	: 7569	78-84-2	: 6370	79-70-9	: 6355	81-30-1	: 725
76-39-1	: 7501	77-75-8	: 7600	78-85-3	: 7696	79-74-3	: 971	81-38-9	: 6594
76-42-6	: 8485	77-76-9	: 3883	78-86-4	: 1926	79-77-6	: 6348	81-48-1	: 154
76-43-7	: 5434	77-77-0	: 4607	78-87-5	: 3271	79-81-2	: 9464	81-49-2	: 294
76-44-8	: 5640	77-78-1	: 4323	78-88-6	: 3284	79-83-4	: 8503	81-54-9	: 10434
76-45-9	: 9281	77-79-2	: 3684	78-89-7	: 2268	79-91-4	: 9693	81-61-8	: 9897
76-49-3	: 10517	77-81-6	: 9672	78-90-0	: 9137	79-92-5	: 1706	81-64-1	: 3692
76-54-0	: 3183	77-83-8	: 5151	78-92-2	: 1432	79-92-5	: 1707	81-77-6	: 3593
76-57-3	: 2443	77-85-0	: 6091	78-93-3	: 1433	79-94-7	: 1077	81-81-2	: 10840
76-59-5	: 1358	77-86-1	: 10693	78-94-4	: 1459	79-95-8	: 9740	81-83-4	: 685
76-60-8	: 1156	77-89-4	: 10343	78-95-5	: 1843	79-97-0	: 1010	81-84-5	: 7980
76-61-9	: 10096	77-90-7	: 10178	78-96-6	: 408	80-00-2	: 2253	81-86-7	: 1289
76-62-0	: 9717	77-92-9	: 2414	78-97-7	: 6181	80-05-7	: 1015	81-88-9	: 9474
76-65-3	: 443	77-93-0	: 10353	78-98-8	: 9142	80-06-8	: 944	81-90-3	: 8764
76-67-5	: 3458	77-94-1	: 1539	78-99-9	: 3270	80-07-9	: 947	81-92-5	: 8765
76-68-6	: 2681	77-95-2	: 9416	79-00-5	: 10229	80-08-0	: 908	82-02-0	: 6571
76-74-4	: 5128	77-99-6	: 10575	79-01-6	: 10231	80-09-1	: 1017	82-05-3	: 594
76-80-2	: 9688	78-00-2	: 9823	79-02-7	: 3065	80-10-4	: 3172	82-12-2	: 5812
76-83-5	: 2339	78-01-3	: 3423	79-03-8	: 9170	80-11-5	: 4161	82-28-0	: 342
76-84-6	: 10654	78-04-6	: 3015	79-04-9	: 1847	80-13-7	: 3074	82-33-7	: 2860
76-86-8	: 2340	78-07-9	: 10334	79-05-0	: 9134	80-15-9	: 6479	82-34-8	: 8041
76-87-9	: 10663	78-08-0	: 10827	79-06-1	: 117	80-18-2	: 7009	82-35-9	: 4365
76-89-1	: 7370	78-09-1	: 9808	79-07-2	: 1839	80-26-2	: 9702	82-38-2	: 6922
76-93-7	: 6149	78-10-4	: 5244	79-08-3	: 1079	80-32-0	: 9636	82-39-3	: 6776
76-94-8	: 7658	78-11-5	: 8575	79-09-4	: 9164	80-33-1	: 2235	82-40-6	: 6634
76-99-3	: 3907	78-13-7	: 9814	79-10-7	: 118	80-35-3	: 9645	82-43-9	: 3083
77-02-1	: 503	78-18-2	: 2590	79-11-8	: 1840	80-38-6	: 2233	82-44-0	: 1859
77-03-2	: 3512	78-26-2	: 7718	79-14-1	: 5584	80-39-7	: 5119	82-45-1	: 236
77-04-3	: 3520	78-27-3	: 5287	79-15-2	: 1078	80-40-0	: 5267	82-46-2	: 3082
77-06-5	: 5532	78-28-4	: 4696	79-16-3	: 6910	80-41-1	: 2021	82-49-5	: 3619
77-07-6	: 6609	78-30-8	: 10302	79-17-4	: 5936	80-42-2	: 9214	82-54-2	: 2472

82-57-5	: 10832	84-54-8	: 6950	85-91-6	: 7447	87-58-1	: 9906	88-89-1	: 10631
82-58-6	: 6644	84-55-9	: 10830	85-94-9	: 2721	87-59-2	: 3927	88-95-9	: 617
82-62-2	: 10262	84-58-2	: 3154	85-95-0	: 677	87-60-5	: 2101	88-96-0	: 623
82-66-6	: 4453	84-60-6	: 3695	85-97-2	: 1920	87-61-6	: 10204	88-98-2	: 2597
82-68-8	: 8540	84-61-7	: 3328	85-98-3	: 3415	87-62-7	: 3930	88-99-3	: 9029
82-71-3	: 10625	84-62-8	: 4511	86-00-0	: 8084	87-63-8	: 2100	89-00-9	: 9338
82-75-7	: 368	84-64-0	: 1549	86-08-8	: 101	87-64-9	: 2148	89-01-0	: 9305
82-86-0	: 9	84-65-1	: 487	86-12-4	: 10015	87-65-0	: 3253	89-02-1	: 4374
82-88-2	: 8759	84-66-2	: 3511	86-21-5	: 4256	87-66-1	: 673	89-05-4	: 665
82-92-8	: 2526	84-68-4	: 3106	86-22-6	: 1382	87-68-3	: 5729	89-08-7	: 9662
82-93-9	: 1823	84-69-5	: 3781	86-26-0	: 6811	87-69-4	: 9681	89-24-7	: 8898
83-01-2	: 4471	84-74-2	: 3045	86-28-2	: 4939	87-73-0	: 5536	89-25-8	: 3653
83-07-8	: 448	84-75-3	: 3589	86-29-3	: 8829	87-79-6	: 9585	89-31-6	: 9503
83-12-5	: 8903	84-76-4	: 4415	86-30-6	: 8189	87-81-0	: 9674	89-32-7	: 688
83-13-6	: 3505	84-77-5	: 3337	86-34-0	: 7659	87-82-1	: 5717	89-36-1	: 3651
83-14-7	: 8518	84-79-7	: 6588	86-35-1	: 4817	87-83-2	: 8526	89-37-2	: 4400
83-25-0	: 8992	84-80-0	: 10838	86-39-5	: 2314	87-84-3	: 8525	89-39-4	: 3878
83-26-1	: 9085	84-83-3	: 3689	86-48-6	: 6114	87-85-4	: 5813	89-40-7	: 8119
83-30-7	: 10436	84-85-5	: 6851	86-50-0	: 556	87-86-5	: 8541	89-51-0	: 1764
83-32-9	: 7	84-86-6	: 364	86-52-2	: 2134	87-87-6	: 9736	89-52-1	: 45
83-33-0	: 3630	84-87-7	: 6121	86-53-3	: 7933	87-88-7	: 1818	89-54-3	: 286
83-34-1	: 7388	84-88-8	: 6191	86-54-4	: 5929	87-89-8	: 6262	89-55-4	: 1233
83-38-5	: 3088	84-89-9	: 365	86-55-5	: 7939	87-90-1	: 9670	89-56-5	: 6081
83-40-9	: 6082	84-95-7	: 3491	86-56-6	: 4148	87-91-2	: 3533	89-57-6	: 327
83-41-0	: 4153	84-96-8	: 10470	86-57-7	: 8125	87-92-3	: 3057	89-58-7	: 4157
83-42-1	: 2139	84-97-9	: 8693	86-59-9	: 9431	87-99-0	: 10861	89-59-8	: 2143
83-43-2	: 7681	84-99-1	: 10847	86-60-2	: 367	88-04-0	: 2000	89-60-1	: 2141
83-44-3	: 2800	85-00-7	: 4572	86-65-7	: 359	88-05-1	: 10504	89-61-2	: 3237
83-45-4	: 9609	85-00-7	: 4571	86-68-0	: 6905	88-06-2	: 10271	89-62-3	: 7484
83-46-5	: 9610	85-01-8	: 8743	86-73-7	: 5343	88-09-5	: 4924	89-63-4	: 2175
83-47-6	: 9611	85-02-9	: 731	86-74-8	: 1730	88-10-8	: 3414	89-64-5	: 2192
83-48-7	: 9608	85-19-8	: 2075	86-75-9	: 9440	88-12-0	: 10823	89-65-6	: 4743
83-49-8	: 3715	85-23-4	: 9587	86-80-6	: 3828	88-13-1	: 10072	89-68-9	: 2127
83-53-4	: 2963	85-29-0	: 2236	86-81-7	: 10477	88-14-2	: 5483	89-69-0	: 10260
83-54-5	: 7775	85-32-5	: 5617	86-84-0	: 6416	88-15-3	: 10036	89-71-4	: 7448
83-55-6	: 371	85-34-7	: 10207	86-86-2	: 7929	88-16-4	: 2331	89-72-5	: 1640
83-56-7	: 7951	85-38-1	: 6130	86-87-3	: 7930	88-17-5	: 10400	89-73-6	: 3702
83-60-3	: 9455	85-40-5	: 9881	86-88-4	: 7973	88-18-6	: 1641	89-74-7	: 4249
83-66-9	: 1602	85-41-6	: 6433	86-89-5	: 8677	88-19-7	: 7007	89-75-8	: 3118
83-67-0	: 10017	85-42-7	: 5807	86-93-1	: 3666	88-20-0	: 7010	89-79-2	: 7464
83-72-7	: 6116	85-43-8	: 9863	86-95-3	: 6192	88-21-1	: 251	89-82-7	: 9292
83-73-8	: 3773	85-44-9	: 9030	86-98-6	: 3289	88-24-4	: 922	89-83-8	: 10095
83-74-9	: 6832	85-46-1	: 7967	87-00-3	: 5920	88-27-7	: 3014	89-84-9	: 3742
83-79-4	: 9488	85-47-2	: 7965	87-08-1	: 8524	88-29-9	: 10782	89-86-1	: 3705
83-81-8	: 9813	85-52-9	: 756	87-10-5	: 2914	88-32-4	: 1604	89-87-2	: 4158
83-88-5	: 9478	85-55-2	: 7035	87-13-8	: 3455	88-43-7	: 284	89-92-9	: 1262
83-89-6	: 9410	85-61-0	: 2445	87-17-2	: 6147	88-44-8	: 345	89-93-0	: 6992
83-95-4	: 10490	85-64-3	: 6592	87-19-4	: 6390	88-51-7	: 288	89-95-2	: 6997
83-98-7	: 8420	85-66-5	: 9295	87-20-7	: 6454	88-58-4	: 3008	89-96-3	: 2032
84-06-0	: 10076	85-68-7	: 791	87-24-1	: 5121	88-60-8	: 1614	89-97-4	: 1881
84-11-7	: 8744	85-69-8	: 5058	87-25-2	: 4882	88-65-3	: 1103	89-98-5	: 1861
84-15-1	: 9696	85-73-4	: 9032	87-28-5	: 6041	88-67-5	: 6278	90-00-6	: 5200
84-16-2	: 5883	85-74-5	: 362	87-29-6	: 2400	88-68-6	: 242	90-01-7	: 5983
84-17-3	: 3346	85-75-6	: 361	87-33-2	: 6554	88-69-7	: 6527	90-02-8	: 9501
84-21-9	: 126	85-79-0	: 2997	87-39-8	: 9384	88-72-2	: 8206	90-04-0	: 6772
84-26-4	: 9491	85-82-5	: 4248	87-40-1	: 10251	88-73-3	: 2181	90-05-1	: 6855
84-31-1	: 9421	85-83-6	: 9515	87-41-2	: 6366	88-74-4	: 8034	90-11-9	: 1287
84-47-9	: 1509	85-84-7	: 8822	87-42-3	: 2286	88-75-5	: 8132	90-12-0	: 7470
84-48-0	: 3620	85-85-8	: 9377	87-44-5	: 1789	88-82-4	: 10449	90-13-1	: 2169
84-50-4	: 3618	85-86-9	: 8825	87-47-8	: 9390	88-85-7	: 4416	90-14-2	: 6319
84-51-5	: 4891	85-87-0	: 9374	87-51-4	: 6244	88-87-9	: 2011	90-15-3	: 7975
84-52-6	: 2722	85-90-5	: 7025	87-52-5	: 4118	88-88-0	: 2337	90-19-7	: 9467

Registry No.	Page	Registry No.	Page	Registry No.	Page	Registry No.	Page	Registry No.	Page
96-41-3	: 2669	97-99-4	: 9860	98-95-3	: 8047	99-93-4	: 6155	100-72-1	: 9885
96-43-5	: 2312	98-00-0	: 5503	98-96-4	: 9303	99-94-5	: 10121	100-74-3	: 5159
96-45-7	: 6219	98-01-1	: 5502	98-97-5	: 9304	99-96-7	: 5992	100-75-4	: 8199
96-47-9	: 7797	98-02-2	: 5488	98-98-6	: 9329	99-97-8	: 7204	100-76-5	: 539
96-48-0	: 1688	98-03-3	: 10069	99-02-5	: 2238	99-98-9	: 3952	100-79-8	: 4044
96-49-1	: 5017	98-04-4	: 9011	99-03-6	: 397	99-99-0	: 8208	100-80-1	: 7785
96-50-4	: 10029	98-05-5	: 603	99-04-7	: 10120	100-00-5	: 2183	100-81-2	: 6993
96-53-7	: 10034	98-06-6	: 1515	99-05-8	: 468	100-01-6	: 8036	100-83-4	: 5968
96-54-8	: 7754	98-07-7	: 10253	99-06-9	: 5991	100-02-7	: 8134	100-84-5	: 6945
96-64-0	: 9582	98-08-8	: 10404	99-07-0	: 3915	100-06-1	: 52	100-86-7	: 8932
96-69-5	: 10046	98-09-9	: 663	99-08-1	: 8207	100-07-2	: 6809	100-87-8	: 833
96-70-8	: 1570	98-10-2	: 661	99-09-2	: 8035	100-09-4	: 6801	100-88-9	: 2640
96-76-4	: 3040	98-11-3	: 662	99-10-5	: 3709	100-10-7	: 3900	100-92-5	: 10511
96-77-5	: 5981	98-13-5	: 10277	99-11-6	: 2413	100-11-8	: 1268	100-97-0	: 5822
96-80-0	: 3804	98-15-7	: 2332	99-12-7	: 4156	100-12-9	: 5169	100-99-2	: 10454
96-81-1	: 107	98-16-8	: 10401	99-14-9	: 9154	100-13-0	: 10813	101-01-9	: 10651
96-82-2	: 5515	98-17-9	: 10410	99-16-1	: 156	100-14-1	: 2138	101-02-0	: 10658
96-83-3	: 6349	98-19-1	: 1555	99-17-2	: 9091	100-15-2	: 7487	101-05-3	: 464
96-88-8	: 6706	98-27-1	: 1616	99-20-7	: 10126	100-16-3	: 8152	101-10-0	: 2227
96-91-3	: 304	98-28-2	: 1945	99-24-1	: 7837	100-17-4	: 8039	101-14-4	: 901
96-96-8	: 6854	98-29-3	: 1516	99-28-5	: 2965	100-18-5	: 3799	101-18-8	: 8814
96-97-9	: 6131	98-37-3	: 323	99-30-9	: 3233	100-19-6	: 8150	101-20-2	: 10215
96-99-1	: 2189	98-46-4	: 8210	99-31-0	: 246	100-20-9	: 619	101-21-3	: 2279
97-00-7	: 2008	98-47-5	: 8067	99-32-1	: 8478	100-21-0	: 9694	101-25-7	: 4408
97-02-9	: 4361	98-48-6	: 636	99-33-2	: 4378	100-22-1	: 9925	101-27-9	: 570
97-05-2	: 6195	98-49-7	: 3800	99-34-3	: 4377	100-23-2	: 4152	101-31-5	: 6204
97-08-5	: 2186	98-50-0	: 394	99-35-4	: 10624	100-25-4	: 4371	101-38-2	: 3141
97-09-6	: 2185	98-51-1	: 1608	99-36-5	: 7449	100-26-5	: 9340	101-39-3	: 7654
97-16-5	: 3260	98-52-2	: 1545	99-45-6	: 134	100-27-6	: 8060	101-40-6	: 4019
97-17-6	: 3063	98-52-2	: 1546	99-48-9	: 7467	100-28-7	: 6419	101-41-7	: 7618
97-18-7	: 1057	98-53-3	: 1547	99-50-3	: 3708	100-29-8	: 4856	101-42-8	: 4259
97-23-4	: 3249	98-54-4	: 1646	99-51-4	: 4154	100-32-3	: 1040	101-43-9	: 2633
97-24-5	: 994	98-56-6	: 2333	99-52-5	: 7481	100-36-7	: 3452	101-48-4	: 3864
97-25-6	: 10234	98-57-7	: 2164	99-54-7	: 3235	100-37-8	: 3384	101-53-1	: 820
97-30-3	: 7278	98-58-8	: 1101	99-55-8	: 7482	100-39-0	: 1247	101-54-2	: 8833
97-39-2	: 4596	98-59-9	: 10118	99-56-9	: 8053	100-40-3	: 10799	101-55-3	: 1316
97-50-7	: 1986	98-60-2	: 1890	99-57-0	: 378	100-41-4	: 4894	101-57-5	: 8811
97-51-8	: 6129	98-61-3	: 6277	99-59-2	: 6853	100-42-5	: 9621	101-60-0	: 9092
97-52-9	: 6852	98-62-4	: 254	99-60-5	: 2187	100-43-6	: 10822	101-61-1	: 964
97-53-0	: 196	98-64-6	: 1888	99-61-6	: 8043	100-44-7	: 2106	101-63-3	: 4383
97-54-1	: 6430	98-66-8	: 1889	99-62-7	: 3798	100-46-9	: 781	101-64-4	: 6876
97-56-3	: 3899	98-67-9	: 5989	99-63-8	: 618	100-47-0	: 715	101-68-8	: 4499
97-59-6	: 157	98-68-0	: 6796	99-64-9	: 3903	100-48-1	: 9319	101-70-2	: 6836
97-61-0	: 7559	98-69-1	: 4901	99-65-0	: 4370	100-49-2	: 2584	101-72-4	: 6530
97-62-1	: 5153	98-71-5	: 5939	99-66-1	: 9254	100-50-5	: 2594	101-73-5	: 6466
97-63-2	: 5100	98-72-6	: 8144	99-71-8	: 1645	100-51-6	: 780	101-75-7	: 8820
97-64-3	: 5096	98-73-7	: 1523	99-73-0	: 1119	100-52-7	: 585	101-76-8	: 946
97-65-4	: 9180	98-74-8	: 8071	99-75-2	: 7450	100-53-8	: 647	101-77-9	: 2855
97-69-8	: 43	98-79-3	: 9389	99-76-3	: 7369	100-54-9	: 9318	101-80-4	: 2854
97-72-3	: 7692	98-80-6	: 604	99-77-4	: 5171	100-55-0	: 9353	101-81-5	: 4498
97-74-5	: 9978	98-81-7	: 1376	99-79-6	: 6350	100-56-1	: 8913	101-82-6	: 826
97-77-8	: 4577	98-82-8	: 6478	99-82-1	: 7405	100-59-4	: 8912	101-83-7	: 3321
97-78-9	: 7538	98-83-9	: 6463	99-83-2	: 8740	100-60-7	: 2634	101-84-8	: 4485
97-85-8	: 6391	98-84-0	: 7040	99-84-3	: 7246	100-61-8	: 6941	101-86-0	: 8928
97-86-9	: 6394	98-85-1	: 6996	99-85-4	: 9700	100-63-0	: 8890	101-90-6	: 974
97-88-1	: 1599	98-86-2	: 39	99-86-5	: 9699	100-64-1	: 2589	101-91-7	: 6151
97-90-5	: 4781	98-87-3	: 3207	99-87-6	: 6509	100-65-2	: 8894	101-92-8	: 1842
97-93-8	: 10344	98-88-4	: 761	99-88-7	: 6475	100-66-3	: 474	101-97-3	: 5205
97-94-9	: 10351	98-89-5	: 2570	99-89-8	: 6529	100-68-5	: 7808	101-98-4	: 7642
97-95-0	: 4926	98-91-9	: 609	99-90-1	: 1320	100-69-6	: 10820	101-99-5	: 5207
97-96-1	: 4922	98-92-0	: 9327	99-91-2	: 2239	100-70-9	: 9317	102-01-2	: 24
97-97-2	: 1987	98-94-2	: 2619	99-92-3	: 398	100-71-0	: 5231	102-02-3	: 8847

CAS	Ref	CAS	Ref	CAS	Ref	CAS	Ref	CAS	Ref
102-04-5	: 4518	103-50-4	: 2895	104-80-3	: 463	106-19-4	: 4548	106-98-9	: 1441
102-06-7	: 4489	103-55-9	: 3972	104-81-4	: 1264	106-20-7	: 980	106-99-0	: 1393
102-07-8	: 4535	103-63-9	: 1202	104-82-5	: 2131	106-21-8	: 4183	107-00-6	: 1675
102-08-9	: 4532	103-65-1	: 9212	104-83-6	: 1952	106-22-9	: 4195	107-01-7	: 1442
102-09-0	: 4474	103-67-3	: 6995	104-84-7	: 6994	106-22-9	: 4196	107-01-7	: 1443
102-13-6	: 6400	103-69-5	: 4890	104-85-8	: 7022	106-23-0	: 4192	107-02-8	: 116
102-14-7	: 8470	103-70-8	: 8882	104-86-9	: 1883	106-24-1	: 5528	107-03-9	: 9151
102-20-5	: 8879	103-71-9	: 8906	104-87-0	: 6959	106-25-2	: 4168	107-04-0	: 1145
102-24-9	: 10485	103-72-0	: 8909	104-88-1	: 1863	106-26-3	: 4164	107-05-1	: 2275
102-25-0	: 10350	103-73-1	: 4828	104-90-5	: 5216	106-27-4	: 6444	107-06-2	: 3174
102-27-2	: 5115	103-74-2	: 9347	104-91-6	: 8197	106-28-5	: 5307	107-07-3	: 2020
102-28-3	: 392	103-75-3	: 4834	104-92-7	: 1092	106-30-9	: 5042	107-08-4	: 6335
102-36-3	: 3265	103-76-4	: 9056	104-93-8	: 6946	106-31-0	: 1430	107-10-8	: 9205
102-45-4	: 4033	103-79-7	: 8965	104-94-9	: 6774	106-32-1	: 5183	107-11-9	: 171
102-47-6	: 3142	103-80-0	: 602	104-96-1	: 7807	106-33-2	: 5097	107-12-0	: 9147
102-50-1	: 6839	103-81-1	: 597	104-98-3	: 10762	106-35-4	: 5690	107-13-1	: 119
102-51-2	: 6785	103-82-2	: 598	105-05-5	: 3403	106-36-5	: 9260	107-14-2	: 1844
102-52-3	: 9914	103-83-3	: 3971	105-06-6	: 4602	106-37-6	: 2910	107-15-3	: 4770
102-53-4	: 9824	103-84-4	: 18	105-07-7	: 5461	106-38-7	: 1361	107-16-4	: 5960
102-54-5	: 5326	103-85-5	: 9006	105-08-8	: 2575	106-39-8	: 1140	107-18-6	: 170
102-56-7	: 3834	103-88-8	: 1318	105-11-3	: 2564	106-40-1	: 1089	107-19-7	: 9175
102-60-3	: 9902	103-89-9	: 6911	105-13-5	: 6795	106-41-2	: 1314	107-20-0	: 1838
102-69-2	: 10668	103-90-2	: 6145	105-16-8	: 3388	106-42-3	: 10853	107-21-1	: 4772
102-70-5	: 2839	103-95-7	: 7401	105-21-5	: 3667	106-43-4	: 2317	107-22-2	: 5596
102-71-6	: 10330	103-99-1	: 6162	105-30-6	: 7562	106-44-5	: 2479	107-25-5	: 7859
102-76-1	: 10128	104-01-8	: 6783	105-31-7	: 5915	106-45-6	: 7014	107-29-9	: 16
102-77-2	: 7906	104-03-0	: 8050	105-34-0	: 7139	106-46-7	: 3094	107-30-2	: 2132
102-79-4	: 996	104-04-1	: 8139	105-36-2	: 4908	106-47-8	: 1852	107-31-3	: 7264
102-81-8	: 3005	104-06-3	: 10020	105-37-3	: 5224	106-48-9	: 2219	107-34-6	: 9211
102-82-9	: 10181	104-09-6	: 6969	105-38-4	: 10819	106-49-0	: 6940	107-35-7	: 305
102-85-2	: 10187	104-10-9	: 247	105-39-5	: 4940	106-50-3	: 613	107-37-9	: 209
102-86-3	: 10429	104-12-1	: 2242	105-40-8	: 5129	106-51-4	: 733	107-39-1	: 10589
102-87-4	: 10329	104-13-2	: 1504	105-41-9	: 7325	106-52-5	: 7678	107-40-4	: 10591
102-92-1	: 8972	104-15-4	: 10116	105-43-1	: 7560	106-53-6	: 1102	107-41-5	: 7555
102-94-3	: 2398	104-21-2	: 6810	105-45-3	: 6913	106-54-7	: 1893	107-43-7	: 845
102-97-6	: 804	104-22-3	: 782	105-46-4	: 1488	106-57-0	: 9053	107-44-8	: 7669
103-01-5	: 8887	104-28-9	: 2404	105-48-6	: 6486	106-58-1	: 4264	107-45-9	: 10578
103-02-6	: 9169	104-29-0	: 2226	105-50-0	: 3497	106-60-5	: 384	107-46-0	: 5819
103-03-7	: 8891	104-36-9	: 2999	105-53-3	: 3477	106-61-6	: 9155	107-47-1	: 3053
103-05-9	: 3960	104-40-5	: 8260	105-54-4	: 4923	106-63-8	: 6374	107-49-3	: 9825
103-09-3	: 5055	104-42-7	: 4643	105-55-5	: 3537	106-65-0	: 4321	107-51-7	: 8342
103-11-7	: 5056	104-43-8	: 4652	105-56-6	: 4954	106-68-3	: 8361	107-52-8	: 9787
103-12-8	: 2864	104-45-0	: 6896	105-57-7	: 3359	106-69-4	: 5847	107-53-9	: 9778
103-16-2	: 8919	104-46-1	: 6892	105-58-8	: 3416	106-70-7	: 7329	107-56-2	: 3810
103-17-3	: 1819	104-47-2	: 6784	105-59-9	: 1002	106-71-8	: 2510	107-58-4	: 1656
103-19-5	: 1025	104-49-4	: 3784	105-60-2	: 1719	106-72-9	: 4084	107-59-5	: 1532
103-23-1	: 979	104-50-7	: 1553	105-65-7	: 3814	106-73-0	: 7288	107-66-4	: 3043
103-24-2	: 981	104-51-8	: 1513	105-66-8	: 9217	106-74-1	: 4842	107-68-6	: 6932
103-25-3	: 7651	104-52-9	: 2281	105-67-9	: 10855	106-75-2	: 3436	107-70-0	: 6846
103-28-6	: 810	104-53-0	: 650	105-68-0	: 6453	106-79-6	: 4317	107-72-2	: 10266
103-29-7	: 4480	104-54-1	: 8970	105-70-4	: 9157	106-86-5	: 10816	107-73-3	: 2379
103-30-0	: 9613	104-55-2	: 2396	105-74-8	: 3341	106-87-6	: 4711	107-74-4	: 4180
103-32-2	: 785	104-57-4	: 799	105-75-9	: 3025	106-88-7	: 4710	107-75-5	: 6031
103-33-3	: 558	104-61-0	: 3661	105-76-0	: 3029	106-89-8	: 4707	107-80-2	: 2917
103-33-3	: 559	104-63-2	: 783	105-83-9	: 410	106-90-1	: 4715	107-81-3	: 1306
103-34-4	: 4588	104-65-4	: 2402	105-86-2	: 4170	106-91-2	: 4716	107-82-4	: 1252
103-36-6	: 4952	104-66-5	: 4450	105-87-3	: 5530	106-92-3	: 190	107-83-5	: 7552
103-37-7	: 790	104-67-6	: 5708	105-99-7	: 3002	106-93-4	: 2938	107-84-6	: 2111
103-38-8	: 807	104-72-3	: 2779	106-02-5	: 8425	106-94-5	: 1327	107-85-7	: 7063
103-45-7	: 8875	104-75-6	: 5057	106-07-0	: 9819	106-95-6	: 1339	107-86-8	: 7086
103-48-0	: 8878	104-76-7	: 5051	106-14-9	: 6133	106-96-7	: 1345	107-87-9	: 8629
103-49-1	: 2892	104-78-9	: 3514	106-18-3	: 1564	106-97-8	: 1401	107-88-0	: 1406

112-36-7	: 3440	114-25-0	: 869	116-52-9	: 1046	119-13-1	: 10106	120-75-2	: 7027
112-37-8	: 10731	114-26-1	: 9196	116-53-0	: 7074	119-15-3	: 4398	120-78-5	: 6710
112-38-9	: 10745	114-75-0	: 6163	116-54-1	: 7190	119-20-0	: 9473	120-80-9	: 9388
112-39-0	: 7542	114-80-7	: 8008	116-58-5	: 9901	119-26-6	: 4401	120-82-1	: 10205
112-40-3	: 4623	114-83-0	: 22	116-63-2	: 334	119-27-7	: 6819	120-83-2	: 3251
112-41-4	: 4635	114-86-3	: 8877	116-71-2	: 477	119-28-8	: 370	120-85-4	: 10606
112-42-5	: 10732	114-91-0	: 6829	116-81-4	: 267	119-32-4	: 7485	120-87-6	: 3740
112-43-6	: 10746	115-02-6	: 544	116-85-8	: 322	119-33-5	: 7498	120-89-8	: 6220
112-44-7	: 10727	115-07-1	: 9178	117-08-8	: 9765	119-34-6	: 380	120-92-3	: 2670
112-45-8	: 10737	115-10-6	: 4058	117-10-2	: 3694	119-36-8	: 7779	120-93-4	: 6221
112-47-0	: 2753	115-11-7	: 6372	117-11-3	: 282	119-38-0	: 6434	120-94-5	: 7757
112-48-1	: 3000	115-17-3	: 10152	117-12-4	: 3693	119-41-5	: 4679	120-95-6	: 972
112-49-2	: 10359	115-18-4	: 7099	117-14-6	: 3617	119-42-6	: 2636	120-97-8	: 3101
112-50-5	: 10361	115-19-5	: 7115	117-18-0	: 9761	119-44-8	: 10845	121-00-6	: 1603
112-52-7	: 2017	115-20-8	: 10230	117-27-1	: 8087	119-47-1	: 993	121-03-9	: 7489
112-53-8	: 4631	115-21-9	: 10235	117-34-0	: 8828	119-48-2	: 4354	121-06-2	: 10686
112-54-9	: 4621	115-22-0	: 6086	117-37-3	: 6870	119-51-7	: 8959	121-14-2	: 7209
112-55-0	: 4628	115-24-2	: 9660	117-39-5	: 9407	119-52-8	: 3853	121-17-5	: 2197
112-56-1	: 1475	115-25-3	: 8699	117-62-4	: 360	119-53-9	: 714	121-19-7	: 8111
112-57-2	: 9820	115-26-4	: 3819	117-73-7	: 9451	119-56-2	: 2232	121-21-1	: 9309
112-58-3	: 3587	115-27-5	: 1829	117-78-2	: 3616	119-58-4	: 966	121-29-9	: 9310
112-59-4	: 3449	115-28-6	: 1828	117-79-3	: 237	119-59-5	: 909	121-30-2	: 283
112-60-7	: 9815	115-29-7	: 4698	117-80-6	: 3231	119-60-8	: 3326	121-32-4	: 4844
112-61-8	: 7783	115-31-1	: 6368	117-81-7	: 986	119-61-9	: 718	121-33-5	: 6060
112-62-9	: 7528	115-32-2	: 949	117-82-8	: 626	119-64-2	: 9874	121-34-6	: 6065
112-63-0	: 7430	115-37-7	: 10013	117-83-9	: 625	119-65-3	: 6551	121-39-1	: 8941
112-64-1	: 9799	115-38-8	: 6703	117-84-0	: 4422	119-67-5	: 5457	121-43-7	: 10523
112-66-3	: 4638	115-39-9	: 1315	117-89-5	: 10376	119-68-6	: 6926	121-44-8	: 10345
112-67-4	: 5755	115-40-2	: 1157	117-92-0	: 9411	119-75-5	: 8142	121-45-9	: 10605
112-69-6	: 5761	115-43-5	: 8810	117-93-1	: 6126	119-80-2	: 4584	121-46-0	: 8263
112-70-9	: 10315	115-44-6	: 9675	117-96-4	: 2869	119-84-6	: 3602	121-47-1	: 252
112-71-0	: 1353	115-53-7	: 9542	117-99-7	: 6164	119-90-4	: 3847	121-50-6	: 2329
112-72-1	: 9797	115-56-0	: 5110	118-00-3	: 5615	119-91-5	: 898	121-51-7	: 8070
112-73-2	: 3439	115-61-7	: 9673	118-08-1	: 5931	119-93-7	: 10109	121-54-0	: 678
112-76-5	: 8308	115-63-9	: 5885	118-10-5	: 2394	120-07-0	: 8868	121-57-3	: 253
112-79-8	: 4690	115-67-3	: 8509	118-12-7	: 7249	120-12-7	: 479	121-59-5	: 277
112-80-1	: 8410	115-69-5	: 352	118-29-6	: 6100	120-14-9	: 3838	121-60-8	: 44
112-82-3	: 1221	115-70-8	: 317	118-31-0	: 7962	120-15-0	: 9867	121-61-9	: 428
112-84-5	: 4614	115-71-9	: 9506	118-32-1	: 6118	120-18-3	: 7966	121-63-1	: 8480
112-85-6	: 4612	115-76-4	: 3516	118-34-3	: 9671	120-20-7	: 3844	121-65-3	: 4645
112-86-7	: 4616	115-77-5	: 8571	118-41-2	: 10483	120-21-8	: 3381	121-66-4	: 8202
112-88-9	: 8312	115-80-0	: 10339	118-44-5	: 5161	120-22-9	: 3493	121-69-7	: 3936
112-89-0	: 1298	115-84-4	: 1571	118-46-7	: 373	120-23-0	: 7991	121-71-1	: 6154
112-90-3	: 8317	115-86-6	: 10655	118-48-9	: 748	120-25-2	: 4845	121-72-2	: 7203
112-91-4	: 8313	115-88-8	: 8389	118-52-5	: 3167	120-29-6	: 10711	121-73-3	: 2182
112-92-5	: 8307	115-90-2	: 5321	118-55-8	: 8995	120-32-1	: 2434	121-75-5	: 6658
112-95-8	: 4681	115-95-7	: 6613	118-58-1	: 831	120-36-5	: 3258	121-79-9	: 9268
112-96-9	: 8324	115-96-8	: 10680	118-60-5	: 5062	120-40-1	: 998	121-81-3	: 4368
112-99-2	: 4417	116-01-8	: 4813	118-61-6	: 5243	120-43-4	: 5217	121-82-4	: 5810
113-00-8	: 5608	116-06-3	: 147	118-69-4	: 3302	120-46-7	: 4515	121-86-8	: 2142
113-15-5	: 4737	116-09-6	: 6184	118-71-8	: 6106	120-47-8	: 5073	121-87-9	: 2173
113-18-8	: 4805	116-11-0	: 6890	118-74-1	: 5724	120-50-3	: 6379	121-88-0	: 379
113-24-6	: 9570	116-14-3	: 9837	118-75-2	: 9741	120-51-4	: 788	121-89-1	: 8149
113-42-8	: 7251	116-15-4	: 8725	118-78-5	: 423	120-55-8	: 3438	121-90-4	: 8082
113-45-1	: 7605	116-16-5	: 5723	118-79-6	: 10172	120-57-0	: 692	121-91-5	: 6458
113-48-4	: 7872	116-17-6	: 10466	118-82-1	: 7233	120-61-6	: 4331	121-92-6	: 8074
113-52-0	: 6227	116-26-7	: 9498	118-90-1	: 10119	120-62-7	: 9078	121-97-1	: 6880
113-53-1	: 4662	116-29-0	: 9806	118-91-2	: 1896	120-66-1	: 7612	121-98-2	: 7443
113-59-7	: 2351	116-31-4	: 9460	118-92-3	: 467	120-70-7	: 9146	122-00-9	: 6914
113-92-8	: 2348	116-38-1	: 4678	118-93-4	: 6153	120-71-8	: 6838	122-01-0	: 1913
114-03-4	: 6199	116-43-8	: 9629	118-96-7	: 10633	120-72-9	: 6243	122-03-2	: 6477
114-07-8	: 4745	116-44-9	: 415	119-06-2	: 4597	120-73-0	: 9293	122-04-3	: 8083

157-06-2	: 513	283-24-9	: 538	301-02-0	: 8311	320-77-4	: 6406	350-30-1 : 2056
157-40-4	: 9589	285-67-6	: 8424	301-12-2	: 8486	321-14-2	: 2074	350-46-9 : 5407
180-43-8	: 9598	286-08-8	: 854	301-21-3	: 6591	321-28-8	: 5396	350-50-5 : 5400
189-55-9	: 2887	286-20-4	: 8423	302-17-0	: 1812	321-38-0	: 5403	352-11-4 : 2124
189-64-0	: 2886	287-23-0	: 2530	302-27-2	: 111	321-60-8	: 5376	352-32-9 : 5426
191-07-1	: 2465	287-27-4	: 10038	302-40-9	: 575	321-98-2	: 4704	352-33-0 : 2051
191-24-2	: 716	287-92-3	: 2659	302-70-5	: 8108	323-09-1	: 5404	352-34-1 : 5392
191-30-0	: 2888	288-13-1	: 9306	302-72-7	: 142	324-74-3	: 5377	352-70-5 : 5425
192-65-4	: 2885	288-14-2	: 6560	302-79-4	: 9462	326-06-7	: 10416	352-93-2 : 3529
192-97-2	: 730	288-32-4	: 6215	302-84-1	: 9529	326-56-7	: 7018	352-97-6 : 5581
193-39-5	: 6238	288-36-8	: 10147	303-07-1	: 3707	326-62-5	: 5363	353-09-3 : 5613
194-59-2	: 2880	288-39-1	: 10022	303-34-4	: 6590	326-91-0	: 10423	353-36-6 : 5386
194-69-4	: 686	288-42-6	: 8441	303-38-8	: 3704	327-54-8	: 9832	353-42-4 : 1069
195-19-7	: 717	288-47-1	: 10030	303-43-5	: 2374	327-57-1	: 8276	353-49-1 : 1757
196-78-1	: 687	288-88-0	: 10148	303-47-9	: 8287	327-92-4	: 3556	353-50-4 : 1760
198-55-0	: 8736	288-94-8	: 10009	303-81-1	: 8284	327-98-0	: 10259	353-54-8 : 10166
204-02-4	: 8730	288-99-3	: 8426	304-21-2	: 5623	328-38-1	: 6601	353-55-9 : 2923
205-82-3	: 701	289-06-5	: 10023	304-55-2	: 9622	328-39-2	: 6600	353-58-2 : 1168
205-99-2	: 700	289-65-6	: 9296	304-84-7	: 3468	328-42-7	: 8431	353-59-3 : 1142
206-44-0	: 5341	289-80-5	: 9312	304-88-1	: 6148	328-50-7	: 8461	353-61-7 : 5402
207-08-9	: 702	289-95-2	: 9382	305-01-1	: 3712	328-74-5	: 1050	353-66-2 : 3555
208-96-8	: 8	290-37-9	: 9302	305-03-3	: 1813	328-84-7	: 3116	353-83-3 : 10392
213-46-7	: 9035	290-38-0	: 10139	305-62-4	: 2851	329-01-1	: 6421	353-85-5 : 10382
215-58-7	: 746	290-87-9	: 10140	305-84-0	: 1773	329-71-5	: 4394	354-04-1 : 2994
217-54-9	: 2882	290-96-0	: 10007	305-85-1	: 3769	329-98-6	: 646	354-06-3 : 1154
217-59-4	: 10648	291-21-4	: 10707	306-08-1	: 6061	330-13-2	: 8155	354-11-0 : 9754
218-01-9	: 2387	291-64-5	: 2554	306-37-6	: 4108	330-54-1	: 4599	354-12-1 : 10221
222-93-5	: 8633	292-46-6	: 8634	306-40-1	: 9628	330-55-2	: 6617	354-14-3 : 9755
224-41-9	: 2877	292-64-8	: 2646	306-44-5	: 8474	331-39-5	: 3744	354-15-4 : 10223
224-42-0	: 2874	293-55-0	: 2641	306-52-5	: 10298	332-77-4	: 3612	354-21-2 : 10222
224-53-3	: 2875	293-96-9	: 2538	306-67-2	: 912	332-80-9	: 7361	354-23-4 : 3306
225-51-4	: 584	294-62-2	: 2545	306-83-2	: 3307	333-18-6	: 4771	354-25-6 : 2307
226-36-8	: 2873	295-48-7	: 2655	306-94-5	: 8702	333-41-5	: 2871	354-32-5 : 10383
229-87-8	: 8745	297-78-9	: 6365	307-30-2	: 8550	333-49-3	: 422	354-33-6 : 8584
230-07-9	: 8748	297-97-2	: 10061	307-34-6	: 8720	333-93-7	: 1404	354-38-1 : 10378
230-27-3	: 732	298-00-0	: 7543	307-35-7	: 8721	334-25-8	: 130	354-51-8 : 2925
230-46-6	: 8746	298-02-2	: 9018	307-45-9	: 8703	334-48-5	: 2757	354-55-2 : 1304
238-84-6	: 703	298-04-4	: 4578	308-24-7	: 10726	334-56-5	: 5383	354-56-3 : 8539
239-64-5	: 2881	298-06-6	: 3510	309-00-2	: 150	334-88-3	: 2872	354-58-5 : 10294
243-17-4	: 704	298-07-7	: 983	311-28-4	: 9724	335-05-7	: 10399	354-92-7 : 8712
244-69-9	: 1740	298-12-4	: 5598	311-45-5	: 8510	335-36-4	: 8698	354-96-1 : 8705
253-52-1	: 9028	298-45-3	: 1388	311-81-9	: 3160	335-44-4	: 10240	355-02-2 : 8715
253-66-7	: 2403	298-46-4	: 2878	311-89-7	: 10702	335-48-8	: 2967	355-04-4 : 8716
253-72-5	: 7995	298-50-0	: 9172	312-40-3	: 3557	335-57-9	: 8708	355-25-9 : 8696
253-82-7	: 9413	298-81-7	: 6767	312-84-5	: 9530	335-67-1	: 8549	355-42-0 : 8710
254-04-6	: 724	298-96-4	: 10662	313-72-4	: 8718	335-76-2	: 8218	355-63-5 : 8709
254-79-5	: 7994	299-27-4	: 9095	314-13-6	: 5299	336-59-4	: 5660	355-68-0 : 8700
258-31-1	: 5722	299-28-5	: 1698	314-19-2	: 502	338-65-8	: 1981	355-75-9 : 8701
260-94-6	: 113	299-29-6	: 5327	314-40-9	: 1074	338-69-2	: 143	355-80-6 : 8331
261-31-4	: 10082	299-42-3	: 4705	315-18-4	: 7871	338-83-0	: 8729	356-16-1 : 9834
262-12-4	: 2883	299-45-6	: 6202	315-22-0	: 7891	339-43-5	: 274	356-18-3 : 3191
271-44-3	: 6233	299-84-3	: 9487	315-30-0	: 163	343-65-7	: 6572	356-24-1 : 7282
271-89-6	: 705	299-85-4	: 4610	317-52-2	: 5792	343-94-2	: 6250	356-27-4 : 5039
273-13-2	: 734	299-86-5	: 2486	319-84-6	: 5731	344-04-7	: 1303	357-57-3 : 1383
273-53-0	: 749	300-38-9	: 2996	319-85-7	: 5732	344-07-0	: 2206	357-67-5 : 3507
274-07-7	: 689	300-39-0	: 3774	319-86-8	: 5733	345-35-7	: 2123	357-70-0 : 5518
274-09-9	: 691	300-57-2	: 173	319-88-0	: 10293	346-18-9	: 9089	358-72-5 : 3892
274-40-8	: 6254	300-62-9	: 8974	319-89-1	: 9900	348-51-6	: 2049	359-01-3 : 5379
275-51-4	: 568	300-76-5	: 7918	320-51-4	: 2330	348-52-7	: 5391	359-02-4 : 3186
279-23-2	: 853	300-87-8	: 4123	320-60-5	: 3309	348-54-9	: 5355	359-06-8 : 5353
280-57-9	: 10354	300-92-5	: 220	320-67-2	: 541	349-88-2	: 5367	359-10-4 : 1982
281-23-2	: 10309	301-00-8	: 7431	320-72-9	: 3199	350-03-8	: 9368	359-11-5 : 10389

CAS	Page	CAS	Page	CAS	Page	CAS	Page	CAS	Page
359-28-4	: 10238	381-71-5	: 3161	430-66-0	: 10387	456-22-4	: 5370	465-42-9	: 1721
359-35-3	: 9836	381-73-7	: 3547	430-71-7	: 9171	456-41-7	: 1260	465-58-7	: 6558
359-83-1	: 8637	382-21-8	: 8713	430-95-5	: 3189	456-48-4	: 5359	465-65-6	: 7922
360-89-4	: 8697	382-45-6	: 458	431-03-8	: 1415	456-49-5	: 5397	465-74-7	: 9442
360-97-4	: 336	383-63-1	: 5269	431-06-1	: 3158	456-59-7	: 2525	465-99-6	: 5629
361-37-5	: 7861	384-04-3	: 8591	431-07-2	: 2325	457-87-4	: 5118	466-43-3	: 530
362-29-8	: 9193	384-22-5	: 8209	431-47-0	: 7835	458-24-2	: 5316	466-49-9	: 79
363-03-1	: 8864	389-08-2	: 7919	431-63-0	: 5799	458-35-5	: 6185	466-81-9	: 4744
363-24-6	: 9278	389-40-2	: 8586	431-89-0	: 5665	458-36-6	: 6071	466-96-6	: 9285
363-72-4	: 8580	392-12-1	: 8464	432-25-7	: 10539	458-37-7	: 2495	466-97-7	: 8277
366-18-7	: 893	392-56-3	: 5794	433-19-2	: 1052	458-88-8	: 9257	466-99-9	: 5954
366-29-0	: 9929	392-83-6	: 1368	434-03-7	: 4812	459-22-3	: 5364	467-14-1	: 8007
366-70-1	: 9116	392-85-8	: 5428	434-07-1	: 8490	459-56-3	: 5366	467-55-0	: 6193
367-11-3	: 3549	393-52-2	: 5373	434-13-9	: 6016	459-57-4	: 5360	467-60-7	: 4512
367-12-4	: 5410	393-75-9	: 2013	434-16-2	: 2363	459-60-9	: 5398	467-62-9	: 10677
367-21-5	: 2048	394-47-8	: 5371	434-22-0	: 7924	459-67-6	: 2676	467-81-2	: 6585
367-23-7	: 10384	395-28-8	: 6561	434-45-7	: 10417	459-72-3	: 5032	467-82-3	: 6586
367-25-9	: 3548	396-01-0	: 8981	434-64-0	: 8728	459-73-4	: 4881	467-85-6	: 3908
367-47-5	: 3749	398-23-2	: 3552	434-90-2	: 2742	459-80-3	: 4167	467-98-1	: 10014
367-57-7	: 10415	401-56-9	: 4944	435-97-2	: 8794	460-00-4	: 1211	468-28-0	: 6637
367-64-6	: 1668	401-78-5	: 1369	436-05-5	: 2496	460-12-8	: 1396	468-45-1	: 9579
368-16-1	: 7362	401-80-9	: 5429	437-38-7	: 5322	460-13-9	: 5415	468-76-8	: 1791
368-43-4	: 664	402-31-3	: 1051	437-50-3	: 3730	460-19-5	: 2512	469-21-6	: 4667
368-47-8	: 10418	402-43-7	: 1370	438-08-4	: 453	460-35-5	: 2334	469-59-0	: 6564
368-77-4	: 10405	402-44-8	: 5430	438-60-8	: 9282	460-43-5	: 10391	469-61-4	: 1793
370-14-9	: 6937	402-67-5	: 5406	439-14-5	: 10770	461-58-5	: 2517	469-62-5	: 9198
371-40-4	: 5357	403-42-9	: 5414	440-17-5	: 10377	461-72-3	: 6218	470-40-6	: 10092
371-41-5	: 5412	403-43-0	: 5375	442-16-0	: 4821	461-78-9	: 2349	470-44-0	: 165
371-62-0	: 5387	404-72-8	: 5393	442-51-3	: 5625	461-89-2	: 10141	470-67-7	: 7419
371-86-8	: 7877	404-86-4	: 1720	443-48-1	: 7868	461-98-3	: 4307	470-82-6	: 5297
371-90-4	: 3553	405-50-5	: 5362	443-79-8	: 6435	462-02-2	: 10639	470-90-6	: 1830
372-09-8	: 2504	405-99-2	: 10803	443-83-4	: 2055	462-06-6	: 5361	470-99-5	: 10540
372-18-9	: 3550	406-33-7	: 5418	444-27-9	: 10032	462-08-8	: 9314	471-16-9	: 7245
372-19-0	: 5356	406-33-7	: 5419	444-30-4	: 10409	462-18-0	: 10317	471-25-0	: 9274
372-20-3	: 5411	407-25-0	: 10380	445-29-4	: 5368	462-20-4	: 3636	471-35-2	: 9942
372-31-6	: 5270	407-41-0	: 9027	446-34-4	: 5401	462-94-2	: 8609	471-46-5	: 8435
372-38-3	: 10385	420-04-2	: 2500	446-35-5	: 3566	462-95-3	: 3362	471-47-6	: 8434
372-46-3	: 5380	420-05-3	: 2502	446-52-6	: 5358	463-04-7	: 8678	471-77-2	: 8004
372-47-4	: 5423	420-12-2	: 10041	446-72-0	: 5526	463-11-6	: 5408	471-87-4	: 9601
372-48-5	: 5422	420-24-6	: 7195	446-86-6	: 545	463-40-1	: 6616	471-95-4	: 1387
372-64-5	: 1053	420-26-8	: 5416	447-31-4	: 2015	463-49-0	: 158	472-70-8	: 1780
372-75-8	: 2417	420-37-1	: 10577	447-41-6	: 8286	463-51-4	: 6570	472-93-5	: 1776
373-02-4	: 8014	420-45-1	: 3567	447-53-0	: 3656	463-52-5	: 6748	473-06-3	: 10521
373-14-8	: 5390	420-46-2	: 10386	448-71-5	: 5292	463-56-9	: 10050	473-29-0	: 3104
373-44-4	: 8347	420-56-4	: 5431	450-95-3	: 5413	463-58-1	: 1753	473-30-3	: 402
373-49-9	: 5757	421-04-5	: 2324	451-13-8	: 3703	463-71-8	: 1750	473-34-7	: 3208
373-52-4	: 1213	421-06-7	: 1365	451-40-1	: 8798	463-72-9	: 1726	473-54-1	: 9045
373-64-8	: 3565	421-07-8	: 10419	451-46-7	: 5033	463-79-6	: 1744	473-55-2	: 9044
373-74-0	: 10412	421-14-7	: 7836	452-35-7	: 4832	463-82-1	: 8006	473-81-4	: 3750
374-01-6	: 10420	421-17-0	: 10396	452-58-4	: 9334	463-88-7	: 8011	473-98-3	: 6635
374-07-2	: 3294	421-20-5	: 7262	452-71-1	: 5399	464-06-2	: 10524	474-25-9	: 3717
375-01-9	: 5661	421-50-1	: 10381	452-86-8	: 6982	464-07-3	: 3992	474-40-8	: 9543
375-03-1	: 8726	421-83-0	: 10398	453-13-4	: 3568	464-10-8	: 10171	474-62-4	: 4734
375-16-6	: 5662	422-05-9	: 8589	453-20-3	: 9857	464-15-3	: 10515	474-69-1	: 4733
375-22-4	: 5659	422-56-0	: 3245	454-14-8	: 2497	464-17-5	: 10518	474-77-1	: 2370
375-83-7	: 8548	425-75-2	: 5271	454-29-5	: 5923	464-41-5	: 1067	474-86-2	: 6037
375-96-2	: 8719	425-82-1	: 8722	454-31-9	: 4985	464-48-2	: 1711	475-00-3	: 10778
376-27-2	: 7601	428-59-1	: 10424	455-14-1	: 10402	464-49-3	: 1710	475-03-6	: 9896
377-41-3	: 2059	429-41-4	: 9722	455-18-5	: 10406	464-72-2	: 9988	475-20-7	: 6627
378-44-9	: 847	430-40-0	: 4986	455-19-6	: 10403	464-85-7	: 9412	475-25-2	: 5633
378-75-6	: 7550	430-51-3	: 5417	455-32-3	: 765	464-86-8	: 2458	475-31-0	: 5580
379-79-3	: 4738	430-57-9	: 3185	455-38-9	: 5369	465-16-7	: 8407	475-38-7	: 3735

475-67-2	: 6408	482-68-8	: 9513	490-02-8	: 526	495-85-2	: 7862	501-53-1	: 793
475-81-0	: 5535	482-74-6	: 2487	490-03-9	: 6090	495-91-0	: 1807	501-65-5	: 4452
476-28-8	: 6642	482-89-3	: 6239	490-10-8	: 8009	496-03-7	: 5078	501-92-8	: 202
476-32-4	: 1810	483-04-5	: 9450	490-11-9	: 9342	496-10-6	: 8334	501-94-0	: 5982
476-45-9	: 6563	483-17-0	: 10488	490-53-9	: 9846	496-11-7	: 6231	501-97-3	: 5986
476-60-8	: 489	483-18-1	: 4695	490-55-1	: 9000	496-15-1	: 3633	502-39-6	: 7437
476-69-7	: 2470	483-55-6	: 6092	490-64-2	: 10482	496-16-2	: 3599	502-41-0	: 2556
476-70-0	: 1061	483-64-7	: 7444	490-78-8	: 66	496-41-3	: 706	502-42-1	: 2557
477-27-0	: 2448	483-65-8	: 7420	490-79-9	: 3706	496-46-8	: 9862	502-44-3	: 8443
477-32-7	: 10831	483-76-1	: 5805	490-91-5	: 7403	496-49-1	: 7761	502-47-6	: 4194
477-47-4	: 9037	483-78-3	: 4121	491-04-3	: 7415	496-67-3	: 279	502-49-8	: 2648
477-60-1	: 574	484-11-7	: 4243	491-07-6	: 7412	496-72-0	: 6978	502-56-7	: 8246
477-75-8	: 3595	484-20-8	: 6899	491-09-8	: 7456	496-73-1	: 6984	502-59-0	: 7399
477-89-4	: 1790	484-29-7	: 6831	491-31-6	: 726	496-74-2	: 6987	502-61-4	: 5303
477-90-7	: 843	484-31-1	: 3886	491-35-0	: 7768	496-77-5	: 6136	502-65-8	: 1777
478-43-3	: 9471	484-47-9	: 10652	491-37-2	: 3603	496-78-6	: 10595	502-72-7	: 2657
478-61-5	: 10486	484-67-3	: 9757	491-38-3	: 728	497-03-0	: 7085	502-97-6	: 4429
478-94-4	: 6643	484-89-9	: 5477	491-67-8	: 10443	497-04-1	: 2263	502-99-8	: 4189
478-95-5	: 6438	484-93-5	: 4673	491-70-3	: 6638	497-18-7	: 1745	503-01-5	: 4085
479-00-5	: 4724	485-31-4	: 870	491-80-5	: 3728	497-20-1	: 5475	503-17-3	: 1676
479-13-0	: 2474	485-35-8	: 2724	491-88-3	: 6459	497-26-7	: 7218	503-28-6	: 4041
479-18-5	: 4672	485-43-8	: 6353	492-17-1	: 881	497-30-3	: 4740	503-29-7	: 547
479-21-0	: 3729	485-47-2	: 6237	492-22-8	: 10083	497-38-1	: 855	503-30-0	: 8444
479-22-1	: 8028	485-64-3	: 5944	492-27-3	: 6190	497-39-2	: 3033	503-38-8	: 4536
479-23-2	: 3594	485-65-4	: 5945	492-37-5	: 6974	497-59-6	: 6139	503-40-2	: 6743
479-27-6	: 7942	485-71-2	: 2393	492-38-6	: 7231	497-76-7	: 6156	503-60-6	: 2114
479-33-4	: 9986	485-72-3	: 5455	492-39-7	: 312	498-00-0	: 6062	503-64-0	: 1451
479-45-8	: 7798	485-89-2	: 6170	492-61-5	: 5540	498-02-2	: 6069	503-66-2	: 6183
479-59-4	: 9842	486-12-4	: 10666	492-80-8	: 3905	498-15-7	: 1769	503-74-2	: 7075
479-61-8	: 2256	486-25-9	: 5348	492-86-4	: 2072	498-23-7	: 7090	503-87-7	: 10084
480-11-5	: 8419	486-35-1	: 3713	492-88-6	: 4843	498-24-8	: 9179	503-93-5	: 10529
480-15-9	: 2734	486-39-5	: 2441	492-89-7	: 8560	498-59-9	: 4609	504-02-9	: 2579
480-16-0	: 7898	486-47-5	: 4804	492-94-4	: 3569	498-60-2	: 5482	504-03-0	: 4267
480-18-2	: 8593	486-66-8	: 2731	492-97-7	: 1058	498-62-4	: 10070	504-07-4	: 3675
480-40-0	: 3741	486-70-4	: 8338	492-99-9	: 2581	498-66-8	: 856	504-15-4	: 6985
480-41-1	: 8000	486-84-0	: 5624	493-01-6	: 2743	498-67-9	: 3188	504-17-6	: 3686
480-44-4	: 4	486-86-2	: 1792	493-02-7	: 2744	498-94-2	: 9063	504-20-1	: 9020
480-54-6	: 9466	486-89-5	: 451	493-05-0	: 3600	498-95-3	: 9062	504-24-5	: 9315
480-63-7	: 10513	487-06-9	: 3854	493-08-3	: 3601	498-96-4	: 9889	504-29-0	: 9313
480-64-8	: 3732	487-16-1	: 6249	493-09-4	: 3598	499-04-7	: 511	504-31-4	: 9297
480-66-0	: 10446	487-21-8	: 9291	493-49-2	: 3613	499-06-9	: 3967	504-53-0	: 8636
480-68-2	: 8177	487-26-3	: 3663	493-52-7	: 7777	499-12-7	: 9182	504-57-4	: 8223
480-81-9	: 9526	487-27-4	: 9519	493-77-6	: 10664	499-12-7	: 9183	504-60-9	: 8565
480-85-3	: 9465	487-68-3	: 10506	494-03-1	: 1837	499-40-1	: 6439	504-60-9	: 8566
480-91-1	: 3635	487-79-6	: 6566	494-04-2	: 8022	499-44-5	: 6053	504-61-0	: 1456
480-96-6	: 710	487-89-8	: 6247	494-19-9	: 3610	499-75-2	: 6515	504-63-2	: 9231
481-05-0	: 516	487-93-4	: 3910	494-47-3	: 5951	499-80-9	: 9339	504-64-3	: 1754
481-06-1	: 9509	488-10-8	: 7590	494-52-0	: 9075	499-81-0	: 9343	504-66-5	: 3316
481-20-9	: 2366	488-17-5	: 6981	494-55-3	: 5953	499-83-2	: 9341	504-78-9	: 10031
481-21-0	: 2365	488-23-3	: 9921	494-90-6	: 9848	499-89-8	: 10090	504-88-1	: 8165
481-26-5	: 9102	488-43-7	: 292	494-97-3	: 9401	500-05-0	: 8477	505-20-4	: 4579
481-29-8	: 5964	488-70-0	: 9965	494-99-5	: 3871	500-22-1	: 9323	505-22-6	: 4427
481-37-8	: 4674	488-73-3	: 2805	495-08-9	: 6080	500-28-7	: 2357	505-23-7	: 4580
481-39-0	: 6117	488-81-3	: 9477	495-18-1	: 5972	500-38-9	: 8268	505-29-3	: 4581
481-42-5	: 6093	488-84-6	: 9482	495-20-5	: 2456	500-44-7	: 7874	505-32-8	: 9951
481-49-2	: 1803	488-87-9	: 3953	495-40-9	: 8853	500-55-0	: 499	505-34-0	: 1808
481-72-1	: 3724	488-93-7	: 5484	495-41-0	: 8858	500-98-1	: 8801	505-48-6	: 8349
481-74-3	: 3731	488-98-2	: 5427	495-45-4	: 4470	500-99-2	: 3880	505-52-2	: 10312
482-05-3	: 883	489-72-5	: 8479	495-48-7	: 566	501-15-5	: 6933	505-54-4	: 5748
482-28-0	: 2392	489-84-9	: 4120	495-61-4	: 7457	501-24-6	: 8562	505-57-7	: 5860
482-44-0	: 6228	489-86-1	: 5604	495-69-2	: 766	501-30-4	: 6050	505-60-2	: 932
482-66-6	: 3609	489-98-5	: 10623	495-76-1	: 696	501-52-0	: 653	505-65-7	: 4432

533-68-6	: 4911	538-23-8	: 9161	540-97-6	: 4619	543-38-4	: 337	551-01-9	: 9086
533-70-0	: 3759	538-24-9	: 5570	541-01-5	: 5744	543-49-7	: 5686	551-06-4	: 7989
533-73-3	: 674	538-32-9	: 839	541-02-6	: 2747	543-59-9	: 2208	551-09-7	: 7986
533-74-4	: 2737	538-39-6	: 4595	541-05-9	: 5815	543-67-9	: 9251	551-11-1	: 9279
533-75-5	: 6019	538-41-0	: 2849	541-15-1	: 1772	543-75-9	: 3615	551-16-6	: 386
533-98-2	: 2916	538-43-2	: 8788	541-16-2	: 3031	543-82-8	: 7283	551-62-2	: 9830
534-07-6	: 3070	538-44-3	: 4252	541-20-8	: 8594	543-83-9	: 7103	551-76-8	: 10257
534-13-4	: 4345	538-51-2	: 8920	541-22-0	: 2746	543-87-3	: 7112	551-77-9	: 10256
534-15-6	: 3888	538-56-7	: 8968	541-25-3	: 3145	544-00-3	: 3789	551-78-0	: 10255
534-22-5	: 7266	538-58-9	: 4507	541-28-6	: 6314	544-01-4	: 3790	551-84-8	: 10864
534-26-9	: 3647	538-62-5	: 4472	541-31-1	: 7070	544-02-5	: 3792	551-88-2	: 8130
534-52-1	: 7214	538-64-7	: 800	541-33-3	: 3128	544-10-5	: 2066	551-92-8	: 4159
534-59-8	: 1653	538-68-1	: 8666	541-35-5	: 1400	544-13-8	: 8610	552-16-9	: 8073
534-84-9	: 9073	538-74-9	: 2899	541-41-3	: 4945	544-16-1	: 1623	552-22-7	: 10097
534-85-0	: 8832	538-75-0	: 3323	541-42-4	: 6521	544-25-2	: 2558	552-30-7	: 10469
535-11-5	: 4920	538-81-8	: 4463	541-46-8	: 7062	544-35-4	: 5178	552-32-9	: 8137
535-13-7	: 4947	538-86-3	: 808	541-47-9	: 7096	544-40-1	: 3051	552-41-0	: 6068
535-15-9	: 4979	538-90-9	: 9190	541-50-4	: 25	544-62-7	: 8327	552-45-4	: 2129
535-32-0	: 4809	538-93-2	: 6378	541-53-7	: 10057	544-63-8	: 9795	552-58-9	: 4742
535-46-6	: 3745	539-00-4	: 9074	541-58-2	: 4336	544-73-0	: 8310	552-59-0	: 9283
535-75-1	: 9061	539-03-7	: 2231	541-59-3	: 9396	544-76-3	: 5747	552-63-6	: 6077
535-77-3	: 6508	539-08-2	: 4863	541-69-5	: 615	544-77-4	: 6305	552-70-5	: 6952
535-80-8	: 1897	539-12-8	: 9189	541-73-1	: 3093	544-85-4	: 4663	552-72-7	: 6623
535-83-1	: 10428	539-15-1	: 3912	541-85-5	: 7309	544-97-8	: 4352	552-79-4	: 7250
535-86-4	: 6512	539-30-0	: 798	541-85-5	: 8361	545-06-2	: 10198	552-82-9	: 7220
535-87-5	: 2850	539-47-9	: 5499	541-88-8	: 1841	545-26-6	: 5533	552-86-3	: 5505
535-89-7	: 2484	539-52-6	: 7591	541-91-3	: 7165	545-47-1	: 6636	552-89-6	: 8042
536-06-1	: 6480	539-74-2	: 4921	542-05-2	: 8467	545-55-1	: 10362	552-94-3	: 1768
536-08-3	: 3756	539-80-0	: 2559	542-08-5	: 6524	545-93-7	: 1086	553-03-7	: 3677
536-17-4	: 3901	539-82-2	: 5193	542-10-9	: 4774	546-06-5	: 2453	553-17-3	: 6859
536-25-4	: 6934	539-86-6	: 160	542-11-0	: 470	546-43-0	: 146	553-19-5	: 10848
536-33-4	: 5234	539-88-8	: 5098	542-15-4	: 472	546-45-2	: 10620	553-20-8	: 8170
536-38-9	: 1149	539-89-9	: 5085	542-16-5	: 473	546-56-5	: 8364	553-24-2	: 8012
536-50-5	: 3958	539-90-2	: 6380	542-18-7	: 1961	546-68-9	: 9908	553-26-4	: 897
536-57-2	: 7005	539-92-4	: 3779	542-28-9	: 9886	546-80-5	: 10091	553-27-5	: 926
536-59-4	: 7857	540-07-8	: 8676	542-32-5	: 318	546-88-3	: 30	553-39-9	: 6120
536-60-7	: 6480	540-08-9	: 5653	542-37-0	: 4285	546-97-4	: 2452	553-60-6	: 6538
536-66-3	: 6483	540-09-0	: 10301	542-46-1	: 2551	547-25-1	: 10719	553-68-4	: 1465
536-69-6	: 1659	540-10-3	: 5762	542-52-9	: 3010	547-44-4	: 9648	553-69-5	: 9366
536-74-3	: 8800	540-18-1	: 8669	542-54-1	: 7556	547-52-4	: 9647	553-72-0	: 10871
536-75-4	: 5233	540-23-8	: 6943	542-55-2	: 6388	547-58-0	: 7529	553-79-7	: 8169
536-78-7	: 5232	540-36-3	: 3551	542-56-3	: 6398	547-63-7	: 7395	553-82-2	: 3205
536-80-1	: 6343	540-37-4	: 6272	542-58-5	: 2027	547-64-8	: 7429	553-86-6	: 707
536-88-9	: 5156	540-38-5	: 6332	542-59-6	: 4786	547-65-9	: 3642	553-90-2	: 4201
536-90-3	: 6773	540-42-1	: 6401	542-69-8	: 6287	547-81-9	: 4762	553-94-6	: 1183
536-95-8	: 834	540-43-2	: 7284	542-75-6	: 3282	547-91-1	: 6051	553-97-9	: 7144
537-17-7	: 9009	540-47-6	: 2699	542-75-6	: 3283	548-40-3	: 10487	554-00-7	: 3076
537-24-6	: 2996	540-49-8	: 2939	542-76-7	: 2265	548-51-6	: 6089	554-01-8	: 355
537-26-8	: 10710	540-49-8	: 2940	542-78-9	: 6663	548-62-9	: 2488	554-12-1	: 7690
537-29-1	: 8274	540-51-2	: 1193	542-81-4	: 2166	548-73-2	: 4670	554-14-3	: 7815
537-39-3	: 5569	540-54-5	: 2260	542-85-8	: 5095	548-80-1	: 2384	554-35-8	: 5544
537-45-1	: 2988	540-59-0	: 3177	542-88-1	: 937	548-84-5	: 9405	554-57-4	: 6754
537-46-2	: 6738	540-59-0	: 3178	542-90-5	: 5258	548-93-6	: 325	554-68-7	: 10346
537-47-3	: 8892	540-61-4	: 230	542-91-6	: 3523	548-98-1	: 2360	554-70-1	: 10366
537-49-5	: 7842	540-63-6	: 4793	542-92-7	: 2658	550-10-7	: 5949	554-73-4	: 8414
537-55-3	: 105	540-67-0	: 5139	543-18-0	: 5612	550-24-3	: 4694	554-84-7	: 8133
537-64-4	: 1028	540-73-8	: 4107	543-20-4	: 1417	550-44-7	: 7397	554-91-6	: 5527
537-65-5	: 395	540-80-7	: 1624	543-21-5	: 1677	550-60-7	: 8127	554-92-7	: 10472
537-91-7	: 1039	540-82-9	: 5249	543-24-8	: 71	550-74-3	: 9036	554-95-0	: 670
537-92-8	: 7613	540-84-1	: 10580	543-27-1	: 6382	550-82-3	: 9453	555-03-3	: 8038
538-07-8	: 1948	540-88-5	: 1489	543-28-2	: 6381	550-97-0	: 7987	555-10-2	: 8741
538-08-9	: 2830	540-92-1	: 9563	543-29-3	: 6397	550-99-2	: 7925	555-16-8	: 8044

CAS	No.	CAS	No.	CAS	No.	CAS	No.	CAS	No.
589-34-4	: 7327	591-23-1	: 7153	592-99-4	: 8373	595-44-8	: 3241	598-99-2	: 7831
589-35-5	: 7563	591-24-2	: 7157	593-08-8	: 10316	595-46-0	: 4127	599-04-2	: 8502
589-38-8	: 5854	591-27-5	: 390	593-39-5	: 8314	595-49-3	: 4404	599-61-1	: 2857
589-40-2	: 1577	591-31-1	: 6778	593-45-3	: 8304	595-89-1	: 9993	599-64-4	: 7635
589-43-5	: 4090	591-34-4	: 1655	593-49-7	: 5645	595-90-4	: 9995	599-66-6	: 1031
589-44-6	: 328	591-35-5	: 3255	593-50-0	: 10131	596-27-0	: 2480	599-67-7	: 7625
589-53-7	: 7287	591-47-9	: 7161	593-51-1	: 6921	596-43-0	: 1373	599-70-2	: 5213
589-55-9	: 5688	591-48-0	: 7160	593-52-2	: 6951	596-51-0	: 5586	599-79-1	: 9650
589-59-3	: 6395	591-49-1	: 7159	593-53-3	: 5395	596-75-8	: 3425	599-91-7	: 9266
589-62-8	: 8359	591-50-4	: 6275	593-54-4	: 7666	597-09-1	: 5173	600-00-0	: 4916
589-63-9	: 8362	591-54-8	: 9381	593-56-6	: 7372	597-12-6	: 6692	600-05-5	: 2978
589-75-3	: 1629	591-60-6	: 1491	593-57-7	: 3942	597-25-1	: 4137	600-07-7	: 7074
589-81-1	: 7286	591-62-8	: 4935	593-60-2	: 1194	597-31-9	: 6032	600-11-3	: 3248
589-82-2	: 5687	591-68-4	: 1635	593-61-3	: 1084	597-35-3	: 3531	600-14-6	: 8617
589-87-7	: 1237	591-76-4	: 7326	593-63-5	: 1848	597-49-9	: 5194	600-15-7	: 6007
589-90-2	: 4015	591-78-6	: 5853	593-66-8	: 6299	597-50-2	: 10367	600-18-0	: 8459
589-90-2	: 4016	591-80-0	: 8645	593-70-4	: 2054	597-51-3	: 10368	600-22-6	: 7541
589-91-3	: 7154	591-81-1	: 6009	593-71-5	: 2081	597-52-4	: 10372	600-25-9	: 2194
589-91-3	: 7155	591-82-2	: 6393	593-74-8	: 4128	597-63-7	: 9822	600-36-2	: 4225
589-92-4	: 7158	591-84-4	: 6403	593-75-9	: 6423	597-64-8	: 9827	600-40-8	: 4385
589-93-5	: 4301	591-87-7	: 167	593-77-1	: 6055	597-71-7	: 8572	601-34-3	: 2373
589-98-0	: 8358	591-93-5	: 8567	593-79-3	: 4318	597-90-0	: 5044	601-57-0	: 2375
590-01-2	: 1654	591-95-7	: 8564	593-80-6	: 4330	597-96-6	: 7336	601-75-2	: 5223
590-02-3	: 1531	591-96-8	: 8568	593-81-7	: 10501	598-01-6	: 7306	601-76-3	: 4402
590-11-4	: 2939	591-97-9	: 1938	593-88-4	: 10505	598-02-7	: 3467	601-77-4	: 6523
590-12-5	: 2940	591-97-9	: 1939	593-90-8	: 10522	598-03-8	: 4566	602-01-7	: 7208
590-13-6	: 1336	592-13-2	: 4091	593-95-3	: 1755	598-04-9	: 3055	602-03-9	: 4360
590-14-7	: 1336	592-20-1	: 91	593-96-4	: 1144	598-05-0	: 4564	602-09-5	: 873
590-15-8	: 1337	592-27-8	: 7285	593-98-6	: 1146	598-09-4	: 2133	602-38-0	: 4390
590-18-1	: 1442	592-31-4	: 1669	594-07-0	: 1727	598-10-7	: 2693	602-55-1	: 8816
590-19-2	: 1392	592-34-7	: 1534	594-08-1	: 10708	598-14-1	: 4980	602-56-2	: 8804
590-21-6	: 2272	592-35-8	: 1530	594-09-2	: 10604	598-16-3	: 10165	602-60-8	: 8040
590-21-6	: 2273	592-41-6	: 5862	594-10-5	: 10617	598-21-0	: 1083	602-64-2	: 10433
590-26-1	: 3765	592-42-7	: 5779	594-11-6	: 7177	598-23-2	: 7114	602-87-9	: 3659
590-35-2	: 4213	592-43-8	: 5863	594-14-9	: 5611	598-25-4	: 7059	602-92-6	: 2995
590-36-3	: 7565	592-43-8	: 5864	594-15-0	: 10162	598-29-8	: 3771	602-94-8	: 8582
590-42-1	: 1597	592-44-9	: 5774	594-16-1	: 2977	598-31-2	: 1080	603-11-2	: 8057
590-46-5	: 846	592-45-0	: 5777	594-18-3	: 2931	598-32-3	: 1458	603-32-7	: 10644
590-50-1	: 4226	592-45-0	: 5778	594-20-7	: 3273	598-38-9	: 3175	603-33-8	: 10646
590-54-5	: 98	592-46-1	: 5780	594-27-4	: 9975	598-41-4	: 229	603-34-9	: 10643
590-55-6	: 1728	592-46-1	: 5781	594-34-3	: 2962	598-45-8	: 6425	603-35-0	: 10656
590-60-3	: 8670	592-46-1	: 5782	594-36-5	: 2112	598-50-5	: 7850	603-36-1	: 10661
590-66-9	: 4010	592-47-2	: 5865	594-37-6	: 3217	598-52-7	: 7824	603-45-2	: 535
590-67-0	: 7149	592-47-2	: 5866	594-38-7	: 6315	598-53-8	: 6514	603-48-5	: 10683
590-73-8	: 4088	592-48-3	: 5775	594-39-8	: 7064	598-55-0	: 7116	603-50-9	: 900
590-86-3	: 7061	592-48-3	: 5776	594-42-3	: 10246	598-56-1	: 4993	603-54-3	: 4534
590-90-9	: 6012	592-50-7	: 5409	594-44-5	: 4796	598-57-2	: 8121	603-71-4	: 10572
590-92-1	: 1331	592-51-8	: 8643	594-45-6	: 4795	598-58-3	: 7477	603-76-9	: 7386
590-93-2	: 1682	592-55-2	: 1198	594-51-4	: 2960	598-61-8	: 7141	603-83-8	: 7480
591-07-1	: 276	592-57-4	: 2561	594-56-9	: 10526	598-72-1	: 1330	604-32-0	: 2372
591-08-2	: 431	592-62-1	: 6956	594-60-5	: 3991	598-73-2	: 1366	604-35-3	: 2371
591-09-3	: 85	592-65-4	: 3782	594-61-6	: 6102	598-74-3	: 7065	604-44-4	: 2171
591-11-7	: 7275	592-76-7	: 5694	594-65-0	: 10195	598-75-4	: 7080	604-53-5	: 871
591-12-8	: 7274	592-77-8	: 5695	594-70-7	: 8118	598-77-6	: 10278	604-59-1	: 699
591-17-3	: 1360	592-77-8	: 5696	594-71-8	: 2195	598-78-7	: 2266	604-68-2	: 5547
591-18-4	: 1236	592-78-9	: 5697	594-72-9	: 3239	598-79-8	: 2277	604-69-3	: 5548
591-19-5	: 1088	592-78-9	: 5698	594-73-0	: 5720	598-82-3	: 6574	604-75-1	: 8440
591-20-8	: 1313	592-82-5	: 1595	594-82-1	: 9931	598-88-9	: 3160	604-88-6	: 5789
591-21-9	: 4013	592-84-7	: 1576	594-83-2	: 10525	598-88-9	: 3161	605-01-6	: 8576
591-21-9	: 4014	592-88-1	: 2841	595-37-9	: 3988	598-92-5	: 2190	605-02-7	: 8937
591-22-0	: 4304	592-90-5	: 8442	595-40-4	: 6558	598-94-7	: 4350	605-32-3	: 5967
591-23-1	: 7152	592-99-4	: 8372	595-41-5	: 4224	598-98-1	: 7205	605-39-0	: 3977

605-45-8	: 3811	609-36-9	: 9123	612-22-6	: 5168	615-13-4	: 3631	617-48-1	: 6662
605-48-1	: 3081	609-39-2	: 8103	612-23-7	: 2136	615-15-6	: 7016	617-50-5	: 6502
605-50-5	: 3791	609-40-5	: 8205	612-25-9	: 8061	615-16-7	: 3597	617-51-6	: 6503
605-54-9	: 975	609-65-4	: 1911	612-28-2	: 7486	615-18-9	: 1909	617-52-7	: 4130
605-65-2	: 3914	609-66-5	: 1864	612-55-5	: 6320	615-20-3	: 1905	617-54-9	: 4129
605-69-6	: 4391	609-67-6	: 6282	612-57-7	: 2298	615-21-4	: 741	617-65-2	: 5552
605-71-0	: 4389	609-72-3	: 7202	612-58-8	: 7767	615-22-5	: 7809	617-73-2	: 6135
606-17-7	: 6264	609-73-4	: 6321	612-60-2	: 7771	615-28-1	: 614	617-75-4	: 10347
606-20-2	: 7211	609-93-8	: 7215	612-62-4	: 2296	615-36-1	: 1087	617-78-7	: 5191
606-21-3	: 2009	609-99-4	: 6033	612-71-5	: 8999	615-37-2	: 6311	617-79-8	: 4934
606-22-4	: 4363	610-02-6	: 10435	612-75-9	: 3978	615-38-3	: 7150	617-83-4	: 3421
606-23-5	: 6236	610-14-0	: 8081	612-78-2	: 872	615-39-4	: 7151	617-84-5	: 3460
606-27-9	: 7490	610-16-2	: 3902	612-82-8	: 10109	615-41-8	: 2077	617-85-6	: 10373
606-28-0	: 7034	610-17-3	: 4150	612-83-9	: 3108	615-42-9	: 3760	617-86-7	: 10371
606-35-9	: 6906	610-27-5	: 8058	612-94-2	: 8938	615-43-0	: 6270	617-88-9	: 2125
606-37-1	: 4388	610-30-0	: 4375	612-96-4	: 8993	615-54-3	: 10155	617-89-0	: 5486
606-43-9	: 7774	610-39-9	: 7213	613-03-6	: 676	615-57-6	: 2905	617-90-3	: 5480
606-45-1	: 7441	610-48-0	: 6947	613-08-1	: 483	615-58-7	: 2973	617-92-5	: 5239
606-46-2	: 3480	610-50-4	: 492	613-12-7	: 6948	615-59-8	: 2957	617-94-7	: 3959
606-83-7	: 8838	610-60-6	: 7039	613-13-8	: 478	615-60-1	: 1995	618-27-9	: 6177
607-00-1	: 4487	610-66-2	: 8051	613-29-6	: 3006	615-62-3	: 2149	618-31-5	: 2959
607-30-7	: 7988	610-67-3	: 4855	613-31-0	: 3592	615-65-6	: 2099	618-32-6	: 760
607-34-1	: 8178	610-69-5	: 8140	613-33-2	: 3979	615-66-7	: 1876	618-36-0	: 7040
607-35-2	: 8180	610-72-0	: 3964	613-35-4	: 2825	615-67-8	: 1880	618-38-2	: 767
607-42-1	: 482	610-89-9	: 4877	613-37-6	: 6812	615-74-7	: 2147	618-40-6	: 7639
607-56-7	: 4357	610-94-6	: 7051	613-42-3	: 789	615-79-2	: 5013	618-41-7	: 659
607-57-8	: 8100	610-96-8	: 7121	613-45-6	: 3836	615-81-6	: 3808	618-42-8	: 99
607-81-8	: 3406	610-97-9	: 7391	613-46-7	: 7934	615-83-8	: 4918	618-45-1	: 6528
607-85-2	: 6501	610-99-1	: 6167	613-48-9	: 3481	615-93-0	: 3146	618-46-2	: 1912
607-90-9	: 9241	611-01-8	: 3963	613-50-3	: 8179	615-94-1	: 3719	618-51-9	: 6279
607-91-0	: 7915	611-06-3	: 3238	613-69-4	: 4825	615-98-5	: 4561	618-62-2	: 3236
607-97-6	: 5027	611-07-4	: 2193	613-70-7	: 6866	615-99-6	: 2838	618-65-5	: 5542
607-99-8	: 10168	611-08-5	: 8176	613-75-2	: 5487	616-02-4	: 7270	618-68-8	: 786
608-07-1	: 6833	611-10-9	: 4968	613-84-3	: 6076	616-05-7	: 1226	618-76-8	: 6025
608-08-2	: 6255	611-13-2	: 7269	613-89-8	: 396	616-06-8	: 8275	618-80-4	: 3240
608-25-3	: 6983	611-14-3	: 5264	613-90-1	: 8453	616-07-9	: 8416	618-83-7	: 5980
608-27-5	: 3075	611-15-4	: 7784	613-91-2	: 8874	616-12-6	: 7581	618-85-9	: 7210
608-28-6	: 6293	611-17-6	: 1256	613-93-4	: 6962	616-20-6	: 2210	618-87-1	: 4364
608-31-1	: 3078	611-19-8	: 1950	613-94-5	: 711	616-21-7	: 3129	618-89-3	: 7052
608-33-3	: 2974	611-20-1	: 5994	613-97-8	: 5117	616-23-9	: 3275	618-91-7	: 7392
608-45-7	: 505	611-21-2	: 3933	614-00-6	: 7503	616-24-0	: 8606	618-94-0	: 8076
608-66-2	: 5512	611-32-5	: 7772	614-16-4	: 8455	616-25-1	: 8646	618-95-1	: 7491
608-68-4	: 4329	611-33-6	: 2299	614-17-5	: 4893	616-29-5	: 2865	618-98-4	: 5170
608-71-9	: 8527	611-34-7	: 9425	614-18-6	: 5236	616-30-8	: 404	619-04-5	: 3966
608-73-1	: 5730	611-35-8	: 2297	614-21-1	: 8151	616-31-9	: 8623	619-08-9	: 2191
608-73-1	: 5731	611-36-9	: 9435	614-26-6	: 3085	616-38-6	: 4005	619-15-8	: 7212
608-73-1	: 5732	611-69-8	: 6481	614-27-7	: 7033	616-39-7	: 3479	619-17-0	: 374
608-73-1	: 5733	611-70-1	: 7653	614-29-9	: 8835	616-42-2	: 4325	619-18-1	: 4362
608-89-9	: 5252	611-72-3	: 5973	614-30-2	: 7623	616-43-3	: 7756	619-21-6	: 5458
608-93-5	: 8533	611-73-4	: 8451	614-33-5	: 9158	616-44-4	: 7816	619-23-8	: 2137
609-02-9	: 4131	611-74-5	: 3947	614-34-6	: 7627	616-45-5	: 9402	619-24-9	: 8078
609-08-5	: 3487	611-92-7	: 4050	614-39-1	: 9115	616-47-7	: 7379	619-25-0	: 8062
609-09-6	: 3474	611-94-9	: 6878	614-45-9	: 1638	616-55-7	: 3034	619-31-8	: 4151
609-11-0	: 4975	611-95-0	: 757	614-47-1	: 4521	616-68-2	: 10468	619-39-6	: 8387
609-12-1	: 4915	611-97-2	: 3968	614-61-9	: 2222	616-79-5	: 375	619-41-0	: 1277
609-14-3	: 5108	611-99-4	: 3711	614-68-6	: 6412	616-91-1	: 64	619-42-1	: 7053
609-19-8	: 10272	612-00-0	: 4479	614-75-5	: 5974	617-05-0	: 5079	619-44-3	: 7393
609-20-1	: 3096	612-12-4	: 934	614-78-8	: 7663	617-12-9	: 2380	619-45-4	: 6925
609-22-3	: 2961	612-14-6	: 630	614-80-2	: 6143	617-31-2	: 6140	619-50-1	: 7492
609-23-4	: 10451	612-15-7	: 10806	614-96-0	: 7389	617-33-4	: 4974	619-55-6	: 6961
609-26-7	: 5146	612-16-8	: 6793	614-99-3	: 5037	617-35-6	: 5189	619-58-9	: 6280
609-31-4	: 8089	612-17-9	: 3657	615-05-4	: 6786	617-45-8	: 524	619-60-3	: 3916

619-65-8	: 2508	622-16-2	: 4473	623-70-1	: 4930	625-36-5	: 2271	626-77-7	: 9240
619-66-9	: 5459	622-24-2	: 2031	623-71-2	: 4948	625-38-7	: 1453	626-82-4	: 1581
619-67-0	: 5940	622-26-4	: 9066	623-73-4	: 4973	625-44-5	: 6396	626-85-7	: 3054
619-72-7	: 8079	622-29-7	: 8927	623-76-7	: 3541	625-45-6	: 6769	626-86-8	: 5067
619-73-8	: 8063	622-31-1	: 588	623-78-9	: 5028	625-48-9	: 8096	626-87-9	: 2970
619-80-7	: 8046	622-32-2	: 587	623-81-4	: 3530	625-50-3	: 4870	626-88-0	: 1272
619-82-9	: 2573	622-37-7	: 550	623-84-7	: 9139	625-52-5	: 5281	626-89-1	: 7564
619-84-1	: 3904	622-38-8	: 5257	623-87-0	: 5564	625-54-7	: 5092	626-93-7	: 5851
619-86-3	: 4831	622-39-9	: 9261	623-91-6	: 3461	625-55-8	: 6496	626-94-8	: 5878
619-99-8	: 5047	622-40-2	: 7904	623-93-8	: 8242	625-56-9	: 2096	626-95-9	: 8612
620-02-0	: 7268	622-42-4	: 8123	623-96-1	: 4553	625-58-1	: 5165	626-96-0	: 8466
620-05-3	: 6313	622-44-6	: 8899	623-97-2	: 927	625-60-5	: 5255	626-97-1	: 8605
620-08-6	: 6903	622-45-7	: 2606	624-03-3	: 4780	625-65-0	: 4235	627-00-9	: 1930
620-13-3	: 1263	622-47-9	: 6973	624-04-4	: 4778	625-68-3	: 1929	627-03-2	: 4818
620-14-4	: 5265	622-57-1	: 5116	624-13-5	: 9252	625-69-4	: 8615	627-04-3	: 5256
620-16-6	: 2033	622-58-2	: 6414	624-16-8	: 2765	625-71-8	: 6008	627-05-4	: 8088
620-17-7	: 5201	622-60-6	: 4848	624-17-9	: 3494	625-72-9	: 6008	627-06-5	: 9269
620-19-9	: 2130	622-61-7	: 2023	624-18-0	: 616	625-75-2	: 8032	627-08-7	: 7256
620-20-2	: 1951	622-62-8	: 4860	624-20-4	: 2950	625-76-3	: 4387	627-09-8	: 9174
620-22-4	: 7021	622-63-9	: 5260	624-22-6	: 7331	625-77-4	: 42	627-11-2	: 2035
620-23-5	: 6958	622-78-6	: 805	624-24-8	: 7558	625-80-9	: 3812	627-12-3	: 9218
620-24-6	: 5984	622-79-7	: 554	624-29-3	: 4015	625-82-1	: 4310	627-13-4	: 9250
620-32-6	: 2900	622-80-0	: 9210	624-31-7	: 6345	625-84-3	: 4311	627-18-9	: 1332
620-40-6	: 10151	622-85-5	: 8976	624-38-4	: 3762	625-86-5	: 4063	627-19-0	: 8687
620-42-8	: 10705	622-86-6	: 2024	624-41-9	: 7081	625-92-3	: 2986	627-20-3	: 8640
620-45-1	: 3200	622-93-5	: 3395	624-42-0	: 7310	625-95-6	: 6312	627-21-4	: 8688
620-63-3	: 5571	622-96-8	: 5266	624-44-2	: 7339	625-96-7	: 7157	627-22-5	: 1922
620-71-3	: 8957	622-97-9	: 7786	624-45-3	: 7539	625-98-9	: 2050	627-26-9	: 1449
620-73-5	: 8861	622-98-0	: 2034	624-48-6	: 4125	625-99-0	: 2078	627-27-0	: 1457
620-80-4	: 4907	623-00-7	: 1108	624-49-7	: 4062	626-00-6	: 3761	627-30-5	: 2269
620-81-5	: 4481	623-03-0	: 1901	624-51-1	: 8240	626-01-7	: 6271	627-31-6	: 3772
620-83-7	: 7042	623-05-2	: 5985	624-54-4	: 8684	626-02-8	: 6331	627-32-7	: 6338
620-88-2	: 8136	623-08-5	: 3935	624-60-2	: 5112	626-04-0	: 639	627-35-0	: 7709
620-92-8	: 1014	623-10-9	: 6075	624-64-6	: 1443	626-15-3	: 915	627-37-2	: 7698
620-93-9	: 7459	623-12-1	: 1857	624-65-7	: 2285	626-16-4	: 935	627-39-4	: 9133
620-94-0	: 1030	623-15-4	: 5491	624-67-9	: 9271	626-17-5	: 3318	627-40-7	: 6891
620-99-5	: 977	623-17-6	: 5489	624-72-6	: 3559	626-18-6	: 631	627-41-8	: 6898
621-03-4	: 2520	623-19-8	: 5504	624-73-7	: 3764	626-19-7	: 621	627-42-9	: 2093
621-07-8	: 6161	623-22-3	: 9238	624-74-8	: 3758	626-20-0	: 10807	627-44-1	: 3478
621-08-9	: 2901	623-24-5	: 916	624-75-9	: 6268	626-23-3	: 3004	627-45-2	: 5034
621-14-7	: 4528	623-25-6	: 936	624-76-0	: 6298	626-26-6	: 3052	627-48-5	: 4953
621-23-8	: 10479	623-26-7	: 3319	624-78-2	: 5111	626-27-7	: 5684	627-49-6	: 3508
621-29-4	: 6413	623-27-8	: 622	624-79-3	: 5087	626-29-9	: 9796	627-51-0	: 4606
621-32-9	: 4847	623-30-3	: 5497	624-80-6	: 5065	626-33-5	: 7311	627-53-2	: 3522
621-33-0	: 4823	623-33-6	: 5578	624-83-9	: 7396	626-35-7	: 5167	627-54-3	: 3534
621-34-1	: 4859	623-36-9	: 7575	624-84-0	: 5934	626-36-8	: 4885	627-58-7	: 4086
621-36-3	: 6972	623-37-0	: 5852	624-89-5	: 5158	626-38-0	: 8663	627-59-8	: 7335
621-37-4	: 5975	623-39-2	: 6886	624-90-8	: 6955	626-39-1	: 10156	627-63-4	: 1447
621-42-1	: 6144	623-40-5	: 8631	624-91-9	: 7478	626-40-4	: 2906	627-70-3	: 36
621-59-0	: 6059	623-42-7	: 7073	624-92-0	: 4051	626-43-7	: 3080	627-83-8	: 4783
621-62-5	: 1975	623-43-8	: 7093	624-95-3	: 3990	626-48-2	: 7753	627-84-9	: 4784
621-64-7	: 8200	623-46-1	: 3179	625-01-4	: 4950	626-51-7	: 7279	627-90-7	: 5279
621-70-5	: 9160	623-47-2	: 5229	625-04-7	: 347	626-53-9	: 5682	627-91-8	: 7894
621-71-6	: 5568	623-48-3	: 5084	625-06-9	: 4222	626-55-1	: 1347	627-93-0	: 3891
621-72-7	: 787	623-49-4	: 4958	625-08-1	: 6085	626-56-2	: 7675	627-97-4	: 7314
621-77-2	: 10642	623-50-7	: 5071	625-22-9	: 3050	626-58-4	: 7676	627-98-5	: 7332
621-82-9	: 2398	623-51-8	: 5099	625-23-0	: 7333	626-60-8	: 2290	628-02-4	: 5828
621-82-9	: 2399	623-53-0	: 5130	625-25-2	: 7294	626-61-9	: 2291	628-04-6	: 5089
621-87-4	: 8792	623-59-6	: 6930	625-27-4	: 7579	626-62-0	: 6289	628-05-7	: 8129
621-95-4	: 907	623-65-4	: 5752	625-28-5	: 7068	626-64-2	: 9363	628-11-5	: 2282
622-08-2	: 817	623-66-5	: 8355	625-31-0	: 8651	626-67-5	: 7673	628-13-7	: 9348
622-15-1	: 4501	623-68-7	: 1454	625-33-2	: 8653	626-71-1	: 234	628-16-0	: 5787

628-17-1	: 6328	629-63-0	: 9793	634-66-2	: 9732	638-37-9	: 1402	645-96-5	: 658
628-20-6	: 1927	629-64-1	: 3583	634-67-3	: 10200	638-38-0	: 6674	646-01-5	: 7820
628-21-7	: 3763	629-65-2	: 3585	634-74-2	: 9469	638-41-5	: 8671	646-04-8	: 8641
628-28-4	: 1611	629-66-3	: 8222	634-90-2	: 9733	638-45-9	: 6306	646-05-9	: 8655
628-29-5	: 1617	629-70-9	: 5758	634-93-5	: 10202	638-46-0	: 1573	646-06-0	: 4433
628-30-8	: 9246	629-72-1	: 1302	634-95-7	: 3540	638-49-3	: 8674	646-07-1	: 7561
628-32-0	: 5225	629-73-2	: 5756	634-97-9	: 9394	638-53-9	: 10314	646-13-9	: 6402
628-34-2	: 2040	629-74-3	: 5772	635-10-9	: 9927	638-58-4	: 9789	646-14-0	: 8110
628-36-4	: 5937	629-76-5	: 8554	635-21-2	: 285	638-59-5	: 9802	646-19-5	: 5673
628-37-5	: 3502	629-78-7	: 5648	635-22-3	: 2176	638-65-3	: 8305	646-20-8	: 5674
628-41-1	: 2562	629-79-8	: 5749	635-46-1	: 9891	638-66-4	: 9603	646-25-3	: 2751
628-44-4	: 7520	629-80-1	: 5745	635-51-8	: 8850	638-67-5	: 10300	646-30-0	: 8220
628-55-7	: 3780	629-82-3	: 4419	635-65-4	: 868	638-68-6	: 10129	646-31-1	: 9779
628-61-5	: 2203	629-90-3	: 5646	635-67-6	: 633	638-95-9	: 10763	650-51-1	: 9577
628-63-7	: 8662	629-92-5	: 8219	635-78-9	: 6142	639-58-7	: 2341	650-69-1	: 2412
628-66-0	: 9140	629-93-6	: 6325	635-90-5	: 8988	639-81-6	: 4739	652-67-5	: 6553
628-67-1	: 1409	629-94-7	: 5638	635-93-8	: 2071	639-99-6	: 4692	653-03-2	: 1438
628-68-2	: 3437	629-96-9	: 4684	636-04-4	: 8831	640-15-3	: 10059	653-37-2	: 8579
628-71-7	: 5712	629-97-0	: 4611	636-09-9	: 3535	640-19-7	: 5351	654-42-2	: 3954
628-73-9	: 5844	629-99-2	: 8547	636-21-5	: 6942	640-61-9	: 3961	655-48-1	: 4483
628-76-2	: 3247	630-01-3	: 5741	636-28-2	: 9710	641-06-5	: 10127	655-86-7	: 8751
628-77-3	: 3770	630-02-4	: 8297	636-30-6	: 10201	641-36-1	: 500	659-70-1	: 6448
628-80-8	: 7594	630-03-5	: 8217	636-36-2	: 6047	641-70-3	: 8116	660-68-4	: 3379
628-81-9	: 1567	630-04-6	: 5639	636-41-9	: 7755	641-85-0	: 9101	660-88-8	: 387
628-83-1	: 1661	630-05-7	: 10709	636-46-4	: 5979	642-18-2	: 217	661-11-0	: 5389
628-87-5	: 6225	630-06-8	: 5856	636-53-3	: 3471	642-31-9	: 481	661-19-8	: 4613
628-89-7	: 2026	630-07-9	: 8635	636-60-2	: 7787	642-44-4	: 309	661-54-1	: 10422
628-92-2	: 2560	630-08-0	: 1746	636-72-6	: 10074	642-84-2	: 1974	661-95-0	: 2949
628-94-4	: 5831	630-10-4	: 9523	636-73-7	: 9358	643-13-0	: 5472	661-97-2	: 3193
628-96-6	: 4782	630-17-1	: 1186	636-82-8	: 2595	643-22-1	: 4747	662-01-1	: 3194
628-97-7	: 5190	630-18-2	: 4275	636-97-5	: 8077	643-28-7	: 6474	665-66-7	: 233
628-99-9	: 8239	630-19-3	: 4269	636-98-6	: 6323	643-58-3	: 7046	666-99-9	: 6132
629-01-6	: 8344	630-20-6	: 9751	637-03-6	: 8472	643-60-7	: 460	670-54-2	: 9783
629-03-8	: 2951	630-60-4	: 8422	637-07-0	: 2426	643-65-2	: 7649	670-96-2	: 8897
629-04-9	: 1218	630-72-8	: 10622	637-27-4	: 8960	643-79-8	: 620	671-16-9	: 9116
629-05-0	: 8402	630-76-2	: 9991	637-39-8	: 995	643-84-5	: 6670	672-13-9	: 6058
629-06-1	: 2060	631-01-6	: 9409	637-44-5	: 8979	643-93-6	: 7047	672-15-1	: 5926
629-08-3	: 5680	631-36-7	: 9826	637-50-3	: 9187	644-08-6	: 7048	672-65-1	: 2030
629-09-4	: 3767	631-40-3	: 9998	637-53-6	: 9002	644-30-4	: 4104	673-22-3	: 6057
629-11-8	: 5835	631-57-2	: 8475	637-55-8	: 8998	644-31-5	: 56	673-31-4	: 655
629-14-1	: 3360	631-64-1	: 2903	637-59-2	: 1344	644-32-6	: 2891	673-32-5	: 9275
629-15-2	: 4779	631-65-2	: 2113	637-64-9	: 9858	644-35-9	: 9255	673-79-0	: 3536
629-17-4	: 4785	631-67-4	: 4338	637-65-0	: 9856	644-36-0	: 6971	673-84-7	: 4190
629-19-6	: 4554	632-05-3	: 10159	637-69-4	: 10808	644-49-5	: 9244	673-84-7	: 4191
629-20-9	: 2649	632-14-4	: 10621	637-78-5	: 6533	644-64-4	: 4353	674-76-0	: 7583
629-27-6	: 6326	632-15-5	: 4343	637-87-6	: 2079	644-90-6	: 381	674-82-8	: 3816
629-30-1	: 5676	632-16-6	: 4340	637-88-7	: 2580	644-97-3	: 8952	675-09-2	: 4288
629-31-2	: 5669	632-22-4	: 9980	637-89-8	: 6715	645-00-1	: 6322	675-10-5	: 6105
629-33-4	: 5897	632-25-7	: 9657	637-92-3	: 1569	645-09-0	: 8045	675-14-9	: 2523
629-35-6	: 3032	632-32-6	: 4736	638-00-6	: 4341	645-12-5	: 8105	675-20-7	: 9072
629-36-7	: 950	632-46-2	: 3965	638-02-8	: 4342	645-13-6	: 6532	676-58-4	: 7433
629-37-8	: 8128	632-50-8	: 9987	638-04-0	: 4013	645-35-2	: 5919	676-59-5	: 4260
629-39-0	: 8393	632-51-9	: 9989	638-07-3	: 4941	645-36-3	: 3358	676-83-5	: 3216
629-40-3	: 8348	632-79-1	: 9714	638-08-4	: 9605	645-41-0	: 10458	676-99-3	: 7668
629-41-4	: 8351	632-80-4	: 9904	638-10-8	: 5127	645-45-4	: 656	677-21-4	: 10421
629-45-8	: 3016	632-93-9	: 3429	638-11-9	: 6484	645-48-7	: 9004	677-69-0	: 5664
629-46-9	: 8394	632-99-5	: 6654	638-16-4	: 10143	645-49-8	: 9612	678-26-2	: 8723
629-50-5	: 10311	633-03-4	: 1071	638-21-1	: 8947	645-56-7	: 9256	680-31-9	: 5824
629-54-9	: 5746	633-65-8	: 842	638-23-3	: 1767	645-59-0	: 651	681-84-5	: 9974
629-59-4	: 9790	634-03-7	: 8756	638-25-5	: 8680	645-62-5	: 5053	682-01-9	: 9996
629-60-7	: 10313	634-19-5	: 9852	638-28-8	: 2067	645-66-9	: 4630	682-30-4	: 3542
629-62-9	: 8552	634-36-6	: 10478	638-29-9	: 8632	645-92-1	: 2867	683-08-9	: 3488

683-18-1	: 3059	694-80-4	: 1138	717-74-8	: 10463	764-47-6	: 9270	771-99-3	: 8956
683-50-1	: 2259	694-83-7	: 2571	719-22-2	: 3013	764-48-7	: 10818	772-00-9	: 8870
683-68-1	: 2930	694-83-7	: 2572	719-32-4	: 9774	764-85-2	: 8247	772-01-0	: 8869
683-72-7	: 3066	694-85-9	: 7747	719-59-5	: 287	764-93-2	: 2788	772-17-8	: 7003
684-16-2	: 8695	694-92-8	: 7513	719-79-9	: 4465	764-96-5	: 10743	772-33-8	: 1269
684-93-5	: 7504	695-06-7	: 4988	721-50-6	: 9113	764-97-6	: 10744	773-76-2	: 3290
685-63-2	: 5795	695-12-5	: 10797	721-90-4	: 8808	764-99-8	: 1056	773-82-0	: 8583
685-73-4	: 5517	695-34-1	: 7733	722-27-0	: 906	765-03-7	: 4658	775-12-2	: 4526
685-87-0	: 3409	695-53-4	: 4202	723-46-6	: 9644	765-05-9	: 2787	775-56-4	: 3364
685-91-6	: 3371	696-07-1	: 6341	723-61-5	: 5631	765-09-3	: 1363	776-35-2	: 3662
686-50-0	: 6603	696-23-1	: 7493	723-62-6	: 484	765-13-9	: 8563	776-74-9	: 1189
687-38-7	: 5593	696-28-6	: 3259	726-42-1	: 4594	765-30-0	: 2697	776-75-0	: 771
688-71-1	: 10670	696-29-7	: 6489	729-43-1	: 40	765-34-4	: 8448	776-76-1	: 7222
688-73-3	: 10190	696-30-0	: 6537	732-11-6	: 9023	765-42-4	: 7179	777-37-7	: 2198
688-74-4	: 10182	696-44-6	: 3934	732-26-3	: 10184	765-43-5	: 2700	778-22-3	: 4514
688-84-6	: 5063	696-59-3	: 9847	736-30-1	: 1041	765-47-9	: 4034	778-24-5	: 4049
689-11-2	: 1670	696-62-8	: 6310	738-70-5	: 10473	765-48-0	: 4312	778-25-6	: 7223
689-12-3	: 6470	696-63-9	: 6798	741-58-2	: 582	765-85-5	: 7142	778-28-9	: 1664
689-89-4	: 7321	696-71-9	: 2647	742-20-1	: 2680	765-87-7	: 2578	778-66-5	: 4520
689-97-4	: 1464	697-11-0	: 5797	744-45-6	: 4494	766-05-2	: 2567	779-02-2	: 6949
690-02-8	: 4242	697-82-5	: 10593	745-65-3	: 9277	766-07-4	: 2626	779-84-0	: 8900
690-08-4	: 4239	697-91-6	: 3147	747-90-0	: 2362	766-08-5	: 7660	780-25-6	: 8923
690-37-9	: 10585	698-01-1	: 1991	753-59-3	: 7196	766-09-6	: 5218	780-69-8	: 10338
690-39-1	: 5800	698-63-5	: 8104	753-89-9	: 2003	766-39-2	: 4064	781-35-1	: 4517
690-92-6	: 4101	698-69-1	: 1993	753-90-2	: 10390	766-51-8	: 1855	781-43-1	: 3940
690-93-7	: 4102	698-76-0	: 9882	754-05-2	: 10828	766-77-8	: 4258	782-74-1	: 945
691-37-2	: 7578	698-87-3	: 8962	754-10-9	: 4270	766-84-7	: 1900	786-19-6	: 1761
691-38-3	: 7582	699-02-5	: 6991	755-25-9	: 8711	766-85-8	: 6309	787-84-8	: 758
691-60-1	: 6545	699-17-2	: 5490	756-79-6	: 4132	766-90-5	: 9186	789-25-3	: 10659
692-04-6	: 76	699-30-9	: 9833	756-80-9	: 4052	766-92-7	: 7813	790-69-2	: 6626
692-24-0	: 7351	699-98-9	: 9344	757-58-4	: 5791	766-94-9	: 9015	791-28-6	: 10657
692-29-5	: 8460	700-12-9	: 8595	758-96-3	: 4271	766-96-1	: 1208	791-31-1	: 10660
692-42-2	: 3399	700-13-0	: 10510	759-05-7	: 7537	767-00-0	: 5996	793-24-8	: 3998
692-45-5	: 10804	700-16-3	: 8727	759-36-4	: 3473	767-12-4	: 4020	797-63-7	: 8278
692-47-7	: 3018	700-87-8	: 6410	759-73-9	: 5175	767-15-7	: 4306	804-30-8	: 5508
692-50-2	: 5796	700-88-9	: 2683	759-94-4	: 4552	767-54-4	: 10535	804-63-7	: 9420
692-86-4	: 5280	701-57-5	: 8204	760-20-3	: 7577	767-58-8	: 3649	811-93-8	: 7684
693-02-7	: 5910	701-97-3	: 2585	760-21-4	: 4928	767-59-9	: 7383	811-95-0	: 10237
693-05-0	: 6936	701-99-5	: 8771	760-23-6	: 3135	767-60-2	: 7385	811-97-2	: 9835
693-07-2	: 2046	702-23-8	: 6790	760-67-8	: 5052	767-99-7	: 7704	812-00-0	: 7665
693-13-0	: 3802	702-54-5	: 3428	760-78-1	: 8281	768-00-3	: 7705	812-03-3	: 9766
693-16-3	: 8345	703-80-0	: 6256	760-79-2	: 3982	768-22-9	: 3632	812-04-4	: 3308
693-21-0	: 3443	704-01-8	: 9924	760-93-0	: 7701	768-32-1	: 10602	813-78-5	: 4109
693-23-2	: 4625	705-86-2	: 9879	761-65-9	: 3024	768-33-2	: 2001	813-94-5	: 10192
693-36-7	: 4576	706-14-9	: 6023	762-04-9	: 3509	768-49-0	: 7706	814-49-3	: 3420
693-54-9	: 2763	706-31-0	: 2548	762-21-0	: 3412	768-52-5	: 6476	814-67-5	: 9144
693-58-3	: 1297	706-78-5	: 8292	762-42-5	: 4000	768-56-9	: 1463	814-68-6	: 9185
693-61-8	: 10740	706-79-6	: 3192	762-49-2	: 1212	768-66-1	: 9969	814-71-1	: 1702
693-62-9	: 10742	708-06-5	: 6112	762-50-5	: 2053	768-94-5	: 10308	814-75-5	: 1127
693-65-2	: 4441	708-76-9	: 6028	762-62-9	: 4233	769-06-2	: 9920	814-78-8	: 7101
693-67-4	: 1374	709-09-1	: 3877	762-63-0	: 4238	769-68-6	: 4897	814-80-2	: 1700
693-72-1	: 8315	709-50-2	: 7277	762-72-1	: 211	769-92-6	: 1506	814-90-4	: 2382
693-89-0	: 7172	709-98-8	: 9163	762-75-4	: 1578	770-09-2	: 10514	815-17-8	: 4208
693-93-6	: 7531	711-79-5	: 6127	763-29-1	: 7576	770-12-7	: 8953	815-24-7	: 3028
693-95-8	: 7801	712-48-1	: 4458	763-32-6	: 7098	770-35-4	: 8791	815-58-7	: 10286
693-98-1	: 7380	712-50-5	: 762	763-69-9	: 5025	771-51-7	: 6245	816-39-7	: 2982
694-05-3	: 9888	713-36-0	: 7041	764-01-2	: 1683	771-56-2	: 8590	816-66-0	: 7540
694-28-0	: 1965	713-68-8	: 8785	764-13-6	: 4087	771-60-8	: 8578	816-79-5	: 5197
694-48-4	: 4293	713-95-1	: 5711	764-35-2	: 5911	771-61-9	: 8587	816-90-0	: 5454
694-51-9	: 5381	716-39-2	: 7974	764-41-0	: 3138	771-62-0	: 8581	817-09-4	: 10291
694-53-1	: 8996	716-79-0	: 8839	764-41-0	: 3139	771-97-1	: 7943	817-91-4	: 7291
694-59-7	: 9355	717-21-5	: 5494	764-42-1	: 1444	771-98-2	: 2600	817-95-8	: 5024

941-98-0	: 31	998-95-8	: 2063	1066-45-1	: 2336	1079-66-9	: 4509	1119-40-0	: 4066
942-01-8	: 9844	999-21-3	: 2837	1066-51-9	: 351	1079-71-6	: 8332	1119-46-6	: 2211
942-06-3	: 3201	999-52-0	: 2062	1067-08-9	: 5147	1080-16-6	: 559	1119-49-9	: 1486
942-92-7	: 8889	999-55-3	: 169	1067-14-7	: 2322	1080-32-6	: 3407	1119-51-3	: 1310
943-15-7	: 7418	999-61-1	: 6186	1067-20-5	: 3498	1081-34-1	: 9706	1119-60-4	: 5699
943-27-1	: 1650	999-78-0	: 4241	1067-53-4	: 10694	1081-75-0	: 4513	1119-62-6	: 4585
943-88-4	: 6867	999-81-5	: 1835	1068-19-5	: 4078	1081-77-2	: 8255	1119-65-9	: 5713
943-89-5	: 6882	999-97-3	: 5818	1068-57-1	: 29	1083-30-3	: 4516	1119-85-3	: 5867
944-22-9	: 5447	1000-50-6	: 1533	1068-87-7	: 5005	1083-56-3	: 4467	1120-06-5	: 2759
945-51-7	: 4531	1000-82-4	: 6111	1068-90-2	: 3372	1085-98-9	: 3064	1120-16-7	: 4622
946-80-5	: 821	1000-86-8	: 4211	1069-53-0	: 10562	1086-80-2	: 3969	1120-21-4	: 10728
947-02-4	: 9022	1001-53-2	: 308	1070-00-4	: 10635	1087-21-4	: 2836	1120-24-7	: 4039
947-04-6	: 542	1001-89-4	: 2061	1070-01-5	: 10321	1088-11-5	: 8267	1120-28-1	: 7229
947-42-2	: 4527	1002-28-4	: 5914	1070-03-7	: 5059	1090-13-7	: 7927	1120-29-2	: 4618
947-73-9	: 8742	1002-43-3	: 7846	1070-19-5	: 1510	1093-58-9	: 2070	1120-34-9	: 7226
947-84-2	: 8841	1002-69-3	: 1968	1070-32-2	: 7290	1094-61-7	: 8024	1120-36-1	: 9801
947-91-1	: 8827	1002-84-2	: 8553	1070-34-4	: 5069	1095-90-5	: 6735	1120-48-5	: 4418
948-32-3	: 10462	1003-03-8	: 2682	1070-70-8	: 1410	1111-74-6	: 4319	1120-49-6	: 3335
948-65-2	: 8904	1003-09-4	: 1356	1070-71-9	: 2506	1111-97-3	: 2115	1120-56-5	: 7236
949-67-7	: 10723	1003-10-7	: 3685	1070-78-6	: 9767	1112-39-6	: 3859	1120-59-8	: 3682
950-10-7	: 6704	1003-29-8	: 9393	1070-83-3	: 1490	1112-48-7	: 1364	1120-62-3	: 7173
950-37-8	: 6758	1003-30-1	: 5253	1070-87-7	: 9962	1112-55-6	: 10005	1120-64-5	: 7533
952-97-6	: 8160	1003-31-2	: 10066	1071-22-3	: 10289	1112-67-0	: 9721	1120-71-4	: 9150
953-17-3	: 7840	1003-64-1	: 5081	1071-23-4	: 9026	1113-02-6	: 8412	1120-72-5	: 7170
955-83-9	: 4488	1003-67-4	: 7746	1071-26-7	: 4070	1113-12-8	: 2832	1120-73-6	: 7175
957-51-7	: 4447	1003-73-2	: 7745	1071-73-4	: 6141	1113-30-0	: 3906	1120-87-2	: 1348
957-68-6	: 281	1003-78-7	: 4326	1071-81-4	: 9953	1113-60-6	: 6138	1120-93-0	: 9300
958-09-8	: 2797	1004-36-0	: 4289	1071-83-6	: 5599	1113-68-4	: 80	1120-94-1	: 10077
959-26-2	: 1006	1004-38-2	: 9385	1071-93-8	: 5833	1113-83-3	: 5585	1120-97-4	: 7217
959-28-4	: 4469	1004-66-6	: 3938	1071-98-3	: 1678	1114-34-7	: 6650	1121-07-9	: 7758
959-33-1	: 6879	1005-64-7	: 1461	1072-05-5	: 4074	1114-51-8	: 3513	1121-18-2	: 7163
959-36-4	: 5970	1005-67-0	: 1619	1072-16-8	: 4176	1114-71-2	: 8516	1121-22-8	: 2572
960-71-4	: 10647	1007-26-7	: 4283	1072-21-5	: 5830	1114-76-7	: 3410	1121-24-0	: 6714
961-11-5	: 9776	1007-32-5	: 8854	1072-43-1	: 7804	1115-11-3	: 7085	1121-31-9	: 9359
961-29-5	: 10439	1008-65-7	: 5309	1072-47-5	: 7219	1115-12-4	: 1758	1121-37-5	: 3334
961-68-2	: 4399	1008-88-4	: 8984	1072-52-2	: 557	1115-15-7	: 10825	1121-55-7	: 10821
965-90-2	: 5022	1008-89-5	: 8983	1072-67-9	: 7426	1115-30-6	: 3375	1121-58-0	: 7737
968-81-0	: 28	1009-14-9	: 8945	1072-71-5	: 10024	1115-47-5	: 77	1121-60-4	: 9322
968-93-4	: 9707	1009-61-6	: 2824	1072-83-9	: 9403	1115-69-1	: 43	1121-92-2	: 8333
972-02-1	: 4448	1009-93-4	: 5814	1072-85-1	: 1209	1116-24-1	: 4547	1122-42-5	: 6294
973-21-7	: 4411	1011-12-7	: 2628	1072-91-9	: 10609	1116-39-8	: 10456	1122-54-9	: 9369
975-17-7	: 10448	1012-72-2	: 3007	1072-98-6	: 2288	1116-40-1	: 10455	1122-58-3	: 4298
976-71-6	: 1717	1013-08-7	: 9880	1073-06-9	: 1210	1116-54-7	: 8185	1122-60-7	: 8092
979-92-0	: 128	1013-88-3	: 8837	1073-07-0	: 2643	1116-61-6	: 10669	1122-61-8	: 8173
981-18-0	: 9012	1014-69-3	: 2814	1073-23-0	: 4305	1116-70-7	: 10180	1122-62-9	: 9367
982-43-4	: 9110	1017-56-7	: 10692	1073-67-2	: 2303	1116-76-3	: 10636	1122-72-1	: 7741
991-42-4	: 8262	1020-84-4	: 8339	1073-72-9	: 7818	1117-55-1	: 5903	1122-81-2	: 9262
992-59-6	: 723	1022-13-5	: 2097	1074-17-5	: 7710	1117-59-5	: 5906	1122-82-3	: 2631
992-94-9	: 7780	1024-57-3	: 5641	1074-43-7	: 7711	1117-61-9	: 4195	1122-83-4	: 9654
993-00-0	: 2163	1025-15-6	: 10135	1074-55-1	: 7712	1117-71-1	: 7056	1122-91-4	: 1095
993-07-7	: 10614	1031-07-8	: 4699	1074-92-6	: 1606	1117-86-8	: 8350	1123-00-8	: 2660
993-13-5	: 7667	1031-47-6	: 10137	1075-38-3	: 1607	1117-97-1	: 6837	1123-09-7	: 4027
994-05-8	: 6841	1032-65-1	: 2801	1076-38-6	: 5998	1118-12-3	: 1671	1123-27-9	: 6022
994-30-9	: 2323	1038-95-5	: 10698	1076-59-1	: 8910	1118-15-6	: 9945	1123-34-8	: 181
994-49-0	: 5790	1047-16-1	: 2405	1076-97-7	: 2573	1118-46-3	: 7890	1123-63-3	: 1999
996-50-9	: 3538	1048-05-1	: 9990	1077-16-3	: 5890	1118-58-7	: 7546	1123-84-8	: 3292
996-98-5	: 8433	1048-08-4	: 9994	1077-28-7	: 10049	1118-68-9	: 4067	1123-85-9	: 8961
998-29-8	: 10676	1053-73-2	: 124	1077-58-3	: 1521	1118-71-4	: 9949	1123-95-1	: 332
998-30-1	: 10340	1055-23-8	: 851	1078-19-9	: 3637	1118-84-9	: 168	1124-05-6	: 3165
998-40-3	: 10186	1066-17-7	: 2450	1078-21-3	: 344	1118-90-7	: 318	1124-06-7	: 1998
998-41-4	: 10189	1066-27-9	: 5291	1078-71-3	: 5704	1119-33-1	: 5038	1124-11-4	: 9972
998-93-6	: 1220	1066-35-9	: 2005	1079-21-6	: 885	1119-34-2	: 515	1124-33-0	: 8174

125-78-6	: 9878	1185-39-3	: 4219	1205-02-3	: 753	1321-74-0	: 4600	1452-15-9	: 2521
125-80-0	: 7424	1185-55-3	: 10492	1205-64-7	: 7621	1321-74-0	: 4601	1452-77-3	: 9326
126-09-6	: 5219	1185-81-5	: 3009	1205-91-0	: 634	1321-74-0	: 4602	1453-24-3	: 4964
126-46-1	: 7123	1186-53-4	: 9961	1207-69-8	: 1849	1330-20-7	: 10851	1453-58-3	: 7726
126-78-9	: 1507	1187-03-7	: 9829	1207-72-3	: 7606	1330-20-7	: 10852	1453-82-3	: 9328
126-79-0	: 1651	1187-42-4	: 2852	1208-52-2	: 7241	1330-20-7	: 10853	1454-80-4	: 880
127-76-0	: 5162	1187-58-2	: 7683	1210-12-4	: 480	1330-45-6	: 2326	1454-84-8	: 8221
128-08-1	: 7593	1188-01-8	: 5587	1210-35-1	: 3611	1330-61-6	: 6426	1454-85-9	: 5651
129-47-1	: 2635	1188-33-6	: 3355	1210-39-5	: 4519	1330-78-5	: 10302	1455-20-5	: 1662
129-50-6	: 8849	1188-37-0	: 70	1212-29-9	: 3329	1330-78-5	: 10303	1455-21-6	: 8236
129-89-1	: 2549	1189-08-8	: 1412	1214-39-7	: 8933	1330-78-5	: 10304	1455-77-2	: 10149
131-16-4	: 4254	1189-11-3	: 9025	1214-47-7	: 6165	1330-86-5	: 3787	1457-46-1	: 9007
131-18-6	: 7656	1189-71-5	: 9664	1217-45-4	: 485	1331-22-2	: 7158	1458-98-6	: 1284
131-60-8	: 2637	1189-85-1	: 1538	1218-34-4	: 104	1331-28-8	: 2301	1458-99-7	: 2216
131-62-0	: 3881	1189-99-7	: 10554	1220-00-4	: 948	1333-41-1	: 7738	1459-09-2	: 5760
132-21-4	: 3852	1190-22-3	: 3130	1222-98-6	: 8154	1333-41-1	: 7739	1459-10-5	: 9804
132-39-4	: 4525	1190-39-2	: 3030	1223-31-0	: 4384	1333-41-1	: 7740	1459-93-4	: 4119
134-23-2	: 2527	1190-63-2	: 5769	1225-55-4	: 9282	1335-86-0	: 7159	1461-25-2	: 9728
134-35-6	: 3980	1190-76-7	: 1448	1226-42-2	: 1022	1335-86-0	: 7161	1462-03-9	: 7168
134-62-9	: 1621	1190-83-6	: 10553	1226-46-6	: 965	1335-88-2	: 9759	1462-12-0	: 3456
135-12-2	: 784	1191-04-4	: 5868	1229-12-5	: 456	1336-36-3	: 2739	1462-84-6	: 10610
135-66-6	: 6437	1191-08-8	: 1418	1234-35-1	: 8916	1336-36-3	: 3122	1463-10-1	: 7851
136-89-6	: 7992	1191-15-7	: 3777	1241-94-7	: 4486	1336-36-3	: 5642	1464-42-2	: 9521
137-41-3	: 260	1191-25-9	: 6048	1260-17-9	: 1771	1336-36-3	: 5725	1464-53-5	: 875
137-42-4	: 5997	1191-41-9	: 5179	1264-62-6	: 4746	1336-36-3	: 5726	1465-25-4	: 7972
138-52-9	: 3042	1191-43-1	: 5843	1271-19-8	: 953	1336-36-3	: 5727	1466-67-7	: 2902
138-80-3	: 814	1191-80-6	: 6720	1271-28-9	: 8018	1336-36-3	: 8215	1466-76-8	: 3850
1141-38-4	: 7946	1191-95-3	: 2535	1272-44-2	: 764	1336-36-3	: 8291	1467-05-6	: 2121
1141-59-9	: 9378	1191-96-4	: 4970	1291-32-3	: 954	1336-36-3	: 8535	1467-79-4	: 4006
1141-88-4	: 905	1191-99-7	: 3624	1300-21-6	: 3173	1336-36-3	: 8536	1468-39-9	: 7076
1142-19-4	: 942	1192-18-3	: 4029	1300-21-6	: 3174	1336-36-3	: 9737	1468-95-7	: 488
1142-39-8	: 5904	1192-28-5	: 2671	1300-32-9	: 5481	1336-36-3	: 9738	1470-94-6	: 3629
1142-70-7	: 1085	1192-30-9	: 1285	1300-64-7	: 6808	1336-36-3	: 10210	1471-03-0	: 9201
1143-38-0	: 491	1192-37-6	: 7237	1300-64-7	: 6809	1336-36-3	: 10211	1476-11-5	: 3138
1143-72-2	: 10438	1192-62-7	: 5492	1300-71-6	: 10854	1338-23-4	: 1436	1476-23-9	: 193
1144-74-7	: 8153	1192-88-7	: 2593	1300-71-6	: 10855	1338-43-8	: 9584	1477-55-0	: 629
1145-01-3	: 4523	1193-02-8	: 256	1300-71-6	: 10856	1361-49-5	: 9683	1482-91-3	: 8365
1146-98-1	: 1322	1193-18-6	: 7164	1300-71-6	: 10857	1397-89-3	: 446	1484-12-4	: 7118
1146-99-2	: 2241	1193-47-1	: 4021	1300-71-6	: 10858	1401-55-4	: 9677	1484-13-5	: 10796
1148-11-4	: 830	1193-81-3	: 7147	1300-71-6	: 10859	1405-86-3	: 5594	1484-84-0	: 9065
1148-79-4	: 9704	1193-82-4	: 7789	1300-73-8	: 3927	1415-73-2	: 214	1484-85-1	: 694
1149-16-2	: 5597	1194-02-1	: 5372	1300-73-8	: 3928	1420-06-0	: 10374	1486-75-5	: 2550
1149-24-2	: 3431	1194-65-6	: 3114	1300-73-8	: 3929	1420-07-1	: 1561	1487-18-9	: 10805
1151-15-1	: 6807	1194-98-5	: 3700	1300-73-8	: 3930	1420-55-9	: 10040	1489-57-2	: 7143
1152-61-0	: 8917	1195-32-0	: 7462	1300-73-8	: 3931	1421-14-3	: 9162	1489-69-6	: 2691
1153-05-5	: 10645	1195-42-2	: 2630	1300-73-8	: 3932	1421-63-2	: 10444	1491-59-4	: 8489
1154-59-2	: 9773	1195-79-5	: 5315	1300-94-3	: 7595	1421-69-8	: 815	1492-24-6	: 269
1155-00-6	: 1038	1196-57-2	: 9445	1319-77-3	: 2477	1423-60-5	: 1686	1493-02-3	: 5464
1156-19-0	: 10107	1197-18-8	: 346	1319-77-3	: 2478	1435-55-8	: 5956	1493-13-6	: 10397
1159-53-1	: 7049	1197-37-1	: 4829	1319-77-3	: 2479	1436-59-5	: 2571	1493-27-2	: 5405
1162-65-8	: 136	1197-55-3	: 245	1319-91-1	: 1699	1438-16-0	: 432	1495-50-7	: 2515
1163-19-5	: 2738	1198-37-4	: 4313	1321-10-4	: 2146	1441-87-8	: 6001	1497-68-3	: 5031
1165-39-5	: 138	1198-55-6	: 9735	1321-10-4	: 2147	1445-45-0	: 10489	1498-51-7	: 5215
1165-48-6	: 3818	1198-59-0	: 3293	1321-10-4	: 2148	1445-73-4	: 7680	1499-10-1	: 4457
1166-52-5	: 4656	1200-03-9	: 1135	1321-12-6	: 8206	1445-75-6	: 3806	1501-27-5	: 7895
1170-76-9	: 816	1200-14-2	: 1511	1321-12-6	: 8207	1445-79-0	: 10552	1501-60-6	: 7546
1174-72-7	: 9984	1201-26-9	: 6257	1321-12-6	: 8208	1446-61-3	: 228	1502-05-2	: 2540
1176-74-5	: 9718	1201-38-3	: 3830	1321-16-0	: 2593	1449-65-6	: 7276	1502-06-3	: 2541
1184-58-3	: 1989	1201-99-6	: 3268	1321-16-0	: 2594	1450-14-2	: 5816	1502-22-3	: 2601
1184-60-7	: 5420	1202-34-2	: 9371	1321-31-9	: 4822	1450-31-3	: 4500	1503-53-3	: 88
1184-78-7	: 10502	1204-06-4	: 6258	1321-31-9	: 4823	1450-63-1	: 9985	1504-58-1	: 8980
1185-33-7	: 3989	1204-28-0	: 671	1321-31-9	: 4824	1450-72-2	: 6097	1511-62-2	: 1173

1825-82-7	: 10233	1918-77-0	: 10064	2032-35-1	: 1171	2052-14-4	: 1586	2142-01-0	: 803
1829-00-1	: 2421	1921-70-6	: 9959	2032-59-9	: 275	2052-15-5	: 1633	2142-63-4	: 1319
1830-54-2	: 4210	1927-31-7	: 2798	2032-65-7	: 6759	2052-49-5	: 9723	2142-73-6	: 4250
1830-95-1	: 227	1928-38-7	: 7192	2033-24-1	: 4042	2055-40-5	: 6541	2146-38-5	: 4969
1833-51-8	: 2118	1929-73-3	: 1478	2036-41-1	: 7752	2055-46-1	: 9890	2150-02-9	: 3823
1836-75-5	: 8099	1929-77-7	: 10781	2037-31-2	: 1892	2057-43-4	: 8654	2150-38-1	: 7199
1838-59-1	: 188	1929-82-4	: 8029	2038-03-1	: 7903	2057-49-0	: 8977	2150-44-9	: 7198
1838-73-9	: 7305	1932-04-3	: 4537	2039-82-9	: 1381	2065-23-8	: 7608	2150-47-2	: 7197
1840-42-2	: 10628	1936-57-8	: 6935	2039-85-2	: 2302	2065-70-5	: 3229	2152-34-3	: 8519
1843-05-6	: 6137	1937-54-8	: 9581	2039-86-3	: 1380	2067-33-6	: 1309	2155-30-8	: 7429
1845-30-3	: 10520	1937-62-8	: 7514	2039-87-4	: 2301	2068-83-9	: 6103	2155-94-4	: 186
1848-84-6	: 4902	1939-99-7	: 645	2039-88-5	: 1379	2077-13-6	: 2086	2156-97-0	: 4639
1850-14-2	: 9913	1940-42-7	: 1170	2039-93-2	: 8857	2078-54-8	: 9195	2157-01-9	: 8392
1851-07-6	: 8023	1941-30-6	: 9997	2040-95-1	: 1550	2079-95-0	: 9794	2159-75-3	: 563
1852-53-5	: 454	1942-45-6	: 8405	2040-96-2	: 9227	2082-59-9	: 8625	2162-92-7	: 3196
1854-26-8	: 4200	1942-46-7	: 2789	2042-37-7	: 1106	2082-79-3	: 8323	2162-98-3	: 3150
1855-63-6	: 2592	1943-16-4	: 2338	2043-38-1	: 1467	2082-81-7	: 1413	2162-99-4	: 3243
1860-27-1	: 7254	1943-83-5	: 2041	2043-61-0	: 2569	2083-91-2	: 8603	2163-00-0	: 3197
1861-21-8	: 4697	1945-53-5	: 8497	2049-67-4	: 3501	2084-18-6	: 7072	2163-42-0	: 7686
1861-32-1	: 4332	1945-77-3	: 7826	2049-80-1	: 3377	2084-19-7	: 8622	2163-48-6	: 3517
1861-40-1	: 569	1948-33-0	: 1517	2049-92-5	: 8665	2085-88-3	: 7645	2163-69-1	: 2703
1864-92-2	: 4833	1949-78-6	: 6651	2049-94-7	: 6443	2086-83-1	: 841	2163-80-6	: 9567
1864-94-4	: 8883	1954-28-5	: 5294	2049-95-8	: 4282	2090-14-4	: 8301	2164-08-1	: 6598
1866-15-5	: 103	1955-45-9	: 4204	2049-96-9	: 8667	2090-89-3	: 1465	2164-09-2	: 3267
1866-31-5	: 180	1961-96-2	: 8902	2050-01-3	: 6450	2091-29-4	: 5757	2164-17-2	: 4347
1866-39-3	: 7136	1967-16-4	: 1822	2050-08-0	: 8685	2094-98-6	: 561	2167-39-7	: 7534
1868-53-7	: 2944	1973-22-4	: 1203	2050-09-1	: 6452	2095-57-0	: 4987	2168-93-6	: 3056
1871-52-9	: 2650	1974-04-5	: 1219	2050-14-8	: 3697	2097-19-0	: 8997	2169-87-1	: 7945
1871-57-4	: 1954	1975-78-6	: 2755	2050-24-0	: 3483	2100-17-6	: 8638	2170-03-8	: 3641
1871-76-7	: 8830	1977-10-2	: 6631	2050-43-3	: 4246	2101-86-2	: 553	2171-96-2	: 7781
1871-96-1	: 2752	1982-47-4	: 2344	2050-46-6	: 3352	2103-57-3	: 10475	2173-56-0	: 8682
1873-25-2	: 1932	1982-49-6	: 9537	2050-47-7	: 919	2103-64-2	: 5521	2175-91-9	: 7259
1873-29-6	: 6392	1983-10-4	: 10183	2050-60-4	: 3037	2104-64-5	: 5172	2176-62-7	: 8544
1873-88-7	: 5667	1984-59-4	: 3203	2050-67-1	: 3123	2104-96-3	: 1325	2176-98-9	: 9999
1873-92-3	: 199	1984-65-2	: 3204	2050-68-2	: 3124	2107-69-9	: 3868	2177-47-1	: 7384
1877-72-1	: 2507	1986-70-5	: 1704	2050-69-3	: 3221	2107-76-8	: 3733	2179-57-9	: 2833
1877-73-2	: 8049	1986-90-9	: 4446	2050-72-8	: 3225	2108-92-1	: 3148	2179-59-1	: 206
1878-18-8	: 7069	1989-33-9	: 5344	2050-73-9	: 3226	2109-66-2	: 7469	2179-60-4	: 7715
1878-65-5	: 1867	1990-29-0	: 218	2050-74-0	: 3227	2110-78-3	: 7376	2181-42-2	: 10618
1878-66-6	: 1868	1992-48-9	: 6539	2050-75-1	: 3228	2111-75-3	: 7855	2182-66-3	: 4040
1878-68-8	: 1097	1999-33-3	: 5588	2050-76-2	: 3232	2111-75-3	: 7856	2189-60-8	: 8383
1878-82-6	: 2519	2001-32-3	: 8156	2050-77-3	: 6291	2113-51-1	: 6284	2196-13-6	: 9321
1879-09-0	: 1557	2001-95-8	: 10769	2050-87-5	: 2842	2113-57-7	: 1113	2197-37-7	: 6615
1885-14-9	: 8862	2002-24-6	: 4802	2050-92-2	: 4439	2113-58-8	: 8085	2197-37-7	: 8300
1885-29-6	: 257	2004-69-5	: 8658	2051-24-3	: 2739	2114-00-3	: 1324	2198-61-0	: 6446
1885-38-7	: 8967	2004-70-8	: 8566	2051-25-4	: 2740	2114-11-6	: 177	2198-75-6	: 3222
1885-48-9	: 3564	2008-41-5	: 9669	2051-30-1	: 4175	2116-62-3	: 8881	2198-77-8	: 3230
1888-71-7	: 5739	2008-58-4	: 3091	2051-31-2	: 2761	2116-65-6	: 827	2199-44-2	: 5009
1889-67-4	: 4046	2009-83-8	: 2069	2051-49-2	: 5849	2122-70-5	: 5164	2199-51-1	: 5010
1892-29-1	: 997	2016-42-4	: 9803	2051-53-8	: 7400	2124-31-4	: 3896	2199-52-2	: 5011
1896-62-4	: 8859	2016-56-0	: 4641	2051-60-7	: 1916	2124-57-4	: 6696	2199-53-3	: 5012
1897-45-6	: 2309	2016-57-1	: 2778	2051-61-8	: 1917	2128-93-0	: 759	2203-80-7	: 7357
1899-24-7	: 1216	2018-61-3	: 95	2051-62-9	: 1918	2130-56-5	: 679	2207-01-4	: 4011
1904-98-9	: 9294	2019-34-3	: 2248	2051-78-7	: 176	2131-18-2	: 8559	2207-03-6	: 4014
1906-79-2	: 5238	2021-28-5	: 5211	2051-79-8	: 3484	2131-41-1	: 10571	2207-04-7	: 4016
1910-41-4	: 5332	2022-85-7	: 5382	2051-85-6	: 8819	2131-55-7	: 2088	2207-27-4	: 9747
1912-24-9	: 531	2025-40-3	: 4960	2051-90-3	: 3171	2132-80-1	: 3855	2207-50-3	: 3665
1916-07-0	: 7838	2027-17-0	: 6519	2051-95-8	: 8454	2134-29-4	: 6179	2209-86-1	: 939
1918-00-9	: 3206	2028-63-9	: 1685	2051-97-0	: 829	2136-89-2	: 2319	2210-25-5	: 6534
1918-02-1	: 434	2029-94-9	: 3351	2052-01-9	: 1281	2136-99-4	: 8291	2210-28-8	: 9247
1918-16-7	: 9131	2031-62-1	: 3365	2052-06-4	: 1172	2138-22-9	: 1878	2210-79-9	: 7610
1918-18-9	: 7193	2031-67-6	: 10336	2052-07-5	: 1112	2141-62-0	: 4866	2211-67-8	: 3133

2211-68-9	: 1941	2270-20-4	: 648	2379-90-0	: 331	2435-53-2	: 9742	2499-59-4	: 8379
2211-69-0	: 1940	2272-45-9	: 7643	2381-15-9	: 6966	2436-79-5	: 3432	2499-95-8	: 5888
2211-70-3	: 1935	2274-11-5	: 4776	2381-16-0	: 6965	2436-85-3	: 4149	2503-56-2	: 7830
2211-98-5	: 9555	2275-18-5	: 10471	2381-21-7	: 7729	2436-90-0	: 4166	2506-41-4	: 2135
2212-67-1	: 7886	2275-23-2	: 10771	2381-31-9	: 6964	2436-96-6	: 4379	2508-29-4	: 388
2213-23-2	: 4072	2277-23-8	: 3752	2382-43-6	: 6187	2437-25-4	: 4627	2512-29-0	: 2408
2213-32-3	: 4230	2279-76-7	: 2342	2382-96-9	: 750	2437-56-1	: 10318	2512-56-3	: 8400
2213-43-6	: 9057	2283-08-1	: 6113	2385-70-8	: 1125	2437-88-9	: 10331	2512-81-4	: 4266
2213-63-0	: 3291	2288-18-8	: 8848	2385-81-1	: 5501	2437-95-8	: 9047	2516-33-8	: 2694
2215-77-2	: 8781	2292-79-7	: 2843	2385-85-5	: 7878	2437-95-8	: 9048	2516-34-9	: 2529
2216-12-8	: 8135	2294-71-5	: 7629	2386-60-9	: 1420	2438-12-2	: 9701	2516-96-3	: 2188
2216-15-1	: 3492	2295-31-0	: 10033	2387-23-7	: 3330	2438-72-4	: 1480	2517-04-6	: 548
2216-30-0	: 4073	2295-58-1	: 10447	2388-14-9	: 6464	2439-01-2	: 1811	2517-43-3	: 6815
2216-32-2	: 5041	2303-16-4	: 2829	2390-59-2	: 5283	2439-10-3	: 4660	2518-24-3	: 340
2216-33-3	: 7516	2303-17-5	: 10133	2393-23-9	: 6792	2439-35-2	: 3909	2523-37-7	: 7261
2216-34-4	: 7517	2305-13-7	: 6063	2396-60-3	: 6875	2439-54-5	: 7525	2524-03-0	: 4262
2216-51-5	: 7408	2306-88-9	: 8395	2396-63-6	: 5658	2439-99-8	: 5601	2524-04-1	: 3417
2216-52-6	: 7409	2307-10-0	: 9599	2396-65-8	: 8229	2444-29-3	: 7499	2524-37-0	: 4991
2216-68-4	: 7476	2310-17-0	: 9021	2396-78-3	: 7354	2444-36-2	: 1866	2524-52-9	: 5235
2216-69-5	: 6848	2310-98-7	: 1153	2396-83-0	: 5054	2444-37-3	: 7805	2524-64-3	: 4476
2216-92-4	: 5209	2312-35-8	: 9173	2396-84-1	: 5045	2445-76-3	: 5908	2524-67-6	: 7907
2216-94-6	: 5212	2313-65-7	: 7334	2398-37-0	: 1091	2445-83-2	: 7024	2525-16-8	: 9866
2217-06-3	: 1055	2314-97-8	: 10393	2398-81-4	: 9332	2446-69-7	: 5907	2525-62-4	: 5899
2217-07-4	: 4551	2315-36-8	: 1977	2399-48-6	: 9859	2447-54-3	: 9505	2528-61-2	: 5692
2217-15-4	: 3813	2315-68-6	: 9215	2401-73-2	: 3297	2450-71-7	: 9272	2530-85-0	: 10497
2217-41-6	: 9873	2316-64-5	: 1232	2401-85-6	: 2010	2455-24-5	: 9861	2530-87-2	: 2283
2218-94-2	: 8124	2318-18-5	: 9527	2402-06-4	: 4038	2456-27-1	: 4414	2531-80-8	: 5174
2219-66-1	: 2921	2319-97-3	: 6726	2402-78-0	: 3286	2456-28-2	: 3336	2532-58-3	: 4031
2219-82-1	: 1615	2321-07-5	: 5349	2402-79-1	: 9771	2459-05-4	: 5068	2536-91-6	: 7026
2221-95-6	: 5329	2326-89-8	: 4668	2403-55-6	: 7707	2459-09-8	: 7743	2540-82-1	: 5456
2223-67-8	: 8005	2328-24-7	: 6974	2403-88-5	: 6197	2459-10-1	: 10512	2541-69-7	: 6963
2223-82-7	: 4542	2338-12-7	: 8080	2404-35-5	: 1158	2459-24-7	: 7472	2545-59-7	: 1479
2227-17-0	: 8529	2338-37-6	: 9199	2404-44-6	: 8396	2459-25-8	: 7473	2547-61-7	: 10247
2227-79-4	: 608	2344-70-9	: 7004	2408-20-0	: 204	2460-77-7	: 3012	2548-87-0	: 8366
2230-70-8	: 10537	2344-80-1	: 2168	2408-37-9	: 10536	2462-84-2	: 7514	2550-04-1	: 210
2231-57-4	: 1751	2344-83-4	: 2284	2409-52-1	: 3486	2462-84-2	: 7528	2550-06-3	: 10219
2233-00-3	: 10285	2345-26-8	: 5529	2409-55-4	: 1613	2462-94-4	: 4689	2550-26-7	: 8855
2234-13-1	: 8294	2345-27-9	: 9798	2411-89-4	: 2481	2463-53-8	: 8248	2550-28-9	: 5916
2234-16-4	: 3262	2345-28-0	: 8555	2412-73-9	: 2613	2463-63-0	: 5693	2550-36-9	: 1259
2235-12-3	: 5857	2345-34-8	: 87	2415-85-2	: 7647	2463-84-5	: 3061	2550-40-5	: 3324
2235-12-3	: 5858	2346-00-1	: 3655	2416-94-6	: 10594	2464-37-1	: 1831	2550-75-6	: 1820
2235-46-3	: 3373	2346-81-8	: 2068	2417-04-1	: 9930	2465-27-2	: 533	2553-19-7	: 3357
2236-60-4	: 335	2348-81-4	: 358	2417-90-5	: 1329	2465-32-9	: 5565	2554-06-5	: 10006
2237-30-1	: 258	2348-82-5	: 6850	2419-73-0	: 3134	2466-76-4	: 73	2555-49-9	: 5203
2238-07-5	: 3578	2349-67-9	: 429	2421-28-5	: 721	2467-02-9	: 3722	2565-58-4	: 2155
2243-27-8	: 8235	2351-13-5	: 918	2422-79-9	: 6967	2470-68-0	: 3582	2568-33-4	: 7066
2243-35-8	: 90	2356-61-8	: 10414	2424-92-2	: 4682	2473-01-0	: 2199	2568-90-3	: 3001
2243-42-7	: 8779	2358-84-1	: 3441	2425-06-1	: 1722	2475-44-7	: 1023	2570-26-5	: 8558
2243-47-2	: 264	2361-96-8	: 96	2425-25-4	: 41	2475-45-8	: 4575	2576-47-8	: 1200
2243-62-1	: 7941	2362-10-9	: 940	2425-54-9	: 2305	2476-37-1	: 3263	2577-63-1	: 2851
2243-76-7	: 153	2363-89-5	: 8366	2425-74-3	: 1575	2483-57-0	: 7479	2578-45-2	: 2012
2243-83-6	: 7936	2365-48-2	: 7435	2425-79-8	: 1411	2487-90-3	: 10494	2579-20-6	: 2574
2243-98-3	: 10750	2366-36-1	: 10236	2426-02-0	: 9864	2489-77-2	: 10619	2579-22-8	: 8978
2244-07-7	: 10729	2366-52-1	: 5378	2426-08-6	: 1579	2492-26-4	: 739	2581-34-2	: 7497
2244-16-8	: 1788	2367-82-0	: 9831	2426-54-2	: 3386	2493-02-9	: 1239	2586-89-2	: 5714
2244-21-5	: 9094	2370-12-9	: 4220	2431-50-7	: 10214	2495-25-2	: 10325	2587-42-0	: 991
2251-65-2	: 10407	2370-13-0	: 4096	2432-12-4	: 3214	2495-27-4	: 5765	2591-86-8	: 9059
2254-94-6	: 7028	2370-14-1	: 4182	2432-63-5	: 4560	2495-35-4	: 779	2592-95-2	: 6000
2257-52-5	: 6500	2370-88-9	: 9939	2432-74-8	: 319	2495-37-6	: 806	2593-15-9	: 9705
2259-96-3	: 2702	2373-51-5	: 2167	2432-87-3	: 4423	2497-18-9	: 5880	2595-54-2	: 6686
2260-50-6	: 61	2373-76-4	: 7134	2432-90-8	: 3344	2497-21-4	: 5882	2595-97-3	: 164
2269-22-9	: 219	2379-55-7	: 4316	2432-99-7	: 435	2498-66-0	: 593	2597-03-7	: 8795

5388-62-5	: 2007	5617-42-5	: 5707	5930-28-9	: 298	6145-31-9	: 4790	6402-36-4	: 4636
5390-04-5	: 8692	5633-20-5	: 8482	5932-68-3	: 6895	6152-67-6	: 4456	6410-10-2	: 8146
5392-40-5	: 4165	5634-39-9	: 6301	5932-79-6	: 8241	6153-56-6	: 8430	6422-86-2	: 989
5394-83-2	: 1712	5648-29-3	: 976	5934-56-5	: 6210	6156-25-8	: 9911	6423-43-4	: 9233
5396-24-7	: 9219	5650-40-8	: 6166	5949-29-1	: 2415	6158-45-8	: 6518	6436-90-4	: 4906
5396-38-3	: 1601	5655-61-8	: 1065	5951-67-7	: 10809	6159-44-0	: 10754	6443-92-1	: 5695
5399-02-0	: 5649	5666-17-1	: 185	5959-52-4	: 357	6159-55-3	: 10775	6454-84-8	: 7427
5401-94-5	: 8114	5675-51-4	: 4626	5965-66-2	: 6580	6159-56-4	: 10774	6482-24-2	: 1243
5405-41-4	: 5075	5676-58-4	: 3970	5965-83-3	: 6196	6163-58-2	: 10696	6485-40-1	: 1787
5406-12-2	: 7001	5683-33-0	: 4297	5966-51-8	: 967	6163-64-0	: 1618	6493-77-2	: 7162
5407-87-4	: 4296	5685-06-3	: 7803	5973-11-5	: 1251	6163-66-2	: 3021	6538-02-9	: 4729
5408-86-6	: 2919	5687-92-3	: 10039	5978-95-0	: 6209	6164-98-3	: 1827	6553-48-6	: 4604
5414-19-7	: 913	5703-26-4	: 6781	5982-99-0	: 456	6166-86-5	: 8596	6556-12-3	: 5550
5414-21-1	: 1308	5728-52-9	: 877	5984-58-7	: 7283	6168-72-5	: 406	6591-63-5	: 9417
5416-93-3	: 6871	5736-88-9	: 1472	5986-55-0	: 8515	6175-49-1	: 4633	6592-85-4	: 5932
5419-55-6	: 10464	5743-27-1	: 1694	5988-76-1	: 2491	6189-41-9	: 4207	6596-40-3	: 8546
5421-04-5	: 6447	5743-97-5	: 9786	5989-27-5	: 6610	6190-65-4	: 2810	6596-50-5	: 9522
5424-21-5	: 3219	5746-57-6	: 5567	5989-54-8	: 6611	6192-52-5	: 10117	6600-40-4	: 8282
5432-85-9	: 6490	5756-43-4	: 5061	5989-81-1	: 6581	6195-92-2	: 7171	6606-59-3	: 5837
5434-27-5	: 2935	5763-61-1	: 3845	6000-40-4	: 3750	6199-67-3	: 2490	6609-56-9	: 6802
5436-21-5	: 3857	5765-44-6	: 7428	6000-43-7	: 5579	6203-18-5	: 3919	6610-29-3	: 7822
5437-98-9	: 6873	5779-94-2	: 3945	6004-44-0	: 9184	6222-35-1	: 2639	6622-76-0	: 7451
5445-51-2	: 2533	5779-95-3	: 3946	6006-33-3	: 10324	6223-78-5	: 3166	6627-55-0	: 1275
5451-52-5	: 2783	5781-53-3	: 7128	6006-95-7	: 8561	6224-63-1	: 10697	6627-72-1	: 1064
5451-80-9	: 5710	5786-21-0	: 2436	6011-14-9	: 231	6225-06-5	: 4212	6628-21-3	: 4982
5451-92-3	: 8258	5786-96-9	: 1384	6012-97-1	: 9775	6232-88-8	: 1248	6628-77-9	: 6900
5452-35-7	: 2553	5787-31-5	: 1122	6027-13-0	: 5924	6236-88-0	: 5131	6630-33-7	: 1093
5452-37-9	: 2645	5794-03-6	: 1706	6027-89-0	: 5621	6258-66-8	: 1884	6639-30-1	: 10252
5452-75-5	: 4966	5794-04-7	: 1707	6029-84-1	: 9486	6261-22-9	: 8690	6640-27-3	: 2146
5453-80-5	: 858	5794-13-8	: 522	6032-29-7	: 8627	6267-24-9	: 10689	6669-13-2	: 4059
5454-28-4	: 1580	5794-24-1	: 521	6035-50-3	: 5469	6270-16-2	: 8090	6703-98-6	: 5628
5454-79-5	: 7152	5794-88-7	: 266	6038-51-3	: 164	6270-34-4	: 6497	6704-31-0	: 8446
5454-83-1	: 7057	5798-75-4	: 4910	6044-68-4	: 3884	6272-38-4	: 8918	6709-39-3	: 4069
5459-58-5	: 1540	5798-79-8	: 1099	6050-13-1	: 2890	6283-25-6	: 2174	6728-26-3	: 5860
5459-93-8	: 2623	5798-88-9	: 1254	6064-63-7	: 6047	6284-40-8	: 2806	6740-88-1	: 6569
5466-84-2	: 8117	5798-94-7	: 1175	6065-32-3	: 4913	6284-84-0	: 4265	6742-54-7	: 10749
5468-37-1	: 232	5805-76-5	: 5120	6065-82-3	: 4983	6285-05-8	: 2250	6750-03-4	: 8226
5469-69-2	: 2287	5808-22-0	: 9553	6066-82-6	: 6189	6287-38-3	: 3089	6753-98-6	: 5927
5470-18-8	: 2196	5809-59-6	: 6013	6069-98-3	: 7404	6290-05-7	: 3469	6754-13-8	: 5630
5471-51-2	: 6152	5810-88-8	: 985	6074-84-6	: 8577	6290-49-9	: 7440	6765-39-5	: 5654
5500-21-0	: 2689	5813-64-9	: 4281	6078-26-8	: 867	6294-31-1	: 3590	6776-19-8	: 4929
5502-88-5	: 7414	5836-10-2	: 2280	6089-09-4	: 8689	6294-34-4	: 928	6779-09-5	: 5214
5503-12-8	: 7855	5837-78-5	: 5126	6091-44-7	: 9067	6294-89-9	: 7364	6779-87-9	: 3607
5507-44-8	: 10801	5857-36-3	: 10587	6091-52-7	: 6513	6314-28-9	: 744	6780-49-0	: 5208
5510-99-6	: 3039	5858-18-4	: 3105	6094-02-6	: 7340	6315-52-2	: 4773	6781-42-6	: 2823
5522-43-0	: 8171	5870-93-9	: 5705	6094-40-2	: 9052	6315-89-5	: 3835	6789-80-6	: 5861
5524-05-0	: 7454	5878-19-3	: 6889	6106-24-7	: 9575	6317-18-6	: 7248	6789-88-4	: 5892
5534-95-2	: 8592	5881-17-4	: 5181	6108-61-8	: 5780	6320-03-2	: 1891	6802-75-1	: 3472
5536-17-4	: 10786	5882-44-0	: 3937	6111-88-2	: 2335	6325-93-5	: 8066	6812-78-8	: 4197
5550-12-9	: 5618	5891-21-4	: 2213	6114-18-7	: 5180	6334-18-5	: 3086	6829-40-9	: 3391
5560-69-0	: 4978	5894-60-0	: 5770	6117-80-2	: 1445	6346-09-4	: 3354	6843-66-9	: 3860
5581-35-1	: 445	5897-76-7	: 409	6117-91-5	: 1455	6351-10-6	: 3628	6846-11-3	: 7640
5582-82-1	: 7300	5902-51-2	: 9689	6117-91-5	: 1456	6358-53-8	: 2418	6848-13-1	: 1992
5586-15-2	: 4356	5905-52-2	: 5328	6119-92-2	: 4412	6358-85-6	: 2409	6850-57-3	: 6791
5587-89-3	: 6351	5910-87-2	: 8226	6120-13-4	: 1067	6361-21-3	: 2179	6852-54-6	: 8925
5588-20-5	: 1825	5910-89-4	: 4290	6125-21-9	: 4403	6362-80-7	: 7221	6863-58-7	: 3020
5588-33-0	: 6724	5911-04-6	: 7507	6125-24-2	: 8098	6378-11-6	: 4949	6870-67-3	: 6562
5588-84-1	: 10467	5911-08-0	: 2116	6130-75-2	: 10250	6378-65-0	: 5898	6871-44-9	: 4677
5591-45-7	: 10080	5912-86-7	: 6894	6138-79-0	: 10666	6380-23-0	: 3887	6873-15-0	: 7044
5593-70-4	: 9730	5915-41-3	: 9691	6138-90-5	: 1185	6380-34-3	: 8817	6874-98-2	: 9512
5598-13-0	: 2353	5917-47-5	: 4653	6139-84-0	: 1924	6381-92-6	: 5019	6876-23-9	: 4012
5617-41-4	: 5706	5921-54-0	: 1434	6140-65-4	: 2674	6382-13-4	: 8686	6881-94-3	: 3451

13071-79-9 : 9690	13531-52-7 : 316	14290-92-7 : 1574	15503-86-3 : 6362	16543-55-8 : 8196
13073-35-3 : 4810	13532-18-8 : 7461	14320-37-7 : 2678	15521-18-3 : 3921	16561-29-8 : 9800
13100-82-8 : 2714	13552-31-3 : 404	14321-27-8 : 796	15521-65-0 : 970	16587-71-6 : 4284
13116-53-5 : 9769	13573-16-5 : 9452	14324-55-1 : 961	15545-48-9 : 2359	16588-34-4 : 2180
13121-70-5 : 2707	13602-12-5 : 9333	14334-40-8 : 6494	15570-12-4 : 6797	16590-41-3 : 7923
13151-06-9 : 7524	13618-93-4 : 8335	14371-10-9 : 2396	15572-56-2 : 6472	16617-46-2 : 416
13151-17-2 : 7348	13640-28-3 : 9042	14417-88-0 : 6694	15625-89-5 : 10667	16629-19-9 : 10075
13151-34-3 : 7184	13673-92-2 : 3097	14436-32-9 : 2773	15647-08-2 : 5060	16642-79-8 : 8157
13169-00-1 : 6884	13674-87-8 : 10682	14452-30-3 : 6333	15647-11-7 : 356	16672-87-0 : 4806
13171-21-6 : 9024	13679-46-4 : 6842	14459-29-1 : 5635	15679-24-0 : 4295	16714-68-4 : 8542
13176-46-0 : 4909	13679-70-4 : 7817	14473-90-6 : 1959	15686-71-2 : 1798	16746-86-4 : 4097
13190-97-1 : 850	13684-56-5 : 2813	14476-30-3 : 10174	15686-83-6 : 9299	16747-25-4 : 10557
13194-48-4 : 4816	13684-63-4 : 8760	14484-64-1 : 5325	15687-27-1 : 6207	16747-26-5 : 10558
13195-80-7 : 2983	13698-16-3 : 4981	14523-22-9 : 9475	15706-73-7 : 1609	16747-28-7 : 10560
13205-44-2 : 1063	13706-86-0 : 7328	14545-08-5 : 6269	15764-04-2 : 10783	16747-30-1 : 10563
13209-15-9 : 955	13718-94-0 : 5541	14548-46-0 : 8987	15764-16-6 : 3944	16747-31-2 : 10564
13248-54-9 : 2616	13741-18-9 : 6369	14595-35-8 : 4559	15804-19-0 : 3678	16747-32-3 : 5003
13250-46-9 : 75	13748-90-8 : 6095	14596-92-0 : 5775	15825-70-4 : 6679	16747-33-4 : 5004
13254-34-7 : 4080	13838-16-9 : 4702	14602-62-1 : 10173	15840-60-5 : 7350	16747-38-9 : 9963
13258-63-4 : 9346	13862-07-2 : 4505	14609-79-1 : 9964	15862-07-4 : 10210	16747-44-7 : 8599
13269-52-8 : 5866	13889-92-4 : 9223	14620-52-1 : 3861	15867-21-7 : 215	16747-45-8 : 8600
13286-92-5 : 7597	13898-58-3 : 754	14630-40-1 : 1054	15869-80-4 : 5040	16747-50-5 : 5132
13290-74-9 : 2140	13905-48-1 : 1258	14641-93-1 : 6579	15869-85-9 : 7509	16751-59-0 : 5671
13291-61-7 : 2621	13909-09-6 : 2042	14660-52-7 : 4919	15869-86-0 : 5182	16752-77-5 : 6762
13292-46-1 : 9485	13912-77-1 : 8290	14667-55-1 : 10608	15869-87-1 : 4171	16789-46-1 : 5140
13313-91-2 : 9267	13925-00-3 : 5230	14686-13-6 : 5696	15869-89-3 : 4174	16790-49-1 : 1439
13323-81-4 : 6996	13927-77-0 : 8015	14686-14-7 : 5698	15869-92-8 : 4177	16842-03-8 : 2438
13325-10-5 : 273	13928-81-9 : 840	14698-29-4 : 8465	15869-94-0 : 4178	16867-04-2 : 6188
13329-40-3 : 6334	13939-06-5 : 7888	14721-66-5 : 9950	15870-10-7 : 7312	16881-77-9 : 7200
13331-27-6 : 8147	13952-84-6 : 1495	14752-75-1 : 5655	15905-32-5 : 4752	16883-48-0 : 10546
13347-42-7 : 1967	13956-29-1 : 1715	14850-22-7 : 8370	15932-80-6 : 7455	16898-52-5 : 4539
13351-73-0 : 7030	13984-57-1 : 5186	14850-23-8 : 8373	15950-66-0 : 10267	16939-57-4 : 1395
13356-08-6 : 5313	13993-65-2 : 7607	14855-76-6 : 7280	15972-60-8 : 141	16957-70-3 : 7584
13360-45-7 : 1821	14002-51-8 : 879	14857-34-2 : 4836	15980-15-1 : 8439	17015-11-1 : 5852
13360-52-6 : 1796	14007-64-8 : 3389	14882-18-9 : 1037	15988-11-1 : 9010	17021-26-0 : 1705
13360-57-1 : 4322	14024-00-1 : 10772	14901-08-7 : 2524	16009-13-5 : 5637	17071-47-5 : 1590
13360-61-7 : 8557	14024-18-1 : 6360	14919-01-8 : 8371	16033-71-9 : 7536	17082-09-6 : 8972
13360-63-9 : 1565	14024-48-7 : 1043	14929-11-4 : 9540	16034-77-8 : 6263	17082-12-1 : 558
13360-64-0 : 5154	14024-63-6 : 10875	14930-96-2 : 2725	16066-09-4 : 8341	17094-01-8 : 9528
13360-65-1 : 5007	14040-11-0 : 10718	14938-35-3 : 8683	16090-49-6 : 4065	17094-21-2 : 7446
13361-63-2 : 4240	14062-18-1 : 5107	14976-57-9 : 2422	16110-51-3 : 2485	17094-34-7 : 5002
13389-42-9 : 8369	14064-10-9 : 3419	14979-39-6 : 7301	16136-84-8 : 2272	17102-64-6 : 5784
13395-16-9 : 2463	14068-53-2 : 5254	15014-25-2 : 2897	16136-85-9 : 2273	17113-33-6 : 8884
13403-37-7 : 4707	14073-00-8 : 7502	15111-56-5 : 4965	16177-46-1 : 4603	17176-77-1 : 2898
13414-95-4 : 3604	14073-97-3 : 7413	15121-84-3 : 8059	16197-92-5 : 7000	17185-29-4 : 10706
13422-51-0 : 5959	14116-69-9 : 506	15164-44-0 : 6274	16219-75-3 : 5082	17201-43-3 : 1250
13426-91-0 : 978	14165-55-0 : 9807	15203-25-5 : 10344	16245-79-7 : 8382	17202-20-9 : 4190
13444-24-1 : 5221	14167-18-1 : 9500	15206-55-0 : 8452	16262-48-9 : 6465	17256-39-2 : 2002
13463-39-3 : 8017	14167-59-0 : 10004	15220-85-6 : 7700	16301-26-1 : 567	17301-94-9 : 7508
13463-40-6 : 6359	14200-76-1 : 10766	15232-76-5 : 8404	16302-35-5 : 3654	17312-63-9 : 1627
13463-41-7 : 9387	14205-39-1 : 6929	15243-01-3 : 978	16357-59-8 : 5026	17341-93-4 : 1749
13466-40-5 : 8942	14210-25-4 : 2254	15243-33-1 : 9492	16369-21-4 : 9208	17348-59-3 : 1594
13475-76-8 : 1563	14221-47-7 : 439	15299-99-7 : 7996	16409-44-2 : 5530	17356-19-3 : 5289
13475-79-1 : 4122	14222-60-7 : 9263	15301-40-3 : 4867	16409-46-4 : 6700	17372-87-1 : 9713
13475-81-5 : 9952	14239-68-0 : 1691	15321-51-4 : 6357	16420-13-6 : 4003	17376-04-4 : 6300
13475-82-6 : 8597	14255-88-0 : 6630	15356-60-2 : 7407	16423-68-0 : 4752	17397-89-6 : 1805
13476-99-8 : 10773	14261-75-7 : 2427	15358-48-2 : 6054	16448-54-7 : 6358	17406-45-0 : 10123
13491-79-7 : 1544	14267-17-5 : 960	15403-89-1 : 5177	16491-15-9 : 4035	17420-30-3 : 376
13511-13-2 : 182	14268-66-7 : 690	15457-05-3 : 5384	16520-62-0 : 1687	17440-90-3 : 6356
13511-38-1 : 2004	14284-06-1 : 2461	15489-90-4 : 5634	16529-66-1 : 8642	17465-86-0 : 2544
13523-86-9 : 9046	14284-93-6 : 9493	15500-66-0 : 8499	16532-02-8 : 1257	17466-45-4 : 8738
13529-27-6 : 3363	14285-68-8 : 9472	15501-33-4 : 6295	16532-79-9 : 1098	17496-08-1 : 441

CAS	Page	CAS	Page	CAS	Page	CAS	Page	CAS	Page
25168-10-9	: 7982	25812-30-0	: 4245	27134-26-5	: 1850	29232-93-7	: 9083	32986-56-4	: 10103
25229-42-9	: 2615	25915-78-0	: 3156	27134-26-5	: 1851	29400-42-8	: 5632	33021-02-2	: 1591
25245-34-5	: 1176	25953-19-9	: 1795	27134-26-5	: 1852	29548-30-9	: 5308	33069-62-4	: 9684
25265-68-3	: 7797	26002-80-2	: 8768	27134-27-6	: 3075	29606-79-9	: 7466	33089-61-1	: 437
25265-71-8	: 4555	26116-12-1	: 5241	27134-27-6	: 3076	29662-90-6	: 4181	33146-45-1	: 3122
25265-75-2	: 1406	26140-60-3	: 9696	27134-27-6	: 3077	29756-38-5	: 1305	33229-34-4	: 5627
25265-75-2	: 1407	26140-60-3	: 9697	27134-27-6	: 3078	29756-38-5	: 1306	33245-39-5	: 5336
25311-71-1	: 6431	26140-60-3	: 9698	27134-27-6	: 3079	29756-38-5	: 1307	33284-53-6	: 9738
25312-34-9	: 10542	26171-23-3	: 10110	27134-27-6	: 3080	29757-24-2	: 8034	33401-87-5	: 8500
25312-65-6	: 2361	26171-83-5	: 1405	27138-31-4	: 4556	29761-21-5	: 6427	33404-34-1	: 1797
25316-40-9	: 4666	26186-02-7	: 10327	27154-33-2	: 10236	29767-20-2	: 9687	33419-42-0	: 5295
25321-14-6	: 7208	26198-63-0	: 3271	27156-03-2	: 3159	29801-94-3	: 9097	33467-76-4	: 5701
25321-14-6	: 7209	26225-79-6	: 4814	27176-87-0	: 4645	29882-57-3	: 2209	33560-15-5	: 6404
25321-14-6	: 7210	26249-12-7	: 2908	27193-86-8	: 4652	29882-72-2	: 8479	33605-72-0	: 812
25321-14-6	: 7211	26249-12-7	: 2909	27195-67-1	: 4010	29883-15-6	: 449	33629-47-9	: 1485
25321-14-6	: 7212	26249-12-7	: 2910	27236-46-0	: 7576	29943-42-8	: 9887	33665-90-6	: 13
25321-14-6	: 7213	26259-45-0	: 1389	27254-36-0	: 8125	29964-84-9	: 6428	33820-53-0	: 6460
25321-22-6	: 3092	26260-02-6	: 6273	27254-36-0	: 8126	29965-97-7	: 2643	33857-26-0	: 3151
25321-22-6	: 3093	26391-16-2	: 2984	27255-72-7	: 8799	30273-11-1	: 1505	33880-83-0	: 4693
25321-22-6	: 3094	26399-36-0	: 9121	27267-69-2	: 2451	30389-18-5	: 5286	33941-07-0	: 9296
25322-14-9	: 10627	26402-26-6	: 3753	27304-13-8	: 8484	30516-87-1	: 10870	33966-50-6	: 1495
25323-30-2	: 3176	26438-26-6	: 8259	27314-13-2	: 8273	30525-89-4	: 8507	33979-03-2	: 5726
25339-17-7	: 2758	26438-27-7	: 2785	27387-79-7	: 1537	30553-04-9	: 7973	33996-33-7	: 72
25339-53-1	: 2768	26441-05-4	: 6034	27464-82-0	: 4334	30560-19-1	: 11	34014-18-1	: 9686
25340-17-4	: 3401	26446-38-8	: 9632	27554-26-3	: 3788	30586-10-8	: 3246	34077-87-7	: 3306
25340-17-4	: 3402	26446-77-5	: 1327	27598-85-2	: 389	30586-10-8	: 3247	34077-87-7	: 3307
25340-17-4	: 3403	26446-77-5	: 1328	27598-85-2	: 390	30586-10-8	: 3248	34077-87-7	: 3308
25340-18-5	: 10349	26471-56-7	: 4360	27598-85-2	: 391	30586-18-6	: 8597	34256-82-1	: 27
25340-18-5	: 10350	26471-56-7	: 4361	27876-94-4	: 2868	30685-95-1	: 6347	34316-15-9	: 1809
25346-31-0	: 1274	26471-56-7	: 4362	27987-06-0	: 10386	30752-19-3	: 1229	34451-18-8	: 1598
25346-32-1	: 2144	26471-56-7	: 4363	27988-97-2	: 10009	30903-87-8	: 6622	34522-32-2	: 1765
25346-33-2	: 1271	26471-56-7	: 4364	28056-54-4	: 10548	31178-68-4	: 508	34590-94-8	: 4557
25354-97-6	: 5896	26472-41-3	: 5928	28076-73-5	: 958	31178-69-5	: 509	34612-38-9	: 6669
25376-45-8	: 6977	26489-02-1	: 4192	28188-41-2	: 1249	31212-28-9	: 8067	34619-03-9	: 3011
25377-73-5	: 4637	26519-91-5	: 7166	28227-92-1	: 8739	31212-28-9	: 8068	34620-76-3	: 6666
25394-57-4	: 135	26530-20-1	: 8377	28249-77-6	: 10045	31218-83-4	: 9191	34698-41-4	: 6232
25394-75-6	: 216	26531-85-1	: 10862	28260-61-9	: 2337	31252-42-3	: 824	34713-94-5	: 7077
25395-22-6	: 280	26549-25-7	: 5687	28289-54-5	: 9871	31294-92-5	: 6304	34883-39-1	: 3121
25395-31-7	: 9155	26571-79-9	: 10218	28300-74-5	: 495	31367-46-1	: 7295	34885-03-5	: 7148
25395-31-7	: 9157	26571-79-9	: 10219	28322-02-3	: 50	31431-39-7	: 6684	34893-92-0	: 3266
25414-22-6	: 6830	26602-89-1	: 1121	28467-92-7	: 8924	31543-75-6	: 2958	34980-39-7	: 6583
25415-67-2	: 5148	26638-19-7	: 3270	28553-12-0	: 3786	31807-55-3	: 7845	35065-27-1	: 5728
25475-67-6	: 6550	26638-19-7	: 3272	28556-81-2	: 6409	31849-78-2	: 1128	35161-71-8	: 7721
25496-08-6	: 5400	26638-19-7	: 3273	28772-56-7	: 1075	31879-05-7	: 8787	35182-51-5	: 5227
25498-49-1	: 10673	26644-46-2	: 10427	28788-62-7	: 1524	31980-29-7	: 8020	35221-79-5	: 9879
25512-62-3	: 2599	26655-34-5	: 5546	28805-86-9	: 1639	32161-06-1	: 100	35274-05-6	: 5764
25542-62-5	: 4914	26754-48-3	: 9850	28805-86-9	: 1642	32222-06-3	: 3718	35305-13-6	: 7717
25550-58-7	: 4392	26760-64-5	: 7087	28805-86-9	: 1644	32315-10-9	: 1048	35306-33-3	: 5525
25550-58-7	: 4393	26760-64-5	: 7088	28911-01-5	: 10145	32360-05-7	: 8325	35367-38-5	: 3546
25550-58-7	: 4394	26761-40-0	: 3785	28994-41-4	: 819	32388-22-0	: 1160	35400-43-2	: 9666
25550-58-7	: 4395	26762-93-6	: 3800	29003-73-4	: 1551	32449-92-6	: 5551	35408-04-9	: 9045
25550-58-7	: 4396	26787-78-0	: 444	29074-38-2	: 1714	32503-27-8	: 9725	35554-08-6	: 9514
25551-13-7	: 10507	26839-75-8	: 10100	29082-74-4	: 8295	32511-34-5	: 1066	35554-44-0	: 6211
25551-13-7	: 10508	26914-02-3	: 6335	29091-05-2	: 4359	32598-13-3	: 9739	35573-93-4	: 1976
25551-13-7	: 10509	26914-02-3	: 6336	29091-21-2	: 9119	32620-08-9	: 1129	35608-64-1	: 5074
25567-67-3	: 2008	26952-21-6	: 7293	29106-32-9	: 2675	32665-23-9	: 6510	35691-65-7	: 1118
25567-67-3	: 2009	26952-23-8	: 3279	29122-68-7	: 529	32718-56-2	: 6825	35693-92-6	: 10211
25567-68-4	: 2136	26952-42-1	: 10624	29136-19-4	: 8225	32729-21-8	: 5588	35794-11-7	: 4268
25611-78-3	: 8834	27025-41-8	: 5559	29191-52-4	: 6772	32768-54-0	: 3299	35860-31-2	: 10634
25639-42-3	: 7149	27072-47-5	: 2975	29191-52-4	: 6773	32809-16-8	: 9118	35860-51-6	: 4372
25653-16-1	: 10687	27072-47-5	: 2976	29191-52-4	: 6774	32830-97-0	: 2214	36011-19-5	: 2727
25790-55-0	: 1921	27072-47-5	: 2977	29204-02-2	: 4687	32854-75-4	: 6589	36049-78-2	: 6327

36069-05-3 : 9286	39515-51-0 : 8776	52081-33-1 : 7883	56830-75-2 : 3554	64902-72-3 : 2354
36069-45-1 : 7909	39546-32-2 : 9060	52152-71-3 : 6288	56961-20-7 : 10208	65530-93-0 : 7324
36088-22-9 : 8537	39589-98-5 : 4004	52181-51-8 : 2331	56961-84-3 : 3152	66215-27-8 : 2711
36150-73-9 : 4748	40058-87-5 : 6488	52181-51-8 : 2332	57041-67-5 : 9838	66325-11-9 : 5909
36309-01-0 : 3820	40487-42-1 : 8520	52181-51-8 : 2333	57129-69-8 : 3646	66333-88-8 : 7532
36357-38-7 : 7748	40520-25-0 : 2638	52217-02-4 : 6011	57377-32-9 : 6203	66357-35-5 : 9449
36383-15-0 : 1514	40580-59-4 : 5606	52315-07-8 : 2708	57392-55-9 : 6455	66429-31-0 : 1826
36386-49-9 : 5314	40596-69-8 : 6763	52387-50-5 : 5877	57526-28-0 : 7083	66441-23-4 : 5318
36566-80-0 : 7359	41011-01-2 : 1137	52390-72-4 : 5686	57576-09-7 : 7465	66841-25-6 : 10124
36653-82-4 : 5753	41096-46-2 : 5955	52562-19-3 : 7853	57653-85-7 : 5735	67210-45-1 : 8396
36701-01-6 : 5493	41198-08-7 : 9120	52645-53-1 : 8731	57837-19-1 : 6727	67329-04-8 : 6364
36734-19-7 : 6352	41464-40-8 : 9737	52663-71-5 : 5642	57968-70-4 : 7309	67485-29-4 : 5930
36768-62-4 : 9968	41587-36-4 : 2173	52663-77-1 : 8215	57968-71-5 : 3533	67952-57-2 : 7307
36791-04-5 : 9476	41587-36-4 : 2174	52708-03-9 : 8241	58138-08-2 : 10328	68085-85-8 : 2706
36823-84-4 : 8681	41587-36-4 : 2175	52896-87-4 : 6499	58167-01-4 : 3813	68359-37-5 : 2704
36847-51-5 : 4976	41587-36-4 : 2176	52918-63-5 : 2793	58479-61-1 : 1944	68420-93-9 : 7765
36894-69-6 : 6573	41587-36-4 : 2177	53022-14-3 : 1068	58560-75-1 : 6207	68476-85-7 : 1401
36895-62-2 : 6756	41814-78-2 : 10306	53152-98-0 : 7675	58654-67-4 : 7522	68476-85-7 : 9135
37143-54-7 : 6835	42027-23-6 : 8614	53178-20-4 : 1926	59080-40-9 : 5718	68694-11-1 : 10375
37248-47-8 : 10767	42200-33-9 : 7917	53188-23-1 : 5471	59094-77-8 : 5255	68767-14-6 : 6632
37273-91-9 : 6728	42205-08-3 : 10348	53230-10-7 : 6690	59113-36-9 : 8602	68844-77-9 : 527
37275-48-2 : 893	42296-74-2 : 5774	53558-25-1 : 9386	59122-46-2 : 7879	68975-47-3 : 7340
37275-48-2 : 894	42296-74-2 : 5779	53566-37-3 : 7358	59227-89-3 : 6595	69377-81-7 : 5440
37275-48-2 : 895	42389-30-0 : 2184	53631-70-2 : 313	59277-89-3 : 120	69409-94-5 : 5441
37275-48-2 : 896	42397-64-8 : 4406	53714-56-0 : 6604	59669-26-0 : 10051	69655-05-6 : 3339
37275-48-2 : 897	42397-65-9 : 4407	53772-78-4 : 8704	59756-60-4 : 5439	69833-18-7 : 521
37517-30-9 : 5	42474-21-5 : 2945	53778-43-1 : 5137	60168-88-9 : 5311	69833-18-7 : 522
37674-63-8 : 7586	42569-16-4 : 8649	53780-34-0 : 6691	60207-90-1 : 9192	70116-68-6 : 7080
37680-73-2 : 8536	42576-02-3 : 864	53861-57-7 : 1766	60212-29-5 : 10751	70124-77-5 : 5337
37688-96-3 : 2989	42852-95-9 : 6931	53910-25-1 : 8660	60251-57-2 : 1300	70470-05-2 : 3645
37746-78-4 : 4913	42874-03-3 : 8488	54182-58-0 : 9630	60435-70-3 : 7289	70644-46-1 : 2604
37853-59-1 : 1045	42978-66-5 : 10672	54327-10-5 : 7280	60583-39-3 : 1385	70715-19-4 : 7255
37893-02-0 : 5335	43121-43-3 : 573	54363-49-4 : 7548	60583-39-3 : 1385	71563-86-5 : 10058
37924-13-3 : 8694	43133-95-5 : 7552	54527-68-3 : 2028	61336-70-7 : 444	71626-11-4 : 577
38136-29-7 : 7574	43133-95-5 : 7553	54532-97-7 : 1526	61361-99-7 : 8795	71751-41-2 : 537
38237-74-0 : 399	43222-48-6 : 3545	54630-50-1 : 7297	61531-45-1 : 8679	71783-56-7 : 9846
38260-54-7 : 5296	44855-57-4 : 8345	54638-10-7 : 8663	61915-27-3 : 8449	72178-02-0 : 5445
38300-67-3 : 10160	44914-03-6 : 7107	54749-90-5 : 2345	61949-83-5 : 6232	72490-01-8 : 5319
38304-91-5 : 7876	48145-04-6 : 8783	54819-86-2 : 1536	62016-38-0 : 10574	72559-06-9 : 9484
38380-07-3 : 5725	49763-65-7 : 8668	54827-17-7 : 9928	62108-66-1 : 9222	72762-00-6 : 9361
38411-22-2 : 5727	50471-44-8 : 10789	54910-89-3 : 5433	62108-68-3 : 6384	73467-76-2 : 729
38433-80-6 : 6101	50594-66-6 : 109	55066-53-0 : 5080	62108-69-4 : 7109	73467-76-2 : 730
38460-95-6 : 10747	50623-57-9 : 1628	55096-26-9 : 7920	62108-70-7 : 7110	73590-58-6 : 8411
38552-72-6 : 7317	50679-08-8 : 9695	55156-16-6 : 7571	62108-71-8 : 9220	74051-80-2 : 9535
38577-97-8 : 3766	50782-69-9 : 4992	55282-16-1 : 1562	62127-47-3 : 10161	74112-36-0 : 7319
38640-62-9 : 3807	50782-69-9 : 4992	55283-68-6 : 4765	62168-25-6 : 2915	74115-24-5 : 2425
38641-94-0 : 5600	50876-33-0 : 9937	55285-14-8 : 1762	62571-86-2 : 1724	74222-97-2 : 9659
38651-65-9 : 3974	51023-22-4 : 10214	55290-64-7 : 3826	62637-91-6 : 9719	74223-64-6 : 7869
38661-81-3 : 1201	51207-31-9 : 9744	55314-57-3 : 5199	62637-91-6 : 9718	74513-58-9 : 9224
38721-71-0 : 3142	51218-45-2 : 7864	55320-06-4 : 2784	63284-71-9 : 8285	74683-66-2 : 8239
38721-71-0 : 3143	51235-04-2 : 5859	55335-06-3 : 10299	63451-35-4 : 6125	74742-08-8 : 8240
38727-55-8 : 3349	51282-49-6 : 7127	55505-26-5 : 7511	63597-41-1 : 8329	74742-10-2 : 2759
38821-53-3 : 1804	51308-54-4 : 1468	55512-33-9 : 9311	63938-10-3 : 2307	74778-22-6 : 8359
38842-05-6 : 9812	51308-99-7 : 2200	55751-54-7 : 1503	63938-10-3 : 2308	75330-75-5 : 6629
38870-89-2 : 6771	51309-10-5 : 1969	55944-70-2 : 2064	63990-84-1 : 768	76058-33-8 : 5242
38873-84-6 : 9278	51338-27-3 : 3313	56030-54-7 : 9634	64044-51-5 : 6581	76578-14-8 : 9446
39029-41-9 : 1690	51481-61-9 : 2391	56053-84-0 : 1663	64057-79-0 : 1019	76738-62-0 : 8496
39148-24-8 : 5466	51524-84-6 : 6927	56073-10-0 : 1073	64091-91-4 : 8190	76824-35-6 : 5301
39191-07-6 : 2107	51630-58-1 : 5324	56103-67-4 : 4442	64168-11-2 : 1033	76963-41-2 : 8214
39196-18-4 : 10054	51632-16-7 : 1276	56298-90-9 : 7296	64204-55-3 : 4754	77094-11-2 : 303
39201-33-7 : 2509	51707-55-2 : 10035	56392-17-7 : 7866	64257-84-7 : 5320	77175-51-0 : 2424
39227-61-7 : 8537	51756-08-2 : 7258	56688-75-6 : 7160	64653-59-4 : 5030	77501-63-4 : 6577

DIAMAGNETIC SUSCEPTIBILITY OF SELECTED ORGANIC COMPOUNDS

When a material is placed in a magnetic field H, a magnetization M is induced in the material which is related to H by $M = \kappa H$, where κ is called the volume susceptibility. Since H and M have the same dimensions, κ is dimensionless. A more useful parameter is the molar susceptibility χ_m, defined by

$$\chi_m = \kappa V_m = \kappa M/\rho$$

where V_m is the molar volume of the substance, M the molar mass, and ρ the mass density. When the cgs system is used, the customary unit for χ_m is $cm^3\ mol^{-1}$; the corresponding SI unit is $m^3\ mol^{-1}$. Substances with no unpaired electrons are called diamagnetic; they have negative values of χ_m.

This table gives values of the diamagnetic susceptibility for about 400 common organic compounds. All values refer to room temperature and atmospheric pressure and to the physical form that is stable under these conditions. Substances are arranged by molecular formula in Hill order. A more extensive table may be found in Reference 1.

In keeping with customary practice, the molar susceptibility is given here in units appropriate to the cgs system. These values should be multiplied by 4π to obtain values for use in SI equations (where the magnetic field strength H has units of A m^{-1}).

References

1. *Landolt-Börnstein, Numerical Data and Functional Relationships in Science and Technology, New Series, II/16, Diamagnetic Susceptibility,* Gupta, R. R., Ed., Springer-Verlag, Heidelberg, 1986.
2. Barter, C., Meisenheimer, R. G., and Stevenson, D. P., *J. Phys. Chem.* 64, 1312, 1960.
3. Broersma, S., *J. Chem. Phys.* 17, 873, 1949.

Molecular formula	Compound	$-\chi_m/10^{-6}\ cm^3\ mol^{-1}$
$CBrCl_3$	Bromotrichloromethane	73.2
CBr_4	Tetrabromomethane	93.7
$CClF_3$	Chlorotrifluoromethane	45.3
$CClN$	Cyanogen chloride	32.4
CCl_2F_2	Dichlorodifluoromethane	52.2
CCl_2O	Carbonyl chloride	47.9
CCl_3F	Trichlorofluoromethane	58.7
CCl_3NO_2	Trichloronitromethane	75.3
CCl_4	Tetrachloromethane	66.8
$CHBrCl_2$	Bromodichloromethane	66.3
$CHBr_3$	Tribromomethane	82.6
$CHCl_3$	Trichloromethane	58.9
CHI_3	Triiodomethane	117.1
CH_2BrCl	Bromochloromethane	55.1
CH_2Br_2	Dibromomethane	65.1
CH_2Cl_2	Dichloromethane	46.6
CH_2I_2	Diiodomethane	93.1
CH_2N_2	Cyanamide	24.8
CH_2O	Formaldehyde	18.6
CH_2O_2	Formic acid	19.9
CH_3Br	Bromomethane	42.8
CH_3Cl	Chloromethane	32.0
CH_3F	Fluoromethane	17.8
CH_3I	Iodomethane	57.2
CH_3NO	Formamide	23.0
CH_3NO_2	Nitromethane	21.0
CH_4	Methane	17.4
CH_4N_2O	Urea	33.5
CH_4O	Methanol	21.4
CH_5N	Methylamine	27.0
CI_4	Tetraiodomethane	136

Molecular formula	Compound	$-\chi_m/10^{-6}\ cm^3\ mol^{-1}$
CN_4O_8	Tetranitromethane	43.0
C_2ClF_3	Chlorotrifluoroethylene	49.1
C_2Cl_4	Tetrachloroethylene	81.6
C_2Cl_6	Hexachloroethane	112.8
C_2HCl_3	Trichloroethylene	65.8
C_2HCl_3O	Trichloroacetaldehyde	73.0
C_2HCl_3O	Dichloroacetyl chloride	69.0
$C_2HCl_3O_2$	Trichloroacetic acid	73.0
C_2HCl_5	Pentachloroethane	99.1
$C_2HF_3O_2$	Trifluoroacetic acid	43.3
C_2H_2	Acetylene	20.8
$C_2H_2Br_4$	1,1,2,2-Tetrabromoethane	123.4
$C_2H_2Cl_2$	1,1-Dichloroethylene	49.2
$C_2H_2Cl_2$	cis-1,2-Dichloroethylene	51.0
$C_2H_2Cl_2$	trans-1,2-Dichloroethylene	48.9
$C_2H_2Cl_4$	1,1,2,2-Tetrachloroethane	89.8
C_2H_3Cl	Chloroethylene	35.9
C_2H_3ClO	Acetyl chloride	39.3
C_2H_3N	Acetonitrile	27.8
C_2H_4	Ethylene	18.8
$C_2H_4Br_2$	1,2-Dibromoethane	78.9
$C_2H_4Cl_2$	1,1-Dichloroethane	57.4
$C_2H_4Cl_2$	1,2-Dichloroethane	59.6
C_2H_4O	Acetaldehyde	22.2
C_2H_4O	Ethylene oxide	30.5
$C_2H_4O_2$	Acetic acid	31.8
$C_2H_4O_2$	Methyl formate	31.1
C_2H_5Br	Bromoethane	78.8
C_2H_5Cl	Chloroethane	69.9
C_2H_5I	Iodoethane	69.1
C_2H_5NO	Acetamide	33.9

Diamagnetic Susceptibility of Selected Organic Compounds

Molecular formula	Compound	$-\chi_m/10^{-6}$ cm³ mol⁻¹	Molecular formula	Compound	$-\chi_m/10^{-6}$ cm³ mol⁻¹
$C_2H_5NO_2$	Nitroethane	35.4	C_4H_6	1,2-Butadiene	35.6
$C_2H_5NO_2$	Glycine	39.6	C_4H_6	1,3-Butadiene	32.1
C_2H_6	Ethane	26.8	$C_4H_6O_2$	Vinyl acetate	46.4
C_2H_6O	Ethanol	33.7	$C_4H_6O_3$	Acetic anhydride	52.8
C_2H_6O	Dimethyl ether	26.3	$C_4H_6O_4$	Succinic acid	58.0
$C_2H_6O_2$	Ethylene glycol	38.9	$C_4H_6O_4$	Dimethyl oxalate	55.7
C_2H_6S	Ethanethiol	47.0	C_4H_7N	Butanenitrile	50.4
C_2H_6S	Dimethyl sulfide	44.9	C_4H_8	1-Butene	41.0
$C_2H_8N_2$	1,2-Ethanediamine	46.5	C_4H_8	cis-2-Butene	42.6
C_2N_2	Cyanogen	21.6	C_4H_8	trans-2-Butene	43.3
C_3H_4	Allene	25.3	C_4H_8	Isobutene	40.8
$C_3H_4O_2$	Vinyl formate	34.7	C_4H_8	Cyclobutane	40.0
C_3H_5Br	3-Bromopropene	58.6	C_4H_8O	Ethyl vinyl ether	47.9
C_3H_5Cl	2-Chloropropene	47.8	C_4H_8O	1,2-Epoxybutane	54.8
C_3H_5Cl	3-Chloropropene	47.8	C_4H_8O	Butanal	45.9
C_3H_5N	Propanenitrile	38.6	C_4H_8O	2-Butanone	45.6
C_3H_6	Propene	30.7	$C_4H_8O_2$	Butanoic acid	55.2
C_3H_6	Cyclopropane	39.2	$C_4H_8O_2$	2-Methylpropanoic acid	56.1
C_3H_6O	Allyl alcohol	36.7	$C_4H_8O_2$	Propyl formate	55.0
C_3H_6O	Propanal	34.2	$C_4H_8O_2$	Ethyl acetate	54.1
C_3H_6O	Acetone	33.8	$C_4H_8O_2$	Methyl propanoate	54.5
C_3H_6O	Methyloxirane	42.5	$C_4H_8O_2$	1,4-Dioxane	52.2
$C_3H_6O_2$	Propanoic acid	43.2	C_4H_9Br	1-Bromobutane	77.1
$C_3H_6O_2$	Ethyl formate	42.4	C_4H_9Br	1-Bromo-2-methylpropane	79.9
C_3H_7Br	1-Bromopropane	65.6	C_4H_9Cl	1-Chlorobutane	67.1
C_3H_7Br	2-Bromopropane	65.1	C_4H_9Cl	2-Chlorobutane	67.4
C_3H_7Cl	1-Chloropropane	56.0	C_4H_9I	1-Iodobutane	93.6
C_3H_7I	1-Iodopropane	84.3	C_4H_9N	Pyrrolidine	54.8
C_3H_7N	Allylamine	40.1	C_4H_9NO	Morpholine	55.0
$C_3H_7NO_2$	1-Nitropropane	45.0	C_4H_{10}	Butane	50.3
$C_3H_7NO_2$	2-Nitropropane	45.4	C_4H_{10}	Isobutane	50.5
$C_3H_7NO_2$	Ethyl carbamate	57.0	$C_4H_{10}O$	1-Butanol	56.4
C_3H_8	Propane	38.6	$C_4H_{10}O$	2-Butanol	57.6
C_3H_8O	1-Propanol	44.8	$C_4H_{10}O$	2-Methyl-1-propanol	57.6
C_3H_8O	2-Propanol	45.7	$C_4H_{10}O$	2-Methyl-2-propanol	56.6
$C_3H_8O_2$	1,3-Propylene glycol	50.2	$C_4H_{10}O$	Diethyl ether	55.5
$C_3H_8O_2$	Dimethoxymethane	47.3	$C_4H_{10}O_2$	1,3-Butanediol	61.8
$C_3H_8O_3$	Glycerol	57.1	$C_4H_{10}O_2$	1,4-Butanediol	61.8
$C_4H_2O_3$	Maleic anhydride	35.8	$C_4H_{10}S$	1-Butanethiol	70.2
$C_4H_4N_2$	Pyrazine	37.8	$C_4H_{11}N$	Butylamine	58.9
$C_4H_4N_2$	Pyrimidine	43.1	$C_4H_{11}N$	Isobutylamine	59.8
C_4H_4O	Furan	43.1	$C_4H_{11}N$	Diethylamine	56.8
$C_4H_4O_3$	Succinic anhydride	47.5	$C_5H_4O_2$	Furfural	47.2
$C_4H_4O_4$	Maleic acid	49.6	C_5H_5N	Pyridine	48.7
$C_4H_4O_4$	Fumaric acid	49.1	$C_5H_6O_2$	Furfuryl alcohol	61.0
C_4H_4S	Thiophene	57.3	$C_5H_7NO_2$	Ethyl cyanoacetate	67.3
C_4H_5N	Pyrrole	48.6	C_5H_8	2-Methyl-1,3-butadiene	46.0

Molecular formula	Compound	$-\chi_m/10^{-6}$ cm^3 mol^{-1}	Molecular formula	Compound	$-\chi_m/10^{-6}$ cm^3 mol^{-1}
C_5H_8O	Cyclopentanone	51.6	C_6H_6ClN	o-Chloroaniline	79.5
$C_5H_8O_2$	Methyl methacrylate	57.3	C_6H_6ClN	m-Chloroaniline	76.6
$C_5H_8O_2$	2,4-Pentanedione	54.9	C_6H_6ClN	p-Chloroaniline	76.7
C_5H_{10}	1-Pentene	54.6	$C_6H_6N_2O_2$	o-Nitroaniline	67.4
C_5H_{10}	2-Methyl-2-butene	54.7	$C_6H_6N_2O_2$	m-Nitroaniline	69.7
C_5H_{10}	Cyclopentane	56.2	$C_6H_6N_2O_2$	p-Nitroaniline	68.0
$C_5H_{10}O$	Cyclopentanol	64.0	C_6H_6O	Phenol	60.6
$C_5H_{10}O$	Pentanal	57.5	$C_6H_6O_2$	p-Hydroquinone	64.7
$C_5H_{10}O$	2-Pentanone	57.5	$C_6H_6O_2$	Pyrocatechol	68.2
$C_5H_{10}O$	3-Pentanone	57.7	$C_6H_6O_2$	Resorcinol	67.2
$C_5H_{10}O_2$	Pentanoic acid	66.5	C_6H_7N	Aniline	62.4
$C_5H_{10}O_2$	3-Methylbutanoic acid	67.7	C_6H_7N	4-Methylpyridine	59.8
$C_5H_{10}O_2$	Butyl formate	65.8	C_6H_8	1,4-Cyclohexadiene	48.7
$C_5H_{10}O_2$	Isobutyl formate	66.8	$C_6H_8N_2$	o-Phenylenediamine	72.5
$C_5H_{10}O_2$	Propyl acetate	65.9	$C_6H_8N_2$	m-Phenylenediamine	70.4
$C_5H_{10}O_2$	Isopropyl acetate	67.0	$C_6H_8N_2$	p-Phenylenediamine	70.7
$C_5H_{10}O_2$	Ethyl propanoate	66.3	C_6H_{10}	1,5-Hexadiene	55.1
$C_5H_{10}O_2$	Tetrahydrofurfuryl alcohol	69.4	C_6H_{10}	1-Hexyne	64.5
$C_5H_{10}O_3$	Diethyl carbonate	75.4	C_6H_{10}	Cyclohexene	58.0
$C_5H_{11}N$	Piperidine	64.2	$C_6H_{10}O$	Cyclohexanone	62.0
C_5H_{12}	Pentane	61.5	$C_6H_{10}O_3$	Ethyl acetoacetate	71.7
C_5H_{12}	Isopentane	63.0	$C_6H_{10}O_4$	Diethyl oxalate	81.7
C_5H_{12}	Neopentane	63.0	C_6H_{12}	1-Hexene	66.4
$C_5H_{12}O$	1-Pentanol	67.0	C_6H_{12}	2,3-Dimethyl-2-butene	65.9
$C_5H_{12}O$	2-Pentanol	69.1	C_6H_{12}	Cyclohexane	68
$C_5H_{12}O_2$	1,5-Pentanediol	73.5	C_6H_{12}	Methylcyclopentane	70.2
$C_5H_{13}N$	Pentylamine	69.3	$C_6H_{12}O$	Hexanal	69.4
C_6Cl_6	Hexachlorobenzene	147.0	$C_6H_{12}O$	2-Hexanone	69.2
$C_6H_4ClNO_2$	1-Chloro-2-nitrobenzene	75.5	$C_6H_{12}O$	3-Hexanone	69.0
$C_6H_4ClNO_2$	1-Chloro-3-nitrobenzene	77.2	$C_6H_{12}O$	4-Methyl-2-pentanone	69.7
$C_6H_4ClNO_2$	1-Chloro-4-nitrobenzene	74.7	$C_6H_{12}O$	Cyclohexanol	73.4
$C_6H_4Cl_2$	o-Dichlorobenzene	84.4	$C_6H_{12}O_2$	Hexanoic acid	78.1
$C_6H_4Cl_2$	m-Dichlorobenzene	84.1	$C_6H_{12}O_2$	Isopentyl formate	78.4
$C_6H_4Cl_2$	p-Dichlorobenzene	81.7	$C_6H_{12}O_2$	Isobutyl acetate	78.7
$C_6H_4O_2$	p-Benzoquinone	36	$C_6H_{12}O_2$	Propyl propanoate	77.7
C_6H_5Br	Bromobenzene	78.4	$C_6H_{12}O_3$	Paraldehyde	86.1
C_6H_5Cl	Chlorobenzene	69.5	C_6H_{14}	Hexane	74.1
C_6H_5ClO	o-Chlorophenol	77.3	C_6H_{14}	2-Methylpentane	75.3
C_6H_5ClO	m-Chlorophenol	77.6	C_6H_{14}	3-Methylpentane	75.5
C_6H_5ClO	p-Chlorophenol	77.7	C_6H_{14}	2,2-Dimethylbutane	76.2
C_6H_5F	Fluorobenzene	58.4	C_6H_{14}	2,3-Dimethylbutane	76.2
C_6H_5I	Iodobenzene	92.0	$C_6H_{14}O$	1-Hexanol	79.5
$C_6H_5NO_2$	Nitrobenzene	61.9	$C_6H_{14}O$	4-Methyl-2-pentanol	80.4
$C_6H_5NO_3$	o-Nitrophenol	68.9	$C_6H_{14}O$	Dipropyl ether	79.4
$C_6H_5NO_3$	m-Nitrophenol	65.9	$C_6H_{14}O_2$	1,6-Hexanediol	84.3
$C_6H_5NO_3$	p-Nitrophenol	66.9	$C_6H_{14}O_2$	1,1-Diethoxyethane	81.4
C_6H_6	Benzene	54.8	$C_6H_{14}O_6$	D-Glucitol	107.8

Molecular formula	Compound	$-\chi_m/10^{-6}$ cm³ mol⁻¹	Molecular formula	Compound	$-\chi_m/10^{-6}$ cm³ mol⁻¹
$C_6H_{15}N$	Triethylamine	83.3	C_7H_{16}	3,3-Dimethylpentane	89.5
C_7H_5N	Benzonitrile	65.2	$C_7H_{16}O$	1-Heptanol	91.7
C_7H_6O	Benzaldehyde	60.7	$C_7H_{16}O$	4-Heptanol	92.1
$C_7H_6O_2$	Salicylaldehyde	66.8	$C_8H_4O_3$	Phthalic anhydride	66.7
$C_7H_6O_3$	Salicylic acid	75	$C_8H_6O_4$	Phthalic acid	83.6
C_7H_7Br	p-Bromotoluene	88.7	$C_8H_6O_4$	Isophthalic acid	84.6
C_7H_7Cl	o-Chlorotoluene	82.4	$C_8H_6O_4$	Terephthalic acid	83.5
C_7H_7Cl	m-Chlorotoluene	79.7	C_8H_7N	Benzeneacetonitrile	76.9
C_7H_7Cl	p-Chlorotoluene	80.3	C_8H_7N	Indole	85.0
C_7H_7Cl	(Chloromethyl)benzene	81.6	C_8H_8	Styrene	68.2
C_7H_7NO	Benzamide	72.0	C_8H_8O	Acetophenone	72.5
$C_7H_7NO_2$	o-Nitrotoluene	72.2	$C_8H_8O_2$	o-Toluic acid	84.3
$C_7H_7NO_2$	m-Nitrotoluene	72.7	$C_8H_8O_2$	m-Toluic acid	83.0
$C_7H_7NO_2$	p-Nitrotoluene	73.3	$C_8H_8O_2$	p-Toluic acid	82.4
C_7H_8	Toluene	65.6	$C_8H_8O_2$	Benzeneacetic acid	82.4
C_7H_8O	o-Cresol	73.3	$C_8H_8O_2$	Methyl benzoate	81.6
C_7H_8O	m-Cresol	72.2	$C_8H_8O_3$	Methyl salicylate	86.6
C_7H_8O	p-Cresol	72.4	C_8H_{10}	Ethylbenzene	77.3
C_7H_8O	Benzyl alcohol	71.8	C_8H_{10}	o-Xylene	77.7
C_7H_8O	Anisole	72.2	C_8H_{10}	m-Xylene	76.4
C_7H_9N	o-Methylaniline	74.9	C_8H_{10}	p-Xylene	77.0
C_7H_9N	m-Methylaniline	74.6	$C_8H_{10}O$	Phenetole	84.5
C_7H_9N	p-Methylaniline	72.5	$C_8H_{11}N$	N-Ethylaniline	85.6
C_7H_9N	N-Methylaniline	74.1	$C_8H_{11}N$	N,N-Dimethylaniline	85.1
C_7H_9N	2,4-Dimethylpyridine	71.3	$C_8H_{11}N$	2,4,6-Trimethylpyridine	83.1
C_7H_9N	2,6-Dimethylpyridine	72.5	$C_8H_{14}O_4$	Ethyl succinate	105.0
C_7H_9NO	o-Methoxyaniline [o-Anisidine]	79.1	C_8H_{16}	1-Octene	88.8
$C_7H_{12}O_4$	Diethyl malonate	92.6	C_8H_{16}	Cyclooctane	85.3
C_7H_{14}	1-Heptene	77.8	$C_8H_{16}O_2$	Octanoic acid	99.5
C_7H_{14}	Cycloheptane	73.9	$C_8H_{16}O_2$	Hexyl acetate	100.9
C_7H_{14}	Methylcyclohexane	78.9	$C_8H_{17}Cl$	1-Chlorooctane	114.9
$C_7H_{14}O$	1-Heptanal	81.0	C_8H_{18}	Octane	96.6
$C_7H_{14}O$	2-Heptanone	80.5	C_8H_{18}	4-Methylheptane	97.3
$C_7H_{14}O$	3-Heptanone	80.7	C_8H_{18}	3-Ethylhexane	97.8
$C_7H_{14}O$	4-Heptanone	80.5	C_8H_{18}	3,4-Dimethylhexane	99.1
$C_7H_{14}O$	2,4-Dimethyl-3-pentanone	81.1	C_8H_{18}	2,2,4-Trimethylpentane	99.1
$C_7H_{14}O_2$	Heptanoic acid	89.0	C_8H_{18}	2,3,4-Trimethylpentane	99.8
$C_7H_{14}O_2$	Pentyl acetate	88.9	$C_8H_{18}O$	1-Octanol	101.6
$C_7H_{14}O_2$	Isopentyl acetate	89.4	$C_8H_{19}N$	Dibutylamine	103.7
$C_7H_{14}O_2$	Butyl propanoate	89.1	C_9H_7N	Quinoline	86.1
$C_7H_{14}O_2$	Ethyl 3-methylbutanoate	91.1	C_9H_7N	Isoquinoline	83.9
C_7H_{16}	Heptane	85.2	C_9H_8	Indene	83
C_7H_{16}	3-Ethylpentane	86.2	C_9H_{10}	Isopropenylbenzene	80.0
C_7H_{16}	2,2-Dimethylpentane	87.0	$C_9H_{10}O_2$	Ethyl benzoate	93.8
C_7H_{16}	2,3-Dimethylpentane	87.5	$C_9H_{10}O_2$	Benzyl acetate	93.2
C_7H_{16}	2,4-Dimethylpentane	87.5	C_9H_{12}	Propylbenzene	89.1

Diamagnetic Susceptibility of Selected Organic Compounds

Molecular formula	Compound	$-\chi_m/10^{-6}$ cm^3 mol^{-1}	Molecular formula	Compound	$-\chi_m/10^{-6}$ cm^3 mol^{-1}
C_9H_{12}	Isopropylbenzene [Cumene]	89.5	$C_{12}H_9N$	Carbazole	119.9
C_9H_{12}	1,3,5-Trimethylbenzene [Mesitylene]	92.3	$C_{12}H_{10}$	Acenaphthene	109.9
C_9H_{18}	1-Nonene	100.1	$C_{12}H_{10}$	Biphenyl	103.3
$C_9H_{18}O$	2,6-Dimethyl-4-heptanone	104.3	$C_{12}H_{10}N_2$	Azobenzene	106.8
C_9H_{20}	Nonane	108.1	$C_{12}H_{11}N$	Diphenylamine	108.4
$C_{10}H_7Br$	1-Bromonaphthalene	123.6	$C_{12}H_{14}O_4$	Diethyl phthalate	127.5
$C_{10}H_7Cl$	1-Chloronaphthalene	107.6	$C_{12}H_{18}$	Hexamethylbenzene	122.5
$C_{10}H_8$	Naphthalene	91.6	$C_{12}H_{24}O_2$	Dodecanoic acid	113.0
$C_{10}H_8$	Azulene	123.7	$C_{13}H_9N$	Acridine	118.8
$C_{10}H_8O$	1-Naphthol	96.2	$C_{13}H_{10}O$	Benzophenone	109.6
$C_{10}H_8O$	2-Naphthol	96.8	$C_{13}H_{12}$	Diphenylmethane	116.0
$C_{10}H_9N$	1-Naphthalenamine	92.5	$C_{13}H_{28}$	Tridecane	153.7
$C_{10}H_9N$	2-Naphthalenamine	98.0	$C_{14}H_8O_2$	9,10-Anthracenedione	113.0
$C_{10}H_{10}O_2$	Safrole	97.5	$C_{14}H_{10}$	Anthracene	129.8
$C_{10}H_{10}O_4$	Dimethyl terephthalate	101.6	$C_{14}H_{10}$	Phenanthrene	127.6
$C_{10}H_{14}$	Butylbenzene	100.7	$C_{14}H_{10}$	Diphenylacetylene	116
$C_{10}H_{14}$	tert-Butylbenzene	101.8	$C_{14}H_{10}O_2$	Benzil	106.8
$C_{10}H_{14}$	Isobutylbenzene	101.7	$C_{14}H_{12}O_2$	Benzyl benzoate	132.2
$C_{10}H_{14}$	p-Cymene	102.8	$C_{14}H_{14}$	1,2-Diphenylethane	127.8
$C_{10}H_{14}$	1,2,4,5-Tetramethylbenzene	101.2	$C_{14}H_{28}O_2$	Tetradecanoic acid [Myristic acid]	176.0
$C_{10}H_{14}O$	p-tert-Butylphenol	108.0	$C_{14}H_{30}$	Tetradecane	166.2
$C_{10}H_{15}N$	N,N-Diethylaniline	107.9	$C_{16}H_{10}$	Pyrene	147
$C_{10}H_{16}$	d-Limonene	98.0	$C_{16}H_{32}O_2$	Hexadecanoic acid [Palmitic acid]	198.6
$C_{10}H_{16}$	α-Pinene	100.7	$C_{16}H_{34}$	Hexadecane	187.6
$C_{10}H_{16}$	β-Pinene	101.9	$C_{16}H_{34}O$	1-Hexadecanol	183.5
$C_{10}H_{16}O$	Camphor, (+)	103.0	$C_{18}H_{12}$	Chrysene	148.0
$C_{10}H_{18}$	cis-Decahydronaphthalene	107.0	$C_{18}H_{14}$	o-Terphenyl	150.4
$C_{10}H_{18}$	trans-Decahydronaphthalene	107.6	$C_{18}H_{14}$	m-Terphenyl	155.5
$C_{10}H_{22}$	Decane	119.5	$C_{18}H_{14}$	p-Terphenyl	156.0
$C_{11}H_{10}$	1-Methylnaphthalene	102.9	$C_{18}H_{34}O_2$	cis-9-Octadecenoic acid [Oleic acid]	208.5
$C_{11}H_{10}$	2-Methylnaphthalene	102.7	$C_{18}H_{36}O_2$	Octadecanoic acid [Stearic acid]	220.8
$C_{11}H_{24}$	Undecane	131.8	$C_{20}H_{12}$	Perylene	167.5
$C_{12}H_8$	Acenaphthylene	111.6			

Section 4
Properties of the Elements and Inorganic Compounds

THE ELEMENTS

C. R. Hammond

One of the most striking facts about the elements is their unequal distribution and occurrence in nature. Present knowledge of the chemical composition of the universe, obtained from the study of the spectra of stars and nebulae, indicates that hydrogen is by far the most abundant element and may account for more than 90% of the atoms or about 75% of the mass of the universe. Helium atoms make up most of the remainder. All of the other elements together contribute only slightly to the total mass.

The chemical composition of the universe is undergoing continuous change. Hydrogen is being converted into helium, and helium is being changed into heavier elements. As time goes on, the ratio of heavier elements increases relative to hydrogen. Presumably, the process is not reversible.

Burbidge, Burbidge, Fowler, and Hoyle, and more recently, Peebles, Penzias, and others have studied the synthesis of elements in stars. To explain all of the features of the nuclear abundance curve — obtained by studies of the composition of the earth, meteorites, stars, etc. — it is necessary to postulate that the elements were originally formed by at least eight different processes: (1) hydrogen burning, (2) helium burning, (3) χ process, (4) e process, (5) s process, (6) r process, (7) p process, and (8) the X process. The X process is thought to account for the existence of light nuclei such as D, Li, Be, and B. Common metals such as Fe, Cr, Ni, Cu, Ti, Zn, etc. were likely produced early in the history of our galaxy. It is also probable that most of the heavy elements on earth and elsewhere in the universe were originally formed in supernovae, or in the hot interior of stars.

Studies of the solar spectrum have led to the identification of 67 elements in the sun's atmosphere; however, all elements cannot be identified with the same degree of certainty. Other elements may be present in the sun, although they have not yet been detected spectroscopically. The element helium was discovered on the sun before it was found on earth. Some elements such as scandium are relatively more plentiful in the sun and stars than here on earth.

Minerals in lunar rocks brought back from the moon on the Apollo missions consist predominantly of *plagioclase* {(Ca,Na)(Al,Si)O$_4$O$_8$} and *pyroxene* {(Ca,Mg,Fe)$_2$Si$_2$O$_6$} — two minerals common in terrestrial volcanic rock. No new elements have been found on the moon that cannot be accounted for on earth; however, three minerals, *armalcolite* {(Fe,Mg)Ti$_2$O$_5$}, *pyroxferroite* {CaFe$_6$(SiO$_3$)$_7$}, and *tranquillityite* {Fe$_8$(Zr,Y)Ti$_3$Si$_3$O$_2$}, are new. The oldest known terrestrial rocks are about 4 billion years old. One rock, known as the "Genesis Rock," brought back from the Apollo 15 Mission, is about 4.15 billion years old. This is only about one-half billion years younger than the supposed age of the moon and solar system. Lunar rocks appear to be relatively enriched in refractory elements such as chromium, titanium, zirconium, and the rare earths, and impoverished in volatile elements such as the alkali metals, in chlorine, and in noble metals such as nickel, platinum, and gold.

Even older than the "Genesis Rock" are *carbonaceous chondrites*, a type of meteorite that has fallen to earth and has been studied. These are some of the most primitive objects of the solar system yet found. The grains making up these objects probably condensed directly out the gaseous nebula from which the sun and planets were born. Most of the condensation of the grains probably was completed within 50,000 years of the time the disk of the nebula was first formed — about 4.6 billion years ago. It is now thought that this type of meteorite may contain a small percentage of presolar dust grains. The relative abundances of the elements of these meteorites are about the same as the abundances found in the solar chromosphere.

The X-ray fluorescent spectrometer sent with the Viking I spacecraft to Mars shows that the Martian soil contains about 12 to 16% iron, 14 to 15% silicon, 3 to 8% calcium, 2 to 7% aluminum, and one-half to 2% titanium. The gas chromatograph — mass spectrometer on Viking II found no trace of organic compounds that should be present if life ever existed there.

F. W. Clarke and others have carefully studied the composition of rocks making up the crust of the earth. Oxygen accounts for about 47% of the crust, by weight, while silicon comprises about 28% and aluminum about 8%. These elements, plus iron, calcium, sodium, potassium, and magnesium, account for about 99% of the composition of the crust.

Many elements such as tin, copper, zinc, lead, mercury, silver, platinum, antimony, arsenic, and gold, which are so essential to our needs and civilization, are among some of the rarest elements in the earth's crust. These are made available to us only by the processes of concentration in ore bodies. Some of the so-called *rare-earth* elements have been found to be much more plentiful than originally thought and are about as abundant as uranium, mercury, lead, or bismuth. The least abundant rare-earth or *lanthanide* element, thulium, is now believed to be more plentiful on earth than silver, cadmium, gold, or iodine, for example. Rubidium, the 16th most abundant element, is more plentiful than chlorine while its compounds are little known in chemistry and commerce.

It is now thought that at least 24 elements are essential to living matter. The four most abundant in the human body are hydrogen, oxygen, carbon, and nitrogen. The seven next most common, in order of abundance, are calcium, phosphorus, chlorine, potassium, sulfur, sodium, and magnesium. Iron, copper, zinc, silicon, iodine, cobalt, manganese, molybdenum, fluorine, tin, chromium, selenium, and vanadium are needed and play a role in living matter. Boron is also thought essential for some plants, and it is possible that aluminum, nickel, and germanium may turn out to be necessary.

Ninety-one elements occur naturally on earth. Minute traces of plutonium-244 have been discovered in rocks mined in Southern California. This discovery supports the theory that heavy elements were produced during creation of the solar system. While technetium and promethium have not yet been found naturally on earth, they have been found to be present in stars. Technetium has been identified in the spectra of certain "late" type stars, and promethium lines have been identified in the spectra of a faintly visible star HR465 in Andromeda. Promethium must have been made very recently near the star's surface for no known isotope of this element has a half-life longer than 17.7 years.

It has been suggested that californium is present in certain stellar explosions known as supernovae; however, this has not been proved. At present no elements are found elsewhere in the universe that cannot be accounted for here on earth.

All atomic mass numbers from 1 to 238 are found naturally on earth except for masses 5 and 8. About 285 relatively stable and 67 naturally radioactive isotopes occur on earth totaling 352. In addition, the neutron, technetium, promethium, and the transuranic elements (lying beyond uranium) have now been produced artificially. In June 1999, scientists at the Lawrence Berkeley National Laboratory reported that they had found evidence of an isotope of Element 118 and its immediate decay products of Elements 116, 114, and 112. This sequence of events tended to reinforce the theory that was predicted since the 1970s that an "island of stability" existed for nuclei with approximately 114 protons and 184 neutrons. This "island" refers to nuclei in which the decay lasts for a period of time instead of a decay that occurs instantaneously. However, on July 27, 2001, researchers at LBNL reported that their laboratory and the facilities at the GSI Laboratory in Germany and at Japanese laboratories failed to confirm the results of their earlier experiments where the fusion of a krypton atom with a lead target resulted in Element 118, with chains of decay leading to Elements 116, 114, and 112, and on down to Element 106. Therefore, the discovery was reported to be spurious. However, with the announcement it was said that different experiments at the Livermore Laboratory and Joint Institute for Nuclear Research in Dubna, Russia indicated that Element 116 had since been created directly. (See also under Elements 116 and 118.)

Laboratory processes have now extended the radioactive element mass numbers beyond 238 to about 280. Each element from atomic numbers 1 to 110 is known to have at least one radioactive isotope. As of December 2001, about 3286 isotopes and isomers were thought to be known and recognized. Many stable and radioactive isotopes are now produced and distributed by the Oak Ridge National Laboratory, Oak Ridge, Tenn., U.S.A., to customers licensed by the U.S. Department of Energy.

The nucleus of an atom is characterized by the number of protons it contains, denoted by Z, and by the number of neutrons, N. Isotopes of an element have the same value of Z, but different values of N. The *mass number A*, is the sum of Z and N. For example, Uranium-238 has a mass number of 238, and contains 92 protons and 146 neutrons.

There is evidence that the definition of chemical elements must be broadened to include the electron. Several compounds known as *electrides*, have recently been made of alkaline metal elements and electrons. A relatively stable combination of a positron and electron, known as *positronium*, has also been studied.

The well-known proton, neutron, and electron are now thought to be members of a group that includes other fundamental particles that have been discovered or hypothesized by physicists. These very elemental particles, of which all matter is made, are now thought to belong to one of two families: namely, **quarks** or **leptons**. Each of these two families consists of six particles. Also, there are four different force carriers that lead to interactions between particles. The six members or "flavors" of the quark family are called **up**, **charm**, **top**, **down**, **strange**, and **bottom**. The force carriers for the quarks are the **gluon** and the **photon**. The six members of the lepton family are the **e neutrino**, the **mu neutrino**, the **tau neutrino**, the **electron**, the **muon particle**, and the **tau particle**. The force carriers for these are the **w boson** and the **z boson**. Furthermore, it appears that each of these particles has an anti-particle that has an opposite electrical charge from the above particles.

Quarks are not found individually, but are found with other quarks arranged to form composites known as **hadrons**. There are two basic types of hadrons: **baryons**, composed of three quarks, and **mesons**, composed of a quark and an anti-quark. Examples of baryons are the neutron and the proton. Neutrons are made of two down quarks and one up quark. Protons are made of two up quarks and one down quark. An example of the meson is the **pion**. This particle is made of an up quark and a down anti-quark. Such particles are unstable and tend to decay rapidly. The anti-particle of the proton is the anti-proton. The exception to the rule is the electron, whose anti-particle is the **positron**.

In recent years a search has been made for a hypothetical particle known as the **Higgs particle** or **Higgs boson**, suggested in 1966 by Peter Higgs of the University of Edinburgh, which could possibly explain why the carriers of the "electro-weak" field (w and z bosons) have mass. The Higgs particle is thought to be responsible possibly for the mass of objects throughout the universe.

Many physicists now hold that all matter and energy in the universe is controlled by four fundamental forces: the **electromagnetic force**, **gravity**, a **weak nuclear force**, and a **strong nuclear force**. The **gluon** binds quarks together by carrying the strong nuclear force. Each of these natural forces is passed back and forth among the basic particles of matter by the force carriers mentioned above. The electromagnetic force is carried by the photon, the weak nuclear force by the **intermediate vector boson**, and the gravity by the **graviton**.

For more complete information on these fundamental particles, please consult recent articles and books on nuclear or particle physics.

The available evidence leads to the conclusion that elements 89 (actinium) through 103 (lawrencium) are chemically similar to the rare-earth or lanthanide elements (elements 57 to 71, inclusive). These elements therefore have been named *actinides* after the first member of this series. Those elements beyond uranium that have been produced artificially have the following names and symbols: neptunium, 93 (Np); plutonium, 94 (Pu); americium, 95 (Am); curium, 96 (Cm); berkelium, 97 (Bk); californium, 98 (Cf); einsteinium, 99 (Es); fermium, 100 (Fm); mendelevium, 101 (Md); nobelium, 102 (No); and lawrencium, 103 (Lr). It is now claimed that Elements 104 through 112 have been produced and identified. More recently, Elements 118, 116, and 114 were reported found (see Element 118). In August 1997, the International Union of Pure and Applied Chemistry (IUPAC) gave final approval to the following names for Elements 104 to 109: Element 104 — rutherfordium (Rf); Element 105 — dubnium (Db); Element 106 — seaborgium (Sg); Element 107 — bohrium (Bh); Element 108 — hassium (Hs); and Element 109 — meitnerium (Mt). The recently discovered elements 110, 111, 112, 114, etc. have not yet been named, but may carry temporary names as designated by the International Union of Pure and Applied Chemistry. IUPAC recommends that until the existence of a new element is proven to their satisfaction, the elements are to have names and symbols derived according to these precise and simple rules: The name is based on the digits in the element's atomic number. Each digit is replaced with these expressions, with the end using the usual –ium suffix as follows: **0 nil, 1 un, 2 bi, 3 tri, 4 quad, 5 pent, 6 hex, 7 sept, 8 oct, 9 enn**. Double letter i's are not used, as for example Ununbiium, but would be Ununbium. The symbol used would be the first letter of the three main syllables. For example, Element 126 would be Unbihexium, with the symbol Ubh. It is thought there is a possibility of producing elements beyond Element 116 and discovering Elements 117, 115, and 113 by altering the beams of ions and the targets from those now being used.

There are many claims in the literature of the existence of various allotropic modifications of the elements, some of which are based on doubtful or incomplete evidence. Also, the physical properties of an element may change drastically by the presence of small amounts of impurities. With new methods of purification, which are now able to produce elements with 99.9999% purity, it has been necessary to restudy the properties of the elements. For example, the melting point of thorium changes by several hundred degrees by the presence of a small percentage of ThO_2 as an impurity. Ordinary commercial tungsten is brittle and can be worked only with difficulty. Pure tungsten, however, can be cut with a hacksaw, forged, spun, drawn, or extruded. In general, the value of a physical property given here applies to the pure element, when it is known.

Many of the chemical elements and their compounds are toxic and should be handled with due respect and care. In recent years there has been a greatly increased knowledge and awareness of the health hazards associated with chemicals, radioactive materials, and other agents. Anyone working with the elements and certain of their compounds should become thoroughly familiar with the proper safeguards to be taken. Information on specific hazards and recommended exposure limits may also be found in Section 16. Reference should also be made to publications such as the following:

1. *Code of Federal Regulations, Title 29, Labor*. With additions found in issues of the *Federal Register*.
2. *Code of Federal Regulations, Title 10, Energy*. With additions found in issues of the *Federal Register*. (Published by the U.S. Government Printing Office. Supt. of Documents.)
3. *Occupational Safety and Health Reporter* (latest edition with amendments and corrections), Bureau of National Affairs, Washington, D.C.
4. *Atomic Energy Law Reporter*, Commerce Clearing House, Chicago, IL.
5. *Nuclear Regulation Reporter*, Commerce Clearing House, Chicago, IL.
6. *TLVs® Threshold Limit Values for Chemical Substances and Physical Agents* is issued annually be the American Conference of Governmental Industrial Hygienists, Cincinnati, Ohio.
7. *The Sigma Aldrich Library of Regulatory and Safety Data. Vol. 3*, Robert E. Lenga and Kristine L. Volonpal, Sigma Chemical Co. and Aldrich Chemical Co., Inc. 1993.
8. *Hazardous Chemicals Desk Reference*, Richard J. Lewis, Sr., 4th ed., John Wiley & Sons, New York, Dec. 1997.
9. *Sittig's Handbook of Toxic and Hazardous Chemicals and Carcinogens*, 3rd ed., Noyes Publications, 2001/2.
10. *Sax's Dangerous Properties of Industrial Materials*, Richard J. Lewis and N. Irving Sax, John Wiley & Sons, New York, 1999.
11. *World Wide Limits for Toxic and Hazardous Chemicals in Air, Water, and Soil*, Marshall Sittig, Noyes Publishers.

The prices of elements as indicated in this article are intended to be only a rough guide. Prices may vary, over time, widely with supplier, quantity, and purity.

Actinium — (Gr. *aktis, aktinos*, beam or ray), Ac; at. wt. (227); at. no. 89; m.p. 1051°C, b.p. 3200 ± 300°C (est.); sp. gr. 10.07 (calc.). Discovered by Andre Debierne in 1899 and independently by F. Giesel in 1902. Occurs naturally in association with uranium minerals. Thirty four isotopes and isomers are now recognized. All are radioactive. Actinium-227, a decay product of uranium-235, is an alpha and beta emitter with a 21.77-year half-life. Its principal decay products are thorium-227 (18.72-day half-life), radium-223 (11.4-day half-life), and a number of short-lived products including radon, bismuth, polonium, and lead isotopes. In equilibrium with its decay products, it is a powerful source of alpha rays. Actinium metal has been prepared by the reduction of actinium fluoride with lithium vapor at about 1100 to 1300°C. The chemical behavior of actinium is similar to that of the rare earths, particularly lanthanum. Purified actinium comes into equilibrium with

its decay products at the end of 185 days, and then decays according to its 21.77-year half-life. It is about 150 times as active as radium, making it of value in the production of neutrons. Actinium-225, with a purity of 99%, is available from the Oak Ridge National Laboratory to holders of a permit for about $500/millicurie, plus packing charges.

Aluminum — (L. *alumen, alum*), Al; at. wt. 26.981539(5); at. no. 13; f.p. 660.323°C; b.p. 2519°C; sp. gr. 2.6989 (20°C); valence 3. The ancient Greeks and Romans used *alum* in medicine as an astringent, and as a mordant in dyeing. In 1761 de Morveau proposed the name *alumine* for the base in alum, and Lavoisier, in 1787, thought this to be the oxide of a still undiscovered metal. Wohler is generally credited with having isolated the metal in 1827, although an impure form was prepared by Oersted two years earlier. In 1807, Davy proposed the name *alumium* for the metal, undiscovered at that time, and later agreed to change it to *aluminum*. Shortly thereafter, the name *aluminium* was adopted to conform with the "ium" ending of most elements, and this spelling is now in use elsewhere in the world. *Aluminium* was also the accepted spelling in the U.S. until 1925, at which time the American Chemical Society officially decided to use the name *aluminum* thereafter in their publications. The method of obtaining aluminum metal by the electrolysis of alumina dissolved in *cryolite* was discovered in 1886 by Hall in the U.S. and at about the same time by Heroult in France. Cryolite, a natural ore found in Greenland, is no longer widely used in commercial production, but has been replaced by an artificial mixture of sodium, aluminum, and calcium fluorides. *Bauxite*, an impure hydrated oxide ore, is found in large deposits in Jamaica, Australia, Suriname, Guyana, Russia, Arkansas, and elsewhere. The Bayer process is most commonly used today to refine bauxite so it can be accommodated in the Hall-Heroult refining process, used to make most aluminum. Aluminum can now be produced from clay, but the process is not economically feasible at present. Aluminum is the most abundant metal to be found in the earth's crust (8.1%), but is never found free in nature. In addition to the minerals mentioned above, it is found in feldspars, granite, and in many other common minerals. Twenty-two isotopes and isomers are known. Natural aluminum is made of one isotope, ^{27}Al. Pure aluminum, a silvery-white metal, possesses many desirable characteristics. It is light, nontoxic, has a pleasing appearance, can easily be formed, machined, or cast, has a high thermal conductivity, and has excellent corrosion resistance. It is nonmagnetic and nonsparking, stands second among metals in the scale of malleability, and sixth in ductility. It is extensively used for kitchen utensils, outside building decoration, and in thousands of industrial applications where a strong, light, easily constructed material is needed. Although its electrical conductivity is only about 60% that of copper, it is used in electrical transmission lines because of its light weight. Pure aluminum is soft and lacks strength, but it can be alloyed with small amounts of copper, magnesium, silicon, manganese, and other elements to impart a variety of useful properties. These alloys are of vital importance in the construction of modern aircraft and rockets. Aluminum, evaporated in a vacuum, forms a highly reflective coating for both visible light and radiant heat. These coatings soon form a thin layer of the protective oxide and do not deteriorate as do silver coatings. They have found application in coatings for telescope mirrors, in making decorative paper, packages, toys, and in many other uses. The compounds of greatest importance are aluminum oxide, the sulfate, and the soluble sulfate with potassium (alum). The oxide, alumina, occurs naturally as ruby, sapphire, corundum, and emery, and is used in glassmaking and refractories. Synthetic ruby and sapphire have found application in the construction of lasers for producing coherent light. In 1852, the price of aluminum was about $1200/kg, and just before Hall's discovery in 1886, about $25/kg. The price rapidly dropped to 60¢ and has been as low as 33¢/kg. The price in December 2001 was about 64¢/lb or $1.40/kg.

Americium — (the Americas), Am; at. wt. 243; at. no. 95; m.p. 1176°C; b.p. 2011°C; sp. gr. 13.67 (20°C); valence 2, 3, 4, 5, or 6. Americium was the fourth transuranium element to be discovered; the isotope ^{241}Am was identified by Seaborg, James, Morgan, and Ghiorso late in 1944 at the wartime Metallurgical Laboratory of the University of Chicago as the result of successive neutron capture reactions by plutonium isotopes in a nuclear reactor:

$$^{239}Pu(n,\gamma) \rightarrow ^{240}Pu(n,\gamma) \rightarrow ^{241}Pu \xrightarrow{\beta} ^{241}Am$$

Since the isotope ^{241}Am can be prepared in relatively pure form by extraction as a decay product over a period of years from strongly neutron-bombarded plutonium, ^{241}Pu, this isotope is used for much of the chemical investigation of this element. Better suited is the isotope ^{243}Am due to its longer half-life (7.37×10^3 years as compared to 432.2 years for ^{241}Am). A mixture of the isotopes ^{241}Am, ^{242}Am, and ^{243}Am can be prepared by intense neutron irradiation of ^{241}Am according to the reactions ^{241}Am (n, γ) → ^{242}Am (n, γ) → ^{243}Am. Nearly isotopically pure ^{243}Am can be prepared by a sequence of neutron bombardments and chemical separations as follows: neutron bombardment of ^{241}Am yields ^{242}Pu by the reactions ^{241}Am (n, γ) → ^{242}Am → ^{242}Pu, after chemical separation the ^{242}Pu can be transformed to ^{243}Am via the reactions ^{242}Pu (n, γ) → ^{243}Pu → ^{243}Am, and the ^{243}Am can be chemically separated. Fairly pure ^{242}Pu can be prepared more simply by very intense neutron irradiation of ^{239}Pu as the result of successive neutron-capture reactions. Seventeen radioactive isotopes and isomers are now recognized. Americium metal has been prepared by reducing the trifluoride with barium vapor at 1000 to 1200°C or the dioxide by lanthanum metal. The luster of freshly prepared americium metal is white and more silvery than plutonium or neptunium prepared in the same manner. It appears to be more malleable than uranium or neptunium and tarnishes slowly in dry air at room temperature. Americium is thought to exist in two forms: an alpha form which has a double hexagonal close-packed structure and a loose-packed cubic beta form. Americium must be handled with great care to avoid personal contamination. As little as 0.03 µCi of ^{241}Am is the maximum permissible total body burden. The alpha activity from ^{241}Am is about three times that of radium. When gram quantities of ^{241}Am are handled, the intense gamma activity makes exposure a serious problem. Americium dioxide, AmO_2, is the most important oxide. AmF_3, AmF_4, $AmCl_3$, $AmBr_3$, AmI_3, and other compounds have been prepared. The isotope ^{241}Am has been used as a portable source for gamma radiography. It has also been used as a radioactive glass thickness gage for the flat glass industry, and as a source of ionization for smoke detectors. Americium-243 (99%) is available from the Oak Ridge National Laboratory at a cost of about $750/g plus packing charges.

Antimony — (Gr. *anti* plus *monos* - a metal not found alone), Sb; at. wt. 121.760(1); at. no. 51; m.p. 630.63°C; b.p. 1587°C; sp. gr. 6.691 (20°C); valence 0, −3, +3, or +5. Antimony was recognized in compounds by the ancients and was known as a metal at the beginning of the 17th century and possibly much earlier. It is not abundant, but is found in over 100 mineral species. It is sometimes found native, but more frequently as the sulfide, *stibnite* (Sb_2S_3); it is also found as antimonides of the heavy metals, and as oxides. It is extracted from the sulfide by roasting to the oxide, which is reduced by salt and scrap iron; from its oxides it is also prepared by reduction with carbon. Two allotropic forms of antimony exist: the normal stable, metallic form, and the amorphous gray form. The so-called explosive antimony is an ill-defined material always containing an appreciable amount of halogen; therefore, it no longer warrants consideration as a separate allotrope. The yellow form, obtained by oxidation of *stibine*, SbH_3, is probably impure, and is not a distinct form. Natural antimony is made of two stable isotopes, ^{121}Sb and ^{123}Sb. Forty five other radioactive isotopes and isomers are now recognized. Metallic antimony is an extremely brittle metal of a flaky, crystalline texture. It is bluish white and has a metallic luster. It is not acted on by air at room temperature, but burns brilliantly when heated with the formation of white fumes of Sb_2O_3. It is a poor conductor of heat and electricity, and has a hardness of 3 to 3.5. Antimony, available commercially with a purity of 99.999 + %, is finding use in semiconductor technology for making infrared detectors, diodes, and Hall-effect devices. Commercial-grade antimony is widely used in alloys with percentages ranging from 1 to 20. It greatly increases the hardness and mechanical strength of lead. Batteries, antifriction alloys, type metal, small arms and tracer bullets, cable sheathing, and minor products use about half the metal produced. Compounds taking up the other half are oxides, sulfides, sodium antimonate, and antimony trichloride. These are used in manufacturing flame-proofing compounds, paints, ceramic enamels, glass, and pottery. Tartar emetic (hydrated potassium antimonyl tartrate) has been used in medicine. Antimony and many of its compounds are toxic. Antimony costs about $1.30/kg for the commercial metal or about $12/g (99.999%).

Argon — (Gr. *argos,* inactive), Ar; at. wt. 39.948(1); at. no. 18; m.p. −189.35°C; b.p. −185.85°C; t_c −122.28; density 1.7837 g/l. Its presence in air was suspected by Cavendish in 1785, discovered by Lord Rayleigh and Sir William Ramsay in 1894. The gas is prepared by fractionation of liquid air, the atmosphere containing 0.94% argon. The atmosphere of Mars contains 1.6% of ^{40}Ar and 5 p.p.m. of ^{36}Ar. Argon is two and one half times as soluble in water as nitrogen, having about the same solubility as oxygen. It is recognized by the characteristic lines in the red end of the spectrum. It is used in electric light bulbs and in fluorescent tubes at a pressure of about 400 Pa, and in filling photo tubes, glow tubes, etc. Argon is also used as an inert gas shield for arc welding and cutting, as a blanket for the production of titanium and other reactive elements, and as a protective atmosphere for growing silicon and germanium crystals. Argon is colorless and odorless, both as a gas and liquid. It is available in high-purity form. Commercial argon is available at a cost of about 3¢ per cubic foot. Argon is considered to be a very inert gas and is not known to form true chemical compounds, as do krypton, xenon, and radon. However, it does form a hydrate having a dissociation pressure of 105 atm at 0°C. Ion molecules such as $(ArKr)^+$, $(ArXe)^+$, $(NeAr)^+$ have been observed spectroscopically. Argon also forms a clathrate with β-hydroquinone. This clathrate is stable and can be stored for a considerable time, but a true chemical bond does not exist. Van der Waals' forces act to hold the argon. In August 2000, researchers at the University of Helsinki, Finland reported they made a new argon compound HArF by shining UV light on frozen argon that contained a small amount of HF. Naturally occurring argon is a mixture of three isotopes. Seventeen other radioactive isotopes are now known to exist. Commercial argon is priced at about $70/300 cu. ft. or 8.5 cu. meters.

Arsenic — (L. *arsenicum*, Gr. *arsenikon*, yellow orpiment, identified with *arsenikos*, male, from the belief that metals were different sexes; Arabic, *Az-zernikh*, the orpiment from Persian *zerni-zar*, gold), As; at. wt. 74.92160(2); at. no. 33; valence –3, 0, +3 or +5. Elemental arsenic occurs in two solid modifications: yellow, and gray or metallic, with specific gravities of 1.97, and 5.73, respectively. Gray arsenic, the ordinary stable form, has a triple point of 817°C and sublimes at 614°C and has a critical temperature of 1400°C. Several other allotropic forms of arsenic are reported in the literature. It is believed that Albertus Magnus obtained the element in 1250 A.D. In 1649 Schroeder published two methods of preparing the element. It is found native, in the sulfides *realgar* and *orpiment*, as arsenides and sulfarsenides of heavy metals, as the oxide, and as arsenates. *Mispickel*, arsenopyrite, (FeSAs) is the most common mineral, from which on heating the arsenic sublimes leaving ferrous sulfide. The element is a steel gray, very brittle, crystalline, semimetallic solid; it tarnishes in air, and when heated is rapidly oxidized to arsenous oxide (As_2O_3) with the odor of garlic. Arsenic and its compounds are poisonous. Exposure to arsenic and its compounds should not exceed 0.01 mg/m^3 as elemental As during an 8-h work day. Arsenic is also used in bronzing, pyrotechny, and for hardening and improving the sphericity of shot. The most important compounds are white arsenic (As_2O_3), the sulfide, Paris green $3Cu(AsO_2)_2 \cdot Cu(C_2H_3O_2)_2$, calcium arsenate, and lead arsenate; the last three have been used as agricultural insecticides and poisons. Marsh's test makes use of the formation and ready decomposition of arsine (AsH_3). Arsenic is available in high-purity form. It is finding increasing uses as a doping agent in solid-state devices such as transistors. Gallium arsenide is used as a laser material to convert electricity directly into coherent light. Natural arsenic is made of one isotope ^{75}As. Thirty other radioactive isotopes and isomers are known. Arsenic (99%) costs about $75/50g. Purified arsenic (99.9995%) costs about $50/g.

Astatine — (Gr. *astatos*, unstable), At; at. wt. (210); at. no. 85; m.p. 300°C (est.); valence probably 1, 3, 5, or 7. Synthesized in 1940 by D. R. Corson, K. R. MacKenzie, and E. Segre at the University of California by bombarding bismuth with alpha particles. The longest-lived isotope, ^{210}At, has a half-life of only 8.1 hours. Thirty-six other isotopes and isomers are now known. Minute quantities of ^{215}At, ^{218}At, and ^{219}At exist in equilibrium in nature with naturally occurring uranium and thorium isotopes, and traces of ^{217}At are in equilibrium with ^{233}U and ^{239}Np resulting from interaction of thorium and uranium with naturally produced neutrons. The total amount of astatine present in the earth's crust, however, is probably less than 1 oz. Astatine can be produced by bombarding bismuth with energetic alpha particles to obtain the relatively long-lived $^{209–211}$At, which can be distilled from the target by heating it in air. Only about 0.05 μg of astatine has been prepared to date.

The "time of flight" mass spectrometer has been used to confirm that this highly radioactive halogen behaves chemically very much like other halogens, particularly iodine. The interhalogen compounds AtI, AtBr, and AtCl are known to form, but it is not yet known if astatine forms diatomic astatine molecules. HAt and CH$_3$At (methyl astatide) have been detected. Astatine is said to be more metallic that iodine, and, like iodine, it probably accumulates in the thyroid gland.

Barium — (Gr. *barys*, heavy), Ba; at. wt. 137.327(7), at. no. 56; m.p. 727°C; b.p. 1897°C; sp. gr. 3.5 (20°C); valence 2. Baryta was distinguished from lime by Scheele in 1774; the element was discovered by Sir Humphrey Davy in 1808. It is found only in combination with other elements, chiefly in *barite* or *heavy spar* (sulfate) and *witherite* (carbonate) and is prepared by electrolysis of the chloride. Large deposits of barite are found in China, Germany, India, Morocco, and in the U.S. Barium is a metallic element, soft, and when pure is silvery white like lead; it belongs to the alkaline earth group, resembling calcium chemically. The metal oxidizes very easily and should be kept under petroleum or other suitable oxygen-free liquids to exclude air. It is decomposed by water or alcohol. The metal is used as a "getter" in vacuum tubes. The most important compounds are the peroxide (BaO_2), chloride, sulfate, carbonate, nitrate, and chlorate. Lithopone, a pigment containing barium sulfate and zinc sulfide, has good covering power, and does not darken in the presence of sulfides. The sulfate, as permanent white or *blanc fixe*, is also used in paint, in X-ray diagnostic work, and in glassmaking. *Barite* is extensively used as a weighting agent in oilwell drilling fluids, and also in making rubber. The carbonate has been used as a rat poison, while the nitrate and chlorate give green colors in pyrotechny. The impure sulfide phosphoresces after exposure to the light. The compounds and the metal are not expensive. Barium metal (99.2 + % pure) costs about $3/g. All barium compounds that are water or acid soluble are poisonous. Naturally occurring barium is a mixture of seven stable isotopes. Thirty six other radioactive isotopes and isomers are known to exist.

Berkelium — (*Berkeley*, home of the University of California), Bk; at. wt. (247); at. no. 97; m.p. 1050°C; valence 3 or 4; sp. gr. 14 (est.). Berkelium, the eighth member of the actinide transition series, was discovered in December 1949 by Thompson, Ghiorso, and Seaborg, and was the fifth transuranium element synthesized. It was produced by cyclotron bombardment of milligram amounts of ^{241}Am with helium ions at Berkeley, California. The first isotope produced had a mass number of 243 and decayed with a half-life of 4.5 hours. Thirteen isotopes are now known and have been synthesized. The existence of ^{249}Bk, with a half-life of 320 days, makes it feasible to isolate berkelium in weighable amounts so that its properties can be investigated with macroscopic quantities. One of the first visible amounts of a pure berkelium compound, berkelium chloride, was produced in 1962. It weighed 3 billionth of a gram. Berkelium probably has not yet been prepared in elemental form, but it is expected to be a silvery metal, easily soluble in dilute mineral acids, and readily oxidized by air or oxygen at elevated temperatures to form the oxide. X-ray diffraction methods have been used to identify the following compounds: BkO$_2$, BkO$_3$, BkF$_3$, BkCl, and BkOCl. As with other actinide elements, berkelium tends to accumulate in the skeletal system. The maximum permissible body burden of ^{249}Bk in the human skeleton is about 0.0004 μg. Because of its rarity, berkelium presently has no commercial or technological use. Berkelium most likely resembles terbium with respect to chemical properties. Berkelium-249 is available from O.R.N.L. at a cost of $185/µg plus packing charges.

Beryllium — (Gr. *beryllos*, beryl; also called Glucinium or Glucinum, Gr. *glykys*, sweet), Be; at. wt. 9.012182(3); at no. 4; m.p. 1287°C; b.p. 2471°C; sp. gr. 1.848 (20°C); valence 2. Discovered as the oxide by Vauquelin in beryl and in emeralds in 1798. The metal was isolated in 1828 by Wohler and by Bussy independently by the action of potassium on beryllium chloride. Beryllium is found in some 30 mineral species, the most important of which are *bertrandite, beryl, chrysoberyl,* and *phenacite. Aquamarine* and *emerald* are precious forms of *beryl.* Beryllium minerals are found in the U.S., Brazil, Russia, Kazakhstan, and elsewhere. Colombia is known for its emeralds. *Beryl* (3BeO · Al_2O_3 · $6SiO_2$) and *bertrandite* (4BeO · $2SiO_2$ · H_2O) are the most important commercial sources of the element and its compounds. Most of the metal is now prepared by reducing beryllium fluoride with magnesium metal. Beryllium metal did not become readily available to industry until 1957. The metal, steel gray in color, has many desirable properties. It is one of the lightest of all metals, and has one of the highest melting points of the light metals. Its modulus of elasticity is about one third greater than that of steel. It resists attack by concentrated nitric acid, has excellent thermal conductivity, and is nonmagnetic. It has a high permeability to X-rays, and when bombarded by alpha particles, as from radium or polonium, neutrons are produced in the ratio of about 30 neutrons/million alpha particles. At ordinary temperatures beryllium resists oxidation in air, although its ability to scratch glass is probably due to the formation of a thin layer of the oxide. Beryllium is used as an alloying agent in producing beryllium copper which is extensively used for springs, electrical contacts, spot-welding electrodes, and nonsparking tools. It has found application as a structural material for high-speed aircraft, missiles, spacecraft, and communication satellites. It is being used in the windshield frame, brake discs, support beams, and other structural components of the space shuttle. Because beryllium is relatively transparent to X-rays, ultra-thin Be-foil is finding use in X-ray lithography for reproduction of microminiature integrated circuits. Natural beryllium is made of ^9Be and is stable. Eight other radioactive isotopes are known.

Beryllium is used in nuclear reactors as a reflector or moderator for it has a low thermal neutron absorption cross section. It is used in gyroscopes, computer parts, and instruments where lightness, stiffness, and dimensional stability are required. The oxide has a very high melting point and is also used in nuclear work and ceramic applications. Beryllium and its salts are toxic and should be handled with the greatest of care. Beryllium and its compounds should not be tasted to verify the sweetish nature of beryllium (as did early experimenters). The metal, its alloys, and its salts can be handled safely if certain work codes are observed, but no attempt should be made to work with beryllium before becoming familiar with proper safeguards. Beryllium metal is available at a cost of about $5/g (99.5% pure).

Bismuth — (Ger. *Weisse Masse*, white mass; later *Wisuth* and *Bisemutum*), Bi; at. wt. 208.98038(2); at. no. 83; m.p. 271.4°C; b.p. 1564°C; sp. gr. 9.747 (20°C); valence 3 or 5. In early times bismuth was confused with tin and lead. Claude Geoffroy the Younger showed it to be distinct from lead in 1753. It is a white crystalline, brittle metal with a pinkish tinge. It occurs native.

The most important ores are *bismuthinite* or bismuth glance (Bi_2S_3) and *bismite* (Bi_2O_3). Peru, Japan, Mexico, Bolivia, and Canada are major bismuth producers. Much of the bismuth produced in the U.S. is obtained as a by-product in refining lead, copper, tin, silver, and gold ores. Bismuth is the most diamagnetic of all metals, and the thermal conductivity is lower than any metal, except mercury. It has a high electrical resistance, and has the highest Hall effect of any metal (i.e., greatest increase in electrical resistance when placed in a magnetic field). "Bismanol" is a permanent magnet of high coercive force, made of MnBi, by the U.S. Naval Surface Weapons Center. Bismuth expands 3.32% on solidification. This property makes bismuth alloys particularly suited to the making of sharp castings of objects subject to damage by high temperatures. With other metals such as tin, cadmium, etc., bismuth forms low-melting alloys which are extensively used for safety devices in fire detection and extinguishing systems. Bismuth is used in producing malleable irons and is finding use as a catalyst for making acrylic fibers. When bismuth is heated in air it burns with a blue flame, forming yellow fumes of the oxide. The metal is also used as a thermocouple material, and has found application as a carrier for U^{235} or U^{233} fuel in atomic reactors. Its soluble salts are characterized by forming insoluble basic salts on the addition of water, a property sometimes used in detection work. Bismuth oxychloride is used extensively in cosmetics. Bismuth subnitrate and subcarbonate are used in medicine. Natural bismuth contains only one isotope ^{209}Bi. Forty-four isotopes and isomers of bismuth are known. Bismuth metal (99.5%) costs about $250/kg.

Bohrium — (Named after Niels Bohr [1885–1962], Danish atomic and nuclear physicist.) Bh; at.wt. [262]. at.no. 107. Bohrium is expected to have chemical properties similar to rhenium. This element was synthesized and unambiguously identified in 1981 using the Universal Linear Accelerator (UNILAC) at the Gesellschaft für Schwerionenforschung (G.S.I.) in Darmstadt, Germany. The discovery team was led by Armbruster and Münzenberg. The reaction producing the element was proposed and applied earlier by a Dubna Group led by Oganessian in 1976. A target of ^{209}Bi was bombarded by a beam of ^{54}Cr ions. In 1983 experiments at Dubna using the 157-inch cyclotron, produced $^{262}107$ by the reaction $^{209}Bi + ^{54}Cr$. The alpha decay of ^{246}Cf, the sixth member in the decay chain of $^{262}107$, served to establish a 1-neutron reaction channel. The IUPAC adopted the name *Bohrium* with the symbol Bh for Element 107 in August 1997. Five isotopes of bohrium are now recognized. One isotope of bohrium appears to have a relatively long life of 15 seconds. Work on this relatively long-lived isotope has been performed with the 88-inch cyclotron at the Lawrence-Berkeley National Laboratory.

Boron — (Ar. *Buraq*, Pers. *Burah*), B; at. wt. 10.811(7); at. no. 5; m.p. 2075°C; b.p. 4000°C; sp. gr. of crystals 2.34, of amorphous variety 2.37; valence 3. Boron compounds have been known for thousands of years, but the element was not discovered until 1808 by Sir Humphry Davy and by Gay-Lussac and Thenard. The element is not found free in nature, but occurs as orthoboric acid usually in certain volcanic spring waters and as borates in *borax* and *colemanite. Ulexite*, another boron mineral, is interesting as it is nature's own version of "fiber optics." Important sources of boron are the ores *rasorite (kernite)* and *tincal (borax ore)*. Both of these ores are found in the Mojave Desert. *Tincal* is the most important source of boron from the Mojave. Extensive *borax* deposits are also found in

Turkey. Boron exists naturally as 19.9% ^{10}B isotope and 80.1% ^{11}B isotope. Ten other isotopes of boron are known. High-purity crystalline boron may be prepared by the vapor phase reduction of boron trichloride or tribromide with hydrogen on electrically heated filaments. The impure, or amorphous, boron, a brownish-black powder, can be obtained by heating the trioxide with magnesium powder. Boron of 99.9999% purity has been produced and is available commercially. Elemental boron has an energy band gap of 1.50 to 1.56 eV, which is higher than that of either silicon or germanium. It has interesting optical characteristics, transmitting portions of the infrared, and is a poor conductor of electricity at room temperature, but a good conductor at high temperature. Amorphous boron is used in pyrotechnic flares to provide a distinctive green color, and in rockets as an igniter. By far the most commercially important boron compound in terms of dollar sales is $Na_2B_4O_7 \cdot 5H_2O$. This pentahydrate is used in very large quantities in the manufacture of insulation fiberglass and sodium perborate bleach. Boric acid is also an important boron compound with major markets in textile fiberglass and in cellulose insulation as a flame retardant. Next in order of importance is borax ($Na_2B_4O_7 \cdot 10H_2O$) which is used principally in laundry products. Use of borax as a mild antiseptic is minor in terms of dollars and tons. Boron compounds are also extensively used in the manufacture of borosilicate glasses. The isotope boron-10 is used as a control for nuclear reactors, as a shield for nuclear radiation, and in instruments used for detecting neutrons. Boron nitride has remarkable properties and can be used to make a material as hard as diamond. The nitride also behaves like an electrical insulator but conducts heat like a metal. It also has lubricating properties similar to graphite. The hydrides are easily oxidized with considerable energy liberation, and have been studied for use as rocket fuels. Demand is increasing for boron filaments, a high-strength, lightweight material chiefly employed for advanced aerospace structures. Boron is similar to carbon in that it has a capacity to form stable covalently bonded molecular networks. Carboranes, metalloboranes, phosphacarboranes, and other families comprise thousands of compounds. Crystalline boron (99.5%) costs about $6/g. Amorphous boron (94–96%) costs about $1.50/g. Elemental boron and the borates are not considered to be toxic, and they do not require special care in handling. However, some of the more exotic boron hydrogen compounds are definitely toxic and do require care.

Bromine — (Gr. *bromos*, stench), Br; at. wt. 79.904(1); at. no. 35; m.p. −7.2°C; b.p. 58.8°C; t_c 315°C; density of gas 7.59 g/l, liquid 3.12 (20°C); valence 1, 3, 5, or 7. Discovered by Balard in 1826, but not prepared in quantity until 1860. A member of the halogen group of elements, it is obtained from natural brines from wells in Michigan and Arkansas. Little bromine is extracted today from seawater, which contains only about 85 ppm. Bromine is the only liquid nonmetallic element. It is a heavy, mobile, reddish-brown liquid, volatilizing readily at room temperature to a red vapor with a strong disagreeable odor, resembling chlorine, and having a very irritating effect on the eyes and throat; it is readily soluble in water or carbon disulfide, forming a red solution, is less active than chlorine but more so than iodine; it unites readily with many elements and has a bleaching action; when spilled on the skin it produces painful sores. It presents a serious health hazard, and maximum safety precautions should be taken when handling it. Much of the bromine output in the U.S. was used in the production of ethylene dibromide, a lead scavenger used in making gasoline antiknock compounds. Lead in gasoline, however, has been drastically reduced, due to environmental considerations. This will greatly affect future production of bromine. Bromine is also used in making fumigants, flameproofing agents, water purification compounds, dyes, medicinals, sanitizers, inorganic bromides for photography, etc. Organic bromides are also important. Natural bromine is made of two isotopes, ^{79}Br and ^{81}Br. Thirty-four isotopes and isomers are known. Bromine (99.8%) costs about $70/kg.

Cadmium — (L. *cadmia*; Gr. *kadmeia* - ancient name for calamine, zinc carbonate), Cd; at. wt. 112.411(8); at. no. 48; m.p. 321.07°C; b.p. 767°C; sp. gr. 8.65 (20°C); valence 2. Discovered by Stromeyer in 1817 from an impurity in zinc carbonate. Cadmium most often occurs in small quantities associated with zinc ores, such as *sphalerite* (ZnS). *Greenockite* (CdS) is the only mineral of any consequence bearing cadmium. Almost all cadmium is obtained as a by-product in the treatment of zinc, copper, and lead ores. It is a soft, bluish-white metal which is easily cut with a knife. It is similar in many respects to zinc. It is a component of some of the lowest melting alloys; it is used in bearing alloys with low coefficients of friction and great resistance to fatigue; it is used extensively in electroplating, which accounts for about 60% of its use. It is also used in many types of solder, for standard E.M.F. cells, for Ni-Cd batteries, and as a barrier to control atomic fission. The market for Ni-Cd batteries is expected to grow significantly. Cadmium compounds are used in black and white television phosphors and in blue and green phosphors for color TV tubes. It forms a number of salts, of which the sulfate is most common; the sulfide is used as a yellow pigment. Cadmium and solutions of its compounds are toxic. Failure to appreciate the toxic properties of cadmium may cause workers to be unwittingly exposed to dangerous fumes. Some silver solders, for example, contain cadmium and should be handled with care. Serious toxicity problems have been found from long-term exposure and work with cadmium plating baths. Cadmium is present in certain phosphate rocks. This has raised concerns that the long-term use of certain phosphate fertilizers might pose a health hazard from levels of cadmium that might enter the food chain. In 1927 the International Conference on Weights and Measures redefined the meter in terms of the wavelength of the red cadmium spectral line (i.e., 1 m = 1,553,164.13 wavelengths). This definition has been changed (see under Krypton). The current price of cadmium is about 50¢/g (99.5%). It is available in high purity form for about $550/kg. Natural cadmium is made of eight isotopes. Thirty four other isotopes and isomers are now known and recognized.

Calcium — (L. *calx*, lime), Ca; at. wt. 40.078(4); at. no. 20; m.p. 842°C; b.p. 1484°C; sp. gr. 1.55 (20°C); valence 2. Though lime was prepared by the Romans in the first century under the name calx, the metal was not discovered until 1808. After learning that Berzelius and Pontin prepared calcium amalgam by electrolyzing lime in mercury, Davy was able to isolate the impure metal. Calcium is a metallic element, fifth in abundance in the earth's crust, of which it forms more than 3%. It is an essential constituent of leaves, bones, teeth, and shells. Never found in nature uncombined, it occurs abundantly as *limestone* ($CaCO_3$), *gypsum* ($CaSO_4 \cdot 2H_2O$), and *fluorite* (CaF_2); *apatite* is the fluorophosphate or chlorophosphate of calcium. The metal has a silvery color, is rather hard, and is prepared by electrolysis of the fused chloride to which calcium fluoride is added to lower the melting point. Chemically it is one of the alkaline earth elements; it readily forms a white coating of oxide in air, reacts with water, burns with a yellow-red flame, forming largely the oxide. The metal is used as a reducing agent in preparing other metals such as thorium, uranium, zirconium, etc., and is used as a deoxidizer, desulfurizer, and inclusion modifier for various ferrous and nonferrous alloys. It is also used as an alloying agent for aluminum, beryllium, copper, lead, and magnesium alloys, and serves as a "getter" for residual gases in vacuum tubes. Its natural and prepared compounds are widely used. Quicklime (CaO), made by heating limestone and changed into slaked lime by the careful addition of water, is the great cheap base of the chemical industry with countless uses. Mixed with sand it hardens as mortar and plaster by taking up carbon dioxide from the air. Calcium from limestone is an important element in Portland cement. The solubility of the carbonate in water containing carbon dioxide causes the formation of caves with stalactites and stalagmites and is responsible for hardness in water. Other important compounds are the carbide (CaC_2), chloride ($CaCl_2$), cyanamide ($CaCN_2$), hypochlorite ($Ca(OCl)_2$), nitrate ($Ca(NO_3)_2$), and sulfide (CaS). Calcium sulfide is phosphorescent after being exposed to light. Natural calcium contains six isotopes. Sixteen other radioactive isotopes are known. Metallic calcium (99.5%) costs about $200/kg.

Californium — (State and University of California), Cf; at. wt. (251); m.p. 900°C; at. no. 98. Californium, the sixth transuranium element to be discovered, was produced by Thompson, Street, Ghiorso, and Seaborg in 1950 by bombarding microgram quantities of ^{242}Cm with 35 MeV helium ions in the Berkeley 60-inch cyclotron. Californium (III) is the only ion stable in aqueous solutions, all attempts to reduce or oxidize californium (III) having failed. The isotope ^{249}Cf results from the beta decay of ^{249}Bk while the heavier isotopes are produced by intense neutron irradiation by the reactions:

$$^{249}Bk(n,\gamma) \rightarrow ^{250}Bk \xrightarrow{\beta} ^{250}Cf \text{ and } ^{249}Cf(n,\gamma) \rightarrow ^{250}Cf$$

followed by

$$^{250}Cf(n,\gamma) \rightarrow ^{251}Cf(n,\gamma) \rightarrow ^{252}Cf$$

The existence of the isotopes ^{249}Cf, ^{250}Cf, ^{251}Cf, and ^{252}Cf makes it feasible to isolate californium in weighable amounts so that its properties can be investigated with macroscopic quantities. Californium-252 is a very strong neutron emitter. One microgram releases 170 million neutrons per minute, which presents biological hazards. Proper safeguards should be used in handling californium. Twenty isotopes of californium are now recognized. ^{249}Cf and ^{252}Cf have half-lives of 351 years and 900 years, respectively. In 1960 a few tenths of a microgram of californium trichloride, $CfCl_3$, californium oxychloride, CfOCl, and californium oxide, Cf_2O_3, were first prepared. Reduction of californium to its metallic state has not yet been accomplished. Because californium is a very efficient source of neutrons, many new uses are expected for it. It has already found use in neutron moisture gages and in well-logging (the determination of water and oil-bearing layers). It is also being used as a portable neutron source for discovery of metals such as gold or silver by on-the-spot activation analysis. ^{252}Cf is now being offered for sale by the Oak Ridge National Laboratory (O.R.N.L.) at a cost of $60/µg and ^{249}Cf at a cost of $185/µg plus packing charges. It has been suggested that californium

may be produced in certain stellar explosions, called *supernovae*, for the radioactive decay of ^{254}Cf (55-day half-life) agrees with the characteristics of the light curves of such explosions observed through telescopes. This suggestion, however, is questioned. Californium is expected to have chemical properties similar to dysprosium.

Carbon — (L. *carbo*, charcoal), C; at. wt. 12.0107(8); at. no. 6; sublimes at 3642°C; triple point (graphite-liquid-gas), 4492°C at a pressure of 101.325 kPa; sp. gr. amorphous 1.8 to 2.1, graphite 1.9 to 2.3, diamond 3.15 to 3.53 (depending on variety); gem diamond 3.513 (25°C); valence 2, 3, or 4. Carbon, an element of prehistoric discovery, is very widely distributed in nature. It is found in abundance in the sun, stars, comets, and atmospheres of most planets. Carbon in the form of microscopic diamonds is found in some meteorites. Natural diamonds are found in *kimberlite* or *lamporite* of ancient formations called "pipes," such as found in South Africa, Arkansas, and elsewhere. Diamonds are now also being recovered from the ocean floor off the Cape of Good Hope. About 30% of all industrial diamonds used in the U.S. are now made synthetically. The energy of the sun and stars can be attributed at least in part to the well-known carbon-nitrogen cycle. Carbon is found free in nature in three allotropic forms: amorphous, graphite, and diamond. A fourth form, known as "white" carbon, is now thought to exist. Graphite is one of the softest known materials while diamond is one of the hardest. Graphite exists in two forms: alpha and beta. These have identical physical properties, except for their crystal structure. Naturally occurring graphites are reported to contain as much as 30% of the rhombohedral (beta) form, whereas synthetic materials contain only the alpha form. The hexagonal alpha type can be converted to the beta by mechanical treatment, and the beta form reverts to the alpha on heating it above 1000°C. In 1969 a new allotropic form of carbon was produced during the sublimation of pyrolytic graphite at low pressures. Under free-vaporization conditions above ~2550 K, "white" carbon forms as small transparent crystals on the edges of the basal planes of graphite. The interplanar spacings of "white" carbon are identical to those of carbon form noted in the graphitic gneiss from the Ries (meteoritic) Crater of Germany. "White" carbon is a transparent birefringent material. Little information is presently available about this allotrope. Of recent interest is the discovery of all-carbon molecules, known as "buckyballs" or fullerenes, which have a number of unusual properties. These interesting molecules, consisting of 60 or 70 carbon atoms linked together, seem capable of withstanding great pressure and trapping foreign atoms inside their network of carbon. They are said to be capable of magnetism and superconductivity and have potential as a nonlinear optical material. Buckyball films are reported to remain superconductive at temperatures as high as 45 K. In combination, carbon is found as carbon dioxide in the atmosphere of the earth and dissolved in all natural waters. It is a component of great rock masses in the form of carbonates of calcium (limestone), magnesium, and iron. Coal, petroleum, and natural gas are chiefly hydrocarbons. Carbon is unique among the elements in the vast number and variety of compounds it can form. With hydrogen, oxygen, nitrogen, and other elements, it forms a very large number of compounds, carbon atom often being linked to carbon atom. There are close to ten million known carbon compounds, many thousands of which are vital to organic and life processes. Without carbon, the basis for life would be impossible. While it has been thought that silicon might take the place of carbon in forming a host of similar compounds, it is now not possible to form stable compounds with very long chains of silicon atoms. The atmosphere of Mars contains 96.2% CO_2. Some of the most important compounds of carbon are carbon dioxide (CO_2), carbon monoxide (CO), carbon disulfide (CS_2), chloroform ($CHCl_3$), carbon tetrachloride (CCl_4), methane (CH_4), ethylene (C_2H_4), acetylene (C_2H_2), benzene (C_6H_6), ethyl alcohol (C_2H_5OH), acetic acid (CH_3COOH), and their derivatives. Carbon has fifteen isotopes. Natural carbon consists of 98.89% ^{12}C and 1.11% ^{13}C. In 1961 the International Union of Pure and Applied Chemistry adopted the isotope carbon-12 as the basis for atomic weights. Carbon-14, an isotope with a half-life of 5715 years, has been widely used to date such materials as wood, archeological specimens, etc. A new brittle form of carbon, known as "glassy carbon," has been developed. It can be obtained with high purity. It has a high resistance to corrosion, has good thermal stability, and is structurally impermeable to both gases and liquids. It has a randomized structure, making it useful in ultra-high technology applications, such as crystal growing, crucibles for high-temperature use, etc. Glassy carbon is available at a cost of about $35/10g. Fullerene powder is available at a cost of about $55/10mg (99%$C_{10}$). Diamond powder (99.9%) costs about $40/g.

Cerium — (named for the asteroid *Ceres*, which was discovered in 1801 only 2 years before the element), Ce; at. wt. 140.115(4); at. no. 58; m.p. 798°C; b.p. 3424°C; sp. gr. 6.770 (25°C); valence 3 or 4. Discovered in 1803 by Klaproth and by Berzelius and Hisinger; metal prepared by Hillebrand and Norton in 1875. Cerium is the most abundant of the metals of the so-called rare earths. It is found in a number of minerals including *allanite* (also known as *orthite*), *monazite, bastnasite, cerite,* and *samarskite*. Monazite and bastnasite are presently the two most important sources of cerium. Large deposits of monazite found on the beaches of Travancore, India, in river sands in Brazil, and deposits of *allanite* in the western United States, and *bastnasite* in Southern California will supply cerium, thorium, and the other rare-earth metals for many years to come. Metallic cerium is prepared by metallothermic reduction techniques, such as by reducing cerous fluoride with calcium, or by electrolysis of molten cerous chloride or other cerous halides. The metallothermic technique is used to produce high-purity cerium. Cerium is especially interesting because of its variable electronic structure. The energy of the inner 4f level is nearly the same as that of the outer or valence electrons, and only small amounts of energy are required to change the relative occupancy of these electronic levels. This gives rise to dual valency states. For example, a volume change of about 10% occurs when cerium is subjected to high pressures or low temperatures. It appears that the valence changes from about 3 to 4 when it is cooled or compressed. The low temperature behavior of cerium is complex. Four allotropic modifications are thought to exist: cerium at room temperature and at atmospheric pressure is known as γ cerium. Upon cooling to −16°C, γ cerium changes to β cerium. The remaining γ cerium starts to change to α cerium when cooled to −172°C, and the transformation is complete at −269°C. α Cerium has a density of 8.16; δ cerium exists above 726°C. At atmospheric pressure, liquid cerium is more dense than its solid form at the melting point. Cerium is an iron-gray lustrous metal. It is malleable, and oxidizes very readily at room temperature, especially in moist air. Except for europium, cerium is the most reactive of the "rare-earth" metals. It slowly decomposes in cold wa-

ter, and rapidly in hot water. Alkali solutions and dilute and concentrated acids attack the metal rapidly. The pure metal is likely to ignite if scratched with a knife. Ceric salts are orange red or yellowish; cerous salts are usually white. Cerium is a component of misch metal, which is extensively used in the manufacture of pyrophoric alloys for cigarette lighters, etc. Natural cerium is stable and contains four isotopes. Thirty-two other radioactive isotopes and isomers are known. While cerium is not radioactive, the impure commercial grade may contain traces of thorium, which is radioactive. The oxide is an important constituent of incandescent gas mantles and it is emerging as a hydrocarbon catalyst in "self-cleaning" ovens. In this application it can be incorporated into oven walls to prevent the collection of cooking residues. As ceric sulfate it finds extensive use as a volumetric oxidizing agent in quantitative analysis. Cerium compounds are used in the manufacture of glass, both as a component and as a decolorizer. The oxide is finding increased use as a glass polishing agent instead of rouge, for it is much faster than rouge in polishing glass surfaces. Cerium compounds are finding use in automobile exhaust catalysts. Cerium is also finding use in making permanent magnets. Cerium, with other rare earths, is used in carbon-arc lighting, especially in the motion picture industry. It is also finding use as an important catalyst in petroleum refining and in metallurgical and nuclear applications. In small lots, cerium costs about $5/g (99.9%).

Cesium — (L. *caesius*, sky blue), Cs; at. wt. 132.90545(2); at. no. 55; m.p. 28.44°C; b.p. 671°C; sp. gr. 1.873 (20°C); valence 1. Cesium was discovered spectroscopically by Bunsen and Kirchhoff in 1860 in mineral water from Durkheim. Cesium, an alkali metal, occurs in *lepidolite, pollucite* (a hydrated silicate of aluminum and cesium), and in other sources. One of the world's richest sources of cesium is located at Bernic Lake, Manitoba. The deposits are estimated to contain 300,000 tons of pollucite, averaging 20% cesium. It can be isolated by electrolysis of the fused cyanide and by a number of other methods. Very pure, gas-free cesium can be prepared by thermal decomposition of cesium azide. The metal is characterized by a spectrum containing two bright lines in the blue along with several others in the red, yellow, and green. It is silvery white, soft, and ductile. It is the most electropositive and most alkaline element. Cesium, gallium, and mercury are the only three metals that are liquid at room temperature. Cesium reacts explosively with cold water, and reacts with ice at temperatures above −116°C. Cesium hydroxide, the strongest base known, attacks glass. Because of its great affinity for oxygen the metal is used as a "getter" in electron tubes. It is also used in photoelectric cells, as well as a catalyst in the hydrogenation of certain organic compounds. The metal has recently found application in ion propulsion systems. Cesium is used in atomic clocks, which are accurate to 5 s in 300 years. A second of time is now defined as being the duration of 9,192,631,770 periods of the radiation corresponding to the transition between the two hyper-fine levels of the ground state of the cesium-133 atom. Its chief compounds are the chloride and the nitrate. Cesium has 52 isotopes and isomers with masses ranging from 112 to 148. The present price of cesium is about $50/g (99.98%) sealed in a glass ampoule.

Chlorine — (Gr. *chloros*, greenish yellow), Cl; at. wt. 35.4527(9); at. no. 17; m.p. −101.5°C; b.p. −34.04°C; t_c 143.8°C; density 3.214 g/l; sp. gr. 1.56 (−33.6°C); valence 1, 3, 5, or 7. Discovered in 1774 by Scheele, who thought it contained oxygen; named in 1810 by Davy, who insisted it was an element. In nature it is found in the combined state only, chiefly with sodium as common salt (NaCl), *carnallite* (KMgCl$_3$ · 6H$_2$O), and *sylvite* (KCl). It is a member of the halogen (salt-forming) group of elements and is obtained from chlorides by the action of oxidizing agents and more often by electrolysis; it is a greenish-yellow gas, combining directly with nearly all elements. At 10°C one volume of water dissolves 3.10 volumes of chlorine, at 30°C only 1.77 volumes. Chlorine is widely used in making many everyday products. It is used for producing safe drinking water the world over. Even the smallest water supplies are now usually chlorinated. It is also extensively used in the production of paper products, dyestuffs, textiles, petroleum products, medicines, antiseptics, insecticides, foodstuffs, solvents, paints, plastics, and many other consumer products. Most of the chlorine produced is used in the manufacture of chlorinated compounds for sanitation, pulp bleaching, disinfectants, and textile processing. Further use is in the manufacture of chlorates, chloroform, carbon tetrachloride, and in the extraction of bromine. Organic chemistry demands much from chlorine, both as an oxidizing agent and in substitution, since it often brings desired properties in an organic compound when substituted for hydrogen, as in one form of synthetic rubber. Chlorine is a respiratory irritant. The gas irritates the mucous membranes and the liquid burns the skin. As little as 3.5 ppm can be detected as an odor, and 1000 ppm is likely to be fatal after a few deep breaths. It was used as a war gas in 1915. Natural chlorine contains two isotopes. Twenty other isotopes and isomers are known.

Chromium — (Gr. *chroma*, color), Cr; at. wt. 51.9961(6); at. no. 24; m.p. 1907°C; b.p. 2671°C; sp. gr. 7.18 to 7.20 (20°C); valence chiefly 2, 3, or 6. Discovered in 1797 by Vauquelin, who prepared the metal the next year, chromium is a steel-gray, lustrous, hard metal that takes a high polish. The principal ore is *chromite* (FeCr$_2$O$_4$), which is found in Zimbabwe, Russia, South Africa, Turkey, Iran, Albania, Finland, Democratic Republic of Madagascar, the Philippines, and elsewhere. The U.S. has no appreciable chromite ore reserves. The metal is usually produced by reducing the oxide with aluminum. Chromium is used to harden steel, to manufacture stainless steel, and to form many useful alloys. Much is used in plating to produce a hard, beautiful surface and to prevent corrosion. Chromium is used to give glass an emerald green color. It finds wide use as a catalyst. All compounds of chromium are colored; the most important are the chromates of sodium and potassium (K$_2$CrO$_4$) and the dichromates (K$_2$Cr$_2$O$_7$) and the potassium and ammonium chrome alums, as KCr(SO$_4$)$_2$ · 12H$_2$O. The dichromates are used as oxidizing agents in quantitative analysis, also in tanning leather. Other compounds are of industrial value; lead chromate is chrome yellow, a valued pigment. Chromium compounds are used in the textile industry as mordants, and by the aircraft and other industries for anodizing aluminum. The refractory industry has found chromite useful for forming bricks and shapes, as it has a high melting point, moderate thermal expansion, and stability of crystalline structure. Chromium is an essential trace element for human health. Many chromium compounds, however, are acutely or chronically toxic, and some are carcinogenic. They should be handled with proper safeguards. Natural chromium contains four isotopes. Twenty other isotopes are known. Chromium metal (99.95%) costs about $1000/kg. Commercial grade chromium (99%) costs about $75/kg.

Cobalt — (*Kobald*, from the German, goblin or evil spirit, *cobalos*, Greek, mine), Co; at. wt. 58.93320(1); at. no. 27; m.p. 1495°C; b.p. 2927°C; sp. gr. 8.9 (20°C); valence 2 or 3. Discovered by Brandt about 1735. Cobalt occurs in the mineral *cobaltite*, *smaltite*, and *erythrite*, and is often associated with nickel, silver, lead, copper, and iron ores, from which it is most frequently obtained as a by-product. It is also present in meteorites. Important ore deposits are found in Congo-Kinshasa, Australia, Zambia, Russia, Canada, and elsewhere. The U.S. Geological Survey has announced that the bottom of the north central Pacific Ocean may have cobalt-rich deposits at relatively shallow depths in waters close to the Hawaiian Islands and other U.S. Pacific territories. Cobalt is a brittle, hard metal, closely resembling iron and nickel in appearance. It has a magnetic permeability of about two thirds that of iron. Cobalt tends to exist as a mixture of two allotropes over a wide temperature range; the β-form predominates below 400°C, and the α above that temperature. The transformation is sluggish and accounts in part for the wide variation in reported data on physical properties of cobalt. It is alloyed with iron, nickel and other metals to make Alnico, an alloy of unusual magnetic strength with many important uses. Stellite alloys, containing cobalt, chromium, and tungsten, are used for high-speed, heavy-duty, high temperature cutting tools, and for dies. Cobalt is also used in other magnet steels and stainless steels, and in alloys used in jet turbines and gas turbine generators. The metal is used in electroplating because of its appearance, hardness, and resistance to oxidation. The salts have been used for centuries for the production of brilliant and permanent blue colors in porcelain, glass, pottery, tiles, and enamels. It is the principal ingredient in Sevre's and Thenard's blue. A solution of the chloride (CoCl$_2$ · 6H$_2$O) is used as sympathetic ink. The cobalt ammines are of interest; the oxide and the nitrate are important. Cobalt carefully used in the form of the chloride, sulfate, acetate, or nitrate has been found effective in correcting a certain mineral deficiency disease in animals. Soils should contain 0.13 to 0.30 ppm of cobalt for proper animal nutrition. Cobalt is found in Vitamin B-12, which is essential for human nutrition. Cobalt of 99.9+% purity is priced at about $250/kg. Cobalt-60, an artificial isotope, is an important gamma ray source, and is extensively used as a tracer and a radiotherapeutic agent. Single compact sources of Cobalt-60 vary from about $1 to $10/curie, depending on quantity and specific activity. Thirty isotopes and isomers of cobalt are known.

Columbium — See Niobium.

Copper — (L. *cuprum*, from the island of Cyprus), Cu; at. wt. 63.546(3); at. no. 29; f.p. 1084.62 °C; b.p. 2562°C; sp. gr. 8.96 (20°C); valence 1 or 2. The discovery of copper dates from prehistoric times. It is said to have been mined for more than 5000 years. It is one of man's most important metals. Copper is reddish colored, takes on a bright metallic luster, and is malleable, ductile, and a good conductor of heat and electricity (second only to silver in electrical conductivity). The electrical industry is one of the greatest users of copper. Copper occasionally occurs native, and is found in many minerals such as *cuprite, malachite, azurite, chalcopyrite*, and *bornite*. Large copper ore deposits are found in the U.S., Chile, Zambia, Zaire, Peru, and Canada. The most important copper ores are the sulfides, oxides, and carbonates. From these, copper is obtained by smelting, leaching, and by electrolysis. Its alloys, brass and bronze, long used, are still very important; all American coins are now copper alloys; monel and gun metals also contain copper. The most important compounds are the oxide and the sulfate, blue vitriol; the latter has wide use as an agricultural poison and as an algicide in water purification. Copper compounds such as Fehling's solution are widely used in analytical chemistry in tests for sugar. High-purity copper (99.999 + %) is readily available commercially. The price of commercial copper has fluctuated widely. The price of copper in December 2001 was about \$1.50/kg. Natural copper contains two isotopes. Twenty-six other radioactive isotopes and isomers are known.

Curium — (Pierre and Marie Curie), Cm; at. wt. (247); at. no. 96; m.p. 1345°C; sp. gr. 13.51 (calc.); valence 3 and 4. Although curium follows americium in the periodic system, it was actually known before americium and was the third transuranium element to be discovered. It was identified by Seaborg, James, and Ghiorso in 1944 at the wartime Metallurgical Laboratory in Chicago as a result of helium-ion bombardment of ^{239}Pu in the Berkeley, California, 60-inch cyclotron. Visible amounts (30 μg) of ^{242}Cm, in the form of the hydroxide, were first isolated by Werner and Perlman of the University of California in 1947. In 1950, Crane, Wallmann, and Cunningham found that the magnetic susceptibility of microgram samples of CmF_3 was of the same magnitude as that of GdF_3. This provided direct experimental evidence for assigning an electronic configuration to Cm^{+3}. In 1951, the same workers prepared curium in its elemental form for the first time. Sixteen isotopes of curium are now known. The most stable, ^{247}Cm, with a half-life of 16 million years, is so short compared to the earth's age that any primordial curium must have disappeared long ago from the natural scene. Minute amounts of curium probably exist in natural deposits of uranium, as a result of a sequence of neutron captures and β decays sustained by the very low flux of neutrons naturally present in uranium ores. The presence of natural curium, however, has never been detected. ^{242}Cm and ^{244}Cm are available in multigram quantities. ^{248}Cm has been produced only in milligram amounts. Curium is similar in some regards to gadolinium, its rare-earth homolog, but it has a more complex crystal structure. Curium is silver in color, is chemically reactive, and is more electropositive than aluminum. CmO_2, Cm_2O_3, CmF_3, CmF_4, $CmCl_3$, $CmBr_3$, and CmI_3 have been prepared. Most compounds of trivalent curium are faintly yellow in color. ^{242}Cm generates about three watts of thermal energy per gram. This compares to one-half watt per gram of ^{238}Pu. This suggests use for curium as a power source. ^{244}Cm is now offered for sale by the O.R.N.L. at \$185/mg plus packing charges. ^{248}Cm is available at a cost of \$160/μg, plus packing charges, from the O.R.N.L. Curium absorbed into the body accumulates in the bones, and is therefore very toxic as its radiation destroys the red-cell forming mechanism. The maximum permissible total body burden of ^{244}Cm (soluble) in a human being is 0.3 μCi (microcurie).

Darmstadtium — (Darmstadt, city in Germany), Ds. In 1987 Oganessian, et al., at Dubna, claimed discovery of this element. Their experiments indicated the spontaneous fissioning nuclide ^{272}110 with a half-life of 10 ms. More recently a group led by Armbruster at G.S.I. in Darmstadt, Germany, reported evidence of ^{269}110, which was produced by bombarding lead for many days with more than 10^{18} nickel atoms. A detector searched each collision for Element 110's distinct decay sequence. On November 9, 1994, evidence of 110 was detected. In 2003 IUPAC approved the name darmstadtium, symbol Ds, for Element 110. Seven isotopes of Element 110 are now recognized.

Deuterium — an isotope of hydrogen — see Hydrogen.

Dubnium — (named after the Joint Institute of Nuclear Research in Dubna, Russia). Db; at.wt. [262]; at.no. 105. In 1967 G. N. Flerov reported that a Soviet team working at the Joint Institute for Nuclear Research at Dubna may have produced a few atoms of ^{260}105 and ^{261}105 by bombarding ^{243}Am with ^{22}Ne. Their evidence was based on time-coincidence measurements of alpha energies. More recently, it was reported that early in 1970 Dubna scientists synthesized Element 105 and that by the end of April 1970 "had investigated all the types of decay of the new element and had determined its chemical properties." In late April 1970, it was announced that Ghiorso, Nurmia, Harris, K. A. Y. Eskola, and P. L. Eskola, working at the University of California at Berkeley, had positively identified Element 105. The discovery was made by bombarding a target of ^{249}Cf with a beam of 84 MeV nitrogen nuclei in the Heavy Ion Linear Accelerator (HILAC). When a ^{15}N nucleus is absorbed by a ^{249}Cf nucleus, four neutrons are emitted and a new atom of ^{260}105 with a half-life of 1.6 s is formed. While the first atoms of Element 105 are said to have been detected conclusively on March 5, 1970, there is evidence that Element 105 had been formed in Berkeley experiments a year earlier by the method described. Ghiorso and his associates have attempted to confirm Soviet findings by more sophisticated methods without success.

In October 1971, it was announced that two new isotopes of Element 105 were synthesized with the heavy ion linear accelerator by A. Ghiorso and co-workers at Berkeley. Element ^{261}105 was produced both by bombarding ^{250}Cf with ^{15}N and by bombarding ^{249}Bk with ^{16}O. The isotope emits 8.93-MeV α particles and decays to ^{257}Lr with a half-life of about 1.8 s. Element ^{262}105 was produced by bombarding ^{249}Bk with ^{18}O. It emits 8.45 MeV α particles and decays to ^{258}Lr with a half-life of about 40 s. Nine isotopes of Dubnium are now recognized. Soon after the discovery the names *Hahnium* and *Joliotium*, named after Otto Hahn and Jean-Frederic Joliot and Mme. Joliot-Curie, were suggested as names for Element 105. The IUPAC in August 1997 finally resolved the issue, naming Element 105 Dubnium with the symbol Db. Dubnium is thought to have properties similar to tantalum.

Dysprosium — (Gr. *dysprositos*, hard to get at), Dy; at. wt. 162.50(3); at. no. 66; m.p. 1412°C; b.p. 2567°C; sp. gr. 8.551 (25°C); valence 3. Dysprosium was discovered in 1886 by Lecoq de Boisbaudran, but not isolated. Neither the oxide nor the metal was available in relatively pure form until the development of ion-exchange separation and metallographic reduction techniques by Spedding and associates about 1950. Dysprosium occurs along with other so-called rare-earth or lanthanide elements in a variety of minerals such as *xenotime*, *fergusonite*, *gadolinite*, *euxenite*, *polycrase*, and *blomstrandine*. The most important sources, however, are from *monazite* and *bastnasite*. Dysprosium can be prepared by reduction of the trifluoride with calcium. The element has a metallic, bright silver luster. It is relatively stable in air at room temperature, and is readily attacked and dissolved, with the evolution of hydrogen, by dilute and concentrated mineral acids. The metal is soft enough to be cut with a knife and can be machined without sparking if overheating is avoided. Small amounts of impurities can greatly affect its physical properties. While dysprosium has not yet found many applications, its thermal neutron absorption cross-section and high melting point suggest metallurgical uses in nuclear control applications and for alloying with special stainless steels. A dysprosium oxide-nickel cermet has found use in cooling nuclear reactor rods. This cermet absorbs neutrons readily without swelling or contracting under prolonged neutron bombardment. In combination with vanadium and other rare earths, dysprosium has been used in making laser materials. Dysprosium-cadmium chalcogenides, as sources of infrared radiation, have been used for studying chemical reactions. The cost of dysprosium metal has dropped in recent years since the development of ion-exchange and solvent extraction techniques, and the discovery of large ore bodies. Thirty two isotopes and isomers are now known. The metal costs about $6/g (99.9% purity).

Einsteinium — (Albert Einstein [1879–1955]), Es; at. wt. (252); m.p. 860°C (est.); at. no. 99. Einsteinium, the seventh transuranic element of the actinide series to be discovered, was identified by Ghiorso and co-workers at Berkeley in December 1952 in debris from the first large thermonuclear explosion, which took place in the Pacific in November 1952. The isotope produced was the 20-day ^{253}Es isotope. In 1961, a sufficient amount of einsteinium was produced to permit separation of a macroscopic amount of ^{253}Es. This sample weighed about 0.01 µg. A special magnetic-type balance was used in making this determination. ^{253}Es so produced was used to produce mendelevium. About 3 µg of einsteinium has been produced at Oak Ridge National Laboratories by irradiating for several years kilogram quantities of ^{239}Pu in a reactor to produce ^{242}Pu. This was then fabricated into pellets of plutonium oxide and aluminum powder, and loaded into target rods for an initial 1-year irradiation at the Savannah River Plant, followed by irradiation in a HFIR (High Flux Isotopic Reactor). After 4 months in the HFIR the targets were removed for chemical separation of the einsteinium from californium. Nineteen isotopes and isomers of einsteinium are now recognized. ^{254}Es has the longest half-life (276 days). Tracer studies using ^{253}Es show that einsteinium has chemical properties typical of a heavy trivalent, actinide element. Einsteinium is extremely radioactive. Great care must be taken when handling it.

Element 112 — In late February 1996, Siguard Hofmann and his collaborators at GSI Darmstadt announced their discovery of Element 112, having 112 protons and 165 neutrons, with an atomic mass of 277. This element was made by bombarding a lead target with high-energy zinc ions. A single nucleus of Element 112 was detected, which decayed after less than 0.001 sec by emitting an α particle, consisting of two protons and two neutrons. This created Element 110_{273}, which in turn decayed by emitting an α particle to form a new isotope of Element 108 and so on. Evidence indicates that nuclei with 162 neutrons are held together more strongly than nuclei with a smaller or larger number of neutrons. This suggests a narrow "peninsula" of relatively stable isotopes around Element 114. GSI scientists are experimenting to bombard targets with ions heavier than zinc to produce Elements 113 and 114. A name has not been suggested for Element 112, although the IUPAC suggested the temporary name of ununbium, with the symbol of Uub, when the element was discovered. Element 112 is expected to have properties similar to mercury.

Element 113 — (Ununtrium) See Element 115.

Element 114 — (Ununquadium) Symbol Uuq. Element 114 is the first new element to be discovered since 1996. This element was found by a Russian–American team, including Livermore researchers, by bombarding a sheet of plutonium with a rare form of calcium hoping to make the atoms stick together in a new element. Radiation showed that the new element broke into smaller pieces. Data of radiation collected at the Russian Joint Institute for Nuclear Research in November and December 1998, were analyzed in January 1999. It was found that some of the heavy atoms created when 114 decayed lived up to 30 seconds, which was longer than ever seen before, for such a heavy element. This isotope decayed into a previously unknown isotope of Element 112, which itself lasted 15 minutes. That isotope, in turn, decayed to a previously undiscovered isotope of Element 108, which survived 17 minutes. Isotopes of these and those with longer life-times have been predicted for some time by theorists. It appears that these isotopes are on the edge of the "island of stability," and that some of the isotopes in this region might last long enough for studies of their nuclear behavior and for a chemical evaluation to be made. No name has yet been suggested for Element 114; however, the temporary name of ununquadium with symbol Uuq may be used.

Element 115— (Ununpentium) On February 2, 2004, it was reported that Element 115 had been discovered at the Joint Institute for Nuclear Research (JINR) in Dubna, Russia. Four atoms of this element were produced by JINR physicists and collaborators from the Lawrence Livermore (California) Laboratory using a 248-MeV beam of calcium-48 ions striking a target of americium-243 atoms. The nuclei of these atoms are said to have a life of 90 milliseconds. The relatively long lifetime of Element 115 suggests that these experiments might be getting closer to the "island of stability" long sought to exist by some nuclear physicists. These atoms were thought to decay first to Element 113 by the emission of an alpha particle, then decaying further to Element 111 by alpha emission again, and then by three more alpha decay processes to Element 105 (dubnium), which after a long delay from the time of the initial interaciton, fissioned. This experiment entailed separating four atoms from trillions of other atoms. A gas-filled separator, employing chemistry, was important in this experiment. Names for Elements 115, Element 113, and Element 111 have not yet been chosen.

Element 116 — (Ununhexium) Symbol Uuh. As of January 2004 it is questionable if this element has been discovered.

Element 117 — (Ununseptium) Symbol Uus. As of January 2004, this element remains undiscovered.

Element 118 — (Ununoctium) Symbol Uuo. In June 1999 it was announced that Elements 118 and 116 had been discovered at the Lawrence Berkeley National Laboratory. A lead target was bombarded for more than 10 days with roughly 1 quintillion krypton ions. The team reported that three atoms of Element 118 were made, which quickly decayed into Elements 116, 114, and elements of lower atomic mass. It was said that the isotopes of Element 118 lasted only about 200 milliseconds, while the isotope of Element 116 lasted only 1.2 milliseconds. It was hoped that these elements might be members of "an island of stability," which had long been sought. At that time it was hoped that a target of bismuth might be bombarded with krypton ions to make Element 119, which, in turn, would decay into Elements 117, 115, and 113.

On July 27, 2001 researchers at the Lawrence Berkeley Laboratory announced that their discovery of Element 118 was being retracted because workers at the GSI Laboratory in Germany and at Japanese laboratories failed to confirm their results. However, it was reported that different experiments at the Livermore Laboratory and Joint Institute from Nuclear Research in Dubna, Russia indicated that Element 116 had since been created.

Researchers at the Australian National Laboratory suggest that super-heavy elements may be more difficult to make than previously thought. Their data suggest the best way to encourage fusion in making super-heavy elements is to combine the lightest projectiles possible with the heaviest possible targets. This would minimize a so-called "quasi-fission process" in which a projectile nucleus steals protons and neutrons from a target nucleus. In this process the two nuclei are said to fly apart without ever having actually combined.

Erbium — (*Ytterby*, a town in Sweden), Er; at. wt. 167.26(3); at. no. 68; m.p. 1529°C; b.p. 2868°C; sp. gr. 9.066 (25°C); valence 3, Erbium, one of the so-called rare-earth elements of the lanthanide series, is found in the minerals mentioned under dysprosium above. In 1842 Mosander separated "yttria," found in the mineral *gadolinite*, into three fractions which he called *yttria*, *erbia*, and *terbia*. The names *erbia* and *terbia* became confused in this early period. After 1860, Mosander's *terbia* was known as *erbia*, and after 1877, the earlier known *erbia* became *terbia*. The *erbia* of this period was later shown to consist of five oxides, now known as *erbia*, *scandia*, *holmia*, *thulia* and *ytterbia*. By 1905 Urbain and James independently succeeded in isolating fairly pure Er_2O_3. Klemm and Bommer first produced reasonably pure erbium metal in 1934 by reducing the anhydrous chloride with potassium vapor. The pure metal is soft and malleable and has a bright, silvery, metallic luster. As with other rare-earth metals, its properties depend to a certain extent on the impurities present. The metal is fairly stable in air and does not oxidize as rapidly as some of the other rare-earth metals. Naturally occurring erbium is a mixture of six isotopes, all of which are stable. Twenty-seven radioactive isotopes of erbium are also recognized. Recent production techniques, using ion-exchange reactions, have resulted in much lower prices of the rare-earth metals and their compounds in recent years. The cost of 99.9% erbium metal is about $21/g. Erbium is finding nuclear and metallurgical uses. Added to vanadium, for example, erbium lowers the hardness and improves workability. Most of the rare-earth oxides have sharp absorption bands in the visible, ultraviolet, and near infrared. This property, associated with the electronic structure, gives beautiful pastel colors to many of the rare-earth salts. Erbium oxide gives a pink color and has been used as a colorant in glasses and porcelain enamel glazes.

Europium — (Europe), Eu; at. wt. 151.964(1); at. no. 63; m.p. 822°C; b.p. 1596°C; sp. gr. 5.244 (25°C); valence 2 or 3. In 1890 Boisbaudran obtained basic fractions from samarium-gadolinium concentrates which had spark spectral lines not accounted for by samarium or gadolinium. These lines subsequently have been shown to belong to europium. The discovery of europium is generally credited to Demarcay, who separated the rare earth in reasonably pure form in 1901. The pure metal was not isolated until recent years. Europium is now prepared by mixing Eu_2O_3 with a 10%-excess of lanthanum metal and heating the mixture in a tantalum crucible under high vacuum. The element is collected as a silvery-white metallic deposit on the walls of the crucible. As with other rare-earth metals, except for lanthanum, europium ignites in air at about 150 to 180°C. Europium is about as hard as lead and is quite ductile. It is the most reactive of the rare-earth metals,

quickly oxidizing in air. It resembles calcium in its reaction with water. *Bastnasite* and *monazite* are the principal ores containing europium. Europium has been identified spectroscopically in the sun and certain stars. Europium isotopes are good neutron absorbers and are being studied for use in nuclear control applications. Europium oxide is now widely used as a phosphor activator and europium-activated yttrium vanadate is in commercial use as the red phosphor in color TV tubes. Europium-doped plastic has been used as a laser material. With the development of ion-exchange techniques and special processes, the cost of the metal has been greatly reduced in recent years. Natural europium contains two stable isotopes. Thirty five other radioactive isotopes and isomers are known. Europium is one of the rarest and most costly of the rare-earth metals. It is priced at about $60/g (99.9% pure).

Fermium — (Enrico Fermi [1901–1954], nuclear physicist), Fm; at. wt. [257]; at. no. 100; m.p. 1527°C. Fermium, the eighth transuranium element of the actinide series to be discovered, was identified by Ghiorso and co-workers in 1952 in the debris from a thermonuclear explosion in the Pacific in work involving the University of California Radiation Laboratory, the Argonne National Laboratory, and the Los Alamos Scientific Laboratory. The isotope produced was the 20-hour ^{255}Fm. During 1953 and early 1954, while discovery of elements 99 and 100 was withheld from publication for security reasons, a group from the Nobel Institute of Physics in Stockholm bombarded ^{238}U with ^{16}O ions, and isolated a 30-min α-emitter, which they ascribed to $^{250}100$, without claiming discovery of the element. This isotope has since been identified positively, and the 30-min half-life confirmed. The chemical properties of fermium have been studied solely with tracer amounts, and in normal aqueous media only the (III) oxidation state appears to exist. The isotope ^{254}Fm and heavier isotopes can be produced by intense neutron irradiation of lower elements such as plutonium by a process of successive neutron capture interspersed with beta decays until these mass numbers and atomic numbers are reached. Twenty isotopes and isomers of fermium are known to exist. ^{257}Fm, with a half-life of about 100.5 days, is the longest lived. ^{250}Fm, with a half-life of 30 min, has been shown to be a product of decay of Element $^{254}102$. It was by chemical identification of ^{250}Fm that production of Element 102 (nobelium) was confirmed. Fermium would probably have chemical properties resembling erbium.

Fluorine — (L. and F. *fluere*, flow, or flux), F; at. wt. 18.9984032(5); at. no. 9; m.p. −219.62°C (1 atm); b.p. −188.12°C (1 atm); t_c −129.02°C; density 1.696 g/L (0°C, 1 atm); liq. den. at b.p. 1.50 g/cm³; valence 1. In 1529, Georgius Agricola described the use of fluorspar as a flux, and as early as 1670 Schwandhard found that glass was etched when exposed to fluorspar treated with acid. Scheele and many later investigators, including Davy, Gay-Lussac, Lavoisier, and Thenard, experimented with hydrofluoric acid, some experiments ending in tragedy. The element was finally isolated in 1886 by Moisson after nearly 74 years of continuous effort. Fluorine occurs chiefly in *fluorspar* (CaF_2) and *cryolite* (Na_3AlF_6), and is in *topaz* and other minerals. It is a member of the halogen family of elements, and is obtained by electrolyzing a solution of potassium hydrogen fluoride in anhydrous hydrogen fluoride in a vessel of metal or transparent fluorspar. Modern commercial production methods are essentially variations on the procedures first used by Moisson. Fluorine is the most electronegative and reactive of all elements. It is a pale yellow, corrosive gas, which reacts with practically all organic and inorganic substances. Finely divided metals, glass, ceramics, carbon, and even water burn in fluorine with a bright flame. Until World War II, there was no commercial production of elemental fluorine. The atom bomb project and nuclear energy applications, however, made it necessary to produce large quantities. Safe handling techniques have now been developed and it is possible at present to transport liquid fluorine by the ton. Fluorine and its compounds are used in producing uranium (from the hexafluoride) and more than 100 commercial fluorochemicals, including many well-known high-temperature plastics. Hydrofluoric acid is extensively used for etching the glass of light bulbs, etc. Fluorochloro hydrocarbons have been extensively used in air conditioning and refrigeration. However, in recent years the U.S. and other countries have been phasing out ozone-depleting substances, such as the fluorochloro hydrocarbons that have been used in these applications. It has been suggested that fluorine might be substituted for hydrogen wherever it occurs in organic compounds, which could lead to an astronomical number of new fluorine compounds. The presence of fluorine as a soluble fluoride in drinking water to the extent of 2 ppm may cause mottled enamel in teeth, when used by children acquiring permanent teeth; in smaller amounts, however, fluorides are said to be beneficial and used in water supplies to prevent dental cavities. Elemental fluorine has been studied as a rocket propellant as it has an exceptionally high specific impulse value. Compounds of fluorine with rare gases have now been confirmed. Fluorides of xenon, radon, and krypton are among those known. Elemental fluorine and the fluoride ion are highly toxic. The free element has a characteristic pungent odor, detectable in concentrations as low as 20 ppb, which is below the safe working level. The recommended maximum allowable concentration for a daily 8-hour time-weighted exposure is 1 ppm. Fluorine is known to have fourteen isotopes.

Francium — (France), Fr; at. no. 87; at. wt. [223]; m.p. 27°C; valence 1. Discovered in 1939 by Mlle. Marguerite Perey of the Curie Institute, Paris. Francium, the heaviest known member of the alkali metal series, occurs as a result of an alpha disintegration of actinium. It can also be made artificially by bombarding thorium with protons. While it occurs naturally in uranium minerals, there is probably less than an ounce of francium at any time in the total crust of the earth. It has the highest equivalent weight of any element, and is the most unstable of the first 101 elements of the periodic system. Thirty-six isotopes and isomers of francium are recognized. The longest lived ^{223}Fr(Ac, K), a daughter of ^{227}Ac, has a half-life of 21.8 min. This is the only isotope of francium occurring in nature. Because all known isotopes of francium are highly unstable, knowledge of the chemical properties of this element comes from radiochemical techniques. No weighable quantity of the element has been prepared or isolated. The chemical properties of francium most closely resemble cesium. In 1996, researchers Orozco, Sprouse, and co-workers at the State University of New York, Stony Brook, reported that they had produced francium atoms by bombarding ^{18}O atoms at a gold target heated almost to its melting point. Collisions between gold and oxygen nuclei created atoms of francium-210 which had 87 protons and 123 neutrons. This team reported they had generated about 1 million francium-210 ions per second and held 1000 or more atoms at a time for about 20 secs. in a magnetic trap they had devised before the atoms decayed or escaped. Enough francium was trapped so that a video camera could capture the light given off by the atoms as they fluoresced. A cluster of about 10,000 francium atoms appeared as a glowing sphere about 1 mm in diameter. It is thought that the francium atoms could serve as miniature laboratoires for probing interactions between electrons and quarks.

Gadolinium — (*gadolinite*, a mineral named for Gadolin, a Finnish chemist), Gd; at. wt. 157.25(3); at. no. 64; m.p. 1313°C; b.p. 3273°C; sp. gr. 7.901 (25°C); valence 3. Gadolinia, the oxide of gadolinium, was separated by Marignac in 1880 and Lecoq de Boisbaudran independently isolated the element from Mosander's "yttria" in 1886. The element was named for the mineral *gadolinite* from which this rare earth was originally obtained. Gadolinium is found in several other minerals, including *monazite* and *bastnasite*, which are of commercial importance. The element has been isolated only in recent years. With the development of ion-exchange and solvent extraction techniques, the availability and price of gadolinium and the other rare-earth metals have greatly improved. Thirty-one isotopes and isomers of gadolinium are now recognized; seven are stable and occur naturally. The metal can be prepared by the reduction of the anhydrous fluoride with metallic calcium. As with other related rare-earth metals, it is silvery white, has a metallic luster, and is malleable and ductile. At room temperature, gadolinium crystallizes in the hexagonal, close-packed α form. Upon heating to 1235°C, α gadolinium transforms into the β form, which has a body-centered cubic structure. The metal is relatively stable in dry air, but in moist air it tarnishes with the formation of a loosely adhering oxide film which splits off and exposes more surface to oxidation. The metal reacts slowly with water and is soluble in dilute acid. Gadolinium has the highest thermal neutron capture cross-section of any known element (49,000 barns). Natural gadolinium is a mixture of seven isotopes. Two of these, ^{155}Gd and ^{157}Gd, have excellent capture characteristics, but they are present naturally in low concentrations. As a result, gadolinium has a very fast burnout rate and has limited use as a nuclear control rod material. It has been used in making gadolinium yttrium garnets, which have microwave applications. Compounds of gadolinium are used in making phosphors for color TV tubes. The metal has unusual superconductive properties. As little as 1% gadolinium has been found to improve the workability and resistance of iron, chromium, and related alloys to high temperatures and oxidation. Gadolinium ethyl sulfate has extremely low noise characteristics and may find use in duplicating the performance of amplifiers, such as the maser. The metal is ferromagnetic. Gadolinium is unique for its high magnetic moment and for its special Curie temperature (above which ferromagnetism vanishes) lying just at room temperature. This suggests uses as a magnetic component that senses hot and cold. The price of the metal is about $5/g (99.9% purity).

Gallium — (L. *Gallia*, France), Ga; at. wt. 69.723(1); at. no. 31; m.p. 29.76°C; b.p. 2204°C; sp. gr. 5.904 (29.6°C) solid; sp. gr. 6.095 (29.6°C) liquid; valence 2 or 3. Predicted and described by Mendeleev as ekaaluminum, and discovered spectroscopically by Lecoq de Boisbaudran in 1875, who in the same year obtained the free metal by electrolysis of a solution of the hydroxide in KOH. Gallium is often found as a trace element in *diaspore, sphalerite, germanite, bauxite,* and *coal.* Some flue dusts from burning coal have been shown to contain as much as 1.5% gallium. It is the only metal, except for mercury, cesium, and rubidium, which can be liquid near room temperatures; this makes possible its use in high-temperature thermometers. It has one of the longest liquid ranges of any metal and has a low vapor pressure even at high temperatures. There is a strong tendency for gallium to supercool below its freezing point. Therefore, seeding may be necessary to initiate solidification. Ultra-pure gallium has a beautiful, silvery appearance, and the solid metal exhibits a conchoidal fracture similar to glass. The metal expands 3.1% on solidifying; therefore, it should not be stored in glass or metal containers, as they may break as the metal solidifies. Gallium wets glass or porcelain, and forms a brilliant mirror when it is painted on glass. It is widely used in doping semiconductors and producing solid-state devices such as transistors. High-purity gallium is attacked only slowly by mineral acids. Magnesium gallate containing divalent impurities such as Mn^{+2} is finding use in commercial ultraviolet activated powder phosphors. Gallium nitride has been used to produce blue light-emitting diodes such as those used in CD and DVD readers. Gallium has found application in the Gallex Detector Experiment located in the Gran Sasso Underground Laboratory in Italy. This underground facility has been built by the Italian Istituto Nazionale di Fisica Nucleare in the middle of a highway tunnel through the Abruzzese mountains, about 150 km east of Rome. In this experiment, 30.3 tons of gallium in the form of 110 tons of $GaCl_3$-HCl solution are being used to detect solar neutrinos. The production of ^{71}Ge from gallium is being measured. Gallium arsenide is capable of converting electricity directly into coherent light. Gallium readily alloys with most metals, and has been used as a component in low melting alloys. Its toxicity appears to be of a low order, but it should be handled with care until more data are forthcoming. Natural gallium contains two stable isotopes. Twenty-six other isotopes, one of which is an isomer, are known. The metal can be supplied in ultrapure form (99.99999+%). The cost is about $5/g (99.999%).

Germanium — (L. *Germania*, Germany), Ge; at. wt. 72.61(2); at. no. 32; m.p. 938.25°C; b.p. 2833°C; sp. gr. 5.323 (25°C); valence 2 and 4. Predicted by Mendeleev in 1871 as ekasilicon, and discovered by Winkler in 1886. The metal is found in *argyrodite*, a sulfide of germanium and silver; in *germanite*, which contains 8% of the element; in zinc ores; in coal; and in other minerals. The element is frequently obtained commercially from flue dusts of smelters processing zinc ores, and has been recovered from the by-products of combustion of certain coals. Its presence in coal insures a large reserve of the element in the years to come. Germanium can be separated from other metals by fractional distillation of its volatile tetrachloride. The tetrachloride may then be hydrolyzed to give GeO_2; the dioxide can be reduced with hydrogen to give the metal. Recently developed zone-refining techniques permit the production of germanium of ultra-high purity. The element is a gray-white metalloid, and in its pure state is crystalline and brittle, retaining its luster in air at room temperature. It is a very important semiconductor material. Zone-refining techniques have led to production of crystalline germanium for semiconductor use with an impurity of only one part in 10^{10}. Doped with arsenic, gallium, or other elements, it is used as a transistor element in thousands of electronic applications. Its application in fiber optics and infrared optical systems now provides the largest use for germanium. Germanium is also finding many other applications including use as an alloying agent, as a phosphor in fluorescent lamps, and as a catalyst. Germanium and germanium oxide are transparent to the infrared and are used

in infrared spectrometers and other optical equipment, including extremely sensitive infrared detectors. Germanium oxide's high index of refraction and dispersion make it useful as a component of glasses used in wide-angle camera lenses and microscope objectives. The field of organogermanium chemistry is becoming increasingly important. Certain germanium compounds have a low mammalian toxicity, but a marked activity against certain bacteria, which makes them of interest as chemotherapeutic agents. The cost of germanium is about $10/g (99.999% purity). Thirty isotopes and isomers are known, five of which occur naturally.

Gold — (Sanskrit *Jval;* Anglo-Saxon *gold*), Au (L. *aurum*, gold); at. wt. 196.96654(3); at. no. 79; m.p. 1064.18°C; b.p. 2856°C; sp. gr. ~19.3 (20°C); valence 1 or 3. Known and highly valued from earliest times, gold is found in nature as the free metal and in tellurides; it is very widely distributed and is almost always associated with quartz or pyrite. It occurs in veins and alluvial deposits, and is often separated from rocks and other minerals by sluicing and panning operations. About 25% of the world's gold output comes from South Africa, and about two thirds of the total U.S. production now comes from South Dakota and Nevada. The metal is recovered from its ores by cyaniding, amalgamating, and smelting processes. Refining is also frequently done by electrolysis. Gold occurs in sea water to the extent of 0.1 to 2 mg/ton, depending on the location where the sample is taken. As yet, no method has been found for recovering gold from sea water profitably. It is estimated that all the gold in the world, so far refined, could be placed in a single cube 60 ft on a side. Of all the elements, gold in its pure state is undoubtedly the most beautiful. It is metallic, having a yellow color when in a mass, but when finely divided it may be black, ruby, or purple. The Purple of Cassius is a delicate test for auric gold. It is the most malleable and ductile metal; 1 oz. of gold can be beaten out to 300 ft^2. It is a soft metal and is usually alloyed to give it more strength. It is a good conductor of heat and electricity, and is unaffected by air and most reagents. It is used in coinage and is a standard for monetary systems in many countries. It is also extensively used for jewelry, decoration, dental work, and for plating. It is used for coating certain space satellites, as it is a good reflector of infrared and is inert. Gold, like other precious metals, is measured in troy weight; when alloyed with other metals, the term *carat* is used to express the amount of gold present, 24 carats being pure gold. For many years the value of gold was set by the U.S. at $20.67/troy ounce; in 1934 this value was fixed by law at $35.00/troy ounce, 9/10th fine. On March 17, 1968, because of a gold crisis, a two-tiered pricing system was established whereby gold was still used to settle international accounts at the old $35.00/troy ounce price while the price of gold on the private market would be allowed to fluctuate. Since this time, the price of gold on the free market has fluctuated widely. The price of gold on the free market reached a price of $620/troy oz. in January 1980. More recently, the U.K. and other nations, including the I.M.F. have sold or threatened to sell a sizeable portion of their gold reserves. This has caused wide fluctuations in the price of gold. Because this has damaged the economy of some countries, a moratorium for a few years has been declared. This has tended to stabilize temporarily the price of gold. The most common gold compounds are auric chloride ($AuCl_3$) and chlorauric acid ($HAuCl_4$), the latter being used in photography for toning the silver image. Gold has forty-eight recognized isotopes and isomers; ^{198}Au, with a half-life of 2.7 days, is used for treating cancer and other diseases. Disodium aurothiomalate is administered intramuscularly as a treatment for arthritis. A mixture of one part nitric acid with three of hydrochloric acid is called *aqua regia* (because it dissolved gold, the King of Metals). Gold is available commercially with a purity of 99.999+%. For many years the temperature assigned to the freezing point of gold has been 1063.0°C; this has served as a calibration point for the International Temperature Scales (ITS-27 and ITS-48) and the International Practical Temperature Scale (IPTS-48). In 1968, a new International Practical Temperature Scale (IPTS-68) was adopted, which demanded that the freezing point of gold be changed to 1064.43°C. In 1990 a new International Temperature Scale (ITS-90) was adopted bringing the t.p. (triple point) of H_2O (t_{90} (°C)) to 0.01°C and the freezing point of gold to 1064.18°C. The specific gravity of gold has been found to vary considerably depending on temperature, how the metal is precipitated, and cold-worked. As of December 2001, gold was priced at about $275/troy oz. ($8.50/g).

Hafnium — (*Hafnia*, Latin name for Copenhagen), Hf; at. wt. 178.49(2); at. no. 72; m.p. 2233°C; b.p. 4603°C; sp. gr. 13.31 (20°C); valence 4. Hafnium was thought to be present in various minerals and concentrations many years prior to its discovery, in 1923, credited to D. Coster and G. von Hevesey. On the basis of the Bohr theory, the new element was expected to be associated with zirconium. It was finally identified in *zircon* from Norway, by means of X-ray spectroscopic analysis. It was named in honor of the city in which the discovery was made. Most zirconium minerals contain 1 to 5% hafnium. It was originally separated from zirconium by repeated recrystallization of the double ammonium or potassium fluorides by von Hevesey and Jantzen. Metallic hafnium was first prepared by van Arkel and deBoer by passing the vapor of the tetraiodide over a heated tungsten filament. Almost all hafnium metal now produced is made by reducing the tetrachloride with magnesium or with sodium (Kroll Process). Hafnium is a ductile metal with a brilliant silver luster. Its properties are considerably influenced by the impurities of zirconium present. Of all the elements, zirconium and hafnium are two of the most difficult to separate. Their chemistry is almost identical, however, the density of zirconium is about half that of hafnium. Very pure hafnium has been produced, with zirconium being the major impurity. Natural hafnium contains six isotopes, one of which is slightly radioactive. Hafnium has a total of 41 recognized isotopes and isomers. Because hafnium has a good absorption cross section for thermal neutrons (almost 600 times that of zirconium), has excellent mechanical properties, and is extremely corrosion resistant, it is used for reactor control rods. Such rods are used in nuclear submarines. Hafnium has been successfully alloyed with iron, titanium, niobium, tantalum, and other metals. Hafnium carbide is the most refractory binary composition known, and the nitride is the most refractory of all known metal nitrides (m.p. 3310°C). Hafnium is used in gas-filled and incandescent lamps, and is an efficient "getter" for scavenging oxygen and nitrogen. Finely divided hafnium is pyrophoric and can ignite spontaneously in air. Care should be taken when machining the metal or when handling hot sponge hafnium. At 700°C hafnium rapidly absorbs hydrogen to form the composition $HfH_{1.86}$. Hafnium is resistant to concentrated alkalis, but at elevated temperatures reacts with oxygen, nitrogen, carbon, boron, sulfur, and silicon. Halogens react directly to form tetrahalides. The price of the metal is about $2/g. The yearly demand for hafnium in the U.S. is now in excess of 50,000 kg.

Hahnium — A name previously used for Element 105, now named *dubnium*.

Hassium — (named for the German state, Hesse) Hs; at.wt. [265]; at.no. 108. This element was first synthesized and identified in 1964 by the same G.S.I. Darmstadt Group who first identified *Bohrium* and *Meitnerium.* Presumably this element has chemical properties similar to osmium. Isotope $^{265}108$ was produced using a beam of ^{58}Fe projectiles, produced by the Universal Linear Accelerator (UNILAC) to bombard a ^{208}Pb target. Discovery of *Bohrium* and *Meitnerium* was made using detection of isotopes with odd proton and neutron numbers. Elements having even atomic numbers have been thought to be less stable against spontaneous fusion than odd elements. The production of $^{265}108$ in the same reaction as was used at G.S.I. was confirmed at Dubna with detection of the seventh member of the decay chain ^{253}Es. Isotopes of *Hassium* are believed to decay by spontaneous fission, explaining why 109 was produced before 108. Isotope $^{265}108$ and $^{266}108$ are thought to decay to $^{261}106$, which in turn decay to $^{257}104$ and $^{253}102$. The IUPAC adopted the name *Hassium* after the German state of Hesse in September 1997. In June 2001 it was announced that hassium is now the heaviest element to have its chemical properties analyzed. A research team at the UNILAC heavy-ion accelerator in Darmstadt, Germany built an instrument to detect and analyze hassium. Atoms of curium-248 were collided with atoms of magnesium-26, producing about 6 atoms of hassium with a half-life of 9 sec. This was sufficiently long to obtain data showing that hassium atoms react with oxygen to form hassium oxide molecules. These condensed at a temperature consistent with the behavior of Group 8 elements. This experiment appears to confirm hassium's location under osmium in the periodic table.

Helium — (Gr. *helios*, the sun), He; at. wt. 4.002602(2); at. no. 2; m.p. below — 272.2°C (26 atm); b.p. — 268.93°C; t_c -267.96°C; density 0.1785 g/l (0°C, 1 atm); liquid density 7.62 lb/ft³ at. b.p.; valence usually 0. Evidence of the existence of helium was first obtained by Janssen during the solar eclipse of 1868 when he detected a new line in the solar spectrum; Lockyer and Frankland suggested the name *helium* for the new element; in 1895, Ramsay discovered helium in the uranium mineral *cleveite*, and it was independently discovered in cleveite by the Swedish chemists Cleve and Langlet about the same time. Rutherford and Royds in 1907 demonstrated that α particles are helium nuclei. Except for hydrogen, helium is the most abundant element found throughout the universe. Helium is extracted from natural gas; all natural gas contains at least trace quantities of helium. It has been detected spectroscopically in great abundance, especially in the hotter stars, and it is an important component in both the proton-proton reaction and the carbon cycle, which account for the energy of the sun and stars. The fusion of hydrogen into helium provides the energy of the hydrogen bomb. The helium content of the atmosphere is about 1 part in 200,000. It is present in various radioactive minerals as a decay product. Much of the world's supply of helium is obtained from wells in Texas, Colorado, and Kansas. The only other known helium extraction plants, outside the United States, in 1999 were in Poland, Russia, China, Algeria, and India. The cost of helium has fallen from $2500/ft³ in 1915 to about 2.5¢/cu.ft. (.028 cu meters) in 1999. Helium has the lowest melting point of any element and has found wide use in cryogenic research, as its boiling point is close to absolute zero. Its use in the study of superconductivity is vital. Using liquid helium, Kurti and co-workers, and others, have succeeded in obtaining temperatures of a few microkelvins by the adiabatic demagnetization of copper nuclei, starting from about 0.01 K. Liquid helium (He^4) exists in two forms: He^{4}I and He^{4}II, with a sharp transition point at 2.174 K (3.83 cm Hg). He^{4}I (above this temperature) is a normal liquid, but He^{4}II (below it) is unlike any other known substance. It expands on cooling; its conductivity for heat is enormous; and neither its heat conduction nor viscosity obeys normal rules. It has other peculiar properties. Helium is the only liquid that cannot be solidified by lowering the temperature. It remains liquid down to absolute zero at ordinary pressures, but it can readily be solidified by increasing the pressure. Solid ^{3}He and ^{4}He are unusual in that both can readily be changed in volume by more than 30% by application of pressure. The specific heat of helium gas is unusually high. The density of helium vapor at the normal boiling point is also very high, with the vapor expanding greatly when heated to room temperature. Containers filled with helium gas at 5 to 10 K should be treated as though they contained liquid helium due to the large increase in pressure resulting from warming the gas to room temperature. While helium normally has a 0 valence, it seems to have a weak tendency to combine with certain other elements. Means of preparing helium diflouride have been studied, and species such as HeNe and the molecular ions He$^+$ and He^{++} have been investigated. Helium is widely used as an inert gas shield for arc welding; as a protective gas in growing silicon and germanium crystals, and in titanium and zirconium production; as a cooling medium for nuclear reactors, and as a gas for supersonic wind tunnels. A mixture of helium and oxygen is used as an artificial atmosphere for divers and others working under pressure. Different ratios of He/O$_2$ are used for different depths at which the diver is operating. Helium is extensively used for filling balloons as it is a much safer gas than hydrogen. One of the recent largest uses for helium has been for pressuring liquid fuel rockets. A Saturn booster such as used on the Apollo lunar missions required about 13 million ft^3 of helium for a firing, plus more for checkouts. Liquid helium's use in magnetic resonance imaging (MRI) continues to increase as the medical profession accepts and develops new uses for the equipment. This equipment is providing accurate diagnoses of problems where exploratory surgery has previously been required to determine problems. Another medical application that is being developed uses MRI to determine by blood analysis whether a patient has any form of cancer. Lifting gas applications are increasing. Various companies in addition to Goodyear, are now using "blimps" for advertising. The Navy and the Air Force are investigating the use of airships to provide early warning systems to detect low-flying cruise missiles. The Drug Enforcement Agency has used radar-equipped blimps to detect drug smugglers along the southern border of the U.S. In addition, NASA is currently using helium-filled balloons to sample the atmosphere in Antarctica to determine what is depleting the ozone layer that protects Earth from harmful U.V. radiation. Research on and development of materials which become superconductive at temperatures well above the boiling point of helium could have a major impact on the demand for helium. Less costly refrigerants having boiling points considerably higher could replace the present need to cool such superconductive materials to the boiling point of helium. Natural helium contains two stable isotopes ^{3}He and ^{4}He. ^{3}He is present in very small quantities. Six other isotopes of helium are now recognized.

Holmium — (L. *Holmia*, for Stockholm), Ho; at. wt. 164.93032(2); at. no 67; m.p. 1474°C; b.p. 2700°C; sp. gr. 8.795 (25°C); valence + 3. The spectral absorption bands of holmium were noticed in 1878 by the Swiss chemists Delafontaine and Soret, who announced the existence of an "Element X." Cleve, of Sweden, later independently discovered the element while working on erbia earth. The element is named after Cleve's native city. Pure holmia, the yellow oxide, was prepared by Homberg in 1911. Holmium occurs in *gadolinite, monazite,* and in other rare-earth minerals. It is commercially obtained from monazite, occurring in that mineral to the extent of about 0.05%. It has been isolated by the reduction of its anhydrous chloride or fluoride with calcium metal. Pure holmium has a metallic to bright silver luster. It is relatively soft and malleable, and is stable in dry air at room temperature, but rapidly oxidizes in moist air and at elevated temperatures. The metal has unusual magnetic properties. Few uses have yet been found for the element. The element, as with other rare earths, seems to have a low acute toxic rating. Natural holmium consists of one isotope ^{165}Ho, which is not radioactive. Holmium has 49 other isotopes known, all of which are radioactive. The price of 99.9% holmium metal is about $20/g.

Hydrogen — (Gr. *hydro*, water, and *genes,* forming), H; at. wt. 1.00794(7); at. no. 1; m.p. –259.34°C; b.p. –252.87°C; t_c –240.18; density 0.08988 g/l; density (liquid) 70.8 g/l (–253°C); density (solid) 70.6 g/l (–262°C); valence 1. Hydrogen was prepared many years before it was recognized as a distinct substance by Cavendish in 1766. It was named by Lavoisier. Hydrogen is the most abundant of all elements in the universe, and it is thought that the heavier elements were, and still are, being built from hydrogen and helium. It has been estimated that hydrogen makes up more than 90% of all the atoms or three quarters of the mass of the universe. It is found in the sun and most stars, and plays an important part in the proton-proton reaction and carbon-nitrogen cycle, which accounts for the energy of the sun and stars. It is thought that hydrogen is a major component of the planet Jupiter and that at some depth in the planet's interior the pressure is so great that solid molecular hydrogen is converted into solid metallic hydrogen. In 1973, it was reported that a group of Russian experimenters may have produced metallic hydrogen at a pressure of 2.8 Mbar. At the transition the density changed from 1.08 to 1.3 g/cm^3. Earlier, in 1972, a Livermore (California) group also reported on a similar experiment in which they observed a pressure-volume point centered at 2 Mbar. It has been predicted that metallic hydrogen may be metastable; others have predicted it would be a superconductor at room temperature. On earth, hydrogen occurs chiefly in combination with oxygen in water, but it is also present in organic matter such as living plants, petroleum, coal, etc. It is present as the free element in the atmosphere, but only to the extent of less than 1 ppm by volume. It is the lightest of all gases, and combines with other elements, sometimes explosively, to form compounds. Great quantities of hydrogen are required commercially for the fixation of nitrogen from the air in the Haber ammonia process and for the hydrogenation of fats and oils. It is also used in large quantities in methanol production, in hydrodealkylation, hydrocracking, and hydrodesulfurization. It is also used as a rocket fuel, for welding, for production of hydrochloric acid, for the reduction of metallic ores, and for filling balloons. The lifting power of 1 ft^3 of hydrogen gas is about 0.076 lb at 0°C, 760 mm pressure. Production of hydrogen in the U.S. alone now amounts to about 3 billion cubic feet per year. It is prepared by the action of steam on heated carbon, by decomposition of certain hydrocarbons with heat, by the electrolysis of water, or by the displacement from acids by certain metals. It is also produced by the action of sodium or potassium hydroxide on aluminum. Liquid hydrogen is important in cryogenics and in the study of superconductivity, as its melting point is only a 20°C above absolute zero. Hydrogen consists of three isotopes, most of which is ^{1}H. The ordinary isotope of hydrogen, H, is known as *protium*. In 1932, Urey announced the discovery of a stable isotope, deuterium (^{2}H or D) with an atomic weight of 2. Deuterium is present in natural hydrogen to the extent of 0.015%. Two years later an unstable isotope, tritium (^{3}H), with an atomic weight of 3 was discovered. Tritium has a half-life of about 12.32 years. Tritium atoms are also present in natural hydrogen but in much smaller proportion. Tritium is readily produced in nuclear reactors and is used in the production of the hydrogen bomb. It is also used as a radioactive agent in making luminous paints, and as a tracer. On August 27, 2001 Russian, French, and Japanese physicists working at the Joint Institute for Nuclear Research near Moscow reported they had made "super-heavy hydrogen," which had a nucleus with one proton and four neutrons. Using an accelerator, they used a beam of helium-6 nuclei to strike a hydrogen target, which resulted in the occasional production of a hydrogen-5 nucleus plus a helium-2 nucleus. These unstable particles quickly disintegrated. This resulted in two protons from the He-2, a triton, and two neutrons from the H-5 breakup. Deuterium gas is readily available, without permit, at about $1/l. Heavy water, deuterium oxide (D_2O), which is used as a moderator to slow down neutrons, is available without permit at a cost of 6c to $1/g, depending on quantity and purity. About 1000 tons (4,400,000 kg) of deuterium oxide (heavy water) are now in use at the Sudbury (Ontario) Neutrino Observatory. This observatory is taking data to provide new revolutionary insight into the properties of neutrinos and into the core of the sun. The heavy water is on loan from Atomic Energy of Canada, Ltd. (AECL). The observatory and detectors are located 6800 ft (2072 m) deep in the Creighton mine of the International Nickel Co., near Sudbury. The heavy water is contained in an acrylic vessel, 12 m in diameter. Neutrinos react with the heavy water to produce Cherenkov radiation. This light is then detected with 9600 photomultiplier tubes surrounding the vessel. The detector laboratory is immensely clean to reduce background radiation, which otherwise hide the very weak signals from neutrinos. Quite apart from isotopes, it has been shown that hydrogen gas under ordinary conditions is a mixture of two kinds of molecules, known as *ortho-* and *para-* hydrogen, which differ from one another by the spins of their electrons and nuclei. Normal hydrogen at room temperature contains 25% of the *para* form and 75% of the *ortho* form. The *ortho* form cannot be prepared in the pure state. Since the two forms differ in energy, the physical properties also differ. The melting and boiling points of *para*hydrogen are about 0.1°C lower than those of normal hydrogen. Consideration is being given to an entire economy based on solar- and nuclear-generated hydrogen. Located in remote regions, power plants would electrolyze sea water; the hydrogen produced would travel to distant cities by pipelines. Pollution-free hydrogen could replace natural gas, gasoline, etc., and could serve as a reducing agent in metallurgy, chemical processing, refining, etc. It could also be used to convert trash into methane and ethylene. Public acceptance, high capital investment, and the high present cost of hydrogen with respect to present fuels are but a few of the problems facing establishment of such an economy. Hydrogen is being investigated as a substitute for deep-sea diving applications below 300 m. Hydrogen is readily available from air product suppliers.

Indium — (from the brilliant indigo line in its spectrum), In; at. wt. 114.818(3); at. no. 49; m.p. 156.60°C; b.p. 2072°C; sp. gr. 7.31 (20°C); valence 1, 2, or 3. Discovered by Reich and Richter, who later isolated the metal. Indium is most frequently associated with zinc materials, and it is from these that most commercial indium is now obtained; however, it is also found in iron, lead, and copper ores. Until 1924, a gram or so constituted the world's supply of this element in isolated form. It is probably about as abundant as silver. About 4 million troy ounces of indium are now produced annually in the Free World. Canada is presently producing more than 1,000,000 troy ounces annually. The present cost of indium is about $2 to $10/g, depending on quantity and purity. It is available in ultrapure form. Indium is a very soft, silvery-white metal with a brilliant luster. The pure metal gives a high-pitched "cry" when bent. It wets glass, as does gallium. It has found application in making low-melting alloys; an alloy of 24% indium-76% gallium is liquid at room temperature. Indium is used in making bearing alloys, germanium transistors, rectifiers, thermistors, liquid crystal displays, high definition television, batteries, and photoconductors. It can be plated onto metal and evaporated onto glass, forming a mirror as good as that made with silver but with more resistance to atmospheric corrosion. There is evidence that indium has a low order of toxicity; however, care should be taken until further information is available. Seventy isotopes and isomers are now recognized (more than any other element). Natural indium contains two isotopes. One is stable. The other, ^{115}In, comprising 95.71% of natural indium is slightly radioactive with a very long half-life.

Iodine — (Gr. *iodes*, violet), I; at. wt. 126.90447(3); at. no. 53; m.p. 113.7°C; b.p. 184.4°C; t_c 546°C; density of the gas 11.27 g/l; sp. gr. solid 4.93 (20°C); valence 1, 3, 5, or 7. Discovered by Courtois in 1811. Iodine, a halogen, occurs sparingly in the form of iodides in sea water from which it is assimilated by seaweeds, in Chilean saltpeter and nitrate-bearing earth, known as *caliche* in brines from old sea deposits, and in brackish waters from oil and salt wells. Ultrapure iodine can be obtained from the reaction of potassium iodide with copper sulfate. Several other methods of isolating the element are known. Iodine is a bluish-black, lustrous solid, volatilizing at ordinary temperatures into a blue-violet gas with an irritating odor; it forms compounds with many elements, but is less active than the other halogens, which displace it from iodides. Iodine exhibits some metallic-like properties. It dissolves readily in chloroform, carbon tetrachloride, or carbon disulfide to form beautiful purple solutions. It is only slightly soluble in water. Iodine compounds are important in organic chemistry and very useful in medicine. Forty two isotopes and isomers are recognized. Only one stable isotope, ^{127}I is found in nature. The artificial radioisotope ^{131}I, with a half-life of 8 days, has been used in treating the thyroid gland. The most common compounds are the iodides of sodium and potassium (KI) and the iodates (KIO_3). Lack of iodine is the cause of goiter. Iodides, and thyroxin which contains iodine, are used internally in medicine, and a solution of KI and iodine in alcohol is used for external wounds. Potassium iodide finds use in photography. The deep blue color with starch solution is characteristic of the free element. Care should be taken in

The Elements

handling and using iodine, as contact with the skin can cause lesions; iodine vapor is intensely irritating to the eyes and mucous membranes. Elemental iodine costs about 25 to 75¢/g depending on purity and quantity.

Iridium — (L. *iris*, rainbow), Ir; at. wt. 192.217(3); at. no. 77; m.p. 2446°C; b.p. 4428°C; sp. gr. 22.42 (17°C); valence 3 or 4. Discovered in 1803 by Tennant in the residue left when crude platinum is dissolved by aqua regia. The name iridium is appropriate, for its salts are highly colored. Iridium, a metal of the platinum family, is white, similar to platinum, but with a slight yellowish cast. It is very hard and brittle, making it very hard to machine, form, or work. It is the most corrosion-resistant metal known, and was used in making the standard meter bar of Paris, which is a 90% platinum-10% iridium alloy. This meter bar was replaced in 1960 as a fundamental unit of length (see under Krypton). Iridium is not attacked by any of the acids nor by aqua regia, but is attacked by molten salts, such as NaCl and NaCN. Iridium occurs uncombined in nature with platinum and other metals of this family in alluvial deposits. It is recovered as a by-product from the nickel mining industry. The largest reserves and production of the platinum group of metals, which includes iridium, is in South Africa, followed by Russia and Canada. The U.S. has only one active mine, located at Nye, MT. The presence of iridium has recently been used in examining the Cretaceous-Tertiary (K-T) boundary. Meteorites contain small amounts of iridium. Because iridium is found widely distributed at the K-T boundary, it has been suggested that a large meteorite or asteroid collided with the earth, killing the dinosaurs, and creating a large dust cloud and crater. Searches for such a crater point to one in the Yucatan, known as Chicxulub. Iridium has found use in making crucibles and apparatus for use at high temperatures. It is also used for electrical contacts. Its principal use is as a hardening agent for platinum. With osmium, it forms an alloy which is used for tipping pens and compass bearings. The specific gravity of iridium is only very slightly lower than that of osmium, which has been generally credited as being the heaviest known element. Calculations of the densities of iridium and osmium from the space lattices gives values of 22.65 and 22.61 g/cm³, respectively. These values may be more reliable than actual physical measurements. At present, therefore, we know that either iridium or osmium is the densest known element, but the data do not yet allow selection between the two. Natural iridium contains two stable isotopes. Forty-five other isotopes, all radioactive, are now recognized. Iridium (99.9%) costs about $100/g.

Iron — (Anglo-Saxon, *iron*), Fe (L. *ferrum*); at. wt. 55.845(2); at. no. 26; m.p. 1538°C; b.p. 2861°C; sp. gr. 7.874 (20°C); valence 2, 3, 4, or 6. The use of iron is prehistoric. Genesis mentions that Tubal-Cain, seven generations from Adam, was "an instructor of every artificer in brass and iron." A remarkable iron pillar, dating to about A.D. 400, remains standing today in Delhi, India. This solid shaft of wrought iron is about $7^1/_4$ m high by 40 cm in diameter. Corrosion to the pillar has been minimal although it has been exposed to the weather since its erection. Iron is a relatively abundant element in the universe. It is found in the sun and many types of stars in considerable quantity. It has been suggested that the iron we have here on earth may have originated in a supernova. Iron is a very difficult element to produce in ordinary nuclear reactions, such as would take place in the sun. Iron is found native as a principal component of a class of iron-nickel meteorites known as *siderites*, and is

a minor constituent of the other two classes of meteorites. The core of the earth, 2150 miles in radius, is thought to be largely composed of iron with about 10% occluded hydrogen. The metal is the fourth most abundant element, by weight, making up the crust of the earth. The most common ore is *hematite* (Fe_2O_3). Magnetite (Fe_3O_4) is frequently seen as *black sands* along beaches and banks of streams. *Lodestone* is another form of magnetite. *Taconite* is becoming increasingly important as a commercial ore. Iron is a vital constituent of plant and animal life, and appears in hemoglobin. The pure metal is not often encountered in commerce, but is usually alloyed with carbon or other metals. The pure metal is very reactive chemically, and rapidly corrodes, especially in moist air or at elevated temperatures. It has four allotropic forms, or ferrites, known as α, β, γ, and δ, with transition points at 700, 928, and 1530°C. The α form is magnetic, but when transformed into the β form, the magnetism disappears although the lattice remains unchanged. The relations of these forms are peculiar. Pig iron is an alloy containing about 3% carbon with varying amounts of S, Si, Mn, and P. It is hard, brittle, fairly fusible, and is used to produce other alloys, including steel. Wrought iron contains only a few tenths of a percent of carbon, is tough, malleable, less fusible, and has usually a "fibrous" structure. Carbon steel is an alloy of iron with carbon, with small amounts of Mn, S, P, and Si. Alloy steels are carbon steels with other additives such as nickel, chromium, vanadium, etc. Iron is the cheapest and most abundant, useful, and important of all metals. Natural iron contains four isotopes. Twenty-six other isotopes and isomers, all radioactive, are now recognized.

Krypton — (Gr. *kryptos*, hidden), Kr; at. wt. 83.80(1); at. no. 36; m.p. −157.36°C; b.p. −153.22 ± 0.10°C; t_c −63.74°C; density 3.733 g/l (0°C); valence usually 0. Discovered in 1898 by Ramsay and Travers in the residue left after liquid air had nearly boiled away. Krypton is present in the air to the extent of about 1 ppm. The atmosphere of Mars has been found to contain 0.3 ppm of krypton. It is one of the "noble" gases. It is characterized by its brilliant green and orange spectral lines. Naturally occurring krypton contains six stable isotopes. Thirty other unstable isotopes and isomers are now recognized. The spectral lines of krypton are easily produced and some are very sharp. In 1960 it was internationally agreed that the fundamental unit of length, the meter, should be defined in terms of the orange-red spectral line of ^{86}Kr. This replaced the standard meter of Paris, which was defined in terms of a bar made of a platinum-iridium alloy. In October 1983 the meter was again redefined by the International Bureau of Weights and Measures as being the length of path traveled by light in a vacuum during a time interval of 1/299,792,458 of a second. Solid krypton is a white crystalline substance with a face-centered cubic structure which is common to all the rare gases. While krypton is generally thought of as a noble gas that normally does not combine with other elements, the existence of some krypton compounds has been established. Krypton difluoride has been prepared in gram quantities and can be made by several methods. A higher fluoride of krypton and a salt of an oxyacid of krypton also have been prepared. Molecule-ions of $ArKr^+$ and KrH^+ have been identified and investigated, and evidence is provided for the formation of KrXe or $KrXe^+$. Krypton clathrates have been prepared with hydroquinone and phenol. ^{85}Kr has found recent application in chemical analysis. By imbedding the isotope in various solids, *kryptonates* are formed. The activity of these kryptonates is sensitive to chemical reactions at the surface. Estimates of the concentration of reactants are therefore made possible. Krypton is used in certain photographic flash lamps for high-speed photography. Uses thus far have been limited because of its high cost. Krypton gas presently costs about $690/100 L.

Kurchatovium — See Rutherfordium.

Lanthanum — (Gr. *lanthanein*, to lie hidden), La; at. wt. 138.9055(2); at. no. 57; m.p. 918°C; b.p. 3464°C; sp. gr. 6.145 (25°C); valence 3. Mosander in 1839 extracted a new earth *lanthana*, from impure cerium nitrate, and recognized the new element. Lanthanum is found in rare-earth minerals such as *cerite, monazite, allanite,* and *bastnasite*. Monazite and bastnasite are principal ores in which lanthanum occurs in percentages up to 25 and 38%, respectively. Misch metal, used in making lighter flints, contains about 25% lanthanum. Lanthanum was isolated in relatively pure form in 1923. Ion-exchange and solvent extraction techniques have led to much easier isolation of the so-called "rare-earth" elements. The availability of lanthanum and other rare earths has improved greatly in recent years. The metal can be produced by reducing the anhydrous fluoride with calcium. Lanthanum is silvery white, malleable, ductile, and soft enough to be cut with a knife. It is one of the most reactive of the rare-earth metals. It oxidizes rapidly when exposed to air. Cold water attacks lanthanum slowly, and hot water attacks it much more rapidly. The metal reacts directly with elemental carbon, nitrogen, boron, selenium, silicon, phosphorus, sulfur, and with halogens. At 310°C, lanthanum changes from a hexagonal to a face-centered cubic structure, and at 865°C it again transforms into a body-centered cubic structure. Natural lanthanum is mixture of two isotopes, one of which is stable and one of which is radioactive with a very long half-life. Thirty other radioactive isotopes are recognized. Rare-earth compounds containing lanthanum are extensively used in carbon lighting applications, especially by the motion picture industry for studio lighting and projection. This application consumes about 25% of the rare-earth compounds produced. La_2O_3 improves the alkali resistance of glass, and is used in making special optical glasses. Small amounts of lanthanum, as an additive, can be used to produce nodular cast iron. There is current interest in hydrogen sponge alloys containing lanthanum. These alloys take up to 400 times their own volume of hydrogen gas, and the process is reversible. Heat energy is released every time they do so; therefore these alloys have possibilities in energy conservation systems. Lanthanum and its compounds have a low to moderate acute toxicity rating; therefore, care should be taken in handling them. The metal costs about $2/g (99.9%).

Lawrencium — (Ernest O. Lawrence [1901–1958], inventor of the cyclotron), Lr; at. no. 103; at. mass no. [262]; valence + 3(?). This member of the 5f transition elements (actinide series) was discovered in March 1961 by A. Ghiorso, T. Sikkeland, A. E. Larsh, and R. M. Latimer. A 3-μg californium target, consisting of a mixture of isotopes of mass number 249, 250, 251, and 252, was bombarded with either [10]B or [11]B. The electrically charged transmutation nuclei recoiled with an atmosphere of helium and were collected on a thin copper conveyor tape which was then moved to place collected atoms in front of a series of solid-state detectors. The isotope of element 103 produced in this way decayed by emitting an 8.6-MeV alpha particle with a half-life of 8 s. In 1967, Flerov and associates of the Dubna Laboratory reported their inability to detect an alpha emitter with a half-life of 8 s which was assigned by the

Berkeley group to [257]103. This assignment has been changed to [258]Lr or [259]Lr. In 1965, the Dubna workers found a longer-lived lawrencium isotope, [256]Lr, with a half-life of 35 s. In 1968, Ghiorso and associates at Berkeley were able to use a few atoms of this isotope to study the oxidation behavior of lawrencium. Using solvent extraction techniques and working very rapidly, they extracted lawrencium ions from a buffered aqueous solution into an organic solvent, completing each extraction in about 30 s. It was found that lawrencium behaves differently from dipositive nobelium and more like the tripositive elements earlier in the actinide series. Ten isotopes of lawrencium are now recognized.

Lead — (Anglo-Saxon *lead*), Pb (L. *plumbum*); at. wt. 207.2(1); at. no. 82; m.p. 327.46°C; b.p. 1749°C; sp. gr. 11.35 (20°C); valence 2 or 4. Long known, mentioned in Exodus. The alchemists believed lead to be the oldest metal and associated it with the planet Saturn. Native lead occurs in nature, but it is rare. Lead is obtained chiefly from *galena* (PbS) by a roasting process. *Anglesite* ($PbSO_4$), *cerussite* ($PbCO_3$), and *minim* (Pb_3O_4) are other common lead minerals. Lead is a bluish-white metal of bright luster, is very soft, highly malleable, ductile, and a poor conductor of electricity. It is very resistant to corrosion; lead pipes bearing the insignia of Roman emperors, used as drains from the baths, are still in service. It is used in containers for corrosive liquids (such as sulfuric acid) and may be toughened by the addition of a small percentage of antimony or other metals. Natural lead is a mixture of four stable isotopes: [204]Pb (1.4%), [206]Pb (24.1%), [207]Pb (22.1%), and [208]Pb (52.4%). Lead isotopes are the end products of each of the three series of naturally occurring radioactive elements: [206]Pb for the uranium series, [207]Pb for the actinium series, and [208]Pb for the thorium series. Forty-three other isotopes of lead, all of which are radioactive, are recognized. Its alloys include solder, type metal, and various antifriction metals. Great quantities of lead, both as the metal and as the dioxide, are used in storage batteries. Lead is also used for cable covering, plumbing, and ammunition. The metal is very effective as a sound absorber, is used as a radiation shield around X-ray equipment and nuclear reactors, and is used to absorb vibration. Lead, alloyed with tin, is used in making organ pipes. White lead, the basic carbonate, sublimed white lead ($PbSO_4$), chrome yellow ($PbCrO_4$), red lead (Pb_3O_4), and other lead compounds are used extensively in paints, although in recent years the use of lead in paints has been drastically curtailed to eliminate or reduce health hazards. Lead oxide is used in producing fine "crystal glass" and "flint glass" of a high index of refraction for achromatic lenses. The nitrate and the acetate are soluble salts. Lead salts such as lead arsenate have been used as insecticides, but their use in recent years has been practically eliminated in favor of less harmful organic compounds. Care must be used in handling lead as it is a cumulative poison. Environmental concern with lead poisoning led to elimination of lead tetraethyl in gasoline. The U.S. Occupational Safety and Health Administration (OSHA) has recommended that industries limit airborne lead to 50 μg/cu. meter. Lead is priced at about 90¢/kg (99.9%).

Lithium — (Gr. *lithos*, stone), Li; at. wt. 6.941(2); at. no. 3; m.p. 180.5°C; b.p. 1342°C; sp. gr. 0.534 (20°C); valence 1. Discovered by Arfvedson in 1817. Lithium is the lightest of all metals, with a density only about half that of water. It does not occur free in nature; combined it is found in small amounts in nearly all igneous rocks and in the waters of many mineral springs. *Lepidolite, spodumene, petalite,* and *amblygonite* are the more important minerals containing it. Lithium is presently being recovered from brines of Searles Lake, in California, and from Nevada, Chile, and Argentina. Large deposits of spodumene are found in North Carolina. The metal is produced electrolytically from the fused chloride. Lithium is silvery in appearance, much like Na and K, other members of the alkali metal series. It reacts with water, but not as vigorously as sodium. Lithium imparts a beautiful crimson color to a flame, but when the metal burns strongly the flame is a dazzling white. Since World War II, the production of lithium metal and its compounds has increased greatly. Because the metal has the highest specific heat of any solid element, it has found use in heat transfer applications; however, it is corrosive and requires special handling. The metal has been used as an alloying agent, is of interest in synthesis of organic compounds, and has nuclear applications. It ranks as a leading contender as a battery anode material as it has a high electrochemical potential. Lithium is used in special glasses and ceramics. The glass for the 200-inch telescope at Mt. Palomar contains lithium as a minor ingredient. Lithium chloride is one of the most hygroscopic materials known, and it, as well as lithium bromide, is used in air conditioning and industrial drying systems. Lithium stearate is used as an all-purpose and high-temperature lubricant. Other lithium compounds are used in dry cells and storage batteries. Seven isotopes of lithium are recognized. Natural lithium contains two isotopes. The metal is priced at about $1.50/g (99.9%).

Lutetium — (Lutetia, ancient name for Paris, sometimes called *cassiopeium* by the Germans), Lu; at. wt. 174.967(1); at. no. 71; m.p. 1663°C; b.p. 3402°C; sp. gr. 9.841 (25°C); valence 3. In 1907, Urbain described a process by which Marignac's ytterbium (1879) could be separated into the two elements, ytterbium (neoytterbium) and lutetium. These elements were identical with "aldebaranium" and "cassiopeium," independently discovered by von Welsbach about the same time. Charles James of the University of New Hampshire also independently prepared the very pure oxide, *lutecia*, at this time. The spelling of the element was changed from *lutecium* to *lutetium* in 1949. Lutetium occurs in very small amounts in nearly all minerals containing yttrium, and is present in *monazite* to the extent of about 0.003%, which is a commercial source. The pure metal has been isolated only in recent years and is one of the most difficult to prepare. It can be prepared by the reduction of anhydrous $LuCl_3$ or LuF_3 by an alkali or alkaline earth metal. The metal is silvery white and relatively stable in air. While new techniques, including ion-exchange reactions, have been developed to separate the various rare-earth elements, lutetium is still the most costly of all rare earths. It is priced at about $100/g (99.9%). ^{176}Lu occurs naturally (97.41%) with ^{175}Lu (2.59%), which is radioactive with a very long half-life of about 4×10^{10} years. Lutetium has 50 isotopes and isomers that are now recognized. Stable lutetium nuclides, which emit pure beta radiation after thermal neutron activation, can be used as catalysts in cracking, alkylation, hydrogenation, and polymerization. Virtually no other commercial uses have been found yet for lutetium. While lutetium, like other rare-earth metals, is thought to have a low toxicity rating, it should be handled with care until more information is available.

Magnesium — (*Magnesia*, district in Thessaly) Mg; at. wt. 24.3050(6); at. no. 12; m.p. 650°C; b.p. 1090°C; sp. gr. 1.738 (20°C); valence 2. Compounds of magnesium have long been known. Black recognized magnesium as an element in 1755. It was isolated by Davy in 1808, and prepared in coherent form by Bussy in 1831. Magnesium is the eighth most abundant element in the earth's crust. It does not occur uncombined, but is found in large deposits in the form of *magnesite, dolomite*, and other minerals. The metal is now principally obtained in the U.S. by electrolysis of fused magnesium chloride derived from brines, wells, and sea water. Magnesium is a light, silvery-white, and fairly tough metal. It tarnishes slightly in air, and finely divided magnesium readily ignites upon heating in air and burns with a dazzling white flame. It is used in flashlight photography, flares, and pyrotechnics, including incendiary bombs. It is one third lighter than aluminum, and in alloys is essential for airplane and missile construction. The metal improves the mechanical, fabrication, and welding characteristics of aluminum when used as an alloying agent. Magnesium is used in producing nodular graphite in cast iron, and is used as an additive to conventional propellants. It is also used as a reducing agent in the production of pure uranium and other metals from their salts. The hydroxide (*milk of magnesia*), chloride, sulfate (*Epsom salts*), and citrate are used in medicine. Dead-burned magnesite is employed for refractory purposes such as brick and liners in furnaces and converters. Calcined magnesia is also used for water treatment and in the manufacture of rubber, paper, etc. Organic magnesium compounds (Grignard's reagents) are important. Magnesium is an important element in both plant and animal life. Chlorophylls are magnesium-centered porphyrins. The adult daily requirement of magnesium is about 300 mg/day, but this is affected by various factors. Great care should be taken in handling magnesium metal, especially in the finely divided state, as serious fires can occur. Water should not be used on burning magnesium or on magnesium fires. Natural magnesium contains three isotopes. Twelve other isotopes are recognized. Magnesium metal costs about $100/kg (99.8%).

Manganese — (L. *magnes*, magnet, from magnetic properties of pyrolusite; It. *manganese*, corrupt form of *magnesia*), Mn; at. wt. 54.938049(9); at. no. 25; m.p. 1246°C; b.p. 2061°C; sp. gr. 7.21 to 7.44, depending on allotropic form; valence 1, 2, 3, 4, 6, or 7. Recognized by Scheele, Bergman, and others as an element and isolated by Gahn in 1774 by reduction of the dioxide with carbon. Manganese minerals are widely distributed; oxides, silicates, and carbonates are the most common. The discovery of large quantities of manganese nodules on the floor of the oceans holds promise as a source of manganese. These nodules contain about 24% manganese together with many other elements in lesser abundance. Most manganese today is obtained from ores found in Ukraine, Brazil, Australia, Republic of So. Africa, Gabon, China, and India. *Pyrolusite* (MnO_2) and *rhodochrosite* ($MnCO_3$) are among the most common manganese minerals. The metal is obtained by reduction of the oxide with sodium, magnesium, aluminum, or by electrolysis. It is gray-white, resembling iron, but is harder and very brittle. The metal is reactive chemically, and decomposes cold water slowly. Manganese is used to form many important alloys. In steel, manganese improves the rolling and forging qualities, strength, toughness, stiffness, wear resistance, hardness, and hardenability. With aluminum and antimony, especially with small amounts of copper, it forms highly ferromagnetic alloys. Manganese metal is ferromagnetic only after special treatment. The pure metal exists in four allotropic forms. The alpha form is stable at ordinary temperature; gamma manganese, which changes to alpha at ordinary temperatures, is soft, easily cut, and capable of being bent. The dioxide (pyrolusite) is used as a depolarizer in dry cells, and is used to "decolorize" glass that is colored green by impurities of iron. Manganese by itself colors glass an amethyst color, and is responsible for the color of true amethyst. The dioxide is also used in the preparation of oxygen and chlorine, and in drying black paints. The permanganate is a powerful oxidizing agent and is used in quantitative analysis and in medicine. Manganese is widely distributed throughout the animal kingdom. It is an important trace element and may be essential for utilization of vitamin B_1. Twenty-seven isotopes and isomers are known. Manganese metal (99.95%) is priced at about $800/kg. Metal of 99.6% purity is priced at about $80/kg.

Meitnerium — (Lise Meitner [1878–1968], Austrian–Swedish physicist and mathematician), Mt; at. wt [266]; at. no. 109. On August 29, 1992, Element 109 was made and identified by physicists at the Heavy Ion Research Laboratory (G.S.I.), Darmstadt, Germany, by bombarding a target of ^{209}Bi with accelerated nuclei of ^{58}Fe. The production of Element 109 has been extremely small. It took a week of target bombardment (10^{11} nuclear encounters) to produce a single atom of 109. Oganessian and his team at Dubna in 1994 repeated the Darmstadt experiment using a tenfold irradiation dose. One fission event from seven alpha decays of 109 was observed, thus indirectly confirming the existence of isotope ^{266}109. In August 1997, the IUPAC adopted the name *meitnerium* for this element, honoring L. Meitner. Four isotopes of *meitnerium* are now recognized.

Mendelevium — (Dmitri Mendeleev [1834–1907]), Md; at. wt. (258); at. no. 101; m.p. 827°C; valence +2, +3. Mendelevium, the ninth transuranium element of the actinide series to be discovered, was first identified by Ghiorso, Harvey, Choppin, Thompson, and Seaborg early in 1955 as a result of the bombardment of the isotope ^{253}Es with helium ions in the Berkeley 60-inch cyclotron. The isotope produced was ^{256}Md, which has a half-life of 78 min. This first identification was notable in that ^{256}Md was synthesized on a one-atom-at-a-time basis. Nineteen isotopes and isomers are now recognized. ^{258}Md has a half-life of 51.5 days. This isotope has been produced by the bombardment of an isotope of einsteinium with ions of helium. It now appears possible that eventually enough ^{258}Md can be made so that some of its physical properties can be determined. ^{256}Md has been used to elucidate some of the chemical properties of mendelevium in aqueous solution. Experiments seem to show that the element possesses a moderately stable dipositive (II) oxidation state in addition to the tripositive (III) oxidation state, which is characteristic of actinide elements.

Mercury — (Planet *Mercury*), Hg (*hydrargyrum*, liquid silver); at. wt. 200.59(2); at. no. 80; t.p. −38.83°C; b.p. 356.73°C; t_c 1447°C; sp. gr. 13.546 (20°C); valence 1 or 2. Known to ancient Chinese and Hindus; found in Egyptian tombs of 1500 B.C. Mercury is the only common metal liquid at ordinary temperatures. It only rarely occurs free in nature. The chief ore is *cinnabar* (HgS). Spain and China produce about 75% of the world's supply of the metal. The commercial unit for handling mercury is the "flask," which weighs 76 lb (34.46 kg). The metal is obtained by heating cinnabar in a current of air and by condensing the vapor. It is a heavy, silvery-white metal; a rather poor conductor of heat, as compared with other metals, and a fair conductor of electricity. It easily forms alloys with many metals, such as gold, silver, and tin, which are called *amalgams*. Its ease in amalgamating with gold is made use of in the recovery of gold from its ores. The metal is widely used in laboratory work for making thermometers, barometers, diffusion pumps, and many other instruments. It is used in making mercury-vapor lamps and advertising signs, etc. and is used in mercury switches and other electrical apparatus. Other uses are in making pesticides, mercury cells for caustic soda and chlorine production, dental preparations, antifouling paint, batteries, and catalysts. The most important salts are mercuric chloride $HgCl_2$ (corrosive sublimate — a violent poison), mercurous chloride Hg_2Cl_2 (calomel, occasionally still used in medicine), mercury fulminate ($Hg(ONC)_2$), a detonator widely used in explosives, and mercuric sulfide (HgS, vermillion, a high-grade paint pigment). Organic mercury compounds are important. It has been found that an electrical discharge causes mercury vapor to combine with neon, argon, krypton, and xenon. These products, held together with van der Waals' forces, correspond to HgNe, HgAr, HgKr, and HgXe. Mercury is a virulent poison and is readily absorbed through the respiratory tract, the gastrointestinal tract, or through unbroken skin. It acts as a cumulative poison and dangerous levels are readily attained in air. Air saturated with mercury vapor at 20°C contains a concentration that exceeds the toxic limit many times. The danger increases at higher temperatures. *It is therefore important that mercury be handled with care.* Containers of mercury should be securely covered and spillage should be avoided. If it is necessary to heat mercury or mercury compounds, it should be done in a well-ventilated hood. Methyl mercury is a dangerous pollutant and is now widely found in water and streams. The triple point of mercury, −38.8344°C, is a fixed point on the International Temperature Scale (ITS-90). Mercury (99.98%) is priced at about $110/kg. Native mercury contains seven isotopes. Thirty-six other isotopes and isomers are known.

Molybdenum — (Gr. *molybdos*, lead), Mo; at. wt. 95.94(1); at. no. 42; m.p. 2623°C; b.p. 4639°C; sp. gr. 10.22 (20°C); valence 2, 3, 4?, 5?, or 6. Before Scheele recognized molybdenite as a distinct ore of a new element in 1778, it was confused with graphite and lead ore. The metal was prepared in an impure form in 1782 by Hjelm. Molybdenum does not occur native, but is obtained principally from *molybdenite* (MoS_2). *Wulfenite* ($PbMoO_4$) and *powellite* ($Ca(MoW)O_4$) are also minor commercial ores. Molybdenum is also recovered as a by-product of copper and tungsten mining operations. The U.S., Canada, Chile, and China produce most of the world's molybdenum ores. The metal is prepared from the powder made by the hydrogen reduction of purified molybdic trioxide or ammonium molybdate. The metal is silvery white, very hard, but is softer and more ductile than tungsten. It has a high elastic modulus, and only tungsten and tantalum, of the more readily available metals, have higher melting points. It is a valuable alloying agent, as it contributes to the hardenability and toughness of quenched and tempered steels. It also improves the strength of steel at high temperatures. It is used in certain nickel-based alloys, such as the Hastelloys® which are heat-resistant and corrosion-resistant to chemical solutions. Molybdenum oxidizes at elevated temperatures. The metal has found recent application as electrodes for electrically heated glass furnaces and forehearths. It is also used in nuclear energy applications and for missile and aircraft parts. Molybdenum is valuable as a catalyst in the refining of petroleum. It has found application as a filament material in electronic and electrical applications. Molybdenum is an essential trace element in plant nutrition. Some lands are barren for lack of this element in the soil. Molybdenum sulfide is useful as a lubricant, especially at high temperatures where oils would decompose. Almost all ultra-high strength steels with minimum yield points up to 300,000 lb/in.^{2} contain molybdenum in amounts from 0.25 to 8%. Natural molybdenum contains seven isotopes. Thirty other isotopes and isomers are known, all of which are radioactive. Molybdenum metal costs about $1/g (99.999% purity). Molybdenum metal (99.9%) costs about $160/kg.

Neodymium — (Gr. *neos*, new, and *didymos*, twin), Nd; at. wt. 144.24(3); at. no. 60; m.p. 1021°C; b.p. 3074°C; sp. gr. 7.008 (25°C); valence 3. In 1841 Mosander extracted from *cerite* a new rose-colored oxide, which he believed contained a new element. He named the element *didymium*, as it was *an inseparable twin brother of lanthanum*. In 1885 von Welsbach separated didymium into two new elemental components, *neodymia* and *praseodymia*, by repeated fractionation of ammonium didymium nitrate. While the free metal is in *misch metal*, long known and used as a pyrophoric alloy for light flints, the element was not isolated in relatively pure form until 1925. Neodymium is present in misch metal to the extent of about 18%. It is present in the minerals *monazite* and *bastnasite*, which are principal sources of rare-earth metals. The element may be obtained by separating neodymium salts from other rare earths by ion-exchange or solvent extraction techniques, and by reducing anhydrous halides such as NdF$_3$ with calcium metal. Other separation techniques are possible. The metal has a bright silvery metallic luster. Neodymium is one of the more reactive rare-earth metals and quickly tarnishes in air, forming an oxide that splits off and exposes metal to oxidation. The metal, therefore, should be kept under light mineral oil or sealed in a plastic material. Neodymium exists in two allotropic forms, with a transformation from a double hexagonal to a body-centered cubic structure taking place at 863°C. Natural neodymium is a mixture of seven isotopes, one of which has a very long half-life. Twenty seven other radioactive isotopes and isomers are recognized. Didymium, of which neodymium is a component, is used for coloring glass to make welder's goggles. By itself, neodymium colors glass delicate shades ranging from pure violet through wine-red and warm gray. Light transmitted through such glass shows unusually sharp absorption bands. The glass has been used in astronomical work to produce sharp bands by which spectral lines may be calibrated. Glass containing neodymium can be used as a laser material to produce coherent light. Neodymium salts are also used as a colorant for enamels. The element is also being used with iron and boron to produce extremely strong magnets. These are the most compact magnets commercially available. The price of the metal is about $4/g. Neodymium has a low-to-moderate acute toxic rating. As with other rare earths, neodymium should be handled with care.

Neon — (Gr. *neos*, new), Ne; at. wt. 20.1797(6); at. no. 10; t.p. −248.59°C; b.p. −246.08°C; t_c −228.7°C (1 atm); density of gas 0.89990 g/l (1 atm, 0°C); density of liquid at b.p. 1.207 g/cm^3; valence 0. Discovered by Ramsay and Travers in 1898. Neon is a rare gaseous element present in the atmosphere to the extent of 1 part in 65,000 of air. It is obtained by liquefaction of air and separated from the other gases by fractional distillation. Natural neon is a mixture of three isotopes. Fourteen other unstable isotopes are known. It is very inert element; however, it is said to form a compound with fluorine. It is still questionable if true compounds of neon exist, but evidence is mounting in favor of their existence. The following ions are known from optical and mass spectrometric studies: Ne$^+$,

(NeAr)$^+$, (NeH)$^+$, and (HeNe$^+$). Neon also forms an unstable hydrate. In a vacuum discharge tube, neon glows reddish orange. Of all the rare gases, the discharge of neon is the most intense at ordinary voltages and currents. Neon is used in making the common neon advertising signs, which accounts for its largest use. It is also used to make high-voltage indicators, lightning arrestors, wave meter tubes, and TV tubes. Neon and helium are used in making gas lasers. Liquid neon is now commercially available and is finding important application as an economical cryogenic refrigerant. It has over 40 times more refrigerating capacity per unit volume than liquid helium and more than three times that of liquid hydrogen. It is compact, inert, and is less expensive than helium when it meets refrigeration requirements. Neon costs about $800/80 cu. ft. (2265 l).

Neptunium — (Planet *Neptune*), Np; at. wt. (237); at. no. 93; m.p. 644°C; sp. gr. 20.25 (20°C); valence 3, 4, 5, and 6. Neptunium was the first synthetic transuranium element of the actinide series discovered; the isotope ^{239}Np was produced by McMillan and Abelson in 1940 at Berkeley, California, as the result of bombarding uranium with cyclotron-produced neutrons. The isotope ^{237}Np (half-life of 2.14×10^6 years) is currently obtained in gram quantities as a by-product from nuclear reactors in the production of plutonium. Twenty-three isotopes and isomers of neptunium are now recognized. Trace quantities of the element are actually found in nature due to transmutation reactions in uranium ores produced by the neutrons which are present. Neptunium is prepared by the reduction of NpF$_3$ with barium or lithium vapor at about 1200°C. Neptunium metal has a silvery appearance, is chemically reactive, and exists in at least three structural modifications: α-neptunium, orthorhombic, density 20.25 g/cm^3, β-neptunium (above 280°C), tetragonal, density (313°C) 19.36 g/cm^3; γ-neptunium (above 577°C), cubic, density (600°C) 18.0 g/cm^3. Neptunium has four ionic oxidation states in solution: Np^{+3} (pale purple), analogous to the rare earth ion Pm^{+3}, Np^{+4} (yellow green); NpO$^+$ (green blue); and NpO^{++} (pale pink). These latter oxygenated species are in contrast to the rare earths which exhibit only simple ions of the (II), (III), and (IV) oxidation states in aqueous solution. The element forms tri- and tetrahalides such as NpF$_3$, NpF$_4$, NpCl$_4$, NpBr$_3$, NpI$_3$, and oxides of various compositions such as are found in the uranium-oxygen system, including Np$_3$O$_8$ and NpO$_2$.

Nickel — (Ger. *Nickel*, Satan or Old Nick's and from *kupfernickel*, Old Nick's copper), Ni; at. wt. 58.6934(2); at. no. 28; m.p. 1455°C; b.p. 2913°C; sp. gr. 8.902 (25°C); valence 0, 1, 2, 3. Discovered by Cronstedt in 1751 in kupfernickel (*niccolite*). Nickel is found as a constituent in most meteorites and often serves as one of the criteria for distinguishing a meteorite from other minerals. Iron meteorites, or *siderites*, may contain iron alloyed with from 5 to nearly 20% nickel. Nickel is obtained commercially from *pentlandite* and *pyrrhotite* of the Sudbury region of Ontario, a district that produces much of the world's nickel. It is now thought that the Sudbury deposit is the result of an ancient meteorite impact. Large deposits of nickel, cobalt, and copper have recently been developed at Voisey's Bay, Labrador. Other deposits of nickel are found in Russia, New Caledonia, Australia, Cuba, Indonesia, and elsewhere. Nickel is silvery white and takes on a high polish. It is hard, malleable, ductile, somewhat ferromagnetic, and a fair conductor of heat and electricity. It belongs to the iron-cobalt group of metals and is chiefly valuable for the alloys it forms. It is extensively used for making stainless steel and other corrosion-resistant alloys such as Invar®, Monel®, Inconel®, and the Hastelloys®. Tubing made of a copper-nickel alloy is extensively used in making desalination plants for converting sea water into fresh water. Nickel is also now used extensively in coinage and in making nickel steel for armor plate and burglar-proof vaults, and is a component in Nichrome®, Permalloy®, and constantan. Nickel added to glass gives a green color. Nickel plating is often used to provide a protective coating for other metals, and finely divided nickel is a catalyst for hydrogenating vegetable oils. It is also used in ceramics, in the manufacture of Alnico magnets, and in batteries. The sulfate and the oxides are important compounds. Natural nickel is a mixture of five stable isotopes; twenty-five other unstable isotopes are known. Nickel sulfide fume and dust, as well as other nickel compounds, are carcinogens. Nickel metal (99.9%) is priced at about $2/g or less in larger quantities.

Niobium — (*Niobe*, daughter of Tantalus), Nb; or Columbium (*Columbia*, name for America); at. wt. 92.90638(2); at. no. 41; m.p. 2477°C; b.p. 4744°C, sp. gr. 8.57 (20°C); valence 2, 3, 4?, 5. Discovered in 1801 by Hatchett in an ore sent to England more that a century before by John Winthrop the Younger, first governor of Connecticut. The metal was first prepared in 1864 by Blomstrand, who reduced the chloride by heating it in a hydrogen atmosphere. The name *niobium* was adopted by the International Union of Pure and Applied Chemistry in 1950 after 100 years of controversy. Most leading chemical societies and government organizations refer to it by this name. Some metallurgists and commercial producers, however, still refer to the metal as "columbium." The element is found in *niobite* (or *columbite*), *niobite-tantalite*, *pyrochlore*, and *euxenite*. Large deposits of niobium have been found associated with *carbonatites* (carbon-silicate rocks), as a constituent of *pyrochlore*. Extensive ore reserves are found in Canada, Brazil, Congo-Kinshasa, Rwanda, and Australia. The metal can be isolated from tantalum, and prepared in several ways. It is a shiny, white, soft, and ductile metal, and takes on a bluish cast when exposed to air at room temperatures for a long time. The metal starts to oxidize in air at 200°C, and when processed at even moderate temperatures must be placed in a protective atmosphere. It is used in arc-welding rods for stabilized grades of stainless steel. Thousands of pounds of niobium have been used in advanced air frame systems such as were used in the Gemini space program. It has also found use in super-alloys for applications such as jet engine components, rocket subassemblies, and heat-resisting equipment. The element has superconductive properties; superconductive magnets have been made with Nb-Zr wire, which retains its superconductivity in strong magnetic fields. Natural niobium is composed of only one isotope, ^{93}Nb. Forty-seven other isotopes and isomers of niobium are now recognized. Niobium metal (99.9% pure) is priced at about 50¢/g.

Nitrogen — (L. *nitrum*, Gr. *nitron*, native soda; genes, *forming*, N; at. wt. 14.00674(7); at. no. 7; m.p. −210.00°C; b.p. −198.79°C; t_c −146.94°C; density 1.2506 g/l; sp. gr. liquid 0.808 (−195.8°C), solid 1.026 (−252°C); valence 3 or 5. Discovered by Daniel Rutherford in 1772, but Scheele, Cavendish, Priestley, and others about the same time studied "burnt or dephlogisticated air," as air without oxygen was then called. Nitrogen makes up 78% of the air, by volume. The atmosphere of Mars, by comparison, is 2.6% nitrogen. The estimated amount of this element in our atmosphere is more than 4000 trillion tons. From this inexhaustible source it can be obtained by liquefaction and fractional distillation. Nitrogen molecules give the orange-red, blue-green, blue-violet, and deep violet shades to the aurora. The element is so inert that Lavoisier named it *azote*, meaning without life, yet its compounds are so active as to be most important in foods, poisons, fertilizers, and explosives. Nitrogen can be also easily prepared by heating a water solution of ammonium nitrite. Nitrogen, as a gas, is colorless, odorless, and a generally inert element. As a liquid it is also colorless and odorless, and is similar in appearance to water. Two allotropic forms of solid nitrogen exist, with the transition from the α to the β form taking place at −237°C. When nitrogen is heated, it combines directly with magnesium, lithium, or calcium; when mixed with oxygen and subjected to electric sparks, it forms first nitric oxide (NO) and then the dioxide (NO_2); when heated under pressure with a catalyst with hydrogen, ammonia is formed (Haber process). The ammonia thus formed is of the utmost importance as it is used in fertilizers, and it can be oxidized to nitric acid (Ostwald process). The ammonia industry is the largest consumer of nitrogen. Large amounts of gas are also used by the electronics industry, which uses the gas as a blanketing medium during production of such components as transistors, diodes, etc. Large quantities of nitrogen are used in annealing stainless steel and other steel mill products. The drug industry also uses large quantities. Nitrogen is used as a refrigerant both for the immersion freezing of food products and for transportation of foods. Liquid nitrogen is also used in missile work as a purge for components, insulators for space chambers, etc., and by the oil industry to build up great pressures in wells to force crude oil upward. Sodium and potassium nitrates are formed by the decomposition of organic matter with compounds of the metals present. In certain dry areas of the world these saltpeters are found in quantity. Ammonia, nitric acid, the nitrates, the five oxides (N_2O, NO, N_2O_3, NO_2, and N_2O_5), TNT, the cyanides, etc. are but a few of the important compounds. Nitrogen gas prices vary from 2¢ to $2.75 per 100 ft^3 (2.83 cu. meters), depending on purity, etc. Production of elemental nitrogen in the U.S. is more than 9 million short tons per year. Natural nitrogen contains two isotopes, ^{14}N and ^{15}N. Ten other isotopes are known.

Nobelium — (Alfred Nobel [1833–1896], inventor of dynamite), No; at. wt. [259]; at. no. 102; valence +2, +3. Nobelium was unambiguously discovered and identified in April 1958 at Berkeley by A. Ghiorso, T. Sikkeland, J. R. Walton, and G. T. Seaborg, who used a new double-recoil technique. A heavy-ion linear accelerator (HILAC) was used to bombard a thin target of curium (95% ^{244}Cm and 4.5% ^{246}Cm) with ^{12}C ions to produce 102^{254} according to the ^{246}Cm (^{12}C, 4n) reaction. Earlier in 1957 workers of the U.S., Britain, and Sweden announced the discovery of an isotope of Element 102 with a 10-min half-life at 8.5 MeV, as a result of bombarding ^{244}Cm with ^{13}C nuclei. On the basis of this experiment the name *nobelium* was assigned and accepted by the Commission on Atomic Weights of the International Union of Pure and Applied Chemistry. The acceptance of the name was premature, for both Russian and American efforts now completely rule out the possibility of any isotope of Element 102 having a half-life of 10 min in the vicinity of 8.5 MeV. Early work in 1957 on the search for this element, in Russia at the Kurchatov Institute, was marred by the assignment of 8.9 ± 0.4 MeV alpha radiation with a half-life of 2 to 40 sec, which was too indefinite to support claim to discovery. Confirmatory experiments at Berkeley in 1966 have shown the existence of ^{254}102 with a 55-s half-life, ^{252}102 with a 2.3-s half-life, and $^{257}102$ with a 25-s half-life. Twelve isotopes are now recognized, one of which — $^{255}102$ — has a half-life of 3.1 min. In view of the discover's traditional right to name an element, the Berkeley group, in 1967, suggested that the hastily given name *nobelium*, along with the symbol No, be retained.

Osmium — (Gr. *osme*, a smell), Os; at. wt. 190.23(3); at. no. 76; m.p. 3033°C; b.p. 5012°C; sp. gr. 22.57; valence 0 to +8, more usually +3, +4, +6, and +8. Discovered in 1803 by Tennant in the residue left when crude platinum is dissolved by *aqua regia*. Osmium occurs in *iridosmine* and in platinum-bearing river sands of the Urals, North America, and South America. It is also found in the nickel-bearing ores of Sudbury, Ontario, region along with other platinum metals. While the quantity of platinum metals in these ores is very small, the large tonnages of nickel ores processed make commercial recovery possible. The metal is lustrous, bluish white, extremely hard, and brittle even at high temperatures. It has the highest melting point and the lowest vapor pressure of the platinum group. The metal is very difficult to fabricate, but the powder can be sintered in a hydrogen atmosphere at a temperature of 2000°C. The solid metal is not affected by air at room temperature, but the powdered or spongy metal slowly gives off osmium tetroxide, which is a powerful oxidizing agent and has a strong smell. The tetroxide is highly toxic, and boils at 130°C (760 mm). Concentrations in air as low as 10^{-7} g/m^3 can cause lung congestion, skin damage, or eye damage. The tetroxide has been used to detect fingerprints and to stain fatty tissue for microscope slides. The metal is almost entirely used to produce very hard alloys, with other metals of the platinum group, for fountain pen tips, instrument pivots, phonograph needles, and electrical contacts. The price of 99.9% pure osmium powder — the form usually supplied commercially — is about $100/g, depending on quantity and supplier. Natural osmium contains seven isotopes, one of which, ^{186}Os, is radioactive with a very long half-life. Thirty four other isotopes and isomers are known, all of which are radioactive.The measured densities of iridium and osmium seem to indicate that osmium is slightly more dense than iridium, so osmium has generally been credited with being the heaviest known element. Calculations of the density from the space lattice, which may be more reliable for these elements than actual measurements, however, give a density of 22.65 for iridium compared to 22.61 for osmium. At present, therefore, we know either iridium or osmium is the heaviest element, but the data do not allow selection between the two.

Oxygen — (Gr. *oxys*, sharp, acid, and *genes*, forming; acid former), O; at. wt. 15.9994(3); at. no. 8; t.p. –218.79°C; t_c –118.56°C; valence 2. For many centuries, workers occasionally realized air was composed of more than one component. The behavior of oxygen and nitrogen as components of air led to the advancement of the phlogiston theory of combustion, which captured the minds of chemists for a century. Oxygen was prepared by several workers, including Bayen and Borch, but they did not know how to collect it, did not study its properties, and did not recognize it as an elementary substance. Priestley is generally credited with its discovery, although Scheele also discovered it independently. Oxygen is the third most abundant element found in the sun, and it plays a part in the carbon-nitrogen cycle, one process thought to give the sun and stars their energy. Oxygen under excited conditions is responsible for the bright red and yellow-green colors of the aurora. Oxygen, as a gaseous element, forms 21% of the atmosphere by volume from which it can be obtained by liquefaction and fractional distillation. The atmosphere of Mars contains about 0.15% oxygen. The element and its compounds make up 49.2%, by weight, of the earth's crust. About two thirds of the human body and nine tenths of water is oxygen. In the laboratory it can be prepared by the electrolysis of water or by heating potassium chlorate with manganese dioxide as a catalyst. The gas is colorless, odorless, and tasteless. The liquid and solid forms are a pale blue color and are strongly paramagnetic. Ozone (O_3), a highly active compound, is formed by the action of an electrical discharge or ultraviolet light on oxygen. Ozone's presence in the atmosphere (amounting to the equivalent of a layer 3 mm thick at ordinary pressures and temperatures) is of vital importance in preventing harmful ultraviolet rays of the sun from reaching the earth's surface. There has been recent concern that pollutants in the atmosphere may have a detrimental effect on this ozone layer. Ozone is toxic and exposure should not exceed 0.2 mg/m^3 (8-hour time-weighted average — 40-hour work week). Undiluted ozone has a bluish color. Liquid ozone is bluish black, and solid ozone is violet-black. Oxygen is very reactive and capable of combining with most elements. It is a component of hundreds of thousands of organic compounds. It is essential for respiration of all plants and animals and for practically all combustion. In hospitals it is frequently used to aid respiration of patients. Its atomic weight was used as a standard of comparison for each of the other elements until 1961 when the International Union of Pure and Applied Chemistry adopted carbon 12 as the new basis. Oxygen has thirteen recognized isotopes. Natural oxygen is a mixture of three isotopes. Oxygen 18 occurs naturally, is stable, and is available commercially. Water (H_2O with 1.5% ^{18}O) is also available. Commercial oxygen consumption in the U.S. is estimated to be 20 million short tons per year and the demand is expected to increase substantially in the next few years. Oxygen enrichment of steel blast furnaces accounts for the greatest use of the gas. Large quantities are also used in making synthesis gas for ammonia and methanol, ethylene oxide, and for oxy-acetylene welding. Air separation plants produce about 99% of the gas, electrolysis plants about 1%. The gas costs 5¢/ft^3 ($1.75/cu. meter) in small quantities.

Palladium — (named after the asteroid *Pallas*, discovered about the same time; Gr. *Pallas*, goddess of wisdom), Pd; at. wt. 106.42(1) at. no. 46; m.p. 1554.9°C; b.p. 2963°C; sp. gr. 12.02 (20°C); valence 2, 3, or 4. Discovered in 1803 by Wollaston. Palladium is found along with platinum and other metals of the platinum group in deposits of Russia, South Africa, Canada (Ontario), and elsewhere. Natural palladium contains six stable isotopes. Twenty-nine other isotopes are recognized, all of which are radioactive. It is frequently found associated with the nickel-copper deposits such as those found in Ontario. Its separation from the platinum metals depends upon the type of ore in which it is found. It is a steel-white metal, does not tarnish in air, and is the least dense and lowest melting of the platinum group of metals. When annealed, it is soft and ductile; cold working greatly increases its strength and hardness. Palladium is attacked by nitric and sulfuric acid. At room temperatures the metal has the unusual property of absorbing up to 900 times its own volume of hydrogen, possibly forming Pd_2H. It is not yet clear if this a true compound. Hydrogen readily diffuses through heated palladium and this provides a means of purifying the gas. Finely divided palladium is a good catalyst and is used for hydrogenation and dehydrogenation reactions. It is alloyed and used in jewelry trades. White gold is an alloy of gold decolorized by the addition of palladium. Like gold, palladium can be beaten into leaf as thin as 1/250,000 in. The metal is used in dentistry, watchmaking, and in making surgical instruments and electrical contacts. Palladium recently has been substituted for higher priced platinum in catalytic converters by some automobile companies. This has caused a large increase in the cost of palladium. The price of the two metals are now, in 2002, about the same. Palladium, however, is less resistant to poisoning by sulfur and lead, than platinum, but it may prove useful in controlling emissions from diesel vehicles. The metal sells for about $350/tr. oz. ($11/g).

Phosphorus — (Gr. *phosphoros*, light bearing; ancient name for the planet Venus when appearing before sunrise), P; at. wt. 30.973762(4); at. no. 15; m.p. (white) 44.15°C; b.p. 280.5°C; sp. gr. (white) 1.82, (red) 2.20, (black) 2.25 to 2.69; valence 3 or 5. Discovered in 1669 by Brand, who prepared it from urine. Phosphorus exists in four or more allotropic forms: white (or yellow), red, and black (or violet). White phosphorus has two modifications: α and β with a transition temperature at −3.8°C. Never found free in nature, it is widely distributed in combination with minerals. Twenty-one isotopes of phosphorus are recognized. *Phosphate* rock, which contains the mineral *apatite*, an impure tricalcium phosphate, is an important source of the element. Large deposits are found in the Russia, China, Morocco, and in Florida, Tennessee, Utah, Idaho, and elsewhere. Phosphorus in an essential ingredient of all cell protoplasm, nervous tissue, and bones. Ordinary phosphorus is a waxy white solid; when pure it is colorless and transparent. It is insoluble in water, but soluble in carbon disulfide. It takes fire spontaneously in air, burning to the pentoxide. It is very poisonous, 50 mg constituting an approximate fatal dose. Exposure to white phosphorus should not exceed 0.1 mg/m^3 (8-hour time-weighted average — 40-hour work week). White phosphorus should be kept under water, as it is dangerously reactive in air, and it should be handled with forceps, as contact with the skin may cause severe burns. When exposed to sunlight or when heated in its own vapor to 250°C, it is converted to the red variety, which does not phosphoresce in air as does the white variety. This form does not ignite spontaneously and it is not as dangerous as white phosphorus. It should, however, be handled with care as it does convert to the white form at some temperatures and it emits highly toxic fumes of the oxides of phosphorus when heated. The red modification is fairly stable, sublimes with a vapor pressure of 1 atm at 417°C, and is used in the manufacture of safety matches, pyrotechnics, pesticides, incendiary shells, smoke bombs, tracer bullets, etc. White phosphorus may be made by several methods. By one process, tricalcium phosphate, the essential ingredient of phosphate rock, is heated in the presence of carbon and silica in an electric furnace or fuel-fired furnace. Elementary phosphorus is liberated as vapor and may be collected under water. If desired, the phosphorus vapor and carbon monoxide produced by the reaction can be oxidized at once in the presence of moisture to produce phosphoric acid, an important compound in making super-phosphate fertilizers. In recent years, concentrated phosphoric acids, which may contain as much as 70 to 75% P$_2$O$_5$ content, have become of great importance to agriculture and farm production. World-wide demand for fertilizers has caused record phosphate production. Phosphates are used in the production of special glasses, such as those used for sodium lamps. Bone-ash, calcium phosphate, is also used to produce fine chinaware and to produce monocalcium phosphate used in baking powder. Phosphorus is also important in the production of steels, phosphor bronze, and many other products. Trisodium phosphate is important as a cleaning agent, as a water softener, and for preventing boiler scale and corrosion of pipes and boiler tubes. Organic compounds of phosphorus are important. Amorphous (red) phosphorus costs about $70/kg (99%).

Platinum — (It. *platina*, silver), Pt; at. wt. 195.078(2); at. no. 78; m.p. 1768.4°C; b.p. 3825°C; sp. gr. 21.45 (20°C); valence 1?, 2, 3, or 4. Discovered in South America by Ulloa in 1735 and by Wood in 1741. The metal was used by pre-Columbian Indians. Platinum occurs native, accompanied by small quantities of iridium, osmium, palladium, ruthenium, and rhodium, all belonging to the same group of metals. These are found in the alluvial deposits of the Ural mountains and in Columbia. *Sperrylite* (PtAs$_2$), occurring with the nickel-bearing deposits of Sudbury, Ontario, is a source of a considerable amount of metal. The large production of nickel offsets there being only one part of the platinum metals in two million parts of ore. The largest supplier of the platinum group of metals is now South Africa, followed by Russia and Canada. Platinum is a beautiful silvery-white metal, when pure, and is malleable and ductile. It has a coefficient of expansion almost equal to that of soda-lime-silica glass, and is therefore used to make sealed electrodes in glass systems. The metal does not oxidize in air at any temperature, but is corroded by halogens, cyanides, sulfur, and caustic alkalis. It is insoluble in hydrochloric and nitric acid, but dissolves when they are mixed as *aqua regia*, forming chloroplatinic acid (H$_2$PtCl$_6$), an important compound. Natural platinum contains six isotopes, one of which, ^{190}Pt, is radioactive with a long half-life. Thirty-seven other radioactive isotopes and isomers are recognized. The metal is extensively used in jewelry, wire, and vessels for laboratory use, and in many valuable instruments including thermocouple elements. It is also used for electrical contacts, corrosion-resistant apparatus, and in dentistry. Platinum-cobalt alloys have magnetic properties. One such alloy made of 76.7% Pt and 23.3% Co, by weight, is an extremely powerful magnet that offers a B-H (max) almost twice that of Alnico V. Platinum resistance wires are used for constructing high-temperature electric furnaces. The metal is used for coating missile nose cones, jet engine fuel nozzles, etc., which must perform reliably for long periods of time at high temperatures. The metal, like palladium, absorbs large volumes, of hydrogen, retaining it at ordinary temperatures but giving it up at red heat. In the finely divided state platinum is an excellent catalyst, having long been used in the contact process for producing sulfuric acid. It is also used as a catalyst in cracking petroleum products. There is also much current interest in the use of platinum as a catalyst in fuel cells and in its use as antipollution devices for automobiles. Platinum anodes are extensively used in cathodic protection systems for large ships and ocean-going vessels, pipelines, steel piers, etc. Pure platinum wire will glow red hot when placed in the vapor of methyl alcohol. It acts here as a catalyst, converting the alcohol to formaldehyde. This phenomenon has been used commercially to produce cigarette lighters and hand warmers. Hydrogen and oxygen explode in the presence of platinum. The price of platinum has varied widely; more than a century ago it was used to adulterate gold. It was nearly eight times as valuable as gold in 1920. The price in January 2002 was about \$430/troy oz. (\$15/g), higher than the price of gold.

Plutonium — (planet *Pluto*), Pu; at. wt. (244); at. no. 94; sp. gr. (α modification) 19.84 (25°C); m.p. 640°C; b.p. 3228°C; valence 3, 4, 5, or 6. Plutonium was the second transuranium element of the actinide series to be discovered. The isotope ^{238}Pu was produced in 1940 by Seaborg, McMillan, Kennedy, and Wahl by deuteron bombardment of uranium in the 60-inch cyclotron at Berkeley, California. Plutonium also exists in trace quantities in naturally occurring uranium ores. It is formed in much the same manner as neptunium, by irradiation of natural uranium with the neutrons which are present. By far of greatest importance is the isotope Pu^{239}, with a half-life of 24,100 years, produced in extensive quantities in nuclear reactors from natural uranium:

$$^{238}U(n,\gamma) \rightarrow {}^{239}U \xrightarrow{\beta} {}^{239}Np \xrightarrow{\beta} {}^{239}Pu$$

Nineteen isotopes of plutonium are now known. Plutonium has assumed the position of dominant importance among the transuranium elements because of its successful use as an explosive ingredient in nuclear weapons and the place which it holds as a key material in the development of industrial use of nuclear power. One kilogram is equivalent to about 22 million kilowatt hours of heat energy. The complete detonation of a kilogram of plutonium produces an explosion equal to about 20,000 tons of chemical explosive. Its importance depends on the nuclear property of being readily fissionable with neutrons and its availability in quantity. The world's nuclear-power reactors are now producing about 20,000 kg of plutonium/yr. By 1982 it was estimated that about 300,000 kg had accumulated. The various nuclear applications of plutonium are well known. ^{238}Pu has been used in the Apollo lunar missions to power seismic and other equipment on the lunar surface. As with neptunium and uranium, plutonium metal can be prepared by reduction of the trifluoride with alkaline-earth metals. The metal has a silvery appearance and takes on a yellow tarnish when slightly oxidized. It is chemically reactive. A relatively large piece of plutonium is warm to the touch because of the energy given off in alpha decay. Larger pieces will produce enough heat to boil water. The metal readily dissolves in concentrated hydrochloric acid, hydroiodic acid, or perchloric acid with formation of the Pu^{+3} ion. The metal exhibits six allotropic modifications having various crystalline structures. The densities of these vary from 16.00 to 19.86 g/cm^3. Plutonium also exhibits four ionic valence states in aqueous solutions: Pu^{+3}(blue lavender), Pu^{+4} (yellow brown), PuO$^+$ (pink?), and PuO$_2^{+2}$ (pink orange). The ion PuO$^+$ is unstable in aqueous solutions, disproportionating into Pu^{+4} and PuO$_2^{+2}$. The Pu^{+4} thus formed, however, oxidizes the PuO$^+$ into PuO$_2^{+2}$, itself being reduced to Pu^{+3}, giving finally Pu^{+3} and PuO$_2^{+2}$. Plutonium forms binary compounds with oxygen: PuO, PuO$_2$, and intermediate oxides of variable composition; with the halides: PuF$_3$, PuF$_4$, PuCl$_3$, PuBr$_3$, PuI$_3$; with carbon, nitrogen, and silicon: PuC, PuN, PuSi$_2$. Oxyhalides are also well known: PuOCl, PuOBr, PuOI. Because of the high rate of emission of alpha particles and the element being specifically absorbed by bone marrow, plutonium, as well as all of the other transuranium elements except neptunium, are radiological poisons and must be handled with very special equipment and precautions. Plutonium is a very dangerous radiological hazard. Precautions must also be taken to prevent the unintentional formation of a critical mass. Plutonium in liquid solution is more likely to become critical than solid plutonium. The shape of the mass must also be considered where criticality is concerned. Plutonium-239 is available to authorized users from the O.R.N.L. at a cost of about \$4.80/mg (99.9%) plus packing costs.

Polonium — (Poland, native country of Mme. Curie [1867–1934]), Po; at. wt. (209); at. no. 84; m.p. 254°C; b.p. 962°C; sp. gr. (alpha modification) 9.32; valence −2, 0, +2, +3(?), +4, and +6. Polonium was the first element discovered by Mme. Curie in 1898, while seeking the cause of radioactivity of pitchblende from Joachimsthal, Bohemia. The electroscope showed it separating with bismuth. Polonium is also called Radium F. Polonium is a very rare natural element. Uranium ores contain only about 100 µg of the element per ton. Its abundance is only about 0.2% of that of radium. In 1934, it was found that when natural bismuth (^{209}Bi) was bombarded by neutrons, ^{210}Bi, the parent of polonium, was obtained. Milligram amounts of polonium may now be prepared this way, by using the high neutron fluxes of nuclear reactors. Polonium-210 is a low-melting, fairly volatile metal, 50% of which is vaporized in air in 45 hours at 55°C. It is an alpha emitter with a half-life of 138.39 days. A milligram emits as many alpha particles as 5 g of radium. The energy released by its decay is so large (140 W/g) that a capsule containing about half a gram reaches a temperature above 500°C. The capsule also presents a contact gamma-ray dose rate of 0.012 Gy/h. A few curies (1 curie = 3.7×10^{10} Bq) of polonium exhibit a blue glow, caused by excitation of the surrounding gas. Because almost all alpha radiation is stopped within the solid source and its container, giving up its energy, polonium has attracted attention for uses as a lightweight heat source for thermoelectric power in space satellites. Thirty-eight isotopes and isomers of polonium are known, with atomic masses ranging from 192 to 218. All are radioactive. Polonium-210 is the most readily available. Isotopes of mass 209 (half-life 102 years) and mass 208 (half-life 2.9 years) can be prepared by alpha, proton, or deuteron bombardment of lead or bismuth in a cyclotron, but these are expensive to produce. Metallic polonium has been prepared from polonium hydroxide and some other polonium compounds in the presence of concentrated aqueous or anhydrous liquid ammonia. Two allotropic modifications are known to exist. Polonium is readily dissolved in dilute acids, but is only slightly soluble in alkalis. Polonium salts of organic acids char rapidly; halide amines are reduced to the metal. Polonium can be mixed or alloyed with beryllium to provide a source of neutrons. It has been used in devices for eliminating static charges in textile mills, etc.; however, beta sources are more commonly used and are less dangerous. It is also used on brushes for removing dust from photographic films. The polonium for these is carefully sealed and controlled, minimizing hazards to the user. Polonium-210 is very dangerous to handle in even milligram or microgram amounts, and special equipment and strict control is necessary. Damage arises from the complete absorption of the energy of the alpha particle into tissue. The maximum permissible body burden for ingested polonium is only 0.03 µCi, which represents a particle weighing only 6.8×10^{-12} g. Weight for weight it is about 2.5×10^{11} times as toxic as hydrocyanic acid. The maximum allowable concentration for soluble polonium compounds in air is about 2×10^{11} µCi/cm^3. Polonium-209 is available on special order from the Oak Ridge National Laboratory at a cost of \$3600/µCi plus packing costs.

Potassium — (English, *potash* — pot ashes; L. *kalium*, Arab. *qali*, alkali), K; at. wt. 39.0983(1); at. no. 19; m.p. 63.38°C; b.p. 759°C; sp. gr. 0.862 (20°C); valence 1. Discovered in 1807 by Davy, who obtained it from caustic potash (KOH); this was the first metal isolated by electrolysis. The metal is the seventh most abundant and makes up about 2.4% by weight of the earth's crust. Most potassium minerals are insoluble and the metal is obtained from them only with great difficulty. Certain minerals, however, such as *sylvite, carnallite, langbeinite,* and *polyhalite* are found in ancient lake and sea beds and form rather extensive deposits from which potassium and its salts can readily be obtained. Potash is mined in Germany, New Mexico, California, Utah, and elsewhere. Large deposits of potash, found at a depth of some 1000 m in Saskatchewan, promise to be important in coming years. Potassium is also found in the ocean, but is present only in relatively small amounts, compared to sodium. The greatest demand for potash has been in its use for fertilizers. Potassium is an essential constituent for plant growth and it is found in most soils. Potassium is never found free in nature, but is obtained by electrolysis of the hydroxide, much in the same manner as prepared by Davy. Thermal methods also are commonly used to produce potassium (such as by reduction of potassium compounds with CaC_2, C, Si, or Na). It is one of the most reactive and electropositive of metals. Except for lithium, it is the lightest known metal. It is soft, easily cut with a knife, and is silvery in appearance immediately after a fresh surface is exposed. It rapidly oxidizes in air and should be preserved in a mineral oil. As with other metals of the alkali group, it decomposes in water with the evolution of hydrogen. It catches fire spontaneously on water. Potassium and its salts impart a violet color to flames. Twenty one isotopes, one of which is an isomer, of potassium are known. Ordinary potassium is composed of three isotopes, one of which is ^{40}K (0.0117%), a radioactive isotope with a half-life of 1.26×10^9 years. The radioactivity presents no appreciable hazard. An alloy of sodium and potassium (NaK) is used as a heat-transfer medium. Many potassium salts are of utmost importance, including the hydroxide, nitrate, carbonate, chloride, chlorate, bromide, iodide, cyanide, sulfate, chromate, and dichromate. Metallic potassium is available commercially for about $1200/kg (98% purity) or $75/g (99.95% purity).

Praseodymium — (Gr. *prasios*, green, and *didymos*, twin), Pr; at. wt. 140.90765(2); at. no. 59; m.p. 931°C; b.p. 3520°C; sp. gr. 6.773; valence 3. In 1841 Mosander extracted the rare earth *didymia* from *lanthana*; in 1879, Lecoq de Boisbaudran isolated a new earth, *samaria*, from didymia obtained from the mineral *samarskite*. Six years later, in 1885, von Welsbach separated didymia into two others, *praseodymia* and *neodymia*, which gave salts of different colors. As with other rare earths, compounds of these elements in solution have distinctive sharp spectral absorption bands or lines, some of which are only a few Angstroms wide. The element occurs along with other rare-earth elements in a variety of minerals. *Monazite* and *bastnasite* are the two principal commercial sources of the rare-earth metals. Ion-exchange and solvent extraction techniques have led to much easier isolation of the rare earths and the cost has dropped greatly. Thirty-seven isotopes and isomers are now recognized. Praseodymium can be prepared by several methods, such as by calcium reduction of the anhydrous chloride or fluoride. Misch metal, used in making cigarette lighters, contains about 5% praseodymium metal. Praseodymium is soft, silvery, malleable, and ductile. It was prepared in relatively pure form in 1931. It is somewhat more resistant to corrosion in air than europium, lanthanum, cerium, or neodymium, but it does develop a green oxide coating that splits off when exposed to air. As with other rare-earth metals it should be kept under a light mineral oil or sealed in plastic. The rare-earth oxides, including Pr_2O_3, are among the most refractory substances known. Along with other rare earths, it is widely used as a core material for carbon arcs used by the motion picture industry for studio lighting and projection. Salts of praseodymium are used to color glasses and enamels; when mixed with certain other materials, praseodymium produces an intense and unusually clean yellow color in glass. Didymium glass, of which praseodymium is a component, is a colorant for welder's goggles. The metal (99.9% pure) is priced at about $4/g.

Promethium — (*Prometheus*, who, according to mythology, stole fire from heaven), Pm; at. no. 61; at. wt. (145); m.p. 1042°C; b.p. 3000°C (est.); sp. gr. 7.264 (25°C); valence 3. In 1902 Branner predicted the existence of an element between neodymium and samarium, and this was confirmed by Moseley in 1914. Unsuccessful searches were made for this predicted element over two decades, and various investigators proposed the names "illinium," "florentium," and "cyclonium" for this element. In 1941, workers at Ohio State University irradiated neodymium and praseodymium with neutrons, deuterons, and alpha particles, resp., and produced several new radioactivities, which most likely were those of element 61. Wu and Segre, and Bethe, in 1942, confirmed the formation; however, chemical proof of the production of element 61 was lacking because of the difficulty in separating the rare earths from each other at that time. In 1945, Marinsky, Glendenin, and Coryell made the first chemical identification by use of ion-exchange chromatography. Their work was done by fission of uranium and by neutron bombardment of neodymium. These investigators named the newly discovered element. Searches for the element on earth have been fruitless, and it now appears that promethium is completely missing from the earth's crust. Promethium, however, has been reported to be in the spectrum of the star HR^{465} in Andromeda. This element is being formed recently near the star's surface, for no known isotope of promethium has a half-life longer than 17.7 years. Thirty five isotopes and isomers of promethium, with atomic masses from 130 to 158 are now known. Promethium-145, with a half-life of 17.7 years, is the most useful. Promethium-145 has a specific activity of 940 Ci/g. It is a soft beta emitter; although no gamma rays are emitted, X-radiation can be generated when beta particles impinge on elements of a high atomic number, and great care must be taken in handling it. Promethium salts luminesce in the dark with a pale blue or greenish glow, due to their high radioactivity. Ion-exchange methods led to the preparation of about 10 g of promethium from atomic reactor fuel processing wastes in early 1963. Little is yet generally known about the properties of metallic promethium. Two allotropic modifications exist. The element has applications as a beta source for thickness gages, and it can be absorbed by a phosphor to produce light. Light produced in this manner can be used for signs or signals that require dependable operation; it can be used as a nuclear-powered battery by capturing light in photocells which convert it into electric current. Such a battery, using ^{147}Pm, would have a useful life of about 5 years. It is being used for fluorescent lighting starters and coatings for self-luminous watch dials. Promethium shows promise as a portable X-ray source, and it may become useful as a heat source to provide auxiliary power for space probes and satellites. More than 30 promethium compounds have been prepared. Most are colored.

Protactinium — (Gr. *protos*, first), Pa; at. wt. 231.03588(2); at. no. 91; m.p. 1572°C; sp. gr. 15.37 (calc.); valence 4 or 5. The first isotope of element 91 to be discovered was ^{234}Pa, also known as UX$_2$, a short-lived member of the naturally occurring ^{238}U decay series. It was identified by K. Fajans and O. H. Gohring in 1913 and they named the new element *brevium*. When the longer-lived isotope ^{231}Pa was identified by Hahn and Meitner in 1918, the name protoactinium was adopted as being more consistent with the characteristics of the most abundant isotope. Soddy, Cranson, and Fleck were also active in this work. The name *protoactinium* was shortened to *protactinium* in 1949. In 1927, Grosse prepared 2 mg of a white powder, which was shown to be Pa$_2$O$_5$. Later, in 1934, from 0.1 g of pure Pa$_2$O$_5$ he isolated the element by two methods, one of which was by converting the oxide to an iodide and "cracking" it in a high vacuum by an electrically heated filament by the reaction

$$2PaI_5 \rightarrow 2Pa + 5I_2$$

Protactinium has a bright metallic luster which it retains for some time in air. The element occurs in *pitchblende* to the extent of about 1 part ^{231}Pa to 10 million of ore. Ores from Congo-Kinshasa have about 3 ppm. Protactinium has twenty-eight isotopes and isomers, the most common of which is ^{231}Pr with a half-life of 32,500 years. A number of protactinium compounds are known, some of which are colored. The element is superconductive below 1.4 K. The element is a dangerous toxic material and requires precautions similar to those used when handling plutonium. In 1959 and 1961, it was announced that the Great Britain Atomic Energy Authority extracted by a 12-stage process 125 g of 99.9% protactinium, the world's only stock of the metal for many years to come. The extraction was made from 60 tons of waste material at a cost of about $500,000. Protactinium is one of the rarest and most expensive naturally occurring elements.

Radium — (L. *radius*, ray), Ra; at. wt. (226); at. no. 88; m.p. 700°C; sp. gr. 5; valence 2. Radium was discovered in 1898 by M. and Mme. Curie in the *pitchblende* or *uraninite* of North Bohemia (Czech Republic), where it occurs. There is about 1 g of radium in 7 tons of pitchblende. The element was isolated in 1911 by Mme. Curie and Debierne by the electrolysis of a solution of pure radium chloride, employing a mercury cathode; on distillation in an atmosphere of hydrogen this amalgam yielded the pure metal. Originally, radium was obtained from the rich pitchblende ore found at Joachimsthal, Bohemia. The *carnotite* sands of Colorado furnish some radium, but richer ores are found in the Republic of Congo-Kinshasa and the Great Bear Lake region of Canada. Radium is present in all uranium minerals, and could be extracted, if desired, from the extensive wastes of uranium processing. Large uranium deposits are located in Ontario, New Mexico, Utah, Australia, and elsewhere. Radium is obtained commercially as the bromide or chloride; it is doubtful if any appreciable stock of the isolated element now exists. The pure metal is brilliant white when freshly prepared, but blackens on exposure to air, probably due to formation of the nitride. It exhibits luminescence, as do its salts; it decomposes in water and is somewhat more volatile than barium. It is a member of the alkaline-earth group

of metals. Radium imparts a carmine red color to a flame. Radium emits alpha, beta, and gamma rays and when mixed with beryllium produce neutrons. One gram of ^{226}Ra undergoes 3.7×10^{10} disintegrations per s. The *curie (Ci)* is defined as that amount of radioactivity which has the same disintegration rate as 1 g of ^{226}Ra. Thirty-six isotopes are now known; radium 226, the common isotope, has a half-life of 1599 years. One gram of radium produces about 0.0001 mL (stp) of emanation, or radon gas, per day. This is pumped from the radium and sealed in minute tubes, which are used in the treatment of cancer and other diseases. One gram of radium yields about 4186 kJ per year. Radium is used in producing self-luminous paints, neutron sources, and in medicine for the treatment of cancer. Some of the more recently discovered radioisotopes, such as ^{60}Co, are now being used in place of radium. Some of these sources are much more powerful, and others are safer to use. Radium loses about 1% of its activity in 25 years, being transformed into elements of lower atomic weight. Lead is a final product of disintegration. Stored radium should be ventilated to prevent build-up of radon. Inhalation, injection, or body exposure to radium can cause cancer and other body disorders. The maximum permissible burden in the total body for ^{226}Ra is 7400 becquerel.

Radon — (from *radium*; called *niton* at first, L. *nitens*, shining), Rn; at. wt. (222); at. no. 86; m.p. −71°C; b.p. −61.7°C; t_c 104°C; density of gas 9.73 g/l; sp. gr. liquid 4.4 at −62°C, solid 4; valence usually 0. The element was discovered in 1900 by Dorn, who called it *radium emanation*. In 1908 Ramsay and Gray, who named it *niton*, isolated the element and determined its density, finding it to be the heaviest known gas. It is essentially inert and occupies the last place in the zero group of gases in the Periodic Table. Since 1923, it has been called radon. Thirty-seven isotopes and isomers are known. Radon-222, coming from radium, has a half-life of 3.823 days and is an alpha emitter; Radon-220, emanating naturally from thorium and called *thoron*, has a half-life of 55.6 s and is also an alpha emitter. Radon-219 emanates from actinium and is called *actinon*. It has a half-life of 3.9 s and is also an alpha emitter. It is estimated that every square mile of soil to a depth of 6 inches contains about 1 g of radium, which releases radon in tiny amounts to the atmosphere. Radon is present in some spring waters, such as those at Hot Springs, Arkansas. On the average, one part of radon is present to 1×10^{21} part of air. At ordinary temperatures radon is a colorless gas; when cooled below the freezing point, radon exhibits a brilliant phosphorescence which becomes yellow as the temperature is lowered and orange-red at the temperature of liquid air. It has been reported that fluorine reacts with radon, forming radon fluoride. Radon clathrates have also been reported. Radon is still produced for therapeutic use by a few hospitals by pumping it from a radium source and sealing it in minute tubes, called seeds or needles, for application to patients. This practice has now been largely discontinued as hospitals can order the seeds directly from suppliers, who make up the seeds with the desired activity for the day of use. Care must be taken in handling radon, as with other radioactive materials. The main hazard is from inhalation of the element and its solid daughters, which are collected on dust in the air. Good ventilation should be provided where radium, thorium, or actinium is stored to prevent build-up of this element. Radon build-up is a health consideration in uranium mines. Recently radon build-up in homes has been a concern. Many deaths from lung cancer are caused by radon exposure. In the U.S. it is recommended that remedial action be taken if the air from radon in homes exceeds 4 pCi/L.

Rhenium — (L. *Rhenus*, Rhine), Re; at. wt. 186.207(1); at. no. 75; m.p. 3186°C; b.p. 5596°C; sp. gr. 21.02 (20°C); valence −1, +1, 2, 3, 4, 5, 6, 7. Discovery of rhenium is generally attributed to Noddack, Tacke, and Berg, who announced in 1925 they had detected the element in platinum ores and *columbite*. They also found the element in *gadolinite* and *molybdenite.* By working up 660 kg of molybdenite they were able in 1928 to extract 1 g of rhenium. The price in 1928 was $10,000/g. Rhenium does not occur free in nature or as a compound in a distinct mineral species. It is, however, widely spread throughout the earth's crust to the extent of about 0.001 ppm. Commercial rhenium in the U.S. today is obtained from molybdenite roaster-flue dusts obtained from copper-sulfide ores mined in the vicinity of Miami, Arizona, and elsewhere in Arizona and Utah. Some molybdenites contain from 0.002 to 0.2% rhenium. It is estimated that in 1999 about 16,000 kg of rhenium was being produced. The total estimated world reserves of rhenium is 11,000,000 kg. Natural rhenium is a mixture of two isotopes, one of which has a very long half-life. Thirty nine other unstable isotopes are recognized. Rhenium metal is prepared by reducing ammonium perrhenate with hydrogen at elevated temperatures. The element is silvery white with a metallic luster; its density is exceeded only by that of platinum, iridium, and osmium, and its melting point is exceeded only by that of tungsten and carbon. It has other useful properties. The usual commercial form of the element is a powder, but it can be consolidated by pressing and resistance-sintering in a vacuum or hydrogen atmosphere. This produces a compact shape in excess of 90% of the density of the metal. Annealed rhenium is very ductile, and can be bent, coiled, or rolled. Rhenium is used as an additive to tungsten and molybdenum-based alloys to impart useful properties. It is widely used for filaments for mass spectrographs and ion gages. Rhenium-molybdenum alloys are superconductive at 10 K. Rhenium is also used as an electrical contact material as it has good wear resistance and withstands arc corrosion. Thermocouples made of Re-W are used for measuring temperatures up to 2200°C, and rhenium wire has been used in photoflash lamps for photography. Rhenium catalysts are exceptionally resistant to poisoning from nitrogen, sulfur, and phosphorus, and are used for hydrogenation of fine chemicals, hydrocracking, reforming, and disproportionation of olefins. Rhenium has recently become especially important as a catalyst for petroleum refining and in making super-alloys for jet engines. Rhenium costs about $16/g (99.99% pure). Little is known of its toxicity; therefore, it should be handled with care until more data are available.

Rhodium — (Gr. *rhodon*, rose), Rh; at. wt. 102.90550(3); at. no. 45; m.p. 1964°C; b.p. 3695°C; sp. gr. 12.41 (20°C); valence 2, 3, 4, 5, and 6. Wollaston discovered rhodium in 1803-4 in crude platinum ore he presumably obtained from South America. Rhodium occurs native with other platinum metals in river sands of the Urals and in North and South America. It is also found with other platinum metals in the copper-nickel sulfide ores of the Sudbury, Ontario region. Although the quantity occurring here is very small, the large tonnages of nickel processed make the recovery commercially feasible. The annual world production of rhodium in 1999 was only about 9000 kg. The metal is silvery white and at red heat slowly changes in air to the sesquioxide. At higher temperatures it converts back to the element. Rhodium has a higher melting point

and lower density than platinum. Its major use is as an alloying agent to harden platinum and palladium. Such alloys are used for furnace windings, thermocouple elements, bushings for glass fiber production, electrodes for aircraft spark plugs, and laboratory crucibles. It is useful as an electrical contact material as it has a low electrical resistance, a low and stable contact resistance, and is highly resistant to corrosion. Plated rhodium, produced by electroplating or evaporation, is exceptionally hard and is used for optical instruments. It has a high reflectance and is hard and durable. Rhodium is also used for jewelry, for decoration, and as a catalyst. Fifty-two isotopes and isomers are now known. Rhodium metal (powder) costs about $180/g (99.9%).

Roentgenium — (Wilhelm Roentgen, discoverer of x-rays), Rg. On December 20, 1994, scientists at GSI Darmstadt, Germany announced they had detected three atoms of a new element with 111 protons and 161 neutrons. This element was made by bombarding ^{83}Bi with ^{28}Ni. Signals of Element 111 appeared for less than 0.002 s, then decayed into lighter elements including Element ^{268}109 and Element ^{264}107. These isotopes had not previously been observed. In 2004 IUPAC approved the name roentgenium for Element 111. Roentgenium is expected to have properties similar to gold.

Rubidium — (L. *rubidus*, deepest red), Rb; at. wt. 85.4678(3); at. no. 37; m.p. 39.31°C; b.p. 688°C; sp. gr. (solid) 1.532 (20°C), (liquid) 1.475 (39°C); valence 1, 2, 3, 4. Discovered in 1861 by Bunsen and Kirchhoff in the mineral *lepidolite* by use of the spectroscope. The element is much more abundant than was thought several years ago. It is now considered to be the 16th most abundant element in the earth's crust. Rubidium occurs in *pollucite, carnallite, leucite,* and *zinnwaldite,* which contains traces up to 1%, in the form of the oxide. It is found in lepidolite to the extent of about 1.5%, and is recovered commercially from this source. Potassium minerals, such as those found at Searles Lake, California, and potassium chloride recovered from brines in Michigan also contain the element and are commercial sources. It is also found along with cesium in the extensive deposits of *pollucite* at Bernic Lake, Manitoba. Rubidium can be liquid at room temperature. It is a soft, silvery-white metallic element of the alkali group and is the second most electropositive and alkaline element. It ignites spontaneously in air and reacts violently in water, setting fire to the liberated hydrogen. As with other alkali metals, it forms amalgams with mercury and it alloys with gold, cesium, sodium, and potassium. It colors a flame yellowish violet. Rubidium metal can be prepared by reducing rubidium chloride with calcium, and by a number of other methods. It must be kept under a dry mineral oil or in a vacuum or inert atmosphere. Thirty five isotopes and isomers of rubidium are known. Naturally occurring rubidium is made of two isotopes, ^{85}Rb and ^{87}Rb. Rubidium-87 is present to the extent of 27.83% in natural rubidium and is a beta emitter with a half-life of 4.9 × 10^{10} years. Ordinary rubidium is sufficiently radioactive to expose a photographic film in about 30 to 60 days. Rubidium forms four oxides: Rb_2O, Rb_2O_2, Rb_2O_3, Rb_2O_4. Because rubidium can be easily ionized, it has been considered for use in "ion engines" for space vehicles; however, cesium is somewhat more efficient for this purpose. It is also proposed for use as a working fluid for vapor turbines and for use in a thermoelectric generator using the magnetohydrodynamic principle where rubidium ions are formed by heat at high temperature and passed through a magnetic field. These conduct electricity and act like an armature of a generator thereby generating an electric current. Rubidium is used as a getter in vacuum tubes and as a photocell component. It has been used in making special glasses. $RbAg_4I_5$ is important, as it has the highest room-temperature conductivity of any known ionic crystal. At 20°C its conductivity is about the same as dilute sulfuric acid. This suggests use in thin film batteries and other applications. The present cost in small quantities is about $50/g (99.8% pure).

Ruthenium — (L. *Ruthenia*, Russia), Ru; at. wt. 101.07(2); at. no. 44, m.p. 2334°C; b.p. 4150°C; sp. gr. 12.41 (20°C); valence 0, 1, 2, 3, 4, 5, 6, 7, 8. Berzelius and Osann in 1827 examined the residues left after dissolving crude platinum from the Ural mountains in *aqua regia*. While Berzelius found no unusual metals, Osann thought he found three new metals, one of which he named ruthenium. In 1844 Klaus, generally recognized as the discoverer, showed that Osann's ruthenium oxide was very impure and that it contained a new metal. Klaus obtained 6 g of ruthenium from the portion of crude platinum that is insoluble in *aqua regia*. A member of the platinum group, ruthenium occurs native with other members of the group in ores found in the Ural mountains and in North and South America. It is also found along with other platinum metals in small but commercial quantities in *pentlandite* of the Sudbury, Ontario, nickel-mining region, and in *pyroxinite* deposits of South Africa. Natural ruthenium contains seven isotopes. Twenty-eight other isotopes and isomers are known, all of which are radioactive. The metal is isolated commercially by a complex chemical process, the final stage of which is the hydrogen reduction of ammonium ruthenium chloride, which yields a powder. The powder is consolidated by powder metallurgy techniques or by argon-arc welding. Ruthenium is a hard, white metal and has four crystal modifications. It does not tarnish at room temperatures, but oxidizes in air at about 800°C. The metal is not attacked by hot or cold acids or *aqua regia*, but when potassium chlorate is added to the solution, it oxidizes explosively. It is attacked by halogens, hydroxides, etc. Ruthenium can be plated by electrodeposition or by thermal decomposition methods. The metal is one of the most effective hardeners for platinum and palladium, and is alloyed with these metals to make electrical contacts for severe wear resistance. A ruthenium-molybdenum alloy is said to be superconductive at 10.6 K. The corrosion resistance of titanium is improved a hundredfold by addition of 0.1% ruthenium. It is a versatile catalyst. Hydrogen sulfide can be split catalytically by light using an aqueous suspension of CdS particles loaded with ruthenium dioxide. It is thought this may have application to removal of H_2S in oil refining and other industrial processes. Compounds in at least eight oxidation states have been found, but of these, the +2. +3. and +4 states are the most common. Ruthenium tetroxide, like osmium tetroxide, is highly toxic. In addition, it may explode. Ruthenium compounds show a marked resemblance to those of osmium. The metal is priced at about $25/g (99.95% pure).

Rutherfordium — (Ernest Rutherford [1871–1937], New Zealand, Canadian, and British physicist); Rf; at. wt. [261]; at. no. 104. In 1964, workers of the Joint Nuclear Research Institute at Dubna (Russia) bombarded plutonium with accelerated 113 to 115 MeV neon ions. By measuring fission tracks in a special glass with a microscope, they detected an isotope that decays by spontaneous fission. They suggested that this isotope,

which has a half-life of 0.3 ± 0.1 s, might be $^{260}104$, produced by the following reaction:

$$^{242}_{94}Pu + ^{22}_{10}Ne \rightarrow ^{260}104 + 4n$$

Element 104, the first *transactinide* element, is expected to have chemical properties similar to those of hafnium. It would, for example, form a relatively volatile compound with chlorine (a tetrachloride). The Soviet scientists have performed experiments aimed at chemical identification, and have attempted to show that the 0.3-s activity is more volatile than that of the relatively nonvolatile actinide trichlorides. This experiment does not fulfill the test of chemically separating the new element from all others, but it provides important evidence for evaluation. New data, reportedly issued by Soviet scientists, have reduced the half-life of the isotope they worked with from 0.3 to 0.15 s. The Dubna scientists suggest the name *kurchatovium* and symbol *Ku* for Element 104, in honor of Igor Vasilevich Kurchatov (1903–1960), late Head of Soviet Nuclear Research. The Dubna Group also has proposed the name *dubnium* for Element 104. In 1969, Ghiorso, Nurmia, Harris, K. A. Y. Eskola, and P. L. Eskola of the University of California at Berkeley reported they had positively identified two, and possibly three, isotopes of Element 104. The group also indicated that after repeated attempts so far they have been unable to produce isotope $^{260}104$ reported by the Dubna groups in 1964. The discoveries at Berkeley were made by bombarding a target of ^{249}Cf with ^{12}C nuclei of 71 MeV, and ^{13}C nuclei of 69 MeV. The combination of ^{12}C with ^{249}Cf followed by instant emission of four neutrons produced Element $^{257}104$. This isotope has a half-life of 4 to 5 s, decaying by emitting an alpha particle into ^{253}No, with a half-life of 105 s. The same reaction, except with the emission of three neutrons, was thought to have produced $^{258}104$ with a half-life of about 1/100 s. Element $^{259}104$ is formed by the merging of a ^{13}C nuclei with ^{249}Cf, followed by emission of three neutrons. This isotope has a half-life of 3 to 4 s, and decays by emitting an alpha particle into ^{255}No, which has a half-life of 185 s. Thousands of atoms of $^{257}104$ and $^{259}104$ have been detected. The Berkeley group believe their identification of $^{258}104$ was correct. Eleven isotopes of Element 104 have now been identified. The Berkeley group proposed for the new element the name *rutherfordium* (symbol Rf), in honor of Ernest Rutherford. This name was formally adapted by IUPAC in August 1997.

Samarium — (*Samarskite*, a mineral), Sm; at. wt. 150.36(3); at. no. 62; m.p. 1074°C; b.p. 1794°C; sp. gr (α) 7.520 (25°C); valence 2 or 3. Discovered spectroscopically by its sharp absorption lines in 1879 by Lecoq de Boisbaudran in the mineral *samarskite*, named in honor of a Russian mine official, Col. Samarski. Samarium is found along with other members of the rare-earth elements in many minerals, including *monazite* and *bastnasite*, which are commercial sources. The largest producer of rare earth minerals is now China, followed by the U.S., India, and Russia. It occurs in monazite to the extent of 2.8%. While *misch metal* containing about 1% of samarium metal, has long been used, samarium has not been isolated in relatively pure form until recent years. Ion-exchange and solvent extraction techniques have recently simplified separation of the rare earths from one another; more recently, electrochemical deposition, using an electrolytic solution of lithium citrate and a mercury electrode, is said to be a simple, fast, and highly specific way to separate the rare earths. Samarium metal can be produced by reducing the oxide with barium or lanthanum. Samarium has a bright silver luster and is reasonably stable in air. Three crystal modifications of the metal exist, with transformations at 734 and 922°C. The metal ignites in air at about 150°C. Thirty-three isotopes and isomers of samarium are now recognized. Natural samarium is a mixture of seven isotopes, three of which are unstable but have long half-lives. Samarium, along with other rare earths, is used for carbon-arc lighting for the motion picture industry. The sulfide has excellent high-temperature stability and good thermoelectric efficiencies up to 1100°C. $SmCo_5$ has been used in making a new permanent magnet material with the highest resistance to demagnetization of any known material. It is said to have an intrinsic coercive force as high as 2200 kA/m. Samarium oxide has been used in optical glass to absorb the infrared. Samarium is used to dope calcium fluoride crystals for use in optical masers or lasers. Compounds of the metal act as sensitizers for phosphors excited in the infrared; the oxide exhibits catalytic properties in the dehydration and dehydrogenation of ethyl alcohol. It is used in infrared absorbing glass and as a neutron absorber in nuclear reactors. The metal is priced at about $3.50/g (99.9%). Little is known of the toxicity of samarium; therefore, it should be handled carefully.

Scandium — (L. *Scandia*, Scandinavia), Sc; at. wt. 44.955910(8); at. no. 21; m.p. 1541°C; b.p. 2836°C; sp. gr. 2.989 (25°C); valence 3. On the basis of the Periodic System, Mendeleev predicted the existence of *ekaboron*, which would have an atomic weight between 40 of calcium and 48 of titanium. The element was discovered by Nilson in 1878 in the minerals *euxenite* and *gadolinite*, which had not yet been found anywhere except in Scandinavia. By processing 10 kg of euxenite and other residues of rare-earth minerals, Nilson was able to prepare about 2 g of scandium oxide of high purity. Cleve later pointed out that Nilson's scandium was identical with Mendeleev's ekaboron. Scandium is apparently a much more abundant element in the sun and certain stars than here on earth. It is about the 23rd most abundant element in the sun, compared to the 50th most abundant on earth. It is widely distributed on earth, occurring in very minute quantities in over 800 mineral species. The blue color of beryl (aquamarine variety) is said to be due to scandium. It occurs as a principal component in the rare mineral *thortveitite*, found in Scandinavia and Malagasy. It is also found in the residues remaining after the extraction of tungsten from Zinnwald *wolframite*, and in *wiikite* and *bazzite*. Most scandium is presently being recovered from *thortveitite* or is extracted as a by-product from uranium mill tailings. Metallic scandium was first prepared in 1937 by Fischer, Brunger, and Grieneisen, who electrolyzed a eutectic melt of potassium, lithium, and scandium chlorides at 700 to 800°C. Tungsten wire and a pool of molten zinc served as the electrodes in a graphite crucible. Pure scandium is now produced by reducing scandium fluoride with calcium metal. The production of the first pound of 99% pure scandium metal was announced in 1960. Scandium is a silver-white metal which develops a slightly yellowish or pinkish cast upon exposure to air. It is relatively soft, and resembles yttrium and the rare-earth metals more than it resembles aluminum or titanium. It is a very light metal and has a much higher melting point than aluminum, making it of interest to designers of spacecraft. Scandium is not attacked by a 1:1 mixture of conc. HNO_3 and 48% HF. Scandium reacts rapidly with many acids. Twenty-three isotopes and isomers of scandium are recognized. The metal is expensive, costing about $200/g with a purity of about 99.9%. About 20 kg of scandium (as Sc_2O_3) are now being used yearly in the U.S. to produce high-intensity lights, and the radioactive isotope ^{46}Sc is used as a tracing agent in refinery crackers for crude oil, etc. Scandium iodide added to mercury vapor lamps produces a highly efficient light source resembling sunlight, which is important for indoor or night-time color TV. Little is yet known about the toxicity of scandium; therefore, it should be handled with care.

Seaborgium — (Glenn T. Seaborg [1912–1999], American chemist and nuclear physicist). Sg; at. wt. [263]; at no. 106. The discovery of *Seaborgium*, Element 106, took place in 1974 almost simultaneously at the Lawrence-Berkeley Laboratory and at the Joint Institute for Nuclear Research at Dubna, Russia. The Berkeley Group, under direction of Ghiorso, used the Super-Heavy Ion Linear Accelerator (Super HILAC) as a source of heavy ^{18}O ions to bombard a 259-μg target of ^{249}Cf. This resulted in the production and positive identification of ^{263}106, which decayed with a half-life of 0.9 ± 0.2 s by the emission of alpha particles as follows:

$$^{263}106 \xrightarrow{\alpha} {}^{259}104 \xrightarrow{\alpha} {}^{255}No \xrightarrow{\alpha}.$$

The Dubna Team, directed by Flerov and Organessian, produced heavy ions of ^{54}Cr with their 310-cm heavy-ion cyclotron to bombard ^{207}Pb and ^{208}Pb and found a product that decayed with a half-life of 7 ms. They assigned ^{259}106 to this isotope. It is now thought seven isotopes of *Seaborgium* have been identified. Two of the isotopes are believed to have half-lives of about 30 s. *Seaborgium* most likely would have properties resembling tungsten. The IUPAC adopted the name *Seaborgium* in August 1997. Normally the naming of an element is not given until after the death of the person for which the element is named; however, in this case, it was named while Dr. Seaborg was still alive.

Selenium — (Gr. *Selene*, moon), Se; at. wt. 78.96(3); at. no. 34; m.p. (gray) 221°C; b.p. (gray) 685°C; sp. gr. (gray) 4.79, (vitreous) 4.28; valence −2, +4, or +6. Discovered by Berzelius in 1817, who found it associated with tellurium, named for the earth. Selenium is found in a few rare minerals, such as *crooksite* and *clausthalite*. In years past it has been obtained from flue dusts remaining from processing copper sulfide ores, but the anode muds from electrolytic copper refineries now provide the source of most of the world's selenium. Selenium is recovered by roasting the muds with soda or sulfuric acid, or by smelting them with soda and niter. Selenium exists in several allotropic forms. Three are generally recognized, but as many as six have been claimed. Selenium can be prepared with either an amorphous or crystalline structure. The color of amorphous selenium is either red, in powder form, or black, in vitreous form. Crystalline monoclinic selenium is a deep red; crystalline hexagonal selenium, the most stable variety, is a metallic gray. Natural selenium contains six stable isotopes. Twenty-nine other isotopes and isomers have been characterized. The element is a member of the sulfur family and resembles sulfur both in its various forms and in its compounds. Selenium exhibits both photovoltaic action, where light is converted directly into electricity, and photoconductive action, where the electrical resistance decreases with increased illumination. These properties make selenium useful in the production of photocells and exposure meters for photographic use, as well as solar cells. Selenium is also able to convert a.c. electricity to d.c., and is extensively used in rectifiers. Below its melting point selenium is a p-type semiconductor and is finding many uses in electronic and solid-state applications. It is used in Xerography for reproducing and copying documents, letters, etc., but recently its use in this application has been decreasing in favor of certain organic compounds. It is used by the glass industry to decolorize glass and to make ruby-colored glasses and enamels. It is also used as a photographic toner, and as an additive to stainless steel. Elemental selenium has been said to be practically nontoxic and is considered to be an essential trace element; however, hydrogen selenide and other selenium compounds are extremely toxic, and resemble arsenic in their physiological reactions. Hydrogen selenide in a concentration of 1.5 ppm is intolerable to man. Selenium occurs in some soils in amounts sufficient to produce serious effects on animals feeding on plants, such as locoweed, grown in such soils. Selenium (99.5%) is priced at about $250/kg. It is also available in high-purity form at a cost of about $350/kg (99.999%).

Silicon — (L. *silex, silicis,* flint), Si; at. wt. 28.0855(3); at. no. 14; m.p. 1414°C; b.p. 3265°C; sp. gr. 2.33 (25°C); valence 4. Davy in 1800 thought silica to be a compound and not an element; later in 1811, Gay Lussac and Thenard probably prepared impure amorphous silicon by heating potassium with silicon tetrafluoride. Berzelius, generally credited with the discovery, in 1824 succeeded in preparing amorphous silicon by the same general method as used earlier, but he purified the product by removing the fluosilicates by repeated washings. Deville in 1854 first prepared crystalline silicon, the second allotropic form of the element. Silicon is present in the sun and stars and is a principal component of a class of meteorites known as "aerolites." It is also a component of *tektites*, a natural glass of uncertain origin. Natural silicon contains three isotopes. Twenty-four other radioactive isotopes are recognized. Silicon makes up 25.7% of the earth's crust, by weight, and is the second most abundant element, being exceeded only by oxygen. Silicon is not found free in nature, but occurs chiefly as the oxide and as silicates. *Sand, quartz, rock crystal, amethyst, agate, flint, jasper,* and *opal* are some of the forms in which the oxide appears. *Granite, hornblende, asbestos, feldspar, clay mica,* etc. are but a few of the numerous silicate minerals. Silicon is prepared commercially by heating silica and carbon in an electric furnace, using carbon electrodes. Several other methods can be used for preparing the element. Amorphous silicon can be prepared as a brown powder, which can be easily melted or vaporized. Crystalline silicon has a metallic luster and grayish color. The Czochralski process is commonly used to produce single crystals of silicon used for solid-state or semiconductor devices. Hyperpure silicon can be prepared by the thermal decomposition of ultra-pure trichlorosilane in a hydrogen atmosphere, and by a vacuum float zone process. This product can be doped with boron, gallium, phosphorus, or arsenic to produce silicon for use in transistors, solar cells, rectifiers, and other solid-state devices which are used extensively in the electronics and space-age industries. Hydrogenated amorphous silicon has shown promise in producing economical cells for converting solar energy into electricity. Silicon is a relatively inert element, but it is attacked by halogens and dilute alkali. Most acids except hydrofluoric, do not affect it. Silicones are important products of silicon. They may be prepared by hydrolyzing a silicon organic chloride, such as dimethyl silicon chloride. Hydrolysis and condensation of various substituted chlorosilanes can be used to produce a very great number of polymeric products, or silicones, ranging from liquids to hard, glasslike solids with many useful properties. Elemental silicon

transmits more than 95% of all wavelengths of infrared, from 1.3 to 6.7 μm. Silicon is one of man's most useful elements. In the form of sand and clay it is used to make concrete and brick; it is a useful refractory material for high-temperature work, and in the form of silicates it is used in making enamels, pottery, etc. Silica, as sand, is a principal ingredient of glass, one of the most inexpensive of materials with excellent mechanical, optical, thermal, and electrical properties. Glass can be made in a very great variety of shapes, and is used as containers, window glass, insulators, and thousands of other uses. Silicon tetrachloride can be used to iridize glass. Silicon is important in plant and animal life. Diatoms in both fresh and salt water extract silica from the water to build up their cell walls. Silica is present in ashes of plants and in the human skeleton. Silicon is an important ingredient in steel; silicon carbide is one of the most important abrasives and has been used in lasers to produce coherent light of 4560 Å. A remarkable material, first discovered in 1930, is *Aerogel,* developed and now used by NASA in their *Stardust* mission, which is expected to encounter Comet Wild 2 in 2004, returning cometary and interplanet dust to Earth in 2006. *Aerogel* is a highly insulative material that has the lowest density of any known solid. One form of *Aerogel* is 99.9% air and 0.1% SiO_2, by volume. It is 1000 times less dense than glass. It has been called "blue smoke" or "solid smoke." A block of *Aerogel* as large as a person may weigh less than a pound and yet support the weight of 1000 lbs (455 kg). This material is expected to trap cometary particles traveling at speeds of 32 km/sec. *Aerogel* is said to be non-toxic and non-inflammable. It has high thermal insulating qualities that could be used in home insulation. Its light weight may have aircraft applications. Regular grade silicon (99.5%) costs about $160/kg. Silicon (99.9999%) pure costs about $200/kg; hyperpure silicon is available at a higher cost. Miners, stonecutters, and other engaged in work where siliceous dust is breathed in large quantities often develop a serious lung disease known as *silicosis.*

Silver — (Anglo-Saxon, *Seolfor siolfur*), Ag (L. *argentum*), at. wt. 107.8682(2); at. no. 47; m.p. 961.78°C; b.p. 2162°C; sp. gr. 10.50 (20°C); valence 1, 2. Silver has been known since ancient times. It is mentioned in Genesis. Slag dumps in Asia Minor and on islands in the Aegean Sea indicate that man learned to separate silver from lead as early as 3000 B.C. Silver occurs native and in ores such as *argentite* (Ag_2S) and *horn silver* (AgCl); lead, lead-zinc, copper, gold, and copper-nickel ores are principal sources. Mexico, Canada, Peru, and the U.S. are the principal silver producers in the western hemisphere. Silver is also recovered during electrolytic refining of copper. Commercial fine silver contains at least 99.9% silver. Purities of 99.999+% are available commercially. Pure silver has a brilliant white metallic luster. It is a little harder than gold and is very ductile and malleable, being exceeded only by gold and perhaps palladium. Pure silver has the highest electrical and thermal conductivity of all metals, and possesses the lowest contact resistance. It is stable in pure air and water, but tarnishes when exposed to ozone, hydrogen sulfide, or air containing sulfur. The alloys of silver are important. Sterling silver is used for jewelry, silverware, etc. where appearance is paramount. This alloy contains 92.5% silver, the remainder being copper or some other metal. Silver is of utmost importance in photography, about 30% of the U.S. industrial consumption going into this application. It is used for dental alloys. Silver is used in making solder and brazing alloys, electrical contacts, and high capacity silver-zinc and silver-cadmium batteries. Silver paints are used for making printed circuits. It is used in mirror production and may be deposited on glass or metals by chemical deposition, electrodeposition, or by evaporation. When freshly deposited, it is the best reflector of visible light known, but is rapidly tarnishes and loses much of its reflectance. It is a poor reflector of ultraviolet. Silver fulminate ($Ag_2C_2N_2O_2$), a powerful explosive, is sometimes formed during the silvering process. Silver iodide is used in seeding clouds to produce rain. Silver chloride has interesting optical properties as it can be made transparent; it also is a cement for glass. Silver nitrate, or *lunar caustic*, the most important silver compound, is used extensively in photography. While silver itself is not considered to be toxic, most of its salts are poisonous. Natural silver contains two stable isotopes. Fifty-six other radioactive isotopes and isomers are known. Silver compounds can be absorbed in the circulatory system and reduced silver deposited in the various tissues of the body. A condition, known as *argyria*, results with a greyish pigmentation of the skin and mucous membranes. Silver has germicidal effects and kills many lower organisms effectively without harm to higher animals. Silver for centuries has been used traditionally for coinage by many countries of the world. In recent times, however, consumption of silver has at times greatly exceeded the output. In 1939, the price of silver was fixed by the U.S. Treasury at 71¢/troy oz., and at 90.5¢/troy oz. in 1946. In November 1961 the U.S. Treasury suspended sales of nonmonetized silver, and the price stabilized for a time at about $1.29, the melt-down value of silver U.S. coins. The Coinage Act of 1965 authorized a change in the metallic composition of the three U.S. subsidiary denominations to clad or composite type coins. This was the first change in U.S. coinage since the monetary system was established in 1792. Clad dimes and quarters are made of an outer layer of 75% Cu and 25% Ni bonded to a central core of pure Cu. The composition of the one- and five-cent pieces remains unchanged. One-cent coins are 95% Cu and 5% Zn. Five-cent coins are 75% Cu and 25% Ni. Old silver dollars are 90% Ag and 10% Cu. Earlier subsidiary coins of 90% Ag and 10% Cu officially were to circulate alongside the clad coins; however, in practice they have largely disappeared (Gresham's Law), as the value of the silver is now greater than their exchange value. Silver coins of other countries have largely been replaced with coins made of other metals. On June 24, 1968, the U.S. Government ceased to redeem U.S. Silver Certificates with silver. Since that time, the price of silver has fluctuated widely. As of January 2002, the price of silver was about $4.10/troy oz. (13¢/g); however the price has fluctuated considerably due to market instability. The price of silver in 2001 was only about four times the cost of the metal about 150 years ago. This has largely been caused by Central Banks disposing of some of their silver reserves and the development of more productive mines with better refining methods. Also, silver has been displaced by other metals or processes, such as digital photography.

Sodium — (English, *soda*; Medieval Latin, *sodanum*, headache remedy), Na (L. *natrium*); at. wt. 22.989770(2); at. no. 11; m.p. 97.80°C; b.p. 883°C; sp. gr. 0.971 (20°C); valence 1. Long recognized in compounds, sodium was first isolated by Davy in 1807 by electrolysis of caustic soda. Sodium is present in fair abundance in the sun and stars. The D lines of sodium are among the most prominent in the solar spectrum. Sodium is the sixth most abundant element on earth, comprising about 2.6% of the earth's crust; it is the most abundant of the alkali group of metals of which it is a member. The most common compound is sodium chloride, but it occurs in many other minerals, such as *soda niter, cryolite, amphibole, zeolite, sodalite*, etc. It is a very reactive element and is never found free in nature. It is now obtained commercially by the electrolysis of absolutely dry fused sodium chloride. This method is much cheaper than that of electrolyzing sodium hydroxide, as was used several years ago. Sodium is a soft, bright, silvery metal which floats on water, decomposing it with the evolution of hydrogen and the formation of the hydroxide. It may or may not ignite spontaneously on water, depending on the amount of oxide and metal exposed to the water. It normally does not ignite in air at temperatures below 115°C. Sodium should be handled with respect, as it can be dangerous when improperly handled. Metallic sodium is vital in the manufacture of sodamide and esters, and in the preparation of organic compounds. The metal may be used to improve the structure of certain alloys, to descale metal, to purify molten metals, and as a heat transfer agent. An alloy of sodium with potassium, NaK, is also an important heat transfer agent. Sodium compounds are important to the paper, glass, soap, textile, petroleum, chemical, and metal industries. Soap is generally a sodium salt of certain fatty acids. The importance of common salt to animal nutrition has been recognized since prehistoric times. Among the many compounds that are of the greatest industrial importance are common salt ($NaCl$), soda ash (Na_2CO_3), baking soda ($NaHCO_3$), caustic soda ($NaOH$), Chile saltpeter ($NaNO_3$), di- and tri-sodium phosphates, sodium thiosulfate (hypo, $Na_2S_2O_3 \cdot 5H_2O$), and borax ($Na_2B_4O_7 \cdot 10H_2O$). Seventeen isotopes of sodium are recognized. Metallic sodium is priced at about $575/kg (99.95%). On a volume basis, it is the cheapest of all metals. Sodium metal should be handled with great care. It should be kept in an inert atmosphere and contact with water and other substances with which sodium reacts should be avoided.

Strontium — (*Strontian*, town in Scotland), Sr; at. wt. 87.62(1); at. no. 38; m.p. 777°C; b.p. 1382°C; sp. gr. 2.54; valence 2. Isolated by Davey by electrolysis in 1808; however, Adair Crawford in 1790 recognized a new mineral (strontianite) as differing from other barium minerals (baryta). Strontium is found chiefly as *celestite* ($SrSO_4$) and *strontianite* ($SrCO_3$). *Celestite* is found in Mexico, Turkey, Iran, Spain, Algeria, and in the U.K. The U.S. has no active *celestite* mines. The metal can be prepared by electrolysis of the fused chloride mixed with potassium chloride, or is made by reducing strontium oxide with aluminum in a vacuum at a temperature at which strontium distills off. Three allotropic forms of the metal exist, with transition points at 235 and 540°C. Strontium is softer than calcium and decomposes water more vigorously. It does not absorb nitrogen below 380°C. It should be kept under mineral oil to prevent oxidation. Freshly cut strontium has a silvery appearance, but rapidly turns a yellowish color with the formation of the oxide. The finely divided metal ignites spontaneously in air. Volatile strontium salts impart a beautiful crimson color to flames, and these salts are used in pyrotechnics and in the production of flares. Natural strontium is a mixture of four stable isotopes. Thirty-two other unstable isotopes and isomers are known to exist. Of greatest importance is ^{90}Sr with a half-life of 29 years. It is a product of nuclear fallout and presents a health problem. This isotope is one of the best long-lived high-energy beta emitters known, and is used in SNAP (Systems for Nuclear Auxiliary Power) devices. These devices hold promise for use in space vehicles, remote weather stations, navigational buoys, etc., where a lightweight, long-lived, nuclear-electric power source is needed. The major use for strontium at present is in producing glass for color television picture tubes. All color TV and cathode ray tubes sold in the U.S. are required by law to contain strontium in the face plate glass to block X-ray emission. Strontium also improves the brilliance of the glass and the quality of the picture. It has also found use in producing ferrite magnets and in refining zinc. Strontium titanate is an interesting optical material as it has an extremely high refractive index and an optical dispersion greater than that of diamond. It has been used as a gemstone, but it is very soft. It does not occur naturally. Strontium metal (99% pure) costs about $220/kg.

Sulfur — (Sanskrit, *sulvere*; L. *sulphurium*), S; at. wt. 32.066(6); at. no. 16; m.p. 115.21°C; b.p. 444.60°C; t_c 1041°C; sp. gr. (rhombic) 2.07, (monoclinic) 1.957 (20°C); valence 2, 4, or 6. Known to the ancients; referred to in Genesis as *brimstone*. Sulfur is found in meteorites. A dark area near the crater Aristarchus on the moon has been studied by R. W. Wood with ultraviolet light. This study suggests strongly that it is a sulfur deposit. Sulfur occurs native in the vicinity of volcanoes and hot springs. It is widely distributed in nature as *iron pyrites, galena, sphalerite, cinnabar, stibnite, gypsum, Epsom salts, celestite, barite*, etc. Sulfur is commercially recovered from wells sunk into the salt domes along the Gulf Coast of the U.S. It is obtained from these wells by the Frasch process, which forces heated water into the wells to melt the sulfur, which is then brought to the surface. Sulfur also occurs in natural gas and petroleum crudes and must be removed from these products. Formerly this was done chemically, which wasted the sulfur. New processes now permit recovery, and these sources promise to be very important. Large amounts of sulfur are being recovered from Alberta gas fields. Sulfur is a pale yellow, odorless, brittle solid, which is insoluble in water but soluble in carbon disulfide. In every state, whether gas, liquid or solid, elemental sulfur occurs in more than one allotropic form or modification; these present a confusing multitude of forms whose relations are not yet fully understood. Amorphous or "plastic" sulfur is obtained by fast cooling of the crystalline form. X-ray studies indicate that amorphous sulfur may have a helical structure with eight atoms per spiral. Crystalline sulfur seems to be made of rings, each containing eight sulfur atoms, which fit together to give a normal X-ray pattern. Twenty-one isotopes of sulfur are now recognized. Four occur in natural sulfur, none of which is radioactive. A finely divided form of sulfur, known as *flowers of sulfur*, is obtained by sublimation. Sulfur readily forms sulfides with many elements. Sulfur is a component of black gunpowder, and is used in the vulcanization of natural rubber and a fungicide. It is also used extensively is making phosphatic fertilizers. A tremendous tonnage is used to produce sulfuric acid, the most important manufactured chemical. It is used in making sulfite paper and other papers, as a fumigant, and in the bleaching of dried fruits. The element is a good electrical insulator. Organic compounds containing sulfur are very important. Calcium sulfate, ammonium sulfate, carbon disulfide, sulfur dioxide, and hydrogen sulfide are but a few of the many other important compounds of sulfur. Sulfur is essential to life. It is a minor constituent of fats, body fluids, and skeletal minerals. Carbon disulfide, hydrogen sulfide, and sulfur dioxide should be handled carefully. Hydrogen sulfide in small concentrations can be metabolized, but in higher concentrations it quickly can cause death by respiratory paralysis. It is insidious in that it quickly deadens the sense of smell. Sulfur dioxide is a dangerous component in atmospheric pollution. Sulfur (99.999%) costs about $575/kg.

Tantalum — (Gr. *Tantalos*, mythological character, father of *Niobe*), Ta; at. wt. 180.9479(1); at. no. 73; m.p. 3017°C; b.p 5458°C; sp. gr. 16.654; valence 2?, 3, 4?, or 5. Discovered in 1802 by Ekeberg, but many chemists thought niobium and tantalum were identical elements until Rose, in 1844, and Marignac, in 1866, showed that niobic and tantalic acids were two different acids. The early investigators only isolated the impure metal. The first relatively pure ductile tantalum was produced by von Bolton in 1903. Tantalum occurs principally in the mineral *columbite-tantalite* (Fe, Mn)(Nb, Ta)$_2$O$_6$. Tantalum ores are found in Australia, Brazil, Rwanda, Zimbabwe, Congo-Kinshasa, Nigeria, and Canada. Separation of tantalum from niobium requires several complicated steps. Several methods are used to commercially produce the element, including electrolysis of molten potassium fluorotantalate, reduction of potassium fluorotantalate with sodium, or reacting tantalum carbide with tantalum oxide. Thirty four isotopes and isomers of tantalum are known to exist. Natural tantalum contains two isotopes, one of which is radioactive with a very long half-life. Tantalum is a gray, heavy, and very hard metal. When pure, it is ductile and can be drawn into fine wire, which is used as a filament for evaporating metals such as aluminum. Tantalum is almost completely immune to chemical attack at temperatures below 150°C, and is attacked only by hydrofluoric acid, acidic solutions containing the fluoride ion, and free sulfur trioxide. Alkalis attack it only slowly. At high temperatures, tantalum becomes much more reactive. The element has a melting point exceeded only by tungsten and rhenium. Tantalum is used to make a variety of alloys with desirable properties such as high melting point, high strength, good ductility, etc. Scientists at Los Alamos have produced a tantalum carbide graphite composite material, which is said to be one of the hardest materials ever made. The compound has a melting point of 3738°C. Tantalum has good "gettering" ability at high temperatures, and tantalum oxide films are stable and have good rectifying and dielectric properties. Tantalum is used to make electrolytic capacitors and vacuum furnace parts, which account for about 60% of its use. The metal is also widely used to fabricate chemical process equipment, nuclear reactors, and aircraft and missile parts. Tantalum is completely immune to body liquids and is a nonirritating metal. It has, therefore, found wide use in making surgical appliances. Tantalum oxide is used to make special glass with high index of refraction for camera lenses. The metal has many other uses. The price of (99.9%) tantalum is about $2/g.

Technetium — (Gr. *technetos*, artificial), Tc; at. wt. (98); at. no. 43; m.p. 2157°C; b.p. 4265°C; sp. gr. 11.50 (calc.); valence 0, +2, +4, +5, +6, and +7. Element 43 was predicted on the basis of the periodic table, and was erroneously reported as having been discovered in 1925, at which time it was named *masurium*. The element was actually discovered by Perrier and Segre in Italy in 1937. It was found in a sample of molybdenum that was bombarded by deuterons in the Berkeley cyclotron, and which E. Lawrence sent to these investigators. Technetium was the first element to be produced artificially. Since its discovery, searches for the element in terrestrial materials have been made without success. If it does exist, the concentration must be very small. Technetium has been found in the spectrum of S-, M-, and N-type stars, and its presence in stellar matter is leading to new theories of the production of heavy elements in the stars. Forty-three isotopes and isomers of technetium, with mass numbers ranging from 86 to 113, are known. ^{97}Tc has a half-life of 2.6×10^6 years. ^{98}Tc has a half-life of 4.2×10^6 years. The isomeric isotope ^{95m}Tc, with a half-life of 61 days, is useful for tracer work, as it produces energetic gamma rays. Technetium metal has been produced in kilogram quantities. The metal was first prepared by passing hydrogen gas at 1100°C over Tc_2S_7. It is now conveniently prepared by the reduction of ammonium pertechnetate with hydrogen. Technetium is a silvery-gray metal that tarnishes slowly in moist air. Until 1960, technetium was available only in small amounts and the price was as high as $2800/g, but the price is now of the order of $100/g. The chemistry of technetium is similar to that of rhenium. Technetium dissolves in nitric acid, aqua regia, and concentrated sulfuric acid, but is not soluble in hydrochloric acid of any strength. The element is a remarkable corrosion inhibitor for steel. It is reported that mild carbon steels may be effectively protected by as little as 55 ppm of $KTcO_4$ in aerated distilled water at temperatures up to 250°C. This corrosion protection is limited to closed systems, since technetium is radioactive and must be confined. ^{99}Tc has a specific activity of 6.2×10^8 Bq/g. Activity of this level must not be allowed to spread. ^{99}Tc is a contamination hazard and should be handled in a glove box. The metal is an excellent superconductor at 11K and below.

Tellurium — (L. *tellus,* earth), Te; at. wt. 127.60(3); at. no. 52; m.p. 449.51°C; b.p. 988°C; sp. gr. 6.24 (20°C); valence −2, 4, or 6. Discovered by Muller von Reichenstein in 1782; named by Klaproth, who isolated it in 1798. Tellurium is occasionally found native, but is more often found as the telluride of gold (*calaverite*), and combined with other metals. It is recovered commercially from the anode muds produced during the electrolytic refining of blister copper. The U.S., Canada, Peru, and Japan are the largest producers of the element. Crystalline tellurium has a silvery-white appearance, and when pure exhibits a metallic luster. It is brittle and easily pulverized. Amorphous tellurium is formed by precipitating tellurium from a solution of telluric or tellurous acid. Whether this form is truly amorphous, or made of minute crystals, is open to question. Tellurium is a p-type semiconductor, and shows greater conductivity in certain directions, depending on alignment of the atoms. Its conductivity increases slightly with exposure to light. It can be doped with silver, copper, gold, tin, or other elements. In air, tellurium burns with a greenish-blue flame, forming the dioxide. Molten tellurium corrodes iron, copper, and stainless steel. Tellurium and its compounds are probably toxic and should be handled with care. Workmen exposed to as little as 0.01 mg/m³ of air, or less, develop "tellurium breath," which has a garlic-like odor. Forty-two isotopes and isomers of tellurium are known, with atomic masses ranging from 106 to 138. Natural tellurium consists of eight isotopes, two of which are radioactive with very long half-lives. Tellurium improves the machinability of copper and stainless steel, and its addition to lead decreases the corrosive action of sulfuric acid on lead and improves its strength and hardness. Tellurium catalysts are used in the oxidation of organic compounds and are used in hydrogenation and halogenation reactions. Tellurium is also used in electronic and semi-conductor devices. It is also used as a basic ingredient in blasting caps, and is added to cast iron for chill control. Tellurium is used in ceramics. Bismuth telluride has been used in thermoelectric devices. Tellurium costs about 50¢/g, with a purity of about 99.5%. The metal with a purity of 99.9999% costs about $5/g.

Terbium — (*Ytterby*, village in Sweden), Tb; at. wt. 158.92534(2); at. no. 65; m.p. 1356°C; b.p. 3230°C; sp. gr. 8.230; valence 3, 4. Discovered by Mosander in 1843. Terbium is a member of the lanthanide or "rare earth" group of elements. It is found in *cerite, gadolinite*, and other minerals along with other rare earths. It is recovered commercially from *monazite* in which it is present to the extent of 0.03%, from *xenotime*, and from *euxenite*, a complex oxide containing 1% of more of terbia. Terbium has been isolated only in recent years with the development of ion-exchange techniques for separating the rare-earth elements. As with other rare earths, it can be produced by reducing the anhydrous chloride or fluoride with calcium metal in a tantalum crucible. Calcium and tantalum impurities can be removed by vacuum remelting. Other methods of isolation are possible. Terbium is reasonably stable in air. It is a silver-gray metal, and is malleable, ductile, and soft enough to be cut with a knife. Two crystal modifications exist, with a transformation temperature of 1289°C. Forty-two isotopes and isomers are recognized. The oxide is a chocolate or dark maroon color. Sodium terbium borate is used as a laser material and emits coherent light at 0.546 μm. Terbium is used to dope calcium fluoride, calcium tungstate, and strontium molybdate, used in solid-state devices. The oxide has potential application as an activator for green phosphors used in color TV tubes. It can be used with ZrO_2 as a crystal stabilizer of fuel cells which operate at elevated temperature. Few other uses have been found. The element is priced at about $40/g (99.9%). Little is known of the toxicity of terbium. It should be handled with care as with other lanthanide elements.

Thallium — (Gr. *thallos*, a green shoot or twig), Tl; at. wt. 204.3833(2); at. no. 81; m.p. 304°C; b.p. 1473°C; sp. gr. 11.85 (20°C); valence 1, or 3. Thallium was discovered spectroscopically in 1861 by Crookes. The element was named after the beautiful green spectral line, which identified the element. The metal was isolated both by Crookes and Lamy in 1862 about the same time. Thallium occurs in *crooksite, lorandite*, and *hutchinsonite*. It is also present in *pyrites* and is recovered from the roasting of this ore in connection with the production of sulfuric acid. It is also obtained from the smelting of lead and zinc ores. Extraction is somewhat complex and depends on the source of the thallium. Manganese nodules, found on the ocean floor, contain thallium. When freshly exposed to air, thallium exhibits a metallic luster, but soon develops a bluish-gray tinge, resembling lead in appearance. A heavy oxide builds up on thallium if left in air, and in the presence of water the hydroxide is formed. The metal is very soft and malleable. It can be cut with a knife. Forty-seven isotopes of thallium, with atomic masses ranging from 179 to 210 are recognized. Natural thallium is a mixture of two isotopes. The element and its compounds are toxic and should be handled carefully. Contact of the metal with skin is dangerous, and when melting the metal adequate ventilation should be provided. Thallium is suspected of carcinogenic potential for man. Thallium sulfate has been widely employed as a rodenticide and ant killer. It is odorless and tasteless, giving no warning of its presence. Its use, however, has been prohibited in the U.S. since 1975 as a household insecticide and rodenticide. The electrical conductivity of thallium sulfide changes with exposure to infrared light, and this compound is used in photocells. Thallium bromide-iodide crystals have been used as infrared optical materials. Thallium has been used, with sulfur or selenium and arsenic, to produce low melting glasses which become fluid between 125 and 150°C. These glasses have properties at room temperatures similar to ordinary glasses and are said to be durable and insoluble in water. Thallium oxide has been used to produce glasses with a high index of refraction. Thallium has been used in treating ringworm and other skin infections; however, its use has been limited because of the narrow margin between toxicity and therapeutic benefits. A mercury-thallium alloy, which forms a eutectic at 8.5% thallium, is reported to freeze at −60°C, some 20° below the freezing point of mercury. Thallium metal (99.999%) costs about $2/g.

Thorium — (*Thor*, Scandinavian god of war), Th; at. wt. 232.0381(1); at. no. 90; m.p. 1750°C; b.p. 4788°C; sp. gr. 11.72; valence +2(?), +3(?), +4. Discovered by Berzelius in 1828. Thorium occurs in *thorite* ($ThSiO_4$) and in *thorianite* (ThO_2 + UO_2). Large deposits of thorium minerals have been reported in New England and elsewhere, but these have not yet been exploited. Thorium is now thought to be about three times as abundant as uranium and about as abundant as lead or molybdenum. The metal is a source of nuclear power. There is probably more energy available for use from thorium in the minerals of the earth's crust than from both uranium and fossil fuels. Any sizable demand for thorium as a nuclear fuel is still several years in the future. Work has been done in developing thorium cycle converter-reactor systems. Several prototypes, including the HTGR (high-temperature gas-cooled reactor) and MSRE (molten salt converter reactor experiment), have operated. While the HTGR reactors are efficient, they are not expected to become important commercially for many years because of certain operating difficulties. Thorium is recovered commercially from the mineral *monazite*, which contains from 3 to 9% ThO_2 along with rare-earth minerals. Much of the internal heat the earth produces has been attributed to thorium and uranium. Several methods are available for producing thorium metal: it can be obtained by reducing thorium oxide with calcium, by electrolysis of anhydrous thorium chloride in a fused mixture of sodium and potassium chlorides, by calcium reduction of thorium tetrachloride mixed with anhydrous zinc chloride, and by reduction of thorium tetrachloride with an alkali metal. Thorium was originally assigned a position in Group IV of the periodic table. Because of its atomic weight, valence, etc., it is now considered to be the second member of the *actinide* series of elements. When pure, thorium is a silvery-white metal which is air-stable and retains its luster for several months. When contaminated with the oxide, thorium slowly tarnishes in air, becoming gray and finally black. The physical properties of thorium are greatly influenced by the degree of contamination with the oxide. The purest specimens often contain several tenths of a percent of the oxide. High-purity thorium has been made. Pure thorium is soft, very ductile, and can be cold-rolled, swaged, and drawn. Thorium is dimorphic, changing at 1400°C from a cubic to a body-centered cubic structure. Thorium oxide has a melting point of 3300°C, which is the highest of all oxides. Only a few elements, such as tungsten, and a few compounds, such as tantalum carbide, have higher melting points. Thorium is slowly attacked by water, but does not dissolve readily in most common acids, except hydrochloric. Powdered thorium metal is often pyrophoric and should be carefully handled. When heated in air, thorium turnings ignite and burn brilliantly with a white light. The principal use of thorium has been in the preparation of the Welsbach mantle, used for portable gas lights. These mantles, consisting of thorium oxide with about 1% cerium oxide and other ingredients, glow with a dazzling light when heated

in a gas flame. Thorium is an important alloying element in magnesium, imparting high strength and creep resistance at elevated temperatures. Because thorium has a low work-function and high electron emission, it is used to coat tungsten wire used in electronic equipment. The oxide is also used to control the grain size of tungsten used for electric lamps; it is also used for high-temperature laboratory crucibles. Glasses containing thorium oxide have a high refractive index and low dispersion. Consequently, they find application in high quality lenses for cameras and scientific instruments. Thorium oxide has also found use as a catalyst in the conversion of ammonia to nitric acid, in petroleum cracking, and in producing sulfuric acid. Thorium has not found many uses due to its radioactive nature and its handling and disposal problems. Thirty isotopes of thorium are known with atomic masses ranging from 210 to 237. All are unstable. ^{232}Th occurs naturally and has a half-life of 1.4×10^{10} years. It is an alpha emitter. ^{232}Th goes through six alpha and four beta decay steps before becoming the stable isotope ^{208}Pb. ^{232}Th is sufficiently radioactive to expose a photographic plate in a few hours. Thorium disintegrates with the production of "thoron" (^{220}Rn), which is an alpha emitter and presents a radiation hazard. Good ventilation of areas where thorium is stored or handled is therefore essential. Thorium metal (99.8%) costs about $25/g.

Thulium — (*Thule*, the earliest name for Scandinavia), Tm; at. wt. 168.93421(3); at. no. 69; m.p. 1545°C; b.p. 1950°C; sp. gr. 9.321 (25°C); valence 3. Discovered in 1879 by Cleve. Thulium occurs in small quantities along with other rare earths in a number of minerals. It is obtained commercially from *monazite*, which contains about 0.007% of the element. Thulium is the least abundant of the rare earth elements, but with new sources recently discovered, it is now considered to be about as rare as silver, gold, or cadmium. Ion-exchange and solvent extraction techniques have recently permitted much easier separation of the rare earths, with much lower costs. Only a few years ago, thulium metal was not obtainable at any cost; in 1996 the oxide cost $20/g. Thulium metal powder now costs $70/g (99.9%). Thulium can be isolated by reduction of the oxide with lanthanum metal or by calcium reduction of the anhydrous fluoride. The pure metal has a bright, silvery luster. It is reasonably stable in air, but the metal should be protected from moisture in a closed container. The element is silver-gray, soft, malleable, and ductile, and can be cut with a knife. Forty-one isotopes and isomers are known, with atomic masses ranging from 146 to 176. Natural thulium, which is 100% ^{169}Tm, is stable. Because of the relatively high price of the metal, thulium has not yet found many practical applications. ^{169}Tm bombarded in a nuclear reactor can be used as a radiation source in portable X-ray equipment. ^{171}Tm is potentially useful as an energy source. Natural thulium also has possible use in *ferrites* (ceramic magnetic materials) used in microwave equipment. As with other lanthanides, thulium has a low-to-moderate acute toxicity rating. It should be handled with care.

Tin — (Anglo-Saxon, *tin*), Sn (L. *stannum*); at. wt. 118.710(7); at. no. 50; m.p. 231.93°C; b.p. 2602°C; sp. gr. (gray) 5.75, (white) 7.31; valence 2, 4. Known to the ancients. Tin is found chiefly in *cassiterite* (SnO_2). Most of the world's supply comes from China, Indonesia, Peru, Brazil, and Bolivia. The U.S. produces almost none, although occurrences have been found in Alaska and Colorado. Tin is obtained by reducing the ore with coal in a reverberatory furnace. Ordinary tin is composed of ten stable isotopes; thirty-six unstable isotopes and isomers are also known. Ordinary tin is a silver-white metal, is malleable, somewhat ductile, and has a highly crystalline structure. Due to the breaking of these crystals, a "tin cry" is heard when a bar is bent. The element has two allotropic forms at normal pressure. On warming, gray, or α tin, with a cubic structure, changes at 13.2°C into white, or β tin, the ordinary form of the metal. White tin has a tetragonal structure. When tin is cooled below 13.2°C, it changes slowly from white to gray. This change is affected by impurities such as aluminum and zinc, and can be prevented by small additions of antimony or bismuth. This change from the α to β form is called the tin pest. Tin-lead alloys are used to make organ pipes. There are few if any uses for gray tin. Tin takes a high polish and is used to coat other metals to prevent corrosion or other chemical action. Such tin plate over steel is used in the so-called tin can for preserving food. Alloys of tin are very important. Soft solder, type metal, fusible metal, pewter, bronze, bell metal, Babbitt metal, White metal, die casting alloy, and phosphor bronze are some of the important alloys using tin. Tin resists distilled sea and soft tap water, but is attacked by strong acids, alkalis, and acid salts. Oxygen in solution accelerates the attack. When heated in air, tin forms SnO_2, which is feebly acid, forming stannate salts with basic oxides. The most important salt is the chloride ($SnCl_2 \cdot H_2O$), which is used as a reducing agent and as a mordant in calico printing. Tin salts sprayed onto glass are used to produce electrically conductive coatings. These have been used for panel lighting and for frost-free windshields. Most window glass is now made by floating molten glass on molten tin (float glass) to produce a flat surface (Pilkington process). Of recent interest is a crystalline tin-niobium alloy that is superconductive at very low temperatures. This promises to be important in the construction of superconductive magnets that generate enormous field strengths but use practically no power. Such magnets, made of tin-niobium wire, weigh but a few pounds and produce magnetic fields that, when started with a small battery, are comparable to that of a 100 ton electromagnet operated continuously with a large power supply. The small amount of tin found in canned foods is quite harmless. The agreed limit of tin content in U.S. foods is 300 mg/kg. The trialkyl and triaryl tin compounds are used as biocides and must be handled carefully. Over the past 25 years the price of commercial tin has varied from 50¢/lb ($1.10/kg) to about $6/kg. Tin (99.99% pure) costs about $260/kg.

Titanium — (L. *Titans*, the first sons of the Earth, myth.), Ti; at. wt. 47.867(1); at. no. 22; m.p. 1668°C; b.p. 3287°C; sp. gr. 4.54; valence 2, 3, or 4. Discovered by Gregor in 1791; named by Klaproth in 1795. Impure titanium was prepared by Nilson and Pettersson in 1887; however, the pure metal (99.9%) was not made until 1910 by Hunter by heating $TiCl_4$ with sodium in a steel bomb. Titanium is present in meteorites and in the sun. Rocks obtained during the Apollo 17 lunar mission showed presence of 12.1% TiO_2. Analyses of rocks obtained during earlier Apollo missions show lower percentages. Titanium oxide bands are prominent in the spectra of M-type stars. The element is the ninth most abundant in the crust of the earth. Titanium is almost always present in igneous rocks and in the sediments derived from them. It occurs in the minerals *rutile*, *ilmenite*, and *sphene*, and is present in titanates and in many iron ores. Deposits of ilmenite and rutile are found in Florida, California, Tennessee, and New York. Australia, Norway, Malaysia, India, and China are also large suppliers of titanium minerals. Titanium is present in the ash of coal,

in plants, and in the human body. The metal was a laboratory curiosity until Kroll, in 1946, showed that titanium could be produced commercially by reducing titanium tetrachloride with magnesium. This method is largely used for producing the metal today. The metal can be purified by decomposing the iodide. Titanium, when pure, is a lustrous, white metal. It has a low density, good strength, is easily fabricated, and has excellent corrosion resistance. It is ductile only when it is free of oxygen. The metal burns in air and is the only element that burns in nitrogen. Titanium is resistant to dilute sulfuric and hydrochloric acid, most organic acids, moist chlorine gas, and chloride solutions. Natural titanium consists of five isotopes with atomic masses from 46 to 50. All are stable. Eighteen other unstable isotopes are known. The metal is dimorphic. The hexagonal α form changes to the cubic β form very slowly at about 880°C. The metal combines with oxygen at red heat, and with chlorine at 550°C. Titanium is important as an alloying agent with aluminum, molybdenum, manganese, iron, and other metals. Alloys of titanium are principally used for aircraft and missiles where lightweight strength and ability to withstand extremes of temperature are important. Titanium is as strong as steel, but 45% lighter. It is 60% heavier than aluminum, but twice as strong. Titanium has potential use in desalination plants for converting sea water into fresh water. The metal has excellent resistance to sea water and is used for propeller shafts, rigging, and other parts of ships exposed to salt water. A titanium anode coated with platinum has been used to provide cathodic protection from corrosion by salt water. Titanium metal is considered to be physiologically inert; however, titanium powder may be a carcinogenic hazard. When pure, titanium dioxide is relatively clear and has an extremely high index of refraction with an optical dispersion higher than diamond. It is produced artificially for use as a gemstone, but it is relatively soft. Star sapphires and rubies exhibit their asterism as a result of the presence of TiO_2. Titanium dioxide is extensively used for both house paint and artist's paint, as it is permanent and has good covering power. Titanium oxide pigment accounts for the largest use of the element. Titanium paint is an excellent reflector of infrared, and is extensively used in solar observatories where heat causes poor seeing conditions. Titanium tetrachloride is used to iridize glass. This compound fumes strongly in air and has been used to produce smoke screens. The price of titanium metal (99.9%) is about $1100/kg.

Tungsten — (Swedish, *tung sten*, heavy stone); also known as *wolfram* (from *wolframite*, said to be named from *wolf rahm* or *spumi lupi*, because the ore interfered with the smelting of tin and was supposed to devour the tin), W; at. wt. 183.84(1); at. no. 74; m.p. 3422°C; b.p. 5555°C; sp. gr. 19.3 (20°C); valence 2, 3, 4, 5, or 6. In 1779 Peter Woulfe examined the mineral now known as *wolframite* and concluded it must contain a new substance. Scheele, in 1781, found that a new acid could be made from *tung sten* (a name first applied about 1758 to a mineral now known as *scheelite*). Scheele and Berman suggested the possibility of obtaining a new metal by reducing this acid. The de Elhuyar brothers found an acid in *wolframite* in 1783 that was identical to the acid of *tungsten* (tungstic acid) of Scheele, and in that year they succeeded in obtaining the element by reduction of this acid with charcoal. Tungsten occurs in *wolframite*, (Fe, Mn)WO_4; *scheelite*, $CaWO_4$; *huebnerite*, $MnWO_4$; and *ferberite*, $FeWO_4$. Important deposits of tungsten occur in California, Colorado, Bolivia, Russia, and Portugal. China is reported to have about 75% of the world's tungsten resources. Natural tungsten contains five stable isotopes. Thirty two other unstable isotopes and isomers are recognized. The metal is obtained commercially by reducing tungsten oxide with hydrogen or carbon. Pure tungsten is a steel-gray to tin-white metal. Very pure tungsten can be cut with a hacksaw, and can be forged, spun, drawn, and extruded. The impure metal is brittle and can be worked only with difficulty. Tungsten has the highest melting point of all metals, and at temperatures over 1650°C has the highest tensile strength. The metal oxidizes in air and must be protected at elevated temperatures. It has excellent corrosion resistance and is attacked only slightly by most mineral acids. The thermal expansion is about the same as borosilicate glass, which makes the metal useful for glass-to-metal seals. Tungsten and its alloys are used extensively for filaments for electric lamps, electron and television tubes, and for metal evaporation work; for electrical contact points for automobile distributors; X-ray targets; windings and heating elements for electrical furnaces; and for numerous spacecraft and high-temperature applications. High-speed tool steels, Hastelloy˙, Stellite˙, and many other alloys contain tungsten. Tungsten carbide is of great importance to the metal-working, mining, and petroleum industries. Calcium and magnesium tungstates are widely used in fluorescent lighting; other salts of tungsten are used in the chemical and tanning industries. Tungsten disulfide is a dry, high-temperature lubricant, stable to 500°C. Tungsten bronzes and other tungsten compounds are used in paints. Zirconium tungstate has found recent applications (see under Zirconium). Tungsten powder (99.999%) costs about $2900/kg.

Uranium — (Planet *Uranus*), U; at. wt. 238.0289(1); at. no. 92; m.p. 1135°C; b.p. 4131°C; sp. gr. ~18.95; valence 2, 3, 4, 5, or 6. Yellow-colored glass, containing more than 1% uranium oxide and dating back to 79 A.D., has been found near Naples, Italy. Klaproth recognized an unknown element in *pitchblende* and attempted to isolate the metal in 1789. The metal apparently was first isolated in 1841 by Peligot, who reduced the anhydrous chloride with potassium. Uranium is not as rare as it was once thought. It is now considered to be more plentiful than mercury, antimony, silver, or cadmium, and is about as abundant as molybdenum or arsenic. It occurs in numerous minerals such as *pitchblende, uraninite, carnotite, autunite, uranophane, davidite,* and *tobernite.* It is also found in *phosphate rock, lignite, monazite sands,* and can be recovered commercially from these sources. Large deposits of uranium ore occur in Utah, Colorado, New Mexico, Canada, and elsewhere. Uranium can be made by reducing uranium halides with alkali or alkaline earth metals or by reducing uranium oxides by calcium, aluminum, or carbon at high temperatures. The metal can also be produced by electrolysis of KUF_5 or UF_4, dissolved in a molten mixture of $CaCl_2$ and NaCl. High-purity uranium can be prepared by the thermal decomposition of uranium halides on a hot filament. Uranium exhibits three crystallographic modifications as follows:

$$\alpha \xrightarrow{688\ C} \beta \xrightarrow{776\ C} \gamma$$

Uranium is a heavy, silvery-white metal which is pyrophoric when finely divided. It is a little softer than steel, and is attacked by cold water in a finely divided state. It is malleable, ductile, and slightly paramagnetic. In air, the metal becomes coated with a layer of oxide. Acids dissolve the metal, but it is unaffected by alkalis. Uranium has twenty three isotopes,

one of which is an isomer and all of which are radioactive. Naturally occurring uranium contains 99.2745% by weight ^{238}U, 0.720% ^{235}U, and 0.0055% ^{234}U. Studies show that the percentage weight of ^{235}U in natural uranium varies by as much as 0.1%, depending on the source. The U.S.D.O.E. has adopted the value of 0.711 as being their "official" percentage of ^{235}U in natural uranium. Natural uranium is sufficiently radioactive to expose a photographic plate in an hour or so. Much of the internal heat of the earth is thought to be attributable to the presence of uranium and thorium. ^{238}U with a half-life of 4.46×10^9 years, has been used to estimate the age of igneous rocks. The origin of uranium, the highest member of the naturally occurring elements — except perhaps for traces of neptunium or plutonium — is not clearly understood, although it has been thought that uranium might be a decay product of elements of higher atomic weight, which may have once been present on earth or elsewhere in the universe. These original elements may have been formed as a result of a primordial "creation," known as "the big bang," in a supernova, or in some other stellar processes. The fact that recent studies show that most trans-uranic elements are extremely rare with very short half-lives indicates that it may be necessary to find some alternative explanation for the very large quantities of radioactive uranium we find on earth. Studies of meteorites from other parts of the solar system show a relatively low radioactive content, compared to terrestrial rocks. Uranium is of great importance as a nuclear fuel. ^{238}U can be converted into fissionable plutonium by the following reactions:

$$^{238}U(n,\gamma) \rightarrow ^{239}U \xrightarrow{\beta} ^{239}Np \xrightarrow{\beta} ^{239}Pu$$

This nuclear conversion can be brought about in "breeder" reactors where it is possible to produce more new fissionable material than the fissionable material used in maintaining the chain reaction. ^{235}U is of even greater importance, for it is the key to the utilization of uranium. ^{235}U, while occurring in natural uranium to the extent of only 0.72%, is so fissionable with slow neutrons that a self-sustaining fission chain reaction can be made to occur in a reactor constructed from natural uranium and a suitable moderator, such as heavy water or graphite, alone. ^{235}U can be concentrated by gaseous diffusion and other physical processes, if desired, and used directly as a nuclear fuel, instead of natural uranium, or used as an explosive. Natural uranium, slightly enriched with ^{235}U by a small percentage, is used to fuel nuclear power reactors for the generation of electricity. Natural thorium can be irradiated with neutrons as follows to produce the important isotope ^{233}U.

$$^{232}Th(n,\gamma) \rightarrow ^{233}Th \xrightarrow{\beta} ^{233}Pa \xrightarrow{\beta} ^{233}U$$

While thorium itself is not fissionable, ^{233}U is, and in this way may be used as a nuclear fuel. One pound of completely fissioned uranium has the fuel value of over 1500 tons of coal. The uses of nuclear fuels to generate electrical power, to make isotopes for peaceful purposes, and to make explosives are well known. The estimated world-wide production of the 437 nuclear power reactors in operation in 1998 amounted to about 352,000 Megawatt hours. In 1998 the U.S. had about 107 commercial reactors with an output of about 100,000 Megawatt-hours. Some nuclear-powered electric generating plants have recently been closed because of safety concerns. There are also serious problems with nuclear waste disposal that have not been completely resolved. Uranium in the U.S.A.

is controlled by the U.S. Nuclear Regulatory Commission, under the Department of Energy. Uses are being found for the large quantities of "depleted" uranium, now available, where uranium-235 has been lowered to about 0.2%. Depleted uranium has been used for inertial guidance devices, gyrocompasses, counterweights for aircraft control surfaces, ballast for missile reentry vehicles, and as a shielding material for tanks, etc. Concerns, however, have been raised over its low radioactive properties. Uranium metal is used for X-ray targets for production of high-energy X-rays. The nitrate has been used as photographic toner, and the acetate is used in analytical chemistry. Crystals of uranium nitrate are triboluminescent. Uranium salts have also been used for producing yellow "vaseline" glass and glazes. Uranium and its compounds are highly toxic, both from a chemical and radiological standpoint. Finely divided uranium metal, being pyrophoric, presents a fire hazard. The maximum permissible total body burden of natural uranium (based on radiotoxicity) is 0.2 μCi for soluble compounds. Recently, the natural presence of uranium and thorium in many soils has become of concern to homeowners because of the generation of radon and its daughters (see under Radon). Uranium metal is available commercially at a cost of about $6/g (99.7%) in air-tight glass under argon.

Vanadium — (Scandinavian goddess, *Vanadis*), V; at. wt. 50.9415(1); at. no. 23; m.p. 1910°C; b.p. 3407°C; sp. gr. 6.11 (18.7°C); valence 2, 3, 4, or 5. Vanadium was first discovered by del Rio in 1801. Unfortunately, a French chemist incorrectly declared del Rio's new element was only impure chromium; del Rio thought himself to be mistaken and accepted the French chemist's statement. The element was rediscovered in 1830 by Sefstrom, who named the element in honor of the Scandinavian goddess *Vanadis* because of its beautiful multicolored compounds. It was isolated in nearly pure form by Roscoe, in 1867, who reduced the chloride with hydrogen. Vanadium of 99.3 to 99.8% purity was not produced until 1927. Vanadium is found in about 65 different minerals among which *carnotite, roscoelite, vanadinite*, and *patronite* are important sources of the metal. Vanadium is also found in phosphate rock and certain iron ores, and is present in some crude oils in the form of organic complexes. It is also found in small percentages in meteorites. Commercial production from petroleum ash holds promise as an important source of the element. China, South Africa, and Russia supply much of the world's vanadium ores. High-purity ductile vanadium can be obtained by reduction of vanadium trichloride with magnesium or with magnesium-sodium mixtures. Much of the vanadium metal being produced is now made by calcium reduction of V_2O_5 in a pressure vessel, an adaptation of a process developed by McKechnie and Seybolt. Natural vanadium is a mixture of two isotopes, ^{50}V (0.25%) and ^{51}V (99.75%). ^{50}V is slightly radioactive, having a long half-life. Twenty other unstable isotopes are recognized. Pure vanadium is a bright white metal, and is soft and ductile. It has good corrosion resistance to alkalis, sulfuric and hydrochloric acid, and salt water, but the metal oxidizes readily above 660°C. The metal has good structural strength and a low fission neutron cross section, making it useful in nuclear applications. Vanadium is used in producing rust resistant, spring, and high-speed tool steels. It is an important carbide stabilizer in making steels. About 80% of the vanadium now produced is used as ferrovanadium or as a steel additive. Vanadium foil is used as a bonding agent in cladding titanium to steel. Vanadium pentoxide is used in ceramics and as a catalyst. It is also used in producing a super-

conductive magnet with a field of 175,000 gauss. Vanadium and its compounds are toxic and should be handled with care. Ductile vanadium is commercially available. Vanadium metal (99.7%) costs about $3/g.

Wolfram — see Tungsten.

Xenon — (Gr. *xenon*, stranger), Xe; at. wt. 131.29(2); at. no. 54; m.p. −111.79°C; b.p. −108.12°C; t_c 16.62°C; density (gas) 5.887 ± 0.009 g/l, sp. gr (liquid) 3.52 (−109°C); valence usually 0. Discovered by Ramsay and Travers in 1898 in the residue left after evaporating liquid air components. Xenon is a member of the so-called noble or "inert" gases. It is present in the atmosphere to the extent of about one part in twenty million. Xenon is present in the Martian atmosphere to the extent of 0.08 ppm. The element is found in the gases evolved from certain mineral springs, and is commercially obtained by extraction from liquid air. Natural xenon is composed of nine stable isotopes. In addition to these, thirty-five unstable isotopes and isomers have been characterized. Before 1962, it had generally been assumed that xenon and other noble gases were unable to form compounds. However, it is now known that xenon, as well as other members of the zero valence elements, do form compounds. Among the compounds of xenon now reported are xenon hydrate, sodium perxenate, xenon deuterate, difluoride, tetrafluoride, hexafluoride, and $XePtF_6$ and $XeRhF_6$. Xenon trioxide, which is highly explosive, has been prepared. More than 80 xenon compounds have been made with xenon chemically bonded to fluorine and oxygen. Some xenon compounds are colored. Metallic xenon has been produced, using several hundred kilobars of pressure. Xenon in a vacuum tube produces a beautiful blue glow when excited by an electrical discharge. The gas is used in making electron tubes, stroboscopic lamps, bactericidal lamps, and lamps used to excite ruby lasers for generating coherent light. Xenon is used in the atomic energy field in bubble chambers, probes, and other applications where its high molecular weight is of value. The perxenates are used in analytical chemistry as oxidizing agents. ^{133}Xe and ^{135}Xe are produced by neutron irradiation in air-cooled nuclear reactors. ^{133}Xe has useful applications as a radioisotope. The element is available in sealed glass containers for about $20/L of gas at standard pressure. Xenon is not toxic, but its compounds are highly toxic because of their strong oxidizing characteristics.

Ytterbium — (*Ytterby*, village in Sweden), Yb; at. wt. 173.04(3); at. no. 70; m.p. 819°C; b.p. 1196°C; sp. gr (α) 6.903 (β) 6.966; valence 2, 3. Marignac in 1878 discovered a new component, which he called *ytterbia*, in the earth then known as *erbia*. In 1907, Urbain separated ytterbia into two components, which he called *neoytterbia* and *lutecia*. The elements in these earths are now known as *ytterbium* and *lutetium*, respectively. These elements are identical with *aldebaranium* and *cassiopeium*, discovered independently and at about the same time by von Welsbach. Ytterbium occurs along with other rare earths in a number of rare minerals. It is commercially recovered principally from *monazite sand*, which contains about 0.03%. Ion-exchange and solvent extraction techniques developed in recent years have greatly simplified the separation of the rare earths from one another. The element was first prepared by Klemm and Bonner in 1937 by reducing ytterbium trichloride with potassium. Their metal was mixed, however, with KCl. Daane, Dennison, and Spedding prepared a much purer form in 1953 from which the chemical and physical properties of

the element could be determined. Ytterbium has a bright silvery luster, is soft, malleable, and quite ductile. While the element is fairly stable, it should be kept in closed containers to protect it from air and moisture. Ytterbium is readily attacked and dissolved by dilute and concentrated mineral acids and reacts slowly with water. Ytterbium has three allotropic forms with transformation points at −13° and 795°C. The beta form is a room-temperature, face-centered, cubic modification, while the high-temperature gamma form is a body-centered cubic form. Another body-centered cubic phase has recently been found to be stable at high pressures at room temperatures. The beta form ordinarily has metallic-type conductivity, but becomes a semiconductor when the pressure is increased above 16,000 atm. The electrical resistance increases tenfold as the pressure is increased to 39,000 atm and drops to about 80% of its standard temperature-pressure resistivity at a pressure of 40,000 atm. Natural ytterbium is a mixture of seven stable isotopes. Twenty six other unstable isotopes and isomers are known. Ytterbium metal has possible use in improving the grain refinement, strength, and other mechanical properties of stainless steel. One isotope is reported to have been used as a radiation source as a substitute for a portable X-ray machine where electricity is unavailable. Few other uses have been found. Ytterbium metal is available with a purity of about 99.9% for about $10/g. Ytterbium has a low acute toxicity rating.

Yttrium — (*Ytterby*, village in Sweden near Vauxholm), Y; at. wt. 88.90585(2); at. no. 39; m.p. 1522°C; b.p. 3345°C; sp. gr. 4.469 (25°C); valence 3. *Yttria*, which is an earth containing yttrium, was discovered by Gadolin in 1794. *Ytterby* is the site of a quarry which yielded many unusually minerals containing rare earths and other elements. This small town, near Stockholm, bears the honor of giving names to *erbium, terbium*, and *ytterbium* as well as *yttrium*. In 1843 Mosander showed that yttria could be resolved into the oxides (or earths) of three elements. The name yttria was reserved for the most basic one; the others were named *erbia* and *terbia*. Yttrium occurs in nearly all of the rare-earth minerals. Analysis of lunar rock samples obtained during the Apollo missions show a relatively high yttrium content. It is recovered commercially from *monazite sand*, which contains about 3%, and from *bastnasite*, which contains about 0.2%. Wohler obtained the impure element in 1828 by reduction of the anhydrous chloride with potassium. The metal is now produced commercially by reduction of the fluoride with calcium metal. It can also be prepared by other techniques. Yttrium has a silver-metallic luster and is relatively stable in air. Turnings of the metal, however, ignite in air if their temperature exceeds 400°C, and finely divided yttrium is very unstable in air. Yttrium oxide is one of the most important compounds of yttrium and accounts for the largest use. It is widely used in making YVO_4 europium, and Y_2O_3 europium phosphors to give the red color in color television tubes. Many hundreds of thousands of pounds are now used in this application. Yttrium oxide also is used to produce yttrium-iron-garnets, which are very effective microwave filters. Yttrium iron, aluminum, and gadolinium garnets, with formulas such as $Y_3Fe_5O_{12}$ and $Y_3Al_5O_{12}$, have interesting magnetic properties. Yttrium iron garnet is also exceptionally efficient as both a transmitter and transducer of acoustic energy. Yttrium aluminum garnet, with a hardness of 8.5, is also finding use as a gemstone (simulated diamond). Small amounts of yttrium (0.1 to 0.2%) can be used to reduce the grain size in chromium, molybdenum, zirconium, and titanium, and to

increase strength of aluminum and magnesium alloys. Alloys with other useful properties can be obtained by using yttrium as an additive. The metal can be used as a deoxidizer for vanadium and other nonferrous metals. The metal has a low cross section for nuclear capture. ^{90}Y, one of the isotopes of yttrium, exists in equilibrium with its parent ^{90}Sr, a product of atomic explosions. Yttrium has been considered for use as a nodulizer for producing nodular cast iron, in which the graphite forms compact nodules instead of the usual flakes. Such iron has increased ductility. Yttrium is also finding application in laser systems and as a catalyst for ethylene polymerization. It has also potential use in ceramic and glass formulas, as the oxide has a high melting point and imparts shock resistance and low expansion characteristics to glass. Natural yttrium contains but one isotope, ^{89}Y. Forty-three other unstable isotopes and isomers have been characterized. Yttrium metal of 99.9% purity is commercially available at a cost of about $5/g.

Zinc — (Ger. *Zink*, of obscure origin), Zn; at. wt. 65.39(2); at. no. 30; m.p. 419.53°C; b.p. 907°C; sp. gr. 7.133 (25°C); valence 2. Centuries before zinc was recognized as a distinct element, zinc ores were used for making brass. Tubal-Cain, seven generations from Adam, is mentioned as being an "instructor in every artificer in brass and iron." An alloy containing 87% zinc has been found in prehistoric ruins in Transylvania. Metallic zinc was produced in the 13th century A.D. in India by reducing calamine with organic substances such as wool. The metal was rediscovered in Europe by Marggraf in 1746, who showed that it could be obtained by reducing *calamine* with charcoal. The principal ores of zinc are *sphalerite* or *blende* (sulfide), *smithsonite* (carbonate), *calamine* (silicate), and *franklinite* (zinc, manganese, iron oxide). Canada, Japan, Belgium, Germany, and The Netherlands are suppliers of zinc ores. Zinc is also mined in Alaska, Tennessee, Missouri, and elsewhere in the U.S. Zinc can be obtained by roasting its ores to form the oxide and by reduction of the oxide with coal or carbon, with subsequent distillation of the metal. Other methods of extraction are possible. Naturally occurring zinc contains five stable isotopes. Twenty-five other unstable isotopes and isomers are recognized. Zinc is a bluish-white, lustrous metal. It is brittle at ordinary temperatures but malleable at 100 to 150°C. It is a fair conductor of electricity, and burns in air at high red heat with evolution of white clouds of the oxide. The metal is employed to form numerous alloys with other metals. Brass, nickel silver, typewriter metal, commercial bronze, spring brass, German silver, soft solder, and aluminum solder are some of the more important alloys. Large quantities of zinc are used to produce die castings, used extensively by the automotive, electrical, and hardware industries. An alloy called *Prestal*, consisting of 78% zinc and 22% aluminum is reported to be almost as strong as steel but as easy to mold as plastic. It is said to be so plastic that it can be molded into form by relatively inexpensive die casts made of ceramics and cement. It exhibits superplasticity. Zinc is also extensively used to galvanize other metals such as iron to prevent corrosion. Neither zinc nor zirconium is ferromagnetic; but $ZrZn_2$ exhibits ferromagnetism at temperatures below 35 K. Zinc oxide is a unique and very useful material to modern civilization. It is widely used in the manufacture of paints, rubber products, cosmetics, pharmaceuticals, floor coverings, plastics, printing inks, soap, storage batteries, textiles, electrical equipment, and other products. It has unusual electrical, thermal, optical, and solid-state properties that have not yet been fully investigated. Lithopone, a mixture of zinc sulfide and barium sulfate, is an important pigment. Zinc sulfide is used in making luminous dials, X-ray and TV screens, and fluorescent lights. The chloride and chromate are also important compounds. Zinc is an essential element in the growth of human beings and animals. Tests show that zinc-deficient animals require 50% more food to gain the same weight as an animal supplied with sufficient zinc. Zinc is not considered to be toxic, but when freshly formed ZnO is inhaled a disorder known as the *oxide shakes* or *zinc chills* sometimes occurs. It is recommended that where zinc oxide is encountered good ventilation be provided. The commercial price of zinc in January 2002 was roughly 40¢/lb ($90 kg). Zinc metal with a purity of 99.9999% is priced at about $5/g.

Zirconium — (Syriac, *zargun*, color of gold), Zr; at. wt. 91.224(2); at. no. 40; m.p. 1855°C; b.p. 4409°C; sp. gr. 6.506 (20°C); valence +2, +3, and +4. The name *zircon* may have originated from the Syriac word *zargono*, which describes the color of certain gemstones now known as *zircon, jargon, hyacinth, jacinth,* or *ligure.* This mineral, or its variations, is mentioned in biblical writings. These minerals were not known to contain this element until Klaproth, in 1789, analyzed a *jargon* from Sri Lanka and found a new earth, which Werner named zircon (*silex circonius*), and Klaproth called *Zirkonerde (zirconia).* The impure metal was first isolated by Berzelius in 1824 by heating a mixture of potassium and potassium zirconium fluoride in a small iron tube. Pure zirconium was first prepared in 1914. Very pure zirconium was first produced in 1925 by van Arkel and de Boer by an iodide decomposition process they developed. Zirconium is found in abundance in S-type stars, and has been identified in the sun and meteorites. Analyses of lunar rock samples obtained during the various Apollo missions to the moon show a surprisingly high zirconium oxide content, compared with terrestrial rocks. Naturally occurring zirconium contains five isotopes. Thirty-one other radioactive isotopes and isomers are known to exist. *Zircon*, $ZrSiO_4$, the principal ore, is found in deposits in Florida, South Carolina, Australia, South Africa, and elsewhere. *Baddeleyite*, found in Brazil, is an important zirconium mineral. It is principally pure ZrO_2 in crystalline form having a hafnium content of about 1%. Zirconium also occurs in some 30 other recognized mineral species. Zirconium is produced commercially by reduction of

the chloride with magnesium (the Kroll Process), and by other methods. It is a grayish-white lustrous metal. When finely divided, the metal may ignite spontaneously in air, especially at elevated temperatures. The solid metal is much more difficult to ignite. The inherent toxicity of zirconium compounds is low. Hafnium is invariably found in zirconium ores, and the separation is difficult. Commercial-grade zirconium contains from 1 to 3% hafnium. Zirconium has a low absorption cross section for neutrons, and is therefore used for nuclear energy applications, such as for cladding fuel elements. Commercial nuclear power generation now takes more than 90% of zirconium metal production. Reactors of the size now being made may use as much as a half-million lineal feet of zirconium alloy tubing. Reactor-grade zirconium is essentially free of hafnium. *Zircaloy* is an important alloy developed specifically for nuclear applications. Zirconium is exceptionally resistant to corrosion by many common acids and alkalis, by sea water, and by other agents. It is used extensively by the chemical industry where corrosive agents are employed. Zirconium is used as a getter in vacuum tubes, as an alloying agent in steel, in surgical appliances, photoflash bulbs, explosive primers, rayon spinnerets, lamp filaments, etc. It is used in poison ivy lotions in the form of the carbonate as it combines with *urushiol*. With niobium, zirconium is superconductive at low temperatures and is used to make superconductive magnets. Alloyed with zinc, zirconium becomes magnetic at temperatures below 35 K. Zirconium oxide (zircon) has a high index of refraction and is used as a gem material. The impure oxide, zirconia, is used for laboratory crucibles that will withstand heat shock, for linings of metallurgical furnaces, and by the glass and ceramic industries as a refractory material. Its use as a refractory material accounts for a large share of all zirconium consumed. Zirconium tungstate is an unusual material that shrinks, rather than expands, when heated. A few other compounds are known to possess this property, but they tend to shrink in one direction, while they stretch out in others in order to maintain an overall volume. Zirconium tungstate shrinks in all directions over a wide temperature range of from near absolute zero to +777°C. It is being considered for use in composite materials where thermal expansion may be a problem. Zirconium of about 99.5% purity is available at a cost of about $2000/kg or about $4/g.

PHYSICAL CONSTANTS OF INORGANIC COMPOUNDS

The compounds in this table were selected on the basis of their laboratory and industrial importance, as well as their value in illustrating trends in the variation of physical properties with position in the periodic table. An effort has been made to include the most frequently encountered inorganic substances; a limited number of organometallics are also covered. Many, if not most, of the compounds that are solids at ambient temperature can exist in more than one crystalline modification. In the absence of other information, the data given here can be assumed to apply to the most stable or common crystalline form. In many cases however, two or more forms are of practical importance, and separate entries will be found in the table.

Compounds are arranged primarily in alphabetical order by the most commonly used name. However, adjustments are made in many instances so as to bring closely related compounds together. For example, hydrides of elements such as boron, silicon, and germanium are grouped together immediately following the entry for the parent element, since they would otherwise be scattered throughout the table. Likewise, the oxoacids of an element are given in one group whenever a strict alphabetical order would separate them (e.g., sulfuric acid and fluorosulfuric acid). The Formula Index following the table provides another means of locating a compound. There is also an index to CAS Registry Numbers.

The following data fields appear in the table:

- **Name**: Systematic name for the substance. The valence state of a metallic element is indicated by a Roman numeral, e.g., copper in the +1 state is written as copper(I) rather than cuprous, iron in the +3 state is iron(III) rather than ferric.
- **Formula**: The simplest descriptive formula is given, but this does not necessarily specify the actual structure of the compound. For example, aluminum chloride is designated as $AlCl_3$, even though a more accurate representation of the structure in the solid phase (and, under some conditions, in the gas phase) is Al_2Cl_6 A few exceptions are made, such as the use of Hg_2^{+2} for the mercury(I) ion.
- **CAS Registry Number**: Chemical Abstracts Service Registry Number. An asterisk* following the CAS RN for a hydrate indicates that the number refers to the anhydrous compound. In most cases the generic CAS RN for the compound is given rather than the number for a specific crystalline form or mineral.
- **Mol. Weight**: Molecular weight (relative molar mass) as calculated with the 2005 IUPAC Recommended Atomic Weights. The number of decimal places corresponds to the number of places in the atomic weight of the least accurately known element (e.g., one place for lead compounds, two places for compounds of selenium, germanium, etc.);

a maximum of three places is given. For compounds of radioactive elements for which IUPAC makes no recommendation, the mass number of the isotope with longest half-life is used.

- **Physical Form**: The crystal system is given, when available, for compounds that are solid at room temperature, together with color and other descriptive features. Abbreviations are listed below.
- **mp**: Normal melting point in °C. The notation tp indicates the temperature where solid, liquid, and gas are in equilibrium at a pressure greater than one atmosphere (i.e., the normal melting point does not exist). When available, the triple point pressure is listed.
- **bp**: Normal boiling point in °C (referred to 101.325 kPa or 760 mmHg pressure). The notation sp following the number indicates the temperature where the pressure of the vapor in equilibrium with the solid reaches 101.325 kPa. See Reference 8, p. 23, for further discussion of sublimation points and triple points. A notation "sublimes" without a temperature being given indicates that there is a perceptible sublimation pressure above the solid at ambient temperatures.
- **Density**: Density values for solids and liquids are always in units of grams per cubic centimeter and can be assumed to refer to temperatures near room temperature unless otherwise stated. Values for gases are the calculated ideal gas densities in grams per liter at 25°C and 101.325 kPa; the unit is always specified for a gas value.
- **Aqueous Solubility:** Solubility is expressed as the number of grams of the compound (excluding any water of hydration) that will dissolve in 100 grams of water. The temperature in °C is given as a superscript. Solubility at other temperatures can be found for many compounds in the table "Aqueous Solubility of Inorganic Compounds at Various Temperatures" in Section 8.
- **Qualitative Solubility**: Qualitative information on the solubility in other solvents (and in water, if quantitative data are unavailable) is given here. The abbreviations are:
  - i insoluble
  - sl slightly soluble
  - s soluble
  - vs very soluble
  - reac reacts with the solvent

Data were taken from a wide variety of reliable sources, including monographs, treatises, review articles, evaluated compilations and databases, and in many cases the primary literature. Some of the most useful references for the properties covered here are listed below.

List of Abbreviations

Ac - acetyl	blk - black	cub - cubic	exp - explodes, explosive
ace - acetone	brn - brown	cyhex - cyclohexane	extrap - extrapolated
acid - acid solutions	bz - benzene	dec - decomposes	flam - flammable
alk - alkaline solutions	chl - chloroform	dil - dilute	gl - glass, glassy
amorp - amorphous	col - colorless	diox - dioxane	grn - green
anh - anhydrous	conc - concentrated	eth - ethyl ether	hc - hydrocarbon solvents
aq - aqueous	cry - crystals, crystalline	EtOH - ethanol	hex - hexagonal, hexane

hp - heptane
HT - high temperature
hyd - hydrate
hyg - hygroscopic
i - insoluble in
liq - liquid
LT - low temperature
MeOH - methanol
monocl - monoclinic
octahed - octahedral
oran - orange

orth - orthorhombic
os - organic solvents
peth - petroleum ether
pow - powder
prec - precipitate
pur - purple
py - pyridine
reac - reacts with
refrac - refractory
rhom - rhombohedral
r.t. - room temperature

s - soluble in
silv - silvery
sl - slightly soluble in
soln - solution
sp - sublimation point
stab - stable
subl - sublimes
temp - temperature
tetr - tetragonal
thf - tetrahydrofuran
tol - toluene

tp - triple point
trans - transition, transformation
tricl - triclinic
trig - trigonal
unstab - unstable
viol - violet
visc - viscous
vs - very soluble in
wh - white
xyl - xylene
yel - yellow

References

1. Phillips, S. L., and Perry, D.L., *Handbook of Inorganic Compounds*, CRC Press, Boca Raton, FL, 1995.
2. Trotman-Dickenson, A. F., Executive Editor, *Comprehensive Inorganic Chemistry*, Vol. 1-5, Pergamon Press, Oxford, 1973.
3. Greenwood, N. N., and Earnshaw, A., *Chemistry of the Elements, Second Edition*, Butterworth-Heinemann, Oxford, 1997.
4. Wiberg, N., Wiberg, E., and Holleman, H. F., *Inorganic Chemistry, 34th Edition*, Academic Press, San Diego, 2001.
5. *GMELIN Handbook of Inorganic and Organometallic Chemistry*, Springer-Verlag, Heidelberg.
6. Chase, M.W., Davies, C.A., Downey, J.R., Frurip, D. J., McDonald, R.A., and Syverud, A.N.; *JANAF Thermochemical Tables, Third Edition, J. Phys. Chem. Ref. Data*, Vol. 14, Suppl. 1, 1985; Chase, M. W., *NIST-JANAF Thermochemical Tables, Fourth Edition, J. Phys. Chem. Ref. Data*, Monograph No. 9, 1998.
7. *Landolt-Börnstein, Numerical Data and Functional Relationships in Science and Technology*, New Series, IV/19A, "Thermodynamic Properties of Inorganic Materials compiled by SGTE", Springer-Verlag, Heidelberg; Part 1, 1999; Part 2; 1999; Part 3, 2000; Part 4, 2001.
8. Lide, D. R., and Kehiaian, H.V., *CRC Handbook of Thermophysical and Thermochemical Data*, CRC Press, Boca Raton, FL, 1994.
9. *Kirk-Othmer Concise Encyclopedia of Chemical Technology*, Wiley-Interscience, New York, 1985.
10. *Dictionary of Inorganic Compounds*, Chapman & Hall, New York, 1992.
11. Massalski, T. B., ed., *Binary Alloy Phase Diagrams, 2nd Edition*, ASM International, Metals Park, Ohio, 1990.
12. Dinsdale, A.T., "SGTE Data for Pure Elements", *CALPHAD*, 15, 317-425, 1991.
13. Madelung, O., *Semiconductors: Group IV Elements and III-IV Compounds*, Springer-Verlag, Heidelberg, 1991.
14. Lidin, R. A., Andreeva, L. L., and Molochko, V. A., *Constants of Inorganic Substances*, Begell House, New York, 1995.
15. Gurvich, L. V., Veyts, I. V., and Alcock, C. B., *Thermodynamic Properties of Individual Substances, Fourth Edition*, Hemisphere Publishing Corp., New York, 1989.
16. *The Combined Chemical Dictionary on CDROM*, Version 9:1, Chapman & Hall/CRC, Boca Raton, FL, & London, 2005.
17. Macdonald, F., Editor, *Chapman & Hall/CRC Combined Chemical Dictionary*, <http://www.chemnetbase.com/scripts/ccdweb.exe>.
18. Sangeeta, G., and LaGraff, J. R., *Inorganic Materials Chemistry, Second Edition*, CRC Press, Boca Raton, FL, & London, 2005.
19. Stern, K. H., *High Temperature Properties and Thermal Decomposition of Inorganic Salts with Oxyanions*, CRC Press, Boca Raton, FL, & London, 2001.
20. Donnay, J.D.H., and Ondik, H.M., *Crystal Data Determinative Tables, Third Edition, Volumes 2 and 4, Inorganic Compounds*, Joint Committee on Powder Diffraction Standards, Swarthmore, PA, 1973.
21. Robie, R., Bethke, P. M., and Beardsley, K. M., *Selected X-ray Crystallographic Data, Molar Volumes, and Densities of Minerals and Related Substances*, U.S. Geological Survey Bulletin 1248, 1967.
22. Carmichael, R. S., *Practical Handbook of Physical Properties of Rocks and Minerals*, CRC Press, Boca Raton, FL, 1989.
23. Deer, W. A., Howie, R.A., and Zussman, J., An *Introduction to the Rock-Forming Minerals*, 2nd Edition, Longman Scientific & Technical, Harlow, Essex, 1992.
24. Linstrom, P. J., and Mallard, W. G., Editors, *NIST Chemistry WebBook, NIST Standard Reference Database No. 69*, June 2005, National Institute of Standards and Technology, Gaithersburg, MD 20899, <http://webbook.nist.gov>.
25. *Phase Diagrams for Ceramists, Volumes 1-8; ACerS-NIST Phase Equilibrium Diagrams, Volumes 9-13*, American Ceramic Society, Westerville, Ohio, 1964-2001.

No.	Name	Formula	CAS Reg No.	Mol. weight	Physical form	mp/°C	bp/°C	Density g cm⁻³	Solubility g/100 g H₂O	Qualitative solubility
1	Actinium	Ac	7440-34-8	227	silv metal; cub	1050	3198	10		
2	Actinium bromide	AcBr₃	33689-81-5	467	wh hex cry		800 subl	5.85		s H₂O
3	Actinium chloride	AcCl₃	22986-54-5	333	wh hex cry		960 subl	4.81		
4	Actinium fluoride	AcF₃	33689-80-4	284	wh hex cry			7.88		i H₂O
5	Actinium iodide	AcI₃	33689-82-6	608	wh cry					s H₂O
6	Actinium oxide	Ac₂O₃	12002-61-8	502	wh hex cry	1977		9.19		i H₂O
7	Aluminum	Al	7429-90-5	26.982	silv-wh metal; cub cry	660.32	2519	2.70		i H₂O; s acid, alk
8	Aluminum acetate	Al(C₂H₃O₂)₃	139-12-8	204.113	wh hyg solid	dec				s H₂O; sl ace
9	Aluminum diacetate	Al(OH)(C₂H₃O₂)₂	142-03-0	162.078	wh amorp powder					i H₂O
10	Aluminum ammonium sulfate	AlNH₄(SO₄)₂	7784-25-0	237.146	wh powder					sl H₂O; i EtOH
11	Aluminum ammonium sulfate dodecahydrate	AlNH₄(SO₄)₂ · 12H₂O	7784-26-1	453.329	col cry or powder	94.5	>280 dec	1.65		s H₂O; i EtOH
12	Aluminum antimonide	AlSb	25152-52-7	148.742	brn cub cry	1065		4.26		
13	Aluminum arsenide	AlAs	22831-42-1	101.903	oran cub cry; hyg	1740		3.76		
14	Aluminum borate	2Al₂O₃ · B₂O₃	11121-16-7	273.543	needles	≈1050				i H₂O
15	Aluminum borohydride	Al(BH₄)₃	16962-07-5	71.510	flam liq	-64.5	44.5			reac H₂O
16	Aluminum bromate nonahydrate	Al(BrO₃)₃ · 9H₂O	11126-81-1*	572.826	wh hyg cry	62	>100 dec			s H₂O
17	Aluminum bromide	AlBr₃	7727-15-3	266.694	wh-yel monocl cry; hyg	97.5	255	3.2		reac H₂O; s bz, tol

No.	Name	Formula	CAS Reg No.	Mol. weight	Physical form	mp/°C	bp/°C	Density g cm⁻³	Solubility g/100 g H₂O	Qualitative solubility
18	Aluminum bromide hexahydrate	$AlBr_3 \cdot 6H_2O$	7784-11-4	374.785	col-yel hyg cry	93		2.54		s H_2O, EtOH, CS_2
19	Aluminum carbide	Al_4C_3	1299-86-1	143.958	yel hex cry	2100	>2200 dec	2.36		reac H_2O
20	Aluminum chlorate nonahydrate	$Al(ClO_3)_3 \cdot 9H_2O$	15477-33-5	439.473	hyg cry					vs H_2O; s EtOH
21	Aluminum chloride	$AlCl_3$	7446-70-0	133.341	wh hex cry or powder; hyg	192.6	180 sp	2.48	45.1²⁵	s bz, ctc, chl
22	Aluminum chloride hexahydrate	$AlCl_3 \cdot 6H_2O$	7784-13-6	241.432	col hyg cry	100 dec		2.398	45.1²⁵	s EtOH, eth
23	Dichloromethylaluminum	$AlCl_2CH_3$	917-65-7	112.923	cry	72.7	95¹⁰0			s bz, eth, hc
24	Chlorodiethylaluminum	$AlCl(C_2H_5)_2$	96-10-6	120.557	col liq	-74		0.96		reac H_2O
25	Chlorodiisobutylaluminum	$AlCl(C_4H_9)_2$	1779-25-5	176.664	hyg col liq	-40		0.95		s eth, hx
26	Aluminum diboride	AlB_2	12041-50-8	48.604	powder	>920 dec		3.19		s dil HCl
27	Aluminum dodecaboride	AlB_{12}	12041-54-2	156.714	yel-brn prisms	2070		2.55		s hot HNO_3; i acid, alk
28	Aluminum ethanolate	$Al(C_2H_5O)_3$	555-75-9	162.163	liq, condenses to wh solid	140				reac H_2O; sl xyl
29	Aluminum fluoride	AlF_3	7784-18-1	83.977	wh hex cry	2250 tp (220 MPa)	1276 sp	3.10	0.50²⁵	
30	Aluminum fluoride monohydrate	$AlF_3 \cdot H_2O$	32287-65-3	101.992	orth cry			2.17	0.50²⁵	
31	Aluminum fluoride trihydrate	$AlF_3 \cdot 3H_2O$	15098-87-0	138.023	wh hyg cry			1.914	0.50²⁵	
32	Aluminum hexafluorosilicate nonahydrate	$Al_2(SiF_6)_3 \cdot 9H_2O$	17099-70-6	642.329	hex prisms	>500 dec				s H_2O
33	Aluminum hydride	AlH_3	7784-21-6	30.006	col hex cry	>150 dec				reac H_2O
34	Aluminum hydroxide	$Al(OH)_3$	21645-51-2	78.004	wh amorp powder			2.42		i H_2O; s alk, acid
35	Aluminum hydroxychloride	$Al_2(OH)_5Cl \cdot 2H_2O$	1327-41-9	210.483	gl solid					s H_2O
36	Aluminum iodide	AlI_3	7784-23-8	407.695	wh leaflets	188.28	382	3.98		reac H_2O
37	Aluminum iodide hexahydrate	$AlI_3 \cdot 6H_2O$	10090-53-6	515.786	yel hyg cry powder					vs H_2O; s EtOH, eth
38	Aluminum lactate	$Al(C_3H_5O_3)_3$	18917-91-4	294.192	powder					vs H_2O
39	Aluminum molybdate	$Al_2(MoO_4)_3$	15123-80-5	533.78	wh pow	≈950				
40	Aluminum nitrate	$Al(NO_3)_3$	13473-90-0	212.997	wh hyg solid	dec			68.9²⁵	vs EtOH; sl ace
41	Aluminum nitrate nonahydrate	$Al(NO_3)_3 \cdot 9H_2O$	7784-27-2	375.134	wh hyg monocl cry	73	135 dec	1.72	68.9²⁵	vs EtOH; i pyr
42	Aluminum nitride	AlN	24304-00-5	40.989	blue-wh hex cry	3000		3.255		reac H_2O
43	Aluminum oleate	$Al(C_{18}H_{33}O_2)_3$	688-37-9	871.342	yel solid					i H_2O; s EtOH, bz
44	Aluminum oxalate monohydrate	$Al_2(C_2O_4)_3 \cdot H_2O$	814-87-9	336.035	wh pow					i H_2O, EtOH; s acid
45	Aluminum oxide (α)	Al_2O_3	1344-28-1	101.961	wh powder; hex	2054	2977	3.99		i H_2O, os; sl alk
46	Aluminum oxide (γ)	Al_2O_3	1344-28-1	101.961	soft wh pow	trans to corundum 1200		3.97		i H_2O; s acid; sl alk
47	Aluminum oxyhydroxide (boehmite)	$AlO(OH)$	1318-23-6	59.989	wh orth cry	trans to diasphore 227		3.07		i H_2O; s hot acid, alk
48	Aluminum oxyhydroxide (diaspore)	$AlO(OH)$	14457-84-2	59.989	orth cry	dec 450		3.38		i H_2O; s acid, alk
49	Aluminum palmitate	$Al(C_{15}H_{31}COO)_3$	555-35-1	793.230	wh-yel powder					i H_2O, EtOH; s peth
50	Aluminum 2,4-pentanedioate	$Al(CH_3COCHCOCH_3)_3$	13963-57-0	324.306	pale yel prisms	194.6	315	1.27		i H2O; s bz, EtOH; sl hex
51	Aluminum perchlorate	$Al(ClO_4)_3$	14452-39-2	325.334	wh hyg cry				55⁰	s H_2O, eth; i ctc
52	Aluminum perchlorate nonahydrate	$Al(ClO_4)_3 \cdot 9H_2O$	14452-39-2	487.471	wh hyg cry	82 dec		2.0	182.4⁰	
53	Aluminum phosphate	$AlPO_4$	7784-30-7	121.953	wh rhomb plates	>1460		2.56		i H_2O; sl acid
54	Aluminum phosphate dihydrate	$AlPO_4 \cdot 2H_2O$	13477-75-3	157.984	wh rhom cry	dec 1500		2.54		i H_2O
55	Aluminum phosphate trihydroxide	$Al_2(OH)_3PO_4$	12004-29-4	199.957	wh or yel monocl cry			2.7		
56	Aluminum metaphosphate	$Al(PO_3)_3$	32823-06-6	263.898	col powder; tetr	≈1525		2.78		i H_2O
57	Aluminum hypophosphite	$Al(H_2PO_2)_3$	7784-22-7	221.948	cry powder	220 dec				i H_2O; s alk, acid
58	Aluminum phosphide	AlP	20859-73-8	57.956	grn or yel cub cry	2550		2.40		reac H_2O
59	Aluminum selenide	Al_2Se_3	1302-82-5	290.84	yel-brown powder	960		3.437		reac H_2O
60	Aluminum silicate (andalusite)	Al_2SiO_5	12183-80-1	162.046	gray-grn cry			3.145		
61	Aluminum silicate (kyanite)	Al_2SiO_5	1302-76-7	162.046	blue or gray tricl cry	dec 1000		3.68		
62	Aluminum silicate (mullite)	$3Al_2O_3 \cdot 2SiO_2$	1302-93-8	426.052	col orth cry	1750		3.17		
63	Aluminum silicate (sillimanite)	Al_2SiO_5	12141-45-6	162.046	wh orth cry	1816		3.25		
64	Aluminum silicate dihydrate	$Al_2O_3 \cdot 2SiO_2 \cdot 2H_2O$	1332-58-7	258.161	wh-yel powder; tricl			2.59		i H_2O, acid, alk
65	Aluminum stearate	$Al(C_{18}H_{35}O_2)_3$	637-12-7	877.390	wh powder	115		1.070		i H_2O, EtOH, eth; s alk
66	Aluminum monostearate	$Al(OH)_2(C_{18}H_{35}O_2)$	7047-84-9	344.467	yel-wh pow	155		1.02		i H_2O
67	Aluminum distearate	$Al(OH)(C_{18}H_{35}O_2)_2$	300-92-5	610.928	wh pow	145				i H_2O
68	Aluminum sulfate	$Al_2(SO_4)_3$	10043-01-3	342.151	wh cry	1040 dec			38.5²⁵	i EtOH
69	Aluminum sulfate octadecahydrate	$Al_2(SO_4)_3 \cdot 18H_2O$	7784-31-8	666.426	col monocl cry	86 dec		1.69	38.5²⁵	
70	Aluminum sulfide	Al_2S_3	1302-81-4	150.158	yel-gray powder	1100		2.02		
71	Aluminum telluride	Al_2Te_3	12043-29-7	436.76	gray-blk hex cry	≈895		4.5		

No.	Name	Formula	CAS Reg No.	Mol. weight	Physical form	mp/°C	bp/°C	Density g cm^{-3}	Solubility g/100 g H$_2$O	Qualitative solubility
72	Aluminum thiocyanate	Al(SCN)$_3$	538-17-0	201.229	yel powder					s H$_2$O; i EtOH, eth
73	Aluminum titanate	Al$_2$TiO$_5$	12004-39-6	181.827	refrac solid	1860				
74	Aluminum zirconium	Al$_3$Zr	12004-50-1	145.187	metallic solid	1645				
75	Americium	Am	7440-35-9	243	silv metal; hex or cub	1176	2011	12		s acid
76	Americium(III) oxide	Am$_2$O$_3$	12254-64-7	534	tan hex cry			11.77		s acid
77	Americium(III) bromide	AmBr$_3$	14933-38-1	483	wh orth cry			6.85		s H$_2$O
78	Americium(III) chloride	AmCl$_3$	13464-46-5	349	pink hex cry	500		5.87		
79	Americium(III) fluoride	AmF$_3$	13708-80-0	300	pink hex cry	1393		9.53		
80	Americium(III) iodide	AmI$_3$	13813-47-3	624	yel ortho cry	≈950		6.9		
81	Americium(IV) fluoride	AmF$_4$	15947-41-8	319	tan monocl cry			7.23		
82	Americium(IV) oxide	AmO$_2$	12005-67-3	275	blk cub cry	>1000 dec		11.68		s acid
83	Ammonia	NH$_3$	7664-41-7	17.031	col gas	-77.73	-33.33	0.696 g/L		vs H$_2$O; s EtOH, eth
84	Ammonium acetate	NH$_4$C$_2$H$_3$O$_2$	631-61-8	77.083	wh hyg cry	114		1.073	148[4]	s EtOH; sl ace
85	Ammonium azide	NH$_4$N$_3$	12164-94-2	60.059	orth cry; flam	160	exp	1.346	20.2[30]	
86	Ammonium benzoate	NH$_4$C$_7$H$_5$O$_2$	1863-63-4	139.152	wh cry or powder	198		1.26		s H$_2$O; sl EtOH
87	Ammonium bromate	NH$_4$BrO$_3$	13843-59-9	145.941	col hex cry	exp				vs H$_2$O
88	Ammonium bromide	NH$_4$Br	12124-97-9	97.943	wh hyg tetr cry	542 dec	396 sp	2.429	78.3[25]	s EtOH, ace; sl eth
89	Ammonium caprylate	NH$_4$C$_8$H$_{15}$O$_2$	5972-76-9	161.243	hyg monocl cry	≈75				reac H$_2$O; s EtOH; i chl, bz
90	Ammonium carbamate	NH$_2$COONH$_4$	1111-78-0	78.071	cry powder					vs H$_2$O; s EtOH
91	Ammonium carbonate	(NH$_4$)$_2$CO$_3$	506-87-6	96.086	col cry powder	58 dec			100[15]	
92	Ammonium chlorate	NH$_4$ClO$_3$	10192-29-7	101.490	wh cry	102 exp		1.80	28.7[0]	
93	Ammonium chloride	NH$_4$Cl	12125-02-9	53.492	col cub cry	520.1 tp (dec)	338 sp	1.519	39.5[25]	
94	Ammonium chromate	(NH$_4$)$_2$CrO$_4$	7788-98-9	152.071	yel cry	185 dec		1.90	37[25]	sl ace, MeOH; i EtOH
95	Ammonium chromic sulfate dodecahydrate	NH$_4$Cr(SO$_4$)$_2$ · 12H$_2$O	10022-47-6	478.343	blue-viol cry	94 dec		1.72		s H$_2$O; sl EtOH
96	Ammonium cobalt(II) phosphate	CoNH$_4$PO$_4$	14590-13-7	171.943	red-viol powder (hyd)					i H$_2$O; s acid
97	Ammonium cobalt(II) phosphate monohydrate	CoNH$_4$PO$_4$ · H$_2$O	16827-96-6	189.959	red-purp orth plates	dec 450				s acid
98	Ammonium cobalt(II) sulfate hexahydrate	(NH$_4$)$_2$Co(SO$_4$)$_2$ · 6H$_2$O	13586-38-4	395.227	red monocl prisms			1.90		s H$_2$O; i EtOH
99	Ammonium copper(II) chloride	CuCl$_2$ · 2NH$_4$Cl	10060-13-6*	241.435	yel hyg orth cry					s H$_2$O
100	Ammonium copper(II) chloride dihydrate	CuCl$_2$ · 2NH$_4$Cl · 2H$_2$O	10060-13-6	277.465	blue-grn tetr cry	110 dec		1.993		s H$_2$O, EtOH
101	Ammonium cyanide	NH$_4$CN	12211-52-8	44.056	col tetr cry	dec		1.10		vs H$_2$O
102	Ammonium dichromate	(NH$_4$)$_2$Cr$_2$O$_7$	7789-09-5	252.065	oran-red monocl cry; hyg	180 dec		2.155	35.6[20]	
103	Ammonium dihydrogen arsenate	NH$_4$H$_2$AsO$_4$	13462-93-6	158.975	tetr cry	300 dec		2.311	52.7[25]	
104	Ammonium dihydrogen phosphate	NH$_4$H$_2$PO$_4$	7722-76-1	115.026	wh tetr cry	190		1.80	40.4[25]	sl EtOH; i ace
105	Ammonium O,O-diethyldithiophosphate	(C$_2$H$_5$O)$_2$P(S)SNH$_4$	1068-22-0	203.264	cry	165				
106	Ammonium dithiocarbamate	NH$_4$NH$_2$CSS	513-74-6	110.202	yel ortho cry	99 dec		1.45		s H$_2$O
107	Ammonium ferricyanide trihydrate	(NH$_4$)$_3$Fe(CN)$_6$ · 3H$_2$O	14221-48-8*	320.110	red cry					s H$_2$O; i EtOH
108	Ammonium ferrocyanide trihydrate	(NH$_4$)$_4$Fe(CN)$_6$ · 3H$_2$O	14481-29-9*	338.149	yel cry	dec				s H$_2$O; i EtOH
109	Ammonium fluoride	NH$_4$F	12125-01-8	37.037	wh hex cry; hyg	238		1.015	83.5[25]	sl EtOH
110	Ammonium fluorosulfonate	NH$_4$SO$_3$F	13446-08-7	117.100	col needles	245				s H$_2$O, EtOH, MeOH
111	Ammonium formate	NH$_4$CHO$_2$	540-69-2	63.057	hyg cry	116		1.27	143[20]	s EtOH
112	Ammonium heptafluorotantalate	(NH$_4$)$_2$TaF$_7$	12022-02-5	350.014	hyg cry					
113	Ammonium hexabromoosmate(IV)	(NH$_4$)$_2$OsBr$_6$	24598-62-7	705.73	small blk cubes					sl H$_2$O; s glycerol; i EtOH
114	Ammonium hexabromoplatinate(IV)	(NH$_4$)$_2$PtBr$_6$	17363-02-9	710.585	powder	145 dec			0.59[20]	
115	Ammonium hexachloroiridate(III)	(NH$_4$)$_3$IrCl$_6$	15752-05-3	459.050	grn pow					
116	Ammonium hexachloroiridate(IV)	(NH$_4$)$_2$IrCl$_6$	16940-92-4	441.012	blk cry powder	dec		2.856	1.09[25]	
117	Ammonium hexachloroosmate(IV)	(NH$_4$)$_2$OsCl$_6$	12125-08-5	439.03	red cry or powder		subl	2.93		s H$_2$O, EtOH
118	Ammonium hexachloropalladate(IV)	(NH$_4$)$_2$PdCl$_6$	19168-23-1	355.22	red-brn hyg cry	dec		2.418		
119	Ammonium hexachloroplatinate(IV)	(NH$_4$)$_2$PtCl$_6$	16919-58-7	443.879	red-oran cub cry	380 dec		3.065	0.5[20]	i EtOH
120	Ammonium hexachlororuthenate(IV)	(NH$_4$)$_2$RuCl$_6$	18746-63-9	349.87	red cry					
121	Ammonium hexafluoroaluminate	(NH$_4$)$_3$AlF$_6$	7784-19-2	195.087	cub cry			1.78		s H$_2$O
122	Ammonium hexafluorogallate	(NH$_4$)$_3$GaF$_6$	14639-94-2	237.828	col cub cry	>200 dec		2.10		
123	Ammonium hexafluorogermanate	(NH$_4$)$_2$GeF$_6$	16962-47-3	222.71	wh cry	380	subl	2.564		s H$_2$O; i EtOH
124	Ammonium hexafluorophosphate	NH$_4$PF$_6$	16941-11-0	163.003	wh cub cry	58 dec		2.180		vs H$_2$O; s ace, EtOH, MeOH
125	Ammonium hexafluorosilicate	(NH$_4$)$_2$SiF$_6$	16919-19-0	178.153	wh cub or trig cry	dec		2.011	22.7[25]	i EtOH, ace
126	Ammonium hexafluorotitanate	(NH$_4$)$_2$TiF$_6$	16962-40-6	197.934	wh solid					s H$_2$O

No.	Name	Formula	CAS Reg No.	Mol. weight	Physical form	mp/°C	bp/°C	Density g cm^{-3}	Solubility g/100 g H$_2$O	Qualitative solubility
127	Ammonium hexafluorozirconate(IV)	$(NH_4)_2ZrF_6$	16919-31-6	241.291	wh hex cry			1.154		s H$_2$O
128	Ammonium hydrogen arsenate	$(NH_4)_2HAsO_4$	7784-44-3	176.004	wh powder			1.99		s H$_2$O
129	Ammonium hydrogen carbonate	NH_4HCO_3	1066-33-7	79.056	col or wh prisms	107 dec		1.586	24.8[25]	
130	Ammonium hydrogen citrate	$(NH_4)_2HC_6H_5O_7$	3012-65-5	226.184	col cry			1.48		vs H$_2$O; sl EtOH
131	Ammonium hydrogen fluoride	NH_4HF_2	1341-49-7	57.044	wh orth cry	125	240 dec	1.50	60.2[20]	
132	Ammonium hydrogen malate	$NH_4C_4H_5O_5$	5972-71-4	151.118	orth cry	160		1.15		s H$_2$O; sl EtOH
133	Ammonium hydrogen oxalate monohydrate	$NH_4HC_2O_4 \cdot H_2O$	5972-72-5*	125.081	col rhomb cry	dec		1.56		sl H$_2$O, EtOH
134	Ammonium hydrogen phosphate	$(NH_4)_2HPO_4$	7783-28-0	132.055	wh cry	155 dec		1.619	69.5[25]	i EtOH, ace
135	Ammonium hydrogen phosphite monohydrate	$(NH_4)_2HPO_3 \cdot H_2O$	51503-61-8	134.071	hyg cry					s H$_2$O
136	Ammonium hydrogen selenate	NH_4HSeO_4	10294-60-7	162.01	rhom cry	dec		2.162		
137	Ammonium hydrogen sulfate	NH_4HSO_4	7803-63-6	115.110	wh hyg cry	147		1.78	100[20]	i EtOH, ace, py
138	Ammonium hydrogen sulfide	NH_4HS	12124-99-1	51.112	wh tetr or orth cry	dec		1.17	128[0]	sl ace; i bz, eth
139	Ammonium hydrogen sulfite	NH_4HSO_3	10192-30-0	99.110	col cry	dec		2.03	71.8[0]	
140	Ammonium hydrogen tartrate	$NH_4HC_4H_4O_6$	3095-65-6	167.117	wh cry			1.68		sl H$_2$O; s alk; i EtOH
141	Ammonium hydroxide	NH_4OH	1336-21-6	35.046	exists only in soln					
142	Ammonium hypophosphite	$NH_4H_2PO_2$	7803-65-8	83.028	wh hyg cry	dec				vs H$_2$O; sl EtOH; i ace
143	Ammonium iodate	NH_4IO_3	13446-09-8	192.941	wh powder	150		3.3	3.84[25]	
144	Ammonium iodide	NH_4I	12027-06-4	144.943	wh tetr cry; hyg	551 dec	405 sp	2.514	178[25]	sl EtOH, MeOH
145	Ammonium iron(II) sulfate hexahydrate	$(NH_4)_2Fe(SO_4)_2 \cdot 6H_2O$	7783-85-9	392.139	blue-grn monocl cry	≈100 dec		1.86		s H$_2$O; i EtOH
146	Ammonium iron(III) chromate	$NH_4Fe(CrO_4)_2$	7789-08-4	305.871	red powder					i H$_2$O
147	Ammonium iron(III) oxalate trihydrate	$(NH_4)_3Fe(C_2O_4)_3 \cdot 3H_2O$	13268-42-3	428.063	grn monocl cry; hyg	≈160 dec		1.780		vs H$_2$O; i EtOH
148	Ammonium iron(III) sulfate dodecahydrate	$NH_4Fe(SO_4)_2 \cdot 12H_2O$	7783-83-7	482.192	col to viol cry	≈37		1.71		vs H$_2$O; i EtOH
149	Ammonium lactate	$NH_4C_3H_5O_3$	52003-58-4	107.108	col cry	92				s H$_2$O, EtOH; sl MeOH; i ace, eth
150	Ammonium magnesium chloride hexahydrate	$NH_4MgCl_3 \cdot 6H_2O$	39733-35-2	256.794	hyg cry	dec 100		1.46	17[20]	s H$_2$O
151	Ammonium mercuric chloride dihydrate	$(NH_4)_2HgCl_4 \cdot 2H_2O$	33445-15-7*	414.51	powder					s H$_2$O; sl EtOH
152	Ammonium metatungstate hexahydrate	$(NH_4)_6W_7O_{24} \cdot 6H_2O$	12028-48-7	1887.19	wh cry					s H$_2$O; i EtOH
153	Ammonium metavanadate	NH_4VO_3	7803-55-6	116.979	wh-yel cry	200 dec		2.326	4.8[20]	
154	Ammonium molybdate(VI) tetrahydrate	$(NH_4)_6Mo_7O_{24} \cdot 4H_2O$	12054-85-2	1235.86	col or grn-yel cry	90 dec		2.498	43	i EtOH
155	Ammonium dimolybdate	$(NH_4)_2Mo_2O_7$	27546-07-2	339.95	cry					s H$_2$O
156	Ammonium molybdophosphate	$(NH_4)_3PO_4 \cdot 12MoO_3$	12026-66-3	1876.35	grn or yel cry	dec			0.02[20]	sl H$_2$O; s alk
157	Ammonium nitrate	NH_4NO_3	6484-52-2	80.043	wh hyg cry; orth	169.7	dec 200-260	1.72	213[25]	sl MeOH
158	Ammonium nitrite	NH_4NO_2	13446-48-5	64.044	wh-yel cry	60 exp		1.69	221[25]	i eth
159	Ammonium nitroferricyanide	$(NH_4)_2Fe(CN)_5NO$	14402-70-1	252.016	red-brn cry					s H$_2$O, EtOH
160	Ammonium oleate	$NH_4C_{18}H_{33}O_2$	544-60-5	299.493	yel-brn paste	21				s H$_2$O; sl ace
161	Ammonium oxalate	$(NH_4)_2C_2O_4$	1113-38-8	124.096	col sol			1.5	5.20[25]	
162	Ammonium oxalate monohydrate	$(NH_4)_2C_2O_4 \cdot H_2O$	6009-70-7	142.110	wh orth cry	dec		1.50	5.20[25]	sl EtOH
163	Ammonium palmitate	$NH_4C_{15}H_{31}CO_2$	593-26-0	273.455	yel-wh powder	22				s H$_2$O; sl bz, xyl; i ace, EtOH, ctc
164	Ammonium pentaborate tetrahydrate	$NH_4B_5O_8 \cdot 4H_2O$	12007-89-5	272.150	wh cry				7.03[18]	
165	Ammonium pentachlororhodate(III) monohydrate	$(NH_4)_2RhCl_5 \cdot H_2O$	63771-33-5	334.262	red cry	dec 210				
166	Ammonium pentachlorozincate	$(NH_4)_3ZnCl_5$	14639-98-6	296.789	hyg orth cry			1.81		vs H$_2$O
167	Ammonium perchlorate	NH_4ClO_4	7790-98-9	117.490	wh orth cry	dec, exp		1.95	24.5[25]	s MeOH; sl EtOH, ace; i eth
168	Ammonium permanganate	NH_4MnO_4	13446-10-1	136.975	purp rhomb cry	70 dec		2.22	7.9[15]	
169	Ammonium peroxydisulfate	$(NH_4)_2S_2O_8$	7727-54-0	228.202	monocl cry or wh powder	dec		1.982	83.5[25]	
170	Ammonium perrhenate	NH_4ReO_4	13598-65-7	268.244	col powder			3.97	6.23[20]	
171	Ammonium phosphate trihydrate	$(NH_4)_3PO_4 \cdot 3H_2O$	10361-65-6*	203.133	wh prisms				25.0[25]	i ace
172	Ammonium phosphomolybdate monohydrate	$(NH_4)_3PO_4 \cdot 12MoO_3 \cdot H_2O$	54723-94-3	1894.36	yel cry or powder	dec			0.02	
173	Ammonium phosphotungstate dihydrate	$(NH_4)_3PO_4 \cdot 12WO_3 \cdot 2H_2O$	1311-90-6	2967.18	cry powder					sl H$_2$O
174	Ammonium picrate	$NH_4C_6H_2N_3O_7$	131-74-8	246.135	yel orth cry	exp		1.72		sl H$_2$O
175	Ammonium polysulfide	$(NH_4)_2S_x$	9080-17-5		yel unstab soln					reac acids
176	Ammonium salicylate	$NH_4C_7H_7O_3$	528-94-9	155.151	wh cry powder					vs H$_2$O; s EtOH

No.	Name	Formula	CAS Reg No.	Mol. weight	Physical form	mp/°C	bp/°C	Density g cm⁻³	Solubility g/100 g H₂O	Qualitative solubility
177	Ammonium selenate	$(NH_4)_2SeO_4$	7783-21-3	179.04	wh monocl cry	dec		2.194	117^{25}	i EtOH, ace
178	Ammonium selenite	$(NH_4)_2SeO_3$	7783-19-9	163.04	wh or red hyg cry	dec			121^{25}	
179	Ammonium stearate	$NH_4C_{18}H_{35}O_2$	1002-89-7	301.509	yel-wh powder	22		0.89		sl H₂O, bz; s EtOH, MeOH; i ace
180	Ammonium sulfamate	$NH_4NH_2SO_3$	7773-06-0	114.124	wh hyg cry	131	160 dec			vs H₂O; sl EtOH
181	Ammonium sulfate	$(NH_4)_2SO_4$	7783-20-2	132.140	wh or brn orth cry	280 dec		1.77	76.4^{25}	i EtOH, ace
182	Ammonium sulfide	$(NH_4)_2S$	12135-76-1	68.142	yel-oran cry	≈0 dec				s H₂O, EtOH, alk
183	Ammonium sulfite	$(NH_4)_2SO_3$	17026-44-7	116.140	wh hyg cry				64.2^{25}	
184	Ammonium sulfite monohydrate	$(NH_4)_2SO_3 \cdot H_2O$	7783-11-1	134.155	col cry	dec		1.41	64.2^{25}	i EtOH, ace
185	Ammonium tartrate	$(NH_4)_2C_4H_4O_6$	3164-29-2	184.147	wh cry	dec		1.601		s H₂O
186	Ammonium tellurate	$(NH_4)_2TeO_4$	13453-06-0	227.68	wh powder	dec		3.024		
187	Ammonium tetraborate tetrahydrate	$(NH_4)_2B_4O_7 \cdot 4H_2O$	12228-87-4	263.377	wh tetr cry	dec 87				vs H₂O; s HNO₃
188	Ammonium tetrachloroaluminate	NH_4AlCl_4	7784-14-7	186.833	wh hyg solid	304				s H₂O, eth
189	Ammonium tetrachloropalladate(II)	$(NH_4)_2PdCl_4$	13820-40-1	284.31	grn cry or red-brn pow					s H₂O
190	Ammonium tetrachloroplatinate(II)	$(NH_4)_2PtCl_4$	13820-41-2	372.973	red cry	dec		2.936		s H₂O; i EtOH
191	Ammonium tetrachlorozincate	$(NH_4)_2ZnCl_4$	14639-97-5	243.298	wh orth plates; hyg	150 dec		1.879		vs H₂O
192	Ammonium tetrafluoroantimonate	NH_4SbF_4	14972-90-8	215.793	col cry					s H₂O
193	Ammonium tetrafluoroborate	NH_4BF_4	13826-83-0	104.844	wh powder; orth	487 dec		1.871	25^{20}	
194	Ammonium tetrathiocyanodiammonochromate(III) monohydrate	$NH_4[Cr(NH_3)_2(SCN)_4] \cdot H_2O$	13573-16-5	354.440	red cry	270 dec				s H₂O, EtOH, ace; i bz
195	Ammonium tetrathiomolybdate	$(NH_4)_2MoS_4$	15060-55-6	260.28	red cry	100 dec				vs H₂O
196	Ammonium tetrathiotungstate	$(NH_4)_2WS_4$	13862-78-7	348.18	oran cry	dec		2.71		s H₂O
197	Ammonium tetrathiovandate	$(NH_4)_3VS_4$	14693-56-2	233.317	dark viol cry					
198	Ammonium thiocyanate	NH_4SCN	1762-95-4	76.121	col hyg cry	≈149	dec	1.30	181^{25}	vs EtOH; s ace; i chl
199	Ammonium thiosulfate	$(NH_4)_2S_2O_3$	7783-18-8	148.205	wh cry	150 dec		1.678		vs H₂O; i EtOH, eth
200	Ammonium titanium oxalate monohydrate	$(NH_4)_2TiO(C_2O_4)_2 \cdot H_2O$	10580-03-7	293.996	hyg cry					vs H₂O
201	Ammonium tungstate(VI)	$(NH_4)_{10}W_{12}O_{41}$	11120-25-5	3042.44	cry powder			2.3		s H₂O; i EtOH
202	Ammonium tungstate(VI) pentahydrate	$(NH_4)_{10}W_{12}O_{41} \cdot 5H_2O$	1311-93-9	3132.52	cry pow or plates			2.3		vs H₂O; i EtOH
203	Ammonium uranate(VI)	$(NH_4)_2U_2O_7$	7783-22-4	624.131	red-yel amorp powder					i H₂O, alk; s acid
204	Ammonium uranium fluoride	$UO_2(NH_4)_4F_5$	18433-40-4	419.135	grn-yel monocl cry					s H₂O; i EtOH
205	Ammonium valerate	$NH_4C_4H_9CO_2$	42739-38-8	119.163	hyg cry	108				vs H₂O, EtOH; s eth
206	Antimony (gray)	Sb	7440-36-0	121.760	silv metal; hex	630.628	1587	6.68		i dil acid
207	Antimony (black)	Sb	7440-36-0	121.760	blk amorp solid	trans gray 0				
208	Stibine	SbH_3	7803-52-3	124.784	col gas; flam	-88	-17	5.100 g/L		sl H₂O; s EtOH
209	Trimethylstibine	$Sb(CH_3)_3$	594-10-5	166.863	col flam liq	-62	81	1.52		
210	Pentamethylstibine	$Sb(CH_3)_5$	15120-50-0	196.933	col hyg liq	-19	127			reac H₂O
211	Tetramethyldistibine	$[Sb(CH_3)_2]_2$	41422-43-9	303.658	yel flam liq or red solid	17				
212	Antimony arsenide	SbAs	12322-34-8	196.682	hex cry	≈680		6.0		
213	Antimony potassium tartrate trihydrate	$K_2(SbC_4H_4O_6)_2 \cdot 3H_2O$	28300-74-5	669.889	col cry			2.6		sl H₂O
214	Antimony(III) acetate	$Sb(C_2H_3O_2)_3$	3643-76-3	298.891	wh pow					
215	Antimony(III) bromide	$SbBr_3$	7789-61-9	361.472	yel orth cry; hyg	97	288	4.35		reac H₂O; s ace, bz, chl
216	Antimony(III) chloride	$SbCl_3$	10025-91-9	228.119	col orth cry; hyg	73.4	220.3	3.14	987^{25}	s acid, EtOH, bz, ace
217	Antimony(III) fluoride	SbF_3	7783-56-4	178.755	wh orth cry; hyg	287	376	4.38	492^{25}	
218	Antimony(III) iodide	SbI_3	7790-44-5	502.473	red rhomb cry	171	400	4.92		reac H₂O; s EtOH, ace; i ctc
219	Antimony(III) iodide sulfide	SbIS	13816-38-1	280.729	dark red prisms or needles	400				
220	Antimony(III) oxide (senarmontite)	Sb_2O_3	1309-64-4	291.518	col cub cry	570 trans	1425	5.58		sl H₂O; i os
221	Antimony(III) oxide (valentinite)	Sb_2O_3	1309-64-4	291.518	wh orth cry	655	1425	5.7		sl H₂O; i os
222	Antimony(III) oxychloride	SbOCl	7791-08-4	173.212	wh momo cry	170 dec				reac H₂O; i EtOH, eth
223	Antimony(III) phosphate	$SbPO_4$	12036-46-3	216.731	cry pow					reac H₂O
224	Antimony(III) potassium oxalate trihydrate	$K_3Sb(C_2O_4)_3 \cdot 3H_2O$	5965-33-3*	557.158	cry pow					s H₂O
225	Antimony(III) selenide	Sb_2Se_3	1315-05-5	480.40	grn orth cry	611		5.81		sl H₂O
226	Antimony(III) sulfate	$Sb_2(SO_4)_3$	7446-32-4	531.708	wh cry powder; hyg	dec		3.62		sl H₂O
227	Antimony(III) sulfide	Sb_2S_3	1345-04-6	339.715	gray-blk orth cry	550		4.562		i H₂O; s conc HCl
228	Antimony(III) telluride	Sb_2Te_3	1327-50-0	626.32	gray cry	620		6.5		

No.	Name	Formula	CAS Reg No.	Mol. weight	Physical form	mp/°C	bp/°C	Density g cm^{-3}	Solubility g/100 g H$_2$O	Qualitative solubility
229	Antimony(III,V) oxide	Sb_2O_4	1332-81-6	307.518	yel orth cry			6.64		
230	Antimony(V) chloride	$SbCl_5$	7647-18-9	299.025	col or yel liq	4	140 dec	2.34		reac H$_2$O; s chl, ctc
231	Antimony(V) fluoride	SbF_5	7783-70-2	216.752	hyg visc liq	8.3	141	3.10		reac H$_2$O
232	Antimony(V) dichlorotrifluoride	$SbCl_2F_3$	7791-16-4	249.661	visc liq					reac H$_2$O
233	Antimony(V) oxide	Sb_2O_5	1314-60-9	323.517	yel powder; cub	dec		3.78	0.3^{20}	
234	Antimony(V) sulfide	Sb_2S_5	1315-04-4	403.845	oran-yel powder	75 dec		4.120		i H$_2$O; s acid, alk
235	Argon	Ar	7440-37-1	39.948	col gas	-189.36 tp (69 kPa)	-185.847	1.633 g/L		sl H$_2$O
236	Arsenic (gray)	As	7440-38-2	74.922	gray metal; rhomb	817 tp (3.70 MPa)	616 sp	5.75		i H$_2$O
237	Arsenic (black)	As	7440-38-2	74.922	blk amorp solid	trans gray As 270		4.9		
238	Arsenic (yellow)	As	7440-38-2	74.922	soft yel cub cry	trans gray As 358		1.97		s CS$_2$
239	Arsine	AsH_3	7784-42-1	77.946	col gas	-116	-62.5	3.186 g/L		sl H$_2$O
240	Diarsine	As_2H_4	15942-63-9	153.875	unstab liq		≈100			
241	Arsenic acid	H_3AsO_4	7778-39-4	141.944	exists only in soln					
242	Arsenic acid hemihydrate	$H_3AsO_4 \cdot 0.5H_2O$	7778-39-4*	150.951	wh hyg cry	36.1		2.5		vs H$_2$O, EtOH
243	Arsenious acid	H_3AsO_3	13464-58-9	125.944	exists only in soln					
244	Arsenic diiodide	As_2I_4	13770-56-4	657.461	red cry	137				reac H$_2$O; s os
245	Arsenic hemiselenide	As_2Se	1303-35-1	228.80	blk cry					i H$_2$O, os; dec acid, alk
246	Arsenic sulfide	As_4S_4	12279-90-2	427.946	red monocl cry	320	565	3.5		i H$_2$O; sl bz; s alk
247	Arsenic(III) bromide	$AsBr_3$	7784-33-0	314.634	col or yel orth cry; hyg	31.1	221	3.40		reac H$_2$O; s hc, ctc; vs eth, bz
248	Arsenic(III) chloride	$AsCl_3$	7784-34-1	181.281	col liq	-16	130	2.150		reac H$_2$O; vs chl, ctc, eth
249	Arsenic(III) ethoxide	$As(C_2H_5O)_3$	3141-12-6	210.103	liq		166	1.21		
250	Arsenic(III) fluoride	AsF_3	7784-35-2	131.917	col liq	-5.9	57.13	2.7		reac H$_2$O; s EtOH, eth, bz
251	Arsenic(III) iodide	AsI_3	7784-45-4	455.635	red hex cry	141	424	4.73		sl H$_2$O, EtOH, eth; s bz. tol
252	Arsenic(III) oxide (arsenolite)	As_2O_3	1327-53-3	197.841	wh cub cry	274	460	3.86	2.05^{25}	
253	Arsenic(III) oxide (claudetite)	As_2O_3	1327-53-3	197.841	wh monocl cry	314	460	3.74	2.05^{25}	s dil acid, alk; i EtOH
254	Arsenic(III) selenide	As_2Se_3	1303-36-2	386.72	brn-blk solid	260		4.75		i H$_2$O; s alk
255	Arsenic(III) sulfide	As_2S_3	1303-33-9	246.038	yel-oran monocl cry	312	707	3.46		i H$_2$O; s alk
256	Arsenic(III) telluride	As_2Te_3	12044-54-1	532.64	blk monocl cry	621		6.50		
257	Arsenic(V) chloride	$AsCl_5$	22441-45-8	252.187	stab at low temp	≈-50 dec				
258	Arsenic(V) fluoride	AsF_5	7784-36-3	169.914	col gas	-79.8	-52.8	6.945 g/L		reac H$_2$O; s EtOH, bz, eth
259	Arsenic(V) oxide	As_2O_5	1303-28-2	229.840	wh amorp powder	315		4.32	65.8^{20}	vs EtOH
260	Arsenic(V) selenide	As_2Se_5	1303-37-3	544.64	blk solid	dec				i H$_2$O, EtOH, eth; s alk
261	Arsenic(V) sulfide	As_2S_5	1303-34-0	310.168	brn-yel amorp solid	dec				i H$_2$O; s alk
262	Astatine	At	7440-68-8	210	cry	302				s HNO$_3$, os
263	Barium	Ba	7440-39-3	137.327	silv-yel metal; cub	727	1897	3.62		reac H$_2$O; sl EtOH
264	Barium acetate	$Ba(C_2H_3O_2)_2$	543-80-6	255.416	wh powder			2.47	79.2^{25}	
265	Barium acetate monohydrate	$Ba(C_2H_3O_2)_2 \cdot H_2O$	5908-64-5	273.431	wh cry	110 dec		2.19	79.2^{25}	sl EtOH
266	Barium aluminate	$BaAl_2O_4$	12004-04-5	255.288	hex cry	1827				
267	Barium aluminide	$BaAl_4$	12672-79-6	245.253	metallic solid	1097				
268	Barium azide	$Ba(N_3)_2$	18810-58-7	221.367	monocl cry; exp	≈120 dec		2.936	17.3^{20}	sl EtOH; i eth
269	Barium bismuthate	$BaBiO_3$	12785-50-1	394.305	bronze cry	1040 dec				
270	Barium bromate	$Ba(BrO_3)_2$	13967-90-3	393.131	col monocl cry				0.79^{25}	s ace
271	Barium bromate monohydrate	$Ba(BrO_3)_2 \cdot H_2O$	10326-26-8	411.147	wh monocl cry	260 dec		3.99	0.831^{25}	i EtOH
272	Barium bromide	$BaBr_2$	10553-31-8	297.135	wh orth cry	857	1835	4.781	100^{25}	
273	Barium bromide dihydrate	$BaBr_2 \cdot 2H_2O$	7791-28-8	333.166	wh cry	75 dec		3.7	100^{25}	s MeOH; i EtOH, ace, diox
274	Barium calcium tungstate	Ba_3CaWO_6	15552-14-4	594.57	cub cry	1420				
275	Barium carbide	BaC_2	50813-65-5	161.348	gray tetr cry	dec		3.74		reac H$_2$O
276	Barium carbonate	$BaCO_3$	513-77-9	197.336	wh orth cry	1380 dec; 1555 (high pres.)		4.308	0.0014^{20}	s acid
277	Barium chlorate	$Ba(ClO_3)_2$	13477-00-4	304.229	wh cry	414			37.9^{25}	sl EtOH, ace
278	Barium chlorate monohydrate	$Ba(ClO_3)_2 \cdot H_2O$	10294-38-9	322.245	wh monocl cry	120 dec		3.179	37.9^{25}	s acid; sl EtOH, ace
279	Barium chloride	$BaCl_2$	10361-37-2	208.233	wh orth cry; hyg	961	1560	3.9	37.0^{25}	
280	Barium chloride dihydrate	$BaCl_2 \cdot 2H_2O$	10326-27-9	244.264	wh monocl cry	≈120 dec		3.097	37.0^{25}	i EtOH
281	Barium chloride fluoride	BaClF	13718-55-3	191.778	wh cry					
282	Barium chromate(V)	$Ba_3(CrO_4)_2$	12345-14-1	643.968	grn-blk hex cry			5.25		s H$_2$O

No.	Name	Formula	CAS Reg No.	Mol. weight	Physical form	mp/°C	bp/°C	Density g cm^{-3}	Solubility g/100 g H$_2$O	Qualitative solubility
283	Barium chromate(VI)	$BaCrO_4$	10294-40-3	253.321	yel orth cry	1380		4.50	0.00026^{20}	reac acid
284	Barium citrate monohydrate	$Ba_3(C_6H_5O_7)_2 \cdot H_2O$	512-25-4*	808.195	gray-wh cry					s H$_2$O, acid
285	Barium copper yttrium oxide	$BaCuY_2O_5$	82642-06-6	458.682	grn cry; not superconductor					
286	Barium copper yttrium oxide	$Ba_2Cu_3YO_7$	109064-29-1	666.194	blk solid; HT superconductor					
287	Barium copper yttrium oxide	$Ba_2Cu_4YO_8$	114104-80-2	745.739	HT superconductor					
288	Barium copper yttrium oxide	$Ba_4Cu_7Y_2O_{15}$	124365-83-9	1411.933	HT superconductor					
289	Barium cyanide	$Ba(CN)_2$	542-62-1	189.361	wh cry powder					vs H$_2$O; s EtOH
290	Barium dichromate dihydrate	$BaCr_2O_7 \cdot 2H_2O$	10031-16-0	389.346	brn-red needles	dec				reac H$_2$O
291	Barium disilicate	$BaSi_2O_5$	12650-28-1	273.495	wh orth cry	1420		3.70		
292	Barium dithionate dihydrate	$BaS_2O_6 \cdot 2H_2O$	13845-17-5	333.484	wh cry	140 dec		4.54	22.1^{20}	sl EtOH
293	Barium ferrite	$BaFe_{12}O_{19}$	11138-11-7	1111.456	magnetic solid					
294	Barium ferrocyanide hexahydrate	$Ba_2Fe(CN)_6 \cdot 6H_2O$	13821-06-2*	594.694	yel monocl cry	80 dec				i H$_2$O, EtOH
295	Barium fluoride	BaF_2	7787-32-8	175.324	wh cub cry	1368	2260	4.893	0.161^{25}	s H$_2$O; i EtOH
296	Barium formate	$Ba(CHO_2)_2$	541-43-5	227.362	cry			3.21		s H$_2$O; i EtOH
297	Barium hexaboride	BaB_6	12046-08-1	202.193	blk cub cry	2070		4.36		i H$_2$O; s acid; i EtOH
298	Barium hexafluorogermanate	$BaGeF_6$		323.96	wh cry	≈665		4.56		i H$_2$O, EtOH; sl acid
299	Barium hexafluorosilicate	$BaSiF_6$	17125-80-3	279.403	wh orth needles	300 dec		4.29		reac H$_2$O
300	Barium hydride	BaH_2	13477-09-3	139.343	gray orth cry	1200		4.16		
301	Barium hydrogen phosphate	$BaHPO_4$	10048-98-3	233.306	wh cry powder	400 dec		4.16	0.015^{20}	s dil acid
302	Barium hydrosulfide	$Ba(HS)_2$	25417-81-6	203.473	yel hyg cry					s H$_2$O
303	Barium hydrosulfide tetrahydrate	$Ba(HS)_2 \cdot 4H_2O$	12230-74-9	275.534	yel rhomb cry	50 dec				s H$_2$O
304	Barium hydroxide	$Ba(OH)_2$	17194-00-2	171.342	wh powder	408			4.91^{25}	
305	Barium hydroxide monohydrate	$Ba(OH)_2 \cdot H_2O$	22326-55-2	189.357	wh powder			3.743	4.91^{25}	s acid
306	Barium hydroxide octahydrate	$Ba(OH)_2 \cdot 8H_2O$	12230-71-6	315.464	wh monocl cry	78 dec		2.18	4.91^{25}	
307	Barium hypophosphite monohydrate	$Ba(H_2PO_2)_2 \cdot H_2O$	14871-79-5*	285.320	monocl plates			2.90		s H$_2$O; i EtOH
308	Barium iodate	$Ba(IO_3)_2$	10567-69-8	487.132	wh cry powder	476 dec		5.23	0.0396^{25}	
309	Barium iodate monohydrate	$Ba(IO_3)_2 \cdot H_2O$	7787-34-0	505.148	cry	130 dec		5.00	0.0396^{25}	s acid; i EtOH
310	Barium iodide	BaI_2	13718-50-8	391.136	wh orth cry	711		5.15	221^{25}	
311	Barium iodide dihydrate	$BaI_2 \cdot 2H_2O$	7787-33-9	427.167	col cry	740 dec		5.0	221^{25}	s EtOH, ace
312	Barium manganate(VI)	$BaMnO_4$	7787-35-1	256.263	grn-gray hyg cry			4.85	0.00041^{20}	
313	Barium metaborate monohydrate	$Ba(BO_2)_2 \cdot H_2O$	26124-86-7	240.962	wh powder	>900		3.3		sl H$_2$O
314	Barium metaborate dihydrate	$Ba(BO_2)_2 \cdot 2H_2O$	23436-05-7	258.977	wh prec	dec			1.3^{25}	sl H$_2$O
315	Barium metaphosphate	$Ba(PO_3)_2$	13466-20-1	295.271	wh powder	1560				i H$_2$O; sl acid
316	Barium metasilicate	$BaSiO_3$	13255-26-0	213.411	col rhomb powder	1605		4.40		i H$_2$O; s acid
317	Barium molybdate	$BaMoO_4$	7787-37-3	297.27	wh powder	1450		4.975	0.0021^{20}	
318	Barium niobate	$Ba(NbO_3)_2$	12009-14-2	419.136	yel orth cry	1455		5.44		i H$_2$O
319	Barium nitrate	$Ba(NO_3)_2$	10022-31-8	261.336	wh cub cry	590		3.24	10.3^{25}	sl EtOH, ace
320	Barium nitride	Ba_3N_2	12047-79-9	439.994	yel-brn cry	>500 dec		4.78		reac H$_2$O
321	Barium nitrite	$Ba(NO_2)_2$	13465-94-6	229.338	col hex cry	267		3.234	79.5^{25}	
322	Barium nitrite monohydrate	$Ba(NO_2)_2 \cdot H_2O$	7787-38-4	247.353	yel-wh hex cry	217 dec		3.18	79.5^{25}	i EtOH
323	Barium orthovanadate	$Ba_3(VO_4)_2$	39416-30-3	641.859	hex cry	707		5.14		
324	Barium oxalate	BaC_2O_4	516-02-9	225.346	wh powder	400 dec		2.658	0.0075	
325	Barium oxalate monohydrate	$BaC_2O_4 \cdot H_2O$	13463-22-4	243.361	wh cry powder			2.66	0.0075^{20}	s acid
326	Barium oxide	BaO	1304-28-5	153.326	wh-yel powder; cub and hex	1973		5.72(cub)	1.5^{20}	s dil acid, EtOH; i ace
327	Barium 2,4-pentanedioate octahydrate	$Ba(CH_3COCHCOCH_3)_2 \cdot 8H_2O$	12084-29-6*	479.665	col hyg cry	320 (anh)				
328	Barium perchlorate	$Ba(ClO_4)_2$	13465-95-7	336.228	col hex cry	505		3.20	312^{25}	vs EtOH
329	Barium perchlorate trihydrate	$Ba(ClO_4)_2 \cdot 3H_2O$	10294-39-0	390.274	col cry			2.74	312^{25}	s MeOH; sl EtOH, ace; i eth
330	Barium permanganate	$Ba(MnO_4)_2$	7787-36-2	375.198	brn-viol cry	200 dec		3.77	62.5^{20}	reac EtOH
331	Barium peroxide	BaO_2	1304-29-6	169.326	gray-wh tetr cry	450 dec		4.96	0.091^{20}	reac dil acid
332	Barium plumbate	$BaPbO_3$	12047-25-5	392.5	orth cry					
333	Barium potassium chromate	$BaK_2(CrO_4)_2$	27133-66-0	447.511	yel hex cry			3.63		vs H$_2$O
334	Barium pyrophosphate	$Ba_2P_2O_7$	13466-21-2	448.597	wh powder	1430		3.9	0.0088^{20}	s acid
335	Barium selenate	$BaSeO_4$	7787-41-9	280.29	wh rhomb cry	dec		4.75	0.015^{20}	
336	Barium selenide	$BaSe$	1304-39-8	216.29	cub cry powder	1780		5.02		reac H$_2$O
337	Barium selenite	$BaSeO_3$	13718-59-7	264.29	solid					i H$_2$O
338	Barium silicide	$BaSi_2$	1304-40-1	193.498	gray lumps	1180				reac H$_2$O
339	Barium sodium niobate	$Ba_2Na(NbO_3)_5$	12323-03-4	1002.167	wh orth cry	1437		5.40		i H$_2$O
340	Barium stannate	$BaSnO_3$	12009-18-6	304.035	cub cry			7.24		sl H$_2$O
341	Barium stannate trihydrate	$BaSnO_3 \cdot 3H_2O$	12009-18-6*	358.081	wh cry powder					sl H$_2$O; s acid

Physical Constants of Inorganic Compounds

No.	Name	Formula	CAS Reg No.	Mol. weight	Physical form	mp/°C	bp/°C	Density g cm⁻³	Solu g/100
342	Barium stearate	$Ba(C_{18}H_{35}O_2)_2$	6865-35-6	704.266	wh powder	160		1.145	
343	Barium strontium niobate	$BaSr(NbO_3)_4$	37185-09-4	788.57	pale yel solid				
344	Barium strontium tungstate	Ba_2SrWO_6	14871-56-8	642.11	hyg pow	1400			
345	Barium sulfate	$BaSO_4$	7727-43-7	233.390	wh orth cry	1580		4.49	0.00031[20]
346	Barium sulfide	BaS	21109-95-5	169.392	col cub cry or gray powder	2227		4.3	8.94[25]
347	Barium sulfite	$BaSO_3$	7787-39-5	217.390	wh monocl cry	dec		4.44	0.0011[25]
348	Barium tartrate	$BaC_4H_4O_6$	5908-81-6	285.398	wh cry			2.98	
349	Barium tetracyanoplatinate(II) tetrahydrate	$BaPt(CN)_4 \cdot 4H_2O$	13755-32-3	508.543	yel powder or cry			2.076	
350	Barium tetraiodomercurate(II)	$BaHgI_4$	10048-99-4	845.54	yel-red hyg cry				
351	Barium thiocyanate	$Ba(SCN)_2$	2092-17-3	253.491	hyg cry				167[25]
352	Barium thiocyanate dihydrate	$Ba(SCN)_2 \cdot 2H_2O$	2092-17-3*	289.522	hyg wh cry				167[25]
353	Barium thiocyanate trihydrate	$Ba(SCN)_2 \cdot 3H_2O$	68016-36-4	307.537	wh needles; hyg			2.286	167[25]
354	Barium thiosulfate	BaS_2O_3	35112-53-9	249.455	wh cry powder	220 dec			0.2[20]
355	Barium thiosulfate monohydrate	$BaS_2O_3 \cdot H_2O$	7787-40-8	267.471	wh cry powder	dec		3.5	0.2
356	Barium titanate ($BaTiO_3$)	$BaTiO_3$	12047-27-7	233.192	wh tetr cry	1625		6.02	
357	Barium titanate ($BaTi_2O_5$)	$BaTi_2O_5$	12009-27-7	313.058	wh solid				
358	Barium titanate ($BaTi_4O_9$)	$BaTi_4O_9$	12009-31-3	472.790	wh solid				
359	Barium titanium silicate	$BaTi(SiO_3)_3$	15491-35-7	413.446	rhom blue-pur cry				
360	Barium tungstate	$BaWO_4$	7787-42-0	385.17	wh tetr cry	1475	1730	5.04	0.0016[20]
361	Barium uranium oxide	BaU_2O_7	10380-31-1	725.381	oran-yel powder				i H_2O; s acid
362	Barium yttrium tungsten oxide	$Ba_3Y_3WO_9$	37265-86-4	1006.53	cub cry	1470			
363	Barium zirconate	$BaZrO_3$	12009-21-1	276.549	gray-wh cub cry	2500		5.52	i H_2O, alk; sl acid
364	Barium zirconium silicate	$BaO \cdot ZrO_2 \cdot SiO_2$		336.634	wh pow				
365	Berkelium (α form)	Bk	7440-40-6	247	hex cry	trans to · 930		14.78	
366	Berkelium (β form)	Bk	7440-40-6	247	cub cry	986		13.25	
367	Beryllium	Be	7440-41-7	9.012	hex cry	1287	2471	1.85	s acid, alk
368	Beryllium acetate	$Be(C_2H_3O_2)_2$	543-81-7	127.101	wh cry	60 dec			i H_2O, EtOH
369	Beryllium basic acetate	$Be_4O(C_2H_3O_2)_6$	1332-52-1	406.312	wh cry	285	330	1.25	i H_2O; s eth, os
370	Beryllium aluminate	$BeAl_2O_4$	12004-06-7	126.973	orth cry			3.65	
371	Beryllium aluminum metasilicate	$Be_3Al_2(SiO_3)_6$	1302-52-9	537.502	col or grn-yel cry; hex			2.64	
372	Beryllium boride (BeB_2)	BeB_2	12228-40-9	30.634	refrac solid	>1970			
373	Beryllium boride (BeB_6)	BeB_6	12429-94-6	73.878	red solid	2070			
374	Beryllium boride (Be_2B)	Be_2B	12536-51-5	28.835	pink cry	1520			
375	Beryllium boride (Be_4B)	Be_4B	12536-52-6	46.860	refrac solid	1160			
376	Beryllium borohydride	$Be(BH_4)_2$	17440-85-6	36.682	solid	125 dec	subl		reac H_2O
377	Beryllium bromide	$BeBr_2$	7787-46-4	168.820	orth cry; hyg	508	473 sp	3.465	vs H_2O; s EtOH, p
378	Beryllium carbide	Be_2C	506-66-1	30.035	red cub cry	2127		1.90	reac H_2O
379	Beryllium carbonate tetrahydrate	$BeCO_3 \cdot 4H_2O$	60883-64-9	93.085	wh solid	100 dec			0.36[0]
380	Beryllium basic carbonate	$Be_2(OH)_2(CO_3)_2$	66104-24-3	181.069	wh powder				i H_2O; s acid, alk
381	Beryllium chloride	$BeCl_2$	7787-47-5	79.918	wh-yel orth cry; hyg	415	482	1.90	71.5[25]	s EtOH, eth, py; i bz, tol
382	Beryllium fluoride	BeF_2	7787-49-7	47.009	tetr cry or gl; hyg	552	1283	2.1	vs H_2O; sl EtOH
383	Beryllium formate	$Be(CHO_2)_2$	1111-71-3	99.047	powder	>250 dec			reac H_2O; i os
384	Beryllium hydride	BeH_2	7787-52-2	11.028	wh amorp solid	250 dec		0.65	reac H_2O; i eth, to
385	Beryllium hydrogen phosphate	$BeHPO_4$	13598-15-7	104.991	cry				i H_2O
386	Beryllium hydroxide (α)	$Be(OH)_2$	13327-32-7	43.027	wh powder or cry	≈200 dec		1.92	sl H_2O, alk; s acid
387	Beryllium hydroxide (β)	$Be(OH)_2$	13327-32-7	43.027	col tetr cry	dec 138			i H_2O; s, acid, alk
388	Beryllium iodide	BeI_2	7787-53-3	262.821	hyg needles	480	590	4.32	reac H_2O; s EtOH
389	Beryllium nitrate trihydrate	$Be(NO_3)_2 \cdot 3H_2O$	13597-99-4	187.068	yel-wh hyg cry	≈30	dec		107[20]	s EtOH
390	Beryllium nitride	Be_3N_2	1304-54-7	55.050	gray refrac cry; cub	2200		2.71	reac acid, alk
391	Beryllium oxalate trihydrate	$BeC_2O_4 \cdot 3H_2O$	15771-43-4	151.077	rhom cry	dec 320			vs H_2O
392	Beryllium oxide	BeO	1304-56-9	25.011	wh hex cry	2578		3.01	i H_2O; sl acid, alk
393	Beryllium 2,4-pentanedioate	$Be(CH_3COCHCOCH_3)_2$	10210-64-7	207.228	monocl cry powder	108	270	1.168	i H_2O; vs EtOH, eth
394	Beryllium perchlorate tetrahydrate	$Be(ClO_4)_2 \cdot 4H_2O$	7787-48-6	279.975	hyg cry	250 dec			198[25]	
395	Beryllium selenate tetrahydrate	$BeSeO_4 \cdot 4H_2O$	10039-31-3	224.03	orth cry	100 dec		2.03	vs H_2O
396	Beryllium sulfate	$BeSO_4$	13510-49-1	105.075	col tetr cry; hyg	1127		2.5	41.3[25]	
397	Beryllium sulfate dihydrate	$BeSO_4 \cdot 2H_2O$	14215-00-0	141.105	col cry	dec 92			
398	Beryllium sulfate tetrahydrate	$BeSO_4 \cdot 4H_2O$	7787-56-6	177.136	col tetr cry	≈100 dec		1.71	41.3[25]	i EtOH
399	Beryllium sulfide	BeS	13598-22-6	41.077	col cub cry			2.36	reac hot H_2O
400	Bismuth	Bi	7440-69-9	208.980	gray-wh soft metal	271.406	1564	9.79	s acid
401	Bismuth acetate	$Bi(C_2H_3O_2)_3$	22306-37-2	386.111	col tablets	250			i H_2O
402	Bismuth subacetate	$BiOC_2H_3O_2$	5142-76-7	284.023	thin cry plates				i H_2O; s dil acid
403	Bismuth antimonide	BiSb	12323-19-2	330.740	cry	475			

No.	Name	Formula	CAS Reg No.	Mol. weight	Physical form	mp/°C	bp/°C	Density g cm⁻³	Solubility g/100 g H_2O	Qualitative solubility
	Bismuth arsenate	$BiAsO_4$	13702-38-0	347.900	wh monocl cry			7.14		i H_2O; sl conc HNO_3
	Bismuth basic carbonate	$(BiO)_2CO_3$	5892-10-4	509.969	wh powder			6.86		i H_2O; s acid
	Bismuth basic dichromate	$Bi_2O_3 \cdot 2CrO_3$		665.948	red-oran amorp pow					i H_2O; s acid, alk
	Bismuth citrate	$BiC_6H_5O_7$	813-93-4	398.080	wh powder			3.458		i H_2O; sl EtOH
	Bismuth hydride	BiH_3	18288-22-7	212.004	col gas; unstab	-67	≈17	8.665 g/L		
	Bismuth hydroxide	$Bi(OH)_3$	10361-43-0	260.002	wh-yel amorp powder			4.962		i H_2O; s acid
	Bismuth germanium oxide	$2Bi_2O_3 \cdot 3GeO_2$	12233-56-6	1245.84	wh pow	1044				
	Bismuth hexafluoro-2,4-pentanedioate	$Bi(CF_3COCHCOCF_3)_3$	142617-56-9	830.132	powder	96				
	Bismuth molybdate	Bi_2MoO_6	13565-96-3	609.90	yel solid			9.32		
	Bismuth molybdate	$Bi_2(MoO_4)_3$	51898-99-8	897.77	monocl cry			5.95		
	Bismuth nitrate pentahydrate	$Bi(NO_3)_3 \cdot 5H_2O$	10035-06-0	485.071	col tricl cry; hyg	≈75 dec		2.83		reac H_2O; s ace; i EtOH
5	Bismuth subnitrate	$Bi_5O(OH)_9(NO_3)_4$	1304-85-4	1461.987	hyg cry powder	260 dec		4.928		i H_2O, EtOH; s dil acid
16	Bismuth oleate	$Bi(C_{18}H_{33}O_2)_3$	52951-38-9	1053.340	soft yel-brn solid					i H_2O; s eth; sl bz
17	Bismuth oxalate	$Bi_2(C_2O_4)_3$	6591-55-5	682.018	wh powder					i H_2O, EtOH; s dil acid
418	Bismuth oxide	Bi_2O_3	1304-76-3	465.959	yel monocl cry or powder	825	1890	8.9		i H_2O; s acid
419	Bismuth tetroxide	Bi_2O_4	12048-50-9	481.959	red-oran powder	305		5.6		reac H_2O
420	Bismuth oxybromide	BiOBr	7787-57-7	304.883	col tetr cry	560 dec		8.08		i H_2O, EtOH; s acid
421	Bismuth oxychloride	BiOCl	7787-59-9	260.432	wh tetr cry	575 dec		7.72		i H_2O
422	Bismuth oxyiodide	BiOI	7787-63-5	351.883	red tetr cry	300 dec		7.92		i H_2O, EtOH, chl; s HCl
423	Bismuth oxynitrate	$BiONO_3$	10361-46-3	286.985	wh powder	260 dec		4.93		i H_2O, EtOH; s acid
424	Bismuth phosphate	$BiPO_4$	10049-01-1	303.951	monocl cry			6.32		sl H_2O, dil acid; i EtOH
425	Bismuth potassium iodide	BiK_4I_7	41944-01-8	1253.704	red cry					reac H_2O; s alk iodide soln
426	Bismuth selenide	Bi_2Se_3	12068-69-8	654.84	blk hex cry	710 dec		7.5		i H_2O
427	Bismuth stannate pentahydrate	$Bi_2(SnO_3)_3 \cdot 5H_2O$	12777-45-6	1008.162	wh cry					i H_2O
428	Bismuth sulfate	$Bi_2(SO_4)_3$	7787-68-0	706.149	wh needles or powder	405 dec		5.08		reac H_2O, EtOH
429	Bismuth sulfide	Bi_2S_3	1345-07-9	514.156	blk-brn orth cry	850		6.78		i H_2O; s acid
430	Bismuth telluride	Bi_2Te_3	1304-82-1	800.76	gray hex plates	580		7.74		i H_2O; s EtOH
431	Bismuth tribromide	$BiBr_3$	7787-58-8	448.692	yel cub cry	219	462	5.72		reac H_2O; s dil acid, ace; i EtOH
432	Bismuth trichloride	$BiCl_3$	7787-60-2	315.339	col or yel cub cry; hyg	234	441	4.75		reac H_2O; s acid, EtOH, ace
433	Bismuth trifluoride	BiF_3	7787-61-3	265.975	wh-gray cub cry	727	900	8.3		i H_2O
434	Bismuth pentafluoride	BiF_5	7787-62-4	303.972	wh tetr needles; hyg	151.4	230	5.55		reac H_2O
435	Bismuth triiodide	BiI_3	7787-64-6	589.693	blk-brn hex cry	408.6	542	5.778	0.00078[20]	s EtOH
436	Bismuth trimethyl	$Bi(CH_3)_3$	593-91-9	254.083	col flam liq	-86	110	2.3		
437	Bismuth titanate	$Bi_4(TiO_4)_3$	12048-51-0	1171.516	wh orth cry			7.85		
438	Bismuth tungstate	$Bi_2(WO_4)_3$	13595-87-4	1161.47	wh powder					
439	Bismuth vanadate	$BiVO_4$	14059-33-7	323.920	orth cry	trans 500		6.25		i H_2O; s acid
440	Bismuth zirconate	$2Bi_2O_3 \cdot 3ZrO_2$	37306-42-6	1301.587	wh pow					
441	Boron	B	7440-42-8	10.811	blk rhomb cry	2075	4000	2.34		i H_2O
442	Diborane	B_2H_6	19287-45-7	27.670	col gas; flam	-164.85	-92.49	1.131 g/L		reac H_2O
443	Tetraborane(10)	B_4H_{10}	18283-93-7	53.323	unstab col gas	-120	18	2.180 g/L		reac hot H_2O
444	Pentaborane(9)	B_5H_9	19624-22-7	63.126	flam col liq	-46.74	60.10	0.60		reac H_2O
445	Pentaborane(11)	B_5H_{11}	18433-84-6	65.142	col liq; unstab	-122	65			reac hot H_2O
446	Hexaborane(10)	B_6H_{10}	23777-80-2	74.945	col liq	-62.3	108 dec	0.67		reac H_2O
447	Hexaborane(12)	B_6H_{12}	12008-19-4	76.961	col liq	-82.3	≈85			
448	Nonaborane(15)	B_9H_{15}	19465-30-6	112.418	col liq	2.7				
449	Decaborane(14)	$B_{10}H_{14}$	17702-41-9	122.221	wh orth cry	98.78	213	0.94		sl H_2O; s EtOH, bz, CS_2, ctc
450	Decaborane(16)	$B_{10}H_{16}$	71595-75-0	124.237	col cry	≈81	dec 170	subl		
451	Dodecaborane(16)	$B_{12}H_{16}$	89711-39-7	145.859	col cry	65				s bz, hx
452	Tridecaborane(19)	$B_{13}H_{19}$	43093-20-5	159.694	yel cry	44				s hx, CH2Cl2
453	Tetradecaborane(18)	$B_{14}H_{18}$	55606-55-8	169.497	visc yel oil		dec 100			s cyhex, CS_2
454	Hexadecaborane(20)	$B_{16}H_{20}$	28265-11-4	193.135	col cry	≈110				s ctc, cyhex, thf
455	Octadecaborane(22)	$B_{18}H_{22}$	11071-61-7	216.773	yel cry	180				s os
456	Tetrabromodiborane	B_2Br_4	14355-29-4	341.238	col liq	≈1	dec 20			
457	Tetrachlorodiborane	B_2Cl_4	13701-67-2	163.434	col liq; flam	-92.6	66.5			reac H_2O

No.	Name	Formula	CAS Reg No.	Mol. weight	Physical form	mp/°C	bp/°C	Density g cm^{-3}	Solubility g/100 g H$_2$O	Qualitative solubility
458	Tetrafluorodiborane	B$_2$F$_4$	13965-73-6	97.616	col gas; flam	-56	-34.0	3.990 g/L		reac H$_2$O
459	Borane carbonyl	BH$_3$CO	13205-44-2	41.845	col gas	-137	-64	1.710 g/L		reac H$_2$O
460	Borazine	B$_3$N$_3$H$_6$	6569-51-3	80.501	col liq	-58	53	0.824		reac H$_4$O
461	Boric acid	H$_3$BO$_3$	10043-35-3	61.833	col tricl cry	170.9		1.5	5.80^{25}	sl EtOH
462	Metaboric acid (α form)	HBO$_2$	13460-50-9	43.818	col orth cry; hyg	176		1.784		s H$_2$O
463	Metaboric acid (β form)	HBO$_2$	13460-50-9	43.818	col monocl cry; hyg	201		2.045		s H$_2$O
464	Metaboric acid (γ form)	HBO$_2$	13460-50-9	43.818	col cub cry	236		2.487		s H$_2$O
465	Tetrafluoroboric acid	HBF$_4$	16872-11-0	87.813	col liq		130 dec	≈1.8		vs H$_2$O, EtOH
466	Boron arsenide	BAs	12005-69-5	85.733	brn cub cry	1100 dec		5.22		
467	Boron carbide	B$_4$C	12069-32-8	55.255	hard blk cry	2350	>3500	2.50		i H$_2$O, acid
468	Boron nitride	BN	10043-11-5	24.818	wh powder; hex or cub cry	2967		2.18		i H$_2$O, acid
469	Boron oxide	B$_2$O$_3$	1303-86-2	69.620	col gl or hex cry; hyg	450		2.55	2.2^{20}	s EtOH
470	Boron phosphide	BP	20205-91-8	41.785	red cub cry or powder	1125 dec				reac H$_2$O, acid
471	Boron silicide	B$_6$Si	12008-29-6	92.952	blk cry	1980				
472	Boron sulfide	B$_2$S$_3$	12007-33-9	117.817	yel amorp solid	563		≈1.7		
473	Boron tribromide	BBr$_3$	10294-33-4	250.523	col liq; hyg	-46	91.3	2.6		reac H$_2$O, EtOH
474	Boron trichloride	BCl$_3$	10294-34-5	117.170	col liq or gas	-107.3	12.5	4.789 g/L		reac H$_2$O, EtOH
475	Boron trifluoride	BF$_3$	7637-07-2	67.806	col gas	-126.8	-99.9	2.772 g/L		s H$_2$O
476	Boron trifluoride etherate	BF$_3$(C$_2$H$_5$)$_2$O	109-63-7	141.927	liq	-60.4	125.5	1.125^{25}		reac H$_2$O; vs eth, EtOH
477	Boron triiodide	BI$_3$	13517-10-7	391.524	wh needles	49.7	209.5	3.35		i H$_2$O
478	Bromine	Br$_2$	7726-95-6	159.808	red liq	-7.2	58.8	3.1028		sl H$_2$O
479	Bromic acid	HBrO$_3$	7789-31-3	128.910	stab only in aq soln					s H$_2$O
480	Hypobromous acid	HOBr	13517-11-8	96.911	exists aq soln					s H$_2$O
481	Bromine dioxide	BrO$_2$	21255-83-4	111.903	unstab yel cry	≈0 dec				
482	Bromine monoxide	Br$_2$O	21308-80-5	175.807	unstab brn solid	-17.5 dec				
483	Dibromine trioxide	Br$_2$O$_3$	53809-75-9	207.806	oran needles (LT)	dec -40				
484	Dibromine pentoxide	Br$_2$O$_5$	58572-43-3	239.805	col cry (low temp)	-20 dec				
485	Bromine azide	BrN$_3$	13973-87-0	121.924	red cry; exp	≈45	exp			
486	Bromine chloride	BrCl	13863-41-7	115.357	dark red liq (<5˜C)	-66	5 dec			reac H$_2$O; s eth, CS$_2$
487	Bromine fluoride	BrF	13863-59-7	98.902	unstab red-brn gas	≈-33	≈20 dec	4.043 g/L		
488	Bromine trifluoride	BrF$_3$	7787-71-5	136.899	col hyg liq	8.77	125.8	2.803		reac H$_2$O
489	Bromine pentafluoride	BrF$_5$	7789-30-2	174.896	col liq	-60.5	41.3	2.460		reac H$_2$O (exp)
490	Bromosyl trifluoride	BrOF$_3$	61519-37-7	152.898	col liq	-5	dec >20			reac H$_2$O
491	Bromyl fluoride	BrO$_2$F	22585-64-4	130.901	col liq	-9	dec 55			reac H$_2$O
492	Perbromyl fluoride	BrO$_3$F	37265-91-1	146.900	col gas	-110	dec 20			reac H$_2$O
493	Cadmium	Cd	7440-43-9	112.411	silv-wh metal	321.069	767	8.69		i H$_2$O; reac acid
494	Cadmium acetate	Cd(C$_2$H$_3$O$_2$)$_2$	543-90-8	230.500	col cry	255		2.34		s H$_2$O, EtOH
495	Cadmium acetate dihydrate	Cd(C$_2$H$_3$O$_2$)$_2$ · 2H$_2$O	5743-04-4	266.529	wh cry	130 dec		2.01		vs H$_2$O; s EtOH
496	Cadmium antimonide	CdSb	12014-29-8	234.171	orth cry	456		6.92		
497	Cadmium arsenide	Cd$_3$As$_2$	12006-15-4	487.076	gray tetr cry	721		6.25		
498	Cadmium azide	Cd(N$_3$)$_2$	14215-29-3	196.451	yel-wh orth cry; exp	exp		3.24		
499	Cadmium borotungstate octadecahydrate	Cd$_5$(BW$_{12}$O$_{40}$) · 18H$_2$O	1306-26-9	3743.20	yel cry					vs H$_2$O
500	Cadmium bromide	CdBr$_2$	7789-42-6	272.219	wh-yel hex cry; hyg	568	863	5.19	115^{25}	sl ace, eth
501	Cadmium bromide tetrahydrate	CdBr$_2$ · 4H$_2$O	13464-92-1	344.281	wh-yel cry				115^{25}	s ace, EtOH
502	Cadmium carbonate	CdCO$_3$	513-78-0	172.420	wh hex cry	500 dec		5.026		i H$_2$O; s acid
503	Cadmium chlorate dihydrate	Cd(ClO$_3$)$_2$ · 2H$_2$O	22750-54-5*	315.344	col hyg cry	80 dec		2.28	2.64^0	
504	Cadmium chloride	CdCl$_2$	10108-64-2	183.317	rhom cry; hyg	568	964	4.08	120^{25}	s ace; sl EtOH; i eth
505	Cadmium chloride monohydrate	CdCl$_2$ · H$_2$O	34330-64-8	201.332	wh cry				120^{25}	
506	Cadmium chloride hemipentahydrate	CdCl$_2$ · 2.5H$_2$O	7790-78-5	228.354	wh rhomb leaflets			3.327	120^{25}	s ace
507	Cadmium chromate	CdCrO$_4$	14312-00-6	228.405	yel orth cry			4.5		i H$_2$O
508	Cadmium cyanide	Cd(CN)$_2$	542-83-6	164.445	wh cub cry			2.23	1.7^{15}	
509	Cadmium dichromate monohydrate	CdCr$_2$O$_7$ · H$_2$O	69239-51-6	346.414	oran solid					s H$_2$O
510	Cadmium 2-ethylhexanoate	Cd(C$_8$H$_{15}$O$_2$)$_2$	2420-98-6	398.818	powder					
511	Cadmium fluoride	CdF$_2$	7790-79-6	150.408	cub cry	1075	1750	6.33	4.36^{25}	s acid; i EtOH
512	Cadmium hydroxide	Cd(OH)$_2$	21041-95-2	146.426	wh trig or hex cry	130 dec		4.79	0.00015^{20}	s dil acid
513	Cadmium iodate	Cd(IO$_3$)$_2$	7790-81-0	462.216	wh powder			6.48	0.091^{25}	s HNO$_3$

No.	Name	Formula	CAS Reg No.	Mol. weight	Physical form	mp/°C	bp/°C	Density g cm⁻³	Solubility g/100 g H₂O	Qualitative solubility
514	Cadmium iodide	CdI_2	7790-80-9	366.220	col hex flakes	388	744	5.64	86.2[25]	vs H_2O; s EtOH, eth, ace
515	Cadmium metasilicate	$CdSiO_3$	13477-19-5	188.495	grn monocl cry	1252		5.10		
516	Cadmium molybdate	$CdMoO_4$	13972-68-4	272.35	col tetr cry	≈900 dec		5.4		i H_2O; s acid
517	Cadmium niobate	$Cd_2Nb_2O_7$	12187-14-3	522.631	cub cry	≈1410		6.28		i H_2O
518	Cadmium nitrate	$Cd(NO_3)_2$	10325-94-7	236.420	wh cub cry; hyg	360		3.6	156[25]	s EtOH
519	Cadmium nitrate tetrahydrate	$Cd(NO_3)_2 \cdot 4H_2O$	10022-68-1	308.482	col orth cry; hyg	59.5		2.45	156[25]	s EtOH, ace
520	Cadmium oxalate	CdC_2O_4	814-88-0	200.430	wh solid			3.32	0.0060[25]	
521	Cadmium oxalate trihydrate	$CdC_2O_4 \cdot 3H_2O$	20712-42-9	254.476	wh amorp powder	340 dec			0.0060[25]	i EtOH; s dil acid
522	Cadmium oxide	CdO	1306-19-0	128.410	brn cub cry		1559 sp	8.15		i H_2O; s dil acid
523	Cadmium 2,4-pentanedioate	$Cd(CH_3COCHCOCH_3)_2$	14689-45-3	310.627	wh solid or red cry	235				
524	Cadmium perchlorate hexahydrate	$Cd(ClO_4)_2 \cdot 6H_2O$	10326-28-0	419.404	wh hex cry			2.37	191.5[25]	
525	Cadmium phosphate	$Cd_3(PO_4)_2$	13477-17-3	527.176	powder	≈1500				i H_2O
526	Cadmium phosphide	Cd_3P_2	12014-28-7	399.181	grn tetr needles	700		5.96		s dil HCl
527	Cadmium selenate dihydrate	$CdSeO_4 \cdot 2H_2O$	10060-09-0	291.40	orth cry	100 dec		3.62	70.5[25]	
528	Cadmium selenide	$CdSe$	1306-24-7	191.37	wh cub cry	1240		5.81		i H_2O
529	Cadmium selenite	$CdSeO_3$	13814-59-0	239.37	col prisms					
530	Cadmium stearate	$Cd(C_{18}H_{35}O_2)_2$	2223-93-0	679.350	wh cry pow	134		1.21		
531	Cadmium succinate	$CdC_4H_4O_4$	141-00-4	228.484	wh pow or needles				0.37[40]	sl H_2O; i EtOH
532	Cadmium sulfate	$CdSO_4$	10124-36-4	208.474	col orth cry	1000		4.69	76.7[25]	i EtOH
533	Cadmium sulfate monohydrate	$CdSO_4 \cdot H_2O$	7790-84-3	226.489	monocl cry	105		3.79	76.7[25]	
534	Cadmium sulfate octahydrate	$CdSO_4 \cdot 8H_2O$	15244-35-6	352.596	col monocl cry	40 dec		3.08	76.7[25]	
535	Cadmium sulfide	CdS	1306-23-6	144.476	yel-oran hex cry	≈1480		4.826		i H_2O; s acid
536	Cadmium sulfite	$CdSO_3$	13477-23-1	192.474	col prisms	dec ≈400			0.05[20]	sl H_2O
537	Cadmium telluride	$CdTe$	1306-25-8	240.01	brn-blk cub cry	1042		6.2		i H_2O, dil acid
538	Cadmium tellurite	$CdTeO_3$	15851-44-2	288.01	col monocl cry	695	dec 1050			
539	Cadmium tetrafluoroborate	$Cd(BF_4)_2$	14486-19-2	286.020	col hyg liq			1.6		vs H_2O, EtOH
540	Cadmium titanate	$CdTiO_3$	12014-14-1	208.276	orth cry			6.5		
541	Cadmium tungstate	$CdWO_4$	7790-85-4	360.25	wh monocl cry			8.0		i H_2O, acid; s NH_4OH
542	Calcium	Ca	7440-70-2	40.078	silv-wh metal	842	1484	1.54		reac H_2O; i bz
543	Calcium acetate	$Ca(C_2H_3O_2)_2$	62-54-4	158.167	wh hyg cry	160 dec		1.50		s H_2O; sl EtOH
544	Calcium acetate monohydrate	$Ca(C_2H_3O_2)_2 \cdot H_2O$	5743-26-0	176.182	wh needles or powder	≈150 dec				s H_2O; sl EtOH
545	Calcium acetate dihydrate	$Ca(C_2H_3O_2)_2 \cdot 2H_2O$	14977-17-4	194.196	long col needles					s H_2O
546	Calcium aluminate	$CaAl_2O_4$	12042-68-1	158.039	wh monocl cry	1605		2.98		reac H_2O
547	Calcium aluminate (β form)	$Ca_3Al_2O_6$	12042-78-3	270.193	wh cub cry; refr	1535		3.04		i H_2O
548	Calcium arsenate	$Ca_3(AsO_4)_2$	7778-44-1	398.072	wh powder	dec		3.6	0.0036[20]	s dil acid
549	Calcium arsenite (1:1)	$CaAsO_3$	52740-16-6	162.998	wh powder					sl H_2O; s acid
550	Calcium borate hexahydrate	$CaB_4O_7 \cdot 6H_2O$	13701-64-9*	303.409	wh cry pow	1162 (anh)				
551	Calcium boride	CaB_6	12007-99-7	104.944	refrac solid	2235		2.49		
552	Calcium bromate	$Ca(BrO_3)_2$	10102-75-7	295.882	wh pow	180				vs H_2O
553	Calcium bromate monohydrate	$Ca(BrO_3)_2 \cdot H_2O$	10102-75-7*	313.898	wh monocl cry	dec 180		3.33		s EtOH, ace
554	Calcium bromide	$CaBr_2$	7789-41-5	199.886	rhom cry; hyg	742	1815	3.38	156[25]	vs H_2O
555	Calcium bromide dihydrate	$CaBr_2 \cdot 2H_2O$	22208-73-7	235.917	wh cry pow					vs H_2O
556	Calcium bromide hexahydrate	$CaBr_2 \cdot 6H_2O$	13477-28-6	307.977	wh hyg powder	38 dec		2.29	156[25]	
557	Calcium carbide	CaC_2	75-20-7	64.099	gray-blk orth cry	2300		2.22		reac H_2O
558	Calcium carbonate (aragonite)	$CaCO_3$	471-34-1	100.087	wh orth cry or powder	825 dec		2.930	0.00066[20]	s dil acid
559	Calcium carbonate (calcite)	$CaCO_3$	471-34-1	100.087	wh hex cry or powder	1330		2.710	0.00066[20]	s dil acid
560	Calcium carbonate (vaterite)	$CaCO_3$	471-34-1	100.087	col hex cry			2.653	0.0011[25]	s dil acid
561	Calcium chlorate	$Ca(ClO_3)_2$	10137-74-3	206.980	wh cry	340			197[25]	
562	Calcium chlorate dihydrate	$Ca(ClO_3)_2 \cdot 2H_2O$	10035-05-9	243.011	wh monocl cry; hyg	100 dec		2.711	197[25]	s EtOH
563	Calcium chloride	$CaCl_2$	10043-52-4	110.984	wh cub cry or powder; hyg	775	1935	2.15	81.3[25]	vs EtOH
564	Calcium chloride monohydrate	$CaCl_2 \cdot H_2O$	13477-29-7	128.999	wh hyg cry	260 dec		2.24	81.3[25]	s EtOH
565	Calcium chloride dihydrate	$CaCl_2 \cdot 2H_2O$	10035-04-8	147.015	hyg flakes or powder	175 dec		1.85	81.3[25]	vs EtOH
566	Calcium chloride tetrahydrate	$CaCl_2 \cdot 4H_2O$	25094-02-4	183.046	col tricl cry			1.83		
567	Calcium chloride hexahydrate	$CaCl_2 \cdot 6H_2O$	7774-34-7	219.075	wh hex cry; hyg	30 dec		1.71	81.3[25]	
568	Calcium chlorite	$Ca(ClO_2)_2$	14674-72-7	174.982	wh cub cry			2.71		reac H_2O
569	Calcium chromate	$CaCrO_4$	13765-19-0	156.072	yel cry	1000 dec				sl H_2O; i EtOH, ace
570	Calcium chromate dihydrate	$CaCrO_4 \cdot 2H_2O$	10060-08-9	192.102	yel orth cry	dec 200		2.50	13.2[20]	s dil acids
571	Calcium citrate tetrahydrate	$Ca_3(C_6H_5O_7)_2 \cdot 4H_2O$	5785-44-4	570.494	wh needles or pow	dec 100			0.096[23]	i eth
572	Calcium cyanamide	$CaCN_2$	156-62-7	80.102	col hex cry	≈1340	subl	2.29		reac H_2O
573	Calcium cyanide	$Ca(CN)_2$	592-01-8	92.112	wh rhomb cry; hyg					s H_2O, EtOH

No.	Name	Formula	CAS Reg No.	Mol. weight	Physical form	mp/°C	bp/°C	Density g cm^{-3}	Solubility g/100 g H$_2$O	Qualitative solubility
574	Calcium dichromate trihydrate	$CaCr_2O_7 \cdot 3H_2O$	14307-33-6*	310.112	red-oran cry	100 dec		2.37		vs H$_2$O; reac EtOH; i eth, ctc
575	Calcium dihydrogen phosphate monohydrate	$Ca(H_2PO_4)_2 \cdot H_2O$	10031-30-8	252.068	col tricl plates	100 dec		2.220		sl H$_2$O; s dil acid
576	Calcium 2-ethylhexanoate	$Ca(C_8H_{15}O_2)_2$	136-51-6	326.485	powder					
577	Calcium ferrocyanide dodecahydrate	$Ca_2Fe(CN)_6 \cdot 12H_2O$		508.289	yel tricl cry	dec		1.68	87[25]	vs H$_2$O; i EtOH
578	Calcium fluoride	CaF_2	7789-75-5	78.075	wh cub cry or powder	1418	2500	3.18	0.0016[25]	sl acid
579	Calcium fluorophosphate	$Ca_5(PO_4)_3F$	12015-73-5	504.302	col hex cry	1650		3.201		i H$_2$O
580	Calcium fluorophosphate dihydrate	$CaPO_3F \cdot 2H_2O$	37809-19-1	174.079	col monocl cry				0.42[27]	i os
581	Calcium formate	$Ca(CHO_2)_2$	544-17-2	130.113	orth cry	300 dec		2.02	16.6[20]	i EtOH
582	Calcium hexaborate pentahydrate	$2CaO \cdot 3B_2O_3 \cdot 5H_2O$	12291-65-5	411.091	col monocl cry	dec 375 (exp)		2.42	1[25]	sl acid
583	Calcium hexafluoro-2,4-pentanedioate	$Ca(CF_3COCHCOCF_3)_2$	121012-90-6	454.180	powder	135				
584	Calcium hexafluorosilicate dihydrate	$CaSiF_6 \cdot 2H_2O$	16925-39-6	218.185	col tetr cry			2.25	0.52[20]	i ace; reac hot H$_2$O
585	Calcium hydride	CaH_2	7789-78-8	42.094	gray orth cry or powder	1000		1.7		reac H$_2$O, EtOH
586	Calcium hydrogen phosphate	$CaHPO_4$	7757-93-9	136.057	wh tricl cry	dec		2.92	0.02[25]	i EtOH
587	Calcium hydrogen phosphate dihydrate	$CaHPO_4 \cdot 2H_2O$	7789-77-7	172.088	monocl cry	≈100 dec		2.31	0.02[25]	i EtOH; s dil acid
588	Calcium hydrogen sulfite	$CaH_2(SO_3)_2$	13780-03-5	202.220				1.06		s H$_2$O
589	Calcium hydrosulfide hexahydrate	$Ca(HS)_2 \cdot 6H_2O$		214.315	col cry	dec				s H$_2$O, EtOH
590	Calcium hydroxide	$Ca(OH)_2$	1305-62-0	74.093	soft hex cry			≈2.2	0.160[20]	s acid
591	Calcium hydroxide phosphate	$Ca_5(OH)(PO_4)_3$	12167-74-7	502.311	col hex cry	dec >900		3.155		i H$_2$O
592	Calcium hypochlorite	$Ca(OCl)_2$	7778-54-3	142.983	powder	100		2.350		
593	Calcium hypophosphite	$Ca(H_2PO_2)_2$	7789-79-9	170.055	wh monocl cry	300 dec				s H$_2$O; i EtOH
594	Calcium iodate	$Ca(IO_3)_2$	7789-80-2	389.883	wh monocl cry			4.52	0.306[25]	s HNO$_3$; i EtOH
595	Calcium iodide	CaI_2	10102-68-8	293.887	hyg hex cry	783	1100	3.96	215[25]	s MeOH, EtOH, ace; i eth
596	Calcium iodide hexahydrate	$CaI_2 \cdot 6H_2O$	71626-98-7	401.978	wh hex needles or powder	42 dec		2.55	215[25]	vs EtOH
597	Calcium metaborate	$Ca(BO_2)_2$	13701-64-9	125.698	powder				0.13[20]	
598	Calcium metasilicate	$CaSiO_3$	1344-95-2	116.162	wh monocl cry	1540		2.92		i H$_2$O
599	Calcium molybdate	$CaMoO_4$	7789-82-4	200.02	wh tetr cry	1520		4.35	0.0011[20]	i EtOH; s conc acid
600	Calcium nitrate	$Ca(NO_3)_2$	10124-37-5	164.087	wh cub cry; hyg	561		2.5	144[25]	s EtOH, MeOH, ace
601	Calcium nitrate tetrahydrate	$Ca(NO_3)_2 \cdot 4H_2O$	13477-34-4	236.149	wh cry	≈40 dec		1.82	144[25]	s EtOH, ace
602	Calcium nitride	Ca_3N_2	12013-82-0	148.247	red-brn cub cry	1195		2.67		s H$_2$O, acid; i EtOH
603	Calcium nitrite	$Ca(NO_2)_2$	13780-06-8	132.089	wh-yel hex cry; hyg	392		2.23	94.6[25]	sl EtOH
604	Calcium nitrite monohydrate	$Ca(NO_2)_2 \cdot H_2O$	10031-34-2	150.104	col or yel cry	dec 100				vs H$_2$O; sl EtOH
605	Calcium oleate	$Ca(C_{18}H_{33}O_2)_2$	142-17-6	602.985	pale yel solid	dec 140			0.04[25]	sl H$_2$O; s bz; i EtOH, ace, eth
606	Calcium oxalate	CaC_2O_4	563-72-4	128.097	wh cry powder			2.2	0.00061[20]	
607	Calcium oxalate monohydrate	$CaC_2O_4 \cdot H_2O$	5794-28-5	146.112	cub cry	200 dec		2.2	0.00061[20]	s dil acid
608	Calcium oxide	CaO	1305-78-8	56.077	gray-wh cub cry	2613		3.34		reac H$_2$O; s acid
609	Calcium oxide silicate	Ca_3OSiO_4	12168-85-3	228.317	refrac solid	2150				
610	Calcium palmitate	$Ca(C_{16}H_{31}O_2)_2$	542-42-7	550.910	wh-yel pow	dec 155				i H$_2$O, EtOH, eth, ace; sl bz
611	Calcium perborate heptahydrate	$Ca(BO_3)_2 \cdot 7H_2O$		283.803	gray-wh pow					s H$_2$O, acid
612	Calcium 2,4-pentanedioate	$Ca(CH_3COCHCOCH_3)_2$	19372-44-2	238.294	cry	dec 175				
613	Calcium perchlorate	$Ca(ClO_4)_2$	13477-36-6	238.979	wh cry	270 dec		2.65	188[25]	s EtOH, MeOH
614	Calcium perchlorate tetrahydrate	$Ca(ClO_4)_2 \cdot 4H_2O$	15627-86-8	311.041	wh cry					vs H$_2$O
615	Calcium permanganate	$Ca(MnO_4)_2$	10118-76-0	277.949	purp hyg cry			2.4	331[20]	reac EtOH
616	Calcium peroxide	CaO_2	1305-79-9	72.077	wh-yel tetr cry; hyg	≈200 dec		2.9		sl H$_2$O; s acid
617	Calcium phosphate	$Ca_3(PO_4)_2$	7758-87-4	310.177	wh amorp powder	1670		3.14	0.00012[20]	i EtOH; s dil acid
618	Calcium phosphide	Ca_3P_2	1305-99-3	182.182	red-brn hyg cry	≈1600		2.51		reac H$_2$O; i EtOH, eth
619	Calcium phosphonate monohydrate	$CaHPO_3 \cdot H_2O$	25232-60-4	138.073	col monocl cry	dec 150				sl H$_2$O; i EtOH
620	Calcium plumbate	Ca_2PbO_4	12013-69-3	351.4	oran-brn orth cry	dec		5.71		i H$_2$O; s acid
621	Calcium propanoate	$Ca(C_3H_5O_2)_2$	4075-81-4	186.219	monocl cry, powder					s H$_2$O; sl MeOH, EtOH; i ace, bz
622	Calcium pyrophosphate	$Ca_2P_2O_7$	7790-76-3	254.099	wh powder	1353		3.09		i H$_2$O; s dil acid
623	Calcium selenate dihydrate	$CaSeO_4 \cdot 2H_2O$	7790-74-1	219.07	wh monocl cry			2.75	8.3[18]	
624	Calcium selenide	$CaSe$	1305-84-6	119.04	wh-brn cub cry	1400 dec		3.8		reac H$_2$O
625	Calcium silicide (CaSi)	$CaSi$	12013-55-7	68.164	orth cry	1324		2.39		

No.	Name	Formula	CAS Reg No.	Mol. weight	Physical form	mp/°C	bp/°C	Density g cm^{-3}	Solubility g/100 g H_2O	Qualitative solubility
626	Calcium silicide (CaSi$_2$)	CaSi$_2$	12013-56-8	96.249	gray hex cry	1040		2.50		i cold H_2O; reac hot H_2O; s acid
627	Calcium stannate trihydrate	CaSnO$_3$ · 3H$_2$O	12013-46-6*	260.832	wh cry pow	dec ≈350				i H_2O
628	Calcium stearate	Ca(C$_{18}$H$_{35}$O$_2$)$_2$	1592-23-0	607.017	granular powder	180				i H_2O, EtOH
629	Calcium succinate trihydrate	CaC$_4$H$_4$O$_4$ · 3H$_2$O	140-99-8	210.196	needles					sl H_2O; s dil acid; i EtOH
630	Calcium sulfate	CaSO$_4$	7778-18-9	136.141	orth cry	1460		2.96	0.205^{25}	
631	Calcium sulfate hemihydrate	CaSO$_4$ · 0.5H$_2$O	10034-76-1	145.149	wh powder				0.205^{25}	
632	Calcium sulfate dihydrate	CaSO$_4$ · 2H$_2$O	10101-41-4	172.171	monocl cry or powder	150 dec		2.32	0.205^{20}	i os
633	Calcium sulfide	CaS	20548-54-3	72.143	wh-yel cub cry; hyg	2524		2.59		sl H_2O; i EtOH
634	Calcium sulfite dihydrate	CaSO$_3$ · 2H$_2$O	10257-55-3	156.172	wh powder				0.0070^{25}	sl EtOH; s acid
635	Calcium tartrate tetrahydrate	CaC$_4$H$_4$O$_6$ · 4H$_2$O	3164-34-9*	260.210	wh pow				0.04^{10}	s dil acid; sl EtOH
636	Calcium telluride	CaTe	12013-57-9	167.68	wh cub cry	1600 dec		4.87		
637	Calcium tetrahydroaluminate	Ca(AlH$_4$)$_2$	16941-10-9	102.105	gray powder; flam					reac H_2O; s thf; i eth, bz
638	Calcium thiocyanate tetrahydrate	Ca(SCN)$_2$ · 4H$_2$O	2092-16-2	228.304	hyg cry	160 dec				vs H_2O; s EtOH, ace
639	Calcium thiosulfate hexahydrate	CaS$_2$O$_3$ · 6H$_2$O	10124-41-1	260.298	tricl cry	45 dec		1.87		s H_2O; i EtOH
640	Calcium titanate	CaTiO$_3$	12049-50-2	135.943	cub cry	1980		3.98		
641	Calcium tungstate	CaWO$_4$	7790-75-2	287.92	wh tetr cry	1620		6.06	0.2^{18}	s hot acid
642	Calcium zirconate	CaZrO$_3$	12013-47-7	179.300	powder	2550				
643	Californium	Cf	7440-71-3	251	hex or cub metal	900		15.1		
644	Carbon (diamond)	C	7782-40-3	12.011	col cub cry	4440 (12.4 GPa)		3.513		i H_2O
645	Carbon (graphite)	C	7782-42-5	12.011	soft blk hex cry	4489 tp (10.3 MPa)	3825 sp	2.2		i H_2O
646	Carbon black	C	1333-86-4	12.011	fine blk pow					i H_2O
647	Carbon (fullerene-C$_{60}$)	C$_{60}$	99685-96-8	720.642	yel needles or plates	>280				s os
648	Carbon (fullerene-C$_{70}$)	C$_{70}$	115383-22-7	840.749	red-brn solid	>280				s bz, tol
649	Fullerene fluoride	C$_{60}$F$_{60}$	134929-59-2	1860.546	col plates	287				vs ace; s thf; i chl
650	Carbon monoxide	CO	630-08-0	28.010	col gas	-205.02	-191.5	1.145 g/L		sl H_2O; s chl, EtOH
651	Carbon dioxide	CO$_2$	124-38-9	44.010	col gas	-56.558 tp	-78.464 sp	1.799 g/L		s H_2O
652	Carbon suboxide	C$_3$O$_2$	504-64-3	68.031	col gas	-112.5	6.8	2.781 g/L		reac H_2O
653	Carbon disulfide	CS$_2$	75-15-0	76.141	col or yel liq	-112.1	46	1.2632^{20}		i H_2O; vs EtOH, bz, os
654	Carbon subsulfide	C$_3$S$_2$	627-34-9	100.162	red liq	-1	90 dec	1.27		reac H_2O
655	Carbon diselenide	CSe$_2$	506-80-9	169.93	yel liq	-43.7	125.5	2.6823^{20}		i H_2O; vs ctc, tol
656	Carbon oxysulfide	COS	463-58-1	60.075	col gas	-138.8	-50	2.456 g/L		s H_2O, EtOH
657	Carbon oxyselenide	COSe	1603-84-5	106.97	col gas; unstab	-124.4	-21.7	4.372 g/L		reac H_2O
658	Carbon sulfide selenide	CSSe	5951-19-9	123.04	yel liq	-85	84.5	1.99		i H_2O
659	Carbon sulfide telluride	CSTe	10340-06-4	171.68	red-yel liq; unstab	-54	20 dec			reac H_2O
660	Carbonyl bromide	COBr$_2$	593-95-3	187.818	col liq		64.5	2.5		reac H_2O
661	Carbonyl chloride	COCl$_2$	75-44-5	98.916	col gas	-127.78	8	4.043 g/L		sl H_2O; s bz, tol
662	Carbonyl fluoride	COF$_2$	353-50-4	66.007	col gas	-111.26	-84.57	2.698 g/L		reac H_2O
663	Cyanogen	C$_2$N$_2$	460-19-5	52.034	col gas	-27.83	-21.1	2.127 g/L		sl H_2O, eth; s EtOH
664	Cyanogen azide	N$_3$CN	764-05-6	68.038	col oily liq	exp				
665	Cyanogen bromide	BrCN	506-68-3	105.922	wh hyg needles	52	61.5	2.015		s H_2O, EtOH, eth
666	Cyanogen chloride	ClCN	506-77-4	61.471	col vol liq or gas	-6.55	13	2.513 g/L		s H_2O, EtOH, eth
667	Cyanogen fluoride	FCN	1495-50-7	45.016	col gas	-82	-46	1.840 g/L		
668	Cyanogen iodide	ICN	506-78-5	152.922	col needles	146.7		2.84		s H_2O, EtOH, eth
669	Cerium	Ce	7440-45-1	140.116	silv metal; cub or hex	799	3443	6.770		s dil acid
670	Cerium boride	CeB$_6$	12008-02-5	204.982	blue refrac solid; hex	2550		4.87		i H_2O, HCl
671	Cerium carbide	CeC$_2$	12012-32-7	164.137	red hex cry	2250		5.47		reac H_2O
672	Cerium carbide	Ce$_2$C$_3$	12115-63-8	316.264	yel-brn cub cry	1505		6.9		
673	Cerium nitride	CeN	25764-08-3	154.123	refrac cub cry	2557		7.89		
674	Cerium silicide	CeSi$_2$	12014-85-6	196.287	tetr cry	1420		5.31		i H_2O
675	Cerium(II) hydride	CeH$_2$	13569-50-1	142.132	cub cry			5.45		reac H_2O
676	Cerium(II) iodide	CeI$_2$	19139-47-0	393.925	bronze cry	808				
677	Cerium(II) sulfide	CeS	12014-82-3	172.181	yel cub cry	2445		5.9		
678	Cerium(III) acetate sesqihydrate	Ce(C$_2$H$_3$O$_2$)$_3$ · 1.5H$_2$O	17829-82-2		col cry	dec 115			26^{15}	s H_2O
679	Cerium(III) ammonium nitrate tetrahydrate	(NH$_4$)$_2$Ce(NO$_3$)$_5$ · 4H$_2$O	13083-04-0	558.279	col monocl cry	74				vs H_2O
680	Cerium(III) ammonium sulfate tetrahydrate	NH$_4$Ce(SO$_4$)$_2$ · 4H$_2$O	21995-38-0*	422.341	monocl cry					s H_2O
681	Cerium(III) bromide	CeBr$_3$	14457-87-5	379.828	wh hex cry; hyg	732	1457			s H_2O

No.	Name	Formula	CAS Reg No.	Mol. weight	Physical form	mp/°C	bp/°C	Density g cm^{-3}	Solubility g/100 g H$_2$O	Qualitative solubility
682	Cerium(III) bromide heptahydrate	CeBr$_3$ · 7H$_2$O	7789-56-2	505.935	col hyg needles	732				s H$_2$O, EtOH
683	Cerium(III) carbonate	Ce$_2$(CO$_3$)$_3$	537-01-9	460.259	wh pow	dec 500				i H$_2$O; s acid
684	Cerium(III) carbonate pentahydrate	Ce$_2$(CO$_3$)$_3$ · 5H$_2$O	72520-94-6	550.335	wh powder					i H$_2$O; s dil acid
685	Cerium(III) chloride	CeCl$_3$	7790-86-5	246.475	wh hex cry	807		3.97		s H$_2$O, EtOH
686	Cerium(III) chloride heptahydrate	CeCl$_3$ · 7H$_2$O	18618-55-8	372.582	yel orth cry; hyg	90 dec				vs H$_2$O, EtOH
687	Cerium(III) fluoride	CeF$_3$	7758-88-5	197.111	wh hex cry; hyg	1430	2180	6.157		i H$_2$O
688	Cerium(III) hydride	CeH$_3$	13864-02-3	143.140	blk pow or blue-blk cry	dec (flam)				reac H$_2$O
689	Cerium(III) hydroxide	Ce(OH)$_3$	15785-09-8	191.138	wh solid					i H$_2$O; s acid
690	Cerium(III) iodide	CeI$_3$	7790-87-6	520.829	yel orth cry; hyg	760				s H$_2$O
691	Cerium(III) iodide nonahydrate	CeI$_3$ · 9H$_2$O	7790-87-6*	682.967	wh-red cry					vs H$_2$O; s EtOH
692	Cerium(III) nitrate hexahydrate	Ce(NO$_3$)$_3$ · 6H$_2$O	10108-73-3*	434.222	col-red cry	150 dec			176^{25}	s ace
693	Cerium(III) oxalate nonahydrate	Ce$_2$(C$_2$O$_4$)$_3$ · 9H$_2$O	13266-83-6	706.426	wh pow	dec				i H$_2$O, EtOH; s acid
694	Cerium(III) oxide	Ce$_2$O$_3$	1345-13-7	328.230	yel-grn cub cry	2210	3730	6.2		i H$_2$O; s acid
695	Cerium(III) 2,4-pentanedioate trihydrate	Ce(CH$_3$COCHCOCH$_3$)$_3$ · 3H$_2$O	15653-01-7	491.486	yel hyg cry	≈150				vs EtOH
696	Cerium(III) perchlorate hexahydrate	Ce(ClO$_4$)$_3$ · 6H$_2$O	36907-38-7	546.559	hyg col cry	dec 200				s H$_2$O, EtOH
697	Cerium(III) selenate	Ce$_2$(SeO$_4$)$_3$		709.11	rhom cry			4.46		s H$_2$O
698	Cerium(III) sulfate	Ce$_2$(SO$_4$)$_3$	13454-94-9	568.420	col hyg cry	920 dec				s H$_2$O
699	Cerium(III) sulfate octahydrate	Ce$_2$(SO$_4$)$_3$ · 8H$_2$O	13454-94-9	712.542	wh orth cry	≈250 dec		2.87		s H$_2$O
700	Cerium(III) sulfide	Ce$_2$S$_3$	12014-93-6	376.427	red cub cry	2450		5.02		i H$_2$O
701	Cerium(III) tungstate	Ce$_2$(WO$_4$)$_3$	13454-74-5	1023.75	yel tetr cry	1089		6.77		i H$_2$O
702	Cerium(IV) ammonium nitrate	(NH$_4$)$_2$Ce(NO$_3$)$_6$	16774-21-3	548.223	red-oran cry					vs H$_2$O
703	Cerium(IV) ammonium nitrate dihydrate	(NH$_4$)$_4$Ce(SO$_4$)$_4$ · 2H$_2$O	10378-47-9	632.551	cry pow	dec 450				
704	Cerium(IV) fluoride	CeF$_4$	10060-10-3	216.110	wh hyg powder	≈600 dec		4.77		i H$_2$O
705	Cerium(IV) hydroxide	Ce(OH)$_4$	12014-56-1	208.146	yel-wh pow					i H$_2$O; s conc acid
706	Cerium(IV) oxide	CeO$_2$	1306-38-3	172.115	wh-yel powder; cub	2480		7.216		i H$_2$O, dil acid
707	Cerium(IV) sulfate tetrahydrate	Ce(SO$_4$)$_2$ · 4H$_2$O	10294-42-5	404.303	yel-oran orth cry	180 dec		3.91	9.66^{20}	
708	Cesium	Cs	7440-46-2	132.905	silv-wh metal	28.5	671	1.93		reac H$_2$O
709	Cesium acetate	CsC$_2$H$_3$O$_2$	3396-11-0	191.949	hyg lumps	194			10^{11}	
710	Cesium aluminum sulfate dodecahydrate	CsAl(SO$_4$)$_2$ · 12H$_2$O	7784-17-0	568.196	col cub cry	117 dec		1.97		s H$_2$O; i EtOH
711	Cesium amide	CsNH$_2$	22205-57-8	148.928	wh tetr cry			3.70		
712	Cesium azide	CsN$_3$	22750-57-8	174.925	hyg tetr cry; exp	326		≈3.5	22^{40}	
713	Cesium bromate	CsBrO$_3$	13454-75-6	260.807	col hex cry			4.11	3.83^{25}	
714	Cesium bromide	CsBr	7787-69-1	212.809	wh cub cry; hyg	636	≈1300	4.43	123^{25}	s EtOH; i ace
715	Cesium carbonate	Cs$_2$CO$_3$	534-17-8	325.820	wh monocl cry; hyg	793		4.24	261^{15}	s EtOH, eth
716	Cesium chlorate	CsClO$_3$	13763-67-2	216.356	col hex cry	342		3.57	7.78^{25}	sl H$_2$O
717	Cesium chloride	CsCl	7647-17-8	168.358	wh cub cry; hyg	646	1297	3.988	191^{25}	s EtOH
718	Cesium chromate(IV)	Cs$_2$CrO$_4$	56320-90-2	647.616	yel hex cry	982		4.24		vs H$_2$O
719	Cesium cyanide	CsCN	21159-32-0	158.923	wh cub cry; hyg	350		3.34		vs H$_2$O
720	Cesium dibromoiodate	CsIBr$_2$	18278-82-5	419.617	dark oran cry	dec				s H$_2$O
721	Cesium fluoride	CsF	13400-13-0	151.903	wh cub cry; hyg	703		4.64	573^{25}	s MeOH; i diox, py
722	Cesium fluoroborate	CsBF$_4$	18909-69-8	219.710	wh orth cry	555 dec		3.2	1.6^{17}	sl H$_2$O
723	Cesium formate	CsCHO$_2$	3495-36-1	177.923	wh cry			1.017		vs H$_2$O
724	Cesium hexafluorogermanate	Cs$_2$GeF$_6$		452.44	wh cry	≈675		4.10		sl cold H$_2$O; s hot H$_2$O
725	Cesium hydride	CsH	58724-12-2	133.913	wh cub cry; flam	528		3.42		reac H$_2$O
726	Cesium hydrogen carbonate	CsHCO$_3$	15519-28-5	193.922	rhom cry	175 dec			209^{15}	s EtOH
727	Cesium hydrogen fluoride	CsHF$_2$	12280-52-3	171.910	tetr cry	170		3.86		
728	Cesium hydrogen sulfate	CsHSO$_4$	7789-16-4	229.976	col rhom prisms	dec		3.352		s H$_2$O
729	Cesium hydroxide	CsOH	21351-79-1	149.912	wh-yel hyg cry	342.3		3.68	300^{30}	s EtOH
730	Cesium iodate	CsIO$_3$	13454-81-4	307.807	wh monocl cry			4.85	2.6^{25}	
731	Cesium iodide	CsI	7789-17-5	259.809	col cub cry; hyg	632	≈1280	4.51	84.8^{25}	s EtOH, MeOH, ace
732	Cesium metaborate	CsBO$_2$	92141-86-1	175.715	cub cry	732		≈3.7		
733	Cesium molybdate	Cs$_2$MoO$_4$	13597-64-3	425.75	wh cry	956.3			67^{18}	s H$_2$O
734	Cesium nitrate	CsNO$_3$	7789-18-6	194.910	wh hex or cub cry	409		3.66	27.9^{25}	s ace; sl EtOH
735	Cesium nitrite	CsNO$_2$	13454-83-6	178.911	yel cry	406				s H$_2$O
736	Cesium oxide	Cs$_2$O	20281-00-9	281.810	yel-oran hex cry	495		4.65		vs H$_2$O
737	Cesium superoxide	CsO$_2$	12018-61-0	164.904	yel tetr cry	432		3.77		reac H$_2$O
738	Cesium trioxide	Cs$_2$O$_3$	12134-22-4	313.809	brn cry	≈400		4.25		reac H$_2$O
739	Cesium perchlorate	CsClO$_4$	13454-84-7	232.356	wh orth cry; hyg	≈600 dec		3.327	2.00^{25}	
740	Cesium periodate	CsIO$_4$	13478-04-1	323.807	wh rhom prisms			4.26	2.2^{15}	
741	Cesium sulfate	Cs$_2$SO$_4$	10294-54-9	361.874	wh orth cry or hex prisms; hyg	1005		4.24	182^{25}	i EtOH, ace, py

No.	Name	Formula	CAS Reg No.	Mol. weight	Physical form	mp/°C	bp/°C	Density g cm⁻³	Solubility g/100 g H₂O	Qualitative solubility
742	Cesium sulfide	Cs_2S	12214-16-3	297.876	yel orth hyg cry	520				vs H_2O
743	Cesium trifluoroacetate	$Cs(C_2F_3O_2)$	21907-50-6	245.920	hyg solid	115				vs H_2O
744	Chlorine	Cl_2	7782-50-5	70.906	grn-yel gas	-101.5	-34.04	2.898 g/L		sl H_2O
745	Hypochlorous acid	HOCl	7790-92-3	52.460	grn-yel; stable only in aq soln					s H_2O
746	Chloric acid	$HClO_3$	7790-93-4	84.459	exists only in aq soln					vs H_2O
747	Perchloric acid	$HClO_4$	7601-90-3	100.459	col hyg liq	-112	≈90 dec	1.77		s H_2O
748	Chlorine monoxide	Cl_2O	7791-21-1	86.905	yel-brn gas	-120.6	2.2	3.552 g/L		vs H_2O
749	Chlorine dioxide	ClO_2	10049-04-4	67.452	oran-grn gas	-59	11	2.757 g/L		sl H_2O
750	Dichlorine trioxide	Cl_2O_3	17496-59-2	118.904	dark brn solid	exp <25				
751	Dichlorine hexoxide	Cl_2O_6	12442-63-6	166.902	red liq	3.5	≈200			reac H_2O
752	Dichlorine heptoxide	Cl_2O_7	10294-48-1	182.902	col oily liq; exp	-91.5	82	1.9		reac H_2O
753	Chlorine fluoride	ClF	7790-89-8	54.451	col gas	-155.6	-101.1	2.226 g/L		reac H_2O
754	Chlorine trifluoride	ClF_3	7790-91-2	92.448	gas	-76.34	11.75	3.779 g/L		reac H_2O
755	Chlorine pentafluoride	ClF_5	13637-63-3	130.445	col gas	-103	-13.1	5.332 g/L		
756	Chlorosyl trifluoride	$ClOF_3$	30708-80-6	108.447	col liq	-42	27			reac H_2O
757	Chloryl fluoride	ClO_2F	13637-83-7	86.450	col gas	-115	-6	3.534 g/L		reac H_2O
758	Chloryl trifluoride	ClO_2F_3	38680-84-1	124.447	col gas	-81.2	-21.6	5.087 g/L		reac H_2O
759	Perchloryl fluoride	ClO_3F	7616-94-6	102.449	col gas	-147	-46.75	4.187 g/L		
760	Chlorine perchlorate	$ClOClO_3$	27218-16-2	134.904	unstab yel liq	-117	≈45 dec	1.81⁰		
761	Chromium	Cr	7440-47-3	51.996	blue-wh metal; cub	1907	2671	7.15		reac dil acid
762	Chromic acid	H_2CrO_4	7738-94-5	118.010	aq soln only					s H_2O
763	Chromium antimonide	CrSb	12053-12-2	173.756	hex cry	1110		7.11		
764	Chromium arsenide	Cr_2As	12254-85-2	178.914	tetr cry			7.04		
765	Chromium boride (CrB)	CrB	12006-79-0	62.807	refrac orth cry	2100		6.1		
766	Chromium boride (CrB₂)	CrB_2	12007-16-8	73.618	refrac solid; hex	2200		5.22		
767	Chromium boride (Cr₂B)	Cr_2B	12006-80-3	114.803	refrac solid	1875				
768	Chromium boride (Cr₅B₃)	Cr_5B_3	12007-38-4	292.414	tetr cry	1900		6.10		
769	Chromium carbide	Cr_3C_2	12012-35-0	180.009	gray orth cry	1895		6.68		
770	Chromium carbonyl	$Cr(CO)_6$	13007-92-6	220.056	col orth cry	130 dec	subl	1.77		i H_2O, EtOH; s eth, chl
771	Chromium nitride (CrN)	CrN	24094-93-7	66.003	gray cub cry	1080 dec		5.9		
772	Chromium nitride (Cr₂N)	Cr_2N	12053-27-9	117.999	hex cry	1650		6.8		
773	Chromium phosphide	CrP	26342-61-0	82.970	orth cry			5.25		
774	Chromium selenide	CrSe	12053-13-3	130.96	hex cry	≈1500		6.1		
775	Chromium silicide (CrSi₂)	$CrSi_2$	12018-09-6	108.167	gray hex cry	1490		4.91		
776	Chromium silicide (Cr₃Si)	Cr_3Si	12018-36-9	184.074	cub cry	1770		6.4		
777	Chromium(II) acetate monohydrate	$Cr(C_2H_3O_2)_2 \cdot H_2O$	628-52-4*	188.100	red monocl cry			1.79		sl H_2O
778	Chromium(II) bromide	$CrBr_2$	10049-25-9	211.804	wh monocl cry; aq soln blue	842		4.236		s H_2O, EtOH
779	Chromium(II) chloride	$CrCl_2$	10049-05-5	122.902	wh hyg needles; aq soln blue	824	1120	2.88		s H_2O
780	Chromium(II) chloride tetrahydrate	$Cr(H_2O)_4Cl_2 \cdot 4H_2O$	13931-94-7	267.024	blue hyg cry	51 dec				s H_2O
781	Chromium(II) fluoride	CrF_2	10049-10-2	89.993	blue-grn monocl cry	894		3.79		sl H_2O; i EtOH
782	Chromium(II) formate monohydrate	$Cr(CHOO)_2 \cdot H_2O$	4493-37-2	160.046	red needles					s H_2O
783	Chromium(II) iodide	CrI_2	13478-28-9	305.805	red-brn cry; hyg	867		5.1		s H_2O
784	Chromium(II) oxalate monohydrate	$CrC_2O_4 \cdot H_2O$	814-90-4*	158.030	yel-grn powder			2.468		sl H_2O
785	Chromium(II) sulfate pentahydrate	$CrSO_4 \cdot 5H_2O$	13825-86-0	238.135	blue cry				21⁰	s dil acid; sl EtOH; i ace
786	Chromium(II,III) oxide	Cr_3O_4	12018-34-7	219.986	cub cry			6.1		
787	Chromium(III) acetate	$Cr(C_2H_3O_2)_3$	1066-30-4	229.127	blue-grn pwd					sl H_2O
788	Chromium(III) acetate monohydrate	$Cr(C_2H_3O_2)_3 \cdot H_2O$	25013-82-5	247.143	gray-grn pow					sl H_2O; i EtOH
789	Chromium(III) acetate hexahydrate	$Cr(C_2H_3O_2)_3 \cdot 6H_2O$	1066-30-4*	337.220	blue needles					s H_2O
790	Chromium(III) acetate hydroxide	$Cr(C_2H_3O_2)_2(OH)$	39430-51-8	187.092	viol cry pow					vs H_2O
791	Chromium(III) bromide	$CrBr_3$	10031-25-1	291.708	dark grn hex cry	812		4.68		s hot H_2O, bz
792	Chromium(III) bromide hexahydrate (α)	$CrBr_3(H_2O)_4 \cdot 2H_2O$	18721-05-6	399.799	grn hyg cry					s H_2O, EtOH
793	Chromium(III) bromide hexahydrate (β)	$Cr(H_2O)_6Br_3$	10031-25-1*	399.799	viol hyg cry					s H_2O; i EtOH, eth
794	Chromium(III) chloride	$CrCl_3$	10025-73-7	158.355	red-viol cry	1152	1300 dec	2.76		sl H_2O
795	Chromium(III) chloride hexahydrate	$[CrCl_2(H_2O)_4]Cl \cdot 2H_2O$	10060-12-5	266.446	grn monocl cry; hyg					s H_2O, EtOH; sl ace; i eth
796	Chromium(III) fluoride	CrF_3	7788-97-8	108.991	grn needles	1425		3.8		i H_2O, EtOH
797	Chromium(III) fluoride trihydrate	$CrF_3 \cdot 3H_2O$	16671-27-5	163.037	grn hex cry			2.2		sl H_2O
798	Chromium(III) hydroxide sulfate	$Cr(OH)SO_4$	12336-95-7	165.066	grn cry					
799	Chromium(III) fluoride nonahydrate	$Cr(H_2O)_6F_3 \cdot 3H_2O$	102430-09-1	271.129	rhom viol cry					sl H_2O

No.	Name	Formula	CAS Reg No.	Mol. weight	Physical form	mp/°C	bp/°C	Density g cm⁻³	Solubility g/100 g H₂O	Qualitative solubility
800	Chromium(III) hydroxide trihydrate	$Cr(OH)_3 \cdot 3H_2O$	1308-14-1	157.063	blue-grn powder					i H_2O; s acid
801	Chromium(III) iodide	CrI_3	13569-75-0	432.709	dark grn hex cry	500 dec		5.32		sl H_2O
802	Chromium(III) nitrate	$Cr(NO_3)_3$	13548-38-4	238.011	grn hyg powder	>60 dec				vs H_2O
803	Chromium(III) nitrate nonahydrate	$Cr(NO_3)_3 \cdot 9H_2O$	7789-02-8	400.148	grn-blk monocl cry	66.3	>100 dec	1.80		vs H_2O
804	Chromium(III) oxide	Cr_2O_3	1308-38-9	151.990	grn hex cry	2320	≈3000	5.22		i H_2O, EtOH; sl acid, alk
805	Chromium(III) 2,4-pentanedioate	$Cr(CH_3COCHCOCH_3)_3$	21679-31-2	349.320	red monocl cry	208	345	1.34		i H_2O; s bz
806	Chromium(III) perchlorate	$Cr(ClO_4)_3$	27535-70-2	350.348	grn-blue cry				58²⁵	vs H_2O
807	Chromium(III) phosphate	$CrPO_4$	7789-04-0	146.967	blue orth cry	>1800		4.6		i H_2O, acid, aqua regia
808	Chromium(III) phosphate hemiheptahydrate	$CrPO_4 \cdot 3.5H_2O$	84359-31-9	210.021	blue-grn powder			2.15		i H_2O; s acid
809	Chromium(III) phosphate hexahydrate	$CrPO_4 \cdot 6H_2O$	84359-31-9	255.059	viol cry	>500 dec		2.121		i H_2O; s acid, alk
810	Chromium(III) potassium oxalate trihydrate	$K_3Cr(C_2O_4)_3 \cdot 3H_2O$	15275-09-9	487.394	blue-grn monocl cry					s H_2O
811	Chromium(III) potassium sulfate dodecahydrate	$CrK(SO_4)_2 \cdot 12H_2O$	7788-99-0	499.403	viol-blk cub cry	89 dec		1.83		s H_2O; i EtOH
812	Chromium(III) sulfate	$Cr_2(SO_4)_3$	10101-53-8	392.180	red pow	dec >700		3.1	64²⁵	s H_2O; vs acid
813	Chromium(III) sulfate octadecahydrate	$Cr_2(SO_4)_3 \cdot 18H_2O$	10101-53-8*	716.455	viol cry	dec 115		1.7		reac H_2O
814	Chromium(III) sulfide	Cr_2S_3	12018-22-3	200.187	brn-blk hex cry			3.8		
815	Chromium(III) telluride	Cr_2Te_3	12053-39-3	486.79	hex cry	≈1300		7.0		
816	Chromium(IV) chloride	$CrCl_4$	15597-88-3	193.808	gas, stable at HT		>600 dec	7.922 g/L		
817	Chromium(IV) fluoride	CrF_4	10049-11-3	127.990	grn cry	277		2.89		reac H_2O
818	Chromium(IV) oxide	CrO_2	12018-01-8	83.995	brn-blk tetr powder	≈400 dec		4.89		i H_2O; s acid
819	Chromium(V) fluoride	CrF_5	14884-42-5	146.988	red orth cry	34	117			reac H_2O
820	Chromium(V) oxide	Cr_2O_5	12218-36-9	183.989	blk needles	dec 200				
821	Chromium(VI) fluoride	CrF_6	13843-28-2	165.986	yel solid; stable at low temp	-100 dec				
822	Chromium(VI) oxide	CrO_3	1333-82-0	99.994	red orth cry	197	≈250 dec	2.7	169²⁵	
823	Chromium(VI) tetrafluoride oxide	$CrOF_4$	23276-90-6	143.989	dark red solid	55				reac H_2O, ace, dmso
824	Chromium(VI) dichloride dioxide	CrO_2Cl_2	14977-61-8	154.901	red liq	-96.5	117	1.91		reac H_2O; s ctc, chl, bz
825	Chromium(VI) difluoride dioxide	CrO_2F_2	7788-96-7	121.992	red-viol cry	30	subl			reac H_2O
826	Cobalt	Co	7440-48-4	58.933	gray metal; hex or cub	1495	2927	8.86		s dil acid
827	Cobaltocene	$Co(C_5H_5)_2$	1277-43-6	189.119	blk-purp cry	173				
828	Cobalt antimonide	$CoSb$	12052-42-5	180.693	hex cry	1202		8.8		
829	Cobalt arsenic sulfide	$CoAsS$	12254-82-9	165.920	silv-wh solid			≈6.1		
830	Cobalt arsenide (CoAs)	$CoAs$	27016-73-5	133.855	orth cry	1180		8.22		
831	Cobalt arsenide (CoAs₂)	$CoAs_2$	12044-42-7	208.776	monocl cry			7.2		
832	Cobalt arsenide (CoAs₃)	$CoAs_3$	12256-04-1	283.698	cub cry	942		6.84		
833	Cobalt boride (CoB)	CoB	12006-77-8	69.744	refrac solid	1460		7.25		
834	Cobalt boride (Co₂B)	Co_2B	12045-01-1	128.677	refrac solid	1280		8.1		reac H_2O, HNO_3
835	Cobalt carbonyl	$Co_2(CO)_8$	10210-68-1	341.947	oran cry	51 dec		1.78		i H_2O; s EtOH, eth, CS_2
836	Cobalt disulfide	CoS_2	12013-10-4	123.063	cub cry			4.3		
837	Cobalt dodecacarbonyl	$Co_4(CO)_{12}$	17786-31-1	571.854	blk cry	60 dec		2.09		
838	Cobalt phosphide	Co_2P	12134-02-0	148.840	gray needles	1386		6.4		i H_2O; s HNO_3
839	Cobalt silicide	$CoSi_2$	12017-12-8	115.104	gray cub cry	1326		4.9		s hot HCl
840	Cobalt(II) acetate	$Co(C_2H_3O_2)_2$	71-48-7	177.022	pink cry					vs H_2O; s EtOH
841	Cobalt(II) acetate tetrahydrate	$Co(C_2H_3O_2)_2 \cdot 4H_2O$	6147-53-1	249.082	red monocl cry			1.705		s H_2O, EtOH, dil acid
842	Cobalt(II) aluminate	$CoAl_2O_4$	13820-62-7	176.894	blue cub cry			4.37		i H_2O
843	Cobalt(II) arsenate octahydrate	$Co_3(AsO_4)_2 \cdot 8H_2O$	24719-19-5	598.760	red monocl needles	400 dec	1000 dec	3.0		i H_2O; s dil acid
844	Cobalt(II) bromate hexahydrate	$Co(BrO_3)_2 \cdot 6H_2O$	13476-01-2	422.829	viol cry			≈2.5		vs H_2O
845	Cobalt(II) bromide	$CoBr_2$	7789-43-7	218.741	grn hex cry; hyg	678		4.91	113.2²⁰	s MeOH, EtOH, ace
846	Cobalt(II) bromide hexahydrate	$CoBr_2 \cdot 6H_2O$	13762-12-4	326.832	red hyg cry	47 dec	100 dec	2.46	113.2	
847	Cobalt(II) carbonate	$CoCO_3$	513-79-1	118.942	pink rhomb cry	dec 280		4.2	0.00014²⁰	i EtOH
848	Cobalt(II) basic carbonate	$2CoCO_3 \cdot 3Co(OH)_2 \cdot H_2O$	7542-09-8	534.743	red-viol cry	dec				i H_2O; s acid
849	Cobalt(II) chlorate hexahydrate	$Co(ClO_3)_2 \cdot 6H_2O$		333.927	dark red hyg cry	dec 61				s H_2O
850	Cobalt(II) chloride	$CoCl_2$	7646-79-9	129.839	blue hyg leaflets	737	1049	3.36	56.2²⁵	s EtOH, eth, ace, py
851	Cobalt(II) chloride dihydrate	$CoCl_2 \cdot 2H_2O$	16544-92-6	165.870	viol-blue cry			2.477	56.2²⁵	

No.	Name	Formula	CAS Reg No.	Mol. weight	Physical form	mp/°C	bp/°C	Density g cm^{-3}	Solubility g/100 g H$_2$O	Qualitative solubility
852	Cobalt(II) chloride hexahydrate	CoCl$_2$ · 6H$_2$O	7791-13-1	237.930	pink-red monocl cry	87 dec		1.924	56.2^{25}	s EtOH, ace, eth
853	Cobalt(II) chromate	CoCrO$_4$	24613-38-5	174.927	yel-brn orth cry			≈4.0		i H$_2$O; s acid
854	Cobalt(II) chromite	CoCr$_2$O$_4$	13455-25-9	226.923	blue-grn cub cry			5.14		i H$_2$O, conc acid
855	Cobalt(II) citrate dihydrate	Co$_3$(C$_6$H$_5$O$_7$)$_2$ · 2H$_2$O	18727-04-3	265.170	rose red cry	dec 150			0.8^{15}	
856	Cobalt(II) cyanide	Co(CN)$_2$	542-84-7	110.967	blue hyg cry			1.872		i H$_2$O
857	Cobalt(II) cyanide dihydrate	Co(CN)$_2$ · 2H$_2$O	20427-11-6	146.998	pink-brn needles					i H$_2$O, acid
858	Cobalt(II) diiron tetroxide	CoFe$_2$O$_4$	12052-28-7	234.621	blk solid					s hot HCl
859	Cobalt(II) ferricyanide	Co$_3$[Fe(CN)$_6$]$_2$	14049-81-1	600.698	red needles					i H$_2$O, HCl; s NH$_4$OH
860	Cobalt(II) fluoride	CoF$_2$	10026-17-2	96.930	red tetr cry	1127	≈1400	4.46	1.4^{25}	s acid
861	Cobalt(II) fluoride tetrahydrate	CoF$_2$ · 4H$_2$O	13817-37-3	168.992	red orth cry	dec		2.22	1.4^{25}	
862	Cobalt(II) formate dihydrate	Co(CHO$_2$)$_2$ · 2H$_2$O	6424-20-0	184.998	red cry powder	140 dec		2.13	5.03^{20}	i EtOH
863	Cobalt(II) hexafluoro-2,4-pentanedioate	Co(CF$_3$COCHCOCF$_3$)$_2$	19648-83-0	473.035	powder	197				
864	Cobalt(II) hexafluorosilicate hexahydrate	CoSiF$_6$ · 6H$_2$O	12021-68-0	309.100	pale red cry			2.087	76.8^{22}	
865	Cobalt(II) hydroxide	Co(OH)$_2$	21041-93-0	92.948	blue-grn cry	≈160 dec		3.60		sl H$_2$O; s acid
866	Cobalt(II) hydroxide monohydrate	Co(OH)$_2$ · H$_2$O	35340-84-2	110.963	blue solid	136 dec				
867	Cobalt(II) iodate	Co(IO$_3$)$_2$	13455-28-2	408.738	blk-viol needles	200 dec		5.09	0.46^{20}	
868	Cobalt(II) iodide	CoI$_2$	15238-00-3	312.742	blk hex cry; hyg	520		5.60	203^{25}	
869	Cobalt(II) iodide dihydrate	CoI$_2$ · 2H$_2$O	13455-29-3	348.773	hyg grn cry	dec 100				
870	Cobalt(II) iodide hexahydrate	CoI$_2$ · 6H$_2$O	15238-00-3*	420.833	red hex prisms	130 dec		2.90	203^{25}	s EtOH, eth, ace
871	Cobalt(II) molybdate	CoMoO$_4$	13762-14-6	218.87	blk monocl cry	1040		4.7		
872	Cobalt(II) molybdate monohydrate	CoMoO$_4$ · H$_2$O	18601-87-1	236.89	blk pow					
873	Cobalt(II) nitrate	Co(NO$_3$)$_2$	10141-05-6	182.942	pale red powder	100 dec		2.49	103^{25}	
874	Cobalt(II) nitrate hexahydrate	Co(NO$_3$)$_2$ · 6H$_2$O	10026-22-9	291.034	red monocl cry; hyg	≈55		1.88	103^{25}	s EtOH
875	Cobalt(II) nitrite	Co(NO$_2$)$_2$	18488-96-5	150.944					0.49^{25}	
876	Cobalt(II) oleate	Co(C$_{18}$H$_{33}$O$_2$)$_2$	14666-94-5	621.840	brn amorp pow					i H$_2$O; s EtOH, eth
877	Cobalt(II) orthosilicate	Co$_2$SiO$_4$	12017-08-2	209.950	red-viol orth cry	1345		4.63		i H$_2$O; s dil HCl
878	Cobalt(II) oxalate	CoC$_2$O$_4$	814-89-1	146.952	pink powder	250 dec		3.02	0.0037^{20}	s acid, NH$_4$OH
879	Cobalt(II) oxalate dihydrate	CoC$_2$O$_4$ · 2H$_2$O	5965-38-8	182.982	pink needles	dec			0.0037	sl acid; s NH$_4$OH
880	Cobalt(II) oxide	CoO	1307-96-6	74.932	gray cub cry	1830		6.44		i H$_2$O; s acid
881	Cobalt(II) 2,4-pentanedioate	Co(CH$_3$COCHCOCH$_3$)$_2$	14024-48-7	257.149	bl-viol cry	167				
882	Cobalt(II) perchlorate	Co(ClO$_4$)$_2$	13455-31-7	257.834	red needles			3.33	113^{25}	i EtOH, ace
883	Cobalt(II) perchlorate hexahydrate	Co(ClO$_4$)$_2$ · 6H$_2$O	13478-33-6	365.926	dark red cry	dec 170		3.33		vs H$_2$O
884	Cobalt(II) phosphate octahydrate	Co$_3$(PO$_4$)$_2$ · 8H$_2$O	10294-50-5	510.865	pink amorp powder			2.77		i H$_2$O; s acid
885	Cobalt(II) potassium sulfate hexahydrate	CoK$_2$(SO$_4$)$_2$ · 6H$_2$O	10026-20-7	437.347	red monocl cry	75 dec		2.22		vs H$_2$O
886	Cobalt(II) selenate pentahydrate	CoSeO$_4$ · 5H$_2$O	14590-19-3	291.97	red tricl cry	dec		2.51	55^{15}	
887	Cobalt(II) selenide	CoSe	1307-99-9	137.89	yel hex cry	1055		7.65		i H$_2$O, alk; s aqua regia
888	Cobalt(II) selenite dihydrate	CoSeO$_3$ · 2H$_2$O	19034-13-0	221.92	blue-red powder					i H$_2$O
889	Cobalt(II) stannate	Co$_2$SnO$_4$	12139-93-4	300.574	grn-blue cub cry			6.30		i H$_2$O; s alk
890	Cobalt(II) stearate	Co(C$_{18}$H$_{35}$O$_2$)$_2$	1002-88-6	625.872	purp solid	74		1.13		
891	Cobalt(II) sulfate	CoSO$_4$	10124-43-3	154.996	red orth cry	>700		3.71	38.3^{25}	
892	Cobalt(II) sulfate monohydrate	CoSO$_4$ · H$_2$O	13455-34-0	173.011	red monocl cry			3.08	38.3^{25}	
893	Cobalt(II) sulfate heptahydrate	CoSO$_4$ · 7H$_2$O	10026-24-1	281.102	pink monocl cry	41 dec		2.03	38.3^{25}	sl EtOH, MeOH
894	Cobalt(II) sulfide	CoS	1317-42-6	90.998	blk amorp powder	1117		5.45		i H$_2$O; s acid
895	Cobalt(II) telluride	CoTe	12017-13-9	186.53	hex cry			≈8.8		
896	Cobalt(II) thiocyanate	Co(SCN)$_2$	3017-60-5	175.097	yel-brn powder				103^{25}	s EtOH, MeOH, ace, eth
897	Cobalt(II) thiocyanate trihydrate	Co(SCN)$_2$ · 3H$_2$O	97126-35-7	229.143	viol rhomb cry				103^{25}	s EtOH, eth, ace
898	Cobalt(II) titanate	CoTiO$_3$	12017-01-5	154.798	grn rhomb cry			5.0		
899	Cobalt(II) tungstate	CoWO$_4$	12640-47-0	306.77	blue monocl cry			≈7.8		i H$_2$O; s hot conc acid
900	Cobalt(II,III) oxide	Co$_3$O$_4$	1308-06-1	240.798	blk cub cry	900 dec		6.11		i H$_2$O; s acid, alk
901	Cobalt(III) acetate	Co(C$_2$H$_3$O$_2$)$_3$	917-69-1	236.064	grn hyg cry	100 dec				s H$_2$O, EtOH
902	Cobalt(III) ammonium tetranitrodiammine	NH$_4$[Co(NH$_3$)$_2$(NO$_2$)$_4$]	13600-89-0	295.054	red-brn orth cry			1.97		s H$_2$O
903	Cobalt(III) fluoride	CoF$_3$	10026-18-3	115.928	brn hex cry	927		3.88		reac H$_2$O; s EtOH, eth, bz
904	Cobalt(III) fluoride dihydrate	Co$_2$F$_6$ · 2H$_2$O	54496-71-8	267.887	red rhomb cry			2.19		s H$_2$O; i EtOH
905	Cobalt(III) hexammine chloride	Co(NH$_3$)$_6$Cl$_3$	10534-89-1	267.475	red monocl cry			1.71		s H$_2$O; i EtOH
906	Cobalt(III) hydroxide	Co(OH)$_3$	1307-86-4	109.955	brn powder	dec		≈4		i H$_2$O; s acid
907	Cobalt(III) nitrate	Co(NO$_3$)$_3$	15520-84-0	244.948	grn cub cry; hyg			≈3.0		s H$_2$O; reac os
908	Cobalt(III) oxide	Co$_2$O$_3$	1308-04-9	165.864	gray-blk powder	895 dec		5.18		i H$_2$O; s conc acid

No.	Name	Formula	CAS Reg No.	Mol. weight	Physical form	mp/°C	bp/°C	Density g cm⁻³	Solubility g/100 g H₂O	Qualitative solubility
909	Cobalt(III) oxide monohydrate	$Co_2O_3 \cdot H_2O$	12016-80-7	183.880	brn-blk hex cry	150 dec				i H₂O; s acid
910	Cobalt(III) 2,4-pentanedioate	$Co(CH_3COCHCOCH_3)_3$	21679-46-9	356.257	dark grn cry	213				s bz, ace
911	Cobalt(III) potassium nitrite sesquihydrate	$CoK_3(NO_2)_6 \cdot 1.5H_2O$	13782-01-9*	479.284	yel cub cry			2.6		sl H₂O; reac acid; i EtOH
912	Cobalt(III) sulfide	Co_2S_3	1332-71-4	214.061	blk cub cry			4.8		reac acid
913	Cobalt(III) titanate	Co_2TiO_4	12017-38-8	229.731	grn-blk cub cry			5.1		s conc HCl
914	Copper	Cu	7440-50-8	63.546	red metal; cub	1084.62	2562	8.96		sl dil acid
915	Copper arsenide	Cu_3As	12005-75-3	265.560	dark gray solid	827				
916	Copper nitride	Cu_3N	1308-80-1	204.645	cub cry	300 dec		5.84		
917	Copper phosphide	CuP_2	12019-11-3	125.494	monocl cry	≈900		4.20		
918	Copper silicide	Cu_5Si	12159-07-8	345.816	solid	825				
919	Copper(I) acetate	$CuC_2H_3O_2$	598-54-9	122.590	col cry	dec	subl			reac H₂O
920	Copper(I) acetylide	Cu_2C_2	1117-94-8	151.113	red amorp powder; exp					
921	Copper(I) azide	CuN_3	14336-80-2	105.566	tetr cry; exp					
922	Copper(I) bromide	CuBr	7787-70-4	143.450	wh cub cry; hyg	483	1345	4.98	0.0012²⁰	i ace
923	Copper(I) chloride	CuCl	7758-89-6	98.999	wh cub cry	423	1490	4.14	0.0047²⁰	i EtOH, ace
924	Copper(I) cyanide	CuCN	544-92-3	89.564	wh powder or grn orth cry	474	dec	2.9		i H₂O, EtOH; s KCN soln
925	Copper(I) fluoride	CuF	13478-41-6	82.544	cub cry			7.1		
926	Copper(I) hydride	CuH	13517-00-5	64.554	red-brn solid	60 dec				
927	Copper(I) iodide	CuI	7681-65-4	190.450	wh cub cry	591	≈1290	5.67	0.000020²⁰	i dil acid
928	Copper(I) mercury iodide	Cu_2HgI_4	13876-85-2	835.30	red cry powder	trans ≈60 (brn)				i H₂O, EtOH
929	Copper(I) oxide	Cu_2O	1317-39-1	143.091	red-brn cub cry	1244	1800 dec	6.0		i H₂O
930	Copper(I) selenide	Cu_2Se	20405-64-5	206.05	blue-blk tetr cry	1113		6.84		i H₂O; s acid
931	Copper(I) sulfide	Cu_2S	22205-45-4	159.157	blue-blk orth cry	1129		5.6		i H₂O; sl acid
932	Copper(I) sulfite hemihydrate	$Cu_2SO_3 \cdot 0.5H_2O$	13982-53-1*	216.164	wh-yel hex cry					sl H₂O; s acid, alk; i EtOH, eth
933	Copper(I) sulfite monohydrate	$Cu_2SO_3 \cdot H_2O$	35788-00-2	225.171	cry			3.83		sl H₂O; s HCl
934	Copper(I) telluride	Cu_2Te	12019-52-2	254.69	blue hex cry	1127		4.6		
935	Copper(I) thiocyanate	CuSCN	1111-67-7	121.629	wh-yel amorp powder	1084		2.85		i H₂O, dil acid, EtOH, ace; s eth
936	Copper(I,II) sulfite dihydrate	$Cu_2SO_3 \cdot CuSO_3 \cdot 2H_2O$	13814-81-8	386.795	red prisms or powder					i H₂O, EtOH; s HCl
937	Copper(II) acetate	$Cu(C_2H_3O_2)_2$	142-71-2	181.635	blue-grn hyg powder					
938	Copper(II) acetate monohydrate	$Cu(C_2H_3O_2)_2 \cdot H_2O$	6046-93-1	199.650	grn monocl cry	115	240 dec	1.88		s H₂O, EtOH; sl eth
939	Copper(II) acetate metaarsenite	$Cu(C_2H_3O_2)_2 \cdot 3Cu(AsO_2)_2$	12002-03-8	1013.795	grn cry powder					i H₂O; reac acid
940	Copper(II) basic acetate	$Cu(C_2H_3O_2)_2 \cdot CuO \cdot 6H_2O$	52503-64-7	369.271	blue-grn cry or powder					sl H₂O, EtOH; s acid, NH₄OH
941	Copper(II) acetylide	CuC_2	12540-13-5	87.567	brn-blk solid; exp	exp 100				
942	Copper(II) arsenate	$Cu_3(AsO_4)_2$	7778-41-8	468.476	blue-grn cry					i H₂O, EtOH; s dil acid
943	Copper(II) arsenite	$CuHAsO_3$	10290-12-7	187.474	yel-grn powder					i H₂O, EtOH; s acid
944	Copper(II) azide	$Cu(N_3)_2$	14215-30-6	147.586	brn orth cry; exp			≈2.6		
945	Copper(II) borate	$Cu(BO_2)_2$	39290-85-2	149.166	blue-grn powder			3.859		i H₂O; s acid
946	Copper(II) bromide	$CuBr_2$	7789-45-9	223.354	blk monocl cry; hyg	498	900	4.710	126²⁵	vs H₂O; s EtOH, ace; i bz, eth
947	Copper(II) butanoate monohydrate	$Cu(C_4H_7O_2)_2 \cdot H_2O$	540-16-9	255.756	grn monocl plates					s H₂O, diox, bz; sl EtOH
948	Copper(II) carbonate	$CuCO_3$	1184-64-1	123.555	cry					i H₂O
949	Copper(II) carbonate hydroxide	$CuCO_3 \cdot Cu(OH)_2$	12069-69-1	221.116	grn monocl cry	200 dec		4.0		i H₂O, EtOH; s dil acid
950	Copper(II) chlorate hexahydrate	$Cu(ClO_3)_2 \cdot 6H_2O$	14721-21-2	338.540	blue-grn hyg cry	65	100 dec		164¹⁸	vs EtOH
951	Copper(II) chloride	$CuCl_2$	7447-39-4	134.452	yel-brn monocl cry; hyg	598	993	3.4	75.7²⁵	s EtOH, ace
952	Copper(II) chloride dihydrate	$CuCl_2 \cdot 2H_2O$	10125-13-0	170.483	grn-blue orth cry; hyg	100 dec		2.51	75.7²⁰	vs EtOH, MeOH; s ace; i eth
953	Copper(II) chloride hydroxide	$Cu_2(OH)_3Cl$	1332-65-6	213.567	pale grn cry					i H₂O; s acid
954	Copper(II) chromate	$CuCrO_4$	13548-42-0	179.540	red-brn cry					i H₂O; s EtOH
955	Copper(II) basic chromate	$CuCrO_4 \cdot 2Cu(OH)_2$	12433-14-6	374.661	brn pow	dec 260				i H₂O; s HNO₃
956	Copper(II) chromite	$CuCr_2O_4$	12018-10-9	231.536	gray-blk tetr cry			5.4		i H₂O, dil acid
957	Copper(II) citrate hemipentahydrate	$Cu_2C_6H_4O_7 \cdot 2.5H_2O$	10402-15-0	360.221	blue-grn cry	100 dec				sl H₂O; s dil acid
958	Copper(II) cyanide	$Cu(CN)_2$	14763-77-0	115.580	grn powder					i H₂O; s acid, alk
959	Copper(II) cyclohexanebutanoate	$Cu(C_{10}H_{17}O_2)_2$	2218-80-6	402.028	powder	126 dec				
960	Copper(II) dichromate dihydrate	$CuCr_2O_7 \cdot 2H_2O$	13675-47-3	315.565	red-brn tricl cry			2.286		vs H₂O
961	Copper(II) ethanolate	$Cu(C_2H_5O)_2$	2850-65-9	153.667	blue hyg solid	120 dec				i os
962	Copper(II) ethylacetoacetate	$Cu(C_2H_5CO_2CHCOCH_3)_2$	14284-06-1	321.813	grn cry	192				s EtOH, chl

No.	Name	Formula	CAS Reg No.	Mol. weight	Physical form	mp/°C	bp/°C	Density g cm⁻³	Solubility g/100 g H₂O	Qualitative solubility
963	Copper(II) 2-ethylhexanoate	$Cu(C_8H_{15}O_2)_2$	149-11-1	349.953	powder	252 dec				
964	Copper(II) ferrate	$CuFe_2O_4$	12018-79-0	239.234	blk cry					
965	Copper(II) ferrocyanide	$Cu_2Fe(CN)_6$	13601-13-3	339.041	red-br cub cry or powder			2.2		i H₂O, acid, os
966	Copper(II) ferrous sulfide	$CuFeS_2$	1308-56-1	183.521	yel tetr cry	950		4.2		i H₂O, HCl; s HNO₃
967	Copper(II) fluoride	CuF_2	7789-19-7	101.543	wh monocl cry	836	1676	4.23	0.075²⁵	
968	Copper(II) fluoride dihydrate	$CuF_2 \cdot 2H_2O$	13454-88-1	137.574	blue monocl cry	130 dec		2.934	0.075²⁵	
969	Copper(II) formate	$Cu(CHO_2)_2$	544-19-4	153.581	blue cry				12.5²⁰	i os
970	Copper(II) formate tetrahydrate	$Cu(CHO_2)_2 \cdot 4H_2O$	5893-61-8	225.641	blue monocl cry				12.5	sl EtOH; i os
971	Copper(II) gluconate	$CuC_{12}H_{22}O_{14}$	527-09-3	453.841	bl-grn cry	156				sl EtOH; i os
972	Copper(II) hexafluoro-2,4-pentanedioate	$Cu(CF_3COCHCOCF_3)_2$	14781-45-4	477.648	cry	98	220 dec			s MeOH, ace, tol
973	Copper(II) hexafluorosilicate tetrahydrate	$CuSiF_6 \cdot 4H_2O$	12062-24-7	277.684	blue monocl cry	dec		2.56	99.7¹⁷	sl EtOH
974	Copper(II) hydroxide	$Cu(OH)_2$	20427-59-2	97.561	blue-grn powder			3.37		i H₂O; s acid, conc alk
975	Copper(II) iodate	$Cu(IO_3)_2$	13454-89-2	413.351	grn monocl cry	dec		5.241	0.15²⁰	s dil acid
976	Copper(II) iodate monohydrate	$Cu(IO_3)_2 \cdot H_2O$	13454-90-5	431.367	blue tricl cry	248 dec		4.872	0.15²⁰	s dil H₂SO₄
977	Copper(II) molybdate	$CuMoO_4$	13767-34-5	223.48	grn cry	≈500		3.4	0.038	
978	Copper(II) nitrate	$Cu(NO_3)_2$	3251-23-8	187.555	blue-grn orth cry; hyg	255	subl		145²⁵	s diox; reac eth
979	Copper(II) nitrate trihydrate	$Cu(NO_3)_2 \cdot 3H_2O$	10031-43-3	241.602	blue rhomb cry	114	170 dec	2.32	145²⁵	vs EtOH
980	Copper(II) nitrate hexahydrate	$Cu(NO_3)_2 \cdot 6H_2O$	13478-38-1	295.647	blue rhomb cry; hyg			2.07	145²⁵	s EtOH
981	Copper(II) oleate	$Cu(C_{18}H_{33}O_2)_2$	1120-44-1	626.453	blue-grn solid					i H₂O; sl EtOH; s eth
982	Copper(II) oxalate	CuC_2O_4	814-91-5	151.565	blue-wh powder	310 dec			0.0026²⁰	i EtOH, eth; s NH₄OH
983	Copper(II) oxalate hemihydrate	$CuC_2O_4 \cdot 0.5H_2O$	814-91-5*	144.573	blue-wh cry	200 dec			0.0026²⁰	s NH₄OH
984	Copper(II) oxide	CuO	1317-38-0	79.545	blk powder or monocl cry	1227		6.31		i H₂O, EtOH; s dil acid
985	Copper(II) 2,4-pentanedioate	$Cu(CH_3COCHCOCH_3)_2$	13395-16-9	261.762	blue powder	284 dec	subl			sl H₂O; s chl
986	Copper(II) oxychloride hemiheptahydrate	$CuCl_2 \cdot 3CuO \cdot 3.5H_2O$	1332-40-7		blue-grn pow	dec 140				i H₂O; s acid, NH₄OH
987	Copper(II) perchlorate	$Cu(ClO_4)_2$	13770-18-8	262.447	grn hyg cry	130 dec			146³⁰	s eth, diox; i bz, ctc
988	Copper(II) perchlorate hexahydrate	$Cu(ClO_4)_2 \cdot 6H_2O$	10294-46-9	370.539	blue monocl cry; hyg	82	120 dec	2.22	146³⁰	vs EtOH, HOAc, ace; sl eth
989	Copper(II) phosphate	$Cu_3(PO_4)_2$	7798-23-4	380.581	blue-grn tricl cry					i H₂O; s acid, NH₄OH
990	Copper(II) phosphate trihydrate	$Cu_3(PO_4)_2 \cdot 3H_2O$	10031-48-8	434.627	blue-grn orth cry					i H₂O; s acid, NH₄OH
991	Copper(II) phthalocyanine	$CuC_{32}H_{16}N_8$	147-14-8	576.069	bl-purp cry					i H₂O, EtOH; s conc H₂SO₄
992	Copper(II) selenate pentahydrate	$CuSeO_4 \cdot 5H_2O$	10031-45-5	296.58	blue tricl cry	80 dec		2.56	27.4²⁵	s acid, NH₄OH; sl ace; i EtOH
993	Copper(II) selenide	$CuSe$	1317-41-5	142.51	blue-blk needles or plates	550 dec		5.99		reac acid
994	Copper(II) selenite dihydrate	$CuSeO_3 \cdot 2H_2O$	15168-20-4	226.54	blue orth cry			3.31		i H₂O; s acid, NH₄OH
995	Copper(II) silicate dihydrate	$CuSiO_3 \cdot 2H_2O$	26318-99-0	175.661	grn-blue orth cry					
996	Copper(II) stannate	$CuSnO_3$	12019-07-7	230.254	blue pow					
997	Copper(II) stearate	$Cu(C_{18}H_{35}O_2)_2$	660-60-6	630.485	blue-grn amorp powder	≈250				i H₂O, EtOH, eth; s py
998	Copper(II) sulfate	$CuSO_4$	7758-98-7	159.609	wh-grn amorp powder or rhomb cry	560 dec		3.60	22.0²⁵	i EtOH
999	Copper(II) sulfate pentahydrate	$CuSO_4 \cdot 5H_2O$	7758-99-8	249.685	blue tricl cry	110 dec		2.286	22.0²⁵	s MeOH; sl EtOH
1000	Copper(II) sulfate, basic	$Cu_3(OH)_4SO_4$	1332-14-5	354.730	grn rhomb cry			3.88		i H₂O
1001	Copper(II) sulfide	CuS	1317-40-4	95.611	blk hex cry	trans 507		4.76		i H₂O, EtOH, dil acid, alk
1002	Copper(II) tartrate trihydrate	$CuC_4H_4O_6 \cdot 3H_2O$	815-82-7	265.663	blue-grn powder					sl H₂O; s acid, alk
1003	Copper(II) telluride	$CuTe$	12019-23-7	191.15	yel orth cry	trans ≈400		7.09		
1004	Copper(II) tellurite	$CuTeO_3$	13812-58-3	239.14	blk glassy solid					i H₂O
1005	Copper(II) tetrafluoroborate	$Cu(BF_4)_2$	14735-84-3	237.155	solid					s H₂O
1006	Copper(II) titanate	$CuTiO_3$	12019-08-8	159.411	gray pow					
1007	Copper(II) 1,1,1-trifluoro-2,4-pentanedioate	$Cu(CF_3COCHCOCH_3)_2$	14324-82-4	369.705	blue-purp cry	197	dec 260			s EtOH, tol
1008	Copper(II) tungstate	$CuWO_4$	13587-35-4	311.38	yel-brn powder			7.5		
1009	Copper(II) tungstate dihydrate	$CuWO_4 \cdot 2H_2O$	13587-35-4*	347.41	grn powder					i H₂O; sl HOAc; reac conc acid
1010	Copper(II) vanadate	$Cu(VO_3)_2$	12789-09-2	261.425	powder					

No.	Name	Formula	CAS Reg No.	Mol. weight	Physical form	mp/°C	bp/°C	Density g cm^{-3}	Solubility g/100 g H$_2$O	Qualitative solubility
1011	Curium	Cm	7440-51-9	247	silv metal; hex or cub	1345	≈3100	13.51		
1012	Dysprosium	Dy	7429-91-6	162.500	silv metal; hex	1412	2567	8.55		s dil acid
1013	Dysprosium boride	DyB$_4$	12310-43-9	205.744	tetr cry	2500		6.98		
1014	Dysprosium nitride	DyN	12019-88-4	176.507	cub cry			9.93		
1015	Dysprosium silicide	DySi$_2$	12133-07-2	218.671	orth cry	1550		5.2		
1016	Dysprosium(II) bromide	DyBr$_2$	83229-05-4	322.308	blk solid					
1017	Dysprosium(II) chloride	DyCl$_2$	13767-31-2	233.406	blk cry	721 dec				reac H$_2$O
1018	Dysprosium(II) iodide	DyI$_2$	36377-94-3	416.309	purp cry	659				reac H$_2$O
1019	Dysprosium(III) acetate tetrahydrate	Dy(C$_2$H$_3$O$_2$)$_3$ · 4H$_2$O	15280-55-4	411.693	yel needles	dec 120				s H$_2$O; sl EtOH
1020	Dysprosium(III) bromide	DyBr$_3$	14456-48-5	402.212	wh hyg cry	879				s H$_2$O
1021	Dysprosium(III) carbonate tetrahydrate	Dy$_2$(CO$_3$)$_3$ · 4H$_2$O	38245-35-1	577.088	wh cry pow					i H$_2$O
1022	Dysprosium(III) chloride	DyCl$_3$	10025-74-8	268.859	wh or yel cry	718	1530	3.67		s H$_2$O, MeOH
1023	Dysprosium(III) chloride hexahydrate	DyCl$_3$ · 6H$_2$O	15059-52-6	376.950	bright yel cry	dec 162				
1024	Dysprosium(III) fluoride	DyF$_3$	13569-80-7	219.495	grn cry	1157				
1025	Dysprosium(III) hydride	DyH$_3$	13537-09-2	165.524	hex cry			7.1		
1026	Dysprosium(III) hydroxide	Dy(OH)$_3$	1308-85-6	213.522	yel or wh needles	205 dec				i H$_2$O
1027	Dysprosium(III) iodide	DyI$_3$	15474-63-2	543.213	grn cry	978				
1028	Dysprosium(III) nitrate pentahydrate	Dy(NO$_3$)$_3$ · 5H$_2$O	10143-38-1*	438.591	yel cry	88.6			208.4^{25}	
1029	Dysprosium(III) oxide	Dy$_2$O$_3$	1308-87-8	372.998	wh cub cry	2228	3900	7.81		s acid
1030	Dysprosium(III) sulfate octahydrate	Dy$_2$(SO$_4$)$_3$ · 8H$_2$O	10031-50-2	757.310	pale yel cry	110				sl H$_2$O
1031	Dysprosium(III) sulfide	Dy$_2$S$_3$	12133-10-7	421.195	red-brn monocl cry			6.08		
1032	Dysprosium(III) telluride	Dy$_2$Te$_3$	12159-43-2	707.80	solid	≈1550				
1033	Einsteinium	Es	7429-92-7	252	metal; cub	860				
1034	Erbium	Er	7440-52-0	167.259	silv metal; hex	1529	2868	9.07		i H$_2$O; s acid
1035	Erbium boride	ErB$_4$	12310-44-0	210.503	tetr cry	2450		7.0		
1036	Erbium acetate tetrahydrate	Er(C$_2$H$_3$O$_2$)$_3$ · 4H$_2$O	15280-57-6	416.452	pink or wh cry			2.11		s H$_2$O
1037	Erbium bromide	ErBr$_3$	13536-73-7	406.971	viol hyg cry	950	≈1460			s H$_2$O, thf
1038	Erbium bromide hexahydrate	ErBr$_3$ · 6H$_2$O	14890-44-9	515.062	pink cry					s H$_2$O
1039	Erbium chloride	ErCl$_3$	10138-41-7	273.618	viol monocl cry; hyg	776		4.1		s H$_2$O
1040	Erbium chloride hexahydrate	ErCl$_3$ · 6H$_2$O	10025-75-9	381.709	pink hyg cry	dec				s H$_2$O; sl EtOH
1041	Erbium fluoride	ErF$_3$	13760-83-3	224.254	pink orth cry	1146		7.8		i H$_2$O
1042	Erbium hydride	ErH$_3$	13550-53-3	170.283	hex cry			≈7.6		
1043	Erbium hydroxide	Er(OH)$_3$	14646-16-3	218.281	pink solid					i H$_2$O
1044	Erbium iodide	ErI$_3$	13813-42-8	547.972	viol hex cry; hyg	1014		≈5.5		s H$_2$O
1045	Erbium nitrate pentahydrate	Er(NO$_3$)$_3$ · 5H$_2$O	10168-80-6*	443.350	red cry	130 dec			240.8^{25}	s EtOH, ace
1046	Erbium nitride	ErN	12020-21-2	181.266	cub cry			10.6		
1047	Erbium oxide	Er$_2$O$_3$	12061-16-4	382.516	pink powder	2344	3920	8.64		i H$_2$O; s acid
1048	Erbium silicide	ErSi$_2$	12020-28-9	223.430	orth cry			7.26		
1049	Erbium sulfate	Er$_2$(SO$_4$)$_3$	13478-49-4	622.706	hyg powder	dec		3.68	13^{20}	
1050	Erbium sulfate octahydrate	Er$_2$(SO$_4$)$_3$ · 8H$_2$O	10031-52-4	766.828	pink monocl cry	dec		3.20	13^{20}	
1051	Erbium sulfide	Er$_2$S$_3$	12159-66-9	430.713	red-brn monocl cry	1730		6.07		
1052	Erbium telluride	Er$_2$Te$_3$	12020-39-2	717.32	orth cry	1213		7.11		
1053	Europium	Eu	7440-53-1	151.964	soft silv metal; cub	822	1529	5.24		reac H$_2$O
1054	Europium boride	EuB$_6$	12008-05-8	216.830	cub cry	≈2600		4.91		
1055	Europium nitride	EuN	12020-58-5	165.971	cub cry			8.7		
1056	Europium silicide	EuSi$_2$	12434-24-1	208.135	tetr cry	1500		5.46		
1057	Europium(II) bromide	EuBr$_2$	13780-48-8	311.772	wh cry	683				s H$_2$O
1058	Europium(II) chloride	EuCl$_2$	13769-20-5	222.870	wh orth cry	731		4.9		s H$_2$O
1059	Europium(II) fluoride	EuF$_2$	14077-39-5	189.961	grn-yel cub cry	≈1380		6.5		
1060	Europium(II) iodide	EuI$_2$	22015-35-6	405.773	grn cry	580				s H$_2$O
1061	Europium(II) selenide	EuSe	12020-66-5	230.92	brn cub cry			6.45		
1062	Europium(II) sulfate	EuSO$_4$	10031-54-6	248.027	col orth cry			4.99		i H$_2$O
1063	Europium(II) sulfide	EuS	12020-65-4	184.029	cub cry			5.7		
1064	Europium(II) telluride	EuTe	12020-69-8	279.56	blk cub cry	1526		6.48		
1065	Europium(III) bromide	EuBr$_3$	13759-88-1	391.676	gray cry	dec				s H$_2$O
1066	Europium(III) chloride	EuCl$_3$	10025-76-0	258.323	grn-yel needles	623		4.89		
1067	Europium(III) chloride hexahydrate	EuCl$_3$ · 6H$_2$O	13759-92-7	366.414	wh-yel hyg cry	850		4.89		s H$_2$O
1068	Europium(III) fluoride	EuF$_3$	13765-25-8	208.959	wh hyg cry	1276				i H$_2$O
1069	Europium(III) iodide	EuI$_3$	13759-90-5	532.677	col cry; unstab	≈875				
1070	Europium(III) nitrate hexahydrate	Eu(NO$_3$)$_3$ · 6H$_2$O	10031-53-5	446.070	wh-pink hyg cry	85 dec			193^{25}	
1071	Europium(III) oxalate	Eu$_2$(C$_2$O$_4$)$_3$	3269-12-3	567.985	wh solid					i H$_2$O; s acid

No.	Name	Formula	CAS Reg No.	Mol. weight	Physical form	mp/°C	bp/°C	Density g cm⁻³	Solubility g/100 g H₂O	Qualitative solubility
1072	Europium(III) oxide	Eu_2O_3	1308-96-9	351.926	pink powder	2291	3790	7.42		i H₂O; s acid
1073	Europium(III) perchlorate hexahydrate	$Eu(ClO_4)_3 \cdot 6H_2O$	36907-40-1	558.407	wh or pink cry					s H₂O, EtOH
1074	Europium(III) sulfate	$Eu_2(SO_4)_3$	13537-15-0	592.116	pale pink cry			4.99	2.1²⁰	
1075	Europium(III) sulfate octahydrate	$Eu_2(SO_4)_3 \cdot 8H_2O$	10031-52-4	736.238	pink cry	375 dec			2.1²⁰	
1076	Fermium	Fm	7440-72-4	257	metal	1527				
1077	Fluorine	F_2	7782-41-4	37.997	pale yel gas	-219.67 tp	-188.12	1.553 g/L		reac H₂O
1078	Fluorine monoxide	F_2O	7783-41-7	53.996	col gas	-223.8	-144.3	2.207 g/L		sl H₂O
1079	Difluorine dioxide	F_2O_2	7783-44-0	69.996	red-oran solid, unstab gas	-163.5	-57 (extrap)	2.861 g/L		
1080	Fluorine tetroxide	F_2O_4	107782-11-6	101.995	red-brn solid	-191	dec -185			
1081	Fluorine nitrate	FNO_3	7789-26-6	81.003	col gas	-175	-46	3.311 g/L		reac H₂O, EtOH, eth; s ace
1082	Fluorine perchlorate	$FOClO_3$	10049-03-3	118.449	col gas; exp	-167.3	-16	4.841 g/L		reac H₂O
1083	Francium	Fr	7440-73-5	223.000	short-lived alkali metal	27				
1084	Gadolinium	Gd	7440-54-2	157.25	silv metal; hex	1313	3273	7.90		s dil acid
1085	Gadolinium boride	GdB_6	12008-06-9	222.12	blk-brn cub cry	2510		5.31		
1086	Gadolinium nitride	GdN	25764-15-2	171.26	cub cry			9.10		
1087	Gadolinium silicide	$GdSi_2$	12134-75-7	213.42	orth cry	1540		5.9		
1088	Gadolinium(II) iodide	GdI_2	13814-72-7	411.06	bronze cry	831				
1089	Gadolinium(II) selenide	GdSe	12024-81-6	236.21	cub cry	2170		8.1		
1090	Gadolinium(III) acetate tetrahydrate	$Gd(C_2H_3O_2)_3 \cdot 4H_2O$	15280-53-2	406.44	wh tricl cry	dec		1.61		s H₂O
1091	Gadolinium(III) bromide	$GdBr_3$	13818-75-2	396.96	wh monocl cry; hyg	770		4.56		
1092	Gadolinium(III) chloride	$GdCl_3$	10138-52-0	263.61	wh monocl cry; hyg	602		4.52		s H₂O
1093	Gadolinium(III) chloride hexahydrate	$GdCl_3 \cdot 6H_2O$	19423-81-5	371.70	col hyg cry			2.424		s H₂O
1094	Gadolinium(III) fluoride	GdF_3	13765-26-9	214.25	wh cry	1232				
1095	Gadolinium(III) iodide	GdI_3	13572-98-0	537.96	yel cry	930				
1096	Gadolinium(III) nitrate pentahydrate	$Gd(NO_3)_3 \cdot 5H_2O$	52788-53-1	433.34	wh cry	92 dec		2.41	190²⁵	
1097	Gadolinium(III) nitrate hexahydrate	$Gd(NO_3)_3 \cdot 6H_2O$	19598-90-4	451.36	hyg tricl cry	91 dec		2.33	190²⁵	s EtOH
1098	Gadolinium(III) oxalate decahydrate	$Gd_2(C_2O_4)_3 \cdot 10H_2O$	22992-15-0	758.71	wh monocl pow	dec 110				i H₂O; sl acid
1099	Gadolinium(III) oxide	Gd_2O_3	12064-62-9	362.50	wh hyg powder	2339	3900	7.07		i H₂O; s acid
1100	Gadolinium(III) sulfate	$Gd_2(SO_4)_3$	13628-54-1	602.69	col cry	500 dec		4.1	2.60²⁰	sl H₂O
1101	Gadolinium(III) sulfate octahydrate	$Gd_2(SO_4)_3 \cdot 8H_2O$	13450-87-8	746.81	col monocl cry	400 dec		4.14	2.3²⁰	sl H₂O
1102	Gadolinium(III) sulfide	Gd_2S_3	12134-77-9	410.70	yel cub cry			6.1		
1103	Gadolinium(III) telluride	Gd_2Te_3	12160-99-5	697.30	orth cry	1255		7.7		
1104	Gallium	Ga	7440-55-3	69.723	silv liq or gray orth cry	29.7666 tp	2204	5.91		reac alk
1105	Gallium antimonide	GaSb	12064-03-8	191.483	brn cub cry	712		5.6137		
1106	Gallium arsenide	GaAs	1303-00-0	144.645	gray cub cry	1238		5.3176		
1107	Gallium nitride	GaN	25617-97-4	83.730	gray hex cry	>2500		6.1		
1108	Gallium phosphide	GaP	12063-98-8	100.697	yel cub cry	1457		4.138		
1109	Gallium suboxide	Ga_2O	12024-20-3	155.445	brn powder	>660	>800 dec	4.77		
1110	Gallium(II) chloride	$GaCl_2$	24597-12-4	140.629	wh orth cry	172.4	535	2.74		
1111	Gallium(II) selenide	GaSe	12024-11-2	148.68	hex cry	960		5.03		
1112	Gallium(II) sulfide	GaS	12024-10-1	101.788	hex cry	965		3.86		
1113	Gallium(II) telluride	GaTe	12024-14-5	197.32	monocl cry	824		5.44		
1114	Gallium(III) bromide	$GaBr_3$	13450-88-9	309.435	wh orth cry	123	279	3.69		
1115	Gallium(III) chloride	$GaCl_3$	13450-90-3	176.082	col needles or gl solid	77.9	201	2.47		
1116	Gallium(III) fluoride	GaF_3	7783-51-9	126.718	wh powder or col needles	>1000		4.47		i H₂O
1117	Gallium(III) fluoride trihydrate	$GaF_3 \cdot 3H_2O$	22886-66-4	180.764	wh cry	>140 dec				sl H₂O
1118	Gallium(III) hydride	GaH_3	13572-93-5	72.747	visc liq	-15	≈0 dec			
1119	Gallium(III) hydroxide	$Ga(OH)_3$	12023-99-3	120.745	unstab prec					
1120	Gallium(III) iodide	GaI_3	13450-91-4	450.436	monocl cry	212	340	4.5		
1121	Gallium(III) nitrate	$Ga(NO_3)_3$	13494-90-1	255.738	wh cry powder					s H₂O, EtOH, eth
1122	Gallium(III) oxide	Ga_2O_3	12024-21-4	187.444	wh cry	1807		≈6.0		s hot acid
1123	Gallium(III) oxide hydroxide	GaOOH	20665-52-5	102.730	orth cry			5.23		
1124	Gallium(III) 2,4-pentanedioate	$Ga(CH_3COCHCOCH_3)_3$	14405-43-7	367.047	wh powder	193	subl	1.42		
1125	Gallium(III) perchlorate hexahydrate	$Ga(ClO_4)_3 \cdot 6H_2O$	17835-81-3	476.166	cry	dec 175				
1126	Gallium(III) selenide	Ga_2Se_3	12024-24-7	376.33	cub cry	937		4.92		
1127	Gallium(III) sulfate	$Ga_2(SO_4)_3$	13494-91-2	427.634	hex cry					
1128	Gallium(III) sulfate octadecahydrate	$Ga_2(SO_4)_3 \cdot 18H_2O$	13780-42-2	751.909	octahed cry					s H₂O, EtOH
1129	Gallium(III) sulfide	Ga_2S_3	12024-22-5	235.641	monocl cry	1090		3.7		

No.	Name	Formula	CAS Reg No.	Mol. weight	Physical form	mp/°C	bp/°C	Density g cm^{-3}	Solubility g/100 g H$_2$O	Qualitative solubility
1130	Gallium(III) telluride	Ga$_2$Te$_3$	12024-27-0	522.25	cub cry	790		5.57		
1131	Germanium	Ge	7440-56-4	72.64	gray-wh cub cry	938.25	2833	5.3234		i H$_2$O, dil acid, alk
1132	Germane	GeH$_4$	7782-65-2	76.67	col gas; flam	-165	-88.1	3.133 g/L		i H$_2$O
1133	Digermane	Ge$_2$H$_6$	13818-89-8	151.33	col liq; flam	-109	29	1.98^{-109}		
1134	Trigermane	Ge$_3$H$_8$	14691-44-2	225.98	col liq	-105.6	110.5	2.20^{-105}		i H$_2$O
1135	Tetragermane	Ge$_4$H$_{10}$	14691-47-5	300.64	col liq		176.9			i H$_2$O
1136	Pentagermane	Ge$_5$H$_{12}$	15587-39-0	375.30	col liq		234			i H$_2$O
1137	Bromogermane	GeH$_3$Br	13569-43-2	155.57	col liq	-32	52	2.34		reac H$_2$O
1138	Chlorogermane	GeH$_3$Cl	13637-65-5	111.12	col liq	-52	28	1.75		reac H$_2$O
1139	Chlorotrifluorogermane	GeF$_3$Cl	14188-40-0	165.09	gas	-66.2	-20.3	6.747 g/L		
1140	Dibromogermane	GeH$_2$Br$_2$	13769-36-3	234.46	col liq	-15	89	2.80		reac H$_2$O
1141	Dichlorogermane	GeH$_2$Cl$_2$	15230-48-5	145.56	col liq	-68	69.5	1.90		reac H$_2$O
1142	Dichlorodifluorogermane	GeF$_2$Cl$_2$	24422-21-7	181.54	col gas	-51.8	-2.8	7.419 g/L		
1143	Dichlorodimethylgermane	Ge(CH$_3$)$_2$Cl$_2$	1529-48-2	173.62	liq	-22	124	1.49		
1144	Fluorogermane	GeH$_3$F	13537-30-9	94.66	col gas			3.868 g/L		reac H$_2$O
1145	Iodogermane	GeH$_3$I	13573-02-9	202.57	liq	-15	≈90			reac H$_2$O
1146	Methylgermane	GeH$_3$CH$_3$	1449-65-6	90.70	col gas	-158	-23	3.706 g/L		
1147	Tribromogermane	GeHBr$_3$	14779-70-5	313.36	col liq	-25	dec			reac H$_2$O
1148	Trichlorogermane	GeHCl$_3$	1184-65-2	180.01	liq	-71	75.3	1.93		reac H$_2$O
1149	Trichlorofluorogermane	GeCl$_3$F	24422-20-6	198.00	liq	-49.8	37.5			
1150	Germanium(II) bromide	GeBr$_2$	24415-00-7	232.45	yel monocl cry	122	150 dec			reac H$_2$O
1151	Germanium(II) chloride	GeCl$_2$	10060-11-4	143.55	wh-yel hyg powder	dec				reac H$_2$O; s eth, bz
1152	Germanium(II) fluoride	GeF$_2$	13940-63-1	110.64	wh orth cry; hyg	110	130 dec	3.64		reac H$_2$O
1153	Germanium(II) iodide	GeI$_2$	13573-08-5	326.45	oran-yel hex cry	428	550 dec	5.4		reac H$_2$O
1154	Germanium(II) oxide	GeO	20619-16-3	88.64	blk solid	700 dec				
1155	Germanium(II) selenide	GeSe	12065-10-0	151.60	gray orth cry or brn powder	675		5.6		
1156	Germanium(II) sulfide	GeS	12025-32-0	104.71	gray orth cry	658		4.1		
1157	Germanium(II) telluride	GeTe	12025-39-7	200.24	cub cry	724		6.16		i H$_2$O; s conc HNO$_3$
1158	Germanium(IV) bromide	GeBr$_4$	13450-92-5	392.26	wh cry	26.1	186.35	3.132		reac H$_2$O
1159	Germanium(IV) chloride	GeCl$_4$	10038-98-9	214.45	col liq	-51.50	86.55	1.88		reac H$_2$O; s bz, eth, EtOH, ctc
1160	Germanium(IV) fluoride	GeF$_4$	7783-58-6	148.63	col gas	-15 tp	-36.5 sp	6.074 g/L		reac H$_2$O
1161	Germanium(IV) iodide	GeI$_4$	13450-95-8	580.26	red-oran cub cry	146	348	4.322		reac H$_2$O
1162	Germanium(IV) nitride	Ge$_3$N$_4$	12065-36-0	273.95	orth cry	900 dec				i H$_2$O, acid, aqua regia
1163	Germanium(IV) oxide	GeO$_2$	1310-53-8	104.64	wh hex cry	1116		4.25		i H$_2$O
1164	Germanium(IV) selenide	GeSe$_2$	12065-11-1	230.56	yel-oran orth cry	707 dec		4.56		
1165	Germanium(IV) sulfide	GeS$_2$	12025-34-2	136.77	blk orth cry	530		3.01		
1166	Gold	Au	7440-57-5	196.967	soft yel metal	1064.18	2856	19.3		s aqua regia
1167	Bromoauric(III) acid pentahydrate	HAuBr$_4$ · 5H$_2$O	17083-68-0	607.667	red-brn hyg cry	27				s H$_2$O, EtOH
1168	Chloroauric(III) acid tetrahydrate	HAuCl$_4$ · 4H$_2$O	16903-35-8	411.848	yel monocl cry; hyg			≈3.9		vs H$_2$O, EtOH; s eth
1169	Gold(I) bromide	AuBr	10294-27-6	276.871	yel-gray tetr cry	165 dec		8.20		i H$_2$O
1170	Gold(I) chloride	AuCl	10294-29-8	232.420	yel orth cry	289 dec		7.6	0.000031^{20}	
1171	Gold(I) cyanide	AuCN	506-65-0	222.985	yel hex cry	dec		7.2		i H$_2$O, EtOH, eth, dil acid
1172	Gold(I) iodide	AuI	10294-31-2	323.871	yel-grn powder; tetr	120 dec		8.25		i H$_2$O; s CN soln
1173	Gold(I) sulfide	Au$_2$S	1303-60-2	425.998	brn-blk cub cry; unstab	240 dec		≈11		i H$_2$O, acid; s aqua regia
1174	Gold(III) bromide	AuBr$_3$	10294-28-7	436.679	red-br monocl cry	≈160 dec				s H$_2$O, EtOH
1175	Gold(III) chloride	AuCl$_3$	13453-07-1	303.326	red monocl cry	>160 dec		4.7	68^{20}	
1176	Gold(III) cyanide trihydrate	Au(CN)$_3$ · 3H$_2$O	535-37-5*	329.065	wh hyg cry	50 dec				vs H$_2$O; sl EtOH
1177	Gold(III) fluoride	AuF$_3$	14720-21-9	253.962	oran-yel hex cry	>300	subl	6.75		
1178	Gold(III) hydroxide	Au(OH)$_3$	1303-52-2	247.989	brn powder	≈100 dec				i H$_2$O; s acid
1179	Gold(III) iodide	AuI$_3$	31032-13-0	577.680	unstab grn powder	20 dec				
1180	Gold(III) oxide	Au$_2$O$_3$	1303-58-8	441.931	brn powder	≈150 dec				i H$_2$O; s acid
1181	Gold(III) selenate	Au$_2$(SeO$_4$)$_3$	10294-32-3	822.81	yel cry					i H$_2$O; s acid
1182	Gold(III) selenide	Au$_2$Se$_3$	1303-62-4	630.81	blk amorp solid	dec		4.65		s aqua regia
1183	Gold(III) sulfide	Au$_2$S$_3$	1303-61-3	490.128	unstab blk powder	200 dec				
1184	Hafnium	Hf	7440-58-6	178.49	gray metal; hex	2233	4603	13.3		s HF
1185	Hafnium boride	HfB$_2$	12007-23-7	200.11	gray hex cry	3100		10.5		
1186	Hafnium carbide	HfC	12069-85-1	190.50	refrac cub cry	≈3000		12.2		
1187	Hafnium hydride	HfH$_2$	12770-26-2	180.51	refrac tetr cry			11.4		
1188	Hafnium nitride	HfN	25817-87-2	192.50	yel-brn cub cry	3310		13.8		
1189	Hafnium phosphide	HfP	12325-59-6	209.46	hex cry			9.78		

No.	Name	Formula	CAS Reg No.	Mol. weight	Physical form	mp/°C	bp/°C	Density g cm⁻³	Solubility g/100 g H₂O	Qualitative solubility
1190	Hafnium silicide	$HfSi_2$	12401-56-8	234.66	gray orth cry	≈1700		7.6		
1191	Hafnocene dichloride	$Hf(C_5H_5)_2Cl_2$	12116-66-4	379.58	col hyg cry	235				s bz, chl; sl thf, eth; i hex
1192	Hafnium(II) bromide	$HfBr_2$	13782-95-1	338.30	blue-blk cry	dec 400				
1193	Hafnium(II) chloride	$HfCl_2$	13782-92-8	249.40	blk solid	dec 400				
1194	Hafnium(III) bromide	$HfBr_3$	13782-96-2	418.20	blue-blk cry	dec 350				
1195	Hafnium(III) chloride	$HfCl_3$	13782-93-9	284.85	blk solid	dec				
1196	Hafnium(III) iodide	HfI_3	13779-73-2	559.20	blk cry	dec				
1197	Hafnium(IV) bromide	$HfBr_4$	13777-22-5	498.11	wh cub cry	424 tp	323 sp	4.90		reac H₂O
1198	Hafnium(IV) chloride	$HfCl_4$	13499-05-3	320.30	wh monocl cry	432 tp	317 sp			reac H₂O
1199	Hafnium(IV) fluoride	HfF_4	13709-52-9	254.48	wh monocl cry	1025	970 sp	7.1		reac H₂O
1200	Hafnium(IV) iodide	HfI_4	13777-23-6	686.11	yel-oran cub cry	449 tp	394 sp	5.6		reac H₂O
1201	Hafnium(IV) oxide	HfO_2	12055-23-1	210.49	wh cub cry	2800	≈5400	9.68		i H₂O
1202	Hafnium(IV) oxychloride octahydrate	$HfOCl_2 \cdot 8H_2O$	14456-34-9	409.52	wh tetr cry	dec				s H₂O
1203	Hafnium(IV) selenide	$HfSe_2$	12162-21-9	336.41	brn hex cry			7.46		
1204	Hafnium(IV) silicate	$HfSiO_4$	13870-13-8	270.57	tetr cry	2758		7.0		
1205	Hafnium(IV) sulfate	$Hf(SO_4)_2$	15823-43-5	370.62	wh cry	>500 dec				
1206	Hafnium(IV) sulfide	HfS_2	18855-94-2	242.62	purp-brn hex cry			6.03		
1207	Hafnium(IV) titanate	$HfTiO_4$	12055-24-2	290.36	wh pow	1980 dec				
1208	Helium	He	7440-59-7	4.003	col gas		-268.93	0.164 g/L		sl H₂O; i EtOH
1209	Holmium	Ho	7440-60-0	164.930	silv metal; hex	1472	2700	8.80		s dil acid
1210	Holmium acetate	$Ho(C_2H_3O_2)_3$	25519-09-9	342.062	yel cry	dec 327				s H₂O
1211	Holmium bromide	$HoBr_3$	13825-76-8	404.642	yel hyg cry	919	1470			
1212	Holmium chloride	$HoCl_3$	10138-62-2	271.289	yel monocl cry; hyg	720	1500	3.7		s H₂O
1213	Holmium chloride hexahydrate	$HoCl_3 \cdot 6H_2O$	14914-84-2	379.381	hyg yel cry	160 dec				s H₂O
1214	Holmium fluoride	HoF_3	13760-78-6	221.925	pink-yel orth cry; hyg	1143	>2200	7.664		s H₂O
1215	Holmium iodide	HoI_3	13813-41-7	545.643	yel hex cry	994		5.4		
1216	Holmium nitrate pentahydrate	$Ho(NO_3)_3 \cdot 5H_2O$	14483-18-2	441.022	hyg oran cry					s H₂O, EtOH, ace
1217	Holmium nitride	HoN	12029-81-1	178.937	cub cry			10.6		
1218	Holmium oxalate decahydrate	$Ho_2(C_2O_4)_3 \cdot 10H_2O$	28965-57-3	774.070	yel solid	dec 40				
1219	Holmium oxide	Ho_2O_3	12055-62-8	377.859	yel cub cry	2330	3900	8.41		s acid
1220	Holmium silicide	$HoSi_2$	12136-24-2	221.101	hex cry			7.1		
1221	Holmium sulfide	Ho_2S_3	12162-59-3	426.056	yel-oran monocl cry			5.92		
1222	Hydrazine	N_2H_4	302-01-2	32.045	col oily liq	1.54	113.55	1.0036		vs H₂O, EtOH, MeOH
1223	Hydrazine acetate	$N_2H_4 \cdot CH_3COOH$	13255-48-6	92.097	cry	100				
1224	Hydrazine azide	$N_2H_4 \cdot HN_3$	14662-04-5	75.074	hyg wh prism	75 exp				vs H₂O
1225	Hydrazine monohydrate	$N_2H_4 \cdot H_2O$	7803-57-8	50.060	fuming liq	-51.7	119	1.030		vs H₂O, EtOH; i chl, eth
1226	Hydrazine hydrobromide	$N_2H_4 \cdot HBr$	13775-80-9	112.957	wh monocl cry flakes	84	≈190 dec	2.3		s H₂O, EtOH
1227	Hydrazine hydrochloride	$N_2H_4 \cdot HCl$	2644-70-4	68.506	wh orth cry	89	240 dec	1.5		s H₂O; i os
1228	Hydrazine dihydrochloride	$N_2H_4 \cdot 2HCl$	5341-61-7	104.967	wh orth cry	198 dec		1.42		s H₂O; sl EtOH
1229	Hydrazine hydroiodide	$N_2H_4 \cdot HI$	10039-55-1	159.957	hyg cry	125				s H₂O
1230	Hydrazine nitrate	$N_2H_4 \cdot HNO_3$	13464-97-6	95.058	monocl cry; exp	70				vs H₂O
1231	Hydrazine dinitrate	$N_2H_4 \cdot 2HNO_3$	13464-98-7	158.071	needles	104 dec			20³⁵	s H₂O
1232	Hydrazine perchlorate hemihydrate	$N_2H_4 \cdot HClO_4 \cdot 0.5H_2O$	13762-65-7		solid	137	exp	1.94		reac H₂O, s EtOH; i eth, bz
1233	Hydrazine sulfate	$N_2H_4 \cdot H_2SO_4$	10034-93-2	130.124	col orth cry	254		1.378		sl H₂O; i EtOH
1234	Dihydrazine sulfate	$(N_2H_4)_2 \cdot H_2SO_4$	13464-80-7	162.169	hyg wh cry flakes	104	dec >180		200²⁵	vs H₂O; i os
1235	Hydrazoic acid	HN_3	7782-79-8	43.028	col liq; exp	-80	35.7			s H₂O
1236	Hydroxylamine	H_2NOH	7803-49-8	33.030	wh orth flakes or needles	33.1	58	1.21		vs H₂O, MeOH
1237	Hydroxylamine hydrobromide	$H_2NOH \cdot HBr$	41591-55-3	113.942	monocl cry			2.35		s H₂O
1238	Hydroxylamine hydrochloride	$H_2NOH \cdot HCl$	5470-11-1	69.491	col monocl cry	159 dec	exp	1.68	94²⁵	vs H₂O
1239	Hydroxylamine perchlorate	$H_2NOH \cdot HClO_4$		133.489	orth cry	88	dec 120			
1240	Hydroxylamine sulfate	$2(H_2NOH) \cdot H_2SO_4$	10039-54-0	164.138	cry	170				vs H₂O
1241	Hydrogen	H_2	1333-74-0	2.016	col gas; flam	-259.198 tp	-252.762	0.082 g/L		sl H₂O
1242	Hydrogen-d_2	D_2	7782-39-0	4.028	col gas	-254.42	-249.48	0.164 g/L		
1243	Hydrogen-t_2	T_2	10028-17-8	6.032	col gas	-252.53	-248.11	0.246 g/L		
1244	Hydrogen-d_1	HD	13983-20-5	3.022	col gas	-256.55	-251.02	0.123 g/L		
1245	Hydrogen-t_1	HT	14885-60-0	4.024	col gas	-254.7	-249.6			
1246	Hydrogen-d_1,t_1	DT	14885-61-1	5.030	col gas	-253.5	-238.9			
1247	Hydrogen bromide	HBr	10035-10-6	80.912	col gas	-86.80	-66.38	3.307 g/L		vs H₂O; s EtOH
1248	Hydrogen bromide-d	DBr	13536-59-9	81.918	col gas	-87.54	-66.9			s H₂O

No.	Name	Formula	CAS Reg No.	Mol. weight	Physical form	mp/°C	bp/°C	Density g cm⁻³	Solubility g/100 g H₂O	Qualitative solubility
1249	Hydrogen chloride	HCl	7647-01-0	36.461	col gas	-114.17	-85	1.490 g/L		vs H_2O
1250	Hydrogen chloride dihydrate	HCl · 2H₂O	13465-05-9	72.492	col liq	-17.7		1.46		
1251	Hydrogen chloride-d	DCl	7698-05-7	37.467	col gas	-114.72	-84.4			s H_2O
1252	Hydrogen cyanide	HCN	74-90-8	27.026	col liq or gas	-13.29	26	0.6876²⁰		vs H_2O, EtOH; sl eth
1253	Hydrogen fluoride	HF	7664-39-3	20.006	col gas	-83.36	20	0.818 g/L		vs H_2O, EtOH; sl eth
1254	Hydrogen iodide	HI	10034-85-2	127.912	col or yel gas	-50.76	-35.55	5.228 g/L		vs H_2O; s os
1255	Hydrogen iodide-d	DI	14104-45-1	128.918	col gas	-51.93	-36.2			s H_2O
1256	Hydrogen peroxide	H₂O₂	7722-84-1	34.015	col liq	-0.43	150.2	1.44		vs H_2O
1257	Hydrogen selenide	H₂Se	7783-07-5	80.98	col gas; flam	-65.73	-41.25	3.310 g/L		s H_2O
1258	Hydrogen sulfide	H₂S	7783-06-4	34.081	col gas; flam	-85.5	-59.55	1.393 g/L		s H_2O
1259	Hydrogen disulfide	H₂S₂	13465-07-1	66.146	col liq		70.7	1.334		
1260	Hydrogen telluride	H₂Te	7783-09-7	129.62	col gas	-49	-2	5.298 g/L		s H_2O, EtOH, alk
1261	Indium	In	7440-74-6	114.818	soft wh metal	156.60	2072	7.31		s acid
1262	Indium antimonide	InSb	1312-41-0	236.578	blk cub cry	524		5.7747		
1263	Indium arsenide	InAs	1303-11-3	189.740	gray cub cry	942		5.67		i acid
1264	Indium nitride	InN	25617-98-5	128.825	brn hex cry	1100		6.88		
1265	Indium phosphide	InP	22398-80-7	145.792	blk cub cry	1062		4.81		sl acid
1266	Indium(I) bromide	InBr	14280-53-6	194.722	oran-red orth cry	285	656	4.96		reac H_2O
1267	Indium(I) chloride	InCl	13465-10-6	150.271	yel cub cry	225	608	4.19		reac H_2O
1268	Indium(I) iodide	InI	13966-94-4	241.722	orth cry	364.4	712	5.32		
1269	Indium(II) bromide	InBr₂	21264-43-7	274.626	orth cry			4.22		reac H_2O
1270	Indium(II) chloride	InCl₂	13465-11-7	185.724	col orth cry	235		3.64		reac H_2O
1271	Indium(II) sulfide	InS	12030-14-7	146.883	red-brn orth cry	692		5.2		
1272	Indium(III) bromide	InBr₃	13465-09-3	354.530	hyg yel-wh monocl cry	420		4.74	414²⁰	
1273	Indium(III) chloride	InCl₃	10025-82-8	221.177	yel monocl cry; hyg	583		4.0	195.1²²	s EtOH
1274	Indium(III) chloride tetrahydrate	InCl₃ · 4H₂O	22519-64-8	293.239	wh cry					s H_2O
1275	Indium(III) fluoride	InF₃	7783-52-0	171.813	wh hex cry; hyg	1172	>1200	4.39		sl H_2O; s dil acid
1276	Indium(III) fluoride trihydrate	InF₃ · 3H₂O	14166-78-0	225.859	wh cry	100 dec				s H_2O
1277	Indium(III) hydroxide	In(OH)₃	20661-21-6	165.840	cub cry			4.4		
1278	Indium(III) iodide	InI₃	13510-35-5	495.531	yel-red monocl cry; hyg	207		4.69	1308²²	
1279	Indium(III) nitrate trihydrate	In(NO₃)₃ · 3H₂O	13770-61-1	354.879	col cry	dec 100				
1280	Indium(III) oxide	In₂O₃	1312-43-2	277.634	yel cub cry	1912		7.18		i H_2O; s hot acid
1281	Indium(III) perchlorate octahydrate	In(ClO₄)₃ · 8H₂O	13465-15-1	557.292	wh cry	≈80	200 dec			
1282	Indium(III) phosphate	InPO₄	14693-82-4	209.789	wh orth cry			4.9		i H_2O
1283	Indium(III) selenide	In₂Se₃	1312-42-1	466.52	blk hex cry	660		5.8		
1284	Indium(III) sulfate	In₂(SO₄)₃	13464-82-9	517.824	hyg wh powder			3.44	117²⁰	
1285	Indium(III) sulfide	In₂S₃	12030-24-9	325.831	oran cub cry	1050		4.45		
1286	Indium(III) telluride	In₂Te₃	1312-45-4	612.44	blk cub cry	667		5.75		
1287	Iodine	I₂	7553-56-2	253.809	blue-blk plates	113.7	184.4	4.933	0.03²⁰	s bz, EtOH, eth, ctc, chl
1288	Iodic acid	HIO₃	7782-68-5	175.910	col orth cry	110 dec		4.63	308²⁵	i EtOH, eth
1289	Periodic acid dihydrate	HIO₄ · 2H₂O	10450-60-9	227.940	monocl hyg cry	122 dec				s H_2O, EtOH; sl eth
1290	Iodine tetroxide	I₂O₄	12399-08-5	317.807	yel cry	130	dec >85	4.2		sl H_2O
1291	Iodine pentoxide	I₂O₅	12029-98-0	333.806	hyg wh cry	≈300 dec		4.98	253.4²⁰	i EtOH, eth, CS₂
1292	Iodine hexoxide	I₂O₆	65355-99-9	349.805	yel solid	dec 150				reac H_2O
1293	Iodine nonaoxide	I₄O₉	73560-00-6	651.613	hyg yel powder	75 dec				
1294	Iodine bromide	IBr	7789-33-5	206.808	blk orth cry	40	116 dec	4.3		s H_2O, EtOH, eth
1295	Iodine chloride	ICl	7790-99-0	162.357	red cry or oily liq	27.38	94.4 dec	3.24		reac H_2O; s EtOH
1296	Iodine trichloride	ICl₃	865-44-1	233.263	yel tricl cry; hyg	101 tp (16 atm)	64 sp dec	3.2		reac H_2O; s EtOH, bz
1297	Iodine fluoride	IF	13873-84-2	145.902	wh pow (-78°C)	-14 dec				
1298	Iodine trifluoride	IF₃	22520-96-3	183.899	yel solid, stable at low temp	-28 dec				
1299	Iodine pentafluoride	IF₅	7783-66-6	221.896	yel liq	9.43	100.5	3.19		reac H_2O
1300	Iodine heptafluoride	IF₇	16921-96-3	259.893	col gas	6.5 tp	4.8 sp	10.62 g/L		s H_2O
1301	Iodosyl trifluoride	IOF₃	19058-78-7	199.898	hyg col needles	dec >110				reac H_2O
1302	Iodosyl pentafluoride	IOF₅	16056-61-4	237.895	col liq	4.5				
1303	Iodyl trifluoride	IO₂F₃	25402-50-0	215.898	yel solid	41	subl			
1304	Periodyl fluoride	IO₃F	30708-86-2	193.900	col cry	dec >100				
1305	Iridium	Ir	7439-88-5	192.217	silv-wh metal; cub	2446	4428	22.562²⁰		s aqua regia
1306	Iridium carbonyl	Ir₄(CO)₁₂	11065-24-0	1104.989	yel cry	210 dec				
1307	Iridium(III) bromide	IrBr₃	10049-24-8	431.929	red-brn monocl cry			6.82		i H_2O, acid, alk

No.	Name	Formula	CAS Reg No.	Mol. weight	Physical form	mp/°C	bp/°C	Density g cm^{-3}	Solubility g/100 g H$_2$O	Qualitative solubility
1308	Iridium(III) bromide tetrahydrate	IrBr$_3 \cdot$ 4H$_2$O	10049-24-8*	503.991	grn-brn cry					s H$_2$O; i EtOH
1309	Iridium(III) chloride	IrCl$_3$	10025-83-9	298.576	brn monocl cry	763 dec		5.30		i H$_2$O, acid, alk
1310	Iridium(III) fluoride	IrF$_3$	23370-59-4	249.212	blk hex cry	250 dec		≈8.0		i H$_2$O, dil acid
1311	Iridium(III) iodide	IrI$_3$	7790-41-2	572.930	dark brn monocl cry			≈7.4		i H$_2$O, acid, bz, chl; s alk
1312	Iridium(III) oxide	Ir$_2$O$_3$	1312-46-5	432.432	blue-blk cry	1000 dec				i H$_2$O; sl hot HCl
1313	Iridium(III) 2,4-pentanedioate	Ir(CH$_3$COCHCOCH$_3$)$_3$	15635-87-7	489.541	oran-yel cry	270	subl			sl H$_2$O; s tol, chl, ace, MeOH
1314	Iridium(III) sulfide	Ir$_2$S$_3$	12136-42-4	480.629	orth cry			10.2		
1315	Iridium(IV) chloride	IrCl$_4$	10025-97-5	334.029	brn hyg solid	≈700 dec				s H$_2$O, EtOH
1316	Iridium(IV) oxide	IrO$_2$	12030-49-8	224.216	brn tetr cry	1100 dec		11.7		
1317	Iridium(IV) sulfide	IrS$_2$	12030-51-2	256.347	orth cry			9.3		
1318	Iridium(VI) fluoride	IrF$_6$	7783-75-7	306.207	yel cub cry; hyg	44	53.6	4.8		reac H$_2$O
1319	Iron	Fe	7439-89-6	55.845	silv-wh or gray met	1538	2861	7.87		s dil acid
1320	Ferrocene	Fe(C$_5$H$_5$)$_2$	102-54-5	186.031	oran needles	172.5	249			i H$_2$O; s EtOH, eth, bz, dil HNO$_3$
1321	Tetracarbonyldihydroiron	Fe(CO)$_4$H$_2$	12002-28-7	169.902	col liq, stab low temp	-70	dec -20			s alk
1322	Iron pentacarbonyl	Fe(CO)$_5$	13463-40-6	195.896	yel oily liq; flam	-20.5	103	1.46		i H$_2$O; s eth, bz, ace
1323	Iron nonacarbonyl	Fe$_2$(CO)$_9$	15321-51-4	363.781	oran-yel cry	100 dec		2.85		
1324	Iron dodecacarbonyl	Fe$_3$(CO)$_{12}$	12088-65-2	503.656	dark grn cry	140		2.00		
1325	Iron arsenide	FeAs	12044-16-5	130.767	gray orth cry	1030		7.85		
1326	Iron boride (FeB)	FeB	12006-84-7	66.656	refrac solid; orth	1658		≈7		
1327	Iron boride (Fe$_2$B)	Fe$_2$B	12006-86-9	122.501	refrac solid; tetr	1389		7.3		
1328	Iron carbide	Fe$_3$C	12011-67-5	179.546	gray cub cry	1227		7.694		
1329	Iron phosphide (FeP)	FeP	26508-33-8	86.819	rhom cry			6.07		
1330	Iron phosphide (Fe$_2$P)	Fe$_2$P	1310-43-6	142.664	gray hex needles	1370		6.8		i H$_2$O, dil acid, alk
1331	Iron phosphide (Fe$_3$P)	Fe$_3$P	12023-53-9	198.509	gray solid	1100		6.74		i H$_2$O
1332	Iron disulfide	FeS$_2$	1317-66-4	119.975	blk cub cry	>600 dec		5.02		i H$_2$O
1333	Iron silicide	FeSi	12022-95-6	83.931	gray cub cry	1410		6.1		
1334	Iron disilicide	FeSi$_2$	12022-99-0	112.016	gray tetr cry	1220		4.74		
1335	Iron(II) acetate	Fe(C$_2$H$_3$O$_2$)$_2$	3094-87-9	173.934	wh cry	190 dec				s H$_2$O
1336	Iron(II) acetate tetrahydrate	Fe(C$_2$H$_3$O$_2$)$_2 \cdot$ 4H$_2$O	3094-87-9*	245.994	grn cry	dec				s H$_2$O, EtOH
1337	Iron(II) aluminate	Fe(AlO$_2$)$_2$	12068-49-4	173.806	blk cub cry			4.3		
1338	Iron(II) arsenate	Fe$_3$(AsO$_4$)$_2$	10102-50-8	445.373	grn powder					i H$_2$O
1339	Iron(II) arsenate hexahydrate	Fe$_3$(AsO$_4$)$_2 \cdot$ 6H$_2$O	10102-50-8*	553.465	grn amorp powder	dec				i H$_2$O; s acid
1340	Iron(II) bromide	FeBr$_2$	7789-46-0	215.653	yel-brn hex cry; hyg	691	dec	4.636	120^{25}	vs EtOH
1341	Iron(II) bromide hexahydrate	FeBr$_2 \cdot$ 6H$_2$O	13463-12-2	323.744	grn hyg cry	27 dec		4.64	120^{25}	s EtOH
1342	Iron(II) carbonate	FeCO$_3$	563-71-3	115.854	gray-brn hex cry			3.944	0.000062^{20}	
1343	Iron(II) chloride	FeCl$_2$	7758-94-3	126.751	wh hex cry; hyg	677	1023	3.16	65.0^{25}	vs EtOH, ace; sl bz
1344	Iron(II) chloride dihydrate	FeCl$_2 \cdot$ 2H$_2$O	16399-77-2	162.782	wh-grn monocl cry	120 dec		2.39	65.0^{25}	
1345	Iron(II) chloride tetrahydrate	FeCl$_2 \cdot$ 4H$_2$O	13478-10-9	198.813	grn monocl cry	105 dec		1.93	65.0^{25}	s EtOH
1346	Iron(II) chromite	FeCr$_2$O$_4$	1308-31-2	223.835	blk cub cry			5.0		
1347	Iron(II) fluoride	FeF$_2$	7789-28-8	93.842	wh tetr cry	1100		4.09		sl H$_2$O; s dil HF; i EtOH, eth
1348	Iron(II) fluoride tetrahydrate	FeF$_2 \cdot$ 4H$_2$O	13940-89-1	165.904	col hex cry			2.20		
1349	Iron(II) hydroxide	Fe(OH)$_2$	18624-44-7	89.860	wh-grn hex cry			3.4	0.000052^{20}	
1350	Iron(II) iodide	FeI$_2$	7783-86-0	309.654	red-viol hex cry; hyg	594		5.3		s H$_2$O, EtOH, eth
1351	Iron(II) iodide tetrahydrate	FeI$_2 \cdot$ 4H$_2$O	7783-86-0*	381.716	blk hyg leaflets	90 dec		2.87		s H$_2$O, EtOH
1352	Iron(II) molybdate	FeMoO$_4$	13718-70-2	215.78	brn-yel monocl cry	1115		5.6		i H$_2$O
1353	Iron(II) nitrate	Fe(NO$_3$)$_2$	14013-86-6	179.854	grn solid				87.5^{25}	
1354	Iron(II) nitrate hexahydrate	Fe(NO$_3$)$_2 \cdot$ 6H$_2$O	14013-86-6*	287.946	grn solid	60 dec			87.5^{25}	
1355	Iron(II) orthosilicate	Fe$_2$SiO$_4$	10179-73-4	203.774	brn orth cry			4.30		
1356	Iron(II) oxalate dihydrate	FeC$_2$O$_4 \cdot$ 2H$_2$O	6047-25-2	179.894	yel cry	150 dec		2.28	0.078^{25}	s acid
1357	Iron(II) oxide	FeO	1345-25-1	71.844	blk cub cry	1377		6.0		i H$_2$O, alk; s acid
1358	Iron(II) 2,4-pentanedioate	Fe(CH$_3$COCHCOCH$_3$)$_2$	14024-17-0	254.061	oran-brn cry	170	subl			sl bz, tol
1359	Iron(II) perchlorate	Fe(ClO$_4$)$_2$	13933-23-8	254.746	grn-wh hyg needles	>100 dec			210^{25}	
1360	Iron(II) phosphate octahydrate	Fe$_3$(PO$_4$)$_2 \cdot$ 8H$_2$O	14940-41-1	501.600	gray-blue monocl cry; hyg			2.58		i H$_2$O; s acid
1361	Iron(II) selenide	FeSe	1310-32-3	134.81	blk hex cry			6.7		i H$_2$O
1362	Iron(II) sulfate	FeSO$_4$	7720-78-7	151.908	wh orth cry; hyg			3.65	29.5^{25}	
1363	Iron(II) sulfate monohydrate	FeSO$_4 \cdot$ H$_2$O	17375-41-6	169.923	wh-yel monocl cry	300 dec		3.0	29.5^{25}	

No.	Name	Formula	CAS Reg No.	Mol. weight	Physical form	mp/°C	bp/°C	Density g cm⁻³	Solubility g/100 g H₂O	Qualitative solubility
1364	Iron(II) sulfate heptahydrate	$FeSO_4 \cdot 7H_2O$	7782-63-0	278.014	blue-grn monocl cry	≈60 dec		1.895	29.5[25]	i EtOH
1365	Iron(II) sulfide	FeS	1317-37-9	87.910	col hex or tetr cry; hyg	1188	dec	4.7		i H₂O; reac acid
1366	Iron(II) tantalate	$Fe(TaO_3)_2$	12140-41-9	513.737	brn tetr cry			7.33		
1367	Iron(II) tartrate	$FeC_4H_4O_6$		203.916	wh cry				0.88	vs acid; s NH₄OH
1368	Iron(II) telluride	FeTe	12125-63-2	183.45	tetr cry	914		6.8		
1369	Iron(II) thiocyanate trihydrate	$Fe(SCN)_2 \cdot 3H_2O$	6010-09-9	226.055	grn monocl cry					s H₂O, EtOH, eth
1370	Iron(II) titanate	$FeTiO_3$	12168-52-4	151.710	blk rhomb cry	≈1470		4.72		
1371	Iron(II) tungstate	$FeWO_4$	13870-24-1	303.68	monocl cry			7.51		
1372	Iron(II,III) oxide	Fe_3O_4	1317-61-9	231.533	blk cub cry or amorp powder	1597		5.17		i H₂O; s acid
1373	Iron(III) acetate, basic	$FeOH(C_2H_3O_2)_2$	10450-55-2	190.941	brn-red amorp powder					i H₂O; s EtOH, acid
1374	Iron(III) ammonium citrate	$Fe(NH_4)_3(C_6H_5O_7)_2$	1185-57-5	488.160	red or brn pow; hyg					s H₂O; i EtOH
1375	Iron(III) arsenate dihydrate	$FeAsO_4 \cdot 2H_2O$	10102-49-5	230.795	grn-brn powder	dec		3.18		i H₂O; s dil acid
1376	Iron(III) bromide	$FeBr_3$	10031-26-2	295.557	dark red hex cry; hyg	dec		4.5	455[25]	s EtOH, eth
1377	Iron(III) chloride	$FeCl_3$	7705-08-0	162.204	grn hex cry; hyg	307.6	≈316	2.90	91.2[25]	s EtOH, eth, ace
1378	Iron(III) chloride hexahydrate	$FeCl_3 \cdot 6H_2O$	10025-77-1	270.295	yel-oran monocl cry; hyg	37 dec		1.82	91.2[25]	s EtOH, eth, ace
1379	Iron(III) chromate	$Fe_2(CrO_4)_3$	10294-52-7	459.671	yel powder					i H₂O, EtOH; s acid
1380	Iron(III) citrate pentahydrate	$FeC_6H_5O_7 \cdot 5H_2O$	3522-50-7	335.021	red-brn cry					s H₂O; i EtOH
1381	Iron(III) dichromate	$Fe_2(Cr_2O_7)_3$	10294-53-8	759.654	red-brn solid					s H₂O, acid
1382	Iron(III) ferrocyanide	$Fe_4[Fe(CN)_6]_3$	14038-43-8	859.229	dark blue powder			1.80		i H₂O, dil acid, os
1383	Iron(III) fluoride	FeF_3	7783-50-8	112.840	grn hex cry	>1000		3.87	5.92[25]	i EtOH, eth, bz
1384	Iron(III) fluoride trihydrate	$FeF_3 \cdot 3H_2O$	15469-38-2	166.886	yel-brn tetr cry			2.3	5.92[25]	
1385	Iron(III) formate	$Fe(CHO_2)_3$	555-76-0	190.897	red-yel cry powder					s H₂O; sl EtOH
1386	Iron(III) hydroxide	$Fe(OH)_3$	1309-33-7	106.867	yel monocl cry			3.12		
1387	Iron(III) hydroxide oxide	FeO(OH)	20344-49-4	88.852	red-brn orth cry			4.26		i H₂O; s acid
1388	Iron(III) metavanadate	$Fe(VO_3)_3$	65842-03-7	352.665	gray-brn powder					i H₂O, EtOH; s acid
1389	Iron(III) nitrate	$Fe(NO_3)_3$	10421-48-4	241.860	cry				82.5[20]	
1390	Iron(III) nitrate hexahydrate	$Fe(NO_3)_3 \cdot 6H_2O$	13476-08-9	349.951	viol cub cry	35 dec			82.5[20]	
1391	Iron(III) nitrate nonahydrate	$Fe(NO_3)_3 \cdot 9H_2O$	7782-61-8	403.997	viol-gray hyg cry	47 dec		1.68	82.5[20]	vs EtOH, ace
1392	Iron(III) oxalate	$Fe_2(C_2O_4)_3$	19469-07-9	375.747	yel amorp powder	100 dec				s H₂O, acid; i alk
1393	Iron(III) oxide	Fe_2O_3	1309-37-1	159.688	red-brn hex cry	1539		5.25		i H₂O; s acid
1394	Iron(III) 2,4-pentanedioate	$Fe(CH_3COCHCOCH_3)_3$	14024-18-1	353.169	red-oran cry	179		5.24		sl H₂O; s os
1395	Iron(III) perchlorate hexahydrate	$Fe(ClO_4)_3 \cdot 6H_2O$	32963-81-8	462.288	viol cry					
1396	Iron(III) phosphate dihydrate	$FePO_4 \cdot 2H_2O$	10045-86-0	186.847	gray-wh orth cry			2.87		i H₂O; s HCl
1397	Iron(III) pyrophosphate nonahydrate	$Fe_4(P_2O_7)_3 \cdot 9H_2O$	10058-44-3	907.348	yel powder					i H₂O; s acid
1398	Iron(III) hypophosphite	$Fe(H_2PO_2)_3$	7783-84-8	250.811	wh-gray powder					i H₂O
1399	Iron(III) sodium pyrophosphate	$FeNaP_2O_7$	10045-87-1	252.778	wh powder			1.5		i H₂O; s HCl
1400	Iron(III) sulfate	$Fe_2(SO_4)_3$	10028-22-5	399.878	gray-wh rhomb cry; hyg			3.10	440[20]	sl EtOH; i ace
1401	Iron(III) sulfate nonahydrate	$Fe_2(SO_4)_3 \cdot 9H_2O$	13520-56-4	562.015	yel hex cry	400 dec		2.1	440[20]	
1402	Iron(III) thiocyanate	$Fe(SCN)_3$	4119-52-2	230.092	red-viol hyg cry	dec				s H₂O, EtOH, ace; i tol, chl
1403	Krypton	Kr	7439-90-9	83.798	col gas	-157.38 tp (73.2 kPa)	-153.34	3.425 g/L		sl H₂O
1404	Krypton difluoride	KrF_2	13773-81-4	121.795	col tetr cry	≈25 dec		3.24		reac H₂O
1405	Krypton fluoride hexafluoroantimonate	$KrFSb_2F_{11}$	39578-36-4	555.299	wh solid	dec 45				
1406	Lanthanum	La	7439-91-0	138.905	silv metal; hex	920	3464	6.15		s dil acid
1407	Lanthanum aluminum oxide	$LaAlO_3$	12003-65-5	213.885	wh rhom cry	trans cub 500				
1408	Lanthanum boride	LaB_6	12008-21-8	203.771	blk cub cry; refrac	2715		4.76		
1409	Lanthanum bromate nonahydrate	$LaBrO_3 \cdot 9H_2O$		684.749	hex cry	dec 100		5.06		vs H₂O
1410	Lanthanum bromide	$LaBr_3$	13536-79-3	378.617	wh hex cry; hyg	788		5.1		s H₂O
1411	Lanthanum carbide	LaC_2	12071-15-7	162.926	tetr cry	2360		5.29		
1412	Lanthanum carbonate octahydrate	$La_2(CO_3)_3 \cdot 8H_2O$	6487-39-4	601.960	wh cry powder			2.6		i H₂O; s dil acid
1413	Lanthanum chloride	$LaCl_3$	10099-58-8	245.264	wh hex cry; hyg	858		3.84	95.7[25]	
1414	Lanthanum chloride heptahydrate	$LaCl_3 \cdot 7H_2O$	20211-76-1	371.371	wh tricl cry; hyg	91 dec			95.7[25]	s EtOH
1415	Lanthanum fluoride	LaF_3	13709-38-1	195.900	wh hex cry; hyg	1493		5.9		i H₂O, acid
1416	Lanthanum hydride	LaH_3	13864-01-2	141.929	blk cub cry			5.36		
1417	Lanthanum hydroxide	$La(OH)_3$	14507-19-8	189.927	wh amorp solid	dec			0.000020[20]	
1418	Lanthanum iodate	$La(IO_3)_3$	13870-19-4	663.614	col cry				1.7	
1419	Lanthanum iodide	LaI_3	13813-22-4	519.619	wh orth cry; hyg	778		5.6		s H₂O
1420	Lanthanum nitrate hexahydrate	$La(NO_3)_3 \cdot 6H_2O$	10277-43-7	433.011	wh hyg tricl cry	≈40 dec			200[25]	vs EtOH; s ace

No.	Name	Formula	CAS Reg No.	Mol. weight	Physical form	mp/°C	bp/°C	Density g cm⁻³	Solubility g/100 g H₂O	Qualitative solubility
1421	Lanthanum nitride	LaN	25764-10-7	152.912	cub cry			6.73		
1422	Lanthanum oxide	La$_2$O$_3$	1312-81-8	325.809	wh amorp powder	2304	3620	6.51		i H$_2$O; s dil acid
1423	Lanthanum perchlorate hexahydrate	La(ClO$_4$)$_3$ · 6H$_2$O	36907-37-6	475.021	hyg col cry	dec 100				vs H$_2$O; s EtOH
1424	Lanthanum silicide	LaSi$_2$	12056-90-5	195.076	gray tetr cry	1520		5.0		
1425	Lanthanum sulfate	La$_2$(SO$_4$)$_3$	10099-60-2	565.999	hyg wh pow	1150				sl H$_2$O
1426	Lanthanum sulfate octahydrate	La$_2$(SO$_4$)$_3$ · 8H$_2$O	57804-25-8	702.058	col cry	dec		2.82		sl H$_2$O
1427	Lanthanum sulfate nonahydrate	La$_2$(SO$_4$)$_3$ · 9H$_2$O	10294-62-9	728.136	hex cry			2.82	2.7[20]	i EtOH
1428	Lanthanum monosulfide	LaS	12031-30-0	170.970	yel cub cry	2300		5.61		
1429	Lanthanum sulfide	La$_2$S$_3$	12031-49-1	374.006	red cub cry	2110		4.9		
1430	Lawrencium	Lr	22537-19-5	262	metal	1627				
1431	Lead	Pb	7439-92-1	207.2	soft silv-gray metal; cub	327.462	1749	11.3		s conc acid
1432	Plumbane	PbH$_4$	15875-18-0	211.2	unstab col gas		-13			
1433	Lead(II) acetate	Pb(C$_2$H$_3$O$_2$)$_2$	301-04-2	325.3	wh cry	280	dec	3.25	44.3[20]	
1434	Lead(II) acetate trihydrate	Pb(C$_2$H$_3$O$_2$)$_2$ · 3H$_2$O	6080-56-4	427.3	col cry	75 dec		2.55		vs H$_2$O; sl EtOH
1435	Lead(II) acetate, basic	Pb(C$_2$H$_3$O$_2$)$_2$ · 2Pb(OH)$_2$	1335-32-6	807.7	wh powder	dec			6.3[0]	
1436	Lead(II) antimonate	Pb$_3$(SbO$_4$)$_2$	13510-89-9	993.1	oran-yel powder			6.58		i H$_2$O, dil acid
1437	Lead(II) arsenate	Pb$_3$(AsO$_4$)$_2$	3687-31-8	899.4	wh cry	1042 dec		5.8		i H$_2$O; s HNO$_3$
1438	Lead(II) arsenite	Pb(AsO$_2$)$_2$	10031-13-7	421.0	wh powder			5.85		i H$_2$O; s dil HNO$_3$
1439	Lead(II) azide	Pb(N$_3$)$_2$	13424-46-9	291.2	col orth needles; exp	exp ≈350		4.7	0.023[18]	vs HOAc
1440	Lead(II) borate monohydrate	Pb(BO$_2$)$_2$ · H$_2$O	10214-39-8	310.8	wh powder	500 dec		5.6		i H$_2$O; s dil HNO$_3$
1441	Lead(II) bromate monohydrate	Pb(BrO$_3$)$_2$ · H$_2$O	10031-21-7	481.0	col cry	≈180 dec		5.53	1.33[20]	
1442	Lead(II) bromide	PbBr$_2$	10031-22-8	367.0	wh orth cry	371	892	6.69	0.975[25]	i EtOH
1443	Lead(II) butanoate	Pb(C$_4$H$_7$O$_2$)$_2$	819-73-8	381.4	col solid	≈90				i H$_2$O; s dil HNO$_3$
1444	Lead(II) carbonate	PbCO$_3$	598-63-0	267.2	col orth cry	≈315 dec		6.582		i H$_2$O
1445	Lead(II) carbonate, basic	Pb(OH)$_2$ · 2PbCO$_3$	1319-46-6	775.6	wh hex cry	400 dec		≈6.5		i H$_2$O, EtOH; s acid
1446	Lead(II) chlorate	Pb(ClO$_3$)$_2$	10294-47-0	374.1	col hyg cry	230 dec		3.9	144[18]	vs EtOH
1447	Lead(II) chloride	PbCl$_2$	7758-95-4	278.1	wh orth needles or powder	501	951	5.98	1.08[25]	s alk
1448	Lead(II) chloride fluoride	PbClF	13847-57-9	261.7	tetr cry			7.05	0.035[20]	
1449	Lead(II) chlorite	Pb(ClO$_2$)$_2$	13453-57-1	342.1	yel monocl cry	dec 126			0.2[25]	sl H$_2$O; s alk
1450	Lead(II) chromate	PbCrO$_4$	7758-97-6	323.2	yel-oran monocl cry	844		6.12	0.000017[20]	s alk, dil acid
1451	Lead(II) chromate(VI) oxide	PbCrO$_4$ · PbO	18454-12-1	546.4	red powder					i H$_2$O
1452	Lead(II) citrate trihydrate	Pb$_3$(C$_6$H$_5$O$_7$)$_2$ · 3H$_2$O	512-26-5	1053.8	wh cry powder					s H$_2$O; sl EtOH
1453	Lead(II) cyanide	Pb(CN)$_2$	592-05-2	259.2	wh-yel powder					sl H$_2$O; reac acid
1454	Lead(II) 2-ethylhexanoate	Pb(C$_7$H$_{15}$CO$_2$)$_2$	301-08-6	493.6	visc liq			1.56		
1455	Lead(II) fluoride	PbF$_2$	7783-46-2	245.2	wh orth cry	830	1293	8.44	0.0670[25]	
1456	Lead(II) fluoroborate	Pb(BF$_4$)$_2$	13814-96-5	380.8	stab only in aq soln					s H$_2$O
1457	Lead(II) formate	Pb(CHO$_2$)$_2$	811-54-1	297.2	wh prisms or needles	190 dec		4.63	1.6[16]	i EtOH
1458	Lead(II) hexafluoro-2,4-pentanedioate	Pb(CF$_3$COCHCOCF$_3$)$_2$	19648-88-5	621.3	cry	155	210			
1459	Lead(II) hydrogen arsenate	PbHAsO$_4$	7784-40-9	347.1	wh monocl cry	280 dec		5.943		i H$_2$O; s HNO$_3$, alk
1460	Lead(II) hydrogen phosphate	PbHPO$_4$	15845-52-0	303.2	wh monocl cry	dec		5.66		
1461	Lead(II) hydroxide	Pb(OH)$_2$	19783-14-3	241.2	wh powder	145 dec		7.59	0.00012[20]	s acid
1462	Lead(II) iodate	Pb(IO$_3$)$_2$	25659-31-8	557.0	wh orth cry			6.50	0.0025[25]	
1463	Lead(II) iodide	PbI$_2$	10101-63-0	461.0	yel hex cry or powder	410	872 dec	6.16	0.076[25]	i EtOH
1464	Lead(II) lactate	Pb(C$_3$H$_5$O$_3$)$_2$	18917-82-3	385.3	wh cry powder					s H$_2$O, hot EtOH
1465	Lead(II) molybdate	PbMoO$_4$	10190-55-3	367.1	yel tetr cry	≈1060		6.7		i H$_2$O; s HNO$_3$, NaOH
1466	Lead(II) niobate	Pb(NbO$_3$)$_2$	12034-88-7	489.0	rhom or tetr cry	1343		6.6		i H$_2$O
1467	Lead(II) nitrate	Pb(NO$_3$)$_2$	10099-74-8	331.2	col cub cry	470		4.53	59.7[25]	sl EtOH
1468	Lead(II) oleate	Pb(C$_{18}$H$_{33}$O$_2$)$_2$	1120-46-3	770.1	wax-like solid					i H$_2$O; s EtOH, bz, eth
1469	Lead(II) oxalate	PbC$_2$O$_4$	814-93-7	295.2	wh powder	300 dec		5.28	0.00025[20]	s dil HNO$_3$
1470	Lead(II) oxide (litharge)	PbO	1317-36-8	223.2	red tetr cry	trans to massicot 489		9.35		i H$_2$O, EtOH; s dil HNO$_3$
1471	Lead(II) oxide (massicot)	PbO	1317-36-8	223.2	yel orth cry	887		9.64		i H$_2$O, EtOH; s dil HNO$_3$
1472	Lead(II) oxide hydrate	3PbO · H$_2$O	1311-11-1	687.6	wh powder			7.41		i H$_2$O; s dil acid
1473	Lead(II) 2,4-pentanedioate	Pb(CH$_3$COCHCOCH$_3$)$_2$	15282-88-9	405.4	cry	143				
1474	Lead(II) perchlorate	Pb(ClO$_4$)$_2$	13453-62-8	406.1	wh cry				441[25]	
1475	Lead(II) perchlorate trihydrate	Pb(ClO$_4$)$_2$ · 3H$_2$O	13637-76-8	460.1	wh cry	100 dec		2.6	441[25]	s EtOH

No.	Name	Formula	CAS Reg No.	Mol. weight	Physical form	mp/°C	bp/°C	Density g cm^{-3}	Solubility g/100 g H$_2$O	Qualitative solubility
1476	Lead(II) phosphate	Pb$_3$(PO$_4$)$_2$	7446-27-7	811.5	wh hex cry	1014		7.01		i H$_2$O, EtOH
1477	Lead(II) hypophosphite	Pb(H$_2$PO$_2$)$_2$	10294-58-3	337.2	hyg cry powder	dec				sl H$_2$O; i EtOH
1478	Lead(II) metasilicate	PbSiO$_3$	10099-76-0	283.3	wh monocl cry powder	764		6.49		i H$_2$O, os
1479	Lead(II) orthosilicate	Pb$_2$SiO$_4$	13566-17-1	506.5	monocl cry	743		7.60		
1480	Lead(II) hexafluorosilicate dihydrate	PbSiF$_6$ · 2H$_2$O	1310-03-8	385.3	col cry	dec				vs H$_2$O
1481	Lead(II) selenate	PbSeO$_4$	7446-15-3	350.2	orth cry			6.37	0.013^{25}	s conc acid
1482	Lead(II) selenide	PbSe	12069-00-0	286.2	gray cub cry	1078		8.1		i H$_2$O; s HNO$_3$
1483	Lead(II) selenite	PbSeO$_3$	7488-51-9	334.2	wh monocl cry	≈500		7.0		i H$_2$O
1484	Lead(II) sodium thiosulfate	Na$_4$Pb(S$_2$O$_3$)$_3$	10101-94-7	635.5	wh cry					sl H$_2$O
1485	Lead(II) stearate	Pb(C$_{18}$H$_{35}$O$_2$)$_2$	1072-35-1	774.1	wh powder	≈100		1.4		i H$_2$O; s hot EtOH
1486	Lead(II) sulfate	PbSO$_4$	7446-14-2	303.3	orth cry	1087		6.29	0.0044^{25}	i acid; sl alk
1487	Lead(II) sulfide	PbS	1314-87-0	239.3	blk powder or silv cub cry	1113		7.60		i H$_2$O; s acid
1488	Lead(II) sulfite	PbSO$_3$	7446-10-8	287.3	wh powder	dec				i H$_2$O; s HNO$_3$
1489	Lead(II) tantalate	Pb(TaO$_3$)$_2$	12065-68-8	665.1	orth cry			7.9		i H$_2$O
1490	Lead(II) telluride	PbTe	1314-91-6	334.8	gray cub cry	924		8.164		i H$_2$O, acid
1491	Lead(II) thiocyanate	Pb(SCN)$_2$	592-87-0	323.4	wh-yel powder			3.82	0.05^{20}	
1492	Lead(II) thiosulfate	PbS$_2$O$_3$	13478-50-7	319.3	wh cry	dec		5.18		i H$_2$O; s acid
1493	Lead(II) titanate	PbTiO$_3$	12060-00-3	303.1	yel tetr cry			7.9		i H$_2$O; reac HCl
1494	Lead(II) tungstate (stolzite)	PbWO$_4$	7759-01-5	455.0	yel tetr cry	1130		8.24	0.03^{20}	s alk
1495	Lead(II) tungstate (raspite)	PbWO$_4$	7759-01-5	455.0	monocl cry	trans 400		8.46	0.03^{20}	s alk
1496	Lead(II) metavanadate	Pb(VO$_3$)$_2$	10099-79-3	405.1	yel powder					i H$_2$O; reac HNO$_3$
1497	Lead(II) zirconate	PbZrO$_3$	12060-01-4	346.4	col orth cry			≈8		i H$_2$O, alk; s acid
1498	Lead(II,IV) oxide	Pb$_2$O$_3$	1314-27-8	462.4	blk monocl cry or red amorp powder	530 dec		10.05		i H$_2$O; s alk; reac conc HCl
1499	Lead(II,II,IV) oxide	Pb$_3$O$_4$	1314-41-6	685.6	red tetr cry	830		8.92		i H$_2$O, EtOH; s hot HCl
1500	Lead(IV) acetate	Pb(C$_2$H$_3$O$_2$)$_4$	546-67-8	443.4	col monocl cry	≈175		2.23		reac H$_2$O, EtOH; s bz, chl
1501	Lead(IV) bromide	PbBr$_4$	13701-91-2	526.8	unstab liq					
1502	Lead(IV) chloride	PbCl$_4$	13463-30-4	349.0	yel oily liq	-15	≈50 dec			
1503	Lead(IV) fluoride	PbF$_4$	7783-59-7	283.2	wh tetr cry; hyg	≈600		6.7		
1504	Lead(IV) oxide	PbO$_2$	1309-60-0	239.2	red tetr cry or brn powder	290 dec		9.64		
1505	Lithium	Li	7439-93-2	6.941	soft silv-wh metal	180.50	1342	0.534		reac H$_2$O
1506	Lithium acetate	LiC$_2$H$_3$O$_2$	546-89-4	65.985	cry	286			45.0^{25}	vs EtOH
1507	Lithium acetate dihydrate	LiC$_2$H$_3$O$_2$ · 2H$_2$O	6108-17-4	102.016	wh rhomb cry	58 dec		1.3	45.0^{25}	s EtOH
1508	Lithium aluminum hydride	LiAlH$_4$	16853-85-3	37.955	gray-wh monocl cry	>125 dec		0.917		reac H$_2$O, EtOH; s eth, thf
1509	Lithium aluminum silicate	LiAlSi$_2$O$_6$	12068-40-5	186.090	wh monocl cry	1430		3.188		
1510	Lithium amide	LiNH$_2$	7782-89-0	22.964	tetr cry	380		1.18		reac H$_2$O
1511	Lithium arsenate	Li$_3$AsO$_4$	13478-14-3	159.743	col orth cry			3.07		sl H$_2$O; s HOAc
1512	Lithium azide	LiN$_3$	19597-69-4	48.961	hyg monocl cry; exp			1.83		vs H$_2$O
1513	Lithium borohydride	LiBH$_4$	16949-15-8	21.784	wh-gray orth cry or powder	268	380 dec	0.66		s alk, eth, thf
1514	Lithium bromate	LiBrO$_3$	13550-28-2	134.843	hyg col orth cry	260			65.4^{25}	vs H$_2$O
1515	Lithium bromide	LiBr	7550-35-8	86.845	wh cub cry; hyg	550	≈1300	3.464	181^{25}	s EtOH, eth
1516	Lithium bromide monohydrate	LiBr · H$_2$O	23303-71-1	104.860	wh orth cry	trans cub 33		3.46	145^4	vs H$_2$O
1517	Lithium carbide	Li$_2$C$_2$	1070-75-3	37.903	wh hyg cry			1.65		reac H$_2$O; i os
1518	Lithium carbonate	Li$_2$CO$_3$	554-13-2	73.891	wh monocl cry	732	1300 dec	2.11	1.30^{25}	s acid; i EtOH
1519	Lithium chlorate	LiClO$_3$	13453-71-9	90.392	col hyg rhom needles	127.6	300 dec	1.119	459^{25}	vs EtOH; sl ace
1520	Lithium chloride	LiCl	7447-41-8	42.394	wh cub cry or powder; hyg	610	1383	2.07	84.5^{25}	s EtOH, ace, py
1521	Lithium chloride monohydrate	LiCl · H$_2$O	16712-20-2	60.409	hyg wh tetr cry	dec 98		1.78	45.9^{25}	vs H$_2$O
1522	Lithium chromate dihydrate	Li$_2$CrO$_4$ · 2H$_2$O	7789-01-7	165.906	yel orth cry; hyg	75 dec		2.15		vs H$_2$O; s EtOH
1523	Lithium citrate tetrahydrate	Li$_3$C$_6$H$_5$O$_7$ · 4H$_2$O	6680-58-6	281.983	wh cry	210 (anh)			75^{25}	vs H$_2$O; sl EtOH
1524	Lithium cobaltite	LiCoO$_2$	12190-79-3	97.873	dark gray pow					i H$_2$O
1525	Lithium cyanide	LiCN	2408-36-8	32.959	wh orth cry	160				
1526	Lithium hydride-d	LiD	13587-16-1	8.955	hyg wh cry	680		0.82		reac H$_2$O
1527	Lithium dichromate dihydrate	Li$_2$Cr$_2$O$_7$ · 2H$_2$O	10022-48-7	265.901	yel-red hyg cry	130 dec		2.34		vs H$_2$O
1528	Lithium dihydrogen phosphate	LiH$_2$PO$_4$	13453-80-0	103.928	col hyg cry	>100		2.461	126^0	
1529	Lithium diisopropylamide	LiN(C$_3$H$_7$)$_2$	4111-54-0	107.123	hyg col cry	dec				s eth; i hc
1530	Lithium ferrosilicon	LiFeSi	64082-35-5	90.872	dark brittle cry					reac H$_2$O

No.	Name	Formula	CAS Reg No.	Mol. weight	Physical form	mp/°C	bp/°C	Density g cm^{-3}	Solubility g/100 g H$_2$O	Qualitative solubility
1531	Lithium fluoride	LiF	7789-24-4	25.939	wh cub cry or powder	848.2	1673	2.640	0.134[25]	s acid
1532	Lithium formate monohydrate	Li(CHO$_2$) · H$_2$O	6108-23-2	69.974	col-wh cry			1.46		s H$_2$O
1533	Lithium hexafluoroantimonate	LiSbF$_6$	18424-17-4	242.691	hyg pow	dec				
1534	Lithium hexafluoroarsenate	LiAsF$_6$	29935-35-1	310.672	rhom wh cry; hyg					
1535	Lithium hexafluorophosphate	LiPF$_6$	21324-40-3	151.905	wh pow					
1536	Lithium hexafluorosilicate	Li$_2$SiF$_6$	17347-95-4	155.958	col hex cry	dec 350				sl ace
1537	Lithium hexafluorostannate	Li$_2$SnF$_6$	17029-16-2	246.582	wh pow					
1538	Lithium hydride	LiH	7580-67-8	7.949	gray cub cry or powder; hyg	692		0.78		reac H$_2$O, EtOH
1539	Lithium hydrogen carbonate	LiHCO$_3$	5006-97-3	67.958	wh pow					sl H$_2$O
1540	Lithium hydroxide	LiOH	1310-65-2	23.948	col tetr cry	473	1626	1.45	12.5[25]	sl EtOH
1541	Lithium hydroxide monohydrate	LiOH · H$_2$O	1310-66-3	41.964	wh monocl cry or powder			1.51	12.5[25]	sl EtOH
1542	Lithium hypochlorite	LiOCl	13840-33-0	58.393	wh pow					vs H$_2$O
1543	Lithium iodate	LiIO$_3$	13765-03-2	181.843	wh hyg hex cry	450		4.502	77.9[25]	i EtOH
1544	Lithium iodide	LiI	10377-51-2	133.845	wh cub cry; hyg	469	1171	4.06	165[25]	
1545	Lithium iodide trihydrate	LiI · 3H$_2$O	7790-22-9	187.891	wh hyg cry	73		2.38	165[25]	vs EtOH, ace
1546	Lithium manganate	Li$_2$MnO$_3$	12163-00-7	116.818	red-brn monocl cry			3.90		i H$_2$O
1547	Lithium metaborate	LiBO$_2$	13453-69-5	49.751	wh monocl cry; hyg	844		2.18	2.6[20]	sl H$_2$O; s EtOH
1548	Lithium metaborate dihydrate	LiBO$_2$ · 2H$_2$O	15293-74-0	85.782	wh cry pow			1.8		s H$_2$O
1549	Lithium metaphosphate	LiPO$_3$	13762-75-9	85.913	wh cry or gl solid			1.8		i H$_2$O
1550	Lithium metasilicate	Li$_2$SiO$_3$	10102-24-6	89.966	wh orth needles	1201		2.52		i cold H$_2$O; reac dil acid
1551	Lithium molybdate	Li$_2$MoO$_4$	13568-40-6	173.82	hyg wh cry	702		2.66	44.8[25]	s H$_2$O
1552	Lithium niobate	LiNbO$_3$	12031-63-9	147.845	wh hex cry	1240		4.30		
1553	Lithium nitrate	LiNO$_3$	7790-69-4	68.946	col hex cry; hyg	253		2.38	102[25]	s EtOH
1554	Lithium nitride	Li$_3$N	26134-62-3	34.830	red hex cry	813		1.27		reac H$_2$O
1555	Lithium nitrite	LiNO$_2$	13568-33-7	52.947	wh hyg cry	222				vs H$_2$O
1556	Lithium nitrite monohydrate	LiNO$_2$ · H$_2$O	13568-33-7*	70.962	col needles	>100		1.615	139.5[25]	vs H$_2$O, EtOH
1557	Lithium orthosilicate	LiSiO$_4$	13453-84-4	99.025	wh rhom cry	1256		2.39		
1558	Lithium oxalate	Li$_2$C$_2$O$_4$	30903-87-8	101.901	col cry	dec		2.121[17]		s H$_2$O; i EtOH, eth
1559	Lithium phosphate	Li$_3$PO$_4$	10377-52-3	115.794	wh orth cry	1205		2.46	0.027[25]	
1560	Lithium oxide	Li$_2$O	12057-24-8	29.881	wh cub cry	1437		2.013		
1561	Lithium perchlorate	LiClO$_4$	7791-03-9	106.392	wh orth cry or powder	236	430 dec	2.428	58.7[25]	s EtOH, ace, eth
1562	Lithium perchlorate trihydrate	LiClO$_4$ · 3H$_2$O	13453-78-6	160.438	wh hex cry	95 dec		1.84		vs H$_2$O, EtOH, ace; i eth
1563	Lithium peroxide	Li$_2$O$_2$	12031-80-0	45.881	wh hex cry			2.31		s H$_2$O; i EtOH
1564	Lithium selenate monohydrate	Li$_2$SeO$_4$ · H$_2$O	7790-71-8	174.86	monocl cry			2.56		vs H$_2$O
1565	Lithium selenite monohydrate	Li$_2$SeO$_3$ · H$_2$O	15593-51-8	158.86	hyg cry					
1566	Lithium stearate	LiC$_{18}$H$_{35}$O$_2$	4485-12-5	290.411	cry	≈220				
1567	Lithium sulfate	Li$_2$SO$_4$	10377-48-7	109.945	wh monocl cry; hyg	860		2.21	34.2[25]	
1568	Lithium sulfate monohydrate	Li$_2$SO$_4$ · H$_2$O	10102-25-7	127.960	col cry	130 dec		2.06	34.2[25]	sl EtOH
1569	Lithium sulfide	Li$_2$S	12136-58-2	45.947	wh cub cry; hyg	1372		1.64		
1570	Lithium tantalate	LiTaO$_3$	12031-66-2	235.887	wh pow	1650				
1571	Lithium tetraborate	Li$_2$B$_4$O$_7$	12007-60-2	169.122	wh tetr cry	917			2.9[20]	sl H$_2$O
1572	Lithium tetraborate pentahydrate	Li$_2$B$_4$O$_7$ · 5H$_2$O	1303-94-2	259.198	wh cry pow	dec 200				vs H$_2$O; i EtOH
1573	Lithium tetracyanoplatinate pentahydrate	Li$_2$Pt(CN)$_4$ · 5H$_2$O	14402-73-4	403.112	grn-yel cry					sl H$_2$O
1574	Lithium tetrafluoroborate	LiBF$_4$	14283-07-9	93.746	hyg wh pow	dec				vs H$_2$O
1575	Lithium thiocyanate	LiSCN	556-65-0	65.024	wh hyg cry				120[25]	
1576	Lithium titanate	Li$_2$TiO$_3$	12031-82-2	109.747	wh pow	1325				i H$_2$O
1577	Lithium tungstate	Li$_2$WO$_4$	13568-45-1	261.72	wh trig pow	740		3.71		s H$_2$O
1578	Lithium vanadate	LiVO$_3$	15060-59-0	105.881	yel pow	subl 1400				
1579	Lithium zirconate	Li$_2$ZrO$_3$	12031-83-3	153.104	wh solid					
1580	Lutetium	Lu	7439-94-3	174.967	silv metal; hex	1663	3402	9.84		s dil acid
1581	Lutetium boride	LuB$_4$	12688-52-7	218.211	tetr cry	2600		≈7.0		
1582	Lutetium bromide	LuBr$_3$	14456-53-2	414.679	wh hyg cry	1025				vs H$_2$O
1583	Lutetium chloride	LuCl$_3$	10099-66-8	281.326	wh monocl cry; hyg	925		3.98		s H$_2$O
1584	Lutetium chloride hexahydrate	LuCl$_3$ · 6H$_2$O	15230-79-2	389.417	col cry	dec 150				s H$_2$O, EtOH
1585	Lutetium fluoride	LuF$_3$	13760-81-1	231.962	orth cry	1182	2200	8.3		i H$_2$O
1586	Lutetium iodide	LuI$_3$	13813-45-1	555.680	brn hex cry; hyg	1050		≈5.6		vs H$_2$O
1587	Lutetium iron oxide	Lu$_3$Fe$_5$O$_{12}$	12023-71-1	996.119	cry					
1588	Lutetium nitrate	Lu(NO$_3$)$_3$	10099-67-9	360.982	hyg col solid					s H$_2$O, EtOH

No.	Name	Formula	CAS Reg No.	Mol. weight	Physical form	mp/°C	bp/°C	Density g cm^{-3}	Solubility g/100 g H$_2$O	Qualitative solubility
1589	Lutetium nitride	LuN	12125-25-6	188.974	cub cry			11.6		
1590	Lutetium oxide	Lu$_2$O$_3$	12032-20-1	397.932	wh cub cry or powder	2427	3980	9.41		
1591	Lutetium perchlorate hexahydrate	Lu(ClO$_4$)$_3$ · 6H$_2$O	14646-29-8	581.410	col cry	dec 350 (anh)				s H$_2$O, MeOH
1592	Lutetium sulfate	Lu$_2$(SO$_4$)$_3$	14986-89-1	638.122	wh pow	dec >850				vs H$_2$O
1593	Lutetium sulfate octahydrate	Lu$_2$(SO$_4$)$_3$ · 8H$_2$O	13473-77-3	782.244	col cry				42.3^{20}	s H$_2$O
1594	Lutetium sulfide	Lu$_2$S$_3$	12163-20-1	446.129	gray rhomb cry	1750 dec		6.26		
1595	Lutetium telluride	Lu$_2$Te$_3$	12163-22-3	732.73	orth cry			7.8		
1596	Magnesium	Mg	7439-95-4	24.305	silv-wh metal	650	1090	1.74		s dil acid
1597	Magnesium acetate	Mg(C$_2$H$_3$O$_2$)$_2$	142-72-3	142.394	wh orth/monocl cry	323 dec		1.50	65.6^{25}	
1598	Magnesium acetate monohydrate	Mg(C$_2$H$_3$O$_2$)$_2$ · H$_2$O	60582-92-5	160.409	orth cry			1.55		
1599	Magnesium acetate tetrahydrate	Mg(C$_2$H$_3$O$_2$)$_2$ · 4H$_2$O	16674-78-5	214.454	col monocl cry; hyg	80 dec		1.45	65.6^{25}	vs EtOH
1600	Magnesium aluminate	Mg(AlO$_2$)$_2$	12068-51-8	142.266	col cub cry	2105		3.55		i H$_2$O
1601	Magnesium aluminum silicate	Mg$_2$Al$_3$(AlSi$_5$O$_{18}$)	1302-88-1	584.953	blue cry			2.6		
1602	Magnesium amide	Mg(NH$_2$)$_2$	7803-54-5	56.350	wh powder; flam	dec		1.39		reac H$_2$O
1603	Magnesium ammonium phosphate hexahydrate	MgNH$_4$PO$_4$ · 6H$_2$O	13478-16-5	245.407	wh pow	dec		1.71		i H$_2$O, EtOH; s acid
1604	Magnesium antimonide	Mg$_3$Sb$_2$	12057-75-9	316.435	hex cry	1245		3.99		
1605	Magnesium arsenide	Mg$_3$As$_2$	12044-49-4	222.758	solid	≈1200		3.15		i H$_2$O
1606	Magnesium diboride	MgB$_2$	12007-25-9	45.927	hex cry	800 dec		2.57		
1607	Magnesium hexaboride	MgB$_6$	12008-22-9	89.171	refrac solid	1100 dec				i H$_2$O
1608	Magnesium dodecaboride	MgB$_{12}$	12230-32-9	154.037	refrac solid	1300 dec				
1609	Magnesium bromate hexahydrate	Mg(BrO$_3$)$_2$ · 6H$_2$O	7789-36-8	388.201	col cub cry	200 dec		2.29	98^{25}	
1610	Magnesium bromide	MgBr$_2$	7789-48-2	184.113	wh hex cry; hyg	711		3.72	102^{25}	
1611	Magnesium bromide hexahydrate	MgBr$_2$ · 6H$_2$O	13446-53-2	292.204	col monocl cry	165 dec		2.0	102^{25}	s EtOH
1612	Magnesium carbonate	MgCO$_3$	546-93-0	84.314	wh hex cry	990		3.010	0.18^{20}	i EtOH; s acid
1613	Magnesium carbonate dihydrate	MgCO$_3$ · 2H$_2$O	5145-48-2	120.345	col tricl cry			2.8		i H$_2$O, ace, NH$_4$OH
1614	Magnesium carbonate trihydrate	MgCO$_3$ · 3H$_2$O	14457-83-1	138.360	col monocl cry	165		1.8	0.18^{16}	
1615	Magnesium carbonate pentahydrate	MgCO$_3$ · 5H$_2$O	61042-72-6	174.390	wh monocl cry	dec >400		3.04	0.38^{16}	
1616	Magnesium carbonate hydroxide tetrahydrate	4MgCO$_3$ · Mg(OH)$_2$ · 4H$_2$O	39409-82-0	467.636	wh monocl cry			2.3		
1617	Magnesium carbonate hydroxide pentahydrate	4MgCO$_3$ · Mg(OH)$_2$ · 5H$_2$O	56378-72-4	485.652	wh pow	dec 700				i H$_2$O; s dil acid; i EtOH
1618	Magnesium carbonate dihydroxide trihydrate	MgCO$_3$ · Mg(OH)$_2$ · 3H$_2$O	12143-96-3	196.680	wh monocl cry	dec		2.04		
1619	Magnesium chlorate hexahydrate	Mg(ClO$_3$)$_2$ · 6H$_2$O	13446-19-0	299.299	wh hyg cry	≈35 dec		1.80	142^{25}	sl EtOH
1620	Magnesium chloride	MgCl$_2$	7786-30-3	95.211	wh hex leaflets; hyg	714	1412	2.325	56.0^{25}	
1621	Magnesium chloride hexahydrate	MgCl$_2$ · 6H$_2$O	7791-18-6	203.302	wh hyg cry	≈100 dec		1.56	56.0^{25}	s EtOH
1622	Magnesium chromate heptahydrate	MgCrO$_4$ · 7H$_2$O	13423-61-5*	266.405	yel rhom cry			1.695	54.8^{25}	
1623	Magnesium chromite	MgCr$_2$O$_4$	12053-26-8	192.295	deep grn cry	2390		4.4		
1624	Magnesium citrate	Mg$_3$(C$_6$H$_5$O$_7$)$_2$	3344-18-1	451.114	wh cry					sl H$_2$O
1625	Magnesium citrate tetradecahydrate	Mg$_3$(C$_6$H$_5$O$_7$)$_2$ · 14H$_2$O	3344-18-1*	703.328	wh cry pow					sl H$_2$O; s acid
1626	Magnesium fluoride	MgF$_2$	7783-40-6	62.302	wh tetr cry	1263	2227	3.148	0.013^{25}	
1627	Magnesium formate dihydrate	Mg(CHO$_2$)$_2$ · 2H$_2$O	6150-82-9	150.370	wh cry	dec				s H$_2$O; i EtOH
1628	Magnesium germanate	Mg$_2$GeO$_4$	12025-13-7	185.25	wh prec					i H$_2$O
1629	Magnesium germanide	Mg$_2$Ge	1310-52-7	121.25	cub cry	1117		3.09		
1630	Magnesium hydride	MgH$_2$	7693-27-8	26.321	wh tetr cry	327		1.45		reac H$_2$O
1631	Magnesium hydrogen phosphate trihydrate	MgHPO$_4$ · 3H$_2$O	7757-86-0	174.331	wh powder	550 dec		2.13		sl H$_2$O; s dil acid
1632	Magnesium hydroxide	Mg(OH)$_2$	1309-42-8	58.320	wh hex cry	350		2.37	0.00069^{20}	s dil acid
1633	Magnesium iodate tetrahydrate	Mg(IO$_3$)$_2$ · 4H$_2$O	7790-32-1*	446.172	col monocl cry	210 dec		3.3	11.1^{25}	
1634	Magnesium iodide	MgI$_2$	10377-58-9	278.114	wh hex cry; hyg	634		4.43	146^{25}	
1635	Magnesium iodide hexahydrate	MgI$_2$ · 6H$_2$O	66778-21-0	386.205	wh monocl cry			2.35		
1636	Magnesium iodide octahydrate	MgI$_2$ · 8H$_2$O	7790-31-0	422.236	wh orth cry; hyg	41 dec		2.10	146^{25}	s EtOH
1637	Magnesium metaborate octahydrate	Mg(BO$_2$)$_2$ · 8H$_2$O	13703-82-7*	254.047	wh pow	988 (anh)				sl H$_2$O
1638	Magnesium metasilicate	MgSiO$_3$	13776-74-4	100.389	wh monocl cry	≈1550 dec		3.19		i H$_2$O; sl HF
1639	Magnesium metatitanate	MgTiO$_3$	12032-30-3	120.170	col hex cry	1565		3.85		
1640	Magnesium molybdate	MgMoO$_4$	12013-21-7	184.24	wh pow	≈1060		2.2	15.9^{25}	s H$_2$O
1641	Magnesium nitrate	Mg(NO$_3$)$_2$	10377-60-3	148.314	wh cub cry			≈2.3	71.2^{25}	
1642	Magnesium nitrate dihydrate	Mg(NO$_3$)$_2$ · 2H$_2$O	15750-45-5	184.345	wh cry	≈100 dec		1.45	71.2^{25}	s EtOH
1643	Magnesium nitrate hexahydrate	Mg(NO$_3$)$_2$ · 6H$_2$O	13446-18-9	256.406	col monocl cry; hyg	≈95 dec		1.46	71.2^{25}	s EtOH
1644	Magnesium nitride	Mg$_3$N$_2$	12057-71-5	100.928	yel cub cry	≈1500 dec		2.71		
1645	Magnesium nitrite trihydrate	Mg(NO$_2$)$_2$ · 3H$_2$O	15070-34-5	170.362	wh hyg prisms	100 dec			129.9^{25}	s EtOH

No.	Name	Formula	CAS Reg No.	Mol. weight	Physical form	mp/°C	bp/°C	Density g cm⁻³	Solubility g/100 g H₂O	Qualitative solubility
1646	Magnesium orthosilicate	Mg_2SiO_4	26686-77-1	140.694	wh orth cry	1897		3.21		i H_2O
1647	Magnesium orthotitanate	Mg_2TiO_4	12032-52-9	160.475	wh cub cry	1840		3.53		
1648	Magnesium oxalate	MgC_2O_4	547-66-0	112.324	wh powder				0.038²⁵	
1649	Magnesium oxalate dihydrate	$MgC_2O_4 \cdot 2H_2O$	6150-88-5	148.354	wh powder				0.038²⁵	i EtOH; s dil acid
1650	Magnesium oxide	MgO	1309-48-4	40.304	wh cub cry	2825	3600	3.6		sl H_2O; i EtOH
1651	Magnesium perborate heptahydrate	$Mg(BO_3)_2 \cdot 7H_2O$	14635-87-1	268.030	wh pow					sl H_2O
1652	Magnesium perchlorate	$Mg(ClO_4)_2$	10034-81-8	223.206	wh hyg powder	250 dec		2.2	100²⁵	
1653	Magnesium perchlorate hexahydrate	$Mg(ClO_4)_2 \cdot 6H_2O$	13446-19-0	331.298	wh hyg cry	190 dec		1.98	100²⁵	s EtOH
1654	Magnesium permanganate hexahydrate	$Mg(MnO_4)_2 \cdot 6H_2O$	10377-62-5	370.268	blue-blk cry	dec		2.18		s H_2O
1655	Magnesium peroxide	MgO_2	1335-26-8	56.304	wh cub cry	100 dec		≈3.0		i H_2O; s dil acid
1656	Magnesium phosphate pentahydrate	$Mg_3(PO_4)_2 \cdot 5H_2O$	7757-87-1*	352.934	wh cry	400 dec			0.00009²⁰	s dil acid
1657	Magnesium phosphate octahydrate	$Mg_3(PO_4)_2 \cdot 8H_2O$	13446-23-6	406.980	wh monocl cry			2.17	0.00009²⁰	s acid
1658	Magnesium pyrophosphate	$Mg_2P_2O_7$	13446-24-7	222.553	col monocl plates	1395		2.56		
1659	Magnesium pyrophosphate trihydrate	$Mg_2P_2O_7 \cdot 3H_2O$	10102-34-8	276.600	wh powder	100 dec		2.56		i H_2O; s acid
1660	Magnesium phosphide	Mg_3P_2	12057-74-8	134.863	yel cub cry			2.06		reac H_2O
1661	Magnesium selenate hexahydrate	$MgSeO_4 \cdot 6H_2O$	13446-28-1	275.35	wh monocl cry			1.928	55.5²⁵	
1662	Magnesium selenide	MgSe	1313-04-8	103.27	brn cub cry			4.2		reac H_2O
1663	Magnesium selenite hexahydrate	$MgSeO_3 \cdot 6H_2O$	15593-61-0	259.36	col hex cry			2.09		i H_2O; s dil acid
1664	Magnesium hexafluorosilicate hexahydrate	$MgSiF_6 \cdot 6H_2O$	60950-56-3	274.472	wh cry	120 dec		1.79	39.3¹⁸	i EtOH
1665	Magnesium silicide	Mg_2Si	22831-39-6	76.696	gray cub cry	1102		1.99		reac H_2O
1666	Magnesium stannide	Mg_2Sn	1313-08-2	167.320	blue cub cry	771		3.60		s H_2O, dil HCl
1667	Magnesium sulfate	$MgSO_4$	7487-88-9	120.368	col orth cry	1137		2.66	35.7²⁵	
1668	Magnesium sulfate monohydrate	$MgSO_4 \cdot H_2O$	14168-73-1	138.383	col monocl cry	150 dec		2.57	35.7²⁵	
1669	Magnesium sulfate heptahydrate	$MgSO_4 \cdot 7H_2O$	10034-99-8	246.474	col orth cry	150 dec		1.67	35.7²⁵	sl EtOH
1670	Magnesium sulfide	MgS	12032-36-9	56.370	red-brn cub cry	2226		2.68		reac H_2O
1671	Magnesium sulfite trihydrate	$MgSO_3 \cdot 3H_2O$	19086-20-5	158.414	col orth cry			2.12	0.79²⁵	
1672	Magnesium sulfite hexahydrate	$MgSO_3 \cdot 6H_2O$	13446-29-2	212.460	wh hex cry	200 dec		1.72	0.79²⁵	i EtOH
1673	Magnesium tetrahydrogen phosphate dihydrate	$Mg(H_2PO_4)_2 \cdot 2H_2O$	15609-80-0	254.311	wh hyg cry	dec 90				s H_2O; i EtOH
1674	Magnesium thiocyanate tetrahydrate	$Mg(SCN)_2 \cdot 4H_2O$	306-61-6	212.531	wh hyg					vs H_2O, EtOH
1675	Magnesium thiosulfate hexahydrate	$MgS_2O_3 \cdot 6H_2O$	13446-30-5	244.525	col cry	170 dec		1.82	93²⁵	i EtOH
1676	Magnesium trisilicate	$Mg_2Si_3O_8$	14987-04-3	260.862	wh powder					i H_2O, EtOH
1677	Magnesium tungstate	$MgWO_4$	13573-11-0	272.14	wh monocl cry			6.89	0.016²⁰	i EtOH
1678	Magnesium vanadate	$Mg_2V_2O_7$	13568-63-3	262.489	tricl cry			3.1		
1679	Magnesium zirconate	$MgZrO_3$	12032-31-4	163.527	col cry	2060		4.23		
1680	Magnesium zirconium silicate	$MgO \cdot ZrO_2 \cdot SiO_2$	52110-05-1	223.612	wh solid					i H_2O, alk; sl acid
1681	Manganese	Mn	7439-96-5	54.938	hard gray metal	1246	2061	7.3		s dil acids
1682	Manganocene	$Mn(C_5H_5)_2$	1271-27-8	185.124	yel-brn cry	173				s py, thf; sl bz
1683	Manganese antimonide (MnSb)	MnSb	12032-82-5	176.698	hex cry	840		6.9		
1684	Manganese antimonide (Mn₂Sb)	Mn_2Sb	12032-97-2	231.636	tetr cry	948		7.0		
1685	Manganese boride (MnB)	MnB	12045-15-7	65.749	orth cry	1890		6.45		
1686	Manganese boride (MnB₂)	MnB_2	12228-50-1	76.560	hex cry	1827		5.3		
1687	Manganese boride (Mn₄B)	Mn_4B	12045-16-8	120.687	red-brn tetr cry	1580		7.20		
1688	Manganese carbide	Mn_3C	12266-65-8	176.825	refrac solid	1520		6.89		
1689	Manganese carbonyl	$Mn_2(CO)_{10}$	10170-69-1	389.977	yel monocl cry	154		1.75		i H_2O; s os
1690	Manganese pentacarbonyl bromide	$Mn(CO)_5Br$	14516-54-2	274.893	oran-yel cry					s os
1691	Manganese phosphide (MnP)	MnP	12032-78-9	85.912	orth cry	1147		5.49		
1692	Manganese phosphide (Mn₂P)	Mn_2P	12333-54-9	140.850	hex cry	1327		6.0		
1693	Manganese silicide	$MnSi_2$	12032-86-9	111.109	gray solid	1152 dec				
1694	Manganese(II) acetate tetrahydrate	$Mn(C_2H_3O_2)_2 \cdot 4H_2O$	6156-78-1	245.087	red monocl cry	80		1.59		s H_2O, EtOH
1695	Manganese(II) bromide	$MnBr_2$	13446-03-2	214.746	pink hex cry	698		4.385	151²⁵	
1696	Manganese(II) bromide tetrahydrate	$MnBr_2 \cdot 4H_2O$	10031-20-6	286.808	red hyg cry	64 dec			151²⁵	
1697	Manganese(II) carbonate	$MnCO_3$	598-62-9	114.947	pink hex cry	>200 dec		3.70	0.00008²⁰	s dil acid
1698	Manganese(II) chloride	$MnCl_2$	7773-01-5	125.844	pink trig cry; hyg	650	1190	2.977	77.3²⁵	s py, EtOH; i eth
1699	Manganese(II) chloride tetrahydrate	$MnCl_2 \cdot 4H_2O$	13446-34-9	197.906	red monocl cry; hyg	87.5		1.913	77.3²⁵	s EtOH; i eth
1700	Manganese(II) dihydrogen phosphate dihydrate	$Mn(H_2PO_4)_2 \cdot 2H_2O$	18718-07-5	284.944	col hyg cry					s H_2O; i EtOH
1701	Manganese(II) fluoride	MnF_2	7782-64-1	92.935	red tetr cry	900		3.98	1.02²⁵	i EtOH
1702	Manganese(II) hydroxide	$Mn(OH)_2$	18933-05-6	88.953	pink hex cry	dec		3.26	0.00034²⁰	
1703	Manganese(II) hypophosphite monohydrate	$Mn(H_2PO_2)_2 \cdot H_2O$	10043-84-2	202.931	pink cry	>250			15²⁰	s H_2O

No.	Name	Formula	CAS Reg No.	Mol. weight	Physical form	mp/°C	bp/°C	Density g cm^{-3}	Solubility g/100 g H$_2$O	Qualitative solubility
1704	Manganese(II) iodide	MnI$_2$	7790-33-2	308.747	wh hex cry; hyg	638		5.04		s H$_2$O, EtOH
1705	Manganese(II) iodide tetrahydrate	MnI$_2$ · 4H$_2$O	7790-33-2*	380.809	red cry					vs H$_2$O; s EtOH
1706	Manganese(II) metasilicate	MnSiO$_3$	7759-00-4	131.022	red orth cry	1291		3.48		i H$_2$O
1707	Manganese(II) molybdate	MnMoO$_4$	14013-15-1	214.88	yel monocl cry			4.05		
1708	Manganese(II) nitrate	Mn(NO$_3$)$_2$	10377-93-2	178.947	col orth cry; hyg			2.2	161^{25}	s diox, thf
1709	Manganese(II) nitrate tetrahydrate	Mn(NO$_3$)$_2$ · 4H$_2$O	20694-39-7	251.009	pink hyg cry	37.1 dec		2.13	161^{25}	s EtOH
1710	Manganese(II) nitrate hexahydrate	Mn(NO$_3$)$_2$ · 6H$_2$O	10377-66-9	287.039	rose monocl cry	28 dec		1.8	161^{25}	vs EtOH
1711	Manganese(II) orthosilicate	Mn$_2$SiO$_4$	13568-32-6	201.960	orth cry			4.11		i H$_2$O
1712	Manganese(II) oxalate dihydrate	MnC$_2$O$_4$ · 2H$_2$O	6556-16-7	178.987	wh cry powder	150 dec		2.45	0.032^{20}	s acid
1713	Manganese(II) oxide	MnO	1344-43-0	70.937	grn cub cry or powder	1842		5.37		i H$_2$O; s acid
1714	Manganese(II) perchlorate hexahydrate	Mn(ClO$_4$)$_2$ · 6H$_2$O	15364-94-0	361.931	pink hex cry			2.10		
1715	Manganese(II) pyrophosphate	Mn$_2$P$_2$O$_7$	53731-35-4	283.819	wh monocl cry	1196		3.71		i H$_2$O
1716	Manganese(II) selenide	MnSe	1313-22-0	133.90	gray cub cry	1460		5.45		i H$_2$O
1717	Manganese(II) sulfate	MnSO$_4$	7785-87-7	151.001	wh orth cry	700	850 dec	3.25	63.7^{25}	
1718	Manganese(II) sulfate monohydrate	MnSO$_4$ · H$_2$O	10034-96-5	169.016	red monocl cry			2.95	63.7^{25}	i EtOH
1719	Manganese(II) sulfate tetrahydrate	MnSO$_4$ · 4H$_2$O	10101-68-5	223.062	red monocl cry	38 dec		2.26	63.7^{25}	i EtOH
1720	Manganese(II) sulfide (α form)	MnS	18820-29-6	87.003	grn cub cry	1610		4.0		i H$_2$O; s dil acid
1721	Manganese(II) sulfide (β form)	MnS	18820-29-6	87.003	red cub cry			3.3		i H$_2$O; s dil acid
1722	Manganese(II) sulfide (γ form)	MnS	18820-29-6	87.003	red hex cry			≈3.3		i H$_2$O; s dil acid
1723	Manganese(II) telluride	MnTe	12032-88-1	182.54	hex cry	≈1150		6.0		
1724	Manganese(II) tetraborate octahydrate	MnB$_4$O$_7$ · 8H$_2$O	12228-91-0	354.300	red solid					i H$_2$O, EtOH; s dil acid
1725	Manganese(II) titanate	MnTiO$_3$	12032-74-5	150.803	red hex cry	1360		4.55		
1726	Manganese(II) tungstate	MnWO$_4$	13918-22-4	302.78	wh monocl cry			7.2	0.0054^{20}	
1727	Manganese(II,III) oxide	Mn$_3$O$_4$	1317-35-7	228.812	brn tetr cry	1567		4.84		i H$_2$O; s HCl
1728	Manganese(III) acetate dihydrate	Mn(C$_2$H$_3$O$_2$)$_3$ · 2H$_2$O	19513-05-4	268.100	brn cry					s eth, HOAc
1729	Manganese(III) fluoride	MnF$_3$	7783-53-1	111.933	red monocl cry; hyg	>600 dec		3.54		reac H$_2$O
1730	Manganese(III) hydroxide	MnO(OH)	1332-63-4	87.945	blk monocl cry	250 dec		≈4.3		i H$_2$O
1731	Manganese(III) oxide	Mn$_2$O$_3$	1317-34-6	157.874	blk cub cry	1080 dec		≈5.0		i H$_2$O
1732	Manganese(IV) oxide	MnO$_2$	1313-13-9	86.937	blk tetr cry	535 dec		5.08		i H$_2$O, HNO$_3$
1733	Manganese(VII) oxide	Mn$_2$O$_7$	12057-92-0	221.872	grn oil; exp	5.9	95 exp	2.40		vs H$_2$O
1734	Mendelevium	Md	7440-11-1	258	metal	827				
1735	Mercury	Hg	7439-97-6	200.59	heavy silv liq	-38.8290	356.62	13.5336		i H$_2$O
1736	Dimethyl mercury	Hg(CH$_3$)$_2$	593-74-8	230.66	liq		93	3.17		i H$_2$O; vs EtOH, eth
1737	Mercury(I) acetate	Hg$_2$(C$_2$H$_3$O$_2$)$_2$	631-60-7	519.27	col scales	dec				sl H$_2$O; i EtOH, eth
1738	Mercury(I) bromate	Hg$_2$(BrO$_3$)$_2$	13465-33-3	656.98	col cry	dec				i H$_2$O; sl acid
1739	Mercury(I) bromide	Hg$_2$Br$_2$	15385-58-7	560.99	wh tetr cry or powder	345 dec		7.307		i H$_2$O, EtOH, eth
1740	Mercury(I) carbonate	Hg$_2$CO$_3$	6824-78-8	461.19	yel-brn cry	130 dec			0.0000045	i EtOH
1741	Mercury(I) chlorate	Hg$_2$(ClO$_3$)$_2$	10294-44-7	568.08	wh rhom cry	≈250 dec		6.409		sl H$_2$O; s EtOH
1742	Mercury(I) chloride	Hg$_2$Cl$_2$	10112-91-1	472.09	wh tetr cry	525 tp	383 sp	7.16	0.0004^{25}	i EtOH, eth
1743	Mercury(I) chromate	Hg$_2$CrO$_4$	13465-34-4	517.17	brn-red solid					i H$_2$O EtOH; s conc HNO$_3$
1744	Mercury(I) fluoride	Hg$_2$F$_2$	13967-25-4	439.18	yel cub cry	570 dec	subl	8.73		reac H$_2$O
1745	Mercury(I) iodate	Hg$_2$(IO$_3$)$_2$	13465-35-5	750.99	yel-wh pow	dec 175			0.0032^{20}	
1746	Mercury(I) iodide	Hg$_2$I$_2$	15385-57-6	654.99	yel amorp powder	290		7.70		i H$_2$O, EtOH, eth
1747	Mercury(I) nitrate	Hg$_2$(NO$_3$)$_2$	10415-75-5	525.19	cry					sl H$_2$O
1748	Mercury(I) nitrate dihydrate	Hg$_2$(NO$_3$)$_2$ · 2H$_2$O	14836-60-3	561.22	col cry	70 dec		4.8		sl H$_2$O
1749	Mercury(I) nitrite	Hg$_2$(NO$_2$)$_2$	13492-25-6	493.19	yel cry	100 dec		7.3		reac H$_2$O
1750	Mercury(I) oxalate	Hg$_2$C$_2$O$_4$	2949-11-3	489.20	cry					i H$_2$O; sl HNO$_3$
1751	Mercury(I) oxide	Hg$_2$O	15829-53-5	417.18	prob mixture of HgO+Hg	100 dec		9.8		i H$_2$O; s HNO$_3$
1752	Mercury(I) perchlorate tetrahydrate	Hg$_2$(ClO$_4$)$_2$ · 4H$_2$O	65202-12-2	672.14	cry	64			442^{25}	
1753	Mercury(I) sulfate	Hg$_2$SO$_4$	7783-36-0	497.24	wh-yel cry powder			7.56	0.051^{25}	s dil HNO$_3$
1754	Mercury(I) sulfide	Hg$_2$S	51595-71-2	433.25	unstab blk pow	dec				i H$_2$O
1755	Mercury(I) thiocyanate	Hg$_2$(SCN)$_2$	13465-37-7	517.34	col powder	dec			0.03^{25}	s HCl, KCNS
1756	Mercury(I) tungstate	Hg$_2$WO$_4$	38705-19-0	649.02	yel amorp solid	dec				i H$_2$O, EtOH
1757	Mercury(II) acetate	Hg(C$_2$H$_3$O$_2$)$_2$	1600-27-7	318.68	wh-yel cry or powder	179 dec		3.28	25^{10}	s EtOH
1758	Mercury(II) amide chloride	Hg(NH$_2$)Cl	10124-48-8	252.07	wh solid		subl	5.38		i H$_2$O, EtOH; s warm acid
1759	Mercury(II) benzoate monohydrate	Hg(C$_7$H$_5$O$_2$)$_2$ · H$_2$O	32839-04-6	460.83	wh cry	165			1.2^{15}	sl EtOH
1760	Mercury(II) bromate	Hg(BrO$_3$)$_2$	26522-91-8	456.39	cry	130 dec			0.15	s acid
1761	Mercury(II) bromide	HgBr$_2$	7789-47-1	360.40	wh rhomb cry or powder	241	318	6.05	0.61^{25}	sl chl; s EtOH, MeOH
1762	Mercury(II) chlorate	Hg(ClO$_3$)$_2$	13465-30-0	367.49	wh needles	dec		4.998	25	

No.	Name	Formula	CAS Reg No.	Mol. weight	Physical form	mp/°C	bp/°C	Density g cm^{-3}	Solubility g/100 g H$_2$O	Qualitative solubility
1763	Mercury(II) chloride	HgCl$_2$	7487-94-7	271.50	wh orth cry	277	304	5.6	7.31^{25}	sl bz; s EtOH, MeOH, ace, eth
1764	Mercury(II) chromate	HgCrO$_4$	13444-75-2	316.58	red monocl cry			6.06		sl H$_2$O
1765	Mercury(II) cyanide	Hg(CN)$_2$	592-04-1	252.62	col tetr cry	320 dec		4.00	11.4^{25}	s EtOH; sl eth
1766	Mercury(II) dichromate	HgCr$_2$O$_7$	7789-10-8	416.58	red cry powder					i H$_2$O; s acid
1767	Mercury(II) fluoride	HgF$_2$	7783-39-3	238.59	wh cub cry; hyg	645 dec		8.95		reac H$_2$O
1768	Mercury(II) fulminate	Hg(CNO)$_2$	628-86-4	284.62	gray cry	exp		4.42		sl H$_2$O; s EtOH, NH$_4$OH
1769	Mercury(II) hydrogen arsenate	HgHAsO$_4$	7784-37-4	340.52	yel powder					i H$_2$O; s acid
1770	Mercury(II) iodate	Hg(IO$_3$)$_2$	7783-32-6	550.40	wh powder	175 dec				i H$_2$O
1771	Mercury(II) iodide (yellow)	HgI$_2$	7774-29-0	454.40	yel tetr cry or powder	256	351	6.28	0.0055^{25}	sl EtOH, ace, eth
1772	Mercury(II) iodide (red)	HgI$_2$	7774-29-0	454.40	red pow	trans to yel 127			0.006^{25}	sl EtOH, ace, eth, chl
1773	Mercury(II) nitrate	Hg(NO$_3$)$_2$	10045-94-0	324.60	col hyg cry	79		4.3		s H$_2$O; i EtOH
1774	Mercury(II) nitrate monohydrate	Hg(NO$_3$)$_2$ · H$_2$O	7783-34-8	342.62	wh-yel hyg cry			4.3		s H$_2$O, dil acid
1775	Mercury(II) nitrate dihydrate	Hg(NO$_3$)$_2$ · 2H$_2$O	22852-67-1	360.63	monocl cry			4.78		s H$_2$O
1776	Mercury(II) oxalate	HgC$_2$O$_4$	3444-13-1	288.61	powder	165 dec				i H$_2$O
1777	Mercury(II) oxide	HgO	21908-53-2	216.59	red or yel orth cry	500 dec		11.14		i H$_2$O, EtOH; s dil acid
1778	Mercury(II) oxide sulfate	(Hg$_3$O$_2$)SO$_4$	1312-03-4	729.83	yel powder					i H$_2$O; s acid
1779	Mercury(II) oxycyanide	Hg(CN)$_2$ · HgO	1335-31-5	469.21	wh orth cry	exp		4.44	11.4^{25}	
1780	Mercury(II) perchlorate trihydrate	Hg(ClO$_4$)$_2$ · 3H$_2$O	7616-83-3	453.54	cry					
1781	Mercury(II) phosphate	Hg$_3$(PO$_4$)$_2$	7782-66-3	791.71	wh-yel powder					i H$_2$O, EtOH; s acid
1782	Mercury(II) selenide	HgSe	20601-83-6	279.55	gray cub cry		subl	8.21		i H$_2$O
1783	Mercury(II) sulfate	HgSO$_4$	7783-35-9	296.65	wh monocl cry			6.47		reac H$_2$O
1784	Mercury(II) sulfide (black)	HgS	1344-48-5	232.66	blk cub cry or powder	850		7.70		i H$_2$O; s acid, EtOH
1785	Mercury(II) sulfide (red)	HgS	1344-48-5	232.66	red hex cry	trans to blk HgS 344		8.17		i H$_2$O, acid; s aqua regia
1786	Mercury(II) telluride	HgTe	12068-90-5	328.19	gray cub cry	673		8.63		
1787	Mercury(II) thiocyanate	Hg(SCN)$_2$	592-85-8	316.75	monocl cry	≈165 dec		3.71	0.070^{25}	s dil HCl
1788	Mercury(II) tungstate	HgWO$_4$	37913-38-5	448.43	yel cry	dec				i H$_2$O, EtOH
1789	Molybdenum	Mo	7439-98-7	95.94	gray-blk metal; cub	2623	4639	10.2		i H$_2$O, dil acid, alk
1790	Molybdophosphoric acid	H$_3$P(Mo$_3$O$_{10}$)$_4$	51429-74-4	1825.25	bright yel cry					
1791	Molybdenum boride (Mo$_2$B)	Mo$_2$B	12006-99-4	202.69	refrac tetr cry	2000		9.2		
1792	Molybdenum boride (Mo$_2$B$_5$)	Mo$_2$B$_5$	12007-97-5	245.94	refrac hex cry	1600		≈7.2		
1793	Molybdenum carbide (MoC)	MoC	12011-97-1	107.95	refrac solid; cub	2577				
1794	Molybdenum carbide (Mo$_2$C)	Mo$_2$C	12069-89-5	203.89	gray orth cry	2687		9.18		
1795	Molybdenum carbonyl	Mo(CO)$_6$	13939-06-5	264.00	wh orth cry	148	155 dec	1.96		i H$_2$O; s bz; sl eth
1796	Molybdenum nitride (MoN)	MoN	12033-19-1	109.95	hex cry	1750		9.20		
1797	Molybdenum nitride (Mo$_2$N)	Mo$_2$N	12033-31-7	205.89	gray cub cry	790 dec		9.46		
1798	Molybdenum phosphide	MoP	12163-69-8	126.91	blk hex cry			7.34		
1799	Molybdenum silicide (MoSi$_2$)	MoSi$_2$	12136-78-6	152.11	gray tetr cry	≈1900		6.2		i H$_2$O; s HF
1800	Molybdenum(II) bromide	MoBr$_2$	13446-56-5	255.75	yel-red cry	dec 700		4.88		i H$_2$O, EtOH
1801	Molybdenum(II) chloride	MoCl$_2$	13478-17-6	166.85	yel cry	dec 500		3.71		i H$_2$O
1802	Molybdenum(II) iodide	MoI$_2$	14055-74-4	349.75	blk hyg cry	700		5.28		i H$_2$O
1803	Molybdenum(III) bromide	MoBr$_3$	13446-57-6	335.65	grn hex cry	dec 500		4.89		i H$_2$O, EtOH
1804	Molybdenum(III) chloride	MoCl$_3$	13478-18-7	202.30	dark red monocl cry	dec 400		3.74		i H$_2$O, os
1805	Molybdenum(III) fluoride	MoF$_3$	20193-58-2	152.94	yel-brn hex cry	>600		4.64		i H$_2$O
1806	Molybdenum(III) iodide	MoI$_3$	14055-75-5	476.65	blk solid	927				i H$_2$O
1807	Molybdenum(III) oxide	Mo$_2$O$_3$	1313-29-7	239.88	gray-blk powder					i H$_2$O; sl acid
1808	Molybdenum(IV) bromide	MoBr$_4$	13520-59-7	415.56	blk cry	dec 110				reac H$_2$O
1809	Molybdenum(IV) chloride	MoCl$_4$	13320-71-3	237.75	blk cry	317				reac H$_2$O, sl chl; i eth, bz
1810	Molybdenum(IV) fluoride	MoF$_4$	23412-45-5	171.93	grn cry	dec				reac H$_2$O
1811	Molybdenum(IV) iodide	MoI$_4$	14055-76-6	603.56	blk cry	dec 100				i H$_2$O
1812	Molybdenum(IV) oxide	MoO$_2$	18868-43-4	127.94	brn-viol tetr cry	≈1800 dec		6.47		i H$_2$O, acid, alk
1813	Molybdenum(IV) selenide	MoSe$_2$	12058-18-3	253.86	gray hex cry	>1200		6.90		
1814	Molybdenum(IV) sulfide	MoS$_2$	1317-33-5	160.07	blk powder or hex cry	1750		5.06		i H$_2$O; s conc acid
1815	Molybdenum(IV) telluride	MoTe$_2$	12058-20-7	351.14	gray hex cry			7.7		
1816	Molybdenum(V) chloride	MoCl$_5$	10241-05-1	273.21	grn-blk monocl cry; hyg	194	268	2.93		reac H$_2$O; s EtOH, eth
1817	Molybdenum(V) fluoride	MoF$_5$	13819-84-6	190.93	yel monocl cry	67	213.6	3.5		reac H$_2$O
1818	Molybdenum(V) oxytrichloride	MoOCl$_3$	13814-74-9	218.30	blk monocl cry	310	subl			reac H$_2$O
1819	Molybdenum(VI) acid monohydrate	H$_2$MoO$_4$ · H$_2$O	7782-91-4	179.97	wh powder			3.1		sl H$_2$O; s alk

No.	Name	Formula	CAS Reg No.	Mol. weight	Physical form	mp/°C	bp/°C	Density g cm^{-3}	Solubility g/100 g H$_2$O	Qualitative solubility
1820	Molybdenum(VI) dioxydichloride	MoO$_2$Cl$_2$	13637-68-8	198.85	yel-oran solid	176	250	3.31		reac H$_2$O
1821	Molybdenum(VI) dioxydifluoride	MoO$_2$F$_2$	13824-57-2	165.94	pale lilac cry	subl 270		3.5		i MeCN, chl; sl HF
1822	Molybdenum(VI) fluoride	MoF$_6$	7783-77-9	209.93	wh cub cry or col liq; hyg	17.5	34.0	2.54		reac H$_2$O; vs hex, ctc
1823	Molybdenum(VI) metaphosphate	Mo(PO$_3$)$_6$	133863-98-6	569.77	yel powder			3.28		i H$_2$O, acid
1824	Molybdenum(VI) oxide	MoO$_3$	1313-27-5	143.94	wh-yel rhomb cry	802	1155	4.70	0.14^{20}	sl H$_2$O; s alk, acid
1825	Molybdenum(VI) oxytetrachloride	MoOCl$_4$	13814-75-0	253.75	grn hyg powder	105	159			
1826	Molybdenum(VI) oxytetrafluoride	MoOF$_4$	14459-59-7	187.93	volatile solid	97.2	186.0			
1827	Molybdenum(VI) sulfide	MoS$_3$	12033-29-3	192.14	blk solid	350 dec				i H$_2$O, os
1828	Neodymium	Nd	7440-00-8	144.242	silv metal; hex	1016	3074	7.01		
1829	Neodymium boride	NdB$_6$	12008-23-0	209.108	blk cub cry	2610		4.93		
1830	Neodymium nitride	NdN	25764-11-8	158.249	blk cub cry			7.69		
1831	Neodymium(II) acetate	Nd(C$_2$H$_3$O$_2$)$_2$	6192-13-8	321.373	red-purp cry					s H$_2$O
1832	Neodymium(II) chloride	NdCl$_2$	25469-93-6	215.148	grn hyg solid	841				
1833	Neodymium(III) bromate nonahydrate	Nd(BrO$_3$)$_3$ · 9H$_2$O	15162-92-2	690.086	red hex cry	66 dec				
1834	Neodymium(III) bromide	NdBr$_3$	13536-80-6	383.954	viol orth cry; hyg	682	1540	5.3		s H$_2$O
1835	Neodymium(III) chloride	NdCl$_3$	10024-93-8	250.601	viol hex cry	759	1600	4.13	100^{25}	vs EtOH; i eth, chl
1836	Neodymium(III) chloride hexahydrate	NdCl$_3$ · 6H$_2$O	13477-89-9	358.692	purp cry	124 dec		2.3	100^{25}	s EtOH
1837	Neodymium(III) fluoride	NdF$_3$	13709-42-7	201.237	viol hex cry; hyg	1377	2300	6.51		i H$_2$O
1838	Neodymium(III) hydroxide	Nd(OH)$_3$	16469-17-3	195.264	blue solid	dec 210				i H$_2$O
1839	Neodymium(III) iodide	NdI$_3$	13813-24-6	524.955	grn orth cry; hyg	787		5.85		s H$_2$O
1840	Neodymium(III) nitrate	Nd(NO$_3$)$_3$	10045-95-1	330.257	viol hyg. cry				152^{25}	s EtOH
1841	Neodymium(III) nitrate hexahydrate	Nd(NO$_3$)$_3$ · 6H$_2$O	14517-29-4	438.348	purp hyg cry				152^{25}	s EtOH, ace
1842	Neodymium(III) oxide	Nd$_2$O$_3$	1313-97-9	336.482	blue hex cry; hyg	2233	3760	7.24		i H$_2$O; s dil acid
1843	Neodymium(III) sulfate	Nd$_2$(SO$_4$)$_3$	13477-91-3	576.672	pink needles	≈700 dec			7.1^{20}	
1844	Neodymium(III) sulfate octahydrate	Nd$_2$(SO$_4$)$_3$ · 8H$_2$O	13477-91-3	720.794	red cry	350 dec		2.85		sl H$_2$O
1845	Neodymium(III) sulfide	Nd$_2$S$_3$	12035-32-4	384.679	orth cry	2207		5.46		
1846	Neodymium(III) telluride	Nd$_2$Te$_3$	12035-35-7	671.28	gray orth cry	1377		7.0		
1847	Neodymium(III) tris(cyclopentadienyl)	Nd(C$_5$H$_5$)$_3$	1273-98-9	339.522	red-blue cry	380				s thf
1848	Neon	Ne	7440-01-9	20.180	col gas	-248.609 tp (43 kPa)	-246.053	0.825 g/L		sl H$_2$O
1849	Neptunium	Np	7439-99-8	237	silv metal	644		20.2		s HCl
1850	Neptunium(IV) oxide	NpO$_2$	12035-79-9	269	grn cub cry	2547		11.1		
1851	Nickel	Ni	7440-02-0	58.693	wh metal; cub	1455	2913	8.90		i H$_2$O; sl dil acid
1852	Nickelocene	Ni(C$_5$H$_5$)$_2$	1271-28-9	188.879	grn cry	173				
1853	Nickel aluminide (NiAl)	NiAl	12003-78-0	85.675	metallic solid	1638				
1854	Nickel antimonide	NiSb	12035-52-8	180.453	hex cry	1147		8.74		
1855	Nickel arsenide	NiAs	27016-75-7	133.615	hex cry	967		7.77		
1856	Nickel boride (NiB)	NiB	12007-00-0	69.504	grn refrac solid	1035		7.13		
1857	Nickel boride (Ni$_2$B)	Ni$_2$B	12007-01-1	128.198	refrac solid	1125		7.90		
1858	Nickel boride (Ni$_3$B)	Ni$_3$B	12007-02-2	186.891	refrac solid	1166		8.17		
1859	Nickel carbonyl	Ni(CO)$_4$	13463-39-3	170.734	col liq	-19.3	42.1 (exp ≈60)	1.31		i H$_2$O; s EtOH, bz, ace, ctc
1860	Nickel phosphide	Ni$_2$P	12035-64-2	148.361	hex cry	1100		7.33		
1861	Nickel silicide (NiSi$_2$)	NiSi$_2$	12201-89-7	114.864	cub cry	993		4.83		
1862	Nickel silicide (Ni$_2$Si)	Ni$_2$Si	12059-14-2	145.473	orth cry	1255		7.40		
1863	Nickel subsulfide	Ni$_3$S$_2$	12035-72-2	240.210	yel hex cry	789		5.87		
1864	Nickel(II) acetate tetrahydrate	Ni(C$_2$H$_3$O$_2$)$_2$ · 4H$_2$O	6018-89-9	248.842	grn monocl cry	250 dec		1.74	16^{20}	s H$_2$O, EtOH
1865	Nickel(II) ammonium chloride hexahydrate	NH$_4$NiCl$_3$ · 6H$_2$O	16122-03-5*	291.182	grn hyg cry			1.65		s H$_2$O
1866	Nickel(II) ammonium sulfate	Ni(NH$_4$)$_2$(SO$_4$)$_2$	15699-18-0	286.895	blue-grn cry	dec 250				sl H$_2$O
1867	Nickel(II) ammonium sulfate hexahydrate	Ni(NH$_4$)$_2$(SO$_4$)$_2$ · 6H$_2$O	7785-20-8	394.987	blue-grn cry	dec 130		1.92	6.5^{20}	s H$_2$O; i EtOH
1868	Nickel(II) arsenate octahydrate	Ni$_3$(AsO$_4$)$_2$ · 8H$_2$O	7784-48-7	598.040	yel-grn powder	dec		4.98		i H$_2$O; s acid
1869	Nickel(II) bromide	NiBr$_2$	13462-88-9	218.501	yel hex cry; hyg	963	subl	5.10	131^{20}	
1870	Nickel(II) bromide trihydrate	NiBr$_2$ · 3H$_2$O	13462-88-9*	272.547	yel-grn hyg cry	200 dec				vs H$_2$O; s EtOH, eth
1871	Nickel(II) carbonate	NiCO$_3$	3333-67-3	118.702	grn rhomb cry			4.389	0.0043^{20}	s dil acid
1872	Nickel(II) chlorate hexahydrate	Ni(ClO$_3$)$_2$ · 6H$_2$O	13477-94-6	333.687	grn cub cry	dec 80		2.07		vs H$_2$O
1873	Nickel(II) chloride	NiCl$_2$	7718-54-9	129.599	yel hex cry; hyg	1031	985 sp	3.51	67.5^{25}	s EtOH
1874	Nickel(II) chloride hexahydrate	NiCl$_2$ · 6H$_2$O	7791-20-0	237.690	grn monocl cry				67.5^{25}	s EtOH
1875	Nickel(II) chromate	NiCrO$_4$	14721-18-7	174.687	red solid					sl H$_2$O
1876	Nickel(II) cyanide tetrahydrate	Ni(CN)$_2$ · 4H$_2$O	13477-95-7	182.789	grn plates	200 dec				i H$_2$O; sl dil acid; s NH$_4$OH
1877	Nickel(II) fluoride	NiF$_2$	10028-18-9	96.690	yel tetr cry	1380		4.7	2.56^{25}	i EtOH, eth
1878	Nickel(II) fluoride tetrahydrate	NiF$_2$ · 4H$_2$O	13940-83-5	168.752	grn pow					sl H$_2$O

No.	Name	Formula	CAS Reg No.	Mol. weight	Physical form	mp/°C	bp/°C	Density g cm^{-3}	Solubility g/100 g H$_2$O	Qualitative solubility
1879	Nickel(II) hydroxide	Ni(OH)$_2$	12054-48-7	92.708	grn hex cry	230 dec		4.1	0.00015^{20}	
1880	Nickel(II) hydroxide monohydrate	Ni(OH)$_2$ · H$_2$O	36897-37-7	110.723	grn powder				0.00015^{20}	s dil acid
1881	Nickel(II) iodate	Ni(IO$_3$)$_2$	13477-98-0	408.498	yel needles			5.07	1.1^{30}	sl H$_2$O
1882	Nickel(II) iodate tetrahydrate	Ni(IO$_3$)$_2$ · 4H$_2$O	13477-99-1	480.560	yel hex cry	dec 100		5.07		sl H$_2$O
1883	Nickel(II) iodide	NiI$_2$	13462-90-3	312.502	blk hex cry; hyg	800	subl	5.22	154^{25}	
1884	Nickel(II) iodide hexahydrate	NiI$_2$ · 6H$_2$O	7790-34-3	420.593	grn monocl cry; hyg				154^{25}	vs EtOH
1885	Nickel(II) nitrate	Ni(NO$_3$)$_2$	13138-45-9	182.702	grn cry				99.2^{25}	s EtOH
1886	Nickel(II) nitrate hexahydrate	Ni(NO$_3$)$_2$ · 6H$_2$O	13478-00-7	290.794	grn monocl cry; hyg	56 dec		2.05	99.2^{25}	s EtOH
1887	Nickel(II) oxalate dihydrate	NiC$_2$O$_4$ · 2H$_2$O	6018-94-6	182.742	grn-wh solid	dec 150			0.0012^{25}	i H$_2$O; s acid, NH$_4$OH
1888	Nickel(II) oxide	NiO	1313-99-1	74.692	grn cub cry	1957		6.72		i H$_2$O; s acid
1889	Nickel(II) perchlorate hexahydrate	Ni(ClO$_4$)$_2$ · 6H$_2$O	13637-71-3*	365.686	grn hex needles	140			158.8^{25}	s EtOH, ace
1890	Nickel(II) phosphate octahydrate	Ni$_3$(PO$_4$)$_2$ · 8H$_2$O	10381-36-9*	510.145	grn plates					s acid
1891	Nickel(II) selenate hexahydrate	NiSeO$_4$ · 6H$_2$O	15060-62-5*	309.74	grn tetr cry			2.314	35.5^{20}	
1892	Nickel(II) selenide	NiSe	1314-05-2	137.65	yel-grn hex cry	980		7.2		
1893	Nickel(II) stannate dihydrate	NiSnO$_3$ · 2H$_2$O	12035-38-0	261.432	grn pow	dec 120				
1894	Nickel(II) sulfate	NiSO$_4$	7786-81-4	154.756	grn-yel orth cry	840 dec		4.01	40.4^{25}	
1895	Nickel(II) sulfate hexahydrate	NiSO$_4$ · 6H$_2$O	10101-97-0	262.847	blue-grn tetr cry	≈100 dec		2.07	40.4^{25}	sl EtOH
1896	Nickel(II) sulfate heptahydrate	NiSO$_4$ · 7H$_2$O	10101-98-1	280.862	grn orth cry			1.98	40.4^{25}	s EtOH
1897	Nickel(II) sulfide	NiS	16812-54-7	90.758	yel hex cry	976		5.5		i H$_2$O
1898	Nickel(II) thiocyanate	Ni(SCN)$_2$	13689-92-4	174.857	grn pwd				55.0^{25}	
1899	Nickel(II) titanate	NiTiO$_3$	12035-39-1	154.558	brn hex cry			5.0		
1900	Nickel(II,III) sulfide	Ni$_3$S$_4$	12137-12-1	304.340	cub cry	995		4.77		
1901	Nickel(III) oxide	Ni$_2$O$_3$	1314-06-3	165.385	gray-blk cub cry	≈600 dec				i H$_2$O; s hot acid
1902	Niobium	Nb	7440-03-1	92.906	gray metal; cub	2477	4744	8.57		i acid
1903	Niobocene dichloride	Nb(C$_5$H$_5$)$_2$Cl$_2$	12793-14-5	293.998	hyg blk cry					sl tol
1904	Niobium boride (NbB)	NbB	12045-19-1	103.717	gray orth cry	2270		7.5		
1905	Niobium boride (NbB$_2$)	NbB$_2$	12007-29-3	114.528	gray hex cry	3050		6.97		
1906	Niobium carbide (NbC)	NbC	12069-94-2	104.917	gray cub cry	3608	4300	7.82		i H$_2$O, acid
1907	Niobium carbide (Nb$_2$C)	Nb$_2$C	12011-99-3	197.824	refrac hex cry	3080		7.8		i H$_2$O
1908	Niobium nitride	NbN	24621-21-4	106.913	gray cry; cub	2300		8.47		i HCl, acid
1909	Niobium phosphide	NbP	12034-66-1	123.880	tetr cry			6.5		
1910	Niobium silicide	NbSi$_2$	12034-80-9	149.077	gray hex cry	1950		5.7		
1911	Niobium(II) oxide	NbO	12034-57-0	108.905	gray cub cry	1937		7.30		
1912	Niobium(III) bromide	NbBr$_3$	15752-41-7	332.618	dark brn solid		subl 400			
1913	Niobium(III) chloride	NbCl$_3$	13569-59-0	199.265	blk solid					
1914	Niobium(III) fluoride	NbF$_3$	15195-53-6	149.901	blue cub cry			4.2		
1915	Niobium(III) iodide	NbI$_3$	13870-20-7	473.619	blk solid	dec 510				
1916	Niobium(IV) bromide	NbBr$_4$	13842-75-6	412.522	dark brn cry		subl 300	4.72		reac H$_2$O
1917	Niobium(IV) chloride	NbCl$_4$	13569-70-5	234.718	viol-blk monocl cry	dec 800	275 subl	3.2		reac H$_2$O
1918	Niobium(IV) fluoride	NbF$_4$	13842-88-1	168.900	blk tetr cry; hyg	>350 dec		4.01		
1919	Niobium(IV) iodide	NbI$_4$	13870-21-8	600.524	gray orth cry	503		5.6		reac H$_2$O
1920	Niobium(IV) oxide	NbO$_2$	12034-59-2	124.905	wh tetr cry or powder	1901		5.9		
1921	Niobium(IV) selenide	NbSe$_2$	12034-77-4	250.83	gray hex cry	>1300		6.3		
1922	Niobium(IV) sulfide	NbS$_2$	12136-97-9	157.036	blk rhomb cry			4.4		
1923	Niobium(IV) telluride	NbTe$_2$	12034-83-2	348.11	hex cry			7.6		
1924	Niobium(V) bromide	NbBr$_5$	13478-45-0	492.426	oran orth cry	265.2	361.6	4.36		s H$_2$O, EtOH
1925	Niobium(V) chloride	NbCl$_5$	10026-12-7	270.171	yel monocl cry; hyg	205.8	247.4	2.78		reac H$_2$O; s HCl, ctc
1926	Niobium(V) dioxyfluoride	NbO$_2$F	15195-33-2	143.903	wh cub cry			4.0		
1927	Niobium(V) ethoxide	Nb(OC$_2$H$_5$)$_5$	3236-82-6	318.209	col hyg liq	5	203	1.258		reac H$_2$O; s peth
1928	Niobium(V) fluoride	NbF$_5$	7783-68-8	187.898	col monocl cry; hyg	80	234	2.70		reac H$_2$O; sl CS$_2$, chl
1929	Niobium(V) iodide	NbI$_5$	13779-92-5	727.428	yel-blk monocl cry	327		5.32		reac H$_2$O
1930	Niobium(V) oxide	Nb$_2$O$_5$	1313-96-8	265.810	wh orth cry	1500		4.47		i H$_2$O; s HF
1931	Niobium(V) oxybromide	NbOBr$_3$	14459-75-7	348.617	yel-brn cry	≈320 dec	subl			
1932	Niobium(V) oxychloride	NbOCl$_3$	13597-20-1	215.264	wh tetr cry		subl	3.72		
1933	Nitrogen	N$_2$	7727-37-9	28.013	col gas	-210.0	-195.798	1.145 g/L		sl H$_2$O; i EtOH
1934	Nitramide	NO$_2$NH$_2$	7782-94-7	62.028	unstab wh cry	72 dec				s H$_2$O, EtOH, ace, eth; i chl
1935	Nitric acid	HNO$_3$	7697-37-2	63.013	col liq; hyg	-41.6	83	1.5129^{20}		vs H$_2$O
1936	Nitrous acid	HNO$_2$	7782-77-6	47.014	stab only in soln					
1937	Nitrous oxide	N$_2$O	10024-97-2	44.012	col gas	-90.8	-88.48	1.799 g/L		sl H$_2$O; s EtOH, eth
1938	Nitric oxide	NO	10102-43-9	30.006	col gas	-163.6	-151.74	1.226 g/L		sl H$_2$O

No.	Name	Formula	CAS Reg No.	Mol. weight	Physical form	mp/°C	bp/°C	Density g cm^{-3}	Solubility g/100 g H$_2$O	Qualitative solubility
1939	Nitrogen dioxide	NO$_2$	10102-44-0	46.006	brn gas; equil with N$_2$O$_4$		see N$_2$O$_4$	1.880 g/L		reac H$_2$O
1940	Nitrogen trioxide	N$_2$O$_3$	10544-73-7	76.011	blue solid or liq (low temp)	-101.1	≈3 dec	1.4^2		reac H$_2$O
1941	Nitrogen tetroxide	N$_2$O$_4$	10544-72-6	92.011	col liq; equil with NO$_2$	-9.3	21.15	1.45^{20}		reac H$_2$O
1942	Nitrogen pentoxide	N$_2$O$_5$	10102-03-1	108.010	col hex cry		33 sp	2.0		s chl; sl ctc
1943	Nitrogen tribromide	NBr$_3$	15162-90-0	253.719	unstab solid	exp -100				
1944	Nitrogen trichloride	NCl$_3$	10025-85-1	120.366	yel oily liq; exp	-40	71	1.653		i H$_2$O; s CS$_2$, bz, ctc
1945	Nitrogen trifluoride	NF$_3$	7783-54-2	71.002	col gas	-206.79	-128.75	2.902 g/L		i H$_2$O
1946	Nitrogen triiodide	NI$_3$	13444-85-4	394.720	unstab blk cry; exp					
1947	Nitrogen chloride difluoride	NClF$_2$	13637-87-1	87.457	col gas	-195	-67	3.575 g/L		
1948	Chloramine	NH$_2$Cl	10599-90-3	51.476	yel liq	-66				s H$_2$O, EtOH, eth; sl bz, ctc
1949	Fluoramine	NH$_2$F	15861-05-9	35.021	unstab gas	≈-110		1.431 g/L		
1950	Difluoramine	NHF$_2$	10405-27-3	53.012	col gas	-116	-23	2.167 g/L		
1951	cis-Difluorodiazine	N$_2$F$_2$	13812-43-6	66.010	col gas	<-195	-105.75	2.698 g/L		
1952	trans-Difluorodiazine	N$_2$F$_2$	13776-62-0	66.010	col gas	-172	-111.45	2.698 g/L		
1953	Tetrafluorohydrazine	N$_2$F$_4$	10036-47-2	104.007	col gas	-164.5	-74	4.251 g/L		
1954	Nitrosyl bromide	NOBr	13444-87-6	109.910	red gas	-56	≈0	4.492 g/L		reac H$_2$O
1955	Nitrosyl chloride	NOCl	2696-92-6	65.459	yel gas	-59.6	-5.5	2.676 g/L		reac H$_2$O
1956	Nitrosyl fluoride	NOF	7789-25-5	49.004	col gas	-132.5	-59.9	2.003 g/L		
1957	Trifluoroamine oxide	NOF$_3$	13847-65-9	87.001	col gas	-161	-87.5	3.556 g/L		
1958	Nitryl chloride	NO$_2$Cl	13444-90-1	81.459	col gas	-145	-15	3.330 g/L		
1959	Nitryl fluoride	NO$_2$F	10022-50-1	65.004	col gas	-166	-72.4	2.657 g/L		reac H$_2$O
1960	Nitrogen selenide	N$_4$Se$_4$	12033-88-4	371.87	red monocl cry; hyg	exp		4.2		i H$_2$O, eth, EtOH; sl bz, CS$_2$
1961	Nobelium	No	10028-14-5	259.000	metal	827				
1962	Osmium	Os	7440-04-2	190.23	blue-wh metal; hex	3033	5012	22.587^{20}		s aqua regia
1963	Osmocene	Os(C$_5$H$_5$)$_2$	1273-81-0	320.42	col cry	229				
1964	Osmium carbonyl	Os$_3$(CO)$_{12}$	15696-40-9	906.81	yel cry	224		3.48		
1965	Osmium pentacarbonyl	Os(CO)$_5$	16406-49-8	330.28	col liq	-15	dec 100			s os
1966	Osmium nonacarbonyl	Os$_2$(CO)$_9$	28411-13-4	632.55	oran-yel cry	65 dec				s hc
1967	Osmium(II) chloride	OsCl$_2$	13444-92-3	261.14	hyg brn solid	dec >450				s EtOH, eth
1968	Osmium(III) bromide	OsBr$_3$	59201-51-3	429.94	dark gray cry	340 dec				i H$_2$O, os, acid
1969	Osmium(III) chloride	OsCl$_3$	13444-93-4	296.59	gray cub cry	450 dec				i H$_2$O, os; s conc acid
1970	Osmium(IV) chloride	OsCl$_4$	10026-01-4	332.04	red-blk orth cry	323 dec		4.38		reac H$_2$O; i os
1971	Osmium(IV) fluoride	OsF$_4$	54120-05-7	266.22	yel cry	230				reac H$_2$O
1972	Osmium(IV) oxide	OsO$_2$	12036-02-1	222.23	yel-brn tetr cry	dec 500		11.4		i H$_2$O, acid
1973	Osmium(V) fluoride	OsF$_5$	31576-40-6	285.22	hyg blue-grn cry	70	233			reac H$_2$O
1974	Osmium(VI) fluoride	OsF$_6$	13768-38-2	304.22	yel cub cry	33.4	47.5	4.1		reac H$_2$O
1975	Osmium(VI) tetrachloride oxide	OsOCl$_4$	36509-15-6	348.04	dark brn hyg cry	32	200			reac H$_2$O; s hc
1976	Osmium(VIII) oxide	OsO$_4$	20816-12-0	254.23	yel monocl cry	40.6	131.2	5.1	6.44^{20}	sl H$_2$O; s ctc, bz, EtOH, eth
1977	Oxygen	O$_2$	7782-44-7	31.999	col gas	-218.79	-182.953	1.308 g/L		sl H$_2$O, EtOH, os
1978	Ozone	O$_3$	10028-15-6	47.998	blue gas	-193	-111.35	1.962 g/L		sl H$_2$O
1979	Palladium	Pd	7440-05-3	106.42	silv-wh metal; cub	1554.8	2963	12.0		s aqua regia
1980	Palladium(II) acetate	Pd(C$_2$H$_3$O$_2$)$_2$	3375-31-3	224.51	oran-brn cry	205 dec				i H$_2$O; s MeCN, eth, ace
1981	Palladium(II) bromide	PdBr$_2$	13444-94-5	266.23	red-blk monocl cry; hyg	250 dec		≈5.2		i H$_2$O
1982	Palladium(II) chloride	PdCl$_2$	7647-10-1	177.33	red rhomb cry; hyg	679		4.0		s H$_2$O, EtOH, ace
1983	Palladium(II) chloride dihydrate	PdCl$_2$ · 2H$_2$O	7647-10-1*	213.36	brn cry					s H$_2$O, EtOH, ace
1984	Palladium(II) cyanide	Pd(CN)$_2$	2035-66-7	158.45	yel solid	dec				
1985	Palladium(II) fluoride	PdF$_2$	13444-96-7	144.42	viol tetr cry; hyg	952		5.76		reac H$_2$O
1986	Palladium(II) iodide	PdI$_2$	7790-38-7	360.23	blk cry	360 dec		6.0		i H$_2$O, EtOH, eth
1987	Palladium(II) nitrate	Pd(NO$_3$)$_2$	10102-05-3	230.43	brn hyg cry	dec				sl H$_2$O; s dil HNO$_3$
1988	Palladium(II) oxide	PdO	1314-08-5	122.42	grn-blk tetr cry	750 dec		8.3		i H$_2$O, acid; sl aqua regia
1989	Palladium(II) 2,4-pentanedioate	Pd(CH$_3$COCHCOCH$_3$)$_2$	14024-61-4	304.64	oran-yel cry	205 dec				s bz, chl
1990	Palladium(II) sulfate dihydrate	PdSO$_4$ · 2H$_2$O	13566-03-5	238.51	grn-brn cry	dec				
1991	Palladium(II) sulfide	PdS	12125-22-3	138.49	gray tetr cry			6.7		
1992	cis-Dichlorodiamminepalladium(II)	Pd(NH$_3$)$_2$Cl$_2$	15684-18-1	211.39	yel pow				0.025^{25}	
1993	trans-Dichlorodiamminepalladium (II)	Pd(NH$_3$)$_2$Cl$_2$	13782-33-7	211.39	yel solid			2.50		

No.	Name	Formula	CAS Reg No.	Mol. weight	Physical form	mp/°C	bp/°C	Density g cm^{-3}	Solubility g/100 g H$_2$O	Qualitative solubility
1994	Phosphorus (white)	P	7723-14-0	30.974	col waxlike cub cry	44.15	280.5	1.823		i H$_2$O; sl bz, EtOH, chl; s CS$_2$
1995	Phosphorus (red)	P	7723-14-0	30.974	red-viol amorp powder	579.2	431 sp	2.16		i H$_2$O, os
1996	Phosphorus (black)	P	7723-14-0	30.974	blk orth cry or amorp solid	610		2.69		i os
1997	Phosphine	PH$_3$	7803-51-2	33.998	col gas; flam	-133.8	-87.75	1.390 g/L		i H$_2$O; sl EtOH, eth
1998	Diphosphine	P$_2$H$_4$	13445-50-6	65.980	col liq	-99	63.5 dec			reac H$_2$O
1999	Diphosphorus tetrachloride	P$_2$Cl$_4$	13497-91-1	203.760	col oily liq	-28	≈180 dec			
2000	Diphosphorus tetrafluoride	P$_2$F$_4$	13824-74-3	137.942	col gas	-86.5	-6.2	5.638 g/L		
2001	Diphosphorus tetraiodide	P$_2$I$_4$	13455-00-0	569.566	red tricl needles	125.5	dec	3.89		
2002	Phosphonium chloride	PH$_4$Cl	24567-53-1	70.459	gas		-27 sp	2.880 g/L		reac H$_2$O
2003	Phosphonium iodide	PH$_4$I	12125-09-6	161.910	col tetr cry	18.5	62.5	2.86		reac H$_2$O, EtOH
2004	Phosphoric acid	H$_3$PO$_4$	7664-38-2	97.995	col visc liq	42.4	407		548[20]	s EtOH
2005	Phosphotungstic acid	H$_3$PW$_{12}$O$_{40}$	12067-99-1	2880.05	wh-yel cry	89				vs H$_2$O; s EtOH, eth
2006	Phosphonic acid	H$_3$PO$_3$	13598-36-2	81.996	wh hyg cry	74.4	200	1.65	309[0]	vs EtOH
2007	Phosphinic acid	HPH$_2$O$_2$	6303-21-5	65.997	hyg cry or col oily liq	26.5	130	1.49		vs H$_2$O, EtOH, eth
2008	Metaphosphoric acid	HPO$_3$	37267-86-0	79.980	gl solid; hyg					sl H$_2$O; s EtOH
2009	Hypophosphoric acid	H$_4$P$_2$O$_6$	7803-60-3	161.976	col orth cry	73 dec				vs H$_2$O
2010	Diphosphoric acid	H$_4$P$_2$O$_7$	2466-09-3	177.975	wh cry	71.5			709[23]	
2011	Difluorophosphoric acid	HPO$_2$F$_2$	13779-41-4	101.978	col liq	≈-94	110 dec	1.583		reac H$_2$O
2012	Hexafluorophosphoric acid	HPF$_6$	16940-81-1	145.972	col oily liq	25 dec				reac H$_2$O
2013	Fluorophosphonic acid	H$_2$PFO$_3$	13537-32-1	99.986	col visc liq	<-70		1.82		vs H$_2$O
2014	Phosphorus nitride (P$_3$N$_5$)	P$_3$N$_5$	12136-91-3	162.955	yel-brn solid	800 dec				i H$_2$O; s os
2015	Phosphorus sesquisulfide	P$_4$S$_3$	1314-85-8	220.090	yel-grn orth cry	173	407	2.03		i H$_2$O; s bz; vs CS$_2$
2016	Phosphorus heptasulfide	P$_4$S$_7$	12037-82-0	348.350	pale yel monocl cry	308	523	2.19		sl CS$_2$
2017	Phosphonitrilic chloride trimer	(PNCl$_2$)$_3$	940-71-6	347.659	wh hyg cry	128.8		1.98		reac H$_2$O
2018	Phosphorus(III) bromide	PBr$_3$	7789-60-8	270.686	col liq	-41.5	173.2	2.8		reac H$_2$O, EtOH; s ace, CS$_2$
2019	Phosphorus(III) dibromide fluoride	PBr$_2$F	15597-39-4	209.780	col liq	-115	78.5			
2020	Phosphorus(III) bromide difluoride	PBrF$_2$	15597-40-7	148.875	col gas	-133.8	-16.1	6.085 g/L		
2021	Phosphorus(III) chloride	PCl$_3$	7719-12-2	137.333	col liq	-93	76	1.574		reac H$_2$O, EtOH; s bz, chl, eth
2022	Phosphorus(III) dichloride fluoride	PCl$_2$F	15597-63-4	120.878	col gas	-144	13.85	4.941 g/L		
2023	Phosphorus(III) chloride difluoride	PClF$_2$	14335-40-1	104.424	col gas	-164.8	-47.3	4.268 g/L		
2024	Phosphorus(III) fluoride	PF$_3$	7783-55-3	87.969	col gas	-151.5	-101.8	3.596 g/L		reac H$_2$O
2025	Phosphorus(III) iodide	PI$_3$	13455-01-1	411.687	red-oran hex cry; hyg	61.2	227 dec	4.18		reac H$_2$O; s EtOH
2026	Phosphorus(III) oxide	P$_2$O$_3$	1314-24-5	109.946	col monocl cry or liq	23.8	173	2.13		reac H$_2$O
2027	Tetraphosphorus(III) hexoxide	P$_4$O$_6$	12440-00-5	219.891	soft wh cry	23.8	175.4			
2028	Phosphorus(III) selenide	P$_2$Se$_3$	1314-86-9	298.83	oran-red cry	245	≈380	1.31		reac H$_2$O; s bz, ctc, CS$_2$, ace
2029	Phosphorus(III) sulfide	P$_2$S$_3$	12165-69-4	158.143	yel solid	290	490			reac H$_2$O; s EtOH, eth, CS$_2$
2030	Phosphorus(V) bromide	PBr$_5$	7789-69-7	430.494	yel orth cry, hyg	≈100 dec		3.61		reac H$_2$O, EtOH; s CS$_2$, ctc
2031	Phosphorus(V) tetrabromide fluoride	PBr$_4$F		369.588	pale yel cry	87 dec				
2032	Phosphorus(V) dibromide trifluoride	PBr$_2$F$_3$	13445-58-4	247.777	yel-red liq	-20	15 dec			
2033	Phosphorus(V) chloride	PCl$_5$	10026-13-8	208.239	wh-yel tetr cry; hyg	167 tp	160 sp	2.1		reac H$_2$O; s CS$_2$, ctc
2034	Phosphorus(V) tetrachloride fluoride	PCl$_4$F	13498-11-8	191.784	col liq	-59	30 dec			
2035	Phosphorus(V) trichloride difluoride	PCl$_3$F$_2$	13537-23-0	175.330	col liq	-63				
2036	Phosphorus(V) dichloride trifluoride	PCl$_2$F$_3$	13454-99-4	158.875	col gas	-125	7.1	6.494 g/L		
2037	Phosphorus(V) chloride tetrafluoride	PClF$_4$	13498-11-8	142.421	col gas	-132	-43.4	5.821 g/L		
2038	Phosphorus(V) fluoride	PF$_5$	7647-19-0	125.966	col gas	-93.8	-84.6	5.149 g/L		reac H$_2$O
2039	Phosphorus(V) oxide	P$_2$O$_5$	1314-56-3	141.945	wh orth cry; hyg	562	605	2.30		reac H$_2$O, EtOH
2040	Phosphorus(V) selenide	P$_2$Se$_5$	1314-82-5	456.75	blk-purp amorp solid					reac hot H$_2$O, ctc; i CS$_2$
2041	Phosphorus(V) sulfide	P$_2$S$_5$	1314-80-3	222.273	grn-yel hyg cry	285	515	2.03		reac H$_2$O; s CS$_2$
2042	Phosphonic difluoride	POF$_2$H	14939-34-5	85.978	volatile liq	>-120	≈60 (gas unstab)			

No.	Name	Formula	CAS Reg No.	Mol. weight	Physical form	mp/°C	bp/°C	Density g cm⁻³	Solubility g/100 g H₂O	Qualitative solubility
2043	Phosphoryl bromide	$POBr_3$	7789-59-5	286.685	faint oran plates	55	191.7	2.822		reac H_2O; s bz, eth chl
2044	Phosphoryl dibromide chloride	$POBr_2Cl$	13550-31-7	242.234	yel solid	31	165			
2045	Phosphoryl dibromide fluoride	$POBr_2F$	14014-19-8	225.779	col liq	-117.2	110.1			
2046	Phosphoryl bromide dichloride	$POBrCl_2$	13455-03-3	197.783	col liq	11	136.5	2.104^{14}		
2047	Phosphoryl bromide difluoride	$POBrF_2$	14014-18-7	164.874	col liq	-84.8	31.6			
2048	Phosphoryl bromide chloride fluoride	$POBrClF$	14518-81-1	181.328	col liq		79			
2049	Phosphoryl chloride	$POCl_3$	10025-87-3	153.332	col liq	1.18	105.5	1.645		reac H_2O, EtOH
2050	Phosphoryl dichloride fluoride	$POCl_2F$	13769-76-1	136.877	col liq	-80.1	52.9			
2051	Phosphoryl chloride difluoride	$POClF_2$	13769-75-0	120.423	col gas	-96.4	3.1	4.922 g/L		
2052	Phosphoryl fluoride	POF_3	13478-20-1	103.968	col gas	-39.1 tp	-39.7 sp	4.250 g/L		reac H_2O
2053	Phosphoryl iodide	POI_3	13455-04-4	427.686	viol cry	53				
2054	Phosphorothioc tribromide	$PSBr_3$	3931-89-3	302.751	yel cry	37.8	212 dec	2.85		
2055	Phosphorothioc dibromide fluoride	$PSBr_2F$	13706-10-0	241.845	yel liq	-75.2	125.3			
2056	Phosphorothioc bromide difluoride	$PSBrF_2$	13706-09-7	180.940	yel liq	-136.9	35.5			
2057	Phosphorothioc trichloride	$PSCl_3$	3982-91-0	169.398	fuming liq	-36.2	125	1.635		reac H_2O; s bz, ctc, chl, CS_2
2058	Phosphorothioc dichloride fluoride	$PSCl_2F$	155698-29-6	152.943	col liq	-96.0	64.7			
2059	Phosphorothioc chloride difluoride	$PSClF_2$	2524-02-9	136.489	col gas	-155.2	6.3	5.579 g/L		
2060	Phosphorothioc trifluoride	PSF_3	2404-52-6	120.034	col gas	-148.8	-52.25	4.906 g/L		
2061	Phosphorothioc triiodide	PSI_3	63972-04-3	443.752	yel cry	48	dec			
2062	Platinum	Pt	7440-06-4	195.084	silv-gray metal; cub	1768.2	3825	21.5		i acid; s aqua regia
2063	Hexachloroplatinic acid	H_2PtCl_6	16941-12-1	409.818	hyg yel-brn cry	60				s H_2O, EtOH
2064	Hydrogen hexahydroxyplatinate(IV)	$H_2Pt(OH)_6$	51850-20-5	299.144	yel needles	dec 100				s H_2O, acid, dil alk
2065	Platinum(II) bromide	$PtBr_2$	13455-12-4	354.892	red-brn powder	250 dec		6.65		i H_2O
2066	Platinum(II) chloride	$PtCl_2$	10025-65-7	265.990	grn hex cry	581 dec		6.0		i H_2O, EtOH, eth; s HCl
2067	Platinum(II) cyanide	$Pt(CN)_2$	592-06-3	247.118	pale yel cry					i H_2O, acid, alk
2068	Platinum(II) iodide	PtI_2	7790-39-8	448.893	blk powder	325 dec		6.4		i H_2O
2069	Platinum(II) oxide	PtO	12035-82-4	211.083	blk tetr cry	325 dec		14.1		i H_2O, EtOH; s aqua regia
2070	Platinum(II) sulfide	PtS	12038-20-9	227.149	tetr cry			10.25		
2071	Platinum(III) bromide	$PtBr_3$	25985-07-3	434.796	grn-blk cry	200 dec				
2072	Platinum(III) chloride	$PtCl_3$	25909-39-1	301.443	grn-blk cry	435 dec		5.26		
2073	Platinum(IV) bromide	$PtBr_4$	68938-92-1	514.700	brn-blk cry	180 dec			0.41^{20}	sl EtOH, eth
2074	Platinum(IV) chloride	$PtCl_4$	37773-49-2	336.896	red-brn cub cry	327 dec		4.30	142^{25}	
2075	Platinum(IV) chloride pentahydrate	$PtCl_4 \cdot 5H_2O$	13454-96-1	426.972	red cry			2.43		s H_2O, EtOH
2076	Platinum(IV) fluoride	PtF_4	13455-15-7	271.078	red cry	600				
2077	Platinum(IV) iodide	PtI_4	7790-46-7	702.702	brn-blk powder	130 dec				s H_2O
2078	Platinum(IV) oxide	PtO_2	1314-15-4	227.083	blk hex cry	450		11.8		i H_2O; s conc acid, dil alk
2079	Platinum(IV) sulfide	PtS_2	12038-21-0	259.214	hex cry			7.85		
2080	Platinum(VI) fluoride	PtF_6	13693-05-5	309.074	red cub cry	61.3	69.1	≈4.0		
2081	cis-Diamminedichloroplatinum	$Pt(NH_3)_2Cl_2$	15663-27-1	300.051	yel solid	270 dec			0.253^{25}	
2082	trans-Diamminedichloroplatinum	$Pt(NH_3)_2Cl_2$	14913-33-8	300.051	pale yel solid	270 dec			0.036^{25}	s DMF, DMSO
2083	Hexachloroplatinic acid hexahydrate	$H_2PtCl_6 \cdot 6H_2O$	16941-12-1	517.909	brn-yel hyg cry	60		2.43	140^{18}	vs EtOH
2084	Platinum silicide	PtSi	12137-83-6	223.170	orth cry	1229		12.4		
2085	Plutonium	Pu	7440-07-5	244	silv-wh metal; monocl	640	3228	19.7		
2086	Plutonium nitride	PuN	12033-54-4	258	gray cub cry	2550		14.4		
2087	Plutonium(II) oxide	PuO	12035-83-5	260	cub cry			14.0		
2088	Plutonium(III) bromide	$PuBr_3$	15752-46-2	484	grn orth cry	681		6.75		s H_2O
2089	Plutonium(III) chloride	$PuCl_3$	13569-62-5	350	grn hex cry	760		5.71		s H_2O
2090	Plutonium(III) fluoride	PuF_3	13842-83-6	301	purp hex cry	1396		9.33		i H_2O; sl acid
2091	Plutonium(III) iodide	PuI_3	13813-46-2	625	grn orth cry; hyg	777		6.92		s H_2O
2092	Plutonium(III) oxide	Pu_2O_3	12036-34-9	536	blk cub cry	2085		10.5		
2093	Plutonium(IV) fluoride	PuF_4	13709-56-3	320	red-brn monocl cry	1037		7.1		
2094	Plutonium(IV) oxide	PuO_2	12059-95-9	276	yel-brn cub cry	2390		11.5		
2095	Plutonium(VI) fluoride	PuF_6	13693-06-6	358	red-brn orth cry	51.6		5.08		
2096	Polonium	Po	7440-08-6	209	silv metal; cub	254	962	9.20		
2097	Polonium(IV) chloride	$PoCl_4$	10026-02-5	351	yel hyg cry	≈300	390			s H_2O, EtOH, ace
2098	Polonium(IV) oxide	PoO_2	7446-06-2	241	yel cub cry	500 dec		8.9		
2099	Potassium	K	7440-09-7	39.098	soft silv-wh metal; cub	63.5	759	0.89		reac H_2O
2100	Potassium acetate	$KC_2H_3O_2$	127-08-2	98.142	wh hyg cry	309		1.57	269^{25}	s EtOH; i eth

No.	Name	Formula	CAS Reg No.	Mol. weight	Physical form	mp/°C	bp/°C	Density g cm⁻³	Solubility g/100 g H₂O	Qualitative solubility
2101	Potassium aluminate trihydrate	$K_2Al_2O_4 \cdot 3H_2O$	12003-63-3*	250.204	wh orth cry			2.13		vs H_2O; i EtOH
2102	Potassium aluminum silicate	$KAlSi_3O_8$	1327-44-2	278.332	col monocl cry			2.56		i H_2O
2103	Potassium aluminum sulfate	$KAl(SO_4)_2$	10043-67-1	258.205	wh hyg powder				5.9²⁰	
2104	Potassium aluminum sulfate dodecahydrate	$KAl(SO_4)_2 \cdot 12H_2O$	7784-24-9	474.389	col cry	≈100 dec		1.72	5.9²⁰	
2105	Potassium amide	KNH_2	17242-52-3	55.121	wh/yel-grn hyg cry	335				reac H_2O, EtOH
2106	Potassium arsenate	K_3AsO_4	13464-36-3	256.215	col cry			2.8	125²⁵	
2107	Potassium arsenite	$KAsO_2$	13464-35-2	146.019	wh hyg powder					s H_2O; sl EtOH
2108	Potassium azide	KN_3	20762-60-1	81.118	tetr cry; exp			2.04	49.7¹⁷	
2109	Potassium borohydride	KBH_4	13762-51-1	53.941	wh cub cry	≈500 dec		1.11		s H_2O
2110	Potassium bromate	$KBrO_3$	7758-01-2	167.000	wh hex cry	434 dec		3.27	8.17²⁵	i EtOH
2111	Potassium bromide	KBr	7758-02-3	119.002	col cub cry; hyg	734	1435	2.74	67.8²⁵	sl EtOH
2112	Potassium carbonate	K_2CO_3	584-08-7	138.206	wh monocl cry; hyg	899	dec	2.29	111²⁵	i EtOH
2113	Potassium carbonate sesquihydrate	$K_2CO_3 \cdot 1.5H_2O$	6381-79-9	165.229	granular cry				111²⁰	
2114	Potassium chlorate	$KClO_3$	3811-04-9	122.549	wh monocl cry	357	dec	2.34	8.61²⁵	
2115	Potassium chloride	KCl	7447-40-7	74.551	wh cub cry	771		1.988	35.5²⁵	i eth, ace
2116	Potassium chlorochromate	$KCrO_3Cl$	16037-50-6	174.545	oran cry			2.5		reac H_2O; s ace, acid
2117	Potassium chromate	K_2CrO_4	7789-00-6	194.191	yel orth cry	974		2.73	65.0²⁵	
2118	Potassium citrate monohydrate	$K_3C_6H_5O_7 \cdot H_2O$	6100-05-6	324.410	col hyg cry	180 dec		1.98	172²⁰	vs H_2O; sl EtOH
2119	Potassium cobalt(II) selenate hexahydrate	$K_2Co(SeO_4)_2 \cdot 6H_2O$	28041-86-3	531.14	red monocl cry			2.51		
2120	Potassium cyanate	$KCNO$	590-28-3	81.115	wh tetr cry	≈700 dec		2.05	75²⁵	sl EtOH
2121	Potassium cyanide	KCN	151-50-8	65.116	wh cub cry; hyg	622		1.55	69.9²⁰	sl EtOH
2122	Potassium cyanoaurite	$KAu(CN)_2$	13967-50-5	288.099	col cry			3.45	14²⁰	s H_2O; sl EtOH; i eth, ace
2123	Potassium dichromate	$K_2Cr_2O_7$	7778-50-9	294.185	oran-red tricl cry	398	≈500 dec	2.68	15.1²⁵	
2124	Potassium dihydrogen arsenate	KH_2AsO_4	7784-41-0	180.034	col cry	288		2.87	19⁶	i EtOH
2125	Potassium dihydrogen phosphate	KH_2PO_4	7778-77-0	136.085	wh tetr cry	253		2.34	25.0²⁵	sl EtOH
2126	Potassium dihydrogen phosphonate	KH_2PO_3	13977-65-6	120.086	col monocl hyg cry					
2127	Potassium dithionate	$K_2S_2O_6$	13455-20-4	238.323	col hex cry	dec		2.27		sl H_2O; i EtOH
2128	Potassium ferricyanide	$K_3Fe(CN)_6$	13746-66-2	329.244	red cry	dec		1.89	48.8²⁵	
2129	Potassium ferrocyanide trihydrate	$K_4Fe(CN)_6 \cdot 3H_2O$	14459-95-1	422.388	yel monocl cry	60 dec		1.85	36.0²⁵	i EtOH, eth
2130	Potassium fluoride	KF	7789-23-3	58.096	wh cub cry	858	1502	2.48	102²⁵	
2131	Potassium fluoride dihydrate	$KF \cdot 2H_2O$	13455-21-5	94.127	monocl cry	41 dec		2.5	102²⁵	
2132	Potassium fluoroborate	KBF_4	14075-53-7	125.903	col orth cry	530		2.505	0.55²⁵	sl EtOH
2133	Potassium fluorotantalate	K_2TaF_7	16924-00-8	392.134	col cry	730		5.24	0.5⁰	
2134	Potassium formate	$KCHO_2$	590-29-4	84.116	col hyg cry	167		1.91	331¹⁸	
2135	Potassium hexachloroosmate(IV)	K_2OsCl_6	16871-60-6	481.15	red cub cry					vs H_2O; sl EtOH
2136	Potassium hexachloroplatinate	K_2PtCl_6	16921-30-5	485.999	yel-oran cub cry	250 dec		3.50	0.77²⁰	i EtOH
2137	Potassium hexacyanocobaltate	$K_3Co(CN)_6$	13963-58-1	332.332	yel monocl cry	dec		1.91		vs H_2O; i EtOH
2138	Potassium hexafluoromanganate(IV)	K_2MnF_6	16962-31-5	247.125	yel hex cry					reac H_2O
2139	Potassium hexafluorosilicate	K_2SiF_6	16871-90-2	220.273	wh cry	dec		2.27	0.084²⁰	i EtOH
2140	Potassium hexafluorozirconate(IV)	K_2ZrF_6	16923-95-8	283.411	col monocl cry			3.48	0.78²	
2141	Potassium hydride	KH	7693-26-7	40.106	cub cry	619		1.43		reac H_2O
2142	Potassium hydrogen arsenate	K_2HAsO_4	21093-83-4	218.125	col monocl prisms	300 dec			18.7⁶	i EtOH
2143	Potassium hydrogen carbonate	$KHCO_3$	298-14-6	100.115	col monocl cry	≈100 dec		2.17	36.2²⁵	i EtOH
2144	Potassium hydrogen fluoride	KHF_2	7789-29-9	78.103	col tetr cry	238.8		2.37	39.2²⁰	i EtOH
2145	Potassium hydrogen iodate	$KH(IO_3)_2$	13455-24-8	389.911	col cry	dec			1.3¹⁵	sl H_2O; i EtOH
2146	Potassium hydrogen oxalate hemihydrate	$KHC_2O_4 \cdot 0.5H_2O$	127-95-7		wh cry	dec		2.09	2.5²⁰	sl EtOH
2147	Potassium hydrogen phosphate	K_2HPO_4	7758-11-4	174.176	wh hyg cry	dec			168²⁵	s EtOH
2148	Potassium hydrogen phosphite	K_2HPO_3	13492-26-7	158.177	wh hyg powder	dec			170²⁰	i EtOH
2149	Potassium hydrogen selenite	$KHSeO_3$	7782-70-9	167.06	hyg orth cry	>100 dec				s H_2O; sl EtOH
2150	Potassium hydrogen sulfate	$KHSO_4$	7646-93-7	136.169	wh monocl cry; hyg	≈200		2.32	50.6²⁵	
2151	Potassium hydrogen sulfide	KHS	1310-61-8	72.171	wh hex cry; hyg	≈450		1.69		s H_2O, EtOH
2152	Potassium hydrogen sulfide hemihydrate	$KHS \cdot 0.5H_2O$	1310-61-8*	81.179	wh-yel hyg cry	≈175		1.7		vs H_2O, EtOH
2153	Potassium hydrogen sulfite	$KHSO_3$	7773-03-7	120.169	wh cry powder	190 dec			49²⁰	i EtOH
2154	Potassium hydrogen tartrate	$KHC_4H_4O_6$	868-14-4	188.177	wh cry			1.98	0.57²⁰	s acid, alk; i EtOH
2155	Potassium hydroxide	KOH	1310-58-3	56.105	wh rhomb cry; hyg	406	1327	2.044	121²⁵	s EtOH; s MeOH
2156	Potassium hypochlorite	$KOCl$	7778-66-7	90.550	exists only in aq soln					
2157	Potassium phosphinate	KH_2PO_2	7782-87-8	104.087	wh hyg cry	dec				vs H_2O; s EtOH

No.	Name	Formula	CAS Reg No.	Mol. weight	Physical form	mp/°C	bp/°C	Density g cm^{-3}	Solubility g/100 g H$_2$O	Qualitative solubility
2158	Potassium iodate	KIO$_3$	7758-05-6	214.001	wh monocl cry	560 dec		3.89	9.22[25]	
2159	Potassium iodide	KI	7681-11-0	166.003	col cub cry	681	1323	3.12	148[25]	sl EtOH
2160	Potassium iron(III) oxalate trihydrate	K$_3$Fe(C$_2$O$_4$)$_3$ · 3H$_2$O		491.243	grn monocl cry	100	230 dec	2.133	4.7[0]	i EtOH
2161	Potassium manganate	K$_2$MnO$_4$	10294-64-1	197.133	grn cry	190 dec				s H$_2$O; reac HCl
2162	Potassium metaarsenate	KAsO$_3$	19197-73-0	162.018	wh solid	660				
2163	Potassium metabisulfite	K$_2$S$_2$O$_5$	16731-55-8	222.324	wh powder	≈150 dec		2.3	49.5[25]	reac acid; i EtOH
2164	Potassium metaborate	KBO$_2$	13709-94-9	81.908	wh hex cry	947		≈2.3		
2165	Potassium molybdate	K$_2$MoO$_4$	13446-49-6	238.14	wh hyg cry	919		2.3	183[25]	i EtOH
2166	Potassium niobate	KNbO$_3$	12030-85-2	180.002	wh rhomb cry	≈1100		4.64		i H$_2$O
2167	Potassium nitrate	KNO$_3$	7757-79-1	101.103	col orth cry or powder	334	400 dec	2.105	38.3[25]	i EtOH
2168	Potassium nitrite	KNO$_2$	7758-09-0	85.104	wh hyg cry	438	537 exp	1.915	312[25]	sl EtOH
2169	Potassium oxalate	K$_2$C$_2$O$_4$	583-52-8	166.216	wh pwd					sl H$_2$O
2170	Potassium oxalate monohydrate	K$_2$C$_2$O$_4$ · H$_2$O	6487-48-5	184.231	col cry	160 dec		2.13	36.4[20]	
2171	Potassium oxide	K$_2$O	12136-45-7	94.196	gray cub cry	740		2.35		s H$_2$O, EtOH, eth
2172	Potassium perbromate	KBrO$_4$	22207-96-1	183.000	wh cry	275 dec			4.21[25]	
2173	Potassium percarbonate monohydrate	K$_2$C$_2$O$_6$ · H$_2$O	589-97-9	216.230	oran or blue pow				6.5[20]	
2174	Potassium perchlorate	KClO$_4$	7778-74-7	138.549	col orth cry; hyg	525		2.52	2.08[25]	
2175	Potassium periodate	KIO$_4$	7790-21-8	230.001	col tetr cry	582	exp	3.618	0.51[25]	
2176	Potassium permanganate	KMnO$_4$	7722-64-7	158.034	purp orth cry	dec		2.7	7.60[25]	reac EtOH
2177	Potassium peroxide	K$_2$O$_2$	17014-71-0	110.196	yel amorp solid	490				reac H$_2$O
2178	Potassium persulfate	K$_2$S$_2$O$_8$	7727-21-1	270.322	col cry	≈100 dec		2.48	4.7[20]	
2179	Potassium phosphate	K$_3$PO$_4$	7778-53-2	212.266	wh orth cry; hyg	1340		2.564	106[25]	i EtOH
2180	Potassium pyrophosphate	K$_4$P$_2$O$_7$	7320-34-5		wh cry	dec 1300				s H$_2$
2181	Potassium pyrophosphate trihydrate	K$_4$P$_2$O$_7$ · 3H$_2$O	7790-67-2	384.383	col hyg cry	dec 300		2.33		vs H$_2$O; i EtOH
2182	Potassium pyrosulfate	K$_2$S$_2$O$_7$	7790-62-7	254.323	col needles	≈325		2.28		s H$_2$O
2183	Potassium selenate	K$_2$SeO$_4$	7790-59-2	221.16	wh powder			3.07	114[25]	
2184	Potassium selenide	K$_2$Se	1312-74-9	157.16	red cub cry; hyg	800		2.29		s H$_2$O
2185	Potassium selenite	K$_2$SeO$_3$	10431-47-7	205.16	wh hyg cry	875 dec			217[25]	sl EtOH
2186	Potassium silver cyanide	KAg(CN)$_2$	506-61-6	199.000	wh cry					s H$_2$O
2187	Potassium sodium tartrate tetrahydrate	KNaC$_4$H$_4$O$_6$ · 4H$_2$O	6381-59-5	282.220	wh cry	≈70 dec	anh at 130	1.79		vs H$_2$O; i EtOH
2188	Potassium stannate trihydrate	K$_2$SnO$_3$ · 3H$_2$O	12142-33-5*	298.951	col cry			3.20		vs H$_2$O; i EtOH
2189	Potassium stearate	KC$_{18}$H$_{35}$O$_2$	593-29-3	322.568	wh pow					sl cold H$_2$O; s hot H$_2$O, EtOH
2190	Potassium sulfate	K$_2$SO$_4$	7778-80-5	174.260	wh orth cry	1069		2.66	12.0[25]	i EtOH
2191	Potassium sulfide	K$_2$S	1312-73-8	110.262	red-yel cub cry; hyg	948		1.74		s H$_2$O, EtOH; i eth
2192	Potassium sulfide pentahydrate	K$_2$S · 5H$_2$O	37248-34-3	200.338	col rhomb cry	60				vs H$_2$O, EtOH; i eth
2193	Potassium sulfite	K$_2$SO$_3$	10117-38-1	158.260	col hex cry				106[25]	sl EtOH
2194	Potassium sulfite dihydrate	K$_2$SO$_3$ · 2H$_2$O	7790-56-9	194.291	wh monocl cry	dec			107[20]	sl EtOH; dec dil acid
2195	Potassium superoxide	KO$_2$	12030-88-5	71.097	yel tetr cry; hyg	380		2.16		reac H$_2$O
2196	Potassium tellurate(VI) trihydrate	K$_2$TeO$_4$ · 3H$_2$O	15571-91-2*	323.84	wh cry powder					s H$_2$O
2197	Potassium tellurite	K$_2$TeO$_3$	7790-58-1	253.80	wh hyg cry	≈460 dec				vs H$_2$O
2198	Potassium tetraborate pentahydrate	K$_2$B$_4$O$_7$ · 5H$_2$O	1332-77-0	323.513	wh cry powder				16.5[30]	sl EtOH
2199	Potassium tetrachloroaurate dihydrate	KAuCl$_4$ · 2H$_2$O	13682-61-6	413.908	yel monocl cry					s H$_2$O, EtOH, eth
2200	Potassium tetrachloroplatinate	K$_2$PtCl$_4$	10025-99-7	415.093	pink-red tetr cry	500 dec		3.38		s H$_2$O; i EtOH
2201	Potassium tetracyanocadmate	K$_2$Cd(CN)$_4$	14402-75-6	294.678	cub cry	≈450		1.85	25[20]	sl EtOH
2202	Potassium tetracyanonickelate monohydrate	K$_2$[Ni(CN)$_4$] · H$_2$O	14220-17-8*	258.975	red-oran cry	dec 100				
2203	Potassium tetracyanoplatinate(II) trihydrate	K$_2$Pt(CN)$_4$ · 3H$_2$O	562-76-5*	431.397	col rhomb prisms					s H$_2$O
2204	Potassium tetracyanozincate	K$_2$Zn(CN)$_4$	14244-62-3	247.676	cry pow					vs H$_2$O
2205	Potassium tetraiodomercurate(II)	K$_2$HgI$_4$	7783-33-7	786.40	yel hyg cry			4.29		vs H$_2$O; s EtOH, eth, ace
2206	Potassium thiocyanate	KSCN	333-20-0	97.181	col tetr cry; hyg	173	500 dec	1.88	238[25]	s EtOH
2207	Potassium thiosulfate	K$_2$S$_2$O$_3$	10294-66-3	190.325	col hyg cry				165[25]	i EtOH
2208	Potassium titanate	K$_2$TiO$_3$	12030-97-6	174.062	wh orth cry	1515		3.1		reac H$_2$O
2209	Potassium triiodide monohydrate	KI$_3$ · H$_2$O	7790-42-3	437.827	brn monocl cry; hyg	225 dec		3.5		s H$_2$O; reac EtOH, eth
2210	Potassium triiodozincate	KZnI$_3$	7790-43-4	485.221	hyg cry					vs H$_2$O
2211	Potassium thiocarbonate	K$_2$CS$_3$	26750-66-3	186.403	yel-red hyg cry					vs H$_2$O
2212	Potassium tungstate	K$_2$WO$_4$	7790-60-5	326.04	hyg cry	921		3.12		vs H$_2$O; i EtOH
2213	Potassium uranate	K$_2$U$_2$O$_7$	7790-63-8	666.251	oran cub cry			6.12		i H$_2$O; s acid

No.	Name	Formula	CAS Reg No.	Mol. weight	Physical form	mp/°C	bp/°C	Density g cm⁻³	Solubility g/100 g H₂O	Qualitative solubility
2214	Potassium uranyl nitrate	$K(UO_2)(NO_3)_3$	18078-40-5	495.140	grn-yel cry pow					vs H_2O
2215	Potassium uranyl sulfate dihydrate	$K_2(UO_2)(SO_4)_2 \cdot 2H_2O$	27709-53-1	576.381	grn-yel cry pow	dec 120		3.36		vs H_2O
2216	Potassium zinc sulfate hexahydrate	$K_2Zn(SO_4)_2 \cdot 6H_2O$	13932-17-7	443.823	cry					s H_2O
2217	Potassium zirconium sulfate trihydrate	$K_4Zr(SO_4)_4 \cdot 3H_2O$	53608-79-0	685.914	wh cry pow					sl H_2O
2218	Praseodymium	Pr	7440-10-0	140.908	silv metal; hex	931	3520	6.77		
2219	Praseodymium boride	PrB_6	12008-27-4	205.774	blk cub cry	2610		4.84		
2220	Praseodymium nitride	PrN	25764-09-4	154.915	cub cry			7.46		
2221	Praseodymium silicide	$PrSi_2$	12066-83-0	197.079	tetr cry	1712		5.46		
2222	Praseodymium(II) iodide	PrI_2	65530-47-4	394.717	bronze solid	758				
2223	Praseodymium(III) bromate	$Pr(BrO_3)_3$	15162-93-3	524.615	grn cry					vs H_2O
2224	Praseodymium(III) bromide	$PrBr_3$	13536-53-3	380.620	grn hex cry; hyg	693		5.28		s H_2O
2225	Praseodymium(III) carbonate octahydrate	$Pr_2(CO_3)_3 \cdot 8H_2O$	14948-62-0	605.964	grn silky plates	dec 420 (anh)				i H_2O; s acid
2226	Praseodymium(III) chloride	$PrCl_3$	10361-79-2	247.267	grn hex needles; hyg	786		4.0	96.1[25]	s EtOH
2227	Praseodymium(III) chloride heptahydrate	$PrCl_3 \cdot 7H_2O$	10025-90-8	373.374	grn cry	110 dec			96.1[25]	s EtOH
2228	Praseodymium(III) fluoride	PrF_3	13709-46-1	197.903	grn hex cry	1399		6.3		
2229	Praseodymium(III) hydroxide	$Pr(OH)_3$	16469-16-2	191.930	grn solid	dec 220		3.7		i H_2O
2230	Praseodymium(III) iodide	PrI_3	13813-23-5	521.621	orth hyg cry	738		≈5.8		s H_2O
2231	Praseodymium(III) nitrate	$Pr(NO_3)_3$	10361-80-5	326.923	pale grn hyg cry				165[25]	s EtOH
2232	Praseodymium(III) nitrate hexahydrate	$Pr(NO_3)_3 \cdot 6H_2O$	15878-77-0	435.014	grn needles				165[25]	s EtOH, ace
2233	Praseodymium(III) oxide	Pr_2O_3	12036-32-7	329.813	wh hex cry	2183	3760	6.9		
2234	Praseodymium(III) perchlorate hexahydrate	$Pr(ClO_4)_3 \cdot 6H_2O$	13498-07-2*	547.351	hyg grn cry	dec 200				s H_2O, EtOH
2235	Praseodymium(III) sulfate octahydrate	$Pr_2(SO_4)_3 \cdot 8H_2O$	13510-41-3	714.125	grn monocl cry			2.83	17[20]	s H_2O
2236	Praseodymium(III) sulfide	Pr_2S_3	12038-13-0	378.010	cub cry	1765		5.1		
2237	Praseodymium(III) telluride	Pr_2Te_3	12038-12-9	664.62	cub cry	1500		≈7.0		
2238	Praseodymium(IV) fluoride	PrF_4	15192-24-2	216.902	yel-wh solid	dec 90				
2239	Promethium	Pm	7440-12-2	145	silv metal; hex	1042	3000	7.26		
2240	Promethium(III) bromide	$PmBr_3$	14325-78-1	385	red cry	625				s H_2O
2241	Promethium(III) chloride	$PmCl_3$	13779-10-7	251	pale blue hyg cry	655				s H_2O
2242	Promethium(III) fluoride	PmF_3	13709-45-0	202	pink solid	1338				s H_2O
2243	Promethium(III) iodide	PmI_3	13818-73-0	526	red solid	695				
2244	Protactinium	Pa	7440-13-3	231.036	shiny metal; tetr or cub	1572		15.4		
2245	Protactinium(V) chloride	$PaCl_5$	13760-41-3	408.301	yel monocl cry	306		3.74		
2246	Radium	Ra	7440-14-4	226	wh metal; cub	696		5		
2247	Radium bromide	$RaBr_2$	10031-23-9	386	wh orth cry	728		5.79	70.6[20]	s EtOH
2248	Radium carbonate	$RaCO_3$	7116-98-5	286	wh orth cry					i H_2O
2249	Radium chloride	$RaCl_2$	10025-66-8	297	wh orth cry	1000		4.9	24.5[20]	s EtOH
2250	Radium fluoride	RaF_2	20610-49-5	264	wh cub cry			6.7		
2251	Radium nitrate	$Ra(NO_3)_2$	10213-12-4	350	cry				13.9	
2252	Radium sulfate	$RaSO_4$	7446-16-4	322	wh cry					i H_2O, acid
2253	Radon	Rn	10043-92-2	222	col gas	-71	-61.7	9.074 g/L		sl H_2O
2254	Rhenium	Re	7440-15-5	186.207	silv-gray metal	3185	5596	20.8		i HCl
2255	Perrhenic acid	$HReO_4$	13768-11-1	251.213	exists only in soln					vs H_2O, os
2256	Rhenium carbonyl	$Re_2(CO)_{10}$	14285-68-8	652.515	yel-wh cry	170 dec		2.87		s os
2257	Rhenium pentacarbonyl bromide	$Re(CO)_5Br$	14220-21-4	406.162	wh cry	90				
2258	Rhenium pentacarbonyl chloride	$Re(CO)_5Cl$	14099-01-5	361.711	wh cry		subl 140			
2259	Rhenium(III) bromide	$ReBr_3$	13569-49-8	425.919	red-brn monocl cry		500 subl	6.10		s ace, MeOH, EtOH
2260	Rhenium(III) chloride	$ReCl_3$	13569-63-6	292.566	red-blk hyg cry	500 dec		4.81		s H_2O
2261	Rhenium(III) iodide	ReI_3	15622-42-1	566.920	blk solid	dec				
2262	Rhenium(IV) chloride	$ReCl_4$	13569-71-6	328.019	purp-blk cry; hyg	300 dec		4.9		
2263	Rhenium(IV) fluoride	ReF_4	15192-42-4	262.201	blue tetr cry	>300 subl		7.49		
2264	Rhenium(IV) oxide	ReO_2	12036-09-8	218.206	gray orth cry	900 dec		11.4		
2265	Rhenium(IV) selenide	$ReSe_2$	12038-64-1	344.13	tricl cry					
2266	Rhenium(IV) silicide	$ReSi_2$	12038-66-3	242.378	refrac solid	2000				
2267	Rhenium(IV) sulfide	ReS_2	12038-63-0	250.337	tricl cry			7.6		
2268	Rhenium(IV) telluride	$ReTe_2$	12067-00-4	441.41	orth cry			8.50		
2269	Rhenium(V) bromide	$ReBr_5$	30937-53-2	585.727	brn solid	110 dec				
2270	Rhenium(V) chloride	$ReCl_5$	39368-69-9	363.472	brn-blk solid	220		4.9		reac H_2O
2271	Rhenium(V) fluoride	ReF_5	30937-52-1	281.199	yel-grn solid	48	221.3			
2272	Rhenium(V) oxide	Re_2O_5	12165-05-8	452.411	blue-blk tetr cry			≈7		

No.	Name	Formula	CAS Reg No.	Mol. weight	Physical form	mp/°C	bp/°C	Density g cm^{-3}	Solubility g/100 g H$_2$O	Qualitative solubility
2273	Rhenium(VI) chloride	ReCl$_6$	31234-26-1	398.925	red-grn solid	29				
2274	Rhenium(VI) dioxydifluoride	ReO$_2$F$_2$	81155-18-2	256.203	col cry	156				
2275	Rhenium(VI) fluoride	ReF$_6$	10049-17-9	300.197	yel liq or cub cry	18.5	33.8	4.06(cry)		s HNO$_3$
2276	Rhenium(VI) oxide	ReO$_3$	1314-28-9	234.205	red cub cry	400 dec		6.9		i H$_2$O, acid, alk
2277	Rhenium(VI) oxytetrachloride	ReOCl$_4$	13814-76-1	344.018	brn cry	29.3	223			reac H$_2$O
2278	Rhenium(VI) oxytetrafluoride	ReOF$_4$	17026-29-8	278.200	blue solid	108	171.7			
2279	Rhenium(VII) fluoride	ReF$_7$	17029-21-9	319.196	yel cub cry	48.3	73.7	4.32		
2280	Rhenium(VII) oxide	Re$_2$O$_7$	1314-68-7	484.410	yel hyg cry	327	360	6.10		s H$_2$O, EtOH, eth, diox, py
2281	Rhenium(VII) trioxychloride	ReO$_3$Cl	7791-09-5	269.658	col liq	4.5	128	3.87		reac H$_2$O
2282	Rhenium(VII) trioxyfluoride	ReO$_3$F	42246-24-2	253.203	yel solid	147	164			
2283	Rhenium(VII) dioxytrifluoride	ReO$_2$F$_3$	57246-89-6	275.201	yel solid	90	185.4			reac H$_2$O
2284	Rhenium(VII) oxypentafluoride	ReOF$_5$	23377-53-9	297.198	cream solid	43.8	73.0			
2285	Rhenium(VII) sulfide	Re$_2$S$_7$	12038-67-4	596.869	brn-blk tetr cry			4.87		i H$_2$O
2286	Rhodium	Rh	7440-16-6	102.906	silv-wh metal; cub	1964	3695	12.4		i acid, sl aqua regia
2287	Rhodium carbonyl	Rh$_6$(CO)$_{16}$	28407-51-4	1065.594	red-brn cry	220 dec				
2288	Rhodium carbonyl chloride	[Rh(CO)$_2$Cl]$_2$	14523-22-9	388.758	red-oran cry	124				s os
2289	Rhodium dodecacarbonyl	Rh$_4$(CO)$_{12}$	19584-30-6	747.743	red hyg cry	150 dec		2.52		reac H$_2$O
2290	Rhodium(III) bromide	RhBr$_3$	15608-29-4	342.618	dark brn plates	800 dec		5.56		s H$_2$O; i acid, os
2291	Rhodium(III) chloride	RhCl$_3$	10049-07-7	209.265	red monocl cry		717	5.38		i H$_2$O; s alk
2292	Rhodium(III) fluoride	RhF$_3$	60804-25-3	159.901	red hex cry			5.4		
2293	Rhodium(III) iodide	RhI$_3$	15492-38-3	483.619	blk monocl cry; hyg			6.4		
2294	Rhodium(III) nitrate	Rh(NO$_3$)$_3$	10139-58-9	288.921	hyg brn solid	600 dec				i H$_2$O
2295	Rhodium(III) nitrate dihydrate	Rh(NO$_3$)$_3$ · 2H$_2$O	13465-43-5	324.951	blk solid	dec				i H$_2$O; s aqua regia
2296	Rhodium(III) oxide	Rh$_2$O$_3$	12036-35-0	253.809	gray hex cry	1100 dec		8.2		
2297	Rhodium(III) oxide pentahydrate	Rh$_2$O$_3$ · 5H$_2$O	39373-27-8	309.010	yel pow	dec				sl H$_2$O; s acid
2298	Rhodium(III) sulfate	Rh$_2$(SO$_4$)$_3$	10489-46-0	493.999	red-yel solid	>500 dec				
2299	Rhodium(IV) oxide	RhO$_2$	12137-27-8	134.905	blk tetr cry			7.2		
2300	Rhodium(IV) oxide dihydrate	RhO$_2$ · 2H$_2$O	12137-27-8	170.936	grn solid	dec		8.20		i H$_2$O, sol HCl, alk
2301	Rhodium(VI) fluoride	RhF$_6$	13693-07-7	216.896	blk cub cry	≈70		3.1		
2302	Rubidium	Rb	7440-17-7	85.468	soft silv metal; cub	39.30	688	1.53		reac H$_2$O
2303	Rubidium acetate	RbC$_2$H$_3$O$_2$	563-67-7	144.512	wh hyg cry	246				vs H$_2$O
2304	Rubidium aluminum sulfate	RbAl(SO$_4$)$_2$	13530-57-9	304.575	hex cry			≈3.1	1.60^{20}	i EtOH
2305	Rubidium aluminum sulfate dodecahydrate	RbAl(SO$_4$)$_2$ · 12H$_2$O	7784-29-4	520.759	col cub cry	≈100 dec		≈1.9		s H$_2$O; i EtOH
2306	Rubidium azide	RbN$_3$	22756-36-1	127.488	tetr cry; exp	317		2.79	107^{16}	
2307	Rubidium bromate	RbBrO$_3$	13446-70-3	213.370	cub cry	430		3.68	2.95^{25}	
2308	Rubidium bromide	RbBr	7789-39-1	165.372	wh cub cry; hyg	692	1340	3.35	116^{25}	
2309	Rubidium carbonate	Rb$_2$CO$_3$	584-09-8	230.945	col monocl cry; hyg	837			223^{20}	
2310	Rubidium chlorate	RbClO$_3$	13446-71-4	168.919	col cry	324	dec 480	3.19	6.63^{25}	sl H$_2$O
2311	Rubidium chloride	RbCl	7791-11-9	120.921	wh cub cry; hyg	724	1390	2.76	93.9^{25}	sl EtOH
2312	Rubidium chromate	Rb$_2$CrO$_4$	13446-72-5	286.930	yel rhom cry			3.518	76.2^{25}	
2313	Rubidium dichromate	Rb$_2$Cr$_2$O$_7$	13446-73-6	386.924	red tricl or yel monocl cry			3.1		s H$_2$O
2314	Rubidium cyanide	RbCN	19073-56-4	111.486	wh cub cry			2.3		s H$_2$O; i EtOH, eth
2315	Rubidium fluoride	RbF	13446-74-7	104.466	wh cub cry; hyg	795	1410	3.2	300^{20}	i EtOH
2316	Rubidium fluoroborate	RbBF$_4$	18909-68-7	172.273	orth cry	612 dec		2.82		sl H$_2$O
2317	Rubidium formate	RbCHO$_2$	3495-35-0	130.486	wh hyg cry	dec				
2318	Rubidium hexafluorogermanate	Rb$_2$GeF$_6$	16962-48-4	357.57	wh cry	696				s H$_2$O
2319	Rubidium hydride	RbH	13446-75-8	86.476	wh cub cry; flam	≈170 dec		2.60		reac H$_2$O
2320	Rubidium hydrogen carbonate	RbHCO$_3$	19088-74-5	146.485	wh rhomb cry	175 dec			116^{20}	
2321	Rubidium hydrogen fluoride	RbHF$_2$	12280-64-7	124.473	tetr cry	188		3.3		
2322	Rubidium hydrogen sulfate	RbHSO$_4$	15587-72-1	182.539	col monocl cry	208		2.9		s H$_2$O
2323	Rubidium hydroxide	RbOH	1310-82-3	102.475	gray-wh orth cry; hyg	385		3.2	173^{30}	s EtOH
2324	Rubidium iodate	RbIO$_3$	13446-76-9	260.370	monocl or cub cry	dec		4.33	2.44^{25}	vs HCl
2325	Rubidium iodide	RbI	7790-29-6	212.372	wh cub cry	656	1300	3.55	165^{25}	s EtOH
2326	Rubidium molybdate	Rb$_2$MoO$_4$	13718-22-4	330.87	wh cry	958				s H$_2$O
2327	Rubidium nitrate	RbNO$_3$	13126-12-0	147.473	wh hex cry; hyg	310		3.11	65.0^{25}	vs H$_2$O
2328	Rubidium nitrite	RbNO$_2$	13825-25-7	131.474	wh cry	422				vs H$_2$O
2329	Rubidium oxide	Rb$_2$O	18088-11-4	186.935	yel-brn cub cry; hyg	400 dec		4.0		reac H$_2$O
2330	Rubidium perchlorate	RbClO$_4$	13510-42-4	184.919	wh hyg cry	597	dec 900	2.9	1.5^{25}	
2331	Rubidium permanganate	RbMnO$_4$	13465-49-1	204.404	dark purp cry	300 dec		3.24		sl H$_2$O
2332	Rubidium peroxide	Rb$_2$O$_2$	23611-30-5	202.935	wh orth cry	570		3.8		reac H$_2$O
2333	Rubidium selenide	Rb$_2$Se	31052-43-4	249.90	wh cub cry	733		3.22		reac H$_2$O

No.	Name	Formula	CAS Reg No.	Mol. weight	Physical form	mp/°C	bp/°C	Density g cm⁻³	Solubility g/100 g H₂O	Qualitative solubility
2334	Rubidium sulfate	Rb_2SO_4	7488-54-2	266.999	wh orth cry	1066		3.61	50.8[25]	
2335	Rubidium sulfide	Rb_2S	31083-74-6	203.001	wh cub cry	425		2.91		s H_2O
2336	Rubidium superoxide	RbO_2	12137-25-6	117.467	tetr cry	412		≈3.0		
2337	Ruthenium	Ru	7440-18-8	101.07	silv-wh metal; hex	2334	4150	12.1		i acid, aqua regia
2338	Ruthenium dodecacarbonyl	$Ru_3(CO)_{12}$	15243-33-1	639.33	oran cry	150 dec				
2339	Ruthenium pentacarbonyl	$Ru(CO)_5$	16406-48-7	241.12	col liq	-22	dec 50			i H_2O; s EtOH, bz, chl, hc
2340	Ruthenium nonacarbonyl	$Ru_2(CO)_9$	63128-11-0	454.23	stab below -40					s hex
2341	Ruthenium nitrosyl chloride monohydrate	$Ru(NO)Cl_3 \cdot H_2O$	18902-42-6	255.45	hyg red cry					
2342	Hexaammineruthenium(III) chloride	$Ru(NH_3)_6Cl_3$	14282-91-8	309.61	col monocl cry					s H_2O
2343	Ruthenium(III) bromide	$RuBr_3$	14014-88-1	340.78	brn hex cry	dec 500		5.3		i H_2O, acid, EtOH
2344	Ruthenium(III) chloride	$RuCl_3$	10049-08-8	207.43	blk-brn hex cry	≈500 dec		3.1		i H_2O; sl EtOH
2345	Ruthenium(III) fluoride	RuF_3	51621-05-7	158.07	brn rhomb cry	≈600 dec		5.36		i H_2O, dil acid
2346	Ruthenium(III) iodide	RuI_3	13896-65-6	481.78	blk hex cry	dec 300		6.0		sl H_2O
2347	Ruthenium(III) 2,4-pentanedioate	$Ru(CH_3COCHCOCH_3)_3$	14284-93-6	398.39	red-brn cry	230				
2348	Ruthenium(IV) fluoride	RuF_4	71500-16-8	177.06	yel-red cry					reac H_2O
2349	Ruthenium(IV) oxide	RuO_2	12036-10-1	133.07	gray-blk tetr cry	dec 1300		7.05		i H_2O, acid
2350	Ruthenium(V) fluoride	RuF_5	14521-18-7	196.06	grn monocl cry	86.5	227	3.90		
2351	Ruthenium(VI) fluoride	RuF_6	13693-08-8	215.06	dark brn orth cry	54	200 dec	3.54		reac H_2O
2352	Ruthenium(VIII) oxide	RuO_4	20427-56-9	165.07	yel monocl prisms	25.4	40	3.29	2.03[20]	sl H_2O; vs ctc; reac EtOH
2353	Samarium	Sm	7440-19-9	150.36	silv metal; rhomb	1072	1794	7.52		
2354	Samarium boride	SmB_6	12008-30-9	215.23	refrac solid	2580		5.07		
2355	Samarium silicide	$SmSi_2$	12300-22-0	206.53	orth cry			5.14		
2356	Samarium(II) bromide	$SmBr_2$	50801-97-3	310.17	brn cry	669				reac H_2O
2357	Samarium(II) chloride	$SmCl_2$	13874-75-4	221.27	brn cry	855		3.69		reac H_2O
2358	Samarium(II) fluoride	SmF_2	15192-17-3	188.36	purp cry					reac H_2O
2359	Samarium(II) iodide	SmI_2	32248-43-4	404.17	grn cry	520				reac H_2O
2360	Samarium(III) acetate trihydrate	$Sm(C_2H_3O_2)_3 \cdot 3H_2O$	17829-86-6	381.54	hyg yel-wh solid			1.94		s H_2O
2361	Samarium(III) bromate nonahydrate	$Sm(BrO_3)_3 \cdot 9H_2O$	63427-22-5	696.20	pink hex cry	75 dec				vs H_2O; sl EtOH
2362	Samarium(III) bromide	$SmBr_3$	13759-87-0	390.07	yel cry	640				reac H_2O
2363	Samarium(III) carbonate	$Sm_2(CO_3)_3$	5895-47-6	480.75	wh-yel pow	dec >500				
2364	Samarium(III) chloride	$SmCl_3$	10361-82-7	256.72	yel cry	682		4.46	93.8[25]	
2365	Samarium(III) chloride hexahydrate	$SmCl_3 \cdot 6H_2O$	13465-55-9	364.81	yel cry	dec		2.38	93.8[25]	
2366	Samarium(III) fluoride	SmF_3	13765-24-7	207.36	wh cry	1306				reac H_2O
2367	Samarium(III) iodide	SmI_3	13813-25-7	531.07	oran cry	850				reac H_2O
2368	Samarium(III) nitrate	$Sm(NO_3)_3$	10361-83-8	336.38	yel-wh hyg solid				144[25]	s EtOH
2369	Samarium(III) nitrate hexahydrate	$Sm(NO_3)_3 \cdot 6H_2O$	13759-83-6	444.47	pale yel cry	78				s H_2O, MeOH, ace
2370	Samarium(III) oxide	Sm_2O_3	12060-58-1	348.72	yel-wh cub cry	2269	3780	7.6		
2371	Samarium(III) sulfate octahydrate	$Sm_2(SO_4)_3 \cdot 8H_2O$	13465-58-2	733.03	yel cry			2.93	2.67[20]	
2372	Samarium(III) sulfide	Sm_2S_3	12067-22-0	396.92	gray-brn cub cry	1720		5.87		
2373	Samarium(III) telluride	Sm_2Te_3	12040-00-5	683.52	orth cry			7.31		
2374	Scandium	Sc	7440-20-2	44.956	silv metal; hex	1541	2836	2.99		
2375	Scandium boride	ScB_2	12007-34-0	66.578	refrac solid	2250		3.17		
2376	Scandium bromide	$ScBr_3$	13465-59-3	284.668	wh hyg cry	969		9.33		s H_2O
2377	Scandium chloride	$ScCl_3$	10361-84-9	151.315	wh hyg cry	967		2.4		s H_2O; i EtOH
2378	Scandium fluoride	ScF_3	13709-47-2	101.951	wh powder	1552				sl H_2O
2379	Scandium hydroxide	$Sc(OH)_3$	17674-34-9	95.978	col amorp solid					i H_2O; s dil acid
2380	Scandium iodide	ScI_3	14474-33-0	425.669	hyg yel cry	953	subl			s H_2O, EtOH, CCD
2381	Scandium nitrate	$Sc(NO_3)_3$	13465-60-6	230.971	wh cry				169[25]	s EtOH
2382	Scandium oxide	Sc_2O_3	12060-08-1	137.910	wh cub cry	2489		3.864		s conc acid
2383	Scandium sulfate pentahydrate	$Sc_2(SO_4)_3 \cdot 5H_2O$	15292-44-1	468.176	col cry	dec 110				vs H_2O
2384	Scandium sulfide	Sc_2S_3	12166-29-9	186.107	yel orth cry	1775		2.91		
2385	Scandium telluride	Sc_2Te_3	12166-44-8	472.71	blk hex cry			5.29		
2386	Selenium (gray)	Se	7782-49-2	78.96	gray metallic cry; hex	220.8	685	4.809		i H_2O, CS_2
2387	Selenium (α form)	Se	7782-49-2	78.96	red monocl cry	trans gray Se >120	685	4.39		i H_2O, EtOH; sl eth
2388	Selenium (vitreous)	Se	7782-49-2	78.96	blk amorp solid	trans gray Se 180	685	4.28		i H_2O; sl CS_2
2389	Selenic acid	H_2SeO_4	7783-08-6	144.97	wh hyg solid	58	260 dec	2.95		vs H_2O; reac EtOH
2390	Pentafluoroorthoselenic acid	$HOSeF_5$	38989-47-8	190.96	col solid	38	47			
2391	Selenous acid	H_2SeO_3	7783-00-8	128.97	wh hyg cry	70 dec		3.0		vs H_2O; s EtOH
2392	Selenium dioxide	SeO_2	7446-08-4	110.96	wh tetr needles or powder	340 tp	315 sp	3.95	264[22]	s EtOH, MeOH; sl ace
2393	Selenium trioxide	SeO_3	13768-86-0	126.96	wh tetr cry; hyg	118	subl	3.44		s H_2O, os

No.	Name	Formula	CAS Reg No.	Mol. weight	Physical form	mp/°C	bp/°C	Density g cm^{-3}	Solubility g/100 g H$_2$O	Qualitative solubility
2394	Selenium bromide	Se_2Br_2	7789-52-8	317.73	red liq	5	225 dec	3.60		reac H$_2$O; s CS$_2$, chl
2395	Selenium chloride	Se_2Cl_2	10025-68-0	228.83	yel-brn oily liq	-85	127 dec	2.774		reac H$_2$O; s CS$_2$, bz ctc, chl
2396	Selenium tetrabromide	$SeBr_4$	7789-65-3	398.58	oran-red cry	123				reac H$_2$O; s CS$_2$, chl
2397	Selenium tetrachloride	$SeCl_4$	10026-03-6	220.77	wh-yel cry	305 tp	191.4 sp	2.6		reac H$_2$O
2398	Selenium tetrafluoride	SeF_4	13465-66-2	154.95	col liq	-9.5	101.6	2.75		reac H$_2$O; vs EtOH, eth
2399	Selenium hexafluoride	SeF_6	7783-79-1	192.95	col gas	-34.6 tp	-46.6 sp	7.887 g/L		i H$_2$O
2400	Selenium chloride pentafluoride	SeF_5Cl	34979-62-9	209.41	col gas	-19	4.5			
2401	Selenium oxybromide	$SeOBr_2$	7789-51-7	254.77	red-yel solid	41.6	220 dec	3.38		reac H$_2$O; s CS$_2$, bz ctc
2402	Selenium oxychloride	$SeOCl_2$	7791-23-3	165.87	col or yel liq	8.5	177	2.44		reac H$_2$O; s ctc, chl bz, tol
2403	Selenium oxyfluoride	$SeOF_2$	7783-43-9	132.96	col liq	15	125	2.8		reac H$_2$O
2404	Selenium oxytetrafluoride	$SeOF_4$	53319-44-1	170.95	unstab col liq	12	65			
2405	Selenium dioxydifluoride	SeO_2F_2	14984-81-7	148.96	col gas	-99.5	-8.4	6.089 g/L		reac H$_2$O
2406	Selenium monosulfide	SeS	7446-34-6	111.03						
2407	Selenium disulfide	SeS_2	7488-56-4	143.09	red-yel cry	100				i H$_2$O; s acid
2408	Selenium sulfide (Se$_2$S$_6$)	Se_2S_6	75926-26-0	350.31	oran needles	121.5		2.44		s CS$_2$; sl bz
2409	Selenium sulfide (Se$_4$S$_4$)	Se_4S_4	75926-28-2	444.10	red cry	113 dec		3.29		s bz; sl CS$_2$
2410	Selenium sulfide (Se$_6$S$_2$)	Se_6S_2	75926-30-6	537.89	oran cry	121.5				s CS$_2$
2411	Silicon	Si	7440-21-3	28.086	gray cry or brn amorp solid	1414	3265	2.3296		i H$_2$O, acid; s alk
2412	Silane	SiH_4	7803-62-5	32.118	col gas; flam	-185	-111.9	1.313 g/L		reac H$_2$O; i EtOH, bz
2413	Disilane	Si_2H_6	1590-87-0	62.219	col gas; flam	-129.4	-14.8	2.543 g/L		reac H$_2$O, ctc, chl; s EtOH, bz
2414	Trisilane	Si_3H_8	7783-26-8	92.321	flam col liq	-117.4	52.9	0.739		reac H$_2$O
2415	Tetrasilane	Si_4H_{10}	7783-29-1	122.421	col liq; flam	-89.9	108.1	0.792		reac H$_2$O
2416	2-Silyltrisilane	Si_4H_{10}	13597-87-0	122.421	col liq	-99.4	101.7	0.792		reac H$_2$O
2417	Cyclopentasilane	Si_5H_{10}	289-22-5	150.507	col liq	-10.5	194.3	0.963		reac H$_2$O
2418	Pentasilane	Si_5H_{12}	14868-53-2	152.523	col liq	-72.8	153.2	0.827		reac H$_2$O
2419	2-Silyltetrasilane	Si_5H_{12}	14868-54-3	152.523	col liq	-109.9	146.2	0.820		reac H$_2$O
2420	2,2-Disilyltrisilane	Si_5H_{12}	15947-57-6	152.523	col liq	-57.8	134.3	0.815		reac H$_2$O
2421	Cyclohexasilane	Si_6H_{12}	291-59-8	180.608	col liq	16.5	226			reac H$_2$O
2422	Hexasilane	Si_6H_{14}	14693-61-9	182.624	col liq	-44.7	193.6	0.847		reac H$_2$O
2423	2-Silylpentasilane	Si_6H_{14}	14868-55-4	182.624	col liq	-78.4	185.2	0.840		
2424	3-Silylpentasilane	Si_6H_{14}	14868-55-4	182.624	col liq	-69	179.5	0.843		reac H$_2$O
2425	Heptasilane	Si_7H_{16}	14693-65-3	212.726	col liq	-30.1	226.8	0.859		reac H$_2$O
2426	Bromosilane	SiH_3Br	13465-73-1	111.014	col gas	-94	1.9	4.538 g/L		
2427	Dibromosilane	SiH_2Br_2	13768-94-0	189.910	liq	-70.1	66			
2428	Tribromosilane	$SiHBr_3$	7789-57-3	268.806	flam liq	-73	109	2.7		reac H$_2$O
2429	Tetrabromosilane	$SiBr_4$	7789-66-4	347.702	col fuming liq	5.39	154	2.8		reac H$_2$O
2430	Bromotrichlorosilane	$SiBrCl_3$	13465-74-2	214.349	col liq	-62	80.3	1.826		reac H$_2$O
2431	Dibromodichlorosilane	$SiBr_2Cl_2$	13465-75-3	258.800	col liq	-45.5	104	2.172		reac H$_2$O
2432	Tribromochlorosilane	$SiBr_3Cl$	13465-76-4	303.251	col liq	-20.8	127	2.497		reac H$_2$O
2433	Hexabromosilane	Si_2Br_6	13517-13-0	535.595	col cry	95	265			
2434	Octabromotrisilane	Si_3Br_8	54804-32-9	723.489	col liq	46				
2435	Chlorosilane	SiH_3Cl	13465-78-6	66.563	col gas	-118	-30.4	2.721 g/L		
2436	Dichlorosilane	SiH_2Cl_2	4109-96-0	101.008	col gas; flam	-122	8.3	4.129 g/L		reac H$_2$O
2437	Trichlorosilane	$SiHCl_3$	10025-78-2	135.453	fuming liq	-128.2	33	1.331		reac H$_2$O
2438	Tetrachlorosilane	$SiCl_4$	10026-04-7	169.898	col fuming liq	-68.74	57.65	1.5		reac H$_2$O
2439	Chlorotrifluorosilane	$SiClF_3$	14049-36-6	120.534	col gas	-138	-70.0	4.927 g/L		reac H$_2$O
2440	Dichlorodifluorosilane	$SiCl_2F_2$	18356-71-3	136.989	col gas	-44	-32	5.599 g/L		reac H$_2$O
2441	Trichlorofluorosilane	$SiCl_3F$	14965-52-7	153.443	col gas		12.25	6.272 g/L		
2442	Trichloroiodosilane	$SiCl_3I$	13465-85-5	261.349	col liq	-60	113.5			reac H$_2$O
2443	Hexachlorodisilane	Si_2Cl_6	13465-77-5	268.889	col liq	2.5	146			reac H$_2$O
2444	Octachlorotrisilane	Si_3Cl_8	13596-23-1	367.881	col liq	-67	216			
2445	Fluorosilane	SiH_3F	13537-33-2	50.108	col gas		-98.6	2.048 g/L		
2446	Difluorosilane	SiH_2F_2	13824-36-7	68.099	col gas	-122	-77.8	2.783 g/L		
2447	Trifluorosilane	$SiHF_3$	13465-71-9	86.089	col gas	-131	-95	3.519 g/L		
2448	Tetrafluorosilane	SiF_4	7783-61-1	104.080	col gas	-90.2	-86	4.254 g/L		reac H$_2$O
2449	Hexafluorodisilane	Si_2F_6	13830-68-7	170.161	col gas	-18.7 tp @ 780 mmHg	-19.1 sp			reac H$_2$O
2450	Octafluorotrisilane	Si_3F_8	14521-14-3	236.244	col liq	-1.2	42			
2451	Decafluorotetrasilane	Si_4F_{10}	14521-15-4	302.326	col cry	68	85.1			
2452	Iodosilane	SiH_3I	13598-42-0	158.014	col liq	-57	45.6			

No.	Name	Formula	CAS Reg No.	Mol. weight	Physical form	mp/°C	bp/°C	Density g cm⁻³	Solubility g/100 g H₂0	Qualitative solubility
2453	Diiodosilane	SiH_2I_2	13760-02-6	283.911	col liq	-1	150			
2454	Triiodosilane	$SiHI_3$	13465-72-0	409.807	liq	8	220 dec			
2455	Tetraiodosilane	SiI_4	13465-84-4	535.704	wh powder	120.5	287.35	4.1		
2456	Hexaiododisilane	Si_2I_6	13510-43-5	817.598	pale yel cry	250				
2457	Disiloxane	$(SiH_3)_2O$	13597-73-4	78.218	gas	-144	-15.2	3.197 g/L		
2458	Hexachlorodisiloxane	$(SiCl_3)_2O$	14986-21-1	284.888	liq	-28	137			
2459	Methylsilane	SiH_3CH_3	992-94-9	46.145	col gas	-156.5	-57.5			
2460	Metasilicic acid	H_2SiO_3	7699-41-4	78.100	wh amorp powder					i H_2O; s HF
2461	Orthosilicic acid	H_4SiO_4	10193-36-9	96.116	exists only in soln					
2462	Fluorosilicic acid	H_2SiF_6	16961-83-4	144.092	stab only in aq soln					s H_2O
2463	Silicon carbide (hexagonal)	SiC	409-21-2	40.097	hard grn-black hex cry	2830		3.16		i H_2O, EtOH
2464	Silicon nitride (Si_3N_4)	Si_3N_4	12033-89-5	140.284	gray refrac solid; hex	1900		3.17		
2465	Silicon monoxide	SiO	10097-28-6	44.085	blk cub cry, stable >1200			2.18		
2466	Silicon dioxide (α-quartz)	SiO_2	14808-60-7	60.085	col hex cry	trans to beta quartz 573	2950	2.648		i H_2O, acid; s HF
2467	Silicon dioxide (β-quartz)	SiO_2	14808-60-7	60.085	col hex cry	trans to tridymite 867	2950	2.533[600]		i H_2O, acid; s HF
2468	Silicon dioxide (tridymite)	SiO_2	15468-32-3	60.085	col hex cry	trans cristobalite 1470	2950	2.265		i H_2O, acid; s HF
2469	Silicon dioxide (cristobalite)	SiO_2	14464-46-1	60.085	col hex cry	1722	2950	2.334		i H_2O, acid; s HF
2470	Silicon dioxide (vitreous)	SiO_2	60676-86-0	60.085	col amorp solid	1713	2950	2.196		i H_2O, acid; s HF
2471	Silicon monosulfide	SiS	12504-41-5	60.151	yel-red hyg powder	1090	940	1.85		reac H_2O
2472	Silicon disulfide	SiS_2	13759-10-9	92.216	wh rhomb cry	1090	subl	2.04		reac H_2O, EtOH; i bz
2473	Silicon tetraacetate	$Si(C_2H_3O_2)_4$	562-90-3	264.262	wh hyg cry	110				reac H_2O; s ace, bz
2474	Silicon tetraboride	SiB_4	12007-81-7	71.330	gray refrac solid	1870 dec		2.4		
2475	Silicotungstic acid	$H_4SiO_4 \cdot (W_3O_9)_4$	12520-88-6	2878.17	hyg yel cry					vs H_2O, EtOH
2476	Silver	Ag	7440-22-4	107.868	silv metal; cub	961.78	2162	10.5		
2477	Silver azide	AgN_3	13863-88-2	149.888	orth cry; exp	exp ≈250		4.9	0.00081[20]	
2478	Silver subfluoride	Ag_2F	1302-01-8	234.734	yel hex cry	100 dec		8.6		reac H_2O
2479	Silver(I) acetate	$AgC_2H_3O_2$	563-63-3	166.912	wh needles or powder	dec		3.26	1.04[20]	
2480	Silver(I) acetylide	Ag_2C_2	7659-31-6	239.757	wh powder; exp					
2481	Silver(I) acetylide (AgC_2H)	AgC_2H	13092-75-6	132.897	wh powder; exp					
2482	Silver(I) arsenate	Ag_3AsO_4	13510-44-6	462.524	red cub cry	dec		6.657	0.00085	s NH_4OH
2483	Silver(I) benzoate	$Ag(C_6H_5CO_2)$	532-31-0	228.982	powder				30[20]	
2484	Silver(I) bromate	$AgBrO_3$	7783-89-3	235.770	wh tetr cry	360 dec		5.21	0.193[25]	
2485	Silver(I) bromide	AgBr	7785-23-1	187.772	yel cub cry	430	1502	6.47	0.000014[25]	i H_2O, acid, EtOH
2486	Silver(I) carbonate	Ag_2CO_3	534-16-7	275.745	yel monocl cry	218		6.077	0.0036[20]	s acid
2487	Silver(I) chlorate	$AgClO_3$	7783-92-8	191.319	wh tetr cry	230	270 dec	4.430	17.6[25]	sl EtOH
2488	Silver(I) chloride	AgCl	7783-90-6	143.321	wh cub cry	455	1547	5.56	0.00019[25]	
2489	Silver(I) chlorite	$AgClO_2$	7783-91-7	175.320	yel cry	105 exp			0.55[25]	
2490	Silver(I) chromate	Ag_2CrO_4	7784-01-2	331.730	brn-red monocl cry			5.625	0.000014[0]	
2491	Silver(I) citrate	$Ag_3C_6H_5O_7$	126-45-4	512.705	wh cry powder					i H_2O; s HNO_3
2492	Silver(I) cyanide	AgCN	506-64-9	133.886	wh-gray hex cry	320 dec		3.95	0.0000011	i H_2O, dil acid
2493	Silver(I) dichromate	$Ag_2Cr_2O_7$	7784-02-3	431.724	red cry			4.770		sl H_2O
2494	Silver(I) diethyldithiocarbamate	$Ag(C_2H_5)_2NCS_2$	1470-61-7	256.138	powder	173				s py
2495	Silver(I) fluoride	AgF	7775-41-9	126.866	yel-brn cub cry; hyg	435	1159	5.852	172[20]	
2496	Silver(I) hexafluoroantimonate	$AgSbF_6$	26042-64-8	343.618	powder					
2497	Silver(I) hexafluoroarsenate	$AgAsF_6$	12005-82-2	296.780	powder					
2498	Silver(I) hexafluorophosphate	$AgPF_6$	26042-63-7	252.832	powder	102 dec				
2499	Silver(I) hydrogen fluoride	$AgHF_2$	12249-52-4	146.873	hyg cry	dec				
2500	Silver(I) iodate	$AgIO_3$	7783-97-3	282.770	wh orth cry	>200		5.53	0.053[25]	
2501	Silver(I) iodide	AgI	7783-96-2	234.772	yel powder; hex	558	1506	5.68	0.000003	i acid
2502	Silver(I) lactate monohydrate	$AgC_3H_5O_3 \cdot H_2O$	128-00-7	214.954	gray cry powder					sl H_2O, EtOH
2503	Silver(I) metaphosphate	$AgPO_3$	13465-96-8	186.840	grn glass	490		6.37		i H_2O; s HNO_3, NH_4OH
2504	Silver(I) molybdate	Ag_2MoO_4	13765-74-7	375.67	yel cub cry	483		6.18		sl H_2O
2505	Silver(I) nitrate	$AgNO_3$	7761-88-8	169.873	col rhomb cry	210	440 dec	4.35	234[25]	sl EtOH, ace
2506	Silver(I) nitrite	$AgNO_2$	7783-99-5	153.874	yel needles	140 dec		4.453	0.415[25]	i EtOH; reac acid
2507	Silver(I) oxalate	$Ag_2C_2O_4$	533-51-7	303.755	wh cry powder	exp 140		5.03	0.0043[20]	

No.	Name	Formula	CAS Reg No.	Mol. weight	Physical form	mp/°C	bp/°C	Density g cm^{-3}	Solubility g/100 g H$_2$O	Qualitative solubility
2508	Silver(I) oxide	Ag$_2$O	20667-12-3	231.735	brn-blk cub cry	≈200 dec		7.2	0.0025	i EtOH; s acid, alk
2509	Silver(I) perchlorate	AgClO$_4$	7783-93-9	207.319	col cub cry; hyg	486 dec		2.806	558[25]	s bz, py, os
2510	Silver(I) perchlorate monohydrate	AgClO$_4$ · H$_2$O	14242-05-8	225.334	hyg wh cry	43 dec			558[25]	
2511	Silver(I) permanganate	AgMnO$_4$	7783-98-4	226.804	viol monocl cry	dec		4.49	0.91[18]	reac EtOH
2512	Silver(I) phosphate	Ag$_3$PO$_4$	7784-09-0	418.576	yel powder	849		6.37	0.0064	sl dil acid
2513	Silver(I) picrate monohydrate	AgC$_6$H$_2$N$_3$O$_7$ · H$_2$O	146-84-9	353.979	yel cry					sl H$_2$O, EtOH; i chl, eth
2514	Silver(I) selenate	Ag$_2$SeO$_4$	7784-07-8	358.69	orth cry			5.72	0.118[20]	
2515	Silver(I) selenide	Ag$_2$Se	1302-09-6	294.70	gray hex needles	880		8.216		i H$_2$O
2516	Silver(I) selenite	Ag$_2$SeO$_3$	7784-05-6	342.69	needles	530	>550 dec	5.930		sl H$_2$O; s acid
2517	Silver(I) sulfate	Ag$_2$SO$_4$	10294-26-5	311.799	col cry or powder	660		5.45	0.84[25]	
2518	Silver(I) sulfide	Ag$_2$S	21548-73-2	247.801	gray-blk orth powder	825 (high press.)		7.23		i H$_2$O; s acid
2519	Silver(I) sulfite	Ag$_2$SO$_3$	13465-98-0	295.799	wh cry	100 dec			0.00046[20]	s acid, NH$_4$OH
2520	Silver(I) telluride	Ag$_2$Te	12002-99-2	343.34	blk orth cry	955		8.4		
2521	Silver(I) tetraiodomercurate(II)	Ag$_2$HgI$_4$	7784-03-4	923.94	yel tetr cry	trans to red cub ≈40		6.1		i H$_2$O, dil acid
2522	Silver(I) thiocyanate	AgSCN	1701-93-5	165.951	wh powder	dec				i H$_2$O
2523	Silver(I) thiosulfate	Ag$_2$S$_2$O$_3$	23149-52-2	327.864	wh cry	dec				sl H$_2$O; s NH$_4$OH
2524	Silver(II) oxide	AgO	1301-96-8	123.867	gray powder; monocl or cub	>100 dec		7.5	0.0027[25]	s alk; reac acid
2525	Silver(I) tungstate	Ag$_2$WO$_4$	13465-93-5	463.57	yel cry	620			0.015	s HNO$_3$, NH$_4$OH
2526	Silver(II) fluoride	AgF$_2$	7783-95-1	145.865	wh or gray hyg cry	690		4.58		reac H$_2$O
2527	Silver(II) oxide (Ag$_2$O$_2$)	Ag$_2$O$_2$	25455-73-6	247.735	gray-blk cub cry	>100		7.44		i H$_2$O; s acid, NH$_4$OH
2528	Sodium	Na	7440-23-5	22.990	soft silv met; cub	97.794	882.940	0.97		reac H$_2$O
2529	Sodium acetate	NaC$_2$H$_3$O$_2$	127-09-3	82.034	col cry	328.2		1.528	50.4[25]	
2530	Sodium acetate trihydrate	NaC$_2$H$_3$O$_2$ · 3H$_2$O	6131-90-4	136.079	col cry	58 dec		1.45	50.4[25]	sl EtOH
2531	Sodium aluminate	NaAlO$_2$	1302-42-7	81.971	wh orth cry; hyg	1650		4.63		vs H$_2$O; i EtOH
2532	Sodium aluminum hydride	NaAlH$_4$	13770-96-2	54.004	wh hyg solid	174 dec		1.24		i eth; s thf
2533	Sodium aluminum sulfate dodecahydrate	NaAl(SO$_4$)$_2$ · 12H$_2$O	10102-71-3	458.281	col cry	≈60		1.61	39.7[20]	i EtOH
2534	Sodium amide	NaNH$_2$	7782-92-5	39.013	wh-grn orth cry	210	500 dec	1.39		reac H$_2$O
2535	Sodium ammonium phosphate tetrahydrate	NaNH$_4$HPO$_4$ · 4H$_2$O	13011-54-6	209.069	monocl cry	≈80 dec		1.54		s H$_2$O; i EtOH
2536	Sodium arsenate dodecahydrate	Na$_3$AsO$_4$ · 12H$_2$O	7778-43-0	424.072	col monocl prism	86 dec				s H$_2$O; sl EtOH; i eth
2537	Sodium arsenite	NaAsO$_2$	7784-46-5	129.911	wh-gray hyg powder			1.87		vs H$_2$O; i EtOH
2538	Sodium azide	NaN$_3$	26628-22-8	65.010	col hex cry	300 dec		1.846	40.8[20]	sl EtOH; i eth
2539	Sodium borohydride	NaBH$_4$	16940-66-2	37.833	wh cub cry; hyg	≈400 dec		1.07	55[20]	reac EtOH
2540	Sodium bromate	NaBrO$_3$	7789-38-0	150.892	col cub cry	381		3.34	39.4[25]	i EtOH
2541	Sodium bromide	NaBr	7647-15-6	102.894	wh cub cry	747	1390	3.200	94.6[25]	s EtOH
2542	Sodium bromide dihydrate	NaBr · 2H$_2$O	13466-08-5	138.925	wh cry	36 dec		2.18	94.6[25]	sl EtOH
2543	Sodium carbonate	Na$_2$CO$_3$	497-19-8	105.989	wh hyg powder	856		2.54	30.7[25]	i EtOH
2544	Sodium carbonate monohydrate	Na$_2$CO$_3$ · H$_2$O	5968-11-6	124.005	col orth cry	100 dec		2.25	30.7[25]	i EtOH
2545	Sodium carbonate decahydrate	Na$_2$CO$_3$ · 10H$_2$O	6132-02-1	286.142	col cry	34 dec		1.46	30.7[25]	i EtOH
2546	Sodium chlorate	NaClO$_3$	7775-09-9	106.441	col cub cry	248	dec 630	2.5	100[25]	sl EtOH
2547	Sodium chloride	NaCl	7647-14-5	58.443	col cub cry	800.7	1465	2.17	36.0[25]	sl EtOH
2548	Sodium chlorite	NaClO$_2$	7758-19-2	90.442	wh hyg cry	≈180 dec			64[17]	
2549	Sodium chromate	Na$_2$CrO$_4$	7775-11-3	161.974	yel orth cry	794		2.72	87.6[25]	sl EtOH
2550	Sodium chromate tetrahydrate	Na$_2$CrO$_4$ · 4H$_2$O	10034-82-9	234.035	yel hyg cry	dec			87.6[25]	sl EtOH
2551	Sodium citrate dihydrate	Na$_3$C$_6$H$_5$O$_7$ · 2H$_2$O	6132-04-3	294.099	wh cry	150 dec				vs H$_2$O; i EtOH, eth
2552	Sodium citrate pentahydrate	Na$_3$C$_6$H$_5$O$_7$ · 5H$_2$O	6858-44-2	348.145	hyg col cry	dec 150		1.86	92[25]	vs H$_2$O; sl EtOH; i eth
2553	Sodium cyanate	NaCNO	917-61-3	65.007	col needles	550		1.89		s H$_2$O; sl EtOH; i eth
2554	Sodium cyanide	NaCN	143-33-9	49.008	wh cub cry; hyg	562		1.6	58.2[20]	sl EtOH
2555	Sodium cyanoborohydride	NaBH$_3$(CN)	25895-60-7	62.843	wh hyg powder	240 dec		1.12		vs H$_2$O; s thf; sl EtOH; i bz, eth
2556	Sodium dichromate	Na$_2$Cr$_2$O$_7$	10588-01-9	261.968	red hyg cry	357	400 dec		187[25]	
2557	Sodium dichromate dihydrate	Na$_2$Cr$_2$O$_7$ · 2H$_2$O	7789-12-0	297.999	oran-red monocl cry	85 dec		2.35		vs H$_2$O; S HOAc
2558	Sodium dihydrogen phosphate	NaH$_2$PO$_4$	7558-80-7	119.977	col monocl cry	200 dec			94.9[25]	
2559	Sodium dihydrogen phosphate monohydrate	NaH$_2$PO$_4$ · H$_2$O	10049-21-5	137.993	wh hyg cry	100 dec			94.9[25]	i EtOH
2560	Sodium dihydrogen phosphate dihydrate	NaH$_2$PO$_4$ · 2H$_2$O	13472-35-0	156.008	col orth cry	60 dec		1.91	94.9[25]	i EtOH
2561	Sodium dihydrogen hypophosphate hexahydrate	Na$_2$H$_2$P$_2$O$_6$ · 6H$_2$O	7782-95-8	314.031	monocl plates	110 dec		1.849	2.0[25]	i EtOH

No.	Name	Formula	CAS Reg No.	Mol. weight	Physical form	mp/°C	bp/°C	Density g cm^{-3}	Solubility g/100 g H$_2$O	Qualitative solubility
2562	Sodium dihydrogen pyrophosphate	Na$_2$H$_2$P$_2$O$_7$	7758-16-9	221.939	wh powder	220 dec		≈1.9		s H$_2$O
2563	Sodium dithionate	Na$_2$S$_2$O$_4$	7775-14-6	174.108	gray-wh powder	52 dec			24.1[20]	sl EtOH
2564	Sodium dithionate dihydrate	Na$_2$S$_2$O$_6$ · 2H$_2$O	7631-94-9*	242.137	col orth cry	110 dec		2.19	15.1[20]	i EtOH
2565	Sodium ethanolate	NaC$_2$H$_5$O	141-52-6	68.050	wh-yel hyg powder	260 dec				reac H$_2$O; s EtOH
2566	Sodium ferricyanide monohydrate	Na$_3$Fe(CN)$_6$ · H$_2$O	14217-21-1*	298.933	red hyg cry					s H$_2$O; i EtOH
2567	Sodium ferrocyanide decahydrate	Na$_4$Fe(CN)$_6$ · 10H$_2$O	13601-19-9	484.061	yel monocl cry	≈50 dec		1.46	20[20]	i os
2568	Sodium fluoride	NaF	7681-49-4	41.988	col cub or tetr cry	996	1704	2.78	4.13[25]	i EtOH
2569	Sodium fluorophosphate	Na$_2$PO$_3$F	10163-15-2	143.950	powder					
2570	Sodium formate	NaCHO$_2$	141-53-7	68.008	wh hyg cry	257.3	dec	1.92	94.9[25]	sl EtOH
2571	Sodium germanate	Na$_2$GeO$_3$	12025-19-3	166.62	wh monocl hyg cry	1083		3.31		
2572	Sodium gold cyanide	NaAu(CN)$_2$	15280-09-8	271.991	wh-yel cry pow					s H$_2$O, NH$_4$OH
2573	Sodium gold thiosulfate dihydrate	Na$_3$Au(S$_2$O$_3$)$_2$ · 5H$_2$O	10233-88-2	526.223	wh needles or prisms	dec 150		3.09		vs H$_2$O; i EtOH
2574	Sodium hexabromoplatinate(IV) hexahydrate	Na$_2$PtBr$_6$ · 6H$_2$O	39277-13-9	828.579	cry					
2575	Sodium hexachloroiridate(IV) hexahydrate	Na$_2$IrCl$_6$ · 6H$_2$O	19567-78-3	559.006	cry	600 dec				
2576	Sodium hexachloroplatinate(IV)	Na$_2$PtCl$_6$	16923-58-3	453.782	yel hyg cry				53[16]	s EtOH
2577	Sodium hexachloroplatinate(IV) hexahydrate	Na$_2$PtCl$_6$ · 6H$_2$O	16923-58-3	561.873	yel cry	110 dec		2.50	53[16]	s EtOH; i eth
2578	Sodium hexafluoroaluminate	Na$_3$AlF$_6$	13775-53-6	209.941	col monocl cry; trans cub 560	1013		2.97		i H$_2$O
2579	Sodium hexafluoroantimonate	NaSbF$_6$	16925-25-0	258.740	wh cub cry			3.375	129[20]	s EtOH, ace
2580	Sodium hexafluorophosphate monohydrate	NaPF$_6$ · H$_2$O	20644-15-9	185.969	col orth cry			2.369	103[0]	s EtOH, MeOH, ace
2581	Sodium hexafluorosilicate	Na$_2$SiF$_6$	16893-85-9	188.056	wh hex cry	847		2.7	0.67[20]	i EtOH
2582	Sodium hexanitrocobaltate(III)	Na$_3$Co(NO$_2$)$_6$	14649-73-1	403.935	yel-brn cry powder					vs H$_2$O; sl EtOH
2583	Sodium hydride	NaH	7646-69-7	23.998	silv cub cry; flam	425 dec		1.39		reac H$_2$O, EtOH
2584	Sodium hydrogen arsenate	Na$_2$HAsO$_4$	7778-43-0	185.908	wh powder	≈195 dec			51[20]	sl EtOH
2585	Sodium hydrogen arsenate heptahydrate	Na$_2$HAsO$_4$ · 7H$_2$O	10048-95-0	312.014	wh monocl cry	≈50 dec		1.87	51[20]	sl EtOH
2586	Sodium hydrogen carbonate	NaHCO$_3$	144-55-8	84.007	wh monocl cry	≈50 dec		2.20	10.3[25]	i EtOH
2587	Sodium hydrogen fluoride	NaHF$_2$	1333-83-1	61.995	wh hex cry	>160 dec		2.08	3.25[20]	
2588	Sodium hydrogen phosphate	Na$_2$HPO$_4$	7558-79-4	141.959	wh hyg powder			1.7	11.8[25]	
2589	Sodium hydrogen phosphate heptahydrate	Na$_2$HPO$_4$ · 7H$_2$O	7782-85-6	268.066	col cry			≈1.7	11.8[25]	i EtOH
2590	Sodium hydrogen phosphate dodecahydrate	Na$_2$HPO$_4$ · 12H$_2$O	10039-32-4	358.143	col cry	≈35 dec		≈1.5	11.8[25]	i EtOH
2591	Sodium hydrogen sulfate	NaHSO$_4$	7681-38-1	120.061	wh hyg cry	≈315		2.43	28.5[25]	
2592	Sodium hydrogen sulfate monohydrate	NaHSO$_4$ · H$_2$O	10034-88-5	138.076	wh monocl cry			2.10	28.5[25]	reac EtOH
2593	Sodium hydrogen sulfide	NaHS	16721-80-5	56.063	col rhomb cry	350		1.79		s H$_2$O, EtOh, eth
2594	Sodium hydrogen sulfide dihydrate	NaHS · 2H$_2$O	16721-80-5	92.094	yel hyg needles	55 dec				vs H$_2$O, EtOH, eth
2595	Sodium hydrogen sulfite	NaHSO$_3$	7631-90-5	104.061	wh cry			1.48		s H$_2$O; sl EtOH
2596	Sodium hydroxide	NaOH	1310-73-2	39.997	wh orth cry; hyg	323	1388	2.13	100[25]	s EtOH, MeOH
2597	Sodium hypochlorite	NaClO	7681-52-9	74.442	stab in aq soln	anh form exp			79.9[25]	
2598	Sodium hypochlorite pentahydrate	NaOCl · 5H$_2$O	10022-70-5	164.518	pale grn orth cry	18		1.6		s H$_2$O
2599	Sodium iodate	NaIO$_3$	7681-55-2	197.892	wh orth cry	422		4.28	9.47[25]	i EtOH
2600	Sodium iodide	NaI	7681-82-5	149.894	wh cub cry; hyg	661	1304	3.67	184[25]	s EtOH, ace
2601	Sodium iodide dihydrate	NaI · 2H$_2$O	13517-06-1	185.925	hyg col monocl cry	69 dec		2.45	318[0]	vs H$_2$O
2602	Sodium bismuthate	NaBiO$_3$	12232-99-4	279.968	yel-brn hyg cry					i cold H$_2$O, reac acid
2603	Sodium metabisulfite	Na$_2$S$_2$O$_5$	7681-57-4	190.107	wh cry				66.7[25]	sl EtOH
2604	Sodium metaborate	NaBO$_2$	7775-19-1	65.800	wh hex cry	966	1434	2.46		s H$_2$O
2605	Sodium metasilicate	Na$_2$SiO$_3$	6834-92-0	122.064	wh amorp solid; hyg	1089		2.61		s cold H$_2$O; reac hot H$_2$O; i EtOH
2606	Sodium metasilicate pentahydrate	Na$_2$SiO$_3$ · 5H$_2$O	13517-24-3	212.140	wh pow	72 dec				s H$_2$O
2607	Sodium molybdate	Na$_2$MoO$_4$	7631-95-0	205.92	col cub cry	687		≈3.5	65.0[25]	
2608	Sodium molybdate dihydrate	Na$_2$MoO$_4$ · 2H$_2$O	10102-40-6	241.95	cry powder	100 dec		≈3.5	65.0[25]	
2609	Sodium molybdophosphate	Na$_3$PO$_4$·12MoO$_3$	1313-30-0	1891.20	hyg solid			2.83		vs H$_2$O, EtOH
2610	Sodium niobate	NaNbO$_3$	12034-09-2	163.894	rhom cry	1422		4.55		i H$_2$O
2611	Sodium nitrate	NaNO$_3$	7631-99-4	84.995	col hex cry; hyg	306.5		2.261	91.2[25]	sl EtOH, MeOH
2612	Sodium nitrite	NaNO$_2$	7632-00-0	68.996	wh orth cry; hyg	284	>320 dec	2.17	84.8[25]	sl EtOH; reac acid
2613	Sodium nitroferricyanide dihydrate	Na$_2$[Fe(CN)$_5$NO] · 2H$_2$O	13755-38-9	297.949	red cry			1.72	40[16]	sl EtOH
2614	Sodium orthovanadate	Na$_3$VO$_4$	13721-39-6	183.909	col hex prisms	860				s H$_2$O; i EtOH
2615	Sodium oxalate	Na$_2$C$_2$O$_4$	62-76-0	133.999	wh powder	≈250 dec		2.34	3.61[25]	i EtOH

No.	Name	Formula	CAS Reg No.	Mol. weight	Physical form	mp/°C	bp/°C	Density g cm^{-3}	Solubility g/100 g H$_2$O	Qualitative solubility
2616	Sodium oxide	Na$_2$O	1313-59-3	61.979	wh amorp powder	1134		2.27		reac H$_2$O
2617	Sodium perborate tetrahydrate	NaBO$_3$ · 4H$_2$O	7632-04-4	153.861	wh cry	60 dec				reac H$_2$O
2618	Sodium perchlorate	NaClO$_4$	7601-89-0	122.441	wh orth cry; hyg	482 dec		2.52	205^{25}	
2619	Sodium perchlorate monohydrate	NaClO$_4$ · H$_2$O	7791-07-3	140.456	wh hyg cry	≈130 dec		2.02	205^{25}	
2620	Sodium periodate	NaIO$_4$	7790-28-5	213.892	wh tetr cry	≈300 dec		3.86	14.4^{25}	s acid
2621	Sodium periodate trihydrate	NaIO$_4$ · 3H$_2$O	13472-31-6	267.938	wh hex cry	175 dec		3.22	14.4^{25}	
2622	Sodium permanganate trihydrate	NaMnO$_4$ · 3H$_2$O	10101-50-5*	195.972	red-blk hyg cry	170 dec		2.47	144^{20}	reac EtOH
2623	Sodium peroxide	Na$_2$O$_2$	1313-60-6	77.979	yel hyg powder	675		2.805		reac H$_2$O
2624	Sodium perrhenate	NaReO$_4$	13472-33-8	273.195	cry	300		5.39		
2625	Sodium persulfate	Na$_2$S$_2$O$_8$	7775-27-1	238.105	wh hyg cry					vs H$_2$O; reac EtOH
2626	Sodium phosphate	Na$_3$PO$_4$	7601-54-9	163.940	col cry	1583		2.54	14.5^{25}	s H$_2$O
2627	Sodium phosphate dodecahydrate	Na$_3$PO$_4$ · 12H$_2$O	10101-89-0	380.124	col hex cry	≈75		1.62	14.4^{25}	i EtOH
2628	Sodium phosphate, chlorinated	Na$_3$PO$_4$ · NaOCl	56802-99-4	238.383	wh cry				25^{25}	
2629	Sodium phosphide	Na$_3$P	12058-85-4	99.943	red solid	>650				reac H$_2$O
2630	Sodium phosphinate	NaH$_2$PO$_2$	7681-53-0	87.979	wh cry				100^{25}	
2631	Sodium phosphinate monohydrate	NaH$_2$PO$_2$ · H$_2$O	10039-56-2	105.994	col hyg cry	310 dec			100^{25}	s EtOH
2632	Sodium phosphonate pentahydrate	Na$_2$HPO$_3$ · 5H$_2$O	13517-23-2	216.036	wh hex plates	dec 200			429^{20}	vs H$_2$O; i EtOH
2633	Sodium pyrophosphate	Na$_4$P$_2$O$_7$	7722-88-5	265.902	col cry	988		2.53	7.09^{25}	
2634	Sodium selenate	Na$_2$SeO$_4$	13410-01-0	188.94	col orth cry				58.5^{25}	
2635	Sodium selenate decahydrate	Na$_2$SeO$_4$ · 10H$_2$O	10102-23-5	369.09	wh cry			1.61	58.5^{25}	
2636	Sodium selenide	Na$_2$Se	1313-85-5	124.94	amorp solid	>875		2.62		reac H$_2$O
2637	Sodium selenite	Na$_2$SeO$_3$	10102-18-8	172.94	wh tetr cry				89.8^{25}	i EtOH
2638	Sodium selenite pentahydrate	Na$_2$SeO$_3$ · 5H$_2$O	26970-82-1	184.054	wh tetr cry	dec				s H$_2$O; i EtOH
2639	Sodium stannate trihydrate	Na$_2$SnO$_3$ · 3H$_2$O	12209-98-2	266.734	col hex cry	dec 140			61^{15}	vs H$_2$O; i EtOH, ace
2640	Sodium stearate	NaC$_{18}$H$_{35}$O$_2$	822-16-2	306.460	wh powder					sl H$_2$O, EtOH; vs hot H$_2$O
2641	Sodium succinate hexahydrate	Na$_2$C$_4$H$_4$O$_4$ · 6H$_2$O	150-90-3	270.144	cry powder	120 dec			20	i EtOH
2642	Sodium sulfate	Na$_2$SO$_4$	7757-82-6	142.043	wh orth cry or powder	884		2.7	28.1^{25}	i EtOH
2643	Sodium sulfate heptahydrate	Na$_2$SO$_4$ · 7H$_2$O	13472-39-4	204.152	wh cry	dec				vs H$_2$O
2644	Sodium sulfate decahydrate	Na$_2$SO$_4$ · 10H$_2$O	7727-73-3	322.196	col monocl cry	32 dec		1.46	28.1^{25}	i EtOH
2645	Sodium sulfide	Na$_2$S	1313-82-2	78.045	wh cub cry; hyg	1172		1.856	20.6^{25}	sl EtOH; i eth
2646	Sodium sulfide pentahydrate	Na$_2$S · 5H$_2$O	1313-83-3	168.121	col orth cry	120 dec		1.58	20.6^{25}	s EtOH; i eth
2647	Sodium sulfide nonahydrate	Na$_2$S · 9H$_2$O	1313-84-4	240.183	wh-yel hyg cry	≈50 dec		1.43	20.6^{25}	sl EtOH; i eth
2648	Sodium sulfite	Na$_2$SO$_3$	7757-83-7	126.043	wh hex cry	911		2.63	30.7^{25}	i EtOH
2649	Sodium sulfite heptahydrate	Na$_2$SO$_3$ · 7H$_2$O	10102-15-5	252.150	wh monocl cry; unstab			1.56	30.7^{25}	sl EtOH
2650	Sodium superoxide	NaO$_2$	12034-12-7	54.989	yel cub cry	552		2.2		reac H$_2$O
2651	Sodium tartrate dihydrate	Na$_2$C$_4$H$_4$O$_6$ · 2H$_2$O	6106-24-7	230.082				1.545		s H$_2$O; i EtOH
2652	Sodium tellurate	Na$_2$TeO$_4$	10101-83-4	237.58	wh powder				0.8	
2653	Sodium tellurite	Na$_2$TeO$_3$	10102-20-2	221.58	wh rhomb prisms					sl H$_2$O
2654	Sodium tetraborate	Na$_2$B$_4$O$_7$	1330-43-4	201.220	col gl solid; hyg	743	1575	2.4	3.17^{25}	sl MeOH
2655	Sodium tetraborate tetrahydrate	Na$_2$B$_4$O$_7$ · 4H$_2$O	12045-87-3	273.281	wh monocl cry			1.95	3.17^{25}	
2656	Sodium tetraborate pentahydrate	Na$_2$B$_4$O$_7$ · 5H$_2$O	12045-88-4	291.296	hex cry	dec		1.88	3.17^{25}	
2657	Sodium tetraborate decahydrate	Na$_2$B$_4$O$_7$ · 10H$_2$O	1303-96-4	381.373	wh monocl cry	75 dec		1.73	3.17^{25}	i EtOH
2658	Sodium tetrachloroaluminate	NaAlCl$_4$	7784-16-9	191.784	orth cry			2.01		s H$_2$O
2659	Sodium tetrachloroaurate(III) dihydrate	NaAuCl$_4$ · 2H$_2$O	13874-02-7	397.800	oran-yel rhom cry	100 dec			150^{10}	s EtOH, eth
2660	Sodium tetrachloropalladate(II) trihydrate	Na$_2$PdCl$_4$ · 3H$_2$O	13820-53-6	348.26	brn-red hyg cry					vs H$_2$O; s EtOH
2661	Sodium tetrachloroplatinate(II) tetrahydrate	Na$_2$PtCl$_4$ · 4H$_2$O	10026-00-3	454.938	red prisms	100				s H$_2$O, EtOH
2662	Sodium tetrafluoroberyllate	Na$_2$BeF$_4$	13871-27-7	130.986	orth cry	575		2.47		sl H$_2$O
2663	Sodium tetrafluoroborate	NaBF$_4$	13755-29-8	109.795	wh orth prisms	384		2.47	108^{20}	sl EtOH
2664	Sodium thioantimonate nonahydrate	Na$_3$SbS$_4$ · 9H$_2$O	10101-91-4	481.127	yel cry	dec 108		1.8	28^{20}	i EtOH
2665	Sodium thiocyanate	NaSCN	540-72-7	81.073	col hyg cry	287			151^{25}	
2666	Sodium thiophosphate dodecahydrate	Na$_3$PO$_3$S · 12H$_2$O	10101-88-9	396.190	hex hyg leaflets	60				vs hot H$_2$O
2667	Sodium thiosulfate	Na$_2$S$_2$O$_3$	7772-98-7	158.108	col monocl cry	100 dec		1.69	76.4^{25}	i EtOH
2668	Sodium thiosulfate pentahydrate	Na$_2$S$_2$O$_3$ · 5H$_2$O	10102-17-7	248.184	col cry	≈50 dec		1.69	76.4^{25}	i EtOH
2669	Sodium trimetaphosphate	Na$_3$(PO$_3$)$_3$	7785-84-4	305.885	wh cry			2.49	22	
2670	Sodium trimetaphosphate hexahydrate	Na$_3$(PO$_3$)$_3$ · 6H$_2$O	7785-84-4	413.976	tricl-rhom hyg prisms	53		1.786	22	i EtOH
2671	Sodium tripolyphosphate	Na$_5$P$_3$O$_{10}$	7758-29-4	367.864	wh hyg powder	622			20^{25}	
2672	Sodium tungstate	Na$_2$WO$_4$	13472-45-2	293.82	wh rhom cry	695		4.18	74.2^{25}	
2673	Sodium tungstate dihydrate	Na$_2$WO$_4$ · 2H$_2$O	10213-10-2	329.85	wh orth cry	100 dec		3.25	74.2^{25}	i EtOH
2674	Sodium uranate(VI) monohydrate	Na$_2$U$_2$O$_7$ · H$_2$O	13721-34-1	652.049	yel powder					i H$_2$O; s acid

No.	Name	Formula	CAS Reg No.	Mol. weight	Physical form	mp/°C	bp/°C	Density g cm^{-3}	Solubility g/100 g H$_2$O	Qualitative solubility
2675	Sodium vanadate(V)	NaVO$_3$	13718-26-8	121.930	col monocl prisms	630			21^{25}	
2676	Sodium vanadate(V) tetrahydrate	NaVO$_3$ · 4H$_2$O	13718-26-8	193.992	yel-wh cry powder				21^{25}	
2677	Strontium	Sr	7440-24-6	87.62	silv-wh metal; cub	777	1382	2.64		reac H$_2$O; s EtOH
2678	Strontium acetate	Sr(C$_2$H$_3$O$_2$)$_2$	543-94-2	205.71	col hyg cry	dec		2.1	40^{25}	vs H$_2$O
2679	Strontium arsenite tetrahydrate	Sr(AsO$_2$)$_2$ · 4H$_2$O	10378-48-0	373.52	wh powder					sl H$_2$O, EtOH; sol dil acid
2680	Strontium bromate monohydrate	Sr(BrO$_3$)$_2$ · H$_2$O	14519-18-7	361.44	yel hyg monocl cry	120 dec		3.773	39.0^{25}	
2681	Strontium bromide	SrBr$_2$	10476-81-0	247.43	wh tetr cry	657		4.216	107^{25}	
2682	Strontium bromide hexahydrate	SrBr$_2$ · 6H$_2$O	7789-53-9	355.52	col hyg cry	88 dec			107^{25}	s EtOH; i eth
2683	Strontium carbide	SrC$_2$	12071-29-3	111.64	blk tetr cry	>1700		3.19		i H$_2$O
2684	Strontium carbonate	SrCO$_3$	1633-05-2	147.63	wh orth cry; hyg	1494		3.785	0.00034^{20}	s dil acid
2685	Strontium chlorate	Sr(ClO$_3$)$_2$	7791-10-8	254.52	col cry	120 dec		3.15	176^{25}	sl EtOH
2686	Strontium chloride	SrCl$_2$	10476-85-4	158.53	wh cub cry; hyg	874	1250	3.052	54.7^{25}	
2687	Strontium chloride hexahydrate	SrCl$_2$ · 6H$_2$O	10025-70-4	266.62	col hyg cry	100 dec		1.96	54.7^{25}	s EtOH
2688	Strontium chromate	SrCrO$_4$	7789-06-2	203.61	yel monocl cry	dec		3.9	0.106^{20}	s dil acid
2689	Strontium cyanide dihydrate	Sr(CN)$_2$ · 4H$_2$O	52870-08-3	211.72	wh hyg cry	dec				vs H$_2$O
2690	Strontium ferrocyanide pentadecahydrate	SrFe(CN)$_6$ · 15H$_2$O	14654-44-5	569.80	yel monocl cry				50	
2691	Strontium fluoride	SrF$_2$	7783-48-4	125.62	wh cub cry or powder	1477	2460	4.24	0.021^{25}	s dil acid
2692	Strontium formate	Sr(CHO$_2$)$_2$	592-89-2	177.66	wh cry	71.9		2.693	9.1^0	
2693	Strontium formate dihydrate	Sr(CHO$_2$)$_2$ · 2H$_2$O	6160-34-5	213.69	col rhom cry	100 dec		2.25	9.1^{37}	i EtOH, eth
2694	Strontium hexaboride	SrB$_6$	12046-54-7	152.49	blk cub cry	2235		3.39		i H$_2$O; s HNO$_3$
2695	Strontium hydride	SrH$_2$	13598-33-9	89.64	orth cry	1050		3.26		reac H$_2$O
2696	Strontium hydroxide	Sr(OH)$_2$	18480-07-4	121.64	col orth cry; hyg	535	710 dec	3.625	2.25^{25}	
2697	Strontium iodate	Sr(IO$_3$)$_2$	13470-01-4	437.43	tricl cry			5.045	0.165^{25}	
2698	Strontium iodide	SrI$_2$	10476-86-5	341.43	wh hyg cry	538	1773 dec	4.55	177^{25}	
2699	Strontium iodide hexahydrate	SrI$_2$ · 6H$_2$O	73796-25-5	449.52	wh-yel hex cry; hyg	120 dec		4.4	177^{25}	s EtOH
2700	Strontium molybdate	SrMoO$_4$	13470-04-7	247.56	wh cry pow	1040		4.54		i H$_2$O
2701	Strontium niobate	SrNb$_2$O$_6$	12034-89-8	369.43	monocl cry	1225		5.11		i H$_2$O
2702	Strontium nitrate	Sr(NO$_3$)$_2$	10042-76-9	211.63	wh cub cry	570		2.99	80.2^{25}	sl EtOH, ace
2703	Strontium nitride	Sr$_3$N$_2$	12033-82-8	290.87	refrac solid	1200				reac H$_2$O; s HCl
2704	Strontium nitrite	Sr(NO$_2$)$_2$	13470-06-9	179.63	wh-yel hyg needles	240 dec		2.8	72.1^{30}	s H$_2$O
2705	Strontium orthosilicate	Sr$_2$SiO$_4$	13597-55-2	267.32	orth cry			4.5		
2706	Strontium oxalate monohydrate	SrC$_2$O$_4$ · H$_2$O	814-95-9	193.65	cry pow	dec 150			0.00005^{20}	sl dil acid
2707	Strontium oxide	SrO	1314-11-0	103.62	col cub cry	2531		5.1		reac H$_2$O
2708	Strontium perchlorate	Sr(ClO$_4$)$_2$	13450-97-0	286.52	col hyg cry				306^{25}	s EtOH, MeOH
2709	Strontium permanganate trihydrate	Sr(MnO$_4$)$_2$ · 3H$_2$O	14446-13-0	379.54	purp cub cry	175 dec		2.75	250^{18}	
2710	Strontium peroxide	SrO$_2$	1314-18-7	119.62	wh tetr cry; unstab	215 dec		4.78		reac H$_2$O
2711	Strontium phosphate	Sr$_3$(PO$_4$)$_2$	7446-28-8	452.80	wh powder				0.000011^{20}	s acid
2712	Strontium selenate	SrSeO$_4$	7446-21-1	230.58	orth cry			4.25	0.115^{20}	s hot HCl
2713	Strontium selenide	SrSe	1315-07-7	166.58	wh cub cry	1600		4.54		
2714	Strontium silicide	SrSi$_2$	12138-28-2	143.79	silv-gray cub cry	1100		3.35		
2715	Strontium sulfate	SrSO$_4$	7759-02-6	183.68	wh orth cry	1606		3.96	0.0135^{25}	i EtOH; sl acid
2716	Strontium sulfide	SrS	1314-96-1	119.69	gray cub cry	2226		3.70		sl H$_2$O; s acid
2717	Strontium sulfite	SrSO$_3$	13451-02-0	167.68	col cry	dec			0.0015^{25}	s H$_2$SO$_4$, HCl
2718	Strontium telluride	SrTe	12040-08-3	215.22	wh cub cry			4.83		
2719	Strontium thiosulfate pentahydrate	SrS$_2$O$_3$ · 5H$_2$O	15123-90-7	289.82	monocl needles	100 dec		2.17	36.3^{25}	i EtOH
2720	Strontium titanate	SrTiO$_3$	12060-59-2	183.49	wh cub cry	2080		5.1		i H$_2$O
2721	Strontium tungstate	SrWO$_4$	13451-05-3	335.46	col tetr cry	dec		6.187	0.14^{15}	i EtOH
2722	Strontium zirconate	SrZrO$_3$	12036-39-4	226.84	col cry	2600				
2723	Sulfur (rhombic)	S	7704-34-9	32.065	yel orth cry	95.3 (trans to monocl)	444.61	2.07		i H$_2$O; sl EtOH, bz, eth; s CS$_2$
2724	Sulfur (monoclinic)	S	7704-34-9	32.065	yel monocl needles, stable 95.3-120	115.21	444.61	2.07		i H$_2$O; sl EtOH, bz, eth; s CS$_2$
2725	Sulfuric acid	H$_2$SO$_4$	7664-93-9	98.079	col oily liq	10.31	337	1.8302		vs H$_2$O
2726	Peroxysulfuric acid	H$_2$SO$_5$	7722-86-3	114.078	wh cry; unstab	45 dec				vs H$_2$O
2727	Nitrosylsulfuric acid	HNOSO$_4$	7782-78-7	127.077	prisms	73 dec				reac H$_2$O; s H$_2$SO$_4$
2728	Chlorosulfonic acid	SO$_2$(OH)Cl	7790-94-5	116.524	col-yel liq	-80	152	1.75		reac H$_2$O; s py
2729	Fluorosulfonic acid	SO$_2$(OH)F	7789-21-1	100.069	col liq	-89	163	1.726		reac H$_2$O
2730	Sulfurous acid	H$_2$SO$_3$	7782-99-2	82.079	exists only in aq soln					
2731	Sulfamic acid	H$_2$NSO$_3$H	5329-14-6	97.094	orth cry	≈205 dec		2.15	14.7^0	sl ace; i eth
2732	Sulfur dioxide	SO$_2$	7446-09-5	64.064	col gas	-75.5	-10.05	2.619 g/L		s H$_2$O, EtOh, eth, chl

No.	Name	Formula	CAS Reg No.	Mol. weight	Physical form	mp/°C	bp/°C	Density g cm^{-3}	Solubility g/100 g H$_2$O	Qualitative solubility
2733	Sulfur trioxide	SO$_3$	7446-11-9	80.063	wh needles	62.2	subl			reac H$_2$O
2734	Sulfur trioxide (γ-form)	SO$_3$	7446-11-9	80.063	col solid or liq	16.8	44.5	1.90		reac H$_2$O
2735	Sulfur trioxide (β-form)	SO$_3$	7446-11-9	80.063	wh needles	30.5	44.5			reac H$_2$O
2736	Sulfur bromide (SSBr$_2$)	SSBr$_2$	13172-31-1	223.938	red oily liq	-46	>25 dec	2.63		reac H$_2$O
2737	Sulfur chloride (SSCl$_2$)	SSCl$_2$	10025-67-9	135.036	yel-red oily liq	-77	137	1.69		reac H$_2$O; s EtOH, bz, eth, ctc
2738	Sulfur fluoride (SSF$_2$)	SSF$_2$	16860-99-4	102.127	col gas	-164.6	-10.6	4.174 g/L		reac H$_2$O
2739	Sulfur fluoride (FSSF)	FSSF	13709-35-8	102.127	col gas	-133	15	4.174 g/L		reac H$_2$O
2740	Sulfur dichloride	SCl$_2$	10545-99-0	102.971	red visc liq	-122	59.6	1.62		reac H$_2$O
2741	Sulfur tetrafluoride	SF$_4$	7783-60-0	108.059	col gas	-125	-40.45	4.417 g/L		reac H$_2$O
2742	Sulfur hexafluoride	SF$_6$	2551-62-4	146.055	col gas	-49.596 tp	-63.8 sp	5.970 g/L		sl H$_2$O; s EtOH
2743	Sulfur bromide pentafluoride	SF$_5$Br	15607-89-3	206.961	col gas	-79	3.1	8.459 g/L		
2744	Sulfur chloride pentafluoride	SF$_5$Cl	13780-57-9	162.510	col gas	-64	-19.05	6.642 g/L		
2745	Sulfur decafluoride	S$_2$F$_{10}$	5714-22-7	254.114	liq	-52.7	30; dec 150	2.08		i H$_2$O
2746	Sulfuryl amide	(NH$_2$)$_2$SO$_2$	7803-58-9	96.109	orth plates	93	250 dec			vs H$_2$O; sl EtOH
2747	Sulfuryl chloride	SO$_2$Cl$_2$	7791-25-5	134.970	col liq	-51	69.4	1.680		reac H$_2$O; s bz, tol, eth
2748	Sulfuryl fluoride	SO$_2$F$_2$	2699-79-8	102.061	col gas	-135.8	-55.4	4.172 g/L		sl H$_2$O, EtOH; s tol, ctc
2749	Sulfuryl bromide fluoride	SO$_2$BrF	13536-61-3	162.966	col liq	-86	41			reac H$_2$O
2750	Sulfuryl chloride fluoride	SO$_2$ClF	13637-84-8	118.515	col gas	-124.7	7.1	1.62^0		reac H$_2$O
2751	Pyrosulfuryl chloride	S$_2$O$_5$Cl$_2$	7791-27-7	215.033	col fuming liq	-37	151	1.837		reac H$_2$O
2752	Thionyl bromide	SOBr$_2$	507-16-4	207.872	yel liq	-50	140			reac H$_2$O
2753	Thionyl chloride	SOCl$_2$	7719-09-7	118.970	yel fuming liq	-101	75.6	1.631		reac H$_2$O; s bz, ctc, chl
2754	Thionyl fluoride	SOF$_2$	7783-42-8	86.061	col gas	-129.5	-43.8	3.518 g/L		reac H$_2$O; s bz, eth
2755	Sulfur fluoride oxide (SOF$_4$)	SOF$_4$	13709-54-1	124.058	col gas	-99.6	-48.5	1.95^{-82}		reac H$_2$O
2756	Sulfur fluoride hypofluorite	F$_5$SOF	15179-32-5	162.054	col gas	-86	-35.1	6.624 g/L		
2757	Tetrasulfur tetranitride	S$_4$N$_4$	28950-34-7	184.287	yel-oran cry	178.2	subl			i H$_2$O; reac alk, acid
2758	Tantalum	Ta	7440-25-7	180.948	gray metal; cub	3017	5458	16.4		reac HF
2759	Tantalum aluminide	TaAl$_3$	12004-76-1	261.893	gray refrac powder	≈1400		7.02		i H$_2$O, acid, alk
2760	Tantalum boride (TaB)	TaB	12007-07-7	191.759	refrac orth cry	2040		14.2		
2761	Tantalum boride (TaB$_2$)	TaB$_2$	12007-35-1	202.570	blk hex cry	3100		11.2		i H$_2$O, acid, alk
2762	Tantalum carbide (TaC)	TaC	12070-06-3	192.959	gold-brown powder; cub	3880	4780	14.3		s HF-HNO$_3$ mixture
2763	Tantalum carbide (Ta$_2$C)	Ta$_2$C	12070-07-4	373.907	refrac hex cry	3327		15.1		
2764	Tantalum hydride	TaH	13981-95-8	181.956	gray metallic solid			15.1		i acid
2765	Tantalum nitride	TaN	12033-62-4	194.955	blk hex cry	3090		13.7		i H$_2$O; sl aqua regia; reac alk
2766	Tantalum silicide	TaSi$_2$	12039-79-1	237.119	gray powder	2200		9.14		
2767	Tantalum(III) bromide	TaBr$_3$	13842-73-4	420.660	gray-grn solid	dec 220				
2768	Tantalum(III) chloride	TaCl$_3$	13569-67-0	287.307	blk-grn solid	dec 440				s H$_2$O
2769	Tantalum(IV) bromide	TaBr$_4$	13842-76-7	500.564	dark blue solid	392		5.77		reac H$_2$O
2770	Tantalum(IV) chloride	TaCl$_4$	13569-72-7	322.760	dark grn solid	dec 300	subl	4.35		reac H$_2$O
2771	Tantalum(IV) iodide	TaI$_4$	14693-80-2	688.566	gray-blk solid	400 dec				reac H$_2$O
2772	Tantalum(IV) oxide	TaO$_2$	12036-14-5	212.947	tetr cry			10.0		
2773	Tantalum(IV) selenide	TaSe$_2$	12039-55-3	338.87	hex cry			6.7		
2774	Tantalum(IV) sulfide	TaS$_2$	12143-72-5	245.078	blk hex cry	>3000		6.86		i H$_2$O
2775	Tantalum(IV) telluride	TaTe$_2$	12067-66-2	436.15	monocl cry			9.4		
2776	Tantalum(V) bromide	TaBr$_5$	13451-11-1	580.468	yel cry powder	265.8	348.8	4.67		
2777	Tantalum(V) chloride	TaCl$_5$	7721-01-9	358.213	yel-wh monocl cry; hyg	216.6	239	3.68		reac H$_2$O; s EtOH
2778	Tantalum(V) fluoride	TaF$_5$	7783-71-3	275.940	wh monocl cry; hyg	96.9	229.5	4.74		s H$_2$O, eth; sl CS$_2$, ctc
2779	Tantalum(V) iodide	TaI$_5$	14693-81-3	815.470	blk hex cry; hyg	496	543	5.80		reac H$_2$O
2780	Tantalum(V) oxide	Ta$_2$O$_5$	1314-61-0	441.893	wh rhomb cry or powder	1875		8.24		i H$_2$O, EtOH, acid; s HF
2781	Technetium	Tc	7440-26-8	98	hex cry	2157	4265	11		
2782	Technetium(V) fluoride	TcF$_5$	31052-14-9	193	yel solid	50	dec			
2783	Technetium(VI) fluoride	TcF$_6$	13842-93-8	212	yel cub cry	37.4	55.3	3.0		
2784	Tellurium	Te	13494-80-9	127.60	gray-wh rhomb cry	449.51	988	6.232		i H$_2$O, bz, CS$_2$
2785	Telluric(VI) acid	H$_6$TeO$_6$	7803-68-1	229.64	wh monocl cry	136		3.07	50.1^{30}	
2786	Tellurous acid	H$_2$TeO$_3$	10049-23-7	177.61	wh cry	40 dec		3.0		sl H$_2$O; s dil acid, alk
2787	Tellurium dioxide	TeO$_2$	7446-07-3	159.60	wh orth cry	733	1245	5.9		i H$_2$O; s alk, acid
2788	Tellurium trioxide	TeO$_3$	13451-18-8	175.60	yel-oran cry	430		5.07		i H$_2$O

No.	Name	Formula	CAS Reg No.	Mol. weight	Physical form	mp/°C	bp/°C	Density g cm^{-3}	Solubility g/100 g H$_2$O	Qualitative solubility
2789	Tellurium dibromide	TeBr$_2$	7789-54-0	287.41	grn-brn hyg cry	210	339			reac H$_2$O; s eth; sl chl
2790	Tellurium dichloride	TeCl$_2$	10025-71-5	198.51	blk amorp solid; hyg	208	328	6.9		reac H$_2$O; i ctc
2791	Tellurium tetrabromide	TeBr$_4$	10031-27-3	447.22	yel-oran monocl cry	380	≈420 dec	4.3		reac H$_2$O; s eth
2792	Tellurium tetrachloride	TeCl$_4$	10026-07-0	269.41	wh monocl cry; hyg	224	387	3.0		reac H$_2$O; s EtOH, tol
2793	Tellurium tetrafluoride	TeF$_4$	15192-26-4	203.59	col cry	129	195 dec			reac H$_2$O
2794	Tellurium decafluoride	Te$_2$F$_{10}$	53214-07-6	445.18	col liq	-33.7	59			
2795	Tellurium tetraiodide	TeI$_4$	7790-48-9	635.22	blk orth cry	280		5.05		reac H$_2$O; sl ace
2796	Tellurium hexafluoride	TeF$_6$	7783-80-4	241.59	col gas	-37.6 tp	-38.9 sp	9.875 g/L		reac H$_2$O
2797	Terbium	Tb	7440-27-9	158.925	silv metal; hex	1359	3230	8.23		
2798	Terbium nitride	TbN	12033-64-6	172.932	cub cry			9.55		
2799	Terbium silicide	TbSi$_2$	12039-80-4	215.096	orth cry			6.66		
2800	Terbium(III) bromide	TbBr$_3$	14456-47-4	398.637	wh hex cry	830	1490			s H$_2$O
2801	Terbium(III) chloride	TbCl$_3$	10042-88-3	265.284	wh orth cry; hyg	582		4.35		s H$_2$O
2802	Terbium(III) chloride hexahydrate	TbCl$_3$ · 6H$_2$O	13798-24-8	373.375	hyg cry			4.35		vs H$_2$O
2803	Terbium(III) iodide	TbI$_3$	13813-40-6	539.638	hex cry; hyg	955		≈5.2		s H$_2$O
2804	Terbium(III) nitrate	Tb(NO$_3$)$_3$	10043-27-3	344.940	pink hyg solid				157[25]	s EtOH
2805	Terbium(III) nitrate hexahydrate	Tb(NO$_3$)$_3$ · 6H$_2$O	13451-19-9	453.031	col needles	89				s H$_2$O, EtOH, ace
2806	Terbium(III) oxide	Tb$_2$O$_3$	12036-41-8	365.849	wh cub cry	2303		7.91		
2807	Terbium(III) sulfate octahydrate	Tb$_2$(SO$_4$)$_3$ · 8H$_2$O	13842-67-6	750.161	wh cry	dec 360				sl H$_2$O
2808	Terbium(III) sulfide	Tb$_2$S$_3$	12138-11-3	414.046	cub cry			6.35		
2809	Terbium(III) fluoride	TbF$_3$	13708-63-9	215.920	wh solid	1175	2280			i H$_2$O
2810	Terbium(IV) fluoride	TbF$_4$	36781-15-4	234.919	wh monocl cry	dec 300				i H$_2$O
2811	Thallium	Tl	7440-28-0	204.383	soft blue-wh metal	304	1473	11.8		i H$_2$O; reac acid
2812	Thallium(I) acetate	TlC$_2$H$_3$O$_2$	563-68-8	263.427	hyg wh cry	131		3.68		s H$_2$O, EtOH
2813	Thallium(I) azide	TlN$_3$	13847-66-0	246.403	yel cry	334	exp			s H$_2$O
2814	Thallium(I) bromate	TlBrO$_3$	14550-84-6	332.285	col needles	120 dec			0.49[30]	s EtOH
2815	Thallium(I) bromide	TlBr	7789-40-4	284.287	yel cub cry	460	819	7.5	0.059[20]	
2816	Thallium(I) carbonate	Tl$_2$CO$_3$	6533-73-9	468.776	wh monocl cry	273		7.11	4.69[20]	i EtOH
2817	Thallium(I) chlorate	TlClO$_3$	13453-30-0	287.834	col hex cry	dec 500		5.5	3.92[20]	
2818	Thallium(I) chloride	TlCl	7791-12-0	239.836	wh cub cry	431	720	7.0	0.33[20]	i EtOH
2819	Thallium(I) chromate	Tl$_2$CrO$_4$	13473-75-1	524.761	yel cry				0.003[20]	sl acid, alk
2820	Thallium(I) cyanide	TlCN	13453-34-4	230.401	wh hex plates			6.523		s H$_2$O, acid, EtOH
2821	Thallium(I) ethanolate	TlC$_2$H$_5$O	20398-06-5	249.443	cloudy liq	-3	130 dec	3.49		reac H$_2$O
2822	Thallium(I) fluoride	TlF	7789-27-7	223.381	wh orth cry	326	826	8.36	245[25]	
2823	Thallium(I) formate	TlCHO$_2$	992-98-3	249.401	hyg col needles	101		4.97		vs H$_2$O; s MeOH
2824	Thallium(I) hexafluorophosphate	TlPF$_6$	60969-19-9	349.347	wh cub cry			4.6		
2825	Thallium(I) hydroxide	TlOH	12026-06-1	221.390	yel needles	139 dec		7.44	34.3[18]	
2826	Thallium(I) iodate	TlIO$_3$	14767-09-0	379.285	wh needles				0.058	sl HNO$_3$
2827	Thallium(I) iodide	TlI	7790-30-9	331.287	yel cry powder	441.7	824	7.1	0.0085[20]	i EtOH
2828	Thallium(I) molybdate	Tl$_2$MoO$_4$	34128-09-1	568.71	yel-wh cub cry					i H$_2$O
2829	Thallium(I) nitrate	TlNO$_3$	10102-45-1	266.388	wh cry	206	450 dec	5.55	9.55[20]	s H$_2$O; i EtOH
2830	Thallium(I) nitrite	TlNO$_2$	13826-63-6	250.389	yel cub cry	186		5.7	32.1[25]	s H$_2$O
2831	Thallium(I) oxalate	Tl$_2$C$_2$O$_4$	30737-24-7	496.786	wh powder			6.31	1.83[20]	
2832	Thallium(I) oxide	Tl$_2$O	1314-12-1	424.766	blk rhomb cry; hyg	579	≈1080	9.52		s H$_2$O, EtOH
2833	Thallium(I) perchlorate	TlClO$_4$	13453-40-2	303.834	col orth cry	501		4.89	19.7[30]	
2834	Thallium(I) selenate	Tl$_2$SeO$_4$	7446-22-2	551.73	orth cry	>400		6.875	2.8[20]	i EtOH, eth
2835	Thallium(I) selenide	Tl$_2$Se	15572-25-5	487.73	gray plates	340				i H$_2$O, acid
2836	Thallium(I) sulfate	Tl$_2$SO$_4$	7446-18-6	504.830	wh rhomb prisms	632		6.77	5.47[25]	
2837	Thallium(I) sulfide	Tl$_2$S	1314-97-2	440.832	blue-blk cry	457	1367	8.39	0.02[20]	sl alk; s acid
2838	Thallium(III) acetate	Tl(C$_2$H$_3$O$_2$)$_3$	2570-63-0	381.514	hyg wh platelets	182 dec				
2839	Thallium(III) bromide tetrahydrate	TlBr$_3$ · 4H$_2$O	13701-90-1	516.157	yel orth cry			3.65		s H$_2$O, EtOH
2840	Thallium(III) chloride	TlCl$_3$	13453-32-2	310.742	monocl cry	155		4.7		vs H$_2$O, EtOH, eth
2841	Thallium(III) chloride tetrahydrate	TlCl$_3$ · 4H$_2$O	13453-32-2*	382.804	orth cry			3.00		s H$_2$O
2842	Thallium(III) fluoride	TlF$_3$	7783-57-5	261.378	wh orth cry; hyg	550 dec		8.65		reac H$_2$O
2843	Thallium(III) nitrate	Tl(NO$_3$)$_3$	13746-98-0	390.398	col cry					reac H$_2$O
2844	Thallium(III) oxide	Tl$_2$O$_3$	1314-32-5	456.765	brn cub cry	834		10.2		i H$_2$O; reac acid
2845	Thallium(III) sulfate	Tl$_2$(SO$_4$)$_3$	16222-66-5	696.955	col leaflets					reac H$_2$O
2846	Thallium selenide	TlSe	12039-52-0	283.34	blk solid	330				i H$_2$O, acid
2847	Thorium	Th	7440-29-1	232.038	soft gray-wh metal; cub	1750	4788	11.7		s acid
2848	Thorium hydride	ThH$_2$	16689-88-6	234.054	tetr cry			9.5		
2849	Thorium boride	ThB$_6$	12229-63-9	296.904	refrac solid	2450		6.99		
2850	Thorium(IV) bromide	ThBr$_4$	13453-49-1	551.654	wh hyg cry	679			65[20]	

No.	Name	Formula	CAS Reg No.	Mol. weight	Physical form	mp/°C	bp/°C	Density g cm^{-3}	Solubility g/100 g H$_2$O	Qualitative solubility
2851	Thorium carbide	ThC	12012-16-7	244.049	cub cry	2500		10.6		reac H$_2$O
2852	Thorium dicarbide	ThC$_2$	12071-31-7	256.059	yel monocl cry	≈2650		9.0		reac H$_2$O
2853	Thorium(IV) chloride	ThCl$_4$	10026-08-1	373.850	gray-wh tetr needles; hyg	770	921	4.59		s H$_2$O, EtOH
2854	Thorium(IV) fluoride	ThF$_4$	13709-59-6	308.032	wh monocl cry; hyg	1110	1680	6.1		
2855	Thorium(IV) iodide	ThI$_4$	7790-49-0	739.656	wh-yel monocl cry	566	837			
2856	Thorium(IV) nitrate	Th(NO$_3$)$_4$	13823-29-5	480.058	hyg wh plates	55 dec				vs H$_2$O, EtOH
2857	Thorium(IV) nitrate tetrahydrate	Th(NO$_3$)$_4$ · 4H$_2$O	13470-07-0	552.119	wh hyg cry	500 dec			191^{20}	s EtOH
2858	Thorium nitride	ThN	12033-65-7	246.045	refrac cub cry	2820		11.6		reac H$_2$O
2859	Thorium(IV) oxide	ThO$_2$	1314-20-1	264.037	wh cub cry	3350	4400	10.0		i H$_2$O, alk; sl acid
2860	Thorium(IV) selenide	ThSe$_2$	60763-24-8	389.96	orth cry			8.5		
2861	Thorium orthosilicate	ThSiO$_4$	14553-44-7	324.122	brn tetr cry			6.7		
2862	Thorium silicide	ThSi$_2$	12067-54-8	288.209	tetr cry	1850		7.9		
2863	Thorium(IV) sulfate nonahydrate	Th(SO$_4$)$_2$ · 9H$_2$O	10381-37-0	586.301	wh monocl cry	dec		2.8	4.2^{20}	
2864	Thorium(IV) sulfide	ThS$_2$	12138-07-7	296.168	dark brn cry	1905		7.30		i H$_2$O; s acid
2865	Thulium	Tm	7440-30-4	168.934	silv metal; hex	1545	1950	9.32		s dil acid
2866	Thulium(II) bromide	TmBr$_2$	64171-97-7	328.742	dark grn solid	619				
2867	Thulium(II) chloride	TmCl$_2$	22852-11-5	239.840	red or grn cry	718				reac H$_2$O
2868	Thulium(II) iodide	TmI$_2$	60864-26-8	422.743	blk hyg solid	756				reac H$_2$O
2869	Thulium(III) bromide	TmBr$_3$	14456-51-0	408.646	wh hyg cry	954				s H$_2$O
2870	Thulium(III) chloride	TmCl$_3$	13537-18-3	275.293	yel hyg cry	845				s H$_2$O
2871	Thulium(III) chloride heptahydrate	TmCl$_3$ · 7H$_2$O	13778-39-7	401.400	hyg cry					s H$_2$O, EtOH
2872	Thulium(III) fluoride	TmF$_3$	13760-79-7	225.929	wh cry	1158				s H$_2$O
2873	Thulium(III) hydroxide	Tm(OH)$_3$	1311-33-7	219.956	wh or grn prec					i H$_2$O
2874	Thulium(III) iodide	TmI$_3$	13813-43-9	549.647	yel hyg cry	1021				
2875	Thulium(III) nitrate	Tm(NO$_3$)$_3$	14985-19-4	354.949	grn hyg solid				212^{25}	s EtOH
2876	Thulium(III) nitrate pentahydrate	Tm(NO$_3$)$_3$ · 5H$_2$O	36548-87-5	445.025	grn hyg cry					s H$_2$O, EtOH, ace
2877	Thulium(III) oxalate hexahydrate	Tm$_2$(C$_2$O$_4$)$_3$ · 6H$_2$O	26677-68-9	710.016	grn solid	dec 50				s alk oxalates
2878	Thulium(III) oxide	Tm$_2$O$_3$	12036-44-1	385.866	grn-wh cub cry	2341	3945	8.6		sl acid
2879	Tin (gray)	Sn	7440-31-5	118.710	cub cry	trans to wh Sn 13.2	2602	5.769		
2880	Tin (white)	Sn	7440-31-5	118.710	silv tetr cry	231.93	2602	7.287		
2881	Stannane	SnH$_4$	2406-52-2	122.742	unstab col gas	-146	-51.8	5.017 g/L		
2882	Methylstannane	SnH$_3$CH$_3$	1631-78-3	136.769	col gas		1.4	5.590 g/L		reac H$_2$O
2883	(Dimethylamino)trimethystannane	Sn(CH$_3$)$_3$N(CH$_3$)$_2$	993-50-0	207.890	liq	1	126	1.22		reac H$_2$O
2884	Tin monophosphide	SnP	25324-56-5	149.684	dull metallic solid	540				
2885	Tin triphosphide	Sn$_4$P$_3$	12286-33-8	567.761	wh cry	≈550		5.2		
2886	Tin(II) acetate	Sn(C$_2$H$_3$O$_2$)$_2$	638-39-1	236.799	wh orth cry	183	subl	2.31		s dil HCl
2887	Tin(II) bromide	SnBr$_2$	10031-24-0	278.518	yel powder	215	639	5.12	85^0	s EtOH, eth, ace
2888	Tin(II) chloride	SnCl$_2$	7772-99-8	189.616	wh orth cry	247.0	623	3.90	178^{10}	s EtOH, ace, eth; i xyl
2889	Tin(II) chloride dihydrate	SnCl$_2$ · 2H$_2$O	10025-69-1	225.647	wh monocl cry	37 dec		2.71	178^{10}	s EtOH, NaOH; vs HCl
2890	Tin(II) fluoride	SnF$_2$	7783-47-3	156.707	wh monocl cry; hyg	215	850	4.57		s H$_2$O; i EtOH, eth, chl
2891	Tin(II) hexafluorozirconate	SnZrF$_6$	12419-43-1	323.924	cry			4.21		s H$_2$O
2892	Tin(II) hydroxide	Sn(OH)$_2$	12026-24-3	152.725	wh amorp solid					
2893	Tin(II) iodide	SnI$_2$	10294-70-9	372.519	red-oran powder	320	714	5.28	0.98^{20}	s bz, chl, CS$_2$
2894	Tin(II) oxalate	SnC$_2$O$_4$	814-94-8	206.729	wh powder	280 dec		3.56		i H$_2$O; s dil HCl
2895	Tin(II) oxide	SnO	21651-19-4	134.709	blue-blk tetr cry	1080 dec		6.45		i H$_2$O, EtOH; s acid
2896	Tin(II) pyrophosphate	Sn$_2$P$_2$O$_7$	15578-26-4	411.363	wh amorp powder	400 dec		4.009		i H$_2$O; s conc acid
2897	Tin(II) selenide	SnSe	1315-06-6	197.67	gray orth cry	861		6.18		i H$_2$O; s aqua regia
2898	Tin(II) sulfate	SnSO$_4$	7488-55-3	214.773	wh orth cry	378 dec		4.15	18.8^{19}	
2899	Tin(II) sulfide	SnS	1314-95-0	150.775	gray orth cry	881	1210	5.08		i H$_2$O; s conc acid
2900	Tin(II) tartrate	SnC$_4$H$_4$O$_6$	815-85-0	266.781	wh cry powder					s H$_2$O, dil HCl
2901	Tin(II) telluride	SnTe	12040-02-7	246.31	gray cub cry	806		6.5		
2902	Tin(IV) bromide	SnBr$_4$	7789-67-5	438.326	wh cry	29.1	205	3.34		vs H$_2$O; s EtOH
2903	Tin(IV) chloride	SnCl$_4$	7646-78-8	260.522	col fuming liq	-34.07	114.15	2.234		reac H$_2$O; s EtOH, ctc, bz, ace
2904	Tin(IV) chloride pentahydrate	SnCl$_4$ · 5H$_2$O	10026-06-9	350.598	wh-yel cry	56 dec		2.04		vs H$_2$O; s EtOH
2905	Tin(IV) chromate	Sn(CrO$_4$)$_2$	38455-77-5	350.697	brn-yel cry powder	dec				s H$_2$O
2906	Tin(IV) fluoride	SnF$_4$	7783-62-2	194.704	wh tetr cry	442	705 subl	4.78		reac H$_2$O
2907	Tin(IV) iodide	SnI$_4$	7790-47-8	626.328	yel-brn cub cry	143	364.35	4.46		reac H$_2$O; s EtOH, bz, chl, eth
2908	Tin(IV) oxide	SnO$_2$	18282-10-5	150.709	gray tetr cry	1630		6.85		i H$_2$O, EtOH; s hot conc alk
2909	Tin(IV) selenide	SnSe$_2$	20770-09-6	276.63	red-brn cry	650		≈5.0		i H$_2$O; s alk, conc acid

No.	Name	Formula	CAS Reg No.	Mol. weight	Physical form	mp/°C	bp/°C	Density g cm⁻³	Solubility g/100 g H₂O	Qualitative solubility
2910	Tin(IV) selenite	$Sn(SeO_3)_2$	7446-25-5	372.63	cry powder					i H₂O; s hot HCl
2911	Tin(IV) sulfide	SnS_2	1315-01-1	182.840	gold-yel hex cry	600 dec		4.5		i H₂O; s alk, aqua regia
2912	Titanium	Ti	7440-32-6	47.867	gray metal; hex	1668	3287	4.506		
2913	Titanocene dichloride	$Ti(C_5H_5)_2Cl_2$	1271-19-8	248.959	red cry	289		1.60		sl H₂O, bz; s chl, EtOH, tol
2914	Titanium hydride	TiH_2	7704-98-5	49.883	gray-blk powder	≈450 dec		3.75		i H₂O
2915	Titanium boride	TiB_2	12045-63-5	69.489	gray refrac solid; hex	3225		4.38		
2916	Titanium carbide	TiC	12070-08-5	59.878	cub cry	3067		4.93		i H₂O; s HNO₃
2917	Titanium nitride	TiN	25583-20-4	61.874	yel-brn cub cry	2947		5.21		i H₂O; s aqua regia
2918	Titanium phosphide	TiP	12037-65-9	78.841	gray hex cry	1990		4.08		
2919	Titanium silicide	$TiSi_2$	12039-83-7	104.038	blk orth cry	1500		4.0		i H₂O, acid, alk; s HF
2920	Titanium(II) bromide	$TiBr_2$	13783-04-5	207.675	blk powder	dec 400		4.0		reac H₂O
2921	Titanium(II) chloride	$TiCl_2$	10049-06-6	118.773	blk hex cryc	1035	1500	3.13		reac H₂O; s EtOH; i chl, eth
2922	Titanium(II) iodide	TiI_2	13783-07-8	301.676	blk hex cry	dec 400		5.02		reac H₂O
2923	Titanium(II) oxide	TiO	12137-20-1	63.866	yel cub cry	1770	3227	4.95		
2924	Titanium(II) sulfide	TiS	12039-07-5	79.932	brn hex cry	1927		3.85		i H₂O; s conc acid
2925	Titanium(III) bromide	$TiBr_3$	13135-31-4	287.579	viol hex cry	dec 400				s H₂O
2926	Titanium(III) chloride	$TiCl_3$	7705-07-9	154.226	red-viol hex cry; hyg	425 dec	960	2.64		reac H₂O
2927	Titanium(III) fluoride	TiF_3	13470-08-1	104.862	viol hex cry	950 dec		2.98		i H₂O, dil acid, alk
2928	Titanium(III) iodide	TiI_3	13783-08-9	428.580	viol cry	dec 350				
2929	Titanium(III) oxide	Ti_2O_3	1344-54-3	143.732	blk hex cry	1842		4.486		s hot HF
2930	Titanium(III) sulfate	$Ti_2(SO_4)_3$	10343-61-0	383.922	grn cry					i H₂O, EtOH; s dil HCl
2931	Titanium(III) sulfide	Ti_2S_3	12039-16-6	191.929	blk hex cry			3.56		
2932	Titanium(III,IV) oxide	Ti_3O_5	12065-65-5	223.598	blk monocl cry	1777		4.24		
2933	Titanium(IV) bromide	$TiBr_4$	7789-68-6	367.483	yel-oran cub cry; hyg	38.3	233.5	3.37		reac H₂O
2934	Titanium(IV) chloride	$TiCl_4$	7550-45-0	189.679	col or yel liq	-24.12	136.45	1.73		reac H₂O; s EtOH
2935	Titanium(IV) fluoride	TiF_4	7783-63-3	123.861	wh hyg powder	377	subl 284	2.798		reac H₂O; s EtOH, py
2936	Titanium(IV) iodide	TiI_4	7720-83-4	555.485	red hyg powder	155	377	4.3		reac H₂O
2937	Titanium(IV) oxide (anatase)	TiO_2	1317-70-0	79.866	brn tetr cry	1560		3.9		
2938	Titanium(IV) oxide (brookite)	TiO_2	12188-41-9	79.866	wh orth cry			4.17		
2939	Titanium(IV) oxide (rutile)	TiO_2	1317-80-2	79.866	wh tetr cry	1843	≈3000	4.17		i H₂O, dil acid; s conc acid
2940	Titanium(IV) oxysulfate monohydrate	$TiOSO_4 \cdot H_2O$	13825-74-6*	177.944	col orth cry			2.71		reac H₂O
2941	Titanium(IV) sulfate	$Ti(SO_4)_2$	13693-11-3	239.992	wh-yel hyg cry	150 dec				s H₂O
2942	Titanium(IV) sulfide	TiS_2	12039-13-3	111.997	yel-brn hex cry; hyg			3.37		s H₂SO₄
2943	Tungsten	W	7440-33-7	183.84	gray-wh metal; cub	3422	5555	19.3		
2944	Tungstic acid	H_2WO_4	7783-03-1	249.85	yel amorp powder	100 dec		5.5		i H₂O, acid; s alk
2945	Tungsten boride (W₂B)	W_2B	12007-10-2	378.49	refrac blk powder	2670		16.0		i H₂O
2946	Tungsten boride (WB)	WB	12007-09-9	194.65	blk refrac powder	2665		15.2		i H₂O
2947	Tungsten boride (W₂B₅)	W_2B_5	12007-98-6	421.74	refrac solid	2370		11.0		i H₂O
2948	Tungsten carbide (W₂C)	W_2C	12070-13-2	379.69	refrac hex cry	≈2800		14.8		i H₂O
2949	Tungsten carbide (WC)	WC	12070-12-1	195.85	gray hex cry	2785		15.6		i H₂O; s HNO₃/HF
2950	Tungsten carbonyl	$W(CO)_6$	14040-11-0	351.90	wh cry	170 dec	subl	2.65		i H₂O; s os
2951	Tungsten nitride (WN₂)	WN_2	60922-26-1	211.85	hex cry	600 dec		7.7		
2952	Tungsten nitride (W₂N)	W_2N	12033-72-6	381.69	gray cub cry	dec		17.8		
2953	Tungsten silicide (WSi₂)	WSi_2	12039-88-2	240.01	blue-gray tetr cry	2160		9.3		i H₂O
2954	Tungsten silicide (W₅Si₃)	W_5Si_3	12039-95-1	1003.46	blue-gray refrac solid	2320		14.4		
2955	Tungsten(II) bromide	WBr_2	13470-10-5	343.65	yel powder	dec 400				i H₂O
2956	Tungsten(II) chloride	WCl_2	13470-12-7	254.75	gray solid	dec 500		5.44		sl H₂O
2957	Tungsten(II) iodide	WI_2	13470-17-2	437.65	oran-brn cry	dec 800		6.79		i H₂O
2958	Tungsten(III) bromide	WBr_3	15163-24-3	423.55	blk hex cry	dec 180				i H₂O
2959	Tungsten(III) chloride	WCl_3	20193-56-0	290.20	red solid	550 dec	subl			reac H₂O
2960	Tungsten(III) iodide	WI_3	15513-69-6	564.55	blk solid	dec r.t.				i H₂O; s ace; sl EtOH, chl
2961	Tungsten(IV) bromide	WBr_4	14055-81-3	503.46	blk orth cry		240 subl			reac H₂O
2962	Tungsten(IV) chloride	WCl_4	13470-13-8	325.65	blk hyg powder	450 dec		4.62		reac H₂O
2963	Tungsten(IV) fluoride	WF_4	13766-47-7	259.83	red-brn cry	dec 800				reac H₂O; s MeCN; i bz, tol, ctc

No.	Name	Formula	CAS Reg No.	Mol. weight	Physical form	mp/°C	bp/°C	Density g cm⁻³	Solubility g/100 g H₂O	Qualitative solubility
2964	Tungsten(IV) iodide	WI_4	14055-84-6	691.46	blk cry	dec				reac H_2O; s EtOH; i eth chl
2965	Tungsten(IV) oxide	WO_2	12036-22-5	215.84	brn monocl cry	≈1500 dec	1730	10.8		i H_2O, os
2966	Tungsten(IV) selenide	WSe_2	12067-46-8	341.76	gray hex cry			9.2		
2967	Tungsten(IV) sulfide	WS_2	12138-09-9	247.97	gray hex cry	1250 dec		7.6		i H_2O, HCl, alk
2968	Tungsten(IV) telluride	WTe_2	12067-76-4	439.04	gray orth cry	1020		9.43		
2969	Tungsten(V) bromide	WBr_5	13470-11-6	583.36	brn-blk hyg solid	286	333			reac H_2O
2970	Tungsten(V) chloride	WCl_5	13470-14-9	361.11	blk-grn hyg cry	253	286	3.88		reac H_2O
2971	Tungsten(V) ethanolate	$W(C_2H_5O)_5$	62571-53-3	409.14	powder		105(0.05 mmHg)			s EtAc
2972	Tungsten(V) fluoride	WF_5	19357-83-6	278.83	yel solid	dec 20				reac H_2O
2973	Tungsten(V) oxytribromide	$WOBr_3$	20213-56-3	439.55	dark brn tetr cry			≈5.9		
2974	Tungsten(V) oxytrichloride	$WOCl_3$	14249-98-0	306.20	grn tetr cry			≈4.6		
2975	Tungsten(VI) bromide	WBr_6	13701-86-5	663.26	blue-blk cry	309		6.9		reac H_2O
2976	Tungsten(VI) chloride	WCl_6	13283-01-7	396.56	purp hex cry; hyg	282	337	3.52		reac H_2O; s EtOH, os
2977	Tungsten(VI) dioxydibromide	WO_2Br_2	13520-75-7	375.65	red cry		440 subl			
2978	Tungsten(VI) dioxydichloride	WO_2Cl_2	13520-76-8	286.75	yel orth cry	265		4.67		i H_2O
2979	Tungsten(VI) dioxydiiodide	WO_2I_2	14447-89-3	469.65	grn monocl cry	400 dec		6.39		
2980	Tungsten(VI) fluoride	WF_6	7783-82-6	297.83	yel liq or col gas	1.9	17.1	3.44		reac H_2O; vs ctc, cyhex
2981	Tungsten(VI) oxide	WO_3	1314-35-8	231.84	yel powder	1473	≈1700	7.2		i H_2O, os; sl acid; s alk
2982	Tungsten(VI) oxytetrabromide	$WOBr_4$	13520-77-9	519.46	red tetr cry	277	327	≈5.5		reac H_2O
2983	Tungsten(VI) oxytetrachloride	$WOCl_4$	13520-78-0	341.65	red hyg cry	210	230	11.92		reac H_2O; s bz, CS_2
2984	Tungsten(VI) oxytetrafluoride	WOF_4	13520-79-1	275.83	wh monocl cry	105	185.9	5.07		reac H_2O
2985	Tungsten(VI) sulfide	WS_3	12125-19-8	280.04	brn powder					sl H_2O; s alk
2986	Uranium	U	7440-61-1	238.029	silv-wh orth cry	1135	4131	19.1		
2987	Uranium boride (UB_2)	UB_2	12007-36-2	259.651	refrac solid	2430		12.7		
2988	Uranium boride (UB_4)	UB_4	12007-84-0	281.273	refrac solid	2530		9.32		i H_2O
2989	Uranium carbide (UC)	UC	12070-09-6	250.040	gray cub cry	2790				
2990	Uranium carbide (UC_2)	UC_2	12071-33-9	262.050	gray tetr cry	2350	4370	11.3		reac H_2O; sl EtOH
2991	Uranium carbide (U_2C_3)	U_2C_3	12076-62-9	512.090	gray cub cry	≈1700 dec		12.7		
2992	Uranium nitride (UN)	UN	25658-43-9	252.036	gray cub cry	2805		14.3		i H_2O
2993	Uranium nitride (U_2N_3)	U_2N_3	12033-83-9	518.078	cub cry	dec		11.3		
2994	Uranium(III) bromide	UBr_3	13470-19-4	477.741	red hyg cry	727				s H_2O
2995	Uranium(III) chloride	UCl_3	10025-93-1	344.388	grn hyg cry	837		5.51		vs H_2O; i bz, ctc
2996	Uranium(III) fluoride	UF_3	13775-06-9	295.024	blk hex cry	1495		8.9		i H_2O; s acid
2997	Uranium(III) hydride	UH_3	13598-56-6	241.053	gray-blk cub cry			11.1		
2998	Uranium(III) iodide	UI_3	13775-18-3	618.742	blk hyg cry	766				s H_2O
2999	Uranium(IV) bromide	UBr_4	13470-20-7	557.645	brn hyg cry	519				s H_2O, EtOH
3000	Uranium(IV) chloride	UCl_4	10026-10-5	379.841	grn octahed cry	590	791	4.72		reac H_2O; s EtOH
3001	Uranium(IV) fluoride	UF_4	10049-14-6	314.023	grn monocl cry	1036	1417	6.7	0.01²⁵	s conc acid, alk
3002	Uranium(IV) iodide	UI_4	13470-22-9	745.647	blk hyg cry	506				s H_2O, EtOH
3003	Uranium(IV) oxide	UO_2	1344-57-6	270.028	brn cub cry	2847		10.97		i H_2O, dil acid; s conc acid
3004	Uranium(IV,V) oxide	U_4O_9	12037-15-9	1096.111	cub cry			11.2		
3005	Uranium(V) bromide	UBr_5	13775-16-1	637.549	brn hyg cry					reac H_2O
3006	Uranium(V) chloride	UCl_5	13470-21-8	415.294	brn hyg cry	287				reac H_2O
3007	Uranium(V) fluoride	UF_5	13775-07-0	333.021	pale blue tetr cry; hyg	348		5.81		s H_2O
3008	Uranium(V,VI) oxide	U_3O_8	1344-59-8	842.082	grn-blk orth cry	1300 dec		8.38		
3009	Uranium(VI) chloride	UCl_6	13763-23-0	450.747	grn hex cry	177		3.6		
3010	Uranium(VI) fluoride	UF_6	7783-81-5	352.019	wh monocl solid	64.06 tp	56.5 sp	5.09		reac H_2O; s ctc, chl
3011	Uranium(VI) oxide	UO_3	1344-58-7	286.027	oran-yel cry			≈7.3		i H_2O; s acid
3012	Uranium(VI) oxide monohydrate	$UO_3 \cdot H_2O$	12326-21-5	304.043	yel orth cry	570 dec		7.05		
3013	Uranium peroxide dihydrate	$UO_4 \cdot 2H_2O$	19525-15-6	338.057	yel hyg cry	115 dec				i H_2O
3014	Uranyl acetate dihydrate	$UO_2(C_2H_3O_2)_2 \cdot 2H_2O$	6159-44-0	424.146	ye cry (HOAc)	80 dec		2.89		sl EtOH
3015	Uranyl chloride	UO_2Cl_2	7791-26-6	340.934	yel orth cry; hyg	577				vs H_2O; s EtOH, ace; i bz
3016	Uranyl fluoride	UO_2F_2	13536-84-0	308.025	yel hyg solid				64.4²⁰	i bz
3017	Uranyl hydrogen phosphate tetrahydrate	$UO_2HPO_4 \cdot 4H_2O$	18433-48-2	438.068	yel cry pow					i H_2O; s acid
3018	Uranyl nitrate	$UO_2(NO_3)_2$	10102-06-4	394.037	yel cry				127²⁵	s eth
3019	Uranyl nitrate hexahydrate	$UO_2(NO_3)_2 \cdot 6H_2O$	13520-83-7	502.129	yel orth cry; hyg	60	118 dec	2.81	127²⁵	s EtOH, eth
3020	Uranyl sulfate	UO_2SO_4	1314-64-3	366.090	yel cry					
3021	Uranyl sulfate trihydrate	$UO_2SO_4 \cdot 3H_2O$	20910-28-5	420.137	yel cry			3.28	152¹⁶	sl EtOH
3022	Vanadium	V	7440-62-2	50.942	gray-wh metal; cub	1910	3407	6.0		i H_2O; s acid

No.	Name	Formula	CAS Reg No.	Mol. weight	Physical form	mp/°C	bp/°C	Density g cm^{-3}	Solubility g/100 g H$_2$O	Qualitative solubility
3023	Vanadocene	V(C$_5$H$_5$)$_2$	1277-47-0	181.128	viol cry; hyg	167				s bz, thf
3024	Vanadocene dichloride	V(C$_5$H$_5$)$_2$Cl$_2$	12083-48-6	252.034	dark grn cry	205 dec				s H$_2$O, chl, EtOH
3025	Vanadium boride (VB)	VB	12045-27-1	61.753	refrac solid	2250				i H$_2$O
3026	Vanadium boride (VB$_2$)	VB$_2$	12007-37-3	72.564	refrac solid	2450				
3027	Vanadium carbide (VC)	VC	12070-10-9	62.953	refrac blk cry; cub	2810		5.77		i H$_2$O
3028	Vanadium carbide (V$_2$C)	V$_2$C	12012-17-8	113.894	hex cry	2167				
3029	Vanadium carbonyl	V(CO)$_6$	14024-00-1	219.002	blue-grn cry; flam	60 dec	subl			i H$_2$O; s aqua regia
3030	Vanadium nitride	VN	24646-85-3	64.949	blk powder; cub	2050		6.13		s HF
3031	Vanadium silicide (VSi$_2$)	VSi$_2$	12039-87-1	107.113	metallic prisms			4.42		
3032	Vanadium silicide (V$_3$Si)	V$_3$Si	12039-76-8	180.911	cub cry	1935		5.70		
3033	Vanadium(II) bromide	VBr$_2$	14890-41-6	210.750	oran-brn hex cry		800 subl	4.58		reac H$_2$O
3034	Vanadium(II) chloride	VCl$_2$	10580-52-6	121.848	grn hex plates	1350	910 subl	3.23		reac H$_2$O; s EtOH, eth
3035	Vanadium(II) fluoride	VF$_2$	13842-80-3	88.939	blue hyg cry	1490				reac H$_2$O
3036	Vanadium(II) iodide	VI$_2$	15513-84-5	304.751	red-viol hex cry		subl 800	5.44		s acid
3037	Vanadium(II) oxide	VO	12035-98-2	66.941	gray-blk cry	1790		5.758		
3038	Vanadium(II) sulfate heptahydrate	VSO$_4$ · 7H$_2$O	36907-42-3	273.111	viol cry					
3039	Vanadium(III) bromide	VBr$_3$	13470-26-3	290.654	blk-grn hyg cry	dec 500	subl	4.00		reac H$_2$O
3040	Vanadium(III) chloride	VCl$_3$	7718-98-1	157.301	red-viol hex cry; hyg	500 dec		3.00		reac H$_2$O; s EtOH, eth
3041	Vanadium(III) fluoride	VF$_3$	10049-12-4	107.937	yel-grn hex cry	1395	subl	3.363		i H$_2$O, EtOH
3042	Vanadium(III) fluoride trihydrate	VF$_3$ · 3H$_2$O	10049-12-4*	161.983	grn rhomb cry	≈100 dec				sl H$_2$O
3043	Vanadium(III) iodide	VI$_3$	15513-94-7	431.655	brn-blk rhomb cry; hyg	dec 300		5.21		reac H$_2$O
3044	Vanadium(III) oxide	V$_2$O$_3$	1314-34-7	149.881	blk powder	1957	≈3000	4.87		i H$_2$O
3045	Vanadium(III) 2,4-pentanedioate	V(CH$_3$COCHCOCH$_3$)$_3$	13476-99-8	348.266	brn cry	≈185	subl	≈1.0		s MeOH, ace, bz chl
3046	Vanadium(III) sulfate	V$_2$(SO$_4$)$_3$	13701-70-7	390.071	yel powder	≈400 dec				sl H$_2$O
3047	Vanadium(III) sulfide	V$_2$S$_3$	1315-03-3	198.078	grn-blk powder	dec		4.7		i H$_2$O; s hot HCl
3048	Vanadium(IV) bromide	VBr$_4$	13595-30-7	370.558	unstab purp cry	-23 dec				
3049	Vanadium(IV) chloride	VCl$_4$	7632-51-1	192.754	red-brn liq	-28	151	1.816		reac H$_2$O; s EtOH, eth
3050	Vanadium(IV) fluoride	VF$_4$	10049-16-8	126.936	grn hyg powder	325 dec	subl	3.15		vs H$_2$O
3051	Vanadium(IV) oxide	VO$_2$	12036-21-4	82.941	blue-blk powder	1967		4.339		i H$_2$O; s acid, alk
3052	Vanadium(V) fluoride	VF$_5$	7783-72-4	145.934	col liq	19.5	48.3	2.50		reac H$_2$O
3053	Vanadium(V) dioxide fluoride	VO$_2$F	14259-82-6	101.939	brn hyg cry	350 dec				reac H$_2$O
3054	Vanadium(V) dioxide chloride	VO$_2$Cl	13759-30-3	118.394	oran hyg cry	dec 180				s thf
3055	Vanadium(V) oxide	V$_2$O$_5$	1314-62-1	181.880	yel-brn orth cry	681	1750	3.35	0.07^{25}	s conc acid, alk; i EtOH
3056	Vanadium(V) sulfide	V$_2$S$_5$	12138-17-9	262.208	grn-blk pow	dec		3.0		i H$_2$O; s acid, alk
3057	Vanadyl bromide	VOBr	13520-88-2	146.845	viol cry	480 dec				
3058	Vanadyl chloride	VOCl	13520-87-1	102.394	brn orth cry		127	1.72		
3059	Vanadyl dibromide	VOBr$_2$	13520-89-3	226.749	yel-brn cry	180 dec				
3060	Vanadyl dichloride	VOCl$_2$	10213-09-9	137.847	grn hyg cry	380 dec		2.88		reac H$_2$O; s EtOH
3061	Vanadyl difluoride	VOF$_2$	13814-83-0	104.938	yel cry					
3062	Vanadyl selenite hydrate	VOSeO$_3$ · H$_2$O	133578-89-9	211.92	grn tricl plates			3.506		
3063	Vanadyl sulfate dihydrate	VOSO$_4$ · 2H$_2$O	27774-13-6	199.035	blue cry powder					s H$_2$O
3064	Vanadyl tribromide	VOBr$_3$	13520-90-6	306.653	deep red liq	-59	170			reac H$_2$O
3065	Vanadyl trichloride	VOCl$_3$	7727-18-6	173.300	fuming red-yel liq	-79	127	1.829		reac H$_2$O; s MeOH, eth, ace
3066	Vanadyl trifluoride	VOF$_3$	13709-31-4	123.936	yel hyg powder	300	480	2.459		reac H$_2$O
3067	Water	H$_2$O	7732-18-5	18.015	col liq	0.00	99.974	0.9970^{25}		vs EtOH, MeOH, ace
3068	Water-d_2	D$_2$O	7789-20-0	20.027	col liq	3.82	101.42	1.1044^{25}		
3069	Water-t_2	T$_2$O	14940-65-9	22.032	col liq	4.48	101.51	1.2138^{25}		
3070	Xenon	Xe	7440-63-3	131.293	col gas	-111.745 tp (81.6 kPa)	-108.09	5.366 g/L		sl H$_2$O
3071	Xenon trioxide	XeO$_3$	13776-58-4	179.291	col orth cry	exp ≈25		4.55		s H$_2$O
3072	Xenon tetroxide	XeO$_4$	12340-14-6	195.291	yel solid or col gas; exp	-35.9	≈0 dec			
3073	Xenon difluoride	XeF$_2$	13709-36-9	169.290	col tetr cry	129.03 tp	114.35 sp	4.32		sl H$_2$O
3074	Xenon tetrafluoride	XeF$_4$	13709-61-0	207.287	col monocl cry	117.10 tp	115.75 sp	4.04		reac H$_2$O
3075	Xenon hexafluoride	XeF$_6$	13693-09-9	245.283	col monocl cry	49.48	75.6	3.56		reac H$_2$O
3076	Xenon fluoride oxide	XeOF$_2$	13780-64-8	185.289	yel solid, stab <-25	exp ≈0				reac H$_2$O
3077	Xenon oxytetrafluoride	XeOF$_4$	13774-85-1	223.286	col liq	-46.2		3.17^0		reac H$_2$O
3078	Xenon dioxydifluoride	XeO$_2$F$_2$	13875-06-4	201.289	col orth cry	30.8 exp		4.10		
3079	Xenon difluoride trioxide	XeO$_3$F$_2$	15192-14-0	217.288	unstab at r.t.	-54.1	exp			
3080	Xenon pentafluoride hexafluoroarsenate	XeF$_5$AsF$_6$	20328-94-3	415.197	wh monocl cry	130.5		3.51		

No.	Name	Formula	CAS Reg No.	Mol. weight	Physical form	mp/°C	bp/°C	Density g cm^{-3}	Solubility g/100 g H$_2$O	Qualitative solubility
3081	Xenon pentafluoride hexafluororuthenate	XeF$_5$RuF$_6$	39796-98-0	441.35	grn orth cry	152		3.79		
3082	Xenon fluoride hexafluoroantimonate	XeF$_3$SbF$_6$	39797-63-2	424.039	yel-grn monocl cry	≈110		3.92		
3083	Xenon fluoride hexafluoroarsenate	Xe$_2$F$_3$AsF$_6$	50432-32-1	508.494	yel-grn monocl cry	99		3.62		reac H$_2$O
3084	Xenon fluoride hexafluororuthenate	XeFRuF$_6$	22527-13-5	365.35	yel-grn monocl cry	110		3.78		
3085	Xenon fluoride undecafluoroantimonate	XeFSb$_2$F$_{11}$	15364-10-0	602.794	yel monocl cry	63		3.69		
3086	Xenon trifluoride undecafluoroantimonate	XeF$_3$Sb$_2$F$_{11}$	35718-37-7	640.791	yel-grn tricl cry	82		3.98		
3087	Ytterbium	Yb	7440-64-4	173.04	silv metal; cub	824	1196	6.90		s dil acid
3088	Ytterbium silicide	YbSi$_2$	12039-89-3	229.21	hex cry			7.54		
3089	Ytterbium(II) bromide	YbBr$_2$	25502-05-0	332.85	yel cry	673				reac H$_2$O
3090	Ytterbium(II) chloride	YbCl$_2$	13874-77-6	243.95	grn cry	721		5.27		reac H$_2$O
3091	Ytterbium(II) fluoride	YbF$_2$	15192-18-4	211.04	gray solid	1407				i H$_2$O
3092	Ytterbium(II) iodide	YbI$_2$	19357-86-9	426.85	blk cry	772				reac H$_2$O
3093	Ytterbium(III) acetate tetrahydrate	Yb(C$_2$H$_3$O$_2$)$_3$ · 4H$_2$O	15280-58-7	422.23	hyg col cry	dec 70		2.09		vs H$_2$O
3094	Ytterbium(III) bromide	YbBr$_3$	13759-89-2	412.75	col cry	956 dec				s H$_2$O
3095	Ytterbium(III) chloride	YbCl$_3$	10361-91-8	279.40	wh hyg powder	854				s H$_2$O
3096	Ytterbium(III) chloride hexahydrate	YbCl$_3$ · 6H$_2$O	19423-87-1	387.49	grn hyg cry	150 dec		2.57		vs H$_2$O
3097	Ytterbium(III) fluoride	YbF$_3$	13760-80-0	230.04	wh cry	1157		8.2		i H$_2$O
3098	Ytterbium(III) iodide	YbI$_3$	13813-44-0	553.75	yel cry	dec 700				s H$_2$O
3099	Ytterbium(III) nitrate	Yb(NO$_3$)$_3$	13768-67-7	359.06	col hyg solid				239[25]	s EtOH
3100	Ytterbium(III) oxide	Yb$_2$O$_3$	1314-37-0	394.08	col cub cry	2355	4070	9.2		s dil acid
3101	Ytterbium(III) sulfate octahydrate	Yb$_2$(SO$_4$)$_3$ · 8H$_2$O	10034-98-7	778.39	col cry			3.3	38.4[20]	
3102	Yttrium	Y	7440-65-5	88.906	silv metal; hex	1522	3345	4.47		reac H$_2$O; s dil acid
3103	Yttrium aluminum oxide	Y$_3$Al$_5$O$_{12}$	12005-21-9	593.619	grn cub cry			≈4.5		
3104	Yttrium antimonide	YSb	12186-97-9	210.666	cub cry	2310		5.97		
3105	Yttrium arsenide	YAs	12255-48-0	163.828	cub cry			5.59		
3106	Yttrium boride	YB$_6$	12008-32-1	153.772	refrac solid	2600		3.72		
3107	Yttrium bromide	YBr$_3$	13469-98-2	328.618	col hyg cry	904			83.3[30]	
3108	Yttrium carbide	YC$_2$	12071-35-1	112.927	refrac solid	≈2400		4.13		
3109	Yttrium carbonate trihydrate	Y$_2$(CO$_3$)$_3$ · 3H$_2$O	5970-44-5	411.885	red-brn powder					i H$_2$O; s dil acid
3110	Yttrium chloride	YCl$_3$	10361-92-9	195.265	wh monocl cry; hyg	721	1482	2.61	75.1[20]	vs H$_2$O
3111	Yttrium chloride hexahydrate	YCl$_3$ · 6H$_2$O	10025-94-2	303.356	hyg col cry	dec 100				vs H$_2$O; s EtOH
3112	Yttrium fluoride	YF$_3$	13709-49-4	145.901	wh hyg powder	1155		4.0		i H$_2$O
3113	Yttrium hydroxide	Y(OH)$_3$	16469-22-0	139.928	wh prec or pow	dec 190				i H$_2$O
3114	Yttrium iodide	YI$_3$	13470-38-7	469.619	hyg wh-yel cry	997				s H$_2$O, ace, EtOH
3115	Yttrium iron oxide	Y$_3$Fe$_5$O$_{12}$	12063-56-8	737.936	cub cry	1555				
3116	Yttrium nitrate	Y(NO$_3$)$_3$	10361-93-0	274.921	wh hyg solid				149[25]	s EtOH
3117	Yttrium nitrate tetrahydrate	Y(NO$_3$)$_3$ · 4H$_2$O	13773-69-8	346.982	red-wh prisms			2.68	149[25]	
3118	Yttrium nitrate hexahydrate	Y(NO$_3$)$_3$ · 6H$_2$O	13494-98-9	383.012	hyg cry				149[25]	
3119	Yttrium oxide	Y$_2$O$_3$	1314-36-9	225.810	wh cry; cub	2439		5.03		s dil acid
3120	Yttrium phosphide	YP	12294-01-8	119.880	cub cry			≈4.4		
3121	Yttrium sulfate octahydrate	Y$_2$(SO$_4$)$_3$ · 8H$_2$O	7446-33-5	610.122	red monocl cry			2.6	7.47[16]	
3122	Yttrium sulfide	Y$_2$S$_3$	12039-19-9	274.007	yel cub cry	1925		3.87		
3123	Zinc	Zn	7440-66-6	65.409	blue-wh metal; hex	419.53	907	7.134		s acid, alk
3124	Zinc acetate dihydrate	Zn(C$_2$H$_3$O$_2$)$_2$ · 2H$_2$O	5970-45-6	219.527	wh powder	237 dec		1.735	30.0[20]	s EtOH
3125	Zinc ammonium sulfate	Zn(NH$_4$)$_2$(SO$_4$)$_2$	7783-24-6	293.611	wh cry				9.2[20]	
3126	Zinc antimonide	ZnSb	12039-35-9	187.169	silv-wh orth cry	565		6.33		reac H$_2$O
3127	Zinc arsenate	Zn$_3$(AsO$_4$)$_2$	13464-44-3	474.065	wh powder				0.000078[20]	s acid, alk
3128	Zinc arsenate octahydrate	Zn$_3$(AsO$_4$)$_2$ · 8H$_2$O	13464-45-4	618.187	wh monocl cry			3.33	0.000078[20]	s acid, alk
3129	Zinc arsenide	Zn$_3$As$_2$	12006-40-5	346.070	powder	1015		5.528		
3130	Zinc arsenite	Zn(AsO$_2$)$_2$	10326-24-6	279.250	col powder					i H$_2$O; s acid
3131	Zinc borate	3ZnO · 2B$_2$O$_3$	27043-84-1	383.466	wh amorp powder			3.64		sl H$_2$O; s dil acid
3132	Zinc borate hemiheptahydrate	2ZnO · 3B$_2$O$_3$ · 3.5H$_2$O	12513-27-8	434.69	wh cry	980		4.22		i H$_2$O
3133	Zinc borate pentahydrate	2ZnO · 3B$_2$O$_3$ · 5H$_2$O	12536-65-1	461.753	wh powder			3.64	0.007[25]	sl HCl
3134	Zinc bromate hexahydrate	Zn(BrO$_3$)$_2$ · 6H$_2$O	13517-27-6	429.305	wh hyg solid	100		2.57		vs H$_2$O
3135	Zinc bromide	ZnBr$_2$	7699-45-8	225.217	wh hex cry; hyg	402	≈670	4.5	488[25]	vs EtOH; s eth
3136	Zinc caprylate	Zn(C$_8$H$_{15}$O$_2$)$_2$	557-09-5	351.816	wh hyg cry	136				sl H$_2$O
3137	Zinc carbonate	ZnCO$_3$	3486-35-9	125.418	wh hex cry	140 dec		4.434	0.000091[20]	s dil acid, alk
3138	Zinc carbonate hydroxide	3Zn(OH)$_2$ · 2ZnCO$_3$	12070-69-8	549.107	wh powder					
3139	Zinc chlorate	Zn(ClO$_3$)$_2$	10361-95-2	232.311	yel hyg cry	60 dec		2.15	200[20]	
3140	Zinc chloride	ZnCl$_2$	7646-85-7	136.315	wh hyg cry	290	732	2.907	408[25]	vs H$_2$O; s EtOH, ace

No.	Name	Formula	CAS Reg No.	Mol. weight	Physical form	mp/°C	bp/°C	Density g cm⁻³	Solubility g/100 g H_2O	Qualitative solubility
3141	Zinc chromate	$ZnCrO_4$	13530-65-9	181.403	yel prisms	316		3.40	3.08	s acid; i ace
3142	Zinc chromite	$ZnCr_2O_4$	12018-19-8	233.399	grn cub cry			5.29		
3143	Zinc citrate dihydrate	$Zn_3(C_6H_5O_7)_2 \cdot 2H_2O$	546-46-3	610.456	col powder					sl H_2O; s dil acid, alk
3144	Zinc cyanide	$Zn(CN)_2$	557-21-1	117.443	wh powder			1.852	0.00047²⁰	reac acid
3145	Zinc diethyl	$Zn(C_2H_5)_2$	557-20-0	123.531	col liq	-28	118	1.2065		reac H_2O; msc eth, peth, bz
3146	Zinc dithionate	ZnS_2O_4	7779-86-4	193.537	wh amorp solid	200 dec			40²⁰	
3147	Zinc fluoride	ZnF_2	7783-49-5	103.406	wh tetr needles; hyg	872	1500	4.9	1.55²⁵	sl H_2O
3148	Zinc fluoride tetrahydrate	$ZnF_2 \cdot 4H_2O$	13986-18-0	175.468	wh orth cry			2.30	1.55²⁵	
3149	Zinc fluoroborate hexahydrate	$Zn(BF_4)_2 \cdot 6H_2O$	27860-83-9	347.109	hex cry			2.12		vs H_2O; s EtOH
3150	Zinc formate dihydrate	$Zn(CHO_2)_2 \cdot 2H_2O$	5970-62-7	191.474	wh cry			2.207	5.2²⁰	i EtOH
3151	Zinc hexafluorosilicate hexahydrate	$ZnSiF_6 \cdot 6H_2O$	16871-71-9	315.576	wh cry					s H_2O
3152	Zinc hydroxide	$Zn(OH)_2$	20427-58-1	99.424	col orth cry	125 dec		3.05	0.000042²⁰	
3153	Zinc iodate	$Zn(IO_3)_2$	7790-37-6	415.214	wh cry powder				0.64²⁵	
3154	Zinc iodide	ZnI_2	10139-47-6	319.218	wh-yel hyg cry	450	625	4.74	438²⁵	vs H_2O; s EtOH, eth
3155	Zinc laurate	$Zn(C_{12}H_{23}O_2)_2$	2452-01-9	464.029	wh powder	128				sl H_2O
3156	Zinc molybdate	$ZnMoO_4$	13767-32-3	225.35	wh tetr cry	>700		4.3		i H_2O
3157	Zinc nitrate	$Zn(NO_3)_2$	7779-88-6	189.418	wh powder				120²⁵	
3158	Zinc nitrate hexahydrate	$Zn(NO_3)_2 \cdot 6H_2O$	10196-18-6	297.510	col orth cry	36 dec		2.067	120²⁵	vs EtOH
3159	Zinc nitride	Zn_3N_2	1313-49-1	224.240	blue-gray cub cry	700 dec		6.22		i H_2O
3160	Zinc nitrite	$Zn(NO_2)_2$	10102-02-0	157.420	hyg solid					reac H_2O
3161	Zinc oleate	$Zn(C_{18}H_{33}O_2)_2$	557-07-3	628.316	wh powder	70 dec				i H_2O; s EtOH, eth, bz
3162	Zinc oxalate	ZnC_2O_4	547-68-2	153.428	wh pwd				0.0026²⁵	
3163	Zinc oxalate dihydrate	$ZnC_2O_4 \cdot 2H_2O$	4255-07-6	189.458	wh powder	100 dec		2.56	0.0026²⁵	s dil acid
3164	Zinc tartrate dihydrate	$ZnC_4H_4O_6 \cdot 2H_2O$	22570-08-7	249.511	wh cry pow	dec 150			0.022²⁰	
3165	Zinc oxide	ZnO	1314-13-2	81.408	wh powder; hex	1974		5.6		i H_2O; s dil acid
3166	Zinc 2,4-pentanedioate	$Zn(CH_3COCHCOCH_3)_2$	14024-63-6	263.625	cry	137 dec				sl H_2O; s EtOH, DMSO
3167	Zinc pentanoate dihydrate	$Zn(C_5H_9O_2)_2 \cdot 2H_2O$	556-38-7	303.687	scales or powder					sl H_2O; reac acid; s EtOH
3168	Zinc perchlorate hexahydrate	$Zn(ClO_4)_2 \cdot 6H_2O$	10025-64-6	372.402	wh cub cry; hyg	106 dec		2.2	121.3²⁵	s EtOH
3169	Zinc permanganate hexahydrate	$Zn(MnO_4)_2 \cdot 6H_2O$	23414-72-4	411.372	blk orth cry; hyg			2.45		s H_2O; reac EtOH
3170	Zinc peroxide	ZnO_2	1314-22-3	97.408	yel-wh powder	>150 dec	212 exp	1.57		i H_2O; reac acid, EtOH, ace
3171	Zinc phosphate	$Zn_3(PO_4)_2$	7779-90-0	386.170	wh monocl cry	900		4.0		i H_2O
3172	Zinc phosphate tetrahydrate	$Zn_3(PO_4)_2 \cdot 4H_2O$	7543-51-3	458.231	col orth cry			3.04		i H_2O, EtOH; s dil acid, alk
3173	Zinc phosphide	Zn_3P_2	1314-84-7	258.175	gray tetr cry	1160		4.55		i H_2O, EtOH; reac acid; s bz
3174	Zinc pyrophosphate	$Zn_2P_2O_7$	7446-26-6	304.761	wh cry powder			3.75		i H_2O; s dil acid
3175	Zinc selenate pentahydrate	$ZnSeO_4 \cdot 5H_2O$	13597-54-1	298.44	tricl cry	50 dec		2.59	63.4²⁵	
3176	Zinc selenide	$ZnSe$	1315-09-9	144.37	yel-red cub cry	>1100	subl	5.65		i H_2O; s dil acid
3177	Zinc orthosilicate	Zn_2SiO_4	13597-65-4	222.902	wh hex cry	1509		4.1		i H_2O, dil acid
3178	Zinc selenite	$ZnSeO_3$	13597-46-1	192.37	wh powder	621				
3179	Zinc stearate	$Zn(C_{18}H_{35}O_2)_2$	557-05-1	632.348	wh powder	130		1.095		i H_2O, EtOH, eth; s bz
3180	Zinc sulfate	$ZnSO_4$	7733-02-0	161.472	col orth cry	680 dec		3.8	57.7²⁵	
3181	Zinc sulfate monohydrate	$ZnSO_4 \cdot H_2O$	7446-19-7	179.487	wh monocl cry	238 dec		3.20	57.7²⁵	i EtOH
3182	Zinc sulfate heptahydrate	$ZnSO_4 \cdot 7H_2O$	7446-20-0	287.578	col orth cry	100 dec		1.97	57.7²⁵	i EtOH
3183	Zinc sulfide (sphalerite)	ZnS	1314-98-3	97.474	gray-wh cub cry	trans wurtzite 1020		4.04		i H_2O, EtOH; s dil acid
3184	Zinc sulfide (wurtzite)	ZnS	1314-98-3	97.474	wh hex cry	1700	subl	4.09		i H_2O; s dil acid
3185	Zinc sulfite dihydrate	$ZnSO_3 \cdot 2H_2O$	7488-52-0	181.503	wh powder	200 dec			0.224²⁵	i EtOH
3186	Zinc telluride	$ZnTe$	1315-11-3	193.01	red cub cry	1239		5.9		i H_2O
3187	Zinc thiocyanate	$Zn(SCN)_2$	557-42-6	181.573	wh hyg cry					sl H_2O; s EtOH
3188	Zirconium	Zr	7440-67-7	91.224	gray-wh metal; hex	1854.7	4409	6.52		s hot conc acid
3189	Zirconocene dichloride	$Zr(C_5H_5)_2Cl_2$	1291-32-3	292.316	col cry	248	subl 150			
3190	Zirconium boride	ZrB_2	12045-64-6	112.846	gray refrac solid; hex	3050		6.17		
3191	Zirconium carbide	ZrC	12020-14-3	103.235	gray refrac solid; cub	3532		6.73		s HF
3192	Zirconium nitride	ZrN	25658-42-8	105.231	yel cub cry	2952		7.09		s conc HF; sl dil acid
3193	Zirconium phosphide	ZrP_2	12037-80-8	153.172	orth cry			≈5.1		
3194	Zirconium silicide	$ZrSi_2$	12039-90-6	147.395	gray powder	1620		4.88		i H_2O, aqua regia; s HF

No.	Name	Formula	CAS Reg No.	Mol. weight	Physical form	mp/°C	bp/°C	Density g cm^{-3}	Solubility g/100 g H$_2$O	Qualitative solubility
3195	Zirconium(II) bromide	ZrBr$_2$	24621-17-8	251.032	blue-blk cry	dec 400				
3196	Zirconium(II) chloride	ZrCl$_2$	13762-26-0	162.130	blk cry	772		3.16		reac H$_2$O
3197	Zirconium(II) fluoride	ZrF$_2$	13842-94-9	129.221	blk cry	902				
3198	Zirconium(II) hydride	ZrH$_2$	7704-99-6	93.240	gray tetr cry	800 dec		5.6		i H$_2$O
3199	Zirconium(II) iodide	ZrI$_2$	15513-85-6	345.033	blk cry	827				
3200	Zirconium(III) bromide	ZrBr$_3$	24621-18-9	330.936	dark blue cry	dec 300				
3201	Zirconium(III) chloride	ZrCl$_3$	10241-03-9	197.583	dark blue cry	627		3.05		reac H$_2$O
3202	Zirconium(III) fluoride	ZrF$_3$	13814-22-7	148.219	blue-grn cry	927		4.26		i H$_2$O; s acid
3203	Zirconium(III) iodide	ZrI$_3$	13779-87-8	471.937	dark blue cry	727				
3204	Zirconium(IV) acetate hydroxide	Zr(C$_2$H$_3$O$_2$)$_2$(OH)$_2$	14311-93-4	243.327	wh amorp solid					s H$_2$O
3205	Zirconium(IV) ammonium carbonate dihydrate	Zr(NH$_4$)$_3$OH(CO$_3$)$_3$ · 2H$_2$O	12616-24-9*	362.404	prisms; unstab					s H$_2$O
3206	Zirconium(IV) bromide	ZrBr$_4$	13777-25-8	410.840	wh cub cry	450 tp	360 sp	3.98		reac H$_2$O
3207	Zirconium(IV) chloride	ZrCl$_4$	10026-11-6	233.036	wh monocl cry; hyg	437 tp	331 sp	2.80		reac H$_2$O; s EtOH, eth
3208	Zirconium(IV) fluoride	ZrF$_4$	7783-64-4	167.218	wh monocl cry	910	912 sp	4.43	1.5^{25}	
3209	Zirconium(IV) hydroxide	Zr(OH)$_4$	14475-63-9	159.254	wh amorp powder	dec		3.25		i H$_2$O; s acid
3210	Zirconium(IV) iodide	ZrI$_4$	13986-26-0	598.842	yel-oran cub cry	500	431 sp	4.85		vs H$_2$O
3211	Zirconium(IV) nitrate pentahydrate	Zr(NO$_3$)$_4$ · 5H$_2$O	13746-89-9	429.320	wh hyg cry	100 dec				vs H$_2$O; s EtOH
3212	Zirconium(IV) orthosilicate	ZrSiO$_4$	10101-52-7	183.308	wh tetr cry	1540 dec		4.6		i H$_2$O, acid
3213	Zirconium(IV) oxide	ZrO$_2$	1314-23-4	123.223	wh amorp powder	2710	4300	5.68		i H$_2$O; sl acid
3214	Zirconium(IV) pyrophosphate	ZrP$_2$O$_7$	13565-97-4	265.167	wh refrac solid	dec 1550				i H$_2$O, dil acid; s HF
3215	Zirconium(IV) sulfate	Zr(SO$_4$)$_2$	14644-61-2	283.349	wh hyg cry	410 dec		3.22		s H$_2$O; sl EtOH
3216	Zirconium(IV) sulfate tetrahydrate	Zr(SO$_4$)$_2$ · 4H$_2$O	7446-31-3	355.411	wh tetr cry	100 dec		2.80		vs H$_2$O
3217	Zirconium(IV) sulfide	ZrS$_2$	12039-15-5	155.354	red-brn hex cry	1550		3.87		i H$_2$O
3218	Zirconium(IV) tungstate	Zr(WO$_4$)$_2$	16853-74-0	586.92	grn pow					
3219	Zirconyl chloride	ZrOCl$_2$	7699-43-6	178.129	wh solid	250 dec				s H$_2$O, EtOH
3220	Zirconyl chloride octahydrate	ZrOCl$_2$ · 8H$_2$O	13520-92-8	322.252	tetr cry	400 dec		1.91		vs H$_2$O, EtOH

FORMULA INDEX OF INORGANIC COMPOUNDS

Formula	No.	Formula	No.	Formula	No.	Formula	No.
LaN	1421	$Li_2CrO_4 \cdot 2H_2O$	1522	$Mg(C_2H_3O_2)_2 \cdot H_2O$	1598	Mn	1681
$La(NO_3)_3 \cdot 6H_2O$	1420	$Li_2Cr_2O_7 \cdot 2H_2O$	1527	$Mg(C_2H_3O_2)_2 \cdot 4H_2O$	1599	MnB	1685
$La(OH)_3$	1417	Li_2MnO_3	1546	$MgCl_2$	1620	MnB_2	1686
LaS	1428	Li_2MoO_4	1551	$MgCl_2 \cdot 6H_2O$	1621	$MnB_4O_7 \cdot 8H_2O$	1724
$LaSi_2$	1424	Li_2O	1560	$Mg(ClO_3)_2 \cdot 6H_2O$	1619	$MnBr_2$	1695
$La_2(CO_3)_3 \cdot 8H_2O$	1412	Li_2O_2	1563	$Mg(ClO_4)_2$	1652	$MnBr_2 \cdot 4H_2O$	1696
La_2O_3	1422	$Li_2Pt(CN)_4 \cdot 5H_2O$	1573	$Mg(ClO_4)_2 \cdot 6H_2O$	1653	$MnCO_3$	1697
La_2S_3	1429	Li_2S	1569	$MgCrO_4 \cdot 7H_2O$	1622	$MnC_2O_4 \cdot 2H_2O$	1712
$La_2(SO_4)_3$	1425	Li_2SO_4	1567	$MgCr_2O_4$	1623	$Mn(C_2H_3O_2)_2 \cdot 4H_2O$	1694
$La_2(SO_4)_3 \cdot 8H_2O$	1426	$Li_2SO_4 \cdot H_2O$	1568	MgF_2	1626	$Mn(C_2H_3O_2)_3 \cdot 2H_2O$	1728
$La_2(SO_4)_3 \cdot 9H_2O$	1427	$Li_2SeO_3 \cdot H_2O$	1565	$MgHPO_4 \cdot 3H_2O$	1631	$Mn(CO)_5Br$	1690
Li	1505	$Li_2SeO_4 \cdot H_2O$	1564	MgH_2	1630	$Mn(C_5H_5)_2$	1682
$LiAlH_4$	1508	Li_2SiF_6	1536	$Mg(H_2PO_4)_2 \cdot 2H_2O$	1673	$MnCl_2$	1698
$LiAlSi_2O_6$	1509	Li_2SiO_3	1550	MgI_2	1634	$MnCl_2 \cdot 4H_2O$	1699
$LiAsF_6$	1534	Li_2SnF_6	1537	$MgI_2 \cdot 6H_2O$	1635	$Mn(ClO_4)_2 \cdot 6H_2O$	1714
$LiBF_4$	1574	Li_2TiO_3	1576	$MgI_2 \cdot 8H_2O$	1636	MnF_2	1701
$LiBH_4$	1513	Li_2WO_4	1577	$Mg(IO_3)_2 \cdot 4H_2O$	1633	MnF_3	1729
$LiBO_2$	1547	Li_2ZrO_3	1579	$Mg(MnO_4)_2 \cdot 6H_2O$	1654	$Mn(H_2PO_2)_2 \cdot H_2O$	1703
$LiBO_2 \cdot 2H_2O$	1548	Li_3AsO_4	1511	$MgMoO_4$	1640	$Mn(H_2PO_4)_2 \cdot 2H_2O$	1700
$LiBr$	1515	$Li_3C_6H_5O_7 \cdot 4H_2O$	1523	$MgNH_4PO_4 \cdot 6H_2O$	1603	MnI_2	1704
$LiBr \cdot H_2O$	1516	Li_3N	1554	$Mg(NH_2)_2$	1602	$MnI_2 \cdot 4H_2O$	1705
$LiBrO_3$	1514	Li_3PO_4	1559	$Mg(NO_2)_2 \cdot 3H_2O$	1645	$MnMoO_4$	1707
$Li(CHO_2) \cdot H_2O$	1532	Lr	1430	$Mg(NO_3)_2$	1641	$Mn(NO_3)_2$	1708
$LiCN$	1525	Lu	1580	$Mg(NO_3)_2 \cdot 2H_2O$	1642	$Mn(NO_3)_2 \cdot 4H_2O$	1709
$LiC_2H_3O_2$	1506	LuB_4	1581	$Mg(NO_3)_2 \cdot 6H_2O$	1643	$Mn(NO_3)_2 \cdot 6H_2O$	1710
$LiC_2H_3O_2 \cdot 2H_2O$	1507	$LuBr_3$	1582	MgO	1650	MnO	1713
$LiC_{18}H_{35}O_2$	1566	$LuCl_3$	1583	$MgO \cdot ZrO_2 \cdot SiO_2$	1680	MnO_2	1732
$LiCl$	1520	$LuCl_3 \cdot 6H_2O$	1584	MgO_2	1655	$MnO(OH)$	1730
$LiCl \cdot H_2O$	1521	$Lu(ClO_4)_3 \cdot 6H_2O$	1591	$Mg(OH)_2$	1632	$Mn(OH)_2$	1702
$LiClO_3$	1519	LuF_3	1585	MgS	1670	MnP	1691
$LiClO_4$	1561	LuI_3	1586	$MgSO_3 \cdot 3H_2O$	1671	MnS	1720
$LiClO_4 \cdot 3H_2O$	1562	LuN	1589	$MgSO_3 \cdot 6H_2O$	1672	MnS	1721
$LiCoO_2$	1524	$Lu(NO_3)_3$	1588	$MgSO_4$	1667	MnS	1722
LiD	1526	Lu_2O_3	1590	$MgSO_4 \cdot H_2O$	1668	$MnSO_4$	1717
LiF	1531	Lu_2S_3	1594	$MgSO_4 \cdot 7H_2O$	1669	$MnSO_4 \cdot H_2O$	1718
$LiFeSi$	1530	$Lu_2(SO_4)_3$	1592	$Mg(SCN)_2 \cdot 4H_2O$	1674	$MnSO_4 \cdot 4H_2O$	1719
LiH	1538	$Lu_2(SO_4)_3 \cdot 8H_2O$	1593	$MgS_2O_3 \cdot 6H_2O$	1675	$MnSb$	1683
LiH_2PO_4	1528	Lu_2Te_3	1595	$MgSe$	1662	$MnSe$	1716
$LiHCO_3$	1539	$Lu_3Fe_5O_{12}$	1587	$MgSeO_3 \cdot 6H_2O$	1663	$MnSiO_3$	1706
LiI	1544	Md	1734	$MgSeO_4 \cdot 6H_2O$	1661	$MnSi_2$	1693
$LiI \cdot 3H_2O$	1545	Mg	1596	$MgSiF_6 \cdot 6H_2O$	1664	$MnTe$	1723
$LiIO_3$	1543	$Mg(AlO_2)_2$	1600	$MgSiO_3$	1638	$MnTiO_3$	1725
$LiN(C_3H_7)_2$	1529	MgB_2	1606	$MgTiO_3$	1639	$MnWO_4$	1726
$LiNH_2$	1510	$Mg(BO_2)_2 \cdot 8H_2O$	1637	$MgWO_4$	1677	Mn_2B	1687
$LiNO_2$	1555	$Mg(BO_3)_2 \cdot 7H_2O$	1651	$MgZrO_3$	1679	$Mn_2(CO)_{10}$	1689
$LiNO_2 \cdot H_2O$	1556	MgB_6	1607	$Mg_2Al_3(AlSi_5O_{18})$	1601	Mn_2O_3	1731
$LiNO_3$	1553	MgB_{12}	1608	Mg_2Ge	1629	Mn_2O_7	1733
LiN_3	1512	$MgBr_2$	1610	Mg_2GeO_4	1628	Mn_2P	1692
$LiNbO_3$	1552	$MgBr_2 \cdot 6H_2O$	1611	$Mg_2P_2O_7$	1658	$Mn_2P_2O_7$	1715
$LiOCl$	1542	$Mg(BrO_3)_2 \cdot 6H_2O$	1609	$Mg_2P_2O_7 \cdot 3H_2O$	1659	Mn_2Sb	1684
$LiOH$	1540	$MgCO_3$	1612	Mg_2Si	1665	Mn_2SiO_4	1711
$LiOH \cdot H_2O$	1541	$MgCO_3 \cdot 2H_2O$	1613	Mg_2SiO_4	1646	Mn_3C	1688
$LiPF_6$	1535	$MgCO_3 \cdot 3H_2O$	1614	$Mg_2Si_3O_8$	1676	Mn_3O_4	1727
$LiPO_3$	1549	$MgCO_3 \cdot 5H_2O$	1615	Mg_2Sn	1666	Mo	1789
$LiSCN$	1575	$MgCO_3 \cdot Mg(OH)_2 \cdot 3H_2O$	1618	Mg_2TiO_4	1647	$MoBr_2$	1800
$LiSbF_6$	1533	$4MgCO_3 \cdot Mg(OH)_2 \cdot 4H_2O$	1616	$Mg_2V_2O_7$	1678	$MoBr_3$	1803
$LiSiO_4$	1557	$4MgCO_3 \cdot Mg(OH)_2 \cdot 5H_2O$	1617	Mg_3As_2	1605	$MoBr_4$	1808
$LiTaO_3$	1570	$Mg(CHO_2)_2 \cdot 2H_2O$	1627	$Mg_3(C_6H_5O_7)_2$	1624	MoC	1793
$LiVO_3$	1578	MgC_2O_4	1648	$Mg_3(C_6H_5O_7)_2 \cdot 14H_2O$	1625	$Mo(CO)_6$	1795
$Li_2B_4O_7$	1571	$MgC_2O_4 \cdot 2H_2O$	1649	Mg_3N_2	1644	$MoCl_2$	1801
$Li_2B_4O_7 \cdot 5H_2O$	1572	$Mg(C_2H_3O_2)_2$	1597	Mg_3P_2	1660	$MoCl_3$	1804
Li_2CO_3	1518			$Mg_3(PO_4)_2 \cdot 5H_2O$	1656	$MoCl_4$	1809
Li_2C_2	1517			$Mg_3(PO_4)_2 \cdot 8H_2O$	1657	$MoCl_5$	1816
$Li_2C_2O_4$	1558			Mg_3Sb_2	1604	MoF_3	1805

Formula	No.	Formula	No.	Formula	No.	Formula	No.
$SrZrO_3$	2722	ThH_2	2848	$TlOH$	2825	U_4O_9	3004
Sr_2SiO_4	2705	ThI_4	2855	$TlPF_6$	2824	V	3022
Sr_3N_2	2703	ThN	2858	$TlSe$	2846	VB	3025
$Sr_3(PO_4)_2$	2711	$Th(NO_3)_4$	2856	Tl_2CO_3	2816	VB_2	3026
T_2	1243	$Th(NO_3)_4 \cdot 4H_2O$	2857	$Tl_2C_2O_4$	2831	VBr_2	3033
T_2O	3069	ThO_2	2859	Tl_2CrO_4	2819	VBr_3	3039
Ta	2758	ThS_2	2864	Tl_2MoO_4	2828	VBr_4	3048
$TaAl_3$	2759	$Th(SO_4)_2 \cdot 9H_2O$	2863	Tl_2O	2832	VC	3027
TaB	2760	$ThSe_2$	2860	Tl_2O_3	2844	$V(CO)_6$	3029
TaB_2	2761	$ThSiO_4$	2861	Tl_2S	2837	$V(C_5H_5)_2$	3023
$TaBr_3$	2767	$ThSi_2$	2862	Tl_2SO_4	2836	$V(C_5H_5)_2Cl_2$	3024
$TaBr_4$	2769	Ti	2912	$Tl_2(SO_4)_3$	2845	$V(CH_3COCHCOCH_3)_3$	3045
$TaBr_5$	2776	TiI_3	2928	Tl_2Se	2835	VCl_2	3034
TaC	2762	TiB_2	2915	Tl_2SeO_4	2834	VCl_3	3040
$TaCl_3$	2768	$TiBr_2$	2920	Tm	2865	VCl_4	3049
$TaCl_4$	2770	$TiBr_3$	2925	$TmBr_2$	2866	VF_2	3035
$TaCl_5$	2777	$TiBr_4$	2933	$TmBr_3$	2869	VF_3	3041
TaF_5	2778	TiC	2916	$TmCl_2$	2867	$VF_3 \cdot 3H_2O$	3042
TaH	2764	$Ti(C_5H_5)_2Cl_2$	2913	$TmCl_3$	2870	VF_4	3050
TaI_4	2771	$TiCl_2$	2921	$TmCl_3 \cdot 7H_2O$	2871	VF_5	3052
TaI_5	2779	$TiCl_3$	2926	TmF_3	2872	VI_2	3036
TaN	2765	$TiCl_4$	2934	TmI_2	2868	VI_3	3043
TaO_2	2772	TiF_3	2927	TmI_3	2874	VN	3030
TaS_2	2774	TiF_4	2935	$Tm(NO_3)_3$	2875	VO	3037
$TaSe_2$	2773	TiH_2	2914	$Tm(NO_3)_3 \cdot 5H_2O$	2876	$VOBr$	3057
$TaSi_2$	2766	TiI_2	2922	$Tm(OH)_3$	2873	$VOBr_2$	3059
$TaTe_2$	2775	TiI_4	2936	$Tm_2(C_2O_4)_3 \cdot 6H_2O$	2877	$VOBr_3$	3064
Ta_2C	2763	TiN	2917	Tm_2O_3	2878	$VOCl$	3058
Ta_2O_5	2780	TiO	2923	U	2986	$VOCl_2$	3060
Tb	2797	$TiOSO_4 \cdot H_2O$	2940	UB_2	2987	$VOCl_3$	3065
$TbBr_3$	2800	TiO_2	2937	UB_4	2938	VOF_2	3061
$TbCl_3$	2801	TiO_2	2938	UBr_3	2994	VOF_3	3066
$TbCl_3 \cdot 6H_2O$	2802	TiO_2	2939	UBr_4	2999	$VOSO_4 \cdot 2H_2O$	3063
TbF_3	2809	TiP	2918	UBr_5	3005	$VOSeO_3 \cdot H_2O$	3062
TbF_4	2810	TiS	2924	UC	2989	VO_2	3051
TbI_3	2803	TiS_2	2942	UC_2	2990	VO_2Cl	3054
TbN	2798	$Ti(SO_4)_2$	2941	UCl_3	2995	VO_2F	3053
$Tb(NO_3)_3$	2804	$TiSi_2$	2919	UCl_4	3000	$VSO_4 \cdot 7H_2O$	3038
$Tb(NO_3)_3 \cdot 6H_2O$	2805	Ti_2O_3	2929	UCl_5	3006	VSi_2	3031
$TbSi_2$	2799	Ti_2S_3	2931	UCl_6	3009	V_2C	3028
Tb_2O_3	2806	$Ti_2(SO_4)_3$	2930	UF_3	2996	V_2O_3	3044
Tb_2S_3	2808	Ti_3O_5	2932	UF_4	3001	V_2O_5	3055
$Tb_2(SO_4)_3 \cdot 8H_2O$	2807	Tl	2811	UF_5	3007	V_2S_3	3047
Tc	2781	$TlBr$	2815	UF_6	3010	$V_2(SO_4)_3$	3046
TcF_5	2782	$TlBrO_3$	2814	UH_3	2997	V_2S_5	3056
TcF_6	2783	$TlBr_3 \cdot 4H_2O$	2839	UI_3	2998	V_3Si	3032
Te	2784	$TlCHO_2$	2823	UI_4	3002	W	2943
$TeBr_2$	2789	$TlCN$	2820	UN	2992	WB	2946
$TeBr_4$	2791	$TlC_2H_3O_2$	2812	UO_2	3003	WBr_2	2955
$TeCl_2$	2790	TlC_2H_5O	2821	$UO_2(C_2H_3O_2)_2 \cdot 2H_2O$	3014	WBr_3	2958
$TeCl_4$	2792	$Tl(C_2H_3O_2)_3$	2838	UO_2Cl_2	3015	WBr_4	2961
TeF_4	2793	$TlCl$	2818	UO_2F_2	3016	WBr_5	2969
TeF_6	2796	$TlClO_3$	2817	$UO_2HPO_4 \cdot 4H_2O$	3017	WBr_6	2975
TeI_4	2795	$TlClO_4$	2833	$UO_2(NO_3)_2$	3018	WC	2949
TeO_2	2787	$TlCl_3$	2840	$UO_2(NO_3)_2 \cdot 6H_2O$	3019	$W(CO)_6$	2950
TeO_3	2788	$TlCl_3 \cdot 4H_2O$	2841	$UO_2(NH_4)_3F_5$	204	$W(C_2H_5O)_5$	2971
Te_2F_{10}	2794	TlF	2822	UO_2SO_4	3020	WCl_2	2956
Th	2847	TlF_3	2842	$UO_2SO_4 \cdot 3H_2O$	3021	WCl_3	2959
ThB_6	2849	TlI	2827	UO_3	3011	WCl_4	2962
$ThBr_4$	2850	$TlIO_3$	2826	$UO_3 \cdot H_2O$	3012	WCl_5	2970
ThC	2851	$TlNO_2$	2830	$UO_4 \cdot 2H_2O$	3013	WCl_6	2976
ThC_2	2852	$TlNO_3$	2829	U_2C_3	2991	WF_4	2963
$ThCl_4$	2853	TlN_3	2813	U_2N_3	2993	WF_5	2972
ThF_4	2854	$Tl(NO_3)_3$	2843	U_3O_8	3008	WF_6	2980

WI_2	2957	YC_2	3108	$Zn(C_8H_{15}O_2)_2$	3136	$Zn_3(AsO_4)_2$	3127
WI_3	2960	YCl_3	3110	$Zn(CH_3COCHCOCH_3)_2$	3166	$Zn_3(AsO_4)_2 \cdot 8H_2O$	3128
WI_4	2964	$YCl_3 \cdot 6H_2O$	3111	$Zn(C_5H_9O_2)_2 \cdot 2H_2O$	3167	$Zn_3(C_6H_5O_7)_2 \cdot 2H_2O$	3143
WN_2	2951	YF_3	3112	$Zn(C_{12}H_{23}O_2)_2$	3155	Zn_3N_2	3159
$WOBr_3$	2973	YI_3	3114	$Zn(C_{18}H_{33}O_2)_2$	3161	Zn_3P_2	3173
$WOBr_4$	2982	$Y(NO_3)_3$	3116	$Zn(C_{18}H_{35}O_2)_2$	3179	$Zn_3(PO_4)_2$	3171
$WOCl_3$	2974	$Y(NO_3)_3 \cdot 4H_2O$	3117	$ZnCl_2$	3140	$Zn_3(PO_4)_2 \cdot 4H_2O$	3172
$WOCl_4$	2983	$Y(NO_3)_3 \cdot 6H_2O$	3118	$Zn(ClO_3)_2$	3139	Zr	3188
WOF_4	2984	$Y(OH)_3$	3113	$Zn(ClO_4)_2 \cdot 6H_2O$	3168	ZrB_2	3190
WO_2	2965	YP	3120	$ZnCrO_4$	3141	$ZrBr_2$	3195
WO_2Br_2	2977	YSb	3104	$ZnCr_2O_4$	3142	$ZrBr_3$	3200
WO_2Cl_2	2978	$Y_2(CO_3)_3 \cdot 3H_2O$	3109	ZnF_2	3147	$ZrBr_4$	3206
WO_2I_2	2979	Y_2O_3	3119	$ZnF_2 \cdot 4H_2O$	3148	ZrC	3191
WO_3	2981	Y_2S_3	3122	ZnI_2	3154	$Zr(C_2H_3O_2)_2(OH)_2$	3204
WS_2	2967	$Y_2(SO_4)_3 \cdot 8H_2O$	3121	$Zn(IO_3)_2$	3153	$Zr(C_5H_5)_2Cl_2$	3189
WS_3	2985	$Y_3Al_5O_{12}$	3103	$Zn(MnO_4)_2 \cdot 6H_2O$	3169	$ZrCl_2$	3196
WSe_2	2966	$Y_3Fe_5O_{12}$	3115	$ZnMoO_4$	3156	$ZrCl_3$	3201
WSi_2	2953	Yb	3087	$Zn(NH_4)_2(SO_4)_2$	3125	$ZrCl_4$	3207
WTe_2	2968	$YbBr_2$	3089	$Zn(NO_2)_2$	3160	ZrF_2	3197
W_2B	2945	$YbBr_3$	3094	$Zn(NO_3)_2$	3157	ZrF_3	3202
W_2B_5	2947	$Yb(C_2H_3O_2)_3 \cdot 4H_2O$	3093	$Zn(NO_3)_2 \cdot 6H_2O$	3158	ZrF_4	3208
W_2C	2948	$YbCl_2$	3090	ZnO	3165	ZrH_2	3198
W_2N	2952	$YbCl_3$	3095	$3ZnO \cdot 2B_2O_3$	3131	ZrI_2	3199
W_5Si_3	2954	$YbCl_3 \cdot 6H_2O$	3096	$2ZnO \cdot 3B_2O_3 \cdot 3.5H_2O$	3132	ZrI_3	3203
Xe	3070	YbF_2	3091	$2ZnO \cdot 3B_2O_3 \cdot 5H_2O$	3133	ZrI_4	3210
$XeFRuF_6$	3084	YbF_3	3097	ZnO_2	3170	ZrN	3192
$XeFSb_2F_{11}$	3085	YbI_2	3092	$Zn(OH)_2$	3152	$Zr(NH_4)_3OH(CO_3)_3 \cdot$	3205
XeF_2	3073	YbI_3	3098	$3Zn(OH)_2 \cdot 2ZnCO_3$	3138	$2H_2O$	
XeF_3SbF_6	3082	$Yb(NO_3)_3$	3099	ZnS	3183	$Zr(NO_3)_4 \cdot 5H_2O$	3211
$XeF_3Sb_2F_{11}$	3086	$YbSi_2$	3088	ZnS	3184	$ZrOCl_2$	3219
XeF_4	3074	Yb_2O_3	3100	$ZnSO_3 \cdot 2H_2O$	3185	$ZrOCl_2 \cdot 8H_2O$	3220
XeF_5AsF_6	3080	$Yb_2(SO_4)_3 \cdot 8H_2O$	3101	$ZnSO_4$	3180	ZrO_2	3213
XeF_5RuF_6	3081	Zn	3123	$ZnSO_4 \cdot H_2O$	3181	$Zr(OH)_4$	3209
XeF_6	3075	$Zn(AsO_2)_2$	3130	$ZnSO_4 \cdot 7H_2O$	3182	ZrP_2	3193
$XeOF_2$	3076	$Zn(BF_4)_2 \cdot 6H_2O$	3149	$Zn(SCN)_2$	3187	ZrP_2O_7	3214
$XeOF_4$	3077	$ZnBr_2$	3135	ZnS_2O_4	3146	ZrS_2	3217
XeO_2F_2	3078	$Zn(BrO_3)_2 \cdot 6H_2O$	3134	$ZnSb$	3126	$Zr(SO_4)_2$	3215
XeO_3	3071	$ZnCO_3$	3137	$ZnSe$	3176	$Zr(SO_4)_2 \cdot 4H_2O$	3216
XeO_3F_2	3079	$Zn(CHO_2)_2 \cdot 2H_2O$	3150	$ZnSeO_3$	3178	$ZrSiO_4$	3212
XeO_4	3072	$Zn(CN)_2$	3144	$ZnSeO_4 \cdot 5H_2O$	3175	$ZrSi_2$	3194
$Xe_2F_3AsF_6$	3083	ZnC_2O_4	3162	$ZnSiF_6 \cdot 6H_2O$	3151	$Zr(WO_4)_2$	3218
Y	3102	$ZnC_2O_4 \cdot 2H_2O$	3163	$ZnTe$	3186		
YAs	3105	$ZnC_4H_4O_6 \cdot 2H_2O$	3164	$Zn_2P_2O_7$	3174		
YB_6	3106	$Zn(C_2H_3O_2)_2 \cdot 2H_2O$	3124	Zn_2SiO_4	3177		
YBr_3	3107	$Zn(C_2H_5)_2$	3145	Zn_3As_2	3129		

CAS REGISTRY NUMBER INDEX OF INORGANIC COMPOUNDS

62-54-4	543	506-65-0	1171	557-09-5	3136	814-93-7	1469
62-76-0	2615	506-66-1	378	557-19-7	1876	814-94-8	2894
71-48-7	840	506-68-3	665	557-20-0	3145	814-95-9	2706
74-90-8	1252	506-77-4	666	557-21-1	3144	815-82-7	1002
75-15-0	653	506-78-5	668	557-39-1	1627	815-85-0	2900
75-20-7	557	506-80-9	655	557-42-6	3187	819-73-8	1443
75-44-5	661	506-87-6	91	562-76-5	2203	822-16-2	2640
96-10-6	24	507-16-4	2752	562-81-2	349	865-44-1	1296
102-54-5	1320	512-25-4	284	562-90-3	2473	868-14-4	2154
109-63-7	476	512-26-5	1452	563-63-3	2479	917-61-3	2553
124-38-9	651	513-74-6	106	563-67-7	2303	917-65-7	23
126-45-4	2491	513-77-9	276	563-68-8	2812	917-69-1	901
127-08-2	2100	513-78-0	502	563-71-3	1342	940-71-6	2017
127-09-3	2529	513-79-1	847	563-72-4	606	992-94-9	2459
127-95-7	2146	516-02-9	324	583-15-3	1759	992-98-3	2823
128-00-7	2502	527-09-3	971	583-52-8	2169	993-50-0	2883
131-74-8	174	528-94-9	176	584-08-7	2112	1002-88-6	890
136-51-6	576	532-31-0	2483	584-09-8	2309	1002-89-7	179
139-12-8	8	533-51-7	2507	589-97-9	2173	1066-30-4	787
140-99-8	629	534-16-7	2486	590-28-3	2120	1066-30-4	789
141-00-4	531	534-17-8	715	590-29-4	2134	1066-30-4	788
141-52-6	2565	535-37-5	1176	592-01-8	573	1066-33-7	129
141-53-7	2570	537-00-8	678	592-04-1	1765	1068-22-0	105
142-03-0	9	537-01-9	683	592-05-2	1453	1070-75-3	1517
142-17-6	605	538-17-0	72	592-06-3	2067	1072-35-1	1485
142-71-2	937	540-16-9	947	592-85-8	1787	1111-67-7	935
142-71-2	938	540-69-2	111	592-87-0	1491	1111-71-3	383
142-72-3	1597	540-72-7	2665	592-89-2	2692	1111-78-0	90
143-33-9	2554	541-43-5	296	592-89-2	2693	1113-38-8	161
144-23-0	1624	542-42-7	610	593-26-0	163	1117-94-8	920
144-23-0	1625	542-62-1	289	593-29-3	2189	1120-44-1	981
144-55-8	2586	542-83-6	508	593-74-8	1736	1120-46-3	1468
146-84-9	2513	542-84-7	856	593-91-9	436	1184-64-1	948
147-14-8	991	543-80-6	264	593-95-3	660	1184-65-2	1148
149-11-1	963	543-81-7	368	594-10-5	209	1185-57-5	1374
150-90-3	2641	543-90-8	494	598-54-9	919	1271-19-8	2913
151-50-8	2121	543-94-2	2678	598-62-9	1697	1271-27-8	1682
156-62-7	572	544-17-2	581	598-63-0	1444	1271-28-9	1852
289-22-5	2417	544-18-3	862	627-34-9	654	1273-81-0	1963
291-59-8	2421	544-19-4	969	628-52-4	777	1273-98-9	1847
298-14-6	2143	544-60-5	160	628-86-4	1768	1277-43-6	827
300-92-5	67	544-92-3	924	630-08-0	650	1277-47-0	3023
301-04-2	1433	546-46-3	3143	631-60-7	1737	1291-32-3	3189
301-08-6	1454	546-67-8	1500	631-61-8	84	1299-86-1	19
302-01-2	1222	546-89-4	1506	637-12-7	65	1301-96-8	2524
304-59-6	2187	546-93-0	1612	638-39-1	2886	1302-01-8	2478
306-61-6	1674	547-66-0	1648	660-60-6	997	1302-09-6	2515
333-20-0	2206	547-68-2	3162	688-37-9	43	1302-37-0	1509
353-50-4	662	547-68-2	3163	764-05-6	664	1302-42-7	2531
409-21-2	2463	551-64-4	3164	811-54-1	1457	1302-52-9	371
460-19-5	663	554-13-2	1518	813-93-4	407	1302-67-6	1600
463-58-1	656	555-35-1	49	813-94-5	571	1302-74-5	45
471-34-1	558	555-75-9	28	814-87-9	44	1302-76-7	61
471-34-1	559	555-76-0	1385	814-88-0	520	1302-81-4	70
471-34-1	560	556-38-7	3167	814-89-1	878	1302-82-5	59
497-19-8	2543	556-63-8	1532	814-89-1	879	1302-88-1	1601
504-64-3	652	556-65-0	1575	814-90-4	784	1302-93-8	62
506-61-6	2186	557-05-1	3179	814-91-5	982	1303-00-0	1106
506-64-9	2492	557-07-3	3161	814-91-5	983	1303-11-3	1263

1303-24-8	252
1303-28-2	259
1303-33-9	255
1303-34-0	261
1303-35-1	245
1303-36-2	254
1303-37-3	260
1303-52-2	1178
1303-58-8	1180
1303-60-2	1173
1303-61-3	1183
1303-62-4	1182
1303-86-2	469
1303-94-2	1572
1303-96-4	2657
1304-28-5	326
1304-29-6	331
1304-39-8	336
1304-40-1	338
1304-54-7	390
1304-56-9	392
1304-76-3	418
1304-82-1	430
1304-85-4	415
1305-62-0	590
1305-78-8	608
1305-79-9	616
1305-84-6	624
1305-99-3	618
1306-05-4	579
1306-06-5	591
1306-19-0	522
1306-23-6	535
1306-24-7	528
1306-25-8	537
1306-26-9	499
1306-38-3	706
1307-86-4	906
1307-96-6	880
1307-99-9	887
1308-04-9	908
1308-06-1	900
1308-14-1	800
1308-31-2	1346
1308-38-9	804
1308-56-1	966
1308-80-1	916
1308-85-6	1026
1308-87-8	1029
1308-96-9	1072
1309-33-7	1386
1309-37-1	1393
1309-42-8	1632
1309-48-4	1650
1309-55-3	1727
1309-56-4	1814
1309-60-0	1504
1309-64-4	220
1309-64-4	221

1310-03-8	1480	1314-36-9	3119	1332-58-7	64	2551-62-4	2742	5970-44-5	3109
1310-32-3	1361	1314-37-0	3100	1332-63-4	1730	2570-63-0	2838	5970-45-6	3124
1310-43-6	1330	1314-41-6	1499	1332-65-6	953	2644-70-4	1227	5970-62-7	3150
1310-52-7	1629	1314-56-3	2039	1332-71-4	912	2696-92-6	1955	5972-71-4	132
1310-53-8	1163	1314-60-9	233	1332-77-0	2198	2699-79-8	2748	5972-72-5	133
1310-58-3	2155	1314-61-0	2780	1332-81-6	229	2850-65-9	961	5972-76-9	89
1310-61-8	2151	1314-62-1	3055	1333-24-0	549	2949-11-3	1750	6009-70-7	162
1310-61-8	2152	1314-64-3	3020	1333-74-0	1241	3012-65-5	130	6010-09-9	1369
1310-65-2	1540	1314-64-3	3021	1333-82-0	822	3017-60-5	896	6018-89-9	1864
1310-66-3	1541	1314-68-7	2280	1333-83-1	2587	3017-60-5	897	6018-94-6	1887
1310-73-2	2596	1314-80-3	2041	1333-86-4	646	3094-87-9	1335	6046-93-1	938
1310-82-3	2323	1314-82-5	2040	1335-26-8	1655	3094-87-9	1336	6047-25-2	1356
1311-11-1	1472	1314-84-7	3173	1335-30-4	63	3095-65-6	140	6080-56-4	1434
1311-33-7	2873	1314-85-8	2015	1335-31-5	1779	3141-12-6	249	6100-05-6	2118
1311-90-6	173	1314-86-9	2028	1335-32-6	1435	3164-29-2	185	6106-24-7	2651
1311-93-9	202	1314-87-0	1487	1336-21-6	141	3164-34-9	635	6108-17-4	1507
1312-03-4	1778	1314-91-6	1490	1341-49-7	131	3236-82-6	1927	6108-23-2	1532
1312-41-0	1262	1314-95-0	2899	1343-88-0	1638	3251-23-8	978	6131-90-4	2530
1312-42-1	1283	1314-96-1	2716	1343-98-2	2460	3269-12-3	1071	6132-02-1	2545
1312-43-2	1280	1314-97-2	2837	1343-98-2	2461	3333-67-3	1871	6132-04-3	2551
1312-45-4	1286	1314-98-3	3183	1344-28-1	45	3344-18-1	1624	6147-53-1	841
1312-46-5	1312	1314-98-3	3184	1344-28-1	46	3344-18-1	1625	6150-82-9	1627
1312-73-8	2191	1315-01-1	2911	1344-43-0	1713	3375-31-3	1980	6150-88-5	1649
1312-73-8	2192	1315-03-3	3047	1344-48-5	1784	3396-11-0	709	6156-78-1	1694
1312-74-9	2184	1315-04-4	234	1344-48-5	1785	3444-13-1	1776	6159-44-0	3014
1312-81-8	1422	1315-05-5	225	1344-54-3	2929	3486-35-9	3137	6160-34-5	2693
1312-99-8	1639	1315-06-6	2897	1344-57-6	3003	3495-35-0	2317	6192-13-8	1831
1313-04-8	1662	1315-07-7	2713	1344-58-7	3011	3495-36-1	723	6303-21-5	2007
1313-08-2	1666	1315-09-9	3176	1344-59-8	3008	3522-50-7	1380	6381-59-5	2187
1313-13-9	1732	1315-11-3	3186	1344-95-2	598	3643-76-3	214	6381-79-9	2113
1313-22-0	1716	1317-33-5	1814	1345-04-6	227	3687-31-8	1437	6424-20-0	862
1313-27-5	1824	1317-34-6	1731	1345-07-9	429	3811-04-9	2114	6484-52-2	157
1313-29-7	1807	1317-35-7	1727	1345-13-7	694	3931-89-3	2054	6487-39-4	1412
1313-30-0	2609	1317-36-8	1470	1345-25-1	1357	3982-91-0	2057	6487-48-5	2170
1313-49-1	3159	1317-36-8	1471	1449-65-6	1146	4075-81-4	621	6533-73-9	2816
1313-59-3	2616	1317-37-9	1365	1470-61-7	2494	4109-96-0	2436	6556-16-7	1712
1313-60-6	2623	1317-38-0	984	1495-50-7	667	4111-54-0	1529	6569-51-3	460
1313-82-2	2645	1317-39-1	929	1529-48-2	1143	4119-52-2	1402	6591-55-5	417
1313-83-3	2646	1317-40-4	1001	1590-87-0	2413	4255-07-6	3163	6680-58-6	1523
1313-84-4	2647	1317-41-5	993	1592-23-0	628	4485-12-5	1566	6824-78-8	1740
1313-85-5	2636	1317-42-6	894	1600-27-7	1757	4493-37-2	782	6834-92-0	2605
1313-96-8	1930	1317-43-7	1632	1603-84-5	657	5006-97-3	1539	6858-44-2	2552
1313-97-9	1842	1317-58-4	535	1631-78-3	2882	5066-34-2	1021	6865-35-6	342
1313-99-1	1888	1317-61-9	1372	1633-05-2	2684	5142-76-7	402	7047-84-9	66
1314-04-1	1897	1317-66-4	1332	1701-93-5	2522	5145-48-2	1613	7116-98-5	2248
1314-05-2	1892	1317-70-0	2937	1762-95-4	198	5329-14-6	2731	7320-34-5	2180
1314-06-3	1901	1317-80-2	2939	1779-25-5	25	5341-61-7	1228	7428-48-0	1485
1314-08-5	1988	1317-98-2	221	1863-63-4	86	5470-11-1	1238	7429-90-5	7
1314-11-0	2707	1318-23-6	47	2035-66-7	1984	5714-22-7	2745	7429-91-6	1012
1314-12-1	2832	1318-33-8	582	2092-16-2	638	5743-04-4	495	7429-92-7	1033
1314-13-2	3165	1318-74-7	64	2092-17-3	351	5743-26-0	544	7439-88-5	1305
1314-15-4	2078	1319-46-6	1445	2092-17-3	352	5785-44-4	571	7439-89-6	1319
1314-18-7	2710	1319-53-5	949	2092-17-3	353	5794-28-5	607	7439-90-9	1403
1314-20-1	2859	1327-41-9	35	2218-80-6	959	5892-10-4	405	7439-91-0	1406
1314-22-3	3170	1327-44-2	2102	2223-93-0	530	5893-61-8	970	7439-92-1	1431
1314-23-4	3213	1327-50-0	228	2404-52-6	2060	5895-47-6	2363	7439-93-2	1505
1314-24-5	2026	1327-53-3	252	2406-52-2	2881	5908-64-5	265	7439-94-3	1580
1314-27-8	1498	1327-53-3	253	2408-36-8	1525	5908-81-6	348	7439-95-4	1596
1314-28-9	2276	1330-43-4	2654	2420-98-6	510	5951-19-9	658	7439-96-5	1681
1314-32-5	2844	1332-14-5	1000	2452-01-9	3155	5965-33-3	224	7439-97-6	1735
1314-34-7	3044	1332-40-7	986	2466-09-3	2010	5965-38-8	879	7439-98-7	1789
1314-35-8	2981	1332-52-1	369	2524-02-9	2059	5968-11-6	2544	7439-99-8	1849

CAS	No.	CAS	No.	CAS	No.	CAS	No.	CAS	No.
7440-00-8	1828	7440-56-4	1131	7558-80-7	2559	7718-54-9	1873	7773-03-7	2153
7440-01-9	1848	7440-57-5	1166	7558-80-7	2560	7718-98-1	3040	7773-06-0	180
7440-02-0	1851	7440-58-6	1184	7580-67-8	1538	7719-09-7	2753	7774-29-0	1771
7440-03-1	1902	7440-59-7	1208	7580-67-8	1526	7719-12-2	2021	7774-29-0	1772
7440-04-2	1962	7440-60-0	1209	7601-54-9	2626	7720-78-7	1362	7774-34-7	567
7440-05-3	1979	7440-61-1	2986	7601-54-9	2627	7720-83-4	2936	7775-09-9	2546
7440-06-4	2062	7440-62-2	3022	7601-89-0	2618	7721-01-9	2777	7775-11-3	2549
7440-07-5	2085	7440-63-3	3070	7601-90-3	747	7722-64-7	2176	7775-14-6	2563
7440-08-6	2096	7440-64-4	3087	7616-83-3	1780	7722-76-1	104	7775-19-1	2604
7440-09-7	2099	7440-65-5	3102	7616-94-6	759	7722-84-1	1256	7775-27-1	2625
7440-10-0	2218	7440-66-6	3123	7631-86-9	2470	7722-86-3	2726	7775-41-9	2495
7440-11-1	1734	7440-67-7	3188	7631-90-5	2595	7722-88-5	2633	7778-18-9	630
7440-12-2	2239	7440-68-8	262	7631-94-9	2564	7722-88-5	2568	7778-39-4	241
7440-13-3	2244	7440-69-9	400	7631-95-0	2607	7723-14-0	1994	7778-39-4	242
7440-14-4	2246	7440-70-2	542	7631-99-4	2611	7723-14-0	1995	7778-41-8	942
7440-15-5	2254	7440-71-3	643	7632-00-0	2612	7723-14-0	1996	7778-43-0	2536
7440-16-6	2286	7440-72-4	1076	7632-04-4	2617	7726-95-6	478	7778-43-0	2584
7440-17-7	2302	7440-73-5	1083	7632-51-1	3049	7727-15-3	17	7778-43-0	2585
7440-18-8	2337	7440-74-6	1261	7637-07-2	475	7727-18-6	3065	7778-44-1	548
7440-19-9	2353	7446-06-2	2098	7646-69-7	2583	7727-21-1	2178	7778-50-9	2123
7440-20-2	2374	7446-07-3	2787	7646-78-8	2903	7727-37-9	1933	7778-53-2	2179
7440-21-3	2411	7446-08-4	2392	7646-79-9	850	7727-43-7	345	7778-54-3	592
7440-22-4	2476	7446-09-5	2732	7646-85-7	3140	7727-54-0	169	7778-66-7	2156
7440-23-5	2528	7446-10-8	1488	7646-93-7	2150	7727-73-3	2644	7778-74-7	2174
7440-24-6	2677	7446-11-9	2733	7647-01-0	1249	7732-18-5	3067	7778-77-0	2125
7440-25-7	2758	7446-11-9	2734	7647-10-1	1982	7733-02-0	3180	7778-80-5	2190
7440-26-8	2781	7446-11-9	2735	7647-10-1	1983	7738-94-5	762	7779-86-4	3146
7440-27-9	2797	7446-14-2	1486	7647-14-5	2547	7738-94-5	822	7779-88-6	3157
7440-28-0	2811	7446-15-3	1481	7647-15-6	2541	7745-87-1	1657	7779-90-0	3171
7440-29-1	2847	7446-16-4	2252	7647-17-8	717	7757-79-1	2167	7779-90-0	3172
7440-30-4	2865	7446-18-6	2836	7647-18-9	230	7757-82-6	2642	7782-39-0	1242
7440-31-5	2879	7446-19-7	3181	7647-19-0	2038	7757-83-7	2648	7782-40-3	644
7440-31-5	2880	7446-20-0	3182	7659-31-6	2480	7757-86-0	1631	7782-41-4	1077
7440-32-6	2912	7446-21-1	2712	7664-38-2	2004	7757-87-1	1656	7782-42-5	645
7440-33-7	2943	7446-22-2	2834	7664-39-3	1253	7757-93-9	586	7782-44-7	1977
7440-34-8	1	7446-25-5	2910	7664-41-7	83	7757-93-9	587	7782-49-2	2386
7440-35-9	75	7446-26-6	3174	7664-93-9	2725	7758-01-2	2110	7782-49-2	2387
7440-36-0	206	7446-27-7	1476	7681-11-0	2159	7758-02-3	2111	7782-49-2	2388
7440-36-0	207	7446-28-8	2711	7681-38-1	2591	7758-05-6	2158	7782-50-5	744
7440-37-1	235	7446-31-3	3216	7681-49-4	2568	7758-09-0	2168	7782-61-8	1391
7440-38-2	236	7446-32-4	226	7681-52-9	2597	7758-11-4	2147	7782-63-0	1364
7440-38-2	237	7446-33-5	3121	7681-52-9	2598	7758-16-9	2562	7782-64-1	1701
7440-38-2	238	7446-34-6	2406	7681-53-0	2630	7758-19-2	2548	7782-65-2	1132
7440-39-3	263	7446-70-0	21	7681-53-0	2631	7758-23-8	575	7782-66-3	1781
7440-40-6	365	7447-39-4	951	7681-55-2	2599	7758-29-4	2671	7782-68-5	1288
7440-40-6	366	7447-40-7	2115	7681-57-4	2603	7758-87-4	617	7782-70-9	2149
7440-41-7	367	7447-41-8	1520	7681-65-4	927	7758-88-5	687	7782-77-6	1936
7440-42-8	441	7487-88-9	1667	7681-82-5	2600	7758-89-6	923	7782-78-7	2727
7440-43-9	493	7487-94-7	1763	7693-26-7	2141	7758-94-3	1343	7782-79-8	1235
7440-44-0	644	7488-51-9	1483	7693-27-8	1630	7758-95-4	1447	7782-85-6	2589
7440-44-0	645	7488-52-0	3185	7697-37-2	1935	7758-97-6	1450	7782-87-8	2157
7440-44-0	646	7488-54-2	2334	7698-05-7	1251	7758-98-7	998	7782-89-0	1510
7440-45-1	669	7488-55-3	2898	7698-05-7	753	7758-99-8	999	7782-91-4	1819
7440-46-2	708	7488-56-4	2407	7699-41-4	2460	7759-00-4	1706	7782-92-5	2534
7440-47-3	761	7542-09-8	848	7699-43-6	3219	7759-01-5	1494	7782-94-7	1934
7440-48-4	826	7542-09-8	847	7699-45-8	3135	7759-01-5	1495	7782-95-8	2561
7440-50-8	914	7543-51-3	3172	7704-34-9	2723	7759-02-6	2715	7782-99-2	2730
7440-51-9	1011	7550-35-8	1515	7704-34-9	2724	7761-88-8	2505	7783-00-8	2391
7440-52-0	1034	7550-45-0	2934	7704-98-5	2914	7772-98-7	2667	7783-03-1	2944
7440-53-1	1053	7553-56-2	1287	7704-99-6	3198	7772-98-7	2668	7783-06-4	1258
7440-54-2	1084	7558-79-4	2588	7705-07-9	2926	7772-99-8	2888	7783-07-5	1257
7440-55-3	1104	7558-80-7	2558	7705-08-0	1377	7773-01-5	1698	7783-08-6	2389

7783-09-7	1260	7783-92-8	2487	7787-46-4	377	7789-51-7	2401	7790-89-8	753
7783-11-1	184	7783-93-9	2509	7787-47-5	381	7789-52-8	2394	7790-91-2	754
7783-18-8	199	7783-95-1	2526	7787-48-6	394	7789-53-9	2682	7790-92-3	745
7783-19-9	178	7783-96-2	2501	7787-49-7	382	7789-54-0	2789	7790-93-4	746
7783-20-2	181	7783-97-3	2500	7787-52-2	384	7789-56-2	682	7790-94-5	2728
7783-21-3	177	7783-98-4	2511	7787-53-3	388	7789-57-3	2428	7790-98-9	167
7783-22-4	203	7783-99-5	2506	7787-56-6	398	7789-59-5	2043	7790-99-0	1295
7783-24-6	3125	7784-01-2	2490	7787-57-7	420	7789-60-8	2018	7791-03-9	1561
7783-26-8	2414	7784-02-3	2493	7787-58-8	431	7789-61-9	215	7791-07-3	2619
7783-28-0	134	7784-03-4	2521	7787-59-9	421	7789-65-3	2396	7791-08-4	222
7783-29-1	2415	7784-05-6	2516	7787-60-2	432	7789-66-4	2429	7791-09-5	2281
7783-30-4	1746	7784-07-8	2514	7787-61-3	433	7789-67-5	2902	7791-10-8	2685
7783-32-6	1770	7784-09-0	2512	7787-62-4	434	7789-68-6	2933	7791-11-9	2311
7783-33-7	2205	7784-11-4	18	7787-63-5	422	7789-69-7	2030	7791-12-0	2818
7783-34-8	1774	7784-13-6	22	7787-64-6	435	7789-75-5	578	7791-13-1	852
7783-35-9	1783	7784-14-7	188	7787-68-0	428	7789-77-7	587	7791-16-4	232
7783-36-0	1753	7784-16-9	2658	7787-69-1	714	7789-78-8	585	7791-18-6	1621
7783-39-3	1767	7784-17-0	710	7787-70-4	922	7789-79-9	593	7791-20-0	1874
7783-40-6	1626	7784-18-1	29	7787-71-5	488	7789-80-2	594	7791-21-1	748
7783-41-7	1078	7784-19-2	121	7788-96-7	825	7789-82-4	599	7791-23-3	2402
7783-42-8	2754	7784-21-6	33	7788-97-8	796	7790-21-8	2175	7791-25-5	2747
7783-43-9	2403	7784-22-7	57	7788-98-9	94	7790-22-9	1545	7791-26-6	3015
7783-44-0	1079	7784-23-8	36	7788-99-0	811	7790-28-5	2620	7791-27-7	2751
7783-46-2	1455	7784-24-9	2104	7789-00-6	2117	7790-29-6	2325	7791-28-8	273
7783-47-3	2890	7784-25-0	10	7789-01-7	1522	7790-30-9	2827	7798-23-4	989
7783-48-4	2691	7784-26-1	11	7789-02-8	803	7790-31-0	1636	7798-23-4	990
7783-49-5	3147	7784-27-2	41	7789-04-0	807	7790-32-1	1633	7803-49-8	1236
7783-50-8	1383	7784-29-4	2305	7789-04-0	808	7790-33-2	1704	7803-51-2	1997
7783-51-9	1116	7784-30-7	53	7789-04-0	809	7790-33-2	1705	7803-52-3	208
7783-52-0	1275	7784-30-7	54	7789-06-2	2688	7790-34-3	1884	7803-54-5	1602
7783-53-1	1729	7784-31-8	69	7789-08-4	146	7790-37-6	3153	7803-55-6	153
7783-54-2	1945	7784-33-0	247	7789-09-5	102	7790-38-7	1986	7803-57-8	1225
7783-55-3	2024	7784-34-1	248	7789-10-8	1766	7790-39-8	2068	7803-58-9	2746
7783-56-4	217	7784-35-2	250	7789-12-0	2557	7790-41-2	1311	7803-60-3	2009
7783-57-5	2842	7784-36-3	258	7789-16-4	728	7790-42-3	2209	7803-62-5	2412
7783-58-6	1160	7784-37-4	1769	7789-17-5	731	7790-43-4	2210	7803-63-6	137
7783-59-7	1503	7784-40-9	1459	7789-18-6	734	7790-44-5	218	7803-65-8	142
7783-60-0	2741	7784-41-0	2124	7789-19-7	967	7790-46-7	2077	7803-68-1	2785
7783-61-1	2448	7784-42-1	239	7789-20-0	3068	7790-47-8	2907	8011-62-9	346
7783-62-2	2906	7784-44-3	128	7789-21-1	2729	7790-48-9	2795	8014-95-7	2725
7783-63-3	2935	7784-45-4	251	7789-23-3	2130	7790-49-0	2855	9080-17-5	175
7783-64-4	3208	7784-46-5	2537	7789-24-4	1531	7790-56-9	2194	10022-31-8	319
7783-66-6	1299	7784-48-7	1868	7789-25-5	1956	7790-58-1	2197	10022-47-6	95
7783-68-8	1928	7785-20-8	1867	7789-26-6	1081	7790-59-2	2183	10022-48-7	1527
7783-70-2	231	7785-23-1	2485	7789-27-7	2822	7790-60-5	2212	10022-50-1	1959
7783-71-3	2778	7785-24-2	843	7789-28-8	1347	7790-62-7	2182	10022-68-1	519
7783-72-4	3052	7785-84-4	2669	7789-29-9	2144	7790-63-8	2213	10022-70-5	2598
7783-75-7	1318	7785-84-4	2670	7789-30-2	489	7790-67-2	2181	10024-93-8	1835
7783-77-9	1822	7785-87-7	1717	7789-31-3	479	7790-69-4	1553	10024-97-2	1937
7783-77-9	1160	7786-30-3	1620	7789-33-5	1294	7790-71-8	1564	10025-64-6	3168
7783-79-1	2399	7786-81-4	1894	7789-36-8	1609	7790-74-1	623	10025-65-7	2066
7783-80-4	2796	7787-32-8	295	7789-38-0	2540	7790-75-2	641	10025-66-8	2249
7783-81-5	3010	7787-33-9	311	7789-39-1	2308	7790-76-3	622	10025-67-9	2737
7783-82-6	2980	7787-34-0	309	7789-40-4	2815	7790-78-5	506	10025-68-0	2395
7783-83-7	148	7787-35-1	312	7789-41-5	554	7790-79-6	511	10025-69-1	2889
7783-84-8	1398	7787-36-2	330	7789-42-6	500	7790-80-9	514	10025-70-4	2687
7783-85-9	145	7787-37-3	317	7789-43-7	845	7790-81-0	513	10025-71-5	2790
7783-86-0	1350	7787-38-4	322	7789-43-7	846	7790-84-3	533	10025-73-7	794
7783-86-0	1351	7787-39-5	347	7789-45-9	946	7790-85-4	541	10025-74-8	1022
7783-89-3	2484	7787-40-8	355	7789-46-0	1340	7790-86-5	685	10025-75-9	1040
7783-90-6	2488	7787-41-9	335	7789-47-1	1761	7790-87-6	690	10025-76-0	1066
7783-91-7	2489	7787-42-0	360	7789-48-2	1610	7790-87-6	691	10025-77-1	1378

10025-78-2	2437	10034-98-7	3101	10099-59-9	1420	10138-52-0	1092	10326-27-9	280
10025-82-8	1273	10034-99-8	1669	10099-60-2	1425	10138-62-2	1212	10326-28-0	524
10025-83-9	1309	10035-04-8	565	10099-60-2	1427	10139-47-6	3154	10340-06-4	659
10025-85-1	1944	10035-05-9	562	10099-66-8	1583	10139-58-9	2294	10343-61-0	2930
10025-87-3	2049	10035-06-0	414	10099-67-9	1588	10141-00-1	811	10361-37-2	279
10025-90-8	2227	10035-10-6	1247	10099-74-8	1467	10141-05-6	873	10361-43-0	409
10025-91-9	216	10036-47-2	1953	10099-76-0	1478	10143-38-1	1028	10361-44-1	414
10025-93-1	2995	10038-98-9	1159	10099-79-3	1496	10163-15-2	2569	10361-46-3	423
10025-94-2	3111	10039-31-3	395	10101-41-4	632	10168-80-6	1045	10361-65-6	171
10025-97-5	1315	10039-32-4	2590	10101-50-5	2622	10170-69-1	1689	10361-79-2	2226
10025-99-7	2200	10039-54-0	1240	10101-52-7	3212	10179-73-4	1355	10361-80-5	2231
10026-00-3	2661	10039-55-1	1229	10101-53-8	812	10190-55-3	1465	10361-80-5	2232
10026-01-4	1970	10039-56-2	2631	10101-53-8	813	10192-29-7	92	10361-82-7	2364
10026-02-5	2097	10042-76-9	2702	10101-63-0	1463	10192-30-0	139	10361-83-8	2368
10026-03-6	2397	10042-88-3	2801	10101-68-5	1719	10193-36-9	2461	10361-83-8	2369
10026-04-7	2438	10043-01-3	68	10101-83-4	2652	10196-04-0	183	10361-84-9	2377
10026-06-9	2904	10043-11-5	468	10101-88-9	2666	10196-04-0	184	10361-91-8	3095
10026-07-0	2792	10043-27-3	2804	10101-89-0	2627	10196-18-6	3158	10361-92-9	3110
10026-08-1	2853	10043-27-3	2805	10101-91-4	2664	10210-64-7	393	10361-93-0	3116
10026-10-5	3000	10043-35-3	461	10101-94-7	1484	10210-68-1	835	10361-93-0	3117
10026-11-6	3207	10043-52-4	563	10101-97-0	1895	10213-09-9	3060	10361-95-2	3139
10026-12-7	1925	10043-67-1	2103	10101-98-1	1896	10213-09-9	3034	10377-37-4	1518
10026-13-8	2033	10043-84-2	1703	10102-02-0	3160	10213-10-2	2673	10377-48-7	1567
10026-17-2	860	10043-92-2	2253	10102-03-1	1942	10213-12-4	2251	10377-51-2	1544
10026-18-3	903	10045-86-0	1396	10102-05-3	1987	10214-39-8	1440	10377-52-3	1559
10026-20-7	885	10045-87-1	1399	10102-06-4	3018	10214-40-1	994	10377-58-9	1634
10026-22-9	874	10045-94-0	1773	10102-15-5	2649	10233-88-2	2573	10377-60-3	1641
10026-24-1	893	10045-95-1	1840	10102-17-7	2668	10241-03-9	3201	10377-62-5	1654
10028-14-5	1961	10048-95-0	2585	10102-18-8	2637	10241-05-1	1816	10377-66-9	1710
10028-15-6	1978	10048-98-3	301	10102-20-2	2653	10257-55-3	634	10377-93-2	1708
10028-17-8	1243	10048-99-4	350	10102-23-5	2635	10277-43-7	1420	10378-47-9	703
10028-18-9	1877	10049-01-1	424	10102-24-6	1550	10290-12-7	943	10378-48-0	2679
10028-22-5	1400	10049-03-3	1082	10102-25-7	1568	10294-26-5	2517	10380-31-1	361
10031-13-7	1438	10049-04-4	749	10102-34-8	1659	10294-27-6	1169	10381-36-9	1890
10031-16-0	290	10049-05-5	779	10102-40-6	2608	10294-28-7	1174	10381-37-0	2863
10031-18-2	1739	10049-06-6	2921	10102-43-9	1938	10294-29-8	1170	10402-15-0	957
10031-20-6	1696	10049-07-7	2291	10102-44-0	1939	10294-31-2	1172	10405-27-3	1950
10031-21-7	1441	10049-08-8	2344	10102-45-1	2829	10294-32-3	1181	10415-75-5	1747
10031-22-8	1442	10049-10-2	781	10102-49-5	1375	10294-33-4	473	10415-75-5	1748
10031-23-9	2247	10049-11-3	817	10102-50-8	1338	10294-34-5	474	10421-48-4	1389
10031-24-0	2887	10049-12-4	3041	10102-50-8	1339	10294-38-9	278	10421-48-4	1390
10031-25-1	791	10049-12-4	3042	10102-68-8	595	10294-39-0	329	10431-47-7	2185
10031-25-1	793	10049-14-6	3001	10102-71-3	2533	10294-40-3	283	10450-55-2	1373
10031-26-2	1376	10049-16-8	3050	10102-75-7	552	10294-42-5	707	10450-60-9	1289
10031-27-3	2791	10049-17-9	2275	10102-75-7	553	10294-44-7	1741	10476-81-0	2681
10031-30-8	575	10049-21-5	2559	10103-62-5	548	10294-46-9	988	10476-85-4	2686
10031-34-2	604	10049-23-7	2786	10108-64-2	504	10294-47-0	1446	10476-86-5	2698
10031-43-3	979	10049-24-8	1307	10108-73-3	692	10294-48-1	752	10489-46-0	2298
10031-45-5	992	10049-24-8	1308	10112-91-1	1742	10294-50-5	884	10534-89-1	905
10031-48-8	990	10049-25-9	778	10117-38-1	2193	10294-52-7	1379	10544-72-6	1941
10031-50-2	1030	10049-25-9	791	10118-76-0	615	10294-53-8	1381	10544-73-7	1940
10031-52-4	1050	10058-44-3	1397	10124-36-4	532	10294-54-9	741	10545-99-0	2740
10031-52-4	1075	10060-08-9	570	10124-36-4	534	10294-58-3	1477	10553-31-8	272
10031-53-5	1070	10060-09-0	527	10124-37-5	600	10294-60-7	136	10567-69-8	308
10031-54-6	1062	10060-10-3	704	10124-41-1	639	10294-62-9	1427	10567-69-8	309
10034-76-1	631	10060-11-4	1151	10124-43-3	891	10294-64-1	2161	10578-16-2	1951
10034-81-8	1652	10060-12-5	795	10124-48-8	1758	10294-66-3	2207	10578-16-2	1952
10034-82-9	2550	10060-13-6	99	10124-53-5	1675	10294-70-9	2893	10580-03-7	200
10034-85-2	1254	10060-13-6	100	10125-13-0	952	10325-94-7	518	10580-52-6	3034
10034-88-5	2592	10090-53-6	37	10137-74-3	561	10326-21-3	1619	10588-01-9	2556
10034-93-2	1233	10097-28-6	2465	10137-74-3	562	10326-24-6	3130	10599-90-3	1948
10034-96-5	1718	10099-58-8	1413	10138-41-7	1039	10326-26-8	271	11065-24-0	1306

11071-61-7	455	12008-27-4	2219	12020-66-5	1061	12033-62-4	2765	12039-76-8	3032
11120-25-5	201	12008-29-6	471	12020-69-8	1064	12033-64-6	2798	12039-79-1	2766
11121-16-7	14	12008-30-9	2354	12021-68-0	864	12033-65-7	2858	12039-80-4	2799
11126-81-1	16	12008-30-9	2354	12022-02-5	112	12033-72-6	2952	12039-83-7	2919
11138-11-7	293	12008-32-1	3106	12022-71-8	1370	12033-82-8	2703	12039-87-1	3031
12002-03-8	939	12009-14-2	318	12022-95-6	1333	12033-83-9	2993	12039-88-2	2953
12002-28-7	1321	12009-18-6	340	12022-99-0	1334	12033-88-4	1960	12039-89-3	3088
12002-61-8	6	12009-18-6	341	12023-53-9	1331	12033-89-5	2464	12039-90-6	3194
12002-99-2	2520	12009-21-1	363	12023-71-1	1587	12034-09-2	2610	12039-95-1	2954
12003-63-3	2101	12009-27-7	357	12023-99-3	1119	12034-12-7	2650	12040-00-5	2373
12003-65-5	1407	12009-31-3	358	12024-10-1	1112	12034-57-0	1911	12040-02-7	2901
12003-78-0	1853	12011-67-5	1328	12024-11-2	1111	12034-59-2	1920	12040-08-3	2718
12004-04-5	266	12011-97-1	1793	12024-14-5	1113	12034-66-1	1909	12041-50-8	26
12004-06-7	370	12011-99-3	1907	12024-20-3	1109	12034-77-4	1921	12041-54-2	27
12004-29-4	55	12012-16-7	2851	12024-21-4	1122	12034-80-9	1910	12042-68-1	546
12004-39-6	73	12012-17-8	3028	12024-22-5	1129	12034-83-2	1923	12042-78-3	547
12004-50-1	74	12012-32-7	671	12024-24-7	1126	12034-88-7	1466	12043-29-7	71
12004-76-1	2759	12012-35-0	769	12024-27-0	1130	12034-89-8	2701	12044-16-5	1325
12005-21-9	3103	12013-10-4	836	12024-81-6	1089	12035-32-4	1845	12044-30-3	246
12005-67-3	82	12013-21-7	1640	12025-13-7	1628	12035-35-7	1846	12044-42-7	831
12005-69-5	466	12013-46-6	627	12025-19-3	2571	12035-38-0	1893	12044-49-4	1605
12005-75-3	915	12013-47-7	642	12025-32-0	1156	12035-39-1	1899	12044-54-1	256
12005-82-2	2497	12013-55-7	625	12025-34-2	1165	12035-52-8	1854	12045-01-1	834
12006-15-4	497	12013-56-8	626	12025-39-7	1157	12035-64-2	1860	12045-15-7	1685
12006-40-5	3129	12013-57-9	636	12026-06-1	2825	12035-72-2	1863	12045-16-8	1687
12006-77-8	833	12013-69-3	620	12026-24-3	2892	12035-79-9	1850	12045-19-1	1904
12006-79-0	765	12013-82-0	602	12026-66-3	156	12035-82-4	2069	12045-27-1	3025
12006-80-3	767	12014-14-1	540	12026-66-3	172	12035-83-5	2087	12045-63-5	2915
12006-84-7	1326	12014-28-7	526	12027-06-4	144	12035-98-2	3037	12045-64-6	3190
12006-86-9	1327	12014-29-8	496	12027-67-7	154	12036-02-1	1972	12045-87-3	2655
12006-99-4	1791	12014-56-1	705	12028-48-7	152	12036-09-8	2264	12045-88-4	2656
12007-00-0	1856	12014-82-3	677	12029-81-1	1217	12036-10-1	2349	12046-08-1	297
12007-01-1	1857	12014-85-6	674	12029-98-0	1291	12036-14-5	2772	12046-54-7	2694
12007-02-2	1858	12014-93-6	700	12030-14-7	1271	12036-21-4	3051	12047-25-5	332
12007-07-7	2760	12015-73-5	579	12030-24-9	1285	12036-22-5	2965	12047-27-7	356
12007-09-9	2946	12016-80-7	909	12030-49-8	1316	12036-32-7	2233	12047-79-9	320
12007-10-2	2945	12017-01-5	898	12030-51-2	1317	12036-34-9	2092	12048-50-9	419
12007-16-8	766	12017-08-2	877	12030-85-2	2166	12036-35-0	2296	12048-51-0	437
12007-23-7	1185	12017-12-8	839	12030-88-5	2195	12036-39-4	2722	12049-33-1	2958
12007-25-9	1606	12017-13-9	895	12030-97-6	2208	12036-41-8	2806	12049-50-2	640
12007-29-3	1905	12017-38-8	913	12031-30-0	1428	12036-44-1	2878	12050-27-0	496
12007-33-9	472	12018-01-8	818	12031-49-1	1429	12036-46-3	223	12052-28-7	858
12007-34-0	2375	12018-09-6	775	12031-63-9	1552	12037-15-9	3004	12052-42-5	828
12007-35-1	2761	12018-10-9	956	12031-66-2	1570	12037-65-9	2918	12053-12-2	763
12007-36-2	2987	12018-19-8	3142	12031-80-0	1563	12037-80-8	3193	12053-13-3	774
12007-37-3	3026	12018-22-3	814	12031-82-2	1576	12037-82-0	2016	12053-26-8	1623
12007-38-4	768	12018-34-7	786	12031-83-3	1579	12038-12-9	2237	12053-27-9	772
12007-58-8	187	12018-36-9	776	12032-20-1	1590	12038-13-0	2236	12053-39-3	815
12007-60-2	1571	12018-61-0	737	12032-30-3	1639	12038-20-9	2070	12054-48-7	1879
12007-81-7	2474	12018-79-0	964	12032-31-4	1679	12038-21-0	2079	12054-48-7	1880
12007-84-0	2988	12019-07-7	996	12032-36-9	1670	12038-63-0	2267	12054-85-2	154
12007-89-5	164	12019-08-8	1006	12032-52-9	1647	12038-64-1	2265	12055-23-1	1201
12007-97-5	1792	12019-11-3	917	12032-74-5	1725	12038-66-3	2266	12055-24-2	1207
12007-98-6	2947	12019-23-7	1003	12032-78-9	1691	12038-67-4	2285	12055-62-8	1219
12007-99-7	551	12019-52-2	934	12032-82-5	1683	12039-07-5	2924	12056-90-5	1424
12008-02-5	670	12019-88-4	1014	12032-86-9	1693	12039-13-3	2942	12057-24-8	1560
12008-05-8	1054	12020-14-3	3191	12032-88-1	1723	12039-15-5	3217	12057-71-5	1644
12008-06-9	1085	12020-21-2	1046	12032-97-2	1684	12039-16-6	2931	12057-74-8	1660
12008-19-4	447	12020-28-9	1048	12033-19-1	1796	12039-19-9	3122	12057-75-9	1604
12008-21-8	1408	12020-39-2	1052	12033-29-3	1827	12039-35-9	3126	12057-92-0	1733
12008-22-9	1607	12020-58-5	1055	12033-31-7	1797	12039-52-0	2846	12058-18-3	1813
12008-23-0	1829	12020-65-4	1063	12033-54-4	2086	12039-55-3	2773	12058-20-7	1815

12058-85-4	2629	12125-02-9	93	12187-14-3	517	12612-73-6	2989	13446-30-5	1675
12059-14-2	1862	12125-08-5	117	12188-41-9	2938	12612-73-6	2991	13446-34-9	1699
12059-95-9	2094	12125-09-6	2003	12190-79-3	1524	12616-24-9	3205	13446-48-5	158
12060-00-3	1493	12125-19-8	2985	12201-89-7	1861	12640-47-0	899	13446-49-6	2165
12060-01-4	1497	12125-22-3	1991	12209-98-2	2639	12650-28-1	291	13446-53-2	1611
12060-08-1	2382	12125-25-6	1589	12211-52-8	101	12672-79-6	267	13446-56-5	1800
12060-58-1	2370	12125-63-2	1368	12214-16-3	742	12673-76-6	2443	13446-57-6	1803
12060-59-2	2720	12133-07-2	1015	12218-36-9	820	12688-52-7	1581	13446-70-3	2307
12061-16-4	1047	12133-10-7	1031	12228-40-9	372	12770-26-2	1187	13446-71-4	2310
12062-24-7	973	12134-02-0	838	12228-50-1	1686	12777-45-6	427	13446-72-5	2312
12063-56-8	3115	12134-22-4	738	12228-87-4	187	12785-50-1	269	13446-73-6	2313
12063-98-8	1108	12134-75-7	1087	12228-91-0	1724	12789-09-2	1010	13446-74-7	2315
12064-03-8	1105	12134-77-9	1102	12229-63-9	2849	12793-14-5	1903	13446-75-8	2319
12064-62-9	1099	12135-76-1	182	12230-32-9	1608	13005-39-5	2199	13446-76-9	2324
12065-10-0	1155	12136-24-2	1220	12230-71-6	306	13007-92-6	770	13450-87-8	1101
12065-11-1	1164	12136-42-4	1314	12230-74-9	303	13011-54-6	2535	13450-88-9	1114
12065-36-0	1162	12136-44-6	2108	12232-99-4	2602	13083-04-0	679	13450-90-3	1115
12065-65-5	2932	12136-45-7	2171	12233-56-6	410	13092-75-6	2481	13450-91-4	1120
12065-68-8	1489	12136-58-2	1569	12249-52-4	2499	13126-12-0	2327	13450-92-5	1158
12066-83-0	2221	12136-78-6	1799	12254-64-7	76	13135-31-4	2925	13450-95-8	1161
12067-00-4	2268	12136-91-3	2014	12254-82-9	829	13138-45-9	1885	13450-97-0	2708
12067-22-0	2372	12136-97-9	1922	12254-85-2	764	13138-45-9	1886	13451-02-0	2717
12067-46-8	2966	12137-12-1	1900	12255-48-0	3105	13172-31-1	2736	13451-05-3	2721
12067-54-8	2862	12137-20-1	2923	12256-04-1	832	13205-44-2	459	13451-11-1	2776
12067-66-2	2775	12137-25-6	2336	12266-65-8	1688	13255-26-0	316	13451-18-8	2788
12067-76-4	2968	12137-27-8	2299	12279-90-2	246	13255-48-6	1223	13451-19-9	2805
12067-99-1	2005	12137-27-8	2300	12280-52-3	727	13266-83-6	693	13453-06-0	186
12068-40-5	1509	12137-83-6	2084	12280-64-7	2321	13268-42-3	147	13453-07-1	1175
12068-49-4	1337	12138-06-6	3184	12286-33-8	2885	13283-01-7	2976	13453-30-0	2817
12068-51-8	1600	12138-07-7	2864	12291-65-5	582	13320-71-3	1809	13453-32-2	2840
12068-56-3	62	12138-09-9	2967	12294-01-8	3120	13327-32-7	386	13453-32-2	2841
12068-57-4	915	12138-11-3	2808	12298-70-3	2957	13327-32-7	387	13453-34-4	2820
12068-69-8	426	12138-17-9	3056	12300-22-0	2355	13395-16-9	985	13453-40-2	2833
12068-90-5	1786	12138-28-2	2714	12310-43-9	1013	13397-24-5	632	13453-49-1	2850
12069-00-0	1482	12139-93-4	889	12310-44-0	1035	13397-26-7	559	13453-57-1	1449
12069-32-8	467	12140-41-9	1366	12322-34-8	212	13400-13-0	721	13453-62-8	1474
12069-69-1	949	12141-45-6	63	12323-03-4	339	13410-01-0	2634	13453-69-5	1547
12069-85-1	1186	12141-46-7	61	12323-19-2	403	13410-01-0	2635	13453-71-9	1519
12069-89-5	1794	12142-33-5	2188	12325-59-6	1189	13423-61-5	1622	13453-78-6	1562
12069-94-2	1906	12143-72-5	2774	12326-21-5	3012	13424-46-9	1439	13453-80-0	1528
12070-06-3	2762	12143-96-3	1618	12333-54-9	1692	13444-75-2	1764	13453-84-4	1557
12070-07-4	2763	12159-07-8	918	12336-95-7	798	13444-85-4	1946	13454-74-5	701
12070-08-5	2916	12159-43-2	1032	12340-14-6	3072	13444-87-6	1954	13454-75-6	713
12070-09-6	2989	12159-66-9	1051	12345-14-1	282	13444-90-1	1958	13454-81-4	730
12070-10-9	3027	12160-99-5	1103	12393-61-2	255	13444-92-3	1967	13454-83-6	735
12070-12-1	2949	12162-21-9	1203	12399-08-5	1290	13444-93-4	1969	13454-84-7	739
12070-13-2	2948	12162-59-3	1221	12401-56-8	1190	13444-94-5	1981	13454-88-1	968
12070-69-8	3138	12163-00-7	1546	12412-52-1	220	13444-96-7	1985	13454-89-2	975
12071-15-7	1411	12163-20-1	1594	12419-43-1	2891	13445-50-6	1998	13454-90-5	976
12071-29-3	2683	12163-22-3	1595	12429-94-6	373	13445-58-4	2032	13454-94-9	698
12071-31-7	2852	12163-69-8	1798	12433-14-6	955	13446-03-2	1695	13454-94-9	699
12071-33-9	2990	12164-94-2	85	12434-24-1	1056	13446-08-7	110	13454-96-1	2075
12071-35-1	3108	12165-05-8	2272	12440-00-5	2027	13446-09-8	143	13454-99-4	2036
12076-62-9	2991	12165-69-4	2029	12442-63-6	751	13446-10-1	168	13455-00-0	2001
12083-48-6	3024	12166-29-9	2384	12504-41-5	2471	13446-18-9	1643	13455-01-1	2025
12084-29-6	327	12166-44-8	2385	12513-27-8	3132	13446-19-0	1619	13455-03-3	2046
12088-65-2	1324	12167-74-7	591	12520-88-6	2475	13446-19-0	1653	13455-04-4	2053
12115-63-8	672	12168-52-4	1370	12528-75-5	1352	13446-23-6	1657	13455-12-4	2065
12116-66-4	1191	12168-85-3	609	12536-51-5	374	13446-24-7	1658	13455-15-7	2076
12124-97-9	88	12183-80-1	60	12536-52-6	375	13446-24-7	1659	13455-20-4	2127
12124-99-1	138	12185-10-3	1994	12536-65-1	3133	13446-28-1	1661	13455-21-5	2131
12125-01-8	109	12186-97-9	3104	12540-13-5	941	13446-29-2	1672	13455-24-8	2145

CAS No.	No.	CAS No.	No.	CAS No.	No.	CAS No.	No.	CAS No.	No.
13455-25-9	854	13465-95-7	328	13478-20-1	2052	13537-23-0	2035	13598-33-9	2695
13455-28-2	867	13465-96-8	2503	13478-28-9	783	13537-24-1	1395	13598-36-2	2006
13455-29-3	869	13465-98-0	2519	13478-33-6	883	13537-30-9	1144	13598-42-0	2452
13455-31-7	882	13466-08-5	2542	13478-38-1	980	13537-32-1	2013	13598-56-6	2997
13455-34-0	892	13466-20-1	315	13478-41-6	925	13537-33-2	2445	13598-65-7	170
13460-50-9	462	13466-21-2	334	13478-45-0	1924	13548-38-4	802	13600-89-0	902
13460-50-9	463	13469-98-2	3107	13478-49-4	1049	13548-42-0	954	13601-13-3	965
13460-50-9	464	13470-01-4	2697	13478-50-7	1492	13548-43-1	95	13601-19-9	2567
13462-88-9	1869	13470-04-7	2700	13492-25-6	1749	13550-28-2	1514	13628-54-1	1100
13462-88-9	1870	13470-06-9	2704	13492-26-7	2148	13550-31-7	2044	13637-61-1	3168
13462-90-3	1883	13470-07-0	2857	13494-80-9	2784	13550-53-3	1042	13637-63-3	755
13462-93-6	103	13470-08-1	2927	13494-90-1	1121	13565-96-3	412	13637-65-5	1138
13463-10-0	1396	13470-10-5	2955	13494-91-2	1127	13565-97-4	3214	13637-68-8	1820
13463-12-2	1341	13470-11-6	2969	13494-98-9	3118	13566-03-5	1990	13637-71-3	1889
13463-22-4	325	13470-12-7	2956	13497-91-1	1999	13566-17-1	1479	13637-76-8	1475
13463-30-4	1502	13470-13-8	2962	13498-07-2	2234	13568-32-6	1711	13637-83-7	757
13463-39-3	1859	13470-14-9	2970	13498-11-8	2034	13568-33-7	1555	13637-84-8	2750
13463-40-6	1322	13470-17-2	2957	13498-11-8	2037	13568-33-7	1556	13637-87-1	1947
13463-67-7	2937	13470-19-4	2994	13499-05-3	1198	13568-40-6	1551	13675-47-3	960
13463-67-7	2938	13470-20-7	2999	13510-35-5	1278	13568-45-1	1577	13682-61-6	2199
13463-67-7	2939	13470-21-8	3006	13510-41-3	2235	13568-63-3	1678	13689-92-4	1898
13464-35-2	2107	13470-22-9	3002	13510-42-4	2330	13569-43-2	1137	13693-05-5	2080
13464-36-3	2106	13470-26-3	3039	13510-43-5	2456	13569-49-8	2259	13693-06-6	2095
13464-44-3	3127	13470-38-7	3114	13510-44-6	2482	13569-50-1	675	13693-07-7	2301
13464-45-4	3128	13472-31-6	2621	13510-49-1	396	13569-59-0	1913	13693-08-8	2351
13464-46-5	78	13472-33-8	2624	13510-89-9	1436	13569-62-5	2089	13693-09-9	3075
13464-58-9	243	13472-35-0	2560	13517-00-5	926	13569-63-6	2260	13693-11-3	2941
13464-80-7	1234	13472-39-4	2643	13517-06-1	2601	13569-67-0	2768	13701-58-1	560
13464-82-9	1284	13472-45-2	2672	13517-10-7	477	13569-70-5	1917	13701-64-9	550
13464-92-1	501	13473-03-5	253	13517-11-8	480	13569-71-6	2262	13701-64-9	597
13464-97-6	1230	13473-75-1	2819	13517-13-0	2433	13569-72-7	2770	13701-67-2	457
13464-98-7	1231	13473-77-3	1593	13517-23-2	2632	13569-75-0	801	13701-70-7	3046
13465-05-9	1250	13473-90-0	40	13517-24-3	2606	13569-80-7	1024	13701-86-5	2975
13465-07-1	1259	13473-90-0	41	13517-27-6	3134	13572-93-5	1118	13701-90-1	2839
13465-09-3	1272	13476-01-2	844	13520-56-4	1401	13572-98-0	1095	13701-91-2	1501
13465-10-6	1267	13476-08-9	1390	13520-59-7	1808	13573-02-9	1145	13702-38-0	404
13465-11-7	1270	13476-99-8	3045	13520-75-7	2977	13573-08-5	1153	13703-82-7	1637
13465-15-1	1281	13477-00-4	277	13520-76-8	2978	13573-11-0	1677	13706-09-7	2056
13465-30-0	1762	13477-00-4	278	13520-77-9	2982	13573-16-5	194	13706-10-0	2055
13465-33-3	1738	13477-09-3	300	13520-78-0	2983	13586-38-4	98	13708-63-9	2809
13465-34-4	1743	13477-17-3	525	13520-79-1	2984	13587-16-1	1526	13708-80-0	79
13465-35-5	1745	13477-19-5	515	13520-83-7	3019	13587-35-4	1008	13708-85-5	2632
13465-37-7	1755	13477-23-1	536	13520-87-1	3058	13587-35-4	1009	13709-31-4	3066
13465-43-5	2295	13477-28-6	556	13520-88-2	3057	13590-82-4	707	13709-35-8	2739
13465-49-1	2331	13477-29-7	564	13520-89-3	3059	13595-30-7	3048	13709-36-9	3073
13465-55-9	2365	13477-34-4	601	13520-90-6	3064	13595-87-4	438	13709-38-1	1415
13465-58-2	2371	13477-36-6	613	13520-92-8	3220	13596-22-0	885	13709-42-7	1837
13465-59-3	2376	13477-75-3	54	13530-57-9	2304	13596-23-1	2444	13709-45-0	2242
13465-60-6	2381	13477-89-9	1836	13530-57-9	2305	13596-46-8	98	13709-46-1	2228
13465-66-2	2398	13477-91-3	1843	13530-65-9	3141	13597-20-1	1932	13709-47-2	2378
13465-71-9	2447	13477-91-3	1844	13536-53-3	2224	13597-44-9	3185	13709-49-4	3112
13465-72-0	2454	13477-94-6	1872	13536-59-9	1248	13597-46-1	3178	13709-52-9	1199
13465-73-1	2426	13477-95-7	1876	13536-61-3	2749	13597-54-1	3175	13709-54-1	2755
13465-74-2	2430	13477-98-0	1881	13536-73-7	1037	13597-55-2	2705	13709-56-3	2093
13465-75-3	2431	13477-99-1	1882	13536-73-7	1038	13597-64-3	733	13709-59-6	2854
13465-76-4	2432	13478-00-7	1886	13536-79-3	1410	13597-65-4	3177	13709-61-0	3074
13465-77-5	2443	13478-04-1	740	13536-80-6	1834	13597-73-4	2457	13709-94-9	2164
13465-78-6	2435	13478-10-9	1345	13536-84-0	3016	13597-87-0	2416	13718-22-4	2326
13465-84-4	2455	13478-14-3	1511	13537-09-2	1025	13597-95-0	394	13718-26-8	2675
13465-85-5	2442	13478-16-5	1603	13537-15-0	1074	13597-99-4	389	13718-26-8	2676
13465-93-5	2525	13478-17-6	1801	13537-18-3	2870	13598-15-7	385	13718-50-8	310
13465-94-6	321	13478-18-7	1804	13537-22-9	1073	13598-22-6	399	13718-55-3	281

13718-59-7	337	13775-18-3	2998	13819-84-6	1817	13940-89-1	1348	14259-82-6	3053
13718-70-2	1352	13775-53-6	2578	13820-40-1	189	13943-58-3	2129	14280-53-6	1266
13721-34-1	2674	13775-80-9	1226	13820-41-2	190	13963-57-0	50	14282-91-8	2342
13721-39-6	2614	13776-58-4	3071	13820-53-6	2660	13963-58-1	2137	14283-07-9	1574
13746-66-2	2128	13776-62-0	1952	13820-62-7	842	13965-73-6	458	14284-06-1	962
13746-89-9	3211	13776-74-4	1638	13821-06-2	294	13966-94-4	1268	14284-93-6	2347
13746-98-0	2843	13777-22-5	1197	13823-29-5	2856	13967-25-4	1744	14285-68-8	2256
13755-29-8	2663	13777-23-6	1200	13823-29-5	2857	13967-50-5	2122	14291-02-2	2715
13755-32-3	349	13777-25-8	3206	13824-36-7	2446	13967-90-3	270	14307-33-6	574
13755-38-9	2613	13778-39-7	2871	13824-57-2	1821	13972-68-4	516	14311-93-4	3204
13759-10-9	2472	13778-96-6	1621	13824-74-3	2000	13973-87-0	485	14312-00-6	507
13759-30-3	3054	13779-10-7	2241	13825-25-7	2328	13977-65-6	2126	14324-82-4	1007
13759-83-6	2369	13779-41-4	2011	13825-74-6	2940	13981-95-8	2764	14325-78-1	2240
13759-87-0	2362	13779-73-2	1196	13825-76-8	1211	13982-53-1	932	14335-40-1	2023
13759-88-1	1065	13779-87-8	3203	13825-86-0	785	13983-20-5	1244	14336-80-2	921
13759_89-2	3094	13779-92-5	1929	13826-63-6	2830	13986-18-0	3148	14355-29-4	456
13759-90-5	1069	13780-03-5	588	13826-83-0	193	13986-26-0	3210	14373-91-2	1030
13759-92-7	1067	13780-06-8	603	13830-68-7	2449	14013-15-1	1707	14402-70-1	159
13760-02-6	2453	13780-06-8	604	13840-00-5	703	14013-86-6	1353	14402-73-4	1573
13760-41-3	2245	13780-42-2	1128	13840-33-0	1542	14013-86-6	1354	14402-75-6	2201
13760-78-6	1214	13780-48-8	1057	13842-67-6	2807	14014-18-7	2047	14405-43-7	1124
13760-79-7	2872	13780-57-9	2744	13842-73-4	2767	14014-19-8	2045	14446-13-0	2709
13760-80-0	3097	13780-64-8	3076	13842-75-6	1916	14014-88-1	2343	14447-89-3	2979
13760-81-1	1585	13782-01-9	911	13842-76-7	2769	14017-47-1	696	14452-39-2	51
13760-83-3	1041	13782-33-7	1993	13842-80-3	3035	14019-91-1	623	14452-39-2	52
13762-12-4	846	13782-92-8	1193	13842-83-6	2090	14024-00-1	3029	14456-34-9	1202
13762-14-6	871	13782-93-9	1195	13842-88-1	1918	14024-17-0	1358	14456-47-4	2800
13762-26-0	3196	13782-95-1	1192	13842-93-8	2783	14024-18-1	1394	14456-48-5	1020
13762-51-1	2109	13782-96-2	1194	13842-94-9	3197	14024-48-7	881	14456-51-0	2869
13762-65-7	1232	13783-04-5	2920	13843-28-2	821	14024-61-4	1989	14456-53-2	1582
13762-75-9	1549	13783-07-8	2922	13843-59-9	87	14024-63-6	3166	14457-83-1	1614
13763-23-0	3009	13783-08-9	2928	13845-17-5	292	14038-43-8	1382	14457-84-2	48
13763-67-2	716	13798-24-8	2802	13847-22-8	3171	14040-11-0	2950	14457-87-5	681
13765-03-2	1543	13812-43-6	1951	13847-57-9	1448	14049-36-6	2439	14457-87-5	682
13765-19-0	569	13812-58-3	1004	13847-65-9	1957	14049-81-1	859	14459-59-7	1826
13765-24-7	2366	13813-22-4	1419	13847-66-0	2813	14055-74-4	1802	14459-75-7	1931
13765-25-8	1068	13813-23-5	2230	13862-78-7	196	14055-75-5	1806	14459-95-1	2129
13765-26-9	1094	13813-24-6	1839	13863-41-7	486	14055-76-6	1811	14464-46-1	2469
13765-74-7	2504	13813-25-7	2367	13863-59-7	487	14055-81-3	2961	14474-33-0	2380
13766-47-7	2963	13813-40-6	2803	13863-88-2	2477	14055-84-6	2964	14475-63-9	3209
13767-31-2	1017	13813-41-7	1215	13864-01-2	1416	14059-33-7	439	14476-12-1	1697
13767-32-3	3156	13813-42-8	1044	13864-02-3	688	14075-53-7	2132	14481-29-9	108
13767-34-5	977	13813-43-9	2874	13870-13-8	1204	14077-39-5	1059	14483-18-2	1216
13768-11-1	2255	13813-44-0	3098	13870-19-4	1418	14099-01-5	2258	14486-19-2	539
13768-38-2	1974	13813-45-1	1586	13870-20-7	1915	14104-45-1	1255	14507-19-8	1417
13768-67-7	3099	13813-46-2	2091	13870-21-8	1919	14166-78-0	1276	14516-54-2	1690
13768-86-0	2393	13813-47-3	80	13870-24-1	1371	14168-73-1	1668	14517-29-4	1841
13768-94-0	2427	13814-22-7	3202	13871-27-7	2662	14175-02-1	1071	14518-81-1	2048
13769-20-5	1058	13814-59-0	529	13873-84-2	1297	14177-46-9	1726	14519-07-4	3134
13769-36-3	1140	13814-62-5	527	13874-02-7	2659	14188-40-0	1139	14519-18-7	2680
13769-75-0	2051	13814-72-7	1088	13874-75-4	2357	14215-00-0	397	14521-14-3	2450
13769-76-1	2050	13814-74-9	1818	13874-77-6	3090	14215-29-3	498	14521-15-4	2451
13770-18-8	987	13814-75-0	1825	13875-06-4	3078	14215-30-6	944	14521-18-7	2350
13770-56-4	244	13814-76-1	2277	13876-85-2	928	14216-75-2	1885	14523-22-9	2288
13770-61-1	1279	13814-81-8	936	13896-65-6	2346	14217-01-7	810	14542-23-5	578
13770-96-2	2532	13814-83-0	3061	13918-22-4	1726	14217-21-1	2566	14550-84-6	2814
13773-69-8	3117	13814-96-5	1456	13931-94-7	780	14220-17-8	2202	14553-44-7	2861
13773-81-4	1404	13816-38-1	219	13932-17-7	2216	14220-21-4	2257	14567-54-5	998
13774-85-1	3077	13817-37-3	861	13933-23-8	1359	14221-48-8	107	14567-59-0	1494
13775-06-9	2996	13818-73-0	2243	13939-06-5	1795	14242-05-8	2510	14567-59-0	1495
13775-07-0	3007	13818-75-2	1091	13940-63-1	1152	14244-62-3	2204	14590-13-7	96
13775-16-1	3005	13818-89-8	1133	13940-83-5	1878	14249-98-0	2974	14590-19-3	886

14635-87-1	1651	14977-61-8	824	15492-38-3	2293	16689-88-6	2848	17829-82-2	678
14639-94-2	122	14984-81-7	2405	15513-69-6	2960	16712-20-2	1521	17829-86-6	2360
14639-97-5	191	14985-19-4	2875	15513-84-5	3036	16721-80-5	2593	17835-81-3	1125
14639-98-6	166	14985-19-4	2876	15513-85-6	3199	16721-80-5	2594	18078-40-5	2214
14644-61-2	3215	14986-21-1	2458	15513-94-7	3043	16731-55-8	2163	18088-11-4	2329
14644-61-2	3216	14986-89-1	1592	15519-28-5	726	16752-60-6	2039	18278-82-5	720
14646-16-3	1043	14986-91-5	1661	15520-84-0	907	16774-21-3	702	18282-10-5	2908
14646-29-8	1591	14987-04-3	1676	15552-14-4	274	16812-54-7	1897	18283-93-7	443
14649-73-1	2582	15059-52-6	1023	15571-91-2	2196	16827-96-6	97	18288-22-7	408
14654-44-5	2690	15060-55-6	195	15572-25-5	2835	16853-74-0	3218	18356-71-3	2440
14662-04-5	1224	15060-59-0	1578	15578-26-4	2896	16853-85-3	1508	18424-17-4	1533
14666-94-5	876	15060-62-5	1891	15587-39-0	1136	16860-99-4	2738	18433-40-4	204
14674-72-7	568	15070-34-5	1645	15587-72-1	2322	16871-60-6	2135	18433-48-2	3017
14676-93-8	784	15098-87-0	31	15593-51-8	1565	16871-71-9	3151	18433-84-6	445
14689-45-3	523	15120-50-0	210	15593-52-9	1564	16871-90-2	2139	18454-12-1	1451
14691-44-2	1134	15123-69-0	992	15593-61-0	1663	16872-11-0	465	18480-07-4	2696
14691-47-5	1135	15123-80-5	39	15597-39-4	2019	16893-85-9	2581	18488-96-5	875
14693-56-2	197	15123-90-7	2719	15597-40-7	2020	16903-35-8	1168	18601-87-1	872
14693-61-9	2422	15162-90-0	1943	15597-63-4	2022	16919-19-0	125	18618-55-8	686
14693-65-3	2425	15162-92-2	1833	15597-88-3	816	16919-31-6	127	18624-44-7	1349
14693-80-2	2771	15162-93-3	2223	15607-89-3	2743	16919-58-7	119	18718-07-5	1700
14693-81-3	2779	15163-24-3	2958	15608-29-4	2290	16921-30-5	2136	18721-05-6	792
14693-82-4	1282	15168-20-4	994	15609-80-0	1673	16921-96-3	1300	18727-04-3	855
14720-21-9	1177	15179-32-5	2756	15610-76-1	100	16923-58-3	2576	18746-63-9	120
14720-53-7	1440	15192-14-0	3079	15622-42-1	2261	16923-58-3	2577	18810-58-7	268
14721-18-7	1875	15192-17-3	2358	15627-86-8	614	16923-95-8	2140	18820-29-6	1720
14721-21-2	950	15192-18-4	3091	15635-87-7	1313	16924-00-8	2133	18820-29-6	1721
14735-84-3	1005	15192-24-2	2238	15653-01-7	695	16925-25-0	2579	18820-29-6	1722
14758-11-3	1459	15192-26-4	2793	15663-27-1	2081	16925-39-6	584	18855-94-2	1206
14763-77-0	958	15192-42-4	2263	15684-18-1	1992	16940-66-2	2539	18868-43-4	1812
14767-09-0	2826	15195-33-2	1926	15696-40-9	1964	16940-81-1	2012	18902-42-6	2341
14779-70-5	1147	15195-53-6	1914	15699-18-0	1866	16940-92-4	116	18909-68-7	2316
14781-45-4	972	15230-48-5	1141	15750-45-5	1642	16941-10-9	637	18909-69-8	722
14791-73-2	558	15230-79-2	1584	15752-05-3	115	16941-11-0	124	18917-82-3	1464
14808-60-7	2466	15238-00-3	868	15752-41-7	1912	16941-12-1	2063	18917-91-4	38
14808-60-7	2467	15238-00-3	870	15752-46-2	2088	16941-12-1	2083	18933-05-6	1702
14808-60-7	2468	15243-27-3	1694	15771-43-4	391	16949-15-8	1513	19034-13-0	888
14808-60-7	2469	15243-33-1	2338	15785-09-8	689	16961-83-4	2462	19049-40-2	369
14836-60-3	1748	15244-10-7	1401	15823-43-5	1205	16962-07-5	15	19058-78-7	1301
14859-67-7	2253	15244-35-6	534	15829-53-5	1751	16962-31-5	2138	19073-56-4	2314
14868-53-2	2418	15244-38-9	813	15843-48-8	1469	16962-40-6	126	19086-20-5	1671
14868-54-3	2419	15275-09-9	810	15845-52-0	1460	16962-47-3	123	19088-74-5	2320
14868-55-4	2423	15280-09-8	2572	15851-44-2	538	16962-48-4	2318	19138-68-2	1001
14868-55-4	2424	15280-53-2	1090	15861-05-9	1949	17014-71-0	2177	19139-47-0	676
14871-56-8	344	15280-55-4	1019	15875-18-0	1432	17026-29-8	2278	19168-23-1	118
14871-79-5	307	15280-57-6	1036	15878-77-0	2232	17026-44-7	183	19197-73-0	2162
14884-42-5	819	15280-58-7	3093	15942-63-9	240	17029-16-2	1537	19262-94-3	1658
14885-60-0	1245	15282-88-9	1473	15947-41-8	81	17029-21-9	2279	19287-45-7	442
14885-61-1	1246	15292-44-1	2383	15947-57-6	2420	17083-68-0	1167	19357-83-6	2972
14890-41-6	3033	15293-74-0	1548	16037-50-6	2116	17099-70-6	32	19357-86-9	3092
14890-44-9	1038	15293-86-4	936	16056-61-4	1302	17125-80-3	299	19372-44-2	612
14902-95-5	395	15318-60-2	679	16122-03-5	1865	17194-00-2	304	19423-81-5	1093
14913-33-8	2082	15321-51-4	1323	16222-66-5	2845	17242-52-3	2105	19423-87-1	3096
14914-84-2	1213	15364-10-0	3085	16399-77-2	1344	17347-95-4	1536	19465-30-6	448
14933-38-1	77	15364-94-0	1714	16406-48-7	2339	17363-02-9	114	19469-07-9	1392
14939-34-5	2042	15385-57-6	1746	16406-49-8	1965	17375-41-6	1363	19513-05-4	1728
14940-41-1	1360	15385-58-7	1739	16469-16-2	2229	17440-85-6	376	19525-15-6	3013
14940-65-9	3069	15468-32-3	2468	16469-17-3	1838	17440-90-3	1321	19567-78-3	2575
14948-62-0	2225	15469-38-2	1384	16469-22-0	3113	17496-59-2	750	19584-30-6	2289
14965-52-7	2441	15474-63-2	1027	16544-92-6	851	17674-34-9	2379	19597-69-4	1512
14972-90-8	192	15477-33-5	20	16671-27-5	797	17702-41-9	449	19598-90-4	1097
14977-17-4	545	15491-35-7	359	16674-78-5	1599	17786-31-1	837	19624-22-7	444

CAS RN	No.	CAS RN	No.	CAS RN	No.	CAS RN	No.	CAS RN	No.
19648-83-0	863	22519-64-8	1274	25764-09-4	2220	33445-15-7	151	51466-48-9	24
19648-88-5	1458	22520-96-3	1298	25764-10-7	1421	33689-80-4	4	51503-61-8	135
19783-14-3	1461	22527-13-5	3084	25764-11-8	1830	33689-81-5	2	51595-71-2	1754
20193-56-0	2959	22537-19-5	1430	25764-15-2	1086	33689-82-6	5	51621-05-7	2345
20193-58-2	1805	22570-08-7	3164	25808-74-6	1480	34018-28-5	1441	51850-20-5	2064
20205-91-8	470	22585-64-4	491	25817-87-2	1188	34128-09-1	2828	51898-99-8	413
20211-76-1	1414	22750-54-5	503	25895-60-7	2555	34330-64-8	505	52003-58-4	149
20213-56-3	2973	22750-57-8	712	25909-39-1	2072	34979-62-9	2400	52110-05-1	1680
20281-00-9	736	22756-36-1	2306	25985-07-3	2071	35112-53-9	354	52503-64-7	940
20328-94-3	3080	22831-39-6	1665	26042-63-7	2498	35340-84-2	866	52740-16-6	549
20344-49-4	1387	22831-42-1	13	26042-64-8	2496	35405-51-7	622	52788-53-1	1096
20398-06-5	2821	22852-11-5	2867	26124-86-7	313	35718-37-7	3086	52788-54-2	2383
20405-64-5	930	22852-67-1	1775	26134-62-3	1554	35788-00-2	933	52870-08-3	2689
20427-11-6	857	22886-66-4	1117	26318-99-0	995	36377-94-3	1018	52951-38-9	416
20427-56-9	2352	22986-54-5	3	26342-61-0	773	36478-76-9	3018	53214-07-6	2794
20427-58-1	3152	22992-15-0	1098	26499-65-0	631	36509-15-6	1975	53319-44-1	2404
20427-59-2	974	23032-72-6	1389	26500-06-1	3084	36539-19-2	3085	53608-79-0	2217
20548-54-3	633	23149-52-2	2523	26508-33-8	1329	36548-87-5	2876	53731-35-4	1715
20601-83-6	1782	23276-90-6	823	26522-91-8	1760	36781-15-4	2810	53809-75-9	483
20610-49-5	2250	23303-71-1	1516	26628-22-8	2538	36897-37-7	1880	54120-05-7	1971
20619-16-3	1154	23370-59-4	1310	26677-68-9	2877	36907-37-6	1423	54496-71-8	904
20644-15-9	2580	23377-53-9	2284	26686-77-1	1646	36907-38-7	696	54723-94-3	172
20661-21-6	1277	23412-45-5	1810	26750-66-3	2211	36907-40-1	1073	54804-32-9	2434
20665-52-5	1123	23414-72-4	3169	26970-82-1	2638	36907-42-3	3038	55147-94-9	806
20667-12-3	2508	23436-05-7	314	27016-73-5	830	37185-09-4	343	55606-55-8	453
20694-39-7	1709	23436-05-7	313	27016-75-7	1855	37248-04-7	1204	56320-90-2	718
20712-42-9	521	23611-30-5	2332	27043-84-1	3131	37248-34-3	2192	56378-72-4	1617
20762-60-1	2108	23777-80-2	446	27133-66-0	333	37265-86-4	362	56802-99-4	2628
20770-09-6	2909	24094-93-7	771	27218-16-2	760	37265-91-1	492	57246-89-6	2283
20816-12-0	1976	24304-00-5	42	27535-70-2	806	37267-86-0	2008	57804-25-8	1426
20859-73-8	58	24401-69-2	927	27546-07-2	155	37306-42-6	440	58572-43-3	484
20910-28-5	3021	24415-00-7	1150	27709-53-1	2215	37773-49-2	2074	58724-12-2	725
21041-93-0	865	24422-20-6	1149	27774-13-6	3063	37809-19-1	580	59201-51-3	1968
21041-95-2	512	24422-21-7	1142	27860-83-9	3149	37913-38-5	1788	60582-92-5	1598
21056-98-4	619	24567-53-1	2002	28041-86-3	2119	38245-35-1	1021	60616-74-2	1630
21093-83-4	2142	24597-12-4	1110	28265-11-4	454	38455-77-5	2905	60676-86-0	2470
21109-95-5	346	24598-62-7	113	28300-74-5	213	38680-84-1	758	60763-24-8	2860
21159-32-0	719	24613-38-5	853	28407-51-4	2287	38705-19-0	1756	60804-25-3	2292
21255-83-4	481	24621-17-8	3195	28411-13-4	1966	38989-47-8	2390	60864-26-8	2868
21264-43-7	1269	24621-18-9	3200	28950-34-7	2757	39277-13-9	2574	60883-64-9	379
21308-45-2	3083	24621-21-4	1908	28958-26-1	2361	39290-85-2	945	60922-26-1	2951
21308-80-5	482	24646-85-3	3030	28965-57-3	1218	39368-69-9	2270	60950-56-3	1664
21324-40-3	1535	24719-19-5	843	29703-01-3	715	39373-27-8	2297	60969-19-9	2824
21351-79-1	729	25013-82-5	788	29809-42-5	2816	39409-82-0	1616	61027-88-1	1601
21548-73-2	2518	25094-02-4	566	29935-35-1	1534	39416-30-3	323	61042-72-6	1615
21645-51-2	34	25152-52-7	12	30708-80-6	756	39430-51-8	790	61519-37-7	490
21651-19-4	2895	25232-60-4	619	30708-86-2	1304	39578-36-4	1405	62571-53-3	2971
21679-31-2	805	25324-56-5	2884	30737-24-7	2831	39733-35-2	150	63128-11-0	2340
21679-46-9	910	25402-50-0	1303	30903-87-8	1558	39796-98-0	3081	63427-22-5	2361
21907-50-6	743	25417-81-6	302	30937-52-1	2271	39797-63-2	3082	63771-33-5	165
21908-53-2	1777	25455-73-6	2527	30937-53-2	2269	41422-43-9	211	63972-04-3	2061
21995-38-0	680	25469-93-6	1832	31032-13-0	1179	41591-55-3	1237	64082-35-5	1530
22015-35-6	1060	25502-05-0	3089	31052-14-9	2782	41944-01-8	425	64171-97-7	2866
22132-71-4	135	25519-09-9	1210	31052-43-4	2333	42246-24-2	2282	64578-39-8	1358
22205-45-4	931	25519-10-2	1036	31083-74-6	2335	42739-38-8	205	65202-12-2	1752
22205-57-8	711	25583-20-4	2917	31234-26-1	2273	43093-20-5	452	65355-99-9	1292
22207-96-1	2172	25617-97-4	1107	31576-40-6	1973	49756-76-5	3080	65530-47-4	2222
22208-73-7	555	25617-98-5	1264	32248-43-4	2359	50432-32-1	3083	65842-03-7	1388
22306-37-2	401	25658-42-8	3192	32287-65-3	30	50801-97-3	2356	66104-24-3	380
22326-55-2	305	25658-43-9	2992	32823-06-6	56	50813-65-5	275	66778-21-0	1635
22398-80-7	1265	25659-31-8	1462	32839-04-6	1759	50927-81-6	2471	67952-43-6	1872
22441-45-8	257	25764-08-3	673	32963-81-8	1395	51429-74-4	1790	68016-36-4	353

68938-92-1	2073	75926-26-0	2408	89711-39-7	451	107539-20-8	287	124365-83-9	288
69239-51-6	509	75926-28-2	2409	92141-86-1	732	107539-20-8	288	133578-89-9	3062
71500-16-8	2348	75926-30-6	2410	94217-84-2	1503	107782-11-6	1080	133863-98-6	1823
71595-75-0	450	81155-18-2	2274	97126-35-7	897	109064-29-1	286	134929-59-2	649
71626-98-7	596	82642-06-6	285	99685-96-8	647	114104-80-2	287	142617-56-9	411
72520-94-6	684	83229-05-4	1016	102430-09-1	799	115383-22-7	648	155698-29-6	2058
73560-00-6	1293	84359-31-9	808	107539-20-8	285	121012-90-6	583		
73796-25-5	2699	84359-31-9	809	107539-20-8	286	123333-98-2	799		

PHYSICAL PROPERTIES OF THE RARE EARTH METALS

K.A. Gschneidner, Jr.

TABLE 1. Data for the Trivalent Ions of the Rare Earth Elements

Rare earth	Symbol	Atomic no.	Atomic wt.[a]	No. 4f electrons	S	L	J	Spectroscopic ground state symbol
Scandium	Sc	21	44.955910	0	—	—	—	—
Yttrium	Y	39	88.90585	0	—	—	—	—
Lanthanum	La	57	138.9055	0	—	—	—	—
Cerium	Ce	58	140.115	1	1/2	3	5/2	$^2F_{5/2}$
Praseodymium	Pr	59	140.90765	2	1	5	4	$^3H_4$
Neodymium	Nd	60	144.24	3	3/2	6	9/2	$^4I_{9/2}$
Promethium	Pm	61	(145)	4	2	6	4	$^5I_4$
Samarium	Sm	62	150.36	5	5/2	5	5/2	$^6H_{5/2}$
Europium	Eu	63	151.965	6	3	3	0	$^7F_0$
Gadolinium	Gd	64	157.25	7	7/2	0	7/2	$^8S_{7/2}$
Terbium	Tb	65	158.92534	8	3	3	6	$^7F_6$
Dysprosium	Dy	66	162.50	9	5/2	5	15/2	$^6H_{15/2}$
Holmium	Ho	67	164.93032	10	2	6	8	$^5I_8$
Erbium	Er	68	167.26	11	3/2	6	15/2	$^4I_{15/2}$
Thulium	Tm	69	168.93421	12	1	5	6	$^3H_6$
Ytterbium	Yb	70	173.04	13	1/2	3	7/2	$^2F_{7/2}$
Lutetium	Lu	71	174.967	14	—	—	—	—

Note: For additional information, see Goldschmidt, Z.B., in *Handbook on the Physics and Chemistry of Rare Earths*, Vol. 1, Gschneidner, K.A., Jr. and Eyring, L., Eds., North-Holland Physics, Amsterdam, 1978; DeLaeter, J.R., and Heumann, K.G., *J. Phys. Chem. Ref. Data*, 20, 1313 , 1991; *Pure Appl. Chem.*, 66, 2423, 1994.

[a] 1993 standard atomic weights.

TABLE 2. Crystallographic Data for the Rare Earth Metals at 24°C (297 K) or Below

Rare earth metal	Crystal structure[a]	a_o	b_o	c_o	Metallic radius CN = 12 (Å)	Atomic volume (cm³/mol)	Density (g/cm³)
αSc	hcp	3.3088	—	5.2680	1.6406	15.039	2.989
αY	hcp	3.6482	—	5.7318	1.8012	19.893	4.469
αLa	dhcp	3.7740	—	12.171	1.8791	22.602	6.146
αCe[b]	fcc	4.85[b]	—	—	1.72[b]	17.2[b]	8.16[b]
βCe	dhcp	3.6810	—	11.857	1.8321	20.947	6.689
γCe[c]	fcc	5.1610	—	—	1.8247	20.696	6.770
αPr	dhcp	3.6721	—	11.8326	1.8279	20.803	6.773
αNd	dhcp	3.6582	—	11.7966	1.8214	20.583	7.008
αPm	dhcp	3.65	—	11.65	1.811	20.24	7.264
αSm	rhomb[d]	3.6290[d]	—	26.207	1.8041	20.000	7.520
Eu	bcc	4.5827	—	—	2.0418	28.979	5.244
αGd	hcp	3.6336	—	5.7810	1.8013	19.903	7.901
α'Tb[e]	ortho	3.605[e]	6.244[e]	5.706[e]	1.784[e]	19.34[e]	8.219[e]
αTb	hcp	3.6055	—	5.6966	1.7833	19.310	8.230
α'Dy[f]	ortho	3.595[f]	6.184[f]	5.678[f]	1.774[f]	19.00[f]	8.551[f]
αDy	hcp	3.5915	—	5.6501	1.7740	19.004	8.551
Ho	hcp	3.5778	—	5.6178	1.7661	18.752	8.795
Er	hcp	3.5592	—	5.5850	1.7566	18.449	9.066
Tm	hcp	3.5375	—	5.5540	1.7462	18.124	9.321
αYb[g]	hcp	3.8799[g]	—	6.3859[g]	1.9451[g]	25.067[g]	6.903[g]
βYb	fcc	5.4848	—	—	1.9392	24.841	6.966
Lu	hcp	3.5052	—	5.5494	1.7349	17.779	9.841

Note: For additional information, see Gschneidner, K.A., Jr. and Calderwood, F.W., in *Handbook on the Physics and Chemistry of Rare Earths*, Vol. 8, Gschneidner, K.A., Jr. and Eyring, L., Eds., North-Holland Physics, Amsterdam, 1986; Gschneidner, K.A., Jr., Pecharsky, V.K., Cho, Jaephil and Martin, S.W., *Scripta Mater.*, 1996, to be published.

[a] hcp = hexagonal close-packed; P6₃/mmc, hP2, A3, Mg-type; dhcp = double-c hexagonal close-packed; P6₃/mmc, hP4, A3´, αLa-type; fcc = face-centered cubic; Fm$\overline{3}$m, cF4, A1, Cu-type; rhomb = rhombohedral; R$\overline{3}$m, hR3, αSm-type; bcc = body-centered cubic; Im$\overline{3}$m, cI2, A2, W-type; ortho = orthorhombic; Cmcm, oC4, α´ Dy-type.

[b] At 77 K (−196°C).

[c] Equilibrium room temperature (standard state) phase.

[d] Rhombohedral is the primitive cell. Lattice parameters given are for the nonprimitive hexagonal cell.

[d] At 220 K (−53°C).

[f] At 86 K (−187°C).

[g] At 23°C.

TABLE 3. Crystallographic Data for Rare Earth Metals at High Temperature

Rare earth metal	Structure	Lattice parameter (Å)	Temp. (°C)	Metallic radius CN = 8 (Å)	Metallic radius CN = 12 (Å)	Atomic volume (cm³/mol)	Density (g/cm³)
βSc	bcc	3.73 (est.)	1337	1.62	1.66	15.6	2.88
βY	bcc	4.10[a]	1478	1.78	1.83	20.8	4.28
βLa	fcc	5.303	325	—	1.875	22.45	6.187
γLa	bcc	4.26	887	1.84	1.90	23.3	5.97
δCe	bcc	4.12	757	1.78	1.84	21.1	6.65
βPr	bcc	4.13	821	1.79	1.84	21.2	6.64
βNd	bcc	4.13	883	1.79	1.84	21.2	6.80
βPm	bcc	4.10 (est.)	890	1.78	1.83	20.8	6.99
βSm	hcp	$a = 3.6630$ $c = 5.8448$	450[b]	—	1.8176	20.450	7.353
γSm	bcc	4.10 (est.)	922	1.77	1.82	20.8	7.25
βGd	bcc	4.06	1265	1.76	1.81	20.2	7.80
βTb	bcc	4.07[a]	1289	1.76	1.81	20.3	7.82
βDy	bcc	4.03[a]	1381	1.75	1.80	19.7	8.23
γYb	bcc	4.44	763[c]	1.92	1.98	26.4	6.57

Note: The rare earths Eu, Ho, Er, Tm, and Lu are monomorphic. For additional information, see Gschneidner, K.A., Jr. and Calderwood, F.W., in *Handbook on the Physics and Chemistry of Rare Earths*, Vol. 8, Gschneidner, K.A., Jr. and Eyring, L., Eds., North-Holland Physics, Amsterdam, 1986, 1.
[a] Determined by extrapolation to 0% solute of a vs. composition data for R-Mg alloys at 24°C and corrected for thermal expansion to temperature given.
[b] The hcp phase was stabilized by impurities and the temperature of measurement was below the equilibrium transition temperature (see Table 4).
[c] The bcc phase was stabilized by impurities and the temperature of measurement was below the equilibrium transition temperature (see Table 4).

TABLE 4. High Temperature Transition Temperatures and Melting Point of Rare Earth Metals

Rare earth metal	Transition I (α – β)[a] Temp. (°C)	Transition I (α – β)[a] Phases	Transition II (β – γ)[a] Temp. (C°)	Transition II (β – γ)[a] Phases	Melting point (C°)
Sc	1337	hcp ⇌ bcc	—	—	1541
Y	1478	hcp ⇌ bcc	—	—	1522
La[b]	310	dhcp ⇌ fcc	865	fcc ⇌ bcc	918
Ce[c,d]	139	dhcp ⇌ fcc (β - γ)	726	fcc ⇌ bcc (γ - δ)	798
Pr	795	dhcp ⇌ bcc	—	—	931
Nd	863	dhcp ⇌ bcc	—	—	1021
Pm	890	dhcp ⇌ bcc	—	—	1042
Sm[e]	734	rhom ⇌ hcp	922	hcp ⇌ bcc	1074
Eu	—	—	—	—	822
Gd	1235	hcp ⇌ bcc	—	—	1313
Tb	1289	hcp ⇌ bcc	—	—	1356
Dy	1381	hcp ⇌ bcc	—	—	1412
Ho	—	—	—	—	1474
Er	—	—	—	—	1529
Tm	—	—	—	—	1545
Yb	795	fcc ⇌ bcc (β - γ)	—	—	819
Lu	—	—	—	—	1663

Note: For additional information, see Gschneidner, K.A., Jr. and Calderwood, F.W., in *Handbook on the Physics and Chemistry of Rare Earths*, Vol. 8, Gschneidner, K.A., Jr. and Eyring, L., Eds., North-Holland Physics, Amsterdam, 1986; Gschneidner, K.A., Jr., Pecharsky, V.K., Cho, Jaephil and Martin, S.W., *Scripta Mater.*, 34, 1717, 1996.
[a] For all the transformations listed, unless otherwise noted.
[b] On cooling, fcc → dhcp (β → α), 260°C.
[c] The β ⇌ γ equilibrium transition temperature is 10 ± 5°C.
[d] On cooling, fcc → dhcp (γ → β), −16°C.
[e] On cooling, hcp → rhomb (β → α), 727°C.

TABLE 5. Low Temperature Transition Temperatures of the Rare Earth Metals

Rare earth metal	Cooling Transformation	Cooling °C	Cooling K	Rare earth metal	Heating Transformation	Heating °C	Heating K
Ce	γ → β[a]	−16	257	Ce	α → β	−148	125
	γ → α	−172	101		α → β + γ	−104	169
	β → α	−228	45		β → γ[a]	139	412
Tb	α → α'	−53	220	Yb	α → β	7	280
Dy	α → α'	−187	86				
Yb	β → α	−13	260				

Note: For additional information, see Beaudry, B.J. and Gschneidner, K.A., Jr., in *Handbook on the Physics and Chemistry of Rare Earths*, Vol. 1, Gschneidner, K.A., Jr. and Eyring, L., Eds., North-Holland Physics, Amsterdam, 1978, 173; Koskenmaki, D.C. and Gschneidner, K.A., Jr., 1978, in *Handbook on the Physics and Chemistry of Rare Earths*, Vol. 1, Gschneidner, K.A., Jr. and Eyring, L., Eds., North-Holland Physics, Amsterdam, 1978, 337; Gschneidner, K.A., Jr., Pecharsky, V.K., Cho, Jaephil and Martin, S.W., *Scripta Mater.*, 34, 1717, 1996.
[a] The β ⇌ γ equilibrium transition temperature is 10 ± 5°C (283 ± 5K).

TABLE 6. Heat Capacity, Standard Entropy, Heats of Transformation, and Fusion of the Rare Earth Metals

Rare earth metal	Heat capacity at 298 K (J/mol K)	Standard entropy $S°_{298}$ (J/mol K)	Heat of transformation (kJ/mol)				Heat of fusion (kJ/mol)
			trans. 1	ΔH_{tr}^1	trans. 2	ΔH_{tr}^2	
Sc	25.5	34.6	$\alpha \rightleftharpoons \beta$	4.00	—	—	14.1
Y	26.5	44.4	$\alpha \rightleftharpoons \beta$	4.99	—	—	11.4
La	27.1	56.9	$\alpha \rightleftharpoons \beta$	0.36	$\beta \rightleftharpoons \gamma$	3.12	6.20
Ce	26.9	72.0	$\beta \rightleftharpoons \gamma$	0.05	$\gamma \rightleftharpoons \delta$	2.99	5.46
Pr	27.2	73.2	$\alpha \rightleftharpoons \beta$	3.17	—	—	6.89
Nd	27.5	71.5	$\alpha \rightleftharpoons \beta$	3.03	—	—	7.14
Pm	27.3[a]	71.6[a]	$\alpha \rightleftharpoons \beta$	3.0[a]	—	—	7.7[a]
Sm	29.5	69.6	$\alpha \rightleftharpoons \beta$	0.2[a]	$\beta \rightleftharpoons \gamma$	3.11	8.62
Eu	27.7	77.8	—	—	—	—	9.21
Gd	37.0	68.1	$\alpha \rightleftharpoons \beta$	3.91	—	—	10.0
Tb	28.9	73.2	$\alpha \rightleftharpoons \beta$	5.02	—	—	10.79
Dy	27.7	75.6	$\alpha \rightleftharpoons \beta$	4.16	—	—	11.06
Ho	27.2	75.3	—	—	—	—	17.0[a]
Er	28.1	73.2	—	—	—	—	19.9
Tm	27.0	74.0	—	—	—	—	16.8
Yb	26.7	59.9	$\beta \rightleftharpoons \gamma$	1.75	—	—	7.66
Lu	26.9	51.0	—	—	—	—	22[a]

Note: For additional information, see Hultgren, R., Desai, P.D., Hawkins, D.T., Gleiser, M., Kelley, K.K., and Wagman, D.D., *Selected Values of the Thermodynamic Properties of the Elements*, ASM International, Metals Park, Ohio, 1973; Wagman, D.D., Evans, W.H., Parker, V.B., Schumm, R.H., Halow, I., Bailey, S.M., Churney, K.L., and Nuttall, R.L., *The NBS Tables of Chemical Thermodynamic Properties, J. Phys. Chem. Ref. Data*, Vol. 11, Suppl 2, 1982; Amitin, E.B., Bessergenev, W.G., Kovalevskaya, Yu. A., and Paukov, I.E., *J. Chem. Thermodyn.*, 15, 181, 1983; Amitin, E.B., Bessergenev, W.G., Kovalevskaya, Yu. A., and Paukov, I.E., *J. Chem. Thermodyn.*, 15, 181, 1983.

[a] Estimated.

TABLE 7. Vapor Pressures, Boiling Points, and Heats of Sublimation of Rare Earth Metals

Rare earth metal	Temperature in °C[a] for a vapor pressure of				Boiling point[a] (°C)	Heat of sublimation at 25°C (kJ/mol)
	10^{-8} atm (0.001 Pa)	10^{-6} atm (0.101 Pa)	10^{-4} atm (10.1 Pa)	10^{-2} atm (1013 Pa)		
Sc	1036	1243	1533	1999	2836	377.8
Y	1222	1460	1812	2360	3345	424.7
La	1301	1566	1938	2506	3464	431.0
Ce	1290	1554	1926	2487	3443	422.6
Pr	1083	1333	1701	2305	3520	355.6
Nd	955	1175	1500	2029	3074	327.6
Pm	—	—	—	—	3000[b]	348[b]
Sm	508	642	835	1150	1794	206.7
Eu	399	515	685	964	1529	175.3
Gd	1167	1408	1760	2306	3273	397.5
Tb	1124	1354	1698	2237	3230	388.7
Dy	804	988	1252	1685	2567	290.4
Ho	845	1036	1313	1771	2700	300.8
Er	908	1113	1405	1896	2868	317.1
Tm	599	748	964	1300	1950	232.2
Yb	301	400	541	776	1196	152.1
Lu	1241	1483	1832	2387	3402	427.6

Note: For additional information, see Hultgren, R., Desai, P.D., Hawkins, D.T., Gleiser, M., Kelley, K.K., and Wagman, D.D., *Selected Values of the Thermodynamic Properties of the Elements*, ASM International, Metals Park, Ohio, 1973; Beaudry, B.J. and Gschneidner, K.A., Jr., in *Handbook on the Physics and Chemistry of Rare Earths*, Vol. 1, Gschneidner, K.A., Jr. and Eyring, L., Eds., North-Holland Physics, Amsterdam, 1978, 173.

[a] International Temperature Scale of 1990 (ITS-90) values.

[b] Estimated.

TABLE 8. Magnetic Properties of the Rare Earth Metals

Rare earth metal	$\chi_A \times 10^6$ at 298 K (emu/mol)	Effective magnetic moment				Easy axis	Néel temp. T_N (K)		Curie temp. T_C (K)	θ_p (K)		
		Paramagnetic at ~298 K		Ferromagnetic at ~0 K			Hex sites	Cubic sites		‖c	⊥c	Polycryst. or avg.
		Theory[a]	Obs.	Theory[b]	Obs.							
αSc	295.2	—	—	—	—	—	—	—	—	—	—	—
αY	187.7	—	—	—	—	—	—	—	—	—	—	—
αLa	95.9	—	—	—	—	—	—	—	—	—	—	—
βLa	105	—	—	—	—	—	—	—	—	—	—	—
γCe	2,270	2.54	2.52	2.14	—	—	—	14.4	—	—	—	−50
βCe	2,500	2.54	2.61	2.14	—	—	13.7	12.5	—	—	—	−41
αPr	5,530	3.58	3.56	3.20	2.7[c]	a	0.03	—	—	—	—	0
αNd	5,930	3.62	3.45	3.27	2.2[c]	b	19.9	7.5	—	0	5	3.3
αPm	—	2.68	—	2.40	—	—	—	—	—	—	—	—
αSm	1,278[d]	0.85	1.74	0.71	0.5[c]	a	109	14.0	—	—	—	—
Eu	30,900	7.94	8.48	7.0	5.9	<110>	—	90.4	—	—	—	100
αGd	185,000[e]	7.94	7.98	7.0	7.63	30° to c	—	—	293.4	317	317	317
αTb	170,000	9.72	9.77	—	—	—	230.0	—	—	195	239	224
α'Tb	—	—	—	9.0	9.34	b	—	—	219.5	—	—	—
αDy	98,000	10.64	10.83	—	—	—	180.2	—	—	121	169	153
α'Dy	—	—	—	10.0	10.33	a	—	—	90.5[g]	—	—	—
Ho	72,900	10.60	11.2	10.0	10.34	b	132	—	19.5	73.0	88.0	83.0
Er	48,000	9.58	9.9	9.0	9.1	30° to c	85	—	18.7	61.7	32.5	42.2
Tm	24,700	7.56	7.61	7.0	7.14	c	58	—	32.0	41.0	−17.0	2.3
βYb	67[d]	—	—	—	—	—	—	—	—	—	—	—
Lu	182.9	—	—	—	—	—	—	—	—	—	—	—

Note: For additional information, see McEwen, K.A., in *Handbook on the Physics and Chemistry of Rare Earths*, Vol. 1, Gschneidner, K.A., Jr. and Eyring, L., Eds., North-Holland Physics, Amsterdam, 1978, 411; Legvold, S., in *Ferromagnetic Materials*, Vol. 1, Wohlfarth, E.P., Ed., North-Holland Physics, Amsterdam, 1980, 183; Pecharsky, V.K., Gschneidner, K.A., Jr. and Fort, D., *Phys. Rev. B*, 47, 5063, 1993; Pecharsky, V.K., Gschneidner, K.A., Jr. and Fort, D., 1996, to be published; Steward, A.M. and Collocott, S.J., *J. Phys.: Condens. Matter*, 1, 677, 1988.

[a] $g[J(J + 1)]^{1/2}$.
[b] gJ.
[c] At 38 T and 4.2 K.
[d] At 290 K.
[e] At 350 K.
[g] On cooling $T_C = 89.6$ K and on warming $T_C = 91.5$ K.

TABLE 9. Room Temperature Coefficient of Thermal Expansion, Thermal Conductivity, Electrical Resistance, and Hall Coefficient

Rare earth metal	Expansion ($\alpha_i \times 10^6$) (°C^{-1})			Thermal conductivity (W/cm K)	Electrical resistance (μΩ·cm)			Hall coefficient ($R_i \times 10^{12}$) (V·cm/A·Oe)		
	α_a	α_c	α_{poly}		ρ_a	ρ_c	ρ_{poly}	R_a	R_c	R_{poly}
αSc	7.6	15.3	10.2	0.158	70.9	26.9	56.2[a]	—	—	−0.13
αY	6.0	19.7	10.6	0.172	72.5	35.5	59.6	−0.27	−1.6	—
aLa	4.5	27.2	12.1	0.134	—	—	61.5	—	—	−0.35
bCe	—	—	—	—	—	—	82.8	—	—	—
γCe	6.3	—	6.3	0.113	—	—	74.4	—	—	+1.81
αPr	4.5	11.2	6.7	0.125	—	—	70.0	—	—	+0.709
αNd	7.6	13.5	9.6	0.165	—	—	64.3	—	—	+0.971
αPm	9[b]	16[b]	11[b]	0.15[b]	—	—	75[b]	—	—	—
αSm	9.6	19.0	12.7	0.133	—	—	94.0	—	—	−0.21
Eu	35.0	—	35.0	0.139[b]	—	—	90.0	—	—	+24.4
αGd	9.1[c]	10.0[c]	9.4[c]	0.105	135.1	121.7	131.0	−10	−54	−4.48[d]
αTb	9.3	12.4	10.3	0.111	123.5	101.5	115.0	−1.0	−3.7	—
αDy	7.1	15.6	9.9	0.107	111.0	76.6	92.6	−0.3	−3.7	—
Ho	7.0	19.5	11.2	0.162	101.5	60.5	81.4	+0.2	−3.2	—
Er	7.9	20.9	12.2	0.145	94.5	60.3	86.0	+0.3	−3.6	—
Tm	8.8	22.2	13.3	0.169	88.0	47.2	67.6	—	—	−1.8
βYb	26.3	—	26.3	0.385	—	—	25.0	—	—	+3.77
Lu	4.8	20.0	9.9	0.164	76.6	34.7	58.2	+0.45	−2.6	−0.535

Note: For additional information, see Beaudry, B. J. and Gschneidner, K.A., Jr., in *Handbook on the Physics and Chemistry of Rare Earths*, Vol. 1, Gschneidner, K.A., Jr. and Eyring, L., Eds., North-Holland Physics, Amsterdam, 1978, 173; McEwen, K.A., in *Handbook on the Physics and Chemistry of Rare Earths*, Vol. 1, Gschneidner, K.A., Jr. and Eyring, L., Eds., North-Holland Physics, Amsterdam, 1978, 411.

[a] Calculated from single crystal values.
[b] Estimated.
[c] At 100°C.
[d] At 77°C.

TABLE 10. Electronic Specific Heat Constant (γ), Electron-Electron (Coulomb) Coupling Constant (μ^*), Electron-Phonon Coupling Constant (λ), Debye Temperature at 0 K(θ_D), and Superconducting Transition Temperature

Rare earth metal	γ (mJ/mol·K^2)	μ^*	λ	θ_D (K) from Heat capacity	θ_D (K) from Elastic constants	Superconducting temperature (K)
αSc	10.334	0.16	0.30	345.3	—	0.050[a]
αY	7.878	0.15	0.30	244.4	258	1.3[b]
αLa	9.45	0.08	0.76	150	154	5.10
βLa	11.5	—	—	140	—	6.00
αCe	12.8	—	—	179	—	0.022[c]
αPr	20	—	1.07[d]	155[e]	153	—
αNd	f	—	0.86[d]	157[e]	163	—
αPm	—	—	—	159[e]	—	—
αSm	8.1 ± 1.5[g]	—	0.81[d]	162[e,f]	169	—
Eu	f	—	—	f	118	—
αGd	4.48	—	0.30	169	182	—
α'Tb	3.71	—	0.34[d]	169.6	177	—
α'Dy	4.9	—	0.32[d]	192	183	—
Ho	2.1	—	0.30[d]	175[e]	190	—
Er	8.7	—	0.33[d]	176.9	188	—
Tm	f	—	0.36[d]	179[e]	200	—
αYb	3.30	—	—	117.6	118	—
βYb	8.36	—	—	109	—	—
Lu	8.194	0.14	0.31	183.2	185	0.022[h]

Note: For additional information, see Sundström, L.J., in *Handbook on the Physics and Chemistry of Rare Earths*, Vol. 1, Gschneidner, K.A., Jr., and Eyring, L., Eds., North-Holland Physics, Amsterdam, 1978, 379; Scott, T., in *Handbook on the Physics and Chemistry of Rare Earths*, Vol. 1, Gschneidner, K.A., Jr. and Eyring, L., Eds., North-Holland Physics, Amsterdam, 1978, 591; Probst, C. and Wittig, J., in *Handbook on the Physics and Chemistry of Rare Earths*, Vol. 1, Gschneidner, K.A., Jr. and Eyring, L., Eds., North-Holland Physics, Amsterdam, 1978, 749; Tsang, T.-W.E., Gschneidner, K.A., Jr., Schmidt, F.A., and Thome, D.K., *Phys. Rev.*, B, 31, 235, 1985; Collocott, S.J., Hill, R.W. and Stewart, A.M., *J. Phys. F*, 18, L223, 1988; Hill, R.W. and Gschneidner, K.A., Jr., *J. Phys. F*, 18, 2545, 1988; Skriver, H.L. and Mertig, I., *Phys. Rev. B*, 41, 6553, 1990. Collocott, S.J. and Stewart, A.M., *J. Phys.: Condens. Matter*, 4, 6743, 1992; Pecharsky, V.K., Gschneidner, K.A., Jr. and Fort, D., *Phys. Rev. B*, 47, 5063, 1993.

[a] At 18.6 GPa.
[b] At 11 GPa.
[c] At 2.2 GPa.
[d] Calculated value.
[e] Estimated.
[f] Heat capacity results have been reported, but the resultant γ and θ_D values are unreliable because of the presence of impurities and/or there was no reliable procedure or model to correct for the magnetic contribution to the heat capacity.
[g] Based on the values reported for the purer Sm sample (IV).
[h] At 4.5 GPa.

TABLE 11. Room Temperature Elastic Moduli and Mechanical Properties

Rare earth metal	Young's (elastic) modulus	Shear modulus	Bulk modulus	Poisson's ratio	Yield strength 0.2% offset	Ultimate tensile strength	Uniform elongation (%)	Reduction in area (%)	Recryst. temp. (°C)
Sc	74.4	29.1	56.6	0.279	173[a]	255[a]	5.0[a]	8.0[a]	550
Y	63.5	25.6	41.2	0.243	42	129	34.0	—	550
αLa	36.6	14.3	27.9	0.280	126[a]	130	7.9[a]	—	300
βCe	—	—	—	—	86	138	—	24.0	—
γCe	33.6	13.5	21.5	0.24	28	117	22.0	30.0	325
αPr	37.3	14.8	28.8	0.281	73	147	15.4	67.0	400
αNd	41.4	16.3	31.8	0.281	71	164	25.0	72.0	400
αPm	46[b]	18[b]	33[b]	0.28[b]	—	—	—	—	400[b]
αSm	49.7	19.5	37.8	0.274	68	156	17.0	29.5	440
Eu	18.2	7.9	8.3	0.152	—	—	—	—	300
αGd	54.8	21.8	37.9	0.259	15	118	37.0	56.0	500
αTb	55.7	22.1	38.7	0.261	—	—	—	—	500
αDy	61.4	24.7	40.5	0.247	43	139	30.0	30.0	550
Ho	64.8	26.3	40.2	0.231	—	—	—	—	520
Er	69.9	28.3	44.4	0.237	60	136	11.5	11.9	520
Tm	74.0	30.5	44.5	0.213	—	—	—	—	600
βYb	23.9	9.9	30.5	0.207	7	58	43.0	92.0	300
Lu	68.6	27.2	47.6	0.261	—	—	—	—	600

Note: For additional information, see Scott, T., in *Handbook on the Physics and Chemistry of Rare Earths*, Vol. 1, Gschneidner, K.A., Jr. and Eyring, L., Eds., North-Holland Physics, Amsterdam, 1978, 591.

[a] Value is questionable.
[b] Estimated.

TABLE 12. Liquid Metal Properties near the Melting Point

Rare earth metal	Density (g/cm³)	Surface tension (N/m)	Viscosity (centipoise)	Heat capacity (J/mol K)	Thermal conductivity (W/cm K)	Magnetic susceptibility χ × 10⁴ (emu/mol)	Electrical resistivity (μΩ·cm)	ΔV (l→s)[a] (%)	Spectral emittance at λ = 645 nm ε (%)	Temp.
Sc	2.80	0.954	—	44.2[b]	—	—	—	—	36.8	1522–1647
Y	4.24	0.871	—	43.1	—	—	—	—	25.4	920–1287
La	5.96	0.718	2.65	34.3	0.238	1.20	133	−0.6	25.4	920–1287
Ce	6.68	0.706	3.20	37.7	0.210	9.37	130	+1.1	32.2	877–1547
Pr	6.59	0.707	2.85	43.0	0.251	17.3	139	−0.02	28.4	931–1537
Nd	6.72	0.687	—	48.8	0.195	18.7	151	−0.9	39.4	1021–1567
Pm	6.9[b]	0.680[b]	—	50[b]	—	—	160[b]	—	—	—
Sm	7.16	0.431	—	50.2[b]	—	18.3	182	−3.6	43.7	1075
Eu	4.87	0.264	—	38.1	—	97	242	−4.8	—	—
Gd	7.4	0.664	—	37.2	0.149	67	195	−2.0	34.2	1313–1600
Tb	7.65	0.669	—	46.5	—	82	193	−3.1	—	—
Dy	8.2	0.648	—	49.9	0.187	95	210	−4.5	29.7	1412–1437
Ho	8.34	0.650	—	43.9	—	88	221	−7.4	—	—
Er	8.6	0.637	—	38.7	—	69	226	−9.0	37.2	1529–1587
Tm	9.0[b]	—	—	41.4	—	41	235[b]	−6.9	—	—
Yb	6.21	0.320	2.67	36.8	—	—	113	−5.1	—	—
Lu	9.3	0.940	—	47.9[b]	—	—	224	−3.6	—	—

Note: For additional information, see Van Zytveld, J., in *Handbook on the Physics and Chemistry of Rare Earths*, Vol. 12, Gschneidner, K.A., Jr. and Eyring, L., Eds., North-Holland Physics, Amsterdam, 1989, 357. Stretz, L.A. and Bautista, R.G., in *Temperature, Its Measurement and Control in Science and Industry*, Vol. 4, part I, H.H. Plumb, Ed., Instrument Society of America, Pittsburgh, 1972, 489. King, T.S., Baria, D.N., and Bautista, R.G., *Met. Trans. B*, 7, 411, 1976; Baria, D.N., King, T.S., and Bautista, R.G., *Met. Trans. B*, 7, 577, 1976.

[a] Volume change on freezing.
[b] Estimated.

TABLE 13. Ionization Potentials (Electronvolts)

Rare earth	I Neutral atom	II Singly ionized	III Doubly ionized	IV Triply ionized	V Quadruply ionized
Sc	6.56144	12.79967	24.75666	73.4894	91.65
Y	6.217	12.24	20.52	60.597	77.0
La	5.5770	11.060	19.1773	49.95	61.6
Ce	5.5387	10.85	20.198	36.758	65.55
Pr	5.464	10.55	21.624	38.98	57.53
Nd	5.5250	10.73	22.1	40.41	—
Pm	5.554	10.90	22.3	41.1	—
Sm	5.6437	11.07	23.4	41.4	—
Eu	5.6704	11.241	24.92	42.7	—
Gd	6.1500	12.09	20.63	44.0	—
Tb	5.8639	11.52	21.91	39.79	—
Dy	5.9389	11.67	22.8	41.47	—
Ho	6.0216	11.80	22.84	42.5	—
Er	6.1078	11.93	22.74	42.7	—
Tm	6.18431	12.05	23.68	42.7	—
Yb	6.25416	12.1761	25.05	43.56	—
Lu	5.42585	13.9	20.9594	45.25	66.8

Note: For references, see the table "Ionization Potentials of Atoms and Atomic Ions" in Section 10.

TABLE 14. Effective Ionic Radii (Å)[A]

Rare earth ion	R^{2+} CN = 6	R^{2+} CN = 8	R^{3+} CN = 6	R^{3+} CN = 8	R^{3+} CN = 12	R^{4+} CN = 6	R^{4+} CN = 8
Sc	—	—	0.745	0.87	1.116	—	—
Y	—	—	0.900	1.015	1.220	—	—
La	—	—	1.045	1.18	1.320	—	—
Ce	—	—	1.010	1.14	1.290	0.80	0.97
Pr	—	—	0.997	1.14	1.286	0.78	0.96
Nd	—	—	0.983	1.12	1.276	—	—
Pm	—	—	0.97	1.10	1.267	—	—
Sm	1.19	1.27	0.958	1.09	1.260	—	—
Eu	1.17	1.25	0.947	1.07	1.252	—	—
Gd	—	—	0.938	1.06	1.246	—	—
Tb	—	—	0.923	1.04	1.236	0.76	0.88
Dy	—	—	0.912	1.03	1.228	—	—
Ho	—	—	0.901	1.02	1.221	—	—
Er	—	—	0.890	1.00	1.214	—	—
Tm	—	—	0.880	0.99	1.207	—	—
Yb	1.00	1.07	0.868	0.98	1.199	—	—
Lu	—	—	0.861	0.97	1.194	—	—

Note: For additional information, see Shannon, R.D. and Prewitt, C.T., *Acta Cryst.*, 25, 925, 1969 and Shannon, R.D. and Prewitt, C.T., *Acta Cryst.*, 26, 1046, 1970.

[a] Radius of O^{2-} is 1.40 Å for a coordination number (CN) of 6.

MELTING, BOILING, TRIPLE, AND CRITICAL POINT TEMPERATURES
OF THE ELEMENTS

This table summarizes the significant points on the phase diagrams for the elements for which data are available. Values are given for the solid-liquid-gas triple point t_{tp}, normal melting point t_m, normal boiling point t_b, and critical temperature t_c; all are on the ITS-90 scale. An "sp" notation indicates a sublimation point, where the vapor pressure of the solid phase reaches 101.325 kPa (1 atm). Transition temperatures between allotropic forms are included for several elements. The major data sources are listed below; values from Reference 1, which deals with reference points on the ITS-90 scale, were adopted when applicable.

References

1. Bedford, R. E., Bonnier, G., Maas, H., and Pavese, F., *Metrologia* 33, 133, 1996.
2. Dinsdale, A.T., "SGTE Data for Pure Elements," *CALPHAD*, 15, 317-425, 1991.
3. Chase, M.W., Davies, C.A., Downey, J.R., Frurip, D.J., McDonald, R.A., and Syverud, A.N., *JANAF Thermochemical Tables, Third Edition, J. Phys. Chem. Ref. Data*, Vol. 14, Suppl. 1, 1985.
4. Gurvich, L.V., Veyts, I.V., and Alcock, C.B., *Thermodynamic Properties of Individual Substances, Fourth Edition*, Hemisphere Publishing Corp., New York, 1989.
5. Greenwood, N. N., and Earnshaw, A., *Chemistry of the Elements, Second Edition*, Butterworth-Heinemann, Oxford, 1997.

Element	t_{tp}/°C	t_m/°C	t_b/°C	t_c/°C
Actinium		1050	3198	
Aluminum		660.32	2519	
Americium		1176	2011	
Antimony		630.628	1587	
Argon	-189.36 (69 kPa)		-185.847	-122.28
Arsenic (gray)	817 (3.70 MPa)		616 sp	1400
Astatine		302		
Barium		727	1897	
Berkelium (β form)		986		
Beryllium		1287	2471	
Bismuth		271.406	1564	
Boron		2075	4000	
Bromine		-7.2	58.8	315
Cadmium		321.069	767	
Calcium		842	1484	
Californium		900		
Carbon (graphite)	4489 (10.3 MPa)		3825 sp	
Carbon (diamond)		4440 (12.4 GPa)		
Cerium		799	3443	
Cesium		28.5	671	1665
Chlorine		-101.5	-34.04	143.8
Chromium		1907	2671	
Cobalt		1495	2927	
Copper		1084.62	2562	
Curium		1345	⁻3100	
Dysprosium		1412	2567	
Einsteinium		860		
Erbium		1529	2868	
Europium		822	1529	
Fermium		1527		
Fluorine	-219.67		-188.12	-129.02
Francium		27		
Gadolinium		1313	3273	
Gallium	29.7666		2204	
Germanium		938.25	2833	
Gold		1064.18	2856	
Hafnium		2233	4603	
Helium			-268.93	-267.96
Holmium		1472	2700	
Hydrogen	-259.198 (7.2 kPa)	-259.1	-252.762	-240.18
Indium	156.5936	156.60	2072	
Iodine		113.7	184.4	546
Iridium		2446	4428	
Iron		1538	2861	
Krypton	-157.38 (73.2 kPa)		-153.34	-63.74
Lanthanum		920	3464	
Lawrencium		1627		
Lead		327.462	1749	

Element	$t_{tp}/°C$	$t_m/°C$	$t_b/°C$	$t_c/°C$
Lithium		180.50	1342	2950
Lutetium		1663	3402	
Magnesium		650	1090	
Manganese		1246	2061	
Mendelevium		827		
Mercury	-38.837	-38.8290	356.62	1477
Molybdenum		2623	4639	
Neodymium		1016	3074	
Neon	-248.609 (43 kPa)		-246.053	-228.7
Neptunium		644		
Nickel		1455	2913	
Niobium		2477	4744	
Nitrogen	-209.999	-210.0	-195.798	-146.94
Nobelium		827		
Osmium		3033	5012	
Oxygen		-218.79	-182.953	-118.56
Palladium		1554.8	2963	
Phosphorus (white)		44.15	280.5	721
Phosphorus (red)	590		431 sp	721
Phosphorus (black)		610		
Platinum		1768.2	3825	
Plutonium		640	3228	
Polonium		254	962	
Potassium		63.5	759	1950
Praseodymium		931	3520	
Promethium		1042	3000	
Protactinium		1572		
Radium		696		
Radon		-71	-61.7	104
Rhenium		3185	5596	
Rhodium		1964	3695	
Rubidium	39.26	39.30	688	1820
Ruthenium		2034	4150	
Samarium		1072	1794	
Scandium		1541	2836	
Selenium (vitreous)		180 (trans to gray)	685	1493
Selenium (gray)		220.8	685	1493
Silicon		1414	3265	
Silver		961.78	2162	
Sodium		97.794	882.940	2300
Strontium		777	1382	
Sulfur (rhombic)		95.3 (trans to monocl)	444.61	1041
Sulfur (monoclinic)		115.21	444.61	1041
Tantalum		3017	5458	
Technetium		2157	4265	
Tellurium		449.51	988	
Terbium		1359	3230	
Thallium		304	1473	
Thorium		1750	4788	
Thulium		1545	1950	
Tin (gray)		13.2 (trans to white)	2602	
Tin (white)		231.93	2602	
Titanium		1668	3287	
Tungsten		3422	5555	
Uranium		1135	4131	
Vanadium		1910	3407	
Xenon	-111.745 (81.6 kPa)		-108.09	16.62
Ytterbium		824	1196	
Yttrium		1522	3345	
Zinc		419.53	907	
Zirconium		1854.7	4409	

HEAT CAPACITY OF THE ELEMENTS AT 25°C

This table gives the specific heat capacity (c_p) in J/g K and the molar heat capacity (C_p) in J/mol K at a temperature of 25°C and a pressure of 100 kPa (1 bar or 0.987 standard atmospheres) for all the elements for which reliable data are available.

Name	c_p J/g K	C_p J/mol K	Name	c_p J/g K	C_p J/mol K
Actinium	0.120	27.2	Molybdenum	0.251	24.06
Aluminum	0.897	24.20	Neodymium	0.190	27.45
Antimony	0.207	25.23	Neon	1.030	20.786
Argon	0.520	20.786	Nickel	0.444	26.07
Arsenic	0.329	24.64	Niobium	0.265	24.60
Barium	0.204	28.07	Nitrogen (N_2)	1.040	29.124
Beryllium	1.825	16.443	Osmium	0.130	24.7
Bismuth	0.122	25.52	Oxygen (O_2)	0.918	29.378
Boron	1.026	11.087	Palladium	0.246	25.98
Bromine (Br_2)	0.474	75.69	Phosphorus (white)	0.769	23.824
Cadmium	0.232	26.020	Platinum	0.133	25.86
Calcium	0.647	25.929	Potassium	0.757	29.600
Carbon (graphite)	0.709	8.517	Praseodymium	0.193	27.20
Cerium	0.192	26.94	Radon	0.094	20.786
Cesium	0.242	32.210	Rhenium	0.137	25.48
Chlorine (Cl_2)	0.479	33.949	Rhodium	0.243	24.98
Chromium	0.449	23.35	Rubidium	0.363	31.060
Cobalt	0.421	24.81	Ruthenium	0.238	24.06
Copper	0.385	24.440	Samarium	0.197	29.54
Dysprosium	0.173	28.16	Scandium	0.568	25.52
Erbium	0.168	28.12	Selenium	0.321	25.363
Europium	0.182	27.66	Silicon	0.712	19.99
Fluorine (F_2)	0.824	31.304	Silver	0.235	25.350
Gadolinium	0.236	37.03	Sodium	1.228	28.230
Gallium	0.373	26.03	Strontium	0.306	26.79
Germanium	0.320	23.222	Sulfur (rhombic)	0.708	22.70
Gold	0.129	25.418	Tantalum	0.140	25.36
Hafnium	0.144	25.73	Tellurium	0.202	25.73
Helium	5.193	20.786	Terbium	0.182	28.91
Holmium	0.165	27.15	Thallium	0.129	26.32
Hydrogen (H_2)	14.304	28.836	Thorium	0.118	27.32
Indium	0.233	26.74	Thulium	0.160	27.03
Iodine (I_2)	0.214	54.43	Tin (white)	0.227	26.99
Iridium	0.131	25.10	Titanium	0.523	25.060
Iron	0.449	25.10	Tungsten	0.132	24.27
Krypton	0.248	20.786	Uranium	0.116	27.665
Lanthanum	0.195	27.11	Vanadium	0.489	24.89
Lead	0.130	26.84	Xenon	0.158	20.786
Lithium	3.582	24.860	Ytterbium	0.155	26.74
Lutetium	0.154	26.86	Yttrium	0.298	26.53
Magnesium	1.023	24.869	Zinc	0.388	25.390
Manganese	0.479	26.32	Zirconium	0.278	25.36
Mercury	0.140	27.983			

VAPOR PRESSURE OF THE METALLIC ELEMENTS — EQUATIONS

C. B. Alcock

This table gives coefficients in an equation for the vapor pressure of 65 metallic elements in both the solid and liquid state. Vapor pressures in the range 10^{-10} to 10^2 Pa (10^{-15} to 10^{-3} atm) are covered. The equation is:

for p in atmospheres: $\log(p/\text{atm}) = A + BT^{-1} + C\log T + DT^{-3}$
for p in pascals: $\log(p/\text{Pa}) = 5.006 + A + BT^{-1} + C\log T + DT^{-3}$
for p in torr (mmHg): $\log(p/\text{torr}) = 2.881 + A + BT^{-1} + C\log T + DT^{-3}$

where T is the temperature in K.

This equation reproduces the observed vapor pressures to an accuracy of 5% or better. The metals are listed alphabetically by name, and the melting point is included.

The table following this one gives values of the vapor pressure at several temperatures in the 400 K to 2400 K range, as calculated from these equations.

Reprinted with permission of the publisher, Pergamon Press.

Reference

Alcock, C. B., Itkin, V. P., and Horrigan, M. K., *Canadian Metallurgical Quarterly*, 23, 309, 1984.

Element	Phase	A	B	C	D	Range/k	mp/k
Aluminum	Solid	9.459	-17342	-0.7927		298-mp	933
Aluminum	Liquid	5.911	-16211			mp-1800	
Americium	Solid	11.311	-15059	-1.3449		298-mp	1449
Barium	Solid	12.405	-9690	-2.2890		298-mp	1000
Barium	Liquid	4.007	-8163			mp-1200	
Beryllium	Solid	8.042	-17020	-0.4440		298-mp	1560
Beryllium	Liquid	5.786	-15731			mp-1800	
Cadmium	Solid	5.939	-5799			298-mp	594
Cadmium	Liquid	5.242	-5392			mp-650	
Calcium	Solid	10.127	-9517	-1.4030		298-mp	1115
Cerium	Solid	6.139	-21752			298-mp	1071
Cerium	Liquid	5.611	-21200			mp-2450	
Cesium	Solid	4.711	-3999			298-mp	302
Cesium	Liquid	4.165	-3830			mp-550	
Chromium	Solid	6.800	-20733	0.4391	-0.4094	298-2000	2180
Cobalt	Solid	10.976	-22576	-1.0280		298-mp	1768
Cobalt	Liquid	6.488	-20578			mp-2150	
Copper	Solid	9.123	-17748	-0.7317		298-mp	1358
Copper	Liquid	5.849	-16415			mp-1850	
Curium	Solid	8.369	-20364	-0.5770		298-mp	1618
Curium	Liquid	5.223	-18292			mp-2200	
Dysprosium	Solid	9.579	-15336	-1.1114		298-mp	1685
Erbium	Solid	9.916	-16642	-1.2154		298-mp	1802
Erbium	Liquid	4.668	-14380			mp-1900	
Europium	Solid	9.240	-9459	-1.1661		298-mp	1095
Gadolinium	Solid	8.344	-20861	-0.5775		298-mp	1586
Gadolinium	Liquid	5.557	-19389			mp-2250	
Gallium	Solid	6.657	-14208			298-mp	303
Gallium	Liquid	6.754	-13984	-0.3413		mp-1600	
Gold	Solid	9.152	-19343	-0.7479		298-mp	1337
Gold	Liquid	5.832	-18024			mp-2050	
Hafnium	Solid	9.445	-32482	-0.6735		298-mp	2506
Holmium	Solid	9.785	-15899	-1.1753		298-mp	1747
Indium	Solid	5.991	-12548			298-mp	430
Indium	Liquid	5.374	-12276			mp-1500	
Iridium	Solid	10.506	-35099	-0.7500		298-2500	2719
Iron	Solid	7.100	-21723	0.4536	-0.5846	298-mp	1811
Iron	Liquid	6.347	-19574			mp-2100	
Lanthanum	Solid	7.463	-22551	-0.3142		298-mp	1191
Lanthanum	Liquid	5.911	-21855			mp-2450	
Lead	Solid	5.643	-10143			298-mp	600
Lead	Liquid	4.911	-9701			mp-1200	
Lithium	Solid	5.667	-8310			298-mp	454
Lithium	Liquid	5.055	-8023			mp-1000	
Lutetium	Solid	8.793	-22423	-0.6200		298-mp	1936
Lutetium	Liquid	5.648	-20302			mp-2350	

Element	Phase	A	B	C	D	Range/k	mp/k
Magnesium	Solid	8.489	-7813	-0.8253		298-mp	923
Manganese	Solid	12.805	-15097	-1.7896		298-mp	1519
Mercury	Liquid	5.116	-3190			298-400	234
Molybdenum	Solid	11.529	-34626	-1.1331		298-2500	2895
Neodymium	Solid	8.996	-17264	-0.9519		298-mp	1294
Neodymium	Liquid	4.912	-15824			mp-2000	
Neptunium	Solid	19.643	-24886	-3.9991		298-mp	917
Neptunium	Liquid	10.076	-23378	-1.3250		mp-2500	
Nickel	Solid	10.557	-22606	-0.8717		298-mp	1728
Nickel	Liquid	6.666	-20765			mp-2150	
Niobium	Solid	8.882	-37818	-0.2575		298-2500	2750
Osmium	Solid	9.419	-41198	-0.3896		298-2500	3306
Palladium	Solid	9.502	-19813	-0.9258		298-mp	1828
Palladium	Liquid	5.426	-17899			mp-2100	
Platinum	Solid	4.882	-29387	1.1039	-0.4527	298-mp	2041
Platinum	Liquid	6.386	-26856			mp-2500	
Plutonium	Solid	26.160	-19162	-6.6675		298-600	913
Plutonium	Solid	18.858	-18460	-4.4720		500-mp	
Plutonium	Liquid	3.666	-16658			mp-2450	
Potassium	Solid	4.961	-4646			298-mp	337
Potassium	Liquid	4.402	-4453			mp-600	
Praseodymium	Solid	8.859	-18720	-0.9512		298-mp	1204
Praseodymium	Liquid	4.772	-17315			mp-2200	
Protactinium	Solid	10.552	-34869	-1.0075		298-mp	1845
Protactinium	Liquid	6.177	-32874			mp-2500	
Rhenium	Solid	11.543	-40726	-1.1629		298-2500	3459
Rhodium	Solid	10.168	-29010	-0.7068		298-mp	2236
Rhodium	Liquid	6.802	-26792			mp-2500	
Rubidium	Solid	4.857	-4215			298-mp	312
Rubidium	Liquid	4.312	-4040			mp-550	
Ruthenium	Solid	9.755	-34154	-0.4723		298-mp	2606
Samarium	Solid	9.988	-11034	-1.3287		298-mp	1347
Scandium	Solid	6.650	-19721	0.2885	-0.3663	298-mp	1814
Scandium	Liquid	5.795	-17681			mp-2000	
Silver	Solid	9.127	-14999	-0.7845		298-mp	1235
Silver	Liquid	5.752	-13827			mp-1600	
Sodium	Solid	5.298	-5603			298-mp	371
Sodium	Liquid	4.704	-5377			mp-700	
Strontium	Solid	9.226	-8572	-1.1926		298-mp	1050
Tantalum	Solid	16.807	-41346	-3.2152	0.7437	298-2500	3280
Terbium	Solid	9.510	-20457	-0.9247		298-mp	1629
Terbium	Liquid	5.411	-18639			mp-2200	
Thallium	Solid	5.971	-9447			298-mp	577
Thallium	Liquid	5.259	-9037			mp-1100	
Thorium	Solid	8.668	-31483	-0.5288		298-mp	2023
Thorium	Liquid	-18.453	-24569	6.6473		mp-2500	
Thulium	Solid	8.882	-12270	-0.9564		298-1400	1818
Tin	Solid	6.036	-15710			298-mp	505
Tin	Liquid	5.262	-15332			mp-1850	
Titanium	Solid	11.925	-24991	-1.3376		298-mp	1943
Titanium	Liquid	6.358	-22747			mp-2400	
Tungsten	Solid	2.945	-44094	1.3677		298-2350	3687
Tungsten	Solid	54.527	-57687	-12.2231		2200-2500	
Uranium	Solid	0.770	-27729	2.6982	-1.5471	298-mp	1408
Uranium	Liquid	20.735	-28776	-4.0962		mp-2500	
Vanadium	Solid	9.744	-27132	-0.5501		298-mp	2183
Vanadium	Liquid	6.929	-25011			mp-2500	
Ytterbium	Solid	9.111	-8111	-1.0849		298-900	1092
Yttrium	Solid	9.735	-22306	-0.8705		298-mp	1795
Yttrium	Liquid	5.795	-20341			mp-2300	
Zinc	Solid	6.102	-6776			298-mp	693
Zinc	Liquid	5.378	-6286			mp-750	
Zirconium	Solid	10.008	-31512	-0.7890		298-m.p	2127
Zirconium	Liquid	6.806	-30295			mp-2500	

VAPOR PRESSURE OF THE METALLIC ELEMENTS — DATA

The following values of the vapor pressure of metallic elements are calculated from the equations in the preceding table. All values are given in pascals. For conversion, note that 1 Pa $= 7.50$ μmHg $= 9.87 \cdot 10^{-6}$ atm.

Metal	mp/K	400 K	600 K	800 K	1000 K	1200 K	1400 K	1600 K	1800 K	2000 K	2200 K	2400 K
Aluminum	933			3.06×10^{-10}	5.08×10^{-6}	0.00256	0.218	6.10	81.4			
Americium	1449			3.88×10^{-7}	0.00167	0.423	21.35					
Barium	1000		7.97×10^{-6}	0.0450	7.11	162						
Beryllium	1560			3.04×10^{-10}	4.96×10^{-6}	0.00314	0.312	9.12	113			
Cadmium	594	0.000280	18.2									
Calcium	1115		2.36×10^{-5}	0.146	25.5							
Cerium	1071				2.47×10^{-11}	8.91×10^{-8}	2.97×10^{-5}	0.00233	0.0691	1.04	9.56	60.8
Cesium	302	0.394										
Chromium	2180				2.45×10^{-8}	7.59×10^{-5}	0.0239	1.80	52.1	774		
Cobalt	1768				2.09×10^{-10}	1.00×10^{-6}	0.000419	0.0379	1.15	16.0		
Copper	1358			6.60×10^{-11}	1.53×10^{-6}	0.00122	0.135	3.94	54.4			
Curium	1618				1.90×10^{-6}	4.24×10^{-6}	0.00103	0.0629	1.17	12.1	82.1	
Dysprosium	1685			1.54×10^{-8}	8.21×10^{-5}	0.0241	1.362	27.5				
Erbium	1802			3.90×10^{-10}	4.30×10^{-6}	0.00205	0.163	4.23	52.5			
Europium	1095		1.74×10^{-5}	0.109	19.4							
Gadolinium	1586				5.70×10^{-10}	1.54×10^{-6}	0.000429	0.0279	0.618	7.39	56.2	
Gallium	303			1.94×10^{-7}	0.000565	0.114	4.98	84.4				
Gold	1337				3.72×10^{-8}	5.44×10^{-5}	0.00920	0.374	6.68	67.0		
Hafnium	2506						1.35×10^{-11}	9.81×10^{-9}	1.63×10^{-6}	9.69×10^{-5}	0.00272	0.0437
Holmium	1747			3.20×10^{-9}	2.32×10^{-5}	0.00837	0.546	12.3				
Indium	430		8.31×10^{-11}	1.08×10^{-5}	0.0127	1.413	40.9					
Iridium	2719							1.48×10^{-9}	3.72×10^{-7}	3.06×10^{-5}	0.00112	0.0225
Iron	1811				5.54×10^{-9}	2.51×10^{-5}	0.0104	0.961	32.7	36.8		
Lanthanum	1191					5.09×10^{-8}	2.02×10^{-5}	0.00181	0.0596	0.976	9.61	64.7
Lead	601		5.54×10^{-7}	0.00618	1.64	68.1						
Lithium	454	7.90×10^{-11}	0.000489	1.08	109							
Lutetium	1936				3.28×10^{-11}	1.59×10^{-7}	6.79×10^{-5}	0.00628	0.211	3.18	26.7	
Magnesium	923	6.53×10^{-9}	0.0152	21.5								
Manganese	1519			5.55×10^{-7}	0.00221	0.524	24.9					
Mercury	234	140										
Molybdenum	2895							1.83×10^{-9}	4.07×10^{-7}	3.03×10^{-5}	0.00102	0.0189
Neodymium	1294			4.55×10^{-11}	7.62×10^{-7}	0.000483	0.0412	1.07	13.4	101		
Neptunium	917					3.31×10^{-9}	1.63×10^{-6}	0.000168	0.00604	0.105	1.06	7.28
Nickel	1728				2.19×10^{-10}	1.09×10^{-6}	0.000471	0.0438	1.37	19.5		
Niobium	2750							2.32×10^{-11}	9.54×10^{-9}	1.17×10^{-6}	5.98×10^{-5}	0.00158
Osmium	3306								1.85×10^{-10}	3.46×10^{-8}	2.49×10^{-6}	8.75×10^{-5}
Palladium	1828				8.27×10^{-9}	1.40×10^{-5}	0.00277	0.144	3.07	30.4		
Platinum	2041						2.34×10^{-8}	1.14×10^{-5}	0.00143	0.0689	0.153	1.59
Plutonium	913				1.03×10^{-8}	6.17×10^{-6}	0.000594	0.0182	0.262	2.20	12.6	53.8
Potassium	337	0.0188	96.9									
Praseodymium	1204				1.95×10^{-8}	2.16×10^{-5}	0.00257	0.0904	1.44	13.2	80.8	
Protactinium	1845							3.44×10^{-10}	8.06×10^{-8}	5.57×10^{-6}	0.000174	0.00306
Rhenium	3459								1.37×10^{-10}	2.22×10^{-8}	1.41×10^{-6}	4.45×10^{-5}
Rhodium	2236						1.69×10^{-8}	5.99×10^{-6}	0.000571	0.0217	0.422	4.41
Rubidium	312	0.165										
Ruthenium	2606							7.96×10^{-9}	1.77×10^{-6}	0.000133	0.00455	0.0858
Samarium	1347		8.17×10^{-8}	0.00221	0.942	51.0						
Scandium	1814				6.31×10^{-8}	0.000129	0.0300	1.80	43.6	91.3		
Silver	1235			1.27×10^{-7}	0.000603	0.165	7.61	131				
Sodium	371	0.000185	5.60									
Strontium	1050	4.99×10^{-11}	0.000429	1.134	121							
Tantalum	3280									3.36×10^{-10}	1.87×10^{-8}	5.21×10^{-7}
Terbium	1629				1.92×10^{-9}	4.18×10^{-6}	0.000988	0.0585	1.15	12.5	88.0	
Thallium	577		1.59×10^{-5}	0.0931	16.9							
Thorium	2023						3.33×10^{-11}	2.00×10^{-8}	2.89×10^{-6}	0.000154	0.00401	0.0610
Thulium	1818		6.03×10^{-10}	5.94×10^{-5}	0.0561	5.22	130					
Tin	505			1.26×10^{-9}	8.62×10^{-6}	0.00310	0.207	4.85	56.3			
Titanium	1943					9.69×10^{-9}	7.44×10^{-6}	0.00106	0.0493	0.978	10.6	76.9
Tungsten	3687									2.62×10^{-10}	3.01×10^{-8}	1.59×10^{-6}
Uranium	1408					9.47×10^{-10}	2.87×10^{-6}	4.27×10^{-6}	0.000263	0.00678	0.0933	0.803
Vanadium	2183					2.79×10^{-10}	4.35×10^{-7}	0.000107	0.00769	0.233	3.68	32.6
Ytterbium	1092	1.03×10^{-9}	0.00384	6.74								
Yttrium	1795				6.66×10^{-11}	2.96×10^{-7}	0.000117	0.0102	0.316	4.27	35.9	
Zinc	693	1.47×10^{-6}	0.653									
Zirconium	2127						1.05×10^{-10}	6.17×10^{-8}	8.68×10^{-6}	0.000450	0.0110	0.155

DENSITY OF MOLTEN ELEMENTS AND REPRESENTATIVE SALTS

This table lists the liquid density at the melting point, ρ_m, for elements that are solid at room temperature, as well as for some representative salts of these elements. Densities at higher temperatures (up to the t_{max} given in the last column) may be estimated from the equation

$$\rho(t) = \rho_m - k(t - t_m)$$

where t_m is the melting point and k is given in the fifth column of the table. If a value of t_{max} is not given, the equation should not be used to extrapolate more than about 20°C beyond the melting point.

Data for the elements were selected from the primary literature; the assistance of Gernot Lang in compiling these data is gratefully acknowledged. The molten salt data were derived from Reference 1.

References

1. Janz, G. J., Thermodynamic and Transport Properties of Molten Salts: Correlation Equations for Critically Evaluated Density, Surface Tension, Electrical Conductance, and Viscosity Data, *J. Phys. Chem. Ref. Data*, 17, Suppl. 2, 1988.
2. Nasch, P. M., and Steinemann, S. G., *Phys. Chem. Liq.*, 29, 43, 1995.

Formula	Name	t_m/°C	ρ_m/g cm^{-3}	k/g cm^{-3} °C^{-1}	t_{max}
Ag	Silver	961.78	9.320	0.0009	1500
AgBr	Silver(I) bromide	430	5.577	0.001035	667
AgCl	Silver(I) chloride	455	4.83	0.00094	627
AgI	Silver(I) iodide	558	5.58	0.00101	802
AgNO$_3$	Silver(I) nitrate	210	3.970	0.001098	360
Ag$_2$SO$_4$	Silver(I) sulfate	660	4.84	0.001089	770
Al	Aluminum	660.32	2.375	0.000233	1340
AlBr$_3$	Aluminum bromide	97.5	2.647	0.002435	267
AlCl$_3$	Aluminum chloride	192.6	1.302	0.002711	296
AlI$_3$	Aluminum iodide	188.32	3.223	0.0025	240
As	Arsenic	817	5.22	0.000544	
Au	Gold	1064.18	17.31	0.001343	1200
B	Boron	2075	2.08		
Ba	Barium	727	3.338	0.000299	1550
BaBr$_2$	Barium bromide	857	3.991	0.000924	900
BaCl$_2$	Barium chloride	961	3.174	0.000681	1081
BaF$_2$	Barium fluoride	1368	4.14	0.000999	1727
BaI$_2$	Barium iodide	711	4.26	0.000977	975
Be	Beryllium	1287	1.690	0.00011	
BeCl$_2$	Beryllium chloride	415	1.54	0.0011	473
BeF$_2$	Beryllium fluoride	552	1.96	0.000015	850
Bi	Bismuth	271.406	10.05	0.00135	800
BiBr$_3$	Bismuth bromide	219	4.76	0.002637	927
BiCl$_3$	Bismuth chloride	234	3.916	0.0023	350
Ca	Calcium	842	1.378	0.000230	1484
CaBr$_2$	Calcium bromide	742	3.111	0.0005	791
CaCl$_2$	Calcium chloride	775	2.085	0.000422	950
CaF$_2$	Calcium fluoride	1418	2.52	0.000391	2027
CaI$_2$	Calcium iodide	783	3.443	0.000751	1028
Cd	Cadmium	321.069	7.996	0.001218	500
CdBr$_2$	Cadmium bromide	568	4.075	0.00108	720
CdCl$_2$	Cadmium chloride	568	3.392	0.00082	807
CdI$_2$	Cadmium iodide	388	4.396	0.001117	700
Ce	Cerium	799	6.55	0.000710	1460
CeCl$_3$	Cerium(III) chloride	807	3.25	0.00092	950
CeF$_3$	Cerium(III) fluoride	1430	4.659	0.000936	1927
Co	Cobalt	1495	7.75	0.00165	1580
Cr	Chromium	1907	6.3	0.0011	2100
Cs	Cesium	28.44	1.843	0.000556	510
CsBr	Cesium bromide	636	3.133	0.001223	860
CsCl	Cesium chloride	646	2.79	0.001065	906
CsF	Cesium fluoride	703	3.649	0.001282	912
CsI	Cesium iodide	632	3.197	0.001183	907
CsNO$_3$	Cesium nitrate	409	2.820	0.001166	491
Cs$_2$SO$_4$	Cesium sulfate	1005	3.1	0.00095	1530
Cu	Copper	1084.62	8.02	0.000609	1630
CuCl	Copper(I) chloride	423	3.692	0.00076	585

Formula	Name	$t_m/°C$	$\rho_m/\text{g cm}^{-3}$	$k/\text{g cm}^{-3}°C^{-1}$	t_{max}
Dy	Dysprosium	1411	8.37	0.00143	1540
DyCl$_3$	Dysprosium(III) chloride	718	3.62	0.00068	987
Er	Erbium	1529	8.86	0.00157	1700
Eu	Europium	822	5.13	0.0028	980
Fe	Iron	1538	6.98	0.000572	1680
FeCl$_2$	Iron(II) chloride	677	2.348	0.000555	877
Ga	Gallium	29.7666	6.08	0.00062	400
GaBr$_3$	Gallium(III) bromide	123	3.116	0.00246	135
GaCl$_3$	Gallium(III) chloride	77.9	2.053	0.002083	141
GaI$_3$	Gallium(III) iodide	212	3.630	0.002377	252
Gd	Gadolinium	1314	7.4		
GdCl$_3$	Gadolinium(III) chloride	602	3.56	0.000671	1007
GdI$_3$	Gadolinium(III) iodide	930	4.12	0.000908	1032
Ge	Germanium	938.25	5.60	0.00055	1600
Hf	Hafnium	2233	12		
HgBr$_2$	Mercury(II) bromide	241	5.126	0.003233	319
HgCl$_2$	Mercury(II) chloride	277	4.368	0.002862	304
HgI$_2$	Mercury(II) iodide	256	5.222	0.003235	354
Ho	Holmium	1472	8.34		
In	Indium	156.60	7.02	0.000836	500
InBr$_3$	Indium(III) bromide	420	3.121	0.0015	528
InCl$_3$	Indium(III) chloride	583	2.140	0.0021	666
InI$_3$	Indium(III) iodide	207	3.820	0.0015	360
Ir	Iridium	2446	19		
K	Potassium	63.38	0.828	0.000232	500
KBr	Potassium bromide	734	2.127	0.000825	930
KCl	Potassium chloride	771	1.527	0.000583	939
KF	Potassium fluoride	858	1.910	0.000651	1037
KI	Potassium iodide	681	2.448	0.000956	904
KNO$_3$	Potassium nitrate	334	1.865	0.000723	457
La	Lanthanum	920	5.94	0.00061	1600
LaBr$_3$	Lanthanum bromide	788	4.933	0.000096	912
LaCl$_3$	Lanthanum chloride	858	3.209	0.000777	973
LaF$_3$	Lanthanum fluoride	1493	4.589	0.000682	2177
LaI$_3$	Lanthanum iodide	778	4.29	0.001110	907
Li	Lithium	180.5	0.512	0.00052	285
LiBr	Lithium bromide	550	2.528	0.000652	739
LiCl	Lithium chloride	610	1.502	0.000432	781
LiF	Lithium fluoride	848.2	1.81	0.000490	1047
LiI	Lithium iodide	469	3.109	0.000917	667
LiNO$_3$	Lithium nitrate	253	1.781	0.000546	441
Li$_2$SO$_4$	Lithium sulfate	860	2.003	0.000407	1214
Lu	Lutetium	1663	9.3		
Mg	Magnesium	650	1.584	0.000234	900
MgBr$_2$	Magnesium bromide	711	2.62	0.000478	935
MgCl$_2$	Magnesium chloride	714	1.68	0.000271	826
MgI$_2$	Magnesium iodide	634	3.05	0.000651	888
Mn	Manganese	1246	5.95	0.00105	1590
MnCl$_2$	Manganese(II) chloride	650	2.353	0.000437	850
Mo	Molybdenum	2623	9.33		
Na	Sodium	97.794	0.927	0.00023	600
NaBr	Sodium bromide	747	2.342	0.000816	945
Na$_2$CO$_3$	Sodium carbonate	856	1.972	0.000448	1004
NaCl	Sodium chloride	800.7	1.556	0.000543	1027
NaF	Sodium fluoride	996	1.948	0.000636	1097
NaI	Sodium iodide	661	2.742	0.000949	912
NaNO$_3$	Sodium nitrate	306.5	1.90	0.000715	370
Na$_2$SO$_4$	Sodium sulfate	884	2.069	0.000483	1077
Nd	Neodymium	1016	6.89	0.00076	1350
Ni	Nickel	1455	7.81	0.000726	1700
NiCl$_2$	Nickel(II) chloride	1031	2.653	0.00066	1057
Os	Osmium	3033	20		
Pb	Lead	327.462	10.66	0.00122	700
PbBr$_2$	Lead(II) bromide	371	5.73	0.00165	600
PbCl$_2$	Lead(II) chloride	501	4.951	0.0015	710

Formula	Name	t_m/°C	ρ_m/g cm^{-3}	k/g cm^{-3} °C^{-1}	t_{max}
PbI$_2$	Lead(II) iodide	410	5.691	0.001594	697
Pd	Palladium	1554.8	10.38	0.001169	1700
Pr	Praseodymium	931	6.50	0.00093	1460
PrCl$_3$	Praseodymium chloride	786	3.23	0.00074	977
Pt	Platinum	1768.2	19.77	0.0024	2200
Pu	Plutonium	640	16.63	0.001419	950
Rb	Rubidium	39.31	1.46	0.000451	800
RbBr	Rubidium bromide	692	2.715	0.001072	907
Rb$_2$CO$_3$	Rubidium carbonate	837	2.84	0.000640	1007
RbCl	Rubidium chloride	724	2.248	0.000883	923
RbF	Rubidium fluoride	795	2.87	0.00102	1067
RbI	Rubidium iodide	656	2.904	0.001143	902
RbNO$_3$	Rubidium nitrate	310	2.519	0.001068	417
Rb$_2$SO$_4$	Rubidium sulfate	1066	2.56	0.000665	1545
Re	Rhenium	3185	18.9		
Rh	Rhodium	1964	10.7	0.000895	2200
Ru	Ruthenium	2334	10.65		
S	Sulfur	115.21	1.819	0.00080	160
Sb	Antimony	630.628	6.53	0.00067	745
SbCl$_3$	Antimony(III) chloride	73.4	2.681	0.002293	77
SbCl$_5$	Antimony(V) chloride	4	2.37	0.001869	77
SbI$_3$	Antimony(III) iodide	171	4.171	0.002483	322
Sc	Scandium	1541	2.80		
Se	Selenium	220.8	3.99		
Si	Silicon	1414	2.57	0.00036	1500
Sm	Samarium	1072	7.16		
Sn	Tin	231.93	6.99	0.000601	1200
SnCl$_2$	Tin(II) chloride	247	3.36	0.001253	480
SnCl$_4$	Tin(IV) chloride	−33	2.37	0.002687	138
Sr	Strontium	777	6.980		
SrBr$_2$	Strontium bromide	657	3.70	0.000745	1004
SrCl$_2$	Strontium chloride	874	2.727	0.000578	1037
SrF$_2$	Strontium fluoride	1477	3.470	0.000751	1927
SrI$_2$	Strontium iodide	538	4.085	0.000885	1026
Ta	Tantalum	3017	15		
TaCl$_5$	Tantalum(V) chloride	216.6	2.700	0.004316	457
Tb	Terbium	1359	7.65		
Te	Tellurium	449.51	5.70	0.00035	600
ThCl$_4$	Thorium chloride	770	3.363	0.0014	847
ThF$_4$	Thorium fluoride	1110	6.058	0.000759	1378
Ti	Titanium	1668	4.11		
TiCl$_4$	Titanium(IV) chloride	−25	1.807	0.001735	137
Tl	Thallium	304	11.22	0.00144	600
TlBr	Thallium(I) bromide	460	5.98	0.001755	647
TlCl	Thallium(I) chloride	431	5.628	0.0018	642
TlI	Thallium(I) iodide	441.8	6.15	0.001761	737
TlNO$_3$	Thallium(I) nitrate	206	4.91	0.001873	279
Tl$_2$SO$_4$	Thallium(I) sulfate	632	5.62	0.00130	927
Tm	Thulium	1545	8.56	0.00050	1675
U	Uranium	1135	17.3		
UCl$_3$	Uranium(III) chloride	837	4.84	0.007943	1057
UCl$_4$	Uranium(IV) chloride	590	3.572	0.001945	667
UF$_4$	Uranium(IV) fluoride	1036	6.485	0.000992	1341
V	Vanadium	1910	5.5		
W	Tungsten	3422	17.6		
Y	Yttrium	1526	4.24		
YCl$_3$	Yttrium chloride	721	2.510	0.0005	845
Yb	Ytterbium	824	6.21		
Zn	Zinc	419.53	6.57	0.0011	700
ZnBr$_2$	Zinc bromide	402	3.47	0.000959	602
ZnCl$_2$	Zinc chloride	290	2.54	0.00053	557
ZnI$_2$	Zinc iodide	450	3.878	0.00136	588
ZnSO$_4$	Zinc sulfate	680	3.14	0.00047	987
Zr	Zirconium	1854.7	5.8		
ZrCl$_4$	Zirconium chloride	437	1.643	0.007464	492

MAGNETIC SUSCEPTIBILITY OF THE ELEMENTS AND INORGANIC COMPOUNDS

When a material is placed in a magnetic field H, a magnetization (magnetic moment per unit volume) M is induced in the material which is related to H by $M = \kappa H$, where κ is called the volume susceptibility. Since H and M have the same dimensions, κ is dimensionless. A more useful parameter is the molar susceptibility χ_m, defined by

$$\chi_m = \kappa V_m = \kappa M / \rho$$

where V_m is the molar volume of the substance, M the molar mass, and ρ the mass density. When the cgs system is used, the customary units for χ_m are $cm^3 \, mol^{-1}$; the corresponding SI units are $m^3 \, mol^{-1}$.

Substances that have no unpaired electron orbital or spin angular momentum generally have negative values of χ_m and are called diamagnetic. Their molar susceptibility varies only slightly with temperature. Substances with unpaired electrons, which are termed paramagnetic, have positive χ_m and show a much stronger temperature dependence, varying roughly as $1/T$. The net susceptibility of a paramagnetic substance is the sum of the paramagnetic and diamagnetic contributions, but the former almost always dominates.

This table gives values of χ_m for the elements and selected inorganic compounds. All values refer to nominal room temperature (285 to 300 K) unless otherwise indicated. When the physical state (s = solid, l = liquid, g = gas, aq = aqueous solution) is not given, the most common crystalline form is understood. An entry of Ferro. indicates a ferromagnetic substance.

Substances are arranged in alphabetical order by the most common name, except that compounds such as hydrides, oxides, and acids are grouped with the parent element (the same ordering used in the table "Physical Constants of Inorganic Compounds").

In keeping with customary practice, the molar susceptibility is given here in units appropriate to the cgs system. These values should be multiplied by 4π to obtain values for use in SI equations (where the magnetic field strength H has units of $A \, m^{-1}$).

References

1. *Landolt-Börnstein, Numerical Data and Functional Relationships in Science and Technology, New Series,* II/16, *Diamagnetic Susceptibility,* Springer-Verlag, Heidelberg, 1986.
2. *Landolt-Börnstein, Numerical Data and Functional Relationships in Science and Technology, New Series,* III/19, Subvolumes a to i2, *Magnetic Properties of Metals,* Springer-Verlag, Heidelberg, 1986-1992.
3. *Landolt-Börnstein, Numerical Data and Functional Relationships in Science and Technology, New Series,* II/2, II/8, II/10, II/11,and II/12a, *Coordination and Organometallic Transition Metal Compounds,* Springer-Verlag, Heidelberg, 1966-1984.
4. *Tables de Constantes et Données Numérique, Volume 7, Relaxation Paramagnetique,* Masson, Paris, 1957.

Name	Formula	$\chi_m/10^{-6} \, cm^3 \, mol^{-1}$
Aluminum	Al	+16.5
Aluminum trifluoride	AlF$_3$	−13.9
Aluminum oxide	Al$_2$O$_3$	−37
Aluminum sulfate	Al$_2$(SO$_4$)$_3$	−93
Ammonia (g)	NH$_3$	−16.3
Ammonia (aq)	NH$_3$	−18.3
Ammonium acetate	NH$_4$C$_2$H$_3$O$_2$	−41.1
Ammonium bromide	NH$_4$Br	−47
Ammonium carbonate	(NH$_4$)$_2$CO$_3$	−42.5
Ammonium chlorate	NH$_4$ClO$_3$	−42.1
Ammonium chloride	NH$_4$Cl	−36.7
Ammonium fluoride	NH$_4$F	−23
Ammonium iodate	NH$_4$IO$_3$	−62.3
Ammonium iodide	NH$_4$I	−66
Ammonium nitrate	NH$_4$NO$_3$	−33
Ammonium sulfate	(NH$_4$)$_2$SO$_4$	−67
Ammonium thiocyanate	NH$_4$SCN	−48.1
Antimony	Sb	−99
Stibine (g)	SbH$_3$	−34.6
Antimony(III) bromide	SbBr$_3$	−111.4
Antimony(III) chloride	SbCl$_3$	−86.7
Antimony(III) fluoride	SbF$_3$	−46
Antimony(III) iodide	SbI$_3$	−147.2
Antimony(III) oxide	Sb$_2$O$_3$	−69.4
Antimony(III) sulfide	Sb$_2$S$_3$	−86
Antimony(V) chloride	SbCl$_5$	−120.5
Argon (g)	Ar	−19.32
Arsenic (gray)	As	−5.6
Arsenic (yellow)	As	−23.2
Arsine (g)	AsH$_3$	−35.2

Name	Formula	$\chi_m/10^{-6} \, cm^3 \, mol^{-1}$
Arsenic(III) bromide	AsBr$_3$	−106
Arsenic(III) chloride	AsCl$_3$	−72.5
Arsenic(III) iodide	AsI$_3$	−142.2
Arsenic(III) oxide	As$_2$O$_3$	−30.34
Arsenic(III) sulfide	As$_2$S$_3$	−70
Barium	Ba	+20.6
Barium bromide	BaBr$_2$	−92
Barium bromide dihydrate	BaBr$_2$·2H$_2$O	−119.3
Barium carbonate	BaCO$_3$	−58.9
Barium chloride	BaCl$_2$	−72.6
Barium chloride dihydrate	BaCl$_2$·2H$_2$O	−100
Barium fluoride	BaF$_2$	−51
Barium hydroxide	Ba(OH)$_2$	−53.2
Barium iodate	Ba(IO$_3$)$_2$	−122.5
Barium iodide	BaI$_2$	−124.4
Barium iodide dihydrate	BaI$_2$·2H$_2$O	−163
Barium nitrate	Ba(NO$_3$)$_2$	−66.5
Barium oxide	BaO	−29.1
Barium peroxide	BaO$_2$	−40.6
Barium sulfate	BaSO$_4$	−65.8
Beryllium	Be	−9.0
Beryllium chloride	BeCl$_2$	−26.5
Beryllium hydroxide	Be(OH)$_2$	−23.1
Beryllium oxide	BeO	−11.9
Beryllium sulfate	BeSO$_4$	−37
Bismuth	Bi	−280.1
Bismuth tribromide	BiBr$_3$	−147
Bismuth trichloride	BiCl$_3$	−26.5
Bismuth fluoride	BiF$_3$	−61.2
Bismuth hydroxide	Bi(OH)$_3$	−65.8

Name	Formula	$\chi_m/10^{-6} \text{ cm}^3 \text{ mol}^{-1}$	Name	Formula	$\chi_m/10^{-6} \text{ cm}^3 \text{ mol}^{-1}$
Bismuth triiodide	BiI_3	−200.5	Cesium iodide	CsI	−82.6
Bismuth nitrate pentahydrate	$Bi(NO_3)_3 \cdot 5H_2O$	−159	Cesium superoxide	CsO_2	+1534
Bismuth oxide	Bi_2O_3	−83	Cesium sulfate	Cs_2SO_4	−116
Bismuth phosphate	$BiPO_4$	−77	Chlorine (l)	Cl_2	−40.4
Bismuth sulfate	$Bi_2(SO_4)_3$	−199	Chlorine trifluoride (g)	ClF_3	−26.5
Bismuth sulfide	Bi_2S_3	−123	Chromium	Cr	+167
Boron	B	−6.7	Chromium(II) chloride	$CrCl_2$	+7230
Diborane (g)	B_2H_6	−21.0	Chromium(III) chloride	$CrCl_3$	+6350
Boric acid (orthoboric acid)	H_3BO_3	−34.1	Chromium(III) fluoride	CrF_3	+4370
Boron trichloride	BCl_3	−59.9	Chromium(III) oxide	Cr_2O_3	+1960
Boron oxide	B_2O_3	−38.7	Chromium(III) sulfate	$Cr_2(SO_4)_3$	+11800
Bromine (l)	Br_2	−56.4	Chromium(VI) oxide	CrO_3	+40
Bromine (g)	Br_2	−73.5	Cobalt	Co	Ferro.
Bromine trifluoride	BrF_3	−33.9	Cobalt(II) bromide	$CoBr_2$	+13000
Bromine pentafluoride	BrF_5	−45.1	Cobalt(II) chloride	$CoCl_2$	+12660
Cadmium	Cd	−19.7	Cobalt(II) chloride hexahydrate	$CoCl_2 \cdot 6H_2O$	+9710
Cadmium bromide	$CdBr_2$	−87.3	Cobalt(II) cyanide	$Co(CN)_2$	+3825
Cadmium bromide tetrahydrate	$CdBr_2 \cdot 4H_2O$	−131.5	Cobalt(II) fluoride	CoF_2	+9490
Cadmium carbonate	$CdCO_3$	−46.7	Cobalt(II) iodide	CoI_2	+10760
Cadmium chloride	$CdCl_2$	−68.7	Cobalt(II) sulfate	$CoSO_4$	+10000
Cadmium chromate	$CdCrO_4$	−16.8	Cobalt(II) sulfide	CoS	+225
Cadmium cyanide	$Cd(CN)_2$	−54	Cobalt(II,III) oxide	Co_3O_4	+7380
Cadmium fluoride	CdF_2	−40.6	Cobalt(III) fluoride	CoF_3	+1900
Cadmium hydroxide	$Cd(OH)_2$	−41	Cobalt(III) oxide	Co_2O_3	+4560
Cadmium iodate	$Cd(IO_3)_2$	−108.4	Copper	Cu	−5.46
Cadmium iodide	CdI_2	−117.2	Copper(I) bromide	CuBr	−49
Cadmium nitrate	$Cd(NO_3)_2$	−55.1	Copper(I) chloride	CuCl	−40
Cadmium nitrate tetrahydrate	$Cd(NO_3)_2 \cdot 4H_2O$	−140	Copper(I) cyanide	CuCN	−24
Cadmium oxide	CdO	−30	Copper(I) iodide	CuI	−63
Cadmium sulfate	$CdSO_4$	−59.2	Copper(I) oxide	Cu_2O	−20
Cadmium sulfide	CdS	−50	Copper(II) bromide	$CuBr_2$	+685
Calcium	Ca	+40	Copper(II) chloride	$CuCl_2$	+1080
Calcium bromide	$CaBr_2$	−73.8	Copper(II) chloride dihydrate	$CuCl_2 \cdot 2H_2O$	+1420
Calcium carbonate	$CaCO_3$	−38.2	Copper(II) fluoride	CuF_2	+1050
Calcium chloride	$CaCl_2$	−54.7	Copper(II) fluoride dihydrate	$CuF_2 \cdot 2H_2O$	+1600
Calcium fluoride	CaF_2	−28	Copper(II) hydroxide	$Cu(OH)_2$	+1170
Calcium hydroxide	$Ca(OH)_2$	−22	Copper(II) nitrate trihydrate	$Cu(NO_3)_2 \cdot 3H_2O$	+1570
Calcium iodate	$Ca(IO_3)_2$	−101.4	Copper(II) nitrate hexahydrate	$Cu(NO_3)_2 \cdot 6H_2O$	+1625
Calcium iodide	CaI_2	−109	Copper(II) oxide	CuO	+238
Calcium oxide	CaO	−15.0	Copper(II) sulfate	$CuSO_4$	+1330
Calcium sulfate	$CaSO_4$	−49.7	Copper(II) sulfate pentahydrate	$CuSO_4 \cdot 5H_2O$	+1460
Calcium sulfate dihydrate	$CaSO_4 \cdot 2H_2O$	−74	Copper(II) sulfide	CuS	−2.0
Carbon (diamond)	C	−5.9	Dysprosium (α)	Dy	+98000
Carbon (graphite)	C	−6.0	Dysprosium(III) oxide	Dy_2O_3	+89600
Carbon monoxide (g)	CO	−11.8	Dysprosium(III) sulfide	Dy_2S_3	+95200
Carbon dioxide (g)	CO_2	−21.0	Erbium	Er	+48000
Cerium (β)	Ce	+2500	Erbium oxide	Er_2O_3	+73920
Cerium(II) sulfide	CeS	+2110	Erbium sulfate octahydrate	$Er_2(SO_4)_3 \cdot 8H_2O$	+74600
Cerium(III) chloride	$CeCl_3$	+2490	Erbium sulfide	Er_2S_3	+77200
Cerium(III) fluoride	CeF_3	+2190	Europium	Eu	+30900
Cerium(III) sulfide	Ce_2S_3	+5080	Europium(II) bromide	$EuBr_2$	+26800
Cerium(IV) oxide	CeO_2	+26	Europium(II) chloride	$EuCl_2$	+26500
Cerium(IV) sulfate tetrahydrate	$Ce(SO_4)_2 \cdot 4H_2O$	−97	Europium(II) fluoride	EuF_2	+23750
Cesium	Cs	+29	Europium(II) iodide	EuI_2	+26000
Cesium bromate	$CsBrO_3$	−75.1	Europium(II) sulfide	EuS	+23800
Cesium bromide	CsBr	−67.2	Europium(III) oxide	Eu_2O_3	+10100
Cesium carbonate	Cs_2CO_3	−103.6	Europium(III) sulfate	$Eu_2(SO_4)_3$	+10400
Cesium chlorate	$CsClO_3$	−65	Fluorine	F_2	−9.63
Cesium chloride	CsCl	−56.7	Gadolinium (350 K)	Gd	+185000
Cesium fluoride	CsF	−44.5	Gadolinium(III) chloride	$GdCl_3$	+27930

Name	Formula	$\chi_m/10^{-6}$ cm^3 mol^{-1}	Name	Formula	$\chi_m/10^{-6}$ cm^3 mol^{-1}
Gadolinium(III) oxide	Gd_2O_3	+53200	Iron(II) chloride tetrahydrate	$FeCl_2 \cdot 4H_2O$	+12900
Gadolinium(III) sulfate octahydrate	$Gd_2(SO_4)_3 \cdot 8H_2O$	+53280	Iron(II) fluoride	FeF_2	+9500
Gadolinium(III) sulfide	Gd_2S_3	+55500	Iron(II) iodide	FeI_2	+13600
Gallium	Ga	−21.6	Iron(II) oxide	FeO	+7200
Gallium suboxide	Ga_2O	−34	Iron(II) sulfate	$FeSO_4$	+12400
Gallium(II) sulfide	GaS	−23	Iron(II) sulfate monohydrate	$FeSO_4 \cdot H_2O$	+10500
Gallium(III) chloride	$GaCl_3$	−63	Iron(II) sulfate heptahydrate	$FeSO_4 \cdot 7H_2O$	+11200
Gallium(III) sulfide	Ga_2S_3	−80	Iron(II) sulfide	FeS	+1074
Germanium	Ge	−11.6	Iron(III) chloride	$FeCl_3$	+13450
Germane (g)	GeH_4	−29.7	Iron(III) chloride hexahydrate	$FeCl_3 \cdot 6H_2O$	+15250
Germanium(II) oxide	GeO	−28.8	Iron(III) fluoride	FeF_3	+13760
Germanium(II) sulfide	GeS	−40.9	Iron(III) fluoride trihydrate	$FeF_3 \cdot 3H_2O$	+7870
Germanium(IV) chloride	$GeCl_4$	−72	Iron(III) nitrate nonahydrate	$Fe(NO_3)_3 \cdot 9H_2O$	+15200
Germanium(IV) fluoride	GeF_4	−50	Krypton (g)	Kr	−29.0
Germanium(IV) iodide	GeI_4	−171	Lanthanum (α)	La	+95.9
Germanium(IV) oxide	GeO_2	−34.3	Lanthanum oxide	La_2O_3	−78
Germanium(IV) sulfide	GeS_2	−53.9	Lanthanum sulfate nonahydrate	$La_2(SO_4)_3 \cdot 9H_2O$	−262
Gold	Au	−28	Lanthanum sulfide	La_2S_3	−37
Gold(I) bromide	AuBr	−61	Lead	Pb	−23
Gold(I) chloride	AuCl	−67	Lead(II) acetate	$Pb(C_2H_3O_2)_2$	−89.1
Gold(I) iodide	AuI	−91	Lead(II) bromide	$PbBr_2$	−90.6
Gold(III) chloride	$AuCl_3$	−112	Lead(II) carbonate	$PbCO_3$	−61.2
Hafnium	Hf	+71	Lead(II) chloride	$PbCl_2$	−73.8
Hafnium oxide	HfO_2	−23	Lead(II) chromate	$PbCrO_4$	−18
Helium (g)	He	−2.02	Lead(II) fluoride	PbF_2	−58.1
Holmium	Ho	+72900	Lead(II) iodate	$Pb(IO_3)_2$	−131
Holmium oxide	Ho_2O_3	+88100	Lead(II) iodide	PbI_2	−126.5
Hydrazine (l)	N_2H_4	−201	Lead(II) nitrate	$Pb(NO_3)_2$	−74
Hydrogen (l, 20.3 K)	H_2	−5.44	Lead(II) oxide	PbO	−42
Hydrogen (g)	H_2	−3.99	Lead(II) phosphate	$Pb_3(PO_4)_2$	−182
Hydrogen chloride (l)	HCl	−22.6	Lead(II) sulfate	$PbSO_4$	−69.7
Hydrogen chloride (aq)	HCl	−22	Lead(II) sulfide	PbS	−83.6
Hydrogen fluoride (l)	HF	−8.6	Lithium	Li	+14.2
Hydrogen fluoride (aq)	HF	−9.3	Lithium bromide	LiBr	−34.3
Hydrogen iodide (s, 195 K)	HI	−47.3	Lithium carbonate	Li_2CO_3	−27
Hydrogen iodide (l, 233 K)	HI	−48.3	Lithium chloride	LiCl	−24.3
Hydrogen iodide (aq)	HI	−50.2	Lithium fluoride	LiF	−10.1
Hydrogen peroxide (l)	H_2O_2	−17.3	Lithium hydride	LiH	−4.6
Hydrogen sulfide (g)	H_2S	−25.5	Lithium hydroxide (aq)	LiOH	−12.3
Indium	In	−10.2	Lithium iodide	LiI	−50
Indium(I) chloride	InCl	−30	Lithium sulfate	Li_2SO_4	−41.6
Indium(II) chloride	$InCl_2$	−56	Lutetium	Lu	+182.9
Indium(II) sulfide	InS	−28	Magnesium	Mg	+13.1
Indium(III) bromide	$InBr_3$	−107	Magnesium bromide	$MgBr_2$	−72
Indium(III) chloride	$InCl_3$	−86	Magnesium carbonate	$MgCO_3$	−32.4
Indium(III) oxide	In_2O_3	−56	Magnesium chloride	$MgCl_2$	−47.4
Indium(III) sulfide	In_2S_3	−98	Magnesium fluoride	MgF_2	−22.7
Iodine	I_2	−90	Magnesium hydroxide	$Mg(OH)_2$	−22.1
Iodic acid	HIO_3	−48	Magnesium iodide	MgI_2	−111
Iodine pentoxide	I_2O_5	−79.4	Magnesium oxide	MgO	−10.2
Iodine chloride	ICl	−54.6	Magnesium sulfate	$MgSO_4$	−42
Iodine trichloride	ICl_3	−90.2	Magnesium sulfate monohydrate	$MgSO_4 \cdot H_2O$	−61
Iodine pentafluoride	IF_5	−58.1	Magnesium sulfate heptahydrate	$MgSO_4 \cdot 7H_2O$	−135.7
Iridium	Ir	+25	Manganese	Mn	+511
Iridium(III) chloride	$IrCl_3$	−14.4	Manganese(II) bromide	$MnBr_2$	+13900
Iridium(IV) oxide	IrO_2	+224	Manganese(II) carbonate	$MnCO_3$	+11400
Iron	Fe	Ferro.	Manganese(II) chloride	$MnCl_2$	+14350
Iron(II) bromide	$FeBr_2$	+13600	Manganese(II) chloride tetrahydrate	$MnCl_2 \cdot 4H_2O$	+14600
Iron(II) carbonate	$FeCO_3$	+11300	Manganese(II) fluoride	MnF_2	+10700
Iron(II) chloride	$FeCl_2$	+14750	Manganese(II) hydroxide	$Mn(OH)_2$	+13500

Name	Formula	$\chi_m/10^{-6}$ cm^3 mol^{-1}	Name	Formula	$\chi_m/10^{-6}$ cm^3 mol^{-1}
Manganese(II) iodide	MnI_2	+14400	Niobium(V) oxide	Nb_2O_5	−10
Manganese(II) oxide	MnO	+4850	Nitrogen (g)	N_2	−12.0
Manganese(II) sulfate	$MnSO_4$	+13660	Nitric acid (l)	HNO_3	−19.9
Manganese(II) sulfate monohydrate	$MnSO_4 \cdot H_2O$	+14200	Nitrous oxide (g)	N_2O	−18.9
Manganese(II) sulfate tetrahydrate	$MnSO_4 \cdot 4H_2O$	+14600	Nitric oxide (s, 90 K)	NO	+19.8
Manganese(II) sulfide (α form)	MnS	+5630	Nitric oxide (l, 118 K)	NO	+114.2
Manganese(II) sulfide (β form)	MnS	+3850	Nitric oxide (g)	NO	+1461
Manganese(II,III) oxide	Mn_3O_4	+12400	Nitrogen dioxide (g, 408 K)	NO_2	+150
Manganese(III) fluoride	MnF_3	+10500	Nitrogen trioxide (g)	N_2O_3	−16
Manganese(III) oxide	Mn_2O_3	+14100	Nitrogen tetroxide (g)	N_2O_4	−23.0
Manganese(IV) oxide	MnO_2	+2280	Osmium	Os	+11
Mercury (s, 234 K)	Hg	−24.1	Oxygen (s, 54 K)	O_2	+10200
Mercury (l)	Hg	−33.5	Oxygen (l, 90 K)	O_2	+7699
Mercury(I) bromide	Hg_2Br_2	−105	Oxygen (g)	O_2	+3449
Mercury(I) chloride	Hg_2Cl_2	−120	Ozone (l)	O_3	+6.7
Mercury(I) fluoride	Hg_2F_2	−106	Palladium	Pd	+540
Mercury(I) iodide	Hg_2I_2	−166	Palladium(II) chloride	$PdCl_2$	−38
Mercury(I) nitrate	$Hg_2(NO_3)_2$	−121	Phosphorus (white)	P	−26.66
Mercury(I) oxide	Hg_2O	−76.3	Phosphorus (red)	P	−20.77
Mercury(I) sulfate	Hg_2SO_4	−123	Phosphine (g)	PH_3	−26.2
Mercury(II) bromide	$HgBr_2$	−94.2	Phosphoric acid (aq)	H_3PO_4	−43.8
Mercury(II) chloride	$HgCl_2$	−82	Phosphorous acid (aq)	H_3PO_3	−42.5
Mercury(II) cyanide	$Hg(CN)_2$	−67	Phosphorus(III) chloride (l)	PCl_3	−63.4
Mercury(II) fluoride	HgF_2	−57.3	Platinum	Pt	+193
Mercury(II) iodide	HgI_2	−165	Platinum(II) chloride	$PtCl_2$	−54
Mercury(II) nitrate	$Hg(NO_3)_2$	−74	Platinum(III) chloride	$PtCl_3$	−66.7
Mercury(II) oxide	HgO	−46	Platinum(IV) chloride	$PtCl_4$	−93
Mercury(II) sulfate	$HgSO_4$	−78.1	Platinum(IV) fluoride	PtF_4	+445
Mercury(II) sulfide	HgS	−55.4	Plutonium	Pu	+525
Mercury(II) thiocyanate	$Hg(SCN)_2$	−96.5	Plutonium(IV) fluoride	PuF_4	+1760
Molybdenum	Mo	+72	Plutonium(IV) oxide	PuO_2	+730
Molybdenum(III) bromide	$MoBr_3$	+525	Plutonium(VI) fluoride	PuF_6	+173
Molybdenum(III) chloride	$MoCl_3$	+43	Potassium	K	+20.8
Molybdenum(III) oxide	Mo_2O_3	−42.0	Potassium bromate	$KBrO_3$	−52.6
Molybdenum(IV) bromide	$MoBr_4$	+520	Potassium bromide	KBr	−49.1
Molybdenum(IV) chloride	$MoCl_4$	+1750	Potassium carbonate	K_2CO_3	−59
Molybdenum(IV) oxide	MoO_2	+41	Potassium chlorate	$KClO_3$	−42.8
Molybdenum(V) chloride	$MoCl_5$	+990	Potassium chloride	KCl	−38.8
Molybdenum(VI) fluoride	MoF_6	−26.0	Potassium chromate	K_2CrO_4	−3.9
Molybdenum(VI) oxide	MoO_3	+3	Potassium cyanide	KCN	−37
Neodymium (α)	Nd	+5930	Potassium ferricyanide	$K_3Fe(CN)_6$	+2290
Neodymium fluoride	NdF_3	+4980	Potassium ferrocyanide trihydrate		−172.3
Neodymium oxide	Nd_2O_3	+10200	Potassium fluoride	KF	−23.6
Neodymium sulfate	$Nd_2(SO_4)_3$	+9990	Potassium hydrogen sulfate	$KHSO_4$	−49.8
Neodymium sulfide	Nd_2S_3	+5550	Potassium hydroxide (aq)	KOH	−22
Neon (g)	Ne	−6.96	Potassium iodate	KIO_3	−63.1
Neptunium	Np	+575	Potassium iodide	KI	−63.8
Nickel	Ni	Ferro.	Potassium nitrate	KNO_3	−33.7
Nickel(II) bromide	$NiBr_2$	+5600	Potassium nitrite	KNO_2	−23.3
Nickel(II) chloride	$NiCl_2$	+6145	Potassium permanganate	$KMnO_4$	+20
Nickel(II) chloride hexahydrate	$NiCl_2 \cdot 6H_2O$	+4240	Potassium sulfate	K_2SO_4	−67
Nickel(II) fluoride	NiF_2	+2410	Potassium sulfide	K_2S	−60
Nickel(II) hydroxide	$Ni(OH)_2$	+4500	Potassium superoxide	KO_2	+3230
Nickel(II) iodide	NiI_2	+3875	Potassium thiocyanate	$KSCN$	−48
Nickel(II) nitrate hexahydrate	$Ni(NO_3)_2 \cdot 6H_2O$	+4300	Praseodymium (α)	Pr	+5530
Nickel(II) oxide	NiO	+660	Praseodymium chloride	$PrCl_3$	+44.5
Nickel(II) sulfate	$NiSO_4$	+4005	Praseodymium oxide	Pr_2O_3	+8994
Nickel(II) sulfide	NiS	+190	Praseodymium sulfide	Pr_2S_3	+10770
Nickel(III) sulfide	Ni_3S_2	+1030	Protactinium	Pa	+277
Niobium	Nb	+208	Rhenium	Re	+67

Name	Formula	$\chi_m/10^{-6}$ cm³ mol⁻¹
Rhenium(IV) oxide	ReO_2	+44
Rhenium(IV) sulfide	ReS_2	+38
Rhenium(V) chloride	$ReCl_5$	+1225
Rhenium(VI) oxide	ReO_3	+16
Rhenium(VII) oxide	Re_2O_7	−16
Rhodium	Rh	+102
Rhodium(III) chloride	$RhCl_3$	−7.5
Rhodium(III) oxide	Rh_2O_3	+104
Rubidium	Rb	+17
Rubidium bromide	RbBr	−56.4
Rubidium carbonate	Rb_2CO_3	−75.4
Rubidium chloride	RbCl	−46
Rubidium fluoride	RbF	−31.9
Rubidium iodide	RbI	−72.2
Rubidium nitrate	$RbNO_3$	−41
Rubidium sulfate	Rb_2SO_4	−88.4
Rubidium superoxide	RbO_2	+1527
Ruthenium	Ru	+39
Ruthenium(III) chloride	$RuCl_3$	+1998
Ruthenium(IV) oxide	RuO_2	+162
Samarium (α)	Sm	+1278
Samarium(II) bromide	$SmBr_2$	+5337
Samarium(III) bromide	$SmBr_3$	+972
Samarium(III) oxide	Sm_2O_3	+1988
Samarium(III) sulfate octahydrate	$Sm_2(SO_4)_3 \cdot 8H_2O$	+1710
Samarium(III) sulfide	Sm_2S_3	+3300
Scandium (α)	Sc	+295.2
Selenium	Se	−25
Selenium dioxide	SeO_2	−27.2
Selenium bromide	Se_2Br_2	−113
Selenium chloride (l)	Se_2Cl_2	−94.8
Selenium hexafluoride (g)	SeF_6	−51
Silicon	Si	−3.12
Silane (g)	SiH_4	−20.4
Disilane (g)	Si_2H_6	−37.3
Tetramethylsilane (l)	$(CH_3)_4Si$	−74.80
Tetraethylsilane (l)	$(C_2H_5)_4Si$	−120.2
Tetrabromosilane (l)	$SiBr_4$	−126
Tetrachlorosilane (l)	$SiCl_4$	−87.5
Silicon carbide	SiC	−12.8
Silicon dioxide	SiO_2	−29.6
Silver	Ag	−19.5
Silver(I) bromide	AgBr	−61
Silver(I) carbonate	Ag_2CO_3	−80.90
Silver(I) chloride	AgCl	−49
Silver(I) chromate	Ag_2CrO_4	−40
Silver(I) cyanide	AgCN	−43.2
Silver(I) fluoride	AgF	−36.5
Silver(I) iodide	AgI	−80
Silver(I) nitrate	$AgNO_3$	−45.7
Silver(I) nitrite	$AgNO_2$	−42
Silver(I) oxide	Ag_2O	−134
Silver(I) phosphate	Ag_3PO_4	−120
Silver(I) sulfate	Ag_2SO_4	−92.90
Silver(I) thiocyanate	AgSCN	−61.8
Silver(II) oxide	AgO	−19.6
Sodium	Na	+16
Sodium acetate	$NaC_2H_3O_2$	−37.6
Sodium bromate	$NaBrO_3$	−44.2
Sodium bromide	NaBr	−41

Name	Formula	$\chi_m/10^{-6}$ cm³ mol⁻¹
Sodium carbonate	Na_2CO_3	−41
Sodium chlorate	$NaClO_3$	−34.7
Sodium chloride	NaCl	−30.2
Sodium dichromate	$Na_2Cr_2O_7$	+55
Sodium fluoride	NaF	−15.6
Sodium hydrogen phosphate	Na_2HPO_4	−56.6
Sodium hydroxide (aq)	NaOH	−15.8
Sodium iodate	$NaIO_3$	−53
Sodium iodide	NaI	−57
Sodium nitrate	$NaNO_3$	−25.6
Sodium nitrite	$NaNO_2$	−14.5
Sodium oxide	Na_2O	−19.8
Sodium peroxide	Na_2O_2	−28.10
Sodium sulfate	Na_2SO_4	−52
Sodium sulfate decahydrate	$Na_2SO_4 \cdot 10H_2O$	−184
Sodium sulfide	Na_2S	−39
Sodium tetraborate	$Na_2B_4O_7$	−85
Strontium	Sr	+92
Strontium bromide	$SrBr_2$	−86.6
Strontium bromide hexahydrate	$SrBr_2 \cdot 6H_2O$	−160
Strontium carbonate	$SrCO_3$	−47
Strontium chlorate	$Sr(ClO_3)_2$	−73
Strontium chloride	$SrCl_2$	−61.5
Strontium chloride hexahydrate	$SrCl_2 \cdot 6H_2O$	−145
Strontium chromate	$SrCrO_4$	−5.1
Strontium fluoride	SrF_2	−37.2
Strontium hydroxide	$Sr(OH)_2$	−40
Strontium iodate	$Sr(IO_3)_2$	−108
Strontium iodide	SrI_2	−112
Strontium nitrate	$Sr(NO_3)_2$	−57.2
Strontium oxide	SrO	−35
Strontium peroxide	SrO_2	−32.3
Strontium sulfate	$SrSO_4$	−57.9
Sulfur (rhombic)	S	−15.5
Sulfur (monoclinic)	S	−14.9
Sulfuric acid (l)	H_2SO_4	−39
Sulfur dioxide (g)	SO_2	−18.2
Sulfur trioxide (l)	SO_3	−28.54
Sulfur chloride (l)	$SSCl_2$	−62.2
Sulfur dichloride (l)	SCl_2	−49.4
Sulfur hexafluoride (g)	SF_6	−44
Thionyl chloride (l)	$SOCl_2$	−44.3
Tantalum	Ta	+154
Tantalum(V) chloride	$TaCl_5$	+140
Tantalum(V) oxide	Ta_2O_5	−32
Technetium	Tc	+115
Tellurium	Te	−38
Tellurium dibromide	$TeBr_2$	−106
Tellurium dichloride	$TeCl_2$	−94
Tellurium hexafluoride (g)	TeF_6	−66
Terbium (α)	Tb	+170000
Terbium oxide	Tb_2O_3	+78340
Thallium	Tl	−50
Thallium(I) bromate	$TlBrO_3$	−75.9
Thallium(I) bromide	TlBr	−63.9
Thallium(I) carbonate	Tl_2CO_3	−101.6
Thallium(I) chlorate	$TlClO_3$	−65.5
Thallium(I) chloride	TlCl	−57.8
Thallium(I) chromate	Tl_2CrO_4	−39.3
Thallium(I) cyanide	TlCN	−49

Name	Formula	$\chi_m/10^{-6}$ cm³ mol⁻¹	Name	Formula	$\chi_m/10^{-6}$ cm³ mol⁻¹
Thallium(I) fluoride	TlF	−44.4	Uranium(IV) bromide	UBr_4	+3530
Thallium(I) iodate	$TlIO_3$	−86.8	Uranium(IV) chloride	UCl_4	+3680
Thallium(I) iodide	TlI	−82.2	Uranium(IV) fluoride	UF_4	+3530
Thallium(I) nitrate	$TlNO_3$	−56.5	Uranium(IV) oxide	UO_2	+2360
Thallium(I) nitrite	$TlNO_2$	−50.8	Uranium(VI) fluoride	UF_6	+43
Thallium(I) sulfate	Tl_2SO_4	−112.6	Uranium(VI) oxide	UO_3	+128
Thallium(I) sulfide	Tl_2S	−88.8	Vanadium	V	+285
Thorium	Th	+97	Vanadium(II) bromide	VBr_2	+3230
Thorium(IV) oxide	ThO_2	−16	Vanadium(II) chloride	VCl_2	+2410
Thulium	Tm	+24700	Vanadium(III) bromide	VBr_3	+2910
Thulium oxide	Tm_2O_3	+51444	Vanadium(III) chloride	VCl_3	+3030
Tin (gray)	Sn	−37.4	Vanadium(III) fluoride	VF_3	+2757
Tin(II) chloride	$SnCl_2$	−69	Vanadium(III) oxide	V_2O_3	+1976
Tin(II) chloride dihydrate	$SnCl_2 \cdot 2H_2O$	−91.4	Vanadium(III) sulfide	V_2S_3	+1560
Tin(II) oxide	SnO	−19	Vanadium(IV) chloride	VCl_4	+1215
Tin(IV) bromide	$SnBr_4$	−149	Vanadium(IV) oxide	VO_2	+99
Tin(IV) chloride (l)	$SnCl_4$	−115	Vanadium(V) oxide	V_2O_5	+128
Tin(IV) oxide	SnO_2	−41	Water (s, 273 K)	H_2O	−12.63
Titanium	Ti	+151	Water (l, 293 K)	H_2O	−12.96
Titanium(II) bromide	$TiBr_2$	+720	Water (l, 373 K)	H_2O	−13.09
Titanium(II) chloride	$TiCl_2$	+484	Water (g, 373 K))	H_2O	−13.1
Titanium(II) iodide	TiI_2	+1790	Xenon (g)	Xe	−45.5
Titanium(II) sulfide	TiS	+432	Ytterbium (β)	Yb	+67
Titanium(III) bromide	$TiBr_3$	+660	Yttrium (α)	Y	+187.7
Titanium(III) chloride	$TiCl_3$	+1110	Yttrium oxide	Y_2O_3	+44.4
Titanium(III) fluoride	TiF_3	+1300	Yttrium sulfide	Y_2S_3	+100
Titanium(III) oxide	Ti_2O_3	+132	Zinc	Zn	−9.15
Titanium(IV) chloride	$TiCl_4$	−54	Zinc carbonate	$ZnCO_3$	−34
Titanium(IV) oxide	TiO_2	+5.9	Zinc chloride	$ZnCl_2$	−55.33
Tungsten	W	+53	Zinc cyanide	$Zn(CN)_2$	−46
Tungsten carbide	WC	+10	Zinc fluoride	ZnF_2	−34.3
Tungsten(II) chloride	WCl_2	−25	Zinc hydroxide	$Zn(OH)_2$	−67
Tungsten(IV) oxide	WO_2	+57	Zinc iodide	ZnI_2	−108
Tungsten(IV) sulfide	WS_2	+5850	Zinc oxide	ZnO	−27.2
Tungsten(V) bromide	WBr_5	+270	Zinc phosphate	$Zn_3(PO_4)_2$	−141
Tungsten(V) chloride	WCl_5	+387	Zinc sulfate	$ZnSO_4$	−47.8
Tungsten(VI) chloride	WCl_6	−71	Zinc sulfate monohydrate	$ZnSO_4 \cdot H_2O$	−63
Tungsten(VI) fluoride (g)	WF_6	−53	Zinc sulfate heptahydrate	$ZnSO_4 \cdot 7H_2O$	−138
Tungsten(VI) oxide	WO_3	−15.8	Zinc sulfide	ZnS	−25
Uranium	U	+409	Zirconium	Zr	+120
Uranium(III) bromide	UBr_3	+4740	Zirconium carbide	ZrC	−26
Uranium(III) chloride	UCl_3	+3460	Zirconium nitrate pentahydrate	$Zr(NO_3)_4 \cdot 5H_2O$	−77
Uranium(III) hydride	UH_3	+6244	Zirconium(IV) oxide	ZrO_2	−13.8
Uranium(III) iodide	UI_3	+4460			

INDEX OF REFRACTION OF INORGANIC LIQUIDS

This table gives the index of refraction n of several inorganic substances in the liquid state at specified temperatures. The measurements refer to ambient atmospheric pressure except for substances whose normal boiling points are greater than the indicated temperature; in this case the pressure is the saturated vapor pressure of the substance. All values refer to a wavelength of 589 nm unless otherwise indicated. Entries are arranged in alphabetical order by chemical formula as normally written.

Data on the index of refraction at other temperatures and wavelengths may be found in Reference 1.

References

1. Wohlfarth, C., and Wohlfarth, B., *Landolt-Börnstein, Numerical Data and Functional Relationships in Science and Technology, New Series*, III/38A, Martienssen, W., Editor, Springer-Verlag, Heidelberg, 1996.
2. Francis, A.W., *J. Chem. Eng. Data*, 5, 534, 1960.

Formula	Name	$t/°C$	n
Ar	Argon	−188	1.2312
$AsCl_3$	Arsenic(III) chloride	16	1.604
BBr_3	Boron tribromide	16	1.312
BrF_3	Bromine trifluoride	25	1.4536
BrF_5	Bromine pentafluoride	25	1.3529
Br_2	Bromine	15	1.659
COS	Carbon oxysulfide	25	1.3506
CO_2	Carbon dioxide	24	1.6630
CS_2	Carbon disulfide	20	1.62774
C_3O_2	Carbon suboxide	0	1.453
Cl_2	Chlorine	20	1.3834
CrO_2Cl_2	Chromyl chloride	23	1.524
$Fe(CO)_5$	Iron pentacarbonyl	14	1.523
$GeBr_4$	Germanium(IV) bromide	26	1.6269
$GeCl_4$	Germanium(IV) chloride	25	1.4614
HBr	Hydrogen bromide	10	1.325
HCN	Hydrogen cyanide	20	1.26136
HCl	Hydrogen chloride	18	1.3287 [a]
$HClO_4$	Perchloric acid	50	1.3819
HF	Hydrogen fluoride	25	1.1574
HI	Hydrogen iodide	16	1.466
HNO_3	Nitric acid	25	1.393
H_2	Hydrogen	−253	1.1096
H_2O	Water	20	1.33336
H_2O_2	Hydrogen peroxide	28	1.4061
H_2S	Hydrogen sulfide	−80	1.460
		20	1.3682
H_2SO_4	Sulfuric acid	20	1.4183
H_2S_2	Hydrogen disulfide	20	1.630

Formula	Name	$t/°C$	n
He	Helium	−269	1.02451 [c]
Kr	Krypton	−157	1.3032 [c]
NH_3	Ammonia	−77	1.3944 [b]
		20	1.3327
NO	Nitric oxide	−90	1.330
N_2	Nitrogen	−196	1.19876 [b]
N_2H_4	Hydrazine	22	1.470
N_2O	Nitrous oxide	25	1.238
O_2	Oxygen	−183	1.2243 [c]
PBr_3	Phosphorus(III) bromide	25	1.687
PCl_3	Phosphorus(III) chloride	21	1.5122
PH_3	Phosphine	17	1.317
P_2O_3	Phosphorus(III) oxide	27	1.540
S	Sulfur	125	1.9170
SCl_2	Sulfur dichloride	14	1.557
SF_6	Sulfur hexafluoride	25	1.167
$SOCl_2$	Thionyl chloride	10	1.527
SO_2	Sulfur dioxide	25	1.3396
SO_2Cl_2	Sulfuryl chloride	12	1.444
SO_3	Sulfur trioxide	20	1.40965
$SSCl_2$	Sulfur chloride	20	1.671
$SbCl_5$	Antimony(V) chloride	22	1.5925
$SiBr_4$	Tetrabromosilane	31	1.5685
$SiCl_4$	Tetrachlorosilane	25	1.41156
$SnBr_4$	Tin(IV) bromide	31	1.6628
$SnCl_4$	Tin(IV) chloride	25	1.5086
$TiCl_4$	Titanium(IV) chloride	18	1.6076
Xe	Xenon	−112	1.3918 [c]

[a] At 581 nm
[b] At 578 nm
[c] At 546 nm

PHYSICAL AND OPTICAL PROPERTIES OF MINERALS

The chemical formula, crystal system, density, hardness, and index of refraction of some common minerals are given in this table. Entries are arranged alphabetically by mineral name. The columns are:

- **Formula:** Chemical formula for a typical sample of the mineral. Composition often varies considerably with the origin of the sample.
- **Crystal system:** tricl = triclinic; monocl = monoclinic; orth = orthorhombic; tetr = tetragonal; hex = hexagonal; rhomb = rhombohedral; cub = cubic.
- **Density:** Typical density in g/cm³. Individual samples may vary by a few percent.
- **Hardness:** On the Mohs' scale (range of 1 to 10, with talc = 1 and diamond = 10).
- **Index of refraction:** Values are given for the three coordinate axes in the order of least, intermediate, and greatest

index. For cubic crystals there is only a single value. See Reference 1 for details on the axis systems. Variations of several percent, depending on the origin and exact composition of the sample, are common.

References

1. Deer, W. A., Howie, R. A., and Zussman, J., *An Introduction to the Rock-Forming Minerals*, 2nd Edition, Longman Scientific & Technical, Harlow, Essex, 1992.
2. Carmichael, R. S., *Practical Handbook of Physical Properties of Rocks and Minerals*, CRC Press, Boca Raton, FL, 1989.
3. Donnay, J. D. H., and Ondik, H. M., *Crystal Data Determinative Tables, Third Edition, Volume 2, Inorganic Compounds*, Joint Committee on Powder Diffraction Standards, Swarthmore, PA, 1973.

Name	Formula	Crystal system	Density g/cm³	Hardness	Index of refraction n_α	n_β	n_γ
Acanthite	Ag_2S	orth	7.2	2.3			
Actinolite	$Ca_2(Mg,Fe)_5Si_8O_{22}(OH,F)_2$	monocl	3.23	5.5	1.624	1.655	1.664
Aegirine	$NaFe(SiO_3)_2$	monocl	3.58	6	1.763	1.800	1.815
Akermanite	$Ca_2MgSi_2O_7$	tetr	2.94	5.5	1.632	1.640	
Alabandite	MnS	cub	4.0	3.8			
Albite	$NaAlSi_3O_8$	tricl	2.63	6.3	1.527	1.531	1.538
Allanite	$(Ca,Mn,Ce,La,Y,Th)_2(Fe,Ti)(Al,Fe)O\cdot OH$ $(Si_2O_7)(SiO_4)$	monocl	3.8	5.8	1.75	1.78	1.80
Allemontite	$SbAs$	hex	6.0	3.5			
Almandine	$Fe_3Al_2Si_3O_{12}$	cub	4.32	6.8	1.830		
Altaite	$PbTe$	cub	8.16	3			
Aluminite	$Al_2(SO_4)(OH)_4 \cdot 7H_2O$	monocl	1.74	1.5	1.459	1.464	1.470
Alunite	$(K,Na)Al_3(SO_4)_2(OH)_6$	rhomb	2.8	3.8	1.572	1.592	
Alunogen	$Al_2(SO_4)_3 \cdot 18H_2O$	monocl	1.69	1.8	1.467	1.47	1.478
Amblygonite	$(Li,Na)Al(PO_4)(F,OH)$	tricl	3.1	5.8	1.591	1.604	1.613
Analcite	$NaAlSi_2O_6 \cdot H_2O$	cub	2.27	5.5	1.486		
Anatase	TiO_2	tetr	4.23	5.8	2.488	2.561	
Andalusite	Al_2OSiO_4	orth	3.15	7.5	1.635	1.639	1.644
Andesine	$NaAlSi_3O_8 \cdot CaAl_2Si_2O_8$	tricl	2.67	6.3	1.550	1.553	1.557
Andorite	$PbAgSb_3S_6$	rhomb	5.35	3.3			
Andradite	$Ca_3(Fe,Ti)_2Si_3O_{12}$	cub	3.86	6.8	1.887		
Anglesite	$PbSO_4$	orth	6.29	2.8	1.877	1.883	1.894
Anhydrite	$CaSO_4$	orth	2.96	3.5	1.570	1.575	1.614
Ankerite	$Ca(Fe,Mg,Mn)(CO_3)_2$	rhomb	3.0	3.8	1.529	1.720	
Anorthite	$CaAl_2Si_2O_8$	tricl	2.76	6.3	1.577	1.585	1.590
Anorthoclase	$(Na,K)AlSi_3O_8$	tricl	2.58	6	1.523	1.528	1.529
Anthophyllite	$(Mg,Fe)_7Si_8O_{22}(OH,F)_2$	rhomb	3.21	5.8	1.645	1.658	1.668
Apatite	$Ca_5(PO_4)_3(OH,F,Cl)$	hex	3.2	5	1.645	1.648	
Apophyllite	$KFCa_4Si_8O_{20} \cdot 8H_2O$	tetr	2.35	4.8	1.535	1.536	
Aragonite	$CaCO_3$	orth	2.83	3.5	1.531	1.680	1.686
Arcanite	K_2SO_4	orth	2.66		1.494	1.494	1.497
Argentite	Ag_2S	orth	7.2	2.3			
Arsenolite	As_2O_3	cub	3.86	1.5	1.755		
Arsenopyrite	$FeAsS$	monocl	6.1	5.8			
Atacamite	$Cu_2(OH)_3Cl$	rhomb	3.76	3.3	1.831	1.861	1.880
Augelite	$Al_2(PO_4)(OH)_3$	monocl	2.70	4.8	1.574	1.576	1.588
Augite	$(Ca,Mg,Fe,Ti,Al)_2(Si,Al)_2O_6$	monocl	3.38	6	1.703	1.707	1.738
Autunite	$Ca(UO_{22})(PO_4)_2 \cdot 10H2O$	tetr	3.2	2.3	1.553	1.577	
Axinite	$(Ca,Mn,Fe)_3Al_2BO_3Si_4O_{12}(OH)$	tricl	3.31	6.8	1.684	1.691	1.694

Name	Formula	Crystal system	Density g/cm³	Hardness	n_α	n_β	n_γ
Azurite	$Cu_3(OH)_2(CO_3)_2$	monocl	3.77	3.8	1.730	1.758	1.838
Baddeleyite	ZrO_2	monocl	5.7	6.5	2.13	2.19	2.20
Barite	$BaSO_4$	orth	4.49	3.3	1.636	1.637	1.648
Benitoite	$BaTi(SiO_3)_3$	rhomb	3.65	6.3	1.757	1.804	
Bertrandite	$Be_4Si_2O_7(OH)_2$	rhomb	2.6	6	1.589	1.602	1.613
Beryl	$Be_3Al_2(SiO_3)_6$	hex	2.64	7.8	1.582	1.589	
Beryllonite	$NaBe(PO)_4$	monocl	2.81	5.8	1.552	1.558	1.561
Biotite	$K(Mg,Fe)_3AlSi_3O_{10}(OH,F)_2$	monocl	3.0	2.8	1.595	1.651	1.651
Bismuthinite	Bi_2S_3	orth	6.78	2			
Bixbyite	$(Mn,Fe)_2O_3$	cub	4.95	6.3			
Bloedite	$Na_2Mg(SO_4)_2 \cdot 4H_2O$	monocl	2.25	2.8	1.483	1.486	1.487
Boehmite	$AlO(OH)$	orth	3.44	3.8	1.64	1.65	1.66
Boracite	$Mg_3B_7O_{13}Cl$	rhomb	2.94	7.3	1.66	1.66	1.67
Borax	$Na_2B_4O_7 \cdot 10H_2O$	monocl	1.73	2.3	1.447	1.469	1.472
Bornite	Cu_5FeS_4	cub	5.07	3			
Boulangerite	$Pb_5Sb_4S_{11}$	monocl	6.1	2.8			
Bournonite	$PbCuSbS_3$	rhomb	5.83	2.8			
Braggite	PtS	tetr	10.2				
Braunite	$(Mn,Si)_2O_3$	tetr	4.78	6.3			
Bravoite	$(Ni,Fe)S_2$	cub	4.62	5.8			
Breithauptite	$NiSb$	hex	≈8.7	5.5			
Brochantite	$Cu_4(SO_4)(OH)_6$	monocl	3.79	3.8	1.728	1.771	1.800
Bromyrite	$AgBr$	cub	6.47	2.5	2.253		
Brookite	TiO_2	orth	4.23	5.8	2.583	2.584	2.700
Brucite	$Mg(OH)_2$	hex	2.37	2.5	1.575	1.59	
Bunsenite	NiO	cub	6.72	5.5			
Cacoxenite	$Fe_4(PO_4)_3(OH)_3 \cdot 12H_2O$	hex	2.3	3.5	1.580	1.646	
Calcite	$CaCO_3$	hex	2.71	3	1.486	1.658	
Caledonite	$Cu_2Pb_5(SO_4)_3(CO_3)(OH)_6$	rhomb	5.76	2.8	1.818	1.866	1.909
Calomel	Hg_2Cl_2	tetr	7.16	1.5	1.973	2.656	
Cancrinite	$(Na,Ca,K)_7[Al_6Si_6O_{24}](CO_3,SO_4,Cl,OH)_2 \cdot H_2O$	hex	2.42	5.5	1.495	1.509	
Carnalite	$KMgCl_3 \cdot 6H_2O$	rhomb	1.60	2.5	1.466	1.475	1.494
Carnotite	$K_2(UO_2)_2(VO_4)_2 \cdot 3H_2O$	rhomb		1.5	1.75	1.92	1.95
Cassiterite	SnO_2	tetr	6.85	6.5	2.006	2.097	
Celestite	$SrSO_4$	orth	3.96	3.3	1.622	1.624	1.631
Celsian	$BaAl_2Si_2O_8$	monocl	3.25	6.3	1.583	1.588	1.594
Cerargyrite	$AgCl$	cub	5.56	2.5	2.071		
Cerussite	$PbCO_3$	orth	6.6	3.3	1.804	2.076	2.079
Cervantite	Sb_2O_4	orth	6.64	4.5			
Chabazite	$Ca[Al_2Si_4O_{12}] \cdot 6H_2O$	trig	2.08	4.5	1.482		
Chalcanthite	$CuSO_4 \cdot 5H_2O$	tricl	2.29	2.5	1.514	1.537	1.543
Chalcocite	Cu_2S	orth	5.6	2.8			
Chalcopyrite	$CuFeS_2$	tetr	4.2	3.8			
Chiolite	$Na_5Al_3F_{14}$	tetr	3.00	3.8	1.342	1.349	
Chlorite	$(Mg,Al,Fe)_{12}(Si,Al)_8O_{20}(OH)_{16}$	monocl	3.0	2.5	1.61	1.62	1.62
Chloritoid	$FeAl_4O_2(SiO_4)_2(OH)_4$	monocl	3.66	6.5	1.717	1.721	1.726
Chondrodite	$Mg(OH,F)_2 \cdot 2Mg_2SiO_4$	monocl	3.21	6.5	1.604	1.615	1.634
Chromite	$FeCr_2O_4$	cub	5.0	5.5	2.16		
Chrysoberyl	$BeAl_2O_4$	orth	3.65	8.5	1.746	1.748	1.756
Chrysocolla	$CuSiO_3 \cdot 2H_2O$	rhomb	2.4	2	1.575	1.597	1.598
Cinnabar	HgS	hex	8.17	2.3	2.814	3.143	
Claudetite	As_2O_3	monocl	3.74	2.5	1.87	1.92	2.01
Clinohumite	$Mg(OH,F)_2 \cdot 4Mg_2SiO_4$	monocl	3.21	6	1.633	1.647	1.668
Clinozoisite	$Ca_2Al_3Si_3O_{12}(OH)$	monocl	3.30	6.5	1.693	1.700	1.712
Cobaltite	$CoAsS$	cub	≈6.1	5.5			
Colemanite	$Ca_2B_6O_{11} \cdot 5H_2O$	monocl	2.42	4.5	1.586	1.592	1.614
Columbite	$(Fe,Mn)(Nb,Ta)_2O_6$	rhomb	5.20	6			
Connellite	$Cu_{19}(SO_4)Cl_4(OH)_{32} \cdot 3H_2O$	hex	3.36	3	1.731	1.752	
Copiapite	$(Fe,Mg)Fe_4(SO_4)_6(OH)_2 \cdot 20H_2O$	tricl	2.13	2.8	1.52	1.54	1.59
Coquimbite	$Fe_2(SO_4)_3 \cdot 9H_2O$	hex	2.1	2.5	1.54	1.56	

Name	Formula	Crystal system	Density g/cm³	Hardness	Index of refraction n_α	n_β	n_γ
Cordierite	$Al_3(Mg,Fe)_2Si_5AlO_{18}$	rhomb	2.66	7	1.540	1.549	1.553
Corundum	Al_2O_3	hex	3.97	9	1.761	1.769	
Cotunnite	$PbCl_2$	orth	5.98	2.5	2.199	2.217	2.260
Covellite	CuS	hex	4.8	1.8			
Cristobalite	SiO_2	hex	2.33	6.5	1.484	1.487	
Crocoite	$PbCrO_4$	monocl	6.12	2.8	2.29	2.36	2.66
Cryolite	Na_3AlF_6	monocl	2.97	2.5	1.338	1.338	1.339
Cryolithionite	$Na_3Li_3Al_2F_{12}$	cub	2.77	2.8	1.340		
Cubanite	$CuFe_2S_3$	rhomb	4.11	3.5			
Cummingtonite	$(Mg,Fe)_7Si_8O_{22}(OH)_2$	monocl	3.4	5.5	1.650	1.660	1.676
Cuprite	Cu_2O	cub	6.0	3.8			
Danburite	$CaSi_2B_2O_8$	rhomb	3.0	7	1.63	1.63	1.63
Datolite	$CaBSiO_4(OH)$	monocl	2.98	5.3	1.624	1.652	1.668
Daubreelite	Cr_2FeS_4	cub	3.81				
Derbylite	$Fe_6Ti_6Sb_2O_{23}$	rhomb	4.53	5	2.45	2.45	2.51
Diamond	C	cub	3.51	10	2.418		
Diaspore	$AlO(OH)$	orth	3.4	6.8	1.694	1.715	1.741
Digenite	$Cu_{2-x}S$	cub	5.55	2.8			
Diopside	$CaMgSi_2O_6$	monocl	3.30	6	1.680	1.687	1.708
Dioptase	$CuSiO_2(OH)_2$	rhomb	3.5	5	1.65	1.70	
Dolomite	$CaMg(CO_3)_2$	rhomb	2.86	3.5	1.500	1.679	
Douglasite	$K_2FeCl_4\cdot2H_2O$	orth	2.16		1.488	1.500	
Dyscrasite	Ag_3Sb	rhomb	9.74	3.8			
Eddingtonite	$BaAl_2Si_3O_{10}\cdot4H_2O$	rhomb	2.8		1.541	1.553	1.557
Eglestonite	Hg_4OCl_2	cub	8.4	2.5	2.49		
Emplectite	$CuBiS_2$	rhomb	6.38	2			
Enargite	Cu_3AsS_4	rhomb	4.5	3			
Enstatite	$MgSiO_3$	monocl	3.19	5.5	1.656	1.662	1.669
Epidote	$Ca_2Al_2(Al,Fe)OH(SiO_4)_3$	monocl	3.44	6	1.733	1.755	1.765
Epsomite	$MgSO_4\cdot7H_2O$	orth	1.67	2.3	1.433	1.455	1.461
Erythrite	$(Co,Ni)_3(AsO_4)_2\cdot8H_2O$	monocl	3.06	2	1.626	1.661	1.699
Eucairite	$CuAgSe$	orth	7.7	2.5			
Euclasite	$BeAlSiO_4(OH)$	monocl	3.1	7.5	1.651	1.655	1.671
Eudialite	$(Na,Ca,Ce)_5(Fe,Mn)(Zr,Ti)(Si_3O_9)_2\,(OH,Cl)$	hex	3.0	5.5	1.623	1.600	1.615
Eulytite	$Bi_4Si_3O_{12}$	cub	6.6	4.5	2.05		
Euxenite	$(Y,Ca,Ce,U,Th)(Nb,Ta,Ti)_2O_6$	rhomb	5.5	6	2.2		
Fayalite	Fe_2SiO_4	orth	4.30	6.5	1.827	1.869	1.879
Ferberite	$FeWO_4$	monocl	7.51	4.3			
Fergussonite	$(Y,Er,Ce,Fe)(Nb,Ta,Ti)O_4$	tetr	5.7	6	2.1		
Fluorite	CaF_2	cub	3.18	4	1.434		
Forsterite	Mg_2SiO_4	orth	3.21	7	1.635	1.651	1.670
Franklinite	$ZnFe_2O_4$	cub	5.21	6	2.36		
Gahnite	$ZnAl_2O_4$	cub	4.62	7.8	1.805		
Galaxite	$MnAl_2O_4$	cub	4.04	7.8	1.92		
Galena	PbS	cub	7.60	2.5	3.91		
Galenabismuthite	$PbBi_2S_4$	rhomb	7.04	3			
Ganomalite	$(Ca,Pb)_{10}(OH,Cl)_2(Si_2O_7)_3$	hex	5.6	3.5	1.910	1.945	
Gaylussite	$Na_2Ca(CO_3)_2\cdot5H_2O$	monocl	1.99	2.8	1.444	1.516	1.523
Gehlenite	$Ca_2Al_2SiO_7$	tetr	3.04	5.5	1.658	1.669	
Geikielite	$MgTiO_3$	hex	3.85	5.5	1.95	2.31	
Gibbsite	$Al(OH)_3$	monocl	2.42	3	1.57	1.57	1.59
Glauberite	$Na_2Ca(SO_4)_2$	monocl	2.80	2.8	1.515	1.535	1.536
Glauconite	$(K,Na,Ca)_{1.6}(Fe,Al,Mg)_{4.0}Si_{7.3}Al_{0.7}O_{20}(OH)_4$	monocl	2.7	2	1.60	1.63	1.63
Glaucophane	$Na_2Mg_3Al_2Si_8O_{22}(OH)_2$	monocl	3.19	6	1.634	1.645	1.648
Gmelinite	$(Ca,Na_2)[Al_2Si_4O_{12}]\cdot6H_2O$	hex	2.10	4.5	1.477	1.485	
Goethite	$FeO(OH)$	orth	4.3	5.3	2.268	2.401	2.457
Goslarite	$ZnSO_4\cdot7H_2O$	orth	1.97	2.3	1.457	1.480	1.484
Greenockite	CdS	hex	4.8	3.3	2.506	2.529	
Grossularite	$Ca_3Al_2Si_3O_{12}$	cub	3.59	6.8	1.734		
Gummite	$UO_3\cdot H_2O$	orth	7.05	3.8			

Name	Formula	Crystal system	Density g/cm³	Hardness	Index of refraction n_α	n_β	n_γ
Gypsum	$CaSO_4 \cdot 2H_2O$	monocl	2.32	2	1.520	1.525	1.530
Halite	$NaCl$	cub	2.17	2	1.544		
Hambergite	$Be_2(OH)(BO_3)$	rhomb	2.36	7.5	1.56	1.59	1.63
Hanksite	$Na_{22}K(SO_4)_9(CO_3)_2Cl$	hex	2.56	3.3	1.461	1.481	
Harmotome	$Ba[Al_2Si_6O_{16}] \cdot 6H_2O$	monocl	2.44	4.5	1.506	1.507	1.511
Hausmannite	Mn_3O_4	tetr	4.84	5.5	2.15	2.46	
Haüyne	$(Na,Ca)_{4-8}Al_6Si_6O_{24}(SO_4,S)_{1-2}$	cub	2.47	5.8	1.502		
Hedenbergite	$CaFeSi_2O_6$	monocl	3.53	6	1.721	1.727	1.746
Helvite	$Mn_4Be_3Si_3O_{12}S$	cub	3.32	6	1.739		
Hematite	Fe_2O_3	hex	5.25	6	2.91	3.19	
Hemimorphite	$Zn_4Si_2O_7(OH)_2 \cdot H_2O$	rhomb	3.45	5	1.614	1.617	1.636
Hercynite	$Fe(AlO_2)_2$	cub	4.3	7.8	1.835		
Herderite	$CaBe(PO_4)(Fe,OH)$	monocl	2.98	5.3	1.592	1.612	1.621
Hessite	Ag_2Te	orth	8.4	2.5			
Heulandite	$(Ca,Na_2,K_2)[Al_2Si_7O_{18}] \cdot 6H_2O$	monocl	2.2	3.8	1.498	1.498	1.506
Hopeite	$Zn_3(PO_4)_2 \cdot 4H_2O$	orth	3.0	3.2	1.58	1.59	1.59
Hornblende	$Ca_2(Mg,Fe)_4Al(Si_7AlO_{22})(OH)_2$	monocl	3.24	5.5	1.67	1.67	1.69
Huebnerite	$MnWO_4$	monocl	7.2	4.3	2.17	2.22	2.32
Humite	$Mg(OH,F)_2 \cdot 3Mg_2SiO_4$	orth	3.3	6	1.625	1.636	1.657
Huntite	$Mg_3Ca(CO_3)_4$	trig	2.70				
Hydrogrossularite	$Ca_3Al_2Si_3O_8(SiO_4)_{1-m}(OH)_{4m}$	cub	3.4	6.8	1.70		
Hydromagnesite	$3MgCO_3 \cdot Mg(OH)_2 \cdot 3H_2O$	monocl	2.24	3.5	1.523	1.527	1.545
Illite	$KAl_4[Si_7AlO_{20}](OH)_4$	monocl	2.8	1.5	1.56	1.59	1.59
Ilmenite	$FeTiO_3$	rhomb	4.72	5.5			
Iodyrite	AgI	hex	5.68	1.5	2.21	2.22	
Jacobsite	$MnFe_2O_4$	cub	4.87	7.8	2.3		
Jadeite	$NaAlSi_2O_6$	monocl	3.34	6	1.649	1.654	1.663
Jamesonite	$Pb_4FeSb_6S_{14}$	monocl	5.63	2.5			
Jarosite	$KFe_3(SO_4)_2(OH)_6$	rhomb	3.09	3	1.715	1.820	
Kainite	$KMg(SO_4)Cl \cdot 3H_2O$	monocl	2.15	2.8	1.494	1.505	1.516
Kaliophylite	$KAlSiO_4$	hex	2.61	6	1.532	1.537	
Kaolinite	$Al_4Si_4O_{10}(OH)_8$	tricl	2.65	2.3	1.549	1.564	1.565
Kernite	$Na_2B_4O_7 \cdot 4H_2O$	monocl	1.95	2.5	1.454	1.472	1.488
Kieserite	$MgSO_4 \cdot H_2O$	monocl	2.57	3.5	1.520	1.533	1.584
Kyanite	Al_2OSiO_4	tricl	3.59	6.3	1.715	1.722	1.731
Lanarkite	$Pb_2(SO_4)O$	monocl	6.92	2.3	1.928	2.007	2.036
Lanthanite	$(La,Ce)_2(CO_3)_3 \cdot 8H_2O$	rhomb	2.72	2.8	1.52	1.587	1.613
Laumontite	$Ca_4[Al_8Si_{16}O_{48}] \cdot 16H_2O$	monocl	2.3	3.3	1.508	1.517	1.519
Laurionite	$Pb(OH)Cl$	rhomb	6.24	3.3	2.08	2.12	2.16
Lawsonite	$CaAl_2(OH)_2Si_2O_7 \cdot H_2O$	rhomb	3.08	6	1.655	1.675	1.685
Lazulite	$(Mg,Fe)Al_2(PO_4)_2(OH)_2$	monocl	3.23	5.8	1.615	1.64	1.650
Lazurite	$Na_4SSi_3Al_3O_{12}$	cub	2.42	5.3	1.500		
Leadhillite	$Pb_4(SO_4)(CO_3)_2(OH)_2$	monocl	6.55	2.8	1.87	2.00	2.01
Lepidocrocite	$FeO(OH)$	orth	4.26	5	1.94	2.20	2.51
Lepidolite	$K_2(Li,Al)_{5-6}[Si_{6-7}Al_{2-1}O_{20}](OH,F)_4$	monocl	2.85	3.3	1.536	1.565	1.566
Leucite	$KAlSi_2O_6$	tetr	2.49	5.8	1.510		
Levyne	$(Ca,Na_2)Al_2Si_4O_{12} \cdot 6H_2O$	rhomb	2.10	4.5	1.496	1.501	
Litharge	PbO	tetr	9.35	2	2.535	2.665	
Loellingite	$FeAs_2$	rhomb	7.40	5.3			
Maghemite	Fe_2O_3	cub	4.88	7.8	2.63		
Magnesite	$MgCO_3$	hex	3.05	4	1.536	1.741	
Magnetite	Fe_3O_4	cub	5.17	6	2.42		
Malachite	$Cu_2(OH)_2(CO_3)$	monocl	4.05	3.8	1.655	1.875	1.909
Manganite	$MnO(OH)$	monocl	≈4.3	4	2.25	2.25	2.53
Manganosite	MnO	cub	5.37	5.5			
Marcasite	FeS_2	cub	5.02	6.3			
Marialite	$Na_4Al_3Si_9O_{24}Cl$	tetr	2.56	5.5	1.541	1.548	
Marshite	CuI	cub	5.67	2.5	2.346		
Mascagnite	$(NH_4)_2SO_4$	orth	1.77	2.3	1.520	1.523	1.533
Matlockite	$PbClF$	tetr	7.05	2.8	2.006	2.145	

Name	Formula	Crystal system	Density g/cm³	Hardness	Index of refraction n_α	n_β	n_γ
Meionite	$Ca_4Al_6Si_6O_{24}CO_3$	tetr	2.78	5.5	1.559	1.595	
Melanterite	$FeSO_4 \cdot 7H_2O$	monocl	1.89	2	1.47	1.48	1.49
Melilite	$(Ca,Na)_2(Mg,Fe,Al,Si)_3O_7$	tetr	3.00	5.5	1.639	1.645	
Mellite	$Al_2C_{12}O_{12} \cdot 18H_2O$	tetr	1.64	2.3	1.511	1.539	
Mendipite	$Pb_3O_2Cl_2$	rhomb	7.24	2.5	2.24	2.27	2.31
Mesolite	$Na_2Ca_2(Al_2Si_3O_{10})_3 \cdot 8H_2O$	orth	2.26	5	1.506		
Metacinnabar	HgS	cub	7.70	3			
Microcline	$KAlSi_3O_8$	monocl	2.56	6.3	1.522	1.526	1.530
Miersite	AgI	hex	5.68	2.5	2.20		
Millerite	NiS	hex	5.5	3.3			
Mimetite	$Pb_5(AsO_4,PO_4)_3Cl$	hex	7.24	3.8	2.128	2.147	
Minium	Pb_3O_4	tetr	8.9	2.5			
Mirabilite	$Na_2SO_4 \cdot 10H_2O$	monocl	1.46	1.8	1.394	1.396	1.398
Moissanite	SiC	hex	3.16	9.5	2.648	2.691	
Molybdenite	MoS_2	hex	5.06	1.3			
Monazite	$(Ce,La,Th)PO_4$	monocl	5.2	5	1.787	1.789	1.840
Monetite	$CaHPO_4$	tricl	2.92	3.5	1.587	1.61	1.640
Monticellite	$Ca(Mg,Fe)SiO_4$	orth	3.18	5.5	1.647	1.655	1.664
Montmorillonite	$(0.5Ca,Na)_{0.7}(Al,Mg,Fe)_4[(Si,Al)_8O_{20}](OH)_4 \cdot nH_2O$	monocl	2.5	1.5	1.55	1.57	1.57
Montroydite	HgO	orth	11.14	2.5	2.37	2.50	2.65
Mordenite	$(Na,K,Ca)[Al_2Si_{10}O_{24}] \cdot 7H_2O$	orth	2.13	3.5	1.478	1.480	1.482
Muscovite	$KAl_2Si_3AlO_{10}(OH,F)_2$	monocl	2.83	2.8	1.563	1.596	1.602
Nantokite	$CuCl$	cub	4.14	2.5	1.930		
Natrolite	$Na_2Al_2Si_3O_{10} \cdot 2H_2O$	orth	2.23	5	1.478	1.481	1.491
Nepheline	$Na_3KAl_4Si_4O_{16}$	hex	2.61	5.8	1.534	1.538	
Newberyite	$MgHPO_4 \cdot 3H_2O$	orth	2.13	3.3	1.514	1.517	1.533
Niccolite	$NiAs$	hex	7.77	5.3			
Norbergite	$Mg(OH,F)_2 \cdot Mg_2SiO_4$	orth	3.21	6.5	1.565	1.573	1.592
Nosean	$Na_8Al_6Si_6O_{24}SO_4$	cub	2.35	5.5	1.495		
Oldhamite	CaS	cub	2.59	4	2.137		
Oligoclase	$([NaSi]_{0.9-0.7}[CaAl]_{0.1-0.3})AlSi_2O_8$	tricl	2.64	6.3	1.539	1.543	1.547
Olivenite	$Cu_2(AsO_4)(OH)$	rhomb	4.2	3	1.77	1.80	1.85
Olivine	$(Mg,Fe)SiO_4$	rhomb	3.81	6.8	1.73	1.76	1.78
Opal	$SiO_2 \cdot nH_2O$	amorp	1.9	5	1.44		
Orpiment	As_2S_3	monocl	3.46	1.8	2.40	2.81	3.02
Orthoclase	$KAlSi_3O_8$	monocl	2.56	6	1.523	1.527	1.531
Orthopyroxene	$(Mg,Fe)SiO_3$	rhomb	3.6	5.5	1.709	1.712	1.723
Paragonite	$NaAl_2Si_3AlO_{10}(OH)_2$	monocl	2.85	2.5	1.572	1.602	1.605
Parisite	$(Ce,La,Na)FCO_3 \cdot CaCO_3$	hex	4.42	4.5	1.672	1.771	
Pectolite	$Ca_2NaH(SiO_3)_3$	tricl	2.88	4.8	1.603	1.610	1.639
Penfieldite	$Pb_4Cl_6(OH)_2$	hex	6.6		2.13	2.21	
Pentlandite	$(Fe,Ni)_9S_8$	cub	4.8	3.8			
Percylite	$PbCuCl_2(OH)_2$	cub		2.5	2.05		
Periclase	MgO	cub	3.6	5.5	1.735		
Perovskite	$CaTiO_3$	cub	3.98	5.5	2.34		
Petalite	$LiAlSi_4O_{10}$	monocl	2.42	6.5	1.506	1.511	1.519
Pharmacosiderite	$Fe_3(AsO_4)_2(OH)_3 \cdot 5H_2O$	cub	2.80	2.5	1.690		
Phenakite	Be_2SiO_4	rhomb	2.98	7.5	1.654	1.670	
Phillipsite	$K(Ca_{0.5},Na)_2[Al_3Si_5O_{16}] \cdot 6H_2O$	monocl	2.2	4.3	1.494	1.497	1.505
Phlogopite	$KMg_3AlSi_3O_{10}(OH,F)_2$	monocl	2.83	2.3	1.560	1.597	1.598
Phosgenite	$Pb_2(CO_3)Cl_2$	tetr	6.13	2.5	2.118	2.145	
Piemontite	$Ca_2(Mn,Fe,Al)_3O(Si_2O_7)(SiO_4)(OH)$	monocl	3.49	6	1.762	1.773	1.796
Pigeonite	$(Mg,Fe,Ca)(Mg,Fe)Si_2O_6$	monocl	3.38	6	1.702	1.703	1.728
Pollucite	$CsAlSi_2O_6$	tetr	2.9	6.5	1.517		
Polybasite	$(Ag,Cu)_{16}Sb_2S_{11}$	monocl	6.1	2.5			
Powellite	$Ca(Mo,W)O_4$	tetr	4.35	3.8	1.971	1.980	
Prehnite	$Ca_2Al_2Si_3O_{10}(OH)_2$	rhomb	2.93	6.3	1.622	1.628	1.648
Proustite	Ag_3AsS_3	rhomb	5.57	2.3	2.792	3.088	
Pseudobrookite	Fe_2TiO_5	rhomb	4.36	6	2.38	2.39	2.42
Psilomelane	$BaMn_9O_{16}(OH)_4$	rhomb	4.71	5.5			

Name	Formula	Crystal system	Density g/cm³	Hardness	n_α	n_β	n_γ
Pumpellyite	$Ca_2Al_2(Al,Fe,Mg)[Si_2(O,OH)_7](SiO_4)(OH,O)_3$	monocl	3.21	5.5	1.688	1.695	1.705
Pyrargyrite	Ag_3SbS_3	rhomb	5.85	2.5	2.88	3.08	
Pyrite	FeS_2	cub	5.02	6.3			
Pyrochlore	$NaCaNb_2O_6F$	cub	5.3	5.3			
Pyrochroite	$Mn(OH)_2$	hex	3.26	2.5	1.68	1.72	
Pyrolusite	MnO_2	tetr	5.08	6.3			
Pyromorphite	$Pb_5(PO_4,AsO_4)_3Cl$	hex	7.04	3.8	2.048	2.058	
Pyrope	$Mg_3Al_2Si_3O_{12}$	cub	3.58	6.8	1.714		
Pyrophyllite	$Al_2Si_4O_{10}(OH)_2$	monocl	2.78	1.5	1.545	1.579	1.599
Pyrrhotite	Fe_7S_8	hex	4.62	4			
Quartz	SiO_2	hex	2.65	7	1.544	1.553	
Rammelsbergite	$NiAs_2$	orth	7.1	5.8			
Raspite	$PbWO_4$	monocl	8.46	2.8	1.27	1.27	1.30
Realgar	As_4S_4	monocl	3.5	1.8	2.538	2.684	2.704
Rhodochrosite	$MnCO_3$	hex	3.70	3.8	1.597	1.816	
Rhodonite	$(Mn,Fe,Ca)SiO_3$	orth	3.48	6	1.725	1.729	1.737
Riebeckite	$Na_2Fe_5(Si_8O_{22})(OH)_2$	monocl	3.3	5	1.675	1.683	1.694
Rutile	TiO_2	tetr	4.23	6.2	2.609	2.900	
Safflorite	$(Co,Fe)As_2$	rhomb	7.3	4.8			
Samarskite	$(Y,Er,Ce,U,Ca,Fe,Pb,Th)(Nb,Ta,Ti,Sn)_2O_6$	rhomb	5.69	5.5	2.200		
Sapphirine	$(Mg,Fe)_2Al_4O_6SiO_4$	monocl	3.49	7.5	1.709	1.712	1.715
Scapolite	$(Na,Ca)_4Al_3(Al,Si)_3Si_6O_{24}(Cl,F,OH,CO_3,SO_4)$	tetr	2.64	5.5	1.551	1.573	
Scheelite	$CaWO_4$	tetr	6.06	4.8	1.920	1.936	
Scolecite	$CaAl_2Si_3O_{10}\cdot 3H_2O$	monocl	2.27	5	1.510	1.518	1.519
Scorodite	$Fe(AsO_4)\cdot 2H_2O$	rhomb	3.28	3.8	1.784	1.795	1.814
Sellaite	MgF_2	tetr	3.15	5	1.378	1.390	
Senarmontite	Sb_2O_3	cub	5.58	2.3	2.087		
Serpentine	$Mg_3Si_2O_5(OH)_4$	monocl	2.55	3	1.55	1.56	1.56
Siderite	$FeCO_3$	hex	3.9	4.3	1.635	1.875	
Sillimanite	Al_2OSiO_4	rhomb	3.25	7	1.658	1.660	1.660
Skutterudite	$(Co,Ni)As_3$	cub	6.8	5.8			
Smithsonite	$ZnCO_3$	rhomb	4.4	4.3	1.621	1.848	
Sodalite	$Na_8Al_6Si_6O_{24}Cl_2$	cub	2.30	5.8	1.485		
Sperrylite	$PtAs_2$	cub	10.58	6.5			
Spessartite	$Mn_3Al_2Si_3O_{12}$	cub	4.19	6.8	1.800		
Sphalerite	ZnS	cub	4.0	3.8	2.369		
Sphene	$CaTiSiO_4(O,OH,F)$	monocl	3.50	5	1.90	1.95	2.03
Spinel	$MgAl_2O_4$	cub	3.55	7.8	1.719		
Spodumene	$LiAlSi_2O_6$	monocl	3.13	6.8	1.656	1.662	1.671
Stannite	Cu_2FeSn_4	tetr	4.4	4			
Staurolite	$(Fe,Mg,Zn)_2(Al,Fe,Ti)_9O_6[(Si,Al)O_4]_4(O,OH)_2$	monocl	3.79	7.5	1.743	1.747	1.755
Stercorite	$Na(NH_4)H(PO_4)\cdot 4H_2O$	tricl	1.62	2	1.439	1.442	1.469
Stibiotantalite	$Sb(Ta,Nb)O_4$	rhomb	6.6	5.5	2.38	2.41	2.46
Stibnite	Sb_2S_3	orth	4.56	2			
Stilbite	$NaCa_2[Al_5Si_{13}O_{36}]\cdot 14H_2O$	monocl	2.2	3.8	1.492	1.499	1.503
Stilpnomelane	$(K,Na,Ca)_{0.6}(Fe,Mg)_6Si_8Al(O,OH)_{27}\cdot 2H_2O$	monocl	2.8	3.5	1.585	1.665	1.665
Stolzite	$PbWO_4$	tetr	8.2	2.8	2.19	2.27	
Strengite	$FePO_4\cdot 2H_2O$	orth	2.87	4	1.707	1.719	1.741
Strontianite	$SrCO_3$	orth	3.5	3.5	1.518	1.666	1.668
Struvite	$Mg(NH_4)(PO_4)\cdot 6H_2O$	rhomb	1.71	2	1.495	1.496	1.504
Sulfur	S	orth	2.07	2	1.958	2.038	2.245
Sylvanite	$(Ag,Au)Te_2$	monocl	8.16	1.8			
Sylvite	KCl	cub	1.99	2	1.490		
Talc	$Mg_3Si_4O_{10}(OH)_2$	monocl	2.71	1	1.545	1.592	1.595
Tantalite	$(Fe,Mn)(Ta,Nb)_2O_6$	rhomb	7.95	6.5	2.26	2.32	2.43
Tapiolite	$FeTa_2O_6$	tetr	7.9	6.3	2.27	2.42	
Tellurobismuthite	Bi_2Te_3	hex	7.74	1.8			
Terlinguaite	Hg_2OCl	monocl	8.73	2.5	2.35	2.64	2.66
Tetrahedrite	$(Cu,Fe)_{12}Sb_4S_{13}$	cub	4.9	3.8			
Thenardite	Na_2SO_4	orth	2.7	2.8	1.468	1.475	1.483

Name	Formula	Crystal system	Density g/cm³	Hardness	n_α	n_β	n_γ
Thermonatrite	$Na_2CO_3 \cdot H_2O$	orth	2.25	1.3	1.420	1.506	1.524
Thomsenolite	$NaCaAlF_6 \cdot H_2O$	monocl	2.98	2	1.407	1.414	1.415
Thorianite	ThO_2	cub	10.0	6.5	2.200		
Thorite	$ThSiO_4$	tetr	6.7	4.8	1.8		
Topaz	$Al_2SiO_4(OH,F)_2$	rhomb	3.53	8	1.618	1.620	1.627
Torbernite	$Cu(UO_2)_2(PO_4)_2 \cdot 8H_2O$	tetr	3.22	2.3	1.582	1.592	
Tourmaline	$Na(Mg,Fe,Mn,Li,Al)_3Al_6Si_6O_{18}(BO_3)_3$	rhomb	3.14	7	1.62	1.65	
Tremolite	$Ca_2Mg_5Si_8O_{22}(OH,F)_2$	monocl	3.0	5.5	1.599	1.612	1.622
Trevorite	$NiFe_2O_4$	cub	5.33	7.8	2.3		
Tridymite	SiO_2	hex	2.27	7	1.475	1.476	1.479
Triphyllite-Lithiophyllite	$Li(Fe,Mn)PO_4$	rhomb	3.46	4.5	1.68	1.68	1.69
Troegerite	$(UO_2)_3(AsO_4)_2 \cdot 12H_2O$	tetr		2.5	1.59	1.630	
Troilite	FeS	hex	4.7	4			
Trona	$Na_3H(CO_3)_2 \cdot 2H_2O$	monocl	2.14	2.8	1.412	1.492	1.540
Turquois	$Cu(Al,Fe)_6(PO_4)_4(OH)_8 \cdot 4H_2O$	tricl	2.9	5.3	1.70	1.73	1.75
Ullmannite	$NiSbS$	cub	6.65	5.3			
Uraninite	UO_2	cub	11.0	5.5			
Uvarovite	$Ca_3Cr_2Si_3O_{12}$	cub	3.83	6.8	1.865		
Valentinite	Sb_2O_3	orth	5.7	2.8	2.18	2.35	2.35
Vanadinite	$Pb_5(VO_4)_3Cl$	hex	6.8	2.9	2.350	2.416	
Variseite-Strengite	$(Al,Fe)(PO_4) \cdot 2H_2O$	rhomb	2.72	4	1.635	1.654	1.668
Vaterite	$CaCO_3$	hex	2.71		1.550	1.645	
Vermiculite	$(Mg,Ca)_{0.7}(Mg,Fe,Al)_6[(Al,Si)_8O_{20}](OH)_4 \cdot 8H_2O$	monocl	2.3	1.5	1.542	1.556	1.556
Vesuvianite	$Ca_{10}(Mg,Fe)_2Al_4(Si_2O_7)_2(SiO_4)_5(OH,F)_4$	tetr	3.33	6.5	1.72	1.73	
Villiaumite	NaF	cub	2.78	2.3	1.327		
Vivianite	$Fe_3(PO_4)_2 \cdot 8H_2O$	monocl	2.58	1.8	1.598	1.629	1.652
Wagnerite	$Mg_2(PO_4)F$	monocl	3.15	5.3	1.568	1.572	1.582
Wavellite	$Al_3(OH)_3(PO_4)_2 \cdot 5H_2O$	rhomb	2.36	3.6	1.527	1.535	1.553
Whewellite	$CaC_2O_4 \cdot H_2O$	cub	2.2	2.8	1.491	1.554	1.650
Willemite	Zn_2SiO_4	hex	4.1	5.5	1.691	1.719	
Witherite	$BaCO_3$	orth	4.29	3.5	1.529	1.676	1.677
Wolframite	$(Fe,Mn)WO_4$	monocl	7.3	4.3	2.26	2.32	2.42
Wollastonite	$CaSiO_3$	monocl	2.92	4.8	1.628	1.639	1.642
Wulfenite	$PbMoO_4$	tetr	6.7	2.9	2.283	2.403	
Wurtzite	ZnS	hex	4.09	3.8	2.356	2.378	
Xenotime	YPO_4	tetr	4.8	4.5	1.721	1.816	
Zeunerite	$Cu(UO_2)_2(AsO_4)_2 \cdot 10H_2O$	tetr			1.606		
Zincite	ZnO	hex	5.6	4	2.013	2.029	
Zircon	$ZrSiO_4$	tetr	4.6	7.5	1.94	1.99	
Zoisite	$Ca_2Al_3Si_3O_{12}(OH)$	rhomb	3.26	6	1.695	1.699	1.711

CRYSTALLOGRAPHIC DATA ON MINERALS

This table contains x-ray crystallographic data on about 400 common minerals, as well as selected crystalline elements. Entries are arranged alphabetically by mineral name. The columns are:

Name: Common name of the mineral.

Formula: Chemical formula for a typical sample of the mineral. Composition often varies considerably with the origin of the sample.

Crystal system: tricl = triclinic; monocl = monoclinic; orth = orthorhombic; tetr = tetragonal; hex = hexagonal; rhomb = rhombohedral; cubic = cubic.

Structure type: Prototype for the structural arrangement of the crystallographic cell.

Z: Number of formula units per the unit cell.

a, b, c: Lengths of the cell edges in Å (1Å = 10^{-8} cm).

α, β, γ : Angles between cell axes.

References

1. Robie, R.A., Bethke, P.M., and Beardsley, K.M., *U. S. Geological Survey Bulletin 1248*, U. S. Government Printing Office, Washington, D.C.
2. Donnay, J.D.H., and Ondik, H.M., *Crystal Data Determinative Tables, Third Edition, Volume 2, Inorganic Compounds*, Joint Committee on Powder Diffraction Standards, Swarthmore, PA, 1973.
3. Deer, W.A., Howie, R.A., and Zussman, J., *An Introduction to the Rock-Forming Minerals, 2nd Edition*, Longman Scientific & Technical, Harlow, Essex, 1992.

Name	Formula	Crystal system	Structure type	Z	a/Å	b/Å	c/Å	α	β	γ
Acanthite	Ag_2S	monocl		4	4.228	6.928	7.862		99.58°	
Acmite (Aegirine)	$NaFe(SiO_3)_2$	monocl	diopside	4	9.658	8.795	5.294		107.42°	
Akermanite	$Ca_2MgSi_2O_7$	tetr	melilite	2	7.8435		5.010			
Alabandite	MnS	cubic	rock salt	4	5.223					
Almandine (Almandite)	$Fe_3Al_2Si_3O_{12}$	cubic	garnet	8	11.526					
Altaite	PbTe	cubic	rock salt	4	6.4606					
Aluminum	Al	cubic	copper	4	4.049					
Alunite	$KAl_3(SO_4)_2(OH)_6$	rhomb		3	6.982		17.32			
Analcite	$NaAlSi_2O_6 \cdot H_2O$	cubic		16	13.733					
Anatase	TiO_2	tetr		4	3.785		9.514			
Andalusite	Al_2OSiO_4	orth		4	7.7959	7.8983	5.5583			
Andradite	$Ca_3Fe_2Si_3O_{12}$	cubic	garnet	8	12.048					
Anglesite	$PbSO_4$	orth	barite	4	8.480	5.398	6.958			
Anhydrite	$CaSO_4$	orth	anhydrite	4	6.991	6.996	6.238			
Annite	$KFe_3[AlSi_3O_{10}](OH)_2$	monocl	1M mica	2	10.29	9.33	5.39		105.1°	
Anorthite	$CaAl_2Si_2O_8$	tricl	primitive cell	8	8.177	12.877	14.169	93.17°	115.85°	91.22°
Anthophyllite	$Mg_7Si_8O_{22}(OH)_2$	orth		4	18.61	18.01	5.24			
Antimony	Sb	rhomb	arsenic	6	4.2996		11.2516			
Aragonite	$CaCO_3$	orth	aragonite	4	5.741	7.968	4.959			
Arcanite	K_2SO_4	orth	arcanite	4	5.772	10.072	7.483			
Argentite	Ag_2S	cubic		2	4.870					
Argentopyrite	$AgFe_2S_3$	orth		4	6.64	11.47	6.45			
Arsenic	As	rhomb	arsenic	6	3.760		10.555			
Arsenolite	As_2O_3	cubic	diamond	16	11.074					
Arsenopyrite	FeAsS	tricl		4	5.760	5.690	5.785	90.00°	112.23°	90.00°
Azurite	$Cu_3(OH)_2(CO_3)_2$	monocl		2	5.008	5.844	10.336		92.45°	
Baddeleyite	ZrO_2	monocl	baddeleyite	4	5.1454	5.2075	5.3107		99.23°	
Banalsite	$BaNa_2Al_4Si_4O_{16}$	orth		4	8.50	9.97	16.72			
Barite	$BaSO_4$	orth	barite	4	8.878	5.450	7.152			
Berlinite	$AlPO_4$	hex	α-quartz	3	4.942		10.97			
Beryl	$Be_3Al_2(SiO_3)_6$	hex	beryl	2	9.215		9.192			
Berzelianite	Cu_2Se	cubic		4	5.85					
Bismite	Bi_2O_3	monocl	pseudo-orth	4	7.48	8.14	5.83		112.9°	
Bismuth	Bi	rhomb	arsenic	6	4.5367		11.8383			
Bismuthinite	Bi_2S_3	orth	stibnite	4	11.150	11.300	3.981			
Bixbyite	Mn_2O_3	cubic	thallium trioxide	16	9.411					
Boehmite	AlO(OH)	orth	lepidocrocite	4	2.868	12.227	3.700			
Borax	$Na_2B_4O_7 \cdot 10H_2O$	monocl		4	11.858	10.674	12.197		106.68°	

Name	Formula	Crystal system	Structure type	Z	a/Å	b/Å	c/Å	α	β	γ
Bornite (metastable)	Cu_5FeS_4	cubic		8	10.94					
Breithauptite	NiSb	hex	niccolite	2	3.942		5.155			
Brochantite	$Cu_4SO_4(OH)_6$	monocl		4	13.066	9.85	6.022		103.27°	
Bromargyrite	AgBr	cubic	rock salt	4	5.7745					
Bromellite	BeO	hex	zincite	2	2.6979		4.3772			
Brookite	TiO_2	orth		8	5.456	9.182	5.143			
Brucite	$Mg(OH)_2$	hex	cadmium iodide	1	3.147		4.769			
Bunsenite	NiO	cubic	rock salt	4	4.177					
Bustamite	$CaMn(SiO_3)_2$	tricl		6	7.736	7.157	13.824	90.52°	94.58°	103.87°
Cadmium telluride	CdTe	cubic	sphalerite	4	6.4805					
Cadmoselite	CdSe	hex	zincite	2	4.2977		7.0021			
Calcite	$CaCO_3$	rhomb	calcite	6	4.9899		17.064			
Calomel	Hg_2Cl_2	tetr		4	4.478		10.910			
Carbonate-apatite	$Ca_{10}(PO_4)_6CO_3 \cdot H_2O$	hex	apatite	1	9.436		6.883			
Cassiterite	SnO_2	tetr	rutile	2	4.738		3.188			
Cattierite	CoS_2	cubic	pyrite	4	5.5345					
Celestite	$SrSO_4$	orth	barite	4	8.359	5.352	6.866			
Celsian	$BaAl_2Si_2O_8$	monocl		8	8.627	13.045	14.408		115.20°	
Cerianite	CeO_2	cubic	fluorite	4	5.4110					
Cerussite	$PbCO_3$	orth	aragonite	4	6.152	8.436	5.195			
Cervantite	Sb_2O_4	orth		4	5.424	11.76	4.804			
Chalcanthite	$CuSO_4 \cdot 5H_2O$	tricl		2	6.1045	10.72	5.949	97.57°	107.28°	77.43°
Chalcocite	Cu_2S	orth		96	11.881	27.323	13.491			
Chalcopyrite	$CuFeS_2$	tetr		4	5.2988		10.434			
Chlorapatite	$Ca_5(PO_4)_3Cl$	hex	apatite	2	9.629		6.777			
Chlorargyrite	AgCl	cubic	rock salt	4	5.5491					
Chloritoid	$FeAl_4O_2(SiO_4)_2(OH)_4$	monocl		8	9.48	5.48	18.18		101.77°	
Chloromagnesite	$MgCl_2$	rhomb		3	3.632		17.795			
Chondrodite	$2Mg_2SiO_4 \cdot MgF_2$	monocl		2	7.89	4.743	10.29		109.03°	
Chrysoberyl	$BeAl_2O_4$	orth	olivine	4	5.4756	9.4041	4.4267			
Cinnabar	HgS	hex	cinnabar	3	4.149		9.495			
Claudetite	As_2O_3	monocl		4	5.339	12.984	4.5405		94.27°	
Clausthalite	PbSe	cubic	rock salt	4	6.1255					
Clinoenstatite	$MgSiO_3$	monocl		8	9.620	8.825	5.188		108.33°	
Clinoferrosilite	$FeSiO_3$	monocl		8	9.7085	9.0872	5.2284		108.43°	
Clinohumite	$4Mg_2SiO_4 \cdot MgF_2$	monocl		2	13.68	4.75	10.27		100.83°	
Clinozoisite	$Ca_2Al_3(SiO_4)_3OH$	monocl		2	8.887	5.581	10.14		115.93°	
Cobalt olivine	Co_2SiO_4	orth	olivine	4	4.782	10.301	6.003			
Cobalt oxide	CoO	cubic	rock salt	4	4.260					
Cobalt sulfide	CoS	cubic	sphalerite	4	5.339					
Cobalt titanate	$CoTiO_3$	rhomb	ilmenite	6	5.066		13.918			
Cobalticalcite	$CoCO_3$	rhomb	calcite	6	4.6581		14.958			
Cobaltite	CoAsS	cubic	NiSbS	4	5.60					
Coesite	SiO_2	monocl		16	7.152	12.379	7.152		120.00°	
Coffinite	$USiO_4$	tetr	zircon	4	6.995		6.263			
Colemanite	$Ca_2B_6O_{11} \cdot 5H_2O$	monocl		4	8.743	11.264	6.102		110.12°	
Coloradoite	HgTe	cubic	sphalerite	4	6.4600					
Cooperite	PtS	tetr		2	3.4699		6.1098			
Copper	Cu	cubic	face-centered cubic	4	3.6150					
Corundum	Al_2O_3	rhomb	corundum	6	4.7591		12.9894			
Cotunnite	$PbCl_2$	orth		4	4.535	7.62	9.05			
Covellite	CuS	hex		6	3.792		16.34			
Cristobalite (α)	SiO_2	tetr		4	4.971		6.918			
Cristobalite (β)	SiO_2	cubic		8	7.1382					
Cryolite	Na_3AlF_6	monocl		2	5.40	5.60	7.776		90.18°	
Cubanite	$CuFe_2S_3$	orth		4	6.46	11.12	6.23			
Cummingtonite	$(Mg,Fe,Mn)_7(Si_4O_{11})_2(OH)_2$	monocl	tremolite	2	9.522	18.223	5.332		101.92°	
Cuprite	Cu_2O	cubic		2	4.2696					
Danburite	$CaB_2Si_2O_8$	orth		4	8.04	8.77	7.74			

Name	Formula	Crystal system	Structure type	Z	a/Å	b/Å	c/Å	α	β	γ
Datolite	$CaBSiO_4(OH)$	monocl		4	9.62	7.60	4.84		90.15°	
Daubreeite	$FeCr_2S_4$	cubic	spinel	8	9.966					
Diamond	C	cubic	diamond	8	3.5670					
Diaspore	$AlO(OH)$	orth		4	4.401	9.421	2.845			
Dickite	$Al_2Si_2O_5(OH)_4$	monocl		4	5.150	8.940	14.736		103.58°	
Digenite	$Cu_{1.79}S$	cubic	deformed fluorite	4	5.5695					
Diopside	$CaMg(SiO_3)_2$	monocl	diopside	4	9.743	8.923	5.251		105.93°	
Dioptase	$CuSiO_2(OH)_2$	rhomb	phenacite	18	14.61		7.80			
Dolerophanite	$Cu_2O(SO_4)$	monocl		4	8.334	6.312	7.628		108.4°	
Dolomite	$CaMg(CO_3)_2$	rhomb	calcite	3	4.8079		16.010			
Dravite	$NaMg_3Al_6B_3Si_6O_{27}(OH)_4$	rhomb	tourmaline	3	15.942		7.224			
Elbaite	$NaLiAl_{7.67}B_3Si_6O_{27}(OH)_4$	rhomb	tourmaline	3	15.842		7.009			
Enargite	Cu_3AsS_4	orth		2	6.426	7.422	6.144			
Enstatite	$MgSiO_3$	orth		16	8.829	18.22	5.192			
Epidote	$Ca_2Al_2(Al,Fe)OH(SiO_4)_3$	monocl		2	8.89	5.63	10.19		115.40°	
Epsomite	$MgSO_4 \cdot 7H_2O$	orth		4	11.86	11.99	6.858			
Eskolaite	Cr_2O_3	rhomb	corundum	6	4.9607		13.599			
Eucairite	AgCuSe	orth		10	4.105	20.35	6.31			
Euclase	$AlBeSiO_4(OH)$	monocl		4	4.763	14.29	4.618		100.25°	
Famatimite	Cu_3SbS_4	tetr		2	5.384		10.770			
Fayalite	Fe_2SiO_4	orth	olivine	4	4.817	10.477	6.105			
Fe-Cordierite	$Fe_2Al_3(AlSi_5O_{18})$	orth	cordierite	4	9.726	17.065	9.287			
Fe-Gehlenite	$Ca_2Fe_2SiO_7$	tetr	melilite	2	7.54		4.855			
Fe-Indialite	$Fe_2Al_3(AlSi_5O_{18})$	hex	beryl	2	9.860		9.285			
Fe-Leucite	$KFeSi_2O_6$	tetr		16	13.205		13.970			
Fe-Microcline	$KFeSi_3O_8$	tricl		4	8.68	13.10	7.340	90.75°	116.05°	86.23°
Fe-Sanidine	$KFeSi_3O_8$	monocl		4	8.689	13.12	7.319		116.10°	
Fe-Skutterudite	$FeAs_{2.95}$	cubic		8	8.1814					
Ferberite	$FeWO_4$	monocl	wolframite	2	4.732	5.708	4.965		90.00°	
Ferriannite	$KFe_3[FeSi_3O_{10}](OH)_2$	monocl		2	5.430	9.404	10.341		100.07°	
Ferroselite	$FeSe_2$	orth	marcasite	2	4.801	5.778	3.587			
Ferrotremolite	$Ca_2Fe_5[Si_8O_{22}](OH)_2$	monocl	tremolite	2	9.97	18.34	5.30		104.50°	
Fluor-edenite	$NaCa_2Mg_5[AlSi_7O_{22}]F_2$	monocl	tremolite	2	9.847	18.00	5.282		104.83°	
Fluor-humite	$3Mg_2SiO_4 \cdot MgF_2$	orth		4	10.243	20.72	4.735			
Fluor-norbergite	$Mg_2SiO_4 \cdot MgF_2$	orth		4	8.727	10.271	4.709			
Fluor-phlogopite	$KMg_3[AlSi_3O_{10}]F_2$	monocl	1M mica	2	5.299	9.188	10.135		99.92°	
Fluor-richterite	$Na_2CaMg_5[Si_8O_{22}]F_2$	monocl	tremolite	2	9.823	17.96	5.268		104.33°	
Fluor-tremolite	$Ca_2Mg_5[Si_8O_{22}]F_2$	monocl	tremolite	2	9.781	18.01	5.267		104.52°	
Fluorapatite	$Ca_5(PO_4)_3F$	hex	apatite	2	9.3684		6.8841			
Fluorite	CaF_2	cubic	fluorite	4	5.4638					
Forsterite	Mg_2SiO_4	orth	olivine	4	4.758	10.214	5.984			
Frohbergite	$FeTe_2$	orth	marcasite	2	5.265	6.265	3.869			
Gahnite	$ZnAl_2O_4$	cubic	spinel	8	8.0848					
Galaxite	$MnAl_2O_4$	cubic	spinel	8	8.258					
Galena	PbS	cubic	rock salt	4	5.9360					
Gallium oxide	Ga_2O_3	rhomb	corundum	6	4.9793		13.429			
Gehlenite	$Ca_2Al_2SiO_7$	tetr	melilite	2	7.690		5.0675			
Geikielite	$MgTiO_3$	rhomb	ilmenite	6	5.054		13.898			
Gerhardite	$Cu_2(NO_3)(OH)_3$	orth		4	6.075	13.812	5.592			
Gersdorfite	NiAsS	cubic		4	5.693					
Gibbsite	$Al(OH)_3$	monocl		8	9.719	5.0705	8.6412		94.57°	
Glauchroite	$CaMnSiO_4$	orth	olivine	4	4.944	11.19	6.529			
Glaucodot	$(Co,Fe)AsS$	orth		24	6.64	28.39	5.64			
Glaucophane I	$Na_2Mg_3Al_2[Si_8O_{22}](OH)_2$	monocl	tremolite	2	9.748	17.915	5.273		102.78°	
Glaucophane II	$Na_2Mg_3Al_2[Si_8O_{22}](OH)_2$	monocl	tremolite	2	9.663	17.696	5.277		103.67°	
Goethite	$FeO(OH)$	orth		4	4.596	9.957	3.021			
Gold	Au	cubic	face-centered cubic	4	4.0786					
Goldmanite	$Ca_3V_2Si_3O_{12}$	cubic	garnet	8	12.070					
Goslarite	$ZnSO_4 \cdot 7H_2O$	orth	epsomite	4	11.779	12.050	6.822			

Name	Formula	Crystal system	Structure type	Z	a/Å	b/Å	c/Å	α	β	γ
Graphite	C	hex	graphite	4	2.4612		6.7079			
Greenockite	CdS	hex	zincite	2	4.1354		6.7120			
Greigite	Fe_3S_4	cubic	spinel	8	9.876					
Grossularite	$Ca_3Al_2Si_3O_{12}$	cubic	garnet	8	11.851					
Grunerite	$Fe_7[Si_8O_{22}](OH)_2$	monocl	tremolite	2	9.572	18.44	5.342		101.77°	
Gudmundite	FeSbS	monocl		8	10.00	5.93	6.73		90.00°	
Gypsum	$CaSO_4 \cdot 2H_2O$	monocl		4	5.68	15.18	6.29		113.83°	
Hafnia	HfO_2	monocl	baddeleyite	4	5.1156	5.1722	5.2948		99.18°	
Halite	NaCl	cubic	rock salt	4	5.6402					
Hambergite	$Be_2(OH,F)BO_3$	orth		8	9.755	12.201	4.426			
Hardystonite	$Ca_2ZnSi_2O_7$	tetr	melilite	2	7.87		5.01			
Hauerite	MnS_2	cubic	pyrite	4	6.1014					
Hausmannite	Mn_3O_4	tetr		8	8.136		9.422			
Hawleyite	CdS	cubic	sphalerite	4	5.833					
Heazelwoodite	Ni_3S_2	rhomb		3	5.746		7.134			
Hedenbergite	$CaFe(SiO_3)_2$	monocl	diopside	4	9.854	9.024	5.263		104.23°	
Hematite	Fe_2O_3	rhomb	corundum	6	5.025		13.735			
Hemimorphite	$Zn_4(OH)_2Si_2O_7 \cdot H_2O$	orth		2	8.370	10.719	5.120			
Hercynite	$Fe(AlO_2)_2$	cubic	spinel	8	8.150					
Herzenbergite	SnS	orth	germanium sulfide	4	4.328	11.190	3.978			
Hessite	Ag_2Te	monocl		4	8.13	4.48	8.09		111.9°	
Hexahydrite	$MgSO_4 \cdot 6H_2O$	monocl		8	10.110	7.212	24.41		98.30°	
High albite (Analbite)	$NaAlSi_3O_8$	tricl		4	8.160	12.870	7.106	93.54°	116.36°	90.19°
High argentite	Ag_2S	cubic		4	6.269					
High bornite	Cu_5FeS_4	cubic		1	5.50					
High carnegeite	$NaAlSiO_4$	cubic		4	7.325					
High chalcocite	Cu_2S	hex		2	3.961		6.722			
High clinoenstatite	$MgSiO_3$	tricl		8	10.000	8.934	5.170	88.27°	70.03°	91.01°
High digenite	Cu_2S	cubic		4	5.725					
High germania	GeO_2	hex	α–quartz	3	4.987		5.652			
High leucite	$KAlSi_2O_6$	cubic		16	13.43					
High naumanite	Ag_2Se	cubic		2	4.993					
High sanidine	$KAlSi_3O_8$	monocl		4	8.615	13.031	7.177		115.98°	
Huebnerite	$MnWO_4$	monocl	wolframite	2	4.834	5.758	4.999		91.18°	
Huntite	$Mg_3Ca(CO_3)_4$	rhomb	calcite	3	9.498		7.816			
Hydroxylapatite	$Ca_5(PO_4)_3OH$	hex	apatite	2	9.418		6.883			
Ice	H_2O	hex		4	4.5212		7.3666			
Ilmenite	$FeTiO_3$	rhomb	ilmenite	6	5.093		14.055			
Indialite (Cordierite)	$Mg_2Al_3(AlSi_5O_{18})$	hex	beryl	2	9.7698		9.3517			
Iodargyrite	AgI	hex	zincite	2	4.5955		7.5005			
Iron (α)	Fe	cubic	body-centered cubic	2	2.8664					
Jacobsite	$MnFe_2O_4$	cubic	spinel	8	8.499					
Jadeite	$NaAl(SiO_3)_2$	monocl	diopside	4	9.409	8.564	5.220		107.50°	
Jalpaite	$Ag_{1.55}Cu_{0.45}S$	tetr		16	8.673		11.756			
Johannsenite	$CaMn(SiO_3)_2$	monocl	diopside	4	9.83	9.04	5.27		105.00°	
Kaliophilite	$KAlSiO_4$	hex		54	26.930		8.522			
Kalsilite	$KAlSiO_4$	hex		2	5.1597		8.7032			
Kaolinite	$Al_2Si_2O_5(OH)_4$	tricl		2	5.155	8.959	7.407	91.68°	104.87°	89.93°
Karelianite	V_2O_3	rhomb	corundum	6	4.952		14.002			
Keatite	SiO_2	tetr		12	7.456		8.604			
Kernite	$Na_2B_4O_7 \cdot 4H_2O$	monocl		4	7.022	9.151	15.676		108.83°	
Kerschsteinite	$CaFeSiO_4$	orth	olivine	4	4.886	11.146	6.434			
Klockmannite	CuSe	hex	deformed covellite	78	14.206		17.25			
Knebelite	$MnFeSiO_4$	orth	olivine	4	4.854	10.602	6.162			
Kyanite	Al_2OSiO_4	tricl		4	7.123	7.848	5.564	89.92°	101.25°	105.97°
Larnite	Ca_2SiO_4	monocl		4	5.48	6.76	9.28		94.55°	
Laurite	RuS_2	cubic	pyrite	4	5.60					

Name	Formula	Crystal system	Structure type	Z	a/Å	b/Å	c/Å	α	β	γ
Lawrencite	$FeCl_2$	rhomb		3	3.593		17.58			
Lawsonite	$CaAl_2Si_2O_7(OH)_2 \cdot H_2O$	orth		4	8.787	5.836	13.123			
Lead	Pb	cubic	face-centered cubic	4	4.9505					
Leonhardtite	$MgSO_4 \cdot 4H_2O$	monocl		4	5.922	13.604	7.905		90.85°	
Lepidocrocite	FeO(OH)	orth		4	3.868	12.525	3.066			
Lepidolite	$K_2Al_3Li_2AlSi_7O_{20}(OH)_4$	monocl	2M2 mica	2	9.2	5.3	20.0		98.00°	
Leucite	$KAlSi_2O_6$	tetr		16	13.074		13.738			
Lime	CaO	cubic	rock salt	4	4.8108					
Lime olivine	Ca_2SiO_4	orth	olivine	4	5.091	11.371	6.782			
Linnaeite	Co_3S_4	cubic	spinel	8	9.401					
Litharge	PbO	tetr		2	3.9759		5.023			
Loellingite	$FeAs_2$	orth	marcasite	2	5.300	5.981	2.882			
Low albite	$NaAlSi_3O_8$	tricl		4	8.139	12.788	7.160	94.27°	116.57°	87.68°
Low bornite	Cu_5FeS_4	tetr		16	10.94		21.88			
Low cordierite	$Mg_2Al_3(AlSi_5O_{18})$	orth		4	9.721	17.062	9.339			
Low germania	GeO_2	tetr	rutile	2	4.3963		2.8626			
Low nepheline	$NaAlSiO_4$	hex		8	9.986		8.330			
Luzonite	Cu_3AsS_4	tetr		2	5.289		10.440			
Mackinawite	FeS	tetr		2	3.675		5.030			
Magnesioriebeckite	$Na_2Mg_3Fe_2[Si_8O_{22}](OH)_2$	monocl	tremolite	2	9.733	17.946	5.299		103.30°	
Magnesite	$MgCO_3$	rhomb	calcite	6	4.6330		15.016			
Magnetite	Fe_3O_4	cubic	spinel	8	8.3940					
Malachite	$Cu_2(OH)_2CO_3$	monocl		4	9.502	11.974	3.240		98.75°	
Maldonite	Au_2Bi	cubic		8	7.958					
Manganese sulfide (γ)	MnS	hex	zincite	2	3.976		6.432			
Manganese sulfide (β)	MnS	cubic	sphalerite	4	5.611					
Manganosite	MnO	cubic	rock salt	4	4.4448					
Marcasite	FeS_2	orth	marcasite	2	4.443	5.423	3.3876			
Margarite	$CaAl_2[AlSi_2O_{10}](OH)_2$	monocl	2M mica	4	5.13	8.92	19.50		95.00°	
Marialite	$Na_4Al_3Si_9O_{24}Cl$	tetr		2	12.064		7.514			
Marshite	CuI	cubic	sphalerite	4	6.0507					
Mascagnite	$(NH_4)_2SO_4$	orth	arcanite	4	7.782	5.993	10.636			
Massicot	PbO	orth		4	5.489	4.755	5.891			
Matlockite	PbClF	tetr		2	4.106		7.23			
Maucherite	$Ni_{11}As_8$	tetr		4	6.870		21.81			
Meionite	$Ca_4Al_6Si_6O_{24}CO_3$	tetr		2	12.174		7.652			
Melanophlogite	SiO_2	cubic	clathrate type	46	13.402					
Melanterite	$FeSO_4 \cdot 7H_2O$	monocl		4	14.072	6.503	11.041		105.57°	
Melonite	$NiTe_2$	hex	cadmium iodide	1	3.869		5.308			
Metacinnabar	HgS	cubic	sphalerite	4	5.8517					
Miargyrite	$AgSbS_2$	monocl		8	12.862	4.111	13.220		98.63°	
Microcline	$KAlSi_3O_8$	tricl		4	8.582	12.964	7.222	90.62°	115.92°	87.68°
Miersite	AgI	cubic	sphalerite	4	6.4963					
Millerite	NiS	rhomb		9	9.616		3.152			
Minium	Pb_3O_4	tetr		4	8.815		6.565			
Minnesotaite	$Fe_3Si_4O_{10}(OH)_2$	monocl		4	5.4	9.42	19.4		100.00°	
Mirabilite	$Na_2SO_4 \cdot 10H_2O$	monocl		4	11.51	10.38	12.83		107.75°	
Mn-Indialite	$Mn_2Al_3(AlSi_5O_{18})$	hex	beryl	2	9.925		9.297			
Molybdenite	MoS_2	hex	molybdenite	2	3.1604		12.295			
Molybdenum	Mo	cubic		2	3.1653					
Molybdite	MoO_3	orth		4	3.962	13.858	3.697			
Monteponite	CdO	cubic	rock salt	4	4.6953					
Monticellite	$CaMgSiO_4$	orth	olivine	4	4.827	11.084	6.376			
Montroydite	HgO	orth		4	6.608	5.518	3.519			
Mullite (2:1)	$2Al_2O_3 \cdot SiO_2$	orth		6	7.5788	7.6909	2.8883			
Mullite (3:2)	$3Al_2O_3 \cdot 2SiO_2$	orth		3	7.557	7.6876	2.8842			
Muscovite	$KAl_2AlSi_3O_{10}(OH)_2$	monocl	2M2 mica	4	5.203	8.995	20.030		94.47°	

Name	Formula	Crystal system	Structure type	Z	a/Å	b/Å	c/Å	α	β	γ
Nacrite	$Al_2Si_2O_5(OH)_4$	monocl		4	8.909	5.146	15.697		113.70°	
Nantokite	CuCl	cubic	sphalerite	4	5.416					
Natroalunite	$NaAl_3(SO_4)_2(OH)_6$	rhomb		3	6.974		16.69			
Natrolite	$Na_2Al_2Si_3O_{10}\cdot2H_2O$	orth		8	18.30	18.63	6.60			
Neighborite	$NaMgF_3$	orth	perovskite	4	5.363	7.676	5.503			
Ni-Skutterudite	$NiAs_{2.95}$	cubic		8	8.3300					
Niccolite	NiAs	hex	niccolite	2	3.618		5.034			
Nickel	Ni	cubic	face-centered cubic	4	3.5238					
Nickel carbonate	$NiCO_3$	rhomb	calcite	6	4.5975		14.723			
Nickel olivine	Ni_2SiO_4	orth	olivine	4	4.727	10.121	5.915			
Nickel selenide	$NiSe_2$	cubic	pyrite	4	5.9604					
Niter	KNO_3	orth	aragonite	4	6.431	9.164	5.414			
Norsethite	$BaMg(CO_3)_2$	rhomb	calcite	3	5.020		16.75			
Oldhamite	CaS	cubic	rock salt	4	5.689					
Orpiment	As_2S_3	monocl		4	11.49	9.59	4.25		90.45°	
Orthoclase	$KAlSi_3O_8$	monocl		4	8.562	12.996	7.193		116.02°	
Orthoferrosilite	$FeSiO_3$	orth	enstatite	16	9.080	18.431	5.238			
Otavite	$CdCO_3$	rhomb	calcite	6	4.9204		16.298			
Paracelsian	$BaAl_2Si_2O_8$	monocl		4	8.58	9.583	9.08		90.00°	
Paragonite	$NaAl_2AlSi_3O_{10}(OH)_2$	monocl	2M1 mica	4	5.13	8.89	19.32		95.17°	
Pararammelsbergite	$NiAs_2$	orth		8	5.75	5.82	11.428			
Paratellurite	TeO_2	tetr		4	4.810		7.613			
Parawollastonite	$CaSiO_3$	monocl		12	15.417	7.321	7.066		95.40°	
Pectolite	$Ca_2NaH(SiO_3)_3$	tricl		2	7.99	7.04	7.02	90.05°	95.27°	102.47°
Pentlandite	$Fe_{5.25}Ni_{3.75}S_8$	cubic		4	10.196					
Pentlandite	$Fe_{4.75}Ni_{5.25}S_8$	cubic		4	10.095					
Periclase	MgO	cubic	rock salt	4	4.2117					
Perovskite	$CaTiO_3$	orth	perovskite	4	5.3670	7.6438	5.4439			
Petalite	$LiAlSi_4O_{10}$	monocl		2	11.32	5.14	7.62		105.90°	
Petzite	Ag_3AuTe_2	cubic		8	10.38					
Phenacite	Be_2SiO_4	rhomb	phenacite	18	12.472		8.252			
Phlogopite	$KMg_3AlSi_3O_{10}(OH)_2$	monocl	1M mica	2	5.326	9.210	10.311		100.17°	
Picrochromite	$MgCr_2O_4$	cubic	spinel	8	8.333					
Piemontite	$Ca_2Al_{1.5}Mn_{1.5}(SiO_4)_3OH$	monocl		2	8.95	5.70	9.41		115.70°	
Platinum	Pt	cubic	face-centered cubic	4	3.9231					
Polymidite	Ni_3S_4	cubic	spinel	8	9.480					
Portlandite	$Ca(OH)_2$	hex	cadmium iodide	1	3.5933		4.9086			
Powellite	$CaMoO_4$	tetr	scheelite	4	5.226		11.43			
Protoenstatite	$MgSiO_3$	orth		8	9.25	8.74	5.32			
Proustite	Ag_3AsS_3	rhomb		6	10.816		8.6948			
Pseudowollastonite	$CaSiO_3$	tricl		24	6.90	11.78	19.65	90.00°	90.80°	90.00°
Pyrargyrite	Ag_3SbS_3	rhomb		6	11.052		8.7177			
Pyrite	FeS_2	cubic	pyrite	4	5.4175					
Pyrolusite	MnO_2	tetr	rutile	2	4.388		2.865			
Pyrope	$Mg_3Al_2Si_3O_{12}$	cubic	garnet	8	11.459					
Pyrophanite	$MnTiO_3$	rhomb	ilmenite	6	5.155		14.18			
Pyrophyllite	$Al_2Si_4O_{10}(OH)_2$	monocl	2M1 mica	4	5.14	8.90	18.55		99.92°	
Pyroxmangite	$MnFe(SiO_3)_2$	tricl		7	7.56	17.45	6.67	84.00°	94.30°	113.70°
Pyrrhotite	$Fe_{0.980}S$	hex	defect niccolite	2	3.446		5.848			
Pyrrhotite	$Fe_{0.885}S$	hex	defect niccolite	2	3.440		5.709			
Quartz (α)	SiO_2	hex		3	4.9136		5.4051			
Quartz (β)	SiO_2	hex		3	4.999		5.4592			
Rammelsbergite	$NiAs_2$	orth	marcasite	2	4.757	5.797	3.542			
Realgar	AsS	monocl		16	9.29	13.53	6.57		106.55°	
Retgersite	$NiSO_4\cdot4H_2O$	tetr		4	6.782		18.28			
Rhodochrosite	$MnCO_3$	rhomb	calcite	6	4.7771		15.664			

Name	Formula	Crystal system	Structure type	Z	a/Å	b/Å	c/Å	α	β	γ
Rhodonite	$MnSiO_3$	tricl		10	7.682	11.818	6.707	92.36°	93.95°	105.66°
Riebeckite	$Na_2Fe_5FSi_8O_{22}(OH)_2$	monocl	tremolite	2	9.729	18.065	5.334		103.31°	
Rutile	TiO_2	tetr		2	4.5937		2.9618			
Safflorite	$Co_{0.5}Fe_{0.5}As_2$	orth	marcasite	2	5.231	5.953	2.962			
Sanmartinite	$ZnWO_4$	monocl	wolframite	2	4.691	5.720	4.925		89.36°	
Sapphirine	$Mg_2Al_4O_6SiO_4$	monocl		8	9.96	28.60	9.85		110.5°	
Scacchite	$MnCl_2$	rhomb		3	3.711		17.59			
Scheelite	$CaWO_4$	tetr	scheelite	4	5.242		11.372			
Schorl	$NaFe_3Al_6B_3Si_6O_{27}(OH)_4$	rhomb	tourmaline	3	16.032		7.149			
Selenium	Se	hex		3	4.3642		4.9588			
Selenolite	SeO_2	tetr		8	8.35		5.05			
Sellaite	MgF_2	tetr	rutile	2	4.621		3.050			
Senarmontite	Sb_2O_3	cubic	arsenic trioxide	16	11.152					
Shandite	$Ni_3Pb_2S_2$	rhomb		3	5.576		13.658			
Shortite	$Na_2Ca_2(CO_3)_3$	orth		2	4.961	11.03	7.12			
Siderite	$FeCO_3$	rhomb	calcite	6	4.6887		15.373			
Silicon	Si	cubic	diamond	8	5.4305					
Sillimanite	Al_2OSiO_4	orth		4	7.4843	7.6730	5.7711			
Silver	Ag	cubic	face-centered cubic	4	4.0862					
Silver telluride I	Ag_2Te	cubic		2	5.29					
Silver telluride II	Ag_2Te	cubic		4	6.585					
Smithsonite	$ZnCO_3$	rhomb	calcite	6	4.6528		15.025			
Soda niter	$NaNO_3$	rhomb	calcite	6	5.0696		16.829			
Sodium melilite	$NaCaAlSi_2O_7$	tetr	melilite	2	8.511		4.809			
Sperrylite	$PtAs_2$	cubic	pyrite	4	5.968					
Spessartite	$Mn_3Al_2Si_3O_{12}$	cubic	garnet	8	11.621					
Sphalerite	ZnS	cubic	sphalerite	4	5.4093					
Sphene	$CaTiSiO_5$	monocl		4	7.07	8.72	6.56		113.95°	
Spinel	$MgAl_2O_4$	cubic	spinel	8	8.080					
Spodumene	$LiAl(SiO_3)_2$	monocl	diopside	4	9.451	8.387	5.208		110.07°	
Spodumene (β)	$LiAl(SiO_3)_2$	tetr		4	7.5332		9.1540			
Staurolite	$Fe_2Al_9Si_4O_{22}(OH)_2$	monocl		2	7.90	16.65	5.63		90.00°	
Sternbergite	$AgFe_2S_3$	orth		8	11.60	12.675	6.63			
Stibnite	Sb_2S_3	orth	stibnite	4	11.229	11.310	3.8389			
Stilleite	$ZnSe$	cubic	sphalerite	4	5.6685					
Stishovite	SiO_2	tetr	rutile	2	4.1790		2.6649			
Stolzite	$PbWO_4$	tetr	scheelite	4	5.4616		12.046			
Stromeyerite	$Ag_{0.93}Cu_{1.07}S$	orth		4	4.066	6.628	7.972			
Strontianite	$SrCO_3$	orth	aragonite	4	6.029	8.414	5.107			
Sulfur (monoclinic)	S	monocl	S8 ring molecules	48	11.04	10.98	10.92		96.73°	
Sulfur (orthorhombic)	S	orth	S8 ring molecules	128	10.4646	12.8660	24.4860			
Sulfur (rhombohedral)	S	rhomb	S6 ring molecules	18	10.818		4.280			
Sylvite	KCl	cubic	rock salt	4	6.2931					
Syngenite	$K_2Ca(SO_4)_2 \cdot H_2O$	monocl		2	9.775	7.156	6.251		104.00°	
Synthetic anorthite	$CaAl_2Si_2O_8$	hex		2	5.10		14.72			
Synthetic anorthite	$CaAl_2Si_2O_8$	orth		2	8.22	8.60	4.83			
Talc	$Mg_3Si_4O_{10}(OH)_2$	monocl	2M1 mica	4	5.287	9.158	18.95		99.50°	
Tantalum	Ta	cubic	tungsten	2	3.3058					
Teallite	$PbSnS_2$	orth	germanium sulfide	2	4.266	11.419	4.090			
Tellurite	TeO_2	orth	tellurite	8	5.607	12.034	5.463			
Tellurium	Te	hex	selenium	3	4.4570		5.9290			
Tellurobismuthite	Bi_2Te_3	rhomb		3	4.3835		30.487			
Tennantite	$Cu_{12}As_4S_{13}$	cubic	tetrahedrite	2	10.190					
Tenorite	CuO	monocl		4	4.684	3.425	5.129		99.47°	
Tephroite	Mn_2SiO_4	orth	olivine	4	4.871	10.636	6.232			

Name	Formula	Crystal system	Structure type	Z	a/Å	b/Å	c/Å	α	β	γ
Tetrahedrite	$Cu_{12}Sb_4S_{13}$	cubic	tetrahedrite	2	10.327					
Thenardite	Na_2SO_4	orth	thenardite	8	5.863	12.304	9.821			
Thorianite	ThO_2	cubic	fluorite	4	5.5952					
Thorite	$ThSiO_4$	tetr	zircon	4	7.143		6.327			
Tiemannite	$HgSe$	cubic	sphalerite	4	6.0853					
Tin	Sn	tetr		4	5.8315		3.1813			
Titanium	Ti	hex		2	2.953		4.729			
Titanium(III) oxide	Ti_2O_3	rhomb	corundum	6	5.149		13.642			
Topaz	$Al_2SiO_4(OH,F)_2$	orth		4	8.394	8.792	4.649			
Tremolite	$Ca_2Mg_5Si_8O_{22}(OH)_2$	monocl	tremolite	2	9.840	18.052	5.275		104.70°	
Trevorite	$NiFe_2O_4$	cubic	spinel	8	8.339					
Tridymite (β)	SiO_2	hex		4	5.0463		8.2563			
Trogtalite	$CoSe_2$	cubic	pyrite	4	5.8588					
Troilite	FeS	hex	niccolite	2	3.446		5.877			
Tschermakite	$CaAl_2SiO_6$	monocl	diopside	4	9.615	8.661	5.272		106.12°	
Tungsten	W	cubic		2	3.1653					
Tungstenite	WS_2	hex	molybdenite	2	3.154		12.362			
Turquois	$CuAl_6(PO_4)_4(OH)_8 \cdot 4H_2O$	tricl		1	7.424	7.629	9.910	68.61°	69.71°	65.08°
Umangite	Cu_3Se_2	tetr		2	6.402		4.276			
Uraninite	UO_2	cubic	fluorite	4	5.4682					
Ureyite	$NaCr(SiO_3)_2$	monocl	diopside	4	9.550	8.712	5.273		107.44°	
Uvarovite	$Ca_3Cr_2Si_3O_{12}$	cubic	garnet	8	11.999					
Uvite	$CaMg_4Al_5B_3Si_6O_{27}(OH)_4$	rhomb	tourmaline	3	15.86		7.19			
Vaesite	NiS_2	cubic	pyrite	4	5.6873					
Valentinite	Sb_2O_3	orth	antimony trioxide	4	4.914	12.468	5.421			
Vanthoffite	$MgSO_4 \cdot 3Na_2SO_4$	monocl		2	9.797	9.217	8.199		113.50°	
Vaterite	$CaCO_3$	hex		6	7.135		8.524			
Villiaumite	NaF	cubic	rock salt	4	4.6342					
Violarite	$FeNi_2S_4$	cubic	spinel	8	9.464					
Willemite	Zn_2SiO_4	rhomb	phenacite	18	13.94		9.309			
Witherite	$BaCO_3$	orth	aragonite	4	6.430	8.904	5.314			
Wolframite	$Fe_{0.5}Mn_{0.5}WO_4$	monocl	wolframite	2	4.782	5.731	4.982		90.57°	
Wollastonite	$CaSiO_3$	tricl		6	7.94	7.32	7.07	90.03°	95.37°	103.43°
Wulfenite	$PbMoO_4$	tetr	scheelite	4	5.435		12.110			
Wurtzite	ZnS	hex	zincite	2	3.8230		6.2565			
Wustite	$Fe_{0.953}O$	cubic	defect rock salt	4	4.3088					
Xenotime	YPO_4	tetr	zircon	4	6.885		5.982			
Zinc	Zn	hex	hexagonal close pack	2	2.665		4.947			
Zinc telluride	$ZnTe$	cubic	sphalerite	4	6.1020					
Zincite	ZnO	hex	zincite	2	3.2495		5.2069			
Zinkosite	$ZnSO_4$	orth	barite	4	8.588	6.740	4.770			
Zircon	$ZrSiO_4$	tetr	zircon	4	6.604		5.979			
Zoisite	$Ca_2Al_3(SiO_4)_3OH$	orth		4	16.15	5.581	10.06			

Section 5
Thermochemistry, Electrochemistry, and Kinetics

CODATA KEY VALUES FOR THERMODYNAMICS

The Committee on Data for Science and Technology (CODATA) has conducted a project to establish internationally agreed values for the thermodynamic properties of key chemical substances. This table presents the final results of the project. Use of these recommended, internally consistent values is encouraged in the analysis of thermodynamic measurements, data reduction, and preparation of other thermodynamic tables.

The table includes the standard enthalpy of formation at 298.15 K, the entropy at 298.15 K, and the quantity $H°$ (298.15 K)$-H°$ (0). A value of 0 in the $\Delta_f H°$ column for an element indicates the reference state for that element. The standard state pressure is 100000 Pa (1 bar). See the reference for information on the dependence of gas-phase entropy on the choice of standard state pressure.

Substances are listed in alphabetical order of their chemical formulas when written in the most common form.

The table is reprinted with permission of CODATA.

Reference

Cox, J. D., Wagman, D. D., and Medvedev, V. A., *CODATA Key Values for Thermodynamics*, Hemisphere Publishing Corp., New York, 1989.

Substance	State	$\Delta_f H°$ (298.15 K) kJ·mol^{-1}	$S°$ (298.15 K) J·K^{-1}·mol^{-1}	$H°$ (298.15 K)$-H°$ (0) kJ·mol^{-1}
Ag	cr	0	42.55 ± 0.20	5.745 ± 0.020
Ag	g	284.9 ± 0.8	172.997 ± 0.004	6.197 ± 0.001
Ag$^+$	aq	105.79 ± 0.08	73.45 ± 0.40	
AgCl	cr	−127.01 ± 0.05	96.25 ± 0.20	12.033 ± 0.020
Al	cr	0	28.30 ± 0.10	4.540 ± 0.020
Al	g	330.0 ± 4.0	164.554 ± 0.004	6.919 ± 0.001
Al^{+3}	aq	−538.4 ± 1.5	−325 ± 10	
AlF$_3$	cr	−1510.4 ± 1.3	66.5 ± 0.5	11.62 ± 0.04
Al$_2$O$_3$	cr, corundum	−1675.7 ± 1.3	50.92 ± 0.10	10.016 ± 0.020
Ar	g	0	154.846 ± 0.003	6.197 ± 0.001
B	cr, rhombic	0	5.90 ± 0.08	1.222 ± 0.008
B	g	565 ± 5	153.436 ± 0.015	6.316 ± 0.002
BF$_3$	g	−1136.0 ± 0.8	254.42 ± 0.20	11.650 ± 0.020
B$_2$O$_3$	cr	−1273.5 ± 1.4	53.97 ± 0.30	9.301 ± 0.040
Be	cr	0	9.50 ± 0.08	1.950 ± 0.020
Be	g	324 ± 5	136.275 ± 0.003	6.197 ± 0.001
BeO	cr	−609.4 ± 2.5	13.77 ± 0.04	2.837 ± 0.008
Br	g	111.87 ± 0.12	175.018 ± 0.004	6.197 ± 0.001
Br$^-$	aq	−121.41 ± 0.15	82.55 ± 0.20	
Br$_2$	l	0	152.21 ± 0.30	24.52 ± 0.01
Br$_2$	g	30.91 ± 0.11	245.468 ± 0.005	9.725 ± 0.001
C	cr, graphite	0	5.74 ± 0.10	1.050 ± 0.020
C	g	716.68 ± 0.45	158.100 ± 0.003	6.536 ± 0.001
CO	g	−110.53 ± 0.17	197.660 ± 0.004	8.671 ± 0.001
CO$_2$	g	−393.51 ± 0.13	213.785 ± 0.010	9.365 ± 0.003
CO$_2$	aq, undissoc.	−413.26 ± 0.20	119.36 ± 0.60	
CO$_3^{-2}$	aq	−675.23 ± 0.25	−50.0 ± 1.0	
Ca	cr	0	41.59 ± 0.40	5.736 ± 0.040
Ca	g	177.8 ± 0.8	154.887 ± 0.004	6.197 ± 0.001
Ca^{+2}	aq	−543.0 ± 1.0	−56.2 ± 1.0	
CaO	cr	−634.92 ± 0.90	38.1 ± 0.4	6.75 ± 0.06
Cd	cr	0	51.80 ± 0.15	6.247 ± 0.015
Cd	g	111.80 ± 0.20	167.749 ± 0.004	6.197 ± 0.001
Cd^{+2}	aq	−75.92 ± 0.60	−72.8 ± 1.5	
CdO	cr	−258.35 ± 0.40	54.8 ± 1.5	8.41 ± 0.08
CdSO$_4$·8/3H$_2$O	cr	−1729.30 ± 0.80	229.65 ± 0.40	35.56 ± 0.04
Cl	g	121.301 ± 0.008	165.190 ± 0.004	6.272 ± 0.001
Cl$^-$	aq	−167.080 ± 0.10	56.60 ± 0.20	
ClO$_4^-$	aq	−128.10 ± 0.40	184.0 ± 1.5	
Cl$_2$	g	0	223.081 ± 0.010	9.181 ± 0.001
Cs	cr	0	85.23 ± 0.40	7.711 ± 0.020
Cs	g	76.5 ± 1.0	175.601 ± 0.003	6.197 ± 0.001
Cs$^+$	aq	−258.00 ± 0.50	132.1 ± 0.5	
Cu	cr	0	33.15 ± 0.08	5.004 ± 0.008

Substance	State	$\dfrac{\Delta_f H^\circ \,(298.15\ \text{K})}{\text{kJ} \cdot \text{mol}^{-1}}$	$\dfrac{S^\circ \,(298.15\ \text{K})}{\text{J} \cdot \text{K}^{-1} \cdot \text{mol}^{-1}}$	$\dfrac{H^\circ \,(298.15\ \text{K}) - H^\circ \,(0)}{\text{kJ} \cdot \text{mol}^{-1}}$
Cu	g	337.4 ± 1.2	166.398 ± 0.004	6.197 ± 0.001
Cu^{+2}	aq	64.9 ± 1.0	−98 ± 4	
CuSO$_4$	cr	−771.4 ± 1.2	109.2 ± 0.4	16.86 ± 0.08
F	g	79.38 ± 0.30	158.751 ± 0.004	6.518 ± 0.001
F$^-$	aq	−335.35 ± 0.65	−13.8 ± 0.8	
F$_2$	g	0	202.791 ± 0.005	8.825 ± 0.001
Ge	cr	0	31.09 ± 0.15	4.636 ± 0.020
Ge	g	372 ± 3	167.904 ± 0.005	7.398 ± 0.001
GeF$_4$	g	−1190.20 ± 0.50	301.9 ± 1.0	17.29 ± 0.10
GeO$_2$	cr, tetragonal	−580.0 ± 1.0	39.71 ± 0.15	7.230 ± 0.020
H	g	217.998 ± 0.006	114.717 ± 0.002	6.197 ± 0.001
H$^+$	aq	0	0	
HBr	g	−36.29 ± 0.16	198.700 ± 0.004	8.648 ± 0.001
HCO$_3^-$	aq	−689.93 ± 0.20	98.4 ± 0.5	
HCl	g	−92.31 ± 0.10	186.902 ± 0.005	8.640 ± 0.001
HF	g	−273.30 ± 0.70	173.779 ± 0.003	8.599 ± 0.001
HI	g	26.50 ± 0.10	206.590 ± 0.004	8.657 ± 0.001
HPO$_4^{-2}$	aq	−1299.0 ± 1.5	−33.5 ± 1.5	
HS$^-$	aq	−16.3 ± 1.5	67 ± 5	
HSO$_4^-$	aq	−886.9 ± 1.0	131.7 ± 3.0	
H$_2$	g	0	130.680 ± 0.003	8.468 ± 0.001
H$_2$O	l	−285.830 ± 0.040	69.95 ± 0.03	13.273 ± 0.020
H$_2$O	g	−241.826 ± 0.040	188.835 ± 0.010	9.905 ± 0.005
H$_2$PO$_4^-$	aq	−1302.6 ± 1.5	92.5 ± 1.5	
H$_2$S	g	−20.6 ± 0.5	205.81 ± 0.05	9.957 ± 0.010
H$_2$S	aq, undissoc.	−38.6 ± 1.5	126 ± 5	
H$_3$BO$_3$	cr	−1094.8 ± 0.8	89.95 ± 0.60	13.52 ± 0.04
H$_3$BO$_3$	aq, undissoc.	−1072.8 ± 0.8	162.4 ± 0.6	
He	g	0	126.153 ± 0.002	6.197 ± 0.001
Hg	l	0	75.90 ± 0.12	9.342 ± 0.008
Hg	g	61.38 ± 0.04	174.971 ± 0.005	6.197 ± 0.001
Hg^{+2}	aq	170.21 ± 0.20	−36.19 ± 0.80	
HgO	cr, red	−90.79 ± 0.12	70.25 ± 0.30	9.117 ± 0.025
Hg$_2^{+2}$	aq	166.87 ± 0.50	65.74 ± 0.80	
Hg$_2$Cl$_2$	cr	−265.37 ± 0.40	191.6 ± 0.8	23.35 ± 0.20
Hg$_2$SO$_4$	cr	−743.09 ± 0.40	200.70 ± 0.20	26.070 ± 0.030
I	g	106.76 ± 0.04	180.787 ± 0.004	6.197 ± 0.001
I$^-$	aq	−56.78 ± 0.05	106.45 ± 0.30	
I$_2$	cr	0	116.14 ± 0.30	13.196 ± 0.040
I$_2$	g	62.42 ± 0.08	260.687 ± 0.005	10.116 ± 0.001
K	cr	0	64.68 ± 0.20	7.088 ± 0.020
K	g	89.0 ± 0.8	160.341 ± 0.003	6.197 ± 0.001
K$^+$	aq	−252.14 ± 0.08	101.20 ± 0.20	
Kr	g	0	164.085 ± 0.003	6.197 ± 0.001
Li	cr	0	29.12 ± 0.20	4.632 ± 0.040
Li	g	159.3 ± 1.0	138.782 ± 0.010	6.197 ± 0.001
Li$^+$	aq	−278.47 ± 0.08	12.24 ± 0.15	
Mg	cr	0	32.67 ± 0.10	4.998 ± 0.030
Mg	g	147.1 ± 0.8	148.648 ± 0.003	6.197 ± 0.001
Mg^{+2}	aq	−467.0 ± 0.6	−137 ± 4	
MgF$_2$	cr	−1124.2 ± 1.2	57.2 ± 0.5	9.91 ± 0.06
MgO	cr	−601.60 ± 0.30	26.95 ± 0.15	5.160 ± 0.020
N	g	472.68 ± 0.40	153.301 ± 0.003	6.197 ± 0.001
NH$_3$	g	−45.94 ± 0.35	192.77 ± 0.05	10.043 ± 0.010
NH$_4^+$	aq	−133.26 ± 0.25	111.17 ± 0.40	
NO$_3^-$	aq	−206.85 ± 0.40	146.70 ± 0.40	
N$_2$	g	0	191.609 ± 0.004	8.670 ± 0.001
Na	cr	0	51.30 ± 0.20	6.460 ± 0.020
Na	g	107.5 ± 0.7	153.718 ± 0.003	6.197 ± 0.001
Na$^+$	aq	−240.34 ± 0.06	58.45 ± 0.15	

Substance	State	$\Delta_f H°$ (298.15 K) kJ·mol^{-1}	$S°$ (298.15 K) J·K^{-1}·mol^{-1}	$H°$ (298.15 K) − $H°$ (0) kJ·mol^{-1}
Ne	g	0	146.328 ± 0.003	6.197 ± 0.001
O	g	249.18 ± 0.10	161.059 ± 0.003	6.725 ± 0.001
OH$^-$	aq	−230.015 ± 0.040	−10.90 ± 0.20	
O$_2$	g	0	205.152 ± 0.005	8.680 ± 0.002
P	cr, white	0	41.09 ± 0.25	5.360 ± 0.015
P	g	316.5 ± 1.0	163.199 ± 0.003	6.197 ± 0.001
P$_2$	g	144.0 ± 2.0	218.123 ± 0.004	8.904 ± 0.001
P$_4$	g	58.9 ± 0.3	280.01 ± 0.50	14.10 ± 0.20
Pb	cr	0	64.80 ± 0.30	6.870 ± 0.030
Pb	g	195.2 ± 0.8	175.375 ± 0.005	6.197 ± 0.001
Pb^{+2}	aq	0.92 ± 0.25	18.5 ± 1.0	
PbSO$_4$	cr	−919.97 ± 0.40	148.50 ± 0.60	20.050 ± 0.040
Rb	cr	0	76.78 ± 0.30	7.489 ± 0.020
Rb	g	80.9 ± 0.8	170.094 ± 0.003	6.197 ± 0.001
Rb$^+$	aq	−251.12 ± 0.10	121.75 ± 0.25	
S	cr, rhombic	0	32.054 ± 0.050	4.412 ± 0.006
S	g	277.17 ± 0.15	167.829 ± 0.006	6.657 ± 0.001
SO$_2$	g	−296.81 ± 0.20	248.223 ± 0.050	10.549 ± 0.010
SO$_4^{-2}$	aq	−909.34 ± 0.40	18.50 ± 0.40	
S$_2$	g	128.60 ± 0.30	228.167 ± 0.010	9.132 ± 0.002
Si	cr	0	18.81 ± 0.08	3.217 ± 0.008
Si	g	450 ± 8	167.981 ± 0.004	7.550 ± 0.001
SiF$_4$	g	−1615.0 ± 0.8	282.76 ± 0.50	15.36 ± 0.05
SiO$_2$	cr, alpha quartz	−910.7 ± 1.0	41.46 ± 0.20	6.916 ± 0.020
Sn	cr, white	0	51.18 ± 0.08	6.323 ± 0.008
Sn	g	301.2 ± 1.5	168.492 ± 0.004	6.215 ± 0.001
Sn^{+2}	aq	−8.9 ± 1.0	−16.7 ± 4.0	
SnO	cr, tetragonal	−280.71 ± 0.20	57.17 ± 0.30	8.736 ± 0.020
SnO$_2$	cr, tetragonal	−577.63 ± 0.20	49.04 ± 0.10	8.384 ± 0.020
Th	cr	0	51.8 ± 0.5	6.35 ± 0.05
Th	g	602 ± 6	190.17 ± 0.05	6.197 ± 0.003
ThO$_2$	cr	−1226.4 ± 3.5	65.23 ± 0.20	10.560 ± 0.020
Ti	cr	0	30.72 ± 0.10	4.824 ± 0.015
Ti	g	473 ± 3	180.298 ± 0.010	7.539 ± 0.002
TiCl$_4$	g	−763.2 ± 3.0	353.2 ± 4.0	21.5 ± 0.5
TiO$_2$	cr, rutile	−944.0 ± 0.8	50.62 ± 0.30	8.68 ± 0.05
U	cr	0	50.20 ± 0.20	6.364 ± 0.020
U	g	533 ± 8	199.79 ± 0.10	6.499 ± 0.020
UO$_2$	cr	−1085.0 ± 1.0	77.03 ± 0.20	11.280 ± 0.020
UO$_2^{+2}$	aq	−1019.0 ± 1.5	−98.2 ± 3.0	
UO$_3$	cr, gamma	−1223.8 ± 1.2	96.11 ± 0.40	14.585 ± 0.050
U$_3$O$_8$	cr	−3574.8 ± 2.5	282.55 ± 0.50	42.74 ± 0.10
Xe	g	0	169.685 ± 0.003	6.197 ± 0.001
Zn	cr	0	41.63 ± 0.15	5.657 ± 0.020
Zn	g	130.40 ± 0.40	160.990 ± 0.004	6.197 ± 0.001
Zn^{+2}	aq	−153.39 ± 0.20	−109.8 ± 0.5	
ZnO	cr	−350.46 ± 0.27	43.65 ± 0.40	6.933 ± 0.040

STANDARD THERMODYNAMIC PROPERTIES OF CHEMICAL SUBSTANCES

his table gives the standard state chemical thermodynamic
erties of about 2500 individual substances in the crystalline,
d, and gaseous states. Substances are listed by molecular for-
in a modified Hill order; all substances not containing car-
appear first, followed by those that contain carbon. The
erties tabulated are:

$H°$ Standard molar enthalpy (heat) of formation at 298.15 K
in kJ/mol

$G°$ Standard molar Gibbs energy of formation at 298.15 K
in kJ/mol

Standard molar entropy at 298.15 K in J/mol K

Molar heat capacity at constant pressure at 298.15 K in
J/mol K

he standard state pressure is 100 kPa (1 bar). The standard
s are defined for different phases by:

The standard state of a pure gaseous substance is that of
the substance as a (hypothetical) ideal gas at the standard
state pressure.

The standard state of a pure liquid substance is that of the
liquid under the standard state pressure.

The standard state of a pure crystalline substance is that of
the crystalline substance under the standard state pressure.

n entry of 0.0 for $\Delta_f H°$ for an element indicates the reference
e of that element. See References 1 and 2 for further informa-
on reference states. A blank means no value is available.

he data are derived from the sources listed in the references,
n other papers appearing in the *Journal of Physical and Chem-
Reference Data*, and from the primary research literature. We
indebted to M. V. Korobov for providing data on fullerene
pounds.

References

1. Cox, J. D., Wagman, D. D., and Medvedev, V. A., *CODATA Key Values
for Thermodynamics*, Hemisphere Publishing Corp., New York, 1989.
2. Wagman, D. D., Evans, W. H., Parker, V. B., Schumm, R. H., Halow, I.,
Bailey, S. M., Churney, K. L., and Nuttall, R. L., *The NBS Tables of
Chemical Thermodynamic Properties, J. Phys. Chem. Ref. Data*, Vol.
11, Suppl. 2, 1982.
3. Chase, M. W., Davies, C. A., Downey, J. R., Frurip, D. J., McDonald, R.
A., and Syverud, A. N., *JANAF Thermochemical Tables, Third Edition, J. Phys. Chem. Ref. Data*, Vol. 14, Suppl.1, 1985.
4. Chase, M. W., *NIST-JANAF Thermochemical Tables, Fourth Edition,
J. Phys. Chem. Ref. Data*, Monograph 9, 1998.
5. Daubert, T. E., Danner, R. P., Sibul, H. M., and Stebbins, C. C., *Physical and Thermodynamic Properties of Pure Compounds: Data Compilation*, extant 1994 (core with 4 supplements), Taylor & Francis,
Bristol, PA.
6. Pedley, J. B., Naylor, R. D., and Kirby, S. P., *Thermochemical Data of
Organic Compounds, Second Edition*, Chapman & Hall, London,
1986.
7. Pedley, J. B., *Thermochemical Data and Structures of Organic Compounds*, Thermodynamic Research Center, Texas A & M University,
College Station, TX, 1994.
8. Domalski, E. S., and Hearing, E. D., Heat Capacities and Entropies of
Organic Compounds in the Condensed Phase, Volume III, *J. Phys.
Chem. Ref. Data*, 25, 1-525, 1996.
9. Zabransky, M., Ruzicka , V., Majer, V., and Domalski, E. S., *Heat
Capacity of Liquids, J. Phys. Chem. Ref. Data*, Monograph No. 6,
1996.
10. Gurvich, L. V., Veyts, I.V., and Alcock, C. B., *Thermodynamic Properties of Individual Substances, Fourth Edition, Vol. 1*, Hemisphere Publishing Corp., New York, 1989.
11. Gurvich, L. V., Veyts, I.V., and Alcock, C. B., *Thermodynamic Properties of Individual Substances, Fourth Edition, Vol. 3*, CRC Press, Boca
Raton, FL, 1994.
12. *NIST Chemistry Webbook*, <webbook.nist.gov>

		Crystal				Liquid				Gas			
ecular ula	Name	$\Delta_f H°$ kJ/mol	$\Delta_f G°$ kJ/mol	$S°$ J/mol K	C_p J/mol K	$\Delta_f H°$ kJ/mol	$\Delta_f G°$ kJ/mol	$S°$ J/mol K	C_p J/mol K	$\Delta_f H°$ kJ/mol	$\Delta_f G°$ kJ/mol	$S°$ J/mol K	C_p J/mol K
tances not containing carbon:													
	Actinium	0.0		56.5	27.2					406.0	366.0	188.1	20.8
	Silver	0.0		42.6	25.4					284.9	246.0	173.0	20.8
	Silver(I) bromide	-100.4	-96.9	107.1	52.4								
3	Silver(I) bromate	-10.5	71.3	151.9									
	Silver(I) chloride	-127.0	-109.8	96.3	50.8								
3	Silver(I) chlorate	-30.3	64.5	142.0									
4	Silver(I) perchlorate	-31.1											
	Silver(I) fluoride	-204.6											
	Silver(II) fluoride	-360.0											
	Silver(I) iodide	-61.8	-66.2	115.5	56.8								
	Silver(I) iodate	-171.1	-93.7	149.4	102.9								
3	Silver(I) nitrate	-124.4	-33.4	140.9	93.1								
	Disilver									410.0	358.8	257.1	37.0
O4	Silver(I) chromate	-731.7	-641.8	217.6	142.3								
	Silver(I) oxide	-31.1	-11.2	121.3	65.9								
	Silver(II) oxide	-24.3	27.6	117.0	88.0								
	Silver(III) oxide	33.9	121.4	100.0									
S	Silver(I) sulfate	-715.9	-618.4	200.4	131.4								
	Silver(I) sulfide (argentite)	-32.6	-40.7	144.0	76.5								
	Aluminum	0.0		28.3	24.2					330.0	289.4	164.6	21.4

Standard Thermodynamic Properties of Chemical Substances

Molecular formula	Name	Crystal $\Delta_f H°$ kJ/mol	$\Delta_f G°$ kJ/mol	$S°$ J/mol K	C_p J/mol K	Liquid $\Delta_f H°$ kJ/mol	$\Delta_f G°$ kJ/mol	$S°$ J/mol K	C_p J/mol K	Gas $\Delta_f H°$ kJ/mol	$\Delta_f G°$ kJ/mol	$S°$ J/mol K	C_p J/mol K
AlB_3H_{12}	Aluminum borohydride					-16.3	145.0	289.1	194.6	13.0	147.0	379.2	
$AlBr$	Aluminum monobromide									-4.0	-42.0	239.5	35.6
$AlBr_3$	Aluminum bromide	-527.2		180.2	100.6					-425.1			
$AlCl$	Aluminum monochloride									-47.7	-74.1	228.1	35.0
$AlCl_2$	Aluminum dichloride									-331.0			
$AlCl_3$	Aluminum chloride	-704.2	-628.8	109.3	91.1					-583.2			
AlF	Aluminum monofluoride									-258.2	-283.7	215.0	31.9
AlF_3	Aluminum fluoride	-1510.4	-1431.1	66.5	75.1					-1204.6	-1188.2	277.1	62.6
AlF_4Na	Sodium tetrafluoroaluminate									-1869.0	-1827.5	345.7	105.9
AlH	Aluminum monohydride									259.2	231.2	187.9	29.4
AlH_3	Aluminum hydride	-46.0		30.0	40.2								
AlH_4K	Potassium aluminum hydride	-183.7											
AlH_4Li	Lithium aluminum hydride	-116.3	-44.7	78.7	83.2								
AlH_4Na	Sodium aluminum hydride	-115.5											
AlI	Aluminum monoiodide									65.5			36.0
AlI_3	Aluminum iodide	-313.8	-300.8	159.0	98.7					-207.5			
AlN	Aluminum nitride	-318.0	-287.0	20.2	30.1								
AlO	Aluminum monoxide									91.2	65.3	218.4	30.9
AlO_4P	Aluminum phosphate	-1733.8	-1617.9	90.8	93.2								
AlP	Aluminum phosphide	-166.5											
AlS	Aluminum monosulfide									200.9	150.1	230.6	33.4
Al_2	Dialuminum									485.9	433.3	233.2	36.4
Al_2Br_6	Aluminum hexabromide									-970.7			
Al_2Cl_6	Aluminum hexachloride									-1290.8	-1220.4	490.0	
Al_2F_6	Aluminum hexafluoride									-2628.0			
Al_2I_6	Aluminum hexaiodide									-516.7			
Al_2O	Aluminum oxide (Al_2O)									-130.0	-159.0	259.4	45.7
Al_2O_3	Aluminum oxide (corundum)	-1675.7	-1582.3	50.9	79.0								
Al_2S_3	Aluminum sulfide	-724.0		116.9	105.1								
Am	Americium	0.0											
Ar	Argon									0.0		154.8	20.8
As	Arsenic (gray)	0.0		35.1	24.6					302.5	261.0	174.2	20.8
As	Arsenic (yellow)	14.6											
$AsBr_3$	Arsenic(III) bromide	-197.5								-130.0	-159.0	363.9	79.2
$AsCl_3$	Arsenic(III) chloride					-305.0	-259.4	216.3		-261.5	-248.9	327.2	75.7
AsF_3	Arsenic(III) fluoride					-821.3	-774.2	181.2	126.6	-785.8	-770.8	289.1	65.6
$AsGa$	Gallium arsenide	-71.0	-67.8	64.2	46.2								
AsH_3	Arsine									66.4	68.9	222.8	38.1
AsH_3O_4	Arsenic acid	-906.3											
AsI_3	Arsenic(III) iodide	-58.2	-59.4	213.1	105.8							388.3	80.6
$AsIn$	Indium arsenide	-58.6	-53.6	75.7	47.8								
AsO	Arsenic monoxide									70.0			
As_2	Diarsenic									222.2	171.9	239.4	35.0
As_2O_5	Arsenic(V) oxide	-924.9	-782.3	105.4	116.5								
As_2S_3	Arsenic(III) sulfide	-169.0	-168.6	163.6	116.3								
At	Astatine	0.0											
Au	Gold	0.0		47.4	25.4					366.1	326.3	180.5	20.8
$AuBr$	Gold(I) bromide	-14.0											
$AuBr_3$	Gold(III) bromide	-53.3											
$AuCl$	Gold(I) chloride	-34.7											
$AuCl_3$	Gold(III) chloride	-117.6											
AuF_3	Gold(III) fluoride	-363.6											
AuH	Gold hydride									295.0	265.7	211.2	29.2
AuI	Gold(I) iodide	0.0											
Au_2	Digold									515.1			36.9
B	Boron (β-rhombohedral)	0.0		5.9	11.1					565.0	521.0	153.4	20.8
BBr	Bromoborane(1)									238.1	195.4	225.0	32.9
BBr_3	Boron tribromide					-239.7	-238.5	229.7		-205.6	-232.5	324.2	67.8
BCl	Chloroborane(1)									149.5	120.9	213.2	31.7
$BClO$	Chloroxyborane									-314.0			
BCl_3	Boron trichloride					-427.2	-387.4	206.3	106.7	-403.8	-388.7	290.1	62.7
$BCsO_2$	Cesium metaborate	-972.0	-915.0	104.4	80.6								
BF	Fluoroborane(1)									-122.2	-149.8	200.5	29.6
BFO	Fluorooxyborane									-607.0			
BF_3	Boron trifluoride									-1136.0	-1119.4	254.4	
BF_3H_3N	Aminetrifluoroboron	-1353.9											

Molecular formula	Name	Crystal $\Delta_f H°$ kJ/mol	Crystal $\Delta_f G°$ kJ/mol	Crystal $S°$ J/mol K	Crystal C_p J/mol K	Liquid $\Delta_f H°$ kJ/mol	Liquid $\Delta_f G°$ kJ/mol	Liquid $S°$ J/mol K	Liquid C_p J/mol K	Gas $\Delta_f H°$ kJ/mol	Gas $\Delta_f G°$ kJ/mol	Gas $S°$ J/mol K	Gas C_p J/mol K
	Trihydro(phosphorus trifluoride)boron									-854.0			
	Sodium tetrafluoroborate	-1844.7	-1750.1	145.3	120.3								
	Borane(1)									442.7	412.7	171.8	29.2
	Metaboric acid (β, monoclinic)	-794.3	-723.4	38.0						-561.9	-551.0	240.1	42.2
	Borane(3)									89.2	93.3	188.2	36.0
	Boric acid	-1094.3	-968.9	90.0	86.1					-994.1			
	Potassium borohydride	-227.4	-160.3	106.3	96.1								
	Lithium borohydride	-190.8	-125.0	75.9	82.6								
	Sodium borohydride	-188.6	-123.9	101.3	86.8								
	Boron triiodide									71.1	20.7	349.2	70.8
	Potassium metaborate	-981.6	-923.4	80.0	66.7								
	Lithium metaborate	-1032.2	-976.1	51.5	59.8								
	Boron nitride	-254.4	-228.4	14.8	19.7					647.5	614.5	212.3	29.5
	Sodium metaborate	-977.0	-920.7	73.5	65.9								
	Boron monoxide									25.0	-4.0	203.5	29.2
	Boron dioxide									-300.4	-305.9	229.6	43.0
	Rubidium metaborate	-971.0	-913.0	94.3	74.1								
	Boron monosulfide									342.0	288.8	216.2	30.0
	Diboron									830.5	774.0	201.9	30.5
	Tetrachlorodiborane					-523.0	-464.8	262.3	137.7	-490.4	-460.6	357.4	95.4
	Tetrafluorodiborane									-1440.1	-1410.4	317.3	79.1
	Diborane									36.4	87.6	232.1	56.7
	Diboron dioxide									-454.8	-462.3	242.5	57.3
	Boron oxide	-1273.5	-1194.3	54.0	62.8					-843.8	-832.0	279.8	66.9
	Boron sulfide	-240.6		100.0	111.7					67.0			
	Borazine					-541.0	-392.7	199.6					
	Tetraborane(10)									66.1	184.3	280.3	93.2
₇	Sodium tetraborate	-3291.1	-3096.0	189.5	186.8								
	Pentaborane(9)					42.7	171.8	184.2	151.1	73.2	173.6	280.6	99.6
	Pentaborane(11)					73.2				103.3	230.6	321.0	130.3
	Hexaborane(10)					56.3				94.6	211.3	296.8	125.7
	Nonaborane(15)									158.4	357.5	364.9	187.0
	Decaborane(14)									47.3	232.8	350.7	186.1
	Barium	0.0		62.5	28.1					180.0	146.0	170.2	20.8
	Barium bromide	-757.3	-736.8	146.0									
	Barium chloride	-855.0	-806.7	123.7	75.1								
₄O₂	Barium chloride dihydrate	-1456.9	-1293.2	203.0									
	Barium fluoride	-1207.1	-1156.8	96.4	71.2								
	Barium hydride	-177.0	-138.2	63.0	46.0								
	Barium hydroxide	-944.7											
₂	Barium iodide	-602.1											
₄	Barium nitrite	-768.2											
₆	Barium nitrate	-988.0	-792.6	214.0	151.4								
	Barium oxide	-548.0	-520.3	72.1	47.3					-112.0			
	Barium sulfate	-1473.2	-1362.2	132.2	101.8								
	Barium sulfide	-460.0	-456.0	78.2	49.4								
	Beryllium	0.0		9.5	16.4					324.0	286.6	136.3	20.8
	Beryllium bromide	-353.5		108.0	69.4								
	Beryllium chloride	-490.4	-445.6	75.8	62.4								
	Beryllium fluoride	-1026.8	-979.4	53.4	51.8								
₂	Beryllium hydroxide	-902.5	-815.0	45.5	62.1								
	Beryllium iodide	-192.5		121.0	71.1								
	Beryllium oxide	-609.4	-580.1	13.8	25.6								
	Beryllium sulfate	-1205.2	-1093.8	77.9	85.7								
	Beryllium sulfide	-234.3		34.0	34.0								
	Bismuth	0.0		56.7	25.5					207.1	168.2	187.0	20.8
	Bismuth oxychloride	-366.9	-322.1	120.5									
	Bismuth trichloride	-379.1	-315.0	177.0	105.0					-265.7	-256.0	358.9	79.7
₃	Bismuth hydroxide	-711.3											
₃	Bismuth triiodide		-175.3										
	Dibismuth									219.7			36.9
	Bismuth oxide	-573.9	-493.7	151.5	113.5								
S₃	Bismuth sulfate	-2544.3											
	Bismuth sulfide	-143.1	-140.6	200.4	122.2								
	Berkelium	0.0											

Molecular formula	Name	Crystal $\Delta_fH°$ kJ/mol	$\Delta_fG°$ kJ/mol	$S°$ J/mol K	C_p J/mol K	Liquid $\Delta_fH°$ kJ/mol	$\Delta_fG°$ kJ/mol	$S°$ J/mol K	C_p J/mol K	Gas $\Delta_fH°$ kJ/mol	$\Delta_fG°$ kJ/mol	$S°$ J/mol K	C_p J/mol K
Br	Bromine (atomic)									111.9	82.4	175.0	20.8
BrCl	Bromine chloride									14.6	-1.0	240.1	35.0
BrCl₃Si	Bromotrichlorosilane									350.1			90.9
BrCs	Cesium bromide	-405.8	-391.4	113.1	52.9								
BrCu	Copper(I) bromide	-104.6	-100.8	96.1	54.7								
BrF	Bromine fluoride									-93.8	-109.2	229.0	33.0
BrF₃	Bromine trifluoride					-300.8	-240.5	178.2	124.6	-255.6	-229.4	292.5	66.6
BrF₅	Bromine pentafluoride					-458.6	-351.8	225.1		-428.9	-350.6	320.2	99.6
BrGe	Germanium monobromide									235.6			37.1
BrGeH₃	Bromogermane									274.8			56.4
BrH	Hydrogen bromide									-36.3	-53.4	198.7	29.1
BrHSi	Bromosilylene									-464.4			
BrH₃Si	Bromosilane									262.4			52.8
BrH₄N	Ammonium bromide	-270.8	-175.2	113.0	96.0								
BrI	Iodine bromide									40.8	3.7	258.8	36.4
BrIn	Indium(I) bromide	-175.3	-169.0	113.0						-56.9	-94.3	259.5	36.7
BrK	Potassium bromide	-393.8	-380.7	95.9	52.3								
BrKO₃	Potassium bromate	-360.2	-271.2	149.2	105.2								
BrKO₄	Potassium perbromate	-287.9	-174.4	170.1	120.2								
BrLi	Lithium bromide	-351.2	-342.0	74.3									
BrNO	Nitrosyl bromide									82.2	82.4	273.7	45.5
BrNa	Sodium bromide	-361.1	-349.0	86.8	51.4					-143.1	-177.1	241.2	36.3
BrNaO₃	Sodium bromate	-334.1	-242.6	128.9									
BrO	Bromine monoxide									125.8	109.6	233.0	34.2
BrO₂	Bromine dioxide									152.0	155.0	271.1	45.4
BrRb	Rubidium bromide	-394.6	-381.8	110.0	52.8								
BrSi	Bromosilyldyne									209.0			38.6
BrTl	Thallium(I) bromide	-173.2	-167.4	120.5						-37.7			
Br₂	Bromine					0.0		152.2	75.7	30.9	3.1	245.5	36.0
Br₂Ca	Calcium bromide	-682.8	-663.6	130.0									
Br₂Cd	Cadmium bromide	-316.2	-296.3	137.2	76.7								
Br₂Co	Cobalt(II) bromide	-220.9			79.5								
Br₂Cr	Chromium(II) bromide	-302.1											
Br₂Cu	Copper(II) bromide	-141.8											
Br₂Fe	Iron(II) bromide	-249.8	-238.1	140.6									
Br₂H₂Si	Dibromosilane									309.7			65.5
Br₂Hg	Mercury(II) bromide	-170.7	-153.1	172.0									
Br₂Hg₂	Mercury(I) bromide	-206.9	-181.1	218.0									
Br₂Mg	Magnesium bromide	-524.3	-503.8	117.2									
Br₂Mn	Manganese(II) bromide	-384.9											
Br₂Ni	Nickel(II) bromide	-212.1											
Br₂Pb	Lead(II) bromide	-278.7	-261.9	161.5	80.1								
Br₂Pt	Platinum(II) bromide	-82.0											
Br₂S₂	Sulfur bromide					-13.0							
Br₂Se	Selenium dibromide									-21.0			
Br₂Sn	Tin(II) bromide	-243.5											
Br₂Sr	Strontium bromide	-717.6	-697.1	135.1	75.3								
Br₂Ti	Titanium(II) bromide	-402.0											
Br₂Zn	Zinc bromide	-328.7	-312.1	138.5									
Br₃Ce	Cerium(III) bromide	-891.4											
Br₃ClSi	Tribromochlorosilane									377.1			95.3
Br₃Dy	Dysprosium(III) bromide	-836.2											
Br₃Fe	Iron(III) bromide	-268.2											
Br₃Ga	Gallium(III) bromide	-386.6	-359.8	180.0									
Br₃HSi	Tribromosilane					-355.6	-336.4	248.1		-317.6	-328.5	348.6	80.8
Br₃In	Indium(III) bromide	-428.9								-282.0			
Br₃OP	Phosphoric tribromide	-458.6								359.8			89.9
Br₃P	Phosphorus(III) bromide					-184.5	-175.7	240.2		-139.3	-162.8	348.1	76.0
Br₃Pt	Platinum(III) bromide	-120.9											
Br₃Re	Rhenium(III) bromide	-167.0											
Br₃Ru	Ruthenium(III) bromide	-138.0											
Br₃Sb	Antimony(III) bromide	-259.4	-239.3	207.1						-194.6	-223.9	372.9	80.2
Br₃Sc	Scandium bromide	-743.1											
Br₃Ti	Titanium(III) bromide	-548.5	-523.8	176.6	101.7								
Br₄Ge	Germanium(IV) bromide					-347.7	-331.4	280.7		-300.0	-318.0	396.2	101.8
Br₄Pa	Protactinium(IV) bromide	-824.0	-787.8	234.0									

Standard Thermodynamic Properties of Chemical Substances

	Name	Crystal				Liquid				Gas			
		$\Delta_f H°$ kJ/mol	$\Delta_f G°$ kJ/mol	$S°$ J/mol K	C_p J/mol K	$\Delta_f H°$ kJ/mol	$\Delta_f G°$ kJ/mol	$S°$ J/mol K	C_p J/mol K	$\Delta_f H°$ kJ/mol	$\Delta_f G°$ kJ/mol	$S°$ J/mol K	C_p J/mol K
	Platinum(IV) bromide	-156.5											
	Tetrabromosilane					-457.3	-443.9	277.8		-415.5	-431.8	377.9	97.1
	Tin(IV) bromide	-377.4	-350.2	264.4						-314.6	-331.4	411.9	103.4
	Tellurium tetrabromide	-190.4											
	Titanium(IV) bromide	-616.7	-589.5	243.5	131.5					-549.4	-568.2	398.4	100.8
	Vanadium(IV) bromide									-336.8			
	Zirconium(IV) bromide	-760.7											
	Phosphorus(V) bromide	-269.9											
	Tantalum(V) bromide	-598.3											
	Tungsten(VI) bromide	-348.5											
	Calcium	0.0		41.6	25.9					177.8	144.0	154.9	20.8
	Calcium chloride	-795.4	-748.8	108.4	72.9								
	Calcium fluoride	-1228.0	-1175.6	68.5	67.0								
	Calcium hydride	-181.5	-142.5	41.4	41.0								
	Calcium hydroxide	-985.2	-897.5	83.4	87.5								
	Calcium iodide	-533.5	-528.9	142.0									
	Calcium nitrate	-938.2	-742.8	193.2	149.4								
	Calcium oxide	-634.9	-603.3	38.1	42.0								
	Calcium sulfate	-1434.5	-1322.0	106.5	99.7								
	Calcium sulfide	-482.4	-477.4	56.5	47.4								
2	Calcium phosphate	-4120.8	-3884.7	236.0	227.8								
	Cadmium	0.0		51.8	26.0					111.8		167.7	20.8
	Cadmium chloride	-391.5	-343.9	115.3	74.7								
	Cadmium fluoride	-700.4	-647.7	77.4									
	Cadmium hydroxide	-560.7	-473.6	96.0									
	Cadmium iodide	-203.3	-201.4	161.1	80.0								
	Cadmium oxide	-258.4	-228.7	54.8	43.4								
	Cadmium sulfate	-933.3	-822.7	123.0	99.6								
	Cadmium sulfide	-161.9	-156.5	64.9									
	Cadmium telluride	-92.5	-92.0	100.0									
	Cerium (γ, fcc)	0.0		72.0	26.9					423.0	385.0	191.8	23.1
	Cerium(III) chloride	-1060.5	-984.8	151.0	87.4								
	Cerium(III) iodide	-669.3											
	Cerium(IV) oxide	-1088.7	-1024.6	62.3	61.6								
	Cerium(II) sulfide	-459.4	-451.5	78.2	50.0								
	Cerium(III) oxide	-1796.2	-1706.2	150.6	114.6								
	Californium	0.0											
	Chlorine (atomic)									121.3	105.3	165.2	21.8
	Cesium chloride	-443.0	-414.5	101.2	52.5								
	Cesium perchlorate	-443.1	-314.3	175.1	108.3								
	Copper(I) chloride	-137.2	-119.9	86.2	48.5								
	Chlorine fluoride									-50.3	-51.8	217.9	32.1
	Perchloryl fluoride									-23.8	48.2	279.0	64.9
	Chlorine trifluoride					-189.5				-163.2	-123.0	281.6	63.9
	Sulfur chloride pentafluoride					-1065.7							
	Germanium monochloride									155.2	124.2	247.0	36.9
3	Chlorogermane											263.7	54.7
	Hydrogen chloride									-92.3	-95.3	186.9	29.1
	Hypochlorous acid									-78.7	-66.1	236.7	37.2
	Perchloric acid					-40.6							
	Chlorosilane											250.7	51.0
	Ammonium chloride	-314.4	-202.9	94.6	84.1								
04	Ammonium perchlorate	-295.3	-88.8	186.2									
	Phosphonium chloride	-145.2											
	Iodine chloride					-23.9	-13.6	135.1		17.8	-5.5	247.6	35.6
	Indium(I) chloride	-186.2								-75.0			
	Potassium chloride	-436.5	-408.5	82.6	51.3					-214.6	-233.3	239.1	36.5
	Potassium chlorate	-397.7	-296.3	143.1	100.3								
	Potassium perchlorate	-432.8	-303.1	151.0	112.4								
	Lithium chloride	-408.6	-384.4	59.3	48.0								
	Lithium perchlorate	-381.0											
	Nitrosyl chloride									51.7	66.1	261.7	44.7
	Nitryl chloride									12.6	54.4	272.2	53.2
	Sodium chloride	-411.2	-384.1	72.1	50.5								
2	Sodium chlorite	-307.0											
3	Sodium chlorate	-365.8	-262.3	123.4									

Standard Thermodynamic Properties of Chemical Substances

Molecular formula	Name	Crystal $\Delta_f H°$ kJ/mol	$\Delta_f G°$ kJ/mol	$S°$ J/mol K	C_p J/mol K	Liquid $\Delta_f H°$ kJ/mol	$\Delta_f G°$ kJ/mol	$S°$ J/mol K	C_p J/mol K	Gas $\Delta_f H°$ kJ/mol	$\Delta_f G°$ kJ/mol	$S°$ J/mol K	C_p J/mol K
ClNaO$_4$	Sodium perchlorate	-383.3	-254.9	142.3									
ClO	Chlorine oxide									101.8	98.1	226.6	31.5
ClOV	Vanadyl chloride	-607.0	-556.0	75.0									
ClO$_2$	Chlorine dioxide									102.5	120.5	256.8	42.0
ClO$_2$	Chlorine superoxide (ClOO)									89.1	105.0	263.7	46.0
ClO$_4$Rb	Rubidium perchlorate	-437.2	-306.9	161.1									
ClRb	Rubidium chloride	-435.4	-407.8	95.9	52.4								
ClSi	Chlorosilylidyne									189.9			36.9
ClTl	Thallium(I) chloride	-204.1	-184.9	111.3	50.9					-67.8			
Cl$_2$	Chlorine									0.0		223.1	33.9
Cl$_2$Co	Cobalt(II) chloride	-312.5	-269.8	109.2	78.5								
Cl$_2$Cr	Chromium(II) chloride	-395.4	-356.0	115.3	71.2								
Cl$_2$CrO$_2$	Chromyl chloride					-579.5	-510.8	221.8		-538.1	-501.6	329.8	84.5
Cl$_2$Cu	Copper(II) chloride	-220.1	-175.7	108.1	71.9								
Cl$_2$Fe	Iron(II) chloride	-341.8	-302.3	118.0	76.7								
Cl$_2$H$_2$Si	Dichlorosilane									285.7			60.5
Cl$_2$Hg	Mercury(II) chloride	-224.3	-178.6	146.0									
Cl$_2$Hg$_2$	Mercury(I) chloride	-265.4	-210.7	191.6									
Cl$_2$Mg	Magnesium chloride	-641.3	-591.8	89.6	71.4								
Cl$_2$Mn	Manganese(II) chloride	-481.3	-440.5	118.2	72.9								
Cl$_2$Ni	Nickel(II) chloride	-305.3	-259.0	97.7	71.7								
Cl$_2$O	Chlorine monoxide									80.3	97.9	266.2	45.4
Cl$_2$OS	Thionyl chloride					-245.6			121.0	-212.5	-198.3	309.8	66.5
Cl$_2$O$_2$S	Sulfuryl chloride					-394.1			134.0	-364.0	-320.0	311.9	77.0
Cl$_2$O$_2$U	Uranyl chloride	-1243.9	-1146.4	150.5	107.9								
Cl$_2$Pb	Lead(II) chloride	-359.4	-314.1	136.0									
Cl$_2$Pt	Platinum(II) chloride	-123.4											
Cl$_2$S	Sulfur dichloride					-50.0							
Cl$_2$S$_2$	Sulfur chloride					-59.4							
Cl$_2$Sn	Tin(II) chloride	-325.1											
Cl$_2$Sr	Strontium chloride	-828.9	-781.1	114.9	75.6								
Cl$_2$Ti	Titanium(II) chloride	-513.8	-464.4	87.4	69.8								
Cl$_2$Zn	Zinc chloride	-415.1	-369.4	111.5	71.3					-266.1			
Cl$_2$Zr	Zirconium(II) chloride	-502.0											
Cl$_3$Cr	Chromium(III) chloride	-556.5	-486.1	123.0	91.8								
Cl$_3$Dy	Dysprosium(III) chloride	-1000.0											
Cl$_3$Er	Erbium chloride	-998.7			100.0								
Cl$_3$Eu	Europium(III) chloride	-936.0											
Cl$_3$Fe	Iron(III) chloride	-399.5	-334.0	142.3	96.7								
Cl$_3$Ga	Gallium(III) chloride	-524.7	-454.8	142.0									
Cl$_3$Gd	Gadolinium(III) chloride	-1008.0			88.0								
Cl$_3$HSi	Trichlorosilane					-539.3	-482.5	227.6		-513.0	-482.0	313.9	75.8
Cl$_3$Ho	Holmium chloride	-1005.4			88.0								
Cl$_3$In	Indium(III) chloride	-537.2								-374.0			
Cl$_3$Ir	Iridium(III) chloride	-245.6											
Cl$_3$La	Lanthanum chloride	-1072.2			108.8								
Cl$_3$Lu	Lutetium chloride	-945.6								-649.0			
Cl$_3$N	Nitrogen trichloride					230.0							
Cl$_3$Nd	Neodymium chloride	-1041.0			113.0								
Cl$_3$OP	Phosphoric trichloride					-597.1	-520.8	222.5	138.8	-558.5	-512.9	325.5	84.9
Cl$_3$OV	Vanadyl trichloride					-734.7	-668.5	244.3		-695.6	-659.3	344.3	89.9
Cl$_3$Os	Osmium(III) chloride	-190.4											
Cl$_3$P	Phosphorus(III) chloride					-319.7	-272.3	217.1		-287.0	-267.8	311.8	71.8
Cl$_3$Pr	Praseodymium chloride	-1056.9			100.0								
Cl$_3$Pt	Platinum(III) chloride	-182.0											
Cl$_3$Re	Rhenium(III) chloride	-264.0	-188.0	123.8	92.4								
Cl$_3$Rh	Rhodium(III) chloride	-299.2											
Cl$_3$Ru	Ruthenium(III) chloride	-205.0											
Cl$_3$Sb	Antimony(III) chloride	-382.2	-323.7	184.1	107.9								
Cl$_3$Sc	Scandium chloride	-925.1											
Cl$_3$Sm	Samarium(III) chloride	-1025.9											
Cl$_3$Tb	Terbium chloride	-997.0											
Cl$_3$Ti	Titanium(III) chloride	-720.9	-653.5	139.7	97.2								
Cl$_3$Tl	Thallium(III) chloride	-315.1											
Cl$_3$Tm	Thulium chloride	-986.6											
Cl$_3$U	Uranium(III) chloride	-866.5	-799.1	159.0	102.5								

Standard Thermodynamic Properties of Chemical Substances

Name	Crystal $\Delta_f H°$ kJ/mol	$\Delta_f G°$ kJ/mol	$S°$ J/mol K	C_p J/mol K	Liquid $\Delta_f H°$ kJ/mol	$\Delta_f G°$ kJ/mol	$S°$ J/mol K	C_p J/mol K	Gas $\Delta_f H°$ kJ/mol	$\Delta_f G°$ kJ/mol	$S°$ J/mol K	C_p J/mol K
Vanadium(III) chloride	-580.7	-511.2	131.0	93.2								
Yttrium chloride	-1000.0								-750.2			75.0
Ytterbium(III) chloride	-959.8											
Germanium(IV) chloride					-531.8	-462.7	245.6		-495.8	-457.3	347.7	96.1
Hafnium(IV) chloride	-990.4	-901.3	190.8	120.5					-884.5			
Protactinium(IV) chloride	-1043.0	-953.0	192.0									
Lead(IV) chloride					-329.3							
Platinum(IV) chloride	-231.8											
Tetrachlorosilane					-687.0	-619.8	239.7	145.3	-657.0	-617.0	330.7	90.3
Tin(IV) chloride					-511.3	-440.1	258.6	165.3	-471.5	-432.2	365.8	98.3
Tellurium tetrachloride	-326.4			138.5								
Thorium(IV) chloride	-1186.2	-1094.1	190.4	120.3					-964.4	-932.0	390.7	107.5
Titanium(IV) chloride					-804.2	-737.2	252.3	145.2	-763.2	-726.3	353.2	95.4
Uranium(IV) chloride	-1019.2	-930.0	197.1	122.0					-809.6	-786.6	419.0	
Vanadium(IV) chloride					-569.4	-503.7	255.0		-525.5	-492.0	362.4	96.2
Zirconium(IV) chloride	-980.5	-889.9	181.6	119.8								
Niobium(V) chloride	-797.5	-683.2	210.5	148.1					-703.7	-646.0	400.6	120.8
Phosphorus(V) chloride	-443.5								-374.9	-305.0	364.6	112.8
Protactinium(V) chloride	-1145.0	-1034.0	238.0									
Tantalum(V) chloride	-859.0											
Uranium(VI) chloride	-1092.0	-962.0	285.8	175.7					-1013.0	-928.0	431.0	
Tungsten(VI) chloride	-602.5								-513.8			
Curium	0.0											
Cobalt	0.0		30.0	24.8					424.7	380.3	179.5	23.0
Cobalt(II) fluoride	-692.0	-647.2	82.0	68.8								
Cobalt(II) hydroxide	-539.7	-454.3	79.0									
Cobalt(II) iodide	-88.7											
Cobalt(II) nitrate	-420.5											
Cobalt(II) oxide	-237.9	-214.2	53.0	55.2								
Cobalt(II) sulfate	-888.3	-782.3	118.0									
Cobalt(II) sulfide	-82.8											
Cobalt(III) sulfide	-147.3											
Cobalt(II,III) oxide	-891.0	-774.0	102.5	123.4								
Chromium	0.0		23.8	23.4					396.6	351.8	174.5	20.8
Chromium(II) fluoride	-778.0											
Chromium(III) fluoride	-1159.0	-1088.0	93.9	78.7								
Chromium(II) iodide	-156.9											
Chromium(III) iodide	-205.0											
Chromium(IV) oxide	-598.0											
Chromium(VI) oxide									-292.9		266.2	56.0
Lead(II) chromate	-930.9											
Chromium iron oxide	-1444.7	-1343.8	146.0	133.6								
Chromium(III) oxide	-1139.7	-1058.1	81.2	118.7								
Chromium(II,III) oxide	-1531.0											
Cesium	0.0		85.2	32.2					76.5	49.6	175.6	20.8
Cesium fluoride	-553.5	-525.5	92.8	51.1								
Cesium hydrogen fluoride	-923.8	-858.9	135.2	87.3								
Cesium hydride	-54.2											
Cesium hydroxide	-416.2	-371.8	104.2	69.9					-256.0	-256.5	254.8	49.7
Cesium hydrogen sulfate	-1158.1											
Cesium amide	-118.4											
Cesium iodide	-346.6	-340.6	123.1	52.8								
Cesium nitrate	-506.0	-406.5	155.2									
Cesium superoxide	-286.2											
Cesium oxide	-345.8	-308.1	146.9	76.0								
Cesium sulfite	-1134.7											
Cesium sulfate	-1443.0	-1323.6	211.9	134.9								
Cesium sulfide	-359.8											
Copper	0.0		33.2	24.4					337.4	297.7	166.4	20.8
Copper(II) fluoride	-542.7											
Copper(II) hydroxide	-449.8											
Copper(I) iodide	-67.8	-69.5	96.7	54.1								
Copper(II) nitrate	-302.9											
Copper(II) oxide	-157.3	-129.7	42.6	42.3								
Copper(II) sulfate	-771.4	-662.2	109.2									
Copper(II) tungstate	-1105.0											

Molecular formula	Name	Crystal				Liquid				Gas			
		$\Delta_f H°$ kJ/mol	$\Delta_f G°$ kJ/mol	$S°$ J/mol K	C_p J/mol K	$\Delta_f H°$ kJ/mol	$\Delta_f G°$ kJ/mol	$S°$ J/mol K	C_p J/mol K	$\Delta_f H°$ kJ/mol	$\Delta_f G°$ kJ/mol	$S°$ J/mol K	C_p J/mol K
CuS	Copper(II) sulfide	-53.1	-53.6	66.5	47.8								
CuSe	Copper(II) selenide	-39.5											
Cu_2	Dicopper									484.2	431.9	241.6	36.6
Cu_2O	Copper(I) oxide	-168.6	-146.0	93.1	63.6								
Cu_2S	Copper(I) sulfide	-79.5	-86.2	120.9	76.3								
Dy	Dysprosium	0.0		75.6	27.7					290.4	254.4	196.6	20.8
DyI_3	Dysprosium(III) iodide	-620.5											
Dy_2O_3	Dysprosium(III) oxide	-1863.1	-1771.5	149.8	116.3								
Er	Erbium	0.0		73.2	28.1					317.1	280.7	195.6	20.8
ErF_3	Erbium fluoride	-1711.0											
Er_2O_3	Erbium oxide	-1897.9	-1808.7	155.6	108.5								
Es	Einsteinium	0.0											
Eu	Europium	0.0		77.8	27.7					175.3	142.2	188.8	20.8
Eu_2O_3	Europium(III) oxide	-1651.4	-1556.8	146.0	122.2								
Eu_3O_4	Europium(II,III) oxide	-2272.0	-2142.0	205.0									
F	Fluorine (atomic)									79.4	62.3	158.8	22.7
FGa	Gallium monofluoride									-251.9			33.3
FGe	Germanium monofluoride									-33.4			34.7
$FGeH_3$	Fluorogermane											252.8	51.6
FH	Hydrogen fluoride					-299.8				-273.3	-275.4	173.8	
FH_3Si	Fluorosilane											238.4	47.4
FH_4N	Ammonium fluoride	-464.0	-348.7	72.0	65.3								
FI	Iodine fluoride									-95.7	-118.5	236.2	33.4
FIn	Indium(I) fluoride									-203.4			
FK	Potassium fluoride	-567.3	-537.8	66.6	49.0								
FLi	Lithium fluoride	-616.0	-587.7	35.7	41.6								
FNO	Nitrosyl fluoride									-66.5	-51.0	248.1	41.3
FNO_2	Nitryl fluoride											260.4	49.8
FNS	Thionitrosyl fluoride (NSF)											259.8	44.1
FNa	Sodium fluoride	-576.6	-546.3	51.1	46.9								
FO	Fluorine oxide									109.0	105.3	216.4	32.0
FO_2	Fluorine superoxide (FOO)									25.4	39.4	259.5	44.5
FRb	Rubidium fluoride	-557.7											
FSi	Fluorosilylidyne									7.1	-24.3	225.8	32.6
FTl	Thallium(I) fluoride	-324.7								-182.4			
F_2	Fluorine									0.0		202.8	31.3
F_2Fe	Iron(II) fluoride	-711.3	-668.6	87.0	68.1								
F_2HK	Potassium hydrogen fluoride	-927.7	-859.7	104.3	76.9								
F_2HN	Difluoramine											252.8	43.4
F_2HNa	Sodium hydrogen fluoride	-920.3	-852.2	90.9	75.0								
F_2HRb	Rubidium hydrogen fluoride	-922.6	-855.6	120.1	79.4								
F_2Mg	Magnesium fluoride	-1124.2	-1071.1	57.2	61.6								
F_2N	Difluoroamidogen									43.1	57.8	249.9	41.0
F_2N_2	cis-Difluorodiazine									69.5			
F_2N_2	trans-Difluorodiazine									82.0			
F_2Ni	Nickel(II) fluoride	-651.4	-604.1	73.6	64.1								
F_2O	Fluorine monoxide									24.5	41.8	247.5	43.3
F_2OS	Thionyl fluoride											278.7	56.8
F_2O_2	Fluorine dioxide									19.2	58.2	277.2	62.1
F_2O_2S	Sulfuryl fluoride											284.0	66.0
F_2O_2U	Uranyl fluoride	-1653.5	-1557.4	135.6	103.2								
F_2Pb	Lead(II) fluoride	-664.0	-617.1	110.5									
F_2Si	Difluorosilylene									-619.0	-628.0	252.7	43.9
F_2Sr	Strontium fluoride	-1216.3	-1164.8	82.1	70.0								
F_2Zn	Zinc fluoride	-764.4	-713.3	73.7	65.7								
F_3Ga	Gallium(III) fluoride	-1163.0	-1085.3	84.0									
F_3Gd	Gadolinium(III) fluoride									-1297.0			
F_3HSi	Trifluorosilane											271.9	60.5
F_3Ho	Holmium fluoride	-1707.0											
F_3N	Nitrogen trifluoride									-132.1	-90.6	260.8	53.4
F_3Nd	Neodymium fluoride	-1657.0											
F_3OP	Phosphoric trifluoride									-1254.3	-1205.8	285.4	68.8
F_3P	Phosphorus(III) fluoride									-958.4	-936.9	273.1	58.7
F_3Sb	Antimony(III) fluoride	-915.5											
F_3Sc	Scandium fluoride	-1629.2	-1555.6	92.0						-1247.0	-1234.0	300.5	67.8
F_3Sm	Samarium(III) fluoride	-1778.0											

Molecular formula	Name	Crystal				Liquid				Gas			
		$\Delta_f H°$ kJ/mol	$\Delta_f G°$ kJ/mol	$S°$ J/mol K	C_p J/mol K	$\Delta_f H°$ kJ/mol	$\Delta_f G°$ kJ/mol	$S°$ J/mol K	C_p J/mol K	$\Delta_f H°$ kJ/mol	$\Delta_f G°$ kJ/mol	$S°$ J/mol K	C_p J/mol K
	Thorium(III) fluoride									-1166.1	-1160.6	339.2	73.3
	Uranium(III) fluoride	-1502.1	-1433.4	123.4	95.1					-1058.5	-1051.9	331.9	74.3
	Yttrium fluoride	-1718.8	-1644.7	100.0						-1288.7	-1277.8	311.8	70.3
	Germanium(IV) fluoride									-1190.2	-1150.0	301.9	
	Hafnium fluoride	-1930.5	-1830.4	113.0						-1669.8			
	Tetrafluorohydrazine									-8.4	79.9	301.2	79.2
	Lead(IV) fluoride	-941.8											
	Sulfur tetrafluoride									-763.2	-722.0	299.6	77.6
	Tetrafluorosilane									-1615.0	-1572.8	282.8	73.6
	Thorium(IV) fluoride	-2097.8	-2003.4	142.0	110.7					-1759.0	-1724.0	341.7	93.0
	Uranium(IV) fluoride	-1914.2	-1823.3	151.7	116.0					-1598.7	-1572.7	368.0	91.2
	Vanadium(IV) fluoride	-1403.3											
	Xenon tetrafluoride	-261.5											
	Zirconium(IV) fluoride	-1911.3	-1809.9	104.6	103.7								
	Iodine pentafluoride					-864.8				-822.5	-751.7	327.7	99.2
	Niobium(V) fluoride	-1813.8	-1699.0	160.2	134.7					-1739.7	-1673.6	321.9	97.1
	Phosphorus(V) fluoride									-1594.4	-1520.7	300.8	84.8
	Tantalum(V) fluoride	-1903.6											
	Vanadium(V) fluoride					-1480.3	-1373.1	175.7		-1433.9	-1369.8	320.9	98.6
Si	Ammonium hexafluorosilicate	-2681.7	-2365.3	280.2	228.1								
	Iridium(VI) fluoride	-579.7	-461.6	247.7						-544.0	-460.0	357.8	121.1
	Potassium hexafluorosilicate	-2956.0	-2798.6	226.0									
	Molybdenum(VI) fluoride					-1585.5	-1473.0	259.7	169.8	-1557.7	-1472.2	350.5	120.6
i	Sodium hexafluorosilicate	-2909.6	-2754.2	207.1	187.1								
	Osmium(VI) fluoride			246.0								358.1	120.8
	Platinum(VI) fluoride			235.6								348.3	122.8
	Sulfur hexafluoride									-1220.5	-1116.5	291.5	97.0
	Selenium hexafluoride									-1117.0	-1017.0	313.9	110.5
	Hexafluorodisilane	-2427.0	-2299.7	219.1	129.5					-2383.3	-2307.3	391.0	129.9
	Tellurium hexafluoride									-1318.0			
	Uranium(VI) fluoride	-2197.0	-2068.5	227.6	166.8					-2147.4	-2063.7	377.9	129.6
	Tungsten(VI) fluoride					-1747.7	-1631.4	251.5		-1721.7	-1632.1	341.1	119.0
	Iron	0.0		27.3	25.1					416.3	370.7	180.5	25.7
	Iron(II) iodide	-113.0											
	Iron(III) iodide									71.0			
)₄	Iron(II) molybdate	-1075.0	-975.0	129.3	118.5								
	Iron(II) oxide	-272.0											
	Iron(II) sulfate	-928.4	-820.8	107.5	100.6								
	Iron(II) tungstate	-1155.0	-1054.0	131.8	114.6								
	Iron(II) sulfide	-100.0	-100.4	60.3	50.5								
	Iron disulfide	-178.2	-166.9	52.9	62.2								
	Iron(III) oxide	-824.2	-742.2	87.4	103.9								
i	Iron(II) orthosilicate	-1479.9	-1379.0	145.2	132.9								
	Iron(II,III) oxide	-1118.4	-1015.4	146.4	143.4								
	Fermium	0.0											
	Francium	0.0		95.4									
	Gallium	0.0	0.0	40.8	26.1	5.6				272.0	233.7	169.0	25.3
₃	Gallium(III) hydroxide	-964.4	-831.3	100.0									
	Gallium(III) iodide	-238.9		205.0	100.0								
	Gallium nitride	-110.5											
	Gallium monoxide									279.5	253.5	231.1	32.1
	Gallium phosphide	-88.0											
	Gallium antimonide	-41.8	-38.9	76.1	48.5								
	Digallium									438.5			
	Gallium suboxide	-356.0											
	Gallium(III) oxide	-1089.1	-998.3	85.0	92.1								
	Gadolinium	0.0		68.1	37.0					397.5	359.8	194.3	27.5
	Gadolinium(III) oxide	-1819.6			106.7								
	Germanium	0.0		31.1	23.3					372.0	331.2	167.9	30.7
	Iodogermane									283.2			57.5
	Germane									90.8	113.4	217.1	45.0
	Germanium(IV) iodide	-141.8	-144.3	271.1						-56.9	-106.3	428.9	104.1
	Germanium(II) oxide	-261.9	-237.2	50.0						-46.2	-73.2	224.3	30.9
	Germanium(IV) oxide	-580.0	-521.4	39.7	52.1								
	Germanium phosphide	-21.0	-17.0	63.0									
	Germanium(II) sulfide	-69.0	-71.5	71.0						92.0	42.0	234.0	33.7

Standard Thermodynamic Properties of Chemical Substances

Molecular formula	Name	Crystal $\Delta_f H°$ kJ/mol	$\Delta_f G°$ kJ/mol	$S°$ J/mol K	C_p J/mol K	Liquid $\Delta_f H°$ kJ/mol	$\Delta_f G°$ kJ/mol	$S°$ J/mol K	C_p J/mol K	Gas $\Delta_f H°$ kJ/mol	$\Delta_f G°$ kJ/mol	$S°$ J/mol K	C_p J/mol K
GeTe	Germanium(II) telluride	20.0											
Ge_2	Digermanium									473.1	416.3	252.8	35.6
Ge_2H_6	Digermane					137.3				162.3			
Ge_3H_8	Trigermane					193.7				226.8			
H	Hydrogen (atomic)									218.0	203.3	114.7	20.8
HI	Hydrogen iodide									26.5	1.7	206.6	29.2
HIO_3	Iodic acid	-230.1											
HK	Potassium hydride	-57.7											
HKO	Potassium hydroxide	-424.6	-379.4	81.2	68.9					-232.0	-229.7	238.3	49.2
HKO_4S	Potassium hydrogen sulfate	-1160.6	-1031.3	138.1									
HLi	Lithium hydride	-90.5	-68.3	20.0	27.9								
HLiO	Lithium hydroxide	-487.5	-441.5	42.8	49.6					-229.0	-234.2	214.4	46.0
HN	Imidogen									351.5	345.6	181.2	29.2
HNO_2	Nitrous acid									-79.5	-46.0	254.1	45.6
HNO_3	Nitric acid					-174.1	-80.7	155.6	109.9	-133.9	-73.5	266.9	54.1
HN_3	Hydrazoic acid					264.0	327.3	140.6		294.1	328.1	239.0	43.7
HNa	Sodium hydride	-56.3	-33.5	40.0	36.4								
HNaO	Sodium hydroxide	-425.8	-379.7	64.4	59.5					-191.0	-193.9	229.0	48.0
$HNaO_4S$	Sodium hydrogen sulfate	-1125.5	-992.8	113.0									
HNa_2O_4P	Sodium hydrogen phosphate	-1748.1	-1608.2	150.5	135.3								
HO	Hydroxyl									39.0	34.2	183.7	29.9
HORb	Rubidium hydroxide	-418.8	-373.9	94.0	69.0					-238.0	-239.1	248.5	49.5
HOTl	Thallium(I) hydroxide	-238.9	-195.8	88.0									
HO_2	Hydroperoxy									10.5	22.6	229.0	34.9
HO_3P	Metaphosphoric acid	-948.5											
HO_4RbS	Rubidium hydrogen sulfate	-1159.0											
HO_4Re	Perrhenic acid	-762.3	-656.4	158.2									
HRb	Rubidium hydride	-52.3											
HS	Mercapto									142.7	113.3	195.7	32.3
HSi	Silylidyne									361.0			
HTa_2	Tantalum hydride	-32.6	-69.0	79.1	90.8								
H_2	Hydrogen									0.0		130.7	28.8
H_2KN	Potassium amide	-128.9											
H_2KO_4P	Potassium dihydrogen phosphate	-1568.3	-1415.9	134.9	116.6								
H_2LiN	Lithium amide	-179.5											
H_2Mg	Magnesium hydride	-75.3	-35.9	31.1	35.4								
H_2MgO_2	Magnesium hydroxide	-924.5	-833.5	63.2	77.0								
H_2N	Amidogen									184.9	194.6	195.0	33.9
H_2NNa	Sodium amide	-123.8	-64.0	76.9	66.2								
H_2NRb	Rubidium amide	-113.0											
$H_2N_2O_2$	Nitramide	-89.5											
H_2NiO_2	Nickel(II) hydroxide	-529.7	-447.2	88.0									
H_2O	Water					-285.8	-237.1	70.0	75.3	-241.8	-228.6	188.8	33.6
H_2O_2	Hydrogen peroxide					-187.8	-120.4	109.6	89.1	-136.3	-105.6	232.7	43.1
H_2O_2Sn	Tin(II) hydroxide	-561.1	-491.6	155.0									
H_2O_2Sr	Strontium hydroxide	-959.0											
H_2O_2Zn	Zinc hydroxide	-641.9	-553.5	81.2									
H_2O_3Si	Metasilicic acid	-1188.7	-1092.4	134.0									
H_2O_4S	Sulfuric acid					-814.0	-690.0	156.9	138.9				
H_2O_4Se	Selenic acid	-530.1											
H_2S	Hydrogen sulfide									-20.6	-33.4	205.8	34.2
H_2S_2	Hydrogen disulfide					-18.1			84.1	15.5			51.5
H_2Se	Hydrogen selenide									29.7	15.9	219.0	34.7
H_2Sr	Strontium hydride	-180.3											
H_2Te	Hydrogen telluride									99.6			
H_2Th	Thorium hydride	-139.7	-100.0	50.7	36.7								
H_2Zr	Zirconium(II) hydride	-169.0	-128.8	35.0	31.0								
H_3ISi	Iodosilane									270.9			54.4
H_3N	Ammonia									-45.9	-16.4	192.8	35.1
H_3NO	Hydroxylamine	-114.2											
H_3O_2P	Phosphinic acid	-604.6				-595.4							
H_3O_3P	Phosphonic acid	-964.4											
H_3O_4P	Phosphoric acid	-1284.4	-1124.3	110.5	106.1	-1271.7	-1123.6	150.8	145.0				
H_3P	Phosphine									5.4	13.5	210.2	37.1
H_3Sb	Stibine									145.1	147.8	232.8	41.1
H_3U	Uranium(III) hydride	-127.2	-72.8	63.7	49.3								

Molecular Formula	Name	Crystal $\Delta_f H°$ kJ/mol	$\Delta_f G°$ kJ/mol	$S°$ J/mol K	C_p J/mol K	Liquid $\Delta_f H°$ kJ/mol	$\Delta_f G°$ kJ/mol	$S°$ J/mol K	C_p J/mol K	Gas $\Delta_f H°$ kJ/mol	$\Delta_f G°$ kJ/mol	$S°$ J/mol K	C_p J/mol K
	Ammonium iodide	-201.4	-112.5	117.0									
	Hydrazine					50.6	149.3	121.2	98.9	95.4	159.4	238.5	48.4
	Ammonium nitrite	-256.5											
	Ammonium nitrate	-365.6	-183.9	151.1	139.3								
	Ammonium azide	115.5	274.2	112.5									
	Orthosilicic acid	-1481.1	-1332.9	192.0									
	Diphosphoric acid	-2241.0				-2231.7							
	Diphosphine					-5.0				20.9			
	Silane									34.3	56.9	204.6	42.8
	Stannane									162.8	188.3	227.7	49.0
	Ammonium hydroxide					-361.2	-254.0	165.6	154.9				
	Ammonium hydrogen sulfite	-768.6											
	Ammonium hydrogen sulfate	-1027.0											
	Disilane									80.3	127.3	272.7	80.8
S	Ammonium sulfate	-1180.9	-901.7	220.1	187.5								
	Trisilane					92.5				120.9			
P	Ammonium hydrogen phosphate	-1566.9			188.0								
P	Ammonium phosphate	-1671.9											
	Helium									0.0		126.2	20.8
	Hafnium	0.0		43.6	25.7					619.2	576.5	186.9	20.8
	Hafnium oxide	-1144.7	-1088.2	59.3	60.3								
	Mercury					0.0		75.9	28.0	61.4	31.8	175.0	20.8
	Mercury(II) iodide	-105.4	-101.7	180.0									
	Mercury(II) oxide	-90.8	-58.5	70.3	44.1								
	Mercury(II) sulfate	-707.5											
	Mercury(II) sulfide (red)	-58.2	-50.6	82.4	48.4								
	Mercury(II) telluride	-42.0											
	Dimercury									108.8	68.2	288.1	37.4
	Mercury(I) iodide	-121.3	-111.0	233.5									
S	Mercury(I) sulfate	-743.1	-625.8	200.7	132.0								
	Holmium	0.0		75.3	27.2					300.8	264.8	195.6	20.8
	Holmium oxide	-1880.7	-1791.1	158.2	115.0								
	Iodine (atomic)									106.8	70.2	180.8	20.8
	Indium(I) iodide	-116.3	-120.5	130.0						7.5	-37.7	267.3	36.8
	Potassium iodide	-327.9	-324.9	106.3	52.9								
	Potassium iodate	-501.4	-418.4	151.5	106.5								
	Potassium periodate	-467.2	-361.4	175.7									
	Lithium iodide	-270.4	-270.3	86.8	51.0								
	Sodium iodide	-287.8	-286.1	98.5	52.1								
	Sodium iodate	-481.8			92.0								
	Sodium periodate	-429.3	-323.0	163.0									
	Iodine monoxide									126.0	102.5	239.6	32.9
	Rubidium iodide	-333.8	-328.9	118.4	53.2								
	Thallium(I) iodide	-123.8	-125.4	127.6						7.1			
	Iodine (rhombic)	0.0		116.1	54.4					62.4	19.3	260.7	36.9
	Magnesium iodide	-364.0	-358.2	129.7									
	Nickel(II) iodide	-78.2											
	Lead(II) iodide	-175.5	-173.6	174.9	77.4								
	Tin(II) iodide	-143.5											
	Strontium iodide	-558.1			81.6								
	Zinc iodide	-208.0	-209.0	161.1									
	Indium(III) iodide	-238.0								-120.5			
	Lanthanum iodide	-668.9											
	Lutetium iodide	-548.0											
	Phosphorus(III) iodide	-45.6										374.4	78.4
	Ruthenium(III) iodide	-65.7											
	Antimony(III) iodide	-100.4											
	Platinum(IV) iodide	-72.8											
	Tetraiodosilane	-189.5											
	Tin(IV) iodide				84.9							446.1	105.4
	Titanium(IV) iodide	-375.7	-371.5	249.4	125.7					-277.8			
	Vanadium(IV) iodide									-122.6			
	Zirconium(IV) iodide	-481.6											
	Indium	0.0		57.8	26.7					243.3	208.7	173.8	20.8
	Indium monoxide									387.0	364.4	236.5	32.6
	Indium phosphide	-88.7	-77.0	59.8	45.4								

Molecular formula	Name	Crystal				Liquid				Gas			
		$\Delta_f H°$ kJ/mol	$\Delta_f G°$ kJ/mol	$S°$ J/mol K	C_p J/mol K	$\Delta_f H°$ kJ/mol	$\Delta_f G°$ kJ/mol	$S°$ J/mol K	C_p J/mol K	$\Delta_f H°$ kJ/mol	$\Delta_f G°$ kJ/mol	$S°$ J/mol K	C_p J/mol K
InS	Indium(II) sulfide	-138.1	-131.8	67.0						238.0			
InSb	Indium antimonide	-30.5	-25.5	86.2	49.5					344.3			
In_2	Diindium									380.9			
In_2O_3	Indium(III) oxide	-925.8	-830.7	104.2	92.0								
In_2S_3	Indium(III) sulfide	-427.0	-412.5	163.6	118.0								
In_2Te_5	Indium(IV) telluride	-175.3											
Ir	Iridium	0.0		35.5	25.1					665.3	617.9	193.6	20.8
IrO_2	Iridium(IV) oxide	-274.1			57.3								
IrS_2	Iridium(IV) sulfide	-138.0											
Ir_2S_3	Iridium(III) sulfide	-234.0											
K	Potassium	0.0		64.7	29.6					89.0	60.5	160.3	20.8
$KMnO_4$	Potassium permanganate	-837.2	-737.6	171.7	117.6								
KNO_2	Potassium nitrite	-369.8	-306.6	152.1	107.4								
KNO_3	Potassium nitrate	-494.6	-394.9	133.1	96.4								
KNa	Potassium sodium					6.3							
KO_2	Potassium superoxide	-284.9	-239.4	116.7	77.5								
K_2	Dipotassium									123.7	87.5	249.7	37.9
K_2O	Potassium oxide	-361.5											
K_2O_2	Potassium peroxide	-494.1	-425.1	102.1									
K_2O_4S	Potassium sulfate	-1437.8	-1321.4	175.6	131.5								
K_2S	Potassium sulfide	-380.7	-364.0	105.0									
K_3O_4P	Potassium phosphate	-1950.2											
Kr	Krypton									0.0		164.1	20.8
La	Lanthanum	0.0		56.9	27.1					431.0	393.6	182.4	22.8
LaS	Lanthanum monosulfide	-456.0	-451.5	73.2	59.0								
La_2O_3	Lanthanum oxide	-1793.7	-1705.8	127.3	108.8								
Li	Lithium	0.0		29.1	24.8					159.3	126.6	138.8	20.8
$LiNO_2$	Lithium nitrite	-372.4	-302.0	96.0									
$LiNO_3$	Lithium nitrate	-483.1	-381.1	90.0									
Li_2	Dilithium									215.9	174.4	197.0	36.1
Li_2O	Lithium oxide	-597.9	-561.2	37.6	54.1								
Li_2O_2	Lithium peroxide	-634.3											
Li_2O_3Si	Lithium metasilicate	-1648.1	-1557.2	79.8	99.1								
Li_2O_4S	Lithium sulfate	-1436.5	-1321.7	115.1	117.6								
Li_2S	Lithium sulfide	-441.4											
Li_3O_4P	Lithium phosphate	-2095.8											
Lr	Lawrencium	0.0											
Lu	Lutetium	0.0		51.0	26.9					427.6	387.8	184.8	20.9
Lu_2O_3	Lutetium oxide	-1878.2	-1789.0	110.0	101.8								
Md	Mendelevium	0.0											
Mg	Magnesium	0.0		32.7	24.9					147.1	112.5	148.6	20.8
MgN_2O_6	Magnesium nitrate	-790.7	-589.4	164.0	141.9								
MgO	Magnesium oxide	-601.6	-569.3	27.0	37.2								
MgO_4S	Magnesium sulfate	-1284.9	-1170.6	91.6	96.5								
MgO_4Se	Magnesium selenate	-968.5											
MgS	Magnesium sulfide	-346.0	-341.8	50.3	45.6								
Mg_2	Dimagnesium									287.7			
Mg_2O_4Si	Magnesium orthosilicate	-2174.0	-2055.1	95.1	118.5								
Mn	Manganese	0.0		32.0	26.3					280.7	238.5	173.7	20.8
MnN_2O_6	Manganese(II) nitrate	-576.3											
$MnNaO_4$	Sodium permanganate	-1156.0											
MnO	Manganese(II) oxide	-385.2	-362.9	59.7	45.4								
MnO_2	Manganese(IV) oxide	-520.0	-465.1	53.1	54.1								
MnO_3Si	Manganese(II) metasilicate	-1320.9	-1240.5	89.1	86.4								
MnS	Manganese(II) sulfide (α form)	-214.2	-218.4	78.2	50.0								
MnSe	Manganese(II) selenide	-106.7	-111.7	90.8	51.0								
Mn_2O_3	Manganese(III) oxide	-959.0	-881.1	110.5	107.7								
Mn_2O_4Si	Manganese(II) orthosilicate	-1730.5	-1632.1	163.2	129.9								
Mn_3O_4	Manganese(II,III) oxide	-1387.8	-1283.2	155.6	139.7								
Mo	Molybdenum	0.0		28.7	24.1					658.1	612.5	182.0	20.8
$MoNa_2O_4$	Sodium molybdate	-1468.1	-1354.3	159.7	141.7								
MoO_2	Molybdenum(IV) oxide	-588.9	-533.0	46.3	56.0								
MoO_3	Molybdenum(VI) oxide	-745.1	-668.0	77.7	75.0								
MoO_4Pb	Lead(II) molybdate	-1051.9	-951.4	166.1	119.7								
MoS_2	Molybdenum(IV) sulfide	-235.1	-225.9	62.6	63.6								
Mo_3Si	Molybdenum silicide	-125.2	-125.7	106.3	93.1								

Standard Thermodynamic Properties of Chemical Substances

Name	Crystal $\Delta_f H°$ kJ/mol	$\Delta_f G°$ kJ/mol	$S°$ J/mol K	C_p J/mol K	Liquid $\Delta_f H°$ kJ/mol	$\Delta_f G°$ kJ/mol	$S°$ J/mol K	C_p J/mol K	Gas $\Delta_f H°$ kJ/mol	$\Delta_f G°$ kJ/mol	$S°$ J/mol K	C_p J/mol K
Nitrogen (atomic)									472.7	455.5	153.3	20.8
Sodium nitrite	-358.7	-284.6	103.8									
Sodium nitrate	-467.9	-367.0	116.5	92.9								
Nitric oxide									91.3	87.6	210.8	29.9
Nitrogen dioxide									33.2	51.3	240.1	37.2
Rubidium nitrite	-367.4	-306.2	172.0									
Rubidium nitrate	-495.1	-395.8	147.3	102.1								
Thallium(I) nitrate	-243.9	-152.4	160.7	99.5								
Phosphorus nitride	-63.0								171.5	149.4	211.1	29.7
Nitrogen									0.0		191.6	29.1
Nitrous oxide									81.6	103.7	220.0	38.6
Nitrogen trioxide					50.3				86.6	142.4	314.7	72.7
Nitrogen tetroxide					-19.5	97.5	209.2	142.7	11.1	99.8	304.4	79.2
Strontium nitrite	-762.3											
Nitrogen pentoxide	-43.1	113.9	178.2	143.1					13.3	117.1	355.7	95.3
Lead(II) nitrate	-451.9											
Radium nitrate	-992.0	-796.1	222.0									
Strontium nitrate	-978.2	-780.0	194.6	149.9								
Zinc nitrate	-483.7											
Sodium azide	21.7	93.8	96.9	76.6								
Silicon nitride	-743.5	-642.6	101.3									
Sodium	0.0		51.3	28.2					107.5	77.0	153.7	20.8
Sodium superoxide	-260.2	-218.4	115.9	72.1								
Disodium									142.1	103.9	230.2	37.6
Sodium oxide	-414.2	-375.5	75.1	69.1								
Sodium peroxide	-510.9	-447.7	95.0	89.2								
Sodium sulfite	-1100.8	-1012.5	145.9	120.3								
Sodium metasilicate	-1554.9	-1462.8	113.9									
Sodium sulfate	-1387.1	-1270.2	149.6	128.2								
Sodium sulfide	-364.8	-349.8	83.7									
Niobium	0.0		36.4	24.6					725.9	681.1	186.3	30.2
Niobium(II) oxide	-405.8	-378.6	48.1	41.3								
Niobium(IV) oxide	-796.2	-740.5	54.5	57.5								
Niobium(V) oxide	-1899.5	-1766.0	137.2	132.1								
Neodymium	0.0		71.5	27.5					327.6	292.4	189.4	22.1
Neodymium oxide	-1807.9	-1720.8	158.6	111.3								
Neon									0.0		146.3	20.8
Nickel	0.0		29.9	26.1					429.7	384.5	182.2	23.4
Nickel(II) sulfate	-872.9	-759.7	92.0	138.0								
Nickel(II) sulfide	-82.0	-79.5	53.0	47.1								
Nickel(III) oxide	-489.5											
Nobelium	0.0											
Oxygen (atomic)									249.2	231.7	161.1	21.9
Phosphorus monoxide									-28.5	-51.9	222.8	31.8
Lead(II) oxide (massicot)	-217.3	-187.9	68.7	45.8								
Lead(II) oxide (litharge)	-219.0	-188.9	66.5	45.8								
Palladium(II) oxide	-85.4			31.4					348.9	325.9	218.0	
Radium oxide	-523.0											
Rubidium oxide	-339.0											
Rhodium monoxide									385.0			
Sulfur monoxide									6.3	-19.9	222.0	30.2
Selenium monoxide									53.4	26.8	234.0	31.3
Silicon monoxide									-99.6	-126.4	211.6	29.9
Tin(II) oxide	-280.7	-251.9	57.2	44.3					15.1	-8.4	232.1	31.6
Strontium oxide	-592.0	-561.9	54.4	45.0					1.5			
Titanium(II) oxide	-519.7	-495.0	50.0	40.0								
Thallium(I) oxide	-178.7	-147.3	126.0									
Uranium(II) oxide									21.0			
Vanadium(II) oxide	-431.8	-404.2	38.9	45.4								
Zinc oxide	-350.5	-320.5	43.7	40.3								
Oxygen									0.0		205.2	29.4
Phosphorus dioxide									-279.9	-281.6	252.1	39.5
Lead(IV) oxide	-277.4	-217.3	68.6	64.6								
Rubidium superoxide	-278.7											
Rubidium peroxide	-472.0											
Ruthenium(IV) oxide	-305.0											

Standard Thermodynamic Properties of Chemical Substances

5-17

Molecular formula	Name	Crystal $\Delta_f H°$ kJ/mol	$\Delta_f G°$ kJ/mol	$S°$ J/mol K	C_p J/mol K	Liquid $\Delta_f H°$ kJ/mol	$\Delta_f G°$ kJ/mol	$S°$ J/mol K	C_p J/mol K	Gas $\Delta_f H°$ kJ/mol	$\Delta_f G°$ kJ/mol	$S°$ J/mol K	C_p J/mol K
O₂S	Sulfur dioxide					-320.5				-296.8	-300.1	248.2	39.9
O₂Se	Selenium dioxide	-225.4											
O₂Si	Silicon dioxide (α-quartz)	-910.7	-856.3	41.5	44.4					-322.0			
O₂Sn	Tin(IV) oxide	-577.6	-515.8	49.0	52.6								
O₂Te	Tellurium dioxide	-322.6	-270.3	79.5									
O₂Th	Thorium(IV) oxide	-1226.4	-1169.2	65.2	61.8								
O₂Ti	Titanium(IV) oxide	-944.0	-888.8	50.6	55.0								
O₂U	Uranium(IV) oxide	-1085.0	-1031.8	77.0	63.6					-465.7	-471.5	274.6	51.4
O₂W	Tungsten(IV) oxide	-589.7	-533.9	50.5	56.1								
O₂Zr	Zirconium(IV) oxide	-1100.6	-1042.8	50.4	56.2								
O₃	Ozone									142.7	163.2	238.9	39.2
O₃PbS	Lead(II) sulfite	-669.9											
O₃PbSi	Lead(II) metasilicate	-1145.7	-1062.1	109.6	90.0								
O₃Pr₂	Praseodymium oxide	-1809.6			117.4								
O₃Rh₂	Rhodium(III) oxide	-343.0			103.8								
O₃S	Sulfur trioxide	-454.5	-374.2	70.7		-441.0	-373.8	113.8		-395.7	-371.1	256.8	50.7
O₃Sc₂	Scandium oxide	-1908.8	-1819.4	77.0	94.2								
O₃SiSr	Strontium metasilicate	-1633.9	-1549.7	96.7	88.5								
O₃Sm₂	Samarium(III) oxide	-1823.0	-1734.6	151.0	114.5								
O₃Tb₂	Terbium oxide	-1865.2			115.9								
O₃Ti₂	Titanium(III) oxide	-1520.9	-1434.2	78.8	97.4								
O₃Tm₂	Thulium oxide	-1888.7	-1794.5	139.7	116.7								
O₃U	Uranium(VI) oxide	-1223.8	-1145.7	96.1	81.7								
O₃V₂	Vanadium(III) oxide	-1218.8	-1139.3	98.3	103.2								
O₃W	Tungsten(VI) oxide	-842.9	-764.0	75.9	73.8								
O₃Y₂	Yttrium oxide	-1905.3	-1816.6	99.1	102.5								
O₃Yb₂	Ytterbium(III) oxide	-1814.6	-1726.7	133.1	115.4								
O₄Os	Osmium(VIII) oxide	-394.1	-304.9	143.9						-337.2	-292.8	293.8	74.1
O₄PbS	Lead(II) sulfate	-920.0	-813.0	148.5	103.2								
O₄PbSe	Lead(II) selenate	-609.2	-504.9	167.8									
O₄Pb₂Si	Lead(II) orthosilicate	-1363.1	-1252.6	186.6	137.2								
O₄Pb₃	Lead(II,II,IV) oxide	-718.4	-601.2	211.3	146.9								
O₄RaS	Radium sulfate	-1471.1	-1365.6	138.0									
O₄Rb₂S	Rubidium sulfate	-1435.6	-1316.9	197.4	134.1								
O₄Ru	Ruthenium(VIII) oxide	-239.3	-152.2	146.4									
O₄SSr	Strontium sulfate	-1453.1	-1340.9	117.0									
O₄STl₂	Thallium(I) sulfate	-931.8	-830.4	230.5									
O₄SZn	Zinc sulfate	-982.8	-871.5	110.5	99.2								
O₄SiSr₂	Strontium orthosilicate	-2304.5	-2191.1	153.1	134.3								
O₄SiZn₂	Zinc orthosilicate	-1636.7	-1523.2	131.4	123.3								
O₄SiZr	Zirconium(IV) orthosilicate	-2033.4	-1919.1	84.1	98.7								
O₄TiZr	Zirconium titanate	-2024.1	-1915.8	116.7	114.0								
O₅Sb₂	Antimony(V) oxide	-971.9	-829.2	125.1									
O₅Ta₂	Tantalum(V) oxide	-2046.0	-1911.2	143.1	135.1								
O₅Ti₃	Titanium(III,IV) oxide	-2459.4	-2317.4	129.3	154.8								
O₅V₂	Vanadium(V) oxide	-1550.6	-1419.5	131.0	127.7								
O₅V₃	Vanadium(III,IV) oxide	-1933.0	-1803.0	163.0									
O₇Re₂	Rhenium(VII) oxide	-1240.1	-1066.0	207.1	166.1					-1100.0	-994.0	452.0	
O₇U₃	Uranium(IV,VI) oxide	-3427.1	-3242.9	250.5	215.5								
O₈S₂Zr	Zirconium(IV) sulfate	-2217.1			172.0								
O₈U₃	Uranium(V,VI) oxide	-3574.8	-3369.5	282.6	238.4								
O₉U₄	Uranium(IV,V) oxide	-4510.4	-4275.1	334.1	293.3								
Os	Osmium	0.0		32.6	24.7					791.0	745.0	192.6	20.8
P	Phosphorus (white)	0.0		41.1	23.8					316.5	280.1	163.2	20.8
P	Phosphorus (red)	-17.6		22.8	21.2								
P	Phosphorus (black)	-39.3											
P₂	Diphosphorus									144.0	103.5	218.1	32.1
P₄	Tetraphosphorus									58.9	24.4	280.0	67.2
Pa	Protactinium	0.0		51.9						607.0	563.0	198.1	22.9
Pb	Lead	0.0		64.8	26.4					195.2	162.2	175.4	20.8
PbS	Lead(II) sulfide	-100.4	-98.7	91.2	49.5								
PbSe	Lead(II) selenide	-102.9	-101.7	102.5	50.2								
PbTe	Lead(II) telluride	-70.7	-69.5	110.0	50.5								
Pd	Palladium	0.0		37.6	26.0					378.2	339.7	167.1	20.8
PdS	Palladium(II) sulfide	-75.0	-67.0	46.0									
Pm	Promethium	0.0										187.1	24.3

Standard Thermodynamic Properties of Chemical Substances

cular la	Name	Crystal $\Delta_f H°$ kJ/mol	$\Delta_f G°$ kJ/mol	$S°$ J/mol K	C_p J/mol K	Liquid $\Delta_f H°$ kJ/mol	$\Delta_f G°$ kJ/mol	$S°$ J/mol K	C_p J/mol K	Gas $\Delta_f H°$ kJ/mol	$\Delta_f G°$ kJ/mol	$S°$ J/mol K	C_p J/mol K
	Polonium	0.0											
	Praseodymium	0.0		73.2	27.2					355.6	320.9	189.8	21.4
	Platinum	0.0		41.6	25.9					565.3	520.5	192.4	25.5
	Platinum(II) sulfide	-81.6	-76.1	55.1	43.4								
	Platinum(IV) sulfide	-108.8	-99.6	74.7	65.9								
	Plutonium	0.0											
	Radium	0.0		71.0						159.0	130.0	176.5	20.8
	Rubidium	0.0		76.8	31.1					80.9	53.1	170.1	20.8
	Rhenium	0.0		36.9	25.5					769.9	724.6	188.9	20.8
	Rhodium	0.0		31.5	25.0					556.9	510.8	185.8	21.0
	Radon									0.0		176.2	20.8
	Ruthenium	0.0		28.5	24.1					642.7	595.8	186.5	21.5
	Sulfur (rhombic)	0.0		32.1	22.6					277.2	236.7	167.8	23.7
	Sulfur (monoclinic)	0.3											
	Silicon monosulfide									112.5	60.9	223.7	32.3
	Tin(II) sulfide	-100.0	-98.3	77.0	49.3								
	Strontium sulfide	-472.4	-467.8	68.2	48.7								
	Thallium(I) sulfide	-97.1	-93.7	151.0									
	Zinc sulfide (wurtzite)	-192.6											
	Zinc sulfide (sphalerite)	-206.0	-201.3	57.7	46.0								
	Disulfur									128.6	79.7	228.2	32.5
	Antimony	0.0		45.7	25.2					262.3	222.1	180.3	20.8
	Diantimony									235.6	187.0	254.9	36.4
	Scandium	0.0		34.6	25.5					377.8	336.0	174.8	22.1
	Selenium (gray)	0.0		42.4	25.4					227.1	187.0	176.7	20.8
	Selenium (α form)	6.7								227.1			
	Selenium (vitreous)	5.0								227.1			
	Strontium selenide	-385.8											
	Thallium(I) selenide	-59.0	-59.0	172.0									
	Zinc selenide	-163.0	-163.0	84.0									
	Diselenium									146.0	96.2	252.0	35.4
	Silicon	0.0		18.8	20.0					450.0	405.5	168.0	22.3
	Disilicon									594.0	536.0	229.9	34.4
	Samarium	0.0		69.6	29.5					206.7	172.8	183.0	30.4
	Tin (white)	0.0		51.2	27.0					301.2	266.2	168.5	21.3
	Tin (gray)	-2.1	0.1	44.1	25.8								
	Strontium	0.0		55.0	26.8					164.4	130.9	164.6	20.8
	Tantalum	0.0		41.5	25.4					782.0	739.3	185.2	20.9
	Terbium	0.0		73.2	28.9					388.7	349.7	203.6	24.6
	Technetium	0.0								678.0		181.1	20.8
	Tellurium	0.0		49.7	25.7					196.7	157.1	182.7	20.8
	Ditellurium									168.2	118.0	268.1	36.7
	Thorium	0.0		51.8	27.3					602.0	560.7	190.2	20.8
	Titanium	0.0		30.7	25.0					473.0	428.4	180.3	24.4
	Thallium	0.0		64.2	26.3					182.2	147.4	181.0	20.8
	Thulium	0.0		74.0	27.0					232.2	197.5	190.1	20.8
	Uranium	0.0		50.2	27.7					533.0	488.4	199.8	23.7
	Vanadium	0.0		28.9	24.9					514.2	754.4	182.3	26.0
	Tungsten	0.0		32.6	24.3					849.4	807.1	174.0	21.3
	Xenon									0.0		169.7	20.8
	Yttrium	0.0		44.4	26.5					421.3	381.1	179.5	25.9
	Ytterbium	0.0		59.9	26.7					152.3	118.4	173.1	20.8
	Zinc	0.0		41.6	25.4					130.4	94.8	161.0	20.8
	Zirconium	0.0		39.0	25.4					608.8	566.5	181.4	26.7

tances containing carbon:

la	Name	Crystal $\Delta_f H°$ kJ/mol	$\Delta_f G°$ kJ/mol	$S°$ J/mol K	C_p J/mol K	Liquid $\Delta_f H°$ kJ/mol	$\Delta_f G°$ kJ/mol	$S°$ J/mol K	C_p J/mol K	Gas $\Delta_f H°$ kJ/mol	$\Delta_f G°$ kJ/mol	$S°$ J/mol K	C_p J/mol K
	Carbon (graphite)	0.0		5.7	8.5					716.7	671.3	158.1	20.8
	Carbon (diamond)	1.9	2.9	2.4	6.1								
	Silver(I) cyanide	146.0	156.9	107.2	66.7								
3	Silver(I) carbonate	-505.8	-436.8	167.4	112.3								
	Barium carbonate	-1213.0	-1134.4	112.1	86.0								
	Beryllium carbonate	-1025.0		52.0	65.0								
2	Bromochlorodifluoromethane											318.5	74.6
F	Bromodichlorofluoromethane											330.6	80.0
3	Bromotrichloromethane										-41.1		85.3
	Bromotrifluoromethane										-648.3		69.3

Standard Thermodynamic Properties of Chemical Substances

Molecular formula	Name	Crystal ΔfH° kJ/mol	ΔfG° kJ/mol	S° J/mol K	Cp J/mol K	Liquid ΔfH° kJ/mol	ΔfG° kJ/mol	S° J/mol K	Cp J/mol K	Gas ΔfH° kJ/mol	ΔfG° kJ/mol	S° J/mol K	Cp J/mol K
CBrN	Cyanogen bromide	140.5								186.2	165.3	248.3	46.9
CBrN$_3$O$_6$	Bromotrinitromethane					32.5				80.3			
CBr$_2$ClF	Dibromochlorofluoromethane											342.8	82.4
CBr$_2$Cl$_2$	Dibromodichloromethane											347.8	87.1
CBr$_2$F$_2$	Dibromodifluoromethane											325.3	77.0
CBr$_2$O	Carbonyl bromide					-127.2				-96.2	-110.9	309.1	61.8
CBr$_3$Cl	Tribromochloromethane											357.8	89.4
CBr$_3$F	Tribromofluoromethane											345.9	84.4
CBr$_4$	Tetrabromomethane	29.4	47.7	212.5	144.3					83.9	67.0	358.1	91.2
CCaO$_3$	Calcium carbonate (calcite)	-1207.6	-1129.1	91.7	83.5								
CCaO$_3$	Calcium carbonate (aragonite)	-1207.8	-1128.2	88.0	82.3								
CCdO$_3$	Cadmium carbonate	-750.6	-669.4	92.5									
CClFO	Carbonyl chloride fluoride											276.7	52.4
CClF$_3$	Chlorotrifluoromethane									-706.3			66.9
CClN	Cyanogen chloride					112.1				138.0	131.0	236.2	45.0
CClN$_3$O$_6$	Chlorotrinitromethane					-27.1				18.4			
CCl$_2$F$_2$	Dichlorodifluoromethane									-477.4	-439.4	300.8	72.3
CCl$_2$O	Carbonyl chloride									-219.1	-204.9	283.5	57.7
CCl$_3$	Trichloromethyl									59.0			
CCl$_3$F	Trichlorofluoromethane					-301.3	-236.8	225.4	121.6	-268.3			78.1
CCl$_4$	Tetrachloromethane					-128.2			130.7	-95.7			83.3
CCoO$_3$	Cobalt(II) carbonate	-713.0											
CCs$_2$O$_3$	Cesium carbonate	-1139.7	-1054.3	204.5	123.9								
CCuN	Copper(I) cyanide	96.2	111.3	84.5									
CFN	Cyanogen fluoride											224.7	41.8
CF$_2$O	Carbonyl fluoride									-639.8			46.8
CF$_3$	Trifluoromethyl									-477.0	-464.0	264.5	49.6
CF$_3$I	Trifluoroiodomethane									-587.8		307.4	70.9
CF$_4$	Tetrafluoromethane									-933.6		261.6	61.1
CFeO$_3$	Iron(II) carbonate	-740.6	-666.7	92.9	82.1								
CFe$_3$	Iron carbide	25.1	20.1	104.6	105.9								
CH	Methylidyne									595.8			
CHBrClF	Bromochlorofluoromethane											304.3	63.2
CHBrCl$_2$	Bromodichloromethane											316.4	67.4
CHBrF$_2$	Bromodifluoromethane									-424.9		295.1	58.7
CHBr$_2$Cl	Chlorodibromomethane											327.7	69.2
CHBr$_2$F	Dibromofluoromethane											316.8	65.1
CHBr$_3$	Tribromomethane					-22.3	-5.0	220.9	130.7	23.8	8.0	330.9	71.2
CHClF$_2$	Chlorodifluoromethane									-482.6		280.9	55.9
CHCl$_2$F	Dichlorofluoromethane											293.1	60.9
CHCl$_3$	Trichloromethane					-134.1	-73.7	201.7	114.2	-102.7	6.0	295.7	65.7
CHCsO$_3$	Cesium hydrogen carbonate	-966.1											
CHFO	Formyl fluoride											246.6	39.9
CHF$_3$	Trifluoromethane									-695.4		259.7	51.0
CHI$_3$	Triiodomethane	-181.1								251.0		356.2	75.0
CHKO$_2$	Potassium formate	-679.7											
CHKO$_3$	Potassium hydrogen carbonate	-963.2	-863.5	115.5									
CHN	Hydrogen cyanide					108.9	125.0	112.8	70.6	135.1	124.7	201.8	35.9
CHNO	Isocyanic acid (HNCO)											238.0	44.9
CHNS	Isothiocyanic acid									127.6	113.0	247.8	46.9
CHN$_3$O$_6$	Trinitromethane					-32.8				-13.4		435.6	134.1
CHNaO$_2$	Sodium formate	-666.5	-599.9	103.8	82.7								
CHNaO$_3$	Sodium hydrogen carbonate	-950.8	-851.0	101.7	87.6								
CHO	Oxomethyl (HCO)									43.1	28.0	224.7	34.6
CH$_2$	Methylene									390.4	372.9	194.9	33.8
CH$_2$BrCl	Bromochloromethane											287.6	52.7
CH$_2$BrF	Bromofluoromethane											276.3	49.2
CH$_2$Br$_2$	Dibromomethane											293.2	54.7
CH$_2$ClF	Chlorofluoromethane											264.4	47.0
CH$_2$Cl$_2$	Dichloromethane					-124.2		177.8	101.2	-95.4		270.2	51.0
CH$_2$F$_2$	Difluoromethane									-452.3		246.7	42.9
CH$_2$I$_2$	Diiodomethane					68.5	90.4	174.1	134.0	119.5	95.8	309.7	57.7
CH$_2$N$_2$	Diazomethane											242.9	52.5
CH$_2$N$_2$	Cyanamide	58.8											
CH$_2$N$_2$O$_4$	Dinitromethane					-104.9				-61.5		358.1	86.4
CH$_2$O	Formaldehyde									-108.6	-102.5	218.8	35.4

Molecular Formula	Name	Crystal				Liquid				Gas			
		$\Delta_f H°$ kJ/mol	$\Delta_f G°$ kJ/mol	$S°$ J/mol K	C_p J/mol K	$\Delta_f H°$ kJ/mol	$\Delta_f G°$ kJ/mol	$S°$ J/mol K	C_p J/mol K	$\Delta_f H°$ kJ/mol	$\Delta_f G°$ kJ/mol	$S°$ J/mol K	C_p J/mol K
x	Paraformaldehyde	-177.6											
	Formic acid					-425.0	-361.4	129.0	99.0	-378.7			
	Trithiocarbonic acid					24.0							
	Methyl									145.7	147.9	194.2	38.7
	Borane carbonyl									-111.2	-92.9	249.4	59.5
	Bromomethane					-59.8				-35.4	-26.3	246.4	42.4
	Chloromethane									-81.9		234.6	40.8
Si	Methyltrichlorosilane							262.8	163.1	-528.9		351.1	102.4
	Fluoromethane											222.9	37.5
	Iodomethane					-13.6		163.2	126.0	14.4		254.1	44.1
	Formamide					-254.0				-193.9			
2	Nitromethane					-112.6	-14.4	171.8	106.6	-80.8		282.9	55.5
2	Methyl nitrite									-66.1			
3	Methyl nitrate					-156.3	-43.4	217.1	157.3	-122.0		305.8	76.6
	Methane									-74.6	-50.5	186.3	35.7
	Ammonium cyanide	0.4			134.0								
O	Urea	-333.1								-245.8			
S	Thiourea	-89.1								22.9			
O₂	Nitroguanidine	-92.4											
	Methanol					-239.2	-166.6	126.8	81.1	-201.0	-162.3	239.9	44.1
	Methanethiol					-46.7	-7.7	169.2	90.5	-22.9	-9.3	255.2	50.3
	Methylamine					-47.3	35.7	150.2	102.1	-22.5	32.7	242.9	50.1
O₃	Ammonium hydrogen carbonate	-849.4	-665.9	120.9									
	Guanidine	-56.0											
S	Hydrazinecarbothioamide	24.7											
O₂	3-Amino-1-nitroguanidine	22.1											
N	Methylamine hydrochloride	-298.1											
	Methylhydrazine					54.2	180.0	165.9	134.9	94.7	187.0	278.8	71.1
	Methylsilane											256.5	65.9
3	Mercury(I) carbonate	-553.5	-468.1	180.0									
	Cyanogen iodide	166.2	185.0	96.2						225.5	196.6	256.8	48.3
	Tetraiodomethane	-392.9								474.0		391.9	95.9
	Potassium cyanide	-113.0	-101.9	128.5	66.3								
	Potassium thiocyanate	-200.2	-178.3	124.3	88.5								
	Potassium carbonate	-1151.0	-1063.5	155.5	114.4								
8	Lithium carbonate	-1215.9	-1132.1	90.4	99.1								
3	Magnesium carbonate	-1095.8	-1012.1	65.7	75.5								
3	Manganese(II) carbonate	-894.1	-816.7	85.8	81.5								
	Cyanide									437.6	407.5	202.6	29.2
	Sodium cyanide	-87.5	-76.4	115.6	70.4								
O	Sodium cyanate	-405.4	-358.1	96.7	86.6								
	Tetranitromethane					38.4				82.4		503.7	176.1
O₃	Sodium carbonate	-1130.7	-1044.4	135.0	112.3								
	Carbon monoxide									-110.5	-137.2	197.7	29.1
	Carbon oxysulfide									-142.0	-169.2	231.6	41.5
	Carbon dioxide									-393.5	-394.4	213.8	37.1
	Lead(II) carbonate	-699.1	-625.5	131.0	87.4								
O₂	Rubidium carbonate	-1136.0	-1051.0	181.3	117.6								
	Strontium carbonate	-1220.1	-1140.1	97.1	81.4								
2	Thallium(I) carbonate	-700.0	-614.6	155.2									
	Zinc carbonate	-812.8	-731.5	82.4	79.7								
	Carbon monosulfide									280.3	228.8	210.6	29.8
	Carbon disulfide					89.0	64.6	151.3	76.4	116.7	67.1	237.8	45.4
	Carbon diselenide					164.8							
	Silicon carbide (cubic)	-65.3	-62.8	16.6	26.9								
	Silicon carbide (hexagonal)	-62.8	-60.2	16.5	26.7								
	Dicarbon									831.9	775.9	199.4	43.2
5	Bromopentafluoroethane									-1064.4			
ClF₃	1,2-Dibromo-1-chloro-1,2,2-trifluoroethane					-691.7				-656.6			
F₄	1,2-Dibromotetrafluoroethane					-817.7				-789.1			
	Tetrabromoethene											387.1	102.7
	Hexabromoethane											441.9	139.3
	Calcium carbide	-59.8	-64.9	70.0	62.7								
N₂	Calcium cyanide	-184.5											
O₄	Calcium oxalate	-1360.6											

Molecular formula	Name	Crystal				Liquid				Gas			
		$\Delta_f H°$ kJ/mol	$\Delta_f G°$ kJ/mol	$S°$ J/mol K	C_p J/mol K	$\Delta_f H°$ kJ/mol	$\Delta_f G°$ kJ/mol	$S°$ J/mol K	C_p J/mol K	$\Delta_f H°$ kJ/mol	$\Delta_f G°$ kJ/mol	$S°$ J/mol K	C_p J/mol K
C_2ClF_3	Chlorotrifluoroethene					-522.7				-505.5	-523.8	322.1	83.9
C_2ClF_5	Chloropentafluoroethane									-1118.8			184.2
$C_2Cl_2F_4$	1,2-Dichloro-1,1,2,2-tetrafluoroethane					-960.2			111.7	-937.0			
$C_2Cl_2O_2$	Oxalyl chloride					-367.6				-335.8			
$C_2Cl_3F_3$	1,1,2-Trichloro-1,2,2-trifluoroethane					-745.0			170.1	-716.8			
C_2Cl_3N	Trichloroacetonitrile											336.6	96.1
C_2Cl_4	Tetrachloroethene					-50.6	3.0	266.9	143.4	-10.9			
$C_2Cl_4F_2$	1,1,1,2-Tetrachloro-2,2-difluoroethane									-489.9	-407.0	382.9	123.4
$C_2Cl_4F_2$	1,1,2,2-Tetrachloro-1,2-difluoroethane								173.6				
C_2Cl_4O	Trichloroacetyl chloride					-280.8				-239.8			
C_2Cl_6	Hexachloroethane	-202.8		237.3	198.2					-143.6			
C_2F_3N	Trifluoroacetonitrile									-497.9		298.1	77.9
C_2F_4	Tetrafluoroethene	-820.5								-658.9		300.1	80.5
C_2F_6	Hexafluoroethane									-1344.2		332.3	106.7
C_2HBr	Bromoacetylene											253.7	55.7
$C_2HBrClF_3$	1-Bromo-2-chloro-1,1,2-trifluoroethane					-675.3				-644.8			
$C_2HBrClF_3$	2-Bromo-2-chloro-1,1,1-trifluoroethane					-720.0				-690.4			
C_2HCl	Chloroacetylene											242.0	54.3
C_2HClF_2	1-Chloro-2,2-difluoroethene									-315.5	-289.1	303.0	72.1
C_2HCl_2F	1,1-Dichloro-2-fluoroethene											313.9	76.5
$C_2HCl_2F_3$	2,2-Dichloro-1,1,1-trifluoroethane											352.8	102.5
C_2HCl_3	Trichloroethene					-43.6		228.4	124.4	-9.0		324.8	80.3
C_2HCl_3O	Trichloroacetaldehyde					-234.5			151.0	-196.6			
C_2HCl_3O	Dichloroacetyl chloride					-280.4				-241.0			
$C_2HCl_3O_2$	Trichloroacetic acid	-503.3											
C_2HCl_5	Pentachloroethane					-187.6			173.8	-142.0			
C_2HF	Fluoroacetylene											231.7	52.4
C_2HF_3	Trifluoroethene									-490.5			
$C_2HF_3O_2$	Trifluoroacetic acid					-1069.9				-1031.4			
C_2HF_5	Pentafluoroethane									-1100.4			
C_2H_2	Acetylene									227.4	209.9	200.9	44.0
$C_2H_2BrF_3$	2-Bromo-1,1,1-trifluoroethane									-694.5			
$C_2H_2Br_2$	cis-1,2-Dibromoethene											311.3	68.8
$C_2H_2Br_2$	trans-1,2-Dibromoethene											313.5	70.3
$C_2H_2Br_2Cl_2$	1,2-Dibromo-1,2-dichloroethane									-36.9			
$C_2H_2Br_4$	1,1,2,2-Tetrabromoethane								165.7				
$C_2H_2ClF_3$	2-Chloro-1,1,1-trifluoroethane											326.5	89.1
$C_2H_2Cl_2$	1,1-Dichloroethene					-23.9	24.1	201.5	111.3	2.8	25.4	289.0	67.1
$C_2H_2Cl_2$	cis-1,2-Dichloroethene					-26.4		198.4	116.4	4.6		289.6	65.1
$C_2H_2Cl_2$	trans-1,2-Dichloroethene					-24.3	27.3	195.9	116.8	5.0	28.6	290.0	66.7
$C_2H_2Cl_2O$	Chloroacetyl chloride					-283.7				-244.8			
$C_2H_2Cl_2O_2$	Dichloroacetic acid					-496.3							
$C_2H_2Cl_3NO$	2,2,2-Trichloroacetamide	-358.0											
$C_2H_2Cl_4$	1,1,1,2-Tetrachloroethane											356.0	102.7
$C_2H_2Cl_4$	1,1,2,2-Tetrachloroethane					-195.0		246.9	162.3	-149.2		362.8	100.8
$C_2H_2F_2$	1,1-Difluoroethene									-335.0		266.2	60.1
$C_2H_2F_2$	cis-1,2-Difluoroethene											268.3	58.2
$C_2H_2F_3I$	1,1,1-Trifluoro-2-iodoethane									-644.5			
$C_2H_2I_2$	cis-1,2-Diiodoethene									-207.4			
C_2H_2O	Ketene					-67.9				-47.5	-48.3	247.6	51.8
$C_2H_2O_2$	Glyoxal									-212.0	-189.7	272.5	60.6
$C_2H_2O_4$	Oxalic acid	-829.9		109.8	91.0					-731.8	-662.7	320.6	86.2
$C_2H_2O_4Sr$	Strontium formate	-1393.3											
C_2H_2S	Thiirene									300.0	275.8	255.3	54.7
C_2H_3Br	Bromoethene									79.2	81.8	275.8	55.5
C_2H_3BrO	Acetyl bromide					-223.5				-190.4			
$C_2H_3BrO_2$	Bromoacetic acid									-383.5	-338.3	337.0	80.5
C_2H_3Cl	Chloroethene	-94.1			59.4	14.6				37.2	53.6	264.0	53.7
$C_2H_3ClF_2$	1-Chloro-1,1-difluoroethane											307.2	82.5
C_2H_3ClO	Acetyl chloride					-272.9	-208.0	200.8	117.0	-242.8	-205.8	295.1	67.8
$C_2H_3ClO_2$	Chloroacetic acid	-509.7								-427.6	-368.5	325.9	78.8

Molecular formula	Name	Crystal				Liquid				Gas			
		$\Delta_f H°$ kJ/mol	$\Delta_f G°$ kJ/mol	$S°$ J/mol K	C_p J/mol K	$\Delta_f H°$ kJ/mol	$\Delta_f G°$ kJ/mol	$S°$ J/mol K	C_p J/mol K	$\Delta_f H°$ kJ/mol	$\Delta_f G°$ kJ/mol	$S°$ J/mol K	C_p J/mol K
F	1,1-Dichloro-1-fluoroethane											320.2	88.7
3	1,1,1-Trichloroethane					-177.4		227.4	144.3	-144.4		323.1	93.3
3	1,1,2-Trichloroethane					-190.8		232.6	150.9	-151.3		337.2	89.0
	Fluoroethene									-138.8			
	Acetyl fluoride					-467.2				-442.1			
	1,1,1-Trifluoroethane									-744.6		279.9	78.2
	1,1,2-Trifluoroethane									-730.7			
O	2,2,2-Trifluoroethanol					-932.4				-888.4			
	Iodoethene											285.0	57.9
	Acetyl iodide					-163.5				-126.4			
O₂	Potassium acetate	-723.0											
	Acetonitrile					40.6	86.5	149.6	91.5	74.0	91.9	243.4	52.2
	Isocyanomethane					130.8	159.5	159.0		163.5	165.7	246.9	52.9
O	Methyl isocyanate					-92.0							
O₂	Nitroethene									33.3		300.5	73.7
O₃	Oxamic acid	-661.2								-552.3			
S	Methyl isothiocyanate	79.4											
aO₂	Sodium acetate	-708.8	-607.2	123.0	79.9								
	Ethylene									52.4	68.4	219.3	42.9
Cl	1-Bromo-2-chloroethane								130.1				
2	1,1-Dibromoethane					-66.2						327.7	80.8
2	1,2-Dibromoethane					-79.2		223.3	136.0	-37.5			
F	1-Chloro-1-fluoroethane									-313.4			
2	1,1-Dichloroethane					-158.4	-73.8	211.8	126.3	-127.7	-70.8	305.1	76.2
2	1,2-Dichloroethane					-166.8			128.4	-126.4		308.4	78.7
	1,1-Difluoroethane									-497.0		282.5	67.8
	1,2-Diiodoethane	9.3								75.0			
₂O₂	Oxamide	-504.4								-387.1			
₂O₂	Ethanedial dioxime	-90.5											
₂O₄	1,1-Dinitroethane					-148.2							
₂O₄	1,2-Dinitroethane					-165.2							
₂S₂	Ethanedithioamide	-20.8								83.0			
4	1H-1,2,4-Triazol-3-amine	76.8											
	Acetaldehyde					-192.2	-127.6	160.2	89.0	-166.2	-133.0	263.8	55.3
	Oxirane					-78.0	-11.8	153.9	88.0	-52.6	-13.0	242.5	47.9
S	Thioacetic acid					-216.9				-175.1			
2	Acetic acid					-484.3	-389.9	159.8	123.3	-432.2	-374.2	283.5	63.4
2	Methyl formate					-386.1			119.1	-357.4		285.3	64.4
3	Peroxyacetic acid												82.4
3	Glycolic acid									-583.0	-504.9	318.6	87.1
	Thiirane					51.6				82.0	96.8	255.2	53.3
i	Ethynylsilane											269.4	72.6
r	Bromoethane					-90.5	-25.8	198.7	100.8	-61.9	-23.9	286.7	64.5
l	Chloroethane					-136.8	-59.3	190.8	104.3	-112.1	-60.4	276.0	62.8
lO	2-Chloroethanol					-295.4							
	Fluoroethane											264.5	58.6
	Iodoethane					-40.0	14.7	211.7	115.1	-8.1	19.2	306.0	66.9
	Ethyleneimine					91.9				126.5			
O	Acetamide	-317.0		115.0	91.3					-238.3			
O	N-Methylformamide								123.8				
O₂	Nitroethane					-143.9			134.4	-103.8		320.5	79.0
O₂	Glycine	-528.5								-392.1			
O₃	2-Nitroethanol					-350.7							
O₃	Ethyl nitrate					-190.4				-154.1			
S	Thioacetamide	-71.7								11.4			
	Ethane									-84.0	-32.0	229.2	52.5
Cd	Dimethyl cadmium					63.6	139.0	201.9	132.0	101.6	146.9	303.0	
Hg	Dimethyl mercury					59.8	140.3	209.0		94.4	146.1	306.0	83.3
₂O	N-Methylurea	-332.8											
₄O₂	1,2-Hydrazinedicarboxamide	-498.7											
₄O₂	Oxalyl dihydrazide	-295.2											
	Ethanol					-277.6	-174.8	160.7	112.3	-234.8	-167.9	281.6	65.6
	Dimethyl ether					-203.3				-184.1	-112.6	266.4	64.4
S	Dimethyl sulfoxide					-204.2	-99.9	188.3	153.0	-151.3			
₂	Ethylene glycol					-460.0		163.2	148.6	-392.2		303.8	82.7
₂S	Dimethyl sulfone	-450.1	-302.4	142.0						-373.1	-272.7	310.6	100.0

Molecular formula	Name	Crystal				Liquid				Gas			
		$\Delta_f H°$ kJ/mol	$\Delta_f G°$ kJ/mol	$S°$ J/mol K	C_p J/mol K	$\Delta_f H°$ kJ/mol	$\Delta_f G°$ kJ/mol	$S°$ J/mol K	C_p J/mol K	$\Delta_f H°$ kJ/mol	$\Delta_f G°$ kJ/mol	$S°$ J/mol K	C_p J/mol K
$C_2H_6O_3S$	Dimethyl sulfite					-523.6				-483.4			
$C_2H_6O_4S$	Dimethyl sulfate					-735.5				-687.0			
C_2H_6S	Ethanethiol					-73.6	-5.5	207.0	117.9	-46.1	-4.8	296.2	72.7
C_2H_6S	Dimethyl sulfide					-65.3		196.4	118.1	-37.4		286.0	74.1
$C_2H_6S_2$	1,2-Ethanedithiol					-54.3				-9.7			
$C_2H_6S_2$	Dimethyl disulfide					-62.6		235.4	146.1	-24.7			
C_2H_6Zn	Dimethyl zinc					23.4		201.6	129.2	53.0			
C_2H_7N	Ethylamine					-74.1			130.0	-47.5	36.3	283.8	71.5
C_2H_7N	Dimethylamine					-43.9	70.0	182.3	137.7	-18.8	68.5	273.1	70.7
C_2H_7NO	Ethanolamine								195.5				
C_2H_8ClN	Dimethylamine hydrochloride	-289.3											
$C_2H_8N_2$	1,2-Ethanediamine					-63.0			172.6	-18.0			
$C_2H_8N_2$	1,1-Dimethylhydrazine					48.9	206.4	198.0	164.1	84.1			
$C_2H_8N_2$	1,2-Dimethylhydrazine					52.7				92.2			
$C_2H_8N_2O_4$	Ammonium oxalate	-1123.0			226.0								
C_2HgO_4	Mercury(II) oxalate	-678.2											
C_2I_2	Diiodoacetylene											313.1	70.3
C_2I_4	Tetraiodoethene	305.0											
$C_2K_2O_4$	Potassium oxalate	-1346.0											
C_2MgO_4	Magnesium oxalate	-1269.0											
C_2N_2	Cyanogen					285.9				306.7		241.9	56.8
$C_2N_4O_6$	Trinitroacetonitrile					183.7							
$C_2Na_2O_4$	Sodium oxalate									-1318.0			
C_2O_4Pb	Lead(II) oxalate	-851.4	-750.1	146.0	105.4								
C_3F_8	Perfluoropropane									-1783.2			
$C_3H_2N_2$	Malononitrile	186.4								265.5			
$C_3H_2O_2$	2-Propynoic acid					-193.2							
$C_3H_2O_3$	1,3-Dioxol-2-one					-459.9				-418.6			
$C_3H_3Cl_3$	1,2,3-Trichloropropene					-101.8							
$C_3H_3F_3$	3,3,3-Trifluoropropene									-614.2			
C_3H_3N	Acrylonitrile					147.1				180.6			
C_3H_3NO	Oxazole					-48.0				-15.5			
C_3H_3NO	Isoxazole					42.1				78.6			
C_3H_4	Allene									190.5			
C_3H_4	Propyne									184.9			
C_3H_4	Cyclopropene									277.1			
$C_3H_4Cl_2$	2,3-Dichloropropene					-73.3							
$C_3H_4Cl_4$	1,1,1,3-Tetrachloropropane					-208.7							
$C_3H_4Cl_4$	1,2,2,3-Tetrachloropropane					-251.8							
$C_3H_4F_4O$	2,2,3,3-Tetrafluoro-1-propanol					-1114.9				-1061.3			
$C_3H_4N_2$	1H-Pyrazole	105.4			81.0					179.4			
$C_3H_4N_2$	Imidazole	49.8								132.9			
C_3H_4O	Acrolein												71.3
$C_3H_4O_2$	1,2-Propanedione					-309.1				-271.0			
$C_3H_4O_2$	Acrylic acid					-383.8			145.7				
$C_3H_4O_2$	2-Oxetanone					-329.9		175.3	122.1	-282.9			
$C_3H_4O_3$	Ethylene carbonate					-682.8			133.9	-508.4			
C_3H_5Br	cis-1-Bromopropene					7.9				40.8			
C_3H_5Br	3-Bromopropene					12.2				45.2			
C_3H_5BrO	Bromoacetone									-181.0			
C_3H_5Cl	2-Chloropropene									-21.0			
C_3H_5Cl	3-Chloropropene								125.1				
C_3H_5ClO	Epichlorohydrin					-148.4			131.6	-107.8			
$C_3H_5ClO_2$	2-Chloropropanoic acid					-522.5				-475.8			
$C_3H_5ClO_2$	3-Chloropropanoic acid	-549.3											
$C_3H_5ClO_2$	Ethyl chloroformate					-505.3				-462.9			
$C_3H_5ClO_2$	Methyl chloroacetate					-487.0				-444.0			
$C_3H_5Cl_3$	1,2,3-Trichloropropane					-230.6			183.6	-182.9			
C_3H_5I	3-Iodopropene					53.7				91.5			
C_3H_5IO	Iodoacetone									-130.5			
$C_3H_5IO_2$	3-Iodopropanoic acid	-460.0											
C_3H_5N	Propanenitrile					15.5			119.3	51.7			
C_3H_5N	2-Propyn-1-amine					205.7							
C_3H_5N	Ethyl isocyanide					108.6				141.7			
C_3H_5NO	Acrylamide	-212.1			110.6	-224.0				-130.2			
$C_3H_5NO_3$	Nitroacetone					-278.6							

Standard Thermodynamic Properties of Chemical Substances

Molecular Formula	Name	Crystal $\Delta_f H°$ kJ/mol	$\Delta_f G°$ kJ/mol	$S°$ J/mol K	C_p J/mol K	Liquid $\Delta_f H°$ kJ/mol	$\Delta_f G°$ kJ/mol	$S°$ J/mol K	C_p J/mol K	Gas $\Delta_f H°$ kJ/mol	$\Delta_f G°$ kJ/mol	$S°$ J/mol K	C_p J/mol K
4	Methyl nitroacetate					-464.0							
9	Trinitroglycerol					-370.9				-279.1		545.9	234.2
	Propene					4.0				20.0			
	Cyclopropane					35.2				53.3	104.5	237.5	55.6
	1,2-Dibromopropane					-113.6				-71.6			
	1,2-Dichloropropane, (±)					-198.8			149.1	-162.8			
	1,3-Dichloropropane					-199.9				-159.2			
	2,2-Dichloropropane					-205.8				-173.2			
O	2,3-Dichloro-1-propanol					-381.5				-316.3			
O	1,3-Dichloro-2-propanol					-385.3				-318.4			
	1,2-Diiodopropane									35.6			
	1,3-Diiodopropane					-9.0							
O_2	Propanediamide	-546.1											
O_2	N-(Aminocarbonyl)acetamide	-544.2								-441.2			
O_4	1,1-Dinitropropane					-163.2				-100.7			
O_4	1,3-Dinitropropane					-207.1							
O_4	2,2-Dinitropropane					-181.2							
O_6	Hexahydro-1,3,5-trinitro-1,3,5-triazine									192.0		482.4	230.2
	Allyl alcohol					-171.8			138.9	-124.5			
	Propanal					-215.6				-185.6		304.5	80.7
	Acetone					-248.4		199.8	126.3	-217.1	-152.7	295.3	74.5
	Methyloxirane					-123.0		196.5	120.4	-94.7		286.9	72.6
	Oxetane					-110.8				-80.5			
	Propanoic acid					-510.7		191.0	152.8	-455.7			
	Ethyl formate								149.3				
	Methyl acetate					-445.9			141.9	-413.3		324.4	86.0
	1,3-Dioxolane					-333.5			118.0	-298.0			
S	Thiolactic acid					-468.4							
	1,3,5-Trioxane	-522.5		133.0	111.4					-465.9			
	Thietane					24.7		184.9		60.6	107.1	285.0	68.3
	Methylthiirane					11.3				45.8			
	1,2-Dithiolane									0.0	47.7	313.5	86.5
	1,3-Dithiolane									10.0	54.7	323.3	84.7
	1,3,5-Trithiane									80.0	130.4	336.4	111.3
r	1-Bromopropane					-121.9				-87.0			
r	2-Bromopropane					-130.5				-99.4			
l	1-Chloropropane					-160.5				-131.9			
l	2-Chloropropane					-172.3				-144.9			
lO_2	3-Chloro-1,2-propanediol					-525.3							
lO_2	2-Chloro-1,3-propanediol					-517.5							
	1-Fluoropropane									-285.9			
	2-Fluoropropane									-293.5			
	1-Iodopropane					-66.0				-30.0			
	2-Iodopropane					-74.8				-40.3			
	Allylamine					-10.0							
	Cyclopropylamine					45.8		187.7	147.1	77.0			
O	N,N-Dimethylformamide					-239.3			150.6	-192.4			
O	Propanamide	-338.2								-259.0			
O_2	1-Nitropropane					-167.2				-124.3		350.0	104.1
O_2	2-Nitropropane					-180.3			170.3	-138.9			
O_2	Ethyl carbamate	-517.1		156.4		-497.3				-446.3			
O_2	DL-Alanine	-563.6											
O_2	D-Alanine	-561.2											
O_2	L-Alanine	-604.0								-465.9			
O_2	β-Alanine	-558.0								-424.0			
O_2	Sarcosine	-513.3								-367.3			
O_2S	L-Cysteine	-534.1											
O_3	Propyl nitrate					-214.5				-174.1		362.6	123.2
O_3	Isopropyl nitrate					-229.7				-191.0			
O_3	DL-Serine	-739.0											
O_3	L-Serine	-732.7											
	Propane					-120.9				-103.8	-23.4	270.3	73.6
N_2O	N-Ethylurea	-357.8											
N_2O	N,N-Dimethylurea	-319.1											
N_2O	N,N-Dimethylurea	-312.1											

Standard Thermodynamic Properties of Chemical Substances

Molecular formula	Name	Crystal $\Delta_f H°$ kJ/mol	$\Delta_f G°$ kJ/mol	$S°$ J/mol K	C_p J/mol K	Liquid $\Delta_f H°$ kJ/mol	$\Delta_f G°$ kJ/mol	$S°$ J/mol K	C_p J/mol K	Gas $\Delta_f H°$ kJ/mol	$\Delta_f G°$ kJ/mol	$S°$ J/mol K	C_p J/mol K
$C_3H_8N_2O_3$	Oxymethurea	-717.0											
C_3H_8O	1-Propanol					-302.6		193.6	143.9	-255.1		322.6	85.6
C_3H_8O	2-Propanol					-318.1		181.1	156.5	-272.6		309.2	89.3
C_3H_8O	Ethyl methyl ether									-216.4		309.2	93.3
$C_3H_8O_2$	1,2-Propylene glycol					-501.0			190.8	-429.8			
$C_3H_8O_2$	1,3-Propylene glycol					-480.8				-408.0			
$C_3H_8O_2$	Ethylene glycol monomethyl ether								171.1				
$C_3H_8O_2$	Dimethoxymethane					-377.8		244.0	162.0	-348.5			
$C_3H_8O_3$	Glycerol					-669.6		206.3	218.9	-577.9			
C_3H_8S	1-Propanethiol					-99.9		242.5	144.6	-67.8			
C_3H_8S	2-Propanethiol					-105.9		233.5	145.3	-76.2			
C_3H_8S	Ethyl methyl sulfide					-91.6		239.1	144.6	-59.6			
$C_3H_8S_2$	1,3-Propanedithiol					-79.4				-29.8			
C_3H_9Al	Trimethyl aluminum					-136.4	-9.9	209.4	155.6	-74.1			
C_3H_9B	Trimethylborane					-143.1	-32.1	238.9		-124.3	-35.9	314.7	88.5
$C_3H_9BO_3$	Trimethyl borate								189.9				
C_3H_9ClSi	Trimethylchlorosilane					-382.8	-246.4	278.2		-352.8	-243.5	369.1	
C_3H_9N	Propylamine					-101.5			164.1	-70.1	39.9	325.4	91.2
C_3H_9N	Isopropylamine					-112.3		218.3	163.8	-83.7	32.2	312.2	97.5
C_3H_9N	Trimethylamine					-45.7		208.5	137.9	-23.6		287.1	91.8
$C_3H_{10}ClN$	Propylamine hydrochloride	-354.7											
$C_3H_{10}ClN$	Trimethylamine hydrochloride	-282.9											
$C_3H_{10}N_2$	1,2-Propanediamine, (±)					-97.8				-53.6			
$C_3H_{10}Si$	Trimethylsilane											331.0	117.9
$C_3H_{12}BN$	Trimethylamine borane	-142.5	70.7	187.0									
$C_3H_{12}BN$	Aminetrimethylboron	-284.1	-79.3	218.0									
C_4Cl_6	Hexachloro-1,3-butadiene					-24.5							
C_4F_8	Perfluorocyclobutane									-1542.6			
C_4F_{10}	Perfluorobutane								127.2				
$C_4H_2N_2$	trans-2-Butenedinitrile	268.2								340.2			
$C_4H_2O_3$	Maleic anhydride	-469.8								-398.3			
$C_4H_2O_4$	2-Butynedioic acid	-577.3											
$C_4H_3NO_3$	2-Nitrofuran	-104.1								-28.8			
$C_4H_4BrNO_2$	N-Bromosuccinimide	-335.9											
$C_4H_4ClNO_2$	N-Chlorosuccinimide	-357.9											
$C_4H_4N_2$	Succinonitrile	139.7		191.6	145.6					209.7			
$C_4H_4N_2$	Pyrazine	139.8								196.1			
$C_4H_4N_2$	Pyrimidine					145.9				195.7			
$C_4H_4N_2$	Pyridazine					224.9				278.3			
$C_4H_4N_2O_2$	Uracil	-429.4			120.5					-302.9			
$C_4H_4N_2O_3$	Barbituric acid	-634.7											
C_4H_4O	Furan					-62.3		177.0	114.8	-34.8		267.2	65.4
$C_4H_4O_2$	Diketene					-233.1				-190.3			
$C_4H_4O_3$	Succinic anhydride	-608.6								-527.9			
$C_4H_4O_4$	Maleic acid	-789.4		160.8	137.0					-679.4			
$C_4H_4O_4$	Fumaric acid	-811.7		168.0	142.0					-675.8			
C_4H_4S	Thiophene					80.2		181.2	123.8	114.9	126.1	278.8	72.8
C_4H_5N	trans-2-Butenenitrile					95.1				134.3			
C_4H_5N	3-Butenenitrile					117.8				159.7			
C_4H_5N	2-Methylacrylonitrile								126.3				
C_4H_5N	Pyrrole					63.1		156.4	127.7	108.2			
C_4H_5N	Cyclopropanecarbonitrile					140.8				182.8			
$C_4H_5NO_2$	Succinimide	-459.0								-375.4			
C_4H_5NS	4-Methylthiazole					67.9				111.8			
$C_4H_5N_3O$	Cytosine	-221.3			132.6								
C_4H_6	1,2-Butadiene					138.6				162.3			
C_4H_6	1,3-Butadiene					88.5		199.0	123.6	110.0			
C_4H_6	1-Butyne					141.4				165.2			
C_4H_6	2-Butyne					119.1				145.7			
C_4H_6	Cyclobutene									156.7			
$C_4H_6N_2O_2$	2,5-Piperazinedione	-446.5											
C_4H_6O	Divinyl ether					-39.8				-13.6			
C_4H_6O	trans-2-Butenal					-138.7				-100.6			
$C_4H_6O_2$	trans-2-Butenoic acid												
$C_4H_6O_2$	Methacrylic acid								161.1				
$C_4H_6O_2$	Vinyl acetate					-349.2				-314.4			

Standard Thermodynamic Properties of Chemical Substances

Molecular formula	Name	Crystal ΔfH° kJ/mol	ΔfG° kJ/mol	S° J/mol K	Cp J/mol K	Liquid ΔfH° kJ/mol	ΔfG° kJ/mol	S° J/mol K	Cp J/mol K	Gas ΔfH° kJ/mol	ΔfG° kJ/mol	S° J/mol K	Cp J/mol K
	Methyl acrylate					-362.2		239.5	158.8	-333.0			
	γ-Butyrolactone					-420.9			141.4	-366.5			
	Acetic anhydride					-624.4				-572.5			
	Propylene carbonate					-613.2			218.6	-582.5			
	Succinic acid	-940.5		167.3	153.1					-823.0			
	Dimethyl oxalate	-756.3								-708.9			
	2,3-Dihydrothiophene					52.9				90.7	133.5	303.5	79.8
	2,5-Dihydrothiophene					47.0				86.9	131.6	297.1	83.3
O	2-Chloroethyl vinyl ether					-208.1				-170.1			
O₂	2-Chlorobutanoic acid					-575.5							
O₂	3-Chlorobutanoic acid					-556.3							
O₂	4-Chlorobutanoic acid					-566.3							
O₂	Propyl chlorocarbonate					-533.4				-492.7			
	Butanenitrile					-5.8				33.6			
	2-Methylpropanenitrile					-13.8				23.4			
	Acetone cyanohydrin					-120.9							
	2-Pyrrolidone					-286.2							
	2-Methyl-2-oxazoline					-169.5				-130.5			
O₄	Iminodiacetic acid	-932.6											
O₄	Ethyl nitroacetate					-487.1							
O₄	L-Aspartic acid	-973.3											
O	Creatinine	-238.5											
	1-Butene					-20.8		227.0	118.0	0.1			
	cis-2-Butene					-29.8		219.9	127.0	-7.1			
	trans-2-Butene					-33.3				-11.4			
	Isobutene					-37.5				-16.9			
	Cyclobutane					3.7				27.7			
	Methylcyclopropane					1.7							
₂	1,2-Dibromobutane					-142.1				-91.6			
₂	1,3-Dibromobutane					-148.0							
₂	1,4-Dibromobutane					-140.3				-87.8			
₂	2,3-Dibromobutane					-139.6				-102.0			
₂	1,2-Dibromo-2-methylpropane					-156.6				-113.3			
₂	1,3-Dichlorobutane					-237.3				-195.0			
₂	1,4-Dichlorobutane					-229.8				-183.4			
₂O	Bis(2-chloroethyl) ether								220.9				
	1,4-Diiodobutane					-30.0							
₂O₂	Succinamide	-581.2											
₂O₂	Dimethylglyoxime	-199.7											
₂O₃	L-Asparagine	-789.4											
₂O₃	N-Glycylglycine	-747.7											
₂O₄	1,4-Dinitrobutane					-237.5							
₈O₈	Cyclotetramethylenetetranitramine									187.9		568.8	275.5
	Ethyl vinyl ether					-167.4				-140.8			
	1,2-Epoxybutane					-168.9		230.9	147.0				
	Butanal					-239.2		246.6	163.7	-204.8		343.7	103.4
	Isobutanal					-247.3				-215.7			
	2-Butanone					-273.3		239.1	158.7	-238.5		339.9	101.7
	Tetrahydrofuran					-216.2		204.3	124.0	-184.1		302.4	76.3
S	S-Ethyl thioacetate					-268.2				-228.1			
₂	Butanoic acid					-533.8		222.2	178.6	-475.9			
₂	2-Methylpropanoic acid								173.0				
₂	Propyl formate					-500.3				-462.7			
₂	Ethyl acetate					-479.3		257.7	170.7	-443.6			
₂	Methyl propanoate								171.2				
₂	1,3-Dioxane					-379.7			143.9	-340.6			
₂	1,4-Dioxane					-353.9		270.2	152.1	-315.3			
₂	2-Methyl-1,3-dioxolane					-386.9				-352.0			
₂S	Sulfolane								180.0				
S	Tetrahydrothiophene					-72.9				-34.1	45.8	309.6	92.5
₂	1,3-Dithiane									-10.0	72.4	333.5	110.4
₂	1,4-Dithiane									0.0	84.5	326.2	109.7
Br	1-Bromobutane					-143.8				-107.1			
Br	2-Bromobutane, (±)					-154.9				-120.3			
Br	2-Bromo-2-methylpropane					-164.4				-132.4			
Cl	1-Chlorobutane					-188.1				-154.4			

Molecular formula	Name	Crystal $\Delta_f H°$ kJ/mol	$\Delta_f G°$ kJ/mol	$S°$ J/mol K	C_p J/mol K	Liquid $\Delta_f H°$ kJ/mol	$\Delta_f G°$ kJ/mol	$S°$ J/mol K	C_p J/mol K	Gas $\Delta_f H°$ kJ/mol	$\Delta_f G°$ kJ/mol	$S°$ J/mol K	C_p J/mol K
C₄H₉Cl	2-Chlorobutane					-192.8				-161.1			
C₄H₉Cl	1-Chloro-2-methylpropane					-191.1				-159.3			
C₄H₉Cl	2-Chloro-2-methylpropane					-211.3				-182.2			
C₄H₉ClO	2-Chloroethyl ethyl ether					-335.6				-301.3			
C₄H₉I	1-Iodo-2-methylpropane								162.3				
C₄H₉I	2-Iodo-2-methylpropane					-107.5				-72.1			
C₄H₉N	Cyclobutanamine					5.6				41.2			
C₄H₉N	Pyrrolidine					-41.1		204.1	156.6	-3.6			
C₄H₉NO	Butanamide	-364.8								-282.0			
C₄H₉NO	N-Methylpropanamide								179.0				
C₄H₉NO	2-Methylpropanamide	-368.6								-282.6			
C₄H₉NO	N,N-Dimethylacetamide					-278.3			175.6	-228.0			
C₄H₉NO	Morpholine								164.8				
C₄H₉NO₂	1-Nitrobutane					-192.5				-143.9		369.9	115.1
C₄H₉NO₂	2-Nitroisobutane					-217.2				-177.1			
C₄H₉NO₂	Propyl carbamate	-552.6								-471.4			
C₄H₉NO₂	4-Aminobutanoic acid	-581.0								-441.0			
C₄H₉NO₃	3-Nitro-2-butanol					-390.0							
C₄H₉NO₃	2-Methyl-2-nitro-1-propanol	-410.1											
C₄H₉NO₃	DL-Threonine	-758.8											
C₄H₉NO₃	L-Threonine	-807.2											
C₄H₉N₃O₂	Creatine	-537.2											
C₄H₁₀	Butane					-147.3			140.9	-125.7			
C₄H₁₀	Isobutane					-154.2				-134.2			
C₄H₁₀Hg	Diethyl mercury					30.1			182.8	75.3			
C₄H₁₀N₂	Piperazine	-45.6											
C₄H₁₀N₂O	Trimethylurea	-330.5											
C₄H₁₀N₂O₂	N-Nitrodiethylamine					-106.2				-53.0			
C₄H₁₀N₂O₄	L-Asparagine, monohydrate	-1086.6											
C₄H₁₀O	1-Butanol					-327.3		225.8	177.2	-274.9			
C₄H₁₀O	2-Butanol					-342.6		214.9	196.9	-292.8		359.5	112.7
C₄H₁₀O	2-Methyl-1-propanol					-334.7		214.7	181.5	-283.8			
C₄H₁₀O	2-Methyl-2-propanol					-359.2		193.3	218.6	-312.5		326.7	113.6
C₄H₁₀O	Diethyl ether					-279.5		172.4	175.6	-252.1		342.7	119.5
C₄H₁₀O	Methyl propyl ether					-266.0		262.9	165.4	-238.1			
C₄H₁₀O	Isopropyl methyl ether					-278.8		253.8	161.9	-252.0			
C₄H₁₀OS	Diethyl sulfoxide					-268.0				-205.6			
C₄H₁₀O₂	1,2-Butanediol, (±)					-523.6							
C₄H₁₀O₂	1,3-Butanediol					-501.0				-433.2			
C₄H₁₀O₂	1,4-Butanediol					-505.3		223.4	200.1	-428.7			
C₄H₁₀O₂	2,3-Butanediol					-541.5			213.0	-482.3			
C₄H₁₀O₂	2-Methyl-1,2-propanediol					-539.7							
C₄H₁₀O₂	Ethylene glycol monoethyl ether								210.8				
C₄H₁₀O₂	Ethylene glycol dimethyl ether					-376.6			193.3				
C₄H₁₀O₂	Dimethylacetal					-420.6				-389.7			
C₄H₁₀O₂	tert-Butyl hydroperoxide					-293.6				-245.9			
C₄H₁₀O₃	Diethylene glycol					-628.5			244.8	-571.2			
C₄H₁₀O₃S	Diethyl sulfite					-600.7				-552.2			
C₄H₁₀O₄S	Diethyl sulfate					-813.2				-756.3			
C₄H₁₀S	1-Butanethiol					-124.7			171.2	-88.0			
C₄H₁₀S	2-Butanethiol					-131.0				-96.9			
C₄H₁₀S	2-Methyl-1-propanethiol					-132.0				-97.3			
C₄H₁₀S	2-Methyl-2-propanethiol					-140.5				-109.6			
C₄H₁₀S	Diethyl sulfide					-119.4		269.3	171.4	-83.5		368.1	117.0
C₄H₁₀S	Methyl propyl sulfide					-118.5		272.5	171.6	-82.2			
C₄H₁₀S	Isopropyl methyl sulfide					-124.7		263.1	172.4	-90.5			
C₄H₁₀S₂	1,4-Butanedithiol					-105.7				-50.6			
C₄H₁₀S₂	Diethyl disulfide					-120.1		269.3	171.4	-79.4			
C₄H₁₁N	Butylamine					-127.6			179.2	-91.9			
C₄H₁₁N	sec-Butylamine					-137.5				-104.6			
C₄H₁₁N	tert-Butylamine					-150.6			192.1	-121.0			
C₄H₁₁N	Isobutylamine					-132.6			183.2	-98.7			
C₄H₁₁N	Diethylamine					-103.7			169.2	-72.2			
C₄H₁₁NO	N,N-Dimethylethanolamine					-253.7				-203.6			
C₄H₁₁NO₂	Diethanolamine	-493.8		233.5						-397.1			
C₄H₁₁NO₃	Tris(hydroxymethyl)methylamine	-717.8											

Standard Thermodynamic Properties of Chemical Substances

Molecular Formula	Name	Crystal ΔfH° kJ/mol	Crystal ΔfG° kJ/mol	Crystal S° J/mol K	Crystal Cp J/mol K	Liquid ΔfH° kJ/mol	Liquid ΔfG° kJ/mol	Liquid S° J/mol K	Liquid Cp J/mol K	Gas ΔfH° kJ/mol	Gas ΔfG° kJ/mol	Gas S° J/mol K	Gas Cp J/mol K
N	Tetramethylammonium bromide	-251.0											
IN	Diethylamine hydrochloride	-358.6											
IN	Tetramethylammonium chloride	-276.4											
J	Tetramethylammonium iodide	-203.9											
2	2-Methyl-1,2-propanediamine					-133.9				-90.3			
b	Tetramethyl lead					97.9				135.9			
i	Tetramethylsilane					-264.0	-100.0	277.3	204.1	-239.1	-99.9	359.0	143.9
n	Tetramethylstannane					-52.3				-18.8			
3	Bis(2-aminoethyl)amine								254.0				
	2-Butynedinitrile					500.4				529.2			
	Nickel carbonyl					-633.0	-588.2	313.4	204.6	-602.9	-587.2	410.6	145.2
	Iron pentacarbonyl					-774.0	-705.3	338.1	240.6				
O2	Hexafluoroacetylacetone	-2286.7											
O5	5-Nitro-2-furancarboxylic acid	-516.8											
	1H-Purine	169.4											
O	Hypoxanthine	-110.8		145.6	134.5								
O2	Xanthine	-379.6		161.1	151.3								
O3	Uric acid	-618.8		173.2	166.1								
	Furfural					-201.6			163.2	-151.0			
	2-Furancarboxylic acid	-498.4								-390.0			
	3-Methyl-2,5-furandione					-504.5				-447.2			
O2	1,1,1-Trifluoro-2,4-pentanedione					-1040.2				-993.3			
	Pyridine					100.2			132.7	140.4			
O	1H-Pyrrole-2-carboxaldehyde	-106.4											
5	Adenine	96.9		147.0						205.7			
0	Guanine	-183.9											
	cis-3-Penten-1-yne					226.5							
	trans-3-Penten-1-yne					228.2							
	1,3-Cyclopentadiene					105.9				134.3			
2O2	Thymine	-462.8		150.8						-328.7			
2	Furfuryl alcohol					-276.2			204.0	-211.8			
4	trans-1-Propene-1,2-dicarboxylic acid	-824.4											
	2-Methylthiophene					44.6		218.5	149.8	83.5			
	3-Methylthiophene					43.1				82.5			
	trans-3-Pentenenitrile					80.9				125.7			
	Cyclobutanecarbonitrile					103.0				147.4			
	1-Methylpyrrole					62.4				103.1			
	2-Methylpyrrole					23.3				74.0			
	3-Methylpyrrole					20.5				70.2			
O2	Ethyl cyanoacetate								220.2				
	1,2-Pentadiene									140.7			
	cis-1,3-Pentadiene									81.4			
	trans-1,3-Pentadiene									76.1			
	1,4-Pentadiene									105.7			
	2,3-Pentadiene									133.1			
	3-Methyl-1,2-butadiene					101.2							
	2-Methyl-1,3-butadiene					48.2		229.3	152.6	75.5			
	Cyclopentene					4.3		201.2	122.4	34.0			
	Spiropentane					157.5		193.7	134.5	185.2			
	Methylenecyclobutane					93.8				121.6			
N4O12	Pentaerythritol tetranitrate	-538.6								-387.0		614.7	294.8
	Cyclopentanone					-235.9				-192.1			
2	4-Pentenoic acid					-430.6							
2	Allyl acetate								184.1				
2	Ethyl acrylate					-370.6				-354.2			
2	Methyl trans-2-butenoate					-382.9				-341.9			
2	Methyl methacrylate								191.2				
2	2,4-Pentanedione					-423.8				-382.0			
2	Dihydro-4-methyl-2(3H)-furanone					-461.3				-406.5			
2	Tetrahydro-2H-pyran-2-one					-436.7				-379.6			
O3	Methyl acetoacetate					-623.2							
O4	Glutaric acid	-960.0											
ClO2	Propyl chloroacetate					-515.5				-467.0			
N	Pentanenitrile					-33.1				10.5			
N	2,2-Dimethylpropanenitrile					-39.8		232.0	179.4	-2.3			

Molecular formula	Name	Crystal $\Delta_f H°$ kJ/mol	$\Delta_f G°$ kJ/mol	$S°$ J/mol K	C_p J/mol K	Liquid $\Delta_f H°$ kJ/mol	$\Delta_f G°$ kJ/mol	$S°$ J/mol K	C_p J/mol K	Gas $\Delta_f H°$ kJ/mol	$\Delta_f G°$ kJ/mol	$S°$ J/mol K	C_p J/mol K
C_5H_9N	1,2,5,6-Tetrahydropyridine					33.5							
C_5H_9NO	2-Piperidinone	-306.6											
C_5H_9NO	N-Methyl-2-pyrrolidone					-262.2			307.8				
$C_5H_9NO_2$	L-Proline	-515.2								-366.2			
$C_5H_9NO_4$	D-Glutamic acid	-1005.3											
$C_5H_9NO_4$	L-Glutamic acid	-1009.7											
C_5H_{10}	1-Pentene					-46.9		262.6	154.0	-21.1			
C_5H_{10}	cis-2-Pentene					-53.7		258.6	151.7	-27.6			
C_5H_{10}	trans-2-Pentene					-58.2		256.5	157.0	-31.9			
C_5H_{10}	2-Methyl-1-butene					-61.1		254.0	157.2	-35.2			
C_5H_{10}	3-Methyl-1-butene					-51.5		253.3	156.1	-27.5			
C_5H_{10}	2-Methyl-2-butene					-68.6		251.0	152.8	-41.7			
C_5H_{10}	Cyclopentane					-105.1		204.5	128.8	-76.4			
C_5H_{10}	Methylcyclobutane					-44.5							
C_5H_{10}	Ethylcyclopropane					-24.8							
C_5H_{10}	1,1-Dimethylcyclopropane					-33.3				-8.2			
C_5H_{10}	cis-1,2-Dimethylcyclopropane					-26.3							
C_5H_{10}	trans-1,2-Dimethylcyclopropane					-30.7							
$C_5H_{10}Br_2$	2,3-Dibromo-2-methylbutane									-137.6			
$C_5H_{10}N_2O$	N-Nitrosopiperidine					-31.1				16.6			
$C_5H_{10}N_2O_2$	N-Nitropiperidine					-93.0				-44.5			
$C_5H_{10}N_2O_3$	L-Glutamine	-826.4											
$C_5H_{10}O$	Cyclopentanol					-300.1		204.1	182.5	-242.5		362.9	
$C_5H_{10}O$	Pentanal					-267.2				-228.4			
$C_5H_{10}O$	2-Pentanone					-297.3			184.1	-258.8			
$C_5H_{10}O$	3-Pentanone					-296.5		266.0	190.9	-257.9			
$C_5H_{10}O$	3-Methyl-2-butanone					-299.5		268.5	179.9	-262.6			
$C_5H_{10}O$	3,3-Dimethyloxetane					-182.2				-148.2			
$C_5H_{10}O$	Tetrahydropyran					-258.3				-223.4			
$C_5H_{10}OS$	S-Propyl thioacetate					-294.5				-250.4			
$C_5H_{10}O_2$	Pentanoic acid					-559.4		259.8	210.3	-491.9			
$C_5H_{10}O_2$	2-Methylbutanoic acid					-554.5							
$C_5H_{10}O_2$	3-Methylbutanoic acid					-561.6				-510.0			
$C_5H_{10}O_2$	2,2-Dimethylpropanoic acid	-564.5								-491.3			
$C_5H_{10}O_2$	Butyl formate								200.2				
$C_5H_{10}O_2$	Propyl acetate								196.2				
$C_5H_{10}O_2$	Isopropyl acetate					-518.9			199.4	-481.6			
$C_5H_{10}O_2$	Ethyl propanoate					-502.7				-463.4			
$C_5H_{10}O_2$	Methyl butanoate								198.2				
$C_5H_{10}O_2$	(Ethoxymethyl)oxirane					-296.5							
$C_5H_{10}O_2$	4-Methyl-1,3-dioxane					-416.1				-376.9			
$C_5H_{10}O_2$	cis-1,2-Cyclopentanediol	-485.0											
$C_5H_{10}O_2$	trans-1,2-Cyclopentanediol	-490.1											
$C_5H_{10}O_2$	Tetrahydrofurfuryl alcohol					-435.7				-369.1			
$C_5H_{10}O_3$	Diethyl carbonate					-681.5				-637.9			
$C_5H_{10}O_3$	Ethylene glycol monomethyl ether acetate								310.0				
$C_5H_{10}O_3$	Ethyl lactate								254.0				
$C_5H_{10}O_4$	Glycerol 1-acetate, (DL)					-909.2							
$C_5H_{10}O_5$	D-Ribose	-1047.2											
$C_5H_{10}O_5$	D-Xylose	-1057.8											
$C_5H_{10}O_5$	α-D-Arabinopyranose	-1057.9											
$C_5H_{10}S$	Thiacyclohexane					-106.3		218.2	163.3	-63.5	53.1	323.0	109.7
$C_5H_{10}S$	Cyclopentanethiol					-89.5		256.9	165.2	-48.1			
$C_5H_{11}Br$	1-Bromopentane					-170.2				-128.9			
$C_5H_{11}Cl$	1-Chloropentane					-213.2				-174.9			
$C_5H_{11}Cl$	1-Chloro-3-methylbutane					-216.0				-179.7			
$C_5H_{11}Cl$	2-Chloro-2-methylbutane					-235.7				-202.2			
$C_5H_{11}Cl$	2-Chloro-3-methylbutane					-226.6				-185.1			
$C_5H_{11}N$	Cyclopentylamine					-95.1		241.0	181.2	-54.9			
$C_5H_{11}N$	Piperidine					-86.4		210.0	179.9	-47.1			
$C_5H_{11}NO$	Pentanamide	-379.5								-290.2			
$C_5H_{11}NO$	2,2-Dimethylpropanamide	-399.7								-313.1			
$C_5H_{11}NO_2$	1-Nitropentane					-215.4				-164.4		390.9	137.1
$C_5H_{11}NO_2$	DL-Valine	-628.9											
$C_5H_{11}NO_2$	L-Valine	-617.9								-455.1			

Molecular formula	Name	Crystal				Liquid				Gas			
		$\Delta_f H°$ kJ/mol	$\Delta_f G°$ kJ/mol	$S°$ J/mol K	C_p J/mol K	$\Delta_f H°$ kJ/mol	$\Delta_f G°$ kJ/mol	$S°$ J/mol K	C_p J/mol K	$\Delta_f H°$ kJ/mol	$\Delta_f G°$ kJ/mol	$S°$ J/mol K	C_p J/mol K
..O₂	5-Aminopentanoic acid	-604.1								-460.0			
..O₂S	L-Methionine	-577.5								-413.5			
..O₄	2-Ethyl-2-nitro-1,3-propanediol	-606.4											
	Pentane					-173.5			167.2	-146.9			
	Isopentane					-178.4		260.4	164.8	-153.6			
	Neopentane					-190.2				-168.0			
..₂O	Butylurea	-419.5											
..₂O	tert-Butylurea	-417.4											
..₂O	N,N-Diethylurea	-372.2											
..₂O	Tetramethylurea					-262.2							
..₂S	Tetramethylthiourea	-38.1								44.9			
	1-Pentanol					-351.6			208.1	-294.6			
	2-Pentanol					-365.2				-311.0			
	3-Pentanol					-368.9			239.7	-314.9			
	2-Methyl-1-butanol, (±)					-356.6				-301.4			
	3-Methyl-1-butanol					-356.4				-300.7			
	2-Methyl-2-butanol					-379.5			247.1	-329.3			
	3-Methyl-2-butanol, (±)					-366.6				-313.5			
	2,2-Dimethyl-1-propanol					-399.4							
	Butyl methyl ether					-290.6		295.3	192.7	-258.1			
	Methyl tert-butyl ether					-313.6		265.3	187.5	-283.7			
	Ethyl propyl ether					-303.6		295.0	197.2	-272.0			
..O₂	1,5-Pentanediol					-528.8				-450.8			
..O₂	2,2-Dimethyl-1,3-propanediol	-551.2											
..O₂	Diethoxymethane					-450.5				-414.7			
..O₂	1,1-Dimethoxypropane					-443.6							
..O₂	2,2-Dimethoxypropane					-459.4				-429.9			
..O₃	Diethylene glycol monomethyl ether								271.1				
..O₃	2-(Hydroxymethyl)-2-methyl-1,3-propanediol	-744.6											
..O₄	Pentaerythritol	-920.6								-776.7			
..O₅	Xylitol	-1118.5											
..S	1-Pentanethiol					-151.3				-110.0			
..S	2-Methyl-1-butanethiol, (+)					-154.4				-114.9			
..S	3-Methyl-1-butanethiol					-154.4				-114.9			
..S	2-Methyl-2-butanethiol					-162.8		290.1	198.1	-127.1			
..S	3-Methyl-2-butanethiol					-158.8				-121.3			
..S	2,2-Dimethyl-1-propanethiol					-165.4				-129.0			
..₂S	Butyl methyl sulfide					-142.9		307.5	200.9	-102.4			
..₂S	tert-Butyl methyl sulfide					-157.1		276.1	199.9	-121.3			
..₂S	Ethyl propyl sulfide					-144.8		309.5	198.4	-104.8			
..₂S	Ethyl isopropyl sulfide					-156.1				-118.3			
..₃N	Pentylamine								218.0				
..₄N₂	N,N,N',N'-Tetramethylmethanediamine					-51.1				-18.2			
..F₅	Chloropentafluorobenzene	-858.4								-809.3			
..₆	Hexachlorobenzene	-127.6		260.2	201.2					-35.5			
	Hexafluorobenzene					-991.3		280.8	221.6	-955.4			
..0	Perfluorocyclohexene					-1963.5				-1932.7			
..₂	Perfluorocyclohexane					-2406.3				-2370.4			
..Cl₅O	Pentachlorophenol	-292.5		253.2	202.0								
..F₅	Pentafluorobenzene	-852.7				-841.8				-806.5			
..F₅O	Pentafluorophenol	-1024.1				-1007.7							
..₂F₄	1,2,4,5-Tetrafluorobenzene					-683.8							
..₃Cl₃	1,2,3-Trichlorobenzene	-70.8								3.8			
..₃Cl₃	1,2,4-Trichlorobenzene					-63.1				-8.1			
..₃Cl₃	1,3,5-Trichlorobenzene	-78.4								-13.4			
..₃N₃O₆	1,3,5-Trinitrobenzene	-37.0			214.6								
..₃N₃O₇	2,4,6-Trinitrophenol	-217.9			239.7								
..₃N₃O₈	2,4,6-Trinitro-1,3-benzenediol	-467.5											
..₄ClNO₂	1-Chloro-4-nitrobenzene	-48.7			250.2								
..₄Cl₂	o-Dichlorobenzene					-17.5			162.4	30.2			
..₄Cl₂	m-Dichlorobenzene					-20.7				25.7			
..₄Cl₂	p-Dichlorobenzene	-42.3		175.4	147.8					22.5			
..₄Cl₂O	2,4-Dichlorophenol	-226.4								-156.3			
..₄F₂	o-Difluorobenzene					-330.0		222.6	159.0	-293.8			
..₄F₂	m-Difluorobenzene					-343.9		223.8	159.1	-309.2			

Molecular formula	Name	Crystal				Liquid				Gas			
		$\Delta_f H°$ kJ/mol	$\Delta_f G°$ kJ/mol	$S°$ J/mol K	C_p J/mol K	$\Delta_f H°$ kJ/mol	$\Delta_f G°$ kJ/mol	$S°$ J/mol K	C_p J/mol K	$\Delta_f H°$ kJ/mol	$\Delta_f G°$ kJ/mol	$S°$ J/mol K	C_p J/mol K
$C_6H_4F_2$	p-Difluorobenzene					-342.3			157.5	-306.7			
$C_6H_4N_2O_4$	1,2-Dinitrobenzene	-2.0			200.4								
$C_6H_4N_2O_4$	1,3-Dinitrobenzene	-27.0			197.5	-36.0							
$C_6H_4N_2O_4$	1,4-Dinitrobenzene	-38.0			200.0								
$C_6H_4N_2O_5$	2,4-Dinitrophenol	-232.7											
$C_6H_4O_2$	p-Benzoquinone	-185.7			129.0					-128.1			
C_6H_5Br	Bromobenzene					60.9		219.2	154.3	-122.9			
C_6H_5Cl	Chlorobenzene					11.1			150.1	52.0			
C_6H_5ClO	2-Chlorophenol								188.7				
C_6H_5ClO	3-Chlorophenol	-206.4				-189.3							
C_6H_5ClO	4-Chlorophenol	-197.7				-181.3							
$C_6H_5Cl_2N$	3,4-Dichloroaniline	-89.1											
C_6H_5F	Fluorobenzene					-150.6		205.9	146.4	-115.9			
C_6H_5I	Iodobenzene					117.2		205.4	158.7	164.9			
$C_6H_5NO_2$	Nitrobenzene					12.5			185.8	68.5		348.8	120.4
$C_6H_5NO_2$	3-Pyridinecarboxylic acid	-344.9								-221.5			
$C_6H_5NO_3$	2-Nitrophenol	-202.4											
$C_6H_5N_3$	1H-Benzotriazole	236.5								335.5			
$C_6H_5N_3O_4$	2,3-Dinitroaniline	-11.7											
$C_6H_5N_3O_4$	2,4-Dinitroaniline	-67.8											
$C_6H_5N_3O_4$	2,5-Dinitroaniline	-44.3											
$C_6H_5N_3O_4$	2,6-Dinitroaniline	-50.6											
$C_6H_5N_3O_4$	3,5-Dinitroaniline	-38.9											
C_6H_6	1,5-Hexadiyne					384.2							
C_6H_6	Benzene					49.1	124.5	173.4	136.0	82.9	129.7	269.2	82.4
C_6H_6ClN	2-Chloroaniline					-4.6							
C_6H_6ClN	3-Chloroaniline					-20.3		198.7					
C_6H_6ClN	4-Chloroaniline	-33.3			147.3								
$C_6H_6N_2O_2$	2-Nitroaniline	-26.1			166.0	-9.4				63.8			
$C_6H_6N_2O_2$	3-Nitroaniline	-38.3			158.8	-14.4				58.4			
$C_6H_6N_2O_2$	4-Nitroaniline	-42.0			167.0	-20.7				58.8			
C_6H_6O	Phenol	-165.1		144.0	127.4					-96.4			
C_6H_6O	2-Vinylfuran					-10.3				27.8			
$C_6H_6O_2$	p-Hydroquinone	-364.5			136.0					-265.3			
$C_6H_6O_2$	Pyrocatechol	-354.1								-267.5			
$C_6H_6O_2$	Resorcinol	-368.0								-274.7			
$C_6H_6O_3$	1,2,3-Benzenetriol	-551.1								-434.2			
$C_6H_6O_3$	1,2,4-Benzenetriol	-563.8								-444.0			
$C_6H_6O_3$	1,3,5-Benzenetriol	-584.6								-452.9			
$C_6H_6O_3$	3,4-Dimethyl-2,5-furandione	-581.4											
$C_6H_6O_6$	cis-1-Propene-1,2,3-tricarboxylic acid	-1224.4											
$C_6H_6O_6$	trans-1-Propene-1,2,3-tricarboxylic acid	-1232.7											
C_6H_6S	Benzenethiol					63.7		222.8	173.2	111.3			
C_6H_7N	Aniline					31.6		191.9		87.5	-7.0	317.9	107.9
C_6H_7N	2-Methylpyridine					56.7		158.6		99.2			
C_6H_7N	3-Methylpyridine					61.9		216.3	158.7	106.4			
C_6H_7N	4-Methylpyridine					59.2		209.1	159.0	103.8			
C_6H_7N	1-Cyclopentenecarbonitrile					111.5				156.5			
$C_6H_8N_2$	Adiponitrile					85.1			128.7	149.5			
$C_6H_8N_2$	1,2-Benzenediamine	-0.3											
$C_6H_8N_2$	1,3-Benzenediamine	-7.8		154.5	159.6								
$C_6H_8N_2$	1,4-Benzenediamine	3.0											
$C_6H_8N_2$	Phenylhydrazine					141.0		217.0	202.9				
$C_6H_8N_2S$	Bis(2-cyanoethyl) sulfide					96.3							
$C_6H_8O_4$	Dimethyl maleate								263.2				
$C_6H_8O_6$	L-Ascorbic acid	-1164.6											
$C_6H_8O_7$	Citric acid	-1543.8											
$C_6H_9Cl_3O_2$	Butyl trichloroacetate					-545.8				-492.3			
$C_6H_9Cl_3O_2$	Isobutyl trichloroacetate					-553.4				-500.2			
C_6H_9N	Cyclopentanecarbonitrile					0.7				44.1			
C_6H_9N	2,4-Dimethylpyrrole	-422.3											
C_6H_9N	2,5-Dimethylpyrrole					-16.7				39.8			
$C_6H_9NO_3$	Triacetamide	-610.5								-550.1			
$C_6H_9NO_6$	Nitrilotriacetic acid	-1311.9											
$C_6H_9N_3O_2$	L-Histidine	-466.7											

Formula	Name	Crystal $\Delta_f H°$ kJ/mol	$\Delta_f G°$ kJ/mol	$S°$ J/mol K	C_p J/mol K	Liquid $\Delta_f H°$ kJ/mol	$\Delta_f G°$ kJ/mol	$S°$ J/mol K	C_p J/mol K	Gas $\Delta_f H°$ kJ/mol	$\Delta_f G°$ kJ/mol	$S°$ J/mol K	C_p J/mol K
	1,5-Hexadiene					54.1				84.2			
	3,3-Dimethyl-1-butyne					78.4							
	Cyclohexene					-38.5		214.6	148.3	-5.0			
	1-Methylcyclopentene					-36.4				-3.8			
	3-Methylcyclopentene					-23.7				7.4			
	4-Methylcyclopentene					-17.6				14.6			
Cl_2O_2	Butyl dichloroacetate					-550.1				-497.8			
O	Cyclohexanone					-271.2			182.2	-226.1			
O	2-Methylcyclopentanone					-265.2							
O	Mesityl oxide								212.5				
O_2	Ethyl trans-2-butenoate					-420.0				-375.6			
O_2	Methyl cyclobutanecarboxylate					-395.0				-350.2			
O_3	Ethyl acetoacetate								248.0				
O_3	Propanoic anhydride					-679.1				-626.5			
O_4	Adipic acid	-994.3											
O_4	Diethyl oxalate					-805.5				-742.0			
O_4	Ethylene glycol diacetate								310.0				
Cl	Chlorocyclohexane					-207.2				-163.7			
ClO_2	Ethyl 4-chlorobutanoate					-566.5				-513.8			
ClO_2	Propyl 3-chloropropanoate					-537.6				-485.7			
ClO_2	Butyl chloroacetate					-538.4				-487.4			
NO	Caprolactam	-329.4		156.8						-239.6			
NO	1-Methyl-2-piperidinone					-293.0							
	1-Hexene					-74.2		295.2	183.3	-43.5			
	cis-2-Hexene					-83.9				-52.3			
	trans-2-Hexene					-85.5				-53.9			
	cis-3-Hexene					-78.9				-47.6			
	trans-3-Hexene					-86.1				-54.4			
	2-Methyl-1-pentene					-90.0				-59.4			
	3-Methyl-1-pentene					-78.2				-49.5			
	4-Methyl-1-pentene					-80.0				-51.3			
	2-Methyl-2-pentene					-98.5				-66.9			
	3-Methyl-cis-2-pentene					-94.5				-62.3			
	3-Methyl-trans-2-pentene					-94.6				-63.1			
	4-Methyl-cis-2-pentene					-87.0				-57.5			
	4-Methyl-trans-2-pentene					-91.6				-61.5			
	2-Ethyl-1-butene					-87.1				-56.0			
	2,3-Dimethyl-1-butene					-93.2				-62.4			
	3,3-Dimethyl-1-butene					-87.5				-60.3			
	2,3-Dimethyl-2-butene					-101.4		270.2	174.7	-68.1			
	Cyclohexane					-156.4			154.9	-123.4			
	Methylcyclopentane					-137.9				-106.2			
	Ethylcyclobutane					-59.0				-27.5			
	1,1,2-Trimethylcyclopropane					-96.2							
$N_2O_4S_2$	L-Cystine	-1032.7											
N_2S_4	Thiram	40.2		301.7									
O	Butyl vinyl ether					-218.8			232.0	-182.6			
O	Hexanal							280.3	210.4				
O	2-Hexanone					-322.0			213.3	-278.9			
O	3-Hexanone					-320.2		305.3	216.9	-277.6			
O	4-Methyl-2-pentanone								213.3				
O	2-Methyl-3-pentanone					-325.9				-286.0			
O	3,3-Dimethyl-2-butanone					-328.6				-290.6			
$_{12}O$	Cyclohexanol					-348.2			208.2	-286.2			
$_{12}O$	cis-2-Methylcyclopentanol					-345.5							
$_{12}O_2$	Hexanoic acid					-583.8				-511.9			
$_{12}O_2$	Butyl acetate					-529.2			227.8	-485.3			
$_{12}O_2$	tert-Butyl acetate					-554.5			231.0	-516.5			
$_{12}O_2$	Isobutyl acetate								233.8				
$_{12}O_2$	Ethyl butanoate								228.0				
$_{12}O_2$	Methyl pentanoate					-514.2			229.3	-471.1			
$_{12}O_2$	Methyl 2,2-dimethylpropanoate					-530.0			257.9	-491.2			
$_{12}O_2$	Diacetone alcohol								221.3				
$_{12}O_3$	Ethylene glycol monoethyl ether acetate								376.0				
$_{12}O_3$	Paraldehyde					-673.1				-631.7			

Standard Thermodynamic Properties of Chemical Substances

Molecular formula	Name	Crystal				Liquid				Gas			
		$\Delta_f H°$ kJ/mol	$\Delta_f G°$ kJ/mol	$S°$ J/mol K	C_p J/mol K	$\Delta_f H°$ kJ/mol	$\Delta_f G°$ kJ/mol	$S°$ J/mol K	C_p J/mol K	$\Delta_f H°$ kJ/mol	$\Delta_f G°$ kJ/mol	$S°$ J/mol K	C_p J/mol K
$C_6H_{12}O_6$	β-D-Fructose	-1265.6											
$C_6H_{12}O_6$	D-Galactose	-1286.3											
$C_6H_{12}O_6$	α-D-Glucose	-1273.3											
$C_6H_{12}O_6$	D-Mannose	-1263.0											
$C_6H_{12}O_6$	L-Sorbose	-1271.5											
$C_6H_{12}S$	Thiepane									-65.8	79.4	363.5	131.3
$C_6H_{12}S$	Cyclohexanethiol					-140.7		255.6	192.6	-96.2			
$C_6H_{12}S$	Cyclopentyl methyl sulfide					-109.8				-64.7			
$C_6H_{13}Br$	1-Bromohexane					-194.2		453.0	204.0	-148.3			
$C_6H_{13}Cl$	2-Chlorohexane					-246.1				-204.3			
$C_6H_{13}N$	Cyclohexylamine					-147.6				-104.0			
$C_6H_{13}N$	2-Methylpiperidine, (±)					-124.9				-84.4			
$C_6H_{13}NO$	Hexanamide	-423.0								-324.2			
$C_6H_{13}NO$	N-Butylacetamide					-380.9				-305.9			
$C_6H_{13}NO_2$	DL-Leucine	-640.6											
$C_6H_{13}NO_2$	D-Leucine	-637.3											
$C_6H_{13}NO_2$	L-Leucine	-637.4			200.1								
$C_6H_{13}NO_2$	DL-Isoleucine	-635.3											
$C_6H_{13}NO_2$	L-Isoleucine	-637.8								-486.8			
$C_6H_{13}NO_2$	L-Norleucine	-639.1											
$C_6H_{13}NO_2$	6-Aminohexanoic acid	-637.3											
C_6H_{14}	Hexane					-198.7			195.6	-166.9			
C_6H_{14}	2-Methylpentane					-204.6		290.6	193.7	-174.6			
C_6H_{14}	3-Methylpentane					-202.4		292.5	190.7	-171.9			
C_6H_{14}	2,2-Dimethylbutane					-213.8		272.5	191.9	-185.9			
C_6H_{14}	2,3-Dimethylbutane					-207.4		287.8	189.7	-178.1			
$C_6H_{14}N_2$	Azopropane					11.5				51.3			
$C_6H_{14}N_2O_2$	DL-Lysine	-678.7											
$C_6H_{14}N_4O_2$	D-Arginine	-623.5		250.6	232.0								
$C_6H_{14}O$	1-Hexanol					-377.5		287.4	240.4	-315.9			
$C_6H_{14}O$	2-Hexanol					-392.0				-333.5			
$C_6H_{14}O$	3-Hexanol					-392.4			286.2				
$C_6H_{14}O$	2-Methyl-1-pentanol								248.0				
$C_6H_{14}O$	3-Methyl-2-pentanol								275.9				
$C_6H_{14}O$	4-Methyl-2-pentanol					-394.7			273.0				
$C_6H_{14}O$	2-Methyl-3-pentanol					-396.4							
$C_6H_{14}O$	3-Methyl-3-pentanol								293.4				
$C_6H_{14}O$	Dipropyl ether					-328.8		323.9	221.6	-293.0			
$C_6H_{14}O$	Diisopropyl ether					-351.5			216.8	-319.2			
$C_6H_{14}O$	Butyl ethyl ether								159.0				
$C_6H_{14}O$	tert-Butyl ethyl ether									-313.9			
$C_6H_{14}OS$	Dipropyl sulfoxide					-329.4				-254.9			
$C_6H_{14}O_2$	1,2-Hexanediol					-577.1				-490.1			
$C_6H_{14}O_2$	1,6-Hexanediol	-569.9				-548.6				-461.2			
$C_6H_{14}O_2$	2-Methyl-2,4-pentanediol								336.0				
$C_6H_{14}O_2$	Ethylene glycol monobutyl ether								281.0				
$C_6H_{14}O_2$	1,1-Diethoxyethane					-491.4				-453.5			
$C_6H_{14}O_2$	Ethylene glycol diethyl ether					-451.4			259.4	-408.1			
$C_6H_{14}O_3$	Diethylene glycol monoethyl ether								301.0				
$C_6H_{14}O_3$	Diethylene glycol dimethyl ether								274.1				
$C_6H_{14}O_3$	Trimethylolpropane	-750.9											
$C_6H_{14}O_4$	Triethylene glycol					-804.3				-725.0			
$C_6H_{14}O_4S$	Dipropyl sulfate					-859.0				-792.0			
$C_6H_{14}O_6$	Galactitol					-1317.0							
$C_6H_{14}O_6$	D-Mannitol					-1314.5							
$C_6H_{14}S$	1-Hexanethiol					-175.7				-129.9			
$C_6H_{14}S$	2-Methyl-2-pentanethiol					-188.3				-148.3			
$C_6H_{14}S$	2,3-Dimethyl-2-butanethiol					-187.1				-147.9			
$C_6H_{14}S$	Diisopropyl sulfide					-181.6		313.0	232.0	-142.0			
$C_6H_{14}S$	Butyl ethyl sulfide					-172.3				-127.8			
$C_6H_{14}S$	Methyl pentyl sulfide					-167.1				-121.8			
$C_6H_{14}S_2$	Dipropyl disulfide					-171.5				-118.3			
$C_6H_{15}B$	Triethylborane					-194.6	9.4	336.7	241.2	-157.7	16.1	437.8	
$C_6H_{15}N$	Dipropylamine					-156.1				-116.0			
$C_6H_{15}N$	Diisopropylamine					-178.5				-143.8			
$C_6H_{15}N$	Triethylamine					-127.7			219.9	-92.7			

Standard Thermodynamic Properties of Chemical Substances

Molecular formula	Name	Crystal $\Delta_f H^\circ$ kJ/mol	Crystal $\Delta_f G^\circ$ kJ/mol	Crystal S° J/mol K	Crystal C_p J/mol K	Liquid $\Delta_f H^\circ$ kJ/mol	Liquid $\Delta_f G^\circ$ kJ/mol	Liquid S° J/mol K	Liquid C_p J/mol K	Gas $\Delta_f H^\circ$ kJ/mol	Gas $\Delta_f G^\circ$ kJ/mol	Gas S° J/mol K	Gas C_p J/mol K
O	2-Diethylaminoethanol					-305.9							
O_3	Triethanolamine	-664.2		389.0						-558.3			
$_2$	1,6-Hexanediamine	-205.0											
$_3OP$	Hexamethylphosphoric triamide								321.0				
OSi_2	Hexamethyldisiloxane					-815.0	-541.5	433.8	311.4	-777.7	-534.5	535.0	238.5
O_6	Molybdenum hexacarbonyl	-982.8	-877.7	325.9	242.3					-912.1	-856.0	490.0	205.0
	Tetracyanoethene	623.8								705.0			
	Perfluorotoluene					-1311.1		355.5	262.3				
	Perfluoromethylcyclohexane					-2931.1			353.1	-2897.2			
	Perfluoroheptane					-3420.0		561.8	419.0	-3383.6			
$_5$	2,3,4,5,6-Pentafluorotoluene					-883.8		306.4	225.8	-842.7			
l_2O	3-Chlorobenzoyl chloride					-189.7							
$_2O_6$	3,5-Dinitrobenzoic acid	-409.8											
ClO	Benzoyl chloride					-158.0				-103.2			
ClO_2	2-Chlorobenzoic acid	-404.5								-325.0			
ClO_2	3-Chlorobenzoic acid	-424.3								-342.3			
ClO_2	4-Chlorobenzoic acid	-428.9		163.2						-341.0			
$_3$	(Trifluoromethyl)benzene								188.4				
N	Benzonitrile					163.2		209.1	165.2	215.7			
NO	Benzoxazole	-24.2								44.8			
NO_4	2-Nitrobenzoic acid	-378.8											
NO_4	3-Nitrobenzoic acid	-394.7											
NO_4	4-Nitrobenzoic acid	-392.2											
N_3O_6	2,4,6-Trinitrotoluene	-63.2		243.3									
N_2	1H-Benzimidazole	79.5								181.7			
N_2	1H-Indazole	151.9								243.0			
N_2O_4	1-Methyl-2,4-dinitrobenzene	-66.4								33.2			
O	Benzaldehyde					-87.0		221.2	172.0	-36.7			
O_2	Benzoic acid	-385.2		167.6	146.8					-294.0			
O_2	Salicylaldehyde								222.0				
O_2	3-(2-Furanyl)-2-propenal	-182.0								-105.9			
O_3	2-Hydroxybenzoic acid	-589.9								-494.8			
Br	4-Bromotoluene					12.0							
Cl	2-Chlorotoluene								166.8				
Cl	(Chloromethyl)benzene					-32.5				18.9			
F	4-Fluorotoluene					-186.9			171.2	-147.4			
NO	Benzamide	-202.6								-100.9			
NO_2	Aniline-2-carboxylic acid	-401.1								-296.0			
NO_2	Aniline-3-carboxylic acid	-417.3								-283.6			
NO_2	Aniline-4-carboxylic acid	-410.0		177.8						-296.7			
NO_2	2-Nitrotoluene					-9.7							
NO_2	3-Nitrotoluene					-31.5							
NO_2	4-Nitrotoluene	-48.1		172.3						31.0			
NO_2	(Nitromethyl)benzene					-22.8				30.7			
NO_2	Salicylaldoxime	-183.7											
	Toluene					12.4			157.3	50.5			
N_2O	Phenylurea	-218.6											
O	o-Cresol	-204.6		165.4	154.6					-128.6			
O	m-Cresol					-194.0		212.6	224.9	-132.3			
O	p-Cresol	-199.3		167.3	150.2					-125.4			
O	Benzyl alcohol					-160.7		216.7	217.9	-100.4			
O	Anisole					-114.8				-67.9			
N	Benzylamine					34.2			207.2	94.4			
N	2-Methylaniline					-6.3				56.4	167.6	351.0	130.2
N	3-Methylaniline					-8.1				54.6	165.4	352.5	125.5
N	4-Methylaniline	-23.5								55.3	167.7	347.0	126.2
N	N-Methylaniline								207.1				
N	1-Cyclohexenecarbonitrile					48.1				101.6			
N	2,3-Dimethylpyridine					19.4		243.7	189.5	67.1			
N	2,4-Dimethylpyridine					16.1		248.5	184.8	63.6			
N	2,5-Dimethylpyridine					18.7		248.8	184.7	66.5			
N	2,6-Dimethylpyridine					12.7		244.2	185.2	58.1			
N	3,4-Dimethylpyridine					18.3		240.7	191.8	68.8			
N	3,5-Dimethylpyridine					22.5		241.7	184.5	72.0			
$_{10}O_2$	Ethyl 2-pentynoate					-301.8				-250.3			
$_{10}O_2$	Methyl 2-hexynoate					-242.7							

Molecular formula	Name	Crystal				Liquid				Gas			
		$\Delta_f H°$ kJ/mol	$\Delta_f G°$ kJ/mol	$S°$ J/mol K	C_p J/mol K	$\Delta_f H°$ kJ/mol	$\Delta_f G°$ kJ/mol	$S°$ J/mol K	C_p J/mol K	$\Delta_f H°$ kJ/mol	$\Delta_f G°$ kJ/mol	$S°$ J/mol K	C_p J/mol K
$C_7H_{11}Cl_3O_2$	Isopentyl trichloroacetate					-580.9				-523.1			
$C_7H_{11}N$	Cyclohexanecarbonitrile					-47.2				4.8			
C_7H_{12}	Bicyclo[2.2.1]heptane	-95.1			151.0					-54.8			
C_7H_{12}	1-Methylbicyclo(3,1,0)hexane					-33.2				1.7			
C_7H_{12}	Methylenecyclohexane					-61.3				-25.2			
C_7H_{12}	Vinylcyclopentane					-34.8							
C_7H_{12}	1-Ethylcyclopentene					-53.3				-19.8			
$C_7H_{12}O$	2-Methylenecyclohexanol					-277.6							
$C_7H_{12}O_2$	Butyl acrylate					-422.6			251.0	-375.3			
$C_7H_{12}O_4$	Diethyl malonate								285.0				
$C_7H_{13}ClO_2$	Butyl 2-chloropropanoate					-571.7				-517.3			
$C_7H_{13}ClO_2$	Isobutyl 2-chloropropanoate					-603.1				-549.6			
$C_7H_{13}ClO_2$	Butyl 3-chloropropanoate					-557.9				-502.3			
$C_7H_{13}ClO_2$	Isobutyl 3-chloropropanoate					-572.6				-517.3			
$C_7H_{13}ClO_2$	Propyl 2-chlorobutanoate					-630.7				-578.4			
$C_7H_{13}N$	Heptanenitrile					-82.8				-31.0			
C_7H_{14}	1-Heptene					-97.9		327.6	211.8	-62.3			
C_7H_{14}	cis-2-Heptene					-105.1							
C_7H_{14}	trans-2-Heptene					-109.5							
C_7H_{14}	cis-3-Heptene					-104.3							
C_7H_{14}	trans-3-Heptene					-109.3							
C_7H_{14}	5-Methyl-1-hexene					-100.0				-65.7			
C_7H_{14}	cis-3-Methyl-3-hexene					-115.9				-79.4			
C_7H_{14}	trans-3-Methyl-3-hexene					-112.7				-76.8			
C_7H_{14}	2,4-Dimethyl-1-pentene					-117.0				-83.8			
C_7H_{14}	4,4-Dimethyl-1-pentene					-110.6				-81.6			
C_7H_{14}	2,4-Dimethyl-2-pentene					-123.1				-88.7			
C_7H_{14}	cis-4,4-Dimethyl-2-pentene					-105.3				-72.6			
C_7H_{14}	trans-4,4-Dimethyl-2-pentene					-121.7				-88.8			
C_7H_{14}	2-Ethyl-3-methyl-1-butene					-114.1				-79.5			
C_7H_{14}	2,3,3-Trimethyl-1-butene					-117.7				-85.5			
C_7H_{14}	Cycloheptane					-156.6				-118.1			
C_7H_{14}	Methylcyclohexane					-190.1			184.8	-154.7			
C_7H_{14}	Ethylcyclopentane					-163.4		279.9		-126.9			
C_7H_{14}	1,1-Dimethylcyclopentane					-172.1				-138.2			
C_7H_{14}	cis-1,2-Dimethylcyclopentane					-165.3		269.2		-129.5			
C_7H_{14}	trans-1,2-Dimethylcyclopentane					-171.2				-136.6			
C_7H_{14}	cis-1,3-Dimethylcyclopentane					-170.1				-135.8			
C_7H_{14}	trans-1,3-Dimethylcyclopentane					-168.1				-133.6			
C_7H_{14}	1,1,2,2-Tetramethylcyclopropane					-119.8							
$C_7H_{14}Br_2$	1,2-Dibromoheptane					-212.3				-157.9			
$C_7H_{14}O$	1-Heptanal					-311.5		335.4	230.1	-263.8			
$C_7H_{14}O$	2-Heptanone								232.6				
$C_7H_{14}O$	3-Heptanone									-297.1			
$C_7H_{14}O$	4-Heptanone									-298.3			
$C_7H_{14}O$	2,2-Dimethyl-3-pentanone					-356.1				-313.6			
$C_7H_{14}O$	2,4-Dimethyl-3-pentanone					-352.9		318.0	233.7	-311.3			
$C_7H_{14}O$	cis-2-Methylcyclohexanol					-390.2				-327.0			
$C_7H_{14}O$	trans-2-Methylcyclohexanol, (±)					-415.7				-352.5			
$C_7H_{14}O$	cis-3-Methylcyclohexanol, (±)					-416.1				-350.9			
$C_7H_{14}O$	trans-3-Methylcyclohexanol, (±)					-394.4				-329.1			
$C_7H_{14}O$	cis-4-Methylcyclohexanol					-413.2				-347.5			
$C_7H_{14}O$	trans-4-Methylcyclohexanol					-433.3				-367.2			
$C_7H_{14}O_2$	Heptanoic acid					-610.2			265.4	-536.2			
$C_7H_{14}O_2$	Pentyl acetate								261.0				
$C_7H_{14}O_2$	Isopentyl acetate								248.5				
$C_7H_{14}O_2$	Ethyl pentanoate					-553.0				-505.9			
$C_7H_{14}O_2$	Ethyl 3-methylbutanoate					-571.0				-527.0			
$C_7H_{14}O_2$	Ethyl 2,2-dimethylpropanoate					-577.2				-536.0			
$C_7H_{14}O_2$	Methyl hexanoate					-540.2				-492.2			
$C_7H_{14}O_6$	α-Methylglucoside	-1233.3											
$C_7H_{15}Br$	1-Bromoheptane					-218.4				-167.8			
C_7H_{16}	Heptane					-224.2			224.7	-187.6			
C_7H_{16}	2-Methylhexane					-229.5		323.3	222.9	-194.5			
C_7H_{16}	3-Methylhexane					-226.4				-191.3			
C_7H_{16}	3-Ethylpentane					-224.9		314.5	219.6	-189.5			

Standard Thermodynamic Properties of Chemical Substances

Molecular formula	Name	Crystal ΔfH° kJ/mol	Crystal ΔfG° kJ/mol	Crystal S° J/mol K	Crystal Cp J/mol K	Liquid ΔfH° kJ/mol	Liquid ΔfG° kJ/mol	Liquid S° J/mol K	Liquid Cp J/mol K	Gas ΔfH° kJ/mol	Gas ΔfG° kJ/mol	Gas S° J/mol K	Gas Cp J/mol K
	2,2-Dimethylpentane					-238.3		300.3	221.1	-205.7			
	2,3-Dimethylpentane					-233.1				-198.7			
	2,4-Dimethylpentane					-234.6		303.2	224.2	-201.6			
	3,3-Dimethylpentane					-234.2				-201.0			
	2,2,3-Trimethylbutane					-236.5		292.2	213.5	-204.4			
	1-Heptanol					-403.3			272.1	-336.5			
	tert-Butyl isopropyl ether					-392.8				-358.1			
	1,7-Heptanediol					-574.2							
	2,2-Diethoxypropane					-538.9				-506.9			
	1-Heptanethiol					-200.5				-149.9			
	Phthalic anhydride	-460.1		180.0	160.0					-371.4			
O2	1H-Indole-2,3-dione	-268.2											
4	Phthalic acid	-782.0		207.9	188.1								
4	Isophthalic acid	-803.0								-696.3			
4	Terephthalic acid	-816.1								-717.9			
	Benzo[b]thiophene	100.6								166.3			
	1H-Indole	86.6								156.5			
	Styrene					103.8			182.0	147.9			
	Phenyl vinyl ether					-26.2				22.7			
	Acetophenone					-142.5				-86.7			
2	o-Toluic acid	-416.5			174.9								
2	m-Toluic acid	-426.1			163.6								
2	p-Toluic acid	-429.2			169.0								
2	Methyl benzoate					-343.5			221.3	-287.9			
3	Methyl salicylate								249.0				
0	Acetanilide	-209.4			179.3								
	1,7-Octadiyne					334.4							
	Ethylbenzene					-12.3			183.2	29.9			
	o-Xylene					-24.4			186.1	19.1			
	m-Xylene					-25.4			183.0	17.3			
	p-Xylene					-24.4			181.5	18.0			
0	2-Ethylphenol					-208.8				-145.2			
0	3-Ethylphenol					-214.3				-146.1			
0	4-Ethylphenol	-224.4			206.9					-144.1			
0	2,3-Xylenol	-241.1								-157.2			
0	2,4-Xylenol					-228.7				-163.8			
0	2,5-Xylenol	-246.6								-161.6			
0	2,6-Xylenol	-237.4								-162.1			
0	3,4-Xylenol	-242.3								-157.3			
0	3,5-Xylenol	-244.4								-162.4			
0	Benzeneethanol								252.6				
0	Ethoxybenzene					-152.6			228.5	-101.6			
O2	1,2-Dimethoxybenzene					-290.3				-223.3			
N	N-Ethylaniline					8.2				56.3			
N	N,N-Dimethylaniline					46.0				100.5			
N	2,4-Dimethylaniline					-39.2							
N	2,5-Dimethylaniline					-38.9							
N	2,6-Dimethylaniline								238.9				
2	1-Octen-3-yne					140.7							
2	cis-1,2-Divinylcyclobutane					124.3				166.5			
2	trans-1,2-Divinylcyclobutane					101.3				143.5			
2N4	2,2'-Azobis[isobutyronitrile]	246.0			237.6								
2O2	2,2,4,4-Tetramethyl-1,3-cyclobutanedione	-379.9								-307.6			
4	Ethylidenecyclohexane					-103.5				-59.5			
4	Allylcyclopentane					-64.5				-24.1			
4ClN5	Atrazine	-125.4											
4O3	Butanoic anhydride								283.7				
15ClO2	3-Methylbutyl 2-chloropropanoate					-627.3				-575.0			
15ClO2	3-Methylbutyl 3-chloropropanoate					-593.4				-539.4			
15N	Octanenitrile					-107.3				-50.5			
16	1-Octene					-124.5			241.0	-81.3			
16	cis-2-Octene					-135.7			239.0				
16	trans-2-Octene					-135.7			239.0				
16	cis-2,2-Dimethyl-3-hexene					-126.4				-89.3			
16	trans-2,2-Dimethyl-3-hexene					-144.9				-107.7			

Molecular formula	Name	Crystal $\Delta_f H°$ kJ/mol	$\Delta_f G°$ kJ/mol	$S°$ J/mol K	C_p J/mol K	Liquid $\Delta_f H°$ kJ/mol	$\Delta_f G°$ kJ/mol	$S°$ J/mol K	C_p J/mol K	Gas $\Delta_f H°$ kJ/mol	$\Delta_f G°$ kJ/mol	$S°$ J/mol K	C_p J/mol K
C_8H_{16}	3-Ethyl-2-methyl-1-pentene					-137.9				-100.3			
C_8H_{16}	2,4,4-Trimethyl-1-pentene					-145.9				-110.5			
C_8H_{16}	2,4,4-Trimethyl-2-pentene					-142.4				-104.9			
C_8H_{16}	Cyclooctane					-167.7				-124.4			
C_8H_{16}	Ethylcyclohexane					-212.1		280.9	211.8	-171.5			
C_8H_{16}	1,1-Dimethylcyclohexane					-218.7		267.2	209.2	-180.9			
C_8H_{16}	cis-1,2-Dimethylcyclohexane					-211.8		274.1	210.2	-172.1			
C_8H_{16}	trans-1,2-Dimethylcyclohexane					-218.2		273.2	209.4	-179.9			
C_8H_{16}	cis-1,3-Dimethylcyclohexane					-222.9		272.6	209.4	-184.6			
C_8H_{16}	trans-1,3-Dimethylcyclohexane					-215.7		276.3	212.8	-176.5			
C_8H_{16}	cis-1,4-Dimethylcyclohexane					-215.6		271.1	212.1	-176.6			
C_8H_{16}	trans-1,4-Dimethylcyclohexane					-222.4		268.0	210.2	-184.5			
C_8H_{16}	Propylcyclopentane					-188.8		310.8	216.3	-147.7			
C_8H_{16}	1-Ethyl-1-methylcyclopentane					-193.8							
C_8H_{16}	cis-1-Ethyl-2-methylcyclopentane					-190.8							
C_8H_{16}	trans-1-Ethyl-2-methylcyclopentane					-195.1				-156.2			
C_8H_{16}	cis-1-Ethyl-3-methylcyclopentane					-194.4							
C_8H_{16}	trans-1-Ethyl-3-methylcyclopentane					-196.0							
$C_8H_{16}O$	Octanal									-291.9		365.4	
$C_8H_{16}O$	2-Ethylhexanal					-348.5				-299.6			
$C_8H_{16}O$	2-Octanone								273.3				
$C_8H_{16}O$	2,2,4-Trimethyl-3-pentanone					-381.6				-338.3			
$C_8H_{16}O_2$	Octanoic acid					-636.0			297.9	-554.3			
$C_8H_{16}O_2$	2-Ethylhexanoic acid					-635.1				-559.5			
$C_8H_{16}O_2$	Hexyl acetate								282.8				
$C_8H_{16}O_2$	Isobutyl isobutanoate					-587.4				-542.9			
$C_8H_{16}O_2$	Propyl pentanoate					-583.0				-533.6			
$C_8H_{16}O_2$	Isopropyl pentanoate					-592.2				-544.9			
$C_8H_{16}O_2$	Methyl heptanoate					-567.1			285.1	-515.5			
$C_8H_{17}Br$	1-Bromooctane					-245.1				-189.3			
$C_8H_{17}Cl$	1-Chlorooctane					-291.3				-238.9			
$C_8H_{17}NO$	Octanamide	-473.2								-362.7			
C_8H_{18}	Octane					-250.1			254.6	-208.5			
C_8H_{18}	2-Methylheptane					-255.0		356.4	252.0	-215.3			
C_8H_{18}	3-Methylheptane, (S)					-252.3		362.6	250.2	-212.5			
C_8H_{18}	4-Methylheptane					-251.6			251.1	-211.9			
C_8H_{18}	3-Ethylhexane					-250.4				-210.7			
C_8H_{18}	2,2-Dimethylhexane					-261.9				-224.5			
C_8H_{18}	2,3-Dimethylhexane					-252.6				-213.8			
C_8H_{18}	2,4-Dimethylhexane					-257.0				-219.2			
C_8H_{18}	2,5-Dimethylhexane					-260.4			249.2	-222.5			
C_8H_{18}	3,3-Dimethylhexane					-257.5			246.6	-219.9			
C_8H_{18}	3,4-Dimethylhexane					-251.8				-212.8			
C_8H_{18}	3-Ethyl-2-methylpentane					-249.6				-211.0			
C_8H_{18}	3-Ethyl-3-methylpentane					-252.8				-214.8			
C_8H_{18}	2,2,3-Trimethylpentane					-256.9				-220.0			
C_8H_{18}	2,2,4-Trimethylpentane					-259.2			239.1	-224.0			
C_8H_{18}	2,3,3-Trimethylpentane					-253.5			245.6	-216.3			
C_8H_{18}	2,3,4-Trimethylpentane					-255.0		329.3	247.3	-217.3			
C_8H_{18}	2,2,3,3-Tetramethylbutane	-269.0		273.7	239.2					-226.0			
$C_8H_{18}N_2$	Azobutane					-40.1				9.2			
$C_8H_{18}O$	1-Octanol					-426.5			305.2	-355.6			
$C_8H_{18}O$	2-Octanol								330.1				
$C_8H_{18}O$	2-Ethyl-1-hexanol					-432.8		347.0	317.5	-365.3			
$C_8H_{18}O$	Dibutyl ether					-377.9			278.2	-332.8			
$C_8H_{18}O$	Di-sec-butyl ether					-401.5				-360.6			
$C_8H_{18}O$	Di-tert-butyl ether					-399.6			276.1	-362.0			
$C_8H_{18}O$	tert-Butyl isobutyl ether					-409.1				-369.0			
$C_8H_{18}O_2$	1,8-Octanediol	-626.6											
$C_8H_{18}O_2$	2,5-Dimethyl-2,5-hexanediol	-681.7											
$C_8H_{18}O_3$	Diethylene glycol monobutyl ether								354.9				
$C_8H_{18}O_3$	Diethylene glycol diethyl ether								341.4				
$C_8H_{18}O_3S$	Dibutyl sulfite					-693.1				-625.3			
$C_8H_{18}O_5$	Tetraethylene glycol					-981.7			428.8	-883.0			
$C_8H_{18}S$	Dibutyl sulfide					-220.7		405.1	284.3	-167.7			
$C_8H_{18}S$	Di-sec-butyl sulfide					-220.7				-167.7			

Standard Thermodynamic Properties of Chemical Substances

Molecular formula	Name	Crystal $\Delta_f H°$ kJ/mol	$\Delta_f G°$ kJ/mol	$S°$ J/mol K	C_p J/mol K	Liquid $\Delta_f H°$ kJ/mol	$\Delta_f G°$ kJ/mol	$S°$ J/mol K	C_p J/mol K	Gas $\Delta_f H°$ kJ/mol	$\Delta_f G°$ kJ/mol	$S°$ J/mol K	C_p J/mol K
	Di-tert-butyl sulfide					-232.6				-188.8			
	Diisobutyl sulfide					-229.2				-180.5			
S_2	Dibutyl disulfide					-222.9				-160.6			
S_2	Di-tert-butyl disulfide					-255.2				-201.0			
N	Dibutylamine					-206.0			292.9	-156.6			
N	Diisobutylamine					-218.5				-179.2			
BrN	Tetraethylammonium bromide	-342.7											
O_4Si	Ethyl silicate							533.1	364.4				
Pb	Tetraethyl lead					52.7	464.6	307.4		109.6			
Si	Tetraethylsilane							298.1					
$_2O_2$	Toluene-2,4-diisocyanate								287.8				
	Quinoline					141.2				200.5			
	Isoquinoline					144.3		216.0	196.2	204.6			
O	2-Quinolinol	-144.9								-25.5			
O	8-Quinolinol	82.1											
	Indene					110.6		215.3	186.9	163.4			
$_4$	2-(Acetyloxy)benzoic acid	-815.6											
	Cyclopropylbenzene					100.3				150.5			
	Indan					11.5		56.0	190.2	60.3			
Cl_2N_2O	Diuron	-329.0											
N_2	2,2'-Dipyrrolylmethane	126.2											
O_2	Ethyl benzoate								246.0				
O_2	Benzyl acetate								148.5				
NO_2	L-Phenylalanine	-466.9		213.6	203.0					-312.9			
NO_3	L-Tyrosine	-685.1		214.0	216.4								
	Propylbenzene					-38.3		287.8	214.7	7.9			
	Isopropylbenzene					-41.1			210.7	4.0			
	2-Ethyltoluene					-46.4				1.3			
	3-Ethyltoluene					-48.7				-1.8			
	4-Ethyltoluene					-49.8				-3.2			
	1,2,3-Trimethylbenzene					-58.5		267.9	216.4	-9.5			
	1,2,4-Trimethylbenzene					-61.8			215.0	-13.8			
	1,3,5-Trimethylbenzene					-63.4			209.3	-15.9			
$_2O$	2-Isopropylphenol					-233.7				-182.2			
$_2O$	3-Isopropylphenol					-252.5				-196.0			
$_2O$	4-Isopropylphenol	-270.0								-175.3			
$_2O_2$	Isopropylbenzene hydroperoxide					-148.3				-78.4			
$_3NO_2$	Ethyl 3,5-dimethylpyrrole-2-carboxylate	-474.5											
$_3NO_2$	Ethyl 2,4-dimethylpyrrole-3-carboxylate	-463.2											
$_3NO_2$	Ethyl 2,5-dimethylpyrrole-3-carboxylate	-478.7											
$_3NO_2$	Ethyl 4,5-dimethylpyrrole-3-carboxylate	-470.3											
$_4O$	Isophorone								253.5				
$_{14}O_6$	Triacetin					-1330.8		458.3	384.7	-1245.0			
$_{15}N$	3-Ethyl-2,4,5-trimethylpyrrole	-89.2											
$_6$	1-Nonyne					16.3				62.3			
$_{16}O_4$	Nonanedioic acid	-1054.3											
$_{17}NO$	2,2,6,6-Tetramethyl-4-piperidinone	-334.2								-273.4			
$_{18}$	Propylcyclohexane					-237.4		311.9	242.0	-192.3			
$_{18}$	1α,3α,5β-1,3,5-Trimethylcyclohexane									-212.1			
$_{18}O$	2-Nonanone					-397.2				-340.7			
$_{18}O$	5-Nonanone					-398.2		401.4	303.6	-344.9			
$_{18}O$	2,6-Dimethyl-4-heptanone					-408.5			297.3	-357.6			
$_{18}O_2$	Nonanoic acid					-659.7			362.4	-577.3			
$_{18}O_2$	Butyl pentanoate					-613.3				-560.2			
$_{18}O_2$	sec-Butyl pentanoate					-624.2				-573.2			
$_{18}O_2$	Isobutyl pentanoate					-620.0				-568.6			
$_{18}O_2$	Methyl octanoate					-590.3				-533.9			
$_{19}N$	N-Butylpiperidine					-171.8							
$_{19}N$	2,2,6,6-Tetramethylpiperidine					-206.9				-159.9			
$_{20}$	Nonane					-274.7			284.4	-228.2			
$_{20}$	2,2-Dimethylheptane					-288.1							
$_{20}$	2,2,3-Trimethylhexane					-282.7							

Molecular Formula	Name	Crystal				Liquid				Gas			
		$\Delta_f H°$ kJ/mol	$\Delta_f G°$ kJ/mol	$S°$ J/mol K	C_p J/mol K	$\Delta_f H°$ kJ/mol	$\Delta_f G°$ kJ/mol	$S°$ J/mol K	C_p J/mol K	$\Delta_f H°$ kJ/mol	$\Delta_f G°$ kJ/mol	$S°$ J/mol K	C_p J/mol K
C9H20	2,2,4-Trimethylhexane					-282.8							
C9H20	2,2,5-Trimethylhexane					-293.3							
C9H20	2,3,3-Trimethylhexane					-281.1							
C9H20	2,3,5-Trimethylhexane					-284.0				-242.6			
C9H20	2,4,4-Trimethylhexane					-280.2							
C9H20	3,3,4-Trimethylhexane					-277.5							
C9H20	3,3-Diethylpentane					-275.4			278.2	-233.3			
C9H20	3-Ethyl-2,2-dimethylpentane					-272.7							
C9H20	3-Ethyl-2,4-dimethylpentane					-269.7							
C9H20	2,2,3,3-Tetramethylpentane					-278.3			271.5	-237.1			
C9H20	2,2,3,4-Tetramethylpentane					-277.7				-236.9			
C9H20	2,2,4,4-Tetramethylpentane					-280.0			266.3	-241.6			
C9H20	2,3,3,4-Tetramethylpentane					-277.9				-236.1			
C9H20N2O	Tetraethylurea					-380.0				-316.4			
C9H20O	1-Nonanol					-453.4				-376.5			
C9H20O2	1,9-Nonanediol	-657.6											
C9H21N	Tripropylamine					-207.1				-161.0			
C10H6N2	2-Quinolinecarbonitrile	246.5											
C10H6N2	3-Quinolinecarbonitrile	242.3											
C10H6N2O4	1,5-Dinitronaphthalene	29.8											
C10H6N2O4	1,8-Dinitronaphthalene	39.7											
C10H7Cl	1-Chloronaphthalene					54.6			212.6	119.8			
C10H7Cl	2-Chloronaphthalene	55.4								137.4			
C10H7I	1-Iodonaphthalene					161.5				233.8			
C10H7I	2-Iodonaphthalene	144.3								235.1			
C10H7NO2	1-Nitronaphthalene	42.6								111.2			
C10H8	Naphthalene	78.5	201.6	167.4	165.7					150.6	224.1	333.1	131.9
C10H8	Azulene	212.3								289.5			
C10H8O	1-Naphthol	-121.5		166.9						-30.4			149.4
C10H8O	2-Naphthol	-124.1	179.0	172.8						-29.9		366.6	147.8
C10H9N	1-Naphthylamine	67.8								132.8			
C10H9N	2-Naphthylamine	60.2								134.3			
C10H10	1,2-Dihydronaphthalene					71.6							
C10H10	1,4-Dihydronaphthalene					84.2							
C10H10O	1-Tetralone	-209.6											
C10H10O4	Dimethyl phthalate								303.1				
C10H10O4	Dimethyl isophthalate	-730.9											
C10H10O4	Dimethyl terephthalate	-732.6		261.1									
C10H12	1,2,3,4-Tetrahydronaphthalene					-29.2			217.5	26.0			
C10H14	Butylbenzene					-63.2		321.2	243.4	-11.8			
C10H14	sec-Butylbenzene, (±)					-66.4				-18.4			
C10H14	tert-Butylbenzene					-71.9				-23.0			
C10H14	Isobutylbenzene					-69.8				-21.9			
C10H14	1-Isopropyl-2-methylbenzene					-73.3							
C10H14	1-Isopropyl-3-methylbenzene					-78.6							
C10H14	1-Isopropyl-4-methylbenzene					-78.0			236.4				
C10H14	o-Diethylbenzene					-68.5							
C10H14	m-Diethylbenzene					-73.5							
C10H14	p-Diethylbenzene					-72.8							
C10H14	3-Ethyl-1,2-dimethylbenzene					-80.5							
C10H14	4-Ethyl-1,2-dimethylbenzene					-86.0							
C10H14	2-Ethyl-1,3-dimethylbenzene					-80.1							
C10H14	2-Ethyl-1,4-dimethylbenzene					-84.8							
C10H14	1-Ethyl-2,4-dimethylbenzene					-84.1							
C10H14	1-Ethyl-3,5-dimethylbenzene					-87.8							
C10H14	1,2,4,5-Tetramethylbenzene	-119.9		245.6	215.1								
C10H14O	Thymol	-309.7								-218.5			
C10H16	Dipentene					-50.8			249.4	-2.6			
C10H16	d-Limonene					-54.5			249.0				
C10H16	α-Pinene					-16.4				28.3			
C10H16	β-Pinene					-7.7				38.7			
C10H16	α-Terpinene									-20.6			
C10H16	β-Myrcene					14.5							
C10H16	cis, cis-2,6-Dimethyl-2,4,6-octatriene					-24.0							
C10H16N2O8	Ethylenediaminetetraacetic acid	-1759.5											

Name	Crystal $\Delta_f H°$ kJ/mol	$\Delta_f G°$ kJ/mol	$S°$ J/mol K	C_p J/mol K	Liquid $\Delta_f H°$ kJ/mol	$\Delta_f G°$ kJ/mol	$S°$ J/mol K	C_p J/mol K	Gas $\Delta_f H°$ kJ/mol	$\Delta_f G°$ kJ/mol	$S°$ J/mol K	C_p J/mol K
Camphor, (±)	-319.4			271.2					-267.5			
1,1'-Bicyclopentyl					-178.9							
cis-Decahydronaphthalene					-219.4		265.0	232.0	-169.2			
trans-Decahydronaphthalene					-230.6		264.9	228.5	-182.1			
Sebacic acid	-1082.6								-921.9			
Decanenitrile					-158.4				-91.5			
1-Decene					-173.8		425.0	300.8	-123.3			
cis-1,2-Di-tert-butylethene					-163.6							
Butylcyclohexane					-263.1		345.0	271.0	-213.7			
Decanoic acid	-713.7				-684.3				-594.9			
Methyl nonanoate					-616.2				-554.2			
1-Nitrodecane					-351.5							
Decane					-300.9			314.4	-249.5			
2-Methylnonane					-309.8		420.1	313.3	-260.2			
5-Methylnonane					-307.9		423.8	314.4	-258.6			
1-Decanol					-478.1			370.6	-396.6			
Dipentyl ether								250.0				
Diisopentyl ether								379.0				
1,10-Decanediol	-678.9											
Ethylene glycol dibutyl ether								350.0				
1-Decanethiol	-309.9				-276.5		476.1	350.4	-211.5			
Dipentyl sulfide					-266.4				-204.9			
Diisopentyl sulfide					-281.8				-221.5			
Octyldimethylamine					-232.8							
1-Naphthalenecarboxylic acid	-333.5								-223.1			
2-Naphthalenecarboxylic acid	-346.1								-232.5			
1-Methylnaphthalene					56.3		254.8	224.4				
2-Methylnaphthalene	44.9		220.0	196.0					106.7			
L-Tryptophan	-415.3		251.0	238.1								
1,1-Dimethylindan					-53.6				-1.6			
1-tert-Butyl-3-methylbenzene					-109.7							
1-tert-Butyl-4-methylbenzene					-109.7				-57.0			
Pentamethylbenzene	-144.6								-67.2			
Spiro[5.5]undecane					-244.5				-188.3			
1-Undecene								344.9				
Methyl decanoate					-640.5				-573.8			
Undecane					-327.2			344.9	-270.8			
1-Undecanol					-504.8							
Tris(perfluorobutyl)amine								418.4				
Acenaphthylene	186.7			166.4					259.7			
Phenazine	237.0								328.8			
Dibenzofuran	-5.3								83.4			
Dibenzothiophene	120.0								205.1			
Thianthrene	182.0								286.0			
Carbazole	101.7								200.7			
Acenaphthene	70.3		188.9	190.4					156.0			
Biphenyl	99.4		209.4	198.4					181.4			
trans-Azoxybenzene	243.4								342.0			
N-Nitrosodiphenylamine	227.2											
Diphenyl ether	-32.1		233.9	216.6	-14.9				52.0			
1-Naphthaleneacetic acid	-359.2											
2-Naphthaleneacetic acid	-371.9											
Diphenylamine	130.2								219.3			
2-Aminobiphenyl	93.8								184.4			
4-Aminobiphenyl	81.0											
p-Benzidine	70.7											
Diethyl phthalate					-776.6		425.1	366.1	-688.4			
Cyclohexylbenzene					-76.6				-16.7			
Diethyl 3,5-dimethylpyrrole-2,4-dicarboxylate	-916.7											
3,9-Dodecadiyne					197.8							
5,7-Dodecadiyne					181.5							
1-tert-Butyl-3,5-dimethylbenzene					-146.5							
Hexamethylbenzene	-162.4		306.3	245.6					-77.4			
Cyclohexylcyclohexane					-273.7				-215.7			
Dodecanedioic acid	-1130.0								-976.9			

Standard Thermodynamic Properties of Chemical Substances

Molecular formula	Name	Crystal $\Delta_f H°$ kJ/mol	$\Delta_f G°$ kJ/mol	$S°$ J/mol K	C_p J/mol K	Liquid $\Delta_f H°$ kJ/mol	$\Delta_f G°$ kJ/mol	$S°$ J/mol K	C_p J/mol K	Gas $\Delta_f H°$ kJ/mol	$\Delta_f G°$ kJ/mol	$S°$ J/mol K	C_p J/mol K
$C_{12}H_{22}O_{11}$	Sucrose	-2226.1											
$C_{12}H_{22}O_{11}$	β-D-Lactose	-2236.7											
$C_{12}H_{24}$	1-Dodecene					-226.2		484.8	360.7	-165.4			
$C_{12}H_{24}O_2$	Dodecanoic acid	-774.6			404.3	-737.9				-642.0			
$C_{12}H_{24}O_2$	Methyl undecanoate					-665.2				-593.8			
$C_{12}H_{24}O_{12}$	α-Lactose monohydrate	-2484.1											
$C_{12}H_{25}Br$	1-Bromododecane					-344.7				-269.9			
$C_{12}H_{25}Cl$	1-Chlorododecane					-392.3				-321.1			
$C_{12}H_{26}$	Dodecane					-350.9			375.8	-289.4			
$C_{12}H_{26}O$	1-Dodecanol					-528.5			438.1	-436.6			
$C_{12}H_{26}O_3$	Diethylene glycol dibutyl ether								452.0				
$C_{12}H_{27}N$	Tributylamine					-281.6							
$C_{12}H_{27}O_4P$	Tributyl phosphate								379.4				
$C_{13}H_8O_2$	Xanthone	-191.5											
$C_{13}H_9N$	Acridine	179.4								273.9			
$C_{13}H_9N$	Phenanthridine	141.9								240.5			
$C_{13}H_9N$	Benzo[f]quinoline	150.6								233.7			
$C_{13}H_{10}$	9H-Fluorene	89.9		207.3	203.1					175.0			173.1
$C_{13}H_{10}N_2$	9-Acridinamine	159.2											
$C_{13}H_{10}O$	Benzophenone	-34.5			224.8					54.9			
$C_{13}H_{11}N$	9-Methyl-9H-carbazole	105.5								201.0			
$C_{13}H_{12}$	Diphenylmethane	71.5		239.3		89.7				139.0			
$C_{13}H_{13}N$	N-Benzylaniline	101.4											
$C_{13}H_{14}N_2$	4,4'-Diaminodiphenylmethane			270.9									
$C_{13}H_{24}O_4$	Tridecanedioic acid	-1148.3											
$C_{13}H_{26}$	1-Tridecene								391.8				
$C_{13}H_{26}O_2$	Methyl dodecanoate					-693.0				-614.9			
$C_{13}H_{28}$	Tridecane								406.7				
$C_{13}H_{28}O$	1-Tridecanol	-599.4											
$C_{14}H_8O_2$	9,10-Anthracenedione	-188.5								-75.7			
$C_{14}H_8O_2$	9,10-Phenanthrenedione	-154.7								-46.6			
$C_{14}H_8O_4$	1,4-Dihydroxy-9,10-anthracenedione	-595.8								-471.7			
$C_{14}H_{10}$	Anthracene	129.2		207.5	210.5					230.9			
$C_{14}H_{10}$	Phenanthrene	116.2		215.1	220.6					207.5			
$C_{14}H_{10}$	Diphenylacetylene	312.4			225.9								
$C_{14}H_{10}O_2$	Benzil	-153.9								-55.5			
$C_{14}H_{10}O_4$	Benzoyl peroxide	-369.4								-281.7			
$C_{14}H_{12}$	cis-Stilbene					183.3				252.3			
$C_{14}H_{12}$	trans-Stilbene	136.9								236.1			
$C_{14}H_{14}$	1,1-Diphenylethane					48.7							
$C_{14}H_{14}$	1,2-Diphenylethane	51.5								142.9			
$C_{14}H_{22}$	1,3-Di-tert-butylbenzene					-188.8							
$C_{14}H_{22}$	1,4-Di-tert-butylbenzene	-212.0											
$C_{14}H_{23}N_3O_{10}$	Pentetic acid	-2225.2											
$C_{14}H_{27}N$	Tetradecanenitrile					-260.2				-174.9			
$C_{14}H_{28}O_2$	Tetradecanoic acid	-833.5			432.0	-788.8				-693.7			
$C_{14}H_{28}O_2$	Methyl tridecanoate					-717.9				-635.3			
$C_{14}H_{30}O$	1-Tetradecanol	-629.6			388.0	-580.6							
$C_{15}H_{16}O_2$	2,2-Bis(4-hydroxyphenyl)propane	-368.6											
$C_{15}H_{24}$	1,3-Di-tert-butyl-5-methylbenzene	-245.8											
$C_{15}H_{24}O$	2,6-Di-tert-butyl-4-methylphenol	-410.0								-296.9			
$C_{15}H_{30}$	Decylcyclopentane					-367.3							
$C_{15}H_{30}O_2$	Pentadecanoic acid	-861.7			443.3	-811.7				-699.0			
$C_{15}H_{30}O_2$	Methyl tetradecanoate					-743.9				-656.9			
$C_{15}H_{32}O$	1-Pentadecanol	-658.2											
$C_{16}H_{10}$	Fluoranthene	189.9		230.6	230.2					289.0			
$C_{16}H_{10}$	Pyrene	125.5		224.9	229.7					225.7			
$C_{16}H_{22}O_4$	Dibutyl phthalate					-842.6				-750.9			
$C_{16}H_{22}O_{11}$	α-D-Glucose pentaacetate	-2249.4											
$C_{16}H_{22}O_{11}$	β-D-Glucose pentaacetate	-2232.6											
$C_{16}H_{26}$	Decylbenzene					-218.3				-138.6			
$C_{16}H_{32}$	1-Hexadecene					-328.7		587.9	488.9	-248.4			
$C_{16}H_{32}O_2$	Hexadecanoic acid	-891.5		452.4	460.7	-838.1				-737.1			
$C_{16}H_{32}O_2$	Methyl pentadecanoate					-771.0				-680.0			
$C_{16}H_{33}Br$	1-Bromohexadecane					-444.5				-350.2			

Standard Thermodynamic Properties of Chemical Substances

Molecular Formula	Name	Crystal				Liquid				Gas			
		$\Delta_f H°$ kJ/mol	$\Delta_f G°$ kJ/mol	$S°$ J/mol K	C_p J/mol K	$\Delta_f H°$ kJ/mol	$\Delta_f G°$ kJ/mol	$S°$ J/mol K	C_p J/mol K	$\Delta_f H°$ kJ/mol	$\Delta_f G°$ kJ/mol	$S°$ J/mol K	C_p J/mol K
	Hexadecane					-456.1			501.6	-374.8			
)	1-Hexadecanol	-686.5			422.0					-517.0			
N	Tetrabutylammonium iodide	-498.6											
O₂	Heptadecanoic acid	-924.4			475.7	-865.6							
	Benz[a]anthracene	170.8								293.0			
	Chrysene	145.3								269.8			
	o-Terphenyl			298.8	274.8			337.1	369.1				
	p-Terphenyl	163.0		285.6	278.7					279.0			
N	Triphenylamine	234.7								326.8			
O₄P	Triphenyl phosphate			397.5	356.2								
P	Triphenylphosphine				312.5								
	1,3,5-Tri-tert-butylbenzene	-320.0											
O₂	Oleic acid								577.0				
O₄	Dibutyl sebacate								619.0				
O₂	Stearic acid	-947.7			501.5	-884.7				-781.2			
Cl	1-Chlorooctadecane					-544.1				-446.0			
	Octadecane	-567.4		480.2	485.6					-414.6			
N	Trihexylamine					-433.0							
O	Triphenylmethanol	-2.5											
O₂	Methyl oleate					-734.5				-649.9			
O₂	Methyl trans-9-octadecenoate					-737.0							
	Perylene	182.8		264.6	274.9								
	Benzo[a]pyrene												254.8
O₄	Diphenyl phthalate	-489.2											
O₂	Ethyl oleate					-775.8							
O₂	Ethyl trans-9-octadecenoate					-773.3							
O₂	Eicosanoic acid	-1011.9			545.1	-940.0				-812.4			
O₄P	Tri-o-cresyl phosphate			570.0	578.0								283.9
	Dibenz[a,h]anthracene												
O₂	trans-13-Docosenoic acid	-960.7											
O₂	Butyl oleate					-816.9							
O₂	Butyl stearate												
O₄	Bis(2-ethylhexyl) phthalate								704.7				
N	Trioctylamine					-585.0							
	9,10-Diphenylanthracene	308.7								465.6			
	5-Butyldocosane					-713.5				-587.6			
	11-Butyldocosane					-716.0				-593.4			
	9,9'-Bianthracene	326.2								454.3			
	11-Decylheneicosane					-848.0				-705.8			
	Dotriacontane	-968.3								-697.2			
	Carbon (fullerene-C₆₀)	2327.0	2302.0	426.0	520.0					2502.0	2442.0	544.0	512.0
	Carbon (fullerene-C₇₀)	2555.0	2537.0	464.0	650.0					2755.0	2692.0	614.0	585.0

THERMODYNAMIC PROPERTIES AS A FUNCTION OF TEMPERATURE

L. V. Gurvich, V. S. Iorish, V. S. Yungman, and O. V. Dorofeeva

The thermodynamic properties $C_p^\circ(T)$, $S^\circ(T)$, $H^\circ(T)-H^\circ(T_r)$, $-[G^\circ(T)-H^\circ(T_r)]/T$ and formation properties $\Delta_f H^\circ(T)$, $\Delta_f G^\circ(T)$, $\log K_f^\circ(T)$ are tabulated as functions of temperature in the range 298.15 to 1500 K for 80 substances in the standard state. The reference temperature, T_r, is equal to 298.15 K. The standard state pressure is taken as 1 bar (100,000 Pa). The tables are presented in the JANAF Thermochemical Tables format (Reference 2). The numerical data are extracted from IVTANTHERMO databases except for C_2H_4O, C_3H_6O, C_6H_6, C_6H_6O, $C_{10}H_8$, and CH_5N, which are based upon TRC Tables. See the references for information on standard states and other details.

References

1. Gurvich, L. V., Veyts, I. V., and Alcock, C. B., Eds., *Thermodynamic Properties of Individual Substances, 4th ed.*, Hemisphere Publishing Corp., New York, 1989.
2. Chase, M. W., et al., *JANAF Thermochemical Tables, 3rd ed., J. Phys. Chem. Ref. Data*, 14, Suppl. 1, 1985.

Order of Listing of Tables

No.	Formula	Name	State	No.	Formula	Name	State
1	Ar	Argon	g	42	$CuCl_2$	Copper dichloride	g
2	Br	Bromine	g	43	F	Fluorine	g
3	Br_2	Dibromine	g	44	F_2	Difluorine	g
4	BrH	Hydrogen bromide	g	45	FH	Hydrogen fluoride	g
5	C	Carbon (graphite)	cr	46	Ge	Germanium	cr, l
6	C	Carbon (diamond)	cr	47	Ge	Germanium	g
7	C_2	Dicarbon	g	48	GeO_2	Germanium dioxide	cr, l
8	C_3	Tricarbon	g	49	$GeCl_4$	Germanium tetrachloride	g
9	CO	Carbon oxide	g	50	H	Hydrogen	g
10	CO_2	Carbon dioxide	g	51	H_2	Dihydrogen	g
11	CH_4	Methane	g	52	HO	Hydroxyl	g
12	C_2H_2	Acetylene	g	53	H_2O	Water	l
13	C_2H_4	Ethylene	g	54	H_2O	Water	g
14	C_2H_6	Ethane	g	55	I	Iodine	g
15	C_3H_6	Cyclopropane	g	56	I_2	Diiodine	cr, l
16	C_3H_8	Propane	g	57	I_2	Diiodine	g
17	C_6H_6	Benzene	l	58	IH	Hydrogen iodide	g
18	C_6H_6	Benzene	g	59	K	Potassium	cr, l
19	$C_{10}H_8$	Naphthalene	cr, l	60	K	Potassium	g
20	$C_{10}H_8$	Naphthalene	g	61	K_2O	Dipotassium oxide	cr, l
21	CH_2O	Formaldehyde	g	62	KOH	Potassium hydroxide	cr, l
22	CH_4O	Methanol	g	63	KOH	Potassium hydroxide	g
23	C_2H_4O	Acetaldehyde	g	64	KCl	Potassium chloride	cr, l
24	C_2H_6O	Ethanol	g	65	KCl	Potassium chloride	g
25	$C_2H_4O_2$	Acetic acid	g	66	N_2	Dinitrogen	g
26	C_3H_6O	Acetone	g	67	NO	Nitric oxide	g
27	C_6H_6O	Phenol	g	68	NO_2	Nitrogen dioxide	g
28	CF_4	Carbon tetrafluoride	g	69	NH_3	Ammonia	g
29	CHF_3	Trifluoromethane	g	70	O	Oxygen	g
30	$CClF_3$	Chlorotrifluoromethane	g	71	O_2	Dioxygen	g
31	CCl_2F_2	Dichlorodifluoromethane	g	72	S	Sulfur	cr, l
32	$CHClF_2$	Chlorodifluoromethane	g	73	S	Sulfur	g
33	CH_5N	Methylamine	g	74	S_2	Disulfur	g
34	Cl	Chlorine	g	75	S_8	Octasulfur	g
35	Cl_2	Dichlorine	g	76	SO_2	Sulfur dioxide	g
36	ClH	Hydrogen chloride	g	77	Si	Silicon	cr
37	Cu	Copper	cr, l	78	Si	Silicon	g
38	Cu	Copper	g	79	SiO_2	Silicon dioxide	cr
39	CuO	Copper oxide	cr	80	$SiCl_4$	Silicon tetrachloride	g
40	Cu_2O	Dicopper oxide	cr				
41	$CuCl_2$	Copper dichloride	cr, l				

T/K	J/K·mol			kJ/mol			Log K_f
	$C_p°$	$S°$	$-(G°-H°(T_r))/T$	$H°-H°(T_r)$	$\Delta_f H°$	$\Delta_f G°$	

1. ARGON   Ar (g)

T/K	$C_p°$	$S°$	$-(G°-H°(T_r))/T$	$H°-H°(T_r)$	$\Delta_f H°$	$\Delta_f G°$	Log K_f
298.15	20.786	154.845	154.845	0.000	0.000	0.000	0.000
300	20.786	154.973	154.845	0.038	0.000	0.000	0.000
400	20.786	160.953	155.660	2.117	0.000	0.000	0.000
500	20.786	165.591	157.200	4.196	0.000	0.000	0.000
600	20.786	169.381	158.924	6.274	0.000	0.000	0.000
700	20.786	172.585	160.653	8.353	0.000	0.000	0.000
800	20.786	175.361	162.322	10.431	0.000	0.000	0.000
900	20.786	177.809	163.909	12.510	0.000	0.000	0.000
1000	20.786	179.999	165.410	14.589	0.000	0.000	0.000
1100	20.786	181.980	166.828	16.667	0.000	0.000	0.000
1200	20.786	183.789	168.167	18.746	0.000	0.000	0.000
1300	20.786	185.453	169.434	20.824	0.000	0.000	0.000
1400	20.786	186.993	170.634	22.903	0.000	0.000	0.000
1500	20.786	188.427	171.773	24.982	0.000	0.000	0.000

2. BROMINE   Br (g)

T/K	$C_p°$	$S°$	$-(G°-H°(T_r))/T$	$H°-H°(T_r)$	$\Delta_f H°$	$\Delta_f G°$	Log K_f
298.15	20.786	175.017	175.017	0.000	111.870	82.379	−14.432
300	20.786	175.146	175.018	0.038	111.838	82.196	−14.311
400	20.787	181.126	175.833	2.117	96.677	75.460	−9.854
500	20.798	185.765	177.373	4.196	96.910	70.129	−7.326
600	20.833	189.559	179.097	6.277	97.131	64.752	−5.637
700	20.908	192.776	180.827	8.364	97.348	59.338	−4.428
800	21.027	195.575	182.499	10.461	97.568	53.893	−3.519
900	21.184	198.061	184.093	12.571	97.796	48.420	−2.810
1000	21.365	200.302	185.604	14.698	98.036	42.921	−2.242
1100	21.559	202.347	187.034	16.844	98.291	37.397	−1.776
1200	21.752	204.231	188.390	19.010	98.560	31.850	−1.386
1300	21.937	205.980	189.676	21.195	98.844	26.279	−1.056
1400	22.107	207.612	190.900	23.397	99.141	20.686	−0.772
1500	22.258	209.142	192.065	25.615	99.449	15.072	−0.525

3. DIBROMINE   Br$_2$ (g)

T/K	$C_p°$	$S°$	$-(G°-H°(T_r))/T$	$H°-H°(T_r)$	$\Delta_f H°$	$\Delta_f G°$	Log K_f
298.15	36.057	245.467	245.467	0.000	30.910	3.105	−0.544
300	36.074	245.690	245.468	0.067	30.836	2.933	−0.511
332.25	36.340	249.387	245.671	1.235		pressure = 1 bar	
400	36.729	256.169	246.892	3.711	0.000	0.000	0.000
500	37.082	264.406	249.600	7.403	0.000	0.000	0.000
600	37.305	271.188	252.650	11.123	0.000	0.000	0.000
700	37.464	276.951	255.720	14.862	0.000	0.000	0.000
800	37.590	281.962	258.694	18.615	0.000	0.000	0.000
900	37.697	286.396	261.530	22.379	0.000	0.000	0.000
1000	37.793	290.373	264.219	26.154	0.000	0.000	0.000
1100	37.883	293.979	266.763	29.938	0.000	0.000	0.000
1200	37.970	297.279	269.170	33.730	0.000	0.000	0.000
1300	38.060	300.322	271.451	37.532	0.000	0.000	0.000
1400	38.158	303.146	273.615	41.343	0.000	0.000	0.000
1500	38.264	305.782	275.673	45.164	0.000	0.000	0.000

4. HYDROGEN BROMIDE   HBr (g)

T/K	$C_p°$	$S°$	$-(G°-H°(T_r))/T$	$H°-H°(T_r)$	$\Delta_f H°$	$\Delta_f G°$	Log K_f
298.15	29.141	198.697	198.697	0.000	−36.290	−53.360	9.348
300	29.141	198.878	198.698	0.054	−36.333	−53.466	9.309
400	29.220	207.269	199.842	2.971	−52.109	−55.940	7.305
500	29.454	213.811	202.005	5.903	−52.484	−56.854	5.939
600	29.872	219.216	204.436	8.868	−52.844	−57.694	5.023
700	30.431	223.861	206.886	11.882	−53.168	−58.476	4.363
800	31.063	227.965	209.269	14.957	−53.446	−59.214	3.866
900	31.709	231.661	211.555	18.095	−53.677	−59.921	3.478
1000	32.335	235.035	213.737	21.298	−53.864	−60.604	3.166
1100	32.919	238.145	215.816	24.561	−54.012	−61.271	2.909

T/K	C°_p	S°	−(G°−H°(T_r))/T	H°−H°(T_r)	Δ_fH°	Δ_fG°	Log K_f
		J/K·mol			kJ/mol		
1200	33.454	241.032	217.799	27.880	−54.129	−61.925	2.696
1300	33.938	243.729	219.691	31.250	−54.220	−62.571	2.514
1400	34.374	246.261	221.499	34.666	−54.291	−63.211	2.358
1500	34.766	248.646	223.230	38.123	−54.348	−63.846	2.223

5. CARBON (GRAPHITE) C (cr; graphite)

T/K	C°_p	S°	−(G°−H°(T_r))/T	H°−H°(T_r)	Δ_fH°	Δ_fG°	Log K_f
298.15	8.536	5.740	5.740	0.000	0.000	0.000	0.000
300	8.610	5.793	5.740	0.016	0.000	0.000	0.000
400	11.974	8.757	6.122	1.054	0.000	0.000	0.000
500	14.537	11.715	6.946	2.385	0.000	0.000	0.000
600	16.607	14.555	7.979	3.945	0.000	0.000	0.000
700	18.306	17.247	9.113	5.694	0.000	0.000	0.000
800	19.699	19.785	10.290	7.596	0.000	0.000	0.000
900	20.832	22.173	11.479	9.625	0.000	0.000	0.000
1000	21.739	24.417	12.662	11.755	0.000	0.000	0.000
1100	22.452	26.524	13.827	13.966	0.000	0.000	0.000
1200	23.000	28.502	14.968	16.240	0.000	0.000	0.000
1300	23.409	30.360	16.082	18.562	0.000	0.000	0.000
1400	23.707	32.106	17.164	20.918	0.000	0.000	0.000
1500	23.919	33.749	18.216	23.300	0.000	0.000	0.000

6. CARBON (DIAMOND) C (cr; diamond)

T/K	C°_p	S°	−(G°−H°(T_r))/T	H°−H°(T_r)	Δ_fH°	Δ_fG°	Log K_f
298.15	6.109	2.362	2.362	0.000	1.850	2.857	−0.501
300	6.201	2.400	2.362	0.011	1.846	2.863	−0.499
400	10.321	4.783	2.659	0.850	1.645	3.235	−0.422
500	13.404	7.431	3.347	2.042	1.507	3.649	−0.381
600	15.885	10.102	4.251	3.511	1.415	4.087	−0.356
700	17.930	12.709	5.274	5.205	1.361	4.537	−0.339
800	19.619	15.217	6.361	7.085	1.338	4.993	−0.326
900	21.006	17.611	7.479	9.118	1.343	5.450	−0.316
1000	22.129	19.884	8.607	11.277	1.372	5.905	−0.308
1100	23.020	22.037	9.731	13.536	1.420	6.356	−0.302
1200	23.709	24.071	10.842	15.874	1.484	6.802	−0.296
1300	24.222	25.990	11.934	18.272	1.561	7.242	−0.291
1400	24.585	27.799	13.003	20.714	1.646	7.675	−0.286
1500	24.824	29.504	14.047	23.185	1.735	8.103	−0.282

7. DICARBON C_2 (g)

T/K	C°_p	S°	−(G°−H°(T_r))/T	H°−H°(T_r)	Δ_fH°	Δ_fG°	Log K_f
298.15	43.548	197.095	197.095	0.000	830.457	775.116	−135.795
300	43.575	197.365	197.096	0.081	830.506	774.772	−134.898
400	42.169	209.809	198.802	4.403	832.751	755.833	−98.700
500	39.529	218.924	201.959	8.483	834.170	736.423	−76.933
600	37.837	225.966	205.395	12.342	834.909	716.795	−62.402
700	36.984	231.726	208.758	16.078	835.148	697.085	−52.016
800	36.621	236.637	211.943	19.755	835.020	677.366	−44.227
900	36.524	240.943	214.931	23.411	834.618	657.681	−38.170
1000	36.569	244.793	217.728	27.065	834.012	638.052	−33.328
1100	36.696	248.284	220.349	30.728	833.252	618.492	−29.369
1200	36.874	251.484	222.812	34.406	832.383	599.006	−26.074
1300	37.089	254.444	225.133	38.104	831.437	579.596	−23.288
1400	37.329	257.201	227.326	41.824	830.445	560.261	−20.903
1500	37.589	259.785	229.405	45.570	829.427	540.997	−18.839

8. TRICARBON C_3 (g)

T/K	C°_p	S°	−(G°−H°(T_r))/T	H°−H°(T_r)	Δ_fH°	Δ_fG°	Log K_f
298.15	42.202	237.611	237.611	0.000	839.958	774.249	−135.643
300	42.218	237.872	237.611	0.078	839.989	773.841	−134.736
400	43.383	250.164	239.280	4.354	841.149	751.592	−98.147
500	44.883	260.003	242.471	8.766	841.570	729.141	−76.172
600	46.406	268.322	246.104	13.331	841.453	706.659	−61.519
700	47.796	275.582	249.807	18.042	840.919	684.230	−51.057
800	48.997	282.045	253.440	22.884	840.053	661.901	−43.217

	J/K·mol			kJ/mol			
T/K	$C_p°$	$S°$	$-(G°-H°(T_r))/T$	$H°-H°(T_r)$	$\Delta_f H°$	$\Delta_f G°$	$\text{Log } K_f$
900	50.006	287.876	256.948	27.835	838.919	639.698	−37.127
1000	50.844	293.189	260.310	32.879	837.572	617.633	−32.261
1100	51.535	298.069	263.524	37.999	836.059	595.711	−28.288
1200	52.106	302.578	266.593	43.182	834.420	573.933	−24.982
1300	52.579	306.768	269.524	48.417	832.690	552.295	−22.191
1400	52.974	310.679	272.326	53.695	830.899	530.793	−19.804
1500	53.307	314.346	275.006	59.010	829.068	509.421	−17.739

9. CARBON OXIDE CO (g)

298.15	29.141	197.658	197.658	0.000	−110.530	−137.168	24.031
300	29.142	197.838	197.659	0.054	−110.519	−137.333	23.912
400	29.340	206.243	198.803	2.976	−110.121	−146.341	19.110
500	29.792	212.834	200.973	5.930	−110.027	−155.412	16.236
600	30.440	218.321	203.419	8.941	−110.157	−164.480	14.319
700	31.170	223.067	205.895	12.021	−110.453	−173.513	12.948
800	31.898	227.277	208.309	15.175	−110.870	−182.494	11.915
900	32.573	231.074	210.631	18.399	−111.378	−191.417	11.109
1000	33.178	234.538	212.851	21.687	−111.952	−200.281	10.461
1100	33.709	237.726	214.969	25.032	−112.573	−209.084	9.928
1200	34.169	240.679	216.990	28.426	−113.228	−217.829	9.482
1300	34.568	243.430	218.920	31.864	−113.904	−226.518	9.101
1400	34.914	246.005	220.763	35.338	−114.594	−235.155	8.774
1500	35.213	248.424	222.527	38.845	−115.291	−243.742	8.488

10. CARBON DIOXIDE CO_2 (g)

298.15	37.135	213.783	213.783	0.000	−393.510	−394.373	69.092
300	37.220	214.013	213.784	0.069	−393.511	−394.379	68.667
400	41.328	225.305	215.296	4.004	−393.586	−394.656	51.536
500	44.627	234.895	218.280	8.307	−393.672	−394.914	41.256
600	47.327	243.278	221.762	12.909	−393.791	−395.152	34.401
700	49.569	250.747	225.379	17.758	−393.946	−395.367	29.502
800	51.442	257.492	228.978	22.811	−394.133	−395.558	25.827
900	53.008	263.644	232.493	28.036	−394.343	−395.724	22.967
1000	54.320	269.299	235.895	33.404	−394.568	−395.865	20.678
1100	55.423	274.529	239.172	38.893	−394.801	−395.984	18.803
1200	56.354	279.393	242.324	44.483	−395.035	−396.081	17.241
1300	57.144	283.936	245.352	50.159	−395.265	−396.159	15.918
1400	57.818	288.196	248.261	55.908	−395.488	−396.219	14.783
1500	58.397	292.205	251.059	61.719	−395.702	−396.264	13.799

11. METHANE CH_4 (g)

298.15	35.695	186.369	186.369	0.000	−74.600	−50.530	8.853
300	35.765	186.590	186.370	0.066	−74.656	−50.381	8.772
400	40.631	197.501	187.825	3.871	−77.703	−41.827	5.462
500	46.627	207.202	190.744	8.229	−80.520	−32.525	3.398
600	52.742	216.246	194.248	13.199	−82.969	−22.690	1.975
700	58.603	224.821	198.008	18.769	−85.023	−12.476	0.931
800	64.084	233.008	201.875	24.907	−86.693	−1.993	0.130
900	69.137	240.852	205.773	31.571	−88.006	8.677	−0.504
1000	73.746	248.379	209.660	38.719	−88.996	19.475	−1.017
1100	77.919	255.607	213.511	46.306	−89.698	30.358	−1.442
1200	81.682	262.551	217.310	54.289	−90.145	41.294	−1.797
1300	85.067	269.225	221.048	62.630	−90.367	52.258	−2.100
1400	88.112	275.643	224.720	71.291	−90.390	63.231	−2.359
1500	90.856	281.817	228.322	80.242	−90.237	74.200	−2.584

12. ACETYLENE C_2H_2 (g)

298.15	44.036	200.927	200.927	0.000	227.400	209.879	−36.769
300	44.174	201.199	200.927	0.082	227.397	209.770	−36.524
400	50.388	214.814	202.741	4.829	227.161	203.928	−26.630
500	54.751	226.552	206.357	10.097	226.846	198.154	−20.701

T/K	$C_p°$	$S°$	$-(G°-H°(T_r))/T$	$H°-H°(T_r)$	$\Delta_f H°$	$\Delta_f G°$	Log K_f
	(J/K·mol)			(kJ/mol)			
600	58.121	236.842	210.598	15.747	226.445	192.452	−16.754
700	60.970	246.021	215.014	21.704	225.968	186.823	−13.941
800	63.511	254.331	219.418	27.931	225.436	181.267	−11.835
900	65.831	261.947	223.726	34.399	224.873	175.779	−10.202
1000	67.960	268.995	227.905	41.090	224.300	170.355	−8.898
1100	69.909	275.565	231.942	47.985	223.734	164.988	−7.835
1200	71.686	281.725	235.837	55.067	223.189	159.672	−6.950
1300	73.299	287.528	239.592	62.317	222.676	154.400	−6.204
1400	74.758	293.014	243.214	69.721	222.203	149.166	−5.565
1500	76.077	298.218	246.709	77.264	221.774	143.964	−5.013

13. ETHYLENE C_2H_4 (g)

T/K	$C_p°$	$S°$	$-(G°-H°(T_r))/T$	$H°-H°(T_r)$	$\Delta_f H°$	$\Delta_f G°$	Log K_f
298.15	42.883	219.316	219.316	0.000	52.400	68.358	−11.976
300	43.059	219.582	219.317	0.079	52.341	68.457	−11.919
400	53.045	233.327	221.124	4.881	49.254	74.302	−9.703
500	62.479	246.198	224.864	10.667	46.533	80.887	−8.450
600	70.673	258.332	229.441	17.335	44.221	87.982	−7.659
700	77.733	269.770	234.393	24.764	42.278	95.434	−7.121
800	83.868	280.559	239.496	32.851	40.655	103.142	−6.734
900	89.234	290.754	244.630	41.512	39.310	111.036	−6.444
1000	93.939	300.405	249.730	50.675	38.205	119.067	−6.219
1100	98.061	309.556	254.756	60.280	37.310	127.198	−6.040
1200	101.670	318.247	259.688	70.271	36.596	135.402	−5.894
1300	104.829	326.512	264.513	80.599	36.041	143.660	−5.772
1400	107.594	334.384	269.225	91.223	35.623	151.955	−5.669
1500	110.018	341.892	273.821	102.107	35.327	160.275	−5.581

14. ETHANE C_2H_6 (g)

T/K	$C_p°$	$S°$	$-(G°-H°(T_r))/T$	$H°-H°(T_r)$	$\Delta_f H°$	$\Delta_f G°$	Log K_f
298.15	52.487	229.161	229.161	0.000	−84.000	−32.015	5.609
300	52.711	229.487	229.162	0.097	−84.094	−31.692	5.518
400	65.459	246.378	231.379	5.999	−88.988	−13.473	1.759
500	77.941	262.344	235.989	13.177	−93.238	5.912	−0.618
600	89.188	277.568	241.660	21.545	−96.779	26.086	−2.271
700	99.136	292.080	247.835	30.972	−99.663	46.800	−3.492
800	107.936	305.904	254.236	41.334	−101.963	67.887	−4.433
900	115.709	319.075	260.715	52.525	−103.754	89.231	−5.179
1000	122.552	331.628	267.183	64.445	−105.105	110.750	−5.785
1100	128.553	343.597	273.590	77.007	−106.082	132.385	−6.286
1200	133.804	355.012	279.904	90.131	−106.741	154.096	−6.708
1300	138.391	365.908	286.103	103.746	−107.131	175.850	−7.066
1400	142.399	376.314	292.178	117.790	−107.292	197.625	−7.373
1500	145.905	386.260	298.121	132.209	−107.260	219.404	−7.640

15. CYCLOPROPANE C_3H_6 (g)

T/K	$C_p°$	$S°$	$-(G°-H°(T_r))/T$	$H°-H°(T_r)$	$\Delta_f H°$	$\Delta_f G°$	Log K_f
298.15	55.571	237.488	237.488	0.000	53.300	104.514	−18.310
300	55.941	237.832	237.489	0.103	53.195	104.832	−18.253
400	76.052	256.695	239.924	6.708	47.967	122.857	−16.043
500	93.859	275.637	245.177	15.230	43.730	142.091	−14.844
600	108.542	294.092	251.801	25.374	40.405	162.089	−14.111
700	120.682	311.763	259.115	36.854	37.825	182.583	−13.624
800	130.910	328.564	266.755	49.447	35.854	203.404	−13.281
900	139.658	344.501	274.516	62.987	34.384	224.441	−13.026
1000	147.207	359.616	282.277	77.339	33.334	245.618	−12.830
1100	153.749	373.961	289.965	92.395	32.640	266.883	−12.673
1200	159.432	387.588	297.538	108.060	32.249	288.197	−12.545
1300	164.378	400.549	304.967	124.257	32.119	309.533	−12.437
1400	168.689	412.892	312.239	140.915	32.215	330.870	−12.345
1500	172.453	424.662	319.344	157.976	32.507	352.193	−12.264

16. PROPANE C_3H_8 (g)

T/K	$C_p°$	$S°$	$-(G°-H°(T_r))/T$	$H°-H°(T_r)$	$\Delta_f H°$	$\Delta_f G°$	Log K_f
298.15	73.597	270.313	270.313	0.000	−103.847	−23.458	4.110

	J/K·mol			kJ/mol			
T/K	C_p°	S°	$-(G^\circ - H^\circ(T_r))/T$	$H^\circ - H^\circ(T_r)$	$\Delta_f H^\circ$	$\Delta_f G^\circ$	$\text{Log } K_f$
300	73.931	270.769	270.314	0.136	−103.972	−22.959	3.997
400	94.014	294.739	273.447	8.517	−110.33	15.029	−0.657
500	112.591	317.768	280.025	18.872	−115.658	34.507	−3.605
600	128.700	339.753	288.162	30.955	−119.973	64.961	−5.655
700	142.674	360.668	297.039	44.540	−123.384	96.065	−7.168
800	154.766	380.528	306.245	59.427	−126.016	127.603	−8.331
900	165.352	399.381	315.555	75.444	−127.982	159.430	−9.253
1000	174.598	417.293	324.841	92.452	−129.380	191.444	−10.000
1100	182.673	434.321	334.026	110.325	−130.296	223.574	−10.617
1200	189.745	450.526	343.064	128.954	−130.802	255.770	−11.133
1300	195.853	465.961	351.929	148.241	−130.961	287.993	−11.572
1400	201.209	480.675	360.604	168.100	−130.829	320.217	−11.947
1500	205.895	494.721	369.080	188.460	−130.445	352.422	−12.272

17. BENZENE C_6H_6 (l)

298.15	135.950	173.450	173.450	0.000	49.080	124.521	−21.815
300	136.312	174.292	173.453	.252	49.077	124.989	−21.762
400	161.793	216.837	179.082	15.102	48.978	150.320	−19.630
500	207.599	257.048	190.639	33.204	50.330	175.559	−18.340

18. BENZENE C_6H_6 (g)

298.15	82.430	269.190	269.190	0.000	82.880	129.750	−22.731
300	83.020	269.700	269.190	0.153	82.780	130.040	−22.641
400	113.510	297.840	272.823	10.007	77.780	146.570	−19.140
500	139.340	326.050	280.658	22.696	73.740	164.260	−17.160
600	160.090	353.360	290.517	37.706	70.490	182.680	−15.903
700	176.790	379.330	301.360	54.579	67.910	201.590	−15.042
800	190.460	403.860	312.658	72.962	65.910	220.820	−14.418
900	201.840	426.970	324.084	92.597	64.410	240.280	−13.945
1000	211.430	448.740	335.473	113.267	63.340	259.890	−13.575
1100	219.580	469.280	346.710	134.827	62.620	277.640	−13.184
1200	226.540	488.690	357.743	157.137	62.200	299.320	−13.029
1300	232.520	507.070	368.534	180.097	62.000	319.090	−12.821
1400	237.680	524.490	379.056	203.607	61.990	338.870	−12.643
1500	242.140	541.040	389.302	227.607	62.110	358.640	−12.489

19. NAPHTHALENE $C_{10}H_8$ (cr, l)

298.15	165.720	167.390	167.390	0.000	78.530	201.585	−35.316
300	167.001	168.419	167.393	0.308	78.466	202.349	−35.232
353.43	208.722	198.948	169.833	10.290	96.099	224.543	−33.186

PHASE TRANSITION: $\Delta_{trs} H = 18.980$ kJ/mol, $\Delta_{trs} S = 53.702$ J/K·mol, cr–l

353.43	217.200	252.650	169.833	29.270	96.099	224.543	−33.186
400	241.577	280.916	181.124	39.917	96.067	241.475	−31.533
470	276.409	322.712	199.114	58.091	97.012	266.859	−29.658

20. NAPHTHALENE $C_{10}H_8$ (g)

298.15	131.920	333.150	333.150	0.000	150.580	224.100	−39.260
300	132.840	333.970	333.157	0.244	150.450	224.560	−39.098
400	180.070	378.800	338.950	15.940	144.190	250.270	−32.681
500	219.740	423.400	351.400	36.000	139.220	277.340	−28.973
600	251.530	466.380	367.007	59.624	135.350	305.330	−26.581
700	277.010	507.140	384.146	86.096	132.330	333.950	−24.919
800	297.730	545.520	401.935	114.868	130.050	362.920	−23.696
900	314.850	581.610	419.918	145.523	128.430	392.150	−22.759
1000	329.170	615.550	437.806	177.744	127.510	421.700	−22.027
1100	341.240	647.500	455.426	211.281	127.100	450.630	−21.398
1200	351.500	677.650	472.707	245.932	126.960	480.450	−20.913
1300	360.260	706.130	489.568	281.531	127.060	509.770	−20.482
1400	367.780	733.110	506.009	317.941	127.390	539.740	−20.137
1500	374.270	758.720	522.019	355.051	127.920	568.940	−19.812

T/K	$C_p°$	$S°$	$-(G°-H°(T_r))/T$	$H°-H°(T_r)$	$\Delta_f H°$	$\Delta_f G°$	Log K_f
		J/K·mol			kJ/mol		

21. FORMALDEHYDE H_2CO (g)

T/K	$C_p°$	$S°$	$-(G°-H°(T_r))/T$	$H°-H°(T_r)$	$\Delta_f H°$	$\Delta_f G°$	Log K_f
298.15	35.387	218.760	218.760	0.000	−108.700	−102.667	17.987
300	35.443	218.979	218.761	0.066	−108.731	−102.630	17.869
400	39.240	229.665	220.192	3.789	−110.438	−100.340	13.103
500	43.736	238.900	223.028	7.936	−112.073	−97.623	10.198
600	48.181	247.270	226.381	12.534	−113.545	−94.592	8.235
700	52.280	255.011	229.924	17.560	−114.833	−91.328	6.815
800	55.941	262.236	233.517	22.975	−115.942	−87.893	5.739
900	59.156	269.014	237.088	28.734	−116.889	−84.328	4.894
1000	61.951	275.395	240.603	34.792	−117.696	−80.666	4.213
1100	64.368	281.416	244.042	41.111	−118.382	−76.929	3.653
1200	66.453	287.108	247.396	47.655	−118.966	−73.134	3.183
1300	68.251	292.500	250.660	54.392	−119.463	−69.294	2.784
1400	69.803	297.616	253.833	61.297	−119.887	−65.418	2.441
1500	71.146	302.479	256.915	68.346	−120.249	−61.514	2.142

22. METHANOL CH_3OH (g)

T/K	$C_p°$	$S°$	$-(G°-H°(T_r))/T$	$H°-H°(T_r)$	$\Delta_f H°$	$\Delta_f G°$	Log K_f
298.15	44.101	239.865	239.865	0.000	−201.000	−162.298	28.434
300	44.219	240.139	239.866	0.082	−201.068	−162.057	28.216
400	51.713	253.845	241.685	4.864	−204.622	−148.509	19.393
500	59.800	266.257	245.374	10.442	−207.750	−134.109	14.010
600	67.294	277.835	249.830	16.803	−210.387	−119.125	10.371
700	73.958	288.719	254.616	23.873	−212.570	−103.737	7.741
800	79.838	298.987	259.526	31.569	−214.350	−88.063	5.750
900	85.025	308.696	264.455	39.817	−215.782	−72.188	4.190
1000	89.597	317.896	269.343	48.553	−216.916	−56.170	2.934
1100	93.624	326.629	274.158	57.718	−217.794	−40.050	1.902
1200	97.165	334.930	278.879	67.262	−218.457	−23.861	1.039
1300	100.277	342.833	283.497	77.137	−218.936	−7.624	0.306
1400	103.014	350.367	288.007	87.304	−219.261	8.644	−0.322
1500	105.422	357.558	292.405	97.729	−219.456	24.930	−0.868

23. ACETALDEHYDE C_2H_4O (g)

T/K	$C_p°$	$S°$	$-(G°-H°(T_r))/T$	$H°-H°(T_r)$	$\Delta_f H°$	$\Delta_f G°$	Log K_f
298.15	55.318	263.840	263.840	0.000	−166.190	−133.010	23.302
300	55.510	264.180	263.837	0.103	−166.250	−132.800	23.122
400	66.282	281.620	266.147	6.189	−169.530	−121.130	15.818
500	76.675	297.540	270.850	13.345	−172.420	−108.700	11.356
600	85.942	312.360	276.550	21.486	−174.870	−95.720	8.334
700	94.035	326.230	282.667	30.494	−176.910	−82.350	6.145
800	101.070	339.260	288.938	40.258	−178.570	−68.730	4.487
900	107.190	351.520	295.189	50.698	−179.880	−54.920	3.187
1000	112.490	363.100	301.431	61.669	−180.850	−40.930	2.138
1100	117.080	374.040	307.537	73.153	−181.560	−27.010	1.283
1200	121.060	384.400	313.512	85.065	−182.070	−12.860	0.560
1300	124.500	394.230	319.350	97.344	−182.420	1.240	−0.050
1400	127.490	403.570	325.031	109.954	−182.640	15.470	−0.577
1500	130.090	412.460	330.571	122.834	−182.750	29.580	−1.030

24. ETHANOL C_2H_5OH (g)

T/K	$C_p°$	$S°$	$-(G°-H°(T_r))/T$	$H°-H°(T_r)$	$\Delta_f H°$	$\Delta_f G°$	Log K_f
298.15	65.652	281.622	281.622	0.000	−234.800	−167.874	29.410
300	65.926	282.029	281.623	0.122	−234.897	−167.458	29.157
400	81.169	303.076	284.390	7.474	−239.826	−144.216	18.832
500	95.400	322.750	290.115	16.318	−243.940	−119.820	12.517
600	107.656	341.257	297.112	26.487	−247.260	−94.672	8.242
700	118.129	358.659	304.674	37.790	−249.895	−69.023	5.151
800	127.171	375.038	312.456	50.065	−251.951	−43.038	2.810
900	135.049	390.482	320.276	63.185	−253.515	−16.825	0.976
1000	141.934	405.075	328.033	77.042	−254.662	9.539	−0.498
1100	147.958	418.892	335.670	91.543	−255.454	36.000	−1.709
1200	153.232	431.997	343.156	106.609	−255.947	62.520	−2.721

	J/K·mol			kJ/mol			
T/K	$C_p°$	$S°$	$-(G°-H°(T_r))/T$	$H°-H°(T_r)$	$\Delta_f H°$	$\Delta_f G°$	$\text{Log } K_f$
1300	157.849	444.448	350.473	122.168	−256.184	89.070	−3.579
1400	161.896	456.298	357.612	138.160	−256.206	115.630	−4.314
1500	165.447	467.591	364.571	154.531	−256.044	142.185	−4.951

25. ACETIC ACID $C_2H_4O_2$ (g)

298.15	63.438	283.470	283.470	0.000	−432.249	−374.254	65.567
300	63.739	283.863	283.471	0.118	−432.324	−373.893	65.100
400	79.665	304.404	286.164	7.296	−436.006	−353.840	46.206
500	93.926	323.751	291.765	15.993	−438.875	−332.950	34.783
600	106.181	341.988	298.631	26.014	−440.993	−311.554	27.123
700	116.627	359.162	306.064	37.169	−442.466	−289.856	21.629
800	125.501	375.331	313.722	49.287	−443.395	−267.985	17.497
900	132.989	390.558	321.422	62.223	−443.873	−246.026	14.279
1000	139.257	404.904	329.060	75.844	−443.982	−224.034	11.702
1100	144.462	418.429	336.576	90.039	−443.798	−202.046	9.594
1200	148.760	431.189	343.933	104.707	−443.385	−180.086	7.839
1300	152.302	443.240	351.113	119.765	−442.795	−158.167	6.355
1400	155.220	454.637	358.105	135.146	−442.071	−136.299	5.085
1500	157.631	465.432	364.903	150.793	−441.247	−114.486	3.987

26. ACETONE C_3H_6O (g)

298.15	74.517	295.349	295.349	0.000	−217.150	−152.716	26.757
300	74.810	295.809	295.349	0.138	−217.233	−152.339	26.521
400	91.755	319.658	298.498	8.464	−222.212	−129.913	16.962
500	107.864	341.916	304.988	18.464	−226.522	−106.315	11.107
600	122.047	362.836	312.873	29.978	−230.120	−81.923	7.133
700	134.306	382.627	321.470	42.810	−233.049	−56.986	4.252
800	144.934	401.246	330.265	56.785	−235.350	−31.673	2.068
900	154.097	418.860	339.141	71.747	−237.149	−6.109	0.353
1000	162.046	435.513	347.950	87.563	−238.404	19.707	−1.030
1100	168.908	451.286	356.617	104.136	−239.283	45.396	−2.157
1200	174.891	466.265	365.155	121.332	−239.827	71.463	−3.110
1300	180.079	480.491	373.513	139.072	−240.120	97.362	−3.912
1400	184.556	493.963	381.596	157.314	−240.203	123.470	−4.607
1500	188.447	506.850	389.533	175.975	−240.120	149.369	−5.202

27. PHENOL C_6H_6O (g)

298.15	103.220	314.810	314.810	0.000	−96.400	−32.630	5.720
300	103.860	315.450	314.810	0.192	−96.490	−32.230	5.610
400	135.790	349.820	319.278	12.217	−100.870	−10.180	1.330
500	161.910	383.040	328.736	27.152	−104.240	12.970	−1.360
600	182.480	414.450	340.430	44.412	−106.810	36.650	−3.190
700	198.840	443.860	353.134	63.508	−108.800	60.750	−4.530
800	212.140	471.310	366.211	84.079	−110.300	85.020	−5.550
900	223.190	496.950	379.327	105.861	−111.370	109.590	−6.360
1000	232.490	520.960	392.302	128.658	−111.990	134.280	−7.010
1100	240.410	543.500	405.033	152.314	−112.280	158.620	−7.530
1200	247.200	564.720	417.468	176.703	−112.390	183.350	−7.980
1300	253.060	584.740	429.568	201.723	−112.330	208.070	−8.360
1400	258.120	603.680	441.331	227.288	−112.120	233.050	−8.700
1500	262.520	621.650	452.767	253.325	−111.780	257.540	−8.970

28. CARBON TETRAFLUORIDE CF_4 (g)

298.15	61.050	261.455	261.455	0.000	−933.200	−888.518	155.663
300	61.284	261.833	261.456	0.113	−933.219	−888.240	154.654
400	72.399	281.057	264.001	6.822	−933.986	−873.120	114.016
500	80.713	298.153	269.155	14.499	−934.372	−857.852	89.618
600	86.783	313.434	275.284	22.890	−934.490	−842.533	73.348
700	91.212	327.162	281.732	31.801	−934.431	−827.210	61.726
800	94.479	339.566	288.199	41.094	−934.261	−811.903	53.011
900	96.929	350.842	294.542	50.670	−934.024	−796.622	46.234

	J/K·mol			kJ/mol			
T/K	C_p°	S°	$-(G^\circ-H^\circ(T_r))/T$	$H^\circ-H^\circ(T_r)$	$\Delta_f H^\circ$	$\Delta_f G^\circ$	$\log K_f$
1000	98.798	361.156	300.695	60.460	−933.745	−781.369	40.814
1100	100.250	370.643	306.629	70.416	−933.442	−766.146	36.381
1200	101.396	379.417	312.334	80.500	−933.125	−750.952	32.688
1300	102.314	387.571	317.811	90.687	−932.800	−735.784	29.564
1400	103.059	395.181	323.069	100.957	−932.470	−720.641	26.887
1500	103.671	402.313	328.116	111.295	−932.137	−705.522	24.568

29. TRIFLUOROMETHANE CHF_3 (g)

298.15	51.069	259.675	259.675	0.000	−696.700	−662.237	116.020
300	51.258	259.991	259.676	0.095	−696.735	−662.023	115.267
400	61.148	276.113	261.807	5.722	−698.427	−650.186	84.905
500	69.631	290.700	266.149	12.275	−699.715	−637.969	66.647
600	76.453	304.022	271.368	19.593	−700.634	−625.528	54.456
700	81.868	316.230	276.917	27.519	−701.253	−612.957	45.739
800	86.201	327.455	282.542	35.930	−701.636	−600.315	39.196
900	89.719	337.818	288.116	44.732	−701.832	−587.636	34.105
1000	92.617	347.426	293.572	53.854	−701.879	−574.944	30.032
1100	95.038	356.370	298.879	63.240	−701.805	−562.253	26.699
1200	97.084	364.730	304.022	72.849	−701.629	−549.574	23.922
1300	98.833	372.571	308.997	82.647	−701.368	−536.913	21.573
1400	100.344	379.952	313.804	92.607	−701.033	−524.274	19.561
1500	101.660	386.921	318.449	102.709	−700.635	−511.662	17.817

30. CHLOROTRIFLUOROMETHANE $CClF_3$ (g)

298.15	66.886	285.419	285.419	0.000	−707.800	−667.238	116.896
300	67.111	285.834	285.421	0.124	−707.810	−666.986	116.131
400	77.528	306.646	288.187	7.383	−708.153	−653.316	85.313
500	85.013	324.797	293.734	15.532	−708.170	−639.599	66.818
600	90.329	340.794	300.271	24.314	−707.975	−625.901	54.489
700	94.132	355.020	307.096	33.547	−707.654	−612.246	45.686
800	96.899	367.780	313.897	43.106	−707.264	−598.642	39.087
900	98.951	379.317	320.536	52.903	−706.837	−585.090	33.957
1000	100.507	389.827	326.947	62.880	−706.396	−571.586	29.856
1100	101.708	399.465	333.108	72.993	−705.950	−558.126	26.503
1200	102.651	408.357	339.013	83.213	−705.505	−544.707	23.710
1300	103.404	416.604	344.668	93.517	−705.064	−531.326	21.349
1400	104.012	424.290	350.084	103.889	−704.628	−517.977	19.326
1500	104.512	431.484	355.273	114.316	−704.196	−504.660	17.574

31. DICHLORODIFLUOROMETHANE CCl_2F_2 (g)

298.15	72.476	300.903	300.903	0.000	−486.000	−447.030	78.317
300	72.691	301.352	300.905	0.134	−486.002	−446.788	77.792
400	82.408	323.682	303.883	7.919	−485.945	−433.716	56.637
500	89.063	342.833	309.804	16.514	−485.618	−420.692	43.949
600	93.635	359.500	316.729	25.663	−485.136	−407.751	35.497
700	96.832	374.189	323.909	35.196	−484.576	−394.897	29.467
800	99.121	387.276	331.027	44.999	−483.984	−382.126	24.950
900	100.801	399.053	337.942	55.000	−483.388	−369.429	21.441
1000	102.062	409.742	344.596	65.146	−482.800	−356.799	18.637
1100	103.030	419.517	350.969	75.402	−482.226	−344.227	16.346
1200	103.786	428.515	357.061	85.745	−481.667	−331.706	14.439
1300	104.388	436.847	362.882	96.154	−481.121	−319.232	12.827
1400	104.874	444.602	368.445	106.618	−480.588	−306.799	11.447
1500	105.270	451.851	373.767	117.126	−480.065	−294.404	10.252

32. CHLORODIFLUOROMETHANE $CHClF_2$ (g)

298.15	55.853	280.915	280.915	0.000	−475.000	−443.845	77.759
300	56.039	281.261	280.916	0.104	−475.028	−443.652	77.246
400	65.395	298.701	283.231	6.188	−476.390	−432.978	56.540
500	73.008	314.145	287.898	13.123	−477.398	−422.001	44.086
600	78.940	328.003	293.448	20.733	−478.103	−410.851	35.767

T/K	C_p°	S°	$-(G^\circ-H^\circ(T_r))/T$	$H^\circ-H^\circ(T_r)$	$\Delta_f H^\circ$	$\Delta_f G^\circ$	Log K_f
		J/K·mol			kJ/mol		
700	83.551	340.533	299.294	28.867	−478.574	−399.603	29.818
800	87.185	351.936	305.172	37.411	−478.870	−388.299	25.353
900	90.100	362.379	310.956	46.280	−479.031	−376.967	21.878
1000	92.475	371.999	316.586	55.413	−479.090	−365.622	19.098
1100	94.433	380.908	322.033	64.761	−479.068	−354.276	16.823
1200	96.066	389.196	327.289	74.289	−478.982	−342.935	14.927
1300	97.438	396.941	332.352	83.966	−478.843	−331.603	13.324
1400	98.601	404.206	337.228	93.769	−478.661	−320.283	11.950
1500	99.593	411.044	341.923	103.681	−478.443	−308.978	10.759

33. METHYLAMINE   CH$_5$N (g)

T/K	C_p°	S°	$-(G^\circ-H^\circ(T_r))/T$	$H^\circ-H^\circ(T_r)$	$\Delta_f H^\circ$	$\Delta_f G^\circ$	Log K_f
298.15	50.053	242.881	242.881	0.000	−22.529	32.734	−5.735
300	50.227	243.196	242.893	0.091	−22.614	33.077	−5.759
400	60.171	258.986	244.975	5.604	−26.846	52.294	−6.829
500	70.057	273.486	249.244	12.121	−30.431	72.510	−7.575
600	78.929	287.063	254.431	19.579	−33.364	93.382	−8.129
700	86.711	299.826	260.008	27.873	−35.712	114.702	−8.559
800	93.545	311.865	265.749	36.893	−37.548	136.316	−8.900
900	99.573	323.239	271.511	46.555	−38.949	158.138	−9.178
1000	104.886	334.006	277.220	56.786	−39.967	180.098	−9.407
1100	109.576	344.233	282.861	67.509	−40.681	201.822	−9.584
1200	113.708	353.944	288.374	78.685	−41.136	224.240	−9.761
1300	117.341	363.190	293.775	90.239	−41.376	246.364	−9.899
1400	120.542	372.012	299.061	102.131	−41.451	268.504	−10.018
1500	123.353	380.426	304.209	114.326	−41.381	290.639	−10.121

34. CHLORINE   Cl (g)

T/K	C_p°	S°	$-(G^\circ-H^\circ(T_r))/T$	$H^\circ-H^\circ(T_r)$	$\Delta_f H^\circ$	$\Delta_f G^\circ$	Log K_f
298.15	21.838	165.190	165.190	0.000	121.302	105.306	−18.449
300	21.852	165.325	165.190	0.040	121.311	105.207	−18.318
400	22.467	171.703	166.055	2.259	121.795	99.766	−13.028
500	22.744	176.752	167.708	4.522	122.272	94.203	−9.841
600	22.781	180.905	169.571	6.800	122.734	88.546	−7.709
700	22.692	184.411	171.448	9.074	123.172	82.813	−6.179
800	22.549	187.432	173.261	11.337	123.585	77.019	−5.029
900	22.389	190.079	174.986	13.584	123.971	71.175	−4.131
1000	22.233	192.430	176.615	15.815	124.334	65.289	−3.410
1100	22.089	194.542	178.150	18.031	124.675	59.368	−2.819
1200	21.959	196.458	179.597	20.233	124.996	53.416	−2.325
1300	21.843	198.211	180.963	22.423	125.299	47.439	−1.906
1400	21.742	199.826	182.253	24.602	125.587	41.439	−1.546
1500	21.652	201.323	183.475	26.772	125.861	35.418	−1.233

35. DICHLORINE   Cl$_2$ (g)

T/K	C_p°	S°	$-(G^\circ-H^\circ(T_r))/T$	$H^\circ-H^\circ(T_r)$	$\Delta_f H^\circ$	$\Delta_f G^\circ$	Log K_f
298.15	33.949	223.079	223.079	0.000	0.000	0.000	0.000
300	33.981	223.290	223.080	0.063	0.000	0.000	0.000
400	35.296	233.263	224.431	3.533	0.000	0.000	0.000
500	36.064	241.229	227.021	7.104	0.000	0.000	0.000
600	36.547	247.850	229.956	10.736	0.000	0.000	0.000
700	36.874	253.510	232.926	14.408	0.000	0.000	0.000
800	37.111	258.450	235.815	18.108	0.000	0.000	0.000
900	37.294	262.832	238.578	21.829	0.000	0.000	0.000
1000	37.442	266.769	241.203	25.566	0.000	0.000	0.000
1100	37.567	270.343	243.692	29.316	0.000	0.000	0.000
1200	37.678	273.617	246.052	33.079	0.000	0.000	0.000
1300	37.778	276.637	248.290	36.851	0.000	0.000	0.000
1400	37.872	279.440	250.416	40.634	0.000	0.000	0.000
1500	37.961	282.056	252.439	44.426	0.000	0.000	0.000

36. HYDROGEN CHLORIDE   HCl (g)

T/K	C_p°	S°	$-(G^\circ-H^\circ(T_r))/T$	$H^\circ-H^\circ(T_r)$	$\Delta_f H^\circ$	$\Delta_f G^\circ$	Log K_f
298.15	29.136	186.902	186.902	0.000	−92.310	−95.298	16.696
300	29.137	187.082	186.902	0.054	−92.314	−95.317	16.596

T/K	J/K·mol			kJ/mol			
	$C_p°$	$S°$	$-(G°-H°(T_r))/T$	$H°-H°(T_r)$	$\Delta_f H°$	$\Delta_f G°$	Log K_f
400	29.175	195.468	188.045	2.969	−92.587	−96.278	12.573
500	29.304	201.990	190.206	5.892	−92.911	−97.164	10.151
600	29.576	207.354	192.630	8.835	−93.249	−97.983	8.530
700	29.988	211.943	195.069	11.812	−93.577	−98.746	7.368
800	30.500	215.980	197.435	14.836	−93.879	−99.464	6.494
900	31.063	219.604	199.700	17.913	−94.149	−100.145	5.812
1000	31.639	222.907	201.858	21.049	−94.384	−100.798	5.265
1100	32.201	225.949	203.912	24.241	−94.587	−101.430	4.816
1200	32.734	228.774	205.867	27.488	−94.760	−102.044	4.442
1300	33.229	231.414	207.732	30.786	−94.908	−102.645	4.124
1400	33.684	233.893	209.513	34.132	−95.035	−103.235	3.852
1500	34.100	236.232	211.217	37.522	−95.146	−103.817	3.615

37. COPPER   Cu (cr, l)

T/K	$C_p°$	$S°$	$-(G°-H°(T_r))/T$	$H°-H°(T_r)$	$\Delta_f H°$	$\Delta_f G°$	Log K_f
298.15	24.440	33.150	33.150	0.000	0.000	0.000	0.000
300	24.460	33.301	33.150	0.045	0.000	0.000	0.000
400	25.339	40.467	34.122	2.538	0.000	0.000	0.000
500	25.966	46.192	35.982	5.105	0.000	0.000	0.000
600	26.479	50.973	38.093	7.728	0.000	0.000	0.000
700	26.953	55.090	40.234	10.399	0.000	0.000	0.000
800	27.448	58.721	42.322	13.119	0.000	0.000	0.000
900	28.014	61.986	44.328	15.891	0.000	0.000	0.000
1000	28.700	64.971	46.245	18.726	0.000	0.000	0.000
1100	29.553	67.745	48.075	21.637	0.000	0.000	0.000
1200	30.617	70.361	49.824	24.644	0.000	0.000	0.000
1300	31.940	72.862	51.501	27.769	0.000	0.000	0.000
1358	32.844	74.275	52.443	29.647	0.000	0.000	0.000

PHASE TRANSITION: $\Delta_{trs} H$ = 13.141 kJ/mol, $\Delta_{trs} S$ = 9.676 J/K·mol, cr–l

T/K	$C_p°$	$S°$	$-(G°-H°(T_r))/T$	$H°-H°(T_r)$	$\Delta_f H°$	$\Delta_f G°$	Log K_f
1358	32.800	83.951	52.443	42.788	0.000	0.000	0.000
1400	32.800	84.950	53.403	44.166	0.000	0.000	0.000
1500	32.800	87.213	55.583	47.446	0.000	0.000	0.000

38. COPPER   Cu (g)

T/K	$C_p°$	$S°$	$-(G°-H°(T_r))/T$	$H°-H°(T_r)$	$\Delta_f H°$	$\Delta_f G°$	Log K_f
298.15	20.786	166.397	166.397	0.000	337.600	297.873	−52.185
300	20.786	166.525	166.397	0.038	337.594	297.626	−51.821
400	20.786	172.505	167.213	2.117	337.179	284.364	−37.134
500	20.786	177.143	168.752	4.196	336.691	271.215	−28.333
600	20.786	180.933	170.476	6.274	336.147	258.170	−22.475
700	20.786	184.137	172.205	8.353	335.554	245.221	−18.298
800	20.786	186.913	173.874	10.431	334.913	232.359	−15.171
900	20.786	189.361	175.461	12.510	334.219	219.581	−12.744
1000	20.786	191.551	176.963	14.589	333.463	206.883	−10.806
1100	20.788	193.532	178.380	16.667	332.631	194.265	−9.225
1200	20.793	195.341	179.719	18.746	331.703	181.726	−7.910
1300	20.803	197.006	180.986	20.826	330.657	169.270	−6.801
1400	20.823	198.548	182.186	22.907	316.342	157.305	−5.869
1500	20.856	199.986	183.325	24.991	315.146	145.987	−5.084

39. COPPER OXIDE   CuO (cr)

T/K	$C_p°$	$S°$	$-(G°-H°(T_r))/T$	$H°-H°(T_r)$	$\Delta_f H°$	$\Delta_f G°$	Log K_f
298.15	42.300	42.740	42.740	0.000	−162.000	−134.277	23.524
300	42.417	43.002	42.741	0.078	−161.994	−134.105	23.349
400	46.783	55.878	44.467	4.564	−161.487	−124.876	16.307
500	49.190	66.596	47.852	9.372	−160.775	−115.803	12.098
600	50.827	75.717	51.755	14.377	−159.973	−106.883	9.305
700	52.099	83.651	55.757	19.526	−159.124	−98.102	7.320
800	53.178	90.680	59.691	24.791	−158.247	−89.444	5.840
900	54.144	97.000	63.491	30.158	−157.356	−80.897	4.695
1000	55.040	102.751	67.134	35.617	−156.462	−72.450	3.784
1100	55.890	108.037	70.615	41.164	−155.582	−64.091	3.043
1200	56.709	112.936	73.941	46.794	−154.733	−55.812	2.429

T/K	$C_p°$	$S°$	$-(G°-H°(T_r))/T$	$H°-H°(T_r)$	$\Delta_f H°$	$\Delta_f G°$	Log K_f
		J/K·mol			kJ/mol		
1300	57.507	117.507	77.118	52.505	−153.940	−47.601	1.913
1400	58.288	121.797	80.158	58.295	−166.354	−39.043	1.457
1500	59.057	125.845	83.070	64.163	−165.589	−29.975	1.044

40. DICOPPER OXIDE   Cu₂O (cr)

T/K	$C_p°$	$S°$	$-(G°-H°(T_r))/T$	$H°-H°(T_r)$	$\Delta_f H°$	$\Delta_f G°$	Log K_f
298.15	62.600	92.550	92.550	0.000	−173.100	−150.344	26.339
300	62.721	92.938	92.551	0.116	−173.102	−150.203	26.152
400	67.587	111.712	95.078	6.654	−173.036	−142.572	18.618
500	70.784	127.155	99.995	13.580	−172.772	−134.984	14.101
600	73.323	140.291	105.643	20.789	−172.389	−127.460	11.096
700	75.552	151.764	111.429	28.235	−171.914	−120.009	8.955
800	77.616	161.989	117.121	35.894	−171.363	−112.631	7.354
900	79.584	171.245	122.629	43.755	−170.750	−105.325	6.113
1000	81.492	179.729	127.920	51.809	−170.097	−98.091	5.124
1100	83.360	187.584	132.992	60.052	−169.431	−90.922	4.317
1200	85.202	194.917	137.850	68.480	−168.791	−83.814	3.648
1300	87.026	201.808	142.507	77.092	−168.223	−76.756	3.084
1400	88.836	208.324	146.978	85.885	−194.030	−68.926	2.572
1500	90.636	214.515	151.276	94.858	−193.438	−60.010	2.090

41. COPPER DICHLORIDE   CuCl₂ (cr, l)

T/K	$C_p°$	$S°$	$-(G°-H°(T_r))/T$	$H°-H°(T_r)$	$\Delta_f H°$	$\Delta_f G°$	Log K_f
298.15	71.880	108.070	108.070	0.000	−218.000	−173.826	30.453
300	71.998	108.515	108.071	0.133	−217.975	−173.552	30.218
400	76.338	129.899	110.957	7.577	−216.494	−158.962	20.758
500	78.654	147.204	116.532	15.336	−214.873	−144.765	15.123
600	80.175	161.687	122.884	23.282	−213.182	−130.901	11.396
675	81.056	171.183	127.732	29.329	−211.185	−120.693	9.340

PHASE TRANSITION: $\Delta_{trs} H$ = 0.700 kJ/mol, $\Delta_{trs} S$ = 1.037 J/K·mol, crII–crI

T/K	$C_p°$	$S°$	$-(G°-H°(T_r))/T$	$H°-H°(T_r)$	$\Delta_f H°$	$\Delta_f G°$	Log K_f
675	82.400	172.220	127.732	30.029	−211.185	−120.693	9.340
700	82.400	175.216	129.375	32.089	−210.719	−117.350	8.757
800	82.400	186.219	135.808	40.329	−208.898	−104.137	6.799
871	82.400	193.226	140.207	46.179	−192.649	−94.893	5.691

PHASE TRANSITION: $\Delta_{trs} H$ = 15.001 kJ/mol, $\Delta_{trs} S$ = 17.221 J/K·mol, crI–l

T/K	$C_p°$	$S°$	$-(G°-H°(T_r))/T$	$H°-H°(T_r)$	$\Delta_f H°$	$\Delta_f G°$	Log K_f
871	100.000	210.447	140.207	61.180	−192.649	−94.893	5.691
900	100.000	213.723	142.523	64.080	−191.640	−91.655	5.319
1000	100.000	224.259	150.179	74.080	−188.212	−80.730	4.217
1100	100.000	233.790	157.353	84.080	−184.873	−70.144	3.331
1130.75	100.000	236.547	159.470	87.155	−183.867	−66.951	3.093

42. COPPER DICHLORIDE   CuCl₂ (g)

T/K	$C_p°$	$S°$	$-(G°-H°(T_r))/T$	$H°-H°(T_r)$	$\Delta_f H°$	$\Delta_f G°$	Log K_f
298.15	56.814	278.418	278.418	0.000	−43.268	−49.883	8.739
300	56.869	278.769	278.419	0.105	−43.271	−49.924	8.692
400	58.992	295.456	280.679	5.911	−43.428	−52.119	6.806
500	60.111	308.752	285.010	11.871	−43.606	−54.271	5.670
600	60.761	319.774	289.911	17.918	−43.814	−56.385	4.909
700	61.168	329.173	294.865	24.015	−44.060	−58.462	4.362
800	61.439	337.360	299.677	30.147	−44.349	−60.500	3.950
900	61.630	344.608	304.274	36.301	−44.688	−62.499	3.627
1000	61.776	351.109	308.638	42.471	−45.088	−64.457	3.367
1100	61.900	357.003	312.771	48.655	−45.566	−66.372	3.152
1200	62.022	362.394	316.685	54.851	−46.139	−68.239	2.970
1300	62.159	367.364	320.395	61.060	−46.829	−70.053	2.815
1400	62.325	371.976	323.916	67.284	−60.784	−71.404	2.664
1500	62.531	376.283	327.265	73.526	−61.613	−72.133	2.512

43. FLUORINE   F (g)

T/K	$C_p°$	$S°$	$-(G°-H°(T_r))/T$	$H°-H°(T_r)$	$\Delta_f H°$	$\Delta_f G°$	Log K_f
298.15	22.746	158.750	158.750	0.000	79.380	62.280	−10.911
300	22.742	158.891	158.750	0.042	79.393	62.173	−10.825
400	22.432	165.394	159.639	2.302	80.043	56.332	−7.356
500	22.100	170.363	161.307	4.528	80.587	50.340	−5.259

T/K	$C_p°$	J/K·mol $S°$	$-(G°-H°(T_r))/T$	$H°-H°(T_r)$	kJ/mol $\Delta_f H°$	$\Delta_f G°$	Log K_f
600	21.832	174.368	163.161	6.724	81.046	44.246	-3.852
700	21.629	177.717	165.008	8.897	81.442	38.081	-2.842
800	21.475	180.595	166.780	11.052	81.792	31.862	-2.080
900	21.357	183.117	168.458	13.193	82.106	25.601	-1.486
1000	21.266	185.362	170.039	15.324	82.391	19.308	-1.009
1100	21.194	187.386	171.525	17.447	82.654	12.986	-0.617
1200	21.137	189.227	172.925	19.563	82.897	6.642	-0.289
1300	21.091	190.917	174.245	21.675	83.123	0.278	-0.011
1400	21.054	192.479	175.492	23.782	83.335	-6.103	0.228
1500	21.022	193.930	176.673	25.886	83.533	-12.498	0.435

44. DIFLUORINE F_2 (g)

T/K	$C_p°$	$S°$	$-(G°-H°(T_r))/T$	$H°-H°(T_r)$	$\Delta_f H°$	$\Delta_f G°$	Log K_f
298.15	31.304	202.790	202.790	0.000	0.000	0.000	0.000
300	31.337	202.984	202.790	0.058	0.000	0.000	0.000
400	32.995	212.233	204.040	3.277	0.000	0.000	0.000
500	34.258	219.739	206.453	6.643	0.000	0.000	0.000
600	35.171	226.070	209.208	10.117	0.000	0.000	0.000
700	35.839	231.545	212.017	13.669	0.000	0.000	0.000
800	36.343	236.365	214.765	17.279	0.000	0.000	0.000
900	36.740	240.669	217.409	20.934	0.000	0.000	0.000
1000	37.065	244.557	219.932	24.625	0.000	0.000	0.000
1100	37.342	248.103	222.334	28.346	0.000	0.000	0.000
1200	37.588	251.363	224.619	32.093	0.000	0.000	0.000
1300	37.811	254.381	226.794	35.863	0.000	0.000	0.000
1400	38.019	257.191	228.866	39.654	0.000	0.000	0.000
1500	38.214	259.820	230.843	43.466	0.000	0.000	0.000

45. HYDROGEN FLUORIDE   HF (g)

T/K	$C_p°$	$S°$	$-(G°-H°(T_r))/T$	$H°-H°(T_r)$	$\Delta_f H°$	$\Delta_f G°$	Log K_f
298.15	29.137	173.776	173.776	0.000	-273.300	-275.399	48.248
300	29.137	173.956	173.776	0.054	-273.302	-275.412	47.953
400	29.149	182.340	174.919	2.968	-273.450	-276.096	36.054
500	29.172	188.846	177.078	5.884	-273.679	-276.733	28.910
600	29.230	194.169	179.496	8.804	-273.961	-277.318	24.142
700	29.350	198.683	181.923	11.732	-274.277	-277.852	20.733
800	29.549	202.614	184.269	14.676	-274.614	-278.340	18.174
900	29.827	206.110	186.505	17.645	-274.961	-278.785	16.180
1000	30.169	209.270	188.626	20.644	-275.309	-279.191	14.583
1100	30.558	212.163	190.636	23.680	-275.652	-279.563	13.275
1200	30.974	214.840	192.543	26.756	-275.988	-279.904	12.184
1300	31.403	217.336	194.355	29.875	-276.315	-280.217	11.259
1400	31.831	219.679	196.081	33.037	-276.631	-280.505	10.466
1500	32.250	221.889	197.729	36.241	-276.937	-280.771	9.777

46. GERMANIUM   Ge (cr, l)

T/K	$C_p°$	$S°$	$-(G°-H°(T_r))/T$	$H°-H°(T_r)$	$\Delta_f H°$	$\Delta_f G°$	Log K_f
298.15	23.222	31.090	31.090	0.000	0.000	0.000	0.000
300	23.249	31.234	31.090	0.043	0.000	0.000	0.000
400	24.310	38.083	32.017	2.426	0.000	0.000	0.000
500	24.962	43.582	33.798	4.892	0.000	0.000	0.000
600	25.452	48.178	35.822	7.414	0.000	0.000	0.000
700	25.867	52.133	37.876	9.980	0.000	0.000	0.000
800	26.240	55.612	39.880	12.586	0.000	0.000	0.000
900	26.591	58.723	41.804	15.227	0.000	0.000	0.000
1000	26.926	61.542	43.639	17.903	0.000	0.000	0.000
1100	27.252	64.124	45.386	20.612	0.000	0.000	0.000
1200	27.571	66.509	47.048	23.353	0.000	0.000	0.000
1211.4	27.608	66.770	47.232	23.668	0.000	0.000	0.000

PHASE TRANSITION: $\Delta_{trs} H$ = 37.030 kJ/mol, $\Delta_{trs} S$ = 30.568 J/K·mol, cr–l

T/K	$C_p°$	$S°$	$-(G°-H°(T_r))/T$	$H°-H°(T_r)$	$\Delta_f H°$	$\Delta_f G°$	Log K_f
1211.4	27.600	97.338	47.232	60.698	0.000	0.000	0.000
1300	27.600	99.286	50.714	63.143	0.000	0.000	0.000
1400	27.600	101.331	54.258	65.903	0.000	0.000	0.000
1500	27.600	103.236	57.460	68.663	0.000	0.000	0.000

T/K	J/K·mol			kJ/mol			Log K_f
	$C_p°$	$S°$	$-(G°-H°(T_r))/T$	$H°-H°(T_r)$	$\Delta_f H°$	$\Delta_f G°$	

47. GERMANIUM   Ge (g)

298.15	30.733	167.903	167.903	0.000	367.800	327.009	−57.290
300	30.757	168.094	167.904	0.057	367.814	326.756	−56.893
400	31.071	177.025	169.119	3.162	368.536	312.959	−40.868
500	30.360	183.893	171.415	6.239	369.147	298.991	−31.235
600	29.265	189.334	173.965	9.222	369.608	284.914	−24.804
700	28.102	193.758	176.487	12.090	369.910	270.773	−20.205
800	27.029	197.439	178.882	14.845	370.060	256.598	−16.754
900	26.108	200.567	181.122	17.501	370.073	242.414	−14.069
1000	25.349	203.277	183.205	20.072	369.969	228.234	−11.922
1100	24.741	205.664	185.141	22.575	369.763	214.069	−10.165
1200	24.264	207.795	186.941	25.025	369.471	199.928	−8.703
1300	23.898	209.722	188.621	27.432	332.088	188.521	−7.575
1400	23.624	211.483	190.192	29.807	331.704	177.492	−6.622
1500	23.426	213.105	191.666	32.159	331.296	166.491	−5.798

48. GERMANIUM DIOXIDE   GeO$_2$ (cr, l)

298.15	50.166	39.710	39.710	0.000	−580.200	−521.605	91.382
300	50.475	40.021	39.711	0.093	−580.204	−521.242	90.755
400	61.281	56.248	41.850	5.759	−579.893	−501.610	65.503
500	66.273	70.519	46.191	12.164	−579.013	−482.134	50.368
600	69.089	82.872	51.299	18.943	−577.915	−462.859	40.295
700	70.974	93.671	56.597	25.952	−576.729	−443.776	33.115
800	72.449	103.247	61.841	33.125	−575.498	−424.866	27.741
900	73.764	111.857	66.928	40.436	−574.235	−406.113	23.570
1000	75.049	119.696	71.819	47.877	−572.934	−387.502	20.241
1100	76.378	126.910	76.504	55.447	−571.582	−369.024	17.523
1200	77.796	133.616	80.987	63.155	−570.166	−350.671	15.264
1300	79.332	139.903	85.279	71.010	−605.685	−329.732	13.249
1308	79.460	140.390	85.615	71.646	−584.059	−328.034	13.100

PHASE TRANSITION: $\Delta_{trs} H = 21.500$ kJ/mol, $\Delta_{trs} S = 16.437$ J/K·mol, crII–crI

1308	80.075	156.827	85.615	93.146	−584.059	−328.034	13.100
1388	81.297	161.617	89.858	99.601	−565.504	−312.415	11.757

PHASE TRANSITION: $\Delta_{trs} H = 17.200$ kJ/mol, $\Delta_{trs} S = 12.392$ J/K·mol, crI–l

1388	78.500	174.009	89.858	116.801	−565.504	−312.415	11.757
1400	78.500	174.685	90.582	117.743	−565.328	−310.228	11.575
1500	78.500	180.100	96.372	125.593	−563.882	−292.057	10.170

49. GERMANIUM TETRACHLORIDE   GeCl$_4$ (g)

298.15	95.918	348.393	348.393	0.000	−500.000	−461.582	80.866
300	96.041	348.987	348.395	0.178	−499.991	−461.343	80.326
400	100.750	377.342	352.229	10.045	−499.447	−448.540	58.573
500	103.206	400.114	359.604	20.255	−498.845	−435.882	45.536
600	104.624	419.067	367.980	30.652	−498.234	−423.347	36.855
700	105.509	435.266	376.463	41.162	−497.634	−410.914	30.662
800	106.096	449.396	384.715	51.744	−497.057	−398.565	26.023
900	106.504	461.917	392.611	62.375	−496.509	−386.287	22.419
1000	106.799	473.155	400.113	73.041	−495.993	−374.068	19.539
1100	107.020	483.344	407.224	83.733	−495.512	−361.899	17.185
1200	107.189	492.664	413.961	94.444	−495.067	−349.772	15.225
1300	107.320	501.249	420.349	105.169	−531.677	−334.973	13.459
1400	107.425	509.206	426.416	115.907	−531.265	−319.857	11.934
1500	107.509	516.621	432.185	126.654	−530.861	−304.771	10.613

50. HYDROGEN   H (g)

298.15	20.786	114.716	114.716	0.000	217.998	203.276	−35.613
300	20.786	114.845	114.716	0.038	218.010	203.185	−35.377
400	20.786	120.824	115.532	2.117	218.635	198.149	−25.875
500	20.786	125.463	117.071	4.196	219.253	192.956	−20.158

| T/K | $C_p°$ | J/K·mol | | $H°-H°(T_r)$ | kJ/mol | | Log K_f |
		$S°$	$-(G°-H°(T_r))/T$		$\Delta_f H°$	$\Delta_f G°$	
600	20.786	129.252	118.795	6.274	219.867	187.639	−16.335
700	20.786	132.457	120.524	8.353	220.476	182.219	−13.597
800	20.786	135.232	122.193	10.431	221.079	176.712	−11.538
900	20.786	137.680	123.780	12.510	221.670	171.131	−9.932
1000	20.786	139.870	125.282	14.589	222.247	165.485	−8.644
1100	20.786	141.852	126.700	16.667	222.806	159.781	−7.587
1200	20.786	143.660	128.039	18.746	223.345	154.028	−6.705
1300	20.786	145.324	129.305	20.824	223.864	148.230	−5.956
1400	20.786	146.864	130.505	22.903	224.360	142.393	−5.313
1500	20.786	148.298	131.644	24.982	224.835	136.522	−4.754

51. DIHYDROGEN H_2 (g)

T/K	$C_p°$	$S°$	$-(G°-H°(T_r))/T$	$H°-H°(T_r)$	$\Delta_f H°$	$\Delta_f G°$	Log K_f
298.15	28.836	130.680	130.680	0.000	0.000	0.000	0.000
300	28.849	130.858	130.680	0.053	0.000	0.000	0.000
400	29.181	139.217	131.818	2.960	0.000	0.000	0.000
500	29.260	145.738	133.974	5.882	0.000	0.000	0.000
600	29.327	151.078	136.393	8.811	0.000	0.000	0.000
700	29.440	155.607	138.822	11.749	0.000	0.000	0.000
800	29.623	159.549	141.172	14.702	0.000	0.000	0.000
900	29.880	163.052	143.412	17.676	0.000	0.000	0.000
1000	30.204	166.217	145.537	20.680	0.000	0.000	0.000
1100	30.580	169.113	147.550	23.719	0.000	0.000	0.000
1200	30.991	171.791	149.460	26.797	0.000	0.000	0.000
1300	31.422	174.288	151.275	29.918	0.000	0.000	0.000
1400	31.860	176.633	153.003	33.082	0.000	0.000	0.000
1500	32.296	178.846	154.653	36.290	0.000	0.000	0.000

52. HYDROXYL OH (g)

T/K	$C_p°$	$S°$	$-(G°-H°(T_r))/T$	$H°-H°(T_r)$	$\Delta_f H°$	$\Delta_f G°$	Log K_f
298.15	29.886	183.737	183.737	0.000	39.349	34.631	−6.067
300	29.879	183.922	183.738	0.055	39.350	34.602	−6.025
400	29.604	192.476	184.906	3.028	39.384	33.012	−4.311
500	29.495	199.067	187.104	5.982	39.347	31.422	−3.283
600	29.513	204.445	189.560	8.931	39.252	29.845	−2.598
700	29.655	209.003	192.020	11.888	39.113	28.287	−2.111
800	29.914	212.979	194.396	14.866	38.945	26.752	−1.747
900	30.265	216.522	196.661	17.874	38.763	25.239	−1.465
1000	30.682	219.731	198.810	20.921	38.577	23.746	−1.240
1100	31.135	222.677	200.848	24.012	38.393	22.272	−1.058
1200	31.603	225.406	202.782	27.149	38.215	20.814	−0.906
1300	32.069	227.954	204.621	30.332	38.046	19.371	−0.778
1400	32.522	230.347	206.374	33.562	37.886	17.941	−0.669
1500	32.956	232.606	208.048	36.836	37.735	16.521	−0.575

53. WATER H_2O (l)

T/K	$C_p°$	$S°$	$-(G°-H°(T_r))/T$	$H°-H°(T_r)$	$\Delta_f H°$	$\Delta_f G°$	Log K_f
298.15	75.300	69.950	69.950	0.000	−285.830	−237.141	41.546
300	75.281	70.416	69.951	0.139	−285.771	−236.839	41.237
373.21	76.079	86.896	71.715	5.666	−283.454	−225.160	31.513

54. WATER H_2O (g)

T/K	$C_p°$	$S°$	$-(G°-H°(T_r))/T$	$H°-H°(T_r)$	$\Delta_f H°$	$\Delta_f G°$	Log K_f
298.15	33.598	188.832	188.832	0.000	−241.826	−228.582	40.046
300	33.606	189.040	188.833	0.062	−241.844	−228.500	39.785
400	34.283	198.791	190.158	3.453	−242.845	−223.900	29.238
500	35.259	206.542	192.685	6.929	−243.822	−219.050	22.884
600	36.371	213.067	195.552	10.509	−244.751	−214.008	18.631
700	37.557	218.762	198.469	14.205	−245.620	−208.814	15.582
800	38.800	223.858	201.329	18.023	−246.424	−203.501	13.287
900	40.084	228.501	204.094	21.966	−247.158	−198.091	11.497
1000	41.385	232.792	206.752	26.040	−247.820	−192.603	10.060
1100	42.675	236.797	209.303	30.243	−248.410	−187.052	8.882
1200	43.932	240.565	211.753	34.574	−248.933	−181.450	7.898
1300	45.138	244.129	214.108	39.028	−249.392	−175.807	7.064

		J/K·mol			kJ/mol		
T/K	C_p°	S°	$-(G^\circ - H^\circ(T_r))/T$	$H^\circ - H^\circ(T_r)$	$\Delta_f H^\circ$	$\Delta_f G^\circ$	Log K_f
1400	46.281	247.516	216.374	43.599	−249.792	−170.132	6.348
1500	47.356	250.746	218.559	48.282	−250.139	−164.429	5.726

55. IODINE   I (g)

298.15	20.786	180.787	180.787	0.000	106.760	70.172	−12.294
300	20.786	180.915	180.787	0.038	106.748	69.945	−12.178
400	20.786	186.895	181.602	2.117	97.974	58.060	−7.582
500	20.786	191.533	183.142	4.196	75.988	50.202	−5.244
600	20.786	195.323	184.866	6.274	76.190	45.025	−3.920
700	20.786	198.527	186.594	8.353	76.385	39.816	−2.971
800	20.787	201.303	188.263	10.432	76.574	34.579	−2.258
900	20.789	203.751	189.851	12.510	76.757	29.319	−1.702
1000	20.795	205.942	191.352	14.589	76.936	24.038	−1.256
1100	20.806	207.924	192.770	16.669	77.109	18.740	−0.890
1200	20.824	209.735	194.110	18.751	77.277	13.426	−0.584
1300	20.851	211.403	195.377	20.835	77.440	8.098	−0.325
1400	20.889	212.950	196.577	22.921	77.596	2.758	−0.103
1500	20.936	214.392	197.717	25.013	77.745	−2.592	0.090

56. DIIODINE I_2 (cr, l)

298.15	54.440	116.139	116.139	0.000	0.000	0.000	0.000
300	54.518	116.476	116.140	0.101	0.000	0.000	0.000
386.75	61.531	131.039	117.884	5.088	0.000	0.000	0.000

PHASE TRANSITION: $\Delta_{trs} H = 15.665$ kJ/mol, $\Delta_{trs} S = 40.504$ J/K·mol, cr–l

386.75	79.555	171.543	117.884	20.753	0.000	0.000	0.000
400	79.555	174.223	119.706	21.807	0.000	0.000	0.000
457.67	79.555	184.938	127.266	26.395	0.000	0.000	0.000

57. DIIODINE I_2 (g)

298.15	36.887	260.685	260.685	0.000	62.420	19.324	−3.385
300	36.897	260.913	260.685	0.068	62.387	19.056	−3.318
400	37.256	271.584	262.138	3.778	44.391	5.447	−0.711
457.67	37.385	276.610	263.652	5.931		pressure = 1 bar	
500	37.464	279.921	264.891	7.515	0.000	0.000	0.000
600	37.613	286.765	267.983	11.269	0.000	0.000	0.000
700	37.735	292.573	271.092	15.037	0.000	0.000	0.000
800	37.847	297.619	274.099	18.816	0.000	0.000	0.000
900	37.956	302.083	276.965	22.606	0.000	0.000	0.000
1000	38.070	306.088	279.681	26.407	0.000	0.000	0.000
1100	38.196	309.722	282.249	30.220	0.000	0.000	0.000
1200	38.341	313.052	284.679	34.047	0.000	0.000	0.000
1300	38.514	316.127	286.981	37.890	0.000	0.000	0.000
1400	38.719	318.989	289.166	41.751	0.000	0.000	0.000
1500	38.959	321.668	291.245	45.635	0.000	0.000	0.000

58. HYDROGEN IODIDE   HI (g)

298.15	29.157	206.589	206.589	0.000	26.500	1.700	−0.298
300	29.158	206.769	206.589	0.054	26.477	1.546	−0.269
400	29.329	215.176	207.734	2.977	17.093	−6.289	0.821
500	29.738	221.760	209.904	5.928	−5.481	−9.946	1.039
600	30.351	227.233	212.348	8.931	−5.819	−10.806	0.941
700	31.070	231.965	214.820	12.002	−6.101	−11.614	0.867
800	31.807	236.162	217.230	15.145	−6.323	−12.386	0.809
900	32.511	239.950	219.548	18.362	−6.489	−13.133	0.762
1000	33.156	243.409	221.763	21.646	−6.608	−13.865	0.724
1100	33.735	246.597	223.878	24.991	−6.689	−14.586	0.693
1200	34.249	249.555	225.896	28.391	−6.741	−15.302	0.666
1300	34.703	252.314	227.823	31.839	−6.775	−16.014	0.643
1400	35.106	254.901	229.666	35.330	−6.797	−16.723	0.624
1500	35.463	257.336	231.430	38.858	−6.814	−17.432	0.607

		J/K·mol			kJ/mol		
T/K	C_p°	S°	$-(G^\circ - H^\circ(T_r))/T$	$H^\circ - H^\circ(T_r)$	$\Delta_f H^\circ$	$\Delta_f G^\circ$	Log K_f

59. POTASSIUM   K (cr, l)

298.15	29.600	64.680	64.680	0.000	0.000	0.000	0.000
300	29.671	64.863	64.681	0.055	0.000	0.000	0.000
336.86	32.130	68.422	64.896	1.188	0.000	0.000	0.000

PHASE TRANSITION: $\Delta_{trs} H$ = 2.321 kJ/mol, $\Delta_{trs} S$ = 6.891 J/K·mol, cr–l

336.86	32.129	75.313	64.896	3.509	0.000	0.000	0.000
400	31.552	80.784	66.986	5.519	0.000	0.000	0.000
500	30.741	87.734	70.469	8.632	0.000	0.000	0.000
600	30.158	93.283	73.824	11.675	0.000	0.000	0.000
700	29.851	97.905	76.943	14.673	0.000	0.000	0.000
800	29.838	101.887	79.818	17.655	0.000	0.000	0.000
900	30.130	105.415	82.470	20.651	0.000	0.000	0.000
1000	30.730	108.618	84.927	23.691	0.000	0.000	0.000
1039.4	31.053	109.812	85.847	24.908	0.000	0.000	0.000

60. POTASSIUM   K (g)

298.15	20.786	160.340	160.340	0.000	89.000	60.479	−10.596
300	20.786	160.468	160.340	0.038	88.984	60.302	−10.499
400	20.786	166.448	161.155	2.117	85.598	51.332	−6.703
500	20.786	171.086	162.695	4.196	84.563	42.887	−4.480
600	20.786	174.876	164.419	6.274	83.599	34.643	−3.016
700	20.786	178.080	166.148	8.353	82.680	26.557	−1.982
800	20.786	180.856	167.817	10.431	81.776	18.601	−1.215
900	20.786	183.304	169.404	12.510	80.859	10.759	−0.624
1000	20.786	185.494	170.905	14.589	79.897	3.021	−0.158
1039.4	20.786	186.297	171.474	15.408		pressure = 1 bar	
1100	20.786	187.475	172.323	16.667	0.000	0.000	0.000
1200	20.786	189.284	173.662	18.746	0.000	0.000	0.000
1300	20.789	190.948	174.929	20.825	0.000	0.000	0.000
1400	20.793	192.489	176.129	22.904	0.000	0.000	0.000
1500	20.801	193.923	177.268	24.983	0.000	0.000	0.000

61. DIPOTASSIUM OXIDE K_2O (cr, l)

298.15	72.000	96.000	96.000	0.000	−361.700	−321.171	56.267
300	72.130	96.446	96.001	0.133	−361.704	−320.920	55.876
400	79.154	118.158	98.914	7.698	−366.554	−306.416	40.013
500	86.178	136.575	104.647	15.964	−366.043	−291.423	30.444
590	92.500	151.348	110.662	24.005	−364.204	−278.079	24.619

PHASE TRANSITION: $\Delta_{trs} H$ = 0.700 kJ/mol, $\Delta_{trs} S$ = 1.186 J/K·mol, crIII–crII

590	100.000	152.534	110.662	24.705	−364.204	−278.079	24.619
600	100.000	154.215	111.374	25.705	−363.968	−276.621	24.082
645	100.000	161.447	114.618	30.205	−358.901	−270.109	21.874

PHASE TRANSITION: $\Delta_{trs} H$ = 4.000 kJ/mol, $\Delta_{trs} S$ = 6.202 J/K·mol, crII–crI

645	100.000	167.649	114.618	34.205	−358.901	−270.109	21.874
700	100.000	175.832	119.111	39.705	−357.592	−262.592	19.595
800	100.000	189.185	127.054	49.705	−355.224	−249.183	16.270
900	100.000	200.963	134.625	59.705	−352.919	−236.067	13.701
1000	100.000	211.499	141.794	69.705	−350.732	−223.202	11.659
1013	100.000	212.791	142.697	71.005	−323.459	−221.546	11.424

PHASE TRANSITION: $\Delta_{trs} H$ = 27.000 kJ/mol, $\Delta_{trs} S$ = 26.654 J/K·mol, crI–l

1013	100.000	239.444	142.697	98.005	−323.459	−221.546	11.424
1100	100.000	247.684	150.679	106.705	−479.439	−203.633	9.670
1200	100.000	256.385	159.131	116.705	−475.371	−178.740	7.780
1300	100.000	264.389	166.924	126.705	−471.321	−154.185	6.195
1400	100.000	271.800	174.154	136.705	−467.287	−129.941	4.848
1500	100.000	278.699	180.896	146.705	−463.268	−105.986	3.691

		J/K·mol			kJ/mol		
T/K	$C_p°$	$S°$	$-(G°-H°(T_r))/T$	$H°-H°(T_r)$	$\Delta_f H°$	$\Delta_f G°$	$\text{Log } K_f$

62. POTASSIUM HYDROXIDE   KOH (cr, l)

T/K	$C_p°$	$S°$	$-(G°-H°(T_r))/T$	$H°-H°(T_r)$	$\Delta_f H°$	$\Delta_f G°$	$\text{Log } K_f$
298.15	64.900	78.870	78.870	0.000	−424.580	−378.747	66.354
300	65.038	79.272	78.871	0.120	−424.569	−378.463	65.895
400	72.519	99.007	81.512	6.998	−426.094	−362.765	47.372
500	80.000	115.993	86.745	14.624	−424.572	−347.093	36.260
520	81.496	119.159	87.931	16.239	−417.725	−344.002	34.555

PHASE TRANSITION: $\Delta_{trs} H = 6.450$ kJ/mol, $\Delta_{trs} S = 12.404$ J/K·mol, crII–crI

520	79.000	131.563	87.931	22.689	−417.725	−344.002	34.555
600	79.000	142.868	94.520	29.009	−416.274	−332.766	28.969
678	79.000	152.523	100.649	35.171	−405.464	−321.998	24.807

PHASE TRANSITION: $\Delta_{trs} H = 9.400$ kJ/mol, $\Delta_{trs} S = 13.865$ J/K·mol, crI–l

678	83.000	166.388	100.649	44.571	−405.464	−321.998	24.807
700	83.000	169.038	102.757	46.397	−404.981	−319.297	23.826
800	83.000	180.121	111.750	54.697	−402.808	−307.206	20.058
900	83.000	189.897	119.901	62.997	−400.694	−295.383	17.143
1000	83.000	198.642	127.345	71.297	−398.668	−283.791	14.824
1100	83.000	206.553	134.192	79.597	−475.618	−267.780	12.716
1200	83.000	213.775	140.527	87.897	−472.711	−249.014	10.839
1300	83.000	220.418	146.421	96.197	−469.843	−230.490	9.261
1400	83.000	226.569	151.929	104.497	−467.011	−212.184	7.917
1500	83.000	232.296	157.098	112.797	−464.217	−194.080	6.758

63. POTASSIUM HYDROXIDE   KOH (g)

T/K	$C_p°$	$S°$	$-(G°-H°(T_r))/T$	$H°-H°(T_r)$	$\Delta_f H°$	$\Delta_f G°$	$\text{Log } K_f$
298.15	49.184	238.283	238.283	0.000	−227.989	−229.685	40.239
300	49.236	238.588	238.284	0.091	−228.007	−229.696	39.993
400	51.178	253.053	240.243	5.124	−231.377	−229.667	29.991
500	52.178	264.591	243.998	10.296	−232.309	−229.129	23.937
600	52.804	274.163	248.251	15.547	−233.145	−228.413	19.885
700	53.296	282.340	252.551	20.853	−233.934	−227.562	16.981
800	53.758	289.487	256.730	26.206	−234.708	−226.599	14.795
900	54.229	295.846	260.730	31.605	−235.495	−225.538	13.090
1000	54.713	301.585	264.533	37.052	−236.322	−224.388	11.721
1100	55.203	306.823	268.143	42.548	−316.077	−218.535	10.377
1200	55.686	311.647	271.570	48.092	−315.925	−209.674	9.127
1300	56.153	316.122	274.827	53.684	−315.764	−200.826	8.069
1400	56.598	320.300	277.927	59.322	−315.595	−191.991	7.163
1500	57.016	324.220	280.884	65.003	−315.420	−183.169	6.378

64. POTASSIUM CHLORIDE   KCl (cr, l)

T/K	$C_p°$	$S°$	$-(G°-H°(T_r))/T$	$H°-H°(T_r)$	$\Delta_f H°$	$\Delta_f G°$	$\text{Log } K_f$
298.15	51.300	82.570	82.570	0.000	−436.490	−408.568	71.579
300	51.333	82.887	82.571	0.095	−436.481	−408.395	71.107
400	52.977	97.886	84.605	5.312	−438.463	−398.651	52.058
500	54.448	109.867	88.498	10.685	−437.990	−388.749	40.612
600	55.885	119.921	92.919	16.201	−437.332	−378.960	32.991
700	57.425	128.649	97.413	21.865	−436.502	−369.295	27.557
800	59.205	136.430	101.812	27.694	−435.505	−359.760	23.490
900	61.361	143.523	106.058	33.719	−434.337	−350.360	20.334
1000	64.032	150.121	110.138	39.983	−432.981	−341.100	17.817
1044	65.405	152.908	111.882	42.830	−485.450	−336.720	16.847

PHASE TRANSITION: $\Delta_{trs} H = 26.320$ kJ/mol, $\Delta_{trs} S = 25.210$ J/K·mol, cr–l

1044	72.000	178.118	111.882	69.150	−485.450	−336.720	16.847
1100	72.000	181.880	115.351	73.182	−483.633	−328.790	15.613
1200	72.000	188.145	121.160	80.382	−480.393	−314.856	13.705
1300	72.000	193.908	126.537	87.582	−477.158	−301.192	12.102
1400	72.000	199.244	131.542	94.782	−473.928	−287.778	10.737
1500	72.000	204.211	136.223	101.982	−470.704	−274.594	9.562

Thermodynamic Properties as a Function of Temperature

T/K	C_p°	S°	$-(G^\circ-H^\circ(T_r))/T$	$H^\circ-H^\circ(T_r)$	$\Delta_f H^\circ$	$\Delta_f G^\circ$	Log K_f
	J/K·mol			kJ/mol			

65. POTASSIUM CHLORIDE   KCl (g)

T/K	C_p°	S°	$-(G^\circ-H^\circ(T_r))/T$	$H^\circ-H^\circ(T_r)$	$\Delta_f H^\circ$	$\Delta_f G^\circ$	Log K_f
298.15	36.505	239.091	239.091	0.000	−214.575	−233.320	40.876
300	36.518	239.317	239.092	0.068	−214.594	−233.436	40.644
400	37.066	249.904	240.532	3.749	−218.112	−239.107	31.224
500	37.384	258.212	243.267	7.473	−219.287	−244.219	25.513
600	37.597	265.048	246.344	11.222	−220.396	−249.100	21.686
700	37.769	270.857	249.441	14.991	−221.461	−253.799	18.938
800	37.907	275.910	252.441	18.775	−222.509	−258.347	16.868
900	38.041	280.382	255.302	22.572	−223.568	−262.764	15.250
1000	38.162	284.397	258.014	26.383	−224.667	−267.061	13.950
1100	38.279	288.039	260.581	30.205	−304.696	−266.627	12.661
1200	38.401	291.375	263.010	34.039	−304.821	−263.161	11.455
1300	38.518	294.454	265.312	37.885	−304.941	−259.684	10.434
1400	38.639	297.313	267.496	41.743	−305.053	−256.199	9.559
1500	38.761	299.983	269.574	45.613	−305.159	−252.706	8.800

66. DINITROGEN   N$_2$ (g)

T/K	C_p°	S°	$-(G^\circ-H^\circ(T_r))/T$	$H^\circ-H^\circ(T_r)$	$\Delta_f H^\circ$	$\Delta_f G^\circ$	Log K_f
298.15	29.124	191.608	191.608	0.000	0.000	0.000	0.000
300	29.125	191.788	191.608	0.054	0.000	0.000	0.000
400	29.249	200.180	192.752	2.971	0.000	0.000	0.000
500	29.580	206.738	194.916	5.911	0.000	0.000	0.000
600	30.109	212.175	197.352	8.894	0.000	0.000	0.000
700	30.754	216.864	199.812	11.936	0.000	0.000	0.000
800	31.433	221.015	202.208	15.046	0.000	0.000	0.000
900	32.090	224.756	204.509	18.222	0.000	0.000	0.000
1000	32.696	228.169	206.706	21.462	0.000	0.000	0.000
1100	33.241	231.311	208.802	24.759	0.000	0.000	0.000
1200	33.723	234.224	210.801	28.108	0.000	0.000	0.000
1300	34.147	236.941	212.708	31.502	0.000	0.000	0.000
1400	34.517	239.485	214.531	34.936	0.000	0.000	0.000
1500	34.842	241.878	216.275	38.404	0.000	0.000	0.000

67. NITRIC OXIDE   NO (g)

T/K	C_p°	S°	$-(G^\circ-H^\circ(T_r))/T$	$H^\circ-H^\circ(T_r)$	$\Delta_f H^\circ$	$\Delta_f G^\circ$	Log K_f
298.15	29.862	210.745	210.745	0.000	91.277	87.590	−15.345
300	29.858	210.930	210.746	0.055	91.278	87.567	−15.247
400	29.954	219.519	211.916	3.041	91.320	86.323	−11.272
500	30.493	226.255	214.133	6.061	91.340	85.071	−8.887
600	31.243	231.879	216.635	9.147	91.354	83.816	−7.297
700	32.031	236.754	219.168	12.310	91.369	82.558	−6.160
800	32.770	241.081	221.642	15.551	91.386	81.298	−5.308
900	33.425	244.979	224.022	18.862	91.405	80.036	−4.645
1000	33.990	248.531	226.298	22.233	91.426	78.772	−4.115
1100	34.473	251.794	228.469	25.657	91.445	77.505	−3.680
1200	34.883	254.811	230.540	29.125	91.464	76.237	−3.318
1300	35.234	257.618	232.516	32.632	91.481	74.967	−3.012
1400	35.533	260.240	234.404	36.170	91.495	73.697	−2.750
1500	35.792	262.700	236.209	39.737	91.506	72.425	−2.522

68. NITROGEN DIOXIDE   NO$_2$ (g)

T/K	C_p°	S°	$-(G^\circ-H^\circ(T_r))/T$	$H^\circ-H^\circ(T_r)$	$\Delta_f H^\circ$	$\Delta_f G^\circ$	Log K_f
298.15	37.178	240.166	240.166	0.000	34.193	52.316	−9.165
300	37.236	240.397	240.167	0.069	34.181	52.429	−9.129
400	40.513	251.554	241.666	3.955	33.637	58.600	−7.652
500	43.664	260.939	244.605	8.167	33.319	64.882	−6.778
600	46.383	269.147	248.026	12.673	33.174	71.211	−6.199
700	48.612	276.471	251.575	17.427	33.151	77.553	−5.787
800	50.405	283.083	255.107	22.381	33.213	83.893	−5.478
900	51.844	289.106	258.555	27.496	33.334	90.221	−5.236
1000	53.007	294.631	261.891	32.741	33.495	96.534	−5.042
1100	53.956	299.729	265.102	38.090	33.686	102.828	−4.883
1200	54.741	304.459	268.187	43.526	33.898	109.105	−4.749

T/K	C_p° J/K·mol	S°	$-(G^\circ-H^\circ(T_r))/T$	$H^\circ-H^\circ(T_r)$ kJ/mol	$\Delta_f H^\circ$	$\Delta_f G^\circ$	$\text{Log } K_f$
1300	55.399	308.867	271.148	49.034	34.124	115.363	−4.635
1400	55.960	312.994	273.992	54.603	34.360	121.603	−4.537
1500	56.446	316.871	276.722	60.224	34.604	127.827	−4.451

69. AMMONIA   NH₃ (g)

T/K	C_p°	S°	$-(G^\circ-H^\circ(T_r))/T$	$H^\circ-H^\circ(T_r)$	$\Delta_f H^\circ$	$\Delta_f G^\circ$	$\text{Log } K_f$
298.15	35.630	192.768	192.768	0.000	−45.940	−16.407	2.874
300	35.678	192.989	192.769	0.066	−45.981	−16.223	2.825
400	38.674	203.647	194.202	3.778	−48.087	−5.980	0.781
500	41.994	212.633	197.011	7.811	−49.908	4.764	−0.498
600	45.229	220.578	200.289	12.174	−51.430	15.846	−1.379
700	48.269	227.781	203.709	16.850	−52.682	27.161	−2.027
800	51.112	234.414	207.138	21.821	−53.695	38.639	−2.523
900	53.769	240.589	210.516	27.066	−54.499	50.231	−2.915
1000	56.244	246.384	213.816	32.569	−55.122	61.903	−3.233
1100	58.535	251.854	217.027	38.309	−55.589	73.629	−3.496
1200	60.644	257.039	220.147	44.270	−55.920	85.392	−3.717
1300	62.576	261.970	223.176	50.432	−56.136	97.177	−3.905
1400	64.339	266.673	226.117	56.779	−56.251	108.975	−4.066
1500	65.945	271.168	228.971	63.295	−56.282	120.779	−4.206

70. OXYGEN   O (g)

T/K	C_p°	S°	$-(G^\circ-H^\circ(T_r))/T$	$H^\circ-H^\circ(T_r)$	$\Delta_f H^\circ$	$\Delta_f G^\circ$	$\text{Log } K_f$
298.15	21.911	161.058	161.058	0.000	249.180	231.743	−40.600
300	21.901	161.194	161.059	0.041	249.193	231.635	−40.331
400	21.482	167.430	161.912	2.207	249.874	225.677	−29.470
500	21.257	172.197	163.511	4.343	250.481	219.556	−22.937
600	21.124	176.060	165.290	6.462	251.019	213.319	−18.571
700	21.040	179.310	167.067	8.570	251.500	206.997	−15.446
800	20.984	182.115	168.777	10.671	251.932	200.610	−13.098
900	20.944	184.584	170.399	12.767	252.325	194.171	−11.269
1000	20.915	186.789	171.930	14.860	252.686	187.689	−9.804
1100	20.893	188.782	173.372	16.950	253.022	181.173	−8.603
1200	20.877	190.599	174.733	19.039	253.335	174.628	−7.601
1300	20.864	192.270	176.019	21.126	253.630	168.057	−6.753
1400	20.853	193.815	177.236	23.212	253.908	161.463	−6.024
1500	20.845	195.254	178.389	25.296	254.171	154.851	−5.392

71. DIOXYGEN   O₂ (g)

T/K	C_p°	S°	$-(G^\circ-H^\circ(T_r))/T$	$H^\circ-H^\circ(T_r)$	$\Delta_f H^\circ$	$\Delta_f G^\circ$	$\text{Log } K_f$
298.15	29.378	205.148	205.148	0.000	0.000	0.000	0.000
300	29.387	205.330	205.148	0.054	0.000	0.000	0.000
400	30.109	213.873	206.308	3.026	0.000	0.000	0.000
500	31.094	220.695	208.525	6.085	0.000	0.000	0.000
600	32.095	226.454	211.045	9.245	0.000	0.000	0.000
700	32.987	231.470	213.612	12.500	0.000	0.000	0.000
800	33.741	235.925	216.128	15.838	0.000	0.000	0.000
900	34.365	239.937	218.554	19.244	0.000	0.000	0.000
1000	34.881	243.585	220.878	22.707	0.000	0.000	0.000
1100	35.314	246.930	223.096	26.217	0.000	0.000	0.000
1200	35.683	250.019	225.213	29.768	0.000	0.000	0.000
1300	36.006	252.888	227.233	33.352	0.000	0.000	0.000
1400	36.297	255.568	229.162	36.968	0.000	0.000	0.000
1500	36.567	258.081	231.007	40.611	0.000	0.000	0.000

72. SULFUR   S (cr, l)

T/K	C_p°	S°	$-(G^\circ-H^\circ(T_r))/T$	$H^\circ-H^\circ(T_r)$	$\Delta_f H^\circ$	$\Delta_f G^\circ$	$\text{Log } K_f$
298.15	22.690	32.070	32.070	0.000	0.000	0.000	0.000
300	22.737	32.210	32.070	0.042	0.000	0.000	0.000
368.3	24.237	37.030	32.554	1.649	0.000	0.000	0.000

PHASE TRANSITION: $\Delta_{trs} H$ = 0.401 kJ/mol, $\Delta_{trs} S$ = 1.089 J/K·mol, crII–crI

T/K	C_p°	S°	$-(G^\circ-H^\circ(T_r))/T$	$H^\circ-H^\circ(T_r)$	$\Delta_f H^\circ$	$\Delta_f G^\circ$	$\text{Log } K_f$
368.3	24.773	38.119	32.553	2.050	0.000	0.000	0.000
388.36	25.180	39.444	32.875	2.551	0.000	0.000	0.000

T/K	C_p°	S°	$-(G^\circ-H^\circ(T_r))/T$	$H^\circ-H^\circ(T_r)$	$\Delta_f H^\circ$	$\Delta_f G^\circ$	Log K_f
		J/K·mol			kJ/mol		
PHASE TRANSITION: $\Delta_{trs} H$ = 1.722 kJ/mol, $\Delta_{trs} S$ = 4.431 J/K·mol, crI–l							
388.36	31.710	43.875	32.872	4.273	0.000	0.000	0.000
400	32.369	44.824	33.206	4.647	0.000	0.000	0.000
500	38.026	53.578	36.411	8.584	0.000	0.000	0.000
600	34.371	60.116	39.842	12.164	0.000	0.000	0.000
700	32.451	65.278	43.120	15.511	0.000	0.000	0.000
800	32.000	69.557	46.163	18.715	0.000	0.000	0.000
882.38	32.000	72.693	48.496	21.351	0.000	0.000	0.000

73. SULFUR   S (g)

T/K	C_p°	S°	$-(G^\circ-H^\circ(T_r))/T$	$H^\circ-H^\circ(T_r)$	$\Delta_f H^\circ$	$\Delta_f G^\circ$	Log K_f
298.15	23.673	167.828	167.828	0.000	277.180	236.704	−41.469
300	23.669	167.974	167.828	0.044	277.182	236.453	−41.170
400	23.233	174.730	168.752	2.391	274.924	222.962	−29.115
500	22.741	179.860	170.482	4.689	273.286	210.145	−21.953
600	22.338	183.969	172.398	6.942	271.958	197.646	−17.206
700	22.031	187.388	174.302	9.160	270.829	185.352	−13.831
800	21.800	190.314	176.125	11.351	269.816	173.210	−11.309
900	21.624	192.871	177.847	13.522	215.723	162.258	−9.417
1000	21.489	195.142	179.465	15.677	216.018	156.301	−8.164
1100	21.386	197.185	180.985	17.821	216.284	150.317	−7.138
1200	21.307	199.043	182.413	19.955	216.525	144.309	−6.282
1300	21.249	200.746	183.759	22.083	216.743	138.282	−5.556
1400	21.209	202.319	185.029	24.206	216.940	132.239	−4.934
1500	21.186	203.781	186.231	26.325	217.119	126.182	−4.394

74. DISULFUR S_2 (g)

T/K	C_p°	S°	$-(G^\circ-H^\circ(T_r))/T$	$H^\circ-H^\circ(T_r)$	$\Delta_f H^\circ$	$\Delta_f G^\circ$	Log K_f
298.15	32.505	228.165	228.165	0.000	128.600	79.696	−13.962
300	32.540	228.366	228.165	0.060	128.576	79.393	−13.823
400	34.108	237.956	229.462	3.398	122.703	63.380	−8.276
500	35.133	245.686	231.959	6.863	118.296	49.031	−5.122
600	35.815	252.156	234.800	10.413	114.685	35.530	−3.093
700	36.305	257.715	237.686	14.020	111.599	22.588	−1.685
800	36.697	262.589	240.501	17.671	108.841	10.060	−0.657
882.38	36.985	266.200	242.734	20.706	pressure = 1 bar		
900	37.045	266.932	243.201	21.358	0.000	0.000	0.000
1000	37.377	270.852	245.773	25.079	0.000	0.000	0.000
1100	37.704	274.430	248.218	28.833	0.000	0.000	0.000
1200	38.030	277.725	250.541	32.620	0.000	0.000	0.000
1300	38.353	280.781	252.751	36.439	0.000	0.000	0.000
1400	38.669	283.635	254.856	40.290	0.000	0.000	0.000
1500	38.976	286.314	256.865	44.173	0.000	0.000	0.000

75. OCTASULFUR S_8 (g)

T/K	C_p°	S°	$-(G^\circ-H^\circ(T_r))/T$	$H^\circ-H^\circ(T_r)$	$\Delta_f H^\circ$	$\Delta_f G^\circ$	Log K_f
298.15	156.500	432.536	432.536	0.000	101.277	48.810	−8.551
300	156.768	433.505	432.539	0.290	101.231	48.484	−8.442
400	167.125	480.190	438.834	16.542	80.642	32.003	−4.179
500	173.181	518.176	451.022	33.577	66.185	21.409	−2.237
600	177.936	550.180	464.951	51.137	55.101	13.549	−1.180
700	182.441	577.948	479.152	69.157	46.349	7.343	−0.548
800	186.764	602.596	493.071	87.620	39.177	2.263	−0.148
900	190.595	624.821	506.495	106.494	−392.062	6.554	−0.380
1000	193.618	645.067	519.355	125.712	−387.728	50.614	−2.644
1100	195.684	663.625	531.639	145.185	−383.272	94.233	−4.475
1200	196.825	680.707	543.359	164.817	−378.786	137.444	−5.983
1300	197.195	696.480	554.539	184.524	−374.356	180.283	−7.244
1400	196.988	711.089	565.206	204.237	−370.048	222.785	−8.312
1500	196.396	724.662	575.389	223.909	−365.905	264.984	−9.227

76. SULFUR DIOXIDE SO_2 (g)

T/K	C_p°	S°	$-(G^\circ-H^\circ(T_r))/T$	$H^\circ-H^\circ(T_r)$	$\Delta_f H^\circ$	$\Delta_f G^\circ$	Log K_f
298.15	39.842	248.219	248.219	0.000	−296.810	−300.090	52.574
300	39.909	248.466	248.220	0.074	−296.833	−300.110	52.253

T/K	C_p°	S°	$-(G^\circ-H^\circ(T_r))/T$	$H^\circ-H^\circ(T_r)$	$\Delta_f H^\circ$	$\Delta_f G^\circ$	$\text{Log } K_f$
	J/K·mol			kJ/mol			
400	43.427	260.435	249.828	4.243	−300.240	−300.935	39.298
500	46.490	270.465	252.978	8.744	−302.735	−300.831	31.427
600	48.938	279.167	256.634	13.520	−304.699	−300.258	26.139
700	50.829	286.859	260.413	18.513	−306.308	−299.386	22.340
800	52.282	293.746	264.157	23.671	−307.691	−298.302	19.477
900	53.407	299.971	267.796	28.958	−362.075	−295.987	17.178
1000	54.290	305.646	271.301	34.345	−362.012	−288.647	15.077
1100	54.993	310.855	274.664	39.810	−361.934	−281.314	13.358
1200	55.564	315.665	277.882	45.339	−361.849	−273.989	11.926
1300	56.033	320.131	280.963	50.920	−361.763	−266.671	10.715
1400	56.426	324.299	283.911	56.543	−361.680	−259.359	9.677
1500	56.759	328.203	286.735	62.203	−361.605	−252.053	8.777

77. SILICON   Si (cr)

T/K	C_p°	S°	$-(G^\circ-H^\circ(T_r))/T$	$H^\circ-H^\circ(T_r)$	$\Delta_f H^\circ$	$\Delta_f G^\circ$	$\text{Log } K_f$
298.15	19.789	18.810	18.810	0.000	0.000	0.000	0.000
300	19.855	18.933	18.810	0.037	0.000	0.000	0.000
400	22.301	25.023	19.624	2.160	0.000	0.000	0.000
500	23.610	30.152	21.231	4.461	0.000	0.000	0.000
600	24.472	34.537	23.092	6.867	0.000	0.000	0.000
700	25.124	38.361	25.006	9.348	0.000	0.000	0.000
800	25.662	41.752	26.891	11.888	0.000	0.000	0.000
900	26.135	44.802	28.715	14.478	0.000	0.000	0.000
1000	26.568	47.578	30.464	17.114	0.000	0.000	0.000
1100	26.974	50.130	32.138	19.791	0.000	0.000	0.000
1200	27.362	52.493	33.737	22.508	0.000	0.000	0.000
1300	27.737	54.698	35.265	25.263	0.000	0.000	0.000
1400	28.103	56.767	36.728	28.055	0.000	0.000	0.000
1500	28.462	58.719	38.130	30.883	0.000	0.000	0.000

78. SILICON   Si (g)

T/K	C_p°	S°	$-(G^\circ-H^\circ(T_r))/T$	$H^\circ-H^\circ(T_r)$	$\Delta_f H^\circ$	$\Delta_f G^\circ$	$\text{Log } K_f$
298.15	22.251	167.980	167.980	0.000	450.000	405.525	−71.045
300	22.234	168.117	167.980	0.041	450.004	405.249	−70.559
400	21.613	174.416	168.843	2.229	450.070	390.312	−50.969
500	21.316	179.204	170.456	4.374	449.913	375.388	−39.216
600	21.153	183.074	172.246	6.497	449.630	360.508	−31.385
700	21.057	186.327	174.032	8.607	449.259	345.682	−25.795
800	21.000	189.135	175.748	10.709	448.821	330.915	−21.606
900	20.971	191.606	177.375	12.808	448.329	316.205	−18.352
1000	20.968	193.815	178.911	14.904	447.791	301.553	−15.751
1100	20.989	195.815	180.358	17.002	447.211	286.957	−13.626
1200	21.033	197.643	181.723	19.103	446.595	272.416	−11.858
1300	21.099	199.329	183.014	21.209	445.946	257.927	−10.364
1400	21.183	200.895	184.236	23.323	445.268	243.489	−9.085
1500	21.282	202.360	185.396	25.446	444.563	229.101	−7.978

79. SILICON DIOXIDE SiO_2 (cr)

T/K	C_p°	S°	$-(G^\circ-H^\circ(T_r))/T$	$H^\circ-H^\circ(T_r)$	$\Delta_f H^\circ$	$\Delta_f G^\circ$	$\text{Log } K_f$
298.15	44.602	41.460	41.460	0.000	−910.700	−856.288	150.016
300	44.712	41.736	41.461	0.083	−910.708	−855.951	149.032
400	53.477	55.744	43.311	4.973	−910.912	−837.651	109.385
500	60.533	68.505	47.094	10.705	−910.540	−819.369	85.598
600	64.452	79.919	51.633	16.971	−909.841	−801.197	69.749
700	68.234	90.114	56.414	23.590	−908.958	−783.157	58.439
800	76.224	99.674	61.226	30.758	−907.668	−765.265	49.966
848	82.967	104.298	63.533	34.569	−906.310	−756.747	46.613
PHASE TRANSITION: $\Delta_{trs} H = 0.411$ kJ/mol, $\Delta_{trs} S = 0.484$ J/K·mol, crII–crII′							
848	67.446	104.782	63.532	34.980	−906.310	−756.747	46.613
900	67.953	108.811	66.033	38.500	−905.922	−747.587	43.388
1000	68.941	116.021	70.676	45.345	−905.176	−730.034	38.133
1100	69.940	122.639	75.104	52.289	−904.420	−712.557	33.836
1200	70.947	128.768	79.323	59.333	−901.382	−695.148	30.259

T/K	J/K·mol			kJ/mol			Log K_f
	C_p°	S°	$-(G^\circ - H^\circ(T_r))/T$	$H^\circ - H^\circ(T_r)$	$\Delta_f H^\circ$	$\Delta_f G^\circ$	

PHASE TRANSITION: $\Delta_{trs} H$ = 2.261 kJ/mol, $\Delta_{trs} S$ = 1.883 J/K·mol, crII′–crI

T/K	C_p°	S°	$-(G^\circ - H^\circ(T_r))/T$	$H^\circ - H^\circ(T_r)$	$\Delta_f H^\circ$	$\Delta_f G^\circ$	Log K_f
1200	71.199	130.651	79.323	61.594	−901.382	−695.148	30.259
1300	71.743	136.372	83.494	68.742	−900.574	−677.994	27.242
1400	72.249	141.707	87.463	75.941	−899.782	−660.903	24.658
1500	72.739	146.709	91.248	83.191	−899.004	−643.867	22.421

80. SILICON TETRACHLORIDE $SiCl_4$ (g)

T/K	C_p°	S°	$-(G^\circ - H^\circ(T_r))/T$	$H^\circ - H^\circ(T_r)$	$\Delta_f H^\circ$	$\Delta_f G^\circ$	Log K_f
298.15	90.404	331.446	331.446	0.000	−662.200	−622.390	109.039
300	90.562	332.006	331.448	0.167	−662.195	−622.143	108.323
400	96.893	359.019	335.088	9.572	−661.853	−608.841	79.505
500	100.449	381.058	342.147	19.456	−661.413	−595.637	62.225
600	102.587	399.576	350.216	29.616	−660.924	−582.527	50.713
700	103.954	415.500	358.432	39.948	−660.417	−569.501	42.496
800	104.875	429.445	366.455	50.392	−659.912	−556.548	36.338
900	105.523	441.837	374.155	60.914	−659.422	−543.657	31.553
1000	105.995	452.981	381.490	71.491	−658.954	−530.819	27.727
1100	106.349	463.101	388.456	82.109	−658.515	−518.027	24.599
1200	106.620	472.366	395.068	92.758	−658.107	−505.274	21.994
1300	106.834	480.909	401.347	103.431	−657.735	−492.553	19.791
1400	107.003	488.833	407.316	114.123	−657.400	−479.860	17.904
1500	107.141	496.220	413.000	124.830	−657.104	−467.189	16.269

THERMODYNAMIC PROPERTIES OF AQUEOUS SYSTEMS

This table contains standard state thermodynamic properties of ions and neutral species in aqueous solution. It includes enthalpy and Gibbs energy of formation, entropy, and heat capacity, and thus serves as a companion to the preceding table, "Standard Thermodynamic Properties of Chemical Substances". The standard state is the hypothetical ideal solution with molality $m = 1$ mol/kg (mean ionic molality $m_\pm$ in the case of a species which is assumed to dissociate at infinite dilution). Further details on conventions may be found in Reference 1.

Cations are listed by formula in the first part of the table, followed by anions and finally neutral species. All values refer to standard conditions of 25°C and 100 kPa pressure.

References

1. Wagman, D. D., Evans, W. H., Parker, V. B., Schumm, R. H., Halow, I., Bailey, S. M., Churney, K. L., and Nuttall, R. L., *The NBS Tables of Chemical Thermodynamic Properties, J. Phys. Chem. Ref. Data*, Vol. 11, Suppl. 2, 1982.
2. Zemaitis, J. F., Clark, D. M., Rafal, M., and Scrivner, N. C., *Handbook of Aqueous Electrolyte Thermodynamics*, American Institute of Chemical Engineers, New York, 1986.

Species	$\Delta_f H°/$ kJ mol^{-1}	$\Delta_f G°/$ kJ mol^{-1}	$S°/$ J mol^{-1}K^{-1}	$C_p/$ J mol^{-1}K^{-1}
Cations				
Ag$^+$	105.6	77.1	72.7	21.8
Al^{+3}	−531.0	−485.0	−321.7	
AlOH^{+2}		−694.1		
Ba^{+2}	−537.6	−560.8	9.6	
BaOH$^+$		−730.5		
Be^{+2}	−382.8	−379.7	−129.7	
Bi^{+3}		82.8		
BiOH^{+2}		−146.4		
Ca^{+2}	−542.8	−553.6	−53.1	
CaOH$^+$		−718.4		
Cd^{+2}	−75.9	−77.6	−73.2	
CdOH$^+$		−261.1		
Ce^{+3}	−696.2	−672.0	−205.0	
Ce^{+4}	−537.2	−503.8	−301.0	
Co^{+2}	−58.2	−54.4	−113.0	
Co^{+3}	92.0	134.0	−305.0	
Cr^{+2}	−143.5			
Cs$^+$	−258.3	−292.0	133.1	−10.5
Cu$^+$	71.7	50.0	40.6	
Cu^{+2}	64.8	65.5	−99.6	
Dy^{+3}	−699.0	−665.0	−231.0	21.0
Er^{+3}	−705.4	−669.1	−244.3	21.0
Eu^{+2}	−527.0	−540.2	−8.0	
Eu^{+3}	−605.0	−574.1	−222.0	8.0
Fe^{+2}	−89.1	−78.9	−137.7	
Fe^{+3}	−48.5	−4.7	−315.9	
FeOH$^+$	−324.7	−277.4	−29.0	
FeOH^{+2}	−290.8	−229.4	−142.0	
Fe(OH)$_2^+$		−438.0		
Ga^{+2}		−88.0		
Ga^{+3}	−211.7	−159.0	−331.0	
GaOH^{+2}		−380.3		
Ga(OH)$_2^+$		−597.4		
Gd^{+3}	−686.0	−661.0	−205.9	
H$^+$	0	0	0	0
Hg^{+2}	171.1	164.4	−32.2	
Hg$_2^{+2}$	172.4	153.5	84.5	
HgOH$^+$	−84.5	−52.3	71.0	
Ho^{+3}	−705.0	−673.7	−226.8	17.0
In$^+$		−12.1		
In^{+2}		−50.7		
In^{+3}	−105.0	−98.0	−151.0	

Species	$\Delta_f H°/$ kJ mol^{-1}	$\Delta_f G°/$ kJ mol^{-1}	$S°/$ J mol^{-1}K^{-1}	$C_p/$ J mol^{-1}K^{-1}
InOH^{+2}	−370.3	−313.0	−88.0	
In(OH)$_2^+$	−619.0	−525.0	25.0	
K$^+$	−252.4	−283.3	102.5	21.8
La^{+3}	−707.1	−683.7	−217.6	−13.0
Li$^+$	−278.5	−293.3	13.4	68.6
Lu^{+3}	−665.0	−628.0	−264.0	25.0
LuF^{+2}		−931.4		
Mg^{+2}	−466.9	−454.8	−138.1	
MgOH$^+$		−626.7		
Mn^{+2}	−220.8	−228.1	−73.6	50.0
MnOH$^+$	−450.6	−405.0	−17.0	
NH$_4^+$	−132.5	−79.3	113.4	79.9
N$_2$H$_5^+$	−7.5	82.5	151.0	70.3
Na$^+$	−240.1	−261.9	59.0	46.4
Nd^{+3}	−696.2	−671.6	−206.7	−21.0
Ni^{+2}	−54.0	−45.6	−128.9	
NiOH$^+$	−287.9	−227.6	−71.0	
PH$_4^+$		92.1		
Pa^{+4}	−619.0			
Pb^{+2}	−1.7	−24.4	10.5	
PbOH$^+$		−226.3		
Pd^{+2}	149.0	176.5	−184.0	
Po^{+2}		71.0		
Po^{+4}		293.0		
Pr^{+3}	−704.6	−679.1	−209.0	−29.0
Pt^{+2}		254.8		
Ra^{+2}	−527.6	−561.5	54.0	
Rb$^+$	−251.2	−284.0	121.5	
Re$^+$		−33.0		
Sc^{+3}	−614.2	−586.6	−255.0	
ScOH^{+2}	−861.5	−801.2	−134.0	
Sm^{+2}		−497.5		
Sm^{+3}	−691.6	−666.6	−211.7	−21.0
Sn^{+2}	−8.8	−27.2	−17.0	
SnOH$^+$	−286.2	−254.8	50.0	
Sr^{+2}	−545.8	−559.5	−32.6	
SrOH$^+$		−721.3		
Tb^{+3}	−682.8	−651.9	−226.0	17.0
Te(OH)$_3^+$	−608.4	−496.1	111.7	
Th^{+4}	−769.0	−705.1	−422.6	
Th(OH)$^{+3}$	−1030.1	−920.5	−343.0	
Th(OH)$_2^{+2}$	−1282.4	−1140.9	−218.0	
Tl$^+$	5.4	−32.4	125.5	

Species	$\Delta_f H°$ / kJ mol⁻¹	$\Delta_f G°$ / kJ mol⁻¹	$S°$ / J mol⁻¹K⁻¹	C_p / J mol⁻¹K⁻¹
Tl^{+3}	196.6	214.6	−192.0	
$TlOH^{+2}$		−15.9		
$Tl(OH)_2^+$		−244.7		
Tm^{+3}	−697.9	−662.0	−243.0	25.0
U^{+3}	−489.1	−476.2	−188.0	
U^{+4}	−591.2	−531.9	−410.0	
Y^{+3}	−723.4	−693.8	−251.0	
$Y_2(OH)_2^{+4}$		−1780.3		
Yb^{+2}		−527.0		
Yb^{+3}	−674.5	−644.0	−238.0	25.0
$Y(OH)^{+2}$		−879.1		
Zn^{+2}	−153.9	−147.1	−112.1	46.0
$ZnOH^+$		−330.1		
Anions				
AlO_2^-	−930.9	−830.9	−36.8	
$Al(OH)_4^-$	−1502.5	−1305.3	102.9	
AsO_2^-	−429.0	−350.0	40.6	
AsO_4^{-3}	−888.1	−648.4	−162.8	
BF_4^-	−1574.9	−1486.9	180.0	
BH_4^-	48.2	114.4	110.5	
BO_2^-	−772.4	−678.9	−37.2	
$B_4O_7^{-2}$		−2604.8		
BeO_2^{-2}	−790.8	−640.1	−159.0	
Br^-	−121.6	−104.0	82.4	−141.8
BrO^-	−94.1	−33.4	42.0	
BrO_3^-	−67.1	18.6	161.7	
BrO_4^-	13.0	118.1	199.6	
$CHOO^-$	−425.6	−351.0	92.0	−87.9
CH_3COO^-	−486.0	−369.3	86.6	−6.3
$C_2O_4^{-2}$	−825.1	−673.9	45.6	
$C_2O_4H^-$	−818.4	−698.3	149.4	
Cl^-	−167.2	−131.2	56.5	−136.4
ClO^-	−107.1	−36.8	42.0	
ClO_2^-	−66.5	17.2	101.3	
ClO_3^-	−104.0	−8.0	162.3	
ClO_4^-	−129.3	−8.5	182.0	
CN^-	150.6	172.4	94.1	
CO_3^{-2}	−677.1	−527.8	−56.9	
CrO_4^{-2}	−881.2	−727.8	50.2	
$Cr_2O_7^{-2}$	−1490.3	−1301.1	261.9	
F^-	−332.6	−278.8	−13.8	−106.7
$Fe(CN)_6^{-3}$	561.9	729.4	270.3	
$Fe(CN)_6^{-4}$	455.6	695.1	95.0	
$HB_4O_7^-$		−2685.1		
HCO_3^-	−692.0	−586.8	91.2	
HF_2^-	−649.9	−578.1	92.5	
HPO_3F^-		−1198.2		
HPO_4^{-2}	−1292.1	−1089.2	−33.5	
$HP_2O_7^{-3}$	−2274.8	−1972.2	46.0	
HS^-	−17.6	12.1	62.8	
HSO_3^-	−626.2	−527.7	139.7	
HSO_4^-	−887.3	−755.9	131.8	−84.0
$HS_2O_4^-$		−614.5		
HSe^-	15.9	44.0	79.0	
$HSeO_3^-$	−514.6	−411.5	135.1	
$HSeO_4^-$	−581.6	−452.2	149.4	
$H_2AsO_3^-$	−714.8	−587.1	110.5	
$H_2AsO_4^-$	−909.6	−753.2	117.0	
$H_2PO_4^-$	−1296.3	−1130.2	90.4	
$H_2P_2O_7^{-2}$	−2278.6	−2010.2	163.0	

Species	$\Delta_f H°$ / kJ mol⁻¹	$\Delta_f G°$ / kJ mol⁻¹	$S°$ / J mol⁻¹K⁻¹	C_p / J mol⁻¹K⁻¹
I^-	−55.2	−51.6	111.3	−142.3
IO^-	−107.5	−38.5	−5.4	
IO_3^-	−221.3	−128.0	118.4	
IO_4^-	−151.5	−58.5	222.0	
MnO_4^-	−541.4	−447.2	191.2	−82.0
MnO_4^{-2}	−653.0	−500.7	59.0	
MoO_4^{-2}	−997.9	−836.3	27.2	
NO_2^-	−104.6	−32.2	123.0	−97.5
NO_3^-	−207.4	−111.3	146.4	−86.6
N_3^-	275.1	348.2	107.9	
OCN^-	−146.0	−97.4	106.7	
OH^-	−230.0	−157.2	−10.8	−148.5
PO_4^{-3}	−1277.4	−1018.7	−220.5	
$P_2O_7^{-4}$	−2271.1	−1919.0	−117.0	
Re^-	46.0	10.1	230.0	
S^{-2}	33.1	85.8	−14.6	
SCN^-	76.4	92.7	144.3	−40.2
SO_3^{-2}	−635.5	−486.5	−29.0	
SO_4^{-2}	−909.3	−744.5	20.1	−293.0
S_2^{-2}	30.1	79.5	28.5	
$S_2O_3^{-2}$	−652.3	−522.5	67.0	
$S_2O_4^{-2}$	−753.5	−600.3	92.0	
$S_2O_8^{-2}$	−1344.7	−1114.9	244.3	
Se^{-2}		129.3		
SeO_3^{-2}	−509.2	−369.8	13.0	
SeO_4^{-2}	−599.1	−441.3	54.0	
VO_3^-	−888.3	−783.6	50.0	
VO_4^{-3}		−899.0		
WO_4^{-2}	−1075.7			
Neutral species				
$AgBr$	−16.0	−26.9	155.2	−120.1
$AgCl$	−61.6	−54.1	129.3	−114.6
AgF	−227.1	−201.7	59.0	−84.9
AgI	50.4	25.5	184.1	−120.5
$AgNO_3$	−101.8	−34.2	219.2	−64.9
Ag_2SO_4	−698.1	−590.3	165.7	−251.0
$AlBr_3$	−895.0	−799.0	−74.5	
$AlCl_3$	−1033.0	−879.0	−152.3	
AlF_3	−1531.0	−1322.0	−363.2	
AlI_3	−699.0	−640.0	12.1	
$Al_2(SO_4)_3$	−3791.0	−3205.0	−583.2	
$BaBr_2$	−780.7	−768.7	174.5	
$BaCO_3$	−1214.8	−1088.6	−47.3	
$BaCl_2$	−872.0	−823.2	122.6	
BaF_2	−1202.9	−1118.4	−18.0	
$Ba(HCO_3)_2$	−1921.6	−1734.3	192.0	
BaI_2	−648.0	−663.9	232.2	
$Ba(NO_3)_2$	−952.4	−783.3	302.5	
$BaSO_4$	−1446.9	−1305.3	29.7	
$BeSO_4$	−1292.0	−1124.3	−109.6	
CCl_3COOH	−516.3			
$CHCl_2COOH$	−512.1			
$CHOOCs$	−683.8	−643.0	226.0	
$CHOOH$	−425.6	−351.0	92.0	−87.9
$CHOOK$	−677.9	−634.2	192.0	−66.1
$CHOONH_4$	−558.1	−430.4	205.0	−7.9
$CHOONa$	−665.7	−612.9	151.0	−41.4
$CHOORb$	−676.7	−635.1	213.0	
$CH_2ClCOOH$	−501.3			
CH_3COOCs	−744.3	−661.3	219.7	

Species	$\Delta_f H°$/ kJ mol⁻¹	$\Delta_f G°$/ kJ mol⁻¹	$S°$/ J mol⁻¹K⁻¹	C_p/ J mol⁻¹K⁻¹
CH_3COOH	−486.0	−369.3	86.6	−6.3
CH_3COOK	−738.4	−652.6	189.1	15.5
CH_3COONH_4	−618.5	−448.6	200.0	73.6
CH_3COONa	−726.1	−631.2	145.6	40.2
CH_3COORb	−737.2	−653.3	207.9	
$(COOH)_2$	−825.1	−673.9	45.6	
$(CH_3)_3N$	−76.0	93.1	133.5	
$CaBr_2$	−785.9	−761.5	111.7	
$CaCO_3$	−1220.0	−1081.4	−110.0	
$CaCl_2$	−877.1	−816.0	59.8	
CaF_2	−1208.1	−1111.2	−80.8	
CaI_2	−653.2	−656.7	169.5	
$Ca(NO_3)_2$	−957.6	−776.1	239.7	
$CaSO_4$	−1452.1	−1298.1	−33.1	
$CdBr_2$	−319.0	−285.5	91.6	
$CdCl_2$	−410.2	−340.1	39.7	
CdF_2	−741.2	−635.2	−100.8	
CdI_2	−186.3	−180.8	149.4	
$Cd(NO_3)_2$	−490.6	−300.1	219.7	
$CdSO_4$	−985.2	−822.1	−53.1	
$CeCl_3$	−1197.5	−1065.6	−38.0	
$CoBr_2$	−301.2	−262.3	50.0	
$CoCl_2$	−392.5	−316.7		
CoI_2	−168.6	−157.7	109.0	
$Co(NO_3)_2$	−472.8	−276.9	180.0	
$CoSO_4$	−967.3	−799.1	−92.0	
$CsBr$	−379.8	−396.0	215.5	
$CsCl$	−425.4	−423.2	189.5	−146.9
CsF	−590.9	−570.8	119.2	
$CsHCO_3$	−950.3	−878.8	224.3	
$CsHSO_4$	−1145.6	−1047.9	264.8	
CsI	−313.5	−343.6	244.3	−152.7
$CsNO_3$	−465.6	−403.3	279.5	−99.0
Cs_2CO_3	−1193.7	−1111.9	209.2	
Cs_2S	−483.7	−498.3	251.0	
Cs_2SO_4	−1425.8	−1328.6	286.2	
Cs_2Se		−454.8		
$Cu(NO_3)_2$	−350.0	−157.0	193.3	
$CuSO_4$	−844.5	−679.0	−79.5	
$DyCl_3$	−1197.0	−1059.0	−61.9	−389.0
$ErCl_3$	−1207.1	−1062.7	−75.3	−389.0
$EuCl_2$	−862.0			
$EuCl_3$	−1106.2	−967.7	−54.0	−402.0
$FeBr_2$	−332.2	−286.8	27.2	
$FeBr_3$	−413.4	−316.7	−68.6	
$FeCl_2$	−423.4	−341.3	−24.7	
$FeCl_3$	−550.2	−398.3	−146.4	
FeF_2	−754.4	−636.5	−165.3	
FeF_3	−1046.4	−840.9	−357.3	
FeI_2	−199.6	−182.1	84.9	
FeI_3	−214.2	−159.4	18.0	
$Fe(NO_3)_3$	−670.7	−338.3	123.4	
$FeSO_4$	−998.3	−823.4	−117.6	
$Fe_2(SO_4)_3$	−2825.0	−2242.8	−571.5	
$GdCl_3$	−1188.0	−1059.0	−36.8	−410.0
HBr	−121.6	−104.0	82.4	−141.8
HCN	150.6	172.4	94.1	
HCl	−167.2	−131.2	56.5	−136.4
HF	−332.6	−278.8	−13.8	−106.7
HI	−55.2	−51.6	111.3	−142.3
HNO_3	−207.4	−111.3	146.4	−86.6
$HSCN$	76.4	92.7	144.3	−40.2
H_2SO_4	−909.3	−744.5	20.1	−293.0
$HoCl_3$	−1206.7	−1067.3	−57.7	−393.0
KBr	−373.9	−387.2	184.9	−120.1
KCl	−419.5	−414.5	159.0	−114.6
KF	−585.0	−562.1	88.7	−84.9
$KHCO_3$	−944.4	−870.0	193.7	
$KHSO_4$	−1139.7	−1039.2	234.3	−63.0
KI	−307.6	−334.9	213.8	−120.5
KNO_3	−459.7	−394.5	248.9	−64.9
K_2CO_3	−1181.9	−1094.4	148.1	
K_2S	−471.5	−480.7	190.4	
K_2SO_4	−1414.0	−1311.1	225.1	−251.0
K_2Se		−437.2		
$LaCl_3$	−1208.8	−1077.3	−50.0	−423.0
$LiBr$	−400.0	−397.3	95.8	−73.2
$LiCl$	−445.6	−424.6	69.9	−67.8
LiF	−611.1	−571.9	−0.4	−38.1
LiI	−333.7	−344.8	124.7	−73.6
$LiNO_3$	−485.9	−404.5	160.2	−18.0
Li_2CO_3	−1234.1	−1114.6	−29.7	
Li_2SO_4	−1466.2	−1331.2	47.3	−155.6
$LuCl_3$	−1167.0	−1021.0	−96.0	−385.0
$MgBr_2$	−709.9	−662.7	26.8	
$MgCl_2$	−801.2	−717.1	−25.1	
MgI_2	−577.2	−558.1	84.5	
$Mg(NO_3)_2$	−881.6	−677.3	154.8	
$MgSO_4$	−1376.1	−1199.5	−118.0	
$MnBr_2$	−464.0			
$MnCl_2$	−555.1	−490.8	38.9	−222.0
MnI_2	−331.0			
$Mn(NO_3)_2$	−635.5	−450.9	218.0	−121.0
$MnSO_4$	−1130.1	−972.7	−53.6	−243.0
NH_4Br	−254.1	−183.3	195.8	−61.9
NH_4BrO_3	−199.6	−60.7	275.1	
NH_4CN	18.0	93.0	207.5	
NH_4Cl	−299.7	−210.5	169.9	−56.5
NH_4ClO_3	−236.5	−87.3	275.7	
NH_4ClO_4	−261.8	−87.8	295.4	
NH_4F	−465.1	−358.1	99.6	−26.8
NH_4HCO_3	−824.5	−666.1	204.6	
NH_4HS	−150.2	−67.2	176.1	
NH_4HSO_3	−758.7	−607.0	253.1	
NH_4HSO_4	−1019.9	−835.2	245.2	−3.8
NH_4HSeO_4	−714.2	−531.6	262.8	
$NH_4H_2AsO_3$	−847.3	−666.4	223.8	
$NH_4H_2AsO_4$	−1042.1	−832.5	230.5	
$NH_4H_2PO_4$	−1428.8	−1209.6	203.8	
$NH_4H_3P_2O_7$	−2409.1	−2102.6	326.0	
NH_4I	−187.7	−130.9	224.7	−62.3
NH_4IO_3	−354.0	−207.4	231.8	
NH_4NO_2	−237.2	−111.6	236.4	−17.6
NH_4NO_3	−339.9	−190.6	259.8	−6.7
NH_4OH	−362.5	−236.5	102.5	−68.6
NH_4SCN	−56.1	13.4	257.7	39.7
$(NH_4)_2CO_3$	−942.2	−686.4	169.9	
$(NH_4)_2CrO_4$	−1146.2	−886.4	277.0	
$(NH_4)_2Cr_2O_7$	−1755.2	−1459.5	488.7	
$(NH_4)_2HAsO_4$	−1171.4	−873.2	225.1	

Species	$\Delta_f H°/$ kJ mol^{-1}	$\Delta_f G°/$ kJ mol^{-1}	$S°/$ J mol^{-1}K^{-1}	$C_p/$ J mol^{-1}K^{-1}	Species	$\Delta_f H°/$ kJ mol^{-1}	$\Delta_f G°/$ kJ mol^{-1}	$S°/$ J mol^{-1}K^{-1}	$C_p/$ J mol^{-1}K^{-1}
$(NH_4)_2HPO_4$	−1557.2	−1247.8	193.3		$RbHCO_3$	−943.2	−870.8	212.7	
$(NH_4)_2S$	−231.8	−72.6	212.1		$RbHSO_4$	−1138.5	−1039.9	253.1	
$(NH_4)_2SO_3$	−900.4	−645.0	197.5		RbI	−306.4	−335.6	232.6	
$(NH_4)_2SO_4$	−1174.3	−903.1	246.9	−133.1	$RbNO_3$	−458.5	−395.2	267.8	
$(NH_4)_2SeO_4$	−864.0	−599.8	280.7		Rb_2CO_3	−1179.5	−1095.8	186.2	
$(NH_4)_3PO_4$	−1674.9	−1256.6	117.0		Rb_2S	−469.4	−482.0	228.4	
$NaBr$	−361.7	−365.8	141.4	−95.4	Rb_2SO_4	−1411.6	−1312.5	263.2	
$NaCl$	−407.3	−393.1	115.5	−90.0	$SmCl_3$	−1193.3	−1060.2	−42.7	−431.0
NaF	−572.8	−540.7	45.2	−60.2	$SrBr_2$	−788.9	−767.4	132.2	
$NaHCO_3$	−932.1	−848.7	150.2		$SrCO_3$	−1222.9	−1087.3	−89.5	
$NaHSO_4$	−1127.5	−1017.8	190.8	−38.0	$SrCl_2$	−880.1	−821.9	80.3	
NaI	−295.3	−313.5	170.3	−95.8	SrI_2	−656.2	−662.6	190.0	
$NaNO_3$	−447.5	−373.2	205.4	−40.2	$Sr(NO_3)_2$	−960.5	−782.0	260.2	
Na_2CO_3	−1157.4	−1051.6	61.1		$SrSO_4$	−1455.1	−1304.0	−12.6	
Na_2S	−447.3	−438.1	103.3		$TbCl_3$	−1184.1	−1045.5	−59.0	−393.0
Na_2SO_4	−1389.5	−1268.4	138.1	−201.0	$TlBr$	−116.2	−136.4	207.9	
Na_2Se		−394.6			$TlBr_3$	−168.2	−97.1	54.0	
$NdCl_3$	−1197.9	−1065.6	−37.7	−431.0	$TlCl$	−161.8	−163.6	182.0	
$NiBr_2$	−297.1	−253.6	36.0		$TlCl_3$	−305.0	−179.0	−23.0	
$NiCl_2$	−388.3	−307.9	−15.1		TlF	−327.3	−311.2	111.7	
NiF_2	−719.2	−603.3	−156.5		TlI	−49.8	−84.0	236.8	
NiI_2	−164.4	−149.0	93.7		$TlNO_3$	−202.0	−143.7	272.0	
$Ni(NO_3)_2$	−468.6	−268.5	164.0		Tl_2SO_4	−898.6	−809.3	271.1	
$NiSO_4$	−963.2	−790.3	−108.8		$TmCl_3$	−1199.1	−1055.6	−75.0	−385.0
$PbBr_2$	−244.8	−232.3	175.3		UCl_4	−1259.8	−1056.8	−184.0	
$PbCl_2$	−336.0	−286.9	123.4		UO_2CO_3	−1696.6	−1481.5	−154.4	
PbF_2	−666.9	−582.0	−17.2		$UO_2(NO_3)_2$	−1434.3	−1176.0	195.4	
PbI_2	−112.1	−127.6	233.0		UO_2SO_4	−1928.8	−1698.2	−77.4	
$Pb(NO_3)_2$	−416.3	−246.9	303.3		$YbCl_3$	−1176.1	−1037.6	−71.0	−385.0
$PrCl_3$	−1206.2	−1072.7	−42.0	−439.0	$ZnBr_2$	−397.0	−355.0	52.7	−238.0
$RaCl_2$	−861.9	−823.8	167.0		$ZnCl_2$	−488.2	−409.5	0.8	−226.0
$Ra(NO_3)_2$	−942.2	−784.0	347.0		ZnF_2	−819.1	−704.6	−139.7	−167.0
$RaSO_4$	−1436.8	−1306.2	75.0		ZnI_2	−264.3	−250.2	110.5	−238.0
$RbBr$	−372.7	−387.9	203.9		$Zn(NO_3)_2$	−568.6	−369.6	180.7	−126.0
$RbCl$	−418.3	−415.2	178.0		$ZnSO_4$	−1063.2	−891.6	−92.0	−247.0
RbF	−583.8	−562.8	107.5						

HEAT OF COMBUSTION

The heat of combustion of a substance at 25°C can be calculated from the enthalpy of formation ($\Delta_f H°$) data in the table "Standard Thermodynamic Properties of Chemical Substances" in this Section. We can write the general combustion reaction as

$$X + O_2 \rightarrow CO_2(g) + H_2O \text{ (l)} + \text{other products}$$

For a compound containing only carbon, hydrogen, and oxygen, the reaction is simply

$$C_a H_b O_c + \left(a + \frac{1}{4}b - \frac{1}{2}c \right) O_2 \rightarrow a\, CO_2 \text{ (g)} + \frac{1}{2}b\, H_2O \text{ (l)}$$

and the standard heat of combustion $\Delta_c H°$, which is defined as the negative of the enthalpy change for the reaction (i.e., the heat released in the combustion process), is given by

$$\Delta_c H° = -a\Delta_f H°(CO_2, g) - \frac{1}{2}b\Delta_f H°(H_2O, l) + \Delta_f H°(C_a H_b O_c)$$

$$= 393.51a + 142.915b + \Delta_f H°(C_a H_b O_c)$$

This equation applies if the reactants start in their standard states (25°C and one atmosphere pressure) and the products return to the same conditions. The same equation applies to a compound containing another element if that element ends in its standard reference state (e.g., nitrogen, if the product is N_2); in general, however, the exact products containing the other elements must be known in order to calculate the heat of combustion.

The following table gives the standard heat of combustion calculated in this manner for a few representative substances.

Molecular formula	Name	$\Delta_c H°$/kJ mol^{-1}
Inorganic substances		
C	Carbon (graphite)	393.5
CO	Carbon monoxide (g)	283.0
H_2	Hydrogen (g)	285.8
H_3N	Ammonia (g)	382.8
H_4N_2	Hydrazine (g)	667.1
N_2O	Nitrous oxide (g)	82.1
Hydrocarbons		
CH_4	Methane (g)	890.8
C_2H_2	Acetylene (g)	1301.1
C_2H_4	Ethylene (g)	1411.2
C_2H_6	Ethane (g)	1560.7
C_3H_6	Propylene (g)	2058.0
C_3H_6	Cyclopropane (g)	2091.3
C_3H_8	Propane (g)	2219.2
C_4H_6	1,3-Butadiene (g)	2541.5
C_4H_{10}	Butane (g)	2877.6
C_5H_{12}	Pentane (l)	3509.0
C_6H_6	Benzene (l)	3267.6
C_6H_{12}	Cyclohexane (l)	3919.6
C_6H_{14}	Hexane (l)	4163.2
C_7H_8	Toluene (l)	3910.3
C_7H_{16}	Heptane (l)	4817.0
$C_{10}H_8$	Naphthalene (s)	5156.3
Alcohols and ethers		
CH_4O	Methanol (l)	726.1
C_2H_6O	Ethanol (l)	1366.8
C_2H_6O	Dimethyl ether (g)	1460.4
$C_2H_6O_2$	Ethylene glycol (l)	1189.2
C_3H_8O	1-Propanol (l)	2021.3

Molecular formula	Name	$\Delta_c H°$/kJ mol^{-1}
$C_3H_8O_3$	Glycerol (l)	1655.4
$C_4H_{10}O$	Diethyl ether (l)	2723.9
$C_5H_{12}O$	1-Pentanol (l)	3330.9
C_6H_6O	Phenol (s)	3053.5
Carbonyl compounds		
CH_2O	Formaldehyde (g)	570.7
C_2H_2O	Ketene (g)	1025.4
C_2H_4O	Acetaldehyde (l)	1166.9
C_3H_6O	Acetone (l)	1789.9
C_3H_6O	Propanal (l)	1822.7
C_4H_8O	2-Butanone (l)	2444.1
Acids and esters		
CH_2O_2	Formic acid (l)	254.6
$C_2H_4O_2$	Acetic acid (l)	874.2
$C_2H_4O_2$	Methyl formate (l)	972.6
$C_3H_6O_2$	Methyl acetate (l)	1592.2
$C_4H_8O_2$	Ethyl acetate (l)	2238.1
$C_6H_5NO_2$	Nicotinic acid (s)	2731.1
$C_7H_6O_2$	Benzoic acid (s)	3228.2
Nitrogen compounds		
CHN	Hydrogen cyanide (g)	671.5
CH_3NO_2	Nitromethane (l)	709.2
CH_4N_2O	Urea (s)	632.7
CH_5N	Methylamine (g)	1085.6
C_2H_3N	Acetonitrile (l)	1247.2
C_2H_5NO	Acetamide (s)	1184.6
C_3H_9N	Trimethylamine (g)	2443.1
C_5H_5N	Pyridine (l)	2782.3
C_6H_7N	Aniline (l)	3392.8

ELECTRICAL CONDUCTIVITY OF WATER

This table gives the electrical conductivity of highly purified water over a range of temperature and pressure. The first column of conductivity data refers to water at its own vapor pressure. Equations for calculating the conductivity at any temperature and pressure may be found in the reference.

Reference

Marshall, W. L., *J. Chem. Eng. Data* 32, 221, 1987.

Conductivity in µS/cm at the Indicated Pressure

$t/°C$	Sat. vapor	50 MPa	100 MPa	200 MPa	400 MPa	600 MPa
0	0.0115	0.0150	0.0189	0.0275	0.0458	0.0667
25	0.0550	0.0686	0.0836	0.117	0.194	0.291
100	0.765	0.942	1.13	1.53	2.45	3.51
200	2.99	4.08	5.22	7.65	13.1	19.5
300	2.41	4.87	7.80	14.1	28.9	46.5
400		1.17	4.91	14.3	39.2	71.3
600			0.134	4.65	33.8	85.7

ELECTRICAL CONDUCTIVITY OF AQUEOUS SOLUTIONS

The following table gives the electrical conductivity of aqueous solutions of some acids, bases, and salts as a function of concentration. All values refer to 20°C. The conductivity κ (often called specific conductance in older literature) is the reciprocal of the resistivity. The molar conductivity Λ is related to this by $\Lambda = \kappa/c$, where c is the amount-of-substance concentration of the electrolyte. Thus if κ has units of millisiemens per centimeter (mS/cm), as in this table, and c is expressed in mol/L, then Λ has units of S cm^2 mol^{-1}. For these electrolytes the concentration c corresponding to the mass percent values given here can be found in the table "Concentrative Properties of Aqueous Solutions" in Section 7.

References

1. *CRC Handbook of Chemistry, and Physics, 70th Edition*, 1989, p. D-221.
2. Wolf, A. V., *Aqueous Solutions and Body Fluids*, Hoeber, 1996.

Electrical Conductivity κ in mS/cm for the Indicated Concentration in Mass Percent

Name	Formula	0.5%	1%	2%	5%	10%	15%	20%	25%	30%	40%	50%
Acetic acid	CH_3COOH	0.3	0.6	0.8	1.2	1.5	1.7	1.7	1.6	1.4	1.1	0.8
Ammonia	NH_3	0.5	0.7	1.0	1.1	1.0	0.7	0.5	0.4			
Ammonium chloride	NH_4Cl	10.5	20.4	40.3	95.3	180						
Ammonium sulfate	$(NH_4)_2SO_4$	7.4	14.2	25.7	57.4	105	147	185	215			
Barium chloride	$BaCl_2$	4.7	9.1	17.4	40.4	76.7	109.0	137.0				
Calcium chloride	$CaCl_2$	8.1	15.7	29.4	67.0	117	157	177	183	172	106	
Cesium chloride	CsCl	3.8	7.4	13.8	32.9	65.8	102	142				
Citric acid	$H_3C(OH)(COO)_3$	1.2	2.1	3.0	4.7	6.2	7.0	7.2	7.1			
Copper(II) sulfate	$CuSO_4$	2.9	5.4	9.3	19.0	32.2	42.3					
Formic acid	HCOOH	1.4	2.4	3.5	5.6	7.8	9.0	9.9	10.4	10.5	9.9	8.6
Hydrogen chloride	HCl	45.1	92.9	183								
Lithium chloride	LiCl	10.1	19.0	34.9	76.4	127	155	170	165	146		
Magnesium chloride	$MgCl_2$	8.6	16.6	31.2	66.9	108	129	134	122	98		
Magnesium sulfate	$MgSO_4$	4.1	7.6	13.3	27.4	42.7	54.2	51.1	44.1			
Manganese(II) sulfate	$MnSO_4$		6.2	10.6	21.6	34.5	43.7	47.6				
Nitric acid	HNO_3	28.4	56.1	108								
Oxalic acid	$H_2C_2O_4$	14.0	21.8	35.3	65.6							
Phosphoric acid	H_3PO_4	5.5	10.1	16.2	31.5	59.4	88.4	118	146	173	209	
Potassium bromide	KBr	5.2	10.2	19.5	47.7	95.6	144	194				
Potassium carbonate	K_2CO_3	7.0	13.6	25.4	58.0	109	152	188	223			
Potassium chloride	KCl	8.2	15.7	29.5	71.9	143	208					
Potassium dihydrogen phosphate	KH_2PO_4	3.0	5.9	11.0	25.0	44.6						
Potassium hydrogen carbonate	$KHCO_3$	4.6	8.9	17.0	38.8	72.4	101	128				
Potassium hydrogen phosphate	K_2HPO_4	5.2	9.9	18.3	40.3							
Potassium hydroxide	KOH	20.0	38.5	75.0	178							
Potassium iodide	KI	3.8	7.5	14.2	35.2	71.8	110	188	224			
Potassium nitrate	KNO_3	5.5	10.7	20.1	47.0	87.3	124	157	182			
Potassium permanganate	$KMnO_4$	3.5	6.9	13.0	30.5							
Potassium sulfate	K_2SO_4	5.8	11.2	21.0	48.0	88.6						
Silver(I) nitrate	$AgNO_3$	3.1	6.1	12.0	26.7	49.8	72.0	92.8	112	129	162	
Sodium acetate	$NaCH_3COO$	3.9	7.6	14.4	30.9	53.4	64.1	69.3	69.2	64.3		
Sodium bromide	NaBr	5.0	9.7	18.4	44.0	84.6	122	157	191	216		
Sodium carbonate	Na_2CO_3	7.0	13.1	23.3	47.0	74.4	88.6					
Sodium chloride	NaCl	8.2	16.0	30.2	70.1	126	171	204	222			
Sodium citrate	$Na_3C_6H_5O_7$		7.4	12.8	26.2	42.1	52.0	57.1	57.3	53.5		
Sodium dihydrogen phosphate	NaH_2PO_4	2.2	4.4	9.1	21.0	33.2	43.3	49.6	53.1	54.0	46.1	
Sodium hydrogen carbonate	$NaHCO_3$	4.2	8.2	15.0	31.4							
Sodium hydrogen phosphate	Na_2HPO_4	4.6	8.7	15.6	31.4							
Sodium hydroxide	NaOH	24.8	48.6	93.1	206							
Sodium nitrate	$NaNO_3$	5.4	10.6	20.4	46.2	82.6	111	134	152	165	178	
Sodium phosphate	Na_3PO_4	7.3	14.1	22.7	43.5							
Sodium sulfate	Na_2SO_4	5.9	11.2	19.8	42.7	71.3	91.1	109				
Sodium thiosulfate	$Na_2S_2O_3$	5.7	10.7	19.5	43.3	76.7	104	123	134	136	118	
Strontium chloride	$SrCl_2$	5.9	11.4	22.0	49.1	91.5	127	153	168	178		
Sulfuric acid	H_2SO_4	24.3	47.8	92	211							
Trichloroacetic acid	CCl_3COOH	10.3	19.6	37.2	84.7	148	193	221				
Zinc sulfate	$ZnSO_4$	2.8	5.4	10.0	20.5	33.7	43.3					

STANDARD KCl SOLUTIONS FOR CALIBRATING CONDUCTIVITY CELLS

This table presents recommended electrolytic conductivity (κ) values for aqueous potassium chloride solutions with molalities of 0.01 mol/kg, 0.1 mol/kg and 1.0 mol/kg at temperatures from 0°C to 50°C. The values, which are based on measurements at the National Institute of Standards and Technology, provide primary standards for the calibration of conductivity cells. The measurements at 0.01 and 0.1 molal are described in Reference 1, while those at 1.0 molal are in Reference 2. Temperatures are given on the ITS-90 scale. The uncertainty in the conductivity is about 0.03% for the 0.01 molal values and about 0.04% for the 0.1 and 1.0 molal values. The conductivity of water saturated with atmospheric CO_2 is given in the last column. These values were subtracted from the original measurements to give the values in the second, third, and fourth columns. All κ values are given in units of 10^{-4} S/m (numerically equal to μS/cm).

The assistance of Kenneth W. Pratt is appreciated.

	$10^4\ \kappa/\mathrm{S\ m^{-1}}$			
$t/°C$	0.01 m KCl	0.1 m KCl	1.0 m KCl	H_2O (CO_2 sat.)
0	772.92	7 116.85	63 488	0.58
5	890.96	8 183.70	72 030	0.68
10	1 013.95	9 291.72	80 844	0.79
15	1 141.45	10 437.1	89 900	0.89
18	1 219.93	11 140.6	—	0.95
20	1 273.03	11 615.9	99 170	0.99
25	1 408.23	12 824.6	108 620	1.10
30	1 546.63	14 059.2	118 240	1.20
35	1 687.79	15 316.0	127 970	1.30
40	1 831.27	16 591.0	137 810	1.40
45	1 976.62	17 880.6	147 720	1.51
50	2 123.43	19 180.9	157 670	1.61

References

1. Wu, Y. C., Koch, W. F., and Pratt, K. W., *J. Res. Natl. Inst. Stand. Technol.* 96, 191, 1991.
2. Wu, Y. C., Koch, W. F., Feng, D., Holland, L. A., Juhasz, E., Arvay, E., and Tomek, A., *J. Res. Natl. Inst. Stand. Technol.* 99, 241, 1994.
3. Pratt, K. W., Koch, W. F., Wu, Y. C., and Berezansky, P. A., *Pure Appl. Chem.* 73, 1783, 2001.

MOLAR CONDUCTIVITY OF AQUEOUS HF, HCl, HBr, AND HI

The molar conductivity Λ of an electrolyte solution is defined as the conductivity divided by amount-of-substance concentration. The customary unit is S cm²mol⁻¹ (i.e., Ω^{-1} cm²mol⁻¹). The first part of this table gives the molar conductivity of the hydrohalogen acids at 25°C as a function of the concentration in mol/L. The second part gives the temperature dependence of Λ for HCl and HBr. More extensive tables and mathematical representations may be found in the reference.

Reference

Hamer, W. J., and DeWane, H. J., *Electrolytic Conductance and the Conductances of the Hydrohalogen Acids in Water*, Natl. Stand. Ref. Data Sys.- Natl. Bur. Standards (U.S.), No. 33, 1970.

c/mol L⁻¹	HF	HCl	HBr	HI
Inf. dil.	405.1	426.1	427.7	426.4
0.0001		424.5	425.9	424.6
0.0005		422.6	424.3	423.0
0.001		421.2	422.9	421.7
0.005	128.1	415.7	417.6	416.4
0.01	96.1	411.9	413.7	412.8
0.05	50.1	398.9	400.4	400.8
0.10	39.1	391.1	391.9	394.0
0.5	26.3	360.7	361.9	369.8
1.0	24.3	332.2	334.5	343.9
1.5		305.8	307.6	316.4
2.0		281.4	281.7	288.9
2.5		258.9	257.8	262.5
3.0		237.6	236.8	237.9
3.5		218.3	217.5	215.4
4.0		200.0	199.4	195.1
4.5		183.1	182.4	176.8
5.0		167.4	166.5	160.4
5.5		152.9	151.8	145.5
6.0		139.7	138.2	131.7
6.5		127.7	125.7	118.6
7.0		116.9	114.2	105.7
7.5		107.0	103.8	
8.0		98.2	94.4	
8.5		90.3	85.8	
9.0		83.1		
9.5		76.6		
10.0		70.7		

HCl

c/mol L⁻¹	−20°C	−10°C	0°C	10°C	20°C	30°C	40°C	50°C
0.5			228.7	283.0	336.4	386.8	436.9	482.4
1.0			211.7	261.6	312.2	359.0	402.9	445.3
1.5			196.2	241.5	287.5	331.1	371.6	410.8
2.0			182.0	222.7	262.9	303.3	342.4	378.2
2.5		131.7	168.5	205.1	239.8	277.0	315.2	347.6
3.0		120.8	154.6	188.5	219.3	253.3	289.3	319.0
3.5	85.5	111.3	139.6	172.2	201.6	232.9	263.9	292.1
4.0	79.3	102.7	129.2	158.1	185.6	214.2	242.2	268.2
4.5	73.7	94.9	119.5	145.4	170.6	196.6	222.5	246.7
5.0	68.5	87.8	110.3	133.5	156.6	180.2	204.1	226.5
5.5	63.6	81.1	101.7	122.5	143.6	165.0	187.1	207.7
6.0	58.9	74.9	93.7	112.3	131.5	151.0	171.3	190.3
6.5	54.4	69.1	86.2	103.0	120.4	138.2	156.9	174.3
7.0	50.2	63.7	79.3	94.4	110.2	126.4	143.3	159.7
7.5	46.3	58.6	73.0	86.5	100.9	115.7	131.6	146.2
8.0	42.7	54.0	67.1	79.4	92.4	106.1	120.6	134.0
8.5	39.4	49.8	61.7	72.9	84.7	97.3	110.7	123.0
9.0	36.4	45.9	56.8	67.1	77.8	89.4	101.7	112.9
9.5	33.6	42.3	52.3	61.8	71.5	82.3	93.6	103.9
10.0	31.2	39.1	48.2	57.0	65.8	75.9	86.3	95.7
10.5	28.9	36.1	44.5	52.7	60.7	70.1	79.6	88.4
11.0	26.8	33.4	41.1	48.8	56.1	64.9	73.6	81.7
11.5	24.9	31.0	38.0	45.3	51.9	60.1	68.0	75.6
12.0	23.1	28.7	35.3	42.0	48.0	55.6	62.8	70.0
12.5	21.4	26.7	32.7	39.0	44.4	51.4	57.9	64.8

HBr

c/mol L⁻¹	−20°C	−10°C	0°C	10°C	20°C	30°C	40°C	50°C
0.5			240.9	295.9	347.0	398.9	453.6	496.8
1.0			229.6	276.0	329.0	380.4	418.6	465.2
1.5			209.5	254.9	298.9	340.6	381.8	421.4
2.0		150.8	188.6	231.3	271.8	314.1	350.5	387.4
2.5		136.8	171.7	208.3	244.8	281.7	316.0	349.1
3.0		125.7	157.2	189.5	222.2	255.0	287.8	318.6
3.5		116.1	144.1	174.6	203.2	234.4	263.7	291.9
4.0	84.0	107.5	132.3	160.2	186.8	214.2	239.7	266.9
4.5	78.0	99.0	123.0	146.4	171.2	195.1	218.8	242.6
5.0	72.3	91.4	112.6	134.0	155.7	178.2	199.6	221.3
5.5	67.0	84.2	103.1	122.7	142.1	162.8	181.4	201.8
6.0	61.8	77.2	94.3	112.0	129.6	148.0	165.4	183.4
6.5	56.8	70.7	86.0	102.0	118.0	134.1	150.5	166.3
7.0	51.9	64.6	78.4	92.6	107.1	121.4	136.3	150.8

EQUIVALENT CONDUCTIVITY OF ELECTROLYTES IN AQUEOUS SOLUTION

Petr Vanýsek

This table gives the equivalent (molar) conductivity Λ at 25°C for some common electrolytes in aqueous solution at concentrations up to 0.1 mol/L. The units of Λ are 10^{-4} m^2 S mol^{-1}.

For very dilute solutions, the equivalent conductivity for any electrolyte of concentration c can be approximately calculated using the Debye-Hückel-Onsager equation, which can be written for a symmetrical (equal charge on cation and anion) electrolyte as

$$\Lambda = \Lambda° - (A + B\Lambda°)c^{1/2}$$

For a solution at 25°C and both cation and anion with charge $|1|$, the constants are $A = 60.20$ and $B = 0.229$. $\Lambda°$ can be found from the next table, "Ionic Conductivity and Diffusion at Infinite Dilution". The equation is reliable for $c < 0.001$ mol/L; with higher concentration the error increases.

Compound	Infinite dilution $\Lambda°$	Concentration (mol/L)						
		0.0005	0.001	0.005	0.01	0.02	0.05	0.1
					Λ			
$AgNO_3$	133.29	131.29	130.45	127.14	124.70	121.35	115.18	109.09
$1/2BaCl_2$	139.91	135.89	134.27	127.96	123.88	119.03	111.42	105.14
$1/2CaCl_2$	135.77	131.86	130.30	124.19	120.30	115.59	108.42	102.41
$1/2Ca(OH)_2$	258	—	—	233	226	214	—	—
$1/2CuSO_4$	133.6	121.6	115.20	94.02	83.08	72.16	59.02	50.55
HCl	425.95	422.53	421.15	415.59	411.80	407.04	398.89	391.13
KBr	151.9	149.8	148.9	146.02	143.36	140.41	135.61	131.32
KCl	149.79	147.74	146.88	143.48	141.20	138.27	133.30	128.90
$KClO_4$	139.97	138.69	137.80	134.09	131.39	127.86	121.56	115.14
$1/3K_3Fe(CN)_6$	174.5	166.4	163.1	150.7	—	—	—	—
$1/4K_4Fe(CN)_6$	184	—	167.16	146.02	134.76	122.76	107.65	97.82
$KHCO_3$	117.94	116.04	115.28	112.18	110.03	107.17	—	—
KI	150.31	148.2	143.32	144.30	142.11	139.38	134.90	131.05
KIO_4	127.86	125.74	124.88	121.18	118.45	114.08	106.67	98.2
KNO_3	144.89	142.70	141.77	138.41	132.75	132.34	126.25	120.34
$KMnO_4$	134.8	132.7	131.9	—	126.5	—	—	113
KOH	271.5	—	234	230	228	—	219	213
$KReO_4$	128.20	126.03	125.12	121.31	118.49	114.49	106.40	97.40
$1/3LaCl_3$	145.9	139.6	137.0	127.5	121.8	115.3	106.2	99.1
LiCl	114.97	113.09	112.34	109.35	107.27	104.60	100.06	95.81
$LiClO_4$	105.93	104.13	103.39	100.52	98.56	96.13	92.15	88.52
$1/2MgCl_2$	129.34	125.55	124.15	118.25	114.49	109.99	103.03	97.05
NH_4Cl	149.6	147.5	146.7	134.4	141.21	138.25	133.22	128.69
NaCl	126.39	124.44	123.68	120.59	118.45	115.70	111.01	106.69
$NaClO_4$	117.42	115.58	114.82	111.70	109.54	106.91	102.35	98.38
NaI	126.88	125.30	124.19	121.19	119.18	116.64	112.73	108.73
$NaOOCCH_3$	91.0	89.2	88.5	85.68	83.72	81.20	76.88	72.76
NaOH	247.7	245.5	244.6	240.7	237.9	—	—	—
Na picrate	80.45	78.7	78.6	75.7	73.7	—	66.3	61.8
$1/2Na_2SO_4$	129.8	125.68	124.09	117.09	112.38	106.73	97.70	89.94
$1/2SrCl_2$	135.73	131.84	130.27	124.18	120.23	115.48	108.20	102.14
$1/2ZnSO_4$	132.7	121.3	114.47	95.44	84.87	74.20	61.17	52.61

IONIC CONDUCTIVITY AND DIFFUSION AT INFINITE DILUTION

Petr Vanýsek

This table gives the molar (equivalent) conductivity λ for common ions at infinite dilution. All values refer to aqueous solutions at 25°C. It also lists the diffusion coefficient D of the ion in dilute aqueous solution, which is related to λ through the equation

$$D = (RT / F^2)(\lambda / |z|)$$

where R is the molar gas constant, T the temperature, F the Faraday constant, and z the charge on the ion. The variation with temperature is fairly sharp; for typical ions, λ and D increase by 2 to 3% per degree as the temperature increases from 25°C.

The diffusion coefficient for a salt, D_{salt}, may be calculated from the D_+ and D_- values of the constituent ions by the relation

$$D_{salt} = \frac{(z_+ + |z_-|)D_+ D_-}{z_+ D_+ + |z_-| D_-}$$

For solutions of simple, pure electrolytes (one positive and one negative ionic species), such as NaCl, equivalent ionic conductivity $\Lambda°$, which is the molar conductivity per unit concentration of charge, is defined as

$$\Lambda° = \Lambda_+ + \Lambda_-$$

where Λ_+ and Λ_- are equivalent ionic conductivities of the cation and anion. The more general formula is

$$\Lambda° = \nu_+ \Lambda_+ + \nu_- \Lambda_-$$

where ν_+ and ν_- refer to the number of moles of cations and anions to which one mole of the electrolyte gives a rise in the solution.

References

1. Gray, D. E., Ed., *American Institute of Physics Handbook*, McGraw-Hill, New York, 1972, 2–226.
2. Robinson, R. A., and Stokes, R. H., *Electrolyte Solutions*, Butterworths, London, 1959.
3. Lobo, V. M. M., and Quaresma, J. L., *Handbook of Electrolyte Solutions*, Physical Science Data Series 41, Elsevier, Amsterdam, 1989.
4. Conway, B. E., *Electrochemical Data*, Elsevier, Amsterdam, 1952.
5. Milazzo, G., *Electrochemistry: Theoretical Principles and Practical Applications*, Elsevier, Amsterdam, 1963.

Ion	$\Lambda_\pm$ $10^{-4}\,m^2\,S\,mol^{-1}$	D $10^{-5}\,cm^2\,s^{-1}$
Inorganic Cations		
Ag^+	61.9	1.648
$1/3Al^{3+}$	61	0.541
$1/2Ba^{2+}$	63.6	0.847
$1/2Be^{2+}$	45	0.599
$1/2Ca^{2+}$	59.47	0.792
$1/2Cd^{2+}$	54	0.719
$1/3Ce^{3+}$	69.8	0.620
$1/2Co^{2+}$	55	0.732
$1/3[Co(NH_3)_6]^{3+}$	101.9	0.904
$1/3[Co(en)_3]^{3+}$	74.7	0.663
$1/6[Co_2(trien)_3]^{6+}$	69	0.306
$1/3Cr^{3+}$	67	0.595
Cs^+	77.2	2.056
$1/2Cu^{2+}$	53.6	0.714
D^+	249.9	6.655
$1/3Dy^{3+}$	65.6	0.582
$1/3Er^{3+}$	65.9	0.585
$1/3Eu^{3+}$	67.8	0.602
$1/2Fe^{2+}$	54	0.719
$1/3Fe^{3+}$	68	0.604
$1/3Gd^{3+}$	67.3	0.597
H^+	349.65	9.311
$1/2Hg^{2+}$	68.6	0.913
$1/2Hg^{2+}$	63.6	0.847
$1/3Ho^{3+}$	66.3	0.589
K^+	73.48	1.957
$1/3La^{3+}$	69.7	0.619
Li^+	38.66	1.029
$1/2Mg^{2+}$	53.0	0.706
$1/2Mn^{2+}$	53.5	0.712
NH_4^+	73.5	1.957
$N_2H_5^+$	59	1.571

Ion	$\Lambda_\pm$ $10^{-4}\,m^2\,S\,mol^{-1}$	D $10^{-5}\,cm^2\,s^{-1}$
Na^+	50.08	1.334
$1/3Nd^{3+}$	69.4	0.616
$1/2Ni^{2+}$	49.6	0.661
$1/4[Ni_2(trien)_3]^{4+}$	52	0.346
$1/2Pb^{2+}$	71	0.945
$1/3Pr^{3+}$	69.5	0.617
$1/2Ra^{2+}$	66.8	0.889
Rb^+	77.8	2.072
$1/3Sc^{3+}$	64.7	0.574
$1/3Sm^{3+}$	68.5	0.608
$1/2Sr^{2+}$	59.4	0.791
Tl^+	74.7	1.989
$1/3Tm^{3+}$	65.4	0.581
$1/2UO_2^{2+}$	32	0.426
$1/3Y^{3+}$	62	0.550
$1/3Yb^{3+}$	65.6	0.582
$1/2Zn^{2+}$	52.8	0.703
Inorganic Anions		
$Au(CN)_2^-$	50	1.331
$Au(CN)_4^-$	36	0.959
$B(C_6H_5)_4^-$	21	0.559
Br^-	78.1	2.080
Br_3^-	43	1.145
BrO_3^-	55.7	1.483
CN^-	78	2.077
CNO^-	64.6	1.720
$1/2CO_3^{2-}$	69.3	0.923
Cl^-	76.31	2.032
ClO_2^-	52	1.385
ClO_3^-	64.6	1.720
ClO_4^-	67.3	1.792
$1/3[Co(CN)_6]^{3-}$	98.9	0.878
$1/2CrO_4^{2-}$	85	1.132

Ion	$\Lambda_{\pm}$ $10^{-4}\,m^2\,S\,mol^{-1}$	D $10^{-5}\,cm^2\,s^{-1}$	Ion	$\Lambda_{\pm}$ $10^{-4}\,m^2\,S\,mol^{-1}$	D $10^{-5}\,cm^2\,s^{-1}$
F^-	55.4	1.475	Histidyl$^+$	23.0	0.612
$1/4[Fe(CN)_6]^{4-}$	110.4	0.735	Hydroxyethyltrimethylarsonium$^+$	39.4	1.049
$1/3[Fe(CN)_6]^{3-}$	100.9	0.896	Methylammonium$^+$	58.7	1.563
$H_2AsO_4^-$	34	0.905	Octadecylpyridinium$^+$	20	0.533
HCO_3^-	44.5	1.185	Octadecyltributylammonium$^+$	16.6	0.442
HF_2^-	75	1.997	Octadecyltriethylammonium$^+$	17.9	0.477
$1/2HPO_4^{2-}$	57	0.759	Octadecyltrimethylammonium$^+$	19.9	0.530
$H_2PO_4^-$	36	0.959	Octadecyltripropylammonium$^+$	17.2	0.458
$H_2PO_2^-$	46	1.225	Octyltrimethylammonium$^+$	26.5	0.706
HS^-	65	1.731	Pentylammonium$^+$	37	0.985
HSO_3^-	58	1.545	Piperidinium$^+$	37.2	0.991
HSO_4^-	52	1.385	Propylammonium$^+$	40.8	1.086
$H_2SbO_4^-$	31	0.825	Pyrilammonium$^+$	24.3	0.647
I^-	76.8	2.045	Tetrabutylammonium$^+$	19.5	0.519
IO_3^-	40.5	1.078	Tetradecyltrimethylammonium$^+$	21.5	0.573
IO_4^-	54.5	1.451	Tetraethylammonium$^+$	32.6	0.868
MnO_4^-	61.3	1.632	Tetramethylammonium$^+$	44.9	1.196
$1/2MoO_4^{2-}$	74.5	1.984	Tetraisopentylammonium$^+$	17.9	0.477
$N(CN)_2^-$	54.5	1.451	Tetrapentylammmonium$^+$	17.5	0.466
NO_2^-	71.8	1.912	Tetrapropylammonium$^+$	23.4	0.623
NO_3^-	71.42	1.902	Triethylammonium$^+$	34.3	0.913
$NH_2SO_3^-$	48.3	1.286	Triethylsulfonium$^+$	36.1	0.961
N_3^-	69	1.837	Trimethylammonium$^+$	47.23	1.258
OCN^-	64.6	1.720	Trimethylhexylammonium$^+$	34.6	0.921
OD^-	119	3.169	Trimethylsulfonium$^+$	51.4	1.369
OH^-	198	5.273	Tripropylammonium$^+$	26.1	0.695
PF_6^-	56.9	1.515			
$1/2PO_3F^{2-}$	63.3	0.843	**Organic Anions**		
$1/3PO_4^{3-}$	92.8	0.824	Acetate$^-$	40.9	1.089
$1/4P_2O_7^{4-}$	96	0.639	p$-$Anisate$^-$	29.0	0.772
$1/3P_3O_9^{3-}$	83.6	0.742	$1/2$Azelate^{2-}	40.6	0.541
$1/5P_3O_{10}^{5-}$	109	0.581	Benzoate$^-$	32.4	0.863
ReO_4^-	54.9	1.462	Bromoacetate$^-$	39.2	1.044
SCN^-	66	1.758	Bromobenzoate$^-$	30	0.799
$1/2SO_3^{2-}$	72	0.959	Butyrate$^-$	32.6	0.868
$1/2SO_4^{2-}$	80.0	1.065	Chloroacetate$^-$	39.8	1.060
$1/2S_2O_3^{2-}$	85.0	1.132	m$-$Chlorobenzoate$^-$	31	0.825
$1/2S_2O_4^{2-}$	66.5	0.885	o$-$Chlorobenzoate$^-$	30.2	0.804
$1/2S_2O_6^{2-}$	93	1.238	$1/3$Citrate^{3-}	70.2	0.623
$1/2S_2O_8^{2-}$	86	1.145	Crotonate$^-$	33.2	0.884
$Sb(OH)_6^-$	31.9	0.849	Cyanoacetate$^-$	43.4	1.156
$SeCN^-$	64.7	1.723	Cyclohexane carboxylate$^-$	28.7	0.764
$1/2SeO_4^{2-}$	75.7	1.008	$1/2$ 1,1$-$Cyclopropanedicarboxylate^{2-}	53.4	0.711
$1/2WO_4^{2-}$	69	0.919	Decylsulfate$^-$	26	0.692
			Dichloroacetate$^-$	38.3	1.020
Organic Cations			$1/2$Diethylbarbiturate^{2-}	26.3	0.350
Benzyltrimethylammonium$^+$	34.6	0.921	Dihydrogencitrate$^-$	30	0.799
Isobutylammonium$^+$	38	1.012	$1/2$Dimethylmalonate^{2-}	49.4	0.658
Butyltrimethylammonium$^+$	33.6	0.895	3,5$-$Dinitrobenzoate$^-$	28.3	0.754
Decylpyridinium$^+$	29.5	0.786	Dodecylsulfate$^-$	24	0.639
Decyltrimethylammonium$^+$	24.4	0.650	Ethylmalonate$^-$	49.3	1.313
Diethylammonium$^+$	42.0	1.118	Ethylsulfate$^-$	39.6	1.055
Dimethylammonium$^+$	51.8	1.379	Fluoroacetate$^-$	44.4	1.182
Dipropylammonium$^+$	30.1	0.802	Fluorobenzoate$^-$	33	0.879
Dodecylammonium$^+$	23.8	0.634	Formate$^-$	54.6	1.454
Dodecyltrimethylammonium$^+$	22.6	0.602	$1/2$Fumarate^{2-}	61.8	0.823
Ethanolammonium$^+$	42.2	1.124	$1/2$Glutarate^{2-}	52.6	0.700
Ethylammonium$^+$	47.2	1.257	Hydrogenoxalate$^-$	40.2	1.070
Ethyltrimethylammonium$^+$	40.5	1.078	Isovalerate$^-$	32.7	0.871
Hexadecyltrimethylammonium$^+$	20.9	0.557	Iodoacetate$^-$	40.6	1.081
Hexyltrimethylammonium$^+$	29.6	0.788	Lactate$^-$	38.8	1.033

Ion	$\Lambda_{\pm}$ $10^{-4}\,m^2\,S\,mol^{-1}$	D $10^{-5}\,cm^2\,s^{-1}$	Ion	$\Lambda_{\pm}$ $10^{-4}\,m^2\,S\,mol^{-1}$	D $10^{-5}\,cm^2\,s^{-1}$
1/2Malate^{2-}	58.8	0.783	Picrate$^-$	30.37	0.809
1/2Maleate^{2-}	61.9	0.824	Pivalate$^-$	31.9	0.849
1/2Malonate^{2-}	63.5	0.845	Propionate$^-$	35.8	0.953
Methylsulfate$^-$	48.8	1.299	Propylsulfate$^-$	37.1	0.988
Naphthylacetate$^-$	28.4	0.756	Salicylate$^-$	36	0.959
1/2Oxalate^{2-}	74.11	0.987	1/2Suberate^{2-}	36	0.479
Octylsulfate$^-$	29	0.772	1/2Succinate^{2-}	58.8	0.783
Phenylacetate$^-$	30.6	0.815	p–Sulfonate	29.3	0.780
1/2o–Phthalate^{2-}	52.3	0.696	1/2Tartarate^{2-}	59.6	0.794
1/2m–Phthalate^{2-}	54.7	0.728	Trichloroacetate$^-$	35	0.932

ACTIVITY COEFFICIENTS OF ACIDS, BASES, AND SALTS

Petr Vanýsek

This table gives mean activity coefficients at 25°C for molalities in the range 0.1 to 1.0. See the following table for definitions, references, and data over a wider concentration range.

	0.1	0.2	0.3	0.4	0.5	0.6	0.7	0.8	0.9	1.0
$AgNO_3$	0.734	0.657	0.606	0.567	0.536	0.509	0.485	0.464	0.446	0.429
$AlCl_3$	0.337	0.305	0.302	0.313	0.331	0.356	0.388	0.429	0.479	0.539
$Al_2(SO_4)_3$	0.035	0.0225	0.0176	0.0153	0.0143	0.014	0.0142	0.0149	0.0159	0.0175
$BaCl_2$	0.500	0.444	0.419	0.405	0.397	0.391	0.391	0.391	0.392	0.395
$BeSO_4$	0.150	0.109	0.0885	0.0769	0.0692	0.0639	0.0600	0.0570	0.0546	0.0530
$CaCl_2$	0.518	0.472	0.455	0.448	0.448	0.453	0.460	0.470	0.484	0.500
$CdCl_2$	0.2280	0.1638	0.1329	0.1139	0.1006	0.0905	0.0827	0.0765	0.0713	0.0669
$Cd(NO_3)_2$	0.513	0.464	0.442	0.430	0.425	0.423	0.423	0.425	0.428	0.433
$CdSO_4$	0.150	0.103	0.0822	0.0699	0.0615	0.0553	0.0505	0.0468	0.0438	0.0415
$CoCl_2$	0.522	0.479	0.463	0.459	0.462	0.470	0.479	0.492	0.511	0.531
$CrCl_3$	0.331	0.298	0.294	0.300	0.314	0.335	0.362	0.397	0.436	0.481
$Cr(NO_3)_3$	0.319	0.285	0.279	0.281	0.291	0.304	0.322	0.344	0.371	0.401
$Cr_2(SO_4)_3$	0.0458	0.0300	0.0238	0.0207	0.0190	0.0182	0.0181	0.0185	0.0194	0.0208
$CsBr$	0.754	0.694	0.654	0.626	0.603	0.586	0.571	0.558	0.547	0.538
$CsCl$	0.756	0.694	0.656	0.628	0.606	0.589	0.575	0.563	0.553	0.544
CsI	0.754	0.692	0.651	0.621	0.599	0.581	0.567	0.554	0.543	0.533
$CsNO_3$	0.733	0.655	0.602	0.561	0.528	0.501	0.478	0.458	0.439	0.422
$CsOH$	0.795	0.761	0.744	0.739	0.739	0.742	0.748	0.754	0.762	0.771
$CsOAc$	0.799	0.771	0.761	0.759	0.762	0.768	0.776	0.783	0.792	0.802
Cs_2SO_4	0.456	0.382	0.338	0.311	0.291	0.274	0.262	0.251	0.242	0.235
$CuCl_2$	0.508	0.455	0.429	0.417	0.411	0.409	0.409	0.410	0.413	0.417
$Cu(NO_3)_2$	0.511	0.460	0.439	0.429	0.426	0.427	0.431	0.437	0.445	0.455
$CuSO_4$	0.150	0.104	0.0829	0.0704	0.0620	0.0559	0.0512	0.0475	0.0446	0.0423
$FeCl_2$	0.5185	0.473	0.454	0.448	0.450	0.454	0.463	0.473	0.488	0.506
HBr	0.805	0.782	0.777	0.781	0.789	0.801	0.815	0.832	0.850	0.871
HCl	0.796	0.767	0.756	0.755	0.757	0.763	0.772	0.783	0.795	0.809
$HClO_4$	0.803	0.778	0.768	0.766	0.769	0.776	0.785	0.795	0.808	0.823
HI	0.818	0.807	0.811	0.823	0.839	0.860	0.883	0.908	0.935	0.963
HNO_3	0.791	0.754	0.735	0.725	0.720	0.717	0.717	0.718	0.721	0.724
H_2SO_4	0.2655	0.2090	0.1826	—	0.1557	—	0.1417	—	—	0.1316
KBr	0.772	0.722	0.693	0.673	0.657	0.646	0.636	0.629	0.622	0.617
KCl	0.770	0.718	0.688	0.666	0.649	0.637	0.626	0.618	0.610	0.604
$KClO_3$	0.749	0.681	0.635	0.599	0.568	0.541	0.518	—	—	—
K_2CrO_4	0.456	0.382	0.340	0.313	0.292	0.276	0.263	0.253	0.243	0.235
KF	0.775	0.727	0.700	0.682	0.670	0.661	0.654	0.650	0.646	0.645
$K_3Fe(CN)_6$	0.268	0.212	0.184	0.167	0.155	0.146	0.140	0.135	0.131	0.128
$K_4Fe(CN)_6$	0.139	0.0993	0.0808	0.0693	0.0614	0.0556	0.0512	0.0479	0.0454	—
KH_2PO_4	0.731	0.653	0.602	0.561	0.529	0.501	0.477	0.456	0.438	0.421
KI	0.778	0.733	0.707	0.689	0.676	0.667	0.660	0.654	0.649	0.645
KNO_3	0.739	0.663	0.614	0.576	0.545	0.519	0.496	0.476	0.459	0.443
$KOAc$	0.796	0.766	0.754	0.750	0.751	0.754	0.759	0.766	0.774	0.783
KOH	0.798	0.760	0.742	0.734	0.732	0.733	0.736	0.742	0.749	0.756
$KSCN$	0.769	0.716	0.685	0.663	0.646	0.633	0.623	0.614	0.606	0.599
K_2SO_4	0.441	0.360	0.316	0.286	0.264	0.246	0.232	—	—	—
$LiBr$	0.796	0.766	0.756	0.752	0.753	0.758	0.767	0.777	0.789	0.803
$LiCl$	0.790	0.757	0.744	0.740	0.739	0.743	0.748	0.755	0.764	0.774
$LiClO_4$	0.812	0.794	0.792	0.798	0.808	0.820	0.834	0.852	0.869	0.887
LiI	0.815	0.802	0.804	0.813	0.824	0.838	0.852	0.870	0.888	0.910
$LiNO_3$	0.788	0.752	0.736	0.728	0.726	0.727	0.729	0.733	0.737	0.743
$LiOH$	0.760	0.702	0.665	0.638	0.617	0.599	0.585	0.573	0.563	0.554
$LiOAc$	0.784	0.742	0.721	0.709	0.700	0.691	0.689	0.688	0.688	0.689
Li_2SO_4	0.468	0.398	0.361	0.337	0.319	0.307	0.297	0.289	0.282	0.277
$MgCl_2$	0.529	0.489	0.477	0.475	0.481	0.491	0.506	0.522	0.544	0.570
$MgSO_4$	0.150	0.107	0.0874	0.0756	0.0675	0.0616	0.0571	0.0536	0.0508	0.0485

	0.1	0.2	0.3	0.4	0.5	0.6	0.7	0.8	0.9	1.0
$MnCl_2$	0.516	0.469	0.450	0.442	0.440	0.443	0.448	0.455	0.466	0.479
$MnSO_4$	0.150	0.105	0.0848	0.0725	0.0640	0.0578	0.0530	0.0493	0.0463	0.0439
NH_4Cl	0.770	0.718	0.687	0.665	0.649	0.636	0.625	0.617	0.609	0.603
NH_4NO_3	0.740	0.677	0.636	0.606	0.582	0.562	0.545	0.530	0.516	0.504
$(NH_4)_2SO_4$	0.439	0.356	0.311	0.280	0.257	0.240	0.226	0.214	0.205	0.196
$NaBr$	0.782	0.741	0.719	0.704	0.697	0.692	0.689	0.687	0.687	0.687
$NaCl$	0.778	0.735	0.710	0.693	0.681	0.673	0.667	0.662	0.659	0.657
$NaClO_3$	0.772	0.720	0.688	0.664	0.645	0.630	0.617	0.606	0.597	0.589
$NaClO_4$	0.775	0.729	0.701	0.683	0.668	0.656	0.648	0.641	0.635	0.629
Na_2CrO_4	0.464	0.394	0.353	0.327	0.307	0.292	0.280	0.269	0.261	0.253
NaF	0.765	0.710	0.676	0.651	0.632	0.616	0.603	0.592	0.582	0.573
NaH_2PO_4	0.744	0.675	0.629	0.593	0.563	0.539	0.517	0.499	0.483	0.468
NaI	0.787	0.751	0.735	0.727	0.723	0.723	0.724	0.727	0.731	0.736
$NaNO_3$	0.762	0.703	0.666	0.638	0.617	0.599	0.583	0.570	0.558	0.548
$NaOAc$	0.791	0.757	0.744	0.737	0.735	0.736	0.740	0.745	0.752	0.757
$NaOH$	0.766	0.727	0.708	0.697	0.690	0.685	0.681	0.679	0.678	0.678
$NaSCN$	0.787	0.750	—	0.720	0.715	0.712	0.710	0.710	0.711	0.712
Na_2SO_4	0.445	0.365	0.320	0.289	0.266	0.248	0.233	0.221	0.210	0.201
$NiCl_2$	0.522	0.479	0.463	0.460	0.464	0.471	0.482	0.496	0.515	0.563
$NiSO_4$	0.150	0.105	0.0841	0.0713	0.0627	0.0562	0.0515	0.0478	0.0448	0.0425
$Pb(NO_3)_2$	0.395	0.308	0.260	0.228	0.205	0.187	0.172	0.160	0.150	0.141
$RbBr$	0.763	0.706	0.673	0.650	0.632	0.617	0.605	0.595	0.586	0.578
$RbCl$	0.764	0.709	0.675	0.652	0.634	0.620	0.608	0.599	0.590	0.583
RbI	0.762	0.705	0.671	0.647	0.629	0.614	0.602	0.591	0.583	0.575
$RbNO_3$	0.734	0.658	0.606	0.565	0.534	0.508	0.485	0.465	0.446	0.430
$RbOAc$	0.796	0.767	0.756	0.753	0.755	0.759	0.766	0.773	0.782	0.792
Rb_2SO_4	0.451	0.374	0.331	0.301	0.279	0.263	0.249	0.238	0.228	0.219
$SrCl_2$	0.511	0.462	0.442	0.433	0.430	0.431	0.434	0.441	0.449	0.461
$TlClO_4$	0.730	0.652	0.599	0.559	0.527	—	—	—	—	—
$TlNO_3$	0.702	0.606	0.545	0.500	—	—	—	—	—	—
UO_2Cl_2	0.544	0.510	0.520	0.505	0.517	0.532	0.549	0.571	0.595	0.620
UO_2SO_4	0.150	0.102	0.0807	0.0689	0.0611	0.0566	0.0515	0.0483	0.0458	0.0439
$ZnCl_2$	0.515	0.462	0.432	0.411	0.394	0.380	0.369	0.357	0.348	0.339
$Zn(NO_3)_2$	0.531	0.489	0.474	0.469	0.473	0.480	0.489	0.501	0.518	0.535
$ZnSO_4$	0.150	0.140	0.0835	0.0714	0.0630	0.0569	0.0523	0.0487	0.0458	0.0435

MEAN ACTIVITY COEFFICIENTS OF ELECTROLYTES AS A FUNCTION OF CONCENTRATION

The mean activity coefficient γ of an electrolyte X_aY_b is defined as

$$\gamma = (\gamma_+^a \gamma_-^b)^{1/(a+b)}$$

where γ_+ and γ_- are activity coefficients of the individual ions (which cannot be directly measured). This table gives the mean activity coefficients of about 100 electrolytes in aqueous solution as a function of concentration, expressed in molality terms. All values refer to a temperature of 25°C. Substances are arranged in alphabetical order by formula.

References

1. Hamer, W. J., and Wu, Y. C., *J. Phys. Chem. Ref. Data*, 1, 1047, 1972.
2. Staples, B. R., *J. Phys. Chem. Ref. Data*, 6, 385, 1977; 10, 767, 1981; 10, 779, 1981.
3. Goldberg, R. N. et al., *J. Phys. Chem. Ref. Data*, 7, 263, 1978; 8, 923, 1979; 8, 1005, 1979; 10, 1, 1981; 10, 671, 1981.

Mean Activity Coefficient at 25°C

m/mol kg^{-1}	AgNO$_3$	BaBr$_2$	BaCl$_2$	BaI$_2$	CaBr$_2$	CaCl$_2$	CaI$_2$
0.001	0.964	0.881	0.887	0.890	0.890	0.888	0.890
0.002	0.950	0.850	0.849	0.853	0.853	0.851	0.853
0.005	0.924	0.785	0.782	0.792	0.791	0.787	0.791
0.010	0.896	0.727	0.721	0.737	0.735	0.727	0.736
0.020	0.859	0.661	0.653	0.678	0.674	0.664	0.677
0.050	0.794	0.573	0.559	0.600	0.594	0.577	0.600
0.100	0.732	0.517	0.492	0.551	0.540	0.517	0.552
0.200	0.656	0.463	0.436	0.520	0.502	0.469	0.524
0.500	0.536	0.435	0.391	0.536	0.500	0.444	0.554
1.000	0.430	0.470	0.393	0.664	0.604	0.495	0.729
2.000	0.316	0.654		1.242	1.125	0.784	
5.000	0.181				18.7	5.907	
10.000	0.108					43.1	
15.000	0.085						

m/mol kg^{-1}	Cd(NO$_2$)$_2$	Cd(NO$_3$)$_2$	CoBr$_2$	CoCl$_2$	CoI$_2$	Co(NO$_3$)$_2$	CsBr
0.001	0.881	0.888	0.890	0.889	0.887	0.888	0.965
0.002	0.837	0.851	0.854	0.852	0.849	0.850	0.951
0.005	0.759	0.787	0.794	0.789	0.783	0.786	0.925
0.010	0.681	0.728	0.740	0.732	0.724	0.728	0.898
0.020	0.589	0.664	0.681	0.670	0.661	0.663	0.864
0.050	0.451	0.576	0.605	0.586	0.582	0.576	0.806
0.100	0.344	0.515	0.556	0.528	0.540	0.516	0.752
0.200	0.247	0.465	0.523	0.483	0.527	0.469	0.691
0.500	0.148	0.428	0.538	0.465	0.596	0.446	0.605
1.000	0.098	0.437	0.685	0.532	0.845	0.492	0.540
2.000	0.069	0.517	1.421	0.864	2.287	0.722	0.485
5.000	0.054		13.9		55.3	3.338	0.454
10.000					196		

m/mol kg^{-1}	CsCl	CsF	CsI	CsNO$_3$	CsOH	Cs$_2$SO$_4$	CuBr$_2$
0.001	0.965	0.965	0.965	0.964	0.966	0.885	0.889
0.002	0.951	0.952	0.951	0.951	0.953	0.845	0.853
0.005	0.925	0.929	0.925	0.924	0.930	0.775	0.791
0.010	0.898	0.905	0.898	0.897	0.906	0.709	0.735
0.020	0.864	0.876	0.863	0.860	0.878	0.634	0.674
0.050	0.805	0.830	0.804	0.796	0.836	0.526	0.594
0.100	0.751	0.792	0.749	0.733	0.802	0.444	0.541
0.200	0.691	0.755	0.688	0.655	0.772	0.369	0.504
0.500	0.607	0.721	0.601	0.529	0.755	0.285	0.503
1.000	0.546	0.726	0.534	0.421	0.782	0.233	0.591
2.000	0.496	0.803	0.470				0.859
5.000	0.474						
10.000	0.508						

m/mol kg^{-1}	CuCl$_2$	Cu(ClO$_4$)$_2$	Cu(NO$_3$)$_2$	FeCl$_2$	HBr	HCl	HClO$_4$
0.001	0.887	0.890	0.888	0.888	0.966	0.965	0.966
0.002	0.849	0.854	0.851	0.850	0.953	0.952	0.953
0.005	0.783	0.795	0.787	0.785	0.930	0.929	0.929
0.010	0.722	0.741	0.729	0.725	0.907	0.905	0.906
0.020	0.654	0.685	0.664	0.659	0.879	0.876	0.878
0.050	0.561	0.613	0.577	0.570	0.837	0.832	0.836
0.100	0.495	0.572	0.516	0.509	0.806	0.797	0.803
0.200	0.441	0.553	0.466	0.462	0.783	0.768	0.776
0.500	0.401	0.617	0.431	0.443	0.790	0.759	0.769
1.000	0.405	0.892	0.456	0.500	0.872	0.811	0.826
2.000	0.453	2.445	0.615	0.782	1.167	1.009	1.055
5.000	0.601		2.083		3.800	2.380	3.100
10.000					33.4	10.4	30.8
15.000							323

m/mol kg^{-1}	HF	HI	HNO$_3$	H$_2$SO$_4$	KBr	KCNS	KCl
0.001	0.551	0.966	0.965	0.804	0.965	0.965	0.965
0.002	0.429	0.953	0.952	0.740	0.952	0.951	0.951
0.005	0.302	0.931	0.929	0.634	0.927	0.927	0.927
0.010	0.225	0.909	0.905	0.542	0.902	0.901	0.901
0.020	0.163	0.884	0.875	0.445	0.870	0.869	0.869
0.050	0.106	0.847	0.829	0.325	0.817	0.815	0.816
0.100	0.0766	0.823	0.792	0.251	0.771	0.768	0.768
0.200	0.0550	0.811	0.756	0.195	0.772	0.716	0.717
0.500	0.0352	0.845	0.725	0.146	0.658	0.647	0.649
1.000	0.0249	0.969	0.730	0.125	0.617	0.598	0.604
2.000	0.0175	1.363	0.788	0.119	0.593	0.556	0.573
5.000	0.0110	4.760	1.063	0.197	0.626	0.525	0.593
10.000	0.0085	49.100	1.644	0.527			
15.000	0.0077		2.212	1.077			
20.000	0.0075		2.607	1.701			

m/mol kg^{-1}	KClO$_3$	K$_2$CrO$_4$	KF	KH$_2$PO$_4$*	K$_2$HPO$_4$**	KI	KNO$_3$
0.001	0.965	0.886	0.965	0.964	0.886	0.965	0.964
0.002	0.951	0.847	0.952	0.950	0.847	0.952	0.950
0.005	0.926	0.779	0.927	0.924	0.779	0.927	0.924
0.010	0.899	0.715	0.902	0.896	0.715	0.902	0.896
0.020	0.865	0.643	0.870	0.859	0.643	0.871	0.860
0.050	0.805	0.539	0.818	0.793	0.538	0.820	0.797
0.100	0.749	0.460	0.773	0.730	0.457	0.776	0.735
0.200	0.681	0.385	0.726	0.652	0.379	0.731	0.662
0.500	0.569	0.296	0.670	0.529	0.283	0.676	0.546
1.000		0.239	0.645	0.422		0.646	0.444
2.000		0.199	0.658			0.638	0.332
5.000			0.871				
10.000			1.715				
15.000			3.120				

m/mol kg^{-1}	KOH	K$_2$SO$_4$	LiBr	LiCl	LiClO$_4$	LiI	LiNO$_3$
0.001	0.965	0.885	0.965	0.965	0.966	0.966	0.965
0.002	0.952	0.844	0.952	0.952	0.953	0.953	0.952
0.005	0.927	0.772	0.929	0.928	0.931	0.930	0.928
0.010	0.902	0.704	0.905	0.904	0.908	0.908	0.904
0.020	0.871	0.625	0.877	0.874	0.882	0.882	0.874
0.050	0.821	0.511	0.832	0.827	0.843	0.843	0.827
0.100	0.779	0.424	0.797	0.789	0.815	0.817	0.788
0.200	0.740	0.343	0.767	0.756	0.795	0.802	0.753
0.500	0.710	0.251	0.754	0.739	0.806	0.824	0.726
1.000	0.733		0.803	0.775	0.887	0.912	0.743
2.000	0.860		1.012	0.924	1.161	1.197	0.837
5.000	1.697		2.696	2.000			1.298

Mean Activity Coefficients of Electrolytes as a Function of Concentration

m/mol kg^{-1}	KOH	K$_2$SO$_4$	LiBr	LiCl	LiClO$_4$	LiI	LiNO$_3$
10.000	6.110		20.0	9.600			2.500
15.000	19.9		147	30.9			3.960
20.000	46.4		486				4.970

m/mol kg^{-1}	LiOH	Li$_2$SO$_4$	MgBr$_2$	MgCl$_2$	MgI$_2$	MnBr$_2$	MnCl$_2$
0.001	0.964	0.887	0.889	0.889	0.889	0.889	0.888
0.002	0.950	0.847	0.852	0.852	0.853	0.853	0.850
0.005	0.923	0.780	0.790	0.790	0.791	0.791	0.850
0.010	0.895	0.716	0.733	0.734	0.736	0.791	0.786
0.020	0.858	0.645	0.672	0.672	0.677	0.735	0.727
0.050	0.794	0.544	0.593	0.590	0.602	0.674	0.662
0.100	0.735	0.469	0.543	0.535	0.556	0.595	0.574
0.200	0.668	0.400	0.512	0.493	0.535	0.543	0.513
0.500	0.579	0.325	0.540	0.485	0.594	0.508	0.464
1.000	0.522	0.284	0.715	0.577	0.858	0.519	0.437
2.000	0.484	0.270	1.590	1.065	2.326	0.650	0.477
5.000	0.493		36.1	14.40	109.8	1.224	0.661
						6.697	1.539

m/mol kg^{-1}	Mn(ClO$_4$)$_2$	NH$_4$Cl	NH$_4$ClO$_4$	(NH$_4$)$_2$HPO$_4$**	NH$_4$NO$_3$	NaBr	NaBrO$_3$
0.001	0.892	0.965	0.964	0.882	0.964	0.965	0.965
0.002	0.858	0.952	0.950	0.839	0.951	0.952	0.951
0.005	0.801	0.927	0.924	0.763	0.925	0.928	0.926
0.010	0.752	0.901	0.895	0.688	0.897	0.903	0.900
0.020	0.700	0.869	0.859	0.600	0.862	0.873	0.867
0.050	0.637	0.816	0.794	0.469	0.801	0.824	0.811
0.100	0.604	0.769	0.734	0.367	0.744	0.783	0.759
0.200	0.596	0.718	0.663	0.273	0.678	0.742	0.698
0.500	0.686	0.649	0.560	0.171	0.582	0.697	0.605
1.000	1.030	0.603	0.479	0.114	0.502	0.687	0.528
2.000	3.072	0.569	0.399	0.074	0.419	0.730	0.449
5.000		0.563			0.303	1.083	
10.000					0.220		
15.000					0.179		
20.000					0.154		

m/mol kg^{-1}	Na$_2$CO$_3$	NaCl	NaClO$_3$	NaClO$_4$	Na$_2$CrO$_4$	NaF	Na$_2$HPO$_4$*
0.001	0.887	0.965	0.965	0.965	0.887	0.965	0.887
0.002	0.847	0.952	0.952	0.952	0.849	0.951	0.848
0.005	0.780	0.928	0.927	0.928	0.783	0.926	0.780
0.010	0.716	0.903	0.902	0.903	0.722	0.901	0.717
0.020	0.644	0.872	0.870	0.872	0.653	0.868	0.644
0.050	0.541	0.822	0.818	0.821	0.554	0.813	0.539
0.100	0.462	0.779	0.771	0.777	0.479	0.764	0.456
0.200	0.385	0.734	0.719	0.729	0.406	0.710	0.373
0.500	0.292	0.681	0.646	0.668	0.318	0.633	0.266
1.000	0.229	0.657	0.590	0.630	0.261	0.573	0.191
2.000	0.182	0.668	0.537	0.608	0.231		0.133
5.000		0.874		0.648			

m/mol kg^{-1}	NaI	NaNO$_3$	NaOH	Na$_2$SO$_3$	Na$_2$SO$_4$	Na$_2$WO$_4$	NiBr$_2$
0.001	0.965	0.965	0.965	0.887	0.886	0.886	0.889
0.002	0.952	0.951	0.952	0.847	0.846	0.846	0.853
0.005	0.928	0.926	0.927	0.779	0.777	0.777	0.791
0.010	0.904	0.900	0.902	0.716	0.712	0.712	0.735
0.020	0.874	0.866	0.870	0.644	0.637	0.638	0.675
0.050	0.827	0.810	0.819	0.540	0.529	0.534	0.596
0.100	0.789	0.759	0.775	0.462	0.446	0.457	0.546
0.200	0.753	0.701	0.731	0.386	0.366	0.388	0.514
0.500	0.722	0.617	0.685	0.296	0.268	0.320	0.535
1.000	0.734	0.550	0.674	0.237	0.204	0.291	0.692
2.000	0.823	0.480	0.714	0.196	0.155	0.291	1.476

Mean Activity Coefficients of Electrolytes as a Function of Concentration

m/mol kg^{-1}	NaI	NaNO$_3$	NaOH	Na$_2$SO$_3$	Na$_2$SO$_4$	Na$_2$WO$_4$	NiBr$_2$
5.000	1.402	0.388	1.076				
10.000	4.011	0.329	3.258				
15.000			9.796				
20.000			19.410				

m/mol kg^{-1}	NiCl$_2$	Ni(ClO$_4$)$_2$	Ni(NO$_3$)$_2$	Pb(ClO$_4$)$_2$	Pb(NO$_3$)$_2$	RbBr	RbCl
0.001	0.889	0.891	0.889	0.889	0.882	0.965	0.965
0.002	0.852	0.855	0.851	0.851	0.840	0.951	0.951
0.005	0.789	0.797	0.787	0.787	0.764	0.926	0.926
0.010	0.732	0.745	0.730	0.729	0.690	0.900	0.900
0.020	0.669	0.690	0.666	0.666	0.604	0.866	0.867
0.050	0.584	0.621	0.581	0.580	0.476	0.811	0.811
0.100	0.527	0.582	0.524	0.522	0.379	0.760	0.761
0.200	0.482	0.567	0.481	0.476	0.291	0.705	0.707
0.500	0.465	0.639	0.467	0.458	0.195	0.630	0.633
1.000	0.538	0.946	0.528	0.516	0.136	0.578	0.583
2.000	0.915	2.812	0.797	0.799		0.535	0.546
5.000	4.785			4.043		0.514	0.544
10.000				33.8			

m/mol kg^{-1}	RbF	RbI	RbNO$_3$	Rb$_2$SO$_4$	SrBr$_2$	SrCl$_2$	SrI$_2$
0.001	0.965	0.965	0.964	0.886	0.889	0.888	0.890
0.002	0.952	0.951	0.950	0.845	0.852	0.850	0.854
0.005	0.927	0.926	0.924	0.776	0.790	0.785	0.793
0.010	0.902	0.900	0.896	0.710	0.734	0.725	0.740
0.020	0.871	0.866	0.859	0.635	0.673	0.659	0.681
0.050	0.821	0.810	0.795	0.526	0.591	0.569	0.606
0.100	0.780	0.759	0.733	0.443	0.535	0.506	0.557
0.200	0.739	0.703	0.657	0.365	0.492	0.455	0.526
0.500	0.701	0.627	0.536	0.274	0.476	0.421	0.542
1.000	0.697	0.574	0.430	0.217	0.545	0.451	0.686
2.000	0.724	0.532	0.320		0.921	0.650	
5.000		0.517					

m/mol kg^{-1}	UO$_2$Cl$_2$	UO$_2$(NO$_3$)$_2$	ZnBr$_2$	ZnCl$_2$	ZnI$_2$
0.001	0.888	0.888	0.890	0.887	0.893
0.002	0.851	0.849	0.854	0.847	0.859
0.005	0.787	0.784	0.794	0.781	0.804
0.010	0.729	0.726	0.741	0.719	0.757
0.020	0.666	0.663	0.683	0.652	0.708
0.050	0.583	0.583	0.606	0.561	0.644
0.100	0.529	0.535	0.553	0.499	0.601
0.200	0.493	0.509	0.515	0.447	0.574
0.500	0.501	0.532	0.516	0.384	0.635
1.000	0.601	0.673	0.558	0.330	0.836
2.000	0.948	1.223	0.578	0.283	1.062
5.000		3.020	0.788	0.342	1.546
10.000			2.317	0.876	4.698
15.000			5.381	1.914	
20.000			7.965	2.968	

* The anion is H$_2$PO$_4^-$.
** The anion is HPO$_4^{-2}$.

ENTHALPY OF DILUTION OF ACIDS

The quantity given in this table is $-\Delta_{dil}H$, the negative of the enthalpy (heat) of dilution to infinite dilution for aqueous solutions of several common acids; i.e., the negative of the enthalphy change when a solution of molality m at a temperature of 25°C is diluted with an infinite amount of water. The tabulated numbers thus represent the heat produced (or, if the value is negative, the heat absorbed) when the acid is diluted. The initial molality m is given in the first column. The second column gives the dilution ratio, which is the number of moles of water that must be added to one mole of the acid to produce a solution of the molality in the first column.

Reference

Parker, V. B., *Thermal Properties of Aqueous Uni-Univalent Electrolytes*, Natl. Stand. Ref. Data Ser. - Natl. Bur. Stand. (U.S.) 2, U.S. Government Printing Office, 1965.

m	Dil. ratio	HF	HCl	HClO$_4$	HBr	HI	HNO$_3$	CH$_2$O$_2$	C$_2$H$_4$O$_2$
55.506	1.0		45.61		48.83		19.73	0.046	2.167
20	2.775	14.88	19.87	13.81	19.92	21.71	9.498	0.038	2.075
15	3.700	14.34	15.40	7.920	14.29	14.02	6.883	0.109	1.962
10	5.551	13.87	10.24	2.013	8.694	7.615	3.933	0.205	1.824
9	6.167	13.81	9.213	1.280	7.719	6.569	3.368	0.230	1.782
8	6.938	13.77	8.201	0.611	6.786	5.607	2.791	0.255	1.724
7	7.929	13.73	7.217	0.046	5.925	4.728	2.251	0.272	1.648
6	9.251	13.69	6.268	-0.351	5.004	3.975	1.749	0.280	1.540
5.5506	10	13.66	5.841	-0.490	4.590	3.577	1.540	0.285	1.477
5	11.10	13.62	5.318	-0.628	4.113	3.197	1.310	0.289	1.393
4.5	12.33	13.58	4.899	-0.732	3.711	2.828	1.109	0.289	1.310
4	13.88	13.53	4.402	-0.787	3.330	2.460	0.958	0.289	1.218
3.5	15.86	13.47	3.958	-0.820	2.966	2.105	0.791	0.289	1.121
3	18.50	13.45	3.506	-0.782	2.611	1.787	0.665	0.289	1.025
2.5	22.20	13.43	3.063	-0.724	2.301	1.527	0.582	0.285	0.912
2	27.75	13.40	2.623	-0.623	1.996	1.318	0.527	0.276	0.803
1.5	37.00	13.36	2.167	-0.431	1.665	1.125	0.506	0.259	0.678
1	55.51	13.30	1.695	-0.201	1.314	0.933	0.506	0.226	0.544
0.5551	100	13.22	1.234	0.050	0.983	0.736	0.502	0.184	0.423
0.5	111.0	13.20	1.172	0.075	0.941	0.711	0.498	0.176	0.406
0.2	277.5	13.09	0.761	0.247	0.649	0.536	0.439	0.146	0.331
0.1	555.1	12.80	0.556	0.272	0.498	0.439	0.372	0.134	0.289
0.0925	600	12.79	0.540	0.272	0.481	0.427	0.368	0.134	0.285
0.0793	700	12.70	0.502	0.272	0.452	0.402	0.351	0.134	0.285
0.0694	800	12.61	0.473	0.268	0.427	0.385	0.339	0.130	0.280
0.0617	900	12.50	0.448	0.264	0.406	0.368	0.326	0.126	0.276
0.05551	1000	12.42	0.427	0.259	0.385	0.351	0.318	0.121	0.272
0.05	1110	12.24	0.406	0.259	0.372	0.339	0.305	0.121	0.272
0.02775	2000	11.29	0.310	0.226	0.285	0.264	0.247	0.117	0.264
0.01850	3000	10.66	0.251	0.197	0.234	0.218	0.213	0.117	0.259
0.01388	4000	10.25	0.226	0.180	0.205	0.192	0.192	0.113	0.259
0.01110	5000	9.874	0.197	0.167	0.184	0.172	0.176	0.109	0.255
0.00555	10000	8.912	0.142	0.126	0.130	0.121	0.130	0.105	0.243
0.00278	20000	7.531	0.105	0.092	0.092	0.084	0.096	0.096	0.230
0.00111	50000	5.439	0.067	0.059	0.054	0.050	0.063	0.084	0.222
0.000555	100000	3.766	0.042	0.042	0.038	0.038	0.046	0.054	0.209
0.000111	500000	1.255	0.021	0.021	0.021	0.021	0.021	0.038	0.167
0	∞	0	0	0	0	0	0	0	0

Header for data: $-\Delta_{dil}H$ in kJ/mol at 25°C

ENTHALPY OF SOLUTION OF ELECTROLYTES

This table gives the molar enthalpy (heat) of solution at infinite dilution for some common uni-univalent electrolytes. This is the enthalpy change when 1 mol of solute in its standard state is dissolved in an infinite amount of water. Values are given in kilojoules per mole at 25°C.

Reference

Parker, V. B., *Thermal Properties of Uni-Univalent Electrolytes*, Natl. Stand. Ref. Data Series — Natl. Bur. Stand.(U.S.), No.2, 1965.

Solute	State	$\Delta_{sol}H°$ kJ/mol
HF	g	−61.50
HCl	g	−74.84
$HClO_4$	l	−88.76
$HClO_4 \cdot H_2O$	c	−32.95
HBr	g	−85.14
HI	g	−81.67
HIO_3	c	8.79
HNO_3	l	−33.28
HCOOH	l	−0.86
CH_3COOH	l	−1.51
NH_3	g	−30.50
NH_4Cl	c	14.78
NH_4ClO_4	c	33.47
NH_4Br	c	16.78
NH_4I	c	13.72
NH_4IO_3	c	31.80
NH_4NO_2	c	19.25
NH_4NO_3	c	25.69
$NH_4C_2H_3O_2$	c	−2.38
NH_4CN	c	17.57
NH_4CNS	c	22.59
CH_3NH_3Cl	c	5.77
$(CH_3)_3NHCl$	c	1.46
$N(CH_3)_4Cl$	c	4.08
$N(CH_3)_4Br$	c	24.27
$N(CH_3)_4I$	c	42.07
$AgClO_4$	c	7.36
$AgNO_2$	c	36.94
$AgNO_3$	c	22.59
LiOH	c	−23.56
$LiOH \cdot H_2O$	c	−6.69
LiF	c	4.73
LiCl	c	−37.03
$LiCl \cdot H_2O$	c	−19.08
$LiClO_4$	c	−26.55
$LiClO_4 \cdot 3H_2O$	c	32.61
LiBr	c	−48.83
$LiBr \cdot H_2O$	c	−23.26
$LiBr \cdot 2H_2O$	c	−9.41
$LiBrO_3$	c	1.42
LiI	c	−63.30
$LiI \cdot H_2O$	c	−29.66
$LiI \cdot 2H_2O$	c	−14.77
$LiI \cdot 3H_2O$	c	0.59
$LiNO_2$	c	−11.00
$LiNO_2 \cdot H_2O$	c	7.03
$LiNO_3$	c	−2.51
NaOH	c	−44.51
$NaOH \cdot H_2O$	c	−21.41
NaF	c	0.91
NaCl	c	3.88
$NaClO_2$	c	0.33
$NaClO_2 \cdot 3H_2O$	c	28.58
$NaClO_3$	c	21.72
$NaClO_4$	c	13.88
$NaClO_4 \cdot H_2O$	c	22.51
NaBr	c	−0.60
$NaBr \cdot 2H_2O$	c	18.64
$NaBrO_3$	c	26.90
NaI	c	−7.53
$NaI \cdot 2H_2O$	c	16.13
$NaIO_3$	c	20.29
$NaNO_2$	c	13.89
$NaNO_3$	c	20.50
$NaC_2H_3O_2$	c	−17.32
$NaC_2H_3O_2 \cdot 3H_2O$	c	19.66
NaCN	c	1.21
$NaCN \cdot 0.5H_2O$	c	3.31
$NaCN \cdot 2H_2O$	c	18.58
NaCNO	c	19.20
NaCNS	c	6.83
KOH	c	−57.61
$KOH \cdot H_2O$	c	−14.64
$KOH \cdot 1.5H_2O$	c	−10.46
KF	c	−17.73
$KF \cdot 2H_2O$	c	6.97
KCl	c	17.22
$KClO_3$	c	41.38
$KClO_4$	c	51.04
KBr	c	19.87
$KBrO_3$	c	41.13
KI	c	20.33
KIO_3	c	27.74
KNO_2	c	13.35
KNO_3	c	34.89
$KC_2H_3O_2$	c	−15.33
KCN	c	11.72
KCNO	c	20.25
KCNS	c	24.23
$KMnO_4$	c	43.56
RbOH	c	−62.34
$RbOH \cdot H_2O$	c	−17.99
$RbOH \cdot 2H_2O$	c	0.88
RbF	c	−26.11
$RbF \cdot H_2O$	c	−0.42
$RbF \cdot 1.5H_2O$	c	1.34
RbCl	c	17.28
$RbClO_3$	c	47.74
$RbClO_4$	c	56.74
RbBr	c	21.88
$RbBrO_3$	c	48.95
RbI	c	25.10
$RbNO_3$	c	36.48
CsOH	c	−71.55
$CsOH \cdot H_2O$	c	−20.50
CsF	c	−36.86
$CsF \cdot H_2O$	c	−10.46
$CsF \cdot 1.5H_2O$	c	−5.44
CsCl	c	17.78
$CsClO_4$	c	55.44
CsBr	c	25.98
$CsBrO_3$	c	50.46
CsI	c	33.35
$CsNO_3$	c	40.00

CHEMICAL KINETIC DATA FOR STRATOSPHERIC MODELING

The present compilation of kinetic data represents the 12th evaluation prepared by the NASA Panel for Data Evaluation. The Panel was established in 1977 by the NASA Upper Atmosphere Research Program Office for the purpose of providing a critical tabulation of the latest kinetic and photochemical data for use by modelers in computer simulations of stratospheric chemistry. The recommended rate data and cross sections are based on laboratory measurements. The major use of theoretical extrapolation of data is in connection with three–body reactions, in which the required pressure or temperature dependence is sometimes unavailable from laboratory measurements, and can be estimated by use of appropriate theoretical treatment. In the case of important rate constants for which no experimental data are available, the panel may provide estimates of rate constant parameters based on analogy to similar reactions for which data are available.

Rate constants are expressed in the form $k(T) = A \exp(-E/RT)$, where A is the pre–exponential factor, E the activation energy, R the gas constant, and T the absolute temperature. Uncertainties are expressed by the factor f, e.g., a value of 4.2×10^{-10} with $f = 2$ indicates that the true value is believed to lie between 2.1×10^{-10} and 8.4×10^{-10}. The value of f at other temperatures may be calculated from $f(298)$, given in the last column, by:

$$f(T) = f(298) \exp[(\Delta E/R)(1/T - 1/298)] ,$$

where $\Delta E/R$ is the uncertainty in E/R.

Table 1 covers rate constant data on second order reactions, grouped by class, while Table 2 covers association reactions. Relevant equilibrium constant data are given in Table 3. All concentrations are measured in molecules cm^{-3}. Notes on each reaction, as well as related photochemical data, may be found in the reference.

The assistance of Robert Hampson is gratefully acknowledged.

Reference

DeMore, W. B., Sander, S. P., Golden, D. M., Hampson, R. F., Kurylo, M. J., Howard, C. J., Ravishankara, A. R., Kolb, C. E., and Molina, M. J., *Chemical Kinetics and Photochemical Data for Use in Atmospheric Modeling. Evaluation Number 12*, Jet Propulsion Laboratory Publication 97–4, Pasadena CA, 1997.

The report is also available at the World Wide Web site < http://remus.jpl.nasa.gov/pub/jpl97>.

TABLE 1. Rate Constants for Second Order Reactions

Reaction	A cm³ molecule⁻¹ s⁻¹	E/R K	k (298 K) cm³ molecule⁻¹ s⁻¹	$f(298)$
O_x Reactions				
$O + O_3 \rightarrow O_2 + O_2$	8.0×10^{-12}	2060 ± 250	8.0×10^{-15}	1.15
$O(^1D)$ Reactions				
$O(^1D) + O_2 \rightarrow O + O_2$	3.2×10^{-11}	$-(70 \pm 100)$	4.0×10^{-11}	1.2
$O(^1D) + O_3 \rightarrow O_2 + O_2$	1.2×10^{-10}	0 ± 100	1.2×10^{-10}	1.3
$\rightarrow O_2 + O + O$	1.2×10^{-10}	0 ± 100	1.2×10^{-10}	1.3
$O(^1D) + H_2 \rightarrow OH + H$	1.1×10^{-10}	0 ± 100	1.1×10^{-10}	1.1
$O(^1D) + H_2O \rightarrow OH + OH$	2.2×10^{-10}	0 ± 100	2.2×10^{-10}	1.2
$O(^1D) + N_2 \rightarrow O + N_2$	1.8×10^{-11}	$-(110 \pm 100)$	2.6×10^{-11}	1.2
$O(^1D) + N_2O \rightarrow N_2 + O_2$	4.9×10^{-11}	0 ± 100	4.9×10^{-11}	1.3
$\rightarrow NO + NO$	6.7×10^{-11}	0 ± 100	6.7×10^{-11}	1.3
$O(^1D) + NH_3 \rightarrow OH + NH_2$	2.5×10^{-10}	0 ± 100	2.5×10^{-10}	1.3
$O(^1D) + CO_2 \rightarrow O + CO_2$	7.4×10^{-11}	$-(120 \pm 100)$	1.1×10^{-10}	1.2
$O(^1D) + CH_4 \rightarrow$ products	1.5×10^{-10}	0 ± 100	1.5×10^{-10}	1.2
$O(^1D) + HCl \rightarrow$ products	1.5×10^{-10}	0 ± 100	1.5×10^{-10}	1.2
$O(^1D) + HF \rightarrow OH + F$	1.4×10^{-10}	0 ± 100	1.4×10^{-10}	2.0
$O(^1D) + HBr \rightarrow$ products	1.5×10^{-10}	0 ± 100	1.5×10^{-10}	2.0
$O(^1D) + Cl_2 \rightarrow$ products	2.8×10^{-10}	0 ± 100	2.8×10^{-10}	2.0
$O(^1D) + CCl_2O \rightarrow$ products	3.6×10^{-10}	0 ± 100	3.6×10^{-10}	2.0
$O(^1D) + CClFO \rightarrow$ products	1.9×10^{-10}	0 ± 100	1.9×10^{-10}	2.0
$O(^1D) + CF_2O \rightarrow$ products	7.4×10^{-11}	0 ± 100	7.4×10^{-11}	2.0
$O(^1D) + CCl_4 \rightarrow$ products (CFC–10)	3.3×10^{-10}	0 ± 100	3.3×10^{-10}	1.2
$O(^1D) + CH_3Br \rightarrow$ products	1.8×10^{-10}	0 ± 100	1.8×10^{-10}	1.3
$O(^1D) + CH_2Br_2 \rightarrow$ products	2.7×10^{-10}	0 ± 100	2.7×10^{-10}	1.3
$O(^1D) + CHBr_3 \rightarrow$ products	6.6×10^{-10}	0 ± 100	6.6×10^{-10}	1.5
$O(^1D) + CH_3F \rightarrow$ products (HFC–41)	1.5×10^{-10}	0 ± 100	1.5×10^{-10}	1.2
$O(^1D) + CH_2F_2 \rightarrow$ products (HFC–32)	5.1×10^{-11}	0 ± 100	5.1×10^{-11}	1.3
$O(^1D) + CHF_3 \rightarrow$ products (HFC–23)	9.1×10^{-12}	0 ± 100	9.1×10^{-12}	1.2

Reaction	A cm^3 molecule^{-1} s^{-1}	E/R K	k (298 K) cm^3 molecule^{-1} s^{-1}	$f(298)$
$O(^1D) + CHCl_2F \rightarrow$ products (HCFC–21)	1.9×10^{-10}	0 ± 100	1.9×10^{-10}	1.3
$O(^1D) + CHClF_2 \rightarrow$ products (HCFC–22)	1.0×10^{-10}	0 ± 100	1.0×10^{-10}	1.2
$O(^1D) + CCl_3F \rightarrow$ products (CFC–11)	2.3×10^{-10}	0 ± 100	2.3×10^{-10}	1.2
$O(^1D) + CCl_2F_2 \rightarrow$ products (CFC–12)	1.4×10^{-10}	0 ± 100	1.4×10^{-10}	1.3
$O(^1D) + CClF_3 \rightarrow$ products (CFC–13)	8.7×10^{-11}	0 ± 100	8.7×10^{-11}	1.3
$O(^1D) + CClBrF_2 \rightarrow$ products (Halon–1211)	1.5×10^{-10}	0 ± 100	1.5×10^{-10}	1.3
$O(^1D) + CBr_2F_2 \rightarrow$ products (Halon–1202)	2.2×10^{-10}	0 ± 100	2.2×10^{-10}	1.3
$O(^1D) + CBrF_3 \rightarrow$ products (Halon–1301)	1.0×10^{-10}	0 ± 100	1.0×10^{-10}	1.3
$O(^1D) + CF_4 \rightarrow CF_4 + O$ (CFC–14)	–	–	2.0×10^{-14}	1.5
$O(^1D) + CH_3CH_2F \rightarrow$ products (HFC–161)	2.6×10^{-10}	0 ± 100	2.6×10^{-10}	1.3
$O(^1D) + CH_3CHF_2 \rightarrow$ products (HFC–152a)	2.0×10^{-10}	0 ± 100	2.0×10^{-10}	1.3
$O(^1D) + CH_3CCl_2F \rightarrow$ products (HCFC–141b)	2.6×10^{-10}	0 ± 100	2.6×10^{-10}	1.3
$O(^1D) + CH_3CClF_2 \rightarrow$ products (HCFC–142b)	2.2×10^{-10}	0 ± 100	2.2×10^{-10}	1.3
$O(^1D) + CH_3CF_3 \rightarrow$ products (HFC–143a)	1.0×10^{-10}	0 ± 100	1.0×10^{-10}	3.0
$O(^1D) + CH_2ClCClF_2 \rightarrow$ products (HCFC–132b)	1.6×10^{-10}	0 ± 100	1.6×10^{-10}	2.0
$O(^1D) + CH_2ClCF_3 \rightarrow$ products (HCFC–133a)	1.2×10^{-10}	0 ± 100	1.2×10^{-10}	1.3
$O(^1D) + CH_2FCF_3 \rightarrow$ products (HFC–134a)	4.9×10^{-11}	0 ± 100	4.9×10^{-11}	1.3
$O(^1D) + CHCl_2CF_3 \rightarrow$ products (HCFC–123)	2.0×10^{-10}	0 ± 100	2.0×10^{-10}	1.3
$O(^1D) + CHClFCF_3 \rightarrow$ products (HCFC–124)	8.6×10^{-11}	0 ± 100	8.6×10^{-11}	1.3
$O(^1D) + CHF_2CF_3 \rightarrow$ products (HFC–125)	1.2×10^{-10}	0 ± 100	1.2×10^{-10}	2.0
$O(^1D) + CCl_3CF_3 \rightarrow$ products (CFC–113a)	2×10^{-10}	0 ± 100	2×10^{-10}	2.0
$O(^1D) + CCl_2FCClF_2 \rightarrow$ products (CFC–113)	2×10^{-10}	0 ± 100	2×10^{-10}	2.0
$O(^1D) + CCl_2FCF_3 \rightarrow$ products (CFC–114a)	1×10^{-10}	0 ± 100	1×10^{-10}	2.0
$O(^1D) + CClF_2CClF_2 \rightarrow$ products (CFC–114)	1.3×10^{-10}	0 ± 100	1.3×10^{-10}	1.3
$O(^1D) + CClF_2CF_3 \rightarrow$ products (CFC–115)	5×10^{-11}	0 ± 100	5×10^{-11}	1.3
$O(^1D) + CBrF_2CBrF_2 \rightarrow$ products (Halon–2402)	1.6×10^{-10}	0 ± 100	1.6×10^{-10}	1.3
$O(^1D) + CF_3CF_3 \rightarrow O + CF_3CF_3$ (CFC–116)	–	–	1.5×10^{-13}	1.5
$O(^1D) + CHF_2CF_2CF_2CHF_2 \rightarrow$ products (HFC–338pcc)	1.8×10^{-11}	0 ± 100	1.8×10^{-11}	1.5
$O(^1D) + c-C_4F_8 \rightarrow$ products	–	–	8×10^{-13}	1.3
$O(^1D) + CF_3CHFCHFCF_2CF_3 \rightarrow$ products (HFC–43–10mee)	2.1×10^{-10}	0 ± 100	2.1×10^{-10}	4
$O(^1D) + C_5F_{12} \rightarrow$ products (CFC–41–12)	–	–	3.9×10^{-13}	2
$O(^1D) + C_6F_{14} \rightarrow$ products (CFC–51–14)	–	–	1×10^{-12}	2
$O(^1D) + 1,2-(CF_3)_2c-C_4F_6 \rightarrow$ products	–	–	2.8×10^{-13}	2
$O(^1D) + SF_6 \rightarrow$ products	–	–	1.8×10^{-14}	1.5

Singlet O₂ Reactions

Reaction	A	E/R	k (298 K)	$f(298)$
$O_2(^1\Delta) + O \rightarrow$ products	–	–	$<2 \times 10^{-16}$	–
$O_2(^1\Delta) + O_2 \rightarrow$ products	3.6×10^{-18}	220 ± 100	1.7×10^{-18}	1.2
$O_2(^1\Delta) + O_3 \rightarrow O + 2O_2$	5.2×10^{-11}	2840 ± 500	3.8×10^{-15}	1.2
$O_2(^1\Delta) + H_2O \rightarrow$ products	–	–	4.8×10^{-18}	1.5
$O_2(^1\Delta) + N \rightarrow NO + O$	–	–	$<9 \times 10^{-17}$	–
$O_2(^1\Delta) + N_2 \rightarrow$ products	–	–	$<10^{-20}$	–
$O_2(^1\Delta) + CO_2 \rightarrow$ products	–	–	$<2 \times 10^{-20}$	–
$O_2(^1\Sigma) + O \rightarrow$ products	–	–	8×10^{-14}	5.0
$O_2(^1\Sigma) + O_2 \rightarrow$ products	–	–	3.9×10^{-17}	1.5
$O_2(^1\Sigma) + O_3 \rightarrow$ products	2.2×10^{-11}	0 ± 200	2.2×10^{-11}	1.2
$O_2(^1\Sigma) + H_2O \rightarrow$ products	–	–	5.4×10^{-12}	1.3
$O_2(^1\Sigma) + N \rightarrow$ products	–	–	$<10^{-13}$	–
$O_2(^1\Sigma) + N_2 \rightarrow$ products	2.1×10^{-15}	0 ± 200	2.1×10^{-15}	1.2
$O_2(^1\Sigma) + CO_2 \rightarrow$ products	4.2×10^{-13}	0 ± 200	4.2×10^{-13}	1.2

HO$_x$ Reactions

Reaction	A	E/R	k (298 K)	$f(298)$
$O + OH \rightarrow O_2 + H$	2.2×10^{-11}	$-(120 \pm 100)$	3.3×10^{-11}	1.2
$O + HO_2 \rightarrow OH + O_2$	3.0×10^{-11}	$-(200 \pm 100)$	5.9×10^{-11}	1.2
$O + H_2O_2 \rightarrow OH + HO_2$	1.4×10^{-12}	2000 ± 1000	1.7×10^{-15}	2.0
$H + O_3 \rightarrow OH + O_2$	1.4×10^{-10}	470 ± 200	2.9×10^{-11}	1.25
$H + HO_2 \rightarrow$ products	8.1×10^{-11}	0 ± 100	8.1×10^{-11}	1.3
$OH + O_3 \rightarrow HO_2 + O_2$	1.6×10^{-12}	940 ± 300	6.8×10^{-14}	1.3
$OH + H_2 \rightarrow H_2O + H$	5.5×10^{-12}	2000 ± 100	6.7×10^{-15}	1.1

Reaction	A cm^3 molecule^{-1} s^{-1}	E/R K	k (298 K) cm^3 molecule^{-1} s^{-1}	f(298)
OH + HD → products	5.0×10^{-12}	2130 ± 200	4.0×10^{-15}	1.2
OH + OH → H_2O + O	4.2×10^{-12}	240 ± 240	1.9×10^{-12}	1.4
OH + HO_2 → H_2O + O_2	4.8×10^{-11}	$-(250\pm200)$	1.1×10^{-10}	1.3
OH + H_2O_2 → H_2O+ HO_2	2.9×10^{-12}	160 ± 100	1.7×10^{-12}	1.2
HO_2 + O_3 → OH + $2O_2$	1.1×10^{-14}	$500\pm$	2.0×10^{-15}	1.3
HO_2 + HO_2 → H_2O_2 + O_2	2.3×10^{-13}	$-(600\pm200)$	1.7×10^{-12}	1.3
H_2O_2 + O_2	1.7×10^{-33}[M]	$-(1000\pm400)$	4.9×10^{-32}[M]	1.3

NO_x Reactions

Reaction	A cm^3 molecule^{-1} s^{-1}	E/R K	k (298 K) cm^3 molecule^{-1} s^{-1}	f(298)
O + NO_2 → NO + O_2	6.5×10^{-12}	$-(120\pm120)$	9.7×10^{-12}	1.1
O + NO_3 → O_2 + NO_2	1.0×10^{-11}	0 ± 150	1.0×10^{-11}	1.5
O + N_2O_5 → products			$<3.0\times10^{-16}$	
O + HNO_3 → OH + NO_3			$<3.0\times10^{-17}$	
O + HO_2NO_2 → products	7.8×10^{-11}	3400 ± 750	8.6×10^{-16}	3.0
H + NO_2 → OH + NO	4.0×10^{-10}	340 ± 300	1.3×10^{-10}	1.3
OH + NO_3 → products			2.2×10^{-11}	1.5
OH + HONO → H_2O + NO_2	1.8×10^{-11}	$390\pm$	4.5×10^{-12}	1.5
OH + HNO_3 → H_2O + NO_3	See reference	1.3		
OH + HO_2NO_2 → products	1.3×10^{-12}	$-(380\pm)$	4.6×10^{-12}	1.5
OH + NH_3 → H_2O + NH_2	1.7×10^{-12}	710 ± 200	1.6×10^{-13}	1.2
HO_2 + NO → NO_2 + OH	3.5×10^{-12}	$-(250\pm50)$	8.1×10^{-12}	1.15
HO_2 + NO_2 → HONO + O_2	See reference			
HO_2 + NO_3 → products			3.5×10^{-12}	1.5
HO_2 + NH_2 → products			3.4×10^{-11}	2.0
N + O_2 → NO + O	1.5×10^{-11}	3600 ± 400	8.5×10^{-17}	1.25
N + O_3 → NO + O_2			$<2.0\times10^{-16}$	
N + NO → N_2 + O	2.1×10^{-11}	$-(100\pm100)$	3.0×10^{-11}	1.3
N + NO_2 → N_2O + O	5.8×10^{-12}	$-(220\pm100)$	1.2×10^{-11}	1.5
NO + O_3 → NO_2 + O_2	2.0×10^{-12}	1400 ± 200	1.8×10^{-14}	1.1
NO + NO_3 → $2NO_2$	1.5×10^{-11}	$-(170\pm100)$	2.6×10^{-11}	1.3
NO_2 + O_3 → NO_3 + O_2	1.2×10^{-13}	2450 ± 150	3.2×10^{-17}	1.15
NO_2 + NO_3 → NO + NO_2 +O_2	See reference			
NO_3 + NO_3 → $2NO_2$ + O_2	8.5×10^{-13}	2450 ± 500	2.3×10^{-16}	1.5
NH_2 + O_2 → products			$<6.0\times10^{-21}$	
NH_2 + O_3 → products	4.3×10^{-12}	930 ± 500	1.9×10^{-13}	3.0
NH_2 + NO → products	4.0×10^{-12}	$-(450\pm150)$	1.8×10^{-11}	1.3
NH_2 + NO_2 → products	2.1×10^{-12}	$-(650\pm250)$	1.9×10^{-11}	3.0
NH + NO → products	4.9×10^{-11}	0 ± 300	4.9×10^{-11}	1.5
NH + NO_2 → products	3.5×10^{-13}	$-(1140\pm500)$	1.6×10^{-11}	2.0
O_3 + HNO_2 → O_2 + HNO_3			$<5.0\times10^{-19}$	
N_2O_5 + H_2O → $2HNO_3$			$<2.0\times10^{-21}$	
N_2(A,v) + O_2 → products			2.5×10^{-12}, v=0	1.5
N_2(A,v) + O_3 → products			4.1×10^{-11}, v=0	2.0

Reactions of Organic Compounds

Reaction	A cm^3 molecule^{-1} s^{-1}	E/R K	k (298 K) cm^3 molecule^{-1} s^{-1}	f(298)
O + CH_3 → products	1.1×10^{-10}	0 ± 250	1.1×10^{-10}	1.3
O + HCN → products	1.0×10^{-11}	4000 ± 1000	1.5×10^{-17}	10
O + C_2H_2 → products	3.0×10^{-11}	1600 ± 250	1.4×10^{-13}	1.3
O + H_2CO → products	3.4×10^{-11}	1600 ± 250	1.6×10^{-13}	1.25
O + CH_3CHO → CH_3CO + OH	1.8×10^{-11}	1100 ± 200	4.5×10^{-13}	1.25
O_3 + C_2H_2 → products	1.0×10^{-14}	4100 ± 500	1.0×10^{-20}	3
O_3 + C_2H_4 → products	1.2×10^{-14}	2630 ± 100	1.7×10^{-18}	1.25
O_3 + C_3H_6 → products	6.5×10^{-15}	1900 ± 200	1.1×10^{-17}	1.2
OH + CO → products	1.5×10^{-13} x $(1+0.6P_{atm})$	0 ± 300	1.5×10^{-13} x $(1+0.6P_{atm})$	1.3
OH + CH_4 → CH_3 + H_2O	2.45×10^{-12}	1775 ± 100	6.3×10^{-15}	1.1
OH + $^{13}CH_4$ → $^{13}CH_3$ + H_2O	See reference			
OH + CH_3D → products	3.5×10^{-12}	1950 ± 200	5.0×10^{-15}	1.15
OH + H_2CO → H_2O + HCO	1.0×10^{-11}	0 ± 200	1.0×10^{-11}	1.25

Reaction	A cm^3 molecule^{-1} s^{-1}	E/R K	k (298 K) cm^3 molecule^{-1} s^{-1}	$f(298)$
$OH + CH_3OH \rightarrow$ products	6.7×10^{-12}	600 ± 300	8.9×10^{-13}	1.2
$OH + CH_3OOH \rightarrow$ products	3.8×10^{-12}	$-(200 \pm 200)$	7.4×10^{-12}	1.5
$OH + HC(O)OH \rightarrow$ products	4.0×10^{-13}	0 ± 200	4.0×10^{-13}	1.3
$OH + HCN \rightarrow$ products	1.2×10^{-13}	400 ± 150	3.1×10^{-14}	3
$OH + C_2H_6 \rightarrow H_2O + C_2H_5$	8.7×10^{-12}	1070 ± 100	2.4×10^{-13}	1.1
$OH + C_3H_8 \rightarrow H_2O + C_3H_7$	1.0×10^{-11}	660 ± 100	1.1×10^{-12}	1.2
$OH + CH_3CHO \rightarrow CH_3CO + H_2O$	5.6×10^{-12}	$-(270 \pm 200)$	1.4×10^{-11}	1.2
$OH + C_2H_5OH \rightarrow$ products	7.0×10^{-12}	235 ± 100	3.2×10^{-12}	1.3
$OH + CH_3C(O)OH \rightarrow$ products	4.0×10^{-13}	$-(200 \pm 400)$	8.0×10^{-13}	1.3
$OH + CH_3C(O)CH_3 \rightarrow CH_3C(O)CH_2 + H_2O$	2.2×10^{-12}	685 ± 100	2.2×10^{-13}	1.15
$OH + CH_3CN \rightarrow$ products	7.8×10^{-13}	1050 ± 200	2.3×10^{-14}	1.5
$OH + CH_3ONO_2 \rightarrow$ products	5.0×10^{-13}	890 ± 500	2.4×10^{-14}	3
$OH + CH_3C(O)O_2NO_2$ (PAN)$\rightarrow$ products			$< 4 \times 10^{-14}$	
$OH + C_2H_5ONO_2 \rightarrow$ products	8.2×10^{-13}	450 ± 300	1.8×10^{-13}	3
$HO_2 + CH_2O \rightarrow$ adduct	6.7×10^{-15}	$-(600 \pm 600)$	5.0×10^{-14}	5
$HO_2 + CH_3O_2 \rightarrow CH_3OOH + O_2$	3.8×10^{-13}	$-(800 \pm 400)$	5.6×10^{-12}	2
$HO_2 + C_2H_5O_2 \rightarrow C_2H_5OOH + O_2$	7.5×10^{-13}	$-(700 \pm 250)$	8.0×10^{-12}	1.5
$HO_2 + CH_3C(O)O_2 \rightarrow$ products	4.5×10^{-13}	$-(1000 \pm 600)$	1.3×10^{-11}	2
$NO_3 + CO \rightarrow$ products			$< 4.0 \times 10^{-19}$	
$NO_3 + CH_2O \rightarrow$ products			5.8×10^{-16}	1.3
$NO_3 + CH_3CHO \rightarrow$ products	1.4×10^{-12}	1900 ± 300	2.4×10^{-15}	1.3
$CH_3 + O_2 \rightarrow$ products			$< 3.0 \times 10^{-16}$	
$CH_3 + O_3 \rightarrow$ products	5.4×10^{-12}	220 ± 150	2.6×10^{-12}	2
$HCO + O_2 \rightarrow CO + HO_2$	3.5×10^{-12}	$-(140 \pm 140)$	5.5×10^{-12}	1.3
$CH_2OH + O_2 \rightarrow CH_2O + HO_2$	9.1×10^{-12}	0 ± 200	9.1×10^{-12}	1.3
$CH_3O + O_2 \rightarrow CH_2O + HO_2$	3.9×10^{-14}	900 ± 300	1.9×10^{-15}	1.5
$CH_3O + NO \rightarrow CH_2O + HNO$	See reference			
$CH_3O + NO_2 \rightarrow CH_2O + HONO$	1.1×10^{-11}	1200 ± 600	2.0×10^{-13}	5
$CH_3O_2 + O_3 \rightarrow$ products			$< 3.0 \times 10^{-17}$	
$CH_3O_2 + CH_3O_2 \rightarrow$ products	2.5×10^{-13}	$-(190 \pm 190)$	4.7×10^{-13}	1.5
$CH_3O_2 + NO \rightarrow CH_3O + NO_2$	3.0×10^{-12}	$-(280 \pm 60)$	7.7×10^{-12}	1.15
$CH_3O_2 + CH_3C(O)O_2 \rightarrow$ products	1.3×10^{-12}	$-(640 \pm 200)$	1.1×10^{-11}	1.5
$C_2H_5 + O_2 \rightarrow C_2H_4 + HO_2$			$< 2.0 \times 10^{-14}$	
$C_2H_5O + O_2 \rightarrow CH_3CHO + HO_2$	6.3×10^{-14}	550 ± 200	1.0×10^{-14}	1.5
$C_2H_5O_2 + C_2H_5O_2 \rightarrow$ products	6.8×10^{-14}	0 ± 300	6.8×10^{-14}	2
$C_2H_5O_2 + NO \rightarrow$ products	2.6×10^{-12}	$-(365 \pm 150)$	8.7×10^{-12}	1.2
$CH_3C(O)O_2 + CH_3C(O)O_2 \rightarrow$ products	2.9×10^{-12}	$-(500 \pm 150)$	1.5×10^{-11}	1.5
$CH_3C(O)O_2 + NO \rightarrow$ products	5.3×10^{-12}	$-(360 \pm 150)$	1.8×10^{-11}	1.4
FO_x Reactions				
$O + FO \rightarrow F + O_2$	2.7×10^{-11}	0 ± 250	2.7×10^{-11}	3.0
$O + FO_2 \rightarrow FO + O_2$	5.0×10^{-11}	0 ± 250	5.0×10^{-11}	5.0
$OH + CH_3F \rightarrow CH_2F + H_2O$ (HFC–41)	3.0×10^{-12}	1500 ± 300	2.0×10^{-14}	1.1
$OH + CH_2F_2 \rightarrow CHF_2 + H_2O$ (HFC–32)	1.9×10^{-12}	1550 ± 200	1.0×10^{-14}	1.2
$OH + CHF_3 \rightarrow CF_3 + H_2O$ (HFC–23)	1.0×10^{-12}	2440 ± 200	2.8×10^{-16}	1.3
$OH + CF_3OH \rightarrow CF_3O + H_2O$			$< 2 \times 10^{-17}$	
$OH + CH_3CH_2F \rightarrow$ products (HFC–161)	7.0×10^{-12}	1100 ± 300	1.7×10^{-13}	1.4
$OH + CH_3CHF_2 \rightarrow$ products (HFC–152a)	2.4×10^{-12}	1260 ± 200	3.5×10^{-14}	1.2
$OH + CH_2FCH_2F \rightarrow CHFCH_2F$ (HFC–152) + H_2O	1.7×10^{-11}	1500 ± 500	1.1×10^{-13}	2.0
$OH + CH_3CF_3 \rightarrow CH_2CF_3 + H_2O$ (HFC–143a)	1.8×10^{-12}	2170 ± 150	1.2×10^{-15}	1.1
$OH + CH_2FCHF_2 \rightarrow$ products (HFC–143)	4.0×10^{-12}	1650 ± 300	1.6×10^{-14}	1.5
$OH + CH_2FCF_3 \rightarrow CHFCF_3 + H_2O$ (HFC–134a)	1.5×10^{-12}	1750 ± 200	4.2×10^{-15}	1.1
$OH + CHF_2CHF_2 \rightarrow CF_2CHF_2$ (HFC–134) + H_2O	1.6×10^{-12}	1680 ± 300	5.7×10^{-15}	2.0
$OH + CHF_2CF_3 \rightarrow CF_2CF_3 + H_2O$ (HFC–125)	5.6×10^{-13}	1700 ± 300	1.9×10^{-15}	1.3
$OH + CH_3OCHF_2 \rightarrow$ products (HFOC–152a)	6.0×10^{-12}	1530 ± 150	3.5×10^{-14}	1.2
$OH + CF_3OCH_3 \rightarrow CF_3OCH_2 + H_2O$ (HFOC–143a)	1.5×10^{-12}	1450 ± 150	1.2×10^{-14}	1.1
$OH + CF_2HOCF_2H \rightarrow CF_2OCF_2H$ (HFOC–134) + H_2O	1.9×10^{-12}	2000 ± 150	2.3×10^{-15}	1.2
$OH + CF_3OCHF_2 \rightarrow CF_3OCF_2 + H_2O$ (HFOC–125)	4.7×10^{-13}	2100 ± 300	4.1×10^{-16}	1.2

Reaction	A cm^3 molecule^{-1} s^{-1}	E/R K	k (298 K) cm^3 molecule^{-1} s^{-1}	f(298)
OH + CF$_3$CH$_2$CH$_3$ $\rightarrow$ products (HFC–263fb)	–	–	4.2×10^{-14}	1.5
OH + CH$_2$FCF$_2$CHF$_2$ $\rightarrow$ products (HFC–245ca)	2.4×10^{-12}	1660 ± 150	9.1×10^{-15}	1.3
OH + CHF$_2$CHFCHF$_2$ $\rightarrow$ products (HFC–245ea)	–	–	1.6×10^{-14}	2.0
OH + CF$_3$CHFCH$_2$F $\rightarrow$ products (HFC–245eb)	–	–	1.6×10^{-14}	2.0
OH + CHF$_2$CH$_2$CF$_3$ $\rightarrow$ products (HFC–245fa)	–	–	1.5×10^{-14}	2.0
OH + CF$_3$CF$_2$CH$_2$F $\rightarrow$ CF$_3$CF$_2$CHF (HFC–236cb) +H$_2$O	6.1×10^{-13}	1330 ± 150	7.0×10^{-15}	1.2
OH + CF$_3$CF$_2$CH$_2$F $\rightarrow$ CF$_3$CF$_2$CHF (HFC–236cb) +H$_2$O	1.5×10^{-12}	1750 ± 500	4.2×10^{-15}	2.0
OH + CF$_3$CHFCHF$_2$ $\rightarrow$ products (HFC–236ea)	1.1×10^{-12}	1590 ± 150	5.3×10^{-15}	1.1
OH + CF$_3$CH$_2$CF$_3$ $\rightarrow$ CF$_3$CHCF$_3$ (HFC–236fa) +H$_2$O	1.3×10^{-12}	2480 ± 150	3.2×10^{-16}	1.1
OH + CF$_3$CHFCF$_3$ $\rightarrow$ CF$_3$CFCF$_3$+H$_2$O (HFC–227ea)	5.0×10^{-13}	1700 ± 300	1.7×10^{-15}	1.1
OH + CHF$_2$OCH$_2$CF$_3$ $\rightarrow$ products (HFOC–245fa)	2.6×10^{-12}	1610 ± 150	1.2×10^{-14}	2.0
OH + CF$_3$CH$_2$CF$_2$CH$_3$ $\rightarrow$ products (HFC–365mfc)	2.0×10^{-12}	1750 ± 200	5.7×10^{-15}	1.3
OH + CF$_3$CH$_2$CH$_2$CF$_3$ $\rightarrow$ products (HFC–356mff)	3.0×10^{-12}	1800 ± 300	7.1×10^{-15}	1.3
OH + CF$_3$CF$_2$CH$_2$CH$_2$F $\rightarrow$ products (HFC–356mcf)	1.7×10^{-12}	1110 ± 200	4.2×10^{-14}	2.0
OH + CHF$_2$CF$_2$CF$_2$CF$_2$H $\rightarrow$ products (HFC–338pcc)	7.8×10^{-13}	1530 ± 200	4.6×10^{-15}	1.5
OH + CF$_3$CH$_2$CF$_2$CH$_2$CF$_3$ $\rightarrow$ products (HFC–458mfcf)	1.2×10^{-12}	1830 ± 200	2.6×10^{-15}	2.0
OH + CF$_3$CHFCHFCF$_2$CF$_3$ $\rightarrow$ products (HFC–43–10mee)	5.2×10^{-13}	1500 ± 300	3.4×10^{-15}	1.3
OH + CF$_3$CF$_2$CH$_2$CH$_2$CF$_2$CF$_3$ $\rightarrow$ (HFC–55–10–mcff) products	–	–	8.3×10^{-15}	1.5
F + O$_3$ $\rightarrow$ FO + O$_2$	2.2×10^{-11}	230 ± 200	1.0×10^{-11}	1.5
F + H$_2$ $\rightarrow$ HF + H	1.4×10^{-10}	500 ± 200	2.6×10^{-11}	1.2
F + H$_2$O $\rightarrow$ HF + OH	1.4×10^{-11}	0 ± 200	1.4×10^{-11}	1.3
F + HNO$_3$ $\rightarrow$ HF + NO$_3$	6.0×10^{-12}	$-(400\pm200)$	2.3×10^{-11}	1.3
F + CH$_4$ $\rightarrow$ HF + CH$_3$	1.6×10^{-10}	260 ± 200	6.7×10^{-11}	1.4
FO + O$_3$ $\rightarrow$ products			$<1 \times 10^{-14}$	
FO + NO $\rightarrow$ NO$_2$ + F	8.2×10^{-12}	$-(300\pm200)$	2.2×10^{-11}	1.5
FO + FO $\rightarrow$ 2 F + O$_2$	1.0×10^{-11}	0 ± 250	1.0×10^{-11}	1.5
FO$_2$ + O$_3$ $\rightarrow$ products			$<3.4\times10^{-16}$	
FO$_2$ + NO $\rightarrow$ FNO + O$_2$	7.5×10^{-12}	690 ± 400	7.5×10^{-13}	2.0
FO$_2$ + NO$_2$ $\rightarrow$ products	3.8×10^{-11}	2040 ± 500	4.0×10^{-14}	2.0
FO$_2$ + CO $\rightarrow$ products			$<5.1\times10^{-16}$	
FO$_2$ + CH$_4$ $\rightarrow$ products			$<2\times10^{-16}$	
CF$_3$O + O$_2$ $\rightarrow$ FO$_2$ + CF$_2$O	$<3 \times 10^{-11}$	5000	$<1.5 \times 10^{-18}$	
CF$_3$O + O$_3$ $\rightarrow$ CF$_3$O$_2$ + O$_2$	2×10^{-12}	1400 ± 600	1.8×10^{-14}	1.3
CF$_3$O + H$_2$O $\rightarrow$ OH + CF$_3$OH	3×10^{-12}	>3600	$<2 \times 10^{-17}$	
CF$_3$O + NO $\rightarrow$ CF$_2$O + FNO	3.7×10^{-11}	$-(110\pm70)$	5.4×10^{-11}	1.2
CF$_3$O + NO$_2$ $\rightarrow$ products	See reference			
CF$_3$O + CO $\rightarrow$ products			$<2 \times 10^{-15}$	
CF$_3$O + CH$_4$ $\rightarrow$ CH$_3$ + CF$_3$OH	2.6×10^{-12}	1420 ± 200	2.2×10^{-14}	1.1
CF$_3$O + C$_2$H$_6$ $\rightarrow$ C$_2$H$_5$ + CF$_3$OH	4.9×10^{-12}	400 ± 100	1.3×10^{-12}	1.2
CF$_3$O$_2$ + O$_3$ $\rightarrow$ CF$_3$O + 2O$_2$			$<3 \times 10^{-15}$	
CF$_3$O$_2$ + CO $\rightarrow$ CF$_3$O + CO$_2$			$<5 \times 10^{-16}$	
CF$_3$O$_2$ + NO $\rightarrow$ CF$_3$O + NO$_2$	5.4×10^{-12}	$-(320\pm150)$	1.6×10^{-11}	1.1

ClO$_x$ Reactions

Reaction	A cm^3 molecule^{-1} s^{-1}	E/R K	k (298 K) cm^3 molecule^{-1} s^{-1}	f(298)
O + ClO $\rightarrow$ Cl + O$_2$	3.0×10^{-11}	$-(70\pm70)$	3.8×10^{-11}	1.2
O + OClO $\rightarrow$ ClO + O$_2$	2.4×10^{-12}	960 ± 300	1.0×10^{-13}	2.0
O + Cl$_2$O $\rightarrow$ ClO + ClO	2.7×10^{-11}	530 ± 150	4.5×10^{-12}	1.3
O + HCl $\rightarrow$ OH + Cl	1.0×10^{-11}	3300 ± 350	1.5×10^{-16}	2.0
O + HOCl $\rightarrow$ OH + ClO	1.7×10^{-13}	0 ± 300	1.7×10^{-13}	3.0
O + ClONO$_2$ $\rightarrow$ products	2.9×10^{-12}	800 ± 200	2.0×10^{-13}	1.5
O$_3$ + OClO $\rightarrow$ products	2.1×10^{-12}	4700 ± 1000	3.0×10^{-19}	2.5
O$_3$ + Cl$_2$O$_2$ $\rightarrow$ products	–	–	$<1.0\times10^{-19}$	–
OH + Cl$_2$ $\rightarrow$ HOCl + Cl	1.4×10^{-12}	900 ± 400	6.7×10^{-14}	1.2
OH + ClO $\rightarrow$ products	1.1×10^{-11}	$-(120\pm150)$	1.7×10^{-11}	1.5
OH + OClO $\rightarrow$ HOCl + O$_2$	4.5×10^{-13}	$-(800\pm200)$	6.8×10^{-12}	2.0
OH + HCl $\rightarrow$ H$_2$O + Cl	2.6×10^{-12}	350 ± 100	8.0×10^{-13}	1.2
OH + HOCl $\rightarrow$ H$_2$O + ClO	3.0×10^{-12}	500 ± 500	5.0×10^{-13}	3.0
OH + ClNO$_2$ $\rightarrow$ HOCl + NO$_2$	2.4×10^{-12}	1250 ± 300	3.6×10^{-14}	2.0
OH + ClONO$_2$ $\rightarrow$ products	1.2×10^{-12}	330 ± 200	3.9×10^{-13}	1.5

Reaction	A cm³ molecule⁻¹ s⁻¹	E/R K	k (298 K) cm³ molecule⁻¹ s⁻¹	f(298)
$OH + CH_3Cl \rightarrow CH_2Cl + H_2O$	4.0×10^{-12}	1400 ± 250	3.6×10^{-14}	1.2
$OH + CH_2Cl_2 \rightarrow CHCl_2 + H_2O$	3.8×10^{-12}	1050 ± 150	1.1×10^{-13}	1.4
$OH + CHCl_3 \rightarrow CCl_3 + H_2O$	2.0×10^{-12}	900 ± 150	1.0×10^{-13}	1.2
$OH + CCl_4 \rightarrow$ products	$\sim 1.0 \times 10^{-12}$	>2300	$<5.0 \times 10^{-16}$	–
$OH + CFCl_3 \rightarrow$ products (CFC–11)	$\sim 1.0 \times 10^{-12}$	>3700	$<5.0 \times 10^{-18}$	–
$OH + CF_2Cl_2 \rightarrow$ products (CFC–12)	$\sim 1.0 \times 10^{-12}$	>3600	$<6.0 \times 10^{-18}$	–
$OH + CH_2ClF \rightarrow CHClF + H_2O$ (HCFC–31)	2.8×10^{-12}	1270 ± 200	3.9×10^{-14}	1.2
$OH + CHFCl_2 \rightarrow CFCl_2 + H_2O$ (HCFC–21)	1.7×10^{-12}	1250 ± 150	2.6×10^{-14}	1.2
$OH + CHF_2Cl \rightarrow CF_2Cl + H_2O$ (HCFC–22)	1.0×10^{-12}	1600 ± 150	4.7×10^{-15}	1.1
$OH + CH_3OCl \rightarrow$ products	2.4×10^{-12}	360 ± 200	7.2×10^{-13}	3.0
$OH + CH_3CCl_3 \rightarrow CH_2CCl_3 + H_2O$ (HCC–140)	1.8×10^{-12}	1550 ± 150	1.0×10^{-14}	1.1
$OH + C_2HCl_3 \rightarrow$ products	4.9×10^{-13}	$-(450 \pm 200)$	2.2×10^{-12}	1.25
$OH + C_2Cl_4 \rightarrow$ products	9.4×10^{-12}	1200 ± 200	1.7×10^{-13}	1.25
$OH + CCl_3CHO \rightarrow H_2O + CCl_3CO$	8.2×10^{-12}	600 ± 300	1.1×10^{-12}	1.5
$OH + CH_3CFCl_2 \rightarrow CH_2CFCl_2 + H_2O$ (HCFC–141b)	1.7×10^{-12}	1700 ± 150	5.7×10^{-15}	1.2
$OH + CH_3CF_2Cl \rightarrow CH_2CF_2Cl + H_2O$ (HCFC–142b)	1.3×10^{-12}	1800 ± 150	3.1×10^{-15}	1.2
$OH + CH_2ClCF_2Cl \rightarrow CHClCF_2Cl$ (HCFC–132b) $+ H_2O$	3.6×10^{-12}	1600 ± 400	1.7×10^{-14}	2.0
$OH + CHCl_2CF_2Cl \rightarrow CCl_2CF_2Cl$ (HCFC–122) $+ H_2O$	1.0×10^{-12}	900 ± 150	4.9×10^{-14}	1.2
$OH + CHFClCFCl_2 \rightarrow CFClCFCl_2$ (HCFC–122a) $+ H_2O$	1.0×10^{-12}	1250 ± 150	1.5×10^{-14}	1.1
$OH + CH_2ClCF_3 \rightarrow CHClCF_3 + H_2O$ (HCFC–133a)	5.2×10^{-13}	1100 ± 300	1.3×10^{-14}	1.3
$OH + CHCl_2CF_3 \rightarrow CCl_2CF_3 + H_2O$ (HCFC–123)	7.0×10^{-13}	900 ± 150	3.4×10^{-14}	1.2
$OH + CHFClCF_2Cl \rightarrow CFClCF_2Cl$ (HCFC–123a) $+ H_2O$	9.2×10^{-13}	1280 ± 150	1.3×10^{-14}	1.2
$OH + CHFClCF_3 \rightarrow CFClCF_3 + H_2O$ (HCFC–124)	8.0×10^{-13}	1350 ± 150	8.6×10^{-15}	1.2
$OH + CH_3CF_2CFCl_2 \rightarrow$ products (HCFC–243cc)	7.7×10^{-13}	1700 ± 300	2.6×10^{-15}	2.0
$OH + CF_3CF_2CHCl_2 \rightarrow$ products (HCFC–225ca)	1.0×10^{-12}	1100 ± 200	2.5×10^{-14}	1.3
$OH + CF_2ClCF_2CHFCl \rightarrow$ products (HCFC–225cb)	5.5×10^{-13}	1250 ± 200	8.3×10^{-15}	1.3
$HO_2 + Cl \rightarrow HCl + O_2$	1.8×10^{-11}	$-(170 \pm 200)$	3.2×10^{-11}	1.5
$\rightarrow OH + ClO$	4.1×10^{-11}	450 ± 200	9.1×10^{-12}	2.0
$HO_2 + ClO \rightarrow HOCl + O_2$	4.8×10^{-13}	$-(700 \pm)$	5.0×10^{-12}	1.4
$H_2O + ClONO_2 \rightarrow$ products	–	–	$<2.0 \times 10^{-21}$	–
$NO + OClO \rightarrow NO_2 + ClO$	2.5×10^{-12}	600 ± 300	3.4×10^{-13}	2.0
$NO + Cl_2O_2 \rightarrow$ products	–	–	$<2.0 \times 10^{-14}$	–
$NO_3 + HCl \rightarrow HNO_3 + Cl$	–	–	$<5.0 \times 10^{-17}$	–
$HO_2NO_2 + HCl \rightarrow$ products	–	–	$<1.0 \times 10^{-21}$	–
$Cl + O_3 \rightarrow ClO + O_2$	2.9×10^{-11}	260 ± 100	1.2×10^{-11}	1.15
$Cl + H_2 \rightarrow HCl + H$	3.7×10^{-11}	2300 ± 200	1.6×10^{-14}	1.25
$Cl + H_2O_2 \rightarrow HCl + HO_2$	1.1×10^{-11}	980 ± 500	4.1×10^{-13}	1.5
$Cl + NO_3 \rightarrow ClO + NO_2$	2.4×10^{-11}	0 ± 400	2.4×10^{-11}	1.5
$Cl + N_2O \rightarrow ClO + N_2$	See reference			
$Cl + HNO_3 \rightarrow$ products	–	–	$<2.0 \times 10^{-16}$	–
$Cl + CH_4 \rightarrow HCl + CH_3$	1.1×10^{-11}	1400 ± 150	1.0×10^{-13}	1.1
$Cl + CH_3D \rightarrow$ products	–	–	7.4×10^{-14}	2.0
$Cl + H_2CO \rightarrow HCl + HCO$	8.1×10^{-11}	30 ± 100	7.3×10^{-11}	1.15
$Cl + CH_3O_2 \rightarrow$ products	–	–	1.6×10^{-10}	1.5
$Cl + CH_3OH \rightarrow CH_2OH + HCl$	5.4×10^{-11}	0 ± 250	5.4×10^{-11}	1.5
$Cl + C_2H_6 \rightarrow HCl + C_2H_5$	7.7×10^{-11}	90 ± 90	5.7×10^{-11}	1.1
$Cl + C_2H_5O_2 \rightarrow ClO + C_2H_5O$	–	–	7.4×10^{-11}	2.0
$\rightarrow HCl + C_2H_4O_2$	–	–	7.7×10^{-11}	2.0
$Cl + CH_3CN \rightarrow$ products	1.6×10^{-11}	2140 ± 300	1.2×10^{-14}	2.0
$Cl + CH_3CO_3NO_2 \rightarrow$ products	–	–	$<1 \times 10^{-14}$	
$Cl + C_3H_8 \rightarrow HCl + C_3H_7$	1.2×10^{-10}	$-(40 \pm 250)$	1.4×10^{-10}	1.3
$Cl + OClO \rightarrow ClO + ClO$	3.4×10^{-11}	$-(160 \pm 200)$	5.8×10^{-11}	1.25
$Cl + ClOO \rightarrow Cl_2 + O_2$	2.3×10^{-10}	0 ± 250	2.3×10^{-10}	3.0
$\rightarrow ClO + ClO$	1.2×10^{-11}	0 ± 250	1.2×10^{-11}	3.0
$Cl + Cl_2O \rightarrow Cl_2 + ClO$	6.2×10^{-11}	$-(130 \pm 130)$	9.6×10^{-11}	1.2
$Cl + Cl_2O_2 \rightarrow$ products	–	–	1.0×10^{-10}	2.0
$Cl + HOCl \rightarrow$ products	2.5×10^{-12}	130 ± 250	1.6×10^{-12}	1.5
$Cl + ClNO \rightarrow NO + Cl_2$	5.8×10^{-11}	$-(100 \pm 200)$	8.1×10^{-11}	1.5

Reaction	A cm^3 molecule^{-1} s^{-1}	E/R K	k (298 K) cm^3 molecule^{-1} s^{-1}	f(298)
$Cl + ClONO_2 \rightarrow$ products	6.5×10^{-12}	$-(135\pm50)$	1.0×10^{-11}	1.2
$Cl + CH_3Cl \rightarrow CH_2Cl + HCl$	3.2×10^{-11}	1250 ± 200	4.8×10^{-13}	1.2
$Cl + CH_2Cl_2 \rightarrow HCl + CHCl_2$	3.1×10^{-11}	1350 ± 500	3.3×10^{-13}	1.5
$Cl + CHCl_3 \rightarrow HCl + CCl_3$	8.2×10^{-12}	1325 ± 300	9.6×10^{-14}	1.3
$Cl + CH_3F \rightarrow HCl + CH_2F$ (HFC–41)	2.0×10^{-11}	1200 ± 500	3.5×10^{-13}	1.3
$Cl + CH_2F_2 \rightarrow HCl + CHF_2$ (HFC–32)	1.2×10^{-11}	1630 ± 500	5.0×10^{-14}	1.5
$Cl + CF_3H \rightarrow HCl + CF_3$ (HFC–23)	–	–	3.0×10^{-18}	5.0
$Cl + CH_2FCl \rightarrow HCl + CHFCl$ (HCFC–31)	1.2×10^{-11}	1390 ± 500	1.1×10^{-13}	2.0
$Cl + CHFCl_2 \rightarrow HCl + CFCl_2$ (HCFC–21)	5.5×10^{-12}	1675 ± 200	2.0×10^{-14}	1.3
$Cl + CHF_2Cl \rightarrow HCl + CF_2Cl$ (HCFC–22)	5.9×10^{-12}	2430 ± 200	1.7×10^{-15}	1.3
$Cl + CH_3CCl_3 \rightarrow CH_2CCl_3 + HCl$	2.8×10^{-12}	1790 ± 400	7.0×10^{-15}	2.0
$Cl + CH_3CH_2F \rightarrow HCl + CH_3CHF$ (HFC–161)	1.8×10^{-11}	290 ± 500	6.8×10^{-12}	3.0
$\rightarrow HCl + CH_2CH_2F$	1.4×10^{-11}	880 ± 500	7.3×10^{-13}	3.0
$Cl + CH_3CHF_2 \rightarrow HCl + CH_3CF_2$ (HFC–152a)	6.4×10^{-12}	950 ± 500	2.6×10^{-13}	1.3
$\rightarrow HCl + CH_2CHF_2$	7.2×10^{-12}	2390 ± 500	2.4×10^{-15}	3.0
$Cl + CH_2FCH_2F \rightarrow HCl + CHFCH_2F$ (HFC–152)	2.6×10^{-11}	1060 ± 500	7.5×10^{-13}	3.0
$Cl + CH_3CFCl_2 \rightarrow HCl + CH_2CFCl_2$ (HCFC–141b)	1.8×10^{-12}	2000 ± 300	2.2×10^{-15}	1.2
$Cl + CH_3CF_2Cl \rightarrow HCl + CH_2CF_2Cl$ (HCFC–142b)	1.4×10^{-12}	2420 ± 500	4.2×10^{-16}	1.2
$Cl + CH_3CF_3 \rightarrow HCl + CH_2CF_3$ (HFC–143a)	1.2×10^{-11}	3880 ± 500	2.6×10^{-17}	5.0
$Cl + CH_2FCHF_2 \rightarrow HCl + CH_2FCF_2$ (HFC–143)	5.5×10^{-12}	1610 ± 500	2.5×10^{-14}	3.0
$\rightarrow HCl + CHFCHF_2$	7.7×10^{-12}	1720 ± 500	2.4×10^{-14}	3.0
$Cl + CH_2ClCF_3 \rightarrow HCl + CHClCF_3$ (HCFC–133a)	1.8×10^{-12}	1710 ± 500	5.9×10^{-15}	3.0
$Cl + CH_2FCF_3 \rightarrow HCl + CHFCF_3$ (HFC–134a)	–	–	1.5×10^{-15}	1.2
$Cl + CHF_2CHF_2 \rightarrow HCl + CF_2CHF_2$ (HCF–134)	7.5×10^{-12}	2430 ± 500	2.2×10^{-15}	1.5
$Cl + CHCl_2CF_3 \rightarrow HCl + CCl_2CF_3$ (HCFC–123)	4.4×10^{-12}	1750 ± 500	1.2×10^{-14}	1.3
$Cl + CHFClCF_3 \rightarrow HCl + CFClCF_3$ (HCFC–124)	1.1×10^{-12}	1800 ± 500	2.7×10^{-15}	1.3
$Cl + CHF_2CF_3 \rightarrow HCl + CF_2CF_3$ (HFC–125)	–	–	2.4×10^{-16}	1.3
$ClO + O_3 \rightarrow ClOO + O_2$			$<1.4\times10^{-17}$	–
$\rightarrow OClO + O_2$	1.0×10^{-12}	>4000	$<1.0\times10^{-18}$	–
$ClO + H_2 \rightarrow$ products	$\sim1.0\times10^{-12}$	>4800	$<1.0\times10^{-19}$	–
$ClO + NO \rightarrow NO_2 + Cl$	6.4×10^{-12}	$-(290\pm100)$	1.7×10^{-11}	1.15
$ClO + NO_3 \rightarrow ClOO + NO_2$	4.7×10^{-13}	0 ± 400	4.7×10^{-13}	1.5
$ClO + N_2O \rightarrow$ products	$\sim1.0\times10^{-12}$	>4300	$<6.0\times10^{-19}$	–
$ClO + CO \rightarrow$ products	$\sim1.0\times10^{-12}$	>3700	$<4.0\times10^{-18}$	–
$ClO + CH_4 \rightarrow$ products	$\sim1.0\times10^{-12}$	>3700	$<4.0\times10^{-18}$	–
$ClO + H_2CO \rightarrow$ products	$\sim1.0\times10^{-12}$	>2100	$<1.0\times10^{-15}$	–
$ClO + CH_3O_2 \rightarrow$ products	3.3×10^{-12}	115 ± 115	2.2×10^{-12}	1.5
$ClO + ClO \rightarrow Cl_2 + O_2$	1.0×10^{-12}	1590 ± 300	4.8×10^{-15}	1.5
$\rightarrow ClOO + Cl$	3.0×10^{-11}	2450 ± 500	8.0×10^{-15}	1.5
$\rightarrow OClO + Cl$	3.5×10^{-13}	1370 ± 300	3.5×10^{-15}	1.5
$HCl + ClONO_2 \rightarrow$ products	–	–	$<1.0\times10^{-20}$	–
$CH_2ClO + O_2 \rightarrow CHClO + HO_2$	–	–	6×10^{-14}	5
$CH_2ClO_2 + HO_2 \rightarrow CH_2ClO_2H + O_2$	3.3×10^{-13}	$-(820\pm200)$	5.2×10^{-12}	1.5
$CH_2ClO_2 + NO \rightarrow CH_2ClO + NO_2$	7×10^{-12}	$-(300\pm200)$	1.9×10^{-11}	1.5
$CCl_3O_2 + NO \rightarrow CCl_3O + NO_2 + Cl$	7.3×10^{-12}	$-(270\pm200)$	1.8×10^{-11}	1.3
$CCl_2FO_2 + NO \rightarrow CClFO + NO_2 + Cl$	4.5×10^{-12}	$-(350\pm200)$	1.5×10^{-11}	1.3
$CClF_2O_2 + NO \rightarrow CF_2O + NO_2 + Cl$	3.8×10^{-12}	$-(400\pm200)$	1.5×10^{-11}	1.2

BrO_x Reactions

Reaction	A cm^3 molecule^{-1} s^{-1}	E/R K	k (298 K) cm^3 molecule^{-1} s^{-1}	f(298)
$O + BrO \rightarrow Br + O_2$	1.9×10^{-11}	$-(230\pm150)$	4.1×10^{-11}	1.5
$O + HBr \rightarrow OH + Br$	5.8×10^{-12}	1500 ± 200	3.8×10^{-14}	1.3
$O + HOBr \rightarrow OH + BrO$	1.2×10^{-10}	430 ± 300	2.8×10^{-11}	3.0
$OH + Br_2 \rightarrow HOBr + Br$	4.2×10^{-11}	0 ± 600	4.2×10^{-11}	1.3
$OH + BrO \rightarrow$ products	–	–	7.5×10^{-11}	3.0
$OH + HBr \rightarrow H_2O + Br$	1.1×10^{-11}	0 ± 250	1.1×10^{-11}	1.2
$OH + CH_3Br \rightarrow CH_2Br + H_2O$	4.0×10^{-12}	1470 ± 150	2.9×10^{-14}	1.1
$OH + CH_2Br_2 \rightarrow CHBr_2 + H_2O$	2.4×10^{-12}	900 ± 300	1.2×10^{-13}	1.1
$OH + CHBr_3 \rightarrow CBr_3 + H_2O$	1.6×10^{-12}	710 ± 200	1.5×10^{-13}	2.0

Reaction	A cm^3 molecule^{-1} s^{-1}	E/R K	k (298 K) cm^3 molecule^{-1} s^{-1}	$f(298)$
OH + CHF$_2$Br → CF$_2$Br + H$_2$O	1.1×10^{-12}	1400 ± 200	1.0×10^{-14}	1.1
OH + CH$_2$ClBr → CHClBr + H$_2$O	2.3×10^{-12}	930 ± 150	1.0×10^{-13}	1.2
OH + CF$_2$ClBr → products	–	–	$<1.5\times10^{-16}$	–
OH + CF$_2$Br$_2$ → products	–	–	$<5.0\times10^{-16}$	–
OH + CF$_3$Br → products	–	–	$<1.2\times10^{-16}$	–
OH + CH$_2$BrCF$_3$ → CHBrCF$_3$ + H$_2$O	1.4×10^{-12}	1340 ± 200	1.6×10^{-14}	1.3
OH + CHFBrCF$_3$ → CFBrCF$_3$	7.2×10^{-13}	1110 ± 150	1.8×10^{-14}	1.5
OH + CHClBrCF$_3$ → CClBrCF$_3$ + H$_2$O	1.3×10^{-12}	995 ± 150	4.5×10^{-14}	1.5
OH + CF$_2$BrCHFCl → CF$_2$BrCFCl + H$_2$O	9.3×10^{-13}	1250 ± 150	1.4×10^{-14}	1.5
OH + CF$_2$BrCF$_2$Br → products	–	–	$<1.5\times10^{-16}$	–
HO$_2$ + Br → HBr + O$_2$	1.5×10^{-11}	600 ± 600	2.0×10^{-12}	2.0
HO$_2$ + BrO → products	3.4×10^{-12}	$-(540\pm200)$	2.1×10^{-11}	1.5
NO$_3$ + HBr → HNO$_3$ + Br	–	–	$<1.0\times10^{-16}$	–
Cl + CH$_2$ClBr → HCl + CHClBr	4.3×10^{-11}	1370 ± 500	4.3×10^{-13}	3.0
Cl + CH$_3$Br → HCl + CH$_2$Br	1.5×10^{-11}	1060 ± 100	4.3×10^{-13}	1.2
Cl + CH$_2$Br$_2$ → HCl + CHBr$_2$	6.4×10^{-12}	810 ± 100	4.2×10^{-13}	1.2
Br + O$_3$ → BrO + O$_2$	1.7×10^{-11}	800 ± 200	1.2×10^{-12}	1.2
Br + H$_2$O$_2$ → HBr + HO$_2$	1.0×10^{-11}	>3000	$<5.0\times10^{-16}$	–
Br + NO$_3$ → BrO + NO$_2$	–	–	1.6×10^{-11}	2.0
Br + H$_2$CO → HBr + HCO	1.7×10^{-11}	800 ± 200	1.1×10^{-12}	1.3
Br + OClO → BrO + ClO	2.6×10^{-11}	1300 ± 300	3.4×10^{-13}	2.0
Br + Cl$_2$O → BrCl + ClO	2.1×10^{-11}	470 ± 150	4.3×10^{-12}	1.3
Br + Cl$_2$O$_2$ → products	–	–	3.0×10^{-12}	2.0
BrO + O$_3$ → products	$\sim1.0\times10^{-12}$	>3200	$<2.0\times10^{-17}$	–
BrO + NO → NO$_2$ + Br	8.8×10^{-12}	$-(260\pm130)$	2.1×10^{-11}	1.15
BrO + NO$_3$ → products	–	–	1.0×10^{-12}	3.0
BrO + ClO → Br + OClO	1.6×10^{-12}	$-(430\pm200)$	6.8×10^{-12}	1.25
→ Br + ClOO	2.9×10^{-12}	$-(220\pm200)$	6.1×10^{-12}	1.25
→ BrCl + O$_2$	5.8×10^{-13}	$-(170\pm200)$	1.0×10^{-12}	1.25
BrO + BrO → products	1.5×10^{-12}	$-(230\pm150)$	3.2×10^{-12}	1.15
CH$_2$BrO$_2$ + NO → CH$_2$O + NO$_2$ + Br	4×10^{-12}	$-(300\pm200)$	1.1×10^{-11}	1.5

IO_x Reactions

Reaction	A	E/R	k (298 K)	$f(298)$
O + I$_2$ → IO + I	1.4×10^{-10}	0 ± 250	1.4×10^{-10}	1.4
O + IO → O$_2$ + I			1.2×10^{-10}	2.0
OH + I$_2$ → HOI + I			1.8×10^{-10}	2.0
OH + HI → H$_2$O + I			3.0×10^{-11}	2.0
OH + CH$_3$I → H$_2$O + CH$_2$I	3.1×10^{-12}	1120 ± 500	7.2×10^{-14}	3.0
OH + CF$_3$I → HOI + CF$_3$			3.1×10^{-14}	5.0
HO$_2$ + I → HI + O$_2$	1.5×10^{-11}	1090 ± 500	3.8×10^{-13}	2.0
HO$_2$ + IO → HOI + O$_2$			8.4×10^{-11}	1.5
NO$_3$ + HI → HNO$_3$ + I	See reference			
I + O$_3$ → IO + O$_2$	2.3×10^{-11}	870 ± 200	1.2×10^{-12}	1.2
I + BrO → IO + Br	–	–	1.2×10^{-11}	2.0
IO + NO → I + NO$_2$	9.1×10^{-12}	$-(240\pm150)$	2.0×10^{-11}	1.2
IO + ClO → products	5.1×10^{-12}	$-(280\pm200)$	1.3×10^{-11}	2.0
IO + BrO → products	–	–	6.9×10^{-11}	1.5
IO + IO → products	1.5×10^{-11}	$-(500\pm500)$	8.0×10^{-11}	1.5
INO + INO → I$_2$ + 2NO	8.4×10^{-11}	2620 ± 600	1.3×10^{-14}	2.5
INO$_2$ + INO$_2$ → I$_2$ + 2NO$_2$	2.9×10^{-11}	2600 ± 1000	4.7×10^{-15}	3.0

SO_x Reactions

Reaction	A	E/R	k (298 K)	$f(298)$
O + SH → SO + H	–	–	1.6×10^{-10}	5.0
O + CS → CO + S	2.7×10^{-10}	760 ± 250	2.1×10^{-11}	1.1
O + H$_2$S → OH + SH	9.2×10^{-12}	1800 ± 550	2.2×10^{-14}	1.7
O + OCS → CO + SO	2.1×10^{-11}	2200 ± 150	1.3×10^{-14}	1.2
O + CS$_2$ → CS + SO	3.2×10^{-11}	650 ± 150	3.6×10^{-12}	1.2
O + CH$_3$SCH$_3$ → CH$_3$SO + CH$_3$	1.3×10^{-11}	$-(410\pm100)$	5.0×10^{-11}	1.1
O + CH$_3$SSCH$_3$ → CH$_3$SO + CH$_3$S	5.5×10^{-11}	$-(250\pm100)$	1.3×10^{-10}	1.3

Reaction	A cm³ molecule⁻¹ s⁻¹	E/R K	k (298 K) cm³ molecule⁻¹ s⁻¹	f(298)
$O_3 + H_2S \rightarrow$ products	–	–	$<2.0 \times 10^{-20}$	–
$O_3 + CH_3SCH_3 \rightarrow$ products			$<1.0 \times 10^{-18}$	–
$O_3 + SO_2 \rightarrow SO_3 + O_2$	3.0×10^{-12}	>7000	$<2.0 \times 10^{-22}$	–
$OH + H_2S \rightarrow SH + H_2O$	6.0×10^{-12}	75±75	4.7×10^{-12}	1.2
$OH + OCS \rightarrow$ products	1.1×10^{-13}	1200±500	1.9×10^{-15}	2.0
$OH + CS_2 \rightarrow$ products	See reference	–	–	–
$OH + CH_3SH \rightarrow CH_3S + H_2O$	9.9×10^{-12}	$-(360\pm100)$	3.3×10^{-11}	1.2
$OH + CH_3SCH_3 \rightarrow H_2O + CH_2SCH_3$	1.2×10^{-11}	260±100	5.0×10^{-12}	1.15
$OH + CH_3SSCH_3 \rightarrow$ products	6.0×10^{-11}	$-(400\pm200)$	2.3×10^{-10}	1.2
$OH + S \rightarrow H + SO$	–	–	6.6×10^{-11}	3.0
$OH + SO \rightarrow H + SO_2$	–	–	8.6×10^{-11}	2.0
$HO_2 + H_2S \rightarrow$ products	–	–	$<3.0 \times 10^{-15}$	–
$HO_2 + CH_3SH \rightarrow$ products	–	–	$<4.0 \times 10^{-15}$	–
$HO_2 + CH_3SCH_3 \rightarrow$ products	–	–	$<5.0 \times 10^{-15}$	–
$HO_2 + SO_2 \rightarrow$ products	–	–	$<1.0 \times 10^{-18}$	–
$NO_2 + SO_2 \rightarrow$ products	–	–	$<2.0 \times 10^{-26}$	–
$NO_3 + H_2S \rightarrow$ products	–	–	$<8.0 \times 10^{-16}$	–
$NO_3 + OCS \rightarrow$ products	–	–	$<1.0 \times 10^{-16}$	–
$NO_3 + CS_2 \rightarrow$ products	–	–	$<4.0 \times 10^{-16}$	–
$NO_3 + CH_3SH \rightarrow$ products	4.4×10^{-13}	$-(210\pm210)$	8.9×10^{-13}	1.25
$NO_3 + CH_3SCH_3 \rightarrow CH_3SCH_2 + HNO_3$	1.9×10^{-13}	$-(500\pm200)$	1.0×10^{-12}	1.2
$NO_3 + CH_3SSCH_3 \rightarrow$ products	1.3×10^{-12}	270±270	5.3×10^{-13}	1.4
$NO_3 + SO_2 \rightarrow$ products	–	–	$<7.0 \times 10^{-21}$	–
$N_2O_5 + CH_3SCH_3 \rightarrow$ products	–	–	$<1.0 \times 10^{-17}$	–
$CH_3O_2 + SO_2 \rightarrow$ products	–	–	$<5.0 \times 10^{-17}$	-
$F + CH_3SCH_3 \rightarrow$ products	–	–	2.4×10^{-10}	2.0
$Cl + H_2S \rightarrow HCl + SH$	3.7×10^{-11}	$-(210\pm100)$	7.4×10^{-11}	1.25
$Cl + OCS \rightarrow$ products	–	–	$<1.0 \times 10^{-16}$	–
$Cl + CS_2 \rightarrow$ products	–	–	$<4.0 \times 10^{-15}$	–
$Cl + CH_3SH \rightarrow CH_3S + HCl$	1.2×10^{-10}	$-(150\pm50)$	2.0×10^{-10}	1.25
$Cl + CH_3SCH_3 \rightarrow$ products	See reference	–	–	–
$ClO + OCS \rightarrow$ products	–	–	$<2.0 \times 10^{-16}$	–
$ClO + CH_3SCH_3 \rightarrow$ products	–	–	9.5×10^{-15}	2.0
$ClO + SO \rightarrow Cl + SO_2$	2.8×10^{-11}	0±50	2.8×10^{-11}	1.3
$ClO + SO_2 \rightarrow Cl + SO_3$	–	–	$<4.0 \times 10^{-18}$	–
$Br + H_2S \rightarrow HBr + SH$	1.4×10^{-11}	2750±300	1.4×10^{-15}	2.0
$Br + CH_3SH \rightarrow CH_3S + HBr$	9.2×10^{-12}	390±100	2.5×10^{-12}	2.0
$Br + CH_3SCH_3 \rightarrow$ products	See reference			
$BrO + CH_3SCH_3 \rightarrow$ products	1.5×10^{-14}	$-(850\pm200)$	2.6×10^{-13}	1.3
$BrO + SO \rightarrow Br + SO_2$			5.7×10^{-11}	1.4
$IO + CH_3SH \rightarrow$ products			6.6×10^{-16}	2.0
$IO + CH_3SCH_3 \rightarrow$ products			1.2×10^{-14}	1.5
$S + O_2 \rightarrow SO + O$	2.3×10^{-12}	0±200	2.3×10^{-12}	1.2
$S + O_3 \rightarrow SO + O_2$			1.2×10^{-11}	2.0
$SO + O_2 \rightarrow SO_2 + O$	2.6×10^{-13}	2400±500	8.4×10^{-17}	2.0
$SO + O_3 \rightarrow SO_2 + O_2$	3.6×10^{-12}	1100±200	9.0×10^{-14}	1.2
$SO + NO_2 \rightarrow SO_2 + NO$	1.4×10^{-11}	0±50	1.4×10^{-11}	1.2
$SO + OClO \rightarrow SO_2 + ClO$			1.9×10^{-12}	3.0
$SO_3 + H_2O \rightarrow$ products	See reference		–	–
$SO_3 + NO_2 \rightarrow$ products			1.0×10^{-19}	10.0
$SH + O_2 \rightarrow OH + SO$			$<4.0 \times 10^{-19}$	–
$SH + O_3 \rightarrow HSO + O_2$	9.0×10^{-12}	280±200	3.5×10^{-12}	1.3
$SH + H_2O_2 \rightarrow$ products			$<5.0 \times 10^{-15}$	–
$SH + NO_2 \rightarrow HSO + NO$	2.9×10^{-11}	$-(240\pm50)$	6.5×10^{-11}	1.2
$SH + Cl_2 \rightarrow ClSH + Cl$	1.7×10^{-11}	690±200	1.7×10^{-12}	2.0
$SH + BrCl \rightarrow$ products	2.3×10^{-11}	$-(350\pm200)$	7.4×10^{-11}	2.0
$SH + Br_2 \rightarrow BrSH + Br$	6.0×10^{-11}	$-(160\pm160)$	1.0×10^{-10}	2.0
$SH + F_2 \rightarrow FSH + F$	4.3×10^{-11}	1390±200	4.0×10^{-13}	2.0

Reaction	A cm^3 molecule^{-1} s^{-1}	E/R K	k (298 K) cm^3 molecule^{-1} s^{-1}	f(298)
HSO + O$_2$ → products			<2.0×10^{-17}	–
HSO + O$_3$ → products			1.0×10^{-13}	1.3
HSO + NO → products			<1.0×10^{-15}	–
HSO + NO$_2$ → HSO$_2$ + NO			9.6×10^{-12}	2.0
HSO$_2$ + O$_2$ → HO$_2$ + SO$_2$			3.0×10^{-13}	3.0
HOSO$_2$ + O$_2$ → HO$_2$ + SO$_3$	1.3×10^{-12}	330±200	4.4×10^{-13}	1.2
CS + O$_2$ → OCS + O			2.9×10^{-19}	2.0
CS + O$_3$ → OCS + O$_2$			3.0×10^{-16}	3.0
CS + NO$_2$ → OCS + NO			7.6×10^{-17}	3.0
CH$_3$S + O$_2$ → products			<3.0×10^{-18}	–
CH$_3$S + O$_3$ → products	2.0×10^{-12}	–(290±100)	5.3×10^{-12}	1.15
CH$_3$S + NO → products			<1.0×10^{-13}	–
CH$_3$S + NO$_2$ → CH$_3$SO + NO	2.1×10^{-11}	–(320±100)	6.1×10^{-11}	1.15
CH$_2$SH + O$_2$ → products			6.5×10^{-12}	2.0
CH$_2$SH + O$_3$ → products			3.5×10^{-11}	2.0
CH$_2$SH + NO → products			1.9×10^{-11}	2.0
CH$_2$SH + NO$_2$ → products			5.2×10^{-11}	2.0
CH$_3$SO + O$_3$ → products			6.0×10^{-13}	1.5
CH$_3$SO + NO$_2$ → CH$_3$SO$_2$ + NO			1.2×10^{-11}	1.4
CH$_3$SOO + O$_3$ → products			<8.0×10^{-13}	–
CH$_3$SOO + NO → products	1.1×10^{-11}	0±100	1.1×10^{-11}	2.0
CH$_3$SO$_2$ + NO$_2$ → products	2.2×10^{-11}	0±100	2.2×10^{-11}	2.0
CH$_3$SCH$_2$ + NO$_3$ → products			3.0 × 10^{-10}	2.0
CH$_3$SCH$_2$O$_2$ + NO → CH$_3$SCH$_2$O + NO$_2$			1.9 × 10^{-11}	2.0
CH$_3$SS + O$_3$ → products			4.6×10^{-13}	2.0
CH$_3$SS + NO$_2$ → products			1.8×10^{-11}	2.0
CH$_3$SSO + NO$_2$ → products			4.5×10^{-12}	2.0
Metal Reactions				
Na + O$_3$ → NaO + O$_2$	1.0×10^{-9}	95±50	7.3×10^{-10}	1.2
→ NaO$_2$ + O	–	–	<4.0×10^{-11}	–
Na + N$_2$O → NaO + N$_2$	2.8×10^{-10}	1600±400	1.3×10^{-12}	1.2
Na + Cl$_2$ → NaCl + Cl	7.3×10^{-10}	0±200	7.3×10^{-10}	1.3
NaO + O → Na + O$_2$	3.7×10^{-10}	0±400	3.7×10^{-10}	3.0
NaO + O$_3$ → NaO$_2$ + O$_2$	1.1×10^{-9}	570±300	1.6×10^{-10}	1.5
→ Na + 2O$_2$	6.0×10^{-11}	0±800	6.0×10^{-11}	3.0
NaO + H$_2$ → NaOH + H	2.6×10^{-11}	0±600	2.6×10^{-11}	2.0
NaO + H$_2$O → NaOH + OH	2.2×10^{-10}	0±400	2.2×10^{-10}	2.0
NaO + NO → Na + NO$_2$	1.5×10^{-10}	0±400	1.5×10^{-10}	4.0
NaO + HCl → products	2.8×10^{-10}	0±400	2.8×10^{-10}	3.0
NaO$_2$ + O → NaO + O$_2$	2.2×10^{-11}	0±600	2.2×10^{-11}	5.0
NaO$_2$ + NO → NaO + NO$_2$	–	–	<10^{-14}	–
NaO$_2$ + HCl → products	2.3×10^{-10}	0±400	2.3×10^{-10}	3.0
NaOH + HCl → NaCl + H$_2$O	2.8×10^{-10}	0±400	2.8×10^{-10}	3.0

TABLE 2. Rate Constants for Association Reactions

The values quoted are suitable for air as the third body, M. The integer in parentheses is the power of ten.

Reaction	Low pressure limit $k_0(T) = k_0(300) (T/300)^{-n}$ cm^6 molecule^{-2} s^{-1}		High pressure limit $k_\infty(T) = k_\infty(300) (T/300)^{-m}$ cm^3 molecule^{-1} s^{-1}	
	$k_0(300)$	n	$k_\infty(300)$	m
O_x Reactions				
O + O$_2$ → O$_3$	(6.0±0.5) (−34)	2.3±0.5	—	—
$O(^1D)$ Reactions				
O(^{1}D) + N$_2$ → N$_2$O	(3.5±3.0) (−37)	0.6	—	—

Chemical Kinetic Data for Stratospheric Modeling

Reaction	Low pressure limit $k_0(T) = k_0(300) (T/300)^{-n}$ cm^6 molecule^{-2} s^{-1}		High pressure limit $k_\infty(T) = k_\infty(300) (T/300)^{-m}$ cm^3 molecule^{-1} s^{-1}	
	$k_0(300)$	n	$k_\infty(300)$	m
HO$_x$ Reactions				
H + O$_2$ → HO$_2$	(5.7±0.5) (−32)	1.6±0.5	(7.5±4.0) (−11)	0±1.0
OH + OH → H$_2$O$_2$	(6.2±1.2) (−31)	1.0	(2.6±1.0) (−11)	0±0.5
NO$_x$ Reactions				
O + NO → NO$_2$	(9.0±2.0) (−32)	1.5±0.3	(3.0±1.0) (−11)	0±1.0
O + NO$_2$ → NO$_3$	(9.0±1.0) (−32)	2.0±1.0	(2.2±0.3) (−11)	0±1.0
OH + NO → HONO	(7.0±1.0) (−31)	2.6±0.3	(3.6±1.0) (−11)	0.1±0.5
OH + NO$_2$ → HNO$_3$	(2.5±0.1) (−30)	4.4±0.3	(1.6±0.2) (−11)	1.7±0.2
HO$_2$ + NO$_2$ → HO$_2$NO$_2$	(1.8±0.3) (−31)	3.2±0.4	(4.7±1.0) (−12)	1.4±1.4
NO$_2$ + NO$_3$ → N$_2$O$_5$	(2.2±0.5) (−30)	3.9±1.0	(1.5±0.8) (−12)	0.7±0.4
NO$_3$ → NO + O$_2$	See reference			
Hydrocarbon Reactions				
CH$_3$ + O$_2$ → CH$_3$O$_2$	(4.5±1.5) (−31)	3.0±1.0	(1.8±0.2) (−12)	1.7±1.7
C$_2$H$_5$ + O$_2$ → C$_2$H$_5$O$_2$	(1.5±1.0) (−28)	3.0±1.0	(8.0±1.0) (−12)	0±1.0
OH + C$_2$H$_2$ → HOCHCH	(5.5±2.0) (−30)	0.0±0.2	(8.3±1.0) (−13)	−2
OH + C$_2$H$_4$ → HOCH$_2$CH$_2$	(1.0±0.6) (−28)	0.8±2.0	(8.8±0.9) (−12)	0
CH$_3$O + NO → CH$_3$ONO	(1.4±0.5) (−29)	3.8±1.0	(3.6±1.6) (−11)	0.6±1.0
CH$_3$O + NO$_2$ → CH$_3$ONO$_2$	(1.1±0.4) (−28)	4.0±2.0	(1.6±0.5) (−11)	1.0±1.0
C$_2$H$_5$O + NO → C$_2$H$_5$ONO	(2.8±1.0) (−27)	4.0±2.0	(5.0±1.0) (−11)	1.0±1.0
C$_2$H$_5$O + NO$_2$ → C$_2$H$_5$ONO$_2$	(2.0±1.0) (−27)	4.0±2.0	(2.8±0.4) (−11)	1.0±1.0
CH$_3$O$_2$ + NO$_2$ → CH$_3$O$_2$NO$_2$	(1.5±0.8) (−30)	4.0±2.0	(6.5±3.2) (−12)	2.0±2.0
CH$_3$C(O)O$_2$ + NO$_2$ → CH$_3$C(O)O$_2$NO$_2$	(9.7±3.8) (−29)	5.6±2.8	(9.3±0.4)(−12)	1.5±0.3
FO$_x$ Reactions				
F + O$_2$ → FO$_2$	(4.4±0.4) (−33)	1.2±0.5	–	–
F + NO → FNO	(1.8±0.3) (−31)	1.0±10	(2.8±1.4) (−10)	0.0±1.0
F + NO$_2$ → FNO$_2$	(6.3±3.0) (−32)	2.0±2.0	(2.6±1.3) (−10)	0.0±1.0
FO + NO$_2$ → FONO$_2$	(2.6±2.0) (−31)	1.3±1.3	(2.0±1.0) (−11)	1.5±1.5
CF$_3$ + O$_2$ → CF$_3$O$_2$	(3.0±0.3) (−29)	4.0±2.0	(4.0±1.0) (−12)	1.0±1.0
CF$_3$O + NO$_2$ → CF$_3$ONO$_2$	See reference			
CF$_3$O$_2$ + NO$_2$CF$_3$O$_2$NO$_2$	(2.2±0.5) (−29)	5.0±1.0	(6.0±1.0) (−12)	2.5±1.0
CF$_3$O + CO → CF$_3$OCO	(2.5±0.2) (−31)	–	(6.8±0.4) (−14)	−1.2
CF$_3$O → CF$_2$O + F	See reference			
ClO$_x$ Reactions				
Cl + O$_2$ → ClOO	(2.7±1.0) (−33)	1.5±0.5	–	–
Cl + NO → ClNO	(9.0±2.0) (−32)	1.6±0.5	–	–
Cl + NO$_2$ ClONO →	(1.3±0.2) (−30)	2.0±1.0	(1.0±0.5) (−10)	1.0±1.0
ClNO$_2$	(1.8±0.3) (−31)	2.0±1.0	(1.0±0.5) (−10)	1.0±1.0
Cl + CO → ClCO	(1.3±0.5) (−33)	3.8±0.5	–	–
Cl + C$_2$H$_2$ → ClC$_2$H$_2$	((5.9±1.0) (−30)	2.1±1.0	(2.1±0.4) (−10)	1.0±0.5
Cl + C$_2$H$_4$ → ClC$_2$H$_4$	(1.6±1) (−29)	3.3±1.0	(3.1±2) (−10)	1.0±0.5
Cl + C$_2$Cl$_4$ → C$_2$Cl$_5$	(1.4±0.6) (−28)	8.5±1.0	(4.0±1.0) (−11)	1.2±0.5
ClO + NO$_2$ → ClONO$_2$	(1.8±0.3) (−31)	3.4±1.0	(1.5±0.7) (−11)	1.9±1.9
OClO + NO$_3$ → O$_2$ClONO$_2$	See reference			
ClO + ClO → Cl$_2$O$_2$	(2.2±0.4) (−32)	3.1±0.5	(3.5±2) (−12)	1.0±1.0
ClO + OClO → Cl$_2$O$_3$	(6.2±1.0) (−32)	4.7±0.6	(2.4±1.2) (−11)	0±1.0
OClO + O → ClO$_3$	(1.9±0.5) (−31)	1.1±1.0	(3.1±0.8) (−11)	0±1.0
CH$_2$Cl + O$_2$ → CH$_2$ClO$_2$	(1.9±0.1) (−30)	3.2±0.2	(2.9±0.2) (−12)	1.2±0.6
CHCl$_2$ + O$_2$ → CHCl$_2$O$_2$	(1.3±0.1) (−30)	4.0±0.2	(2.8±0.2) (−12)	1.4±0.6
CCl$_3$ + O$_2$ → CCl$_3$O$_2$	(6.9±0.2) (−31)	6.4±0.3	(2.4±0.2) (−12)	2.1±0.6
CFCl$_2$ + O$_2$ → CFCl$_2$O$_2$	(5.0±0.8) (−30)	4.0±2.0	(6.0±1.0) (−12)	1.0±1.0
CF$_2$Cl + O$_2$ → CF$_2$ClO$_2$	(3.0±1.5) (−30)	4.0±2.0	(3±2) (−12)	1.0±1.0
CCl$_3$O$_2$ + NO$_2$ → CCl$_3$O$_2$NO$_2$	(5.0±1.0) (−29)	5.0±1.0	(6.0±1.0) (−12)	2.5±1.0
CFCl$_2$O$_2$ + NO$_2$ → CFCl$_2$O$_2$NO$_2$	(3.5±0.5) (−29)	5.0±1.0	(6.0±1.0) (−12)	2.5±1.0
CF$_2$ClO$_2$ + NO$_2$ → CF$_2$ClO$_2$NO$_2$	(3.3±0.7) (−29)	6.7±1.3	(4.1±1.9) (−12)	2.8±0.7

Reaction	Low pressure limit $k_0(T) = k_0(300)\,(T/300)^{-n}\ \mathrm{cm^6\,molecule^{-2}\,s^{-1}}$		High pressure limit $k_\infty(T) = k_\infty(300)\,(T/300)^{-m}\ \mathrm{cm^3\,molecule^{-1}\,s^{-1}}$	
	$k_0(300)$	n	$k_\infty(300)$	m
BrO_x Reactions				
$Br + NO_2 \rightarrow BrNO_2$	(4.2±0.8) (−31)	2.4±0.5	(2.7±0.5) (−11)	0±1.0
$BrO + NO_2 \rightarrow BrONO_2$	(5.2±0.6) (−31)	3.2±0.8	(6.9±1.0) (−12)	2.9±1.0
IO_x Reactions				
$I + NO \rightarrow INO$	(1.8±0.5) (−32)	1.0±0.5	(1.7±1.0) (−11)	0±1.0
$I + NO_2 \rightarrow INO_2$	(3.0±1.5) (−31)	1.0±1.0	(6.6±5.0) (−11)	0±1.0
$IO + NO_2 \rightarrow IONO_2$	(5.9±2.0) (−31)	3.5±1.0	(9.0±1.0) (−12)	1.5±1.0
SO_x Reactions				
$HS + NO \rightarrow HSNO$	(2.4±0.4) (−31)	3.0±1.0	(2.7±0.5) (−11)	0
$CH_3S + NO \rightarrow CH_3SNO$	(3.2±0.4) (−29)	4.0±1.0	(3.9±0.6) (−11)	2.7±1.0
$O + SO_2 \rightarrow SO_3$	(1.3±) (−33)	−3.6±0.7		
$OH + SO_2 \rightarrow HOSO_2$	(3.0±1.0) (−31)	3.3±1.5	(1.5±0.5) (−12)	0
$CH_3SCH_2 + O_2 \rightarrow CH_3SCH_2O_2$	See reference			
$SO_3 + NH_3 \rightarrow H_3NSO_3$	(3.9±0.8) (−30)	3.0±3.0	(4.7±1.3) (−11)	0±1.0
Metal Reactions				
$Na + O_2 \rightarrow NaO_2$	(3.2±0.3) (−30)	1.4±0.3	(6.0±2.0) (−10)	0±1.0
$NaO + O_2 \rightarrow NaO_3$	(3.5±0.7) (−30)	2.0±2.0	(5.7±3.0) (−10)	0±1.0
$NaO + CO_2 \rightarrow NaCO_3$	(8.7±2.6) (−28)	2.0±2.0	(6.5±3.0) (−10)	0±1.0
$NaOH + CO_2 \rightarrow NaHCO_3$	(1.3±0.3) (−28)	2.0±2.0	(6.8±4.0) (−10)	0±1.0

TABLE 3. Equilibrium Constants

$K(T)/\mathrm{cm^3\,molecule^{-1}} = A \exp(B/T)$ [$200 < T/\mathrm{K} < 300$]

Reaction	$A/\mathrm{cm^3\,molecule^{-1}}$	B/K	K (298 K)	f (298 K)
$HO_2 + NO_2 \rightarrow HO_2NO_2$	2.1×10^{-27}	10900±1000	1.6×10^{-11}	5
$NO + NO_2 \rightarrow N_2O_3$	3.3×10^{-27}	4667±100	2.1×10^{-20}	2
$NO_2 + NO_2 \rightarrow N_2O_4$	5.2×10^{-29}	6643±250	2.5×10^{-19}	2
$NO_2 + NO_3 \rightarrow N_2O_5$	2.7×10^{-27}	11000±500	2.9×10^{-11}	1.3
$CH_3O_2 + NO_2 \rightarrow CH_3O_2NO_2$	1.3×10^{-28}	11200±1000	2.7×10^{-12}	2
$CH_3C(O)O_2 + NO_2 \rightarrow CH_3C(O)O_2NO_2$	9.0×10^{-29}	14000±200	2.3×10^{-8}	2
$F + O_2 \rightarrow FOO$	3.2×10^{-25}	6100±1200	2.5×10^{-16}	1.0
$Cl + O_2 \rightarrow ClOO$	5.7×10^{-25}	2500±750	2.5×10^{-21}	2
$Cl + CO \rightarrow ClCO$	1.6×10^{-25}	4000±500	1.1×10^{-19}	5
$ClO + O_2 \rightarrow ClOO_2$	2.9×10^{-26}	<3700	$<7.2 \times 10^{-21}$	–
$ClO + ClO \rightarrow Cl_2O_2$	1.3×10^{-27}	8744±850	7.2×10^{-15}	1.5
$ClO + OClO \rightarrow Cl_2O_3$	1.1×10^{-24}	5455±300	9.8×10^{-17}	3
$OClO + NO_3 \rightarrow O_2ClONO_2$	1×10^{-28}	9300±1000	3.6×10^{-15}	5
$OH + CS_2 \rightarrow CS_2OH$	4.5×10^{-25}	5140±500	1.4×10^{-17}	1.4
$CH_3S + O_2 \rightarrow CH_3SO_2$	1.8×10^{-27}	5545±300	2.2×10^{-19}	1.4

Section 6
Fluid Properties

THERMODYNAMIC PROPERTIES OF AIR

These tables summarize the thermodynamic properties of air in the liquid and gaseous states, as well as along the saturation line. In the table for the saturation state, P(boil) is the bubble point temperature (i.e., the pressure at which boiling begins as the temperature of the liquid is raised), and P(con) is the dew point temperature (pressure at which condensation begins as the temperature of the gas is lowered). The other properties tabulated are density (ρ), enthalpy (H), entropy (S), and isobaric heat capacity (C_p). More detailed tables may be found in the references.

References

1. Vasserman, A. A., and Rabinovich, V. A., *Thermophysical Properties of Liquid Air and its Components*, Izdatel'stvo Komiteta, Standartov, Moscow, 1968.
2. Vasserman, A. A., et al., *Thermophysical Properties of Air and Air Components*, Izdatel'stvo Nauka, Moscow, 1966.

Properties in the saturation state:

T K	P(boil) bar	P(con) bar	ρ (liq) g/cm³	ρ (gas) g/L
65	0.1468	0.0861	0.939	0.464
70	0.3234	0.2052	0.917	1.033
75	0.6366	0.4321	0.894	2.048
80	1.146	0.8245	0.871	3.709
85	1.921	1.453	0.845	6.258
90	3.036	2.397	0.819	9.980
95	4.574	3.748	0.792	15.21
100	6.621	5.599	0.763	22.39
110	12.59	11.22	0.699	45.15
120	21.61	20.14	0.622	87.34
130	34.16	33.32	0.487	184.33
132.55	37.69	37.69	0.313	312.89

Properties of air in the gaseous state:

P bar	T K	ρ g/L	H J/g	S J/g K	C_p J/g K
1	100	3.556	98.3	5.759	1.032
1	200	1.746	199.7	6.463	1.007
1	300	1.161	300.3	6.871	1.007
1	500	0.696	503.4	7.389	1.030
1	1000	0.348	1046.6	8.138	1.141
10	200	17.835	195.2	5.766	1.049
10	300	11.643	298.3	6.204	1.021
10	500	6.944	502.9	6.727	1.034
10	1000	3.471	1047.2	7.477	1.142
100	200	213.950	148.8	4.949	1.650
100	300	116.945	279.9	5.486	1.158
100	500	66.934	499.0	6.048	1.073
100	1000	33.613	1052.4	6.812	1.151

Properties of liquid air:

P bar	T K	ρ g/cm³	H J/g	S J/g K	C_p J/g K
1	75	0.8935	−131.7	2.918	1.843
5	75	0.8942	−131.4	2.916	1.840
5	80	0.8718	−122.3	3.031	1.868
5	85	0.8482	−112.9	3.143	1.901
5	90	0.8230	−103.3	3.250	1.941
5	95	0.7962	−93.5	3.356	1.991
10	75	0.8952	−131.1	2.913	1.836
10	80	0.8729	−122.0	3.028	1.863
10	90	0.8245	−103.1	3.246	1.932
10	100	0.7695	−83.2	3.452	2.041
50	75	0.9025	−128.2	2.892	1.806
50	100	0.7859	−81.8	3.415	1.939
50	125	0.6222	−28.3	3.889	2.614
50	150	0.1879	91.9	4.764	2.721
100	75	0.9111	−124.5	2.867	1.774
100	100	0.8033	−79.4	3.376	1.852
100	125	0.6746	−31.4	3.805	2.062
100	150	0.4871	32.8	4.271	2.832

PROPERTIES OF WATER IN THE RANGE 0–100 °C

This table summarizes the best available values of the density, specific heat capacity at constant pressure (C_p), vapor pressure, viscosity, thermal conductivity, dielectric constant, and surface tension for liquid water in the range 0 – 100 °C. All values (except vapor pressure) refer to a pressure of 100 kPa (1 bar). The temperature scale is IPTS-68.

References

1. L. Harr, J. S. Gallagher, and G. S. Kell, *NBS/NRC Steam Tables*, Hemisphere Publishing Corp., New York, 1984.

2. K. N. Marsh, Ed., *Recommended Reference Materials for the Realization of Physicochemical Properties*, Blackwell Scientific Publications, Oxford, 1987.
3. J. V. Sengers and J. T. R. Watson, Improved international formulations for the viscosity and thermal conductivity of water substance, *J. Phys. Chem. Ref. Data*, 15, 1291, 1986.
4. D. G. Archer and P. Wang, The dielectric constant of water and Debye-Hückel limiting law slopes, *J. Phys. Chem. Ref. Data*, 19, 371, 1990.
5. N. B. Vargaftik, et al., International tables of the surface tension of water, *J. Phys. Chem. Ref. Data*, 12, 817, 1983.

t °C	Density g/cm³	C_p J/g K	Vap. pres. kPa	Visc. µPa s	Ther. cond. mW/K m	Diel. const.	Surf. ten. mN/m
0	0.99984	4.2176	0.6113	1793	561.0	87.90	75.64
10	0.99970	4.1921	1.2281	1307	580.0	83.96	74.23
20	0.99821	4.1818	2.3388	1002	598.4	80.20	72.75
30	0.99565	4.1784	4.2455	797.7	615.4	76.60	71.20
40	0.99222	4.1785	7.3814	653.2	630.5	73.17	69.60
50	0.98803	4.1806	12.344	547.0	643.5	69.88	67.94
60	0.98320	4.1843	19.932	466.5	654.3	66.73	66.24
70	0.97778	4.1895	31.176	404.0	663.1	63.73	64.47
80	0.97182	4.1963	47.373	354.4	670.0	60.86	62.67
90	0.96535	4.2050	70.117	314.5	675.3	58.12	60.82
100	0.95840	4.2159	101.325	281.8	679.1	55.51	58.91
Ref.	1–3	2	1, 3	3	3	4	5

ENTHALPY OF VAPORIZATION OF WATER

The enthalpy (heat) of vaporization of water is tabulated as a function of temperature on the IPTS-68 scale.

Reference

Marsh, K. N., Ed., *Recommended Reference Materials for the Realization of Physicochemical Properties*, Blackwell, Oxford, 1987.

t °C	$\Delta_{vap}H$ kJ/mol	t °C	$\Delta_{vap}H$ kJ/mol
0	45.054	200	34.962
25	43.990	220	33.468
40	43.350	240	31.809
60	42.482	260	29.930
80	41.585	280	27.795
100	40.657	300	25.300
120	39.684	320	22.297
140	38.643	340	18.502
160	37.518	360	12.966
180	36.304	374	2.066

FIXED POINT PROPERTIES OF H₂O AND D₂O

Temperatures are given on the IPTS-68 scale.

References

1. Haar, L., Gallagher, J. S., and Kell, G. S., *NBS/NRC Steam Tables*, Hemisphere Publishing Corp., New York, 1984.
2. Levelt Sengers, J. M. H., Straub, J., Watanabe, K., and Hill, P. G., Assessment of critical parameter values for H_2O and D_2O, *J. Phys. Chem. Ref. Data*, 14, 193, 1985.

3. Kestin, J. et. al., Thermophysical properties of fluid D_2O, *J. Phys. Chem. Ref. Data*, 13, 601, 1984.
4. Kestin, J. et. al., Thermophysical properties of fluid H_2O, *J. Phys. Chem. Ref. Data*, 13, 175, 1984.
5. Hill, P. G., MacMillan, R. D. C., and Lee, V., A fundamental equation of state for heavy water, *J. Phys. Chem. Ref. Data*, 11, 1, 1982.

	Unit	H₂O	D₂O
Molar mass	g/mol	18.01528	20.02748
Melting point(101.325 kPa)	°C	0.00	3.82
Boiling point(101.325 kPa)	°C	100.00	101.42
Triple point temperature	°C	0.01	3.82
Triple point pressure	Pa	611.73	661
Triple point density(l)	g/cm³	0.99978	1.1055
Triple point density(g)	mg/L	4.885	5.75
Critical temperature	°C	373.99	370.74
Critical pressure	MPa	22.064	21.671
Critical density	g/cm³	0.322	0.356
Critical specific volume	cm³/g	3.11	2.81
Maximum density(saturated liquid)	g/cm³	0.99995	1.1053
Temperature of maximum density	°C	4.0	11.2

THERMAL CONDUCTIVITY OF SATURATED H₂O AND D₂O

This table gives the thermal conductivity λ for water (H_2O or D_2O) in equilibrium with its vapor. Values for the liquid (λ_l) and vapor (λ_v) are listed, as well as the vapor pressure.

References

1. Sengers, J.V. and Watson, J.T.R., Improved international formulations for the viscosity and thermal conductivity of water substance, *J. Phys. Chem. Ref. Data*, 15, 1291, 1986.
2. Matsunaga, N. and Nagashima, A., Transport properties of liquid and gaseous D_2O over a wide range of temperature and pressure, *J. Phys. Chem. Ref. Data*, 12, 933, 1983.

	H₂O			D₂O		
t/°C	P/kPa	λ_l/(mW/K m)	λ_v/(mW/K m)	P/kPa	λ_l/(mW/K m)	λ_v/(mW/K m)
0	0.6	561.0	16.49			
10	1.2	580.0	17.21	1.0	575	17.0
20	2.3	598.4	17.95	2.0	589	17.8
30	4.2	615.4	18.70	3.7	600	18.5
40	7.4	630.5	19.48	6.5	610	19.3
50	12.3	643.5	20.28	11.1	618	20.2
60	19.9	654.3	21.10	18.2	625	21.0
70	31.2	663.1	21.96	28.8	629	21.9
80	47.4	670.0	22.86	44.2	633	22.8
90	70.1	675.3	23.80	66.1	635	23.8
100	101.3	679.1	24.79	96.2	636	24.8
150	476	682.1	30.77	465	625	30.8
200	1555	663.4	39.10	1546	592	39.0
250	3978	621.4	51.18	3995	541	52.0
300	8593	547.7	71.78	8688	473	75.2
350	16530	447.6	134.59	16820	391	143.0

STANDARD DENSITY OF WATER

This table gives the density ρ of water in the temperature range from 0°C to 100°C at a pressure of 101325 Pa (one standard atmosphere). From 0°C to 40°C the values are taken from the IUPAC publication in Reference 1 and refer to standard mean ocean water (SMOW), free from dissolved salts and gases. SMOW is a standard water sample of high purity and known isotopic composition. Methods of correcting for different isotopic compositions are discussed in Reference 1. The remaining values are taken from the NIST Chemistry WebBook, Reference 2.

Note that the IUPAC values refer to the IPTS-68 temperature scale, while the NIST values are based on the ITS-90 scale (where the normal boiling point is 99.974°C). The conversion between these scales can be found in Sec. 1. The difference between the scales leads to a difference in the density of water of about 20 ppm in the neighborhood of 100°C and much less at lower temperatures.

References

1. Marsh, K. N., Ed., *Recommended Reference Materials for the Realization of Physicochemical Properties*, Blackwell Scientific Publications, Oxford, 1987.
2. Lemmon, E.W., McLinden, M.O., and Friend, D.G., "Thermophysical Properties of Fluid Systems" in NIST Chemistry WebBook, NIST Standard Reference Database Number 69, Eds. P.J. Linstrom and W.G. Mallard, June 2005, National Institute of Standards and Technology, Gaithersburg MD, 20899 (http://webbook.nist.gov).
3. Wagner, W., and Pruss, A., "The IAPWS formulation 1995 for the thermodynamic properties of ordinary water substance for general and scientific use", J. Phys. Chem. Ref. Data 31, 387-535, 2002.
4. Saul, A., and Wagner, W., "A Fundamental Equation for Water Covering the Range From the Melting Line to 1273 K at Pressures up to 25000 MPa", J. Phys. Chem. Ref. Data 18, 1537-1564, 1989.

$t/°C$	$\rho/\text{g cm}^{-3}$	$t/°C$	$\rho/\text{g cm}^{-3}$	$t/°C$	$\rho/\text{g cm}^{-3}$	$t/°C$	$\rho/\text{g cm}^{-3}$	$t/°C$	$\rho/\text{g cm}^{-3}$
0.1	0.9998493	4.3	0.9999742	8.5	0.9998189	12.7	0.9994167	16.9	0.9987942
0.2	0.9998558	4.4	0.9999736	8.6	0.9998121	12.8	0.9994043	17.0	0.9987769
0.3	0.9998622	4.5	0.9999728	8.7	0.9998051	12.9	0.9993918	17.1	0.9987595
0.4	0.9998683	4.6	0.9999719	8.8	0.9997980	13.0	0.9993792	17.2	0.9987419
0.5	0.9998743	4.7	0.9999709	8.9	0.9997908	13.1	0.9993665	17.3	0.9987243
0.6	0.9998801	4.8	0.9999696	9.0	0.9997834	13.2	0.9993536	17.4	0.9987065
0.7	0.9998857	4.9	0.9999683	9.1	0.9997759	13.3	0.9993407	17.5	0.9986886
0.8	0.9998912	5.0	0.9999668	9.2	0.9997682	13.4	0.9993276	17.6	0.9986706
0.9	0.9998964	5.1	0.9999651	9.3	0.9997604	13.5	0.9993143	17.7	0.9986525
1.0	0.9999015	5.2	0.9999632	9.4	0.9997525	13.6	0.9993010	17.8	0.9986343
1.1	0.9999065	5.3	0.9999612	9.5	0.9997444	13.7	0.9992875	17.9	0.9986160
1.2	0.9999112	5.4	0.9999591	9.6	0.9997362	13.8	0.9992740	18.0	0.9985976
1.3	0.9999158	5.5	0.9999568	9.7	0.9997279	13.9	0.9992602	18.1	0.9985790
1.4	0.9999202	5.6	0.9999544	9.8	0.9997194	14.0	0.9992464	18.2	0.9985604
1.5	0.9999244	5.7	0.9999518	9.9	0.9997108	14.1	0.9992325	18.3	0.9985416
1.6	0.9999284	5.8	0.9999490	10.0	0.9997021	14.2	0.9992184	18.4	0.9985228
1.7	0.9999323	5.9	0.9999461	10.1	0.9996932	14.3	0.9992042	18.5	0.9985038
1.8	0.9999360	6.0	0.9999430	10.2	0.9996842	14.4	0.9991899	18.6	0.9984847
1.9	0.9999395	6.1	0.9999398	10.3	0.9996751	14.5	0.9991755	18.7	0.9984655
2.0	0.9999429	6.2	0.9999365	10.4	0.9996658	14.6	0.9991609	18.8	0.9984462
2.1	0.9999461	6.3	0.9999330	10.5	0.9996564	14.7	0.9991463	18.9	0.9984268
2.2	0.9999491	6.4	0.9999293	10.6	0.9996468	14.8	0.9991315	19.0	0.9984073
2.3	0.9999519	6.5	0.9999255	10.7	0.9996372	14.9	0.9991166	19.1	0.9983877
2.4	0.9999546	6.6	0.9999216	10.8	0.9996274	15.0	0.9991016	19.2	0.9983680
2.5	0.9999571	6.7	0.9999175	10.9	0.9996174	15.1	0.9990864	19.3	0.9983481
2.6	0.9999595	6.8	0.9999132	11.0	0.9996074	15.2	0.9990712	19.4	0.9983282
2.7	0.9999616	6.9	0.9999088	11.1	0.9995972	15.3	0.9990558	19.5	0.9983081
2.8	0.9999636	7.0	0.9999043	11.2	0.9995869	15.4	0.9990403	19.6	0.9982880
2.9	0.9999655	7.1	0.9998996	11.3	0.9995764	15.5	0.9990247	19.7	0.9982677
3.0	0.9999672	7.2	0.9998948	11.4	0.9995658	15.6	0.9990090	19.8	0.9982474
3.1	0.9999687	7.3	0.9998898	11.5	0.9995551	15.7	0.9989932	19.9	0.9982269
3.2	0.9999700	7.4	0.9998847	11.6	0.9995443	15.8	0.9989772	20.0	0.9982063
3.3	0.9999712	7.5	0.9998794	11.7	0.9995333	15.9	0.9989612	20.1	0.9981856
3.4	0.9999722	7.6	0.9998740	11.8	0.9995222	16.0	0.9989450	20.2	0.9981649
3.5	0.9999731	7.7	0.9998684	11.9	0.9995110	16.1	0.9989287	20.3	0.9981440
3.6	0.9999738	7.8	0.9998627	12.0	0.9994996	16.2	0.9989123	20.4	0.9981230
3.7	0.9999743	7.9	0.9998569	12.1	0.9994882	16.3	0.9988957	20.5	0.9981019
3.8	0.9999747	8.0	0.9998509	12.2	0.9994766	16.4	0.9988791	20.6	0.9980807
3.9	0.9999749	8.1	0.9998448	12.3	0.9994648	16.5	0.9988623	20.7	0.9980594
4.0	0.9999750	8.2	0.9998385	12.4	0.9994530	16.6	0.9988455	20.8	0.9980380
4.1	0.9999748	8.3	0.9998321	12.5	0.9994410	16.7	0.9988285	20.9	0.9980164
4.2	0.9999746	8.4	0.9998256	12.6	0.9994289	16.8	0.9988114	21.0	0.9979948

$t/°C$	$\rho/\text{g cm}^{-3}$	$t/°C$	$\rho/\text{g cm}^{-3}$	$t/°C$	$\rho/\text{g cm}^{-3}$	$t/°C$	$\rho/\text{g cm}^{-3}$	$t/°C$	$\rho/\text{g cm}^{-3}$
21.1	0.9979731	26.1	0.9967604	31.1	0.9953139	36.1	0.9936531	51.0	0.98758
21.2	0.9979513	26.2	0.9967337	31.2	0.9952827	36.2	0.9936178	52.0	0.98712
21.3	0.9979294	26.3	0.9967069	31.3	0.9952514	36.3	0.9935825	53.0	0.98665
21.4	0.9979073	26.4	0.9966800	31.4	0.9952201	36.4	0.9935470	54.0	0.98617
21.5	0.9978852	26.5	0.9966530	31.5	0.9951887	36.5	0.9935115	55.0	0.98569
21.6	0.9978630	26.6	0.9966259	31.6	0.9951572	36.6	0.9934759	56.0	0.98521
21.7	0.9978406	26.7	0.9965987	31.7	0.9951255	36.7	0.9934403	57.0	0.98471
21.8	0.9978182	26.8	0.9965714	31.8	0.9950939	36.8	0.9934045	58.0	0.98421
21.9	0.9977957	26.9	0.9965441	31.9	0.9950621	36.9	0.9933687	59.0	0.98371
22.0	0.9977730	27.0	0.9965166	32.0	0.9950302	37.0	0.9933328	60.0	0.98320
22.1	0.9977503	27.1	0.9964891	32.1	0.9949983	37.1	0.9932968	61.0	0.98268
22.2	0.9977275	27.2	0.9964615	32.2	0.9949663	37.2	0.9932607	62.0	0.98216
22.3	0.9977045	27.3	0.9964337	32.3	0.9949342	37.3	0.9932246	63.0	0.98163
22.4	0.9976815	27.4	0.9964059	32.4	0.9949020	37.4	0.9931884	64.0	0.98109
22.5	0.9976584	27.5	0.9963780	32.5	0.9948697	37.5	0.9931521	65.0	0.98055
22.6	0.9976351	27.6	0.9963500	32.6	0.9948373	37.6	0.9931157	66.0	0.98000
22.7	0.9976118	27.7	0.9963219	32.7	0.9948049	37.7	0.9930793	67.0	0.97945
22.8	0.9975883	27.8	0.9962938	32.8	0.9947724	37.8	0.9930428	68.0	0.97890
22.9	0.9975648	27.9	0.9962655	32.9	0.9947397	37.9	0.9930062	69.0	0.97833
23.0	0.9975412	28.0	0.9962371	33.0	0.9947071	38.0	0.9929695	70.0	0.97776
23.1	0.9975174	28.1	0.9962087	33.1	0.9946743	38.1	0.9929328	71.0	0.97719
23.2	0.9974936	28.2	0.9961801	33.2	0.9946414	38.2	0.9928960	72.0	0.97661
23.3	0.9974697	28.3	0.9961515	33.3	0.9946085	38.3	0.9928591	73.0	0.97603
23.4	0.9974456	28.4	0.9961228	33.4	0.9945755	38.4	0.9928221	74.0	0.97544
23.5	0.9974215	28.5	0.9960940	33.5	0.9945423	38.5	0.9927850	75.0	0.97484
23.6	0.9973973	28.6	0.9960651	33.6	0.9945092	38.6	0.9927479	76.0	0.97424
23.7	0.9973730	28.7	0.9960361	33.7	0.9944759	38.7	0.9927107	77.0	0.97364
23.8	0.9973485	28.8	0.9960070	33.8	0.9944425	38.8	0.9926735	78.0	0.97303
23.9	0.9973240	28.9	0.9959778	33.9	0.9944091	38.9	0.9926361	79.0	0.97241
24.0	0.9972994	29.0	0.9959486	34.0	0.9943756	39.0	0.9925987	80.0	0.97179
24.1	0.9972747	29.1	0.9959192	34.1	0.9943420	39.1	0.9925612	81.0	0.97116
24.2	0.9972499	29.2	0.9958898	34.2	0.9943083	39.2	0.9925236	82.0	0.97053
24.3	0.9972250	29.3	0.9958603	34.3	0.9942745	39.3	0.9924860	83.0	0.96990
24.4	0.9972000	29.4	0.9958306	34.4	0.9942407	39.4	0.9924483	84.0	0.96926
24.5	0.9971749	29.5	0.9958009	34.5	0.9942068	39.5	0.9924105	85.0	0.96861
24.6	0.9971497	29.6	0.9957712	34.6	0.9941728	39.6	0.9923726	86.0	0.96796
24.7	0.9971244	29.7	0.9957413	34.7	0.9941387	39.7	0.9923347	87.0	0.96731
24.8	0.9970990	29.8	0.9957113	34.8	0.9941045	39.8	0.9922966	88.0	0.96664
24.9	0.9970735	29.9	0.9956813	34.9	0.9940703	39.9	0.9922586	89.0	0.96598
25.0	0.9970480	30.0	0.9956511	35.0	0.9940359	40.0	0.9922204	90.0	0.96531
25.1	0.9970223	30.1	0.9956209	35.1	0.9940015	41.0	0.99183	91.0	0.96463
25.2	0.9969965	30.2	0.9955906	35.2	0.9939671	42.0	0.99144	92.0	0.96396
25.3	0.9969707	30.3	0.9955602	35.3	0.9939325	43.0	0.99104	93.0	0.96327
25.4	0.9969447	30.4	0.9955297	35.4	0.9938978	44.0	0.99063	94.0	0.96258
25.5	0.9969186	30.5	0.9954991	35.5	0.9938631	45.0	0.99021	95.0	0.96189
25.6	0.9968925	30.6	0.9954685	35.6	0.9938283	46.0	0.98979	96.0	0.96119
25.7	0.9968663	30.7	0.9954377	35.7	0.9937934	47.0	0.98936	97.0	0.96049
25.8	0.9968399	30.8	0.9954069	35.8	0.9937585	48.0	0.98893	98.0	0.95978
25.9	0.9968135	30.9	0.9953760	35.9	0.9937234	49.0	0.98848	99.0	0.95907
26.0	0.9967870	31.0	0.9953450	36.0	0.9936883	50.0	0.98804	99.974	0.95837

PROPERTIES OF ICE AND SUPERCOOLED WATER

The common form of ice at ambient temperature and pressure is hexagonal ice, designated as ice I_h (see phase diagram in Section 12). The data given here refer to that form. Data have been taken from the references indicated; values have been interpolated and smoothed in some cases. All properties are sensitive to the method of preparation of the sample, since air or other gases are sometimes occluded. For this reason there is often disagreement among values found in the literature.

Density values (except at 0°C) and the thermal expansion coefficient were calculated from the temperature variation in the crystal lattice constants of ice (see Ref. 1). The thermal expansion coefficient appears to become negative around −200°C, but there is considerable scatter in the data.

Density of ice I_h and supercooled water in g cm^{-3}

$t/°C$	ρ (ice)	ρ (supercooled water)
0	0.9167	0.9998
−10	0.9187	0.9982
−20	0.9203	0.9935
−30	0.9216	0.9839
−40	0.9228	
−50	0.9240	
−60	0.9252	
−80	0.9274	
−100	0.9292	
−120	0.9305	
−140	0.9314	
−160	0.9331	
−180	0.9340	
Ref.	1	8

Phase transition properties:

$\Delta_{fus}H(0°C) = 333.6$ J/g (Ref. 2)
$\Delta_{sub}H(0°C) = 2838$ J/g (Ref. 2)

Other properties of ice I_h:

α_V: cubic thermal expansion coefficient, $\alpha_V = -(1/V)(\partial V/\partial t)_p$
κ: adiabatic compressibility, $\kappa = -(1/V)(\partial V/\partial p)_s$
ε: relative permittivity (dielectric constant)
k: thermal conductivity
c_p: specific heat capacity at constant pressure

$t/°C$	$\alpha_V/10^{-6}\,°C^{-1}$	$\kappa/10^{-5}\,MPa^{-1}$	ε	$k/W\,cm^{-1}\,°C^{-1}$	$c_p/J\,g^{-1}\,°C^{-1}$
0	159	13.0	91.6	0.0214	2.11
−10	155	12.8	94.4	0.023	2.03
−20	149	12.7	97.5	0.024	1.96
−30	143	12.5	99.7	0.025	1.88
−40	137	12.4	101.9	0.026	1.80
−50	130	12.2	106.9	0.028	1.72
−60	122	12.1	119.5	0.030	1.65
−80	105	11.9		0.033	1.50
−100	85	11.6		0.037	1.36
−120	77	11.4		0.042	1.23
−140	60	11.3		0.049	1.10
−160	45	11.2		0.057	0.97
−180	30	11.1		0.070	0.83
−200		11.0		0.087	0.67
−220		10.9		0.118	0.50
−240		10.9		0.20	0.29
−250		10.9		0.32	0.17
Ref.	1,2,3,5	1,5	6	7	1

References

1. Eisenberg, D., and Kauzmann, W., *The Structure and Properties of Water*, Oxford University Press, Oxford, 1969.
2. *Landolt-Börnstein, Numerical Data and Functional Relationships in Science and Technology, New Series*, V/1b, Springer-Verlag, Heidelberg, 1982.
3. LaPlaca, S., and Post, B., *Acta Cryst.*, 13, 503, 1960. [Thermal expansion of lattice]
4. Brill, R., and Tippe, A., *Acta Cryst.*, 23, 343, 1967. [Thermal expansion of lattice]
5. Leadbetter, A. J., *Proc. Roy. Soc. A* 287, 403, 1965. [Compressibility and thermal expansion]
6. Auty, R. P., and Cole, R. H., *J. Chem. Phys.*, 20, 1309, 1952. [Dielectric constant]
7. Slack, G. A., *Phys. Rev. B*, 22, 3065, 1980. [Thermal conductivity]
8. Hare, D. E., and Sorensen, C. M., *J. Chem. Phys.*, 87, 4840, 1987. [Supercooled water]
9. Hobbs, P. V., *Ice Physics*, Clarendon Press, Oxford, 1974.

VOLUMETRIC PROPERTIES OF AQUEOUS SODIUM CHLORIDE SOLUTIONS

This table gives the following properties of aqueous solutions of NaCl as a function of temperature and concentration:

Specific volume v (reciprocal of density) in cm³/g
Isothermal compressibility $\kappa_T = -(1/v)(\partial v/\partial P)_T$ in GPa⁻¹
Cubic expansion coefficient $\alpha_v = (1/v)(\partial v/\partial T)_p$ in kK⁻¹

All data refer to a pressure of 100 kPa (1 bar). The reference gives properties over a wider range of temperature and pressure.

Reference

Rogers, P. S. Z., and Pitzer, K. S., *J. Phys. Chem. Ref. Data*, 11, 15, 1982.

t/°C	Molality in mol/kg								
	0.100	0.250	0.500	0.750	1.000	2.000	3.000	4.000	5.000
Specific volume v in cm³/g									
0	0.995732	0.989259	0.978889	0.968991	0.959525	0.925426	0.896292	0.870996	0.848646
10	0.995998	0.989781	0.979804	0.970256	0.961101	0.927905	0.899262	0.874201	0.851958
20	0.997620	0.991564	0.981833	0.972505	0.963544	0.930909	0.902565	0.877643	0.855469
25	0.998834	0.992832	0.983185	0.973932	0.965038	0.932590	0.904339	0.879457	0.857301
30	1.000279	0.994319	0.984735	0.975539	0.966694	0.934382	0.906194	0.881334	0.859185
40	1.003796	0.997883	0.988374	0.979243	0.970455	0.938287	0.910145	0.885276	0.863108
50	1.008064	1.002161	0.992668	0.983551	0.974772	0.942603	0.914411	0.889473	0.867241
60	1.0130	1.0071	0.9976	0.9885	0.9797	0.9474	0.9191	0.8940	0.8716
70	1.0186	1.0127	1.0031	0.9939	0.9851	0.9526	0.9240	0.8987	0.8762
80	1.0249	1.0188	1.0092	0.9999	0.9909	0.9581	0.9293	0.9037	0.8809
90	1.0317	1.0256	1.0157	1.0063	0.9972	0.9640	0.9348	0.9089	0.8858
100	1.0391	1.0329	1.0228	1.0133	1.0040	0.9703	0.9406	0.9144	0.8910
Compressibility κ_T in GPa⁻¹									
0	0.503	0.492	0.475	0.459	0.443	0.389	0.346	0.315	0.294
10	0.472	0.463	0.449	0.436	0.423	0.377	0.341	0.313	0.294
20	0.453	0.446	0.433	0.422	0.411	0.371	0.338	0.313	0.294
25	0.447	0.440	0.428	0.417	0.407	0.369	0.337	0.313	0.294
30	0.443	0.436	0.425	0.414	0.404	0.367	0.337	0.313	0.294
40	0.438	0.432	0.421	0.411	0.401	0.367	0.338	0.315	0.296
50	0.438	0.431	0.421	0.411	0.402	0.369	0.340	0.317	0.299
60	0.44	0.44	0.43	0.42	0.41	0.38	0.35	0.32	0.30
70	0.45	0.44	0.43	0.42	0.42	0.38	0.36	0.33	0.31
80	0.46	0.45	0.44	0.43	0.43	0.39	0.37	0.34	0.32
90	0.47	0.47	0.46	0.45	0.44	0.41	0.38	0.35	0.33
100	0.49	0.48	0.47	0.46	0.45	0.42	0.39	0.37	0.34
Cubic expansion coefficient α_v in kK⁻¹									
0	−0.058	−0.026	0.024	0.069	0.110	0.237	0.313	0.355	
10	0.102	0.123	0.156	0.186	0.213	0.297	0.349	0.380	
20	0.218	0.232	0.254	0.274	0.292	0.349	0.384	0.406	
25	0.267	0.278	0.296	0.312	0.327	0.373	0.401	0.420	
30	0.311	0.320	0.334	0.347	0.359	0.395	0.418	0.433	
40	0.389	0.394	0.402	0.410	0.417	0.438	0.451	0.460	
50	0.458	0.460	0.464	0.467	0.470	0.479	0.484	0.486	
60	0.52	0.52	0.52	0.52	0.52	0.52	0.52	0.52	
70	0.58	0.58	0.58	0.57	0.57	0.56	0.55	0.54	
80	0.64	0.63	0.63	0.62	0.61	0.60	0.58	0.56	
90	0.69	0.68	0.67	0.67	0.66	0.63	0.61	0.59	
100	0.74	0.73	0.72	0.71	0.70	0.66	0.64	0.61	

DENSITY OF D$_2$O

Density of liquid D$_2$O in g/cm^3 at a pressure of 100 kPa (1 bar).

Reference

Kirillin, V.A., Ed., *Heavy Water: Thermophysical Properties*, Gosudarstvennoe Energeticheskoe Izdatel'stvo, Moscow, 1963.

t/°C	3.8	5	10	15	20	25	30
Density	1.1053	1.1055	1.1057	1.1056	1.105	1.1044	1.1034

t/°C	35	40	45	50	55	60	65
Density	1.1019	1.1001	1.0979	1.0957	1.0931	1.0905	1.0875

t/°C	70	75	80	85	90	95	100
Density	1.0847	1.0815	1.0783	1.0748	1.0712	1.0673	1.0635

VAPOR PRESSURE OF ICE

The values of the vapor (sublimation) pressure of ice in this table were calculated from the equation recommended by the International Association for the Properties of Steam (IAPS) in 1993. Temperature values correspond to the ITS-90 temperature scale. The uncertainty in the pressure is estimated to be 0.1% for $t > -25$°C and 0.5% for $t < -25$°C. The first entry in the table is the triple point of water.

Reference

Wagner, W., Saul, A., and Pruss, A., *J. Phys. Chem. Ref. Data*, 23, 515, 1994.

t/°C	p/Pa	t/°C	p/Pa	t/°C	p/Pa
0.01	611.657	−16	150.68	−33	27.71
0	611.15	−17	137.25	−34	24.90
−1	562.67	−18	124.92	−35	22.35
−2	517.72	−19	113.62	−36	20.04
−3	476.06	−20	103.26	−37	17.96
−4	437.47	−21	93.77	−38	16.07
−5	401.76	−22	85.10	−39	14.37
−6	368.73	−23	77.16	−40	12.84
−7	338.19	−24	69.91	−45	7.202
−8	309.98	−25	63.29	−50	3.936
−9	283.94	−26	57.25	−55	2.093
−10	259.90	−27	51.74	−60	1.080
−11	237.74	−28	46.73	−65	0.540
−12	217.32	−29	42.16	−70	0.261
−13	198.52	−30	38.01	−75	0.122
−14	181.22	−31	34.24	−80	0.055
−15	165.30	−32	30.82		

VAPOR PRESSURE OF WATER FROM 0 TO 370°C

This table gives the vapor pressure of water at intervals of 1° C from the melting point to the critical point.

Reference

Haar, L., Gallagher, J. S., and Kell, G. S., *NBS/NRC Steam Tables*, Hemisphere Publishing Corp., New York, 1984.

t/°C	P/kPa	t/°C	P/kPa	t/°C	P/kPa	t/°C	P/kPa
0	0.61129	52	13.623	104	116.67	156	557.32
1	0.65716	53	14.303	105	120.79	157	571.94
2	0.70605	54	15.012	106	125.03	158	586.87
3	0.75813	55	15.752	107	129.39	159	602.11
4	0.81359	56	16.522	108	133.88	160	617.66
5	0.87260	57	17.324	109	138.50	161	633.53
6	0.93537	58	18.159	110	143.24	162	649.73
7	1.0021	59	19.028	111	148.12	163	666.25
8	1.0730	60	19.932	112	153.13	164	683.10
9	1.1482	61	20.873	113	158.29	165	700.29
10	1.2281	62	21.851	114	163.58	166	717.83
11	1.3129	63	22.868	115	169.02	167	735.70
12	1.4027	64	23.925	116	174.61	168	753.94
13	1.4979	65	25.022	117	180.34	169	772.52
14	1.5988	66	26.163	118	186.23	170	791.47
15	1.7056	67	27.347	119	192.28	171	810.78
16	1.8185	68	28.576	120	198.48	172	830.47
17	1.9380	69	29.852	121	204.85	173	850.53
18	2.0644	70	31.176	122	211.38	174	870.98
19	2.1978	71	32.549	123	218.09	175	891.80
20	2.3388	72	33.972	124	224.96	176	913.03
21	2.4877	73	35.448	125	232.01	177	934.64
22	2.6447	74	36.978	126	239.24	178	956.66
23	2.8104	75	38.563	127	246.66	179	979.09
24	2.9850	76	40.205	128	254.25	180	1001.9
25	3.1690	77	41.905	129	262.04	181	1025.2
26	3.3629	78	43.665	130	270.02	182	1048.9
27	3.5670	79	45.487	131	278.20	183	1073.0
28	3.7818	80	47.373	132	286.57	184	1097.5
29	4.0078	81	49.324	133	295.15	185	1122.5
30	4.2455	82	51.342	134	303.93	186	1147.9
31	4.4953	83	53.428	135	312.93	187	1173.8
32	4.7578	84	55.585	136	322.14	188	1200.1
33	5.0335	85	57.815	137	331.57	189	1226.9
34	5.3229	86	60.119	138	341.22	190	1254.2
35	5.6267	87	62.499	139	351.09	191	1281.9
36	5.9453	88	64.958	140	361.19	192	1310.1
37	6.2795	89	67.496	141	371.53	193	1338.8
38	6.6298	90	70.117	142	382.11	194	1368.0
39	6.9969	91	72.823	143	392.92	195	1397.6
40	7.3814	92	75.614	144	403.98	196	1427.8
41	7.7840	93	78.494	145	415.29	197	1458.5
42	8.2054	94	81.465	146	426.85	198	1489.7
43	8.6463	95	84.529	147	438.67	199	1521.4
44	9.1075	96	87.688	148	450.75	200	1553.6
45	9.5898	97	90.945	149	463.10	201	1586.4
46	10.094	98	94.301	150	475.72	202	1619.7
47	10.620	99	97.759	151	488.61	203	1653.6
48	11.171	100	101.32	152	501.78	204	1688.0
49	11.745	101	104.99	153	515.23	205	1722.9
50	12.344	102	108.77	154	528.96	206	1758.4
51	12.970	103	112.66	155	542.99	207	1794.5

t/°C	P/kPa	t/°C	P/kPa	t/°C	P/kPa	t/°C	P/kPa
208	1831.1	250	3973.6	292	7657.2	334	13528
209	1868.4	251	4041.2	293	7768.6	335	13701
210	1906.2	252	4109.6	294	7881.3	336	13876
211	1944.6	253	4178.9	295	7995.2	337	14053
212	1983.6	254	4249.1	296	8110.3	338	14232
213	2023.2	255	4320.2	297	8226.8	339	14412
214	2063.4	256	4392.2	298	8344.5	340	14594
215	2104.2	257	4465.1	299	8463.5	341	14778
216	2145.7	258	4539.0	300	8583.8	342	14964
217	2187.8	259	4613.7	301	8705.4	343	15152
218	2230.5	260	4689.4	302	8828.3	344	15342
219	2273.8	261	4766.1	303	8952.6	345	15533
220	2317.8	262	4843.7	304	9078.2	346	15727
221	2362.5	263	4922.3	305	9205.1	347	15922
222	2407.8	264	5001.8	306	9333.4	348	16120
223	2453.8	265	5082.3	307	9463.1	349	16320
224	2500.5	266	5163.8	308	9594.2	350	16521
225	2547.9	267	5246.3	309	9726.7	351	16725
226	2595.9	268	5329.8	310	9860.5	352	16931
227	2644.6	269	5414.3	311	9995.8	353	17138
228	2694.1	270	5499.9	312	10133	354	17348
229	2744.2	271	5586.4	313	10271	355	17561
230	2795.1	272	5674.0	314	10410	356	17775
231	2846.7	273	5762.7	315	10551	357	17992
232	2899.0	274	5852.4	316	10694	358	18211
233	2952.1	275	5943.1	317	10838	359	18432
234	3005.9	276	6035.0	318	10984	360	18655
235	3060.4	277	6127.9	319	11131	361	18881
236	3115.7	278	6221.9	320	11279	362	19110
237	3171.8	279	6317.0	321	11429	363	19340
238	3228.6	280	6413.2	322	11581	364	19574
239	3286.3	281	6510.5	323	11734	365	19809
240	3344.7	282	6608.9	324	11889	366	20048
241	3403.9	283	6708.5	325	12046	367	20289
242	3463.9	284	6809.2	326	12204	368	20533
243	3524.7	285	6911.1	327	12364	369	20780
244	3586.3	286	7014.1	328	12525	370	21030
245	3648.8	287	7118.3	329	12688	371	21283
246	3712.1	288	7223.7	330	12852	372	21539
247	3776.2	289	7330.2	331	13019	373	21799
248	3841.2	290	7438.0	332	13187	373.98	22055
249	3907.0	291	7547.0	333	13357		

BOILING POINT OF WATER AT VARIOUS PRESSURES

Data are based on the equation of state recommended by the International Association for the Properties of Steam in 1984, as presented in Haar, Gallagher, and Kell, *NBS-NRC Steam Tables* (Hemisphere Publishing Corp., New York, 1984). The temperature scale is IPTS-68.

Note that: 1 mbar = 100 Pa = 0.000986923 atmos = 0.750062 mmHg.

P/mbar	T/°C	P/mbar	T/°C	P/mbar	T/°C	P/mbar	T/°C
50	32.88	915	97.17	1013.25	100.00	1200	104.81
100	45.82	920	97.32	1015	100.05	1250	105.99
150	53.98	925	97.47	1020	100.19	1300	107.14
200	60.07	930	97.62	1025	100.32	1350	108.25
250	64.98	935	97.76	1030	100.46	1400	109.32
300	69.11	940	97.91	1035	100.60	1450	110.36
350	72.70	945	98.06	1040	100.73	1500	111.38
400	75.88	950	98.21	1045	100.87	1550	112.37
450	78.74	955	98.35	1050	101.00	1600	113.33
500	81.34	960	98.50	1055	101.14	1650	114.26
550	83.73	965	98.64	1060	101.27	1700	115.18
600	85.95	970	98.78	1065	101.40	1750	116.07
650	88.02	975	98.93	1070	101.54	1800	116.94
700	89.96	980	99.07	1075	101.67	1850	117.79
750	91.78	985	99.21	1080	101.80	1900	118.63
800	93.51	990	99.35	1085	101.93	1950	119.44
850	95.15	995	99.49	1090	102.06	2000	120.24
900	96.71	1000	99.63	1095	102.19	2050	121.02
905	96.87	1005	99.77	1100	102.32	2100	121.79
910	97.02	1010	99.91	1150	103.59	2150	122.54

MELTING POINT OF ICE AS A FUNCTION OF PRESSURE

This table gives values of the melting temperature of ice at various pressures, as calculated from the equation for the ice I - liquid water phase boundary recommended by the International Association for the Properties of Steam (IAPS). Temperatures are on the ITS-90 scale. See the Reference for information on forms of ice that exist at higher pressures. The transition points for transformations of the various forms of ice (in each case in equilibrium with liquid water) are:

ice I - ice III	209.9 MPa	−21.985°C
ice III - ice V	350.1	−16.986
ice V - ice VI	632.4	0.16
ice VI - ice VII	2216	82

Reference

Wagner, W., Saul, A., and Pruss, A., *J. Phys. Chem. Ref. Data*, 23, 515, 1994.

p/MPa	t/°C	p/MPa	t/°C	p/MPa	t/°C
0.1	0.00	40	−3.15	130	−12.07
1	−0.06	50	−4.02	140	−13.22
2	−0.14	60	−4.91	150	−14.40
3	−0.21	70	−5.83	160	−15.62
4	−0.29	80	−6.79	170	−16.85
5	−0.36	90	−7.78	180	−18.11
10	−0.74	100	−8.80	190	−19.39
20	−1.52	110	−9.86	200	−20.69
30	−2.32	120	−10.95	210	−22.00

PROPERTIES OF WATER AND STEAM AS A FUNCTION OF TEMPERATURE AND PRESSURE

This table gives properties of compressed water and superheated steam at selected pressures and temperatures. The properties included are density ρ, enthalpy H, entropy S, heat capacity at constant pressure C_p, and static dielectric constant (relative permittivity). The table was generated from the formulation approved by the International Association for the Properties of Water and Steam for general and scientific use. The reference state for this table is the liquid at the triple point, at which the internal energy and entropy are taken as zero. A duplicate entry in the temperature column indicates a phase transition (liquid–vapor) at that temperature; property values are then given for both phases. In the 100 MPa section of the table, an entry is given at the critical temperature, 647.10 K. Temperatures refer to the ITS-90 scale, on which the normal boiling point of water is 373.12 K (99.97°C).

References

1. Release on the IAPWS Formulation 1995 for the Thermodynamic Properties of Ordinary Water Substance for General and Scientific Use, September 1996; available from Executive Secretary of IAPWS, Electric Power Research Institute, 3412 Hillview Ave., Palo Alto, CA 94304-1395.
2. NIST Chemistry WebBook, NIST Standard Reference Database Number 69, Mallard, W. G., and Linstrom, P. J., Eds., March 1998, National Institute of Standards and Technology, Gaithersburg, MD, 20899 (http://webbook.nist.gov).
3. Pruss, A. and Wagner, W., to be published.
4. Fernandez, D. P., Goodwin, A. R. H., Lemmon, E. W., Levelt Sengers, J. M. H., and Williams, R. C., *J. Phys. Chem. Ref. Data*, 26, 1125, 1997. [Dielectric constant]

p/MPa	T/K	ρ/kg m^{-3}	H/J g^{-1}	S/J g^{-1}K^{-1}	C_p/J g^{-1}K^{-1}	Diel. const.
0.1	273.16	999.84	0.10	0.0000	4.2194	87.90
0.1	300	996.56	112.65	0.3931	4.1806	77.75
0.1	325	987.19	217.15	0.7276	4.1819	69.32
0.1	350	973.73	321.84	1.0380	4.1945	61.79
0.1	372.76	958.63	417.50	1.3028	4.2152	55.61
0.1	372.76	0.59034	2674.9	7.3588	2.0784	1.006
0.1	375	0.58653	2679.6	7.3713	2.0686	1.006
0.1	400	0.54761	2730.4	7.5025	2.0078	1.005
0.1	450	0.48458	2829.7	7.7365	1.9752	1.004
0.1	500	0.43514	2928.6	7.9447	1.9813	1.003
0.1	550	0.39507	3028.1	8.1344	2.0010	1.003
0.1	600	0.36185	3128.8	8.3096	2.0268	1.002
0.1	650	0.33384	3230.8	8.4730	2.0557	1.002
0.1	700	0.30988	3334.4	8.6264	2.0867	1.002
0.1	750	0.28915	3439.5	8.7715	2.1191	1.002
0.1	800	0.27102	3546.3	8.9093	2.1525	1.001
0.1	850	0.25504	3654.8	9.0408	2.1868	1.001
0.1	900	0.24085	3765.0	9.1668	2.2216	1.001
0.1	950	0.22815	3876.9	9.2879	2.2568	1.001
0.1	1000	0.21673	3990.7	9.4045	2.2921	1.001
0.1	1050	0.20640	4106.1	9.5172	2.3273	1.001
0.1	1100	0.19701	4223.4	9.6263	2.3621	1.001
0.1	1150	0.18844	4342.3	9.7321	2.3965	1.001
0.1	1200	0.18058	4463.0	9.8348	2.4302	1.001
1	273.16	1000.3	1.02	0.0000	4.2150	87.93
1	300	996.96	113.48	0.3928	4.1781	77.78
1	325	987.58	217.93	0.7272	4.1798	69.36
1	350	974.13	322.56	1.0374	4.1925	61.82
1	375	957.43	427.64	1.3274	4.2158	55.09
1	400	937.87	533.47	1.6005	4.2535	49.06
1	450	890.39	749.20	2.1086	4.3924	38.81
1	453.03	887.13	762.51	2.1381	4.4045	38.23
1	453.03	5.1450	2777.1	6.5850	2.7114	1.042
1	500	4.5323	2891.2	6.8250	2.2795	1.034
1	550	4.0581	3001.8	7.0359	2.1647	1.028
1	600	3.6871	3109.0	7.2224	2.1292	1.024
1	650	3.3843	3215.2	7.3925	2.1254	1.020
1	700	3.1305	3321.7	7.5504	2.1368	1.017
1	750	2.9140	3429.0	7.6984	2.1566	1.015
1	800	2.7265	3537.5	7.8384	2.1816	1.013
1	850	2.5624	3647.3	7.9715	2.2098	1.012
1	900	2.4174	3758.5	8.0986	2.2402	1.011

p/MPa	T/K	ρ/kg m⁻³	H/J g⁻¹	S/J g⁻¹K⁻¹	C_p/J g⁻¹K⁻¹	Diel. const.
1	950	2.2882	3871.3	8.2206	2.2721	1.010
1	1000	2.1723	3985.7	8.3380	2.3048	1.009
1	1050	2.0678	4101.8	8.4512	2.3380	1.008
1	1100	1.9729	4219.5	8.5608	2.3713	1.007
1	1150	1.8865	4338.9	8.6669	2.4044	1.007
1	1200	1.8074	4460.0	8.7699	2.4371	1.006
10	273.16	1004.8	10.1	0.000	4.173	88.30
10	300	1001.0	121.7	0.390	4.153	78.11
10	325	991.46	225.6	0.723	4.160	69.67
10	350	978.09	329.7	1.031	4.173	62.13
10	375	961.62	434.4	1.320	4.195	55.40
10	400	942.42	539.6	1.592	4.230	49.39
10	450	896.16	753.9	2.096	4.355	39.17
10	500	838.02	977.1	2.566	4.602	30.79
10	550	761.82	1218	3.027	5.140	23.53
10	584.15	688.42	1408	3.360	6.123	18.70
10	584.15	55.463	2725	5.616	7.140	1.404
10	600	49.773	2820	5.775	5.136	1.365
10	650	40.479	3022	6.100	3.396	1.267
10	700	35.355	3177	6.330	2.874	1.214
10	750	31.810	3314	6.520	2.645	1.179
10	800	29.107	3443	6.686	2.531	1.154
10	850	26.933	3568	6.838	2.473	1.134
10	900	25.123	3691	6.978	2.445	1.118
10	950	23.580	3813	7.110	2.436	1.105
10	1000	22.241	3935	7.235	2.439	1.095
10	1050	21.063	4057	7.354	2.450	1.086
10	1100	20.017	4180	7.469	2.466	1.078
10	1150	19.078	4304	7.579	2.485	1.072
10	1200	18.230	4429	7.685	2.507	1.066
100	273.16	1045.3	95.4	−0.008	3.905	91.83
100	300	1037.2	201.4	0.362	3.979	81.22
100	325	1026.6	301.3	0.682	4.008	72.58
100	350	1013.6	401.7	0.979	4.025	64.95
100	375	998.59	502.6	1.258	4.040	58.19
100	400	981.82	603.7	1.518	4.056	52.20
100	450	943.51	807.8	1.999	4.110	42.15
100	500	899.21	1015	2.436	4.196	34.15
100	550	848.78	1228	2.842	4.323	27.67
100	600	791.49	1448	3.225	4.501	22.29
100	647.10	730.24	1665	3.573	4.733	17.97
100	650	726.21	1679	3.595	4.750	17.72
100	700	651.77	1925	3.958	5.083	13.75
100	750	568.52	2188	4.322	5.449	10.34
100	800	482.23	2466	4.681	5.610	7.562
100	850	404.66	2742	5.016	5.380	5.571
100	900	343.61	3000	5.310	4.887	4.284
100	950	298.61	3231	5.560	4.382	3.477
100	1000	265.45	3440	5.774	3.978	2.956
100	1050	240.32	3631	5.961	3.683	2.601
100	1100	220.62	3809	6.127	3.471	2.347
100	1150	204.71	3979	6.278	3.319	2.158
100	1200	191.53	4142	6.417	3.209	2.011

PERMITTIVITY (DIELECTRIC CONSTANT) OF WATER AS A FUNCTION OF TEMPERATURE AND PRESSURE

The following table summarizes the relative permittivity (static dielectric constant) of liquid water and steam over a wide range of temperature and pressure. Values are given from slightly above the freezing point to 1000 K and at pressures from normal atmospheric to 1000 MPa (about 10000 atm). The values are generated from an equation that correlates the best experimental measurements from a large number of sources. The correlating equation and full details of the formulation may be found in Reference 1.

Temperatures are given on the ITS-90 scale. Liquid–vapor boundaries are indicated by horizontal lines.

Reference

Fernandez, D. P., Goodwin, A. R. H., Lemmon, E. W., Levelt Sengers, J. M. H., and Williams, R. C., *J. Phys. Chem. Ref. Data*, 26, 1125, 1997.

					Pressure in MPa						
T/K	0.1	1	2	5	10	20	50	100	200	500	1000
275	87.16	87.20	87.24	87.36	87.57	87.97	89.16	91 05	94.55	103.7	
280	85.19	85.23	85.27	85.39	85.59	85.98	87.14	88.98	92.38	101.3	
285	83.27	83.30	83.34	83.46	83.65	84.04	85.17	86.96	90.27	98.91	
290	81.39	81.42	81.46	81.57	81.76	82.14	83.24	84.99	88.22	96.64	
295	79.55	79.58	79.62	79 73	79.92	80.29	81.37	83.08	86.24	94.44	
300	77.75	77.78	77.82	77.93	78.11	78.48	79.54	81.22	84.31	92.31	
305	75.99	76.02	76.06	76.17	76.35	76.71	77.75	79.40	82.43	90.25	101.3
310	74.27	74.30	74.33	74.44	74.62	74.98	76.01	77.63	80.61	88.26	99.06
315	72.58	72.61	72.65	72.76	72.93	73.28	74.30	75.90	78.84	86.34	96.87
320	70.93	70.97	71.00	71.11	71.28	71.63	72.64	74.22	77.11	84.48	94.76
340	64.70	64.73	64.77	64.87	65.04	65.38	66.36	67.89	70.65	77.58	87.07
360	59.00	59.03	59 07	59.17	59.34	59.68	60.65	62.15	64.83	71.45	80.36
380	1.006	53.83	53.86	53.97	54.14	54.48	55.45	56.94	59.57	65.95	74.43
400	1.005	49.06	49.10	49.21	49.39	49.73	50.71	52.20	54.80	61.00	69.12
420	1.005	44.70	44.74	44.85	45.04	45.39	46.39	47.90	50.48	56.53	64.35
440	1.004	40.70	40.74	40 85	41.05	41.42	42.45	43.98	46.55	52.48	60.03
460	1.004	1.041	37.04	37.17	37.37	37.76	38.84	40.40	42.99	48.81	56.11
480	1.004	1.038	33.61	33.75	33.97	34.39	35.53	37.14	39.75	45.47	52.55
500	1.003	1.034	1.074	30.55	30.79	31.25	32.47	34.15	36.79	42.44	49.30
550	1.003	1.028	1.059	1.177	23.53	24.18	25.73	27.67	30.46	35.99	42.38
600	1.002	1.024	1.049	1.137	1.365	17.50	19.90	22.29	25.34	30.82	36.82
650	1.002	1.020	1.041	1.112	1.267	2.066	14.50	17.72	21.12	26.62	32.31
700	1.002	1.017	1.036	1.095	1.214	1.603	8.963	13.75	17.60	23.17	28.60
750	1.002	1.015	1.031	1.082	1.179	1.452	4.424	10.34	14.65	20.30	25.51
800	1.001	1.013	1.027	1.071	1.154	1.365	2.844	7.562	12.17	17.88	22.91
850	1.001	1.012	1.024	1.063	1.134	1.307	2.269	5.571	10.10	15.83	20.70
900	1.001	1.011	1.022	1.056	1.118	1.265	1.975	4.284	8.416	14.08	18.80
950	1.001	1.010	1.020	1.050	1.105	1.232	1.793	3.477	7.066	12.57	17.15
1000	1.001	1.009	1.018	1.046	1.095	1.206	1.668	2.956	6.003	11.27	15.72
1050	1.001	1.008	1.016	1.041	1.086	1.184	1.576	2.601	5.172	10.14	14.45
1100	1.001	1.007	1.015	1.038	1.078	1.167	1.505	2.347	4.523	9.160	13.34
1150	1.001	1.007	1.014	1.035	1.072	1.151	1.449	2.158	4.012	8.309	12.35
1200	1.001	1.006	1.013	1.032	1.066	1.139	1.403	2.011	3.606	7.569	11.47

PERMITTIVITY (DIELECTRIC CONSTANT) OF WATER AT VARIOUS FREQUENCIES

The permittivity of liquid water in the radiofrequency and microwave regions can be represented by the Debye equation (References 1 and 2):

$$\varepsilon' = \varepsilon_\infty + \frac{\varepsilon_s - \varepsilon_\infty}{1 + \omega^2 \tau^2}$$

$$\varepsilon'' = \frac{(\varepsilon_s - \varepsilon_\infty)\omega\tau}{1 + \omega^2 \tau^2}$$

where $\varepsilon = \varepsilon' + i\,\varepsilon''$ is the (complex) relative permittivity (i.e., the absolute permittivity divided by the permittivity of free space $\varepsilon_0 = 8.854 \cdot 10^{-12}$ F m^{-1}). Here ε_s is the static permittivity (see Reference 3 and the table "Properties of Water in the Range 0–100°C" in this Section); ε_∞ is a parameter describing the permittivity in the high frequency limit; τ is the relaxation time for molecular orientation; and $\omega = 2\pi f$ is the angular frequency. The values in this table have been calculated from parameters given in Reference 2:

Other useful quantities that can be calculated from the values in the table are the loss tangent:

$$\tan\delta = \varepsilon''/\varepsilon'$$

and the absorption coefficient α which describes the power attenuation per unit length ($P = P_0\,e^{-\alpha l}$):

$$\alpha = \frac{\pi f \varepsilon''}{c\sqrt{\varepsilon'}}$$

and c is the speed of light. The last equation is valid when $\varepsilon''/\varepsilon' \ll 1$.

References

1. Fernendez, D. P., Mulev, Y., Goodwin, A. R. H., and Levelt Sengers, J. M. H., *J. Phys. Chem. Ref. Data*, 24, 33, 1995.
2. Kaatze, U., *J. Chem. Eng. Data*, 34, 371, 1989.
3. Archer, D. G., and Wang, P., *J. Phys. Chem. Ref. Data*, 12, 817, 1983.

	0°C	25°C	50°C
ε_∞	5.7	5.2	4.0
τ/ps	17.67	8.27	4.75

Frequency	0°C ε'	0°C ε''	25°C ε'	25°C ε''	50°C ε'	50°C ε''
0	87.90	0.00	78.36	0.00	69.88	0.00
1 kHz	87.90	0.00	78.36	0.00	69.88	0.00
1 MHz	87.90	0.01	78.36	0.00	69.88	0.00
10 MHz	87.90	0.09	78.36	0.04	69.88	0.02
100 MHz	87.89	0.91	78.36	0.38	69.88	0.20
200 MHz	87.86	1.82	78.35	0.76	69.88	0.39
500 MHz	87.65	4.55	78.31	1.90	69.87	0.98
1 GHz	86.90	9.01	78.16	3.79	69.82	1.96
2 GHz	84.04	17.39	77.58	7.52	69.65	3.92
3 GHz	79.69	24.64	76.62	11.13	69.36	5.85
4 GHz	74.36	30.49	75.33	14.58	68.95	7.75
5 GHz	68.54	34.88	73.73	17.81	68.45	9.62
10 GHz	42.52	40.88	62.81	29.93	64.49	18.05
20 GHz	19.56	30.78	40.37	36.55	52.57	28.99
30 GHz	12.50	22.64	26.53	33.25	40.57	32.74
40 GHz	9.67	17.62	18.95	28.58	31.17	32.43
50 GHz	8.28	14.34	14.64	24.53	24.42	30.47

THERMOPHYSICAL PROPERTIES OF FLUIDS

These tables give thermodynamic and transport properties of some important fluids, as generated from the equations of state presented in the references below. The properties tabulated are density (ρ), energy (E), enthalpy (H), entropy (S), isochoric heat capacity (C_v), isobaric heat capacity (C_p), speed of sound (v_s), viscosity (η), thermal conductivity (λ), and dielectric constant (D). All extensive properties are given on a molar basis. Not all properties are included for every substance. The references should be consulted for information on the uncertainties and the reference states for E, H, and S.

Values are given as a function of temperature for several isobars. The phase can be determined by noting the sharp decrease in density between two successive temperature entries; all lines above

this point refer to the liquid phase, and all lines below refer to the gas phase. If there is no sharp discontinuity in density, all data in the table refer to the supercritical region (i.e., the isobar is above the critical pressure).

References

1. Younglove, B. A., Thermophysical Properties of Fluids. Part I, *J. Phys. Chem. Ref. Data,* 11, Suppl. 1, 1982.
2. Younglove, B. A., and Ely, J. F., Thermophysical Properties of Fluids. Part II, *J. Phys. Chem. Ref. Data,* 16, 577, 1987.
3. McCarty, R. D., Thermodynamic Properties of Helium, *J. Phys. Chem. Ref. Data,* 2, 923, 1973.

NITROGEN (N₂)

T K	ρ mol/L	E J/mol	H J/mol	S J/mol K	C_v J/mol K	C_p J/mol K	η μPa s	λ mW/m K	D
P = 0.1 MPa (1 bar)									
70	30.017	−3828	−3824	73.8	28.5	57.2	203.9	143.5	1.45269
77.25	28.881	−3411	−3407	79.5	27.8	57.8	152.2	133.8	1.43386
77.25	0.163	1546	2161	151.6	21.6	31.4	5.3	7.6	1.00215
100	0.123	2041	2856	159.5	21.1	30.0	6.8	9.6	1.00162
200	0.060	4140	5800	179.9	20.8	29.2	12.9	18.4	1.00079
300	0.040	6223	8717	191.8	20.8	29.2	18.0	25.8	1.00053
400	0.030	8308	11635	200.2	20.9	29.2	22.2	32.3	1.00040
500	0.024	10414	14573	206.7	21.2	29.6	26.1	38.5	1.00032
600	0.020	12563	17554	212.2	21.8	30.1	29.5	44.5	1.00026
700	0.017	14770	20593	216.8	22.4	30.7	32.8	50.5	1.00023
800	0.015	17044	23698	221.0	23.1	31.4	35.8	56.3	1.00020
900	0.013	19383	26869	224.7	23.7	32.0	38.7	62.0	1.00017
1000	0.012	21786	30103	228.1	24.3	32.6	41.5	67.7	1.00016
1500	0.008	34530	47004	241.8	26.4	34.7	54.0	93.3	1.00010
P = 1 MPa									
70	30.070	−3838	−3805	73.6	28.9	56.9	205.9	144.1	1.45355
80	28.504	−3267	−3232	81.3	27.8	57.7	139.5	130.7	1.42760
90	26.721	−2685	−2648	88.2	26.7	59.4	100.1	115.3	1.39824
100	24.634	−2073	−2032	94.6	26.2	64.4	73.1	98.5	1.36417
103.75	23.727	−1828	−1786	97.1	26.2	67.8	64.8	91.8	1.34947
103.75	1.472	1788	2467	138.1	24.1	45.0	7.6	12.5	1.01954
200	0.614	4048	5675	160.3	21.0	30.4	13.2	19.3	1.00812
300	0.402	6171	8661	172.5	20.9	29.6	18.1	26.3	1.00529
400	0.300	8273	11609	180.9	20.9	29.5	22.4	32.7	1.00395
500	0.240	10389	14563	187.5	21.3	29.7	26.1	38.8	1.00315
600	0.200	12544	17554	193.0	21.8	30.2	29.6	44.8	1.00262
700	0.171	14756	20600	197.7	22.4	30.8	32.8	50.7	1.00224
800	0.150	17032	23709	201.8	23.1	31.4	35.9	56.5	1.00196
900	0.133	19374	26884	205.6	23.7	32.1	38.8	62.2	1.00174
1000	0.120	21778	30121	209.0	24.3	32.7	41.5	67.8	1.00157
1500	0.080	34527	47029	222.7	26.4	34.8	54.0	93.4	1.00104
P = 10 MPa									
65.32	31.120	−4176	−3855	68.6	31.8	53.8	275.7	153.8	1.47067
100	26.201	−2328	−1946	92.0	27.4	56.3	90.2	112.3	1.38942
200	7.117	3037	4442	136.4	22.7	45.5	17.6	30.4	1.09698
300	3.989	5667	8174	151.7	21.4	33.4	20.1	31.9	1.05347
400	2.898	7941	11392	161.0	21.3	31.3	23.7	36.7	1.03860

T K	ρ mol/L	E J/mol	H J/mol	S J/mol K	C_v J/mol K	C_p J/mol K	η μPa s	λ mW/m K	D
500	2.302	10148	14492	167.9	21.5	30.8	27.1	42.0	1.03055
600	1.918	12361	17575	173.5	21.9	30.9	30.4	47.4	1.02538
700	1.647	14613	20683	178.3	22.5	31.3	33.5	53.0	1.02175
800	1.445	16919	23837	182.5	23.2	31.8	36.4	58.6	1.01904
900	1.288	19283	27046	186.3	23.8	32.4	39.3	64.1	1.01694
1000	1.162	21705	30308	189.8	24.4	32.9	42.0	69.6	1.01526
1500	0.783	34504	47283	203.5	26.5	34.8	54.3	94.7	1.01020

OXYGEN (O_2)

T K	ρ mol/L	E J/mol	H J/mol	S J/mol K	C_v J/mol K	C_p J/mol K	η μPa s	λ mW/m K	D
P = 0.1 MPa (1 bar)									
60	40.049	−5883	−5880	72.4	34.9	53.4	425.2	188.2	1.55619
80	37.204	−4814	−4812	87.7	31.0	53.6	251.7	166.1	1.51114
100	0.123	2029	2840	172.9	21.4	30.5	7.5	9.3	1.00146
120	0.102	2458	3442	178.4	21.0	29.8	9.0	11.2	1.00121
140	0.087	2881	4035	182.9	20.9	29.5	10.5	13.1	1.00103
160	0.076	3301	4624	186.9	20.9	29.4	11.9	15.0	1.00090
180	0.067	3720	5210	190.3	20.8	29.3	13.3	16.7	1.00080
200	0.060	4138	5796	193.4	20.8	29.3	14.6	18.4	1.00072
220	0.055	4556	6381	196.2	20.8	29.3	15.9	20.1	1.00065
240	0.050	4974	6966	198.8	20.9	29.3	17.2	21.7	1.00060
260	0.046	5393	7552	201.1	20.9	29.3	18.4	23.2	1.00055
280	0.043	5812	8138	203.3	21.0	29.4	19.5	24.8	1.00051
300	0.040	6234	8726	205.3	21.1	29.4	20.6	26.3	1.00048
320	0.038	6657	9316	207.2	21.2	29.5	21.7	27.8	1.00045
340	0.035	7082	9908	209.0	21.3	29.7	22.8	29.3	1.00042
360	0.033	7510	10503	210.7	21.5	29.8	23.8	30.8	1.00040
380	0.032	7941	11100	212.3	21.6	30.0	24.8	32.2	1.00038
P = 1 MPa									
60	40.084	−5887	−5863	72.3	34.9	53.3	428.5	188.4	1.55674
80	37.254	−4822	−4795	87.6	31.0	53.5	253.8	166.4	1.51192
100	34.153	−3741	−3712	99.7	28.5	55.2	155.6	137.9	1.46381
120	1.198	2163	2997	156.7	24.0	40.6	9.4	13.9	1.01429
140	0.950	2683	3735	162.4	22.2	34.4	10.8	14.9	1.01133
160	0.802	3151	4398	166.8	21.5	32.2	12.2	16.3	1.00955
180	0.698	3598	5030	170.5	21.2	31.2	13.5	17.7	1.00831
200	0.620	4035	5647	173.8	21.1	30.6	14.8	19.3	1.00738
220	0.559	4466	6255	176.7	21.0	30.3	16.1	20.8	1.00665
240	0.509	4894	6858	179.3	21.0	30.1	17.3	22.3	1.00606
260	0.468	5321	7458	181.7	21.0	29.9	18.5	23.8	1.00556
280	0.433	5748	8056	183.9	21.1	29.9	19.6	25.2	1.00515
300	0.403	6174	8654	186.0	21.1	29.9	20.7	26.7	1.00479
320	0.377	6602	9252	187.9	21.2	29.9	21.8	28.2	1.00448
340	0.355	7032	9851	189.7	21.4	30.0	22.8	29.6	1.00421
360	0.335	7463	10452	191.4	21.5	30.1	23.9	31.1	1.00397
380	0.317	7898	11056	193.1	21.7	30.2	24.9	32.6	1.00376
P = 10 MPa									
60	40.419	−5931	−5684	71.5	35.1	53.0	461.8	189.9	1.56210
80	37.727	−4893	−4628	86.7	31.6	52.7	274.4	168.6	1.51936
100	34.881	−3856	−3570	98.5	29.1	53.4	171.0	141.2	1.47500
120	31.721	−2796	−2481	108.4	27.3	55.9	113.0	115.1	1.42677
140	27.890	−1662	−1304	117.5	26.2	62.9	76.3	91.8	1.36972
160	22.379	−322	125	127.0	26.1	84.8	48.6	71.2	1.29037
180	13.232	1489	2245	139.5	26.6	105.9	26.2	46.8	1.16560
200	8.666	2681	3835	147.9	24.0	60.6	21.2	34.0	1.10650

T K	ρ mol/L	E J/mol	H J/mol	S J/mol K	C_v J/mol K	C_p J/mol K	η μPa s	λ mW/m K	D
220	6.868	3424	4880	152.9	22.6	46.4	20.5	30.8	1.08380
240	5.836	4029	5742	156.6	22.0	40.6	20.8	30.1	1.07090
260	5.134	4573	6521	159.7	21.8	37.6	21.4	30.2	1.06219
280	4.613	5086	7254	162.5	21.6	35.8	22.1	30.8	1.05575
300	4.205	5581	7959	164.9	21.6	34.7	22.9	31.6	1.05073
320	3.874	6063	8645	167.1	21.7	33.9	23.7	32.6	1.04667
340	3.598	6538	9318	169.1	21.8	33.4	24.6	33.7	1.04329
360	3.363	7009	9982	171.0	21.9	33.0	25.4	34.9	1.04043
380	3.161	7477	10641	172.8	22.0	32.8	26.3	36.1	1.03796

HYDROGEN (H_2)

T K	ρ mol/L	E J/mol	H J/mol	S J/mol K	C_v J/mol K	C_p J/mol K	v_s m/s	D
$P = 0.1$ MPa (1 bar)								
15	37.738	−605	−603	11.2	9.7	14.4	1319	1.24827
20	35.278	−524	−521	15.8	11.3	19.1	1111	1.23093
40	0.305	491	818	75.6	12.5	21.3	521	1.00186
60	0.201	748	1244	84.3	13.1	21.6	636	1.00122
80	0.151	1030	1694	90.7	15.3	23.7	714	1.00091
100	0.120	1370	2202	96.4	18.7	27.1	773	1.00073
120	0.100	1777	2776	101.6	21.8	30.2	827	1.00061
140	0.086	2237	3401	106.4	23.8	32.2	883	1.00052
160	0.075	2723	4054	110.8	24.6	33.0	940	1.00046
180	0.067	3216	4714	114.7	24.6	32.9	998	1.00041
200	0.060	3703	5367	118.1	24.1	32.4	1054	1.00037
220	0.055	4179	6009	121.2	23.4	31.8	1110	1.00033
240	0.050	4641	6638	123.9	22.8	31.2	1163	1.00030
260	0.046	5093	7256	126.4	22.3	30.6	1214	1.00028
280	0.043	5535	7865	128.6	21.9	30.2	1263	1.00026
300	0.040	5970	8466	130.7	21.6	29.9	1310	1.00024
400	0.030	8093	11421	139.2	21.0	29.3	1518	1.00018
$P = 1$ MPa								
15	38.109	−609	−583	10.9	10.1	14.1	1315	1.25089
20	35.852	−532	−504	15.5	11.4	18.4	1155	1.23496
40	3.608	399	676	54.1	12.9	28.4	498	1.02209
60	2.098	697	1173	64.3	13.2	23.5	635	1.01280
80	1.523	994	1651	71.1	15.4	24.7	719	1.00928
100	1.204	1343	2174	77.0	18.8	27.7	779	1.00733
120	0.999	1756	2758	82.3	21.9	30.6	835	1.00608
140	0.854	2219	3390	87.1	23.9	32.5	891	1.00520
160	0.747	2709	4048	91.5	24.7	33.2	949	1.00454
180	0.663	3204	4712	95.4	24.6	33.1	1006	1.00404
200	0.597	3693	5368	98.9	24.1	32.5	1063	1.00363
220	0.543	4170	6012	102.0	23.5	31.9	1118	1.00330
240	0.498	4634	6643	104.7	22.9	31.2	1171	1.00303
260	0.460	5087	7263	107.2	22.3	30.7	1222	1.00279
280	0.427	5530	7873	109.5	21.9	30.3	1271	1.00259
300	0.399	5966	8475	111.5	21.6	30.0	1317	1.00242
400	0.299	8091	11433	120.1	21.0	29.4	1525	1.00182
$P = 10$ MPa								
20	39.669	−568	−316	13.0	10.9	15.0	1458	1.26198
40	31.344	−209	110	27.3	13.2	27.0	1171	1.20354
60	21.273	255	725	39.7	13.8	32.5	931	1.13527
80	14.830	686	1360	48.8	15.9	31.1	886	1.09303
100	11.417	1110	1986	55.8	19.3	31.9	904	1.07109
120	9.357	1571	2640	61.8	22.4	33.5	941	1.05801

T K	ρ mol/L	*E* J/mol	*H* J/mol	*S* J/mol K	*C*$_v$ J/mol K	*C*$_p$ J/mol K	*v*$_s$ m/s	*D*
140	7.969	2068	3323	67.0	24.3	34.6	989	1.04925
160	6.963	2583	4020	71.7	25.0	34.9	1042	1.04294
180	6.195	3099	4713	75.7	24.9	34.4	1096	1.03814
200	5.588	3604	5393	79.3	24.4	33.6	1150	1.03436
220	5.094	4094	6057	82.5	23.7	32.8	1203	1.03129
240	4.683	4569	6704	85.3	23.1	32.0	1254	1.02874
260	4.336	5030	7336	87.8	22.6	31.3	1302	1.02659
280	4.038	5481	7958	90.1	22.1	30.8	1349	1.02475
300	3.780	5924	8570	92.3	21.8	30.4	1394	1.02315
400	2.869	8073	11559	100.9	21.2	29.6	1592	1.01753

HELIUM (He-4)

T K	ρ mol/L	*E* J/mol	*H* J/mol	*S* J/mol K	*C*$_v$ J/mol K	*C*$_p$ J/mol K	*v*$_s$ m/s	η μPa s	*D*
P = 0.1 MPa (1 bar)									
3	35.794	−39	−36	9.8	7.6	9.4	222	3.85	1.05646
4	32.477	−27	−24	13.3	9.1	16.3	185	3.33	1.05114
5	2.935	52	86	39.1	12.7	27.1	120	1.39	1.00456
10	1.238	120	201	55.2	12.5	21.7	185	2.26	1.00192
20	0.602	247	413	69.9	12.5	21.0	264	3.58	1.00093
50	0.240	623	1039	89.0	12.5	20.8	417	6.36	1.00037
100	0.120	1247	2079	103.4	12.5	20.8	589	9.78	1.00019
200	0.060	2494	4158	117.8	12.5	20.8	833	15.14	1.00009
300	0.040	3741	6237	126.3	12.5	20.8	1020	19.93	1.00006
400	0.030	4988	8315	132.3	12.5	20.8	1177	24.29	1.00005
500	0.024	6236	10394	136.9	12.5	20.8	1316	28.36	1.00004
600	0.020	7483	12472	140.7	12.5	20.8	1441	32.22	1.00003
700	0.017	8730	14551	143.9	12.5	20.8	1557	35.89	1.00003
800	0.015	9977	16630	146.7	12.5	20.8	1664	39.43	1.00002
900	0.013	11224	18708	149.1	12.5	20.8	1765	42.85	1.00002
1000	0.012	12471	20787	151.3	12.5	20.8	1861	46.16	1.00002
1500	0.008	18707	31179	159.7	12.5	20.8	2279	61.55	1.00001
P = 1 MPa									
3	39.703	−42	−16	8.6	7.1	7.8	300	5.63	1.06274
4	38.210	−34	−7	11.2	8.3	10.9	290	5.01	1.06034
5	35.818	−22	6	14.0	9.7	15.1	269	4.38	1.05650
10	15.378	78	143	32.2	12.3	30.5	198	3.07	1.02402
20	6.067	228	393	49.8	12.6	22.9	274	3.94	1.00943
50	2.353	617	1042	69.8	12.5	21.1	428	6.53	1.00365
100	1.186	1245	2089	84.3	12.5	20.9	597	9.89	1.00184
200	0.597	2495	4170	98.7	12.5	20.8	838	15.21	1.00093
300	0.399	3742	6249	107.1	12.5	20.8	1024	19.96	1.00062
400	0.300	4990	8327	113.1	12.5	20.8	1180	24.32	1.00046
500	0.240	6237	10406	117.8	12.5	20.8	1319	28.38	1.00037
600	0.200	7485	12484	121.5	12.5	20.8	1444	32.23	1.00031
700	0.172	8732	14562	124.7	12.5	20.8	1559	35.91	1.00027
800	0.150	9979	16641	127.5	12.5	20.8	1666	39.44	1.00023
900	0.133	11227	18719	130.0	12.5	20.8	1767	42.86	1.00021
1000	0.120	12474	20798	132.2	12.5	20.8	1862	46.17	1.00019
1500	0.080	18710	31190	140.6	12.5	20.8	2280	61.55	1.00012
P = 10 MPa									
4	51.978	−24	169	6.7	6.0	7.3	586	24.27	1.08262
5	51.118	−18	177	8.5	7.9	9.3	576	18.16	1.08122
10	46.872	23	236	16.6	11.0	14.5	546	9.31	1.07432
20	37.092	154	423	29.5	12.6	20.7	498	6.99	1.05854
50	19.192	572	1093	49.9	12.9	22.4	541	8.07	1.03003

T K	ρ mol/L	E J/mol	H J/mol	S J/mol K	C_v J/mol K	C_p J/mol K	v_s m/s	η μPa s	D
100	10.525	1231	2181	65.0	12.8	21.3	674	10.93	1.01640
200	5.605	2500	4284	79.6	12.6	20.9	889	15.82	1.00871
300	3.829	3755	6367	88.0	12.6	20.8	1063	20.25	1.00595
400	2.908	5006	8445	94.0	12.6	20.8	1212	24.54	1.00452
500	2.344	6256	10522	98.6	12.5	20.8	1346	28.56	1.00364
600	1.963	7505	12599	102.4	12.5	20.8	1467	32.38	1.00305
700	1.689	8754	14676	105.6	12.5	20.8	1580	36.04	1.00262
800	1.481	10003	16753	108.4	12.5	20.8	1685	39.56	1.00230
900	1.320	11252	18830	110.9	12.5	20.8	1784	42.96	1.00205
1000	1.189	12500	20907	113.0	12.5	20.8	1877	46.26	1.00185
1500	0.797	18742	31294	121.5	12.5	20.8	2289	61.62	1.00124

ARGON (Ar)

T K	ρ mol/L	E J/mol	H J/mol	S J/mol K	C_v J/mol K	C_p J/mol K	v_s m/s	η μPa s	λ mW/m K
P = 0.1 MPa (1 bar)									
85	35.243	−4811	−4808	53.6	23.1	44.7	820	278.8	132.4
90	0.138	1077	1802	129.4	13.1	22.5	174	7.5	6.0
100	0.123	1211	2024	131.8	12.9	21.9	184	8.2	6.6
120	0.102	1471	2456	135.7	12.6	21.4	203	9.8	7.8
140	0.087	1727	2881	139.0	12.6	21.1	220	11.4	9.0
160	0.076	1980	3302	141.8	12.5	21.0	235	13.0	10.2
180	0.067	2232	3722	144.3	12.5	21.0	250	14.5	11.4
200	0.060	2483	4141	146.5	12.5	20.9	263	16.0	12.5
220	0.055	2734	4559	148.5	12.5	20.9	276	17.5	13.7
240	0.050	2984	4976	150.3	12.5	20.9	289	18.9	14.8
260	0.046	3234	5394	152.0	12.5	20.9	300	20.3	15.8
280	0.043	3484	5811	153.5	12.5	20.8	312	21.6	16.9
300	0.040	3734	6227	155.0	12.5	20.8	323	22.9	17.9
320	0.038	3984	6644	156.3	12.5	20.8	333	24.2	18.9
340	0.035	4234	7060	157.6	12.5	20.8	344	25.4	19.9
360	0.033	4484	7477	158.7	12.5	20.8	354	26.6	20.8
380	0.032	4734	7893	159.9	12.5	20.8	363	27.8	21.7
P = 1 MPa									
85	35.307	−4820	−4792	53.5	23.1	44.6	823	281.3	133.0
90	34.542	−4598	−4569	56.1	21.6	44.7	808	242.7	124.2
100	32.909	−4145	−4115	60.9	19.9	46.2	753	185.0	109.2
120	1.181	1210	2057	114.3	14.7	30.1	189	10.3	9.3
140	0.945	1544	2603	118.5	13.5	25.4	212	11.8	10.1
160	0.799	1838	3089	121.8	13.0	23.6	231	13.3	11.1
180	0.697	2116	3551	124.5	12.8	22.7	247	14.8	12.1
200	0.619	2384	3999	126.9	12.7	22.2	262	16.3	13.2
220	0.559	2648	4438	128.9	12.6	21.8	275	17.7	14.2
240	0.509	2908	4873	130.8	12.6	21.6	288	19.1	15.3
260	0.468	3167	5304	132.6	12.6	21.5	301	20.4	16.3
280	0.433	3423	5732	134.2	12.6	21.4	312	21.8	17.3
300	0.403	3679	6159	135.6	12.5	21.3	324	23.1	18.3
320	0.377	3934	6583	137.0	12.5	21.2	334	24.3	19.2
340	0.355	4188	7007	138.3	12.5	21.2	345	25.5	20.2
360	0.335	4441	7429	139.5	12.5	21.1	355	26.7	21.1
380	0.317	4694	7851	140.6	12.5	21.1	365	27.9	22.0
P = 10 MPa									
90	35.208	−4694	−4410	55.0	21.9	43.2	846	265.2	129.5
100	33.744	−4271	−3974	59.6	20.4	44.0	800	205.0	115.1
120	30.525	−3396	−3069	67.8	18.8	46.9	672	131.2	92.1
140	26.609	−2447	−2072	75.5	17.6	54.1	526	85.9	71.7

T K	ρ mol/L	E J/mol	H J/mol	S J/mol K	C_v J/mol K	C_p J/mol K	ν_s m/s	η μPa s	λ mW/m K
160	20.816	−1279	−799	83.9	17.4	78.6	357	51.3	52.8
180	12.296	228	1042	94.8	17.3	83.6	257	27.8	32.0
200	8.442	1118	2302	101.4	15.3	48.6	268	23.3	23.6
220	6.776	1661	3137	105.4	14.2	36.8	284	22.8	21.6
240	5.787	2087	3815	108.4	13.7	31.6	300	23.2	21.3
260	5.105	2458	4416	110.8	13.4	28.8	314	23.9	21.4
280	4.596	2798	4974	112.9	13.2	27.1	327	24.8	21.8
300	4.195	3119	5503	114.7	13.1	25.9	339	25.7	22.3
320	3.869	3427	6012	116.3	13.0	25.0	350	26.7	22.9
340	3.596	3726	6506	117.8	13.0	24.4	361	27.7	23.5
360	3.364	4017	6989	119.2	12.9	23.9	372	28.7	24.2
380	3.164	4303	7464	120.5	12.9	23.5	381	29.7	24.9

METHANE (CH₄)

T K	ρ mol/L	E J/mol	H J/mol	S J/mol K	C_v J/mol K	C_p J/mol K	η μPa s	λ mW/m K	D
P = 0.1 MPa (1 bar)									
100	27.370	−5258	−5254	73.0	33.4	54.1	156.3	208.1	1.65504
125	0.099	3026	4039	156.5	25.4	34.6	5.0	13.4	1.00193
150	0.081	3667	4896	162.7	25.2	34.0	5.9	16.2	1.00159
175	0.069	4301	5743	168.0	25.2	33.8	6.9	19.1	1.00136
200	0.061	4935	6587	172.5	25.3	33.8	7.8	21.9	1.00119
225	0.054	5571	7434	176.5	25.5	34.0	8.7	24.8	1.00105
250	0.048	6216	8288	180.1	26.0	34.4	9.6	27.8	1.00095
275	0.044	6875	9156	183.4	26.6	35.0	10.4	30.9	1.00086
300	0.040	7552	10042	186.4	27.5	35.9	11.2	34.1	1.00079
325	0.037	8252	10951	189.4	28.5	36.9	12.0	37.6	1.00073
350	0.034	8979	11887	192.1	29.7	38.0	12.8	41.2	1.00068
375	0.032	9737	12853	194.8	30.9	39.3	13.5	45.1	1.00063
400	0.030	10528	13852	197.4	32.3	40.7	14.3	49.1	1.00059
425	0.028	11354	14886	199.9	33.7	42.1	15.0	53.3	1.00056
450	0.027	12215	15956	202.3	35.2	43.5	15.7	57.6	1.00053
500	0.024	14047	18204	207.1	38.0	46.4	17.0	66.5	1.00047
600	0.020	18111	23101	216.0	42.9	51.3	19.4	84.1	1.00039
P = 1 MPa									
100	27.413	−5268	−5231	72.9	33.4	54.0	158.1	208.9	1.65617
125	25.137	−3882	−3842	85.3	32.4	57.4	89.2	168.2	1.59261
150	0.969	3282	4315	140.9	27.9	45.2	6.2	18.4	1.01911
175	0.765	4041	5348	147.3	26.4	38.9	7.1	20.6	1.01507
200	0.644	4736	6289	152.3	25.9	36.8	8.0	23.1	1.01268
225	0.560	5410	7197	156.6	25.9	36.0	8.9	25.8	1.01102
250	0.497	6081	8093	160.4	26.2	35.8	9.7	28.7	1.00979
275	0.448	6758	8991	163.8	26.8	36.1	10.6	31.7	1.00882
300	0.408	7449	9901	167.0	27.6	36.7	11.4	34.9	1.00803
325	0.375	8160	10829	169.9	28.6	37.6	12.1	38.3	1.00738
350	0.347	8897	11781	172.8	29.7	38.6	12.9	41.9	1.00683
375	0.323	9662	12760	175.5	31.0	39.8	13.6	45.7	1.00636
400	0.302	10460	13770	178.1	32.4	41.1	14.4	49.6	1.00595
425	0.284	11291	14814	180.6	33.8	42.4	15.1	53.8	1.00559
450	0.268	12157	15892	183.1	35.2	43.8	15.7	58.1	1.00527
500	0.241	13997	18153	187.8	38.1	46.6	17.0	66.9	1.00474
600	0.200	18073	23070	196.8	43.0	51.4	19.5	84.5	1.00394
P = 10 MPa									
100	27.815	−5362	−5003	72.0	33.8	53.2	175.4	217	1.66668
125	25.754	−4036	−3648	84.1	32.7	55.3	100.4	178.8	1.60895
150	23.441	−2655	−2229	94.4	31.4	58.6	65.7	144.6	1.54553

T K	ρ mol/L	E J/mol	H J/mol	S J/mol K	C_v J/mol K	C_p J/mol K	η μPa s	λ mW/m K	D
175	20.613	−1175	−689	103.9	30.3	65.5	44.9	113.4	1.47021
200	16.602	542	1144	113.6	30.1	84.7	29.4	85.8	1.36789
225	10.547	2680	3628	125.3	30.8	102.2	17.6	61.0	1.22352
250	7.013	4289	5714	134.1	29.3	67.4	14.3	47.6	1.14481
275	5.530	5387	7195	139.8	28.7	53.4	13.8	44.1	1.11297
300	4.685	6320	8454	144.2	28.9	48.0	13.9	44.6	1.09513
325	4.115	7192	9622	147.9	29.6	45.8	14.3	46.6	1.08322
350	3.695	8047	10753	151.3	30.5	44.9	14.7	49.2	1.07450
375	3.366	8903	11874	154.4	31.7	44.8	15.2	52.3	1.06773
400	3.101	9774	12999	157.3	32.9	45.2	15.8	55.7	1.06227
425	2.880	10666	14138	160.0	34.3	46.0	16.3	59.4	1.05775
450	2.692	11584	15298	162.7	35.7	46.9	16.9	63.3	1.05392
500	2.389	13507	17692	167.7	38.5	48.9	18.0	71.6	1.04775
600	1.963	17700	22795	177.0	43.3	52.9	20.2	88.3	1.03911

ETHANE (C_2H_6)

T K	ρ mol/L	E J/mol	H J/mol	S J/mol K	C_v J/mol K	C_p J/mol K	v_s m/s	D
P = 0.1 MPa (1 bar)								
95	21.50	−14555	−14550	80.2	47.2	68.7	1970	1.93480
100	21.32	−14210	−14205	83.8	47.1	69.3	1943	1.92500
125	20.41	−12468	−12463	99.3	45.0	69.8	1775	1.87634
150	19.47	−10717	−10712	112.1	43.4	70.4	1587	1.82726
175	18.49	−8938	-8933	123.1	42.7	72.1	1396	1.77671
200	0.062	5503	7123	210.1	34.5	43.8	258	1.00208
225	0.054	6401	8238	215.4	36.5	45.5	273	1.00183
250	0.049	7349	9401	220.3	38.9	47.7	287	1.00164
275	0.044	8360	10624	224.9	41.6	50.2	300	1.00148
300	0.040	9439	11914	229.4	44.5	53.1	312	1.00136
325	0.037	10592	13278	233.8	47.6	56.1	324	1.00125
350	0.035	11823	14719	238.1	50.7	59.2	335	1.00116
375	0.032	13133	16240	242.3	54.0	62.4	345	1.00108
400	0.030	14525	17841	246.4	57.2	65.6	355	1.00101
450	0.027	17548	21282	254.5	63.6	72.0	375	1.00090
500	0.024	20883	25035	262.4	69.7	78.1	393	1.00081
600	0.020	28429	33415	277.6	80.9	89.3	428	1.00067
P = 1 MPa								
95	21.514	−14562	−14515	80.2	47.3	68.7	1972	1.93537
100	21.334	−14217	−14170	83.7	47.2	69.3	1946	1.92560
125	20.427	−12478	−12429	99.2	45.0	69.8	1778	1.87709
150	19.494	−10731	−10679	112.0	43.4	70.3	1592	1.82823
175	18.515	−8957	−8903	123.0	42.7	72.0	1402	1.77800
200	17.464	−7127	−7070	132.7	42.9	74.9	1209	1.72513
225	16.288	−5199	−5137	141.8	43.8	80.2	1008	1.66733
250	0.564	6762	8534	198.7	41.6	57.5	260	1.01909
275	0.489	7902	9949	204.1	43.2	56.2	280	1.01650
300	0.435	9063	11363	209.0	45.5	57.2	297	1.01467
325	0.393	10273	12815	213.7	48.3	59.1	311	1.01327
350	0.360	11546	14321	218.1	51.3	61.5	325	1.01214
375	0.333	12889	15893	222.5	54.4	64.2	337	1.01121
400	0.310	14306	17534	226.7	57.5	67.1	349	1.01043
450	0.272	17367	21038	234.9	63.8	73.0	370	1.00917
500	0.244	20730	24836	242.9	69.9	78.9	390	1.00819
600	0.201	28313	33278	258.3	81.0	89.8	427	1.00677
P = 10 MPa								
95	21.624	−14626	−14163	79.5	47.4	68.5	2000	1.94104

T K	ρ mol/L	E J/mol	H J/mol	S J/mol K	C_v J/mol K	C_p J/mol K	v_s m/s	D
100	21.448	−14286	−13819	83.0	47.4	69.1	1974	1.93146
125	20.570	−12572	−12086	98.5	45.5	69.3	1814	1.88436
150	19.678	−10858	−10350	111.1	43.9	69.6	1637	1.83753
175	18.758	−9130	−8596	121.9	43.3	70.8	1459	1.79010
200	17.793	−7363	−6801	131.5	43.5	73.0	1284	1.74134
225	16.760	−5535	−4938	140.3	44.3	76.4	1110	1.69017
250	15.620	−3609	−2969	148.6	45.8	81.5	935	1.63488
275	14.301	−1539	−839	156.7	47.9	89.4	758	1.57249
300	12.666	757	1547	165.0	50.8	102.7	577	1.49740
325	10.398	3443	4404	174.1	54.7	129.1	399	1.39745
350	7.292	6643	8015	184.8	58.8	150.1	290	1.26832
375	5.182	9419	11349	194.1	60.0	115.7	289	1.18570
400	4.182	11577	13968	200.8	61.4	96.9	310	1.14797
450	3.204	15379	18500	211.5	65.8	87.5	347	1.11193
500	2.677	19135	22870	220.7	71.2	88.0	378	1.09288
600	2.076	27160	31978	237.3	81.8	94.7	427	1.07142

PROPANE (C$_3$H$_8$)

T K	ρ mol/L	E J/mol	H J/mol	S J/mol K	C_v J/mol K	C_p J/mol K	v_s m/s	D
P = 0.1 MPa (1 bar)								
90	16.526	−21486	−21426	87.3	59.2	84.5	2126	2.07988
100	16.295	−20639	−20577	96.2	59.6	85.2	2041	2.05806
125	15.726	−18495	−18432	115.4	59.2	86.5	1856	2.00674
150	15.156	−16319	−16253	131.3	58.9	88.0	1685	1.95796
175	14.577	−14096	−14028	145.0	59.5	90.3	1521	1.91036
200	13.982	−11806	−11735	157.3	61.0	93.5	1359	1.86300
225	13.339	−9395	−9387	168.5	63.4	97.9	1197	1.81487
250	0.050	9194	11213	257.6	57.2	66.8	228	1.00238
275	0.045	10691	12930	264.1	61.6	70.7	239	1.00215
300	0.041	12297	14752	270.5	66.2	75.1	249	1.00195
325	0.037	14019	16689	276.7	71.1	79.8	259	1.00179
350	0.035	15862	18744	282.8	76.0	84.6	269	1.00166
375	0.032	17827	20921	288.8	80.9	89.5	278	1.00154
400	0.030	19912	23217	294.7	85.7	94.3	286	1.00144
450	0.027	24441	28166	306.4	95.2	103.6	303	1.00128
500	0.024	29428	33573	317.7	104.1	112.6	318	1.00115
600	0.020	40677	45658	339.7	120.4	128.8	347	1.00095
P = 1 MPa								
90	16.526	−21486	−21426	87.2	59.3	84.5	2128	2.08034
100	16.295	−20639	−20577	96.2	59.7	85.2	2043	2.05856
125	15.726	−18495	−18432	115.3	59.2	86.4	1859	2.00736
150	15.156	−16319	−16253	131.2	59.0	88.0	1690	1.95873
175	14.577	−14096	−14028	144.9	59.6	90.2	1526	1.91132
200	13.982	−11806	−11735	157.2	61.1	93.4	1365	1.86421
225	13.361	−9424	−9349	168.4	63.4	97.7	1205	1.81642
250	12.696	−6919	−6840	179.0	66.4	103.3	1045	1.76672
275	11.962	−4252	−4169	189.1	70.0	110.8	881	1.71316
300	11.102	−1360	−1270	199.2	74.1	121.9	708	1.65216
325	0.428	13278	15614	255.2	74.1	89.6	233	1.02067
350	0.383	15259	17869	261.9	78.0	91.2	248	1.01846
375	0.349	17318	20183	268.3	82.2	94.2	261	1.01678
400	0.322	19472	22582	274.4	86.7	97.8	272	1.01544
450	0.279	24092	27672	286.4	95.7	105.9	293	1.01337
500	0.248	29137	33172	298.0	104.4	114.1	312	1.01184
600	0.203	40455	45374	320.2	120.5	129.7	344	1.00968

T K	ρ mol/L	E J/mol	H J/mol	S J/mol K	C_v J/mol K	C_p J/mol K	v_s m/s	D
P = 10 MPa								
90	16.590	−21553	−20951	86.5	59.9	84.4	2146	2.08489
100	16.364	−20714	−20103	95.4	60.1	85.1	2068	2.06350
125	15.810	−18595	−17962	114.5	59.6	86.1	1895	2.01342
150	15.259	−16448	−15793	130.3	59.3	87.5	1733	1.96617
175	14.705	−14261	−13581	144.0	59.9	89.5	1577	1.92048
200	14.141	−12016	−11309	156.1	61.4	92.4	1425	1.87557
225	13.562	−9692	−8955	167.2	63.7	96.1	1277	1.83076
250	12.960	−7268	−6496	177.5	66.7	100.7	1133	1.78529
275	12.322	−4721	−3909	187.4	70.2	106.4	991	1.73826
300	11.631	−2027	−1167	196.9	74.1	113.2	851	1.68849
325	10.860	843	1764	206.3	78.4	121.5	715	1.63437
350	9.973	3924	4927	215.7	82.9	132.0	582	1.57361
375	8.905	7270	8393	225.2	87.7	146.1	455	1.50271
400	7.561	10957	12279	235.3	93.0	165.7	339	1.41671
450	4.614	18845	21013	255.8	101.8	167.8	249	1.24060
500	3.241	25567	28652	272.0	107.8	142.7	276	1.16439
600	2.242	38131	42591	297.4	121.7	140.5	332	1.11122

VIRIAL COEFFICIENTS OF SELECTED GASES

Henry V. Kehiaian

This table gives second virial coefficients of about 110 inorganic and organic gases as a function of temperature. Selected data from the literature have been fitted by least squares to the equation

$$B / cm^3 mol^{-1} = \sum_{i=1}^{n} a(i)[(T_o / T) - 1]^{i-1}$$

where T_o = 298.15 K. The table gives the coefficients $a(i)$ and values of B at fixed temperature increments, as calculated from this smoothing equation.

The equation may be used with the tabulated coefficients for interpolation within the indicated temperature range. It should not be used for extrapolation beyond this range.

Compounds are listed in the modified Hill order (see Introduction), with carbon-containing compounds following those compounds not containing carbon.

A useful compilation of virial coefficient data from the literature may be found in:

J. H. Dymond and E. B. Smith, *The Virial Coefficients of Pure Gases and Mixtures, A Critical Compilation*, Oxford University Press, Oxford, 1980.

Compounds Not Containing Carbon

Mol. form.	Name		T/K	B/cm³ mol⁻¹
Ar	Argon		100	−184
			120	−131
			140	−98
		a(1) = −16	160	−76
		a(2) = −60	80	−60
		a(3) = −10	200	−48
			300	−16
			400	−1
			500	7
			600	12
			700	15
			800	18
			900	20
			1000	22
BF₃	Boron trifluoride		200	−338
			240	−202
			280	−129
		a(1) = −106	320	−85
		a(2) = −330	360	−56
		a(3) = −251	400	−37
		a(4) = −80	440	−23
ClH	Hydrogen chloride		190	−451
			230	−269
			270	−181
		a(1) = −144	310	−132
		a(2) = −325	350	−102
		a(3) = −277	390	−81
		a(4) = −170	430	−66
			470	−54
Cl₂	Chlorine		210	−508
			220	−483
			230	−457
		a(1) = −303	240	−432
		a(2) = −555	250	−407
		a(3) = 9	260	−383
		a(4) = 329	270	−360
		a(5) = 68	280	−339
			290	−318
			300	−299
			350	−221
			400	−166
			450	−126

Mol. form.	Name		T/K	B/cm³ mol⁻¹
			500	−97
			600	−59
			700	−36
			800	−22
			900	−12
F₂	Fluorine		80	−386
			110	−171
			140	−113
		a(1) = −25	170	−73
		a(2) = 21	200	−47
		a(3) = −185	230	−32
		a(4) = 113	260	−25
F₄Si	Silicon tetrafluoride		210	−268
			240	−213
			270	−170
		a(1) = −138	300	−136
		a(2) = −312	330	−108
			360	−84
			390	−64
			420	−47
			450	−32
F₅I	Iodine pentafluoride		320	−2540
			330	−2344
			340	−2172
		a(1) = −3077	350	−2021
		a(2) = −8474	360	−1890
		a(3) = −9116	370	−1775
			380	−1674
			390	−1587
			400	−1510
			410	−1443
F₅P	Phosphorus pentafluoride		320	−162
			340	−143
			360	−127
		a(1) = −186	380	−112
		a(2) = −345	400	−98
			420	−86
			440	−75
			460	−64
F₆Mo	Molybdenum hexafluoride		300	−896
			310	−810
			320	−737

Mol. form.	Name		T/K	B/cm³ mol⁻¹
		$a(1) = -914$	330	−677
		$a(2) = -2922$	340	−627
		$a(3) = -4778$	350	−586
			360	−553
			370	−527
			380	−506
			390	−491
F_6S	Sulfur hexafluoride		200	−685
			250	−416
			300	−275
		$a(1) = -279$	350	−190
		$a(2) = -647$	400	−135
		$a(3) = -335$	450	−96
		$a(4) = -72$	500	−68
F_6U	Uranium hexafluoride		320	−1030
			340	−905
			360	−805
		$a(1) = -1204$	380	−724
		$a(2) = -2690$	400	−658
		$a(3) = -2144$	420	−604
			440	−560
F_6W	Tungsten hexafluoride		320	−641
			340	−578
			360	−523
		$a(1) = -719$	380	−473
		$a(2) = -1143$	400	−428
			420	−387
			440	−350
			460	−317
H_2	Hydrogen		15	−230
			20	−151
			25	−108
		$a(1) = 15.4$	30	−82
		$a(2) = -9.0$	35	−64
		$a(3) = -0.2$	40	−52
			45	−42
			50	−35
			60	−24
			70	−16
			80	−11
			90	−7
			100	−3
			200	11
			300	15
			400	18
H_2O	Water		300	−1126
			320	−850
			340	−660
		$a(1) = -1158$	360	−526
		$a(2) = -5157$	380	−428
		$a(3) = -10301$	400	−356
		$a(4) = -10597$	420	−301
		$a(5) = -4415$	440	−258
			460	−224
			480	−197
			500	−175
			600	−104
			700	−67
			800	−44
			900	−30
			1000	−20

Mol. form.	Name		T/K	B/cm³ mol⁻¹
			1100	−14
			1200	−11
H_3N	Ammonia		290	−302
			300	−265
			310	−236
		$a(1) = -271$	320	−213
		$a(2) = -1022$	330	−194
		$a(3) = -2715$	340	−179
		$a(4) = -4189$	350	−166
			360	−154
			370	−144
			380	−135
			400	−118
			420	−101
H_3P	Phosphine		190	−457
			200	−404
			210	−364
		$a(1) = -146$	220	−332
		$a(2) = -733$	230	−305
		$a(3) = 1022$	240	−281
		$a(4) = -1220$	250	−258
			260	−235
			270	−213
			280	−190
			290	−166
He	Helium		2	−172
			6	−48
			10	−24
		$a(1) = 12$	14	−13
		$a(2) = -1$	18	−7
			22	−3
			26	−1
			30	1
			50	6
			70	8
			90	10
			110	10
			150	11
			250	12
			650	13
			700	13
Kr	Krypton		110	−363
			120	−307
			130	−263
		$a(1) = -51$	140	−229
		$a(2) = -118$	150	−201
		$a(3) = -29$	160	−178
		$a(4) = -5$	170	−159
			180	−143
			190	−129
			200	−117
			250	−75
			300	−51
			400	−23
			500	−8
			600	2
			700	8
NO	Nitric oxide		120	−232
			130	−176
			140	−138
		$a(1) = -12$	150	−113

Virial Coefficients of Selected Gases

Mol. form.	Name	T/K	$B/cm^3\ mol^{-1}$
	$a(2) = -119$	160	-96
	$a(3) = 89$	170	-83
	$a(4) = -73$	180	-73
		190	-65
		200	-58
		210	-52
		230	-42
		250	-32
		270	-24
N_2	Nitrogen	75	-274
		100	-161
		125	-104
	$a(1) = -4$	150	-71
	$a(2) = -56$	175	-49
	$a(3) = -12$	200	-34
		225	-24
		250	-15
		300	-4
		400	9
		500	16
		600	21
		700	24
N_2O	Nitrous oxide	240	-219
		260	-181
		280	-151
	$a(1) = -130$	300	-128
	$a(2) = -307$	320	-110
	$a(3) = -248$	340	-96
		360	-85
		380	-76
		400	-68
Ne	Neon	60	-25
		80	-13
		100	-6
	$a(1) = 10.8$	120	-1
	$a(2) = -7.5$	140	2
	$a(3) = 0.4$	160	4
		180	6
		200	7
		300	11
		400	13
		500	14
		600	15
O_2	Oxygen	90	-241
		110	-161
		130	-117
	$a(1) = -16$	150	-88
	$a(2) = -62$	170	-69
	$a(3) = -8$	190	-55
	$a(4) = -3$	210	-44
		230	-36
		250	-29
		270	-23
		290	-18
		310	-14
		330	-10
		350	-7
		400	-1
O_2S	Sulfur dioxide	290	-465
		320	-354
		350	-276

Mol. form.	Name	T/K	$B/cm^3\ mol^{-1}$
	$a(1) = -430$	380	-221
	$a(2) = -1193$	410	-181
	$a(3) = -1029$	440	-153
		470	-132
Xe	Xenon	160	-421
		170	-377
		180	-340
	$a(1) = -130$	190	-307
	$a(2) = -262$	200	-280
	$a(3) = -87$	210	-255
		220	-234
		230	-215
		240	-199
		250	-184
		300	-129
		350	-93
		400	-69
		500	-39
		600	-21
		650	-14

Compounds Containing Carbon

Mol. form.	Name	T/K	$B/cm^3\ mol^{-1}$
$CClF_3$	Chlorotrifluoromethane	240	-369
		290	-237
		340	-165
	$a(1) = -223$	390	-119
	$a(2) = -504$	440	-86
	$a(3) = -340$	490	-60
	$a(4) = -291$	540	-39
CCl_2F_2	Dichlorodifluoromethane	250	-769
		280	-570
		310	-441
	$a(1) = -486$	340	-353
	$a(2) = -1217$	370	-289
	$a(3) = -1188$	400	-241
	$a(4) = -698$	430	-204
		460	-174
CCl_3F	Trichlorofluoromethane	240	-1140
		280	-879
		320	-689
	$a(1) = -786$	360	-545
	$a(2) = -1428$	400	-431
	$a(3) = -142$	440	-340
		480	-265
CCl_4	Tetrachloromethane	320	-1345
		340	-1171
		360	-1040
	$a(1) = -1600$	380	-942
	$a(2) = -4059$	400	-868
	$a(3) = -4653$	420	-814
CF_4	Tetrafluoromethane	250	-137
		300	-87
		350	-55
	$a(1) = -88$	400	-32
	$a(2) = -238$	450	-16
	$a(3) = -70$	500	-4
		600	14
		700	25

Mol. form.	Name	T/K	B/cm³ mol⁻¹	Mol. form.	Name	T/K	B/cm³ mol⁻¹
		800	33	CH_3Cl	Chloromethane	280	−466
$CHClF_2$	Chlorodifluoromethane	300	−343			300	−402
		325	−298			320	−348
		350	−257		a(1) = −407	340	−304
	a(1) = −347	375	−221		a(2) = −887	360	−266
	a(2) = −575	400	−188		a(3) = −385	380	−234
	a(3) = 187	425	−158			400	−206
$CHCl_2F$	Dichlorofluoromethane	250	−728			420	−182
		275	−634			440	−161
		300	−557			460	−142
	a(1) = −562	325	−491			480	−126
	a(2) = −862	350	−434			500	−112
		375	−385			600	−58
		400	−343	CH_3F	Fluoromethane	280	−244
		425	−305			300	−205
		450	−271			320	−174
$CHCl_3$	Trichloromethane	320	−1001		a(1) = −209	340	−150
		330	−926		a(2) = −525	360	−129
		340	−858		a(3) = −365	380	−112
	a(1) = −1193	350	−797			400	−99
	a(2) = −2936	360	−740			420	−87
	a(3) = −1751	370	−689	CH_3I	Iodomethane	310	−725
		380	−642			320	−646
		390	−599			330	−582
		400	−559		a(1) = −844	340	−531
CHF_3	Trifluoromethane	200	−433		a(2) = −3353	350	−492
		220	−350		a(3) = −6590	360	−462
		240	−288			370	−441
	a(1) = −177	260	−241			380	−427
	a(2) = −399	280	−204	CH_4	Methane	110	−328
	a(3) = −250	300	−174			120	−276
		320	−151			130	−237
		340	−132		a(1) = −43	140	−206
		360	−116		a(2) = −114	150	−181
		380	−103		a(3) = −19	160	−160
		400	−91		a(4) = −7	170	−143
CH_2Cl_2	Dichloromethane	320	−706			180	−128
		330	−634			190	−116
		340	−574			200	−105
	a(1) = −913	350	−524			250	−66
	a(2) = −3371	360	−482			300	−43
	a(3) = −5013	370	−447			350	−27
		380	−420			400	−16
		400	−380			500	0
		420	−357			600	10
CH_2F_2	Difluoromethane	280	−375	CH_4O	Methanol	320	−1431
		290	−343			330	−1299
		300	−316			340	−1174
	a(1) = −321	310	−294		a(1) = −1752	350	−1056
	a(2) = −754	320	−275		a(2) = −4694	360	−945
	a(3) = −1300	330	−260			370	−840
		340	−248			380	−741
		350	−238			390	−646
CH_3Br	Bromomethane	280	−645			400	−557
		290	−596	CH_5N	Methylamine	300	−451
		300	−551			325	−367
		310	−509			350	−304
	a(1) = −559	320	−469		a(1) = −459	375	−257
	a(2) = −1324	340	−396		a(2) = −1191	400	−220
		360	−332		a(3) = −995	425	−192
		380	−274			450	−170

Virial Coefficients of Selected Gases

Mol. form.	Name	T/K	B/cm³ mol⁻¹
		500	−140
		550	−122
CO	Carbon monoxide	210	−36
		240	−24
		270	−15
	a(1) = −9	300	−8
	a(2) = −58	330	−3
	a(3) = −18	360	1
		420	7
		480	11
CO_2	Carbon dioxide	220	−244
		240	−204
		260	−172
	a(1) = −127	280	−146
	a(2) = −288	300	−126
	a(3) = −118	320	−108
		340	−94
		360	−81
		380	−71
		400	−62
		500	−30
		600	−13
		700	−1
		800	7
		900	12
		1000	16
		1100	19
CS_2	Carbon disulfide	280	−932
		310	−740
		340	−603
	a(1) = −807	370	−504
	a(2) = −1829	400	−431
	a(3) = −1371	430	−375
$C_2Cl_2F_4$	1,2-Dichloro-1,1,2,2-tetrafluoroethane	300	−801
		320	−695
		340	−608
	a(1) = −812	360	−536
	a(2) = −1773	380	−475
	a(3) = −963	400	−423
		420	−379
		440	−341
		460	−307
		480	−279
		500	−253
$C_2Cl_3F_3$	1,1,2-Trichloro-1,2,2-trifluoroethane	290	−1041
		310	−943
		330	−856
	a(1) = −999	350	−780
	a(2) = −1479	370	−712
		390	−651
		410	−596
		430	−546
		450	−500
C_2H_2	Ethyne	200	−573
		210	−500
		220	−440
	a(1) = −216	230	−390
	a(2) = −375	240	−349
	a(3) = −716	250	−315
		260	−287

Mol. form.	Name	T/K	B/cm³ mol⁻¹
		270	−263
C_2H_3N	Ethanenitrile	330	−3468
		340	−2971
		350	−2563
	a(1) = −5840	360	−2233
	a(2) = −29175	370	−1970
	a(3) = −47611	380	−1765
		390	−1610
		400	−1499
		410	−1425
C_2H_4	Ethene	240	−218
		270	−172
		300	−139
	a(1) = −140	330	−113
	a(2) = −296	360	−92
	a(3) = −101	390	−76
		420	−63
		450	−52
$C_2H_4Cl_2$	1,2-Dichloroethane	370	−812
		390	−716
		410	−635
	a(1) = −1362	430	−566
	a(2) = −3240	450	−508
	a(3) = −2100	470	−458
		490	−416
		510	−379
		530	−347
		550	−319
		570	−295
C_2H_4O	Ethanal	290	−1352
		320	−927
		350	−654
	a(1) = −1217	380	−482
	a(2) = −4647	410	−375
	a(3) = −5725	440	−314
		470	−283
$C_2H_4O_2$	Methyl methanoate	320	−821
		330	−744
		340	−677
	a(1) = −1035	350	−620
	a(2) = −3425	360	−571
	a(3) = −4203	370	−528
		380	−492
		390	−461
		400	−435
C_2H_5Cl	Chloroethane	320	−634
		360	−450
		400	−330
	a(1) = −777	440	−249
	a(2) = −2205	480	−195
	a(3) = −1764	520	−157
		560	−131
		600	−114
C_2H_6	Ethane	200	−409
		220	−337
		240	−284
	a(1) = −184	260	−242
	a(2) = −376	280	−209
	a(3) = −143	300	−181
	a(4) = −54	320	−159
		340	−140

Mol. form.	Name		T/K	B/cm^3 mol^{-1}
			360	−123
			380	−109
			400	−96
			500	−52
			600	−24
C$_2$H$_6$O	Ethanol		320	−2710
			330	−2135
			340	−1676
		a(1) = −4475	350	−1317
		a(2) = −29719	360	−1043
		a(3) = −56716	370	−843
			380	−705
			390	−622
C$_2$H$_6$O	Dimethyl ether		275	−536
			280	−517
			285	−499
		a(1) = −455	290	−482
		a(2) = −965	295	−465
			300	−449
			305	−433
			310	−418
C$_2$H$_7$N	Dimethylamine		310	−606
			320	−563
			330	−523
		a(1) = −662	340	−487
		a(2) = −1504	350	−454
		a(3) = −667	360	−423
			370	−395
			380	−369
			390	−345
			400	−322
C$_2$H$_7$N	Ethylamine		300	−773
			310	−710
			320	−654
		a(1) = −785	330	−604
		a(2) = −2012	340	−558
		a(3) = −1397	350	−517
			360	−480
			370	−447
			380	−416
			390	−389
			400	−363
C$_3$H$_6$	Cyclopropane		300	−383
			310	−356
			320	−332
		a(1) = −388	330	−310
		a(2) = −861	340	−290
		a(3) = −538	350	−272
			360	−256
			370	−241
			380	−227
			390	−215
			400	−204
C$_3$H$_6$	Propene		280	−395
			300	−342
			320	−299
		a(1) = −347	340	−262
		a(2) = −727	360	−232
		a(3) = −325	380	−205
			400	−183
			420	−163

Mol. form.	Name		T/K	B/cm^3 mol^{-1}
			440	−146
			460	−131
			480	−118
			500	−106
C$_3$H$_6$O	2-Propanone		300	−1996
			320	−1522
			340	−1198
		a(1) = −2051	360	−971
		a(2) = −8903	380	−806
		a(3) = −18056	400	−683
		a(4) = −16448	420	−586
			440	−506
			460	−437
			480	−375
C$_3$H$_6$O	Ethyl methanoate		330	−1003
			340	−916
			350	−839
		a(1) = −1371	360	−771
		a(2) = −4231	370	−712
		a(3) = −4312	380	−660
			390	−614
C$_3$H$_6$O	Methyl ethanoate		320	−1320
			330	−1186
			340	−1074
		a(1) = −1709	350	−980
		a(2) = −6348	360	−903
		a(3) = −9650	370	−840
			380	−789
			390	−749
C$_3$H$_7$Cl	1-Chloropropane		310	−1001
			340	−772
			370	−614
		a(1) = −1121	400	−501
		a(2) = −3271	430	−417
		a(3) = −3786	460	−352
		a(4) = −1974	490	−302
			520	−261
			550	−227
			580	−198
C$_3$H$_8$	Propane		240	−641
			260	−527
			280	−444
		a(1) = −386	300	−381
		a(2) = −844	320	−331
		a(3) = −720	340	−292
		a(4) = −574	360	−259
			380	−232
			400	−208
			440	−169
			480	−138
			520	−112
			560	−90
C$_3$H$_8$O	1-Propanol		380	−873
			385	−826
			390	−783
		a(1) = −2690	395	−744
		a(2) = −12040	400	−709
		a(3) = −16738	405	−679
			410	−651
			415	−627
			420	−606

Virial Coefficients of Selected Gases

Mol. form.	Name		T/K	B/cm^3 mol^{-1}
C$_3$H$_8$O	2-Propanol		380	−821
			385	−766
			390	−717
		a(1) = −3165	395	−674
		a(2) = −16092	400	−636
		a(3) = −24197	405	−604
			410	−576
			415	−552
			420	−533
C$_3$H$_9$N	Trimethylamine		310	−675
			320	−628
			330	−585
		a(1) = −737	340	−547
		a(2) = −1669	350	−512
		a(3) = −986	360	−480
			370	−450
C$_4$H$_8$	1-Butene		300	−624
			320	−539
			340	−470
		a(1) = −633	360	−413
		a(2) = −1442	380	−366
		a(3) = −932	400	−327
			420	−294
C$_4$H$_8$O	2-Butanone		310	−2056
			320	−1878
			330	−1712
		a(1) = −2282	340	−1555
		a(2) = −5907	350	−1407
			360	−1267
			370	−1135
C$_4$H$_8$O$_2$	Propyl methanoate		330	−1496
			340	−1354
			350	−1231
		a(1) = −2118	360	−1126
		a(2) = −7299	370	−1035
		a(3) = −8851	380	−957
			390	−890
			400	−834
C$_4$H$_8$O$_2$	Ethyl ethanoate		330	−1543
			340	−1385
			350	−1254
		a(1) = −2272	360	−1144
		a(2) = −8818	370	−1055
		a(3) = −13130	380	−982
			390	−923
			400	−878
C$_4$H$_8$O$_2$	Methyl propanoate		330	−1588
			340	−1444
			350	−1319
		a(1) = −2216	360	−1211
		a(2) = −7339	370	−1117
		a(3) = −8658	380	−1037
			390	−968
			400	−908
C$_4$H$_9$Cl	1-Chlorobutane		330	−1224
			370	−898
			410	−691
		a(1) = −1643	450	−551
		a(2) = −4897	490	−449
		a(3) = −6178	530	−371
		a(4) = −3718	570	−309

Mol. form.	Name		T/K	B/cm^3 mol^{-1}
C$_4$H$_{10}$	Butane		250	−1170
			280	−863
			310	−668
		a(1) = −735	340	−536
		a(2) = −1835	370	−442
		a(3) = −1922	400	−371
		a(4) = −1330	430	−315
			460	−270
			490	−232
			520	−199
			550	−171
C$_4$H$_{10}$	2-Methylpropane		270	−900
			300	−697
			330	−553
		a(1) = −707	360	−450
		a(2) = −1719	390	−374
		a(3) = −1282	420	−317
			450	−273
			480	−240
			510	−215
C$_4$H$_{10}$O	1-Butanol		350	−1693
			360	−1544
			370	−1402
		a(1) = −2629	380	−1268
		a(2) = −6315	390	−1141
			400	−1021
			420	−796
			440	−593
C$_4$H$_{10}$O	2-Methyl-1-propanol		390	−1076
			400	−979
			410	−887
		a(1) = −2269	420	−800
		a(2) = −5065	430	−716
			440	−636
C$_4$H$_{10}$O	2-Butanol		380	−1110
			390	−1005
			400	−906
		a(1) = −2232	410	−811
		a(2) = −5209	420	−721
C$_4$H$_{10}$O	2-Methyl-2-propanol		380	−924
			390	−827
			400	−736
		a(1) = −1952	410	−649
		a(2) = −4775	420	−567
C$_4$H$_{10}$O	Diethyl ether		280	−1550
			300	−1199
			320	−954
		a(1) = −1226	340	−776
		a(2) = −4458	360	−638
		a(3) = −7746	380	−525
		a(4) = −10005	400	−428
			420	−340
C$_4$H$_{11}$N	Diethylamine		320	−1228
			330	−1134
			340	−1056
		a(1) = −1522	350	−988
		a(2) = −5204	360	−926
		a(3) = −15047	370	−868
		a(4) = −28835	380	−812
			390	−755

Mol. form.	Name	T/K	B/cm³ mol⁻¹
		400	−697
C_5H_5N	Pyridine	350	−1257
		360	−1176
		370	−1099
a(1) = −1765		380	−1026
a(2) = −3431		390	−957
		400	−892
		420	−770
		440	−659
C_5H_{10}	Cyclopentane	300	−1049
		305	−1015
		310	−981
a(1) = −1062		315	−949
a(2) = −2116		320	−918
C_5H_{10}	1-Pentene	310	−966
		320	−898
		330	−836
a(1) = −1055		340	−780
a(2) = −2377		350	−729
a(3) = −1189		360	−681
		370	−638
		380	−598
		390	−561
		400	−527
		410	−495
$C_5H_{10}O$	2-Pentanone	330	−2850
		340	−2420
		350	−2076
a(1) = −4962		360	−1804
a(2) = −26372		370	−1595
a(3) = −46537		380	−1440
		390	−1332
C_5H_{12}	Pentane	300	−1234
		310	−1130
		320	−1038
a(1) = −1254		330	−957
a(2) = −3345		340	−884
a(3) = −2726		350	−818
		400	−579
		450	−436
		500	−348
		550	−294
C_5H_{12}	2-Methylbutane	280	−1263
		290	−1166
		300	−1079
a(1) = −1095		310	−1001
a(2) = −2503		320	−931
a(3) = −1534		330	−867
		340	−810
		350	−757
		400	−557
		450	−424
C_5H_{12}	2,2-Dimethylpropane	300	−916
		310	−843
		320	−780
a(1) = −931		330	−724
a(2) = −2387		340	−674
a(3) = −2641		350	−629
a(4) = −1810		360	−590
		370	−554
		380	−521

Mol. form.	Name	T/K	B/cm³ mol⁻¹
		390	−492
		400	−464
		450	−357
		500	−279
		550	−218
C_6H_6	Benzene	290	−1588
		300	−1454
		310	−1335
a(1) = −1477		320	−1231
a(2) = −3851		330	−1139
a(3) = −3683		340	−1056
a(4) = −1423		350	−983
		400	−712
		450	−542
		500	−429
		550	−349
		600	−291
C_6H_7N	2-Methylpyridine	360	−1656
		370	−1523
		380	−1404
a(1) = −2940		390	−1297
a(2) = −8813		400	−1202
a(3) = −7809		410	−1117
		420	−1040
		430	−972
C_6H_7N	3-Methylpyridine	380	−1819
		390	−1612
		400	−1448
a(1) = −6304		410	−1322
a(2) = −30415		420	−1230
a(3) = −44549		430	−1166
C_6H_7N	4-Methylpyridine	380	−1787
		390	−1578
		400	−1417
a(1) = −6553		410	−1297
a(2) = −32873		420	−1214
a(3) = −49874		430	−1163
C_6H_{12}	Cyclohexane	300	−1698
		320	−1391
		340	−1170
a(1) = −1733		360	−1007
a(2) = −5618		380	−883
a(3) = −9486		400	−786
a(4) = −7936		420	−707
		440	−641
		460	−584
		480	−534
		500	−488
		520	−446
		540	−406
		560	−368
C_6H_{12}	Methylcyclopentane	305	−1447
		315	−1357
		325	−1272
a(1) = −1512		335	−1192
a(2) = −2910		345	−1117
C_6H_{14}	Hexane	300	−1920
		310	−1724
		320	−1561
a(1) = −1961		330	−1424
a(2) = −6691		340	−1309

Virial Coefficients of Selected Gases

Mol. form.	Name	T/K	B/cm³ mol⁻¹
	a(3) = −13167	350	−1209
	a(4) = −15273	360	−1123
		370	−1046
		380	−978
		390	−916
		400	−859
		410	−806
		430	−707
		450	−616
$C_6H_{15}N$	Triethylamine	330	−1562
		340	−1444
		350	−1340
	a(1) = −2061	360	−1249
	a(2) = −5735	370	−1169
	a(3) = −5899	380	−1099
		390	−1037
		400	−983
C_7H_8	Toluene	350	−1641
		360	−1511
		370	−1394
	a(1) = −2620	380	−1289
	a(2) = −7548	390	−1195
	a(3) = −6349	400	−1110
		410	−1034
		420	−965
		430	−903
C_7H_{14}	1-Heptene	340	−1781
		350	−1651
		360	−1532
	a(1) = −2491	370	−1424
	a(2) = −6230	380	−1324
	a(3) = −3780	390	−1233
		400	−1150
		410	−1073
C_7H_{16}	Heptane	300	−2782
		320	−2297
		340	−1928
	a(1) = −2834	360	−1641
	a(2) = −8523	380	−1415
	a(3) = −10068	400	−1233
	a(4) = −5051	420	−1085
		440	−963
		460	−862
		480	−775

Mol. form.	Name	T/K	B/cm³ mol⁻¹
		500	−702
		540	−583
		580	−490
		620	−416
		660	−355
		700	−304
C_8H_{10}	1,2-Dimethylbenzene	380	−2046
		390	−1848
		400	−1681
	a(1) = −5632	410	−1543
	a(2) = −22873	420	−1428
	a(3) = −28900	430	−1335
		440	−1261
C_8H_{10}	1,3-Dimethylbenzene	380	−2082
		390	−1865
		400	−1679
	a(1) = −5808	410	−1521
	a(2) = −23244	420	−1388
	a(3) = −27607	430	−1276
		440	−1184
C_8H_{10}	1,4-Dimethylbenzene	380	−2043
		390	−1851
		400	−1680
	a(1) = −4921	410	−1529
	a(2) = −16843	420	−1395
	a(3) = −16159	430	−1276
		440	−1171
C_8H_{16}	1-Octene	360	−2147
		370	−2000
		380	−1861
	a(1) = −3273	390	−1729
	a(2) = −6557	400	−1604
		410	−1485
C_8H_{18}	Octane	300	−4042
		350	−2511
		400	−1704
	a(1) = −4123	450	−1234
	a(2) = −13120	500	−936
	a(3) = −16408	550	−732
	a(4) = −8580	600	−583
		650	−468
		700	−375

VAN DER WAALS CONSTANTS FOR GASES

The van der Waals equation of state for a real gas is

$$(P + n^2 a/V^2)(V - nb) = nRT$$

where P is the pressure, V the volume, T the temperature, n the amount of substance (in moles), and R the gas constant. The van der Waals constants a and b are characteristic of the substance and are independent of temperature. They are related to the critical temperature and pressure, T_c and P_c, by

$$a = 27R^2T_c^2/64P_c \quad b = RT_c/8P_c$$

This table gives values of a and b for some common gases. Most of the values have been calculated from the critical temperature and pressure values given in the table "Critical Constants" in this section. Van der Waals constants for other gases may easily be calculated from the data in that table.

To convert the van der Waals constants to SI units, note that 1 bar L^2/mol^2 = 0.1 Pa m^6/mol^2 and 1 L/mol = 0.001 m^3/mol.

Reference

Reid, R. C, Prausnitz, J. M., and Poling, B. E., *The Properties of Gases and Liquids, Fourth Edition*, McGraw-Hill, New York, 1987.

Substance	a bar L^2/mol^2	b L/mol
Acetic acid	17.71	0.1065
Acetone	16.02	0.1124
Acetylene	4.516	0.0522
Ammonia	4.225	0.0371
Aniline	29.14	0.1486
Argon	1.355	0.0320
Benzene	18.82	0.1193
Bromine	9.75	0.0591
Butane	13.89	0.1164
1-Butanol	20.94	0.1326
2-Butanone	19.97	0.1326
Carbon dioxide	3.658	0.0429
Carbon disulfide	11.25	0.0726
Carbon monoxide	1.472	0.0395
Chlorine	6.343	0.0542
Chlorobenzene	25.80	0.1454
Chloroethane	11.66	0.0903
Chloromethane	7.566	0.0648
Cyclohexane	21.92	0.1411
Cyclopropane	8.34	0.0747
Decane	52.74	0.3043
1-Decanol	59.51	0.3086
Diethyl ether	17.46	0.1333
Dimethyl ether	8.690	0.0774
Dodecane	69.38	0.3758
1-Dodecanol	75.70	0.3750
Ethane	5.580	0.0651
Ethanol	12.56	0.0871
Ethylene	4.612	0.0582
Fluorine	1.171	0.0290
Furan	12.74	0.0926
Helium	0.0346	0.0238
Heptane	31.06	0.2049
1-Heptanol	38.17	0.2150
Hexane	24.84	0.1744
1-Hexanol	31.79	0.1856
Hydrazine	8.46	0.0462
Hydrogen	0.2452	0.0265
Hydrogen bromide	4.500	0.0442
Hydrogen chloride	3.700	0.0406
Hydrogen cyanide	11.29	0.0881
Hydrogen fluoride	9.565	0.0739
Hydrogen iodide	6.309	0.0530

Substance	a bar L^2/mol^2	b L/mol
Hydrogen sulfide	4.544	0.0434
Isobutane	13.32	0.1164
Krypton	5.193	0.0106
Methane	2.303	0.0431
Methanol	9.476	0.0659
Methylamine	7.106	0.0588
Neon	0.208	0.0167
Neopentane	17.17	0.1411
Nitric oxide	1.46	0.0289
Nitrogen	1.370	0.0387
Nitrogen dioxide	5.36	0.0443
Nitrogen trifluoride	3.58	0.0545
Nitrous oxide	3.852	0.0444
Octane	37.88	0.2374
1-Octanol	44.71	0.2442
Oxygen	1.382	0.0319
Ozone	3.570	0.0487
Pentane	19.09	0.1449
1-Pentanol	25.88	0.1568
Phenol	22.93	0.1177
Propane	9.39	0.0905
1-Propanol	16.26	0.1079
2-Propanol	15.82	0.1109
Propene	8.442	0.0824
Pyridine	19.77	0.1137
Pyrrole	18.82	0.1049
Silane	4.38	0.0579
Sulfur dioxide	6.865	0.0568
Sulfur hexafluoride	7.857	0.0879
Tetrachloromethane	20.01	0.1281
Tetrachlorosilane	20.96	0.1470
Tetrafluoroethylene	6.954	0.0809
Tetrafluoromethane	4.040	0.0633
Tetrafluorosilane	5.259	0.0724
Tetrahydrofuran	16.39	0.1082
Thiophene	17.21	0.1058
Toluene	24.86	0.1497
1,1,1-Trichloroethane	20.15	0.1317
Trichloromethane	15.34	0.1019
Trifluoromethane	5.378	0.0640
Trimethylamine	13.37	0.1101
Water	5.537	0.0305
Xenon	4.192	0.0516

MEAN FREE PATH AND RELATED PROPERTIES OF GASES

In the simplest version of the kinetic theory of gases, molecules are treated as hard spheres of diameter d which make binary collisions only. In this approximation the mean distance traveled by a molecule between successive collisions, the mean free path l, is related to the collision diameter by:

$$l = \frac{kT}{\pi\sqrt{2}Pd^2}$$

where P is the pressure, T the absolute temperature, and k the Boltzmann constant. At standard conditions ($P = 100\,000$ Pa and $T = 298.15$ K) this relation becomes:

$$l = \frac{9.27 \cdot 10^{27}}{d^2}$$

where l and d are in meters.

Using the same model and the same standard pressure, the collision diameter can be calculated from the viscosity η by the kinetic theory relation:

$$\eta = \frac{2.67 \cdot 10^{-20}(MT)^{1/2}}{d^2}$$

where η is in units of μPa s and M is the molar mass in g/mol. Kinetic theory also gives a relation for the mean velocity $\bar{v}$ of molecules of mass m:

$$\bar{v} = \left(\frac{8kT}{\pi m}\right)^{1/2} = 145.5(T/M)^{1/2}\,\text{m/s}$$

Finally, the mean time τ between collisions can be calculated from the relation $\tau\bar{v} = l$.

The table below gives values of l, $\bar{v}$, and τ for some common gases at 25°C and atmospheric pressure, as well as the value of d, all calculated from measured gas viscosities (see References 2 and 3 and the table "Viscosity of Gases" in this section). It is seen from the above equations that the mean free path varies directly with T and inversely with P, while the mean velocity varies as the square root of T and, in this approximation, is independent of P.

A more accurate model, in which molecular interactions are described by a Lennard-Jones potential, gives mean free path values about 5% lower than this table (see Reference 4).

References

1. Reid, R. C., Prausnitz, J. M., and Poling, B. E., *The Properties of Gases and Liquids, Fourth Edition*, McGraw-Hill, New York, 1987.
2. Lide, D. R., and Kehiaian, H. V., *CRC Handbook of Thermophysical and Thermochemical Data*, CRC Press, Boca Raton, FL, 1994.
3. Vargaftik, N. B., *Tables of Thermophysical Properties of Liquids and Gases, Second Edition*, John Wiley, New York, 1975.
4. Kaye, G. W. C., and Laby, T. H., *Tables of Physical and Chemical Constants, 15th Edition*, Longman, London, 1986.

Gas	d	l	$\bar{v}$	τ
Air	$3.66 \cdot 10^{-10}$ m	$6.91 \cdot 10^{-8}$ m	467 m/s	148 ps
Ar	3.58	7.22	397	182
CO_2	4.53	4.51	379	119
H_2	2.71	12.6	1769	71
He	2.15	20.0	1256	159
Kr	4.08	5.58	274	203
N_2	3.70	6.76	475	142
NH_3	4.32	4.97	609	82
Ne	2.54	14.3	559	256
O_2	3.55	7.36	444	166
Xe	4.78	4.05	219	185

INFLUENCE OF PRESSURE ON FREEZING POINTS

This table illustrates the variation of the freezing point of representative types of liquids with pressure. Substances are listed in alphabetical order. Note that 1 MPa = 0.01 kbar = 9.87 atm.

References

1. Isaacs, N. S., *Liquid Phase High Pressure Chemistry*, John Wiley, New York, 1981.
2. Merrill, L., *J. Phys. Chem. Ref. Data*, 6, 1205, 1977; 11, 1005, 1982.

Substance	Molecular formula	Freezing point in °C at:		
		0.1 MPa	100 MPa	1000 MPa
Acetic acid	$C_2H_4O_2$	16.6	37	
Acetophenone	C_8H_8O	20.0	41.2	
Aniline	C_6H_7N	−6.0	13.5	140
Benzene	C_6H_6	5.5	33.4	
Benzonitrile	C_7H_5N	−12.8	7.6	
Benzyl alcohol	C_7H_8O	−15.2	0.2	
Bromobenzene	C_6H_5Br	−30.6	−12	108
Bromoethane	C_2H_5Br	−118.6	−108	
1-Bromonaphthalene	$C_{10}H_7Br$	−1.8	6.1	
1-Bromopropane	C_3H_7Br	−110	−98	
p-Bromotoluene	C_7H_7Br	28.0	56.7	
Butanoic acid	$C_4H_8O_2$	−5.7	13.8	
1-Butanol	$C_4H_{10}O$	−89.8	−77.2	
Carbon disulfide	CS_2	−111.5	−98	
Chlorobenzene	C_6H_5Cl	−45.2	−28	84
p-Chlorotoluene	C_7H_7Cl	6.9	33.1	
o-Cresol	C_7H_8O	29.8	47.7	
m-Cresol	C_7H_8O	11.8	25.6	
p-Cresol	C_7H_8O	35.8	56.2	
Cyclohexane	C_6H_{12}	6.6	32.5	
Cyclohexanol	$C_6H_{12}O$	25.5	62.3	
1,2-Dibromoethane	$C_2H_4Br_2$	9.9	34.0	
p-Dichlorobenzene	$C_6H_4Cl_2$	52.7	79.1	
Dichloromethane	CH_2Cl_2	−95.1	−83	
N,N-Dimethylaniline	$C_8H_{11}N$	2.5	26.3	
1,4-Dioxane	$C_4H_8O_2$	11	23	
Ethanol	C_2H_6O	−114.1	−108	
Formamide	CH_3NO	−15.5	10.8	
Formic acid	CH_2O_2	8.3	20.6	
Furan	C_4H_4O	−85.6	−73	
Hexamethyldisiloxane	$C_6H_{18}OSi_2$	−66	−37	
Menthol	$C_{10}H_{20}O$	42	60	
Methyl benzoate	$C_8H_8O_2$	−15	31.8	
2-Methyl-2-butanol	$C_5H_{12}O$	−8.8	13.4	
2-Methyl-2-propanol	$C_4H_{10}O$	25.4	58.1	
Naphthalene	$C_{10}H_8$	78.2	115.7	
Nitrobenzene	$C_6H_5NO_2$	5.7	13.5	
m-Nitrotoluene	$C_7H_7NO_2$	15.5	40.6	
Pentachloroethane	C_2HCl_5	−29.0	−6.3	
Potassium	K	63.7	78	170
Potassium chloride	ClK	771		945
Propanoic acid	$C_3H_6O_2$	−20.7	−1.2	
Silver chloride	AgCl	455		545
Sodium	Na	97.8	106	167
Sodium chloride	ClNa	800.7		997
Sodium fluoride	FNa	996		1115
Tetrachloromethane	CCl_4	−23.0	14.2	
Tribromomethane	$CHBr_3$	8.1	31.5	
Trichloromethane	$CHCl_3$	−63.6	−45.2	
Water	H_2O	0.0	−9.0	
o-Xylene	C_8H_{10}	−25.2	−3.5	
m-Xylene	C_8H_{10}	−47.8	−25.2	
p-Xylene	C_8H_{10}	13.2	46.0	

CRITICAL CONSTANTS

The parameters of the liquid–gas critical point are important constants in determining the behavior of fluids. This table lists the critical temperature, pressure, and molar volume, as well as the normal boiling point, for approximately 850 inorganic and organic substances. The properties and their units are:

T_b: Normal boiling point in kelvins at a pressure of 101.325 kPa (1 atmosphere); an "s" following the value indicates a sublimation point (temperature at which the solid is in equilibrium with the gas at a pressure of 101.325 kPa)

T_c: Critical temperature in kelvins

P_c: Critical pressure in megapascals

V_c: Critical molar volume in cm^3/mol

The number of digits given for T_b, T_c, and P_c indicates the estimated accuracy of these quantities; however, values of T_c greater than 750 K may be in error by 10 K or more. Although most V_c values are given to three figures, they cannot be assumed accurate to better than a few percent. All values are experimentally determined except for a few values, indicated by an asterisk*, which are based on extrapolations. Methods of measurement are described and critiqued in Reference 1.

Many of the critical constants in this table are taken from reviews produced by the IUPAC Commission on Thermodynamics (References 1– 8). Compounds are listed by molecular formula in modified Hill order, with compounds not containing carbon preceding those that do contain carbon.

The assistance of Douglas Ambrose is gratefully acknowledged.

References

1. Ambrose, D., and Young, C. L., *J. Chem. Eng. Data*, 40, 345, 1995. [IUPAC Part 1]
2. Ambrose, D., and Tsonopoulos, C., *J. Chem. Eng. Data*, 40, 531, 1995. [IUPAC Part 2]
3. Tsonopoulos, C., and Ambrose, D., *J. Chem. Eng. Data*, 40, 547, 1995. [IUPAC Part 3]
4. Gude, M., and Teja, A. S., *J. Chem. Eng. Data*, 40, 1025, 1995. [IUPAC Part 4]
5. Daubert, T. E., *J. Chem. Eng. Data*, 41, 365, 1996. [IUPAC Part 5]
6. Tsonopoulos, C., and Ambrose, D., *J. Chem. Eng. Data*, 41, 645, 1996. [IUPAC Part 6]
7. Kudcharker, A. P., Ambrose, D., and Tsonopoulos, C., *J. Chem. Eng. Data*, 46, 457, 2001. [IUPAC Part 7]
8. Tsonopoulos, C., and Ambrose, D., *J. Chem. Eng. Data*, 46, 480, 2001. [IUPAC Part 8]
9. Ambrose, D., "Vapor-Liquid Constants of Fluids", in Stevenson, R. M., and Malanowski, S., Eds., *Handbook of the Thermodynamics of Organic Compounds*, Elsevier, New York, 1987.
10. Das, A., Frenkel, M., Gadalla, N. A. M., Kudchadker, S., Marsh, K. N., Rodgers, A. S., and Wilhoit, R. C., *J. Phys. Chem. Ref. Data*, 22, 659, 1993.
11. Wilson, L. C., Wilson, H. L., Wilding, W. V., and Wilson, G. M., *J. Chem. Eng. Data*, 41, 1252, 1996.
12. Daubert, T. E., Danner, R. P., Sibul, H. M., and Stebbins, C. C., *Physical and Thermodynamic Properties of Pure Compounds: Data Compilation*, extant 2002 (core with supplements), Taylor & Francis, Bristol, PA.
13. Morton, D. W., Lui, M. P. W., Tran, C. A., and Young, C. L., *J. Chem. Eng. Data*, 45, 437, 2000.
14. VonNiederhausern, D. M., Wilson, L. C., Giles, N. F., and Wilson, G. M., *J. Chem. Eng. Data*, 45, 154, 2000.
15. VonNiederhausern, D. M., Wilson, G. M., and Giles, N. F., *J. Chem. Eng. Data*, 45, 157, 2000.
16. Nikitin, E. D., Popov, A. P., Bogatishcheva, N. S., and Yatluk, Y. G., *J. Chem. Eng. Data*, 47, 1012, 2002.
17. Wilson, G. M., VonNiederhausern, D. M., and Giles, N. F., *J. Chem. Eng. Data*, 47, 761, 2002.
18. Wang, B. H., Adcock, J. L., Mathur, S. B., and Van Hook, W. A., *J. Chem. Thermodynamics*, 23, 699, 1991.
19. Chae, H. B., Schmidt, J. W., and Moldover, M. R., *J. Phys. Chem.*, 94, 8840, 1990.
20. Dillon, I. G., Nelson, P. A., and Swanson, B. S., *J. Chem. Phys.*, 44, 4229, 1966.
21. *Physical Constants of Hydrocarbon and Non-Hydrocarbon Compounds*, ASTM Data Series DS 4B, ASTM, Philadelphia, 1988.
22. Nowak, P., Tielkes, T., Kleinraum, R., and Wagner, W., *J. Chem. Thermodynamics*, 29, 885, 1997.
23. Steele, W. V., Chirico, R. D., Nguyen, A., and Knipmeyer, S. E., *J. Chem. Thermodynamics*, 27, 311, 1995
24. Duan, Y. Y., Shi, L., Zhu, M. S., and Han, L. Z., *J. Chem. Eng. Data*, 44, 501, 1999.
25. Weber, L. A., and Defibaugh, D. R., *J. Chem. Eng. Data*, 41, 382, 1996.
26. Duarte-Garza, H. A., Hwang, C. A., Kellerman, S. A., Miller, R. C., Hall, K. R., and Holste, J. C., *J. Chem. Eng. Data*, 42, 497, 1997.
27. Weber, L. A., and Defibaugh, D. R., *J. Chem. Eng. Data*, 41, 1477, 1996.
28. Fujiwara, K., Nakamura, S., and Noguchi, M., *J. Chem. Eng. Data*, 43, 55, 1998.
29. Widiatmo, J. V., Morimoto, Y., and Watanabe, K., *J. Chem. Eng. Data*, 47, 1246, 2002.
30. Duarte-Garza, H. A., Stouffer, C. E., Hall, K. R., Holste, J. C., Marsh, K. N., and Gammon, B. E., *J. Chem. Eng. Data*, 42, 745, 1997.
31. Nikitin, E. D., Pavlov, P. A., Popov, A. P., and Nikitina, H. E., *J. Chem. Thermodynamics*, 27, 945, 1995.
32. Sako, T., Sato, M., Nakazawa, N., Oowa, M., Yasumoto, M., Ito, H., and Yamashita, S., *J. Chem. Eng. Data*, 41, 802, 1996.
33. Zhang, H-L, Sato, H., and Watanabe, K., *J. Chem. Eng. Data*, 40, 1281, 1995.
34. Sifner, O., and Klomfar, J., *J. Phys. Chem. Ref. Data*, 23, 63, 1994.
35. Younglove, B. A., and McLinden, M. O., *J. Phys. Chem. Ref. Data*, 23, 731, 1994.
36. Tillner-Roth, R., and Baehr, H. D., *J. Phys. Chem. Ref. Data*, 23, 657, 1994.
37. Xiang, H. W., *J. Phys. Chem. Ref. Data*, 30, 1161, 2001.
38. Goodwin, A. H. R., Defibaugh, D. R., and Weber, L. A., *J. Chem. Eng. Data*, 43, 846, 1998.
39. Lim, J. S., Park, K. H., Lee, B. G., and Kim, J-D., *J. Chem. Eng. Data*, 46, 1580, 2001.
40. Linstrom, P. J., and Mallard, W. G., Eds., *NIST Chemistry WebBook*, NIST Standard Reference Database No. 69, July 2001, National Institute of Standards and Technology, Gaithersburg, MD 20899, http://webbook.nist.gov.
41. *ASHRAE Fundamentals Handbook 2001*, Chapter 19. Refrigerants, American Society of Heating, Refrigerating, and Air-Conditioning Engineers, Atlanta, GA, 2001.
42. Fialho, P. S., and Nieto de Castro, C. A., *Int. J. Thermophys.*, 21, 385, 2000.
43. Vargaftik, N. B., *Int. J. Thermophys.*, 11, 467, 1990
44. Vargaftik, N.B., Vinogradov, Y. K., and Yargin, V. S., *Handbook of Physical Properties of Liquids and Gases, Third Edition*, Begell House, New York, 1996.
45. Schmidt, J. W., Carrillo-Nava, E., and Moldover, M. R., *Fluid Phase Equilibria*, 122, 187, 1996.
46. Defibaugh, D. R., Gillis, K. A., Moldover, M. R., Morrison, G., and Schmidt, J. W., *Fluid Phase Equilibria*, 81, 285, 1992.
47. Salvi-Narkhede, M., Wang, B-H., Adcock, J. L., and Van Hook, W. A., *J. Chem. Thermodynamics*, 24, 1065, 1992.
48. Nikitin, E. D., Pavlov, P. A., and Pavlov, A. P., *Fluid Phase Equilib.*, 189, 151–161, 2001.

Molecular formula	Name	T_b/K	T_c/K	P_c/MPa	$V_c/cm^3 mol^{-1}$	Ref.
$AlBr_3$	Aluminum bromide	528	763	2.89	310	9
$AlCl_3$	Aluminum chloride	453 s	620	2.63	257	9
AlI_3	Aluminum iodide	655	983		408	9
Ar	Argon	87.30	150.87	4.898	75	9
As	Arsenic	876	1673		35	9
$AsCl_3$	Arsenic(III) chloride	403	654		252	9
AsH_3	Arsine	210.7	373.1			9
BBr_3	Boron tribromide	364	581		272	9
BCl_3	Boron trichloride	285.80	455	3.87	239	9
BF_3	Boron trifluoride	172	260.8	4.98	115	9
BI_3	Boron triiodide	482.7	773		356	9
B_2H_6	Diborane	180.8	289.8	4.05		9
$BiBr_3$	Bismuth tribromide	726	1220		301	9
$BiCl_3$	Bismuth trichloride	720	1179	12.0	261	9
BrH	Hydrogen bromide	206.77	363.2	8.55		9
BrI	Iodine bromide	389	719		139	9
Br_2	Bromine	332.0	588	10.34	127	9
Br_2Hg	Mercury(II) bromide	595	1012			9
Br_3Ga	Gallium(III) bromide	552	806.7		303	9
Br_3HSi	Tribromosilane	382	610.0		305	9
Br_3P	Phosphorus(III) bromide	446.4	711		300	9
Br_3Sb	Antimony(III) bromide	553	904		300	9
Br_4Ge	Germanium(IV) bromide	459.50	718		392	9
Br_4Hf	Hafnium(IV) bromide	596 s	746		415	9
Br_4Si	Tetrabromosilane	427	663		382	9
Br_4Sn	Tin(IV) bromide	478	744		417	9
Br_4Ti	Titanium(IV) bromide	503	795.7		391	9
Br_4Zr	Zirconium(IV) bromide	633 s	805		424	9
Br_5Ta	Tantalum(V) bromide	622	974		461	9
$ClFO_3$	Perchloryl fluoride	226.40	368.4	5.37	161	9
ClF_2N	Nitrogen chloride difluoride	206	337.5	5.15		9
ClF_2P	Phosphorus(III) chloride difluoride	225.9	362.4	4.52		9
ClF_2PS	Phosphorothioc chloride difluoride	279.5	439.2	4.14		9
ClF_3Si	Chlorotrifluorosilane	203.2	307.7	3.46		9
ClF_5	Chlorine pentafluoride	260.1	416	5.27	233	9
ClF_5S	Sulfur chloride pentafluoride	254.10	390.9			9
ClH	Hydrogen chloride	188	324.7	8.31	81	9
ClH_4N	Ammonium chloride	611 s	1155	163.5		9
ClH_4P	Phosphonium chloride	246 s	322.3	7.37		9
$ClNO$	Nitrosyl chloride	267.7	440			9
$ClOV$	Vanadyl chloride	400	636		171	9
Cl_2	Chlorine	239.11	416.9	7.991	123	9
Cl_2FP	Phosphorus(III) dichloride fluoride	287.00	463.0	4.96		9
Cl_2F_2Si	Dichlorodifluorosilane	241	369.0	3.5		9
Cl_2Hg	Mercury(II) chloride	577	973		174	9
Cl_2OSe	Selenium oxychloride	450	730	7.09	235	9
Cl_3FSi	Trichlorofluorosilane	285.40	438.6	3.58		9
Cl_3Ga	Gallium(III) chloride	474	694		263	9
Cl_3HSi	Trichlorosilane	306	479		268	9
Cl_3P	Phosphorus(III) chloride	349.3	563		264	9
Cl_3Sb	Antimony(III) chloride	493.5	794		272	9
Cl_4Ge	Germanium(IV) chloride	359.70	553.2	3.861	330	9
Cl_4Hf	Hafnium(IV) chloride	590 s	725.7	5.42	314	9
Cl_4ORe	Rhenium(VI) oxytetrachloride	496	781		362	9
Cl_4OW	Tungsten(VI) oxytetrachloride	500.70	782		338	9
Cl_4Si	Tetrachlorosilane	330.80	508.1	3.593	326	9
Cl_4Sn	Tin(IV) chloride	387.30	591.9	3.75	351	9
Cl_4Te	Tellurium tetrachloride	660	1002	8.56	310	9
Cl_4Ti	Titanium(IV) chloride	409.60	638	4.66	339	9
Cl_4Zr	Zirconium(IV) chloride	604 s	778	5.77	319	9

Molecular formula	Name	T_b/K	T_c/K	P_c/MPa	V_c/cm^3 mol^{-1}	Ref.
Cl$_5$Mo	Molybdenum(V) chloride	541	850		369	9
Cl$_5$Nb	Niobium(V) chloride	527.2	803.5	4.88	397	9
Cl$_5$P	Phosphorus(V) chloride	433 s	646			9
Cl$_5$Ta	Tantalum(V) chloride	512.50	767		402	9
Cl$_6$W	Tungsten(VI) chloride	619.90	923		422	9
Cs	Cesium	944	1938	9.4	341	43
FH	Hydrogen fluoride	293	461	6.48	69	9
FNO$_2$	Nitryl fluoride	200.8	349.5			9
F$_2$	Fluorine	85.03	144.13	5.172	66	9
F$_2$HN	Difluoramine	250	403			9
F$_2$N$_2$	cis-Difluorodiazine	167.40	272	7.09		9
F$_2$N$_2$	trans-Difluorodiazine	161.70	260	5.57		9
F$_2$O	Fluorine monoxide	128.40	215			9
F$_2$Xe	Xenon difluoride	387.50	631	9.32	148	9
F$_3$N	Nitrogen trifluoride	144.40	234.0	4.46	126	9
F$_3$NO	Trifluoramine oxide	185.7	303	6.43	147	9
F$_3$P	Phosphorus(III) fluoride	171.4	271.2	4.33		9
F$_3$PS	Phosphorothioc trifluoride	220.90	346.0	3.82		9
F$_4$N$_2$	Tetrafluorohydrazine	199	309	3.75		9
F$_4$S	Sulfur tetrafluoride	232.70	364			9
F$_4$Si	Tetrafluorosilane	187	259.0	3.72		9
F$_4$Xe	Xenon tetrafluoride	388.90	612	7.04	188	9
F$_5$Nb	Niobium(V) fluoride	502	737	6.28	155	9
F$_6$Mo	Molybdenum(VI) fluoride	307.2	473	4.75	226	9
F$_6$S	Sulfur hexafluoride	209.35	318.69	3.77	199	9
F$_6$Se	Selenium hexafluoride	226.55	345.5			9
F$_6$Te	Tellurium hexafluoride	234.25	356			9
F$_6$U	Uranium(VI) fluoride	329.65	505.8	4.66	250	9
F$_6$W	Tungsten(VI) fluoride	290.3	444	4.34	233	9
GaI$_3$	Gallium(III) iodide	613	951		395	9
GeH$_4$	Germane	185.1	312.2	4.95	147	9
GeI$_4$	Germanium(IV) iodide	650	973		500	9
HI	Hydrogen iodide	237.60	424.0	8.31		9
H$_2$	Hydrogen	20.28	32.97	1.293	65	9
H$_2$O	Water	373.2	647.14	22.06	56	9
H$_2$O$_2$	Hydrogen peroxide	423.4	728*	22*		31
H$_2$S	Hydrogen sulfide	213.60	373.2	8.94	99	9
H$_2$Se	Hydrogen selenide	231.90	411	8.92		9
H$_3$N	Ammonia	239.82	405.5	11.35	72	9
H$_3$P	Phosphine	185.40	324.5	6.54		9
H$_4$N$_2$	Hydrazine	386.70	653	14.7		9
He	Helium	4.22	5.19	0.227	57	9
HfI$_4$	Hafnium iodide	667 s	916		528	9
Hg	Mercury	629.88	1750	172.00	43	9
HgI$_2$	Mercury(II) iodide	627	1072			9
I$_2$	Iodine	457.6	819		155	9
I$_3$Sb	Antimony(III) iodide	674	1102			9
I$_4$Si	Tetraiodosilane	560.50	944		558	9
I$_4$Sn	Tin(IV) iodide	637.50	968		531	9
I$_4$Ti	Titanium(IV) iodide	650	1040		505	9
I$_4$Zr	Zirconium(IV) iodide	704 s	960		530	9
K	Potassium	1032	2223*	16*	209*	20
Kr	Krypton	119.93	209.41	5.50	91	9
Li	Lithium	1615	3223*	67*	66*	20
NO	Nitric oxide	121.41	180	6.48	58	9
N$_2$	Nitrogen	77.36	126.21	3.39	90	9
N$_2$O	Nitrous oxide	184.67	309.57	7.255	97	9
N$_2$O$_4$	Nitrogen tetroxide	294.30	431	10.1	167	9
Na	Sodium	1156	2573*	35*	116*	20
Ne	Neon	27.07	44.4	2.76	42	9

Molecular formula	Name	T_b/K	T_c/K	P_c/MPa	$V_c/cm^3\,mol^{-1}$	Ref.
O_2	Oxygen	90.20	154.59	5.043	73	9
O_2S	Sulfur dioxide	263.10	430.8	7.884	122	9
O_3	Ozone	161.80	261.1	5.57	89	9
O_3S	Sulfur trioxide	318	491.0	8.2	127	9
O_4Os	Osmium(VIII) oxide	408	678			9
O_7Re_2	Rhenium(VII) oxide	633	942		334	9
P	Phosphorus	553.7	994			9
Rb	Rubidium	961	2093*	16*	247*	20
Rn	Radon	211.5	377	6.28		9
S	Sulfur	717.75	1314	20.7		9
Se	Selenium	958	1766	27.2		9
Xe	Xenon	165.03	289.77	5.841	118	34
$CBrClF_2$	Bromochlorodifluoromethane	269.5	426.88	4.254	246	9
$CBrF_3$	Bromotrifluoromethane	215.4	340.2	3.97	196	9
CBr_2F_2	Dibromodifluoromethane	295.91	471.3	4.45		9
$CClF_3$	Chlorotrifluoromethane	191.8	302	3.870	180	9
CCl_2F_2	Dichlorodifluoromethane	243.4	384.95	4.136	217	9
CCl_2O	Carbonyl chloride [Phosgene]	281	455	5.67	190	18
CCl_3F	Trichlorofluoromethane	296.9	471.1	4.47	247	9
CCl_4	Tetrachloromethane	350.0	556.6	4.516	276	9
CF_3I	Trifluoroiodomethane	250.7	396.44	3.953	226	24
CF_4	Tetrafluoromethane	145.2	227.6	3.74	140	9
$CHBrF_2$	Bromodifluoromethane	258.6	411.98	5.132	275	47
$CHClF_2$	Chlorodifluoromethane	232.5	369.5	5.035	164	18
$CHCl_2F$	Dichlorofluoromethane	282.1	451.58	5.18	196	9
$CHCl_3$	Trichloromethane	334.32	536.4	5.47	239	9
CHF_3	Trifluoromethane	191.1	298.98	4.82	133	42
CHN	Hydrogen cyanide	299	456.7	5.39	139	9
CH_2ClF	Chlorofluoromethane	264.1	427	5.70		37
CH_2Cl_2	Dichloromethane	313	510	6.10		9
CH_2F_2	Difluoromethane	221.6	351.56	5.83	123	42
CH_2O_2	Formic acid	374	588			7
CH_3Cl	Chloromethane	249.06	416.25	6.679	139	9
CH_3Cl_3Si	Methyltrichlorosilane	338.8	517	3.28	348	9
CH_3F	Fluoromethane	194.8	317.8	5.88	113	9
CH_3I	Iodomethane	315.58	528			9
CH_3NO_2	Nitromethane	374.34	588	5.87	173	9
CH_4	Methane	111.67	190.56	4.599	98.60	2
CH_4O	Methanol	337.8	512.5	8.084	117	4
CH_4S	Methanethiol	279.1	470	7.23	147	8
CH_5ClSi	Chloromethylsilane	280	517.8			9
CH_5N	Methylamine	266.83	430.7	7.614		9
CH_6N_2	Methylhydrazine	360.7	567	8.24	271	9
CH_6Si	Methylsilane	215.7	352.4			8
CO	Carbon monoxide	81.7	132.91	3.499	93	9
COS	Carbon oxysulfide	223	375	5.88	137	9
CO_2	Carbon dioxide	194.6 s	304.13	7.375	94	22
CS_2	Carbon disulfide	319	552	7.90	173	9
$C_2Br_2ClF_3$	1,2-Dibromo-1-chloro-1,2,2-trifluoroethane	366	560.7	3.61	368	9
$C_2Br_2F_4$	1,2-Dibromotetrafluoroethane	320.50	487.8	3.393	341	9
C_2ClF_3	Chlorotrifluoroethene	245.4	379	4.05	212	9
C_2ClF_5	Chloropentafluoroethane	234.1	353.2	3.229	252	9
$C_2Cl_2F_4$	1,1-Dichloro-1,2,2,2-tetrafluoroethane	276.6	418.6	3.30	294	9
$C_2Cl_2F_4$	1,2-Dichloro-1,1,2,2-tetrafluoroethane	276.7	418.78	3.252	297	42
$C_2Cl_3F_3$	1,1,1-Trichloro-2,2,2-trifluoroethane	318.7	482.9			40
$C_2Cl_3F_3$	1,1,2-Trichloro-1,2,2-trifluoroethane	320.9	487.3	3.42	325	9
C_2Cl_4	Tetrachloroethene	394.5	620.2			9
$C_2Cl_4F_2$	1,1,2,2-Tetrachloro-1,2-difluoroethane	366.0	551			9
C_2Cl_6	Hexachloroethane	457.85	695	3.34*	412*	12
C_2F_3N	Trifluoroacetonitrile	204.4	311.11	3.618	202	9

Molecular formula	Name	T_b/K	T_c/K	P_c/MPa	$V_c/cm^3 mol^{-1}$	Ref.
C_2F_4	Tetrafluoroethene	197.3	306.5	3.94	172	9
C_2F_6	Hexafluoroethane	195.1	293		222	9
C_2HClF_2	1-Chloro-2,2-difluoroethene	254.7	400.6	4.46	197	9
C_2HClF_4	1-Chloro-1,1,2,2-tetrafluoroethane	261.5	399.9	3.72	244	9
C_2HClF_4	1-Chloro-1,2,2,2-tetrafluoroethane	261	395.65	3.643	244	42
$C_2HCl_2F_3$	1,2-Dichloro-1,1,2-trifluoroethane	302.7	461.6		278	19
$C_2HCl_2F_3$	2,2-Dichloro-1,1,1-trifluoroethane	300.97	456.83	3.661	278	35
C_2HCl_3	Trichloroethene	360.36	544.2	5.02		9
$C_2HF_3O_2$	Trifluoroacetic acid	346	491.3	3.258	204	9
C_2HF_5	Pentafluoroethane	225.1	339.17	3.620	208.0	29,30
C_2HF_5O	Trifluoromethyl difluoromethyl ether	235	354.0	3.33	192	25,45,46
C_2H_2	Acetylene	188.45	308.3	6.138	112.2	6
$C_2H_2ClF_3$	2-Chloro-1,1,1-trifluoroethane	279.3	425.01		228	40
$C_2H_2Cl_2$	cis-1,2-Dichloroethene	333.3	544.2			9
$C_2H_2Cl_2$	trans-1,2-Dichloroethene	321.9	516.5	5.51		9
$C_2H_2Cl_4$	1,1,2,2-Tetrachloroethane	418.4	661.15			9
$C_2H_2F_2$	1,1-Difluoroethene	187.5	302.9	4.46	154	9
$C_2H_2F_4$	1,1,1,2-Tetrafluoroethane	247.07	374.18	4.065	198	36
$C_2H_2F_4$	1,1,2,2-Tetrafluoroethane	253.3	391.74	4.615	191	19,42
$C_2H_2F_4O$	Bis(difluoromethyl) ether	275	420.25	4.228	223	46
C_2H_3Cl	Chloroethene [Vinyl chloride]	259.4	432	5.67	179	12
$C_2H_3ClF_2$	1-Chloro-1,1-difluoroethane	264.1	410.34	4.048	225	19,32
$C_2H_3Cl_2F$	1,1-Dichloro-1-fluoroethane	305.2	477.5	4.194	255	26,42
$C_2H_3Cl_3$	1,1,1-Trichloroethane	347.24	545	4.30		9
$C_2H_3Cl_3$	1,1,2-Trichloroethane	387.0	602*	4.48*	281*	12
C_2H_3F	Fluoroethene	201	327.9	5.24	144	9
$C_2H_3F_3$	1,1,1-Trifluoroethane	225.90	345.86	3.764	194	27,28
$C_2H_3F_3$	1,1,2-Trifluoroethane	276.9	429.8	5.241	207	40
$C_2H_3F_3O$	Methyl trifluoromethyl ether	249.49	378.02	3.588	228	47
C_2H_3N	Acetonitrile	354.80	545.6	4.884	173	14
C_2H_4	Ethylene [Ethene]	169.38	282.34	5.041	131	6
$C_2H_4Br_2$	1,2-Dibromoethane	404.8	583.0	7.2		9
$C_2H_4Cl_2$	1,1-Dichloroethane	330.5	523	5.07	236	9
$C_2H_4Cl_2$	1,2-Dichloroethane	356.7	561	5.4	225	9
$C_2H_4F_2$	1,1-Difluoroethane	249.10	386.7	4.50	181	9,19
C_2H_4O	Acetaldehyde	293.3	466		154	7
C_2H_4O	Oxirane [Ethylene oxide]	283.8	469	7.2	142	7
$C_2H_4O_2$	Acetic acid	391.1	590.7	5.78	171	7
$C_2H_4O_2$	Methyl formate	304.9	487.2	6.00	172	7
C_2H_5Br	Bromoethane	311.7	503.9	6.23	215	9
C_2H_5Cl	Chloroethane	285.5	460.4	5.3		9
C_2H_5F	Fluoroethane	235.5	375.31	5.028		9
C_2H_6	Ethane	184.6	305.32	4.872	145.5	2
$C_2H_6Cl_2Si$	Dichlorodimethylsilane	343.5	520.4	3.49	350	9
C_2H_6O	Ethanol	351.44	514.0	6.137	168	4
C_2H_6O	Dimethyl ether	248.4	400.2	5.34	168	7
$C_2H_6O_2$	1,2-Ethanediol	470.5	720	8		7,14
C_2H_6S	Ethanethiol	308.2	499	5.49	207	8
C_2H_6S	Dimethyl sulfide	310.48	503	5.53	203.7	8
$C_2H_6S_2$	Dimethyl disulfide	382.89	615			8
C_2H_7N	Ethylamine	289.7	456	5.62	182	9
C_2H_7N	Dimethylamine	280.03	437.22	5.340		9
$C_2H_8N_2$	1,2-Ethanediamine	390	613.1	6.707		11,12
C_2N_2	Cyanogen	252.1	400	5.98		9
C_3ClF_5O	Chloropentafluoroacetone	281	410.6	2.878		9
$C_3Cl_2F_6$	1,3-Dichloro-1,1,2,2,3,3-hexafluoropropane	308.9	453	2.753		41
C_3F_6O	Perfluoroacetone	245.8	357.14	2.84	329	9
C_3F_6O	Perfluorooxetane	244.8	361.8	3.03	272	18,47
C_3F_8	Perfluoropropane	236.6	345.1	2.680	299	9
$C_3F_8O_2$	Perfluorodimethoxymethane	263	372.3	2.333	363	18

Molecular formula	Name	T_b/K	T_c/K	P_c/MPa	$V_c/cm^3\ mol^{-1}$	Ref.
C_3HF_7	1,1,1,2,3,3,3-Heptafluoropropane	256.8	374.89	2.929	274	39,47
C_3HF_7O	Trifluoromethyl 1,1,2,2-tetrafluoroethyl ether	270	387.78	2.293	337	18,47
$C_3H_2F_6$	1,1,1,2,3,3-Hexafluoropropane	279.3	412.38	3.412	269	33
$C_3H_2F_6$	1,1,1,3,3,3-Hexafluoropropane	272.2	398.07			45
$C_3H_2F_6O$	1,2,2,2-Tetrafluoroethyl difluoromethyl ether	296.50	428.95	3.050	315	32
$C_3H_3F_3$	3,3,3-Trifluoropropene	256	376.2	3.80	211	9
$C_3H_3F_5$	1,1,1,3,3-Pentafluoropropane	288.5	427.20			45
$C_3H_3F_5$	1,1,1,2,2-Pentafluoropropane	255.8	380.11	3.137	273	9
$C_3H_3F_5$	1,1,2,2,3-Pentafluoropropane	298.2	447.57			45
$C_3H_3F_5O$	Methyl pentafluoroethyl ether	278.74	406.80	2.887	301	32
$C_3H_3F_5O$	Difluoromethyl 2,2,2-trifluoroethyl ether	302.39	443.99			38
C_3H_3N	Acrylonitrile	350.5	540	4.660		11,12
C_3H_3NO	Isoxazole	368	552.0			9
C_3H_4	Allene	238.8	394	5.25		6
C_3H_4	Propyne	250.0	402.4	5.63	163.5	6
C_3H_5Cl	3-Chloropropene	318.3	514			9
$C_3H_5F_3O$	2,2,2-Trifluoroethyl methyl ether	304.77	448.98	3.513	277	32
C_3H_5N	Propanenitrile	370.29	561.3	4.26	229	9
C_3H_6	Propene	225.46	364.9	4.60	185	6
C_3H_6	Cyclopropane	240.34	398.0	5.54	162	5
$C_3H_6Cl_2$	1,2-Dichloropropane	369.6	578.5			13
$C_3H_6Cl_2$	1,3-Dichloropropane	394.1	614.6			13
C_3H_6O	Allyl alcohol	370.2	545.1			4
C_3H_6O	Propanal	321	505	5.26	204	7
C_3H_6O	Acetone	329.20	508.1	4.700	213	7
C_3H_6O	Methyloxirane [1,2-Propylene oxide]	308	485	5.2	190	7
$C_3H_6O_2$	Propanoic acid	414.30	598.5	4.67	233	7
$C_3H_6O_2$	Ethyl formate	327.6	508.54	4.74	229	7
$C_3H_6O_2$	Methyl acetate	330.02	506.5	4.750	228	7
$C_3H_6O_3$	Dimethyl carbonate	363.7	557	4.80	252	7
C_3H_7Cl	1-Chloropropane	319.7	503	4.58		9
C_3H_7Cl	2-Chloropropane	308.9	484			13
C_3H_7NO	N,N-Dimethylformamide	426	649.6		262	9
C_3H_8	Propane	231.1	369.83	4.248	200	2
C_3H_8O	1-Propanol	370.4	536.8	5.169	218	4
C_3H_8O	2-Propanol	355.5	508.3	4.764	222	4
C_3H_8O	Ethyl methyl ether	280.6	437.9	4.38	222	7
$C_3H_8O_2$	1,2-Propylene glycol	460.8	676.4	5.941		7,14
$C_3H_8O_2$	1,3-Propylene glycol [Trimethylene glycol]	487.6	718.2	6.55		14,17
$C_3H_8O_2$	2-Methoxyethanol [Ethylene glycol monomethyl ether]	397.3	597.6	5.285	263	7,11,12
$C_3H_8O_2$	Dimethoxymethane [Methylal]	315	491	3.96	213	7
$C_3H_8O_3$	Glycerol	563	850	7.5		7
C_3H_8S	1-Propanethiol	341.0	537	4.6	286	8
C_3H_8S	Ethyl methyl sulfide	339.9	533	4.25		8
$C_3H_9BO_3$	Trimethyl borate	340.7	501.7	3.59		9
C_3H_9ClSi	Trimethylchlorosilane	333	497.8	3.20	366	9
C_3H_9N	Propylamine	320.37	497.0	4.72		9
C_3H_9N	Isopropylamine	304.91	471.8	4.54	221	9
C_3H_9N	Trimethylamine	276.02	432.79	4.087	254	9
$C_4Br_2F_8$	1,4-Dibromooctafluorobutane	370	532.5	2.39		9
$C_4Cl_2F_6$	1,2-Dichloro-1,2,3,3,4,4-hexafluorocyclobutane	332.7	497*	2.73*	386*	12
C_4F_8	Perfluorocyclobutane	267.3	388.46	2.784	324	9
C_4F_{10}	Perfluorobutane	271.3	386.4	2.323	378	9
C_4F_{10}	Perfluoroisobutane	273	395.4			9
$C_4H_2F_8O$	Perfluoroethyl 2,2,2-trifluoroethyl ether	301.04	421.68	2.330	409	32
$C_4H_3F_7O$	Perfluoropropyl methyl ether	307.38	437.70	2.481	377	32
$C_4H_3F_7O$	Perfluoroisopropyl methyl ether	302.49	433.30	2.553	369	32
C_4H_4O	Furan	304.7	490.2	5.3	218	7
$C_4H_4O_4$	Maleic acid		620			7
C_4H_4S	Thiophene	357.2	580	5.70	219	8

Molecular formula	Name	T_b/K	T_c/K	P_c/MPa	V_c/cm³ mol⁻¹	Ref.
$C_4H_5F_5O$	Perfluoroethyl ethyl ether	301.26	431.23	2.533	366	32
C_4H_5N	Pyrrole	402.94	639.7	6.34	200	10
C_4H_6	1,3-Butadiene	268.74	425	4.32	221	6
C_4H_6	1-Butyne	281.23	440	4.60	208	6
C_4H_6	2-Butyne	300.1	488.7			9
$C_4H_6O_2$	Vinyl acetate	346.0	519.2	4.185		7
$C_4H_6O_2$	γ-Butyrolactone	477	731	5.13		7,11
$C_4H_6O_3$	Acetic anhydride	412.7	606	4.0		7
$C_4H_6O_3$	Propylene carbonate	515	762.7	4.14		17
C_4H_7N	Butanenitrile	390.8	585.4	3.88		9
C_4H_8	1-Butene	266.89	419.5	4.02	240.8	6
C_4H_8	cis-2-Butene	276.86	435.5	4.21	233.8	6
C_4H_8	trans-2-Butene	274.03	428.6	4.10	237.7	6
C_4H_8	Isobutene	266.3	417.9	4.000	238.8	6
C_4H_8	Cyclobutane	285.8	460.0	4.98	210	9
C_4H_8O	Ethyl vinyl ether	308.7	475	4.07		7
C_4H_8O	Butanal	348.0	537	4.32	258	7
C_4H_8O	Isobutanal	337.7	544	5.1		7
C_4H_8O	2-Butanone [Methyl ethyl ketone]	352.74	536.7	4.207	267	7
C_4H_8O	Tetrahydrofuran	338	540.5	5.19	224	7
C_4H_8OS	S-Ethyl thioacetate	389.6	590.55	4.075	319	11,12
$C_4H_8O_2$	Butanoic acid	436.90	615.2	4.06	292	7
$C_4H_8O_2$	2-Methylpropanoic acid	427.60	605.0	3.70	290	7
$C_4H_8O_2$	Propyl formate	354.1	538.0	4.06	285	7
$C_4H_8O_2$	Isopropyl formate	341.4	535	3.95		7
$C_4H_8O_2$	Ethyl acetate	350.26	523.3	3.87	286	7
$C_4H_8O_2$	Methyl propanoate	353.0	530.7	4.00	282	7
$C_4H_8O_2$	1,4-Dioxane	374.7	588	5.21	238	7
C_4H_8S	Tetrahydrothiophene	394.3	632	5.4		8
C_4H_9Cl	1-Chlorobutane	351.6	539.2			13
C_4H_9Cl	2-Chlorobutane	341.4	518.6			13
C_4H_9Cl	2-Chloro-2-methylpropane	324.1	500			13
C_4H_9N	Pyrrolidine	359.71	568	6.00	238	10
C_4H_{10}	Butane	272.7	425.12	3.796	255	2
C_4H_{10}	Isobutane	261.42	407.8	3.640	259	5
$C_4H_{10}O$	1-Butanol	390.88	563.0	4.414	274	4
$C_4H_{10}O$	2-Butanol [sec-Butyl alcohol]	372.66	536.2	4.202	269	4
$C_4H_{10}O$	2-Methyl-1-propanol [Isobutyl alcohol]	381.04	547.8	4.295	274	4
$C_4H_{10}O$	2-Methyl-2-propanol [tert-Butyl alcohol]	355.6	506.2	3.972	275	4
$C_4H_{10}O$	Diethyl ether	307.7	466.7	3.644	281	7
$C_4H_{10}O$	Methyl propyl ether	312.3	476.2	3.801		7
$C_4H_{10}O$	Isopropyl methyl ether	303.92	464.4	3.762		7
$C_4H_{10}O_2$	2-Methyl-1,3-propanediol	484.8	708.0	5.35		17
$C_4H_{10}O_2$	1,2-Butanediol	463.7	680	5.21	303	7,23
$C_4H_{10}O_2$	1,3-Butanediol	480.7	676	4.02	305	7,23
$C_4H_{10}O_2$	1,4-Butanediol [Tetramethylene glycol]	508	723.8	5.52		17
$C_4H_{10}O_2$	1,2-Dimethoxyethane [Ethylene glycol dimethyl ether]	357.7	540	3.90	308	7
$C_4H_{10}O_2$	1,2-Propylene glycol monomethyl ether	392	579.8	4.113		11,12
$C_4H_{10}O_3$	Diethylene glycol	519.0	750	4.7		7
$C_4H_{10}S$	1-Butanethiol	371.7	570	4.0	324	8
$C_4H_{10}S$	Diethyl sulfide	365.3	557.8	3.90	317.6	8,15
$C_4H_{10}S_2$	Diethyl disulfide	427.2	642			8
$C_4H_{11}N$	Butylamine	350.15	531.9	4.25	277	10
$C_4H_{11}N$	sec-Butylamine	335.88	514.3	4.20	278	10
$C_4H_{11}N$	tert-Butylamine	317.19	483.9	3.84	292	10
$C_4H_{11}N$	Isobutylamine	340.90	519	4.07	278	10
$C_4H_{11}N$	Diethylamine	328.7	499.99	3.758		9
$C_4H_{12}N_2O$	N-(2-Aminoethyl)ethanolamine	512	739.2	4.65		17
$C_4H_{12}Si$	Tetramethylsilane	299.8	448.6	2.821	361.6	8
$C_4H_{12}Sn$	Tetramethylstannane	351	521.8	2.981		8

Molecular formula	Name	T_b/K	T_c/K	P_c/MPa	V_c/cm^3 mol^{-1}	Ref.
$C_4H_{13}N_3$	Bis(2-aminoethyl)amine	480	709.8	4.38		14,17
C_5F_{12}	Perfluoropentane	302.4	420.59	2.045	473	9
$C_5H_2F_6O_2$	Hexafluoroacetylacetone	327.30	485.1	2.767		9
$C_5H_4O_2$	Furfural	434.9	670*	5.89*		7
C_5H_5N	Pyridine	388.38	620.0	5.67	243	10
$C_5H_6N_2$	2-Methylpyrazine	410	634.3	5.01	283	9
C_5H_6O	2-Methylfuran	337.9	528	4.7	247	7
C_5H_7N	1-Methylpyrrole	385.96	596.0	4.86	271	10
C_5H_7N	2-Methylpyrrole	420.8	654	5.08	266	10
C_5H_7N	3-Methylpyrrole	416.1	647	5.08	266	10
C_5H_8	1-Pentyne	313.3	493.5			9
C_5H_8	Cyclopentene	317.4	506.5	4.80	245	6
C_5H_8O	Cyclopentanone	403.72	624	4.60		7
C_5H_8O	3,4-Dihydro-2H-pyran	359	562	4.56	268	7
C_5H_9N	Pentanenitrile	414.5	610.3	3.58		9
C_5H_9NO	N-Methyl-2-pyrrolidone	475	721.8		311	9
C_5H_{10}	1-Pentene	303.11	464.8	3.56	298.4	6
C_5H_{10}	cis-2-Pentene	310.08	475	3.69		6
C_5H_{10}	trans-2-Pentene	309.49	471	3.52		9
C_5H_{10}	2-Methyl-1-butene	304.4	470	3.8		9
C_5H_{10}	3-Methyl-1-butene	293.3	452.7	3.53	304.9	6
C_5H_{10}	2-Methyl-2-butene	311.71	470	3.42		6
C_5H_{10}	Cyclopentane	322.5	511.7	4.51	259	5
$C_5H_{10}O$	Cyclopentanol	413.57	619.5	4.9		4
$C_5H_{10}O$	Allyl ethyl ether	340.8	518			7
$C_5H_{10}O$	Pentanal	376	567	3.97	313	7
$C_5H_{10}O$	2-Pentanone [Methyl propyl ketone]	375.41	561.1	3.683	321	7
$C_5H_{10}O$	3-Pentanone [Diethyl ketone]	374.9	561.4	3.729	331	7
$C_5H_{10}O$	3-Methyl-2-butanone	367.48	553.0	3.80	308	7
$C_5H_{10}O$	Tetrahydropyran	361	572	4.77	263	7
$C_5H_{10}O$	2-Methyltetrahydrofuran	351	537	3.76	267	7
$C_5H_{10}O_2$	Pentanoic acid	459.3	637.2	3.63	346	7
$C_5H_{10}O_2$	3-Methylbutanoic acid	449.7	629	3.40		7
$C_5H_{10}O_2$	Isobutyl formate	371.4	551	3.88	355	7
$C_5H_{10}O_2$	Propyl acetate	374.69	549.7	3.36	345	7
$C_5H_{10}O_2$	Isopropyl acetate	361.8	531.0	3.31	344	7
$C_5H_{10}O_2$	Ethyl propanoate	372.3	546.7	3.45	342	7
$C_5H_{10}O_2$	Methyl butanoate	376.0	554.5	3.47	340	7
$C_5H_{10}O_2$	Methyl isobutanoate	365.7	540.7	3.43	339	7
$C_5H_{10}O_3$	2-Methoxyethyl acetate	416	630.0			7
$C_5H_{11}Cl$	1-Chloropentane	381.6	571.2			13
$C_5H_{11}Cl$	2-Chloro-2-methylbutane	358.8	509.1			13
$C_5H_{11}N$	Piperidine	379.37	594	4.94	288	10
C_5H_{12}	Pentane	309.21	469.7	3.370	311	2
C_5H_{12}	Isopentane	301.03	460.4	3.38	306	5
C_5H_{12}	Neopentane	282.63	433.8	3.196	307	5
$C_5H_{12}O$	1-Pentanol	411.13	588.1	3.897	326	4
$C_5H_{12}O$	2-Pentanol	392.5	560.3	3.675	329	4
$C_5H_{12}O$	3-Pentanol	389.40	559.6		325	4
$C_5H_{12}O$	2-Methyl-1-butanol	400.7	575.4	3.94		4
$C_5H_{12}O$	3-Methyl-1-butanol	404.3	577.2	3.93		4
$C_5H_{12}O$	2-Methyl-2-butanol	375.6	543.7	3.71		4
$C_5H_{12}O$	3-Methyl-2-butanol	386.1	556.1	3.87		4
$C_5H_{12}O$	Butyl methyl ether	343.31	512.7	3.37	329	7
$C_5H_{12}O$	Methyl tert-butyl ether	328.2	497.1	3.430		7
$C_5H_{12}O$	Ethyl propyl ether	336.36	500.2	3.370	339	7
$C_5H_{12}O_2$	2-Propoxyethanol	423.0	615	3.65	364	7
$C_5H_{12}O_2$	Diethoxymethane	361	531.7			7
$C_5H_{12}O_2$	1,2-Dimethoxypropane	369	543.0			7
$C_5H_{12}O_2$	2,2-Dimethoxypropane	356	510			7

Molecular formula	Name	T_b/K	T_c/K	P_c/MPa	$V_c/cm^3 mol^{-1}$	Ref.
$C_5H_{12}O_3$	Diethylene glycol monomethyl ether	466	672	3.67		11,12
$C_5H_{12}S$	3-Methyl-1-butanethiol	389	594			8
C_6BrF_5	Bromopentafluorobenzene	410	601	3.0		9
C_6ClF_5	Chloropentafluorobenzene	391.11	570.81	3.238	376	9
$C_6Cl_2F_4$	1,2-Dichloro-3,4,5,6-tetrafluorobenzene	430.9	626	5.32		9
$C_6Cl_3F_3$	1,3,5-Trichloro-2,4,6-trifluorobenzene	471.6	684.8	3.27	448	9
C_6F_6	Hexafluorobenzene	353.41	516.73	3.273	335	9
C_6F_{10}	Perfluorocyclohexene	325.2	461.8			9
C_6F_{12}	Perfluoro-1-hexene	330.2	454.4			9
C_6F_{12}	Perfluorocyclohexane	325.95	457.2			9
C_6F_{14}	Perfluorohexane	329.8	448.77	2.43		9
C_6F_{14}	Perfluoro-2-methylpentane	330.8	455.3	1.868	606	9
C_6F_{14}	Perfluoro-3-methylpentane	331.6	450	1.923	532	9
C_6F_{14}	Perfluoro-2,3-dimethylbutane	333.0	463	1.69		9
C_6HF_5	Pentafluorobenzene	358.89	530.97	1.87	525	9
C_6HF_5O	Pentafluorophenol	418.8	609	3.531	324	9
C_6HF_{11}	Undecafluorocyclohexane	335.2	477.7	4.0	348	9
$C_6H_2F_4$	1,2,3,4-Tetrafluorobenzene	367.5	550.83			9
$C_6H_2F_4$	1,2,3,5-Tetrafluorobenzene	357.6	535.25	3.791	313	9
$C_6H_2F_4$	1,2,4,5-Tetrafluorobenzene	363.4	543.35	3.747		9
$C_6H_3ClF_2$	1-Chloro-2,4-difluorobenzene	400	609.6	3.801		13
$C_6H_3ClF_2$	1-Chloro-2,5-difluorobenzene	401	612.5			13
$C_6H_3ClF_2$	1-Chloro-3,4-difluorobenzene	400	609.2			13
$C_6H_3ClF_2$	1-Chloro-3,5-difluorobenzene	391.7	592.0			13
$C_6H_3F_3$	1,2,3-Trifluorobenzene	368	560.3			13
$C_6H_3F_3$	1,2,4-Trifluorobenzene	363	551.1			13
$C_6H_3F_3$	1,3,5-Trifluorobenzene	348.7	530.9			13
C_6H_4BrF	1-Bromo-2-fluorobenzene	427	669.6			13
C_6H_4BrF	1-Bromo-3-fluorobenzene	423	652.0			13
C_6H_4BrF	1-Bromo-4-fluorobenzene	424.7	654.8			13
C_6H_4ClF	1-Chloro-2-fluorobenzene	410.8	633.8			13
C_6H_4ClF	1-Chloro-3-fluorobenzene	400.8	615.9			13
C_6H_4ClF	1-Chloro-4-fluorobenzene	403	620.1			13
$C_6H_4Cl_2$	m-Dichlorobenzene	446	685.7			13
$C_6H_4F_2$	o-Difluorobenzene	367	566.0			13
$C_6H_4F_2$	m-Difluorobenzene	355.8	548.4			13
$C_6H_4F_2$	p-Difluorobenzene	362	556.9			13
C_6H_5Br	Bromobenzene	429.21	670	4.40		9,13
C_6H_5Cl	Chlorobenzene	404.87	633.4	4.52	324	9
C_6H_5F	Fluorobenzene	357.88	560.09	4.52	308	9,13
C_6H_5I	Iodobenzene	461.6	721	4.551	269	9
C_6H_6	Benzene	353.24	562.05	4.52	351	9
C_6H_6O	Phenol	455.02	694.2	4.895	256	3
C_6H_7N	Aniline	457.32	699	5.93		7
C_6H_7N	2-Methylpyridine [2-Picoline]	402.53	621.0	4.89	287	9
C_6H_7N	3-Methylpyridine [3-Picoline]	417.29	645.0	4.60	292	10
C_6H_7N	4-Methylpyridine [4-Picoline]	418.51	645.7	4.65	288	10
C_6H_{10}	1,5-Hexadiene	332.6	508	4.70	292	10
C_6H_{10}	Cyclohexene	356.13	560.4			6
$C_6H_{10}O$	Cyclohexanone	428.58	665	4.6		6
$C_6H_{10}O$	2-Methylcyclopentanone	412.7	631			7
$C_6H_{10}O$	Mesityl oxide	403	605	4.00		7
$C_6H_{10}O_2$	Ethyl trans-2-butenoate	411	599		353	7
$C_6H_{10}S$	Diallyl sulfide	411.8	653			8
$C_6H_{11}Cl$	Chlorocyclohexane	415	586			13
$C_6H_{11}N$	Hexanenitrile	436.80	633.8	3.30		9
C_6H_{12}	1-Hexene	336.63	504.0	3.21	355.1	6
C_6H_{12}	Cyclohexane	353.88	553.8	4.08	308	5
C_6H_{12}	Methylcyclopentane	345.0	532.7	3.79	318	5
$C_6H_{12}O$	Butyl vinyl ether	367	540	3.20	384	7

Molecular formula	Name	T_b/K	T_c/K	P_c/MPa	V_c/cm^3 mol^{-1}	Ref.
$C_6H_{12}O$	Hexanal	404	592	3.46	378	7
$C_6H_{12}O$	2-Hexanone [Butyl methyl ketone]	400.8	587.1	3.30	377	7
$C_6H_{12}O$	3-Hexanone [Ethyl propyl ketone]	396.7	583.0	3.320	378	7
$C_6H_{12}O$	4-Methyl-2-pentanone [Isobutyl methyl ketone]	389.7	574.6	3.270		7
$C_6H_{12}O$	3,3-Dimethyl-2-butanone	379.3	570.9	3.43	382	7
$C_6H_{12}O$	Cyclohexanol	433.99	647.1	4.401		11,12
$C_6H_{12}O_2$	Hexanoic acid	478.4	655	3.38	413	7
$C_6H_{12}O_2$	Pentyl formate	403.6	576	3.46	412	7
$C_6H_{12}O_2$	Isopentyl formate	396.7	578			7
$C_6H_{12}O_2$	Butyl acetate	399.3	575.6	3.14		7
$C_6H_{12}O_2$	sec-Butyl acetate	385	571	3.01		7
$C_6H_{12}O_2$	Isobutyl acetate	389.7	561	2.99	401	7
$C_6H_{12}O_2$	Propyl propanoate	395.7	570	3.06		7
$C_6H_{12}O_2$	Ethyl butanoate	394.5	568.8	3.1	415	7
$C_6H_{12}O_2$	Ethyl 2-methylpropanoate	383.3	554	3.1	415	7
$C_6H_{12}O_2$	Methyl pentanoate	400.6	590	3.20	422	7
$C_6H_{12}O_3$	1,2-Propylene glycol monomethyl ether acetate	420	597.8	3.01	432	7
$C_6H_{12}O_3$	2-Ethoxyethyl acetate	429.6	608.0	3.17	443	7,23
$C_6H_{12}O_3$	Paraldehyde	397.5	563			7
$C_6H_{12}S$	Cyclohexanethiol	432.0	684		401	8
$C_6H_{13}Cl$	1-Chlorohexane	408.3	599			13
$C_6H_{13}Cl$	3-Chloro-3-methylpentane	389	528			13
C_6H_{14}	Hexane	341.88	507.6	3.025	368	2
C_6H_{14}	2-Methylpentane	333.41	497.7	3.04	368	5
C_6H_{14}	3-Methylpentane	336.42	504.6	3.12	368	5
C_6H_{14}	2,2-Dimethylbutane	322.88	489.0	3.10	358	5
C_6H_{14}	2,3-Dimethylbutane	331.08	500.0	3.15	361	5
$C_6H_{14}O$	2-Methoxy-2-methylbutane	359.3	535	3.20	374	7
$C_6H_{14}O$	1-Hexanol	430.8	610.3	3.417	387	4
$C_6H_{14}O$	2-Hexanol	413	585.9	3.31	384	4
$C_6H_{14}O$	3-Hexanol	408	582.4	3.36	383	4
$C_6H_{14}O$	2-Methyl-1-pentanol	422	604.4	3.45		4
$C_6H_{14}O$	4-Methyl-1-pentanol	425.1	603.5			4
$C_6H_{14}O$	2-Methyl-2-pentanol	394.3	559.5			4
$C_6H_{14}O$	4-Methyl-2-pentanol	404.8	574.4			4
$C_6H_{14}O$	2-Methyl-3-pentanol	399.7	576.0	3.46		4
$C_6H_{14}O$	3-Methyl-3-pentanol	395.6	575.6	3.52		4
$C_6H_{14}O$	Dipropyl ether	363.23	530.6	3.028		7
$C_6H_{14}O$	Diisopropyl ether	341.6	500.3	2.832	386	7
$C_6H_{14}O$	tert-Butyl ethyl ether	345.8	509.4	2.934	395	7
$C_6H_{14}O$	Methyl pentyl ether	372	546.5	3.042	391	7
$C_6H_{14}O_2$	1-Propoxy-2-propanol	423	605.1	3.051		14
$C_6H_{14}O_2$	2-Butoxyethanol	441.6	634	3.27	424	7
$C_6H_{14}O_2$	1,1-Diethoxyethane [Acetal]	375.40	540	3.22		7
$C_6H_{14}O_2$	1,2-Diethoxyethane [Ethylene glycol diethyl ether]	394.4	542			7
$C_6H_{14}O_3$	Diethylene glycol monoethyl ether [Carbitol]	469	670	3.167		11,12
$C_6H_{14}O_3$	Diethylene glycol dimethyl ether	435	617			7
$C_6H_{14}O_4$	Triethylene glycol	558	780	3.3		7
$C_6H_{15}N$	Dipropylamine	382.5	555.8	3.63		9
$C_6H_{15}N$	Diisopropylamine	357.1	523.1	3.02		9
$C_6H_{15}N$	Triethylamine	362	535.6	3.032	389	9
C_7F_8	Perfluorotoluene	377.7	534.47	2.705	428	9
C_7F_{14}	Perfluoro-1-heptene	354.2	478.2			9
C_7F_{14}	Perfluoromethylcyclohexane	349.5	485.91	2.019	570	9
C_7F_{16}	Perfluoroheptane	355.7	474.8	1.62	664	9
C_7HF_{15}	1H-Pentadecafluoroheptane	369.2	495.8			9
$C_7H_3F_5$	2,3,4,5,6-Pentafluorotoluene	390.7	566.52	3.126	384	9
$C_7H_4BrF_3$	1-Bromo-2-(trifluoromethyl)benzene	440.7	656.5			13
$C_7H_4BrF_3$	1-Bromo-3-(trifluoromethyl)benzene	424.7	627.1			13
$C_7H_4BrF_3$	1-Bromo-4-(trifluoromethyl)benzene	433	629.8			13

Molecular formula	Name	T_b/K	T_c/K	P_c/MPa	V_c/cm³ mol⁻¹	Ref.
C_7H_5N	Benzonitrile	464.3	699.4	4.21		9
$C_7H_6F_2$	2,4-Difluorotoluene	390	581.4			13
$C_7H_6F_2$	2,5-Difluorotoluene	391	587.8			13
$C_7H_6F_2$	2,6-Difluorotoluene	385	581.8			13
$C_7H_6F_2$	3,4-Difluorotoluene	385	598.5			13
C_7H_6O	Benzaldehyde	452.0	695	4.7		7
C_7H_7F	2-Fluorotoluene	388	591.2			13
C_7H_7F	3-Fluorotoluene	388	591.8			13
C_7H_7F	4-Fluorotoluene	389.8	592.1			13
C_7H_8	Toluene	383.78	591.80	4.110	316	3,15
C_7H_8O	o-Cresol	464.19	697.6	4.17		7
C_7H_8O	m-Cresol	475.42	705.8	4.36		7
C_7H_8O	p-Cresol	475.13	704.6	4.07		7
C_7H_8O	Benzyl alcohol	478.46	715	4.3		9
C_7H_8O	Anisole [Methoxybenzene]	426.9	646.5	4.24	341	7,11,12
C_7H_9N	2-Methylaniline	473.5	707	4.37		9
C_7H_9N	3-Methylaniline	476.5	707	4.28		9
C_7H_9N	4-Methylaniline	473.6	706	4.58		9
C_7H_9N	N-Methylaniline	469.4	701	5.20		9
C_7H_9N	2,3-Dimethylpyridine	434.27	655.4	4.10	356	23
C_7H_9N	2,4-Dimethylpyridine	431.53	647	3.95	361	23
C_7H_9N	2,5-Dimethylpyridine	430.13	645	3.85	361	23
C_7H_9N	2,6-Dimethylpyridine	417.16	624	3.85	361	23
C_7H_9N	3,4-Dimethylpyridine	452.25	684	4.20	355	23
C_7H_9N	3,5-Dimethylpyridine	444.99	668	4.05	361	23
C_7H_{14}	1-Heptene	366.79	537.3	2.92	409	6
C_7H_{14}	Cycloheptane	391.6	604.2	3.82	353	5
C_7H_{14}	Methylcyclohexane	374.08	572.1	3.48	369	5
C_7H_{14}	Ethylcyclopentane	376.7	569.5	3.40	375	5
C_7H_{14}	1,1-Dimethylcyclopentane	360.7	547	3.45		21
C_7H_{14}	cis-1,2-Dimethylcyclopentane	372.7	565	3.45		21
C_7H_{14}	trans-1,2-Dimethylcyclopentane	365.1	553	3.45		21
C_7H_{14}	cis-1,3-Dimethylcyclopentane	364.0	551	3.45		21
C_7H_{14}	trans-1,3-Dimethylcyclopentane	364.9	553	3.45		21
$C_7H_{14}O$	Heptanal	426.0	617	3.16	434	7
$C_7H_{14}O$	2-Heptanone [Methyl pentyl ketone]	424.20	611.4	2.97	436	7
$C_7H_{14}O$	3-Heptanone [Ethyl butyl ketone]	420	606.6		433	7
$C_7H_{14}O$	4-Heptanone	417	602.0		434	7
$C_7H_{14}O$	5-Methyl-2-hexanone [Methyl isopentyl ketone]	417	604.1			7
$C_7H_{14}O$	2-Methyl-3-hexanone	408	593.3			7
$C_7H_{14}O_2$	Heptanoic acid	495.4	678	3.16		7
$C_7H_{14}O_2$	Pentyl acetate	422.4	599	2.73	470	7,23
$C_7H_{14}O_2$	Isopentyl acetate	415.7	586.1	2.76		7
$C_7H_{14}O_2$	Butyl propanoate	420.0	594.5			7
$C_7H_{14}O_2$	Isobutyl propanoate	410	584			7
$C_7H_{14}O_2$	Propyl butanoate	416.2	593.1	2.72		7
$C_7H_{14}O_2$	Propyl isobutanoate	408.6	579.4			7
$C_7H_{14}O_2$	Ethyl pentanoate	419.3	593.3			7
$C_7H_{14}O_2$	Ethyl 3-methylbutanoate	408.2	582.4			7
$C_7H_{14}O_3$	Ethyl 3-ethoxypropanoate	439	621.0	2.66	458	7
$C_7H_{15}Cl$	1-Chloroheptane	433.6	614			13
C_7H_{16}	Heptane	371.6	540.2	2.74	428	2
C_7H_{16}	2-Methylhexane	363.19	530.4	2.74	421	5
C_7H_{16}	3-Methylhexane	365	535.2	2.81	404	5
C_7H_{16}	3-Ethylpentane	366.7	540.6	2.89	416	5
C_7H_{16}	2,2-Dimethylpentane	352.4	520.5	2.77	416	5
C_7H_{16}	2,3-Dimethylpentane	362.93	537.3	2.91	393	5
C_7H_{16}	2,4-Dimethylpentane	353.64	519.8	2.74	418	5
C_7H_{16}	3,3-Dimethylpentane	359.21	536.4	2.95	414	5
C_7H_{16}	2,2,3-Trimethylbutane	354.01	531.1	2.95	398	5

Molecular formula	Name	T_b/K	T_c/K	P_c/MPa	V_c/cm³ mol⁻¹	Ref.
$C_7H_{16}O$	2-Ethoxy-2-methylbutane	375	546	2.935	463	7
$C_7H_{16}O$	1-Heptanol	449.60	632.6	3.058	435	4
$C_7H_{16}O$	2-Heptanol	432	608.3	3.021	442	4
$C_7H_{16}O$	3-Heptanol, (S)	430	605.4		434	4
$C_7H_{16}O$	4-Heptanol	429	602.6		432	4
$C_7H_{16}O_2$	1-Butoxy-2-propanol	444.7	624.9	2.739		14
$C_7H_{16}O_2$	1-tert-Butoxy-2-methoxyethane		574			7
$C_7H_{16}O_2$	2,2-Diethoxypropane	387	510.7			7
$C_7H_{16}O_3$	Diethylene glycol monopropyl ether	486	680	3.00	489	7
$C_7H_{20}Si_2$	Bis(trimethylsilyl)methane	406	573.9	1.99		8
$C_8F_{16}O$	Perfluoro-2-butyltetrahydrofuran	375.8	500.2	1.607	588	9
C_8F_{18}	Perfluorooctane	379.1	502	1.66		9
C_8H_6S	Benzo[b]thiophene	494	764	4.76	379	8
C_8H_7N	4-Methylbenzonitrile	490.2	723			9
C_8H_7N	1H-Indole	526.8	794	4.8	356	10
C_8H_8	Styrene	418	635.2	3.87		15
C_8H_8O	Acetophenone	475	709.6	4.01	388	7,23
$C_8H_8O_2$	Phenyl acetate	469	685.7	3.59		17
$C_8H_8O_3$	Methyl salicylate	496.1	709			7
C_8H_{10}	Ethylbenzene	409.34	617.15	3.609	374	3,15
C_8H_{10}	o-Xylene	417.7	630.3	3.732	370	3
C_8H_{10}	m-Xylene	412.27	617.0	3.541	375	3
C_8H_{10}	p-Xylene	411.52	616.2	3.511	378	3
$C_8H_{10}O$	2-Ethylphenol	477.7	703.0			7
$C_8H_{10}O$	3-Ethylphenol	491.6	716.4			7
$C_8H_{10}O$	4-Ethylphenol	491.1	716.4			7
$C_8H_{10}O$	2,3-Xylenol	490.1	722.8			7
$C_8H_{10}O$	2,4-Xylenol	484.13	707.6			7
$C_8H_{10}O$	2,5-Xylenol	484.3	706.9			7
$C_8H_{10}O$	2,6-Xylenol	474.22	701.0			7
$C_8H_{10}O$	3,4-Xylenol	500	729.8			7
$C_8H_{10}O$	3,5-Xylenol	494.89	715.6			7
$C_8H_{10}O$	α-Methylbenzenemethanol	478	699	3.77		14
$C_8H_{10}O$	Ethoxybenzene	442.96	647	3.4		7
$C_8H_{10}O$	2-Methylanisole	444	662.0			7
$C_8H_{10}O$	3-Methylanisole	448.7	665.3			7
$C_8H_{10}O$	4-Methylanisole	448.7	666			7
$C_8H_{11}N$	N-Ethylaniline	476.2	698			9
$C_8H_{11}N$	N,N-Dimethylaniline	467.30	687	3.63		9
$C_8H_{14}O_4$	Diethyl succinate	490.9	663			7
$C_8H_{15}N$	Octanenitrile	478.40	674.4	2.85		9
C_8H_{16}	1-Octene	394.44	567.0	2.68	468	6
C_8H_{16}	Cyclooctane	422	647.2	3.56	410	5
C_8H_{16}	Ethylcyclohexane	405.1	609	3.04		21
C_8H_{16}	cis-1,2-Dimethylcyclohexane	403.0	606	2.95		21
C_8H_{16}	trans-1,2-Dimethylcyclohexane	396.7	596	2.94		21
C_8H_{16}	cis-1,3-Dimethylcyclohexane	393.3	591	2.94		21
C_8H_{16}	trans-1,3-Dimethylcyclohexane	397.7	598	2.94		21
C_8H_{16}	trans-1,4-Dimethylcyclohexane	392.6	587.7			5
$C_8H_{16}O$	Octanal	444	639	2.96	488	7
$C_8H_{16}O$	2-Octanone [Hexyl methyl ketone]	445.7	632.7		497	7
$C_8H_{16}O$	3-Octanone [Ethyl amyl ketone]	440.7	627.7		497	7
$C_8H_{16}O$	4-Octanone [Butyl propyl ketone]	436	623.8		497	7
$C_8H_{16}O$	2-Methyl-3-heptanone [Butyl isopropyl ketone]	431	614.9			7
$C_8H_{16}O$	5-Methyl-3-heptanone	434	619.0			7
$C_8H_{16}O_2$	Octanoic acid	512	693	2.87	519	7
$C_8H_{16}O_2$	2-Ethylhexanoic acid	501	674	2.78	528	7
$C_8H_{16}O_2$	Hexyl acetate	444.7	618.4			7
$C_8H_{16}O_2$	Isopentyl propanoate	433.4	611			7
$C_8H_{16}O_2$	Butyl butanoate	439	612.1			7

Molecular formula	Name	T_b/K	T_c/K	P_c/MPa	V_c/cm^3 mol^{-1}	Ref.
$C_8H_{16}O_2$	Isobutyl butanoate	430.1	611			7
$C_8H_{16}O_2$	Isobutyl isobutanoate	421.8	602			7
$C_8H_{16}O_2$	Propyl 3-methylbutanoate	429.1	609			7
$C_8H_{16}O_2$	Ethyl hexanoate	440	615.2			7
$C_8H_{16}O_2$	Methyl heptanoate	447	628			7
$C_8H_{16}O_3$	2-Butoxyethyl acetate	465	640.7	2.694	549	7
$C_8H_{16}O_4$	Diethylene glycol monoethyl ether acetate	491.7	673.5	2.59		17
$C_8H_{17}Cl$	1-Chlorooctane	456.7	643			13
C_8H_{18}	Octane	398.82	568.7	2.49	492	2
C_8H_{18}	2-Methylheptane	390.81	559.7	2.50	488	5
C_8H_{18}	3-Methylheptane	392.1	563.6	2.55	464	5
C_8H_{18}	4-Methylheptane	390.87	561.7	2.54	476	5
C_8H_{18}	3-Ethylhexane	391.8	565.5	2.61	455	5
C_8H_{18}	2,2-Dimethylhexane	380.01	549.8	2.53	478	5
C_8H_{18}	2,3-Dimethylhexane	388.77	563.5	2.63	468	5
C_8H_{18}	2,4-Dimethylhexane	382.7	553.5	2.56	472	5
C_8H_{18}	2,5-Dimethylhexane	382.27	550.0	2.49	482	5
C_8H_{18}	3,3-Dimethylhexane	385.12	562.0	2.65	443	5
C_8H_{18}	3,4-Dimethylhexane	390.88	568.8	2.69	466	5
C_8H_{18}	3-Ethyl-2-methylpentane	388.81	567.1	2.70	442	5
C_8H_{18}	3-Ethyl-3-methylpentane	391.42	576.5	2.81	455	5
C_8H_{18}	2,2,3-Trimethylpentane	383	563.5	2.73	436	5
C_8H_{18}	2,2,4-Trimethylpentane [Isooctane]	372.37	543.8	2.57	468	5
C_8H_{18}	2,3,3-Trimethylpentane	388.0	573.5	2.82	455	5
C_8H_{18}	2,3,4-Trimethylpentane	386.7	566.4	2.73	460	5
C_8H_{18}	2,2,3,3-Tetramethylbutane	379.60	567.8	2.87	461	9
$C_8H_{18}O$	1-Octanol	468.31	652.5	2.777	497	4
$C_8H_{18}O$	2-Octanol	452.5	629.6	2.754	519	4
$C_8H_{18}O$	3-Octanol	444	628.5		515	4
$C_8H_{18}O$	4-Octanol	449.5	625.1		515	4
$C_8H_{18}O$	4-Methyl-3-heptanol	443	623.5			4
$C_8H_{18}O$	5-Methyl-3-heptanol	445	621.2			4
$C_8H_{18}O$	2-Ethyl-1-hexanol	457.8	640.6	2.8		4
$C_8H_{18}O$	Dibutyl ether	413.43	584	3.0		7
$C_8H_{18}O$	Di-tert-butyl ether	380.38	550			9
$C_8H_{18}O_2$	1-tert-Butoxy-2-ethoxyethane	421.2	585			7
$C_8H_{18}O_3$	Diethylene glycol monobutyl ether	504	692	2.79		7
$C_8H_{18}O_3$	Diethylene glycol diethyl ether	461	612			7
$C_8H_{18}O_5$	Tetraethylene glycol	601	800	3.2		7
$C_8H_{18}S$	1-Octanethiol	472.3	667		504	8
$C_8H_{18}S$	Dibutyl sulfide	458	650	2.48		8
$C_8H_{19}N$	Dibutylamine	432.8	607.5	3.11		9
$C_8H_{19}N$	Diisobutylamine	412.8	584.4	3.20		9
$C_8H_{20}Si$	Tetraethylsilane	427.9	605	2.50	587	8
C_9F_{20}	Perfluorononane	398.5	524	1.56		9
C_9H_7N	Quinoline	510.31	782	4.86	371	10
C_9H_7N	Isoquinoline	516.37	803	5.10	374	10
C_9H_{10}	Indan	451.12	684.9	3.95		3
C_9H_{12}	Propylbenzene	432.39	638.35	3.200	440	3
C_9H_{12}	Isopropylbenzene [Cumene]	425.56	631.0	3.209		3
C_9H_{12}	2-Ethyltoluene	438.4	651	3.38		21
C_9H_{12}	3-Ethyltoluene	434.5	637	3.25		21
C_9H_{12}	4-Ethyltoluene	435	640.2	3.23		3
C_9H_{12}	1,2,3-Trimethylbenzene	449.27	664.5	3.454		3
C_9H_{12}	1,2,4-Trimethylbenzene	442.53	649.1	3.232		3
C_9H_{12}	1,3,5-Trimethylbenzene [Mesitylene]	437.89	637.3	3.127		3
$C_9H_{12}O$	2-Methoxy-1,4-dimethylbenzene	467	677.3			7
$C_9H_{12}O$	1-Methoxy-2,4-dimethylbenzene	465	682			7
$C_9H_{13}N$	2-Methyl-N,N-dimethylaniline	467.3	668	3.12		9
C_9H_{18}	1-Nonene	420.1	594.0		526	6

Molecular formula	Name	T_b/K	T_c/K	P_c/MPa	V_c/cm^3 mol^{-1}	Ref.
C_9H_{18}	Cyclononane	451.6	682	3.34		21
C_9H_{18}	1α,3α,5β-1,3,5-Trimethylcyclohexane	413.7	602.2			5
$C_9H_{18}O$	Nonanal	464	659	2.68	543	7
$C_9H_{18}O$	2-Nonanone [Heptyl methyl ketone]	468.5	652.2	2.48	560	7,11,12
$C_9H_{18}O$	3-Nonanone [Ethyl hexyl ketone]	463	648.1		560	7
$C_9H_{18}O$	4-Nonanone [Pentyl propyl ketone]	460.7	643.7		560	7
$C_9H_{18}O$	5-Nonanone [Dibutyl ketone]	461.60	641.4	2.32	560	7
$C_9H_{18}O_2$	Nonanoic acid	527.7	712	2.35		7
$C_9H_{18}O_2$	Isopentyl butanoate	452	619			7
$C_9H_{18}O_2$	Isobutyl 3-methylbutanoate	441.7	621			7
$C_9H_{18}O_2$	Ethyl heptanoate	460	634			7
C_9H_{20}	Nonane	423.97	594.6	2.29	555	2
C_9H_{20}	2-Methyloctane	416.4	582.8	2.31		5
C_9H_{20}	2,2-Dimethylheptane	405.9	576.7	2.35		5
C_9H_{20}	2,2,5-Trimethylhexane	397.24	569.8			5
C_9H_{20}	2,2,3,3-Tetramethylpentane	413.4	607.5	2.74		5
C_9H_{20}	2,2,3,4-Tetramethylpentane	406.2	592.6	2.60		5
C_9H_{20}	2,2,4,4-Tetramethylpentane	395.44	574.6	2.49		5
C_9H_{20}	2,3,3,4-Tetramethylpentane	414.7	607.5	2.72		5
$C_9H_{20}O$	1-Nonanol	486.52	670.7	2.528	572	4
$C_9H_{20}O$	2-Nonanol	466.7	649.6	2.53	575	4
$C_9H_{20}O$	3-Nonanol	468	648.0		577	4
$C_9H_{20}O$	4-Nonanol	465.7	645.1		575	4
$C_{10}F_8$	Perfluoronaphthalene	482	673.1			9
$C_{10}F_{18}$	Perfluorodecalin	415	566	1.52		9
$C_{10}F_{22}$	Perfluorodecane	417.4	542	1.45		9
$C_{10}H_8$	Naphthalene	491.1	748.4	4.05	407	3
$C_{10}H_9N$	1-Naphthylamine	573.9	850	5.0	438	10
$C_{10}H_9N$	2-Naphthylamine	579.4	850	4.9	438	10
$C_{10}H_{12}$	1,2,3,4-Tetrahydronaphthalene [Tetralin]	480.8	720	3.65	408	4
$C_{10}H_{14}$	Butylbenzene	456.46	660.5	2.89	497	3
$C_{10}H_{14}$	Isobutylbenzene	445.94	650	3.05		4
$C_{10}H_{14}$	1-Isopropyl-4-methylbenzene [p-Cymene]	450.3	652	2.8		3
$C_{10}H_{14}$	p-Diethylbenzene	456.9	657.9	2.803		3
$C_{10}H_{14}$	1,2,4,5-Tetramethylbenzene [Durene]	470.0	676	2.9		3
$C_{10}H_{14}O$	Thymol	505.7	698			7
$C_{10}H_{16}$	d-Limonene	451	653		470	6
$C_{10}H_{16}$	α-Pinene	429.4	644		454	6
$C_{10}H_{16}$	3-Carene, (+)	444	658		487	6
$C_{10}H_{18}$	1,3-Decadiene	442	615			9
$C_{10}H_{18}$	cis-Decahydronaphthalene	469.0	702.3	3.20		9
$C_{10}H_{18}$	trans-Decahydronaphthalene	460.5	687.1			9
$C_{10}H_{20}$	1-Decene	443.7	617	2.22	584	6
$C_{10}H_{20}O$	Decanal	481.7	674	2.60	599	7
$C_{10}H_{20}O$	2-Decanone [Methyl octyl ketone]	483	671.8		625	7
$C_{10}H_{20}O$	3-Decanone [Ethyl heptyl ketone]	476	667.6		628	7
$C_{10}H_{20}O$	4-Decanone [Hexyl propyl ketone]	479.7	662.9		628	7
$C_{10}H_{20}O$	5-Decanone [Butyl pentyl ketone]	477	661.0		628	7
$C_{10}H_{20}O$	5-Methyl-2-isopropylcyclohexanol [Menthol]	489	694			9
$C_{10}H_{20}O_2$	Decanoic acid [Capric acid]	541.9	722	2.10	638	7
$C_{10}H_{20}O_2$	2-Ethylhexyl acetate	472	642	2.09	681	7
$C_{10}H_{20}O_2$	Ethyl octanoate	481.7	649			7
$C_{10}H_{20}O_4$	Diethylene glycol monobutyl ether acetate	518	693.9	2.15		17
$C_{10}H_{22}$	Decane	447.30	617.7	2.11	624	2
$C_{10}H_{22}$	3,3,5-Trimethylheptane	428.9	609.5	2.32		5
$C_{10}H_{22}$	2,2,3,3-Tetramethylhexane	433.5	623.0	2.51		5
$C_{10}H_{22}$	2,2,5,5-Tetramethylhexane	410.6	581.4	2.19		5
$C_{10}H_{22}O$	1-Decanol	504.3	687.3	2.315	649	4
$C_{10}H_{22}O$	2-Decanol	484	668.6		646	4
$C_{10}H_{22}O$	3-Decanol	486	666.1		643	4

Molecular formula	Name	T_b/K	T_c/K	P_c/MPa	$V_c/cm^3\,mol^{-1}$	Ref.
$C_{10}H_{22}O$	4-Decanol	483.7	663.7		643	4
$C_{10}H_{22}O$	5-Decanol	474	663.2		646	4
$C_{10}H_{22}S$	Diisopentyl sulfide	484	664			8
$C_{11}H_{10}$	1-Methylnaphthalene	517.9	772	3.60		3
$C_{11}H_{10}$	2-Methylnaphthalene	514.3	761			3
$C_{11}H_{16}$	Pentylbenzene	478.6	675	2.58		16
$C_{11}H_{22}O$	2-Undecanone	504.7	688		692	7
$C_{11}H_{22}O$	3-Undecanone	500	685		692	7
$C_{11}H_{22}O$	4-Undecanone		681		692	7
$C_{11}H_{22}O$	5-Undecanone	500	679		692	7
$C_{11}H_{22}O$	6-Undecanone	501	678		692	7
$C_{11}H_{22}O_2$	Undecanoic acid		728	2.13		48
$C_{11}H_{22}O_2$	Ethyl nonanoate	500.2	664			7
$C_{11}H_{24}$	Undecane	469.1	639	1.98	689	2
$C_{11}H_{24}O$	1-Undecanol	518	703.6	2.147	718	4
$C_{12}H_8$	Acenaphthylene	553	792	3.20		21
$C_{12}H_8O$	Dibenzofuran	560	824	3.64	495	7
$C_{12}H_8S$	Dibenzothiophene	605.7	897	3.86	512	8
$C_{12}H_9N$	Carbazole	627.84	901.8	3.13	454	10
$C_{12}H_{10}$	Biphenyl	529.3	773	3.38	497	3
$C_{12}H_{10}O$	Diphenyl ether	531.2	767			7
$C_{12}H_{12}$	2,7-Dimethylnaphthalene	538	775	3.23	601	3
$C_{12}H_{18}$	Hexylbenzene	499.3	695	2.35		16
$C_{12}H_{18}$	Hexamethylbenzene	536.6	758			3
$C_{12}H_{20}O$	[1,1'-Bicyclohexyl]-2-one	537	787			7
$C_{12}H_{24}$	1-Dodecene	487.0	658	1.93		6
$C_{12}H_{24}O$	2-Dodecanone	519.7	702		752	7
$C_{12}H_{24}O$	3-Dodecanone		701		752	7
$C_{12}H_{24}O$	4-Dodecanone		697		759	7
$C_{12}H_{24}O$	5-Dodecanone		695		759	7
$C_{12}H_{24}O$	6-Dodecanone		694		762	7
$C_{12}H_{24}O_2$	Dodecanoic acid		743	1.93		48
$C_{12}H_{26}$	Dodecane	489.47	658	1.82	754	2
$C_{12}H_{26}O$	1-Dodecanol	533	719.4	1.994		4
$C_{13}H_9N$	Acridine	618.01	891.1	3.21	548	10
$C_{13}H_9N$	Phenanthridine	622.1	895	3.6	548	10
$C_{13}H_{10}O$	Benzophenone	578.6	830	3.35	568	7
$C_{13}H_{11}N$	9-Methyl-9H-carbazole	616.79	890	3.38	572	10
$C_{13}H_{12}$	Diphenylmethane	538.2	760	2.71	563	3
$C_{13}H_{20}$	Heptylbenzene	513	708	2.14		16
$C_{13}H_{26}O$	2-Tridecanone	536	717		820	7
$C_{13}H_{26}O$	3-Tridecanone		716		823	7
$C_{13}H_{26}O$	4-Tridecanone		712		823	7
$C_{13}H_{26}O$	5-Tridecanone		710		826	7
$C_{13}H_{26}O$	6-Tridecanone		709		826	7
$C_{13}H_{26}O$	7-Tridecanone	534	708		830	7
$C_{14}H_{28}O_2$	Tridecanoic acid		763	1.64		48
$C_{13}H_{26}O_2$	Methyl dodecanoate	540	712			7
$C_{13}H_{28}$	Tridecane	508.62	675	1.68	823	2
$C_{13}H_{28}O$	1-Tridecanol	547	734	1.935		9
$C_{14}H_{10}$	Anthracene	613.1	869.3		554	9
$C_{14}H_{10}$	Phenanthrene	613	869			4
$C_{14}H_{22}$	Octylbenzene	537	725	1.98		16
$C_{14}H_{28}O$	2-Tetradecanone		728		896	7
$C_{14}H_{28}O$	3-Tetradecanone		727		896	7
$C_{14}H_{28}O$	4-Tetradecanone		725		900	7
$C_{14}H_{28}O$	7-Tetradecanone		723		904	7
$C_{14}H_{28}O_2$	Tetradecanoic acid		763	1.64		48
$C_{14}H_{30}$	Tetradecane	526.73	693	1.57	894	2
$C_{14}H_{30}O$	1-Tetradecanol	560	747	1.81		9

Molecular formula	Name	T_b/K	T_c/K	P_c/MPa	$V_c/cm^3\,mol^{-1}$	Ref.
$C_{15}H_{30}O_2$	Pentadecanoic acid		777	1.57		48
$C_{15}H_{32}$	Pentadecane	543.8	708	1.48	966	2
$C_{16}H_{26}$	Decylbenzene	566	752	1.72		16
$C_{16}H_{32}O_2$	Hexadecanoic acid		785	1.49		48
$C_{16}H_{34}$	Hexadecane	560.01	723	1.40	1034	2
$C_{16}H_{34}$	2,2,4,4,6,8,8-Heptamethylnonane	519.5	692			5
$C_{16}H_{34}O$	1-Hexadecanol	585	770	1.61		9
$C_{17}H_{28}$	Undecylbenzene	589	763	1.64		16
$C_{17}H_{34}O_2$	Heptadecanoic acid		792	1.37		48
$C_{17}H_{36}$	Heptadecane	575.2	736	1.34	1103	2
$C_{17}H_{36}O$	1-Heptadecanol	597	780	1.50		9
$C_{18}H_{14}$	o-Terphenyl	605	857	2.99	731	3
$C_{18}H_{14}$	m-Terphenyl	636	883	2.48	724	3
$C_{18}H_{14}$	p-Terphenyl	649	908	2.99	729	3
$C_{18}H_{36}O_2$	Octadecanoic acid		803	1.33		48
$C_{18}H_{38}$	Octadecane	589.5	747	1.29	1189	2
$C_{18}H_{38}O$	1-Octadecanol	608	790	1.44		9
$C_{19}H_{32}$	Tridecylbenzene	619	790	1.54		16
$C_{19}H_{40}$	Nonadecane	603.1	755	1.16		3
$C_{20}H_{40}O_2$	Eicosanoic acid		820	1.20		48
$C_{20}H_{42}$	Eicosane	616	768	1.07		3
$C_{20}H_{42}O$	1-Eicosanol [Arachic alcohol]	629	809	1.30		9
$C_{21}H_{44}$	Heneicosane	629.7	778	1.03		2
$C_{22}H_{44}O_2$	Docosanoic acid		837	1.11		48
$C_{22}H_{46}$	Docosane	641.8	786	0.98		2
$C_{23}H_{48}$	Tricosane	653	790	0.92		2
$C_{24}H_{50}$	Tetracosane	664.5	800	0.87		2
$C_{30}H_{50}$	Squalene	694.5	795.9	0.59		15

SUBLIMATION PRESSURE OF SOLIDS

This table gives the sublimation (vapor) pressure of some representative solids as a function of temperature. Entries include simple inorganic and organic substances in their solid phase below room temperature, as well as polycyclic organic compounds which show measurable sublimation pressure only at elevated temperatures. Substances are listed by molecular formula in the Hill order. Values marked by * represent the solid–liquid–gas triple point. Note that some pressure values are in pascals (Pa) and others are in kilopascals (kPa). For conversion, 1 kPa = 7.506 mmHg = 0.0098692 atm.

References

1. Lide, D. R. and Kehiaian, H. V., CRC *Handbook of Thermophysical and Thermochemical Data*, CRC Press, Boca Raton, FL, 1994.
2. *TRC Thermodynamic Tables*, Thermodynamic Research Center, Texas A&M University, College Station, TX.
3. Oja, V. and Suuberg, E. M., *J. Chem. Eng. Data*, 43, 486, 1998.

Substance									
Ar Argon	T/K	55	60	65	70	75	80	83.81*	
	p/kPa	0.2	0.8	2.8	7.7	18.7	40.7	68.8*	
BrH Hydrogen bromide	T/K	135	140	150	160	170	180	185.1*	
	p/kPa	0.1	0.3	1.1	3.3	8.7	20.1	27.4*	
Br_2 Bromine	T/K	170	180	190	200	210	220	230	240*
	p/Pa	0.069	0.416	2.04	8.45	30.3	96.0	273	710*
ClH Hydrogen chloride	T/K	120	130	140	150	155	159.0*		
	p/kPa	0.1	0.5	1.9	5.8	9.5	13.5*		
Cl_2 Chlorine	T/K	120	130	140	150	160	170*		
	p/Pa	0.144	1.52	11.2	63.1	283	1054*		
F_4Si Tetrafluorosilane	T/K	130	140	150	160	170	175	180	186.3*
	p/kPa	0.2	0.9	3.9	14.0	43.8	74.2	122.4	220.8*
F_6S Sulfur hexafluoride	T/K	150	165	180	190	200	210	220	223.1*
	p/kPa	0.4	2.6	11.3	25.9	54.5	106.1	195.1	232.7*
HI Hydrogen iodide	T/K	160	170	180	190	200	210	220	222.4*
	p/kPa	0.2	0.8	2.2	5.3	11.7	23.6	44.1	49.3*
H_2O Water	T/K	190	210	225	240	250	260	270	273.16*
	p/Pa	0.032	0.702	4.942	27.28	76.04	195.8	470.1	611.66*
H_2S Hydrogen sulfide	T/K	140	150	160	165	170	175	180	187.6*
	p/kPa	0.2	0.6	1.9	3.2	5.2	8.3	12.7	22.7*
H_3N Ammonia	T/K	160	170	180	190	195	195.4*		
	p/kPa	0.1	0.4	1.2	3.5	5.8	6.12*		
I_2 Iodine	T/K	240	250	260	270	280	290	300	310*
	p/Pa	0.081	0.297	0.971	2.89	7.92	20.1	47.9	107*
Kr Krypton	T/K	80	90	95	100	105	110	115.8*	
	p/kPa	0.4	2.7	6.0	12.1	22.8	40.4	73.1*	
NO Nitric oxide	T/K	85	90	95	100	105	109.5*		
	p/kPa	0.1	0.4	1.3	3.8	10.0	21.9*		
Xe Xenon	T/K	110	120	130	140	150	155	160	161.4*
	p/kPa	0.3	1.5	4.9	14.0	34.2	51.1	74.2	81.7*
CHN Hydrogen cyanide	T/K	200	210	220	230	240	250	255	259.83*
	p/kPa	0.2	0.4	1.0	2.2	4.8	9.7	13.6	18.62*
CH_4 Methane	T/K	65	70	75	80	85	90.69*		
	p/kPa	0.1	0.3	0.8	2.1	4.9	11.70*		
CO Carbon monoxide	T/K	50	55	60	65	68.13*			
	p/kPa	0.1	0.6	2.6	8.2	15.4*			
CO_2 Carbon dioxide	T/K	130	140	155	170	185	194.7	205	216.58*
	p/kPa	0.032	0.187	1.674	9.987	44.02	101.3	227.1	518.0*
C_2Cl_6 Hexachloroethane	T/K	275	300	325	350	375	400	425	459.9*
	p/Pa	0.004	0.056	0.383	1.62	5.30	14.8	36.4	107.4*
C_2H_2 Acetylene	T/K	130	140	150	160	170	180	190	192.4*
	p/kPa	0.2	0.7	2.6	7.8	20.6	49.0	106.3	126.0*
$C_2H_4O_2$ Acetic acid	T/K	250	260	270	280	289.7*			
	p/kPa	0.092	0.199	0.406	0.79	1.29*			
C_5H_{12} Neopentane	T/K	200	210	220	230	240	250	255	256.58*
	p/kPa	0.7	1.6	3.6	7.3	13.9	24.8	32.4	35.8*

$C_6H_6Cl_6$ 1,2,3,4,5,6–Hexa–chlorocyclohexane (Lindane)	T/K	300	320	330	340	350	360	370	380
	p/Pa	0.01	0.13	0.39	1.04	2.66	6.42	14.8	32.7
$C_6H_6O_2$ Resorcinol	T/K	330	340	350	360	370	380		
	p/Pa	1.03	2.78	7.09	17.2	39.6	87.6		
$C_6H_6O_2$ p–Hydroquinone	T/K	350	360	370	380	390	400		
	p/Pa	1.20	3.18	7.96	19.0	43.4	95.1		
$C_{10}H_8$ Naphthalene	T/K	250	270	280	290	300	310	330	353.43*
	p/Pa	0.036	0.514	1.662	4.918	13.43	34.15	182.9	999.6*
$C_{12}H_8N_2$ Phenazine	T/K	290	300	310	320				
	p/Pa	0.0013	0.0046	0.0150	0.0448				
$C_{12}H_8O$ Dibenzofuran	T/K	300	310	320	330	340	350		
	p/Pa	0.408	1.21	3.35	8.71	21.4	50.0		
$C_{12}H_9N$ Carbazole	T/K	350	355	360					
	p/Pa	0.086	0.140	0.245					
$C_{13}H_7NO_2$ Benz[g]isoquinoline–5,10–dione	T/K	330	340	350	360	370	380		
	p/Pa	0.006	0.018	0.053	0.148	0.394	0.994		
$C_{13}H_8O$ 1H–Phenalen–1–one	T/K	330	340	350					
	p/Pa	0.040	0.113	0.302					
$C_{13}H_8O_2$ 3–Hydroxy–1H–phenalen–1–one	T/K	400	410	420	430				
	p/Pa	0.006	0.018	0.053	0.144				
$C_{13}H_9N$ Acridine	T/K	290	300	310	320				
	p/Pa	0.0024	0.0085	0.0278	0.0845				
$C_{13}H_9N$ Phenanthridine	T/K	310	320	330	340				
	p/Pa	0.020	0.066	0.206	0.603				
$C_{14}H_{10}$ Anthracene	T/K	320	330	340	350	360	370	380	390
	p/Pa	0.014	0.043	0.125	0.342	1.01	2.38	5.35	11.5
$C_{14}H_{10}$ Phenanthrene	T/K	300	310	320	330	340	350	360	
	p/Pa	0.025	0.085	0.270	0.796	2.02	4.89	11.2	
$C_{16}H_{10}$ Pyrene	T/K	320	330	340	350	360	370	380	390
	p/Pa	0.008	0.024	0.073	0.208	0.556	1.32	2.86	6.30
$C_{16}H_{10}O$ 1–Pyrenol	T/K	360	370	380	390	400			
	p/Pa	0.005	0.016	0.047	0.135	0.364			
$C_{16}H_{12}S$ Benzo[b]naphtho–(2,1–d)thiophene	T/K	330	340	350	360	370	380	390	
	p/Pa	0.001	0.004	0.012	0.036	0.098	0.255	0.631	
$C_{17}H_{12}$ 11H–Benzo[b]fluorene	T/K	340	350	360	370	380	390	400	
	p/Pa	0.003	0.009	0.029	0.085	0.235	0.619	1.55	
$C_{18}H_{10}O_4$ 6,11–Dihydroxy–5,12–naphthacenedione	T/K	420	430	440	450				
	p/Pa	0.008	0.022	0.055	0.131				
$C_{18}H_{12}$ Chrysene	T/K	390	400	410	420				
	p/Pa	0.087	0.221	0.539	1.26				
$C_{18}H_{12}$ Naphthacene	T/K	390	400	410	420	430	440	450	460
	p/Pa	0.005	0.014	0.035	0.084	0.194	0.432	0.928	1.929
$C_{20}H_{12}$ Perylene	T/K	390	400	410	420	430			
	p/Pa	0.006	0.015	0.040	0.102	0.246			
$C_{22}H_{14}$ Pentacene	T/K	450	460	470	480	490			
	p/Pa	0.002	0.006	0.013	0.031	0.069			
$C_{24}H_{12}$ Coronene	T/K	430	440	450	460	470	480	490	500
	p/Pa	0.004	0.010	0.021	0.046	0.097	0.197	0.389	0.747

VAPOR PRESSURE

This table gives vapor pressure data for about 1800 inorganic and organic substances. In order to accommodate elements and compounds ranging from refractory to highly volatile in a single table, the temperature at which the vapor pressure reaches specified pressure values is listed. The pressure values run in decade steps from 1 Pa (about 7.5 μm Hg) to 100 kPa (about 750 mm Hg). All temperatures are given in °C.

The data used in preparing the table came from a large number of sources; the main references used for each substance are indicated in the last column. Since the data were refit in most cases, values appearing in this table may not be identical with values in the source cited. The temperature entry in the 100 kPa column is close to, but not identical with, the normal boiling point (which is defined as the temperature at which the vapor pressure reaches 101.325 kPa). Although some temperatures are quoted to 0.1°C, uncertainties of several degrees should generally be assumed. Values followed by an "e" were obtained by extrapolating (usually with an Antoine equation) beyond the region for which experimental measurements were available and are thus subject to even greater uncertainty.

Compounds are listed by molecular formula following the Hill convention. Substances not containing carbon are listed first, followed by those that contain carbon. To locate an organic compound by name or CAS Registry Number when the molecular formula is not known, use the table "Physical Constants of Organic Compounds" in Section 3 and its indexes to determine the molecular formula. The indexes to "Physical Constants of Inorganic Compounds" in Section 4 can be used in a similar way.

More extensive and detailed vapor pressure data on selected important substances appear in other tables in this section of the *Handbook*. These substances are flagged by a symbol following the name as follows:

*   See "Vapor Pressure of Fluids below 300 K"
**   See "IUPAC Recommended Data for Vapor Pressure Calibration"
***   See "Vapor Pressure of Ice" and "Vapor Pressure of Water from 0 to 370°C"

The following notations appear after individual temperature entries:

s — Indicates the substance is a solid at this temperature.
e — Indicates an extrapolation beyond the region where experimental measurements exist.
i — Indicates the value was calculated from ideal gas hermodynamic functions, such as those in the *JANAF Thermochemical Tables* (see Reference 8).

References

1. Lide, D.R., and Kehiaian, H.V., CRC Handbook of *Thermophysical and Thermochemical Data*, CRC Press, Boca Raton, FL, 1994.
2. Stull, D., in *American Institute of Physics Handbook, Third Edition*, Gray, D.E., Ed., McGraw Hill, New York, 1972.
3. Hultgren, R., Desai, P.D., Hawkins, D.T., Gleiser, M., Kelley, K.K., and Wagman, D.D., *Selected Values of Thermodynamic Properties of the Elements*, American Society for Metals, Metals Park, OH, 1973.
4. Stull, D., *Ind. Eng. Chem.*, 39, 517, 1947.
5. *TRCVP, Vapor Pressure Database, Version 2.2P*, Thermodynamic Research Center, Texas A&M University, College Station, TX.
6. *TRC Thermodynamic Tables*, Thermodynamic Research Center, Texas A&M University, College Station, TX.
7. Ohe, S., *Computer Aided Data Book of Vapor Pressure*, Data Book Publishing Co., Tokyo, 1976.
8. Chase, M.W., Davies, C.A., Downey, J.R., Frurip, D.J., McDonald, R.A., and Syverud, A.N., *JANAF Thermochemical Tables, Third Edition, J. Phys. Chem. Ref. Data*, Vol. 14, Suppl. 1, 1985.
9. Barin, I., *Thermochemical Data of Pure Substances*, VCH Publishers, New York, 1993.
10. Jacobsen, R.T., et. al, *International Thermodynamic Tables of the Fluid State, No. 10. Ethylene*, Blackwell Scientific Publications, Oxford, 1988.
11. Wakeham, W.A., *International Thermodynamic Tables of the Fluid State, No. 12. Methanol*, Blackwell Scientific Publications, Oxford, 1993.
12. Janz, G.J., *Molten Salts Handbook*, Academic Press, New York, 1967.
13. Ohse, R.W. *Handbook of Thermodynamic and Transport Properties of Alkali Metals*, Blackwell Scientific Publications, Oxford, 1994.
14. Gschneidner, K.A., in *CRC Handbook of Chemistry and Physics, 77th Edition*, p. 4–112, CRC Press, Boca Raton, FL, 1996.
15. Leider, H.R., Krikorian, O.H., and Young, D.A., *Carbon*, 11, 555, 1973.
16. Ruzicka, K., and Majer, V., *J. Phys. Chem. Ref. Data*, 23, 1, 1994.
17. Tillner–Roth, R., and Baehr, H.D., *J. Phys. Chem. Ref. Data*, 23, 657, 1994.
18. Younglove, B.A., and McLinden, M.O., *J. Phys. Chem. Ref. Data*, 23, 731, 1994.
19. Outcalt, S.L., and McLinden, M.O., *J. Phys. Chem. Ref. Data*, 25, 605, 1996.
20. Weber, L.A., and Defibaugh, D.R., *J. Chem. Eng. Data*, 41, 382, 1996.
21. Rodrigues, M.F., and Bernardo-Gil, M.G., *J. Chem. Eng. Data*, 41, 581, 1996.
22. Piacente, V., Gigli, G., Scardala, P., and Giustini, A., *J. Phys. Chem.*, 100, 9815, 1996.
23. Barton, J.L., and Bloom, H., *J. Phys. Chem.*, 60, 1413, 1956.
24. Sense, K.A., Alexander, C.A., Bowman, R.E., and Filbert, R.B., *J. Phys. Chem.*, 61, 337, 1957.
25. Ewing, C.T., and Stern, K.H., *J. Phys. Chem.* 78, 1998, 1974.
26. Cady, G.H., and Hargreaves, G.B., *J. Chem. Soc.*, 1563, 1961; 1568, 1961.
27. Skudlarski, K., Dudek, J., and Kapala, J., *J. Chem. Thermodynamics*, 19, 857, 1987.
28. Wagner, W., and de Reuck, K.M., *International Thermodynamic Tables of the Fluid State, No. 9. Oxygen*, Blackwell Scientific Publications, Oxford, 1987.
29. Marsh, K.N., Editor, *Recommended Reference Materials for the Realization of Physicochemical Properties*, Blackwell Scientific Publications, Oxford, 1987.
30. Alcock, C.B., Itkin, V.P., and Horrigan, M.K., *Canadian Metallurgical Quarterly*, 23, 309, 1984.
31. Stewart, R.B., and Jacobsen, R.T., *J. Phys. Chem. Ref. Data*, 18, 639, 1989.
32. Sifner, O., and Klomfar, J., *J. Phys. Chem. Ref. Data*, 23, 63, 1994.
33. Bah, A., and Dupont–Pavlovsky, N., *J. Chem. Eng. Data*, 40, 869, 1995.
34. Behrens, R.G., and Rosenblatt, G., *J. Chem. Thermodynamics*, 4, 175, 1972.
35. Behrens, R.G., and Rosenblatt, G., *J. Chem. Thermodynamics*, 5, 173, 1973.
36. Haar, L., Gallagher, J.S., and Kell, G.S., *NBS/NRC Steam Tables*, Hemisphere Publishing Corp., New York, 1984.
37. Wagner, W., Saul, A., and Pruss, A., *J. Phys. Chem. Ref. Data*, 23, 515, 1994.
38. Behrens, R.G., Lemons, R.S., and Rosenblatt, G., *J. Chem. Thermodynamics*, 6, 457, 1974.
39. Boublik, T., Fried, V., and Hala, E., *The Vapor Pressure of Pure Substances, Second Edition*, Elsevier, Amsterdam, 1984.
40. Goodwin, R.D., *J. Phys. Chem. Ref. Data*, 14, 849, 1985.
41. Younglove, B.A., and Ely, J.F., *J. Phys. Chem. Ref. Data*, 16, 577, 1987.

Mol. form.	Name	Temperature in °C for the indicated pressure						Ref.
		1 Pa	10 Pa	100 Pa	1 kPa	10 kPa	100 kPa	

Substances not containing carbon:

Mol. form.	Name	1 Pa	10 Pa	100 Pa	1 kPa	10 kPa	100 kPa	Ref.
Ag	Silver	1010	1140	1302	1509	1782	2160	2
AgBr	Silver(I) bromide	569 i	656 i	765 i	905 i	1093 i	1359 i	9
AgCl	Silver(I) chloride	670	769	873	1052	1264	1561	4
AgI	Silver(I) iodide	594	686	803	959	1177	1503	4
Al	Aluminum	1209	1359	1544	1781	2091	2517	2
AlB_3H_{12}	Aluminum borohydride				−46.8	−9.4	45.5	4
$AlCl_3$	Aluminum trichloride	58.4 s	76.5 s	97.1 s	120.7 s	148.2 s	180.5 s	4
AlF_3	Aluminum trifluoride	744 s	819 s	906 s	1008 s	1130 s	1276 s	8
AlI_3	Aluminum triiodide				218	285	385	4
Al_2O_3	Aluminum oxide			2122	2351	2629	2975	4
Ar	Argon*		−226.4 s	−220.3 s	−212.4 s	−201.7 s	−186.0	1,5,31
As	Arsenic	280 s	323 s	373 s	433 s	508 s	601 s	3
$AsCl_3$	Arsenic(III) chloride		−8 e	21.3	63.1	129.4		1
AsF_3	Arsenic(III) fluoride				8.1	56.0		4
AsI_3	Arsenic(III) iodide			187	261	367 e		7
As_2O_3	Arsenic(III) oxide (arsenolite)	133.7 s	163.0 s	196.8 s	236.2 s	283.0		34
At	Astatine	88 s	119 s	156 s	202 s	258 s	334	2
Au	Gold	1373	1541	1748	2008	2347	2805	2
B	Boron	2075	2289	2549	2868	3272	3799	2
BBr_3	Boron tribromide		−45 e	−15 e	27.5	90.4		1
BCl_3	Boron trichloride*		−94.0	−70.5	−37.4	12.3		4
BF_3	Boron trifluoride*	−173.9 s	−166.0 s	−156.0 s	−143.0 s	−125.9	−101.1	4
B_2F_4	Tetrafluorodiborane						−34	1
B_2H_6	Diborane			−162 e	−147.0	−125.8	−92.6	1
B_5H_9	Pentaborane(9)				−34.8	3.8	57.6	4
Ba	Barium	638 s	765	912	1115	1413	1897	9
Be	Beryllium	1189 s	1335	1518	1750	2054	2469	2
$BeBr_2$	Beryllium bromide	203 s	240 s	283 s	335 s	397 s	473 s	4
$BeCl_2$	Beryllium chloride	196 s	237 s	284 s	339 s	402 s	487	4
BeF_2	Beryllium fluoride		686 e	767 e	869	999	1172 e	7
BeI_2	Beryllium iodide	188 s	229 s	276 s	333 s	402 s	487	4
Bi	Bismuth	668	768	892	1052	1265	1562	2
$BiBr_3$	Bismuth tribromide			217 s	273 i	348 i	455 i	4,9
$BiCl_3$	Bismuth trichloride				248.9	328.6	438.7	1,4
BrCs	Cesium bromide	531 s	601 s	701 i	834 i	1019 i	1293 e	9
BrH	Hydrogen bromide*		−153.3 s	−140.4 s	−123.8 s	−101.5 s	−67.0	5
BrH_3Si	Bromosilane				−81.0	−47.3	2.2	4
BrH_4N	Ammonium bromide	121 s	154 s	195 s	246 s	310.4 s	395.1 s	5
BrK	Potassium bromide	597 s	674 s	773				25
BrLi	Lithium bromide		630	733	868	1049	1308	4
BrNa	Sodium bromide			791	931	1120	1389	4
BrRb	Rubidium bromide			766	903	1087	1350	4
BrTl	Thallium(I) bromide				509	635	817	4
Br_2	Bromine*	−87.7 s	−71.8 s	−52.7 s	−29.3 s	2.5	58.4	1
Br_2Cd	Cadmium bromide	373 s	435 s	509 s				27
Br_2Hg	Mercury(II) bromide	71 s	98 s	132 s	174 s	227 s	318	4
Br_2OS	Thionyl bromide	−49 e	−29 e	−5 e	27.8	72.9	139.6	5
Br_2Pb	Lead(II) bromide	374	431	502	597	726	914	4
Br_2S_2	Sulfur bromide	−7 e	15 e	42 e	78.4	128.1	200.9	5
Br_3In	Indium(III) bromide			304.6 s	328.7 s	364.8 s		1
Br_3OP	Phosphorus(V) oxybromide				64 e	115.5	191.4	5
Br_3P	Phosphorus(III) bromide		−23 e	5 e	42.3	94.6	172.6	5
Br_3Sb	Antimony(III) bromide				136.5	196.9	286.5	1
Br_4Ge	Germanium(IV) bromide				51	105	188	4
Br_4Sn	Tin(IV) bromide				67	122	204	4
Br_4Zr	Zirconium(IV) bromide	136 s	167 s	203 s	245 s	295 s	356 s	4
Br_5P	Phosphorus(V) bromide		−19 s	4 s	31 s	65.5 s	110.1	5
Ca	Calcium	591 s	683 s	798 s	954	1170	1482	2
Cd	Cadmium	257 s	310 s	381	472	594	767	2

Mol. form.	Name	Temperature in °C for the indicated pressure						Ref.
		1 Pa	10 Pa	100 Pa	1 kPa	10 kPa	100 kPa	
$CdCl_2$	Cadmium chloride	412 s	471 s	541 s	634	768	959	23, 27
CdF_2	Cadmium fluoride				1257	1461	1742	4
CdI_2	Cadmium iodide	296 s	344 s	406	498	622	795	4,27
CdO	Cadmium oxide	770 s	866 s	983 s	1128 s	1314 s	1558 s	4
Ce	Cerium	1719	1921	2169	2481	2886	3432	14
ClCs	Cesium chloride			730	864	1043	1297	4
ClCu	Copper(I) chloride		459	543	675	914	1477	4
ClF	Chlorine fluoride*				−144.4	−122.6	−90.2	5
ClF_2P	Phosphorus(III) chloride difluoride				−119.5	−91.1	−47.6	5
ClF_3	Chlorine trifluoride				−63.7	−33.0	11.4	5
ClF_5	Chlorine pentafluoride				−88 e	−59	−14	7
ClH	Hydrogen chloride*				−138.2 s	−118.0	−85.2	1,5
$ClHO_3S$	Chlorosulfonic acid	−40 e	−20 e	5 e	38.7	85.0	153.6	5
ClH_4N	Ammonium chloride	91 s	121 s	159 s	204.7 s	263.1 s	339.5 s	5
ClK	Potassium chloride	625 s	704 s	804	945	1137	1411	23,25
ClLi	Lithium chloride		649 i	761 i	905 i	1101 i	1381 i	8
ClNO	Nitrosyl chloride		−116 s	−100 s	−78.7 s	−50.2	−5.7	5
$ClNO_2$	Nitryl chloride	−121 e	−113 e	−102 e	−86.1	−60.9	−15.7	5
ClNa	Sodium chloride	653 s	733 s	835	987	1182	1461	23,25
ClO_2	Chlorine dioxide*					−34.3	10.5	5
ClRb	Rubidium chloride			777	916	1105	1379	4
ClTl	Thallium(I) chloride				504	626	806	4
Cl_2	Chlorine*	−145 s	−133.7 s	−120.2 s	−103.6 s	−76.1	−34.2	1
Cl_2Co	Cobalt(II) chloride					818	1048	4
Cl_2FP	Phosphorus(III) dichloride fluoride				−71.1	−37.4	13.5	5
Cl_2F_3P	Phosphorus(V) dichloride trifluoride		−120 e	−101 e	−77.1	−44.3	3 e	7
Cl_2Fe	Iron(II) chloride				685	821	1025	4
Cl_2Hg	Mercury(II) chloride	64.4 s	94.7 s	130.8 s	174.5 s	228.5 s	304.0	4
Cl_2Mg	Magnesium chloride			762	908	1111	1414	4
Cl_2Mn	Manganese(II) chloride				760	933	1189	4
Cl_2Ni	Nickel(II) chloride	534 s	592 s	662 s	747 s	852 s	985 s	4
Cl_2OS	Thionyl chloride	−99 e	−81 e	−58 e	−27.1	14.6	75.2	5
Cl_2O_2S	Sulfuryl chloride				−27 e	11.8	69.0	5
Cl_2Pb	Lead(II) chloride			541 e	637	765	949	23
Cl_2S	Sulfur dichloride	−76 e	−61 e	−41 e	−16.7	15.3	58.7	5
Cl_2S_2	Sulfur chloride	−55 e	−36 e	−12 e	21.0	67.2	137.1	5
Cl_2Sn	Tin(II) chloride		253	308	381	479	622	4
Cl_2Zn	Zinc chloride	305 i	356 i	419 i	497 i	596 i	726 i	4,9,12
Cl_3Fe	Iron(III) chloride	118 s	153 s	190 s	229 s	268 s	319	4
Cl_3HSi	Trichlorosilane			−81 e	−56 e	−21 e	31.6	7
Cl_3N	Nitrogen trichloride				−25 e	13.2	70.6	5
Cl_3OP	Phosphorus(V) oxychloride				39.9	105.0		5
Cl_3P	Phosphorus(III) chloride	−93 e	−77 e	−55 e	−26.0	14.5	75.7	5
Cl_4Po	Polonium(IV) chloride				300.6	389.4		5
Cl_4Se	Selenium tetrachloride	23 s	45 s	71 s	102 s	141.4 s	191.1 s	5
Cl_4Si	Tetrachlorosilane*				−39 e	0 e	57.3	1
Cl_4Te	Tellurium tetrachloride				237 e	299.4	387.8	5
Cl_4Zr	Zirconium(IV) chloride	117 s	146 s	181 s	222 s	272 s	336 s	9
Cl_5P	Phosphorus(V) chloride	−2 s	19 s	44 s	74 s	111.4 s	158.9 s	5
Co	Cobalt	1517	1687	1892	2150	2482	2925	2
Cr	Chromium	1383 s	1534 s	1718 s	1950	2257	2669	2
Cs	Cesium	144.5	195.6	260.9	350.0	477.1	667.0	13,30
CsF	Cesium fluoride				825	999	1249	4
CsI	Cesium iodide	523 s	595 s	692	854	1029	1278	4,25
Cu	Copper	1236	1388	1577	1816	2131	2563	2
CuI	Copper(I) iodide				636	864	1331	4
Dy	Dysprosium	1105 s	1250 s	1431 i	1681 i	2031 i	2558 i	3
Er	Erbium	1231 s	1390 s	1612 i	1890 i	2279 i	2859 i	3

Mol. form.	Name	Temperature in °C for the indicated pressure						Ref.
		1 Pa	10 Pa	100 Pa	1 kPa	10 kPa	100 kPa	
Eu	Europium	590 s	684 s	799 s	961	1179	1523	14
FH	Hydrogen fluoride*				−71.1	−33.7	19.2	1,5
FHO$_3$S	Fluorosulfonic acid	−14 e	4 e	28 e	59.1	101.3	162.2	5
FK	Potassium fluoride			869	1017	1216	1499	4
FLi	Lithium fluoride	801 s	896	1024	1188	1395	1672	4,12,25
FNO	Nitrosyl fluoride			−131 e	−116.1	−94.3	−60.1	5
FNO$_2$	Nitryl fluoride		−156 e	−144 e	−128.1	−106.0	−72.6	5
FNO$_3$	Fluorine nitrate	−160 e	−149 e	−135 e	−115.1	−87.4	−45.0	5
FNa	Sodium fluoride		920 s	1058	1218	1426	1702	4,12,24
FRb	Rubidium fluoride			910	1001	1145	1409	4,12
F$_2$	Fluorine*	−235 s	−229.5 s	−222.9 s	−214.8	−204.3	−188.3	1,5
F$_2$O	Fluorine monoxide*	−211.7	−204.7	−195.9	−184.2	−168.2	−144.9	5
F$_2$OS	Thionyl fluoride			−124 e	−106.5	−81.5	−44.1	5
F$_2$O$_2$Re	Rhenium(VI) dioxydifluoride				89.2	131.9	185 e	26
F$_2$Pb	Lead(II) fluoride				865	1054	1292	4
F$_2$Xe	Xenon difluoride			2.9 s	31.8 s	67.9 s	114 s	1,5
F$_2$Zn	Zinc fluoride	731 s	813 s	911 i	1048 i	1237 i	1503 i	9
F$_3$N	Nitrogen trifluoride*	−201 e	−194 e	−185 e	−172.8	−155.5	−129.2	5
F$_3$OP	Phosphorus(V) oxyfluoride	−124 s	−113 s	−100 s	−83.7 s	−64.1 s	−39.7 s	5
F$_3$P	Phosphorus(III) fluoride*				−152 e	−132.6	−101.4	5
F$_4$MoO	Molybdenum(VI) oxytetrafluoride	−21 s	3 s	33 s	69.3 s	117.3	184.1	26
F$_4$ORe	Rhenium(VI) oxytetrafluoride	5 s	26 s	50.7 s	80.1 s	117.1	171.2	26
F$_4$OW	Tungsten(VI) oxytetrafluoride	2 s	25 s	52.1 s	84.3 s	126.7	185.4	26
F$_4$S	Sulfur tetrafluoride				−110.0	−82.1	−40.3	5
F$_4$Se	Selenium tetrafluoride				13.6	51.6	104.7	5
F$_4$Si	Tetrafluorosilane*	−166 s	−157 s	−145.6 s	−132.3 s	−115.7 s	−94.9 s	4,7
F$_5$Mo	Molybdenum(V) fluoride				86.6	140.3	213 e	26
F$_5$Nb	Niobium(V) fluoride				80	140	224	4
F$_5$ORe	Rhenium(VII) oxypentafluoride	−103 s	−84 s	−59 s	−28 s	13.7 s	72.8	26
F$_5$Os	Osmium(V) fluoride			74.1	113.2	162.3	226 e	26
F$_5$P	Phosphorus(V) fluoride	−157 s	−148 s	−137 s	−124.5 s	−108.6 s	−84.8 s	5
F$_5$Re	Rhenium(V) fluoride			58.8	99.5	152 e	221 e	26
F$_5$Ta	Tantalum(V) fluoride					119	229	4
F$_6$Ir	Iridium(VI) fluoride	−88 s	−71 s	−51 s	−27 s	3.8 s	53.1	26
F$_6$Mo	Molybdenum(VI) fluoride	−98 s	−82 s	−64 s	−41.2 s	−13.4 s	33.5	26
F$_6$Os	Osmium(VI) fluoride	−89 s	−73 s	−54 s	−30.6 s	−1.7 s	47.4	26
F$_6$Re	Rhenium(VI) fluoride	−97 s	−82 s	−63 s	−40.2 s	−11.9 s	33.4	26
F$_6$S	Sulfur hexafluoride*	−158 s	−147 s	−133.6 s	−116.6 s	−94.4 s	−64.1 s	5
F$_6$Se	Selenium hexafluoride	−143 s	−132 s	−118 s	−100.7 s	−77.8 s	−46.5 s	5
F$_6$Te	Tellurium hexafluoride	−142 s	−130 s	−115 s	−96 s	−71.8 s	−39.1 s	5
F$_6$W	Tungsten(VI) fluoride	−107 s	−92 s	−74 s	−52.1 s	−24.8 s	16.9	26
F$_{10}$S$_2$	Sulfur decafluoride					−22.0	28.5	5
Fe	Iron	1455 s	1617	1818	2073	2406	2859	2
Fr	Francium	131 e	181 e	246 e	335 e	465 e	673 e	2
Ga	Gallium	1037	1175	1347	1565	1852	2245	2
Gd	Gadolinium	1563 i	1755 i	1994 i	2300 i	2703 i	3262 i	3
Ge	Germanium	1371	1541	1750	2014	2360	2831	2
HI	Hydrogen iodide*	−146 s	−135.2 s	−120.8 s	−101.9 s	−75.9 s	−35.9	5
HKO	Potassium hydroxide	520 e	601 e	704	842	1035	1325	4
HNO$_3$	Nitric acid			−37 e	−9 e	28.4	82.2	5
HN$_3$	Hydrazoic acid			−79 e	−54 e	−18.0	35.7	5
HNaO	Sodium hydroxide	513	605	722	874	1080	1377	4
H$_2$	Hydrogen*					−258.6	−252.8	1
H$_2$I$_2$Si	Diiodosilane				11.8	70.5	149.4	4
H$_2$O	Water***	−60.7 s	−42.2 s	−20.3 s	7.0	45.8	99.6	36,37
H$_2$O$_2$	Hydrogen peroxide			13 e	45 e	89.0	149.8	5
H$_2$O$_4$S	Sulfuric acid	72	103	140	187	248	330	4
H$_2$S	Hydrogen sulfide*		−149 s	−136 s	−118.9 s	−95.9 s	−60.5	1,5
H$_2$S$_2$	Hydrogen disulfide				−27 e	12.2	70.7	5

Mol. form.	Name	Temperature in °C for the indicated pressure						Ref.
		1 Pa	10 Pa	100 Pa	1 kPa	10 kPa	100 kPa	
H_2Se	Hydrogen selenide	−145 s	−134 s	−120 s	−102.8 s	−78.9 s	−41.5	5
H_2Te	Hydrogen telluride					−46.6	−2.3	5
H_3ISi	Iodosilane				−47.7	−10.1	45.2	4
H_3N	Ammonia*	−139 s	−127 s	−112 s	−94.5 s	−71.3	−33.6	1,5,6
H_3NO	Hydroxylamine				43.7	73.3	109.8	4
H_3P	Phosphine*	−182 s	−173 s	−161 s	−145 s	−122.7	−88.0	5
H_4IN	Ammonium iodide	125 s	159 s	201 s	253 s	318.4 s	405.2 s	5
H_4N_2	Hydrazine				14.7	55.6	113 e	5
H_4Si	Silane*			−181	−165.4	−143.7	−111.8	4
He	Helium*					−270.6	−268.9	2
Hf	Hafnium	2416	2681	3004	3406	3921	4603	9
Hg	Mercury**	42.0	76.6	120.0	175.6	250.3	355.9	29,30
HgI_2	Mercury(II) iodide	85.1 s	115.6 s	152.4 s	197.8 s	255.1 s	353.6	4
Ho	Holmium	1159 s	1311 s	1502 i	1767 i	2137 i	2691 i	3
IK	Potassium iodide			731	866	1052	1322	4
ILi	Lithium iodide	545	619	710	824	972	1170	4
INa	Sodium iodide			753	883	1058	1301	4
IRb	Rubidium iodide			733	866	1045	1302	4
ITl	Thallium(I) iodide				520	644	821	4
I_2	Iodine (rhombic)	−12.8 s	9.3 s	35.9 s	68.7 s	108 s	184.0	1,2
I_2Pb	Lead(II) iodide			470	558	682	869	4
I_2Zn	Zinc iodide	301 s	351 s	409 s	488 i	598 i	750 i	9
I_3Sb	Antimony(III) iodide				214.9	292.0	401.2	4
I_4Sn	Tin(IV) iodide				167.1	242.7	347.7	4
I_4Zr	Zirconium(IV) iodide	187 s	220 s	259 s	305 s	361 s	430 s	4
In	Indium	923	1052	1212	1417	1689	2067	2
Ir	Iridium	2440 s	2684	2979	3341	3796	4386	2
K	Potassium	200.2	256.5	328	424	559	756.2	13,30
Kr	Krypton*	−214.0 s	−208.0 s	−199.4 s	−188.9 s	−174.6 s	−153.6	5
La	Lanthanum	1732 i	1935 i	2185 i	2499 i	2905 i	3453 i	3
Li	Lithium	524.3	612.3	722.1	871.2	1064.3	1337.1	13,30
Lu	Lutetium	1633 s	1829.8	2072.8	2380 i	2799 i	3390 i	3
Mg	Magnesium	428 s	500 s	588 s	698	859	1088	2
Mn	Manganese	955 s	1074 s	1220 s	1418	1682	2060	2
Mo	Molybdenum	2469 s	2721	3039	3434	3939	4606	2
MoO_3	Molybdenum(VI) oxide				801	935	1151	4
NO	Nitric oxide*	−201 s	−195 s	−188 s	−179.3 s	−168.1 s	−151.9	5
N_2	Nitrogen*	−236 s	−232 s	−226.8 s	−220.2 s	−211.1 s	−195.9	1,5
N_2O	Nitrous oxide*	−167 s	−157 s	−145.4 s	−131.1 s	−112.9 s	−88.7	5
N_2O_4	Nitrogen tetroxide	−92 s	−78 s	−61 s	−41.1 s	−16.6 s	28.7	5
N_2O_5	Nitrogen pentoxide	−71 s	−56 s	−40 s	−19.9 s	3.9 s	33.2	5
Na	Sodium	280.6	344.2	424.3	529	673	880.2	13,30
Nb	Niobium	2669	2934	3251	3637	4120	4740	2
Nd	Neodymium	1322.3	1501.2	1725.3	2023 i	2442 i	3063 i	3
Ne	Neon*	−261 s	−260 s	−258 s	−255 s	−252 s	−246.1	2
Ni	Nickel	1510	1677	1881	2137	2468	2911	2
OPb	Lead(II) oxide	724	816	928	1065	1241	1471	4
OSr	Strontium oxide	1789 s	1903 s	2047 s	2235 s	2488 s		4
O_2	Oxygen*				−211.9	−200.5	−183.1	1,28
O_2S	Sulfur dioxide*			−98 s	−80 s	−52.2	−10.3	1,5
O_2Se	Selenium dioxide	124.5 s	153.9 s	188 s	228 s	275 s	315 s	38
O_2Si	Silicon dioxide	1966 i	2149 i	2368 i				8
O_3	Ozone*	−189 e	−182 e	−172 e	−158 e	−139.7	−111.5	5
O_3P_2	Phosphorus(III) oxide				47.3	100.3	172.8	4
O_3S	Sulfur trioxide				−20 s	6.6 s	44.5	5
O_3Sb_2	Antimony(III) oxide (valentinite)	426.1 s	478 s	539 s	610 s	907	1420	4,35
O_5P_2	Phosphorus(V) oxide	285 s	328 s	377.5 s	434.4 s	500.5 s	591	4
O_7Re_2	Rhenium(VII) oxide	147 s	176 s	208 s	244 s	284 s	362	4
Os	Osmium	2887 s	3150	3478	3875	4365	4983	2

		Temperature in °C for the indicated pressure						
Mol. form.	Name	1 Pa	10 Pa	100 Pa	1 kPa	10 kPa	100 kPa	Ref.
P	Phosphorus (white)	6 s	34 s	69	115	180	276	3,9
P	Phosphorus (red)	182 s	216 s	256 s	303 s	362 s	431 s	2,3
Pb	Lead	705	815	956	1139	1387	1754	2
PbS	Lead(II) sulfide	656 s	741 s	838 s	953 s	1088 s	1280	4
Pd	Palladium	1448 s	1624	1844	2122	2480	2961	2
Po	Polonium				573 e	730.2	963.3	5
Pr	Praseodymium	1497.7	1699.4	1954 i	2298 i	2781 i	3506 i	3
Pt	Platinum	2057	2277 e	2542	2870	3283	3821	2
Pu	Plutonium	1483	1680	1925	2238	2653	3226	2
Ra	Radium	546 s	633 s	764	936	1173	1526	2
Rb	Rubidium	160.4	212.5	278.9	368	496.1	685.3	13,30
Re	Rhenium	3030 s	3341	3736	4227	4854	5681	2
Rh	Rhodium	2015	2223	2476	2790	3132	3724	2
Rn	Radon*	−163 s	−152 s	−139 s	−121.4 s	−97.6 s	−62.3	5
Ru	Ruthenium	2315 s	2538	2814	3151	3572	4115	2
S	Sulfur	102 s	135	176	235	318	444	3
Sb	Antimony	534 s	603 s	738	946	1218	1585	2,3
Sc	Scandium	1372 s	1531 s	1733 i	1993 i	2340 i	2828 i	3
Se	Selenium	227	279	344	431	540	685	3
Si	Silicon	1635	1829	2066	2363	2748	3264	2
Sm	Samarium	728 s	833 s	967 s	1148 i	1402 i	1788 i	3
Sn	Tin	1224	1384	1582	1834	2165	2620	2
Sr	Strontium	523 s	609 s	717 s	866	1072	1373	2
Ta	Tantalum	3024	3324	3684	4122	4666	5361	2
Tb	Terbium	1516.1	1706.1	1928 i	2232 i	2640 i	3218 i	3
Tc	Technetium	2454 e	2725 e	3051 e	3453 e	3961 e	4621 e	2
Te	Tellurium			502 e	615 e	768.8	992.4	5
Th	Thorium	2360	2634	2975	3410	3986	4782	2
Ti	Titanium	1709	1898	2130 e	2419	2791	3285	2
Tl	Thallium	609	704	824	979	1188	1485	2
Tm	Thulium	844 s	962 s	1108 s	1297 s	1548 i	1944 i	3
U	Uranium	2052	2291	2586	2961	3454	4129	2
V	Vanadium	1828 s	2016	2250	2541	2914	3406	2
W	Tungsten	3204 s	3500	3864	4306	4854	5550	2
Xe	Xenon*	−190 s	−181 s	−170 s	−155.8 s	−136.6 s	−108.4	5,32
Y	Yttrium	1610.1	1802.3	2047 i	2354 i	2763 i	3334 i	3
Yb	Ytterbium	463 s	540 s	637 s	774 s	993 i	1192 i	3
Zn	Zinc	337 s	397 s	477	579	717	912 e	2
Zr	Zirconium	2366	2618	2924	3302	3780	4405	2
Substances containing carbon:								
C	Carbon (graphite)		2566 s	2775 s	3016 s	3299 s	3635 s	15
CBrClF$_2$	Bromochloro-difluoromethane	−136 e	−123 e	−106 e	−83.4	−51.8	−4.3	1
CBrCl$_3$	Bromotrichloromethane				−6 e	38.9	104.4	5
CBrF$_3$	Bromotrifluoromethane*	−168 e	−156 e	−142 e	−122.8	−96.6	−58.1	5
CBrN	Cyanogen bromide				−13 s	17.7 s	61.0	1
CBr$_2$F$_2$	Dibromodifluoromethane		−110 e	−91 e	−66 e	−30 e	22.5	1
CBr$_4$	Tetrabromomethane			25.6 s	65.8 s	111.6	188.9	5
CClF$_3$	Chlorotrifluoromethane	−176 e	−167 e	−155 e	−139 e	−116 e	−81.7	5
CClN	Cyanogen chloride		−94.6 s	−78.1 s	−57 s	−29 s	13.0	5
CCl$_2$F$_2$	Dichlorodifluoromethane*	−150 e	−138 e	−122 e	−101.8	−73.1	−30.0	5
CCl$_2$O	Carbonyl chloride	−127 e	−113 e	−96 e	−73 e	−40.6	7.2	5
CCl$_3$F	Trichlorofluoromethane*		−107 e	−89 e	−63 e	−28.5	23.3	1,5
CCl$_3$NO$_2$	Trichloronitromethane		−59 e	−30 e	4.4	47.8	112.0	5
CCl$_4$	Tetrachloromethane*	−79.4 s	−70.8 s	−53.5 s	−24.4 s	15.8	76.2	1,5
CFN	Cyanogen fluoride		−135 s	−121.2 s	−104.1 s	−82.8 s	−46.2	1,5
CF$_4$	Tetrafluoromethane*	−199.9 s	−193 s	−183.9 s	−171.6	−153.9	−128.3	1,5
CHBrF$_2$	Bromodifluoromethane		−128 s	−111.4 s	−89.7 s	−59.7 s	−16 s	5
CHBr$_3$	Tribromomethane				30.5	78.3	148.8	1
CHClF$_2$	Chlorodifluoromethane*	−152 e	−141 e	−126 e	−107.1	−80.5	−41.1	5
CHCl$_2$F	Dichlorofluoromethane	−76 e	−70 e	−61 e	−49 e	−28.7	8.6	1

Mol. form.	Name	Temperature in °C for the indicated pressure						Ref.
		1 Pa	10 Pa	100 Pa	1 kPa	10 kPa	100 kPa	
CHCl$_3$	Trichloromethane*			−61 e	−34 e	4.3	60.8	1
CHF$_3$	Trifluoromethane*			−152 e	−136 e	−114.4	−82.3	1
CHI$_3$	Triiodomethane	51.1 s	82.7 s	121 e			218.0	5
CHN	Hydrogen cyanide*			−77 s	−52.6 s	−22.7 s	25.4	1,5
CHNO	Cyanic acid			−81.1	−56.8	−23.9	23 e	5
CH$_2$BrCl	Bromochloromethane	−83 e	−69 e	−50 e	−25 e	11.4	67.7	1
CH$_2$Br$_2$	Dibromomethane			−37 e	−7 e	35.2	96.5	5
CH$_2$ClF	Chlorofluoromethane		−124 e	−108 e	−86.2	−55.7	−9.4	5
CH$_2$Cl$_2$	Dichloromethane*		−92 e	−73 e	−48 e	−12.5	39.3	1
CH$_2$F$_2$	Difluoromethane*	−156.7	−145.8	−131.9	−113.6	−88.6	−51.9	1
CH$_2$I$_2$	Diiodomethane			17 e	55 e	106.1	181.6	5
CH$_2$O	Formaldehyde*				−91	−61.7	−19.3	1
CH$_2$O$_2$	Formic acid	−56 s	−40.4 s	−22.3 s	−0.8 s	37.0	100.2	1,5
CH$_3$AsF$_2$	Methyldifluoroarsine				−15 e	22.1	76.1	5
CH$_3$BO	Borane carbonyl				−124	−99	−64	4
CH$_3$Br	Bromomethane				−77	−44.3	3.3	1
CH$_3$Cl	Chloromethane*	−140.2 s	−128.6 s	−114.7 s	−96 e	−67.1	−24.4	1,33
CH$_3$Cl$_3$Si	Methyltrichlorosilane		−83 e	−61 e	−33 e	7 e	65.7	1
CH$_3$F	Fluoromethane*				−130 e	−111 e	−78.6	1
CH$_3$I	Iodomethane				−49 e	−12.4	42.1	1
CH$_3$NO	Formamide		22 e	53 e	93 e	145.0	218 e	5
CH$_3$NO$_2$	Nitromethane				−2 e	40 e	100.8	1
CH$_3$NO$_3$	Methyl nitrate		−75 e	−55 e	−27 e	9.8	63 e	5
CH$_4$	Methane*	−220 s	−214.2 s	−206.8 s	−197 s	−183.6 s	−161.7	5,41
CH$_4$Cl$_2$Si	Dichloromethylsilane			−77 e	−51 e	−14 e	40.5	1
CH$_4$O	Methanol*	−87 e	−69 e	−47.5	−20.4	15.2	64.2	11
CH$_4$S	Methanethiol		−115 e	−97 e	−74 e	−41.7	5.7	1
CH$_5$ClSi	Chloromethylsilane	−129 e	−115 e	−97.9	−74.4	−41.5	8.3	5
CH$_5$N	Methylamine				−76.7	−48.1	−6.6	1
CH$_6$N$_2$	Methylhydrazine			−31 e	−4.7	32.9	91 e	1
CH$_6$OSi	Methyl silyl ether				−90.2	−61.8	−18 e	1
CH$_6$Si	Methylsilane			−144 e	−124.6	−97.5	−57.5	5
CIN	Cyanogen iodide						153.8	5
CNNa	Sodium cyanide		672 e	798	961	1182	1497	4
CN$_4$O$_8$	Tetranitromethane				18.0	61.8	124 e	5
CO	Carbon monoxide*			−223 s	−216.5 s	−207.2 s	−191.7	40
COS	Carbon oxysulfide*			−136 e	−117 e	−90.0	−50.4	1
COSe	Carbon oxyselenide			−120	−98	−67	−22	4
CO$_2$	Carbon dioxide*	−159.1 s	−148.9 s	−136.7 s	−121.6 s	−103.1 s	−78.6 s	5
CS$_2$	Carbon disulfide		−96 e	−76 e	−49 e	−10.9	45.9	1
CSe$_2$	Carbon diselenide			−24 e	9.4	56.2	127 e	1
C$_2$Br$_2$ClF$_3$	1,2-Dibromo-1-chloro-1,2,2-trifluoroethane						92.3	5
C$_2$Br$_2$F$_4$	1,2-Dibromotetrafluoroethane		−97 e	−75 e	−46 e	−7.2	47.1	5
C$_2$Br$_4$	Tetrabromoethylene		−54.5 s	−31.7 s	−3.5 s	32.2 s	226.0	5
C$_2$ClF$_3$	Chlorotrifluoroethylene	−146 e	−134 e	−119 e	−99 e	−71 e	−28.4	1
C$_2$ClF$_5$	Chloropentafluoroethane					−80.3	−39.4	1
C$_2$Cl$_2$F$_4$	1,1-Dichlorotetrafluoroethane					−45.4	2.7	5
C$_2$Cl$_2$F$_4$	1,2-Dichlorotetrafluoroethane				−76.8	−44.9	3.2	5
C$_2$Cl$_3$F$_3$	1,1,1-Trichlorotrifluoroethane						45.6	1,5
C$_2$Cl$_3$F$_3$	1,1,2-Trichlorotrifluoroethane					−8.2	47.3	1,5
C$_2$Cl$_3$N	Trichloroacetonitrile				−16 e	25.3	85.1	1
C$_2$Cl$_4$	Tetrachloroethylene			−22 e	10 e	54.4	120.7	1
C$_2$Cl$_4$F$_2$	1,1,1,2-Tetrachloro-2,2-difluoroethane				−7 e	31.0	91.1	5
C$_2$Cl$_4$F$_2$	1,1,2,2-Tetrachloro-1,2-difluoroethane					32.3	92.5	1
C$_2$Cl$_4$O	Trichloroacetyl chloride			−25 e	7 e	51.7	117.8	1,5
C$_2$Cl$_6$	Hexachloroethane	−7.6 s	9.9 s	33.6 s	67.7 s	116.9 s	184.2 s	5
C$_2$F$_3$N	Trifluoroacetonitrile				−126.1	−102.5	−67.8	1
C$_2$F$_4$	Tetrafluoroethylene				−132.3	−109.7	−75.8	1
C$_2$F$_4$N$_2$O$_4$	1,1,2,2-Tetrafluoro-1,2-dinitroethane				−30 e	6.4	59.5	5

Mol. form.	Name	Temperature in °C for the indicated pressure						Ref.
		1 Pa	10 Pa	100 Pa	1 kPa	10 kPa	100 kPa	
C_2F_6	Hexafluoroethane**			−155.2 s	−137.5 s	−113.4 s	−78.4 s	1,5
$C_2HBrClF_3$	2-Bromo-2-chloro-1,1,1-trifluoroethane				−41.4	−4.8	49.8	1
C_2HBr_3O	Tribromoacetaldehyde			15.0	52.7	103.0	173.5	5
C_2HClF_4	1-Chloro-1,1,2,2-tetrafluoroethane			−110 e	−87.6	−57.0	−12.1	5
$C_2HCl_2F_3$	2,2-Dichloro-1,1,1-trifluoroethane		−101.0	−82.2	−57.4	−23.3	26.7	18
C_2HCl_3	Trichloroethylene	−74 e	−59 e	−39 e	−12 e	26.7	86.8	1
C_2HCl_3O	Trichloroacetaldehyde			−41.6	−9.8	33.8	97.4	5
$C_2HCl_3O_2$	Trichloroacetic acid				83.8	130.0	197.2	1,5
C_2HCl_5	Pentachloroethane		−23 e	3 e	37.4	86.0	159.4	1
$C_2HF_3O_2$	Trifluoroacetic acid					16.8	71.4	1,5
C_2HF_5O	Trifluoromethyl difluoromethyl ether	−147 e	−136 e	−121 e	−102 e	−75.0	−35.4	20
C_2H_2	Acetylene*			−146.6 s	−130.7 s	−110.6 s	−84.8 s	5
$C_2H_2Br_2$	cis-1,2-Dibromoethylene		−45 e	−21 e	10 e	52.2	114.8	1
$C_2H_2Br_2$	trans-1,2–Dibromoethylene				−4 e	42.2	107.4	5
$C_2H_2Br_2Cl_2$	1,2-Dibromo-1,1-dichloroethane					103.6	177.8	5
$C_2H_2Br_2Cl_2$	1,2-Dibromo-1,2-dichloroethane		−11 e	22 e	64.1	119 e	193 e	5
$C_2H_2Br_4$	1,1,2,2-Tetrabromoethane	14 e	38 e	69 e	109 e	163.7	242.9	5
$C_2H_2Cl_2$	1,1-Dichloroethylene	−116 e	−101 e	−82 e	−57 e	−21.4	31.2	1
$C_2H_2Cl_2$	cis-1,2-Dichloroethylene			−62 e	−34 e	3.8	60.3	1
$C_2H_2Cl_2$	trans-1,2-Dichloroethylene				−44 e	−7.5	47.3	1
$C_2H_2Cl_2F_2$	1,2-Dichloro-1,1-difluoroethane	−101 e	−87 e	−68 e	−42.2	−6.8	46.3	5
$C_2H_2Cl_2O$	Chloroacetyl chloride			−23.7	5.6	46.1	105.6	5
$C_2H_2Cl_4$	1,1,1,2-Tetrachloroethane	−58 e	−40 e	−15 e	17 e	62.2	129.7	1
$C_2H_2Cl_4$	1,1,2,2-Tetrachloroethane		−22 e	1 e	32.4	76.9	144.7	1
$C_2H_2F_4$	1,1,1,2-Tetrafluoroethane				−94.3	−66.8	−26.4	17
$C_2H_2F_4$	1,1,2,2-Tetrafluoroethane				−96.0	−66.9	−23.3	5
C_2H_2O	Ketene		−151 e	−135 e	−115 e	−88.2	−50.0	1
C_2H_3Br	Bromoethylene	−124 e	−110 e	−92 e	−68 e	−34.5	15.4	5
C_2H_3BrO	Acetyl bromide	−78 e	−65 e	−49 e	−25 e	13.9	84 e	5
$C_2H_3Br_3$	1,1,2-Tribromoethane	−18 e	4 e	32 e	68 e	117.1	188.4	5
C_2H_3Cl	Chloroethylene	−139 e	−127 e	−110 e	−89 e	−59.0	−14.1	1
$C_2H_3ClF_2$	1-Chloro-1,1-difluoroethane		−123 e	−107 e	−85.3	−55.4	−10.5	5
C_2H_3ClO	Acetyl chloride	−100 e	−85 e	−66 e	−40 e	−3.6	50.4	1
$C_2H_3ClO_2$	Chloroacetic acid				78.4	123.9	188.9	1
$C_2H_3Cl_2F$	1,1-Dichloro-1-fluoroethane		−101 e	−83 e	−57.9	−22.7	31.4	5
$C_2H_3Cl_2F$	1,2-Dichloro-1-fluoroethane			−50 e	−23.8	14.1	73.4	5
$C_2H_3Cl_3$	1,1,1-Trichloroethane				−25.3	14.2	73.7	5
$C_2H_3Cl_3$	1,1,2-Trichloroethane			−23 e	7 e	49.9	113.4	1
C_2H_3F	Fluoroethylene			−153.3	−135.2	−109.9	−72.2	5
C_2H_3FO	Acetyl fluoride					−64.1	17.0	5
$C_2H_3F_3$	1,1,1-Trifluoroethane				−113 e	−86.6	−47.8	1
$C_2H_3F_3O$	2,2,2-Trifluoroethanol			−33 e	−8 e	26.0	74 e	5
C_2H_3I	Iodoethylene				−41 e	−3 e	55.6	5
C_2H_3IO	Acetyl iodide				−0.6	47 e	107.0	5
C_2H_3N	Acetonitrile				−20 e	21.4	81.2	1
C_2H_3NO	Methylisocyanate				−43.5	−10.2	38.8	1
C_2H_3NS	Methyl thiocyanate			−18.4	16.2	63.5	132.5	5
C_2H_4	Ethylene*				−155.6	−135.1	−104.0	1,10
C_2H_4BrCl	1-Bromo-2-chloroethane				−0.4	41.7	105.7	6
$C_2H_4Br_2$	1,1-Dibromoethane		−49 e	−26 e	5 e	46.4	107.6	5
$C_2H_4Br_2$	1,2-Dibromoethane				18 e	62.2	130.9	1
C_2H_4ClF	1-Chloro-1-fluoroethane				−69.9	−36.1	15.8	5
$C_2H_4Cl_2$	1,1-Dichloroethane		−84 e	−64 e	−36.7	1.0	56.9	1
$C_2H_4Cl_2$	1,2-Dichloroethane				−16.4	23.7	83.1	1
$C_2H_4F_2$	1,1-Difluoroethane			−115.2	−94.6	−66.1	−24.3	19
$C_2H_4N_2O_6$	Ethylene glycol dinitrate	4 e	25.6	51.0	81 e	117 e	162 e	5
C_2H_4O	Acetaldehyde		−105 e	−87 e	−62.8	−29.4	20.0	5
C_2H_4O	Ethylene oxide		−111 e	−93 e	−70 e	−37.0	10.2	1
$C_2H_4O_2$	Acetic acid	−42.8 s	−26.7 s	−8 s	14.2 s	55.9	117.5	1,5
$C_2H_4O_2$	Methyl formate			−95 e	−76 e	−51.8	−18.1	31.4	5

Mol. form.	Name	Temperature in °C for the indicated pressure						Ref.
		1 Pa	10 Pa	100 Pa	1 kPa	10 kPa	100 kPa	
$C_2H_4O_3$	Peroxyacetic acid				14.4	55.3	109.7	5
$C_2H_4O_3$	Glycolic acid						99.9	5
$C_2H_5AsF_2$	Ethyldifluoroarsine			−36 e	−6.0	35.0	93.1	5
C_2H_5Br	Bromoethane	−111 e	−96 e	−77 e	−51.3	−15.5	38.0	5
C_2H_5Cl	Chloroethane	−126 e	−112 e	−94 e	−70 e	−37.0	12.0	1
C_2H_5ClO	2-Chloroethanol	−61 e	−39 e	−12 e	23 e	67.1	127.3	5
C_2H_5ClO	Chloromethyl methyl ether	−96 e	−80 e	−59 e	−32 e	6 e	61 e	5
$C_2H_5Cl_3OSi$	Trichloroethoxysilane	−78 e	−60 e	−36.0	−4.6	38.7	102.0	5
$C_2H_5Cl_3Si$	Trichloroethylsilane	−79 e	−61 e	−38 e	−8 e	34.9	98.7	5
C_2H_5F	Fluoroethane		−142 e	−127 e	−106.3	−78.7	−37.9	1
C_2H_5FO	2-Fluoroethanol		−22 e	8.3	47.5	99 e		5
C_2H_5I	Iodoethane	−94 e	−78 e	−56 e	−27.9	11.9	71.9	5
C_2H_5N	Ethyleneimine		−74 e	−55 e	−30 e	4.1	55 e	5
C_2H_5NO	Acetamide	16.7 s	39.1 s	65.2 s	102.8	150.8	218.2	5
C_2H_5NO	N-Methylformamide		13 e	41 e	78 e	127.9	199.1	1
$C_2H_5NO_2$	Nitroethane	−61 e	−44 e	−21 e	8.3	50.1	113.5	5
$C_2H_5NO_3$	Ethyl nitrate	−81 e	−63 e	−41 e	−12 e	28.2	87 e	1
C_2H_6	Ethane*	−183.3 s	−173.2	−161.3	−145.3	−122.8	−88.8	41
$C_2H_6Cl_2Si$	Dichlorodimethylsilane					11.1	70.1	5
C_2H_6Hg	Dimethyl mercury				−13.5	29.0	92.1	5
$C_2H_6N_2O$	N-Nitrosodimethylamine				30.7	80.5	149.8	5
C_2H_6O	Ethanol	−73 e	−56 e	−34 e	−7 e	29.2	78.0	1,5
C_2H_6O	Dimethyl ether*		−135 e	−118 e	−96.8	−67.6	−25.1	1,5
C_2H_6OS	Dimethyl sulfoxide			27.4	65.0	115.9	188.6	1
$C_2H_6O_2$	Ethylene glycol	2 e	24 e	51.1	86.1	132.5	196.9	1
$C_2H_6O_2$	Ethyl hydroperoxide	−70 e	−49 e	−25 e	6.8	47.0	101 e	5
$C_2H_6O_2S$	Dimethyl sulfone				109 e	166.8	248.9	5
C_2H_6S	Ethanethiol	−112 e	−97 e	−78 e	−53 e	−18 e	34.7	1
C_2H_6S	Dimethyl sulfide		−96 e	−77 e	−51.2	−16.0	37.0	1,5
$C_2H_6S_2$	Dimethyl disulfide	−71 e	−53 e	−29 e	1.7	45.0	109.3	5
$C_2H_7BO_2$	Dimethoxyborane	−116 e	−101.9	−83.5	−59.2	−25.4	25 e	5
C_2H_7N	Ethylamine			−71 e	−53 e	−27 e	16.4	1
C_2H_7N	Dimethylamine			−88 e	−66.9	−37.2	6.6	1
C_2H_7NO	Ethanolamine		11 e	35 e	66.2	109.0	170.6	1
$C_2H_8N_2$	1,2-Ethanediamine				17.0	57.5	116.6	1,5
$C_2H_8N_2$	1,1-Dimethylhydrazine			−52 e	−25.6	10.5	63 e	5
$C_2H_8N_2$	1,2-Dimethylhydrazine		−49 e	−33 e	−9 e	26.4	88 e	1
C_2N_2	Cyanogen	−127 s	−114.1 s	−98.5 s	−79.2 s	−54.9 s	−21.4	5
C_3ClF_5O	Chloropentafluoroacetone	−122 e	−109 e	−93 e	−71 e	−39.4	7.4	5
C_3Cl_6	Hexachloropropene	−12 e	11 e	40 e	79 e	132.8	213.6	5
C_3F_6	Perfluoropropene	−150 e	−138 e	−122 e	−101 e	−72 e	−30.6	5
C_3F_6O	Perfluoroacetone			−113 e	−94 e	−67.8	−27.6	5
C_3F_8	Perfluoropropane		−139 e	−124 e	−105 e	−77.5	−37.0	1
C_3HN	Cyanoacetylene			−58.7 s	−35.6 s	−7 s	42.0	5
$C_3H_2F_6O$	1,1,1,3,3,3-Hexafluoro-2-propanol					12.7	57.1	5
$C_3H_3F_5$	1,1,1,2,2−Pentafluoropropane					−60 e	−17.9	5
C_3H_3N	2-Propenenitrile		−72 e	−50 e	−22 e	17.7	77.0	1
C_3H_3NS	Thiazole					54.4	117.8	5
C_3H_4	Allene*		−129 e	−118 e	−101.4	−76.7	−34.7	5
C_3H_4	Propyne				−94 e	−65.3	−23.2	1
$C_3H_4ClF_3$	3-Chloro-1,1,1-trifluoropropane	−102 e	−87 e	−68 e	−43 e	−8 e	45.3	5
$C_3H_4Cl_2O$	1,1-Dichloroacetone				1 e	47.8	118.0	5
$C_3H_4Cl_2O_2$	Methyl dichloroacetate	−44 e	−25 e	0 e	33 e	77.7	142.3	5
$C_3H_4Cl_4$	1,1,1,2-Tetrachloropropane	−48 e	−28 e	−2 e	32 e	79.1	149.5	5
$C_3H_4F_4O$	2,2,3,3-Tetrafluoro-1-propanol			−10 e	17 e	53.9	107.2	5
C_3H_4O	Acrolein		−87 e	−67 e	−40 e	−3.0	52.8	1
$C_3H_4O_2$	Propenoic acid				35 e	78.0	140.7	1
$C_3H_4O_2$	Vinyl formate			−58 e	−34 e	−1.6	46.2	1
$C_3H_4O_2$	2-Oxetanone		−21 e	8 e	45.5	93.8	159.3	5
$C_3H_4O_3$	Ethylene carbonate	12.7 s	37 e				247	5

		Temperature in °C for the indicated pressure						
Mol. form.	Name	1 Pa	10 Pa	100 Pa	1 kPa	10 kPa	100 kPa	Ref.
C$_3$H$_5$Br	cis-1-Bromopropene	−100 e	−84 e	−64 e	−37 e	1.0	57.4	5
C$_3$H$_5$Br	2-Bromopropene	−112 e	−95 e	−75 e	−47 e	−9 e	48.0	5
C$_3$H$_5$Br	3-Bromopropene	−98 e	−80 e	−58 e	−28 e	12 e	69.6	5
C$_3$H$_5$Cl	cis-1-Chloropropene	−114 e	−100 e	−81 e	−55 e	−20.1	32.4	5
C$_3$H$_5$Cl	trans-1-Chloropropene		−97 e	−77 e	−52 e	−16.2	37.0	5
C$_3$H$_5$Cl	2-Chloropropene	−120 e	−106 e	−87 e	−63 e	−28.7	22.3	5
C$_3$H$_5$Cl	3-Chloropropene	−107 e	−92 e	−72.4	−46.3	−9.8	44.6	5
C$_3$H$_5$ClO	Epichlorohydrin			−21 e	11 e	53.8	115.5	5
C$_3$H$_5$ClO$_2$	Methyl chloroacetate		−28 e	−5 e	25 e	66.9	129.1	5
C$_3$H$_5$Cl$_3$	1,1,3-Trichloropropane	−51 e	−31 e	−5 e	28 e	75.3	145.1	5
C$_3$H$_5$Cl$_3$	1,2,3-Trichloropropane			2 e	37 e	84.9	156.3	5
C$_3$H$_5$Cl$_3$Si	Trichloro-2-propenylsilane					53.0	116.5	5
C$_3$H$_5$I	3-Iodopropene	−80 e	−62 e	−39 e	−8 e	36 e	101.5	5
C$_3$H$_5$N	Propanenitrile	−69.4	−55.3	−36.0	−7.9	35.2	97.4	1,5
C$_3$H$_5$NO	Acrylamide			109.6	161 e			5
C$_3$H$_5$NO	3-Hydroxypropanenitrile	−11 e	18 e	53 e	96.1	150.3	220.8	5
C$_3$H$_5$NS	Ethyl thiocyanate	−39 e	−20 e	4 e	35 e	79.1	143.4	5
C$_3$H$_5$NS	Ethyl isothiocyanate				17.4	66 e	136 e	5
C$_3$H$_5$N$_3$O$_9$	Trinitroglycerol	48.6	75.7	118 e	191 e	353 e	1007 e	5
C$_3$H$_6$	Propene*	−160.6	−149.0	−134.3	−114.9	−88.2	−47.9	1,5
C$_3$H$_6$	Cyclopropane			−124 e	−104 e	−75.7	−33.1	1
C$_3$H$_6$BrCl	1-Bromo-3-chloropropane	−51 e	−31 e	−6 e	28 e	74.1	142.9	5
C$_3$H$_6$Br$_2$	1,2-Dibromopropane	−46 e	−26 e	−2 e	31 e	75.3	139.5	5
C$_3$H$_6$Br$_2$	1,3-Dibromopropane	−30 e	−9 e	17 e	52 e	98.7	166.8	5
C$_3$H$_6$Cl$_2$	1,1-Dichloropropane				−14 e	27.0	87.7	5
C$_3$H$_6$Cl$_2$	1,2-Dichloropropane	−78 e	−61 e	−38.1	−8.1	33.7	95.9	5
C$_3$H$_6$Cl$_2$	1,3-Dichloropropane	−65 e	−46 e	−22 e	10 e	54.0	119.9	5
C$_3$H$_6$Cl$_2$	2,2-Dichloropropane				−28 e	10.8	68.9	5
C$_3$H$_6$Cl$_2$O	1,3-Dichloro-2-propanol			21.8	59.0	107.6	173.9	5
C$_3$H$_6$N$_2$O$_4$	1,1-Dinitropropane	−9 e	12 e	39 e	73.2	120 e	187 e	5
C$_3$H$_6$O	Allyl alcohol	−63 e	−48 e	−21.9	6.8	44.5	96.2	5
C$_3$H$_6$O	Methyl vinyl ether			−114 e	−89 e	−52.7	4.6	1
C$_3$H$_6$O	Propanal			−69 e	−42 e	−6 e	47.7	1
C$_3$H$_6$O	Acetone	−95	−81.8	−62.8	−35.6	1.3	55.7	1,5
C$_3$H$_6$O	Methyloxirane	−109 e	−95 e	−76 e	−51.5	−17.2	33.9	5
C$_3$H$_6$O$_2$	Propanoic acid			0 e	35.1	79.9	140.8	1,5
C$_3$H$_6$O$_2$	Ethyl formate		−80 e	−61 e	−35 e	1 e	54.0	1
C$_3$H$_6$O$_2$	Methyl acetate	−95 e	−79 e	−59 e	−33 e	3.3	56.6	1
C$_3$H$_6$O$_2$	1,3-Dioxolane		−72 e	−50 e	−22 e	17.0	75.3	1
C$_3$H$_6$O$_3$	1,3,5-Trioxane					53 e	113.7	1
C$_3$H$_6$S	Thietane		−62 e	−40 e	−9 e	32.5	94.5	5
C$_3$H$_7$Br	1-Bromopropane	−95 e	−78 e	−57 e	−28 e	11.6	70.6	1
C$_3$H$_7$Br	2-Bromopropane		−84 e	−65 e	−39.6	−1.7	59.1	1,5
C$_3$H$_7$Cl	1-Chloropropane	−106 e	−90 e	−71 e	−44.5	−8.1	46.2	1
C$_3$H$_7$Cl	2-Chloropropane		−91 e	−74 e	−51.1	−17.8	35.4	1,5
C$_3$H$_7$ClO	2-Chloro-1-propanol			23 e	63.8		125.7	5
C$_3$H$_7$F	1-Fluoropropane	−133 e	−120 e	−103 e	−80.7	−49.4	−2.8	5
C$_3$H$_7$I	1-Iodopropane	−78 e	−60 e	−37 e	−6 e	36.9	102.0	5
C$_3$H$_7$I	2-Iodopropane	−89 e	−71 e	−47 e	−16.3	26.5	89.2	5
C$_3$H$_7$N	Allylamine		−88 e	−65 e	−37 e	0.4	52 e	5
C$_3$H$_7$NO	N,N-Dimethylformamide	−39 e	−20 e	5 e	38.0	83.9	152.6	1
C$_3$H$_7$NO	N-Methylacetamide	−13.3 s	13 s	43 e	83.8	136.1	206.3	5
C$_3$H$_7$NO$_2$	1-Nitropropane	−56 e	−37 e	−13 e	20 e	64.8	130.8	1
C$_3$H$_7$NO$_2$	2-Nitropropane		−48 e	−22 e	10.7	55.6	119.8	1
C$_3$H$_7$NO$_3$	Propyl nitrate			−23.9	6.1	48.1	111 e	5
C$_3$H$_8$	Propane*	−156.9	−145.6	−130.9	−111.4	−83.8	−42.3	1,41
C$_3$H$_8$O	1-Propanol	−54 e	−38 e	−16 e	10 e	47 e	96.9	1,5
C$_3$H$_8$O	2-Propanol	−65 e	−49 e	−28 e	−1.3	33.6	82.0	1,5
C$_3$H$_8$O	Ethyl methyl ether	−98 e	−89 e	−77 e	−60 e	−34.8	7.0	5
C$_3$H$_8$O$_2$	1,2-Propylene glycol	−11 e	13 e	42 e	78 e	125.0	187.2	5

Mol. form.	Name	1 Pa	10 Pa	100 Pa	1 kPa	10 kPa	100 kPa	Ref.
		\multicolumn{6}{c}{Temperature in °C for the indicated pressure}		

Mol. form.	Name	1 Pa	10 Pa	100 Pa	1 kPa	10 kPa	100 kPa	Ref.
$C_3H_8O_2$	1,3-Propylene glycol	4 e	30 e	62 e	101 e	149.9	214.0	5
$C_3H_8O_2$	Ethylene glycol monomethyl ether	−57 e	−37 e	−12 e	21 e	63.8	124.3	1
$C_3H_8O_2$	Dimethoxymethane	−93 e	−81 e	−64 e	−42 e	−9.3	41.7	5
$C_3H_8O_3$	Glycerol	96 e	113 e	136 e	168 e	213.4	287 e	1
C_3H_8S	1-Propanethiol	−94 e	−78 e	−57 e	−29.1	9.6	67.4	1,5
C_3H_8S	2-Propanethiol	−102 e	−87 e	−67 e	−41 e	−3 e	52.2	1
C_3H_8S	Ethyl methyl sulfide	−94 e	−78 e	−57 e	−29.7	8.8	66.3	1
$C_3H_8S_2$	1,3–Propanedithiol	−53 e	−28 e	3 e	43 e	97 e	172.4	5
C_3H_9As	Trimethylarsine			−74 e	−45 e	−5.4	52.0	5
$C_3H_9BO_3$	Trimethyl borate				−14 e	15.6	67.9	5
C_3H_9BS	Methyl dimethylthioborane			−62 e	−30.4	11.4	70.7	5
C_3H_9ClSi	Trimethylchlorosilane				−37.8	0.4	57.3	5
C_3H_9N	Propylamine		−81 e	−63 e	−38.3	−4.1	46.9	1,5
C_3H_9N	Isopropylamine		−91 e	−74 e	−50.4	−17.6	31.5	1,5
C_3H_9N	Trimethylamine		−114 e	−97 e	−75.0	−43.8	2.6	1,5
C_3H_9NO	1-Amino-2-propanol			18 e	53.2	98.2	157.9	5
$C_3H_9O_4P$	Trimethyl phosphate	−31 e	−7 e	23.6	62.8	116.0	192.0	5
C_3H_9P	Trimethylphosphine			−81 e	−53 e	−15.0	37.1	5
C_3H_9Sb	Trimethylstibine			−56 e	−23.8	19 e	80 e	5
$C_3H_{10}N_2$	1,2-Propanediamine		−35.4	−12.0	18.8	61 e	119 e	5
C_3N_2O	Carbonyl dicyanide				−21.7	15.3	65.2	5
C_4Cl_6	Hexachloro-1,3-butadiene	−1 e	22 e	50 e	86.7	137.0	209.7	5
$C_4F_6O_3$	Trifluoroacetic acid anhydride			−63 e	−39 e	−7.1	38.8	5
C_4F_8	Perfluorocyclobutane						−6.2	1
C_4F_{10}	Perfluorobutane		−122 e	−105 e	−82 e	−49.8	−2.5	1,5
$C_4H_2Cl_2O_2$	trans-2-Butenedioyl dichloride			8.0	45.6	94.3	159.8	5
$C_4H_2Cl_2S$	2,5-Dichlorothiophene			−20 e	22 e	81.4	171 e	5
$C_4H_2O_3$	Maleic anhydride				73.7	127.9	201.7	5
C_4H_3ClS	2-Chlorothiophene		−62 e	−35 e	2 e	51.8	123 e	5
C_4H_3IS	2-Iodothiophene			−25 e	23 e	94.9	181.0	5
C_4H_4	1-Buten-3-yne			−96.1	−73.4	−41.8	4.9	5
$C_4H_4N_2$	Succinonitrile	24.8 s					266.0	5
C_4H_4O	Furan			−78 e	−54 e	−20 e	31.0	1
$C_4H_4O_2$	Diketene				19.3	63.3	126 e	5
$C_4H_4O_3$	Succinic anhydride				121 e	180.8	260.8	5
$C_4H_4O_4$	Fumaric acid	123.9 s	150 s	180 s				5
C_4H_4S	Thiophene				−17 e	23.7	83.7	5
C_4H_5Cl	2-Chloro-1,3-butadiene	−113 e	−95 e	−71 e	−41 e	0.3	59.0	5
C_4H_5ClO	2-Methyl-2-propenoyl chloride		−57 e	−35 e	−5 e	36.4	98.2	5
$C_4H_5Cl_3O_2$	Ethyl trichloroacetate			15.3	51.9	100.1	166.6	5
C_4H_5N	3-Butenenitrile	−67 e	−48 e	−23.1	9.3	53.7	118.4	5
C_4H_5N	Methylacrylonitrile				−12 e	29.0	89.8	5
C_4H_5N	Pyrrole			−8 e	24 e	66.7	129.4	1
$C_4H_5NO_2$	Methyl cyanoacetate	−3 e	19 e	48 e	84 e	134.0	204.6	5
C_4H_5NS	Allyl isothiocyanate	−45 e	−27 e	−3 e	32.1	89 e	198 e	5
C_4H_5NS	4-Methylthiazole						67.0	5
C_4H_6	1,2-Butadiene	−132 e	−117 e	−98 e	−72.8	−38.9	10.5	5
C_4H_6	1,3-Butadiene*			−106 e	−83 e	−51.9	−4.7	1
C_4H_6	1-Butyne	−125 e	−111 e	−94 e	−71.2	−39.4	7.8	1
C_4H_6	2-Butyne		−89.2 s	−73.8 s	−53.5 s	−23.9	26.6	5
$C_4H_6Cl_2O_2$	Ethyl dichloroacetate			2.6	40.1	89.1	156.3	5
C_4H_6O	Divinyl ether		−99 e	−80 e	−56 e	−22.1	28.0	5
C_4H_6O	trans-2-Butenal	−74 e	−56 e	−33 e	−3 e	39.7	102.4	5
C_4H_6O	3-Buten-2-one					21 e	81.0	5
C_4H_6O	Cyclobutanone			−34 e	−4 e	37.1	97 e	5
$C_4H_6O_2$	cis-Crotonic acid			30 e	63 e	106.7	168.9	5
$C_4H_6O_2$	trans-Crotonic acid				74 e	120.8	184.9	5
$C_4H_6O_2$	3-Butenoic acid	−19 e	2 e	27 e	61 e	105.6	168.6	5
$C_4H_6O_2$	Methacrylic acid			22 e	56 e	99.9	161.5	5
$C_4H_6O_2$	Vinyl acetate	−88 e	−71 e	−50 e	−22 e	16.2	72.2	1

		Temperature in °C for the indicated pressure						
Mol. form.	Name	1 Pa	10 Pa	100 Pa	1 kPa	10 kPa	100 kPa	Ref.
$C_4H_6O_2$	Methyl acrylate		−71 e	−48 e	−18 e	22 e	79.9	5
$C_4H_6O_2$	2,3-Butanedione					30.7	84.8	5
$C_4H_6O_2$	gamma-Butyrolactone		−17 e	24 e	72 e	130.2	203 e	5
$C_4H_6O_3$	Acetic anhydride	−44 e	−25 e	−1 e	31 e	75.1	139.7	1
$C_4H_6O_3$	Propylene carbonate	−40 e	−5 e	43 e	112 e	220 e	410 e	5
$C_4H_6O_4$	Dimethyl oxalate				50.5	98.1	163.0	5
C_4H_7Br	trans-1-Bromo-1-butene	−87 e	−68 e	−43.3	−11.4	31.9	94.4	5
C_4H_7Br	2-Bromo-1-butene	−87 e	−70 e	−48 e	−20 e	20.7	80.6	5
C_4H_7Br	cis-2-Bromo-2-butene	−90 e	−72 e	−49.0	−18.5	23.5	85.2	5
C_4H_7Br	trans-2-Bromo-2-butene	−86 e	−67 e	−43.4	−12.0	31.0	93.5	5
$C_4H_7Br_3$	1,2,3-Tribromobutane	0 e	23 e	53 e	91 e	143.7	219.5	5
$C_4H_7Br_3$	1,2,4-Tribromobutane	−3 e	20 e	49 e	87 e	139.4	214.5	5
C_4H_7Cl	3-Chloro-1-butene			−64 e	−36 e	4 e	63.6	5
C_4H_7Cl	cis-2-Chloro-2-butene	−100 e	−83 e	−62 e	−34 e	6 e	66.4	5
C_4H_7Cl	trans-2-Chloro-2-butene	−102 e	−86 e	−65 e	−37 e	3 e	62.2	5
C_4H_7Cl	3-Chloro-2-methylpropene		−75 e	−54 e	−25 e	13.8	71.5	5
$C_4H_7ClO_2$	Ethyl chloroacetate			−2.6	32.6	79.1	143.8	5
C_4H_7N	Butanenitrile	−67 e	−48 e	−24 e	8 e	52.3	117.2	1
C_4H_8	1-Butene	−139.0	−125.2	−107.8	−85.3	−53.7	−6.6	1,5
C_4H_8	cis-2-Butene	−131.2	−117.4	−99.8	−76.7	−44.8	3.4	1,5
C_4H_8	trans-2-Butene			−102 e	−80 e	−47.6	0.6	1
C_4H_8	Isobutene	−139.1	−125.5	−108.2	−85.5	−54.5	−7.3	1,5
C_4H_8	Cyclobutane			−71.8	−38.1	12.1	5
C_4H_8	Methylcyclopropane	−130 e	−116 e	−99.3	−76.3	−44.2	4.2	5
$C_4H_8Br_2$	1,2-Dibromobutane	−54 e	−30 e	0.4	39.6	92.1	166.1	5
$C_4H_8Br_2$	1,4-Dibromobutane	−13 e	9 e	37 e	74 e	124.0	196.5	5
$C_4H_8Cl_2$	1,1-Dichlorobutane			−25 e	6 e	49.3	113.4	5
$C_4H_8Cl_2$	1,2-Dichlorobutane			−28.4	5.8	53.1	123.1	5
$C_4H_8Cl_2$	1,4-Dichlorobutane		−26 e	0 e	35 e	82.4	153.4	5
$C_4H_8Cl_2$	2,2-Dichlorobutane		−58 e	−35 e	−5 e	37.8	102.1	5
$C_4H_8Cl_2O$	Bis(2-chloroethyl) ether	−32 e	−9 e	19.8	56.9	106.9	177.9	5
C_4H_8O	Ethyl vinyl ether		−102 e	−81 e	−53.1	−16.5	34.7	5
C_4H_8O	1,2-Epoxybutane	−135 e	−114 e	−87 e	−53 e	−5.5	62.1	5
C_4H_8O	Butanal	−88 e	−72 e	−50 e	−22 e	16.6	74.5	1,5
C_4H_8O	Isobutanal			−56 e	−29 e	8 e	63.8	1
C_4H_8O	2-Butanone	−85 e	−68 e	−46 e	−18.1	21.2	79.2	1
C_4H_8O	Tetrahydrofuran	−94 e	−78 e	−57.3	−29.8	9 e	65.6	1
$C_4H_8O_2$	Butanoic acid			12.9	52.2	101.4	163.3	1,5
$C_4H_8O_2$	2-Methylpropanoic acid	−30.1	−8.2	18.1	50.5	92.9	154.0	5
$C_4H_8O_2$	Propyl formate	−78 e	−62 e	−42 e	−15.1	23.0	80.4	1,5
$C_4H_8O_2$	Isopropyl formate	−80 e	−65 e	−47 e	−22.2	13.2	67.7	5
$C_4H_8O_2$	Ethyl acetate	−83 e	−66 e	−45 e	−18 e	20.4	76.8	1
$C_4H_8O_2$	Methyl propanoate	−80 e	−64 e	−43 e	−15.8	22.2	79.0	1
$C_4H_8O_2$	cis-2-Butene-1,4-diol	17 e	44 e	77 e	117.4	168.5	234.9	5
$C_4H_8O_2$	1,3-Dioxane			−37 e	−3 e	43.4	106.0	5
$C_4H_8O_2$	1,4-Dioxane					39.6	101.0	1
$C_4H_8O_2S$	Sulfolane		49 e	87 e	135 e	198.0	283.5	5
C_4H_8S	Tetrahydrothiophene	−66 e	−47 e	−23 e	9.4	54.1	120.5	1
C_4H_9Br	1-Bromobutane	−68.4	−53.9	−34.1	−5.4	37.6	101.1	1,5
C_4H_9Br	2-Bromobutane	−86 e	−68 e	−46 e	−16 e	26.6	90.7	5
C_4H_9Br	1-Bromo-2-methylpropane	−85 e	−68 e	−46 e	−16 e	26.8	91.1	5
C_4H_9Br	2-Bromo-2-methylpropane					11.7	72.4	1,5
C_4H_9Cl	1-Chlorobutane	−87 e	−71 e	−49 e	−21 e	18.4	78.1	1
C_4H_9Cl	2-Chlorobutane	−96 e	−80 e	−59 e	−31.0	8.5	67.9	1
C_4H_9Cl	1-Chloro-2-methylpropane	−94 e	−78 e	−56.6	−28.7	10.2	68.5	5
C_4H_9Cl	2-Chloro-2-methylpropane					−4.2	50.3	5
$C_4H_9Cl_3Si$	Butyltrichlorosilane					77.2	148.4	5
C_4H_9F	1-Fluorobutane	−114 e	−99 e	−80 e	−55 e	−20.0	32.1	5
C_4H_9F	2-Fluorobutane	−117 e	−103 e	−85 e	−60.7	−26.7	24.7	5
C_4H_9I	1-Iodobutane	−62 e	−43 e	−19 e	14 e	60.5	130.0	5

Mol. form.	Name	Temperature in °C for the indicated pressure						Ref.
		1 Pa	10 Pa	100 Pa	1 kPa	10 kPa	100 kPa	
C_4H_9I	2-Iodobutane	−70 e	−51 e	−27 e	5 e	50 e	119.5	5
C_4H_9I	1-Iodo-2-methylpropane		−47 e	−21.4	12.0	56.8	120.0	5
C_4H_9I	2-Iodo-2-methylpropane	−75.1 s	−58.8 s	−39.5 s	-5.2	41 e	100.0	5
C_4H_9N	Pyrrolidine		−59 e	−38 e	−10 e	28.5	86.2	1
C_4H_9NO	N-Methylpropanamide				81.1	105 e		5
C_4H_9NO	N,N-Dimethylacetamide	−8 e	8 e	28.0	56.4	98.2	165.7	1
C_4H_9NO	2-Butanone oxime		−18 e	7 e	38.9	81.9	142.9	5
C_4H_9NO	Morpholine				21 e	64.5	128.5	1
$C_4H_9NO_3$	Isobutyl nitrate			−18 e	15.1	59.2	123.0	5
C_4H_{10}	Butane*	−134.3	−121.0	−103.9	−81.1	−49.1	−0.8	1,41
C_4H_{10}	Isobutane*		−129.0	−113.0	−90.9	−59.4	−12.0	1,41
$C_4H_{10}O$	1-Butanol	−37 e	−20 e	0 e	28 e	64 e	117.4	1
$C_4H_{10}O$	2-Butanol	−50 e	−34 e	−14 e	12.6	48.2	99.2	1,5
$C_4H_{10}O$	2-Methyl-1-propanol	−39 e	−24 e	−5 e	20.9	56.0	107.6	1,5
$C_4H_{10}O$	2-Methyl-2-propanol					34.4	82.1	1,5
$C_4H_{10}O$	Diethyl ether	−111 e	−96 e	−77 e	−52.6	−17.8	34.1	1
$C_4H_{10}O$	Methyl propyl ether				−40 e	−11.3	38.7	5
$C_4H_{10}O$	Isopropyl methyl ether				−56 e	−21.2	30.4	5
$C_4H_{10}O_2$	1,3-Butanediol	−4 e	23 e	55 e	94 e	142.9	206.1	5
$C_4H_{10}O_2$	1,4-Butanediol		45 e	77 e	116 e	164.7	227.6	5
$C_4H_{10}O_2$	2,3-Butanediol		15 e	43 e	77 e	121.2	180.3	5
$C_4H_{10}O_2$	Ethylene glycol monoethyl ether	−49 e	−29 e	−3 e	30 e	73.6	135.3	1
$C_4H_{10}O_2$	Ethylene glycol dimethyl ether			−44 e	−15 e	25.2	85.2	1
$C_4H_{10}O_2$	Dimethylacetal	−89 e	−74 e	−55 e	−29 e	7.7	64.1	5
$C_4H_{10}O_2$	Diethylperoxide				−39 e	3.6	65.0	5
$C_4H_{10}O_2S$	Bis(2-hydroxyethyl) sulfide			31 e	114.2		282.0	5
$C_4H_{10}O_3$	Diethylene glycol	35 e	58 e	86 e	123 e	173.6	245.2	1
$C_4H_{10}O_4S$	Diethyl sulfate		3 e	36 e	79 e	134 e	208.3	5
$C_4H_{10}S$	1-Butanethiol	−77 e	−59 e	−37 e	−6 e	35.4	98.0	5
$C_4H_{10}S$	2-Butanethiol	−86 e	−69 e	−47 e	−17 e	23.4	84.5	5
$C_4H_{10}S$	2-Methyl-1-propanethiol		−66 e	−44 e	−15 e	26.5	88.1	5
$C_4H_{10}S$	2-Methyl-2-propanethiol					5.8	63.8	5
$C_4H_{10}S$	Diethyl sulfide	−80 e	−62 e	−40 e	−10.8	30.3	91.7	1
$C_4H_{10}S$	Methyl propyl sulfide	−78 e	−61 e	−38 e	−8 e	33.1	95.1	5
$C_4H_{10}S$	Isopropyl methyl sulfide	−85 e	−68 e	−46 e	−17 e	23.4	84.3	5
$C_4H_{10}S_2$	1,4-Butanedithiol	−17 e	5 e	32 e	69.1	119.9	195.1	5
$C_4H_{10}S_2$	Diethyl disulfide	−46 e	−26 e	0 e	35 e	82.4	153.5	5
$C_4H_{11}N$	Butylamine			−46 e	−18.1	20.0	75.9	5
$C_4H_{11}N$	sec-Butylamine			−55 e	−29.1	7.5	62.3	5
$C_4H_{11}N$	tert-Butylamine			−67 e	−42.4	−8.1	43.7	5
$C_4H_{11}N$	Isobutylamine	−85 p	−70 p	−50 e	−24.5	12.0	67.3	5
$C_4H_{11}N$	Diethylamine			−46 e	−26 e	5 e	55.2	1
$C_4H_{11}NO$	N,N-Dimethylethanolamine	−52 e	−31 e	−6 e	27 e	70.9	133 e	5
$C_4H_{11}NO_2$	Diethanolamine	53 e	77 e	107 e	146 e	197.3	268 e	5
$C_4H_{12}BN$	(Dimethylamino)dimethyl-borane		−81 e	−60.1	−31.9	7.0	64.2	5
$C_4H_{12}Cl_2OSi_2$	1,3-Dichloro-1,1,3,3-tetramethyldisiloxane		−33 e	−9 e	23.8	69.1	136.5	5
$C_4H_{12}O_4Si$	Tetramethyl silicate				14.4	59.3	119.7	5
$C_4H_{12}Si$	Tetramethylsilane			−83 e	−59 e	−25 e	26.7	5
$C_4H_{12}Sn$	Tetramethylstannane			−55.0	−25.6	16.6	77.7	5
$C_4H_{13}N_3$	Diethylenetriamine	−10 e	13 e	43 e	80 e	129.6	198 e	5
C_4NiO_4	Nickel carbonyl					−12	42	4
C_5F_{12}	Perfluoropentane				−54.7	−20.9	28.6	5
C_5FeO_5	Iron pentacarbonyl				0	44	105	4
C_5H_4ClN	2-Chloropyridine			7.4	45.8	97.3	169.9	5
$C_5H_4O_2$	Furfural	−26 e	−8 e	16 e	47 e	92.4	161.4	1
C_5H_5N	Pyridine			−23 e	8 e	51.0	114.9	1
C_5H_6	1,3-Cyclopentadiene			−77 e	−51 e	−14 e	39.8	5
$C_5H_6N_2$	Pentanedinitrile	24.1	52 e	85 e	126 e	178 e	245 e	5
C_5H_6O	2-Methylfuran			−66 e	−35 e	6 e	64.5	1

		Temperature in °C for the indicated pressure						
Mol. form.	Name	1 Pa	10 Pa	100 Pa	1 kPa	10 kPa	100 kPa	Ref.
$C_5H_6O_2$	Furfuryl alcohol	−30 e	−5 e	25 e	62.6	109.3	169.7	5
C_5H_6S	2-Methylthiophene		−58 e	−32 e	2 e	47.9	112.2	1
C_5H_6S	3-Methylthiophene		−53 e	−28 e	6 e	50.6	115.1	1
C_5H_7N	1-Methylpyrrole				8 e	49.9	112.3	5
$C_5H_7NO_2$	Ethyl cyanoacetate	16 e	39 e	67.0	102.1	146.7	205.6	5
C_5H_8	1,2-Pentadiene	−109 e	−93 e	−73 e	−46.1	−9.7	44.5	5
C_5H_8	cis-1,3-Pentadiene	−109 e	−93 e	−73 e	−47.0	−10.5	43.7	1,5
C_5H_8	trans-1,3-Pentadiene			−75 e	−49.0	−13 e	42 e	1
C_5H_8	1,4-Pentadiene	−120 e	−105 e	−86 e	−60.9	−26.2	25.6	5
C_5H_8	2,3-Pentadiene	−106 e	−90 e	−70 e	−42.9	−6.3	47.9	5
C_5H_8	3-Methyl-1,2-butadiene	−111 e	−95 e	−75 e	−49.2	−13.1	40.4	5
C_5H_8	2-Methyl-1,3-butadiene	−115 e	−100 e	−81 e	−55.4	−19.7	33.7	1,5
C_5H_8	1-Pentyne			−75 e	−49.1	−13.5	39.9	5
C_5H_8	2-Pentyne	−100 e	−85 e	−65 e	−37.9	−0.5	55.7	5
C_5H_8	3-Methyl-1-butyne			−82 e	−57.5	−23.1	28.6	5
C_5H_8	Cyclopentene	−109 e	−94 e	−74 e	−48 e	−11.1	43.8	5
C_5H_8	Spiropentane	−110 e	−95 e	−76 e	−51 e	−15 e	38.6	5
C_5H_8O	3-Methyl-3-buten-2-one			−35 e	−5 e	36.0	97.3	5
C_5H_8O	Cyclopropyl methyl ketone		−57 e	−31 e	3 e	49 e	112 e	5
C_5H_8O	Cyclopentanone		−39 e	−14 e	19 e	64 e	130.3	1
C_5H_8O	3,4-Dihydro-2H-pyran				−22 e	22.0	84.9	5
$C_5H_8O_2$	4-Pentenoic acid	0 e	19 e	44 e	77 e	122.0	187.5	5
$C_5H_8O_2$	Vinyl propanoate					31.2	94 e	5
$C_5H_8O_2$	Ethyl acrylate		−55 e	−32.7	−2.8	38.5	99.2	5
$C_5H_8O_2$	Methyl methacrylate			−31 e	−1 e	39.7	100.0	1
$C_5H_8O_2$	2,4-Pentanedione			−5 e	24.7	67.8	137.4	1
$C_5H_8O_2$	Tetrahydro-2H-pyran-2-one		5 e	35.1	74.4	128.3	207.0	5
$C_5H_8O_3$	Methyl acetoacetate				50.1	101.1	171.3	5
$C_5H_8O_4$	Glutaric acid		121 e	153.2	191.9	240.3	302.5	5
$C_5H_8O_4$	Dimethyl malonate	−22 e	1 e	30.0	66.7	114.7	180.2	5
$C_5H_9ClO_2$	Ethyl 2-chloropropanoate			1.4	36.4	82.5	146.0	5
$C_5H_9ClO_2$	Isopropyl chloroacetate			−2 e	35.0	83.3	148.1	5
C_5H_9N	Pentanenitrile	−54 e	−34 e	−8 e	26 e	72.2	140.9	1
C_5H_9N	2,2-Dimethylpropanenitrile					41.1	104.8	5
C_5H_9NO	N-Methyl-2-pyrrolidone	1 e	24 e	53.1	92.3	147.2	229 e	5
C_5H_{10}	1-Pentene	−118.9	−103.4	−84.0	−58.8	−23.3	29.6	1,5
C_5H_{10}	cis-2-Pentene	−113.8	−98.1	−78.4	−52.7	−16.8	36.6	1,5
C_5H_{10}	trans-2-Pentene	−114.5	−98.9	−79.1	−53.3	−17.5	36.0	1,5
C_5H_{10}	2-Methyl-1-butene	−117.7	−102.2	−82.7	−57.2	−21.9	30.8	1,5
C_5H_{10}	3-Methyl-1-butene	−125.0	−110.1	−91.2	−66.7	−32.2	19.7	1,5
C_5H_{10}	2-Methyl-2-butene	−113.4	−97.6	−77.7	−51.6	−15.8	38.2	1,5
C_5H_{10}	Cyclopentane			−77.0	−45.4	−7.1	48.8	5
C_5H_{10}	Ethylcyclopropane	−118 e	−102 e	−83 e	−57 e	−20 e	35.5	5
C_5H_{10}	cis-1,2-Dimethylcyclo-propane	−118 e	−103 e	−83 e	−57 e	−20 e	36.6	5
C_5H_{10}	trans-1,2-Dimethylcyclo-propane	−122 e	−108 e	−89 e	−63 e	−27 e	27.8	5
$C_5H_{10}Br_2$	1,5-Dibromopentane	1 e	25 e	54 e	93 e	145.6	221.8	5
$C_5H_{10}Cl_2$	1,2-Dichloropentane			30 e	77.4	147.8		5
$C_5H_{10}Cl_2$	1,5-Dichloropentane	−31 e	−10 e	17 e	54 e	104.1	178.9	5
$C_5H_{10}N_2$	3-(Dimethylamino)-propanenitrile				51.1	101.8	171.4	5
$C_5H_{10}O$	Cyclopentanol		−13 e	11.5	42.2	82.5	140.0	5
$C_5H_{10}O$	Allyl ethyl ether			−56 e	−28.7	9.8	67.2	5
$C_5H_{10}O$	Pentanal	−71 e	−53 e	−31 e	−1 e	40.8	102.6	5
$C_5H_{10}O$	2-Pentanone				−1 e	40.3	101.9	1,5
$C_5H_{10}O$	3-Pentanone			−31 e	−1 e	40 e	101.6	1
$C_5H_{10}O$	3-Methyl-2-butanone	−69 e	−54 e	−34 e	−6.9	32.2	94.0	1,5
$C_5H_{10}O$	Tetrahydropyran				−15 e	26.0	88 e	5
$C_5H_{10}O$	2-Methyltetrahydrofuran				−20 e	19.7	79.8	5
$C_5H_{10}O_2$	Pentanoic acid	−7.4	15.3	42.7	76.3	122.1	185.7	5
$C_5H_{10}O_2$	2-Methylbutanoic acid	−10 e	10 e	36 e	69 e	112.8	175.2	5
$C_5H_{10}O_2$	3-Methylbutanoic acid	−15.8	4 e	30.0	64.7	110.6	176.1	5

Mol. form.	Name	1 Pa	10 Pa	100 Pa	1 kPa	10 kPa	100 kPa	Ref.
$C_5H_{10}O_2$	Butyl formate			−29 e	2 e	44.4	105.7	5
$C_5H_{10}O_2$	Isobutyl formate	−69 e	−53 e	−31 e	−3 e	37.4	97.6	5
$C_5H_{10}O_2$	Propyl acetate	−69 e	−51 e	−29 e	0 e	40.9	101.2	1
$C_5H_{10}O_2$	Isopropyl acetate		−61 e	−40 e	−11 e	29.8	88.2	5
$C_5H_{10}O_2$	Ethyl propanoate	−69 e	−52 e	−30 e	−1 e	38.9	98.7	1
$C_5H_{10}O_2$	Methyl butanoate	−68 e	−50 e	−28 e	0.9	41.7	102.3	5
$C_5H_{10}O_2$	Methyl isobutanoate	−83 e	−65 e	−41 e	−11 e	31 e	92.1	5
$C_5H_{10}O_2$	Tetrahydrofurfuryl alcohol	−40 e	−16 e	15 e	55 e	106 e	176.8	5
$C_5H_{10}O_3$	Diethyl carbonate		−42 e	−17 e	17 e	61.6	125.9	5
$C_5H_{10}O_3$	Ethylene glycol monomethyl ether acetate	−47 e	−26 e	0 e	34 e	79.4	144.1	5
$C_5H_{10}S$	Thiacyclohexane				24 e	71.1	141.2	5
$C_5H_{10}S$	Cyclopentanethiol				18 e	64 e	131.7	5
$C_5H_{11}Br$	1-Bromopentane	−60 e	−41 e	−16 e	16 e	61.5	129.1	5
$C_5H_{11}Br$	2-Bromopentane	−69 e	−51 e	−27 e	5 e	49.7	116.9	5
$C_5H_{11}Br$	3-Bromopentane	−68 e	−50 e	−26 e	6 e	50.8	118.1	5
$C_5H_{11}Br$	1-Bromo-3-methylbutane	−67 e	−49 e	−25 e	8 e	52.4	119.9	5
$C_5H_{11}Cl$	1-Chloropentane	−73 e	−55 e	−32 e	−1 e	42.5	107.9	5
$C_5H_{11}Cl$	2-Chloropentane	−80 e	−62 e	−39 e	−9 e	33.2	96.1	5
$C_5H_{11}Cl$	3-Chloropentane	−77 e	−60 e	−37 e	−7 e	34.9	97.3	5
$C_5H_{11}Cl$	2-Chloro-2-methylbutane			−52 e	−21 e	21.8	85.2	5
$C_5H_{11}Cl$	1-Chloro-2,2-dimethyl-propane				−17 e	23.5	83.9	5
$C_5H_{11}F$	1-Fluoropentane	−97 e	−80 e	−60 e	−32 e	5.7	62.4	5
$C_5H_{11}I$	1-Iodopentane	−47 e	−27 e	−1 e	34 e	83.0	156.5	5
$C_5H_{11}I$	1-Iodo-3-methylbutane		−34 e	−6.6	28.8	77.3	147.8	5
$C_5H_{11}N$	Cyclopentylamine	−66 e	−48 e	−26 e	4 e	45.8	108 e	5
$C_5H_{11}N$	Piperidine				2 e	43.3	105.8	5
$C_5H_{11}N$	N-Methylpyrrolidine				−23 e	18.5	78 e	5
$C_5H_{11}NO_3$	3-Methylbutyl nitrate		−26 e	1.0	35.5	81.7	147.0	5
C_5H_{12}	Pentane**	−115.5	−99.8	−80.0	−54.0	−18.1	35.7	16
C_5H_{12}	Isopentane	−119 e	−105 e	−86 e	−61 e	−26 e	27.5	1
C_5H_{12}	Neopentane*		−107.5 s	−90.8 s	−68.8 s	−38.5 s	9.2	1,5
$C_5H_{12}N_2O$	Tetramethylurea			20.7	58.0	106.7	179.5	5
$C_5H_{12}O$	1-Pentanol	−27 e	−10 e	12 e	41 e	79.8	137.4	5
$C_5H_{12}O$	2-Pentanol	−35 e	−19 e	1 e	28.0	64.9	118.7	1
$C_5H_{12}O$	3-Pentanol	−41 e	−25 e	−4 e	24 e	61.1	114.9	5
$C_5H_{12}O$	2-Methyl-1-butanol	−27 e	−11 e	9 e	36.2	73.4	128.3	1
$C_5H_{12}O$	3-Methyl-1-butanol	−22 e	−7 e	13 e	39.1	75.7	130.1	5
$C_5H_{12}O$	2-Methyl-2-butanol			−5 e	17.7	50.6	101.7	1,5
$C_5H_{12}O$	3-Methyl-2-butanol			−3 e	22.7	58.2	111.1	5
$C_5H_{12}O$	2,2-Dimethyl-1-propanol					59.2	112.7	5
$C_5H_{12}O$	Butyl methyl ether			−54 e	−27 e	12 e	69.8	1
$C_5H_{12}O$	Methyl tert-butyl ether			−66 e	−39 e	−2 e	54.8	1
$C_5H_{12}O$	Ethyl propyl ether	−92 e	−77 e	−57 e	−30.5	6.7	63.4	1,5
$C_5H_{12}O_2$	1,5-Pentanediol	25 e	52 e	85 e	125 e	175.1	238.9	5
$C_5H_{12}O_2$	Ethylene glycol monopropyl ether				40 e	85.6	149.3	5
$C_5H_{12}O_2$	Diethoxymethane		−65 e	−43 e	−14 e	27.3	87.7	5
$C_5H_{12}O_3$	Diethylene glycol monomethyl ether		12 e	40 e	76 e	124.2	193.7	1
$C_5H_{12}S$	1-Pentanethiol	−60 e	−41 e	−17 e	15 e	60 e	126.2	1
$C_5H_{12}S$	2-Pentanethiol	−70 e	−52 e	−28 e	3 e	46.6	111.9	5
$C_5H_{12}S$	3-Pentanethiol	−70 e	−51 e	−28 e	4 e	47.7	113.4	5
$C_5H_{12}S$	2-Methyl-1-butanethiol				8.0	52.3	118.5	5
$C_5H_{12}S$	3-Methyl-1-butanethiol				7.8	51.9	117.9	5
$C_5H_{12}S$	2-Methyl-2-butanethiol				−8.0	34.6	98.7	5
$C_5H_{12}S$	Butyl methyl sulfide		−43 e	−19 e	13 e	57 e	123.0	1
$C_5H_{12}S$	tert-Butyl methyl sulfide				−7.8	34.7	98.4	5
$C_5H_{12}S$	Ethyl propyl sulfide	−64 e	−46 e	−23 e	9 e	52.7	118.0	5
$C_5H_{12}S$	Ethyl isopropyl sulfide	−72 e	−54 e	−31 e	0 e	42.7	106.9	5
$C_5H_{13}N$	Pentylamine		−52 e	−29 e	1 e	42.8	104.0	5
C_6BrF_5	Bromopentafluorobenzene		−10 e	23 e	68 e	136.0		5
C_6ClF_5	Chloropentafluorobenzene		−44 e	−21 e	11 e	53.8	117.6	1

		Temperature in °C for the indicated pressure						
Mol. form.	Name	1 Pa	10 Pa	100 Pa	1 kPa	10 kPa	100 kPa	Ref.
C₆Cl₃F₃	1,3,5-Trichloro-2,4,6-trifluorobenzene	−19 e	4 e	32 e	70 e	121.7	197.9	1
C₆F₆	Hexafluorobenzene		−56.9 s	−36 s	−11.5 s	22.6	79.9	1,5
C₆F₁₂	Perfluorocyclohexane				−46.2 s	−7.6 s	48.9 s	5
C₆F₁₄	Perfluorohexane		−75 e	−57 e	−32 e	2.8	56.8	5
C₆F₁₄	Perfluoro-2-methylpentane				−33 e	2.9	57.1	5
C₆F₁₄	Perfluoro-3-methylpentane	−95 e	−80 e	−60 e	−34 e	2.8	57.9	5
C₆F₁₄	Perfluoro-2,3-dimethylbutane					4.3	59.3	5
C₆HF₅	Pentafluorobenzene			−41 e	−13 e	27 e	85.3	5
C₆HF₅O	Pentafluorophenol				39 e	82 e	145.2	5
C₆H₂F₄	1,2,3,4-Tetrafluorobenzene			−36 e	−7 e	33.8	94.0	1
C₆H₂F₄	1,2,3,5-Tetrafluorobenzene			−43 e	−14 e	25.5	84.1	1
C₆H₂F₄	1,2,4,5-Tetrafluorobenzene					30.7	89.9	1
C₆H₃Cl₃O	2,4,6-Trichlorophenol			71.8	114.0	169.5	245.7	5
C₆H₃F₃	1,3,5-Trifluorobenzene					18.2	75.0	5
C₆H₄Br₂	m-Dibromobenzene	−7 e	16 e	44 e	83 e	137.0	218.2	5
C₆H₄ClNO₂	1-Chloro-4-nitrobenzene	15.4 s	35.8 s		97 e	156.0	238 e	5
C₆H₄Cl₂	o-Dichlorobenzene		−13 e	16.3	53.9	104.6	180.0	1,5
C₆H₄Cl₂	m-Dichlorobenzene		−22 e	8.0	46.7	97.8	172.5	1,5
C₆H₄Cl₂	p-Dichlorobenzene	−45.5 s	−21.8 s	8 s	46.7 s	99.0	173.6	1,5
C₆H₄O₂	p-Benzoquinone	−4.1 s	17.8 s	43.5 s	74.3 s	111.6 s		5
C₆H₅AsCl₂	Dichlorophenylarsine	6.9	35.2	70 e	113 e	170 e	245 e	5
C₆H₅Br	Bromobenzene		−25 e	1 e	34.9	83.1	155.4	1
C₆H₅Cl	Chlorobenzene		−43 e	−17 e	16.8	62.9	131.3	1,5
C₆H₅ClO	o-Chlorophenol				45.8	97.9	173.9	5
C₆H₅ClO	m-Chlorophenol			39.7	80.2	135.1	213.4	5
C₆H₅ClO	p-Chlorophenol			45.0	86.5	142.0	219.9	5
C₆H₅Cl₃Si	Trichlorophenylsilane			33 e	70.2	122.6	201 e	5
C₆H₅F	Fluorobenzene				−16.9	24.2	84.4	1
C₆H₅I	Iodobenzene	−30 e	−7 e	20.9	58.5	110.6	187.8	1
C₆H₅NO₂	Nitrobenzene		10 e	40 e	78 e	132 e	210.3	1
C₆H₅NO₃	p-Nitrophenol	72.6 s	97.4 s					5
C₆H₆	1,5-Hexadien-3-yne	−82 e	−66 e	−44.3	−16.0	23.7	83.6	5
C₆H₆	Benzene**			−40 s	−15.1 s	20.0	79.7	1,5
C₆H₆ClN	o-Chloroaniline		10 e	39.0	75.2	131.4	208.3	5
C₆H₆ClN	m-Chloroaniline	−5 e	19.7	49.4	94.2	162 e	1069 e	5
C₆H₆N₂O₂	p-Nitroaniline	87.8 s			192.0	252.6	331.2	5
C₆H₆O	Phenol	−9.7 s	9.6 s	34.1 s	68.9	113.7	181.4	1,5
C₆H₆O₃	1,2,3-Benzenetriol				162.0	222.8	308.3	5
C₆H₆S	Benzenethiol		−15 e	12 e	47 e	96.0	168.6	5
C₆H₇N	Aniline		−2.5	26.7	63.5	112.5	183.5	1,5
C₆H₇N	2-Methylpyridine	−56.5	−37.8	−13.9	18.3	62.9	129.0	1,5
C₆H₇N	3-Methylpyridine			−5 e	28.8	75.2	143.7	1
C₆H₇N	4-Methylpyridine	−58.2 s	−43.1 s	−3.9 s	29.6	76.1	144.9	1,5
C₆H₈	cis-1,3,5-Hexatriene					21 e	78 e	5
C₆H₈	1,3-Cyclohexadiene	−88 e	−71 e	−50 e	−21 e	19 e	79.9	5
C₆H₈	1,4-Cyclohexadiene				−15 e	27.3	85.0	5
C₆H₈N₂	Adiponitrile	30 e	61 e	100 e	148.6	211.8	297 e	5
C₆H₈N₂	m-Phenylenediamine			94.5	140.2	200.8	285.0	5
C₆H₈N₂	Phenylhydrazine		38 e	69 e	109 e	163.9	242.5	5
C₆H₈O₄	Dimethyl maleate		5 e	36 e	76 e	127.3	197 e	5
C₆H₈S	2,5-Dimethylthiophene		−43 e	−16 e	20 e	67.5	134.8	5
C₆H₁₀	trans-1,3-Hexadiene	−86 e	−70 e	−51 e	−24 e	14 e	72 e	5
C₆H₁₀	trans-1,4-Hexadiene	−98 e	−81 e	−60 e	−33 e	7 e	65 e	5
C₆H₁₀	1,5-Hexadiene	−99 e	−84 e	−64 e	−37 e	0.9	59.2	5
C₆H₁₀	cis,cis-2,4-Hexadiene					18 e	79.6	5
C₆H₁₀	trans,cis-2,4-Hexadiene	−89 e	−73 e	−52 e	−23 e	18 e	79.6	5
C₆H₁₀	trans,trans-2,4-Hexadiene				−23 e	18 e	79.6	5
C₆H₁₀	trans-2-Methyl-1,3-pentadiene	−92 e	−75 e	−54 e	−26 e	14 e	75.6	5
C₆H₁₀	2,3-Dimethyl-1,3-butadiene			−59 e	−30 e	9.7	68.1	5
C₆H₁₀	1-Hexyne	−91 e	−75 e	−54 e	−26 e	12.8	71.0	5

Mol. form.	Name	1 Pa	10 Pa	100 Pa	1 kPa	10 kPa	100 kPa	Ref.
C_6H_{10}	2-Hexyne	−84 e	−67 e	−46 e	−17 e	23.6	84.1	5
C_6H_{10}	3-Hexyne	−86 e	−69 e	−48 e	−19.1	21.0	81.0	1,5
C_6H_{10}	4-Methyl-1-pentyne	−97 e	−81 e	−61 e	−34 e	4.1	60.7	5
C_6H_{10}	4-Methyl-2-pentyne	−91 e	−74 e	−54 e	−26 e	13.8	72.7	5
C_6H_{10}	Cyclohexene	−87 e	−70 e	−49 e	−19 e	21 e	82.6	1
$C_6H_{10}Cl_2$	1,1-Dichlorocyclohexane	−39 e	−19 e	8 e	43 e	93.5	170.5	5
$C_6H_{10}Cl_2$	cis-1,2-Dichlorocyclohexane			27 e	69 e	125.7	206.2	5
$C_6H_{10}O$	4-Methyl-4-penten-2-one	−59 e	−41 e	−17 e	14 e	57.0	121.0	5
$C_6H_{10}O$	Cyclohexanone		−25 e	1 e	36 e	84 e	155.2	1
$C_6H_{10}O$	Mesityl oxide	−56 e	−37 e	−13 e	19 e	63.5	129.3	5
$C_6H_{10}O_2$	Vinyl butanoate					53 e	114.5	5
$C_6H_{10}O_2$	Ethyl methacrylate				8 e	53.2	116.8	5
$C_6H_{10}O_2$	Allyl glycidyl ether				40.1	85.7	152.8	5
$C_6H_{10}O_3$	Ethyl acetoacetate	−25 e	−3 e	25.7	62.3	111.3	180.2	5
$C_6H_{10}O_3$	Propanoic anhydride	−32 e	−15 e	6 e	36 e	77.6	142.9	5
$C_6H_{10}O_4$	Diethyl oxalate	−5 e	18 e	44.9	79.4	124.3	185.2	5
$C_6H_{10}O_4$	Dimethyl succinate			30 e	70.4	123.3	195.4	5
$C_6H_{10}O_4$	Ethylene glycol diacetate	−17 e	6 e	35.0	71.9	121.1	190.0	5
$C_6H_{10}S$	Diallylsulfide	−58 e	−38 e	−12.4	21.7	68.8	138.1	5
$C_6H_{11}Cl$	Chlorocyclohexane		−35 e	−9 e	25 e	71.6	142.1	5
$C_6H_{11}N$	Hexanenitrile	−40 e	−19 e	8 e	43 e	91.5	163.2	1,5
$C_6H_{11}N$	4-Methylpentanenitrile		−50 e	−20 e	20 e	75.2	155.2	5
$C_6H_{11}NO$	Caprolactam	36.8 s	58.9 s	86.6 s			270	5
C_6H_{12}	1-Hexene	−99.8	−82.8	−61.4	−33.7	5.2	63.1	1,5
C_6H_{12}	cis-2-Hexene	−97 e	−80 e	−58 e	−30 e	9.9	68.5	5
C_6H_{12}	trans-2-Hexene	−94 e	−78 e	−57 e	−30 e	9.3	67.5	5
C_6H_{12}	cis-3-Hexene	−96 e	−79 e	−59 e	−30.8	7.9	66.0	5
C_6H_{12}	trans-3-Hexene	−95 e	−79 e	−58 e	−30.0	8.8	66.7	5
C_6H_{12}	2-Methyl-1-pentene	−98 e	−82 e	−62 e	−34.2	4.1	61.7	5
C_6H_{12}	3-Methyl-1-pentene	−104 e	−88 e	−68 e	−41.5	−3.6	53.8	5
C_6H_{12}	4-Methyl-1-pentene	−105 e	−89 e	−69 e	−41.6	−3.6	53.5	5
C_6H_{12}	2-Methyl-2-pentene	−95 e	−78 e	−58 e	−30 e	9.0	66.9	5
C_6H_{12}	3-Methyl-cis-2-pentene	−95 e	−79 e	−58 e	−30 e	8.9	67.3	5
C_6H_{12}	3-Methyl-trans-2-pentene	−93 e	−77 e	−55 e	−27.4	11.7	70.0	5
C_6H_{12}	4-Methyl-cis-2-pentene	−102 e	−86 e	−66 e	−38.7	−0.9	56.0	5
C_6H_{12}	4-Methyl-trans-2-pentene	−100 e	−84 e	−64 e	−36.8	1.2	58.2	5
C_6H_{12}	2-Ethyl-1-butene	−98 e	−81 e	−60 e	−32 e	6.6	64.3	5
C_6H_{12}	2,3-Dimethyl-1-butene	−103 e	−87 e	−67 e	−39.9	−1.9	55.2	5
C_6H_{12}	3,3-Dimethyl-1-butene	−110 e	−95 e	−76 e	−50.8	−14.5	40.8	5
C_6H_{12}	2,3-Dimethyl-2-butene		−75 e	−54 e	−25 e	14 e	72.9	1
C_6H_{12}	Cyclohexane	−85.6 s	−68.9 s	−47.6 s	−19.8 s	19.3	80.4	1,5
C_6H_{12}	Methylcyclopentane	−97 e	−80 e	−58 e	−28.8	11.6	71.4	1,5
C_6H_{12}	Ethylcyclobutane	−99 e	−82 e	−61 e	−32 e	9 e	70.2	5
C_6H_{12}	Isopropylcyclopropane	−104 e	−88 e	−68 e	−40 e	−1 e	57.9	5
C_6H_{12}	1-Ethyl-1-methylcyclopropane	−105 e	−89 e	−69 e	−41 e	−3 e	56.3	5
C_6H_{12}	1,1,2-Trimethylcyclopropane	−109 e	−94 e	−73 e	−46 e	−7 e	52.0	5
$C_6H_{12}Cl_2$	1,2-Dichlorohexane				49 e	98.1	171.7	5
$C_6H_{12}Cl_2O$	2,2′-Dichlorodiisopropyl ether		−1 e	27.3	63.4	112.3	182.1	5
$C_6H_{12}O$	Butyl vinyl ether	−87 e	−67 e	−42 e	−9.3	33.6	93.2	5
$C_6H_{12}O$	Isobutyl vinyl ether	−87 e	−68 e	−44 e	−13 e	26.5	80.7	5
$C_6H_{12}O$	Hexanal	−56 e	−37 e	−13 e	19 e	62.6	127.8	5
$C_6H_{12}O$	2-Hexanone	−43 e	−21 e	4.2	34.5	61.9	127.2	1,5
$C_6H_{12}O$	3-Hexanone		−40 e	−16 e	15 e	58.5	123.1	1
$C_6H_{12}O$	3-Methyl-2-pentanone				8.5	52.7	117.0	5
$C_6H_{12}O$	4-Methyl-2-pentanone	−61 e	−43 e	−21 e	9 e	51.5	116.1	5
$C_6H_{12}O$	2-Methyl-3-pentanone					50.2	113.0	5
$C_6H_{12}O$	3,3-Dimethyl-2-butanone			−30 e	0 e	42.5	105.7	1
$C_6H_{12}O$	Cyclohexanol			34 e	61 e	99.2	160.7	1
$C_6H_{12}O_2$	Hexanoic acid		33 e	59 e	93 e	139.3	204.5	1
$C_6H_{12}O_2$	4-Methylpentanoic acid	36 e	49 e	67.1	92.9	133.6	206.8	5

		Temperature in °C for the indicated pressure						
Mol. form.	Name	1 Pa	10 Pa	100 Pa	1 kPa	10 kPa	100 kPa	Ref.
$C_6H_{12}O_2$	Diethylacetic acid	–9 e	16 e	46 e	83 e	130.7	192.5	5
$C_6H_{12}O_2$	Isopentyl formate	–60 e	–41 e	–17 e	15 e	59.1	124 e	5
$C_6H_{12}O_2$	Butyl acetate	–63 e	–43 e	–19 e	14 e	61.0	125.6	1,5
$C_6H_{12}O_2$	Isobutyl acetate	–63 e	–45 e	–21 e	10 e	53.4	116 e	5
$C_6H_{12}O_2$	Propyl propanoate	–62 e	–42 e	–18 e	14 e	58.3	122.0	5
$C_6H_{12}O_2$	Ethyl butanoate	–49 e	–34 e	–14 e	14.3	55.2	121.1	5
$C_6H_{12}O_2$	Ethyl 2-methylpropanoate	–65 e	–47 e	–24.6	5.4	47.3	109.8	5
$C_6H_{12}O_2$	Methyl pentanoate				19.2	63.7	127.4	5
$C_6H_{12}O_2$	Methyl isopentanoate					53.3	116.3	5
$C_6H_{12}O_2$	Diacetone alcohol	–41 e	–17 e	13 e	50.1	98.5	164 e	5
$C_6H_{12}O_3$	Ethylene glycol monoethyl ether acetate	–25 e	–8 e	14 e	44.6	88.0	155.6	5
$C_6H_{12}O_3$	Paraldehyde				17 e	62.2	124 e	5
$C_6H_{12}S$	Cyclohexanethiol					84.8	158.3	5
$C_6H_{12}S$	cis-Tetrahydro-2,5-dimethylthiophene	–53 e	–34 e	–8 e	25 e	72.0	142.1	5
$C_6H_{12}S$	Tetrahydro-3-methyl-2H-thiopyran	–48 e	–27 e	0 e	35 e	84.1	157.5	5
$C_6H_{13}Br$	1-Bromohexane	–45 e	–25 e	2 e	36 e	83.7	154.8	5
$C_6H_{13}Cl$	1-Chlorohexane	–55 e	–36 e	–11 e	21 e	66.7	134.6	5
$C_6H_{13}F$	1-Fluorohexane	–80 e	–62 e	–40 e	–11 e	30.4	91.1	5
$C_6H_{13}I$	1-Iodohexane	–33 e	–11 e	16 e	53 e	104.0	180.8	5
$C_6H_{13}N$	Cyclohexylamine			–9 e	22 e	66.6	133.5	1
C_6H_{14}	Hexane	–96.4 s	–79.2	–57.6	–29.3	9.8	68.3	16
C_6H_{14}	2-Methylpentane	–100 e	–84 e	–64 e	–36 e	2 e	59.9	1
C_6H_{14}	3-Methylpentane	–99 e	–83 e	–62 e	–34.3	4.6	62.9	1
C_6H_{14}	2,2-Dimethylbutane		–90 e	–71.5	–45.5	–7.7	49.4	1
C_6H_{14}	2,3-Dimethylbutane	–103 e	–87 e	–66 e	–39.0	–0.4	57.6	1
$C_6H_{14}O$	1-Hexanol		5 e	28 e	56.8	97.3	157.1	1
$C_6H_{14}O$	2-Hexanol	–28 e	–10 e	12 e	41.4	81.5	139.6	1
$C_6H_{14}O$	3-Hexanol	–43 e	–23 e	1 e	33 e	75.4	135.1	1
$C_6H_{14}O$	2-Methyl-1-pentanol			14 e	45.9	88.3	147.6	5
$C_6H_{14}O$	4-Methyl-1-pentanol			24 e	53 e	92.4	151.4	5
$C_6H_{14}O$	2-Methyl-2-pentanol	–29 e	–15 e	3 e	27.1	63.0	120.9	5
$C_6H_{14}O$	3-Methyl-2-pentanol				36.5	76.1	133.8	5
$C_6H_{14}O$	4-Methyl-2-pentanol	–43 e	–24 e	0 e	30 e	71.9	131.3	5
$C_6H_{14}O$	2-Methyl-3-pentanol				29.8	68.8	126.0	5
$C_6H_{14}O$	3-Methyl-3-pentanol		–23 e	–4 e	22.9	61.1	121.1	5
$C_6H_{14}O$	2-Ethyl-1-butanol		–5 e	17 e	46 e	85.7	146.1	5
$C_6H_{14}O$	3,3-Dimethyl-1-butanol	–37 e	–16 e	9 e	42 e	84.3	142.5	5
$C_6H_{14}O$	2,3-Dimethyl-2-butanol			–5 e	23 e	61.3	118.2	5
$C_6H_{14}O$	Dipropyl ether	–80 e	–63 e	–41 e	–12 e	28.8	89.7	1
$C_6H_{14}O$	Diisopropyl ether		–76 e	–55 e	–28 e	11 e	68.1	1
$C_6H_{14}O$	Butyl ethyl ether	–78 e	–61 e	–39 e	–10 e	31.0	91.9	1
$C_6H_{14}O$	tert-Butyl ethyl ether	–90 e	–74 e	–53 e	–24.6	14.4	72.6	5
$C_6H_{14}O_2$	2-Methyl-2,4-pentanediol	–8 e	17 e	48 e	86 e	134.4	197.5	5
$C_6H_{14}O_2$	Ethylene glycol monobutyl ether	–31 e	–8 e	20 e	55 e	103.2	170.2	5
$C_6H_{14}O_2$	1,1-Diethoxyethane	–68 e	–49 e	–26 e	3.7	44.2	101.9	5
$C_6H_{14}O_2$	Ethylene glycol diethyl ether		–59 e	–35.3	–2.8	44.4	118.8	5
$C_6H_{14}O_3$	1,2,6-Hexanetriol	92 e	114.8	146.0	191 e			5
$C_6H_{14}O_3$	Dipropylene glycol				110 e	162.6	231.4	5
$C_6H_{14}O_3$	Diethylene glycol monoethyl ether			40 e	80.3	132.4	201.4	5
$C_6H_{14}O_3$	Diethylene glycol dimethyl ether	–42 e	–20 e	8.3	44.3	92.3	159.4	5
$C_6H_{14}O_3$	Trimethylolpropane	73 e	98 e	128 e	167.8	220.5	295 e	5
$C_6H_{14}O_4$	Triethylene glycol	44 e	74 e	109.0	152.6	207.2	277.9	5
$C_6H_{14}S$	1-Hexanethiol	–45 e	–25 e	1 e	35 e	81.7	152.2	5
$C_6H_{14}S$	2-Hexanethiol	–50 e	–32 e	–8 e	25 e	69.9	138.4	5
$C_6H_{14}S$	Dipropyl sulfide	–50 e	–30 e	–6 e	28 e	73.6	142.4	5
$C_6H_{14}S$	Diisopropyl sulfide	–65 e	–47 e	–23 e	9 e	53.1	119.6	5
$C_6H_{14}S$	Isopropyl propyl sulfide				18.5	63.8	131.6	5
$C_6H_{14}S$	Butyl ethyl sulfide	–49 e	–30 e	–5 e	29 e	74.8	143.8	5
$C_6H_{15}N$	Hexylamine			–10 e	22 e	66.0	130.6	5
$C_6H_{15}N$	Butylethylamine				6.1	47.7	107.0	5

Mol. form.	Name	1 Pa	10 Pa	100 Pa	1 kPa	10 kPa	100 kPa	Ref.
$C_6H_{15}N$	Dipropylamine		−48 e	−25 e	6 e	47.5	108.8	5
$C_6H_{15}N$	Diisopropylamine			−47 e	−17.5	23.5	84.0	5
$C_6H_{15}N$	Triethylamine	−58 e	−45 e	−29 e	−5 e	29.9	88.5	1
$C_6H_{15}NO$	2-Diethylaminoethanol					97 e	160.6	5
$C_6H_{15}NO_3$	Triethanolamine	75 e	108 e	148 e	196 e	256.7	334 e	5
$C_6H_{15}O_4P$	Triethyl phosphate			34	76	132	211	4
$C_6H_{16}N_2$	Hexamethylenediamine				76.0	128.2	199.0	5
$C_6H_{16}O_2Si$	Diethoxydimethylsilane	−62 e	−44 e	−21.2	9.1	51.0	113.0	5
$C_6H_{18}Cl_2O_2Si_3$	1,5-Dichloro-1,1,3,3,5,5-hexamethyltrisiloxane	−29 e	−7 e	22.2	59.7	110.5	183.4	5
$C_6H_{18}OSi_2$	Hexamethyldisiloxane		−56 e	−34 e	−5 e	37.1	100.1	5
C_6MoO_6	Molybdenum hexacarbonyl		17.4 s	42.8 s	73.1 s	109.9 s	155.4 s	5
C_7F_{14}	Perfluoromethylcyclohexane				−21 e	18 e	75.9	1
C_7F_{16}	Perfluoroheptane		−62 e	−41 e	−14 e	24.7	82.1	1
C_7HF_{15}	1H-Pentadecafluoroheptane			−7 e	35.9	96.0		5
$C_7H_3ClF_3NO_2$	1-Chloro-2-nitro-4-trifluoromethyl)benzene	3 e	26 e	55 e	92.8	145.2	222.0	5
$C_7H_3F_5$	2,3,4,5,6-Pentafluorotoluene			−20 e	11 e	53.6	117.0	5
$C_7H_4ClF_3$	1-Chloro-2-(trifluoromethyl) benzene			1 e	34.5	81.8	151.8	5
$C_7H_4ClF_3$	1-Chloro-3-(trifluoromethyl) benzene	−53 e	−34 e	−9 e	24.2	69.8	137.2	5
$C_7H_4ClF_3$	1-Chloro-4-(trifluoromethyl) benzene			−9 e	24.2	70.4	138.1	5
$C_7H_4Cl_2O$	o-Chlorobenzoyl chloride				93 e	149 e	237.0	5
$C_7H_4Cl_2O$	m-Chlorobenzoyl chloride				87.8	147 e	225.0	5
$C_7H_4F_3NO_2$	1-Nitro-3-(trifluoromethyl) benzene		11 e	39 e	76.2	127.3	202.2	5
$C_7H_4F_4$	1-Fluoro-4-(trifluoromethyl) benzene			−38 e	−6 e	38.6	102.3	5
C_7H_5BrO	Benzoyl bromide	−15 e	11 e	42.6	83.9	139.5	218.0	5
C_7H_5ClO	Benzoyl chloride			27.5	67.0	120.4	196.7	5
$C_7H_5Cl_3$	(Trichloromethyl)benzene		9 e	40.6	81.5	136.2	213.0	5
$C_7H_5F_3$	(Trifluoromethyl)benzene			−3 e	39 e	101.6		5
C_7H_5N	Benzonitrile		−6 e	23.9	63.1	115.7	190.0	5
C_7H_5NS	Phenyl isothiocyanate			79.4	105 e	117 e		5
$C_7H_6Cl_2$	2,4-Dichlorotoluene		6 e	33 e	68.3	119.5	199.1	5
$C_7H_6Cl_2$	3,4-Dichlorotoluene	−13 e	9 e	38 e	76 e	129.3	208.4	5
$C_7H_6Cl_2$	(Dichloromethyl)benzene			31	72	130	213	4
C_7H_6O	Benzaldehyde		−9 e	19 e	54.6	104.6	178.3	1
$C_7H_6O_2$	Salicylaldehyde		−1 e	29 e	68 e	120.7	196.2	5
C_7H_7Br	o-Bromotoluene		−10 e	17 e	54 e	104.8	181.1	5
C_7H_7Br	m-Bromotoluene	−34 e	−11 e	19.4	58.1	109.9	183.1	5
C_7H_7Br	p-Bromotoluene				57 e	107.8	183.8	5
C_7H_7Br	(Bromomethyl)benzene			25.4	66.8	121.7	198.3	5
C_7H_7Cl	o-Chlorotoluene		−24 e	3 e	38 e	86.3	158.7	1,5
C_7H_7Cl	m-Chlorotoluene	−41 e	−21 e	6 e	41 e	89 e	161.8	5
C_7H_7Cl	p-Chlorotoluene				40 e	88.9	161.5	1,5
C_7H_7Cl	(Chloromethyl)benzene	−34 e	−11 e	17.7	55.4	106.3	178.9	5
C_7H_7ClO	1-Chloro-2-methoxy-benzene	−22 e	2 e	33 e	72 e	125.2	201 e	5
C_7H_7F	o-Fluorotoluene		−50 e	−26 e	5 e	49.0	113.9	5
C_7H_7F	m-Fluorotoluene	−67 e	−48 e	−25 e	7 e	51.0	116.1	5
C_7H_7F	p-Fluorotoluene		−48 e	−24 e	7 e	51 e	116.2	5
$C_7H_7NO_2$	o-Nitrotoluene	23 e	40 e	62 e	94 e	141.9	221.9	5
$C_7H_7NO_2$	m-Nitrotoluene			45 e	89.7	148.7	231.3	5
$C_7H_7NO_3$	2-Nitroanisole	15 e	45 e	82 e	129 e	189.4	271.8	5
C_7H_8	Toluene	−78.1	−57.1	−31.3	1.5	45.2	110.1	5
C_7H_8	Bicyclo[2.2.1]hepta-2,5-diene				−15 e	27.4	91 e	5
$C_7H_8Cl_2Si$	Dichloromethylphenylsilane			32.4	71.8	126.0	205.0	5
C_7H_8O	o-Cresol	−6.4 s	12.8 s	40.2	72.3	120.3	190.5	1,5
C_7H_8O	m-Cresol	20.8	33.6	52.4	82.6	130.6	201.8	1,5
C_7H_8O	p-Cresol	−0.2 s	20.7 s	52.7	83.1	130.7	201.5	1,5
C_7H_8O	Benzyl alcohol	8 e	28 e	54 e	88 e	134.7	204.9	1
C_7H_8O	Anisole		−21 e	4 e	38 e	84 e	153.2	1,5
C_7H_8S	3-Methylbenzenethiol		0 e	29 e	66 e	117.9	194.6	5
C_7H_9N	Benzylamine			25.6	62.6	112.7	183.9	5

		Temperature in °C for the indicated pressure						
Mol. form.	Name	1 Pa	10 Pa	100 Pa	1 kPa	10 kPa	100 kPa	Ref.
C₇H₉N	o-Methylaniline	1.0	18.8	42.6	76.1	125.6	199.9	1,5
C₇H₉N	m-Methylaniline	3.8	22.0	46.2	80.1	128.8	202.9	1,5
C₇H₉N	p-Methylaniline				77.1	126.2	199.9	5
C₇H₉N	N-Methylaniline	−16 e	6 e	34 e	70.3	121.1	195.8	1
C₇H₉N	2-Ethylpyridine	−46 e	−26 e	−1 e	33 e	79.3	149.0	5
C₇H₉N	3-Ethylpyridine	−38 e	−17 e	9 e	44 e	92.7	166.5	5
C₇H₉N	4-Ethylpyridine	−35 e	−15 e	11 e	46 e	94.4	168.6	5
C₇H₉N	2,3-Dimethylpyridine				42 e	89.9	160.6	5
C₇H₉N	2,4-Dimethylpyridine		−25 e	3.7	40.0	87.5	157.9	1,5
C₇H₉N	2,5-Dimethylpyridine			4 e	39 e	86.2	156.6	1
C₇H₉N	2,6-Dimethylpyridine			−3 e	29.9	75.8	143.6	1
C₇H₉N	3,4-Dimethylpyridine		−9 e	19 e	55 e	104.8	178.6	5
C₇H₉N	3,5-Dimethylpyridine			11 e	48 e	98 e	171.5	1
C₇H₁₀N₂	Toluene-2,4-diamine			100.4	145.3	202.9	279.5	5
C₇H₁₂	1-Heptyne	−75 e	−57 e	−35 e	−5 e	37.1	99.5	5
C₇H₁₂	2-Heptyne		−51 e	−27 e	4 e	46.9	111.5	5
C₇H₁₂	3-Heptyne	−71 e	−53 e	−31 e	0 e	42.7	106.4	5
C₇H₁₂	5-Methyl-1-hexyne	−80 e	−62 e	−40 e	−11 e	30.1	91.4	5
C₇H₁₂	5-Methyl-2-hexyne	−75 e	−57 e	−34 e	−4 e	38.6	102.0	5
C₇H₁₂	2-Methyl-3-hexyne	−78 e	−61 e	−39 e	−9 e	32.6	94.8	5
C₇H₁₂	4,4-Dimethyl-1-pentyne		−73 e	−52 e	−24 e	15.9	75.6	5
C₇H₁₂	4,4-Dimethyl-2-pentyne		−70 e	−48 e	−19 e	21.4	82.6	5
C₇H₁₂	Bicyclo[4.1.0]heptane					49.9	116.3	5
C₇H₁₂	Cycloheptene			−30.0	3.4	47.5	108 e	5
C₇H₁₂	1-Methylbicyclo(3,1,0)hexane					29.8	92.6	5
C₇H₁₂	Methylenecyclohexane	−76 e	−58 e	−35 e	−5 e	38	103.0	5
C₇H₁₂	1-Methylcyclohexene	−72 e	−53 e	−30 e	1 e	45 e	109.8	5
C₇H₁₂	4-Methylcyclohexene	−76 e	−59 e	−36 e	−5 e	37.9	102.3	5
C₇H₁₂	1-Ethylcyclopentene	−75 e	−57 e	−34 e	−3 e	40.7	105.8	5
C₇H₁₂	1,2-Dimethylcyclopentene	−75 e	−57 e	−34 e	−3 e	40.2	105.3	5
C₇H₁₂	1,5-Dimethylcyclopentene	−77 e	−59 e	−36 e	−5.5	37.3	101.5	5
C₇H₁₂O	Cycloheptanone			18 e	53.7	104.0	178.7	5
C₇H₁₂O₂	Butyl acrylate	−52 e	−31 e	−4.5	30.4	78.0	146.9	5
C₇H₁₂O₂	Propyl methacrylate				26 e	73.8	139.7	5
C₇H₁₂O₃	Ethyl levulinate		17 e	45.3	82.6	133.2	205.7	5
C₇H₁₂O₄	Diethyl malonate	−23 e	4 e	36.0	76.4	128.5	198.3	5
C₇H₁₂O₄	Dimethyl glutarate	−11 e	15 e	47 e	87.7	139.8	209.5	5
C₇H₁₃ClO	Heptanoyl chloride	−17 e	4 e	29.4	59.7	96.9	144.0	5
C₇H₁₄	1-Heptene	−82.1	−63.8	−40.6	−10.7	31.1	93.2	1,5
C₇H₁₄	cis-2-Heptene	−79 e	−61 e	−38 e	−8 e	34.3	98.0	5
C₇H₁₄	trans-2-Heptene	−79 e	−61 e	−39 e	−8 e	34.0	97.5	5
C₇H₁₄	cis-3-Heptene	−80 e	−62 e	−40 e	−10 e	32.3	95.3	5
C₇H₁₄	trans-3-Heptene	−80 e	−62 e	−40 e	−10 e	32.2	95.2	5
C₇H₁₄	2-Methyl-1-hexene	−81 e	−64 e	−42 e	−12 e	29.3	91.6	5
C₇H₁₄	4-Methyl-1-hexene	−84 e	-67 e	−45 e	−16 e	25.3	86.3	5
C₇H₁₄	2-Methyl-2-hexene	−80 e	−63 e	−40 e	−10 e	32.0	95.0	5
C₇H₁₄	cis-3-Methyl-2-hexene	−79 e	−62 e	−39 e	−9 e	33.4	96.8	5
C₇H₁₄	trans-4-Methyl-2-hexene	−83 e	−66 e	−44 e	−15 e	25.9	87.1	5
C₇H₁₄	trans-5-Methyl-2-hexene	−83 e	−66 e	−44 e	−15 e	26.3	87.7	5
C₇H₁₄	trans-2-Methyl-3-hexene	−84 e	−67 e	−45 e	−16 e	24.6	85.5	5
C₇H₁₄	3-Ethyl-1-pentene	−85 e	−68 e	−46 e	−17 e	23.2	83.7	5
C₇H₁₄	2,3-Dimethyl-1-pentene	−85 e	−68 e	−46 e	−17 e	23.4	83.8	5
C₇H₁₄	2,4-Dimethyl-1-pentene	−88 e	−71 e	−50 e	−21 e	20.0	81.2	5
C₇H₁₄	3,3-Dimethyl-1-pentene	−87 e	−71 e	−50 e	−21 e	18.1	77.1	5
C₇H₁₄	4,4-Dimethyl-1-pentene	−94 e	−78 e	−57 e	−28 e	11.5	72.1	5
C₇H₁₄	2,3-Dimethyl-2-pentene	−79 e	−62 e	−39 e	−9 e	33.5	96.9	5
C₇H₁₄	2,4-Dimethyl-2-pentene	−84 e	−68 e	−46 e	−18 e	22.6	82.9	5
C₇H₁₄	cis-3,4-Dimethyl-2-pentene	−83 e	−65 e	−43 e	−14 e	27.2	88.8	5
C₇H₁₄	trans-3,4-Dimethyl-2-pentene	−82 e	−64 e	−42 e	−13 e	29.0	91.1	5
C₇H₁₄	cis-4,4-Dimethyl-2-pentene	−90 e	−73 e	−51 e	−22 e	18.6	80.0	5

Mol. form.	Name	Temperature in °C for the indicated pressure						Ref.
		1 Pa	10 Pa	100 Pa	1 kPa	10 kPa	100 kPa	
C_7H_{14}	trans-4,4-Dimethyl-2-pentene	−90 e	−73 e	−52 e	−23 e	16.6	76.3	5
C_7H_{14}	2,3,3-Trimethyl-1-butene	−91 e	−75 e	−53 e	−24.2	16.3	77.5	5
C_7H_{14}	Cycloheptane				6 e	51.1	118.4	1
C_7H_{14}	Methylcyclohexane	−79 e	−62 e	−39 e	−7.9	35.5	100.5	1
C_7H_{14}	Ethylcyclopentane	−76 e	−59 e	−35 e	−5 e	38.4	103.0	5
C_7H_{14}	1,1-Dimethylcyclopentane		−69 e	−47 e	−17 e	24.8	87.4	5
C_7H_{14}	cis-1,2-Dimethylcyclopentane			−38 e	−8 e	34.9	99.0	5
C_7H_{14}	trans-1,2-Dimethylcyclopentane	−83 e	−66 e	−43 e	−13 e	28.4	91.4	5
C_7H_{14}	cis-1,3-Dimethylcyclopentane	−84 e	−66 e	−44 e	−14 e	28.2	91.1	5
C_7H_{14}	trans-1,3-Dimethylcyclopentane	−84 e	−67 e	−44 e	−14 e	27.4	90.3	5
$C_7H_{14}O$	1-Heptanal	−41 e	−21 e	4 e	37 e	83.7	152.3	5
$C_7H_{14}O$	2-Heptanone		−22 e	3 e	36 e	82.2	150.6	1
$C_7H_{14}O$	3-Heptanone		−28 e	0 e	36 e	83.2	147.0	5
$C_7H_{14}O$	4-Heptanone	−27 e	−6 e	18.8	50.2	90.3	143.4	5
$C_7H_{14}O$	5-Methyl-2-hexanone		−27 e	−2 e	31.0	76.6	144.4	5
$C_7H_{14}O$	2,4-Dimethyl-3-pentanone	−61 e	−42 e	−18 e	14 e	58.5	124.8	1
$C_7H_{14}O_2$	Heptanoic acid	24 e	46 e	72 e	107 e	154.6	222.6	5
$C_7H_{14}O_2$	Pentyl acetate	−58 e	−39 e	−14 e	20 e	70.1	149 e	5
$C_7H_{14}O_2$	Isopentyl acetate	−51 e	−30 e	−4 e	30.3	76.2	141.4	5
$C_7H_{14}O_2$	Isobutyl propanoate	−35 e	−19 e	2 e	31 e	72.0	136.1	5
$C_7H_{14}O_2$	Propyl butanoate	−35 e	−19 e	3 e	32.0	74.9	142.8	5
$C_7H_{14}O_2$	Propyl isobutanoate		−28 e	−5.7	24.5	67.5	133.3	5
$C_7H_{14}O_2$	Isopropyl isobutanoate		−44 e	−19.7	12.2	56.0	120.1	5
$C_7H_{14}O_2$	Ethyl 3-methylbutanoate	−57 e	−36 e	−10 e	23.9	69.5	134.4	5
$C_7H_{14}O_2$	Methyl hexanoate	−47 e	−26 e	2 e	36.6	83.3	149 e	5
$C_7H_{14}O_2$	4-Methoxy-4-methyl-2-pentanone				43 e	89.8	160 e	5
$C_7H_{15}Br$	1-Bromoheptane	−30 e	−9 e	18 e	54 e	104.4	178.4	5
$C_7H_{15}Cl$	1-Chloroheptane	−39 e	−19 e	7 e	41 e	88.6	159.9	5
$C_7H_{15}F$	1-Fluoroheptane	−64 e	−45 e	−22 e	10 e	53.3	117.4	5
$C_7H_{15}I$	1-Iodoheptane	−19 e	3 e	32 e	71 e	123.8	203.4	5
C_7H_{16}	Heptane	−78.6	−60.2	−37.0	−6.6	35.4	98.0	16
C_7H_{16}	2-Methylhexane	−82 e	−65 e	−43 e	−13 e	27.8	89.7	1
C_7H_{16}	3-Methylhexane	−81 e	−64 e	−42 e	−12 e	29.2	91.5	1
C_7H_{16}	3-Ethylpentane	−81 e	−63 e	−41 e	−11 e	30.5	93.1	1
C_7H_{16}	2,2-Dimethylpentane	−90 e	−73 e	−52 e	−22.9	17.6	78.8	1
C_7H_{16}	2,3-Dimethylpentane	−87 e	−68.4	−45.3	−14.9	26.8	89.3	5
C_7H_{16}	2,4-Dimethylpentane	−89 e	−72 e	−50 e	−21.3	19.2	80.1	1
C_7H_{16}	3,3-Dimethylpentane	−88 e	−71 e	−49 e	−18.8	22.9	85.6	1
C_7H_{16}	2,2,3-Trimethylbutane				−23.2	18.1	80.4	5
$C_7H_{16}O$	1-Heptanol		17 e	40 e	70.1	112.5	176 e	1
$C_7H_{16}O$	2-Heptanol	−9 e	7 e	27 e	55.0	95.2	158.7	5
$C_7H_{16}O$	3-Heptanol	−8 e	7 e	27 e	54.5	93.9	156.3	5
$C_7H_{16}O$	4-Heptanol	−16 e	1 e	22 e	51 e	91.9	154.6	5
$C_7H_{16}O$	2,2-Dimethyl-3-pentanol			9 e	35 e	73.1	135.5	5
$C_7H_{16}S$	1-Heptanethiol	−30 e	−9 e	18 e	53 e	102.7	176.4	5
$C_7H_{17}N$	Heptylamine		5 e	39 e	86.7	156.4		5
$C_7H_{18}N_2$	N,N-Diethyl-1,3-propanediamine			50.1	99.9	167.7		5
C_8F_{18}	Perfluorooctane			5 e	45.0	105.6		5
$C_8H_4O_3$	Phthalic anhydride	48.2 s	72.4 s			192.7	284.2	5
C_8H_6O	Benzofuran		−16 e	12 e	47.9	97.7	170.7	5
C_8H_7Cl	o-Chlorostyrene	−33 e	−10 e	20 e	58 e	110.8	188 e	5
C_8H_7N	2-Methylbenzonitrile		1 e	32.1	72.2	126.6	204.7	5
C_8H_7N	4-Methylbenzonitrile			40.1	78.7	134.3	221.3	5
C_8H_7N	Benzeneacetonitrile	−3 e	23 e	55.3	97.4	153.7	233.1	5
$C_8H_7NO_4$	Indole	20.6 s	44.5 s				254.0	5
C_8H_8	Methyl 2-nitrobenzoate	17 e	49 e	89 e	140 e	208 e	302 e	5
C_8H_8	Styrene		−31 e	−5 e	28.6	75.4	144.7	1
C_8H_8	1,3,5,7-Cyclooctatetraene				24.3	71.0	140.1	5
C_8H_8O	Acetophenone			36 e	73 e	125.3	201.5	5
$C_8H_8O_2$	Phenyl acetate		3 e	33.1	72.2	123.9	195.5	5

Mol. form.	Name	1 Pa	10 Pa	100 Pa	1 kPa	10 kPa	100 kPa	Ref.
$C_8H_8O_2$	Methyl benzoate		−1 e	29 e	68 e	121.2	198.9	5
$C_8H_8O_2$	4-Methoxybenzaldehyde	9 e	35 e	68.1	110.8	167.9	248.5	5
$C_8H_8O_3$	Methyl salicylate	−1 e	22 e	51 e	88.8	141.8	219.9	5
C_8H_9Cl	1-Chloro-2-ethylbenzene	−30 e	−9 e	18 e	54 e	103.7	177.9	5
C_8H_9Cl	1-Chloro-4-ethylbenzene	−27 e	−6 e	22 e	58 e	108.7	183.9	5
$C_8H_9NO_2$	1-Ethyl-4-nitrobenzene	10 e	36 e	69 e	111.6	168 e	245 e	5
C_8H_{10}	Ethylbenzene	−56.2	−36.8	−12.0	21.1	67.1	135.7	1
C_8H_{10}	o-Xylene		−7 e	27 e	27 e	74.2	143.9	1
C_8H_{10}	m-Xylene		−35 e	−10 e	23.4	69.8	138.7	1
C_8H_{10}	p-Xylene				22.4	68.9	137.9	1
$C_8H_{10}O$	o-Ethylphenol		16.9	44.5	81.1	130.9	204.0	5
$C_8H_{10}O$	m-Ethylphenol	5.6	29.2	57.5	91.9	144.8	217.9	5
$C_8H_{10}O$	p-Ethylphenol			60 e	95.5	144.6	217.5	5
$C_8H_{10}O$	2,3-Xylenol	14.3 s	34.3 s	57.2 s	91.4	141.7	216.4	1,5
$C_8H_{10}O$	2,4-Xylenol			50.2	85.5	137.2	210.5	1,5
$C_8H_{10}O$	2,5-Xylenol	13.4 s	33.2 s	55.9 s	87.4	137.0	210.6	5
$C_8H_{10}O$	2,6-Xylenol	−3.1 s	16.7 s	39.6 s	75.3	125.9	200.6	1,5
$C_8H_{10}O$	3,4-Xylenol	19.7 s	40.2 s	63.7 s	102.1	152.3	226.4	1,5
$C_8H_{10}O$	3,5-Xylenol	16.5 s	37.2 s	61.1 s	98.0	147.9	221.3	1,5
$C_8H_{10}O$	Benzeneethanol	2 e	25 e	54 e	92 e	143.6	217.7	5
$C_8H_{10}O$	Phenetole		−9 e	17 e	51 e	99 e	169.3	5
$C_8H_{10}O_2$	2-Phenoxyethanol	21 e	46 e	75.9	115.4	168.7	244.8	5
$C_8H_{10}O_2$	1,3-Dimethoxybenzene	18 e	34 e	56 e	86.7	135.5	223 e	5
$C_8H_{11}N$	p-Ethylaniline	−2 e	21 e	49 e	87 e	139.4	216.7	5
$C_8H_{11}N$	N-Ethylaniline	−15 e	8 e	38 e	76.4	128.8	204.2	5
$C_8H_{11}N$	N,N-Dimethylaniline			28 e	66 e	118.1	193.6	1
$C_8H_{11}N$	2,4-Xylidine	−2 e	21 e	51 e	88 e	139.1	210.9	5
$C_8H_{11}N$	2,6-Xylidine			37 e	80 e	137.7	217.7	5
$C_8H_{11}N$	5-Ethyl-2-picoline	−33 e	−9.3	20 e			178.0	5
$C_8H_{11}NO$	o-Phenetidine	0 e	27 e	60 e	102.2	156.0	228.1	5
C_8H_{12}	1,5-Cyclooctadiene		−37 e	−8 e	30 e	80.2	150 e	5
C_8H_{12}	4-Vinylcyclohexene	−62 e	−43 e	−19 e	14.1	59.9	129 e	5
$C_8H_{12}O_4$	Diethyl maleate	−6 e	20 e	52.2	93.5	148.4	224.8	5
C_8H_{14}	2,5-Dimethyl-1,5-hexadiene	−38 e	−26 e	−10 e	14 e	50.8	115.1	5
C_8H_{14}	1-Octyne	−59 e	−40 e	−16 e	16 e	60.3	125.8	1
C_8H_{14}	2-Octyne	−52 e	−33 e	−8 e	25 e	70.6	137.8	1
C_8H_{14}	3-Octyne	−55 e	−35 e	−11 e	22 e	66.8	132.8	1
C_8H_{14}	4-Octyne	−56 e	−36 e	−12 e	21 e	65.6	131.4	1
C_8H_{14}	1-Ethylcyclohexene	−55 e	−35 e	−11 e	22 e	68 e	136.5	5
$C_8H_{14}O_2$	Cyclohexyl acetate					103.1	172.9	5
$C_8H_{14}O_2$	Butyl methacrylate				47 e	93.3	159.0	5
$C_8H_{14}O_3$	Butanoic anhydride	−28 e	−2 e	30 e	71 e	123.8	196.5	5
$C_8H_{14}O_4$	Ethyl succinate	−6 e	20 e	51.0	91.1	143.7	216.1	5
$C_8H_{14}O_4$	Dipropyl oxalate	−4 e	20 e	49.9	88.6	140.4	213.0	5
$C_8H_{14}O_4$	Dimethyl adipate		28 e	61 e	103 e	156.1	227.3	5
$C_8H_{15}Br$	(2-Bromoethyl)cyclohexane	−14 e	8 e	36.9	75.3	129.7	212.5	5
$C_8H_{15}ClO$	Octanoyl chloride	1 e	22 e	46 e	74.7	109 e	150 e	5
$C_8H_{15}N$	Octanenitrile	−15 e	8 e	37 e	75 e	127.7	204.4	5
C_8H_{16}	1-Octene	−65.7	−46.1	−21.4	10.5	54.9	120.9	1,5
C_8H_{16}	cis-2-Octene	−59 e	−41 e	−17 e	15 e	59 e	125.2	5
C_8H_{16}	trans-2-Octene	−59 e	−41 e	−17 e	14 e	59 e	124.5	5
C_8H_{16}	cis-3-Octene	−65 e	−46 e	−22 e	10 e	55.1	122.4	5
C_8H_{16}	trans-3-Octene	−61 e	−43 e	−19 e	13 e	57 e	122.8	5
C_8H_{16}	cis-4-Octene	−63 e	−44 e	−20 e	11 e	56 e	122.1	5
C_8H_{16}	trans-4-Octene	−65 e	−46 e	−22 e	10 e	54.6	121.8	5
C_8H_{16}	2-Methyl-1-heptene	−66 e	−48 e	−24 e	8 e	52.3	118.7	5
C_8H_{16}	2,2-Dimethyl-cis-3-hexene	−74 e	−56 e	−33 e	−3 e	40.1	105.0	5
C_8H_{16}	2,3-Dimethyl-2-hexene	−65 e	−47 e	−23 e	10 e	54.3	121.3	5
C_8H_{16}	2,3,3-Trimethyl-1-pentene		−53 e	−30 e	1 e	43.8	107.9	5
C_8H_{16}	2,4,4-Trimethyl-1-pentene	−79 e	−61 e	−38 e	−7 e	36.2	101.0	5

Mol. form.	Name	Temperature in °C for the indicated pressure						Ref.
		1 Pa	10 Pa	100 Pa	1 kPa	10 kPa	100 kPa	
C_8H_{16}	2,3,4-Trimethyl-2-pentene	−68 e	−49 e	−26 e	6 e	50.0	115.8	5
C_8H_{16}	2,4,4-Trimethyl-2-pentene	−73 e	−56 e	−33 e	−2 e	40.4	104.5	5
C_8H_{16}	Cyclooctane				30 e	78 e	150.7	1
C_8H_{16}	Ethylcyclohexane	−61 e	−42 e	−17 e	15.8	61.9	131.3	5
C_8H_{16}	1,1-Dimethylcyclohexane			−27 e	5 e	50.6	119.1	5
C_8H_{16}	cis-1,2-Dimethylcyclohexane		−44 e	−20 e	14 e	59.7	129.2	5
C_8H_{16}	trans-1,2-Dimethylcyclohexane	−68 e	−49 e	−25 e	8 e	53.9	122.9	5
C_8H_{16}	cis-1,3-Dimethylcyclohexane	−68 e	−48 e	−23 e	10 e	55.6	123.1	5
C_8H_{16}	trans-1,3-Dimethylcyclohexane	−62 e	−45 e	−23 e	8 e	51.5	120.9	5
C_8H_{16}	cis-1,4-Dimethylcyclohexane	−66 e	−47 e	−23 e	10 e	55.3	123.8	5
C_8H_{16}	trans-1,4-Dimethylcyclohexane			−27 e	5 e	50.6	118.9	5
C_8H_{16}	Propylcyclopentane	−60 e	−41 e	−16 e	16.5	62.1	130.5	5
C_8H_{16}	Isopropylcyclopentane	−65 e	−46 e	−21 e	12 e	57.3	125.9	5
C_8H_{16}	1-Ethyl-1-methylcyclopentane	−67 e	−49 e	−24 e	8 e	53.2	121.0	5
C_8H_{16}	cis-1-Ethyl-2-methylcyclopentane	−63 e	−44 e	−19 e	13.3	59.1	127.6	5
C_8H_{16}	1,1,2-Trimethylcyclopentane				2 e	46.2	113.2	5
C_8H_{16}	1,1,3-Trimethylcyclopentane	−77 e	−59 e	−36 e	−5 e	38.7	104.4	5
C_8H_{16}	1′,2′,4a-1,2,4-Trimethylcyclo-pentane	−70 e	−52 e	−28 e	4 e	48.9	116.2	5
C_8H_{16}	1′,2a,4′-1,2,4-Trimethylcyclo-pentane	−74 e	−56 e	−33 e	−1 e	42.8	108.8	5
$C_8H_{16}O$	1-Propylcyclopentanol	9 e	24 e	43 e	69.0	108.4	173.5	5
$C_8H_{16}O$	Octanal				6 e	45.7	97.8	170.2	5
$C_8H_{16}O$	2-Octanone		−3 e	23 e	57 e	103.8	172.1	5
$C_8H_{16}O$	3-Octanone			8 e	47.7	97 e	161 e	5
$C_8H_{16}O$	2,2,4-Trimethyl-3-pentanone			11.3	42.1	81.7	134.6	5
$C_8H_{16}O_2$	Octanoic acid	37 e	58 e	85 e	120 e	165.5	238.4	1,5
$C_8H_{16}O_2$	2-Ethylhexanoic acid				108 e	159.6	226.6	5
$C_8H_{16}O_2$	Hexyl acetate	−37 e	−13 e	16 e	52.8	100.4	164 e	5
$C_8H_{16}O_2$	Isopentyl propanoate			3.1	40.7	90.6	159.8	5
$C_8H_{16}O_2$	Isobutyl isobutanoate	−47 e	−26 e	0.4	34.8	81.1	147.0	5
$C_8H_{16}O_2$	Propyl 3-methylbutanoate			1.8	38.9	87.9	155.6	5
$C_8H_{16}O_2$	Ethyl hexanoate	−31 e	−9 e	18.7	53.9	100.7	166.2	5
$C_8H_{16}O_2$	Methyl heptanoate	−30 e	−9 e	19 e	54.2	102.4	172 e	5
$C_8H_{16}O_4$	Diethylene glycol monoethyl ether acetate	−16 e	10.6	43.9	86.2	141.3	216.6	5
$C_8H_{17}Br$	1-Bromooctane	−17 e	6 e	34 e	72 e	123.8	200.3	5
$C_8H_{17}Cl$	1-Chlorooctane	−25 e	−4 e	23 e	59 e	108.8	182.9	5
$C_8H_{17}Cl$	3-(Chloromethyl)heptane					100.3	172.4	5
$C_8H_{17}F$	1-Fluorooctane				29 e	74.6	141.8	5
$C_8H_{17}I$	1-Iodooctane	−6 e	18 e	48 e	87 e	142.5	224.5	5
C_8H_{18}	Octane		−42.6	−17.9	14.4	58.9	125.3	16
C_8H_{18}	2-Methylheptane	−69 e	−49.1	−24.5	7.6	51.6	117.2	1,5
C_8H_{18}	3-Methylheptane	−67 e	−48.1	−23.6	8.5	52.7	118.5	1,5
C_8H_{18}	4-Methylheptane	−65 e	−47 e	−24 e	7.8	51.6	117.2	5
C_8H_{18}	3-Ethylhexane				8 e	52.1	118.1	5
C_8H_{18}	2,2-Dimethylhexane	−73 e	−55 e	−32 e	−1.5	41.6	106.4	5
C_8H_{18}	2,3-Dimethylhexane				5 e	49.2	115.1	5
C_8H_{18}	2,4-Dimethylhexane				0.6	43.9	109.0	5
C_8H_{18}	2,5-Dimethylhexane	−71 e	−53 e	−30 e	0.7	43.8	108.6	5
C_8H_{18}	3,3-Dimethylhexane	−72 e	−54 e	−30 e	1.4	45.4	111.5	5
C_8H_{18}	3,4-Dimethylhexane				7 e	50.9	117.3	5
C_8H_{18}	3-Ethyl-2-methylpentane	−69 e	−50 e	−27 e	5 e	48.9	115.2	5
C_8H_{18}	3-Ethyl-3-methylpentane	−70 e	−51 e	−27 e	5 e	50.2	117.8	5
C_8H_{18}	2,2,3-Trimethylpentane	−74 e	−56 e	−32 e	−0.8	43.1	109.4	5
C_8H_{18}	2,2,4-Trimethylpentane	−81.9	−63.4	−39.8	−8.9	34.0	98.8	5
C_8H_{18}	2,3,3-Trimethylpentane	−72 e	−54 e	−30 e	2.1	46.9	114.3	5
C_8H_{18}	2,3,4-Trimethylpentane	−74 e	−54.5	−30.0	2.2	46.7	113.1	1,5
C_8H_{18}	2,2,3,3-Tetramethylbutane	−62.5 s	−44 s	−20.9 s	8.9 s	48.8 s	105.8	5
$C_8H_{18}O$	1-Octanol	12 e	30 e	53 e	84 e	128.2	194.8	1,39
$C_8H_{18}O$	2-Octanol			40 e	69.9	112.5	179.4	1,39
$C_8H_{18}O$	3-Octanol	12 e	24 e	40 e	64 e	102.8	174.1	1

		Temperature in °C for the indicated pressure						
Mol. form.	Name	1 Pa	10 Pa	100 Pa	1 kPa	10 kPa	100 kPa	Ref.
$C_8H_{18}O$	4-Octanol			40 e	66.9	107.3	176.0	1,39
$C_8H_{18}O$	4-Methyl-3-heptanol	−52 e	−28 e	1 e	39 e	87.6	155.0	5
$C_8H_{18}O$	5-Methyl-3-heptanol	−35 e	−16 e	8 e	40 e	84.8	153.0	5
$C_8H_{18}O$	4-Methyl-4-heptanol	−17 e	1 e	24 e	55 e	97.2	160.7	5
$C_8H_{18}O$	2-Ethyl-1-hexanol			45 e	75 e	118.3	184.2	1
$C_8H_{18}O$	2-Ethyl-2-hexanol	−13 e	4 e	26 e	55 e	96.3	160.3	5
$C_8H_{18}O$	2,4,4-Trimethyl-2-pentanol		−7 e	13 e	40 e	79.8	146.1	5
$C_8H_{18}O$	2,2,4-Trimethyl-3-pentanol	−2 e	9 e	24 e	47 e	82.6	150.4	5
$C_8H_{18}O$	Dibutyl ether	−55 e	−35 e	−8 e	26 e	73.0	141.2	5
$C_8H_{18}O$	Di-sec-butyl ether			−19 e	12.1	55.4	120.6	5
$C_8H_{18}O$	Di-tert-butyl ether			−33 e	−2 e	41.7	106.8	1
$C_8H_{18}O_2$	Ethylene glycol monohexyl ether	−13 e	14 e	46 e	86 e	137.7	206.9	5
$C_8H_{18}O_2$	1,2-Dipropoxyethane			−44.2	−2.0	63.6	179.2	5
$C_8H_{18}O_2$	Di-tert-butyl peroxide			−26 e	4.3	46.6	110.5	5
$C_8H_{18}O_3$	Diethylene glycol monobutyl ether	14 e	37 e	66.8	104.9	153 e	230.4	5
$C_8H_{18}O_3$	Diethylene glycol diethyl ether	−32 e	−7 e	25 e	64.9	117.1	189 e	5
$C_8H_{18}O_5$	Tetraethylene glycol	89 e	117 e	151.1	192.2	242.9	307.3	5
$C_8H_{18}S$	1-Octanethiol	−15 e	6 e	34 e	71 e	122.1	198.5	5
$C_8H_{18}S$	Dibutyl sulfide	−22 e	0 e	27 e	63 e	113.5	188.4	5
$C_8H_{19}N$	Dibutylamine	−37 e	−16 e	10 e	44 e	90.8	159.1	5
$C_8H_{19}N$	Diisobutylamine	−57 e	−36 e	−9.0	25.5	72.2	139.0	5
$C_8H_{20}O_4Si$	Ethyl silicate	−77 e	−52 e	−21 e	21.6	80.5	164.1	5
$C_8H_{20}Si$	Tetraethylsilane			−6.5	30.5	80.6	152.6	5
C_9F_{20}	Perfluorononane					40 e	114.7	5
$C_9H_6N_2O_2$	Toluene-2,4-diisocyanate		39 e	72 e	113.9	169.7	247 e	5
C_9H_7N	Quinoline	−1.3	23.7	55.4	96.8	153.4	236.5	1,5
C_9H_7N	Isoquinoline		30.2	60.7	101.3	157.9	242.7	1,5
C_9H_8	Indene			12 e	53.0	106.8	181.0	5
C_9H_{10}	cis-1-Propenylbenzene	−38 e	−15.4	13.3	51.4	103.7	178.4	5
C_9H_{10}	trans-1-Propenylbenzene		−16 e	13.3	51.6	103.7	178.4	5
C_9H_{10}	Isopropenylbenzene			3.2	41.5	92.8	164.9	5
C_9H_{10}	Indan	−33 e	−12 e	16 e	52 e	102.3	177.5	1
$C_9H_{10}O$	2,4-Dimethylbenzaldehyde	−3 e	23 e	54 e	93.2	144.6	214.5	5
$C_9H_{10}O_2$	Ethyl benzoate	−18 e	8 e	39 e	80.1	135.1	212.8	5
$C_9H_{10}O_2$	Benzyl acetate	−11 e	15 e	46.6	86.9	139.5	211 e	5
$C_9H_{11}Br$	1-Bromo-4-isopropylbenzene	−8 e	15 e	45 e	84 e	138.1	218.5	5
$C_9H_{11}Cl$	1-Chloro-2-isopropylbenzene	−23 e	−1 e	27 e	64 e	114.6	190.5	5
$C_9H_{11}Cl$	1-Chloro-4-isopropylbenzene		3 e	31 e	69 e	120.5	197.8	5
C_9H_{12}	Propylbenzene	−43 e	−23 e	4 e	38 e	86.7	158.8	1
C_9H_{12}	Isopropylbenzene	−46 e	−26 e	−1 e	33 e	80.9	152.0	1
C_9H_{12}	o-Ethyltoluene	−40 e	−19 e	8 e	43 e	92.1	164.7	5
C_9H_{12}	m-Ethyltoluene	−42 e	−21 e	5 e	40.4	88.9	160.8	5
C_9H_{12}	p-Ethyltoluene	−41 e	−21 e	6 e	41 e	89.2	161.5	5
C_9H_{12}	1,2,3-Trimethylbenzene		−12 e	15 e	52 e	101.5	175.6	1
C_9H_{12}	1,2,4-Trimethylbenzene	−37 e	−16 e	11 e	47 e	95.9	168.9	1
C_9H_{12}	1,3,5-Trimethylbenzene	−39 e	−18 e	9 e	43.7	92.4	164.3	1
$C_9H_{12}O$	Benzyl ethyl ether		−10 e	20.4	59.3	111.3	184.5	5
$C_9H_{12}O$	Phenyl propyl ether		−10 e	21 e	61 e	113.9	189.3	5
$C_9H_{12}O$	Phenyl isopropyl ether	−20 e	−1 e	23 e	56 e	103.7	176.9	5
$C_9H_{13}N$	2,4,6-Trimethylaniline	12 e	36 e	66 e	104.1	154.9	226 e	5
$C_9H_{13}N$	N,N-Dimethyl-o-toluidine	−25 e	−3 e	24.4	60.6	110.7	184.5	5
$C_9H_{13}N$	Amphetamine			33 e	70.1	118 e	202.0	5
$C_9H_{14}O$	Isophorone		1 e	33.1	75.1	132.4	215.1	5
$C_9H_{14}O_6$	Triacetin	37.6	62 e	90 e	124 e	165 e	214 e	5
$C_9H_{16}O_4$	Diethyl glutarate	−1 e	26 e	60.2	103.3	159.6	236.5	5
$C_9H_{17}N$	Nonanenitrile	−3 e	21 e	50.9	90.7	145.4	225.1	5
C_9H_{18}	1-Nonene	−50.1	−29.4	−3.3	30.4	77.1	146.4	1,5
C_9H_{18}	2-Methyl-1-octene	−53 e	−34 e	−9 e	25 e	72 e	144.1	5
C_9H_{18}	Butylcyclopentane	−45 e	−24 e	1 e	36 e	84 e	156.1	5
C_9H_{18}	Propylcyclohexane	−46 e	−26 e	0 e	35.1	83.6	156.2	5

Mol. form.	Name	\multicolumn{6}{c}{Temperature in °C for the indicated pressure}						Ref.
		1 Pa	10 Pa	100 Pa	1 kPa	10 kPa	100 kPa	
C_9H_{18}	Isopropylcyclohexane	−48 e	−28 e	−2 e	33 e	81.3	154.0	5
C_9H_{18}	trans-1-Ethyl-4-methylcyclo-hexane	−53 e	−33 e	−8 e	25 e	71.8	141.5	5
C_9H_{18}	1,1,2-Trimethylcyclohexane			−12 e	23 e	71.5	145.5	5
C_9H_{18}	1,1,3-Trimethylcyclohexane	−60 e	−41 e	−16 e	18 e	65.2	136.1	5
C_9H_{18}	1′,2a,4a-1,2,4-Trimethylcyclo-hexane	−71 e	−50 e	−22 e	15 e	65.7	140.7	5
C_9H_{18}	1′,3′,5′-1,3,5-Trimethylcyclo-hexane	−72 e	−50 e	−22 e	14 e	65.1	140.0	5
C_9H_{18}	Isobutylcyclopentane	−105 e	−88 e	−64 e	−28 e	31 e	147.0	5
C_9H_{18}	cis-1-Methyl-2-propylcyclo-pentane	−52 e	−33 e	−7 e	28 e	77 e	152.0	5
C_9H_{18}	trans-1-Methyl-2-propylcyclo-pentane	−56 e	−36 e	−11 e	23 e	72 e	145.8	5
C_9H_{18}	1,1,3,3-Tetramethylcyclo-pentane	−72 e	−54 e	−30 e	2 e	47 e	117.4	5
$C_9H_{18}O$	Nonanal		−3 e	27.4	65.5	115.6	184.6	5
$C_9H_{18}O$	2-Nonanone		8 e	35 e	71 e	121.0	194.0	5
$C_9H_{18}O$	5-Nonanone			−1 e	39.1	94 e	188 e	5
$C_9H_{18}O$	2,6–Dimethyl-4-heptanone	−32 e	−12 e	14 e	48 e	96.2	167.7	5
$C_9H_{18}O_2$	Nonanoic acid	48 e	69 e	97 e	133 e	182.7	255.1	5
$C_9H_{18}O_2$	Heptyl acetate	−16 e	6 e	34 e	70 e	119.9	191.9	5
$C_9H_{18}O_2$	Isopentyl butanoate				55 e	105.6	178.4	5
$C_9H_{18}O_2$	Isobutyl 3-methylbutanoate			11.3	48.3	97.9	168.3	5
$C_9H_{18}O_2$	Propyl hexanoate	−26 e	−2 e	28 e	65.1	113.4	178 e	5
$C_9H_{18}O_2$	Methyl octanoate	−26 e	−9 e	13 e	40 e	76 e	127.9	5
$C_9H_{19}Cl$	1-Chlorononane	−11 e	11 e	39 e	76 e	127.8	204.7	5
C_9H_{20}	Nonane	−46.8	−26.0	0.0	34.0	80.8	150.3	16
C_9H_{20}	2-Methyloctane	−49 e	−30 e	−5 e	28 e	73.9	142.8	5
C_9H_{20}	3-Methyloctane	−49 e	−29 e	−5 e	29 e	74.7	143.7	5
C_9H_{20}	4-Methyloctane	−50 e	−30 e	−6 e	27 e	73.2	141.9	5
C_9H_{20}	2,2-Dimethylheptane	−58 e	−39 e	−15 e	18 e	63.6	132.3	5
C_9H_{20}	2,3-Dimethylheptane	−53 e	−33 e	−9 e	25 e	70.8	140.0	5
C_9H_{20}	2,6-Dimethylheptane	−55 e	−36 e	−12 e	21 e	66.4	134.7	5
C_9H_{20}	3-Ethyl-4-methylhexane			−9 e	24 e	70.6	139.9	5
C_9H_{20}	2,2,4-Trimethylhexane	−66.1	−46.4	−21.3	11.8	57.7	126.0	5
C_9H_{20}	2,2,5-Trimethylhexane	−65.1	−45.8	−21.2	11.2	56.2	123.7	1,5
C_9H_{20}	2,3,3-Trimethylhexane	−58 e	−38 e	−13 e	20 e	66.7	137.2	5
C_9H_{20}	2,3,5-Trimethylhexane	−60 e	−41 e	−16 e	17 e	62.3	130.9	5
C_9H_{20}	2,4,4-Trimethylhexane	−62 e	−43 e	−18 e	15 e	61.0	130.2	5
C_9H_{20}	3,3,4-Trimethylhexane	−53 e	−33 e	−7 e	28 e	76.3	148.9	5
C_9H_{20}	3,3-Diethylpentane			−9 e	26 e	73.7	145.7	1
C_9H_{20}	3-Ethyl-2,4-dimethylpentane	−58 e	−38 e	−13 e	20 e	66.7	136.2	5
C_9H_{20}	2,2,3,3-Tetramethylpentane				21 e	68.5	139.8	1
C_9H_{20}	2,2,3,4-Tetramethylpentane	−61 e	−42 e	−17 e	16 e	62.5	132.6	1
C_9H_{20}	2,2,4,4-Tetramethylpentane		−49 e	−25 e	8 e	53.2	121.8	1
C_9H_{20}	2,3,3,4-Tetramethylpentane	−57 e	−37 e	−12 e	22 e	69.7	141.1	1
$C_9H_{20}O$	1-Nonanol		40 e	64 e	96.9	141.0	213.0	5,39
$C_9H_{20}O$	3-Nonanol		24 e	47 e	78 e	123.0	194.2	5
$C_9H_{20}O$	4-Nonanol			45 e	76.4	121.3	192.0	5
$C_9H_{20}O$	5-Nonanol	13 e	31 e	54 e	84.5	128.1	194.7	5
$C_9H_{20}O$	2,2,4,4-Tetramethyl-3-pentanol				58	100	167	5
$C_9H_{20}S$	1-Nonanethiol	−2 e	21 e	49 e	87 e	140.4	219.2	5
$C_9H_{21}BO_3$	Triisopropyl borate					73.1	139.0	5
$C_9H_{21}N$	Nonylamine		9 e	37 e	75 e	126.2	202.1	5
$C_9H_{21}N$	Tripropylamine	−39 e	−18 e	8 e	42 e	88.2	156.0	5
$C_{10}F_8$	Perfluoronaphthalene	5.2 s	25.1 s	48.1 s				5
$C_{10}F_{22}$	Perfluorodecane					52 e	132.9	5
$C_{10}H_7Br$	1-Bromonaphthalene	17 e	45 e	80.3	126.7	189.8	280.5	5
$C_{10}H_7Cl$	1-Chloronaphthalene	14 e	39 e	70.5	112.8	171.6	258.6	5
$C_{10}H_8$	Naphthalene**	3.2 s	24.1 s	49.3 s	80.7	135.6	217.5	1,5
$C_{10}H_8$	Azulene	24.1 s	46 s	71.5 s	103.3	162.6	244.0	5
$C_{10}H_8O$	1-Naphthol				137.2	196.7	281.8	5
$C_{10}H_8O$	2-Naphthol				140.7	200.5	286.8	5
$C_{10}H_9N$	1-Naphthalenamine		62 e	99.0	146.9	210.7	300.1	5
$C_{10}H_9N$	2-Naphthalenamine	36.3 s	65.9 s	103 s	150.9	215.1	305.5	5

Mol. form.	Name	\multicolumn{6}{c}{Temperature in °C for the indicated pressure}					Ref.
		1 Pa	10 Pa	100 Pa	1 kPa	10 kPa	100 kPa	
$C_{10}H_9N$	2-Methylquinoline	5.3	31.9	63.8	102.9	165.8	247.2	5
$C_{10}H_9N$	4-Methylquinoline	29 e	54 e	85 e	127 e	183.0	265.1	5
$C_{10}H_9N$	6-Methylquinoline	27 e	51 e	81 e	122 e	179.2	264.5	5
$C_{10}H_9N$	8-Methylquinoline	15 e	40 e	70 e	111 e	166.1	247.3	5
$C_{10}H_{10}$	m-Divinylbenzene	−29 e	−4 e	27.1	67.6	122.1	199 e	5
$C_{10}H_{10}O_4$	Dimethyl phthalate	27 e	56 e	92.7	137.8	195.8	272.7	5
$C_{10}H_{10}O_4$	Dimethyl isophthalate			85 e	129.5	189.2	273 e	5
$C_{10}H_{10}O_4$	Dimethyl terephthalate	56.6 s	79.4 s	106.1 s	137.9 s	197.9	282 e	5
$C_{10}H_{12}$	1,2,3,4-Tetrahydronaphthalene	−21 e	3 e	33.2	74.1	127.4	207.8	5
$C_{10}H_{12}$	2-Ethylstyrene	−31 e	−8 e	21 e	60 e	111.7	187 e	5
$C_{10}H_{12}$	3-Ethylstyrene	−28 e	−5.3	24.1	62.6	116 e	193 e	5
$C_{10}H_{12}$	4-Ethylstyrene	−31 e	−8.2	21.3	60.5	115 e	196 e	5
$C_{10}H_{12}O$	Estragole			48.5	88.0	140.7	214.6	5
$C_{10}H_{12}O$	4-Isopropylbenzaldehyde			54.1	96.0	152.2	231.5	5
$C_{10}H_{12}O_2$	4-Allyl-2-methoxyphenol	9 e	37 e	72 e	115.9	173.8	252.9	5
$C_{10}H_{12}O_2$	2-Phenylethyl acetate	−4 e	22 e	54 e	96 e	152.3	232.0	5
$C_{10}H_{12}O_2$	Propyl benzoate	−8 e	18 e	50.2	92.3	149.2	230.5	5
$C_{10}H_{12}O_2$	Ethyl phenylacetate	−9 e	19 e	52 e	95 e	150.2	225 e	5
$C_{10}H_{12}O_2$	Isoeugenol				125 e	185.3	267.1	5
$C_{10}H_{14}$	Butylbenzene	−28 e	−7 e	21 e	56.9	107.6	182.8	1,5
$C_{10}H_{14}$	sec-Butylbenzene	−35 e	−14 e	13 e	48 e	98.3	172.8	5
$C_{10}H_{14}$	tert-Butylbenzene	−37 e	−16 e	10 e	46 e	94.9	168.6	5
$C_{10}H_{14}$	Isobutylbenzene	−36 e	−15 e	12 e	47.9	97.8	172.3	5
$C_{10}H_{14}$	o-Cymene	−39 e	−16 e	13 e	51 e	103.1	177.8	5
$C_{10}H_{14}$	m-Cymene	−34 e	−13 e	14 e	50 e	99.9	174.6	5
$C_{10}H_{14}$	p-Cymene	−33 e	−12 e	16 e	52 e	102.2	176.6	5
$C_{10}H_{14}$	o-Diethylbenzene	−28 e	−6 e	21 e	58 e	107.9	182.9	5
$C_{10}H_{14}$	m-Diethylbenzene	−28 e	−7 e	20 e	56 e	106.2	180.6	5
$C_{10}H_{14}$	p-Diethylbenzene	−28 e	−6 e	21 e	57 e	108.1	183.3	5
$C_{10}H_{14}$	3-Ethyl-1,2-dimethylbenzene	−22 e	0 e	28 e	66 e	117.2	193.4	5
$C_{10}H_{14}$	4-Ethyl-1,2-dimethylbenzene	−24 e	−2 e	26 e	63 e	113.6	189.2	5
$C_{10}H_{14}$	2-Ethyl-1,3-dimethylbenzene		−2 e	26 e	63 e	113.7	189.5	5
$C_{10}H_{14}$	2-Ethyl-1,4-dimethylbenzene	−27 e	−5 e	23 e	60 e	110.6	186.4	5
$C_{10}H_{14}$	1-Ethyl-2,4-dimethylbenzene	−25 e	−4 e	24 e	61 e	112.2	187.9	5
$C_{10}H_{14}$	1-Ethyl-3,5-dimethylbenzene	−28 e	−6 e	21 e	58 e	108.3	183.2	5
$C_{10}H_{14}$	1-Methyl-2-propylbenzene	−27 e	−6 e	22 e	58.2	108.9	184.3	5
$C_{10}H_{14}$	1-Methyl-3-propylbenzene	−29 e	−8 e	20 e	56.1	106.5	181.3	5
$C_{10}H_{14}$	1-Methyl-4-propylbenzene	−29 e	−7 e	20 e	56.6	107.4	182.8	5
$C_{10}H_{14}$	1,2,3,4-Tetramethylbenzene		7 e	36 e	74 e	126.6	204.5	5
$C_{10}H_{14}$	1,2,3,5-Tetramethylbenzene	−19 e	3 e	32 e	69 e	120.9	197.5	5
$C_{10}H_{14}$	1,2,4,5-Tetramethylbenzene					119.9	196.3	5
$C_{10}H_{14}O$	2-Butylphenol	7 e	31 e	61 e	101 e	155.2	234.4	5
$C_{10}H_{14}O$	Butyl phenyl ether	−16 e	8 e	38 e	77 e	131.3	209.7	5
$C_{10}H_{14}O$	Thymol	18.9 s	37.9 s	59.5	101.2	155.0	230.4	5
$C_{10}H_{15}N$	2-Methyl-5-isopropylaniline	19 e	43 e	72 e	107.4	150 e	204 e	5
$C_{10}H_{15}N$	N-Butylaniline	11 e	35 e	66 e	106 e	160.9	241.0	5
$C_{10}H_{15}N$	N,N-Diethylaniline	−11 e	14 e	44.3	84.2	138.4	216.3	5
$C_{10}H_{16}$	Dipentene	−42 e	−19 e	10.6	48.7	100.2	173.9	5
$C_{10}H_{16}$	d-Limonene	−45 e	−21 e	9.1	48.0	100.4	174.5	5
$C_{10}H_{16}$	l-Limonene	−33 e	−12 e	16 e	52.0	102.3	177.0	21
$C_{10}H_{16}$	β-Myrcene			9.4	47.3	98.3	171.0	5
$C_{10}H_{16}$	α-Pinene	−48 e	−27 e	−1 e	33.6	82.2	155.1	21
$C_{10}H_{16}$	β-Pinene	−43 e	−22 e	5.0	40.6	90.5	165.5	21
$C_{10}H_{16}$	Camphene					90.7	160.1	4
$C_{10}H_{16}$	Terpinolene			26.5	64.9	115.4	184.6	5
$C_{10}H_{16}$	β-Phellandrene			16 e	53.2	104 e	171.0	5
$C_{10}H_{16}O$	(+)-Camphor	−15.8 s	10 s	41.5 s	80.8 s	131.4 s	207.6	5
$C_{10}H_{16}O$	Pulegone	37 e	49.1	66.4	92.2	135.1	220.2	5
$C_{10}H_{18}$	1-Decyne	−34 e	−13 e	14 e	51 e	100.3	173.5	5
$C_{10}H_{18}$	cis-Decahydronaphthalene	−26 e	−4 e	24 e	62.4	115.5	195.3	1

Mol. form.	Name	Temperature in °C for the indicated pressure						Ref.
		1 Pa	10 Pa	100 Pa	1 kPa	10 kPa	100 kPa	
$C_{10}H_{18}$	trans-Decahydronaphthalene		−10 e	18 e	55.3	107.9	186.8	1
$C_{10}H_{18}O$	α-Terpineol			48	89	142	217	4
$C_{10}H_{18}O$	Eucalyptol			10.6	48.5	100.3	175.4	5
$C_{10}H_{18}O$	trans-Geraniol	4 e	31 e	63.2	104.3	157.7	229.6	5
$C_{10}H_{18}O_4$	Sebacic acid	125.9 s						5
$C_{10}H_{18}O_4$	Dipropyl succinate	11 e	38 e	72.1	115.4	172.3	250.4	5
$C_{10}H_{18}O_4$	Diethyl adipate	4 e	35 e	72 e	116.6	171.2	239.5	5
$C_{10}H_{19}N$	Decanenitrile	13 e	36 e	66 e	105.8	160.6	241.6	5
$C_{10}H_{20}$	1-Decene	−35.5	−13.7	13.7	49.0	97.9	170.1	1,5
$C_{10}H_{20}$	Cyclodecane			29 e	68 e	121.3	201.8	1
$C_{10}H_{20}$	Butylcyclohexane	−31 e	−9 e	18 e	54 e	104.7	180.4	5
$C_{10}H_{20}$	Isobutylcyclohexane	−37 e	−16 e	10 e	46 e	95.9	170.8	5
$C_{10}H_{20}$	tert-Butylcyclohexane	−39 e	−18 e	9 e	45 e	95.3	171.1	5
$C_{10}H_{20}O$	Decanal		16 e	47.2	86.3	137.7	208.0	5
$C_{10}H_{20}O_2$	Decanoic acid	58 e	80 e	108 e	145 e	195.2	269.5	5
$C_{10}H_{20}O_2$	Octyl acetate	−26 e	−3 e	27 e	66.3	120.0	198.2	5
$C_{10}H_{20}O_2$	2-Ethylhexyl acetate	−11 e	5 e	26 e	57.6	107.1	197.2	5
$C_{10}H_{20}O_2$	Isopentyl isopentanoate			22 e	62.8	116.9	193.6	5
$C_{10}H_{20}O_2$	Ethyl octanoate	−17 e	9 e	41 e	81.4	133.2	203 e	5
$C_{10}H_{20}O_4$	Diethylene glycol monobutyl ether acetate	6 e	34 e	69 e	112.6	169.2	245.4	5
$C_{10}H_{21}Br$	1-Bromodecane	9 e	33 e	63 e	104 e	159.2	240.0	5
$C_{10}H_{21}Cl$	1-Chlorodecane	2 e	25 e	54 e	92 e	145.7	225.3	5
$C_{10}H_{21}F$	1-Fluorodecane	−22 e	0 e	27 e	64 e	113.3	185.7	5
$C_{10}H_{22}$	Decane		−10.6	16.7	52.3	101.1	173.7	16
$C_{10}H_{22}$	2-Methylnonane	−34 e	−14 e	12 e	47 e	94.8	166.5	5
$C_{10}H_{22}$	3-Methylnonane	−34 e	−14 e	12 e	47 e	95.1	167.3	5
$C_{10}H_{22}$	4-Methylnonane	−36 e	−16 e	10 e	45 e	93.1	165.2	5
$C_{10}H_{22}$	5-Methylnonane	−36 e	−16 e	10 e	45 e	92.6	164.6	5
$C_{10}H_{22}$	2,4-Dimethyloctane				38 e	84.9	155.4	5
$C_{10}H_{22}$	2,7-Dimethyloctane	−39 e	−19 e	7 e	41 e	88.4	159.4	5
$C_{10}H_{22}$	2,2,6-Trimethylheptane	−46 e	−27 e	−2 e	32 e	78.5	148.4	5
$C_{10}H_{22}$	3,3,5-Trimethylheptane			0 e	35 e	82.7	155.2	5
$C_{10}H_{22}$	2,2,3,3-Tetramethylhexane	−46 e	−25 e	1 e	36 e	85.6	159.8	5
$C_{10}H_{22}$	2,2,5,5-Tetramethylhexane			−10 e	22 e	68.3	137.0	5
$C_{10}H_{22}$	2,4-Dimethyl-3-isopropylpentane	−46 e	−26 e	0 e	35 e	83.2	156.5	5
$C_{10}H_{22}$	2,2,3,3,4-Pentamethylpentane		−24 e	3 e	39 e	89.1	165.5	5
$C_{10}H_{22}$	2,2,3,4,4-Pentamethylpentane		−29 e	−3 e	33 e	82.8	158.7	5
$C_{10}H_{22}O$	1-Decanol	30 e	50 e	75 e	109 e	157.3	230.6	1,39
$C_{10}H_{22}O$	4-Decanol	18 e	37 e	61 e	93 e	139 e	210 e	5
$C_{10}H_{22}O$	Dipentyl ether	−31 e	−8 e	22 e	60 e	111.6	186.2	5
$C_{10}H_{22}O$	Diisopentyl ether			14.0	51.5	101.8	172.8	5
$C_{10}H_{22}O_2$	Ethylene glycol dibutyl ether	0 e	20 e	44 e	78.4	127.1	202.9	5
$C_{10}H_{22}O_5$	Tetraethylene glycol dimethyl ether				138 e	200.9	275.3	5
$C_{10}H_{22}S$	1-Decanethiol	11 e	34 e	64 e	103 e	157.5	238.6	5
$C_{10}H_{22}S$	Diisopentylsulfide			7 e	82 e	118 e	139 e	5
$C_{10}H_{23}N$	Dipentylamine				77 e	127.7	202.0	5
$C_{10}H_{30}O_3Si_4$	Decamethyltetrasiloxane	−31 e	−6 e	26 e	66.8	118.8	193.9	5
$C_{10}H_{30}O_5Si_5$	Decamethylcyclopentasiloxane	−2 e	19 e	46 e	82 e	132.9	210.4	5
$C_{11}H_8O_2$	1-Naphthalenecarboxylic acid				191.9	239.3	299.6	5
$C_{11}H_8O_2$	2-Naphthalenecarboxylic acid				197.9	246.0	308.1	5
$C_{11}H_{10}$	1-Methylnaphthalene	5 e	29 e	60 e	102 e	159.1	244.1	1
$C_{11}H_{10}$	2-Methylnaphthalene			57 e	99 e	156.0	240.5	1
$C_{11}H_{12}O_2$	Ethyl trans-cinnamate			79	125	187	271	4
$C_{11}H_{12}O_3$	Myristicin	23 e	53 e	88.9	135.2	196.0	279.4	5
$C_{11}H_{14}$	4-Isopropylstyrene	−25 e	−1 e	30.2	70.3	124.5	202.1	5
$C_{11}H_{14}$	1,2,3,4-Tetrahydro-5-methylnaphthalene	9 e	31 e	60 e	99 e	153.1	233.8	5
$C_{11}H_{14}$	1,2,3,4-Tetrahydro-6-methylnaphthalene	17 e	36 e	62 e	97 e	147.8	228.5	5
$C_{11}H_{14}O_2$	Butyl benzoate	6 e	34 e	67.9	110.3	165 e	237 e	5
$C_{11}H_{16}$	Pentylbenzene	−14 e	8 e	37 e	74 e	126.7	204.9	5

		Temperature in °C for the indicated pressure						
Mol. form.	Name	1 Pa	10 Pa	100 Pa	1 kPa	10 kPa	100 kPa	Ref.
C$_{11}$H$_{16}$	*p-tert*-Butyltoluene	−24 e	−2 e	27 e	64.1	115.5	190.8	5
C$_{11}$H$_{16}$	1,3-Diethyl-5-methylbenzene	−26 e	−1 e	29.5	69.5	123.5	200.2	5
C$_{11}$H$_{16}$	2-Ethyl-1,3,5-trimethylbenzene		6 e	36 e	75.7	129.6	207.6	5
C$_{11}$H$_{16}$	1-Ethyl-2,4,5-trimethylbenzene	−13 e	11 e	40 e	79.4	132.1	207.7	5
C$_{11}$H$_{20}$	1-Undecyne	−22 e	0 e	29 e	67 e	118.5	194.5	5
C$_{11}$H$_{20}$	2-Undecyne	−17 e	6 e	35 e	74 e	127.4	205.4	5
C$_{11}$H$_{20}$O$_2$	10-Undecenoic acid	35 e	67 e	105 e	150.0	205.4	274.5	5
C$_{11}$H$_{20}$O$_4$	Ethyl diethylmalonate			74 e	105 e	149.4	219 e	5
C$_{11}$H$_{21}$N	Undecanenitrile			78.6	120.3	177.3	259.9	5
C$_{11}$H$_{22}$	1-Undecene	−21.6	1.2	29.7	66.4	117.1	192.2	5
C$_{11}$H$_{22}$	*cis*-2-Undecene	−14 e	7 e	34 e	70.2	120.6	196 e	5
C$_{11}$H$_{22}$	*trans*-2-Undecene	−14 e	7 e	33 e	69.3	119.6	195 e	5
C$_{11}$H$_{22}$	*cis*-4-Undecene	−19 e	3 e	30 e	66.6	117.1	192 e	5
C$_{11}$H$_{22}$	*trans*-4-Undecene	−17 e	4 e	31 e	67.1	117.4	193 e	5
C$_{11}$H$_{22}$	*cis*-5-Undecene	−19 e	2 e	30 e	66.2	116.7	191 e	5
C$_{11}$H$_{22}$	*trans*-5-Undecene	−18 e	3 e	31 e	67.0	117.4	192 e	5
C$_{11}$H$_{22}$	Pentylcyclohexane	−17 e	6 e	34 e	72 e	124.2	202.7	5
C$_{11}$H$_{22}$	Hexylcyclopentane	−15 e	7 e	36 e	73 e	125.0	202.5	5
C$_{11}$H$_{22}$O	2-Undecanone	17 e	37 e	64.3	103.0	153.6	232.6	1,5
C$_{11}$H$_{22}$O	6-Undecanone		28 e	57 e	95 e	148.4	226.9	1
C$_{11}$H$_{22}$O$_2$	Undecanoic acid	68 e	90 e	118 e	156 e	207.2	283.6	5
C$_{11}$H$_{22}$O$_2$	Heptyl butanoate	2 e	29 e	62 e	102.6	155.1	224.7	5
C$_{11}$H$_{22}$O$_2$	Propyl octanoate	−2 e	23 e	55 e	94.0	145.2	215 e	5
C$_{11}$H$_{22}$O$_2$	Methyl decanoate	10 e	33 e	62 e	100.9	154.0	232 e	5
C$_{11}$H$_{24}$	Undecane	−18.4	4.3	32.6	69.5	120.2	195.4	16
C$_{11}$H$_{24}$	2-Methyldecane	−20 e	1 e	28 e	64 e	114.0	188.7	5
C$_{11}$H$_{24}$	3-Methyldecane	−35 e	−10 e	22 e	61.9	115.6	190.4	5
C$_{11}$H$_{24}$	4-Methyldecane	−38 e	−12 e	20 e	60.8	113.9	186.4	5
C$_{11}$H$_{24}$	2,4,7-Trimethyloctane				43 e	94 e	170.4	5
C$_{11}$H$_{24}$O	1-Undecanol	52.2	80.0	82 e	118 e	167.6	244.1	5
C$_{11}$H$_{24}$S	1-Undecanethiol	23 e	47 e	77 e	118 e	173.6	256.8	5
C$_{12}$F$_{27}$N	Trinonafluorobutylamine		3 e	29.0	63.3	109.9	176.8	5
C$_{12}$H$_8$	Acenaphthylene	24 s	49.8 s	80.6 s				5
C$_{12}$H$_9$N	Carbazole					254.7	354.0	5
C$_{12}$H$_{10}$	Acenaphthene				126.2	187 e	276 e	1
C$_{12}$H$_{10}$	Biphenyl			69.0	111.1	169.5	254.7	1
C$_{12}$H$_{10}$N$_2$	Azobenzene			98.1	144.8	206.7	292.7	4
C$_{12}$H$_{10}$O	Diphenyl ether		44 e	75 e	116 e	173 e	257.4	5
C$_{12}$H$_{10}$O	1-Acetonaphthone	37 e	69 e	107.0	154.6	215.2	294.9	5
C$_{12}$H$_{10}$O	2-Acetonaphthone	48.3 s		118.7	163.0	221.1	300.3	5
C$_{12}$H$_{10}$S	Diphenyl sulfide	20 e	51 e	88.7	137.5	202.2	291.8	5
C$_{12}$H$_{11}$N	Diphenylamine	48 s		102.8	150.5	213.7	301.4	5
C$_{12}$H$_{12}$	1-Ethylnaphthalene	16 e	41 e	72 e	114 e	171.8	257.7	5
C$_{12}$H$_{12}$	2-Ethylnaphthalene	14 e	39 e	71 e	113 e	171.2	257.3	5
C$_{12}$H$_{12}$	1,2-Dimethylnaphthalene	26 e	51 e	82 e	123 e	180.5	265.7	5
C$_{12}$H$_{12}$	2,7-Dimethylnaphthalene	31.5 s	53.1 s	78.8 s	115.9	175 e	260 e	5
C$_{12}$H$_{14}$O$_4$	Diethyl phthalate	12 e	51 e	96 e	150.5	215.9	296.2	5
C$_{12}$H$_{16}$	*p*-Isopropenylisopropylbenzene	−11 e	15 e	46 e	87 e	142.4	221 e	5
C$_{12}$H$_{16}$	Cyclohexylbenzene		28 e	58 e	98 e	154.7	239.5	5
C$_{12}$H$_{16}$O$_2$	3-Methylbutyl benzoate			66 e	115.0	177.7	261.4	5
C$_{12}$H$_{18}$	Hexylbenzene	−2 e	22 e	51 e	90 e	144.5	225.5	5
C$_{12}$H$_{18}$	1,2-Diisopropylbenzene	−14 e	9 e	37 e	74 e	125.9	203.2	5
C$_{12}$H$_{18}$	1,3-Diisopropylbenzene	−14 e	8 e	36 e	74 e	125.5	202.6	5
C$_{12}$H$_{18}$	1,4-Diisopropylbenzene	−6 e	18 e	49 e	90 e	148.8	238 e	5
C$_{12}$H$_{18}$	Hexamethylbenzene	46.3 s	72.5 s	81.7 s	121.8 s	178.3	263.7	5
C$_{12}$H$_{18}$	1,5,9-Cyclododecatriene	−14 e	11 e	44 e	87 e	145.0	229.8	5
C$_{12}$H$_{20}$O$_2$	Geranyl acetate			67.7	110.8	166.9	242.9	5
C$_{12}$H$_{20}$O$_4$	Dibutyl maleate	12.3	50.4	94.0	144.2	203 e	272 e	5
C$_{12}$H$_{22}$	1-Dodecyne	−11 e	13 e	43 e	82 e	135.8	214.4	5
C$_{12}$H$_{22}$	Cyclohexylcyclohexane		20 e	53.1	96.0	154.1	237.2	5

Mol. form.	Name	Temperature in °C for the indicated pressure						Ref.
		1 Pa	10 Pa	100 Pa	1 kPa	10 kPa	100 kPa	
$C_{12}H_{22}O_2$	Methyl 10-undecenoate	10 e	38 e	73 e	116 e	172.2	247.1	5
$C_{12}H_{22}O_4$	Dimethyl sebacate		53 e	97	150	214	293	4
$C_{12}H_{23}N$	Dodecanenitrile	36 e	60 e	92 e	133 e	190.5	275.5	5
$C_{12}H_{24}$	1-Dodecene	−8.3	15.2	44.8	82.9	135.4	212.8	5
$C_{12}H_{24}$	Hexylcyclohexane	−3 e	20 e	50 e	89 e	143.1	224.2	5
$C_{12}H_{24}$	Heptylcyclopentane	−1 e	22 e	51 e	90 e	143.5	223.5	5
$C_{12}H_{24}O$	Dodecanal				70 e	116.2	175.9	256.6	5
$C_{12}H_{24}O_2$	Dodecanoic acid	78 e	100 e	128 e	166 e	219.1	298.1	5
$C_{12}H_{24}O_2$	Decyl acetate	12 e	40 e	74 e	115.1	168.1	238 e	5
$C_{12}H_{24}O_2$	Ethyl decanoate	8 e	35 e	69 e	111.8	166.1	238 e	5
$C_{12}H_{25}Br$	1-Bromododecane	31 e	57 e	90 e	132 e	190.8	275.3	5
$C_{12}H_{25}Cl$	1-Chlorododecane	27 e	51 e	81 e	122 e	178.7	262.6	5
$C_{12}H_{26}$	Dodecane	−5.4	18.2	47.6	85.8	138.2	215.8	16
$C_{12}H_{26}O$	1-Dodecanol				133 e	185.0	264.1	1
$C_{12}H_{26}O_3$	Diethylene glycol dibutyl ether	5 e	34.4	70.2	115.3	174.1	253.8	5
$C_{12}H_{27}N$	Tributylamine	−26 e	1 e	35 e	77.7	134.5	213.4	5
$C_{12}H_{27}N$	Triisobutylamine		1 e	28.9	64.9	112.5	178.5	5
$C_{12}H_{27}O_4P$	Tributyl phosphate					205 e	288.3	5
$C_{12}H_{36}O_6Si_6$	Dodecamethylcyclohexasiloxane	18 e	41 e	69 e	108 e	162.2	244.7	5
$C_{13}H_9N$	Acridine			124.4	176.2	246.0	345.4	5
$C_{13}H_9N$	Phenanthridine	79 s						5
$C_{13}H_{10}$	Fluorene	48.4 s			137.4	205.4	295 e	5
$C_{13}H_{10}O_2$	Phenyl benzoate			102.3	151.4	217.9	313.3	5
$C_{13}H_{10}O_3$	Phenyl salicylate				166.0	224.8	312.4	5
$C_{13}H_{12}$	Diphenylmethane		45 e	77 e	119.3	177.7	263.6	1,5
$C_{13}H_{13}N$	Methyldiphenylamine	35 e	63 e	98.4	143.1	201.6	281.6	5
$C_{13}H_{14}$	1-Isopropylnaphthalene	27 e	51 e	82 e	123.2	180.8	267.3	5
$C_{13}H_{20}$	Heptylbenzene	12 e	36 e	66 e	107 e	162.7	246.2	5
$C_{13}H_{24}O_2$	Ethyl 10-undecenoate	32 e	55 e	86 e	125.2	179.5	258.4	5
$C_{13}H_{26}$	1-Tridecene	4.1	28.5	59.0	98.3	152.5	232.3	5
$C_{13}H_{26}$	Heptylcyclohexane	11 e	34 e	65 e	105 e	160.9	244.3	5
$C_{13}H_{26}$	Octylcyclopentane	13 e	36 e	66 e	106 e	160.9	243.1	5
$C_{13}H_{26}O_2$	Tridecanoic acid	87 e	109 e	138 e	176 e	230.3	311.5	5
$C_{13}H_{26}O_2$	Methyl dodecanoate	38 e	61 e	90 e	130 e	184.9	269 e	5
$C_{13}H_{28}$	Tridecane	7.2	31.5	61.8	101.1	155.1	234.9	16
$C_{13}H_{28}O$	1-Tridecanol	71.6	101.0	103 e	140 e	192.3	273.1	5
$C_{14}H_{10}$	Anthracene	89.2 s	125.9 s	151.5 s	165 s	238.8	340.2	1,5
$C_{14}H_{10}$	Phenanthrene	53 s	83 s	120.8	170.4	238.4	337.7	5
$C_{14}H_{10}O_2$	Benzil			123	175	246	346	4
$C_{14}H_{12}$	cis-Stilbene	26 e	54 e	88 e	130.4	183 e	253 e	5
$C_{14}H_{12}$	trans-Stilbene				155.6	218.1	305.8	5
$C_{14}H_{12}O_2$	Benzoin				181	248	342	4
$C_{14}H_{14}$	1,1-Diphenylethane	19 e	47 e	82.0	125.3	181 e	254 e	5
$C_{14}H_{15}N$	Dibenzylamine	48 e	77 e	113.1	158.9	218.5	299.4	5
$C_{14}H_{16}$	1-Butylnaphthalene	67 e	82 e	103 e	135 e	186.7	288.6	5
$C_{14}H_{16}$	2-Butylnaphthalene	44 e	67 e	98 e	139 e	197.5	287.4	5
$C_{14}H_{22}$	Octylbenzene	20.1	46.2	79.1	121.9	178.1	263.8	5
$C_{14}H_{26}O_4$	Diethyl sebacate		83 e	120	166	225	305	4
$C_{14}H_{27}N$	Tetradecanenitrile	52 e	79 e	114.0	159.0	219.7	306.3	5
$C_{14}H_{28}$	1-Tetradecene	16.1	41.3	72.7	113.2	168.7	250.6	5
$C_{14}H_{28}$	Octylcyclohexane	16.9	44.3	77.8	120.0	177.6	263.2	5
$C_{14}H_{28}$	Nonylcyclopentane	25 e	49 e	80 e	120 e	177.2	261.5	5
$C_{14}H_{28}O_2$	Tetradecanoic acid	96 e	118 e	147 e	186 e	241.3	325.6	5
$C_{14}H_{30}$	Tetradecane	19.1	44.1	75.3	115.7	171.1	253.0	16
$C_{14}H_{30}O$	1-Tetradecanol	80.0	110.5	149.6	152 e	205.3	286.7	5
$C_{14}H_{31}N$	Tetradecylamine			104 e	147 e	206.1	290.9	5
$C_{14}H_{42}O_5Si_6$	Tetradecamethylhexasiloxane	6 e	36 e	72 e	117 e	176.0	259.1	5
$C_{15}H_{18}$	1-Pentylnaphthalene	34 e	62 e	96 e	141.3	202.2	289 e	5
$C_{15}H_{24}$	Nonylbenzene	33.0	58.9	92.0	135.4	193.7	281.4	5
$C_{15}H_{30}$	Nonylcyclohexane	35 e	60 e	92 e	134 e	193.4	280.9	5

Mol. form.	Name	Temperature in °C for the indicated pressure						Ref.
		1 Pa	10 Pa	100 Pa	1 kPa	10 kPa	100 kPa	
$C_{15}H_{30}$	Decylcyclopentane	37 e	61 e	93 e	134 e	192.5	278.8	5
$C_{15}H_{30}O_2$	Methyl tetradecanoate		75 e	110	155	214	295	4
$C_{15}H_{32}$	Pentadecane	30.5	56.1	88.1	129.6	186.3	270.1	16
$C_{16}H_{22}O_4$	Dibutyl phthalate		104.0	142.7	191.5	254.5	339.4	4
$C_{16}H_{32}$	1-Hexadecene	38.4	65.0	98.1	140.5	198.8	284.3	5
$C_{16}H_{32}O_2$	Hexadecanoic acid		136 e	165 e	205 e	261.9	350.2	5
$C_{16}H_{34}$	Hexadecane	41.1	67.4	100.3	142.7	200.7	286.3	16
$C_{16}H_{34}O$	1-Hexadecanol	99.5	130.6	171.9	175 e	229.0	311.7	5
$C_{16}H_{35}N$	Hexadecylamine	63 e	91 e	126 e	171 e	232.6	320.5	5
$C_{17}H_{10}O$	Benzanthrone		184 e	229.3	290.3	377.2	511 e	5
$C_{17}H_{34}O_2$	Methyl hexadecanoate	65 e	93	129	177			4
$C_{17}H_{36}$	Heptadecane	51.5	78.5	112.0	155.3	214.5	302 e	16
$C_{17}H_{36}O$	1-Heptadecanol	94 e	117 e	146 e	185 e	240.1	323.3	5
$C_{18}H_{14}$	o-Terphenyl	66 e	94 e	129 e	176 e	241.3	336.3	5
$C_{18}H_{14}$	m-Terphenyl	87 e	118 e	156 e	206.6	275.3	374.6	5
$C_{18}H_{14}$	p-Terphenyl	127.1 s	154.7 s		217.2	284.0	383.0	5
$C_{18}H_{30}$	Hexaethylbenzene				144.1	206.8	297.5	5
$C_{18}H_{34}O_2$	Oleic acid	94 e	126 e	165.5	214.5	277.0	359.7	5
$C_{18}H_{34}O_2$	Elaidic acid		124 e	166	216	280	361	4
$C_{18}H_{36}O$	Stearaldehyde			142 e	186 e	246.9	336.7	5
$C_{18}H_{36}O_2$	Stearic acid		153 e	183 e	223 e	281.6	374.5	5
$C_{18}H_{38}$	Octadecane	61.5	89.0	123.1	167.3	227.6	316 e	16
$C_{18}H_{38}O$	1-Octadecanol	106 e	130 e	160 e	200.5	257.3	343.0	5
$C_{19}H_{16}$	Triphenylmethane	81 s		112 e	175 e	254.6	360.0	5
$C_{19}H_{36}O_2$	Methyl oleate	85 e	114 e	149.7	195.6	256 e	340 e	5
$C_{19}H_{40}$	Nonadecane	71.1	99.1	133.8	178.8	240.1	330 e	16
$C_{20}H_{42}$	Eicosane	80.4	108.9	144.2	189.8	252.1	344 e	16
$C_{20}H_{42}O$	1-Eicosanol	119 e	143 e	173 e	213 e	270.0	355.1	5
$C_{20}H_{60}O_8Si_9$	Eicosamethylnonasiloxane			141 e	183.1	236.7	307.1	5
$C_{21}H_{21}O_4P$	Tri-o-cresyl phosphate	119.0	156.1	201.0	256.3	326.3	418 e	5
$C_{21}H_{21}O_4P$	Tri-m-cresyl phosphate	147.8	177.3	211.4	251.3	298 e	355 e	5
$C_{21}H_{21}O_4P$	Tri-p-cresyl phosphate	140.6	174 e	214 e	262 e	320 e	392 e	5
$C_{21}H_{44}$	Heneicosane	82.3	113.5	152.2	201.6	263.8	355.9	5
$C_{22}H_{42}O_2$	Brassidic acid	134 e	166 e	203.6	249.8	307.6	382.0	5
$C_{22}H_{42}O_2$	Erucic acid	126 e	160 e	199.4	247.4	306.5	381.1	5
$C_{22}H_{42}O_2$	Butyl oleate	95.5	124.2	158 e	198 e	245 e	304 e	5
$C_{22}H_{44}O_2$	Behenic acid	145.4	176.5	213.7	259.3	316.2	390 e	5
$C_{22}H_{44}O_2$	Butyl stearate	99.6	128 e	162 e	201 e	249 e	307 e	5
$C_{22}H_{46}$	Docosane	83.5	115.0	154.0	203.6	274.8	368.0	5
$C_{23}H_{48}$	Tricosane	102.9	135.1	174.8	221 e	285.3	379.5	5
$C_{24}H_{38}O_4$	Dioctyl phthalate	130 e	163.7	203.8	252 e	311 e	385 e	5
$C_{24}H_{38}O_4$	Bis(2-ethylhexyl) phthalate	122.0	153.2	189.2	231.3	281.1	341.1	5
$C_{24}H_{50}$	Tetracosane	115.0	148.1	188.5	239.1	295.4	390.6	5
$C_{25}H_{52}$	Pentacosane	119.7	152.7	193.2	244.4	305.0	401.1	5
$C_{26}H_{54}$	Hexacosane	125.1	158.8	200.1	252.1	314.3	411.3	5
$C_{27}H_{56}$	Heptacosane	136.7	168.8	206.5	255.8	323.3	421.2	5
$C_{28}H_{58}$	Octacosane	136.5	169.8	210.9	263.1	332.0	430.6	5
$C_{29}H_{60}$	Nonacosane	148.2	182.8	221.2	271.5	340.2	439.7	5
$C_{30}H_{62}$	Squalane	66 e	84 e	105.8	131.9	163.7	203.2	5
C_{70}	Carbon (fullerene-C_{70})	598 s	662 s					22

VAPOR PRESSURE OF FLUIDS AT TEMPERATURES BELOW 300 K

This table gives vapor pressures of 67 important fluids in the temperature range 2 to 300 K. Helium (^{4}He), hydrogen (H$_2$), and neon (Ne) are covered on this page. The remaining fluids are listed on subsequent pages by molecular formula in the Hill order (see Introduction). The data have been taken from evaluated sources; references are listed at the end of the table.

Pressures are given in kilopascals (kPa). Note that:
  1 kPa = 7.50062 Torr
  100 kPa = 1 bar
  101.325 kPa = 1 atmos
"s" following an entry indicates that the compound is solid at that temperature.

Helium		Hydrogen		Neon		Helium		Hydrogen		Neon	
T/K	*P*/kPa	*T*/K	*P*/kPa	*T*/K	*P*/kPa	*T*/K	*P*/kPa	*T*/K	*P*/kPa	*T*/K	*P*/kPa
2.2	5.3	14.0	7.90	25.0	51.3	4.2	99.0	24.0	264.2		
2.3	6.7	14.5	10.38	26.0	71.8	4.3	108.7	24.5	295.1		
2.4	8.3	15.0	13.43	27.0	98.5	4.4	119.0	25.0	328.5		
2.5	10.2	15.5	17.12	28.0	132.1	4.5	129.9	25.5	364.3		
2.6	12.4	16.0	21.53	29.0	173.5	4.6	141.6	26.0	402.9		
2.7	14.8	16.5	26.74	30.0	223.8	4.7	153.9	26.5	444.3		
2.8	17.5	17.0	32.84	31.0	284.0	4.8	167.0	27.0	488.5		
2.9	20.6	17.5	39.92	32.0	355.2	4.9	180.8	27.5	535.7		
3.0	24.0	18.0	48.08	33.0	438.6	5.0	195.4	28.0	586.1		
3.1	27.8	18.5	57.39	34.0	535.2	5.1	210.9	28.5	639.7		
3.2	32.0	19.0	67.96	35.0	646.2			29.0	696.7		
3.3	36.5	19.5	79.89	36.0	772.8			29.5	757.3		
3.4	41.5	20.0	93.26	37.0	916.4			30.0	821.4		
3.5	47.0	20.5	108.2	38.0	1078			30.5	889.5		
3.6	52.9	21.0	124.7	39.0	1260			31.0	961.5		
3.7	59.3	21.5	143.1	40.0	1462			31.5	1038.0		
3.8	66.1	22.0	163.2	41.0	1688			32.0	1119.0		
3.9	73.5	22.5	185.3	42.0	1939			32.5	1204.0		
4.0	81.5	23.0	209.4	43.0	2216	Ref.	17,18		1		13
4.1	90.0	23.5	235.7	44.0	2522						

| | Ar | | BCl$_3$ | BF$_3$ | BrH | | Br$_2$ | ClF | ClH | |
T/K	Argon		Boron trichloride	Boron trifluoride	Hydrogen bromide		Bromine	Chlorine fluoride	Hydrogen chloride	
50	0.1	s								
55	0.2	s								
60	0.8	s								
65	2.8	s								
70	7.7	s								
75	18.7	s								
80	40.7	s								
85	79.0									
90	134									
95	213									
100	324									
105	473									
110	666									
115	910							0.1		
120	1214							0.3	0.1	s
125	1584							0.6	0.3	s
130	2027							1.2	0.5	s
135	2553				0.1	s		2.1	1.0	s
140	3170				0.3	s		3.6	1.9	s
145	3892			7.7	0.6	s		6.0	3.4	s
150	4736			13.4	1.1	s		9.5	5.8	s
155				22.3	1.9	s		14.6	9.5	s
160				35.2	3.3	s		21.8	14.7	
165				53.7	5.4	s		31.7	22.0	
170				79.1	8.7	s		44.8	31.9	
175				113	13.4	s		62.0	45.1	

T/K	Ar Argon	BCl$_3$ Boron trichloride	BF$_3$ Boron trifluoride	BrH Hydrogen bromide		Br$_2$ Bromine		ClF Chlorine fluoride	ClH Hydrogen chloride
180		0.1	157	20.1	s			84.2	62.5
185		0.2	214	29.5	s			112	84.7
190		0.3	285	37.9				147	113
195		0.5	372	51.8				190	148
200		0.8	479	69.5				242	190
205		1.2	608	91.8				304	242
210		1.8	762	119				378	304
215		2.6	944	153				464	377
220		3.8	1160	194		0.1	s	564	463
225		5.2	1413	242		0.2	s	680	563
230		7.2	1709	299		0.3	s	812	678
235		9.7	2056	366		0.4	s	961	811
240		12.9	2460	443		0.7	s	1130	961
245		17.0	2913	532		1.1	s	1319	1132
250		22.0	3481	633		1.7	s	1529	1325
255		28.1	4123	748		2.6	s	1762	1542
260		35.6	4874	878		3.8	s	2019	1784
265		44.5		1023		5.5	s	2301	2054
270		55.1		1185		7.3		2608	2354
275		67.6		1364		9.5		2941	2686
280		82.2		1562		12.3		3303	3053
285		99.1		1780		15.6		3693	3457
290		119		2018		19.7		4111	3901
295		141		2278		24.6		4560	4388
300		166		2561		30.5		5039	4921
Ref.	8,15	12	12	12		12		12	12

T/K	ClO$_2$ Chlorine dioxide	Cl$_2$ Chlorine	Cl$_4$Si Silicon tetrachloride	FH Hydrogen fluoride	F$_2$ Fluorine	F$_2$O Difluorine oxide	F$_3$N Nitrogen trifluoride
50							
55					0.4		
60					1.5		
65					4.8		
70					12.3		
75					27.6	0.1	
80					55.3	0.2	
85					101	0.5	0.1
90					172	1.2	0.2
95					276	2.6	0.4
100					420	5.3	0.9
105					615	10.1	2.0
110					870	18.0	4.0
115					1196	30.5	7.3
120					1605	49.3	12.8
125					2108	76.7	21.1
130					2721	115	33.5
135					3458	168	51.1
140					4339	237	75.4
145						328	108
150						444	150
155						588	205
160						766	273
165						981	357
170						1238	459
175		1.8				1541	581
180		2.8				1895	726
185		4.2				2303	896
190		6.1		0.3		2771	1092
195	0.1	8.7		0.5		3302	1319
200	0.3	12.3		0.8		3899	1578
205	0.5	16.9		1.2		4567	1871
210	0.9	22.9	0.1	1.7		5308	2203

T/K	ClO_2 Chlorine dioxide	Cl_2 Chlorine	Cl_4Si Silicon tetrachloride	FH Hydrogen fluoride	F_2 Fluorine	F_2O Difluorine oxide	F_3N Nitrogen trifluoride
215	1.4	30.5	0.2	2.3			2577
220	2.3	40.1	0.3	3.2			2995
225	3.5	51.9	0.5	4.4			3464
230	5.3	66.4	0.7	5.9			3991
235	7.6	84.0	1.0	7.9			
240	10.8	105	1.5	10.3			
245	14.9	130	2.0	13.4			
250	20.1	160	2.8	17.2			
255	26.6	194	3.8	21.8			
260	34.6	234	5.0	27.4			
265	44.4	280	6.6	34.2			
270	56.1	332	8.6	42.2			
275	69.9	392	11.1	51.8			
280	86.2	459	14.2	63.1			
285	105	535	17.9	76.3			
290	127	619	22.3	91.7			
295	151	714	27.7	110			
300	179	818	34.0	130			
Ref.	12	5	12	12	12	12	1

T/K	F_3P Phosphorus trifluoride	F_4Si Silicon tetrafluoride		F_6S Sulfur hexafluoride		HI Hydrogen iodide		H_2S Hydrogen sulfide		H_3N Ammonia		H_3P Phosphine
105	0.1											
110	0.2											0.1
115	0.5											0.2
120	1.0											0.4
125	1.9	0.1	s									0.7
130	3.5	0.2	s									1.3
135	5.9	0.4	s					0.1	s			2.3
140	9.5	0.9	s	0.1	s			0.2	s			3.9
145	14.9	1.9	s	0.2	s			0.3	s			6.2
150	22.5	3.8	s	0.4	s			0.6	s			9.6
155	33.1	7.5	s	0.8	s	0.1	s	1.1	s			14.5
160	47.3	14.0	s	1.5	s	0.2	s	1.9	s	0.1	s	21.1
165	66.0	25.2	s	2.6	s	0.4	s	3.2	s	0.2	s	30.0
170	90.1	43.8	s	4.4	s	0.8	s	5.2	s	0.3	s	41.6
175	121	74.2	s	7.1	s	1.3	s	8.3	s	0.6	s	56.6
180	159	122	s	11.3	s	2.2	s	12.7	s	1.2	s	75.6
185	206	197	s	17.3	s	3.4	s	18.9	s	2.1	s	99.2
190	262	280		25.9	s	5.3	s	26.6		3.5	s	128
195	330	376		38.0	s	8.0	s	36.7		5.8	s	163
200	410	488		54.4	s	11.7	s	49.8		8.7		205
205	503	618		76.6	s	16.8	s	66.4		12.6		254
210	611	766		106	s	23.6	s	87.1		17.9		312
215	736	932		145	s	32.5	s	113		24.9		379
220	877	1117		195	s	44.0	s	144		34.1		456
225	1037	1324		249		56.2		182		45.9		544
230	1217	1555		305		71.4		227		60.8		644
235	1418	1816		371		89.7		281		79.6		756
240	1640	2111		448		112		344		103		881
245	1885	2449		536		137		416		131		1019
250	2154	2841		636		168		500		165		1172
255	2448	3301		750		203		597		207		1341
260	2767			878		244		706		256		1525
265	3112			1021		290		830		313		1725
270				1181		343		969		381		1942
275				1358		404		1124		460		2176
280				1554		472		1297		552		2428
285				1768		548		1488		655		2699
290				2003		633		1698		774		2987
295				2258		727		1929		909		3295
300				2534		831		2181		1062		3621
Ref.	12	12		12,15		12		12,15		11		12

T/K	H₄Si Silane	Kr Krypton		NO Nitric oxide		N₂ Nitrogen		N₂O Nitrous oxide	O₂ Oxygen	O₂S Sulfur dioxide
50						0.4	s			
55						1.8	s		0.2	
60						6.3	s		0.7	
65						17.4			2.3	
70						38.6			6.3	
75		0.1	s			76.1			14.5	
80		0.4	s			137			30.1	
85		1.1	s	0.1	s	229			56.8	
90		2.7	s	0.4	s	361			99.3	
95	0.1	6.0	s	1.3	s	541			163	
100	0.2	12.1	s	3.8	s	779			254	
105	0.4	22.8	s	10.0	s	1084			379	
110	1.0	40.4	s	23.5		1467			543	
115	1.9	68.0	s	46.8		1939		0.1	756	
120	3.5	103		86.5		2513		0.1	1022	
125	6.1	150		151		3209		0.3	1351	
130	10.0	211		248				0.7	1749	
135	15.8	290		391				1.3	2225	
140	24.1	388		592				2.5	2788	
145	35.3	509		867				4.3	3448	
150	50.3	655		1231				7.1	4219	
155	69.8	830		1703				11.4		
160	94.6	1037		2302				17.6		
165	126	1278		3050				26.4		
170	164	1557		3971				38.5		0.1
175	210	1877		5089				54.7		0.2
180	265	2241		6433				75.9		0.3
185	331	2655						103		0.5
190	408	3120						138		0.8
195	498	3641						181		1.3
200	602	4223						234		2.0
205	722	4870						298		3.0
210	859							374		4.4
215	1017							465		6.3
220	1196							571		9.0
225	1398							694		12.6
230	1628							835		17.3
235	1888							996		23.3
240	2180							1179		31.1
245	2509							1385		40.9
250	2880							1615		53.2
255	3296							1870		68.3
260	3763							2152		86.7
265	4288							2462		109
270								2802		136
275								3172		168
280								3573		205
285								4006		249
290								4473		300
295								4973		359
300								5508		426
Ref.	12	13, 15		12, 15		1		12	3	12

T/K	O₃ Ozone	Rn Radon	Xe Xenon		CBrF₃ Bromotri-fluoromethane	CClF₃ Chlorotri-fluoromethane	CCl₂F₂ Dichlorodi-fluoromethane	CCl₃F Trichloro-fluoromethane
100	0.1		0.1	s				
105	0.2		0.1	s				
110	0.4		0.3	s				
115	1.0		0.7	s		0.1		
120	2.0		1.5	s		0.2		
125	3.8		2.7	s		0.3		
130	6.8	0.1	4.9	s		0.6		

T/K	O₃ Ozone	Rn Radon	Xe Xenon		CBrF₃ Bromotri-fluoromethane	CClF₃ Chlorotri-fluoromethane	CCl₂F₂ Dichlorodi-fluoromethane	CCl₃F Trichloro-fluoromethane
135	11.5	0.3	8.5	s	0.1	1.1		
140	18.7	0.5	14.0	s	0.3	2.0		
145	29.1	0.9	22.2	s	0.5	3.3		
150	43.7	1.5	34.2	s	0.9	5.3		
155	63.6	2.4	51.1	s	1.5	8.3	0.1	
160	89.9	3.8	74.2	s	2.5	12.6	0.3	
165	124	5.8	101		3.9	18.6	0.5	
170	168	8.6	134		5.9	26.8	0.8	
175	222	12.5	173		8.8	37.6	1.3	
180	289	17.7	222		12.8	51.7	2.1	
185	367	24.5	280		18.1	69.7	3.2	
190	468	33.2	348		25.1	92.3	4.8	0.2
195	584	44.4	428		34.1	120	6.9	0.3
200	721	58.2	521		45.6	155	9.9	0.4
205	881	75.3	628		60.0	196	13.7	0.6
210	1068	96	750		77.8	246	18.8	1.0
215	1285	121	889		99.5	304	25.2	1.4
220	1536	151	1045		126	372	33.3	2.0
225	1824	185	1220		157	451	43.3	2.9
230	2155		1416		194	542	55.5	4.1
235	2534		1633		237	646	70.4	5.6
240	2968		1872		287	763	88.1	7.6
245	3464		2136		344	896	109	10.1
250	4031		2425		410	1044	134	13.3
255	4678		2742		485	1210	163	17.2
260	5417		3087		570	1394	196	22.1
265			3462		665	1598	234	28.0
270			3869		771	1823	278	35.1
275			4310		889	2071	327	43.7
280			4786		1021	2343	383	53.8
285			5299		1166	2641	445	65.7
290					1325	2968	515	79.6
295					1501	3325	593	95.6
300					1692	3716	679	114.1
Ref.	12	15	12,13		12	12	12	12

T/K	CCl₄ Tetrachloro-methane	CF₄ Tetrafluoro-methane	CO Carbon monoxide		COS Carbon oxysulfide	CO₂ Carbon dioxide		CHClF₂ Chlorodifluoro-methane	CHCl₃ Trichloro-methane
50			0.1	s					
55			0.6	s					
60			2.6	s					
65			8.2	s					
70			21.0						
75			44.4						
80			83.7						
85			147						
90		0.1	239						
95		0.3	371						
100		0.8	545						
105		1.7	771						
110		3.4	1067						
115		6.5	1428						
120		11.5	1877						
125		19.3	2400						
130		30.8	3064						
135		47.4				0.1	s		
140		70.2			0.1	0.2	s		
145		101			0.2	0.4	s		
150		141			0.4	0.8	s	0.1	
155		191			0.8	1.7	s	0.3	

T/K	CCl_4 Tetrachloro-methane	CF_4 Tetrafluoro-methane	CO Carbon monoxide	COS Carbon oxysulfide	CO_2 Carbon dioxide		$CHClF_2$ Chlorodifluo-romethane	$CHCl_3$ Trichloro-methane
160		254		1.3	3.1	s	0.5	
165		332		2.2	5.7	s	0.8	
170		425		3.4	9.9	s	1.4	
175		537		5.2	16.8	s	2.3	
180		669		7.8	27.6	s	3.6	
185		824		11.3	44.0	s	5.5	
190		1005		15.9	68.4	s	8.1	
195		1216		22.1	104	s	11.8	
200		1460		30.0	155	s	16.7	
205		1743		40.1	227	s	23.1	
210		2073		52.7	327	s	31.5	
215		2457		68.2	465	s	42.1	0.1
220		2907		87.2	599		55.3	0.2
225		3438		110	734		71.7	0.3
230				137	893		91.6	0.4
235				169	1075		116	0.7
240				207	1283		144	1.0
245				250	1519		178	1.4
250				301	1785		218	2.0
255	1.5			358	2085		264	2.7
260	2.1			423	2419		317	3.7
265	2.8			497	2790		377	5.0
270	3.7			580	3203		446	6.6
275	4.9			673	3658		525	8.7
280	6.4			777	4161		613	11.3
285	8.2			892	4714		711	14.4
290	10.5			1019	5318		821	18.3
295	13.2			1159	5984		944	22.9
300	16.5			1313	6713		1080	28.5
Ref.	12	12	9	12	6, 19		12	12

T/K	CHF_3 Trifluoro-methane	CHN Hydrogen cyanide		CH_2Cl_2 Dichloro-methane	CH_2F_2 Difluoro-methane	CH_2O Formaldehyde	CH_3Cl Chloromethane	CH_3F Fluoromethane
120	0.1							
125	0.2							
130	0.4							
135	0.7							0.6
140	1.4				0.1			1.2
145	2.5				0.2			2.1
150	4.3				0.3			3.6
155	7.1				0.6			5.9
160	11.1				1.0			9.3
165	17.0				1.7			14.1
170	25.3				2.8			20.9
175	36.5				4.4			29.9
180	51.4				6.8			42.0
185	70.9				10.2	1.3	2.1	57.6
190	95.8				14.8	2.0	3.1	77.4
195	127				21.2	3.0	4.6	102
200	166	0.1	s	0.1	29.5	4.4	6.7	133
205	214	0.2	s	0.2	40.5	6.4	9.5	171
210	271	0.4	s	0.3	54.5	9.1	13.1	216
215	340	0.6	s	0.4	72.1	12.7	17.9	270
220	421	1	s	0.6	94.1	17.4	24.0	333
225	516	1.5	s	0.9	121	23.4	31.8	408
230	626	2.2	s	1.4	154	31.0	41.4	495
235	754	3.3	s	2.0	193	40.6	53.3	595
240	900	4.7	s	2.8	240	52.5	67.7	711
245	1067	6.8	s	3.8	295	67.0	85.1	843
250	1257	9.7	s	5.3	360	84.6	106	993
255	1472	13.6	s	7.1	434	106	131	1163
260	1713	18.8		9.5	521	131	159	1355

T/K	CHF_3 Trifluoro-methane	CHN Hydrogen cyanide	CH_2Cl_2 Dichloro-methane	CH_2F_2 Difluoro-methane	CH_2O Formaldehyde	CH_3Cl Chloromethane	CH_3F Fluoromethane
265	1984	24.1	12.4	620	161	193	1571
270	2287	30.5	16.1	732	196	232	1813
275	2624	38.3	20.7	860	236	277	2084
280	3000	47.7	26.3	1004	283	327	2387
285	3418	58.8	33.0	1165	337	385	2724
290	3881	72.1	41.1	1346	399	450	3099
295	4393	87.6	50.8	1547	470	524	3516
300		105.9	62.1	1770	549	606	3978
Ref.	12	12,16	12	12	12	12	12

T/K	CH_4 Methane	CH_4O Methanol	C_2H_2 Acetylene		C_2H_4 Ethylene	C_2H_6 Ethane	C_2H_6O Dimethyl ether	C_3H_4 Propadiene
65	0.1							
70	0.3							
75	0.8							
80	2.1							
85	4.9							
90	10.6							
95	20.0							
100	34.5							
105	57.0							
110	88.4				0.3			
115	133				0.8	0.1		
120	192				1.4	0.4		
125	269				2.7	0.7		
130	368		0.1	s	4.5	1.3		
135	491		0.3	s	7.7	2.2		
140	642		0.7	s	11.9	3.8		
145	824		1.3	s	18.3	6.0		
150	1041		2.6	s	27.5	9.7		0.1
155	1297		4.6	s	39.9	15.0	0.1	0.2
160	1594		7.8	s	56.4	21.5	0.2	0.3
165	1937		12.8	s	77.9	31.0	0.3	0.6
170	2331		20.6	s	105	42.9	0.5	1.0
175	2779		32.2	s	140	59.0	0.9	1.7
180	3288		49.0	s	182	78.7	1.4	2.7
185	3865		72.9	s	234	104	2.1	4.1
190	4520		106	s	296	135	3.2	6.1
195			146		369	172	4.7	8.9
200			190		456	217	6.8	12.5
205			244		557	271	9.6	17.4
210			309		673	334	13.3	23.7
215			385		806	407	18.1	31.6
220			475		958	492	24.3	41.4
225			579		1128	590	32.1	53.5
230		0.1	699		1321	700	41.9	68.2
235		0.2	837		1535	826	53.9	85.8
240		0.4	993		1774	967	68.6	107
245		0.5	1170		2039	1125	86.3	131
250		0.8	1370		2331	1301	108	160
255		1.2	1593		2652	1496	133	193
260		1.7	1843		3005	1712	162	230
265		2.4	2121		3391	1949	197	273
270		3.3	2429		3813	2210	237	322
275		4.5	2771		4275	2495	283	376
280		6.2	3150			2806	335	438
285		8.3	3567			3146	395	506
290		11	4028			3515	463	582
295		14.4	4535			3917	538	666
300		18.7	5093			4355	623	759
Ref.	2,16	12	12,16		4	2	12	12

T/K	C_3H_6 Propylene	C_3H_8 Propane	C_4H_6 Buta-1,3-diene	C_4H_{10} Butane	C_4H_{10} Isobutane	C_5H_{12} Pentane	C_5H_{12} Neopentane	
140	0.1							
145	0.2							
150	0.4							
155	0.7							
160	1.2	0.8			0.1			
165	2.0	1.4			0.1			
170	3.1	2.2	0.1	0.1	0.3			
175	4.7	3.3	0.2	0.2	0.4			
180	7.0	5.0	0.4	0.3	0.7			
185	10.1	7.3	0.6	0.5	1.1		0.1	s
190	14.2	10.5	1.0	0.8	1.7		0.2	s
195	19.7	15.0	1.5	1.3	2.5		0.4	s
200	26.9	20.1	2.3	1.9	3.7		0.7	s
205	35.9	27.0	3.4	2.8	5.3		1.1	s
210	47.3	36.0	4.8	4.0	7.4		1.6	s
215	61.3	47.0	6.7	5.7	10.2		2.4	s
220	78.5	60.0	9.2	7.8	13.8	1.0	3.6	s
225	99.2	77.0	12.5	10.6	18.3	1.5	5.2	s
230	124	97.0	16.7	14.1	24.0	2.1	7.3	s
235	153	120	21.9	18.5	31.1	3.0	10.2	s
240	188	148	28.4	24.1	39.8	4.2	13.9	s
245	228	180	36.3	30.9	50.3	5.7	18.7	s
250	274	218	46.0	39.1	62.9	7.6	24.8	s
255	327	261	57.6	49.1	77.8	10.0	32.4	s
260	387	311	71.3	61.0	95.4	13.0	41.6	
265	456	367	87.6	75.0	116	16.6	51.4	
270	533	431	107	91.5	140	21.1	63.0	
275	619	502	129	111	167	26.6	76.6	
280	715	582	154	133	198	33.1	92.3	
285	822	671	184	159	234	40.8	111	
290	940	769	217	188	274	50.0	131	
295	1069	878	255	221	319	60.7	155	
300	1212	998	297	258	370	73.2	182	
Ref.	7	2	12	2	2	14	12,16	

References

1. B. A. Younglove, Thermophysical properties of fluids. I. Ethylene, parahydrogen, nitrogen trifluoride, and oxygen, *J. Phys. Chem. Ref. Data*, 11, Supp. 1, 1982.
2. B. A. Younglove and J. F. Ely, Thermophysical properties of fluids. II. Methane, ethane, propane, isobutane, and normal butane, *J. Phys. Chem. Ref. Data*, 16, 577, 1987.
3. W. Wagner, et al., *International Tables for the Fluid State: Oxygen*, Blackwell Scientific Publications, Oxford, 1987.
4. R. T. Jacobsen, et al., *International Tables for the Fluid State: Ethylene*, Blackwell Scientific Publications, Oxford, 1988.
5. S. Angus, et al., *International Tables for the Fluid State: Chlorine*, Pergamon Press, Oxford, 1985.
6. S. Angus, et al., *International Tables for the Fluid State: Carbon Dioxide*, Pergamon Press, Oxford, 1976.
7. S. Angus, et al., *International Tables for the Fluid State: Propylene*, Pergamon Press, Oxford, 1980.
8. R. B. Stewart and R. T. Jacobsen, Thermophysical properties of argon, *J. Phys. Chem. Ref. Data*, 18, 639, 1989.
9. R. D. Goodwin, Carbon monoxide thermophysical properties, *J. Phys. Chem. Ref. Data*, 14, 849, 1985.
10. R. D. Goodwin, Methanol thermophysical properties, *J. Phys. Chem. Ref. Data*, 16, 799, 1987.
11. L. Haar, Thermodynamic properties of ammonia, *J. Phys. Chem. Ref. Data*, 7, 635, 1978.
12. DIPPR Data Compilation of Pure Compound Properties, Design Institute for Physical Properties Data, American Institute of Chemical Engineers, 1987.
13. V. A. Rabinovich, et al., *Thermophysical Properties of Neon, Argon, Krypton, and Xenon*, Hemisphere Publishing Corp., New York, 1987.
14. K. N. Marsh, *Recommended Reference Methods for the Realization of Physicochemical Properties*, Blackwell Scientific Publications, Oxford, 1987.
15. TRC Thermodynamic Tables: Non-Hydrocarbons, Thermodynamic Research Center, Texas A & M University, College Station, Texas, 1985.
16. R. M. Stevenson and S. Malanowski, *Handbook of the Thermodynamics of Organic Compounds*, Elsevier, New York, 1987.
17. S. Angus and K. M. de Reuck, *International Tables of the Fluid State: Helium-4*, Pergamon Press, Oxford, 1977.
18. R. D. McCarty, *J. Phys. Chem. Ref. Data*, 2, 923, 1973.
19. R. Span and W. Wagner, *J. Phys. Chem. Ref. Data*, 25, 1509, 1996.

VAPOR PRESSURE OF SATURATED SALT SOLUTIONS

This table gives the vapor pressure of water above saturated solutions of some common salts at ambient temperatures. Data on pure water are given on the last line for comparison.

The references provide additional information on water activity, osmotic coefficient, and enthalpy of vaporization.

References

1. Apelblat, A., *J. Chem. Thermodynamics*, 24, 619, 1992.
2. Apelblat, A., *J. Chem. Thermodynamics*, 25, 63, 1993.
3. Apelblat, A., *J. Chem. Thermodynamics*, 25, 1513, 1993.
4. Apelblat, A. and Korin, E., *J. Chem. Thermodynamics*, 30, 59, 1998.

	Vapor Pressure in kPa							
Salt	10°C	15°C	20°C	25°C	30°C	35°C	40°C	Ref.
$BaCl_2$	0.971	1.443	2.073	2.887	3.903	5.133	6.576	1
$Ca(NO_3)_2$	0.701	1.015	1.381	1.772	2.154	2.487		1
$CuSO_4$	1.113	1.574	2.189	2.996	4.037	5.363		3
$FeSO_4$	0.978	1.516	2.208	3.035	3.950	4.884		3
KBr	0.953	1.338	1.853	2.533	3.419	4.563		3
KIO_3	1.100	1.564	2.177	2.970	3.979	5.236	6.778	4
K_2CO_3	0.541	0.802	1.134	1.536	1.997	2.499	3.016	1
$LiCl$	0.128	0.193	0.279	0.384				2
$Mg(NO_3)_2$	0.726	0.999	1.339	1.749	2.231	2.782	3.397	1
$MnCl_2$	0.697	1.064	1.515	2.020	2.535	3.002		3
NH_4Cl	0.971	1.328	1.836	2.481				2
NH_4NO_3	0.853	1.152	1.524	1.972				2
$(NH_4)_2SO_4$	0.901	1.319	1.871	2.573	3.439	4.474		3
$NaBr$	0.722	1.004	1.376	1.858	2.475	3.255	4.229	4
$NaCl$	0.921	1.285	1.768	2.401	3.218	4.262	5.581	4
$NaNO_2$	0.703	0.994	1.381	1.888	2.540	3.368	4.403	4
$NaNO_3$	0.884	1.244	1.719	2.335	3.121	4.109	5.333	4
$RbCl$	0.862	1.215	1.684	2.298	3.088	4.089	5.343	4
$ZnSO_4$	0.945	1.401	1.986	2.698	3.523	4.431	5.382	1
Water	1.228	1.706	2.339	3.169	4.246	5.627	7.381	

IUPAC RECOMMENDED DATA FOR VAPOR PRESSURE CALIBRATION

These precise vapor pressure values are recommended as secondary standards. Values are given in kPa (1 kPa = 0.0098692 atm = 7.5006 Torr). Reprinted by permission of IUPAC.

References

1. Marsh, K.N., Ed., *Recommended Reference Materials for the Realization of Physicochemical Properties*, Blackwell Scientific Publications, Oxford, 1987.
2. Ruzicka, K., Fulem, M., and Ruzicka, V., "Recommended Vapor Pressure of Solid Naphthalene", *J. Chem. Eng. Data*, 50, 1956–1970, 2005.

T/K	$CO_2(s)$	$H_2O(s)$	$C_{10}H_8(s)$	$n\text{-}C_5H_{12}$	C_6H_6	C_6F_6	H_2O	Hg
180	27.62		$3.04 \cdot 10^{-11}$					
190	68.44		$4.33 \cdot 10^{-10}$					
200	155.11	0.0002	$4.69 \cdot 10^{-9}$					
210	327.17	0.0007	$4.02 \cdot 10^{-8}$					
220		0.0026	$2.82 \cdot 10^{-7}$					
230		0.0089	$1.66 \cdot 10^{-6}$					
240		0.0273	$8.37 \cdot 10^{-6}$					
250		0.0760	$3.69 \cdot 10^{-5}$	7.60				
260		0.1958	$1.446 \cdot 10^{-4}$	12.98				
270		0.4701	$5.09 \cdot 10^{-4}$	21.15			0.485	
280			0.00163	33.11	5.148	4.322	0.991	
290			0.004798	50.01	8.606	7.463	1.919	
300			0.01308	73.17	13.816	12.328	3.535	
310			0.03328	104.07	21.389	19.576	6.228	
320			0.07956	144.3	32.054	30.009	10.540	
330			0.1797	195.7	46.656	44.578	17.202	
340			0.3854	260.1	66.152	64.380	27.167	
350			0.7884	339.4	91.609	90.664	41.647	
360				435.9	124.192	124.816	62.139	
370				551.5	165.2	168.4	90.453	
380				688.8	215.9	223.0	128.74	
390				850.2	277.7	290.4	179.48	
400				1038	353.2	372.6	245.54	0.138
410				1256	441.0	471.5	330.15	0.215
420				1507	545.5	589.3	436.89	0.329
430				1793	667.6	728.3	569.73	0.493
440				2120	808.8	890.9	732.99	0.724
450				2490	971.1	1080	931.36	1.045
460				2910	1156	1297	1169.9	1.485
470					1366	1547	1453.9	2.078
480					1602	1833	1789.0	2.866
490					1868	2159	2181.4	3.899
500					2164	2530	2637.3	5.239
510					2494	2954	3163.3	6.955
520					2861		3766.4	9.131
530					3267		4453.9	11.861
540					3717		5233.5	15.256
550					4216		6113.4	19.438
560					4770		7102.0	25.547
570							8208.6	30.74
580							9443.0	38.19
590							10816	47.09
600							12339	57.64

ENTHALPY OF VAPORIZATION

The molar enthalpy (heat) of vaporization $\Delta_{vap}H$, which is defined as the enthalpy change in the conversion of one mole of liquid to gas at constant temperature, is tabulated here for approximately 850 inorganic and organic compounds. Values are given, when available, both at the normal boiling point t_b, referred to a pressure of 101.325 kPa (760 mmHg), and at 25°C. Substances are listed by molecular formula in the modified Hill order (see Preface).

The values in this table were measured either by calorimetric techniques or by application of the Claperyon equation to the variation of vapor pressure with temperature. See Reference 1 for a discussion of the accuracy of different experimental techniques and for methods of estimating enthalpy of vaporization at other temperatures.

References

1. Majer, V. and Svoboda, V., *Enthalpies of Vaporization of Organic Compounds*, Blackwell Scientific Publications, Oxford, 1985.
2. Chase, M. W., Davies, C. A., Downey, J. R., Frurip, D. J., McDonald, R. A., and Syverud, A. N., *JANAF Thermochemical Tables, Third Edition, J. Phys. Chem. Ref. Data*, Vol. 14, Suppl. 1, 1985.
3. *Landolt-Börnstein, Numerical Data and Functional Relationships in Science and Technology, Sixth Edition*, II/4, *Caloric Quantities of State*, Springer-Verlag, Heidelberg, 1961.
4. Daubert, T. E., Danner, R. P., Sibul, H. M., and Stebbins, C. C., *Physical and Thermodynamic Properties of Pure Compounds: Data Compilation*, extant 1994 (core with 4 supplements), Taylor & Francis, Bristol, PA.
5. Ruzicka, K. and Majer, V., Simultaneous Treatment of Vapor Pressures and Related Thermal Data Between the Triple and Normal Boiling Temperatures for *n*–Alkanes $C_5 - C_{20}$, *J. Phys. Chem. Ref. Data*, 23, 1, 1994.
6. Verevkin, S. P., Thermochemistry of Amines: Experimental Standard Molar Enthalpies of Formation of Some Aliphatic and Aromatic Amines, *J. Chem. Thermodynamics*, 29, 891, 1997.
7. Cady, G. H. and Hargreaves, G. B., The Vapor Pressure of Some Heavy Transition Metal Hexafluorides, *J. Chem. Soc.*, 1563, 1961; 1578, 1961.
8. Steele, W. V., Chirico, R. D., Knipmeyer, S. E., and Nguyen, A., *J. Chem. Eng. Data*, 41, 1255, 1996.

Mol. form.	Name	t_b/°C	$\Delta_{vap}H(t_b)$ kJ/mol	$\Delta_{vap}H(25°C)$ kJ/mol
AgBr	Silver(I) bromide	1502	198	
AgCl	Silver(I) chloride	1547	199	
AgI	Silver(I) iodide	1506	143.9	
Al	Aluminum	2519	294	
AlB_3H_{12}	Aluminum borohydride	44.5	30	
$AlBr_3$	Aluminum tribromide	255	23.5	
AlI_3	Aluminum triiodide	382	32.2	
Ar	Argon	−185.85	6.43	
$AsBr_3$	Arsenic(III) bromide	221	41.8	
$AsCl_3$	Arsenic(III) chloride	130	35.01	
AsF_3	Arsenic(III) fluoride	57.8	29.7	
AsF_5	Arsenic(V) fluoride	−52.8	20.8	
AsH_3	Arsine	−62.5	16.69	
AsI_3	Arsenic(III) iodide	424	59.3	
Au	Gold	2856	324	
B	Boron	4000	480	
BBr_3	Boron tribromide	91	30.5	
BCl_3	Boron trichloride	12.65	23.77	23.1
BF_3	Boron trifluoride	−101	19.33	
BI_3	Boron triiodide	210	40.5	
B_2F_4	Tetrafluorodiborane	−34	28	
B_2H_6	Diborane	−92.4	14.28	
B_4H_{10}	Tetraborane	18	27.1	
B_5H_{11}	Pentaborane(11)	63	31.8	
Ba	Barium	1897	140	
$BeCl_2$	Beryllium chloride	482	105	
BeI_2	Beryllium iodide	487	70.5	
Bi	Bismuth	1564	151	
$BiBr_3$	Bismuth tribromide	453	75.4	
$BiCl_3$	Bismuth trichloride	447	72.61	
BrF	Bromine fluoride	20	25.1	
BrF_3	Bromine trifluoride	125.8	47.57	
BrF_5	Bromine pentafluoride	40.76	30.6	
BrH	Hydrogen bromide	−66.38		12.69
BrH_3Si	Bromosilane	1.9	24.4	
BrIn	Indium(I) bromide	656	92	
BrTl	Thallium(I) bromide	819	99.56	

Mol. form.	Name	t_b/°C	$\Delta_{vap}H(t_b)$ kJ/mol	$\Delta_{vap}H$(25°C) kJ/mol
Br_2	Bromine	58.8	29.96	30.91
Br_2Cd	Cadmium bromide	844	115	
Br_2H_2Si	Dibromosilane	66	31	
Br_2Hg	Mercury(II) bromide	322	58.89	
Br_2Pb	Lead(II) bromide	892	133	
Br_2Sn	Tin(II) bromide	639	102	
Br_2Zn	Zinc bromide	697	118	
Br_3Ga	Gallium(III) bromide	279	38.9	
Br_3HSi	Tribromosilane	109	34.8	
Br_3OP	Phosphorus(V) oxybromide	191.7	38	
Br_3P	Phosphorus(III) bromide	172.95	38.8	
Br_3Sb	Antimony(III) bromide	280	59	
Br_4Ge	Germanium(IV) bromide	186.35	41.4	
Br_4Si	Tetrabromosilane	154	37.9	
Br_4Sn	Tin(IV) bromide	205	43.5	
Br_4Ti	Titanium(IV) bromide	230	44.37	
Br_5Ta	Tantalum(V) bromide	349	62.3	
Cd	Cadmium	767	99.87	
$CdCl_2$	Cadmium chloride	960	124.3	
CdF_2	Cadmium fluoride	1748	214	
CdI_2	Cadmium iodide	742	115	
ClF	Chlorine fluoride	−101.1	24	
$ClFO_3$	Perchloryl fluoride	−46.75	19.33	
ClF_2P	Phosphorus(III) chloride difluoride	−47.25	17.6	
ClF_3	Chlorine trifluoride	11.75	27.53	
ClF_3Si	Chlorotrifluorosilane	−70.0	18.7	
ClH	Hydrogen chloride	−85	16.15	9.08
ClH_3Si	Chlorosilane	−30.4	21	
$ClNO$	Nitrosyl chloride	−5.5	25.78	
$ClNO_2$	Nitryl chloride	−15	25.7	
ClO_2	Chlorine dioxide	11	30	
$ClTl$	Thallium(I) chloride	720	102.2	
Cl_2	Chlorine	−34.04	20.41	17.65
Cl_2Cr	Chromium(II) chloride	1300	197	
Cl_2CrO_2	Chromyl chloride	117	35.1	
Cl_2FP	Phosphorus(III) dichloride fluoride	14	24.9	
Cl_2F_2Si	Dichlorodifluorosilane	−32	21.2	
Cl_2H_2Si	Dichlorosilane	8.3	25	24.2
Cl_2Hg	Mercury(II) chloride	304	58.9	
Cl_2O	Chlorine monoxide	2.2	25.9	
Cl_2OS	Thionyl chloride	75.6	31.7	31
Cl_2O_2S	Sulfuryl chloride	69.4	31.4	30.1
Cl_2Pb	Lead(II) chloride	951	127	
Cl_2Sn	Tin(II) chloride	623	86.8	
Cl_2Ti	Titanium(II) chloride	1500	232	
Cl_2Zn	Zinc chloride	732	126	
Cl_3Ga	Gallium(III) chloride	201	23.9	
Cl_3HSi	Trichlorosilane	33		25.7
Cl_3OP	Phosphorus(V) oxychloride	105.5	34.35	38.6
Cl_3OV	Vanadyl trichloride	127	36.78	
Cl_3P	Phosphorus(III) chloride	75.95	30.5	32.1
Cl_3Sb	Antimony(III) chloride	220.3	45.19	
Cl_3Ti	Titanium(III) chloride	960	124	
Cl_4Ge	Germanium(IV) chloride	86.55	27.9	
Cl_4OW	Tungsten(VI) oxytetrachloride	227.55	67.8	
Cl_4Si	Tetrachlorosilane	57.65	28.7	29.7
Cl_4Sn	Tin(IV) chloride	114.15	34.9	
Cl_4Te	Tellurium tetrachloride	387	77	
Cl_4Th	Thorium(IV) chloride	921	146.4	
Cl_4Ti	Titanium(IV) chloride	136.45	36.2	

Mol. form.	Name	t_b/°C	$\Delta_{vap}H(t_b)$ kJ/mol	$\Delta_{vap}H(25°C)$ kJ/mol
Cl_4V	Vanadium(IV) chloride	148	41.4	42.5
Cl_5Mo	Molybdenum(V) chloride	268	62.8	
Cl_5Nb	Niobium(V) chloride	254.0	52.7	
Cl_5Ta	Tantalum(V) chloride	239.35	54.8	
Cl_6W	Tungsten(VI) chloride	346.75	52.7	
FH_3Si	Fluorosilane	−98.6	18.8	
FLi	Lithium fluoride	1673	147	
FNO	Nitrosyl fluoride	−59.9	19.28	
FNO_2	Nitryl fluoride	−72.4	18.05	
FNS	Thionitrosyl fluoride (NSF)	4.8	22.2	
F_2	Fluorine	−188.12	6.62	
F_2H_2Si	Difluorosilane	−77.8	16.3	
F_2O	Fluorine monoxide	−144.75	11.09	
F_2OS	Thionyl fluoride	−43.8	21.8	
F_2O_2	Fluorine dioxide	−57	19.1	
F_2Pb	Lead(II) fluoride	1293	160.4	
F_2Zn	Zinc fluoride	1500	190.1	
F_3HSi	Trifluorosilane	−95	16.2	
F_3N	Nitrogen trifluoride	−128.75	11.56	
F_3O_2Re	Rhenium(VII) dioxytrifluoride	185.4	65.7	
F_3P	Phosphorus(III) fluoride	−101.5	16.5	
F_3PS	Phosphorus(V) sulfide trifluoride	−52.25	19.6	
F_4MoO	Molybdenum(VI) oxytetrafluoride	186.0	50.6	
F_4N_2	Tetrafluorohydrazine	−74	13.27	
F_4ORe	Rhenium(VI) oxytetrafluoride	171.7	61.0	
F_4OW	Tungsten(VI) oxytetrafluoride	185.9	59.5	
F_4S	Sulfur tetrafluoride	−40.45	26.44	
F_4Se	Selenium tetrafluoride	106	47.2	
F_4Th	Thorium(IV) fluoride	1680	258	
F_5I	Iodine pentafluoride	100.5	41.3	
F_5Mo	Molybdenum(V) fluoride	213.6	51.8	
F_5Nb	Niobium(V) fluoride	229	52.3	
F_5Os	Osmium(V) fluoride	225.9	65.6	
F_5P	Phosphorus(V) fluoride	−84.6	17.2	
F_5Re	Rhenium(V) fluoride	221.3	58.1	
F_5Ta	Tantalum(V) fluoride	229.2	56.9	
F_5V	Vanadium(V) fluoride	48.3	44.52	
F_6Ir	Iridium(VI) fluoride	53.6	30.9	
F_6Mo	Molybdenum(VI) fluoride	34.0	29.0	
F_6Os	Osmium(VI) fluoride	47.5	28.1	
F_6Re	Rhenium(VI) fluoride	33.8	28.7	
F_6S	Sulfur hexafluoride			8.99
F_6W	Tungsten(VI) fluoride	17.1	26.5	
Ga	Gallium	2204	254	
GaI_3	Gallium(III) iodide	340	56.5	
Ge	Germanium	2833	334	
GeH_4	Germane	−88.1	14.06	
Ge_2H_6	Digermane	30.8	25.1	
Ge_3H_8	Trigermane	110.5	32.2	
HI	Hydrogen iodide	−35.55	19.76	17.36
HLiO	Lithium hydroxide	1626	188	
HNO_3	Nitric acid	83		39.1
HN_3	Hydrazoic acid	35.7	30.5	
HNaO	Sodium hydroxide	1388	175	
H_2	Hydrogen	−252.87	0.90	
H_2O	Water	100.0	40.657	43.990
H_2O_2	Hydrogen peroxide	150.2		51.6
H_2S	Hydrogen sulfide	−59.55	18.67	14.08
H_2S_2	Hydrogen disulfide	70.7		33.78
H_2Se	Hydrogen selenide	−41.25	19.7	

Mol. form.	Name	t_b/°C	$\Delta_{vap}H(t_b)$ kJ/mol	$\Delta_{vap}H(25°C)$ kJ/mol
H_2Te	Hydrogen telluride	−2	19.2	
H_3N	Ammonia	−33.33	23.33	19.86
H_3P	Phosphine	−87.75	14.6	
H_3Sb	Stibine	−17	21.3	
H_4N_2	Hydrazine	113.55	41.8	44.7
H_4P_2	Diphosphine	63.5	28.8	
H_4Si	Silane	−111.9	12.1	
H_4Sn	Stannane	−51.8	19.05	
H_6Si_2	Disilane	−14.3	21.2	
H_8Si_3	Trisilane	52.9	28.5	
He	Helium	−268.93	0.08	
Hg	Mercury	356.73	59.11	
HgI_2	Mercury(II) iodide	354	59.2	
IIn	Indium(I) iodide	712	90.8	
ITl	Thallium(I) iodide	824	104.7	
I_2	Iodine	184.4	41.57	
I_2Pb	Lead(II) iodide	872	104	
I_2Sn	Tin(II) iodide	714	105	
I_3P	Phosphorus(III) iodide	227	43.9	
I_3Sb	Antimony(III) iodide	401	68.6	
I_4Si	Tetraiodosilane	287.35	50.2	
I_4Sn	Tin(IV) iodide	364.35	56.9	
I_4Ti	Titanium(IV) iodide	377	58.4	
Kr	Krypton	−153.22	9.08	
MoO_3	Molybdenum(VI) oxide	1155	138	
NO	Nitric oxide	−151.74	13.83	
N_2	Nitrogen	−195.79	5.57	
N_2O	Nitrous oxide	−88.48	16.53	
N_2O_4	Nitrogen tetroxide	21.15	38.12	
Ne	Neon	−246.08	1.71	
O_2	Oxygen	−182.95	6.82	
O_2S	Sulfur dioxide	−10.05	24.94	22.92
O_3S	Sulfur trioxide	45	40.69	43.14
P	Phosphorus	280.5	12.4	14.2
Pb	Lead	1749	179.5	
S	Sulfur	444.60	45	
STl_2	Thallium(I) sulfide	1367	154	
Se	Selenium	685	95.48	
Te	Tellurium	988	114.1	
Xe	Xenon	−108.11	12.57	
$CClF_3$	Chlorotrifluoromethane	−81.4	15.8	
CCl_2F_2	Dichlorodifluoromethane	−29.8	20.1	
CCl_3F	Trichlorofluoromethane	23.7	25.1	
CCl_4	Tetrachloromethane	76.8	29.82	32.43
$CHBr_3$	Tribromomethane	149.1	39.66	46.05
$CHClF_2$	Chlorodifluoromethane	−40.7	20.2	
$CHCl_2F$	Dichlorofluoromethane	8.9	25.2	
$CHCl_3$	Trichloromethane	61.17	29.24	31.28
CH_2BrCl	Bromochloromethane	68.0	30.0	
CH_2Br_2	Dibromomethane	97	32.92	36.97
CH_2Cl_2	Dichloromethane	40	28.06	28.82
CH_2I_2	Diiodomethane	182	42.5	
CH_2O_2	Formic acid	101	22.69	20.10
CH_3Br	Bromomethane	3.5	23.91	22.81
CH_3Cl	Chloromethane	−24.09	21.40	18.92
CH_3I	Iodomethane	42.55	27.34	27.97
CH_3NO	Formamide	220		60.15
CH_3NO_2	Nitromethane	101.19	33.99	38.27
CH_4	Methane	−161.48	8.19	
CH_4O	Methanol	64.6	35.21	37.43

Mol. form.	Name	t_b/°C	$\Delta_{vap}H(t_b)$ kJ/mol	$\Delta_{vap}H(25°C)$ kJ/mol
CH_5N	Methylamine	−6.32	25.60	23.37
CH_6N_2	Methylhydrazine	87.5	36.12	40.37
CN_4O_8	Tetranitromethane	126.1	40.74	49.93
CO	Carbon monoxide	−191.5	6.04	
CS_2	Carbon disulfide	46	26.74	27.51
$C_2Br_2ClF_3$	1,2-Dibromo-1-chloro-1,2,2-trifluoroethane	93	31.17	35.04
$C_2Br_2F_4$	1,2-Dibromotetrafluoroethane	47.35	27.03	28.39
C_2ClF_5	Chloropentafluoroethane	−37.95	19.41	
$C_2Cl_2F_4$	1,2-Dichlorotetrafluoroethane	3.8	23.3	
$C_2Cl_3F_3$	1,1,1-Trichlorotrifluoroethane	46.1	26.85	28.08
$C_2Cl_3F_3$	1,1,2-Trichloro-1,2,2-trifluoroethane	47.7	27.04	28.40
C_2Cl_4	Tetrachloroethylene	121.3	34.68	39.68
C_2F_6	Hexafluoroethane	−78.1	16.15	
$C_2HBrClF_3$	2-Bromo-2-chloro-1,1,1-trifluoroethane	50.2	28.08	29.61
C_2HCl_3	Trichloroethylene	87.21	31.40	34.54
C_2HCl_5	Pentachloroethane	159.8	36.9	
$C_2HF_3O_2$	Trifluoroacetic acid	73	33.3	
$C_2H_2Br_4$	1,1,2,2-Tetrabromoethane	243.5	48.7	
$C_2H_2Cl_2$	1,1-Dichloroethylene	31.6	26.14	26.48
$C_2H_2Cl_2$	cis-1,2-Dichloroethylene	60.1	30.2	
$C_2H_2Cl_2$	trans-1,2-Dichloroethylene	48.7	28.9	
$C_2H_2Cl_4$	1,1,2,2-Tetrachloroethane	146.5	37.64	45.71
C_2H_3Br	Bromoethylene	15.8	23.4	
C_2H_3Cl	Chloroethylene	−13.3	20.8	
$C_2H_3Cl_2F$	1,1-Dichloro-1-fluoroethane	32.0	26.06	26.48
$C_2H_3Cl_3$	1,1,1-Trichloroethane	74.09	29.86	32.50
$C_2H_3Cl_3$	1,1,2-Trichloroethane	113.8	34.82	40.24
$C_2H_3F_3$	1,1,1-Trifluoroethane	−47.25	18.99	
C_2H_3N	Acetonitrile	81.65	29.75	32.94
C_2H_4	Ethylene	−103.77	13.53	
$C_2H_4Br_2$	1,2-Dibromoethane	131.6	34.77	41.73
$C_2H_4Cl_2$	1,1-Dichloroethane	57.4	28.85	30.62
$C_2H_4Cl_2$	1,2-Dichloroethane	83.5	31.98	35.16
$C_2H_4F_2$	1,1-Difluoroethane	−24.95	21.56	19.08
C_2H_4O	Acetaldehyde	20.1	25.76	25.47
C_2H_4O	Ethylene oxide	10.6	25.54	24.75
$C_2H_4O_2$	Acetic acid	117.9	23.70	23.36
$C_2H_4O_2$	Methyl formate	31.7	27.92	28.35
C_2H_5Br	Bromoethane	38.5	27.04	28.03
C_2H_5Cl	Chloroethane	12.3	24.65	
C_2H_5ClO	2-Chloroethanol	128.6	41.4	
C_2H_5I	Iodoethane	72.5	29.44	31.93
C_2H_5NO	N-Methylformamide	199.51		56.19
$C_2H_5NO_2$	Nitroethane	114.0	38.0	
C_2H_6	Ethane	−88.6	14.69	5.16
C_2H_6O	Ethanol	78.29	38.56	42.32
C_2H_6O	Dimethyl ether	−24.8	21.51	18.51
C_2H_6OS	Dimethyl sulfoxide	189	43.1	
$C_2H_6O_2$	Ethylene glycol	197.3	50.5	
C_2H_6S	Ethanethiol	35.1	26.79	27.30
C_2H_6S	Dimethyl sulfide	37.33	27.0	27.65
$C_2H_6S_2$	1,2-Ethanedithiol	146.1	37.93	44.68
$C_2H_6S_2$	Dimethyl disulfide	109.8	33.78	37.86
C_2H_7N	Dimethylamine	6.88	26.40	25.05
C_2H_7NO	Ethanolamine	171	49.83	
$C_2H_8N_2$	1,2-Ethanediamine	117	37.98	44.98
$C_2H_8N_2$	1,1-Dimethylhydrazine	63.9	32.55	35.0
C_2N_2	Cyanogen	−21.1	23.33	19.75
$C_3Cl_2F_6$	1,2-Dichlorohexafluoropropane	34.1	26.28	26.93
$C_3H_3Cl_3O_2$	Methyl trichloroacetate	153.8		48.33

Mol. form.	Name	$t_b/°C$	$\Delta_{vap}H(t_b)$ kJ/mol	$\Delta_{vap}H(25°C)$ kJ/mol
C_3H_3N	Acrylonitrile	77.3	32.6	
$C_3H_4Cl_2O_2$	Methyl dichloroacetate	142.9	39.28	47.72
C_3H_4O	Acrolein	52.6	28.3	
$C_3H_4O_2$	2-Oxetanone	162		47.03
C_3H_5Br	3-Bromopropene	70.1	30.24	32.73
C_3H_5Cl	3-Chloropropene	45.1	29.0	
$C_3H_5ClO_2$	Methyl chloroacetate	129.5	39.23	46.73
$C_3H_5Cl_3$	1,2,3-Trichloropropane	157	37.1	
C_3H_5N	Propanenitrile	97.14	31.81	36.03
C_3H_6	Propene	−47.69	18.42	14.24
C_3H_6	Cyclopropane	−32.81	20.05	16.93
$C_3H_6Br_2$	1,2-Dibromopropane	141.9	35.61	41.67
$C_3H_6Br_2$	1,3-Dibromopropane	167.3		47.45
$C_3H_6Cl_2$	1,3-Dichloropropane	120.9	35.18	40.75
C_3H_6O	Allyl alcohol	97.0	40.0	
C_3H_6O	Propanal	48	28.31	29.62
C_3H_6O	Acetone	56.05	29.10	30.99
C_3H_6O	Methyloxirane	35	27.35	27.89
C_3H_6O	Oxetane	47.6	28.67	29.85
$C_3H_6O_2$	Propanoic acid	141.15		32.14
$C_3H_6O_2$	Ethyl formate	54.4	29.91	31.96
$C_3H_6O_2$	Methyl acetate	56.87	30.32	32.29
C_3H_6S	Thietane	95	32.32	35.97
C_3H_7Br	1-Bromopropane	71.1	29.84	32.01
C_3H_7Br	2-Bromopropane	59.5	28.33	30.17
C_3H_7Cl	1-Chloropropane	46.5	27.18	28.35
C_3H_7Cl	2-Chloropropane	35.7	26.30	26.90
C_3H_7I	1-Iodopropane	102.6	32.08	36.25
C_3H_7I	2-Iodopropane	89.5	30.68	34.06
C_3H_7NO	N-Ethylformamide	198		58.44
C_3H_7NO	N,N-Dimethylformamide	153		46.89
$C_3H_7NO_2$	1-Nitropropane	131.1	38.5	
$C_3H_7NO_2$	2-Nitropropane	120.2	36.8	
C_3H_8	Propane	−42.1	19.04	14.79
C_3H_8O	1-Propanol	97.2	41.44	47.45
C_3H_8O	2-Propanol	82.3	39.85	45.39
$C_3H_8O_2$	1,2-Propylene glycol	187.6	52.4	
$C_3H_8O_2$	1,3-Propylene glycol	214.4	57.9	
$C_3H_8O_2$	Ethylene glycol monomethyl ether	124.1	37.54	45.17
$C_3H_8O_3$	Glycerol	290	61.0	
C_3H_8S	1-Propanethiol	67.8	29.54	31.89
C_3H_8S	2-Propanethiol	52.6	27.91	29.45
C_3H_8S	Ethyl methyl sulfide	66.7	29.53	31.85
$C_3H_8S_2$	1,3-Propanedithiol	172.9		49.66
C_3H_9N	Propylamine	47.22	29.55	31.27
C_3H_9N	Isopropylamine	31.76	27.83	28.36
C_3H_9N	Trimethylamine	2.87	22.94	21.66
$C_3H_{10}N_2$	1,3-Propanediamine	139.8	40.85	50.16
C_4F_8	Perfluorocyclobutane	−5.9	23.2	
C_4F_{10}	Perfluorobutane	−1.9	22.9	
$C_4H_4N_2$	Succinonitrile	266	48.5	
$C_4H_4N_2$	Pyrimidine	123.8	43.09	49.79
$C_4H_4N_2$	Pyridazine	208		53.47
C_4H_4O	Furan	31.5	27.10	27.45
$C_4H_4O_2$	Diketene	126.1	36.80	42.89
C_4H_4S	Thiophene	84.0	31.48	34.70
$C_4H_5Cl_3O_2$	Ethyl trichloroacetate	167.5		50.97
C_4H_5N	2-Methylacrylonitrile	90.3	31.8	
C_4H_5N	Pyrrole	129.79	38.75	45.09
C_4H_5N	Cyclopropanecarbonitrile	135.1	35.55	41.94

Mol. form.	Name	$t_b/°C$	$\Delta_{vap}H(t_b)$ kJ/mol	$\Delta_{vap}H(25°C)$ kJ/mol
$C_4H_5NO_2$	Methyl cyanoacetate	200.5	48.2	
C_4H_5NS	4-Methylthiazole	133.3	37.58	43.85
C_4H_6	1,2-Butadiene	10.9	24.02	23.21
C_4H_6	1,3-Butadiene	−4.41	22.47	20.86
C_4H_6	1-Butyne	8.08	24.52	23.35
$C_4H_6Cl_2O_2$	Ethyl dichloroacetate	155		50.60
$C_4H_6O_2$	Vinyl acetate	72.5	34.6	
$C_4H_6O_2$	Methyl acrylate	80.7	33.1	
$C_4H_6O_2$	γ-Butyrolactone	204	52.2	
$C_4H_6O_3$	Acetic anhydride	139.5	38.2	
C_4H_6S	2,3-Dihydrothiophene	112.1	33.24	37.74
C_4H_6S	2,5-Dihydrothiophene	122.4	34.83	39.95
$C_4H_7ClO_2$	Ethyl chloroacetate	144.3	40.43	49.47
C_4H_7N	Butanenitrile	117.6	33.68	39.33
C_4H_7N	2-Methylpropanenitrile	103.9	32.39	37.13
C_4H_8	1-Butene	−6.26	22.07	20.22
C_4H_8	cis-2-Butene	3.71	23.34	22.16
C_4H_8	trans-2-Butene	0.88	22.72	21.40
C_4H_8	Cyclobutane	12.6	24.19	23.51
$C_4H_8Br_2$	1,4-Dibromobutane	197		53.09
$C_4H_8Cl_2$	1,2-Dichlorobutane	124.1	33.90	39.58
$C_4H_8Cl_2$	1,4-Dichlorobutane	161		46.36
$C_4H_8Cl_2O$	Bis(2-chloroethyl) ether	178.5	45.2	
C_4H_8O	Ethyl vinyl ether	35.5	26.2	
C_4H_8O	1,2-Epoxybutane	63.4	30.3	
C_4H_8O	Butanal	74.8	31.5	
C_4H_8O	2-Butanone	79.59	31.30	34.79
C_4H_8O	Tetrahydrofuran	65	29.81	31.99
$C_4H_8O_2$	Butanoic acid	163.75		40.45
$C_4H_8O_2$	2-Methylpropanoic acid	154.45		35.30
$C_4H_8O_2$	Propyl formate	80.9	33.61	37.53
$C_4H_8O_2$	Ethyl acetate	77.11	31.94	35.60
$C_4H_8O_2$	Methyl propanoate	79.8	32.24	35.85
$C_4H_8O_2$	1,3-Dioxane	106.1	34.37	39.09
$C_4H_8O_2$	1,4-Dioxane	101.5	34.16	38.60
C_4H_8S	Tetrahydrothiophene	121.0	34.66	39.43
C_4H_9Br	1-Bromobutane	101.6	32.51	36.64
C_4H_9Br	2-Bromobutane	91.3	30.77	34.41
C_4H_9Br	1-Bromo-2-methylpropane	91.1	31.33	34.82
C_4H_9Br	2-Bromo-2-methylpropane	73.3	29.23	31.81
C_4H_9Cl	1-Chlorobutane	78.6	30.39	33.51
C_4H_9Cl	2-Chlorobutane	68.2	29.17	31.53
C_4H_9Cl	1-Chloro-2-methylpropane	68.5	29.22	31.67
C_4H_9Cl	2-Chloro-2-methylpropane	50.9	27.55	28.98
C_4H_9I	1-Iodobutane	130.6	34.66	40.63
C_4H_9I	2-Iodobutane	120.1	33.27	38.46
C_4H_9I	1-Iodo-2-methylpropane	121.1	33.54	38.83
C_4H_9I	2-Iodo-2-methylpropane	100.1	31.43	35.41
C_4H_9N	Pyrrolidine	86.56	33.01	37.52
C_4H_9NO	N-Ethylacetamide	205		64.89
C_4H_9NO	N,N-Dimethylacetamide	165		50.24
C_4H_9NO	Morpholine	128	37.1	
C_4H_{10}	Butane	−0.5	22.44	21.02
C_4H_{10}	Isobutane	−11.73	21.30	19.23
$C_4H_{10}O$	1-Butanol	117.73	43.29	52.35
$C_4H_{10}O$	2-Butanol	99.51	40.75	49.72
$C_4H_{10}O$	2-Methyl-1-propanol	107.89	41.82	50.82
$C_4H_{10}O$	2-Methyl-2-propanol	82.4	39.07	46.69
$C_4H_{10}O$	Diethyl ether	34.5	26.52	27.10
$C_4H_{10}O$	Methyl propyl ether	39.1	26.75	27.60

Mol. form.	Name	$t_b/°C$	$\Delta_{vap}H(t_b)$ kJ/mol	$\Delta_{vap}H(25°C)$ kJ/mol
$C_4H_{10}O$	Isopropyl methyl ether	30.77	26.05	26.41
$C_4H_{10}O_2$	1,2-Butanediol	190.5	52.84	71.55
$C_4H_{10}O_2$	1,3-Butanediol	207.5	54.31	74.46
$C_4H_{10}O_2$	Ethylene glycol monoethyl ether	135	39.22	48.21
$C_4H_{10}O_2$	Ethylene glycol dimethyl ether	85	32.42	36.39
$C_4H_{10}O_3$	Diethylene glycol	245.8	52.3	
$C_4H_{10}S$	1-Butanethiol	98.5	32.23	36.63
$C_4H_{10}S$	2-Butanethiol	85	30.59	33.99
$C_4H_{10}S$	2-Methyl-1-propanethiol	88.5	31.01	34.63
$C_4H_{10}S$	2-Methyl-2-propanethiol	64.3	28.45	30.78
$C_4H_{10}S$	Diethyl sulfide	92.1	31.77	35.80
$C_4H_{10}S$	Methyl propyl sulfide	95.6	32.08	36.24
$C_4H_{10}S$	Isopropyl methyl sulfide	84.8	30.71	34.15
$C_4H_{10}S_2$	1,4-Butanedithiol	195.5		55.10
$C_4H_{10}S_2$	Diethyl disulfide	154.1	37.58	45.18
$C_4H_{11}N$	Butylamine	77.00	31.81	35.72
$C_4H_{11}N$	sec-Butylamine	62.73	29.92	32.85
$C_4H_{11}N$	tert-Butylamine	44.04	28.27	29.64
$C_4H_{11}N$	Isobutylamine	67.75	30.61	33.85
$C_4H_{11}N$	Diethylamine	55.5	29.06	31.31
$C_4H_{11}N$	Isopropylmethylamine	50.4	28.71	30.69
$C_4H_{11}NO$	2-Amino-2-methyl-1-propanol	165.5	50.6	
$C_4H_{11}NO_2$	Diethanolamine	268.8	65.2	
$C_5H_2F_6O_2$	Hexafluoroacetylacetone	54.15	27.05	30.58
$C_5H_4O_2$	Furfural	161.7	43.2	
C_5H_5N	Pyridine	115.23	35.09	40.21
$C_5H_6O_2$	Furfuryl alcohol	171	53.6	
C_5H_6S	2-Methylthiophene	112.6	33.90	38.87
C_5H_6S	3-Methylthiophene	115.5	34.24	39.43
C_5H_7N	trans-3-Pentenenitrile	142.6	37.09	44.77
C_5H_7N	Cyclobutanecarbonitrile	149.6	36.88	44.34
C_5H_8	Spiropentane	39	26.76	27.49
C_5H_8O	Cyclopropyl methyl ketone	111.3	34.07	39.41
C_5H_8O	Cyclopentanone	130.57	36.35	42.72
$C_5H_8O_2$	Methyl cyclopropanecarboxylate	114.9	35.25	41.27
$C_5H_8O_2$	Allyl acetate	103.5	36.3	
$C_5H_8O_2$	Ethyl acrylate	99.4	34.7	
$C_5H_8O_2$	Methyl methacrylate	100.5	36.0	
$C_5H_8O_2$	2,4-Pentanedione	138	34.30	41.77
C_5H_9N	Pentanenitrile	141.3	36.09	43.60
C_5H_9N	3-Methylbutanenitrile	127.5	35.10	41.64
C_5H_9N	2,2-Dimethylpropanenitrile	106.1	32.40	37.35
C_5H_{10}	1-Pentene	29.96	25.20	25.47
C_5H_{10}	cis-2-Pentene	36.93		26.86
C_5H_{10}	trans-2-Pentene	36.34		26.76
C_5H_{10}	2-Methyl-1-butene	31.2	25.50	25.92
C_5H_{10}	3-Methyl-1-butene	20.1		23.77
C_5H_{10}	2-Methyl-2-butene	38.56	26.31	27.06
C_5H_{10}	Cyclopentane	49.3	27.30	28.52
$C_5H_{10}Cl_2$	1,2-Dichloropentane	148.3	36.45	43.89
$C_5H_{10}Cl_2$	1,5-Dichloropentane	179		50.71
$C_5H_{10}O$	Cyclopentanol	140.42		57.05
$C_5H_{10}O$	2-Pentanone	102.26	33.44	38.40
$C_5H_{10}O$	3-Pentanone	101.96	33.45	38.52
$C_5H_{10}O$	3-Methyl-2-butanone	94.33	32.35	36.78
$C_5H_{10}O$	3,3-Dimethyloxetane	80.6	30.85	33.94
$C_5H_{10}O$	Tetrahydropyran	88	31.17	34.58
$C_5H_{10}O_2$	Pentanoic acid	186.1	44.1	
$C_5H_{10}O_2$	2-Methylbutanoic acid	177		46.91
$C_5H_{10}O_2$	Butyl formate	106.1	36.58	41.11

Mol. form.	Name	$t_b/°C$	$\Delta_{vap}H(t_b)$ kJ/mol	$\Delta_{vap}H(25°C)$ kJ/mol
$C_5H_{10}O_2$	Isobutyl formate	98.2	33.6	
$C_5H_{10}O_2$	Propyl acetate	101.54	33.92	39.72
$C_5H_{10}O_2$	Isopropyl acetate	88.6	32.93	37.20
$C_5H_{10}O_2$	Ethyl propanoate	99.1	33.88	39.21
$C_5H_{10}O_2$	Methyl butanoate	102.8	33.79	39.28
$C_5H_{10}O_2$	Methyl isobutanoate	92.5	32.61	37.32
$C_5H_{10}O_2$	Tetrahydrofurfuryl alcohol	178	45.2	
$C_5H_{10}O_3$	Diethyl carbonate	126		43.60
$C_5H_{10}O_3$	Ethylene glycol monomethyl ether acetate	143	43.9	
$C_5H_{10}S$	Thiacyclohexane	141.8	35.96	42.58
$C_5H_{10}S$	Cyclopentanethiol	132.1	35.32	41.42
$C_5H_{11}Br$	1-Bromopentane	129.8	35.01	41.28
$C_5H_{11}Cl$	1-Chloropentane	107.8	33.15	38.24
$C_5H_{11}Cl$	2-Chloropentane	97.0	31.79	36.03
$C_5H_{11}Cl$	1-Chloro-3-methylbutane	98.9	32.02	36.24
$C_5H_{11}I$	1-Iodopentane	155		45.27
$C_5H_{11}N$	Piperidine	106.22		39.29
C_5H_{12}	Pentane	36.06	25.79	26.43
C_5H_{12}	Isopentane	27.88	24.69	24.85
C_5H_{12}	Neopentane	9.48	22.74	21.84
$C_5H_{12}O$	1-Pentanol	137.98	44.36	57.02
$C_5H_{12}O$	2-Pentanol	119.3	41.40	54.21
$C_5H_{12}O$	3-Pentanol	116.25		54.0
$C_5H_{12}O$	2-Methyl-1-butanol	128		55.16
$C_5H_{12}O$	3-Methyl-1-butanol	131.1	44.07	55.61
$C_5H_{12}O$	2-Methyl-2-butanol	102.4	39.04	50.10
$C_5H_{12}O$	3-Methyl-2-butanol	112.9		53.0
$C_5H_{12}O$	Butyl methyl ether	70.16	29.55	32.37
$C_5H_{12}O$	sec-Butyl methyl ether	59.1	28.09	30.23
$C_5H_{12}O$	Methyl tert-butyl ether	55.2	27.94	29.82
$C_5H_{12}O$	Isobutyl methyl ether	58.6	28.02	30.13
$C_5H_{12}O$	Ethyl propyl ether	63.21	28.94	31.43
$C_5H_{12}O$	Ethyl isopropyl ether	54.1	28.21	30.08
$C_5H_{12}O_2$	1-Ethoxy-2-methoxyethane	102.1	34.33	39.83
$C_5H_{12}O_2$	1,5-Pentanediol	239	60.7	
$C_5H_{12}O_2$	Ethylene glycol monopropyl ether	149.8	41.40	52.12
$C_5H_{12}O_2$	Diethoxymethane	88	31.33	35.65
$C_5H_{12}O_3$	Diethylene glycol monomethyl ether	193	46.6	
$C_5H_{12}S$	1-Pentanethiol	126.6	34.88	41.24
$C_5H_{12}S$	2-Methyl-1-butanethiol	119.1	33.79	39.45
$C_5H_{12}S$	2-Methyl-2-butanethiol	99.1	31.37	35.67
$C_5H_{12}S$	Butyl methyl sulfide	123.5	34.47	40.46
$C_5H_{12}S$	tert-Butyl methyl sulfide	99	31.47	35.84
$C_5H_{12}S$	Ethyl propyl sulfide	118.6	34.24	39.97
$C_5H_{12}S$	Ethyl isopropyl sulfide	107.5	32.74	37.78
$C_5H_{13}N$	Pentylamine	104.3	34.01	40.08
$C_5H_{13}N$	Ethylisopropylamine	69.6	29.94	33.13
C_6ClF_5	Chloropentafluorobenzene	117.96	34.76	41.07
C_6F_6	Hexafluorobenzene	80.26	31.66	35.71
C_6HF_5	Pentafluorobenzene	85.74	32.15	36.27
$C_6H_4Cl_2$	o-Dichlorobenzene	180	39.66	50.21
$C_6H_4Cl_2$	m-Dichlorobenzene	173	38.62	48.58
$C_6H_4Cl_2$	p-Dichlorobenzene	174	38.79	49.0
$C_6H_4F_2$	o-Difluorobenzene	94	32.21	36.18
$C_6H_4F_2$	m-Difluorobenzene	82.6	31.10	34.59
$C_6H_4F_2$	p-Difluorobenzene	89	31.77	35.54
C_6H_5Br	Bromobenzene	156.06		44.54
C_6H_5Cl	Chlorobenzene	131.72	35.19	40.97
C_6H_5F	Fluorobenzene	84.73	31.19	34.58
C_6H_5I	Iodobenzene	188.4	39.5	

Mol. form.	Name	t_b/°C	$\Delta_{vap}H(t_b)$ kJ/mol	$\Delta_{vap}H(25°C)$ kJ/mol
$C_6H_5NO_2$	Nitrobenzene	210.8		55.01
C_6H_6	Benzene	80.09	30.72	33.83
C_6H_6ClN	o-Chloroaniline	208.8	44.4	
C_6H_6O	Phenol	181.87	45.69	57.82
C_6H_6S	Benzenethiol	169.1	39.93	47.56
C_6H_7N	Aniline	184.17	42.44	55.83
C_6H_7N	2-Methylpyridine	129.38	36.17	42.48
C_6H_7N	3-Methylpyridine	144.14	37.35	44.44
C_6H_7N	4-Methylpyridine	145.36	37.51	44.56
C_6H_7N	1-Cyclopentenecarbonitrile			44.98
C_6H_9N	Cyclopentanecarbonitrile			43.43
$C_6H_9NO_3$	Triacetamide			60.41
C_6H_{10}	Cyclohexene	82.98	30.46	33.47
$C_6H_{10}O$	Cyclohexanone	155.43		45.06
$C_6H_{10}O$	Mesityl oxide	130	36.1	
$C_6H_{10}O_2$	Methyl cyclobutanecarboxylate	135.5	37.13	44.72
$C_6H_{10}O_3$	Propanoic anhydride	170	41.7	
$C_6H_{10}O_4$	Diethyl oxalate	185.7	42.0	
$C_6H_{10}O_4$	Ethylene glycol diacetate	190		61.44
$C_6H_{11}N$	Hexanenitrile	163.65		47.91
C_6H_{12}	1-Hexene	63.48		30.61
C_6H_{12}	cis-2-Hexene	68.8		32.19
C_6H_{12}	trans-2-Hexene	67.9		31.60
C_6H_{12}	cis-3-Hexene	66.4		31.23
C_6H_{12}	trans-3-Hexene	67.1		31.55
C_6H_{12}	2-Methyl-1-pentene	62.1		30.48
C_6H_{12}	3-Methyl-1-pentene	54.2		28.62
C_6H_{12}	4-Methyl-1-pentene	53.9		28.71
C_6H_{12}	2-Methyl-2-pentene	67.3		31.60
C_6H_{12}	3-Methyl-cis-2-pentene	67.7		32.09
C_6H_{12}	3-Methyl-trans-2-pentene	70.4		31.35
C_6H_{12}	4-Methyl-cis-2-pentene	56.3		29.48
C_6H_{12}	4-Methyl-trans-2-pentene	58.6		29.97
C_6H_{12}	2-Ethyl-1-butene	64.7		31.13
C_6H_{12}	2,3-Dimethyl-1-butene	55.6		29.18
C_6H_{12}	3,3-Dimethyl-1-butene	41.2		26.61
C_6H_{12}	2,3-Dimethyl-2-butene	73.3	29.64	32.51
C_6H_{12}	Cyclohexane	80.73	29.97	33.01
C_6H_{12}	Methylcyclopentane	71.8	29.08	31.64
C_6H_{12}	Ethylcyclobutane	70.8	28.67	31.24
$C_6H_{12}Cl_2$	1,2-Dichlorohexane	173		48.16
$C_6H_{12}O$	Butyl vinyl ether	94	31.58	36.17
$C_6H_{12}O$	2-Hexanone	127.6	36.35	43.14
$C_6H_{12}O$	3-Hexanone	123.5	35.36	42.47
$C_6H_{12}O$	3-Methyl-2-pentanone	117.5	34.16	40.53
$C_6H_{12}O$	4-Methyl-2-pentanone	116.5	34.49	40.61
$C_6H_{12}O$	2-Methyl-3-pentanone	113.5	33.84	39.79
$C_6H_{12}O$	3,3-Dimethyl-2-butanone	106.1	33.39	37.91
$C_6H_{12}O$	Cyclohexanol	160.84		62.01
$C_6H_{12}O_2$	Butyl acetate	126.1	36.28	43.86
$C_6H_{12}O_2$	tert-Butyl acetate	95.1	33.07	38.03
$C_6H_{12}O_2$	Isobutyl acetate	116.5	35.9	
$C_6H_{12}O_2$	Propyl propanoate	122.5	35.54	43.45
$C_6H_{12}O_2$	Ethyl butanoate	121.5	35.47	42.68
$C_6H_{12}O_2$	Ethyl 2-methylpropanoate	110.1	33.67	39.83
$C_6H_{12}O_2$	Methyl pentanoate	127.4	35.36	43.10
$C_6H_{12}O_2$	Methyl 2,2-dimethylpropanoate	101.1	33.42	38.76
$C_6H_{12}O_3$	Ethylene glycol monoethyl ether acetate	156.4	40.76	52.61
$C_6H_{12}S$	Cyclohexanethiol	158.9	37.06	44.57
$C_6H_{13}Br$	1-Bromohexane	155.3		45.89

Mol. form.	Name	$t_b/°C$	$\Delta_{vap}H(t_b)$ kJ/mol	$\Delta_{vap}H(25°C)$ kJ/mol
$C_6H_{13}Cl$	1-Chlorohexane	135	35.67	42.83
$C_6H_{13}I$	1-Iodohexane	181		49.75
$C_6H_{13}N$	Cyclohexylamine	134	36.14	43.67
C_6H_{14}	Hexane	68.73	28.85	31.56
C_6H_{14}	2-Methylpentane	60.26	27.79	29.89
C_6H_{14}	3-Methylpentane	63.27	28.06	30.28
C_6H_{14}	2,2-Dimethylbutane	49.73	26.31	27.68
C_6H_{14}	2,3-Dimethylbutane	57.93	27.38	29.12
$C_6H_{14}N_2$	Azopropane	114		39.88
$C_6H_{14}O$	1-Hexanol	157.6	44.50	61.61
$C_6H_{14}O$	2-Hexanol	140	41.01	58.46
$C_6H_{14}O$	2-Methyl-1-pentanol	149	50.2	
$C_6H_{14}O$	4-Methyl-1-pentanol	151.9	44.46	60.47
$C_6H_{14}O$	2-Methyl-2-pentanol	121.1	39.59	54.77
$C_6H_{14}O$	4-Methyl-2-pentanol	131.6	44.2	
$C_6H_{14}O$	2-Ethyl-1-butanol	147	43.2	
$C_6H_{14}O$	Dipropyl ether	90.08	31.31	35.69
$C_6H_{14}O$	Diisopropyl ether	68.51	29.10	32.12
$C_6H_{14}O$	Butyl ethyl ether	92.3	31.63	36.32
$C_6H_{14}O$	Methyl pentyl ether	99	32.02	36.85
$C_6H_{14}O_2$	2-Methyl-2,4-pentanediol	197.1	57.3	
$C_6H_{14}O_2$	Ethylene glycol monobutyl ether	168.4		56.59
$C_6H_{14}O_2$	1,1-Diethoxyethane	102.25	36.28	43.20
$C_6H_{14}O_2$	Ethylene glycol diethyl ether	119.4	36.28	43.20
$C_6H_{14}O_3$	Bis(ethoxymethyl) ether	140.6	36.17	44.69
$C_6H_{14}O_3$	Diethylene glycol monoethyl ether	196	47.5	
$C_6H_{14}O_3$	Diethylene glycol dimethyl ether	162	36.17	44.69
$C_6H_{14}O_4$	Triethylene glycol	285	71.4	
$C_6H_{14}S$	Dipropyl sulfide	142.9	36.60	44.21
$C_6H_{14}S$	Diisopropyl sulfide	120.1	33.80	39.60
$C_6H_{14}S$	Isopropyl propyl sulfide	132.1	35.11	41.78
$C_6H_{14}S$	Butyl ethyl sulfide	144.3	37.01	44.51
$C_6H_{14}S$	Methyl pentyl sulfide	145.1	37.41	45.24
$C_6H_{15}N$	Hexylamine	132.8	36.54	45.10
$C_6H_{15}N$	Butylethylamine	107.5	33.97	40.15
$C_6H_{15}N$	Dipropylamine	109.3	33.47	40.04
$C_6H_{15}N$	Diisopropylamine	83.9	30.40	34.61
$C_6H_{15}N$	Isopropylpropylamine	96.9	32.14	37.23
$C_6H_{15}N$	Triethylamine	89	31.01	34.84
C_6MoO_6	Molybdenum hexacarbonyl	701	72.51	
$C_7H_3F_5$	2,3,4,5,6-Pentafluorotoluene	117.5	34.75	41.12
$C_7H_5F_3$	(Trifluoromethyl)benzene	102.1	32.63	37.60
C_7H_5N	Benzonitrile	191.1	45.9	
C_7H_6O	Benzaldehyde	179.0	42.5	
$C_7H_6O_2$	Salicylaldehyde	197	38.2	
C_7H_7Cl	o-Chlorotoluene	159.0	37.5	
C_7H_7Cl	p-Chlorotoluene	162.4	38.7	
C_7H_7F	o-Fluorotoluene	115	35.4	
C_7H_7F	p-Fluorotoluene	116.6	34.08	39.42
C_7H_8	Toluene	110.63	33.18	38.01
C_7H_8O	o-Cresol	191.04	45.19	
C_7H_8O	m-Cresol	202.27	47.40	61.71
C_7H_8O	p-Cresol	201.98	47.45	
C_7H_8O	Benzyl alcohol	205.31	50.48	
C_7H_8O	Anisole	153.7	38.97	46.90
C_7H_9N	Benzylamine	185		60.16
C_7H_9N	o-Methylaniline	200.3	44.6	
C_7H_9N	m-Methylaniline	203.3	44.9	
C_7H_9N	p-Methylaniline	200.4	44.3	
C_7H_9N	1-Cyclohexenecarbonitrile			53.55

Mol. form.	Name	t_b/°C	$\Delta_{vap}H(t_b)$ kJ/mol	$\Delta_{vap}H(25°C)$ kJ/mol
C_7H_9N	2,3-Dimethylpyridine	161.12	39.08	47.82
C_7H_9N	2,4-Dimethylpyridine	158.38	38.53	47.49
C_7H_9N	2,5-Dimethylpyridine	156.98	38.68	47.04
C_7H_9N	2,6-Dimethylpyridine	144.01	37.46	45.34
C_7H_9N	3,4-Dimethylpyridine	179.10	39.99	50.50
C_7H_9N	3,5-Dimethylpyridine	171.84	39.46	49.33
$C_7H_{10}O$	Dicyclopropyl ketone	161		53.70
$C_7H_{11}N$	Cyclohexanecarbonitrile			51.92
C_7H_{12}	1-Methylbicyclo(3,1,0)hexane	93.1	31.07	34.77
$C_7H_{12}O_4$	Diethyl malonate	200	54.8	
C_7H_{14}	1-Heptene	93.64		35.49
C_7H_{14}	cis-2-Heptene	98.4		36.26
C_7H_{14}	trans-2-Heptene	98		36.27
C_7H_{14}	cis-3-Heptene	95.8		35.81
C_7H_{14}	trans-3-Heptene	95.7		35.84
C_7H_{14}	cis-3-Methyl-3-hexene	95.4		36.31
C_7H_{14}	trans-3-Methyl-3-hexene	93.5		35.70
C_7H_{14}	2,4-Dimethyl-1-pentene	81.6		33.03
C_7H_{14}	4,4-Dimethyl-1-pentene	72.5		31.13
C_7H_{14}	2,4-Dimethyl-2-pentene	83.4		34.19
C_7H_{14}	cis-4,4-Dimethyl-2-pentene	80.4		32.56
C_7H_{14}	trans-4,4-Dimethyl-2-pentene	76.7		32.81
C_7H_{14}	2-Ethyl-3-methyl-1-butene	89		34.35
C_7H_{14}	2,3,3-Trimethyl-1-butene	77.9		32.09
C_7H_{14}	Methylcyclohexane	100.93	31.27	35.36
C_7H_{14}	Ethylcyclopentane	103.5	31.96	36.40
C_7H_{14}	cis-1,3-Dimethylcyclopentane	90.8	30.40	34.20
$C_7H_{14}O$	2-Heptanone	151.05		47.24
$C_7H_{14}O$	2,2-Dimethyl-3-pentanone	125.6	36.09	42.34
$C_7H_{14}O$	2,4-Dimethyl-3-pentanone	125.4	34.64	41.51
$C_7H_{14}O$	1-Methylcyclohexanol	155	79.0	
$C_7H_{14}O$	cis-2-Methylcyclohexanol	165	48.5	
$C_7H_{14}O$	trans-2-Methylcyclohexanol	167.5	53.0	
$C_7H_{14}O_2$	Pentyl acetate	149.2	38.42	48.56
$C_7H_{14}O_2$	Isopentyl acetate	142.5	37.5	
$C_7H_{14}O_2$	Ethyl pentanoate	146.1	36.96	47.01
$C_7H_{14}O_2$	Ethyl 3-methylbutanoate	135.0	37.0	
$C_7H_{14}O_2$	Ethyl 2,2-dimethylpropanoate	118.4	34.51	41.25
$C_7H_{14}O_2$	Methyl hexanoate	149.5	38.55	48.04
$C_7H_{15}Br$	1-Bromoheptane	179		50.60
$C_7H_{15}Cl$	1-Chloroheptane	159		47.66
C_7H_{16}	Heptane	98.5	31.77	36.57
C_7H_{16}	2-Methylhexane	90.04	30.62	34.87
C_7H_{16}	3-Methylhexane	92	30.9	
C_7H_{16}	3-Ethylpentane	93.5	31.12	35.22
C_7H_{16}	2,2-Dimethylpentane	79.2	29.23	32.42
C_7H_{16}	2,3-Dimethylpentane	89.78	30.46	34.26
C_7H_{16}	2,4-Dimethylpentane	80.49	29.55	32.88
C_7H_{16}	3,3-Dimethylpentane	86.06	29.62	33.03
C_7H_{16}	2,2,3-Trimethylbutane	80.86	28.90	32.05
$C_7H_{16}O$	Hexyl methyl ether	126.1	34.93	42.07
$C_7H_{16}O$	1-Heptanol	176.45		66.81
$C_7H_{16}O$	3-Heptanol	157	42.5	
$C_7H_{16}O$	Butyl propyl ether	118.1	33.72	40.22
$C_7H_{16}O$	Ethyl pentyl ether	117.6	34.41	41.01
$C_7H_{17}N$	Heptylamine	156		49.96
C_8F_{18}	Perfluorooctane	105.9	33.38	41.13
C_8H_8	Styrene	145	38.7	
C_8H_8O	Acetophenone	202	43.98	55.40
$C_8H_8O_2$	Methyl benzoate	199		55.57

Mol. form.	Name	t_b/°C	$\Delta_{vap}H(t_b)$ kJ/mol	$\Delta_{vap}H(25°C)$ kJ/mol
$C_8H_8O_3$	Methyl salicylate	222.9	46.7	
C_8H_{10}	Ethylbenzene	136.19	35.57	42.24
C_8H_{10}	o-Xylene	144.5	36.24	43.43
C_8H_{10}	m-Xylene	139.12	35.66	42.65
C_8H_{10}	p-Xylene	138.37	35.67	42.40
$C_8H_{10}O$	2,4-Xylenol	210.98		64.96
$C_8H_{10}O$	2,5-Xylenol	211.1	46.9	
$C_8H_{10}O$	2,6-Xylenol	201.07		75.31
$C_8H_{10}O$	3,4-Xylenol	227		85.03
$C_8H_{10}O$	3,5-Xylenol	221.74		82.01
$C_8H_{10}O$	Phenetole	169.81		51.04
$C_8H_{11}N$	N-Ethylaniline	203.0		58.3
$C_8H_{11}N$	N,N-Dimethylaniline	194.15		52.83
$C_8H_{11}N$	2,4-Dimethylaniline	214		61.3
$C_8H_{11}N$	2,5-Dimethylaniline	214		61.7
$C_8H_{11}N$	2,3,6-Trimethylpyridine	171.6	39.95	50.61
$C_8H_{11}N$	2,4,6-Trimethylpyridine	170.6	39.87	50.33
C_8H_{14}	1-Octyne	126.3	35.83	42.30
C_8H_{14}	2-Octyne	137.6	37.26	44.49
C_8H_{14}	3-Octyne	133.1	36.94	43.92
C_8H_{14}	4-Octyne	131.6	36.0	42.73
$C_8H_{14}O_3$	Butanoic anhydride	200	50.0	
$C_8H_{15}N$	Octanenitrile	205.25		56.80
C_8H_{16}	1-Octene	121.29	34.07	40.34
C_8H_{16}	cis-2,2-Dimethyl-3-hexene	105.5		36.86
C_8H_{16}	trans-2,2-Dimethyl-3-hexene	100.8		37.03
C_8H_{16}	3-Ethyl-2-methyl-1-pentene	109.5		37.27
C_8H_{16}	2,4,4-Trimethyl-1-pentene	101.4		35.59
C_8H_{16}	2,4,4-Trimethyl-2-pentene	104.9		37.23
C_8H_{16}	Ethylcyclohexane	131.9	34.04	40.56
C_8H_{16}	1,1-Dimethylcyclohexane	119.6	32.51	37.92
C_8H_{16}	cis-1,2-Dimethylcyclohexane	129.8	33.47	39.70
C_8H_{16}	trans-1,2-Dimethylcyclohexane	123.5	32.96	38.36
C_8H_{16}	cis-1,3-Dimethylcyclohexane	120.1	32.91	38.26
C_8H_{16}	trans-1,3-Dimethylcyclohexane	124.5	33.39	39.16
C_8H_{16}	cis-1,4-Dimethylcyclohexane	124.4	33.28	39.02
C_8H_{16}	trans-1,4-Dimethylcyclohexane	119.4	32.56	37.90
C_8H_{16}	Propylcyclopentane	131	34.70	41.08
C_8H_{16}	Isopropylcyclopentane	126.5	33.56	39.44
C_8H_{16}	1-Ethyl-1-methylcyclopentane	121.6	33.20	38.85
$C_8H_{16}O$	2,2,4-Trimethyl-3-pentanone	135.1	35.64	43.30
$C_8H_{16}O_2$	Octanoic acid	239	58.5	
$C_8H_{16}O_2$	2-Ethylhexanoic acid	228		75.60
$C_8H_{16}O_2$	Isobutyl isobutanoate	148.6	38.2	
$C_8H_{16}O_2$	Ethyl hexanoate	167		51.72
$C_8H_{16}O_2$	Methyl heptanoate	174		51.62
$C_8H_{17}Br$	1-Bromooctane	200		55.77
$C_8H_{17}Cl$	1-Chlorooctane	181.5		52.42
$C_8H_{17}F$	1-Fluorooctane	142.4	40.43	49.65
C_8H_{18}	Octane	125.67	34.41	41.49
C_8H_{18}	2-Methylheptane	117.66	33.26	39.67
C_8H_{18}	3-Methylheptane	118.9	33.66	39.83
C_8H_{18}	4-Methylheptane	117.72	33.35	39.69
C_8H_{18}	3-Ethylhexane	118.6	33.59	39.64
C_8H_{18}	2,2-Dimethylhexane	106.86	32.07	37.28
C_8H_{18}	2,3-Dimethylhexane	115.62	33.17	38.78
C_8H_{18}	2,4-Dimethylhexane	109.5	32.51	37.76
C_8H_{18}	2,5-Dimethylhexane	109.12	32.54	37.85
C_8H_{18}	3,3-Dimethylhexane	111.97	32.31	37.53
C_8H_{18}	3,4-Dimethylhexane	117.73	33.24	38.97

Mol. form.	Name	$t_b/°C$	$\Delta_{vap}H(t_b)$ kJ/mol	$\Delta_{vap}H(25°C)$ kJ/mol
C_8H_{18}	3-Ethyl-2-methylpentane	115.66	32.93	38.52
C_8H_{18}	3-Ethyl-3-methylpentane	118.27	32.78	37.99
C_8H_{18}	2,2,3-Trimethylpentane	110	31.94	36.91
C_8H_{18}	2,2,4-Trimethylpentane	99.22	30.79	35.14
C_8H_{18}	2,3,3-Trimethylpentane	114.8	32.12	37.27
C_8H_{18}	2,3,4-Trimethylpentane	113.5	32.36	37.75
C_8H_{18}	2,2,3,3-Tetramethylbutane	106.45		42.90
$C_8H_{18}N_2$	Azobutane			49.31
$C_8H_{18}O$	1-Octanol	195.16		70.98
$C_8H_{18}O$	2-Octanol	180	44.4	
$C_8H_{18}O$	2-Ethyl-1-hexanol	184.6	54.2	
$C_8H_{18}O$	Dibutyl ether	140.28	36.49	44.97
$C_8H_{18}O$	Di-sec-butyl ether	121.1	34.06	40.84
$C_8H_{18}O$	Di-tert-butyl ether	107.23	32.15	37.61
$C_8H_{18}O_2$	1,2-Dipropoxyethane			50.62
$C_8H_{18}O_3$	Diethylene glycol diethyl ether	188		58.40
$C_8H_{18}S$	Dibutyl sulfide	185		52.96
$C_8H_{18}S$	Di-tert-butyl sulfide	149.1	33.26	43.76
$C_8H_{18}S$	Diisobutyl sulfide	171		48.71
$C_8H_{19}N$	Dibutylamine	159.6	38.44	49.45
$C_8H_{19}N$	2-Ethylhexylamine	169.2	40.0	
C_9H_7N	Quinoline	237.16	49.7	59.30
C_9H_7N	Isoquinoline	243.22	49.0	60.26
C_9H_{10}	Cyclopropylbenzene	173.6		50.22
C_9H_{10}	Indan	177.97	39.63	48.79
$C_9H_{10}O_2$	Benzyl acetate	213	49.4	
C_9H_{12}	Propylbenzene	159.24		46.22
C_9H_{12}	Isopropylbenzene	152.41		45.13
C_9H_{12}	1,2,3-Trimethylbenzene	176.12		49.05
C_9H_{12}	1,2,4-Trimethylbenzene	169.38		47.93
C_9H_{12}	1,3,5-Trimethylbenzene	164.74		47.50
$C_9H_{14}O_6$	Triacetin	259		85.74
C_9H_{18}	Butylcyclopentane	156.6	36.16	45.89
C_9H_{18}	Propylcyclohexane	156.7		45.08
C_9H_{18}	Isopropylcyclohexane	154.8		44.02
$C_9H_{18}O$	2-Nonanone	195.3		56.44
$C_9H_{18}O$	5-Nonanone	188.45		53.30
$C_9H_{18}O$	2,6-Dimethyl-4-heptanone	169.4		50.92
$C_9H_{18}O_2$	Methyl octanoate	192.9		56.41
C_9H_{20}	Nonane	150.82	37.18	46.55
C_9H_{20}	2,2,5-Trimethylhexane	124.09	33.65	40.16
C_9H_{20}	2,3,5-Trimethylhexane	131.4	34.43	41.41
C_9H_{20}	3,3-Diethylpentane	146.3	34.61	42.0
C_9H_{20}	2,2,4,4-Tetramethylpentane	122.29	32.51	38.49
$C_9H_{20}O$	1-Nonanol	213.37		76.86
$C_{10}H_7Br$	1-Bromonaphthalene	281	39.3	
$C_{10}H_7Cl$	1-Chloronaphthalene	259	52.1	
$C_{10}H_8$	Naphthalene	217.9	43.2	
$C_{10}H_9N$	2-Methylquinoline	246.5		66.1
$C_{10}H_9N$	4-Methylquinoline	262		67.6
$C_{10}H_9N$	6-Methylquinoline	258.6		67.7
$C_{10}H_9N$	8-Methylquinoline	247.5		65.7
$C_{10}H_{12}$	1,2,3,4-Tetrahydronaphthalene	207.6	43.9	
$C_{10}H_{14}$	Butylbenzene	183.31	38.87	51.36
$C_{10}H_{14}$	sec-Butylbenzene	173.3		47.98
$C_{10}H_{14}$	tert-Butylbenzene	169.1		47.71
$C_{10}H_{14}$	Isobutylbenzene	172.79		47.86
$C_{10}H_{14}$	1-Isopropyl-4-methylbenzene	177.1	38.2	
$C_{10}H_{16}O$	(+)-Camphor	207.4	59.5	
$C_{10}H_{18}$	cis-Decahydronaphthalene	195.8	41.0	

Mol. form.	Name	t_b/°C	$\Delta_{vap}H(t_b)$ kJ/mol	$\Delta_{vap}H(25°C)$ kJ/mol
$C_{10}H_{18}$	*trans*-Decahydronaphthalene	187.3	40.2	
$C_{10}H_{19}N$	Decanenitrile	243		66.84
$C_{10}H_{20}$	1-Decene	170.5		50.43
$C_{10}H_{20}$	Butylcyclohexane	180.9		49.36
$C_{10}H_{20}O_2$	2-Ethylhexyl acetate	199	43.5	
$C_{10}H_{20}O_2$	Isopentyl isopentanoate	190.4	45.9	
$C_{10}H_{22}$	Decane	174.15	39.58	51.42
$C_{10}H_{22}$	2-Methylnonane	167.1	38.23	49.63
$C_{10}H_{22}$	3-Methylnonane	167.9	38.26	49.71
$C_{10}H_{22}$	5-Methylnonane	165.1	38.14	49.34
$C_{10}H_{22}$	2,4-Dimethyloctane	156	36.47	47.13
$C_{10}H_{22}O$	1-Decanol	231.1		81.50
$C_{10}H_{22}O$	Diisopentyl ether	172.5	35.1	
$C_{10}H_{22}S$	1-Decanethiol	240.6		65.48
$C_{11}H_{10}$	1-Methylnaphthalene	244.7	45.5	
$C_{11}H_{21}N$	Undecanenitrile	253		71.14
$C_{11}H_{22}$	Pentylcyclohexane	203.7		53.88
$C_{11}H_{24}$	Undecane	195.9	41.91	56.58
$C_{11}H_{24}$	2-Methyldecane	189.3	40.25	54.28
$C_{11}H_{24}$	4-Methyldecane	187	40.70	53.76
$C_{11}H_{24}$	2,4,7-Trimethyloctane	168.1	38.22	49.91
$C_{12}F_{27}N$	Tris(perfluorobutyl)amine	178	46.4	
$C_{12}H_{10}O$	Diphenyl ether	258.0	48.2	
$C_{12}H_{16}$	Cyclohexylbenzene	240.1		59.94
$C_{12}H_{22}$	Cyclohexylcyclohexane	238		57.98
$C_{12}H_{23}N$	Dodecanenitrile	277		76.12
$C_{12}H_{24}$	1-Dodecene	213.8		60.78
$C_{12}H_{26}$	2,2,4,6,6-Pentamethylheptane	177.8		48.97
$C_{12}H_{26}$	Dodecane	216.32	44.09	61.52
$C_{12}H_{26}O$	1-Dodecanol	259		91.96
$C_{12}H_{27}BO_3$	Tributyl borate	234	56.1	
$C_{12}H_{27}N$	Tributylamine	216.5	46.9	
$C_{13}H_{13}N$	*N*-Benzylaniline	306.5		79.6
$C_{13}H_{26}O_2$	Methyl dodecanoate	267		77.17
$C_{13}H_{28}$	Tridecane	235.47	46.20	66.68
$C_{14}H_{10}$	Phenanthrene	340		75.50
$C_{14}H_{12}O_2$	Benzyl benzoate	323.5	53.6	
$C_{14}H_{27}N$	Tetradecanenitrile			85.29
$C_{14}H_{30}$	Tetradecane	253.58	48.16	71.73
$C_{14}H_{30}O$	1-Tetradecanol	289		102.20
$C_{15}H_{32}$	Pentadecane	270.6	50.08	76.77
$C_{16}H_{22}O_4$	Dibutyl phthalate	340	79.2	
$C_{16}H_{32}$	1-Hexadecene	284.9		80.25
$C_{16}H_{34}$	Hexadecane	286.86	51.84	81.35
$C_{17}H_{36}$	Heptadecane	302.0	53.58	86.47
$C_{18}H_{34}O_2$	Oleic acid	360	67.4	
$C_{18}H_{38}$	Octadecane	316.3	55.23	91.44
$C_{19}H_{40}$	Nonadecane	329.9	56.93	96.4
$C_{20}H_{42}$	Eicosane	343	58.49	101.81

ENTHALPY OF FUSION

This table lists the molar enthalpy (heat) of fusion, $\Delta_{fus}H$, of over 1100 inorganic and organic compounds. All values refer to the enthalpy change at equilibrium between the liquid phase and the most stable solid phase at the phase transition temperature. Most values of $\Delta_{fus}H$ are given at the normal melting point t_m. However, a "t" following the entry in the melting point column indicate a triple-point temperature, where the solid, liquid, and gas phases are in equilibrium. Temperatures are given on the ITS-90 scale.

A * following an entry indicates that the value includes the enthalpy of transition between crystalline phases whose transformation occurs within 1°C of the melting point.

Substances are listed by name, either an IUPAC systematic name or, in the case of drugs and other complex compounds, a common synonym. Inorganic compounds, including metal salts of organic acids, are listed first, followed by organic compounds. The molecular formula in the Hill convention is included.

References

1. Chase, M. W., Davies, C. A., Downey, J. R., Frurip, D. J., McDonald, R. A., and Syverud, A. N., *JANAF Thermochemical Tables, Third Edition, J. Phys. Chem. Ref. Data*, Vol. 14, Suppl. 1, 1985.
2. Chase, M. W., *NIST-JANAF Thermochemical Tables, Fourth Edition, J. Phys. Chem. Ref. Data*, Monograph No. 9, 1998.
3. Gurvich, L. V., Veyts, I. V., and Alcock, C. B., *Thermodynamic Properties of Individual Substances, Fourth Edition*; Vol. 2, Hemisphere Publishing Corp., New York, 1991; Vol. 3, CRC Press, Boca Raton, FL, 1994.
4. Dinsdale, A. T., "SGTE Data for Pure Elements", *CALPHAD*, 15, 317-425, 1991.
5. *Landolt-Börnstein, Numerical Data and Functional Relationships in Science and Technology, New Series*, IV/8A, "Enthalpies of Fusion and Transition of Organic Compounds", Springer-Verlag, Heidelberg, 1995.
6. *Landolt-Börnstein, Numerical Data and Functional Relationships in Science and Technology, New Series*, IV/19A, "Thermodynamic Properties of Inorganic Materials compiled by SGTE", Springer-Verlag, Heidelberg; Part 1, 1999; Part 2; 1999; Part 3, 2000; Part 4, 2001.
7. Janz, G. J., et al., *Physical Properties Data Compilations Relevant to Energy Storage. II. Molten Salts*, Nat. Stand. Ref. Data Sys.- Nat. Bur. Standards (U.S.), No. 61, Part 2, 1979.
8. Dirand, M., Bouroukba, M., Chevallier, V., Petitjean, D., Behar, E., and Ruffier-Meray, V., "Normal Alkanes, Multialkane Synthetic Model Mixtures, and Real Petroleum Waxes: Crystallographic Structures, Thermodynamic Properties, and Crystallization", *J. Chem. Eng. Data*, 47, 115-143, 2002.
9. Linstrom, P. J., and Mallard, W. G., Editors, *NIST Chemistry WebBook*, NIST Standard Reference Database No. 69, June 2005, National Institute of Standards and Technology, Gaithersburg, MD 20899, <http://webbook.nist.gov>.
10. Thermodynamic Research Center, National Institute of Standards and Technology, *TRC Thermodynamic Tables*, <http://trc.nist.gov>.
11. Sangster, J., "Phase Diagrams and Thermodynamic Properties of Binary Systems of Drugs", J. Phys. Chem. Ref. Data 28, 889, 1999.

Name	Molecular formula	t_m/°C	$\Delta_{fus}H$/kJ mol^{-1}
Inorganic compounds (including salts of organic acids)			
Actinium	Ac	1050	12.0
Aluminum	Al	660.32	10.71
Aluminum bromide	AlBr$_3$	97.5	11.25
Aluminum chloride	AlCl$_3$	192.6	35.35
Aluminum fluoride	AlF$_3$	2250 t	0.56
Aluminum iodide	AlI$_3$	188.28	15.90
Aluminum oxide (α)	Al$_2$O$_3$	2054	111.1
Aluminum sulfide	Al$_2$S$_3$	1100	66
Americium	Am	1176	14.39
Ammonia	H$_3$N	-77.73	5.66
Ammonium chloride	ClH$_4$N	520.1	10.6
Ammonium fluoride	FH$_4$N	238	12.6
Ammonium iodide	H$_4$IN	551	21
Ammonium nitrate	H$_4$N$_2$O$_3$	169.7	5.86
Antimony (gray)	Sb	630.628	19.79
Antimony(III) bromide	Br$_3$Sb	97	14.6
Antimony(III) chloride	Cl$_3$Sb	73.4	12.97
Antimony(III) fluoride	F$_3$Sb	287	22.8
Antimony(III) iodide	I$_3$Sb	171	22.8
Antimony(III) oxide (valentinite)	O$_3$Sb$_2$	655	54
Antimony(III) sulfide	S$_3$Sb$_2$	550	47.9
Argon	Ar	-189.36	1.18
Arsenic (gray)	As	817	24.44
Arsenic(III) bromide	AsBr$_3$	31.1	11.7
Arsenic(III) chloride	AsCl$_3$	-16	10.1
Arsenic(III) fluoride	AsF$_3$	-5.9	10.4
Arsenic(III) iodide	AsI$_3$	141	21.8
Arsenic(III) oxide (claudetite)	As$_2$O$_3$	314	18
Arsenic(V) oxide	As$_2$O$_5$	730	60
Arsenic(III) selenide	As$_2$Se$_3$	377	40.8
Arsenic(III) sulfide	As$_2$S$_3$	312	28.7
Arsenic sulfide	As$_4$S$_4$	307	25.4
Arsenic(III) telluride	As$_2$Te$_3$	375	46.0
Barium	Ba	727	7.12
Barium bromide	BaBr$_2$	857	32.2
Barium carbonate	CBaO$_3$	1555 (high pres.)	40
Barium chloride	BaCl$_2$	961	15.85
Barium fluoride	BaF$_2$	1368	23.36
Barium hydride	BaH$_2$	1200	25
Barium hydroxide	BaH$_2$O$_2$	408	16
Barium iodide	BaI$_2$	711	26.5
Barium oxide	BaO	1973	46
Barium sulfate	BaO$_4$S	1580	40
Barium sulfide	BaS	2227	63
Beryllium	Be	1287	7.895
Beryllium bromide	BeBr$_2$	508	18
Beryllium carbide	CBe$_2$	2127	75.3
Beryllium chloride	BeCl$_2$	415	8.66
Beryllium fluoride	BeF$_2$	552	4.77
Beryllium iodide	BeI$_2$	480	20.92
Beryllium nitride	Be$_3$N$_2$	2200	111
Beryllium oxide	BeO	2578	86
Beryllium sulfate	BeO$_4$S	1127	6
Bismuth	Bi	271.406	11.106
Bismuth oxide	Bi$_2$O$_3$	825	14.7

Enthalpy of Fusion

Name	Molecular formula	$t_m/°C$	$\Delta_{fus}H/kJ\ mol^{-1}$	Name	Molecular formula	$t_m/°C$	$\Delta_{fus}H/kJ\ mol^{-1}$
Bismuth sulfide	Bi_2S_3	777	78.2	Chromium(II) fluoride	CrF_2	894	34
Bismuth tribromide	$BiBr_3$	219	21.7	Chromium(III) fluoride	CrF_3	1425	66
Bismuth trichloride	$BiCl_3$	234	23.6	Chromium(II) iodide	CrI_2	867	46
Bismuth trifluoride	BiF_3	649	21.6	Chromium(III) iodide	CrI_3	857	61
Bismuth triiodide	BiI_3	408.6	39.1	Chromium(III) oxide	Cr_2O_3	2432	125
Boric acid	BH_3O_3	170.9	22.3	Chromium(VI) oxide	CrO_3	197	14.2
Boron	B	2075	50.2	Chromium(II) sulfide	CrS	1567	25.5
Boron nitride	BN	2967	81	Cobalt	Co	1495	16.20
Boron oxide	B_2O_3	450	24.56	Cobalt(II) bromide	Br_2Co	678	43
Boron sulfide	B_2S_3	563	48.12	Cobalt(II) chloride	Cl_2Co	737	46.0
Boron trichloride	BCl_3	-107.3	2.10	Cobalt(II) fluoride	CoF_2	1127	58.1
Boron trifluoride	BF_3	-126.8	4.20	Cobalt(II) iodide	CoI_2	520	35
Bromine	Br_2	-7.2	10.57	Cobalt(II) selenite	CoO_3Se	659	16.3
Bromine pentafluoride	BrF_5	-60.5	5.67	Cobalt(II) sulfide	CoS	1117	30
Cadmium	Cd	321.069	6.21	Copper	Cu	1084.62	13.26
Cadmium bromide	Br_2Cd	568	33.35	Copper(I) bromide	BrCu	483	5.1
Cadmium chloride	$CdCl_2$	568	48.58	Copper(I) chloride	ClCu	423	7.08
Cadmium fluoride	CdF_2	1075	22.6	Copper(II) chloride	Cl_2Cu	598	15.0
Cadmium iodide	CdI_2	388	15.3	Copper(II) fluoride	CuF_2	836	55
Cadmium nitrate	CdN_2O_6	360	18.3	Copper(I) iodide	CuI	591	7.93
Calcium	Ca	842	8.54	Copper(I) oxide	Cu_2O	1244	65.6
Calcium bromide	Br_2Ca	742	29.1	Copper(II) oxide	CuO	1227	49
Calcium carbonate (calcite)	$CCaO_3$	1330	36	Copper(I) sulfide	Cu_2S	1129	9.62
Calcium chloride	$CaCl_2$	775	28.05	Curium	Cm	1345	14.64
Calcium fluoride	CaF_2	1418	30	Decaborane(14)	$B_{10}H_{14}$	98.78	21.97
Calcium hydride	CaH_2	1000	6.7	Dysprosium	Dy	1412	11.35
Calcium iodide	CaI_2	783	41.8	Dysprosium(III) fluoride	DyF_3	1157	58.6
Calcium nitrate	CaN_2O_6	561	23.4	Dysprosium(III) oxide	Dy_2O_3	2408	120
Calcium oxide	CaO	2613	80	Einsteinium	Es	860	9.41
Calcium sulfate	CaO_4S	1460	28	Erbium	Er	1529	19.90
Calcium sulfide	CaS	2524	70	Erbium chloride	Cl_3Er	776	32.6
Carbon (graphite)	C	4489	117.4	Erbium fluoride	ErF_3	1146	28.2
Cerium	Ce	799	5.460	Erbium oxide	Er_2O_3	2418	130
Cerium(III) bromide	Br_3Ce	732	51.9	Europium	Eu	822	9.21
Cerium(III) chloride	$CeCl_3$	807	53.1	Europium(II) bromide	Br_2Eu	683	25.1
Cerium(III) fluoride	CeF_3	1430	55.6	Europium(III) chloride	Cl_3Eu	623	33.1
Cerium(III) iodide	CeI_3	760	51.0	Europium(III) fluoride	EuF_3	647	6.40
Cerium(III) oxide	Ce_2O_3	2250	120	Europium(II) oxide	EuO	1967	40
Cerium(IV) oxide	CeO_2	2480	80	Europium(III) oxide	Eu_2O_3	2350	117
Cesium	Cs	28.5	2.09	Fluorine	F_2	-219.67	0.51
Cesium carbonate	CCs_2O_3	793	31	Gadolinium	Gd	1313	9.67
Cesium chloride	ClCs	646	20.4	Gadolinium(III) bromide	Br_3Gd	785	38.1
Cesium chromate	$CrCs_2O_4$	963	35.3	Gadolinium(III) chloride	Cl_3Gd	602	40.6
Cesium fluoride	CsF	703	21.7	Gadolinium(III) fluoride	F_3Gd	1232	52.4
Cesium hydride	CsH	528	15	Gadolinium(III) iodide	GdI_3	930	54.0
Cesium hydroxide	CsHO	342.3	7.78	Gadolinium(III) oxide	Gd_2O_3	2425	60
Cesium iodide	CsI	632	25.7	Gallium	Ga	29.7666	5.585
Cesium metaborate	$BCsO_2$	732	27	Gallium antimonide	GaSb	712	25.1
Cesium molybdate	Cs_2MoO_4	956.3	31.8	Gallium arsenide	AsGa	1238	87.64
Cesium nitrate	$CsNO_3$	409	13.8	Gallium(III) bromide	Br_3Ga	123	11.7
Cesium nitrite	$CsNO_2$	406	10.9	Gallium(III) chloride	Cl_3Ga	77.9	11.51
Cesium oxide	Cs_2O	495	20	Gallium(III) iodide	GaI_3	212	12.9
Cesium peroxide	Cs_2O_2	594	22	Gallium(III) oxide	Ga_2O_3	1807	100
Cesium sulfate	Cs_2O_4S	1005	35.7	Germanium	Ge	938.25	36.94
Chlorine	Cl_2	-101.5	6.40	Germanium(IV) bromide	Br_4Ge	26.1	12
Chromium	Cr	1907	21.00	Germanium(II) iodide	GeI_2	428	33.3
Chromium(II) bromide	Br_2Cr	842	45	Germanium(IV) iodide	GeI_4	146	19.1
Chromium(III) bromide	Br_3Cr	812	60	Germanium(IV) oxide	GeO_2	1116	12.6
Chromium(II) chloride	Cl_2Cr	824	45.0	Germanium(II) selenide	GeSe	675	24.7
Chromium(III) chloride	Cl_3Cr	827	60	Germanium(II) sulfide	GeS	658	21.3

Name	Molecular formula	$t_m/°C$	$\Delta_{fus}H$/kJ mol^{-1}
Germanium(IV) sulfide	GeS$_2$	840	16.3
Germanium(II) telluride	GeTe	724	47.3
Gold	Au	1064.18	12.55
Hafnium	Hf	2233	27.20
Hafnium nitride	HfN	3310	62.8
Hafnium(IV) oxide	HfO$_2$	2800	96
Holmium	Ho	1472	11.76
Holmium bromide	Br$_3$Ho	919	50.1
Holmium chloride	Cl$_3$Ho	720	30.5
Holmium fluoride	F$_3$Ho	1143	56.3
Holmium oxide	Ho$_2$O$_3$	2415	130
Hydrazine	H$_4$N$_2$	1.54	12.66
Hydrogen	H$_2$	-259.198 t	0.12
Hydrogen bromide	BrH	-86.80	2.41
Hydrogen chloride	ClH	-114.17	2.00
Hydrogen fluoride	FH	-83.36	4.58
Hydrogen iodide	HI	-50.76	2.87
Hydrogen peroxide	H$_2$O$_2$	-0.43	12.50
Hydrogen sulfide	H$_2$S	-85.5	2.38
Indium	In	156.60	3.291
Indium antimonide	InSb	524	47.7
Indium arsenide	AsIn	942	77.0
Indium(I) bromide	BrIn	285	24.3
Indium(III) bromide	Br$_3$In	420	26
Indium(I) chloride	ClIn	225	9.20
Indium(III) chloride	Cl$_3$In	583	27
Indium(III) fluoride	F$_3$In	1172	64
Indium(I) iodide	IIn	364.4	17.26
Indium(II) iodide	I$_2$In	155	1.29
Indium(III) iodide	I$_3$In	207	18.48
Indium(III) oxide	In$_2$O$_3$	1912	105
Indium(II) sulfide	InS	692	36.0
Iodine	I$_2$	113.7	15.52
Iodine chloride	ClI	27.38	11.6
Iridium	Ir	2446	41.12
Iridium(VI) fluoride	F$_6$Ir	44	8.40
Iron	Fe	1538	13.81
Iron boride (FeB)	BFe	1658	62.66
Iron(II) bromide	Br$_2$Fe	691	43.0
Iron(II) chloride	Cl$_2$Fe	677	42.83
Iron(III) chloride	Cl$_3$Fe	307.6	40
Iron(II) fluoride	F$_2$Fe	1100	50
Iron(III) fluoride	F$_3$Fe	367	0.58
Iron(II) iodide	FeI$_2$	594	39
Iron(II) oxide	FeO	1377	24.1
Iron(II,III) oxide	Fe$_3$O$_4$	1597	138
Iron(III) oxide	Fe$_2$O$_3$	1539	87
Iron sodium oxide	FeNaO$_2$	1347	49.4
Iron(II) sulfide	FeS	1188	31.5
Krypton	Kr	-157.38	1.64
Lanthanum	La	920	6.20
Lanthanum bromide	Br$_3$La	788	54.0
Lanthanum chloride	Cl$_3$La	858	54.4
Lanthanum fluoride	F$_3$La	1493	50.2
Lanthanum iodide	I$_3$La	778	56.1
Lead	Pb	327.462	4.774
Lead(II) bromide	Br$_2$Pb	371	16.44
Lead(II) chloride	Cl$_2$Pb	501	21.88
Lead(II) fluoride	F$_2$Pb	830	14.7
Lead(II) iodide	I$_2$Pb	410	23.4

Name	Molecular formula	$t_m/°C$	$\Delta_{fus}H$/kJ mol^{-1}
Lead(II) oxide (massicot)	OPb	887	25.6
Lead(II) sulfate	O$_4$PbS	1087	40.2
Lead(II) sulfide	PbS	1113	49.4
Lithium	Li	180.50	3.00
Lithium aluminate	AlLiO$_2$	1610	87.9
Lithium bromide	BrLi	550	17.66
Lithium carbonate	CLi$_2$O$_3$	732	44.8
Lithium chloride	ClLi	610	19.8
Lithium chromate	CrLi$_2$O$_4$	482	30.5
Lithium fluoride	FLi	848.2	27.09
Lithium hexafluoroaluminate	AlF$_6$Li$_3$	785	86.19
Lithium hydride	HLi	692	21.8
Lithium hydride-d	DLi	694	22
Lithium hydroxide	HLiO	473	20.9
Lithium iodide	ILi	469	14.6
Lithium metasilicate	Li$_2$O$_3$Si	1201	28
Lithium nitrate	LiNO$_3$	253	26.7
Lithium nitrite	LiNO$_2$	222	9.2
Lithium oxide	Li$_2$O	1437	35.6
Lithium perchlorate	ClLiO$_4$	236	29.3
Lithium sulfate	Li$_2$O$_4$S	860	9.00
Lutetium	Lu	1663	18.65
Lutetium oxide	Lu$_2$O$_3$	2490	133
Magnesium	Mg	650	8.48
Magnesium bromide	Br$_2$Mg	711	39.3
Magnesium carbonate	CMgO$_3$	990	59
Magnesium chloride	Cl$_2$Mg	714	43.1
Magnesium fluoride	F$_2$Mg	1263	58.7
Magnesium hydride	H$_2$Mg	327	14
Magnesium iodide	I$_2$Mg	634	26
Magnesium orthosilicate	Mg$_2$O$_4$Si	1897	71
Magnesium oxide	MgO	2825	77
Magnesium phosphate	Mg$_3$O$_8$P$_2$	1348	121
Magnesium sulfate	MgO$_4$S	1137	14.6
Magnesium sulfide	MgS	2226	63
Magnesium tetraboride	B$_4$Mg	727	0.0
Manganese	Mn	1246	12.91
Manganese(II) bromide	Br$_2$Mn	698	33.5
Manganese(II) chloride	Cl$_2$Mn	650	30.7
Manganese(II) fluoride	F$_2$Mn	900	30
Manganese(II) iodide	I$_2$Mn	638	41.8
Manganese(II) oxide	MnO	1842	43.9
Manganese(II) sulfide (α form)	MnS	1530	26.1
Mercury	Hg	-38.829	2.295
Mercury(II) bromide	Br$_2$Hg	241	17.9
Mercury(II) chloride	Cl$_2$Hg	277	19.41
Mercury(II) fluoride	F$_2$Hg	645	23.0
Mercury(I) iodide	Hg$_2$I$_2$	290	31.4
Mercury(II) iodide (yellow)	HgI$_2$	256	15.6
Mercury(II) sulfide (black)	HgS	820	40
Metaboric acid (γ form)	BHO$_2$	236	14.3
Molybdenum	Mo	2623	37.48
Molybdenum boride (Mo$_2$B$_5$)	B$_5$Mo$_2$	2210	226
Molybdenum(IV) chloride	Cl$_4$Mo	317	16.7
Molybdenum(V) chloride	Cl$_5$Mo	194	19
Molybdenum(VI) dioxydichloride	Cl$_2$MoO$_2$	176	17.0
Molybdenum(V) fluoride	F$_5$Mo	45.67	6.1
Molybdenum(VI) fluoride	F$_6$Mo	17.5	4.33
Molybdenum monoboride	BMo	2600	55.23
Molybdenum(VI) oxide	MoO$_3$	802	48.7

Name	Molecular formula	$t_m/°C$	$\Delta_{fus}H/kJ\ mol^{-1}$	Name	Molecular formula	$t_m/°C$	$\Delta_{fus}H/kJ\ mol^{-1}$
Molybdenum(VI) oxytetrachloride	Cl_4MoO	105	14.3	Plutonium(III) iodide	I_3Pu	777	50.2
Molybdenum(VI) oxytetrafluoride	F_4MoO	97.2	4	Plutonium(III) oxide	O_3Pu_2	2085	113
Molybdenum(V) oxytrichloride	Cl_3MoO	310	22	Plutonium(IV) oxide	O_2Pu	2390	67
Molybdenum(III) sulfide	Mo_2S_3	1807	0.13	Polonium	Po	254	10.0
Neodymium	Nd	1016	7.14	Potassium	K	63.5	2.335
Neodymium(III) bromide	Br_3Nd	682	45.3	Potassium aluminate	$AlKO_2$	1713	82
Neodymium(III) chloride	Cl_3Nd	759	48.5	Potassium bromide	BrK	734	25.52
Neodymium(III) fluoride	F_3Nd	1377	54.8	Potassium carbonate	CK_2O_3	899	27.6
Neodymium(III) iodide	I_3Nd	787	41.5	Potassium chloride	ClK	771	26.28
Neon	Ne	-248.609	0.328	Potassium chromate	CrK_2O_4	974	33.0
Neptunium	Np	644	3.20	Potassium cyanide	CKN	622	14.6
Nickel	Ni	1455	17.48	Potassium fluoride	FK	858	27.2
Nickel boride (Ni_2B)	BNi_2	1125	42.15	Potassium fluoroborate	BF_4K	570	17.66
Nickel boride (Ni_3B)	BNi_3	1166	72.28	Potassium hydride	HK	619	21
Nickel(II) bromide	Br_2Ni	963	56	Potassium hydrogen fluoride	F_2HK	238.8	6.62
Nickel(II) chloride	Cl_2Ni	1031	77.9	Potassium hydroxide	HKO	406	7.90
Nickel(II) fluoride	F_2Ni	1380	69	Potassium iodide	IK	681	24.0
Nickel(II) iodide	I_2Ni	800	48	Potassium metaborate	BKO_2	947	31.38
Nickel(II) oxide	NiO	1957	50.7	Potassium nitrate	KNO_3	334	9.6
Nickel(II) sulfide	NiS	976	30.1	Potassium nitrite	KNO_2	438	16.7
Nickel disulfide	NiS_2	1007	65.7	Potassium oxide	K_2O	740	27
Nickel subsulfide	Ni_3S_2	789	19.7	Potassium peroxide	K_2O_2	545	20.5
Niobium	Nb	2477	30	Potassium sulfate	K_2O_4S	1069	36.6
Niobium(V) bromide	Br_5Nb	254	24.0	Potassium sulfide	K_2S	948	16.15
Niobium(V) chloride	Cl_5Nb	205.8	33.9	Potassium superoxide	KO_2	535	20.6
Niobium(V) fluoride	F_5Nb	80	12.2	Praseodymium	Pr	931	6.89
Niobium(V) iodide	I_5Nb	327	37.7	Praseodymium(III) bromide	Br_3Pr	693	47.3
Niobium nitride	NNb	2050	46.0	Praseodymium(III) chloride	Cl_3Pr	786	50.6
Niobium(II) oxide	NbO	1937	85.4	Praseodymium(III) fluoride	F_3Pr	1399	57.3
Niobium(IV) oxide	NbO_2	1901	92	Praseodymium(III) iodide	I_3Pr	738	53.1
Niobium(V) oxide	Nb_2O_5	1512	104.3	Protactinium	Pa	1572	12.34
Nitric acid	HNO_3	-41.6	10.5	Radium	Ra	696	7.7
Nitric oxide	NO	-163.6	2.30	Rhenium	Re	3185	34.08
Nitrogen	N_2	-210.0	0.71	Rhenium(VII) oxide	O_7Re_2	327	65.7
Nitrogen tetroxide	N_2O_4	-9.3	14.65	Rhodium	Rh	1964	26.59
Nitrous oxide	N_2O	-90.8	6.54	Rubidium	Rb	39.30	2.19
Osmium	Os	3033	57.85	Rubidium bromide	BrRb	692	23.3
Osmium(VIII) oxide	O_4Os	40.6	14.3	Rubidium carbonate	CO_3Rb_2	873	30
Oxygen	O_2	-218.79	0.44	Rubidium chloride	ClRb	724	24.4
Palladium	Pd	1554.8	16.74	Rubidium fluoride	FRb	795	25.8
Palladium(II) chloride	Cl_2Pd	679	18.41	Rubidium hydride	HRb	585	22
Phosphinic acid	H_3O_2P	26.5	9.7	Rubidium hydroxide	HORb	385	8.0
Phosphonic acid	H_3O_3P	74.4	12.8	Rubidium iodide	IRb	656	22.1
Phosphoric acid	H_3O_4P	42.4	13.4	Rubidium metaborate	BO_2Rb	860	31
Phosphorus (white)	P	44.15	0.659	Rubidium nitrate	NO_3Rb	310	4.6
Phosphorus (red)	P	579.2	18.54	Rubidium nitrite	NO_2Rb	422	12.1
Phosphorus(III) chloride	Cl_3P	-93	7.10	Rubidium oxide	ORb_2	505	20
Phosphorus heptasulfide	P_4S_7	308	36.6	Rubidium peroxide	O_2Rb_2	570	21
Phosphorus(V) oxide	O_5P_2	562	27.2	Rubidium sulfate	O_4Rb_2S	1066	37.3
Phosphorus sesquisulfide	P_4S_3	173	20.1	Rubidium superoxide	O_2Rb	540	21
Phosphoryl chloride	Cl_3OP	1.18	13.1	Ruthenium	Ru	2334	38.59
Platinum	Pt	1768.2	22.175	Ruthenium(V) fluoride	F_5Ru	101	74.5
Plutonium	Pu	640	2.824	Samarium	Sm	1072	8.62
Plutonium(III) bromide	Br_3Pu	681	58.6	Samarium(III) oxide	O_3Sm_2	2335	119
Plutonium(III) chloride	Cl_3Pu	760	63.6	Scandium	Sc	1541	14.10
Plutonium(III) fluoride	F_3Pu	1396	59.8	Scandium chloride	Cl_3Sc	967	67.4
Plutonium(IV) fluoride	F_4Pu	1037	42.7	Scandium fluoride	F_3Sc	1552	62.6
Plutonium(VI) fluoride	F_6Pu	51.6	18.6	Scandium oxide	O_3Sc_2	2489	127
				Selenium (gray)	Se	220.8	6.69
				Selenium dioxide	O_2Se	360	17.6

Name	Molecular formula	$t_m/°C$	$\Delta_{fus}H$/kJ mol^{-1}	Name	Molecular formula	$t_m/°C$	$\Delta_{fus}H$/kJ mol^{-1}
Silicon	Si	1414	50.21	Tantalum(V) oxide	O_5Ta_2	1875	120
Silicon dioxide (cristobalite)	O_2Si	1722	9.6	Technetium	Tc	2157	33.29
Silicon monosulfide	SSi	1090	31	Tellurium	Te	449.51	17.38
Silver	Ag	961.78	11.30	Tellurium dioxide	O_2Te	733	28.9
Silver(I) bromide	AgBr	430	9.163	Tellurium tetrabromide	Br_4Te	380	24.7
Silver(I) chloride	AgCl	455	13.054	Tellurium tetrachloride	Cl_4Te	224	18.9
Silver(I) iodide	AgI	558	9.414	Terbium	Tb	1359	10.15
Silver(I) nitrate	$AgNO_3$	210	11.72	Terbium(III) bromide	Br_3Tb	830	31.5
Silver(I) oxide	Ag_2O	827	15	Terbium(III) chloride	Cl_3Tb	582	19.5
Silver(I) sulfate	Ag_2O_4S	660	17.99	Tetrachlorosilane	Cl_4Si	-68.74	7.60
Silver(I) sulfide	Ag_2S	836	7.9	Tetraiodosilane	I_4Si	120.5	19.7
Sodium	Na	97.794	2.60	Thallium	Tl	304	4.142
Sodium bromate	$BrNaO_3$	381	28.11	Thallium(I) bromide	BrTl	460	16.4
Sodium bromide	BrNa	747	26.23	Thallium(I) carbonate	CO_3Tl_2	273	18
Sodium carbonate	CNa_2O_3	856	29.7	Thallium(I) chloride	ClTl	431	15.56
Sodium chlorate	$ClNaO_3$	248	22.6	Thallium(I) fluoride	FTl	326	13.87
Sodium chloride	ClNa	800.7	28.16	Thallium(I) formate	CHO_2Tl	101	10.9
Sodium chromate	$CrNa_2O_4$	794	24.7	Thallium(I) iodide	ITl	441.7	14.7
Sodium cyanide	CNNa	562	8.79	Thallium(I) nitrate	NO_3Tl	206	9.6
Sodium fluoride	FNa	996	33.35	Thallium(I) oxide	OTl_2	579	30.3
Sodium formate	$CHNaO_2$	257.3	17.7	Thallium(III) oxide	O_3Tl_2	834	53
Sodium hexafluoroaluminate	AlF_6Na_3	1013	114.4	Thallium(I) sulfate	O_4STl_2	632	23.8
Sodium hexafluorosilicate	F_6Na_2Si	847	99.6	Thallium(I) sulfide	STl_2	457	23.0
Sodium hydride	HNa	638	26	Thorium	Th	1750	13.81
Sodium hydroxide	HNaO	323	6.60	Thorium(IV) bromide	Br_4Th	679	54.4
Sodium iodate	$INaO_3$	422	35.1	Thorium(IV) chloride	Cl_4Th	770	43.9
Sodium iodide	INa	661	23.7	Thorium(IV) fluoride	F_4Th	1110	41.8
Sodium metaborate	$BNaO_2$	966	36.2	Thorium(IV) iodide	I_4Th	566	48.1
Sodium metasilicate	Na_2O_3Si	1089	51.8	Thorium(IV) oxide	O_2Th	3350	90
Sodium nitrate	$NNaO_3$	306.5	15.5	Thulium	Tm	1545	16.84
Sodium nitrite	$NNaO_2$	284	14.9	Thulium(III) chloride	Cl_3Tm	845	34.9
Sodium oxide	Na_2O	1134	47.7	Thulium(III) fluoride	F_3Tm	1158	28.9
Sodium peroxide	Na_2O_2	675	24.5	Tin (white)	Sn	231.93	7.148
Sodium sulfate	Na_2O_4S	884	23.85	Tin(II) bromide	Br_2Sn	232	18.0
Sodium sulfide	Na_2S	1172	19	Tin(IV) bromide	Br_4Sn	29.1	12.2
Sodium sulfite	Na_2O_3S	911	25.9	Tin(II) chloride	Cl_2Sn	247.0	14.52
Strontium	Sr	777	7.43	Tin(IV) chloride	Cl_4Sn	-34.07	9.20
Strontium bromide	Br_2Sr	657	10.5	Tin(II) fluoride	F_2Sn	215	10.5
Strontium carbonate	CO_3Sr	1494	40	Tin(IV) fluoride	F_4Sn	442	27.6
Strontium chloride	Cl_2Sr	874	16.22	Tin(II) iodide	I_2Sn	320	18.0
Strontium fluoride	F_2Sr	1477	29.7	Tin(IV) iodide	I_4Sn	402	0.16
Strontium hydride	H_2Sr	1050	23	Tin(II) oxide	OSn	977	27.7
Strontium hydroxide	H_2O_2Sr	535	23	Tin(IV) oxide	O_2Sn	1630	23.4
Strontium iodide	I_2Sr	538	19.7	Tin(II) sulfide	SSn	881	31.6
Strontium nitrate	N_2O_6Sr	570	44.6	Tin(II) telluride	SnTe	806	45.2
Strontium oxide	OSr	2531	81	Titanium	Ti	1668	14.15
Strontium sulfate	O_4SSr	1606	36	Titanium boride	B_2Ti	2920	100.4
Strontium sulfide	SSr	2226	63	Titanium(IV) bromide	Br_4Ti	38.3	12.9
Sulfur (monoclinic)	S	115.21	1.721	Titanium(II) chloride	Cl_2Ti	1035	34.3
Sulfur hexafluoride	F_6S	-49.596	5.02	Titanium(IV) chloride	Cl_4Ti	-24.12	9.97
Sulfuric acid	H_2O_4S	10.31	10.71	Titanium(IV) fluoride	F_4Ti	377	41
Sulfur trioxide (γ-form)	O_3S	16.8	8.60	Titanium(IV) iodide	I_4Ti	155	19.8
Tantalum	Ta	3017	36.57	Titanium nitride	NTi	2947	66.9
Tantalum boride (TaB$_2$)	B_2Ta	3100	83.68	Titanium(III) oxide	O_3Ti_2	1842	104.6
Tantalum(V) bromide	Br_5Ta	240	37.7	Titanium(IV) oxide (rutile)	O_2Ti	1912	68
Tantalum(V) chloride	Cl_5Ta	216.6	35.1	Titanium(II) sulfide	STi	1927	32
Tantalum(V) fluoride	F_5Ta	96.9	12	Tungsten	W	3422	52.31
Tantalum(V) iodide	I_5Ta	496	7.74	Tungsten boride (WB)	BW	2800	80
Tantalum nitride (TaN)	NTa	3090	6.7	Tungsten boride (W$_2$B)	BW_2	2740	117
Tantalum nitride (Ta$_2$N)	NTa_2	2727	92.0	Tungsten boride (W$_2$B$_5$)	B_5W_2	2370	240

Name	Molecular formula	$t_m/°C$	$\Delta_{fus}H$/kJ mol^{-1}	Name	Molecular formula	$t_m/°C$	$\Delta_{fus}H$/kJ mol^{-1}
Tungsten(V) bromide	Br$_5$W	286	17.2	Zirconium(II) iodide	I$_2$Zr	827	28
Tungsten(V) chloride	Cl$_5$W	253	20.6	Zirconium(III) iodide	I$_3$Zr	727	33
Tungsten(VI) chloride	Cl$_6$W	282	6.69	Zirconium(IV) iodide	I$_4$Zr	500	32
Tungsten(VI) fluoride	F$_6$W	1.9	4.10	Zirconium nitride	NZr	2952	67.4
Tungsten(VI) oxide	O$_3$W	1473	73	Zirconium(IV) oxide	O$_2$Zr	2710	90
Tungsten(VI) oxytetrachloride	Cl$_4$OW	210	18.8	Zirconium(IV) sulfide	S$_2$Zr	1550	45
Tungsten(VI) oxytetrafluoride	F$_4$OW	105	6				
Uranium	U	1135	9.14	*Organic compounds*			
Uranium(III) bromide	Br$_3$U	727	43.9	Acenaphthene	C$_{12}$H$_{10}$	93.4	21.49
Uranium(IV) bromide	Br$_4$U	519	55.2	Acenaphthylene	C$_{12}$H$_8$	91.8	6.9
Uranium(IV) chloride	Cl$_4$U	590	44.8	Acetaldehyde	C$_2$H$_4$O	-123.37	2.31
Uranium(III) fluoride	F$_3$U	1495	36.8	Acetamide	C$_2$H$_5$NO	80.16	15.59
Uranium(IV) fluoride	F$_4$U	1036	47	Acetaminophen	C$_8$H$_9$NO$_2$	169.3	30.5
Uranium(V) fluoride	F$_5$U	348	35	Acetanilide	C$_8$H$_9$NO	114.3	21.3
Uranium(VI) fluoride	F$_6$U	64.06	19.2	Acetic acid	C$_2$H$_4$O$_2$	16.64	11.73
Uranium(IV) iodide	I$_4$U	506	42.1	Acetic anhydride	C$_4$H$_6$O$_3$	-74.1	10.5
Uranium(IV) oxide	O$_2$U	2847	74.2	Acetone	C$_3$H$_6$O	-94.7	5.77
Uranyl chloride	Cl$_2$O$_2$U	577	44.06	Acetonitrile	C$_2$H$_3$N	-43.82	8.16
Vanadium	V	1910	21.5	Acrylic acid	C$_3$H$_4$O$_2$	12.5	9.51
Vanadium(II) chloride	Cl$_2$V	1350	35.0	Acrylonitrile	C$_3$H$_3$N	-83.48	6.23
Vanadium(IV) chloride	Cl$_4$V	-28	2.30	Allene	C$_3$H$_4$	-136.6	4.40
Vanadium(II) fluoride	F$_2$V	1490	44	Allobarbital	C$_{10}$H$_{12}$N$_2$O$_3$	172	32.3
Vanadium(III) fluoride	F$_3$V	1395	57	2-Aminobenzoic acid	C$_7$H$_7$NO$_2$	146	20.5
Vanadium(V) fluoride	F$_5$V	19.5	49.96	4-Aminobenzoic acid	C$_7$H$_7$NO$_2$	188.2	22.5
Vanadium(II) oxide	OV	1790	50	3-Amino-1-propanol	C$_3$H$_9$NO	12.4	19.7
Vanadium(III) oxide	O$_3$V$_2$	1957	140	Aminopyrine	C$_{13}$H$_{17}$N$_3$O	107.5	27.6
Vanadium(IV) oxide	O$_2$V	1545	56.0	Ampyrone	C$_{11}$H$_{13}$N$_3$O	109	24.9
Vanadium(V) oxide	O$_5$V$_2$	681	64	Aniline	C$_6$H$_7$N	-6.02	10.54
Water	H$_2$O	0.00	6.01	Anisole	C$_7$H$_8$O	-37.13	12.9
Xenon	Xe	-111.745 t	2.27	Anthracene	C$_{14}$H$_{10}$	215.76	29.4
Xenon difluoride	F$_2$Xe	129.03	16.8	Antipyrine	C$_{11}$H$_{12}$N$_2$O	112	27.3
Xenon tetrafluoride	F$_4$Xe	117.1	16.3	trans-Azobenzene	C$_{12}$H$_{10}$N$_2$	67.88	22.52
Xenon hexafluoride	F$_6$Xe	49.48	5.74	trans-Azoxybenzene	C$_{12}$H$_{10}$N$_2$O	34.6	17.9
Ytterbium	Yb	824	7.66	Barbital	C$_8$H$_{12}$N$_2$O$_3$	190	24.7
Ytterbium(III) chloride	Cl$_3$Yb	854	35.4	Benzaldehyde	C$_7$H$_6$O	-57.1	9.32
Yttrium	Y	1522	11.39	Benzamide	C$_7$H$_7$NO	127.3	19.5
Yttrium chloride	Cl$_3$Y	721	31.5	Benz[a]anthracene	C$_{18}$H$_{12}$	160.5	21.4
Yttrium fluoride	F$_3$Y	1155	27.9	Benzene	C$_6$H$_6$	5.49	9.87
Yttrium oxide	O$_3$Y$_2$	2439	81	Benzeneacetic acid	C$_8$H$_8$O$_2$	76.5	16.3
Zinc	Zn	419.53	7.068	1,2-Benzenediamine	C$_6$H$_8$N$_2$	102.1	23.1
Zinc bromide	Br$_2$Zn	402	15.7	1,3-Benzenediamine	C$_6$H$_8$N$_2$	66.0	15.57
Zinc chloride	Cl$_2$Zn	325	10.30	1,4-Benzenediamine	C$_6$H$_8$N$_2$	141.1	23.8
Zinc fluoride	F$_2$Zn	872	40	Benzenethiol	C$_6$H$_6$S	-14.93	11.48
Zinc iodide	I$_2$Zn	450	17	p-Benzidine	C$_{12}$H$_{12}$N$_2$	127	19.1
Zinc oxide	OZn	1974	70	Benzil	C$_{14}$H$_{10}$O$_2$	94.87	23.5
Zinc phosphide (ZnP$_2$)	P$_2$Zn	980	92.9	Benzocaine	C$_9$H$_{11}$NO$_2$	89.7	22.3
Zinc selenite	O$_3$SeZn	621	46.4	Benzoic acid	C$_7$H$_6$O$_2$	122.35	18.02
Zinc sulfide (wurtzite)	SZn	1827	30	Benzonitrile	C$_7$H$_5$N	-13.99	9.1
Zinc telluride	TeZn	1295	63	Benzo[c]phenanthrene	C$_{18}$H$_{12}$	68	16.3
Zirconium	Zr	1854.7	21.00	Benzophenone	C$_{13}$H$_{10}$O	47.9	18.19
Zirconium boride	B$_2$Zr	3050	104.6	Benzo[a]pyrene	C$_{20}$H$_{12}$	181.1	17.3
Zirconium(II) bromide	Br$_2$Zr	827	28	Benzo[e]pyrene	C$_{20}$H$_{12}$	181.4	16.6
Zirconium(III) bromide	Br$_3$Zr	727	33	p-Benzoquinone	C$_6$H$_4$O$_2$	115	18.5
Zirconium(IV) bromide	Br$_4$Zr	450		Benzoyl chloride	C$_7$H$_5$ClO	-0.4	19.2
Zirconium(II) chloride	Cl$_2$Zr	722	27.0	Benzyl alcohol	C$_7$H$_8$O	-15.4	8.97
Zirconium(III) chloride	Cl$_3$Zr	627	30	2,2'-Binaphthalene	C$_{20}$H$_{14}$	187.9	38.9
Zirconium(IV) chloride	Cl$_4$Zr	437	29	Biphenyl	C$_{12}$H$_{10}$	68.93	18.57
Zirconium(II) fluoride	F$_2$Zr	902	37.7	Bromobenzene	C$_6$H$_5$Br	-30.72	10.70
Zirconium(III) fluoride	F$_3$Zr	927	50	1-Bromobutane	C$_4$H$_9$Br	-112.6	9.23
Zirconium(IV) fluoride	F$_4$Zr	910	61	2-Bromobutane	C$_4$H$_9$Br	-112.65	6.89

Name	Molecular formula	t_m/°C	$\Delta_{fus}H$/kJ mol^{-1}	Name	Molecular formula	t_m/°C	$\Delta_{fus}H$/kJ mol^{-1}
Bromoethane	C_2H_5Br	-118.6	7.47	2-Chlorophenol	C_6H_5ClO	9.4	13.0
Bromoethene	C_2H_3Br	-139.54	5.12	3-Chlorophenol	C_6H_5ClO	32.6	14.9
1-Bromoheptane	$C_7H_{15}Br$	-56.1	21.8	4-Chlorophenol	C_6H_5ClO	42.8	14.1
1-Bromohexane	$C_6H_{13}Br$	-83.7	18.1	1-Chloropropane	C_3H_7Cl	-122.9	5.54
Bromomethane	CH_3Br	-93.68	5.98	2-Chloropropane	C_3H_7Cl	-117.18	7.39
1-Bromonaphthalene	$C_{10}H_7Br$	6.1	15.2	2-Chlorotoluene	C_7H_7Cl	-35.8	9.6
2-Bromonaphthalene	$C_{10}H_7Br$	55.9	14.4	Chlorotrifluoroethene	C_2ClF_3	-158.2	5.55
1-Bromooctane	$C_8H_{17}Br$	-55.0	24.7	Chrysene	$C_{18}H_{12}$	255.5	26.2
1-Bromopentane	$C_5H_{11}Br$	-88.0	14.37	Coronene	$C_{24}H_{12}$	437.4	19.2
1-Bromopropane	C_3H_7Br	-110.3	6.44	o-Cresol	C_7H_8O	31.03	15.82
2-Bromopropane	C_3H_7Br	-89.0	6.53	m-Cresol	C_7H_8O	12.24	10.71
Bromotrichloromethane	$CBrCl_3$	-5.65	2.53	p-Cresol	C_7H_8O	34.77	12.71
1,2-Butadiene	C_4H_6	-136.2	6.96	Cyanamide	CH_2N_2	45.56	7.27
1,3-Butadiene	C_4H_6	-108.91	7.98	Cyanogen	C_2N_2	-27.83	8.11
Butanal	C_4H_8O	-96.86	10.77	Cyclobutane	C_4H_8	-90.7	1.09
Butane	C_4H_{10}	-138.3	4.66	Cycloheptane	C_7H_{14}	-8.46	1.88
1,4-Butanediol	$C_4H_{10}O_2$	20.4	18.70	Cycloheptanol	$C_7H_{14}O$	7.2	1.60
1-Butanethiol	$C_4H_{10}S$	-115.7	10.46	Cyclohexane	C_6H_{12}	6.59	2.68
Butanoic acid	$C_4H_8O_2$	-5.1	11.59	Cyclohexanol	$C_6H_{12}O$	25.93	1.78
1-Butanol	$C_4H_{10}O$	-88.6	9.37	Cyclohexanone	$C_6H_{10}O$	-27.9	1.328
2-Butanol	$C_4H_{10}O$	-88.5	5.97	Cyclohexene	C_6H_{10}	-103.5	3.29
2-Butanone	C_4H_8O	-86.64	8.39	Cyclohexylamine	$C_6H_{13}N$	-17.8	17.5
1-Butene	C_4H_8	-185.34	3.96	Cyclohexylbenzene	$C_{12}H_{16}$	7.07	15.6
cis-2-Butene	C_4H_8	-138.88	7.31	Cyclooctane	C_8H_{16}	14.59	2.41
trans-2-Butene	C_4H_8	-105.52	9.76	Cyclopentane	C_5H_{10}	-93.4	0.61
cis-2-Butenoic acid	$C_4H_6O_2$	15	12.6	Cyclopentanol	$C_5H_{10}O$	-17.5	1.535
trans-2-Butenoic acid	$C_4H_6O_2$	71.5	13.0	Cyclopentene	C_5H_8	-135.0	3.36
tert-Butylamine	$C_4H_{11}N$	-66.94	0.882	Cyclopentylamine	$C_5H_{11}N$	-82.7	8.31
Butylbenzene	$C_{10}H_{14}$	-87.85	11.22	Cyclopropane	C_3H_6	-127.58	5.44
Butylcyclohexane	$C_{10}H_{20}$	-74.73	14.16	Cyclopropylamine	C_3H_7N	-35.39	13.18
Butyl methyl ether	$C_5H_{12}O$	-115.7	10.85	cis-Decahydronaphthalene	$C_{10}H_{18}$	-42.9	9.49
1-Butyne	C_4H_6	-125.7	6.03	trans-Decahydronaphthalene	$C_{10}H_{18}$	-30.4	14.41
2-Butyne	C_4H_6	-32.2	9.23	Decanal	$C_{10}H_{20}O$	-4.0	34.5
γ-Butyrolactone	$C_4H_6O_2$	-43.61	9.57	Decane	$C_{10}H_{22}$	-29.6	28.72
Caffeine	$C_8H_{10}N_4O_2$	236.3	22.0	Decanoic acid	$C_{10}H_{20}O_2$	31.4	27.8
Carbazole	$C_{12}H_9N$	246.3	24.1	1-Decanol	$C_{10}H_{22}O$	6.9	43
Carbon dioxide	CO_2	-56.558	9.02	1-Decene	$C_{10}H_{20}$	-66.3	13.81
Carbon diselenide	CSe_2	-43.7	6.36	1,2-Dibromoethane	$C_2H_4Br_2$	9.84	10.89
Carbon disulfide	CS_2	-112.1	4.39	1,2-Dibromopropane	$C_3H_6Br_2$	-55.49	8.94
Carbon monoxide	CO	-205.02	0.833	1,3-Dibromopropane	$C_3H_6Br_2$	-34.5	14.6
Carbon oxysulfide	COS	-138.8	4.73	1,2-Dibromotetrafluoroethane	$C_2Br_2F_4$	-110.32	7.04
Carbonyl chloride	CCl_2O	-127.78	5.74	o-Dichlorobenzene	$C_6H_4Cl_2$	-17.0	12.4
Chloroacetic acid	$C_2H_3ClO_2$	63	12.28	m-Dichlorobenzene	$C_6H_4Cl_2$	-24.8	12.6
2-Chloroaniline	C_6H_6ClN	-1.9	11.9	p-Dichlorobenzene	$C_6H_4Cl_2$	53.09	18.19
3-Chloroaniline	C_6H_6ClN	-10.28	10.15	1,1-Dichloroethane	$C_2H_4Cl_2$	-96.9	7.87
4-Chloroaniline	C_6H_6ClN	70.5	20.0	1,2-Dichloroethane	$C_2H_4Cl_2$	-35.7	8.84
Chlorobenzene	C_6H_5Cl	-45.31	9.6	1,1-Dichloroethene	$C_2H_2Cl_2$	-122.56	6.51
2-Chlorobenzoic acid	$C_7H_5ClO_2$	140.2	25.6	cis-1,2-Dichloroethene	$C_2H_2Cl_2$	-80.0	7.2
Chlorocyclohexane	$C_6H_{11}Cl$	-43.81	2.043	Dichloromethane	CH_2Cl_2	-97.2	4.60
Chlorodifluoromethane	$CHClF_2$	-157.42	4.12	1,2-Dichloropropane	$C_3H_6Cl_2$	-100.53	6.40
Chloroethane	C_2H_5Cl	-138.4	4.45	2,2-Dichloropropane	$C_3H_6Cl_2$	-33.9	2.30
Chloroethene	C_2H_3Cl	-153.84	4.92	1,2-Dichloro-1,1,2,2-tetrafluoroethane	$C_2Cl_2F_4$	-92.53	1.51
Chloromethane	CH_3Cl	-97.7	6.43	Diethyl ether	$C_4H_{10}O$	-116.2	7.19
2-Chloro-2-methylpropane	C_4H_9Cl	-25.60	2.07	3,3-Diethylpentane	C_9H_{20}	-33.1	10.09
1-Chloronaphthalene	$C_{10}H_7Cl$	-2.5	12.9	Diethyl sulfide	$C_4H_{10}S$	-103.91	10.90
2-Chloronaphthalene	$C_{10}H_7Cl$	58.0	14.0	o-Difluorobenzene	$C_6H_4F_2$	-47.1	11.05
1-Chloro-2-nitrobenzene	$C_6H_4ClNO_2$	32.1	17.9	m-Difluorobenzene	$C_6H_4F_2$	-69.12	8.58
1-Chloro-3-nitrobenzene	$C_6H_4ClNO_2$	44.4	19.4	Diisopropyl ether	$C_6H_{14}O$	-85.4	12.04
1-Chloro-4-nitrobenzene	$C_6H_4ClNO_2$	82	14.1	1,2-Dimethoxyethane	$C_4H_{10}O_2$	-69.20	12.6
Chloropentafluoroethane	C_2ClF_5	-99.4	1.86	Dimethoxymethane	$C_3H_8O_2$	-105.1	8.33

Name	Molecular formula	$t_m/°C$	$\Delta_{fus}H/kJ\ mol^{-1}$	Name	Molecular formula	$t_m/°C$	$\Delta_{fus}H/kJ\ mol^{-1}$
Dimethylamine	C_2H_7N	-92.18	5.94	Formamide	CH_3NO	2.49	8.44
2,2-Dimethylbutane	C_6H_{14}	-98.8	0.58	Formic acid	CH_2O_2	8.3	12.68
2,3-Dimethylbutane	C_6H_{14}	-128.10	0.79	Furan	C_4H_4O	-85.61	3.80
2,3-Dimethyl-2-butene	C_6H_{12}	-74.19	6.45	Furfural	$C_5H_4O_2$	-38.1	14.37
1,1-Dimethylcyclohexane	C_8H_{16}	-33.3	2.07	Furfuryl alcohol	$C_5H_6O_2$	-14.6	13.13
cis-1,2-Dimethylcyclohexane	C_8H_{16}	-49.8	1.64	Glycerol	$C_3H_8O_3$	18.1	18.3
trans-1,2-Dimethylcyclohexane	C_8H_{16}	-88.15	10.49	Heneicosane	$C_{21}H_{44}$	40.01	45.21
cis-1,3-Dimethylcyclohexane	C_8H_{16}	-75.53	10.82	Heptacosane	$C_{27}H_{56}$	59.23	61.9
trans-1,3-Dimethylcyclohexane	C_8H_{16}	-90.07	9.87	Heptadecane	$C_{17}H_{36}$	22.0	40.16
cis-1,4-Dimethylcyclohexane	C_8H_{16}	-87.39	9.31	Heptanal	$C_7H_{14}O$	-43.4	23.2
trans-1,4-Dimethylcyclohexane	C_8H_{16}	-36.93	12.33	Heptane	C_7H_{16}	-90.55	14.03
Dimethyl disulfide	$C_2H_6S_2$	-84.67	9.19	Heptanoic acid	$C_7H_{14}O_2$	-7.17	15.13
Dimethyl ether	C_2H_6O	-141.5	4.94	1-Heptanol	$C_7H_{16}O$	-33.2	18.17
N,N-Dimethylformamide	C_3H_7NO	-60.48	7.90	1-Heptene	C_7H_{14}	-118.9	12.41
1,1-Dimethylhydrazine	$C_2H_8N_2$	-57.20	10.07	Hexachlorobenzene	C_6Cl_6	228.83	25.2
1,2-Dimethylhydrazine	$C_2H_8N_2$	-8.9	13.64	Hexachloroethane	C_2Cl_6	186.8t	9.75
Dimethyl oxalate	$C_4H_6O_4$	54.8	21.1	Hexacontane	$C_{60}H_{122}$	99.3	193.2
2,2-Dimethylpentane	C_7H_{16}	-123.7	5.82	Hexacosane	$C_{26}H_{54}$	56.1	60.0
2,4-Dimethylpentane	C_7H_{16}	-119.2	6.85	Hexadecane	$C_{16}H_{34}$	18.12	53.36
3,3-Dimethylpentane	C_7H_{16}	-134.4	6.85	Hexadecanoic acid	$C_{16}H_{32}O_2$	62.5	53.7
Dimethyl sulfide	C_2H_6S	-98.24	7.99	1-Hexadecanol	$C_{16}H_{34}O$	49.2	33.6
Dimethyl sulfone	$C_2H_6O_2S$	108.9	18.30	Hexafluorobenzene	C_6F_6	5.03	11.59
Dimethyl sulfoxide	C_2H_6OS	17.89	14.37	Hexafluoroethane	C_2F_6	-100.05	2.69
N,N-Dimethylurea	$C_3H_8N_2O$	182.1	23.0	Hexamethylbenzene	$C_{12}H_{18}$	165.5	20.6
N,N'-Dimethylurea	$C_3H_8N_2O$	106.6	13.0	Hexanal	$C_6H_{12}O$	-56	13.3
Dimethyl zinc	C_2H_6Zn	-43.0	6.83	Hexane	C_6H_{14}	-95.35	13.08
1,4-Dioxane	$C_4H_8O_2$	11.85	12.84	1,6-Hexanedioic acid	$C_6H_{10}O_4$	152.5	36.3
1,3-Dioxolane	$C_3H_6O_2$	-97.22	6.57	1,6-Hexanediol	$C_6H_{14}O_2$	41.5	22.2
Diphenylamine	$C_{12}H_{11}N$	53.2	18.5	1-Hexanol	$C_6H_{14}O$	-47.4	15.38
Diphenyl ether	$C_{12}H_{10}O$	26.864	17.22	2-Hexanone	$C_6H_{12}O$	-55.5	14.9
Diphenylmethane	$C_{13}H_{12}$	25.4	18.6	3-Hexanone	$C_6H_{12}O$	-55.4	13.49
Dipropyl ether	$C_6H_{14}O$	-114.8	10.8	Hexatetracontane	$C_{46}H_{94}$	87.6	176.0
Divinyl ether	C_4H_6O	-100.6	7.9	Hexatriacontane	$C_{36}H_{74}$	75.8	87.7
Docosane	$C_{22}H_{46}$	43.6	48.8	1-Hexene	C_6H_{12}	-139.76	9.35
Dodecane	$C_{12}H_{26}$	-9.57	36.8	cis-2-Hexene	C_6H_{12}	-141.11	8.88
Dodecanoic acid	$C_{12}H_{24}O_2$	43.8	36.3	Hydrogen cyanide	CHN	-13.29	8.41
1-Dodecanol	$C_{12}H_{26}O$	23.9	40.2	p-Hydroquinone	$C_6H_6O_2$	172.4	26.8
1-Dodecene	$C_{12}H_{24}$	-35.2	19.9	2-Hydroxybenzoic acid	$C_7H_6O_3$	159.0	14.2
Dotriacontane	$C_{32}H_{66}$	69.4	75.8	Imidazole	$C_3H_4N_2$	89.5	12.82
Eicosane	$C_{20}H_{42}$	36.6	69.9	Indan	C_9H_{10}	-51.38	8.60
1-Eicosanol	$C_{20}H_{42}O$	65.4	42	Indene	C_9H_8	-1.5	10.20
Estradiol benzoate	$C_{25}H_{28}O_3$	193	41.8	Indomethacin	$C_{19}H_{16}ClNO_4$	160	36.9
Ethane	C_2H_6	-182.79	2.72*	Iodobenzene	C_6H_5I	-31.3	9.75
1,2-Ethanediamine	$C_2H_8N_2$	11.14	22.58	Isobutane	C_4H_{10}	-159.4	4.54
1,2-Ethanediol	$C_2H_6O_2$	-12.69	9.96	Isobutene	C_4H_8	-140.7	5.92
Ethanethiol	C_2H_6S	-147.88	4.98	Isopentane	C_5H_{12}	-159.77	5.15
Ethanol	C_2H_6O	-114.14	4.931	Isopropylamine	C_3H_9N	-95.13	7.33
Ethinylestradiol	$C_{20}H_{24}O_2$	183.5	27.9	Isopropylbenzene	C_9H_{12}	-96.02	7.33
Ethyl acetate	$C_4H_8O_2$	-83.8	10.48	1-Isopropyl-4-methylbenzene	$C_{10}H_{14}$	-67.94	9.66
Ethylbenzene	C_8H_{10}	-94.96	9.18	Isoquinoline	C_9H_7N	26.47	13.54
Ethylcyclohexane	C_8H_{16}	-111.3	8.33	Khellin	$C_{14}H_{12}O_5$	154	32.3
Ethylene	C_2H_4	-169.15	3.35	Maleic anhydride	$C_4H_2O_3$	52.56	13.60
Ethyl methyl sulfide	C_3H_8S	-105.93	9.76	Methane	CH_4	-182.47	0.94
3-Ethylpentane	C_7H_{16}	-118.55	9.55	Methanethiol	CH_4S	-123	5.91
2-Ethyltoluene	C_9H_{12}	-79.83	9.96	Methanol	CH_4O	-97.53	3.215
3-Ethyltoluene	C_9H_{12}	-95.6	7.6	Methyl acetate	$C_3H_6O_2$	-98.25	7.49
4-Ethyltoluene	C_9H_{12}	-62.35	12.7	Methylamine	CH_5N	-93.5	6.13
Fluoranthene	$C_{16}H_{10}$	110.19	18.69	2-Methylaniline	C_7H_9N	-14.41	11.66
9H-Fluorene	$C_{13}H_{10}$	114.77	19.58	3-Methylaniline	C_7H_9N	-31.3	7.9
Fluorobenzene	C_6H_5F	-42.18	11.31	4-Methylaniline	C_7H_9N	43.6	18.9

Name	Molecular formula	$t_m/°C$	$\Delta_{fus}H$/kJ mol^{-1}	Name	Molecular formula	$t_m/°C$	$\Delta_{fus}H$/kJ mol^{-1}
Methyl benzoate	$C_8H_8O_2$	-12.4	9.74	1-Octene	C_8H_{16}	-101.7	15.31
2-Methyl-1,3-butadiene	C_5H_8	-145.9	4.93	2-Oxepanone	$C_6H_{10}O_2$	-1.0	13.83
2-Methyl-2-butanol	$C_5H_{12}O$	-9.1	4.46	Oxetane	C_3H_6O	-97	6.5
3-Methyl-2-butanone	$C_5H_{10}O$	-93.1	9.34	Oxirane	C_2H_4O	-112.5	5.17
2-Methyl-1-butene	C_5H_{10}	-137.53	7.91	4-Oxopentanoic acid	$C_5H_8O_3$	33	9.22
3-Methyl-1-butene	C_5H_{10}	-168.43	5.36	Paraldehyde	$C_6H_{12}O_3$	12.6	13.5
2-Methyl-2-butene	C_5H_{10}	-133.72	7.60	Pentachloroethane	C_2HCl_5	-28.78	11.3
Methyl tert-butyl ether	$C_5H_{12}O$	-108.6	7.60	Pentacontane	$C_{50}H_{102}$	92.1	162.4
Methylcyclohexane	C_7H_{14}	-126.6	6.75	Pentacosane	$C_{25}H_{52}$	53.93	56.9
Methylcyclopentane	C_6H_{12}	-142.42	6.93	Pentadecane	$C_{15}H_{32}$	9.95	34.6
Methylcyclopropane	C_4H_8	-177.6	2.8	cis-1,3-Pentadiene	C_5H_8	-140.8	5.64
2-Methylfuran	C_5H_6O	-91.3	8.55	trans-1,3-Pentadiene	C_5H_8	-87.4	7.14
2-Methylheptane	C_8H_{18}	-109.02	11.92	1,4-Pentadiene	C_5H_8	-148.2	6.12
3-Methylheptane	C_8H_{18}	-120.48	11.69	Pentaerythritol	$C_5H_{12}O_4$	258	4.8
4-Methylheptane	C_8H_{18}	-121.0	10.8	Pentafluorobenzene	C_6HF_5	-47.4	10.87
2-Methylhexane	C_7H_{16}	-118.2	9.19	Pentafluorophenol	C_6HF_5O	37.5	16.41
Methylhydrazine	CH_6N_2	-52.36	10.42	2,3,4,5,6-Pentafluorotoluene	$C_7H_3F_5$	-29.78	13.1
Methyl methacrylate	$C_5H_8O_2$	-47.55	14.4	Pentane	C_5H_{12}	-129.67	8.40
1-Methylnaphthalene	$C_{11}H_{10}$	-30.43	6.95	Pentanedioic acid	$C_5H_8O_4$	97.8	20.3
2-Methylnaphthalene	$C_{11}H_{10}$	34.6	12.13	Pentanenitrile	C_5H_9N	-96.2	9
Methyl nitrate	CH_3NO_3	-83.0	8.24	1-Pentanethiol	$C_5H_{12}S$	-75.65	17.53
Methyloxirane	C_3H_6O	-111.9	6.53	Pentanoic acid	$C_5H_{10}O_2$	-33.6	14.16
2-Methylpentane	C_6H_{14}	-153.6	6.27	1-Pentanol	$C_5H_{12}O$	-77.6	10.50
3-Methylpentane	C_6H_{14}	-162.90	5.30	2-Pentanone	$C_5H_{10}O$	-76.8	10.63
2-Methyl-1-propanol	$C_4H_{10}O$	-101.9	6.32	3-Pentanone	$C_5H_{10}O$	-39	11.59
2-Methyl-2-propanol	$C_4H_{10}O$	25.69	6.70	Pentatriacontane	$C_{35}H_{72}$	74.6	86.3
2-Methylpyridine	C_6H_7N	-66.68	9.72	1-Pentene	C_5H_{10}	-165.12	5.94
3-Methylpyridine	C_6H_7N	-18.14	14.18	cis-2-Pentene	C_5H_{10}	-151.36	7.11
4-Methylpyridine	C_6H_7N	3.67	12.58	trans-2-Pentene	C_5H_{10}	-140.21	8.35
N-Methylurea	$C_2H_6N_2O$	104.9	14.0	Perfluoroacetone	C_3F_6O	-125.45	8.38
Morpholine	C_4H_9NO	-4.8	14.5	Perfluorobutane	C_4F_{10}	-129.1	7.66
Naphthalene	$C_{10}H_8$	80.26	19.01	Perfluorocyclobutane	C_4F_8	-40.19	2.77
1-Naphthol	$C_{10}H_8O$	95.0	23.1	Perfluoroheptane	C_7F_{16}	-51.2	6.95
2-Naphthol	$C_{10}H_8O$	121.5	18.1	Perfluorohexane	C_6F_{14}	-88.2	6.84
Neopentane	C_5H_{12}	-16.4	3.10	Perfluoropropane	C_3F_8	-147.70	0.477
Niacinamide	$C_6H_6N_2O$	130	23.2	Perfluorotoluene	C_7F_8	-65.49	11.54
2-Nitroaniline	$C_6H_6N_2O_2$	71.0	16.1	Perylene	$C_{20}H_{12}$	277.76	31.9
3-Nitroaniline	$C_6H_6N_2O_2$	113.4	23.6	Phenacetin	$C_{10}H_{13}NO_2$	134	33.0
4-Nitroaniline	$C_6H_6N_2O_2$	147.5	21.2	Phenanthrene	$C_{14}H_{10}$	99.24	16.46
Nitrobenzene	$C_6H_5NO_2$	5.7	12.12	Phenobarbital	$C_{12}H_{12}N_2O_3$	174.0	27.8
Nitroethane	$C_2H_5NO_2$	-89.5	9.85	Phenol	C_6H_6O	40.89	11.51
Nitromethane	CH_3NO_2	-28.38	9.70	α-Phenylbenzeneacetic acid	$C_{14}H_{12}O_2$	147.29	31.3
2-Nitrophenol	$C_6H_5NO_3$	44.8	17.7	Phenylbutazone	$C_{19}H_{20}N_2O_2$	105	27.7
3-Nitrophenol	$C_6H_5NO_3$	96.8	20.6	Phenylhydrazine	$C_6H_8N_2$	20.6	14.05
4-Nitrophenol	$C_6H_5NO_3$	113.6	18.8	Piperidine	$C_5H_{11}N$	-11.02	14.85
Nitrosobenzene	C_6H_5NO	67	31.0	Potassium acetate	$C_2H_3KO_2$	309	7.65
4-Nitrotoluene	$C_7H_7NO_2$	51.63	16.81	Propane	C_3H_8	-187.63	3.50
Nonacosane	$C_{29}H_{60}$	63.7	66.9	1,3-Propanediol	$C_3H_8O_2$	-27.7	7.1
Nonadecane	$C_{19}H_{40}$	32.0	45.8	Propanenitrile	C_3H_5N	-92.78	5.03
Nonanal	$C_9H_{18}O$	-19.3	30.5	1-Propanethiol	C_3H_8S	-113.13	5.48
Nonane	C_9H_{20}	-53.46	15.47	2-Propanethiol	C_3H_8S	-130.5	5.74
Nonanoic acid	$C_9H_{18}O_2$	12.4	19.82	Propanoic acid	$C_3H_6O_2$	-20.5	10.66
5-Nonanone	$C_9H_{18}O$	-3.8	24.93	1-Propanol	C_3H_8O	-124.39	5.37
Octacosane	$C_{28}H_{58}$	61.1	65.1	2-Propanol	C_3H_8O	-87.9	5.41
Octadecane	$C_{18}H_{38}$	28.2	61.7	Propene	C_3H_6	-185.24	3.003
1-Octadecanol	$C_{18}H_{38}O$	57.9	45	Propylamine	C_3H_9N	-84.75	10.97
Octane	C_8H_{18}	-56.82	20.73	Propylbenzene	C_9H_{12}	-99.6	9.27
Octanoic acid	$C_8H_{16}O_2$	16.5	21.35	Propylcyclohexane	C_9H_{18}	-94.9	10.37
1-Octanol	$C_8H_{18}O$	-14.8	23.7	Pyrazine	$C_4H_4N_2$	51.0	12.9
Octatriacontane	$C_{38}H_{78}$	78.6	133.2	1H-Pyrazole	$C_3H_4N_2$	70.7	14.0

Name	Molecular formula	$t_m/°C$	$\Delta_{fus}H$/kJ mol^{-1}	Name	Molecular formula	$t_m/°C$	$\Delta_{fus}H$/kJ mol^{-1}
Pyrene	$C_{16}H_{10}$	150.62	17.36	Thiazole	C_3H_3NS	-33.62	9.57
Pyridine	C_5H_5N	-41.70	8.28	Thietane	C_3H_6S	-73.24	8.25
Pyrocatechol	$C_6H_6O_2$	104.6	22.8	Thiophene	C_4H_4S	-38.21	5.07
Pyrrole	C_4H_5N	-23.39	7.91	Thiourea	CH_4N_2S	178	14.0
Pyrrolidine	C_4H_9N	-57.79	8.58	Thymol	$C_{10}H_{14}O$	49.5	21.3
Quinoline	C_9H_7N	-14.78	10.66	Toluene	C_7H_8	-94.95	6.64
Resorcinol	$C_6H_6O_2$	109.4	20.4	o-Toluic acid	$C_8H_8O_2$	103.5	19.5
Sebacic acid	$C_{10}H_{18}O_4$	130.9	40.8	m-Toluic acid	$C_8H_8O_2$	109.9	15.7
Sodium acetate	$C_2H_3NaO_2$	328.2	17.9	p-Toluic acid	$C_8H_8O_2$	179.6	22.7
Sodium hydrogen carbonate	$CHNaO_3$	527	25	Triacontane	$C_{30}H_{62}$	65.1	68.3
Spiro[2.2]pentane	C_5H_8	-107.0	6.43	1,3,5-Triazine	$C_3H_3N_3$	80.3	14.56
Stearic acid	$C_{18}H_{36}O_2$	69.3	61.2	Tribromomethane	$CHBr_3$	8.69	11.05
trans-Stilbene	$C_{14}H_{12}$	124.2	27.7	Trichloroacetic acid	$C_2HCl_3O_2$	59.2	5.90
Styrene	C_8H_8	-30.65	10.9	1,2,3-Trichlorobenzene	$C_6H_3Cl_3$	51.3	17.9
Succinic acid	$C_4H_6O_4$	187.9	32.4	1,2,4-Trichlorobenzene	$C_6H_3Cl_3$	16.92	16.4
Succinic anhydride	$C_4H_4O_3$	119	20.4	1,3,5-Trichlorobenzene	$C_6H_3Cl_3$	62.8	18.1
Succinonitrile	$C_4H_4N_2$	58.06	3.70	1,1,1-Trichloroethane	$C_2H_3Cl_3$	-30.01	2.35
Sulfacetamide	$C_8H_{10}N_2O_3S$	183	22.4	1,1,2-Trichloroethane	$C_2H_3Cl_3$	-36.3	11.46
Sulfadiazine	$C_{10}H_{10}N_4O_2S$	258	42.6	Trichloroethene	C_2HCl_3	-84.7	8.45
Sulfamerazine	$C_{11}H_{12}N_4O_2S$	236	38.7	Trichlorofluoromethane	CCl_3F	-110.44	6.89
Sulfamethoxazole	$C_{10}H_{11}N_3O_3S$	170	32.2	Trichloromethane	$CHCl_3$	-63.41	9.5
Sulfamethoxypyridazine	$C_{11}H_{12}N_4O_3S$	182.5	31.3	1,1,2-Trichloro-1,2,2-trifluoroethane	$C_2Cl_3F_3$	-36.22	2.47
Sulfapyridine	$C_{11}H_{11}N_3O_2S$	192	34.4				
Sulfathiazole	$C_9H_9N_3O_2S_2$	202	26.4	Tricosane	$C_{23}H_{48}$	47.76	50.86
Sulfisoxazole	$C_{11}H_{13}N_3O_3S$	196	30.2	Tridecane	$C_{13}H_{28}$	-5.4	28.50
o-Terphenyl	$C_{18}H_{14}$	56.20	17.19	1-Tridecanol	$C_{13}H_{28}O$	31.7	41.4
p-Terphenyl	$C_{18}H_{14}$	213.9	35.3	1,1,1-Trifluoroethane	$C_2H_3F_3$	-111.3	6.19
Tetrabromomethane	CBr_4	92.3	3.76	Trifluoromethane	CHF_3	-155.2	4.06
1,1,2,2-Tetrachloro-1,2-difluoroethane	$C_2Cl_4F_2$	24.8	3.67	Triiodomethane	CHI_3	121.2	16.44
				Trimethoprim	$C_{14}H_{18}N_4O_3$	199	49.4
1,1,2,2-Tetrachloroethane	$C_2H_2Cl_4$	-42.4	9.17	Trimethylamine	C_3H_9N	-117.1	7
Tetrachloroethene	C_2Cl_4	-22.3	10.88	1,2,3-Trimethylbenzene	C_9H_{12}	-25.4	8.18
Tetrachloromethane	CCl_4	-22.62	2.56	1,2,4-Trimethylbenzene	C_9H_{12}	-43.77	13.19
Tetracontane	$C_{40}H_{82}$	81.5	135.5	1,3,5-Trimethylbenzene	C_9H_{12}	-44.72	9.51
Tetracosane	$C_{24}H_{50}$	50.4	54.4	2,2,3-Trimethylbutane	C_7H_{16}	-24.6	2.26
Tetradecane	$C_{14}H_{30}$	5.82	45.07	2,2,4-Trimethylpentane	C_8H_{18}	-107.3	9.20
Tetradecanoic acid	$C_{14}H_{28}O_2$	54.2	45.1	1,3,5-Trinitrobenzene	$C_6H_3N_3O_6$	122.9	15.4
1-Tetradecanol	$C_{14}H_{30}O$	38.2	25.1*	Trinitroglycerol	$C_3H_5N_3O_9$	13.5	21.87
1,2,3,5-Tetrafluorobenzene	$C_6H_2F_4$	-46.25	6.36	2,4,6-Trinitrotoluene	$C_7H_5N_3O_6$	80.5	22.9
1,2,4,5-Tetrafluorobenzene	$C_6H_2F_4$	3.88	15.05	1,3,5-Trioxane	$C_3H_6O_3$	60.29	15.11
Tetrafluoroethene	C_2F_4	-131.15	7.72	Triphenylamine	$C_{18}H_{15}N$	126.5	24.9
Tetrafluoromethane	CF_4	-183.60	0.704	Triphenylene	$C_{18}H_{12}$	197.8	24.74
Tetrahydrofuran	C_4H_8O	-108.44	8.54	Tritriacontane	$C_{33}H_{68}$	71.2	79.5
Tetrahydropyran	$C_5H_{10}O$	-49.1	1.8	Undecane	$C_{11}H_{24}$	-25.5	22.2
Tetrahydrothiophene	C_4H_8S	-96.2	7.35	Urea	CH_4N_2O	133.3	13.9
1,2,4,5-Tetramethylbenzene	$C_{10}H_{14}$	79.3	21	o-Xylene	C_8H_{10}	-25.2	13.6
Tetramethyl lead	$C_4H_{12}Pb$	-30.2	10.80	m-Xylene	C_8H_{10}	-47.8	11.6
2,2,3,3-Tetramethylpentane	C_9H_{20}	-9.75	2.33	p-Xylene	C_8H_{10}	13.25	17.12
2,2,4,4-Tetramethylpentane	C_9H_{20}	-66.54	9.74	2,3-Xylenol	$C_8H_{10}O$	72.5	21.0
Tetramethylsilane	$C_4H_{12}Si$	-99.06	6.87	2,5-Xylenol	$C_8H_{10}O$	74.8	23.4
Tetramethylstannane	$C_4H_{12}Sn$	-55.1	9.30	2,6-Xylenol	$C_8H_{10}O$	45.8	18.9
Tetratetracontane	$C_{44}H_{90}$	85.6	149.6	3,4-Xylenol	$C_8H_{10}O$	65.1	18.1
Tetratriacontane	$C_{34}H_{70}$	72.5	79.4	3,5-Xylenol	$C_8H_{10}O$	63.4	17.4
1H-Tetrazole	CH_2N_4	157.3	18.2				

PRESSURE AND TEMPERATURE DEPENDENCE OF LIQUID DENSITY

This table gives data on the variation of the density of some common liquids with pressure and temperature. The pressure dependence is described to first order by the isothermal compressibility coefficient κ defined as

$$\kappa = -(1/V)\,(\partial V/\partial P)_T$$

where V is the volume, and the temperature dependence by the cubic expansion coefficient α,

$$\alpha = (1/V)\,(\partial V/\partial T)_P$$

Substances are listed by molecular formula in the Hill order. More precise data on the variation of density with temperature over a wide temperature range can be found in Reference 1.

References

1. Lide, D. R., and Kehiaian, H. V., *CRC Handbook of Thermophysical and Thermochemical Data*, CRC Press, Boca Raton, FL, 1994.
2. Le Neindre, B., *Effets des Hautes et Très Hautes Pressions*, in *Techniques de l'Ingénieur*, Paris, 1991.
3. *Landolt-Börnstein, Numerical Data and Functional Relationships in Science and Technology, New Series*, IV/4, *High-Pressure Properties of Matter*, Springer–Verlag, Heidelberg, 1980.
4. Riddick, J.A., Bunger, W.B., and Sakano, T.K., *Organic Solvents, Fourth Edition*, John Wiley & Sons, New York, 1986.
5. Isaacs, N. S., *Liquid Phase High Pressure Chemistry*, John Wiley, New York, 1981.

Molecular formula	Name	Isothermal compressibility		Cubic thermal expansion	
		$t/°C$	$\kappa \times 10^4/MPa^{-1}$	$t/°C$	$\alpha \times 10^3/°C^{-1}$
Cl_3P	Phosphorus trichloride	20	9.45	20	1.9
H_2O	Water	20	4.591	20	0.206
		25	4.524	25	0.256
		30	4.475	30	0.302
Hg	Mercury	20	0.401	20	0.1811
CCl_4	Tetrachloromethane	20	10.50	20	1.14
		40	12.20	40	1.21
		70	15.6	70	1.33
$CHBr_3$	Tribromomethane	50	8.76	25	0.91
$CHCl_3$	Trichloromethane	20	9.96	20	1.21
		50	12.9	50	1.33
CH_2Br_2	Dibromomethane	27	6.85		
CH_2Cl_2	Dichloromethane	25	10.3	25	1.39
CH_3I	Iodomethane	27	10.3	25	1.26
CH_4O	Methanol	20	12.14	20	1.49
		40	13.83	40	1.59
CS_2	Carbon disulfide	20	9.38	20	1.12
		40	10.6	35	1.16
C_2Cl_4	Tetrachloroethylene	25	7.56	25	1.02
C_2HCl_3	Trichloroethylene	25	8.57	25	1.17
$C_2H_2Cl_2$	*trans*-1,2-Dichloroethylene	25	11.2	25	1.36
$C_2H_4Cl_2$	1,1-Dichloroethane	20	7.97	25	0.93
$C_2H_4Cl_2$	1,2-Dichloroethane	30	8.46	20	1.14
$C_2H_4O_2$	Acetic acid	20	9.08	20	1.08
		80	13.7	80	1.38
C_2H_5Br	Bromoethane	20	11.53	20	1.31
C_2H_5I	Iodoethane	20	9.82	25	1.17
C_2H_6O	Ethanol	20	11.19	20	1.40
		70	15.93	70	1.67
$C_2H_6O_2$	Ethylene glycol	20	3.64	20	0.626
C_3H_6O	Acetone	20	12.62	20	1.46
		40	15.6	40	1.57
C_3H_7Br	1-Bromopropane	0	10.22	25	1.2
C_3H_7Cl	1-Chloropropane	0	12.09	20	1.4
C_3H_7I	1-Iodopropane	0	10.22	25	1.09
C_3H_8O	1-Propanol	0	8.43	0	1.22
C_3H_8O	2-Propanol	40	13.32	40	1.55
$C_3H_8O_2$	1,2-Propanediol	0	4.45	20	0.695
$C_3H_8O_2$	1,3-Propanediol	0	4.09	20	0.61
$C_3H_8O_3$	Glycerol	0	2.54	20	0.520
$C_4H_8O_2$	Ethyl acetate	20	11.32	20	1.35
		60	16.2	60	1.54

Molecular formula	Name	Isothermal compressibility		Cubic thermal expansion	
		$t/°C$	$\kappa \times 10^4/MPa^{-1}$	$t/°C$	$\alpha \times 10^3/°C^{-1}$
C_4H_9Br	1-Bromobutane	25	10.26	20	1.13
C_4H_9I	1-Iodobutane	0	7.73	25	1.02
$C_4H_{10}O$	1-Butanol	0	8.10	0	1.12
$C_4H_{10}O$	Diethyl ether	20	18.65	20	1.65
		30	20.85	30	1.72
$C_4H_{10}O_3$	Diethylene glycol	0	3.34	20	0.635
C_5H_{10}	Cyclopentane	20	13.31	20	1.35
$C_5H_{11}Br$	1-Bromopentane	0	8.42	25	1.04
$C_5H_{11}I$	1-Iodopentane	0	7.56		
C_5H_{12}	Pentane	25	21.80	25	1.64
$C_5H_{12}O$	1-Pentanol	0	7.71	0	1.02
C_6H_5Br	Bromobenzene	20	6.46	20	0.86
C_6H_5Cl	Chlorobenzene	20	7.45	20	0.94
$C_6H_5NO_2$	Nitrobenzene	20	4.93	25	0.833
C_6H_6	Benzene	25	9.66	25	1.14
		45	11.28	45	1.21
C_6H_6O	Phenol	60	6.05	60	0.82
C_6H_7N	Aniline	20	4.53	20	0.81
		80	6.32	80	0.91
C_6H_{12}	Cyclohexane	20	11.30	20	1.15
		60	15.2	60	1.29
C_6H_{14}	Hexane	25	16.69	25	1.41
		45	20.27	45	1.52
C_6H_{14}	2-Methylpentane	0	13.97	25	1.43
C_6H_{14}	3-Methylpentane	0	14.57	25	1.40
C_6H_{14}	2,3-Dimethylbutane	20	17.97	25	1.39
$C_6H_{14}O$	1-Hexanol	25	8.24	25	1.03
$C_6H_{15}NO_3$	Triethanolamine	0	3.61	55	0.53
C_7H_8	Toluene	20	8.96	20	1.05
		50	11.0	50	1.13
C_7H_8O	Anisole	20	6.60	20	0.951
C_7H_{14}	Cycloheptane	20	9.22		
C_7H_{16}	Heptane	25	14.38	25	1.26
C_8H_{10}	o-Xylene	25	8.10	25	0.96
C_8H_{10}	m-Xylene	20	8.46	20	0.99
C_8H_{10}	p-Xylene	25	8.59	25	1.00
C_8H_{16}	Cyclooctane	20	8.03		
C_8H_{18}	Octane	25	12.82	25	1.16
		45	15.06	45	1.23
$C_8H_{18}O$	1-Octanol	25	7.64	25	0.827
C_9H_{12}	Mesitylene	25	8.14	25	0.94
$C_9H_{14}O_6$	Triacetin	0	4.49	25	0.94
C_9H_{20}	Nonane	25	11.75	25	1.08
$C_{10}H_{22}$	Decane	25	10.94	25	1.02
$C_{11}H_{24}$	Undecane	25	10.31	25	0.97
$C_{12}H_{26}$	Dodecane	25	9.88	25	0.93
$C_{13}H_{28}$	Tridecane	25	9.48	25	0.90
$C_{14}H_{30}$	Tetradecane	25	9.10	25	0.87
$C_{15}H_{32}$	Pentadecane	25	8.82		
$C_{16}H_{22}O_4$	Butyl phthalate	0	5.0	25	0.86
$C_{16}H_{34}$	Hexadecane	25	8.57		
		45	9.78		
$C_{19}H_{36}O_2$	Methyl oleate	0	6.18	60	0.85

PROPERTIES OF CRYOGENIC FLUIDS

This table gives physical and thermodynamic properties of eight cryogenic fluids. The properties are:

M	Molar mass in grams per mole
T_t	Triple point temperature in kelvins
P_t	Triple point pressure in kilopascals
$\rho_t\,(l)$	Liquid density at the triple point in grams per milliliter
$\Delta_{fus}H$ @ T_t	Enthalpy of fusion at the triple point in joules per gram
T_b	Normal boiling point in kelvins at a pressure of 101325 pascals (760 mmHg)
$\Delta_{vap}H$ @ T_b	Enthalpy of vaporization at the normal boiling point in joules per gram
$\rho\,(l)$ @ T_b	Liquid density at the normal boiling point in grams per milliliter
$\rho\,(g)$ @ T_b	Vapor density at the normal boiling point in grams per liter
$C_p\,(l)$ @ T_b	Liquid heat capacity at constant pressure at the normal boiling point in joules per gram kelvin
$C_p\,(g)$ @ T_b	Vapor heat capacity at constant pressure at the normal boiling point in joules per gram kelvin
T_c	Critical temperature in kelvins
P_c	Critical pressure in megapascals
ρ_c	Critical density in grams per milliliter

In the case of air, the value given for the triple point temperature is the incipient solidification temperature, and the normal boiling point value is the incipient boiling (bubble) point. See Reference 3 for more details.

References

1. Younglove, B. A., *J. Phys. Chem. Ref. Data*, 11, Suppl. 1, 1982.
2. Daubert, T. E., Danner, R. P., Sibul, H. M., and Stebbins, C. C., *Physical and Thermodynamic Properties of Pure Compounds: Data Compilation*, extant 1994 (core with 4 supplements), Taylor & Francis, Bristol, PA (also available as database).
3. Sytchev, V. V., et al., *Thermodynamic Properties of Air*, Hemisphere Publishing, New York, 1987.
4. Jacobsen, R. T., Stewart, R. B., and Jahangiri, M., *J. Phys. Chem. Ref. Data*, 15, 735, 1986. [Nitrogen]
5. Stewart, R. B., Jacobsen, R. T., and Wagner, W., *J. Phys. Chem. Ref. Data*, 20, 917, 1991. [Oxygen]
6. McCarty, R. D., *J. Phys. Chem. Ref. Data*, 2, 923, 1973. [Helium] Also, Donnelly, R. J., private communication.
7. Stewart, R. B. and Jacobsen, R. T., *J. Phys. Chem. Ref. Data*, 18, 639, 1989. [Argon]
8. Setzmann, U. and Wagner, W., *J. Phys. Chem. Ref. Data*, 20, 1061, 1991. [Methane]
9. Vargaftik, N. B., *Thermophysical Properties of Liquids and Gases*, 2nd ed., John Wiley, New York, 1975.

Property	Units	Air	N_2	O_2	H_2	He	Ne	Ar	Kr	Xe	CH_4
M	g/mol	28.96	28.014	31.999	2.0159	4.0026	20.180	39.948	83.800	131.290	16.043
T_t	K	59.75	63.15	54.3584	13.8		24.5561	83.8058	115.8	161.4	90.694
P_t	kPa		12.463	0.14633	7.042		50	68.95	72.92	81.59	11.696
$\rho_t\,(l)$	g/mL	0.959	0.870	1.306	0.0770		1.251	1.417	2.449	2.978	0.4515
$\Delta_{fus}H$ @ T_t	J/g		25.3	13.7	59.5		16.8	28.0	16.3	13.8	58.41
T_b	K	78.67	77.35	90.188	20.28	4.2221	27.07	87.293	119.92	165.10	111.668
$\Delta_{vap}H$ @ T_b	J/g	198.7	198.8	213.1	445	20.7	84.8	161.0	108.4	96.1	510.83
$\rho\,(l)$ @ T_b	g/mL	0.8754	0.807	1.141	0.0708	0.124901	1.204	1.396	2.418	2.953	0.4224
$\rho\,(g)$ @ T_b	g/L	3.199	4.622	4.467	1.3390	16.89	9.51	5.79	8.94		1.816
$C_p\,(l)$ @ T_b	J/g K	1.865	2.042	1.699	9.668	4.545	1.877	1.078	0.533	0.340	3.481
$C_p\,(g)$ @ T_b	J/g K		1.341	0.980	12.24	9.78		0.570	0.248	0.158	2.218
T_c	K	132.5	126.20	154.581	32.98	5.1953	44.40	150.663	209.40	289.73	190.56
P_c	MPa	3.766	3.390	5.043	1.293	0.227460	2.760	4.860	5.500	5.840	4.592
ρ_c	g/mL	0.316	0.313	0.436	0.031	0.06964	0.484	0.531	0.919	1.110	0.1627

PROPERTIES OF LIQUID HELIUM

The following data were obtained by a critical evaluation of all existing experimental measurements on liquid helium, using a fitting procedure described in the reference. All values refer to liquid helium at saturated vapor pressure; temperatures are on the ITS–90 scale. Several properties show a singularity at the lambda point (2.1768 K).

p : vapor pressure
ρ : density
C_s : molar heat capacity
$\Delta_{vap}H$: molar enthalpy of vaporization

ε : relative permittivity (dielectric constant)
σ : surface tension
α : coefficient of linear expansion
η : viscosity
λ : thermal conductivity

Reference

Donnelly, R. J., and Barenghi, C. F., *J. Phys. Chem. Reference Data*, 27, 1217, 1998.

T/K	p/kPa	ρ/g cm^{-3}	C_s/J mol^{-1}K^{-1}	$\Delta_{vap}H$/J mol^{-1}	ε	σ /mN m^{-1}	$10^3\alpha$/K^{-1}	η/μPa s	λ /W cm^{-1}K^{-1}
0.0		0.1451397	0	59.83	1.057255		0.000		
0.5		0.1451377	0.010	70.24	1.057254	0.3530	0.107		
1.0	0.01558	0.1451183	0.415	80.33	1.057246	0.3471	0.309	3.873	
1.5	0.4715	0.1451646	4.468	89.35	1.057265	0.3322	−2.36	1.346	
2.0	3.130	0.1456217	21.28	93.07	1.057449	0.3021	−12.2	1.468	
2.5	10.23	0.1448402	9.083	92.50	1.057135	0.2623	39.4	3.259	0.1497
3.0	24.05	0.1412269	9.944	94.11	1.055683	0.2161	61.5	3.517	0.1717
3.5	47.05	0.1360736	12.37	92.84	1.053615	0.1626	88.7	3.509	0.1868
4.0	81.62	0.1289745	15.96	87.00	1.050770	0.1095	129	3.319	0.1965
4.5	130.3	0.1188552	21.8	75.86	1.046725	0.0609	211		
5.0	196.0		44.7	47.67		0.0157			

PROPERTIES OF REFRIGERANTS

This table gives physical properties of compounds that have been used as working fluids in traditional refrigeration systems or are under consideration as replacements in newer systems. Some are also used as solvents and blowing agents. Many of the compounds listed are believed to be less harmful to the environment than the traditional halocarbons refrigerants.

Compounds are listed by their ASHRAE standard refrigerant designations (Reference 1), which appear in the first column. These codes are often prefixed by symbols such as CFC– (for chlorofluorocarbon), HCFC– (for hydrochlorofluorocarbon), or simply R– (for refrigerant). The molecular formula and CAS Registry Number are also given. The properties tabulated are:

t_m normal melting point in °C
t_b normal boiling point in °C (at 101.325 kPa or 760 mmHg)
t_c critical temperature in °C
TLV   Threshold Limit Value, which is the maximum safe concentration in air in the workplace, expressed as the time–weighted average (TWA) in parts per million by volume over an 8–hr workday and 40–hr workweek. A value followed by C is an absolute ceiling limit.

References

1. *ASHRAE Standard 34–1997*, Number Designation and Safety Classification of Refrigerants.
2. *ASHRAE Fundamentals Handbook 2001*, Chapter 19. Refrigerants, American Society of Heating, Refrigerating, and Air-Conditioning Engineers, Atlanta, GA, 2001.
3. Platzer, B., Polt, A., and Mauer, G., *Thermophysical Properties of Refrigerants*, Springer, Berlin, 1990.
4. Sako, T., Sato, M., Nakazawa, N., Oowa, M., Yasumoto, M., Ito, H., and Yamashita, S., *J. Chem. Eng. Data*, 41, 802, 1996.
5. Schmidt, J. W., Carrillo-Nava, E., and Moldover, M. R., *Fluid Phase Equilibria*, 122, 187, 1996.
6. Salvi–Narkhede, M., Wang, B.-H., Adcock, J. L., and Van Hook, W. A., *J. Chem. Thermodynamics*, 24, 1065, 1992.
7. Fialho, P. S., and Nieto de Castro, C. A., *Int. J. Thermophys.*, 21, 385, 2000.
8. Daubert, T. E., Danner, R. P., Sibul, H. M., and Stebbins, C. C., *Physical and Thermodynamic Properties of Pure Compounds: Data Compilation*, extant 2002 (core with supplements), Taylor & Francis, Bristol, PA.

Further references and additional data on the critical properties may be found in the table "Critical Constants" in this section.

Code	Name	Molecular formula	CAS Reg. No.	t_m/°C	t_b/°C	t_c/°C	TLV
10	Tetrachloromethane	CCl_4	56–23–5	−22.62	76.8	283.4	5
11	Trichlorofluoromethane	CCl_3F	75–69–4	−110.44	23.7	197.9	1000C
12	Dichlorodifluoromethane	CCl_2F_2	75–71–8	−158	−29.8	111.80	1000
12B1	Bromochlorodifluoromethane	$CBrClF_2$	353–59–3	−159.5	−3.7	153.73	
12B2	Dibromodifluoromethane	CBr_2F_2	75–61–6	−110.1	22.76	198.1	100
13	Chlorotrifluoromethane	$CClF_3$	75–72–9	−181	−81.4	29	
13B1	Bromotrifluoromethane	$CBrF_3$	75–63–8	−172	−57.8	67.0	1000
14	Tetrafluoromethane	CF_4	75–73–0	−183.60	−128.0	−45.5	
20	Trichloromethane	$CHCl_3$	67–66–3	−63.41	61.17	263.2	10
21	Dichlorofluoromethane	$CHCl_2F$	75–43–4	−135	8.9	178.43	10
22	Chlorodifluoromethane	$CHClF_2$	75–45–6	−157.42	−40.7	96.3	1000
22B1	Bromodifluoromethane	$CHBrF_2$	1511–62–2	−145	−14.6	138.83	
23	Trifluoromethane	CHF_3	75–46–7	−155.2	−82.1	25.83	
30	Dichloromethane	CH_2Cl_2	75–09–2	−97.2	40	237	50
31	Chlorofluoromethane	CH_2ClF	593–70–4	−135.1	−9.1	154	
32	Difluoromethane	CH_2F_2	75–10–5	−136.8	−51.6	78.41	
40	Chloromethane	CH_3Cl	74–87–3	−97.7	−24.09	143.10	50
41	Fluoromethane	CH_3F	593–53–3	−141.8	−78.4	44.6	
50	Methane	CH_4	74–82–8	−182.47	−161.48	−82.59	1000
110	Hexachloroethane	C_2Cl_6	67–72–1	186.8	184.7 sp	422	1
111	Pentachlorofluoroethane	C_2Cl_5F	354–56–3	101.3	138		
112	1,1,2,2-Tetrachloro-1,2-difluoroethane	$C_2Cl_4F_2$	76–12–0	24.8	92.8	278	500
112a	1,1,1,2-Tetrachloro-2,2-difluoroethane	$C_2Cl_4F_2$	76–11–9	41.0	92.8		500
113	1,1,2-Trichloro-1,2,2-trifluoroethane	$C_2Cl_3F_3$	76–13–1	−36.22	47.7	214.1	1000
113a	1,1,1-Trichloro-2,2,2-trifluoroethane	$C_2Cl_3F_3$	354–58–5	14.37	45.5	209.7	
114	1,2-Dichloro-1,1,2,2-tetrafluoroethane	$C_2Cl_2F_4$	76–14–2	−92.53	3.5	145.63	1000
114a	1,1-Dichloro-1,2,2,2-tetrafluoroethane	$C_2Cl_2F_4$	374–07–2	−56.6	3.4	145.4	
114B2	1,2-Dibromotetrafluoroethane	$C_2Br_2F_4$	124–73–2	−110.32	47.35	214.6	
115	Chloropentafluoroethane	C_2ClF_5	76–15–3	−99.4	−39.1	80.0	1000
116	Hexafluoroethane	C_2F_6	76–16–4	−100.05	−78.1	20	
120	Pentachloroethane	C_2HCl_5	76–01–7	−28.78	162.0		
121	1,1,2,2-Tetrachloro-1-fluoroethane	C_2HCl_4F	354–14–3	−82.6	116.7		
121a	1,1,1,2-Tetrachloro-2-fluoroethane	C_2HCl_4F	354–11–0	−95.3	117.1		
122	1,2,2-Trichloro-1,1-difluoroethane	$C_2HCl_3F_2$	354–21–2	−140	71.9		

Code	Name	Molecular formula	CAS Reg. No.	t_m/°C	t_b/°C	t_c/°C	TLV
122a	1,2,2-Trichloro-1,2-difluoroethane	$C_2HCl_3F_2$	354–15–4	−174	72.5		
122b	1,1,1-Trichloro-2,2-difluoroethane	$C_2HCl_3F_2$	354–12–1		73		
123	2,2-Dichloro-1,1,1-trifluoroethane	$C_2HCl_2F_3$	306–83–2	−107	27.82	183.68	
123a	1,2-Dichloro-1,1,2-trifluoroethane	$C_2HCl_2F_3$	354–23–4	−78	29.5	188.4	
124	1-Chloro-1,2,2,2-tetrafluoroethane	C_2HClF_4	2837–89–0		−12	122.50	
124a	1-Chloro-1,1,2,2-tetrafluoroethane	C_2HClF_4	354–25–6	−117	−11.7	126.7	
125	Pentafluoroethane	C_2HF_5	354–33–6	−103	−48.1	66.02	
E125	Trifluoromethyl difluoromethyl ether	C_2HF_5O	3822–68–2	−157	−38	80.8	
130	1,1,2,2-Tetrachloroethane	$C_2H_2Cl_4$	79–34–5	−42.4	145.2	388.00	1
131	1,1,2-Trichloro-2-fluoroethane	$C_2H_2Cl_3F$	359–28–4		102.4		
132	1,2-Dichloro-1,2-difluoroethane	$C_2H_2Cl_2F_2$	431–06–1	−101.2	59.6		
132b	1,2-Dichloro-1,1-difluoroethane	$C_2H_2Cl_2F_2$	1649–08–7	−101.2	46.2		
133	1-Chloro-1,2,2-trifluoroethane	$C_2H_2ClF_3$	431–07–2		17.3		
133a	2-Chloro-1,1,1-trifluoroethane	$C_2H_2ClF_3$	75–88–7	−105.5	6.1	151.86	
133b	1-Chloro-1,1,2-trifluoroethane	$C_2H_2ClF_3$	421–04–5		12		
134	1,1,2,2-Tetrafluoroethane	$C_2H_2F_4$	359–35–3	−89	−19.9	118.59	
134a	1,1,1,2-Tetrafluoroethane	$C_2H_2F_4$	811–97–2	−103.3	−26.08	101.03	
E134	Bis(difluoromethyl) ether	$C_2H_2F_4O$	1691–17–4		2	147.10	
140	1,1,2-Trichloroethane	$C_2H_3Cl_3$	79–00–5	−36.3	113.8	329	10
140a	1,1,1-Trichloroethane	$C_2H_3Cl_3$	71–55–6	−30.01	74.09	272	350
141	1,2-Dichloro-1-fluoroethane	$C_2H_3Cl_2F$	430–57–9	−60	73.8		
141b	1,1-Dichloro-1-fluoroethane	$C_2H_3Cl_2F$	1717–00–6	−103.5	32.0	204.1	
142	1-Chloro-2,2-difluoroethane	$C_2H_3ClF_2$	338–65–8		35.1		
142b	1-Chloro-1,1-difluoroethane	$C_2H_3ClF_2$	75–68–3	−130.8	−9.1	137.19	
143	1,1,2-Trifluoroethane	$C_2H_3F_3$	430–66–0	−84	3.7	156.6	
143a	1,1,1-Trifluoroethane	$C_2H_3F_3$	420–46–2	−111.3	−47.25	72.71	
143m	Methyl trifluoromethyl ether	$C_2H_3F_3O$	421–14–7	−149	−23.66	104.87	
E143a	2,2,2-Trifluoroethyl methyl ether	$C_3H_5F_3O$	460–43–5		31.62	175.83	
150	1,2-Dichloroethane	$C_2H_4Cl_2$	107–06–2	−35.7	83.5	288	10
150a	1,1-Dichloroethane	$C_2H_4Cl_2$	75–34–3	−96.9	57.3	250	100
151	1-Chloro-2-fluoroethane	C_2H_4ClF	762–50–5		52.8		
151a	1-Chloro-1-fluoroethane	C_2H_4ClF	1615–75–4		16.2		
152	1,2-Difluoroethane	$C_2H_4F_2$	624–72–6		26		
152a	1,1-Difluoroethane	$C_2H_4F_2$	75–37–6	−117	−24.05	113.5	
160	Chloroethane	C_2H_5Cl	75–00–3	−138.4	12.3	187.2	100
161	Fluoroethane	C_2H_5F	353–36–6	−143.2	−37.7	102.16	
170	Ethane	C_2H_6	74–84–0	−182.79	−88.6	32.17	1000
216ca	1,3-Dichloro-1,1,2,2,3,3-hexafluoropropane	$C_3Cl_2F_6$	662–01–1	−125.4	35.7	180	
218	Perfluoropropane	C_3F_8	76–19–7	−147.70	−36.6	71.9	
227ca2	Trifluoromethyl 1,1,2,2-tetrafluoroethyl ether	C_3HF_7O	2356–61–8	−141	−3	114.63	
227ea	1,1,1,2,3,3,3-Heptafluoropropane	C_3HF_7	431–89–0	−131	−16.4	101.74	
227me	Trifluoromethyl 1,2,2,2-tetrafluoroethyl ether	C_3HF_7O	2356–62–9		−9.6		
236ea	1,1,1,2,3,3-Hexafluoropropane	$C_3H_2F_6$	431–63–0		6.1	139.23	
236fa	1,1,1,3,3,3-Hexafluoropropane	$C_3H_2F_6$	690–39–1	−93.6	−1.0	124.92	
236me	1,2,2,2-Tetrafluoroethyl difluoromethyl ether	$C_3H_2F_6O$	57041–67–5		23.35	155.80	
245ca	1,1,2,2,3-Pentafluoropropane	$C_3H_3F_5$	679–86–7		25.0	174.42	
245cb	1,1,1,2,2-Pentafluoropropane	$C_3H_3F_5$	1814–88–6		−17.4	106.96	
245fa	1,1,1,3,3-Pentafluoropropane	$C_3H_3F_5$	460–73–1		15.3	154.05	
245mc	Methyl pentafluoroethyl ether	$C_3H_3F_5O$	22410–44–2		5.59	133.65	
245mf	Difluoromethyl 2,2,2-trifluoroethyl ether	$C_3H_3F_5O$	1885–48–9		29.24	170.84	
245qc	Difluoromethyl 1,1,2-trifluoroethyl ether	$C_3H_3F_5O$	69948–24–9		43.1		
254pc	Methyl 1,1,2,2-tetrafluoroethyl ether	$C_3H_4F_4O$	425–88–7	−107	37.1		
290	Propane	C_3H_8	74–98–6	−187.63	−42.1	96.68	1000
C316	1,2-Dichloro-1,2,3,3,4,4-hexafluorocyclobutane	$C_4Cl_2F_6$	356–18–3	−24.2	59.5	224	
C317	1-Chloro-1,2,2,3,3,4,4-heptafluorocyclobutane	C_4ClF_7	377–41–3	−39.1	25		
C318	Perfluorocyclobutane	C_4F_8	115–25–3	−40.19	−5.9	115.31	
347mcc	Perfluoropropyl methyl ether	$C_4H_3F_7O$	375–03–1		34.23	164.55	
347mmy	Perfluoroisopropyl methyl ether	$C_4H_3F_7O$	22052–84–2		29.34	160.15	
600	Butane	C_4H_{10}	106–97–8	−138.3	−0.5	151.97	1000
600a	Isobutane	C_4H_{10}	75–28–5	−159.4	−11.73	134.6	1000

Code	Name	Molecular formula	CAS Reg. No.	t_m/°C	t_b/°C	t_c/°C	TLV
610	Diethyl ether	$C_4H_{10}O$	60–29–7	−116.2	34.5	193.5	400
611	Methyl formate	$C_2H_4O_2$	107–31–3	−99	31.7	214.0	100
717	Ammonia	H_3N	7664–41–7	−77.73	−33.33	132.3	25
744	Carbon dioxide	CO_2	124–38–9	−56.56	−78.5 sp	30.98	5000
764	Sulfur dioxide	O_2S	7446–09–5	−75.5	−10.05	157.6	2
1112a	1,1-Dichloro-2,2-difluoroethene	$C_2Cl_2F_2$	79–35–6	−116	19		
1113	Chlorotrifluoroethene	C_2ClF_3	79–38–9	−158.2	−27.8	106	
1114	Tetrafluoroethene	C_2F_4	116–14–3	−131.15	−75.9	33.3	2
1120	Trichloroethene	C_2HCl_3	79–01–6	−84.7	87.21	271.0	50
1130	*trans*-1,2-Dichloroethene	$C_2H_2Cl_2$	156–60–5	−49.8	48.7	243.3	200
1132a	1,1-Difluoroethene	$C_2H_2F_2$	75–38–7	−144	−85.7	29.7	500
1140	Chloroethene	C_2H_3Cl	75–01–4	−153.84	−13.8	159	
1141	Fluoroethene	C_2H_3F	75–02–5	−160.5	−72	54.7	1
1150	Ethylene	C_2H_4	74–85–1	−169.15	−103.77	9.19	
1270	Propene	C_3H_6	115–07–1	−185.24	−47.69	91.7	

DENSITY AND SPECIFIC VOLUME OF MERCURY

The data in this table have been adjusted to the ITS–90 temperature scale. The uncertainty in density values is 0.0003 g/mL between −20 and −10°C; 0.0001 or less between −10 and 200°C; and 0.0002 between 200 and 300°C.

Reference

Ambrose, D., *Metrologia*, 27, 245, 1990.

$t/°C$	$\rho/(g/mL)$	$v/(mL/kg)$	$t/°C$	$\rho/(g/mL)$	$v/(mL/kg)$	$t/°C$	$\rho/(g/mL)$	$v/(mL/kg)$
−20	13.64461	73.2890	30	13.52134	73.9572	80	13.39971	74.6285
−19	13.64212	73.3024	31	13.51889	73.9705	81	13.39729	74.6420
−18	13.63964	73.3157	32	13.51645	73.9839	82	13.39487	74.6554
−17	13.63716	73.3291	33	13.51400	73.9973	83	13.39245	74.6689
−16	13.63468	73.3424	34	13.51156	74.0107	84	13.39003	74.6824
−15	13.63220	73.3558	35	13.50911	74.0241	85	13.38762	74.6959
−14	13.62972	73.3691	36	13.50667	74.0375	86	13.38520	74.7094
−13	13.62724	73.3824	37	13.50422	74.0509	87	13.38278	74.7229
−12	13.62476	73.3958	38	13.50178	74.0643	88	13.38037	74.7364
−11	13.62228	73.4091	39	13.49934	74.0777	89	13.37795	74.7498
−10	13.61981	73.4225	40	13.49690	74.0911	90	13.37554	74.7633
−9	13.61733	73.4358	41	13.49446	74.1045	91	13.37313	74.7768
−8	13.61485	73.4492	42	13.49202	74.1179	92	13.37071	74.7903
−7	13.61238	73.4625	43	13.48958	74.1313	93	13.36830	74.8038
−6	13.60991	73.4759	44	13.48714	74.1447	94	13.36589	74.8173
−5	13.60743	73.4892	45	13.48470	74.1581	95	13.36347	74.8308
−4	13.60496	73.5026	46	13.48226	74.1715	96	13.36106	74.8443
−3	13.60249	73.5160	47	13.47982	74.1850	97	13.35865	74.8579
−2	13.60002	73.5293	48	13.47739	74.1984	98	13.35624	74.8714
−1	13.59755	73.5427	49	13.47495	74.2118	99	13.35383	74.8849
0	13.59508	73.5560	50	13.47251	74.2252	100	13.35142	74.8984
1	13.59261	73.5694	51	13.47008	74.2386	110	13.3273	75.0337
2	13.59014	73.5827	52	13.46765	74.2520	120	13.3033	75.1693
3	13.58768	73.5961	53	13.46521	74.2655	130	13.2793	75.3052
4	13.58521	73.6095	54	13.46278	74.2789	140	13.2553	75.4413
5	13.58275	73.6228	55	13.46035	74.2923	150	13.2314	75.5778
6	13.58028	73.6362	56	13.45791	74.3057	160	13.2075	75.7147
7	13.57782	73.6495	57	13.45548	74.3192	170	13.1836	75.8519
8	13.57535	73.6629	58	13.45305	74.3326	180	13.1597	75.9895
9	13.57289	73.6763	59	13.45062	74.3460	190	13.1359	76.1274
10	13.57043	73.6896	60	13.44819	74.3594	200	13.1120	76.2659
11	13.56797	73.7030	61	13.44576	74.3729	210	13.0882	76.4047
12	13.56551	73.7164	62	13.44333	74.3863	220	13.0644	76.5440
13	13.56305	73.7297	63	13.44090	74.3998	230	13.0406	76.6838
14	13.56059	73.7431	64	13.43848	74.4132	240	13.0167	76.8241
15	13.55813	73.7565	65	13.43605	74.4266	250	12.9929	76.9650
16	13.55567	73.7698	66	13.43362	74.4401	260	12.9691	77.1064
17	13.55322	73.7832	67	13.43120	74.4535	270	12.9453	77.2484
18	13.55076	73.7966	68	13.42877	74.4670	280	12.9214	77.3909
19	13.54831	73.8100	69	13.42635	74.4804	290	12.8975	77.5341
20	13.54585	73.8233	70	13.42392	74.4939	300	12.8736	77.6779
21	13.54340	73.8367	71	13.42150	74.5073			
22	13.54094	73.8501	72	13.41908	74.5208			
23	13.53849	73.8635	73	13.41665	74.5342			
24	13.53604	73.8769	74	13.41423	74.5477			
25	13.53359	73.8902	75	13.41181	74.5612			
26	13.53114	73.9036	76	13.40939	74.5746			
27	13.52869	73.9170	77	13.40697	74.5881			
28	13.52624	73.9304	78	13.40455	74.6016			
29	13.52379	73.9438	79	13.40213	74.6150			

THERMAL PROPERTIES OF MERCURY

Lev R. Fokin

The first of these tables gives the molar heat capacity at constant pressure of liquid and gaseous mercury as a function of temperature. To convert to specific heat in units of J/g K, divide these values by 200.59, the atomic weight of mercury.

Reference

Douglas, T. B., Ball, A. T., and Ginnings, D. C., *J. Res. Natl. Bur. Stands.*, 46, 334, 1951.

$t/°C$	$C_p/$(J/mol K) Liquid	Gas	$t/°C$	$C_p/$(J/mol K) Liquid	Gas	$t/°C$	$C_p/$(J/mol K) Liquid	Gas
−38.84	28.2746	20.786	140	27.3675	20.786	340	27.1500	20.836
−20	28.1466	20.786	160	27.3090	20.786	356.73	27.1677	20.849
0	28.0190	20.786	180	27.2588	20.790	360	27.1709	20.853
20	27.9002	20.786	200	27.2169	20.790	380	27.1981	20.870
25	27.8717	20.786	220	27.1834	20.794	400	27.2324	20.891
40	27.7897	20.786	240	27.1583	20.794	420	27.2738	20.916
60	27.6880	20.786	260	27.1412	20.799	440	27.3207	20.941
80	27.5952	20.786	280	27.1320	20.807	460	27.3742	20.974
100	27.5106	20.786	300	27.1303	20.815	480	27.4332	21.008
120	27.4349	20.786	320	27.1366	20.824	500	27.4985	21.046

The second table gives the molar heat capacity of solid mercury in its rhombohedral (α–mercury) form.

References

1. Busey and Giaque, *J. Am. Chem. Soc.*, 75, 806, 1953.
2. Amitin, Lebedeva, and Paukov, *Rus. J. Phys. Chem.*, 2666, 1979.

$t/°C$	$C_p/$(J/mol K)	$t/°C$	$C_p/$(J/mol K)	$t/°C$	$C_p/$(J/mol K)	$t/°C$	$C_p/$(J/mol K)
−268.99	0.99*	−248.15	12.74	−193.15	23.16	−93.15	26.69
−268.99	0.97**	−243.15	14.78	−183.15	23.76	−73.15	27.28
−268.15	1.6	−233.15	17.90	−173.15	24.24	−53.15	27.96
−263.15	4.6	−223.15	19.94	−153.15	25.00	−38.87	28.5
−258.15	7.6	−213.15	21.40	−133.15	25.61	* Superconducting state	
−253.15	10.33	−203.15	22.42	−113.15	26.15	**Normal state	

The final table gives the cubic thermal expansion coefficient α, the isothermal compressibility coefficient κ_T, and the speed of sound U for liquid mercury as a function of temperature. These properties are defined as follows:

$$\alpha = \frac{1}{v}\left(\frac{\partial v}{\partial T}\right)_p \qquad \kappa_T = -\frac{1}{v}\left(\frac{\partial v}{\partial P}\right)_T \qquad U^2 = \left(\frac{\partial P}{\partial \rho}\right)_s \qquad \rho = v^{-1}$$

where v is the specific volume (given in the table on the preceding page).

Reference

Vukalovich, M. P., et al., *Thermophysical Properties of Mercury*, Moscow Standard Press, 1971.

$t/°C$	$\alpha \times 10^4/K^{-1}$	$\kappa_T \times 10^6/bar^{-1}$ At 1 bar	At 1000 bar	$U/m\,s^{-1}$	$t/°C$	$\alpha \times 10^4/K^{-1}$	$\kappa_T \times 10^6/bar^{-1}$ At 1 bar	At 1000 bar	$U/m\,s^{-1}$
−20	1.818	3.83		1470	120	1.8058	4.513	4.33	1404.7
0	1.8144	3.918	3.78	1460.8	140	1.8074	4.622		1395.4
20	1.8110	4.013	3.87	1451.4	160	1.8100	4.731	4.53	1386.1
40	1.8083	4.109	3.96	1442.0	180	1.8136	4.844		1376.7
60	1.8064	4.207		1432.7	200	1.818	4.96		1367
80	1.8053	4.308	4.14	1423.4	250	1.834	5.26		1344
100	1.8051	4.410		1414.1	300	1.856	5.59		1321

VAPOR PRESSURE OF MERCURY

The following table gives the vapor pressure of mercury in kilopascals (100 kPa = 1 bar) from 0°C to the critical temperature. The values through 800°C come from References 1 and 2. Those from 900°C through 1400°C are taken from Reference 3; they should only be considered approximations. The final set refers to the critical point.

References

1. Vukalovich, M. P., and Fokin, L. R., *Thermophysical Properties of Mercury*, Standards Press (USSR), 1972.
2. Vargaftik, N. B., Vinogradov, Y. K., and Yargin, V. S., *Handbook of Physical Properties of Liquids and Gases, Third Edition*, Begell House, New York, 1996.
3. Hensel, F., and Frank, E. U., *Rev. Mod. Phys.* 40, 697, 1968.

t/°C	p/kPa	t/°C	p/kPa	t/°C	p/kPa	t/°C	p/kPa
0	2.728×10^{-5}	220	4.304	440	380.1	660	3953
10	7.101×10^{-5}	230	5.758	450	436.2	670	4285
20	1.729×10^{-4}	240	7.614	460	498.6	680	4636
30	3.680×10^{-4}	250	9.959	470	567.9	690	5009
40	8.626×10^{-4}	260	12.892	480	644.6	700	5403
50	1.786×10^{-3}	270	16.527	490	729.2	710	5820
60	3.536×10^{-3}	280	20.993	500	822.2	720	6259
70	6.724×10^{-3}	290	26.435	510	924.2	730	6722
80	0.01232	300	33.015	520	1035.8	740	7210
90	0.02128	310	40.910	530	1157.6	750	7722
100	0.03745	320	50.320	540	1290.1	760	8260
110	0.06247	330	61.460	550	1434.0	770	8825
120	0.1015	340	74.567	560	1589.9	780	9417
130	0.1608	350	89.896	570	1758.4	790	10037
140	0.2491	360	107.72	580	1940.3	800	10685
150	0.3778	370	128.34	590	2136	900	*19300*
160	0.5618	380	152.07	600	2346	1000	*32000*
170	0.8204	390	179.25	610	2572	1100	*50000*
180	1.178	400	210.24	620	2814	1200	*73000*
190	1.664	410	245.4	630	3072	1300	*103000*
200	2.315	420	285.2	640	3347	1400	*139000*
210	3.177	430	329.9	650	3641	1477	172000

SURFACE TENSION OF COMMON LIQUIDS

The surface tension γ of about 200 liquids is tabulated here as a function of temperature. Values of γ are given in units of millinewtons per meter (mN/m), which is equivalent to dyn/cm in cgs units. The values refer to a nominal pressure of one atmosphere (about 100 kPa) except in cases where the indicated temperature is above the normal boiling point of the substance; in those cases, the applicable pressure is the saturation vapor pressure at the temperature in question.

The uncertainty of the values is 0.1 to 0.2 mN/m or less in most cases. Values at temperatures between the points tabulated can be obtained by linear interpolation to a good approximation.

Substances are listed by molecular formula in the modified Hill order, with substances not containing carbon appearing before those that do contain carbon. A more extensive compilation of surface tension may be found in Reference 1.

References

1. Jasper, J. J., *J. Phys. Chem. Ref. Data*, 1, 841, 1972.
2. Kahl, H., Wadewitz, T., and Winkelmann, J., *J. Chem. Eng. Data*, 48, 580, 2003.

Mol. formula	Name	γ in mN/m				
		10°C	25°C	50°C	75°C	100°C
Br_2	Bromine	43.68	40.95	36.40		
Cl_2O_2S	Sulfuryl chloride		28.78			
Cl_3OP	Phosphoryl chloride		32.03	28.85	25.66	
Cl_3P	Phosphorus trichloride		27.98	24.81		
Cl_4Si	Silicon tetrachloride	19.78	18.29	15.80		
H_2O	Water	74.23	71.99	67.94	63.57	58.91
H_4N_2	Hydrazine		66.39			
Hg	Mercury	488.55	485.48	480.36	475.23	470.11
CCl_4	Tetrachloromethane		26.43	23.37	20.31	17.25
CS_2	Carbon disulfide	33.81	31.58	27.87		
$CHBr_3$	Tribromomethane		44.87	41.60	38.33	
$CHCl_3$	Trichloromethane		26.67	23.44	20.20	
CH_2Br_2	Dibromomethane		39.05	35.33	31.61	
CH_2Cl_2	Dichloromethane		27.20			
CH_2O_2	Formic acid		37.13	34.38	31.64	
CH_3I	Iodomethane	32.19	30.34			
CH_3NO	Formamide		57.03	54.92	52.82	50.71
CH_3NO_2	Nitromethane	39.04	36.53	32.33		
CH_4O	Methanol	23.23	22.07	20.14		
CH_5N	Methylamine		19.15			
C_2HCl_5	Pentachloroethane		34.15	31.20	28.26	
$C_2HF_3O_2$	Trifluoroacetic acid		13.53	11.42		
$C_2H_2Cl_4$	1,1,2,2-Tetrachloroethane		35.58	32.41	29.24	26.07
$C_2H_3Cl_3$	1,1,1-Trichloroethane		25.18	22.07		
$C_2H_3Cl_3$	1,1,2-Trichloroethane		34.02	30.65	27.27	23.89
C_2H_3N	Acetonitrile		28.66	25.51		
$C_2H_4Br_2$	1,2-Dibromoethane		39.55	36.25	32.95	
$C_2H_4Cl_2$	1,1-Dichloroethane		24.07			
$C_2H_4Cl_2$	1,2-Dichloroethane		31.86	28.29	24.72	
C_2H_4O	Acetaldehyde	22.54	20.50	17.10		
$C_2H_4O_2$	Acetic acid		27.10	24.61	22.13	
$C_2H_4O_2$	Methyl formate	26.72	24.36	20.43	16.50	12.57
C_2H_5Br	Bromoethane	25.36	23.62			
C_2H_5I	Iodoethane	30.38	28.46	25.24		
$C_2H_5NO_2$	Nitroethane	34.02	32.13	29.00		
C_2H_6O	Ethanol	23.22	21.97	19.89		
C_2H_6OS	Dimethyl sulfoxide		42.92	40.06		
$C_2H_6O_2$	Ethylene glycol		47.99	45.76	43.54	41.31
C_2H_6S	Dimethyl sulfide	25.27	24.06			
C_2H_6S	Ethanethiol		23.08			
$C_2H_6S_2$	Dimethyl disulfide		33.39	30.04		
C_2H_7N	Dimethylamine		26.34			
C_2H_7N	Ethylamine		19.20			
C_2H_7NO	Ethanolamine		48.32	45.53	42.73	
C_3H_5Br	3-Bromopropene		26.31	23.17		

Mol. formula	Name	γ in mN/m				
		10°C	25°C	50°C	75°C	100°C
C₃H₅Cl	3-Chloropropene		23.14			
C₃H₅ClO	Epichlorohydrin	38.40	36.36	32.96	29.56	26.16
C₃H₅N	Propanenitrile		26.75	23.87		
C₃H₆Cl₂	1,2-Dichloropropane		28.32	25.22	22.12	
C₃H₆O	Acetone	24.57	22.72	19.65		
C₃H₆O	Allyl alcohol	26.63	25.28	23.02	20.77	
C₃H₆O₂	Ethyl formate	25.16	23.18			
C₃H₆O₂	Methyl acetate	26.66	24.73	21.51		
C₃H₆O₂	Propanoic acid		26.20	23.72	21.23	
C₃H₇Br	1-Bromopropane	27.08	25.26	22.21		
C₃H₇Br	2-Bromopropane	25.03	23.25	20.30		
C₃H₇Cl	1-Chloropropane	23.16	21.30			
C₃H₇Cl	2-Chloropropane	20.49	19.16			
C₃H₇NO	N,N-Dimethylformamide	37.56	35.74	32.70	29.66	26.62
C₃H₇NO₂	2-Nitropropane	31.02	29.29	26.39		
C₃H₈O	1-Propanol	24.48	23.32	21.38	19.43	
C₃H₈O	2-Propanol	22.11	20.93	18.96	16.98	
C₃H₈O₂	2-Methoxyethanol	32.32	30.84	28.38	25.92	23.46
C₃H₈S	1-Propanethiol		24.20	21.02		
C₃H₈S	2-Propanethiol		21.33	18.39		
C₃H₉N	Propylamine		21.75			
C₃H₉N	Trimethylamine		13.41			
C₄H₄N₂	Pyridazine	49.51	47.96	45.37	42.78	40.19
C₄H₄N₂	Pyrimidine		30.33	27.80	25.28	22.75
C₄H₄S	Thiophene		30.68	27.36		
C₄H₅N	Pyrrole	38.71	37.06	34.31		
C₄H₆O₃	Acetic anhydride	34.08	31.93	28.34	24.75	21.16
C₄H₇N	Butanenitrile		26.92	24.33	21.73	
C₄H₈O	2-Butanone		23.97	21.16		
C₄H₈O₂	1,4-Dioxane		32.75	29.28	25.80	22.32
C₄H₈O₂	Ethyl acetate	25.13	23.39	20.49	17.58	14.68
C₄H₈O₂	Methyl propanoate	26.32	24.44	21.29		
C₄H₈O₂	Butanoic acid		26.05	23.75	21.45	
C₄H₉Br	1-Bromobutane	27.58	25.90	23.08	20.27	17.45
C₄H₉Cl	1-Chlorobutane	24.85	23.18	20.39		
C₄H₉I	1-Iodobutane	29.79	28.24	25.67	23.09	20.51
C₄H₉N	Pyrrolidine	30.58	29.23	26.98		
C₄H₁₀O	1-Butanol	26.28	24.93	22.69	20.44	18.20
C₄H₁₀O	2-Butanol	23.74	22.54	20.56	18.57	16.58
C₄H₁₀O	2-Methyl-2-propanol		19.96	17.71		
C₄H₁₀O	Diethyl ether		16.65			
C₄H₁₀O₂	2-Ethoxyethanol		28.35	26.11	23.86	21.62
C₄H₁₀O₃	Diethylene glycol		44.77	42.57	40.37	38.17
C₄H₁₀S	Diethyl sulfide	26.22	24.57	21.80		
C₄H₁₁N	Butylamine		23.44	20.63		
C₄H₁₁N	Isobutylamine		21.75	19.02		
C₄H₁₁N	tert-Butylamine		16.87			
C₄H₁₁N	Diethylamine		19.85			
C₅H₄O₂	Furfural	45.08	43.09	39.78	36.46	33.14
C₅H₅N	Pyridine		36.56	33.29	30.03	
C₅H₈	Cyclopentene	24.45	22.20			
C₅H₈O	Cyclopentanone	34.45	32.80	30.05	27.30	24.55
C₅H₉NO	N-Methyl-2-pyrrolidinone	41.94	40.21	37.33	34.45	31.57
C₅H₁₀	1-Pentene	17.10	15.45			
C₅H₁₀	2-Methyl-2-butene	18.61	17.15			
C₅H₁₀	Cyclopentane	24.07	21.88	18.22		
C₅H₁₀O	2-Pentanone		23.25	21.62		
C₅H₁₀O	3-Pentanone		24.74	22.13		
C₅H₁₀O	Pentanal	26.95	25.44	22.91		
C₅H₁₀O₂	Butyl formate	26.05	24.52	21.95	19.39	16.82

Mol. formula	Name	γ in mN/m				
		10°C	25°C	50°C	75°C	100°C
$C_5H_{10}O_2$	Propyl acetate	25.48	23.80	21.00	18.20	15.40
$C_5H_{10}O_2$	Isopropyl acetate	23.37	21.76	19.08	16.40	
$C_5H_{10}O_2$	Ethyl propanoate	25.55	23.80	20.88	17.96	
$C_5H_{10}O_2$	Methyl butanoate	26.34	24.62	21.76	18.89	16.03
$C_5H_{11}Cl$	1-Chloropentane	26.01	24.40	21.71	19.02	16.33
$C_5H_{11}N$	Piperidine	30.64	28.91	26.03	23.14	20.26
C_5H_{12}	Pentane	17.15	15.49			
$C_5H_{12}O$	1-Pentanol	26.67	25.36	23.17	20.99	18.80
$C_5H_{12}O$	2-Pentanol	24.96	23.45	20.94	18.43	15.92
$C_5H_{12}O$	3-Methyl-1-butanol	24.94	23.71	21.66	19.61	17.56
$C_5H_{13}N$	Pentylamine		24.69	22.14	19.58	
$C_6H_4Cl_2$	m-Dichlorobenzene	37.15	35.43	32.57	29.70	26.83
C_6H_5Br	Bromobenzene	36.98	35.24	32.34	29.44	26.54
C_6H_5Cl	Chlorobenzene	34.78	32.99	30.02	27.04	24.06
C_6H_5ClO	o-Chlorophenol		39.70	36.89	34.09	31.28
C_6H_5ClO	m-Chlorophenol		41.18	38.66	36.13	33.61
C_6H_5F	Fluorobenzene	28.47	26.66	23.65	20.64	
C_6H_5I	Iodobenzene	40.40	38.71	35.91	33.10	30.29
$C_6H_5NO_2$	Nitrobenzene			40.56	37.66	34.77
C_6H_6	Benzene		28.22	25.00	21.77	
C_6H_6O	Phenol			38.20	35.53	32.86
C_6H_7N	Aniline		42.12	39.41	36.69	
C_6H_7N	2-Methylpyridine		33.00	29.90	26.79	
$C_6H_8N_2$	Adiponitrile		45.45	43.02	40.58	
C_6H_{10}	Cyclohexene	28.01	26.17	23.12		
$C_6H_{10}O$	Cyclohexanone	36.43	34.57	31.46	28.36	25.25
$C_6H_{11}N$	Hexanenitrile		27.37	25.11	22.84	
C_6H_{12}	Cyclohexane	25.91	24.16	21.26	15.44	
C_6H_{12}	Methylcyclopentane	23.47	21.72	18.82		
C_6H_{12}	1-Hexene	19.44	17.90	15.33		
$C_6H_{12}O$	Cyclohexanol		32.92	30.50	28.09	25.67
$C_6H_{12}O$	2-Hexanone		25.45	22.72		
$C_6H_{12}O_2$	Butyl acetate	26.48	24.88	22.21	19.54	16.87
$C_6H_{12}O_2$	Isobutyl acetate	24.58	23.06	20.53	17.99	15.46
$C_6H_{12}O_2$	Ethyl butanoate	25.51	23.94	21.33	18.71	16.10
$C_6H_{12}O_3$	Paraldehyde	27.22	25.63	22.97	20.32	17.66
$C_6H_{13}Cl$	1-Chlorohexane	27.28	25.73	23.13	20.54	17.94
$C_6H_{13}N$	Cyclohexylamine		31.22	28.25	25.28	
C_6H_{14}	Hexane	19.42	17.89	15.33		
C_6H_{14}	2-Methylpentane	18.37	16.88	14.39		
C_6H_{14}	3-Methylpentane	19.20	17.61	14.96		
$C_6H_{14}O$	Diisopropyl ether		17.27	14.65		
$C_6H_{14}O$	1-Hexanol		25.81	23.81	21.80	19.80
$C_6H_{14}O_2$	1,1-Diethoxyethane		20.89	18.31	15.74	
$C_6H_{14}O_2$	2-Butoxyethanol	27.36	26.14	24.10	22.06	20.02
$C_6H_{15}N$	Triethylamine		20.22	17.74		
$C_6H_{15}N$	Dipropylamine		22.31	19.75	17.20	
$C_6H_{15}N$	Diisopropylamine		19.14	16.45		
C_7H_5N	Benzonitrile		38.79	35.90	33.00	
C_7H_6O	Benzaldehyde	39.63	38.00	35.27	32.55	29.82
C_7H_8	Toluene	29.46	27.73	24.85	21.98	19.10
C_7H_8O	o-Cresol		36.90	34.38	31.85	29.32
C_7H_8O	m-Cresol		35.69	33.38	31.07	28.76
C_7H_8O	Benzyl alcohol				27.89	24.44
C_7H_8O	Anisole		35.10	32.09	29.08	
C_7H_9N	N-Methylaniline		36.90	34.47	32.05	
C_7H_9N	2,3-Dimethylpyridine		32.71	30.04	27.36	
C_7H_9N	Benzylamine		39.30	36.27	33.23	
C_7H_{14}	Methylcyclohexane	24.98	23.29	20.46		
C_7H_{14}	1-Heptene	21.29	19.80	17.33	14.85	

Mol. formula	Name	γ in mN/m				
		10°C	25°C	50°C	75°C	100°C
C$_7$H$_{14}$O	2-Heptanone		26.12	23.48		
C$_7$H$_{14}$O$_2$	Pentyl acetate	26.67	25.17	22.69	20.20	17.72
C$_7$H$_{14}$O$_2$	Heptanoic acid		27.76	25.64		
C$_7$H$_{16}$	Heptane	21.14	19.66	17.19	14.73	
C$_7$H$_{16}$	3-Methylhexane	20.76	19.31	16.88	14.46	
C$_8$H$_8$O	Acetophenone		39.04	36.15	33.27	
C$_8$H$_8$O$_2$	Methyl benzoate		37.17	34.25	31.32	
C$_8$H$_8$O$_3$	Methyl salicylate	40.98	39.22	36.28	33.35	30.41
C$_8$H$_{10}$	Ethylbenzene	30.39	28.75	26.01	23.28	20.54
C$_8$H$_{10}$	o-Xylene	31.41	29.76	27.01	24.25	21.50
C$_8$H$_{10}$	m-Xylene	30.13	28.47	25.71	22.95	20.19
C$_8$H$_{10}$	p-Xylene		28.01	25.32	22.64	19.95
C$_8$H$_{10}$O	Phenetole		32.41	29.65	26.89	
C$_8$H$_{11}$N	N,N-Dimethylaniline		35.52	32.90	30.27	
C$_8$H$_{11}$N	N-Ethylaniline		36.33	33.65	30.98	
C$_8$H$_{16}$	Ethylcyclohexane	26.73	25.15	22.51		
C$_8$H$_{18}$	Octane	22.57	21.14	18.77	16.39	14.01
C$_8$H$_{18}$	2,5-Dimethylhexane	20.77	19.40	17.12	14.84	12.56
C$_8$H$_{18}$O	1-Octanol	28.30	27.10	25.12		
C$_8$H$_{19}$N	Dibutylamine		24.12	21.74	19.36	
C$_8$H$_{19}$N	Diisobutylamine		21.72	19.44	17.16	
C$_9$H$_7$N	Quinoline	44.19	42.59	39.94	37.28	34.62
C$_9$H$_{12}$	Cumene	29.27	27.69	25.05	22.42	19.78
C$_9$H$_{12}$	1,2,4-Trimethylbenzene	30.74	29.20	26.64	24.07	21.51
C$_9$H$_{12}$	Mesitylene	28.89	27.55	25.31	23.07	20.82
C$_9$H$_{18}$O	5-Nonanone		26.28	23.85		
C$_9$H$_{20}$	Nonane	23.79	22.38	20.05	17.71	15.37
C$_9$H$_{20}$O	1-Nonanol	29.03	27.89	26.00	24.10	22.20
C$_{10}$H$_{12}$	1,2,3,4-Tetrahydronaphthalene		33.17	30.78	28.40	
C$_{10}$H$_{22}$	Decane	24.75	23.37	21.07	18.77	16.47
C$_{10}$H$_{22}$O	1-Decanol	29.61	28.51	26.68	24.85	23.02
C$_{11}$H$_{24}$	Undecane	25.56	24.21	21.96	19.70	17.45
C$_{12}$H$_{10}$O	Diphenyl ether		26.75	24.80		
C$_{12}$H$_{27}$N	Tributylamine		24.39	22.32	20.24	
C$_{13}$H$_{28}$	Tridecane	26.86	25.55	23.37	21.19	19.01
C$_{14}$H$_{12}$O$_2$	Benzyl benzoate	44.47	42.82	40.06	37.31	34.55
C$_{14}$H$_{30}$	Tetradecane	27.43	26.13	23.96	21.78	19.61
C$_{16}$H$_{34}$	Hexadecane		27.05	24.91	22.78	20.64
C$_{18}$H$_{38}$	Octadecane		27.87	25.77	23.66	21.55

SURFACE TENSION OF AQUEOUS MIXTURES

The composition dependence of the surface tension of binary mixtures of several compounds with water is given in this table. The data are tabulated as a function of the mass percent of the non-aqueous component. Data for methanol, ethanol, 1-propanol, and 2-propanol are taken from Reference 1, which also gives values at other temperatures.

References

1. Vazquez, G., Alvarez, E., and Navaza, J. M., *J. Chem. Eng. Data*, 40, 611, 1995.
2. *Landolt-Börnstein, Numerical Data and Functional Relationships in Science and Technology, New Series*, IV/16, *Surface Tension*, Springer-Verlag, Heidelberg, 1997.

Surface Tension in mN/m for the Specified Mass %

Compound	t/°C	0%	10%	20%	30%	40%	50%	60%	70%	80%	90%	100%
Acetic acid	30	71.2	51.4	43.3	41.2	38.2	37.4	36.1	33.5	31.5	30.2	26.3
Acetone	25	72.0	44.9	40.5	36.7	33.0	30.1	29.4	29.4	27.6	24.5	23.1
Acetonitrile	20	72.8	48.5	40.2	34.1	31.6	30.6	30.0	29.6	29.1	28.7	28.4
1,2-Butanediol	25	72.0	66.1	60.4	55.1	50.1	45.6	43.3	41.9	40.8	39.2	35.8
1,3-Butanediol	30	71.2	58.1	51.6	48.7	45.8	43.9	42.4	41.2	40.0	39.0	37.0
1,4-Butanediol	30	71.2	61.2	56.9	54.2	52.0	50.7	49.5	47.9	46.6	45.2	43.8
Butanoic acid	30	71.2	42.4	37.5	35.5	34.8	32.2	30.8	29.2	27.4	26.3	25.5
2-Butanone	20	72.8	41.6	32.2			25.2					24.6
γ-Butyrolactone	30	71.2	64	58	53	50	48	46	45	44	42.8	42.7
Chloroacetic acid	25	72.0	59.8	53.6	51.3	49.7	48.3	47.5	46.1			
Diethanolamine	25	72.0	66.8	63.2	60.7	58.8	57.2	55.7	54.3	52.7	50.6	47.2
N,N-Dimethylacetamide	25	72.0	72.0	72.0	72.4	73.5	74.9	75.4	73.0	65.7	54.7	36.4
N,N-Dimethylformamide	25	72.0	65.4	59.2	53.8	49.6	47.3	46.9	44.9	42.3	38.4	35.2
1,4-Dioxane	25	72.0					41.2	39.6	37.9	36.2	34.5	33.7
Ethanol	25	72.01	47.53	37.97	32.98	30.16	27.96	26.23	25.01	23.82	22.72	21.82
Ethylene glycol	20	72.8	68.5	64.9	61.9		57.0					48.2
Formic acid	20	72.8	66	60	55.7	52.2	50.3	48.8	47.1	44.7	40.9	38.0
Glycerol	25	72.0	70.5	69.5	68.5	67.9	67.4	66.9	66.5	65.7	64.5	62.5
Methanol	25	72.01	56.18	47.21	41.09	36.51	32.86	29.83	27.48	25.54	23.93	22.51
Morpholine	20	72.8	65.1	60.7	58.9	56.7	53.0	49.6	47.0	43.7	41.8	38.7
Nitric acid	20	72.8	71.9	70.7	68.9	66.6	63.8	60.6	56.8	52.6	47.9	42.6
Propanoic acid	30	71.2	46.6	42.2	37.7	35.6	33.1	31.7	30.2	28.2	27.4	25.8
1-Propanol	25	72.01	34.32	27.84	25.98	25.26	24.80	24.49	24.08	23.86	23.59	23.28
2-Propanol	25	72.01	40.42	30.57	26.82	25.27	24.26	23.51	22.68	22.14	21.69	21.22
1,2-Propylene glycol	30	71.2	60.5	54.9	50.7	47.2	44.5	41.5	38.6	37.6	36.3	35.5
1,3-Propylene glycol	30	71.2	62.6	58.8	55.7	53.8	52.8	51.7	50.8	49.6	48.2	47.0
Pyridine	25	72.0	52.8	51.2	48.0	46.8	46.6	45.8	45.0	43.6	40.9	37.0
Sulfolane	20	72.8					62.5	61.6	59.6	57.1	54.9	50.9
Sulfuric acid	50	67.9	73.5	75.1	73.6	71.2	68.0	64.1	60.0	56.4	53.6	51.7
Trichloroacetaldehyde	25	72.0	56.7	51.0	46.7	44.1	43.0	42.5	41.5	38.9	34.7	29.4
Trichloroacetic acid	25	72.0	55.8	46.5	42.8	41.6	40.6	39.4	38.3	37.4	36.5	

PERMITTIVITY (DIELECTRIC CONSTANT) OF LIQUIDS

Christian Wohlfarth

The permittivity of a substance (often called the dielectric constant) is the ratio of the electric displacement D to the electric field strength E when an external field is applied to the substance. The quantity tabulated here is the relative permittivity, which is the ratio of the actual permittivity to the permittivity of a vacuum; it is a dimensionless number.

The table gives the static relative permittivity ε_r, i.e., the relative permittivity measured in static fields or at low frequencies where no relaxation effects occur. The fourth column of the table lists the value of ε_r at the temperature specified in the third column, usually 293.15 or 298.15 K. Otherwise, the temperature closest to 293.15 K was chosen, or (as it is the case for many of the substances included here) ε_r is given at the only temperature for which data are available.

The static permittivity refers to nominal atmospheric pressure as long as the corresponding temperature is below the normal boiling point. Otherwise, at temperatures above the normal boiling point, the pressure is understood to be the saturated vapor pressure of the substance considered.

For substances where information on the temperature dependence of the permittivity is available, the table gives the coefficients of a simple polynomial fitting of permittivity to temperature with an equation of the form

$$\varepsilon_r(T) = a + bT + cT^2 + dT^3$$

where T is the absolute temperature in K. Since the parameter d was used in only a few cases where the quadratic fit was not satisfactory, only a, b, and c are listed as columns in the table, while the d values are given at the end of this introduction. For all other substances, $d = 0$. The temperature range of the fit is given in the last column. The coefficients of the fitting equation can be used to calculate dielectric constants within the fitted temperature range but should not be used for extrapolation outside this range. The user who needs dielectric constant data with more accuracy than can be provided by this equation is referred to Reference 1, which gives the original data together with their literature source.

Substances are listed by molecular formula in modified Hill order, with substances not containing carbon preceding those that do contain carbon.

* Indicates that the isomer was not specified in the original reference.

** Indicates a compound for which the cubic term is needed:

Ethanol	$d = -0.15512E-05$
N-Methylacetamide	$d = -0.12998E-04$
1,2-Propylene glycol	$d = -0.32544E-05$
1-Butanol	$d = -0.48841E-06$
2-Butanol	$d = -0.89512E-06$
2-Methyl-1-propanol	$d = -0.45229E-06$
2-Methyl-2-propanol	$d = -0.25968E-05$
N-Butylacetamide	$d = -0.48716E-05$

References

1. Wohlfarth, Ch., Static Dielectric Constants of Pure Liquids and Binary Liquid Mixtures, *Landolt-Börnstein, Numerical Data and Functional Relationships in Science and Technology, New Series*, Editor in Chief, O. Madelung, Group IV, Macroscopic and Technical Properties of Matter, Volume 6, Springer-Verlag, Berlin, 1991.
2. Marsh, K. N., Ed., *Recommended Reference Materials for the Realization of Physicochemical Properties*, Blackwell Scientific Publications, Oxford, 1987.

Mol. form.	Name	T/K	ε_r	a	b	c	Range/K
AlBr$_3$	Aluminum tribromide	373.2	3.38				
Ar	Argon	140.00	1.3247	0.12408E+01	0.68755E-02	−0.45344E-04	87-149
AsH$_3$	Arsine	200.9	2.40	0.37674E+01	−0.97454E-02	0.14537E-04	157-201
BBr$_3$	Boron tribromide	273.2	2.58				
B$_2$H$_6$	Diborane	180.66	1.8725	0.23848E+01	−0.29501E-02	0.64189E-06	108-181
B$_5$H$_9$	Pentaborane(9)	298.2	21.1	0.40952E+03	−0.24414E+01	0.38225E-02	226-298
BrF$_3$	Bromine trifluoride	298.2	106.8				
BrF$_5$	Bromine pentafluoride	297.7	7.91	0.11428E+02	−0.11822E-01		262-298
BrH	Hydrogen bromide	186.8	8.23				
BrNO	Nitrosyl bromide	288.4	13.4				
Br$_2$	Bromine	297.9	3.1484	0.32701E+01	−0.12535E-03		273-327
Br$_2$OS	Thionyl bromide	293.2	9.06				
Br$_3$OV	Vanadyl tribromide	298.2	3.6	0.61112E+01	−0.84211E-02		203-298
Br$_4$Ge	Germanium(IV) bromide	299.9	2.955	0.34450E+01	−0.16083E-02		300-316
Br$_4$Sn	Tin(IV) bromide	303.45	3.169	0.50001E+01	−0.60383E-02		304-316
ClFO$_3$	Perchloryl fluoride	150.2	2.194	0.23808E+01	−0.38629E-03	−0.57143E-05	125-150
ClF$_3$	Chlorine trifluoride	293.2	4.394	0.96716E+01	−0.18000E-01		273-313
ClF$_5$	Chlorine pentafluoride	193.2	4.28	0.78192E+01	−0.20860E-01	0.13132E-04	193-256
ClH	Hydrogen chloride	158.9	14.3	0.47316E+02	−0.28455E+00	0.48650E-03	159-258
ClNO	Nitrosyl chloride	285.2	18.2				
Cl$_2$	Chlorine	208.0	2.147	0.29440E+01	−0.44649E-02	0.30388E-05	208-240
Cl$_2$F$_3$P	Phosphorus(V) dichloride trifluoride	228.63	2.8129	0.46501E+01	−0.80358E-02		172-229
Cl$_2$OS	Thionyl chloride	298.2	8.675				
Cl$_2$OSe	Selenium oxychloride	293.2	46.2				
Cl$_2$O$_2$S	Sulfuryl chloride	293.2	9.1				

Mol. form.	Name	T/K	ε_r	a	b	c	Range/K
Cl_2S	Sulfur dichloride	298.2	2.915				
Cl_2S_2	Sulfur chloride	288.2	4.79				
Cl_3F_2P	Phosphorus(V) trichloride difluoride	268.0	2.3752	0.28905E+01	−0.19228E−02		215-268
Cl_3OP	Phosphorus(V) oxychloride	293.2	14.1				
Cl_3OV	Vanadyl trichloride	298.2	3.4				
Cl_3P	Phosphorus(III) chloride	290.2	3.498	0.59098E+01	−0.83322E−02		290-333
Cl_3PS	Phosphorus(V) sulfide trichloride	298.2	4.94				
Cl_4FP	Phosphorus(V) tetrachloride fluoride	272.64	2.6499	0.33503E+01	−0.29651E−02		244-273
Cl_4Ge	Germanium(IV) chloride	273.2	2.463	−0.55078E+01	0.64881E−01	−0.13091E−03	246-273
Cl_4Pb	Lead(IV) chloride	293.2	2.78				
Cl_4Si	Tetrachlorosilane	273.2	2.248	0.58041E+01	−0.27129E−01	0.51678E−04	207-273
Cl_4Sn	Tin(IV) chloride	273.2	3.014	0.43951E+01	−0.48805E−02		234-273
Cl_4Ti	Titanium(IV) chloride	257.4	2.843	0.33668E+01	−0.19675E−02		237-257
Cl_4V	Vanadium(IV) chloride	298.2	3.05				
Cl_5P	Phosphorus(V) chloride	433.2	2.85				
Cl_5Sb	Antimony(V) chloride	293.0	3.222	0.45413E+01	−0.45078E−02		276-320
FH	Hydrogen fluoride	273.2	83.6	0.50352E+03	−0.19297E+01	0.14372E−02	200-273
F_2	Fluorine	53.48	1.4913	0.14144E+01	0.26387E−02	−0.28356E−04	54-144
F_5I	Iodine pentafluoride	293.2	37.13	0.95184E+02	−0.19800E+00		273-313
F_6S	Sulfur hexafluoride	223.2	1.81				
F_6Xe	Xenon hexafluoride	328.2	4.10				
F_7I	Iodine heptafluoride	298.2	1.75				
$F_{10}S_2$	Sulfur decafluoride	293.2	2.0202				
HI	Hydrogen iodide	220.2	3.87	0.51557E+03	−0.44552E+01	0.96795E−02	220-236
H_2	Hydrogen	13.52	1.2792	0.13327E+01	−0.51946E−02		14-19
H_2O	Water	293.2	80.100	0.24921E+03	−0.79069E+00	0.72997E−03	273-372
H_2O_2	Hydrogen peroxide	290.2	74.6	0.48511E+03	−0.23145E+01	0.31020E−02	233-303
H_2S	Hydrogen sulfide	283.2	5.93	0.14736E+02	−0.33675E−01	0.96740E−05	212-363
H_3N	Ammonia	293.2	16.61	0.66756E+02	−0.24696E+00	0.25913E−03	238-323
H_4N_2	Hydrazine	298.2	51.7	0.22061E+03	−0.89633E+00	0.11066E−02	278-323
He	Helium	2.055	1.0555	0.10640E+01	−0.35584E−02		2-4
I_2	Iodine	391.25	11.08	0.64730E+02	−0.29266E+00	0.39759E−03	391-441
Kr	Krypton	119.80	1.664				
Mn_2O_7	Manganese(VII) oxide	293.2	3.28	0.37655E+01	−0.16463E−02		283-312
NO	Nitric oxide	1.997					
N_2	Nitrogen	63.15	1.4680	0.12550E+01	0.67949E−02	−0.56704E−04	63-126
N_2O_3	Nitrogen trioxide	203.2	31.13	0.92287E+02	−0.43306E+00	0.65000E−03	203-243
N_2O_4	Nitrogen tetroxide	293.2	2.44	0.28212E+01	−0.13000E−02		253-293
Ne	Neon	26.11	1.1907	0.12667E+01	−0.29064E−02		26-29
O_2	Oxygen	54.478	1.5684	0.15434E+01	0.14615E−02	−0.21964E−04	55-154
O_2S	Sulfur dioxide	298.2	16.3	0.52045E+02	−0.16125E+00	0.11042E−03	213-449
O_3	Ozone	90.2	4.75	0.86344E+01	−0.54807E−01	0.12596E−03	90-185
O_3S	Sulfur trioxide	291.2	3.11				
P	Phosphorus	307.2	4.096	0.79018E+00	0.23911E−01	−0.42826E−04	307-358
S	Sulfur	407.2	3.4991	0.51651E+01	−0.77381E−02	0.89120E−05	407-479
Se	Selenium	510.65	5.44	0.67569E+01	−0.25829E−02		511-575
Xe	Xenon	161.35	1.880				
$CBrClF_2$	Bromochlorodifluoromethane	123.2	3.920	0.52442E+01	−0.11000E−01		123-223
$CBrCl_3$	Bromotrichloromethane	293.2	2.405	0.29249E+01	−0.17650E−02		273-333
$CBrF_3$	Bromotrifluoromethane	123.2	3.730	0.54154E+01	−0.13680E−01		123-173
CBr_2Cl_2	Dibromodichloromethane	298.2	2.542	0.32330E+01	−0.23162E−02		298-333
CBr_2F_2	Dibromodifluoromethane	273.2	2.939	0.67296E+01	−0.22133E−01	0.30213E−04	139-273
CBr_3Cl	Tribromochloromethane	333.2	2.601				
CBr_3F	Tribromofluoromethane	293.2	3.00	0.53203E+01	−0.11061E−01	0.10688E−04	206-323
CBr_3NO_2	Tribromonitromethane	298.2	9.034	0.16079E+02	−0.23630E−01		298-328
$CClF_3$	Chlorotrifluoromethane	123.2	3.010	0.43677E+01	−0.11020E−01		123-173
CCl_2F_2	Dichlorodifluoromethane	123.2	3.500	0.46984E+01	−0.97600E−02		123-223
CCl_2O	Carbonyl chloride	295.2	4.30				
CCl_3D	Trichloromethane-d	298.2	4.67				
CCl_3F	Trichlorofluoromethane	293.2	3.00	0.53203E+01	−0.11061E−01	0.10688E−04	206-323
CCl_3NO_2	Trichloronitromethane	293.2	7.319	0.14403E+02	−0.24178E−01		276-333
CCl_4	Tetrachloromethane	293.2	2.2379	0.28280E+01	−0.20339E−02	0.71795E−07	283-333
CF_4	Tetrafluoromethane	126.3	1.685	0.20350E+01	−0.27616E−02		126-142

Mol. form.	Name	T/K	ε_r	a	b	c	Range/K
$CHBr_3$	Tribromomethane	283.2	4.404	0.71707E+01	−0.98000E-02		283-343
$CHCl_3$	Trichloromethane	293.2	4.8069	0.15115E+02	−0.51830E-01	0.56803E-04	218-323
CHF_3	Trifluoromethane	294.0	5.2	0.11442E+03	−0.75600E+00	0.13562E-02	130-263
CHN	Hydrogen cyanide	293.2	114.9	0.37331E+04	−0.23180E+02	0.36963E-01	258-299
CH_2Br_2	Dibromomethane	283.2	7.77	0.18060E+02	−0.36333E-01		283-313
CH_2Cl_2	Dichloromethane	298.0	8.93	0.40452E+02	−0.17748E+00	0.23942E-03	184-306
CH_2F_2	Difluoromethane	152.2	53.74	0.19428E+03	−0.12939E+01	0.24280E-02	152-224
CH_2I_2	Diiodomethane	298.2	5.32				
CH_2O_2	Formic acid	298.2	51.1	0.14040E+03	−0.24673E+00	−0.17151E-03	287-358
CH_3Br	Bromomethane	275.7	9.71	0.40580E+02	−0.18418E+00	0.26219E-03	195-276
CH_3Cl	Chloromethane	295.2	10.0	0.42775E+02	−0.16175E+00	0.17108E-03	190-392
CH_3ClO_2S	Methanesulfonyl chloride	293.2	34.0	0.10384E+03	−0.33838E+00	0.34156E-03	293-373
CH_3DO	Methan-d_1-ol	297.5	31.68	0.20839E+03	−0.10318E+01	0.14740E-02	176-298
CH_3F	Fluoromethane	131.0	51.0	0.11338E+03	−0.63979E+00	0.96983E-03	150-299
CH_3I	Iodomethane	293.2	6.97	0.24264E+02	−0.93914E-01	0.11926E-03	223-303
CH_3NO	Formamide	293.2	111.0	0.26076E+03	−0.61145E+00	0.34296E-03	278-333
CH_3NO_2	Nitromethane	293.2	37.27	0.11227E+03	−0.35591E+00	0.34206E-03	288-343
CH_3NO_2	Methyl nitrite	200.0	20.77	0.11071E+03	−0.73428E+00	0.14054E-02	110-260
CH_3NO_3	Methyl nitrate	293.2	23.9				
CH_4	Methane	91.0	1.6761	0.15996E+01	0.27434E-02	−0.22086E-04	91-184
CH_4O	Methanol	293.2	33.0	0.19341E+03	−0.92211E+00	0.12839E-02	177-293
CH_5N	Methylamine	215.2	16.7	0.34398E+02	−0.73630E-01	−0.41279E-04	198-258
CN_4O_8	Tetranitromethane	293.2	2.317				
COS	Carbon oxysulfide	185.0	4.47	0.84702E+01	−0.21488E-01		143-185
$COSe$	Carbon oxyselenide	283.2	3.47	0.48740E+01	−0.49425E-02		219-283
CO_2	Carbon dioxide	295.0	1.4492	0.79062E+00	0.10639E-01	−0.28510E-04	220-300
CS_2	Carbon disulfide	293.2	2.6320	0.45024E+01	−0.12054E-01	0.19147E-04	154-319
$C_2Br_2F_4$	1,2-Dibromotetrafluoroethane	298.2	2.34				
$C_2Cl_2F_4$	1,2-Dichlorotetrafluoroethane	273.2	2.4842	0.36663E+01	−0.42271E-02	−0.36255E-06	193-273
$C_2Cl_2O_2$	Oxalyl chloride	294.35	3.470				
C_2Cl_3N	Trichloroacetonitrile	292.2	7.85				
C_2Cl_4	Tetrachloroethylene	303.2	2.268				
$C_2Cl_4F_2$	1,1,2,2-Tetrachloro-1,2-difluoroethane	308.2	2.52				
C_2HBr_3O	Tribromoacetaldehyde	293.2	7.6				
C_2HCl_3	Trichloroethylene	301.5	3.390	0.58319E+01	−0.80828E-02		302-338
$C_2HCl_3F_2$	1,2,2-Trichloro-1,1-difluoroethane	303.2	4.01	0.75423E+01	−0.11667E-01		303-333
C_2HCl_3O	Trichloroacetaldehyde	298.2	6.8				
$C_2HCl_3O_2$	Trichloroacetic acid	333.2	4.34	0.13412E+01	0.90000E-02	−0.24130E-14	333-393
C_2HCl_5	Pentachloroethane	298.2	3.716	0.65972E+01	−0.96800E-02		298-338
$C_2HF_3O_2$	Trifluoroacetic acid	293.2	8.42	0.21652E+02	−0.68146E-01	0.78571E-04	263-323
C_2H_2	Acetylene	195.0	2.4841				
$C_2H_2Br_2$	cis-1,2-Dibromoethylene	298.2	7.08				
$C_2H_2Br_2$	trans-1,2-Dibromoethylene	298.2	2.88				
$C_2H_2Br_4$	1,1,2,2-Tetrabromoethane	303.2	6.72	0.16246E+02	−0.31500E-01		303-333
$C_2H_2Cl_2$	1,1-Dichloroethylene	293.2	4.60				
$C_2H_2Cl_2$	cis-1,2-Dichloroethylene	298.2	9.20				
$C_2H_2Cl_2$	trans-1,2-Dichloroethylene	293.2	2.14				
$C_2H_2Cl_2O_2$	Dichloroacetic acid	293.2	8.33	0.11014E+02	−0.10859E-01	0.49242E-05	284-363
$C_2H_2Cl_4$	1,1,1,2-Tetrachloroethane	207.2	9.22	0.19606E+02	−0.49847E-01		207-233
$C_2H_2Cl_4$	1,1,2,2-Tetrachloroethane	293.2	8.50				
$C_2H_2I_2$	cis-1,2-Diiodoethylene	345.65	4.46				
C_2H_3ClO	Acetyl chloride	295.2	15.8				
$C_2H_3ClO_2$	Chloroacetic acid	338.2	12.35	0.17310E+02	−0.14674E-01		338-393
$C_2H_3Cl_2NO_2$	1,1-Dichloro-1-nitroethane	303.2	16.3	0.37576E+02	−0.70400E-01		303-333
$C_2H_3Cl_3$	1,1,1-Trichloroethane	293.2	7.243	0.27705E+02	−0.10621E+00	0.12424E-03	258-318
$C_2H_3Cl_3$	1,1,2-Trichloroethane	298.2	7.1937	0.17147E+02	−0.33371E-01		288-318
$C_2H_3F_3O$	2,2,2-Trifluoroethanol	293.2	27.68	0.90593E+02	−0.21421E+00		293-318
C_2H_3N	Acetonitrile	293.2	36.64	0.29724E+03	−0.15508E+01	0.22591E-02	288-333
C_2H_3NO	Methyl isocyanate	288.7	21.75				
C_2H_4	Ethylene	270.0	1.4833	0.13546E+01	0.62614E-02	−0.21374E-04	200-270
C_2H_4BrCl	1-Bromo-2-chloroethane	283.2	7.41	0.19493E+02	−0.59054E-01	0.58036E-04	263-363
$C_2H_4Br_2$	1,2-Dibromoethane	293.2	4.9612	0.67142E+01	−0.59800E-02		293-313
$C_2H_4Cl_2$	1,1-Dichloroethane	298.2	10.10	0.24429E+02	−0.48000E-01		288-318

Mol. form.	Name	T/K	ε_r	a	b	c	Range/K
$C_2H_4Cl_2$	1,2-Dichloroethane	293.2	10.42	0.24404E+02	−0.47892E−01		293-343
$C_2H_4Cl_2O$	Bis(chloromethyl) ether	293.2	3.51				
$C_2H_4N_2O_6$	Ethylene glycol dinitrate	293.2	28.26				
C_2H_4O	Acetaldehyde	291.2	21.0				
C_2H_4O	Ethylene oxide	293.2	12.42	0.52661E+02	−0.21337E+00	0.25947E−03	293-243
C_2H_4OS	Thioacetic acid	298.2	14.30				
$C_2H_4O_2$	Acetic acid	293.2	6.20	−0.15731E+02	0.12662E+00	−0.17738E−03	293-363
$C_2H_4O_2$	Methyl formate	288.2	9.20	0.19699E+02	−0.36429E−01		288-302
$C_2H_4O_3S$	Ethylene glycol sulfite	298.2	39.6	0.85483E+02	−0.15400E+00		298-328
C_2H_5Br	Bromoethane	298.2	9.01	0.28473E+02	−0.85495E−01	0.67971E−04	243-308
C_2H_5Cl	Chloroethane	293.2	9.45	0.60693E+02	−0.31290E+00	0.47154E−03	237-293
C_2H_5ClO	2-Chloroethanol	293.2	25.80	0.11155E+03	−0.30149E+00		140-175
C_2H_5I	Iodoethane	293.2	7.82	0.25598E+02	−0.94367E−01	0.11424E−03	183-343
C_2H_5N	Ethyleneimine	298.2	18.3	0.61405E+02	−0.14474E+00		273-298
C_2H_5NO	Acetamide	363.7	67.6	−0.20055E+03	0.15515E+01	−0.22392E−02	364-448
C_2H_5NO	N-Methylformamide	293.2	189.0	0.10383E+04	−0.43165E+01	0.48398E−02	276-353
C_2H_5NO	Acetaldoxime	298.2	4.70				
$C_2H_5NO_2$	Nitroethane	288.2	29.11	0.57406E+02	−0.97657E−01		276-333
$C_2H_5NO_2$	Methyl carbamate	328.2	18.48	0.36773E+02	−0.55700E−01		328-368
$C_2H_5NO_3$	Ethyl nitrate	293.2	19.7				
C_2H_6	Ethane	95.0	1.9356	0.20185E+01	−0.51493E−03	−0.48148E−05	95-295
C_2H_6O	Ethanol	293.2	25.3	0.15145E+03	−0.87020E+00	0.19570E−02	163-523
C_2H_6O	Dimethyl ether	258.0	6.18	0.22389E+02	−0.86524E−01	0.91291E−04	155-258
C_2H_6OS	Dimethyl sulfoxide	293.2	47.24	0.38478E+02	0.16939E+00	−0.47423E−03	288-343
$C_2H_6O_2$	Ethylene glycol	293.2	41.4	0.14355E+03	−0.48573E+00	0.46703E−03	293-423
$C_2H_6O_2S$	Dimethyl sulfone	383.2	47.39	0.10830E+03	−0.15900E+00		383-398
$C_2H_6O_4S$	Dimethyl sulfate	298.2	55.0				
C_2H_6S	Ethanethiol	298.2	6.667				
C_2H_6S	Dimethyl sulfide	294.2	6.70				
$C_2H_6S_2$	1,2-Ethanedithiol	293.2	7.26	0.11228E+02	−0.13500E−01		293-333
$C_2H_6S_2$	Dimethyl disulfide	298.2	9.6	0.19109E+02	−0.32000E−01		298-323
C_2H_7N	Ethylamine	273.2	8.7	0.30163E+02	−0.79000E−01		233-273
C_2H_7NO	Ethanolamine	293.2	31.94	0.14890E+03	−0.62491E+00	0.77143E−03	253-293
$C_2H_8N_2$	1,2-Ethanediamine	293.2	13.82	0.48922E+02	−0.17021E+00	0.17262E−03	273-333
C_3Cl_6O	Hexachloroacetone	291.9	3.925	0.76423E+01	−0.15838E−01	0.10618E−04	269-303
C_3F_6O	Perfluoroacetone	202.2	2.104	0.34809E+01	−0.92883E−02	0.12282E−04	151-238
C_3HN	Cyanoacetylene	291.9	72.3	0.91803E+03	−0.49149E+01	0.69104E−02	281-314
$C_3H_2F_6O$	1,1,1,3,3,3-Hexafluoro-2-propanol	293.2	16.70				
$C_3H_3ClO_3$	4-Chloro-1,3-dioxolan-2-one	313.2	62.0				
C_3H_3N	Acrylonitrile	293.2	33.0	0.11109E+03	−0.36806E+00	0.34879E−03	233-413
$C_3H_3NO_2$	Cyanoacetic acid	277.2	33.4				
C_3H_4	Allene	269.0	2.025	0.26049E+01	−0.44147E−03	−0.63420E−05	156-269
C_3H_4	Propyne	246.0	3.218	0.60871E+01	−0.11730E−01		185-246
$C_3H_4ClF_3$	3-Chloro-1,1,1-trifluoropropane	295.2	7.32	0.22361E+02	−0.68840E−01	0.60594E−04	275-313
C_3H_4ClNO	2-Chloroethyl isocyanate	288.2	29.1	0.64311E+02	−0.12217E+00		288-403
$C_3H_4Cl_2O$	1,1-Dichloroacetone	293.2	14.6				
$C_3H_4F_4O$	2,2,3,3-Tetrafluoro-1-propanol	298.2	21.03				
C_3H_4O	Propargyl alcohol	293.2	20.8	0.99895E+02	−0.38911E+00	0.40776E−03	213-293
$C_3H_4O_3$	Ethylene carbonate	313.2	89.78	0.20746E+03	−0.37610E+00		313-343
C_3H_5Br	3-Bromopropene	293.2	7.0				
$C_3H_5BrO_2$	2-Bromopropanoic acid	294.2	11.0				
$C_3H_5Br_3$	1,2,3-Tribromopropane	303.2	6.00	0.11024E+02	−0.16596E−01		303-358
C_3H_5Cl	2-Chloropropene	299.25	8.92				
C_3H_5Cl	3-Chloropropene	293.2	8.2				
$C_3H_5ClN_2O_6$	3-Chloro-1,2-propanediol dinitrate	293.2	17.50				
C_3H_5ClO	Epichlorohydrin	293.2	22.6				
$C_3H_5ClO_2$	Ethyl chloroformate	308.7	9.736	0.15356E+02	−0.18250E−01		309-349
$C_3H_5ClO_2$	Methyl chloroacetate	293.2	12.0				
$C_3H_5Cl_3$	1,2,3-Trichloropropane	293.2	7.5				
C_3H_5I	3-Iodopropene	292.2	6.1				
C_3H_5N	Propanenitrile	293.2	29.7	0.82222E+02	−0.22937E+00	0.17424E−03	213-473
C_3H_5NO	Ethyl isocyanate	293.2	19.7				
C_3H_5NS	Ethyl isothiocyanate	293.2	19.6				

Mol. form.	Name	T/K	ε_r	a	b	c	Range/K
$C_3H_5N_3O_9$	Trinitroglycerol	293.2	19.25				
C_3H_6	Propene	220.0	2.1365	0.29623E+01	−0.37564E-02		220-250
$C_3H_6Br_2$	1,2-Dibromopropane	283.2	4.60	0.54973E+01	−0.31695E-02		283-333
$C_3H_6Br_2$	1,3-Dibromopropane	293.2	9.482	0.29193E+02	−0.94450E-01	0.92800E-04	293-368
$C_3H_6ClNO_2$	2-Chloro-2-nitropropane	250.4	31.90				
$C_3H_6Cl_2$	1,2-Dichloropropane	293.2	8.37	0.18915E+02	−0.35907E-01		281-323
$C_3H_6Cl_2$	1,3-Dichloropropane	303.2	10.27	0.21609E+02	−0.37333E-01		303-333
$C_3H_6Cl_2$	2,2-Dichloropropane	293.2	11.37	0.32421E+02	−0.72188E-01		245-293
$C_3H_6N_2O_4$	2,2-Dinitropropane	325.1	42.4				
C_3H_6O	Allyl alcohol	293.2	19.7	0.62714E+02	−0.14771E+00	0.37879E-05	213-303
C_3H_6O	Propanal	290.2	18.5				
C_3H_6O	Acetone	293.2	21.01	0.88157E+02	−0.34300E+00	0.38925E-03	273-323
$C_3H_6O_2$	Propanoic acid	298.2	3.44	0.18793E+01	0.46841E-02	0.19983E-05	289-408
$C_3H_6O_2$	Ethyl formate	288.2	8.57	0.15884E+02	−0.25333E-01		288-318
$C_3H_6O_2$	Methyl acetate	288.2	7.07	0.13190E+02	−0.21226E-01		276-318
$C_3H_6O_3$	3-Hydroxypropanoic acid	296.2	30.0				
$C_3H_6O_3$	Dimethyl carbonate	298.2	3.087				
$C_3H_6O_3$	1,3,5-Trioxane	338.2	15.55				
C_3H_7Br	1-Bromopropane	293.2	8.09	0.17769E+02	−0.32599E-01		274-328
C_3H_7Br	2-Bromopropane	293.2	9.46	0.26195E+02	−0.72995E-01	0.55454E-04	186-328
C_3H_7Cl	1-Chloropropane	293.2	8.588	0.21214E+02	−0.43130E-01		273-313
C_3H_7ClO	3-Chloro-1-propanol	215.2	36.0	0.12436E+03	−0.60841E+00	0.92060E-03	145-215
C_3H_7ClO	1-Chloro-2-propanol	153.2	59.0	−0.19169E+02	0.13605E+01	−0.55567E-02	153-177
$C_3H_7ClO_2$	3-Chloro-1,2-propanediol	293.2	31.0				
C_3H_7I	1-Iodopropane	293.2	7.07	0.13744E+02	−0.22745E-01		293-323
C_3H_7I	2-Iodopropane	298.2	8.19				
C_3H_7NO	N-Ethylformamide	298.2	102.7	0.64764E+03	−0.28499E+01	0.34286E-02	298-338
C_3H_7NO	N,N-Dimethylformamide	293.2	38.25	0.15364E+03	−0.60367E+00	0.71505E-03	213-353
C_3H_7NO	N-Methylacetamide	303.2	179.0	0.15975E+04	−0.90451E+01	0.18345E-01	303-473
$C_3H_7NO_2$	1-Nitropropane	288.2	24.70	0.94999E+02	−0.38358E+00	0.48480E-03	276-333
$C_3H_7NO_2$	2-Nitropropane	288.2	26.74	0.60138E+02	−0.11566E+00		276-303
$C_3H_7NO_2$	Propyl nitrite	250.0	12.35	0.70552E+02	−0.40362E+00	0.66687E-03	110-310
$C_3H_7NO_2$	Isopropyl nitrite	260.0	13.92	0.74578E+02	−0.38283E+00	0.57071E-03	150-300
$C_3H_7NO_2$	Ethyl carbamate	328.2	14.14	0.32431E+02	−0.65097E-01	0.28571E-04	328-368
C_3H_8	Propane	293.19	1.6678	0.22883E+01	−0.23276E-02	0.84710E-06	90-300
C_3H_8O	1-Propanol	293.2	20.8	0.98045E+02	−0.36860E+00	0.36422E-03	193-493
C_3H_8O	2-Propanol	293.2	20.18	0.10416E+03	−0.41011E+00	0.42049E-03	193-493
$C_3H_8O_2$	1,2-Propylene glycol	303.2	27.5	0.24546E+03	−0.15738E+01	0.38068E-02	193-403
$C_3H_8O_2$	1,3-Propylene glycol	293.2	35.1	0.11365E+03	−0.36680E+00	0.33766E-03	288-328
$C_3H_8O_2$	Ethylene glycol monomethyl ether	298.2	17.2	0.11803E+03	−0.58000E+00	0.81001E-03	254-318
$C_3H_8O_2$	Dimethoxymethane	293.2	2.644	0.25877E+01	−0.93019E-03	0.38472E-05	171-293
$C_3H_8O_3$	Glycerol	293.2	46.53	0.77503E+02	−0.37984E-01	−0.23107E-03	288-343
C_3H_8S	1-Propanethiol	288.2	5.937	0.11602E+02	−0.19580E-01		273-318
C_3H_8S	2-Propanethiol	298.2	5.952				
$C_3H_8S_2$	1,2-Propanedithiol	293.2	7.24	0.14667E+02	−0.32660E-01	0.25000E-04	293-333
$C_3H_8S_2$	1,3-Propanedithiol	303.2	8.11	0.66607E+01	0.31310E-01	−0.87500E-04	303-343
$C_3H_9BO_3$	Trimethyl borate	293.2	2.2762				
C_3H_9ClSi	Trimethylchlorosilane	273.2	10.21	−0.19492E+02	0.29806E+00	−0.69284E-03	223-273
C_3H_9N	Propylamine	296.2	5.08	0.17719E+02	−0.59022E-01	0.54780E-04	204-296
C_3H_9N	Isopropylamine	293.2	5.6268	0.40429E+02	−0.21441E+00	0.32634E-03	213-298
C_3H_9N	Trimethylamine	298.2	2.440	0.39745E+01	−0.51331E-02		273-298
$C_3H_9O_4P$	Trimethyl phosphate	293.2	20.6				
C_4Cl_6	Hexachloro-1,3-butadiene	293.2	2.55				
$C_4Cl_6O_3$	Trichloroacetic anhydride	298.2	5.0				
$C_4F_6O_3$	Trifluoroacetic acid anhydride	298.2	2.7				
$C_4H_2Cl_4O_3$	Dichloroacetic anhydride	298.2	15.8				
$C_4H_2O_3$	Maleic anhydride	326.2	52.75				
$C_4H_3F_7O$	2,2,3,3,4,4,4-Heptafluoro-1-butanol	298.2	14.4				
$C_4H_4N_2$	Succinonitrile	298.2	62.6	0.17724E+03	−0.54654E+00	0.54046E-03	236-351
$C_4H_4N_2$	Pyrazine	323.2	2.80				
C_4H_4O	Furan	277.1	2.88	0.13636E+01	0.12864E-01	−0.22701E-04	188-277
C_4H_4S	Thiophene	293.2	2.739	0.32941E+01	−0.19019E-02		253-293
C_4H_5Cl	2-Chloro-1,3-butadiene	293.2	4.914				

Mol. form.	Name	T/K	ε_r	a	b	c	Range/K
$C_4H_5Cl_3O_2$	Ethyl trichloroacetate	293.2	8.428				
C_4H_5N	Pyrrole	293.0	8.00	0.12672E+02	−0.14075E-01	−0.62671E-05	293-357
C_4H_5NO	Allyl isocynate	288.2	15.15	0.34299E+02	−0.66444E-01		288-333
C_4H_6	1,3-Butadiene	265.0	2.050	0.27674E+01	−0.26738E-02		185-265
C_4H_6O	Divinyl ether	288.2	3.94				
C_4H_6O	Ethoxyacetylene	298.2	8.05				
C_4H_6O	Cyclobutanone	298.2	14.27	0.43974E+02	−0.15712E+00	0.19264E-03	220-317
$C_4H_6O_2$	Methyl acrylate	303.2	7.03	0.11968E+02	−0.16500E-01		303-333
$C_4H_6O_2$	2,3-Butanedione	298.2	4.04	0.46907E+01	−0.22302E-02		278-348
$C_4H_6O_2$	γ-Butyrolactone	293.2	39.0				
$C_4H_6O_3$	Acetic anhydride	293.2	22.45				
$C_4H_6O_3$	Propylene carbonate	293.0	66.14	0.15940E+03	−0.39530E+00	0.26284E-03	273-333
C_4H_7Br	cis-2-Bromo-2-butene	293.2	5.38				
C_4H_7Br	trans-2-Bromo-2-butene	293.2	6.76				
$C_4H_7BrO_2$	2-Bromobutanoic acid	293.2	7.2				
$C_4H_7BrO_2$	Ethyl bromoacetate	303.2	9.75	0.15627E+02	−0.19600E-01		303-333
$C_4H_7BrO_2$	Methyl 3-bromopropanoate	303.2	5.81	0.36001E+01	0.72500E-02		303-343
$C_4H_7ClO_2$	Propyl chlorocarbonate	293.2	11.2				
$C_4H_7ClO_2$	Methyl 2-chloropropanoate	303.2	11.45	0.22449E+02	−0.36250E-01		303-343
C_4H_7N	Butanenitrile	293.2	24.83	0.53884E+02	−0.99257E-01		293-333
C_4H_7N	2-Methylpropanenitrile	293.2	24.42	0.52554E+02	−0.96000E-01		293-313
C_4H_7NO	2-Pyrrolidone	298.2	28.18	0.11054E+03	−0.47945E+00	0.68182E-03	298-338
C_4H_8	1-Butene	220.0	2.2195	0.29354E+01	−0.32580E-02		220-250
C_4H_8	cis-2-Butene	296.0	1.960	0.28802E+01	−0.31064E-02		197-296
C_4H_8	Isobutene	288.7	2.1225	0.33701E+01	−0.43295E-02		220-289
$C_4H_8Br_2$	1,2-Dibromobutane	293.2	4.74	0.11199E+03	−0.63334E+00	0.91250E-03	293-333
$C_4H_8Br_2$	1,3-Dibromobutane	293.2	9.14	0.34031E+02	−0.13254E+00	0.16250E-03	293-333
$C_4H_8Br_2$	1,4-Dibromobutane	303.2	8.68	0.20944E+02	−0.55620E-01	0.50000E-04	303-333
$C_4H_8Br_2$	2,3-Dibromobutane	298.2	6.245	0.23849E+02	−0.96300E-01	0.12500E-03	293-333
$C_4H_8Br_2$	1,2-Dibromo-2-methylpropane	293.2	4.1				
$C_4H_8Cl_2$	1,2-Dichlorobutane	293.2	7.74	0.31925E+02	−0.13232E+00	0.17007E-03	293-356
$C_4H_8Cl_2$	1,4-Dichlorobutane	308.2	9.30	0.59766E+01	0.49300E-01	−0.12500E-03	308-338
$C_4H_8Cl_2$	1,2-Dichloro-2-methylpropane	296.0	7.15	0.39429E+02	−0.20028E+00	0.30917E-03	165-296
$C_4H_8Cl_2O$	Bis(2-chloroethyl) ether	293.2	21.20				
C_4H_8O	Butanal	298.2	13.45				
C_4H_8O	2-Butanone	293.2	18.56	0.15457E+02	0.90152E-01	−0.27100E-03	293-333
C_4H_8O	Tetrahydrofuran	295.2	7.52	0.30739E+02	−0.12946E+00	0.17195E-03	224-295
$C_4H_8O_2$	Butanoic acid	287.2	2.98	0.15010E+01	0.50046E-02		287-403
$C_4H_8O_2$	2-Methylpropanoic acid	293.2	2.58				
$C_4H_8O_2$	Propyl formate	303.2	6.92				
$C_4H_8O_2$	Ethyl acetate	293.2	6.0814	0.15646E+02	−0.44066E-01	0.39137E-04	293-433
$C_4H_8O_2$	Methyl propanoate	293.2	6.200	0.12798E+02	−0.22540E-01		293-333
$C_4H_8O_2$	1,4-Dioxane	293.2	2.2189	0.27299E+01	−0.17440E-02		293-313
$C_4H_8O_3$	2-Hydroxybutanoic acid	296.2	37.7				
$C_4H_8O_3$	3-Hydroxybutanoic acid	296.2	31.5				
$C_4H_8O_3$	Ethyl methyl carbonate	293.2	2.985				
$C_4H_8O_3$	Ethylene glycol monoacetate	303.2	12.95				
C_4H_9Br	1-Bromobutane	283.2	7.315	0.22542E+02	−0.79306E-01	0.89867E-04	183-363
C_4H_9Br	2-Bromobutane	298.2	8.64	0.18461E+02	−0.32933E-01		274-328
C_4H_9Br	1-Bromo-2-methylpropane	273.2	7.70	0.37558E+02	−0.20571E+00	0.35496E-03	112-273
C_4H_9Br	2-Bromo-2-methylpropane	293.0	10.98	0.35085E+02	−0.14075E+00	0.19960E-03	258-293
C_4H_9Cl	1-Chlorobutane	293.2	7.276	0.13565E+02	−0.10161E-01	−0.38750E-04	273-323
C_4H_9Cl	2-Chlorobutane	293.2	8.564	0.30376E+02	−0.11377E+00	0.13429E-03	273-323
C_4H_9Cl	1-Chloro-2-methylpropane	293.2	7.027	0.14945E+02	−0.33747E-01	0.23036E-04	273-323
C_4H_9Cl	2-Chloro-2-methylpropane	293.2	9.663	0.35077E+02	−0.12867E+00	0.14304E-03	273-323
C_4H_9I	1-Iodobutane	293.2	6.27	0.16493E+02	−0.50262E-01	0.52485E-04	293-323
C_4H_9I	2-Iodobutane	293.2	7.873	0.10883E+02	−0.14680E-02	−0.30000E-04	293-323
C_4H_9I	2-Iodo-2-methylpropane	283.2	6.65	0.76780E+01	0.69900E-02	−0.37500E-04	283-323
C_4H_9N	Pyrrolidine	293.0	8.30	0.38191E+02	−0.15462E+00	0.17941E-03	274-333
C_4H_9NO	N-Methylpropanamide	293.2	170.0				
C_4H_9NO	N-Ethylacetamide	293.2	135.0	0.74494E+03	−0.31400E+01	0.36131E-02	213-353
C_4H_9NO	N,N-Dimethylacetamide	294.2	38.85	0.15420E+03	−0.57506E+00	0.61911E-03	294-433
C_4H_9NO	2-Butanone oxime	293.2	3.4				

Mol. form.	Name	T/K	ε_r	a	b	c	Range/K
C_4H_9NO	Morpholine	298.2	7.42				
$C_4H_9NO_2$	tert-Butyl nitrite	298.2	11.47				
$C_4H_9NO_2$	Propyl carbamate	338.2	12.06	0.24356E+02	−0.36400E−01		338-378
$C_4H_9NO_2$	Ethyl-N-methyl carbamate	298.2	21.10	0.11477E+03	−0.47568E+00	0.54127E−03	298-373
$C_4H_9NO_2$	N-Acetylethanolamine	298.2	96.6	0.37016E+03	−0.13113E+01	0.13214E−02	298-348
$C_4H_9NO_3$	Butyl nitrate	293.2	13.10				
C_4H_{10}	Butane	295.0	1.7697	0.22379E+01	−0.13884E−02	−0.66711E−06	135-303
C_4H_{10}	Isobutane	295.0	1.7518	0.23295E+01	−0.19953E−02	0.14197E−06	115-303
$C_4H_{10}O$	1-Butanol	293.2	17.84	0.10578E+03	−0.50587E+00	0.84733E−03	193-553
$C_4H_{10}O$	2-Butanol	293.2	17.26	0.13850E+03	−0.75146E+00	0.14086E−02	172-533
$C_4H_{10}O$	2-Methyl-1-propanol	293.2	17.93	0.10762E+03	−0.51398E+00	0.83702E−03	173-533
$C_4H_{10}O$	2-Methyl-2-propanol	298.2	12.47	0.22541E+03	−0.14990E+01	0.34050E−02	298-503
$C_4H_{10}O$	Diethyl ether	293.2	4.2666	0.79725E+01	−0.12519E−01		283-301
$C_4H_{10}O_2$	1,2-Butanediol	298.2	22.4	0.63702E+02	−0.13807E+00		278-323
$C_4H_{10}O_2$	1,3-Butanediol	298.2	28.8	0.72883E+02	−0.14770E+00		278-323
$C_4H_{10}O_2$	1,4-Butanediol	298.2	31.9	0.13079E+03	−0.46985E+00	0.46320E−03	288-328
$C_4H_{10}O_2$	Ethylene glycol monoethyl ether	298.2	13.38				
$C_4H_{10}O_2$	Ethylene glycol dimethyl ether	296.7	7.30	0.48832E+02	−0.24218E+00	0.34413E−03	256-318
$C_4H_{10}O_2S$	Bis(2-hydroxyethyl) sulfide	293.2	28.61	0.13128E+03	−0.52719E+00	0.60465E−03	253-333
$C_4H_{10}O_3$	Diethylene glycol	293.2	31.82	0.13973E+03	−0.54725E+00	0.61149E−03	288-343
$C_4H_{10}O_3S$	Diethyl sulfite	293.2	15.6				
$C_4H_{10}O_4$	1,2,3,4-Butanetetrol	393.2	28.2				
$C_4H_{10}O_4S$	Diethyl sulfate	293.2	29.2				
$C_4H_{10}S$	1-Butanethiol	288.2	5.204	0.11201E+02	−0.20767E−01		273-318
$C_4H_{10}S$	2-Butanethiol	288.2	5.645	0.10866E+02	−0.17993E−01		273-318
$C_4H_{10}S$	2-Methyl-1-propanethiol	298.2	4.961				
$C_4H_{10}S$	2-Methyl-2-propanethiol	293.2	5.475	0.10597E+02	−0.17500E−01		283-313
$C_4H_{10}S$	Diethyl sulfide	298.2	5.723				
$C_4H_{11}N$	Butylamine	293.2	4.71	0.13322E+02	−0.44176E−01	0.50250E−04	223-333
$C_4H_{11}N$	Diethylamine	293.2	3.680	0.26462E+02	−0.13750E+00	0.20373E−03	243-323
$C_4H_{11}NO_2$	Diethanolamine	293.2	25.75	0.73435E+02	−0.21377E+00	0.17500E−03	273-323
$C_4H_{12}O_2Si$	Dimethoxydimethylsilane	298.2	3.663				
$C_4H_{12}O_3Si$	Trimethoxymethylsilane	298.2	4.9				
$C_4H_{12}O_4Si$	Tetramethyl silicate	293.2	6.0				
$C_4H_{12}Si$	Diethylsilane	293.2	2.544				
$C_4H_{12}Si$	Tetramethylsilane	293.2	1.921				
$C_4H_{13}N_3$	Diethylenetriamine	293.2	12.62	0.57840E+02	−0.23873E+00	0.28841E−03	213-333
C_5FeO_5	Iron pentacarbonyl	293.2	2.602				
C_5H_4BrN	2-Bromopyridine	298.2	23.18	0.73391E+02	−0.23678E+00	0.22930E−03	298-398
C_5H_4ClN	2-Chloropyridine	298.2	27.32	0.98702E+02	−0.34237E+00	0.34502E−03	298-398
$C_5H_4F_8O$	2,2,3,3,4,4,5,5-Octafluoro-1-pentanol	298.2	15.30				
$C_5H_4O_2$	Furfural	293.2	42.1				
C_5H_5N	Pyridine	293.2	13.260	0.43991E+02	−0.15150E+00	0.15925E−03	293-323
C_5H_5NO	Pyridine-1-oxide	343.0	35.94	0.20878E+02	0.16450E+00	−0.35269E−03	343-398
C_5H_6O	2-Methylfuran	293.2	2.76				
$C_5H_6O_2$	Furfuryl alcohol	298.2	16.85				
$C_5H_7Cl_3O_2$	Propyl trichloroacetate	298.2	8.32				
$C_5H_7NO_2$	Ethyl cyanoacetate	263.2	31.62				
C_5H_8	1,3-Pentadiene*	298.2	2.319				
C_5H_8	1,4-Pentadiene	294.0	2.054	0.29994E+01	−0.34578E−02	0.85300E−06	178-294
C_5H_8	2-Methyl-1,3-butadiene	293.2	2.098	0.28170E+01	−0.23147E−02	−0.43975E−06	198-293
C_5H_8	Cyclopentene	295.0	2.083	0.28177E+01	−0.27597E−02	0.89346E−06	171-319
C_5H_8O	Cyclopentanone	298.2	13.58	0.24083E+02	−0.30286E−01	−0.16802E−04	219-298
$C_5H_8O_2$	Ethyl acrylate	303.2	6.05	0.47827E+02	−0.24394E+00	0.35000E−03	303-343
$C_5H_8O_2$	Methyl trans-2-butenoate	293.2	6.6645				
$C_5H_8O_2$	Methyl methacrylate	303.2	6.32	0.32098E+02	−0.14568E+00	0.20000E−03	303-343
$C_5H_8O_2$	2,4-Pentanedione	303.2	26.524				
$C_5H_8O_4$	Dimethyl malonate	293.2	9.82	0.26470E+02	−0.76656E−01	0.67888E−04	293-433
$C_5H_9BrO_2$	Ethyl 2-bromopropanoate	293.2	9.4				
$C_5H_9ClO_2$	Isobutyl chlorocarbonate	293.2	9.1				
$C_5H_9ClO_2$	Ethyl 2-chloropropanoate	303.2	11.95	0.25965E+02	−0.46250E−01		303-343
$C_5H_9ClO_2$	Ethyl 3-chloropropanoate	303.2	10.19	0.21951E+02	−0.38750E−01		303-343
$C_5H_9ClO_2$	Methyl 4-chlorobutanoate	303.2	9.51	0.17127E+02	−0.25000E−01		303-343

Mol. form.	Name	T/K	ε_r	a	b	c	Range/K
C_5H_9N	Pentanenitrile	293.2	20.04	0.55793E+02	−0.15750E+00	0.12432E−03	183-333
C_5H_9N	2,2-Dimethylpropanenitrile	293.2	21.1	0.58418E+02	−0.16884E+00	0.14131E−03	293-453
C_5H_9NO	Isobutyl isocyanate	293.2	11.638	0.38026E+02	−0.12714E+00	0.12679E−03	293-353
C_5H_9NO	N-Methyl-2-pyrrolidone	293.2	32.55				
C_5H_{10}	1-Pentene	293.2	2.011	−0.11438E+01	0.25420E−01	−0.50000E−04	273-293
C_5H_{10}	2-Methyl-1-butene	293.2	2.180				
C_5H_{10}	2-Methyl-2-butene	296.0	1.979	0.26064E+01	−0.19578E−02	−0.53908E−06	225-296
C_5H_{10}	Cyclopentane	293.2	1.9687	0.24287E+01	−0.15304E−02	−0.13095E−06	278-313
C_5H_{10}	Ethylcyclopropane	293.2	1.933				
$C_5H_{10}Br_2$	1,2-Dibromopentane	298.2	4.39				
$C_5H_{10}Br_2$	1,4-Dibromopentane	293.2	9.05	0.26443E+02	−0.88640E−01	0.10000E−03	293-333
$C_5H_{10}Br_2$	1,5-Dibromopentane	303.2	9.14	0.38192E+02	−0.15648E+00	0.20000E−03	303-333
$C_5H_{10}Cl_2$	1,2-Dichloropentane	293.2	6.89	0.19016E+02	−0.57954E−01	0.56801E−04	293-356
$C_5H_{10}Cl_2$	1,5-Dichloropentane	298.2	9.92				
$C_5H_{10}O$	Cyclopentanol	288.2	18.5	0.10565E+03	−0.44244E+00	0.48657E−03	258-323
$C_5H_{10}O$	Pentanal	293.2	10.00				
$C_5H_{10}O$	2,2-Dimethylpropanal	293.2	9.051	0.18645E+02	−0.32395E−01	−0.16157E−05	280-333
$C_5H_{10}O$	2-Pentanone	293.2	15.45	0.40893E+02	−0.10423E+00	0.60557E−04	204-353
$C_5H_{10}O$	3-Pentanone	293.2	17.00	0.12690E+02	0.95177E−01	−0.27321E−03	233-353
$C_5H_{10}O$	3-Methyl-2-butanone	293.2	10.37	0.30695E+02	−0.10962E+00	0.13810E−03	293-328
$C_5H_{10}O$	Tetrahydropyran	293.2	5.66	0.19793E+02	−0.76071E−01	0.94852E−04	234-333
$C_5H_{10}O$	2-Methyltetrahydrofuran	298.2	6.97				
$C_5H_{10}O_2$	Pentanoic acid	294.4	2.661	0.33491E+01	−0.75156E−02	0.17820E−04	250-344
$C_5H_{10}O_2$	Butyl formate	303.2	6.10	0.21532E+02	−0.84106E−01	0.10952E−03	288-323
$C_5H_{10}O_2$	Isobutyl formate	293.2	6.41				
$C_5H_{10}O_2$	Propyl acetate	293.2	5.62	0.17677E+02	−0.61404E−01	0.69196E−04	253-353
$C_5H_{10}O_2$	Ethyl propanoate	293.2	5.76				
$C_5H_{10}O_2$	Methyl butanoate	301.2	5.48	0.38604E+02	−0.19171E+00	0.27128E−03	301-343
$C_5H_{10}O_2$	Tetrahydrofurfuryl alcohol	303.2	13.48				
$C_5H_{10}O_2S$	3-Methyl sulfolane	298.2	29.4	0.53158E+02	−0.93730E−01	0.47275E−04	298-398
$C_5H_{10}O_3$	Diethyl carbonate	297.2	2.820				
$C_5H_{10}O_3$	Ethyl lactate	303.2	15.4	0.31225E+02	−0.43531E−01	−0.28571E−04	273-373
$C_5H_{10}O_4$	1,2,3-Propanetriol-1-acetate	242.2	38.57	0.10653E+03	−0.26439E+00	−0.62371E−04	215-242
$C_5H_{11}Br$	2-Bromo-2-methylbutane	298.2	9.21				
$C_5H_{11}Br$	1-Bromopentane	299.2	6.31	0.20954E+02	−0.78743E−01	0.98908E−04	183-328
$C_5H_{11}Br$	3-Bromopentane	298.2	8.37				
$C_5H_{11}Br$	1-Bromo-3-methylbutane	291.5	6.33	0.27743E+02	−0.13927E+00	0.22627E−03	123-292
$C_5H_{11}Cl$	1-Chloropentane	293.2	6.654	0.18626E+02	−0.54719E−01	0.47143E−04	273-323
$C_5H_{11}Cl$	1-Chloro-3-methylbutane	292.0	6.10	0.22228E+02	−0.93189E−01	0.12991E−03	171-297
$C_5H_{11}Cl$	2-Chloro-2-methylbutane	222.75	12.31	0.55104E+02	−0.29866E+00	0.47840E−03	201-223
$C_5H_{11}F$	1-Fluoropentane	293.2	3.931				
$C_5H_{11}I$	1-Iodopentane	293.2	5.78	0.15753E+02	−0.50543E−01	0.56401E−04	293-323
$C_5H_{11}I$	3-Iodopentane	293.2	7.432				
$C_5H_{11}I$	1-Iodo-3-methylbutane	292.2	5.6				
$C_5H_{11}I$	2-Iodo-2-methylbutane	293.2	8.192				
$C_5H_{11}N$	Piperidine	293.0	4.33	0.82317E+01	−0.11229E−01	−0.71429E−05	293-333
$C_5H_{11}N$	N-Methylpyrrolidine	298.2	32.2				
$C_5H_{11}NO$	2,2-Dimethylpropanamide	298.2	20.13	0.10400E+03	−0.46017E+00	0.60000E−03	298-328
$C_5H_{11}NO$	N,N-Diethylformamide	293.2	29.6				
$C_5H_{11}NO$	2-Pentanone oxime	293.2	3.3				
$C_5H_{11}NO_2$	Pentyl nitrite	298.2	7.21				
C_5H_{12}	Pentane	293.2	1.8371				
C_5H_{12}	Isopentane	293.2	1.845	0.22384E+01	−0.12985E−02	−0.16182E−06	143-293
C_5H_{12}	Neopentane	296.0	1.769	0.10949E+02	−0.63057E−01	0.10835E−03	251-296
$C_5H_{12}N_2O$	Tetramethylurea	293.2	23.10				
$C_5H_{12}O$	1-Pentanol	298.2	15.13	0.73397E+02	−0.28165E+00	0.28427E−03	213-513
$C_5H_{12}O$	2-Pentanol	298.2	13.71	0.16437E+03	−0.86506E+00	0.11955E−02	273-323
$C_5H_{12}O$	3-Pentanol	298.2	13.35	0.12838E+03	−0.60980E+00	0.75000E−03	288-318
$C_5H_{12}O$	2-Methyl-1-butanol	298.2	15.63	0.14020E+02	0.13948E+00	−0.45000E−03	288-318
$C_5H_{12}O$	3-Methyl-1-butanol	293.2	15.63	0.79733E+02	−0.31272E+00	0.32014E−03	173-513
$C_5H_{12}O$	2-Methyl-2-butanol	298.2	5.78	0.11662E+03	−0.69756E+00	0.10920E−02	268-318
$C_5H_{12}O$	3-Methyl-2-butanol	298.2	12.1				
$C_5H_{12}O$	2,2-Dimethyl-1-propanol	333.2	8.35	0.92350E+02	−0.41870E+00	0.50000E−03	333-373

Mol. form.	Name	T/K	ε_r	a	b	c	Range/K
$C_5H_{12}O_2$	1,2-Pentanediol	296.8	17.31	0.18436E+03	−0.10682E+01	0.17037E-02	197-297
$C_5H_{12}O_2$	1,4-Pentanediol	295.7	26.74	0.13568E+03	−0.59198E+00	0.75398E-03	193-318
$C_5H_{12}O_2$	1,5-Pentanediol	293.2	26.2	0.11858E+03	−0.45920E+00	0.49341E-03	243-343
$C_5H_{12}O_2$	2,3-Pentanediol	296.9	17.37	0.95876E+02	−0.46463E+00	0.67434E-03	238-297
$C_5H_{12}O_2$	2,4-Pentanediol	294.2	24.69	0.11914E+03	−0.52569E+00	0.69607E-03	224-294
$C_5H_{12}O_2$	Diethoxymethane	293.2	2.527	0.25294E+01	0.73988E-04	−0.28331E-06	227-293
$C_5H_{12}O_4$	Tetramethoxymethane	293.2	2.40				
$C_5H_{12}O_5$	Xylitol	293.2	40.0				
$C_5H_{12}S$	1-Pentanethiol	293.2	4.847	0.71131E+01	−0.30228E-02	−0.16414E-04	273-333
$C_5H_{12}S$	2-Methyl-2-butanethiol	293.2	5.087	0.15116E+02	−0.50700E-01	0.56250E-04	273-333
$C_5H_{12}S_4$	Tetrakis(methylthio)methane	343.2	2.818				
$C_5H_{13}N$	Pentylamine	293.2	4.27	0.11274E+02	−0.34965E-01	0.37706E-04	223-353
$C_5H_{13}N_3$	1,1,3,3-Tetramethylguanidine	298.2	11.5				
$C_5H_{14}OSi$	Ethoxytrimethylsilane	298.2	3.013				
C_6F_6	Hexafluorobenzene	298.2	2.029	0.24041E+01	−0.83086E-03	−0.14286E-05	298-338
C_6F_{14}	Perfluorohexane	298.2	1.76				
$C_6H_3N_3O_7$	2,4,6-Trinitrophenol	294.2	4.0				
C_6H_4BrF	1-Bromo-2-fluorobenzene	298.2	4.72				
C_6H_4BrF	1-Bromo-3-fluorobenzene	298.2	4.85				
C_6H_4BrF	1-Bromo-4-fluorobenzene	298.2	2.60				
$C_6H_4BrNO_2$	1-Bromo-3-nitrobenzene	328.2	20.2	0.81413E+02	−0.27645E+00	0.27367E-03	328-413
$C_6H_4Br_2$	o-Dibromobenzene	293.2	7.86	−0.81849E-02	0.62671E-01	−0.12222E-03	293-353
$C_6H_4Br_2$	m-Dibromobenzene	293.2	4.81	0.93214E+01	−0.20273E-01	0.16667E-04	293-353
$C_6H_4Br_2$	p-Dibromobenzene	368.2	2.57				
C_6H_4ClF	1-Chloro-2-fluorobenzene	298.2	6.10				
C_6H_4ClF	1-Chloro-3-fluorobenzene	298.2	4.96				
C_6H_4ClF	1-Chloro-4-fluorobenzene	298.2	3.34				
$C_6H_4ClNO_2$	1-Chloro-2-nitrobenzene	323.2	37.7	0.16800E+03	−0.59708E+00	0.59957E-03	323-436
$C_6H_4ClNO_2$	1-Chloro-3-nitrobenzene	323.2	20.9	0.77193E+02	−0.25118E+00	0.23798E-03	323-433
$C_6H_4ClNO_2$	1-Chloro-4-nitrobenzene	393.2	8.09				
$C_6H_4Cl_2$	o-Dichlorobenzene	293.2	10.12	0.13629E+02	0.10622E-02	−0.44444E-04	293-353
$C_6H_4Cl_2$	m-Dichlorobenzene	293.2	5.02	0.77565E+01	−0.93333E-02	−0.26880E-14	293-353
$C_6H_4Cl_2$	p-Dichlorobenzene	328.2	2.3943	0.26999E+01	−0.35325E-03	−0.17619E-05	328-363
C_6H_4FI	1-Fluoro-2-iodobenzene	298.2	8.22				
C_6H_4FI	1-Fluoro-4-iodobenzene	298.2	3.12				
$C_6H_4F_2$	o-Difluorobenzene	301.2	13.38	0.59107E+02	−0.23611E+00	0.27987E-03	273-323
$C_6H_4F_2$	m-Difluorobenzene	301.2	5.01	0.14448E+02	−0.46982E-01	0.51948E-04	273-323
$C_6H_4I_2$	o-Diiodobenzene	323.2	5.41	0.31150E+02	−0.14428E+00	0.20000E-03	323-353
$C_6H_4I_2$	m-Diiodobenzene	323.2	4.11				
$C_6H_4I_2$	p-Diiodobenzene	393.2	2.88				
$C_6H_4N_2$	2-Pyridinecarbonitrile	303.2	93.77	0.45596E+03	−0.17746E+01	0.19105E-02	303-398
$C_6H_4N_2$	3-Pyridinecarbonitrile	323.2	20.54	0.60484E+02	−0.17280E+00	0.15218E-03	323-398
$C_6H_4N_2$	4-Pyridinecarbonitrile	353.2	5.23	0.12533E+02	−0.30115E-01	0.26674E-04	353-398
$C_6H_4N_2O_4$	1,3-Dinitrobenzene	365.2	22.9	0.10406E+03	−0.34133E+00	0.32609E-03	365-413
C_6H_5Br	Bromobenzene	293.2	5.45	0.94100E+01	−0.12537E-01	−0.31127E-05	234-333
C_6H_5Cl	Chlorobenzene	293.2	5.6895	0.19471E+02	−0.70786E-01	0.82466E-04	293-430
C_6H_5ClO	o-Chlorophenol	296.2	7.40	0.29755E+02	−0.11256E+00	0.12390E-03	296-448
C_6H_5ClO	m-Chlorophenol	293.2	6.255				
C_6H_5ClO	p-Chlorophenol	314.2	11.18	0.31997E+02	−0.94241E-01	0.88392E-04	314-453
$C_6H_5ClO_2S$	Benzenesulfonyl chloride	323.2	28.90	0.83886E+02	−0.23405E+00	0.19713E-03	323-473
C_6H_5ClS	4-Chlorobenzenethiol	338.2	3.59				
C_6H_5F	Fluorobenzene	293.2	5.465				
C_6H_5I	Iodobenzene	293.2	4.59	0.89442E+01	−0.20008E-01	0.17641E-04	243-323
C_6H_5NOS	N-Sulfinylaniline	298.2	6.97				
$C_6H_5NO_2$	Nitrobenzene	293.0	35.6	0.11212E+03	−0.35211E+00	0.31128E-03	279-533
$C_6H_5NO_3$	o-Nitrophenol	323.2	16.50	0.33827E+02	−0.62123E-01	0.26774E-04	323-453
$C_6H_5NO_3$	m-Nitrophenol	373.2	35.45	0.18967E+03	−0.66144E+00	0.66532E-03	373-458
$C_6H_5NO_3$	p-Nitrophenol	393.2	42.20	0.22901E+03	−0.74264E+00	0.68006E-03	393-463
C_6H_6	Benzene	293.2	2.2825	0.26706E+01	−0.91648E-03	−0.14257E-05	293-513
C_6H_6BrN	m-Bromoaniline	293.2	13.0				
C_6H_6ClN	o-Chloroaniline	293.2	13.40				
C_6H_6ClN	m-Chloroaniline	293.2	13.3				
$C_6H_6N_2O_2$	o-Nitroaniline	353.0	47.3	0.18900E+03	−0.56977E+00	0.47484E-03	353-468

Mol. form.	Name	T/K	ε_r	a	b	c	Range/K
$C_6H_6N_2O_2$	*m*-Nitroaniline	398.0	35.6	0.20352E+03	−0.66582E+00	0.61310E-03	398-468
$C_6H_6N_2O_2$	*p*-Nitroaniline	428.0	78.5	0.48673E+03	−0.15040E+01	0.12857E-02	428-468
C_6H_6O	Phenol	303.2	12.40	0.63391E+02	−0.24988E+00	0.26930E-03	303-433
$C_6H_6O_2$	Pyrocatechol	388.2	17.57	0.74930E+02	−0.22142E+00	0.18919E-03	388-463
$C_6H_6O_2$	Resorcinol	393.2	13.55	0.30252E+02	−0.56443E-01	0.35578E-04	393-463
C_6H_6S	Benzenethiol	303.2	4.26	0.57155E+01	−0.70336E-02	0.73617E-05	303-358
C_6H_7N	Aniline	293.2	7.06	0.89534E+01	0.38990E-02	−0.36310E-04	293-413
C_6H_7N	2-Methylpyridine	293.2	10.18	0.34560E+02	−0.11980E+00	0.12500E-03	293-333
C_6H_7N	3-Methylpyridine	303.0	11.10	0.19643E+03	−0.11167E+01	0.16667E-02	303-333
C_6H_7N	4-Methylpyridine	293.0	12.2	0.33765E+02	−0.10113E+00	0.93860E-04	274-333
C_6H_7NO	2-Methylpyridine-1-oxide	323.2	36.4	0.11705E+03	−0.35301E+00	0.32000E-03	323-398
C_6H_7NO	3-Methylpyridine-1-oxide	318.2	28.26	0.59851E+02	−0.12682E+00	0.86622E-04	318-398
C_6H_8	1,3-Cyclohexadiene	184.2	2.68				
C_6H_8	1,4-Cyclohexadiene	296.0	2.211	0.27459E+01	−0.16975E-02	−0.36461E-06	232-356
$C_6H_8N_2$	Phenylhydrazine	293.2	7.15				
$C_6H_8N_2$	2,5-Dimethylpyrazine	293.2	2.436				
$C_6H_8N_2$	2,6-Dimethylpyrazine	308.2	2.653				
$C_6H_8O_2$	1,4-Cyclohexanedione	351.2	4.40				
$C_6H_9Cl_3O_2$	Butyl trichloroacetate	293.2	7.480				
$C_6H_9Cl_3O_2$	Isobutyl trichloroacetate	293.2	7.667				
C_6H_9N	Cyclopentanecarbonitrile	293.2	22.68	0.69830E+02	−0.25303E+00	0.31491E-03	201-293
C_6H_{10}	1,5-Hexadiene	294.0	2.125	0.30014E+01	−0.28668E-02	−0.31026E-06	151-294
C_6H_{10}	*cis,cis*-2,4-Hexadiene	297.0	2.163	0.27284E+01	−0.17178E-02	−0.62926E-06	234-351
C_6H_{10}	*trans,trans*-2,4-Hexadiene	297.0	2.123	0.26774E+01	−0.16977E-02	−0.55637E-06	232-353
C_6H_{10}	2-Methyl-1,3-pentadiene*	298.2	2.422				
C_6H_{10}	3-Methyl-1,3-pentadiene	298.2	2.426				
C_6H_{10}	4-Methyl-1,3-pentadiene	293.2	2.599	0.51328E+01	−0.12774E-01	0.14215E-04	198-323
C_6H_{10}	2,3-Dimethyl-1,3-butadiene	293.2	2.102	0.26258E+01	−0.17990E-02	0.12035E-06	223-323
C_6H_{10}	1-Hexyne	296.0	2.621	0.58591E+01	−0.17099E-01	0.20856E-04	184-296
C_6H_{10}	Cyclohexene	293.2	2.2176	0.30598E+01	−0.39841E-02	0.37554E-05	141-313
$C_6H_{10}O$	Butoxyacetylene	298.2	6.62				
$C_6H_{10}O$	Cyclohexanone	293.0	16.1	0.41577E+02	−0.11463E+00	0.92454E-04	253-423
$C_6H_{10}O$	Mesityl oxide	273.2	15.6				
$C_6H_{10}O_2$	Ethyl 2-butenoate	293.2	5.4				
$C_6H_{10}O_2$	Ethyl methacrylate	303.2	5.68	0.40962E+02	−0.20520E+00	0.29286E-03	303-343
$C_6H_{10}O_3$	Ethyl acetoacetate	293.2	14.0				
$C_6H_{10}O_3$	Propanoic anhydride	293.2	18.30				
$C_6H_{10}O_4$	Monomethyl glutarate	293.2	8.37	0.16779E+02	−0.39839E-01	0.38095E-04	293-363
$C_6H_{10}O_4$	Diethyl oxalate	293.2	8.266	0.21938E+02	−0.66226E-01	0.66800E-04	293-368
$C_6H_{10}O_4$	Dimethyl succinate	293.2	7.19	0.13551E+02	−0.23109E-01	0.55440E-05	293-433
$C_6H_{10}O_4$	Ethylene glycol diacetate	290.2	7.7	0.25093E+02	−0.95171E-01	0.12224E-03	223-290
$C_6H_{11}Br$	Bromocyclohexane	303.2	8.0026				
$C_6H_{11}BrO_2$	Ethyl 2-bromobutanoate	303.2	8.57	0.49005E+02	−0.23193E+00	0.32500E-03	303-333
$C_6H_{11}BrO_2$	Ethyl 2-bromo-2-methylpropanoate	303.2	8.55	0.77044E+02	−0.40784E+00	0.60000E-03	303-333
$C_6H_{11}Cl$	Chlorocyclohexane	303.2	7.9505				
$C_6H_{11}N$	Hexanenitrile	298.2	17.26				
$C_6H_{11}N$	4-Methylpentanenitrile	295.2	17.5				
$C_6H_{11}NO$	Cyclohexanone oxime	362.2	3.04				
C_6H_{12}	1-Hexene	294.0	2.077	0.31476E+01	−0.50003E-02	0.46673E-05	149-294
C_6H_{12}	*trans*-2-Hexene	295.0	1.978	0.24338E+01	−0.11323E-02	−0.13720E-05	157-295
C_6H_{12}	*cis*-3-Hexene	296.0	2.069	0.30691E+01	−0.45458E-02	0.39898E-05	155-296
C_6H_{12}	*trans*-3-Hexene	293.2	1.954				
C_6H_{12}	Cyclohexane	293.2	2.0243	0.24293E+01	−0.12095E-02	−0.58741E-06	283-333
C_6H_{12}	Methylcyclopentane	293.2	1.9853	0.21587E+01	−0.22450E-03	−0.12500E-05	293-323
C_6H_{12}	Ethylcyclobutane	293.2	1.965				
$C_6H_{12}Br_2$	1,6-Dibromohexane	298.2	8.52	−0.55185E+01	0.11746E+00	−0.23658E-03	274-328
$C_6H_{12}Br_2$	3,4-Dibromohexane	298.2	6.732				
$C_6H_{12}Cl_2$	1,6-Dichlorohexane	308.2	8.60	0.11277E+02	0.67200E-02	−0.50000E-04	308-338
$C_6H_{12}O$	1-Methylcyclopentanol	310.1	7.11	0.75444E+02	−0.36617E+00	0.47021E-03	310-333
$C_6H_{12}O$	Isobutyl vinyl ether	293.2	3.34	0.48060E+01	−0.50000E-02	−0.41495E-14	293-323
$C_6H_{12}O$	2-Hexanone	293.2	14.56	0.70378E+02	−0.29385E+00	0.35289E-03	243-293
$C_6H_{12}O$	4-Methyl-2-pentanone	293.2	13.11	0.36341E+02	−0.97119E-01	0.61896E-04	204-373
$C_6H_{12}O$	3,3-Dimethyl-2-butanone	293.2	12.73	0.66857E+02	−0.28552E+00	0.34422E-03	243-293

Mol. form.	Name	T/K	ε_r	a	b	c	Range/K
$C_6H_{12}O$	Cyclohexanol	293.2	16.40	0.10173E+03	−0.43072E+00	0.47926E−03	293-423
$C_6H_{12}O_2$	Hexanoic acid	298.2	2.600	0.21730E+01	0.14840E−02	−0.16526E−06	298-433
$C_6H_{12}O_2$	2-Ethylbutanoic acid	296.2	2.72				
$C_6H_{12}O_2$	tert-Butylacetic acid	296.2	2.85				
$C_6H_{12}O_2$	Pentyl formate	292.2	5.7				
$C_6H_{12}O_2$	Isopentyl formate	288.2	5.44	0.29257E+02	−0.14028E+00	0.20000E−03	288-323
$C_6H_{12}O_2$	Butyl acetate	293.2	5.07	0.13825E+02	−0.43994E−01	0.48214E−04	253-353
$C_6H_{12}O_2$	sec-Butyl acetate	293.2	5.135	0.12427E+02	−0.32035E−01	0.24286E−04	273-323
$C_6H_{12}O_2$	tert-Butyl acetate	293.2	5.672	0.55435E+02	−0.30494E+00	0.46107E−03	273-323
$C_6H_{12}O_2$	Isobutyl acetate	293.2	5.068	0.14323E+02	−0.46048E−01	0.49286E−04	273-323
$C_6H_{12}O_2$	Propyl propanoate	293.2	5.249				
$C_6H_{12}O_2$	Ethyl butanoate	301.2	5.18	0.48698E+02	−0.25660E+00	0.37237E−03	301-343
$C_6H_{12}O_2$	Methyl pentanoate	293.2	4.992				
$C_6H_{12}O_2$	Diacetone alcohol	298.2	18.2				
$C_6H_{12}O_3$	Ethylene glycol monoethyl ether acetate	303.2	7.567	0.23290E+02	−0.71566E−01	0.65000E−04	303-323
$C_6H_{12}S$	Cyclohexanethiol	298.2	5.420				
$C_6H_{13}Br$	1-Bromohexane	298.2	5.82	0.15233E+02	−0.44385E−01	0.43039E−04	274-328
$C_6H_{13}Cl$	1-Chlorohexane	293.2	6.104	0.15994E+02	−0.43647E−01	0.33393E−04	273-323
$C_6H_{13}ClO$	6-Chloro-1-hexanol	242.2	21.6	−0.73364E+01	0.46377E+00	−0.14202E−02	195-242
$C_6H_{13}I$	1-Iodohexane	293.3	5.35	0.16685E+02	−0.61309E−01	0.77262E−04	293-323
$C_6H_{13}N$	Cyclohexylamine	293.2	4.547				
$C_6H_{13}NO$	N-Propylpropanamide	298.2	118.1	0.58846E+03	−0.22012E+01	0.20870E−02	298-328
$C_6H_{13}NO$	N-Butylacetamide	293.2	104.0	0.70739E+03	−0.37369E+01	0.71585E−02	253-493
$C_6H_{13}NO$	N,N-Diethylacetamide	293.2	32.1				
C_6H_{14}	Hexane	293.2	1.8865	0.19768E+01	0.70933E−03	−0.34470E−05	293-473
C_6H_{14}	2-Methylpentane	293.2	1.886	0.20745E+01	0.50871E−03	−0.39286E−05	273-323
C_6H_{14}	3-Methylpentane	293.2	1.886	0.24739E+01	−0.23190E−02	0.10714E−05	273-323
C_6H_{14}	2,2-Dimethylbutane	293.2	1.869	0.22740E+01	−0.96229E−03	−0.14286E−05	273-313
C_6H_{14}	2,3-Dimethylbutane	293.2	1.889	0.24305E+01	−0.20081E−02	0.53571E−06	273-323
$C_6H_{14}O$	1-Hexanol	293.2	13.03	0.62744E+02	−0.24214E+00	0.24704E−03	233-513
$C_6H_{14}O$	2-Hexanol	298.2	11.06				
$C_6H_{14}O$	3-Hexanol	298.2	9.66				
$C_6H_{14}O$	3-Methyl-1-pentanol	298.2	15.2				
$C_6H_{14}O$	3-Methyl-3-pentanol	293.2	4.322				
$C_6H_{14}O$	2-Ethyl-1-butanol	362.2	6.19				
$C_6H_{14}O$	2,2-Dimethyl-1-butanol	293.2	10.5	0.14054E+03	−0.72925E+00	0.97821E−03	243-393
$C_6H_{14}O$	Dipropyl ether	297.0	3.38	0.14600E+02	−0.72670E−01	0.11742E−03	161-297
$C_6H_{14}O$	Diisopropyl ether	303.2	3.805				
$C_6H_{14}OS$	Dipropyl sulfoxide	303.2	30.37	0.84868E+02	−0.23486E+00	0.18198E−03	303-373
$C_6H_{14}O_2$	2-Methyl-2,4-pentanediol	293.2	25.86	0.14531E+03	−0.65285E+00	0.83503E−03	203-333
$C_6H_{14}O_2$	Ethylene glycol diethyl ether	293.2	3.90	0.99099E+01	−0.33403E−01	0.44048E−04	223-303
$C_6H_{14}O_2S$	Dipropyl sulfone	303.2	32.62	0.70195E+02	−0.15008E+00	0.86506E−04	303-398
$C_6H_{14}O_3$	1,2,6-Hexanetriol	285.3	31.5	0.26127E+03	−0.14552E+01	0.22765E−02	261-285
$C_6H_{14}O_3$	Diethylene glycol dimethyl ether	298.2	7.23	0.28291E+02	−0.11236E+00	0.14000E−03	298-333
$C_6H_{14}O_4$	Triethylene glycol	293.2	23.69	0.91845E+02	−0.33827E+00	0.36062E−03	253-333
$C_6H_{14}O_6$	D-Glucitol	353.2	35.5				
$C_6H_{14}O_6$	D-Mannitol	443.2	24.6				
$C_6H_{14}S$	1-Hexanethiol	293.2	4.436	0.11774E+02	−0.37298E−01	0.41875E−04	273-333
$C_6H_{15}B$	Triethylborane	293.2	1.974				
$C_6H_{15}N$	Hexylamine	293.2	4.08	0.80244E+01	−0.16627E−01	0.10874E−04	253-373
$C_6H_{15}N$	Dipropylamine	293.2	2.923	0.11376E+02	−0.49796E−01	0.71792E−04	243-323
$C_6H_{15}N$	Triethylamine	293.2	2.418	0.29205E+01	−0.14007E−02	−0.13469E−05	233-323
$C_6H_{15}OP$	Triethylphosphine oxide	323.2	35.5				
$C_6H_{15}O_4P$	Triethyl phosphate	298.2	13.20	0.61230E+02	−0.26047E+00	0.33333E−03	298-333
$C_6H_{15}PS$	Triethylphosphine sulfide	371.2	39.0				
$C_6H_{16}O_2Si$	Diethoxydimethylsilane	298.2	3.216				
$C_6H_{16}Si$	Triethylsilane	293.2	2.323				
$C_6H_{18}N_3OP$	Hexamethylphosphoric triamide	293.2	31.3	0.95666E+02	−0.29769E+00	0.26407E−03	283-363
$C_6H_{18}N_4$	N,N'-Bis(2-aminoethyl)-1,2-ethanediamine	293.2	10.76	0.50699E+02	−0.21730E+00	0.27582E−03	213-333
$C_6H_{18}OSi_2$	Hexamethyldisiloxane	293.2	2.179	0.34537E+01	−0.61530E−02	0.61544E−05	213-313
$C_6H_{18}O_3Si_3$	Hexamethylcyclotrisiloxane	343.2	2.139				

Mol. form.	Name	T/K	ε_r	a	b	c	Range/K
$C_6H_{19}NSi_2$	Hexamethyldisilazane	294.2	2.273	0.23358E+01	0.16127E-02	−0.62078E-05	294-333
C_7F_{14}	Perfluoromethylcyclohexane	298.2	1.82				
C_7F_{16}	Perfluoroheptane	289.2	1.847				
$C_7H_3Cl_5$	2,3,4,5,6-Pentachlorotoluene	293.2	4.8				
C_7H_4ClNO	4-Chlorophenyl isocyanate	288.2	3.177	0.40896E+01	−0.31667E-02		288-348
C_7H_5BrO	Benzoyl bromide	293.2	21.33	0.84231E+02	−0.31089E+00	0.32857E-03	283-313
C_7H_5ClO	Benzoyl chloride	293.2	23.0				
C_7H_5FO	Benzoyl fluoride	293.2	22.7				
$C_7H_5F_3$	(Trifluoromethyl)benzene	298.2	9.22				
C_7H_5N	Benzonitrile	293.2	25.9	0.57605E+02	−0.13354E+00	0.87767E-04	273-453
C_7H_5NO	Phenyl isocyanate	293.2	8.940	0.17541E+02	−0.29790E-01	0.15476E-05	293-353
$C_7H_6ClNO_2$	4-Chloro-3-nitrotoluene	301.2	28.07				
$C_7H_6Cl_2$	2,4-Dichlorotoluene	301.2	5.68				
$C_7H_6Cl_2$	2,6-Dichlorotoluene	301.2	3.36				
$C_7H_6Cl_2$	3,4-Dichlorotoluene	301.2	9.39				
$C_7H_6Cl_2$	(Dichloromethyl)benzene	293.2	6.9				
C_7H_6O	Benzaldehyde	293.2	17.85	0.35046E+02	−0.61271E-01	0.16222E-04	301-346
$C_7H_6O_2$	Salicylaldehyde	293.2	18.35	0.51315E+02	−0.15379E+00	0.14111E-03	289-453
C_7H_7Br	o-Bromotoluene	293.2	4.641	0.10229E+02	−0.25050E-01	0.20357E-04	273-323
C_7H_7Br	m-Bromotoluene	293.2	5.566	0.11522E+02	−0.24946E-01	0.15714E-04	273-323
C_7H_7Br	p-Bromotoluene	293.2	5.503	0.10014E+02	−0.13918E-01	−0.50000E-05	273-293
C_7H_7Br	(Bromomethyl)benzene	293.2	6.658	0.18482E+02	−0.57207E-01	0.57321E-04	273-323
C_7H_7BrO	o-Bromoanisole	303.2	8.96	0.12023E+02	−0.59116E-02	−0.13787E-04	303-358
C_7H_7BrO	p-Bromoanisole	303.2	7.40	0.74367E+01	0.12648E-01	−0.42128E-04	303-358
C_7H_7Cl	o-Chlorotoluene	293.2	4.721	0.11507E+02	−0.31148E-01	0.27143E-04	273-323
C_7H_7Cl	m-Chlorotoluene	293.2	5.763	0.13921E+02	−0.37186E-01	0.31786E-04	273-323
C_7H_7Cl	p-Chlorotoluene	293.2	6.25	0.20265E+01	0.40060E-01	−0.87500E-04	293-333
C_7H_7Cl	(Chloromethyl)benzene	293.2	6.854	0.17108E+02	−0.45285E-01	0.35000E-04	273-323
C_7H_7ClO	p-Chloroanisole	293.2	7.84	0.64019E+01	0.30560E-01	−0.87500E-04	293-333
$C_7H_7ClO_2S$	p-Toluenesulfonyl chloride	343.2	22.6				
$C_7H_7ClO_3S$	4-Methoxybenzenesulfonyl chloride	314.2	27.2				
C_7H_7F	o-Fluorotoluene	298.2	4.23				
C_7H_7F	m-Fluorotoluene	298.2	5.41				
C_7H_7F	p-Fluorotoluene	298.2	5.88				
C_7H_7I	p-Iodotoluene	308.2	4.4				
C_7H_7N	2-Vinylpyridine	293.2	9.126				
C_7H_7N	4-Vinylpyridine	293.2	10.50				
$C_7H_7NO_2$	Benzyl nitrite	298.2	7.78				
$C_7H_7NO_2$	o-Nitrotoluene	293.0	26.26	0.10420E+03	−0.41726E+00	0.51607E-03	273-323
$C_7H_7NO_2$	m-Nitrotoluene	303.2	24.95	0.62492E+02	−0.16235E+00	0.12844E-03	303-403
$C_7H_7NO_2$	p-Nitrotoluene	331.2	22.2				
$C_7H_7NO_2S$	4-Nitrothioanisole	346.0	21.7				
$C_7H_7NO_3$	2-Nitroanisole	293.2	45.75	0.16684E+03	−0.58196E+00	0.57382E-03	293-423
$C_7H_7NO_3$	3-Nitroanisole	318.2	25.7	0.65402E+02	−0.16460E+00	0.12560E-03	318-443
$C_7H_7NO_3$	4-Nitroanisole	338.2	26.95	0.59811E+02	−0.10955E+00	0.36042E-04	338-443
C_7H_8	Toluene	296.35	2.379	0.32584E+01	−0.34410E-02	0.15937E-05	207-316
C_7H_8O	o-Cresol	298.2	6.76	0.21633E+02	−0.71069E-01	0.70590E-04	298-453
C_7H_8O	m-Cresol	298.2	12.44	0.81716E+02	−0.35039E+00	0.39878E-03	274-463
C_7H_8O	p-Cresol	298.2	13.05	0.70253E+02	−0.28870E+00	0.31979E-03	298-453
C_7H_8O	Benzyl alcohol	303.2	11.916	0.13661E+03	−0.72127E+00	0.10225E-02	303-333
C_7H_8O	Anisole	294.2	4.30	0.10887E+02	−0.32372E-01	0.33629E-04	294-413
$C_7H_8O_2$	2-Methoxyphenol	298.2	11.95	0.31751E+02	−0.88173E-01	0.72953E-04	291-448
$C_7H_8O_2$	3-Methoxyphenol	298.2	11.59	0.37279E+02	−0.12113E+00	0.11698E-03	298-433
$C_7H_8O_2$	4-Methoxyphenol	333.7	11.05	0.39483E+02	−0.12142E+00	0.10841E-03	334-453
$C_7H_8O_2S$	Ethyl thiophene-2-carboxylate	293.2	6.18				
$C_7H_8O_2S$	Methyl phenyl sulfone	373.2	37.9				
C_7H_8S	Benzenemethanethiol	298.2	4.705	0.16628E+02	−0.68276E-01	0.94636E-04	298-358
C_7H_8S	4-Methylbenzenethiol	323.2	4.74	0.87052E+01	−0.15347E-01	0.95238E-05	323-358
C_7H_8S	(Methylthio)benzene	303.2	4.88	0.21841E+02	−0.97630E-01	0.13750E-03	303-343
C_7H_9N	Benzylamine	293.2	5.18				
C_7H_9N	o-Methylaniline	298.2	6.138	0.10988E+02	−0.18976E-01	0.91958E-05	298-398
C_7H_9N	m-Methylaniline	298.2	5.816	0.13477E+02	−0.35551E-01	0.33135E-04	298-398
C_7H_9N	p-Methylaniline	333.2	5.058	0.78897E+01	−0.10196E-01	0.51190E-05	333-403

Mol. form.	Name	T/K	ε_r	a	b	c	Range/K
C_7H_9N	N-Methylaniline	293.2	5.96				
C_7H_9N	2-Ethylpyridine	293.2	8.33	0.36397E+02	−0.15070E+00	0.18750E-03	293-333
C_7H_9N	4-Ethylpyridine	293.2	10.98	−0.73831E+01	0.14326E+00	−0.27500E-03	293-333
C_7H_9N	2,4-Dimethylpyridine	293.2	9.60	0.25895E+02	−0.73900E-01	0.62500E-04	293-333
C_7H_9N	2,6-Dimethylpyridine	293.2	7.33	0.17714E+02	−0.39080E-01	0.12500E-04	293-333
C_7H_9NO	2,6-Dimethylpyridine-1-oxide	298.2	46.11	0.22765E+03	−0.90760E+00	0.10011E-02	298-398
C_7H_9NO	o-Methoxyaniline	303.2	5.230	0.79911E+01	−0.92183E-02	0.37879E-06	303-393
C_7H_9NO	m-Methoxyaniline	298.2	8.76	0.28179E+02	−0.97840E-01	0.11027E-03	289-393
C_7H_9NO	p-Methoxyaniline	333.2	7.85	0.30149E+02	−0.10523E+00	0.11467E-03	333-453
$C_7H_{10}N_2$	1-Methyl-1-phenylhydrazine	292.2	7.3				
$C_7H_{11}Cl_3O_2$	Isopentyl trichloroacetate	293.2	7.287				
C_7H_{12}	1,6-Heptadiene	293.0	2.161	0.30815E+01	−0.36095E-02	0.16354E-05	184-293
C_7H_{12}	Cycloheptene	295.0	2.265	0.32309E+01	−0.42373E-02	0.32572E-05	227-363
$C_7H_{12}O$	Cycloheptanone	298.2	13.16	0.17511E+03	−0.11221E+01	0.19417E-02	258-298
$C_7H_{12}O$	2-Methylcyclohexanone	293.2	14.0				
$C_7H_{12}O$	3-Methylcyclohexanone	293.2	12.4				
$C_7H_{12}O$	4-Methylcyclohexanone	293.2	12.35				
$C_7H_{12}O_2$	Cyclohexanecarboxylic acid	304.2	2.67				
$C_7H_{12}O_2$	Cyclohexyl formate	293.2	6.47				
$C_7H_{12}O_2$	Butyl acrylate	301.2	5.25	0.38296E+02	−0.19109E+00	0.27006E-03	301-343
$C_7H_{12}O_4$	Monomethyl adipate	293.2	6.69	0.11962E+02	−0.23973E-01	0.20608E-04	293-433
$C_7H_{12}O_4$	Diethyl malonate	304.2	7.550	0.14809E+02	−0.31207E-01	0.24066E-04	304-393
$C_7H_{12}O_4$	Dimethyl glutarate	293.2	7.87	0.20697E+02	−0.57794E-01	0.48405E-04	293-433
$C_7H_{12}O_5$	1,2,3-Propanetriol-1,3-diacetate	288.2	9.80	0.28321E+02	−0.89073E-01	0.86891E-04	258-374
C_7H_{14}	1-Heptene	293.2	2.092	0.21755E+01	0.13896E-02	−0.57049E-05	273-323
C_7H_{14}	2-Methyl-2-hexene	293.2	2.962				
C_7H_{14}	3-Ethyl-2-pentene	293.2	2.051				
C_7H_{14}	Cycloheptane	293.2	2.0784	0.25136E+01	−0.15089E-02	0.84915E-07	278-333
C_7H_{14}	Methylcyclohexane	293.2	2.024				
$C_7H_{14}Br_2$	1,2-Dibromoheptane	298.2	3.77				
$C_7H_{14}Br_2$	2,3-Dibromoheptane	298.2	5.08				
$C_7H_{14}Br_2$	3,4-Dibromoheptane	298.2	4.70				
$C_7H_{14}Cl_2$	1,7-Dichloroheptane	298.2	8.34				
$C_7H_{14}O$	1-Heptanal	295.2	9.07				
$C_7H_{14}O$	2-Heptanone	293.2	11.95	0.38348E+02	−0.12531E+00	0.12005E-03	253-413
$C_7H_{14}O$	3-Heptanone	293.2	12.7				
$C_7H_{14}O$	4-Heptanone	293.2	12.60	0.41520E+02	−0.13839E+00	0.13497E-03	253-393
$C_7H_{14}O$	5-Methyl-2-hexanone	293.2	13.53	0.52353E+02	−0.17695E+00	0.15195E-03	293-333
$C_7H_{14}O$	Cyclohexanemethanol	333.2	9.70	0.10164E+03	−0.45839E+00	0.54762E-03	333-368
$C_7H_{14}O$	2-Methylcyclohexanol*	293.2	9.375	0.17315E+03	−0.98794E+00	0.14634E-02	273-323
$C_7H_{14}O$	3-Methylcyclohexanol*	293.2	13.79	0.65896E+02	−0.21954E+00	0.14107E-03	273-323
$C_7H_{14}O$	4-Methylcyclohexanol*	293.2	13.45	0.65021E+02	−0.22896E+00	0.17946E-03	273-323
$C_7H_{14}O_2$	Heptanoic acid	288.2	3.04	0.36423E+01	−0.31996E-02	0.39362E-05	288-423
$C_7H_{14}O_2$	Pentyl acetate	293.2	4.79	0.12091E+02	−0.36536E-01	0.39732E-04	253-353
$C_7H_{14}O_2$	Isopentyl acetate	293.2	4.72				
$C_7H_{14}O_2$	Butyl propanoate	293.2	4.838				
$C_7H_{14}O_2$	Propyl butanoate	293.2	4.3				
$C_7H_{14}O_2$	Ethyl pentanoate	291.2	4.71				
$C_7H_{14}O_2$	Ethyl 3-methylbutanoate	293.2	4.71				
$C_7H_{14}O_2$	Methyl hexanoate	293.2	4.615				
$C_7H_{15}Br$	1-Bromoheptane	303.2	5.255	0.15289E+02	−0.50621E-01	0.57753E-04	203-343
$C_7H_{15}Br$	2-Bromoheptane	295.2	6.46				
$C_7H_{15}Br$	4-Bromoheptane	295.2	6.81				
$C_7H_{15}Cl$	1-Chloroheptane	293.2	5.521	0.14279E+02	−0.39431E-01	0.32321E-04	273-323
$C_7H_{15}Cl$	2-Chloroheptane	295.2	6.52				
$C_7H_{15}Cl$	3-Chloroheptane	295.2	6.70				
$C_7H_{15}Cl$	4-Chloroheptane	295.2	6.54				
$C_7H_{15}I$	1-Iodoheptane	298.2	4.92	0.11856E+02	−0.33493E-01	0.34368E-04	294-323
$C_7H_{15}I$	3-Iodoheptane	295.2	6.39				
C_7H_{16}	Heptane	293.2	1.9209	0.24740E+01	−0.22577E-02	0.12428E-05	273-373
C_7H_{16}	2-Methylhexane	293.2	1.9221	0.24759E+01	−0.22535E-02	0.12500E-05	293-323
C_7H_{16}	3-Methylhexane	293.2	1.920	0.27089E+01	−0.37908E-02	0.37500E-05	273-323
C_7H_{16}	3-Ethylpentane	293.2	1.942	0.23771E+01	−0.15140E-02	0.10093E-06	163-363

Mol. form.	Name	T/K	ε_r	a	b	c	Range/K
C_7H_{16}	2,2-Dimethylpentane	293.2	1.915	0.23414E+01	−0.14362E-02	−0.51322E-07	153-353
C_7H_{16}	2,3-Dimethylpentane	293.2	1.929	0.25637E+01	−0.26328E-02	0.16071E-05	273-323
C_7H_{16}	2,4-Dimethylpentane	293.2	1.902	0.23979E+01	−0.17436E-02	0.17857E-06	273-323
C_7H_{16}	3,3-Dimethylpentane	291.3	1.9419	0.24007E+01	−0.16802E-02	0.36069E-06	291-322
C_7H_{16}	2,2,3-Trimethylbutane	293.2	1.930				
$C_7H_{16}O$	1-Heptanol	293.2	11.75	0.60662E+02	−0.24049E+00	0.25155E-03	239-513
$C_7H_{16}O$	2-Heptanol	293.7	9.72	0.10050E+03	−0.49793E+00	0.64504E-03	207-365
$C_7H_{16}O$	3-Heptanol	296.1	7.07	0.19586E+03	−0.11465E+01	0.17175E-02	248-349
$C_7H_{16}O$	4-Heptanol	296.2	6.18	0.28995E+03	−0.18499E+01	0.30109E-02	270-301
$C_7H_{16}O$	2-Methyl-2-hexanol	297.0	3.257				
$C_7H_{16}O$	3-Methyl-2-hexanol	297.2	4.990	0.59724E+02	−0.32417E+00	0.47058E-03	244-372
$C_7H_{16}O$	3-Methyl-3-hexanol	298.2	3.248				
$C_7H_{16}O$	3-Ethyl-3-pentanol	293.2	3.158				
$C_7H_{16}O$	2,2-Dimethyl-1-pentanol	293.2	6.020	0.37318E+02	−0.17095E+00	0.22022E-03	283-393
$C_7H_{16}O$	Ethyl pentyl ether	296.2	3.6				
$C_7H_{16}O$	Ethyl isopentyl ether	293.2	3.955	0.66541E+01	−0.55450E-02	−0.12500E-04	293-323
$C_7H_{16}O_3$	Triethoxymethane	293.2	4.779				
$C_7H_{16}S$	1-Heptanethiol	293.2	4.194	0.71333E+01	−0.97320E-02	−0.12500E-05	273-333
$C_7H_{17}N$	Heptylamine	293.2	3.81	0.87794E+01	−0.24363E-01	0.25325E-04	253-373
$C_7H_{18}O_3Si$	Triethoxymethylsilane	298.2	3.845				
$C_8H_4F_6$	1,3-Bis(trifluoromethyl)benzene	303.2	5.98				
C_8H_6	Phenylacetylene	298.2	2.98				
$C_8H_6Cl_2$	2,5-Dichlorostyrene	298.2	2.58				
$C_8H_6Cl_4$	1,2,3,4-Tetrachloro-5,6-dimethylbenzene	293.2	8.0				
$C_8H_6Cl_4$	1,2,3,5-Tetrachloro-4,6-dimethylbenzene	293.2	5.4				
C_8H_6O	Phenoxyacetylene	298.2	4.76				
C_8H_7N	Benzeneacetonitrile	299.2	17.87	0.82175E+02	−0.37416E+00	0.53220E-03	299-343
$C_8H_7NO_2$	4-Methoxyphenyl isocyanate	333.2	10.26	0.20780E+02	−0.31571E-01		333-403
$C_8H_7NO_4$	Methyl 2-nitrobenzoate	300.1	27.76				
C_8H_8	Styrene	293.2	2.4737	0.44473E+01	−0.11422E-01	0.16000E-04	293-313
C_8H_8O	Acetophenone	298.2	17.44	0.26099E+02	0.64048E-02	−0.11905E-03	298-333
$C_8H_8O_2$	Benzeneacetic acid	353.2	3.47	0.24104E+01	0.30000E-02		353-393
$C_8H_8O_2$	Benzyl formate	303.2	6.34	0.26162E+02	−0.11026E+00	0.14787E-03	303-358
$C_8H_8O_2$	Phenyl acetate	298.2	5.403	0.11327E+02	−0.26707E-01	0.22938E-04	298-404
$C_8H_8O_2$	Methyl benzoate	302.7	6.642	0.17486E+02	−0.51027E-01	0.50222E-04	303-393
$C_8H_8O_2$	(Hydroxyacetyl)benzene	298.2	21.33	0.42286E+02	−0.69215E-01	−0.35714E-05	298-368
$C_8H_8O_2$	4-Methoxybenzaldehyde	303.2	22.0				
$C_8H_8O_3$	Methyl salicylate	314.4	8.80	0.20501E+02	−0.39045E-01	0.68298E-05	223-398
C_8H_9Br	1-Bromo-2-ethylbenzene	298.2	5.55				
C_8H_9Br	1-Bromo-3-ethylbenzene	298.2	5.56				
C_8H_9Br	1-Bromo-4-ethylbenzene	298.2	5.42				
C_8H_9BrO	1-Bromo-2-ethoxybenzene	313.2	7.04	0.23146E+02	−0.75753E-01	0.77778E-04	313-358
C_8H_9Cl	1-Chloro-2-ethylbenzene	298.2	4.36				
C_8H_9Cl	1-Chloro-3-ethylbenzene	298.2	5.18				
C_8H_9Cl	1-Chloro-4-ethylbenzene	298.2	5.16				
$C_8H_9NO_2$	1-Ethyl-2-nitrobenzene	273.4	21.9				
$C_8H_9NO_2$	Methyl 2-aminobenzoate	298.2	21.9				
$C_8H_9NO_2$	Ethyl 4-pyridinecarboxylate	293.2	8.95				
C_8H_{10}	Ethylbenzene	293.2	2.4463	0.35969E+01	−0.53169E-02	0.47500E-05	293-323
C_8H_{10}	o-Xylene	293.2	2.562	0.36163E+01	−0.40177E-02	0.14286E-05	273-323
C_8H_{10}	m-Xylene	293.2	2.359	0.28421E+01	−0.10191E-02	−0.21429E-05	273-323
C_8H_{10}	p-Xylene	293.2	2.2735	0.23140E+01	0.97221E-03	−0.37500E-05	293-363
$C_8H_{10}O$	2,3-Xylenol	343.2	4.81	0.14399E+02	−0.41438E-01	0.39244E-04	343-433
$C_8H_{10}O$	2,4-Xylenol	303.2	5.060	0.22125E+02	−0.85543E-01	0.96548E-04	303-363
$C_8H_{10}O$	2,5-Xylenol	338.2	5.36	0.18049E+02	−0.54991E-01	0.51656E-04	338-455
$C_8H_{10}O$	2,6-Xylenol	313.2	4.90	0.12284E+02	−0.32996E-01	0.29867E-04	313-453
$C_8H_{10}O$	3,4-Xylenol	333.2	9.02	0.54423E+02	−0.21153E+00	0.22508E-03	333-453
$C_8H_{10}O$	3,5-Xylenol	323.2	9.06	0.54251E+02	−0.21647E+00	0.23542E-03	323-453
$C_8H_{10}O$	Benzeneethanol	293.2	12.31	0.12170E+03	−0.63124E+00	0.87776E-03	278-333
$C_8H_{10}O$	1-Phenylethanol	293.2	8.77	0.32971E+02	−0.12042E+00	0.12809E-03	293-423
$C_8H_{10}O$	Phenetole	293.2	4.216	−0.15043E+02	0.13752E+00	−0.24500E-03	293-313

Mol. form.	Name	T/K	ε_r	a	b	c	Range/K
$C_8H_{10}O$	2-Methylanisole	293.2	3.502	0.50825E+01	−0.62297E-02	0.28571E-05	293-333
$C_8H_{10}O$	3-Methylanisole	293.2	3.967	0.12830E+02	−0.49701E-01	0.66429E-04	293-333
$C_8H_{10}O$	4-Methylanisole	293.2	3.914	0.86608E+01	−0.23510E-01	0.25000E-04	293-333
$C_8H_{10}O_2$	1,2-Dimethoxybenzene	293.2	4.45	0.74604E+01	−0.13445E-01	0.10737E-04	293-443
$C_8H_{10}O_2$	1,3-Dimethoxybenzene	298.2	5.363	0.11911E+02	−0.30804E-01	0.29643E-04	298-358
$C_8H_{10}O_2$	1,4-Dimethoxybenzene	333.7	5.60	0.11289E+02	−0.20765E-01	0.11987E-04	334-463
$C_8H_{10}O_2S$	Ethyl phenyl sulfone	348.2	39.0				
$C_8H_{10}S$	(Ethylthio)benzene	298.2	4.95				
$C_8H_{11}N$	p-Ethylaniline	298.2	4.84				
$C_8H_{11}N$	N-Ethylaniline	293.2	5.87				
$C_8H_{11}N$	N,N-Dimethylaniline	298.2	4.90	0.84052E+01	−0.13549E-01	0.62835E-05	289-453
$C_8H_{11}N$	2,4,6-Trimethylpyridine	298.2	7.807	0.20990E+02	−0.57419E-01	0.44286E-04	298-358
$C_8H_{11}NO$	4-Ethoxyaniline	298.2	7.43				
$C_8H_{12}N_2O_2$	Hexamethylene diisocyanate	288.2	14.41	0.26715E+02	−0.42696E-01		288-403
$C_8H_{12}O_4$	Diethyl maleate	298.2	7.560	0.13953E+02	−0.21969E-01	0.17817E-05	298-343
$C_8H_{12}O_4$	Diethyl fumarate	296.2	6.56				
C_8H_{14}	1,7-Octadiene	293.0	2.186	0.28376E+01	−0.17442E-02	−0.16141E-05	214-293
C_8H_{14}	cis-Cyclooctene	296.0	2.306	0.31115E+01	−0.32058E-02	0.16713E-05	269-406
C_8H_{14}	1,2-Dimethylcyclohexene	296.0	2.144	0.26443E+01	−0.17973E-02	0.35815E-06	211-374
C_8H_{14}	1,3-Dimethylcyclohexene	296.0	2.182	0.29951E+01	−0.34615E-02	0.24026E-05	213-373
$C_8H_{14}O_2$	Methyl cyclohexanecarboxylate	293.2	4.87				
$C_8H_{14}O_2$	Cyclohexyl acetate	293.2	5.08				
$C_8H_{14}O_3$	Butanoic anhydride	293.2	12.8				
$C_8H_{14}O_3$	2-Methylpropanoic anhydride	292.2	13.6				
$C_8H_{14}O_4$	Diisopropyl oxalate	293.2	6.403	0.10709E+02	−0.16328E-01	0.56000E-05	293-368
$C_8H_{14}O_4$	Diethyl succinate	293.2	6.098	0.80213E+01	0.11810E-02	−0.26400E-04	293-343
$C_8H_{14}O_4$	Dimethyl adipate	293.2	6.84	0.11739E+02	−0.17281E-01	0.11447E-05	293-433
$C_8H_{15}N$	Octanenitrile	293.2	13.90				
C_8H_{16}	1-Octene	293.2	2.113	0.24348E+01	0.34200E-03	−0.50000E-05	273-323
C_8H_{16}	cis-3-Octene	298.2	2.062				
C_8H_{16}	trans-3-Octene	298.2	2.002				
C_8H_{16}	cis-4-Octene	298.2	2.053				
C_8H_{16}	trans-4-Octene	298.2	2.004				
C_8H_{16}	3-Methyl-2-heptene*	293.2	2.436				
C_8H_{16}	2,5-Dimethyl-2-hexene	293.2	2.431				
C_8H_{16}	2,4,4-Trimethyl-1-pentene	298.2	2.0908				
C_8H_{16}	Cyclooctane	295.0	2.116	0.25036E+01	−0.12460E-02	−0.23175E-06	295-411
$C_8H_{16}Br_2$	1,8-Dibromooctane	298.2	7.43	0.94117E+00	0.61520E-01	−0.13333E-03	298-328
$C_8H_{16}Cl_2$	1,8-Dichlorooctane	298.2	7.64				
$C_8H_{16}O$	2-Octanone	293.2	9.51	−0.16219E+02	0.18799E+00	−0.34156E-03	293-333
$C_8H_{16}O$	3-Octanone	303.2	10.50				
$C_8H_{16}O_2$	Octanoic acid	288.2	2.85	0.29391E+01	−0.38721E-03		288-423
$C_8H_{16}O_2$	2-Ethylhexanoic acid	296.2	2.64				
$C_8H_{16}O_2$	Hexyl acetate	293.2	4.42				
$C_8H_{16}O_2$	Pentyl propanoate	293.2	4.552				
$C_8H_{16}O_2$	Isopentyl propanoate	273.2	5.21	0.17665E+02	−0.71718E-01	0.95635E-04	273-373
$C_8H_{16}O_2$	Butyl butanoate	298.2	4.39	0.79684E+01	−0.12000E-01	0.15266E-13	298-318
$C_8H_{16}O_2$	Propyl pentanoate	292.2	4.0				
$C_8H_{16}O_2$	Ethyl hexanoate	293.2	4.45	0.11007E+02	−0.32800E-01	0.35714E-04	253-353
$C_8H_{16}O_2$	Methyl heptanoate	293.2	4.355				
$C_8H_{16}O_3$	Isopentyl lactate	273.2	11.2	0.48649E+02	−0.21253E+00	0.27619E-03	273-373
$C_8H_{17}Br$	1-Bromooctane	293.2	5.0957	0.12404E+02	−0.35050E-01	0.34542E-04	283-353
$C_8H_{17}Br$	2-Bromooctane	293.2	5.44				
$C_8H_{17}Cl$	1-Chlorooctane	298.2	5.05	0.11346E+02	−0.25120E-01	0.13450E-04	274-328
$C_8H_{17}Cl$	2-Chlorooctane	293.2	5.42				
$C_8H_{17}F$	1-Fluorooctane	293.2	3.89				
$C_8H_{17}I$	1-Iodooctane	293.2	4.67	0.12452E+02	−0.41229E-01	0.50108E-04	233-313
$C_8H_{17}NO_2$	1-Nitrooctane	293.2	11.46				
C_8H_{18}	Octane	293.2	1.948	0.22590E+01	−0.84212E-03	−0.75758E-06	233-393
C_8H_{18}	2-Methylheptane	293.2	1.9519				
C_8H_{18}	3-Ethylhexane	293.2	1.9617				
C_8H_{18}	2,2-Dimethylhexane	293.2	1.9498				
C_8H_{18}	2,5-Dimethylhexane	293.95	1.9619	0.25821E+01	−0.26804E-02	0.19404E-05	294-324

Mol. form.	Name	T/K	ε_r	a	b	c	Range/K
C_8H_{18}	3,3-Dimethylhexane	293.2	1.9645				
C_8H_{18}	3,4-Dimethylhexane	292.1	1.9814	0.26849E+01	−0.33712E-02	0.32949E-05	292-324
C_8H_{18}	3-Ethyl-3-methylpentane	291.49	1.9869	0.25983E+01	−0.28027E-02	0.24195E-05	292-324
C_8H_{18}	2,2,3-Trimethylpentane	293.2	1.960				
C_8H_{18}	2,2,4-Trimethylpentane	293.2	1.943	0.23677E+01	−0.14768E-02	0.94261E-07	173-373
C_8H_{18}	2,3,3-Trimethylpentane	293.2	1.9780				
C_8H_{18}	2,3,4-Trimethylpentane	293.2	1.9738				
$C_8H_{18}O$	1-Octanol	293.2	10.30	0.51647E+02	−0.20371E+00	0.21320E-03	258-513
$C_8H_{18}O$	2-Octanol	293.2	8.13	0.63760E+02	−0.27643E+00	0.31075E-03	213-513
$C_8H_{18}O$	3-Octanol	293.2	5.55	0.12505E+03	−0.70646E+00	0.10245E-02	223-383
$C_8H_{18}O$	4-Octanol	293.2	4.48	0.51049E+02	−0.26664E+00	0.37280E-03	243-403
$C_8H_{18}O$	2-Methyl-1-heptanol	293.1	5.16	0.61698E+02	−0.33647E+00	0.49066E-03	236-328
$C_8H_{18}O$	3-Methyl-1-heptanol	290.3	2.884	0.84687E+01	−0.33712E-01	0.49793E-04	241-316
$C_8H_{18}OS$	4-Methyl-1-heptanol	290.6	4.63	0.48612E+02	−0.26773E+00	0.39972E-03	237-332
$C_8H_{18}O$	5-Methyl-1-heptanol	290.4	7.68	0.54581E+02	−0.24772E+00	0.29734E-03	235-328
$C_8H_{18}O$	6-Methyl-1-heptanol	290.3	10.54	0.57997E+02	−0.23517E+00	0.24663E-03	265-328
$C_8H_{18}O$	2-Methyl-2-heptanol	292.2	3.43				
$C_8H_{18}O$	3-Methyl-2-heptanol	289.6	7.47	0.39178E+02	−0.17976E+00	0.24218E-03	229-329
$C_8H_{18}O$	4-Methyl-2-heptanol	290.0	3.59	0.39715E+02	−0.23115E+00	0.36771E-03	240-333
$C_8H_{18}O$	5-Methyl-2-heptanol	278.5	7.5	0.68568E+02	−0.40706E+00	0.67433E-03	230-279
$C_8H_{18}O$	6-Methyl-2-heptanol	290.1	6.41	0.77520E+02	−0.41724E+00	0.59448E-03	239-329
$C_8H_{18}O$	2-Methyl-3-heptanol	293.2	3.260	−0.59739E+01	0.56700E-01	−0.83125E-04	343-403
$C_8H_{18}O$	3-Methyl-3-heptanol	293.2	3.013	−0.38440E+01	0.42327E-01	−0.61250E-04	343-403
$C_8H_{18}O$	4-Methyl-3-heptanol	293.2	3.312	−0.48003E+01	0.50740E-01	−0.75000E-04	343-403
$C_8H_{18}O$	5-Methyl-3-heptanol	293.2	3.832	0.61967E+01	−0.63750E-02		343-383
$C_8H_{18}O$	6-Methyl-3-heptanol	293.2	4.992	0.23037E+02	−0.98029E-01	0.12479E-03	283-383
$C_8H_{18}O$	2-Methyl-4-heptanol	296.3	3.338	0.42102E+00	0.10427E-01	−0.20438E-05	230-333
$C_8H_{18}O$	3-Methyl-4-heptanol	290.0	7.46	0.33354E+02	−0.14077E+00	0.17750E-03	230-330
$C_8H_{18}O$	4-Methyl-4-heptanol	296.2	2.902				
$C_8H_{18}O$	2-Ethyl-1-hexanol	298.2	7.58	0.86074E+02	−0.42636E+00	0.55078E-03	208-318
$C_8H_{18}O$	2,2-Dimethyl-1-hexanol	293.2	4.50	0.91244E+01	−0.21785E-01	0.21018E-04	283-393
$C_8H_{18}O$	Dibutyl ether	293.2	3.0830	0.65383E+01	−0.16172E-01	0.14969E-04	293-314
$C_8H_{18}OS$	Dibutyl sulfoxide	313.2	24.73	0.67156E+02	−0.16448E+00	0.92275E-04	313-393
$C_8H_{18}O_2$	2-Ethyl-1,3-hexanediol	293.2	18.73	0.57919E+02	−0.17128E+00	0.12949E-03	233-333
$C_8H_{18}O_2S$	Dibutyl sulfone	323.2	25.72	0.66248E+02	−0.16417E+00	0.12001E-03	323-398
$C_8H_{18}O_4$	Triethylene glycol dimethyl ether	298.2	7.62				
$C_8H_{18}O_5$	Tetraethylene glycol	293.2	20.44	0.83547E+02	−0.31691E+00	0.34689E-03	253-333
$C_8H_{18}S$	1-Octanethiol	293.2	3.949	0.63667E+01	−0.87920E-02	0.18750E-05	273-333
$C_8H_{18}S$	Dibutyl sulfide	298.2	4.29				
$C_8H_{19}N$	Octylamine	293.2	3.58	0.77931E+01	−0.20015E-01	0.19347E-04	273-373
$C_8H_{19}N$	Dibutylamine	293.2	2.765	0.52504E+01	−0.10538E-01	0.71485E-05	243-323
$C_8H_{20}O_4Si$	Ethyl silicate	293.2	2.50				
$C_8H_{20}Si$	Tetraethylsilane	293.2	2.090				
$C_8H_{20}Sn$	Tetraethylstannane	293.2	2.241				
$C_8H_{23}N_5$	Tetraethylenepentamine	293.2	9.40	0.40553E+02	−0.16681E+00	0.20659E-03	213-333
$C_8H_{24}O_4Si_4$	Octamethylcyclotetrasiloxane	296.2	2.390	0.36286E+01	−0.56885E-02	0.50874E-05	296-333
$C_9H_6N_2O_2$	Toluene-2,4-diisocyanate	293.2	8.433	0.22174E+02	−0.66982E-01	0.68571E-04	293-353
$C_9H_6O_2$	2H-1-Benzopyran-2-one	343.2	34.04	0.11311E+03	−0.33804E+00	0.31324E-03	343-423
C_9H_7N	Quinoline	293.2	9.16	0.33432E+02	−0.13497E+00	0.17788E-03	258-323
C_9H_7N	Isoquinoline	298.2	11.0	0.14412E+03	−0.79935E+00	0.11839E-02	298-323
C_9H_8O	Cinnamaldehyde	305.8	17.72	0.41837E+02	−0.11060E+00	0.10401E-03	306-354
$C_9H_8O_4$	2-(Acetyloxy)benzoic acid	333.2	6.55	0.69994E+01	−0.14553E-02		333-416
C_9H_{10}	1-Propenylbenzene	293.2	2.73				
C_9H_{10}	Allylbenzene	293.2	2.63				
C_9H_{10}	Isopropenylbenzene	293.2	2.28				
$C_9H_{10}OS$	4-Acetylthioanisole	355.2	11.34				
$C_9H_{10}O_2$	Ethyl benzoate	293.2	6.20	0.18216E+02	−0.62361E-01	0.72884E-04	288-343
$C_9H_{10}O_2$	Methyl 4-methylbenzoate	306.2	4.3				
$C_9H_{10}O_2$	Benzyl acetate	303.2	5.34	0.11727E+02	−0.30869E-01	0.32340E-04	303-358
$C_9H_{10}O_2$	Phenyl propanoate	293.2	4.77				
$C_9H_{10}O_2$	4-Acetylanisole	313.2	17.3				
$C_9H_{10}O_3$	Ethyl salicylate	308.2	8.48	0.18910E+02	−0.35623E-01	0.46529E-05	225-321
$C_9H_{10}O_3$	Methyl 2-methoxybenzoate	294.2	7.7				

Mol. form.	Name	T/K	ε_r	a	b	c	Range/K
$C_9H_{11}Br$	(3-Bromopropyl)benzene	302.2	5.41	0.11360E+02	−0.27471E-01	0.25775E-04	302-358
$C_9H_{11}NO$	N-Ethylbenzamide	352.7	42.6	−0.20109E+03	0.17866E+01	−0.31065E-02	353-389
$C_9H_{11}NO$	N,N-Dimethylbenzamide	318.2	20.77	0.76725E+02	−0.26908E+00	0.29409E-03	318-443
$C_9H_{11}NO_2$	Ethyl 2-aminobenzoate	298.2	4.14				
C_9H_{12}	Propylbenzene	293.2	2.370	0.26933E+01	0.21679E-03	−0.44643E-05	273-323
C_9H_{12}	Isopropylbenzene	293.2	2.381	0.31149E+01	−0.30801E-02	0.19643E-05	273-323
C_9H_{12}	o-Ethyltoluene	293.2	2.595				
C_9H_{12}	m-Ethyltoluene	293.2	2.365				
C_9H_{12}	p-Ethyltoluene	293.2	2.265				
C_9H_{12}	1,2,3-Trimethylbenzene	293.2	2.656	0.76006E+01	−0.29118E-01	0.41786E-04	273-323
C_9H_{12}	1,2,4-Trimethylbenzene	293.2	2.377	0.31517E+01	−0.30634E-02	0.14286E-05	273-323
C_9H_{12}	1,3,5-Trimethylbenzene	293.2	2.279	0.38998E+01	−0.88072E-02	0.11149E-04	288-358
$C_9H_{12}O$	Benzenepropanol	293.2	11.97	0.94482E+02	−0.45540E+00	0.59307E-03	213-303
$C_9H_{12}O$	α-Ethylbenzenemethanol	293.2	6.68	0.44520E+02	−0.21505E+00	0.29443E-03	233-373
$C_9H_{12}O$	α,α-Dimethylbenzenemethanol	303.2	5.61	0.57072E+01	0.86568E-02	−0.29580E-04	303-373
$C_9H_{12}O$	1-Phenyl-2-propanol	293.2	9.35	0.10762E+03	−0.56026E+00	0.76915E-03	233-373
$C_9H_{12}O$	Benzyl ethyl ether	298.2	3.90				
$C_9H_{12}O$	2,6-Dimethylanisole	293.2	3.780	0.76700E+01	−0.18298E-01	0.17143E-04	293-333
$C_9H_{12}O$	3,5-Dimethylanisole	293.2	3.711	0.54981E+01	−0.56651E-02	−0.14286E-05	293-333
$C_9H_{12}O_2S$	Butyl thiophene-2-carboxylate	293.2	6.40				
$C_9H_{12}S$	Benzenepropanethiol	303.2	4.36	0.82411E+01	−0.15034E-01	0.73617E-05	303-358
$C_9H_{13}N$	Benzylethylamine	293.2	4.3				
$C_9H_{13}N$	N-Propylaniline	293.2	5.48				
$C_9H_{13}N$	2-Methyl-N,N-dimethylaniline	293.2	3.4				
$C_9H_{13}N$	4-Methyl-N,N-dimethylaniline	293.2	3.9				
$C_9H_{14}OSi$	Trimethylphenoxysilane	298.2	3.3953				
$C_9H_{14}O_6$	Triacetin	293.6	7.11	0.17819E+02	−0.53656E-01	0.57759E-04	219-304
$C_9H_{14}Si$	Trimethylphenylsilane	298.2	2.3533	0.21463E+01	0.32711E-02	−0.86264E-05	288-323
$C_9H_{16}O_2$	2-Nonenoic acid	296.2	2.5				
$C_9H_{16}O_2$	Cyclohexyl propanoate	293.2	4.82				
$C_9H_{16}O_2$	Ethyl cyclohexanecarboxylate	293.2	4.64				
$C_9H_{16}O_4$	Diethyl glutarate	303.2	6.659				
$C_9H_{17}N$	Nonanenitrile	293.2	12.08				
C_9H_{18}	1-Nonene	293.2	2.180	0.22710E+01	0.15797E-02	−0.64286E-05	273-323
$C_9H_{18}Br_2$	1,9-Dibromononane	293.2	7.153	0.18931E+02	−0.57764E-01	0.60000E-04	293-343
$C_9H_{18}O$	2-Nonanone	295.2	9.14				
$C_9H_{18}O$	5-Nonanone	293.2	10.6				
$C_9H_{18}O$	Di-tert-butyl ketone	287.65	10.0				
$C_9H_{18}O$	2,6-Dimethyl-4-heptanone	293.2	9.91	0.33178E+02	−0.11290E+00	0.11454E-03	273-393
$C_9H_{18}O_2$	Nonanoic acid	294.9	2.475	0.25039E+01	0.67274E-03	−0.24180E-05	295-365
$C_9H_{18}O_2$	2-Methyloctanoic acid	293.2	2.39				
$C_9H_{18}O_2$	2-Ethylheptanoic acid	293.2	1.98				
$C_9H_{18}O_2$	Heptyl acetate	293.2	4.2				
$C_9H_{18}O_2$	Pentyl butanoate	301.2	4.08	0.59029E+01	−0.49905E-02	−0.34292E-05	301-343
$C_9H_{18}O_2$	Isopentyl butanoate	293.2	4.0				
$C_9H_{18}O_2$	Isobutyl pentanoate	292.2	3.8				
$C_9H_{18}O_2$	Methyl octanoate	293.2	4.101				
$C_9H_{19}Br$	1-Bromononane	298.2	4.74	0.79870E+01	−0.10488E-01	−0.13450E-05	274-328
$C_9H_{19}Cl$	1-Chlorononane	293.2	4.803	0.95528E+01	−0.16200E-01	−0.16365E-13	293-323
$C_9H_{19}NO$	N,N-Dibutylformamide	293.2	18.4				
C_9H_{20}	Nonane	293.2	1.9722	0.23894E+01	−0.14830E-02	0.14881E-06	253-393
C_9H_{20}	2-Methyloctane	293.2	1.967				
C_9H_{20}	4-Methyloctane	293.2	1.967				
C_9H_{20}	2,4-Dimethylheptane	293.2	1.89				
C_9H_{20}	2,5-Dimethylheptane	293.2	1.89				
C_9H_{20}	2,6-Dimethylheptane	293.2	1.987				
$C_9H_{20}N_2O$	Tetraethylurea	296.8	14.29	0.52820E+02	−0.18790E+00	0.19580E-03	205-411
$C_9H_{20}O$	1-Nonanol	293.2	8.83	0.97467E+02	−0.51103E+00	0.71429E-03	288-343
$C_9H_{20}O$	2-Nonanol	298.2	6.66	0.10136E+03	−0.55612E+00	0.80000E-03	288-308
$C_9H_{20}O$	3-Nonanol	298.2	4.49	0.55214E+02	−0.31920E+00	0.50000E-03	288-308
$C_9H_{20}O$	4-Nonanol	298.2	3.69	0.27954E+01	0.30000E-02	−0.52375E-13	288-308
$C_9H_{20}O$	5-Nonanol	298.2	3.54	−0.25463E+01	0.35320E-01	−0.50000E-04	288-308
$C_9H_{21}B$	Tripropylborane	293.2	2.026				

Mol. form.	Name	T/K	ε_r	a	b	c	Range/K
$C_9H_{21}N$	Nonylamine	293.2	3.42	0.53575E+01	−0.71982E-02	0.19481E-05	293-373
$C_9H_{21}N$	Tripropylamine	293.2	2.380	0.33380E+01	−0.86332E-02	0.18322E-04	243-293
$C_9H_{21}O_4P$	Tripropyl phosphate	293.2	10.93	0.33166E+02	−0.10514E+00	0.10000E-03	293-373
$C_{10}H_7Br$	1-Bromonaphthalene	298.2	4.768	0.10561E+02	−0.27671E-01	0.27655E-04	293-323
$C_{10}H_7Cl$	1-Chloronaphthalene	298.2	5.04	0.84861E+01	−0.12357E-01	0.26899E-05	274-328
$C_{10}H_7NO_2$	1-Nitronaphthalene	333.2	19.68	0.36267E+02	−0.41283E-01	−0.25595E-04	333-403
$C_{10}H_8$	Naphthalene	363.2	2.54				
$C_{10}H_8O$	1-Naphthol	373.0	5.03	0.16489E+02	−0.46700E-01	0.42857E-04	373-453
$C_{10}H_8O$	2-Naphthol	413.0	4.95	0.92865E+01	−0.10500E-01	0.42501E-15	413-453
$C_{10}H_9N$	1-Naphthylamine	333.2	5.20	0.10577E+02	−0.22114E-01	0.17857E-04	333-453
$C_{10}H_9N$	2-Naphthylamine	393.0	5.26	0.19722E+02	−0.60679E-01	0.60714E-04	393-473
$C_{10}H_9N$	2-Methylquinoline	293.2	7.24	0.11688E+02	−0.78400E-02	−0.25000E-04	293-333
$C_{10}H_9N$	4-Methylquinoline	293.2	9.31	0.17788E+02	−0.32580E-01	0.12500E-04	293-333
$C_{10}H_9N$	6-Methylquinoline	293.2	8.48	0.21696E+02	−0.63400E-01	0.62500E-04	293-333
$C_{10}H_9N$	8-Methylquinoline	293.2	6.58	0.19356E+02	−0.61900E-01	0.62500E-04	293-333
$C_{10}H_{10}O_4$	Methyl 2-(acetyloxy)benzoate	328.9	5.31	0.19579E+02	−0.69970E-01	0.80889E-04	329-371
$C_{10}H_{10}O_4$	Dimethyl phthalate	293.2	8.66				
$C_{10}H_{12}$	1,2,3,4-Tetrahydronaphthalene	298.2	2.771	0.29172E+01	0.12832E-02	−0.59453E-05	298-343
$C_{10}H_{12}$	4-Ethylstyrene	298.2	3.350				
$C_{10}H_{12}$	Dicyclopentadiene	313.2	2.43	0.30564E+01	−0.20000E-02	0.82443E-15	313-373
$C_{10}H_{12}O$	Tetrahydro-2-naphthol*	293.2	11.70	0.98978E+02	−0.48267E+00	0.63008E-03	293-363
$C_{10}H_{12}O$	4-Isopropylbenzaldehyde	288.2	10.68				
$C_{10}H_{12}O_2$	4-Allyl-2-methoxyphenol	293.2	9.55	0.52377E+02	−0.24380E+00	0.33333E-03	273-323
$C_{10}H_{12}O_2$	2-Phenylethyl acetate	297.2	4.93				
$C_{10}H_{12}O_2$	Benzyl propanoate	303.0	5.11	0.42301E+01	0.13962E-01	−0.36426E-04	303-358
$C_{10}H_{12}O_2$	Phenyl butanoate	293.2	4.48				
$C_{10}H_{12}O_2$	Propyl benzoate	303.2	5.78	0.10927E+02	−0.20535E-01	0.11745E-04	303-358
$C_{10}H_{12}O_2$	Ethyl phenylacetate	293.2	5.320				
$C_{10}H_{14}$	Butylbenzene	293.2	2.359				
$C_{10}H_{14}$	sec-Butylbenzene	293.2	2.357	0.28348E+01	−0.68586E-03	−0.32143E-05	273-323
$C_{10}H_{14}$	tert-Butylbenzene	293.2	2.359	0.27924E+01	−0.38350E-03	−0.37500E-05	273-323
$C_{10}H_{14}$	Isobutylbenzene	293.2	2.318	0.28055E+01	−0.92614E-03	−0.25000E-05	273-323
$C_{10}H_{14}$	1-Isopropyl-4-methylbenzene	298.2	2.2322	0.25266E+01	−0.25121E-03	−0.24867E-05	277-333
$C_{10}H_{14}$	o-Diethylbenzene	293.2	2.594				
$C_{10}H_{14}$	m-Diethylbenzene	293.2	2.369				
$C_{10}H_{14}$	p-Diethylbenzene	293.2	2.259				
$C_{10}H_{14}$	1-Ethyl-3,5-dimethylbenzene	293.2	2.275				
$C_{10}H_{14}$	1,2,3,4-Tetramethylbenzene	296.0	2.538	0.33822E+01	−0.33630E-02	0.17475E-05	273-412
$C_{10}H_{14}$	1,2,4,5-Tetramethylbenzene	356.0	2.223	0.26834E+01	−0.10327E-02	−0.73533E-06	356-430
$C_{10}H_{14}N_2$	L-Nicotine	293.2	8.937	0.21347E+02	−0.57177E-01	0.50655E-04	293-363
$C_{10}H_{14}O$	1-Phenyl-2-methyl-2-propanol	298.2	5.71	0.21922E+02	−0.84231E-01	0.99475E-04	298-423
$C_{10}H_{14}O$	Butyl phenyl ether	293.2	3.734				
$C_{10}H_{14}O$	Thymol	333.2	4.259				
$C_{10}H_{15}N$	N,N-Diethylaniline	303.2	5.15	0.50773E+01	0.15399E-01	−0.50000E-04	303-328
$C_{10}H_{16}$	γ-Terpinene	298.2	2.2738				
$C_{10}H_{16}$	d-Limonene	298.2	2.3746				
$C_{10}H_{16}$	l-Limonene	298.2	2.3738				
$C_{10}H_{16}$	Terpinolene	298.2	2.2918				
$C_{10}H_{16}$	α-Pinene	298.2	2.1787				
$C_{10}H_{16}$	β-Pinene	298.2	2.4970				
$C_{10}H_{16}$	α-Terpinene	298.2	2.4526				
$C_{10}H_{16}$	β-Myrcene	298.2	2.3				
$C_{10}H_{16}O$	Carvenone	293.2	18.8				
$C_{10}H_{16}O$	d-Fenchone	294.2	12.8				
$C_{10}H_{17}Cl$	2-Chlorobornane	368.2	5.21				
$C_{10}H_{18}$	Pinane	298.2	2.1456				
$C_{10}H_{18}$	cis-Decahydronaphthalene	293.2	2.219	0.25410E+01	−0.11420E-02	0.15092E-06	293-373
$C_{10}H_{18}$	trans-Decahydronaphthalene	293.2	2.184	0.26615E+01	−0.21241E-02	0.16864E-05	293-373
$C_{10}H_{18}O$	Eucalyptol	298.2	4.57				
$C_{10}H_{18}O_2$	Cyclohexyl butanoate	293.2	4.58				
$C_{10}H_{18}O_4$	Diethyl adipate	293.2	6.109	0.14824E+02	−0.40749E-01	0.37600E-04	293-343
$C_{10}H_{20}$	1-Decene	293.2	2.136	0.19091E+01	0.33442E-02	−0.87500E-05	273-323
$C_{10}H_{20}$	cis-5-Decene	298.2	2.071				

Mol. form.	Name	T/K	ε_r	a	b	c	Range/K
$C_{10}H_{20}$	trans-5-Decene	298.2	2.030				
$C_{10}H_{20}$	5-Methyl-4-nonene	293.2	2.175				
$C_{10}H_{20}$	2,4,6-Trimethyl-3-heptene	293.2	2.293				
$C_{10}H_{20}Br_2$	1,10-Dibromodecane	303.2	6.56	0.17350E+02	−0.50328E−01	0.48633E−04	303-368
$C_{10}H_{20}Cl_2$	1,10-Dichlorodecane	308.2	6.68	−0.57423E+01	0.94220E−01	−0.17500E−03	308-338
$C_{10}H_{20}O$	2-Decanone	287.2	8.3				
$C_{10}H_{20}O$	Menthol	309.3	3.90	0.68202E+01	−0.15894E−01	0.20837E−04	309-358
$C_{10}H_{20}O_2$	2,2-Dimethyloctanoic acid	296.2	2.8				
$C_{10}H_{20}O_2$	Octyl acetate	288.2	4.18	−0.34691E+01	0.58106E−01	−0.10952E−03	288-323
$C_{10}H_{20}O_2$	2-Methylheptyl acetate	288.2	4.27	0.23285E+02	−0.11538E+00	0.17143E−03	288-323
$C_{10}H_{20}O_2$	Pentyl pentanoate	305.6	4.076	0.77641E+01	−0.14335E−01	0.73740E−05	306-393
$C_{10}H_{20}O_2$	Isopentyl pentanoate	292.2	3.6				
$C_{10}H_{20}O_2$	Isopentyl isopentanoate	288.2	4.39	0.14698E+02	−0.57726E−01	0.76190E−04	288-323
$C_{10}H_{20}O_2$	Methyl nonanoate	293.2	3.943				
$C_{10}H_{21}Br$	1-Bromodecane	298.2	4.44	0.11202E+02	−0.33491E−01	0.36314E−04	274-328
$C_{10}H_{21}Cl$	1-Chlorodecane	293.2	4.581	0.68741E+01	−0.12210E−02	−0.22500E−04	293-323
$C_{10}H_{21}NO$	N,N-Dibutylacetamide	293.2	19.1				
$C_{10}H_{22}$	Decane	293.2	1.9853	0.24054E+01	−0.15445E−02	0.44643E−06	253-393
$C_{10}H_{22}$	2,7-Dimethyloctane	293.2	1.98				
$C_{10}H_{22}$	4-Propylheptane	293.2	1.9955				
$C_{10}H_{22}O$	1-Decanol	293.2	7.93	0.47195E+02	−0.20740E+00	0.24942E−03	293-343
$C_{10}H_{22}O$	2-Decanol	298.2	5.82	0.13621E+03	−0.81000E+00	0.12500E−02	288-308
$C_{10}H_{22}O$	3-Decanol	298.2	4.05	0.52090E+02	−0.31020E+00	0.50000E−03	288-308
$C_{10}H_{22}O$	4-Decanol	298.2	3.42	−0.11260E+02	0.93960E−01	−0.15000E−03	288-308
$C_{10}H_{22}O$	5-Decanol	298.2	3.24	−0.25832E+01	0.31456E−01	−0.40000E−04	288-308
$C_{10}H_{22}O$	2,2-Dimethyl-1-octanol	293.2	7.86	0.69536E+02	−0.34596E+00	0.46250E−03	293-333
$C_{10}H_{22}O$	Dipentyl ether	298.2	2.798				
$C_{10}H_{22}O$	Diisopentyl ether	293.2	2.817	0.44690E+01	−0.63710E−02	0.25000E−05	293-323
$C_{10}H_{22}OS$	Dipentyl sulfoxide	348.2	18.8				
$C_{10}H_{22}O_5$	Tetraethylene glycol dimethyl ether	298.2	7.68				
$C_{10}H_{22}S$	Dipentyl sulfide	298.2	3.826				
$C_{10}H_{23}N$	Decylamine	293.2	3.31	0.61497E+01	−0.12801E−01	0.10606E−04	293-373
$C_{10}H_{30}O_3Si_4$	Decamethyltetrasiloxane	293.2	2.370				
$C_{10}H_{30}O_5Si_5$	Decamethylcyclopentasiloxane	293.2	2.50				
$C_{11}H_{10}$	1-Methylnaphthalene	293.2	2.915	0.45126E+01	−0.76480E−02	0.75000E−05	293-333
$C_{11}H_{10}$	2-Methylnaphthalene	313.2	2.747				
$C_{11}H_{10}O$	1-Methoxynaphthalene	293.2	4.020	0.71885E+01	−0.14838E−01	0.13750E−04	293-333
$C_{11}H_{10}O$	2-Methoxynaphthalene	353.2	3.563	0.56702E+01	−0.69754E−02	0.28571E−05	353-373
$C_{11}H_{12}O_2$	Ethyl trans-cinnamate	293.2	5.63				
$C_{11}H_{12}O_3$	Ethyl benzoylacetate	303.2	13.50	0.93644E+01	0.74280E−01	−0.20000E−03	303-323
$C_{11}H_{14}O_2$	Benzyl butanoate	301.2	4.55				
$C_{11}H_{14}O_2$	Phenyl pentanoate	293.2	4.30				
$C_{11}H_{14}O_2$	Butyl benzoate	303.2	5.52	0.77854E+01	−0.34972E−02	−0.13149E−04	303-358
$C_{11}H_{14}O_2$	Isobutyl benzoate	291.2	5.39				
$C_{11}H_{16}$	1,3-Diethyl-5-methylbenzene	293.2	2.264				
$C_{11}H_{16}$	Pentamethylbenzene	334.0	2.358	0.30196E+01	−0.22619E−02	0.83831E−06	334-413
$C_{11}H_{22}$	1-Undecene	293.2	2.137	0.22132E+01	0.13121E−02	−0.53571E−05	273-323
$C_{11}H_{22}O$	2-Undecanone	285.3	8.3				
$C_{11}H_{22}O_2$	Nonyl acetate	293.2	3.87				
$C_{11}H_{22}O_2$	Pentyl hexanoate	288.2	4.22	0.83503E+01	−0.18449E−01	0.14286E−04	288-323
$C_{11}H_{23}Br$	1-Bromoundecane	272.6	4.61				
$C_{11}H_{24}$	Undecane	293.2	1.9972	0.23637E+01	−0.12500E−02	−0.85869E−16	283-363
$C_{11}H_{24}O$	1-Undecanol	313.2	5.98				
$C_{11}H_{25}N$	Undecylamine	293.2	3.25	0.54945E+01	−0.96161E−02	0.66017E−05	293-373
$C_{12}F_{27}N$	Tris(perfluorobutyl)amine	293.2	2.15				
$C_{12}H_8O$	Dibenzofuran	373.2	3.00				
$C_{12}H_{10}$	Biphenyl	348.2	2.53	0.26869E+01	0.63072E−03	−0.30995E−05	348-428
$C_{12}H_{10}N_2O$	trans-Azoxybenzene	311.2	5.2				
$C_{12}H_{10}O$	Diphenyl ether	283.2	3.726				
$C_{12}H_{10}O$	2-Acetonaphthone	333.2	13.03	0.14538E+03	−0.73040E+00	0.10000E−02	333-363
$C_{12}H_{10}OS$	Diphenyl sulfoxide	344.7	16.6				
$C_{12}H_{10}O_2S$	Diphenyl sulfone	406.2	21.1				
$C_{12}H_{10}S$	Diphenyl sulfide	298.2	5.43				

Mol. form.	Name	T/K	ε_r	a	b	c	Range/K
$C_{12}H_{11}N$	Diphenylamine	323.2	3.73				
$C_{12}H_{11}NO$	N-1-Naphthylenylacetamide	433.2	24.3	0.84739E+02	−0.12391E+00	−0.35714E-04	433-533
$C_{12}H_{12}$	1,6-Dimethylnaphthalene	293.2	2.7250				
$C_{12}H_{12}O$	1-Ethoxynaphthalene	292.2	3.3				
$C_{12}H_{14}O_2$	Propyl cinnamate	293.2	5.45				
$C_{12}H_{14}O_4$	Diethyl phthalate	293.2	7.86				
$C_{12}H_{16}O$	2-Cyclohexylphenol	328.2	3.97				
$C_{12}H_{16}O$	4-Cyclohexylphenol	404.2	4.42				
$C_{12}H_{16}O_2$	Pentyl benzoate	293.2	5.07				
$C_{12}H_{16}O_3$	Pentyl salicylate	301.2	6.25				
$C_{12}H_{16}O_3$	Isopentyl salicylate	293.12	7.26	0.13129E+02	−0.19190E-01	−0.36060E-05	225-397
$C_{12}H_{17}NO$	N-Butyl-N-phenylacetamide	298.2	11.66				
$C_{12}H_{18}$	Hexylbenzene	293.2	2.3				
$C_{12}H_{18}$	1,3,5-Triethylbenzene	293.2	2.256				
$C_{12}H_{18}$	Hexamethylbenzene	449.0	2.172	0.35710E+01	−0.46912E-02	0.35088E-05	449-489
$C_{12}H_{20}O_2$	l-Bornyl acetate	303.2	4.46	0.60791E+01	0.98200E-02	−0.50000E-04	303-323
$C_{12}H_{22}O$	Dicyclohexyl ether	293.2	3.45	0.95324E+01	−0.31740E-01	0.37500E-04	293-333
$C_{12}H_{22}O$	Cyclododecanone	303.2	11.4	0.39327E+02	−0.13248E+00	0.13298E-03	303-423
$C_{12}H_{22}O_6$	Dibutyl tartrate	314.2	9.4				
$C_{12}H_{24}$	1-Dodecene	293.2	2.152	0.22581E+01	0.11106E-02	−0.50000E-05	273-323
$C_{12}H_{24}O_2$	Decyl acetate	293.2	3.75				
$C_{12}H_{24}O_2$	Ethyl decanoate	293.2	3.75	0.70969E+01	−0.15080E-01	0.12500E-04	293-353
$C_{12}H_{24}O_2$	Methyl undecanoate	293.2	3.671				
$C_{12}H_{25}Br$	1-Bromododecane	298.2	4.07	0.86103E+01	−0.20891E-01	0.18994E-04	274-328
$C_{12}H_{25}Cl$	1-Chlorododecane	298.2	4.17	0.10002E+02	−0.27798E-01	0.27559E-04	274-328
$C_{12}H_{25}I$	1-Iodododecane	298.2	3.91	0.34641E+01	0.97404E-02	−0.27602E-04	293-323
$C_{12}H_{26}$	Dodecane	293.2	2.0120	0.23697E+01	−0.12200E-02	−0.36375E-16	283-363
$C_{12}H_{26}O$	1-Dodecanol	303.2	5.82	0.18518E+02	−0.44859E-01	0.99900E-05	303-358
$C_{12}H_{26}O$	2-Butyl-1-octanol	363.2	3.28				
$C_{12}H_{27}BO_3$	Tributyl borate	293.2	2.23				
$C_{12}H_{27}N$	Dodecylamine	303.2	3.07	0.27999E+01	0.44810E-02	−0.11905E-04	303-373
$C_{12}H_{27}N$	Tributylamine	293.2	2.340	0.19846E+01	0.28108E-02	−0.54545E-05	233-293
$C_{12}H_{27}O_4P$	Tributyl phosphate	293.2	8.34	0.26304E+02	−0.88480E-01	0.92857E-04	293-373
$C_{12}H_{28}O_4Si$	Tetrapropoxysilane	298.2	3.21				
$C_{12}H_{28}Sn$	Tetrapropylstannane	293.2	2.267				
$C_{12}H_{30}OSi_2$	Hexaethyldisiloxane	298.2	2.259	0.36559E+01	−0.72406E-02	0.85714E-05	298-333
$C_{13}H_{10}O$	Benzophenone	300.2	12.62	0.34130E+02	−0.10249E+00	0.10268E-03	300-420
$C_{13}H_{10}O_3$	Phenyl salicylate	290.2	6.92	0.26545E+02	−0.11180E+00	0.15220E-03	290-358
$C_{13}H_{12}$	Diphenylmethane	303.2	2.540	0.30638E+01	−0.17286E-02		303-333
$C_{13}H_{12}O$	Benzyl phenyl ether	313.2	3.748				
$C_{13}H_{18}O_2$	Hexyl benzoate	293.2	4.80				
$C_{13}H_{20}$	Heptylbenzene	293.2	2.26				
$C_{13}H_{20}O$	α-Ionone*	292.4	10.78				
$C_{13}H_{20}O$	β-Ionone*	297.65	11.66				
$C_{13}H_{24}O_4$	Diethyl nonanedioate	303.2	5.133				
$C_{13}H_{26}$	1-Tridecene	293.2	2.139	0.14154E+01	0.66514E-02	−0.14286E-04	273-323
$C_{13}H_{26}O$	7-Tridecanone	303.2	7.6				
$C_{13}H_{26}O_2$	Ethyl undecanoate	293.2	3.55				
$C_{13}H_{26}O_2$	Methyl dodecanoate	293.2	3.539				
$C_{13}H_{27}Br$	1-Bromotridecane	281.15	4.19				
$C_{13}H_{28}$	Tridecane	293.2	2.0213	0.23731E+01	−0.12000E-02	−0.21841E-15	283-363
$C_{13}H_{28}$	5-Butylnonane	293.2	2.0319				
$C_{13}H_{28}O$	1-Tridecanol	333.2	4.02				
$C_{14}H_{10}$	Anthracene	502.0	2.649	0.20571E+02	−0.69169E-01	0.66667E-04	502-516
$C_{14}H_{10}$	Phenanthrene	383.2	2.72				
$C_{14}H_{10}O_2$	Benzil	368.2	13.04	−0.23599E+02	0.22715E+00	−0.34667E-03	368-393
$C_{14}H_{12}O_2$	Benzyl benzoate	303.2	5.26	0.76856E+01	−0.80000E-02	−0.80361E-15	303-358
$C_{14}H_{12}O_3$	Benzyl salicylate	301.2	4.12				
$C_{14}H_{14}$	1,2-Diphenylethane	331.2	2.47	0.31178E+01	−0.21572E-02	0.59800E-06	331-451
$C_{14}H_{14}O$	Dibenzyl ether	293.2	3.821	0.80154E+01	−0.20536E-01	0.21250E-04	293-333
$C_{14}H_{15}N$	Dibenzylamine	293.2	3.446				
$C_{14}H_{16}O_2Si$	Dimethyldiphenoxysilane	298.2	3.500	0.51669E+01	−0.77001E-02	0.70156E-05	283-353
$C_{14}H_{18}O_2$	Pentyl cinnamate	293.2	4.89				

Mol. form.	Name	T/K	ε_r	a	b	c	Range/K
$C_{14}H_{22}$	Octylbenzene	293.2	2.26				
$C_{14}H_{26}O_4$	Diisobutyl adipate	293.2	5.19				
$C_{14}H_{26}O_4$	Diethyl sebacate	303.2	4.995	0.39143E+02	−0.20965E+00	0.32000E-03	303-313
$C_{14}H_{28}O_2$	Dodecyl acetate	293.2	3.6				
$C_{14}H_{28}O_2$	Ethyl laurate	273.2	3.94				
$C_{14}H_{28}O_2$	Methyl tridecanoate	293.2	3.442				
$C_{14}H_{29}Br$	1-Bromotetradecane	293.2	3.84	0.10058E+02	−0.33905E-01	0.43528E-04	274-328
$C_{14}H_{30}$	Tetradecane	293.2	2.0343	0.23832E+01	−0.11900E-02	−0.51229E-16	283-363
$C_{14}H_{30}O$	1-Tetradecanol	318.2	4.42	0.12272E+02	−0.24667E-01	−0.13168E-13	318-358
$C_{14}H_{31}N$	Tetradecylamine	312.55	2.90				
$C_{15}H_{12}O_4$	Phenyl 2-(acetyloxy)benzoate	384.2	4.33				
$C_{15}H_{26}O_6$	Tributyrin	282.8	5.72	0.13152E+02	−0.36684E-01	0.36795E-04	199-283
$C_{15}H_{30}O_2$	Methyl tetradecanoate	293.2	3.352				
$C_{15}H_{31}Br$	1-Bromopentadecane	293.35	3.88				
$C_{15}H_{32}$	Pentadecane	293.2	2.0391	0.23792E+01	−0.11600E-02	−0.71069E-16	283-363
$C_{15}H_{32}O$	1-Pentadecanol	333.2	3.70				
$C_{15}H_{33}N$	Pentadecylamine	313.25	2.85				
$C_{15}H_{33}N$	Triisopentylamine	294.2	2.29				
$C_{16}H_{22}O_4$	Dibutyl phthalate	293.2	6.58	0.12444E+02	−0.20000E-01		293-333
$C_{16}H_{32}O_2$	Hexadecanoic acid	338.2	2.417				
$C_{16}H_{32}O_2$	Ethyl myristate	293.2	3.50	0.52642E+01	−0.60000E-02	−0.47358E-15	293-353
$C_{16}H_{32}O_2$	Methyl pentadecanoate	293.2	3.296				
$C_{16}H_{33}Br$	1-Bromohexadecane	298.2	3.68	0.58668E+01	−0.73333E-02	−0.52666E-14	298-328
$C_{16}H_{33}I$	1-Iodohexadecane	293.2	3.57	0.79531E+01	−0.22859E-01	0.26955E-04	293-323
$C_{16}H_{34}$	Hexadecane	293.2	2.0460	0.23861E+01	−0.11600E-02	0.25555E-15	293-363
$C_{16}H_{34}O$	1-Hexadecanol	333.2	3.69	0.85935E+01	−0.14714E-01	−0.45533E-13	333-363
$C_{16}H_{35}N$	Hexadecylamine	328.35	2.71				
$C_{16}H_{36}Sn$	Tetrabutylstannane	293.2	9.74	0.56115E+02	−0.24812E+00	0.30682E-03	293-313
$C_{17}H_{12}O_3$	2-Naphthyl salicylate	293.0	6.30	0.11229E+02	−0.18857E-01	0.70332E-05	293-353
$C_{17}H_{34}O$	9-Heptadecanone	328.2	5.43	0.44176E+02	−0.21183E+00	0.28571E-03	328-363
$C_{17}H_{34}O_2$	Methyl palmitate	313.2	3.124				
$C_{17}H_{36}$	Heptadecane	293.2	2.0578	0.23627E+01	−0.10400E-02	−0.10397E-12	293-308
$C_{17}H_{36}O$	1-Heptadecanol	333.2	3.41				
$C_{18}H_{26}O_4$	Dipentyl phthalate	293.2	6.00				
$C_{18}H_{28}O_2$	Phenyl laurate	293.2	3.28				
$C_{18}H_{30}O_2$	Linolenic acid	293.2	2.825	0.33867E+01	−0.19181E-02		274-368
$C_{18}H_{30}O_4$	Dicyclohexyl adipate	308.2	4.84				
$C_{18}H_{32}O_2$	Linoleic acid	293.2	2.754	0.32073E+01	−0.15477E-02		275-368
$C_{18}H_{34}O_2$	Oleic acid	293.2	2.336	0.25385E+01	−0.69448E-03		275-368
$C_{18}H_{34}O_4$	Dibutyl sebacate	293.2	4.54				
$C_{18}H_{36}O_2$	Stearic acid	293.2	2.314	0.27159E+01	−0.13300E-02		293-373
$C_{18}H_{36}O_2$	Hexadecyl acetate	308.2	3.19	0.47310E+01	−0.50000E-02	0.41338E-14	308-348
$C_{18}H_{36}O_2$	Ethyl palmitate	303.2	3.07	0.57938E+01	−0.12294E-01	0.10919E-04	303-455
$C_{18}H_{36}O_2$	Methyl heptadecanoate	313.2	3.07				
$C_{18}H_{37}Br$	1-Bromooctadecane	303.35	3.53	0.46790E+01	−0.30355E-02	−0.24798E-05	303-332
$C_{18}H_{38}O$	1-Octadecanol	333.2	3.38	0.73784E+01	−0.12000E-01	−0.22871E-13	333-363
$C_{18}H_{39}BO_3$	Trihexyl borate	293.2	2.22				
$C_{18}H_{39}N$	Octadecylamine	326.35	2.67				
$C_{19}H_{16}$	Triphenylmethane	367.2	2.46	0.40201E+01	−0.66507E-02	0.65329E-05	367-448
$C_{19}H_{18}O_3Si$	Methyltriphenoxysilane	298.2	3.628				
$C_{19}H_{32}O_2$	Methyl linolenate	293.2	3.355				
$C_{19}H_{34}O_2$	Methyl linoleate	293.2	3.466				
$C_{19}H_{36}O_2$	Methyl oleate	293.2	3.211				
$C_{19}H_{38}O$	10-Nonadecanone	353.2	5.37				
$C_{19}H_{38}O_2$	Methyl stearate	313.2	3.021				
$C_{19}H_{40}$	Nonadecane	293.2	2.0706				
$C_{20}H_{30}O_4$	Dihexyl phthalate	293.2	5.62				
$C_{20}H_{38}O_2$	Ethyl oleate	301.2	3.17	0.57033E+01	−0.11223E-01	0.93447E-05	301-423
$C_{20}H_{40}O_2$	Octadecyl acetate	308.2	3.07	0.44569E+01	−0.45000E-02	0.33923E-14	308-348
$C_{20}H_{40}O_2$	Ethyl stearate	313.2	2.958	0.70930E+01	−0.19081E-01	0.19555E-04	331-440
$C_{20}H_{40}O_2$	Methyl nonadecanoate	313.2	2.982				
$C_{20}H_{42}O$	1-Eicosanol	338.2	3.13	0.21700E+01	0.12497E-01	−0.28571E-04	338-363
$C_{20}H_{42}O$	Didecyl ether	293.2	2.644	0.41465E+01	−0.62240E-02	0.37500E-05	293-333

Mol. form.	Name	T/K	ε_r	a	b	c	Range/K
$C_{20}H_{60}O_8Si_9$	Eicosamethylnonasiloxane	293.2	2.645	0.57840E+01	−0.16568E-01	0.20000E-04	293-323
$C_{21}H_{21}O_4P$	Tricresyl phosphate*	298.2	6.7				
$C_{21}H_{38}O_6$	1,2,3-Propanetriyl hexanoate	293.2	4.476				
$C_{22}H_{42}O_2$	Butyl oleate	298.2	4.00				
$C_{22}H_{44}O_2$	Butyl stearate	298.2	3.120	0.73894E+02	−0.46261E+00	0.75500E-03	298-343
$C_{22}H_{46}$	Docosane	293.2	2.0840				
$C_{22}H_{46}O$	1-Docosanol	348.2	2.94	0.82062E+01	−0.25069E-01	0.28571E-04	348-373
$C_{24}H_{20}O_4Si$	Tetraphenoxysilane	333.2	3.4915				
$C_{24}H_{38}O_4$	Dioctyl phthalate	293.2	5.22				
$C_{26}H_{50}O_4$	Dioctyl sebacate	299.2	4.01				
$C_{27}H_{50}O_6$	1,2,3-Propanetriyl octanoate	293.2	3.931				
$C_{30}H_{58}O_4$	Ethylene glycol ditetradecanoate	343.2	2.98				
$C_{30}H_{62}$	Triacontane	373.2	1.9112				
$C_{30}H_{62}$	2,6,10,15,19,23-Hexamethyltetracosane	373.2	1.9106				
$C_{34}H_{66}O_4$	Ethylene glycol dipalmitate	348.2	2.89				
$C_{34}H_{68}O_2$	Hexadecyl stearate	333.2	2.61				
$C_{38}H_{74}O_4$	Ethylene glycol distearate	353.2	2.79				
$C_{39}H_{74}O_6$	Glycerol trilaurate	313.2	3.287				
$C_{51}H_{98}O_6$	Glycerol tripalmitate	328.2	2.901	−0.29131E+01	0.32206E-01	−0.44154E-04	328-393
$C_{57}H_{104}O_6$	Glycerol trioleate	293.2	3.109				
$C_{57}H_{104}O_6$	Glycerol trielaidate	313.2	2.980				
$C_{57}H_{110}O_6$	Glycerol tristearate	353.2	2.740				

PERMITTIVITY (DIELECTRIC CONSTANT) OF GASES

This table gives the relative permittivity ε (often called the dielectric constant) of some common gases at a temperature of 20°C and pressure of one atmosphere (101.325 kPa). Values of the permanent dipole moment μ in Debye Units (1 D = 3.33564 × 10^{-30} C m) are also included.

The density dependence of the permittivity is given by the equation

$$\frac{\varepsilon-1}{\varepsilon-2} = \rho_m \left(\frac{4\pi N\alpha}{3} + \frac{4\pi N \mu^2}{9kT} \right)$$

where ρ_m is the molar density, N is Avogadro's number, k is the Boltzmann constant, T is the temperature, and α is the molecular polarizability. Therefore, in regions where the gas can be considered ideal, $\varepsilon - 1$ is approximately proportional to the pressure at constant temperature. For nonpolar gases ($\mu = 0$), $\varepsilon - 1$ is inversely proportional to temperature at constant pressure.

The number of significant figures indicates the accuracy of the values given. The values of ε for air, Ar, H$_2$, He, N$_2$, O$_2$, and CO$_2$ are recommended as reference values; these are accurate to 1 ppm or better.

The second part of the table gives the permittivity of water vapor in equilibrium with liquid water as a function of temperature (derived from Reference 4).

References

1. A. A. Maryott and F. Buckley, *Table of Dielectric Constants and Electric Dipole Moments of Substances in the Gaseous State*, National Bureau of Standards Circular 537, 1953.
2. B. A. Younglove, *J. Phys. Chem. Ref. Data*, 11, Suppl. 1, 1982; 16, 577, 1987 (for data on N$_2$, H$_2$, O$_2$, and hydrocarbons over a range of pressure and temperature).
3. *Landolt-Börnstein, Numerical Data and Functional Relationships in Science and Technology*, New Series, Group IV, Vol. 4, Springer-Verlag, Heidelberg, 1980 (for data at high pressures).
4. G. Birnbaum and S. K. Chatterjee, *J. Appl. Phys.*, 23, 220, 1952 (for data on water vapor).

Mol. form.	Name	ε	μ/D
Compounds not containing carbon			
	Air (dry, CO$_2$ free)	1.0005364	
Ar	Argon	1.0005172	0
BF$_3$	Boron trifluoride	1.0011	0
BrH	Hydrogen bromide	1.00279	0.827
ClH	Hydrogen chloride	1.00390	1.109
F$_3$N	Nitrogen trifluoride	1.0013	0.235
F$_6$S	Sulfur hexafluoride	1.00200	0
HI	Hydrogen iodide	1.00214	0.448
H$_2$	Hydrogen	1.0002538	0
H$_2$S	Hydrogen sulfide	1.00344	0.97
H$_3$N	Ammonia	1.00622	1.471
He	Helium	1.0000650	0
Kr	Krypton	1.00078	0
NO	Nitric oxide	1.00060	0.159
N$_2$	Nitrogen	1.0005480	0
N$_2$O	Nitrous oxide	1.00104	0.161
Ne	Neon	1.00013	0
O$_2$	Oxygen	1.0004947	0
O$_2$S	Sulfur dioxide	1.00825	1.633
O$_3$	Ozone	1.0017	0.534
Xe	Xenon	1.00126	0

Mol. form.	Name	ε	μ/D
Compounds containing carbon			
CF$_4$	Tetrafluoromethane	1.00121	0
CO	Carbon monoxide	1.00065	0.110
CO$_2$	Carbon dioxide	1.000922	0
CH$_3$Br	Bromomethane	1.01028	1.822
CH$_3$Cl	Chloromethane	1.01080	1.892
CH$_3$F	Fluoromethane	1.00973	1.858
CH$_3$I	Iodomethane	1.00914	1.62
CH$_4$	Methane	1.00081	0
C$_2$H$_2$	Acetylene	1.00124	0
C$_2$H$_3$Cl	Chloroethylene	1.0075	1.45
C$_2$H$_4$	Ethylene	1.00134	0
C$_2$H$_5$Cl	Chloroethane	1.01325	2.05
C$_2$H$_6$	Ethane	1.00140	0
C$_2$H$_6$O	Dimethyl ether	1.0062	1.30
C$_3$H$_6$	Propene	1.00228	0.366
C$_3$H$_6$	Cyclopropane	1.00178	0
C$_3$H$_8$	Propane	1.00200	0.084
C$_4$H$_{10}$	Butane	1.00258	0
C$_4$H$_{10}$	Isobutane	1.00260	0.132

Permittivity of Saturated Water Vapor

t/°C	ε	t/°C	ε
0	1.00007	60	1.00144
10	1.00012	70	1.00213
20	1.00022	80	1.00305
30	1.00037	90	1.00428
40	1.00060	100	1.00587
50	1.00095		

AZEOTROPIC DATA FOR BINARY MIXTURES

J. Gmehling, J. Menke, J. Krafczyk, K. Fischer, J.-C. Fontaine, and H. V. Kehiaian

Binary homogeneous (single-phase) liquid mixtures having an extremum (maximum or minimum) vapor pressure P at constant temperature T, as a function of composition, are called azeotropic mixtures, or simply azeotropes. The composition is usually expressed as mole fractions, where x_1 for component 1 in the liquid phase and y_1 for component 1 in the vapor phase are identical. Mixtures that do not show a maximum or minimum are called zeotropic. A maximum (minimum) of the $P(x_1)$ or $P(y_1)$ curves corresponds to a minimum (maximum) of the boiling temperature T at constant P, plotted as a function of x_1 or y_1 [see $T(x_1)$ and $T(y_1)$ curves, Types I and III, in Fig.1]. Azeotropes in which the pressure is a maximum (temperature is a minimum) are often called positive azeotropes, while pressure-minimum (temperature-maximum) azeotropes are called negative azeotropes. The coordinates of an azeotropic point are the azeotropic temperature T_{Az}, pressure P_{Az}, and the vapor-phase composition $y_{1,Az}$, which is the same as the liquid-phase composition $x_{1,Az}$.

In the two-phase liquid-liquid region of partially miscible (heterogeneous) mixtures, the vapor pressure at constant T (or the boiling temperature at constant P) is independent of the global composition x_1 of the two coexisting liquid phases between the equilibrium compositions x_1' and x_1'' $(x_1' < x_1'')$.

The constant vapor pressure (boiling temperature) above the two-phase region of certain partially miscible mixtures is usually larger (smaller) than the vapor pressure (boiling temperature) at any other liquid-phase composition in the homogeneous region. In this case, the vapor-phase composition is inside the miscibility gap. Mixtures of this type are called heteroazeotropic mixtures, or simply heteroazeotropes. (Fig. 1, Type II), as opposed to the other types of azeotropes, called homoazeotropes.

Only in a few cases partially miscible mixtures present a positive or negative azeotropic point in the single-phase region, outside the miscibility gap, similar to the azeotropic points of homogeneous mixtures (Fig. 1, Types IV and VI).

A few binary mixtures, for example the system perfluorobenzene + benzene, may present two azeotropic points at constant temperature (pressure), a positive and a negative one. They are called double azeotropic mixtures, or simply double azeotropes. (Fig. 1, Type V).

The knowledge of the occurrence of azeotropic points in binary and higher systems is of special importance for the design of distillation processes. The number of theoretical stages of a distillation column required for the separation depends on the separation factor α_{12}, i.e. the ratio of the K_i-factors ($K_i = y_i/x_i$) of the components i $(i = 1, 2)$. The required separation factor can be calculated with the following simplified relation (Reference 1):

$$\alpha_{12} = K_1/K_2 = (y_1/x_1)/(y_2/x_2) = (\gamma_1 P_1^s)/(\gamma_2 P_2^s) \qquad (1)$$

where γ_i is the activity coefficient of component i in the liquid phase and P_i^s is the vapor pressure of the pure component i.

In distillation processes, only the difference between the separation factor and unity $(\alpha_{12} - 1)$ can be exploited for the separation. If the separation factor is close to unity, a large number of theoretical stages is required for the separation. If the binary system to be separated shows an azeotropic point ($\alpha_{12} = 1$), the separation is impossible by ordinary distillation, even with an infinitely large number of stages.

Following eq. (1) azeotropic behavior will always occur in homogeneous binary systems when the vapor pressure ratio P_1^s/P_2^s is equal to the ratio of the activity coefficients γ_2/γ_1.

Various thermodynamic methods based on g^E-models (Wilson, NRTL, UNIQUAC) or group contribution methods (UNIFAC, modified UNIFAC, ASOG, PSRK) can be used for either calculating or predicting the required activity coefficients for the components under given conditions of temperature and composition (Reference 2).

Because of the importance of azeotropic data for the design of distillation processes, compilations have been available in book form for quite some time (References 3-7). The most recent printed data collection was published in 1994 (Reference 8). A revised and extended version appeared in 2004 (Reference 9).

A collection of approximately 47,400 zeotropic and azeotropic data sets, compiled from 6600 references, are stored in a comprehensive computerized data bank (Reference 10). The references from the above-mentioned compilations and from the vapor-liquid equilibrium part of the Dortmund Data Bank (Reference 11) were supplemented by references found from CAS online searches, private communications, data from industry, etc.. Over 24,000 zeotropic data and over 20,000 azeotropic data are available for binary systems. Nearly 90% of the binary azeotropic data show a pressure maximum. In most cases (ca. 90%) these are homogeneous azeotropes, and in approximately 7–8% of the cases heterogeneous azeotropes are reported. Less than 10% of the data stored show a pressure minimum. Approximately 21,000 of the data sets stored were published after 1970.

The table below provides information about azeotropes for 808 selected binary systems. Compounds are listed in the modified Hill order, with carbon-containing compounds following those compounds not containing carbon. In columns 1 and 2 are the molecular formulas of components 1 and 2 written in the Hill convention. In column 3 the names of the components are given, either a systematic IUPAC name or a name in ubiquitous use. Columns 4, 5, and 6 contain the azeotropic coordinates of the mixtures: temperature T_{Az}, pressure P_{Az}, and vapor-phase composition $y_{1,Az}$. The explanation of the type of azeotrope (column 7) is given by the following codes:

O: homogeneous azeotrope in a completely miscible system
L: homogeneous azeotrope in a partially miscible system
E: heterogeneous azeotrope
X: pressure maximum
N: pressure minimum
D: double azeotrope
C: system contains a supercritical compound

References

1. Gmehling, J. and Brehm, A., *Grundoperationen*, Thieme-Verlag, Stuttgart, 1996.
2. Gmehling, J. and Kolbe, B., *Thermodynamik*, VCH-Verlag, Weinheim, 1992.
3. Lecat, M., *Doctoral Dissertation*, 1908.

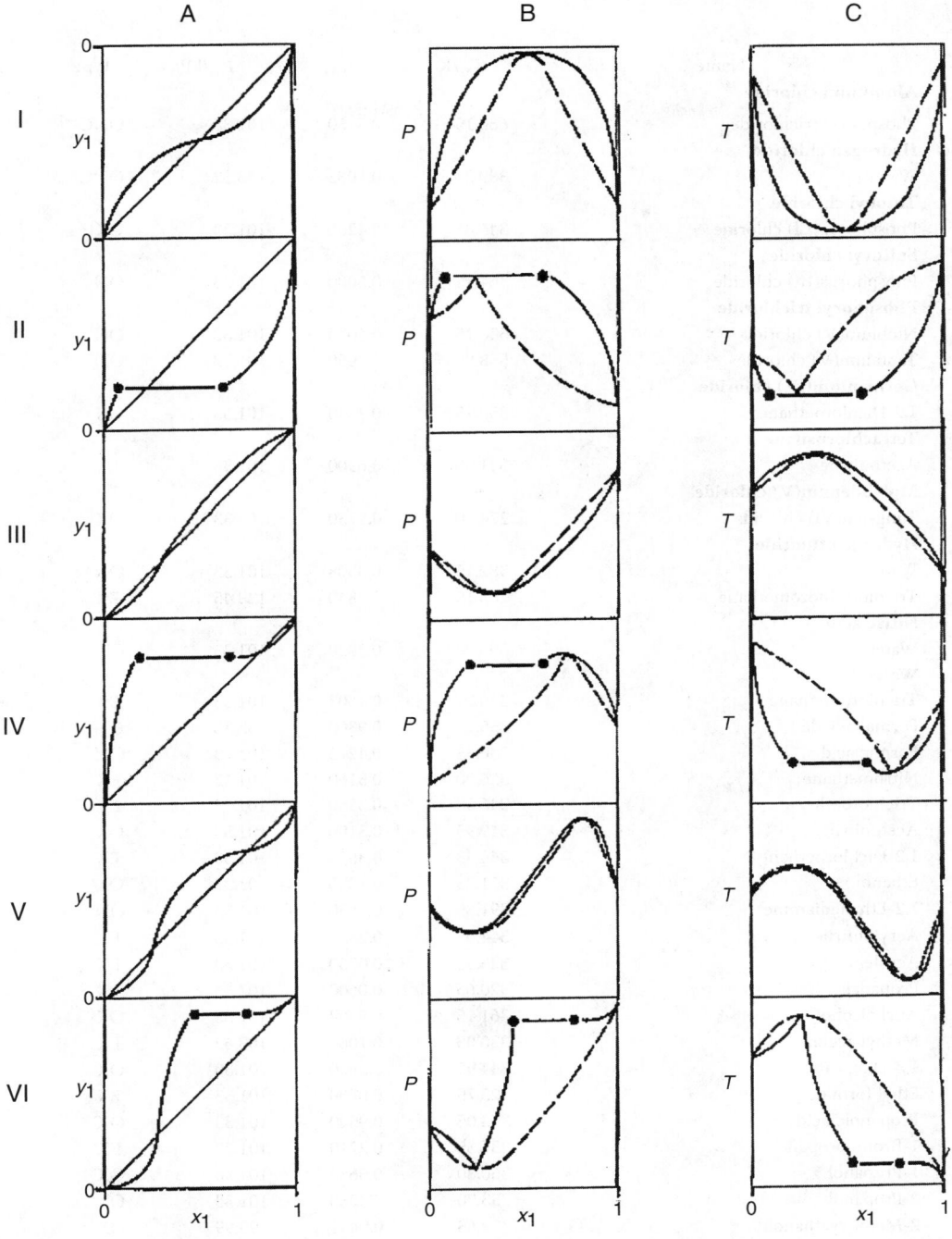

Figure 1 Different types of binary azeotropic systems: I - homogeneous pressure-maximum azeotrope in a completely miscible system (OX); II - heterogeneous pressure-maximum azeotrope (EX); III - homogeneous pressure-minimum azeotrope in a completely miscible system (ON); IV - homogeneous pressure-maximum azeotrope in a partially miscible system (LX); V - D: double azeotrope (OND, OXD); VI - homogeneous pressure-minimum azeotrope in a partially miscible system (LN). A - $y_1(x_1)$; B - $P(x_1)$ and $P(y_1)$; C - $T(x_1)$ and $T(y_1)$. Continuous line - (x_1); Dashed line - (y_1).

4. Lecat, M., *L'Azeotropisme*, Monograph, L'Auteur, Brussel, 1918.
5. Lecat, M., *Tables Azeotropiques*, Monograph, Lamertin, Brussel 1949.
6. Ogorodnikov, S. K., Lesteva, T. M., and Kogan V. B., *Azeotropic Mixtures*, Khimia, Leningrad, 1971.
7. Horsley, L. H., *Azeotropic Data III*, American Chemical Society, Washington, 1973.

8. Gmehling, J., Menke, J., Krafczyk, J., and Fischer, K., *Azeotropic Data*, 2 Volumes, VCH Verlag, Weinheim, 1994.
9. Gmehling, J., Menke, J., Krafczyk, J., and Fischer, K., *Azeotropic Data*, 2nd Ed., 3 Volumes, VCH Verlag, Weinheim, 2004.
10. Gmehling, J., Menke, J., Krafczyk, J., and Fischer, K., *A Data Bank for Azeotropic Data, Status and Applications, Fluid Phase Equilib.* 103, 51, 1995.
11. Dortmund Data Bank, www.ddbst.de

Molecular formula						
Comp. 1	Comp. 2	Name	T_{Az}/K	$y_{1,Az}$	P_{Az}/kPa	Type
$AlCl_3$		**Aluminum chloride**				
	Cl_3OP	Phosphoryl trichloride	660.15	0.5150	101.33	ONC
ClH		**Hydrogen chloride**				
	H_2O	Water	389.34	0.1083	133.32	ONC
Cl_2OS		**Thionyl chloride**				
	Cl_3P	Phosphorus(III) chloride	345.85	0.4200	101.33	OX
Cl_2O_2S		**Sulfuryl chloride**				
	Cl_3P	Phosphorus(III) chloride	364.15	0.5000	101.33	ON
Cl_3OP		**Phosphoryl trichloride**				
	Cl_5Nb	Niobium(V) chloride	536.15	0.4020	101.33	ON
	Cl_5Ta	Tantalum(V) chloride	558.85	0.4650	101.33	ON
Cl_4Ge		**Germanium(IV) chloride**				
	$C_2H_4Cl_2$	1,2-Dichloroethane	350.75	0.4630	101.33	OX
Cl_4Si		**Tetrachlorosilane**				
	C_2H_3N	Acetonitrile	321.05	0.6900	101.33	EX
Cl_5Mo		**Molybdenum(V) chloride**				
	Cl_6W	Tungsten(VI) chloride	274.70	0.9750	101.33	OX
FH		**Hydrogen fluoride**				
	H_2O	Water	382.15	0.3508	101.33	ON
	CCl_3F	Trichlorofluoromethane	283.15	0.7840	129.45	EX
HNO_3		**Nitric acid**				
	H_2O	Water	393.20	0.3820	101.33	ON
H_2O		**Water**				
	$CHCl_3$	Trichloromethane	329.27	0.1603	101.33	EX
	CH_2O	Formaldehyde	355.75	0.9300	53.33	OX
	CH_2O_2	Formic acid	380.35	0.4272	101.33	ON
	CH_3NO_2	Nitromethane	356.90	0.5160	101.33	EX
	C_2HCl_3	Trichloroethene	346.55	0.3560	101.33	EX
	C_2H_3N	Acetonitrile	349.95	0.3100	101.33	OX
	$C_2H_4Cl_2$	1,2-Dichloroethane	345.43	0.3570	101.33	EX
	C_2H_6O	Ethanol	351.25	0.1030	101.33	OX
	$C_2H_8N_2$	1,2-Ethanediamine	391.85	0.4450	101.33	ON
	C_3H_3N	Acrylonitrile	344.05	0.2850	101.33	EX
	C_3H_4O	Acrolein	325.45	0.0730	101.33	LX
	C_3H_6O	Propanal	320.65	0.0600	101.33	LX
	C_3H_6O	Allyl alcohol	361.15	0.5562	101.33	OX
	$C_3H_6O_2$	Methyl acetate	330.05	0.1060	103.62	LX
	$C_3H_6O_2$	1,3-Dioxolane	344.95	0.2520	101.30	OX
	$C_3H_6O_2$	Ethyl formate	325.75	0.0700	101.33	EX
	$C_3H_6O_2$	Propanoic acid	373.05	0.9500	101.33	OX
	C_3H_7Br	1-Bromopropane	336.35	0.2210	101.33	EX
	C_3H_8O	1-Propanol	360.80	0.5680	101.33	OX
	C_3H_8O	2-Propanol	353.70	0.3260	101.33	OX
	$C_3H_8O_2$	2-Methoxyethanol	372.65	0.9441	99.99	OX
	$C_3H_8O_2$	Dimethoxymethane	315.05	0.0269	101.38	LX
	C_4H_5N	cis-2-Butenenitrile	358.45	0.3832	101.33	EX
	C_4H_5N	trans-2-Butenenitrile	363.05	0.6843	101.33	EX
	C_4H_5N	Pyrrole	348.15	0.7514	50.13	EX
	$C_4H_6O_2$	Methacrylic acid	372.25	0.9464	98.93	OX
	C_4H_8O	2-Butanone	346.54	0.3480	101.33	LX
	C_4H_8O	Tetrahydrofuran	336.67	0.1828	101.33	OX
	C_4H_8O	Isobutanal	332.80	0.1698	100.99	EX
	$C_4H_8O_2$	Ethyl acetate	343.55	0.2990	101.33	EX
	$C_4H_8O_2$	Butanoic acid	372.95	0.9559	101.33	OX
	$C_4H_8O_2$	1,4-Dioxane	360.65	0.5280	101.33	OX
	$C_4H_8O_2$	Propyl formate	344.85	0.3090	101.33	EX
	$C_4H_8O_2$	Methyl propanoate	344.75	0.3050	101.33	EX
	C_4H_9Br	1-Bromobutane	353.95	0.4950	101.33	EX

Molecular formula						
Comp. 1	Comp. 2	Name	T_{Az}/K	$y_{1,Az}$	P_{Az}/kPa	Type
	C_4H_9Br	1-Bromo-2-methylpropane	348.45	0.3730	101.33	EX
	C_4H_9Cl	1-Chloro-2-methylpropane	333.95	0.1970	101.33	LX
	$C_4H_{10}O$	1-Butanol	365.45	0.7540	101.33	EX
	$C_4H_{10}O$	2-Butanol	360.50	0.6200	101.33	LX
	$C_4H_{10}O$	2-Methyl-2-propanol	353.00	0.4011	101.33	OX
	$C_4H_{11}N$	Butylamine	349.85	0.0700	101.33	OX
	C_5H_5N	Pyridine	367.30	0.7500	101.33	OX
	C_5H_8	2-Methyl-1,3-butadiene	305.85	0.0520	101.33	EX
	C_5H_8	Methylenecyclobutane	313.15	0.0212	101.30	EX
	C_5H_8O	Cyclopropyl methyl ketone	361.65	0.7060	101.19	EX
	$C_5H_8O_2$	Methyl methacrylate	354.45	0.4996	101.33	EX
	C_5H_{10}	2-Methyl-2-butene	309.75	0.0650	101.33	EX
	$C_5H_{10}O$	3-Methyl-2-buten-1-ol	369.55	0.9141	101.33	EX
	$C_5H_{10}O$	3-Methyl-3-buten-1-ol	333.15	0.8680	101.33	EX
	$C_5H_{10}O$	2-Methyl-3-buten-2-ol	359.25	0.5770	101.33	LX
	$C_5H_{10}O$	3-Pentanone	356.05	0.4750	101.33	EX
	$C_5H_{10}O_2$	Isopropyl acetate	349.75	0.3960	101.33	EX
	$C_5H_{10}O_2$	Propyl acetate	355.91	0.5228	101.33	EX
	$C_5H_{10}O_2$	Butyl formate	356.95	0.5360	101.33	EX
	$C_5H_{10}O_2$	Isobutyl formate	352.75	0.4460	101.33	EX
	$C_5H_{12}O$	3-Methyl-1-butanol	367.97	0.8265	101.33	EX
	$C_5H_{12}O$	2-Methyl-2-butanol	360.85	0.6355	101.75	EX
	$C_5H_{12}O$	1-Pentanol	369.08	0.8633	101.33	EX
	$C_5H_{12}O$	2-Pentanol	363.15	0.7550	92.49	EX
	C_6H_6	Benzene	342.35	0.2980	101.33	EX
	C_6H_7N	Aniline	372.55	0.9580	101.33	EX
	C_6H_7N	4-Methylpyridine	370.50	0.8972	101.33	OX
	C_6H_{10}	Cyclohexene	343.95	0.3090	101.33	EX
	$C_6H_{10}O$	Cyclohexanone	369.45	0.8694	101.33	EX
	$C_6H_{10}O$	Methyldihydropyran (unspecified isomer)	360.75	0.5841	100.93	EX
	$C_6H_{10}O_2$	4-Vinyl-1,3-dioxane	367.65	0.8955	101.33	EX
	C_6H_{12}	1-Hexene	318.15	0.1510	63.35	EX
	$C_6H_{12}O_2$	Butyl acetate	363.35	0.7013	101.33	EX
	$C_6H_{12}O_2$	Isobutyl acetate	361.05	0.6440	101.33	EX
	$C_6H_{12}O_2$	4,4-Dimethyl-1,3-dioxane	366.00	0.7779	101.33	EX
	$C_6H_{12}O_2$	4,5-Dimethyl-1,3-dioxane (unspecified isomer)	365.05	0.7966	101.50	EX
	$C_6H_{12}O_2$	4-Ethyl-1,3-dioxane	365.75	0.7257	101.30	EX
	$C_6H_{12}O_2$	Diacetone alcohol	370.00	0.9900	90.79	OX
	$C_6H_{12}O_2$	Propyl propanoate	362.05	0.6600	101.33	EX
	$C_6H_{13}N$	Cyclohexylamine	369.55	0.8692	101.33	OX
	C_6H_{14}	Hexane	334.75	0.2110	101.33	EX
	$C_6H_{14}O$	Butyl ethyl ether	349.85	0.4070	101.33	EX
	$C_6H_{14}O$	1-Hexanol	367.89	0.9432	101.33	EX
	$C_6H_{14}O_3$	Di(ethylene glycol) dimethyl ether	372.70	0.9679	101.33	OX
	$C_6H_{15}N$	Diisopropylamine	347.25	0.3654	101.33	EX
	$C_6H_{15}N$	Dipropylamine	359.00	0.6046	101.33	EX
	C_7H_8	Toluene	357.25	0.5230	101.33	EX
	C_7H_8O	Benzyl alcohol	373.05	0.9840	101.33	EX
	C_7H_9N	2,6-Dimethylpyridine	369.17	0.8647	101.33	EX
	$C_7H_{12}O_4$	1,2-Propanediol diacetate	358.15	0.9740	59.41	EX
	C_7H_{14}	1-Heptene	350.20	0.4100	101.33	EX
	$C_7H_{14}O_2$	Isopentyl acetate	367.05	0.7990	101.46	EX
	$C_7H_{14}O_2$	Butyl propanoate	367.95	0.8340	101.33	EX
	C_7H_{16}	Heptane	352.35	0.4510	101.33	EX
	$C_7H_{16}O$	1-Heptanol	371.99	0.9703	101.33	EX
	C_8H_8	Styrene	367.15	0.8000	101.33	EX
	C_8H_8O	Acetophenone	371.15	0.9675	101.19	EX
	C_8H_{10}	m-Xylene	365.15	0.7667	101.33	EX

Molecular formula

Comp. 1	Comp. 2	Name	T_{Az}/K	$y_{1,Az}$	P_{Az}/kPa	Type
	C_8H_{10}	p-Xylene	365.15	0.7450	101.33	EX
	C_8H_{10}	Ethylbenzene	364.15	0.7221	101.33	EX
	$C_8H_{16}O_2$	Butyl butanoate	369.85	0.9110	101.33	EX
	C_8H_{18}	Octane	362.75	0.6850	101.33	EX
	C_8H_{18}	2,2,4-Trimethylpentane	351.95	0.4420	101.33	EX
	$C_8H_{18}O$	Dibutyl ether	368.65	0.7628	101.33	EX
	$C_8H_{18}O$	1-Octanol	372.75	0.9820	101.33	EX
	$C_8H_{19}N$	Dibutylamine	370.05	0.8850	101.33	EX
	C_9H_{10}	Isopropenylbenzene	369.95	0.8880	101.33	EX
	C_9H_{12}	Isopropylbenzene	368.15	0.8340	101.33	EX
	$C_9H_{12}O$	2-Phenyl-2-propanol	371.25	0.9718	101.33	EX
	C_9H_{20}	Nonane	367.95	0.8280	101.33	EX
	$C_9H_{20}O$	1-Nonanol	373.00	0.9846	101.33	EX
	$C_{10}H_{22}$	Decane	370.75	0.9180	101.33	EX
	$C_{10}H_{22}O$	1-Decanol	373.13	0.9865	101.33	EX
	$C_{12}H_{27}N$	Tributylamine	372.80	0.9762	101.46	EX
CCl_4		**Tetrachloromethane**				
	C_2H_6O	Ethanol	338.19	0.6140	101.33	OX
	C_3H_6O	Acetone	341.25	0.0337	149.93	OX
	C_3H_8O	1-Propanol	346.28	0.8032	101.33	OX
	C_3H_8O	2-Propanol	341.83	0.6686	101.33	OX
	C_4H_6O	2-Butenal	348.15	0.6500	97.86	OX
	C_4H_6O	2-Methylpropenal	339.15	0.6000	97.86	OX
	C_4H_8O	2-Butanone	346.99	0.6630	101.33	OX
	$C_4H_8O_2$	Ethyl acetate	347.95	0.5700	101.33	OX
	$C_4H_{10}O$	1-Butanol	349.71	0.9500	101.33	OX
	$C_4H_{10}O$	2-Methyl-1-propanol	348.95	0.9080	101.33	OX
	$C_5H_{10}O$	2-Methyl-3-buten-2-ol	348.45	0.9009	101.06	OX
CS_2		**Carbon disulfide**				
	CH_4O	Methanol	310.65	0.7000	101.33	LX
$CHCl_3$		**Trichloromethane**				
	CH_4O	Methanol	328.15	0.6480	107.99	OX
	C_2H_6O	Ethanol	332.45	0.8410	101.33	OX
	C_3H_6O	Acetone	337.58	0.6398	101.33	ON
	$C_3H_6O_2$	Methyl acetate	337.51	0.6760	101.33	ON
	C_3H_8O	2-Propanol	334.15	0.9500	101.33	OX
	C_4H_6O	2-Butenal	329.15	0.9950	97.86	OX
	C_6H_{12}	2-Methyl-1-pentene	333.95	0.6235	101.19	OX
	C_6H_{14}	Hexane	333.45	0.7840	101.33	OX
CHN		**Hydrogen cyanide**				
	C_3H_5Cl	3-Chloropropene	296.45	0.8016	101.33	OX
CH_2Cl_2		**Dichloromethane**				
	C_2H_6O	Ethanol	312.05	0.9600	101.33	OX
CH_2O_2		**Formic acid**				
	$C_2H_4Cl_2$	1,2-Dichloroethane	350.17	0.4275	101.33	OX
	$C_5H_{10}O_2$	Butyl formate	372.15	0.8700	101.33	OX
	C_8H_{10}	m-Xylene	365.95	0.8545	101.33	EX
CH_3NO_2		**Nitromethane**				
	C_2H_6O	Ethanol	333.15	0.2850	53.61	OX
	C_3H_7Br	1-Bromopropane	343.25	0.1020	99.82	OX
	$C_4H_8O_2$	1,4-Dioxane	373.25	0.4101	101.48	OX
	C_5H_{10}	2-Methyl-2-butene	311.15	0.0570	101.33	LX
	C_7H_{14}	Methylcyclohexane	354.85	0.5123	101.33	EX
	C_7H_{16}	Heptane	353.25	0.4790	101.33	EX
	C_8H_{18}	Octane	363.38	0.6964	99.73	EX
	C_9H_{20}	Nonane	369.29	0.8403	99.73	EX
	$C_{10}H_{22}$	Decane	371.96	0.9239	99.73	EX
	$C_{11}H_{24}$	Undecane	373.16	0.9619	99.73	EX

Molecular formula						
Comp. 1	Comp. 2	Name	T_{Az}/K	$y_{1,Az}$	P_{Az}/kPa	Type
	$C_{12}H_{26}$	Dodecane	373.75	0.9846	99.73	EX
CH$_4$O		**Methanol**				
	$C_2HBrClF_3$	2-Bromo-2-chloro-1,1,1-trifluoroethane	317.25	0.1890	93.33	OX
	C_2H_5Br	Bromoethane	308.05	0.1610	101.33	OX
	C_3H_5Cl	3-Chloropropene	312.15	0.2570	100.39	OX
	C_3H_6O	Acetone	328.29	0.2400	101.33	OX
	$C_3H_6O_2$	Methyl acetate	328.15	0.3480	107.19	OX
	$C_3H_6O_2$	1,3-Dioxolane	334.66	0.6910	101.30	OX
	$C_3H_6O_2$	Ethyl formate	318.15	0.3000	81.34	OX
	$C_3H_6O_3$	Dimethyl carbonate	337.25	0.8504	102.52	OX
	C_3H_7Cl	1-Chloropropane	313.35	0.2500	101.59	OX
	$C_4H_4F_6O$	Bis(2,2,2-trifluoroethyl) ether	326.28	0.4450	101.30	OX
	$C_4H_6O_2$	Vinyl acetate	332.05	0.6182	101.33	OX
	C_4H_8O	2-Butanone	323.15	0.8020	58.80	OX
	C_4H_8O	Tetrahydrofuran	332.75	0.5040	101.33	OX
	$C_4H_8O_2$	Ethyl acetate	335.66	0.7120	101.33	OX
	$C_4H_{10}O$	Diethyl ether	305.15	0.0500	93.33	OX
	$C_4H_{10}O_2$	Dimethylacetal	330.35	0.4700	101.33	OX
	$C_5H_3F_9O$	1,1,1,2,3,3-Hexafluoro-3-(2,2,2-trifluoroethoxy)propane	330.67	0.5600	101.30	OX
	C_5H_6	1,3-Cyclopentadiene	309.05	0.2120	101.33	OX
	C_5H_8	2-Methyl-1,3-butadiene	303.55	0.1670	101.33	OX
	C_5H_8	Methylenecyclobutane	309.05	0.2190	101.33	OX
	C_5H_8	1-Methylcyclobutene	304.85	0.1900	101.33	OX
	C_5H_8	cis-1,3-Pentadiene	311.10	0.2300	101.33	OX
	C_5H_8	trans-1,3-Pentadiene	309.65	0.2110	101.33	OX
	C_5H_{10}	2-Methyl-1-butene	300.55	0.1720	101.33	OX
	C_5H_{10}	3-Methyl-1-butene	291.05	0.0890	101.33	OX
	C_5H_{10}	2-Methyl-2-butene	306.25	0.2160	101.33	OX
	C_5H_{10}	1-Pentene	300.05	0.1469	102.47	OX
	$C_5H_{10}O$	2,3-Epoxy-2-methylbutane	334.95	0.6590	101.33	OX
	C_5H_{12}	Isopentane	297.05	0.0930	101.33	OX
	C_5H_{12}	Pentane	303.20	0.1930	101.30	OX
	$C_5H_{12}O$	Butyl methyl ether	330.00	0.5515	100.08	OX
	$C_5H_{12}O$	Methyl tert-butyl ether	325.00	0.3140	103.15	OX
	$C_5H_{12}O$	Ethyl propyl ether	330.00	0.4050	112.25	OX
	$C_5H_{12}O_2$	Diethoxymethane	336.03	0.8127	101.52	OX
	$C_5H_{12}O_2$	2,2-Dimethoxypropane	334.15	0.7250	100.00	OX
	$C_5H_{14}N_2$	N,N,N',N'-Tetramethylmethanediamine	335.15	0.7670	101.33	OX
	C_6F_6	Hexafluorobenzene	318.15	0.6100	61.73	OX
	C_6H_5F	Fluorobenzene	333.35	0.6625	101.62	OX
	C_6H_6	Benzene	331.56	0.6090	101.33	OX
	C_6H_{12}	Cyclohexane	328.75	0.6090	106.66	OX
	C_6H_{12}	2-Methyl-1-pentene	330.00	0.4517	141.80	OX
	C_6H_{14}	2,3-Dimethylbutane	313.15	0.3620	85.50	OX
	C_6H_{14}	Hexane	333.15	0.5160	149.64	OX
	$C_6H_{14}O$	tert-Butyl ethyl ether	330.95	0.6002	101.54	OX
	$C_6H_{14}O$	Diisopropyl ether	330.00	0.5390	101.61	OX
	$C_6H_{14}O$	Butyl ethyl ether	335.00	0.8010	98.84	OX
	$C_6H_{14}O$	2-Methoxy-2-methylbutane	335.55	0.7735	101.69	OX
	C_7H_8	Toluene	336.65	0.8820	101.33	OX
	C_7H_{14}	Methylcyclohexane	333.15	0.7520	102.87	EX
	C_7H_{16}	Heptane	331.95	0.7279	101.33	OX
	$C_7H_{16}O$	2-Ethoxy-2-methylbutane	335.15	0.8736	97.28	OX
	C_8H_{18}	Octane	335.55	0.8830	101.33	LX
	C_9H_{20}	Nonane	337.25	0.9526	101.33	OX
C$_2$Cl$_3$F$_3$		**1,1,2-Trichloro-1,2,2-trifluoroethane**				
	$C_2H_3F_3O$	2,2,2-Trifluoroethanol	316.58	0.7770	101.33	EX

Molecular formula						
Comp. 1	Comp. 2	Name	T_{Az}/K	$y_{1,Az}$	P_{Az}/kPa	Type
	C_2H_6O	Ethanol	317.75	0.8456	101.42	OX
	C_3H_8O	2-Propanol	319.35	0.9159	100.95	OX
	$C_4H_{10}O$	2-Methyl-2-propanol	319.95	0.9426	101.09	OX
C_2Cl_4		**Tetrachloroethene**				
	$C_2H_3Cl_3$	1,1,2-Trichloroethane	385.95	0.2115	101.33	OX
	C_8H_{16}	1-Octene	393.15	0.5900	101.33	OX
	C_8H_{16}	cis-4-Octene	393.65	0.7100	101.33	OX
	C_8H_{16}	trans-4-Octene	393.45	0.6700	101.33	OX
	C_8H_{18}	Octane	371.90	0.8781	53.44	OX
$C_2Cl_4F_2$		**1,1,2,2-Tetrachloro-1,2-difluoroethane**				
	$C_2H_4Cl_2$	1,2-Dichloroethane	353.80	0.2700	101.33	OX
$C_2HBrClF_3$		**2-Bromo-2-chloro-1,1,1-trifluoroethane**				
	$C_4H_{10}O$	Diethyl ether	323.65	0.7200	93.33	ON
C_2HCl_3		**Trichloroethene**				
	$C_2H_4Cl_2$	1,2-Dichloroethane	355.35	0.3324	101.36	OX
	C_2H_6O	Ethanol	343.85	0.4741	101.33	OX
	C_4H_6O	2-Butenal	360.15	0.9000	97.86	OX
	C_6H_{12}	Cyclohexane	353.40	0.0975	101.32	OX
$C_2H_2Cl_2$		**trans-1,2-Dichloroethene**				
	$C_5H_3F_9O$	1,1,1,2,3,3-Hexafluoro-3-(2,2,2-trifluoroethoxy)propane	318.50	0.8390	101.30	OX
C_2H_3N		**Acetonitrile**				
	C_3H_8O	2-Propanol	348.15	0.5287	100.81	OX
	$C_4H_6O_2$	Vinyl acetate	344.65	0.1948	98.33	OX
	C_4H_8O	2-Butanone	352.15	0.3195	101.15	OX
	C_4H_8O	Tetrahydrofuran	338.95	0.0784	101.13	OX
	$C_4H_{10}O$	2-Methyl-2-propanol	333.15	0.6200	56.93	OX
	C_5H_8	2-Methyl-1,3-butadiene	306.75	0.0410	101.33	OX
	C_5H_8	Methylenecyclobutane	312.45	0.1450	101.33	OX
	$C_5H_8O_2$	Methyl methacrylate	355.25	0.9866	102.07	OX
	C_5H_{10}	2-Methyl-2-butene	308.95	0.1320	101.33	OX
	C_5H_{10}	1-Pentene	301.85	0.0830	101.33	OX
	C_5H_{12}	Isopentane	298.45	0.1040	101.33	EX
	C_6H_6	Benzene	328.15	0.4560	54.65	OX
	$C_6H_{14}O$	2-Methoxy-2-methylbutane	346.13	0.5835	100.56	OX
	$C_7H_{16}O$	2-Ethoxy-2-methylbutane	348.85	0.7219	98.99	OX
	$C_{10}H_{20}$	1-Decene	354.55	0.9924	100.51	OX
$C_2H_4Cl_2$		**1,1-Dichloroethane**				
	C_3H_8O	2-Propanol	329.55	0.8928	101.60	OX
	C_6H_{14}	Hexane	329.30	0.8025	101.21	OX
$C_2H_4Cl_2$		**1,2-Dichloroethane**				
	C_3H_8O	2-Propanol	347.25	0.5258	100.32	OX
	$C_4H_{10}O$	2-Methyl-1-propanol	356.05	0.9173	101.26	OX
	$C_4H_{10}O$	2-Methyl-2-propanol	349.45	0.5336	101.43	OX
	C_7H_{14}	Methylcyclohexane	354.65	0.8036	101.21	OX
	C_8H_{18}	2,2,4-Trimethylpentane	343.15	0.7600	73.13	OX
C_2H_4O		**Acetaldehyde**				
	C_4H_6	1,3-Butadiene	268.15	0.0520	101.33	OX
	C_5H_8	2-Methyl-1,3-butadiene	292.23	0.8140	101.33	OX
$C_2H_4O_2$		**Acetic acid**				
	C_5H_5N	Pyridine	411.25	0.5780	101.33	ON
	$C_5H_{12}O$	3-Methyl-2-butanol	392.65	0.7210	101.33	ON
	C_6H_7N	2-Methylpyridine	417.27	0.5120	101.33	ON
	$C_6H_{10}O_2$	Vinyl butanoate	386.45	0.5750	101.33	OX
	C_6H_{14}	Hexane	341.40	0.0839	101.33	OX
	C_7H_9N	2,4-Dimethylpyridine	435.45	0.3022	101.33	ON
	C_7H_{16}	Heptane	364.95	0.4490	101.33	OX
	C_8H_{10}	o-Xylene	389.75	0.8640	101.33	OX

Molecular formula						
Comp. 1	Comp. 2	Name	T_{Az}/K	$y_{1,Az}$	P_{Az}/kPa	Type
	C_8H_{10}	p-Xylene	388.40	0.8200	101.33	OX
	C_8H_{18}	Octane	378.85	0.6870	101.33	OX
	C_9H_{20}	Nonane	386.05	0.8250	101.33	OX
	$C_{10}H_{22}$	Decane	390.05	0.9250	101.33	OX
	$C_{11}H_{24}$	Undecane	391.15	0.9720	101.33	OX
$C_2H_4O_2$		**Methyl formate**				
	C_2H_5Br	Bromoethane	303.05	0.7360	101.33	OX
	$C_4H_{10}O$	Diethyl ether	301.55	0.6030	101.33	OX
	C_5H_8	2-Methyl-1,3-butadiene	298.90	0.5150	101.33	OX
	C_5H_{10}	2-Methyl-2-butene	297.75	0.5760	101.33	OX
	C_5H_{12}	Isopentane	291.55	0.4920	101.33	OX
	C_5H_{12}	Pentane	294.85	0.5740	101.33	OX
	C_6H_{14}	Hexane	302.65	0.8490	101.33	OX
C_2H_5Br		**Bromoethane**				
	C_5H_{10}	2-Methyl-2-butene	308.55	0.5110	101.33	OX
	C_5H_{12}	Isopentane	300.55	0.2180	101.33	OX
$C_2H_5NO_2$		**Nitroethane**				
	$C_4H_{10}O$	2-Methyl-1-propanol	375.81	0.4080	101.33	OX
	C_7H_{16}	Heptane	362.95	0.3520	101.33	OX
C_2H_6O		**Ethanol**				
	C_3H_3N	Acrylonitrile	343.95	0.4440	101.33	OX
	$C_3H_6O_2$	Methyl acetate	329.79	0.0362	101.33	OX
	$C_4H_3F_7O$	1,1,2,2-Tetrafluoroethyl 1,1,1-trifluoroethyl ether	326.67	0.2000	101.30	OX
	$C_4H_4F_6O$	Bis(2,2,2-trifluoroethyl) ether	331.90	0.2840	101.30	OX
	C_4H_8O	Butanal	345.45	0.3690	101.33	OX
	C_4H_8O	2-Butanone	347.15	0.5080	101.33	OX
	C_4H_8O	Tetrahydrofuran	344.95	0.1290	125.00	OX
	$C_4H_8O_2$	Ethyl acetate	344.85	0.4590	101.33	OX
	$C_4H_8O_2$	1,4-Dioxane	351.33	0.9480	101.33	OX
	$C_4H_8O_2$	Methyl propanoate	346.30	0.5140	103.91	OX
	$C_4H_{11}N$	Butylamine	354.99	0.5900	101.33	ON
	$C_5H_3F_9O$	1,1,1,2,3,3-Hexafluoro-3-(2,2,2-trifluoroethoxy)propane	337.88	0.3980	101.30	OX
	C_5H_8	2-Methyl-1,3-butadiene	305.95	0.1500	101.33	OX
	C_5H_8	Cyclopentene	323.40	0.1440	134.00	OX
	C_5H_{10}	2-Methyl-2-butene	309.79	0.0795	101.33	OX
	C_5H_{10}	Cyclopentane	323.44	0.1800	121.00	OX
	$C_5H_{10}O$	2,3-Epoxy-2-methylbutane	343.45	0.2930	101.33	OX
	$C_5H_{10}O$	3-Methyl-2-butanone	350.85	0.8250	101.33	OX
	$C_5H_{10}O$	2-Pentanone	351.15	0.9779	100.50	OX
	$C_5H_{10}O$	3-Pentanone	351.33	0.9590	101.33	OX
	$C_5H_{10}O_2$	Isopropyl acetate	349.85	0.7010	101.33	OX
	$C_5H_{10}O_2$	Methyl butanoate	346.30	0.8800	83.88	OX
	C_5H_{12}	Isopentane	299.95	0.0540	101.33	OX
	C_5H_{12}	Pentane	307.15	0.0537	101.33	OX
	$C_5H_{12}O$	Methyl tert-butyl ether	327.75	0.0380	101.33	OX
	$C_5H_{12}O_2$	Diethoxymethane	348.30	0.6497	102.35	OX
	C_6H_5F	Fluorobenzene	343.85	0.4752	101.54	OX
	C_6H_6	Benzene	341.25	0.4600	101.33	OX
	C_6H_{12}	Cyclohexane	337.95	0.4540	102.26	OX
	C_6H_{14}	Hexane	331.65	0.3410	101.33	OX
	$C_6H_{14}O$	tert-Butyl ethyl ether	339.95	0.3728	101.72	OX
	$C_6H_{14}O$	2-Methoxy-2-methylbutane	346.81	0.5820	101.32	OX
	C_7H_8	Toluene	349.75	0.8152	101.33	OX
	$C_7H_{16}O$	2-Ethoxy-2-methylbutane	349.35	0.7644	101.54	OX
	C_8H_{18}	Octane	349.85	0.8250	101.33	OX
	C_8H_{18}	2,2,4-Trimethylpentane	344.42	0.6450	101.33	OX

Molecular formula						
Comp. 1	**Comp. 2**	**Name**	T_{Az}/K	$y_{1,Az}$	P_{Az}/kPa	**Type**
	C_9H_{20}	Nonane	351.35	0.9400	101.33	OX
$C_2H_6O_2$		**1,2-Ethanediol**				
	$C_5H_{12}O_3$	Di(ethylene glycol) monomethyl ether	463.95	0.4388	101.33	OX
	$C_6H_{14}O_3$	Di(ethylene glycol) monoethyl ether	467.15	0.6480	101.33	OX
	C_7H_8O	o-Cresol	462.67	0.3797	101.33	OX
	$C_7H_{16}O_3$	Di(ethylene glycol) monoisopropyl ether	466.35	0.6964	101.33	OX
	$C_7H_{16}O_3$	Di(ethylene glycol) monopropyl ether	468.55	0.8448	101.33	OX
	$C_7H_{16}O_3$	Di(propylene glycol) monomethyl ether (unspecified isomer)	457.65	0.3500	101.33	OX
	$C_8H_{11}N$	2,4,6-Trimethylpyridine	443.65	0.1734	101.33	OX
	$C_8H_{18}O_3$	Di(ethylene glycol) monobutyl ether	469.15	0.9102	101.33	OX
	$C_8H_{18}O_3$	Di(ethylene glycol) monoisobutyl ether	467.55	0.8355	101.33	OX
	$C_8H_{18}O_3$	Di(propylene glycol) monoethyl ether (unspecified isomer)	458.65	0.4800	101.33	OX
	$C_9H_{20}O_3$	Di(propylene glycol) monopropyl ether (unspecified isomer)	463.15	0.6590	101.33	OX
	$C_{10}H_{22}O_3$	Di(propylene glycol) monobutyl ether (unspecified isomer)	465.75	0.8130	101.33	OX
C_3H_3N		**Acrylonitrile**				
	C_5H_8	Methylenecyclobutane	313.80	0.1275	101.33	OX
	C_6H_6	Benzene	347.45	0.5575	101.46	OX
	C_6H_{12}	Cyclohexane	337.75	0.4836	101.94	OX
	C_6H_{14}	Hexane	330.90	0.4048	101.05	OX
C_3H_4O		**Acrolein**				
	C_5H_8	2-Methyl-1,3-butadiene	306.45	0.1980	101.33	OX
C_3H_6O		**Propanal**				
	C_5H_8	2-Methyl-1,3-butadiene	306.35	0.1700	101.33	OX
	C_5H_8	Methylenecyclobutane	311.30	0.3600	101.33	OX
C_3H_6O		**Acetone**				
	$C_3H_6O_2$	Methyl acetate	328.85	0.6470	101.33	OX
	C_3H_7Br	1-Bromopropane	328.75	0.9915	99.75	OX
	C_4H_8O	Tetrahydrofuran	328.85	0.9603	100.35	OX
	C_4H_9Cl	2-Chloro-2-methylpropane	322.05	0.1944	102.11	OX
	C_5H_8	2-Methyl-1,3-butadiene	306.95	0.0610	101.33	OX
	C_5H_8	Methylenecyclobutane	311.25	0.2800	101.33	OX
	C_5H_8	1-Methylcyclobutene	307.75	0.2220	101.33	OX
	C_5H_{10}	2-Methyl-1-butene	303.25	0.1400	101.33	OX
	C_5H_{10}	2-Methyl-2-butene	308.75	0.2440	101.33	OX
	C_5H_{12}	Isopentane	298.75	0.1730	101.33	OX
	$C_5H_{12}O$	Methyl tert-butyl ether	324.35	0.4824	102.19	OX
	C_6H_{12}	Cyclohexane	330.05	0.7590	109.32	OX
	C_6H_{12}	1-Hexene	323.35	0.5973	101.40	OX
	C_6H_{12}	2-Methyl-1-pentene	333.40	0.5793	140.60	OX
	C_6H_{14}	Hexane	322.95	0.6480	101.33	OX
	$C_6H_{14}O$	Diisopropyl ether	327.10	0.7424	100.17	OX
	$C_6H_{15}N$	Triethylamine	318.15	0.9800	68.13	OX
	C_7H_{14}	Methylcyclohexane	318.15	0.9500	68.66	OX
C_3H_6O		**Allyl alcohol**				
	$C_5H_{10}O_2$	Ethyl propanoate	367.65	0.5597	99.79	OX
	C_6H_6	Benzene	349.90	0.2203	101.33	OX
	C_6H_{12}	Cyclohexane	333.15	0.2790	63.98	OX
$C_3H_6O_2$		**Methyl acetate**				
	C_3H_7Br	1-Bromopropane	329.60	0.9727	99.56	OX
	C_6H_{10}	Cyclohexene	330.35	0.9121	102.87	OX
	C_6H_{12}	Cyclohexane	328.65	0.8000	101.33	OX
	C_6H_{12}	Methylcyclopentane	325.85	0.6917	99.50	OX
	C_6H_{12}	1-Hexene	323.15	0.6340	92.08	OX
	C_6H_{12}	2-Methyl-1-pentene	325.15	0.5931	100.38	OX

Molecular formula						
Comp. 1	Comp. 2	Name	T_{Az}/K	$y_{1,Az}$	P_{Az}/kPa	Type
	C_6H_{14}	Hexane	326.65	0.6590	106.66	OX
	C_7H_{16}	Heptane	323.15	0.9570	79.48	OX
$C_3H_6O_2$		**Ethyl formate**				
	C_3H_7Br	2-Bromopropane	326.15	0.7090	101.33	OX
	C_6H_{12}	Cyclohexane	323.15	0.8210	91.46	OX
$C_3H_6O_2$		**Propanoic acid**				
	C_5H_5N	Pyridine	421.75	0.6860	101.33	ON
$C_3H_6O_3$		**Dimethyl carbonate**				
	$C_5H_{12}O_2$	Diethoxymethane	358.71	0.4437	100.42	OX
	C_6H_6	Benzene	353.50	0.1366	100.48	OX
	C_6H_{12}	Cyclohexane	346.95	0.3780	101.49	OX
	C_6H_{12}	Methylcyclopentane	342.35	0.2680	103.46	OX
	C_6H_{14}	Hexane	338.15	0.2540	98.46	OX
	$C_6H_{14}O$	Dipropyl ether	356.45	0.5044	100.73	OX
	C_7H_{16}	Heptane	355.15	0.5930	99.67	OX
C_3H_7Br		**1-Bromopropane**				
	C_3H_8O	2-Propanol	339.15	0.7349	99.97	OX
	C_6H_{12}	Cyclohexane	343.35	0.9219	98.84	OX
C_3H_7NO		N,N-**Dimethylformamide**				
	C_7H_{16}	Heptane	370.15	0.0800	101.33	OX
	$C_{10}H_{16}$	1,4-Dimethyl-4-vinylcyclohexene	415.65	0.5880	101.33	OX
	$C_{10}H_{16}$	1-Methyl-3-(1-methylethylidene)cyclohexene	419.05	0.7250	101.33	OX
$C_3H_7NO_2$		**1-Nitropropane**				
	C_7H_{16}	Heptane	369.25	0.1630	101.33	OX
$C_3H_7NO_2$		**2-Nitropropane**				
	C_7H_{16}	Heptane	367.55	0.2920	101.33	OX
C_3H_8O		**1-Propanol**				
	$C_4H_3F_7O$	1,1,2,2-Tetrafluoroethyl 1,1,1-trifluoroethyl ether	329.23	0.0350	101.30	OX
	$C_4H_4F_6O$	Bis(2,2,2-trifluoroethyl) ether	336.22	0.1100	101.30	OX
	$C_4H_6O_2$	2,3-Butanedione	359.30	0.3600	100.67	OX
	$C_4H_8O_2$	1,4-Dioxane	365.30	0.6418	101.30	OX
	$C_5H_{10}O_2$	Propyl acetate	367.88	0.6190	101.33	OX
	$C_5H_{12}O_2$	Diethoxymethane	359.01	0.2320	99.43	OX
	C_6H_6	Benzene	350.20	0.2060	101.33	OX
	C_6H_{12}	Cyclohexane	347.68	0.2490	101.33	OX
	C_6H_{12}	Methylcyclopentane	340.85	0.1729	101.19	OX
	$C_6H_{12}O_2$	4,4-Dimethyl-1,3-dioxane	368.20	0.9597	101.30	OX
	C_6H_{14}	Hexane	348.15	0.1900	137.23	OX
	C_7H_8	Toluene	365.35	0.6770	101.33	OX
	C_7H_{16}	Heptane	357.65	0.4830	101.33	OX
	C_8H_8	Styrene	369.08	0.9884	98.13	OX
	C_8H_{10}	o-Xylene	369.85	0.9886	98.66	OX
	C_8H_{10}	m-Xylene	369.90	0.9531	99.06	OX
	C_8H_{10}	p-Xylene	369.60	0.9531	99.99	OX
	C_8H_{14}	1-Octyne	369.00	0.8600	101.33	OX
	C_8H_{18}	Octane	366.85	0.7483	101.33	OX
	C_8H_{18}	2,2,4-Trimethylpentane	357.89	0.4580	101.30	OX
	C_9H_{20}	Nonane	369.95	0.9225	101.33	OX
C_3H_8O		**2-Propanol**				
	$C_4H_4F_6O$	Bis(2,2,2-trifluoroethyl) ether	334.16	0.2230	101.30	OX
	$C_4H_6O_2$	2,3-Butanedione	350.85	0.6454	100.95	OX
	C_4H_8O	2-Butanone	350.55	0.3830	101.33	OX
	$C_4H_{10}O$	2-Methyl-2-propanol	343.05	0.5551	60.27	ON
	$C_5H_3F_9O$	1,1,1,2,3,3-Hexafluoro-3-(2,2,2-trifluoroethoxy)propane	341.23	0.3420	101.30	OX
	C_5H_8	2-Methyl-1,3-butadiene	307.05	0.0150	101.33	OX

Molecular formula

Comp. 1	Comp. 2	Name	T_{Az}/K	$y_{1,Az}$	P_{Az}/kPa	Type
	C_5H_{10}	2-Methyl-2-butene	310.95	0.0460	101.33	OX
	$C_5H_{10}O$	2,3-Epoxy-2-methylbutane	346.10	0.1400	101.33	OX
	$C_5H_{10}O$	3-Methyl-2-butanone	354.75	0.8500	101.33	OX
	C_5H_{12}	Isopentane	298.15	0.1370	101.33	OX
	$C_5H_{12}O_2$	Diethoxymethane	351.45	0.6107	98.61	OX
	C_6H_5F	Fluorobenzene	347.75	0.4666	101.25	OX
	C_6H_6	Benzene	345.03	0.3960	101.33	OX
	C_6H_{10}	Cyclohexene	344.65	0.4271	101.40	OX
	C_6H_{12}	Cyclohexane	342.75	0.4050	101.33	OX
	C_6H_{12}	Methylcyclopentane	336.45	0.2900	98.14	OX
	C_6H_{14}	Hexane	338.15	0.2900	112.66	OX
	$C_6H_{14}O$	Diisopropyl ether	340.00	0.2050	103.36	OX
	$C_6H_{15}N$	Diisopropylamine	352.94	0.4890	101.33	OX
	C_7H_8	Toluene	354.65	0.8370	101.33	OX
	C_7H_{14}	Methylcyclohexane	350.85	0.6530	101.33	OX
	C_7H_{16}	Heptane	349.55	0.6023	101.33	OX
	$C_7H_{16}O$	*tert*-Butyl isopropyl ether	349.95	0.5306	102.70	OX
	C_8H_{18}	Octane	354.63	0.8990	101.33	OX
	C_8H_{18}	2,2,4-Trimethylpentane	349.58	0.6350	101.30	OX
$C_3H_8O_2$		**2-Methoxyethanol**				
	C_8H_8	Styrene	393.95	0.7787	98.93	OX
	C_8H_{10}	*o*-Xylene	392.65	0.7127	98.79	OX
	C_8H_{10}	*m*-Xylene	392.15	0.6397	99.73	OX
	C_8H_{10}	*p*-Xylene	392.65	0.6303	99.99	OX
	C_8H_{16}	1-Octene	380.75	0.4700	101.33	OX
	C_8H_{16}	*cis*-4-Octene	381.25	0.4900	101.33	OX
	C_8H_{16}	*trans*-4-Octene	381.05	0.4900	101.33	OX
$C_3H_8O_2$		**Dimethoxymethane**				
	C_5H_6	1,3-Cyclopentadiene	313.65	0.3350	101.33	OX
	C_5H_8	2-Methyl-1,3-butadiene	306.80	0.0160	101.33	OX
	C_5H_8	Methylenecyclobutane	310.35	0.4630	101.33	OX
	C_5H_8	1-Methylcyclobutene	309.05	0.2900	101.33	OX
$C_3H_8O_2$		**1,2-Propanediol**				
	$C_7H_{16}O_3$	Di(propylene glycol) monomethyl ether (unspecified isomer)	456.85	0.5691	101.33	OX
	$C_8H_{18}O_3$	Di(propylene glycol) monoethyl ether (unspecified isomer)	458.75	0.7778	101.33	OX
	$C_9H_{20}O_3$	Di(propylene glycol) monoisopropyl ether (unspecified isomer)	458.95	0.8130	101.33	OX
	$C_9H_{20}O_3$	Di(propylene glycol) monopropyl ether (unspecified isomer)	458.95	0.9010	101.33	OX
	$C_{10}H_{22}O_3$	Di(propylene glycol) monobutyl ether (unspecified isomer)	459.65	0.9721	101.33	OX
	$C_{10}H_{22}O_3$	Di(propylene glycol) monoisobutyl ether (unspecified isomer)	459.05	0.9255	101.33	OX
$C_3H_8O_2$		**1,3-Propanediol**				
	$C_5H_{12}O_3$	Di(ethylene glycol) monomethyl ether	455.25	0.6300	101.33	OX
	$C_6H_{14}O_3$	Di(ethylene glycol) monoethyl ether	459.25	0.9350	101.33	OX
C_4H_6		**1,3-Butadiene**				
	C_4H_8	2-Butene (unspecified isomer)	267.59	0.7650	101.33	OX
C_4H_6O		**2-Butenal**				
	C_7H_8	Toluene	374.15	0.5950	97.86	OX
	C_8H_{18}	Octane	353.15	0.4950	97.86	OX
$C_4H_6O_2$		**Vinyl acetate**				
	C_6H_{12}	Cyclohexane	340.45	0.6200	101.33	OX
	C_6H_{14}	Hexane	335.25	0.4450	101.33	OX
$C_4H_6O_2$		**2,3-Butanedione**				
	C_7H_8	Toluene	362.70	0.9513	101.34	OX

Molecular formula						
Comp. 1	Comp. 2	Name	T_{Az}/K	$y_{1,Az}$	P_{Az}/kPa	Type
$C_4H_6O_3$		**Acetic anhydride**				
	C_8H_{16}	1-Octene	367.53	0.2840	53.88	OX
	C_8H_{18}	Octane	397.65	0.3500	129.80	OX
C_4H_8O		**Butanal**				
	C_6H_{12}	2-Methyl-1-pentene	334.15	0.2293	101.48	OX
C_4H_8O		**2-Butanone**				
	$C_4H_8O_2$	Ethyl acetate	349.55	0.1700	101.33	OX
	C_6H_6	Benzene	351.53	0.4790	101.33	OX
	C_6H_{10}	Cyclohexene	343.29	0.5110	89.35	OX
	C_6H_{12}	1-Hexene	334.75	0.1760	100.58	OX
	C_6H_{14}	Hexane	337.15	0.3280	101.33	OX
	$C_6H_{14}O$	Diisopropyl ether	340.55	0.1938	101.56	OX
	$C_6H_{14}O$	Dipropyl ether	351.40	0.7785	100.88	OX
	C_7H_{14}	Methylcyclohexane	350.50	0.7984	98.93	OX
	C_7H_{16}	Heptane	350.15	0.7670	101.33	OX
C_4H_8O		**Tetrahydrofuran**				
	C_6H_{12}	2-Methyl-1-pentene	334.65	0.2867	101.29	OX
	C_6H_{14}	Hexane	323.15	0.5900	65.83	OX
$C_4H_8O_2$		**Ethyl acetate**				
	$C_4H_{10}O$	2-Methyl-2-propanol	349.75	0.7778	101.28	OX
	C_6H_6	Benzene	350.55	0.9453	102.45	OX
	C_6H_{10}	Cyclohexene	347.45	0.6183	100.87	OX
	C_6H_{12}	Cyclohexane	345.00	0.5390	102.45	OX
	C_6H_{12}	1-Hexene	333.15	0.1230	91.47	OX
	C_6H_{14}	Hexane	338.00	0.3430	101.32	OX
	C_7H_{14}	Methylcyclohexane	349.90	0.9001	101.83	OX
$C_4H_8O_2$		**Butanoic acid**				
	C_5H_5N	Pyridine	436.35	0.9117	101.33	ON
	$C_8H_{16}O_2$	Butyl butanoate	434.60	0.6532	93.33	OXD
	$C_8H_{16}O_2$	Butyl butanoate	434.78	0.8639	93.33	OND
	$C_{11}H_{24}$	Undecane	435.55	0.9060	101.33	OX
$C_4H_8O_2$		**1,4-Dioxane**				
	$C_4H_{10}O$	2-Butanol	371.75	0.4732	100.77	OX
	$C_5H_{10}O_2$	Propyl acetate	373.35	0.6334	101.13	OX
	$C_5H_{12}O$	2-Methyl-2-butanol	373.75	0.8119	99.62	OX
	C_6H_{10}	Cyclohexene	355.75	0.1065	101.44	OX
	C_6H_{12}	Methylcyclopentane	343.85	0.0538	99.79	OX
	$C_6H_{15}N$	Triethylamine	343.15	0.2500	56.80	OX
	C_7H_{16}	Heptane	364.30	0.4868	101.06	OX
	$C_7H_{16}O$	2-Ethoxy-2-methylbutane	369.15	0.5452	100.27	OX
$C_4H_8O_2$		**Propyl formate**				
	C_6H_6	Benzene	343.15	0.3770	76.08	OX
$C_4H_8O_2$		**Methyl propanoate**				
	C_7H_{14}	Methylcyclohexane	352.45	0.8956	101.33	OX
C_4H_9Cl		**1-Chlorobutane**				
	C_6H_{12}	Cyclohexane	348.31	0.5800	95.85	OX
C_4H_9NO		***N,N*-Dimethylacetamide**				
	C_8H_{10}	o-Xylene	416.95	0.0591	103.40	OX
	C_8H_{10}	Ethylbenzene	408.95	0.0037	101.70	OX
$C_4H_{10}O$		**1-Butanol**				
	C_5H_5N	Pyridine	392.00	0.7050	101.33	ON
	$C_5H_{10}O_3$	Diethyl carbonate	370.85	0.6346	53.20	OX
	C_6H_5Cl	Chlorobenzene	388.25	0.6950	101.33	OX
	C_6H_{12}	Cyclohexane	352.68	0.0787	101.33	OX
	$C_6H_{12}O_2$	Butyl acetate	389.97	0.7700	101.33	OX
	$C_6H_{12}O_2$	Isobutyl acetate	387.15	0.5980	101.33	OX
	C_6H_{14}	Hexane	341.35	0.0370	101.33	OX
	C_7H_8	Toluene	378.85	0.3320	101.33	OX

Molecular formula						
Comp. 1	Comp. 2	Name	T_{Az}/K	$y_{1,Az}$	P_{Az}/kPa	Type
	C_7H_{12}	3-Ethylcyclopentene	367.65	0.1900	101.33	OX
	C_7H_{16}	Heptane	366.55	0.2272	101.38	OX
	C_8H_8	Styrene	388.71	0.8923	98.39	OX
	C_8H_{10}	o-Xylene	388.05	0.8671	100.13	OX
	C_8H_{10}	m-Xylene	387.75	0.7865	101.46	OX
	C_8H_{10}	p-Xylene	387.85	0.7823	99.73	OX
	C_8H_{14}	1-Octyne	386.50	0.6200	101.33	OX
	C_8H_{14}	2-Octyne	398.30	0.7910	101.33	OX
	C_8H_{16}	1-Octene	363.45	0.4530	53.33	OX
	C_8H_{16}	cis-4-Octene	382.35	0.5300	101.33	OX
	C_8H_{16}	trans-4-Octene	382.15	0.5310	101.33	OX
	C_8H_{18}	Octane	383.15	0.5500	102.79	OX
	$C_8H_{18}O$	Dibutyl ether	390.59	0.8754	101.33	OX
	C_9H_{16}	1-Butylcyclopentene	356.70	0.8450	79.99	OX
	C_9H_{16}	1-Nonyne	390.60	0.9400	101.33	OX
	C_9H_{20}	Nonane	389.05	0.8128	101.33	OX
$C_4H_{10}O$		**2-Butanol**				
	$C_5H_{10}O$	3-Pentanone	370.50	0.6075	99.98	OX
	C_6H_{10}	Cyclohexene	352.75	0.2046	101.25	OX
	C_6H_{12}	Cyclohexane	349.90	0.1892	101.02	OX
	C_6H_{14}	Hexane	348.15	0.1010	128.66	OX
	$C_6H_{14}O$	2-Methoxy-2-methylbutane	359.15	0.0991	102.12	OX
	C_7H_8	Toluene	353.44	0.5550	56.67	OX
	C_7H_{16}	Heptane	361.95	0.4116	102.70	OX
	$C_7H_{16}O$	2-Ethoxy-2-methylbutane	367.75	0.4931	102.89	OX
	C_8H_{10}	m-Xylene	369.85	0.9717	101.06	OX
	C_8H_{10}	p-Xylene	369.55	0.9646	101.46	OX
	C_8H_{18}	Octane	371.05	0.8001	101.30	OX
$C_4H_{10}O$		**Diethyl ether**				
	C_5H_{12}	Pentane	306.85	0.5500	101.33	OX
$C_4H_{10}O$		**2-Methyl-1-propanol**				
	$C_5H_{10}O_2$	Isobutyl formate	370.90	0.1930	101.33	OX
	C_6H_6	Benzene	352.45	0.0780	101.33	OX
	C_6H_{10}	Cyclohexene	353.75	0.1363	100.31	OX
	C_6H_{12}	Cyclohexane	351.35	0.1325	101.45	OX
	C_6H_{12}	Methylcyclopentane	343.15	0.0567	100.35	OX
	C_7H_8	Toluene	374.35	0.4941	101.33	OX
	C_8H_{10}	m-Xylene	380.35	0.9300	101.33	OX
	C_8H_{10}	p-Xylene	380.30	0.9200	101.33	OX
	C_8H_{18}	Octane	376.58	0.6700	101.30	OX
$C_4H_{10}O$		**2-Methyl-2-propanol**				
	C_5H_8	Methylenecyclobutane	314.65	0.0150	101.33	OX
	C_6H_{10}	Cyclohexene	346.00	0.4172	99.61	OX
	C_6H_{12}	Methylcyclopentane	339.35	0.2559	99.93	OX
	C_6H_{12}	1-Hexene	333.25	0.2650	101.30	OX
	C_6H_{14}	Hexane	337.70	0.2502	101.30	OX
	$C_6H_{14}O$	tert-Butyl ethyl ether	342.85	0.2512	101.44	OX
	$C_6H_{14}O$	Diisopropyl ether	340.45	0.1058	101.72	OX
	$C_6H_{14}O$	2-Methoxy-2-methylbutane	353.20	0.5617	101.80	OX
	C_7H_8	Toluene	353.44	0.9200	93.61	OX
	$C_7H_{16}O$	tert-Butyl isopropyl ether	350.90	0.5390	102.94	OX
	C_8H_{18}	Octane	343.15	0.9680	61.18	OX
	C_8H_{18}	2,2,4-Trimethylpentane	339.28	0.6040	59.49	OX
$C_4H_{10}O_2$		**1,4-Butanediol**				
	$C_{15}H_{32}O$	1-Pentadecanol	502.75	0.9980	101.33	OX
$C_4H_{10}O_2$		**1,2-Dimethoxyethane**				
	C_7H_{14}	Methylcyclohexane	350.00	0.8190	79.42	OX
$C_4H_{10}O_2$		**2-Ethoxyethanol**				

	Molecular formula						
Comp. 1	Comp. 2		Name	T_{Az}/K	$y_{1,Az}$	P_{Az}/kPa	Type
	C_8H_8		Styrene	405.75	0.6438	101.33	OX
	C_8H_{10}		o-Xylene	404.95	0.5965	101.36	OX
	C_8H_{10}		m-Xylene	401.75	0.5159	101.33	OX
	C_8H_{10}		p-Xylene	402.55	0.5042	102.19	OX
	C_8H_{10}		Ethylbenzene	401.05	0.4632	100.94	OX
$C_4H_{10}O_3$			**Di(ethylene glycol)**				
	$C_9H_{20}O$		1-Nonanol	486.65	0.0095	101.33	OX
$C_4H_{11}N$			**Butylamine**				
	C_6H_6		Benzene	343.15	0.7000	80.89	OX
C_5H_5N			**Pyridine**				
	C_7H_8		Toluene	383.19	0.2250	101.33	OX
	C_7H_{16}		Heptane	368.61	0.3002	101.33	OX
	C_9H_{20}		Nonane	388.15	0.9350	101.33	OX
C_5H_6			**2-Methyl-1-buten-3-yne**				
	C_5H_8		2-Methyl-1,3-butadiene	305.88	0.7210	101.33	OX
	C_5H_{10}		2-Methyl-1-butene	303.15	0.3450	101.33	OX
	C_5H_{12}		Isopentane	299.35	0.3620	101.33	OX
C_5H_6			**1,3-Cyclopentadiene**				
	C_5H_{10}		2-Methyl-2-butene	310.85	0.3000	101.33	OX
	C_5H_{12}		Pentane	307.75	0.1959	101.30	OX
C_5H_8			**2-Methyl-1,3-butadiene**				
	C_5H_{12}		Pentane	310.55	0.7421	114.66	OX
	$C_6F_{15}N$		Tris(perfluoroethyl)amine	303.35	0.8200	101.33	EX
C_5H_8			**3-Methyl-1-butyne**				
	C_5H_{12}		Isopentane	297.15	0.5650	101.33	OX
C_5H_8			**1-Pentyne**				
	C_5H_{10}		2-Methyl-2-butene	310.95	0.3300	101.33	OX
	C_5H_{12}		Pentane	307.55	0.3050	101.33	OX
C_5H_8O			**Cyclopentanone**				
	$C_5H_{12}O$		3-Methyl-1-butanol	402.02	0.5944	101.33	OX
	$C_5H_{12}O$		1-Pentanol	403.84	0.9196	101.33	OX
$C_5H_8O_2$			**Methyl methacrylate**				
	C_7H_{16}		Heptane	366.35	0.4597	99.94	OX
	C_8H_{18}		Octane	373.70	0.9651	100.16	OX
C_5H_{10}			**2-Methyl-1-butene**				
	$C_6F_{15}N$		Tris(perfluoroethyl)amine	301.95	0.8450	101.33	OX
C_5H_{10}			**2-Methyl-2-butene**				
	$C_6F_{15}N$		Tris(perfluoroethyl)amine	307.65	0.8170	101.33	OX
$C_5H_{10}O$			**3-Methyl-3-buten-1-ol**				
	C_6H_{12}		Cyclohexane	352.65	0.0215	101.10	OX
	$C_6H_{12}O_2$		4,4-Dimethyl-1,3-dioxane	403.05	0.7590	102.26	OX
	C_7H_8		Toluene	381.55	0.2391	101.60	OX
	C_7H_{16}		Heptane	370.00	0.2100	101.30	OX
$C_5H_{10}O$			**2-Methyl-3-buten-2-ol**				
	C_6H_{12}		Cyclohexane	350.15	0.1904	101.20	OX
	C_6H_{12}		1-Hexene	336.55	0.0479	101.30	OX
	C_7H_8		Toluene	366.55	0.7788	101.20	OX
$C_5H_{10}O$			**3-Pentanone**				
	C_7H_{14}		Methylcyclohexane	366.95	0.4441	99.82	OX
	$C_7H_{16}O$		2-Ethoxy-2-methylbutane	371.15	0.4764	100.21	OX
$C_5H_{10}O_2$			**Propyl acetate**				
	C_6H_{12}		Cyclohexane	353.15	0.0598	100.43	OX
	C_7H_{14}		Methylcyclohexane	368.40	0.4746	100.90	OX
	C_7H_{16}		Heptane	366.75	0.4215	101.38	OX
	$C_7H_{16}O$		2-Ethoxy-2-methylbutane	370.95	0.6529	100.03	OX
C_5H_{12}			**Isopentane**				
	$C_6F_{15}N$		Tris(perfluoroethyl)amine	299.65	0.9020	101.33	OX
$C_5H_{12}O$			**2-Methyl-1-butanol**				

Molecular formula						
Comp. 1	Comp. 2	Name	T_{Az}/K	$y_{1,Az}$	P_{Az}/kPa	Type
	C_8H_{10}	o-Xylene	402.05	0.7417	101.87	OX
	C_8H_{10}	m-Xylene	400.65	0.6316	101.85	OX
	C_8H_{10}	p-Xylene	400.15	0.6273	101.07	OX
	C_8H_{10}	Ethylbenzene	398.75	0.5657	99.46	OX
$C_5H_{12}O$		**3-Methyl-1-butanol**				
	$C_6H_{10}O$	Cyclohexanone	404.87	0.9094	101.33	OX
	C_7H_8	Toluene	383.15	0.1250	101.33	OX
	$C_7H_{14}O_2$	Isopentyl acetate	403.95	0.9900	101.33	OX
	C_7H_{16}	Heptane	368.15	0.1016	95.06	OX
$C_5H_{12}O$		**2-Methyl-2-butanol**				
	C_6H_6	Benzene	352.35	0.1500	101.33	OX
	C_6H_{12}	Cyclohexane	351.95	0.1100	101.33	OX
	C_6H_{12}	Methylcyclopentane	344.75	0.0551	101.80	OX
	C_6H_{14}	Hexane	339.06	0.0436	93.55	OX
	C_7H_{14}	Methylcyclohexane	366.60	0.3965	99.87	OX
	C_7H_{16}	Heptane	348.15	0.3140	56.83	OX
	$C_7H_{16}O$	2-Ethoxy-2-methylbutane	369.85	0.3904	100.52	OX
$C_5H_{12}O$		**1-Pentanol**				
	$C_6H_{10}O$	Cyclohexanone	392.37	0.9748	53.32	OX
	$C_7H_{14}O_2$	Isopentyl acetate	407.45	0.6000	101.33	OX
	C_7H_{16}	Heptane	371.45	0.0576	101.33	OX
	C_8H_{18}	Octane	393.15	0.2847	101.33	OX
	C_9H_{20}	Nonane	404.45	0.6242	101.33	OX
	$C_{10}H_{22}$	Decane	410.65	0.9221	101.33	OX
$C_5H_{12}O$		**3-Pentanol**				
	C_7H_{16}	Heptane	368.15	0.2001	98.62	OX
$C_5H_{12}O_2$		**Diethoxymethane**				
	C_6H_{12}	Cyclohexane	353.21	0.1774	101.39	OX
	C_6H_{14}	Hexane	361.27	0.9101	102.30	OX
C_6F_6		**Hexafluorobenzene**				
	C_6H_6	Benzene	353.60	0.7600	101.33	OND
	C_6H_6	Benzene	352.50	0.1832	101.33	OXD
$C_6F_{15}N$		**Tris(perfluoroethyl)amine**				
	C_6H_6	Benzene	329.95	0.5900	101.33	EX
	C_6H_{12}	Cyclohexane	329.35	0.5690	101.33	EX
	C_6H_{14}	Hexane	327.65	0.4840	101.33	OX
C_6H_5Br		**Bromobenzene**				
	$C_6H_{12}O$	Cyclohexanol	403.15	0.7390	52.45	OX
C_6H_6		**Benzene**				
	C_6H_{12}	Cyclohexane	353.15	0.5460	109.18	OX
	C_6H_{12}	Methylcyclopentane	333.15	0.1390	69.93	OX
	C_6H_{14}	Hexane	341.45	0.0500	101.33	OX
	C_7H_{16}	Heptane	353.25	0.9922	101.32	OX
	C_8H_{18}	2,2,4-Trimethylpentane	353.25	0.9751	101.32	OX
C_6H_6O		**Phenol**				
	C_6H_7N	Aniline	459.09	0.3884	101.33	ON
	C_6H_7N	2-Methylpyridine	458.33	0.7852	101.32	ON
	C_6H_7N	3-Methylpyridine	462.93	0.6918	101.32	ON
	C_7H_5N	Benzonitrile	465.11	0.2345	101.33	ON
	C_7H_6O	Benzaldehyde	447.00	0.6001	73.00	ON
	C_7H_9N	2,6-Dimethylpyridine	459.32	0.7539	101.32	ON
	C_8H_{18}	Octane	398.17	0.0690	101.32	OX
	C_9H_{12}	Propylbenzene	428.15	0.1150	91.85	OX
	C_9H_{12}	1,2,3-Trimethylbenzene	443.45	0.3936	101.33	OX
	C_9H_{12}	1,2,4-Trimethylbenzene	440.65	0.2409	101.33	OX
	C_9H_{12}	1,3,5-Trimethylbenzene	436.95	0.1828	101.33	OX
	C_9H_{18}	1-Nonene	413.15	0.1297	86.57	OX
	C_9H_{20}	Nonane	419.18	0.2180	101.32	OX

	Molecular formula						
Comp. 1	**Comp. 2**	**Name**	T_{Az}/K	$y_{1,Az}$	P_{Az}/kPa	**Type**	
	C$_{10}$H$_{12}$	1,2,3,4-Tetrahydronaphthalene	448.15	0.9031	84.25	OX	
	C$_{10}$H$_{14}$	Butylbenzene	447.05	0.5535	101.33	OX	
	C$_{10}$H$_{14}$	sec-Butylbenzene	441.15	0.3129	101.33	OX	
	C$_{10}$H$_{14}$	tert-Butylbenzene	439.95	0.2773	101.33	OX	
	C$_{10}$H$_{14}$	Diethylbenzene (unspecified isomer)	446.45	0.4705	101.33	OX	
	C$_{10}$H$_{14}$	o-Diethylbenzene	447.15	0.5565	101.33	OX	
	C$_{10}$H$_{14}$	m-Diethylbenzene	445.95	0.5152	101.33	OX	
	C$_{10}$H$_{14}$	1-Ethyl-3,5-dimethylbenzene	447.95	0.5840	101.33	OX	
	C$_{10}$H$_{14}$	2-Ethyl-1,4-dimethylbenzene	447.95	0.5840	101.33	OX	
	C$_{10}$H$_{14}$	Isobutylbenzene	441.75	0.3522	101.33	OX	
	C$_{10}$H$_{14}$	1-Isopropyl-2-methylbenzene	443.75	0.4643	101.33	OX	
	C$_{10}$H$_{14}$	1-Isopropyl-3-methylbenzene	443.05	0.4027	101.33	OX	
	C$_{10}$H$_{14}$	1-Isopropyl-4-methylbenzene	443.65	0.4430	101.33	OX	
	C$_{10}$H$_{14}$	1-Methyl-2-propylbenzene	447.75	0.5801	101.33	OX	
	C$_{10}$H$_{14}$	1-Methyl-3-propylbenzene	446.35	0.5264	101.33	OX	
	C$_{10}$H$_{14}$	1-Methyl-4-propylbenzene	447.15	0.5575	101.33	OX	
	C$_{10}$H$_{14}$	1,2,3,5-Tetramethylbenzene	454.25	0.7957	101.33	OX	
	C$_{10}$H$_{14}$	1,2,4,5-Tetramethylbenzene	453.36	0.7857	101.33	OX	
	C$_{10}$H$_{18}$	trans-Decahydronaphthalene	443.15	0.5419	99.85	OX	
	C$_{10}$H$_{22}$	Decane	434.15	0.4450	101.32	OX	
	C$_{11}$H$_{16}$	1-Butyl-2-methylbenzene	455.55	0.8504	101.33	OX	
	C$_{11}$H$_{16}$	1-Butyl-3-methylbenzene	454.15	0.8099	101.33	OX	
	C$_{11}$H$_{16}$	1-Butyl-4-methylbenzene	454.55	0.8230	101.33	OX	
	C$_{11}$H$_{22}$	1-Undecene	443.15	0.6426	92.31	OX	
	C$_{12}$H$_{26}$	Dodecane	450.73	0.7900	101.32	OX	
	C$_{14}$H$_{30}$	Tetradecane	452.48	0.9650	101.32	OX	
C$_6$H$_7$N		**Aniline**					
	C$_7$H$_8$O	o-Cresol	464.29	0.0953	101.33	ON	
	C$_9$H$_{12}$	1,2,3-Trimethylbenzene	444.65	0.3331	101.33	OX	
	C$_9$H$_{12}$	1,2,4-Trimethylbenzene	441.80	0.1850	101.33	OX	
	C$_9$H$_{12}$	1,3,5-Trimethylbenzene	437.68	0.1071	101.33	OX	
	C$_9$H$_{20}$	Nonane	422.35	0.1770	101.33	OX	
	C$_{10}$H$_{14}$	Butylbenzene	448.65	0.4993	101.33	OX	
	C$_{10}$H$_{14}$	sec-Butylbenzene	443.15	0.3021	101.33	OX	
	C$_{10}$H$_{14}$	tert-Butylbenzene	438.25	0.2104	101.33	OX	
	C$_{10}$H$_{14}$	o-Diethylbenzene	448.75	0.5024	101.33	OX	
	C$_{10}$H$_{14}$	m-Diethylbenzene	447.45	0.4584	101.33	OX	
	C$_{10}$H$_{14}$	p-Diethylbenzene	448.85	0.5086	101.33	OX	
	C$_{10}$H$_{14}$	1-Ethyl-3,5-dimethylbenzene	449.65	0.5310	101.33	OX	
	C$_{10}$H$_{14}$	2-Ethyl-1,4-dimethylbenzene	449.65	0.5310	101.33	OX	
	C$_{10}$H$_{14}$	Isobutylbenzene	442.75	0.2890	101.33	OX	
	C$_{10}$H$_{14}$	1-Isopropyl-2-methylbenzene	445.95	0.4052	101.33	OX	
	C$_{10}$H$_{14}$	1-Isopropyl-3-methylbenzene	444.15	0.3419	101.33	OX	
	C$_{10}$H$_{14}$	1-Isopropyl-4-methylbenzene	445.35	0.3829	101.33	OX	
	C$_{10}$H$_{14}$	1-Methyl-2-propylbenzene	449.45	0.5270	101.33	OX	
	C$_{10}$H$_{14}$	1-Methyl-3-propylbenzene	447.85	0.4711	101.33	OX	
	C$_{10}$H$_{14}$	1-Methyl-4-propylbenzene	448.75	0.5035	101.33	OX	
	C$_{10}$H$_{14}$	1,2,3,5-Tetramethylbenzene	456.55	0.7504	101.33	OX	
	C$_{10}$H$_{14}$	1,2,4,5-Tetramethylbenzene	455.36	0.7349	101.33	OX	
	C$_{10}$H$_{22}$	Decane	440.43	0.4660	101.33	OX	
	C$_{11}$H$_{16}$	1-Butyl-2-methylbenzene	458.15	0.8091	101.33	OX	
	C$_{11}$H$_{16}$	1-Butyl-3-methylbenzene	456.55	0.7661	101.33	OX	
	C$_{11}$H$_{16}$	1-Butyl-4-methylbenzene	457.05	0.7807	101.33	OX	
	C$_{11}$H$_{24}$	Undecane	449.05	0.6970	101.33	OX	
	C$_{12}$H$_{26}$	Dodecane	453.52	0.8220	101.33	OX	
	C$_{13}$H$_{28}$	Tridecane	456.22	0.9300	101.33	OX	
	C$_{14}$H$_{30}$	Tetradecane	457.05	0.9770	101.33	OX	
C$_6$H$_7$N		**2-Methylpyridine**					

Molecular formula						
Comp. 1	Comp. 2	Name	T_{Az}/K	$y_{1,Az}$	P_{Az}/kPa	Type
	C_8H_{18}	Octane	394.27	0.4610	101.33	OX
	C_9H_{20}	Nonane	402.35	0.8790	101.33	OX
C_6H_7N		**3-Methylpyridine**				
	C_7H_8O	*m*-Cresol	477.01	0.1556	101.32	ON
	C_7H_9N	2,6-Dimethylpyridine	416.64	0.2940	101.33	OX
C_6H_7N		**4-Methylpyridine**				
	C_7H_8O	*m*-Cresol	477.74	0.1822	101.32	ON
	C_7H_9N	2,6-Dimethylpyridine	417.08	0.2000	101.32	OX
$C_6H_{10}O$		**Methyldihydropyran (unspecified isomer)**				
	C_7H_8	Toluene	381.85	0.0207	101.30	OX
$C_6H_{10}O$		**4-Methylenetetrahydropyran**				
	C_7H_8	Toluene	381.15	0.5253	101.20	OX
$C_6H_{12}O$		**Cyclohexanol**				
	C_8H_{10}	*o*-Xylene	415.95	0.1426	101.33	OX
	C_8H_{10}	*m*-Xylene	411.85	0.0503	101.33	OX
	C_8H_{10}	*p*-Xylene	410.95	0.0505	101.33	OX
	C_9H_{20}	Nonane	410.20	0.3350	79.99	OX
$C_6H_{12}O_2$		**Butyl acetate**				
	C_8H_{16}	1-Octene	393.00	0.3030	101.33	OX
$C_6H_{12}O_2$		**4,4-Dimethyl-1,3-dioxane**				
	C_8H_{10}	*o*-Xylene	404.65	0.9662	101.30	OX
	C_8H_{18}	Octane	393.95	0.3343	101.20	OX
	C_9H_{20}	Nonane	402.15	0.8864	101.30	OX
	$C_{10}H_{22}$	Decane	405.35	0.9999	100.60	OX
$C_6H_{14}O$		**1-Hexanol**				
	C_8H_{18}	Octane	398.55	0.0886	101.33	OX
	C_9H_{20}	Nonane	416.95	0.3649	101.33	OX
	$C_{10}H_{22}$	Decane	427.05	0.7123	101.33	OX
$C_6H_{14}O_2$		**2,2-Dimethoxybutane**				
	C_7H_8	Toluene	380.15	0.9180	101.44	OX
C_7F_{16}		**Perfluoroheptane**				
	C_7H_{16}	Heptane	328.16	0.6100	53.60	OX
C_7H_5N		**Benzonitrile**				
	C_7H_8O	*o*-Cresol	468.91	0.5100	101.33	ON
	C_7H_8O	*m*-Cresol	476.10	0.1441	101.33	ON
	C_7H_8O	*p*-Cresol	476.95	0.0898	101.33	ON
	$C_8H_{10}O$	2,6-Xylenol	477.15	0.0807	101.33	ON
C_7H_8O		***o*-Cresol**				
	$C_8H_{11}N$	2,4,6-Trimethylpyridine	470.35	0.6561	101.33	ON
	$C_{10}H_{14}$	*sec*-Butylbenzene	444.65	0.0938	101.33	OX
	$C_{10}H_{14}$	Diethylbenzene (unspecified isomer)	453.10	0.2694	101.33	OX
	$C_{10}H_{14}$	1,2,4,5-Tetramethylbenzene	462.37	0.6273	101.33	OX
	$C_{10}H_{22}$	Decane	433.15	0.3100	78.71	OX
	$C_{11}H_{22}$	1-Undecene	448.15	0.5516	83.07	OX
	$C_{11}H_{24}$	Undecane	433.15	0.5800	56.40	OX
	$C_{12}H_{26}$	Dodecane	458.15	0.8466	93.55	OX
C_7H_8O		***m*-Cresol**				
	C_7H_9N	2,6-Dimethylpyridine	475.66	0.9869	101.32	ON
	C_9H_7N	Quinoline	511.20	0.0356	101.33	ON
	C_9H_{20}	Nonane	413.15	0.0400	76.54	OX
	$C_{10}H_8$	Naphthalene	474.65	0.9680	101.33	OX
	$C_{10}H_{12}$	1,2,3,4-Tetrahydronaphthalene	468.45	0.5900	93.10	OX
	$C_{10}H_{14}$	*sec*-Butylbenzene	445.85	0.0136	101.33	OX
	$C_{10}H_{14}$	Diethylbenzene (unspecified isomer)	454.10	0.1010	101.33	OX
	$C_{10}H_{14}$	1,2,4,5-Tetramethylbenzene	466.87	0.3591	101.33	OX
	$C_{10}H_{22}$	Decane	433.15	0.2170	75.85	OX
C_7H_8O		***p*-Cresol**				
	$C_{10}H_8$	Naphthalene	474.55	0.9414	101.33	OX

Molecular formula						
Comp. 1	Comp. 2	Name	T_{Az}/K	$y_{1,Az}$	P_{Az}/kPa	Type
	$C_{10}H_{14}$	sec-Butylbenzene	446.05	0.0186	101.33	OX
	$C_{10}H_{14}$	Diethylbenzene (unspecified isomer)	454.50	0.1105	101.33	OX
C_7H_8O		**Benzyl alcohol**				
	$C_{10}H_{22}$	Decane	445.75	0.2490	101.33	OX
C_7H_9N		**2-Methylaniline**				
	$C_{10}H_{22}$	Decane	446.91	0.1770	101.33	OX
	$C_{11}H_{24}$	Undecane	461.40	0.4930	101.33	OX
	$C_{12}H_{26}$	Dodecane	468.90	0.7650	101.33	OX
	$C_{13}H_{28}$	Tridecane	472.55	0.9070	101.33	OX
$C_7H_{14}O_2$		**Pentyl acetate**				
	C_9H_{20}	Nonane	419.20	0.5380	101.32	OX
$C_7H_{16}O$		**1-Heptanol**				
	C_9H_{20}	Nonane	423.45	0.1071	101.33	OX
	$C_{10}H_{22}$	Decane	438.75	0.4308	101.33	OX
	$C_{11}H_{24}$	Undecane	447.85	0.8014	101.33	OX
C_8H_{10}		**o-Xylene**				
	C_9H_{20}	Nonane	417.40	0.8498	101.33	OX
$C_8H_{10}O$		**2,6-Xylenol**				
	C_9H_7N	Quinoline	511.00	0.0890	101.33	ON
	$C_{10}H_8$	Naphthalene	475.70	0.9381	101.33	OX
	$C_{10}H_{14}$	1,2,4,5-Tetramethylbenzene	468.85	0.3480	101.33	OX
$C_8H_{10}O$		**2,3-Xylenol**				
	C_9H_7N	Quinoline	513.30	0.2684	101.33	ON
	$C_{10}H_8$	Naphthalene	485.45	0.4123	101.33	OX
$C_8H_{10}O$		**2,4-Xylenol**				
	C_9H_7N	Quinoline	512.30	0.1717	101.33	ON
	$C_{10}H_8$	Naphthalene	481.25	0.6435	101.33	OX
	$C_{10}H_{14}$	1,2,4,5-Tetramethylbenzene	474.05	0.1869	101.33	OX
$C_8H_{10}O$		**2,5-Xylenol**				
	C_9H_7N	Quinoline	512.30	0.1717	101.33	ON
	$C_{10}H_8$	Naphthalene	481.25	0.6435	101.33	OX
	$C_{10}H_{14}$	1,2,4,5-Tetramethylbenzene	474.35	0.1763	101.33	OX
$C_8H_{10}O$		**3,4-Xylenol**				
	C_9H_7N	Isoquinoline	519.75	0.2955	101.33	ON
	C_9H_7N	Quinoline	514.77	0.3907	101.33	ON
	$C_{10}H_8$	Naphthalene	490.95	0.1158	101.33	OX
	$C_{10}H_9N$	3-Methylisoquinoline	524.35	0.0811	101.33	ON
	$C_{10}H_9N$	2-Methylquinoline	521.17	0.1647	101.33	ON
	$C_{10}H_9N$	3-Methylquinoline	523.60	0.1152	101.33	ON
	$C_{10}H_9N$	7-Methylquinoline	525.85	0.0466	101.33	ON
	$C_{11}H_{11}N$	2,3-Dimethylquinoline	521.60	0.2113	101.33	ON
$C_8H_{10}O$		**3,5-Xylenol**				
	C_9H_7N	Isoquinoline	518.05	0.1915	101.33	ON
	C_9H_7N	Quinoline	513.58	0.3287	101.33	ON
	$C_{10}H_8$	Naphthalene	489.33	0.2601	101.33	OX
	$C_{10}H_9N$	2-Methylquinoline	520.65	0.0094	101.33	ON
	$C_{11}H_{11}N$	2,3-Dimethylquinoline	520.70	0.0530	101.33	ON
$C_8H_{10}O$		**2-Ethylphenol**				
	C_9H_7N	Quinoline	511.75	0.1041	101.33	ON
	$C_{10}H_8$	Naphthalene	478.35	0.8005	101.33	OX
	$C_{10}H_{14}$	1,2,4,5-Tetramethylbenzene	471.45	0.3707	101.33	OX
$C_8H_{10}O$		**3-Ethylphenol**				
	C_9H_7N	Quinoline	512.70	0.2089	101.33	ON
	$C_{10}H_8$	Naphthalene	483.45	0.5551	101.33	OX
	$C_{10}H_{14}$	1,2,4,5-Tetramethylbenzene	475.95	0.1249	101.33	OX
$C_8H_{10}O$		**4-Ethylphenol**				
	C_9H_7N	Quinoline	513.45	0.2832	101.33	ON
	$C_{10}H_8$	Naphthalene	486.10	0.3762	101.33	OX

Molecular formula						
Comp. 1	Comp. 2	Name	T_{Az}/K	$y_{1,Az}$	P_{Az}/kPa	Type
$C_8H_{11}N$		**2,4-Dimethylaniline**				
	$C_{11}H_{24}$	Undecane	468.13	0.1490	101.33	OX
	$C_{12}H_{26}$	Dodecane	482.95	0.4520	101.33	OX
	$C_{13}H_{28}$	Tridecane	488.43	0.7880	101.33	OX
	$C_{14}H_{30}$	Tetradecane	490.53	0.9840	101.33	OX
$C_8H_{18}O$		**1-Octanol**				
	$C_{10}H_{22}$	Decane	446.45	0.1029	101.33	OX
	$C_{11}H_{24}$	Undecane	460.05	0.4772	101.33	OX
	$C_{12}H_{26}$	Dodecane	466.95	0.8836	101.33	OX
C_9H_7N		**Isoquinoline**				
	$C_{11}H_{10}$	2-Methylnaphthalene	513.90	0.2074	101.33	OX
C_9H_7N		**Quinoline**				
	$C_9H_{12}O$	3-Isopropylphenol	514.70	0.6109	101.33	ON
	$C_9H_{12}O$	2-Isopropylphenol	512.75	0.8015	101.33	ON
	$C_9H_{12}O$	2-Propylphenol	513.60	0.7243	101.33	ON
	$C_9H_{12}O$	3-Propylphenol	514.70	0.6109	101.33	ON
	$C_9H_{12}O$	4-Propylphenol	515.35	0.5451	101.33	ON
	$C_{10}H_{14}O$	2-Butylphenol	515.70	0.5350	101.33	ON
	$C_{10}H_{14}O$	2-tert-Butylphenol	513.70	0.7299	101.33	ON
	$C_{10}H_{14}O$	3-tert-Butylphenol	517.05	0.4315	101.33	ON
	$C_{10}H_{14}O$	4-Isobutylphenol	515.95	0.5061	101.33	ON
	$C_{10}H_{14}O$	2-sec-Butylphenol	514.70	0.6339	101.33	ON
	$C_{10}H_{14}O$	4-sec-Butylphenol	516.45	0.4551	101.33	ON
	$C_{11}H_{10}$	2-Methylnaphthalene	511.05	0.9213	101.33	OX
	$C_{11}H_{16}O$	2-tert-Butyl-5-methylphenol	515.45	0.5854	101.33	ON
	$C_{11}H_{16}O$	2-sec-Butyl-4-methylphenol	516.10	0.5139	101.33	ON
C_9H_{12}		**1,2,3-Trimethylbenzene**				
	$C_{10}H_{22}$	Decane	433.35	0.4010	72.54	OX
C_9H_{12}		**1,2,4-Trimethylbenzene**				
	$C_{10}H_{22}$	Decane	433.35	0.8600	80.25	OX
$C_9H_{12}O$		**2-Ethyl-4-methylphenol**				
	$C_{10}H_8$	Naphthalene	488.20	0.2218	101.33	OX
$C_9H_{12}O$		**2-Ethyl-5-methylphenol**				
	$C_{10}H_8$	Naphthalene	489.45	0.1710	101.33	OX
$C_9H_{12}O$		**2-Isopropylphenol**				
	$C_{10}H_8$	Naphthalene	483.15	0.5102	101.33	OX
	$C_{10}H_{14}$	1,2,4,5-Tetramethylbenzene	476.25	0.1036	101.33	OX
$C_9H_{12}O$		**2,4,6-Trimethylphenol**				
	$C_{10}H_8$	Naphthalene	486.70	0.3161	101.33	OX
$C_9H_{20}O$		**1-Nonanol**				
	$C_{11}H_{24}$	Undecane	468.45	0.0925	101.33	OX
	$C_{12}H_{26}$	Dodecane	480.65	0.5235	101.33	OX
$C_{10}H_{22}O$		**1-Decanol**				
	$C_{12}H_{26}$	Dodecane	489.25	0.1068	101.33	OX

VISCOSITY OF GASES

The following table gives the viscosity of some common gases as a function of temperature. Unless otherwise noted, the viscosity values refer to a pressure of 100 kPa (1 bar). The notation $P=0$ indicates the low pressure limiting value is given. The difference between the viscosity at 100 kPa and the limiting value is generally less than 1%. Viscosity is given in units of µPa s; note that 1 µPa s = 10^{-5} poise. Substances are listed in the modified Hill order (see Introduction).

		Viscosity in micropascal seconds (µPa s)						
		100 K	200 K	300 K	400 K	500 K	600 K	Ref.
	Air	7.1	13.3	18.6	23.1	27.1	30.8	1
Ar	Argon	8.0	15.9	22.9	28.8	34.2	39.0	2,8
BF_3	Boron trifluoride		12.3	17.1	21.7	26.1	30.2	13
ClH	Hydrogen chloride			14.6	19.7	24.3		13
F_6S	Sulfur hexafluoride ($P=0$)			15.3	19.8	23.9	27.7	10
H_2	Hydrogen ($P=0$)	4.2	6.8	9.0	10.9	12.7	14.4	4
D_2	Deuterium ($P=0$)	5.9	9.6	12.6	15.4	17.9	20.3	11
H_2O	Water			10.0	13.3	17.3	21.4	6
D_2O	Deuterium oxide			11.1	13.7	17.7	22.0	7
He	Helium ($P=0$)	9.7	15.3	20.0	24.4	28.4	32.3	8
Kr	Krypton ($P=0$)	8.8	17.1	25.6	33.1	39.8	45.9	8
NO	Nitric oxide		13.8	19.2	23.8	28.0	31.9	13
N_2	Nitrogen ($P=0$)		12.9	17.9	22.2	26.1	29.6	12
N_2O	Nitrous oxide		10.0	15.0	19.4	23.6	27.4	13
Ne	Neon ($P=0$)	14.4	24.3	32.1	38.9	45.0	50.8	8
O_2	Oxygen ($P=0$)	7.5	14.6	20.8	26.1	30.8	35.1	12
O_2S	Sulfur dioxide		8.6	12.9	17.5	21.7		13
Xe	Xenon ($P=0$)	8.3	15.4	23.2	30.7	37.6	44.0	8
CO	Carbon monoxide	6.7	12.9	17.8	22.1	25.8	29.1	13
CO_2	Carbon dioxide		10.0	15.0	19.7	24.0	28.0	9,10
$CHCl_3$	Chloroform			10.2	13.7	16.9	20.1	13
CH_4	Methane		7.7	11.2	14.3	17.0	19.4	10
CH_4O	Methanol			13.2	16.5	19.6		13
C_2H_2	Acetylene			10.4	13.5	16.5		13
C_2H_4	Ethylene		7.0	10.4	13.6	16.5	19.1	3
C_2H_6	Ethane		6.4	9.5	12.3	14.9	17.3	5
C_2H_6O	Ethanol			11.6	14.5	17.0		13
C_3H_8	Propane			8.3	10.9	13.4	15.8	5
C_4H_{10}	Butane			7.5	10.0	12.3	14.6	5
C_4H_{10}	Isobutane			7.6	10.0	12.3	14.6	5
$C_4H_{10}O$	Diethyl ether			7.6	10.1	12.4		13
C_5H_{12}	Pentane			6.7	9.2	11.4	13.4	13
C_6H_{14}	Hexane				8.6	10.8	12.8	13

References

1. K. Kadoya, N. Matsunaga, and A. Nagashima, Viscosity and thermal conductivity of dry air in the gaseous phase, *J. Phys. Chem. Ref. Data*, 14, 947, 1985.

2. B. A. Younglove and H. J. M. Hanley, The viscosity and thermal conductivity coefficients of gaseous and liquid argon, *J. Phys. Chem. Ref. Data*, 15, 1323, 1986.

3. P. M. Holland, B. E. Eaton, and H. J. M. Hanley, A correlation of the viscosity and thermal conductivity data of gaseous and liquid ethylene, *J. Phys. Chem. Ref. Data*, 12, 917, 1983.

4. M. J. Assael, S. Mixafendi, and W. A. Wakeham, The viscosity and thermal conductivity of normal hydrogen in the limit zero density, *J. Phys. Chem. Ref. Data*, 15, 1315, 1986.

5. B. A. Younglove and J. F. Ely, Thermophysical properties of fluids. II. Methane, ethane, propane, isobutane, and normal butane, *J. Phys. Chem. Ref. Data*, 16, 577, 1987.

6. J. V. Sengers and J. T. R. Watson, Improved international formulations for the viscosity and thermal conductivity of water substance, *J. Phys. Chem. Ref. Data*, 15, 1291, 1986.

7. N. Matsunaga and A. Nagashima, Transport properties of liquid and gaseous D_2O over a wide range of temperature and pressure, *J. Phys. Chem. Ref. Data*, 12, 933, 1983.

8. J. Kestin, et al., Equilibrium and transport properties of the noble gases and their mixtures at low density, *J. Phys. Chem. Ref. Data*, 13, 299, 1984.

9. V. Vescovic, et al., The transport properties of carbon dioxide, *J. Phys. Chem. Ref. Data*, 19, 1990.

10. R. D. Trengove and W. A. Wakeham, The viscosity of carbon dioxide, methane, and sulfur hexafluoride in the limit of zero density, *J. Phys. Chem. Ref. Data*, 16, 175, 1987.

11. M. J. Assael, S. Mixafendi, and W. A. Wakeham, The viscosity of normal deuterium in the limit of zero density, *J. Phys. Chem. Ref. Data*, 16, 189, 1987.

12. W. A. Cole and W. A. Wakeham, The viscosity of nitrogen, oxygen, and their binary mixtures in the limit of zero density, *J. Phys. Chem. Ref. Data*, 14, 209, 1985.

13. C. Y. Ho, Ed., *Properties of Inorganic and Organic Fluids, CINDAS Data Series on Materials Properties*, Vol. V-1, Hemisphere Publishing Corp., New York, 1988.

VISCOSITY OF LIQUIDS

The absolute viscosity of some common liquids at temperatures between −25 and 100°C is given in this table. Values were derived by fitting experimental data to suitable expressions for the temperature dependence. The substances are arranged by molecular formula in the modified Hill order (see Preface). All values are given in units of millipascal seconds (mPa s); this unit is identical to centipoise (cp).

Viscosity values correspond to a nominal pressure of 1 atmosphere. If a value is given at a temperature above the normal boiling point, the applicable pressure is understood to be the vapor pressure of the liquid at that temperature. A few values are given at a temperature slightly below the normal freezing point; these refer to the supercooled liquid.

The accuracy ranges from 1% in the best cases to 5 to 10% in the worst cases. Additional significant figures are included in the table to facilitate interpolation.

References

1. Viswanath, D. S. and Natarajan, G., *Data Book on the Viscosity of Liquids*, Hemisphere Publishing Corp., New York, 1989.
2. Daubert, T. E., Danner, R. P., Sibul, H. M., and Stebbins, C. C., *Physical and Thermodynamic Properties of Pure Compounds: Data Compilation*, extant 1994 (core with 4 supplements), Taylor & Francis, Bristol, PA (also available as database).
3. Ho, C. Y., Ed., *CINDAS Data Series on Material Properties*, Vol. V-1, *Properties of Inorganic and Organic Fluids*, Hemisphere Publishing Corp., New York, 1988.
4. Stephan, K. and Lucas, K., *Viscosity of Dense Fluids*, Plenum Press, New York, 1979.
5. Vargaftik, N. B., *Tables of Thermophysical Properties of Liquids and Gases*, 2nd ed., John Wiley, New York, 1975.

Molecular formula	Name	Viscosity in mPa s					
		−25°C	0°C	25°C	50°C	75°C	100°C
Compounds not containing carbon							
Br_2	Bromine		1.252	0.944	0.746		
Cl_3HSi	Trichlorosilane		0.415	0.326			
Cl_3P	Phosphorous trichloride	0.870	0.662	0.529	0.439		
Cl_4Si	Tetrachlorosilane			99.4	96.2		
H_2O	Water		1.793	0.890	0.547	0.378	0.282
H_4N_2	Hydrazine			0.876	0.628	0.480	0.384
Hg	Mercury			1.526	1.402	1.312	1.245
NO_2	Nitrogen dioxide		0.532	0.402			
Compounds containing carbon							
CCl_3F	Trichlorofluoromethane	0.740	0.539	0.421			
CCl_4	Tetrachloromethane		1.321	0.908	0.656	0.494	
CS_2	Carbon disulfide		0.429	0.352			
$CHBr_3$	Tribromomethane			1.857	1.367	1.029	
$CHCl_3$	Trichloromethane	0.988	0.706	0.537	0.427		
CHN	Hydrogen cyanide		0.235	0.183			
CH_2Br_2	Dibromomethane	1.948	1.320	0.980	0.779	0.652	
CH_2Cl_2	Dichloromethane	0.727	0.533	0.413			
CH_2O_2	Formic acid			1.607	1.030	0.724	0.545
CH_3I	Iodomethane		0.594	0.469			
CH_3NO	Formamide		7.114	3.343	1.833		
CH_3NO_2	Nitromethane	1.311	0.875	0.630	0.481	0.383	0.317
CH_4O	Methanol	1.258	0.793	0.544			
CH_5N	Methylamine	0.319	0.231				
$C_2Cl_3F_3$	1,1,2-Trichlorotrifluoro-ethane	1.465	0.945	0.656	0.481		
C_2Cl_4	Tetrachloroethylene		1.114	0.844	0.663	0.535	0.442
C_2HCl_3	Trichloroethylene		0.703	0.545	0.444	0.376	
C_2HCl_5	Pentachloroethane		3.761	2.254	1.491	1.061	
$C_2HF_3O_2$	Trifluoroacetic acid			0.808	0.571		
$C_2H_2Cl_2$	*cis*-1,2-Dichloroethylene	0.786	0.575	0.445			
$C_2H_2Cl_2$	*trans*-1,2-Dichloroethylene	0.522	0.398	0.317	0.261		
$C_2H_2Cl_4$	1,1,1,2-Tetrachloroethane	3.660	2.200	1.437	1.006	0.741	0.570
$C_2H_3ClF_2$	1-Chloro-1,1-difluoro-ethane	0.477	0.376				
C_2H_3ClO	Acetyl chloride			0.368	0.294		
$C_2H_3Cl_3$	1,1,1-Trichloroethane	1.847	1.161	0.793	0.578	0.428	
C_2H_3N	Acetonitrile		0.400	0.369	0.284	0.234	
$C_2H_4Br_2$	1,2-Dibromoethane			1.595	1.116	0.837	0.661

Molecular formula	Name	Viscosity in mPa s					
		−25°C	0°C	25°C	50°C	75°C	100°C
$C_2H_4Cl_2$	1,1-Dichloroethane			0.464	0.362		
$C_2H_4Cl_2$	1,2-Dichloroethane		1.125	0.779	0.576	0.447	
$C_2H_4O_2$	Acetic acid			1.056	0.786	0.599	0.464
$C_2H_4O_2$	Methyl formate		0.424	0.325			
C_2H_5Br	Bromoethane	0.635	0.477	0.374			
C_2H_5Cl	Chloroethane	0.416	0.319				
C_2H_5I	Iodoethane		0.723	0.556	0.444	0.365	
C_2H_5NO	N-Methylformamide		2.549	1.678	1.155	0.824	0.606
$C_2H_5NO_2$	Nitroethane	1.354	0.940	0.688	0.526	0.415	0.337
C_2H_6O	Ethanol	3.262	1.786	1.074	0.694	0.476	
C_2H_6OS	Dimethyl sulfoxide			1.987	1.290		
$C_2H_6O_2$	Ethylene glycol			16.1	6.554	3.340	1.975
C_2H_6S	Dimethyl sulfide		0.356	0.284			
C_2H_6S	Ethanethiol		0.364	0.287			
C_2H_7N	Dimethylamine	0.300	0.232				
C_2H_7NO	Ethanolamine			21.1	8.560	3.935	1.998
C_3H_5Br	3-Bromopropene		0.620	0.471	0.373		
C_3H_5Cl	3-Chloropropene		0.408	0.314			
C_3H_5ClO	Epichlorohydrin	2.492	1.570	1.073	0.781	0.597	0.474
C_3H_5N	Propanenitrile			0.294	0.240	0.202	
C_3H_6O	Acetone	0.540	0.395	0.306	0.247		
C_3H_6O	Allyl alcohol			1.218	0.759	0.505	
C_3H_6O	Propanal			0.321	0.249		
$C_3H_6O_2$	Ethyl formate		0.506	0.380	0.300		
$C_3H_6O_2$	Methyl acetate		0.477	0.364	0.284		
$C_3H_6O_2$	Propanoic acid		1.499	1.030	0.749	0.569	0.449
C_3H_7Br	1-Bromopropane		0.645	0.489	0.387		
C_3H_7Br	2-Bromopropane		0.612	0.458	0.359		
C_3H_7Cl	1-Chloropropane		0.436	0.334			
C_3H_7Cl	2-Chloropropane		0.401	0.303			
C_3H_7I	1-Iodopropane		0.970	0.703	0.541	0.436	0.363
C_3H_7I	2-Iodopropane		0.883	0.653	0.506	0.407	
C_3H_7NO	N,N-Dimethylformamide		1.176	0.794	0.624		
$C_3H_7NO_2$	1-Nitropropane	1.851	1.160	0.798	0.589	0.460	0.374
C_3H_8O	1-Propanol	8.645	3.815	1.945	1.107	0.685	
C_3H_8O	2-Propanol		4.619	2.038	1.028	0.576	
$C_3H_8O_2$	1,2-Propylene glycol		248	40.4	11.3	4.770	2.750
$C_3H_8O_3$	Glycerol			934	152	39.8	14.8
C_3H_8S	1-Propanethiol		0.503	0.385			
C_3H_8S	2-Propanethiol		0.477	0.357	0.280		
C_3H_9N	Propylamine			0.376			
C_3H_9N	Isopropylamine		0.454	0.325			
C_4H_4O	Furan	0.661	0.475	0.361			
C_4H_5N	Pyrrole		2.085	1.225	0.828	0.612	
$C_4H_6O_3$	Acetic anhydride		1.241	0.843	0.614	0.472	0.377
C_4H_7N	Butanenitrile			0.553	0.418	0.330	0.268
C_4H_8O	2-Butanone	0.720	0.533	0.405	0.315	0.249	
C_4H_8O	Tetrahydrofuran	0.849	0.605	0.456	0.359		
$C_4H_8O_2$	1,4-Dioxane			1.177	0.787	0.569	
$C_4H_8O_2$	Ethyl acetate		0.578	0.423	0.325	0.259	
$C_4H_8O_2$	Methyl propionate		0.581	0.431	0.333	0.266	
$C_4H_8O_2$	Propyl formate		0.669	0.485	0.370	0.293	
$C_4H_8O_2$	Butanoic acid		2.215	1.426	0.982	0.714	0.542
$C_4H_8O_2$	2-Methylpropanoic acid		1.857	1.226	0.863	0.639	0.492
$C_4H_8O_2S$	Sulfolane				6.280	3.818	2.559
C_4H_8S	Tetrahydrothiophene			0.973	0.912		
C_4H_9Br	1-Bromobutane		0.815	0.606	0.471	0.379	
C_4H_9Cl	1-Chlorobutane		0.556	0.422	0.329	0.261	
C_4H_9N	Pyrrolidine	1.914	1.071	0.704	0.512		
C_4H_9NO	N,N-Dimethylacetamide			1.927			

Molecular formula	Name	\multicolumn{6}{c}{Viscosity in mPa s}					
		−25°C	0°C	25°C	50°C	75°C	100°C
C_4H_9NO	Morpholine			2.021	1.247	0.850	0.627
$C_4H_{10}O$	1-Butanol	12.19	5.185	2.544	1.394	0.833	0.533
$C_4H_{10}O$	2-Butanol			3.096	1.332	0.698	0.419
$C_4H_{10}O$	2-Methyl-2-propanol			4.312	1.421	0.678	
$C_4H_{10}O$	Diethyl ether		0.283	0.224			
$C_4H_{10}O_3$	Diethylene glycol			30.200	11.130	4.917	2.505
$C_4H_{10}S$	Diethyl sulfide		0.558	0.422	0.331	0.267	
$C_4H_{11}N$	Butylamine		0.830	0.574	0.409	0.298	
$C_4H_{11}N$	Isobutylamine		0.770	0.571	0.367		
$C_4H_{11}N$	Diethylamine			0.319	0.239		
$C_4H_{11}NO_2$	Diethanolamine				109.5	28.7	9.100
$C_5H_4O_2$	Furfural		2.501	1.587	1.143	0.906	0.772
C_5H_5N	Pyridine		1.361	0.879	0.637	0.497	0.409
C_5H_{10}	1-Pentene	0.313	0.241	0.195			
C_5H_{10}	2-Methyl-2-butene		0.255	0.203			
C_5H_{10}	Cyclopentane		0.555	0.413	0.321		
$C_5H_{10}O$	Mesityl oxide	1.291	0.838	0.602	0.465	0.381	0.326
$C_5H_{10}O$	2-Pentanone		0.641	0.470	0.362	0.289	0.238
$C_5H_{10}O$	3-Pentanone		0.592	0.444	0.345	0.276	0.227
$C_5H_{10}O_2$	Butyl formate		0.937	0.644	0.472	0.362	0.289
$C_5H_{10}O_2$	Propyl acetate		0.768	0.544	0.406	0.316	0.255
$C_5H_{10}O_2$	Ethyl propanoate		0.691	0.501	0.380	0.299	0.242
$C_5H_{10}O_2$	Methyl butanoate		0.759	0.541	0.406	0.318	0.257
$C_5H_{10}O_2$	Methyl isobutanoate		0.672	0.488	0.373	0.296	
$C_5H_{11}N$	Piperidine			1.573	0.958	0.649	0.474
C_5H_{12}	Pentane	0.351	0.274	0.224			
C_5H_{12}	Isopentane	0.376	0.277	0.214			
$C_5H_{12}O$	1-Pentanol	25.4	8.512	3.619	1.820	1.035	0.646
$C_5H_{12}O$	2-Pentanol			3.470	1.447	0.761	0.465
$C_5H_{12}O$	3-Pentanol			4.149	1.473	0.727	0.436
$C_5H_{12}O$	2-Methyl-1-butanol			4.453	1.963	1.031	0.612
$C_5H_{12}O$	3-Methyl-1-butanol		8.627	3.692	1.842	1.031	0.631
$C_5H_{13}N$	Pentylamine		1.030	0.702	0.493	0.356	
C_6F_6	Hexafluorobenzene			2.789	1.730	1.151	
$C_6H_4Cl_2$	o-Dichlorobenzene		1.958	1.324	0.962	0.739	0.593
$C_6H_4Cl_2$	m-Dichlorobenzene		1.492	1.044	0.787	0.628	0.525
C_6H_5Br	Bromobenzene		1.560	1.074	0.798	0.627	0.512
C_6H_5Cl	Chlorobenzene	1.703	1.058	0.753	0.575	0.456	0.369
C_6H_5ClO	o-Chlorophenol			3.589	1.835	1.131	0.786
C_6H_5ClO	m-Chlorophenol				4.041		
C_6H_5F	Fluorobenzene		0.749	0.550	0.423	0.338	
C_6H_5I	Iodobenzene		2.354	1.554	1.117	0.854	0.683
$C_6H_5NO_2$	Nitrobenzene		3.036	1.863	1.262	0.918	0.704
C_6H_6	Benzene			0.604	0.436	0.335	
C_6H_6ClN	o-Chloroaniline			3.316	1.913	1.248	0.887
C_6H_6O	Phenol				3.437	1.784	1.099
C_6H_7N	Aniline			3.847	2.029	1.247	0.850
$C_6H_8N_2$	Phenylhydrazine			13.0	4.553	1.850	0.848
C_6H_{10}	Cyclohexene		0.882	0.625	0.467	0.364	
$C_6H_{10}O$	Cyclohexanone			2.017	1.321	0.919	0.671
$C_6H_{11}N$	Hexanenitrile			0.912	0.650	0.488	0.382
C_6H_{12}	Cyclohexane			0.894	0.615	0.447	
C_6H_{12}	Methylcyclopentane	0.927	0.653	0.479	0.364		
C_6H_{12}	1-Hexene	0.441	0.326	0.252	0.202		
$C_6H_{12}O$	Cyclohexanol			57.5	12.3	4.274	1.982
$C_6H_{12}O$	2-Hexanone	1.300	0.840	0.583	0.429	0.329	0.262
$C_6H_{12}O$	4-Methyl-2-pentanone			0.545	0.406		
$C_6H_{12}O_2$	Butyl acetate		1.002	0.685	0.500	0.383	0.305
$C_6H_{12}O_2$	Isobutyl acetate			0.676	0.493	0.370	0.286
$C_6H_{12}O_2$	Ethyl butanoate			0.639	0.453		

Molecular formula	Name	Viscosity in mPa s					
		−25°C	0°C	25°C	50°C	75°C	100°C
$C_6H_{12}O_2$	Diacetone alcohol	28.7	6.621	2.798	1.829	1.648	
$C_6H_{12}O_3$	Paraldehyde			1.079	0.692	0.485	0.362
$C_6H_{13}N$	Cyclohexylamine			1.944	1.169	0.782	0.565
C_6H_{14}	Hexane		0.405	0.300	0.240		
C_6H_{14}	2-Methylpentane		0.372	0.286	0.226		
C_6H_{14}	3-Methylpentane		0.395	0.306			
$C_6H_{14}O$	Dipropyl ether		0.542	0.396	0.304	0.242	
$C_6H_{14}O$	1-Hexanol			4.578	2.271	1.270	0.781
$C_6H_{15}N$	Triethylamine		0.455	0.347	0.273	0.221	
$C_6H_{15}N$	Dipropylamine		0.751	0.517	0.377	0.288	0.228
$C_6H_{15}N$	Diisopropylamine			0.393	0.300	0.237	
$C_6H_{15}NO_3$	Triethanolamine			609	114	31.5	11.7
C_7H_5N	Benzonitrile			1.267	0.883	0.662	0.524
C_7H_7Cl	o-Chlorotoluene		1.390	0.964	0.710	0.547	0.437
C_7H_7Cl	m-Chlorotoluene		1.165	0.823	0.616	0.482	0.391
C_7H_7Cl	p-Chlorotoluene			0.837	0.621	0.483	0.390
C_7H_8	Toluene	1.165	0.778	0.560	0.424	0.333	0.270
C_7H_8O	o-Cresol				3.035	1.562	0.961
C_7H_8O	m-Cresol			12.9	4.417	2.093	1.207
C_7H_8O	Benzyl alcohol			5.474	2.760	1.618	1.055
C_7H_8O	Anisole			1.056	0.747	0.554	0.427
C_7H_9N	N-Methylaniline		4.120	2.042	1.222	0.825	0.606
C_7H_9N	o-Methyl aniline		10.3	3.823	1.936	1.198	0.839
C_7H_9N	m-Methyl aniline		8.180	3.306	1.679	1.014	0.699
C_7H_9N	Benzylamine			1.624	1.080	0.769	0.577
C_7H_{14}	Methylcyclohexane		0.991	0.679	0.501	0.390	0.316
C_7H_{14}	1-Heptene		0.441	0.340	0.273	0.226	
$C_7H_{14}O$	2-Heptanone			0.714	0.407	0.297	
$C_7H_{14}O_2$	Heptanoic acid			3.840	2.282	1.488	1.041
C_7H_{16}	Heptane	0.757	0.523	0.387	0.301	0.243	
C_7H_{16}	3-Methylhexane			0.350			
$C_7H_{16}O$	1-Heptanol			5.810	2.603	1.389	0.849
$C_7H_{16}O$	2-Heptanol			3.955	1.799	0.987	0.615
$C_7H_{16}O$	3-Heptanol				1.957	0.976	0.584
$C_7H_{16}O$	4-Heptanol			4.207	1.695	0.882	0.539
$C_7H_{17}N$	Heptylamine			1.314	0.865	0.600	0.434
C_8H_8	Styrene		1.050	0.695	0.507	0.390	0.310
C_8H_8O	Acetophenone			1.681			0.634
$C_8H_8O_2$	Methyl benzoate			1.857			
$C_8H_8O_3$	Methyl salicylate					1.102	0.815
C_8H_{10}	Ethylbenzene		0.872	0.631	0.482	0.380	0.304
C_8H_{10}	o-Xylene		1.084	0.760	0.561	0.432	0.345
C_8H_{10}	m-Xylene		0.795	0.581	0.445	0.353	0.289
C_8H_{10}	p-Xylene			0.603	0.457	0.359	0.290
$C_8H_{10}O$	Phenetole			1.197	0.817	0.594	0.453
$C_8H_{11}N$	N,N-Dimethylaniline		1.996	1.300	0.911	0.675	0.523
$C_8H_{11}N$	N-Ethylaniline		3.981	2.047	1.231	0.825	0.596
C_8H_{16}	Ethylcyclohexane		1.139	0.784	0.579		
$C_8H_{16}O_2$	Octanoic acid			5.020	2.656	1.654	1.147
C_8H_{18}	Octane		0.700	0.508	0.385	0.302	0.243
$C_8H_{18}O$	1-Octanol			7.288	3.232	1.681	0.991
$C_8H_{18}O$	4-Methyl-3-heptanol		1.904	1.085	0.702	0.497	0.375
$C_8H_{18}O$	5-Methyl-3-heptanol		2.052	1.178	0.762	0.536	0.401
$C_8H_{18}O$	2-Ethyl-1-hexanol		20.7	6.271	2.631	1.360	0.810
$C_8H_{18}O$	Dibutyl ether	1.417	0.918	0.637	0.466	0.356	0.281
$C_8H_{19}N$	Dibutylamine		1.509	0.918	0.619	0.449	0.345
$C_8H_{19}N$	Diisobutylamine		1.115	0.723	0.511	0.384	0.303
C_9H_7N	Quinoline			3.337	1.892	1.201	0.833
C_9H_{10}	Indane		2.230	1.357	0.931	0.692	0.545
C_9H_{12}	Cumene		1.075	0.737	0.547		

Molecular formula	Name	Viscosity in mPa s					
		−25°C	0°C	25°C	50°C	75°C	100°C
$C_9H_{14}O$	Isophorone		4.201	2.329	1.415	0.923	0.638
$C_9H_{18}O$	5-Nonanone			1.199	0.834	0.619	0.484
$C_9H_{18}O_2$	Nonanoic acid			7.011	3.712	2.234	1.475
C_9H_{20}	Nonane		0.964	0.665	0.488	0.375	0.300
$C_9H_{20}O$	1-Nonanol			9.123	4.032		
$C_{10}H_{10}O_4$	Dimethyl phthalate		63.2	14.4	5.309	2.824	1.980
$C_{10}H_{14}$	Butylbenzene			0.950	0.683	0.515	
$C_{10}H_{18}$	cis-Decahydronaphthalene	12.8	5.645	3.042	1.875	1.271	0.924
$C_{10}H_{18}$	trans-Decahydronaphthalene	6.192	3.243	1.948	1.289	0.917	0.689
$C_{10}H_{20}O_2$	Decanoic acid				4.327	2.651	
$C_{10}H_{22}$	Decane	2.188	1.277	0.838	0.598	0.453	0.359
$C_{10}H_{22}O$	1-Decanol			10.9	4.590		
$C_{11}H_{24}$	Undecane		1.707	1.098	0.763	0.562	0.433
$C_{12}H_{10}O$	Diphenyl ether				2.130	1.407	1.023
$C_{12}H_{26}$	Dodecane		2.277	1.383	0.930	0.673	0.514
$C_{13}H_{12}$	Diphenylmethane					1.265	0.929
$C_{13}H_{28}$	Tridecane		2.909	1.724	1.129	0.796	0.594
$C_{14}H_{30}$	Tetradecane			2.128	1.376	0.953	0.697
$C_{16}H_{22}O_4$	Dibutyl phthalate	483	66.4	16.6	6.470	3.495	2.425
$C_{16}H_{34}$	Hexadecane			3.032	1.879	1.260	0.899
$C_{18}H_{38}$	Octadecane				2.487	1.609	1.132

VISCOSITY OF CARBON DIOXIDE ALONG THE SATURATION LINE

The table below gives the viscosity of gas and liquid CO_2 along the liquid–vapor saturation line.

References

1. Fenghour, A., Wakeham, W. A., and Vesovic, V., *J. Phys. Chem. Ref. Data*, 27, 31, 1998.
2. Angus, S., et al., *International Tables for the Fluid State: Carbon Dioxide*, Pergamon Press, Oxford, 1976.

T/K	P/kPa	Gas η/μPa s	Liquid η/μPa s
205	227	10.33	
210	327	10.60	
215	465	10.87	
220	600	11.13	241.68
225	735	11.41	221.72
230	894	11.69	203.75
235	1075	11.98	187.48
240	1283	12.27	172.67
245	1519	12.58	159.13
250	1786	12.90	146.69
255	2085	13.24	135.20
260	2419	13.61	124.30
265	2790	14.02	114.63
270	3203	14.47	105.21
275	3658	14.99	96.44
280	4160	15.61	87.89
285	4712	16.37	79.64
290	5315	17.36	71.47
295	5984	18.79	63.01
300	6710	21.29	53.33
302	6997	23.52	48.30

VISCOSITY AND DENSITY OF AQUEOUS HYDROXIDE SOLUTIONS

The viscosity and density of aqueous hydroxide solutions at 25°C is tabulated here as a function of concentration. Viscosity is given in millipascal second, which is equal to the c.g.s. unit centipoise (cP). The last entry in each column refers to the saturated solution.

Reference

Sipos, P. M., Hefter, G., and May, P. M., *J. Chem. Eng. Data*, 45, 613, 2000.

Viscosity in mPa s

c/mol L^{-1}	LiOH	NaOH	KOH	CsOH	(CH$_3$)$_4$NOH
0.5	1.017	0.997	0.937	0.91	1.017
1.0	1.169	1.116	0.990	0.94	1.186
1.5	1.340	1.248	1.050	0.97	1.430
2.0	1.537	1.396	1.116	1.03	1.762
3.0	2.050	1.754	1.269	1.19	3.031
4.0	2.734	2.228	1.448	1.41	7.238
5.0		2.867	1.657	1.67	
6.0		3.727	1.902	1.98	
7.0		4.869	2.196	2.40	
8.0		6.351	2.554	3.09	
9.0		8.230	3.005	4.31	
10.0		10.554	3.581	6.46	
11.0		13.362	4.328		
12.0		16.677	5.303		
13.0		20.503	6.577		
14.0		24.826	8.235		
15.0		29.604			
16.0		34.767			
17.0		40.212			
18.0		45.800			
19.0		51.354			
Sat.	3.311	51.911	8.526		8.850

Density in g/cm^3

c/mol L^{-1}	LiOH	NaOH	KOH	CsOH	(CH$_3$)$_4$NOH
0.5	1.012	1.019	1.022	1.063	0.999
1.0	1.025	1.040	1.045	1.128	1.002
1.5	1.038	1.059	1.068	1.193	1.005
2.0	1.050	1.078	1.090	1.257	1.009
3.0	1.072	1.115	1.133	1.383	1.019
4.0	1.093	1.149	1.174	1.508	1.030
5.0		1.182	1.214	1.632	
6.0		1.213	1.253	1.755	
7.0		1.243	1.290	1.876	
8.0		1.271	1.326	1.997	
9.0		1.299	1.362	2.117	
10.0		1.325	1.396	2.236	
11.0		1.350	1.429	2.354	
12.0		1.374	1.462	2.471	
13.0		1.397	1.494	2.587	
14.0		1.419	1.524	2.703	
15.0		1.441			
16.0		1.461			
17.0		1.481			
18.0		1.499			
19.0		1.517			
Sat.	1.109	1.519	1.529	2.800	1.032

VISCOSITY OF LIQUID METALS

This table gives the viscosity of several liquid metals as a function of temperature. Experimental data from some of the references were smoothed to produce the table. Viscosity is given in millipascal second (mPa s), which equals the c.g.s. unit centipoise (cP).

References

1. Shpil'rain, E. E., Yakimovich, K. A., Fomin, V. A., Skovorodjko, S. N., and Mozgovoi, A. G., in *Handbook of Thermodynamic and Transport Properties of the Alkali Metals*, Ohse, R. H., Ed., Blackwell Scientific Publishers, Oxford, 1985. [Li, Na, K, Rb, Cs]
2. Rothwell, E., *J. Inst. Metals*, 90, 389, 1961. [Al]
3. Culpin, M. F., *Proc. Phys. Soc.*, 70, 1079, 1957. [Ca]
4. *Landolt-Börnstein, Numerical Data and Functional Relationships in Science and Technology, Sixth Edition*, II/5a, *Transport Phenomena I (Viscosity and Diffusion)*, Springer–Verlag, Heidelberg, 1961 [Co, Au, Mg, Ni, Ag]
5. Spells, K. E., *Proc. Phys. Soc.*, 48, 299, 1936. [Ga]
6. Walsdorfer, H., Arpshofen, I., and Predel, B., *Z. Met.*, 79, 503, 1988. [In]

$t/°C$	Lithium	Sodium	Potassium	Rubidium	Cesium	Gallium
				Viscosity in mPa s		
50				0.542	0.598	1.921
100		0.687	0.441	0.435	0.469	1.608
150		0.542	0.358	0.365	0.389	1.397
200	0.566	0.451	0.303	0.316	0.334	1.245
250	0.503	0.387	0.263	0.280	0.294	1.130
300	0.453	0.341	0.234	0.252	0.264	1.040
350	0.412	0.306	0.211	0.230	0.240	0.968
400	0.379	0.278	0.193	0.212	0.221	0.909
450	0.352	0.255	0.178	0.197	0.206	0.859
500	0.328	0.237	0.166	0.185	0.192	0.817
550	0.308	0.221	0.155	0.174	0.181	0.781
600	0.290	0.208	0.146	0.165	0.171	0.750
650	0.275	0.196	0.138	0.157	0.163	0.722
700	0.261	0.186	0.132	0.150	0.156	0.698
750	0.249	0.177	0.126	0.143	0.149	0.677
800	0.238	0.170	0.120	0.138	0.143	0.657
850	0.228	0.163	0.115	0.133	0.138	0.640
900	0.219	0.156	0.111	0.128	0.134	0.624
950	0.211	0.151	0.107	0.124	0.129	0.610
1000	0.204	0.146	0.104	0.120	0.125	0.597
1050	0.197	0.141	0.101	0.117	0.122	0.585
1100	0.191	0.137	0.098	0.114	0.119	0.574
1150	0.185	0.133	0.095	0.111	0.116	
1200	0.180	0.129	0.092	0.108	0.113	
1250	0.175	0.126	0.090	0.105	0.110	
1300	0.170	0.123	0.088	0.103	0.108	
1350	0.166	0.120	0.086	0.101	0.106	
1400	0.162	0.117	0.084	0.099	0.104	
1450	0.158	0.115	0.082	0.097	0.102	
1500	0.155	0.113	0.081	0.095	0.100	
1550	0.151	0.110	0.079	0.093	0.098	
1600	0.148	0.108	0.078	0.092	0.097	
1650	0.145	0.106	0.076	0.090	0.095	
1700	0.142	0.105	0.075		0.094	
1750	0.139	0.103	0.074		0.092	
1800	0.137	0.101			0.091	
1850	0.135	0.100			0.090	
1900	0.132	0.098			0.089	
1950	0.130	0.097			0.088	
2000	0.128	0.096			0.086	

			Viscosity in mPa s					
t/°C	Aluminum	Calcium	Cobalt	Gold	Indium	Magnesium	Nickel	Silver
250					1.35			
300					1.22			
350					1.12			
400					1.04			
450					0.98			
700	1.289					1.10		
750	1.200					0.96		
800	1.115					0.84		
850	1.028	1.107				0.74		
900		0.959				0.67		
1000								3.80
1050								3.56
1100				5.130				3.31
1150				4.874				3.06
1200				4.640				2.82
1250				4.429				2.61
1300				4.240				2.42
1350								2.28
1400								2.20
1450								2.19
1500			4.15				4.35	
1550			3.89				4.09	
1600			3.64				3.87	
1650			3.41				3.67	
1700			3.20				3.49	
1750			2.99				3.32	

THERMAL CONDUCTIVITY OF GASES

This table gives the thermal conductivity of several gases as a function of temperature. Unless otherwise noted, the values refer to a pressure of 100 kPa (1 bar) or to the saturation vapor pressure if that is less than 100 kPa. The notation $P = 0$ indicates the low pressure limiting value is given. In general, the $P = 0$ and $P = 100$ kPa values differ by less than 1%. Units are milliwatts per meter kelvin. Substances are listed in the modified Hill order.

MF	Name	Thermal conductivity in mW/m K						Ref.
		100 K	200 K	300 K	400 K	500 K	600 K	
	Air	9.4	18.4	26.2	33.3	39.7	45.7	1
Ar	Argon	6.2	12.4	17.9	22.6	26.8	30.6	2,8
BF_3	Boron trifluoride			19.0	24.6			11
ClH	Hydrogen chloride		9.2	14.5	19.5	24.0	28.1	11
F_6S	Sulfur hexafluoride ($P = 0$)			13.0	20.6	27.5	33.8	16
H_2	Hydrogen ($P = 0$)	68.6	131.7	186.9	230.4			4
H_2O	Water			18.7	27.1	35.7	47.1	6
	Deuterium oxide				27.0	36.5	47.6	7
H_2S	Hydrogen sulfide			14.6	20.5	26.4	32.4	11
H_3N	Ammonia			24.4	37.4	51.6	66.8	11
He	Helium ($P = 0$)	75.5	119.3	156.7	190.6	222.3	252.4	8
Kr	Krypton ($P = 0$)	3.3	6.4	9.5	12.3	14.8	17.1	8
NO	Nitric oxide		17.8	25.9	33.1	39.6	46.2	11
N_2	Nitrogen	9.8	18.7	26.0	32.3	38.3	44.0	12
N_2O	Nitrous oxide		9.8	17.4	26.0	34.1	41.8	11
Ne	Neon ($P = 0$)	22.3	37.6	49.8	60.3	69.9	78.7	8
O_2	Oxygen	9.3	18.4	26.3	33.7	41.0	48.1	10
O_2S	Sulfur dioxide			9.6	14.3	20.0	25.6	11
Xe	Xenon ($P = 0$)	2.0	3.6	5.5	7.3	8.9	10.4	8
CCl_2F_2	Dichlorodifluoromethane			9.9	15.0	20.1	25.2	13
CF_4	Tetrafluoromethane ($P = 0$)			16.0	24.1	32.2	39.9	16
CO	Carbon monoxide ($P = 0$)			25.0	32.3	39.2	45.7	14
CO_2	Carbon dioxide		9.6	16.8	25.1	33.5	41.6	9
$CHCl_3$	Trichloromethane			7.5	11.1	15.1		11
CH_4	Methane		22.5	34.1	49.1	66.5	84.1	5,15
CH_4O	Methanol				26.2	38.6	53.0	11
$C_2Cl_2F_4$	1,2-Dichlorotetrafluoroethane			10.25	15.7	21.1		13
$C_2Cl_3F_3$	1,1,2-Trichlorotrifluoroethane			9.0	13.6	18.3		13
C_2H_2	Acetylene			21.4	33.3	45.4	56.8	11
C_2H_4	Ethylene		11.1	20.5	34.6	49.9	68.6	3
C_2H_6	Ethane		11.0	21.3	35.4	52.2	70.5	5
C_2H_6O	Ethanol			14.4	25.8	38.4	53.2	11
C_3H_6O	Acetone			11.5	20.2	30.6	42.7	11
C_3H_8	Propane			18.0	30.6	45.5	61.9	5
C_4F_8	Perfluorocyclobutane			12.5	19.5			13
C_4H_{10}	Butane			16.4	28.4	43.0	59.1	5
C_4H_{10}	Isobutane			16.1	27.9	42.1	57.6	5
$C_4H_{10}O$	Diethyl ether			15.1	25.0	37.1		11
C_5H_{12}	Pentane			14.4	24.9	37.8	52.7	11
C_6H_{14}	Hexane				23.4	35.4	48.7	11

References

1. Kadoya, K. Matsunaga, N., and Nagashima, A., Viscosity and thermal conductivity of dry air in the gaseous phase, *J. Phys. Chem. Ref. Data*, 14, 947, 1985.
2. Younglove, B. A. and Hanley, H. J. M., The viscosity and thermal conductivity coefficients of gaseous and liquid argon, *J. Phys. Chem. Ref. Data*, 15, 1323, 1986.
3. Holland, P. M., Eaton, B. E., and Hanley, H. J. M., A correlation of the viscosity and thermal conductivity data of gaseous and liquid ethylene, *J. Phys. Chem. Ref. Data*, 12, 917, 1983.
4. Assael, M. J., Mixafendi, S., and Wakeham, W. A., The viscosity and thermal conductivity of normal hydrogen in the limit of zero density, *J. Phys. Chem. Ref. Data*, 15, 1315, 1986.
5. Younglove, B. A. and Ely, J. F., Thermophysical properties of fluids. II. Methane, ethane, propane, isobutane, and normal butane, *J. Phys. Chem. Ref. Data*, 16, 577, 1987.

6. Sengers, J. V. and Watson, J. T. R., Improved international formulations for the viscosity and thermal conductivity of water substance, *J. Phys. Chem. Ref. Data*, 15, 1291, 1986.

7. Matsunaga, N. and Nagashima, A., Transport properties of liquid and gaseous D_2O over a wide range of temperature and pressure, *J. Phys. Chem. Ref. Data*, 12, 933, 1983.

8. Kestin, J. et al., Equilibrium and transport properties of the noble gases and their mixtures at low density, *J. Phys. Chem. Ref. Data*, 13, 229, 1984.

9. Vescovic, V. et al., The transport properties of carbon dioxide, *J. Phys. Chem. Ref. Data*, 19, 1990.

10. Younglove, B. A., Thermophysical properties of fluids. I. Argon, ethylene, parahydrogen, nitrogen, nitrogen trifluoride, and oxygen, *J. Phys. Chem. Ref. Data*, 11, Suppl. 1, 1982.

11. Ho, C. Y., Ed., *Properties of Inorganic and Organic Fluids, CINDAS Data Series on Materials Properties*, Volume V-1, Hemisphere Publishing Corp., New York, 1988.

12. Stephen, K., Krauss, R., and Laesecke, A., Viscosity and thermal conductivity of nitrogen for a wide range of fluid states, *J. Phys. Chem. Ref. Data*, 16, 993, 1987.

13. Krauss, R. and Stephan, K., Thermal conductivity of refrigerants in a wide range of temperature and pressure, *J. Phys. Chem. Ref. Data*, 18, 43, 1989.

14. Millat, J. and Wakeham, W. A., The thermal conductivity of nitrogen and carbon monoxide in the limit of zero density, *J. Phys. Chem. Ref. Data*, 18, 565, 1989.

15. Friend, D. G., Ely, J. F., and Ingham, H., Thermophysical properties of methane, *J. Phys. Chem. Ref. Data*, 18, 583, 1989.

16. Uribe, F. J., Mason, E. A., and Kestin, J., Thermal conductivity of nine polyatomic gases at low density, *J. Phys. Chem. Ref. Data*, 19, 1123, 1990.

THERMAL CONDUCTIVITY OF LIQUIDS

This table gives the thermal conductivity of about 275 liquids at temperatures between −25 and 100°C. Values refer to nominal atmospheric pressure; when an entry is given for a temperature above the normal boiling point of the liquid, the pressure is understood to be the saturation vapor pressure at that temperature. Reference 1 contains data on many of these liquids at high pressures. Data on halocarbon refrigerants over a wide range of temperature and pressure may be found in Reference 6.

Values given to three decimal places (i.e., to 0.001 W/m K) have an uncertainty of 2% to 5%. Values given to 0.0001 W/m K should be accurate to 1% or better.

Substances are arranged by molecular formula in Hill order, except that compounds not containing carbon precede those that do contain carbon.

References

1. Vargaftik, N. B., Filippov, L. P., Tarzimanov, A. A., and Totskii, E. E., *Handbook of Thermal Conductivity of Liquids and Gases*, CRC Press, Boca Raton FL, 1994.
2. Daubert, T. E., Danner, R. P., Sibul, H. M., and Stebbins, C. C., *Physical and Thermodynamic Properties of Pure Compounds: Data Compilation*, extant 1994 (core with four supplements), Taylor and Francis, Bristol, PA (also available as a database).
3. Watanabe, H., *J. Chem. Eng. Data* 48, 124, 2003.
4. Watanabe, H., and Seong, D. J., *Int. J. Thermophys.* 23, 337, 2002.
5. Nieto de Castro, C. A., Li, S. F. Y., Nagashima, A., Trengove, R. D., and Wakeham, W. A., *J. Phys. Chem. Ref. Data* 15, 1073, 1986.
6. Krauss, R., and Stephan, K., *J. Phys. Chem. Ref. Data* 18, 43, 1989.
7. Assael, M. J., Ramires, M. L. V., Nieto de Castro, C. A., and Wakeham, W. A., *J. Phys. Chem. Ref. Data* 19, 113, 1990.
8. Ramires, M. L. V., Nieto de Castro, C. A., Nagasaka, Y., Nagashima, A., Assael, M. J., and Wakeham, W. A., *J. Phys. Chem. Ref. Data* 24, 1377, 1995.
9. Ramires, M. L. V., Nieto de Castro, C. A., Perkins, R. A., Nagasaka, Y., Nagashima, A., Assael, M. J., and Wakeham, W. A., *J. Phys. Chem. Ref. Data* 29, 133, 2000.
10. Marsh, K. N., Ed., *Recommended Reference Materials for the Realization of Physicochemical Properties*, Blackwell Scientific Publications, Oxford, 1987.
11. Beaton, C. F., and Hewitt, G. F., *Physical Property Data for the Design Engineer*, Hemisphere Publishing Corp., New York, 1989.

Molecular formula	Name	Thermal conductivity in W/m K						
		−25°C	0°C	25°C	50°C	75°C	100°C	Ref.
Cl₄Ge	Germanium(IV) chloride	0.111	0.105	0.100	0.095	0.090	0.084	1
Cl₄Si	Tetrachlorosilane			0.099	0.096			2
Cl₄Sn	Tin(IV) chloride	0.123	0.117	0.112	0.106	0.101	0.095	1
Cl₄Ti	Titanium(IV) chloride		0.143	0.138	0.134	0.129	0.124	1
H₂O	Water		0.5562	0.6062	0.6423	0.6643	0.6729	8
Hg	Mercury	7.85	8.175	8.514	8.842	9.161	9.475	11
CCl₃F	Trichlorofluoromethane	0.102	0.096	0.089	0.083	0.076	0.070	1
CCl₄	Tetrachloromethane		0.109	0.103	0.098	0.092	0.087	1
CHCl₃	Trichloromethane	0.127	0.122	0.117	0.112	0.107	0.102	2
CH₂Br₂	Dibromomethane	0.120	0.114	0.108	0.103	0.097		2
CH₂Cl₂	Dichloromethane	0.158	0.149	0.140	0.133	0.128	0.127	1
CH₂I₂	Diiodomethane			0.098	0.093	0.088	0.083	1
CH₂O₂	Formic acid			0.267	0.265	0.263	0.261	1
CH₃NO₂	Nitromethane	0.226	0.215	0.204	0.193	0.182	0.171	1
CH₄O	Methanol	0.218	0.210	0.202	0.195	0.189	0.182	1
CS₂	Carbon disulfide		0.154	0.149				2
C₂Br₂F₄	1,2-Dibromotetrafluoroethane	0.071	0.066	0.061	0.057	0.053	0.049	1
C₂Cl₃F₃	1,1,2-Trichloro-1,2,2-trifluoroethane	0.0847	0.0790	0.0736	0.0683			6
C₂Cl₄	Tetrachloroethene		0.117	0.110	0.104	0.098	0.093	1
C₂Cl₄F₂	1,1,2,2-Tetrachloro-1,2-difluoroethane			0.082	0.078	0.074	0.069	1
C₂HCl₃	Trichloroethene	0.128	0.121	0.114	0.106	0.098	0.090	1
C₂H₂Cl₄	1,1,2,2-Tetrachloroethane	0.124	0.118	0.111	0.104	0.098	0.091	1
C₂H₃Cl₃	1,1,1-Trichloroethane		0.106	0.101	0.096			2
C₂H₃N	Acetonitrile	0.208	0.198	0.188	0.178	0.168		2
C₂H₄Br₂	1,2-Dibromoethane			0.100	0.096	0.092	0.088	1
C₂H₄Cl₂	1,2-Dichloroethane	0.144	0.139	0.133	0.128	0.122	0.117	1
C₂H₄O₂	Acetic acid			0.158	0.153	0.149	0.144	2
C₂H₄O₂	Methyl formate		0.194	0.187				1
C₂H₅Br	Bromoethane	0.107	0.104	0.101				1
C₂H₅Cl	Chloroethane	0.145	0.132	0.119	0.106	0.093		2
C₂H₅I	Iodoethane		0.091	0.087	0.083	0.079		1
C₂H₅NO	N-Methylformamide			0.203	0.201	0.199	0.196	2
C₂H₅NO₂	Nitroethane			0.173	0.161	0.149		1
C₂H₆O	Ethanol	0.181	0.174	0.167	0.160	0.153	0.148	1
C₂H₆O₂	1,2-Ethanediol		0.248	0.254	0.258	0.261	0.261	1
C₂H₇NO	Ethanolamine			0.240	0.238	0.236		1

Molecular formula	Name	Thermal conductivity in W/m K						
		−25°C	0°C	25°C	50°C	75°C	100°C	Ref.
C_3F_8	Perfluoropropane	0.062	0.056	0.051	0.046	0.041	0.035	1
C_3H_3N	Acrylonitrile	0.186	0.176	0.166	0.156	0.146	0.136	1
C_3H_5ClO	Epichlorohydrin	0.142	0.137	0.131	0.125	0.119	0.114	2
C_3H_6O	Allyl alcohol			0.162				1
C_3H_6O	Acetone		0.169	0.161				2
C_3H_6O	Methyloxirane			0.181	0.171			1
$C_3H_6O_2$	Propanoic acid		0.147	0.144	0.141	0.139	0.136	1
$C_3H_6O_2$	Ethyl formate	0.181	0.171	0.160	0.149	0.138		1
$C_3H_6O_2$	Methyl acetate	0.174	0.164	0.153	0.143	0.133	0.122	2
C_3H_7Br	1-Bromopropane	0.108	0.104	0.099	0.094			1
C_3H_7Cl	1-Chloropropane	0.129	0.123	0.116	0.110	0.104	0.098	1
C_3H_7I	1-Iodopropane	0.096	0.092	0.087	0.083	0.078	0.074	1
C_3H_7I	2-Iodopropane	0.089	0.085	0.082	0.078	0.074	0.071	1
C_3H_7NO	N,N-Dimethylformamide			0.183	0.175	0.167	0.159	1
$C_3H_7NO_2$	1-Nitropropane			0.152	0.144	0.137		1
C_3H_8O	1-Propanol	0.162	0.158	0.154	0.149	0.145	0.141	2
C_3H_8O	2-Propanol	0.146	0.141	0.135	0.129	0.124	0.118	2
$C_3H_8O_2$	1,2-Propanediol	0.199	0.200	0.200	0.200	0.199	0.197	1
$C_3H_8O_2$	2-Methoxyethanol			0.190	0.180	0.170		1
$C_3H_8O_3$	Glycerol			0.285	0.288	0.292	0.296	1
C_3H_9N	Trimethylamine	0.143	0.133					2
C_4F_8	Perfluorocyclobutane	0.082	0.072	0.063	0.053	0.044	0.034	1
C_4H_4O	Furan	0.142	0.134	0.126				2
C_4H_4S	Thiophene			0.199	0.195	0.191	0.186	2
C_4H_6	1,2-Butadiene	0.147	0.134					1
C_4H_6	2-Butyne	0.137	0.129	0.121				2
$C_4H_6O_2$	Vinyl acetate			0.151	0.141	0.131	0.120	1
$C_4H_6O_3$	Acetic anhydride		0.170	0.164	0.158	0.152	0.146	1
C_4H_8O	Butanal		0.155	0.147	0.140	0.132		1
C_4H_8O	2-Butanone	0.158	0.151	0.145	0.139	0.133		2
C_4H_8O	Tetrahydrofuran	0.132	0.126	0.120	0.114			2
$C_4H_8O_2$	Propyl formate		0.151	0.144	0.137	0.130		1
$C_4H_8O_2$	Ethyl acetate		0.151	0.144	0.136			1
$C_4H_8O_2$	Methyl propanoate			0.141	0.137			1
$C_4H_8O_2$	1,4-Dioxane			0.159	0.147	0.135	0.123	2
C_4H_9Br	1-Bromobutane	0.112	0.107	0.103	0.098	0.093	0.088	1
C_4H_9I	1-Iodobutane		0.094	0.090	0.085	0.081	0.077	1
C_4H_9NO	N,N-Dimethylacetamide			0.175	0.172	0.168		1
$C_4H_{10}O$	1-Butanol		0.158	0.153	0.147	0.142	0.137	1
$C_4H_{10}O$	2-Methyl-2-propanol			0.112	0.110	0.109	0.108	1
$C_4H_{10}O$	Diethyl ether	0.150	0.140	0.130	0.120	0.110	0.100	2
$C_4H_{10}O_2$	2-Ethoxyethanol			0.190	0.182	0.174	0.165	1
C_5H_5N	Pyridine		0.171	0.166	0.162	0.157	0.153	1
$C_5H_6O_2$	Furfuryl alcohol			0.179				1
C_5H_8	2-Methyl-1,3-butadiene	0.141	0.130	0.119				1
C_5H_8	1-Pentyne	0.144	0.136	0.127	0.119			1
C_5H_8	Cyclopentene	0.143	0.136	0.129				2
$C_5H_8O_2$	Methyl methacrylate		0.156	0.147	0.137	0.127	0.117	1
$C_5H_8O_2$	2,4-Pentanedione			0.154	0.150	0.146	0.143	1
C_5H_9NO	N-Methyl-2-pyrrolidone			0.167	0.162	0.157		1
C_5H_{10}	1-Pentene	0.131	0.124	0.116				2
C_5H_{10}	Cyclopentane	0.140	0.133	0.126				2
$C_5H_{10}O$	Pentanal		0.146	0.139	0.133	0.127	0.121	1
$C_5H_{10}O$	2-Pentanone		0.149	0.142	0.135	0.128	0.121	1
$C_5H_{10}O$	3-Pentanone		0.151	0.144	0.137	0.129	0.122	1
$C_5H_{10}O_2$	Pentanoic acid			0.140	0.137	0.133	0.130	1
$C_5H_{10}O_2$	Butyl formate			0.136	0.130	0.123	0.117	1
$C_5H_{10}O_2$	Propyl acetate		0.146	0.140	0.135	0.130	0.124	1
$C_5H_{10}O_2$	Ethyl propanoate				0.133	0.121		1
$C_5H_{10}O_2$	Methyl butanoate			0.140				1
$C_5H_{11}Br$	1-Bromopentane	0.113	0.109	0.105	0.101	0.097	0.093	1
$C_5H_{11}Cl$	1-Chloropentane		0.125	0.120	0.115	0.109		1

Molecular formula	Name	Thermal conductivity in W/m K						
		−25°C	0°C	25°C	50°C	75°C	100°C	Ref.
$C_5H_{11}I$	1-Iodopentane		0.096	0.092	0.088	0.084	0.081	1
C_5H_{12}	Pentane	0.130	0.1207	0.1113	0.1018	0.0923	0.083	4
C_5H_{12}	Isopentane			0.111				1
$C_5H_{12}O$	1-Pentanol	0.159	0.155	0.150	0.145	0.141	0.136	1
$C_5H_{12}O$	2-Methyl-2-butanol		0.119	0.116	0.113	0.109	0.106	1
$C_5H_{12}O_2$	1,5-Pentanediol		0.221	0.222				1
$C_5H_{12}O_3$	Diethylene glycol monomethyl ether			0.190	0.185	0.180	0.175	1
C_6F_6	Hexafluorobenzene				0.083			1
C_6F_{14}	Perfluorohexane		0.067	0.065	0.064			1
$C_6H_3Cl_3$	1,2,3-Trichlorobenzene				0.110	0.108	0.106	1
$C_6H_3Cl_3$	1,2,4-Trichlorobenzene			0.112	0.109	0.106		1
$C_6H_4Cl_2$	o-Dichlorobenzene		0.125	0.121	0.117	0.113	0.109	1
$C_6H_4Cl_2$	m-Dichlorobenzene		0.120	0.116	0.113	0.109		1
$C_6H_4Cl_2$	p-Dichlorobenzene				0.112	0.108	0.105	1
C_6H_5Br	Bromobenzene	0.119	0.115	0.111	0.107	0.103	0.099	1
C_6H_5Cl	Chlorobenzene	0.137	0.132	0.127	0.123	0.118	0.113	1
C_6H_5F	Fluorobenzene			0.136	0.131	0.126		1
C_6H_5I	Iodobenzene	0.106	0.103	0.101	0.098	0.095	0.092	1
$C_6H_5NO_2$	Nitrobenzene			0.149	0.145	0.142	0.139	1
C_6H_6	Benzene			0.1411	0.1329	0.1247		7
C_6H_6ClN	2-Chloroaniline			0.148				1
C_6H_6O	Phenol				0.153	0.149	0.147	1
C_6H_7N	Aniline		0.175					1
$C_6H_8N_2$	Hexanedinitrile			0.174	0.168			1
C_6H_{10}	Cyclohexene	0.142	0.136	0.130	0.124	0.118		2
$C_6H_{10}O$	Cyclohexanone			0.138	0.134	0.130	0.126	1
$C_6H_{10}O$	Mesityl oxide	0.170	0.163	0.156	0.149	0.142	0.134	2
$C_6H_{10}O_3$	Ethyl acetoacetate			0.155	0.152	0.148	0.144	1
$C_6H_{10}O_4$	Diethyl oxalate			0.157				1
C_6H_{12}	1-Hexene	0.138	0.129	0.121	0.113			1
C_6H_{12}	Cyclohexane			0.123	0.117	0.111		2
$C_6H_{12}O$	2-Hexanone		0.156	0.145	0.134	0.124	0.115	1
$C_6H_{12}O$	Cyclohexanol			0.138	0.134	0.130	0.126	1
$C_6H_{12}O_2$	Hexanoic acid		0.148	0.142	0.137	0.131		1
$C_6H_{12}O_2$	Butyl acetate		0.143	0.136	0.130	0.123	0.116	1
$C_6H_{12}O_2$	Propyl propanoate				0.133			1
$C_6H_{12}O_2$	Ethyl butanoate		0.143	0.137	0.131	0.126		1
$C_6H_{12}O_2$	Methyl pentanoate		0.143	0.138	0.132	0.127		1
$C_6H_{12}O_3$	Paraldehyde			0.130				1
$C_6H_{13}Br$	1-Bromohexane	0.115	0.111	0.108	0.104	0.101	0.097	1
$C_6H_{13}I$	1-Iodohexane		0.098	0.095	0.091	0.088	0.084	1
C_6H_{14}	Hexane	0.133	0.1250	0.1167	0.1083	0.0999	0.092	4
C_6H_{14}	2-Methylpentane	0.120	0.1127	0.1050	0.0972	0.0894	0.082	3
C_6H_{14}	3-Methylpentane	0.122	0.1142	0.1064	0.0986	0.0909	0.083	3
C_6H_{14}	2,2-Dimethylbutane	0.108	0.1006	0.0934	0.0861	0.0788	0.072	3
C_6H_{14}	2,3-Dimethylbutane	0.115	0.1076	0.1003	0.0930	0.0857	0.078	3
$C_6H_{14}O$	1-Hexanol	0.161	0.157	0.152	0.147	0.142	0.137	1
$C_6H_{14}O$	Dipropyl ether		0.137	0.130	0.123	0.117		1
$C_6H_{14}O_2$	1,2-Diethoxyethane			0.140	0.133	0.125		1
$C_6H_{14}O_3$	Diethylene glycol monoethyl ether			0.188	0.184	0.180		1
$C_6H_{14}O_4$	Triethylene glycol		0.193	0.195	0.196	0.196	0.196	1
$C_6H_{15}N$	Triethylamine	0.146	0.139	0.132	0.125	0.118	0.111	1
C_7F_{16}	Perfluoroheptane	0.068	0.064	0.060	0.056	0.053		1
C_7H_5N	Benzonitrile			0.148	0.142	0.136	0.130	1
C_7H_6O	Benzaldehyde			0.153	0.148	0.143	0.139	1
C_7H_8	Toluene	0.1455	0.1385	0.1310	0.1235	0.1162	0.1095	9
C_7H_8O	o-Cresol							1
C_7H_8O	m-Cresol			0.149	0.147	0.145		1
C_7H_8O	Benzyl alcohol			0.159	0.158	0.156	0.154	1
C_7H_8O	Anisole			0.145	0.142	0.139	0.136	1
C_7H_9N	2-Methylaniline			0.162				1
C_7H_9N	3-Methylaniline			0.161				1

Molecular formula	Name	Thermal conductivity in W/m K						
		−25°C	0°C	25°C	50°C	75°C	100°C	Ref.
C_7H_{14}	1-Heptene	0.139	0.132	0.125	0.118	0.111		1
C_7H_{14}	Cycloheptane			0.123	0.118	0.112	0.108	1
$C_7H_{14}O$	Heptanal			0.140				1
$C_7H_{14}O$	3-Heptanone		0.143	0.137	0.131	0.125	0.119	1
$C_7H_{14}O$	4-Heptanone			0.136	0.131	0.125	0.120	1
$C_7H_{14}O_2$	Hexyl formate			0.141	0.133	0.126	0.119	1
$C_7H_{14}O_2$	Heptanoic acid			0.140	0.137	0.133		1
$C_7H_{14}O_2$	Pentyl acetate		0.141	0.134	0.126	0.120	0.113	1
$C_7H_{14}O_2$	Butyl propanoate			0.139	0.133	0.126	0.121	1
$C_7H_{14}O_2$	Ethyl pentanoate			0.132				1
$C_7H_{14}O_2$	Methyl hexanoate			0.136	0.131	0.126	0.121	1
C_7H_{16}	Heptane	0.1378	0.1303	0.1228	0.1152	0.1077		5
C_7H_{16}	2-Methylhexane	0.125	0.1177	0.1105	0.1033	0.0961	0.089	3
C_7H_{16}	3-Methylhexane	0.126	0.1184	0.1112	0.1040	0.0968	0.090	3
C_7H_{16}	3-Ethylpentane	0.128	0.1203	0.1128	0.1053	0.0978	0.090	3
C_7H_{16}	2,2-Dimethylpentane	0.111	0.1046	0.0980	0.0913	0.0847	0.078	3
C_7H_{16}	2,3-Dimethylpentane	0.120	0.1127	0.1059	0.0990	0.0922	0.085	3
C_7H_{16}	2,4-Dimethylpentane	0.116	0.1089	0.1020	0.0951	0.0882	0.081	3
C_7H_{16}	3,3-Dimethylpentane	0.113	0.1068	0.1001	0.0934	0.0867	0.080	3
C_7H_{16}	2,2,3-Trimethylbutane	0.107	0.1011	0.0950	0.0889	0.0828	0.077	3
$C_7H_{16}O$	1-Heptanol	0.160	0.158	0.153	0.149	0.144	0.139	1
C_8F_{18}	Perfluorooctane		0.066	0.062	0.059	0.055	0.052	1
C_8H_8	Styrene	0.148	0.142	0.137	0.131	0.126	0.120	2
C_8H_8O	Acetophenone			0.147	0.146	0.144	0.142	1
$C_8H_8O_2$	Methyl benzoate			0.147				1
C_8H_{10}	Ethylbenzene	0.143	0.137	0.130	0.123	0.116	0.110	1
C_8H_{10}	o-Xylene			0.131	0.126	0.120	0.114	2
C_8H_{10}	m-Xylene			0.130	0.124	0.118	0.113	2
C_8H_{10}	p-Xylene			0.130	0.124	0.118	0.112	2
$C_8H_{10}O$	Ethoxybenzene	0.151	0.145	0.140	0.135	0.130		1
$C_8H_{10}O_2$	2-Phenoxyethanol			0.169	0.168	0.166	0.165	1
$C_8H_{11}N$	N-Ethylaniline			0.150				1
$C_8H_{11}N$	N,N-Dimethylaniline			0.122	0.119	0.115		1
C_8H_{16}	1-Octene	0.139	0.133	0.126	0.120	0.114	0.107	1
$C_8H_{16}O$	2-Octanone		0.141	0.135	0.129	0.124	0.118	1
$C_8H_{16}O_2$	Heptyl formate		0.141	0.137	0.132	0.128	0.123	1
$C_8H_{16}O_2$	Octanoic acid			0.146	0.143	0.139	0.135	1
$C_8H_{16}O_2$	Hexyl acetate			0.135	0.129	0.123	0.118	1
$C_8H_{16}O_2$	Pentyl propanoate			0.138	0.132			1
$C_8H_{16}O_2$	Ethyl hexanoate		0.142	0.137	0.133	0.128	0.123	1
$C_8H_{17}Cl$	1-Chlorooctane		0.130	0.127	0.124	0.121	0.119	1
C_8H_{18}	Octane	0.139	0.1317	0.1244	0.1171	0.1097	0.102	4
C_8H_{18}	2-Methylheptane	0.127	0.1206	0.1139	0.1072	0.1005	0.094	3
C_8H_{18}	3-Methylheptane	0.128	0.1216	0.1149	0.1081	0.1014	0.095	3
C_8H_{18}	2,2,4-Trimethylpentane	0.107	0.1007	0.0948	0.0888	0.0829	0.077	3
C_8H_{18}	2,3,4-Trimethylpentane	0.115	0.1093	0.1035	0.0976	0.0918	0.086	3
$C_8H_{18}O$	Ethyl hexyl ether		0.131	0.126	0.120	0.114	0.109	1
$C_8H_{18}O$	1-Octanol		0.162	0.158	0.153	0.148	0.143	1
$C_8H_{18}O$	Dibutyl ether		0.139	0.132	0.125	0.118	0.112	1
$C_8H_{18}O_3$	Diethylene glycol monobutyl ether			0.163	0.158	0.153	0.148	1
$C_8H_{18}O_4$	Triethylene glycol dimethyl ether			0.169	0.158	0.147		1
$C_8H_{18}O_5$	Tetraethylene glycol			0.191	0.192			1
C_9H_7N	Quinoline			0.147	0.144	0.141	0.138	1
C_9H_{10}	Indan			0.135				1
$C_9H_{10}O_2$	Ethyl benzoate			0.141				1
C_9H_{12}	Propylbenzene	0.134	0.130	0.125	0.120	0.115	0.109	1
C_9H_{12}	Isopropylbenzene	0.132	0.128	0.123	0.118	0.112	0.107	1
C_9H_{12}	1,2,4-Trimethylbenzene			0.129	0.124	0.118	0.114	1
C_9H_{12}	1,3,5-Trimethylbenzene	0.143	0.139	0.134	0.129	0.123	0.117	1
C_9H_{18}	1-Nonene	0.136	0.130	0.123	0.116	0.110	0.104	1
$C_9H_{18}O_2$	Nonanoic acid			0.150	0.146	0.142	0.138	1
$C_9H_{18}O_2$	Heptyl acetate			0.135	0.128	0.122	0.116	1

Molecular formula	Name	Thermal conductivity in W/m K						
		−25°C	0°C	25°C	50°C	75°C	100°C	Ref.
$C_9H_{19}Br$	1-Bromononane		0.116	0.112	0.109	0.106	0.103	1
$C_9H_{19}Cl$	1-Chlorononane		0.132	0.128	0.124	0.120	0.115	1
$C_9H_{19}I$	1-Iodononane		0.105	0.102	0.099	0.095	0.092	1
C_9H_{20}	Nonane	0.141	0.1337	0.1269	0.1201	0.1133	0.106	4
$C_9H_{20}O$	1-Nonanol		0.164	0.159	0.155	0.150	0.145	1
$C_{10}H_7Br$	1-Bromonaphthalene			0.110	0.109	0.108	0.106	1
$C_{10}H_7Cl$	1-Chloronaphthalene			0.126				1
$C_{10}H_{10}O_4$	Dimethyl phthalate			0.1473	0.1443	0.1409	0.1373	10
$C_{10}H_{12}$	1,2,3,4-Tetrahydronaphthalene			0.131	0.129	0.128	0.126	1
$C_{10}H_{14}$	Butylbenzene			0.126	0.121	0.116	0.111	1
$C_{10}H_{14}$	sec-Butylbenzene, (±)-		0.129	0.124	0.119	0.114	0.108	1
$C_{10}H_{14}$	tert-Butylbenzene			0.117	0.114	0.110	0.106	1
$C_{10}H_{14}$	1-Isopropyl-4-methylbenzene	0.132	0.127	0.122	0.117	0.112	0.107	2
$C_{10}H_{14}$	o-Diethylbenzene		0.133	0.127	0.122	0.116	0.111	1
$C_{10}H_{18}$	trans-Decahydronaphthalene			0.113				1
$C_{10}H_{20}$	1-Decene	0.138	0.132	0.126	0.120	0.114	0.109	1
$C_{10}H_{20}O$	Decanal		0.149	0.144	0.139	0.134	0.129	1
$C_{10}H_{20}O_2$	Heptyl propanoate			0.137	0.132	0.127	0.122	1
$C_{10}H_{20}O_2$	Hexyl butanoate			0.137	0.132	0.127	0.121	1
$C_{10}H_{20}O_2$	Decanoic acid				0.148	0.144	0.140	1
$C_{10}H_{22}$	Decane	0.142	0.1360	0.1296	0.1232	0.1167	0.110	4
$C_{10}H_{22}O$	1-Decanol			0.162	0.159	0.155	0.151	1
$C_{10}H_{22}O$	Dipentyl ether			0.131	0.125	0.121	0.116	1
$C_{10}H_{22}O_2$	1,2-Dibutoxyethane			0.140	0.134	0.127	0.120	1
$C_{11}H_{16}$	Pentylbenzene		0.135	0.130	0.125	0.120	0.115	1
$C_{11}H_{22}$	1-Undecene			0.126	0.118	0.114	0.108	1
$C_{11}H_{22}O$	6-Undecanone			0.137	0.132	0.127		1
$C_{11}H_{22}O_2$	Undecanoic acid				0.153	0.149		1
$C_{11}H_{22}O_2$	Octyl propanoate			0.135	0.130	0.125	0.120	1
$C_{11}H_{22}O_2$	Heptyl butanoate			0.139	0.134	0.129	0.123	1
$C_{11}H_{24}$	Undecane			0.136	0.128	0.122	0.116	1
$C_{11}H_{24}O$	1-Undecanol			0.169	0.165	0.161	0.158	1
$C_{12}H_{10}O$	Diphenyl ether			0.139	0.135	0.131		2
$C_{12}H_{14}O_4$	Diethyl phthalate			0.172	0.169	0.166		1
$C_{12}H_{16}$	Cyclohexylbenzene			0.121	0.119	0.117		1
$C_{12}H_{18}$	Hexylbenzene		0.141	0.137	0.132	0.128	0.124	1
$C_{12}H_{24}O_2$	Decyl acetate			0.146	0.136	0.126		1
$C_{12}H_{24}O_2$	Octyl butanoate			0.139	0.134	0.129	0.125	1
$C_{12}H_{26}$	Dodecane			0.135	0.130	0.124	0.119	1
$C_{12}H_{26}O$	1-Dodecanol				0.167	0.163	0.159	1
$C_{12}H_{26}O_3$	Diethylene glycol dibutyl ether		0.150	0.146	0.143	0.139	0.135	1
$C_{12}H_{27}N$	Tributylamine			0.129				1
$C_{13}H_{26}$	1-Tridecene			0.130	0.125	0.120	0.115	1
$C_{13}H_{28}$	Tridecane			0.130	0.125	0.120	0.115	1
$C_{14}H_{28}$	1-Tetradecene			0.136	0.131	0.126	0.121	1
$C_{14}H_{30}$	Tetradecane			0.139	0.134	0.129	0.124	1
$C_{14}H_{30}O$	1-Tetradecanol				0.167	0.162	0.157	2
$C_{16}H_{22}O_4$	Dibutyl phthalate		0.139	0.136	0.134	0.131	0.129	1
$C_{16}H_{34}$	Hexadecane			0.140	0.135	0.130	0.125	2
$C_{18}H_{38}$	Octadecane				0.146	0.142	0.137	2
$C_{20}H_{40}O_2$	Butyl palmitate			0.151	0.148	0.144	0.140	1
$C_{22}H_{42}O_2$	Butyl oleate			0.157	0.153	0.149	0.145	1
$C_{22}H_{42}O_4$	Dioctyl hexanedioate			0.157	0.153	0.149	0.145	1

DIFFUSION IN GASES

This table gives binary diffusion coefficients D_{12} for a number of common gases as a function of temperature. Values refer to atmospheric pressure. The diffusion coefficient is inversely proportional to pressure as long as the gas is in a regime where binary collisions dominate. See Reference 1 for a discussion of the dependence of D_{12} on temperature and composition.

The first part of the table gives data for several gases in the presence of a large excess of air. The remainder applies to equimolar mixtures of gases. Each gas pair is ordered alphabetically according to the most common way of writing the formula. The listing of pairs then follows alphabetical order by the first constituent.

References

1. Marrero, T. R., and Mason, E. A., *J. Phys. Chem. Ref. Data*, 1, 1, 1972.
2. Kestin, J., et al., *J. Phys. Chem. Ref. Data*, 13, 229, 1984.

$D_{12}/\text{cm}^2\,\text{s}^{-1}$ for $p = 101.325$ kPa and the Specified T/K

System	200	273.15	293.15	373.15	473.15	573.15	673.15
Large Excess of Air							
Ar-air		0.167	0.189	0.289	0.437	0.612	0.810
CH_4-air			0.210	0.321	0.485	0.678	0.899
CO-air			0.208	0.315	0.475	0.662	0.875
CO_2-air			0.160	0.252	0.390	0.549	0.728
H_2-air		0.668	0.756	1.153	1.747	2.444	3.238
H_2O-air			0.242	0.399	0.638	0.873	1.135
He-air		0.617	0.697	1.057	1.594	2.221	2.933
SF_6-air				0.150	0.233	0.329	0.438
Equimolar Mixture							
Ar-CH_4				0.306	0.467	0.657	0.876
Ar-CO		0.168	0.190	0.290	0.439	0.615	0.815
Ar-CO_2		0.129	0.148	0.235	0.365	0.517	0.689
Ar-H_2		0.698	0.794	1.228	1.876	2.634	3.496
Ar-He	0.381	0.645	0.726	1.088	1.617	2.226	2.911
Ar-Kr	0.064	0.117	0.134	0.210	0.323	0.456	0.605
Ar-N_2		0.168	0.190	0.290	0.439	0.615	0.815
Ar-Ne	0.160	0.277	0.313	0.475	0.710	0.979	1.283
Ar-O_2		0.166	0.187	0.285	0.430	0.600	0.793
Ar-SF_6				0.128	0.202	0.290	0.389
Ar-Xe	0.052	0.095	0.108	0.171	0.264	0.374	0.498
CH_4-H_2			0.708	1.084	1.648	2.311	3.070
CH_4-He			0.650	0.992	1.502	2.101	2.784
CH_4-N_2			0.208	0.317	0.480	0.671	0.890
CH_4-O_2			0.220	0.341	0.523	0.736	0.978
CH_4-SF_6				0.167	0.257	0.363	0.482
CO-CO_2			0.162	0.250	0.384		
CO-H_2	0.408	0.686	0.772	1.162	1.743	2.423	3.196
CO-He	0.365	0.619	0.698	1.052	1.577	2.188	2.882
CO-Kr		0.131	0.149	0.227	0.346	0.485	0.645
CO-N_2	0.133	0.208	0.231	0.336	0.491	0.673	0.878
CO-O_2			0.202	0.307	0.462	0.643	0.849
CO-SF_6				0.144	0.226	0.323	0.432
CO_2-C_3H_8			0.084	0.133	0.209		
CO_2-H_2	0.315	0.552	0.627	0.964	1.470	2.066	2.745
CO_2-H_2O			0.162	0.292	0.496	0.741	1.021
CO_2-He	0.300	0.513	0.580	0.878	1.321		
CO_2-N_2			0.160	0.253	0.392	0.553	0.733
CO_2-N_2O	0.055	0.099	0.113	0.177	0.276		
CO_2-Ne	0.131	0.227	0.258	0.395	0.603	0.847	
CO_2-O_2			0.159	0.248	0.380	0.535	0.710
CO_2-SF_6				0.099	0.155		
D_2-H_2	0.631	1.079	1.219	1.846	2.778	3.866	5.103
H_2-He	0.775	1.320	1.490	2.255	3.394	4.726	6.242
H_2-Kr	0.340	0.601	0.682	1.053	1.607	2.258	2.999

System	200	273.15	293.15	373.15	473.15	573.15	673.15
H_2-N_2	0.408	0.686	0.772	1.162	1.743	2.423	3.196
H_2-Ne	0.572	0.982	1.109	1.684	2.541	3.541	4.677
H_2-O_2		0.692	0.782	1.188	1.792	2.497	3.299
H_2-SF_6			0.412	0.649	0.998	1.400	1.851
H_2-Xe		0.513	0.581	0.890	1.349	1.885	2.493
H_2O-N_2			0.242	0.399			
H_2O-O_2			0.244	0.403	0.645	0.882	1.147
He-Kr	0.330	0.559	0.629	0.942	1.404	1.942	2.550
He-N_2	0.365	0.619	0.698	1.052	1.577	2.188	2.882
He-Ne	0.563	0.948	1.066	1.592	2.362	3.254	4.262
He-O_2		0.641	0.723	1.092	1.640	2.276	2.996
He-SF_6			0.400	0.592	0.871	1.190	1.545
He-Xe	0.282	0.478	0.538	0.807	1.201	1.655	2.168
Kr-N_2		0.131	0.149	0.227	0.346	0.485	0.645
Kr-Ne	0.131	0.228	0.258	0.392	0.587	0.812	1.063
Kr-Xe	0.035	0.064	0.073	0.116	0.181	0.257	0.344
N_2-Ne			0.317	0.483	0.731	1.021	1.351
N_2-O_2			0.202	0.307	0.462	0.643	0.849
N_2-SF_6			0.148	0.231	0.328	0.436
N_2-Xe		0.107	0.122	0.188	0.287	0.404	0.539
Ne-Xe	0.111	0.193	0.219	0.332	0.498	0.688	0.901
O_2-SF_6			0.097	0.154	0.238	0.334	0.441

DIFFUSION OF GASES IN WATER

This table gives values of the diffusion coefficient, D, for diffusion of several common gases in water at various temperatures. For simple one-dimensional transport, the diffusion coefficient describes the time-rate of change of concentration, dc/dt, through the equation

$$dc/dt = D\, d^2c/dx^2$$

where x is, for example, the perpendicular distance from a gas–liquid interface. The values below have been selected from the references indicated; in some cases data have been refitted to permit interpolation in temperature.

Gas–liquid diffusion coefficients are difficult to measure, and large differences are found between values obtained by different authors and through different experimental methods. See References 1 and 2 for a discussion of measurement techniques.

References

1. Jähne, B., Heinz, G., and Dietrich, W., *J. Geophys. Res.*, 92, 10767, 1987.
2. Himmelblau, D. M., *Chem. Rev.*, 64, 527, 1964.
3. Boerboom, A. J. H., and Kleyn, G., *J. Chem. Phys.*, 50, 1086, 1969.
4. O'Brien, R. N., and Hyslop, W. F., *Can. J. Chem.*, 55, 1415, 1977.
5. Maharajh, D. M., and Walkley, J., *Can. J. Chem.*, 51, 944, 1973.
6. *Landolt-Börnstein, Numerical Data and Functional Relationships in Science and Technology, Sixth Edition*, II/5a, *Transport Phenomena I (Viscosity and Diffusion)*, Springer–Verlag, Heidelberg, 1969.

$$D/10^{-5}\,\text{cm}^2\,\text{s}^{-1}$$

	10°C	15°C	20°C	25°C	30°C	35°C	Ref.
Ar				2.5			3,4
$CHCl_2F$				1.80			5
CH_3Br				1.35			5
CH_3Cl				1.40			5
CH_4	1.24	1.43	1.62	1.84	2.08	2.35	1
CO_2	1.26	1.45	1.67	1.91	2.17	2.47	1
C_2H_2	1.43	1.59	1.78	1.99	2.23		2
Cl_2		1.13	1.5	1.89			2,6
HBr				3.15			6
HCl				3.07			6
H_2	3.62	4.08	4.58	5.11	5.69	6.31	1
H_2S				1.36			2,6
He	5.67	6.18	6.71	7.28	7.87	8.48	1,3
Kr	1.20	1.39	1.60	1.84	2.11	2.40	1,3
NH_3		1.3	1.5				2
NO_2			1.23	1.4	1.59		2,6
N_2				2.0			2
N_2O		1.62	2.11	2.57			2,6
Ne	2.93	3.27	3.64	4.03	4.45	4.89	1,3
O_2		1.67	2.01	2.42			2,6
Rn	0.81	0.96	1.13	1.33	1.55	1.80	1
SO_2		1.62	1.83	2.07	2.32		2
Xe	0.93	1.08	1.27	1.47	1.70	1.95	1,3

DIFFUSION COEFFICIENTS IN LIQUIDS AT INFINITE DILUTION

This table lists diffusion coefficients D_{AB} at infinite dilution for some binary liquid mixtures. Although values are given to two decimal places, measurements in the literature are often in poor agreement. Therefore most values in the table cannot be relied upon to better than 10%.

Solvents are listed in alphabetical order, as are the solutes within each solvent group.

Reference

Landolt-Börnstein, *Numerical Data and Functional Relationships in Science and Technology*, Sixth Edition, Vol. II/5a, Springer–Verlag, Heidelberg, 1969.

Solute	Solvent	$t/°C$	D_{AB} $10^{-5}\,\mathrm{cm^2\,s^{-1}}$	Solute	Solvent	$t/°C$	D_{AB} $10^{-5}\,\mathrm{cm^2\,s^{-1}}$
Acetic acid	Acetone	25	3.31	Acetone	Tetrachloromethane	25	1.75
Benzoic acid	Acetone	25	2.62	Benzene	Tetrachloromethane	25	1.42
Formic acid	Acetone	25	3.77	Cyclohexane	Tetrachloromethane	25	1.30
Nitrobenzene	Acetone	20	2.94	Ethanol	Tetrachloromethane	25	1.90
Tetrachloromethane	Acetone	25	3.29	Iodine	Tetrachloromethane	30	1.63
Trichloromethane	Acetone	25	3.64	Trichloromethane	Tetrachloromethane	25	1.66
Water	Acetone	25	4.56	Acetic acid	Toluene	25	2.26
Acetic acid	Benzene	25	2.09	Benzene	Toluene	25	2.54
Aniline	Benzene	25	1.96	Benzoic acid	Toluene	25	1.49
Benzoic acid	Benzene	25	1.38	Cyclohexane	Toluene	25	2.42
Bromobenzene	Benzene	8	1.45	Formic acid	Toluene	25	2.65
2-Butanone	Benzene	30	2.09	Water	Toluene	25	6.19
Chloroethylene	Benzene	8	1.77	Acetone	Trichloromethane	25	2.55
Cyclohexane	Benzene	25	2.25	Benzene	Trichloromethane	25	2.89
Ethanol	Benzene	25	3.02	2-Butanone	Trichloromethane	25	2.13
Formic acid	Benzene	25	2.28	Butyl acetate	Trichloromethane	25	1.71
Heptane	Benzene	25	1.78	Diethyl ether	Trichloromethane	25	2.15
Methanol	Benzene	25	3.80	Ethanol	Trichloromethane	15	2.20
Toluene	Benzene	25	1.85	Ethyl acetate	Trichloromethane	25	2.02
1,2,4-Trichlorobenzene	Benzene	8	1.34	Acetic acid	Water	25	1.29
Trichloromethane	Benzene	25	2.26	Acetone	Water	25	1.28
Adipic acid	1-Butanol	30	0.40	Acetonitrile	Water	15	1.26
Benzene	1-Butanol	25	1.00	Alanine	Water	25	0.91
Biphenyl	1-Butanol	25	0.63	Allyl alcohol	Water	15	0.90
Butyric acid	1-Butanol	30	0.51	Aniline	Water	20	0.92
p-Dichlorobenzene	1-Butanol	25	0.82	Arabinose	Water	20	0.69
Methanol	1-Butanol	30	0.59	Benzene	Water	20	1.02
Oleic acid	1-Butanol	30	0.25	1-Butanol	Water	25	0.56
Propane	1-Butanol	25	1.57	Caprolactam	Water	25	0.87
Water	1-Butanol	25	0.56	Chloroethylene	Water	25	1.34
Benzene	Cyclohexane	25	1.41	Cyclohexane	Water	20	0.84
Tetrachloromethane	Cyclohexane	25	1.49	Diethylamine	Water	20	0.97
Toluene	Cyclohexane	25	1.57	Ethanol	Water	25	1.24
Allyl alcohol	Ethanol	20	0.98	Ethanolamine	Water	25	1.08
Benzene	Ethanol	25	1.81	Ethyl acetate	Water	20	1.00
Iodine	Ethanol	25	1.32	Ethylbenzene	Water	20	0.81
Iodobenzene	Ethanol	20	1.00	Ethylene glycol	Water	25	1.16
3-Methyl-1-butanol	Ethanol	20	0.81	Glucose	Water	25	0.67
Pyridine	Ethanol	20	1.10	Glycerol	Water	25	1.06
Tetrachloromethane	Ethanol	25	1.50	Glycine	Water	25	1.05
Water	Ethanol	25	1.24	Lactose	Water	15	0.38
Acetic acid	Ethyl acetate	20	2.18	Maltose	Water	15	0.38
Acetone	Ethyl acetate	20	3.18	Mannitol	Water	15	0.50
2-Butanone	Ethyl acetate	30	2.93	Methane	Water	25	1.49
Ethyl benzoate	Ethyl acetate	20	1.85	Methanol	Water	15	1.28
Nitrobenzene	Ethyl acetate	20	2.25	3-Methyl-1-butanol	Water	10	0.69
Water	Ethyl acetate	25	3.20	Methylcyclopentane	Water	20	0.85
Benzene	Heptane	25	3.91	Phenol	Water	20	0.89
Toluene	Heptane	25	3.72	1-Propanol	Water	15	0.87
Bromobenzene	Hexane	8	2.60	Propene	Water	25	1.44
2-Butanone	Hexane	30	3.74	Pyridine	Water	25	0.58
Dodecane	Hexane	25	2.73	Raffinose	Water	15	0.33
Iodine	Hexane	25	4.45	Sucrose	Water	25	0.52
Methane	Hexane	25	0.09	Toluene	Water	20	0.85
Propane	Hexane	25	4.87	Urea	Water	25	1.38
Tetrachloromethane	Hexane	25	3.70	Urethane	Water	15	0.80
Toluene	Hexane	25	4.21				

Section 7
Biochemistry

PROPERTIES OF AMINO ACIDS

This table gives selected properties of some important amino acids and closely related compounds. The first part of the table lists the 20 "standard" amino acids that are the basic constituents of proteins. The second part includes other amino acids and related compounds of biochemical importance. Within each part of the table the compounds are listed by name in alphabetical order. Structures are given in the following table.

Symbol : Three-letter symbol for the standard amino acids
M_r: Molecular weight
t_m: Melting point
pK_a, pK_b, pK_c, pK_d : Negative of the logarithm of the acid dissociation constants for the COOH and NH_2 groups (and, in some cases, other groups) in the molecule (at 25°C)
pI: pH at the isoelectric point
S: Solubility in water in units of grams of compound per kilogram of water; a temperature of 25°C is understood unless otherwise stated in a superscript. When quantitative data are not available, the notations sl.s. (for slightly soluble), s. (for soluble), and v.s. (for very soluble) are used.
V_2^0: Partial molar volume in aqueous solution at infinite dilution (at 25°C)

Data on the enthalpy of formation of many of these compounds are included in the table "Standard Thermodynamic Properties of Chemical Substances" in Section 5 of this Handbook. Absorption spectra and optical rotation data can be found in Reference 3. Partial molar volume is taken from Reference 5; other thermodynamic properties, including solubility as a function of temperature, are given in References 3 and 5. Most of the pK values come from References 1, 6, and 7.

References

1. Dawson, R. M. C., Elliott, D. C., Elliott, W. H., and Jones, K. M., *Data for Biochemical Research*, Third Edition, Clarendon Press, Oxford, 1986.
2. O'Neil, Maryadele J., Ed., *The Merck Index, Thirteenth Edition*, Merck & Co., Rahway, NJ, 2001.
3. Sober, H. A., Ed., *CRC Handbook of Biochemistry. Selected Data for Molecular Biology*, CRC Press, Boca Raton, FL, 1968.
4. Voet, D., and Voet, J. G., *Biochemistry, Second Edition*, John Wiley & Sons, New York, 1995.
5. Hinz, H. J., Ed., *Thermodynamic Data for Biochemistry and Biotechnology*, Springer–Verlag, Heidelberg, 1986.
6. Fasman, G. D., Ed. *Practical Handbook of Biochemistry and Molecular Biology*, CRC Press, Boca Raton, FL, 1989.
7. Smith, R. M., and Martell, A. E., *NIST Standard Reference Database 46: Critically Selected Stability Constants of Metal Complexes Database*, Version 3.0, National Institute of Standards and Technology, Gaithersburg, MD, 1997.
8. Ramasami, P., *J. Chem. Eng. Data*, 47, 1164, 2002.

Symbol	Name	Mol. form	M_r	t_m/°C	pK_a	pK_b	pK_c	pK_d	pI	S/g kg^{-1}	V_2^0/cm^3 mol^{-1}
Ala	L-Alanine	$C_3H_7NO_2$	89.09	297	2.33	9.71			6.00	166.9	60.54
Arg	L-Arginine	$C_6H_{14}N_4O_2$	174.20	244	2.03	9.00	12.10		10.76	182.6	127.42
Asn	L-Asparagine	$C_4H_8N_2O_3$	132.12	235	2.16	8.73			5.41	25.1	78.0
Asp	L-Aspartic acid	$C_4H_7NO_4$	133.10	270	1.95	9.66	3.71		2.77	5.04	74.8
Cys	L-Cysteine	$C_3H_7NO_2S$	121.16	240	1.91	10.28	8.14		5.07	v.s.	73.45
Gln	L-Glutamine	$C_5H_{10}N_2O_3$	146.14	185	2.18	9.00			5.65	42.5	
Glu	L-Glutamic acid	$C_5H_9NO_4$	147.13	160	2.16	9.58	4.15		3.22	8.6	89.85
Gly	Glycine	$C_2H_5NO_2$	75.07	290	2.34	9.58			5.97	250.2	43.26
His	L-Histidine	$C_6H_9N_3O_2$	155.15	287	1.70	9.09	6.04		7.59	43.5	98.3
Ile	L-Isoleucine	$C_6H_{13}NO_2$	131.17	284	2.26	9.60			6.02	34.2	105.80
Leu	L-Leucine	$C_6H_{13}NO_2$	131.17	293	2.32	9.58			5.98	22.0	107.77
Lys	L-Lysine	$C_6H_{14}N_2O_2$	146.19	224	2.15	9.16	10.67		9.74	5.8	108.5
Met	L-Methionine	$C_5H_{11}NO_2S$	149.21	281	2.16	9.08			5.74	56	105.57
Phe	L-Phenylalanine	$C_9H_{11}NO_2$	165.19	283	2.18	9.09			5.48	27.9	121.5
Pro	L-Proline	$C_5H_9NO_2$	115.13	221	1.95	10.47			6.30	1622	82.76
Ser	L-Serine	$C_3H_7NO_3$	105.09	228	2.13	9.05			5.68	250	60.62
Thr	L-Threonine	$C_4H_9NO_3$	119.12	256	2.20	8.96			5.60	98.1	76.90
Trp	L-Tryptophan	$C_{11}H_{12}N_2O_2$	204.23	289	2.38	9.34			5.89	13.2	143.8
Tyr	L-Tyrosine	$C_9H_{11}NO_3$	181.19	343	2.24	9.04	10.10		5.66	0.46	
Val	L-Valine	$C_5H_{11}NO_2$	117.15	315	2.27	9.52			5.96	88	90.75

Name	Mol. form	M_r	t_m/°C	pK_a	pK_b	pK_c	pK_d	pI	S/g kg^{-1}	V_2^0/cm^3 mol^{-1}
N-Acetylglutamic acid	$C_7H_{11}NO_5$	189.17	199						s.	
N6-Acetyl-L-lysine	$C_8H_{16}N_2O_3$	188.22	265	2.12	9.51					
β-Alanine	$C_3H_7NO_2$	89.09	200	3.51	10.08				723.6	58.28
2-Aminoadipic acid	$C_6H_{11}NO_4$	161.16	207	2.14	4.21	9.77		3.18	2.2^{40}	
DL-2-Aminobutanoic acid	$C_4H_9NO_2$	103.12	304	2.30	9.63			6.06	210	75.6
DL-3-Aminobutanoic acid	$C_4H_9NO_2$	103.12	194.3	3.43	10.05			7.30	1250	76.3
4-Aminobutanoic acid	$C_4H_9NO_2$	103.12	203	4.02	10.35				971	73.2
10-Aminodecanoic acid	$C_{10}H_{21}NO_2$	187.28	188.5							167.3

Name	Mol. form	M_r	$t_m/°C$	pK_a	pK_b	pK_c	pK_d	pI	S/g kg^{-1}	V_2^0/cm^3 mol^{-1}
7-Aminoheptanoic acid	$C_7H_{15}NO_2$	145.20	195						v.s.	120.0
6-Aminohexanoic acid	$C_6H_{13}NO_2$	131.17	205					7.29	863	104.2
L-3-Amino-2-methylpropanoic acid	$C_4H_9NO_2$	103.12	185						s.	
2-Amino-2-methylpropanoic acid	$C_4H_9NO_2$	103.12	335	2.36	10.21			5.72	137	77.55
9-Aminononanoic acid	$C_9H_{19}NO_2$	173.26	191							151.3
8-Aminooctanoic acid	$C_8H_{17}NO_2$	159.23	192							136.1
5-Amino-4-oxopentanoic acid	$C_5H_9NO_3$	131.13	118	4.05	8.90					
5-Aminopentanoic acid	$C_5H_{11}NO_2$	117.15	157 dec						s.	87.6
o-Anthranilic acid	$C_7H_7NO_2$	137.14	146	2.05	4.95				3.5^{14}	
Azaserine	$C_5H_7N_3O_4$	173.13	150		8.55				v.s.	
Canavanine	$C_5H_{12}N_4O_3$	176.17	172	2.50	6.60	9.25		7.93	v.s.	
L-γ-Carboxyglutamic acid	$C_6H_9NO_6$	191.14	167	1.70	9.90	4.75	3.20			
Carnosine	$C_9H_{14}N_4O_3$	226.23	260	2.51	9.35	6.76			322	
Citrulline	$C_6H_{13}N_3O_3$	175.19	222	2.32	9.30			5.92	s.	
Creatine	$C_4H_9N_3O_2$	131.13	303	2.63	14.30				16	
L-Cysteic acid	$C_3H_7NO_5S$	169.16	260	1.89	8.70	1.30			v.s.	
L-Cystine	$C_6H_{12}N_2O_4S_2$	240.30	260	1.50	8.80	2.05	8.03		0.11	
2,4-Diaminobutanoic acid	$C_4H_{10}N_2O_2$	118.13	118.1	1.85	8.24	10.44		9.27	s.	
3,5-Dibromo-L-tyrosine	$C_9H_9Br_2NO_3$	338.98	245						2.72	
3,5-Dichloro-L-tyrosine	$C_9H_9Cl_2NO_3$	250.08	247						1.97	
3,5-Diiodo-L-tyrosine	$C_9H_9I_2NO_3$	432.98	213	2.12	9.10	6.16			0.62	
Dopamine	$C_8H_{11}NO_2$	153.18			10.36	8.88			s.	
L-Ethionine	$C_6H_{13}NO_2S$	163.24	273	2.18	9.05	13.10				
N-Glycylglycine	$C_4H_8N_2O_3$	132.12	263	3.13	8.10				225	
Guanidinoacetic acid	$C_3H_7N_3O_2$	117.11	282	2.82					5	
Histamine	$C_5H_9N_3$	111.15	83		9.83	6.11			v.s.	
L-Homocysteine	$C_4H_9NO_2S$	135.19	232	2.15	8.57	10.38		5.55	s.	
Homocystine	$C_8H_{16}N_2O_4S_2$	268.35	264	1.59	9.44	2.54	8.52		0.2	
L-Homoserine	$C_4H_9NO_3$	119.12	203	2.27	9.28			6.17	1100	
3-Hydroxy-DL-glutamic acid	$C_5H_9NO_5$	163.13	209					3.28		
5-Hydroxylysine	$C_6H_{14}N_2O_3$	162.19		2.13	8.85	9.83		9.15		
trans-4-Hydroxy-L-proline	$C_5H_9NO_3$	131.13	274	1.82	9.47			5.74	361	84.49
L-3-Iodotyrosine	$C_9H_{10}INO_3$	307.08	205	2.20	9.10	8.70			sl.s.	
L-Kynurenine	$C_{10}H_{12}N_2O_3$	208.21	194						sl.s.	
L-Lanthionine	$C_6H_{12}N_2O_4S$	208.24	294						1.5	
Levodopa	$C_9H_{11}NO_4$	197.19	277	2.32	8.72	9.96	11.79		5^{20}	
L-1-Methylhistidine	$C_7H_{11}N_3O_2$	169.18	249	1.69	8.85	6.48			200	
L-Norleucine	$C_6H_{13}NO_2$	131.17	301	2.31	9.68			6.09	15	107.7
L-Norvaline	$C_5H_{11}NO_2$	117.15	307	2.31	9.65				107	91.8
L-Ornithine	$C_5H_{12}N_2O_2$	132.16	140	1.94	8.78	10.52		9.73	v.s.	
O-Phosphoserine	$C_3H_8NO_6P$	185.07	166	2.14	9.80	5.70				
L-Pyroglutamic acid	$C_5H_7NO_3$	129.12	162	3.32						
Sarcosine	$C_3H_7NO_2$	89.09	212	2.18	9.97				428	
Taurine	$C_2H_7NO_3S$	125.15	328	-0.3	9.06				105	
L-Thyroxine	$C_{15}H_{11}I_4NO_4$	776.87	235	2.20	10.01	6.45			sl.s.	

STRUCTURES OF COMMON AMINO ACIDS

L-Alanine (Ala)

L-Arginine (Arg)

L-Asparagine (Asn)

L-Aspartic acid (Asp)

L-Cysteine (Cys)

L-Glutamine (Gln)

L-Glutamic acid (Glu)

Glycine (Gly)

L-Histidine (His)

L-Isoleucine (Ile)

L-Leucine (Leu)

L-Lysine (Lys)

L-Methionine (Met)

L-Phenylalanine (Phe)

L-Proline (Pro)

L-Serine (Ser)

L-Threonine (Thr)

L-Tryptophan (Trp)

L-Tyrosine (Tyr)

L-Valine (Val)

N-Acetylglutamic acid

N6-Acetyl-L-lysine

β-Alanine

2-Aminoadipic acid

DL-2-Aminobutanoic acid

DL-3-Aminobutanoic acid

4-Aminobutanoic acid

6-Aminohexanoic acid

L-3-Amino-2-methylpropa-
noic acid

2-Amino-2-methylpropanoic
acid

5-Amino-4-oxopentanoic acid

5-Aminopentanoic acid

Azaserine

Canavanine

L-γ-Carboxyglutamic acid

Carnosine

Citrulline

Creatine

L-Cysteic acid

L-Cystine

2,4-Diaminobutanoic acid

3,5-Dibromo-L-tyrosine

3,5-Diiodo-L-tyrosine

Dopamine

L-Ethionine

N-Glycylglycine

Guanidinoacetic acid

Histamine

L-Homocysteine

Homocystine

L-Homoserine

trans-4-Hydroxy-L-proline

L-3-Iodotyrosine

L-Kynurenine

L-Lanthionine

Levodopa

L-1-Methylhistidine

L-Norleucine

L-Norvaline

L-Ornithine

O-Phosphoserine

L-Pyroglutamic acid

Sarcosine

Taurine

L-Thyroxine

PROPERTIES OF PURINE AND PYRIMIDINE BASES

This table lists some of the important purine and pyrimidine bases that occur in nucleic acids. The pK_a values (negative logarithm of the acid dissociation constant) are given for each ionization stage. The last column gives the aqueous solubility S at the indicated temperature in units of grams per 100 grams of solution. The numbering system in the rings is:

Purine

Pyrimidine

References

1. Dawson, R. M. C., et al., *Data for Biochemical Research*, 3rd ed., Clarendon Press, Oxford, 1986.
2. O'Neil, M. J., Ed., *The Merck Index*, 13th ed., Merck and Co., Rahway, NJ, 2001.

Common name	Systematic name	Mol. form.	Mol. wt.	pK_a values			S/mass % (temp.)
Pyrimidines							
Cytosine	4-Amino-2-hydroxypyrimidine	$C_4H_5N_3O$	111.10	4.60	12.16		0.73 (25°C)
5-Methylcytosine	4-Amino-2-hydroxy-5-methylpyrimidine	$C_5H_7N_3O$	125.13	4.6	12.4		0.45 (25°C)
5-Hydroxymethylcytosine	4-Amino-2-hydroxy-5-hydroxy-methylpyrimidine	$C_5H_7N_3O_2$	141.13	4.3	13		
Uracil	2,4-Dihydroxypyrimidine	$C_4H_4N_2O_2$	112.09	0.5	9.5	>13	0.27 (25°C)
Thymine	5-Methyluracil	$C_5H_6N_2O_2$	126.11	9.94	>13		0.35 (25°C)
Orotic acid	Uracil-6-carboxylic acid	$C_5H_4N_2O_4$	156.10	2.4	9.5	>13	0.18 (18°C)
Purines							
Adenine	6-Aminopurine	$C_5H_5N_5$	135.14	<1	4.3	9.83	0.104 (25°C)
Guanine	2-Amino-6-hydroxypurine	$C_5H_5N_5O$	151.13	3.3	9.2	12.3	0.0068 (40°C)
7-Methylguanine	7-Methyl-2-amino-6-hydroxypurine	$C_6H_7N_5O$	165.16	3.5	9.9		
Isoguanine	6-Amino-2-hydroxypurine	$C_5H_5N_5O$	151.13	4.5	9.0		0.006 (25°C)
Xanthine	2,6-Dioxopurine	$C_5H_4N_4O_2$	152.11	0.8	7.4	11.1	0.05 (20°C)
Hypoxanthine	6-Hydroxypurine	$C_5H_4N_4O$	136.11	2.0	8.9	12.1	0.07 (19°C)
Uric acid	2,6,8-Trihydroxypurine	$C_5H_4N_4O_3$	168.11	5.4	11.3		0.002 (20°C)

THE GENETIC CODE

This table gives the correspondence between a messenger RNA codon and the amino acid which it specifies. The symbols for bases in the codon are:

U: uracil
C: cytosine
A: adenine
G: guanine

The amino acid symbols are given in the table entitled "Structures of Common Amino Acids". A chain-initiating codon is indicated by **init** and a chain-terminating codon by **term**.

Example: UCA codes for **Ser**, UAC codes for **Tyr**, etc.

First position	Second position				Third position
	U	C	A	G	
U	Phe	Ser	Tyr	Cys	U
	Phe	Ser	Tyr	Cys	C
	Leu	Ser	**term**	**term**	A
	Leu	Ser	**term**	Trp	G
C	Leu	Pro	His	Arg	U
	Leu	Pro	His	Arg	C
	Leu	Pro	Gln	Arg	A
	Leu	Pro	Gln	Arg	G
A	Ile	Thr	Asn	Ser	U
	Ile	Thr	Asn	Ser	C
	Ile	Thr	Lys	Arg	A
	Met (**init**)	Thr	Lys	Arg	G
G	Val	Ala	Asp	Gly	U
	Val	Ala	Asp	Gly	C
	Val	Ala	Glu	Gly	A
	Val (**init**)	Ala	Glu	Gly	G

PROPERTIES OF FATTY ACIDS

This table gives the systematic names and selected properties of some of the more important fatty acids of five or more carbon atoms. Compounds are listed first by degree of saturation and, secondly, by number of carbon atoms. The following data are included:

M_r: Molecular weight
S: Aqueous solubility at 20°C in units of grams of solute per 100 grams of water
t_m: Melting point in °C

References

1. Dawson, R. M. C., Elliott, D. C., Elliott, W. H., and Jones, K. M., *Data for Biochemical Research*, 3rd ed., Clarendon Press, Oxford, 1986.
2. Fasman, G. D., Ed., *Practical Handbook of Biochemistry and Molecular Biology*, CRC Press, Boca Raton, FL, 1989.

Common name	Systematic name	Mol. form.	M_r	t_m/°C	S
Saturated					
Valeric acid	Pentanoic acid	$C_5H_{10}O_2$	102.13	−33.6	2.5
Isovaleric acid	3-Methylbutanoic acid	$C_5H_{10}O_2$	102.13	−29.3	4.3
Caproic acid	Hexanoic acid	$C_6H_{12}O_2$	116.16	−3	0.967
Enanthic acid	Heptanoic acid	$C_7H_{14}O_2$	130.19	−7.17	0.24
Caprylic acid	Octanoic acid	$C_8H_{16}O_2$	144.21	16.5	0.080
Pelargonic acid	Nonanoic acid	$C_9H_{18}O_2$	158.24	12.4	0.0284
Capric acid	Decanoic acid	$C_{10}H_{20}O_2$	172.27	31.4	0.015
Lauric acid	Dodecanoic acid	$C_{12}H_{24}O_2$	200.32	43.8	0.0055
Tridecylic acid	Tridecanoic acid	$C_{13}H_{26}O_2$	214.35	41.5	0.0033
Myristic acid	Tetradecanoic acid	$C_{14}H_{28}O_2$	228.38	54.2	0.0020
Pentadecylic acid	Pentadecanoic acid	$C_{15}H_{30}O_2$	242.40	52.3	0.0012
Palmitic acid	Hexadecanoic acid	$C_{16}H_{32}O_2$	256.43	62.5	0.00072
Margaric acid	Heptadecanoic acid	$C_{17}H_{34}O_2$	270.46	61.3	0.00042
Stearic acid	Octadecanoic acid	$C_{18}H_{36}O_2$	284.48	69.3	0.00029
Arachidic acid	Eicosanoic acid	$C_{20}H_{40}O_2$	312.54	76.5	
Phytanic acid	3,7,11,15-Tetramethylhexadecanoic acid	$C_{20}H_{40}O_2$	312.54	-65	
Behenic acid	Docosanoic acid	$C_{22}H_{44}O_2$	340.59	81.5	
Lignoceric acid	Tetracosanoic acid	$C_{24}H_{48}O_2$	368.64	87.5	
Cerotic acid	Hexacosanoic acid	$C_{26}H_{52}O_2$	396.70	88.5	
Montanic acid	Octacosanoic acid	$C_{28}H_{56}O_2$	424.75	90.9	
Monounsaturated					
Caproleic acid	9-Decenoic acid	$C_{10}H_{18}O_2$	170.25	26.5	
Palmitoleic acid	*cis*-9-Hexadecenoic acid	$C_{16}H_{30}O_2$	254.41	0.5	
Oleic acid	*cis*-9-Octadecenoic acid	$C_{18}H_{34}O_2$	282.47	13.4	
Elaidic acid	*trans*-9-Octadecenoic acid	$C_{18}H_{34}O_2$	282.47	45	
Vaccenic acid	*trans*-11-Octadecenoic acid	$C_{18}H_{34}O_2$	282.47	44	
Erucic acid	*cis*-13-Docosenoic acid	$C_{22}H_{42}O_2$	338.57	34.7	
Brassidic acid	*trans*-13-Docosenoic acid	$C_{22}H_{42}O_2$	338.57	61.9	
Nervonic acid	*cis*-15-Tetracosenoic acid	$C_{24}H_{46}O_2$	366.63	43	
Diunsaturated					
Linoleic acid	*cis,cis*-9,12-Octadecadienoic acid	$C_{18}H_{32}O_2$	280.45	−7	
Triunsaturated					
cis-Eleostearic acid	*trans,cis,trans*-9,11,13-Octadecatrienoic acid	$C_{18}H_{30}O_2$	278.44	49	
trans-Eleostearic acid	*trans,trans,trans*-9,11,13-Octadecatrienoic acid	$C_{18}H_{30}O_2$	278.44	71.5	
Linolenic acid	*cis,cis,cis*-9,12,15-Octadecatrienoic acid	$C_{18}H_{30}O_2$	278.44	−11	
Tetraunsaturated					
Arachidonic acid	5,8,11,14-Eicosatetraenoic acid, (all-*trans*)	$C_{20}H_{32}O_2$	304.47	−49.5	

CARBOHYDRATE NAMES AND SYMBOLS

The following table lists the systematic names and symbols for selected carbohydrates and some of their derivatives. The symbols for monosaccharide residues and derivatives are recommended by IUPAC for use in describing the structures of oligosaccharide chains. A more complete list can be found in the reference.

Reference

McNaught, A. D., *Pure Appl. Chem.*, 68, 1919-2008, 1996.

Common name	Symbol	Systematic name
Abequose	Abe	3,6-Dideoxy-D-*xylo*-hexose
N-Acetyl-2-deoxyneur-2-enaminic acid	Neu2en5Ac	
N-Acetylgalactosamine	GalNAc	
N-Acetylglucosamine	GlcNAc	
N-Acetylneuraminic acid	Neu5Ac	
Allose	All	*allo*-Hexose
Altrose	Alt	*altro*-Hexose
Apiose	Api	3-*C*-(Hydroxymethyl)-*glycero*-tetrose
Arabinitol	Ara-ol	Arabinitol
Arabinose	Ara	*arabino*-Pentose
Arcanose		2,6-Dideoxy-3-*C*-methyl-3-*O*-methyl-*xylo*-hexose
Ascarylose		3,6-Dideoxy-L-*arabino*-hexose
Boivinose		2,6-Dideoxy-D-gulose
Chalcose		4,6-Dideoxy-3-*O*-methyl-D-*xylo*-hexose
Cladinose		2,6-Dideoxy-3-*C*-methyl-3-*O*-methyl-L-*ribo*-hexose
Colitose		3,6-Dideoxy-L-*xylo*-hexose
Cymarose		6-Deoxy-3-*O*-methyl-*ribo*-hexose
3-Deoxy-D-*manno*-oct-2-ulosonic acid	Kdo	
2-Deoxyribose	dRib	2-Deoxy-*erythro*-pentose
2,3-Diamino-2,3-dideoxy-D-glucose	GlcN3N	
Diginose		2,6-Dideoxy-3-*O*-methyl-*lyxo*-hexose
Digitalose		6-Deoxy-3-*O*-methyl-D-galactose
Digitoxose		2,6-Dideoxy-D-*ribo*-hexose
3,4-Di-*O*-methylrhamnose	Rha3,4Me$_2$	
Ethyl glucopyranuronate	Glc*p*A6Et	
Evalose		6-Deoxy-3-*C*-methyl-D-mannose
Fructose	Fru	*arabino*-Hex-2-ulose
Fucitol	Fuc-ol	6-Deoxy-D-galactitol
Fucose	Fuc	6-Deoxygalactose
β-D-Galactopyranose 4-sulfate	β-D-Gal*p*4*S*	
Galactosamine	GalN	2-Amino-2-deoxygalactose
Galactose	Gal	*galacto*-Hexose
Glucitol	Glc-ol	
Glucosamine	GlcN	2-Amino-2-deoxyglucose
Glucose	Glc	*gluco*-Hexose
Glucuronic acid	GlcA	
N-Glycoloylneuraminic acid	Neu5Gc	
Gulose	Gul	*gulo*-Hexose
Hamamelose		2-*C*-(Hydroxymethyl)-D-ribose
Idose	Ido	*ido*-Hexose
Iduronic acid	IdoA	
Lactose	Lac	β-D-Galactopyranosyl-(1→4)-D-glucose
Lyxose	Lyx	*lyxo*-Pentose
Maltose		α-D-Glucopyranosyl-(1→4)-D-glucose
Mannose	Man	*manno*-Hexose
2-*C*-Methylxylose	Xyl2*C*Me	
Muramic acid	Mur	2-Amino-3-*O*-[(R)-1-carboxyethyl]-2-deoxy-D-glucose
Mycarose		2,6-Dideoxy-3-*C*-methyl-L-*ribo*-hexose
Mycinose		6-Deoxy-2,3-di-*O*-methyl-D-allose

Common name	Symbol	Systematic name
Neuraminic acid	Neu	5-Amino-3,5-dideoxy-D-*glycero*-D-*galacto*-non-2-ulosonic acid
Panose		α-D-Glucopyranosyl-(1→6)-α-D-glucopyranosyl-(1→4)-D-glucose
Paratose		3,6-Dideoxy-D-*ribo*-hexose
Primeverose		β-D-Xylopyranosyl-(1→6)-D-glucose
Psicose	Psi	*ribo*-Hex-2-ulose
Quinovose	Qui	6-Deoxyglucose
Raffinose		β-D-Fructofuranosyl-α-D-galactopyranosyl-(1→6)-α-D-glucopyranoside
Rhamnose	Rha	6-Deoxymannose
Rhodinose		2,3,6-Trideoxy-L-*threo*-hexose
Ribose	Rib	*ribo*-Pentose
Ribose 5-phosphate	Rib5*P*	
Ribulose	Ribulo (Rul)	*erythro*-Pent-2-ulose
Rutinose		α-L-Rhamnopyranosyl-(1→6)-D-glucose
Sarmentose		2,6-Dideoxy-3-*O*-methyl-D-*xylo*-hexose
Sedoheptulose		D-*altro*-Hept-2-ulose
Sorbose	Sor	*xylo*-Hex-2-ulose
Streptose		5-Deoxy-3-*C*-formyl-L-lyxose
Sucrose		β-D-Fructofuranosyl-α-D-glucopyranoside
Tagatose	Tag	*lyxo*-Hex-2-ulose
Talose	Tal	*talo*-Hexose
Turanose		α-D-Glucopyranosyl-(1→3)-D-fructose
Tyvelose	Tyv	3,6-Dideoxy-D-*arabino*-hexose
Xylose	Xyl	*xylo*-Pentose
Xylulose	Xylulo (Xul)	*threo*-Pent-2-ulose

STANDARD TRANSFORMED GIBBS ENERGIES OF FORMATION FOR BIOCHEMICAL REACTANTS

Robert N. Goldberg and Robert A. Alberty

This table contains values of the standard transformed Gibbs energies of formation $\Delta_f G'^\circ$ for 130 biochemical reactants. Values of $\Delta_f G'^\circ$ are given at pH 7.0, the temperature 298.15 K, and the pressure 100 kPa for three ionic strengths: $I = 0$, $I = 0.1$ mol/L and $I = 0.25$ mol/L. The table can be used for calculating apparent equilibrium constants K' and standard apparent reduction potentials E'° for biochemical reactions. Such a listing is more compact than tabulating the actual apparent equilibrium constants or standard apparent reduction potentials, which would require a very large number of reactant-product combinations. In the table, all reactants are in aqueous solution unless indicated otherwise.

A biochemical reactant is a sum of species. For example, ATP consists of an equilibrium mixture of the aqueous species ATP^{4-}, $HATP^{3-}$, H_2ATP^{2-}, $MgATP^{2-}$, etc. Similarly, phosphate refers to the equilibrium mixture of the aqueous species PO_4^{3-}, HPO_4^{2-}, $H_2PO_4^{-}$, H_3PO_4, $MgHPO_4$, etc. Biochemical reactions are written using biochemical reactants in terms of an apparent equilibrium constant K', which is distinct from the standard equilibrium constant K. This subject is discussed in an IUPAC report (see Reference 1 below).

The apparent equilibrium constant K' and the standard transformed Gibbs energy change $\Delta_r G'^\circ$ for a biochemical reaction can be calculated from the $\Delta_f G'^\circ$ values by using the relationship

$$-RT \ln K' = \Delta_r G'^\circ = \Sigma v'_i \Delta_f G'^\circ,$$

where the summation is over all of the biochemical reactants. The quantity v'_i is the stoichiometric number of reactant i (v'_i is positive for reactants on the right side of the equation and negative for reactants on the left side); R is the gas constant. As an example, the hydrolysis reaction of ATP is

$$ATP + H_2O(l) = ADP + phosphate.$$

At pH 7.00 and $I = 0.25$ M, $\Delta_r G'^\circ$ and K' are calculated as follows:

$$\Delta_r G'^\circ = \{-1424.70 - 1059.49 - (-2292.50 - 155.66)\} \cdot (kJ\ mol^{-1}) = -36.03\ kJ\ mol^{-1}$$

$$K' = \exp[-(-36030\ J\ mol^{-1})/\{(8.3145\ J\ mol^{-1}\ K^{-1}) \cdot (298.15\ K)\}] = 2.05 \cdot 10^6$$

An example involving a biochemical half-cell reaction is

$$acetaldehyde(aq) + 2\ e^- = ethanol(aq).$$

At 298.15 K, pH 7.00, and $I = 0$, the standard apparent reduction potential E'° can be calculated as follows

$$E'^\circ = -(1/nF) \cdot \{\Delta_f G'^\circ(ethanol) - \Delta_f G'^\circ(acetaldehyde)\},$$

where n is the number of electrons in the half-cell reaction and F is the Faraday constant. Then,

$$E'^\circ = [-1/(2 \cdot 9.6485 \cdot 10^4\ C\ mol^{-1})] \cdot (58.10 \cdot 10^3\ J\ mol^{-1} - 20.83 \cdot 10^3\ J\ mol^{-1}) = -0.193\ V$$

References

1. Alberty, R.A., Cornish-Bowden, A., Gibson, Q.H., Goldberg, R.N., Hammes, G., Jencks, W., Tipton, K.F, Veech, R., Westerhoff, H.V., and Webb, E.C. *Pure Appl. Chem.* 66, 1641-1666, 1994.
2. Alberty, R.A., *Arch. Biochem. Biophys.*, 353, 116-130, 1998; 358, 25-39, 1998.
3. Alberty, R.A., *Thermodynamics of Biochemical Reactions*, Wiley-Interscience, New York, 2003.
4. Alberty, R.A., *BasicBiochemData2: Data and Programs for Biochemical Thermodynamics*, <http://library.wolfram.com/infocenter/MathSource/797>.

Reactant	$\Delta_f G'^\circ (I = 0)$ kJ mol^{-1}	$\Delta_f G'^\circ (I = 0.1\ M)$ kJ mol^{-1}	$\Delta_f G'^\circ (I = 0.25\ M)$ kJ mol^{-1}
Acetaldehyde	20.83	23.27	24.06
Acetate	-249.46	-248.23	-247.83
Acetone	80.04	83.71	84.90
Acetyl Coenzyme A	-60.49	-58.65	-58.06
Acetylphosphate	-1109.34	-1107.57	-1107.02
cis-Aconitate	-797.26	-800.93	-802.12
Adenine	510.45	513.51	514.50
Adenosine	324.93	332.89	335.46
Adenosine 5'-diphosphate (ADP)	-1428.93	-1425.55	-1424.70
Adenosine 5'-monophosphate (AMP)	-562.04	-556.53	-554.83
Adenosine 5'-triphosphate (ATP)	-2292.61	-2292.16	-2292.50
D-Alanine	-91.31	-87.02	-85.64
Ammonia	80.50	82.34	82.93
D-Arabinose	-342.67	-336.55	-334.57
L-Asparagine	-206.28	-201.38	-199.80
L-Aspartate	-456.14	-453.08	-452.09
1,3-Biphosphoglycerate	-2202.06	-2205.69	-2207.30
Butanoate	-72.94	-69.26	-68.08
1-Butanol	227.72	233.84	235.82

Reactant	$\Delta_f G'^{\circ}(I = 0)$ kJ mol^{-1}	$\Delta_f G'^{\circ}(I = 0.1\ M)$ kJ mol^{-1}	$\Delta_f G'^{\circ}(I = 0.25\ M)$ kJ mol^{-1}
Citrate	-963.46	-965.49	-966.23
Isocitrate	-956.82	-958.84	-959.58
Coenzyme A (CoA)	-7.98	-7.43	-7.26
CO(aq)	-119.90	-119.90	-119.90
CO(g)	-137.17	-137.17	-137.17
CO_2(aq)[total]	-547.33	-547.15	-547.10
CO_2(g)	-394.36	-394.36	-394.36
Creatine	100.41	105.92	107.69
Creatinine	256.55	260.84	262.22
L-Cysteine	-59.23	-55.01	-53.65
L-Cystine	-187.03	-179.69	-177.32
Cytochrome c [oxidized]	0.00	-5.51	-7.29
Cytochrome c [reduced]	-24.51	-26.96	-27.75
Dihydroxyacetone phosphate	-1096.60	-1095.91	-1095.70
Ethanol	58.10	61.77	62.96
Ethyl acetate	-18.00	-13.10	-11.52
Ferredoxin [oxidized]	0.00	-0.61	-0.81
Ferredoxin [reduced]	38.07	38.07	38.07
Flavine adenine dinucleotide (FAD) [oxidized]	1238.65	1255.17	1260.51
Flavine adenine dinucleotide (FAD) [reduced]	1279.68	1297.43	1303.16
Flavin adenine dinucleotide-enzyme (FADenz) [oxidized]	1238.65	1255.17	1260.51
Flavin adenine dinucleotide-enzyme (FADenz) [reduced]	1229.96	1247.71	1253.44
Flavin mononucleotide (FMN) [oxidized]	759.17	768.35	771.32
Flavin mononucleotide (FMN) [reduced]	800.20	810.61	813.97
Formate	-311.04	-311.04	-311.04
D-Fructose	-436.03	-428.69	-426.32
D-Fructose 1,6-diphosphate	-2202.84	-2205.66	-2206.78
D-Fructose 6-phosphate	-1321.71	-1317.16	-1315.74
Fumarate	-521.97	-523.19	-523.58
D-Galactose	-429.45	-422.11	-419.74
α-D-Galactose 1-phosphate	-1317.50	-1313.01	-1311.60
D-Glucose	-436.42	-429.08	-426.71
α-D-Glucose 1-phosphate	-1318.03	-1313.34	-1311.89
D-Glucose 6-phosphate	-1325.00	-1320.37	-1318.92
Glutamate	-377.82	-373.54	-372.16
D-Glutamine	-128.46	-122.34	-120.36
Glutathione [oxidized]	1198.69	1214.60	1219.74
Glutathione [reduced]	625.75	634.76	637.62
Glutathione-coenzyme A	563.49	572.06	574.83
D-Glyceraldehyde 3-phosphate	-1088.94	-1088.25	-1088.04
Glycerol	-177.83	-172.93	-171.35
sn-Glycerol 3-phosphate	-1080.22	-1077.83	-1077.13
Glycine	-180.13	-177.07	-176.08
Glycolate	-411.08	-409.86	-409.46
Glycylglycine	-200.55	-195.65	-194.07
Glyoxylate	-428.64	-428.64	-428.64
H_2(aq)	97.51	98.74	99.13
H_2(g)	79.91	81.14	81.53
H_2O(l)	-157.28	-156.05	-155.66
H_2O_2(aq)	-54.12	-52.89	-52.50
3-Hydroxypropanoate	-318.62	-316.17	-315.38
Hypoxanthine	249.33	251.77	252.56
Indole	503.49	507.78	509.16
Lactate	-316.94	-314.49	-313.70
Lactose	-688.29	-674.83	-670.48
L-Leucine	167.18	175.14	177.71
L-Isoleucine	175.53	183.49	186.06
D-Lyxose	-349.58	-343.46	-341.48
Malate	-682.88	-682.85	-682.85
Maltose	-695.65	-682.19	-677.84

Reactant	$\Delta_f G'^{\circ}(I = 0)$ kJ mol^{-1}	$\Delta_f G'^{\circ}(I = 0.1$ M$)$ kJ mol^{-1}	$\Delta_f G'^{\circ}(I = 0.25$ M$)$ kJ mol^{-1}
D-Mannitol	-383.22	-374.65	-371.89
Mannose	-430.52	-423.18	-420.81
Methane(aq)	125.50	127.94	128.73
Methane(g)	109.11	111.55	112.34
Methanol	-15.48	-13.04	-12.25
L-Methionine	-63.40	-56.67	-54.49
N_2(aq)	18.70	18.70	18.70
N_2(g)	0.00	0.00	0.00
Nicotinamide Adenine Dinucleotide (NAD) [oxidized]	1038.86	1054.17	1059.11
Nicotinamide Adenine Dinucleotide (NAD) [reduced]	1101.47	1115.55	1120.09
Nicotinamide Adenine Dinucleotide Phosphate (NADP) [oxidized]	163.73	173.52	176.68
Nicotinamide Adenine Dinucleotide Phosphate (NADP) [reduced]	229.67	235.79	237.77
O_2(aq)	16.40	16.40	16.40
O_2(g)	0.00	0.00	0.00
Oxalate	-673.90	-676.35	-677.14
Oxaloacetate	-713.38	-714.60	-715.00
Oxalosuccinate	-979.05	-979.05	-979.05
2-Oxoglutarate	-633.58	-633.58	-633.58
Palmitate	979.25	997.61	1003.54
L-Phenylalanine	232.42	239.15	241.33
Phosphate	-1058.56	-1059.17	-1059.49
2-Phospho-D-glycerate	-1340.72	-1341.32	-1341.79
3-Phospho-D-glycerate	-1346.38	-1347.19	-1347.73
Phosphoenolpyruvate	-1185.46	-1188.53	-1189.73
1-Propanol	143.84	148.74	150.32
2-Propanol	134.42	139.32	140.90
Pyrophosphate	-1934.95	-1939.13	-1940.66
Pyruvate	-352.40	-351.18	-350.78
Retinal	1118.78	1135.91	1141.45
Retinol	1170.78	1189.14	1195.07
Ribose	-339.23	-333.11	-331.13
Ribose 1-phosphate	-1215.87	-1212.24	-1211.14
Ribose 5-phosphate	-1223.95	-1220.32	-1219.22
Ribulose	-336.38	-330.26	-328.28
L-Serine	-231.18	-226.89	-225.51
Sorbose	-432.47	-425.13	-422.76
Succinate	-530.72	-530.65	-530.64
Succinyl Coenzyme A	-349.90	-348.06	-347.47
Sucrose	-685.66	-672.20	-667.85
Thioredoxin [oxidized]	0.00	0.00	0.00
Thioredoxin [reduced]	54.32	55.41	55.74
L-Tryptophan	364.78	372.12	374.49
L-Tyrosine	68.82	75.55	77.73
Ubiquinone [oxidized]	3596.07	3651.15	3668.94
Ubiquinone [reduced]	3586.06	3642.37	3660.55
Urate	-206.03	-204.81	-204.41
Urea	-42.97	-40.53	-39.74
Uric acid	-197.07	-194.63	-193.84
L-Valine	80.87	87.60	89.78
D-Xylose	-350.93	-344.81	-342.83
D-Xylulose	-346.59	-340.47	-338.49

THERMODYNAMIC QUANTITIES FOR THE IONIZATION REACTIONS OF BUFFERS IN WATER

Robert N. Goldberg, Nand Kishore, and Rebecca M. Lennen

This table contains selected values for the pK, standard molar enthalpy of reaction $\Delta_r H°$, and standard molar heat-capacity change $\Delta_r C_p°$ for the ionization reactions of 64 buffers many of which are relevant to biochemistry and to biology.[1] The values pertain to the temperature $T = 298.15$ K and the pressure $p = 0.1$ MPa. The standard state is the hypothetical ideal solution of unit molality. These data permit one to calculate values of the pK and of $\Delta_r H°$ at temperatures in the vicinity {$T \approx (274$ K to 350 K)} of the reference temperature $\theta = 298.15$ K by using the following equations[2]

$$\Delta_r G°_T = -RT \ln K_T = \ln(10) \cdot RT \cdot pK_T, \tag{1}$$

$$R \ln K_T = -(\Delta_r G°_\theta /\theta) + \Delta_r H°_\theta \{(1/\theta) - (1/T)\} + \Delta_r C°_{p\theta}\{(\theta /T) - 1 + \ln(T/\theta)\}, \tag{2}$$

$$\Delta_r H°_T = \Delta_r H°_\theta + \Delta_r C°_{p\theta}(T - \theta). \tag{3}$$

Here, $\Delta_r G°$ is the standard molar Gibbs energy change and K is the equilibrium constant for a reaction; R is the gas constant (8.314 472 J K^{-1} mol^{-1}). The subscripts T and θ denote the temperature to which a quantity pertains, the subscript p denotes constant pressure, and the subscript r denotes that the quantity refers to a reaction. Combination of equations (1) and (2) yields the following equation that gives pK as a function of temperature:

$$pK_T = -\{R \cdot \ln(10)\}^{-1}[-\{\ln(10) \cdot RT \cdot pK_\theta /\theta\} + \Delta_r H°_\theta \{(1/\theta) - (1/T)\} + \Delta_r C°_{p\theta}\{(\theta /T) - 1 + \ln(T/\theta)\}]. \tag{4}$$

The above equations neglect higher order terms that involve temperature derivatives of $\Delta_r C_p°$. Also, it is important to recognize that the values of pK and $\Delta_r H°$ effectively pertain to ionic strength $I = 0$. However, the values of pK and $\Delta_r H°$ are almost always dependent on the ionic strength and the actual composition of the solution. These issues are discussed in Reference 1 which also gives an approximate method for making appropriate corrections.

References

1. Goldberg, R. N., Kishore, N., and Lennen, R. M., "Thermodynamic Quantities for the Ionization Reactions of Buffers," *J. Phys. Chem. Ref. Data*, in press.
2. Clarke, E. C. W., and Glew, D. N., *Trans. Faraday Soc.*, 62, 539-547, 1966.

Selected Values of Thermodynamic Quantities for the Ionization Reactions of Buffers in Water at $T = 298.15$ K and $p = 0.1$ MPa

Buffer	Reaction	pK	$\Delta_r H°$ kJ mol^{-1}	$\Delta_r C_p°$ J K^{-1} mol^{-1}
ACES	$HL^{\pm} = H^+ + L^-$, $(HL = C_4H_{10}N_2O_4S)$	6.847	30.43	−49
Acetate	$HL = H^+ + L^-$, $(HL = C_2H_4O_2)$	4.756	−0.41	−142
ADA	$H_3L^+ = H^+ + H_2L^\pm$, $(H_2L = C_6H_{10}N_2O_5)$	1.59		
	$H_2L^\pm = H^+ + HL^-$	2.48	16.7	
	$HL^- = H^+ + L^{2-}$	6.844	12.23	−144
2-Amino-2-methyl-1,3-propanediol	$HL^+ = H^+ + L$, $(L = C_4H_{11}NO_2)$	8.801	49.85	−44
2-Amino-2-methyl-1-propanol	$HL^+ = H^+ + L$, $(L = C_4H_{11}NO)$	9.694	54.05	$\approx$−21
3-Amino-1-propanesulfonic acid	$HL = H^+ + L^-$, $(HL = C_3H_9NO_3S)$	10.2		
Ammonia	$NH_4^+ = H^+ + NH_3$	9.245	51.95	8
AMPSO	$HL^\pm = H^+ + L^-$, $(HL = C_7H_{17}NO_5S)$	9.138	43.19	−61
Arsenate	$H_3AsO_4 = H^+ + H_2AsO_4^-$	2.31	−7.8	
	$H_2AsO_4^- = H^+ + HAsO_4^{2-}$	7.05	1.7	
	$HAsO_4^{2-} = H^+ + AsO_4^{3-}$	11.9	15.9	
Barbital	$H_2L = H^+ + HL^-$, $(H_2L = C_8H_{12}N_2O_3)$	7.980	24.27	−135
	$HL^- = H^+ + L^{2-}$	12.8		
BES	$HL^\pm = H^+ + L^-$, $(HL = C_6H_{15}NO_5S)$	7.187	24.25	−2
Bicine	$H_2L^+ = H^+ + HL^\pm$, $(HL = C_6H_{13}NO_4)$	2.0		
	$HL^\pm = H^+ + L^-$	8.334	26.34	0
Bis-tris	$H_3L^+ = H^+ + H_2L^+$, $(H_2L = C_8H_{19}NO_5)$	6.484	28.4	27
Bis-tris propane	$H_2L^{2+} = H^+ + HL^+$, $(L = C_{11}H_{26}N_2O_6)$	6.65		
	$HL^+ = H^+ + L$	9.10		
Borate	$H_3BO_3 = H^+ + H_2BO_3^-$	9.237	13.8	$\approx$−240
Cacodylate	$H_2L^+ = H^+ + HL$, $(HL = C_2H_6AsO_2)$	1.78	−3.5	
	$HL = H^+ + L^-$	6.28	−3.0	−86
CAPS	$HL^\pm = H^+ + L^-$, $(HL = C_9H_{19}NO_3S)$	10.499	48.1	57
CAPSO	$HL^\pm = H^+ + L^-$, $(HL = C_9H_{19}NO_4S)$	9.825	46.67	21
Carbonate	$H_2CO_3 = H^+ + HCO_3^-$	6.351	9.15	−371
	$HCO_3^- = H^+ + CO_3^{2-}$	10.329	14.70	−249
CHES	$HL^\pm = H^+ + L^-$, $(HL = C_8H_{17}NO_3S)$	9.394	39.55	9

Buffer	Reaction	pK	$\Delta_r H°$ kJ mol^{-1}	$\Delta_r C_p°$ J K^{-1} mol^{-1}
Citrate	$H_3L = H^+ + H_2L^-$, $(H_3L = C_6H_8O_7)$	3.128	4.07	−131
	$H_2L^- = H^+ + HL^{2-}$	4.761	2.23	−178
	$HL^{2-} = H^+ + L^{3-}$	6.396	−3.38	−254
L-Cysteine	$H_3L^+ = H^+ + H_2L$, $(H_2L = C_3H_7NO_2S)$	1.71	≈−0.6	
	$H_2L = H^+ + HL^-$	8.36	36.1	≈−66
	$HL^- = H^+ + L^{2-}$	10.75	34.1	≈−204
Diethanolamine	$HL^+ = H^+ + L$, $(L = C_4H_{11}NO_2)$	8.883	42.08	36
Diglycolate	$H_2L = H^+ + HL^-$, $(H_2L = C_4H_6O_5)$	3.05	−0.1	≈−142
	$HL^- = H^+ + L^{2-}$	4.37	−7.2	≈−138
3,3-Dimethylglutarate	$H_2L = H^+ + HL^-$, $(H_2L = C_7H_{12}O_4)$	3.70		
	$HL^- = H^+ + L^{2-}$	6.34		
DIPSO	$HL^± = H^+ + L^-$, $(HL = C_7H_{17}NO_5S)$	7.576	30.18	42
Ethanolamine	$HL^+ = H^+ + L$, $(L = C_2H_7NO)$	9.498	50.52	26
N-Ethylmorpholine	$HL^+ = H^+ + L$, $(L = C_6H_{13}NO)$	7.77	27.4	
Glycerol 2-phosphate	$H_2L = H^+ + HL^-$, $(H_2L = C_3H_9NO_6P)$	1.329	−12.2	−330
	$HL^- = H^+ + L^{2-}$	6.650	−1.85	−212
Glycine	$H_2L^+ = H^+ + HL^±$, $(HL = C_2H_5NO_2)$	2.351	4.00	−139
	$HL^± = H^+ + L^-$	9.780	44.2	−57
Glycine amide	$HL^+ = H^+ + L$, $(L = C_2H_6N_2O)$	8.04	42.9	
Glycylglycine	$H_2L^+ = H^+ + HL^±$, $(HL = C_4H_8N_2O_3)$	3.140	0.11	−128
	$HL^± = H^+ + L^-$	8.265	43.4	−16
Glycylglycylglycine	$H_2L^+ = H^+ + HL^±$, $(HL = C_6H_{11}N_3O_4)$	3.224	0.84	
	$HL^± = H^+ + L^-$	8.090	41.7	
HEPES	$H_2L^+ = H^+ + HL^±$, $(HL = C_8H_{18}N_2O_4S)$	≈3.0		
	$HL^± = H^+ + L^-$	7.564	20.4	47
HEPPS	$HL^± = H^+ + L^-$, $(HL = C_9H_{20}N_2O_4S)$	7.957	21.3	48
HEPPSO	$HL^± = H^+ + L^-$, $(HL = C_9H_{20}N_2O_5S)$	8.042	23.70	47
L-Histidine	$H_3L^{2+} = H^+ + H_2L^+$, $(HL = C_6H_9N_3O_2)$	1.5$_4$	3.6	
	$H_2L^+ = H^+ + HL$	6.07	29.5	176
	$HL = H^+ + L^-$	9.34	43.8	−233
Hydrazine	$H_2L^{2+} = H^+ + HL^+$, $(L = H_4N_2)$	−0.99	38.1	
	$HL^+ = H^+ + L$	8.02	41.7	
Imidazole	$HL^+ = H^+ + L$, $(L = C_3H_4N_2)$	6.993	36.64	−9
Maleate	$H_2L = H^+ + HL^-$, $(H_2L = C_4H_4O_4)$	1.92	1.1	≈−21
	$HL^- = H^+ + L^{2-}$	6.27	−3.6	≈−31
2-Mercaptoethanol	$HL = H^+ + L^-$, $(HL = C_2H_6OS)$	9.7$_5$	26.2	
MES	$HL^± = H^+ + L^-$, $(HL = C_6H_{13}NO_4S)$	6.270	14.8	5
Methylamine	$HL^+ = H^+ + L$, $(L = CH_5N)$	10.645	55.34	33
2-Methylimidazole	$HL^+ = H^+ + L$, $(L = C_4H_6N_2)$	8.0$_1$	36.8	
MOPS	$HL^± = H^+ + L^-$, $(HL = C_7H_{15}NO_4S)$	7.184	21.1	25
MOPSO	$H_2L^+ = H^+ + HL^±$, $(HL = C_7H_{15}NO_5S)$	0.060		
	$HL^± = H^+ + L^-$	6.90	25.0	≈38
Oxalate	$H_2L = H^+ + HL^-$, $(H_2L = C_2H_2O_4)$	1.27	−3.9	≈−231
	$HL^- = H^+ + L^{2-}$	4.266	7.00	−231
Phosphate	$H_3PO_4 = H^+ + H_2PO_4^-$	2.148	−8.0	−141
	$H_2PO_4^- = H^+ + HPO_4^{2-}$	7.198	3.6	−230
	$HPO_4^{2-} = H^+ + PO_4^{3-}$	12.35	16.0	−242
Phthalate	$H_2L = H^+ + HL^-$, $(H_2L = C_8H_6O_4)$	2.950	−2.70	−91
	$HL^- = H^+ + L^{2-}$	5.408	−2.17	−295
Piperazine	$H_2L^{2+} = H^+ + HL^+$, $(L = C_4H_{10}N_2)$	5.333	31.11	86
	$HL^+ = H^+ + L$	9.731	42.89	75
PIPES	$HL^± = H^+ + L^-$, $(HL = C_8H_{18}N_2O_6S_2)$	7.141	11.2	22
POPSO	$HL^± = H^+ + L^-$, $(HL = C_{10}H_{22}N_2O_8S_2)$	≈8.0		
Pyrophosphate	$H_4P_2O_7 = H^+ + H_3P_2O_7^-$	0.83	−9.2	≈−90
	$H_3P_2O_7^- = H^+ + H_2P_2O_7^{2-}$	2.26	−5.0	≈−130
	$H_2P_2O_7^{2-} = H^+ + HP_2O_7^{3-}$	6.72	0.5	−136
	$HP_2O_7^{3-} = H^+ + P_2O_7^{4-}$	9.46	1.4	−141
Succinate	$H_2L = H^+ + HL^-$, $(H_2L = C_4H_6O_4)$	4.207	3.0	−121
	$HL^- = H^+ + L^{2-}$	5.636	−0.5	−217
Sulfate	$HSO_4^- = H^+ + SO_4^{2-}$	1.987	−22.4	−258

Buffer	Reaction	pK	$\Delta_r H°$ kJ mol^{-1}	$\Delta_r C_p°$ J K^{-1} mol^{-1}
Sulfite	$H_2SO_3 = H^+ + HSO_3^-$	1.857	−17.80	−272
	$HSO_3^- = H^+ + SO_3^{2-}$	7.172	−3.65	−262
TAPS	$HL^{\pm} = H^+ + L^-$, (HL = $C_7H_{17}NO_6S$)	8.44	40.4	15
TAPSO	$HL^{\pm} = H^+ + L^-$, (HL = $C_7H_{17}NO_7S$)	7.635	39.09	−16
L(+)-Tartaric acid	$H_2L = H^+ + HL^-$, ($H_2L = C_4H_6O_6$)	3.036	3.19	−147
	$HL^- = H^+ + L^{2-}$	4.366	0.93	−218
TES	$HL^{\pm} = H^+ + L^-$, (HL = $C_6H_{15}NO_6S$)	7.550	32.13	0
Tricine	$H_2L^+ = H^+ + HL^{\pm}$, (HL = $C_6H_{13}NO_5$)	2.023	5.85	−196
	$HL^{\pm} = H^+ + L^-$	8.135	31.37	−53
Triethanolamine	$HL^+ = H^+ + L$, (L = $C_6H_{15}NO_3$)	7.762	33.6	50
Triethylamine	$HL^+ = H^+ + L$, (L = $C_6H_{15}N$)	10.72	43.13	151
Tris	$HL^+ = H^+ + L$, (L = $C_4H_{11}NO_3$)	8.072	47.45	−59

BIOLOGICAL BUFFERS

This table of frequently used buffers gives the pK_a value at 25°C and the useful pH range of each buffer. The buffers are listed in order of increasing pH.

The table is reprinted with permission of Sigma Chemical Company, St. Louis, Mo.

Acronym	Name	Mol. wt.	pK_a	Useful pH range
MES	2-(N-Morpholino)ethanesulfonic acid	195.2	6.1	5.5–6.7
BIS TRIS	Bis(2-hydroxyethyl)iminotris(hydroxymethyl)methane	209.2	6.5	5.8–7.2
ADA	N-(2-Acetamido)-2-iminodiacetic acid	190.2	6.6	6.0–7.2
ACES	2-[(2-Amino-2-oxoethyl)amino]ethanesulfonic acid	182.2	6.8	6.1–7.5
PIPES	Piperazine-N,N´-bis(2-ethanesulfonic acid)	302.4	6.8	6.1–7.5
MOPSO	3-(N-Morpholino)-2-hydroxypropanesulfonic acid	225.3	6.9	6.2–7.6
BIS TRISPROPANE	1,3-Bis[tris(hydroxymethyl)methylamino]propane	282.3	6.8[a]	6.3–9.5
BES	N,N-Bis(2-hydroxyethyl)-2-aminoethanesulfonic acid	213.2	7.1	6.4–7.8
MOPS	3-(N-Morpholino)propanesulfonic acid	209.3	7.2	6.5–7.9
HEPES	N-(2-Hydroxyethyl)piperazine-N´-(2-ethanesulfonic acid)	238.3	7.5	6.8–8.2
TES	N-Tris(hydroxymethyl)methyl-2-aminoethanesulfonic acid	229.2	7.5	6.8–8.2
DIPSO	3-[N,N-Bis(2-hydroxyethyl)amino]-2-hydroxypropanesulfonic acid	243.3	7.6	7.0–8.2
TAPSO	3-[N-Tris(hydroxymethyl)methylamino]-2-hydroxypropanesulfonic acid	259.3	7.6	7.0–8.2
TRIZMA	Tris(hydroxymethyl)aminomethane	121.1	8.1	7.0–9.1
HEPPSO	N-(2-hydroxyethyl)piperazine-N´-(2-hydroxypropanesulfonic acid)	268.3	7.8	7.1–8.5
POPSO	Piperazine-N,N´-bis(2-hydroxypropanesulfonic acid)	362.4	7.8	7.2–8.5
EPPS	N-(2-Hydroxyethyl)piperazine-N´-(3-propanesulfonic acid)	252.3	8.0	7.3–8.7
TEA	Triethanolamine	149.2	7.8	7.3–8.3
TRICINE	N-Tris(hydroxymethyl)methylglycine	179.2	8.1	7.4–8.8
BICINE	N,N-Bis(2-hydroxyethyl)glycine	163.2	8.3	7.6–9.0
TAPS	N-Tris(hydroxymethyl)methyl-3-aminopropanesulfonic acid	243.3	8.4	7.7–9.1
AMPSO	3-[(1,1-Dimethyl-2-hydroxyethyl)amino]-2-hydroxypropanesulfonic acid	227.3	9.0	8.3–9.7
CHES	2-(N-Cyclohexylamino)ethanesulfonic acid	207.3	9.3	8.6–10.0
CAPSO	3-(Cyclohexylamino)-2-hydroxy-1-propanesulfonic acid	237.3	9.6	8.9–10.3
AMP	2-Amino-2-methyl-1-propanol	89.1	9.7	9.0–10.5
CAPS	3-(Cyclohexylamino)-1-propanesulfonic acid	221.3	10.4	9.7–11.1

[a] pK_a = 9.0 for the second dissociation stage.

TYPICAL pH VALUES OF BIOLOGICAL MATERIALS AND FOODS

This table gives typical pH ranges for various biological fluids and common foods. All values refer to 25°C.

Biological Materials	
Blood, human	7.35–7.45
Blood, dog	6.9–7.2
Spinal fluid, human	7.3–7.5
Saliva, human	6.5–7.5
Gastric contents, human	1.0–3.0
Duodenal contents, human	4.8–8.2
Feces, human	4.6–8.4
Urine, human	4.8–8.4
Milk, human	6.6–7.6
Bile, human	6.8–7.0

Foods	
Apples	2.9–3.3
Apricots	3.6–4.0
Asparagus	5.4–5.8
Bananas	4.5–4.7
Beans	5.0–6.0
Beers	4.0–5.0
Beets	4.9–5.5
Blackberries	3.2–3.6
Bread, white	5.0–6.0
Butter	6.1–6.4
Cabbage	5.2–5.4
Carrots	4.9–5.3
Cheese	4.8–6.4
Cherries	3.2–4.0
Cider	2.9–3.3
Corn	6.0–6.5
Crackers	6.5–8.5
Dates	6.2–6.4
Eggs, fresh white	7.6–8.0
Flour, wheat	5.5–6.5
Gooseberries	2.8–3.0
Grapefruit	3.0–3.3
Grapes	3.5–4.5

Hominy (lye)	6.8–8.0
Jams, fruit	3.5–4.0
Jellies, fruit	2.8–3.4
Lemons	2.2–2.4
Limes	1.8–2.0
Maple syrup	6.5–7.0
Milk, cows	6.3–6.6
Olives	3.6–3.8
Oranges	3.0–4.0
Oysters	6.1–6.6
Peaches	3.4–3.6
Pears	3.6–4.0
Peas	5.8–6.4
Pickles, dill	3.2–3.6
Pickles, sour	3.0–3.4
Pimento	4.6–5.2
Plums	2.8–3.0
Potatoes	5.6–6.0
Pumpkin	4.8–5.2
Raspberries	3.2–3.6
Rhubarb	3.1–3.2
Salmon	6.1–6.3
Sauerkraut	3.4–3.6
Shrimp	6.8–7.0
Soft drinks	2.0–4.0
Spinach	5.1–5.7
Squash	5.0–5.4
Strawberries	3.0–3.5
Sweet potatoes	5.3–5.6
Tomatoes	4.0–4.4
Tuna	5.9–6.1
Turnips	5.2–5.6
Vinegar	2.4–3.4
Water, drinking	6.5–8.0
Wines	2.8–3.8

CHEMICAL COMPOSITION OF THE HUMAN BODY

The elemental composition of the "standard man" of mass 70 kg is given below.

References

1. Padikal, T. N., and Fivozinsky, S. P., *Medical Physics Data Book*, *National Bureau of Standards Handbook 138*, U. S. Government Printing Office, Washington, DC, 1981.
2. Snyde, W. S., et al., *Reference Man: Anatomical, Physiological, and Metabolic Characteristics*, Pergamon, New York, 1975.

Element	Amount (g)	Percent of total body mass
Oxygen	43,000	61
Carbon	16,000	23
Hydrogen	7000	10
Nitrogen	1800	2.6
Calcium	1000	1.4
Phosphorus	780	1.1
Sulfur	140	0.20
Potassium	140	0.20
Sodium	100	0.14
Chlorine	95	0.12
Magnesium	19	0.027
Silicon	18	0.026
Iron	4.2	0.006
Fluorine	2.6	0.0037
Zinc	2.3	0.0033
Rubidium	0.32	0.00046
Strontium	0.32	0.00046
Bromine	0.20	0.00029
Lead	0.12	0.00017
Copper	0.072	0.00010
Aluminum	0.061	0.00009
Cadmium	0.050	0.00007
Boron	<0.048	0.00007
Barium	0.022	0.00003
Tin	<0.017	0.00002
Manganese	0.012	0.00002
Iodine	0.013	0.00002
Nickel	0.010	0.00001
Gold	<0.010	0.00001
Molybdenum	<0.0093	0.00001
Chromium	<0.0018	0.000003
Cesium	0.0015	0.000002
Cobalt	0.0015	0.000002
Uranium	0.00009	0.0000001
Beryllium	0.000036	
Radium	$3.1 \cdot 10^{-11}$	

Section 8
Analytical Chemistry

PREPARATION OF SPECIAL ANALYTICAL REAGENTS

Aluminon (qualitative test for aluminum). Aluminon is a trade name for the ammonium salt of aurintricarboxylic acid. Dissolve 1 g of the salt in 1 L of distilled water. Shake the solution well to insure thorough mixing.

Bang's reagent (for glucose estimation). Dissolve 100 g of K_2CO_3, 66 g of KCl and 160 g of $KHCO_3$ in the order given in about 700 mL of water at 30°C. Add 4.4 g of $CuSO_4$ and dilute to 1 L after the CO_2 is evolved. This solution should be shaken only in such a manner as not to allow entry of air. After 24 hours 300 mL are diluted to 1 L with saturated KCl solution, shaken gently and used after 24 hours; 50 mL is equivalent to 10 mg glucose.

Barfoed's reagent (test for glucose). See Cupric acetate.

Baudisch's reagent. See Cupferron.

Benedict's solution (qualitative reagent for glucose). With the aid of heat, dissolve 173 g of sodium citrate and 100 g of Na_2CO_3 in 800 mL of water. Filter, if necessary, and dilute to 850 mL. Dissolve 17.3 g of $CuSO_4 \cdot 5H_2O$ in 100 mL of water. Pour the latter solution, with constant stirring, into the carbonate-citrate solution, and dilute to 1 L.

Benzidine hydrochloride solution (for sulfite determination). Make a paste of 8 g of benzidine hydrochloride ($C_{12}H_8(NH_3)_2 \cdot 2HCl$) and 20 mL of water, add 20 mL of HCl (sp. gr. 1.12) and dilute to 1 L with water. Each mL of this solution is equivalent to 0.00357 g of H_2SO_4.

Bertrand's reagent (glucose estimation). Consists of the following solutions:
1. Dissolve 200 g of Rochelle salt and 150 g of NaOH in sufficient water to make 1 L of solution.
2. Dissolve 40 g of $CuSO_4$ in enough water to make 1 L of solution.
3. Dissolve 50 g of $Fe_2(SO_4)_3$ and 200 g of H_2SO_4 (sp. gr. 1.84) in sufficient water to make 1 L of solution.
4. Dissolve 5 g of $KMnO_4$ in sufficient water to make 1 L of solution.

Bial's reagent (for pentose). Dissolve 1 g of orcinol (5-methyl-1,3-benzenediol) in 500 mL of 30% HCl to which 30 drops of a 10% solution of $FeCl_3$ has been added.

Boutron — Boudet soap solution:
1. Dissolve 100 g of pure castile soap in about 2.5 L of 56% ethanol.
2. Dissolve 0.59 g of $Ba(NO_3)_2$ in 1 L of water.

Adjust the castile soap solution so that 2.4 mL of it will give a permanent lather with 40 mL of solution (b). When adjusted, 2.4 mL of soap solution is equivalent to 220 parts per million of hardness (as $CaCO_3$) for a 40 mL sample. See also Soap solution.

Brucke's reagent (protein precipitation). See Potassium iodide-mercuric iodide.

Clarke's soap solution (estimation of hardness in water).
1. Dissolve 100 g of pure powdered castile soap in 1 L of 80% ethanol and allow to stand over night.
2. Prepare a solution of $CaCl_2$ by dissolving 0.5 g of $CaCO_3$ in HCl (sp. gr. 1.19), neutralize with NH_4OH and make slightly alkaline to litmus, and dilute to 500 mL. One mL is equivalent to 1 mg of $CaCO_3$.

Titrate (1) against (2) and dilute (1) with 80% ethanol until 1 mL of the resulting solution is equivalent to 1 mL of (2) after making allowance for the lather factor (the amount of standard soap solution required to produce a permanent lather in 50 mL of distilled water). One mL of the adjusted solution after subtracting the lather factor is equivalent to 1 mg of $CaCO_3$. See also Soap solution.

Cobalticyanide paper (Rinnmann's test for Zn). Dissolve 4 g of $K_3Co(CN)_6$ and 1 g of $KClO_3$ in 100 mL of water. Soak filter paper in solution and dry at 100°C. Apply drop of zinc solution and burn in an evaporating dish. A green disk is obtained if zinc is present.

Cochineal. Extract 1 g of cochineal for 4 days with 20 mL of alcohol and 60 mL of distilled water. Filter.

Congo red. Dissolve 0.5 g of congo red in 90 mL of distilled water and 10 mL of alcohol.

Cupferron (Baudisch's reagent for iron analysis). Dissolve 6 g of the ammonium salt of *N*-hydroxy-*N*-nitrosoaniline (cupferron) in 100 mL of H_2O. Reagent good for 1 week only and must be kept in the dark.

Cupric acetate (Barfoed's reagent for reducing monosaccharides). Dissolve 66 g of cupric acetate and 10 mL of glacial acetic acid in water and dilute to 1 L.

Cupric oxide, ammoniacal; Schweitzer's reagent (dissolves cotton, linen, and silk, but not wool).
1. Dissolve 5 g of cupric sulfate in 100 mL of boiling water, and add sodium hydroxide until precipitation is complete. Wash the precipitate well, and dissolve it in a minimum quantity of ammonium hydroxide.
2. Bubble a slow stream of air through 300 mL of strong ammonium hydroxide containing 50 g of fine copper turnings. Continue for 1 hour.

Cupric sulfate in glycerin-potassium hydroxide (reagent for silk). Dissolve 10 g of cupric sulfate, $CuSO_4 \cdot 5H_2O$, in 100 mL of water and add 5 g of glycerol. Add KOH solution slowly until a deep blue solution is obtained.

Cupron (precipitates copper). Dissolve 5 g of benzoinoxime in 100 mL of 95% ethanol.

Cuprous chloride, acidic (reagent for CO in gas analysis).
1. Cover the bottom of a 2-L flask with a layer of cupric oxide about 0.5 inch deep, suspend a coil of copper wire so as to reach from the bottom to the top of the solution, and fill the flask with hydrochloric acid (sp. gr. 1.10). Shake occasionally. When the solution becomes nearly colorless, transfer to reagent bottles, which should also contain copper wire. The stock bottle may be refilled with dilute hydrochloric acid until either the cupric oxide or the copper wire is used up. Copper sulfate may be substituted for copper oxide in the above procedure.
2. Dissolve 340 g of $CuCl_2 \cdot 2H_2O$ in 600 mL of conc. HCl and reduce the cupric chloride by adding 190 mL of a saturated solution of stannous chloride or until the solution is colorless. The stannous chloride is prepared by treating 300 g of metallic tin in a 500 mL flask with conc. HCl until no more tin goes into solution.
3. (Winkler method). Add a mixture of 86 g of CuO and 17 g of finely divided metallic Cu, made by the reduction of CuO with hydrogen, to a solution of HCl, made by dilut-

ing 650 mL of conc. HCl with 325 mL of water. After the mixture has been added slowly and with frequent stirring, a spiral of copper wire is suspended in the bottle, reaching all the way to the bottom. Shake occasionally, and when the solution becomes colorless, it is ready for use.

Cuprous chloride, ammoniacal (reagent for CO in gas analysis).
1. The acid solution of cuprous chloride as prepared above is neutralized with ammonium hydroxide until an ammonia odor persists. An excess of metallic copper must be kept in the solution.
2. Pour 800 mL of acidic cuprous chloride, prepared by the Winkler method, into about 4 L of water. Transfer the precipitate to a 250 mL graduate. After several hours, siphon off the liquid above the 50 mL mark and refill with 7.5% NH_4OH solution which may be prepared by diluting 50 mL of conc. NH_4OH with 150 mL of water. The solution is well shaken and allowed to stand for several hours. It should have a faint odor of ammonia.

Dichlorofluorescein indicator. Dissolve 1 g in 1 L of 70% alcohol or 1 g of the sodium salt in 1 L of water.

Dimethyglyoxime, 0.01 N. Dissolve 0.6 g of dimethylglyoxime (2,3-butanedione oxime) in 500 mL of 95% ethanol. This is an especially sensitive test for nickel, a very definite crimson color being produced.

Diphenylamine (reagent for rayon). Dissolve 0.2 g in 100 mL of concentrated sulfuric acid.

Diphenylamine sulfonate (for titration of iron with $K_2Cr_2O_7$). Dissolve 0.32 g of the barium salt of diphenylamine sulfonic acid in 100 mL of water, add 0.5 g of sodium sulfate and filter off the precipitate of $BaSO_4$.

Diphenylcarbazide. Dissolve 0.2 g of diphenylcarbazide in 10 mL of glacial acetic acid and dilute to 100 mL with 95% ethanol.

Esbach's reagent (estimation of protein). To a water solution of 10 g of picric acid and 20 g of citric acid, add sufficient water to make 1 L of solution.

Eschka's compound. Two parts of calcined ("light") magnesia are thoroughly mixed with 1 part of anhydrous sodium carbonate.

Fehling's solution (reagent for reducing sugars.)
1. Copper sulfate solution. Dissolve 34.66 g of $CuSO_4 \cdot 5H_2O$ in water and dilute to 500 mL.
2. Alkaline tartrate solution. Dissolve 173 g of potassium sodium tartrate (Rochelle salt, $KNaC_4H_4O_6 \cdot 4H_2O$) and 50 g of NaOH in water and dilute when cold to 500 mL.

Mix equal volumes of the two solutions at the time of using.

Ferric-alum indicator. Dissolve 140 g of ferric ammonium sulfate crystals in 400 mL of hot water. When cool, filter, and make up to a volume of 500 mL with dilute nitric acid.

Folin's mixture (for uric acid). To 650 mL of water add 500 g of $(NH_4)_2SO_4$, 5 g of uranium acetate, and 6 g of glacial acetic acid. Dilute to 1 L.

Formaldehyde — sulfuric acid (Marquis' reagent for alkaloids). Add 10 mL of formaldehyde solution to 50 mL of sulfuric acid.

Froehde's reagent. See Sulfomolybdic acid.

Fuchsin (reagent for linen). Dissolve 1 g of fuchsin in 100 mL of alcohol.

Fuchsin — sulfurous acid (Schiff's reagent for aldehydes). Dissolve 0.5 g of fuchsin and 9 g of sodium bisulfite in 500 mL of water, and add 10 mL of HCl. Keep in well-stoppered bottles and protect from light.

Gunzberg's reagent (detection of HCl in gastric juice). Prepare as needed a solution containing 4 g of phloroglucinol (1,3,5-benzenetriol) and 2 g of vanillin in 100 mL of absolute ethanol.

Hager's reagent. See Picric acid.

Hanus solution (for iodine number). Dissolve 13.2 g of resublimed iodine in 1 L of glacial acetic acid which will pass the dichromate test for reducible matter. Add sufficient bromine to double the halogen content, determined by titration (3 mL is about the proper amount). The iodine may be dissolved by the aid of heat, but the solution should be cold when the bromine is added.

Iodine, tincture of. To 50 mL of water add 70 g of I_2 and 50 g of KI. Dilute to 1 L with alcohol.

Iodo-potassium iodide (Wagner's reagent for alkaloids). Dissolve 2 g of iodine and 6 g of KI in 100 mL of water.

Litmus (indicator). Extract litmus powder three times with boiling alcohol, each treatment consuming an hour. Reject the alcoholic extract. Treat residue with an equal weight of cold water and filter; then exhaust with five times its weight of boiling water, cool and filter. Combine the aqueous extracts.

Magnesia mixture (reagent for phosphates and arsenates). Dissolve 55 g of magnesium chloride and 105 g of ammonium chloride in water, barely acidify with hydrochloric acid, and dilute to 1 L. The ammonium hydroxide may be omitted until just previous to use. The reagent, if completely mixed and stored for any period of time, becomes turbid.

Magnesium uranyl acetate. Dissolve 100 g of $UO_2(C_2H_3O_2)_2 \cdot 2H_2O$ in 60 mL of glacial acetic acid and dilute to 500 mL. Dissolve 330 g of $Mg(C_2H_3O_2)_2 \cdot 4H_2O$ in 60 mL of glacial acetic acid and dilute to 200 mL. Heat solutions to the boiling point until clear, pour the magnesium solution into the uranyl solution, cool and dilute to 1 L. Let stand over night and filter if necessary.

Marme's reagent. See Potassium-cadmium iodide.

Marquis' reagent. See Formaldehyde-sulfuric acid.

Mayer's reagent (white precipitate with most alkaloids in slightly acid solutions). Dissolve 1.358 g of $HgCl_2$ in 60 mL of water and pour into a solution of 5 g of KI in 10 mL of H_2O. Add sufficient water to make 100 mL.

Methyl orange indicator. Dissolve 1 g of methyl orange in 1 L of water. Filter, if necessary.

Methyl orange, modified. Dissolve 2 g of methyl orange and 2.8 g of xylene cyanole FF in 1 L of 50% alcohol.

Methyl red indicator. Dissolve 1 g of methyl red in 600 mL of alcohol and dilute with 400 mL of water.

Methyl red, modified. Dissolve 0.50 g of methyl red and 1.25 g of xylene cyanole FF in 1 L of 90% alcohol. Or, dissolve 1.25 g of methyl red and 0.825 g of methylene blue in 1 L of 90% alcohol.

Millon's reagent (for albumins and phenols). Dissolve 1 part of mercury in 1 part of cold fuming nitric acid. Dilute with twice the volume of water and decant the clear solution after several hours.

Molisch's reagent. See 1-Naphthol.

1-Naphthol (Molisch's reagent for wool). Dissolve 15 g of 1-naphthol in 100 mL of alcohol or chloroform.

Nessler's reagent (for ammonia). Dissolve 50 g of KI in the smallest possible quantity of cold water (50 mL). Add a saturated solution of mercuric chloride (about 22 g in 350 mL of water will be needed) until an excess is indicated by the formation of a precipitate. Then add 200 mL of 5 N NaOH and dilute to 1 L. Let settle, and draw off the clear liquid.

Nickel oxide, ammoniacal (reagent for silk). Dissolve 5 g of nickel sulfate in 100 mL of water, and add sodium hydroxide solution until nickel hydroxide is completely precipitated. Wash the precipitate well and dissolve in 25 mL of concentrated ammonium hydroxide and 25 mL of water.

Nitron (detection of nitrate radical). Dissolve 10 g of nitron (1,4-diphenyl-3-(phenylamino)-1,2,4-triazolium hydroxide) in 5 mL of glacial acetic acid and 95 mL of water. The solution may be filtered with slight suction through an alumdum crucible and kept in a dark bottle.

1-Nitroso-2-naphthol. Make a saturated solution in 50% acetic acid (1 part of glacial acetic acid with 1 part of water). Does not keep well.

Nylander's solution (carbohydrates). Dissolve 20 g of bismuth subnitrate and 40 g of Rochelle salt in 1 L of 8% NaOH solution. Cool and filter.

Obermayer's reagent (for indoxyl in urine). Dissolve 4 g of $FeCl_3$ in 1 L of HCl (sp. gr. 1.19).

Oxine. Dissolve 14 g of 8-hydroxyquinoline in 30 mL of glacial acetic acid. Warm slightly, if necessary. Dilute to 1 L.

Oxygen absorbent. Dissolve 300 g of ammonium chloride in 1 L of water and add 1 L of concentrated ammonium hydroxide solution. Shake the solution thoroughly. For use as an oxygen absorbent, a bottle half full of copper turnings is filled nearly full with the NH_4Cl-NH_4OH solution and the gas passed through.

Pasteur's salt solution. To 1 L of distilled water add 2.5 g of potassium phosphate, 0.25 g of calcium phosphate, 0.25 g of magnesium sulfate, and 12.00 g of ammonium tartrate.

Pavy's solution (glucose reagent). To 120 mL of Fehling's solution, add 300 mL of NH_4OH (sp. gr. 0.88) and dilute to 1 L with water.

Phenanthroline ferrous ion indicator. Dissolve 1.485 g of 1,10-phenanthroline monohydrate in 100 mL of 0.025 M ferrous sulfate solution.

Phenolphthalein. Dissolve 1 g of phenolphthalein in 50 mL of alcohol and add 50 mL of water.

Phenolsulfonic acid (determination of nitrogen as nitrate). Dissolve 25 g of phenol in 150 mL of conc. H_2SO_4, add 75 mL of fuming H_2SO_4 (15% SO_3), stir well and heat for 2 hours at 100°C.

Phloroglucinol solution (pentosans). Make a 3% phloroglucinol (1,3,5-benzenetriol) solution in alcohol. Keep in a dark bottle.

Phosphomolybdic acid (Sonnenschein's reagent for alkaloids).
1. Prepare ammonium phosphomolybdate and after washing with water, boil with nitric acid and expel NH_3; evaporate to dryness and dissolve in 2 M nitric acid.
2. Dissolve ammonium molybdate in HNO_3 and treat with phosphoric acid. Filter, wash the precipitate, and boil with aqua regia until the ammonium salt is decomposed. Evaporate to dryness. The residue dissolved in 10% HNO_3 constitutes Sonnenschein's reagent.

Phosphoric acid — sulfuric acid mixture. Dilute 150 mL of conc. H_2SO_4 and 100 mL of conc. H_3PO_4 (85%) with water to a volume of 1 L.

Phosphotungstic acid (Schcibicr's reagent for alkaloids).
1. Dissolve 20 g of sodium tungstate and 15 g of sodium phosphate in 100 mL of water containing a little nitric acid.
2. The reagent is a 10% solution of phosphotungstic acid in water. Thc.phosphotungstic acid is prepared by evaporating a mixture of 10 g of sodium tungstate dissolved in 5 g of phosphoric acid (sp. gr. 1.13) and enough boiling water to effect solution. Crystals of phosphotungstic acid separate.

Picric acid (Hager's reagent for alkaloids, wool and silk). Dissolve 1 g of picric acid in 100 mL of water.

Potassium antimonate (reagent for sodium). Boil 22 g of potassium antimonate with 1 L of water until nearly all of the salt has dissolved, cool quickly, and add 35 mL of 10% potassium hydroxide. Filter after standing overnight.

Potassium-cadmium iodide (Marme's reagent for alkaloids). Add 2 g of CdI_2 to a boiling solution of 4 g of KI in 12 mL of water, and then mix with 12 mL of saturated KI solution.

Potassium hydroxide (for CO_2 absorption). Dissolve 360 g of KOH in water and dilute to 1 L.

Potassium iodide — mercuric iodide (Brucke's reagent for proteins). Dissolve 50 g of KI in 500 mL of water, and saturate with mercuric iodide (about 120 g). Dilute to 1 L.

Potassium pyrogallate (for oxygen absorption). For mixtures of gases containing less than 28% oxygen, add 100 mL of KOH solution (50 g of KOH to 100 mL of water) to 5 g of pyrogallol. For mixtures containing more than 28% oxygen the KOH solution should contain 120 g of KOH to 100 mL of water.

Pyrogallol, alkaline.
1. Dissolve 75 g of pyrogallic acid in 75 mL of water.
2. Dissolve 500 g of KOH in 250 mL of water. When cool, adjust until sp. gr. is 1.55.
For use, add 270 mL of solution (2) to 30 mL of solution (1).

Rosolic acid (indicator). Dissolve 1 g of rosolic acid in 10 mL of alcohol and add 100 mL of water.

Scheibler's reagent. See Phosphotungstic acid.

Schiff's reagent. See Fuchsin-sulfurous acid.

Schweitzer's reagent. See Cupric oxide, ammoniacal.

Soap solution (reagent for hardness in water). Dissolve 100 g of dry castile soap in 1 L of 80% alcohol (5 parts alcohol to 1 part water). Allow to stand several days and dilute with 70% to 80% alcohol until 6.4 mL produces a permanent lather with 20 mL of standard calcium solution. The latter solution is made by dissolving 0.2 g of $CaCO_3$ in a small amount of dilute HCl, evaporating to dryness and making up to 1 L.

Sodium bismuthate (oxidation of manganese). Heat 20 parts of NaOH nearly to redness in an iron or nickel crucible and add slowly 10 parts of basic bismuth nitrate which has been previously dried. Add 2 parts of sodium peroxide, and pour the brownish-yellow fused mass onto an iron plate to cool. When cold, break up in a mortar, extract with water, and collect on an asbestos filter.

Sodium hydroxide (for CO_2 absorption). Dissolve 330 g of NaOH in water and dilute to 1 L.

Sodium nitroprusside (reagent for hydrogen sulfide and wool). Use a freshly prepared solution of 1 g of sodium nitroferricyanide in 10 mL of water.

Sodium oxalate (primary standard). Dissolve 30 g of the commercial salt in 1 L of water, make slightly alkaline with sodium hydroxide, and let stand until perfectly clear. Filter and evaporate the filtrate to 100 mL. Cool and filter. Pulverize the residue and wash it several times with small volumes of water. The procedure is repeated until the mother liquor is free from sulfate and is neutral to phenolphthalein.

Sodium plumbite (reagent for wool). Dissolve 5 g of sodium hydroxide in 100 mL of water. Add 5 g of litharge (PbO) and boil until dissolved.

Sodium polysulfide. Dissolve 480 g of $Na_2S \cdot 9H_2O$ in 500 mL of water, add 40 g of NaOH and 18 g of sulfur. Stir thoroughly and dilute to 1 L with water.

Sonnenschein's reagent. See Phosphomolybdic acid.

Starch solution.

1. Make a paste with 2 g of soluble starch and 0.01 g of HgI_2 with a small amount of water. Add the mixture slowly to 1 L of boiling water and boil for a few minutes. Keep in a glass stoppered bottle. If other than soluble starch is used, the solution will not clear on boiling; it should be allowed to stand and the clear liquid decanted.

2. A solution of starch which keeps indefinitely is made as follows: Mix 500 mL of saturated NaCl solution (filtered), 80 mL of glacial acetic acid, 20 mL of water and 3 g of starch. Bring slowly to a boil and boil for 2 minutes.

3. Make a paste with 1 g of soluble starch and 5 mg of HgI_2, using as little cold water as possible. Then pour about 200 mL of boiling water on the paste and stir immediately. This will give a clear solution if the paste is prepared correctly and the water actually boiling. Cool and add 4 g of KI. Starch solution decomposes on standing due to bacterial action, but this solution will keep well if stored under a layer of toluene.

Stoke's reagent. Dissolve 30 g of $FeSO_4$ and 20 g of tartaric acid in water and dilute to 1 L. Just before using, add concentrated NH_4OH until the precipitate first formed is redissolved.

Sulfanilic acid (reagent for nitrites). Dissolve 0.5 g of sulfanilic acid in a mixture of 15 mL of glacial acetic acid and 135 mL of recently boiled water.

Sulfomolybdic acid (Froehde's reagent for alkaloids and glucosides). Dissolve 10 g of molybdic acid or sodium molybdate in 100 mL of conc. H_2SO_4.

Tannic acid (reagent for albumin, alkaloids, and gelatin). Dissolve 10 g of tannic acid in 10 mL of alcohol and dilute with water to 100 mL.

Titration mixture (residual chlorine in water analyasis). Prepare 1 L of dilute HCl (100 mL of HCl (sp. gr. 1.19) in sufficient water to make 1 L). Dissolve 1 g of o-tolidine in 100 mL of the dilute HCl and dilute to 1 L with dilute HCl solution.

Trinitrophenol solution. See Picric acid.

Turmeric tincture (reagent for borates). Digest ground turmeric root with several quantities of water which are discarded. Dry the residue and digest it several days with six times its weight of alcohol. Filter.

Uffelmann's reagent (turns yellow in presence of lactic acid). To a 2% solution of pure phenol in water, add a water solution of $FeCl_3$ until the phenol solution becomes violet in color.

Wagner's reagent. See Iodo-potassium iodide.

Wagner's solution (used in phosphate rock analysis to prevent precipitation of iron and aluminum). Dissolve 25 g of citric acid and 1 g of salicylic acid in water and dilute to 1 L. Use 50 mL of the reagent.

Wij's iodine monochloride solution (for iodine number). Dissolve 13 g of resublimed iodine in 1 L of glacial acetic acid which will pass the dichromate test for reducible matter. Set aside 25 mL of this solution. Pass into the remainder of the solution dry chlorine gas (dried and washed by passing through H_2SO_4 (sp. gr. 1.84)) until the characteristic color of free iodine has been discharged. Now add the iodine solution which was reserved, until all free chlorine has been destroyed. A slight excess of iodine does little or no harm, but an excess of chlorine must be avoided. Preserve in well stoppered, amber colored bottles. Avoid use of solutions which have been prepared for more than 30 days.

Wij's special solution (for iodine number). To 200 mL of glacial acetic acid that will pass the dichromate test for reducible matter, add 12 g of dichloramine T (N,N-dichloro-4-methyl-benzenesulfonamide), and 16.6 g of dry KI (in small quantities with continual shaking until all the KI has dissolved). Make up to 1 L with the same quality of acetic acid used above and preserve in a dark colored bottle.

Zimmermann-Reinhardt reagent (determination of iron). Dissolve 70 g of $MnSO_4 \cdot 4H_2O$ in 500 mL of water, add 125 mL of conc. H_2SO_4 and 125 mL of 85% H_3PO_4, and dilute to 1 L.

Zinc chloride solution, basic (reagent for silk). Dissolve 1000 g of zinc chloride in 850 mL of water, and add 40 g of zinc oxide. Heat until solution is complete.

Zinc uranyl acetate (reagent for sodium). Dissolve 10 g of $UO_2(C_2H_3O_2)_2 \cdot 2H_2O$ in 6 g of 30% acetic acid with heat, if necessary, and dilute to 50 mL. Dissolve 30 g of $Zn(C_2H_3O_2)_2 \cdot H_2O$ in 3 g of 30% acetic acid and dilute to 50 mL. Mix the two solutions, add 50 mg of NaCl, allow to stand overnight and filter.

STANDARD SOLUTIONS OF ACIDS, BASES, AND SALTS

For each compound listed, the last column of this table gives the mass in grams which is contained in 1 liter of a solution whose amount-of-substance concentration divided by the equivalence factor of the compound equals 0.1 mol/L. In the older literature such a solution is often referred to as a "decinormal solution" (0.1 N).

Reference

Compendium of Analytical Nomenclature (IUPAC), Pergamon Press, Oxford, 1978.

Name	Formula	Atomic or molecular weight	Equivalence factor	Mass in grams
Acetic acid	$HC_2H_3O_2$	60.0530	1	6.0053
Ammonia	NH_3	17.0306	1	1.7031
Ammonium ion	NH_4^+	18.0386	1	1.8039
Ammonium chloride	NH_4Cl	53.4916	1	5.3492
Ammonium sulfate	$(NH_4)_2SO_4$	132.1388	1/2	6.6069
Ammonium thiocyanate	NH_4CNS	76.1204	1	7.6120
Barium	Ba	137.34	1/2	6.867
Barium carbonate	$BaCO_3$	197.3494	1/2	9.8675
Barium chloride hydrate	$BaCl_2 \cdot 2H_2O$	244.2767	1/2	12.2138
Barium hydroxide	$Ba(OH)_2$	171.3547	1/2	8.5677
Barium oxide	BaO	153.3394	1/2	7.6670
Bromine	Br	79.909	1	7.9909
Calcium	Ca	40.08	1/2	2.004
Calcium carbonate	$CaCO_3$	100.0894	1/2	5.0045
Calcium chloride	$CaCl_2$	110.9860	1/2	5.5493
Calcium chloride hydrate	$CaCl_2 \cdot 6H_2O$	219.0150	1/2	10.9508
Calcium hydroxide	$Ca(OH)_2$	74.0947	1/2	3.7047
Calcium oxide	CaO	56.0794	1/2	2.8040
Chlorine	Cl	35.453	1	3.5453
Citric acid	$C_6H_8O_7 \cdot H_2O$	210.1418	1/3	7.0047
Cobalt	Co	58.9332	1/2	2.9466
Copper	Cu	63.54	1/2	3.177
Copper oxide (cupric)	CuO	79.5394	1/2	3.9770
Copper sulfate hydrate	$CuSO_4 \cdot 5H_2O$	249.6783	1/2	12.4839
Hydrochloric acid	HCl	36.4610	1	3.6461
Hydrocyanic acid	HCN	27.0258	1	2.7026
Iodine	I	126.9044	1	12.6904
Lactic acid	$C_3H_6O_3$	90.0795	1	9.0080
Malic acid	$C_4H_6O_5$	134.0894	1/2	6.7045
Magnesium	Mg	24.312	1/2	1.2156
Magnesium carbonate	$MgCO_3$	84.3214	1/2	4.2161
Magnesium chloride	$MgCl_2$	95.2180	1/2	4.7609
Magnesium chloride hydrate	$MgCl_2 \cdot 6H_2O$	203.2370	1/2	10.1623
Magnesium oxide	MgO	40.3114	1/2	2.0156
Manganese	Mn	54.938	1/2	2.7469
Manganese sulfate	$MnSO_4$	150.9996	1/2	7.5500
Mercuric chloride	$HgCl_2$	271.4960	1/2	13.5748
Nickel	Ni	58.71	1/2	2.9356
Nitric acid	HNO_3	63.0129	1	6.3013
Oxalic acid	$H_2C_2O_4$	90.0358	1/2	4.5018
Oxalic acid hydrate	$H_2C_2O_4 \cdot 2H_2O$	126.0665	1/2	6.3033
Oxalic acid anhydride	C_2O_3	72.0205	1/2	3.6010
Phosphoric acid	H_3PO_4	97.9953	1/3	3.2665
Potassium	K	39.102	1	3.9102
Potassium bicarbonate	$KHCO_3$	100.1193	1	10.0119
Potassium carbonate	K_2CO_3	138.2134	1/2	6.9106
Potassium chloride	KCl	74.5550	1	7.4555
Potassium cyanide	KCN	65.1199	1	6.5120
Potassium hydroxide	KOH	56.1094	1	5.6109

Name	Formula	Atomic or molecular weight	Equivalence factor	Mass in grams
Potassium oxide	K_2O	94.2034	1/2	4.7102
Potassium tartrate	$K_2H_4C_4O_6$	226.2769	1/2	11.3139
Silver	Ag	107.87	1	10.787
Silver nitrate	$AgNO_3$	169.8749	1	16.9875
Sodium	Na	22.9898	1	2.2990
Sodium bicarbonate	$NaHCO_3$	84.0071	1	8.4007
Sodium carbonate	Na_2CO_3	105.9890	1/2	5.2995
Sodium chloride	NaCl	58.4428	1	5.8443
Sodium hydroxide	NaOH	39.9972	1	3.9997
Sodium oxide	Na_2O	61.9790	1/2	3.0990
Sodium sulfide	Na_2S	78.0436	1/2	3.9022
Succinic acid	$H_2C_4H_4O_4$	118.0900	1/2	5.9045
Sulfuric acid	H_2SO_4	98.0775	1/2	4.9039
Tartaric acid	$C_4H_6O_6$	150.0888	1/2	7.5044
Zinc	Zn	65.37	1/2	3.269
Zinc sulfate hydrate	$ZnSO_4 \cdot 7H_2O$	287.5390	1/2	14.3769

STANDARD SOLUTIONS OF OXIDATION AND REDUCTION REAGENTS

For each reagent listed, the last column of this table gives the mass in grams which is contained in a solution whose amount-of-substance concentration divided by the equivalence factor of the compound equals 0.1 mol/L. The equivalence factor given refers to the most common reactions of the reagent. In the older literature such a solution is often called a "decinormal solution" (0.1 N).

Reference

Compendium of Analytical Nomenclature (IUPAC), Pergamon Press, Oxford, 1978.

Name	Formula	Atomic or molecular weight	Equivalence factor	Mass in grams
Antimony	Sb	121.75	1/2	6.0875
Arsenic	As	74.9216	1/2	3.7461
Arsenic trisulfide	As_2S_3	246.0352	1/4	6.1509
Arsenous oxide	As_2O_3	197.8414	1/4	4.9460
Barium peroxide	BaO_2	169.3388	1/2	8.4669
Barium peroxide hydrate	$BaO_2 \cdot 8H_2O$	313.4615	1/2	15.6730
Calcium	Ca	40.08	1/2	2.004
Calcium carbonate	$CaCO_3$	100.0894	1/2	5.0045
Calcium hypochlorite	$Ca(OCl)_2$	142.9848	1/4	3.5746
Calcium oxide	CaO	56.0794	1/2	2.8040
Chlorine	Cl	35.453	1	3.5453
Chromium trioxide	CrO_3	99.9942	1/3	3.3331
Ferrous ammonium sulfate	$FeSO_4(NH_4)SO_4 \cdot 6H_2O$	392.0764	1	39.2076
Hydroferrocyanic acid	$H_4Fe(CN)_6$	215.9860	1	21.5986
Hydrogen peroxide	H_2O_2	34.0147	1/2	1.7007
Hydrogen sulfide	H_2S	34.0799	1/2	1.7040
Iodine	I	126.9044	1	12.6904
Iron	Fe	55.847	1	5.5847
Iron oxide (ferrous)	FeO	71.8464	1	7.1846
Iron oxide (ferric)	Fe_2O_3	159.6922	1/2	7.9846
Lead peroxide	PbO_2	239.1888	1/2	11.9594
Manganese dioxide	MnO_2	86.9368	1/2	4.3468
Nitric acid	HNO_3	63.0129	1/3	2.1004
Nitrogen trioxide	N_2O_3	76.0116	1/4	1.9002
Nitrogen pentoxide	N_2O_5	108.0104	1/6	1.8001
Oxalic acid	$C_2H_2O_4$	90.0358	1/2	4.5018
Oxalic acid hydrate	$C_2H_2O_4 \cdot 2H_2O$	126.0665	1/2	6.3033
Oxygen	O	15.9994	1/2	0.8000
Potassium dichromate	$K_2Cr_2O_7$	294.1918	1/6	4.9032
Potassium chlorate	$KClO_3$	122.5532	1/6	2.0425
Potassium chromate	K_2CrO_4	194.1076	1/3	6.4733
Potassium ferrocyanide	$K_4Fe(CN)_6$	368.3621	1	36.8362
Potassium ferrocyanide hydrate	$K_4Fe(CN)_6 \cdot 3H_2O$	422.4081	1	42.2408
Potassium iodide	KI	166.0064	1	16.6006
Potassium nitrate	KNO_3	101.1069	1/3	3.3702
Potassium perchlorate	$KClO_4$	138.5526	1/8	1.7319
Potassium permanganate	$KMnO_4$	158.0376	1/5	3.1608
Sodium chlorate	$NaClO_3$	106.4410	1/6	1.7740
Sodium nitrate	$NaNO_3$	84.9947	1/3	2.8332
Sodium thiosulfate hydrate	$Na_2S_2O_3 \cdot 5H_2O$	248.1825	1	24.8183
Stannous chloride	$SnCl_2$	189.5960	1/2	9.4798
Stannous oxide	SnO	134.6894	1/2	6.7345
Sulfur dioxide	SO_2	64.0628	1/2	3.2031
Tin	Sn	118.69	1/2	5.935

ORGANIC ANALYTICAL REAGENTS FOR THE DETERMINATION OF INORGANIC SUBSTANCES

G. Ackermann, L. Sommer, and D. Thorburn Burns

Determination	Reagents	Ref.
Aluminium	Alizarin Red S	Onishi, Part II a, p 28. (5), Snell, *Metals I*, p 587. (7)
	Aluminon	Fries/Getrost, p 16. (2), Onishi, IIa, p 21. (5), Snell, *Metals I*, p 590. (7)
	Aluminon + Cetyltrimethylammonium bromide	Huaxue Shiji, *8*, 85, (1986)
	Chrome Azurol S	Onishi, Part IIa, p 26. (5), Snell, *Metals I*, p 605. (7)
	Chrome Azurol S + Cetyltrimethylammonium bromide	Marczenko, p 133. (3), Snell, *Metals I*, p 606. (7)
	Chromazol KS + Cetylpyridinium bromide	*Analyst*, *107*, 428, (1982).
	Eriochrome Cyanine R	Fries/Getrost, p 19. (2), Onishi, Part IIa, p 25. (5), Snell, *Metals I*, p 611. (7)
	Eriochrome Cyanine R + Cetyltrimethylammonium bromide	Snell, *Metals I*, p 613. (7), *Analyst*, *107*, 1431, (1982).
	8-Hydroxyquinoline	Fries/Getrost, p 22. (2), Marczenko, p 131. (3), Onishi, Part IIa, p 31. (5), Snell, *Metals I* , p 622. (7)
Ammonia	Phenol + Sodium hypochlorite	Boltz, p 210 (1), Marczenko, p 413. (3), Snell, *Nonmetals*, p 604. (9)
Antimony	Brilliant Green	Onishi, Part IIa, p 102. (5), Snell, *Metals I*, p 384. (7)
	Bromopyrogallol Red	*Talanta*, *13*, 507, (1966).
	Rhodamine B	Fries/Getrost, p 32. (2), Marczenko, p 141. (3), Onishi, Part IIa, p 93. (5), Snell, *Metals I*, p 404. (7)
	Silver diethyldithiocarbamate	Fries/Getrost, p 36. (2)
Arsenic	Silver diethyldithiocarbamate	Fries/Getrost, p 41. (2), Marczenko, p 153. (3), Onishi, Part IIa, p 153. (5), Snell, *Metals I*, p 370. (7)
Barium	Sulfonazo III	Fries/Getrost, p 46. (2), Snell, *Metals II*, p 1782. (8), Onishi, Part IIa, p 202. (5)
Beryllium	Beryllon II	Snell, *Metals I*, p 667. (7)
	Chrome Azurol S	Marczenko, p 163. (3), Snell, *Metals I*, p 672. (7)
	Chrome Azurol S + Cetyltrimethylammonium bromide	Marczenko, p 164. (3), Snell, *Metals I*, p 673. (7)
	Eriochrome Cyanine R	Snell, *Metals I*, p 675. (7), *Talanta*, *31*, 249, (1984).
	Eriochrome Cyanine R + Cetyltrimethylammonium bromide	Zh. *Anal. Khim.*, *33*, 1298, (1978).
Bismuth	Dithizone	Onishi, Part IIa, p 262. (5), Snell, *Metals I*, p 303. (7)
	Pyrocatechol Violet	Fres. *Z. Anal. Chem.*, *186*, 418, (1962).
	Pyrocatechol Violet + Cetyltrimethylammonium bromide	Zh. *Anal. Khim.*, *38*, 216, (1983).
	Thiourea	Onishi, Part IIa, p 260. (5), Snell, *Metals I*, p 317. (7)
	Xylenol Orange	Friez/Getrost, p 57. (2), Marczenko, p 172. (3), Snell, *Metals I*, p 320. (7)
Boron	Azomethine H	Snell, Nonmetals, p 165. (9)
	Carminic acid	Boltz, p 14. (1), Fries/Getrost, p 65. (2), Snell, *Nonmetals*, p 170. (9), Williams, p 35. (11)
	Curcumin	Boltz, p 8. (1), Fries/Getrost, p 68. (2), Marczenko, p 180. (3), Snell, *Nonmetals*, p 180. (9), Fres. *Z. Anal. Chem.*, *323*, 266, (1986).
	Methylene Blue	Boltz, p 21. (1), Marczenko, p 183. (3), Snell, *Nonmetals*, p 205. (9), *Talanta*, 31, 547, (1984).
Bromide	Fluorescein	Boltz, p 48. (1), Snell, *Nonmetals*, p 276., Fres. *Z. Anal. Chem.*, *301*, 28 (1980).
	Phenol Red	Boltz, p 44. (1), Marczenko, p 190. (3), Snell, Nonmetals, p 28. (9)
Cadmium	2-(5-Bromo-2-pyridylazo)-5-diethylaminophenol	Marczenko, p 197. (3)
	Cadion	Onishi, Part IIa, p 323. (5)
	Dithizone	Fries/Getrost, p 78. (2), Onishi, Part IIa, p 315. (5), Snell, *Metals I*, p 279. (7), West, p 25. (10).
	4–(2-Pyridylazo)resorcinol	Fres. *Z. Anal. Chem.*, *310*, 51, (1982).
Calcium	Chlorophosphonazo III	Marczenko, p 207. (3), Snell, *Metals II*, p 1744. (8)
	Glyoxal-bis(2-hydroxyanil)	Fries/Getrost, p 86. (2), Onishi, Part IIa, p 352. (5), Snell, *Metals I*, p 1762. (8)
	Murexide	Onishi, Part IIa, p 357. (5), Snell, *Metals II*, p 1769. (8)
	Phthalein Purple	*Anal. Chim. Acta*, *34*, 71 (1966).
Cerium	*N*-benzoyl-*N*-phenylhydroxylamine	*Anal. Chim. Acta*, *48*, 155, (1969).

Determination	Reagents	Ref.
	8-Hydroxyquinoline	Fries/Getrost, p 93. (2), Marczenko, p 220. (3), Onishi, Part IIa, p 383. (7)
Chlorine	N,N-Diethyl-1,4-phenylenediamine	Boltz, p 92. (1), Fries/Getrost, p 101. (2), Snell, Nonmetals, p 225. (9), *Analyst, 90,* 187, (1965).
Chromium	1,5-Diphenylcarbazide	Fries/Getrost, p 105. (2), Onishi, Part IIa, p 412. (5), Snell, *Metals I,* p 714. (7), West, p 12. (10)
	4-(2-Pyridylazo)resorcinol	Snell, *Metals I,* p 736. (7), West, p 17. (10)
	4-(2-Pyridylazo)resorcinol + Tetradecyldimethyl-benzylammonium chloride	West, p 17. (10), A*nal. Chim. Acta, 67,* 297, (1973).
	4-(2-Pyridylazo)resorcinol + Hydrogen peroxide	Fres. *Z. Anal. Chem., 304,* 382, (1980).
Cobalt	Nitroso-R salt	Fries/Getrost, p 118. (2), Onishi, Part IIa, p 454. (5), Snell, *Metals I,* p 953. (7)
	1-Nitroso-2-naphthol	Fries/Getrost, p 111. (2), Marczenko, p 246. (3), Snell, *Metals I,* p 947. (5)
	2-Nitroso-1-naphthol	Fries/Getrost, p 113. (2), Onishi, Part IIa, p 459. (5), Snell, *Metals I,* p 949. (7), West, p 45. (10)
	4-(2-Pyridylazo)resorcinol	Snell, *Metals I,* p 969. (7), West, p 44. (10)
	4-(2-Pyridylazo)resorcinol + Diphenylguanidine	Zh. *Anal. Khim., 35,* 1306, (1980).
Copper	Bathocuproine	Fries/Getrost, p 135. (2), Snell, *Metals I,* p 148. (7)
	Bathocuproine disulfonic acid	Fries/Getrost, p 137. (2), West, p 52. (10)
	Dithizone	Marczenko, p 258. (3), Onishi, Part IIa, p 529. (5), Snell, *Metals I,* p 199. (7)
	Neocuproine	Snell, *Metals I,* p 217. (5), West, p 51. (10)
	Cuprizone	Onishi, Part IIa, p 534. (5), Snell, *Metals I,* p 157. (7), West, p 53. (10)
	4-(2-pyridylazo)resorcinol + Tetradecyldimethyl-benzylammonium chloride	*Anal. Chim. Acta, 138,* 321, (1982).
Cyanide	Barbituric Acid + Pyridine	Fries/Getrost, p 153. (2), Snell, *Nonmetals,* p 653. (9)
	Barbituric Acid + Pyridine-4-carboxylic acid	*Anal. Chim. Acta, 99,* 197, (1978).
Fluoride	Alizarin Fluorine blue + Lanthanum(III) ion	Boltz, p 129. (1), Fries/Getrost, p 158. (2), Snell, *Nonmetals,* p 333. (9), Williams, p 354. (11)
	Eriochrome Cyanine R + Zirconium(IV) ion	Boltz, p 119. (1), Snell, *Nonmetals,* p 359. (2), Williams, p 357. (11)
Gallium	Pyrocatechol violet + Diphenylguanidine	Snell, *Metals I,* p 500. (7)
	8-Hydroxyquinoline	Onishi Pt IIa, p 582. (5), Snell, *Metals I,* p 505. (7)
	1-(2-Pyridylazo)-2-naphthol	Snell, *Metals I,* p 512. (7)
	4-(2-Pyridylazo)resorcinol	Snell, *Metals I,* p 513. (7)
	Rhodamine B	Marczenko, p 284. (3), Onishi, Part IIa, p 578. (5), Snell, *Metals I,* p 515. (7)
	Xylenol Orange	Fries/Getrost, p 166. (2), Snell, *Metals I,* p 523. (7)
	Xylenol Orange + 8-Hydroxyquinoline	*Zh. Anal. Khim., 26,* 75, (1971).
Germanium	Brilliant Green + Molybdate	Snell, *Metals I,* p 562. (7)
	Phenylfluorone	Fries/Getrost, p 168. (2), Marczenko, p 292. (3), Onishi, Part IIa, p 607. (5), Snell, *Metals I,* p 570. (7)
Gold	5-(4-Diethylaminobenzylidene) rhodanine	Fries/Getrost, p 173. (2), Onishi, Part IIa, p 631. (5), Snell, *Metals II,* p 1516. (8)
	Rhodamine B	Fries/Getrost, p 175. (2), Marczenko, p 301. (3), Onishi, Part IIa, p 637. (5), Snell, *Metals II,* p 513. (8)
Hafnium	Arsenazo III	Snell, *Metals II,* p 1184. (8), *Talanta, 19,* 807, (1972).
Indium	Bromopyrogallol Red	Snell, *Metals I,* p 469. (7)
	Chrome Azurol S	Snell, *Metals I,* p 474. (7)
	Chrome Azurol S + Cetyltrimethylammonium bromide	*Anal. Chim. Acta, 67,* 107, (1973).
	Dithizone	Fries/Getrost, p 179. (2), Onishi, Part IIa, p 672. (5), Snell, *Metals I,* p 474. (7)
	8-Hydroxyquinoline	Onishi, Part IIa, p 670. (5), Snell, *Metals I,* p 475. (7)
	1-(2-Pyridylazo)-2-naphthol	Snell, *Metals I,* p 480. (7)
	4-(2-Pyridylazo)resorcinol	Marczenko, p 309. (3), Snell, *Metals* I, p 480. (7)
Iodide	Neocuproine + Copper(II)	*Anal. Chim. Acta, 69,* 321, (1974).
Iodine	Starch	Boltz, p 162. (1), Marczenko, p 316. (3), Snell, *Nonmetals,* p 307. (9)
Iridium	Rhodamine 6G + Tin(II)	Marczenko, p 323. (3)
	N,N-Dimethyl-4-nitrosoaniline	*Anal. Chem., 27,* 1776, (1955).
Iron	Bathophenanthroline	Fries/Getrost, p 189. (2), Onishi, Part IIa, p 729. (5), Snell, *Metals I,* p 763. (7)
	Bathophenanthroline disulfonic acid	Fries/Getrost, p 191. (2), Snell, *Metals I,* p 772. (7)

Determination	Reagents	Ref.
	2,2´-Bipyridyl	Snell, *Metals I*, p 750. (7)
	Chrome Azurol S + Cetyltrimethylammonium bromide	Snell, *Metals I*, p 757. (7), *Coll. Czech. Chem. Comm.*, 45, 2656, (1980).
	1,10-Phenanthroline	Fries/Getrost, p 199. (2), Marczenko, p 331. (3), Onishi, Part IIa, p 725. (5), Snell, *Metals I*, p 795. (7)
	1,10-Phenanthroline + Bromothymol Blue	*Zh. Anal. Khim.*, 25, 1348, (1970).
	Ferrozine	Onishi, Part IIa, p 730. (5), Snell, *Metals I*, p 783. (7)
Lanthanum	Arsenazo III	Marczenko, p 468. (3), Snell, *Metals II*, p 1910. (8)
Lead	Dithizone	Fries/Getrost, p 207. (2), Onishi, Part IIa, p 824. (5), Snell, *Metals I*, p 2. (7), West, p 34. (10)
	Sodium diethyldithiocarbamate	Fries/Getrost, p 214. (2), Snell, *Metals I*, p 27. (7)
	4-(2-Pyridylazo)resorcinol	Fries/Getrost, p 220. (2), Marczenko, p 347. (3), Snell, *Metals I*, p 34. (7)
Lithium	Thoron	Onishi, Part IIa, p 863. (5), Snell, *Metals II*, p 1726. (8), *Talanta*, 30, 587, (1983).
Magnesium	Eriochrome Black T	Fries/Getrost, p 226. (2), Marczenko, p 355. (3), Onishi, Part IIb, p 13. (6), Snell, *Metals II*, p 1932. (8)
	8-Hydroxyquinoline	Onishi, Part IIb, p 11. (6), Snell, *Metals II*, p 1938. (8)
	8-Hydroxyquinoline + Butylamine	Fries/Getrost, p 228. (2), Snell, *Metals II*, p 1938. (8)
	Titan Yellow	Fries/Getrost, p 234. (2), Marczenko, p 352. (3), Snell, *Metals II*, p 1945. (8)
	Xylidyl Blue	Fries/Getrost, p 231. (2), Onishi, Part IIb, p 14. (6), Snell, *Metals II*, p 1950. (8)
Manganese	Formaldoxime	Fries/Getrost, p 236. (2), Marczenko, p 364. (3), Onishi Part IIb, p 38. (6), Snell, *Metals II*, 1010. (8)
Mercury	Dithizone	Fries/Getrost, p 243. (2), Marczenko, p 373. (3), Onishi, Part IIb, p 66. (6), Snell, *Metals I*, p 107. (7), West, p 29. (10)
	Michler's thioketone	Marczenko, p 375. (3), Snell, *Metals I*, p 126. (7)
	Xylenol Orange	*Talanta*, 16, 1023, (1969)
Molybdenum	Bromopyrogallol Red + Cetylpyridium chloride	West, p 58. (10)
	Phenylfluorone	Snell, *Metals II*, p 1311. (8), *Microchem. J.*, 31, 56, (1985).
	Toluene-3,4-dithiol	Fries/Getrost, p 251. (2), Marczenko, p 384. (3), Onishi, Part IIb, p 96. (6), Snell, *Metals II*, p 1301. (8)
Nickel	2-(5-Bromo-2-pyridylazo)-5-diethylaminophenol	Marczenko, p 397. (3), *Talanta 28*, 189, (1981).
	Dimethylglyoxime	Fries/Getrost, p 263. (2), Marczenko, p 393. (3), Onishi, Part IIb, p 125. (6), Snell, *Metals I*, p 887. (7)
	Dimethylglyoxime + Oxidant	Fries/Getrost, p 263. (2), Onishi, Part IIb, p 125. (6), Snell, *Metals I*, p 887. (7)
	2,2´-Furildioxime	Marczenko, p 396. (3), Snell, *Metals I*, p 904. (7)
	2-(2-Pyridylazo)-2-naphthol	Snell, *Metals I*, p 910. (7)
	4-(2-Pyridylazo)resorcinol	Snell, *Metals I*, p 911. (7), West, p 39. (10), *Anal. Chim. Acta, 82*, 431, (1976).
Niobium	*N*-Benzoyl-*N*-phenylhydroxylamine	Snell, *Metals II*, p 1425. (8)
	Pyrocatechol + EDTA or 2,2´Bipyridyl or 1-(2-thenoyl)-3,3,3,-trifluoroacetone	Snell, *Metals II*, p 1427. (8)
	Bromopyrogallol red	Marczenko, p 407. (3), Snell, *Metals II*, p 1426. (8)
	Bromopyrogallol red + Cetylpyridinium chloride	*Talanta, 32*, 189, (1985).
	4-(2-Pyridylazo)resorcinol	Fries/Getrost, p 274. (2), Marczenko, p 406. (3), Onishi, Part IIb, p 160. (7), Snell, *Metals II*, p 1447. (8)
	Sulfochlorophenol S	Onishi, Part IIb, p 161. (7), Snell, *Metals II*, p 1430. (8)
	Xylenol Orange	Onishi, Part IIb, p 164. (7)
Nitrate	Brucine	Boltz, p 227. (1), Fries/Getrost, p 280. (2), Snell, *Nonmetals*, p 546. (9)
	Chromotropic acid	Boltz, p 229. (1), Fries/Getrost, p 281. (2), Snell, *Nonmetals*, p 548. (9), Williams, p 132. (11), *Fres. Z. Anal. Chem., 320*, 490, (1985).
	Sulfanilamide + *N*-(1-Naphthyl)ethylenediamine dihydrochloride	Fries/Getrost, p 279. (2), Snell, *Nonmetals*, p 559. (9)
Nitrite	Sulfanilamide + *N*-(1-Naphthyl)ethylenediamine dihydrochlorine	Boltz, p 241. (1), Snell, *Nonmetals, p* 585. (8), *Analyst, 109*, 1281, (1984).
	Sulfanilic acid + 1-Naphthylamine	Boltz, p 237. (1), Fries/Getrost, p 285. (2), Marczenko, p 419. (3), Snell, *Nonmetals*, p 586. (9)
Osmium	1,5-Diphenylcarbazide	Marczenko, p 428. (3)
Palladium	2-(5 Bromo-2-pyridylazo)-5-diethylaminophenol	*Talanta, 33*, 939, (1986).

Determination	Reagents	Ref.
	Dithizone	Marczenko, p 440. (3), Onishi, Part IIb, p 227. (6), Snell, *Metals II*, p 1577. (8)
	2-Nitroso-1-naphthol	Fries/Getrost, p 294. (2), Onishi, Part IIb, p 226. (6), Snell, *Metals II*, p 1581. (8)
	4-(2-Pyridylazo)resorcinol	Snell, *Metals II*, p 1583. (8) *Analyst*, *107*, 708, (1982).
Phosphate	Rhodamine B + Molybdate	Snell, *Nonmetals*, p 103. (9)
	Malachite Green + Molybdate	Snell, *Nonmetals*, p 12. (9), Analyst, *108*, 361, (1983).
Platinum	Sulfochlorophenolazorhodamine	Onishi, Part IIb, p 253. (6), *Talanta*, *34*, 87, (1987).
	Dithizone	Fries/Getrost, p 300. (2), Onishi, Part IIb, p 253. (6), Snell, *Metals II*, p 1534. (8)
	2-Mercaptobenzothiazole	Fries/Getrost, p 302. (2), *Zh. Anal. Khim.*, *24*, 1172, (1969).
Rare Earths	Arsenazo I	Marczenko, p 470. (3), Onishi, Part IIa, p 785. (5), Snell, *Metals II*, p 1857. (8)
	Arsenazo III	Fries/Getrost, p 309. (2), Marczenko, p 468. (3), Onishi, Part IIa, p 786. (5), Snell, *Metals II*, p 1862. (8)
	Xylenol Orange	Onishi, Part IIa, p 787. (5), Snell, *Metals II*, p 1874. (8)
Rhenium	2,2′-Furildioxime	Fries/Getrost, p 310. (2), Marczenko, p 481. (3), Onishi, Part IIb, p 288. (6), Snell, *Metals II*, p 1659. (8)
Rhodium	1-(2-Pyridylazo)-2-naphthol	Fries/Getrost, p 311. (2), Snell, *Metals II*, p 1553. (8)
Ruthenium	1,10-Phenanthroline	Onishi, Part IIb, p 331. (6), Snell, *Metals II*, p 1623. (8)
	Thiourea	Fries/Getrost, p 318. (2), Onishi, Part IIb, 329. (6), Snell, *Metals II*, p 1626. (8)
	1,4-Diphenylthiosemicarbazide	Marczenko, p 493. (3), Onishi, Part IIb, p 330. (8)
Scandium	Alizarin red S	Fries/Getrost, p 319. (2), Onishi, Part IIb, p 360. (6), Snell, *Metals I*, p 536. (7)
	Arsenazo III	Onishi, Part IIb, p 359. (6), Snell, *Metals I*, p 539. (7)
	Chrome Azurol S	Snell, *Metals I*, p 551. (7), *Anal. Chim. Acta*, *159*, 309, (1984).
	Xylenol Orange	Marczenko, p 501. (3), Onishi, Part IIb, p 357. (6), Snell, *Metals I*, p 547. (7)
Selenium	3,3′-Diaminobenzidine	Boltz, p 391. (1), Fries/Getrost, p 323. (2), Marczenko, p 508. (3), Snell, *Nonmetals*, p 490. (9), West, p 4. (10).
	2,3-Diaminonaphthaline	Snell, *Nonmetals*, p 501. (9)
Silver	Dithizone	Fries/Getrost, p 328. (2), Marczenko, p 524. (3), Onishi, Part IIb, p 379. (6), Snell, *Metals I*, p 82. (7)
	Eosin + 1,10-Phenanthroline	Snell, *Metals I*, p 93. (7)
Sulfate	Methylthymol blue + Barium (II)	Snell, *Nonmetals*, p 457. (9)
Sulfide	*N,N,*-Dimethyl-1,4-phenylenediamine	Boltz, p 483. (1), Fries/Getrost, p 344. (2), Snell, *Nonmetals*, p 400. (9), Williams, p 578. (11)
Sulfite	Pararosaniline + Formaldehyde	Boltz, p 478. (1), Marczenko, p 540. (3), Snell, *Nonmetals*, p 430. (9), Williams, p 591. (11)
Tantalum	Methyl Violet	Marczenko, p 551. (3), Snell, *Metals II*, p 1485. (8)
	4-(2-Pyridylazo)resorcinol	Snell, *Metals II*, p 1488. (8)
	Phenylfluorone	Onishi, Part IIb, p 166. (6), Snell, *Metals II*, p 1486. (8)
Tellurium	Diethyldithiocarbamate	Boltz, p 402. (1), Fries/Getrost, p 348. (2), Snell, *Nonmetals*, p 533. (9), Williams, p 220. (10)
	Bismuthiol II	Boltz, p 401. (1), Marczenko, p 557. (3), Snell, *Nonmetals*, p 524. (9)
Thallium	Brilliant green	Fries/Getrost, p 352. (2), Marczenko, p 567. (3), Onishi, Part IIb, p 426. (6), Snell, *Metals I*, p 45. (7)
	Dithizone	Fries/Getrost, p 355. (2), Onishi, Part IIb, p 426. (6), Snell, *Metals I*, p 54. (7)
	Rhodamine B	Fries/Getrost, p 354. (2), Marczenko, p 566. (3), Onishi, Part IIb, p 424. (6), Snell, *Metals I*, p 63. (7)
Thorium	Arsenazo III	Fries/Getrost, p 360. (2), Marczenko, p 575. (3), Onishi, Part IIb, p 460. (6), Snell, *Metals II*, p 1820. (8)
	Thoron	Marczenko, p 574. (3), Onishi, Part IIb, p 463. (6), Snell, *Metals I*, p 1835. (7)
	Xylenol Orange	Snell, *Metals I*, p 1852. (7)
	Xylenol Orange + Cetyltrimethylammonium bromide	*Talanta*, *26*, 499, (1979).
Tin	Pyrocatechol violet (and + Cetyltrimethylammonium bromide)	Marczenko, p 585. (3), Onishi, Part IIb, p 501. (6), Snell, *Metals I*, p 422. (7)
	Gallein	Onishi, Part IIb, p 507, 510. (6), Snell, *Metals I*, p 432. (7)
	Phenylfluorone	Fries/Getrost, p 368. (2), Marczenko, p 582. (3), Onishi, Part IIb, p 497. (6), Snell, *Metals I*, p 444. (7)

Determination	Reagents	Ref.
	Toluene-3,4-dithiol + Dispersant	Fries/Getrost, p 366. (2), Onishi, Part IIb, p 502. (6), Snell, *Metals I*, p 427. (7)
Titanium	Chromotropic acid	Marczenko, p 593. (3), Onishi, Part IIb, p 551. (6), Snell, *Metals II*, p 1080. (8)
	Diantipyrinylmethane	Onishi, Part IIb, p 545. (6), Snell, *Metals II*, 1085. (8)
	Tiron	Fries/Getrost, p 376. (2), Onishi, Part IIb, p 549. (6), Snell, *Metals II*, p 1114. (8)
Tungsten	Pyrocatechol Violet	Snell, *Metals II*, p 1265. (8)
	Tetraphenylarsonium chloride + Thiocyanate	Onishi, Part IIb, p 596. (6), Snell, *Metals II*, p 1278. (8)
	Toluene-3,5-dithiol	Marczenko, p 605. (3), Onishi, Part IIb, p 590. (6), Snell, *Metals II*, p 1267. (8)
Uranium	Arsenazo III	Marczenko, p 611. (3), Onishi, Part IIb, p 627. (6), Snell, *Metals II*, p 1356. (8)
	2-(5-Bromo-2-pyridylazo)diethylaminophenol	Fries/Getrost, p 388. (2), Onishi, Part IIb, p 625. (6)
	Chlorophosphonazo III	Snell, *Metals II*, p 1367. (8), *Fres. Z. Anal. Chem.*, *306*, 110, (1981).
	1-(2-Pyridylazo)-2-naphthol	Fries/Getrost, p 386. (2), Onishi, Part IIb, p 625. (6), Snell, *Metals II*, p 1387. (8)
Vanadium	N-Benzoyl-N-phenylhydroxylamine	Fries/Getrost, p 395. (2), Marczenko, p 625. (3), Snell, *Metals II*, p 1196. (8)
	8-Hydroxyquinoline	Marczenko, p 623. (3), Snell, *Metals II*, p 1209. (8)
	4-(2-pyridylazo)resorcinol	Fries/Getrost, p 404. (2), Marczenko, p 628. (3), Onishi, Part IIb, p 625. (6), Snell, *Metals II*, p 1226. (8)
Yttrium	Alizarin Red S	Fries/Getrost, p 406. (2), Onishi, Part IIa, p 784. (5), Snell, *Metals II*, p 1919. (8)
	Arsenazo III	Marczenko, p 468. (3), Onishi, Part IIa, p 786. (5), Snell, *Metals II*, p 1921. (8)
	Xylenol Orange	Fries/Getrost, p 406. (2), Onishi, Part IIa, p 787. (5), Snell, *Metals II*, p 1923. (8)
Zinc	Dithizone	Fries/Getrost, p 408. (2), Marczenko, p 637. (3), Onishi, Part IIb, p 708. (6), Snell, *Metals II*, p 1042. (8)
	1-(2-Pyridylazo)-2-naphthol	Marczenko, p 639. (3), Onishi, Part IIb, p 719. (6), Snell, *Metals II*, p 1056. (8)
	Xylenol Orange	Fries/Getrost, p 417. (2), Snell, *Metals II*, p 1062. (8), Talanta, *26*, 693, (1979).
	Zircon	Fries/Getrost, p 412. (2), Onishi, Part IIb, p 719. (6), Snell, *Metals II*, p 1063. (8), West, p 23. (10)
Zirconium	Alizarin Red S	Fries/Getrost, p 421. (2), Marczenko, p 647. (3), Onishi, Part IIb, p 763. (6), Snell, *Metals II*, p 1136. (8)
	Arsenazo III	Fries/Getrost, p 421. (2), Onishi, Part IIb, p 770. (6), Snell, *Metals II*, p 1143. (8)
	Pyrocatechol Violet	Onishi, Part IIb, p 771. (6), Snell, *Metals II*, p 1149. (8)
	Morin	Fries/Getrost, p 424. (2), Onishi, Part IIb, p 765. (6), Snell, *Metals II*, p 1158. (8)
	Xylenol Orange	Fries/Getrost, p 419. (2), Marczenko, p 648. (3), Onishi, Part IIb, p 767. (6), Snell, *Metals II*, p 1167. (8)

Reviews

Sommer, L, Ackermann, G., Thorburn Burns, D., and Savvin, S. B., *Pure and Applied Chem.*, 62, 2147, 1990.

Sommer, L., Ackermann, G., and Thorburn Burns, D., *Pure and Applied Chem.*, 62, 2323, 1990)

Sommer, L., Komarek, J., and Thorburn Burns, D., *Pure and Applied Chem.*, 64, 213, 1992.

Savvin, S. B., *Crit. Rev. Anal. Chem.*, 8, 55, 1979.

Monographs

1. Boltz, D. F., and Howell, J. A., *Colorimetric Determination of Nonmetals*, 2nd ed, Wiley, New York, 1978.
2. Fries, J. and Getrost, H., *Organic Reagents for Trace Analysis*, E Merck, Darmstadt, 1977.
3. Marczenko, Z., *Separation and Spectrophotometric Determination of Elements*, Ellis Horwood, Chichester, 1986.

4. Sandell, E. B. and Onishi, H., *Photometric Determination of Traces of Metals. General Aspects, Part I*, 4th ed, J. Wiley, New York, 1978.
5. Onishi, H., *Photometric Determination of Traces of Metals. Part IIa: Individual Metals, Aluminium to Lithium*, 4th ed, J. Wiley, New York, 1986.
6. Onishi, H., *Photometric Determination of Traces of Metals. Part IIb: Individual Metals, Magnesium to Zinc*, 4th ed, J. Wiley, New York, 1989.
7. Snell, F. D., *Photometric and Fluorimetric Methods of Analysis, Metals Part 1*, J. Wiley, New York, 1978.
8. Snell, F. D., *Photometric and Fluorimetric Methods of Analysis, Metals Part 2*, J. Wiley, New York, 1978.
9. Snell, F. D., *Photometric and Fluorimetric Methods of Analysis, Nonmetals*, J. Wiley, New York, 1981.
10. West, T. S. and Nürnberg, H. W., Eds., *The Determination of Trace Metals in Natural Waters*, Blackwell, Oxford, 1988.
11. Williams, W. J., *Handbook of Anion Determination*, Butterworth, London, 1979.
12. Townshend, A., Burns, D. T., Guilbault, G. G., Lobinski, R., Marczenko, Z., Newman, E., and Onishi, H., *Dictionary of Analytical Reagents*, Chapman & Hall, London, 1993.

FLAME AND BEAD TESTS

Flame Colorations

Violet

Potassium compounds. Purple red through blue glass. Easily obscured by sodium flame. Bluish-green through green glass. Rubidium and cesium compounds impart same flame as potassium compounds.

Blues

Azure — Copper chloride. Copper bromide gives azure blue followed by green. Other copper compounds give same coloration when moistened with hydrochloric acid.

Light blue — Lead, arsenic, selenium.

Greens

Emerald — Copper compounds except the halides, and when not moistened with hydrochloric acid.

Pure green — Compounds of thallium and tellurium.

Yellowish — Barium compounds. Some molybdenum compounds. Borates, especially when treated with sulfuric acid or when burned with alcohol.

Bluish — Phosphates with sulfuric acid.

Feeble — Antimony compounds. Ammonium compounds.

Whitish — Zinc.

Reds

Carmine — Lithium compounds. Violet through blue glass. Invisible through green glass. Masked by barium flame.

Scarlet — Strontium compounds. Violet through blue glass. Yellowish through green glass. Masked by barium flame.

Yellowish — Calcium compounds. Greenish through blue glass. Green through green glass. Masked by barium flame.

Yellow

Yellow — All sodium compounds. Invisible with blue glass.

Bead Tests

Abbreviations employed: s = saturated; ss = supersaturated; ns = not saturated; h = hot; c = cold

Borax Beads

Substance	Oxidizing flame	Reducing flame
Aluminum	Colorless (h, c, ns); opaque (ss)	Colorless; opaque (s)
Antimony	Colorless; yellow or brownish (h, ss)	Gray and opaque
Barium	Colorless (ns)	
Bismuth	Colorless; yellow or brownish (h, ss)	Gray and opaque
Cadmium	Colorless	Gray and opaque
Calcium	Colorless (ns)	
Cerium	Red (h)	Colorless (h, c)
Chromium	Green (c)	Green
Cobalt	Blue (h, c)	Blue (h, c)
Copper	Green (h); blue (c)	Red (c); opaque (ss); colorless (h)
Iron	Yellow or brownish red (h, ns)	Green (ss)
Lead	Colorless; yellow or brownish (h, ss)	Gray and opaque
Magnesium	Colorless (ns)	
Manganese	Violet (h, c)	Colorless (h, c)
Molybdenum	Colorless	Yellow or brown (h)
Nickel	Brown; red (c)	Gray and opaque
Silicon	Colorless (h, c); opaque (ss)	Colorless; opaque (s)
Silver	Colorless (ns)	Gray and opaque
Strontium	Colorless (ns)	
Tin	Colorless (h, c); opaque (ss)	Colorless; opaque (s)
Titanium	Colorless	Yellow (h); violet (c)
Tungsten	Colorless	Brown
Uranium	Yellow or brownish (h, ns)	Green
Vanadium	Colorless	Green

Beads of Microcosmic Salt
$NaNH_4HPO_4$

Substance	Oxidizing flame	Reducing flame
Aluminum	Colorless; opaque (s)	Colorless; not clear (ss)
Antimony	Colorless (ns)	Gray and opaque
Barium	Colorless; opaque (s)	Colorless; not clear (ss)
Bismuth	Colorless (ns)	Gray and opaque
Cadmium	Colorless (ns)	Gray and opaque
Calcium	Colorless; opaque (s)	Colorless; not clear (ss)

Beads of Microcosmic Salt
$NaNH_4HPO_4$

Substance	Oxidizing flame	Reducing flame
Cerium	Yellow or brownish red (h, s)	Colorless
Chromium	Red (h, s); green (c)	Green (c)
Cobalt	Blue (h, c)	Blue (h, c)
Copper	Blue (c); green (h)	Red and opaque (c)
Iron	Yellow or brown (h, s)	Colorless; yellow or brownish (h)
Lead	Colorless (ns)	Gray and opaque
Magnesium	Colorless; opaque (s)	Colorless; not clear (ss)
Manganese	Violet (h, c)	Colorless
Molybdenum	Colorless; green (h)	Green (h)
Nickel	Yellow (c); red (h, s)	Yellow (c); red (h); gray and opaque
Silver		Gray and opaque
Strontium	Colorless; opaque (s)	Colorless; not clear (ss)
Tin	Colorless; opaque (s)	Colorless
Titanium	Colorless (ns)	Violet (c); yellow or brownish (h)
Uranium	Green; yellow or brownish	Green (h) (h, s)
Vanadium	Yellow	Green
Zinc	Colorless (ns)	Gray and opaque

Sodium Carbonate Bead

Substance	Oxidizing flame	Reducing flame
Manganese	Green	Colorless

ACID-BASE INDICATORS

A. K. Covington

The first part of this table lists some common acid-base indicators in alphabetical order along with the approximate pH range(s) at which a color change occurs. Following this is a table of the same indicators ordered by pH range, which includes the nature of the color change, instructions on preparation of the indicator solution, and the acid dissociation constant pK, when available.

The color code is:

C = colorless	A = amber	B/G = blue-green
Pk = pink	Y = yellow	V = violet
R = red	B = blue	P = purple
O = orange		

Reference

Bishop, E., Ed., *Indicators*, Pergamon, Oxford, 1972.

Indicator	pH Range
Alizarin	5.6-7.2; 11.0-12.4
Alizarin Red S	4.6-6.0
Alizarin Yellow R	10.1-12.0
Benzopurpurine 4B	2.2-4.2
4,4'-Bis(2-amino-1-naphthylazo)-2,2'-stilbenedisulfonic acid	3.0-4.0
4,4'-Bis(4-amino-1-naphthylazo)-2,2'-stilbenedisulfonic acid	8.0-9.0
Brilliant Yellow	6.6-7.8
Bromocresol Green	3.8-5.4
Bromocresol Purple	5.2-6.8
Bromophenol Blue	3.0-4.6
Bromothymol Blue	6.0-7.6
Chlorophenol Red	5.2-6.8
Clayton Yellow	12.2-13.2
Congo Red	3.0-5.0
o-Cresolphthalein	8.2-9.8
Cresol Red	0.0-1.0; 7.0-8.8
Crystal Violet	0.0-1.8
Curcumin (Turmaric)	7.4-8.6
p-(2,4-Dihydroxyphenylazo) benzenesulfonic acid, sodium salt	11.4-12.6
p-Dimethylaminoazobenzene	2.8-4.4
4-(4-Dimethylamino-1-naphylazo)-3-methoxybenzenesulfonic acid	3.5-4.8
2-(p-Dimethylamino-phenylazo)pyridine	0.2-1.8; 4.4-5.6
N,N-Dimethyl-p-(m-tolylazo)aniline	2.6-4.8
2,4-Dinitrophenol	2.0-4.7
2-(2,4 Dinitrophenylazo)-1-naphthol-3,6-disulfonic acid, disodium salt	6.0-7.0
6,8-Dinitro-2,4-(1H)quinazolinedione	6.4-8.0
Erythrosin, disodium salt	2.2-3.6
4-(p-Ethoxyphenylazo)-m-phenylene-diamine monohydrochloride	4.4-5.8

Indicator	pH Range
Ethyl bis(2,4-dimethylphenyl)ethanoate	8.4-9.6
Ethyl Orange	3.4-4.8
Ethyl Red	4.0-5.8
Ethyl Violet	0.0-2.4
5,5'-Indigodisulfonic acid, disodium salt	11.4-13.0
Malachite Green	0.2-1.8
Metacresol Purple	1.2-2.8; 7.4-9.0
Metanil Yellow	1.2-2.4
Methyl Green	0.2-1.8
Methyl Orange	3.2-4.4
Methyl Red	4.8-6.0
Methyl Violet	0.0-1.6
p-Naphtholbenzein	8.2-10.0
Neutral Red	6.8-8.0
p-Nitrophenol	5.4-6.6
m-Nitrophenol	6.8-8.6
Orange IV	1.4-2.8
Paramethyl Red	1.0-3.0
Phenolphthalein	8.2-10.0
Phenol Red	6.6-8.0
4-Phenylazodiphenylamine	1.2-2.6
4-Phenylazo-1-naphthylamine	4.0-5.6
Propyl Red	4.8-6.6
Quinaldine Red	1.4-3.2
Resazurin	3.8-6.4
Resorcin Blue	4.4-6.2
Tetrabromophenolphthalein ethyl ester, potassium salt	3.0-4.2
Thymol Blue	1.2-2.8; 8.0-9.6
Thymolphthalein	9.4-10.6
4-o-Tolylazo-o-toluidine	1.4-2.8
1,3,5-Trinitrobenzene	12.0-14.0
2,4,6-Trinitrotoluene	11.5-13.0
Turmaric	7.4-8.6

pH range	Color change	Indicator	pK	Preparation
0.0-1.0	R-Y	Cresol Red		0.1 g in 26.2 mL 0.01 M NaOH + 223.8 mL water
0.0-1.6	Y-B	Methyl Violet		0.01-0.05% in water
0.0-1.8	Y-B	Crystal Violet		0.02% in water
0.0-2.4	Y-B	Ethyl Violet		0.1 g in 50 mL 50% v/v methanol-water
0.2-1.8	Y-B/G	Malachite Green	1.3	water
0.2-1.8	Y-B	Methyl Green		0.1% in water
0.2-1.8	Y-R	2-(p-Dimethylaminophenylazo)pyridine		0.1% in ethanol
1.0-3.0	R-Y	Paramethyl Red		ethanol
1.2-2.4	R-Y	Metanil Yellow		0.01% in water
1.2-2.6	R-Y	4-Phenylazodiphenylamine		0.01 g in 1 mL 1 M HCl + 50 mL ethanol + 49 mL water
1.2-2.8	R-Y	Thymol Blue	1.65	0.1 g in 21.5 mL 0.01 M NaOH + 228.5 mL water
1.2-2.8	R-Y	Metacresol Purple	1.51	0.1 g in 26.2 mL 0.01 M NaOH + 223.8 mL water
1.4-2.8	R-Y	Orange IV		0.01% in water
1.4-2.8	O-Y	4-o-Tolylazo-o-toluidine		water
1.4-3.2	C-R	Quinaldine Red	2.63	1% in ethanol
2.0-4.7	C-Y	2,4-Dinitrophenol	3.96	sat. solution in water
2.2-3.6	O-R	Erythrosin, disodium salt		0.1% in water
2.2-4.2	V-R	Benzopurpurine 4B		0.1% in water
2.6-4.8	R-Y	N,N-Dimethyl-p-(m-tolylazo)aniline		0.1% in water
2.8-4.4	R-Y	p-Dimethylaminoazobenzene		0.1 g in 100 mL 90% v/v ethanol-water
3.0-4.0	P-R	4,4'-Bis(2-amino-1-naphthylazo)-2,2'-stilbenedisulfonic acid		0.1 g in 5.9 mL 0.05 M NaOH + 94.1 mL water
3.0-4.2	Y-B	Tetrabromophenolphthalein ethyl ester, potassium salt		0.1% in ethanol
3.0-4.6	Y-B	Bromophenol Blue	4.10	0.1 g in 14.9 mL 0.01 M NaOH + 235.1 mL water
3.0-5.0	B-R	Congo Red		0.1% in water
3.2-4.4	R-Y	Methyl Orange	3.46	0.1% in water
3.4-4.8	R-Y	Ethyl Orange	4.34	0.05-0.2% in water or aqueous ethanol
3.5-4.8	V-Y	4-(4-Dimethylamino-1-naphylazo)-3-methoxybenzenesulfonic acid		0.1% in 60% ethanol-water
3.8-5.4	Y-B	Bromocresol Green	4.90	0.1 g in 14.3 mL 0.01 M NaOH + 235.7 mL water
3.8-6.4	O-V	Resazurin		water
4.0-5.6	R-Y	4-Phenylazo-1-naphthylamine		0.1% in ethanol
4.0-5.8	C-R	Ethyl Red	5.42	0.1 g in 100 mL 50% v/v methanol-water 0.1% in ethanol
4.4-5.6	R-Y	2-(p-Dimethylaminophenylazo)pyridine		0.1% in ethanol
4.4-5.8	O-Y	4-(p-Ethoxyphenylazo)-m-phenylene-diamine monohydrochloride		0.1% in water
4.4-6.2	R-B	Resorcin Blue		0.2% in ethanol
4.6-6.0	Y-R	Alizarin Red S		water
4.8-6.0	R-Y	Methyl Red	5.00	0.02 g in 100 mL 60% v/v ethanol-water
4.8-6.6	R-Y	Propyl Red	5.48	ethanol
5.2-6.8	Y-P	Bromocresol Purple	6.40	0.1 g in 18.5 mL 0.01 M NaOH + 231.5 mL water
5.2-6.8	Y-R	Chlorophenol Red	6.25	0.1 g in 23.6 mL 0.01 M NaOH + 226.4 mL water
5.4-6.6	C-Y	p-Nitrophenol	7.15	0.1% in water
5.6-7.2	Y-R	Alizarin		0.1% in methanol
6.0-7.0	Y-B	2-(2,4-Dinitrophenylazo)-1-naphthol-3,6-disulfonic acid, disodium salt		0.1% in water
6.0-7.6	Y-B	Bromothymol Blue	7.30	0.1 g in 16 mL 0.01 M NaOH + 234 mL water
6.4-8.0	C-Y	6,8-Dinitro-2,4-(1H)quinazolinedione		25 g in 115 mL 1 M NaOH + 50 mL water at 100°C
6.6-7.8	Y-R	Brilliant Yellow		1% in water
6.6-8.0	Y-R	Phenol Red	8.00	0.1 g in 28.2 mL 0.01 M NaOH + 221.8 mL water
6.8-8.0	R-A	Neutral Red		0.01 g in 100 mL 50% v/v ethanol-water
6.8-8.6	C-Y	m-Nitrophenol	8.28	0.3% in water
7.0-8.8	Y-R	Cresol Red	8.46	0.1 g in 26.2 mL 0.01 M NaOH + 223.8 mL water
7.4-8.6	Y-R	Turmaric (Curcumin)		ethanol
7.4-9.0	Y-P	Metacresol Purple	8.3	0.1 g in 26.2 mL 0.01 M NaOH + 223.8 mL water
8.0-9.0	B-R	4,4'-Bis(4-amino-1-naphthylazo)-2,2'-stilbenedisulfonic acid		0.1 g in 5.9 mL 0.05 M NaOH + 94.1 mL water
8.0-9.6	Y-B	Thymol Blue	9.20	0.1 g in 21.5 mL 0.01 M NaOH + 228.5 mL water
8.2-10.0	O-B	p-Naphtholbenzein		1% in dil. alkali

pH range	Color change	Indicator	pK	Preparation
8.2-10.0	C-Pk	Phenolphthalein	9.5	0.5 g in 100 mL 50% v/v ethanol-water
8.2-9.8	C-R	o-Cresolphthalein		0.04% in ethanol
8.4-9.6	C-B	Ethyl bis(2,4-dimethylphenyl)ethanoate		sat. solution in 50% acetone-ethanol
9.4-10.6	C-B	Thymolphthalein		0.04 g in 100 mL 50% v/v ethanol-water
10.1-12.0	Y-R	Alizarin Yellow R		0.01% in water
11.0-12.4	R-P	Alizarin		0.1% in methanol
11.4-12.6	Y-O	p-(2,4-Dihydroxyphenylazo) benzenesulfonic acid, sodium salt		0.1% in water
11.4-13.0	B-Y	5,5'-Indigodisulfonic acid, disodium salt		water
11.5-13.0	C-O	2,4,6-Trinitrotoluene		0.1-0.5% in ethanol
12.0-14.0	C-O	1,3,5-Trinitrobenzene		0.1-0.5% in ethanol
12.2-13.2	Y-A	Clayton Yellow		0.1% in water

FLUORESCENT INDICATORS

Jack DeMent

Fluorescent indicators are substances which show definite changes in fluorescence with change in pH. Some fluorescent materials are not suitable for indicators since their change in fluorescence is too gradual. Fluorescent indicators find greatest utility in the titration of opaque, highly turbid or deeply colored solutions. A long wavelength ultraviolet ("black light") lamp in a dimly lighted room provides the best environment for titrations involving fluorescent indictors, although bright daylight is sometimes sufficient to evoke a response in the bright green, yellow and orange fluorescent indicators. Titrations are carried out in non-fluorescent glassware. One should check the glassware prior to use to make certain that it does not fluoresce due to the wavelengths of light involved in the titration. The meniscus of the liquid in the burette can be followed when a few particles of an insoluble fluorescent solid are dropped onto its surface.

In this table the indicators are arranged by approximate pH range covered. In the case of some of the dyestuffs the end point may vary slightly with the source or manufacturer.

Indicator	C.I.	From pH	To pH
pH 0 to 2			
Benzoflavine	—	0.3, yellow fl.	1.7, green fl.
3,6-Dioxyphthalimide	—	0, blue fl.	2.4, green fl.
Eosine YS	768	0, yellow colored	3.0, yellow fl.
Erythrosine	772	0, yellow colored	3.6, yellow fl.
Esculin	—	1.5, colorless	2, blue fl.
4-Ethoxyacridone	—	1.2, green fl.	3.2, blue fl.
3,6-Tetramethyldiaminooxanthone	—	1.2, green fl.	3.4, blue fl.
pH 2 to 4			
Chromotropic acid	—	3.5, colorless	4.5, blue fl
Fluorescein	766	4, colorless	4.5, green fl.
Magdala Red	—	3.0, purple colored	4.0, fl
α-Naphthylamine	—	3.4, colorless	4.8, blue fl.
β-Naphthylamine	—	2.8, colorless	4.4, violet fl.
Phloxine	774	3.4, colorless	5.0, bright yellow fl.
Salicylic acid	—	2.5, colorless	3.5, blue fl.
pH 4 to 6			
Acridine	788	4.9, green fl.	5.1, violet colored
Dichlorofluorescein	—	4.0, colorless	5.0, green fl.
3,6-Dioxyxanthone	—	5.4, colorless	7.6, blue-violet fl.
Erythrosine	772	4.0, colorless	4.5, yellow-green fl.
β-Methylesculetin	—	4.0, colorless	6.2, blue fl
Neville-Winther acid	—	6.0, colorless	6.5, blue fl.
Resorufin	—	4.4, yellow fl	6.4, weak orange fl.
Quininic acid	—	4.4, yellow colored	5.0, blue fl.
Quinine [first end point]	—	5.0, blue fl.	6.1, violet fl.
pH 6 to 8			
Acid R Phosphine	—	(claimed for range pH 6.0–7.0)	
Brilliant Diazol Yellow	—	6.5, colorless	7.5, violet, fl.
Cleves acid	—	6.5, colorless	7.5, green fl.
Coumaric acid	—	7.2, colorless	9.0, green fl.
3,6-Dioxyphthalic dinitrile	—	5.8, blue fl.	8.2, green fl.
Magnesium 8-hydrozyquinolinate	—	6.5, colorless	7.5, golden fl.
β-Methylumbelliferone	—	7.0, colorless	7.5, blue fl.
1-Naphthol-4-sulfonic acid	—	6.0, colorless	6.5, blue fl.
Orcinaurine	—	6.5, colorless	8.0, green fl.
Patent Phosphine	789	(for the range pH 6.0–7.0, green-yellow fl.)	
Thioflavine	816	(for the region pH 6.5–7.0, yellow fl.)	
Umbelliferone	—	6.5, colorless	7.6, blue fl.
pH 8 to 10			
Acridine Orange	788	8.4, orange colored	10.4, green fl.
Ethoxyphenylnaphthostilbazonium chloride	—	9, green fl.	11, non-fl.

Name			
G Salt	—	9.0, dull blue fl.	9.5, bright blue fl.
Naphthazol derivatives	—	8.2, colorless	10.0, yellow or green fl.
α-Naphthionic acid	—	9, blue fl.	11, green fl.
2-Naphthol-3,6-disulfonic acid	—	9.5, dark blue fl	Light blue fl. at higher pH
β-Naphthol	—	8.6, colorless	Blue fl. at higher pH
α-Naphtholsulfonic acid	—	8.0, dark blue fl.	9.0, bright violet fl.
1,4-Naphtholsulfonic acid	—	8.2, dark blue fl.	Light blue fl. at higher pH
Orcinsulfonphthalein	—	8.6, yellow colored	10.0 fl.
Quinine [second end point]	—	9.5, violet fl.	10.0, colorless
R-Salt	—	9.0, dull blue fl.	9.5, bright blue fl.
Sodium 1-naphthol-2-sulfonate	—	9.0, dark blue fl.	10.0, bright violet fl.

pH 10 to 12

Name			
Courmarin	—	9.8, deep green fl.	12, light green fl.
Eosine BN	771	10.5, colorless	14.0, yellow fl.
Papaverine (permanganate oxidized)	—	9.5, yellow fl.	11.0, blue fl.
Schaffers Salt	—	5.0, violet fl.	11.0, green-blue fl.
SS-Acid (sodium salt)	—	10.0, violet fl.	12.0, yellow-colored

pH 12 to 14

Name			
Cotarnine	—	12.0, yellow fl.	13.0, white fl.
α-Naphthionic acid	—	12, blue fl.	13, green fl.
β-Naphthionic acid	—	12, blue fl.	13, violet fl.

CONVERSION FORMULAS FOR CONCENTRATION OF SOLUTIONS

A = Weight percent of solute
B = Molecular weight of solvent
E = Molecular weight of solute
F = Grams of solute per liter of solution

G = Molality
M = Molarity
N = Mole fraction
R = Density of solution in grams per milliliter

Concentration of solute—SOUGHT	Concentration of solute—GIVEN				
	A	N	G	M	F
A	—	$\dfrac{100N \times E}{N \times E + (1-N)B}$	$\dfrac{100G \times E}{1000 + G \times E}$	$\dfrac{M \times E}{10R}$	$\dfrac{F}{10R}$
N	$\dfrac{\dfrac{A}{E}}{\dfrac{A}{E} + \dfrac{100-A}{B}}$	—	$\dfrac{B \times G}{B \times G + 1000}$	$\dfrac{B \times M}{M(B-E) + 1000R}$	$\dfrac{B \times F}{F(B-E) + 1000R \times E}$
G	$\dfrac{1000A}{E(100-A)}$	$\dfrac{1000N}{B - N \times B}$	—	$\dfrac{1000M}{1000R - (M \times E)}$	$\dfrac{1000F}{E(1000R - F)}$
M	$\dfrac{10R \times A}{E}$	$\dfrac{1000R \times N}{N \times E + (1-N)B}$	$\dfrac{1000R \times G}{1000 + E \times G}$	—	$\dfrac{F}{E}$
F	$10AR$	$\dfrac{1000R \times N \times E}{N \times E + (1-N)B}$	$\dfrac{1000R \times G \times E}{1000 + G \times E}$	$M \times E$	—

ELECTROCHEMICAL SERIES

Petr Vanýsek

There are three tables for this electrochemical series. Each table lists standard reduction potentials, $E°$ values, at 298.15 K (25°C), and at a pressure of 101.325 kPa (1 atm). Table 1 is an alphabetical listing of the elements, according to the symbol of the elements. Thus, data for silver (Ag) precedes those for aluminum (Al). Table 2 lists only those reduction reactions which have $E°$ values positive in respect to the standard hydrogen electrode. In Table 2, the reactions are listed in the order of increasing positive potential, and they range from 0.0000 V to + 3.4 V. Table 3 lists only those reduction potentials which have $E°$ negative with respect to the standard hydrogen electrode. In Table 3, the reactions are listed in the order of decreasing potential and range from 0.0000 V to −4.10 V. The reliability of the potentials is not the same for all the data. Typically, the values with fewer significant figures have lower reliability. The values of reduction potentials, in particular those of less common reactions, are not definite; they are subject to occasional revisions.

Abbreviations: ac = acetate; bipy = 2,2´-dipyridine, or bipyridine; en = ethylenediamine; phen = 1,10-phenanthroline.

References

1. Milazzo, G., Caroli, S., and Sharma, V. K. *Tables of Standard Electrode Potentials*, Wiley, Chichester, 1978.
2. Bard, A. J., Parsons, R., and Jordan, J. *Standard Potentials in Aqueous Solutions*, Marcel Dekker, New York, 1985.
3. Bratsch, S. G. *J. Phys. Chem. Ref. Data*, 18, 1–21, 1989.

TABLE 1. Alphabetical Listing

Reaction	$E°/V$	Reaction	$E°/V$
$Ac^{3+} + 3\,e \rightleftharpoons Ac$	−2.20	$Am^{4+} + e \rightleftharpoons Am^{3+}$	2.60
$Ag^+ + e \rightleftharpoons Ag$	0.7996	$Am^{2+} + 2\,e \rightleftharpoons Am$	−1.9
$Ag^{2+} + e \rightleftharpoons Ag^+$	1.980	$Am^{3+} + 3\,e \rightleftharpoons Am$	−2.048
$Ag(ac) + e \rightleftharpoons Ag + (ac)^-$	0.643	$Am^{3+} + e \rightleftharpoons Am^{2+}$	−2.3
$AgBr + e \rightleftharpoons Ag + Br^-$	0.07133	$As + 3\,H^+ + 3\,e \rightleftharpoons AsH_3$	−0.608
$AgBrO_3 + e \rightleftharpoons Ag + BrO_3^-$	0.546	$As_2O_3 + 6\,H^+ + 6\,e \rightleftharpoons 2\,As + 3\,H_2O$	0.234
$Ag_2C_2O_4 + 2\,e \rightleftharpoons 2\,Ag + C_2O_4^{2-}$	0.4647	$HAsO_2 + 3\,H^+ + 3\,e \rightleftharpoons As + 2\,H_2O$	0.248
$AgCl + e \rightleftharpoons Ag + Cl^-$	0.22233	$AsO_2^- + 2\,H_2O + 3\,e \rightleftharpoons As + 4\,OH^-$	−0.68
$AgCN + e \rightleftharpoons Ag + CN^-$	−0.017	$H_3AsO_4 + 2\,H^+ + 2\,e^- \rightleftharpoons HAsO_2 + 2\,H_2O$	0.560
$Ag_2CO_3 + 2\,e \rightleftharpoons 2\,Ag + CO_3^{2-}$	0.47	$AsO_4^{3-} + 2\,H_2O + 2\,e \rightleftharpoons AsO_2^- + 4\,OH^-$	−0.71
$Ag_2CrO_4 + 2\,e \rightleftharpoons 2\,Ag + CrO_4^{2-}$	0.4470	$At_2 + 2\,e \rightleftharpoons 2\,At^-$	0.3
$AgF + e \rightleftharpoons Ag + F^-$	0.779	$Au^+ + e \rightleftharpoons Au$	1.692
$Ag_4[Fe(CN)_6] + 4\,e \rightleftharpoons 4\,Ag + [Fe(CN)_6]^{4-}$	0.1478	$Au^{3+} + 2\,e \rightleftharpoons Au^+$	1.401
$AgI + e \rightleftharpoons Ag + I^-$	−0.15224	$Au^{3+} + 3\,e \rightleftharpoons Au$	1.498
$AgIO_3 + e \rightleftharpoons Ag + IO_3^-$	0.354	$Au^{2+} + e^- \rightleftharpoons Au^+$	1.8
$Ag_2MoO_4 + 2\,e \rightleftharpoons 2\,Ag + MoO_4^{2-}$	0.4573	$AuOH^{2+} + H^+ + 2\,e \rightleftharpoons Au^+ + H_2O$	1.32
$AgNO_2 + e \rightleftharpoons Ag + 2\,NO_2^-$	0.564	$AuBr_2^- + e \rightleftharpoons Au + 2\,Br^-$	0.959
$Ag_2O + H_2O + 2\,e \rightleftharpoons 2\,Ag + 2\,OH^-$	0.342	$AuBr_4^- + 3\,e \rightleftharpoons Au + 4\,Br^-$	0.854
$Ag_2O_3 + H_2O + 2\,e \rightleftharpoons 2\,AgO + 2\,OH^-$	0.739	$AuCl_4^- + 3\,e \rightleftharpoons Au + 4\,Cl^-$	1.002
$Ag^{3+} + 2\,e \rightleftharpoons Ag^+$	1.9	$Au(OH)_3 + 3\,H^+ + 3\,e \rightleftharpoons Au + 3\,H_2O$	1.45
$Ag^{3+} + e \rightleftharpoons Ag^{2+}$	1.8	$H_2BO_3^- + 5\,H_2O + 8\,e \rightleftharpoons BH_4^- + 8\,OH^-$	−1.24
$Ag_2O_2 + 4\,H^+ + e \rightleftharpoons 2\,Ag + 2\,H_2O$	1.802	$H_2BO_3^- + H_2O + 3\,e \rightleftharpoons B + 4\,OH^-$	−1.79
$2\,AgO + H_2O + 2\,e \rightleftharpoons Ag_2O + 2\,OH^-$	0.607	$H_3BO_3 + 3\,H^+ + 3\,e \rightleftharpoons B + 3\,H_2O$	−0.8698
$AgOCN + e \rightleftharpoons Ag + OCN^-$	0.41	$B(OH)_3 + 7\,H^+ + 8\,e \rightleftharpoons BH_4^- + 3\,H_2O$	−0.481
$Ag_2S + 2\,e \rightleftharpoons 2\,Ag + S^{2-}$	−0.691	$Ba^{2+} + 2\,e \rightleftharpoons Ba$	−2.912
$Ag_2S + 2\,H^+ + 2\,e \rightleftharpoons 2\,Ag + H_2S$	−0.0366	$Ba^{2+} + 2\,e \rightleftharpoons Ba(Hg)$	−1.570
$AgSCN + e \rightleftharpoons Ag + SCN^-$	0.08951	$Ba(OH)_2 + 2\,e \rightleftharpoons Ba + 2\,OH^-$	−2.99
$Ag_2SeO_3 + 2\,e \rightleftharpoons 2\,Ag + SeO_3^{2-}$	0.3629	$Be^{2+} + 2\,e \rightleftharpoons Be$	−1.847
$Ag_2SO_4 + 2\,e \rightleftharpoons 2\,Ag + SO_4^{2-}$	0.654	$Be_2O_3^{2-} + 3\,H_2O + 4\,e \rightleftharpoons 2\,Be + 6\,OH^-$	−2.63
$Ag_2WO_4 + 2\,e \rightleftharpoons 2\,Ag + WO_4^{2-}$	0.4660	$p-$benzoquinone $+ 2\,H^+ + 2\,e \rightleftharpoons$ hydroquinone	0.6992
$Al^{3+} + 3\,e \rightleftharpoons Al$	−1.662	$Bi^+ + e \rightleftharpoons Bi$	0.5
$Al(OH)_3 + 3\,e \rightleftharpoons Al + 3\,OH^-$	−2.31	$Bi^{3+} + 3\,e \rightleftharpoons Bi$	0.308
$Al(OH)_4^- + 3\,e \rightleftharpoons Al + 4\,OH^-$	−2.328	$Bi^{3+} + 2\,e \rightleftharpoons Bi^+$	0.2
$H_2AlO_3^- + H_2O + 3\,e \rightleftharpoons Al + 4\,OH^-$	−2.33	$Bi + 3\,H^+ + 3\,e \rightleftharpoons BiH_3$	−0.8
$AlF_6^{3-} + 3\,e \rightleftharpoons Al + 6\,F^-$	−2.069		

Reaction	$E°/V$
$BiCl_4^- + 3\,e \rightleftharpoons Bi + 4\,Cl^-$	0.16
$Bi_2O_3 + 3\,H_2O + 6\,e \rightleftharpoons 2\,Bi + 6\,OH^-$	−0.46
$Bi_2O_4 + 4\,H^+ + 2\,e \rightleftharpoons 2\,BiO^+ + 2\,H_2O$	1.593
$BiO^+ + 2\,H^+ + 3\,e \rightleftharpoons Bi + H_2O$	0.320
$BiOCl + 2\,H^+ + 3\,e \rightleftharpoons Bi + Cl^- + H_2O$	0.1583
$Bk^{4+} + e \rightleftharpoons Bk^{3+}$	1.67
$Bk^{2+} + 2\,e \rightleftharpoons Bk$	−1.6
$Bk^{3+} + e \rightleftharpoons Bk^{2+}$	−2.8
$Br_2(aq) + 2\,e \rightleftharpoons 2\,Br^-$	1.0873
$Br_2(l) + 2\,e \rightleftharpoons 2\,Br^-$	1.066
$HBrO + H^+ + 2\,e \rightleftharpoons Br^- + H_2O$	1.331
$HBrO + H^+ + e \rightleftharpoons 1/2\,Br_2(aq) + H_2O$	1.574
$HBrO + H^+ + e \rightleftharpoons 1/2\,Br_2(l) + H_2O$	1.596
$BrO^- + H_2O + 2\,e \rightleftharpoons Br^- + 2\,OH^-$	0.761
$BrO_3^- + 6\,H^+ + 5\,e \rightleftharpoons 1/2\,Br_2 + 3\,H_2O$	1.482
$BrO_3^- + 6\,H^+ + 6\,e \rightleftharpoons Br^- + 3\,H_2O$	1.423
$BrO_3^- + 3\,H_2O + 6\,e \rightleftharpoons Br^- + 6\,OH^-$	0.61
$(CN)_2 + 2\,H^+ + 2\,e \rightleftharpoons 2\,HCN$	0.373
$2\,HCNO + 2\,H^+ + 2\,e \rightleftharpoons (CN)_2 + 2\,H_2O$	0.330
$(CNS)_2 + 2\,e \rightleftharpoons 2\,CNS^-$	0.77
$CO_2 + 2\,H^+ + 2\,e \rightleftharpoons HCOOH$	−0.199
$Ca^+ + e \rightleftharpoons Ca$	−3.80
$Ca^{2+} + 2\,e \rightleftharpoons Ca$	−2.868
$Ca(OH)_2 + 2\,e \rightleftharpoons Ca + 2\,OH^-$	−3.02
Calomel electrode, 1 molal KCl	0.2800
Calomel electrode, 1 molar KCl (NCE)	0.2801
Calomel electrode, 0.1 molar KCl	0.3337
Calomel electrode, saturated KCl (SCE)	0.2412
Calomel electrode, saturated NaCl (SSCE)	0.2360
$Cd^{2+} + 2\,e \rightleftharpoons Cd$	−0.4030
$Cd^{2+} + 2\,e \rightleftharpoons Cd(Hg)$	−0.3521
$Cd(OH)_2 + 2\,e \rightleftharpoons Cd(Hg) + 2\,OH^-$	−0.809
$CdSO_4 + 2\,e \rightleftharpoons Cd + SO_4^{2-}$	−0.246
$Cd(OH)_4^{2-} + 2\,e \rightleftharpoons Cd + 4\,OH^-$	−0.658
$CdO + H_2O + 2\,e \rightleftharpoons Cd + 2\,OH^-$	−0.783
$Ce^{3+} + 3\,e \rightleftharpoons Ce$	−2.336
$Ce^{3+} + 3\,e \rightleftharpoons Ce(Hg)$	−1.4373
$Ce^{4+} + e \rightleftharpoons Ce^{3+}$	1.72
$CeOH^{3+} + H^+ + e \rightleftharpoons Ce^{3+} + H_2O$	1.715
$Cf^{4+} + e \rightleftharpoons Cf^{3+}$	3.3
$Cf^{3+} + e \rightleftharpoons Cf^{2+}$	−1.6
$Cf^{3+} + 3\,e \rightleftharpoons Cf$	−1.94
$Cf^{2+} + 2\,e \rightleftharpoons Cf$	−2.12
$Cl_2(g) + 2\,e \rightleftharpoons 2\,Cl^-$	1.35827
$HClO + H^+ + e \rightleftharpoons 1/2\,Cl_2 + H_2O$	1.611
$HClO + H^+ + 2\,e \rightleftharpoons Cl^- + H_2O$	1.482
$ClO^- + H_2O + 2\,e \rightleftharpoons Cl^- + 2\,OH^-$	0.81
$ClO_2 + H^+ + e \rightleftharpoons HClO_2$	1.277
$HClO_2 + 2\,H^+ + 2\,e \rightleftharpoons HClO + H_2O$	1.645
$HClO_2 + 3\,H^+ + 3\,e \rightleftharpoons 1/2\,Cl_2 + 2\,H_2O$	1.628
$HClO_2 + 3\,H^+ + 4\,e \rightleftharpoons Cl^- + 2\,H_2O$	1.570
$ClO_2^- + H_2O + 2\,e \rightleftharpoons ClO^- + 2\,OH^-$	0.66
$ClO_2^- + 2\,H_2O + 4\,e \rightleftharpoons Cl^- + 4\,OH^-$	0.76
$ClO_2(aq) + e \rightleftharpoons ClO_2^-$	0.954

Reaction	$E°/V$
$ClO_3^- + 2\,H^+ + e \rightleftharpoons ClO_2 + H_2O$	1.152
$ClO_3^- + 3\,H^+ + 2\,e \rightleftharpoons HClO_2 + H_2O$	1.214
$ClO_3^- + 6\,H^+ + 5\,e \rightleftharpoons 1/2\,Cl_2 + 3\,H_2O$	1.47
$ClO_3^- + 6\,H^+ + 6\,e \rightleftharpoons Cl^- + 3\,H_2O$	1.451
$ClO_3^- + H_2O + 2\,e \rightleftharpoons ClO_2^- + 2\,OH^-$	0.33
$ClO_3^- + 3\,H_2O + 6\,e \rightleftharpoons Cl^- + 6\,OH^-$	0.62
$ClO_4^- + 2\,H^+ + 2\,e \rightleftharpoons ClO_3^- H_2O$	1.189
$ClO_4^- + 8\,H^+ + 7\,e \rightleftharpoons 1/2\,Cl_2 + 4\,H_2O$	1.39
$ClO_4^- + 8\,H^+ + 8\,e \rightleftharpoons Cl^- + 4\,H_2O$	1.389
$ClO_4^- + H_2O + 2\,e \rightleftharpoons ClO_3^- + 2\,OH^-$	0.36
$Cm^{4+} + e \rightleftharpoons Cm^{3+}$	3.0
$Cm^{3+} + 3\,e \rightleftharpoons Cm$	−2.04
$Co^{2+} + 2\,e \rightleftharpoons Co$	−0.28
$Co^{3+} + e \rightleftharpoons Co^{2+}$	1.92
$[Co(NH_3)_6]^{3+} + e \rightleftharpoons [Co(NH_3)_6]^{2+}$	0.108
$Co(OH)_2 + 2\,e \rightleftharpoons Co + 2\,OH^-$	−0.73
$Co(OH)_3 + e \rightleftharpoons Co(OH)_2 + OH^-$	0.17
$Cr^{2+} + 2\,e \rightleftharpoons Cr$	−0.913
$Cr^{3+} + e \rightleftharpoons Cr^{2+}$	−0.407
$Cr^{3+} + 3\,e \rightleftharpoons Cr$	−0.744
$Cr_2O_7^{2-} + 14\,H^+ + 6\,e \rightleftharpoons 2\,Cr^{3+} + 7\,H_2O$	1.232
$CrO_2^- + 2\,H_2O + 3\,e \rightleftharpoons Cr + 4\,OH^-$	−1.2
$HCrO_4^- + 7\,H^+ + 3\,e \rightleftharpoons Cr^{3+} + 4\,H_2O$	1.350
$CrO_2 + 4\,H^+ + e \rightleftharpoons Cr^{3+} + 2\,H_2O$	1.48
$Cr(V) + e \rightleftharpoons Cr(IV)$	1.34
$CrO_4^{2-} + 4\,H_2O + 3\,e \rightleftharpoons Cr(OH)_3 + 5\,OH^-$	−0.13
$Cr(OH)_3 + 3\,e \rightleftharpoons Cr + 3\,OH^-$	−1.48
$Cs^+ + e \rightleftharpoons Cs$	−3.026
$Cu^+ + e \rightleftharpoons Cu$	0.521
$Cu^{2+} + e \rightleftharpoons Cu^+$	0.153
$Cu^{2+} + 2\,e \rightleftharpoons Cu$	0.3419
$Cu^{2+} + 2\,e \rightleftharpoons Cu(Hg)$	0.345
$Cu^{3+} + e \rightleftharpoons Cu^{2+}$	2.4
$Cu_2O_3 + 6\,H^+ + 2\,e \rightleftharpoons 2\,Cu^{2+} + 3\,H_2O$	2.0
$Cu^{2+} + 2\,CN^- + e \rightleftharpoons [Cu(CN)_2]^-$	1.103
$CuI_2^- + e \rightleftharpoons Cu + 2\,I^-$	0.00
$Cu_2O + H_2O + 2\,e \rightleftharpoons 2\,Cu + 2\,OH^-$	−0.360
$Cu(OH)_2 + 2\,e \rightleftharpoons Cu + 2\,OH^-$	−0.222
$2\,Cu(OH)_2 + 2\,e \rightleftharpoons Cu_2O + 2\,OH^- + H_2O$	−0.080
$2\,D^+ + 2\,e \rightleftharpoons D_2$	−0.013
$Dy^{2+} + 2\,e \rightleftharpoons Dy$	−2.2
$Dy^{3+} + 3\,e \rightleftharpoons Dy$	−2.295
$Dy^{3+} + e \rightleftharpoons Dy^{2+}$	−2.6
$Er^{2+} + 2\,e \rightleftharpoons Er$	−2.0
$Er^{3+} + 3\,e \rightleftharpoons Er$	−2.331
$Er^{3+} + e \rightleftharpoons Er^{2+}$	−3.0
$Es^{3+} + e \rightleftharpoons Es^{2+}$	−1.3
$Es^{3+} + 3\,e \rightleftharpoons Es$	−1.91
$Es^{2+} + 2\,e \rightleftharpoons Es$	−2.23
$Eu^{2+} + 2\,e \rightleftharpoons Eu$	−2.812
$Eu^{3+} + 3\,e \rightleftharpoons Eu$	−1.991
$Eu^{3+} + e \rightleftharpoons Eu^{2+}$	−0.36
$F_2 + 2\,H^+ + 2\,e \rightleftharpoons 2\,HF$	3.053

Reaction	$E°/V$
$F_2 + 2\,e \rightleftharpoons 2\,F^-$	2.866
$F_2O + 2\,H^+ + 4\,e \rightleftharpoons H_2O + 2\,F^-$	2.153
$Fe^{2+} + 2\,e \rightleftharpoons Fe$	−0.447
$Fe^{3+} + 3\,e \rightleftharpoons Fe$	−0.037
$Fe^{3+} + e \rightleftharpoons Fe^{2+}$	0.771
$2\,HFeO_4^- + 8\,H^+ + 6\,e \rightleftharpoons Fe_2O_3 + 5\,H_2O$	2.09
$HFeO_4^- + 4\,H^+ + 3\,e \rightleftharpoons FeOOH + 2\,H_2O$	2.08
$HFeO_4^- + 7\,H^+ + 3\,e \rightleftharpoons Fe^{3+} + 4\,H_2O$	2.07
$Fe_2O_3 + 4\,H^+ + 2\,e \rightleftharpoons 2\,FeOH^+ + H_2O$	0.16
$[Fe(CN)_6]^{3-} + e \rightleftharpoons [Fe(CN)_6]^{4-}$	0.358
$FeO_4^{2-} + 8\,H^+ + 3\,e \quad Fe^{3+} + 4\,H_2O$	2.20
$[Fe(bipy)_2]^{3+} + e \rightleftharpoons Fe(bipy)_2]^{2+}$	0.78
$[Fe(bipy)_3]^{3+} + e \rightleftharpoons Fe(bipy)_3]^{2+}$	1.03
$Fe(OH)_3 + e \rightleftharpoons Fe(OH)_2 + OH^-$	−0.56
$[Fe(phen)_3]^{3+} + e \rightleftharpoons [Fe(phen)_3]^{2+}$	1.147
$[Fe(phen)_3]^{3+} + e \rightleftharpoons [Fe(phen)_3]^{2+}$ (1 molar H_2SO_4)	1.06
$[Ferricinium]^+ + e \rightleftharpoons$ ferrocene	0.400
$Fm^{3+} + e \rightleftharpoons Fm^{2+}$	−1.1
$Fm^{3+} + 3\,e \rightleftharpoons Fm$	−1.89
$Fm^{2+} + 2\,e \rightleftharpoons Fm$	−2.30
$Fr^+ + e \rightleftharpoons Fr$	−2.9
$Ga^{3+} + 3\,e \rightleftharpoons Ga$	−0.549
$Ga^+ + e \rightleftharpoons Ga$	−0.2
$GaOH^{2+} + H^+ + 3\,e \rightleftharpoons Ga + H_2O$	−0.498
$H_2GaO_3^- + H_2O + 3\,e \rightleftharpoons Ga + 4\,OH^-$	−1.219
$Gd^{3+} + 3\,e \rightleftharpoons Gd$	−2.279
$Ge^{2+} + 2\,e \rightleftharpoons Ge$	0.24
$Ge^{4+} + 4\,e \rightleftharpoons Ge$	0.124
$Ge^{4+} + 2\,e \rightleftharpoons Ge^{2+}$	0.00
$GeO_2 + 2\,H^+ + 2\,e \rightleftharpoons GeO + H_2O$	−0.118
$H_2GeO_3 + 4\,H^+ + 4\,e \rightleftharpoons Ge + 3\,H_2O$	−0.182
$2\,H^+ + 2\,e \_ H_2$	0.00000
$H_2 + 2\,e \rightleftharpoons 2\,H^-$	−2.23
$HO_2 + H^+ + e \rightleftharpoons H_2O_2$	1.495
$2\,H_2O + 2\,e \rightleftharpoons H_2 + 2\,OH^-$	−0.8277
$H_2O_2 + 2\,H^+ + 2\,e \rightleftharpoons 2\,H_2O$	1.776
$Hf^{4+} + 4\,e \rightleftharpoons Hf$	−1.55
$HfO^{2+} + 2\,H^+ + 4\,e \rightleftharpoons Hf + H_2O$	−1.724
$HfO_2 + 4\,H^+ + 4\,e \rightleftharpoons Hf + 2\,H_2O$	−1.505
$HfO(OH)_2 + H_2O + 4\,e \rightleftharpoons Hf + 4\,OH^-$	−2.50
$Hg^{2+} + 2\,e \rightleftharpoons Hg$	0.851
$2\,Hg^{2+} + 2\,e \rightleftharpoons Hg_2^{2+}$	0.920
$Hg_2^{2+} + 2\,e \rightleftharpoons 2\,Hg$	0.7973
$Hg_2(ac)_2 + 2\,e \rightleftharpoons 2\,Hg + 2(ac)^-$	0.51163
$Hg_2Br_2 + 2\,e \rightleftharpoons 2\,Hg + 2\,Br^-$	0.13923
$Hg_2Cl_2 + 2\,e \rightleftharpoons 2\,Hg + 2\,Cl^-$	0.26808
$Hg_2HPO_4 + 2\,e \rightleftharpoons 2\,Hg + HPO_4^{2-}$	0.6359
$Hg_2I_2 + 2\,e \rightleftharpoons 2\,Hg + 2\,I^-$	−0.0405
$Hg_2O + H_2O + 2\,e \rightleftharpoons 2\,Hg + 2\,OH^-$	0.123
$HgO + H_2O + 2\,e \rightleftharpoons Hg + 2\,OH^-$	0.0977
$Hg(OH)_2 + 2\,H^+ + 2\,e \rightleftharpoons Hg + 2\,H_2O$	1.034
$Hg_2SO_4 + 2\,e \rightleftharpoons 2\,Hg + SO_4^{2-}$	0.6125

Reaction	$E°/V$
$Ho^{2+} + 2\,e \rightleftharpoons Ho$	−2.1
$Ho^{3+} + 3\,e \rightleftharpoons Ho$	−2.33
$Ho^{3+} + e \rightleftharpoons Ho^{2+}$	−2.8
$I_2 + 2\,e \rightleftharpoons 2\,I^-$	0.5355
$I_3^- + 2\,e \rightleftharpoons 3\,I^-$	0.536
$H_3IO_6^{2-} + 2\,e \rightleftharpoons IO_3^- + 3\,OH^-$	0.7
$H_5IO_6 + H^+ + 2\,e \rightleftharpoons IO_3^- + 3\,H_2O$	1.601
$2\,HIO + 2\,H^+ + 2\,e \rightleftharpoons I_2 + 2\,H_2O$	1.439
$HIO + H^+ + 2\,e \rightleftharpoons I^- + H_2O$	0.987
$IO^- + H_2O + 2\,e \rightleftharpoons I^- + 2\,OH^-$	0.485
$2\,IO_3^- + 12\,H^+ + 10\,e \rightleftharpoons I_2 + 6\,H_2O$	1.195
$IO_3^- + 6\,H^+ + 6\,e \rightleftharpoons I^- + 3\,H_2O$	1.085
$IO_3^- + 2\,H_2O + 4\,e \rightleftharpoons IO^- + 4\,OH^-$	0.15
$IO_3^- + 3\,H_2O + 6\,e \rightleftharpoons IO^- + 6\,OH^-$	0.26
$In^+ + e \rightleftharpoons In$	−0.14
$In^{2+} + e \rightleftharpoons In^+$	−0.40
$In^{3+} + e \rightleftharpoons In^{2+}$	−0.49
$In^{3+} + 2\,e \rightleftharpoons In^+$	−0.443
$In^{3+} + 3\,e \rightleftharpoons In$	−0.3382
$In(OH)_3 + 3\,e \rightleftharpoons In + 3\,OH^-$	−0.99
$In(OH)_4^- + 3\,e \rightleftharpoons In + 4\,OH^-$	−1.007
$In_2O_3 + 3\,H_2O + 6\,e \rightleftharpoons 2\,In + 6\,OH^-$	−1.034
$Ir^{3+} + 3\,e \rightleftharpoons Ir$	1.156
$[IrCl_6]^{2-} + e \rightleftharpoons [IrCl_6]^{3-}$	0.8665
$[IrCl_6]^{3-} + 3\,e \rightleftharpoons Ir + 6\,Cl^-$	0.77
$Ir_2O_3 + 3\,H_2O + 6\,e \rightleftharpoons 2\,Ir + 6\,OH^-$	0.098
$K^+ + e \rightleftharpoons K$	−2.931
$La^{3+} + 3\,e \rightleftharpoons La$	−2.379
$La(OH)_3 + 3\,e \rightleftharpoons La + 3\,OH^-$	−2.90
$Li^+ + e \rightleftharpoons Li$	−3.0401
$Lr^{3+} + 3\,e \rightleftharpoons Lr$	−1.96
$Lu^{3+} + 3\,e \rightleftharpoons Lu$	−2.28
$Md^{3+} + e \rightleftharpoons Md^{2+}$	−0.1
$Md^{3+} + 3\,e \rightleftharpoons Md$	−1.65
$Md^{2+} + 2\,e \rightleftharpoons Md$	−2.40
$Mg^+ + e \rightleftharpoons Mg$	−2.70
$Mg^{2+} + 2\,e \rightleftharpoons Mg$	−2.372
$Mg(OH)_2 + 2\,e \rightleftharpoons Mg + 2\,OH^-$	−2.690
$Mn^{2+} + 2\,e \rightleftharpoons Mn$	−1.185
$Mn^{3+} + e \rightleftharpoons Mn^{2+}$	1.5415
$MnO_2 + 4\,H^+ + 2\,e \rightleftharpoons Mn^{2+} + 2\,H_2O$	1.224
$MnO_4^- + e \rightleftharpoons MnO_4^{2-}$	0.558
$MnO_4^- + 4\,H^+ + 3\,e \rightleftharpoons MnO_2 + 2\,H_2O$	1.679
$MnO_4^- + 8\,H^+ + 5\,e \rightleftharpoons Mn^{2+} + 4\,H_2O$	1.507
$MnO_4^- + 2\,H_2O + 3\,e \rightleftharpoons MnO_2 + 4\,OH^-$	0.595
$MnO_4^{2-} + 2\,H_2O + 2\,e \rightleftharpoons MnO_2 + 4\,OH^-$	0.60
$Mn(OH)_2 + 2\,e \rightleftharpoons Mn + 2\,OH^-$	−1.56
$Mn(OH)_3 + e \rightleftharpoons Mn(OH)_2 + OH^-$	0.15
$Mn_2O_3 + 6\,H^+ + e \rightleftharpoons 2\,Mn^{2+} + 3\,H_2O$	1.485
$Mo^{3+} + 3\,e \rightleftharpoons Mo$	−0.200
$MoO_2 + 4\,H^+ + 4\,e \rightleftharpoons Mo + 4\,H_2O$	−0.152
$H_3Mo_7O_{24}^{3-} + 45\,H^+ + 42\,e \rightleftharpoons 7\,Mo + 24\,H_2O$	0.082

Reaction	$E°/V$
$MoO_3 + 6 H^+ + 6 e \rightleftharpoons Mo + 3 H_2O$	0.075
$N_2 + 2 H_2O + 6 H^+ + 6 e \rightleftharpoons 2 NH_4OH$	0.092
$3 N_2 + 2 H^+ + 2 e \rightleftharpoons 2 HN_3$	−3.09
$N_5^+ + 3 H^+ + 2 e \rightleftharpoons 2 NH_4^+$	1.275
$N_2O + 2 H^+ + 2 e \rightleftharpoons N_2 + H_2O$	1.766
$H_2N_2O_2 + 2 H^+ + 2 e \rightleftharpoons N_2 + 2 H_2O$	2.65
$N_2O_4 + 2 e \rightleftharpoons 2 NO_2^-$	0.867
$N_2O_4 + 2 H^+ + 2 e \rightleftharpoons 2 HNO_2$	1.065
$N_2O_4 + 4 H^+ + 4 e \rightleftharpoons 2 NO + 2 H_2O$	1.035
$2 NH_3OH^+ + H^+ + 2 e \rightleftharpoons N_2H_5^+ + 2 H_2O$	1.42
$2 NO + 2 H^+ + 2 e \rightleftharpoons N_2O + H_2O$	1.591
$2 NO + H_2O + 2 e \rightleftharpoons N_2O + 2 OH^-$	0.76
$HNO_2 + H^+ + e \rightleftharpoons NO + H_2O$	0.983
$2 HNO_2 + 4 H^+ + 4 e \rightleftharpoons H_2N_2O_2 + 2 H_2O$	0.86
$2 HNO_2 + 4 H^+ + 4 e \rightleftharpoons N_2O + 3 H_2O$	1.297
$NO_2^- + H_2O + e \rightleftharpoons NO + 2 OH^-$	−0.46
$2 NO_2^- + 2 H_2O + 4 e \rightleftharpoons N_2O_2^{2-} + 4 OH^-$	−0.18
$2 NO_2^- + 3 H_2O + 4 e \rightleftharpoons N_2O + 6 OH^-$	0.15
$NO_3^- + 3 H^+ + 2 e \rightleftharpoons HNO_2 + H_2O$	0.934
$NO_3^- + 4 H^+ + 3 e \rightleftharpoons NO + 2 H_2O$	0.957
$2 NO_3^- + 4 H^+ + 2 e \rightleftharpoons N_2O_4 + 2 H_2O$	0.803
$NO_3^- + H_2O + 2 e \rightleftharpoons NO_2^- + 2 OH^-$	0.01
$2 NO_3^- + 2 H_2O + 2 e \rightleftharpoons N_2O_4 + 4 OH^-$	−0.85
$Na^+ + e \rightleftharpoons Na$	−2.71
$Nb^{3+} + 3 e \rightleftharpoons Nb$	−1.099
$NbO_2 + 2 H^+ + 2 e \rightleftharpoons NbO + H_2O$	−0.646
$NbO_2 + 4 H^+ + 4 e \rightleftharpoons Nb + 2 H_2O$	−0.690
$NbO + 2 H^+ + 2 e \rightleftharpoons Nb + H_2O$	−0.733
$Nb_2O_5 + 10 H^+ + 10 e \rightleftharpoons 2 Nb + 5 H_2O$	−0.644
$Nd^{3+} + 3 e \rightleftharpoons Nd$	−2.323
$Nd^{2+} + 2 e \rightleftharpoons Nd$	−2.1
$Nd^{3+} + e \rightleftharpoons Nd^{2+}$	−2.7
$Ni^{2+} + 2 e \rightleftharpoons Ni$	−0.257
$Ni(OH)_2 + 2 e \rightleftharpoons Ni + 2 OH^-$	−0.72
$NiO_2 + 4 H^+ + 2 e \rightleftharpoons Ni^{2+} + 2 H_2O$	1.678
$NiO_2 + 2 H_2O + 2 e \rightleftharpoons Ni(OH)_2 + 2 OH^-$	−0.490
$No^{3+} + e \rightleftharpoons No^{2+}$	1.4
$No^{3+} + 3 e \rightleftharpoons No$	−1.20
$No^{2+} + 2 e \rightleftharpoons No$	−2.50
$Np^{3+} + 3 e \rightleftharpoons Np$	−1.856
$Np^{4+} + e \rightleftharpoons Np^{3+}$	0.147
$NpO_2 + H_2O + H^+ + e \rightleftharpoons Np(OH)_3$	−0.962
$O_2 + 2 H^+ + 2 e \rightleftharpoons H_2O_2$	0.695
$O_2 + 4 H^+ + 4 e \rightleftharpoons 2 H_2O$	1.229
$O_2 + H_2O + 2 e \rightleftharpoons HO_2^- + OH^-$	−0.076
$O_2 + 2 H_2O + 2 e \rightleftharpoons H_2O_2 + 2 OH^-$	−0.146
$O_2 + 2 H_2O + 4 e \rightleftharpoons 4 OH^-$	0.401
$O_3 + 2 H^+ + 2 e \rightleftharpoons O_2 + H_2O$	2.076
$O_3 + H_2O + 2 e \rightleftharpoons O_2 + 2 OH^-$	1.24
$O(g) + 2 H^+ + 2 e \rightleftharpoons H_2O$	2.421
$OH + e \rightleftharpoons OH^-$	2.02
$HO_2^- + H_2O + 2 e \rightleftharpoons 3 OH^-$	0.878
$OsO_4 + 8 H^+ + 8 e \rightleftharpoons Os + 4 H_2O$	0.838

Reaction	$E°/V$
$OsO_4 + 4 H^+ + 4 e \rightleftharpoons OsO_2 + 2 H_2O$	1.02
$[Os(bipy)_2]^{3+} + e \rightleftharpoons [Os(bipy)_2]^{2+}$	0.81
$[Os(bipy)_3]^{3+} + e \rightleftharpoons [Os(bipy)_3]^{2+}$	0.80
$P(red) + 3 H^+ + 3 e \rightleftharpoons PH_3(g)$	−0.111
$P(white) + 3 H^+ + 3 e \rightleftharpoons PH_3(g)$	−0.063
$P + 3 H_2O + 3 e \rightleftharpoons PH_3(g) + 3 OH^-$	−0.87
$H_2P_2^- + e \rightleftharpoons P + 2 OH^-$	−1.82
$H_3PO_2 + H^+ + e \rightleftharpoons P + 2 H_2O$	−0.508
$H_3PO_3 + 2 H^+ + 2 e \rightleftharpoons H_3PO_2 + H_2O$	−0.499
$H_3PO_3 + 3 H^+ + 3 e \rightleftharpoons P + 3 H_2O$	−0.454
$HPO_3^{2-} + 2 H_2O + 2 e \rightleftharpoons H_2PO_2^- + 3 OH^-$	−1.65
$HPO_3^{2-} + 2 H_2O + 3 e \rightleftharpoons P + 5 OH^-$	−1.71
$H_3PO_4 + 2 H^+ + 2 e \rightleftharpoons H_3PO_3 + H_2O$	−0.276
$PO_4^{3-} + 2 H_2O + 2 e \rightleftharpoons HPO_3^{2-} + 3 OH^-$	−1.05
$Pa^{3+} + 3 e \rightleftharpoons Pa$	−1.34
$Pa^{4+} + 4 e \rightleftharpoons Pa$	−1.49
$Pa^{4+} + e \rightleftharpoons Pa^{3+}$	−1.9
$Pb^{2+} + 2 e \rightleftharpoons Pb$	−0.1262
$Pb^{2+} + 2 e \rightleftharpoons Pb(Hg)$	−0.1205
$PbBr_2 + 2 e \rightleftharpoons Pb + 2 Br^-$	−0.284
$PbCl_2 + 2 e \rightleftharpoons Pb + 2 Cl^-$	−0.2675
$PbF_2 + 2 e \rightleftharpoons Pb + 2 F^-$	−0.3444
$PbHPO_4 + 2 e \rightleftharpoons Pb + HPO_4^{2-}$	−0.465
$PbI_2 + 2 e \rightleftharpoons Pb + 2 I^-$	−0.365
$PbO + H_2O + 2 e \rightleftharpoons Pb + 2 OH^-$	−0.580
$PbO_2 + 4 H^+ + 2 e \rightleftharpoons Pb^{2+} + 2 H_2O$	1.455
$HPbO_2^- + H_2O + 2 e \rightleftharpoons Pb + 3 OH^-$	−0.537
$PbO_2 + H_2O + 2 e \rightleftharpoons PbO + 2 OH^-$	0.247
$PbO_2 + SO_4^{2-} + 4 H^+ + 2 e \rightleftharpoons PbSO_4 + 2 H_2O$	1.6913
$PbSO_4 + 2 e \rightleftharpoons Pb + SO_4^{2-}$	−0.3588
$PbSO_4 + 2 e \rightleftharpoons Pb(Hg) + SO_4^{2-}$	−0.3505
$Pd^{2+} + 2 e \rightleftharpoons Pd$	0.951
$[PdCl_4]^{2-} + 2 e \rightleftharpoons Pd + 4 Cl^-$	0.591
$[PdCl_6]^{2-} + 2 e \rightleftharpoons [PdCl_4]^{2-} + 2 Cl^-$	1.288
$Pd(OH)_2 + 2 e \rightleftharpoons Pd + 2 OH^-$	0.07
$Pm^{2+} + 2 e \rightleftharpoons Pm$	−2.2
$Pm^{3+} + 3 e \rightleftharpoons Pm$	−2.30
$Pm^{3+} + e \rightleftharpoons Pm^{2+}$	−2.6
$Po^{4+} + 2 e \rightleftharpoons Po^{2+}$	0.9
$Po^{4+} + 4 e \rightleftharpoons Po$	0.76
$Pr^{4+} + e \rightleftharpoons Pr^{3+}$	3.2
$Pr^{2+} + 2 e \rightleftharpoons Pr$	−2.0
$Pr^{3+} + 3 e \rightleftharpoons Pr$	−2.353
$Pr^{3+} + e \rightleftharpoons Pr^{2+}$	−3.1
$Pt^{2+} + 2 e \rightleftharpoons Pt$	1.18
$[PtCl_4]^{2-} + 2 e \rightleftharpoons Pt + 4 Cl^-$	0.755
$[PtCl_6]^{2-} + 2 e \rightleftharpoons [PtCl_4]^{2-} + 2 Cl^-$	0.68
$Pt(OH)_2 + 2 e \rightleftharpoons Pt + 2 OH^-$	0.14
$PtO_3 + 2 H^+ + 2 e \rightleftharpoons PtO_2 + H_2O$	1.7
$PtO_3 + 4 H^+ + 2 e \rightleftharpoons Pt(OH)_2^{2+} + H_2O$	1.5
$PtOH^+ + H^+ + 2 e \rightleftharpoons Pt + H_2O$	1.2
$PtO_2 + 2 H^+ + 2 e \rightleftharpoons PtO + H_2O$	1.01

Reaction	$E°/V$
$PtO_2 + 4\,H^+ + 4\,e \rightleftharpoons Pt + 2\,H_2O$	1.00
$Pu^{3+} + 3\,e \rightleftharpoons Pu$	−2.031
$Pu^{4+} + e \rightleftharpoons Pu^{3+}$	1.006
$Pu^{5+} + e \rightleftharpoons Pu^{4+}$	1.099
$PuO_2(OH)_2 + 2\,H^+ + 2\,e \rightleftharpoons Pu(OH)_4$	1.325
$PuO_2(OH)_2 + H^+ + e \rightleftharpoons PuO_2OH + H_2O$	1.062
$Ra^{2+} + 2\,e \rightleftharpoons Ra$	−2.8
$Rb^+ + e \rightleftharpoons Rb$	−2.98
$Re^{3+} + 3\,e \rightleftharpoons Re$	0.300
$ReO_4^- + 4\,H^+ + 3\,e \rightleftharpoons ReO_2 + 2\,H_2O$	0.510
$ReO_2 + 4\,H^+ + 4\,e \rightleftharpoons Re + 2\,H_2O$	0.2513
$ReO_4^- + 2\,H^+ + e \rightleftharpoons ReO_3 + H_2O$	0.768
$ReO_4^- + 4\,H_2O + 7\,e \rightleftharpoons Re + 8\,OH^-$	−0.584
$ReO_4^- + 8\,H^+ + 7\,e \_ Re + 4\,H_2O$	0.368
$Rh^+ + e \rightleftharpoons Rh$	0.600
$Rh^+ + 2e \rightleftharpoons Rh$	0.600
$Rh^{3+} + 3\,e \rightleftharpoons Rh$	0.758
$[RhCl_6]^{3-} + 3\,e \rightleftharpoons Rh + 6\,Cl^-$	0.431
$RhOH^{2+} + H^+ + 3\,e \rightleftharpoons Rh + H_2O$	0.83
$Ru^{2+} + 2\,e \rightleftharpoons Ru$	0.455
$Ru^{3+} + e \rightleftharpoons Ru^{2+}$	0.2487
$RuO_2 + 4\,H^+ + 2\,e \rightleftharpoons Ru^{2+} + 2\,H_2O$	1.120
$RuO_4^- + e \rightleftharpoons RuO_4^{2-}$	0.59
$RuO_4 + e \rightleftharpoons RuO_4^-$	1.00
$RuO_4 + 6\,H^+ + 4\,e \rightleftharpoons Ru(OH)_2^{2+} + 2\,H_2O$	1.40
$RuO_4 + 8\,H^+ + 8\,e \rightleftharpoons Ru + 4\,H_2O$	1.038
$[Ru(bipy)_3]^{3+} + e^- \rightleftharpoons [Ru(bipy)_3]^{2+}$	1.24
$[Ru(H_2O)_6]^{3+} + e^- \rightleftharpoons [Ru(H_2O)_6]^{2+}$	0.23
$[Ru(NH_3)_6]^{3+} + e^- \rightleftharpoons [Ru(NH_3)_6]^{2+}$	0.10
$[Ru(en)_3]^{3+} + e^- \rightleftharpoons [Ru(en)_3]^{2+}$	0.210
$[Ru(CN)_6]^{3-} + e^- \rightleftharpoons [Ru(CN)_6]^{4-}$	0.86
$S + 2\,e \rightleftharpoons S^{2-}$	−0.47627
$S + 2H^+ + 2\,e \rightleftharpoons H_2S(aq)$	0.142
$S + H_2O + 2\,e \rightleftharpoons SH^- + OH^-$	−0.478
$2\,S + 2\,e \rightleftharpoons S_2^{2-}$	−0.42836
$S_2O_6^{2-} + 4\,H^+ + 2\,e \rightleftharpoons 2\,H_2SO_3$	0.564
$S_2O_8^{2-} + 2\,e \rightleftharpoons 2\,SO_4^{2-}$	2.010
$S_2O_8^{2-} + 2\,H^+ + 2\,e \rightleftharpoons 2\,HSO_4^-$	2.123
$S_4O_6^{2-} + 2\,e \rightleftharpoons 2\,S_2O_3^{2-}$	0.08
$2\,H_2SO_3 + H^+ + 2\,e \rightleftharpoons HS_2O_4^- + 2\,H_2O$	−0.056
$H_2SO_3 + 4\,H^+ + 4\,e \rightleftharpoons S + 3\,H_2O$	0.449
$2\,SO_3^{2-} + 2\,H_2O + 2\,e \rightleftharpoons S_2O_4^{2-} + 4\,OH^-$	−1.12
$2\,SO_3^{2-} + 3\,H_2O + 4\,e \rightleftharpoons S_2O_3^{2-} + 6\,OH^-$	−0.571
$SO_4^{2-} + 4\,H^+ + 2\,e \rightleftharpoons H_2SO_3 + H_2O$	0.172
$2\,SO_4^{2-} + 4\,H^+ + 2\,e \rightleftharpoons S_2O_6^{2-} + H_2O$	−0.22
$SO_4^{2-} + H_2O + 2\,e \rightleftharpoons SO_3^{2-} + 2\,OH^-$	−0.93
$Sb + 3\,H^+ + 3\,e \rightleftharpoons SbH_3$	−0.510
$Sb_2O_3 + 6\,H^+ + 6\,e \rightleftharpoons 2\,Sb + 3\,H_2O$	0.152
$Sb_2O_5 \text{ (senarmontite)} + 4\,H^+ + 4\,e \rightleftharpoons Sb_2O_3 + 2\,H_2O$	0.671
$Sb_2O_5 \text{ (valentinite)} + 4\,H^+ + 4\,e \rightleftharpoons Sb_2O_3 + 2\,H_2O$	0.649
$Sb_2O_5 + 6\,H^+ + 4\,e \rightleftharpoons 2\,SbO^+ + 3\,H_2O$	0.581

Reaction	$E°/V$
$SbO^+ + 2\,H^+ + 3\,e \rightleftharpoons Sb + 2\,H_2O$	0.212
$SbO_2^- + 2\,H_2O + 3\,e \rightleftharpoons Sb + 4\,OH^-$	−0.66
$SbO_3^- + H_2O + 2\,e \rightleftharpoons SbO_2^- + 2\,OH^-$	−0.59
$Sc^{3+} + 3\,e \rightleftharpoons Sc$	−2.077
$Se + 2\,e \rightleftharpoons Se^{2-}$	−0.924
$Se + 2\,H^+ + 2\,e \rightleftharpoons H_2Se(aq)$	−0.399
$H_2SeO_3 + 4\,H^+ + 4\,e \rightleftharpoons Se + 3\,H_2O$	0.74
$Se + 2\,H^+ + 2\,e \rightleftharpoons H_2Se$	−0.082
$SeO_3^{2-} + 3\,H_2O + 4\,e \rightleftharpoons Se + 6\,OH^-$	−0.366
$SeO_4^{2-} + 4\,H^+ + 2\,e \rightleftharpoons H_2SeO_3 + H_2O$	1.151
$SeO_4^{2-} + H_2O + 2\,e \rightleftharpoons SeO_3^{2-} + 2\,OH^-$	0.05
$SiF_6^{2-} + 4\,e \rightleftharpoons Si + 6\,F^-$	−1.24
$SiO + 2\,H^+ + 2\,e \rightleftharpoons Si + H_2O$	−0.8
$SiO_2 \text{ (quartz)} + 4\,H^+ + 4\,e \rightleftharpoons Si + 2\,H_2O$	0.857
$SiO_3^{2-} + 3\,H_2O + 4\,e \rightleftharpoons Si + 6\,OH^-$	−1.697
$Sm^{3+} + e \rightleftharpoons Sm^{2+}$	−1.55
$Sm^{3+} + 3\,e \rightleftharpoons Sm$	−2.304
$Sm^{2+} + 2\,e \rightleftharpoons Sm$	−2.68
$Sn^{2+} + 2\,e \rightleftharpoons Sn$	−0.1375
$Sn^{4+} + 2\,e \_ Sn^{2+}$	0.151
$Sn(OH)_3^+ + 3\,H^+ + 2\,e \rightleftharpoons Sn^{2+} + 3\,H_2O$	0.142
$SnO_2 + 4\,H^+ + 2\,e^- \rightleftharpoons Sn^{2+} + 2\,H_2O$	−0.094
$SnO_2 + 4\,H^+ + 4\,e \rightleftharpoons Sn + 2\,H_2O$	−0.117
$SnO_2 + 3\,H^+ + 2\,e \rightleftharpoons SnOH^+ + H_2O$	−0.194
$SnO_2 + 2\,H_2O + 4\,e \rightleftharpoons Sn + 4\,OH^-$	−0.945
$HSnO_2^- + H_2O + 2\,e \rightleftharpoons Sn + 3\,OH^-$	−0.909
$Sn(OH)_6^{2-} + 2\,e \rightleftharpoons HSnO_2^- + 3\,OH^- + H_2O$	−0.93
$Sr^+ + e \rightleftharpoons Sr$	−4.10
$Sr^{2+} + 2\,e \rightleftharpoons Sr$	−2.899
$Sr^{2+} + 2\,e \rightleftharpoons Sr(Hg)$	−1.793
$Sr(OH)_2 + 2\,e \rightleftharpoons Sr + 2\,OH^-$	−2.88
$Ta_2O_5 + 10\,H^+ + 10\,e \rightleftharpoons 2\,Ta + 5\,H_2O$	−0.750
$Ta^{3+} + 3\,e \rightleftharpoons Ta$	−0.6
$Tc^{2+} + 2\,e \rightleftharpoons Tc$	0.400
$TcO_4^- + 4\,H^+ + 3\,e \rightleftharpoons TcO_2 + 2\,H_2O$	0.782
$Tc^{3+} + e \rightleftharpoons Tc^{2+}$	0.3
$TcO_4^- + 8\,H^+ + 7\,e \rightleftharpoons Tc + 4\,H_2O$	0.472
$Tb^{4+} + e \rightleftharpoons Tb^{3+}$	3.1
$Tb^{3+} + 3\,e \rightleftharpoons Tb$	−2.28
$Te + 2\,e \rightleftharpoons Te^{2-}$	−1.143
$Te + 2\,H^+ + 2\,e \rightleftharpoons H_2Te$	−0.793
$Te^{4+} + 4\,e \rightleftharpoons Te$	0.568
$TeO_2 + 4\,H^+ + 4\,e \rightleftharpoons Te + 2\,H_2O$	0.593
$TeO_3^{2-} + 3\,H_2O + 4\,e \rightleftharpoons Te + 6\,OH^-$	−0.57
$TeO_4^- + 8\,H^+ + 7\,e \rightleftharpoons Te + 4\,H_2O$	0.472
$H_6TeO_6 + 2\,H^+ + 2\,e \rightleftharpoons TeO_2 + 4\,H_2O$	1.02
$Th^{4+} + 4\,e \rightleftharpoons Th$	−1.899
$ThO_2 + 4\,H^+ + 4\,e \rightleftharpoons Th + 2\,H_2O$	−1.789
$Th(OH)_4 + 4\,e \rightleftharpoons Th + 4\,OH^-$	−2.48
$Ti^{2+} + 2\,e \rightleftharpoons Ti$	−1.630
$Ti^{3+} + e \rightleftharpoons Ti^{2+}$	−0.9
$TiO_2 + 4\,H^+ + 2\,e \rightleftharpoons Ti^{2+} + 2\,H_2O$	−0.502
$Ti^{3+} + 3\,e \rightleftharpoons Ti$	−1.37

Reaction	$E°/V$
$TiOH^{3+} + H^+ + e \rightleftharpoons Ti^{3+} + H_2O$	−0.055
$Tl^+ + e \rightleftharpoons Tl$	−0.336
$Tl^+ + e \rightleftharpoons Tl(Hg)$	−0.3338
$Tl^{3+} + 2 e \rightleftharpoons Tl^+$	1.252
$Tl^{3+} + 3 e \rightleftharpoons Tl$	0.741
$TlBr + e \rightleftharpoons Tl + Br^-$	−0.658
$TlCl + e \rightleftharpoons Tl + Cl^-$	−0.5568
$TlI + e \rightleftharpoons Tl + I^-$	−0.752
$Tl_2O_3 + 3 H_2O + 4 e \rightleftharpoons 2 Tl^+ + 6 OH^-$	0.02
$TlOH + e \rightleftharpoons Tl + OH^-$	−0.34
$Tl(OH)_3 + 2 e \rightleftharpoons TlOH + 2 OH^-$	−0.05
$Tl_2SO_4 + 2 e \rightleftharpoons Tl + SO_4^{2-}$	−0.4360
$Tm^{3+} + e \rightleftharpoons Tm^{2+}$	−2.2
$Tm^{3+} + 3 e \rightleftharpoons Tm$	−2.319
$Tm^{2+} + 2 e \rightleftharpoons Tm$	−2.4
$U^{3+} + 3 e \rightleftharpoons U$	−1.798
$U^{4+} + e \rightleftharpoons U^{3+}$	−0.607
$UO_2^+ + 4 H^+ + e \rightleftharpoons U^{4+} + 2 H_2O$	0.612
$UO_2^{2+} + e \rightleftharpoons UO^+_2$	0.062
$UO_2^{2+} + 4 H^+ + 2 e \rightleftharpoons U^{4+} + 2 H_2O$	0.327
$UO_2^{2+} + 4 H^+ + 6 e \rightleftharpoons U + 2 H_2O$	−1.444
$V^{2+} + 2 e \rightleftharpoons V$	−1.175
$V^{3+} + e \rightleftharpoons V^{2+}$	−0.255
$VO^{2+} + 2 H^+ + e \rightleftharpoons V^{3+} + H_2O$	0.337
$VO_2^+ + 2 H^+ + e \rightleftharpoons VO^{2+} + H_2O$	0.991
$V_2O_5 + 6 H^+ + 2 e \rightleftharpoons 2 VO^{2+} + 3 H_2O$	0.957
$V_2O_5 + 10 H^+ + 10 e \rightleftharpoons 2 V + 5 H_2O$	−0.242

Reaction	$E°/V$
$V(OH)_4^+ + 2 H^+ + e \rightleftharpoons VO^{2+} + 3 H_2O$	1.00
$V(OH)_4^+ + 4 H^+ + 5 e \rightleftharpoons V + 4 H_2O$	−0.254
$[V(phen)_3]^{3+} + e \rightleftharpoons [V(phen)_3]^{2+}$	0.14
$W^{3+} + 3 e \rightleftharpoons W$	0.1
$W_2O_5 + 2 H^+ + 2 e \rightleftharpoons 2 WO_2 + H_2O$	−0.031
$WO_2 + 4 H^+ + 4 e \rightleftharpoons W + 2 H_2O$	−0.119
$WO_3 + 6 H^+ + 6 e \rightleftharpoons W + 3 H_2O$	−0.090
$WO_3 + 2 H^+ + 2 e \rightleftharpoons WO_2 + H_2O$	0.036
$2 WO_3 + 2 H^+ + 2 e \rightleftharpoons W_2O_5 + H_2O$	−0.029
$H_4XeO_6 + 2 H^+ + 2 e \rightleftharpoons XeO_3 + 3 H_2O$	2.42
$XeO_3 + 6 H^+ + 6 e \rightleftharpoons Xe + 3 H_2O$	2.10
$XeF + e \rightleftharpoons Xe + F^-$	3.4
$Y^{3+} + 3 e \rightleftharpoons Y$	−2.372
$Yb^{3+} + e \rightleftharpoons Yb^{2+}$	−1.05
$Yb^{3+} + 3 e \rightleftharpoons Yb$	−2.19
$Yb^{2+} + 2 e \rightleftharpoons Yb$	−2.76
$Zn^{2+} + 2 e \rightleftharpoons Zn$	−0.7618
$Zn^{2+} + 2 e \rightleftharpoons Zn(Hg)$	−0.7628
$ZnO_2^{2-} + 2 H_2O + 2 e \rightleftharpoons Zn + 4 OH^-$	−1.215
$ZnSO_4 \cdot 7 H_2O + 2 e = Zn(Hg) + SO_4^{2-} + 7 H_2O$ (Saturated $ZnSO_4$)	−0.7993
$ZnOH^+ + H^+ + 2 e \rightleftharpoons Zn + H_2O$	−0.497
$Zn(OH)_4^{2-} + 2 e \rightleftharpoons Zn + 4 OH^-$	−1.199
$Zn(OH)_2 + 2 e \rightleftharpoons Zn + 2 OH^-$	−1.249
$ZnO + H_2O + 2 e \rightleftharpoons Zn + 2 OH^-$	−1.260
$ZrO_2 + 4 H^+ + 4 e \rightleftharpoons Zr + 2 H_2O$	−1.553
$ZrO(OH)_2 + H_2O + 4 e \rightleftharpoons Zr + 4 OH^-$	−2.36
$Zr^{4+} + 4 e \rightleftharpoons Zr$	−1.45

TABLE 2. Reduction Reactions Having $E°$ Values More Positive than That of the Standard Hydrogen Electrode

Reaction	$E°/V$
$2 H^+ + 2 e \rightleftharpoons H_2$	0.00000
$CuI_2^- + e \rightleftharpoons Cu + 2 I^-$	0.00
$Ge^{4+} + 2 e \rightleftharpoons Ge^{2+}$	0.00
$NO_3^- + H_2O + 2 e \rightleftharpoons NO_2^- + 2 OH^-$	0.01
$Tl_2O_3 + 3 H_2O + 4 e \rightleftharpoons 2 Tl^+ + 6 OH^-$	0.02
$SeO_4^{2-} + H_2O + 2 e \rightleftharpoons SeO_3^{2-} + 2 OH^-$	0.05
$WO_3 + 2 H^+ + 2 e \rightleftharpoons WO_2 + H_2O$	0.036
$UO_2^{2+} + e = UO_2^+$	0.062
$Pd(OH)_2 + 2 e \rightleftharpoons Pd + 2 OH^-$	0.07
$AgBr + e \rightleftharpoons Ag + Br^-$	0.07133
$MoO_3 + 6 H^+ + 6 e \rightleftharpoons Mo + 3 H_2O$	0.075
$S_4O_6^{2-} + 2 e \rightleftharpoons 2 S_2O_3^{2-}$	0.08
$H_3Mo_7O_{24}^{3-} + 45 H^+ + 42 e \rightleftharpoons 7 Mo + 24 H_2O$	0.082
$AgSCN + e \rightleftharpoons Ag + SCN^-$	0.8951
$N_2 + 2 H_2O + 6 H^+ + 6 e \rightleftharpoons 2 NH_4OH$	0.092
$HgO + H_2O + 2 e \rightleftharpoons Hg + 2 OH^-$	0.0977
$Ir_2O_3 + 3 H_2O + 6 e \rightleftharpoons 2 Ir + 6 OH^-$	0.098
$2 NO + 2 e \rightleftharpoons N_2O_2^{2-}$	0.10
$[Ru(NH_3)_6]^{3+} + e \rightleftharpoons [Ru(NH_3)_6]^{2+}$	0.10
$W^{3+} + 3 e \rightleftharpoons W$	0.1
$[Co(NH_3)_6]^{3+} + e \rightleftharpoons [Co(NH_3)_6]^{2+}$	0.108
$Hg_2O + H_2O + 2 e \rightleftharpoons 2 Hg + 2 OH^-$	0.123

Reaction	$E°/V$
$Ge^{4+} + 4 e \rightleftharpoons Ge$	0.124
$Hg_2Br_2 + 2 e \rightleftharpoons 2 Hg + 2 Br^-$	0.13923
$Pt(OH)_2 + 2 e \rightleftharpoons Pt + 2 OH^-$	0.14
$[V(phen)_3]^{3+} + e \rightleftharpoons [V(phen)_3]^{2+}$	0.14
$S + 2 H^+ + 2 e \rightleftharpoons H_2S(aq)$	0.142
$Sn(OH)_3^+ + 3 H^+ + 2 e \rightleftharpoons Sn^{2+} + 3 H_2O$	0.142
$Np^{4+} + e \rightleftharpoons Np^{3+}$	0.147
$Ag_4[Fe(CN)_6] + 4 e \rightleftharpoons 4 Ag + [Fe(CN)_6]^{4-}$	0.1478
$IO_3^- + 2 H_2O + 4 e \rightleftharpoons IO^- + 4 OH^-$	0.15
$Mn(OH)_3 + e \rightleftharpoons Mn(OH)_2 + OH^-$	0.15
$2 NO_2^- + 3 H_2O + 4 e \rightleftharpoons N_2O + 6 OH^-$	0.15
$Sn^{4+} + 2 e \rightleftharpoons Sn^{2+}$	0.151
$Sb_2O_3 + 6 H^+ + 6 e \rightleftharpoons 2 Sb + 3 H_2O$	0.152
$Cu^{2+} + e \rightleftharpoons Cu^+$	0.153
$BiOCl + 2 H^+ + 3 e \rightleftharpoons Bi + Cl^- + H_2O$	0.1583
$BiCl_4^- + 3 e \rightleftharpoons Bi + 4 Cl^-$	0.16
$Fe_2O_3 + 4 H^+ + 2 e \rightleftharpoons 2 FeOH^+ + H_2O$	0.16
$Co(OH)_3 + e \rightleftharpoons Co(OH)_2 + OH^-$	0.17
$SO_4^{2-} + 4 H^+ + 2 e \rightleftharpoons H_2SO_3 + H_2O$	0.172
$Bi^{3+} + 2 e \rightleftharpoons Bi^+$	0.2
$[Ru(en)_3]^{3+} + e \rightleftharpoons [Ru(en)_3]^{2+}$	0.210
$SbO^+ + 2 H^+ + 3 e \rightleftharpoons Sb + 2 H_2O$	0.212

Reaction	$E°$/V
$AgCl + e \rightleftharpoons Ag + Cl^-$	0.22233
$[Ru(H_2O)_6]^{3+} + e \rightleftharpoons [Ru(H_2O)_6]^{2+}$	0.23
$As_2O_3 + 6 H^+ + 6 e \rightleftharpoons 2 As + 3 H_2O$	0.234
Calomel electrode, saturated NaCl (SSCE)	0.2360
$Ge^{2+} + 2 e \rightleftharpoons Ge$	0.24
$Ru^{3+} + e \rightleftharpoons Ru^{2+}$	0.24
Calomel electrode, saturated KCl	0.2412
$PbO_2 + H_2O + 2 e \rightleftharpoons PbO + 2 OH^-$	0.247
$HAsO_2 + 3 H^+ + 3 e \rightleftharpoons As + 2 H_2O$	0.248
$Ru^{3+} + e \rightleftharpoons Ru^{2+}$	0.2487
$ReO_2 + 4 H^+ + 4 e \rightleftharpoons Re + 2 H_2O$	0.2513
$IO_3^- + 3 H_2O + 6 e \rightleftharpoons I^- + OH^-$	0.26
$Hg_2Cl_2 + 2 e \rightleftharpoons 2 Hg + 2 Cl^-$	0.26808
Calomel electrode, 1 molal KCl	0.2800
Calomel electrode, 1 molar KCl (NCE)	0.2801
$At_2 + 2 e \rightleftharpoons 2 At^-$	0.3
$Re^{3+} + 3 e \rightleftharpoons Re$	0.300
$Tc^{3+} + e \rightleftharpoons Tc^{2+}$	0.3
$Bi^{3+} + 3 e \rightleftharpoons Bi$	0.308
$BiO^+ + 2 H^+ + 3 e \rightleftharpoons Bi + H_2O$	0.320
$UO_2^{2+} + 4 H^+ + 2 e \rightleftharpoons U^{4+} + 2 H_2O$	0.327
$ClO_3^- + H_2O + 2 e \rightleftharpoons ClO_2^- + 2 OH^-$	0.33
$2 HCNO + 2 H^+ + 2 e \rightleftharpoons (CN)_2 + 2 H_2O$	0.330
Calomel electrode, 0.1 molar KCl	0.3337
$VO^{2+} + 2 H^+ + e \rightleftharpoons V^{3+} + H_2O$	0.337
$Cu^{2+} + 2 e \rightleftharpoons Cu$	0.3419
$Ag_2O + H_2O + 2 e \rightleftharpoons 2 Ag + 2 OH^-$	0.342
$Cu^{2+} + 2 e \rightleftharpoons Cu(Hg)$	0.345
$AgIO_3 + e \rightleftharpoons Ag + IO_3^-$	0.354
$[Fe(CN)_6]^{3-} + e \rightleftharpoons [Fe(CN)_6]^{4-}$	0.358
$ClO_4^- + H_2O + 2 e \rightleftharpoons ClO_3^- + 2 OH^-$	0.36
$Ag_2SeO_3 + 2 e \rightleftharpoons 2 Ag + SeO_3^{2-}$	0.3629
$ReO_4^- + 8 H^+ + 7 e \rightleftharpoons Re + 4 H_2O$	0.368
$(CN)_2 + 2 H^+ + 2 e \rightleftharpoons 2 HCN$	0.373
$[Ferricinium]^+ + e \rightleftharpoons ferrocene$	0.400
$Tc^{2+} + 2 e \rightleftharpoons Tc$	0.400
$O_2 + 2 H_2O + 4 e \rightleftharpoons 4 OH^-$	0.401
$AgOCN + e \rightleftharpoons Ag + OCN^-$	0.41
$[RhCl_6]^{3-} + 3 e \rightleftharpoons Rh + 6 Cl^-$	0.431
$Ag_2CrO_4 + 2 e \rightleftharpoons 2 Ag + CrO_4^{2-}$	0.4470
$H_2SO_3 + 4 H^+ + 4 e \rightleftharpoons S + 3 H_2O$	0.449
$Ru^{2+} + 2 e \rightleftharpoons Ru$	0.455
$Ag_2MoO_4 + 2 e \rightleftharpoons 2 Ag + MoO_4^{2-}$	0.4573
$Ag_2C_2O_4 + 2 e \rightleftharpoons 2 Ag + C_2O_4^{2-}$	0.4647
$Ag_2WO_4 + 2 e \rightleftharpoons 2 Ag + WO_4^{2-}$	0.4660
$Ag_2CO_3 + 2 e \rightleftharpoons 2 Ag + CO_3^{2-}$	0.47
$TcO_4^- + 8 H^+ + 7 e \rightleftharpoons Tc + 4 H_2O$	0.472
$TeO_4^- + 8 H^+ + 7 e \rightleftharpoons Te + 4 H_2O$	0.472
$IO^- + H_2O + 2 e \rightleftharpoons I^- + 2 OH^-$	0.485
$NiO_2 + 2 H_2O + 2 e \rightleftharpoons Ni(OH)_2 + 2 OH^-$	0.490
$Bi^+ + e \rightleftharpoons Bi$	0.5
$ReO_4^- + 4 H^+ + 3 e \rightleftharpoons ReO_2 + 2 H_2O$	0.510
$Hg_2(ac)_2 + 2 e \rightleftharpoons 2 Hg + 2(ac)^-$	0.51163

Reaction	$E°$/V
$Cu^+ + e \rightleftharpoons Cu$	0.521
$I_2 + 2 e \rightleftharpoons 2 I^-$	0.5355
$I_3^- + 2 e \rightleftharpoons 3 I^-$	0.536
$AgBrO_3 + e \rightleftharpoons Ag + BrO_3^-$	0.546
$MnO_4^- + e \rightleftharpoons MnO_4^-$	0.558
$H_3AsO_4 + 2 H^+ + 2 e \rightleftharpoons HAsO_2 + 2 H_2O$	0.560
$S_2O_6^{2-} + 4 H^+ + 2 e \rightleftharpoons 2 H_2SO_3$	0.564
$AgNO_2 + e \rightleftharpoons Ag + NO_2^-$	0.564
$Te^{4+} + 4 e \rightleftharpoons Te$	0.568
$Sb_2O_5 + 6 H^+ + 4 e \rightleftharpoons 2 SbO^+ + 3 H_2O$	0.581
$RuO_4^- + e \rightleftharpoons RuO_4^{2-}$	0.59
$[PdCl_4]^{2-} + 2 e \rightleftharpoons Pd + 4 Cl^-$	0.591
$TeO_2 + 4 H^+ + 4 e \rightleftharpoons Te + 2 H_2O$	0.593
$MnO_4^- + 2 H_2O + 3 e \rightleftharpoons MnO_2 + 4 OH^-$	0.595
$Rh^{2+} + 2 e \rightleftharpoons Rh$	0.600
$Rh^+ + e \rightleftharpoons Rh$	0.600
$MnO_4^{2-} + 2 H_2O + 2 e \rightleftharpoons MnO_2 + 4 OH^-$	0.60
$2 AgO + H_2O + 2 e \rightleftharpoons Ag_2O + 2 OH^-$	0.607
$BrO_3^- + 3 H_2O + 6 e \rightleftharpoons Br^- + 6 OH^-$	0.61
$UO_2^+ + 4 H^+ + e \rightleftharpoons U^{4+} + 2 H_2O$	0.612
$Hg_2SO_4 + 2 e \rightleftharpoons 2 Hg + SO_4^{2-}$	0.6125
$ClO_3^- + 3 H_2O + 6 e \rightleftharpoons Cl^- + 6 OH^-$	0.62
$Hg_2HPO_4 + 2 e \rightleftharpoons 2 Hg + HPO_4^{2-}$	0.6359
$Ag(ac) + e \rightleftharpoons Ag + (ac)^-$	0.643
$Sb_2O_5(valentinite) + 4 H^+ + 4 e \rightleftharpoons Sb_2O_3 + 2 H_2O$	0.649
$Ag_2SO_4 + 2 e \rightleftharpoons 2 Ag + SO_4^{2-}$	0.654
$ClO_2^- + H_2O + 2 e \rightleftharpoons ClO^- + 2 OH^-$	0.66
$Sb_2O_5(senarmontite) + 4 H^+ + 4 e \rightleftharpoons Sb_2O_5 + 2 H_2O$	0.671
$[PtCl_6]^{2-} + 2 e \rightleftharpoons [PtCl_4]^{2-} + 2 Cl^-$	0.68
$O_2 + 2 H^+ + 2 e \rightleftharpoons H_2O_2$	0.695
$p-benzoquinone + 2 H^+ + 2 e \rightleftharpoons hydroquinone$	0.6992
$H_3IO_6^{2-} + 2 e \rightleftharpoons IO_3^- + 3 OH^-$	0.7
$Ag_2O_3 + H_2O + 2 e \rightleftharpoons 2 AgO + 2 OH^-$	0.739
$Tl^{3+} + 3 e \rightleftharpoons Tl$	0.741
$[PtCl_4]^{2-} + 2 e \rightleftharpoons Pt + 4 Cl^-$	0.755
$Rh^{3+} + 3 e \rightleftharpoons Rh$	0.758
$ClO_2^- + 2 H_2O + 4 e \rightleftharpoons Cl^- + 4 OH^-$	0.76
$2 NO + H_2O + 2 e \rightleftharpoons N_2O + 2 OH^-$	0.76
$Po^{4+} + 4 e \rightleftharpoons Po$	0.76
$BrO^- + H_2O + 2 e \rightleftharpoons Br^- + 2 OH^-$	0.761
$ReO_4^- + 2 H^+ + e \rightleftharpoons ReO_3 + H_2O$	0.768
$(CNS)_2 + 2 e \rightleftharpoons 2 CNS^-$	0.77
$[IrCl_6]^{3-} + 3 e \rightleftharpoons Ir + 6 Cl^-$	0.77
$Fe^{3+} + e \rightleftharpoons Fe^{2+}$	0.771
$AgF + e \rightleftharpoons Ag + F^-$	0.779
$[Fe(bipy)_2]^{3+} + e \rightleftharpoons [Fe(bipy)_2]^{2+}$	0.78
$TcO_4^- + 4 H^+ + 3 e \rightleftharpoons TcO_2 + 2 H_2O$	0.782
$Hg_2^{2+} + 2 e \rightleftharpoons 2 Hg$	0.7973
$Ag^+ + e \rightleftharpoons Ag$	0.7996
$[Os(bipy)_3]^{3+} + e \rightleftharpoons [Os(bipy)_3]^{2+}$	0.80
$2 NO_3^- + 4 H^+ + 2 e \rightleftharpoons N_2O_4 + 2 H_2O$	0.803

Reaction	$E°/V$	Reaction	$E°/V$
$[Os(bipy)_2]^{3+} + e \rightleftharpoons [Os(bipy)_2]^{2+}$	0.81	$ClO_3^- + 3\,H^+ + 2\,e \rightleftharpoons HClO_2 + H_2O$	1.214
$RhOH^{2+} + H^+ + 3\,e \rightleftharpoons Rh + H_2O$	0.83	$MnO_2 + 4\,H^+ + 2\,e \rightleftharpoons Mn^{2+} + 2\,H_2O$	1.224
$OsO_4 + 8\,H^+ + 8\,e \rightleftharpoons Os + 4\,H_2O$	0.838	$O_2 + 4\,H^+ + 4\,e \rightleftharpoons 2\,H_2O$	1.229
$ClO^- + H_2O + 2\,e \rightleftharpoons Cl^- + 2\,OH^-$	0.841	$Cr_2O_7^{2-} + 14\,H^+ + 6\,e \rightleftharpoons 2\,Cr^{3+} + 7\,H_2O$	1.232
$Hg^{2+} + 2\,e \rightleftharpoons Hg$	0.851	$O_3 + H_2O + 2\,e \rightleftharpoons O_2 + 2\,OH^-$	1.24
$AuBr_4^- + 3\,e \rightleftharpoons Au + 4\,Br^-$	0.854	$[Ru(bipy)_3]^{3+} + e \rightleftharpoons [Ru(bipy)_3]^{2+}$	1.24
$SiO_2(quartz) + 4\,H^+ + 4\,e \rightleftharpoons Si + 2\,H_2O$	0.857	$Tl^{3+} + 2\,e \rightleftharpoons Tl^+$	1.252
$2\,HNO_2 + 4\,H^+ + 4\,e \rightleftharpoons H_2N_2O_2 + H_2O$	0.86	$N_2H_5^+ + 3\,H^+ + 2\,e \rightleftharpoons 2\,NH_4^+$	1.275
$[Ru(CN)_6]^{3-} + e^- \rightleftharpoons [Ru(CN)_6]^{4-}$	0.86	$ClO_2 + H^+ + e \rightleftharpoons HClO_2$	1.277
$[IrCl_6]^{2-} + e \rightleftharpoons [IrCl_6]^{3-}$	0.8665	$[PdCl_6]^{2-} + 2\,e \rightleftharpoons [PdCl_4]^{2-} + 2\,Cl^-$	1.288
$N_2O_4 + 2\,e \rightleftharpoons 2\,NO_2^-$	0.867	$2\,HNO_2 + 4\,H^+ + 4\,e \rightleftharpoons N_2O + 3\,H_2O$	1.297
$HO_2^- + H_2O + 2\,e \rightleftharpoons 3\,OH^-$	0.878	$AuOH^{2+} + H^+ + 2\,e \rightleftharpoons Au^+ + H_2O$	1.32
$Po^{4+} + 2\,e \rightleftharpoons Po^{2+}$	0.9	$PuO_2(OH)_2 + 2\,H^- + 2\,e \rightleftharpoons Pu(OH)_4$	1.325
$2\,Hg^{2+} + 2\,e \rightleftharpoons Hg_2^{2+}$	0.920	$HBrO + H^+ + 2\,e \rightleftharpoons Br^- + H_2O$	1.331
$NO_3^- + 3\,H^+ + 2\,e \rightleftharpoons HNO_2 + H_2O$	0.934	$Cr(V) + e \rightleftharpoons Cr(IV)$	1.34
$Pd^{2+} + 2\,e \rightleftharpoons Pd$	0.951	$HCrO_4^- + 7\,H^+ + 3\,e \rightleftharpoons Cr^{3+} + 4\,H_2O$	1.350
$ClO_2(aq) + e \rightleftharpoons ClO_2^-$	0.954	$Cl_2(g) + 2\,e \rightleftharpoons 2Cl^-$	1.35827
$NO_3^- + 4\,H^+ + 3\,e \rightleftharpoons NO + 2\,H_2O$	0.957	$ClO_4^- + 8\,H^+ + 8\,e \rightleftharpoons Cl^- + 4\,H_2O$	1.389
$V_2O_5 + 6\,H^+ + 2\,e \rightleftharpoons 2\,VO^{2+} + 3\,H_2O$	0.957	$ClO_4^- + 8\,H^+ + 7\,e \rightleftharpoons 1/2\,Cl_2 + 4\,H_2O$	1.39
$AuBr_2^- + e \rightleftharpoons Au + 2\,Br^-$	0.959	$No^{3+} + e \rightleftharpoons No^{2+}$	1.4
$HNO_2 + H^+ + e \rightleftharpoons NO + H_2O$	0.983	$RuO_4 + 6\,H^+ + 4\,e \rightleftharpoons Ru(OH)_2^{2+} + 2\,H_2O$	1.40
$HIO + H^+ + 2\,e \rightleftharpoons I^- + H_2O$	0.987	$Au^{3+} + 2\,e \rightleftharpoons Au^+$	1.401
$VO_2^+ + 2\,H^+ + e \rightleftharpoons VO^{2+} + H_2O$	0.991	$2\,NH_3OH^+ + H^+ + 2\,e \rightleftharpoons N_2H_5^+ + 2\,H_2O$	1.42
$PtO_2 + 4\,H^+ + 4\,e \rightleftharpoons Pt + 2\,H_2O$	1.00	$BrO_3^- + 6\,H^+ + 6\,e \rightleftharpoons Br^- + 3\,H_2O$	1.423
$RuO_4 + e \rightleftharpoons RuO_4^-$	1.00	$2\,HIO + 2\,H^+ + 2\,e \rightleftharpoons I_2 + 2\,H_2O$	1.439
$V(OH)_4^+ + 2\,H^+ + e \rightleftharpoons VO^{2+} + 3\,H_2O$	1.00	$Au(OH)_3 + 3\,H^+ + 3\,e \rightleftharpoons Au^- + 3\,H_2O$	1.45
$AuCl_4^- + 3\,e \rightleftharpoons Au + 4\,Cl^-$	1.002	$3IO_3^- + 6\,H^+ + 6\,e \rightleftharpoons Cl^- + 3\,H_2O$	1.451
$Pu^{4+} + e \rightleftharpoons Pu^{3+}$	1.006	$PbO_2 + 4\,H^+ + 2\,e \rightleftharpoons Pb^{2+} + 2\,H_2O$	1.455
$PtO_2 + 2\,H^+ + 2\,e \rightleftharpoons PtO + H_2O$	1.01	$ClO_3^- + 6\,H^+ + 5\,e \rightleftharpoons 1/2\,Cl_2 + 3\,H_2O$	1.47
$OsO_4 + 4\,H^+ + 4\,e \rightleftharpoons OsO_2 + 2\,H_2O$	1.02	$CrO_2 + 4\,H^+ + e \rightleftharpoons Cr^{3+} + 2\,H_2O$	1.48
$H_6TeO_6 + 2\,H^+ + 2\,e \rightleftharpoons TeO_2 + 4\,H_2O$	1.02	$BrO_3^- + 6\,H^+ + 5\,e \rightleftharpoons 1/2\,Br_2 + 3\,H_2O$	1.482
$[Fe(bipy)_3]^{3+} + e \rightleftharpoons [Fe(bipy)_3]^{2+}$	1.03	$HClO + H^+ + 2\,e \rightleftharpoons Cl^- + H_2O$	1.482
$Hg(OH)_2 + 2\,H^+ + 2\,e \rightleftharpoons Hg + 2\,H_2O$	1.034	$Mn_2O_3 + 6\,H^+ + e \rightleftharpoons 2\,Mn^{2+} + 3\,H_2O$	1.485
$N_2O_4 + 4\,H^+ + 4\,e \rightleftharpoons 2\,NO + 2\,H_2O$	1.035	$HO_2 + H^+ + e \rightleftharpoons H_2O_2$	1.495
$RuO_4 + 8\,H^+ + 8\,e \rightleftharpoons Ru + 4H_2O$	1.038	$Au^{3+} + 3\,e \rightleftharpoons Au$	1.498
$[Fe(phen)_3]^{3+} + e \rightleftharpoons [Fe(phen)_3]^{2+}$ (1 molar H_2SO_4)	1.06	$PtO_3 + 4\,H^+ + 2\,e \rightleftharpoons Pt(OH)_2^{2+} + H_2O$	1.5
$PuO_2(OH)_2 + H^+ + e \rightleftharpoons PuO_2OH + H_2O$	1.062	$MnO_4^- + 8\,H^+ + 5\,e \rightleftharpoons Mn^{2+} + 4\,H_2O$	1.507
$N_2O_4 + 2\,H^+ + 2\,e \rightleftharpoons 2\,HNO_2$	1.065	$Mn^{3+} + e \rightleftharpoons Mn^{2-}$	1.5415
$Br_2(l) + 2\,e \rightleftharpoons 2Br^-$	1.066	$HClO_2 + 3\,H^+ + 4\,e \rightleftharpoons Cl^- + 2\,H_2O$	1.570
$IO_3^- + 6\,H^+ + 6\,e \rightleftharpoons I^- + 3\,H_2O$	1.085	$HBrO + H^+ + e \rightleftharpoons 1/2\,Br_2(aq) + H_2O$	1.574
$Br_2(aq) + 2\,e \rightleftharpoons 2Br^-$	1.0873	$2\,NO + 2\,H^+ + 2\,e \rightleftharpoons N_2O + H_2O$	1.591
$Pu^{5+} + e \rightleftharpoons Pu^{4+}$	1.099	$Bi_2O_4 + 4\,H^+ + 2\,e \rightleftharpoons 2\,BiO^+ + 2\,H_2O$	1.593
$Cu^{2+} + 2\,CN^- + e \rightleftharpoons [Cu(CN)_2]^-$	1.103	$HBrO + H^+ + e \rightleftharpoons 1/2\,Br_2(l) + H_2O$	1.596
$RuO_2 + 4\,H^+ + 2\,e \rightleftharpoons Ru^{2+} + 2\,H_2O$	1.120	$H_5IO_6 + H^+ + 2\,e \rightleftharpoons IO_3^- + 3\,H_2O$	1.601
$[Fe(phen)_3]^{3+} + e \rightleftharpoons [Fe(phen)_3]^{2+}$	1.147	$HClO + H^+ + e \rightleftharpoons 1/2\,Cl_2 + H_2O$	1.611
$SeO_4^{2-} + 4\,H^+ + 2\,e \rightleftharpoons H_2SeO_3 + H_2O$	1.151	$HClO_2 + 3\,H^+ + 3\,e \rightleftharpoons 1/2\,Cl_2 + 2\,H_2O$	1.628
$ClO_3^- + 2\,H^+ + e \rightleftharpoons ClO_2 + H_2O$	1.152	$HClO_2 + 2\,H^+ + 2\,e \rightleftharpoons HClO + H_2O$	1.645
$Ir^{3+} + 3\,e \rightleftharpoons Ir$	1.156	$Bk^{4+} + e \rightleftharpoons Bk^{3+}$	1.67
$Pt^{2+} + 2\,e \rightleftharpoons Pt$	1.18	$NiO_2 + 4\,H^+ + 2\,e \rightleftharpoons Ni^{2+} + 2\,H_2O$	1.678
$ClO_4^- + 2\,H^+ + 2\,e \rightleftharpoons ClO_3^- + H_2O$	1.189	$MnO_4^- + 4\,H^+ + 3\,e \rightleftharpoons MnO_2 + 2\,H_2O$	1.679
$2\,IO_3^- + 12\,H^+ + 10\,e \rightleftharpoons I_2 + 6\,H_2O$	1.195	$PbO_2 + SO_4^{2-} + 4\,H^+ + 2\,e \rightleftharpoons PbSO_4 + 2\,H_2O$	1.6913
$PtOH^+ + H^+ + 2\,e \rightleftharpoons Pt + H_2O$	1.2	$Au^+ + e \rightleftharpoons Au$	1.692

Reaction	$E°/V$	Reaction	$E°/V$
$PtO_3 + 2\,H^+ + 2\,e \rightleftharpoons PtO_2 + H_2O$	1.7	$2\,HFeO_4^- + 8\,H^+ + 6\,e \rightleftharpoons Fe_2O_3 + 5\,H_2O$	2.09
$CeOH^{3+} + H^+ + e \rightleftharpoons Ce^{3+} + H_2O$	1.715	$XeO_3 + 6\,H^+ + 6\,e \rightleftharpoons Xe + 3\,H_2O$	2.10
$Ce^{4+} + e \rightleftharpoons Ce^{3+}$	1.72	$S_2O_8^{2-} + 2\,H^+ + 2\,e \rightleftharpoons 2\,HSO_4^-$	2.123
$N_2O + 2\,H^+ + 2\,e \rightleftharpoons N_2 + H_2O$	1.766	$F_2O + 2\,H^+ + 4\,e \rightleftharpoons H_2O + 2\,F^-$	2.153
$H_2O_2 + 2\,H^+ + 2\,e \rightleftharpoons 2\,H_2O$	1.776	$FeO_4^{2-} + 8\,H^+ + 3\,e \rightleftharpoons Fe^{3+} + 4\,H_2O$	2.20
$Ag^{3+} + e \rightleftharpoons Ag^{2+}$	1.8	$Cu^{3+} + e \rightleftharpoons Cu^{2+}$	2.4
$Au^{2+} + e^- \rightleftharpoons Au^+$	1.8	$H_4XeO_6 + 2\,H^+ + 2\,e \rightleftharpoons XeO_3 + 3\,H_2O$	2.42
$Ag_2O_2 + 4\,H^+ + e \rightleftharpoons 2\,Ag + 2\,H_2O$	1.802	$O(g) + 2\,H^+ + 2\,e \rightleftharpoons H_2O$	2.421
$Co^{3+} + e \rightleftharpoons Co^{2-}(\text{2 molar } H_2SO_4)$	1.83	$Am^{4+} + e \rightleftharpoons Am^{3+}$	2.60
$Ag^{3+} + 2\,e \rightleftharpoons Ag^+$	1.9	$H_2N_2O_2 + 2\,H^+ + 2\,e \rightleftharpoons N_2 + 2\,H_2O$	2.65
$Co^{3+} + e \rightleftharpoons Co^{2+}$	1.92	$F_2 + 2\,e \rightleftharpoons 2\,F^-$	2.866
$Ag^{2+} + e \rightleftharpoons Ag^+$	1.980	$Cm^{4+} + e \rightleftharpoons Cm^{3+}$	3.0
$Cu_2O_3 + 6\,H^+ + 2\,e \rightleftharpoons 2\,Cu^{2+} + 3\,H_2O$	2.0	$F_2 + 2\,H^+ + 2\,e \rightleftharpoons 2\,HF$	3.053
$S_2O_8^{2-} + 2\,e \rightleftharpoons 2\,SO_4^{2-}$	2.010	$Tb^{4+} + e \rightleftharpoons Tb^{3+}$	3.1
$OH + e \rightleftharpoons OH^-$	2.02	$Pr^{4+} + e \rightleftharpoons Pr^{3+}$	3.2
$HFeO_4^- + 7\,H^+ + 3\,e \rightleftharpoons Fe^{3+} + 4\,H_2O$	2.07	$Cf^{4+} + e \rightleftharpoons Cf^{3+}$	3.3
$O_3 + 2\,H^+ + 2\,e \rightleftharpoons O_2 + H_2O$	2.076	$XeF + e \rightleftharpoons Xe + F^-$	3.4
$HFeO_4^- + 4\,H^+ + 3\,e \rightleftharpoons FeOOH + 2\,H_2O$	2.08		

TABLE 3. Reduction Reactions Having $E°$ Values More Negative than That of the Standard Hydrogen Electrode

Reaction	$E°/V$	Reaction	$E°/V$
$2\,H^+ + 2\,e \rightleftharpoons H_2$	0.00000	$H_2GeO_3 + 4\,H^+ + 4\,e \rightleftharpoons Ge + 3\,H_2O$	−0.182
$2\,D^+ + 2\,e \rightleftharpoons D_2$	−0.013	$SnO_2 + 3\,H^+ + 2\,e \rightleftharpoons SnOH^+ + H_2O$	−0.194
$AgCN + e \rightleftharpoons Ag + CN^-$	−0.017	$CO_2 + 2\,H^+ + 2\,e \rightleftharpoons HCOOH$	−0.199
$2\,WO_3 + 2\,H^+ + 2\,e \rightleftharpoons W_2O_5 + H_2O$	−0.029	$Mo^{3+} + 3\,e \rightleftharpoons Mo$	−0.200
$W_2O_5 + 2\,H^+ + 2\,e \rightleftharpoons 2\,WO_2 + H_2O$	−0.031	$Ga^+ + e \rightleftharpoons Ga$	−0.2
$Ag_2S + 2\,H^+ + 2\,e \rightleftharpoons 2\,Ag + H_2S$	−0.0366	$2\,SO_3^{2-} + 4\,H^+ + 2\,e \rightleftharpoons S_2O_6^{2-} + H_2O$	−0.22
$Fe^{3+} + 3\,e \rightleftharpoons Fe$	−0.037	$Cu(OH)_2 + 2\,e \rightleftharpoons Cu + 2\,OH^-$	−0.222
$Hg_2I_2 + 2\,e \rightleftharpoons 2\,Hg + 2\,I^-$	−0.0405	$V_2O_5 + 10\,H^+ + 10\,e \rightleftharpoons 2\,V + 5\,H_2O$	−0.242
$Tl(OH)_3 + 2\,e \rightleftharpoons TlOH + 2\,OH^-$	−0.05	$CdSO_4 + 2\,e \rightleftharpoons Cd + SO_4^{2-}$	−0.246
$TiOH^{3+} + H^+ + e \rightleftharpoons Ti^{3+} + H_2O$	−0.055	$V(OH)_4^+ + 4\,H^+ + 5\,e \rightleftharpoons V + 4\,H_2O$	−0.254
$2\,H_2SO_3 + H^+ + 2\,e \rightleftharpoons HS_2O_4^- + 2\,H_2O$	−0.056	$V^{3+} + e \rightleftharpoons V^{2+}$	−0.255
$P(white) + 3\,H^+ + 3\,e \rightleftharpoons PH_3(g)$	−0.063	$Ni^{2+} + 2\,e \rightleftharpoons Ni$	−0.257
$O_2 + H_2O + 2\,e \rightleftharpoons HO_2^- + OH^-$	−0.076	$PbCl_2 + 2\,e \rightleftharpoons Pb + 2\,Cl^-$	−0.2675
$2\,Cu(OH)_2 + 2\,e \rightleftharpoons Cu_2O + 2\,OH^- + H_2O$	−0.080	$H_3PO_4 + 2\,H^+ + 2\,e \rightleftharpoons H_3PO_3 + H_2O$	−0.276
$Se + 2\,H^+ + 2\,e \rightleftharpoons H_2Se$	−0.082	$Co^{2+} + 2\,e \rightleftharpoons Co$	−0.28
$WO_3 + 6\,H^+ + 6\,e \rightleftharpoons W + 3\,H_2O$	−0.090	$PbBr_2 + 2\,e \rightleftharpoons Pb + 2\,Br^-$	−0.284
$SnO_2 + 4\,H^+ + 2\,e \rightleftharpoons Sn^{2+} + 2\,H_2O$	−0.094	$Tl^+ + e \rightleftharpoons Tl(Hg)$	−0.3338
$Md^{3+} + e \rightleftharpoons Md^{2+}$	−0.1	$Tl^+ + e \rightleftharpoons Tl$	−0.336
$P(red) + 3\,H^+ + 3\,e \rightleftharpoons PH_3(g)$	−0.111	$In^{3+} + 3\,e \rightleftharpoons In$	−0.3382
$SnO_2 + 4\,H^+ + 4\,e \rightleftharpoons Sn + 2\,H_2O$	−0.117	$TlOH + e \rightleftharpoons Tl + OH^-$	−0.34
$GeO_2 + 2\,H^+ + 2\,e \rightleftharpoons GeO + H_2O$	−0.118	$PbF_2 + 2\,e \rightleftharpoons Pb + 2\,F^-$	−0.3444
$WO_2 + 4\,H^+ + 4\,e \rightleftharpoons W + 2\,H_2O$	−0.119	$PbSO_4 + 2\,e \rightleftharpoons Pb(Hg) + SO_4^{2-}$	−0.3505
$Pb^{2+} + 2\,e \rightleftharpoons Pb(Hg)$	−0.1205	$Cd^{2+} + 2\,e \rightleftharpoons Cd(Hg)$	−0.3521
$Pb^{2+} + 2\,e \rightleftharpoons Pb$	−0.1262	$PbSO_4 + 2\,e \rightleftharpoons Pb + SO_4^{2-}$	−0.3588
$CrO_4^{2-} + 4\,H_2O + 3\,e \rightleftharpoons Cr(OH)_3 + 5\,OH^-$	−0.13	$Cu_2O + H_2O + 2\,e \rightleftharpoons 2\,Cu + 2\,OH^-$	−0.360
$Sn^{2-} + 2\,e \rightleftharpoons Sn$	−0.1375	$Eu^{3+} + e \rightleftharpoons Eu^{2+}$	−0.36
$In^+ + e \rightleftharpoons In$	−0.14	$PbI_2 + 2\,e \rightleftharpoons Pb + 2\,I^-$	−0.365
$O_2 + 2\,H_2O + 2\,e \rightleftharpoons H_2O_2 + 2\,OH^-$	−0.146	$SeO_3^{2-} + 3\,H_2O + 4\,e \rightleftharpoons Se + 6\,OH^-$	−0.366
$MoO_2 + 4\,H^+ + 4\,e \rightleftharpoons Mo + 4\,H_2O$	−0.152	$Se + 2\,H^+ + 2\,e \rightleftharpoons H_2Se(aq)$	−0.399
$AgI + e \rightleftharpoons Ag + I^-$	−0.15224	$In^{2+} + e \rightleftharpoons In^+$	−0.40
$2\,NO_2^- + 2\,H_2O + 4\,e \rightleftharpoons N_2O_2^{2-} + 4\,OH^-$	−0.18	$Cd^{2+} + 2\,e \rightleftharpoons Cd$	−0.4030

Reaction	$E°/V$	Reaction	$E°/V$
$Cr^{3+} + e \rightleftharpoons Cr^{2+}$	-0.407	$ZnSO_4 7H_2O + 2\,e \rightleftharpoons Zn(Hg) + SO_4^{2-} + 7\,H_2O$	-0.7993
$2\,S + 2\,e \rightleftharpoons S_2^{2-}$	-0.42836	(Saturated $ZnSO_4$)	
$Tl_2SO_4 + 2\,e \rightleftharpoons Tl + SO_4^{2-}$	-0.4360	$Bi + 3\,H^+ + 3\,e \rightleftharpoons BiH_3$	-0.8
$In^{3+} + 2\,e \rightleftharpoons In^+$	-0.443	$SiO + 2\,H^+ + 2\,e \rightleftharpoons Si + H_2O$	-0.8
$Fe^{2+} + 2\,e \rightleftharpoons Fe$	-0.447	$Cd(OH)_2 + 2\,e \rightleftharpoons Cd(Hg) + 2\,OH^-$	-0.809
$H_3PO_3 + 3\,H^+ + 3\,e \rightleftharpoons P + 3\,H_2O$	-0.454	$2\,H_2O + 2\,e \rightleftharpoons H_2 + 2\,OH^-$	-0.8277
$Bi_2O_3 + 3\,H_2O + 6\,e \rightleftharpoons 2\,Bi + 6\,OH^-$	-0.46	$2\,NO_2^- + 2\,H_2O + 2\,e \rightleftharpoons N_2O_4 + 4\,OH^-$	-0.85
$NO_2^- + H_2O + e \rightleftharpoons NO + 2\,OH$	-0.46	$H_3BO_3 + 3\,H^+ + 3\,e \rightleftharpoons B + 3\,H_2O$	-0.8698
$PbHPO_4 + 2\,e \rightleftharpoons Pb + HPO_4^{2-}$	-0.465	$P + 3\,H_2O + 3\,e \rightleftharpoons PH_3(g) + 3\,OH^-$	-0.87
$S + 2\,e \rightleftharpoons S^{2-}$	-0.47627	$Ti^{3+} + e \rightleftharpoons Ti^{2+}$	-0.9
$S + H_2O + 2\,e \rightleftharpoons HS^- + OH^-$	-0.478	$HSnO_2^- + H_2O + 2\,e \rightleftharpoons Sn + 3\,OH^-$	-0.909
$B(OH)_3 + 7\,H^+ + 8\,e \rightleftharpoons BH_4^- + 3\,H_2O$	-0.481	$Cr^{2+} + 2\,e \rightleftharpoons Cr$	-0.913
$In^{3+} + e \rightleftharpoons In^{2+}$	-0.49	$Se + 2\,e \rightleftharpoons Se^{2-}$	-0.924
$ZnOH^+ + H^+ + 2\,e \rightleftharpoons Zn + H_2O$	-0.497	$SO_4^{2-} + H_2O + 2\,e \rightleftharpoons SO_3^{2-} + 2\,OH^-$	-0.93
$GaOH^{2+} + H^+ + 3\,e \rightleftharpoons Ga + H_2O$	-0.498	$Sn(OH)_6^{2-} + 2\,e \rightleftharpoons HSnO_2^- + 3\,OH^- + H_2O$	-0.93
$H_3PO_3 + 2\,H^+ + 2\,e \rightleftharpoons H_3PO_2 + H_2O$	-0.499	$SnO_2 + 2\,H_2O + 4\,e \rightleftharpoons Sn + 4\,OH^-$	-0.945
$TiO_2 + 4\,H^+ + 2\,e \rightleftharpoons Ti^{2+} + 2\,H_2O$	-0.502	$In(OH)_3 + 3\,e \rightleftharpoons In + 3\,OH^-$	-0.99
$H_3PO_2 + H^+ + e \rightleftharpoons P + 2\,H_2O$	-0.508	$NpO_2 + H_2O + H^+ + e \rightleftharpoons Np(OH)_3$	-0.962
$Sb + 3\,H^+ + 3\,e \rightleftharpoons SbH_3$	-0.510	$In(OH)_4^- + 3\,e \rightleftharpoons In + 4\,OH^-$	-1.007
$HPbO_2^- + H_2O + 2\,e \rightleftharpoons Pb + 3\,OH^-$	-0.537	$In_2O_3 + 3\,H_2O + 6\,e \rightleftharpoons 2\,In + 6\,OH^-$	-1.034
$Ga^{3+} + 3\,e \rightleftharpoons Ga$	-0.549	$PO_4^{3-} + 2\,H_2O + 2\,e \rightleftharpoons HPO_3^{2-} + 3\,OH^-$	-1.05
$TlCl + e \rightleftharpoons Tl + Cl^-$	-0.5568	$Yb^{3+} + e \rightleftharpoons Yb^{2+}$	-1.05
$Fe(OH)_3 + e \rightleftharpoons Fe(OH)_2 + OH^-$	-0.56	$Nb^{3+} + 3\,e \rightleftharpoons Nb$	-1.099
$TeO_3^{2-} + 3\,H_2O + 4\,e \rightleftharpoons Te + 6\,OH^-$	-0.57	$Fm^{3+} + e \rightleftharpoons Fm^{2+}$	-1.1
$2\,SO_3^{2-} + 3\,H_2O + 4\,e \rightleftharpoons S_2O_3^{2-} + 6\,OH^-$	-0.571	$2\,SO_3^{2-} + 2\,H_2O + 2\,e \rightleftharpoons S_2O_4^{2-} + 4\,OH^-$	-1.12
$PbO + H_2O + 2\,e \rightleftharpoons Pb + 2\,OH^-$	-0.580	$Te + 2\,e \rightleftharpoons Te^{2-}$	-1.143
$ReO_2 + 4\,H_2O + 7\,e \rightleftharpoons Re + 8\,OH^-$	-0.584	$V^{2+} + 2\,e \rightleftharpoons V$	-1.175
$SbO_3^- + H_2O + 2\,e \rightleftharpoons SbO_2^- + 2\,OH^-$	-0.59	$Mn^{2+} + 2\,e \rightleftharpoons Mn$	-1.185
$Ta^{3+} + 3\,e \rightleftharpoons Ta$	-0.6	$Zn(OH)_4^{2-} + 2\,e \rightleftharpoons Zn + 4\,OH^-$	-1.199
$U^{4+} + e \rightleftharpoons U^{3+}$	-0.607	$CrO_2 + 2\,H_2O + 3\,e \rightleftharpoons Cr + 4\,OH^-$	-1.2
$As + 3\,H^+ + 3\,e \rightleftharpoons AsH_3$	-0.608	$No^{3+} + 3\,e \rightleftharpoons No$	-1.20
$Nb_2O_5 + 10\,H^+ + 10\,e \rightleftharpoons 2\,Nb + 5\,H_2O$	-0.644	$ZnO_2^- + 2\,H_2O + 2\,e \rightleftharpoons Zn + 4\,OH^-$	-1.215
$NbO_2 + 2\,H^+ + 2\,e \rightleftharpoons NbO + H_2O$	-0.646	$H_2GaO_3^- + H_2O + 3\,e \rightleftharpoons Ga + 4\,OH^-$	-1.219
$Cd(OH)_4^{2-} + 2\,e \rightleftharpoons Cd + 4\,OH^-$	-0.658	$H_2BO_3^- + 5\,H_2O + 8\,e \rightleftharpoons BH_4^- + 8\,OH^-$	-1.24
$TlBr + e \rightleftharpoons Tl + Br^-$	-0.658	$SiF_6^{2-} + 4\,e \rightleftharpoons Si + 6\,F^-$	-1.24
$SbO_2^- + 2\,H_2O + 3\,e \rightleftharpoons Sb + 4\,OH^-$	-0.66	$Zn(OH)_2 + 2\,e \rightleftharpoons Zn + 2\,OH^-$	-1.249
$AsO_2^- + 2\,H_2O + 3\,e \rightleftharpoons As + 4\,OH^-$	-0.68	$ZnO + H_2O + 2\,e \rightleftharpoons Zn + 2\,OH^-$	-1.260
$NbO_2 + 4\,H^+ + 4\,e \rightleftharpoons Nb + 2\,H_2O$	-0.690	$Es^{3+} + e \rightleftharpoons Es^{2+}$	-1.3
$Ag_2S + 2\,e \rightleftharpoons 2\,Ag + S^{2-}$	-0.691	$Pa^{3+} + 3\,e \rightleftharpoons Pa$	-1.34
$AsO_4^{3-} + 2\,H_2O + 2\,e \rightleftharpoons AsO_2^- + 4\,OH^-$	-0.71	$Ti^{3+} + 3\,e \rightleftharpoons Ti$	-1.37
$Ni(OH)_2 + 2\,e \rightleftharpoons Ni + 2\,OH^-$	-0.72	$Ce^{3+} + 3\,e \rightleftharpoons Ce(Hg)$	-1.4373
$Co(OH)_2 + 2\,e \rightleftharpoons Co + 2\,OH^-$	-0.73	$UO_2^{2+} + 4\,H^+ + 6\,e \rightleftharpoons U + 2\,H_2O$	-1.444
$NbO + 2\,H^+ + 2\,e \rightleftharpoons Nb + H_2O$	-0.733	$Zr^{4+} + 4\,e \rightleftharpoons Zr$	-1.45
$H_2SeO_3 + 4\,H^+ + 4\,e \rightleftharpoons Se + 3\,H_2O$	-0.74	$Cr(OH)_3 + 3\,e \rightleftharpoons Cr + 3\,OH^-$	-1.48
$Cr^{3+} + 3\,e \rightleftharpoons Cr$	-0.744	$Pa^{4+} + 4\,e \rightleftharpoons Pa$	-1.49
$Ta_2O_5 + 10\,H^+ + 10\,e \rightleftharpoons 2\,Ta + 5\,H_2O$	-0.750	$HfO_2 + 4\,H^+ + 4\,e \rightleftharpoons Hf + 2\,H_2O$	-1.505
$TlI + e \rightleftharpoons Tl + I^-$	-0.752	$Hf^{4+} + 4\,e \rightleftharpoons Hf$	-1.55
$Zn^{2+} + 2\,e \rightleftharpoons Zn$	-0.7618	$Sm^{3+} + e \rightleftharpoons Sm^{2+}$	-1.55
$Zn^{2+} + 2\,e \rightleftharpoons Zn(Hg)$	-0.7628	$ZrO_2 + 4\,H^+ + 4\,e \rightleftharpoons Zr + 2\,H_2O$	-1.553
$CdO + H_2O + 2\,e \rightleftharpoons Cd + 2\,OH^-$	-0.783	$Mn(OH)_2 + 2\,e \rightleftharpoons Mn + 2\,OH^-$	-1.56
$Te + 2\,H^+ + 2\,e \rightleftharpoons H_2Te$	-0.793	$Ba^{2+} + 2\,e \rightleftharpoons Ba(Hg)$	-1.570

Reaction	$E°/V$	Reaction	$E°/V$
$Bk^{2+} + 2\,e \rightleftharpoons Bk$	−1.6	$Al(OH)_3 + 3\,e \rightleftharpoons Al + 3\,OH^-$	−2.31
$Cf^{3+} + e \rightleftharpoons Cf^{2+}$	−1.6	$Tm^{3+} + 3\,e \rightleftharpoons Tm$	−2.319
$Ti^{2+} + 2\,e \rightleftharpoons Ti$	−1.630	$Nd^{3+} + 3\,e \rightleftharpoons Nd$	−2.323
$Md^{3+} + 3\,e \rightleftharpoons Md$	−1.65	$Al(OH)^- + 3\,e \rightleftharpoons Al + 4\,OH^-$	−2.328
$HPO_3^{2-} + 2\,H_2O + 2\,e \rightleftharpoons H_2PO_2^- + 3\,OH^-$	−1.65	$H_2AlO_3^- + H_2O + 3\,e \rightleftharpoons Al + 4\,OH^-$	−2.33
$Al^{3+} + 3\,e \rightleftharpoons Al$	−1.662	$Ho^{3+} + 3\,e \rightleftharpoons Ho$	−2.33
$SiO_3^{2-} + H_2O + 4\,e \rightleftharpoons Si + 6\,OH^-$	−1.697	$Er^{3+} + 3\,e \rightleftharpoons Er$	−2.331
$HPO_3^{2-} + 2\,H_2O + 3\,e \rightleftharpoons P + 5\,OH^-$	−1.71	$Ce^{3+} + 3\,e \rightleftharpoons Ce$	−2.336
$HfO^{2+} + 2\,H^+ + 4\,e \rightleftharpoons Hf + H_2O$	−1.724	$Pr^{3+} + 3\,e \rightleftharpoons Pr$	−2.353
$ThO_2 + 4\,H^+ + 4\,e \rightleftharpoons Th + 2\,H_2O$	−1.789	$ZrO(OH)_2 + H_2O + 4\,e \rightleftharpoons Zr + 4\,OH^-$	−2.36
$H_2BO_3^- + H_2O + 3\,e \rightleftharpoons B + 4\,OH^-$	−1.79	$Mg^{2+} + 2\,e \rightleftharpoons Mg$	−2.372
$Sr^{2+} + 2\,e \rightleftharpoons Sr(Hg)$	−1.793	$Y^{3+} + 3\,e \rightleftharpoons Y$	−2.372
$U^{3+} + 3\,e \rightleftharpoons U$	−1.798	$La^{3+} + 3\,e \rightleftharpoons La$	−2.379
$H_2PO_2^- + e \rightleftharpoons P + 2\,OH^-$	−1.82	$Tm^{2+} + 2\,e \rightleftharpoons Tm$	−2.4
$Be^{2+} + 2\,e \rightleftharpoons Be$	−1.847	$Md^{2+} + 2\,e \rightleftharpoons Md$	−2.40
$Np^{3+} + 3\,e \rightleftharpoons Np$	−1.856	$Th(OH)_4 + 4\,e \rightleftharpoons Th + 4\,OH^-$	−2.48
$Fm^{3+} + 3\,e \rightleftharpoons Fm$	−1.89	$HfO(OH)_2 + H_2O + 4\,e \rightleftharpoons Hf + 4\,OH^-$	−2.50
$Th^{4+} + 4\,e \rightleftharpoons Th$	−1.899	$No^{2+} + 2\,e \rightleftharpoons No$	−2.50
$Am^{2+} + 2\,e \rightleftharpoons Am$	−1.9	$Dy^{3+} + e \rightleftharpoons Dy^{2+}$	−2.6
$Pa^{4+} + e \rightleftharpoons Pa^{3+}$	−1.9	$Pm^{3+} + e \rightleftharpoons Pm^{2+}$	−2.6
$Es^{3+} + 3\,e \rightleftharpoons Es$	−1.91	$Be_2O_3^{2-} + 3\,H_2O + 4\,e \rightleftharpoons 2\,Be + 6\,OH^-$	−2.63
$Cf^{3+} + 3\,e \rightleftharpoons Cf$	−1.94	$Sm^{2+} + 2\,e \rightleftharpoons Sm$	−2.68
$Lr^{3+} + 3\,e \rightleftharpoons Lr$	−1.96	$Mg(OH)_2 + 2\,e \rightleftharpoons Mg + 2\,OH^-$	−2.690
$Eu^{3+} + 3\,e \rightleftharpoons Eu$	−1.991	$Nd^{3+} + e \rightleftharpoons Nd^{2+}$	−2.7
$Er^{2+} + 2\,e \rightleftharpoons Er$	−2.0	$Mg^+ + e \rightleftharpoons Mg$	−2.70
$Pr^{2+} + 2\,e \rightleftharpoons Pr$	−2.0	$Na^+ + e \rightleftharpoons Na$	−2.71
$Pu^{3+} + 3\,e \rightleftharpoons Pu$	−2.031	$Yb^{2+} + 2\,e \rightleftharpoons Yb$	−2.76
$Cm^{3+} + 3\,e \rightleftharpoons Cm$	−2.04	$Bk^{3+} + e \rightleftharpoons Bk^{2+}$	−2.8
$Am^{3+} + 3\,e \rightleftharpoons Am$	−2.048	$Ho^{3+} + e \rightleftharpoons Ho^{2+}$	−2.8
$AlF_6^{3-} + 3\,e \rightleftharpoons Al + 6\,F^-$	−2.069	$Ra^{2+} + 2\,e \rightleftharpoons Ra$	−2.8
$Sc^{3+} + 3\,e \rightleftharpoons Sc$	−2.077	$Eu^{2+} + 2\,e \rightleftharpoons Eu$	−2.812
$Ho^{2+} + 2\,e \rightleftharpoons Ho$	−2.1	$Ca^{2+} + 2\,e \rightleftharpoons Ca$	−2.868
$Nd^{2+} + 2\,e \rightleftharpoons Nd$	−2.1	$Sr(OH)_2 + 2\,e \rightleftharpoons Sr + 2\,OH^-$	−2.88
$Cf^{2+} + 2\,e \rightleftharpoons Cf$	−2.12	$Sr^{2+} + 2\,e \rightleftharpoons Sr$	−2.899
$Yb^{3+} + 3\,e \rightleftharpoons Yb$	−2.19	$Fr^+ + e \rightleftharpoons Fr$	−2.9
$Ac^{3+} + 3\,e \rightleftharpoons Ac$	−2.20	$La(OH)_3 + 3\,e \rightleftharpoons La + 3\,OH^-$	−2.90
$Dy^{2+} + 2\,e \rightleftharpoons Dy$	−2.2	$Ba^{2+} + 2\,e \rightleftharpoons Ba$	−2.912
$Tm^{3+} + e \rightleftharpoons Tm^{2+}$	−2.2	$K^+ + e \rightleftharpoons K$	−2.931
$Pm^{2+} + 2\,e \rightleftharpoons Pm$	−2.2	$Rb^+ + e \rightleftharpoons Rb$	−2.98
$Es^{2+} + 2\,e \rightleftharpoons Es$	−2.23	$Ba(OH)_2 + 2\,e \rightleftharpoons Ba + 2\,OH^-$	−2.99
$H_2 + 2\,e \rightleftharpoons 2\,H^-$	−2.23	$Er^{3+} + e \rightleftharpoons Er^{2+}$	−3.0
$Gd^{3+} + 3\,e \rightleftharpoons Gd$	−2.279	$Ca(OH)_2 + 2\,e \rightleftharpoons Ca + 2\,OH^-$	−3.02
$Tb^{3+} + 3\,e \rightleftharpoons Tb$	−2.28	$Cs^+ + e \rightleftharpoons Cs$	−3.026
$Lu^{3+} + 3\,e \rightleftharpoons Lu$	−2.28	$Li^+ + e \rightleftharpoons Li$	−3.0401
$Dy^{3+} + 3\,e \rightleftharpoons Dy$	−2.295	$3\,N_2 + 2\,H^+ + 2\,e \rightleftharpoons 2\,HN_3$	−3.09
$Am^{3+} + e \rightleftharpoons Am^{2+}$	−2.3	$Pr^{3+} + e \rightleftharpoons Pr^{2+}$	−3.1
$Fm^{2+} + 2\,e \rightleftharpoons Fm$	−2.30	$Ca^+ + e \rightleftharpoons Ca$	−3.80
$Pm^{3+} + 3\,e \rightleftharpoons Pm$	−2.30	$Sr^+ + e \rightleftharpoons Sr$	−4.10
$Sm^{3+} + 3\,e \rightleftharpoons Sm$	−2.304		

REDUCTION AND OXIDATION POTENTIALS FOR CERTAIN ION RADICALS

Petr Vanýsek

There are two tables for ion radicals. The first table lists reduction potentials for organic compounds which produce anion radicals during reduction, a process described as $A + e^- \rightleftharpoons A^-$. The second table lists oxidation potentials for organic compounds which produce cation radicals during oxidation, a process described as $A \rightleftharpoons A^{+\cdot} + e^-$. To obtain reduction potential for a reverse reaction, the sign for the potential is changed.

Unlike the table of the Electrochemical Series, which lists *standard* potentials, values for radicals are experimental values with experimental conditions given in the second column. Since the measurements leading to potentials for ion radicals are very dependent on conditions, an attempt to report standard potentials for radicals would serve no useful purpose. For the same reason, the potentials are also reported as experimental values, usually a half-wave potential ($E_{1/2}$ in polarography) or a peak potential (E_p in cyclic voltammetry). Unless otherwise stated, the values are reported vs. SCE (saturated calomel electrode). To obtain a value vs.

normal hydrogen electrode, 0.241 V has to be added to the SCE values. All the ion radicals chosen for inclusion in the tables result from electrochemically reversible reactions. More detailed data on ion radicals can be found in the *Encyclopedia of Electrochemistry of Elements*, (A. J. Bard, Ed.), Vol. XI and XII in particular, Marcel Dekker, New York, 1978.

Abbreviations are: CV — cyclic voltammetry; DMF — *N,N*-Dimethylformamide; E swp — potential sweep; $E°$ — standard potential; E_p — peak potential; $E_{p/2}$ — half-peak potential; $E_{1/2}$ — half wave potential; M — mol/L; MeCN — acetonitrile; pol — polarography; rot Pt dsk — rotated Pt disk; SCE — saturated calomel electrode; TBABF$_4$ — tetrabutylammonium tetrafluoroborate; TBAI — tetrabutylammonium iodide; TBAP — tetrabutylammonium perchlorate; TEABr — tetraethylammonium bromide; TEAP — tetraethylammonium perchlorate; THF — tetrahydrofuran; TPACF$_3$SO$_3$ — tetrapropylammonium trifluoromethanesulfite; TPAP — tetrapropylammonium perchlorate; and wr — wire.

Reduction Potentials (Products Are Anion Radicals)

Substance	Conditions/electrode/technique	Potential V (vs. SCE)
Acetone	DMF, 0.1 M TEABr/Hg/pol	$E_{1/2} = -2.84$
1-Naphthyphenylacetylene	DMF, 0.03 M TBAI/Hg/pol	$E_{1/2} = -1.91$
1-Naphthalenecarboxyaldehyde	-/Hg/pol	$E_{1/2} = -0.91$
2-Naphthalenecarboxyaldehyde	-/Hg/pol	$E_{1/2} = -0.96$
2-Phenanthrenecarboxaldehyde	-/Hg/pol	$E_{1/2} = -1.00$
3-Phenanthrenecarboxaldehyde	-/Hg/pol	$E_{1/2} = -0.94$
9-Phenanthrenecarboxaldehyde	-/Hg/pol	$E_{1/2} = -0.83$
1-Anthracenecarboxaldehyde	-/Hg/pol	$E_{1/2} = -0.75$
1-Pyrenecarboxaldehyde	-/Hg/pol	$E_{1/2} = -0.76$
2-Pyrenecarboxaldehyde	-/Hg/pol	$E_{1/2} = -1.00$
Anthracene	DMF, 0.1 M TBAP/Pt dsk/CV	$E_p = -2.00$
	DMF, 0.5 M TBABF$_4$/Hg/CV	$E_{1/2} = -1.93$
	MeCN, 0.1 M TEAP/Hg/CV	$E_{1/2} = -2.07$
	DMF, 0.1 M TBAI/Hg/pol	$E_{1/2} = -1.92$
9,10-Dimethylanthracene	DMF, 0.1 M TBAP/Pt/CV	$E_p = -2.08$
	MeCN, 0.1 M TBAP/Pt/CV	$E_p = -2.10$
1-Phenylanthracene	DMF, 0.5 M TBABF$_2$/Hg/CV	$E_{1/2} = -1.91$
	DMF, 0.1 M TBAI/Hg/pol	$E_{1/2} = -1.878$
2-Phenylanthracene	DMF, 0.1 M TBAI/Hg/pol	$E_{1/2} = -1.875$
8-Phenylanthracene	DMF, 0.5 M TBABF$_4$/Hg/CV	$E_{1/2} = -1.91$
9-Phenylanthracene	DMF, 0.5 M TBABF$_4$/Hg/CV	$E_{1/2} = -1.93$
	DMF, 0.1 M TBAI/Hg/pol	$E_{1/2} = -1.863$
1,8-Diphenylanthracene	DMF, 0.5 M TBABF$_4$/Hg/CV	$E_{1/2} = -1.88$
1,9-Diphenylanthracene	DMF, 0.1 M TBAI/Hg/pol	$E_{1/2} = -1.846$
1,10-Diphenylanthracene	DMF, 0.1 M TBAI/Hg/pol	$E_{1/2} = -1.786$
8,9-Diphenylanthracene	DMF, 0.5 M TBABF$_4$/Hg/CV	$E_{1/2} = -1.90$
9,10-Diphenylanthracene	MeCN, 0.1 M TBAP/rot Pt/E swp	$E_{1/2} = -1.83$
	DMF, 0.1 M TBAI/Hg/pol	$E_{1/2} = -1.835$
1,8,9-Triphenylanthracene	DMF, 0.5 M TBABF$_4$/Hg/CV	$E_{1/2} = -1.85$
1,8,10-Triphenylanthracene	DMF, 0.5 M TBABF$_4$/Hg/CV	$E_{1/2} = -1.81$
9,10-Dibiphenylanthracene	MeCN, 0.1 M TBAP/rot Pt/E swp	$E_{1/2} = -1.94$
Benz(a)anthracene	MeCN, 0.1 M TEAP/Hg/CV	$E_{1/2} = -2.11$
	MeCN, 0.1 M TEAP/Hg/pol	$E_{1/2} = -2.40^a$
Azulene	DMF, 0.1 M TBAI/Hg/pol	$E_{1/2} = -1.10^c$
Annulene	DMF, 0.5 M TBAP 0°C/Hg/pol	$E_{1/2} = -1.23$
Benzaldehyde	DMF, 0.1 M TBAP/Hg/pol	$E_{1/2} = -1.67$
Benzil	DMSO, 0.1 M TBAP/Hg/pol	$E_{1/2} = -1.04$
Benzophenone	-/Hg/pol	$E_{1/2} = -1.80$
	DMF/Pt dsk/CV	$E° = -1.72$
Chrysene	MeCN, 0.1 M TEAP/Hg/pol	$E_{1/2} = -2.73^a$
Fluoranthrene	DMF, 0.1 M TBAP/Pt dsk/CV	$E_p = -1.76$
Cyclohexanone	DMF, 0.1 M TEABr/Hg/pol	$E_{1/2} = -2.79$

Substance	Conditions/electrode/technique	Potential V (vs. SCE)
5,5-Dimethyl-3-phenyl-2-cyclohexen-1-one	DMF, 0.5 M/Hg/pol	$E_{1/2} = -1.71$
1,2,3-Indanetrione hydrate (ninhydrin)	DMF, 0.2 M NaNO$_3$/Hg/pol	$E_{1/2} = -0.039$
Naphthacene	DMF, 0.1 M TBAI/Hg/pol	$E_{1/2} = -1.53$
Naphthalene	DMF, 0.1 M TBAP/Pt dsk/CV	$E_p = -2.55$
	DMF, 0.5 M TBABF$_4$/Hg/CV	$E_{1/2} = -2.56$
	DMF, MeCN, 0.1 M TEAP/Hg/CV	$E_{1/2} = -2.63$
	DMF, 0.1 M TBAI/Hg/pol	$E_{1/2} = -2.50$
1-Phenylnaphthalene	DMF, 0.5 M TBABF$_4$/Hg/CV	$E_{1/2} = -2.36$
1,2-Diphenylnaphthalene	DMF, 0.5 M TBABF$_4$/Hg/CV	$E_{1/2} = -2.25$
Cyclopentanone	DMF, 0.1 M TEABr/Hg/pol	$E_{1/2} = -2.82$
Phenanthrene	MeCN, 0.1 M TBAP/Pt wr/CV	$E_{1/2} = -2.47$
	MeCN, 0.1 M TEAP/Hg/pol	$E_{1/2} = -2.88^a$
Pentacene	THF, 0.1 M TBAP/rot Pt dsk/E swp	$E_{1/2} = -1.40$
Perylene	MeCN, 0.1 M TEAP/Hg/CV	$E_{1/2} = -1.73$
1,3-Diphenyl-1,3-propanedione	DMSO, 0.2 M TBAP/Hg/CV	$E_{1/2} = -1.42$
2,2-Dimethyl-1,3-diphenyl-1,3 propanedione	DMSO, TBAP/Hg/CV	$E_{1/2} = -1.80$
Pyrene	DMF, 0.1 M TBAP/Pt/CV	$E_p = -2.14$
	MeCN, 0.1 M TEAP/Hg/pol	$E_{1/2} = -2.49^a$
Diphenylsulfone	DMF, TEABr	$E_{1/2} = -2.16$
Triphenylene	MeCN, 0.1 M TEAP/Hg/pol	$E_{1/2} = -2.87^a$
9,10-Anthraquinone	DMF, 0.5 M TBAP, 20°/Pt dsk/CV	$E_{1/2} = -1.01$
1,4-Benzoquinone	MeCN, 0.1 M TEAP/Pt/CV	$E_p = -0.54$
1,4-Naphthohydroquinone, dipotassium salt	DMF, 0.5 M TBAP, 20°/Pt dsk/CV	$E_{1/2} = -1.55$
Rubrene	DMF, 0.1 M TBAP/Pt dsk/CV	$E_p = -1.48$
	DMF, 0.1 M TBAI/Hg/pol	$E_{1/2} = -1.410$
Benzocyclooctatetraene	THF, 0.1 M TBAP/Hg/pol	$E_{1/2} = -2.13$
sym-Dibenzocyclooctatetraene	THF, 0.1 M TBAP/Hg/pol	$E_{1/2} = -2.29$
Ubiquinone-6	MeCN, 0.1 M TEAP/Pt/CV	$E_p = -1.05^e$
(9-Phenyl-fluorenyl)$^+$	10.2 M H$_2$SO$_4$/Hg/CV	$E_p = -0.01^b$
(Triphenylcyclopropenyl)$^+$	MeCN, 0.1 M TEAP/Hg/CV	$E_p = -1.87$
(Triphenylmethyl)$^+$	MeCN, 0.1 M TBAP/Hg/pol	$E_{1/2} = 0.27$
	H$_2$SO$_4$, 10.2 M/Hg/CV	$E_p = -0.58^b$
(Tribiphenylmethyl)$^+$	MeCN, 0.1 M TBAP/Hg/pol	$E_{1/2} = 0.19$
(Tri-4-t-butyl-5-phenylmethyl)$^+$	MeCN, 0.1 M TBAP/Hg/pol	$E_{1/2} = 0.13$
(Tri-4-isopropylphenylmethyl)$^+$	MeCN, 0.1 M TBAP/Hg/pol	$E_{1/2} = 0.07$
(Tri-4-methylphenylmethyl)$^+$	MeCN, 0.1 M TBAP/Hg/pol	$E_{1/2} = 0.05$
(Tri-4-cyclopropylphenylmethyl)$^+$	MeCN, 0.1 M TBAP/Hg/pol	$E_{1/2} = 0.01$
(Tropylium)$^+$	MeCN, 0.1 M TBAP/Hg/pol	$E_{1/2} = -0.17$
	DMF, 0.15 M TBAI/Hg/pol	$E_{1/2} = -1.55$
	DMF, 0.15 M TBAI/Hg/pol	$E_{1/2} = -1.55$
	DMF, 0.15 M TBAI/Hg/pol	$E_{1/2} = -1.57$
	DMF, 0.15 M TBAI/Hg/pol	$E_{1/2} = -1.60$
	DMF, 0.15 M TBAI/Hg/pol	$E_{1/2} = -1.87$
	DMF, 0.15 M TBAI/Hg/pol	$E_{1/2} = -1.96$
	DMF, 0.15 M TBAI/Hg/pol	$E_{1/2} = -2.05$

Oxidation Potentials (Products Are Cation Radicals)

Substance	Conditions/electrode/technique	Potential V (vs. SCE)
Anthracene	CH$_2$Cl$_2$, 0.2 M TBABF$_4$, −70°C/Pt dsk/CV	$E_p = +0.73^d$
9,10-Dimethylanthracene	MeCN, 0.1 M LiClO$_4$/Pt wr/CV	$E_p = +1.0$
9,10-Dipropylanthracene	MeCN, 0.1 M TEAP/Pt/CV	$E_p = +1.08$
1,8-Diphenylanthracene	CH$_2$Cl$_2$, 0.2 M TPrACF$_3$SO$_3$/rot Pt wr/E swp	$E_{1/2} = +1.34$
8,9-Diphenylanthracene	CH$_2$Cl$_2$, 0.2 M TPrACF$_3$SO$_3$/rot Pt wr/E swp	$E_{1/2} = +1.30$
9,10-Diphenylanthracene	MeCN/Pt/CV	$E_p = +1.22$
Perylene	MeCN, 0.1 M TBAP/Pt/CV	$E_p = +1.34$
Pyrene	DMF, 0.1 M TBAP/Pt dsk/CV	$E_p = +1.25$
Rubrene	DMF, 0.1 M TBAP/Pt dsk/CV	$E_p = +1.10$
Tetracene	CH$_2$Cl$_2$, 0.2 M TBABF$_4$, −70°C/Pt wr/CV	$E_p = +0.35^d$
1,4-Dithiabenzene	MeCN, 0.1 M TEAP/Pt dsk/rot	$E_{1/2} = +0.69$
1,4-Dithianaphthalene	MeCN, 0.1 M TEAP/Pt dsk/rot	$E_{1/2} = +0.80$
Thianthrene	0.1 M TPAP/Pt/CV	$E_{1/2} = +1.28$

^{a} vs 0.01 M Ag/AgClO$_4$
^{b} vs. Hg/Hg$_2$SO$_4$, 17 M H$_2$SO$_4$
^{c} vs Hg pool
^{d} vs Ag/saturated AgNO$_3$
^{e} vs Ag/0.01 M Ag+

pH SCALE FOR AQUEOUS SOLUTIONS

A.K. Covington

A Working Party of IUPAC, after extensive considerations over five years, has recently produced a report (1) which sets pH firmly within the International System of Units (SI). A summary of these important developments is given below.

The concept of pH is unique amongst the commonly encountered physicochemical quantities in that, in terms of its definition,

$$pH = -\lg a_H \qquad (1)$$

it involves a single ion quantity, the activity of the hydrogen ion, which is immeasurable by any thermodynamically valid method and requires a convention for its evaluation.

pH was originally defined by Sørensen (2) in terms of the concentration of hydrogen ions (in modern nomenclature) as pH = $-\lg (c_H/c^o)$ where c_H is the hydrogen ion concentration in mol dm^{-3}, and $c^o = 1$ mol dm^{-3} is the standard amount concentration. Subsequently (3), it was accepted as more satisfactory to define pH in terms of the relative activity of hydrogen ions in solution

$$pH = -\lg a_H = -\lg (m_H \gamma_H / m^o) \qquad (2)$$

where a_H is the relative (molality basis) activity and γ_H is the molal activity coefficient of the hydrogen ion H$^+$ at the molality m_H, and m^o the standard molality. The quantity pH is intended to be a measure of the activity of hydrogen ions in solution. However, since it is defined in terms of a quantity that cannot be measured by a thermodynamically valid method, eqn.(2) can only be considered a *notional definition* of pH.

pH being a single ion quantity, it is not determinable in terms of a fundamental (or base) unit of any measurement system, and there is difficulty providing a proper basis for the traceability of pH measurements. A satisfactory approach is now available in that pH determinations can be incorporated into the International System (SI) if they can be traced to measurements made using a method that fulfils the definition of a 'primary method of measurement' (4).

The essential feature of a primary method is that it must operate according to a well-defined measurement equation in which all of the variables can be determined experimentally in terms of SI units. Any limitation in the determination of the experimental variables, or in the theory, must be included within the estimated uncertainty of the method if traceability to the SI is to be established. If a convention were used without an estimate of its uncertainty, true traceability to SI would not be established. The electrochemical cell without liquid junction, known as the Harned cell (5), fulfils the definition of a primary method for the measurement of the acidity function, p($a_H \gamma_{Cl}$), and subsequently of the pH of buffer solutions.

The Harned cell is written as

$$Pt \mid H_2 \mid buffer\ S,\ Cl^- \mid AgCl \mid Ag \qquad Cell\ I$$

and contains a standard buffer, S, with chloride ions, as potassium or sodium chloride, added in order to use the silver-silver chloride electrode as reference electrode. The application of the Nernst equation to the spontaneous cell reaction of Cell I:

$$\tfrac{1}{2}\ H_2 + AgCl \rightarrow Ag(s) + H^+ + Cl^-$$

yields the potential difference E_I of the cell (corrected to 1 atm (101.325 kPa), the partial pressure of hydrogen gas used in electrochemistry in preference to 100 kPa) as

$$E_I = E^o - (RT/F)\ln 10\ \lg [(m_H\gamma_H/m^o)(m_{Cl}\gamma_{Cl}/m^o)] \qquad (3)$$

which can be rearranged, since $a_H = m_H\gamma_H/m^o$, to give the acidity function

$$p(a_H\gamma_{Cl}) = -\lg(a_H\gamma_{Cl}) = (E_I - E^o)/[(RT/F)\ln10] + \lg(m_{Cl}/m^o) \qquad (4)$$

where E^o is the standard potential difference of the cell, and hence of the silver-silver chloride electrode, and γ_{Cl} is the activity coefficient of the chloride ion.

The standard potential difference of the silver/silver chloride electrode, E^o, is determined from a Harned cell in which only HCl is present at a fixed molality (e.g. $m = 0.01$ mol kg^{-1})

$$Pt \mid H_2 \mid HCl\ (m) \mid AgCl \mid Ag \qquad Cell\ Ia$$

The application of the Nernst equation to the HCl cell (Ia) gives

$$E_{Ia} = E^o - (2RT/F)\ln 10\ \lg[(m_{HCl}/m^o)(\gamma_{\pm HCl})] \qquad (5)$$

where E_{Ia} has been corrected to 1 atmosphere partial pressure of hydrogen gas (101.325 kPa) and $\gamma_{\pm HCl}$ is the mean ionic activity coefficient of HCl.

Values of the activity coefficient ($\gamma_{\pm HCl}$) at molality 0.01 mol kg^{-1} and various temperatures were given by Bates and Robinson (6). The standard potential difference depends on the method of preparation of the electrodes, but individual determinations of the activity coefficient of HCl at 0.01 mol kg^{-1} are more uniform than values of E^o. Hence the practical determination of the potential difference of the cell with HCl at 0.01 mol kg^{-1} is recommended at 298.15 K at which the mean ionic activity coefficient is 0.904. (It is unnecessary to repeat the measurement of E^o at other temperatures but simply to correct published smoothed values by the observed difference in E^o at 298.15 K)

In national metrology institutes (NMIs), measurements of Cells I and Ia are often done simultaneously in a thermostat bath. Subtracting eqn.(5) from (3) gives

$$\Delta E = E_I - E_{Ia} = -(RT/F)\ln 10\{\lg[(m_H\gamma_H/m^o)(m_{Cl}\gamma_{Cl}/m^o)] - \lg[(m_{HCl}/m^o)^2\gamma^2_{\pm HCl}]\} \qquad (6)$$

which is independent of the standard potential difference. Therefore, the subsequently calculated pH does not depend on the standard potential difference and hence does not depend on the assumption that the standard potential of the hydrogen electrode is zero at all temperatures. Therefore, the Harned cell gives an exact comparison between hydrogen ion activities at different temperatures.

The quantity p($a_H\gamma_{Cl}$) = $-\lg (a_H\gamma_{Cl})$, on the left hand side of (4), is called the acidity function (5). To obtain the quantity pH according to eqn. (2) from the acidity function, it is necessary to evaluate $\lg \gamma_{Cl}$ independently. This is done in two steps: (i) the

ue of lg $(a_H\gamma_{Cl})$ at zero chloride molality, lg $(a_H\gamma_{Cl})^\circ$, is evaluated
 (ii) a value for the activity of the chloride ion γ°_{Cl}, at zero chlo-
e molality (sometimes referred to as the limiting or 'trace'
vity coefficient) is calculated using the Bates-Guggenheim
vention (7). The value of lg $(a_H\gamma_{Cl})^\circ$ corresponding to zero
oride molality is determined by linear extrapolation of mea-
ements using Harned cells with at least three added molalities
odium or potassium chloride ($I < 0.1$ mol kg^{-1}).
he value of lg $(a_H\gamma_{Cl})^\circ$ corresponding to zero chloride molality
determined by linear extrapolation of measurements using
ned cells with at least three added molalities of sodium or
assium chloride ($I < 0.1$ mol kg^{-1}) in accord with eqn. (7):

$$-lg\,(a_H\gamma_{Cl}) = -lg\,(a_H\gamma_{Cl})^\circ + Sm_{Cl} \tag{7}$$

re S is an empirical, temperature dependent, constant.
he Bates-Guggenheim convention (7) assumes that the trace
vity coefficient of the chloride ion γ°_{Cl} is given by

$$lg\,\gamma^\circ_{Cl} = -A\,I^{1/2}\,/\,(\,1 + Ba\,I^{1/2}) \tag{8}$$

re A is the Debye-Hückel temperature dependent constant (lim-
; slope), a is the *mean* distance of closest approach of the ions
 size parameter), Ba is set equal to 1.5 (mol kg^{-1})$^{-1/2}$ at all temper-
es in the range 5-50 °C, and I is the ionic strength of the buffer
ich for its evaluation requires knowledge of appropriate acid dis-
ation constants).

he various stages in the assignment of primary standard pH
es are combined in eqn. (9), which is derived from eqns. (4),
nd (8)

$$pH(PS) = \lim m_{Cl \to 0}\{(E_I - E^\circ)/[(RT/F)\ln 10] + lg\,(m_{Cl}/m^\circ)\} \\ - AI^{1/2}/[1 + 1.5\,(I/m^\circ)^{1/2}] \tag{9}$$

 order for a particular buffer solution to be considered a pri-
y buffer solution, it must be of the "highest metrological"
ity (4) in accordance with the definition of a primary stan-
. It is recommended that it have the following attributes (9):

. High buffer value in the range 0.016-0.07 (mol OH$^-$)/
  pH.
. Small dilution value at half concentration (change in
  pH with change in buffer concentration) in the range
  0.01-0.20.
. Small dependence of pH on temperature less than ±
  0.01 K^{-1}.
. Low residual liquid junction potential < 0.01 in pH.
. Ionic strength ≤0.1 mol kg^{-1} to permit applicability of
  Bates-Guggenheim convention.
. NMI certificate for specific batch.
. Reproducible purity of preparation (lot to lot differ-
  ences of |ΔpH(PS)| < 0.003).
. Long term stability of stored solid material.

lues for the above and other important parameters for the
ary and secondary buffer materials are given in Table 1.

Primary Standard Buffers

 there can be significant variations in the purity of samples of
fer of the same nominal chemical composition, it is essential
 the primary buffer material used has been certified with val-
hat have been measured with Cell I. The Harned cell is used
by many national metrological institutes for accurate measurements of pH of buffer solutions.

Typical values of the pH(PS) of the seven solutions from the six
accepted primary standard reference buffers, which meet the conditions stated above, are listed in Table 2. Batch-to-batch variations in
purity can result in changes in the pH value of samples of at most
0.003. The typical values in Table 2 should not be used in place of
the certified value (from a Harned cell measurement) for a specific
batch of buffer material.

The required attributes listed above effectively limit the range of
primary buffers available to between pH 3 and 10 (at 25 °C). Calcium hydroxide and potassium tetraoxalate are excluded because
the contribution of hydroxide or hydrogen ions to the ionic strength
is significant. Also excluded are the nitrogen bases of the type BH$^+$
(such as tris(hydroxymethyl)aminomethane and piperazine phosphate) and the zwitterionic buffers (e.g. HEPES and MOPS (10)).
These do not comply because either the Bates-Guggenheim convention is not applicable, or the liquid junction potentials are high.
This means the choice of primary standards is restricted to buffers
derived from oxy-carbon, -phosphorus, -boron and mono, di- and
tri-protic carboxylic acids. The uncertainties (11) associated with
Harned cell measurements are calculated (1) to be 0.004 in pH at
NMIs, with typical variation between batches of primary standard
buffers of 0.003.

Secondary Standards

Substances that do not fulfil all the criteria for primary standards, but to which pH values can be assigned using Cell I are
considered to be secondary standards (Table 3). Reasons for their
exclusion as primary standards include difficulties in achieving
consistent and suitable chemical quality (e.g. acetic acid is a liquid), suspected high liquid junction potential, or inappropriate-
ness of the Bates-Guggenheim convention (e.g. other charge-type
buffers). The uncertainty is higher (e.g. 0.01) for biological buffers. Certain other substances, which cannot be used in cells containing hydrogen gas electrodes, are also classed as secondary
standards.

Calibration Procedures

(a) One-point calibration
   A single point calibration is insufficient to determine
   both slope and one-point parameters. The theoretical value for the slope can be assumed but the practical slope may be up to 5% lower. Alternatively, a value
   for the practical slope can be assumed from the manufacturer's prior calibration. The one-point calibration therefore yields only an estimate of pH(X). Since
   both parameters may change with age of the electrodes, this is not a reliable procedure.

(b) Two-point calibration [target uncertainty: 0.02-0.03
    at 25 °C]
   In the majority of practical applications, glass electrodes cells are calibrated by a two-point calibration,
   or bracketing, procedure using two standard buffer
   solutions, with pH values, pH(S$_1$) and pH(S$_2$), bracketing the unknown pH(X). Bracketing is often taken
   to mean that the pH(S$_1$) and pH(S$_2$) buffers selected
   from Table 2 should be those that are immediately
   above and below pH(X). This may not be appropriate
   in all situations and choice of a wider range may be
   better.

(c) Multi-point calibration [target uncertainty: 0.01-0.03 at 25 °C].

Multi-point calibration is carried out using up to five standard buffers . The use of more than five points yields no significant improvement in the statistical information obtainable.

Details of uncertainty computations (11) have been given (1).

Measurement of pH and choice of pH Standard Solutions

1a) If pH is not required to better than ±0.05 any pH standard solution may be selected.

1b) If pH is required to ±0.002 and interpretation in terms of hydrogen ion concentration or activity is desired, choose a standard solution, pH(PS), to match X as closely as possible in terms of pH, composition and ionic strength.

2) Alternatively, a bracketing procedure may be adopted whereby two standard solutions are chosen whose pH values, pH(S1), pH(S2) are on either side of pH(X).   Then if the corresponding potential difference measurements are $E(S1)$, $E(S2)$, $E(X)$, then pH(X) is obtained from

$$pH(X) = pH(S1) + [E(X) - E(S1)]/ \%k$$

where $\%k = 100[E(S2) - E(S1)]/[pH(S2) - pH(S1)]$ is the apparent percentage slope.   This procedure is very easily done on some pH meters simply by adjusting downwards the slope factor control with the electrodes in S2.   The purpose of the bracketing procedure is to compensate for deficiencies in the electrodes and measuring system.

Information to be given about the measurement of pH(X)

The standard solutions selected for calibration of the pH meter system should be reported with the measurement as follows:

System calibrated with pH(S) = ... at ...K.

System calibrated with two primary standards, pH(PS1) =.... and pH(PS2) =....at ....K.

System calibrated with n standards, pH(S1) =.... , pH(S2) =.... etc. at....K.

Interpretation of pH(X) in terms of hydrogen ion concentration

The defined pH has no simple interpretation in terms of hydrogen ion concentration but the mean ionic activity coefficient of a typical 1:1 electrolyte can be used to obtain hydrogen ion concentration subject to an uncertainty of 3.9% in concentration, corresponding to 0.02 in pH.

References

1. Buck, R.P., Rondinini, S., Covington, A.K., Baucke, F.G.K., Brett, C.M.A., Camoes, M.F.C., Milton, M.J.T., Mussini, T., Naumann, R., Pratt, K.W., Spitzer, P., and Wilson, G.S. *Pure Appl. Chem.*, 74, 2105, 2002.
2. Sørensen, S.P.L. *Comp. Rend.Trav. Lab. Carlsberg*, 8, 1, 1909.
3. Sørensen, S.P.L., and Linderstrøm-Lang, K.L., *Comp. Rend. Trav. Lab. Carlsberg*, 15, 1924.
4. BIPM, *Com. Cons. Quantité de Matière* 4, 1998. See also M.J.T. Milton and T.J. Quinn, *Metrologia* 38, 289, 2001.
5. Harned H.S., and Owen, B.B., *The Physical Chemistry of Electrolytic Solutions*, Ch 14, Reinhold, New York, 1958.
6. Bates R.G., and Robinson, R.A., *J. Soln. Chem.* 9, 455, 1980.
7. Bates R.G., and Guggenheim, E.A., *Pure Appl. Chem.* 1, 163, 1960.
8. *International Vocabulary of Basic and General Terms in Metrology* (VIM), Beuth, Berlin, 2nd. Edn. 1994.
9. Bates, R.G. *Determination of pH*, Wiley, New York, 1973.
10. Good, N.E. et al., *Biochem. J.* 5, 467, 1966.
11. *Guide to the Expression of Uncertainty* (GUM), BIPM, IEC, IFCC, ISO, IUPAC, IUPAP, OIML, 1993.

TABLE 1. Summary of Useful Properties of Some Primary and Secondary Standard Buffer Substances and Solutions

Salt or solid substance	Formula	Molality/ mol kg^{-1}	Molar mass/ g mol^{-1}	Density/ g/mL	Amount conc. at 20°C/ mol dm^{-3}	Mass/g to make 1 dm^3	Dilution value $\Delta pH_{1/2}$	Buffer value (β)/ mol OH$^-$ dm^{-3}	pH Temperature coefficient/ K^{-1}
Potassium tetroxalate dihydrate	$KH_3C_4O_8 \cdot 2H_2O$	0.1	254.191	1.0091	0.09875	25.101			
Potassium tetraoxalate dihydrate	$KH_3C_4O_8 \cdot 2H_2O$	0.05	254.191	1.0032	0.04965	12.620	0.186	0.070	0.001
Potassium hydrogen tartrate (sat at 25°C)	$KHC_4H_4O_6$	0.0341	188.18	1.0036	0.034	6.4	0.049	0.027	- 0.0014
Potassium dihydrogen citrate	$KH_2C_6H_5O_7$	0.05	230.22	1.0029	0.04958	11.41	0.024	0.034	- 0.022
Potassium hydrogen phthalate	$KHC_8H_4O_4$	0.05	204.44	1.0017	0.04958	10.12	0.052	0.016	0.00012
Disodium hydrogen orthophosphate +	Na_2HPO_4	0.025	141.958	1.0038	0.02492	3.5379	0.080	0.029	- 0.0028

36   **pH Scale for Aqueous Solutions**

Salt or solid substance	Formula	Molality/ mol kg⁻¹	Molar mass/ g mol⁻¹	Density/ g/mL	Amount conc. at 20°C/ mol dm⁻³	Mass/g to make 1 dm³	Dilution value $\Delta pH_{1/2}$	Buffer value (β)/ mol OH⁻ dm⁻³	pH Temperature coefficient/ K⁻¹
potassium dihydrogen orthophosphate	KH_2PO_4	0.025	136.085			3.3912			
Disodium hydrogen orthophosphate +	Na_2HPO_4	0.03043	141.959	1.0020	0.08665	4.302	0.07	0.016	− 0.0028
potassium dihydrogen orthophosphate	KH_2PO_4	0.00869	136.085		0.03032	1.179			
Disodium tetraborate decahydrate	$Na_2B_4O_7\cdot10H_2O$	0.05	381.367	1.0075	0.04985	19.012			
Disodium tetraborate decahydrate	$Na_2B_4O_7\cdot10H_2O$	0.01	381.367	1.0001	0.00998	3.806	0.01	0.020	− 0.0082
Sodium hydrogen carbonate +	$NaHCO_3$	0.025	84.01	1.0013	0.02492	2.092	0.079	0.029	−0.0096
sodium carbonate	Na_2CO_3	0.025	105.99			2.640			
Calcium hydroxide (sat. at 25°C)	$Ca(OH)_2$	0.0203	74.09	0.9991	0.02025	1.5	-0.28	0.09	-0.033

TABLE 2. Typical Values of pH(PS) for Primary Standards at 0–50°C

Primary standards (PS)	Temperature in °C										
	0	5	10	15	20	25	30	35	37	40	50
Sat. potassium hydrogen tartrate (at 25°C)						3.557	3.552	3.549	3.548	3.547	3.549
0.05 mol kg⁻¹ potassium dihydrogen citrate	3.863	3.840	3.820	3.802	3.788	3.776	3.766	3.759	3.756	3.754	3.749
0.05 mol kg⁻¹ potassium hydrogen phthalate	4.000	3.998	3.997	3.998	4.000	4.005	4.011	4.018	4.022	4.027	4.050
0.025 mol kg⁻¹ disodium hydrogen phosphate + 0.025 mol kg⁻¹ potassium dihydrogen phosphate	6.984	6.951	6.923	6.900	6.881	6.865	6.853	6.844	6.841	6.838	6.833
0.03043 mol kg⁻¹ disodium hydrogen phosphate + 0.008695 mol kg⁻¹ potassium dihydrogen phosphate	7.534	7.500	7.472	7.448	7.429	7.413	7.400	7.389	7.386	7.380	7.367
0.01 mol kg⁻¹ disodium tetraborate	9.464	9.395	9.332	9.276	9.225	9.180	9.139	9.102	9.088	9.068	9.011
0.025 mol kg⁻¹ sodium hydrogen carbonate + 0.025 mol kg⁻¹ sodium carbonate	10.317	10.245	10.179	10.118	10.062	10.012	9.966	9.926	9.910	9.889	9.828

TABLE 3. Values of pH(SS) of Some Secondary Standards from Harned Cell I Measurements

Secondary standards	0	5	10	15	20	25	30	37	40	50
					Temperature in °C					
0.05 mol kg^{-1} potassium tetroxalate[a]	1.67	1.67	1.67	1.67	1.68	1.68	1.68	1.69	1.69	1.71
0.05 mol kg^{-1} sodium hydrogen diglycolate[b]		3.47	3.47	3.48	3.48	3.49	3.50	3.52	3.53	3.56
0.1 mol dm^{-3} acetic acid + 0.1 mol dm^{-3} sodium acetate	4.68	4.67	4.67	4.66	4.66	4.65	4.65	4.66	4.66	4.68
mol dm^{-3} acetic acid + 0.1 mol dm^{-3} sodium acetate	4.74	4.73	4.73	4.72	4.72	4.72	4.72	4.73	4.73	4.75
0.02 mol kg^{-1} piperazine phosphate[c]	6.58	6.51	6.45	6.39	6.34	6.29	6.24	6.16	6.14	6.06
0.05 mol kg^{-1} tris hydrochloride + 0.01667 mol kg^{-1} tris[c]	8.47	8.30	8.14	7.99	7.84	7.70	7.56	7.38	7.31	7.07
0.05 mol kg^{-1} disodium tetraborate	9.51	9.43	9.36	9.30	9.25	9.19	9.15	9.09	9.07	9.01
Saturated (at 25 °C) calcium hydroxide	13.42	13.21	13.00	12.81	12.63	12.45	12.29	12.07	11.98	11.71

[a] Potassium trihydrogen dioxalate (KH$_3$C$_4$O$_8$)
[b] Sodium hydrogen 2,2'-oxydiacetate
[c] 2-Amino-2-(hydroxymethyl)-1,3 propanediol or tris(hydroxymethyl)aminomethane

PRACTICAL pH MEASUREMENTS ON NATURAL WATERS

A. K. Covington and W. Davison

(1) Dilute solutions and freshwater including 'acid-rain' samples ($I < 0.02$ mol kg^{-1})

Major problems could be encountered due to errors associated with the liquid junction. It is recommended that either a free diffusion junction is used or it is verified that the junction is working correctly using dilute solutions as follows. For commercial electrodes calibrated with IUPAC aqueous RVS or PS standards, the pH(X) of dilute solutions should be within ±0.02 of those given in Table 1. The difference in determined pH(X) between a stirred and unstirred dilute solution should be < 0.02. The characteristics of glass electrodes are such that below pH 5 the readings should be stable within 2 min, but for pH 5 to 8.8 or so minutes may be necessary to attain stability. Interpretation of pH(X) measured in this way in terms of activity of hydrogen ion, a_{H+} is subject[1] to an uncertainty of ±0.02 in pH.

(2) Seawater

Measurements made by calibration of electrodes with IUPAC aqueous RVS or PS standards to obtain pH(X) are perfectly valid. However, the interpretation of pH(X) in terms of the activity of hydrogen ion is complicated by the non zero residual liquid junction potential as well as by systematic differences between electrode pairs, principally attributable to the reference electrode. For 35‰ salinity seawater ($S = 0.035$) a_{H+} calculated from pH(X) is typically 12% too low. Special seawater pH scales have been devised to overcome this problem:

(i) The total hydrogen ion scale, pH_T, is defined in terms of the sum of free and complexed (total) hydrogen ion concentrations, where

$$^TC_H = [H^+] + [HSO_4^-] + [HF].$$

$$\text{So, } pH_T = -\log {}^TC_H$$

Calibration of the electrodes with a buffer having a composition similar to that of seawater, to which pH_T has been assigned, results in values of pHT(X) (Tables 2, 3) which are accurately interpretable in terms of ^TC_H.

(ii) The free hydrogen ion scale, pH_F, is defined, and fully interpretable, in terms of the concentration of free hydrogen ions.

$$pH_F = -\log [H^+]$$

Values of pH_F as a function of temperature have been assigned to the same set of pH_T seawater buffers, and so alternatively can be used for calibration (Tables 2, 3) [2,3]

(3) Estuarine water

Prescriptions for seawater scale buffers are available for a range of salinities. Reliable estuarine pH measurements can be made by calibrating with a buffer of the same salinity as the sample. However, these buffers are difficult to prepare and their use presumes prior knowledge of salinity of the sample. Interpretable measurements of estuarine pH can be made by calibration with IUPAC aqueous RVS or PS standards if the electrode pair is additionally calibrated using a 20‰ salinity seawater buffer.[4] The difference between the assigned pH_{SWS} of the seawater buffer and its measured pH(X) value using RVS or PS standards is

$$\Delta pH = pH_{SWS} - pH(X)$$

Values of ΔpH should be in the range of 0.08 to 0.18. It empirically corrects for differences between the two pH scales and for measurement errors associated with the electrode pair. The pH(X) of samples measured using IUPAC aqueous buffers, can be converted to pH_T or pH_F using the appropriate measured ΔpH:

$$pH_T = pH(X) - \Delta pH$$

$$\text{or } pH_F = pH(X) - \Delta pH$$

This simple procedure is appropriate to pH measurement at salinities from 2‰ to 35‰. For salinities lower than 2‰ the procedures for freshwaters should be adopted.

References

1. Davison, W. and Harbinson, T. R., *Analyst*, 113, 709, 1988.
2. Culberson, C. H., in *Marine Electrochemistry*, Whitfield, M. and Jagner, D., Eds., Wiley, 1981.
3. Millero, F. J., *Limnol. Oceanogr.*, 31, 839, 1986.
4. Covington, A. K., Whalley, P. D., Davison, W., and Whitfield, M., in *The Determination of Trace Metals in Natural Waters*, West, T. S. and Nurnberg, H. W., Eds., Blackwell, Oxford, 1988.
5. Koch, W. F., Marinenko, G., and Paule, R. C., *J. Res. NBS*, 91, 33, 1986.

TABLE 1. pH of Dilute Solutions at 25°C, Degassed and Equilibrated with Air, Suitable as Quality Control Standards

	Ionic strength mmol kg^{-1}	Concentration(x) mmol kg^{-1}	pH $p_{CO_2} = 0$	pH p_{CO_2} = air
Potassium hydrogen phthalate	10.7	10	4.12	4.12
	1.1	1	4.33	4.33
xKH$_2$PO$_4$ + xNa$_2$HPO$_4$	9.9	2.5	7.07	7.05
xKH$_2$PO$_4$ + 3.5xNa$_2$HPO$_4$	10	0.87	7.61	7.58
Na$_2$B$_4$O$_7 \cdot$ 10H$_2$O	10	5	9.20	—
HCl	0.1	0.1	4.03	4.03
SRM2694-I[a]	—	—	4.30	—
SRM2694-II[a]	—	—	3.59	—

Note: The pH of solutions near to pH 4 is virtually independent of temperature over the range of 5 to 30°C.

[a] Simulated rainwater samples are available (Reference 5) from NIST containing sulfate, nitrate, chloride, fluoride, sodium, potassium, calcium and magnesium.

TABLE 2. Composition of Seawater Buffer of Salinity S = 35‰ at 25°C (Reference 3)

Solute	mol dm^{-3}	mol kg^{-1}	g kg^{-1}	g dm^{-3}
NaCl	0.3666	0.3493	20.416	20.946
Na$_2$SO$_4$	0.02926	0.02788	3.96	4.063
KCl	0.01058	0.01008	0.752	0.772
CaCl$_2$	0.01077	0.01026	1.139	1.169
MgCl$_2$	0.05518	0.05258	5.006	5.139
Tris	0.06	0.05717	6.926	7.106
Tris $\cdot$ HCl	0.06	0.05717	9.010	9.244

Tris = tris(hydroxymethyl)aminomethane (HOCH$_2$)$_3$CNH$_2$.
A 20‰ buffer is made by diluting the 35‰ in the ratio 20:35.

TABLE 3. Assigned Values of 20‰ and 35‰ Buffers on Free and Total Hydrogen Ion Scales. Calculated from Equations Provided by Millero (Reference 3)

Temp (°C)	pH$_T$ $S = 20‰$	pH$_T$ $S = 35‰$	pH$_F$ $S = 20‰$	pH$_F$ $S = 35‰$
5	8.683	8.718	8.759	8.81
10	8.513	8.542	8.597	8.647
15	8.351	8.374	8.442	8.491
20	8.195	8.212	8.292	8.341
25	8.045	8.057	8.149	8.197
30	7.901	7.908	8.011	8.059
35	7.762	7.764	7.879	7.926

BUFFER SOLUTIONS GIVING ROUND VALUES OF pH AT 25°C

A pH	A x	B pH	B x	C pH	C x	D pH	D x	E pH	E x
1.00	67.0	2.20	49.5	4.10	1.3	5.80	3.6	7.00	46.6
1.10	52.8	2.30	45.8	4.20	3.0	5.90	4.6	7.10	45.7
1.20	42.5	2.40	42.2	4.30	4.7	6.00	5.6	7.20	44.7
1.30	33.6	2.50	38.8	4.40	6.6	6.10	6.8	7.30	43.4
1.40	26.6	2.60	35.4	4.50	8.7	6.20	8.1	7.40	42.0
1.50	20.7	2.70	32.1	4.60	11.1	6.30	9.7	7.50	40.3
1.60	16.2	2.80	28.9	4.70	13.6	6.40	11.6	7.60	38.5
1.70	13.0	2.90	25.7	4.80	16.5	6.50	13.9	7.70	36.6
1.80	10.2	3.00	22.3	4.90	19.4	6.60	16.4	7.80	34.5
1.90	8.1	3.10	18.8	5.00	22.6	6.70	19.3	7.90	32.0
2.00	6.5	3.20	15.7	5.10	25.5	6.80	22.4	8.00	29.2
2.10	5.10	3.30	12.9	5.20	28.8	6.90	25.9	8.10	26.2
2.20	3.9	3.40	10.4	5.30	31.6	7.00	29.1	8.20	22.9
		3.50	8.2	5.40	34.1	7.10	32.1	8.30	19.9
		3.60	6.3	5.50	36.6	7.20	34.7	8.40	17.2
		3.70	4.5	5.60	38.8	7.30	37.0	8.50	14.7
		3.80	2.9	5.70	40.6	7.40	39.1	8.60	12.2
		3.90	1.4	5.80	42.3	7.50	40.9	8.70	10.3
		4.00	0.1	5.90	43.7	7.60	42.4	8.80	8.5
						7.70	43.5	8.90	7.0
						7.80	44.5	9.00	5.7
						7.90	45.3		
						8.00	46.1		

F pH	F x	G pH	G x	H pH	H x	I pH	I x	J pH	J x
8.00	20.5	9.20	0.9	9.60	5.0	10.90	3.3	12.00	6.0
8.10	19.7	9.30	3.6	9.70	6.2	11.00	4.1	12.10	8.0
8.20	18.8	9.40	6.2	9.80	7.6	11.10	5.1	12.20	10.2
8.30	17.7	9.50	8.8	9.90	9.1	11.20	6.3	12.30	12.8
8.40	16.6	9.60	11.1	10.00	10.7	11.30	7.6	12.40	16.2
8.50	15.2	9.70	13.1	10.10	12.2	11.40	9.1	12.50	20.4
8.60	13.5	9.80	15.0	10.20	13.8	11.50	11.1	12.60	25.6
8.70	11.6	9.90	16.7	10.30	15.2	11.60	13.5	12.70	32.2
8.80	9.6	10.00	18.3	10.40	16.5	11.70	16.2	12.80	41.2
8.90	7.1	10.10	19.5	10.50	17.8	11.80	19.4	12.90	53.0
9.00	4.6	10.20	20.5	10.60	19.1	11.90	23.0	13.00	66.0
9.10	2.0	10.30	21.3	10.70	20.2	12.00	26.9		
		10.40	22.1	10.80	21.2				
		10.50	22.7	10.90	22.0				
		10.60	23.3	11.00	22.7				
		10.70	23.8						
		10.80	24.25						

A. 25 ml of 0.2 molar KCl + x ml of 0.2 molar HCl.
B. 50 ml of 0.1 molar potassium hydrogen phthalate + x ml of 0.1 molar HCl.
C. 50 ml of 0.1 molar potassium hydrogen phthalate + x ml of 0.1 molar NaOH.
D. 50 ml of 0.1 molar potassium dihydrogen phosphate + x ml of 0.1 molar NaOH.
E. 50 ml of 0.1 molar tris(hydroxymethyl)aminomethane + x ml of 0.1 M HCl.
F. 50 ml of 0.025 molar borax + x ml of 0.1 molar HCl.
G. 50 ml of 0.025 molar borax + x ml of 0.1 molar NaOH.
H. 50 ml of 0.05 molar sodium bicarbonate + x ml of 0.1 molar NaOH.
I. 50 ml of 0.05 molar disodium hydrogen phosphate + x ml of 0.1 molar NaOH.
J. 25 ml of 0.2 molar KCl + x ml of 0.2 molar NaOH.

Final volume of mixtures = 100 ml.

References

1. Bower, V. E., and Bates, R. G., *J. Res. Natl. Bur. Stand.*, 55, 197, 1955 (A–D).
2. Bates, R. G., and Bower, V. E., *Anal. Chem.*, 28, 1322, 1956 (E–J).

DISSOCIATION CONSTANTS OF INORGANIC ACIDS AND BASES

The data in this table are presented as values of pK_a, defined as the negative logarithm of the acid dissociation constant K_a for the reaction

$$BH \rightleftharpoons B^- + H^+$$

Thus $pK_a = -\log K_a$, and the hydrogen ion concentration $[H^+]$ can be calculated from

$$K_a = \frac{H^+ \; B^-}{[BH]}$$

In the case of bases, the entry in the table is for the conjugate acid; e.g., ammonium ion for ammonia. The OH^- concentration in the system

$$NH_3 + H_2O \rightleftharpoons NH_4^+ + OH^-$$

can be calculated from the equation

$$K_b = K_{water}/K_a = \frac{OH^- \; NH_4^+}{[NH_3]}$$

where $K_{water} = 1.01 \times 10^{-14}$ at 25 °C. Note that $pK_a + pK_b = pK_{water}$.

All values refer to dilute aqueous solutions at zero ionic strength at the temperature indicated. The table is arranged alphabetically by compound name.

Reference

1. Perrin, D. D., *Ionization Constants of Inorganic Acids and Bases in Aqueous Solution, Second Edition*, Pergamon, Oxford, 1982.

Name	Formula	Step	$t/°C$	pK_a
Aluminum(III) ion	Al^{+3}		25	5.0
Ammonia	NH_3		25	9.25
Arsenic acid	H_3AsO_4	1	25	2.26
		2	25	6.76
		3	25	11.29
Arsenious acid	H_2AsO_3		25	9.29
Barium(II) ion	Ba^{+2}		25	13.4
Boric acid	H_3BO_3	1	20	9.27
		2	20	>14
Calcium(II) ion	Ca^{+2}		25	12.6
Carbonic acid	H_2CO_3	1	25	6.35
		2	25	10.33
Chlorous acid	$HClO_2$		25	1.94
Chromic acid	H_2CrO_4	1	25	0.74
		2	25	6.49
Cyanic acid	$HCNO$		25	3.46
Germanic acid	H_2GeO_3	1	25	9.01
		2	25	12.3
Hydrazine	N_2H_4		25	8.1
Hydrazoic acid	HN_3		25	4.6
Hydrocyanic acid	HCN		25	9.21
Hydrofluoric acid	HF		25	3.20
Hydrogen peroxide	H_2O_2		25	11.62
Hydrogen selenide	H_2Se	1	25	3.89
		2	25	11.0
Hydrogen sulfide	H_2S	1	25	7.05
		2	25	19
Hydrogen telluride	H_2Te	1	18	2.6
		2	25	11
Hydroxylamine	NH_2OH		25	5.94
Hypobromous acid	$HBrO$		25	8.55
Hypochlorous acid	$HClO$		25	7.40
Hypoiodous acid	HIO		25	10.5
Iodic acid	HIO_3		25	0.78
Lithium ion	Li^+		25	13.8
Magnesium(II) ion	Mg^{+2}		25	11.4
Nitrous acid	HNO_2		25	3.25
Perchloric acid	$HClO_4$		20	-1.6
Periodic acid	HIO_4		25	1.64
Phosphoric acid	H_3PO_4	1	25	2.16

Name	Formula	Step	$t/°C$	pK_a
		2	25	7.21
		3	25	12.32
Phosphorous acid	H_3PO_3	1	20	1.3
		2	20	6.70
Pyrophosphoric acid	$H_4P_2O_7$	1	25	0.91
		2	25	2.10
		3	25	6.70
		4	25	9.32
Selenic acid	H_2SeO_4	2	25	1.7
Selenious acid	H_2SeO_3	1	25	2.62
		2	25	8.32
Silicic acid	H_4SiO_4	1	30	9.9
		2	30	11.8
		3	30	12
		4	30	12
Sodium ion	Na^+		25	14.8
Strontium(II) ion	Sr^{+2}		25	13.2
Sulfamic acid	NH_2SO_3H		25	1.05
Sulfuric acid	H_2SO_4	2	25	1.99
Sulfurous acid	H_2SO_3	1	25	1.85
		2	25	7.2
Telluric acid	H_2TeO_4	1	18	7.68
		2	18	11.0
Tellurous acid	H_2TeO_3	1	25	6.27
		2	25	8.43
Tetrafluoroboric acid	HBF_4		25	0.5
Thiocyanic acid	$HSCN$		25	−1.8
Water	H_2O		25	13.995

DISSOCIATION CONSTANTS OF ORGANIC ACIDS AND BASES

This table lists the dissociation (ionization) constants of over 1070 organic acids, bases, and amphoteric compounds. All data apply to dilute aqueous solutions and are presented as values of pK_a, which is defined as the negative of the logarithm of the equilibrium constant K_a for the reaction

$$HA \rightleftharpoons H^+ + A^-$$

i.e.,

$$K_a = [H^+][A^-]/[HA]$$

where $[H^+]$, etc. represent the concentrations of the respective species in mol/L. It follows that $pK_a = pH + \log[HA] - \log[A^-]$, so that a solution with 50% dissociation has pH equal to the pK_a of the acid.

Data for bases are presented as pK_a values for the conjugate acid, i.e., for the reaction

$$BH^+ \rightleftharpoons H^+ + B$$

In older literature, an ionization constant K_b was used for the reaction $B + H_2O \rightleftharpoons BH^+ + OH^-$. This is related to K_a by

$$pK_a + pK_b = pK_{water} = 14.00 \quad (\text{at } 25°C)$$

Compounds are listed by molecular formula in Hill order.

References

1. Perrin, D. D., *Dissociation Constants of Organic Bases in Aqueous Solution*, Butterworths, London, 1965; Supplement, 1972.
2. Serjeant, E. P., and Dempsey, B., *Ionization Constants of Organic Acids in Aqueous Solution*, Pergamon, Oxford, 1979.
3. Albert, A., "Ionization Constants of Heterocyclic Substances", in Katritzky, A. R., Ed., *Physical Methods in Heterocyclic Chemistry*, Academic Press, New York, 1963.
4. Sober, H.A., Ed., *CRC Handbook of Biochemistry*, CRC Press, Boca Raton, FL, 1968.
5. Perrin, D. D., Dempsey, B., and Serjeant, E. P., pK_a *Prediction for Organic Acids and Bases*, Chapman and Hall, London, 1981.
6. Albert, A., and Serjeant, E. P., *The Determination of Ionization Constants, Third Edition*, Chapman and Hall, London, 1984.
7. Budavari, S., Ed., *The Merck Index, Twelfth Edition*, Merck & Co., Whitehouse Station, NJ, 1996.

Mol. form.	Name	Step	t/°C	pK_a
CHNO	Cyanic acid		25	3.7
CH_2N_2	Cyanamide		29	1.1
CH_2O	Formaldehyde		25	13.27
CH_2O_2	Formic acid		25	3.75
CH_3NO_2	Nitromethane		25	10.21
CH_3NS_2	Carbamodithioic acid		25	2.95
CH_4N_2O	Urea		25	0.10
CH_4N_2S	Thiourea		25	-1
CH_4O	Methanol		25	15.5
CH_4S	Methanethiol		25	10.33
CH_5N	Methylamine		25	10.66
CH_5NO	O-Methylhydroxylamine			12.5
CH_5N_3	Guanidine		25	13.6
C_2HCl_3O	Trichloroacetaldehyde		25	10.04
$C_2HCl_3O_2$	Trichloroacetic acid		20	0.66
$C_2HF_3O_2$	Trifluoroacetic acid		25	0.52
$C_2H_2Cl_2O_2$	Dichloroacetic acid		25	1.35
$C_2H_2O_3$	Glyoxylic acid		25	3.18
$C_2H_2O_4$	Oxalic acid	1	25	1.25
		2	25	3.81
$C_2H_3BrO_2$	Bromoacetic acid		25	2.90
$C_2H_3ClO_2$	Chloroacetic acid		25	2.87
$C_2H_3Cl_3O$	2,2,2-Trichloroethanol		25	12.24
$C_2H_3FO_2$	Fluoroacetic acid		25	2.59
$C_2H_3F_3O$	2,2,2-Trifluoroethanol		25	12.37
$C_2H_3IO_2$	Iodoacetic acid		25	3.18
$C_2H_3NO_4$	Nitroacetic acid		24	1.48
$C_2H_3N_3$	1H-1,2,3-Triazole		20	1.17
$C_2H_3N_3$	1H-1,2,4-Triazole		20	2.27
$C_2H_4N_2$	Aminoacetonitrile		25	5.34
C_2H_4O	Acetaldehyde		25	13.57
C_2H_4OS	Thioacetic acid		25	3.33
$C_2H_4O_2$	Acetic acid		25	4.756
$C_2H_4O_2S$	Thioglycolic acid		25	3.68
$C_2H_4O_3$	Glycolic acid		25	3.83
C_2H_5N	Ethyleneimine		25	8.04

Mol. form.	Name	Step	t/°C	pK_a
C_2H_5NO	Acetamide		25	15.1
$C_2H_5NO_2$	Acetohydroxamic acid			8.70
$C_2H_5NO_2$	Nitroethane		25	8.46
$C_2H_5NO_2$	Glycine	1	25	2.35
		2	25	9.78
$C_2H_6N_2$	Ethanimidamide		25	12.1
C_2H_6O	Ethanol		25	15.5
C_2H_6OS	2-Mercaptoethanol		25	9.72
$C_2H_6O_2$	Ethyleneglycol		25	15.1
$C_2H_7AsO_2$	Dimethylarsinic acid	1	25	1.57
		2	25	6.27
C_2H_7N	Ethylamine		25	10.65
C_2H_7N	Dimethylamine		25	10.73
C_2H_7NO	Ethanolamine		25	9.50
$C_2H_7NO_3S$	2-Aminoethanesulfonic acid	1	25	1.5
		2	25	9.06
C_2H_7NS	Cysteamine	1	25	8.27
		2	25	10.53
$C_2H_7N_5$	Biguanide	1		11.52
		2		2.93
$C_2H_8N_2$	1,2-Ethanediamine	1	25	9.92
		2	25	6.86
$C_2H_8O_7P_2$	1-Hydroxy-1,1-diphosphonoethane	1		1.35
		2		2.87
		3		7.03
		4		11.3
$C_3H_2O_2$	2-Propynoic acid		25	1.84
C_3H_3NO	Oxazole		33	0.8
C_3H_3NO	Isoxazole		25	-2.0
$C_3H_3NO_2$	Cyanoacetic acid		25	2.47
C_3H_3NS	Thiazole		25	2.52
$C_3H_3N_3O_3$	Cyanuric acid	1		6.88
		2		11.40
		3		13.5
$C_3H_4N_2$	1H-Pyrazole		25	2.49
$C_3H_4N_2$	Imidazole		25	6.99

Mol. form.	Name	Step	$t/°C$	pK_a
$C_3H_4N_2S$	2-Thiazolamine		20	5.36
C_3H_4O	Propargyl alcohol		25	13.6
$C_3H_4O_2$	Acrylic acid		25	4.25
$C_3H_4O_3$	Pyruvic acid		25	2.39
$C_3H_4O_4$	Malonic acid	1	25	2.85
		2	25	5.70
$C_3H_4O_5$	Hydroxypropanedioic acid	1		2.42
		2		4.54
$C_3H_5BrO_2$	3-Bromopropanoic acid		25	4.00
$C_3H_5ClO_2$	2-Chloropropanoic acid		25	2.83
$C_3H_5ClO_2$	3-Chloropropanoic acid		25	3.98
$C_3H_6N_2$	3-Aminopropanenitrile		20	7.80
$C_3H_6N_6$	1,3,5-Triazine-2,4,6-triamine		25	5.00
C_3H_6O	Allyl alcohol		25	15.5
$C_3H_6O_2$	Propanoic acid		25	4.87
$C_3H_6O_2S$	(Methylthio)acetic acid		25	3.66
$C_3H_6O_3$	Lactic acid		25	3.86
$C_3H_6O_3$	3-Hydroxypropanoic acid		25	4.51
$C_3H_6O_4$	Glyceric acid		25	3.52
C_3H_7N	Allylamine		25	9.49
C_3H_7N	Azetidine		25	11.29
C_3H_7NO	2-Propanone oxime		25	12.42
$C_3H_7NO_2$	L-Alanine	1	25	2.34
		2	25	9.87
$C_3H_7NO_2$	β-Alanine	1	25	3.55
		2	25	10.24
$C_3H_7NO_2$	Sarcosine	1	25	2.21
		2	25	10.1
$C_3H_7NO_2S$	L-Cysteine	1	25	1.5
		2	25	8.7
		3	25	10.2
$C_3H_7NO_3$	L-Serine	1	25	2.19
		2	25	9.21
$C_3H_7NO_5S$	DL-Cysteic acid	1	25	1.3
		2	25	1.9
		3	25	8.70
$C_3H_7N_3O_2$	Glycocyamine		25	2.82
$C_3H_8O_2$	Ethylene glycol monomethyl ether		25	14.8
$C_3H_8O_3$	Glycerol		25	14.15
C_3H_9N	Propylamine		25	10.54
C_3H_9N	Isopropylamine		25	10.63
C_3H_9N	Trimethylamine		25	9.80
C_3H_9NO	2-Methoxyethylamine		25	9.40
C_3H_9NO	Trimethylamine oxide		20	4.65
$C_3H_{10}N_2$	1,2-Propanediamine, (±)	1	25	9.82
		2	25	6.61
$C_3H_{10}N_2$	1,3-Propanediamine	1	25	10.55
		2	25	8.88
$C_3H_{10}N_2O$	1,3-Diamino-2-propanol	1	20	9.69
		2	20	7.93
$C_3H_{11}N_3$	1,2,3-Triaminopropane	1	20	9.59
		2	20	7.95
$C_4H_4FN_3O$	Flucytosine			3.26
$C_4H_4N_2$	Pyrazine		20	0.65
$C_4H_4N_2$	Pyrimidine		20	1.23
$C_4H_4N_2$	Pyridazine		20	2.24
$C_4H_4N_2O_2$	Uracil		25	9.45
$C_4H_4N_2O_3$	Barbituric acid		25	4.01
$C_4H_4N_2O_5$	Alloxanic acid		25	6.64

Mol. form.	Name	Step	$t/°C$	pK_a
$C_4H_4N_4O_2$	5-Nitropyrimidinamine		20	0.35
$C_4H_4O_2$	2-Butynoic acid		25	2.62
$C_4H_4O_4$	Maleic acid	1	25	1.92
		2	25	6.23
$C_4H_4O_4$	Fumaric acid	1	25	3.02
		2	25	4.38
$C_4H_4O_5$	Oxaloacetic acid	1	25	2.55
		2	25	4.37
		3	25	13.03
C_4H_5N	Pyrrole		25	-3.8
$C_4H_5NO_2$	Succinimide		25	9.62
$C_4H_5N_3$	2-Pyrimidinamine		20	3.45
$C_4H_5N_3$	4-Pyrimidinamine		20	5.71
$C_4H_5N_3O$	Cytosine	1		4.60
		2		12.16
$C_4H_5N_3O_2$	6-Methyl-1,2,4-triazine-3,5(2H,4H)-dione			7.6
$C_4H_6N_2$	1-Methylimidazol		25	6.95
$C_4H_6N_4O_3$	Allantoin		25	8.96
$C_4H_6N_4O_3S_2$	Acetazolamide			7.2
$C_4H_6O_2$	trans-Crotonic acid		25	4.69
$C_4H_6O_2$	3-Butenoic acid		25	4.34
$C_4H_6O_2$	Cyclopropanecarboxylic acid		25	4.83
$C_4H_6O_3$	2-Oxobutanoic acid		25	2.50
$C_4H_6O_3$	Acetoacetic acid		25	3.6
$C_4H_6O_4$	Succinic acid	1	25	4.21
		2	25	5.64
$C_4H_6O_4$	Methylmalonic acid	1	25	3.07
		2	25	5.76
$C_4H_6O_5$	Malic acid	1	25	3.40
		2	25	5.11
$C_4H_6O_6$	DL-Tartaric acid	1	25	3.03
		2	25	4.37
$C_4H_6O_6$	meso-Tartaric acid	1	25	3.17
		2	25	4.91
$C_4H_6O_6$	L-Tartaric acid	1	25	2.98
		2	25	4.34
$C_4H_6O_8$	Dihydroxytartaric acid		25	1.92
$C_4H_7ClO_2$	2-Chlorobutanoic acid			2.86
$C_4H_7ClO_2$	3-Chlorobutanoic acid			4.05
$C_4H_7ClO_2$	4-Chlorobutanoic acid			4.52
$C_4H_7NO_2$	4-Cyanobutanoic acid		25	2.42
$C_4H_7NO_3$	N-Acetylglycine		25	3.67
$C_4H_7NO_4$	Iminodiacetic acid	1		2.98
		2		9.89
$C_4H_7NO_4$	L-Aspartic acid	1	25	1.99
		2	25	3.90
		3	25	9.90
$C_4H_7N_3O$	Creatinine	1	25	4.8
		2		9.2
$C_4H_7N_5$	2,4,6-Pyrimidinetriamine		20	6.84
$C_4H_8N_2O_3$	L-Asparagine	1	20	2.1
		2	20	8.80
$C_4H_8N_2O_3$	N-Glycylglycine	1	25	3.14
		2		8.17
$C_4H_8O_2$	Butanoic acid		25	4.83
$C_4H_8O_2$	2-Methylpropanoic acid		20	4.84
$C_4H_8O_3$	3-Hydroxybutanoic acid, (±)		25	4.70
$C_4H_8O_3$	4-Hydroxybutanoic acid		25	4.72
$C_4H_8O_3$	Ethoxyacetic acid		18	3.65
C_4H_9N	Pyrrolidine		25	11.31

Mol. form.	Name	Step	$t/°C$	pK_a
C_4H_9NO	Morpholine		25	8.50
$C_4H_9NO_2$	2-Methylalanine	1	25	2.36
		2	25	10.21
$C_4H_9NO_2$	N,N-Dimethylglycine		25	9.89
$C_4H_9NO_2$	DL-2-Aminobutanoic acid	1	25	2.29
		2	25	9.83
$C_4H_9NO_2$	4-Aminobutanoic acid	1	25	4.031
		2	25	10.556
$C_4H_9NO_2S$	DL-Homocysteine	1	25	2.22
		2	25	8.87
		3	25	10.86
$C_4H_9NO_3$	L-Threonine	1	25	2.09
		2	25	9.10
$C_4H_9NO_3$	L-Homoserine	1	25	2.71
		2	25	9.62
$C_4H_9N_3O_2$	Creatine	1	25	2.63
		2	25	14.3
$C_4H_{10}N_2$	Piperazine	1	25	9.73
		2	25	5.33
$C_4H_{10}N_2O_2$	2,4-Diaminobutanoic acid	1	25	1.85
		2	25	8.24
		3	25	10.44
$C_4H_{10}O_4$	1,2,3,4-Butanetetrol			13.9
$C_4H_{11}N$	Butylamine		25	10.60
$C_4H_{11}N$	sec-Butylamine		25	10.56
$C_4H_{11}N$	tert-Butylamine		25	10.68
$C_4H_{11}N$	Diethylamine		25	10.84
$C_4H_{11}NO_3$	Tris(hydroxymethyl) methylamine		20	8.3
$C_4H_{12}N_2$	1,4-Butanediamine	1	25	10.80
		2	25	9.63
C_5H_4BrN	3-Bromopyridine		25	2.84
C_5H_4ClN	2-Chloropyridine		25	0.49
C_5H_4ClN	3-Chloropyridine		25	2.81
C_5H_4ClN	4-Chloropyridine		25	3.83
C_5H_4FN	2-Fluoropyridine		25	-0.44
$C_5H_4N_2O_2$	4-Nitropyridine		25	1.61
$C_5H_4N_4$	1H-Purine	1	20	2.30
		2	20	8.96
$C_5H_4N_4O$	Hypoxanthine		25	8.7
$C_5H_4N_4O$	Allopurinol			10.2
$C_5H_4N_4O_3$	Uric acid		12	3.89
$C_5H_4N_4S$	1,7-Dihydro-6H-purine-6-thione	1		7.77
		2		11.17
$C_5H_4O_2S$	2-Thiophenecarboxylic acid		25	3.49
$C_5H_4O_2S$	3-Thiophenecarboxylic acid		25	4.1
$C_5H_4O_3$	2-Furancarboxylic acid		25	3.16
$C_5H_4O_3$	3-Furancarboxylic acid		25	3.9
C_5H_5N	Pyridine		25	5.23
C_5H_5NO	2-Pyridinol	1	20	0.75
		2	20	11.65
C_5H_5NO	3-Pyridinol	1	20	4.79
		2	20	8.75
C_5H_5NO	4-Pyridinol	1	20	3.20
		2	20	11.12
C_5H_5NO	2(1H)-Pyridinone	1	20	0.75
		2	20	11.65
C_5H_5NO	Pyridine-1-oxide		24	0.79
$C_5H_5NO_2$	1H-Pyrrole-2-carboxylic acid		20	4.45

Mol. form.	Name	Step	$t/°C$	pK_a
$C_5H_5NO_2$	1H-Pyrrole-3-carboxylic acid		20	5.00
$C_5H_5N_3O$	Pyrazinecarboxamide			0.5
$C_5H_5N_5$	Adenine	1		4.3
		2		9.83
$C_5H_5N_5O$	Guanine		40	9.92
$C_5H_6N_2$	2-Pyridinamine		20	6.82
$C_5H_6N_2$	3-Pyridinamine		25	6.04
$C_5H_6N_2$	4-Pyridinamine		25	9.11
$C_5H_6N_2$	2-Methylpyrazine		27	1.45
$C_5H_6N_2O_2$	Thymine		25	9.94
$C_5H_6O_4$	1,1-Cyclopropanedicarboxylic acid	1	25	1.82
		2	25	7.43
$C_5H_6O_4$	trans-1-Propene-1,2-dicarboxylic acid	1	25	3.09
		2	25	4.75
$C_5H_6O_4$	1-Propene-2,3-dicarboxylic acid	1	25	3.85
		2	25	5.45
$C_5H_6O_5$	2-Oxoglutaric acid	1	25	2.47
		2	25	4.68
$C_5H_7NO_3$	5,5-Dimethyl-2,4-oxazolidinedione		37	6.13
$C_5H_7NO_3$	L-Pyroglutamic acid		25	3.32
$C_5H_7N_3$	2,5-Pyridinediamine		20	6.48
$C_5H_7N_3$	Methylaminopyrazine		25	3.39
$C_5H_7N_3O_4$	Azaserine			8.55
$C_5H_8N_2$	2,4-Dimethylimidazole		25	8.36
$C_5H_8N_4O_3S_2$	Methazolamide			7.30
$C_5H_8O_2$	trans-3-Pentenoic acid		25	4.51
$C_5H_8O_4$	Dimethylmalonic acid		25	3.15
$C_5H_8O_4$	Glutaric acid	1	18	4.32
		2	25	5.42
$C_5H_8O_4$	Methylsuccinic acid	1	25	4.13
		2	25	5.64
$C_5H_9NO_2$	L-Proline	1	25	1.95
		2	25	10.64
$C_5H_9NO_3$	5-Amino-4-oxopentanoic acid	1	25	4.05
		2	25	8.90
$C_5H_9NO_3$	trans-4-Hydroxyproline	1	25	1.82
		2	25	9.66
$C_5H_9NO_4$	L-Glutamic acid	1	25	2.13
		2	25	4.31
		3		9.67
$C_5H_9N_3$	Histamine	1	25	6.04
		2	25	9.75
$C_5H_{10}N_2O_3$	Glycylalanine		25	3.15
$C_5H_{10}N_2O_3$	L-Glutamine	1	25	2.17
		2	25	9.13
$C_5H_{10}N_2O_4$	Glycylserine	1	25	2.98
		2	25	8.38
$C_5H_{10}O_2$	Pentanoic acid		20	4.83
$C_5H_{10}O_2$	2-Methylbutanoic acid		25	4.80
$C_5H_{10}O_2$	3-Methylbutanoic acid		25	4.77
$C_5H_{10}O_2$	2,2-Dimethylpropanoic acid		20	5.03
$C_5H_{10}O_4$	D-2-Deoxyribose		25	12.61
$C_5H_{10}O_5$	L-Ribose		25	12.22
$C_5H_{10}O_5$	D-Xylose		18	12.14
$C_5H_{11}N$	Piperidine		25	11.123
$C_5H_{11}N$	N-Methylpyrrolidine		25	10.46
$C_5H_{11}NO$	4-Methylmorpholine		25	7.38
$C_5H_{11}NO_2$	L-Valine	1	25	2.29
		2	25	9.74

Mol. form.	Name	Step	$t/°C$	pK_a	Mol. form.	Name	Step	$t/°C$	pK_a
$C_5H_{11}NO_2$	DL-Norvaline	1		2.36			3	20	9.31
		2		9.72	C_6H_6BrN	2-Bromoaniline		25	2.53
$C_5H_{11}NO_2$	L-Norvaline	1	25	2.32	C_6H_6BrN	3-Bromoaniline		25	3.53
		2	25	9.81	C_6H_6BrN	4-Bromoaniline		25	3.89
$C_5H_{11}NO_2$	N-Propylglycine	1	25	2.35	C_6H_6ClN	2-Chloroaniline		25	2.66
		2	25	10.19	C_6H_6ClN	3-Chloroaniline		25	3.52
$C_5H_{11}NO_2$	5-Aminopentanoic acid	1	25	4.27	C_6H_6ClN	4-Chloroaniline		25	3.98
		2	25	10.77	C_6H_6FN	2-Fluoroaniline		25	3.20
$C_5H_{11}NO_2$	Betaine	0		1.83	C_6H_6FN	3-Fluoroaniline		25	3.59
$C_5H_{11}NO_2S$	L-Methionine	1	25	2.13	C_6H_6FN	4-Fluoroaniline		25	4.65
		2	25	9.27	C_6H_6IN	2-Iodoaniline		25	2.54
$C_5H_{12}N_2O$	Tetramethylurea			2	C_6H_6IN	3-Iodoaniline		25	3.58
$C_5H_{12}N_2O_2$	L-Ornithine	1	25	1.71	C_6H_6IN	4-Iodoaniline		25	3.81
		2	25	8.69	$C_6H_6N_2O$	3-Pyridinecarboxamide		20	3.3
		3	25	10.76	$C_6H_6N_2O$	2-Pyridinecarbox-	1	20	3.59
$C_5H_{13}N$	Pentylamine		25	10.63		aldehyde oxime	2	20	10.18
$C_5H_{13}N$	3-Pentanamine		17	10.59	$C_6H_6N_2O_2$	2-Nitroaniline		25	-0.25
$C_5H_{13}N$	3-Methyl-1-butanamine		25	10.60	$C_6H_6N_2O_2$	3-Nitroaniline		25	2.46
$C_5H_{13}N$	2-Methyl-2-butanamine		19	10.85	$C_6H_6N_2O_2$	4-Nitroaniline		25	1.02
$C_5H_{13}N$	2,2-Dimethylpropylamine		25	10.15	C_6H_6O	Phenol		25	9.99
$C_5H_{13}N$	Diethylmethylamine		25	10.35	$C_6H_6O_2$	p-Hydroquinone	1	25	9.85
$C_5H_{14}NO$	Choline		25	13.9			2	25	11.4
$C_5H_{14}N_2$	1,5-Pentanediamine	1	25	10.05	$C_6H_6O_2$	Pyrocatechol	1	25	9.34
		2	25	10.93			2	25	12.6
$C_6H_3Cl_3N_2O_2$	4-Amino-3,5,6-trichloro-			3.6	$C_6H_6O_2$	Resorcinol	1	25	9.32
	2-pyridinecarboxylic acid						2	25	11.1
$C_6H_3N_3O_7$	2,4,6-Trinitrophenol		24	0.42	$C_6H_6O_2S$	Benzenesulfinic acid		20	1.3
$C_6H_4Cl_2O$	2,3-Dichlorophenol		25	7.44	$C_6H_6O_3S$	Benzenesulfonic acid		25	0.70
$C_6H_4N_2O_5$	2,4-Dinitrophenol		25	4.07	$C_6H_6O_4$	5-Hydroxy-2-(hydroxy-			7.9
$C_6H_4N_2O_5$	2,5-Dinitrophenol		15	5.15		methyl)-4H-pyran-4-one			
$C_6H_4N_4$	Pteridine		20	4.05	$C_6H_6O_4S$	3-Hydroxybenzene-		25	9.07
C_6H_5BrO	2-Bromophenol		25	8.45		sulfonic acid			
C_6H_5BrO	3-Bromophenol		25	9.03	$C_6H_6O_4S$	4-Hydroxybenzene-		25	9.11
C_6H_5BrO	4-Bromophenol		25	9.37		sulfonic acid			
$C_6H_5Br_2N$	3,5-Dibromoaniline		25	2.34	$C_6H_6O_6$	cis-1-Propene-1,2,3-		25	1.95
C_6H_5ClO	2-Chlorophenol		25	8.56		tricarboxylic acid			
C_6H_5ClO	3-Chlorophenol		25	9.12	$C_6H_6O_6$	trans-1-Propene-1,2,3-	1	25	2.80
C_6H_5ClO	4-Chlorophenol		25	9.41		tricarboxylic acid	2	25	4.46
$C_6H_5Cl_2N$	2,4-Dichloroaniline		22	2.05	C_6H_6S	Benzenethiol		25	6.62
C_6H_5FO	2-Fluorophenol		25	8.73	$C_6H_7BO_2$	Benzeneboronic acid			8.83
C_6H_5FO	3-Fluorophenol		25	9.29	C_6H_7N	Aniline		25	4.87
C_6H_5FO	4-Fluorophenol		25	9.89	C_6H_7N	2-Methylpyridine		25	6.00
C_6H_5IO	2-Iodophenol		25	8.51	C_6H_7N	3-Methylpyridine		25	5.70
C_6H_5IO	3-Iodophenol		25	9.03	C_6H_7N	4-Methylpyridine		25	5.99
C_6H_5IO	4-Iodophenol		25	9.33	C_6H_7NO	2-Aminophenol	1	20	4.78
C_6H_5NO	2-Pyridinecarboxaldehyde		25	12.68			2	20	9.97
C_6H_5NO	4-Pyridinecarboxaldehyde		30	12.05	C_6H_7NO	3-Aminophenol	1	20	4.37
$C_6H_5NO_2$	Nitrobenzene		0	3.98			2	20	9.82
$C_6H_5NO_2$	2-Pyridinecarboxylic acid	1	20	0.99	C_6H_7NO	4-Aminophenol	1	25	5.48
		2	20	5.39			2	25	10.30
$C_6H_5NO_2$	3-Pyridinecarboxylic acid	1	25	2.00	C_6H_7NO	2-Methoxypyridine		20	3.28
		2	25	4.82	C_6H_7NO	3-Methoxypyridine		25	4.78
$C_6H_5NO_2$	4-Pyridinecarboxylic acid	1	25	1.77	C_6H_7NO	4-Methoxypyridine		25	6.58
		2	25	4.84	$C_6H_7NO_3S$	2-Aminobenzenesulfonic		25	2.46
$C_6H_5NO_3$	2-Nitrophenol		25	7.23		acid			
$C_6H_5NO_3$	3-Nitrophenol		25	8.36	$C_6H_7NO_3S$	3-Aminobenzenesulfonic		25	3.74
$C_6H_5NO_3$	4-Nitrophenol		25	7.15		acid			
$C_6H_5N_3$	1H-Benzotriazole		20	1.6	$C_6H_7NO_3S$	4-Aminobenzenesulfonic		25	3.23
$C_6H_5N_5O$	2-Amino-4-	1	20	2.27		acid			
	hydroxypteridine	2	20	7.96	$C_6H_8N_2$	N-Methylpyridinamine		20	9.65
$C_6H_5N_5O_2$	Xanthopterin	2	20	6.59	$C_6H_8N_2$	o-Phenylenediamine	1	20	4.57

Mol. form.	Name	Step	$t/°C$	pK_a	Mol. form.	Name	Step	$t/°C$	pK_a
		2	20	0.80	$C_6H_{13}NO_2$	L-Leucine	1	25	2.33
$C_6H_8N_2$	m-Phenylenediamine	1	20	5.11			2	25	9.74
		2	20	2.50	$C_6H_{13}NO_2$	L-Isoleucine	1	25	2.32
$C_6H_8N_2$	p-Phenylenediamine	1	20	6.31			2	25	9.76
		2	20	2.97	$C_6H_{13}NO_2$	L-Norleucine	1	25	2.34
$C_6H_8N_2$	Phenylhydrazine		15	8.79			2	25	9.83
$C_6H_8O_2$	2,4-Hexadienoic acid		25	4.76	$C_6H_{13}NO_2$	6-Aminohexanoic acid	1	25	4.37
$C_6H_8O_2$	1,3-Cyclohexanedione		25	5.26			2	25	10.80
$C_6H_8O_4$	2,2-Dimethyl-1,3-dioxane-4,6-dione			5.1	$C_6H_{13}NO_4$	N,N-Bis(2-hydroxy-ethyl)glycine	2	20	8.35
$C_6H_8O_6$	L-Ascorbic acid	1	25	4.04	$C_6H_{13}N_3O_3$	Citrulline	1	25	2.43
		2	16	11.7			2	25	9.69
$C_6H_8O_7$	Citric acid	1	25	3.13	$C_6H_{14}N_2$	cis-1,2-Cyclohexane-diamine	1	20	9.93
		2	25	4.76			2	20	6.13
		3	25	6.40	$C_6H_{14}N_2$	trans-1,2-Cyclohexane-diamine	1	20	9.94
$C_6H_8O_7$	Isocitric acid	1	25	3.29			2	20	6.47
		2	25	4.71	$C_6H_{14}N_2$	cis-2,5-Dimethyl-piperazine	1	25	9.66
		3	25	6.40			2	25	5.20
$C_6H_9NO_6$	Nitrilotriacetic acid	1	20	3.03	$C_6H_{14}N_2O_2$	L-Lysine	1	25	2.16
		2	20	3.07			2	25	9.06
		3	20	10.70			3	25	10.54
$C_6H_9NO_6$	L-γ-Carboxyglutamic acid	1	25	1.7	$C_6H_{14}N_4O_2$	L-Arginine	1	25	1.82
		2	25	3.2			2	25	8.99
		3	25	4.75			3	25	12.5
		4	25	9.9	$C_6H_{14}O_6$	D-Mannitol		18	13.5
$C_6H_9N_3$	4,6-Dimethylpyrimi-dinamine		20	4.82	$C_6H_{15}N$	Hexylamine		25	10.56
					$C_6H_{15}N$	Diisopropylamine		25	11.05
$C_6H_9N_3O_2$	L-Histidine	1	25	1.80	$C_6H_{15}N$	Triethylamine		25	10.75
		2	25	6.04	$C_6H_{15}NO_3$	Triethanolamine		25	7.76
		3	25	9.33	$C_6H_{16}N_2$	1,6-Hexanediamine	1	0	11.86
$C_6H_{10}O_2$	Cyclopentanecarboxylic acid		25	4.99			2	0	10.76
$C_6H_{10}O_3$	Ethyl acetoacetate		25	10.68	$C_6H_{16}N_2$	N,N,N',N'-Tetramethyl-1,2-ethanediamine	1	25	10.40
$C_6H_{10}O_4$	3-Methylglutaric acid		25	4.24			2	25	8.26
$C_6H_{10}O_4$	Adipic acid	1	18	4.41	$C_6H_{19}NSi_2$	Hexamethyldisilazane			7.55
		2	18	5.41	$C_7HF_5O_2$	Pentafluorobenzoic acid			1.75
$C_6H_{11}NO_2$	2-Piperidinecarboxylic acid	1	25	2.28	$C_7H_3Br_2NO$	3,5-Dibromo-4-hydroxybenzonitrile			4.06
		2	25	10.72	$C_7H_3N_3O_8$	2,4,6-Trinitrobenzoic acid		25	0.65
$C_6H_{11}NO_3$	Adipamic acid		25	4.63	$C_7H_4Cl_3NO_3$	Triclopyr			2.68
$C_6H_{11}NO_4$	2-Aminoadipic acid	1	25	2.14	$C_7H_4N_2O_6$	2,4-Dinitrobenzoic acid		25	1.43
		2	25	4.21	$C_7H_5BrO_2$	2-Bromobenzoic acid		25	2.85
		3	25	9.77	$C_7H_5BrO_2$	3-Bromobenzoic acid		25	3.81
$C_6H_{11}N_3O_4$	N-(N-Glycylglycyl)glycine	1	25	3.225	$C_7H_5BrO_2$	4-Bromobenzoic acid		25	3.96
		2	25	8.09	$C_7H_5ClO_2$	2-Chlorobenzoic acid		25	2.90
$C_6H_{11}N_3O_4$	Glycylasparagine	1	25	2.942	$C_7H_5ClO_2$	3-Chlorobenzoic acid		25	3.84
		2	18	8.44	$C_7H_5ClO_2$	4-Chlorobenzoic acid		25	4.00
$C_6H_{12}N_2$	Triethylenediamine	1		3.0	$C_7H_5FO_2$	2-Fluorobenzoic acid		25	3.27
		2		8.7	$C_7H_5FO_2$	3-Fluorobenzoic acid		25	3.86
$C_6H_{12}N_2O_4S_2$	L-Cystine	1		1	$C_7H_5FO_2$	4-Fluorobenzoic acid		25	4.15
		2		2.1	$C_7H_5F_3O$	2-(Trifluoromethyl)phenol		25	8.95
		3		8.02	$C_7H_5F_3O$	3-(Trifluoromethyl)phenol		25	8.68
		4		8.71	$C_7H_5IO_2$	2-Iodobenzoic acid		25	2.86
$C_6H_{12}O_2$	Hexanoic acid		25	4.85	$C_7H_5IO_2$	3-Iodobenzoic acid		25	3.87
$C_6H_{12}O_2$	4-Methylpentanoic acid		18	4.84	$C_7H_5IO_2$	4-Iodobenzoic acid		25	4.00
$C_6H_{12}O_6$	β-D-Fructose		25	12.27	C_7H_5NO	2-Hydroxybenzonitrile		25	6.86
$C_6H_{12}O_6$	α-D-Glucose		25	12.46	C_7H_5NO	3-Hydroxybenzonitrile		25	8.61
$C_6H_{12}O_6$	D-Mannose		25	12.08	C_7H_5NO	4-Hydroxybenzonitrile		25	7.97
$C_6H_{13}N$	Cyclohexylamine		25	10.64	$C_7H_5NO_3S$	Saccharin		18	11.68
$C_6H_{13}N$	1-Methylpiperidine		25	10.38	$C_7H_5NO_4$	2-Nitrobenzoic acid		25	2.17
$C_6H_{13}N$	1,2-Dimethylpyrrolidine		26	10.20	$C_7H_5NO_4$	3-Nitrobenzoic acid		25	3.46
$C_6H_{13}NO$	N-Ethylmorpholine		25	7.67	$C_7H_5NO_4$	4-Nitrobenzoic acid		25	3.43

Mol. form.	Name	Step	$t/°C$	pK_a
$C_7H_5NO_4$	2,3-Pyridinedicarboxylic acid	1	25	2.43
		2	25	4.78
$C_7H_5NO_4$	2,4-Pyridinedicarboxylic acid	1	25	2.15
$C_7H_5NO_4$	2,6-Pyridinedicarboxylic acid	1	25	2.16
		2	25	4.76
$C_7H_5NO_4$	3,5-Pyridinedicarboxylic acid	1	25	2.80
	Chlorothiazide	1		6.85
$C_7H_6ClN_3O_4S_2$		2		9.45
$C_7H_6F_3N$	3-(Trifluoromethyl)aniline		25	3.49
$C_7H_6F_3N$	4-(Trifluoromethyl)aniline		25	2.45
$C_7H_6N_2$	1H-Benzimidazole		25	5.53
$C_7H_6N_2$	2-Aminobenzonitrile		25	0.77
$C_7H_6N_2$	3-Aminobenzonitrile		25	2.75
$C_7H_6N_2$	4-Aminobenzonitrile		25	1.74
C_7H_6O	Benzaldehyde		25	14.90
$C_7H_6O_2$	Benzoic acid		25	4.204
$C_7H_6O_2$	Salicylaldehyde		25	8.37
$C_7H_6O_2$	3-Hydroxybenzaldehyde		25	8.98
$C_7H_6O_2$	4-Hydroxybenzaldehyde		25	7.61
$C_7H_6O_3$	2-Hydroxybenzoic acid	1	20	2.98
		2	20	13.6
$C_7H_6O_3$	3-Hydroxybenzoic acid	1	25	4.08
		2	19	9.92
$C_7H_6O_3$	4-Hydroxybenzoic acid	1	25	4.57
		2	25	9.46
$C_7H_6O_4$	2,4-Dihydroxybenzoic acid	1	25	3.11
		2	25	8.55
		3	25	14.0
$C_7H_6O_4$	2,5-Dihydroxybenzoic acid	1	25	2.97
$C_7H_6O_4$	3,4-Dihydroxybenzoic acid	1	25	4.48
		2	25	8.83
		3	25	12.6
$C_7H_6O_4$	3,5-Dihydroxybenzoic acid	1	25	4.04
$C_7H_6O_5$	2,4,6-Trihydroxybenzoic acid		25	1.68
$C_7H_6O_5$	3,4,5-Trihydroxybenzoic acid		25	4.41
C_7H_7NO	Benzamide		25	~13
$C_7H_7NO_2$	Aniline-2-carboxylic acid	1	25	2.17
		2	25	4.85
$C_7H_7NO_2$	Aniline-3-carboxylic acid	1	25	3.07
		2	25	4.79
$C_7H_7NO_2$	Aniline-4-carboxylic acid	1	25	2.50
		2	25	4.87
$C_7H_7NO_3$	4-Amino-2-hydroxy-benzoic acid			3.25
$C_7H_8ClN_3O_4S_2$	Hydrochlorothiazide	1		7.9
		2		9.2
$C_7H_8N_4O_2$	Theobromine		18	7.89
$C_7H_8N_4O_2$	Theophylline	1	25	8.77
C_7H_8O	o-Cresol		25	10.29
C_7H_8O	m-Cresol		25	10.09
C_7H_8O	p-Cresol		25	10.26
C_7H_8OS	4-(Methylthio)phenol		25	9.53
$C_7H_8O_2$	2-Methoxyphenol		25	9.98
$C_7H_8O_2$	3-Methoxyphenol		25	9.65
$C_7H_8O_2$	4-Methoxyphenol		25	10.21
C_7H_8S	Benzenemethanethiol		25	9.43
C_7H_9N	Benzylamine		25	9.34

Mol. form.	Name	Step	$t/°C$	pK_a
C_7H_9N	2-Methylaniline		25	4.45
C_7H_9N	3-Methylaniline		25	4.71
C_7H_9N	4-Methylaniline		25	5.08
C_7H_9N	N-Methylaniline		25	4.85
C_7H_9N	2-Ethylpyridine		25	5.89
C_7H_9N	2,3-Dimethylpyridine		25	6.57
C_7H_9N	2,4-Dimethylpyridine		25	6.99
C_7H_9N	2,5-Dimethylpyridine		25	6.40
C_7H_9N	2,6-Dimethylpyridine		25	6.65
C_7H_9N	3,4-Dimethylpyridine		25	6.46
C_7H_9N	3,5-Dimethylpyridine		25	6.15
C_7H_9NO	2-Methoxyaniline		25	4.53
C_7H_9NO	3-Methoxyaniline		25	4.20
C_7H_9NO	4-Methoxyaniline		25	5.36
C_7H_9NS	2-(Methylthio)aniline		25	3.45
C_7H_9NS	4-(Methylthio)aniline		25	4.35
$C_7H_9N_5$	2-Dimethylaminopurine	1	20	4.00
		2	20	10.24
$C_7H_{11}N_3O_2$	L-1-Methylhistidine	1	25	1.69
		2	25	6.48
		3	25	8.85
$C_7H_{11}N_3O_2$	L-3-Methylhistidine	1	25	1.92
		2	25	6.56
		3	25	8.73
$C_7H_{12}O_2$	Cyclohexanecarboxylic acid		25	4.91
$C_7H_{12}O_4$	Heptanedioic acid	1	25	4.71
		2	25	5.58
$C_7H_{12}O_4$	Butylpropanedioic acid	1	5	2.96
$C_7H_{13}NO_4$	α-Ethylglutamic acid	1	25	3.846
		2	25	7.838
$C_7H_{14}O_2$	Heptanoic acid		25	4.89
$C_7H_{14}O_6$	α-Methylglucoside		25	13.71
$C_7H_{15}N$	1-Ethylpiperidine		23	10.45
$C_7H_{15}N$	1,2-Dimethylpiperidine,(±)		25	10.22
$C_7H_{15}NO_3$	Carnitine		25	3.80
$C_7H_{17}N$	Heptylamine		25	10.67
$C_7H_{17}N$	2-Heptanamine		19	10.7
$C_8H_5NO_2$	3-Cyanobenzoic acid		25	3.60
$C_8H_5NO_2$	4-Cyanobenzoic acid		25	3.55
$C_8H_6N_2$	Cinnoline		20	2.37
$C_8H_6N_2$	Quinazoline		29	3.43
$C_8H_6N_2$	Quinoxaline		20	0.56
$C_8H_6N_2$	Phthalazine		20	3.47
$C_8H_6N_4O_5$	Nitrofurantoin			7.2
$C_8H_6O_3$	3-Formylbenzoic acid		25	3.84
$C_8H_6O_3$	4-Formylbenzoic acid		25	3.77
$C_8H_6O_4$	Phthalic acid	1	25	2.943
		2	25	5.432
$C_8H_6O_4$	Isophthalic acid	1	25	3.70
		2	25	4.60
$C_8H_6O_4$	Terephthalic acid	1	25	3.54
		2	25	4.34
$C_8H_7ClO_2$	2-Chlorobenzeneacetic acid		25	4.07
$C_8H_7ClO_2$	3-Chlorobenzeneacetic acid		25	4.14
$C_8H_7ClO_2$	4-Chlorobenzeneacetic acid		25	4.19
$C_8H_7ClO_3$	2-Chlorophenoxyacetic acid		25	3.05
$C_8H_7ClO_3$	3-Chlorophenoxyacetic acid		25	3.10
$C_8H_7NO_4$	2-Nitrobenzeneacetic acid		25	4.00
$C_8H_7NO_4$	3-Nitrobenzeneacetic acid		25	3.97
$C_8H_7NO_4$	4-Nitrobenzeneacetic acid		25	3.85
$C_8H_8F_3N_3O_4S_2$	Hydroflumethiazide	1		8.9

Mol. form.	Name	Step	t/°C	pKₐ
		2		9.7
$C_8H_8N_2$	2-Methyl-1H-benzimidazole	1	25	6.19
$C_8H_8O_2$	o-Toluic acid		25	3.91
$C_8H_8O_2$	m-Toluic acid		25	4.25
$C_8H_8O_2$	p-Toluic acid		25	4.37
$C_8H_8O_2$	Benzeneacetic acid		25	4.31
$C_8H_8O_2$	1-(2-Hydroxyphenyl)ethanone		25	10.06
$C_8H_8O_2$	1-(3-Hydroxyphenyl)ethanone		25	9.19
$C_8H_8O_2$	1-(4-Hydroxyphenyl)ethanone		25	8.05
$C_8H_8O_3$	2-Methoxybenzoic acid		25	4.08
$C_8H_8O_3$	3-Methoxybenzoic acid		25	4.10
$C_8H_8O_3$	4-Methoxybenzoic acid		25	4.50
$C_8H_8O_3$	Phenoxyacetic acid		25	3.17
$C_8H_8O_3$	Mandelic acid		25	3.37
$C_8H_8O_4$	2,5-Hydroxybenzeneacetic acid		25	4.40
C_8H_9NO	Acetanilide		25	0.5
$C_8H_9NO_2$	2-(Methylamino)benzoic acid		25	5.34
$C_8H_9NO_2$	3-(Methylamino)benzoic acid		25	5.10
$C_8H_9NO_2$	4-(Methylamino)benzoic acid		25	5.04
$C_8H_9NO_2$	N-Phenylglycine	1	25	1.83
		2		4.39
$C_8H_{10}BrN$	4-Bromo-N,N-dimethylaniline		25	4.23
$C_8H_{10}ClN$	3-Chloro-N,N-dimethylaniline		20	3.83
$C_8H_{10}ClN$	4-Chloro-N,N-dimethylaniline		20	4.39
$C_8H_{10}N_2O_2$	N,N-Dimethyl-3-nitroaniline		25	2.62
$C_8H_{11}N$	N-Ethylaniline		25	5.12
$C_8H_{11}N$	N,N-Dimethylaniline		25	5.07
$C_8H_{11}N$	2,6-Dimethylaniline		25	3.89
$C_8H_{11}N$	Benzeneethanamine		25	9.83
$C_8H_{11}N$	2,4,6-Trimethylpyridine		25	7.43
$C_8H_{11}NO$	2-Ethoxyaniline		28	4.43
$C_8H_{11}NO$	3-Ethoxyaniline		25	4.18
$C_8H_{11}NO$	4-Ethoxyaniline		28	5.20
$C_8H_{11}NO$	4-(2-Aminoethyl)phenol	1	25	9.74
		2	25	10.52
$C_8H_{11}NO$	2-(2-Methoxyethyl)pyridine			5.5
$C_8H_{11}NO_2$	Dopamine	1	25	8.9
		2	25	10.6
$C_8H_{11}NO_3$	Norepinephrine	1	25	8.64
		2	25	9.70
$C_8H_{11}N_3O_6$	6-Azauridine			6.70
$C_8H_{11}N_5$	Phenylbiguanide	1		10.76
		2		2.13
$C_8H_{12}N_2O_3$	Barbital		25	7.43
$C_8H_{12}O_2$	5,5-Dimethyl-1,3-cyclohexanedione		25	5.15
$C_8H_{13}NO_2$	Arecoline			6.84
$C_8H_{14}O_2S_2$	Thioctic acid			5.4
$C_8H_{14}O_4$	Octanedioic acid	1	25	4.52
$C_8H_{15}NO$	Tropine		15	3.80
$C_8H_{15}NO$	Pseudotropine		15	3.80
$C_8H_{16}N_2O_3$	N-Glycylleucine		25	3.18
$C_8H_{16}N_2O_3$	N-Leucylglycine	1	25	3.25

Mol. form.	Name	Step	t/°C	pKₐ
		2	25	8.2
$C_8H_{16}N_2O_4S_2$	Homocystine	1	25	1.59
		2	25	2.54
		3	25	8.52
		4	25	9.44
$C_8H_{16}O_2$	Octanoic acid		25	4.89
$C_8H_{16}O_2$	2-Propylpentanoic acid			4.6
$C_8H_{17}N$	2-Propylpiperidine,(S)			10.9
$C_8H_{17}N$	2,2,4-Trimethylpiperidine		30	11.04
$C_8H_{17}NO$	$trans$-6-Propyl-3-piperidinol,(3S)			10.3
$C_8H_{19}N$	Octylamine		25	10.65
$C_8H_{19}N$	N-Methyl-2-heptanamine		17	10.99
$C_8H_{19}N$	Dibutylamine		21	11.25
$C_8H_{20}N_2$	1,8-Octanediamine	1	20	11.00
		2	20	10.1
C_9H_6BrN	3-Bromoquinoline		25	2.69
$C_9H_7ClO_2$	$trans$-o-Chlorocinnamic acid		25	4.23
$C_9H_7ClO_2$	$trans$-m-Chlorocinnamic acid		25	4.29
$C_9H_7ClO_2$	$trans$-p-Chlorocinnamic acid		25	4.41
C_9H_7N	Quinoline		20	4.90
C_9H_7N	Isoquinoline		20	5.40
C_9H_7NO	2-Quinolinol	1	20	-0.31
		2	20	11.76
C_9H_7NO	3-Quinolinol	1	20	4.28
		2	20	8.08
C_9H_7NO	4-Quinolinol	1	20	2.23
		2	20	11.28
C_9H_7NO	6-Quinolinol	1	20	5.15
		2	20	8.90
C_9H_7NO	8-Quinolinol	1	25	4.91
		2	25	9.81
C_9H_7NO	7-Isoquinolinol	1	20	5.68
		2	20	8.90
$C_9H_7NO_3$	2-Cyanophenoxyacetic acid		25	2.98
$C_9H_7NO_3$	3-Cyanophenoxyacetic acid		25	3.03
$C_9H_7NO_3$	4-Cyanophenoxyacetic acid		25	2.93
$C_9H_7N_7O_2S$	Azathioprine			8.2
$C_9H_8N_2$	2-Quinolinamine		20	7.34
$C_9H_8N_2$	3-Quinolinamine		20	4.91
$C_9H_8N_2$	4-Quinolinamine		20	9.17
$C_9H_8N_2$	1-Isoquinolinamine		20	7.62
$C_9H_8N_2$	3-Isoquinolinamine		20	5.05
$C_9H_8O_2$	cis-Cinnamic acid		25	3.88
$C_9H_8O_2$	$trans$-Cinnamic acid		25	4.44
$C_9H_8O_2$	α-Methylenebenzene-acetic acid			4.35
$C_9H_8O_4$	2-(Acetyloxy)benzoic acid		25	3.48
$C_9H_9Br_2NO_3$	3,5-Dibromo-L-tyrosine	1		2.17
		2		6.45
		3		7.60
$C_9H_9ClO_2$	3-(2-Chlorophenyl)-propanoic acid		25	4.58
$C_9H_9ClO_2$	3-(3-Chlorophenyl)-propanoic acid		25	4.59
$C_9H_9ClO_2$	3-(4-Chlorophenyl)-propanoic acid		25	4.61
$C_9H_9I_2NO_3$	L-3,5-Diiodotyrosine	1	25	2.12

Mol. form.	Name	Step	$t/°C$	pK_a
		2	25	5.32
		3	25	9.48
$C_9H_9NO_3$	N-Benzoylglycine		25	3.62
$C_9H_9NO_4$	3-(2-Nitrophenyl)-propanoic acid		25	4.50
$C_9H_9NO_4$	3-(4-Nitrophenyl)-propanoic acid		25	4.47
$C_9H_9N_3O_2$	Carbendazim			4.48
$C_9H_9N_3O_2S_2$	Sulfathiazole			7.2
$C_9H_{10}INO_3$	L-3-Iodotyrosine	1	25	2.2
		2	25	8.7
		3	25	9.1
$C_9H_{10}N_2$	2-Ethylbenzimidazole		25	6.18
$C_9H_{10}O_2$	3,5-Dimethylbenzoic acid		25	4.32
$C_9H_{10}O_2$	Benzenepropanoic acid		25	4.66
$C_9H_{10}O_2$	α-Methylbenzeneacetic acid		25	4.64
$C_9H_{10}O_3$	α-Hydroxy-α-methyl-benezeneacetic acid		25	3.47
$C_9H_{11}Cl_2N_3O_4S_2$	Methylclothiazide			9.4
$C_9H_{11}N$	N-Allylaniline		25	4.17
$C_9H_{11}N$	1-Indanamine		22	9.21
$C_9H_{11}NO_2$	4-(Dimethylamino)-benzoic acid	1		6.03
		2		11.49
$C_9H_{11}NO_2$	Ethyl 4-aminobenzoate			2.5
$C_9H_{11}NO_2$	L-Phenylalanine	1	25	2.20
		2	25	9.31
$C_9H_{11}NO_3$	L-Tyrosine	1	25	2.20
		2	25	9.11
		3	25	10.1
$C_9H_{11}NO_4$	Levodopa	1	25	2.32
		2	25	8.72
		3	25	9.96
		4	25	11.79
$C_9H_{12}N_2O_2$	Tyrosineamide		25	7.33
$C_9H_{13}N$	N-Isopropylaniline		25	5.77
$C_9H_{13}NO_3$	Epinephrine	1	25	8.66
		2	25	9.95
$C_9H_{13}N_2O_9P$	5'-Uridylic acid	1		6.4
		2		9.5
$C_9H_{13}N_3O_5$	Cytidine	1		4.22
		2		12.5
$C_9H_{14}ClNO$	Phenylpropanolamine hydrochloride			9.44
$C_9H_{14}N_2O_3$	Metharbital			8.45
$C_9H_{14}N_3O_8P$	3'-Cytidylic acid	1		0.8
		2		4.28
		3		6.0
$C_9H_{14}N_4O_3$	Carnosine	1	20	2.73
		2	20	6.87
		3	20	9.73
$C_9H_{15}NO_3S$	Captopril	1		3.7
		2		9.8
$C_9H_{15}N_5O$	Minoxidil			4.61
$C_9H_{16}O_4$	Nonanedioic acid	1	25	4.53
		2	25	5.33
$C_9H_{18}O_2$	Nonanoic acid		25	4.96
$C_9H_{19}N$	N-Butylpiperidine		23	10.47
$C_9H_{19}N$	2,2,6,6-Tetramethyl-piperidine		25	11.07
$C_9H_{21}N$	Nonylamine		25	10.64
$C_{10}H_7NO_2$	8-Quinolinecarboxylic acid		25	1.82
$C_{10}H_8O$	1-Naphthol		25	9.39
$C_{10}H_8O$	2-Naphthol		25	9.63
$C_{10}H_9N$	1-Naphthylamine		25	3.92
$C_{10}H_9N$	2-Naphthylamine		25	4.16
$C_{10}H_9N$	2-Methylquinoline		20	5.83
$C_{10}H_9N$	4-Methylquinoline		20	5.67
$C_{10}H_9N$	5-Methylquinoline		20	5.20
$C_{10}H_9NO$	5-Amino-1-naphthol		25	3.97
$C_{10}H_9NO$	6-Methoxyquinoline		20	5.03
$C_{10}H_9NO_2$	1H-Indole-3-acetic acid			4.75
$C_{10}H_{10}O_2$	o-Methylcinnamic acid		25	4.50
$C_{10}H_{10}O_2$	m-Methylcinnamic acid		25	4.44
$C_{10}H_{10}O_2$	p-Methylcinnamic acid		25	4.56
$C_{10}H_{12}N_2$	Tryptamine		25	10.2
$C_{10}H_{12}N_2O$	5-Hydroxytryptamine	1	25	9.8
		2	25	11.1
$C_{10}H_{12}N_2O_5$	Dinoseb			4.62
$C_{10}H_{12}N_4O_3$	Dideoxyinosine			9.12
$C_{10}H_{12}O$	5,6,7,8-Tetrahydro-2-naphthalenol		25	10.48
$C_{10}H_{12}O_2$	Benzenebutanoic acid		25	4.76
$C_{10}H_{12}O_5$	Propyl 3,4,5-trihydroxy-benzoate			8.11
$C_{10}H_{13}N_5O_4$	Adenosine	1	25	3.6
		2	25	12.4
$C_{10}H_{14}N_2$	L-Nicotine	1		8.02
		2		3.12
$C_{10}H_{14}N_5O_7P$	5'-Adenylic acid	1		3.8
		2		6.2
$C_{10}H_{14}O$	2-tert-Butylphenol		25	10.62
$C_{10}H_{14}O$	3-tert-Butylphenol		25	10.12
$C_{10}H_{14}O$	4-tert-Butylphenol		25	10.23
$C_{10}H_{15}N$	N-tert-Butylaniline		25	7.00
$C_{10}H_{15}N$	N,N-Diethylaniline		25	6.57
$C_{10}H_{15}NO$	d-Ephedrine		10	10.139
$C_{10}H_{15}NO$	l-Ephedrine		10	9.958
$C_{10}H_{17}N_3O_6S$	l-Glutathione	1	25	2.12
		2	25	3.59
		3	25	8.75
		4	25	9.65
$C_{10}H_{18}N_4O_5$	L-Argininosuccinic acid	1	25	1.62
		2	25	2.70
		3	25	4.26
		4	25	9.58
$C_{10}H_{18}O_4$	Sebacic acid	1	25	4.59
		2	25	5.59
$C_{10}H_{19}N$	Bornylamine		25	10.17
$C_{10}H_{19}N$	Neobornylamine		25	10.01
$C_{10}H_{21}N$	Butylcyclohexylamine		25	11.23
$C_{10}H_{21}N$	1,2,2,6,6-Pentamethyl-piperidine		30	11.25
$C_{10}H_{23}N$	Decylamine		25	10.64
$C_{11}H_8N_2$	1H-Perimidine		20	6.35
$C_{11}H_8O_2$	1-Naphthalenecarboxylic acid		25	3.69
$C_{11}H_8O_2$	2-Naphthalenecarboxylic acid		25	4.16
$C_{11}H_{11}N$	Methyl-1-naphthylamine		27	3.67
$C_{11}H_{12}I_3NO_2$	Iopanoic acid			4.8
$C_{11}H_{12}N_2O_2$	L-Tryptophan	1	25	2.46
		2	25	9.41

Dissociation Constants of Organic Acids and Bases

Mol. form.	Name	Step	t/°C	pK_a
$C_{11}H_{12}N_4O_3S$	Sulfamethoxypyridazine			6.7
$C_{11}H_{13}F_3N_2O_3S$	Mefluidide			4.6
$C_{11}H_{13}NO_3$	Hydrastinine			11.38
$C_{11}H_{13}N_3O_3S$	Sulfisoxazole			5
$C_{11}H_{14}N_2O$	Cytisine	1		6.11
		2		13.08
$C_{11}H_{14}O_2$	2-tert-Butylbenzoic acid		25	3.54
$C_{11}H_{14}O_2$	3-tert-Butylbenzoic acid		25	4.20
$C_{11}H_{14}O_2$	4-tert-Butylbenzoic acid		25	4.38
$C_{11}H_{16}N_2O_2$	Pilocarpine	1	25	1.6
		2	25	6.9
$C_{11}H_{16}N_4O_4$	Pentostatin			5.2
$C_{11}H_{17}N$	N,N-Diethyl-2-methyl-aniline		25	7.24
$C_{11}H_{17}NO_3$	Isoproterenol			8.64
$C_{11}H_{17}N_3O_8$	Tetrodotoxin			8.76
$C_{11}H_{18}ClNO_3$	Methoxamine hydrochloride		25	9.2
$C_{11}H_{18}N_2O_3$	Amobarbital		25	8.0
$C_{11}H_{25}N$	Undecylamine		25	10.63
$C_{11}H_{26}NO_2PS$	Methylphosphonothioic acid S[2-[bis(1-isopropyl)amino]ethyl], O-ethylester			7.9
$C_{12}H_6Cl_4O_2S$	Bithionol	1		4.82
		2		10.50
$C_{12}H_8N_2$	1,10-Phenanthroline		25	4.84
$C_{12}H_8N_2$	Phenazine		20	1.20
$C_{12}H_{10}O$	2-Hydroxybiphenyl		25	10.01
$C_{12}H_{10}O$	3-Hydroxybiphenyl		25	9.64
$C_{12}H_{10}O$	4-Hydroxybiphenyl		25	9.55
$C_{12}H_{11}N$	Diphenylamine		25	0.79
$C_{12}H_{11}N$	2-Aminobiphenyl		25	3.83
$C_{12}H_{11}N$	3-Aminobiphenyl		18	4.25
$C_{12}H_{11}N$	4-Aminobiphenyl		18	4.35
$C_{12}H_{11}N$	2-Benzylpyridine		25	5.13
$C_{12}H_{11}N_3$	4-Aminoazobenzene		25	2.82
$C_{12}H_{12}N_2$	p-Benzidine	1	20	4.65
		2	20	3.43
$C_{12}H_{12}N_2O_3$	Phenobarbital	1		7.3
		2		11.8
$C_{12}H_{13}I_3N_2O_3$	Iocetamic acid			4
$C_{12}H_{13}N$	N,N-Dimethyl-1-naphthylamine		25	4.83
$C_{12}H_{13}N$	N,N-Dimethyl-2-naphthylamine		25	4.566
$C_{12}H_{14}N_4O_2S$	Sulfamethazine	1		7.4
		2		2.65
$C_{12}H_{14}N_4O_3S$	Sulfacytine			6.9
$C_{12}H_{17}N_3O_4$	Agaritine	1		3.4
		2		8.86
$C_{12}H_{20}N_2O_2$	Aspergillic acid			5.5
$C_{12}H_{21}N_5O_2S_2$	Nizatidine	1		2.1
		2		6.8
$C_{12}H_{22}O_{11}$	Sucrose		25	12.7
$C_{12}H_{22}O_{11}$	α-Maltose		21	12.05
$C_{12}H_{23}N$	Dicyclohexylamine			10.4
$C_{12}H_{27}N$	Dodecylamine		25	10.63
$C_{13}H_9N$	Acridine		20	5.58
$C_{13}H_9N$	Phenanthridine		20	5.58
$C_{13}H_{10}N_2$	9-Acridinamine		20	9.99
$C_{13}H_{10}N_2$	2-Phenylbenzimidazole	1	25	5.23
		2	25	11.91
$C_{13}H_{10}O_2$	2-Phenylbenzoic acid		25	3.46
$C_{13}H_{10}O_3$	2-Phenoxybenzoic acid		25	3.53
$C_{13}H_{10}O_3$	3-Phenoxybenzoic acid		25	3.95
$C_{13}H_{10}O_3$	4-Phenoxybenzoic acid		25	4.57
$C_{13}H_{11}N_3$	3,6-Acridinediamine		20	9.65
$C_{13}H_{12}Cl_2O_4$	Ethacrynic acid			3.50
$C_{13}H_{12}N_2O$	Harmine			7.70
$C_{13}H_{12}N_2O_3S$	Sulfabenzamide		25	4.57
$C_{13}H_{13}N$	4-Benzylaniline		25	2.17
$C_{13}H_{14}N_2O_{13}$	Harmaline			4.2
$C_{13}H_{15}N_3O_3$	Imazapyr	1		1.9
		2		3.6
$C_{13}H_{16}ClNO$	Ketamine			7.5
$C_{13}H_{19}NO_4S$	4-[(Dipropylamino)-sulfonyl]benzoic acid			5.8
$C_{13}H_{21}N$	2,6-Di-tert-butylpyridine			3.58
$C_{13}H_{29}N$	(Tridecyl)amine		25	10.63
$C_{14}H_{12}F_3NO_4S_2$	Perfluidone			2.5
$C_{14}H_{12}O_2$	α-Phenylbenzeneacetic acid		25	3.94
$C_{14}H_{12}O_3$	α-Hydroxy-α-phenyl-benzeneacetic acid		25	3.04
$C_{14}H_{18}N_4O_3$	Trimethoprim			6.6
$C_{14}H_{19}NO_2$	Methylphenidate			8.9
$C_{14}H_{21}N_3O_3S$	Tolazamide		25	3.6
$C_{14}H_{22}N_2O_3$	Atenolol			9.6
$C_{14}H_{31}N$	Tetradecylamine		25	10.62
$C_{15}H_{10}ClN_3O_3$	Clonazepam	1		1.5
		2		10.5
$C_{15}H_{11}I_4NO_4$	L-Thyroxine	1	25	2.2
		2	25	6.45
		3	25	10.1
$C_{15}H_{14}O_3$	Fenoprofen			7.3
$C_{15}H_{15}NO_2$	Mefenamic acid			4.2
$C_{15}H_{15}N_3O_2$	Methyl Red	1		2.5
		2		9.5
$C_{15}H_{17}ClN_4$	NeutralRed			6.7
$C_{15}H_{19}NO_2$	Tropacocaine		15	4.32
$C_{15}H_{19}N_3O_3$	Imazethapyr	1		2.1
		2		3.9
$C_{15}H_{21}N_3O_2$	Physostigmine	1		6.12
		2		12.24
$C_{15}H_{26}N_2$	Sparteine	1	20	2.24
		2	20	9.46
$C_{15}H_{33}N$	Pentadecylamine		25	10.61
$C_{16}H_{13}ClN_2O$	Valium			3.4
$C_{16}H_{14}ClN_3O$	Chlorodiazepoxide			4.8
$C_{16}H_{16}N_2O_2$	Lysergic acid	1		3.44
		2		7.68
$C_{16}H_{17}N_3O_4S$	Cephalexin	1		5.2
		2		7.3
$C_{16}H_{19}N_3O_4S$	Cephradine	1		2.63
		2		7.27
$C_{16}H_{22}N_2$	Lycodine	1		3.97
		2		8.08
$C_{16}H_{35}N$	Hexadecylamine		25	10.61
$C_{17}H_{17}NO_2$	Apomorphine	1		7.0
		2		8.92
$C_{17}H_{19}NO_3$	Piperine		18	12.22
$C_{17}H_{19}NO_3$	Morphine	1	25	8.21
		2	20	9.85
$C_{17}H_{20}N_4O_6$	Riboflavin	1		1.7

Mol. form.	Name	Step	$t/°C$	pK_a
		2	25	9.69
$C_{17}H_{20}O_6$	Mycophenolic acid			4.5
$C_{17}H_{23}NO_3$	Hyoscyamine		21	9.7
$C_{17}H_{27}NO_4$	Nadolol			9.67
$C_{18}H_{19}ClN_4$	Clozapine	1		3.70
		2		7.60
$C_{18}H_{21}NO_3$	Codeine			8.21
$C_{18}H_{21}N_3O$	Dibenzepin			8.25
$C_{18}H_{32}O_2$	Linoleic acid			7.6
$C_{18}H_{33}ClN_2O_5S$	Clindamycin			7.6
$C_{18}H_{39}N$	Octadecylamine		25	10.60
$C_{19}H_{10}Br_4O_5S$	Bromophenol Blue			4.0
$C_{19}H_{14}O_5S$	Phenol Red			7.9
$C_{19}H_{16}ClNO_4$	Indomethacin			4.5
$C_{19}H_{17}N_3O_4S_2$	Cephaloridine			3.2
$C_{19}H_{20}N_2O_2$	Phenylbutazone			4.5
$C_{19}H_{21}N$	Protriptyline			8.2
$C_{19}H_{21}NO_3$	Thebaine		15	6.05
$C_{19}H_{22}N_2O$	Cinchonine	1		5.85
		2		9.92
$C_{19}H_{22}N_2O$	Cinchonidine	1		5.80
		2		10.03
$C_{19}H_{22}N_2O_2$	Cupreine			6.57
$C_{19}H_{22}O_6$	Gibberellic acid			4.0
$C_{19}H_{23}N_3O_2$	Ergometrinine			7.3
$C_{19}H_{23}N_3O_2$	Ergonovine			6.8
$C_{20}H_{14}O_4$	Phenolphthalein		25	9.7
$C_{20}H_{21}NO_4$	Papaverine			6.4
$C_{20}H_{23}N$	Amitriptyline			9.4
$C_{20}H_{23}N_7O_7$	Folinic acid	1		3.1
		2		4.8
		3		10.4
$C_{20}H_{24}N_2O_2$	Quinine	1	25	8.52
		2	25	4.13
$C_{20}H_{24}N_2O_2$	Quinidine	1	20	5.4
		2	20	10.0
$C_{20}H_{26}N_2O_2$	Hydroquinine			5.33
$C_{21}H_{14}Br_4O_5S$	Bromocresol Green			4.7
$C_{21}H_{16}Br_2O_5S$	Bromocresol Purple			6.3
$C_{21}H_{18}O_5S$	CresolRed			8.3
$C_{21}H_{21}NO_6$	Hydrastine			7.8
$C_{21}H_{22}N_2O_2$	Strychnine		25	8.26

Mol. form.	Name	Step	$t/°C$	pK_a
$C_{21}H_{23}ClFNO_2$	Haloperidol			8.3
$C_{21}H_{31}NO_4$	Furethidine			7.48
$C_{21}H_{35}N_3O_7$	Lisinopril	1		2.5
		2		4.0
		3		6.7
		4		10.1
$C_{22}H_{18}O_4$	o-Cresolphthalein			9.4
$C_{22}H_{22}FN_3O_2$	Droperidol			7.64
$C_{22}H_{23}NO_7$	Noscapine			7.8
$C_{22}H_{25}NO_6$	Colchicine		20	12.36
$C_{22}H_{25}N_3O$	Benzpiperylon	1		6.73
		2		9.13
$C_{22}H_{33}NO_2$	Atisine			12.2
$C_{23}H_{26}N_2O_4$	Brucine	1		6.04
		2		11.07
$C_{24}H_{40}O_4$	Deoxycholic acid			6.58
$C_{24}H_{40}O_5$	Cholic acid			6.4
$C_{25}H_{29}I_2NO_3$	Amiodarone		25	6.56
$C_{25}H_{41}NO_9$	Aconine			9.52
$C_{26}H_{43}NO_6$	Glycocholic acid			4.4
$C_{26}H_{45}NO_7S$	Taurocholic acid			1.4
$C_{27}H_{28}Br_2O_5S$	Bromothymol Blue			7.0
$C_{27}H_{38}N_2O_4$	Verapamil			8.6
$C_{29}H_{32}O_{13}$	Etoposide			9.8
$C_{29}H_{40}N_2O_4$	Emetine	1		5.77
		2		6.64
$C_{30}H_{23}BrO_4$	Bromadiolone		21	4.04
$C_{30}H_{48}O_3$	Oleanolic acid			2.52
$C_{31}H_{36}N_2O_{11}$	Novobiocin	1		4.3
		2		9.1
$C_{32}H_{32}O_{13}S$	Teniposide			10.13
$C_{33}H_{40}N_2O_9$	Reserpine			6.6
$C_{34}H_{47}NO_{11}$	Aconitine			5.88
$C_{36}H_{51}NO_{11}$	Veratridine			9.54
$C_{37}H_{67}NO_{13}$	Erythromycin			8.8
$C_{43}H_{58}N_4O_{12}$	Rifampin	1		1.7
		2		7.9
$C_{45}H_{73}NO_{15}$	Solanine		15	6.66
$C_{46}H_{56}N_4O_{10}$	Vincristine			5.4
$C_{46}H_{58}N_4O_9$	Vinblastine	1		5.4
		2		7.4

CONCENTRATIVE PROPERTIES OF AQUEOUS SOLUTIONS: DENSITY, REFRACTIVE INDEX, FREEZING POINT DEPRESSION, AND VISCOSITY

This table gives properties of aqueous solutions of 66 substances as a function of concentration. All data refer to a temperature of 20°C. The properties are:

Mass %: Mass of solute divided by total mass of solution, expressed as percent.

m Molality (moles of solute per kg of water).

c Molarity (moles of solute per liter of solution).

ρ Density of solution in g/cm³.

n Index of refraction, relative to air, at a wavelength of 589 nm (sodium D line); the index of pure water at 20°C is 1.3330.

Δ Freezing point depression in °C relative to pure water.

η Absolute (dynamic) viscosity in mPa s (equal to centipoise, cP); the viscosity of pure water at 20°C is 1.002 mPa s.

Density data for aqueous solutions over a wider range of temperatures and pressures (and for other compounds) may be found in Reference 2. Solutes are listed in the following order:

Acetic acid	Lithium chloride	2-Propanol
Acetone	Magnesium chloride	Silver nitrate
Ammonia	Magnesium sulfate	Sodium acetate
Ammonium chloride	Maltose	Sodium bicarbonate
Ammonium sulfate	Manganese(II) sulfate	Sodium bromide
Barium chloride	D-Mannitol	Sodium carbonate
Calcium chloride	Methanol	Sodium chloride
Cesium chloride	Nitric acid	Sodium citrate
Citric acid	Oxalic acid	Sodium hydroxide
Copper sulfate	Phosphoric acid	Sodium nitrate
Disodium ethylenediamine	Potassium bicarbonate	Sodium phosphate
tetraacetate (EDTA sodium)	Potassium bromide	Sodium hydrogen phosphate
Ethanol	Potassium carbonate	Sodium dihydrogen phosphate
Ethylene glycol	Potassium chloride	Sodium sulfate
Ferric chloride	Potassium hydroxide	Sodium thiosulfate
Formic acid	Potassium iodide	Strontium chloride
D-Fructose	Potassium nitrate	Sucrose
D-Glucose	Potassium permanganate	Sulfuric acid
Glycerol	Potassium hydrogen phosphate	Trichloroacetic acid
Hydrochloric acid	Potassium dihydrogen phosphate	Tris(hydroxymethyl)methylamine
Lactic acid	Potassium sulfate	Urea
Lactose	1-Propanol	Zinc sulfate

References

1. Wolf, A. V., *Aqueous Solutions and Body Fluids*, Hoeber, 1966.
2. Söhnel, O., and Novotny, P., *Densities of Aqueous Solutions of Inorganic Substances*, Elsevier, Amsterdam, 1985.

Solute	Mass %	m/mol kg⁻¹	c/mol L⁻¹	ρ/g cm⁻³	n	Δ/°C	η/mPa s
Acetic acid	0.5	0.084	0.083	0.9989	1.3334	0.16	1.012
CH_3COOH	1.0	0.168	0.166	0.9996	1.3337	0.32	1.022
	2.0	0.340	0.333	1.0011	1.3345	0.63	1.042
	3.0	0.515	0.501	1.0025	1.3352	0.94	1.063
	4.0	0.694	0.669	1.0038	1.3359	1.26	1.084
	5.0	0.876	0.837	1.0052	1.3366	1.58	1.105
	6.0	1.063	1.006	1.0066	1.3373	1.90	1.125
	7.0	1.253	1.175	1.0080	1.3381	2.23	1.143
	8.0	1.448	1.345	1.0093	1.3388	2.56	1.162
	9.0	1.647	1.515	1.0107	1.3395	2.89	1.186
	10.0	1.850	1.685	1.0121	1.3402	3.23	1.210
	12.0	2.271	2.028	1.0147	1.3416	3.91	1.253
	14.0	2.711	2.372	1.0174	1.3430	4.61	1.298
	16.0	3.172	2.718	1.0200	1.3444	5.33	1.341
	18.0	3.655	3.065	1.0225	1.3458	6.06	1.380
	20.0	4.163	3.414	1.0250	1.3472	6.81	1.431
	22.0	4.697	3.764	1.0275	1.3485	7.57	1.478
	24.0	5.259	4.116	1.0299	1.3498	8.36	1.525

Solute	Mass %	m/mol kg^{-1}	c/mol L^{-1}	ρ/g cm^{-3}	n	Δ/°C	η/mPa s
	26.0	5.851	4.470	1.0323	1.3512	9.17	1.572
	28.0	6.476	4.824	1.0346	1.3525	10.00	1.613
	30.0	7.137	5.180	1.0369	1.3537	10.84	1.669
	32.0	7.837	5.537	1.0391	1.3550	11.70	1.715
	34.0	8.579	5.896	1.0413	1.3562	12.55	1.762
	36.0	9.367	6.255	1.0434	1.3574	13.38	1.812
	38.0	10.207	6.615	1.0454	1.3586		1.852
	40.0	11.102	6.977	1.0474	1.3598		1.912
	50.0	16.653	8.794	1.0562	1.3653		2.158
	60.0	24.979	10.620	1.0629	1.3700		2.409
	70.0	38.857	12.441	1.0673	1.3738		2.629
	80.0	66.611	14.228	1.0680	1.3767		2.720
	90.0	149.875	15.953	1.0644	1.3771		2.386
	92.0	191.507	16.284	1.0629	1.3766		2.240
	94.0	260.894	16.602	1.0606	1.3759		2.036
	96.0	399.667	16.911	1.0578	1.3748		1.813
	98.0	815.987	17.198	1.0538	1.3734		1.535
	100.0		17.447	1.0477	1.3716		1.223
Acetone	0.5	0.087	0.086	0.9975	1.3334	0.16	1.013
$(CH_3)_2CO$	1.0	0.174	0.172	0.9968	1.3337	0.32	1.024
	2.0	0.351	0.343	0.9954	1.3344	0.65	1.047
	3.0	0.533	0.513	0.9940	1.3352	0.97	1.072
	4.0	0.717	0.684	0.9926	1.3359	1.30	1.099
	5.0	0.906	0.853	0.9912	1.3366	1.63	1.125
	6.0	1.099	1.023	0.9899	1.3373	1.96	1.150
	7.0	1.296	1.191	0.9886	1.3381	2.29	1.174
	8.0	1.497	1.360	0.9874	1.3388	2.62	1.198
	9.0	1.703	1.528	0.9861	1.3395	2.95	1.221
	10.0	1.913	1.696	0.9849	1.3402	3.29	1.244
Ammonia	0.5	0.295	0.292	0.9960	1.3332	0.55	1.009
NH_3	1.0	0.593	0.584	0.9938	1.3335	1.14	1.015
	2.0	1.198	1.162	0.9895	1.3339	2.32	1.029
	3.0	1.816	1.736	0.9853	1.3344	3.53	1.043
	4.0	2.447	2.304	0.9811	1.3349	4.78	1.057
	5.0	3.090	2.868	0.9770	1.3354	6.08	1.071
	6.0	3.748	3.428	0.9730	1.3359	7.43	1.085
	7.0	4.420	3.983	0.9690	1.3365	8.95	1.099
	8.0	5.106	4.533	0.9651	1.3370	10.34	1.113
	9.0	5.807	5.080	0.9613	1.3376	11.90	1.127
	10.0	6.524	5.622	0.9575	1.3381	13.55	1.141
	12.0	8.007	6.695	0.9502	1.3393	17.13	1.169
	14.0	9.558	7.753	0.9431	1.3404	21.13	1.195
	16.0	11.184	8.794	0.9361	1.3416	25.63	1.218
	18.0	12.889	9.823	0.9294	1.3428	30.70	1.237
	20.0	14.679	10.837	0.9228	1.3440	36.42	1.254
	22.0	16.561	11.838	0.9164	1.3453	43.36	1.268
	24.0	18.542	12.826	0.9102	1.3465	51.38	1.280
	26.0	20.630	13.801	0.9040	1.3477	60.77	1.288
	28.0	22.834	14.764	0.8980	1.3490	71.66	
	30.0	25.164	15.713	0.8920	1.3502	84.06	
Ammonium	0.5	0.094	0.093	0.9998	1.3340	0.32	0.999
chloride	1.0	0.189	0.187	1.0014	1.3349	0.64	0.996
NH_4Cl	2.0	0.382	0.376	1.0045	1.3369	1.27	0.992
	3.0	0.578	0.565	1.0076	1.3388	1.91	0.988
	4.0	0.779	0.756	1.0107	1.3407	2.57	0.985
	5.0	0.984	0.948	1.0138	1.3426	3.25	0.982
	6.0	1.193	1.141	1.0168	1.3445	3.94	0.979
	7.0	1.407	1.335	1.0198	1.3464	4.66	0.976

Solute	Mass %	m/mol kg^{-1}	c/mol L^{-1}	ρ/g cm^{-3}	n	Δ/°C	η/mPa s
	8.0	1.626	1.529	1.0227	1.3483	5.40	0.974
	9.0	1.849	1.726	1.0257	1.3502	6.16	0.972
	10.0	2.077	1.923	1.0286	1.3521	6.95	0.970
	12.0	2.549	2.320	1.0344	1.3559	8.60	0.969
	14.0	3.043	2.722	1.0401	1.3596		0.969
	16.0	3.561	3.128	1.0457	1.3634		0.971
	18.0	4.104	3.537	1.0512	1.3671		0.973
	20.0	4.674	3.951	1.0567	1.3708		0.978
	22.0	5.273	4.368	1.0621	1.3745		0.986
	24.0	5.903	4.789	1.0674	1.3782		0.996
Ammonium sulfate $(NH_4)_2SO_4$	0.5	0.038	0.038	1.0012	1.3338	0.17	1.008
	1.0	0.076	0.076	1.0042	1.3346	0.33	1.014
	2.0	0.154	0.153	1.0101	1.3363	0.63	1.027
	3.0	0.234	0.231	1.0160	1.3379	0.92	1.041
	4.0	0.315	0.309	1.0220	1.3395	1.21	1.057
	5.0	0.398	0.389	1.0279	1.3411	1.49	1.073
	6.0	0.483	0.469	1.0338	1.3428	1.77	1.090
	7.0	0.570	0.551	1.0397	1.3444	2.05	1.108
	8.0	0.658	0.633	1.0456	1.3460	2.33	1.127
	9.0	0.748	0.716	1.0515	1.3476	2.61	1.147
	10.0	0.841	0.800	1.0574	1.3492	2.89	1.168
	12.0	1.032	0.971	1.0691	1.3523	3.47	1.210
	14.0	1.232	1.145	1.0808	1.3555	4.07	1.256
	16.0	1.441	1.323	1.0924	1.3586	4.69	1.305
	18.0	1.661	1.504	1.1039	1.3616		1.359
	20.0	1.892	1.688	1.1154	1.3647		1.421
	22.0	2.134	1.876	1.1269	1.3677		1.490
	24.0	2.390	2.067	1.1383	1.3707		1.566
	26.0	2.659	2.262	1.1496	1.3737		1.650
	28.0	2.943	2.460	1.1609	1.3766		1.743
	30.0	3.243	2.661	1.1721	1.3795		1.847
	32.0	3.561	2.866	1.1833	1.3824		1.961
	34.0	3.898	3.073	1.1945	1.3853		2.086
	36.0	4.257	3.284	1.2056	1.3881		2.222
	38.0	4.638	3.499	1.2166	1.3909		2.371
	40.0	5.045	3.716	1.2277	1.3938		2.530
Barium chloride $BaCl_2$	0.5	0.024	0.024	1.0026	1.3337	0.12	1.009
	1.0	0.049	0.048	1.0070	1.3345	0.23	1.016
	2.0	0.098	0.098	1.0159	1.3360	0.46	1.026
	3.0	0.149	0.148	1.0249	1.3375	0.69	1.037
	4.0	0.200	0.199	1.0341	1.3391	0.93	1.049
	5.0	0.253	0.251	1.0434	1.3406	1.18	1.062
	6.0	0.307	0.303	1.0528	1.3422	1.44	1.075
	7.0	0.361	0.357	1.0624	1.3438	1.70	1.087
	8.0	0.418	0.412	1.0721	1.3454	1.98	1.101
	9.0	0.475	0.468	1.0820	1.3470	2.27	1.114
	10.0	0.534	0.524	1.0921	1.3487	2.58	1.129
	12.0	0.655	0.641	1.1128	1.3520	3.22	1.161
	14.0	0.782	0.763	1.1342	1.3555	3.92	1.195
	16.0	0.915	0.889	1.1564	1.3591	4.69	1.234
	18.0	1.054	1.019	1.1793	1.3627		1.277
	20.0	1.201	1.156	1.2031	1.3664		1.325
	22.0	1.355	1.297	1.2277	1.3703		1.378
	24.0	1.517	1.444	1.2531	1.3741		1.437
	26.0	1.687	1.597	1.2793	1.3781		1.503
Calcium chloride $CaCl_2$	0.5	0.045	0.045	1.0024	1.3342	0.22	1.015
	1.0	0.091	0.091	1.0065	1.3354	0.44	1.028
	2.0	0.184	0.183	1.0148	1.3378	0.88	1.050

Solute	Mass %	m/mol kg^{-1}	c/mol L^{-1}	ρ/g cm^{-3}	n	Δ/°C	η/mPa s
	3.0	0.279	0.277	1.0232	1.3402	1.33	1.078
	4.0	0.375	0.372	1.0316	1.3426	1.82	1.110
	5.0	0.474	0.469	1.0401	1.3451	2.35	1.143
	6.0	0.575	0.567	1.0486	1.3475	2.93	1.175
	7.0	0.678	0.667	1.0572	1.3500	3.57	1.208
	8.0	0.784	0.768	1.0659	1.3525	4.28	1.242
	9.0	0.891	0.872	1.0747	1.3549	5.04	1.279
	10.0	1.001	0.976	1.0835	1.3575	5.86	1.319
	12.0	1.229	1.191	1.1014	1.3625	7.70	1.408
	14.0	1.467	1.413	1.1198	1.3677	9.83	1.508
	16.0	1.716	1.641	1.1386	1.3730	12.28	1.625
	18.0	1.978	1.878	1.1579	1.3784	15.11	1.764
	20.0	2.253	2.122	1.1775	1.3839	18.30	1.930
	22.0	2.541	2.374	1.1976	1.3895	21.70	2.127
	24.0	2.845	2.634	1.2180	1.3951	25.30	2.356
	26.0	3.166	2.902	1.2388	1.4008	29.70	2.645
	28.0	3.504	3.179	1.2600	1.4066	34.70	3.000
	30.0	3.862	3.464	1.2816	1.4124	41.00	3.467
	32.0	4.240	3.759	1.3036	1.4183	49.70	4.035
	34.0	4.642	4.062	1.3260	1.4242		4.820
	36.0	5.068	4.375	1.3488	1.4301		5.807
	38.0	5.522	4.698	1.3720	1.4361		7.321
	40.0	6.007	5.030	1.3957	1.4420		8.997
Cesium	0.5	0.030	0.030	1.0020	1.3334	0.10	1.000
chloride	1.0	0.060	0.060	1.0058	1.3337	0.20	0.997
CsCl	2.0	0.121	0.120	1.0135	1.3345	0.40	0.992
	3.0	0.184	0.182	1.0214	1.3353	0.61	0.988
	4.0	0.247	0.245	1.0293	1.3361	0.81	0.984
	5.0	0.313	0.308	1.0374	1.3369	1.02	0.980
	6.0	0.379	0.373	1.0456	1.3377	1.22	0.977
	7.0	0.447	0.438	1.0540	1.3386	1.43	0.974
	8.0	0.516	0.505	1.0625	1.3394	1.64	0.971
	9.0	0.587	0.573	1.0711	1.3403	1.85	0.969
	10.0	0.660	0.641	1.0798	1.3412	2.06	0.966
	12.0	0.810	0.782	1.0978	1.3430	2.51	0.961
	14.0	0.967	0.928	1.1163	1.3448	2.97	0.955
	16.0	1.131	1.079	1.1355	1.3468	3.46	0.950
	18.0	1.304	1.235	1.1552	1.3487	3.96	0.945
	20.0	1.485	1.397	1.1756	1.3507	4.49	0.939
	22.0	1.675	1.564	1.1967	1.3528		0.934
	24.0	1.876	1.737	1.2185	1.3550		0.930
	26.0	2.087	1.917	1.2411	1.3572		0.926
	28.0	2.310	2.103	1.2644	1.3594		0.924
	30.0	2.546	2.296	1.2885	1.3617		0.922
	32.0	2.795	2.497	1.3135	1.3641		0.922
	34.0	3.060	2.705	1.3393	1.3666		0.924
	36.0	3.341	2.921	1.3661	1.3691		0.926
	38.0	3.640	3.146	1.3938	1.3717		0.930
	40.0	3.960	3.380	1.4226	1.3744		0.934
	42.0	4.301	3.624	1.4525	1.3771		0.940
	44.0	4.667	3.877	1.4835	1.3800		0.947
	46.0	5.060	4.142	1.5158	1.3829		0.956
	48.0	5.483	4.418	1.5495	1.3860		0.967
	50.0	5.940	4.706	1.5846	1.3892		0.981
	60.0	8.910	6.368	1.7868	1.4076		1.120
	64.0	10.560	7.163	1.8842	1.4167		1.238
Citric acid	0.5	0.026	0.026	1.0002	1.3336	0.05	1.013
(HO)C(COOH)$_3$	1.0	0.053	0.052	1.0022	1.3343	0.11	1.024
	2.0	0.106	0.105	1.0063	1.3356	0.21	1.048

Solute	Mass %	m/mol kg^{-1}	c/mol L^{-1}	ρ/g cm^{-3}	n	Δ/°C	η/mPa s
	3.0	0.161	0.158	1.0105	1.3368	0.32	1.073
	4.0	0.217	0.211	1.0147	1.3381	0.43	1.098
	5.0	0.274	0.265	1.0189	1.3394	0.54	1.125
	6.0	0.332	0.320	1.0232	1.3407	0.65	1.153
	7.0	0.392	0.374	1.0274	1.3420	0.76	1.183
	8.0	0.453	0.430	1.0316	1.3433	0.88	1.214
	9.0	0.515	0.485	1.0359	1.3446	1.00	1.247
	10.0	0.578	0.541	1.0402	1.3459	1.12	1.283
	12.0	0.710	0.655	1.0490	1.3486	1.38	1.357
	14.0	0.847	0.771	1.0580	1.3514	1.66	1.436
	16.0	0.991	0.889	1.0672	1.3541	1.95	1.525
	18.0	1.143	1.008	1.0764	1.3569	2.26	1.625
	20.0	1.301	1.130	1.0858	1.3598	2.57	1.740
	22.0	1.468	1.254	1.0953	1.3626	2.88	1.872
	24.0	1.644	1.380	1.1049	1.3655	3.21	2.017
	26.0	1.829	1.508	1.1147	1.3684	3.55	2.178
	28.0	2.024	1.639	1.1246	1.3714	3.89	2.356
	30.0	2.231	1.772	1.1346	1.3744	4.25	2.549
Copper	0.5	0.031	0.031	1.0033	1.3339	0.08	1.017
sulfate	1.0	0.063	0.063	1.0085	1.3348	0.14	1.036
$CuSO_4$	2.0	0.128	0.128	1.0190	1.3367	0.26	1.084
	3.0	0.194	0.194	1.0296	1.3386	0.37	1.129
	4.0	0.261	0.261	1.0403	1.3405	0.48	1.173
	5.0	0.330	0.329	1.0511	1.3424	0.59	1.221
	6.0	0.400	0.399	1.0620	1.3443	0.70	1.276
	7.0	0.472	0.471	1.0730	1.3462	0.82	1.336
	8.0	0.545	0.543	1.0842	1.3481	0.93	1.400
	9.0	0.620	0.618	1.0955	1.3501	1.05	1.469
	10.0	0.696	0.694	1.1070	1.3520	1.18	1.543
	12.0	0.854	0.850	1.1304	1.3560	1.45	1.701
	14.0	1.020	1.013	1.1545	1.3601	1.75	1.889
	16.0	1.193	1.182	1.1796	1.3644		2.136
	18.0	1.375	1.360	1.2059	1.3689		2.449
Disodium	0.5	0.015	0.015	1.0009	1.3339	0.07	1.017
ethylenediamine	1.0	0.030	0.030	1.0036	1.3348	0.14	1.032
tetraacetate	1.5	0.045	0.045	1.0062	1.3356	0.21	1.046
(EDTA sodium)	2.0	0.061	0.060	1.0089	1.3365	0.27	1.062
$Na_2C_{10}H_{14}N_2O_8$	2.5	0.076	0.075	1.0115	1.3374	0.33	1.077
	3.0	0.092	0.090	1.0142	1.3383	0.40	1.093
	3.5	0.108	0.106	1.0169	1.3392	0.46	1.109
	4.0	0.124	0.121	1.0196	1.3400	0.52	1.125
	4.5	0.140	0.137	1.0223	1.3409	0.58	1.142
	5.0	0.157	0.152	1.0250	1.3418	0.65	1.160
	5.5	0.173	0.168	1.0277	1.3427	0.71	1.178
	6.0	0.190	0.184	1.0305	1.3436	0.77	1.197
Ethanol	0.5	0.109	0.108	0.9973	1.3333	0.20	1.023
CH_3CH_2OH	1.0	0.219	0.216	0.9963	1.3336	0.40	1.046
	2.0	0.443	0.432	0.9945	1.3342	0.81	1.095
	3.0	0.671	0.646	0.9927	1.3348	1.23	1.140
	4.0	0.904	0.860	0.9910	1.3354	1.65	1.183
	5.0	1.142	1.074	0.9893	1.3360	2.09	1.228
	6.0	1.385	1.286	0.9878	1.3367	2.54	1.279
	7.0	1.634	1.498	0.9862	1.3374	2.99	1.331
	8.0	1.887	1.710	0.9847	1.3381	3.47	1.385
	9.0	2.147	1.921	0.9833	1.3388	3.96	1.442
	10.0	2.412	2.131	0.9819	1.3395	4.47	1.501
	12.0	2.960	2.551	0.9792	1.3410	5.56	1.627
	14.0	3.534	2.967	0.9765	1.3425	6.73	1.761

Solute	Mass %	m/mol kg^{-1}	c/mol L^{-1}	ρ/g cm^{-3}	n	Δ/°C	η/mPa s
	16.0	4.134	3.382	0.9739	1.3440	8.01	1.890
	18.0	4.765	3.795	0.9713	1.3455	9.40	2.019
	20.0	5.427	4.205	0.9687	1.3469	10.92	2.142
	22.0	6.122	4.613	0.9660	1.3484	12.60	2.259
	24.0	6.855	5.018	0.9632	1.3498	14.47	2.370
	26.0	7.626	5.419	0.9602	1.3511	16.41	2.476
	28.0	8.441	5.817	0.9571	1.3524	18.43	2.581
	30.0	9.303	6.212	0.9539	1.3535	20.47	2.667
	32.0	10.215	6.601	0.9504	1.3546	22.44	2.726
	34.0	11.182	6.987	0.9468	1.3557	24.27	2.768
	36.0	12.210	7.370	0.9431	1.3566	25.98	2.803
	38.0	13.304	7.747	0.9392	1.3575	27.62	2.829
	40.0	14.471	8.120	0.9352	1.3583	29.26	2.846
	42.0	15.718	8.488	0.9311	1.3590	30.98	2.852
	44.0	17.055	8.853	0.9269	1.3598	32.68	2.850
	46.0	18.490	9.213	0.9227	1.3604	34.36	2.843
	48.0	20.036	9.568	0.9183	1.3610	36.04	2.832
	50.0	21.706	9.919	0.9139	1.3616	37.67	2.813
	60.0	32.559	11.605	0.8911	1.3638	44.93	2.547
	70.0	50.648	13.183	0.8676	1.3652		2.214
	80.0	86.824	14.649	0.8436	1.3658		1.881
	90.0	195.355	15.980	0.8180	1.3650		1.542
	92.0	249.620	16.225	0.8125	1.3646		1.475
	94.0	340.062	16.466	0.8070	1.3642		1.407
	96.0	520.946	16.697	0.8013	1.3636		1.342
	98.0		16.920	0.7954	1.3630		1.273
	100.0		17.133	0.7893	1.3614		1.203
Ethylene	0.5	0.081	0.080	0.9988	1.3335	0.15	1.010
glycol	1.0	0.163	0.161	0.9995	1.3339	0.30	1.020
$(CH_2OH)_2$	2.0	0.329	0.322	1.0007	1.3348	0.61	1.048
	3.0	0.498	0.484	1.0019	1.3358	0.92	1.074
	4.0	0.671	0.646	1.0032	1.3367	1.24	1.099
	5.0	0.848	0.809	1.0044	1.3377	1.58	1.125
	6.0	1.028	0.972	1.0057	1.3386	1.91	1.153
	7.0	1.213	1.136	1.0070	1.3396	2.26	1.182
	8.0	1.401	1.299	1.0082	1.3405	2.62	1.212
	9.0	1.593	1.464	1.0095	1.3415	2.99	1.243
	10.0	1.790	1.628	1.0108	1.3425	3.37	1.277
	12.0	2.197	1.959	1.0134	1.3444	4.16	1.348
	14.0	2.623	2.292	1.0161	1.3464	5.01	1.424
	16.0	3.069	2.626	1.0188	1.3484	5.91	1.500
	18.0	3.537	2.962	1.0214	1.3503	6.89	1.578
	20.0	4.028	3.300	1.0241	1.3523	7.93	1.661
	24.0	5.088	3.981	1.0296	1.3564	10.28	1.843
	28.0	6.265	4.669	1.0350	1.3605	13.03	2.047
	32.0	7.582	5.364	1.0405	1.3646	16.23	2.280
	36.0	9.062	6.067	1.0460	1.3687	19.82	2.537
	40.0	10.741	6.776	1.0514	1.3728	23.84	2.832
	44.0	12.659	7.491	1.0567	1.3769	28.32	3.166
	48.0	14.872	8.212	1.0619	1.3811	33.30	3.544
	52.0	17.453	8.939	1.0670	1.3851	38.81	3.981
	56.0	20.505	9.671	1.0719	1.3892	44.83	4.475
	60.0	24.166	10.406	1.0765	1.3931	51.23	5.026
Ferric	0.5	0.031	0.031	1.0025	1.3344	0.21	1.024
chloride	1.0	0.062	0.062	1.0068	1.3358	0.39	1.047
$FeCl_3$	2.0	0.126	0.125	1.0153	1.3386	0.75	1.093
	3.0	0.191	0.189	1.0238	1.3413	1.15	1.139
	4.0	0.257	0.255	1.0323	1.3441	1.56	1.187
	5.0	0.324	0.321	1.0408	1.3468	2.00	1.238

Solute	Mass %	m/mol kg^{-1}	c/mol L^{-1}	ρ/g cm^{-3}	n	Δ/°C	η/mPa s
	6.0	0.394	0.388	1.0493	1.3496	2.48	1.292
	7.0	0.464	0.457	1.0580	1.3524	2.99	1.350
	8.0	0.536	0.526	1.0668	1.3552	3.57	1.412
	9.0	0.610	0.597	1.0760	1.3581	4.19	1.480
	10.0	0.685	0.669	1.0853	1.3611	4.85	1.553
	12.0	0.841	0.817	1.1040	1.3670	6.38	1.707
	14.0	1.004	0.969	1.1228	1.3730	8.22	1.879
	16.0	1.174	1.126	1.1420		10.45	2.080
	18.0	1.353	1.289	1.1615		13.08	2.311
	20.0	1.541	1.457	1.1816		16.14	2.570
	24.0	1.947	1.810	1.2234		23.79	3.178
	28.0	2.398	2.189	1.2679		33.61	4.038
	32.0	2.901	2.595	1.3153		49.16	5.274
	36.0	3.468	3.030	1.3654			7.130
	40.0	4.110	3.496	1.4176			9.674
Formic acid	0.5	0.109	0.109	0.9994	1.3333	0.21	1.006
HCOOH	1.0	0.219	0.217	1.0006	1.3336	0.42	1.011
	2.0	0.443	0.436	1.0029	1.3342	0.82	1.017
	3.0	0.672	0.655	1.0053	1.3348	1.24	1.195
	4.0	0.905	0.876	1.0077	1.3354	1.67	1.032
	5.0	1.143	1.097	1.0102	1.3359	2.10	1.039
	6.0	1.387	1.320	1.0126	1.3365	2.53	1.046
	7.0	1.635	1.544	1.0150	1.3371	2.97	1.052
	8.0	1.889	1.768	1.0175	1.3376	3.40	1.058
	9.0	2.149	1.994	1.0199	1.3382	3.84	1.064
	10.0	2.414	2.221	1.0224	1.3387	4.27	1.070
	12.0	2.962	2.678	1.0273	1.3397	5.19	1.082
	14.0	3.537	3.139	1.0322	1.3408	6.11	1.094
	16.0	4.138	3.605	1.0371	1.3418	7.06	1.106
	18.0	4.769	4.074	1.0419	1.3428	8.08	1.119
	20.0	5.431	4.548	1.0467	1.3437	9.11	1.132
	28.0	8.449	6.481	1.0654	1.3475	13.10	1.179
	36.0	12.220	8.477	1.0839	1.3511	17.65	1.227
	44.0	17.070	10.529	1.1015	1.3547	22.93	1.281
	52.0	23.535	12.633	1.1183	1.3581	29.69	1.340
	60.0	32.587	14.813	1.1364	1.3612	38.26	1.410
	68.0	46.166	17.054	1.1544	1.3641		1.490
D-Fructose	0.5	0.028	0.028	1.0002	1.3337	0.05	1.015
$C_6H_{12}O_6$	1.0	0.056	0.056	1.0021	1.3344	0.10	1.028
	2.0	0.113	0.112	1.0061	1.3358	0.21	1.054
	3.0	0.172	0.168	1.0101	1.3373	0.32	1.080
	4.0	0.231	0.225	1.0140	1.3387	0.43	1.106
	5.0	0.292	0.283	1.0181	1.3402	0.54	1.134
	6.0	0.354	0.340	1.0221	1.3417	0.66	1.165
	7.0	0.418	0.399	1.0262	1.3431	0.78	1.198
	8.0	0.483	0.458	1.0303	1.3446	0.90	1.232
	9.0	0.549	0.517	1.0344	1.3461	1.03	1.270
	10.0	0.617	0.576	1.0385	1.3476	1.16	1.309
	12.0	0.757	0.697	1.0469	1.3507	1.43	1.391
	14.0	0.904	0.820	1.0554	1.3538	1.71	1.483
	16.0	1.057	0.945	1.0640	1.3569	2.01	1.587
	18.0	1.218	1.072	1.0728	1.3601	2.32	1.703
	20.0	1.388	1.201	1.0816	1.3634	2.64	1.837
	22.0	1.566	1.332	1.0906	1.3667	3.05	1.986
	24.0	1.753	1.465	1.0996	1.3700	3.43	2.154
	26.0	1.950	1.600	1.1089	1.3734	3.82	2.348
	28.0	2.159	1.738	1.1182	1.3768	4.20	2.562
	30.0	2.379	1.878	1.1276	1.3803		2.817
	32.0	2.612	2.020	1.1372	1.3839		3.112

Solute	Mass %	m/mol kg^{-1}	c/mol L^{-1}	ρ/g cm^{-3}	n	Δ/°C	η/mPa s
	34.0	2.859	2.164	1.1469	1.3874		3.462
	36.0	3.122	2.312	1.1568	1.3911		3.899
	38.0	3.402	2.461	1.1668	1.3948		4.418
	40.0	3.700	2.613	1.1769	1.3985		5.046
	42.0	4.019	2.767	1.1871	1.4023		5.773
	44.0	4.361	2.925	1.1975	1.4062		6.644
	46.0	4.728	3.084	1.2080	1.4101		7.753
	48.0	5.124	3.247	1.2187	1.4141		9.060
D-Glucose	0.5	0.028	0.028	1.0001	1.3337	0.05	1.010
C$_6$H$_{12}$O$_6$	1.0	0.056	0.056	1.0020	1.3344	0.11	1.021
	2.0	0.113	0.112	1.0058	1.3358	0.21	1.052
	3.0	0.172	0.168	1.0097	1.3373	0.32	1.083
	4.0	0.231	0.225	1.0136	1.3387	0.43	1.113
	5.0	0.292	0.282	1.0175	1.3402	0.55	1.145
	6.0	0.354	0.340	1.0214	1.3417	0.67	1.179
	7.0	0.418	0.398	1.0254	1.3432	0.79	1.214
	8.0	0.483	0.457	1.0294	1.3447	0.91	1.250
	9.0	0.549	0.516	1.0334	1.3462	1.04	1.289
	10.0	0.617	0.576	1.0375	1.3477	1.17	1.330
	12.0	0.757	0.697	1.0457	1.3508	1.44	1.416
	14.0	0.904	0.819	1.0540	1.3539	1.73	1.512
	16.0	1.057	0.944	1.0624	1.3571	2.03	1.625
	18.0	1.218	1.070	1.0710	1.3603	2.35	1.757
	20.0	1.388	1.199	1.0797	1.3635	2.70	1.904
	22.0	1.566	1.329	1.0884	1.3668	3.07	2.063
	24.0	1.753	1.462	1.0973	1.3702	3.48	2.242
	26.0	1.950	1.597	1.1063	1.3736	3.90	2.458
	28.0	2.159	1.734	1.1154	1.3770	4.34	2.707
	30.0	2.379	1.873	1.1246	1.3805	4.79	2.998
	32.0	2.612	2.014	1.1340	1.3840		3.324
	34.0	2.859	2.158	1.1434	1.3876		3.704
	36.0	3.122	2.304	1.1529	1.3912		4.193
	38.0	3.402	2.452	1.1626	1.3949		4.786
	40.0	3.700	2.603	1.1724	1.3986		5.493
	42.0	4.019	2.756	1.1823	1.4024		6.288
	44.0	4.361	2.912	1.1924	1.4062		7.235
	46.0	4.728	3.071	1.2026	1.4101		8.454
	48.0	5.124	3.232	1.2130	1.4141		9.883
	50.0	5.551	3.396	1.2235	1.4181		11.884
	52.0	6.013	3.562	1.2342	1.4222		14.489
	54.0	6.516	3.732	1.2451	1.4263		17.916
	56.0	7.064	3.905	1.2562	1.4306		22.886
	58.0	7.665	4.081	1.2676	1.4349		29.389
	60.0	8.326	4.261	1.2793	1.4394		37.445
Glycerol	0.5	0.055	0.054	0.9994	1.3336	0.07	1.011
CH$_2$OHCHOHCH$_2$OH	1.0	0.110	0.109	1.0005	1.3342	0.18	1.022
	2.0	0.222	0.218	1.0028	1.3353	0.41	1.048
	3.0	0.336	0.327	1.0051	1.3365	0.63	1.074
	4.0	0.452	0.438	1.0074	1.3376	0.85	1.100
	5.0	0.572	0.548	1.0097	1.3388	1.08	1.127
	6.0	0.693	0.659	1.0120	1.3400	1.32	1.157
	7.0	0.817	0.771	1.0144	1.3412	1.56	1.188
	8.0	0.944	0.883	1.0167	1.3424	1.81	1.220
	9.0	1.074	0.996	1.0191	1.3436	2.06	1.256
	10.0	1.207	1.109	1.0215	1.3448	2.32	1.291
	12.0	1.481	1.337	1.0262	1.3472	2.88	1.365
	14.0	1.768	1.568	1.0311	1.3496	3.47	1.445
	16.0	2.068	1.800	1.0360	1.3521	4.09	1.533
	18.0	2.384	2.035	1.0409	1.3547	4.76	1.630

Solute	Mass %	m/mol kg^{-1}	c/mol L^{-1}	ρ/g cm^{-3}	n	Δ/°C	η/mPa s
	20.0	2.715	2.271	1.0459	1.3572	5.46	1.737
	24.0	3.429	2.752	1.0561	1.3624	7.01	1.988
	28.0	4.223	3.242	1.0664	1.3676	8.77	2.279
	32.0	5.110	3.742	1.0770	1.3730	10.74	2.637
	36.0	6.108	4.252	1.0876	1.3785	12.96	3.088
	40.0	7.239	4.771	1.0984	1.3841	15.50	3.653
	44.0	8.532	5.300	1.1092	1.3897		4.443
	48.0	10.024	5.838	1.1200	1.3954		5.413
	52.0	11.764	6.385	1.1308	1.4011		6.666
	56.0	13.820	6.944	1.1419	1.4069		8.349
	60.0	16.288	7.512	1.1530	1.4129		10.681
	64.0	19.305	8.092	1.1643	1.4189		13.657
	68.0	23.075	8.680	1.1755	1.4249		18.457
	72.0	27.923	9.277	1.1866	1.4310		27.625
	76.0	34.387	9.884	1.1976	1.4370		40.571
	80.0	43.436	10.498	1.2085	1.4431		59.900
	84.0	57.009	11.121	1.2192	1.4492		84.338
	88.0	79.632	11.753	1.2299	1.4553		147.494
	92.0	124.878	12.392	1.2404	1.4613		384.467
	96.0	260.615	13.039	1.2508	1.4674		780.458
	100.0		13.694	1.2611	1.4735		
Hydrochloric	0.5	0.138	0.137	1.0007	1.3341	0.49	1.008
acid	1.0	0.277	0.275	1.0031	1.3353	0.99	1.015
HCl	2.0	0.560	0.553	1.0081	1.3376	2.08	1.029
	3.0	0.848	0.833	1.0130	1.3399	3.28	1.044
	4.0	1.143	1.117	1.0179	1.3422	4.58	1.059
	5.0	1.444	1.403	1.0228	1.3445	5.98	1.075
	6.0	1.751	1.691	1.0278	1.3468	7.52	1.091
	7.0	2.064	1.983	1.0327	1.3491	9.22	1.108
	8.0	2.385	2.277	1.0377	1.3515	11.10	1.125
	9.0	2.713	2.574	1.0426	1.3538	13.15	1.143
	10.0	3.047	2.873	1.0476	1.3561	15.40	1.161
	12.0	3.740	3.481	1.0576	1.3607	20.51	1.199
	14.0	4.465	4.099	1.0676	1.3653		1.239
	16.0	5.224	4.729	1.0777	1.3700		1.282
	18.0	6.020	5.370	1.0878	1.3746		1.326
	20.0	6.857	6.023	1.0980	1.3792		1.374
	22.0	7.736	6.687	1.1083	1.3838		1.426
	24.0	8.661	7.362	1.1185	1.3884		1.483
	26.0	9.636	8.049	1.1288	1.3930		1.547
	28.0	10.666	8.748	1.1391	1.3976		1.620
	30.0	11.754	9.456	1.1492	1.4020		1.705
	32.0	12.907	10.175	1.1594	1.4066		1.799
	34.0	14.129	10.904	1.1693	1.4112		1.900
	36.0	15.427	11.642	1.1791	1.4158		2.002
	38.0	16.810	12.388	1.1886	1.4204		2.105
	40.0	18.284	13.140	1.1977	1.4250		
Lactic acid	0.5	0.056	0.055	0.9992	1.3335	0.10	1.014
CH$_3$CHOHCOOH	1.0	0.112	0.111	1.0002	1.3340	0.19	1.027
	2.0	0.227	0.223	1.0023	1.3350	0.38	1.056
	3.0	0.343	0.334	1.0043	1.3360	0.57	1.084
	4.0	0.463	0.447	1.0065	1.3370	0.76	1.110
	5.0	0.584	0.560	1.0086	1.3380	0.95	1.138
	6.0	0.709	0.673	1.0108	1.3390	1.16	1.167
	7.0	0.836	0.787	1.0131	1.3400	1.36	1.198
	8.0	0.965	0.902	1.0153	1.3410	1.57	1.229
	9.0	1.098	1.017	1.0176	1.3420	1.79	1.262
	10.0	1.233	1.132	1.0199	1.3430	2.02	1.296
	12.0	1.514	1.365	1.0246	1.3450	2.49	1.366

Solute	Mass %	m/mol kg^{-1}	c/mol L^{-1}	ρ/g cm^{-3}	n	Δ/°C	η/mPa s
	14.0	1.807	1.600	1.0294	1.3470	2.99	1.441
	16.0	2.115	1.837	1.0342	1.3491	3.48	1.522
	18.0	2.437	2.076	1.0390	1.3511	3.96	1.607
	20.0	2.775	2.318	1.0439	1.3532	4.44	1.699
	24.0	3.506	2.807	1.0536	1.3573		1.902
	28.0	4.317	3.305	1.0632	1.3615		2.136
	32.0	5.224	3.811	1.0728	1.3657		2.414
	36.0	6.244	4.325	1.0822	1.3700		2.730
	40.0	7.401	4.847	1.0915	1.3743		3.114
	44.0	8.722	5.377	1.1008	1.3786		3.566
	48.0	10.247	5.917	1.1105	1.3828		4.106
	52.0	12.026	6.466	1.1201	1.3871		4.789
	56.0	14.129	7.023	1.1297	1.3914		5.579
	60.0	16.652	7.588	1.1392	1.3958		6.679
	64.0	19.736	8.161	1.1486	1.4001		8.024
	68.0	23.590	8.741	1.1579	1.4045		9.863
	72.0	28.546	9.328	1.1670	1.4088		12.866
	76.0	35.154	9.922	1.1760	1.4131		16.974
	80.0	44.405	10.522	1.1848	1.4173		22.164
Lactose	0.5	0.015	0.015	1.0002	1.3337	0.03	1.013
$C_{12}H_{22}O_{11}$	1.0	0.030	0.029	1.0021	1.3345	0.06	1.026
	2.0	0.060	0.059	1.0061	1.3359	0.11	1.058
	3.0	0.090	0.089	1.0102	1.3375	0.17	1.089
	4.0	0.122	0.119	1.0143	1.3390	0.23	1.120
	5.0	0.154	0.149	1.0184	1.3406	0.29	1.154
	6.0	0.186	0.179	1.0225	1.3421	0.35	1.191
	7.0	0.220	0.210	1.0267	1.3437	0.42	1.232
	8.0	0.254	0.241	1.0308	1.3453	0.50	1.276
	9.0	0.289	0.272	1.0349	1.3468		1.321
	10.0	0.325	0.304	1.0390	1.3484		1.370
	12.0	0.398	0.367	1.0473	1.3515		1.476
	14.0	0.476	0.432	1.0558	1.3548		1.593
	16.0	0.556	0.498	1.0648	1.3582		1.724
	18.0	0.641	0.565	1.0746	1.3619		1.869
Lithium	0.5	0.119	0.118	1.0012	1.3341	0.42	1.019
chloride	1.0	0.238	0.237	1.0041	1.3351	0.84	1.037
LiCl	2.0	0.481	0.476	1.0099	1.3373	1.72	1.072
	3.0	0.730	0.719	1.0157	1.3394	2.68	1.108
	4.0	0.983	0.964	1.0215	1.3415	3.73	1.146
	5.0	1.241	1.211	1.0272	1.3436	4.86	1.185
	6.0	1.506	1.462	1.0330	1.3457	6.14	1.226
	7.0	1.775	1.715	1.0387	1.3478	7.56	1.269
	8.0	2.051	1.971	1.0444	1.3499	9.11	1.313
	9.0	2.333	2.230	1.0502	1.3520	10.79	1.360
	10.0	2.621	2.491	1.0560	1.3541	12.61	1.411
	12.0	3.217	3.022	1.0675	1.3583	16.59	1.522
	14.0	3.840	3.564	1.0792	1.3625	21.04	1.647
	16.0	4.493	4.118	1.0910	1.3668		1.787
	18.0	5.178	4.683	1.1029	1.3711		1.942
	20.0	5.897	5.260	1.1150	1.3755		2.128
	22.0	6.653	5.851	1.1274	1.3799		2.341
	24.0	7.449	6.453	1.1399	1.3844		2.600
	26.0	8.288	7.069	1.1527	1.3890		2.925
	28.0	9.173	7.700	1.1658	1.3936		3.318
	30.0	10.109	8.344	1.1791	1.3983		3.785
Magnesium	0.5	0.053	0.053	1.0022	1.3343	0.26	1.024
chloride	1.0	0.106	0.106	1.0062	1.3356	0.52	1.046
$MgCl_2$	2.0	0.214	0.213	1.0144	1.3381	1.06	1.091

Solute	Mass %	m/mol kg^{-1}	c/mol L^{-1}	ρ/g cm^{-3}	n	Δ/°C	η/mPa s
	3.0	0.325	0.322	1.0226	1.3406	1.65	1.139
	4.0	0.438	0.433	1.0309	1.3432	2.30	1.188
	5.0	0.553	0.546	1.0394	1.3457	3.01	1.241
	6.0	0.670	0.660	1.0479	1.3483		1.298
	7.0	0.791	0.777	1.0564	1.3508		1.358
	8.0	0.913	0.895	1.0651	1.3534		1.423
	9.0	1.039	1.015	1.0738	1.3560		1.493
	10.0	1.167	1.137	1.0826	1.3587		1.570
	12.0	1.432	1.387	1.1005	1.3641		1.745
	14.0	1.710	1.645	1.1189	1.3695		1.956
	16.0	2.001	1.911	1.1372	1.3749		2.207
	18.0	2.306	2.184	1.1553	1.3804		2.507
	20.0	2.626	2.467	1.1742	1.3859		2.867
	22.0	2.962	2.758	1.1938	1.3915		3.323
	24.0	3.317	3.060	1.2140	1.3972		3.917
	26.0	3.690	3.371	1.2346	1.4030		4.694
	28.0	4.085	3.692	1.2555	1.4089		5.709
	30.0	4.501	4.022	1.2763	1.4148		7.017
Magnesium sulfate $MgSO_4$	0.5	0.042	0.042	1.0033	1.3340	0.10	1.027
	1.0	0.084	0.084	1.0084	1.3350	0.19	1.054
	2.0	0.170	0.169	1.0186	1.3371	0.36	1.112
	3.0	0.257	0.256	1.0289	1.3391	0.52	1.177
	4.0	0.346	0.345	1.0392	1.3411	0.69	1.249
	5.0	0.437	0.436	1.0497	1.3431	0.87	1.328
	6.0	0.530	0.528	1.0602	1.3451	1.05	1.411
	7.0	0.625	0.623	1.0708	1.3471	1.24	1.498
	8.0	0.722	0.719	1.0816	1.3492	1.43	1.593
	9.0	0.822	0.817	1.0924	1.3512	1.64	1.702
	10.0	0.923	0.917	1.1034	1.3532	1.85	1.829
	12.0	1.133	1.122	1.1257	1.3572	2.31	2.104
	14.0	1.352	1.336	1.1484	1.3613	2.86	2.412
	16.0	1.582	1.557	1.1717	1.3654	3.67	2.809
	18.0	1.824	1.788	1.1955	1.3694		3.360
	20.0	2.077	2.027	1.2198	1.3735		4.147
	22.0	2.343	2.275	1.2447	1.3776		5.199
	24.0	2.624	2.532	1.2701	1.3817		6.498
	26.0	2.919	2.800	1.2961	1.3858		8.066
Maltose $C_{12}H_{22}O_{11}$	0.5	0.015	0.015	1.0003	1.3337	0.03	1.016
	1.0	0.030	0.029	1.0023	1.3345	0.06	1.030
	2.0	0.060	0.059	1.0063	1.3359	0.11	1.060
	3.0	0.090	0.089	1.0104	1.3374	0.17	1.092
	4.0	0.122	0.119	1.0144	1.3389	0.23	1.126
	5.0	0.154	0.149	1.0184	1.3404	0.29	1.162
	6.0	0.186	0.179	1.0224	1.3420	0.35	1.200
	7.0	0.220	0.210	1.0265	1.3435	0.42	1.239
	8.0	0.254	0.241	1.0305	1.3450	0.48	1.281
	9.0	0.289	0.272	1.0345	1.3466	0.55	1.325
	10.0	0.325	0.303	1.0385	1.3482	0.62	1.372
	12.0	0.398	0.367	1.0465	1.3513	0.77	1.474
	14.0	0.476	0.431	1.0545	1.3546	0.92	1.588
	16.0	0.556	0.497	1.0629	1.3578	1.08	1.715
	18.0	0.641	0.564	1.0716	1.3612	1.25	1.859
	20.0	0.730	0.631	1.0801	1.3644	1.43	2.021
	22.0	0.824	0.700	1.0894	1.3678	1.64	2.216
	24.0	0.923	0.770	1.0984	1.3714	1.85	2.463
	26.0	1.026	0.842	1.1080	1.3749	2.08	2.753
	28.0	1.136	0.914	1.1171	1.3785	2.34	3.066
	30.0	1.252	0.988	1.1269	1.3821	2.62	3.427
	40.0	1.948	1.375	1.1769	1.4013	4.41	6.926

Solute	Mass %	m/mol kg^{-1}	c/mol L^{-1}	ρ/g cm^{-3}	n	Δ/°C	η/mPa s
	50.0	2.921	1.797	1.2304	1.4217		17.786
	52.0	3.165	1.886	1.2416	1.4260		22.034
	54.0	3.429	1.976	1.2528	1.4308		28.757
	56.0	3.718	2.068	1.2638	1.4350		38.226
	58.0	4.034	2.159	1.2740	1.4394		49.298
	60.0	4.382	2.253	1.2855	1.4440		
Manganese(II)	1.0	0.067	0.067	1.0080	1.3348	0.16	1.046
sulfate	2.0	0.135	0.135	1.0178	1.3366	0.31	1.090
MnSO$_4$	3.0	0.205	0.204	1.0277	1.3384	0.44	1.137
	4.0	0.276	0.275	1.0378	1.3402	0.57	1.187
	5.0	0.349	0.347	1.0480	1.3420	0.70	1.242
	6.0	0.423	0.421	1.0583	1.3438	0.84	1.301
	7.0	0.498	0.495	1.0688	1.3457	0.98	1.363
	8.0	0.576	0.572	1.0794	1.3475	1.12	1.431
	9.0	0.655	0.650	1.0902	1.3494	1.28	1.505
	10.0	0.736	0.729	1.1012	1.3513	1.44	1.587
	12.0	0.903	0.893	1.1236	1.3551	1.80	1.779
	14.0	1.078	1.063	1.1467	1.3589	2.21	2.005
	16.0	1.261	1.240	1.1705	1.3629	2.67	2.272
	18.0	1.454	1.424	1.1950	1.3668	3.19	2.580
	20.0	1.656	1.616	1.2203	1.3708	3.80	2.938
D-Mannitol	0.5	0.028	0.027	1.0000	1.3337	0.05	1.019
CH$_2$(CHOH)$_4$CH$_2$OH	1.0	0.055	0.055	1.0017	1.3345	0.10	1.032
	2.0	0.112	0.110	1.0053	1.3359	0.21	1.057
	3.0	0.170	0.166	1.0088	1.3374	0.32	1.081
	4.0	0.229	0.222	1.0124	1.3389	0.43	1.107
	5.0	0.289	0.279	1.0159	1.3403	0.54	1.135
	6.0	0.350	0.336	1.0195	1.3418	0.66	1.166
	7.0	0.413	0.393	1.0230	1.3433	0.77	1.200
	8.0	0.477	0.451	1.0266	1.3447	0.90	1.236
	9.0	0.543	0.509	1.0302	1.3462	1.02	1.275
	10.0	0.610	0.567	1.0338	1.3477	1.15	1.314
	11.0	0.678	0.626	1.0375	1.3491	1.28	1.355
	12.0	0.749	0.686	1.0412	1.3506	1.41	1.398
	13.0	0.820	0.746	1.0450	1.3521	1.55	1.443
	14.0	0.894	0.806	1.0489	1.3536	1.69	1.489
	15.0	0.969	0.867	1.0529	1.3552	1.84	1.537
Methanol	0.5	0.157	0.156	0.9973	1.3331	0.28	1.022
CH$_3$OH	1.0	0.315	0.311	0.9964	1.3332	0.56	1.040
	2.0	0.637	0.621	0.9947	1.3334	1.14	1.070
	3.0	0.965	0.930	0.9930	1.3336	1.75	1.100
	4.0	1.300	1.238	0.9913	1.3339	2.37	1.131
	5.0	1.643	1.544	0.9896	1.3341	3.02	1.163
	6.0	1.992	1.850	0.9880	1.3343	3.71	1.196
	7.0	2.349	2.155	0.9864	1.3346	4.41	1.229
	8.0	2.714	2.459	0.9848	1.3348	5.13	1.264
	9.0	3.087	2.762	0.9832	1.3351	5.85	1.297
	10.0	3.468	3.064	0.9816	1.3354	6.60	1.329
	12.0	4.256	3.665	0.9785	1.3359	8.14	1.389
	14.0	5.081	4.262	0.9755	1.3365	9.72	1.446
	16.0	5.945	4.856	0.9725	1.3370	11.36	1.501
	18.0	6.851	5.447	0.9695	1.3376	13.13	1.554
	20.0	7.803	6.034	0.9666	1.3381	15.02	1.604
	22.0	8.803	6.616	0.9636	1.3387	16.98	1.652
	24.0	9.856	7.196	0.9606	1.3392	19.04	1.697
	26.0	10.966	7.771	0.9576	1.3397	21.23	1.735
	28.0	12.138	8.341	0.9545	1.3402	23.59	1.769
	30.0	13.376	8.908	0.9514	1.3407	25.91	1.795

Solute	Mass %	$m/\text{mol kg}^{-1}$	$c/\text{mol L}^{-1}$	$\rho/\text{g cm}^{-3}$	n	$\Delta/°\text{C}$	$\eta/\text{mPa s}$
	32.0	14.688	9.470	0.9482	1.3411	28.15	1.814
	34.0	16.078	10.028	0.9450	1.3415	30.48	1.827
	36.0	17.556	10.580	0.9416	1.3419	32.97	1.835
	38.0	19.129	11.127	0.9382	1.3422	35.60	1.839
	40.0	20.807	11.669	0.9347	1.3425	38.60	1.837
	50.0	31.211	14.288	0.9156	1.3431	54.50	1.761
	60.0	46.816	16.749	0.8944	1.3426	74.50	1.600
	70.0	72.826	19.040	0.8715	1.3411		1.368
	80.0	124.844	21.144	0.8468	1.3385		1.128
	90.0	280.899	23.045	0.8204	1.3348		0.861
	100.0		24.710	0.7917	1.3290		0.586
Nitric acid	0.5	0.080	0.079	1.0009	1.3336	0.28	1.004
HNO_3	1.0	0.160	0.159	1.0037	1.3343	0.56	1.005
	2.0	0.324	0.320	1.0091	1.3356	1.12	1.007
	3.0	0.491	0.483	1.0146	1.3368	1.70	1.010
	4.0	0.661	0.648	1.0202	1.3381	2.32	1.014
	5.0	0.835	0.814	1.0257	1.3394	2.96	1.018
	6.0	1.013	0.982	1.0314	1.3407	3.63	1.022
	7.0	1.194	1.152	1.0370	1.3421	4.33	1.027
	8.0	1.380	1.324	1.0427	1.3434	5.05	1.032
	9.0	1.570	1.498	1.0485	1.3447	5.81	1.038
	10.0	1.763	1.673	1.0543	1.3460	6.60	1.044
	12.0	2.164	2.030	1.0660	1.3487	8.27	1.058
	14.0	2.583	2.395	1.0780	1.3514	10.08	1.075
	16.0	3.023	2.768	1.0901	1.3541	12.04	1.094
	18.0	3.484	3.149	1.1025	1.3569	14.16	1.116
	20.0	3.967	3.539	1.1150	1.3596		1.141
	22.0	4.476	3.937	1.1277	1.3624		1.169
	24.0	5.011	4.344	1.1406	1.3652		1.199
	26.0	5.576	4.760	1.1536	1.3680		1.233
	28.0	6.172	5.185	1.1668	1.3708		1.271
	30.0	6.801	5.618	1.1801	1.3736		1.311
	32.0	7.468	6.060	1.1934	1.3763		1.354
	34.0	8.175	6.512	1.2068	1.3790		1.400
	36.0	8.927	6.971	1.2202	1.3817		1.450
	38.0	9.727	7.439	1.2335	1.3842		1.504
	40.0	10.580	7.913	1.2466	1.3867		1.561
Oxalic acid	0.5	0.056	0.056	1.0006	1.3336	0.16	1.013
$(COOH)_2$	1.0	0.112	0.111	1.0030	1.3342	0.30	1.023
	1.5	0.169	0.167	1.0054	1.3347	0.44	1.033
	2.0	0.227	0.224	1.0079	1.3353	0.57	1.044
	2.5	0.285	0.281	1.0103	1.3359	0.71	1.055
	3.0	0.343	0.337	1.0126	1.3364	0.84	1.065
	3.5	0.403	0.395	1.0150	1.3370	0.97	1.076
	4.0	0.463	0.452	1.0174	1.3375	1.09	1.086
	4.5	0.523	0.510	1.0197	1.3381		1.097
	5.0	0.585	0.568	1.0220	1.3386		1.108
	6.0	0.709	0.684	1.0265	1.3397		1.129
	7.0	0.836	0.802	1.0310	1.3407		1.150
	8.0	0.966	0.920	1.0355	1.3418		1.172
Phosphoric	0.5	0.051	0.051	1.0010	1.3335	0.12	1.010
acid	1.0	0.103	0.102	1.0038	1.3340	0.24	1.020
H_3PO_4	2.0	0.208	0.206	1.0092	1.3349	0.46	1.050
	3.0	0.316	0.311	1.0146	1.3358	0.69	1.079
	4.0	0.425	0.416	1.0200	1.3367	0.93	1.108
	5.0	0.537	0.523	1.0254	1.3376	1.16	1.138
	6.0	0.651	0.631	1.0309	1.3385	1.38	1.169
	7.0	0.768	0.740	1.0363	1.3394	1.62	1.200

Solute	Mass %	m/mol kg^{-1}	c/mol L^{-1}	ρ/g cm^{-3}	n	Δ/°C	η/mPa s
	8.0	0.887	0.850	1.0418	1.3403	1.88	1.232
	9.0	1.009	0.962	1.0474	1.3413	2.16	1.267
	10.0	1.134	1.075	1.0531	1.3422	2.45	1.303
	12.0	1.392	1.304	1.0647	1.3441	3.01	1.382
	14.0	1.661	1.538	1.0765	1.3460	3.76	1.469
	16.0	1.944	1.777	1.0885	1.3480	4.45	1.565
	18.0	2.240	2.022	1.1009	1.3500	5.25	1.671
	20.0	2.551	2.273	1.1135	1.3520	6.23	1.788
	22.0	2.878	2.529	1.1263	1.3540	7.38	1.914
	24.0	3.223	2.791	1.1395	1.3561	8.69	2.049
	26.0	3.585	3.059	1.1528	1.3582	10.12	2.198
	28.0	3.968	3.333	1.1665	1.3604	11.64	2.365
	30.0	4.373	3.614	1.1804	1.3625	13.23	2.553
	32.0	4.802	3.901	1.1945	1.3647	14.94	2.766
	34.0	5.257	4.194	1.2089	1.3669	16.81	3.001
	36.0	5.740	4.495	1.2236	1.3691	18.85	3.260
	38.0	6.254	4.803	1.2385	1.3713	21.09	3.544
	40.0	6.803	5.117	1.2536	1.3735	23.58	3.856
Potassium	0.5	0.050	0.050	1.0014	1.3335	0.18	1.009
bicarbonate	1.0	0.101	0.100	1.0046	1.3341	0.34	1.015
KHCO$_3$	2.0	0.204	0.202	1.0114	1.3353	0.67	1.027
	3.0	0.309	0.305	1.0181	1.3365	0.98	1.040
	4.0	0.416	0.409	1.0247	1.3376	1.29	1.053
	5.0	0.526	0.515	1.0310	1.3386	1.60	1.067
	6.0	0.638	0.622	1.0379	1.3397	1.91	1.081
	7.0	0.752	0.730	1.0446	1.3409	2.22	1.096
	8.0	0.869	0.840	1.0514	1.3419	2.53	1.112
	9.0	0.988	0.951	1.0581	1.3430	2.84	1.128
	10.0	1.110	1.064	1.0650	1.3441	3.16	1.145
	12.0	1.362	1.293	1.0788	1.3462	3.79	1.183
	14.0	1.626	1.528	1.0929	1.3484	4.41	1.224
	16.0	1.903	1.770	1.1073	1.3506		1.270
	18.0	2.193	2.017	1.1221	1.3528		1.319
	20.0	2.497	2.272	1.1372	1.3550		1.373
	22.0	2.817	2.533	1.1527	1.3572		1.432
	24.0	3.154	2.801	1.1685	1.3595		1.497
Potassium	0.5	0.042	0.042	1.0018	1.3336	0.15	1.000
bromide	1.0	0.085	0.084	1.0054	1.3342	0.29	0.998
KBr	2.0	0.171	0.170	1.0127	1.3354	0.59	0.994
	3.0	0.260	0.257	1.0200	1.3366	0.88	0.990
	4.0	0.350	0.345	1.0275	1.3379	1.18	0.985
	5.0	0.442	0.435	1.0350	1.3391	1.48	0.981
	6.0	0.536	0.526	1.0426	1.3403	1.78	0.977
	7.0	0.633	0.618	1.0503	1.3416	2.10	0.974
	8.0	0.731	0.711	1.0581	1.3429	2.42	0.970
	9.0	0.831	0.806	1.0660	1.3441	2.74	0.967
	10.0	0.934	0.903	1.0740	1.3454	3.07	0.964
	12.0	1.146	1.099	1.0903	1.3481	3.76	0.958
	14.0	1.368	1.302	1.1070	1.3507	4.49	0.953
	16.0	1.601	1.512	1.1242	1.3535	5.25	0.949
	18.0	1.845	1.727	1.1419	1.3562	6.04	0.946
	20.0	2.101	1.950	1.1601	1.3591	6.88	0.944
	22.0	2.370	2.179	1.1788	1.3620	7.76	0.943
	24.0	2.654	2.416	1.1980	1.3650	8.70	0.943
	26.0	2.952	2.661	1.2179	1.3680	9.68	0.944
	28.0	3.268	2.914	1.2383	1.3711	10.72	0.947
	30.0	3.601	3.175	1.2593	1.3743	11.82	0.952
	32.0	3.954	3.445	1.2810	1.3776	12.98	0.959
	34.0	4.329	3.724	1.3033	1.3809		0.968

Solute	Mass %	m/mol kg^{-1}	c/mol L^{-1}	ρ/g cm^{-3}	n	Δ/°C	η/mPa s
	36.0	4.727	4.012	1.3263	1.3843		0.979
	38.0	5.150	4.311	1.3501	1.3878		0.993
	40.0	5.602	4.620	1.3746	1.3914		1.010
Potassium	0.5	0.036	0.036	1.0027	1.3339	0.18	1.013
carbonate	1.0	0.073	0.073	1.0072	1.3347	0.34	1.025
K$_2$CO$_3$	2.0	0.148	0.147	1.0163	1.3365	0.66	1.048
	3.0	0.224	0.223	1.0254	1.3382	0.99	1.071
	4.0	0.301	0.299	1.0345	1.3399	1.32	1.094
	5.0	0.381	0.378	1.0437	1.3416	1.67	1.119
	6.0	0.462	0.457	1.0529	1.3433	2.03	1.146
	7.0	0.545	0.538	1.0622	1.3450	2.40	1.174
	8.0	0.629	0.620	1.0715	1.3467	2.77	1.204
	9.0	0.716	0.704	1.0809	1.3484	3.17	1.235
	10.0	0.804	0.789	1.0904	1.3501	3.57	1.269
	12.0	0.987	0.963	1.1095	1.3535	4.45	1.339
	14.0	1.178	1.144	1.1291	1.3569	5.39	1.414
	16.0	1.378	1.330	1.1490	1.3603	6.42	1.497
	18.0	1.588	1.523	1.1692	1.3637	7.55	1.594
	20.0	1.809	1.722	1.1898	1.3671	8.82	1.707
	24.0	2.285	2.139	1.2320	1.3739	11.96	1.978
	28.0	2.814	2.584	1.2755	1.3807	16.01	2.331
	32.0	3.405	3.057	1.3204	1.3874	21.46	2.834
	36.0	4.070	3.559	1.3665	1.3940	28.58	3.503
	40.0	4.824	4.093	1.4142	1.4006	37.55	4.360
	50.0	7.236	5.573	1.5404	1.4168		9.369
Potassium	0.5	0.067	0.067	1.0014	1.3337	0.23	1.000
chloride	1.0	0.135	0.135	1.0046	1.3343	0.46	0.999
KCl	2.0	0.274	0.271	1.0110	1.3357	0.92	0.999
	3.0	0.415	0.409	1.0174	1.3371	1.38	0.998
	4.0	0.559	0.549	1.0239	1.3384	1.85	0.997
	5.0	0.706	0.691	1.0304	1.3398	2.32	0.996
	6.0	0.856	0.835	1.0369	1.3411	2.80	0.994
	7.0	1.010	0.980	1.0434	1.3425	3.29	0.992
	8.0	1.166	1.127	1.0500	1.3438	3.80	0.990
	9.0	1.327	1.276	1.0566	1.3452	4.30	0.989
	10.0	1.490	1.426	1.0633	1.3466	4.81	0.988
	12.0	1.829	1.733	1.0768	1.3493	5.88	0.990
	14.0	2.184	2.048	1.0905	1.3521		0.994
	16.0	2.555	2.370	1.1043	1.3549		0.999
	18.0	2.944	2.701	1.1185	1.3577		1.004
	20.0	3.353	3.039	1.1328	1.3606		1.012
	22.0	3.783	3.386	1.1474	1.3635		1.024
	24.0	4.236	3.742	1.1623	1.3665		1.040
Potassium	0.5	0.090	0.089	1.0025	1.3340	0.30	1.010
hydroxide	1.0	0.180	0.179	1.0068	1.3350	0.61	1.019
KOH	2.0	0.364	0.362	1.0155	1.3369	1.24	1.038
	3.0	0.551	0.548	1.0242	1.3388	1.89	1.058
	4.0	0.743	0.736	1.0330	1.3408	2.57	1.079
	5.0	0.938	0.929	1.0419	1.3427	3.36	1.102
	6.0	1.138	1.124	1.0509	1.3445	4.14	1.126
	7.0	1.342	1.322	1.0599	1.3464	4.92	1.151
	8.0	1.550	1.524	1.0690	1.3483		1.177
	9.0	1.763	1.729	1.0781	1.3502		1.205
	10.0	1.980	1.938	1.0873	1.3520		1.233
	12.0	2.431	2.365	1.1059	1.3558		1.294
	14.0	2.902	2.806	1.1246	1.3595		1.361
	16.0	3.395	3.261	1.1435	1.3632		1.436
	18.0	3.913	3.730	1.1626	1.3670		1.521

Solute	Mass %	m/mol kg^{-1}	c/mol L^{-1}	ρ/g cm^{-3}	n	Δ/°C	η/mPa s
	20.0	4.456	4.213	1.1818	1.3707		1.619
	22.0	5.027	4.711	1.2014	1.3744		1.732
	24.0	5.629	5.223	1.2210	1.3781		1.861
	26.0	6.262	5.750	1.2408	1.3818		2.006
	28.0	6.931	6.293	1.2609	1.3854		2.170
	30.0	7.639	6.851	1.2813	1.3889		2.357
	40.0	11.882	9.896	1.3881	1.4068		3.879
	50.0	17.824	13.389	1.5024	1.4247		7.892
Potassium	0.5	0.030	0.030	1.0019	1.3337	0.11	0.999
iodide	1.0	0.061	0.061	1.0056	1.3343	0.22	0.997
KI	2.0	0.123	0.122	1.0131	1.3357	0.43	0.991
	3.0	0.186	0.184	1.0206	1.3370	0.64	0.986
	4.0	0.251	0.248	1.0282	1.3384	0.86	0.981
	5.0	0.317	0.312	1.0360	1.3397	1.08	0.976
	6.0	0.385	0.377	1.0438	1.3411	1.30	0.969
	7.0	0.453	0.443	1.0517	1.3425	1.53	0.963
	8.0	0.524	0.511	1.0598	1.3440	1.77	0.957
	9.0	0.596	0.579	1.0679	1.3454	2.01	0.951
	10.0	0.669	0.648	1.0762	1.3469	2.26	0.946
	12.0	0.821	0.790	1.0931	1.3498	2.77	0.937
	14.0	0.981	0.937	1.1105	1.3529	3.30	0.929
	16.0	1.147	1.088	1.1284	1.3560	3.87	0.921
	18.0	1.322	1.244	1.1469	1.3593	4.46	0.915
	20.0	1.506	1.405	1.1659	1.3626	5.09	0.910
	22.0	1.699	1.571	1.1856	1.3661	5.76	0.905
	24.0	1.902	1.744	1.2060	1.3696	6.46	0.901
	26.0	2.117	1.922	1.2270	1.3733	7.21	0.898
	28.0	2.343	2.106	1.2487	1.3771	8.01	0.895
	30.0	2.582	2.297	1.2712	1.3810	8.86	0.892
	32.0	2.835	2.495	1.2944	1.3851	9.76	0.891
	34.0	3.103	2.700	1.3185	1.3893	10.72	0.890
	36.0	3.388	2.913	1.3434	1.3936	11.73	0.890
	38.0	3.692	3.134	1.3692	1.3981	12.81	0.893
	40.0	4.016	3.364	1.3959	1.4027	13.97	0.897
Potassium	0.5	0.050	0.050	1.0014	1.3335	0.17	0.999
nitrate	1.0	0.100	0.099	1.0045	1.3339	0.33	0.996
KNO$_3$	2.0	0.202	0.200	1.0108	1.3349	0.64	0.990
	3.0	0.306	0.302	1.0171	1.3358	0.94	0.986
	4.0	0.412	0.405	1.0234	1.3368	1.22	0.983
	5.0	0.521	0.509	1.0298	1.3377	1.50	0.980
	6.0	0.631	0.615	1.0363	1.3386	1.76	0.977
	7.0	0.744	0.722	1.0428	1.3396	2.02	0.975
	8.0	0.860	0.830	1.0494	1.3405	2.27	0.973
	9.0	0.978	0.940	1.0560	1.3415	2.52	0.971
	10.0	1.099	1.051	1.0627	1.3425	2.75	0.970
	12.0	1.349	1.277	1.0762	1.3444		0.970
	14.0	1.610	1.509	1.0899	1.3463		0.972
	16.0	1.884	1.747	1.1039	1.3482		0.976
	18.0	2.171	1.991	1.1181	1.3502		0.982
	20.0	2.473	2.240	1.1326	1.3521		0.990
	22.0	2.790	2.497	1.1473	1.3541		0.999
	24.0	3.123	2.759	1.1623	1.3561		1.010
Potassium	0.5	0.032	0.032	1.0017		0.11	1.001
permanganate	1.0	0.064	0.064	1.0051		0.22	1.000
KMnO$_4$	1.5	0.096	0.096	1.0085		0.32	0.999
	2.0	0.129	0.128	1.0118		0.43	0.998
	3.0	0.196	0.193	1.0186			0.995
	4.0	0.264	0.260	1.0254			0.992

Solute	Mass %	m/mol kg^{-1}	c/mol L^{-1}	ρ/g cm^{-3}	n	Δ/°C	η/mPa s
	5.0	0.333	0.327	1.0322			0.989
	6.0	0.404	0.394	1.0390			0.985
Potassium	0.5	0.029	0.029	1.0025	1.3338	0.13	1.013
hydrogen	1.0	0.058	0.058	1.0068	1.3345	0.25	1.023
phosphate	1.5	0.087	0.087	1.0110	1.3353	0.37	1.034
K_2HPO_4	2.0	0.117	0.117	1.0153	1.3361	0.49	1.046
	2.5	0.147	0.146	1.0195	1.3368	0.61	1.057
	3.0	0.178	0.176	1.0238	1.3376	0.73	1.069
	3.5	0.208	0.207	1.0281	1.3384	0.86	1.081
	4.0	0.239	0.237	1.0324	1.3392	0.97	1.094
	4.5	0.271	0.268	1.0368	1.3399	1.10	1.107
	5.0	0.302	0.299	1.0412	1.3407	1.22	1.120
	6.0	0.366	0.362	1.0500	1.3422	1.46	1.147
	7.0	0.432	0.426	1.0590	1.3438	1.70	1.177
	8.0	0.499	0.491	1.0680	1.3453	1.95	1.209
Potassium	0.5	0.037	0.037	1.0018	1.3336	0.13	1.010
dihydrogen	1.0	0.074	0.074	1.0053	1.3342	0.25	1.019
phosphate	1.5	0.112	0.111	1.0089	1.3348	0.37	1.028
KH_2PO_4	2.0	0.150	0.149	1.0125	1.3354	0.49	1.038
	3.0	0.227	0.225	1.0197	1.3365	0.72	1.060
	4.0	0.306	0.302	1.0269	1.3377	0.96	1.083
	5.0	0.387	0.380	1.0342	1.3388	1.19	1.108
	6.0	0.469	0.459	1.0414	1.3400	1.41	1.133
	7.0	0.553	0.539	1.0486	1.3411	1.63	1.160
	8.0	0.639	0.621	1.0558	1.3422	1.84	1.187
	9.0	0.727	0.703	1.0630	1.3434	2.04	1.215
	10.0	0.816	0.786	1.0703	1.3445	2.23	1.245
Potassium	0.5	0.029	0.029	1.0022	1.3336	0.14	1.006
sulfate	1.0	0.058	0.058	1.0062	1.3343	0.26	1.011
K_2SO_4	2.0	0.117	0.116	1.0143	1.3355	0.50	1.021
	3.0	0.177	0.176	1.0224	1.3368	0.73	1.033
	4.0	0.239	0.237	1.0306	1.3380	0.95	1.045
	5.0	0.302	0.298	1.0388	1.3393	1.17	1.058
	6.0	0.366	0.360	1.0470	1.3405		1.072
	7.0	0.432	0.424	1.0553	1.3417		1.087
	8.0	0.499	0.488	1.0637	1.3428		1.102
	9.0	0.568	0.554	1.0721	1.3440		1.117
	10.0	0.638	0.620	1.0806	1.3452		1.132
1-Propanol	1.0	0.168	0.166	0.9963	1.3339	0.31	1.051
$CH_3CH_2CH_2OH$	2.0	0.340	0.331	0.9946	1.3348	0.61	1.100
	3.0	0.515	0.496	0.9928	1.3357	0.93	1.152
	4.0	0.693	0.660	0.9911	1.3366	1.24	1.208
	5.0	0.876	0.823	0.9896	1.3376	1.57	1.267
	6.0	1.062	0.987	0.9882	1.3385	1.91	1.325
	7.0	1.252	1.149	0.9868	1.3394	2.26	1.387
	8.0	1.447	1.312	0.9855	1.3404	2.61	1.449
	9.0	1.646	1.474	0.9842	1.3414	2.99	1.514
	10.0	1.849	1.635	0.9829	1.3423	3.36	1.577
	12.0	2.269	1.958	0.9804	1.3442	4.09	1.710
	14.0	2.709	2.278	0.9779	1.3460	4.91	1.849
	16.0	3.169	2.595	0.9749	1.3477	5.78	1.986
	18.0	3.652	2.911	0.9719	1.3494	6.67	2.106
	20.0	4.160	3.223	0.9686	1.3510	7.76	2.218
	24.0	5.254	3.838	0.9612	1.3539	9.12	2.432
	28.0	6.471	4.441	0.9533	1.3566	10.17	2.612
	32.0	7.830	5.033	0.9452	1.3592	10.66	2.765
	36.0	9.359	5.613	0.9370	1.3614		2.900

Solute	Mass %	m/mol kg^{-1}	c/mol L^{-1}	ρ/g cm^{-3}	n	Δ/°C	η/mPa s
	40.0	11.093	6.182	0.9288	1.3635		3.010
	60.0	24.958	8.860	0.8875	1.3734		3.186
	80.0	66.556	11.275	0.8470	1.3812		2.822
	100.0		13.368	0.8034	1.3852		2.227
2-Propanol	1.0	0.168	0.166	0.9960	1.3338	0.30	1.056
$CH_3CHOHCH_3$	2.0	0.340	0.331	0.9939	1.3346	0.60	1.112
	3.0	0.515	0.495	0.9920	1.3355	0.93	1.166
	4.0	0.693	0.659	0.9902	1.3364	1.26	1.225
	5.0	0.876	0.822	0.9884	1.3373	1.61	1.287
	6.0	1.062	0.985	0.9871	1.3382	1.96	1.352
	7.0	1.252	1.148	0.9855	1.3392	2.32	1.417
	8.0	1.447	1.310	0.9843	1.3400	2.68	1.485
	9.0	1.646	1.472	0.9831	1.3410	3.06	1.553
	10.0	1.849	1.633	0.9816	1.3420	3.48	1.629
	12.0	2.269	1.955	0.9793	1.3439	4.43	1.794
	14.0	2.709	2.276	0.9772	1.3459	5.29	1.970
	16.0	3.169	2.596	0.9751	1.3478	6.36	2.160
	18.0	3.652	2.913	0.9725	1.3496	7.40	2.352
	20.0	4.160	3.227	0.9696	1.3514	8.52	2.550
	40.0	11.093	6.191	0.9302	1.3642		
	60.0	24.958	8.809	0.8824	1.3717		
	80.0	66.556	11.103	0.8341	1.3742		
	100.0		13.058	0.7848	1.3776		
Silver	0.5	0.030	0.030	1.0027	1.3336	0.10	1.003
nitrate	1.0	0.059	0.059	1.0070	1.3342	0.20	1.005
$AgNO_3$	2.0	0.120	0.120	1.0154	1.3352	0.40	1.009
	3.0	0.182	0.181	1.0239	1.3363	0.59	1.013
	4.0	0.245	0.243	1.0327	1.3374	0.78	1.016
	5.0	0.310	0.307	1.0417	1.3385	0.96	1.020
	6.0	0.376	0.371	1.0506	1.3396	1.15	1.024
	7.0	0.443	0.437	1.0597	1.3407	1.33	1.027
	8.0	0.512	0.503	1.0690	1.3419	1.51	1.031
	9.0	0.582	0.571	1.0785	1.3431	1.69	1.035
	10.0	0.654	0.641	1.0882	1.3443	1.87	1.039
	12.0	0.803	0.783	1.1079	1.3467	2.21	1.049
	14.0	0.958	0.930	1.1284	1.3493	2.55	1.060
	16.0	1.121	1.083	1.1496	1.3519	2.86	1.072
	18.0	1.292	1.241	1.1715	1.3546		1.086
	20.0	1.472	1.406	1.1942	1.3574		1.101
	22.0	1.660	1.577	1.2177	1.3602		1.117
	24.0	1.859	1.755	1.2420	1.3632		1.135
	26.0	2.068	1.940	1.2672	1.3662		1.154
	28.0	2.289	2.132	1.2933	1.3694		1.176
	30.0	2.523	2.332	1.3204	1.3726		1.200
	32.0	2.770	2.541	1.3487	1.3760		1.227
	34.0	3.033	2.758	1.3780	1.3795		1.257
	36.0	3.311	2.985	1.4087	1.3832		1.290
	38.0	3.608	3.223	1.4407	1.3871		1.326
	40.0	3.925	3.472	1.4743	1.3911		1.366
Sodium	0.5	0.061	0.061	1.0008	1.3337	0.22	1.021
acetate	1.0	0.123	0.122	1.0034	1.3344	0.43	1.040
CH_3COONa	2.0	0.249	0.246	1.0085	1.3358	0.88	1.080
	3.0	0.377	0.371	1.0135	1.3372	1.34	1.124
	4.0	0.508	0.497	1.0184	1.3386	1.82	1.171
	5.0	0.642	0.624	1.0234	1.3400	2.32	1.222
	6.0	0.778	0.752	1.0283	1.3414	2.85	1.278
	7.0	0.918	0.882	1.0334	1.3428	3.40	1.337
	8.0	1.060	1.013	1.0386	1.3442	3.98	1.401

Solute	Mass %	m/mol kg^{-1}	c/mol L^{-1}	ρ/g cm^{-3}	n	Δ/°C	η/mPa s
	9.0	1.206	1.145	1.0440	1.3456	4.57	1.468
	10.0	1.354	1.279	1.0495	1.3470		1.539
	12.0	1.662	1.552	1.0607	1.3498		1.688
	14.0	1.984	1.829	1.0718	1.3526		1.855
	16.0	2.322	2.112	1.0830	1.3554		2.054
	18.0	2.676	2.400	1.0940	1.3583		2.284
	20.0	3.047	2.694	1.1050	1.3611		2.567
	22.0	3.438	2.993	1.1159	1.3639		2.948
	24.0	3.849	3.297	1.1268	1.3666		3.400
	26.0	4.283	3.606	1.1377	1.3693		3.877
	28.0	4.741	3.921	1.1488	1.3720		4.388
	30.0	5.224	4.243	1.1602	1.3748		4.940
Sodium bicarbonate NaHCO$_3$	0.5	0.060	0.060	1.0018	1.3337	0.20	1.015
	1.0	0.120	0.120	1.0054	1.3344	0.40	1.028
	1.5	0.181	0.180	1.0089	1.3351	0.59	1.042
	2.0	0.243	0.241	1.0125	1.3357	0.78	1.057
	2.5	0.305	0.302	1.0160	1.3364	0.98	1.071
	3.0	0.368	0.364	1.0196	1.3370	1.16	1.086
	3.5	0.432	0.426	1.0231	1.3377	1.35	1.102
	4.0	0.496	0.489	1.0266	1.3383	1.54	1.118
	4.5	0.561	0.552	1.0301	1.3390	1.72	1.134
	5.0	0.627	0.615	1.0337	1.3396	1.90	1.151
	5.5	0.693	0.679	1.0372	1.3403	2.08	1.168
	6.0	0.760	0.743	1.0408	1.3409	2.26	1.185
Sodium bromide NaBr	0.5	0.049	0.049	1.0021	1.3337	0.17	1.004
	1.0	0.098	0.098	1.0060	1.3344	0.34	1.007
	2.0	0.198	0.197	1.0139	1.3358	0.69	1.012
	3.0	0.301	0.298	1.0218	1.3372	1.04	1.017
	4.0	0.405	0.400	1.0298	1.3386	1.39	1.022
	5.0	0.512	0.504	1.0380	1.3401	1.76	1.028
	6.0	0.620	0.610	1.0462	1.3415	2.14	1.034
	7.0	0.732	0.717	1.0546	1.3430	2.53	1.040
	8.0	0.845	0.826	1.0630	1.3445	2.93	1.046
	9.0	0.961	0.937	1.0716	1.3460	3.34	1.053
	10.0	1.080	1.050	1.0803	1.3475	3.77	1.060
	12.0	1.325	1.281	1.0981	1.3506	4.67	1.077
	14.0	1.582	1.519	1.1164	1.3538	5.65	1.096
	16.0	1.851	1.765	1.1352	1.3570	6.74	1.119
	18.0	2.133	2.020	1.1546	1.3604		1.144
	20.0	2.430	2.283	1.1745	1.3638		1.174
	22.0	2.741	2.555	1.1951	1.3673		1.207
	24.0	3.069	2.837	1.2163	1.3708		1.244
	26.0	3.415	3.129	1.2382	1.3745		1.287
	28.0	3.780	3.431	1.2608	1.3783		1.336
	30.0	4.165	3.744	1.2842	1.3822		1.395
	32.0	4.574	4.069	1.3083	1.3862		1.465
	34.0	5.007	4.406	1.3333	1.3903		1.546
	36.0	5.467	4.755	1.3592	1.3946		1.639
	38.0	5.957	5.119	1.3860	1.3990		1.745
	40.0	6.479	5.496	1.4138	1.4035		1.866
Sodium carbonate Na$_2$CO$_3$	0.5	0.047	0.047	1.0034	1.3341	0.22	1.025
	1.0	0.095	0.095	1.0086	1.3352	0.43	1.049
	2.0	0.193	0.192	1.0190	1.3375	0.75	1.102
	3.0	0.292	0.291	1.0294	1.3397	1.08	1.159
	4.0	0.393	0.392	1.0398	1.3419	1.42	1.222
	5.0	0.497	0.495	1.0502	1.3440	1.77	1.292
	6.0	0.602	0.600	1.0606	1.3462	2.13	1.367
	7.0	0.710	0.707	1.0711	1.3483		1.448

Solute	Mass %	m/mol kg^{-1}	c/mol L^{-1}	ρ/g cm^{-3}	n	Δ/°C	η/mPa s
	8.0	0.820	0.816	1.0816	1.3504		1.538
	9.0	0.933	0.927	1.0922	1.3525		1.638
	10.0	1.048	1.041	1.1029	1.3547		1.754
	11.0	1.166	1.156	1.1136	1.3568		1.884
	12.0	1.287	1.273	1.1244	1.3589		2.028
	13.0	1.410	1.392	1.1353	1.3610		2.186
	14.0	1.536	1.514	1.1463	1.3631		2.361
	15.0	1.665	1.638	1.1574	1.3652		2.551
Sodium	0.5	0.086	0.086	1.0018	1.3339	0.30	1.011
chloride	1.0	0.173	0.172	1.0053	1.3347	0.59	1.020
NaCl	2.0	0.349	0.346	1.0125	1.3365	1.19	1.036
	3.0	0.529	0.523	1.0196	1.3383	1.79	1.052
	4.0	0.713	0.703	1.0268	1.3400	2.41	1.068
	5.0	0.901	0.885	1.0340	1.3418	3.05	1.085
	6.0	1.092	1.069	1.0413	1.3435	3.70	1.104
	7.0	1.288	1.256	1.0486	1.3453	4.38	1.124
	8.0	1.488	1.445	1.0559	1.3470	5.08	1.145
	9.0	1.692	1.637	1.0633	1.3488	5.81	1.168
	10.0	1.901	1.832	1.0707	1.3505	6.56	1.193
	12.0	2.333	2.229	1.0857	1.3541	8.18	1.250
	14.0	2.785	2.637	1.1008	1.3576	9.94	1.317
	16.0	3.259	3.056	1.1162	1.3612	11.89	1.388
	18.0	3.756	3.486	1.1319	1.3648	14.04	1.463
	20.0	4.278	3.928	1.1478	1.3684	16.46	1.557
	22.0	4.826	4.382	1.1640	1.3721	19.18	1.676
	24.0	5.403	4.847	1.1804	1.3757		1.821
	26.0	6.012	5.326	1.1972	1.3795		1.990
Sodium	1.0	0.039	0.039	1.0049	1.3348	0.20	1.043
citrate	2.0	0.079	0.078	1.0120	1.3366	0.39	1.081
(HO)C(COONa)$_3$	3.0	0.120	0.118	1.0186	1.3383	0.59	1.122
	4.0	0.161	0.159	1.0260	1.3401	0.79	1.166
	5.0	0.204	0.200	1.0331	1.3419	0.97	1.210
	6.0	0.247	0.242	1.0405	1.3437	1.17	1.263
	7.0	0.292	0.284	1.0482	1.3455	1.36	1.314
	8.0	0.337	0.327	1.0557	1.3473	1.57	1.371
	9.0	0.383	0.371	1.0632	1.3491	1.77	1.427
	10.0	0.431	0.415	1.0708	1.3509	1.96	1.499
	12.0	0.528	0.505	1.0861	1.3546	2.38	1.649
	14.0	0.631	0.598	1.1019	1.3583	2.82	1.832
	16.0	0.738	0.693	1.1173	1.3618	3.27	2.045
	18.0	0.851	0.790	1.1327	1.3656	3.82	2.290
	20.0	0.969	0.891	1.1492	1.3693	4.39	2.596
	24.0	1.224	1.099	1.1813	1.3767		3.409
	28.0	1.507	1.318	1.2151	1.3845		4.586
	32.0	1.823	1.548	1.2487	1.3923		6.541
	36.0	2.180	1.792	1.2843	1.4001		9.788
Sodium	0.5	0.126	0.125	1.0039	1.3344	0.43	1.027
hydroxide	1.0	0.253	0.252	1.0095	1.3358	0.86	1.054
NaOH	2.0	0.510	0.510	1.0207	1.3386	1.74	1.112
	3.0	0.773	0.774	1.0318	1.3414	2.64	1.176
	4.0	1.042	1.043	1.0428	1.3441	3.59	1.248
	5.0	1.316	1.317	1.0538	1.3467	4.57	1.329
	6.0	1.596	1.597	1.0648	1.3494	5.60	1.416
	7.0	1.882	1.883	1.0758	1.3520	6.69	1.510
	8.0	2.174	2.174	1.0869	1.3546	7.87	1.616
	9.0	2.473	2.470	1.0979	1.3572	9.12	1.737
	10.0	2.778	2.772	1.1089	1.3597	10.47	1.882
	12.0	3.409	3.393	1.1309	1.3648	13.42	2.201

Solute	Mass %	m/mol kg^{-1}	c/mol L^{-1}	ρ/g cm^{-3}	n	Δ/°C	η/mPa s
	14.0	4.070	4.036	1.1530	1.3697	16.76	2.568
	15.0	4.412	4.365	1.1640	1.3722		2.789
	16.0	4.762	4.701	1.1751	1.3746		3.043
	18.0	5.488	5.387	1.1971	1.3793		3.698
	20.0	6.250	6.096	1.2192	1.3840		4.619
	22.0	7.052	6.827	1.2412	1.3885		5.765
	24.0	7.895	7.579	1.2631	1.3929		7.100
	26.0	8.784	8.352	1.2848	1.3971		8.744
	28.0	9.723	9.145	1.3064	1.4012		10.832
	30.0	10.715	9.958	1.3277	1.4051		13.517
	32.0	11.766	10.791	1.3488	1.4088		16.844
	34.0	12.880	11.643	1.3697	1.4123		20.751
	36.0	14.064	12.512	1.3901	1.4156		25.290
	38.0	15.324	13.398	1.4102	1.4186		30.461
	40.0	16.668	14.300	1.4299	1.4215		36.312
Sodium	0.5	0.059	0.059	1.0016	1.3336	0.20	1.004
nitrate	1.0	0.119	0.118	1.0050	1.3341	0.40	1.007
NaNO$_3$	2.0	0.240	0.238	1.0117	1.3353	0.79	1.012
	3.0	0.364	0.359	1.0185	1.3364	1.18	1.018
	4.0	0.490	0.483	1.0254	1.3375	1.56	1.025
	5.0	0.619	0.607	1.0322	1.3387	1.94	1.032
	6.0	0.751	0.734	1.0392	1.3398	2.32	1.040
	7.0	0.886	0.862	1.0462	1.3409	2.70	1.049
	8.0	1.023	0.991	1.0532	1.3421	3.08	1.059
	9.0	1.164	1.123	1.0603	1.3432	3.46	1.069
	10.0	1.307	1.256	1.0674	1.3443	3.84	1.081
	12.0	1.604	1.527	1.0819	1.3466	4.60	1.107
	14.0	1.915	1.806	1.0967	1.3489	5.37	1.138
	18.0	2.583	2.387	1.1272	1.3536	6.98	1.215
	20.0	2.941	2.689	1.1429	1.3559	7.81	1.263
	30.0	5.042	4.326	1.2256	1.3678		1.609
	40.0	7.844	6.200	1.3175	1.3802		2.226
Sodium	0.5	0.031	0.031	1.0042	1.3343	0.19	1.033
phosphate	1.0	0.062	0.062	1.0100	1.3356	0.37	1.064
Na$_3$PO$_4$	1.5	0.093	0.093	1.0158	1.3369	0.53	1.094
	2.0	0.124	0.125	1.0216	1.3381	0.67	1.126
	2.5	0.156	0.157	1.0275	1.3394	0.79	1.161
	3.0	0.189	0.189	1.0335	1.3406		1.198
	3.5	0.221	0.222	1.0395	1.3419		1.238
	4.0	0.254	0.255	1.0456	1.3432		1.281
	4.5	0.287	0.289	1.0517	1.3444		1.327
	5.0	0.321	0.323	1.0579	1.3457		1.375
	5.5	0.355	0.357	1.0642	1.3470		1.426
	6.0	0.389	0.392	1.0705	1.3482		1.480
	6.5	0.424	0.427	1.0768	1.3495		1.538
	7.0	0.459	0.462	1.0832	1.3507		1.598
	7.5	0.495	0.498	1.0896	1.3519		1.662
	8.0	0.530	0.535	1.0961	1.3532		1.729
Sodium hydrogen	0.5	0.035	0.035	1.0032	1.3340	0.17	1.021
phosphate	1.0	0.071	0.071	1.0082	1.3349	0.32	1.042
Na$_2$HPO$_4$	1.5	0.107	0.107	1.0131	1.3358	0.46	1.064
	2.0	0.144	0.143	1.0180	1.3368		1.088
	2.5	0.181	0.180	1.0229	1.3377		1.113
	3.0	0.218	0.217	1.0279	1.3386		1.138
	3.5	0.255	0.255	1.0328	1.3396		1.165
	4.0	0.293	0.292	1.0378	1.3405		1.193
	4.5	0.332	0.331	1.0428	1.3414		1.223
	5.0	0.371	0.369	1.0478	1.3424		1.254

Solute	Mass %	m/mol kg^{-1}	c/mol L^{-1}	ρ/g cm^{-3}	n	Δ/°C	η/mPa s
	5.5	0.410	0.408	1.0528	1.3433		1.286
Sodium	0.5	0.042	0.042	1.0019	1.3336	0.14	1.018
dihydrogen	1.0	0.084	0.084	1.0056	1.3343	0.28	1.035
phosphate	1.5	0.127	0.126	1.0094	1.3349	0.42	1.051
NaH_2PO_4	2.0	0.170	0.169	1.0131	1.3356	0.56	1.068
	2.5	0.214	0.212	1.0168	1.3362	0.70	1.085
	3.0	0.258	0.255	1.0206	1.3369	0.84	1.103
	3.5	0.302	0.299	1.0244	1.3375	0.98	1.121
	4.0	0.347	0.343	1.0281	1.3382	1.12	1.140
	4.5	0.393	0.387	1.0319	1.3388	1.25	1.160
	5.0	0.439	0.432	1.0358	1.3395	1.39	1.180
	6.0	0.532	0.522	1.0434	1.3408	1.65	1.223
	7.0	0.627	0.613	1.0511	1.3421	1.89	1.270
	8.0	0.725	0.706	1.0589	1.3434	2.12	1.319
	9.0	0.824	0.800	1.0668	1.3447	2.35	1.371
	10.0	0.926	0.896	1.0747	1.3460	2.58	1.428
	12.0	1.137	1.091	1.0907	1.3486	3.06	1.552
	14.0	1.357	1.292	1.1070	1.3512	3.53	1.694
	16.0	1.588	1.499	1.1236	1.3538	4.03	1.861
	18.0	1.830	1.711	1.1404	1.3565	4.55	2.050
	20.0	2.084	1.930	1.1576	1.3592	5.10	2.283
	22.0	2.351	2.155	1.1752	1.3618		2.550
	24.0	2.632	2.387	1.1931	1.3646		2.850
	26.0	2.929	2.625	1.2113	1.3673		3.214
	28.0	3.242	2.870	1.2299	1.3700		3.682
	30.0	3.572	3.123	1.2488	1.3728		4.300
	32.0	3.923	3.383	1.2682	1.3756		5.079
	34.0	4.294	3.650	1.2879	1.3784		6.008
	36.0	4.689	3.925	1.3080	1.3812		7.098
	38.0	5.109	4.208	1.3285	1.3840		8.363
	40.0	5.557	4.499	1.3493	1.3869		9.814
Sodium	0.5	0.035	0.035	1.0027	1.3338	0.17	1.013
sulfate	1.0	0.071	0.071	1.0071	1.3345	0.32	1.026
Na_2SO_4	2.0	0.144	0.143	1.0161	1.3360	0.61	1.058
	3.0	0.218	0.217	1.0252	1.3376	0.87	1.091
	4.0	0.293	0.291	1.0343	1.3391	1.13	1.126
	5.0	0.371	0.367	1.0436	1.3406	1.36	1.163
	6.0	0.449	0.445	1.0526	1.3420	1.56	1.202
	7.0	0.530	0.523	1.0619	1.3435		1.244
	8.0	0.612	0.603	1.0713	1.3449		1.289
	9.0	0.696	0.685	1.0808	1.3464		1.337
	10.0	0.782	0.768	1.0905	1.3479		1.390
	12.0	0.960	0.938	1.1101	1.3509		1.508
	14.0	1.146	1.114	1.1301	1.3539		1.646
	16.0	1.341	1.296	1.1503	1.3567		1.812
	18.0	1.545	1.483	1.1705	1.3595		2.005
	20.0	1.760	1.677	1.1907	1.3620		2.227
	22.0	1.986	1.875	1.2106	1.3643		2.481
Sodium	0.5	0.032	0.032	1.0024	1.3340	0.14	1.012
thiosulfate	1.0	0.064	0.064	1.0065	1.3351	0.28	1.023
$Na_2S_2O_3$	2.0	0.129	0.128	1.0148	1.3371	0.57	1.044
	3.0	0.196	0.194	1.0231	1.3392	0.84	1.066
	4.0	0.264	0.261	1.0315	1.3413	1.09	1.090
	5.0	0.333	0.329	1.0399	1.3434	1.34	1.115
	6.0	0.404	0.398	1.0483	1.3454	1.59	1.141
	7.0	0.476	0.468	1.0568	1.3475	1.83	1.169
	8.0	0.550	0.539	1.0654	1.3496	2.06	1.199
	9.0	0.626	0.611	1.0740	1.3517	2.30	1.231

Concentrative Properties of Aqueous Solutions

Table 8-75

Solute	Mass %	m/mol kg^{-1}	c/mol L^{-1}	ρ/g cm^{-3}	n	Δ/°C	η/mPa s
	10.0	0.703	0.685	1.0827	1.3538	2.55	1.267
	12.0	0.862	0.835	1.1003	1.3581	3.06	1.345
	14.0	1.030	0.990	1.1182	1.3624	3.60	1.435
	16.0	1.205	1.150	1.1365	1.3667	4.17	1.537
	18.0	1.388	1.315	1.1551	1.3711	4.76	1.657
	20.0	1.581	1.485	1.1740	1.3756	5.37	1.798
	30.0	2.711	2.417	1.2739	1.3987		2.903
	40.0	4.216	3.498	1.3827	1.4229		5.758
Strontium chloride SrCl$_2$	0.5	0.032	0.032	1.0027	1.3339	0.16	1.012
	1.0	0.064	0.064	1.0071	1.3348	0.31	1.021
	2.0	0.129	0.128	1.0161	1.3366	0.62	1.039
	3.0	0.195	0.194	1.0252	1.3384	0.93	1.057
	4.0	0.263	0.261	1.0344	1.3402	1.26	1.076
	5.0	0.332	0.329	1.0437	1.3421	1.61	1.096
	6.0	0.403	0.399	1.0532	1.3440	1.98	1.116
	7.0	0.475	0.469	1.0628	1.3459	2.38	1.136
	8.0	0.549	0.541	1.0726	1.3478	2.80	1.157
	9.0	0.624	0.615	1.0825	1.3498	3.25	1.180
	10.0	0.701	0.689	1.0925	1.3518	3.74	1.204
	12.0	0.860	0.843	1.1131	1.3558	4.81	1.258
	14.0	1.027	1.002	1.1342	1.3599	6.03	1.317
	16.0	1.202	1.167	1.1558	1.3641	7.41	1.383
	18.0	1.385	1.338	1.1780	1.3684	8.98	1.460
	20.0	1.577	1.515	1.2008	1.3728	10.74	1.549
	22.0	1.779	1.699	1.2241	1.3772	12.74	1.650
	24.0	1.992	1.890	1.2481	1.3817	14.99	1.765
	26.0	2.216	2.087	1.2728	1.3864		1.897
	28.0	2.453	2.293	1.2983	1.3911		2.056
	30.0	2.703	2.507	1.3248	1.3961		2.245
	32.0	2.968	2.730	1.3523	1.4013		2.527
	34.0	3.250	2.962	1.3811	1.4067		2.846
	36.0	3.548	3.205	1.4114	1.4124		3.206
Sucrose C$_{12}$H$_{22}$O$_{11}$	0.5	0.015	0.015	1.0002	1.3337	0.03	1.015
	1.0	0.030	0.029	1.0021	1.3344	0.06	1.028
	2.0	0.060	0.059	1.0060	1.3359	0.11	1.055
	3.0	0.090	0.089	1.0099	1.3373	0.17	1.084
	4.0	0.122	0.118	1.0139	1.3388	0.23	1.114
	5.0	0.154	0.149	1.0178	1.3403	0.29	1.146
	6.0	0.186	0.179	1.0218	1.3418	0.35	1.179
	7.0	0.220	0.210	1.0259	1.3433	0.42	1.215
	8.0	0.254	0.241	1.0299	1.3448	0.49	1.254
	9.0	0.289	0.272	1.0340	1.3463	0.55	1.294
	10.0	0.325	0.303	1.0381	1.3478	0.63	1.336
	12.0	0.398	0.367	1.0465	1.3509	0.77	1.429
	14.0	0.476	0.431	1.0549	1.3541	0.93	1.534
	16.0	0.556	0.497	1.0635	1.3573	1.10	1.653
	18.0	0.641	0.564	1.0722	1.3606	1.27	1.790
	20.0	0.730	0.632	1.0810	1.3639	1.47	1.945
	22.0	0.824	0.700	1.0899	1.3672	1.67	2.124
	24.0	0.923	0.771	1.0990	1.3706	1.89	2.331
	26.0	1.026	0.842	1.1082	1.3741	2.12	2.573
	28.0	1.136	0.914	1.1175	1.3776	2.37	2.855
	30.0	1.252	0.988	1.1270	1.3812	2.64	3.187
	32.0	1.375	1.063	1.1366	1.3848	2.94	3.762
	34.0	1.505	1.139	1.1464	1.3885	3.27	4.052
	36.0	1.643	1.216	1.1562	1.3922	3.63	4.621
	38.0	1.791	1.295	1.1663	1.3960	4.02	5.315
	40.0	1.948	1.375	1.1765	1.3999	4.45	6.162
	42.0	2.116	1.456	1.1868	1.4038	4.93	7.234

Solute	Mass %	m/mol kg^{-1}	c/mol L^{-1}	ρ/g cm^{-3}	n	Δ/°C	η/mPa s
	44.0	2.295	1.539	1.1972	1.4078		8.596
	46.0	2.489	1.623	1.2079	1.4118		10.301
	48.0	2.697	1.709	1.2186	1.4159		12.515
	50.0	2.921	1.796	1.2295	1.4201		15.431
	60.0	4.382	2.255	1.2864	1.4419		58.487
	70.0	6.817	2.755	1.3472	1.4654		481.561
	80.0	11.686	3.299	1.4117	1.4906		
Sulfuric acid	0.5	0.051	0.051	1.0016	1.3336	0.21	1.010
H_2SO_4	1.0	0.103	0.102	1.0049	1.3342	0.42	1.019
	2.0	0.208	0.206	1.0116	1.3355	0.80	1.036
	3.0	0.315	0.311	1.0183	1.3367	1.17	1.059
	4.0	0.425	0.418	1.0250	1.3379	1.60	1.085
	5.0	0.537	0.526	1.0318	1.3391	2.05	1.112
	6.0	0.651	0.635	1.0385	1.3403	2.50	1.136
	7.0	0.767	0.746	1.0453	1.3415	2.95	1.159
	8.0	0.887	0.858	1.0522	1.3427	3.49	1.182
	9.0	1.008	0.972	1.0591	1.3439	4.08	1.206
	10.0	1.133	1.087	1.0661	1.3451	4.64	1.230
	12.0	1.390	1.322	1.0802	1.3475	5.93	1.282
	14.0	1.660	1.563	1.0947	1.3500	7.49	1.337
	16.0	1.942	1.810	1.1094	1.3525	9.26	1.399
	18.0	2.238	2.064	1.1245	1.3551	11.29	1.470
	20.0	2.549	2.324	1.1398	1.3576	13.64	1.546
	22.0	2.876	2.592	1.1554	1.3602	16.48	1.624
	24.0	3.220	2.866	1.1714	1.3628	19.85	1.706
	26.0	3.582	3.147	1.1872	1.3653	24.29	1.797
	28.0	3.965	3.435	1.2031	1.3677	29.65	1.894
	30.0	4.370	3.729	1.2191	1.3701	36.21	2.001
	32.0	4.798	4.030	1.2353	1.3725	44.76	2.122
	34.0	5.252	4.339	1.2518	1.3749	55.28	2.255
	36.0	5.735	4.656	1.2685	1.3773		2.392
	38.0	6.249	4.981	1.2855	1.3797		2.533
	40.0	6.797	5.313	1.3028	1.3821		2.690
	42.0	7.383	5.655	1.3205	1.3846		2.872
	44.0	8.011	6.005	1.3386	1.3870		3.073
	46.0	8.685	6.364	1.3570	1.3895		3.299
	48.0	9.411	6.734	1.3759	1.3920		3.546
	50.0	10.196	7.113	1.3952	1.3945		3.826
	52.0	11.045	7.502	1.4149	1.3971		4.142
	54.0	11.969	7.901	1.4351	1.3997		4.499
	56.0	12.976	8.312	1.4558	1.4024		4.906
	58.0	14.080	8.734	1.4770	1.4050		5.354
	60.0	15.294	9.168	1.4987	1.4077		5.917
	70.0	23.790	11.494	1.6105			
	80.0	40.783	14.088	1.7272			
	90.0	91.762	16.649	1.8144			
	92.0	117.251	17.109	1.8240			
	94.0	159.734	17.550	1.8312			
	96.0	244.698	17.966	1.8355			
	98.0	499.592	18.346	1.8361			
	100.0		18.663	1.8305			
Trichloroacetic	0.5	0.031	0.031	1.0008	1.3337	0.11	1.011
acid	1.0	0.062	0.061	1.0034	1.3343	0.21	1.021
CCl_3COOH	2.0	0.125	0.123	1.0083	1.3356	0.42	1.044
	3.0	0.189	0.186	1.0133	1.3369	0.64	1.069
	4.0	0.255	0.249	1.0182	1.3381	0.86	1.096
	5.0	0.322	0.313	1.0230	1.3394	1.08	1.123
	6.0	0.391	0.377	1.0279	1.3406	1.30	1.150
	7.0	0.461	0.442	1.0328	1.3418	1.53	1.177

Solute	Mass %	m/mol kg^{-1}	c/mol L^{-1}	ρ/g cm^{-3}	n	Δ/°C	η/mPa s
	8.0	0.532	0.508	1.0378	1.3431	1.76	1.204
	9.0	0.605	0.574	1.0428	1.3444	1.99	1.233
	10.0	0.680	0.641	1.0479	1.3456	2.23	1.263
	12.0	0.835	0.777	1.0583	1.3483	2.73	1.326
	14.0	0.996	0.916	1.0692	1.3510	3.26	1.393
	16.0	1.166	1.058	1.0806	1.3539	3.82	1.462
	18.0	1.343	1.203	1.0921	1.3568		1.533
	20.0	1.530	1.351	1.1035	1.3597		1.608
	24.0	1.933	1.654	1.1260	1.3652		1.768
	28.0	2.380	1.968	1.1485	1.3705		1.935
	32.0	2.880	2.294	1.1713	1.3759		2.118
	36.0	3.443	2.632	1.1947	1.3813		2.320
	40.0	4.080	2.984	1.2188	1.3868		1.543
	44.0	4.809	3.349	1.2435	1.3923		2.797
	48.0	5.650	3.726	1.2682	1.3977		3.076
Tris	0.5	0.041	0.041	0.9994	1.3337	0.08	1.014
(hydroxymethyl)-	1.0	0.083	0.083	1.0006	1.3344	0.16	1.027
methylamine	2.0	0.168	0.166	1.0030	1.3359	0.31	1.054
$H_2NC(CH_2OH)_3$	3.0	0.255	0.249	1.0054	1.3374	0.47	1.083
	4.0	0.344	0.333	1.0078	1.3388	0.64	1.115
	5.0	0.434	0.417	1.0103	1.3403	0.80	1.148
	6.0	0.527	0.502	1.0128	1.3418	0.97	1.182
	7.0	0.621	0.587	1.0153	1.3433	1.15	1.218
	8.0	0.718	0.672	1.0179	1.3448	1.33	1.256
	9.0	0.816	0.758	1.0204	1.3463	1.51	1.295
	10.0	0.917	0.844	1.0230	1.3478	1.70	1.337
	12.0	1.126	1.019	1.0282	1.3508	2.08	1.427
	14.0	1.344	1.194	1.0335	1.3539	2.47	1.527
	16.0	1.572	1.372	1.0389	1.3570	2.90	1.642
	18.0	1.812	1.552	1.0443	1.3601	3.36	1.772
	20.0	2.064	1.733	1.0498	1.3633	3.85	1.920
	30.0	3.538	2.670	1.0781	1.3797		2.998
	40.0	5.503	3.657	1.1076	1.3970		5.208
Urea	0.5	0.084	0.083	0.9995	1.3337	0.16	1.007
$(NH_2)_2CO$	1.0	0.168	0.167	1.0007	1.3344	0.31	1.010
	2.0	0.340	0.334	1.0033	1.3358	0.62	1.012
	3.0	0.515	0.502	1.0058	1.3372	0.93	1.017
	4.0	0.694	0.672	1.0085	1.3387	1.24	1.025
	5.0	0.876	0.842	1.0111	1.3401	1.55	1.033
	6.0	1.063	1.013	1.0138	1.3416	1.88	1.041
	7.0	1.253	1.185	1.0165	1.3431	2.22	1.049
	8.0	1.448	1.358	1.0192	1.3446	2.56	1.057
	9.0	1.647	1.531	1.0220	1.3461	2.91	1.065
	10.0	1.850	1.706	1.0248	1.3476	3.26	1.074
	12.0	2.270	2.059	1.0304	1.3506	3.95	1.091
	14.0	2.710	2.415	1.0360	1.3537	4.66	1.109
	16.0	3.171	2.775	1.0417	1.3568	5.40	1.130
	18.0	3.655	3.139	1.0473	1.3599	6.19	1.153
	20.0	4.163	3.506	1.0530	1.3629	7.00	1.178
	22.0	4.696	3.878	1.0586	1.3661	7.81	1.205
	24.0	5.258	4.253	1.0643	1.3692	8.64	1.235
	26.0	5.850	4.632	1.0699	1.3723	9.52	1.266
	28.0	6.475	5.014	1.0756	1.3754	10.45	1.298
	30.0	7.136	5.401	1.0812	1.3785	11.40	1.332
	32.0	7.835	5.791	1.0869	1.3817	12.34	1.371
	34.0	8.577	6.185	1.0926	1.3848	13.27	1.413
	36.0	9.366	6.584	1.0984	1.3881	14.20	1.459
	38.0	10.205	6.988	1.1044	1.3913	15.11	1.509
	40.0	11.100	7.397	1.1106	1.3947	15.99	1.565

Solute	Mass %	m/mol kg^{-1}	c/mol L^{-1}	ρ/g cm^{-3}	n	Δ/°C	η/mPa s
	42.0	12.057	7.812	1.1171	1.3982	16.83	1.629
	44.0	13.082	8.234	1.1239	1.4018	17.62	1.700
	46.0	14.183	8.665	1.1313	1.4056		1.780
Zinc sulfate	0.5	0.031	0.031	1.0034	1.3339	0.08	1.021
$ZnSO_4$	1.0	0.063	0.062	1.0085	1.3348	0.15	1.040
	2.0	0.126	0.126	1.0190	1.3366	0.28	1.081
	3.0	0.192	0.191	1.0296	1.3384	0.41	1.126
	4.0	0.258	0.258	1.0403	1.3403	0.53	1.175
	5.0	0.326	0.326	1.0511	1.3421	0.65	1.227
	6.0	0.395	0.395	1.0620	1.3439	0.77	1.283
	7.0	0.466	0.465	1.0730	1.3457	0.89	1.341
	8.0	0.539	0.537	1.0842	1.3475	1.01	1.403
	9.0	0.613	0.611	1.0956	1.3494	1.14	1.470
	10.0	0.688	0.686	1.1071	1.3513	1.27	1.545
	12.0	0.845	0.840	1.1308	1.3551	1.55	1.716
	14.0	1.008	1.002	1.1553	1.3590	1.89	1.918
	16.0	1.180	1.170	1.1806	1.3630	2.31	2.152

ION PRODUCT OF WATER SUBSTANCE

William L. Marshall and E. U. Franck

Pressure (bars)	Temperature (°C)								
	0	25	50	75	100	150	200	250	300
Saturated vapor	14.938	13.995	13.275	12.712	12.265	11.638	11.289	11.191	11.406
250	14.83	13.90	13.19	12.63	12.18	11.54	11.16	11.01	11.14
500	14.72	13.82	13.11	12.55	12.10	11.45	11.05	10.85	10.86
750	14.62	13.73	13.04	12.48	12.03	11.36	10.95	10.72	10.66
1,000	14.53	13.66	12.96	12.41	11.96	11.29	10.86	10.60	10.50
1,500	14.34	13.53	12.85	12.29	11.84	11.16	10.71	10.43	10.26
2,000	14.21	13.40	12.73	12.18	11.72	11.04	10.57	10.27	10.08
2,500	14.08	13.28	12.62	12.07	11.61	10.92	10.45	10.12	9.91
3,000	13.97	13.18	12.53	11.98	11.53	10.83	10.34	9.99	9.76
3,500	13.87	13.09	12.44	11.90	11.44	10.74	10.24	9.88	9.63
4,000	13.77	13.00	12.35	11.82	11.37	10.66	10.16	9.79	9.52
5,000	13.60	12.83	12.19	11.66	11.22	10.52	10.00	9.62	9.34
6,000	13.44	12.68	12.05	11.53	11.09	10.39	9.87	9.48	9.18
7,000	13.31	12.55	11.93	11.41	10.97	10.27	9.75	9.35	9.04
8,000	13.18	12.43	11.82	11.30	10.86	10.17	9.64	9.24	8.93
9,000	13.04	12.31	11.71	11.20	10.77	10.07	9.54	9.13	8.82
10,000	12.91	12.21	11.62	11.11	10.68	9.98	9.45	9.04	8.71

Pressure (bars)	Temperature (°C)								
	350	400	450	500	600	700	800	900	1000
Saturated vapor	12.30	—	—	—	—	—	—	—	—
250	11.77	19.43	21.59	22.40	23.27	23.81	24.23	24.59	24.93
500	11.14	11.88	13.74	16.13	18.30	19.29	19.92	20.39	20.80
750	10.79	11.17	11.89	13.01	15.25	16.55	17.35	17.93	18.39
1,000	10.54	10.77	11.19	11.81	13.40	14.70	15.58	16.22	16.72
1,500	10.22	10.29	10.48	10.77	11.59	12.50	13.30	13.97	14.50
2,000	9.98	9.98	10.07	10.23	10.73	11.36	11.98	12.54	12.97
2,500	9.79	9.74	9.77	9.86	10.18	10.63	11.11	11.59	12.02
3,000	9.61	9.54	9.53	9.57	9.78	10.11	10.49	10.89	11.24
3,500	9.47	9.37	9.33	9.34	9.48	9.71	10.02	10.35	10.62
4,000	9.34	9.22	9.16	9.15	9.23	9.41	9.65	9.93	10.13
5,000	9.13	8.99	8.90	8.85	8.85	8.95	9.11	9.30	9.42
6,000	8.96	8.80	8.69	8.62	8.57	8.61	8.72	8.86	8.97
7,000	8.81	8.64	8.51	8.42	8.34	8.34	8.40	8.51	8.64
8,000	8.68	8.50	8.36	8.25	8.13	8.10	8.13	8.21	8.38
9,000	8.57	8.37	8.22	8.10	7.95	7.89	7.89	7.95	8.12
10,000	8.46	8.25	8.09	7.96	7.78	7.70	7.68	7.70	7.85

Data in this table were calculated from the equation, $\log_{10} K_w^* = A + B/T + C/T^2 + D/T^3 + (E + F/T + G/T^2) \log_{10} \rho_w^*$, where $K_w^* = K_w/(\text{mol kg}^{-1})$, and $\rho_w^* = \rho_w/(\text{g cm}^{-3})$. The parameters are:

$A = -4.098$
$B = -3245.2$ K
$C = +2.2362 \times 10^5$ K^2
$D = -3.984 \times 10^7$ K^3

$E = +13.957$
$F = 1262.3$ K
$G = +8.5641 \times 10^5$ K^2

Reprinted with permission from W. L. Marshall and E. U. Franck, *J. Phys. Chem. Ref. Data*, 10, 295, 1981.

IONIZATION CONSTANT OF NORMAL AND HEAVY WATER

This table gives the ionization constant in molality terms for H_2O and D_2O at temperatures from 0 to 100°C at the saturated vapor pressure. The quantity tabulated is $-\log K_w$, where K_w is defined by

$$K_w = m_+ \times m_-$$

and m_+ and m_- are the molalities, in mol/kg of water, for H^+ and OH^-, respectively.

References

1. Marshall W. L., and Franck, E. U., *J. Phys. Chem. Ref. Data*, 10, 295, 1981.
2. Mesmer R. E., and Herting, D. L., *J. Solution Chem.*, 7, 901, 1978.

$t/°C$	$-\log K_w$	
	H_2O	D_2O
0	14.938	15.972
5	14.727	15.743
10	14.528	15.527
15	14.340	15.324
20	14.163	15.132
25	13.995	14.951
30	13.836	14.779
35	13.685	14.616
40	13.542	14.462
45	13.405	14.316
50	13.275	14.176
55	13.152	14.044
60	13.034	13.918
65	12.921	13.798
70	12.814	13.683
75	12.712	13.574
80	12.613	13.470
85	12.520	13.371
90	12.428	13.276
95	12.345	13.186
100	12.265	13.099

SOLUBILITY OF SELECTED GASES IN WATER

L. H. Gevantman

The values in this table are taken almost exclusively from the International Union of Pure and Applied Chemistry "Solubility Data Series". Unless noted, they comprise evaluated data fitted to a smoothing equation. The data at each temperature are then derived from the smoothing equation which expresses the mole fraction solubility X_1 of the gas in solution as:

$$\ln X_1 = A + B/T^* + C \ln T^*$$

where

$$T^* = T/100\ \text{K}$$

All values refer to a partial pressure of the gas of 101.325 kPa (one atmosphere).

The equation constants, the standard deviation for $\ln X_1$ (except where noted), and the temperature range over which the equation applies are given in the column headed Equation constants. There are two exceptions. The equation for methane has an added term, DT^*. The equation for H_2Se and H_2S takes the form,

$$\ln X_1 = A + B/T + C \ln T + DT$$

where T is the temperature in kelvin.

Solubilities given for those gases which react with water, namely ozone, nitrogen oxides, chlorine and its oxides, carbon dioxide, hydrogen sulfide, hydrogen selenide and sulfur dioxide, are recorded as bulk solubilities; i.e., all chemical species of the gas and its reaction products with water are included.

Gas	T/K	Solubility (X_1)	Equation constants	Ref.
Hydrogen (H_2)	288.15	1.510×10^{-5}	$A = -48.1611$	1
$M_r = 2.01588$	293.15	1.455×10^{-5}	$B = 55.2845$	
	298.15	1.411×10^{-5}	$C = 16.8893$	
	303.15	1.377×10^{-5}	Std. dev. = ± 0.54%	
	308.15	1.350×10^{-5}	Temp. range = 273.15—353.15	
Deuterium (D_2)	283.15	$1.675 \times 10^{-5} \pm 0.57\%$	Averaged experimental values	1
$M_r = 4.0282$	288.15	$1.595 \times 10^{-5} \pm 0.57\%$		
	293.15	$1.512 \times 10^{-5} \pm 0.78\%$	Temp. range = 278.15—303.15	
	298.15	$1.460 \times 10^{-5} \pm 0.52\%$		
	303.15	$1.395 \times 10^{-5} \pm 0.37\%$		
Helium (He)	288.15	7.123×10^{-6}	$A = -41.4611$	2
$A_r = 4.0026$	293.15	7.044×10^{-6}	$B = 42.5962$	
	298.15	6.997×10^{-6}	$C = 14.0094$	
	303.15	6.978×10^{-6}	Std. dev. = ±0.54%	
	308.15	6.987×10^{-6}	Temp. range = 273.15—348.15	
Neon (Ne)	288.15	8.702×10^{-6}	$A = -52.8573$	2
$A_r = 20.1797$	293.15	8.395×10^{-6}	$B = 61.0494$	
	298.15	8.152×10^{-6}	$C = 18.9157$	
	303.15	7.966×10^{-6}	Std. dev. = ±0.47%	
	308.15	7.829×10^{-6}	Temp. range = 273.15—348.15	
Argon (Ar)	288.15	3.025×10^{-5}	$A = -57.6661$	3
$A_r = 39.948$	293.15	2.748×10^{-5}	$B = 74.7627$	
	298.15	2.519×10^{-5}	$C = 20.1398$	
	303.15	2.328×10^{-5}	Std. dev. = ±0.26%	
	308.15	2.169×10^{-5}	Temp. range = 273.15—348.15	
Krypton (Kr)	288.15	5.696×10^{-5}	$A = -66.9928$	4
$A_r = 83.80$	293.15	5.041×10^{-5}	$B = 91.0166$	
	298.15	4.512×10^{-5}	$C = 24.2207$	
	303.15	4.079×10^{-5}	Std. dev. = ±0.32%	
	308.15	3.725×10^{-5}	Temp. range = 273.15—353.15	
Xenon (Xe)	288.15	10.519×10^{-5}	$A = -74.7398$	4
$A_r = 131.29$	293.15	9.051×10^{-5}	$B = 105.210$	
	298.15	7.890×10^{-5}	$C = 27.4664$	
	303.15	6.961×10^{-5}	Std. dev. = ±0.35%	
	308.15	6.212×10^{-5}	Temp. range = 273.15—348.15	

Gas	T/K	Solubility (X_1)	Equation constants	Ref.
Radon-222(^{222}Rn)	288.15	2.299×10^{-4}	$A = -90.5481$	
$A_r = 222$	293.15	1.945×10^{-4}	$B = 130.026$	
	298.15	1.671×10^{-4}	$C = 35.0047$	
	303.15	1.457×10^{-4}	Std. dev. = ±1.02%	
	308.15	1.288×10^{-4}	Temp. range = 273.15—373.15	
Oxygen (O_2)	288.15	2.756×10^{-5}	$A = -66.7354$	5
$M_r = 31.9988$	293.15	2.501×10^{-5}	$B = 87.4755$	
	298.15	2.293×10^{-5}	$C = 24.4526$	
	303.15	2.122×10^{-5}	Std. dev. = ±0.36%	
	308.15	1.982×10^{-5}	Temp. range = 273.15—348.15	
Ozone (O_3)	293.15	1.885×10^{-6} ± 10%	Experimental value derived	5
$M_r = 47.9982$		pH = 7.0	from Henry's Law Constant Equation	
Nitrogen (N_2)	288.15	1.386×10^{-5}	$A = -67.3877$	6
$M_r = 28.0134$	293.15	1.274×10^{-5}	$B = 86.3213$	
	298.15	1.183×10^{-5}	$C = 24.7981$	
	303.15	1.108×10^{-5}	Std. dev. = ±0.72%	
	308.15	1.047×10^{-5}	Temp. range = 273.15—348.15	
Nitrous oxide (N_2O)	288.15	5.948×10^{-4}	$A = -60.7467$	7
$M_r = 44.0129$	293.15	5.068×10^{-4}	$B = 88.8280$	
	298.15	4.367×10^{-4}	$C = 21.2531$	
	303.15	3.805×10^{-4}	Std. dev. = ±1.2%	
	308.15	3.348×10^{-4}	Temp. range = 273.15—313.15	
Nitric oxide (NO)	288.15	4.163×10^{-5}	$A = -62.8086$	7
$M_r = 30.0061$	293.15	3.786×10^{-5}	$B = 82.3420$	
	298.15	3.477×10^{-5}	$C = 22.8155$	
	303.15	3.222×10^{-5}	Std. dev. = ±0.76%	
	308.15	3.012×10^{-5}	Temp. range = 273.15—358.15	
Carbon monoxide (CO)	288.15	2.095×10^{-5}	Derived from Henry's	8
$M_r = 28.0104$	293.15	1.918×10^{-5}	Law Constant Equation	
	298.15	1.774×10^{-5}	Std. dev. = ±0.043%	
	303.15	1.657×10^{-5}	Temp. range = 273.15—328.15	
	308.15	1.562×10^{-5}		
Carbon dioxide (CO_2)	288.15	8.21×10^{-4}	Derived from Henry's	9
$M_r = 44.0098$	293.15	7.07×10^{-4}	Law Constant Equation	
	298.15	6.15×10^{-4}	Std. dev. = ±1.1%	
	303.15	5.41×10^{-4}	Temp. range = 273.15—353.15	
	308.15	4.80×10^{-4}		
Hydrogen selenide (H_2Se)	288.15	1.80×10^{-3}	$A = 9.15$	10
$M_r = 80.976$	298.15	1.49×10^{-3}	$B = 974$	
	308.15	1.24×10^{-3}	$C = -3.542$	
			$D = 0.0042$	
			Std. dev. = ±2.3×10^{-5}	
			Temp. range = 288.15—343.15	
Hydrogen sulfide (H_2S)	288.15	2.335×10^{-3}	$A = -24.912$	10
$M_r = 34.082$	293.15	2.075×10^{-3}	$B = 3477$	
	298.15	1.85×10^{-3}	$C = 0.3993$	
	303.15	1.66×10^{-3}	$D = 0.0157$	
	308.15	1.51×10^{-3}	Std. dev. = ±6.5×10^{-5}	
			Temp. range = 283.15—603.15	
Sulfur dioxide (SO_2)	288.15	3.45×10^{-2}	$A = -25.2629$	11
$M_r = 64.0648$	293.15	2.90×10^{-2}	$B = 45.7552$	

Gas	T/K	Solubility (X_1)	Equation constants	Ref.
	298.15	2.46×10^{-2}	$C = 5.6855$	
	303.15	2.10×10^{-2}	Std. dev. $= \pm1.8\%$	
	308.15	1.80×10^{-2}	Temp. range $= 278.15—328.15$	
Chlorine (Cl_2)	283.15	$2.48 \times 10^{-3} \pm 2\%$	Experimental data	11
$M_r = 70.9054$	293.15	$1.88 \times 10^{-3} \pm 2\%$	Temp. range $= 283.15—333.15$	
	303.15	$1.50 \times 10^{-3} \pm 2\%$		
	313.15	$1.23 \times 10^{-3} \pm 2\%$		
Chlorine monoxide (Cl_2O)	273.15	$5.25 \times 10^{-1} \pm 1\%$	Experimental data	11
$M_r = 86.9048$	276.61	$4.54 \times 10^{-1} \pm 1\%$	Temp. range $= 273.15—293.15$	
	283.15	$4.273 \times 10^{-1} \pm 1\%$		
	293.15	$3.353 \times 10^{-1} \pm 1\%$		
Chlorine dioxide (ClO_2)	288.15	2.67×10^{-2}	$A = 7.9163$	11
$M_r = 67.4515$	293.15	2.20×10^{-2}	$B = 0.4791$	
	298.15	1.823×10^{-2}	$C = 11.0593$	
	303.15	1.513×10^{-2}	Std. dev. $= \pm4.6\%$	
	308.15	1.259×10^{-2}	Temp. range $= 283.15—333.15$	
Methane (CH_4)	288.15	3.122×10^{-5}	$A = -115.6477$	12
$M_r = 16.0428$	293.15	2.806×10^{-5}	$B = 155.5756$	
	298.15	2.552×10^{-5}	$C = 65.2553$	
	303.15	2.346×10^{-5}	$D = -6.1698$	
	308.15	2.180×10^{-5}	Std. dev. $= \pm0.056\%$	
			Temp. range $= 273.15—328.15$	
Ethane (C_2H_6)	288.15	4.556×10^{-5}	$A = -90.8225$	13
$M_r = 30.0696$	293.15	3.907×10^{-5}	$B = 126.9559$	
	298.15	3.401×10^{-5}	$C = 34.7413$	
	303.15	3.002×10^{-5}	Std. dev. $= \pm0.13\%$	
	308.15	2.686×10^{-5}	Temp. range $= 273.15—323.15$	
Propane (C_3H_8)	288.15	3.813×10^{-5}	$A = -102.044$	14
$M_r = 44.097$	293.15	3.200×10^{-5}	$B = 144.345$	
	298.15	2.732×10^{-5}	$C = 39.4740$	
	303.15	2.370×10^{-5}	Std. dev. $= \pm0.012\%$	
	308.15	2.088×10^{-5}	Temp. range $= 273.15—347.15$	
Butane (C_4H_{10})	288.15	3.274×10^{-5}	$A = -102.029$	14
$M_r = 58.123$	293.15	2.687×10^{-5}	$B = 146.040$	
	298.15	2.244×10^{-5}	$C = 38.7599$	
	303.15	1.906×10^{-5}	Std. dev. $= \pm0.026\%$	
	308.15	1.645×10^{-5}	Temp. range $= 273.15—349.15$	
2-Methyl propane (Isobutane)	288.15	2.333×10^{-5}	$A = -129.714$	14
(C_4H_{10})	293.15	1.947×10^{-5}	$B = 183.044$	
$M_r = 58.123$	298.15	1.659×10^{-5}	$C = 53.4651$	
	303.15	1.443×10^{-5}	Std. dev. $= \pm0.034\%$	
	308.15	1.278×10^{-5}	Temp. range $= 278.15—318.15$	

References

1. C. L. Young, Ed., *IUPAC Solubility Data Series*, Vol. 5/6, Hydrogen and Deuterium, Pergamon Press, Oxford, England, 1981.
2. H. L. Clever, Ed., *IUPAC Solubility Data Series*, Vol. 1, Helium and Neon, Pergamon Press, Oxford, England, 1979.
3. H. L. Clever, Ed., *IUPAC Solubility Data Series*, Vol. 4, Argon, Pergamon Press, Oxford, England, 1980.
4. H. L. Clever, Ed., *IUPAC Solubility Data Series*, Vol. 2, Krypton, Xenon and Radon, Pergamon Press, Oxford, England, 1979.
5. R. Battino, Ed., *IUPAC Solubility Data Series*, Vol. 7, Oxygen and Ozone, Pergamon Press, Oxford, England, 1981.
6. R. Battino, Ed., *IUPAC Solubility Data Series*, Vol. 10, Nitrogen and Air, Pergamon Press, Oxford, England, 1982.
7. C. L. Young, Ed., *IUPAC Solubility Data Series*, Vol. 8, Oxides of Nitrogen, Pergamon Press, Oxford, England, 1981.
8. R. W. Cargill, Ed., *IUPAC Solubility Data Series*, Vol. 43, Carbon Monoxide, Pergamon Press, Oxford, England, 1990.
9. R. Crovetto, Evaluation of Solubility Data for the System CO2-H2O, *J. Phys. Chem. Ref. Data*, 20, 575, 1991.

10. P. G. T. Fogg and C. L. Young, Eds., *IUPAC Solubility Data Series*, Vol. 32, Hydrogen Sulfide, Deuterium Sulfide, and Hydrogen Selenide, Pergamon Press, Oxford, England, 1988.

11. C. L. Young, Ed., *IUPAC Solubility Data Series*, Vol. 12, Sulfur Dioxide, Chlorine, Fluorine and Chlorine Oxides, Pergamon Press, Oxford, England, 1983.

12. H. L. Clever and C. L. Young, Eds., *IUPAC Solubility Data Series*, Vol. 27/28, Methane, Pergamon Press, Oxford, England, 1987.

13. W. Hayduk, Ed., *IUPAC Solubility Data Series*, Vol. 9, Ethane, Pergamon Press, Oxford, England, 1982.

14. W. Hayduk, Ed., *IUPAC Solubility Data Series*, Vol. 24, Propane, Butane and 2-Methylpropane, Pergamon Press, Oxford, England, 1986.

SOLUBILITY OF CARBON DIOXIDE IN WATER AT VARIOUS TEMPERATURES AND PRESSURES

The solubility of CO_2 in water, expressed as mole fraction of CO_2 in the liquid phase, is given for pressures up to atmospheric and temperatures of 0 to 100°C. Note that 1 standard atmosphere equals 101.325 kPa. The references give data over a wider range of temperature and pressure. The estimated accuracy is about 2%.

References

1. Carroll, J. J., Slupsky, J. D., and Mather, A. E., *J. Phys. Chem. Ref. Data*, 20, 1201, 1991.
2. Fernandez-Prini, R. and Crovetto, R., *J. Phys. Chem. Ref. Data*, 18, 1231, 1989.
3. Crovetto, R., *J. Phys. Chem. Ref. Data*, 20, 575,1991.

	1000 × mole fraction of CO_2 in liquid phase						
	Partial pressure of CO_2 in kPa						
$t/°C$	5	10	20	30	40	50	100
0	0.067	0.135	0.269	0.404	0.538	0.671	1.337
5	0.056	0.113	0.226	0.338	0.451	0.564	1.123
10	0.048	0.096	0.191	0.287	0.382	0.477	0.950
15	0.041	0.082	0.164	0.245	0.327	0.409	0.814
20	0.035	0.071	0.141	0.212	0.283	0.353	0.704
25	0.031	0.062	0.123	0.185	0.247	0.308	0.614
30	0.027	0.054	0.109	0.163	0.218	0.271	0.541
35	0.024	0.048	0.097	0.145	0.193	0.242	0.481
40	0.022	0.043	0.087	0.130	0.173	0.216	0.431
45	0.020	0.039	0.078	0.117	0.156	0.196	0.389
50	0.018	0.036	0.071	0.107	0.142	0.178	0.354
55	0.016	0.033	0.065	0.098	0.131	0.163	0.325
60	0.015	0.030	0.060	0.090	0.121	0.150	0.300
65	0.014	0.028	0.056	0.084	0.112	0.140	0.279
70	0.013	0.026	0.052	0.079	0.105	0.131	0.261
75	0.012	0.025	0.049	0.074	0.099	0.123	0.245
80	0.012	0.023	0.047	0.070	0.093	0.116	0.232
85	0.011	0.022	0.044	0.067	0.089	0.111	0.221
90	0.011	0.021	0.042	0.064	0.085	0.106	0.211
95	0.010	0.020	0.041	0.061	0.082	0.102	0.203
100	0.010	0.020	0.039	0.059	0.079	0.098	0.196

AQUEOUS SOLUBILITY AND HENRY'S LAW CONSTANTS OF ORGANIC COMPOUNDS

The solubility in water of over 1200 organic compounds, including many compounds of environmental interest, is tabulated here. Values are given at ambient temperature (usually 20°C - 25°C) and at higher temperatures when data are available. Solids, liquids, and gases are included; additional data on gases can be found in the table "Solubility of Selected Gases in Water" in Section 8.

Solubility of solids is defined as the concentration of the compound in a solution that is in equilibrium with the solid phase at the specified temperature and one atmosphere pressure. For liquids whose water mixtures separate into two phases, the solubility given here is the concentration of the specified compound in the water-rich phase at equilibrium. In the case of gases (i.e., compounds whose vapor pressure at the specified temperature exceeds one atmosphere) the solubility is defined here as the concentration in the water phase when the partial pressure of the compound above the solution is 101.325 kPa (1 atm). Values for gases are marked with [a].

The primary solubility values in this tables are expressed as mass percent of solute, $S = 100w_2$, where the mass fraction w_2 is defined as

$$w_2 = m_2/(m_1 + m_2),$$

where m_2 is the mass of solute and m_1 the mass of water. For convenience, the solubility expressed in grams of solute per liter of solution is tabulated in the adjacent column to mass percent. The conversion between these two measures involves the density of the solution, which usually is not readily available. For compounds with low solubility (say, $S < 1\%$), the error from approximating the density is generally less than the uncertainty in the experimental solubility measurement, so that little accuracy is lost in the conversion. However, this may not be true for more soluble compounds; for that reason, some values in the table are indicated as approximate (~).

The mass fraction w_2 is related to other common measures of solubility as follows:

molality: $\qquad m_2 = 1000\, w_2/M_2(1-w_2)$

molarity: $\qquad c_2 = 1000\rho w_2/M_2$

mole fraction: $x_2 = (w_2/M_2)/\{(w_2/M_2) + (1-w_2)/M_1\}$

mass of solute per 100 g of H_2O: $\quad r_2 = 100w_2/(1-w_2)$

mass of solute per liter of solution: $1000\rho w_2$

where M_2 is the molar mass of the solute, $M_1 = 18.015$ g/mol is the molar mass of water, and ρ is the density of the solution in g/mL. Data have been selected from evaluated sources wherever possible, in particular the *IUPAC Solubility Data Series* (References 1,3,4,35,36,38,39). Many values come from experimental measurements reported in the *Journal of Chemical and Engineering Data* and the *Journal of Chemical Thermodynamics*, as well as critical review papers in the *Journal of Physical and Chemical Reference Data*. The primary source for each value is listed in the column following the solubility values. The user is cautioned that

wide variations of data are found in the literature for the lower solubility compounds.

The table also contains values of the Henry's Law constant k_H, which provides a measure of the partition of a substance between the atmosphere and the aqueous phase. Here k_H is defined as the limit of p_2/c_2 as the concentration approaches zero, where p_2 is the partial pressure of the solute above the solution and c_2 is the concentration in the solution at equilibrium (other formulations of Henry's Law are often used; see Reference 5). The values of k_H listed here are based on direct experimental measurement whenever available, but many of them are simply calculated as the ratio of the pure compound vapor pressure to the solubility. This approximation is reliable only for compounds of very low solubility. In fact, values of k_H found in the literature frequently differ by a factor of two or three, and variations over an order of magnitude are not unusual (Reference 5). Therefore the data given here should be taken only as a rough indication of the true Henry's Law constant, which is difficult to measure precisely.

All values of k_H refer to 25°C. If the vapor pressure of the compound at 25°C is greater than one atmosphere, it can be assumed that the k_H value has been calculated as $101.325/c_2$ kPa m^3/mol. The source of the Henry's Law data is given in the last column. The air-water partition coefficient (i.e., ratio of air concentration to water concentration when both are expressed in the same units) is equal to k_H/RT or $k_H/2.48$ in the units used here.

Compounds are listed by molecular formula following the Hill convention. To locate a compound by name or CAS Registry Number when the molecular formula is not known, use the table "Physical Constants of Organic Compounds" in Section 3 and its indexes to determine the molecular formula.

References

1. *Solubility Data Series, International Union of Pure and Applied Chemistry, Vol. 15*, Pergamon Press, Oxford, 1982.
2. *Solubility Data Series, International Union of Pure and Applied Chemistry, Vol. 20*, Pergamon Press, Oxford, 1985.
3. *Solubility Data Series, International Union of Pure and Applied Chemistry, Vol. 37*, Pergamon Press, Oxford, 1988.
4. *Solubility Data Series, International Union of Pure and Applied Chemistry, Vol. 38*, Pergamon Press, Oxford, 1988.
5. Mackay, D., and Shiu, W. Y., *J. Phys. Chem. Ref. Data*, 10, 1175, 1981.
6. Pearlman, R. S., and Yalkowsky, S. H., *J. Phys. Chem. Ref. Data*, 13, 975, 1984.
7. Shiu, W. Y., and Mackay, D., *J. Phys. Chem. Ref. Data*, 15, 911, 1986.
8. Varhanickova, D., Lee, S. C., Shiu, W. Y., and Mackay, D., *J. Chem. Eng. Data*, 40, 620, 1995.
9. Miller, M. M., Ghodbane, S., Wasik, S. P., Tewari, Y. B., and Martire, D. E., *J. Chem. Eng. Data*, 29, 184, 1984.
10. Riddick, J. A., Bunger, W. B., and Sakano, T. K., *Organic Solvents, Fourth Edition*, John Wiley & Sons, New York, 1986.
11. Mackay, D., Shiu, W. Y., and Ma, K. C., *Illustrated Handbook of Physical-Chemical Properties and Environmental Fate for Organic Chemicals, Vol. I*, Lewis Publishers/CRC Press, Boca Raton, FL, 1992.
12. Mackay, D., Shiu, W. Y., and Ma, K. C., *Illustrated Handbook of Physical-Chemical Properties and Environmental Fate for Organic Chemicals, Vol. II*, Lewis Publishers/CRC Press, Boca Raton, FL, 1992.
13. Mackay, D., Shiu, W. Y., and Ma, K. C., *Illustrated Handbook of Physical-Chemical Properties and Environmental Fate for Organic Chemicals, Vol. III*, Lewis Publishers/CRC Press, Boca Raton, FL, 1993.

4. Horvath, A. L., *Halogenated Hydrocarbons*, Marcel Dekker, New York, 1982.
5. Howard, P. H., *Handbook of Environmental Fate and Exposure Data for Organic Chemicals, Vol. I*, Lewis Publishers/CRC Press, Boca Raton, FL, 1989.
6. Howard, P. H., *Handbook of Environmental Fate and Exposure Data for Organic Chemicals, Vol. II*, Lewis Publishers/CRC Press, Boca Raton, FL, 1990.
7. Banergee, S., Yalkowsky, S. H., and Valvani, S. C., *Environ. Sci. Technol.*, 14, 1227, 1980.
8. Gevantman, L., in *CRC Handbook of Chemistry and Physics, 84th Edition*, p. 8-88, CRC Press, Boca Raton, FL, 2003.
9. Wilhelm, E., Battino, R., and Wilcock, R. J., *Chem. Rev.* 77, 219, 1977.
0. Stephenson, R. M., *J. Chem. Eng. Data*, 37, 80, 1992.
1. Stephenson, R. M., Stuart, J., and Tabak, M., *J. Chem. Eng. Data*, 29, 287, 1984.
2. Shiu, W.-Y., and Ma, K.-C, *J. Phys. Chem. Ref. Data*, 29, 41, 2000.
3. Lun, R., Varhanickova, D., Shiu, W.-Y., and Mackay, D., *J. Chem. Eng. Data*, 42, 951 (1997).
4. Huang, G.-L., Xiao, H., Chi, J., Shiu, W.-Y., and Mackay, D., *J. Chem. Eng. Data*, 45, 411, 2000.
5. Horvath, A. L., Getzen, F. W., and Maczynska, Z., *J. Phys. Chem. Ref. Data*, 28, 395, 2000 [IUPAC No. 67].
6. Dawson, R. M. C., Elliott, D. C., Elliott, W. H., and Jones, K. M., *Data for Biochemical Research*, Third Edition, Clarendon Press, Oxford, 1986.
7. Stephen, H., and Stephen, T., *Solubilities of Organic and Inorganic Compounds*, MacMillan, New York, 1963.
28. Shiu, W.-Y., and Mackay, D., *J. Chem. Eng. Data*, 42, 27, 1997.
29. Hinz, H.-J., ed., *Thermodynamic Data for Biochemistry and Biotechnology*, Springer-Verlag, Berlin, 1986.
30. Budavari, S., ed., *The Merck Index, Twelfth Edition*, Merck & Co., Rahway, NJ, 1996.
31. Bamford, H. A., Poster, D. L., and Baker, J. E., *J. Chem. Eng. Data*, 45, 1069, 2000.
32. Lide, D. R., and Milne, G. W. A., *Handbook of Data on Organic Compounds, Third Edition*, CRC Press, Boca Raton, FL, 1994.
33. Apelblat, A., and Manzurola, E., *J. Chem. Thermodynamics* 21, 1005, 1989.
34. Apelblat, A., and Manzurola, E., *J. Chem. Thermodynamics* 22, 289, 1990.
35. Horvath, A. L., and Getzen, F. W., *J. Phys. Chem. Ref. Data* 28, 649, 1999 [IUPAC No. 68].
36. Sazonov, V. P., Marsh, K. N., and Hefter, G. T., *J. Phys. Chem. Ref. Data* 29, 1165, 2000 [IUPAC No. 71].
37. Verbruggen, E. M. J., Hermens, J. L. M., and Tolls, J., *J. Phys. Chem. Ref. Data* 29, 1435, 2000.
38. Sazonov, V. P., Shaw, D. G., and Marsh, K. N., *J. Phys. Chem. Ref. Data* 31, 1, 2002 [IUPAC No. 77].
39. Sazonov, V. P., and Shaw, D. G., *J. Phys. Chem. Ref. Data* 31, 989, 2002 [IUPAC No. 78].
40. Yalkowsky, S. H., and He, Y., *Handbook of Aqueous Solubility Data*, CRC Press, Boca Raton, FL, 2003.
41. Shiu, W.-Y., and Ma, K.-C., *J. Phys. Chem. Ref. Data* 29, 387, 2000.

Mol. formula	Name	Mol. wt.	$t/°C$	Mass%	g/L	Ref.	k_H kPa m³mol⁻¹	Ref.
$CBrF_3$	Bromotrifluoromethane	148.910	25	0.032[a]	0.32[a]	14		
CBr_3F	Tribromofluoromethane	270.721	25	0.040	0.40	14		
CBr_4	Tetrabromomethane	331.627	30	0.024	0.24	14		
$CClF_3$	Chlorotrifluoromethane	104.459	25	0.009[a]	0.09[a]	10	6.9	13
$CClN$	Cyanogen chloride	61.471	0	5.7	60	40		
CCl_2F_2	Dichlorodifluoromethane	120.914	20	0.028[a]	0.28[a]	5	41	13
CCl_3F	Trichlorofluoromethane	137.368	20	0.11	1.1	5	10.2	13
CCl_3NO_2	Trichloronitromethane	164.376	0	0.227	2.27	40		
			25	0.162	1.62	40		
CCl_4	Tetrachloromethane	153.823	25	0.065	0.65	20	2.99	13
			75	0.115	11.5	20	2.99	13
CF_4	Tetrafluoromethane	88.005	25	0.00187[a]	0.0187[a]	19		
$CHBrCl_2$	Bromodichloromethane	163.829	30	0.300	3.00	40		
$CHBr_2Cl$	Chlorodibromomethane	208.280	30	0.251	2.51	40		
$CHBr_3$	Tribromomethane	252.731	25	0.30	3.0	5	0.047	13
$CHClF_2$	Chlorodifluoromethane	86.469	25	0.30[a]	3.0[a]	10	3.0	13
$CHCl_2F$	Dichlorofluoromethane	102.923	25	0.95[a]	9.5[a]	10		
$CHCl_3$	Trichloromethane	119.378	25	0.80	8.0	20	0.43	13
			59	0.79	7.9	20	0.43	13
CHF_3	Trifluoromethane	70.014	25	0.09[a]	0.9[a]	14		
CHI_3	Triiodomethane	393.732	25	0.012	0.12	14		
CH_2BrCl	Bromochloromethane	129.384	25	1.7	17	10	0.18	13
CH_2Br_2	Dibromomethane	173.835	20	1.28	11.5	20	0.086	13
			90	1.51	15.3	20		
CH_2ClF	Chlorofluoromethane	68.478	25	1.05[a]	10.6[a]	14		
CH_2Cl_2	Dichloromethane	84.933	25	1.73	17.6	20	0.30	13
CH_2I_2	Diiodomethane	267.836	30	0.124	1.24	10	0.032	13
CH_3Br	Bromomethane	94.939	20	1.80[a]	18.3[a]	5	0.63	13
CH_3Cl	Chloromethane	50.488	25	0.535[a]	5.35[a]	5	0.98	13
CH_3F	Fluoromethane	34.033	30	0.177[a]	1.77[a]	5		
CH_3I	Iodomethane	141.939	20	1.4	14	10	0.54	13
CH_3NO_2	Nitromethane	61.041	0	9.2	101	36		

Aqueous Solubility and Henry's Law Constants of Organic Compounds

Mol. formula	Name	Mol. wt.	t/°C	Solubility S Mass%	g/L	Ref.	Henry Const. k_H kPa m³mol⁻¹	Ref.
			25	11.0	~125	36		
			50	14.8	~175	36		
CH_4	Methane	16.043	25	0.00227[a]	0.0227[a]	18	67.4	5
CH_4N_2O	Urea	60.055	5	44		26		
			25	54.4		26		
CH_4N_2S	Thiourea	76.121	20	10.6	~120	40		
			80	~37	580	40		
$CH_4N_4O_2$	Nitroguanidine	104.069	25	1.2	12	40		
CO	Carbon monoxide	28.010	25	0.00276[a]	0.0276[a]	18		
CO_2	Carbon dioxide	44.010	25	0.1501	1.501	18		
CS_2	Carbon disulfide	76.141	20	0.210	2.10	10		
$C_2Br_2F_4$	1,2-Dibromotetrafluoroethane	259.823	25	0.00030	0.0030	25		
C_2ClF_5	Chloropentafluoroethane	154.466	25	0.006[a]	0.06[a]	10	260	13
$C_2Cl_2F_4$	1,2-Dichloro-1,1,2,2-tetrafluoroethane	170.921	25	0.013[a]	0.13[a]	10	127	13
$C_2Cl_3F_3$	1,1,2-Trichloro-1,2,2-trifluoroethane	187.375	25	0.017	0.17	25	32	13
C_2Cl_4	Tetrachloroethene	165.833	0	0.0273	0.24	20		
			20	0.0286	0.21	20	1.73	13
			80	0.0380	0.20	20		
$C_2Cl_4F_2$	1,1,2,2-Tetrachloro-1,2-difluoroethane	203.830	27	0.016	0.16	25		
C_2Cl_6	Hexachloroethane	236.739	25	0.005	0.05	25	0.85	13
C_2F_4	Tetrafluoroethene	100.015	25	0.0158[a]	0.158[a]	19		
$C_2HBrClF_3$	2-Bromo-2-chloro-1,1,1-trifluoroethane	197.381	10	0.52	5.2	25		
			25	0.41	4.1	25		
			40	0.40	4.0	25		
$C_2HCl_2F_3$	2,2-Dichloro-1,1,1-trifluoroethane	152.930	25	0.46	4.6	25		
C_2HCl_3	Trichloroethene	131.388	0	0.145	1.45	25		
			25	0.128	1.28	25	1.03	13
			60	0.133	1.33	25		
C_2HCl_3O	Trichloroacetaldehyde	147.387	25	~39	650	40		
$C_2HCl_3O_2$	Trichloroacetic acid	163.387	25	92.3		27		
C_2HCl_5	Pentachloroethane	202.294	25	0.049	0.49	25	0.25	13
C_2H_2	Acetylene	26.037	25	0.1081[a]	1.081[a]	19		
$C_2H_2Br_2Cl_2$	1,2-Dibromo-1,2-dichloroethane	256.751	20	0.070	0.70	25		
$C_2H_2Br_4$	1,1,2,2-Tetrabromoethane	345.653	0	0.052	0.52	25		
			25	0.068	0.68	25		
			50	0.106	1.06	25		
			100	0.307	3.07	25		
$C_2H_2Cl_2$	1,1-Dichloroethene	96.943	5	0.310	3.10	25		
			25	0.242	2.42	25	2.62	13
			50	0.225	2.25	25		
			90	0.355	3.55	25		
$C_2H_2Cl_2$	cis-1,2-Dichloroethene	96.943	10	0.76	7.6	25		
			25	0.64	6.4	25	0.46	13
			40	0.66	6.6	25		
$C_2H_2Cl_2$	trans-1,2-Dichloroethene	96.943	10	0.53	5.3	25		
			25	0.45	4.5	25	0.96	13
			40	0.41	4.1	25		
$C_2H_2Cl_2F_2$	1,2-Dichloro-1,1-difluoroethane	134.940	24	0.49	4.9	25		
$C_2H_2Cl_4$	1,1,1,2-Tetrachloroethane	167.849	0	0.120	1.20	25		
			25	0.107	1.07	25	0.24	13
			50	0.123	1.23	25		
$C_2H_2Cl_4$	1,1,2,2-Tetrachloroethane	167.849	5	0.302	3.02	25		
			25	0.283	2.83	25	0.026	13
			50	0.318	3.18	25		
$C_2H_2I_2$	cis-1,2-Diiodoethene	279.846	25	0.046	0.46	25		
$C_2H_2I_2$	trans-1,2-Diiodoethene	279.846	25	0.015	0.15	25		
$C_2H_2O_4$	Oxalic acid	90.035	20	8.69	95.1	27		
			80	45.8		27		

Mol. formula	Name	Mol. wt.	$t/°C$	Solubility S			Henry Const. k_H	
				Mass%	g/L	Ref.	kPa m³mol⁻¹	Ref.
$C_2H_3Br_2Cl$	1,2-Dibromo-1-chloroethane	222.306	20	0.060	0.60	25		
$C_2H_3Br_3$	1,1,2-Tribromoethane	266.757	20	0.050	0.50	25		
C_2H_3Cl	Chloroethene	62.498	25	0.27[a]	2.7[a]	5	2.68	13
$C_2H_3Cl_2F$	1,1-Dichloro-1-fluoroethane	116.949	25	0.042	0.42	25		
$C_2H_3Cl_3$	1,1,1-Trichloroethane	133.404	0	0.134	1.34	25		
			25	0.129	1.29	25	1.76	13
			50	0.138	1.38	25		
$C_2H_3Cl_3$	1,1,2-Trichloroethane	133.404	0	0.425	4.25	25		
			25	0.459	4.59	25	0.092	13
			50	0.536	5.36	25		
C_2H_3N	Acetonitrile	41.052	-3	40.5		39		
			-10	31.7		39		
C_2H_3NS	Methyl isothiocyanate	73.117	20	0.75	7.6	40		
C_2H_4	Ethylene	28.053	25	0.01336[a]	0.1336[a]	19	21.7	5
C_2H_4BrCl	1-Bromo-2-chloroethane	143.410	30	0.683	6.83	25		
$C_2H_4Br_2$	1,2-Dibromoethane	187.861	20	0.412	3.1	20		
			50	0.493	3.9	20	0.066	13
			80	0.572	5.4	20		
$C_2H_4Cl_2$	1,1-Dichloroethane	98.959	0	0.62	6.2	25		
			25	0.50	5.0	25	0.63	13
			50	0.50	5.0	25		
$C_2H_4Cl_2$	1,2-Dichloroethane	98.959	0	0.92	9.2	25		
			25	0.86	8.6	25	0.14	13
			50	1.05	10.6	25		
			100	2.17	22.1	25		
$C_2H_4N_4$	Cyanoguanidine	84.080	25	3.8	40	40		
$C_2H_4N_4$	1H-1,2,4-Triazol-3-amine	84.080	23	22		26		
$C_2H_4O_2$	Methyl formate	60.052	25	23		10		
$C_2H_4O_3$	Glycolic acid	76.051	25	71.21		34		
			55	77.95		34		
C_2H_5Br	Bromoethane	108.965	0	1.05	10.6	25		
			25	0.90	9.0	25	1.23	13
C_2H_5Cl	Chloroethane	64.514	0	0.45	4.5	25		
			25	0.67	6.7	25	1.02	13
C_2H_5F	Fluoroethane	48.059	25	0.216[a]	2.16[a]	14		
C_2H_5I	Iodoethane	155.965	0	0.44	4.4	25		
			25	0.40	4.0	25	0.52	13
C_2H_5N	Ethyleneimine	43.068	20	0.90	9.12	40		
C_2H_5NO	Acetamide	59.067	20	40.8		10		
$C_2H_5NO_2$	Nitroethane	75.067	25	4.4	46	38		
			50	5.3	56	38		
$C_2H_5NO_2$	Methyl carbamate	75.067	15	69		27		
$C_2H_5NO_2$	Glycine	75.067	25	20.06		26		
C_2H_5NS	Thioacetamide	75.133	25	12.3	140	40		
$C_2H_5N_3O_2$	N-Methyl-N-nitrosourea	103.080	14	2.3	24	40		
$C_2H_5N_3O_2$	Imidodicarbonic diamide	103.080	15	1.5	15	40		
C_2H_6	Ethane	30.069	25	0.00568[a]	0.0568[a]	18	50.6	5
C_2H_6O	Dimethyl ether	46.068	24	35.3[a]		10	0.077	13
C_2H_6OS	Dimethyl sulfoxide	78.133	25	25.3		10		
$C_2H_6O_4S$	Dimethyl sulfate	126.132	18	2.7	28	27		
C_2H_6S	Dimethyl sulfide	62.134	25	2	20	10		
$C_2H_7AsO_2$	Dimethylarsinic acid	137.998	25	~41	684	40		
C_2N_2	Cyanogen	52.034	25	0.8	8	30		
$C_3Br_2F_6$	1,2-Dibromo-1,1,2,3,3,3-hexafluoropropane	309.830	21	0.0068	0.068	35		
$C_3Cl_2F_6$	1,2-Dichloro-1,1,2,3,3,3-hexafluoropropane	220.928	21	0.0096	0.096	35		
$C_3Cl_3F_5$	1,1,1-Trichloro-2,2,3,3,3-pentafluoropropane	237.383	21	0.0058	0.058	35		

| Mol. formula | Name | Mol. wt. | $t/°C$ | Solubility S | | | Henry Const. k_H | |
				Mass%	g/L	Ref.	kPa m³mol⁻¹	Ref.
$C_3Cl_4F_4$	1,1,1,3-Tetrachloro-2,2,3,3-tetrafluoropropane	253.838	21	0.0052	0.052	35		
C_3Cl_6	Hexachloropropene	248.750	20	0.00118	0.0118	35		
C_3F_6	Perfluoropropene	150.022	25	0.0194[a]	0.194[a]	14		
C_3F_8	Perfluoropropane	188.019	15	0.0015[a]	0.015[a]	14		
$C_3H_2N_2$	Malononitrile	66.061	20	10.6	118	40		
C_3H_3N	Acrylonitrile	53.063	20	7.35	79.3	10		
$C_3H_3NOS_2$	2-Thioxo-4-thiazolidinone	133.192	25	0.225	2.25	40		
$C_3H_3N_3O_3$	Cyanuric acid	129.074	25	0.259	2.59	40		
C_3H_4	Propyne	40.064	25	0.364[a]	3.64[a]	5	1.11	5
$C_3H_4ClF_3$	3-Chloro-1,1,1-trifluoropropane	132.512	20	0.133	1.33	35		
$C_3H_4Cl_2$	cis-1,3-Dichloropropene	110.970	20	0.27	2.7	5	0.24	5
$C_3H_4Cl_2$	trans-1,3-Dichloropropene	110.970	20	0.28	2.8	5	0.18	5
$C_3H_4Cl_2$	2,3-Dichloropropene	110.970	25	0.215	2.15	5	0.36	5
$C_3H_4N_2O$	2-Cyanoacetamide	84.076	20	11.5	130	40		
$C_3H_4N_2O_2$	2,4-Imidazolidinedione	100.076	25	3.93	40.9	29		
C_3H_4O	Acrolein	56.063	20	20.8		10		
$C_3H_4O_4$	Malonic acid	104.062	0	37.9		26		
			20	42.4		26		
			50	48.1		26		
C_3H_5Br	3-Bromopropene	120.976	25	0.38	3.8	35		
$C_3H_5Br_2Cl$	1,2-Dibromo-3-chloropropane	236.333	20	0.123	1.23	35		
C_3H_5Cl	3-Chloropropene	76.525	25	0.40	4.0	35	1.10	5
			50	0.13	1.3	35		
C_3H_5ClO	Epichlorohydrin	92.524	20	6.58	70.4	10	0.003	13
			65	7.2	77.9	40		
$C_3H_5Cl_3$	1,2,3-Trichloropropane	147.431	10	0.14	1.4	35		
			25	0.20	2.0	35	0.038	13
C_3H_5N	Propanenitrile	55.079	25	10.3	~115	10		
C_3H_5NO	Acrylamide	71.078	20	~27	371	40		
$C_3H_5N_3O_9$	Trinitroglycerol	227.087	25	0.13	1.3	40		
			80	0.34	3.4	40		
C_3H_6	Propene	42.080	25	0.0200[a]	0.200[a]	5	21.3	5
C_3H_6	Cyclopropane	42.080	25	0.0484[a]	0.484[a]	19		
C_3H_6BrCl	1-Bromo-3-chloropropane	157.437	25	0.223	2.23	35		
$C_3H_6Br_2$	1,2-Dibromopropane	201.888	25	0.143	1.43	10		
$C_3H_6Br_2$	1,3-Dibromopropane	201.888	25	0.169	1.69	35		
$C_3H_6Cl_2$	1,2-Dichloropropane, (±)-	112.986	5	0.270	2.70	35		
			25	0.274	2.74	35	0.29	13
			40	0.297	2.97	35		
$C_3H_6Cl_2$	1,3-Dichloropropane	112.986	5	0.218	2.18	35		
			25	0.280	2.80	35		
$C_3H_6N_6$	1,3,5-Triazine-2,4,6-triamine	126.120	20	0.323	3.23	40		
			95	4.2	44	40		
$C_3H_6N_6O_6$	Hexahydro-1,3,5-trinitro-1,3,5-triazine	222.116	25	0.0060	0.060	17		
C_3H_6O	Propanal	58.079	25	30.6		10		
C_3H_6O	Methyloxirane	58.079	20	40.5		10	0.0087	13
$C_3H_6O_2$	Ethyl formate	74.079	25	11.8	~135	10		
$C_3H_6O_2$	Methyl acetate	74.079	20	24.5		10		
$C_3H_6O_3$	1,3,5-Trioxane	90.078	25	17.4	~210	30		
C_3H_7Br	1-Bromopropane	122.992	0	0.298	2.98	35		
			25	0.234	2.34	35	3.8	13
C_3H_7Br	2-Bromopropane	122.992	20	0.32	3.2	35	1.27	13
C_3H_7Cl	1-Chloropropane	78.541	25	0.250	2.50	35	1.41	13
C_3H_7Cl	2-Chloropropane	78.541	0	0.44	4.4	35		
			20	0.30	3.0	35		
C_3H_7ClO	1-Chloro-2-methoxyethane	94.540	20	7.79	84	20		
			70	6.31	67	20		

Mol. formula	Name	Mol. wt.	t/°C	Solubility S			Henry Const. k_H	
				Mass%	g/L	Ref.	kPa m³mol⁻¹	Ref.
C_3H_7F	1-Fluoropropane	62.086	14	0.386[a]	3.86[a]	14		
C_3H_7F	2-Fluoropropane	62.086	15	0.366	3.66	14		
C_3H_7I	1-Iodopropane	169.992	0	0.114	1.14	35		
			20	0.100	1.00	35	0.93	13
C_3H_7I	2-Iodopropane	169.992	0	0.167	1.67	35		
			20	0.140	1.40	35		
$C_3H_7NO_2$	1-Nitropropane	89.094	25	1.54	15.6	38		
			90	2.29	23	20		
$C_3H_7NO_2$	2-Nitropropane	89.094	25	1.75	17.8	38		
			90	2.36	24	20		
$C_3H_7NO_2$	Ethyl carbamate	89.094	15	48		27		
$C_3H_7NO_2$	L-Alanine	89.094	25	14.30	~170	26		
$C_3H_7NO_2$	β-Alanine	89.094	25	47.1		26		
$C_3H_7NO_2$	Sarcosine	89.094	25	30.0		26		
$C_3H_7NO_3$	L-Serine	105.093	25	20	~200	26		
$C_3H_7N_3O_2$	N-Ethyl-N-nitrosourea	117.107	20	1.3	13	40		
$C_3H_7N_3O_2$	Guanidinoacetic acid	117.107	25	0.5	5	26		
C_3H_8	Propane	44.096	25	0.00669[a]	0.0669[a]	18	71.6	5
$C_3H_8NO_5P$	Glyphosate	169.074	25	1.2	12	32		
$C_3H_8O_2$	Dimethoxymethane	76.095	16	24.4		10		
$C_3H_9O_4P$	Trimethyl phosphate	140.074	25	~33	500	40		
C_4Cl_6	Hexachloro-1,3-butadiene	260.761	25	0.41	4.1	35		
C_4F_8	Perfluorocyclobutane	200.030	21	0.014[a]	0.14[a]	14		
$C_4H_3FN_2O_2$	5-Fluorouracil	130.077	22	1.10	11.1	40		
$C_4H_4N_2$	Succinonitrile	80.088	25	11.5	~130	10		
$C_4H_4N_2O_2$	Uracil	112.087	25	0.27	2.7	29		
C_4H_4O	Furan	68.074	25	1	10	10	0.54	13
$C_4H_4O_4$	Maleic acid	116.073	25	44.1		26		
C_4H_5N	2-Methylacrylonitrile	67.090	20	2.57	26.3	10		
C_4H_5N	Pyrrole	67.090	25	4.5	47	10		
$C_4H_5N_3O$	Cytosine	111.102	25	0.73	7.3	29		
C_4H_6	1,3-Butadiene	54.091	25	0.0735[a]	0.735[a]	5	20.7	13
C_4H_6	1-Butyne	54.091	25	0.287[a]	2.87[a]	5	1.91	5
$C_4H_6N_2O_2$	2,5-Piperazinedione	114.103	25	1.64	16.6	29		
$C_4H_6N_4O_3S_2$	Acetazolamide	222.246	30	0.10	1.0	40		
C_4H_6O	$trans$-2-Butenal	70.090	20	15.6	~185	10		
$C_4H_6O_2$	$trans$-2-Butenoic acid	86.090	20	7.1	76	26		
$C_4H_6O_2$	Methacrylic acid	86.090	20	8.9	98	10		
$C_4H_6O_2$	Vinyl acetate	86.090	20	2.0	20	10		
$C_4H_6O_2$	Methyl acrylate	86.090	25	4.94	52.0	10		
$C_4H_6O_2$	2,3-Butanedione	86.090	20	31.7		20		
			80	21.8		20		
$C_4H_6O_4$	Succinic acid	118.089	25	7.71	83.5	27		
			100	55		27		
$C_4H_6O_4$	Methylmalonic acid	118.089	0	30.1		26		
			20	40		26		
$C_4H_6O_4$	Dimethyl oxalate	118.089	20	5.82	61.8	27		
$C_4H_6O_5$	Diglycolic acid	134.088	24	40.03		34		
$C_4H_6O_5$	Malic acid	134.088	26	59		26		
			50	59.9		34		
$C_4H_6O_6$	DL-Tartaric acid	150.087	0	8.95	98	26		
			20	17.1	~200	26		
			100	65		26		
$C_4H_6O_6$	L-Tartaric acid	150.087	20	58		26		
			100	77		26		
C_4H_7Br	4-Bromo-1-butene	135.003	25	0.076	0.76	35		
C_4H_7Cl	1-Chloro-2-methylpropene	90.552	25	0.916	9.16	5	0.12	5
C_4H_7ClO	3-Chloro-2-butanone	106.551	19	2.80	29	20		

Mol. formula	Name	Mol. wt.	t/°C	Solubility S			Henry Const. k_H kPa m³mol⁻¹	
				Mass%	g/L	Ref.	kPa m³mol⁻¹	Ref.
			92	3.38	35	20		
C_4H_7N	Butanenitrile	69.106	20	3.3	34	10		
$C_4H_7NO_4$	Iminodiacetic acid	133.104	5	2.32	23.7	40		
$C_4H_7NO_4$	L-Aspartic acid	133.104	25	0.501	5.01	26		
C_4H_8	1-Butene	56.107	25	0.0222[a]	0.222[a]	5	25.6	13
C_4H_8	Isobutene	56.107	25	0.0263[a]	0.263[a]	5	21.6	13
$C_4H_8Br_2$	1,4-Dibromobutane	215.915	25	0.035	0.35	35		
$C_4H_8Cl_2$	1,1-Dichlorobutane	127.013	25	0.050	0.50	35		
$C_4H_8Cl_2$	1,4-Dichlorobutane	127.013	25	0.16	1.6	35		
$C_4H_8Cl_2$	2,3-Dichlorobutane, (±)-	127.013	20	0.056	0.56	35		
$C_4H_8Cl_2O$	Bis(2-chloroethyl) ether	143.012	20	1.04	10.4	20	0.003	13
			81	1.26	12.8	20		
$C_4H_8N_2O_2$	Succinamide	116.119	50	18.4	~225	27		
$C_4H_8N_2O_2$	Dimethylglyoxime	116.119	20	0.06	0.6	40		
$C_4H_8N_2O_3$	L-Asparagine	132.118	25	2.45	25.1	26		
$C_4H_8N_2O_3$	N-Glycylglycine	132.118	25	18.4	~225	29		
C_4H_8O	cis-2-Buten-1-ol	72.106	20	16.6	~200	10		
C_4H_8O	Ethyl vinyl ether	72.106	20	0.9	9	10		
C_4H_8O	Butanal	72.106	25	7.1	76	10		
C_4H_8O	Isobutanal	72.106	20	9.1	100	10		
C_4H_8O	2-Butanone	72.106	25	25.9		20		
			70	18.1	~220	20		
$C_4H_8O_2$	2-Methylpropanoic acid	88.106	20	22.8		10		
$C_4H_8O_2$	Propyl formate	88.106	22	2.05	20.9	10		
$C_4H_8O_2$	Ethyl acetate	88.106	25	8.08	87.9	10		
$C_4H_8O_2$	Methyl propanoate	88.106	25	6	6	30		
C_4H_9Br	1-Bromobutane	137.018	25	0.087	0.87	35	1.2	13
C_4H_9Br	1-Bromo-2-methylpropane	137.018	18	0.051	0.51	35		
C_4H_9Cl	1-Chlorobutane	92.567	1	0.062	0.62	35		
			25	0.087	0.87	35	1.54	13
C_4H_9Cl	2-Chlorobutane	92.567	0	0.107	1.07	35		
			25	0.092	0.92	35		
C_4H_9Cl	1-Chloro-2-methylpropane	92.567	25	0.92	9.2	35		
C_4H_9Cl	2-Chloro-2-methylpropane	92.567	15	0.29	2.9	35		
C_4H_9I	1-Iodobutane	184.018	17	0.021	0.21	10	1.87	13
C_4H_9NO	Butanamide	87.120	25	~19	230	40		
$C_4H_9NO_2$	Ethyl N-methylcarbamate	103.120	15	69		27		
$C_4H_9NO_2$	2-Methylalanine	103.120	25	12.0	~135	26		
$C_4H_9NO_2$	DL-2-Aminobutanoic acid	103.120	25	17.4	~210	26		
$C_4H_9NO_2$	DL-3-Aminobutanoic acid	103.120	25	55.6		26		
$C_4H_9NO_3$	L-Threonine	119.119	25	8.93	98.0	26		
$C_4H_9NO_3$	L-Homoserine	119.119	25	52.4		26		
$C_4H_9N_3O_2$	Creatine	131.133	25	1.6	16	26		
C_4H_{10}	Butane	58.122	25	0.00724[a]	0.0724[a]	18	95.9	5
C_4H_{10}	Isobutane	58.122	25	0.00535[a]	0.0535[a]	18	120	5
$C_4H_{10}NO_3PS$	Acephate	183.166	20	~28	394	40		
$C_4H_{10}N_2O$	N-Nitrosodiethylamine	102.134	24	9.6	106	40		
$C_4H_{10}O$	1-Butanol	74.121	0	10.4	~115	1		
			25	7.4	80	1		
			50	6.4	68	1		
$C_4H_{10}O$	2-Butanol	74.121	10	23.9		1		
			25	18.1	~220	1		
			50	14.0	~165	1		
$C_4H_{10}O$	2-Methyl-1-propanol	74.121	0	11.5	~130	1		
			25	8.1	88	1	0.00273	28
			50	6.5	70	1		
$C_4H_{10}O$	Diethyl ether	74.121	25	6.04	64.2	10	0.088	13
$C_4H_{10}O$	Methyl propyl ether	74.121	25	3.5	36	30		

Mol. formula	Name	Mol. wt.	$t/°C$	Mass%	g/L	Ref.	kPa $m^3 mol^{-1}$	Ref.
				Solubility S			Henry Const. k_H	
$C_4H_{10}O_4$	1,2,3,4-Butanetetrol	122.120	20	38.0		27		
$C_4H_{10}S$	1-Butanethiol	90.187	20	0.0597	0.597	10		
$C_4H_{10}S$	Diethyl sulfide	90.187	25	0.307	3.07	40		
$C_4H_{11}NO_2$	Diethanolamine	105.136	20	95.4		10		
$C_4H_{11}NO_3$	Tris(hydroxymethyl)methylamine	121.135	25	~41	699	40		
$C_4H_{12}Si$	Tetramethylsilane	88.224	25	0.00196	0.0196	10		
C_5Cl_8	Octachloro-1,3-pentadiene	343.678	20	0.000020	0.00020	35		
C_5F_{12}	Perfluoropentane	288.035	25	0.00012	0.0012	35		
$C_5H_4N_2O_4$	Orotic acid	156.097	18	0.18	1.8	26		
$C_5H_4N_4O$	Allopurinol	136.112	25	0.057	0.57	40		
$C_5H_4N_4O$	Hypoxanthine	136.112	25	0.070	0.70	29		
$C_5H_4N_4O_2$	Xanthine	152.112	20	0.05	0.5	26		
$C_5H_4N_4O_3$	Uric acid	168.111	20	0.002	0.02	26		
$C_5H_4N_4S$	1,7-Dihydro-6H-purine-6-thione	152.178	25	0.0124	0.124	40		
$C_5H_4O_2$	Furfural	96.085	20	8.2	89	10		
$C_5H_4O_3$	2-Furancarboxylic acid	112.084	25	4.758	50.0	33		
			50	25.16		33		
$C_5H_5N_5$	Adenine	135.128	25	0.104	1.04	29		
$C_5H_5N_5O$	Guanine	151.127	25	0.0068	0.068	29		
$C_5H_5N_5O$	6-Amino-1,3-dihydro-2H-purin-2-one	151.127	25	0.006	0.06	26		
C_5H_6	1,3-Cyclopentadiene	66.102	25	0.068	0.68	3		
$C_5H_6Cl_2N_2O_2$	1,3-Dichloro-5,5-dimethyl hydantoin	197.019	20	0.050	0.50	40		
$C_5H_6N_2OS$	Methylthiouracil	142.179	25	0.0533	0.533	40		
$C_5H_6N_2O_2$	Thymine	126.114	25	0.35	3.5	29		
$C_5H_6O_4$	1-Propene-2,3-dicarboxylic acid	130.100	20	7.7	83	26		
$C_5H_7NO_2$	Ethyl cyanoacetate	113.116	20	25.9		10		
$C_5H_7N_3O$	4-Amino-5-methyl-2(1H)-pyrimidinone	125.129	25	0.45	4.5	26		
C_5H_8	1,4-Pentadiene	68.118	25	0.056	0.56	3	12	5
C_5H_8	2-Methyl-1,3-butadiene	68.118	25	0.061	0.61	3	7.78	5
			50	0.076[a]	0.76[a]	3		
C_5H_8	1-Pentyne	68.118	25	0.157	1.57	3	2.5	5
C_5H_8	Cyclopentene	68.118	25	0.054	0.54	3	6.56	13
$C_5H_8N_4O_3S_2$	Methazolamide	236.273	15	0.0472	0.472	40		
$C_5H_8N_4O_{12}$	Pentaerythritol tetranitrate	316.138	20	0.0002	0.002	40		
C_5H_8O	Cyclopentanone	84.117	20	31.0		20		
			80	24.8		20		
C_5H_8O	3,4-Dihydro-2H-pyran	84.117	20	1.04	10.5	20		
			82	2.26	23	20		
$C_5H_8O_2$	Ethyl acrylate	100.117	25	1.50	15.2	10		
$C_5H_8O_2$	Methyl methacrylate	100.117	20	1.56	15.9	10		
$C_5H_8O_2$	2,4-Pentanedione	100.117	20	16.1	~200	20		
			80	32.2		20		
$C_5H_8O_3$	4-Oxopentanoic acid	116.116	10	63.6		34		
			25	83.97		34		
$C_5H_8O_4$	Pentanedioic acid	132.116	25	58.3		33		
			50	78.06		33		
$C_5H_8O_4$	Dimethyl malonate	132.116	19	14.9	~175	20		
			90	29.8		20		
C_5H_9ClO	5-Chloro-2-pentanone	120.577	22	4.7	49	20		
			71	13.5	~155	20		
$C_5H_9NO_2$	L-Proline	115.131	25	61.9		26		
$C_5H_9NO_3$	trans-4-Hydroxy-L-proline	131.130	25	26.5		26		
$C_5H_9NO_4$	DL-Glutamic acid	147.130	25	2.30	23.5	29		
$C_5H_9NO_4$	L-Glutamic acid	147.130	25	0.85	8.5	26		
C_5H_{10}	1-Pentene	70.133	25	0.0148	0.148	3	40.3	5
C_5H_{10}	cis-2-Pentene	70.133	25	0.0203	0.203	3	22.8	5
C_5H_{10}	3-Methyl-1-butene	70.133	25	0.013[a]	0.13[a]	3	54.7	5
C_5H_{10}	2-Methyl-2-butene	70.133	25	0.041	0.41	3		

Aqueous Solubility and Henry's Law Constants of Organic Compounds

Mol. formula	Name	Mol. wt.	$t/°C$	Solubility S			Henry Const. k_H	
				Mass%	g/L	Ref.	kPa m^{3}mol^{-1}	Ref.
C_5H_{10}	Cyclopentane	70.133	25	0.0157	0.157	3	19.1	13
$C_5H_{10}Cl_2$	1,2-Dichloropentane	141.038	25	0.029	0.29	35		
$C_5H_{10}Cl_2$	1,5-Dichloropentane	141.038	19	0.02	0.2	35		
$C_5H_{10}Cl_2$	2,3-Dichloropentane	141.038	25	0.029	0.29	35		
$C_5H_{10}Cl_2$	2,3-Dichloro-2-methylbutane	141.038	25	0.029	0.29	35		
$C_5H_{10}N_2O_2S$	Methomyl	162.210	25	5.5	58	40		
$C_5H_{10}N_2O_3$	L-Glutamine	146.144	25	4.0	42	26		
$C_5H_{10}N_2S_2$	Dazomet	162.276	25	0.12	1.2	40		
$C_5H_{10}O$	Pentanal	86.132	25	1.2	12	40		
$C_5H_{10}O$	2-Pentanone	86.132	25	5.5	58	20	0.00847	28
			80	3.8	40	20	0.00847	28
$C_5H_{10}O$	3-Pentanone	86.132	25	4.72	49.5	20		
			80	3.16	33	20		
$C_5H_{10}O$	3-Methyl-2-butanone	86.132	18	6.7	72	20		
			80	3.9	41	20		
$C_5H_{10}O$	Tetrahydropyran	86.132	20	8.57	87.1	20		
			81	4.29	45	20		
$C_5H_{10}O$	2-Methyltetrahydrofuran	86.132	19	14.4	~160	20	0.67	13
			71	6.0	64	20		
$C_5H_{10}O_2$	Pentanoic acid	102.132	16	3.6	37	26		
$C_5H_{10}O_2$	3-Methylbutanoic acid	102.132	20	4.0	42	26		
$C_5H_{10}O_2$	Isobutyl formate	102.132	22	1.0	10	10		
$C_5H_{10}O_2$	Propyl acetate	102.132	20	2.3	34	10		
$C_5H_{10}O_2$	Isopropyl acetate	102.132	20	2.9	30	10		
$C_5H_{10}O_2$	Ethyl propanoate	102.132	20	1.92	19.6	10		
$C_5H_{10}O_2$	Methyl butanoate	102.132		1.6	16	30		
$C_5H_{10}O_3$	Diethyl carbonate	118.131	20	1.8	18	40		
$C_5H_{10}O_5$	D-Xylose	150.130	25	~30	432	40		
$C_5H_{11}Br$	1-Bromopentane	151.045	25	0.0127	0.127	35		
$C_5H_{11}Br$	1-Bromo-3-methylbutane	151.045	16	0.020	0.20	35		
$C_5H_{11}Cl$	1-Chloropentane	106.594	5	0.020	0.20	35		
			25	0.0201	0.201	35	2.37	13
$C_5H_{11}Cl$	3-Chloropentane	106.594	25	0.025	0.25	35		
$C_5H_{11}NO_2$	L-Valine	117.147	25	8.13	88.4	26		
$C_5H_{11}NO_2$	L-Norvaline	117.147	25	9.7	107	26		
$C_5H_{11}NO_2S$	L-Methionine	149.212	25	5.3	56	26		
C_5H_{12}	Pentane	72.149	25	0.0041	0.041	3	128	13
C_5H_{12}	Isopentane	72.149	25	0.00485	0.0485	3	479	13
C_5H_{12}	Neopentane	72.149	25	0.00332[a]	0.0332[a]	3	220	13
$C_5H_{12}NO_3PS_2$	Cygon	229.258	20	2.6	27	40		
$C_5H_{12}O$	1-Pentanol	88.148	0	3.1	32	1		
			25	2.20	22.4	1		
			50	1.8	18	1		
$C_5H_{12}O$	2-Pentanol	88.148	25	4.3	45	21		
$C_5H_{12}O$	3-Pentanol	88.148	25	5.6	59	21		
$C_5H_{12}O$	2-Methyl-1-butanol, (±)-	88.148	25	3.0	31	3		
$C_5H_{12}O$	3-Methyl-1-butanol	88.148	25	2.7	28	1		
$C_5H_{12}O$	2-Methyl-2-butanol	88.148	25	11.0	~125	1		
$C_5H_{12}O$	3-Methyl-2-butanol, (±)-	88.148	25	5.6	59	1		
$C_5H_{12}O$	2,2-Dimethyl-1-propanol	88.148	25	3.5	36	1		
$C_5H_{12}O$	Methyl tert-butyl ether	88.148	0	8.3	37.6	20	0.070	13
			20	4.2	44	20		
			49	1.9	19	20		
$C_5H_{12}O_4$	Pentaerythritol	136.147	15	5.3	56	30		
$C_6Cl_4O_2$	2,3,5,6-Tetrachloro-2,5-cyclohexadiene-1,4-dione	245.875	20	0.025	0.25	40		
$C_6Cl_5NO_2$	Pentachloronitrobenzene	295.335	20	0.000044	0.00044	40		
C_6Cl_6	Hexachlorobenzene	284.782	25	0.0000005	0.000005	41	0.131	11

Mol. formula	Name	Mol. wt.	$t/°C$	Solubility S			Henry Const. k_H	
				Mass%	g/L	Ref.	kPa m^{3}mol^{-1}	Ref.
C_6F_{14}	Perfluorohexane	338.042	25	0.0000098	0.000098	35		
C_6F_{14}	Perfluoro-2-methylpentane	338.042	25	0.000017	0.00017	35		
C_6HCl_5	Pentachlorobenzene	250.337	25	0.000050	0.00050	41	0.085	11
C_6HCl_5O	Pentachlorophenol	266.336	25	0.0013	0.013	24		
$C_6H_2Br_4$	1,2,4,5-Tetrabromobenzene	393.696	25	0.00000434	0.0000434	2		
$C_6H_2ClN_3O_6$	2-Chloro-1,3,5-trinitrobenzene	247.549	15	0.018	0.18	40		
$C_6H_2Cl_4$	1,2,3,4-Tetrachlorobenzene	215.892	25	0.0007	0.007	41	0.144	11
$C_6H_2Cl_4$	1,2,3,5-Tetrachlorobenzene	215.892	25	0.00035	0.0035	41	0.59	11
$C_6H_2Cl_4$	1,2,4,5-Tetrachlorobenzene	215.892	25	0.000060	0.00060	41	0.122	11
$C_6H_2Cl_4O$	2,3,4,6-Tetrachlorophenol	231.891	25	0.017	0.17	24		
$C_6H_2Cl_4O_2$	3,4,5,6-Tetrachloro-1,2-benzenediol	247.891	25	0.071	0.71	8		
$C_6H_3Br_3$	1,2,4-Tribromobenzene	314.800	25	0.0010	0.010	2		
$C_6H_3Br_3$	1,3,5-Tribromobenzene	314.800	25	0.0000789	0.000789	2		
$C_6H_3Br_3O$	2,4,6-Tribromophenol	330.799	15	0.0007	0.007	2		
$C_6H_3ClN_2O_4$	1-Chloro-2,4-dinitrobenzene	202.552	25	0.00092	0.0092	40		
$C_6H_3Cl_2NO_2$	1,2-Dichloro-4-nitrobenzene	192.000	20	0.0121	0.121	40		
$C_6H_3Cl_2NO_2$	Clopyralid	192.000	20	0.1	1	40		
$C_6H_3Cl_3$	1,2,3-Trichlorobenzene	181.447	25	0.0021	0.021	41	0.242	11
$C_6H_3Cl_3$	1,2,4-Trichlorobenzene	181.447	25	0.0040	0.040	41	0.277	11
$C_6H_3Cl_3$	1,3,5-Trichlorobenzene	181.447	25	0.0008	0.008	41	1.1	11
$C_6H_3Cl_3O$	2,4,5-Trichlorophenol	197.446	25	0.1	1	2		
$C_6H_3Cl_3O$	2,4,6-Trichlorophenol	197.446	25	0.050	0.50	24		
$C_6H_3Cl_3O_2$	3,4,5-Trichloro-1,2-benzenediol	213.446	25	0.051	0.51	8		
$C_6H_3Cl_4N$	Nitrapyrin	230.907	20	0.0040	0.040	40		
$C_6H_3N_3O_6$	1,3,5-Trinitrobenzene	213.104	15	0.028	0.28	40		
$C_6H_3N_3O_7$	2,4,6-Trinitrophenol	229.104	25	1.25	12.7	40		
			90	4.9	51	40		
C_6H_4BrCl	1-Bromo-2-chlorobenzene	191.453	25	0.0124	0.124	2		
C_6H_4BrCl	1-Bromo-3-chlorobenzene	191.453	25	0.0118	0.118	2		
C_6H_4BrCl	1-Bromo-4-chlorobenzene	191.453	25	0.00442	0.0442	2		
C_6H_4BrI	1-Bromo-4-iodobenzene	282.904	25	0.000794	0.00794	2		
$C_6H_4Br_2$	o-Dibromobenzene	235.904	25	0.00748	0.0748	2		
$C_6H_4Br_2$	m-Dibromobenzene	235.904	25	0.0064	0.064	2		
$C_6H_4Br_2$	p-Dibromobenzene	235.904	25	0.0020	0.020	2		
$C_6H_4Br_2O$	2,4-Dibromophenol	251.903	25	0.2	2	2		
C_6H_4ClF	1-Chloro-2-fluorobenzene	130.547	25	0.0502	0.502	40		
C_6H_4ClI	1-Chloro-2-iodobenzene	238.453	25	0.00689	0.0689	2		
C_6H_4ClI	1-Chloro-3-iodobenzene	238.453	25	0.00674	0.0674	2		
C_6H_4ClI	1-Chloro-4-iodobenzene	238.453	25	0.00311	0.0311	2		
$C_6H_4ClNO_2$	1-Chloro-2-nitrobenzene	157.555	20	0.0441	0.441	40		
$C_6H_4ClNO_2$	1-Chloro-3-nitrobenzene	157.555	20	0.0273	0.273	40		
$C_6H_4ClNO_2$	1-Chloro-4-nitrobenzene	157.555	20	0.0453	0.453	40		
$C_6H_4Cl_2$	o-Dichlorobenzene	147.002	0	0.0142	0.142	2		
			25	0.0147	0.147	2	0.195	28
			50	0.0212	0.212	2		
$C_6H_4Cl_2$	m-Dichlorobenzene	147.002	10	0.0103	0.103	2		
			25	0.0120	0.120	41	0.376	11
			50	0.0165	0.165	2		
$C_6H_4Cl_2$	p-Dichlorobenzene	147.002	10	0.00512	0.0512	2		
			25	0.0080	0.080	41	0.244	28
			50	0.0167	0.167	2		
$C_6H_4Cl_2O$	2,3-Dichlorophenol	163.001	25	0.82	8.22	40		
$C_6H_4Cl_2O$	2,4-Dichlorophenol	163.001	20	0.49	4.9	24		
$C_6H_4Cl_2O$	2,6-Dichlorophenol	163.001	25	0.262	2.62	40		
$C_6H_4Cl_2O_2$	3,5-Dichloro-1,2-benzenediol	179.001	25	0.78	7.8	8		
$C_6H_4Cl_2O_2$	4,5-Dichloro-1,2-benzenediol	179.001	25	1.19	12.0	8		
$C_6H_4F_2$	o-Difluorobenzene	114.093	25	0.114	1.14	2		
$C_6H_4F_2$	m-Difluorobenzene	114.093	25	0.114	1.14	2		

| Mol. formula | Name | Mol. wt. | $t/°C$ | Solubility S | | | Henry Const. k_H | |
				Mass%	g/L	Ref.	kPa m³mol⁻¹	Ref.
$C_6H_4F_2$	p-Difluorobenzene	114.093	25	0.122	1.22	2		
$C_6H_4I_2$	o-Diiodobenzene	329.905	25	0.00192	0.0192	2		
$C_6H_4I_2$	m-Diiodobenzene	329.905	25	0.000185	0.00185	2		
$C_6H_4I_2$	p-Diiodobenzene	329.905	25	0.000893	0.00893	2		
$C_6H_4N_2O_4$	1,2-Dinitrobenzene	168.107	20	0.21	2.1	27		
$C_6H_4N_2O_4$	1,3-Dinitrobenzene	168.107	20	2.09	21.3	27		
$C_6H_4N_2O_4$	1,4-Dinitrobenzene	168.107	20	1.30	13.1	27		
$C_6H_4N_2O_5$	2,4-Dinitrophenol	184.106	25	0.046	0.46	40		
			100	1.4	14	40		
$C_6H_4O_2$	p-Benzoquinone	108.095	25	1.36	13.8	27		
C_6H_5Br	Bromobenzene	157.008	10	0.0387	0.387	2		
			25	0.0445	0.445	2	0.250	28
			40	0.0516	0.516	2		
C_6H_5BrO	4-Bromophenol	173.007	25	1.86	19.0	2		
C_6H_5Cl	Chlorobenzene	112.557	10	0.0387	0.387	2		
			25	0.0484	0.484	41	0.32	28
			50	0.0882	0.882	2		
C_6H_5ClO	2-Chlorophenol	128.556	25	2.0	20	2		
C_6H_5ClO	3-Chlorophenol	128.556	25	2.2	22	2		
C_6H_5ClO	4-Chlorophenol	128.556	25	2.7	28	2		
C_6H_5F	Fluorobenzene	96.102	19	0.170	1.70	20	0.70	11
			80	0.188	1.88	20	0.70	11
C_6H_5I	Iodobenzene	204.008	10	0.0193	0.193	2		
			25	0.0226	0.226	2	0.078	11
			45	0.0279	0.279	2		
$C_6H_5NO_2$	Nitrobenzene	123.110	25	0.21	2.1	17		
$C_6H_5NO_2$	3-Pyridinecarboxylic acid	123.110	20	1.8	18	40		
$C_6H_5NO_3$	2-Nitrophenol	139.109	20	0.21	2.1	27		
$C_6H_5NO_3$	3-Nitrophenol	139.109	20	2.14	21.9	27		
$C_6H_5NO_3$	4-Nitrophenol	139.109	20	1.32	13.3	27		
$C_6H_5N_3O_4$	2,4-Dinitroaniline	183.122	25	0.0078	0.078	40		
C_6H_6	Benzene	78.112	10	0.178	1.78	3		
			25	0.178	1.78	22	0.557	22
			50	0.208	2.08	3		
C_6H_6ClN	2-Chloroaniline	127.572	25	0.876	8.76	10		
C_6H_6ClN	3-Chloroaniline	127.572	20	0.54	5.44	40		
C_6H_6ClN	4-Chloroaniline	127.572	20	0.275	2.75	40		
$C_6H_6Cl_6$	1,2,3,4,5,6-Hexachlorocyclohexane, (1α,2α,3β,4α,5α,6β)	290.830	25	0.00073	0.0073	40		
$C_6H_6Cl_6$	1,2,3,4,5,6-Hexachlorocyclohexane, (1α,2α,3β,4α,5β,6β)	290.830	25	0.00020	0.0020	40		
$C_6H_6Cl_6$	1,2,3,4,5,6-Hexachlorocyclohexane, (1α,2β,3α,4β,5α,6β)	290.830	25	0.000024	0.00024	40		
$C_6H_6N_2O$	3-Pyridinecarboxamide	122.124	20	~33	500	40		
$C_6H_6N_2O_2$	2-Nitroaniline	138.124	30	1.47	14.9	27		
$C_6H_6N_2O_2$	3-Nitroaniline	138.124	30	0.121	1.21	27		
$C_6H_6N_2O_2$	4-Nitroaniline	138.124	30	0.073	0.73	27		
$C_6H_6N_4O_4$	Nitrofurazone	198.137	20	0.0238	0.238	40		
C_6H_6O	Phenol	94.111	16	6.3	67	26		
			25	8.66	94.8	10		
$C_6H_6O_2$	p-Hydroquinone	110.111	25	7.42	80.1	27		
$C_6H_6O_2$	Pyrocatechol	110.111	20	31.1		27		
$C_6H_6O_2$	Resorcinol	110.111	20	63.7		27		
$C_6H_6O_3$	1,2,3-Benzenetriol	126.110	25	38.5		27		
$C_6H_6O_3$	1,3,5-Benzenetriol	126.110	20	1.12	11.3	27		
$C_6H_6O_6$	trans-1-Propene-1,2,3-tricarboxylic acid	174.108	25	20.9		26		
			90	52.5		26		
C_6H_7N	Aniline	93.127	25	3.38	35.0	10	14	15

| Mol. formula | Name | Mol. wt. | $t/°C$ | Solubility S | | | Henry Const. k_H | |
				Mass%	g/L	Ref.	kPa m³mol⁻¹	Ref.
C_6H_7NO	2-Aminophenol	109.126	20	1.92	19.6	40		
C_6H_7NO	3-Aminophenol	109.126	20	2.56	26.3	40		
			70	~24	319	40		
C_6H_7NO	4-Aminophenol	109.126	20	1.55	15.7	40		
$C_6H_7NO_3S$	4-Aminobenzenesulfonic acid	173.190	7	0.59	5.9	27		
$C_6H_7N_3O$	Isoniazid	137.139	25	11.0	123	40		
C_6H_8	1,4-Cyclohexadiene	80.128	25	0.08	0.8	3	1.03	13
C_6H_8ClN	Aniline hydrochloride	129.588	15	15.1	~180	27		
$C_6H_8N_2$	Hexanedinitrile	108.141	20	0.80	8.0	16		
$C_6H_8N_2$	1,2-Benzenediamine	108.141	20	3.02	31.1	40		
$C_6H_8N_2$	1,3-Benzenediamine	108.141	20	3.48	36.1	40		
$C_6H_8N_2$	1,4-Benzenediamine	108.141	24	3.45	35.7	40		
$C_6H_8N_2O_2S$	4-Aminobenzenesulfonamide	172.205	20	0.71	7.14	40		
$C_6H_8N_2O_8$	Isosorbide dinitrate	236.136	25	0.055	0.55	40		
$C_6H_8O_4$	Dimethyl maleate	144.126	25	8.0	87	10		
$C_6H_8O_6$	L-Ascorbic acid	176.124	25	25.22		33		
			50	41.00		33		
$C_6H_8O_7$	Citric acid	192.124	20	59		26		
$C_6H_9N_3O_2$	L-Histidine	155.154	25	4.17	43.5	26		
$C_6H_9N_3O_3$	Metronidazole	171.153	20	0.93	9.4	40		
C_6H_{10}	1,5-Hexadiene	82.143	25	0.017	0.17	3		
C_6H_{10}	1-Hexyne	82.143	25	0.036	0.36	3	4.14	13
C_6H_{10}	Cyclohexene	82.143	25	0.016	0.16	3	4.57	13
$C_6H_{10}O$	Cyclohexanone	98.142	25	8.8	96	20		
			80	6.8	73	20		
$C_6H_{10}O$	Mesityl oxide	98.142	20	2.89	29.8	10		
$C_6H_{10}O_3$	Ethyl acetoacetate	130.141	25	12	~135	10		
$C_6H_{10}O_4$	1,6-Hexanedioic acid	146.141	15	1.48	15.0	26		
			100	61.5		26		
$C_6H_{10}O_4$	Dimethyl succinate	146.141	21	12.4	~140	20		
			92	17.1	~205	20		
$C_6H_{10}O_4$	1,2-Ethanediol, diacetate	146.141	25	13.3	153	40		
$C_6H_{10}O_8$	Galactaric acid	210.138	14	0.33	3.3	40		
$C_6H_{11}NO$	Cyclohexanone oxime	113.157	25	1.57	15.9	40		
$C_6H_{11}NO$	Caprolactam	113.157	25	84.0		10		
$C_6H_{11}N_2O_4PS_3$	Methidathion	302.330	20	0.0187	0.187	40		
C_6H_{12}	1-Hexene	84.159	25	0.0053	0.053	3	41.8	5
C_6H_{12}	trans-2-Hexene	84.159	25	0.0067	0.067	3		
C_6H_{12}	2-Methyl-1-pentene	84.159	25	0.0078	0.078	3	28.1	5
C_6H_{12}	4-Methyl-1-pentene	84.159	25	0.0048	0.048	3	63.2	5
C_6H_{12}	2,3-Dimethyl-1-butene	84.159	30	0.046	0.46	3		
C_6H_{12}	Cyclohexane	84.159	25	0.0058	0.058	3	19.4	13
C_6H_{12}	Methylcyclopentane	84.159	25	0.0043	0.043	3	36.7	5
$C_6H_{12}N_2O_3$	Daminozide	160.170	25	9.1	100	40		
$C_6H_{12}N_2O_4S$	L-Lanthionine	208.235	25	0.15	1.5	26		
$C_6H_{12}N_2O_4S_2$	L-Cystine	240.300	25	0.011	0.11	26		
$C_6H_{12}N_2S_4$	Thiram	240.432	20	0.003	0.03	40		
$C_6H_{12}N_2S_4Zn$	Ziram	305.841	20	0.0065	0.065	40		
$C_6H_{12}N_4$	Hexamethylenetetramine	140.186	12	44.8		27		
$C_6H_{12}O$	1-Hexen-3-ol	100.158	25	2.52	25.9	1		
$C_6H_{12}O$	4-Hexen-2-ol	100.158	25	3.81	39.6	1		
$C_6H_{12}O$	Butyl vinyl ether	100.158	20	0.3	3	10		
$C_6H_{12}O$	2-Hexanone	100.158	20	1.51	17.8	20		
			81	1.15	11.6	20		
$C_6H_{12}O$	4-Methyl-2-pentanone	100.158	19	1.92	17	20		
			90	1.22	12.4	20		
$C_6H_{12}O$	3,3-Dimethyl-2-butanone	100.158	19	1.97	18.4	20		
			90	1.14	11.5	20		

Mol. formula	Name	Mol. wt.	$t/°C$	Solubility S			Henry Const. k_H	
				Mass%	g/L	Ref.	kPa m³mol⁻¹	Ref.
$C_6H_{12}O$	Cyclohexanol	100.158	10	4.62	48.4	1		
			25	3.8	40	1		
			40	3.30	34.1	1		
$C_6H_{12}O_2$	Hexanoic acid	116.158	20	0.96	9.6	26		
			60	1.16	11.7	26		
$C_6H_{12}O_2$	Isopentyl formate	116.158	22	0.3	3	27		
$C_6H_{12}O_2$	Butyl acetate	116.158	20	0.68	6.8	10		
$C_6H_{12}O_2$	sec-Butyl acetate	116.158	20	0.62	6.2	10		
$C_6H_{12}O_2$	Isobutyl acetate	116.158	20	0.63	6.3	10		
$C_6H_{12}O_2$	Propyl propanoate	116.158	25	0.6	6	27		
$C_6H_{12}O_2$	Ethyl butanoate	116.158	20	0.49	4.9	10		
$C_6H_{12}O_3$	2-Ethoxyethyl acetate	132.157		14	~165	30		
$C_6H_{12}O_3$	Tetrahydro-2,5-dimethoxyfuran	132.157	21	32		20		
			90	19	~235	20		
$C_6H_{12}O_3$	Paraldehyde	132.157	25	11	~125	30		
$C_6H_{12}O_6$	β-D-Fructose	180.155	20	~31	444	40		
$C_6H_{12}O_6$	D-Galactose	180.155	20	40.6		27		
$C_6H_{12}O_6$	α-D-Glucose	180.155	15	45.0		27		
			30	54.6		27		
			80	81.5		27		
$C_6H_{12}O_6$	L-Sorbose	180.155	17	~26	355	40		
$C_6H_{13}Br$	1-Bromohexane	165.071	25	0.00258	0.0258	35		
$C_6H_{13}Cl$	1-Chlorohexane	120.620	5	0.0047	0.047	35		
			25	0.0064	0.064	35		
$C_6H_{13}NO_2$	L-Leucine	131.173	25	2.15	22.0	26		
$C_6H_{13}NO_2$	L-Isoleucine	131.173	25	3.31	34.2	26		
$C_6H_{13}NO_2$	L-Norleucine	131.173	25	1.5	15	26		
$C_6H_{13}NO_2$	6-Aminohexanoic acid	131.173	25	46		29		
$C_6H_{13}NO_2$	Ethyl N-propylcarbamate	131.173	15	7.70	83.4	27		
C_6H_{14}	Hexane	86.175	25	0.0011	0.011	3	183	13
			60	0.00136	0.0136	3		
C_6H_{14}	2-Methylpentane	86.175	25	0.00137	0.0137	3	176	13
C_6H_{14}	3-Methylpentane	86.175	25	0.00129	0.0129	3	170	13
C_6H_{14}	2,2-Dimethylbutane	86.175	25	0.0021	0.021	3	199	13
C_6H_{14}	2,3-Dimethylbutane	86.175	25	0.0021	0.021	3	144	13
$C_6H_{14}N_2O_2$	L-Lysine	146.187	25	0.58	5.8	26		
$C_6H_{14}N_4O_2$	L-Arginine	174.201	25	15.44	~185	26		
$C_6H_{14}O$	2-Methoxy-2-methylbutane	102.174	20	1.10	12.7	20		
			79	0.36	3.6	20		
$C_6H_{14}O$	1-Hexanol	102.174	0	0.79	7.9	1		
			25	0.60	6.0	1		
			50	0.51	5.1	1		
$C_6H_{14}O$	2-Hexanol	102.174	25	1.4	14	1		
$C_6H_{14}O$	3-Hexanol	102.174	25	1.6	16	1		
$C_6H_{14}O$	2-Methyl-1-pentanol	102.174	25	0.81	8.1	1		
$C_6H_{14}O$	4-Methyl-1-pentanol	102.174	25	0.76	7.6	1		
$C_6H_{14}O$	2-Methyl-2-pentanol	102.174	25	3.2	33	1		
$C_6H_{14}O$	3-Methyl-2-pentanol	102.174	25	1.9	19	1		
$C_6H_{14}O$	4-Methyl-2-pentanol	102.174	27	1.5	15	1		
$C_6H_{14}O$	2-Methyl-3-pentanol	102.174	25	2.0	20	1		
$C_6H_{14}O$	3-Methyl-3-pentanol	102.174	25	4.3	45	1		
$C_6H_{14}O$	2-Ethyl-1-butanol	102.174	25	1.0	10	1		
$C_6H_{14}O$	2,2-Dimethyl-1-butanol	102.174	25	0.8	8	1		
$C_6H_{14}O$	2,3-Dimethyl-2-butanol	102.174	25	4.2	44	1		
$C_6H_{14}O$	3,3-Dimethyl-2-butanol, (±)-	102.174	25	2.4	25	1		
$C_6H_{14}O$	Dipropyl ether	102.174	25	0.49	4.9	10	0.26	13
$C_6H_{14}O$	Diisopropyl ether	102.174	20	0.79	12	20	0.26	13
			61	0.22	2.2	20		

Mol. formula	Name	Mol. wt.	$t/°C$	Solubility S			Henry Const. k_H	
				Mass%	g/L	Ref.	kPa m³mol⁻¹	Ref.
$C_6H_{14}O$	Butyl ethyl ether	102.174	20	0.65	6.5	20		
			70	0.39	3.9	20		
$C_6H_{14}O_2$	1,1-Diethoxyethane	118.174	25	5	5	10		
$C_6H_{14}O_2$	1,2-Diethoxyethane	118.174	20	21.0		10		
$C_6H_{14}O_6$	D-Glucitol	182.171	20	~41	~689	40		
$C_6H_{14}O_6$	D-Mannitol	182.171	25	17.7	~215	27		
$C_6H_{15}N$	Dipropylamine	101.190	20	2.5	26	10		
$C_6H_{15}N$	Triethylamine	101.190	20	5.5	58	10		
$C_6H_{16}ClN$	Triethylamine hydrochloride	137.651	25	57.8		27		
$C_6H_{16}N_2$	1,6-Hexanediamine	116.204	5	~42	~711	40		
C_7F_{16}	Perfluoroheptane	388.049	25	0.0000013	0.000013	35		
$C_7H_3Br_2NO$	3,5-Dibromo-4-hydroxybenzonitrile	276.913	25	0.013	0.13	40		
$C_7H_3N_3O_8$	2,4,6-Trinitrobenzoic acid	257.114	23	1.97	20.1	40		
$C_7H_4ClNO_4$	3-Chloro-2-nitrobenzoic acid	201.565	25	0.047	0.47	27		
$C_7H_4ClNO_4$	5-Chloro-2-nitrobenzoic acid	201.565	25	0.96	9.6	27		
$C_7H_4Cl_4O$	2,3,4,6-Tetrachloro-5-methylphenol	245.918	25	0.00061	0.0061	2		
$C_7H_4N_2O_6$	3,5-Dinitrobenzoic acid	212.116	25	0.134	1.34	27		
$C_7H_4O_6$	4-Oxo-4H-pyran-2,6-dicarboxylic acid	184.103	25	1.45	14.7	27		
$C_7H_4O_7$	3-Hydroxy-4-oxo-4H-pyran-2,6-dicarboxylic acid	200.103	25	0.84	8.4	27		
$C_7H_5BrO_2$	2-Bromobenzoic acid	201.018	25	0.185	1.85	27		
$C_7H_5BrO_2$	3-Bromobenzoic acid	201.018	25	0.040	0.40	27		
$C_7H_5BrO_2$	4-Bromobenzoic acid	201.018	25	0.0056	0.056	27		
$C_7H_5ClO_2$	2-Chlorobenzoic acid	156.567	25	0.209	2.09	27		
$C_7H_5ClO_2$	3-Chlorobenzoic acid	156.567	25	0.040	0.40	27		
$C_7H_5ClO_2$	4-Chlorobenzoic acid	156.567	25	0.072	0.72	27		
$C_7H_5Cl_2NO_2$	3-Amino-2,5-dichlorobenzoic acid	206.027	25	0.070	0.70	40		
$C_7H_5Cl_3$	(Trichloromethyl)benzene	195.474	5	0.0053	0.053	10		
$C_7H_5Cl_3O$	2,4,6-Trichloro-3-methylphenol	211.473	25	0.0112	0.112	2		
$C_7H_5FO_2$	2-Fluorobenzoic acid	140.112	25	0.72	7.2	27		
$C_7H_5FO_2$	3-Fluorobenzoic acid	140.112	25	0.15	1.5	27		
$C_7H_5FO_2$	4-Fluorobenzoic acid	140.112	25	0.12	1.2	27		
$C_7H_5IO_2$	2-Iodobenzoic acid	248.018	25	0.095	0.95	27		
$C_7H_5IO_2$	3-Iodobenzoic acid	248.018	25	0.016	0.16	27		
$C_7H_5IO_2$	4-Iodobenzoic acid	248.018	25	0.0027	0.027	27		
C_7H_5N	Benzonitrile	103.122	25	0.2	2	10		
C_7H_5NO	Benzoxazole	119.121	20	0.834	8.34	6		
$C_7H_5NO_3$	3-Nitrobenzaldehyde	151.120	25	0.16	1.6	27		
$C_7H_5NO_3$	4-Nitrobenzaldehyde	151.120	25	0.23	2.3	27		
$C_7H_5NO_3S$	Saccharin	183.185	25	0.40	4.0	27		
			100	4.0	42	27		
$C_7H_5NO_4$	2-Nitrobenzoic acid	167.120	25	0.55	5.58	40		
$C_7H_5NO_4$	3-Nitrobenzoic acid	167.120	25	0.256	2.56	40		
$C_7H_5NO_4$	4-Nitrobenzoic acid	167.120	25	0.0422	0.422	40		
$C_7H_5N_3O_6$	2,4,6-Trinitrotoluene	227.131	20	0.012	0.12	40		
			100	0.15	1.5	40		
$C_7H_5N_5O_8$	N-Methyl-N,2,4,6-tetranitroaniline	287.144	20	0.0074	0.074	40		
$C_7H_6ClN_3O_4S_2$	Chlorothiazide	295.724	25	0.0283	0.283	40		
$C_7H_6Cl_2$	(Dichloromethyl)benzene	161.029	30	0.025	0.25	10		
$C_7H_6Cl_2O$	2,4-Dichloro-6-methylphenol	177.028	25	0.0283	0.283	2		
$C_7H_6Cl_2O$	2,6-Dichloro-4-methylphenol	177.028	25	0.0673	0.673	2		
$C_7H_6N_2$	1H-Benzimidazole	118.136	20	0.201	2.01	6		
$C_7H_6N_2$	1H-Indazole	118.136	20	0.0827	0.827	6		
$C_7H_6N_2O_4$	1-Methyl-2,4-dinitrobenzene	182.134	20	0.0271	0.271	40		
$C_7H_6N_2O_5$	2-Methyl-4,6-dinitrophenol	198.133	15	0.0130	0.130	40		
C_7H_6O	Benzaldehyde	106.122	20	0.3	3	10		
$C_7H_6O_2$	Benzoic acid	122.122	25	0.34	3.4	27		
			95	6.4	68	26		

Mol. formula	Name	Mol. wt.	$t/°C$	Solubility S Mass%	g/L	Ref.	Henry Const. k_H kPa $m^3 mol^{-1}$	Ref.
$C_7H_6O_2$	Salicylaldehyde	122.122	86	1.68	17.0	10		
$C_7H_6O_2$	4-Hydroxybenzaldehyde	122.122	30	1.27	12.9	40		
$C_7H_6O_3$	2-Hydroxybenzoic acid	138.121	25	0.2550	2.550	33		
			50	0.5665	5.665	33		
$C_7H_6O_3$	4-Hydroxybenzoic acid	138.121	15	0.8	8	26		
			75	2.5	26	27		
$C_7H_6O_4$	3,4-Dihydroxybenzoic acid	154.121	14	1.8	18	26		
			80	21.3		26		
$C_7H_6O_5$	3,4,5-Trihydroxybenzoic acid	170.120	15	0.94	9.4	27		
			100	25.0		27		
C_7H_7Br	4-Bromotoluene	171.035	25	0.011	0.11	2		
C_7H_7Cl	4-Chlorotoluene	126.584	20	0.0137	0.137	40		
C_7H_7Cl	(Chloromethyl)benzene	126.584	20	0.0493	0.493	10		
C_7H_7ClO	2-Chloro-6-methylphenol	142.583	25	0.36	3.6	2		
C_7H_7ClO	4-Chloro-2-methylphenol	142.583	25	0.68	6.8	2		
C_7H_7ClO	4-Chloro-3-methylphenol	142.583	25	0.40	4.0	2		
C_7H_7NO	Benzamide	121.137	12	0.577	5.77	27		
$C_7H_7NO_2$	2-Hydroxybenzamide	137.137	15	0.145	1.45	40		
$C_7H_7NO_2$	Aniline-2-carboxylic acid	137.137	20	0.349	3.49	40		
$C_7H_7NO_2$	Aniline-4-carboxylic acid	137.137	25	0.54	5.39	40		
$C_7H_7NO_2$	2-Nitrotoluene	137.137	30	0.065	0.65	27		
$C_7H_7NO_2$	3-Nitrotoluene	137.137	30	0.050	0.50	27		
$C_7H_7NO_2$	4-Nitrotoluene	137.137	30	0.044	0.44	27		
$C_7H_7NO_3$	4-Amino-2-hydroxybenzoic acid	153.136	20	0.20	2.0	40		
$C_7H_7NO_3$	2-Nitroanisole	153.136	30	0.169	1.69	10		
$C_7H_7NO_3$	4-Nitroanisole	153.136	30	0.059	0.59	27		
C_7H_8	Toluene	92.139	25	0.053	0.531	22	0.660	22
			90	0.12	1.2	22		
C_7H_8	1,3,5-Cycloheptatriene	92.139	25	0.064	0.64	3	0.47	13
C_7H_8	1,6-Heptadiyne	92.139	25	0.125	1.25	3		
$C_7H_8ClN_3O_4S_2$	Hydrochlorothiazide	297.740	25	0.007	0.07	40		
$C_7H_8N_2S$	Phenylthiourea	152.217	25	2.55	26.1	27		
$C_7H_8N_4O_2$	Theophylline	180.165	20	0.52	5.2	29		
C_7H_8O	o-Cresol	108.138	40	3.08	31.8	10		
C_7H_8O	m-Cresol	108.138	40	2.51	25.8	10		
C_7H_8O	p-Cresol	108.138	40	2.26	23.1	10		
C_7H_8O	Benzyl alcohol	108.138	20	0.08	0.8	10		
C_7H_8O	Anisole	108.138	20	0.203	2.0	20	0.025	13
			81	0.294	2.9	20	0.025	13
$C_7H_8O_2$	4-Methoxyphenol	124.138	20	2.51	25.7	40		
$C_7H_8O_3S$	p-Toluenesulfonic acid	172.202	40	~33	500	40		
C_7H_9N	2-Methylaniline	107.153	20	1.66	16.9	10		
C_7H_9N	4-Methylaniline	107.153	21	7.35	79.3	10		
C_7H_9N	N-Methylaniline	107.153	25	0.56	5.62	40		
C_7H_9NO	2-Methoxyaniline	123.152	25	1.24	12.6	40		
C_7H_9NO	4-Methoxyaniline	123.152	20	1.14	11.5	40		
$C_7H_9NO_2S$	2-Methylbenzenesulfonamide	171.217	25	0.162	1.62	27		
$C_7H_9NO_2S$	3-Methylbenzenesulfonamide	171.217	25	0.78	7.8	27		
$C_7H_9NO_2S$	4-Methylbenzenesulfonamide	171.217	25	0.316	3.16	27		
$C_7H_{10}N_2OS$	2,3-Dihydro-6-propyl-2-thioxo-4(1H)-pyrimidinone	170.231	25	0.120	1.20	40		
$C_7H_{10}O_5$	Shikimic acid	174.151		15	~175	26		
C_7H_{12}	1-Heptyne	96.170	25	0.0094	0.094	3	4.47	13
C_7H_{12}	Cycloheptene	96.170	25	0.0066	0.066	3	4.9	13
C_7H_{12}	1-Methylcyclohexene	96.170	25	0.0052	0.052	3		
$C_7H_{12}O$	Cycloheptanone	112.169	20	3.61	37	20		
			92	2.82	29	20		
$C_7H_{12}O$	2-Methylcyclohexanone, (±)-	112.169	20	1.98	20.2	20		

Mol. formula	Name	Mol. wt.	$t/°C$	Solubility S			Henry Const. k_H	
				Mass%	g/L	Ref.	kPa $m^3 mol^{-1}$	Ref.
			90	1.54	15.6	20		
$C_7H_{12}O$	4-Methylcyclohexanone	112.169	20	2.43	25	20		
			80	1.95	19.9	20		
$C_7H_{12}O_2$	Cyclohexanecarboxylic acid	128.169	15	0.201	2.01	27		
$C_7H_{12}O_4$	Heptanedioic acid	160.168	25	6.347	67.77	33		
			50	42.80		33		
$C_7H_{12}O_4$	Diethyl malonate	160.168	20	2.26	23.2	20		
			91	2.47	25	20		
$C_7H_{12}O_6$	Quinic acid	192.166	9	29		26		
$C_7H_{13}N_3O_3S$	Oxamyl	219.261	25	~21	280	40		
C_7H_{14}	1-Heptene	98.186	25	0.032	0.32	3	40.3	13
C_7H_{14}	trans-2-Heptene	98.186	25	0.015	0.15	3	42.2	13
C_7H_{14}	Cycloheptane	98.186	25	0.0030	0.030	3	9.59	13
C_7H_{14}	Methylcyclohexane	98.186	25	0.00151	0.0151	3	43.3	13
			50	0.0019	0.019	3		
C_7H_{14}	Ethylcyclopentane	98.186	20	0.012	0.12	3		
$C_7H_{14}N_2O_2S$	Aldicarb	190.263	20	0.60	6.02	40		
$C_7H_{14}O$	Heptanal	114.185	11	0.124	1.24	27		
$C_7H_{14}O$	2-Heptanone	114.185	25	0.435	4.3	20	0.0171	28
			90	0.353	3.53	20	0.0171	28
$C_7H_{14}O$	3-Heptanone	114.185	20	0.479	4.8	20		
			90	0.309	3.1	20		
$C_7H_{14}O$	4-Heptanone	114.185	20	0.457	4.57	20		
			90	0.316	3.16	20		
$C_7H_{14}O$	5-Methyl-2-hexanone	114.185	19	0.537	5.40	20		
			90	0.417	4.19	20		
$C_7H_{14}O$	5-Methyl-3-hexanone	114.185	20	0.47	4.7	20		
			81	0.32	3.2	20		
$C_7H_{14}O$	2,4-Dimethyl-3-pentanone	114.185	20	0.52	5.9	20		
			90	0.30	3.0	20		
$C_7H_{14}O_2$	Ethyl 2-methylbutanoate, (+)	130.185	19	0.257	2.58	20		
			91	0.151	1.51	20		
$C_7H_{14}O_2$	Heptanoic acid	130.185	15	0.24	2.4	27		
$C_7H_{14}O_2$	Pentyl acetate	130.185	20	0.17	1.7	10		
$C_7H_{14}O_2$	Isopentyl acetate	130.185	20	0.2	2	10		
$C_7H_{14}O_2$	sec-Pentyl acetate (S)-	130.185	25	0.2	2	27		
$C_7H_{14}O_2$	Butyl propanoate	130.185	22	0.572	5.72	27		
$C_7H_{14}O_2$	Isobutyl propanoate	130.185	19	0.225	2.26	20		
			91	0.142	1.42	20		
$C_7H_{14}O_2$	Propyl butanoate	130.185	17	0.162	1.62	27		
$C_7H_{14}O_2$	Ethyl pentanoate	130.185	25	0.3	3	27		
$C_7H_{14}O_2$	Ethyl 3-methylbutanoate	130.185	20	0.2	2	10		
$C_7H_{15}Br$	1-Bromoheptane	179.098	25	0.00067	0.0067	35		
$C_7H_{15}Cl$	1-Chloroheptane	134.647	25	0.00136	0.0136	35		
$C_7H_{15}I$	1-Iodoheptane	226.098	25	0.00035	0.0035	35		
C_7H_{16}	Heptane	100.202	0	0.0003	0.003	3		
			25	0.00024	0.0024	3	209	13
			40	0.00025	0.0025	3		
C_7H_{16}	2-Methylhexane	100.202	25	0.00025	0.0025	3	346	5
C_7H_{16}	3-Methylhexane	100.202	25	0.00026	0.0026	3	249	13
C_7H_{16}	2,2-Dimethylpentane	100.202	25	0.00044	0.0044	3	318	5
C_7H_{16}	2,3-Dimethylpentane	100.202	25	0.00052	0.0052	3	175	5
C_7H_{16}	2,4-Dimethylpentane	100.202	25	0.00042	0.0042	3	323	13
C_7H_{16}	3,3-Dimethylpentane	100.202	25	0.00059	0.0059	3	186	5
$C_7H_{16}O$	1-Heptanol	116.201	10	0.25	2.5	1		
			25	0.174	1.74	1	0.00562	28
			50	0.12	1.2	1		
$C_7H_{16}O$	2-Heptanol, (±)-	116.201	30	0.33	3.3	1		

Aqueous Solubility and Henry's Law Constants of Organic Compounds

Mol. formula	Name	Mol. wt.	$t/^\circ C$	Solubility S			Henry Const. k_H	
				Mass%	g/L	Ref.	kPa m³mol⁻¹	Ref.
$C_7H_{16}O$	3-Heptanol, (S)-	116.201	25	0.43	4.3	1		
$C_7H_{16}O$	4-Heptanol	116.201	25	0.47	4.7	1		
$C_7H_{16}O$	2-Methyl-2-hexanol	116.201	25	1.0	10	1		
$C_7H_{16}O$	5-Methyl-2-hexanol	116.201	25	0.49	4.9	1		
$C_7H_{16}O$	3-Methyl-3-hexanol	116.201	25	1.2	12	1		
$C_7H_{16}O$	3-Ethyl-3-pentanol	116.201	25	1.7	17	1		
$C_7H_{16}O$	2,3-Dimethyl-2-pentanol	116.201	25	1.5	15	1		
$C_7H_{16}O$	2,4-Dimethyl-2-pentanol	116.201	25	1.3	13	1		
$C_7H_{16}O$	2,2-Dimethyl-3-pentanol	116.201	25	0.82	8.2	1		
$C_7H_{16}O$	2,3-Dimethyl-3-pentanol	116.201	25	1.6	16	1		
$C_7H_{16}O$	2,4-Dimethyl-3-pentanol	116.201	25	0.70	7.0	1		
$C_7H_{16}O$	2,3,3-Trimethyl-2-butanol	116.201	40	2.2	22	1		
$C_8Cl_4N_2$	Chlorothalonil	265.911	25	0.00006	0.0006	40		
C_8F_{18}	Perfluorooctane	438.057	25	0.00000017	0.0000017	35		
$C_8H_4F_6$	1,3-Bis(trifluoromethyl)benzene	214.108	25	0.0041	0.041	2		
$C_8H_4O_3$	Phthalic anhydride	148.116	27	0.62	6.20	40		
$C_8H_5Cl_3O_3$	2,4,5-Trichlorophenoxyacetic acid	255.483	25	0.028	0.28	40		
$C_8H_5NO_2$	1H-Isoindole-1,3(2H)-dione	147.132	25	0.036	0.36	40		
$C_8H_6Cl_2O_3$	(2,4-Dichlorophenoxy)acetic acid	221.038	25	0.07	0.7	40		
$C_8H_6Cl_2O_3$	3,6-Dichloro-2-methoxybenzoic acid	221.038	25	0.45	4.5	40		
$C_8H_6N_2$	Quinoxaline	130.147	50	54		6		
$C_8H_6N_4O_5$	Nitrofurantoin	238.158	30	0.011	0.11	40		
$C_8H_6O_3$	1,3-Benzodioxole-5-carboxaldehyde	150.132	20	0.35	3.5	40		
$C_8H_6O_4$	Phthalic acid	166.132	25	0.6977	6.977	33		
			65	3.575	37.08	33		
$C_8H_6O_4$	Isophthalic acid	166.132	25	0.013	0.13	27		
C_8H_6S	Benzo[b]thiophene	134.199	20	0.0130	0.130	6		
$C_8H_7ClO_3$	2-Chloro-4-hydroxy-5-methoxybenzaldehyde	186.593	25	0.013	0.13	8		
$C_8H_7ClO_3$	3-Chloro-4-hydroxy-5-methoxybenzaldehyde	186.593	25	0.093	0.93	8		
$C_8H_7Cl_3O$	2,4,6-Trichloro-3,5-dimethylphenol	225.500	25	0.00050	0.0050	2		
C_8H_7N	1H-Indole	117.149	20	0.187	1.87	6		
C_8H_8	Styrene	104.150	25	0.0321	0.321	22	0.286	22
			50	0.046	0.46	4	0.30	13
$C_8H_8Cl_3O_3PS$	Ronnel	321.546	20	0.00011	0.0011	40		
$C_8H_8F_3N_3O_4S_2$	Hydroflumethiazide	331.293	37	0.068	0.68	40		
$C_8H_8HgO_2$	Mercury(II) phenyl acetate	336.74	20	0.2	2	30		
$C_8H_8N_2$	2-Methyl-1H-benzimidazole	132.163	20	0.145	1.45	6		
$C_8H_8N_2O_2$	1,2-Benzenedicarboxamide	164.162	30	0.59	5.9	40		
C_8H_8O	Acetophenone	120.149	25	0.55	5.5	28	0.00108	28
			80	1.204	12.2	20		
$C_8H_8O_2$	o-Toluic acid	136.149	25	0.118	1.18	27		
$C_8H_8O_2$	m-Toluic acid	136.149	25	0.098	0.98	27		
$C_8H_8O_2$	p-Toluic acid	136.149	25	0.345	3.45	27		
$C_8H_8O_2$	Benzeneacetic acid	136.149	25	1.71	17.4	27		
$C_8H_8O_2$	Benzyl formate	136.149	20	1.07	10.8	20		
			80	1.43	14.5	20		
$C_8H_8O_2$	Phenyl acetate	136.149	20	0.59	5.9	20		
			91	0.91	9.2	20		
$C_8H_8O_2$	Methyl benzoate	136.149	20	0.21	2.1	10		
$C_8H_8O_2$	4-Methoxybenzaldehyde	136.149	25	0.429	4.29	40		
$C_8H_8O_3$	Methyl 4-hydroxybenzoate	152.148	25	0.24	2.4	40		
$C_8H_8O_3$	4-Methoxybenzoic acid	152.148	25	0.023	0.23	27		
$C_8H_8O_3$	α-Hydroxybenzeneacetic acid, (±)-	152.148	25	11.3	~125	27		
$C_8H_8O_3$	Methyl salicylate	152.148	30	0.74	7.4	10		
$C_8H_8O_3$	4-Hydroxy-3-methoxybenzaldehyde	152.148	25	0.247	2.47	8		
C_8H_9ClO	4-Chloro-2,5-dimethylphenol	156.609	25	0.89	8.9	2		

Mol. formula	Name	Mol. wt.	$t/°C$	Solubility S			Henry Const. k_H	
				Mass%	g/L	Ref.	kPa m³mol⁻¹	Ref.
C_8H_9ClO	4-Chloro-2,6-dimethylphenol	156.609	25	0.52	5.2	2		
C_8H_9ClO	4-Chloro-3,5-dimethylphenol	156.609	25	0.34	3.4	2		
C_8H_9NO	Acetanilide	135.163	20	0.52	5.2	27		
			70	2.7	28	27		
$C_8H_9NO_2$	N-(4-Hydroxyphenyl)acetamide	151.163	25	1.3	13	40		
C_8H_{10}	Ethylbenzene	106.165	0	0.020	0.20	4		
			25	0.0161	0.161	22	0.843	22
			40	0.0200	0.200	4		
C_8H_{10}	o-Xylene	106.165	25	0.0171	0.171	22	0.551	22
			45	0.021	0.21	4		
C_8H_{10}	m-Xylene	106.165	0	0.0203	0.203	4		
			25	0.0161	0.161	22	0.730	22
			40	0.022	0.22	4		
C_8H_{10}	p-Xylene	106.165	0	0.0160	0.160	4		
			25	0.0181	0.181	22	0.690	22
			40	0.022	0.22	4		
$C_8H_{10}NO_5PS$	Methyl parathion	263.208	10	0.00218	0.0218	40		
			20	0.00380	0.0380	40		
			30	0.0059	0.059	40		
$C_8H_{10}N_4O_2$	Caffeine	194.191	25	2.12	21.7	29		
$C_8H_{10}O$	4-Ethylphenol	122.164	20	0.59	5.9	40		
$C_8H_{10}O$	2,3-Xylenol	122.164	25	0.457	4.57	40		
$C_8H_{10}O$	2,4-Xylenol	122.164	25	0.787	7.87	10		
$C_8H_{10}O$	2,5-Xylenol	122.164	25	0.354	3.54	40		
$C_8H_{10}O$	2,6-Xylenol	122.164	25	0.60	6.05	40		
$C_8H_{10}O$	3,4-Xylenol	122.164	25	0.477	4.77	40		
$C_8H_{10}O$	3,5-Xylenol	122.164	29	0.62	6.2	10		
$C_8H_{10}O$	Benzeneethanol	122.164	25	1.72	17.5	40		
$C_8H_{10}O$	Ethoxybenzene	122.164	25	0.12	1.2	10		
$C_8H_{10}O_2$	2-Phenoxyethanol	138.164	20	2.53	26.0	40		
$C_8H_{10}O_2$	1,2-Dimethoxybenzene	138.164	20	0.716	7.21	20		
			92	1.073	10.9	20		
$C_8H_{11}N$	N,N-Dimethylaniline	121.180	25	0.111	1.11	40		
$C_8H_{11}N$	2,5-Dimethylaniline	121.180	20	0.66	6.6	27		
$C_8H_{11}NO$	4-(2-Aminoethyl)phenol	137.179	15	1.03	10.4	40		
C_8H_{12}	4-Vinylcyclohexene	108.181	25	0.005	0.05	4		
$C_8H_{12}O_2$	1,2-Epoxy-4-(epoxyethyl)cyclohexane	140.180	20	13.4	155	40		
$C_8H_{12}O_4$	Diethyl maleate	172.179	20	1.56	15.9	20		
			91	1.75	17.8	20		
C_8H_{14}	1-Octyne	110.197	25	0.0024	0.024	4	7.87	13
$C_8H_{14}ClN_5$	Atrazine	215.684	25	0.007	0.07	26		
$C_8H_{14}O_4$	Octanedioic acid	174.195	25	0.2416	2.416	34		
			50	0.5570	5.570	34		
$C_8H_{14}O_4$	Diethyl succinate	174.195	20	0.19	1.9	40		
$C_8H_{15}N_3O_7$	Streptozotocin	265.221	25	0.50	5.07	40		
C_8H_{16}	1-Octene	112.213	25	0.00027	0.0027	4	96.3	13
C_8H_{16}	Cyclooctane	112.213	25	0.00079	0.0079	4	10.7	13
C_8H_{16}	Ethylcyclohexane	112.213	40	0.00066	0.0066	4		
C_8H_{16}	cis-1,2-Dimethylcyclohexane	112.213	25	0.00060	0.0060	4	36	5
C_8H_{16}	trans-1,4-Dimethylcyclohexane	112.213	25	0.000384	0.00384	4	88.2	5
C_8H_{16}	Propylcyclopentane	112.213	25	0.00020	0.0020	4	90.2	5
C_8H_{16}	1,1,3-Trimethylcyclopentane	112.213	25	0.00037	0.0037	4	159	5
$C_8H_{16}N_2O_4S_2$	Homocystine	268.354	25	0.02	0.2	26		
$C_8H_{16}O$	2-Octanone	128.212	25	0.113	1.13	10		
			91	0.094	0.94	20		
$C_8H_{16}O$	3-Octanone	128.212	20	0.137	1.37	20		
			91	0.106	1.06	20		
$C_8H_{16}O$	5-Methyl-3-heptanone	128.212	20	0.192	1.92	20		

Mol. formula	Name	Mol. wt.	t/°C	Solubility S			Henry Const. k_H	
				Mass%	g/L	Ref.	kPa m³mol⁻¹	Ref.
			90	0.131	1.31	20		
$C_8H_{16}O_2$	Octanoic acid	144.212	25	0.080	0.80	26		
$C_8H_{16}O_2$	Hexyl acetate	144.212	20	0.02	0.2	10		
$C_8H_{16}O_2$	sec-Hexyl acetate	144.212	20	0.13	1.3	10		
$C_8H_{16}O_2$	Pentyl propanoate	144.212	20	0.1	1	27		
$C_8H_{16}O_2$	Isobutyl isobutanoate	144.212	20	0.5	5	10		
$C_8H_{16}O_2$	Ethyl hexanoate	144.212	20	0.063	0.63	27		
$C_8H_{17}Br$	1-Bromooctane	193.125	25	0.000167	0.00167	35		
$C_8H_{17}Cl$	1-Chlorooctane	148.674	25	0.0345	0.345	35		
$C_8H_{17}Cl$	3-(Chloromethyl)heptane	148.674	20	0.01	0.1	10		
C_8H_{18}	Octane	114.229	25	0.000071	0.00071	4	311	13
			50	0.00010	0.0010	4		
C_8H_{18}	3-Methylheptane, (S)-	114.229	25	0.000079	0.00079	4	376	5
C_8H_{18}	2,2,4-Trimethylpentane	114.229	25	0.00022	0.0022	4	307	13
C_8H_{18}	2,3,4-Trimethylpentane	114.229	25	0.00018	0.0018	4	206	13
$C_8H_{18}O$	1-Octanol	130.228	25	0.054	0.54	1		
$C_8H_{18}O$	2-Octanol	130.228	25	0.4	4	1		
$C_8H_{18}O$	2-Methyl-2-heptanol	130.228	30	0.25	2.5	1		
$C_8H_{18}O$	2-Ethyl-1-hexanol	130.228	25	0.01	0.1	1		
$C_8H_{18}O$	Dibutyl ether	130.228	20	0.023	0.3	20	0.48	13
			90	0.010	0.10	20		
$C_8H_{19}N$	Dibutylamine	129.244	20	0.47	4.7	10		
$C_8H_{19}N$	2-Ethylhexylamine	129.244	20	0.25	2.5	10		
$C_8H_{20}Si$	Tetraethylsilane	144.331	25	0.0000325	0.000325	10		
$C_9H_4Cl_3NO_2S$	Folpet	296.558	20	0.00010	0.0010	40		
$C_9H_5Cl_3N_4$	Anilazine	275.522	20	0.001	0.01	40		
$C_9H_6Cl_2N_2O_3$	Methazole	261.061	24	0.00015	0.0015	40		
$C_9H_6O_2$	2H-1-Benzopyran-2-one	146.143	20	0.190	1.90	40		
			60	0.69	6.95	40		
$C_9H_7BrO_4$	2-(Acetyloxy)-5-bromobenzoic acid	259.054		0.07	0.7	30		
$C_9H_7Cl_3O_3$	Silvex	269.509	25	0.014	0.14	40		
C_9H_7N	Quinoline	129.159	20	0.633	6.33	6		
C_9H_7N	Isoquinoline	129.159	20	0.452	4.52	6		
C_9H_7NO	8-Quinolinol	145.158	25	0.065	0.65	40		
$C_9H_8Cl_2O_3$	2-(2,4-Dichlorophenoxy)propanoic acid	235.064	25	0.083	0.83	40		
$C_9H_8Cl_3NO_2S$	Captan	300.590	20	0.00005	0.0005	40		
C_9H_8O	trans-Cinnamaldehyde	132.159	25	0.135	1.35	40		
$C_9H_8O_2$	trans-Cinnamic acid	148.159	20	0.1	1	26		
			98	0.59	5.9	26		
$C_9H_8O_4$	2-(Acetyloxy)benzoic acid	180.158	25	0.25	2.5	27		
$C_9H_9ClO_3$	(4-Chloro-2-methylphenoxy)acetic acid	200.618	25	0.117	1.17	40		
$C_9H_9Cl_2NO$	Propanil	218.079	20	0.013	0.13	40		
$C_9H_9I_2NO_3$	3,5-Diiodo-L-tyrosine	432.981	25	0.062	0.62	26		
C_9H_9N	3-Methyl-1H-indole	131.174	20	0.050	0.50	6		
$C_9H_9NO_3$	N-Benzoylglycine	179.172	25	0.37	3.7	29		
$C_9H_9N_3O_2S_2$	Sulfathiazole	255.316	20	0.048	0.48	40		
C_9H_{10}	Isopropenylbenzene	118.175	20	0.0116	0.116	40		
C_9H_{10}	Indan	118.175	25	0.010	0.10	4		
$C_9H_{10}Cl_2N_2O$	Diuron	233.093	25	0.0042	0.042	40		
$C_9H_{10}Cl_2N_2O_2$	Linuron	249.093	25	0.0075	0.075	40		
$C_9H_{10}O$	1-Phenyl-1-propanone	134.174	19	0.32	3.2	20		
			80	0.24	2.4	20		
$C_9H_{10}O_2$	Ethyl benzoate	150.174	25	0.083	0.83	20		
$C_9H_{10}O_2$	Benzyl acetate	150.174	25	0.150	1.50	40		
$C_9H_{10}O_3$	Ethyl 4-hydroxybenzoate	166.173	25	0.0080	0.080	40		
$C_9H_{11}ClN_2O$	N'-(4-Chlorophenyl)-N,N-dimethylurea	198.648	25	0.023	0.23	26		
$C_9H_{11}Cl_2N_3O_4S_2$	Methyclothiazide	360.237	20	0.005	0.05	40		
$C_9H_{11}Cl_3NO_3PS$	Chlorpyrifos	350.586	20	0.000073	0.00073	40		

Mol. formula	Name	Mol. wt.	$t/°C$	Solubility S Mass%	g/L	Ref.	Henry Const. k_H kPa m³mol⁻¹	Ref.
$C_9H_{11}NO_2$	DL-Phenylalanine	165.189	25	1.40	14.2	29		
$C_9H_{11}NO_2$	L-Phenylalanine	165.189	25	2.71	27.9	26		
$C_9H_{11}NO_3$	DL-Tyrosine	181.188	25	0.35	3.5	30		
$C_9H_{11}NO_3$	L-Tyrosine	181.188	25	0.046	0.46	26		
$C_9H_{11}NO_4$	Levodopa	197.188	25	62.3		26		
C_9H_{12}	1,8-Nonadiyne	120.191	25	0.0125	0.125	4		
C_9H_{12}	Propylbenzene	120.191	25	0.0052	0.052	22	1.041	22
C_9H_{12}	Isopropylbenzene	120.191	25	0.0050	0.050	22	1.466	22
C_9H_{12}	2-Ethyltoluene	120.191	25	0.0093	0.093	5	0.529	13
C_9H_{12}	4-Ethyltoluene	120.191	25	0.0094	0.094	5	0.500	13
C_9H_{12}	1,2,3-Trimethylbenzene	120.191	25	0.0070	0.070	22	0.343	22
C_9H_{12}	1,2,4-Trimethylbenzene	120.191	25	0.0057	0.057	22	0.569	22
C_9H_{12}	1,3,5-Trimethylbenzene	120.191	25	0.0050	0.050	22	0.781	22
$C_9H_{12}N_2O$	N,N-Dimethyl-N'-phenylurea	164.203	25	0.32	3.2	40		
$C_9H_{13}BrN_2O_2$	Bromacil	261.115	25	0.082	0.82	40		
$C_9H_{13}ClN_2O_2$	Terbacil	216.664	25	0.071	0.71	40		
$C_9H_{13}ClN_6$	Cyanazine	240.692	25	0.0171	0.171	40		
$C_9H_{13}NO_3$	Epinephrine	183.204	20	0.018	0.18	40		
$C_9H_{14}N_4O_3$	Carnosine	226.232	25	24.4		26		
$C_9H_{14}O$	Isophorone	138.206	20	1.57	16.0	20		
			80	1.27	12.9	20		
$C_9H_{14}O_6$	Glycerol triacetate	218.203	25	5.8	62	10		
C_9H_{16}	1-Nonyne	124.223	25	0.00072	0.0072	4		
$C_9H_{16}ClN_5$	Propazine	229.710	20	0.00086	0.0086	40		
$C_9H_{16}N_4OS$	Tebuthiuron	228.314	20	0.23	2.3	40		
$C_9H_{16}O_4$	Nonanedioic acid	188.221	25	0.1780	1.780	34		
			65	1.322	13.40	34		
$C_9H_{16}O_4$	Diethyl glutarate	188.221	30	1.20	12.2	20		
			91	0.91	9.2	20		
$C_9H_{17}N_5S$	Ametryn	227.330	20	0.0190	0.190	40		
C_9H_{18}	1-Nonene	126.239	25	0.000112	0.00112	40		
C_9H_{18}	1,1,3-Trimethylcyclohexane	126.239	25	0.000177	0.00177	4	105	13
$C_9H_{18}FeN_3S_6$	Ferbam	416.494	20	0.013	0.13	40		
$C_9H_{18}N_2O_4$	2-Methyl-2-propyl-1,3-propanediol dicarbamate	218.250	25	0.33	3.3	40		
$C_9H_{18}O$	2-Nonanone	142.238	20	0.038	0.38	20		
			70	0.034	0.34	20		
$C_9H_{18}O$	3-Nonanone	142.238	30	0.056	0.56	20		
			80	0.046	0.46	20		
$C_9H_{18}O$	5-Nonanone	142.238	20	0.054	0.54	20		
			80	0.029	0.29	20		
$C_9H_{18}O$	2,6-Dimethyl-4-heptanone	142.238	21	0.045	0.43	20		
			91	0.037	0.37	20		
$C_9H_{18}O_2$	Nonanoic acid	158.238	20	0.0284	0.284	26		
$C_9H_{18}O_2$	Ethyl heptanoate	158.238	20	0.029	0.29	27		
C_9H_{20}	Nonane	128.255	25	0.000017	0.00017	4	333	13
			50	0.000022	0.00022	4		
C_9H_{20}	4-Methyloctane	128.255	25	0.0000115	0.000115	4	1000	5
C_9H_{20}	2,2,5-Trimethylhexane	128.255	25	0.00008	0.0008	4	246	13
$C_9H_{20}O$	3,5-Dimethyl-4-heptanol	144.254	15	0.072	0.72	1		
$C_9H_{20}O$	1-Nonanol	144.254	25	0.014	0.14	1		
$C_9H_{20}O$	2-Nonanol, (±)-	144.254	15	0.026	0.26	1		
$C_9H_{20}O$	3-Nonanol, (±)-	144.254	15	0.032	0.32	1		
$C_9H_{20}O$	4-Nonanol	144.254	15	0.0026	0.026	1		
$C_9H_{20}O$	5-Nonanol	144.254	15	0.0032	0.032	1		
$C_{10}Cl_{10}O$	Kepone	490.636	100	0.4	4	40		
$C_{10}Cl_{12}$	Mirex	545.543	25	0.0000085	0.000085	40		
$C_{10}F_{22}$	Perfluorodecane	538.072	20	0.000031	0.00031	35		

Mol. formula	Name	Mol. wt.	$t/°C$	Solubility S			Henry Const. k_H	
				Mass%	g/L	Ref.	kPa m³mol⁻¹	Ref.
$C_{10}H_4Cl_2O_2$	2,3-Dichloro-1,4-naphthalenedione	227.044	25	0.00001	0.0001	40		
$C_{10}H_5Cl_7$	Heptachlor	373.318	25	0.0000056	0.000056	40		
$C_{10}H_6Cl_4O_4$	Dimethyl tetrachloroterephthalate	331.965	25	0.00005	0.0005	40		
$C_{10}H_6Cl_8$	Chlordane	409.779	25	0.000185	0.00185	40		
$C_{10}H_7Cl$	1-Chloronaphthalene	162.616	25	0.00224	0.0224	5	0.0363	28
$C_{10}H_7Cl$	2-Chloronaphthalene	162.616	25	0.00117	0.0117	5	0.0335	28
$C_{10}H_7NO_2$	1-Nitronaphthalene	173.169	18	0.005	0.05	40		
$C_{10}H_8$	Naphthalene	128.171	10	0.0019	0.019	4		
			25	0.00316	0.0316	22	0.043	22
			50	0.0082	0.082	4		
$C_{10}H_8N_2$	2,2'-Bipyridine	156.184	25	0.61	6.1	40		
$C_{10}H_8O$	1-Naphthol	144.170	20	0.111	1.11	40		
$C_{10}H_8O$	2-Naphthol	144.170	20	0.064	0.64	40		
			80	0.67	6.7	40		
$C_{10}H_9Cl_4NO_2S$	Captafol	349.061	20	0.000142	0.00142	40		
$C_{10}H_9N$	1-Naphthylamine	143.185	20	0.17	1.7	40		
$C_{10}H_9N$	2-Naphthylamine	143.185	20	0.0189	0.189	40		
$C_{10}H_9N$	3-Methylisoquinoline	143.185	20	0.092	0.92	6		
$C_{10}H_{10}Cl_2O_3$	4-(2,4-Dichlorophenoxy)butanoic acid	249.090	25	0.0046	0.046	40		
$C_{10}H_{10}O_4$	Dimethyl phthalate	194.184	25	0.40	4.0	15		
$C_{10}H_{10}O_4$	Dimethyl terephthalate	194.184	25	0.00328	0.0328	40		
$C_{10}H_{11}F_3N_2O$	N,N-Dimethyl-N'-[3-(trifluoromethyl)phenyl]urea	232.201	20	0.0105	0.105	40		
$C_{10}H_{11}N_3O_3S$	Sulfamethoxazole	253.277	25	0.0281	0.281	40		
$C_{10}H_{12}$	1,2,3,4-Tetrahydronaphthalene	132.202	20	0.0045	0.045	40		
$C_{10}H_{12}ClNO_2$	Chloropropham	213.661	25	0.0080	0.080	40		
$C_{10}H_{12}N_2O_3S$	Bentazon	240.278	20	0.050	0.50	40		
$C_{10}H_{12}N_3O_3PS_2$	Azinphos-methyl	317.324	20	0.00209	0.0209	40		
$C_{10}H_{12}N_4O_5$	Inosine	268.226	20	1.6	16	29		
$C_{10}H_{12}O_3$	Propyl 4-hydroxybenzoate	180.200	25	0.04	0.4	40		
$C_{10}H_{12}O_4$	Cantharidin	196.200	20	0.003	0.03	40		
$C_{10}H_{13}NO_2$	Isopropyl phenylcarbamate	179.216	20	0.01	0.1	40		
$C_{10}H_{13}NO_2$	N-(4-Ethoxyphenyl)acetamide	179.216	25	0.0502	0.502	40		
$C_{10}H_{13}N_5O_3$	2'-Deoxyadenosine	251.242	25	0.67	6.7	29		
$C_{10}H_{13}N_5O_4$	Adenosine	267.242	25	0.51	5.1	29		
$C_{10}H_{13}N_5O_5$	Guanosine	283.241	25	0.0500	0.500	29		
$C_{10}H_{14}$	Butylbenzene	134.218	25	0.00138	0.0138	22	1.33	22
$C_{10}H_{14}$	sec-Butylbenzene, (±)-	134.218	25	0.0014	0.014	4	1.89	11
$C_{10}H_{14}$	$tert$-Butylbenzene	134.218	25	0.0032	0.032	4	1.28	11
$C_{10}H_{14}$	Isobutylbenzene	134.218	25	0.0010	0.010	4	3.32	11
$C_{10}H_{14}$	1-Isopropyl-4-methylbenzene	134.218	25	0.0051	0.051	23	0.80	5
$C_{10}H_{14}$	o-Diethylbenzene	134.218	20	0.0071	0.071	40		
$C_{10}H_{14}$	p-Diethylbenzene	134.218	20	0.0025	0.025	40		
$C_{10}H_{14}$	1,2,4,5-Tetramethylbenzene	134.218	25	0.000348	0.00348	4	2.55	11
$C_{10}H_{14}NO_5PS$	Parathion	291.261	20	0.00129	0.0129	40		
$C_{10}H_{14}N_2O_5$	Thymidine	242.228	25	5.1	54	29		
$C_{10}H_{14}O$	4-$tert$-Butylphenol	150.217	25	0.058	0.58	40		
$C_{10}H_{14}O$	Carvone, (±)-	150.217	15	0.13	1.3	27		
$C_{10}H_{14}O$	Thymol	150.217		0.1	1	30		
$C_{10}H_{15}NO$	l-Ephedrine	165.232	25	0.57	5.69	40		
$C_{10}H_{15}N_5O_5$	Vidarabine	285.257	20	0.051	0.51	40		
$C_{10}H_{16}$	d-Limonene	136.234	0	0.00097	0.0097	4		
			25	0.00138	0.0138	4		
$C_{10}H_{16}N_2O_3S$	Biotin	244.310	25	0.035	0.35	40		
$C_{10}H_{16}O$	Camphor, (+)	152.233	20	0.01	0.1	10		
$C_{10}H_{16}O$	Carvenone, (S)-	152.233	15	0.22	2.2	27		
$C_{10}H_{16}O_4$	$trans$-Camphoric acid, (±)-	200.232	25	0.8	8	27		
$C_{10}H_{18}$	cis-Decahydronaphthalene	138.250	25	0.000089	0.00089	37		

Mol. formula	Name	Mol. wt.	$t/°C$	Solubility S			Henry Const. k_H	
				Mass%	g/L	Ref.	kPa m³mol⁻¹	Ref.
$C_{10}H_{18}$	trans-Decahydronaphthalene	138.250	25	0.000089	0.00089	4	3	13
$C_{10}H_{18}O$	Borneol, (±)-	154.249	25	0.074	0.74	27		
$C_{10}H_{18}O$	α-Terpineol	154.249	15	0.20	2.0	27		
$C_{10}H_{18}O$	Eucalyptol	154.249	21	0.379	3.79	40		
			50	0.170	1.70	40		
$C_{10}H_{18}O_2$	Cyclohexyl butanoate	170.249	20	0.11	1.1	20		
			90	0.09	0.90	20		
$C_{10}H_{18}O_4$	Sebacic acid	202.248	20	0.10	1.0	40		
$C_{10}H_{19}N_5O$	Prometone	225.291	20	0.075	0.75	40		
$C_{10}H_{19}N_5S$	Prometryn	241.357	20	0.0048	0.048	32		
$C_{10}H_{20}$	1-Decene	140.266	25	0.00057	0.0057	4		
$C_{10}H_{20}$	Pentylcyclopentane	140.266	25	0.0000115	0.000115	4	185	5
$C_{10}H_{20}O$	4-Decanone	156.265	20	0.0238	0.238	20		
			80	0.0064	0.064	20		
$C_{10}H_{20}O_2$	Nonyl formate	172.265	10	0.012	0.12	20		
$C_{10}H_{20}O_2$	Hexyl butanoate	172.265	29	0.021	0.21	20		
			90	0.039	0.39	20		
$C_{10}H_{20}O_2$	Decanoic acid	172.265	20	0.015	0.15	26		
$C_{10}H_{20}O_2$	Octyl acetate	172.265	19	0.020	0.20	20		
			92	0.012	0.12	20		
$C_{10}H_{20}O_2$	Ethyl octanoate	172.265	20	0.007	0.07	27		
$C_{10}H_{22}$	Decane	142.282	0	0.0000015	0.000015	4	479	13
$C_{10}H_{22}O$	1-Decanol	158.281	25	0.0037	0.037	1		
$C_{10}H_{22}O$	Diisopentyl ether	158.281	20	0.02	0.2	10		
$C_{11}H_8O_2$	2-Methyl-1,4-naphthalenedione	172.181	25	0.016	0.16	40		
$C_{11}H_8O_2$	1-Naphthalenecarboxylic acid	172.181	25	0.0058	0.058	27		
$C_{11}H_{10}$	1-Methylnaphthalene	142.197	25	0.00281	0.0281	22	0.045	22
$C_{11}H_{10}$	2-Methylnaphthalene	142.197	25	0.0025	0.025	4	0.051	12
$C_{11}H_{10}N_2S$	1-Naphthalenylthiourea	202.275	20	0.06	0.6	40		
$C_{11}H_{12}Cl_2N_2O_5$	Chloramphenicol	323.129	25	0.38	3.8	40		
$C_{11}H_{12}I_3NO_2$	Iopanoic acid	570.932	37	0.034	0.34	40		
$C_{11}H_{12}NO_4PS_2$	Phosmet	317.321	25	0.0025	0.025	40		
$C_{11}H_{12}N_2O_2$	L-Tryptophan	204.225	25	1.30	13.1	26		
$C_{11}H_{13}NO_4$	Bendiocarb	223.226	25	0.004	0.04	40		
$C_{11}H_{13}N_3O_3S$	Sulfisoxazole	267.304	37	0.03	0.3	40		
$C_{11}H_{14}ClNO$	Propachlor	211.688	20	0.07	0.7	40		
$C_{11}H_{14}N_2O$	Cytisine	190.241	16	~30	439	40		
$C_{11}H_{14}O_3$	Butyl 4-hydroxybenzoate	194.227	25	0.020	0.20	40		
$C_{11}H_{15}NO_2$	Butyl 4-aminobenzoate	193.243	25	0.018	0.18	40		
$C_{11}H_{15}NO_3$	Propoxur	209.242	20	0.193	1.93	40		
$C_{11}H_{16}$	Pentylbenzene	148.245	25	0.00105	0.0105	5	1.69	11
$C_{11}H_{16}O$	4-(1,1-Dimethylpropyl)phenol	164.244	25	0.017	0.17	40		
$C_{11}H_{17}N_3O_3S$	4-Amino-N-[(butylamino)carbonyl] benzenesulfonamide	271.336	37	0.053	0.53	40		
$C_{11}H_{18}N_2O_3$	Amobarbital	226.272	25	0.06	0.6	40		
$C_{11}H_{22}O_2$	Heptyl butanoate	186.292	20	0.028	0.28	20		
			80	0.020	0.20	20		
$C_{11}H_{22}O_2$	Ethyl nonanoate	186.292	20	0.003	0.03	27		
$C_{11}H_{24}$	Undecane	156.309	25	0.0000004	0.000004	37		
$C_{12}Br_{10}O$	Decabromobiphenyl ether	959.167	25	0.0000025	0.000025	40		
$C_{12}Cl_{10}$	Decachlorobiphenyl	498.658	25	0.00000000012	0.0000000012	7	0.0208	7
$C_{12}F_{26}$	Hexacosafluorododecane	638.086	20	0.00000096	0.0000096	35		
$C_{12}HCl_9$	2,2',3,3',4,5,5',6,6'-Nonachlorobiphenyl	464.213	25	0.0000000018	0.000000018	7		
$C_{12}H_2Cl_8$	2,2',3,3',5,5',6,6'-Octachlorobiphenyl	429.768	25	0.00000015	0.0000015	41	0.0381	7
$C_{12}H_3Cl_7$	2,2',3,3',4,4',6-Heptachlorobiphenyl	395.323	25	0.0000002	0.000002	7	0.0054	7
$C_{12}H_4Cl_4O_2$	2,3,7,8-Tetrachlorodibenzo-p-dioxin	321.971	22	0.0000000019	0.000000019	40		
$C_{12}H_4Cl_6$	2,2',3,3',4,4'-Hexachlorobiphenyl	360.878	25	0.00000006	0.0000006	7	0.0354	31
$C_{12}H_4Cl_6$	2,2',4,4',6,6'-Hexachlorobiphenyl	360.878	25	0.0000003	0.000003	41	0.818	7

Mol. formula	Name	Mol. wt.	$t/°C$	Solubility S			Henry Const. k_H	
				Mass%	g/L	Ref.	kPa m³mol⁻¹	Ref.
$C_{12}H_4Cl_6$	2,2',3,3',6,6'-Hexachlorobiphenyl	360.878	25	0.0000004	0.000004	41		
$C_{12}H_5Cl_5$	2,3,4,5,6-Pentachlorobiphenyl	326.433	25	0.0000008	0.000008	7		
$C_{12}H_5Cl_5$	2,2',4,5,5'-Pentachlorobiphenyl	326.433	25	0.000001	0.00001	7	0.0421	31
$C_{12}H_5N_7O_{12}$	2,4,6-Trinitro-N-(2,4,6-trinitrophenyl) aniline	439.208	17	0.0060	0.060	40		
$C_{12}H_6Cl_4$	2,3,4,5-Tetrachlorobiphenyl	291.988	25	0.000002	0.00002	7		
$C_{12}H_6Cl_4$	2,2',4,5-Tetrachlorobiphenyl	291.988	25	0.0000016	0.000016	9		
$C_{12}H_7Cl_2NO_3$	Nitrofen	284.095	22	0.00095	0.0095	40		
$C_{12}H_7Cl_3$	2,4,5-Trichlorobiphenyl	257.543	25	0.000014	0.00014	7	0.0379	31
$C_{12}H_7Cl_3$	2,4,6-Trichlorobiphenyl	257.543	25	0.00002	0.0002	7	0.0495	7
$C_{12}H_8$	Acenaphthylene	152.192	20	0.0016	0.016	28	0.012	28
$C_{12}H_8Cl_2$	2,5-Dichlorobiphenyl	223.098	25	0.0002	0.002	7	0.0201	7
$C_{12}H_8Cl_2$	2,6-Dichlorobiphenyl	223.098	25	0.00014	0.0014	7		
$C_{12}H_8Cl_6$	Aldrin	364.910	25	0.00002	0.0002	40		
$C_{12}H_8Cl_6O$	Dieldrin	380.909	25	0.000020	0.00020	40		
$C_{12}H_8Cl_6O$	Endrin	380.909	25	0.000026	0.00026	40		
$C_{12}H_8O$	Dibenzofuran	168.191	25	0.000475	0.00475	41	0.011	12
$C_{12}H_8O_4$	Methoxsalen	216.190	30	0.0048	0.048	40		
$C_{12}H_8S$	Dibenzothiophene	184.257	25	0.000103	0.00103	6		
$C_{12}H_9Cl$	2-Chlorobiphenyl	188.652	25	0.00055	0.0055	7	0.0701	7
$C_{12}H_9ClF_3N_3O$	Norflurazon	303.666	25	0.0028	0.028	40		
$C_{12}H_9Cl_2NO_3$	Vinclozolin	286.110	20	0.1	1	32		
$C_{12}H_9N$	Carbazole	167.206	22	0.000120	0.00120	6		
$C_{12}H_9NS$	10H-Phenothiazine	199.271	25	0.00016	0.0016	40		
$C_{12}H_{10}$	Acenaphthene	154.207	0	0.00015	0.0015	4		
			25	0.000380	0.00380	22	0.01217	22
			50	0.00092	0.0092	4		
$C_{12}H_{10}$	Biphenyl	154.207	0	0.000272	0.00272	4		
			25	0.00072	0.0072	22	0.0280	22
			50	0.0022	0.022	4		
$C_{12}H_{10}Cl_2N_2$	3,3'-Dichloro-p-benzidine	253.126	25	0.00031	0.0031	40		
$C_{12}H_{10}N_2$	trans-Azobenzene	182.220	20	0.03	0.3	27		
$C_{12}H_{10}N_2O$	N-Nitrosodiphenylamine	198.219	25	0.0035	0.035	17		
$C_{12}H_{10}O$	2-Hydroxybiphenyl	170.206	25	0.07	0.7	40		
$C_{12}H_{10}O$	4-Hydroxybiphenyl	170.206	25	0.0056	0.056	40		
$C_{12}H_{10}O$	Diphenyl ether	170.206	25	0.00180	0.0180	6	0.027	13
$C_{12}H_{10}O_2$	1-Naphthaleneacetic acid	186.206	25	0.0415	0.415	40		
$C_{12}H_{11}Cl_2NO$	Propyzamide	256.127	25	0.0015	0.015	32		
$C_{12}H_{11}N$	Diphenylamine	169.222	20	0.0055	0.055	40		
			50	0.0058	0.058	40		
$C_{12}H_{11}NO_2$	Carbaryl	201.221	20	0.0102	0.102	40		
$C_{12}H_{11}N_3$	4-Aminoazobenzene	197.235	25	0.0030	0.030	40		
			97	0.068	0.68	40		
$C_{12}H_{11}N_3$	1,3-Diphenyl-1-triazene	197.235	20	0.050	0.50	40		
$C_{12}H_{12}$	1-Ethylnaphthalene	156.223	25	0.00101	0.0101	4	0.039	12
$C_{12}H_{12}$	2-Ethylnaphthalene	156.223	25	0.00080	0.0080	4	0.078	12
$C_{12}H_{12}$	1,3-Dimethylnaphthalene	156.223	25	0.0008	0.008	4		
$C_{12}H_{12}$	1,4-Dimethylnaphthalene	156.223	25	0.00114	0.0114	4		
$C_{12}H_{12}$	1,5-Dimethylnaphthalene	156.223	25	0.00031	0.0031	4	0.036	28
$C_{12}H_{12}$	2,3-Dimethylnaphthalene	156.223	25	0.00025	0.0025	4		
$C_{12}H_{12}$	2,6-Dimethylnaphthalene	156.223	25	0.00017	0.0017	4		
$C_{12}H_{12}ClN_5O_4S$	Chlorsulfuron	357.773	25	2.71	27.9	32		
$C_{12}H_{12}N_2$	p-Benzidine	184.236	24	0.0360	0.360	40		
$C_{12}H_{12}N_2O_2S$	Bis(4-aminophenyl) sulfone	248.300	25	0.016	0.16	40		
$C_{12}H_{12}N_2O_3$	Phenobarbital	232.234	25	0.12	1.2	40		
			45	0.26	2.6	40		
$C_{12}H_{13}NO_2S$	Carboxin	235.302	25	0.017	0.17	40		
$C_{12}H_{14}N_4O_2S$	Sulfamethazine	278.330	20	0.053	0.53	40		

Mol. formula	Name	Mol. wt.	$t/°C$	Solubility S			Henry Const. k_H	
				Mass%	g/L	Ref.	kPa m³mol⁻¹	Ref.
$C_{12}H_{14}O_4$	Diethyl phthalate	222.237	25	0.12	1.2	40		
$C_{12}H_{15}ClNO_4PS_2$	Phosalone	367.808	20	0.00026	0.0026	40		
$C_{12}H_{15}NO_3$	Carbofuran	221.252	20	0.032	0.32	40		
$C_{12}H_{18}$	Hexylbenzene	162.271	25	0.00021	0.0021	4		
$C_{12}H_{18}N_2O_3S$	Tolbutamide	270.347	25	0.011	0.11	40		
$C_{12}H_{18}N_4O_6S$	Oryzalin	346.359	25	0.00024	0.0024	40		
$C_{12}H_{18}O_2$	4-Hexyl-1,3-benzenediol	194.270	18	0.05	0.5	40		
$C_{12}H_{19}ClNO_3P$	Crufomate	291.711	20	0.50	5.0	40		
$C_{12}H_{20}N_4O_2$	Hexazinone	252.313	25	3.2	33	40		
$C_{12}H_{22}O_4$	Dodecanedioic acid	230.301	20	0.004	0.04	40		
$C_{12}H_{22}O_{11}$	Sucrose	342.296	20	67.1		27		
			50	72.3		27		
			100	83.0		27		
$C_{12}H_{22}O_{11}$	α-Maltose	342.296	20	51.9		27		
$C_{12}H_{24}O$	2,6,8-Trimethyl-4-nonanone	184.318	10	0.012	0.12	20		
			80	0.014	0.14	20		
$C_{12}H_{24}O_2$	Dodecanoic acid	200.318	20	0.0055	0.055	26		
$C_{12}H_{24}O_2$	Ethyl decanoate	200.318	20	0.0015	0.015	27		
$C_{12}H_{26}$	Dodecane	170.334	25	0.00000037	0.0000037	4	750	5
$C_{12}H_{26}O$	1-Dodecanol	186.333	25	0.0004	0.004	1		
$C_{12}H_{26}O$	Dihexyl ether	186.333	20	0.019	0.19	20		
			90	0.019	0.19	20		
$C_{12}H_{27}N$	Tributylamine	185.349	25	0.0142	0.142	40		
$C_{12}H_{27}O_4P$	Tributyl phosphate	266.313	25	0.039	0.39	10		
$C_{13}H_9N$	Acridine	179.217	25	0.00466	0.0466	6		
$C_{13}H_9N$	Benzo[f]quinoline	179.217	25	0.0079	0.079	6		
$C_{13}H_{10}$	9H-Fluorene	166.218	0	0.00007	0.0007	4		
			25	0.00019	0.0019	22	0.00787	22
			50	0.00063	0.0063	4		
$C_{13}H_{10}Cl_2O_2$	Dichlorophene	269.123	25	0.003	0.03	40		
$C_{13}H_{10}O$	Benzophenone	182.217	20	0.0075	0.075	40		
$C_{13}H_{11}ClO$	Clorophene	218.678	20	0.42	4.2	40		
$C_{13}H_{12}$	Diphenylmethane	168.234	25	0.000141	0.00141	4	0.001	12
$C_{13}H_{12}N_2O$	N,N'-Diphenylurea	212.246	20	0.015	0.15	40		
$C_{13}H_{13}Cl_2N_3O_3$	Iprodione	330.166	20	0.0013	0.013	40		
$C_{13}H_{14}$	1,4,5-Trimethylnaphthalene	170.250	25	0.00021	0.0021	4		
$C_{13}H_{17}N_3O$	Aminopyrine	231.293	25	4.8	50	40		
$C_{13}H_{18}Cl_2N_2O_2$	Melphalan	305.200	30	0.44	4.4	40		
$C_{13}H_{18}O_2$	Ibuprofen	206.281	25	0.0011	0.011	40		
			60	0.0048	0.048	40		
$C_{13}H_{19}N_3O_4$	Pendimethalin	281.308	20	0.00003	0.0003	40		
$C_{13}H_{22}NO_3PS$	Fenamiphos	303.358	20	0.0329	0.329	40		
$C_{13}H_{26}O_2$	Tridecanoic acid	214.344	20	0.0033	0.033	26		
$C_{13}H_{28}$	Tridecane	184.361	25	0.00000003	0.0000003	37		
$C_{14}H_8O_2$	9,10-Anthracenedione	208.213	25	0.00014	0.0014	40		
$C_{14}H_8O_4$	1,4-Dihydroxy-9,10-anthracenedione	240.212	25	0.0000096	0.000096	40		
$C_{14}H_9ClF_2N_2O_2$	Diflubenzuron	310.683	20	0.00002	0.0002	40		
$C_{14}H_9Cl_5$	1,1,1-Trichloro-2,2-bis(4-chlorophenyl) ethane	354.486	25	0.0000001	0.000001	40		
$C_{14}H_9Cl_5O$	1,1-Bis(4-chlorophenyl)-2,2,2-trichloroethanol	370.485	25	0.00013	0.0013	40		
$C_{14}H_9NO_2$	2-Amino-9,10-anthracenedione	223.227	25	0.000016	0.00016	40		
$C_{14}H_{10}$	Anthracene	178.229	0	0.0000022	0.000022	4		
			25	0.0000045	0.000045	22	0.00396	22
$C_{14}H_{10}$	Phenanthrene	178.229	10	0.000050	0.00050	4		
			25	0.00011	0.0011	22	0.00324	22
			50	0.00041	0.0041	4		
$C_{14}H_{10}Cl_4$	1,1-Dichloro-2,2-bis(p-chlorophenyl) ethane	320.041	25	0.000009	0.00009	40		

Mol. formula	Name	Mol. wt.	t/°C	Solubility S Mass%	g/L	Ref.	Henry Const. k_H kPa m³mol⁻¹	Ref.
			45	0.000024	0.00024	40		
$C_{14}H_{10}O_4$	Benzoyl peroxide	242.227	20	0.000016	0.00016	40		
$C_{14}H_{12}$	trans-Stilbene	180.245	25	0.000029	0.00029	4	0.040	12
$C_{14}H_{12}O_2$	Benzoin	212.244	25	0.03	0.3	40		
$C_{14}H_{14}$	1,2-Diphenylethane	182.261	25	0.00044	0.0044	6	0.017	12
$C_{14}H_{14}NO_4PS$	Ethyl p-nitrophenyl benzenethiophosphate	323.304	22	0.00031	0.0031	40		
$C_{14}H_{14}O$	Dibenzyl ether	198.260	35	0.0040	0.040	10		
$C_{14}H_{14}O_3$	2-Pivaloyl-1,3-indandione	230.259	25	0.0018	0.018	40		
$C_{14}H_{15}N_3$	4-(Dimethylamino)azobenzene	225.289	20	0.00014	0.0014	40		
$C_{14}H_{15}N_3$	2′,3-Dimethyl-4-aminoazobenzene	225.289	37	0.0007	0.007	40		
$C_{14}H_{16}ClN_3O_2$	Bayleton	293.749	20	0.026	0.26	40		
$C_{14}H_{16}N_2$	o-Tolidine	212.290	25	0.13	1.3	40		
$C_{14}H_{16}N_2O_2$	3,3′-Dimethoxybenzidine	244.289	25	0.006	0.06	40		
$C_{14}H_{18}N_4O_3$	Trimethoprim	290.318	25	0.04	0.4	40		
$C_{14}H_{20}ClNO_2$	Alachlor	269.768	23	0.024	0.24	40		
$C_{14}H_{21}N_3O_3S$	Tolazamide	311.400	30	0.0065	0.065	40		
$C_{14}H_{22}N_2O$	2-(Diethylamino)-N-(2,6-dimethylphenyl)acetamide	234.337	25	0.38	3.8	40		
$C_{14}H_{28}O_2$	Tetradecanoic acid	228.371	20	0.0020	0.020	26		
$C_{14}H_{29}Cl$	1-Chlorotetradecane	232.833	25	0.0232	0.232	35		
$C_{14}H_{30}$	Tetradecane	198.388	25	0.000012	0.00012	5		
$C_{14}H_{30}O$	1-Tetradecanol	214.387	25	0.000031	0.00031	1		
$C_{15}H_{12}$	2-Methylanthracene	192.256	25	0.00003	0.0003	22		
$C_{15}H_{12}$	9-Methylanthracene	192.256	25	0.000026	0.00026	4		
$C_{15}H_{12}$	1-Methylphenanthrene	192.256	25	0.0000269	0.000269	4		
$C_{15}H_{12}N_2O_2$	Phenytoin	252.268	37	0.0038	0.038	40		
$C_{15}H_{15}NO_2$	Mefenamic acid	241.286	20	0.0026	0.026	40		
$C_{15}H_{16}O_2$	2,2-Bis(4-hydroxyphenyl)propane	228.287	20	0.035	0.35	40		
$C_{15}H_{20}N_2O_4S$	Acetohexamide	324.396	37	0.0013	0.013	40		
$C_{15}H_{24}O$	4-Nonylphenol	220.351	25	0.000636	0.00636	40		
$C_{15}H_{26}O_6$	Tributyrin	302.363	20	0.010	0.10	40		
$C_{15}H_{30}O_2$	Pentadecanoic acid	242.398	20	0.0012	0.012	26		
$C_{15}H_{32}O$	1-Pentadecanol	228.414	25	0.000010	0.00010	1		
$C_{16}H_{10}$	Fluoranthene	202.250	25	0.000026	0.00026	22	0.00096	22
$C_{16}H_{10}$	Pyrene	202.250	25	0.000013	0.00013	22	0.00092	22
			50	0.00009	0.0009	4		
$C_{16}H_{13}ClN_2O$	Valium	284.739	25	0.005	0.05	40		
$C_{16}H_{14}$	9,10-Dimethylanthracene	206.282	25	0.0000056	0.000056	4		
$C_{16}H_{14}ClN_3O$	Chlorodiazepoxide	299.754	20	0.2	2	40		
$C_{16}H_{14}Cl_2O_3$	Chlorobenzilate	325.186	20	0.001	0.01	32		
$C_{16}H_{14}Cl_2O_4$	Diclofop-methyl	341.186	20	0.0003	0.003	32		
$C_{16}H_{14}N_2O$	2-Methyl-3-(2-methylphenyl)-4(3H)-quinazolinone	250.294	23	0.03	0.3	40		
$C_{16}H_{15}Cl_3O_2$	Methoxychlor	345.648	25	0.000005	0.00005	40		
$C_{16}H_{15}NO_3$	N-Benzoyl-L-phenylalanine	269.295	25	0.085	0.85	29		
$C_{16}H_{16}N_2O_4$	Phenmedipham	300.309	25	0.00047	0.0047	32		
$C_{16}H_{17}NO$	Diphenamid	239.312	27	0.026	0.26	32		
$C_{16}H_{17}N_3O_4S$	Cephalexin	347.389	25	1.2	12	40		
$C_{16}H_{22}O_4$	Dibutyl phthalate	278.344	25	0.00112	0.0112	15		
$C_{16}H_{32}O_2$	Hexadecanoic acid	256.424	20	0.00072	0.0072	26		
$C_{16}H_{34}$	Hexadecane	226.441	25	0.0000006	0.000006	37		
$C_{16}H_{34}O$	1-Hexadecanol	242.440	25	0.000003	0.00003	1		
$C_{17}H_{12}$	11H-Benzo[a]fluorene	216.277	25	0.0000045	0.000045	4		
$C_{17}H_{12}$	11H-Benzo[b]fluorene	216.277	25	0.0000002	0.000002	4		
$C_{17}H_{17}NO_2$	Apomorphine	267.323	25	2.0	20	40		
$C_{17}H_{17}N_3O_3$	Imazaquin	311.335	20	0.009	0.09	32		
$C_{17}H_{19}NO_3$	Morphine	285.338	20	0.015	0.15	27		
$C_{17}H_{19}NO_4$	Fenoxycarb	301.338	20	0.0006	0.006	32		

Mol. formula	Name	Mol. wt.	$t/°C$	Solubility S Mass%	g/L	Ref.	Henry Const. k_H kPa m^{3}mol^{-1}	Ref.
$C_{17}H_{20}N_2O$	N,N,N',N'-Tetramethyl-4,4'-diaminobenzophenone	268.353	20	0.04	0.4	40		
$C_{17}H_{20}N_4O_6$	Riboflavin	376.364	25	0.0075	0.075	40		
$C_{17}H_{21}NO_4$	Cocaine	303.354	25	0.17	1.7	27		
$C_{17}H_{23}NO_3$	Atropine	289.370	20	0.3	3	40		
$C_{17}H_{23}NO_3$	Hyoscyamine	289.370	20	0.36	3.6	40		
$C_{17}H_{34}O_2$	Heptadecanoic acid	270.451	20	0.00042	0.0042	26		
$C_{18}H_{12}$	Benz[a]anthracene	228.288	25	0.0000011	0.000011	22	0.00058	22
$C_{18}H_{12}$	Chrysene	228.288	25	0.0000002	0.000002	22	0.000065	22
$C_{18}H_{12}$	Naphthacene	228.288	25	0.00000006	0.0000006	4	0.000004	12
$C_{18}H_{12}$	Triphenylene	228.288	25	0.0000041	0.000041	4	0.00001	12
$C_{18}H_{12}N_2$	2,2'-Biquinoline	256.301	24	0.000102	0.00102	6		
$C_{18}H_{14}$	o-Terphenyl	230.304	25	0.00012	0.0012	40		
$C_{18}H_{14}$	m-Terphenyl	230.304	25	0.00015	0.0015	40		
$C_{18}H_{14}$	p-Terphenyl	230.304		0.0000018	0.000018	40		
$C_{18}H_{15}O_4P$	Triphenyl phosphate	326.283	24	0.000073	0.00073	40		
$C_{18}H_{16}OSn$	Triphenyltin hydroxide	367.029	20	0.0001	0.001	32		
$C_{18}H_{20}O_2$	trans-Diethylstilbestrol	268.351	20	0.01	0.1	40		
$C_{18}H_{21}NO_3$	Codeine	299.365	25	0.79	7.9	27		
$C_{18}H_{32}O_{16}$	Raffinose	504.437	20	12.5	~145	27		
$C_{18}H_{34}O_4$	Dibutyl sebacate	314.461	20	0.004	0.04	10		
$C_{18}H_{36}O_2$	Stearic acid	284.478	20	0.00029	0.0029	26		
$C_{18}H_{38}$	Octadecane	254.495	25	0.0000006	0.000006	37		
$C_{18}H_{38}O$	1-Octadecanol	270.494	34	0.000011	0.00011	1		
$C_{19}H_{14}$	9-Methylbenz[a]anthracene	242.314	27	0.0000066	0.000066	4		
$C_{19}H_{14}$	10-Methylbenz[a]anthracene	242.314	25	0.0000055	0.000055	4		
$C_{19}H_{14}$	5-Methylchrysene	242.314	27	0.0000062	0.000062	4		
$C_{19}H_{16}ClNO_4$	Indomethacin	357.788	25	0.001	0.01	40		
$C_{19}H_{16}O_4$	Warfarin	308.328	20	0.004	0.04	40		
$C_{19}H_{19}N_7O_6$	Folic acid	441.397	0	0.001	0.01	26		
			100	0.05	0.5	26		
$C_{19}H_{20}N_2O_2$	Phenylbutazone	308.374	25	0.0034	0.034	40		
$C_{19}H_{28}O_2$	17-Hydroxyandrost-4-en-3-one, (17β)	288.424	25	0.0024	0.024	40		
$C_{20}H_{12}$	Perylene	252.309	25	0.00000004	0.0000004	4	0.000003	12
$C_{20}H_{12}$	Benzo[a]pyrene	252.309	25	0.0000003	0.000003	22	0.0000465	22
$C_{20}H_{12}$	Benzo[e]pyrene	252.309	20	0.0000005	0.000005	22	0.0000467	22
$C_{20}H_{12}$	Benzo[k]fluoranthene	252.309		0.00000008	0.0000008	40		
$C_{20}H_{12}$	Benzo[b]fluoranthene	252.309	20	0.0000002	0.000002	40		
$C_{20}H_{12}O_5$	Fluorescein	332.306	20	0.005	0.05	27		
$C_{20}H_{13}N$	13H-Dibenzo[a,i]carbazole	267.324	24	0.00000104	0.0000104	6		
$C_{20}H_{14}$	1,2-Dihydrobenz[j]aceanthrylene	254.325	25	0.00000036	0.0000036	6		
$C_{20}H_{14}O_4$	Phenolphthalein	318.323	20	0.018	0.18	27		
$C_{20}H_{14}O_4$	Diphenyl phthalate	318.323	24	0.000008	0.00008	40		
$C_{20}H_{21}NO_4$	Papaverine	339.386	37	0.0037	0.037	40		
$C_{20}H_{23}N$	Amitriptyline	277.404	24	0.00097	0.0097	40		
$C_{20}H_{24}N_2O_2$	Quinine	324.417	25	0.057	0.57	27		
$C_{20}H_{24}N_2O_2$	Quinidine	324.417	20	0.020	0.20	27		
$C_{20}H_{26}O_2$	Norethisterone	298.419	25	0.00063	0.0063	40		
$C_{20}H_{30}O_2$	17-Methyltestosterone	302.451	25	0.0033	0.033	40		
$C_{20}H_{42}$	Eicosane	282.547	25	0.00000019	0.0000019	4		
$C_{21}H_{13}N$	Dibenz[a,j]acridine	279.335	25	0.000016	0.00016	6		
$C_{21}H_{16}$	1,2-Dihydro-3-methylbenz[j]aceanthrylene	268.352	25	0.00000022	0.0000022	6		
$C_{21}H_{20}Cl_2O_3$	Permethrin	391.288	20	0.00002	0.0002	32		
$C_{21}H_{21}O_4P$	Tri-p-cresyl phosphate	368.363	25	0.00004	0.0004	40		
$C_{21}H_{22}N_2O_2$	Strychnine	334.412	20	0.013	0.13	27		
$C_{21}H_{23}ClFNO_2$	Haloperidol	375.865	30	0.0003	0.003	40		
$C_{21}H_{26}O_5$	17,21-Dihydroxypregna-1,4-diene-3,11,20-trione	358.428	25	0.012	0.12	40		

Mol. formula	Name	Mol. wt.	$t/°C$	Solubility S Mass%	g/L	Ref.	Henry Const. k_H kPa m^3mol^{-1}	Ref.
$C_{21}H_{28}O_5$	Prednisolone	360.444	25	0.03	0.3	40		
$C_{21}H_{28}O_5$	17,21-Dihydroxypregn-4-ene-3,11,20-trione	360.444	25	0.028	0.28	30		
$C_{21}H_{30}O_2$	Progesterone	314.462	25	0.00088	0.0088	40		
			41	0.00206	0.0206	40		
$C_{21}H_{30}O_5$	Hydrocortisone	362.460	25	0.029	0.29	40		
$C_{21}H_{33}NO_7$	Lasiocarpine	411.490	20	0.67	6.75	40		
$C_{22}H_{12}$	Indeno[1,2,3-cd]pyrene	276.330	20	0.00000002	0.0000002	40		
$C_{22}H_{12}$	Benzo[ghi]perylene	276.330	25	0.000000026	0.00000026	4	0.000075	12
$C_{22}H_{14}$	Benzo[b]triphenylene	278.346	25	0.0000027	0.000027	4		
$C_{22}H_{14}$	Dibenz[a,h]anthracene	278.346	25	0.00000006	0.0000006	4		
$C_{22}H_{14}$	Dibenz[a,j]anthracene	278.346	25	0.0000012	0.000012	4		
$C_{22}H_{14}$	Picene	278.346	25	0.00000025	0.0000025	4		
$C_{22}H_{18}Cl_2FNO_3$	Cyfluthrin	434.287	20	0.0000002	0.000002	32		
$C_{22}H_{19}Cl_2NO_3$	Cypermethrin	416.297	20	0.000001	0.00001	32		
$C_{22}H_{20}O_{13}$	Carminic acid	492.386	20	0.13	1.3	40		
$C_{22}H_{22}FN_3O_2$	Droperidol	379.427	30	0.00041	0.0041	40		
$C_{22}H_{23}NO_7$	Noscapine	413.421	25	0.03	0.3	40		
$C_{22}H_{25}NO_6$	Colchicine	399.437	20	4	4	26		
$C_{22}H_{29}FO_5$	Dexamethasone	392.460	25	0.009	0.09	40		
$C_{22}H_{30}O_5$	Methylprednisolone	374.470	25	0.012	0.12	40		
$C_{22}H_{44}O_2$	Butyl stearate	340.583	25	0.2	2	10		
$C_{22}H_{46}$	Docosane	310.600	22	0.0000006	0.000006	37		
$C_{23}H_{19}ClF_3NO_3$	Cyhalothrin	449.850	20	0.0000005	0.000005	32		
$C_{23}H_{22}ClF_3O_2$	Bifenthrin	422.868	25	0.00001	0.0001	32		
$C_{23}H_{22}O_6$	Rotenone	394.417	25	0.000017	0.00017	40		
$C_{23}H_{26}N_2O_4$	Brucine	394.463	20	0.012	0.12	27		
$C_{23}H_{27}NO_8$	Narceine	445.462	13	0.078	0.78	27		
$C_{24}H_{12}$	Coronene	300.352	25	0.000000014	0.00000014	4		
$C_{24}H_{38}O_4$	Bis(2-ethylhexyl) phthalate	390.557	25	0.000027	0.00027	40		
$C_{24}H_{40}O_4$	3,12-Dihydroxycholan-24-oic acid, $(3\alpha,5\beta,12\alpha)$	392.573	20	0.001	0.01	40		
$C_{24}H_{40}O_5$	Cholic acid	408.572	20	0.028	0.28	26		
$C_{24}H_{50}$	Tetracosane	338.654	22	0.0000004	0.000004	37		
$C_{25}H_{24}F_6N_4$	Hydramethylnon	494.476	20	0.0000006	0.000006	32		
$C_{26}H_{54}$	Hexacosane	366.707	25	0.0000002	0.000002	37		
$C_{28}H_{58}$	Octacosane	394.761	22	0.0000006	0.000006	37		
$C_{29}H_{32}O_{13}$	Etoposide	588.556	20	0.02	0.2	40		
$C_{29}H_{40}N_2O_4$	Emetine	480.639	15	0.096	0.96	40		
$C_{29}H_{44}O_{12}$	Ouabain	584.652	25	1.3	13	40		
$C_{33}H_{40}N_2O_9$	Reserpine	608.679	30	0.0073	0.073	40		
$C_{36}H_{74}$	Hexatriacontane	506.973	25	0.0000002	0.000002	37		
$C_{37}H_{67}NO_{13}$	Erythromycin	733.927	30	0.12	1.2	40		
			80	0.04	0.4	40		
$C_{41}H_{64}O_{13}$	Digitoxin	764.939	25	0.0004	0.004	40		
$C_{41}H_{64}O_{14}$	Digoxin	780.939	25	0.0059	0.059	40		
$C_{45}H_{73}NO_{15}$	Solanine	868.060	15	0.0026	0.026	40		
$C_{60}H_{78}OSn_2$	Fenbutatin oxide	1052.68	23	0.0000005	0.000005	32		

[a] Indicates a value of S for a gas at a partial pressure of 101.325 kPa (1 atm) in equilibrium with the solution.

AQUEOUS SOLUBILITY OF INORGANIC COMPOUNDS AT VARIOUS TEMPERATURES

The solubility of over 300 common inorganic compounds in water is tabulated here as a function of temperature. Solubility is defined as the concentration of the compound in a solution that is in equilibrium with a solid phase at the specified temperature. In this table the solid phase is generally the most stable crystalline phase at the temperature in question. An asterisk * on solubility values in adjacent columns indicates that the solid phase changes between those two temperatures (usually from one hydrated phase to another or from a hydrate to the anhydrous solid). In such cases the slope of the solubility vs. temperature curve may show a discontinuity.

All solubility values are expressed as mass percent of solute, $100 \cdot w_2$, where

$$w_2 = m_2/(m_1 + m_2)$$

and m_2 is the mass of solute and m_1 the mass of water. This quantity is related to other common measures of solubility as follows:

Molarity: $c_2 = 1000 \rho w_2/M_2$
Molality: $m_2 = 1000 w_2/M_2(1-w_2)$
Mole fraction: $x_2 = (w_2/M_2)/\{(w_2/M_2) + (1-w_2)/M_1\}$
Mass of solute per 100 g of H_2O: $r_2 = 100 w_2/(1-w_2)$

Here M_2 is the molar mass of the solute and $M_1 = 18.015$ g/mol is the molar mass of water. ρ is the density of the solution in g cm^{-3}.

The data in the table have been derived from the references indicated; in many cases the data have been refitted or interpolated in order to present solubility at rounded values of temperature. Where available, values were taken from the IUPAC *Solubility Data Series* (Reference 1) or the related papers in the *Journal of Physical and Chemical Reference Data* (References 2 to 5), which present carefully evaluated data.

The solubility of sparingly soluble compounds that do not appear in this table may be calculated from the data in the table "Solubility Product Constants". Solubility of inorganic gases may be found in the table "Solubility of Selected Gases in Water".

Compounds are listed alphabetically by chemical formula in the most commonly used form (e.g., NaCl, NH_4NO_3, etc.).

References

1. *Solubility Data Series*, International Union of Pure and Applied Chemistry. Volumes 1 to 53 were published by Pergamon Press, Oxford, from 1979 to 1994; subsequent volumes were published by Oxford University Press, Oxford. The number following the colon is the volume number in the series.
2. Clever, H. L., and Johnston, F. J., *J. Phys. Chem. Ref. Data*, 9, 751, 1980.
3. Marcus, Y., *J. Phys. Chem. Ref. Data*, 9, 1307, 1980.
4. Clever, H. L., Johnson, S. A., and Derrick, M. E., *J. Phys. Chem. Ref. Data*, 14, 631, 1985.
5. Clever, H. L., Johnson, S. A., and Derrick, M. E., *J. Phys. Chem. Ref. Data*, 21, 941, 1992.
6. Söhnel, O., and Novotny, P., *Densities of Aqueous Solutions of Inorganic Substances*, Elsevier, Amsterdam, 1985.
7. Krumgalz, B.S., *Mineral Solubility in Water at Various Temperatures*, Israel Oceanographic and Limnological Research Ltd., Haifa, 1994.
8. Potter, R. W., and Clynne, M. A., *J. Research U.S. Geological Survey*, 6, 701, 1978; Clynne, M. A., and Potter, R. W., *J. Chem. Eng. Data*, 24, 338, 1979.
9. Marshal, W. L., and Slusher, R., *J. Phys. Chem.*, 70, 4015, 1966; Knacke, O., and Gans, W., *Zeit. Phys. Chem.*, NF, 104, 41, 1977.
10. Stephen, H., and Stephen, T., *Solubilities of Inorganic and Organic Compounds, Vol. 1*, Macmillan, New York, 1963.

Compound	0°C	10°C	20°C	25°C	30°C	40°C	50°C	60°C	70°C	80°C	90°C	100°C	Ref.
$AgBrO_3$				0.193							1.32		7
$AgClO_2$	0.17	0.31	0.47	0.55	0.64	0.82	1.02	1.22	1.44	1.66	1.88	2.11	7
$AgClO_3$				15									7
$AgClO_4$	81.6	83.0	84.2	84.8	85.3	86.3	86.9	87.5	87.9	88.3	88.6	88.8	6
$AgNO_2$	0.155			0.413									7
$AgNO_3$	55.9	62.3	67.8	70.1	72.3	76.1	79.2	81.7	83.8	85.4	86.7	87.8	6
Ag_2SO_4	0.56	0.67	0.78	0.83	0.88	0.97	1.05	1.13	1.20	1.26	1.32	1.39	7
$AlCl_3$	30.84	30.91	31.03	31.10	31.18	31.37	31.60	31.87	32.17	32.51	32.90	33.32	7
$Al(ClO_4)_3$	54.9										64.4		7
AlF_3	0.25	0.34	0.44	0.50	0.56	0.68	0.81	0.96	1.11	1.28	1.45	1.64	7
$Al(NO_3)_3$	37.0	38.2	39.9	40.8	42.0	44.5	47.3	50.4	53.8*			61.5*	6
$Al_2(SO_4)_3$	27.5			27.8	28.2	29.2	30.7	32.6	34.9	37.6	40.7	44.2	7
As_2O_3	1.19	1.48	1.80	2.01	2.27	2.86	3.43	4.11	4.89	5.77	6.72	7.71	10
$BaBr_2$	47.6	48.5	49.5	50.0	50.4	51.4	52.5	53.5	54.5	55.5	56.6	57.6	6
$Ba(BrO_3)_2$	0.285	0.442	0.656	0.788	0.935	1.30	1.74	2.27	2.90	3.61	4.40	5.25	1:14
$Ba(C_2H_3O_2)_2$	37.0			44.2									7
$BaCl_2$	23.30	24.88	26.33	27.03	27.70	29.00	30.27	31.53	32.81	34.14	35.54	37.05	8
$Ba(ClO_2)_2$	30.5			31.3							44.7		7
$Ba(ClO_3)_2$	16.90	21.23	23.66	27.50	29.43	33.16	36.69	40.05	43.04	45.90	48.70	51.17	1:14
$Ba(ClO_4)_2$	67.30	70.96	74.30	75.75	77.05	79.23	80.92	82.21	83.16	83.88	84.43	84.90	7
BaF_2		0.158		0.161									7
BaI_2	62.5	64.7	67.3	68.8	69.1	69.5	70.1	70.7	71.3	72.0	72.7	73.4	6

Compound	0°C	10°C	20°C	25°C	30°C	40°C	50°C	60°C	70°C	80°C	90°C	100°C	Ref.
$Ba(IO_3)_2$	0.0182	0.0262	0.0342	0.0396	0.045*	0.058*	0.073	0.090	0.109	0.131	0.156	0.182	1:14
$Ba(NO_2)_2$	31.1	36.6	41.8	44.3	46.8	51.6	56.2	60.5	64.6	68.5	72.1	75.6	10
$Ba(NO_3)_2$	4.7	6.3	8.2	9.3	10.2	12.4	14.7	17.0	19.3	21.5	23.5	25.5	6
$Ba(OH)_2$	1.67			4.68	8.4	19	33	52	74	100			7
BaS	2.79	4.78	6.97	8.21	9.58	12.67	16.18	20.05	24.19	28.55	33.04	37.61	7
$Ba(SCN)_2$				62.6									7
$BaSO_3$				0.0011									1:26
$BeCl_2$	40.5			41.7									7
$Be(ClO_4)_2$				59.5									7
$BeSO_4$	26.69	27.58	28.61	29.22	29.90	31.51	33.39	35.50	37.78	40.21	42.72	45.28	7
$CaBr_2$	55	56	59	61	63	68	71	73					10
$CaCl_2$	36.70	39.19	42.13	44.83*	49.12*	52.85*	56.05*	56.73	57.44	58.21	59.04	59.94	8
$Ca(ClO_3)_2$	63.2	64.2	65.5	66.3	67.2	69.0	71.0	73.2	75.5*	77.4*	77.7	78.0	1:14
$Ca(ClO_4)_2$				65.3									7
CaF_2	0.0013			0.0016									10
CaI_2	64.6	66.0	67.6	68.3	69.0	70.8	72.4	74.0	76.0	78.0	79.6	81.0	7
$Ca(IO_3)_2$	0.082	0.155	0.243	0.305	0.384*	0.517*	0.590	0.652	0.811*	0.665*	0.668		1:14
$Ca(NO_2)_2$	38.6	39.5	44.5	48.6									7
$Ca(NO_3)_2$	50.1	53.1	56.7	59.0	60.9	65.4	77.8	78.1	78.2	78.3	78.4	78.5	6
$CaSO_3$			0.0059	0.0054	0.0049	0.0041	0.0035	0.0030	0.0026	0.0023	0.0020	0.0019	1:26
$CaSO_4$	0.174	0.191	0.202	0.205	0.208	0.210	0.207	0.201	0.193	0.184	0.173	0.163	9
$CdBr_2$	36.0	43.0	49.9	53.4	56.4	60.3*	60.3*	60.5	60.7	60.9	61.3	61.6	6
CdC_2O_4				0.0060									5
$CdCl_2$	47.2	50.1	53.2	54.6	56.3*	57.3*	57.5	57.8	58.1	58.51	58.98	59.5	6
$Cd(ClO_4)_2$				58.7								66.9	7
CdF_2		5.82	4.65	4.18	3.76								5
CdI_2	44.1	44.9	45.8	46.3	46.8	47.9	49.0	50.2	51.5	52.7	54.1	55.4	6
$Cd(IO_3)_2$				0.091									5
$Cd(NO_3)_2$	55.4	57.1	59.6	61.0	62.8	66.5	70.6	86.1	86.5	86.8	87.1	87.4	6
$CdSO_4$	43.1	43.1	43.2	43.4	43.6	44.1	43.5	42.5	41.4	40.2	38.5	36.7	6
$CdSeO_4$	42.04	40.59	39.02	38.18	37.29	35.35	33.15	30.65	27.84	24.69	21.24	17.49	5
$Ce(NO_3)_3$	57.99	59.80	61.89	63.05	64.31*	67.0*	68.6	71.1*	74.9*	79.2	80.9	83.1	1:13
$CoCl_2$	30.30	32.60	34.87	35.99	37.10	39.27	41.38	43.46	45.50	47.51	49.51	51.50	7
$Co(ClO_4)_2$	50.0			53.0									7
CoF_2				1.4									7
CoI_2	58.00	61.78	65.35	66.99	68.51	71.17	73.41	75.29	76.89	78.28	79.52	80.70	7
$Co(NO_2)_2$	0.076			0.49									7
$Co(NO_3)_2$	45.5	47.0	49.4	50.8	52.4	56.0	60.1	62.6	64.9	67.7			6
$CoSO_4$	19.9	23.0	26.1	27.7	29.2	32.3	34.4	35.9	35.5	33.2	30.6	27.8	6
$Co(SCN)_2$				50.7									7
CrO_3	62.2	62.3	62.6	62.8	63.0	63.5	64.1	64.7	65.5	66.2	67.1	67.9	6
$CsBr$				55.2									7
$CsBrO_3$	1.16	1.93	3.01	3.69	4.46	6.32	8.60	11.32	14.45	17.96	21.83	25.98	1:30
$CsCl$	61.83	63.48	64.96	65.64	66.29	67.50	68.60	69.61	70.54	71.40	72.21	72.96	1:47
$CsClO_3$	2.40	3.87	5.94	7.22	8.69	12.15	16.33	21.14	26.45	32.10	37.89	43.42	1:30
$CsClO_4$	0.79	1.01	1.51	1.96	2.57	4.28	6.55	9.29	12.41	15.80	19.39	23.07	7
CsI	30.9	37.2	43.2	45.9	48.6	53.3	57.3	60.7	63.6	65.9	67.7	69.2	6
$CsIO_3$	1.08	1.58	2.21	2.59	3.02	3.96	5.06	6.29	7.70	9.20	10.79	12.45	1:30
$CsNO_3$	8.46	13.0	18.6	21.8	25.1	32.0	39.0	45.7	51.9	57.3	62.1	66.2	6
$CsOH$				75									7
Cs_2SO_4	62.6	63.4	64.1	64.5	64.8	65.5	66.1	66.7	67.3	67.8	68.3	68.8	6
$CuBr_2$				55.8									7
$CuCl_2$	40.8	41.7	42.6	43.1	43.7	44.8	46.0	47.2	48.5	49.9	51.3	52.7	6
$Cu(ClO_4)_2$	54.3			59.3									7
CuF_2				0.075									7
$Cu(NO_3)_2$	45.2	49.8	56.3	59.2	61.1	62.0	63.1	64.5	65.9	67.5	69.2	71.0	6
$CuSO_4$	12.4	14.4	16.7	18.0	19.3	22.2	25.4	28.8	32.4	36.3	40.3	43.5	6
$CuSeO_4$	10.6			16.0									7
$Dy(NO_3)_3$	58.79	59.99	61.49	62.35	63.29	65.43	68.04	71.58					1:13
$Er(NO_3)_3$	61.58	63.15	64.84	65.75	66.69	68.70	70.96	73.64	77.75				1:13
$Eu(NO_3)_3$	55.2	56.7	58.5	59.4	60.4	62.5	64.6						1:13

Compound	0°C	10°C	20°C	25°C	30°C	40°C	50°C	60°C	70°C	80°C	90°C	100°C	Ref.
$FeBr_2$				54.6								64.8*	7
$FeCl_2$	33.2*			39.4*								48.7*	7
$FeCl_3$	42.7	44.9	47.9	47.7	51.6	74.8	76.7	84.6	84.3	84.3	84.4	84.7	6
$Fe(ClO_4)_2$	63.39			67.76									7
FeF_3				5.59									7
$Fe(NO_3)_3$	40.15			46.57									7
$Fe(NO_3)_2$	41.44			46.67									7
$FeSO_4$	13.5	17.0	20.8	22.8	24.8	28.8	32.8	35.5	33.6	30.4	27.1	24.0	6
$Gd(NO_3)_3$	56.3	57.7	59.2	60.1	61.0	62.9	65.2	67.9	71.5				1:13
HIO_3	73.45	74.10	74.98	75.48	76.03	77.20	78.46	79.78	81.13	82.48	83.82	85.14	1:30
H_3BO_3	2.61	3.57	4.77	5.48	6.27	8.10	10.3	12.9	15.9	19.3	23.1	27.3	6
$HgBr_2$	0.26	0.37	0.52	0.61	0.72	0.96	1.26	1.63	2.08	2.61	3.23	3.95	4
$Hg(CN)_2$	6.57	7.83	9.33	10.2	11.1	13.1	15.5	18.2	21.2	24.6	28.3	32.3	6
$HgCl_2$	4.24	5.05	6.17	6.81	7.62	9.53	12.02	15.18	19.16	24.06	29.90	36.62	4
HgI_2			0.0041	0.0055	0.0072	0.0122	0.0199						4
$Hg(SCN)_2$				0.070									4
Hg_2Cl_2				0.0004									3
$Hg_2(ClO_4)_2$	73.8			79.8*								85.3*	7
Hg_2SO_4	0.038	0.043	0.048	0.051	0.054	0.059	0.065	0.070	0.076	0.082	0.088	0.093	4
$Ho(NO_3)_3$				63.8									1:13
KBF_4	0.28	0.34	0.45	0.55	0.75	1.38	2.09	2.82	3.58	4.34	5.12	5.90	10
KBr	35.0	37.3	39.4	40.4	41.4	43.2	44.8	46.2	47.6	48.8	49.8	50.8	6
$KBrO_3$	2.97	4.48	6.42	7.55	8.79	11.57	14.71	18.14	21.79	25.57	29.42	33.28	1:30
$KC_2H_3O_2$	68.40	70.29	72.09	72.92	73.70	75.08	76.27	77.31	78.22	79.04	79.80	80.55	7
KCl	21.74	23.61	25.39	26.22	27.04	28.59	30.04	31.40	32.66	33.86	34.99	36.05	1:47
$KClO_3$	3.03	4.67	6.74	7.93	9.21	12.06	15.26	18.78	22.65	26.88	31.53	36.65	1:30
$KClO_4$	0.70	1.10	1.67	2.04	2.47	3.54	4.94	6.74	8.99	11.71	14.94	18.67	6
KF	30.90	39.8	47.3	50.41	53.2				60.0				7
$KHCO_3$	18.62	21.73	24.92	26.6	28.13	31.32	34.46	37.51	40.45				6
$KHSO_4$	27.1	29.7	32.3	33.6	35.0	37.8	40.5	43.4	46.2	49.02	51.82	54.6	6
KH_2PO_4	11.74	14.91	18.25	19.97	21.77	25.28	28.95	32.76	36.75	40.96	45.41	50.12	1:31
KI	56.0	57.6	59.0	59.7	60.4	61.6	62.8	63.8	64.8	65.7	66.6	67.4	6
KIO_3	4.53	5.96	7.57	8.44	9.34	11.09	13.22	15.29	17.41	19.58	21.78	24.03	1:30
KIO_4	0.16	0.22	0.37	0.51	0.70	1.24	1.96	2.83	3.82	4.89	6.02	7.17	7
$KMnO_4$	2.74	4.12	5.96	7.06	8.28	11.11	14.42	18.16					6
KNO_2	73.7	74.6	75.3	75.7	76.0	76.7	77.4	78.0	78.5	79.1	79.6	80.1	6
KNO_3	12.0	17.6	24.2	27.7	31.3	38.6	45.7	52.2	58.0	63.0	67.3	70.8	6
KOH	48.7	50.8	53.2	54.7	56.1	57.9	58.6	59.5	60.6	61.8	63.1	64.6	6
$KSCN$	63.8	66.4	69.1	70.4	71.6	74.1	76.5	78.9	81.1	83.3	85.3	87.3	6
K_2CO_3	51.3	51.7	52.3	52.7	53.1	54.0	54.9	56.0	57.2	58.4	59.6	61.0	6
K_2CrO_4	37.1	38.1	38.9	39.4	39.8	40.5	41.3	41.9	42.6	43.2	43.8	44.3	6
$K_2Cr_2O_7$	4.30	7.12	10.9	13.1	15.5	20.8	26.3	31.7	36.9	41.5	45.5	48.9	6
K_2HAsO_4	48.5*			63.6*								79.8*	7
K_2HPO_4	57.0	59.1	61.5	62.7	64.1	67.7*		72.7*					1:31
K_2MoO_4				64.7							66.5		7
K_2SO_3	51.30	51.39	51.49	51.55	51.62	51.76	51.93	52.11	52.32	52.54	52.79	53.06	1:26
K_2SO_4	7.11	8.46	9.95	10.7	11.4	12.9	14.2	15.5	16.7	17.7	18.6	19.3	6
$K_2S_2O_3$	49.0*			62.3*							75.7*		7
$K_2S_2O_5$	22.1	26.7	31.1	33.1	35.2	39.0	42.6	46.0	49.1	52.0	54.6		1:26
K_2SeO_3	68.4*			68.5*								68.5*	7
K_2SeO_4	52.70	52.93	53.17	53.30	53.43	53.70	53.99	54.30	54.61	54.94	55.26	55.60	7
K_3AsO_4	51.5*			55.6*								73*	7
$K_3Fe(CN)_6$	23.9	27.6	31.1	32.8	34.3	37.2	39.6	41.7	43.5	45.0	46.1	47.0	6
K_3PO_4	44.3			51.4									7
$K_4Fe(CN)_6$	12.5	17.3	22.0	23.9	25.6	29.2	32.5	35.5	38.2	40.6	41.4	43.1	6
$LaCl_3$	49.0	48.5	48.6	48.9	49.3	50.5	52.1	54.0	56.3	58.9	61.7		6
$La(NO_3)_3$	55.0	56.9	58.9	60.0	61.1	63.6	66.3	69.9*	74.1*				1:13
$LiBr$	58.4	60.1	62.7	64.4	65.9	67.8	68.3	69.0	69.8	70.7	71.7	72.8	6
$LiBrO_3$	61.03	62.62	64.44	65.44	66.51	68.90	71.68*	73.24*	74.43	75.66	76.93	78.32	1:30
$LiC_2H_3O_2$	23.76	26.49	29.42	31.02	32.72	36.48	40.65	45.15	49.93	54.91	60.04	65.26	7
$LiCl$	40.45	42.46*	45.29*	45.81	46.25	47.30	48.47	49.78	51.27	52.98	54.98*	56.34*	1:47

Compound	0°C	10°C	20°C	25°C	30°C	40°C	50°C	60°C	70°C	80°C	90°C	100°C	Ref.
$LiClO_3$	73.2	75.6*	80.8*	82.1	83.4	85.9*	87.1*	88.2	89.6	91.3	93.4	95.7	1:30
$LiClO_4$	30.1	32.6	35.5	37.0	38.6	41.9	45.5	49.2	53.2	57.2	61.3	71.4	6
LiF	0.120	0.126	0.131	0.134									7
LiH_2PO_4	55.8												7
LiI	59.4	60.5	61.7	62.3	63.0	64.3	65.8	67.3	68.8	81.3	81.7	82.6	6
$LiIO_3$				43.8									1:30
$LiNO_2$	41	45	49	51	53	56	60	63	66	68			10
$LiNO_3$	34.8	37.6	42.7	50.5	57.9	60.1	62.2	64.0	65.7	67.2	68.5	69.7	6
$LiOH$	10.8	10.8	11.0	11.1	11.3	11.7	12.2	12.7	13.4	14.2	15.1	16.1	6
$LiSCN$				54.5									7
Li_2CO_3	1.54	1.43	1.33	1.28	1.24	1.15	1.07	0.99	0.92	0.85	0.78	0.72	7
$Li_2C_2O_4$				5.87									7
Li_2HPO_3	9.07	8.40	7.77	7.47	7.18	6.64	6.16	5.71	5.30	4.91	4.53	4.16	7
Li_2SO_4	26.3	25.9	25.6	25.5	25.3	25.0	24.8	24.5	24.3	24.0	23.8	23.6	6
Li_3PO_4				0.027									1:31
$Lu(NO_3)_3$				71.1									1:13
$MgBr_2$	49.3	49.8	50.3	50.6	50.9	51.5	52.1	52.8	53.5	54.2	55.0	55.7	6
$Mg(BrO_3)_2$	43.0	45.2	48.0	49.4	51.0	54.3	57.9	61.6	65.3	69.0*	70.9*	71.7	1:14
$Mg(C_2H_3O_2)_2$	36.18	37.55	38.92	39.61									7
MgC_2O_4				0.038									7
$MgCl_2$	33.96	34.85	35.58	35.90	36.20	36.77	37.34	37.97	38.71	39.62	40.75	42.15	8
$Mg(ClO_3)_2$	53.35	54.40	56.81	58.66	60.91*	65.46*	67.33	69.27	71.01	72.44	73.48		1:14
$Mg(ClO_4)_2$	47.8	48.7	49.6	50.1	50.5	51.3	52.1						6
$MgCrO_4$	32.06*			35.39*									7
$MgCr_2O_7$				58.9						67.0			7
MgF_2				0.013									7
MgI_2	54.7	56.1	58.2	59.4	60.8	63.9	65.0	65.0	65.0	65.0	65.1	65.2	6
$Mg(IO_3)_2$	3.19*	6.70*	7.92	8.52	9.11	10.45	11.99	13.7	15.6	17.6	19.6		1:14
$Mg(NO_2)_2$				47									7
$Mg(NO_3)_2$	38.4	39.5	40.8	41.6	42.4	44.1	45.9	47.9	50.0	52.2	70.6	72.0	6
$MgSO_3$	0.32	0.37	0.46	0.52	0.61	0.87*	0.85*	0.76	0.69	0.64	0.62	0.60	1:26
$MgSO_4$	18.2	21.7	25.1	26.3	28.2	30.9	33.4	35.6	36.9	35.9	34.7	33.3	6
MgS_2O_3	30.7			34.1									7
$MgSeO_4$	31.4*			35.7*							47*		7
$MnBr_2$	56.00	57.72	59.39	60.19	60.96	62.41	63.75	65.01	66.19	67.32	68.42	69.50	7
$MnCl_2$	38.7	40.6	42.5	43.6	44.7	47.0	49.4	54.1	54.7	55.2	55.7	56.1	6
MnF_2	0.80*			1.01*								0.48	7
$Mn(IO_3)_2$				0.27							0.34		7
$Mn(NO_3)_2$	50.5			61.7									7
$MnSO_4$	34.6	37.3	38.6	38.9	38.9	37.7	36.3	34.6	32.8	30.8	28.8	26.7	6
NH_4Br	37.5	40.2	42.7	43.9	45.1	47.3	49.4	51.3	53.0	54.6	56.1	57.4	7
NH_4Cl	22.92	25.12	27.27	28.34	29.39	31.46	33.50	35.49	37.46	39.40	41.33	43.24	1:47
NH_4ClO_4	10.8	14.1	17.8	19.7	21.7	25.8	29.8	33.6	37.3	40.7	43.8	46.6	6
NH_4F	41.7	43.2	44.7	45.5	46.3	47.8	49.3	50.9	52.5	54.1			7
NH_4HCO_3	10.6	13.7	17.6	19.9	22.4	27.9	34.2	41.4	49.3	58.1	67.6	78.0	7
$NH_4H_2AsO_4$	25.2	29.0	32.7	34.5	36.3	39.7	43.1	46.2	49.3	52.2	55.0		7
$NH_4H_2PO_4$	17.8	22.0	26.4	28.8	31.2	36.2	41.6	47.2	53.0	59.2	65.7	72.4	7
NH_4I	60.7	62.1	63.4	64.0	64.6	65.8	66.8	67.8	68.7	69.6	70.4	71.1	6
NH_4IO_3				3.70	4.20	5.64	7.63						1:30
NH_4NO_2	55.7	59.0	64.9	68.8									7
NH_4NO_3	54.0	60.1	65.5	68.0	70.3	74.3	77.7	80.8	83.4	85.8	88.2	90.3	6
NH_4SCN				64.4						81.1			7
$(NH_4)_2C_2O_4$	2.31	3.11	4.25	4.94	5.73	7.56	9.73	12.2	15.1	18.3	21.8	25.7	7
$(NH_4)_2HPO_4$	36.4	38.2	40.0	41.0	42.0	44.1	46.2	48.5	50.9	53.3	55.9	58.6	7
$(NH_4)_2S_2O_5$	65.5	67.9	69.8	70.5	71.3	72.3	72.9	73.1					1:26
$(NH_4)_2S_2O_8$	37.00	40.45	43.84	45.49	47.11	50.25	53.28	56.23	59.13	62.00			7
$(NH_4)_2SO_3$	32.2	34.9	37.7	39.1	40.6	43.7	47.0	50.6	54.5	58.9			1:26
$(NH_4)_2SO_4$	41.3	42.1	42.9	43.3	43.8	44.7	45.6	46.6	47.5	48.5	49.5	50.5	6
$(NH_4)_2SeO_3$	49.0	51.1	53.4	54.7	56.0	58.9	62.0	65.4	69.1				7
$(NH_4)_2SeO_4$				54.02									7
$(NH_4)_3PO_4$				15.5									7

Compound	0°C	10°C	20°C	25°C	30°C	40°C	50°C	60°C	70°C	80°C	90°C	100°C	Ref.
NaBr	44.4	45.9	47.7	48.6	49.6	51.6	53.7	54.1	54.3	54.5	54.7	54.9	6
NaBrO$_3$	20.0	23.22	26.65	28.28	29.86	32.83	35.55	38.05	40.37	42.52			1:30
NaCHO$_2$	30.8	37.9	45.7	48.7	50.6	52.0	53.5	55.0					6
NaC$_2$H$_3$O$_2$	26.5	28.8	31.8	33.5	35.5	39.9	45.1	58.3	59.3	60.5	61.7	62.9	6
NaCl	26.28	26.32	26.41	26.45	26.52	26.67	26.84	27.03	27.25	27.50	27.78	28.05	1:47
NaClO	22.7			44.4									7
NaClO$_2$				97.0*				95.3*					7
NaClO$_3$	44.27	46.67	49.3	50.1	51.2	53.6	55.5	57.0	58.5	60.5	63.3	67.1	1:30
NaClO$_4$	61.9	64.1	66.2	67.2	68.3	70.4	72.5	74.1	74.7	75.4	76.1	76.7	6
NaF	3.52	3.72	3.89	3.97	4.05	4.20	4.34	4.46	4.57	4.66	4.75	4.82	6
NaHCO$_3$	6.48	7.59	8.73	9.32	9.91	11.13	12.40	13.70	15.02	16.37	17.73	19.10	7
NaHSO$_4$				22.2								33.3	10
NaH$_2$PO$_4$	36.54	41.07	46.00	48.68	51.54	57.89*	61.7*	62.3*	65.9	68.7			1:31
NaI	61.2	62.4	63.9	64.8	65.7	67.7	69.8	72.0	74.7	74.8	74.9	75.1	6
NaIO$_3$	2.43	4.40	7.78*	8.65*	9.60	11.67	13.99	16.52	19.25*	21.1*	22.9	24.7	1:30
NaIO$_4$				12.62									7
NaNO$_2$	41.9	43.4	45.1	45.9	46.8	48.7	50.7	52.8	55.0	57.2	59.5	61.8	6
NaNO$_3$	42.2	44.4	46.6	47.7	48.8	51.0	53.2	55.3	57.5	59.6	61.7	63.8	6
NaOH	30	39	46	50	53	58	63	67	71	74	76	79	10
NaSCN		52.9	57.1	60.2	62.7	63.5	64.2	65.0	65.9	66.9	67.9	69.0	6
Na$_2$B$_4$O$_7$	1.23	1.71	2.50	3.07	3.82	6.02	9.7	14.9	17.1	19.9	23.5	28.0	6
Na$_2$CO$_3$	6.44	10.8	17.9	23.5	28.7	32.8	32.2	31.7	31.3	31.1	30.9	30.9	6
Na$_2$C$_2$O$_4$	2.62	2.95	3.30	3.48	3.65	4.00	4.36	4.71	5.06	5.41	5.75	6.08	6
Na$_2$CrO$_4$	22.6	32.3	44.6	46.7	46.9	48.9	51.0	53.4	55.3	55.5	55.8	56.1	6
Na$_2$Cr$_2$O$_7$	62.1	63.1	64.4	65.2	66.1	68.0	70.1	72.3	74.6	77.0	79.6	80.7	6
Na$_2$HAsO$_4$	5.6*			29.3*								67*	7
Na$_2$HPO$_4$	1.66	4.19	7.51	10.55	16.34*	35.17*	44.64*	45.20	46.81	48.78	50.52	51.53	1:31
Na$_2$MoO$_4$	30.6	38.8	39.4	39.4	39.8	40.3	41.0	41.7	42.6	43.5	44.5	45.5	6
Na$_2$S	11.1	13.2	15.7	17.1	18.6	22.1	26.7	28.1	30.2	33.0	36.4	41.0	6
Na$_2$SO$_3$	12.0	16.1	20.9	23.5	26.3*	27.3*	25.9	24.8	23.7	22.8	22.1	21.5	1:26
Na$_2$SO$_4$			16.13	21.94	29.22*	32.35*	31.55	30.90	30.39	30.02	29.79	29.67	8
Na$_2$S$_2$O$_3$	33.1	36.3	40.6	43.3	45.9	52.0	62.3	65.7	68.8	69.4	70.1	71.0	6
Na$_2$S$_2$O$_5$		38.4	39.5	40.0	40.6	41.8	43.0	44.2	45.5	46.8	48.1	49.5	1:26
Na$_2$SeO$_3$				47.3*								45*	7
Na$_2$SeO$_4$	11.7			36.9*								42.1*	7
Na$_2$WO$_4$	41.6	41.9	42.3	42.6	42.9	43.6	44.4	45.3	46.2	47.3	48.4	49.5	6
Na$_3$PO$_4$	4.28	7.30	10.8	12.6	14.1	16.6	22.9	28.4	32.4	37.6	40.4	43.5	6
Na$_4$P$_2$O$_7$	2.23	3.28	4.81	6.62	7.00	10.10	14.38	20.07	27.31	36.03	32.37	30.67	6
NdCl$_3$	49.0	49.3	49.7	50.0	50.4	51.2	52.2	53.3	54.5	55.8	57.1	58.5	6
Nd(NO$_3$)$_3$	55.76	57.49	59.37	60.38	61.43	63.69	66.27	69.47					1:13
NiCl$_2$	34.7	36.1	38.5	40.3	41.7	42.1	43.2	45.0	46.1	46.2	46.4	46.6	6
Ni(ClO$_4$)$_2$	51.1			52.8									7
NiF$_2$				2.50							2.52		7
NiI$_2$	55.40	57.68	59.78	60.69	61.50	62.80	63.73	64.38	64.80	65.09	65.30		7
Ni(NO$_3$)$_2$	44.1	46.0	48.4	49.8	51.3	54.6	58.3	61.0	63.1	65.6	67.9	69.0	6
NiSO$_4$	21.4	24.4	27.4	28.8	30.3*	32.0*	34.1	35.8	37.7	39.9	42.3	44.8	6
Ni(SCN)$_2$				35.48									7
NiSeO$_4$	21.6		26.2*								45.6*		7
PbBr$_2$	0.449	0.620	0.841	0.966	1.118	1.46	1.89						2
PbCl$_2$	0.66	0.81	0.98	1.07	1.17	1.39	1.64	1.93	2.24	2.60	2.99	3.42	2
Pb(ClO$_4$)$_2$				81.5									7
PbF$_2$		0.0603	0.0649	0.0670	0.0693								2
PbI$_2$	0.041	0.052	0.067	0.076	0.086	0.112	0.144	0.187	0.243	0.315			2
Pb(IO$_3$)$_2$				0.0025									7
Pb(NO$_3$)$_2$	28.46	32.13	35.67	37.38	39.05	42.22	45.17	47.90	50.42	52.72	54.82	56.75	2
PbSO$_4$	0.0033	0.0038	0.0042	0.0044	0.0047	0.0052	0.0058						2
PrCl$_3$	48.0	48.1	48.6	49.0	49.5	50.8	52.3	54.1	56.1	58.3			6
Pr(NO$_3$)$_3$	57.50	59.20	61.16	62.24	63.40*	65.7*	67.8	70.2	73.4				1:13
RbBr	47.4	50.1	52.6	53.8	54.9	57.0	58.8	60.6	62.1	63.5	64.8	65.9	6
RbBrO$_3$	0.97	1.55	2.36	2.87	3.45	4.87	6.64	8.78	11.29	14.15	17.32	20.76	1:30
RbCl	43.58	45.65	47.53	48.42	49.27	50.86	52.34	53.67	54.92	56.08	57.16	58.15	1:47

Compound	0°C	10°C	20°C	25°C	30°C	40°C	50°C	60°C	70°C	80°C	90°C	100°C	Ref.
$RbClO_3$	2.10	3.38	5.14	6.22	7.45	10.35	13.85	17.93	22.53	27.57	32.96	38.60	1:30
$RbClO_4$	1			1.5								17	7
RbF			75										7
$RbHCO_3$			53.7										7
RbI	55.8	58.6	61.1	62.3	63.4	65.4	67.2	68.8	70.3	71.6	72.7	73.8	6
$RbIO_3$	1.09	1.53	2.07	2.38	2.74	3.52	4.41	5.42	6.52	7.74	9.00	10.36	1:30
$RbNO_3$	16.4	25.0	34.6	39.4	44.2	53.1	60.8	67.2	72.2	76.1	79.0	81.2	6
$RbOH$					63.4								7
Rb_2CrO_4	38.27			43.26									7
Rb_2SO_4	27.3	30.0	32.5	33.7	34.8	36.9	38.7	40.3	41.8	43.0	44.1	44.9	6
$SbCl_3$	85.7			90.8									7
SbF_3	79.4			83.1									7
$Sc(NO_3)_3$	57.0	59.3	61.6	62.8	63.9	66.2	68.5						1:13
$Sm(NO_3)_3$	54.83	56.33	58.08	59.05	60.08	62.38	65.05*	68.1*	70.8	74.2			1:13
$SmCl_3$		48.0	48.2	48.4	48.6	49.2	50.0						6
$SnCl_2$	46	64											7
SnI_2			0.97									3.87	7
$SrBr_2$	46.0	48.3	50.6	51.7	52.9	55.2	57.6	59.9	62.3	64.6	66.8	69.0	6
$Sr(BrO_3)_2$	18.53	22.00	25.39	27.02	28.59	31.55	34.21	36.57	38.64*	40.2*	40.8	41.0	1:14
$SrCl_2$	31.94	32.93	34.43	35.37	36.43	38.93	41.94	45.44*	46.81*	47.69	48.70	49.87	8
$Sr(ClO_2)_2$	13.0	13.6	14.1	14.3	14.5	14.9	15.3	15.6	15.9				7
$Sr(ClO_3)_2$	63.29	63.42	63.64	63.77	63.93	64.29	64.70	65.16	65.65	66.18	66.74	67.31	1:14
$Sr(ClO_4)_2$	70.04*			75.35*		78.44*							7
SrF_2	0.011			0.021									7
SrI_2	62.5	62.8	63.5	63.9	64.5	65.8	67.3	69.0	70.8	72.7	74.7	79.2	6
$Sr(IO_3)_2$	0.102	0.126	0.152	0.165	0.179	0.206	0.233	0.259	0.284	0.307	0.328	0.346	1:14
$Sr(MnO_4)_2$	2.5												7
$Sr(NO_2)_2$					41.9	44.3						58.6	7
$Sr(NO_3)_2$	28.2	34.6	41.0	44.5	47.0	47.4	47.9	48.4	48.9	49.5	50.1	50.7	6
$Sr(OH)_2$	0.9			2.2									7
$SrSO_3$				0.0015									1:26
$SrSO_4$				0.0135									7
SrS_2O_3	8.8	13.2	17.7	20.0	22.2	26.8							7
$Tb(NO_3)_3$			60.6	61.02									1:13
Tl_2SO_4	2.65	3.56	4.61	5.19	5.80	7.09	8.46	9.89	11.33	12.77	14.18	15.53	6
$Tm(NO_3)_3$				67.9									1:13
$UO_2(NO_3)_2$	49.52	51.82	54.42	55.85	57.55	61.59	67.07						1:55
$Y(NO_3)_3$	55.57	56.93	58.75	59.86	61.11*	63.3*	64.9	67.9	72.5				1:13
$Yb(NO_3)_3$				70.5									1:13
$ZnBr_2$	79.3	80.1	81.8	83.0	84.1	85.6	85.8	86.1	86.3	86.6	86.8	87.1	6
ZnC_2O_4		0.0010	0.0019	0.0026									5
$ZnCl_2$		76.6	79.0	80.3	81.4	81.8	82.4	83.0	83.7	84.4	85.2	86.0	6
$Zn(ClO_4)_2$	44.29*			46.27*			48.70						7
ZnF_2				1.53									5
ZnI_2	81.1	81.2	81.3	81.4	81.5	81.7	82.0	82.3	82.6	83.0	83.3	83.7	6
$Zn(IO_3)_2$		0.58	0.64	0.69	0.77	0.82							5
$Zn(NO_3)_2$	47.8	50.8	54.4	54.6	58.5	79.1	80.1	87.5	89.9				6
$ZnSO_3$			0.1786	0.1790	0.1794	0.1803	0.1812						5
$ZnSO_4$	29.1	32.0	35.0	36.6	38.2	41.3	43.0	42.1	41.0	39.9	38.8	37.6	6
$ZnSeO_4$	33.06	34.98	37.38	38.79	40.34								5

SOLUBILITY PRODUCT CONSTANTS

The solubility product constant K_{sp} is a useful parameter for calculating the aqueous solubility of sparingly soluble compounds under various conditions. It may be determined by direct measurement or calculated from the standard Gibbs energies of formation $\Delta_f G°$ of the species involved at their standard states. Thus if $K_{sp} = [M^+]^m$, $[A^-]^n$ is the equilibrium constant for the reaction

$$M_m A_n(s) \rightleftharpoons mM^+(aq) + nA^-(aq),$$

where $M_m A_n$ is the slightly soluble substance and M^+ and A^- are the ions produced in solution by the dissociation of $M_m A_n$, then the Gibbs energy change is

$$\Delta G° = m \, \Delta_f G° \, (M^+, aq) + n \, \Delta_f G° \, (A^-, aq) - \Delta_f G° \, (M_m A_n, s)$$

The solubility product constant is calculated from the equation

$$\ln K_{sp} = -\Delta G°/RT$$

The first table below gives selected values of K_{sp} at 25°C. Many of these have been calculated from standard state thermodynamic data in References 1 and 2; other values are taken from publications of the IUPAC Solubility Data Project (References 3 to 7).

The above formulation is not convenient for treating sulfides because the S^{-2} ion is usually not present in significant concentrations (see Reference 8). This is due to the hydrolysis reaction

$$S^{-2} + H_2O \rightleftharpoons HS^- + OH^-$$

which is strongly shifted to the right except in very basic solutions. Furthermore, the equilibrium constant for this reaction, which depends on the second ionization constant of H_2S, is poorly known. Therefore it is more useful in the case of sulfides to define a different solubility product K_{spa} based on the reaction

$$M_m S_n(s) + 2H^+ \rightleftharpoons mM^+ + nH_2S \, (aq)$$

Values of K_{spa}, taken from Reference 8, are given for several sulfides in the auxiliary table following the main table. Additional discussion of sulfide equilibria may be found in References 7 and 9.

References

1. Wagman, D. D., Evans, W. H., Parker, V. B., Schumm, R. H., Halow, I., Bailey, S. M., Churney, K. L., and Nuttall, R. L., *The NBS Tables of Chemical Thermodynamic Properties*, J. Phys. Chem. Ref. Data, Vol. 11, Suppl. 2, 1982.
2. Garvin, D., Parker, V. B., and White, H. J., *CODATA Thermodynamic Tables*, Hemisphere, New York, 1987.
3. *Solubility Data Series* (53 Volumes), International Union of Pure and Applied Chemistry, Pergamon Press, Oxford, 1979–1992.
4. Clever, H. L., and Johnston, F. J., *J. Phys. Chem. Ref. Data*, 9, 751, 1980.
5. Marcus, Y., *J. Phys. Chem. Ref. Data*, 9, 1307, 1980.
6. Clever, H. L., Johnson, S. A., and Derrick, M. E., *J. Phys. Chem. Ref. Data*, 14, 631, 1985.
7. Clever, H. L., Johnson, S. A., and Derrick, M. E., *J. Phys. Chem. Ref. Data*, 21, 941, 1992.
8. Myers, R. J., *J. Chem. Educ.*, 63, 687, 1986.
9. Licht, S., *J. Electrochem. Soc.*, 135, 2971, 1988.

Compound	Formula	K_{sp}
Aluminum phosphate	$AlPO_4$	$9.84 \cdot 10^{-21}$
Barium bromate	$Ba(BrO_3)_2$	$2.43 \cdot 10^{-4}$
Barium carbonate	$BaCO_3$	$2.58 \cdot 10^{-9}$
Barium chromate	$BaCrO_4$	$1.17 \cdot 10^{-10}$
Barium fluoride	BaF_2	$1.84 \cdot 10^{-7}$
Barium hydroxide octahydrate	$Ba(OH)_2 \cdot 8H_2O$	$2.55 \cdot 10^{-4}$
Barium iodate	$Ba(IO_3)_2$	$4.01 \cdot 10^{-9}$
Barium iodate monohydrate	$Ba(IO_3)_2 \cdot H_2O$	$1.67 \cdot 10^{-9}$
Barium molybdate	$BaMoO_4$	$3.54 \cdot 10^{-8}$
Barium nitrate	$Ba(NO_3)_2$	$4.64 \cdot 10^{-3}$
Barium selenate	$BaSeO_4$	$3.40 \cdot 10^{-8}$
Barium sulfate	$BaSO_4$	$1.08 \cdot 10^{-10}$
Barium sulfite	$BaSO_3$	$5.0 \cdot 10^{-10}$
Beryllium hydroxide	$Be(OH)_2$	$6.92 \cdot 10^{-22}$
Bismuth arsenate	$BiAsO_4$	$4.43 \cdot 10^{-10}$
Bismuth iodide	BiI_3	$7.71 \cdot 10^{-19}$
Cadmium arsenate	$Cd_3(AsO_4)_2$	$2.2 \cdot 10^{-33}$
Cadmium carbonate	$CdCO_3$	$1.0 \cdot 10^{-12}$
Cadmium fluoride	CdF_2	$6.44 \cdot 10^{-3}$
Cadmium hydroxide	$Cd(OH)_2$	$7.2 \cdot 10^{-15}$
Cadmium iodate	$Cd(IO_3)_2$	$2.5 \cdot 10^{-8}$
Cadmium oxalate trihydrate	$CdC_2O_4 \cdot 3H_2O$	$1.42 \cdot 10^{-8}$
Cadmium phosphate	$Cd_3(PO_4)_2$	$2.53 \cdot 10^{-33}$
Calcium carbonate (calcite)	$CaCO_3$	$3.36 \cdot 10^{-9}$
Calcium fluoride	CaF_2	$3.45 \cdot 10^{-11}$
Calcium hydroxide	$Ca(OH)_2$	$5.02 \cdot 10^{-6}$

Compound	Formula	K_{sp}
Calcium iodate	$Ca(IO_3)_2$	$6.47 \cdot 10^{-6}$
Calcium iodate hexahydrate	$Ca(IO_3)_2 \cdot 6H_2O$	$7.10 \cdot 10^{-7}$
Calcium molybdate	$CaMoO_4$	$1.46 \cdot 10^{-8}$
Calcium oxalate monohydrate	$CaC_2O_4 \cdot H_2O$	$2.32 \cdot 10^{-9}$
Calcium phosphate	$Ca_3(PO_4)_2$	$2.07 \cdot 10^{-33}$
Calcium sulfate	$CaSO_4$	$4.93 \cdot 10^{-5}$
Calcium sulfate dihydrate	$CaSO_4 \cdot 2H_2O$	$3.14 \cdot 10^{-5}$
Calcium sulfite hemihydrate	$CaSO_3 \cdot 0.5H_2O$	$3.1 \cdot 10^{-7}$
Cesium perchlorate	$CsClO_4$	$3.95 \cdot 10^{-3}$
Cesium periodate	$CsIO_4$	$5.16 \cdot 10^{-6}$
Cobalt(II) arsenate	$Co_3(AsO_4)_2$	$6.80 \cdot 10^{-29}$
Cobalt(II) hydroxide (blue)	$Co(OH)_2$	$5.92 \cdot 10^{-15}$
Cobalt(II) iodate dihydrate	$Co(IO_3)_2 \cdot 2H_2O$	$1.21 \cdot 10^{-2}$
Cobalt(II) phosphate	$Co_3(PO_4)_2$	$2.05 \cdot 10^{-35}$
Copper(I) bromide	$CuBr$	$6.27 \cdot 10^{-9}$
Copper(I) chloride	$CuCl$	$1.72 \cdot 10^{-7}$
Copper(I) cyanide	$CuCN$	$3.47 \cdot 10^{-20}$
Copper(I) iodide	CuI	$1.27 \cdot 10^{-12}$
Copper(I) thiocyanate	$CuSCN$	$1.77 \cdot 10^{-13}$
Copper(II) arsenate	$Cu_3(AsO_4)_2$	$7.95 \cdot 10^{-36}$
Copper(II) iodate monohydrate	$Cu(IO_3)_2 \cdot H_2O$	$6.94 \cdot 10^{-8}$
Copper(II) oxalate	CuC_2O_4	$4.43 \cdot 10^{-10}$
Copper(II) phosphate	$Cu_3(PO_4)_2$	$1.40 \cdot 10^{-37}$
Europium(III) hydroxide	$Eu(OH)_3$	$9.38 \cdot 10^{-27}$
Gallium(III) hydroxide	$Ga(OH)_3$	$7.28 \cdot 10^{-36}$
Iron(II) carbonate	$FeCO_3$	$3.13 \cdot 10^{-11}$

Compound	Formula	K_{sp}
Iron(II) fluoride	FeF_2	$2.36 \cdot 10^{-6}$
Iron(II) hydroxide	$Fe(OH)_2$	$4.87 \cdot 10^{-17}$
Iron(III) hydroxide	$Fe(OH)_3$	$2.79 \cdot 10^{-39}$
Iron(III) phosphate dihydrate	$FePO_4 \cdot 2H_2O$	$9.91 \cdot 10^{-16}$
Lanthanum iodate	$La(IO_3)_3$	$7.50 \cdot 10^{-12}$
Lead(II) bromide	$PbBr_2$	$6.60 \cdot 10^{-6}$
Lead(II) carbonate	$PbCO_3$	$7.40 \cdot 10^{-14}$
Lead(II) chloride	$PbCl_2$	$1.70 \cdot 10^{-5}$
Lead(II) fluoride	PbF_2	$3.3 \cdot 10^{-8}$
Lead(II) hydroxide	$Pb(OH)_2$	$1.43 \cdot 10^{-20}$
Lead(II) iodate	$Pb(IO_3)_2$	$3.69 \cdot 10^{-13}$
Lead(II) iodide	PbI_2	$9.8 \cdot 10^{-9}$
Lead(II) selenate	$PbSeO_4$	$1.37 \cdot 10^{-7}$
Lead(II) sulfate	$PbSO_4$	$2.53 \cdot 10^{-8}$
Lithium carbonate	Li_2CO_3	$8.15 \cdot 10^{-4}$
Lithium fluoride	LiF	$1.84 \cdot 10^{-3}$
Lithium phosphate	Li_3PO_4	$2.37 \cdot 10^{-11}$
Magnesium carbonate	$MgCO_3$	$6.82 \cdot 10^{-6}$
Magnesium carbonate trihydrate	$MgCO_3 \cdot 3H_2O$	$2.38 \cdot 10^{-6}$
Magnesium carbonate pentahydrate	$MgCO_3 \cdot 5H_2O$	$3.79 \cdot 10^{-6}$
Magnesium fluoride	MgF_2	$5.16 \cdot 10^{-11}$
Magnesium hydroxide	$Mg(OH)_2$	$5.61 \cdot 10^{-12}$
Magnesium oxalate dihydrate	$MgC_2O_4 \cdot 2H_2O$	$4.83 \cdot 10^{-6}$
Magnesium phosphate	$Mg_3(PO_4)_2$	$1.04 \cdot 10^{-24}$
Manganese(II) carbonate	$MnCO_3$	$2.24 \cdot 10^{-11}$
Manganese(II) iodate	$Mn(IO_3)_2$	$4.37 \cdot 10^{-7}$
Manganese(II) oxalate dihydrate	$MnC_2O_4 \cdot 2H_2O$	$1.70 \cdot 10^{-7}$
Mercury(I) bromide	Hg_2Br_2	$6.40 \cdot 10^{-23}$
Mercury(I) carbonate	Hg_2CO_3	$3.6 \cdot 10^{-17}$
Mercury(I) chloride	Hg_2Cl_2	$1.43 \cdot 10^{-18}$
Mercury(I) fluoride	Hg_2F_2	$3.10 \cdot 10^{-6}$
Mercury(I) iodide	Hg_2I_2	$5.2 \cdot 10^{-29}$
Mercury(I) oxalate	$Hg_2C_2O_4$	$1.75 \cdot 10^{-13}$
Mercury(I) sulfate	Hg_2SO_4	$6.5 \cdot 10^{-7}$
Mercury(I) thiocyanate	$Hg_2(SCN)_2$	$3.2 \cdot 10^{-20}$
Mercury(II) bromide	$HgBr_2$	$6.2 \cdot 10^{-20}$
Mercury(II) iodide	HgI_2	$2.9 \cdot 10^{-29}$
Neodymium carbonate	$Nd_2(CO_3)_3$	$1.08 \cdot 10^{-33}$
Nickel(II) carbonate	$NiCO_3$	$1.42 \cdot 10^{-7}$
Nickel(II) hydroxide	$Ni(OH)_2$	$5.48 \cdot 10^{-16}$
Nickel(II) iodate	$Ni(IO_3)_2$	$4.71 \cdot 10^{-5}$
Nickel(II) phosphate	$Ni_3(PO_4)_2$	$4.74 \cdot 10^{-32}$
Palladium(II) thiocyanate	$Pd(SCN)_2$	$4.39 \cdot 10^{-23}$
Potassium hexachloroplatinate	K_2PtCl_6	$7.48 \cdot 10^{-6}$
Potassium perchlorate	$KClO_4$	$1.05 \cdot 10^{-2}$
Potassium periodate	KIO_4	$3.71 \cdot 10^{-4}$
Praseodymium hydroxide	$Pr(OH)_3$	$3.39 \cdot 10^{-24}$
Radium iodate	$Ra(IO_3)_2$	$1.16 \cdot 10^{-9}$
Radium sulfate	$RaSO_4$	$3.66 \cdot 10^{-11}$
Rubidium perchlorate	$RbClO_4$	$3.00 \cdot 10^{-3}$
Scandium fluoride	ScF_3	$5.81 \cdot 10^{-24}$
Scandium hydroxide	$Sc(OH)_3$	$2.22 \cdot 10^{-31}$
Silver(I) acetate	$AgCH_3COO$	$1.94 \cdot 10^{-3}$
Silver(I) arsenate	Ag_3AsO_4	$1.03 \cdot 10^{-22}$
Silver(I) bromate	$AgBrO_3$	$5.38 \cdot 10^{-5}$
Silver(I) bromide	$AgBr$	$5.35 \cdot 10^{-13}$
Silver(I) carbonate	Ag_2CO_3	$8.46 \cdot 10^{-12}$
Silver(I) chloride	$AgCl$	$1.77 \cdot 10^{-10}$
Silver(I) chromate	Ag_2CrO_4	$1.12 \cdot 10^{-12}$
Silver(I) cyanide	$AgCN$	$5.97 \cdot 10^{-17}$
Silver(I) iodate	$AgIO_3$	$3.17 \cdot 10^{-8}$
Silver(I) iodide	AgI	$8.52 \cdot 10^{-17}$
Silver(I) oxalate	$Ag_2C_2O_4$	$5.40 \cdot 10^{-12}$
Silver(I) phosphate	Ag_3PO_4	$8.89 \cdot 10^{-17}$
Silver(I) sulfate	Ag_2SO_4	$1.20 \cdot 10^{-5}$
Silver(I) sulfite	Ag_2SO_3	$1.50 \cdot 10^{-14}$
Silver(I) thiocyanate	$AgSCN$	$1.03 \cdot 10^{-12}$
Strontium arsenate	$Sr_3(AsO_4)_2$	$4.29 \cdot 10^{-19}$
Strontium carbonate	$SrCO_3$	$5.60 \cdot 10^{-10}$
Strontium fluoride	SrF_2	$4.33 \cdot 10^{-9}$
Strontium iodate	$Sr(IO_3)_2$	$1.14 \cdot 10^{-7}$
Strontium iodate monohydrate	$Sr(IO_3)_2 \cdot H_2O$	$3.77 \cdot 10^{-7}$
Strontium iodate hexahydrate	$Sr(IO_3)_2 \cdot 6H_2O$	$4.55 \cdot 10^{-7}$
Strontium sulfate	$SrSO_4$	$3.44 \cdot 10^{-7}$
Thallium(I) bromate	$TlBrO_3$	$1.10 \cdot 10^{-4}$
Thallium(I) bromide	$TlBr$	$3.71 \cdot 10^{-6}$
Thallium(I) chloride	$TlCl$	$1.86 \cdot 10^{-4}$
Thallium(I) chromate	Tl_2CrO_4	$8.67 \cdot 10^{-13}$
Thallium(I) iodate	$TlIO_3$	$3.12 \cdot 10^{-6}$
Thallium(I) iodide	TlI	$5.54 \cdot 10^{-8}$
Thallium(I) thiocyanate	$TlSCN$	$1.57 \cdot 10^{-4}$
Thallium(III) hydroxide	$Tl(OH)_3$	$1.68 \cdot 10^{-44}$
Tin(II) hydroxide	$Sn(OH)_2$	$5.45 \cdot 10^{-27}$
Yttrium carbonate	$Y_2(CO_3)_3$	$1.03 \cdot 10^{-31}$
Yttrium fluoride	YF_3	$8.62 \cdot 10^{-21}$
Yttrium hydroxide	$Y(OH)_3$	$1.00 \cdot 10^{-22}$
Yttrium iodate	$Y(IO_3)_3$	$1.12 \cdot 10^{-10}$
Zinc arsenate	$Zn_3(AsO_4)_2$	$2.8 \cdot 10^{-28}$
Zinc carbonate	$ZnCO_3$	$1.46 \cdot 10^{-10}$
Zinc carbonate monohydrate	$ZnCO_3 \cdot H_2O$	$5.42 \cdot 10^{-11}$
Zinc fluoride	ZnF_2	$3.04 \cdot 10^{-2}$
Zinc hydroxide	$Zn(OH)_2$	$3 \cdot 10^{-17}$
Zinc iodate dihydrate	$Zn(IO_3)_2 \cdot 2H_2O$	$4.1 \cdot 10^{-6}$
Zinc oxalate dihydrate	$ZnC_2O_4 \cdot 2H_2O$	$1.38 \cdot 10^{-9}$
Zinc selenide	$ZnSe$	$3.6 \cdot 10^{-26}$
Zinc selenite monohydrate	$ZnSeO_3 \cdot H_2O$	$1.59 \cdot 10^{-7}$

Sulfides

Compound	Formula	K_{spa}
Cadmium sulfide	CdS	$8 \cdot 10^{-7}$
Copper(II) sulfide	CuS	$6 \cdot 10^{-16}$
Iron(II) sulfide	FeS	$6 \cdot 10^{2}$
Lead(II) sulfide	PbS	$3 \cdot 10^{-7}$
Manganese(II) sulfide (green)	MnS	$3 \cdot 10^{7}$
Mercury(II) sulfide (red)	HgS	$4 \cdot 10^{-33}$
Mercury(II) sulfide (black)	HgS	$2 \cdot 10^{-32}$
Silver(I) sulfide	Ag$_2$S	$6 \cdot 10^{-30}$
Tin(II) sulfide	SnS	$1 \cdot 10^{-5}$
Zinc sulfide (sphalerite)	ZnS	$2 \cdot 10^{-4}$
Zinc sulfide (wurtzite)	ZnS	$3 \cdot 10^{-2}$

SOLUBILITY OF COMMON SALTS AT AMBIENT TEMPERATURES

This table gives the aqueous solubility of selected salts at temperatures from 10°C to 40°C. Values are given in molality terms.

References

1. Apelblat, A., *J. Chem. Thermodynamics*, 24, 619, 1992.
2. Apelblat, A., *J. Chem. Thermodynamics*, 25, 63, 1993.
3. Apelblat, A., *J. Chem. Thermodynamics*, 25, 1513, 1993.
4. Apelblat, A. and Korin, E., *J. Chem. Thermodynamics*, 30, 59, 1998.

Salt	10°C	15°C	20°C	25°C	30°C	35°C	40°C	Ref.
$BaCl_2$	1.603	1.659	1.716	1.774	1.834	1.895	1.958	1
$Ca(NO_3)_2$	6.896	7.398	7.986	8.675	9.480	10.421		1
$CuSO_4$	1.055	1.153	1.260	1.376	1.502	1.639		3
$FeSO_4$	1.352	1.533	1.729	1.940	2.165	2.405		3
KBr	5.002	5.237	5.471	5.703	5.932	6.157		3
KIO_3	0.291	0.333	0.378	0.426	0.478	0.534	0.593	4
K_2CO_3	7.756	7.846	7.948	8.063	8.191	8.331	8.483	1
$LiCl$	19.296	19.456	19.670	19.935				2
$Mg(NO_3)_2$	4.403	4.523	4.656	4.800	4.958	5.130	5.314	1
$MnCl_2$	5.421	5.644	5.884	6.143	6.422	6.721		3
NH_4Cl	6.199	6.566	6.943	7.331				2
NH_4NO_3	18.809	21.163	23.721	26.496				2
$(NH_4)_2SO_4$	5.494	5.589	5.688	5.790	5.896	6.005		3
$NaBr$	8.258	8.546	8.856	9.191	9.550	9.937	10.351	4
$NaCl$	6.110	6.121	6.136	6.153	6.174	6.197	6.222	4
$NaNO_2$	11.111	11.484	11.883	12.310	12.766	13.253	13.772	4
$NaNO_3$	9.395	9.819	10.261	10.723	11.204	11.706	12.230	4
$RbCl$	6.911	7.180	7.449	7.717	7.986	8.253	8.520	4
$ZnSO_4$	2.911	3.116	3.336	3.573	3.827	4.099	4.194	1

SOLUBILITY CHART

Abbreviations: **W**, soluble in water; **A**, insoluble in water but soluble in acids; **w**, sparingly soluble in water but soluble in acids; **a**, insoluble in water and only sparingly soluble in acids; **I**, insoluble in water and acids; **d**, decomposes in water. * Indicates two modifications of the salt.

No.		Al	NH₄	Sb	Ba	Bi	Cd	Ca	Cr	Co	Cu	Au (I)	Au (II)	H	Fe (II)	Fe (III)
1	Acetate	W	W		W	W	W	W	W	W	W	W	W	W		
	—(C₂H₃O₂)	Al(—)₃	NH₄(—)		Ba(—)₂	Bi(—)₃	Cd(—)₂	Ca(—)₂	Cr(—)₃	Co(—)₂	Cu(—)₂			C₂H₄O₂	Fe(—)₂	Fe₂(—)₆
2	Arsenate	a	W	A	w	A	A	w		A	A			W	A	A
	—(AsO₄)	Al(—)	(NH₄)₃(—)	Sb(—)	Ba₃(—)₂	Bi(—)	Cd₃(—)₂	Ca₃(—)₂		Co₃(—)₂	Cu₃(—)₂			H₃AsO₄	Fe₃(—)₂	Fe(—)
3	Arsenite		W	A				w		A	A					
	—(AsO₃)		NH₄AsO₂	Sb(—)				Ca₃(—)₂		Co₃H₆(—)₄	CuH(—)					
4	Benzoate		W		W	A	W	W		W	W			W	W	A
	—(C₇H₅O₂)		NH₄(—)		Ba(—)₂	Bi(—)₃	Cd(—)₂	Ca(—)₂		Co(—)₂	Cu(—)₂			C₇H₆O₂	Fe(—)₂	Fe₂(—)₆
5	Bromide	W	W	d	W	d	W	W	W(I)*	W	W	w	W	W	W	W
		AlBr₂	NH₄Br	SbBr₃	BaBr₂	BiBr₃	CdBr₂	CaBr₂	CrBr₃	CoBr₂	CuBr₂	AuBr	AuBr₃	HBr	FeBr₂	FeBr₃
6	Carbonate		W	w			A	w	W	A					w	
			(NH₄)₂CO₃	BaCO₃			CdCO₃	CaCO₃	CrCO₃	CoCO₃					FeCO₃	
7	Chlorate	W	W		W	W	W	W		W	W			W	W	W
	—(ClO₃)	Al(—)₃	NH₄(—)		Ba(—)₂	Bi(—)	Cd(—)₂	Ca(—)₂		Co(—)₂	Cu(—)₂			HClO₃	Fe(—)₂	Fe(—)₃
8	Chloride	W	W	W	W	d	W	W	W	W	W	W	W	W	W	W
		AlCl₃	NH₄Cl	SbCl₃	BaCl₂	BiCl₃	CdCl₂	CaCl₂	CrCl₃	CoCl₂	CuCl₂	AuCl	AuCl₃	HCl	FeCl₂	FeCl₃
9	Chromate		W		A		A	W		A						A
	—(CrO₄)		(NH₄)₂(—)		Ba(—)		Cd(—)	Ca(—)		Co(—)						Fe₂(—)₃
10	Citrate	W	W		w	A	A	w		w				W		W
	—(C₆H₅O₇)	Al(—)	(NH₄)₃(—)		Ba₃(—)₂	Bi(—)	Cd₃(—)₂	Ca₃(—)₂		Co₃(—)₂				C₆H₈O₇		Fe(—)
11	Cyanide		W		W	w	W	W	A	A	A	w	W	W	a	
			NH₄CN		Ba(CN)₂	Bi(CN)₃	Cd(CN)₂	Ca(CN)₂	Cr(CN)₃	Co(CN)₂	Cu(CN)₂	AuCN	Au(CN)₃	HCN	Fe(CN)₂	
12	Ferricy'de		W		w		A	W		I	I			W	I	
	—(Fe(CN)₆)		(NH₄)₃(—)		Ba₃(—)₂		Cd₃(—)₂	Ca₃(—)₂		Co₃(—)₂	Cu₃(—)₂			H₃(—)	Fe₃(—)₂	
13	Ferrocy'de	w	W		W		A	W		I	I			W	A	a
	—(Fe(CN)₆)	Al₄(—)₃	(NH₄)₄(—)		Ba₂(—)		Cd₂(—)	Ca₂(—)		Co₂(—)	Cu₂(—)			H₄(—)	Fe₂(—)	Fe₄(—)₃
14	Fluoride	W	W	W	w	W	W	w	W(a)*	W	w			W	w	w
		AlF₃	NH₄F	SbF₃	BaF₂	BiF₃	CdF₂	CaF₂	CrF₃	CoF₂	CuF₂			HF	FeF₂	FeF₃
15	Formate	W	W		W	W	W	W		W	W			W	W	W
	—(CHO₂)	Al(—)₃	NH₄(—)		Ba(—)₂	Bi(—)₃	Cd(—)₂	Ca(—)₂		Co(—)₂	Cu(—)₂			CH₂O₂	Fe(—)₂	Fe(—)₃
16	Hydroxide	A	W		W	A	A	W	A	A	A	W	A		A	A
		Al(OH)₃	NH₄OH		Ba(OH)₂	Bi(OH)₃	Cd(OH)₂	Ca(OH)₂	Cr(OH)₃	Co(OH)₂	Cu(OH)₂	AuOH	Au(OH)₃		Fe(OH)₂	Fe(OH)₃
17	Iodide	W	W	d	W	A	W	W	W	W	W	a	a	W	W	W
		AlI₃	NH₄I	SbI₃	BaI₂	BiI₃	CdI₂	CaI₂	CrI₃	CoI₂	CuI	AuI	AuI₃	HI	FeI₂	FeI₃
18	Nitrate	W	W		W	d	W	W	W	W	W			W	W	W
		Al(NO₃)₃	NH₄NO₃		Ba(NO₃)₂	Bi(NO₃)₃	Cd(NO₃)₂	Ca(NO₃)₂	Cr(NO₃)₃	Co(NO₃)₂	Cu(NO₃)₂			HNO₃	Fe(NO₃)₂	Fe(NO₃)₃
19	Oxalate	W	W		W	A	W	A	W	W	W			W	A	A
	—(C₂O₄)	Al₂(—)₃	(NH₄)₂(—)		Ba(—)	Bi₂(—)₃	Cd(—)	Ca(—)	Cr(—)	Co(—)	Cu(—)			C₂H₂O₄	Fe(—)	Fe₂(—)₃
20	Oxide	a		w	W	A	A	w	a	A	A	A	A		A	A
		Al₂O₃		Sb₂O₃	BaO	Bi₂O₃	CdO	CaO	Cr₂O₃	CoO	CuO	Au₂O	Au₂O₃	H₂O₂	FeO	Fe₂O₃
21	Phosphate	A	W		A	A	A	w	w	A	A			W	A	w
	—(PO₄)	AlPO₄	NH₄H₂PO₄		Ba₃(PO₄)₂	BiPO₄	Cd₃(PO₄)₂	Ca₃(PO₄)₂	Cr₂(PO₄)₂	Co₃(PO₄)₂	Cu₃(PO₄)₂			H₃PO₄	Fe₃(PO₄)₂	FePO₄
22	Silicate,	I			W		A	w		A	A			I		
	—(SiO₃)	Al₂(—)₃			Ba(—)		Cd(—)	Ca(—)		Co₂SiO₄	Cu(—)			H₂SiO₃		
23	Sulfate	W	W	A	a	d	W	w	W(I)*	W	W			W	W	w
		Al₂(SO₄)₃	(NH₄)₂SO₄	Sb₂(SO₄)₃	BaSO₄	Bi₂(SO₄)₃	CdSO₄	CaSO₄	Cr₂(SO₄)₃	CoSO₄	CuSO₄			H₂SO₄	FeSO₄	Fe(SO₄)₃
24	Sulfide	d	W	A	d	A	A	w	d	A	A	I	I	W	A	w
		Al₂S₃	(NH₄)₂S	Sb₂S₃	BaS	Bi₂S₃	CdS	CaS	Cr₂S₃	CoS	CuS	Au₂S	Au₂S₃	H₂S	FeS	Fe₂S₃
25	Tartrate	w	W	W	w	A	A	A	d	A	A			W	A	d
	—(C₄H₄O₆)	Al₂(—)₃	(NH₄)₂(—)	Sb₂(—)₃	Ba(—)	Bi₂(—)₃	Cd(—)	Ca(—)		Co(—)	Cu(—)			C₄H₆O₆	Fe(—)	Fe₂(—)₃
26	Thiocy'te		W		W			W		W	d			W	W	W
			NH₄CNS		Ba(CNS)₂			Ca(CNS)		Co(CNS)₂	CuCNS			CNSH	Fe(CNS)₂	Fe(CNS)₃

No.	Ion	Pb	Mg	Mn	Hg (I)	Hg (II)	Ni	K	Pt	Ag	Na	Sn (IV)	Sn (II)	Sr	Zn
1	Acetate	W	W	W	w	W	W	W		w	W	W	d	W	W
	—($C_2H_3O_2$)	$Pb(—)_2$	$Mg(—)_2$	$Mn(—)_2$	$Hg(—)$	$Hg(—)_2$	$Ni(—)_2$	$K(—)$		$Ag(—)$	$Na(—)$	$Sn(—)_4$	$Sn(—)_2$	$Sr(—)_2$	$Zn(—)_2$
2	Arsenate	A	A	w	A	w	A	W		A	W			w	A
	—(AsO_4)	$PbH(—)$	$Mg_3(—)$	$MnH(—)$	$Hg_3(—)$	$Hg_3(—)_2$	$Ni_3(—)_2$	$K_3(—)$		$Ag_3(—)$	$Na_3(—)$			$SrH(—)$	$Zn_3(—)_2$
3	Arsenite		W	A	A	A	A	W		A	W		A	w	
	—(AsO_3)		$Mg_3(—)_2$	$Mn_3H_6(—)_4$	$Hg_3(—)$	$Hg_3(—)$	$Ni_3H_6(—)_4$	K_3AsO_3		$Ag_3(—)$	$Na_2H(—)$		$Sn_3(—)_2$	$Sr_3(—)_2$	
4	Benzoate	w	W	W	A	w	w	W		w	W			W	W
	—($C_7H_5O_2$)	$Pb(—)_2$	$Mg(—)_2$	$Mn(—)_2$	$Hg_2(—)_2$	$Hg(—)_2$	$Ni(—)_2$	$K(—)$		$Ag(—)$	$Na(—)$				$Zn(—)_2$
5	Bromide	W	W	W	W	A	W	W	w	a	W	W	W	W	W
		$PbBr_2$	$MgBr_2$	$MnBr_2$	$HgBr$	$HgBr_2$	$NiBr_2$	KBr	$PtBr_4$	$AgBr$	$NaBr$	$SnBr_4$	$SnBr_2$	$SrBr_2$	$ZnBr_2$
6	Carbonate	A	w	w	A		w	W		A	W			w	w
		$PbCO_3$	$MgCO_3$	$MnCO_3$	Hg_2CO_3		$NiCO_3$	K_2CO_3		Ag_2CO_3	Na_2CO_3			$SrCO_3$	$ZnCO_3$
7	Chlorate	W	W	W	W	W	W	W		W	W		W	W	W
	—(ClO_3)	$Pb(—)_2$	$Mg(—)_2$	$Mn(—)_2$	$Hg(—)$	$Hg(—)_2$	$Ni(—)_2$	$K(—)$		$Ag(—)$	$Na(—)$		$Sn(—)_2$	$Sr(—)_2$	$Zn(—)_2$
8	Chloride	W	W	W	a	W	W	W	W	a	W	W	W	W	W
		$PbCl_2$	$MgCl_2$	$MnCl_2$	$HgCl$	$HgCl_2$	$NiCl_2$	KCl	$PtCl_4$	$AgCl$	$NaCl$	$SnCl_4$	$SnCl_2$	$SrCl_2$	$ZnCl_2$
9	Chromate				w	w	A	A		W	W	W	A	W	w
	—(CrO_4)	$Pb(—)$	$Mg(—)$		$Hg_2(—)$	$Hg(—)$	$Ni(—)$	$K_2(—)$		$Ag_2(—)$	$Na_2(—)$	$Sn(—)_2$	$Sn(—)$	$Sr(—)$	$Zn(—)$
10	Citrate	W	W	w			W	W		W	W			A	W
	—($C_6H_5O_7$)	$Pb_3(—)_2$	$Mg_3(—)_2$	$MnH(—)$	$Hg_3(—)$		$Ni_3(—)_2$	$K_3(—)$		$Ag_3(—)$	$Na_3(—)$			$SrH(—)$	$Zn_3(—)_2$
11	Cyanide	w	W		A	W	a	W	I	a	W			W	A
		$Pb(CN)_2$	$Mg(CN)_2$		$HgCN$	$Hg(CN)_2$	$Ni(CN)_2$	KCN	$Pt(CN)_2$	$AgCN$	$NaCN$			$Sr(CN)_2$	$Zn(CN)_2$
12	Ferricy'de	w	W			A	I	W		I			A		
	—$Fe(CN)_6$	$Pb_3(—)_2$	$Mg_3(—)_2$			$Hg_3(—)_2$	$Ni_3(—)_2$	$K_3(—)$		$Ag_3(—)$	$Na_3(—)$		$Sn_3(—)_2$	$Sr_3(—)_2$	$Zn_3(—)_2$
13	Ferrocy'de	a	W	A		I	I	W		I	W		a	W	I
	—$Fe(CN)_6$	$Pb_2(—)$	$Mg_2(—)$	$Mn_2(—)$		$Hg_2(—)$	$Ni_2(—)$	$K_4(—)$		$Ag_4(—)$	$Na_4(—)$		$Sn_2(—)$	$Sr_2(—)$	$Zn_2(—)$
14	Fluoride	w	w	A	d	d	w	W	W	W	W	W	W	w	w
		PbF_2	MgF_2	MnF_2	HgF	HgF_2	NiF_2	KF	PtF_4	AgF	NaF	SnF_4	SnF_2	SrF_2	ZnF_2
15	Formate	W	W	W	w	W	W	W		w	W			W	w
	—(CHO_2)	$Pb(—)_2$	$Mg(—)_2$	$Mn(—)_2$	$Hg(—)$	$Hg(—)_2$	$Ni(—)_2$	$K(—)$		$Ag(—)$	$Na(—)$			$Sr(—)_2$	$Zn(—)_2$
16	Hydroxide	w	A	A		A	w	W	A		W	w	A	W	A
		$Pb(OH)_2$	$Mg(OH)_2$	$Mn(OH)_2$		$Hg(OH)_2$	$Ni(OH)_2$	KOH	$Pt(OH)_4$		$NaOH$	$Sn(OH)_4$	$Sn(OH)_2$	$Sr(OH)_2$	$Zn(OH)_2$
17	Iodide	w	w	w	A	A	W	W	I	I	W	W	W	W	W
		PbI_2	MgI_2	MnI_2	HgI	HgI_2	NiI_2	KI	PtI_2	AgI	NaI	SnI_4	SnI_2	SrI_2	ZnI_2
18	Nitrate	W	W	W	W	W	W	W	W	W	W		d	W	W
		$Pb(NO_3)_2$	$Mg(NO_3)_2$	$Mn(NO_3)_2$	$HgNO_3$	$Hg(NO_3)_2$	$Ni(NO_3)_2$	KNO_3	$Pt(NO_3)_4$	$AgNO_3$	$NaNO_3$		$Sn(NO_3)_2$	$Sr(NO_3)_2$	$Zn(NO_3)_2$
19	Oxalate	A	w	w	a	A	A	W		a	W		A	A	A
	—(C_2O_4)	$Pb(—)$	$Mg(—)$	$Mn(—)$	$Hg_2(—)$	$Hg(—)$	$Ni(—)$	$K_2(—)$		$Ag_2(—)$	$Na_2(—)$		$Sn(—)$	$Sr(—)$	$Zn(—)$
20	Oxide	w	A	A	A	A	w	W		w	d	A	A	W	w
		PbO	MgO	MnO	Hg_2O	HgO	NiO	K_2O	PtO	Ag_2O	Na_2O	SnO_2	SnO	SrO	ZnO
21	Phosphate	A	w	w	A	A	A	W		A	W		A	A	A
		$Pb_3(PO_4)_2$	$Mg_3(PO_4)_2$	$Mn_3(PO_4)_2$	Hg_3PO_4	$Hg_3(PO_4)_2$	$Ni_3(PO_4)_2$	K_3PO_4		Ag_3PO_4	Na_3PO_4		$Sn_3(PO_4)$	$Sr_3(PO_4)_2$	$Zn_3(PO_4)_2$
22	Silicate	A	A	I				W			W			A	A
	—(SiO_3)	$Pb(—)$	$Mg(—)$	$Mn(—)$				$K_2(—)$			$Na_2(—)$			$Sr(—)$	$Zn(—)$
23	Sulfate	w	W	W	w	d	W	W	W	w	W	W	W	w	W
		$PbSO_4$	$MgSO_4$	$MnSO_4$	Hg_2SO_4	$HgSO_4$	$NiSO_4$	K_2SO_4	$Pt(SO_4)_2$	Ag_2SO_4	Na_2SO_4	$Sn(SO_4)_2$	$SnSO_4$	$SrSO_4$	$ZnSO_4$
24	Sulfide	A	d	A	I	I	A	W	I	A	W	A	A	W	A
		PbS	MgS	MnS	Hg_2S	HgS	NiS	K_2S	PtS	Ag_2S	Na_2S	SnS_2	SnS	SrS	ZnS
25	Tartrate	A	w	w	I	I	A	W	I	A	W		A	W	A
	—($C_4H_4O_6$)	$Pb(—)$	$Mg(—)$	$Mn(—)$	$Hg_2(—)$		$Ni(—)$	$K_2(—)$		$Ag_2(—)$	$Na_2(—)$		$Sn(—)$	$Sr(—)$	$Zn(—)$
26	Thiocy'te	w	W	W	A	w			I		W			W	W
		$Pb(CNS)_2$	$Mg(CNS)_2$	$Mn(CNS)_2$	$HgCNS$	$Hg(CNS)_2$		$KCNS$		$AgCNS$	$NaCNS$			$Sr(CNS)_2$	$Zn(CNS)_2$

REDUCTION OF WEIGHINGS IN AIR TO VACUO

When the mass M of a body is determined in air, a correction is necessary for the buoyancy of the air. The corrected mass is given by $M + kM/1000$, where k is a function of the material used for the weights, given by

$$k = 1000\rho_{air}(1/\rho_{body} - 1/\rho_{weight})$$

and ρ is density. The table below is computed for an air density of 0.0012 g/cm^3 and for densities of three common weights: platinum-iridium (21.6 g/cm^3), brass (8.5 g/cm^3), and aluminum or quartz (2.65 g/cm^3).

References

1. Kaye, G. W. C., and Laby, T. H., *Tables of Physical and Chemical Constants*, 16th Edition, pp. 25-28, Longman, London, 1995.
2. Giacomo, P., *Metrologia*, 18, 33, 1982.
3. Davis, R. S., *Metrologia*, 29, 67, 1992.

Density of body (g/cm^3)	Value of k for weights of: Pt-Ir	Brass	Quartz or Al
0.5	2.34	2.26	1.95
0.6	1.94	1.86	1.55
0.7	1.66	1.57	1.26
0.8	1.44	1.36	1.05
0.9	1.28	1.19	0.88
1.0	1.14	1.06	0.75
1.1	1.04	0.95	0.64
1.2	0.94	0.86	0.55
1.3	0.87	0.78	0.47
1.4	0.80	0.72	0.40
1.5	0.74	0.66	0.35
1.6	0.69	0.61	0.30
1.7	0.65	0.56	0.25

Density of body (g/cm^3)	Value of k for weights of: Pt-Ir	Brass	Quartz or Al
1.8	0.61	0.53	0.21
1.9	0.58	0.49	0.18
2.0	0.54	0.46	0.15
2.5	0.42	0.34	0.03
3.0	0.34	0.26	-0.05
4.0	0.24	0.16	-0.15
6.0	0.14	0.06	-0.25
8.0	0.09	0.01	-0.30
10.0	0.06	-0.02	-0.33
15.0	0.02	-0.06	-0.37
20.0	0.00	-0.08	-0.39
22.0	0.00	-0.09	-0.40

For a more accurate calculation, use the following values of the density of air (assuming 50% relative humidity and 0.04% CO_2):

P/kPa	Air temperature 10°C	20°C	30°C
85	0.001043	0.001005	0.000968
90	0.001105	0.001065	0.001025
95	0.001166	0.001124	0.001083
100	0.001228	0.001184	0.001140
105	0.001290	0.001243	0.001198

Formulas for calculating the density of air over more extended ranges of temperature, pressure, and humidity may be found in the references.

VOLUME OF ONE GRAM OF WATER

The following table, which is designed for gravimetric calibration of volumetric apparatus, gives the specific volume of water at standard atmospheric pressure as a function of temperature.

Reference

Marsh, K. N., Ed., *Recommended Reference Materials for the Realization of Physicochemical Properties*, pp. 25-27, Blackwell Scientific Publications, Oxford, 1987.

t/°C	Volume of 1 g H₂O in cm³
10	1.0002980
11	1.0003928
12	1.0005007
13	1.0006212
14	1.0007542
15	1.0008992
16	1.0010561
17	1.0012246
18	1.0014044
19	1.0015952
20	1.0017969
21	1.0020092
22	1.0022320
23	1.0024649
24	1.0027079
25	1.0029607
26	1.0032234
27	1.0034956
28	1.0037771
29	1.0040679
30	1.0043679

PROPERTIES OF CARRIER GASES FOR GAS CHROMATOGRAPHY

The following is a list of carrier gases sometimes used in gas chromatography, with properties relevant to the design of chromatographic systems. All data refer to normal atmospheric pressure (101.325 kPa).

M_r: Molecular weight (relative molar mass)
ρ_{25}: Density at 25°C in g/L
λ: Thermal conductivity in mW/m °C
η: Viscosity in µPa s (equal to 10^{-3} cp)
c_p: Specific heat at 25°C in J/g °C

References

1. Lide, D. R., and Kehiaian, H. V., *CRC Handbook of Thermophysical and Thermochemical Data*, CRC Press, Boca Raton, FL, 1994.
2. Bruno, T. J., and Svoronos, P. D. N., *CRC Handbook of Basic Tables for Chemical Analysis*, CRC Press, Boca Raton, FL, 1989.

Gas	M_r	ρ_{25} g L^{-1}	At 25°C λ mW/m °C	At 25°C η µPa s	At 250°C λ mW/m °C	At 250°C η µPa s	c_p(25 °C) J/g °C
Hydrogen	2.016	0.0824	185.9	8.9	280	13.1	14.3
Helium	4.003	0.1636	154.6	19.9	230	29.5	5.20
Argon	39.95	1.6329	17.8	22.7	27.7	35.3	0.521
Nitrogen	28.01	1.1449	25.9	17.9	39.6	26.8	1.039
Oxygen	32.00	1.3080	26.2	20.7	42.6	31.8	0.919
Carbon monoxide	28.01	1.1449	24.8	17.8	40.7	26.5	1.039
Carbon dioxide	44.01	1.7989	16.7	14.9	35.5	24.9	0.843
Sulfur hexafluoride	146.05	5.9696	13.1	28.1	15.3	24.8	0.664
Methane	16.04	0.6556	34.5	11.1	75.0	17.6	2.23
Ethane	30.07	1.2291	20.9	9.4	57.7	15.5	1.75
Ethylene	28.05	1.1465	20.5	10.3	53.8	17.2	1.53
Propane	44.10	1.8025	17.9	8.3	49.2	14.0	1.67

SOLVENTS FOR ULTRAVIOLET SPECTROPHOTOMETRY

This table lists some solvents commonly used for sample preparation for ultraviolet spectrophotometry. The properties given are:

λ_c: cutoff wavelength, below which the solvent absorption becomes excessive.

ε: dielectric constant (relative permittivity); the temperature in °C is given as a superscript.

t_b: normal boiling point.

References

1. Bruno, T. J., and Svoronos, P. D. N., *CRC Handbook of Basic Tables for Chemical Analysis*, CRC Press, Boca Raton, FL, 1989.
2. *Landolt–Börnstein, Numerical Data and Functional Relationships in Science and Technology*, New Series, IV/6, *Static Dielectric Constants of Pure Liquids and Binary Liquid Mixtures*, Springer–Verlag, Heidelberg, 1991.

Name	λ_c/nm	ε	t_b/°C
Acetic acid	260	6.20^{20}	117.9
Acetone	330	21.01^{20}	56.0
Acetonitrile	190	36.64^{20}	81.6
Benzene	280	2.28^{20}	80.0
2-Butanol	260	17.26^{20}	99.5
Butyl acetate	254	5.07^{20}	126.1
Carbon disulfide	380	2.63^{20}	46
Carbon tetrachloride	265	2.24^{20}	76.8
1-Chlorobutane	220	7.28^{20}	78.6
Chloroform	245	4.81^{20}	61.1
Cyclohexane	210	2.02^{20}	80.7
1,2-Dichloroethane	226	10.42^{20}	83.5
Dichloromethane	235	8.93^{25}	40
Diethyl ether	218	4.27^{20}	34.5
N,N-Dimethylacetamide	268	38.85^{21}	165
N,N-Dimethylformamide	270	38.25^{20}	153
Dimethyl sulfoxide	265	47.24^{20}	189
1,4-Dioxane	215	2.22^{20}	101.5
Ethanol	210	25.3^{20}	78.2
Ethyl acetate	255	6.08^{20}	77.1
Ethylene glycol dimethyl ether	240	7.30^{24}	85
Ethylene glycol monoethyl ether	210	13.38^{25}	135
Ethylene glycol monomethyl ether	210	17.2^{25}	124.1
Glycerol	207	46.53^{20}	290
Heptane	197	1.92^{20}	98.5
Hexadecane	200	2.05^{20}	286.8
Hexane	210	1.89^{20}	68.7
Methanol	210	33.0^{20}	64.6
Methylcyclohexane	210	2.02^{20}	100.9
Methyl ethyl ketone	330	18.56^{20}	79.5
Methyl isobutyl ketone	335	13.11^{20}	116.5
2-Methyl-1-propanol	230	17.93^{20}	107.8
N-Methyl-2-pyrrolidone	285	32.55^{20}	202
Nitromethane	380	37.27^{20}	101.1
Pentane	210	1.84^{20}	36.0
Pentyl acetate	212	4.79^{20}	149.2
1-Propanol	210	20.8^{20}	97.2
2-Propanol	210	20.18^{20}	82.3
Pyridine	330	13.26^{20}	115.2
Tetrachloroethylene	290	2.27^{30}	121.3
Tetrahydrofuran	220	7.52^{22}	65
Toluene	286	2.38^{23}	110.6
1,1,2-Trichloro-1,2,2-trifluoroethane	231	2.41^{25}	47.7
2,2,4-Trimethylpentane	215	1.94^{20}	99.2
Water	191	80.10^{20}	100.0
o-Xylene	290	2.56^{20}	144.5
m-Xylene	290	2.36^{20}	139.1
p-Xylene	290	2.27^{20}	138.3

¹³C CHEMICAL SHIFTS OF USEFUL NMR SOLVENTS

The following table gives the expected carbon-13 chemical shifts, relative to tetramethylsilane, for various useful NMR solvents. In some solvents, slight changes can occur with change of concentration.[2,3]

References

1. Bruno, T. J., and Svoronos, P. D. N., *CRC Handbook of Basic Tables for Chemical Analysis*, CRC Press, Boca Raton, FL, 1989.

2. Silverstein, R. M., Bassler, G. C., and Morrill, T. C., *Spectrometric Identification of Organic Compounds*, John Wiley & Sons, Now York, 1981.

3. Rahman, A. U., *Nuclear Magnetic Resonance. Basic Principles*, Springer-Verlag, New York, 1986.

4. Pretsch, E., Clerc, T., Seibl, J., and Simon, W., *Spectral Data for Structure Determination of Organic Compounds, Second Edition*, Springer-Verlag, Heidelberg, 1989.

Solvent	Formula	Chemical shift (ppm)
Acetic acid-d_4	CD_3COOD	20.0 (CD_3) 205.8 (C=O)
Acetone	$(CH_3)_2C=O$	30.7 (CH_3) 206.7 (C=O)
Acetone-d_6	$(CD_3)_2C=O$	29.2 (CD_3) 204.1 (C=O)
Acetonitrile-d_3	$CD_3C\equiv N$	1.3 (CD_3) 117.1 (C$\equiv$N)
Benzene	C_6H_6	128.5
Benzene-d_6	C_6D_6	128.4
Carbon disulfide	CS_2	192.3
Carbon tetrachloride	CCl_4	96.0
Chloroform	$CHCl_3$	77.2
Chloroform-d_3	$CDCl_3$	77.05
Cyclohexane-d_{12}	C_6D_{12}	27.5
Dichloromethane-d_2	CD_2Cl_2	53.6
Dimethylformamide-d_7	$(CD_3)_2NCDO$	31 (CD_3) 36 (CD_3) 162.4 (C=O)
Dimethylsulfoxide-d_6	$(CD_3)_2S=O$	39.6
Dioxane-d_8	$C_4D_8O_2$	67.4
Formic acid-d_2	DCOOD	165.5
Methanol-d_4	CD_3OD	49.3
Nitromethane-d_3	CD_3NO_2	57.3
Pyridine	C_5H_5N	123.6 (C_3) 135.7 (C_4) 149.8 (C_2)
Pyridine-d_5	C_5D_5N	123.9 (C_3) 135.9 (C_4) 150.2 (C_2)
1,1,2,2-Tetrachloroethane-d_2	$CDCl_2CDCl_2$	75.5
Tetrahydrofuran-d_8	C_4D_8O	25.8 (C_2) 67.9 (C_1)
Trichlorofluoromethane	$CFCl_3$	117.6

MASS SPECTRAL PEAKS OF COMMON ORGANIC SOLVENTS

The strongest peaks in the mass spectra of 200 important solvents are listed in this table. The e/m value for each peak is followed by the relative intensity in parentheses, with the strongest peak assigned an intensity of 100. The peaks for each compound are listed in order of decreasing intensity. Solvents are listed in alphabetical order by common name.

Data on the physical properties of the same compounds may be found in Section 15 in the table "Properties of Common Laboratory Solvents".

References

1. NIST/EPA/NIH Mass Spectral Database, National Institute of Standards and Technology, Gaithersburg, MD, 20899.
2. Lide, D. R., and Milne, G. W. A., Eds., *Handbook of Data on Organic Compounds, Third Edition*, CRC Press, Boca Raton, FL, 1994. (Also available as a CD ROM database.)

Compound					e/m (intensity)					
Acetal (1,1-Diethoxyethane)	44(100)	43(92)	29(77)	31(76)	45(74)	27(52)	72(48)	73(23)	28(17)	46(15)
Acetic acid	43(100)	45(87)	60(57)	15(42)	42(14)	29(13)	14(13)	28(7)	18(6)	16(6)
Acetone	43(100)	15(34)	58(23)	27(9)	14(9)	42(8)	26(7)	29(5)	28(5)	39(4)
Acetonitrile	41(100)	40(46)	39(13)	14(9)	38(6)	28(4)	26(4)	25(3)	42(2)	27(2)
Acetylacetone	43(100)	85(31)	100(20)	27(12)	42(10)	29(10)	41(7)	39(7)	31(5)	26(5)
Acrylonitrile	53(100)	26(85)	52(79)	51(34)	27(13)	50(8)	25(7)	38(5)	54(3)	37(3)
Adiponitrile	41(100)	68(50)	54(42)	40(21)	55(20)	27(17)	39(16)	28(13)	52(7)	42(6)
Allyl alcohol	57(100)	31(34)	29(32)	28(31)	58(25)	39(22)	27(20)	30(16)	32(14)	26(11)
Allylamine	30(100)	56(80)	28(76)	57(33)	39(21)	29(20)	27(18)	26(13)	41(8)	18(8)
2-Aminoisobutanol	58(100)	41(18)	18(17)	42(13)	28(11)	56(10)	30(10)	29(8)	43(6)	59(5)
Benzal chloride	125(100)	127(32)	160(14)	89(13)	162(9)	63(9)	126(8)	62(7)	105(5)	39(5)
Benzaldehyde	51(100)	77(81)	50(55)	106(44)	105(43)	52(26)	78(16)	39(13)	27(10)	74(8)
Benzene	78(100)	77(20)	52(19)	51(17)	50(15)	39(12)	79(6)	76(5)	74(4)	38(4)
Benzonitrile	103(100)	76(34)	50(13)	104(9)	75(7)	51(7)	77(5)	52(4)	39(4)	74(3)
Benzyl chloride	91(100)	126(20)	65(14)	92(9)	39(9)	63(8)	128(6)	45(6)	89(5)	125(3)
Bromochloromethane	49(100)	130(67)	128(52)	51(31)	93(23)	81(20)	79(20)	95(17)	132(16)	47(8)
Bromoform (Tribromomethane)	173(100)	171(50)	175(49)	93(22)	91(22)	79(18)	81(17)	94(13)	92(13)	254(11)
Butyl acetate	43(100)	56(34)	41(17)	27(16)	29(15)	73(11)	61(10)	28(7)	55(6)	39(6)
Butyl alcohol	31(100)	56(81)	41(62)	43(60)	27(50)	42(31)	29(31)	28(17)	39(16)	55(12)
sec-Butyl alcohol	45(100)	31(22)	27(22)	59(20)	29(18)	43(13)	41(12)	44(8)	18(8)	28(5)
tert-Butyl alcohol	59(100)	31(33)	41(22)	43(18)	29(13)	27(11)	57(10)	42(4)	60(3)	28(3)
Butylamine	30(100)	73(10)	28(5)	41(3)	27(3)	18(3)	44(2)	42(2)	31(2)	29(2)
tert-Butylamine	58(100)	41(21)	42(15)	18(9)	30(8)	15(8)	39(7)	57(6)	28(6)	59(4)
Butyl methyl ketone	43(100)	58(60)	57(17)	100(16)	29(15)	41(13)	85(8)	27(8)	71(7)	59(5)
p-tert-Butyltoluene	133(100)	105(38)	41(23)	148(18)	93(16)	91(14)	115(13)	134(11)	39(11)	116(10)
γ-Butyrolactone	28(100)	42(74)	29(48)	27(33)	41(27)	56(25)	86(24)	26(18)	85(10)	39(10)
Caprolactam	55(100)	113(87)	30(81)	56(66)	84(60)	85(57)	42(51)	41(33)	28(26)	43(17)
Carbon disulfide	76(100)	32(22)	44(17)	78(9)	38(6)	28(5)	77(3)	64(1)	46(1)	39(1)
Carbon tetrachloride	117(100)	119(98)	121(31)	82(24)	47(23)	84(16)	35(14)	49(8)	28(8)	36(6)
1-Chloro-1,1-difluoroethane	65(100)	45(31)	85(14)	31(10)	64(8)	44(7)	35(6)	26(6)	87(5)	81(4)
Chlorobenzene	112(100)	77(63)	114(33)	51(29)	50(14)	75(8)	113(7)	78(5)	76(5)	28(4)
Chloroform	83(100)	85(64)	47(35)	35(19)	48(16)	49(12)	87(10)	37(6)	50(5)	84(4)
Chloropentafluoroethane	85(100)	69(61)	31(38)	87(32)	50(17)	35(8)	119(6)	66(4)	100(3)	47(3)
Cumene (Isopropylbenzene)	105(100)	120(25)	77(13)	51(12)	79(10)	106(9)	39(9)	27(8)	103(6)	91(5)
Cyclohexane	56(100)	84(71)	41(70)	27(37)	55(36)	39(35)	42(30)	69(23)	28(18)	43(14)
Cyclohexanol	57(100)	44(68)	41(68)	39(51)	32(40)	43(38)	31(32)	42(22)	67(18)	82(16)
Cyclohexanone	55(100)	42(85)	41(34)	27(33)	98(31)	39(27)	69(26)	70(20)	43(14)	28(14)
Cyclohexylamine	56(100)	43(23)	28(17)	99(10)	70(8)	57(6)	30(6)	93(5)	54(4)	41(4)
Cyclopentane	42(100)	70(30)	55(29)	41(29)	39(22)	27(15)	40(7)	29(5)	28(4)	43(3)
Cyclopentanone	55(100)	28(50)	84(42)	41(38)	56(29)	27(24)	39(19)	42(15)	26(9)	29(7)
p-Cymene (1-Methyl-4-isopropylbenzene)	119(100)	91(42)	134(33)	39(27)	41(20)	117(18)	65(18)	77(17)	27(16)	120(15)
cis-Decalin	67(100)	81(87)	41(81)	138(67)	96(62)	82(62)	39(50)	55(45)	27(44)	95(42)
trans-Decalin	41(100)	68(91)	67(88)	82(67)	27(65)	96(61)	95(55)	138(51)	81(51)	29(51)
Diacetone alcohol	43(100)	59(41)	58(17)	101(10)	41(9)	31(9)	83(6)	56(6)	55(6)	29(6)
1,2-Dibromoethane	27(100)	107(77)	109(72)	26(24)	28(10)	81(5)	79(5)	25(5)	95(4)	93(4)
Dibromofluoromethane	111(100)	113(98)	192(29)	43(16)	41(16)	190(15)	194(14)	81(9)	79(9)	122(7)
Dibromomethane	174(100)	93(96)	95(84)	172(53)	176(50)	91(11)	81(9)	79(9)	94(5)	65(5)
1,2-Dibromotetrafluoroethane	179(100)	181(97)	129(34)	131(33)	100(17)	31(13)	260(12)	50(8)	69(7)	262(6)
Dibutylamine	86(100)	72(52)	30(48)	44(40)	29(31)	57(24)	41(21)	73(15)	28(15)	43(13)

Compound	e/m (intensity)									
o-Dichlorobenzene	146(100)	148(64)	111(38)	75(23)	113(12)	74(12)	50(11)	150(10)	73(9)	147(7)
1,1-Dichloroethane (Ethylidene dichloride)	63(100)	27(71)	65(31)	26(19)	83(11)	85(7)	61(7)	35(6)	98(5)	62(5)
1,2-Dichloroethane (Ethylene dichloride)	62(100)	27(91)	49(40)	64(32)	26(31)	63(19)	98(14)	51(13)	61(12)	100(9)
1,1-Dichloroethylene	61(100)	96(61)	98(38)	63(32)	26(16)	60(15)	62(7)	25(7)	100(6)	35(6)
cis-1,2-Dichloroethylene	61(100)	96(73)	98(47)	63(32)	26(30)	60(21)	25(13)	35(12)	62(9)	100(8)
trans-1,2-Dichloroethylene	61(100)	96(67)	98(43)	26(34)	63(32)	60(24)	25(15)	62(10)	100(7)	47(7)
Dichloroethyl ether	93(100)	63(74)	27(38)	95(32)	65(24)	31(9)	49(4)	28(4)	94(3)	62(3)
Dichloromethane (Methylene chloride)	49(100)	84(64)	86(39)	51(31)	47(14)	48(8)	88(6)	50(3)	85(2)	83(2)
1,2-Dichloropropane	63(100)	62(71)	27(57)	41(49)	39(32)	65(31)	76(27)	64(25)	49(13)	77(12)
1,2-Dichlorotetrafluoroethane	85(100)	135(52)	87(33)	137(17)	101(9)	31(9)	103(6)	100(6)	50(5)	69(4)
Diethanolamine	30(100)	74(82)	28(77)	56(69)	18(50)	42(46)	29(36)	27(34)	45(30)	43(19)
Diethylamine	30(100)	58(81)	44(28)	73(18)	29(18)	28(17)	72(12)	42(11)	27(11)	59(4)
Diethyl carbonate	29(100)	45(70)	31(53)	27(39)	91(24)	28(15)	63(11)	26(10)	30(6)	43(5)
Diethylene glycol	45(100)	75(23)	31(20)	44(16)	27(14)	76(12)	29(12)	43(11)	42(9)	41(4)
Diethylene glycol dimethyl ether (Diglyme)	59(100)	58(43)	31(34)	29(32)	45(28)	28(19)	89(15)	43(9)	27(5)	60(4)
Diethylene glycol monoethyl ether (Carbitol)	45(100)	59(56)	72(37)	73(22)	60(14)	31(13)	75(11)	44(9)	104(8)	103(7)
Diethylene glycol monoethyl ether acetate	43(100)	29(51)	31(42)	45(40)	59(24)	72(18)	44(10)	73(9)	42(9)	30(6)
Diethylene glycol monomethyl ether	45(100)	31(42)	59(41)	29(38)	28(32)	58(21)	43(14)	27(13)	44(11)	32(10)
Diethylenetriamine	44(100)	73(59)	30(35)	19(18)	56(16)	28(16)	27(16)	42(11)	99(8)	43(8)
Diethyl ether	31(100)	29(63)	59(40)	27(35)	45(33)	74(23)	15(17)	43(9)	28(9)	26(9)
Diisobutyl ketone (Isovalerone)	57(100)	85(82)	41(46)	43(39)	58(33)	28(30)	26(30)	39(22)	42(12)	142(11)
Diisopropyl ether	45(100)	43(39)	87(15)	41(12)	59(10)	27(8)	39(4)	69(3)	42(3)	31(3)
N,N-Dimethylacetamide	44(100)	87(69)	43(46)	45(23)	42(19)	72(15)	15(11)	30(8)	28(5)	88(4)
Dimethylamine	44(100)	45(81)	18(32)	28(30)	43(19)	42(15)	15(9)	46(5)	41(5)	27(5)
Dimethyl disulfide	94(100)	45(63)	79(59)	46(38)	47(26)	15(18)	48(14)	61(12)	64(11)	96(9)
N,N-Dimethylformamide	73(100)	44(86)	42(36)	30(22)	28(20)	29(8)	43(7)	72(6)	58(5)	74(4)
Dimethyl sulfoxide	63(100)	78(70)	15(40)	45(35)	29(16)	61(13)	46(12)	31(11)	48(10)	47(10)
1,4-Dioxane	28(100)	29(37)	88(31)	58(24)	31(17)	15(17)	27(15)	30(13)	43(11)	26(9)
1,3-Dioxolane	73(100)	29(56)	44(53)	45(28)	28(21)	43(20)	27(13)	31(7)	74(5)	42(3)
Dipentene	68(100)	93(50)	67(44)	94(22)	39(22)	107(18)	92(18)	53(18)	136(16)	79(16)
Epichlorohydrin	57(100)	27(39)	29(32)	49(25)	31(22)	62(18)	28(16)	92(1)		
Ethanolamine (Glycinol)	30(100)	18(30)	28(15)	42(7)	31(6)	17(6)	61(5)	15(5)	43(3)	29(3)
Ethyl acetate	43(100)	29(46)	27(33)	45(32)	61(28)	28(25)	42(18)	73(11)	88(10)	70(10)
Ethyl acetoacetate	43(100)	29(24)	88(18)	28(16)	85(14)	27(12)	42(11)	60(9)	130(6)	45(6)
Ethyl alcohol	31(100)	45(44)	46(18)	27(18)	29(15)	43(14)	30(6)	42(3)	19(3)	14(3)
Ethylamine	30(100)	28(32)	44(20)	45(19)	27(13)	15(10)	42(9)	29(8)	41(5)	40(5)
Ethylbenzene	91(100)	106(31)	51(14)	39(10)	77(8)	65(8)	105(7)	92(7)	78(7)	27(6)
Ethyl bromide (Bromoethane)	108(100)	110(97)	29(62)	27(51)	28(35)	26(14)	93(6)	32(6)	95(5)	81(5)
Ethyl chloride (Cloroethane)	64(100)	28(91)	29(84)	27(75)	66(32)	26(28)	49(25)	51(8)	63(6)	65(4)
Ethylene carbonate	29(100)	44(62)	43(54)	88(40)	30(16)	28(11)	45(7)	58(6)	42(6)	73(4)
Ethylenediamine (1,2-Ethane-diamine)	30(100)	18(13)	42(6)	43(5)	27(5)	44(4)	29(4)	17(4)	15(4)	41(3)
Ethylene glycol	31(100)	33(35)	29(13)	32(11)	43(6)	27(5)	28(4)	62(3)	30(3)	44(2)
Ethylene glycol diethyl ether	31(100)	59(71)	29(58)	45(43)	27(33)	74(27)	43(15)	15(14)	28(12)	44(10)
Ethylene glycol dimethyl ether	45(100)	60(13)	29(13)	90(7)	58(6)	31(5)	28(5)	43(4)	59(3)	46(2)
Ethylene glycol monobutyl ether	57(100)	45(38)	29(35)	41(31)	87(16)	27(12)	56(11)	31(9)	75(7)	28(7)
Ethylene glycol monoethyl ether (Cellosolve)	31(100)	29(52)	59(50)	27(27)	45(26)	72(14)	43(14)	15(14)	28(8)	26(6)
Ethylene glycol monoethyl ether acetate	43(100)	31(34)	59(31)	72(28)	44(25)	29(24)	45(12)	27(11)	15(11)	87(7)
Ethylene glycol monomethyl ether	45(100)	31(15)	29(14)	28(11)	47(9)	76(6)	43(6)	58(4)	46(4)	27(4)
Ethylene glycol monomethyl ether acetate	43(100)	45(48)	58(42)	29(10)	42(4)	31(4)	73(3)	27(3)	59(2)	26(2)
Ethyl formate	31(100)	28(73)	27(51)	29(38)	45(34)	26(17)	74(11)	43(9)	47(8)	56(4)
Furan	68(100)	39(64)	40(9)	38(9)	42(6)	29(6)	37(5)	69(4)	34(2)	67(1)
Furfural	39(100)	96(55)	95(52)	38(38)	29(35)	37(29)	40(11)	97(9)	50(7)	42(7)
Furfuryl alcohol	98(100)	41(65)	39(59)	81(55)	53(53)	97(51)	42(49)	69(39)	70(36)	29(28)
Glycerol	61(100)	43(90)	31(57)	44(54)	29(38)	18(32)	27(12)	42(11)	60(10)	45(10)
Heptane	43(100)	41(56)	29(49)	57(47)	27(46)	71(45)	56(27)	42(26)	39(23)	70(18)
1-Heptanol	41(100)	70(87)	56(86)	31(78)	43(72)	29(70)	55(67)	27(65)	42(54)	69(41)
Hexane	57(100)	43(78)	41(77)	29(61)	27(57)	56(45)	42(39)	39(27)	28(16)	86(14)
1-Hexanol (Caproyl alcohol)	56(100)	43(78)	31(74)	41(71)	27(64)	29(59)	55(58)	42(53)	39(37)	69(27)
Hexylene glycol	59(100)	43(61)	56(25)	45(17)	41(16)	57(13)	42(13)	85(11)	61(10)	31(10)

Compound	e/m (intensity)								
Hexyl methyl ketone	43(100)	58(79)	41(56)	59(52)	71(49)	27(46)	29(36)	39(27)	57(18)	55(17)
Isobutyl acetate	43(100)	56(26)	73(15)	41(10)	29(5)	71(3)	57(3)	39(3)	27(3)	86(2)
Isobutyl alcohol	43(100)	33(73)	31(72)	41(66)	42(60)	27(43)	29(18)	39(17)	28(8)	74(6)
Isobutylamine	30(100)	28(9)	41(6)	73(5)	27(5)	39(4)	29(3)	15(3)	58(2)	56(2)
Isopentyl acetate	43(100)	70(49)	55(38)	61(15)	42(15)	41(14)	27(12)	87(11)	29(10)	73(9)
Isophorone	82(100)	39(20)	138(17)	54(13)	27(12)	41(10)	53(8)	83(7)	29(7)	55(6)
Isopropyl acetate	43(100)	61(17)	41(14)	87(9)	59(8)	27(8)	42(7)	39(4)	45(3)	44(2)
Isopropyl alcohol	45(100)	43(19)	27(17)	29(12)	41(7)	31(6)	19(6)	42(5)	44(4)	59(3)
Isoquinoline	129(100)	102(26)	51(20)	128(18)	50(11)	130(10)	75(10)	76(9)	103(8)	74(7)
d-Limonene (Citrene)	68(100)	93(50)	67(49)	41(22)	94(21)	79(21)	39(21)	136(20)	53(19)	121(16)
2,6-Lutidine (2,6-Dimethyl-pyridine)	107(100)	39(39)	106(29)	66(22)	92(18)	65(18)	38(12)	27(11)	79(9)	63(9)
Mesitylene (1,3,5-Trimethyl-benzene)	105(100)	120(64)	119(15)	77(13)	39(11)	106(9)	91(9)	51(8)	27(7)	121(6)
Mesityl oxide	55(100)	83(89)	43(73)	29(42)	98(36)	39(32)	27(28)	53(11)	41(10)	56(5)
Methyl acetate	43(100)	74(52)	28(38)	42(19)	59(17)	44(8)	32(8)	29(6)	31(4)	75(2)
Methylal (Dimethoxymethane)	45(100)	75(61)	29(59)	31(13)	30(6)	15(6)	47(5)	76(2)	46(2)	44(2)
Methyl alcohol	31(100)	29(72)	32(67)	15(42)	28(12)	14(10)	30(9)	13(6)	12(3)	16(2)
Methylamine	30(100)	31(87)	28(56)	29(19)	32(15)	15(12)	27(9)			
Methyl benzoate	105(100)	77(81)	51(45)	136(24)	50(18)	106(8)	78(6)	28(6)	39(5)	27(5)
Methylcyclohexane	83(100)	55(82)	41(60)	98(44)	42(35)	56(30)	27(29)	39(27)	69(23)	70(22)
Methyl ethyl ketone	43(100)	72(24)	29(19)	27(12)	57(7)	42(5)	26(4)	28(3)	44(2)	39(2)
N-Methylformamide	59(100)	30(54)	28(34)	29(13)	58(8)	15(7)	60(3)	41(3)	27(3)	31(2)
Methyl formate	31(100)	29(63)	32(34)	60(28)	30(7)	28(7)	44(2)	18(2)	61(1)	59(1)
Methyl iodide (Iodomethane)	142(100)	127(38)	141(14)	15(13)	139(5)	140(4)	128(3)	14(1)	13(1)	71(0)
Methyl isobutyl ketone	43(100)	58(84)	29(65)	41(56)	57(44)	27(42)	39(31)	85(19)	100(14)	42(14)
Methyl isopentyl ketone	43(100)	58(34)	27(14)	41(13)	15(13)	57(11)	39(9)	71(8)	59(8)	29(8)
2-Methylpentane	43(100)	42(53)	41(35)	27(31)	71(29)	39(20)	29(18)	57(11)	15(10)	70(7)
4-Methyl-2-pentanol	45(100)	43(47)	69(30)	41(27)	27(19)	39(13)	29(12)	87(11)	84(10)	57(10)
Methyl pentyl ketone	43(100)	58(60)	71(14)	41(11)	27(11)	59(9)	39(8)	29(8)	42(5)	114(4)
Methyl propyl ketone	43(100)	41(17)	86(12)	42(12)	27(11)	39(8)	71(7)	58(7)	45(7)	44(3)
N-Methyl-2-pyrrolidone	99(100)	44(89)	98(80)	42(60)	41(38)	43(17)	28(17)	71(13)	39(11)	70(10)
Morpholine	57(100)	29(100)	87(69)	28(69)	30(38)	56(33)	86(28)	31(28)	27(12)	15(7)
Nitrobenzene	77(100)	51(59)	123(42)	50(25)	30(15)	65(14)	39(10)	93(9)	74(7)	78(6)
Nitroethane	29(100)	30(12)	28(11)	26(9)	27(8)	43(5)	41(5)	14(5)	15(3)	46(2)
Nitromethane	30(100)	61(64)	46(39)	28(30)	45(8)	27(8)	44(7)	29(7)	60(5)	43(4)
1-Nitropropane	43(100)	27(93)	41(90)	39(34)	30(25)	44(20)	42(20)	26(20)	28(13)	54(12)
2-Nitropropane	43(100)	41(73)	27(71)	39(30)	30(18)	15(11)	42(9)	28(8)	26(8)	38(6)
Octane	43(100)	57(30)	85(25)	41(25)	71(19)	29(17)	56(14)	70(10)	42(10)	27(10)
1-Octanol	41(100)	56(85)	43(82)	55(81)	31(69)	27(69)	29(68)	42(62)	70(53)	69(48)
Pentachloroethane	167(100)	165(91)	117(90)	119(89)	83(58)	169(54)	130(43)	132(42)	60(40)	85(37)
Pentamethylene glycol (1,5-Pentanediol)	31(100)	56(85)	41(67)	57(59)	55(51)	44(45)	29(37)	43(31)	68(29)	27(26)
Pentane	43(100)	42(55)	41(45)	27(42)	29(26)	39(19)	57(13)	28(9)	15(9)	72(8)
1-Pentanol (Amyl alcohol)	42(100)	70(72)	55(65)	41(56)	31(47)	29(41)	27(26)	57(22)	28(22)	43(21)
Pentyl acetate (Amyl acetate)	43(100)	70(90)	42(52)	28(51)	61(50)	55(41)	73(21)	41(20)	29(14)	69(11)
2-Picoline (2-Methylpyridine)	93(100)	66(41)	39(31)	92(20)	78(19)	51(19)	65(16)	38(13)	50(12)	52(11)
α-Pinene	93(100)	92(30)	39(24)	41(23)	77(22)	91(21)	27(21)	79(18)	121(13)	53(10)
β-Pinene	93(100)	41(64)	69(47)	39(33)	27(31)	79(20)	77(18)	53(14)	94(13)	91(13)
Piperidine (Hexahydropyridine)	84(100)	85(53)	56(46)	57(43)	28(41)	29(37)	44(34)	42(30)	30(30)	43(25)
Propanenitrile	28(100)	54(63)	26(20)	27(17)	52(11)	55(10)	51(9)	15(9)	53(7)	25(7)
Propyl acetate	43(100)	61(19)	31(18)	27(15)	42(11)	73(9)	41(9)	29(9)	59(5)	39(5)
Propyl alcohol	31(100)	27(19)	29(18)	59(11)	42(9)	60(7)	41(7)	28(7)	43(3)	32(3)
Propylamine	30(100)	28(13)	59(8)	27(7)	41(5)	42(3)	39(3)	29(3)	26(3)	18(3)
Propylbenzene	91(100)	120(21)	92(10)	38(10)	65(9)	78(6)	51(6)	27(5)	63(4)	105(3)
1,2-Propylene glycol	45(100)	18(46)	29(21)	43(19)	31(18)	27(17)	28(11)	19(8)	44(6)	61(5)
Pseudocumene (1,2,4-Trimethyl-benzene)	105(100)	120(56)	119(17)	77(15)	39(15)	51(11)	91(10)	27(10)	106(9)	79(7)
Pyridine	79(100)	52(62)	51(31)	50(19)	78(11)	53(7)	39(7)	80(6)	27(3)	77(2)
Pyrrole	67(100)	41(58)	39(58)	40(51)	28(42)	38(20)	37(12)	66(7)	68(5)	27(3)
Pyrrolidine	43(100)	28(52)	70(33)	71(26)	42(22)	41(20)	27(16)	39(15)	29(10)	30(9)
2-Pyrrolidone	85(100)	42(43)	41(36)	28(33)	30(29)	56(16)	84(14)	40(12)	27(12)	29(9)
Quinoline	129(100)	51(28)	76(25)	128(24)	44(24)	50(20)	32(19)	75(18)	74(12)	103(11)
Styrene	104(100)	103(41)	78(32)	51(28)	77(23)	105(12)	50(12)	52(11)	39(11)	102(10)
Sulfolane	41(100)	28(94)	56(82)	55(72)	120(37)	27(32)	39(19)	29(17)	26(11)	48(5)
α-Terpinene	121(100)	93(85)	136(43)	91(40)	77(34)	39(33)	27(33)	79(27)	41(26)	43(18)
1,1,1-Tetrachloro-2,2-difluoroethane	167(100)	169(96)	117(85)	119(82)	171(31)	85(29)	121(26)	82(14)	47(14)	101(13)
Tetrachloro-1,2-difluoroethane	101(100)	103(64)	167(54)	169(52)	117(19)	119(18)	171(17)	105(11)	31(11)	132(9)

Compound	e/m (intensity)									
1,1,1,2-Tetrachloroethane	131(100)	133(96)	117(76)	119(73)	95(34)	135(31)	121(23)	97(23)	61(19)	60(18)
1,1,2,2-Tetrachloroethane	83(100)	85(63)	95(11)	87(10)	168(8)	133(8)	131(8)	96(8)	61(8)	60(8)
Tetrachloroethylene	166(100)	164(82)	131(71)	129(71)	168(45)	94(38)	47(31)	96(24)	133(20)	59(17)
Tetraethylene glycol	45(100)	89(10)	44(8)	43(6)	31(6)	29(6)	27(6)	101(5)	75(5)	28(5)
Tetrahydrofuran	42(100)	41(52)	27(33)	72(29)	71(27)	39(24)	43(22)	29(22)	40(13)	15(10)
1,2,3,4-Tetrahydronaphthalene	104(100)	132(53)	91(43)	51(17)	39(17)	131(15)	117(15)	115(14)	78(13)	77(13)
Tetrahydropyran	41(100)	28(64)	56(57)	45(57)	29(51)	27(49)	85(47)	86(42)	39(28)	55(23)
Tetramethylsilane	73(100)	43(14)	45(12)	74(8)	29(7)	15(5)	75(4)	44(4)	42(4)	31(4)
Toluene	91(100)	92(73)	39(20)	65(14)	63(11)	51(11)	50(7)	27(6)	93(5)	90(5)
o-Toluidine	106(100)	107(83)	77(17)	79(13)	39(12)	53(10)	52(10)	54(9)	51(9)	28(9)
Triacetin	43(100)	103(44)	145(34)	116(17)	115(13)	44(10)	86(9)	28(8)	73(7)	42(7)
Tributylamine	142(100)	100(19)	143(11)	29(8)	185(7)	57(6)	44(6)	41(6)	30(5)	86(4)
1,1,1-Trichloroethane	97(100)	99(64)	61(58)	26(31)	27(24)	117(19)	63(19)	119(18)	35(17)	62(11)
1,1,2-Trichloroethane	97(100)	83(95)	99(62)	85(60)	61(58)	26(23)	96(21)	63(19)	27(17)	98(15)
Trichloroethylene	95(100)	130(90)	132(85)	60(65)	97(64)	35(40)	134(27)	47(26)	62(21)	59(13)
Trichlorofluoromethane	101(100)	103(66)	66(13)	105(11)	35(11)	47(9)	31(8)	82(4)	68(4)	37(4)
1,1,2-Trichlorotrifluoroethane	101(100)	151(68)	103(64)	85(45)	31(45)	153(44)	35(20)	66(19)	47(18)	87(14)
Triethanolamine	118(100)	56(69)	45(60)	42(56)	44(27)	43(25)	41(14)	116(8)	57(8)	86(7)
Triethylamine	86(100)	30(68)	58(37)	28(24)	29(23)	27(19)	44(18)	101(17)	42(16)	56(8)
Triethylene glycol	45(100)	58(11)	89(9)	31(8)	29(8)	75(7)	44(7)	43(7)	27(7)	28(5)
Triethyl phosphate	99(100)	81(71)	155(56)	82(45)	45(45)	109(44)	127(41)	43(24)	125(16)	111(14)
Trimethylamine	58(100)	59(47)	30(29)	42(26)	44(17)	15(14)	28(10)	18(10)	43(8)	57(7)
Trimethylene glycol (1,3-Propanediol)	28(100)	58(93)	31(76)	57(70)	29(40)	27(26)	45(24)	43(23)	19(18)	30(17)
Trimethyl phosphate	110(100)	109(35)	79(34)	95(25)	80(23)	15(20)	140(18)	47(10)	31(7)	139(5)
Veratrole	138(100)	95(65)	77(48)	123(44)	52(42)	41(33)	65(30)	51(29)	39(19)	63(17)
o-Xylene	91(100)	106(40)	39(21)	105(17)	51(17)	77(15)	27(12)	65(10)	92(8)	79(8)
m-Xylene	91(100)	106(65)	105(29)	39(18)	51(15)	77(14)	27(10)	92(8)	79(8)	78(8)
p-Xylene	91(100)	106(62)	105(30)	51(16)	39(16)	77(13)	27(11)	92(7)	78(7)	65(7)

Section 9
Molecular Structure and Spectroscopy

BOND LENGTHS IN CRYSTALLINE ORGANIC COMPOUNDS

The following table gives average interatomic distances for bonds between the elements H, B, C, N, O, F, Si, P, S, Cl, As, Se, Br, Te, and I as determined from X-ray and neutron diffraction measurements on organic crystals. The table has been derived from an analysis of high-precision structure data on about 10,000 crystals contained in the 1985 version of the Cambridge Structural Database, which is maintained by the Cambridge Crystallographic Data Center. The explanation of the columns is:

Column 1: Specification of elements in the bond, with coordination number given in parentheses, and bond type (single, double, etc.). For carbon, the hybridization state is given.

Column 2: Substructure in which the bond is found. The target bond is set in boldface. Where X is not specified, it denotes any element type. C# indicates any sp^3 carbon atom, and C* denotes an sp^3 carbon whose bonds, in addition to those specified in the linear formulation, are to C and H atoms only.

Column 3: d is the unweighted mean in Å units of all the values for that bond length found in the sample.

Column 4: m is the median in Å units of all values.

Column 5: σ is the standard deviation in the sample.

Column 6: q_1 is the lower quartile for the sample (i.e., 25% of values are less than q_1 and 75% exceed it).

Column 7: q_u is the upper quartile for the sample.

Column 8: n is number of observations in the sample.

Column 9: Notes refer to the footnotes in Appendix 1.

References to special cases are given in a shorthand form and listed in Appendix 2. Further information on the method of analysis of the data may be found in the reference cited below.

The table is reprinted with permission of the authors, the Royal Society of Chemistry, and the International Union of Crystallography.

Reference

Frank H. Allen, Olga Kennard, David G. Watson, Lee Brammer, A. Guy Orpen, and Robin Taylor, *J. Chem. Soc. Perkin Trans.* II, S1–S19, 1987.

Bond	Substructure	d	m	σ	q_1	q_u	n	Note
As(3)–As(3)	X$_2$–**As–As**–X$_2$	2.459	2.457	0.011	2.456	2.466	8	
As–B	see CUDLOC (2.065), CUDLUI (2.041)							
As–BR	see CODDEE, CODDII (2.346–3.203)							
As(4)–C	**X$_3$–As–CH$_3$**	1.903	1.907	0.016	1.893	1.916	12	
	(X)$_2$(C,O,S=)**As–C**sp^3	1.927	1.929	0.017	1.921	1.937	16	
	As–Car in Ph$_4$As$^+$	1.905	1.909	0.012	1.897	1.912	108	
	(X)$_2$(C,O,S=)**As–C**ar	1.922	1.927	0.016	1.908	1.934	36	
As(3)–C	X$_2$–**As–C**sp^3	1.963	1.965	0.017	1.948	1.978	6	
	X$_2$–**As–C**ar	1.956	1.956	0.015	1.944	1.964	41	
As(3)–Cl	X$_2$–**As–Cl**	2.268	2.256	0.039	2.247	2.281	10	
As(6)–F	in **AsF$_6^-$**	1.678	1.676	0.020	1.659	1.695	36	
As(3)–I	see OPIMAS (2.579, 2.590)							
As(3)–N(3)	X$_2$–**As–N**–X$_2$	1.858	1.858	0.029	1.839	1.873	19	
As(4)–N(2)	see TPASSN (1.837)							
As(4)–O	(X)$_2$(O=)**As–OH**	1.710	1.712	0.017	1.695	1.726	6	
As(3)–O	see ASAZOC, PHASOC01 (1.787–1.845)							
As(4)=O	**X$_3$–As=O**	1.661	1.661	0.016	1.652	1.667	9	
As(3)=P(3)	see BELNIP (2.350, 2.362)							†
As(3)–P(3)	see BUTHAZ10 (2.124)							†
As(3)–S	X$_2$–**As–S**	2.275	2.266	0.032	2.247	2.298	14	
As(4)=S	**X$_3$–As=S**	2.083	2.082	0.004	2.080	2.086	9	
As(3)–Se(2)	see COSDIX, ESEARS (2.355–2.401)							†
As(3)–Si(4)	see BICGEZ, MESIAD (2.351–2.365)							†
As(3)–Te(2)	see ETEARS (2.571, 2.576)							†
B(n)–B(n)	n = 5–7 in boron cages	1.775	1.773	0.031	1.763	1.786	688	
B(4)–B(4)	see CETTAW (2.041)							
B(4)–B(3)	see COFVOI (1.698)							
B(3)–B(3)	X$_2$–**B–B**–X$_2$	1.701	1.700	0.014	1.691	1.712	8	
B(6)–BR		1.967	1.971	0.014	1.954	1.979	7	†
B(4)–BR		2.017	2.008	0.031	1.990	2.044	15	†
B(n)–C	n = 5–7: **B–C** in cages	1.716	1.717	0.020	1.707	1.728	96	
	n = 3–4: **B–C**sp^3 not cages	1.597	1.599	0.022	1.585	1.611	29	1
	n = 4: **B–C**ar	1.606	1.607	0.012	1.596	1.615	41	
	n = 4: **B–C**ar in Ph$_4$B$^-$	1.643	1.643	0.006	1.641	1.645	16	
B(n)–C	n = 3: **B–C**ar	1.556	1.552	0.015	1.546	1.566	24	
B(n)–Cl	**B(5)–Cl** and **B(3)–Cl**	1.751	1.751	0.011	1.743	1.761	14	

Bond	Substructure	d	m	σ	q_l	q_u	n	Note
	B(4)–Cl	1.833	1.833	0.013	1.821	1.843	22	
B(4)–F	**B–F** (B neutral)	1.366	1.368	0.017	1.356	1.375	25	
	B⁻–F in BF₄⁻	1.365	1.372	0.029	1.352	1.390	84	
B(4)–I	see TMPBTI (2.220, 2.253)							
B(4)–N(3)	X₃–**B–N**(=C)(X)	1.611	1.617	0.013	1.601	1.625	8	
	in pyrazaboles	1.549	1.552	0.015	1.536	1.560	10	
B(3)–N(3)	X₂–**B–N**–C₂: all coplanar	1.404	1.404	0.014	1.389	1.408	40	2
	for τ(BN) > 30° see BOGSUL, BUSHAY, CILRUK							
	(1.434–1.530)							
	S₂–**B–N**–X₂	1.447	1.443	0.013	1.435	1.470	14	
B(4)–O	**B⁻–O** in BO₄⁻	1.468	1.468	0.022	1.453	1.479	24	
	for neutral B–O see Note 3							3
B(3)–O(2)	X₂–**B–O**–X	1.367	1.367	0.024	1.349	1.382	35	
B(n)–P	n = 4: **B–P**	1.922	1.927	0.027	1.900	1.954	10	
	n = 3: see BUPSIB10 (1.892, 1.893)							
B(4)–S	**B(4)–S**(3)	1.930	1.927	0.009	1.925	1.934	10	
	B(4)–S(2)	1.896	1.896	0.004	1.893	1.899	6	
B(3)–S	N–**B–S**₂	1.806	1.806	0.010	1.799	1.816	28	
	(=X–)(N–)**B–S**	1.851	1.854	0.013	1.842	1.859	10	
Br–Br	see BEPZEB, TPASTB	2.542	2.548	0.015	2.526	2.551	4	
Br–C	**Br–C***	1.966	1.967	0.029	1.951	1.983	100	4
	Br–Csp³ (cyclopropane)	1.910	1.910	0.010	1.900	1.914	8	
	Br–Csp²	1.883	1.881	0.015	1.874	1.894	31	4
	Br–Car (mono-Br + m.p-Br₂)	1.899	1.899	0.012	1.892	1.906	119	4
	Br–Car (o-Br₂)	1.875	1.872	0.011	1.864	1.884	8	4
⁻Br(2)–Cl	see TEACBR (2.362–2.402)							†
Br–I	see DTHIBR10 (2.646), TPHOSI (2.695)							
Br–N	see NBBZAM (1.843)							
Br–O	see CIYFOF	1.581	1.581	0.007	1.574	1.587	4	
Br–P	see CISTED (2.366)							
Br–S(2)	see BEMLIO (2.206)							†
Br–S(3)	see CIWYIQ (2.435, 2.453)							†
Br–S(3)⁺	see THINBR (2.321)							†
Br–SE	see CIFZUM (2.508, 2.619)							
Br–Si	see BIZJAV (2.284)							
Br–Te	In **Br₆Te**²⁻ see CUGBAH (2.692–2.716)							
	Br–Te(4) see BETUTE10 (3.079, 3.015)							
	Br–Te(3) see BTUPTE (2.835)							
Csp³–Csp³	C#–**CH₂–CH₃**	1.513	1.514	0.014	1.507	1.523	192	
	(C#)₂–**CH–CH₃**	1.524	1.526	0.015	1.518	1.534	226	
	(C#)₃–**C–CH₃**	1.534	1.534	0.011	1.527	1.541	825	
	C#–**CH₂–CH₂**–C#	1.524	1.524	0.014	1.516	1.532	2459	
	(C#)₂–**CH–CH₂**–C#	1.531	1.531	0.012	1.524	1.538	1217	
	(C#)₃–**C–CH₂**–C#	1.538	1.539	0.010	1.533	1.544	330	
	(C#)₂–**CH–CH**–(C#)₂	1.542	1.542	0.011	1.536	1.549	321	
	(C#)₃–**C–CH**–(C#)₂	1.556	1.556	0.011	1.549	1.562	215	
	(C#)₃–**C–C**–(C#)₃	1.588	1.580	0.025	1.566	1.610	21	
	C*–C* (overall)	1.530	1.530	0.015	1.521	1.539	5777	5,6
	in cyclopropane (any subst.)	1.510	1.509	0.026	1.497	1.523	888	7
	in cyclobutane (any subst.)	1.554	1.553	0.021	1.540	1.567	679	8
	in cyclopentane (C,H-subst.)	1.543	1.543	0.018	1.532	1.554	1641	
	in cyclohexane (C,H-subst.)	1.535	1.535	0.016	1.525	1.545	2814	
	cyclopropyl-C* (exocyclic)	1.518	1.518	0.019	1.505	1.531	366	7
	cyclobutyl-C* (exocyclic)	1.529	1.529	0.016	1.519	1.539	376	8
	cyclopentyl-C* (exocyclic)	1.540	1.541	0.017	1.527	1.549	956	
	cyclohexyl-C* (exocyclic)	1.539	1.538	0.016	1.529	1.549	2682	
	in cyclobutene (any subst.)	1.573	1.574	0.017	1.566	1.586	25	8
	in cyclopentene (C,H-subst.)	1.541	1.539	0.015	1.532	1.549	208	
	in cyclohexene (C,H-subst.)	1.541	1.541	0.020	1.528	1.554	586	
	in oxirane (epoxide)	1.466	1.466	0.015	1.458	1.474	249	9
	in aziridine	1.480	1.481	0.021	1.465	1.496	67	9

Bond	Substructure	d	m	σ	q_1	q_u	n	Note
	in oxetane	1.541	1.541	0.019	1.527	1.557	16	
	in azetidine	1.548	1.543	0.018	1.536	1.558	22	
	oxiranyl-C* (exocyclic)	1.509	1.507	0.018	1.497	1.519	333	9
	aziridinyl-C* (exocyclic)	1.512	1.512	0.018	1.496	1.526	13	9
Csp^3–Csp^2	CH₃–C=C	1.503	1.504	0.011	1.497	1.509	215	
	C#–CH₂–C=C	1.502	1.502	0.013	1.494	1.510	483	
	(C#)₂–CH–C=C	1.510	1.510	0.014	1.501	1.518	564	
	(C#)₃–C–C=C	1.522	1.522	0.016	1.511	1.533	193	
Csp^3–Csp^2	C*–C=C (overall)	1.507	1.507	0.015	1.499	1.517	1456	5
	C*–C=C (endocyclic)							
	in cyclopropene	1.509	1.508	0.016	1.500	1.516	20	10
	in cyclobutene	1.513	1.512	0.018	1.500	1.525	50	8
	in cyclopentene	1.512	1.512	0.014	1.502	1.521	208	
	in cyclohexene	1.506	1.505	0.016	1.495	1.516	391	
	in cyclopentadiene	1.502	1.503	0.019	1.490	1.515	18	
	in cyclohexa-1,3-diene	1.504	1.504	0.017	1.491	1.517	56	
	C*–C=C (exocyclic):							
	cyclopropenyl-C*	1.478	1.475	0.012	1.470	1.485	7	10
	cyclobutenyl-C*	1.489	1.483	0.015	1.479	1.496	11	8
	cyclopentenyl-C*	1.504	1.506	0.012	1.495	1.512	115	
	cyclohexenyl-C*	1.511	1.511	0.013	1.502	1.519	292	
	C*CH=O in aldehydes	1.510	1.510	0.008	1.501	1.518	7	
	(C*)₂–C=O							
	in ketones	1.511	1.511	0.015	1.501	1.521	952	11
	in cyclobutanone	1.529	1.530	0.016	1.514	1.545	18	
	in cyclopentanone	1.514	1.514	0.016	1.505	1.523	312	
	acyclic and 6 + rings	1.509	1.509	0.016	1.499	1.519	626	
	C*–COOH in carboxylic acids	1.502	1.502	0.014	1.495	1.510	176	
	C*–COO⁻ in carboxylate anions	1.520	1.521	0.011	1.516	1.528	57	
	C*–C(=O)(–OC*)							
	in acyclic esters	1.497	1.496	0.018	1.484	1.509	553	12
	in β-lactones	1.519	1.519	0.020	1.500	1.538	4	13
	in γ-lactones	1.512	1.512	0.015	1.501	1.521	110	12
	in δ-lactones	1.504	1.502	0.013	1.495	1.517	27	12
	cyclopropyl (C)–C=O in ketones, acids and esters	1.486	1.485	0.018	1.474	1.497	105	7
	C*–C(=O)(–NH₂) in acyclic amides	1.514	1.512	0.016	1.506	1.526	32	14
	C*–C(=O)(–NHC*) in acyclic amides	1.506	1.505	0.012	1.498	1.515	78	14
	C*–C(=O)[–N(C*)₂] in acyclic amides	1.505	1.505	0.011	1.496	1.517	15	14
Csp^3–Car	CH₃–Car	1.506	1.507	0.011	1.501	1.513	454	
	C#–CH₂–Car	1.510	1.510	0.009	1.505	1.516	674	
	(C#)₂–CH–Car	1.515	1.515	0.011	1.508	1.522	363	
	(C#)₃–C–Car	1.527	1.530	0.016	1.517	1.539	308	
	C*–Car (overall)	1.513	1.513	0.014	1.505	1.521	1813	
	cyclopropyl (C)–Car	1.490	1.490	0.015	1.479	1.503	90	7
Csp^3–Csp^1	C*–C≡C	1.466	1.465	0.010	1.460	1.469	21	15
	C#–C≡C	1.472	1.472	0.012	1.464	1.481	88	15
	C*–C≡N	1.470	1.469	0.013	1.463	1.479	106	7b
	cyclopropyl (C)–C≡N	1.444	1.447	0.010	1.436	1.451	38	7
Csp^2–Csp^2	C=C–C=C							
	(conjugated)	1.455	1.455	0.011	1.447	1.463	30	16,18
	(unconjugated)	1.478	1.476	0.012	1.470	1.479	8	17,18
	(overall)	1.460	1.460	0.015	1.450	1.470	38	
	C=C–C=C–C=C	1.443	1.445	0.013	1.431	1.454	29	18
	C=C–C=C (endocyclic in TCNQ)	1.432	1.433	0.012	1.424	1.441	280	19
	C=C–C(=O)(–C*)							
	(conjugated)	1.464	1.462	0.018	1.453	1.476	211	16,18
	(unconjugated)	1.484	1.486	0.017	1.475	1.497	14	17,18
	(overall)	1.465	1.462	0.018	1.453	1.478	226	
	C=C–C(=O)–C=C							
	in benzoquinone (C,H-subst. only)	1.478	1.476	0.011	1.469	1.488	28	
	in benzoquinone (any subst.)	1.478	1.478	0.031	1.464	1.498	172	

Bond	Substructure	d	m	σ	q_l	q_u	n	Note
	non-quinonoid	1.456	1.455	0.012	1.447	1.464	28	
	C=**C**–COOH	1.475	1.476	0.015	1.461	1.488	22	
	C=**C**–COOC*	1.488	1.489	0.014	1.478	1.497	113	
	C=**C**–COO⁻	1.502	1.499	0.017	1.488	1.510	11	
	HOO**C**–**C**OOH	1.538	1.537	0.007	1.535	1.541	9	
	HOO**C**–**C**OO⁻	1.549	1.552	0.009	1.546	1.553	13	
	⁻OO**C**–**C**OO⁻	1.564	1.559	0.022	1.554	1.568	9	
	formal Csp^2–Csp^2 single bond in selected non-fused heterocycles:							
	in 1*H*-pyrrole (C3–C4)	1.412	1.410	0.016	1.401	1.427	29	
	in furan (C3–C4)	1.423	1.423	0.016	1.412	1.433	62	
	in thiophene (C3–C4)	1.424	1.425	0.015	1.415	1.433	40	
	in pyrazole (C3–C4)	1.410	1.412	0.016	1.400	1.418	20	
	in isoxazole (C3–C4)	1.425	1.425	0.016	1.413	1.438	9	
	in furazan (C3–C4)	1.428	1.427	0.007	1.422	1.435	6	
	in furoxan (C3–C4)	1.417	1.417	0.006	1.412	1.422	14	
Csp^2–C*ar*	C=**C**–**C***ar*							
	(conjugated)	1.470	1.470	0.015	1.463	1.480	37	16,18
Csp^2–C*ar*		1.488	1.490	0.012	1.480	1.496	87	17,18
	(overall)	1.483	1.483	0.015	1.472	1.494	124	
	cyclopropenyl (C=**C**)–**C***ar*	1.447	1.448	0.006	1.441	1.452	8	10
	C*ar*–**C**(=O)–C*	1.488	1.489	0.016	1.478	1.500	84	
	C*ar*–**C**(=O)–**C***ar*	1.480	1.481	0.017	1.468	1.494	58	
	C*ar*–**C**OOH	1.484	1.485	0.014	1.474	1.491	75	
	C*ar*–**C**(=O)(–OC*)	1.487	1.487	0.012	1.480	1.494	218	
	C*ar*–**C**OO⁻	1.504	1.509	0.014	1.495	1.512	26	
	C*ar*–**C**(–O)–NH₂	1.500	1.503	0.020	1.498	1.510	19	
	C*ar*–**C**=N–C#							
	(conjugated)	1.476	1.478	0.014	1.466	1.486	27	16
	(unconjugated)	1.491	1.490	0.008	1.485	1.496	48	17
	(overall)	1.485	1.487	0.013	1.481	1.493	75	
	in indole (C3–C3a)	1.434	1.434	0.011	1.428	1.439	40	
Csp^2–Csp^1	C=**C**–**C**≡C	1.431	1.427	0.014	1.425	1.441	11	7b
	C=**C**–**C**≡N in TCNQ	1.427	1.427	0.010	1.420	1.433	280	19
C*ar*–C*ar*	in biphenyls (*ortho* subst. all **H**)	1.487	1.488	0.007	1.484	1.493	30	
	(≥1 non-**H** *ortho*-subst.)	1.490	1.491	0.010	1.486	1.495	212	
C*ar*–Csp^1	**C***ar*–**C**≡C	1.434	1.436	0.006	1.430	1.437	37	
	C*ar*–**C**≡N	1.443	1.444	0.008	1.436	1.448	31	
Csp^1–Csp^1	C≡**C**–**C**=C	1.377	1.378	0.012	1.374	1.384	21	
Csp^2=Csp^2	C*–**CH**=**CH**₂	1.299	1.300	0.027	1.280	1.311	42	
	(**C***)₂–**C**=**CH**₂	1.321	1.321	0.013	1.313	1.328	77	
	C*–**CH**=**CH**–C*							
	(*cis*)	1.317	1.318	0.013	1.310	1.323	106	
	(*trans*)	1.312	1.311	0.011	1.304	1.320	19	
	(overall)	1.316	1.317	0.015	1.309	1.323	127	
	(**C***)₂–**C**=**CH**–C*	1.326	1.328	0.011	1.319	1.334	168	
	(**C***₂–**C**=**C**–(**C***)₂	1.331	1.330	0.009	1.326	1.334	89	
	(**C***,H)₂–**C**=**C**–(**C***,H)₂ (overall)	1.322	1.323	0.014	1.315	1.331	493	5
	in cyclopropene (any subst.)	1.294	1.288	0.017	1.284	1.302	10	10
	in cyclobutene (any subst.)	1.335	1.335	0.019	1.324	1.347	25	8
	in cyclopentene (C,H-subst.)	1.323	1.324	0.013	1.314	1.331	104	
	in cyclohexene (C,H-subst.)	1.326	1.325	0.012	1.318	1.334	196	
	C=**C**=C (allenes, any subst.)	1.307	1.307	0.005	1.303	1.310	18	
	C=**C**–**C**=C (C,**H** subst., conjugated)	1.330	1.330	0.014	1.322	1.338	76	16
	C=**C**–**C**=**C**–C=C (C,**H** subst., conjugated)	1.345	1.345	0.012	1.337	1.350	58	16
	C=**C**–**C***ar* (C,H subst., conjugated)	1.339	1.340	0.011	1.334	1.346	124	16
	C=**C** in cyclopenta-1,3-diene (any subst.)	1.341	1.341	0.017	1.328	1.356	18	
	C=**C** in cyclohexa-1,3-diene (any subst.)	1.332	1.332	0.013	1.323	1.341	56	
	in **C**=**C**–C=O							
	(C,**H** subst., conjugated)	1.340	1.340	0.013	1.332	1.348	211	16,18
	(C,**H** subst., unconjugated)	1.331	1.330	0.008	1.326	1.339	14	17,18

Bond	Substructure	d	m	σ	q_l	q_u	n	Note
	(C,H subst., overall)	1.340	1.339	0.013	1.332	1.348	226	
	in cyclohexa-2,5-dien-1-ones	1.329	1.327	0.011	1.321	1.335	28	
	in p-benzoquinones							
	(C*,H subst.)	1.333	1.337	0.011	1.325	1.338	14	
	(any subst.)	1.349	1.339	0.030	1.330	1.364	86	
	in TCNQ							
	(endocyclic)	1.352	1.353	0.010	1.345	1.358	142	19
	(exocyclic)	1.392	1.391	0.017	1.379	1.405	139	19
	C=**C**–OH in enol tautomers	1.362	1.360	0.020	1.349	1.370	54	
	in heterocycles (any subst.):							
	1H-pyrrole (C2–C3, C4–C5)	1.375	1.377	0.018	1.361	1.388	58	
	furan (C2–C3, C4–C5)	1.341	1.342	0.021	1.329	1.351	125	
	thiophene (C2–C3, C4–C5)	1.362	1.359	0.025	1.346	1.377	60	
	pyrazole (C4–C5)	1.369	1.372	0.019	1.362	1.383	20	
	imidazole (C4–C5)	1.360	1.361	0.014	1.352	1.367	44	
	isoxazole (C4–C5)	1.341	1.336	0.012	1.331	1.355	9	
	indole (C2–C3)	1.364	1.363	0.012	1.355	1.371	40	
$Car \simeq Car$	in phenyl rings with C*, H subst. only							
	H–**C** $\simeq$ **C**–H	1.380	1.381	0.013	1.372	1.388	2191	
	C*–**C** $\simeq$ **C**–H	1.387	1.388	0.010	1.382	1.393	891	
	C*–**C** $\simeq$ **C**–C*	1.397	1.397	0.009	1.392	1.403	182	
	C $\simeq$ **C** (overall)	1.384	1.384	0.013	1.375	1.391	3264	
	F–**C** $\simeq$ **C**–F	1.372	1.374	0.011	1.366	1.380	84	4
	Cl–**C** $\simeq$ **C**–Cl	1.388	1.389	0.014	1.380	1.398	152	4
	in naphthalene (D_{2h}, any subst.)							
	C1–C2	1.364	1.364	0.014	1.356	1.373	440	
	C2–C3	1.406	1.406	0.014	1.397	1.415	218	
	C1–C8a	1.420	1.419	0.012	1.412	1.426	440	
	C4a–C8a	1.422	1.424	0.011	1.417	1.429	109	
$Car \simeq Car$	in anthracene (D_{2h}, any subst.)							
	C1–C2	1.356	1.356	0.009	1.350	1.360	56	
	C2–C3	1.410	1.410	0.010	1.401	1.416	34	
	C1–C9a	1.430	1.430	0.006	1.426	1.434	56	
	C4a–C9a	1.435	1.436	0.007	1.429	1.440	34	
	C9–C9a	1.400	1.402	0.009	1.395	1.406	68	
	in pyridine (C,H subst.)	1.379	1.381	0.012	1.371	1.387	276	20
	(any subst.)	1.380	1.380	0.015	1.371	1.389	537	20
	in pyridinium cation							
	(N^+ –H; C,H subst. on C)							
	C2–C3	1.373	1.375	0.012	1.368	1.380	30	
	C3–C4	1.379	1.380	0.011	1.371	1.388	30	
	(N^+ –X; C,H subst. on C)							
	C2–C3	1.373	1.372	0.019	1.362	1.382	151	
	C3–C4	1.383	1.385	0.019	1.372	1.394	151	
	in pyrazine (H subst. on C)	1.379	1.377	0.010	1.370	1.388	10	
	(any subst. on C)	1.405	1.405	0.024	1.388	1.420	60	
	in pyrimidine (C,H subst. on C)	1.387	1.389	0.018	1.379	1.400	28	
$Csp^1 \equiv Csp^1$	X–**C**≡**C**–X	1.183	1.183	0.014	1.174	1.193	119	15
	C,H–**C**≡**C**–C,H	1.181	1.181	0.014	1.173	1.192	104	15
	in **C**≡**C**–C(sp^2,ar)	1.189	1.193	0.010	1.181	1.195	38	15
	in **C**≡**C**–**C**≡**C**	1.192	1.192	0.010	1.187	1.197	42	15
	in **CH**≡**C**–C#	1.174	1.174	0.011	1.167	1.180	42	15
Csp^3–Cl	Omitting 1,2-dichlorides:							
	C–**CH$_2$**–**Cl**	1.790	1.790	0.007	1.783	1.795	13	4
	C$_2$–**CH**–**Cl**	1.803	1.802	0.003	1.800	1.807	8	4
	C$_3$–**C**–**Cl**	1.849	1.856	0.011	1.837	1.858	5	4
	X–CH$_2$–Cl (X = C,H,N,O)	1.790	1.791	0.011	1.783	1.797	37	4
	X$_2$–CH–Cl (X = C,H,N,O)	1.805	1.803	0.014	1.800	1.812	26	4
	X$_3$–**C**–Cl (X = C,H,N,O)	1.843	1.838	0.014	1.835	1.858	7	4
	X$_2$–**C**–Cl$_2$ (X = C,H,N,O)	1.779	1.776	0.015	1.769	1.790	18	4
	X–**C**–Cl$_3$ (X = C,H,N,O)	1.768	1.765	0.011	1.761	1.776	33	4

Bond	Substructure	d	m	σ	q_1	q_u	n	Note
	Cl–CH(–C)–CH(–C)–Cl	1.793	1.793	0.013	1.786	1.800	66	4
	Cl–C(–C$_2$)–C(–C$_2$)–Cl	1.762	1.760	0.010	1.757	1.765	54	4
	cyclopropyl–Cl	1.755	1.756	0.011	1.749	1.763	64	
Csp^2–Cl	C=C–Cl (C,H,N,O subst. on C)	1.734	1.729	0.019	1.719	1.748	63	4
	C=C–Cl$_2$ (C,H,N,O subst. on C)	1.720	1.716	0.013	1.708	1.729	20	4
	Cl–C=C–Cl	1.713	1.711	0.011	1.705	1.720	80	4
Car–Cl	Car–Cl (mono–Cl + m,p-Cl$_2$)	1.739	1.741	0.010	1.734	1.745	340	4
	Car–Cl (o–Cl$_2$)	1.720	1.720	0.010	1.713	1.717	364	4
Csp^1Cl	see HCLENE10 (1.634, 1.646)							
Csp^3–F	Omitting 1,2-difluorides							
	C–CH$_2$–F and C$_2$–CH–F	1.399	1.399	0.017	1.389	1.408	25	4
	C$_3$–C–F	1.428	1.431	0.009	1.421	1.435	11	4
	(C*,H)$_2$–C–F$_2$	1.349	1.347	0.012	1.342	1.356	58	4
	C*–C–F$_3$	1.336	1.334	0.007	1.330	1.344	12	4
	F–C*–C*–F	1.371	1.374	0.007	1.362	1.375	26	4
	X$_3$–C–F (X = C,H,N,O)	1.386	1.389	0.033	1.373	1.408	70	4
	X$_2$–C–F$_2$ (X = C,H,N,O)	1.351	1.349	0.013	1.342	1.356	58	4
	X–C–F$_3$ (X = C,H,N,O)	1.322	1.323	0.015	1.314	1.332	309	4
	F–C(–X)$_2$–C(–X)$_2$–F (X = C,H,N,O)	1.373	1.374	0.009	1.362	1.377	30	4
	F–C(–X)$_2$–NO$_2$ (X = any subst.)	1.320	1.319	0.009	1.312	1.327	18	
Csp^2–F	C=C–F (C,H,N,O subst. on C)	1.340	1.340	0.013	1.334	1.346	34	4
Car–F	Car–F (mono-F + m,p-F$_2$)	1.363	1.362	0.008	1.357	1.368	38	4
	Car–F (o-F$_2$)	1.340	1.340	0.009	1.336	1.344	167	4
Csp^3–H	C–C–H$_3$ (methyl)	1.059	1.061	0.030	1.039	1.083	83	21
	C$_2$–C–H$_2$ (primary)	1.092	1.095	0.013	1.088	1.099	100	21
	C$_3$–C–H (secondary)	1.099	1.097	0.004	1.095	1.103	14	21
	C$_{2,3}$–C–H (primary and secondary)	1.093	1.095	0.012	1.089	1.100	118	21
	X–C–H$_3$ (methyl)	1.066	1.074	0.028	1.049	1.087	160	21
	X$_2$–C–H$_2$ (primary)	1.092	1.095	0.012	1.088	1.099	230	21
	X$_3$–C–H (secondary)	1.099	1.099	0.007	1.095	1.103	117	21
	X$_{2,3}$–C–H (primary and secondary)	1.094	1.096	0.011	1.091	1.100	348	21
Csp^2–H	C–C=C–H	1.077	1.079	0.012	1.074	1.085	14	21
Car–H	Car–H	1.083	1.083	0.011	1.080	1.087	218	21
Csp^3–I	C*–I	2.162	2.159	0.015	2.149	2.179	15	4
Car–I	Car–I	2.095	2.095	0.015	2.089	2.104	51	4
Csp^3–N(4)	C*–NH$_3^+$	1.488	1.488	0.013	1.482	1.495	298	
	(C*)$_2$–NH$_2^+$	1.494	1.493	0.016	1.484	1.503	249	
	(C*)$_3$–NH$^+$	1.502	1.502	0.015	1.491	1.512	509	
	(C*)$_4$–N$^+$	1.510	1.509	0.020	1.496	1.523	319	
	C*–N$^+$ (overall)	1.499	1.498	0.018	1.488	1.510	1370	
Csp^3–N(3)	C*–N$^+$ in N-subst. pyridinium	1.485	1.484	0.009	1.477	1.490	32	
	C*–NH$_2$ (Nsp^3: pyramidal)	1.469	1.470	0.010	1.462	1.474	19	22
	(C*)$_2$–NH (Nsp^3: pyramidal)	1.469	1.467	0.012	1.461	1.477	152	5,22
	(C*)$_3$–N (Nsp^3: pyramidal)	1.469	1.468	0.014	1.460	1.476	1042	5,22
	C*–Nsp^3 (overall)	1.469	1.468	0.014	1.460	1.476	1201	
	Csp^3–Nsp^3							
	in aziridine	1.472	1.471	0.016	1.464	1.482	134	
	in azetidine	1.484	1.481	0.018	1.472	1.495	21	
	in tetrahydropyrrole	1.475	1.473	0.016	1.464	1.483	66	
	in piperidine	1.473	1.473	0.013	1.460	1.479	240	
	Csp^3–Nsp^2 (N planar) in:							23
	acyclic amides C*–NH–C=O	1.454	1.451	0.011	1.446	1.461	78	14
	β-lactams C*–N(–X)–C=O (endo)	1.464	1.465	0.012	1.458	1.475	23	13
	γ-lactams							
	C*–NH–C=O (endo)	1.457	1.458	0.011	1.449	1.465	20	13
	C*–N(–C*)–C=O (endo)	1.462	1.461	0.010	1.453	1.466	15	13
	C*–N(–C*)–C=O (exo)	1.458	1.456	0.014	1.448	1.465	15	13
	δ-lactams							
	C*–NH–C=O (endo)	1.478	1.472	0.016	1.467	1.491	6	14
	C*–N(–C*)–C=O (endo)	1.479	1.476	0.007	1.475	1.482	15	14
	C*–N(–C*)–C=O (exo)	1.468	1.471	0.009	1.462	1.477	15	14

Bond	Substructure	d	m	σ	q_1	q_u	n	Note
	nitro compounds (1,2-dinitro omitted):							
	C–**CH$_2$–NO$_2$**	1.485	1.483	0.020	1.478	1.502	8	
	C$_2$–**CH–NO$_2$**	1.509	1.509	0.011	1.502	1.511	12	
	C$_3$–**C–NO$_2$**	1.533	1.533	0.013	1.530	1.539	17	
	C$_2$–**C–(NO$_2$)$_2$**	1.537	1.536	0.016	1.525	1.550	19	
	1,2-dinitro: **NO$_2$–C*–C*–NO$_2$**	1.552	1.550	0.023	1.536	1.572	32	
Csp^3–N(2)	C#–**N**=N	1.493	1.493	0.020	1.477	1.506	54	
	C*–N=C–Car	1.465	1.468	0.011	1.461	1.472	75	
Csp^2–N(3)	C=**C–NH$_2$** Nsp^2 planar	1.336	1.344	0.017	1.317	1.348	10	23
	C=**C–NH**–C# Nsp^2 planar	1.339	1.340	0.016	1.327	1.351	17	23
	C=**C–N**–(C#)$_2$							
	Nsp^2 planar	1.355	1.358	0.014	1.341	1.363	22	23
	Nsp^3 pyramidal	1.416	1.418	0.018	1.397	1.432	18	22
	Csp^2–Nsp^2 (N planar) in:							23
	acyclic amides							
	NH$_2$–C=O	1.325	1.323	0.009	1.318	1.331	32	14
	C*–NH–C=O	1.334	1.333	0.011	1.326	1.343	78	14
	(C*)$_2$–N–C=O	1.346	1.342	0.011	1.339	1.356	5	14
	β-lactams **C*–NH–C**=O	1.385	1.388	0.019	1.374	1.396	23	13
	γ-lactams							
	C*–NH–C=O	1.331	1.331	0.011	1.326	1.337	20	13
	C*–N(–C*)–**C**=O	1.347	1.344	0.014	1.335	1.359	15	13
	δ-lactams							
	C*–NH–C=O	1.334	1.334	0.006	1.330	1.339	6	14
	(C*)–**N**(–C*)–**C**=O	1.352	1.353	0.010	1.344	1.356	15	14
	peptides C#–**N**(–X)–**C**(–C#)(=O)	1.333	1.334	0.013	1.326	1.340	380	24
	ureas							
	(**NH$_2$**)$_2$–**C**=O	1.334	1.334	0.008	1.329	1.339	48	25,26
	(C#–**NH**)$_2$–**C**=O	1.347	1.345	0.010	1.341	1.354	26	25
	[(C#)$_n$–**N**]$_2$–**C**=O	1.363	1.359	0.014	1.354	1.370	40	25,27
	thioureas	1.346	1.343	0.023	1.328	1.361	192	
	(X$_2$**N**)$_2$–**C**=S							
	imides							
	[C#–**C**(=O)]$_2$–**NH**	1.376	1.377	0.012	1.369	1.383	64	
	[C#–**C**(=O)]$_2$–**N**–C#	1.389	1.383	0.017	1.376	1.404	38	
	[Csp^2–**C**(=O)]$_2$–**N**–C#	1.396	1.396	0.010	1.389	1.403	46	
	[Csp^2–**C**(=O)]$_2$–**N**–Csp^2	1.409	1.406	0.020	1.391	1.419	28	
	guanidinium [**C**–(**NH$_2$**)$_3$]$^+$ (unsubst.)	1.321	1.320	0.008	1.314	1.327	39	
	(any subst.)	1.328	1.325	0.015	1.317	1.333	140	
	in heterocyclic systems (any subst.)							
	1H-pyrrole (N1–C2, N1–C5)	1.372	1.374	0.016	1.363	1.384	58	
	indole (N1–C2)	1.370	1.370	0.012	1.364	1.377	40	
	pyrazole (N1–C5)	1.357	1.359	0.012	1.347	1.365	20	
	imidazole (N1–C2)	1.349	1.349	0.018	1.338	1.358	44	
	imidazole (N1–C5)	1.370	1.370	0.010	1.365	1.377	44	
Csp^2–N(2)	in imidazole (N3–C4)	1.376	1.377	0.011	1.369	1.384	44	
Car–N(4)	**Car–N$^+$**–(C,H)$_3$	1.465	1.466	0.007	1.461	1.470	23	
Car–N(3)	**Car–NH$_2$**							
	(Nsp^2: planar)	1.355	1.360	0.020	1.340	1.372	33	23
	(Nsp^3: pyramidal)	1.394	1.396	0.011	1.385	1.403	25	22
	(overall)	1.375	1.377	0.025	1.363	1.394	98	28
Car–N(3)	**Car–NH**–C#							
	(Nsp^2: planar)	1.353	1.353	0.007	1.347	1.359	16	23
	(Nsp^3: pyramidal)	1.419	1.423	0.017	1.412	1.432	8	22
	(overall)	1.380	1.364	0.032	1.353	1.412	31	28
	Car–N–(C#)$_2$							
	(Nsp^2: planar)	1.371	1.370	0.016	1.363	1.382	41	23
	(Nsp^3: pyramidal)	1.426	1.425	0.011	1.421	1.431	22	22
	(overall)	1.390	1.385	0.030	1.366	1.420	69	28
	in indole (N1–C7a)	1.372	1.372	0.007	1.367	1.376	40	
	Car–NO$_2$	1.468	1.469	0.014	1.460	1.476	556	

Bond	Substructure	d	m	σ	q_1	q_u	n	Note
Car–N(2)	**Car–N=N**	1.431	1.435	0.020	1.422	1.442	26	
Csp^2=N(3)	in furoxan ($^+$N2=C3)	1.316	1.316	0.009	1.311	1.324	14	
Csp^2=N(2)	Car–**C=N**–C#	1.279	1.279	0.008	1.275	1.285	75	
	(C,H)$_2$–**C=N**–OH in oximes	1.281	1.280	0.013	1.273	1.288	67	
	S–**C=N**–X	1.302	1.302	0.021	1.285	1.319	36	
	in pyrazole (N2=C3)	1.329	1.331	0.014	1.315	1.339	20	
	in imidazole (C2=N3)	1.313	1.314	0.011	1.307	1.319	44	
	in isoxazole (N2=C3)	1.314	1.315	0.009	1.305	1.320	9	
	in furazan (N2=C3, C4=N5)	1.298	1.299	0.006	1.294	1.303	12	
	in furoxan (C4=N5)	1.304	1.306	0.008	1.300	1.308	14	
$Car \simeq$ N(3)	**C** $\simeq$ **N$^+$**–H (pyrimidinium)	1.335	1.334	0.015	1.325	1.342	30	
	C $\simeq$ **N$^+$**–C* (pyrimidinium)	1.346	1.346	0.010	1.340	1.352	64	
	C $\simeq$ **N$^+$**–O$^-$ (pyrimidinium)	1.362	1.359	0.013	1.353	1.369	56	
$Car \simeq$ N(2)	**C** $\simeq$ **N** (pyridine)	1.337	1.338	0.012	1.330	1.344	269	
	C $\simeq$ **N** (pyrazine)	1.336	1.335	0.022	1.319	1.347	120	
	C $\simeq$ **N** $\simeq$ **C** (pyrimidine)	1.339	1.338	0.015	1.333	1.342	28	
	N $\simeq$ **C** $\simeq$ **N** (pyrimidine)	1.333	1.335	0.013	1.326	1.337	28	
	C $\simeq$ **N** (pyrimidine) (overall)	1.336	1.337	0.014	1.331	1.339	56	
	in any 6-membered N-containing aromatic ring:							
	H–C $\simeq$ **N** $\simeq$ **C–H**	1.334	1.334	0.014	1.327	1.341	146	
	H–C $\simeq$ **N** $\simeq$ **C–C***	1.339	1.341	0.013	1.336	1.345	38	
	C*–C $\simeq$ **N** $\simeq$ **C–C***	1.345	1.345	0.008	1.342	1.348	24	
	C $\simeq$ **N** $\simeq$ **C** (overall)	1.336	1.337	0.014	1.329	1.344	204	
$Csp^1 \equiv$ N(2)	X–S–**N$\equiv$C**$^-$ (isothiocyanide)	1.144	1.147	0.006	1.140	1.148	6	
$Csp^1 \equiv$ N(1)	C*–**C$\equiv$N**	1.136	1.137	0.010	1.131	1.142	140	
	C=C–**C$\equiv$N** in TCNQ	1.144	1.144	0.008	1.139	1.149	284	19
	Car–**C$\equiv$N**	1.138	1.138	0.007	1.133	1.143	31	
	X–**C$\equiv$N**	1.144	1.141	0.012	1.138	1.151	10	
	(S–**C$\equiv$N**)$^-$	1.155	1.156	0.012	1.147	1.165	14	
Csp^3–O(2)	in alcohols							
	CH$_3$–OH	1.413	1.414	0.018	1.395	1.425	17	
	C–**CH$_2$–OH**	1.426	1.426	0.011	1.420	1.431	75	
	C$_2$–**CH–OH**	1.432	1.431	0.011	1.425	1.439	266	
	C$_3$–**C–OH**	1.440	1.440	0.012	1.432	1.449	106	
	C*–OH (overall)	1.432	1.431	0.013	1.424	1.441	464	
	in dialkyl ethers							29
	CH$_3$–O–C*	1.416	1.418	0.016	1.405	1.426	110	
	C–**CH$_2$–O**–C*	1.426	1.424	0.011	1.418	1.435	34	
	C$_2$–**CH–O**–C*	1.429	1.430	0.010	1.420	1.437	53	
	C$_3$–**C–O**–C*	1.452	1.450	0.011	1.445	1.458	39	
	C*–O–C* (overall)	1.426	1.425	0.019	1.414	1.437	236	5
	in aryl alkyl ethers							29
	CH$_3$–O–Car	1.424	1.424	0.012	1.417	1.431	616	
	C–**CH$_2$–O**–Car	1.431	1.430	0.013	1.422	1.438	188	
	C$_2$–**CH–O**–Car	1.447	1.446	0.020	1.435	1.466	58	
	C$_3$–**C–O**–Car	1.470	1.469	0.018	1.456	1.483	55	
	C*–O–Car (overall)	1.429	1.427	0.018	1.419	1.436	917	
	in alkyl esters of carboxylic acids							12,29
	CH$_3$–O–C(=O)–C*	1.448	1.449	0.010	1.442	1.455	200	
	C–**CH$_2$–O**–C(=O)–C*	1.452	1.453	0.009	1.445	1.458	32	
	C$_2$–**CH–O**–C(=O)–C*	1.460	1.460	0.010	1.454	1.465	78	
	C$_3$–**C–O**–C(=O)–C*	1.477	1.475	0.008	1.472	1.484	6	
	C*–O–C(=O)–C* (overall)	1.450	1.451	0.014	1.442	1.459	314	
	in alkyl esters of α,β-unsaturated acids:							
	C*–O–C(=O)–C=C (overall)	1.453	1.452	0.013	1.444	1.459	112	
	in alkyl esters of benzoic acid							
	C*–O–C(=O)–C(phenyl) (overall)	1.454	1.454	0.012	1.446	1.463	219	
	in ring systems							
	oxirane (epoxides) (any subst.)	1.446	1.446	0.014	1.438	1.456	498	9
	oxetane (any subst.)	1.463	1.460	0.015	1.451	1.474	16	
	tetrahydrofuran (C,H subst.)	1.442	1.441	0.017	1.430	1.451	154	

Bond	Substructure	d	m	σ	q_1	q_u	n	Note
Csp^3–O(2)	tetrahydropyran (C,H subst.)	1.441	1.442	0.015	1.431	1.451	22	
	β-lactones: C^*–O–C(=O)	1.492	1.494	0.010	1.481	1.501	4	16
	γ-lactones: C^*–O–C(=O)	1.464	1.464	0.012	1.455	1.473	110	12
	δ-lactones: C^*–O–C(=O)	1.461	1.464	0.017	1.452	1.473	27	12
	O–C–O system in *gem*-diols, and pyranose and furanose sugars:							30,31
	HO–C^*–OH	1.397	1.401	0.012	1.388	1.405	18	
	C_5–O_5–C_1–O_1H in pyranoses							
	O_1 axial (α):							
	C_5–O_5	1.439	1.440	0.008	1.432	1.445	29	
	O_5–C_1	1.427	1.426	0.012	1.421	1.432	29	
	C_1–O_1	1.403	1.400	0.012	1.391	1.412	29	
	O_1 equatorial (β):							
	C_5–O_5	1.435	1.436	0.008	1.429	1.440	17	
	O_5–C_1	1.430	1.431	0.010	1.424	1.436	17	
	C_1–O_1	1.393	1.393	0.007	1.386	1.399	17	
	α + β (overall):							
	C_5–O_5	1.439	1.440	0.008	1.432	1.446	60	
	O_5–C_1	1.430	1.429	0.012	1.421	1.436	60	
	C_1–O_1	1.401	1.399	0.011	1.392	1.407	60	
	C_4–O_4–C_1–O_1H in furanoses (overall values)							
	C_4–O_4	1.442	1.446	0.012	1.436	1.449	18	
	O_4–C_1	1.432	1.432	0.012	1.421	1.443	18	
	C_1–O_1	1.404	1.405	0.013	1.397	1.409	18	
	C_5–O_5–C_1–O_1–C^* in pyranoses							
	O_1 axial (α):							
	C_5–O_5	1.439	1.438	0.010	1.433	1.446	67	
	O_5–C_1	1.417	1.417	0.009	1.410	1.424	67	
	C_1–O_1	1.409	1.409	0.014	1.401	1.417	67	
	O_1–C^*	1.435	1.435	0.013	1.427	1.443	67	
	O_1 equatorial (β):							
	C_5–O_5	1.434	1.435	0.006	1.429	1.439	39	
	O_5–C_1	1.424	1.424	0.008	1.418	1.431	39	
	C_1–O_1	1.390	1.390	0.011	1.381	1.400	39	
	O_1–C^*	1.437	1.438	0.013	1.428	1.445	39	
	α + β (overall):							
	C_5–O_5	1.436	1.436	0.009	1.431	1.442	126	
	O_5–C_1	1.419	1.419	0.011	1.412	1.426	126	
	C_1–O_1	1.402	1.403	0.016	1.391	1.413	126	
	O_1–C^*	1.436	1.436	0.013	1.428	1.445	126	
	C_4–O_4–C_1–O_1–C^* in furanoses (overall values)							
	C_4–O_4	1.443	1.445	0.013	1.429	1.453	23	
	O_4–C_1	1.421	1.418	0.012	1.413	1.431	23	
	C_1–O_1	1.410	1.409	0.014	1.401	1.420	23	
	O_1–C^*	1.439	1.437	0.014	1.429	1.449	23	
	Miscellaneous:							
	$C\#$–O–SiX_3	1.416	1.416	0.017	1.405	1.428	29	
	C^*–O–SO_2–C	1.465	1.461	0.014	1.454	1.475	33	
Csp^2–O(2)	in enols: C=C–OH	1.333	1.331	0.017	1.324	1.342	53	
	in enol esters: C=C–O–C^*	1.354	1.353	0.016	1.341	1.363	40	
	in acids:							
	C^*–C(=O)–OH	1.308	1.311	0.019	1.298	1.320	174	
	C=C–C(=O)–OH	1.293	1.295	0.019	1.279	1.307	22	
	Car–C(=O)–OH	1.305	1.311	0.020	1.291	1.317	75	
	in esters:							
	C^*–C(=O)–O–C^*	1.336	1.337	0.014	1.328	1.346	551	12,29
	C=C–C(=O)–O–C^*	1.332	1.331	0.011	1.324	1.339	112	
	Car–C(=O)–O–C^*	1.337	1.335	0.013	1.329	1.344	219	12
	C^*–C(=O)–O–C=C	1.362	1.359	0.018	1.351	1.374	26	
	C^*–C(=O)–O–C=C	1.407	1.405	0.017	1.394	1.420	26	

Bond	Substructure	d	m	σ	q_1	q_u	n	Note
	C*–**C**(=O)–O–C*ar*	1.360	1.359	0.011	1.355	1.367	40	12
	in anhydrides: O=**C**–O–C=O	1.386	1.386	0.011	1.379	1.393	70	
	in ring systems:							
	furan (O1–C2, O1–C5)	1.368	1.369	0.015	1.359	1.377	125	
	isoxazole (O1–C5)	1.354	1.354	0.010	1.345	1.360	9	
	β-lactones: C*–**C**(=O)–O–C*	1.359	1.359	0.013	1.348	1.371	4	13
	γ-lactones: C*–**C**(=O)–O–C*	1.350	1.349	0.012	1.342	1.359	110	12
	δ-lactones: C*–**C**(=O)–O–C*	1.339	1.339	0.016	1.332	1.347	27	12
C*ar*–O(2)	in phenols: **C**ar–**OH**	1.362	1.364	0.015	1.353	1.373	551	
	in aryl alkyl ethers: **C**ar–**O**–C*	1.370	1.370	0.011	1.363	1.377	920	29,32
C*ar*–O(2)	in diaryl ethers: **C**ar–**O**–**C**ar	1.384	1.381	0.014	1.375	1.391	132	
	in esters: **C**ar–**O**–C(=O)–C*	1.401	1.401	0.010	1.394	1.408	40	12
Csp^2=O(1)	in aldehydes and ketones:							
	C*–**CH**=**O**	1.192	1.192	0.005	1.188	1.197	7	
	(C*)$_2$–**C**=**O**	1.210	1.210	0.008	1.206	1.215	474	5
	(C#)$_2$–**C**=**O**							
	in cyclobutanones	1.198	1.198	0.007	1.194	1.204	12	
	in cyclopentanones	1.208	1.208	0.007	1.203	1.212	155	
	in cyclohexanones	1.211	1.211	0.009	1.207	1.216	312	
	C=C–**C**=**O**	1.222	1.222	0.010	1.216	1.229	225	
	(C=C)$_2$–**C**=**O**	1.233	1.229	0.010	1.226	1.242	28	
	C*ar*–**C**=**O**	1.221	1.218	0.014	1.212	1.229	85	
	(C*ar*)$_2$–**C**=**O**	1.230	1.226	0.015	1.220	1.238	66	
	C=**O** in benzoquinones	1.222	1.220	0.013	1.211	1.231	86	
	delocalized double bonds in carboxylate anions:							
	H–**C** $\simeq$ **O**$_2^-$ (formate)	1.242	1.243	0.012	1.234	1.252	24	
	C*–**C** $\simeq$ **O**$_2^-$	1.254	1.253	0.010	1.247	1.261	114	
	C=C–**C** $\simeq$ **O**$_2^-$	1.250	1.248	0.017	1.238	1.261	52	
	C*ar*–**C** $\simeq$ **O**$_2^-$	1.255	1.253	0.010	1.249	1.262	22	
	HOO**C**–**C** $\simeq$ **O**$_2^-$ (hydrogen oxalate)	1.243	1.247	0.015	1.232	1.256	26	
	$^-$**O**$_2$ $\simeq$ **C**–**C** $\simeq$ **O**$_2^-$ (oxalate)	1.251	1.251	0.007	1.248	1.254	18	
	in carboxylic acids (X–COOH)							
	C*–**C**(=**O**)–OH	1.214	1.214	0.019	1.203	1.224	175	
	C=C–**C**(=**O**)–OH	1.229	1.226	0.017	1.218	1.237	22	
	C*ar*–**C**(=**O**)–OH	1.226	1.223	0.020	1.211	1.241	75	
	in esters:							
	C*–**C**(=**O**)–O–C*	1.196	1.196	0.010	1.190	1.202	551	12
	C=C–**C**(=**O**)–O–C*	1.199	1.198	0.009	1.193	1.203	113	
	C*ar*–**C**(=**O**)–O–C*	1.202	1.201	0.009	1.196	1.207	218	12
	C*–**C**(=**O**)–O–C=C	1.190	1.190	0.014	1.184	1.198	26	
	C*–**C**(=**O**)–O–C*ar*	1.187	1.188	0.011	1.181	1.195	40	12
	in anhydrides: O=**C**–O–**C**=O	1.187	1.187	0.010	1.184	1.193	70	
	in β-lactones: C*–**C**(=**O**)–O–C*	1.193	1.193	0.006	1.187	1.198	4	13
	γ-lactones: C*–**C**(=**O**)–O–C*	1.201	1.202	0.009	1.196	1.206	109	12
	δ-lactones: C*–**C**(=**O**)–O–C*	1.205	1.207	0.008	1.201	1.209	27	12
	in amides:							
	NH$_2$–**C**(–C*)=**O**	1.234	1.233	0.012	1.225	1.243	32	14
	(C*–)(C*,H–)N–**C**(–C*)=**O**	1.231	1.231	0.012	1.224	1.238	378	14
	β-lactams: C*–NH–**C**=**O**	1.198	1.200	0.012	1.193	1.204	23	13
	γ-lactams:							
	C*–NH–**C**=**O**	1.235	1.235	0.008	1.232	1.240	20	13
	C*–N(–C*)–**C**=O	1.225	1.226	0.011	1.217	1.233	15	13
	δ-lactams:							
	C*–NH–**C**=**O**	1.240	1.241	0.003	1.237	1.243	6	14
	C*–N(–C*)–**C**=**O**	1.233	1.233	0.007	1.229	1.239	15	14
	in ureas:							
	(NH$_2$)$_2$–**C**=**O**	1.256	1.256	0.007	1.249	1.261	24	25,26
	(C#–NH)$_2$–**C**=**O**	1.241	1.237	0.011	1.235	1.245	13	25
	[(C#)$_n$–N]$_2$–**C**=**O**	1.230	1.230	0.007	1.224	1.234	20	25,27
Csp^3–P(4)	C$_3$–**P**$^+$–**C***	1.800	1.802	0.015	1.790	1.812	35	33
	C$_2$–**P**(=O)–CH$_3$	1.791	1.790	0.006	1.786	1.795	10	

Bond	Substructure	d	m	σ	q_1	q_u	n	Note
	$C_2-P(=O)-CH_2-C$	1.806	1.806	0.009	1.801	1.813	45	
	$C_2-P(=O)-CH-C_2$	1.821	1.821	0.009	1.815	1.828	15	
	$C_2-P(=O)-C-C_3$	1.841	1.842	0.008	1.835	1.847	14	
	$C_2-P(=O)-C^*$ (overall)	1.813	1.811	0.017	1.800	1.822	84	
$Csp^3-P(3)$	C_2-P-C^*	1.855	1.857	0.019	1.840	1.870	23	
$Car-P(4)$	C_3-P^+-Car	1.793	1.792	0.011	1.786	1.800	276	
	$C_2-P(=O)-Car$	1.801	1.802	0.011	1.796	1.807	98	
	$Ph_3-P=N^+=P-Ph_3$	1.795	1.795	0.008	1.789	1.800	197	
$Car-P(3)$	$C_2-P-Car$	1.836	1.837	0.010	1.830	1.844	102	
	$(N\simeq)_2P-Car(P \simeq N$ aromatic$)$	1.795	1.793	0.011	1.788	1.803	43	
$Csp^3-S(4)$	C^*-SO_2-C ($C^* = CH_3$ excluded)	1.786	1.782	0.018	1.774	1.797	75	
	C^*-SO_2-C (overall)	1.779	1.778	0.020	1.764	1.790	94	
	C^*-SO_2-O-X	1.745	1.744	0.009	1.738	1.754	7	34
	$C^*-SO_2-N-X_2$	1.758	1.756	0.018	1.746	1.773	17	34
$Csp^3-S(3)$	$C^*-S(=O)-C$ ($C^* = CH_3$ excluded)	1.818	1.814	0.024	1.802	1.829	69	
	$C^*-S(=O)-C$ (overall)	1.809	1.806	0.025	1.793	1.820	88	
	$CH_3-S^+-X_2$	1.786	1.787	0.007	1.779	1.792	21	
	$C^*-S^+-X_2$ ($C^* = CH_3$ excluded)	1.823	1.820	0.016	1.812	1.834	18	
	$C^*-S^+-X_2$ (overall)	1.804	1.794	0.025	1.788	1.820	41	
$Csp^3-S(2)$	C^*-SH	1.808	1.805	0.010	1.800	1.819	6	
	CH_3-S-C^*	1.789	1.787	0.008	1.784	1.794	9	
$Csp^3-S(2)$	$C-CH_2-S-C^*$	1.817	1.816	0.013	1.808	1.824	92	
	$C_2-CH-S-C^*$	1.819	1.819	0.011	1.811	1.825	32	
	$C_3-C-S-C^*$	1.856	1.860	0.011	1.854	1.863	26	
	C^*-S-C^* (overall)	1.819	1.817	0.019	1.809	1.827	242	
	in thiirane	1.834	1.835	0.025	1.810	1.858	4	9
	in thiirane: see ZCMXSP (1.817, 1.844)							
	in tetrahydrothiophene	1.827	1.826	0.018	1.811	1.837	20	
	in tetrahydrothiopyran	1.823	1.821	0.014	1.812	1.832	24	
	$C-CH_2-S-S-X$	1.823	1.820	0.014	1.813	1.832	41	
	$C_3-C-S-S-X$	1.863	1.865	0.015	1.848	1.878	11	
	$C^*-S-S-X$ (overall)	1.833	1.828	0.022	1.818	1.848	59	
$Csp^2-S(2)$	$C=C-S-C^*$	1.751	1.755	0.017	1.740	1.764	61	
	$C=C-S-C=C$ (in tetrathiafulvalene)	1.741	1.741	0.011	1.733	1.750	88	
	$C=C-S-C=C$ (in thiophene)	1.712	1.712	0.013	1.703	1.722	60	
	$O=C-S-C\#$	1.762	1.759	0.018	1.747	1.778	20	
$Car-S(4)$	$Car-SO_2-C$	1.763	1.764	0.009	1.756	1.769	96	
	$Car-SO_2-O-X$	1.752	1.750	0.008	1.749	1.756	27	
	$Car-SO_2-N-X_2$	1.758	1.759	0.013	1.749	1.765	106	35
$Car-S(3)$	$Car-S(=O)-C$	1.790	1.790	0.010	1.783	1.798	41	
	$Car-S^+-X_2$	1.778	1.779	0.010	1.771	1.787	10	
$Car-S(2)$	$Car-S-C^*$	1.773	1.774	0.009	1.765	1.779	44	
	$Car-S-Car$	1.768	1.767	0.010	1.762	1.774	158	
	$Car-S-Car$ (in phenothiazine)	1.764	1.764	0.008	1.760	1.769	48	
	$Car-S-S-X$	1.777	1.777	0.012	1.767	1.785	47	
$Csp^1-S(2)$	$N\equiv C-S-X$	1.679	1.683	0.026	1.645	1.698	10	
$Csp^1-S(1)$	$(N\equiv C-S)^-$	1.630	1.630	0.014	1.619	1.641	14	
$Csp^2=S(1)$	$(C^*)_2-C=S$: see IPMUDS (1.599)							
	$(Car)_2-C=S$: see CELDOM (1.611)							
	$(X)_2-C=S$ ($X = C,N,O,S$)	1.671	1.675	0.024	1.656	1.689	245	
	$X_2N-C(=S)-S-X$	1.660	1.660	0.016	1.648	1.674	38	
	$(X_2N)_2-C=S$ (thioureas)	1.681	1.684	0.020	1.669	1.693	96	
	$N-C(\simeq S)_2$	1.720	1.721	0.012	1.709	1.731	20	
Csp^3-Se	$C\#-Se$	1.970	1.967	0.032	1.948	1.998	21	
$Csp^2-Se(2)$	$C=C-Se-C=C$ (in tetraselenafulvalene)	1.893	1.895	0.013	1.882	1.902	32	
$Car-Se(3)$	Ph_3-Se^+	1.930	1.929	0.006	1.924	1.936	13	
$Csp^3-Si(5)$	$C\#-Si^--X_4$	1.874	1.876	0.015	1.859	1.884	9	
$Csp^3-Si(4)$	CH_3-Si-X_3	1.857	1.857	0.018	1.848	1.869	552	
	C^*-Si-X_3 ($C^* = CH_3$ excluded)	1.888	1.887	0.023	1.872	1.905	124	
	C^*-Si-X_3 (overall)	1.863	1.861	0.024	1.850	1.875	681	
$Car-Si(4)$	$Car-Si-X_3$	1.868	1.868	0.014	1.857	1.878	178	

Bond	Substructure	d	m	σ	q_1	q_u	n	Note
Csp^1–Si(4)	C≡**C**–**Si**–X$_3$	1.837	1.840	0.012	1.824	1.849	8	
Csp^3–Te	**C#**–**Te**	2.158	2.159	0.030	2.128	2.177	13	
Car–Te	Car–**Te**	2.116	2.115	0.020	2.104	2.130	72	
Csp^2=Te	see CEDCUJ (2.044)							
Cl–Cl	see PHASCL (2.306, 2.227)							
Cl–I	see CMBIDZ (2.563), HXPASC (2.541, 2.513), METAMM (2.552), BQUINI (2.416, 2.718)							
Cl–N	see BECTAE (1.743–1.757), BOGPOC (1.705)							
Cl–O(1)	in ClO$_4^-$	1.414	1.419	0.026	1.403	1.431	252	
Cl–P	(N≃)$_2$**P**–**Cl** (N ≃ P aromatic)	1.997	1.994	0.015	1.989	2.004	46	
	Cl–**P** (overall)	2.008	2.001	0.035	1.986	2.028	111	
Cl–S	**Cl**–**S** (overall)	2.072	2.079	0.023	2.047	2.091	6	
	see also longer bonds in CILSAR (2.283), BIHXIZ (2.357), CANLUY (2.749)							
Cl–Se	see BIRGUE10, BIRHAL10, CTCNSE (2.234–2.851)							
Cl–Si(4)	**Cl**–**Si**–X$_3$ (monochloro)	2.072	2.075	0.009	2.066	2.078	5	
	Cl$_2$–**Si**–X$_2$ and **Cl**$_3$–**Si**–X	2.020	2.012	0.015	2.007	2.036	5	
Cl–Te	Cl–Te in range 2.34–2.60	2.520	2.515	0.034	2.493	2.537	22	36
	see also longer bonds in BARRIV, BOJPUL, CETUTE, EPHTEA, OPNTEC10 (2.73–2.94)							
F–N(3)	**F**–**N**–C$_2$ and **F**$_2$–**N**–C	1.406	1.404	0.016	1.395	1.416	9	
F–P(6)	in hexafluorophosphate, PF$_6^-$	1.579	1.587	0.025	1.563	1.598	72	
F–P(3)	(N≃)$_2$**P**–**F**(N ≃ P aromatic)	1.495	1.497	0.016	1.481	1.510	10	
F–S	43 observations in range 1.409–1.770 in a wide variety of environments; F–S(6) in F$_2$–SO$_2$–C$_2$ (see FPSULF10, BETJOZ)	1.640	1.646	0.011	1.626	1.649	6	
	F–**S**(4) in F$_2$–**S**(=O)–N (see BUDTEZ)	1.527	1.528	0.004	1.524	1.530	24	37
F–Si(6)	in SiF$_6^{2-}$	1.694	1.701	0.013	1.677	1.703	6	
F–Si(5)	**F**–**Si**–X$_4$	1.636	1.639	0.035	1.602	1.657	10	
F–Si(4)	**F**–**Si**–X$_3$	1.588	1.587	0.014	1.581	1.599	24	
F–Te	see CUCPIZ (F–Te(6) = 1.942, 1.937), FPHTEL(F–Te(4) = 2.006)							
H–N(4)	X$_3$–**N$^+$**–**H**	1.033	1.036	0.022	1.026	1.045	87	21
H–N(3)	X$_3$–**N**–**H**	1.009	1.010	0.019	0.997	1.023	95	21
H–O(2)	in alcohols C*–**O**–**H**	0.967	0.969	0.010	0.959	0.974	63	21
	C#–**O**–**H**	0.967	0.970	0.010	0.959	0.974	73	21
	in acids O=C–**O**–**H**	1.015	1.017	0.017	1.001	1.031	16	21,38
I–I	in I$_3^-$	2.917	2.918	0.011	2.907	2.927	6	
I–N	see BZPRIB, CMBIDZ, HMTITI, HMTNTI, IFORAM, IODMAM (2.042–2.475)							
I–O	X–**I**–**O**(see BZPRIB, CAJMAB, IBZDAC11) for IO$_6^-$ see BOVMEE (1.829–1.912)	2.144	2.144	0.028	2.127	2.164	6	
I–P(3)	see CEHKAB (2.490–2.493)							†
I–S	sec DTHIBR10 (2.687), ISUREA10 (2.629), BZTPPI (3.251)							
I–Te(4)	**I**–**Te**–X$_3$	2.926	2.928	0.026	2.902	2.944	8	
N(4)–N(3)	X$_3$–**N$^+$**–**N^0**–X$_2$ (N^0 planar)	1.414	1.414	0.005	1.412	1.418	13	
N(3)–N(3)	(C)(C,H)–**N$_a$**–**N$_b$**(C)(C,H)							5,39
	N$_a$, N$_b$ pyramidal	1.454	1.452	0.021	1.444	1.457	44	40
	N$_a$ pyramidal, N$_b$ planar	1.420	1.420	0.015	1.407	1.433	68	40
	N$_a$, N$_b$ planar	1.401	1.401	0.018	1.384	1.418	40	40
	overall	1.425	1.425	0.027	1.407	1.443	139	
N(3)–N(2)	in pyrazole (N1–N2)	1.366	1.366	0.019	1.350	1.375	20	
	in pyridaznium (N1$^+$≃N2)	1.350	1.349	0.010	1.345	1.361	7	
N(2) ≃ N(2)	**N** ≃ **N** (aromatic) in pyridazine							
	with C,H as *ortho* substituents	1.304	1.300	0.019	1.287	1.326	6	
	with N,Cl as *ortho* substituents	1.368	1.373	0.011	1.362	1.375	9	

Bond	Substructure	d	m	σ	q_1	q_u	n	Note
N(2)=N(2)	C#–**N**=**N**–C#							
	cis	1.245	1.244	0.009	1.239	1.252	21	
	trans	1.222	1.222	0.006	1.218	1.227	6	
	(overall)	1.240	1.241	0.012	1.230	1.251	27	
	Car–**N**=**N**–Car	1.255	1.253	0.016	1.247	1.262	13	
	X–**N**=**N**=N (azides)	1.216	1.226	0.028	1.202	1.237	19	
N(2)=N(1)	X–N=**N**=**N** (azides)	1.124	1.128	0.015	1.114	1.137	19	
N(3)–O(2)	(C,H)$_2$–**N**–**OH** (Nsp^2: planar)	1.396	1.394	0.012	1.390	1.401	28	
	C$_2$–**N**–**O**–C							
	(Nsp^3: pyramidal)	1.463	1.465	0.012	1.457	1.468	22	
	(Nsp^2: planar)	1.397	1.394	0.011	1.388	1.409	12	
	in furoxan (N2–O1)	1.438	1.436	0.009	1.430	1.447	14	
N(3)–O(1)	(C$\simeq$)$_2$**N**$^+$–**O**$^-$ in pyridine N-oxides	1.304	1.299	0.015	1.291	1.316	11	
	in furoxan ($^+$N2–O6$^-$)	1.234	1.234	0.008	1.228	1.240	14	
N(2)–O(2)	in oximes							
	(C#)$_2$–C=**N**–**OH**	1.416	1.418	0.006	1.416	1.420	7	
	(H)(Csp^2)–C=**N**–**OH**	1.390	1.390	0.011	1.380	1.401	20	
	(C#)(Csp^2)–C=**N**–**OH**	1.402	1.403	0.010	1.393	1.410	18	
	(Csp^2)$_2$–C=**N**–**OH**	1.378	1.377	0.017	1.365	1.393	16	
	(C,H)$_2$–C=**N**–**OH** (overall)	1.394	1.395	0.018	1.379	1.408	67	
	in furazan (O1–N2, O1–N5)	1.385	1.383	0.013	1.378	1.392	12	
	in furoxan (O1–N5)	1.380	1.380	0.011	1.370	1.388	14	
	in isoxazole (O1–N2)	1.425	1.425	0.010	1.417	1.434	9	
N(3)=O(1)	in nitrate ions **NO**$_3^-$	1.239	1.240	0.020	1.227	1.251	105	
	in nitro groups							
	C*–**NO**$_2$	1.212	1.214	0.012	1.206	1.221	84	
	C#–**NO**$_2$	1.210	1.210	0.011	1.203	1.218	251	
	Car–**NO**$_2$	1.217	1.218	0.011	1.211	1.215	1116	
	C–NO$_2$ (overall)	1.218	1.219	0.013	1.210	1.226	1733	
N(3)–P(4)	X$_2$–**P**(=X)–**N**X$_2$							
	Nsp^2: planar	1.652	1.651	0.024	1.634	1.670	205	
	Nsp^3: pyramidal	1.683	1.683	0.005	1.680	1.686	6	
	(overall)	1.662	1.662	0.029	1.639	1.682	358	
	subsets of this group are:							
	O$_2$–**P**(=S)–**N**X$_2$	1.628	1.624	0.015	1.615	1.634	9	
	C–**P**(=S)–(**N**X$_2$)$_2$	1.691	1.694	0.018	1.678	1.703	28	
	O–**P**(=S)–(**N**X$_2$)$_2$	1.652	1.654	0.014	1.642	1.664	28	
	P(=O)–(**N**X$_2$)$_3$	1.663	1.668	0.026	1.640	1.679	78	
N(3)–P(3)	–**N**X–**P**(–X)–**N**X–**P**(–X)–(P$_2$N$_2$ ring)	1.730	1.721	0.017	1.716	1.748	20	
	–**N**X–**P**(=S)–**N**X–**P**(=S)–(P$_2$N$_2$ ring)	1.697	1.697	0.015	1.690	1.703	44	
	in P-substituted phosphazenes:							
	(N$\simeq$)$_2$**P**–**N** (amino)	1.637	1.638	0.014	1.625	1.651	16	
	(aziridinyl)	1.672	1.674	0.010	1.665	1.676	15	
N(2)=P(4)	Ph$_3$–**P**=**N**$^+$=P–Ph$_3$	1.571	1.573	0.013	1.563	1.580	66	
N(2)=P(3)	Ph$_3$–**P**=**N**–C,S	1.599	1.597	0.018	1.580	1.615	7	
N(2)$\simeq$P(3)	**N** $\simeq$ **P** aromatic							
	in phosphazenes	1.582	1.582	0.019	1.571	1.594	126	
	in P $\simeq$ N $\simeq$ S	1.604	1.606	0.009	1.594	1.612	36	
N(3)–S(4)	C–SO$_2$–**NH**$_2$	1.600	1.601	0.012	1.591	1.610	14	35
	C–SO$_2$–**NH**–C#	1.633	1.633	0.019	1.615	1.652	47	35
	C–SO$_2$–**N**–C(#)$_2$	1.642	1.641	0.024	1.623	1.659	38	35
N(3)–S(2)	C–S–**N**X$_2$ Nsp^2: planar	1.710	1.707	0.019	1.698	1.722	22	23
	(for Nsp^3 pyramidal see MODIAZ: 1.765)							
	X–S–**N**X$_2$ Nsp^2: planar	1.707	1.705	0.012	1.699	1.715	30	23
N(2)–S(2)	C=**N**–**S**–X	1.656	1.663	0.027	1.632	1.677	36	
N(2)$\simeq$S(2)	**N** $\simeq$ **S** aromatic in **P** $\simeq$ **N** $\simeq$ **S**	1.560	1.558	0.011	1.554	1.563	37	
N(2)=S(2)	**N=S** in **N=S**=N and **N=S**=S	1.541	1.546	0.022	1.521	1.558	37	
N(3)–SE	see COJCUZ (1.830), DSEMOR10 (1.846, 1.852), MORTRS10 (1.841)							
N(2)–Se	see SEBZQI (1.805), NAPSEZ10 (1.809, 1.820)							
N(2)=Se	see CISMUM (1.790, 1.791)							

Bond	Substructure	d	m	σ	q_l	q_u	n	Note
N(3)–Si(5)	see DMESIP01, BOJLER, CASSAQ, CASYOK, CECXEN, CINTEY, CIPBUY, FMESIB, MNPSIL, PNPOSI (1.973–2.344)							
N(3)–Si(4)	X_3–**Si**–NX_2 (overall)	1.748	1.746	0.022	1.735	1.757	170	
	subsets of this group are:							
	X_3–**Si**–NHX	1.714	1.719	0.014	1.702	1.727	16	
	X_3–**Si**–NX–Si–X_3 acyclic	1.743	1.744	0.016	1.731	1.755	45	
	N–**Si**–**N** in 4-membered rings	1.742	1.742	0.009	1.735	1.748	53	
	N–**Si**–**N** in 5-membered rings	1.741	1.742	0.019	1.726	1.749	33	
N(2)–Si(4)	X_3–**Si**–N^-–Si–X_3	1.711	1.712	0.019	1.693	1.729	15	
N–Te	see ACLTEP (2.402), BIBLAZ (1.980), CESSAU (2.023)							
O(2)–O(2)	C^*–**O**–**O**–C^*,H							
	τ(OO) = 70–85°	1.464	1.464	0.009	1.458	1.472	12	
	τ(OO) *ca.* 180°	1.482	1.480	0.005	1.478	1.486	5	
	overall	1.469	1.471	0.012	1.461	1.478	17	
	O=C–**O**–**O**–C=O see ACBZPO01 (1.446), CEYLUN (1.452), CIMHIP (1.454)							
	Si–**O**–**O**–Si	1.496	1.499	0.005	1.490	1.499	10	
O(2)–P(5)	X–**P**–$(OX)_4$							41
	trigonal bipyramidal:							
	axial	1.689	1.685	0.024	1.675	1.712	20	
	equatorial	1.619	1.622	0.024	1.604	1.628	20	
	square pyramidal	1.662	1.661	0.020	1.649	1.673	28	
O(2)–P(4)	C–**O**–**P**$(\simeq O)_3{}^{2-}$	1.621	1.622	0.007	1.615	1.628	12	
	$(H-O)_2$–**P**$(\simeq O)_2{}^-$	1.560	1.561	0.009	1.555	1.566	16	
	$(C-O)_2$–**P**$(\simeq O)_2{}^-$	1.608	1.607	0.013	1.599	1.615	16	
	$(C\#-O)_3$–**P**=O	1.558	1.554	0.011	1.550	1.564	30	
	$(Car-O)_3$–**P**=O	1.587	1.588	0.014	1.572	1.599	19	
	X–**O**–**P**(=O)–$(C,N)_2$	1.590	1.585	0.016	1.577	1.601	33	
	$(X-O)_2$–**P**(=O)–(C,N)	1.571	1.572	0.013	1.563	1.579	70	
O(2)–P(3)	$(N\simeq)_2$**P**–**O**–C (N $\simeq$ P aromatic)	1.573	1.573	0.011	1.563	1.584	16	
O(1)=P(4)	C–O–**P**$(\simeq O)_3{}^{2-}$ (delocalized)	1.513	1.512	0.008	1.508	1.518	42	
	$(H-O)_2$–**P**$(\simeq O)_2{}^-$ (delocalized)	1.503	1.503	0.005	1.499	1.508	16	
	$(C-O)_2$–**P**$(\simeq O)_2{}^-$ (delocalized)	1.483	1.485	0.008	1.474	1.490	16	
	$(C-O)_3$–**P**=O	1.449	1.448	0.007	1.446	1.452	18	
	C_3–**P**=O	1.489	1.486	0.010	1.481	1.496	72	
	N_3–**P**=O	1.461	1.462	0.014	1.449	1.470	26	
	$(C)_2(N)$–**P**=O	1.487	1.489	0.007	1.479	1.493	5	
	$(C,N)_2(O)$–**P**=O	1.467	1.462	0.007	1.462	1.472	33	
	$(C,N)(O)_2$–**P**=O	1.457	1.458	0.009	1.454	1.462	35	
O(2)–S(4)	C–**O**–SO_2–C	1.577	1.576	0.015	1.566	1.584	41	
	C–**O**–SO_2–CH_3	1.569	1.569	0.013	1.556	1.582	7	
	C–**O**–SO_2–Car	1.580	1.578	0.015	1.571	1.588	27	
O(1)=S(4)	C–SO_2–C	1.436	1.437	0.010	1.431	1.442	316	42
	X–SO_2–NX_2	1.428	1.428	0.010	1.422	1.434	326	
	C–SO_2–N–$(C,H)_2$	1.430	1.430	0.009	1.425	1.435	206	
	C–SO_2–O–C	1.423	1.423	0.008	1.418	1.428	82	
	in $SO_4{}^{2-}$	1.472	1.473	0.013	1.463	1.481	104	
O(1)=S(3)	C–**S**(=O)–C	1.497	1.498	0.013	1.489	1.505	90	5
O–Se	see BAPPAJ, BIRGUE10, BIRHAL10, CXMSEO, DGLYSE, SPSEBU (1.597 for **O**=**Se** to 1.974 for **O**–**Se**)							
O(2)–Si(5)	$(X-O)_3$–**Si**–(N)(C)	1.663	1.658	0.023	1.650	1.665	21	
O(2)–Si(4)	X_3–**Si**–**O**–X (overall)	1.631	1.630	0.022	1.617	1.646	191	
O(2)–Si(4)	subsets of this group are:							
	X_3–**Si**–**O**–C#	1.645	1.647	0.012	1.634	1.652	29	
	X_3–**Si**–**O**–Si–X_3	1.622	1.625	0.014	1.614	1.631	70	
	X_3–**Si**–**O**–O–Si–X_3	1.680	1.676	0.008	1.673	1.688	10	
O(2)–Te(6)	$(X-O)_6$–**Te**	1.927	1.927	0.020	1.908	1.942	16	
O(2)–Te(4)	$(X-O)_2$–**Te**–X_2	2.133	2.136	0.054	2.078	2.177	12	
P(4)–P(4)	X_3–**P**–**P**–X_3	2.256	2.259	0.025	2.243	2.277	6	

Bond	Substructure	d	m	σ	q_1	q_u	n	Note
P(4)–P(3)	see CECHEX (2.197), COZPIQ (2.249)							
P(3)–P(3)	X_2–**P**–**P**–X_2	2.214	2.210	0.022	2.200	2.224	41	
P(4)=P(4)	see BUTSUE (2.054)							
P(3)=P(3)	see BALXOB (2.034)							
P(4)=S(1)	C_3–**P**=**S**	1.954	1.952	0.005	1.950	1.957	13	
	$(N,O)_2(C)$–**P**=**S**	1.922	1.924	0.014	1.913	1.927	26	
	$(N,O)_3$–**P**=**S**	1.913	1.914	0.014	1.906	1.921	50	
P(4)=Se(1)	X_3–**P**=**Se**	2.093	2.099	0.019	2.075	2.108	12	
P(3)–Si(4)	X_2–**P**–**Si**–X_3: 3- and 4-rings	2.264	2.260	0.019	2.249	2.283	22	
	excluded (see BOPFER, BOPFIV, CASTOF10, COZVIW: 2.201–2.317)							
P(4)=Te(1)	see MOPHTE (2.356), TTEBPZ (2.327)							
S(2)–S(2)	C–**S**–**S**–C							
	$\tau(SS) = 75–105°$	2.031	2.029	0.015	2.021	2.038	46	
	$\tau(SS) = 0–20°$	2.070	2.068	0.022	2.057	2.077	28	
	(overall)	2.048	2.045	0.026	2.028	2.068	99	
	in polysulphide chain–**S**–**S**–**S**–	2.051	2.050	0.022	2.037	2.065	126	
S(2)–S(1)	X–N=**S**–**S**	1.897	1.896	0.012	1.887	1.908	5	
S–Se(4)	see BUWZUO (2.264, 2.269)							
S–Se(2)	X–**Se**–**S** (any)	2.193	2.195	0.015	2.174	2.207	9	
S(2)–Si(4)	X_3–**Si**–**S**–X	2.145	2.138	0.020	2.130	2.158	19	
S(2)–Te	X–**S**–**Te** (any)	2.405	2.406	0.022	2.383	2.424	10	
	X=**S**–**Te** (any)	2.682	2.686	0.035	2.673	2.694	28	
Se(2)–Se(2)	X–**Se**–**Se**–X	2.340	2.340	0.024	2.315	2.361	15	
Se(2)–Te(2)	see BAWFUA, BAWGAH (2.524–2.561)							†
Si(4)–Si(4)	X_3–**Si**–**Si**–X_3 3–membered rings excluded: see CIHRAM (2.511)	2.359	2.359	0.012	2.349	2.366	42	
Te–Te	see CAHJOK (2.751, 2.704)							

Appendix 1. (Footnotes to Table)

1. Sample dominated by B–CH_3. For longer bonds in B⁻–CH_3 see LITMEB10 [B(4)–CH_3 = 1.621–1.644Å].
2. p(π)–p(π) Bonding with Bsp^2 and Nsp^2 coplanar (τBN = 0 ± 15°) predominates. See G. Schmidt, R. Boese, and D. Bläser, *Z. Naturforsch.*, 1982, **37b**, 1230.
3. 84 observations range from 1.38 to 1.61 Å and individual values depend on substituents on B and O. For a discussion of borinic acid adducts see S. J. Rettig and J. Trotter, *Can. J. Chem.*, 1982, **60**, 2957.
4. See M. Kaftory in 'The Chemistry of Functional Groups. Supplement D: The Chemistry of Halides, Pseudohalides, and Azides', S. Patai and Z. Rappoport, Eds., Wiley: New York, 1983, Part 2, ch. 24.
5. Bonds which are endocyclic or exocyclic to any 3- or 4-membered rings have been omitted from all averages in this section.
6. The overall average given here is for Csp^3–Csp^3 bonds which carry only C or H substituents. The value cited reflects the relative abundance of each 'substitution' group. The 'mean of means' for the 9 subgroups is 1.538 (σ = 0.022) Å.
7. See F. H. Allen, (a) *Acta Crystallogr.*, 1980, **B36**, 81; (b) 1981, **B37**, 890.
8. See F. H. Allen, *Acta Crystallogr.*, 1984, **B40**, 64.
9. See F. H. Allen, *Tetrahedron*, 1982, **38**, 2843.
10. See F. H. Allen, *Tetrahedron*, 1982, **38**, 645.
11. Cyclopropanones and cyclobutanones excluded.
12. See W. B. Schweizer and J. D. Dunitz, *Helv. Chim. Acta*, 1982, **65**, 1547.
13. See L. Norskov-Lauritsen, H.-B. Bürgi, P. Hoffmann, and H. R. Schmidt, *Helv. Chim. Acta*, 1985, **68**, 76.
14. See P. Chakrabarti and J. D. Dunitz, *Helv. Chim. Acta*, 1982, **65**, 1555.
15. See J. L. Hencher in 'The Chemistry of the C≡C Triple Bond', S. Patai, Ed., Wiley, New York, 1978, ch. 2.
16. Conjugated: torsion angle about central C–C single bond is 0 ± 20° (*cis*) or 180 ± 20° (*trans*).
17. Unconjugated: torsion angle about central C–C single bond is 20–160°.
18. Other conjugative substituents excluded.
19. TCNQ is tetracyanoquinodimethane.
20. No difference detected between C2 ≃ C3 and C3 ≃ C4 bonds.
21. Derived from neutron diffraction results only.
22. Nsp^3: pyramidal; mean valence angle at N is in range 108–114°.
23. Nsp^2: planar; mean valence angle at N is ≥ 117.5°.
24. Cyclic and acyclic peptides.
25. See R. H. Blessing, *J. Am. Chem. Soc.*, 1983, **105**, 2776.
26. See L. Lebioda, *Acta Crystallogr.*, 1980, **B36**, 271.
27. n = 3 or 4, *i.e.* tri- or tetra-substituted ureas.
28. Overall value also includes structures with mean valence angle at N in the range 115–118°.
29. See F. H. Allen and A. J. Kirby, *J. Am. Chem. Soc.*, 1984, **106**, 6197.
30. See A. J. Kirby, 'The Anomeric Effect and Related Stereoelectronic Effects at Oxygen', Springer, Berlin, 1983.
31. See B. Fuchs, L. Schleifer, and E. Tartakovsky, *Nouv. J. Chim.*, 1984, **8**, 275.
32. See S. C. Nyburg and C. H. Faerman, *J. Mol. Struct.*, 1986, **140**, 347.
33. Sample dominated by P–CH_3 and P–CH_2–C.
34. Sample dominated by C* = methyl.
35. See A. Kalman, M. Czugler, and G. Argay, *Acta Crystallogr.*, 1981, **B37**, 868.
36. Bimodal distribution resolved into 22 'short' bonds and 5 longer outliers.
37. All 24 observations come from BUDTEZ.
38. 'Long' O–H bonds in centrosymmetric O---H---O H–bonded dimers are excluded.
39. N–N bond length also dependent on torsion angle about N–N bond and on nature of substituent C atoms; these effects are ignored here.
40. N pyramidal has average angle at N in range 100–113.5°; N planar has average angle of ≥ 117.5°.
41. See R. R. Holmes and J. A. Deiters, *J. Amer. Chem. Soc.*, 1977, **99**, 3318.
42. No detectable variation in S=O bond length with type of C-substituent.

Appendix 2

Short-form references to individual CSD entries cited by reference code in the Table. A full list of CSD bibliographic entries is given in SUP 56701.

ACBZPO01	*J. Am. Chem. Soc.*, 1975, **97**, 6729.	CIWYIQ	*Inorg. Chem.*, 1984, **23**, 1946.
ACLTEP	*J. Organomet. Chem.*, 1980, **184**, 417.	CIYFOF	*Inorg. Chem.*, 1984, **23**, 1790.
ASAZOC	*Dokl. Akad. Nauk SSSR*, 1979, **249**, 120.	CMBIDZ	*J. Org. Chem.*, 1979, **44**, 1447.
BALXOB	*J. Am. Chem. Soc.*, 1981, **103**, 4587.	CODDEE	*Z. Naturforsch., Teil B*, 1984, **39**, 1257.
BAPPAJ	*Inorg. Chem.*, 1981, **20**, 3071.	CODDII	*Z. Naturforsch., Teil B*, 1984, **39**, 1257.
BARRIV	*Acta Chem. Scand., Ser. A*, 1981, **35**, 443.	COFVOI	*Z. Naturforsch., Teil B*, 1984, **39**, 1027.
BAWFUA	*Cryst. Struct. Commun.*, 1981, **10**, 1345.	COJCUZ	*Chem. Ber.*, 1984, **117**, 2686.
BAWGAH	*Cryst. Struct. Commun.*, 1981, **10**, 1353.	COSDIX	*Z. Naturforsch., Teil B*, 1984, **39**, 1344.
BECTAE	*J. Org. Chem.*, 1981, **46**, 5048, 1981.	COZPIQ	*Chem. Ber.*, 1984, **117**, 2063.
BELNIP	*Z. Naturforsch., Teil B*, 1982, **37**, 299.	COZVIW	*Z. Anorg. Allg. Chem.*, 1984, **515**, 7.
BEMLIO	*Chem. Ber.*, 1982, **115**, 1126.	CTCNSE	*J. Am. Chem. Soc.*, 1980, **102**, 5430.
BEPZEB	*Cryst. Struct. Commun.*, 1982, **11**, 175.	CUCPIZ	*J. Am. Chem. Soc.*, 1984, **106**, 7529.
BETJOZ	*J. Am. Chem. Soc.*, 1982, **104**, 1683.	CUDLOC	*J. Cryst. Spectrosc.*, 1985, **15**, 53.
BETUTE10	*Acta Chem. Scand., Ser. A*, 1976, **30**, 719.	CUDLUI	*J. Cryst. Spectrosc.*, 1985, **15**, 53.
BIBLAZ	*Zh. Strukt. Khim.*, 1981, **22**, 118.	CUGBAH	*Acta Crystallogr., Sect. C*, 1985, **41**, 476.
BICGEZ	*Z. Anorg. Allg. Chem.*, 1982, **486**, 90.	CXMSEO	*Acta Crystallogr., Sect. B*, 1973, **29**, 595.
BIHXIZ	*J. Chem. Soc., Chem. Commun.*, 1982, 982.	DGLYSE	*Acta Crystallogr., Sect. B*, 1975, **31**, 1785.
BIRGUE10	*Z. Naturforsch., Teil B*, 1983, **38**, 20.	DMESIP01	*Acta Crystallogr., Sect. C*, 1984, **40**, 895.
BIRHAL10	*Z. Naturforsch., Teil B*, 1982, **37**, 1410.	DSEMOR10	*J. Chem. Soc., Dalton Trans.*, 1980, 628.
BIZJAV	*J. Organomet. Chem.*, 1982, **238**, C1.	DTHIBR10	*Inorg. Chem.*, 1971, **10**, 697.
BOGPOC	*Z. Naturforsch., Teil B*, 1982, **37**, 1402.	EPHTEA	*Inorg. Chem.*, 1980, **19**, 2487.
BOGSUL	*Z. Naturforsch., Teil B*, 1982, **37**, 1230.	ESEARS	*J. Chem. Soc. C*, 1971, 1511.
BOJLER	*Z. Anorg. Allg. Chem.*, 1982, **493**, 53.	ETEARS	*J. Chem. Soc. C*, 1971, 1511.
BOJPUL	*Acta Chem. Scand., Ser. A*, 1982, **36**, 829.	FMESIB	*J. Organomet. Chem.*, 1980, **197**, 275.
BOPFER	*Chem. Ber.*, 1983, **116**, 146.	FPHTEL	*J. Chem. Soc., Dalton Trans.*, 1980, 2306.
BOPFIV	*Chem. Ber.*, 1983, **116**, 146.	FPSULF10	*J. Am. Chem. Soc.*, 1982, **104**, 1683.
BOVMEE	*Acta Crystallogr., Sect. B*, 1982, **38**, 1048.	HCLENE10	*Acta Crystallogr., Sect. B*, 1982, **38**, 3139.
BQUINI	*Acta Crystallogr., Sect. B*, 1979, **35**, 1930.	HMTITI	*Acta Crystallogr., Sect. B*, 1975, **31**, 1505.
BTUPTE	*Acta Chem. Scand., Ser. A*, 1975, **29**, 738.	HMTNTI	*Z. Anorg. Allg. Chem.*, 1974, **409**, 237.
BUDTEZ	*Z. Naturforsch., Teil B*, 1983, **38**, 454.	HXPASC	*J. Chem. Soc., Dalton Trans.*, 1975, 1381.
BUPSIB10	*Z. Anorg. Allg. Chem.*, 1981, **474**, 31.	IBZDAC11	*J. Chem. Soc., Dalton Trans.*, 1979, 854.
BUSHAY	*Z. Naturforsch., Teil. B*, 1983, **38**, 692.	IFORAM	*Monatsh. Chem.*, 1974, **105**, 621.
BUTHAZ10	*Inorg. Chem.*, 1984, **23**, 2582.	IODMAM	*Acta Crystallogr., Sect. B*, 1977, **33**, 3209.
BUTSUE	*J. Chem. Soc., Chem. Commun.*, 1983, 862.	IPMUDS	*Acta Crystallogr., Sect. B*, 1973, **29**, 2128.
BUWZUO	*Acta Chem. Scand., Ser A*, 1983, **37**, 219.	ISUREA10	*Acta Crystallogr., Sect. B*, 1972, **28**, 643.
BZPRIB	*Z. Naturforsch., Teil B*, 1981, **36**, 922.	LITMEB10	*J. Am. Chem. Soc.*, 1975, **97**, 6401.
BZTPPI	*Inorg. Chem.*, 1978, **17**, 894.	MESIAD	*Z. Naturforsch., Teil B*, 1980, **35**, 789.
CAHJOK	*Inorg. Chem.*, 1983, **22**, 1809.	METAMM	*Acta Crystallogr.*, 1964, **17**, 1336.
CAJMAB	*Chem. Z*, 1983, **107**, 169.	MNPSIL	*J. Am. Chem. Soc.*, 1969, **91**, 4134.
CANLUY	*Tetrahedron Lett.*, 1983, **24**, 4337.	MODIAZ	*J. Heterocycl. Chem.*, 1980, **17**, 1217.
CASSAQ	*J. Struct. Chem.*, 1983, **2**, 101.	MOPHTE	*Acta Chem. Scand., Ser. A*, 1980, **34**, 333.
CASTOF10	*Acta Crystallogr., Sect. C*, 1984, **40**, 1879.	MORTRS10	*J. Chem. Soc., Dalton Trans.*, 1980, 628.
CASYOK	*J. Struct. Chem.*, 1983, **2**, 107.	NAPSEZ10	*J. Am. Chem. Soc.*, 1980, **102**, 5070.
CECHEX	*Z. Anorg. Allg. Chem.*, 1984, **508**, 61.	NBBZAM	*Z. Naturforsch., Teil B*, 1977, **32**, 1416.
CECXEN	*J. Struct. Chem.*, 1983, **2**, 207.	OPIMAS	*Aust. J. Chem.*, 1977, **30**, 2417.
CEDCUJ	*J. Org. Chem.*, 1983, **48**, 5149.	OPNTEC10	*J. Chem. Soc., Dalton Trans.*, 1982, 251.
CEHKAB	*Z. Naturforsch., Teil B*, 1984, **39**, 139.	PHASCL	*Acta Crystallogr., Sect. B*, 1981, **37**, 1357.
CELDOM	*Acta Crystallogr., Sect. C*, 1984, **40**, 556.	PHASOC01	*Aust. J. Chem.*, 1975, **28**, 15.
CESSAU	*Acta Crystallogr., Sect. C*, 1984, **40**, 653.	PNPOSI	*J. Am. Chem. Soc.*, 1968, **90**, 5102.
CETTAW	*Chem. Ber.*, 1984, **117**, 1089.	SEBZQI	*J. Chem. Soc., Chem. Commun.*, 1977, 325.
CETUTE	*Acta Chem. Scand., Ser A*, 1975, **29**, 763.	SPSEBU	*Acta Chem. Scand., Ser. A*, 1979, **33**, 403.
CEYLUN	*Izv. Akad. Nauk SSSR, Ser. Khim.*, 1983, 2744.	TEACBR	*Cryst. Struct. Commun.*, 1974, **3**, 753.
CIFZUM	*Acta Chem. Scand., Ser A*, 1984, **38**, 289.	THINBR	*J. Am. Chem. Soc.*, 1970, **92**, 4002.
CIHRAM	*Angew. Chem., Int. Ed. Engl.*, 1984, **23**, 302.	TMPBTI	*Acta Crystallogr., Sect. B*, 1975, **31**, 1116.
CILRUK	*J. Chem. Soc., Chem. Commun.*, 1984, 1023.	TPASSN	*J. Chem. Soc., Dalton Trans.*, 1977, 514.
CILSAR	*J. Chem. Soc., Chem. Commun.*, 1984, 1021.	TPASTB	*Cryst. Struct. Commun.*, 1976, **5**, 39.
CIMHIP	*Acta Crystallogr., C*, 1984, **40**, 1458.	TPHOSI	*Z. Naturforsch., Teil B*, 1979, **34**, 1064.
CINTEY	*Dokl. Akad. Nauk SSSR*, 1984, **274**, 615.	TTEBPZ	*Z. Naturforsch., Teil B*, 1979, **34**, 256.
CIPBUY	*J. Struct. Chem.*, 1983, **2**, 281.	ZCMXSP	*Cryst. Struct. Commun.*, 1977, **6**, 93.
CISMUM	*Z. Naturforsch., Teil B*, 1984, **39**, 485.		
CISTED	*Z. Anorg. Allg. Chem.*, 1984, **511**, 95.		

BOND LENGTHS IN ORGANOMETALLIC COMPOUNDS

This table summarizes the average values of interatomic distances of representative metal–ligand bonds. Sigma bonds between d- and f-block metals and the elements C, N, O, P, S, and As are included. The values are extracted from a much larger list in Reference 1. The tabulated values are the unweighted means of reported measurements on compounds in each category. If four or more measurements are available, the standard deviation is given in parentheses. All values are in Ångstrom units (10^{-10} m).

The first part of the table covers metal-carbon bonds in different ligand categories, while the second part covers metal bonds to other elements. R stands for any alkyl group; Me for a CH_3 group; C_6R_5 indicates an aryl group; and C(=O)R an acyl group. Metals are listed in atomic number order.

Reference

1. Orpen, A. G., Brammer, L., Allen, F.H., Kennard, O., Watson, D. G., and Taylor, R., *J. Chem. Soc. Dalton Trans.*, 1989, S1-S83.

M	M-CH$_3$	M-CH$_2$R	M-CR=CR$_2$	M-C$_6$R$_5$	M-C(=O)R
Ti		2.167	2.215(0.042)	2.148	
V				2.114(0.012)	
Cr	2.168		2.035(0.009)	2.075(0.019)	
Mn	2.095(0.030)	2.176(0.024)	2.007	2.064(0.021)	2.044
Fe	2.074	2.091(0.030)	1.991(0.039)	2.031(0.062)	1.997(0.033)
Co	2.014(0.023)	2.039(0.032)	1.934(0.019)	1.974	1.990
Ni	2.029	1.964	1.892(0.017)	1.917(0.038)	1.850(0.059)
Cu				2.020	
Zn		1.964			
Zr	2.292(0.049)		2.257		
Nb	2.336	1.319			
Mo	2.254(0.065)	2.250(0.061)	2.204(0.049)	2.193(0.054)	2.109
Ru	2.179(0.045)	2.036(0.010)	2.063	2.092(0.057)	2.091
Rh	2.092(0.027)	2.100	2.040(0.054)	2.011(0.026)	1.995(0.031)
Pd		2.028	2.000(0.024)	1.981(0.032)	1.982(0.029)
Hf	2.275(0.049)		2.205		
Ta	2.217(0.035)	2.225(0.056)		2.199(0.073)	
W	2.189(0.039)	2.175	2.224		
Re	2.173(0.051)	2.290		2.027	2.190(0.027)
Os		2.221	2.052	2.090(0.032)	2.161
Ir	2.175		2.071(0.044)	2.070(0.038)	2.019
Pt	2.083(0.045)	2.062(0.031)	2.024(0.037)	2.049(0.046)	1.991(0.025)
Au	2.066(0.045)		2.042	2.059(0.024)	
Hg	2.072(0.026)	2.125		2.086(0.040)	
Th	2.567				

M	M-NH$_3$	M-OH$_2$	M-PMe$_3$	M-SR	M-AsR$_3$
Ti		2.066(0.052)		2.369	2.686
V		2.129(0.131)	2.510(0.010)	2.378(0.007)	
Cr	2.069(0.008)	1.997(0.070)	2.389(0.069)	2.362	2.460(0.040)
Mn		2.189(0.040)	2.455(0.164)	2.366(0.054)	2.400(0.013)
Fe		2.085(0.066)	2.246(0.042)	2.271(0.028)	2.352(0.043)
Co	1.965(0.021)	2.085(0.064)	2.217(0.043)	2.254(0.025)	2.323(0.021)
Ni	2.074(0.093)	2.079(0.038)	2.204(0.031)	2.187(0.007)	2.333(0.035)
Cu	1.987(0.017)	2.186(0.215)			2.367(0.016)
Zn	2.044	2.090(0.061)		2.295	
Y		2.398(0.068)			
Zr			2.692		
Nb		2.248(0.137)			2.741(0.008)
Mo	2.217	2.201(0.094)	2.462(0.046)	2.401(0.050)	2.582(0.036)
Ru	2.126(0.024)	2.074(0.051)	2.307(0.050)		2.446(0.031)
Rh	2.114(0.018)	2.190(0.096)	2.266(0.036)		2.416(0.039)
Pd	2.032	2.200	2.287(0.018)		2.386(0.052)
Ag		2.350			
Cd		2.318(0.065)		2.444	
La		2.556(0.062)			
Ce		2.565(0.063)			
Pr		2.518(0.038)			
Nd		2.533(0.058)			
Sm		2.459(0.050)			
Eu		2.441(0.055)			
Gd		2.443(0.074)			
Tb		2.455			
Dy		2.409(0.074)			
Ho		2.407(0.069)			
Er		2.404(0.083)			
Yb		2.353(0.066)			
Lu		2.404(0.116)			
Ta			2.589(0.044)		
W		2.115(0.065)	2.485(0.039)		
Re	2.253	2.199(0.091)	2.369(0.065)		2.575(0.006)
Os	2.136	2.166	2.328(0.029)		
Ir	2.050(0.021)		2.323(0.028)	2.461	
Pt			2.295(0.036)	2.320(0.015)	2.366(0.058)
Au		2.157		2.293	
Hg		2.690(0.083)		2.402(0.065)	
Th		2.483(0.032)			
U		2.455(0.047)			

BOND LENGTHS AND ANGLES IN GAS-PHASE MOLECULES

This table is reprinted from *Kagaku Benran, 3rd Edition*, Vol. II, pp. 649–661 (1984), with permission of the publisher, Maruzen Company, LTD. (Copyright 1984 by the Chemical Society of Japan). Translation was carried out by Kozo Kuchitsu.

Internuclear distances and bond angles are represented in units of Å (1 Å = 10^{-10} m) and degrees, respectively. The same but inequivalent atoms are discriminated by subscripts a, b, etc. In some molecules ax for axial and eq for equatorial are also used. All measurements were made in the gas phase. The methods used are abbreviated as follows. UV: ultraviolet (including visible) spectroscopy; IR: infrared spectroscopy; R: Raman spectroscopy; MW: microwave spectroscopy; ED: electron diffraction; NMR: nuclear magnetic resonance; LMR: laser magnetic resonance; EPR: electron paramagnetic resonance; MBE: molecular beam electric resonance. If two methods were used jointly for structure determination, they are listed together, as (ED, MW). If the numerical values listed refer to the equilibrium values, they are specified by r_e and θ_e. In other cases the listed values represent various average values in vibrational states; it is frequently the case that they represent the r_s structure derived from several isotopic species for MW or the r_g structure (i.e., the average internuclear distances at thermal equilibrium) for ED. These internuclear distances for the same atom pair with different definitions may sometimes differ as much

as 0.01 Å. Appropriate comments are made on the symmetry and conformation in the equilibrium structure.

In general, the numerical values listed in the following tables contain uncertainties in the last digits. However, for certain molecules such as diatomic molecules, with experimental uncertainties of the order of 10^{-5} Å or smaller, numerical values are listed to four decimal places.

References

1. L. E. Sutton, Ed., *Tables of Interatomic Distances and Configuration in Molecules and Ions,* The Chemical Society Special Publication, No. 11, 18, The Chemical Society (London) (1958, 1965).
2. K.-H. Hellwege, Ed., *Landolt-Börnstein Numerical Data and Functional Relations in Science and Technology,* New Series, II/7, J. H. Callomon, E. Hirota, K. Kuchitsu, W. J. Lafferty, A. G. Maki, C. S. Pote, with assistance of I. Buck and B. Starck, *Structure Data of Free Polyatomic Molecules,* Springer-Verlag (1976).
3. K. P. Huber and G. Herzberg, *Molecular Spectra and Molecular Structure IV. Constants of Diatomic Molecules,* Van Nostrand Reinhold Co., London (1979).
4. B. Starck, *Microwave Catalogue and Supplements.*
5. B. Starck, *Electron Diffraction Catalogue and Supplements.*

STRUCTURES OF ELEMENTS AND INORGANIC COMPOUNDS
Compounds are arranged in alphabetical order by their chemical formulas
(Lengths in Å and Angles in Degrees)

Compound	Structure				Method
AgBr	Ag—Br (r_e)	2.3931			MW
AgCl	Ag—Cl (r_e)	2.2808			MW
AgF	Ag—F (r_e)	1.9832			MW
AgH	Ag—H (r_e)	1.617			UV
AgI	Ag—I (r_e)	2.5446			MW
AgO	Ag—O (r_e)	2.0030			UV
AlBr	Al—Br (r_e)	2.295			UV
AlCl	Al—Cl (r_e)	2.1301			MW
AlF	Al—F (r_e)	1.6544			MW
AlH	Al—H (r_e)	1.6482			UV
AlI	Al—I (r_e)	2.5371			MW
AlO	Al—O (r_e)	1.6176			UV
Al_2Br_6	Br$_a$, Br$_b$, Br$_a$ / Al Al / Br$_a$, Br$_b$, Br$_a$		Al—Br$_a$	2.22	ED
			Al—Br$_b$	2.38	
			∠Br$_b$AlBr$_b$	82	
			∠Br$_a$AlBr$_a$	118	
			(D$_{2h}$)		
Al_2Cl_6	Cl$_a$, Cl$_b$, Cl$_a$ / Al Al / Cl$_a$, Cl$_b$, Cl$_a$		Al—Cl$_a$	2.04	ED
			Al—Cl$_b$	2.24	
			∠Cl$_b$AlCl$_b$	87	
			∠Cl$_a$AlCl$_a$	122	
			(D$_{2h}$)		
$AsBr_3$	As—Br	2.324	∠BrAsBr	99.6	ED
$AsCl_3$	As—Cl	2.165	∠ClAsCl	98.6	ED, MW
AsF_3	As—F	1.710	∠FAsF	95.9	ED, MW
AsF_5	F$_a$, F$_b$ / F$_b$—As, F$_a$, F$_b$	As—F$_a$ 1.711	As—F$_b$	1.656	
			(D$_{3h}$)		

Compound			Structure				Method
AsH_3	As—H (r_e)	1.511		∠HAsH (θ_e)	92.1		MW, IR
AsI_3	As—I	2.557		∠IAsI	100.2		ED
AuH	Au—H (r_e)	1.5237					UV
BBr_3	B—Br	1.893	(D_{3h})				ED
BCl_3	B—Cl	1.742	(D_{3h})				ED
BF	B—F (r_e)	1.2626					UV
BF_2H	B—H	1.189	B—F	1.311	∠FBF	118.3	MW
BF_2OH	B—F	1.32	B—O	1.34	O—H	0.941	MW
	∠FBF	118	∠FBO	123	∠BOH	114.1	
BF_3	B—F	1.313	(D_{3h})				ED, IR
BH	B—H (r_e)	1.2325					UV
BH_3PH_3	B—P	1.937	B—H	1.212	P—H	1.399	MW
	∠PBH	103.6	∠BPH	116.9	∠HBH	114.6	
	∠HPH	101.3	staggered form				
BI_3	B—I	2.118	(D_{3h})				ED
BN	B—N (r_e)	1.281					UV
BO	B—O (r_e)	1.2045					EPR
BO_2	B—O	1.265	linear				UV
BS	B—S	1.6091					UV

B_2H_6

			B—H_a	1.19			IR, ED
			B—H_b	1.33			
			B···B	1.77			
			∠H_aBH_a	122			
			∠H_bBH_b	97			

Compound			Structure				Method
$B_3H_3O_3$	B—O	1.376		∠BOB≅∠OBO	120		ED
$B_3H_6N_3$	B—N	1.435	B—H	1.26	N—H	1.05	ED
	∠NBN	118	∠BNB	121	(C_2)		
BaH	Ba—H (r_e)	2.2318					UV
BaO	Ba—O (r_e)	1.9397					MW
BaS	Ba—S (r_e)	2.5074					MBE
BeF	Be—F (r_e)	1.3609					UV
BeH	Be—H (r_e)	1.3431					UV
BeO	Be—O (r_e)	1.3308					UV
BiBr	Bi—Br (r_e)	2.6095					MW
$BiBr_3$	Bi—Br	2.63		∠BrBiBr	90	(C_{3v})	ED
BiCl	Bi—Cl (r_e)	2.4716					MW
$BiCl_3$	Bi—Cl	2.423		∠ClBiCl	100	(C_{3v})	ED
BiF	Bi—F (r_e)	2.0516					MW
BiH	Bi—H (r_e)	1.805					UV
BiI	Bi—I (r_e)	2.8005					MW
BiO	Bi—O (r_e)	1.934					UV
BrCN	C—N (r_e)	1.157		C—Br (r_e)	1.790		IR
BrCl	Br—Cl (r_e)	2.1361					MW
BrF	Br—F (r_e)	1.7590					MW
BrF_3	F_a—Br—F_a		Br—F_a	1.810	Br—F_b	1.721	MW
		(vertical F_b)	∠F_aBrF_b	86.2	(C_{2v})		
BrF_5	Br—F (average)	1.753					ED, MW
	(Br—F_eq) − (Br—F_ax) = 0.069						
	∠F_axBrF_eq	85.1	(C_{4v})				
BrO	Br—O (r_e)	1.7172					MW
Br_2	Br—Br (r_e)	2.2811					R
CBr_4	C—Br	1.935	(T_d)				ED
CCl	C—Cl	1.6512					UV
$CClF_3$	C—Cl	1.752	C—F	1.325	∠FCF	108.6	ED, MW
CCl_3F	C—Cl	1.754	C—F	1.362	∠ClCCl	111	MW
					(C_{3v})		
CCl_4	C—Cl	1.767	(T_d)				ED
CF	C—F (r_e)	1.2718					EPR
CF_3I	C—I	2.138	C—F	1.330	∠FCF	108.1	ED, MW
CF_4	C—F	1.323	(T_d)				ED

Compound	Structure						Method
CH	C—H (r_e)	1.1199					UV
CI$_4$	C—I	2.15	(T$_d$)				ED
CN	C—N (r_e)	1.1718					MW
CO	C—O (r_e)	1.1283					MW
COBr$_2$	C—O	1.178			C—Br	1.923	ED, MW
	∠BrCBr	112.3					
COClF	C—F	1.334	C—O	1.173	C—Cl	1.725	ED, MW
	∠FCCl	108.8	∠ClCO	127.5			
COCl$_2$	C—O	1.179			C—Cl	1.742	ED, MW
	∠ClCCl	111.8					
COF$_2$	C—F	1.3157			C—O	1.172	ED, MW
	∠FCF	107.71					
CO$_2$	C—O (r_e)	1.1600					IR
CP	C—P (r_e)	1.562					UV
CS	C—S (r_e)	1.5349					MW
CS$_2$	C—S (r_e)	1.5526					IR
C$_2$	C—C (r_e)	1.2425					UV
C$_3$O$_2$	C—O	1.163			C—C	1.289	ED
	linear (large-amplitude bending vibration)						
CaH	Ca—H (r_e)	2.002					UV
CaO	Ca—O (r_e)	1.8221					UV
CaS	Ca—S (r_e)	2.3178					UV
CdH	Cd—H (r_e)	1.781					EPR
CdBr$_2$	Cd—Br	2.35	linear				ED
CdCl$_2$	Cd—Cl	2.24	linear				ED
CdI$_2$	Cd—I	2.56	linear				ED
ClCN	C—Cl (r_e)	1.629			C—N (r_e)	1.160	MW
ClF	Cl—F (r_e)	1.6283					MW
ClF$_3$	F$_a$—Cl—F$_a$		Cl—F$_a$	1.698	Cl—F$_b$	1.598	MW
	$\mid$		∠F$_a$ClF$_b$	87.5	(C$_{2v}$)		
	F$_b$						
ClO	Cl—O (r_e)	1.5696					MW, UV
ClOH	O—Cl	1.690	O—H	0.975	∠HOCl	102.5	MW, IR
ClO$_2$	Cl—O	1.470			∠OClO	117.38	MW
ClO$_3$(OH)	O$_a$—Cl	1.407			O$_b$—Cl	1.639	ED
	∠O$_a$ClO$_a$		114.3		∠O$_a$ClO$_b$	104.1	

Compound	Structure						Method
Cl$_2$	Cl—Cl (r_e)	1.9878					UV
Cl$_2$O	Cl—O	1.6959			∠ClOCl	110.89	MW
CoH	Co—H (r_e)	1.542					UV
Cr (CO)$_6$	C—O	1.16			Cr—C	1.92	ED
	∠CrCO	180					
CrO	Cr—O (r_e)	1.615					UV
CsBr	Cs—Br (r_e)	3.0723					MW
CsCl	Cs—Cl (r_e)	2.9063					MW
CsF	Cs—F (r_e)	2.3454					MW
CsH	Cs—H (r_e)	2.4938					UV
CsI	Cs—I (r_e)	3.3152					MW
CsOH	Cs—O (r_e)	2.395			O—H (r_e)	0.97	MW
CuBr	Cu—Br (r_e)	2.1734					MW
CuCl	Cu—Cl (r_e)	2.0512					MW
CuF	Cu—F (r_e)	1.7449					MW
CuH	Cu—H (r_e)	1.4626					UV
CuI	Cu—I (r_e)	2.3383					MW
FCN	C—F	1.262			C—N	1.159	MW
FOH	O—H	0.96	O—F	1.442	∠HOF	97.2	MW
F$_2$	F—F (r_e)	1.4119					R

Compound			Structure					Method
$Fe(CO)_5$	Fe—C (average)	1.821						ED
	$(Fe—C)_{eq} - (Fe—C)_{ax}$		0.020					
	C—O (average)	1.153			(D_{3h})			
GaBr	Ga—Br (r_e)	2.3525						MW
GaCl	Ga—Cl (r_e)	2.2017						MW
GaF	Ga—F (r_e)	1.7744						MW
GaF_3	Ga—F	1.88			(D_{3h})			ED
GaI	Ga—I (r_e)	2.5747						MW
GaI_3	Ga—I	2.458			(D_{3h})			ED
GdI_3	Gd—I	2.841			∠IGdI	108	(C_{3v})	ED
$GeBrH_3$	Ge—H	1.526			Ge—Br	2.299		MW, IR
	∠HGeH	106.2						
$GeBr_4$	Ge—Br	2.272			(T_d)			ED
$GeClH_3$	Ge—H	1.537			Ge—Cl	2.150		IR, MW
	∠HGeH	111.0						
$GeCl_2$	Ge—Cl	2.183			∠ClGeCl	100.3		ED
$GeCl_4$	Ge—Cl	2.113			(T_d)			ED
$GeFH_3$	Ge—H	1.522			Ge—F	1.732		MW, IR
	∠HGeH	113.0						
GeF_2	Ge—F (r_e)	1.7321			∠FGeF (θ_e)	97.17		MW
GeH	Ge—H (r_e)	1.5880						UV
GeH_4	Ge—H	1.5251			(T_d)			IR, R
GeO	Ge—O (r_e)	1.6246						MW
GeS	Ge—S (r_e)	2.0121						MW
GeSe	Ge—Se (r_e)	2.1346						MW
GeTe	Ge—Te (r_e)	2.3402						MW
Ge_2H_6	Ge—H	1.541			Ge—Ge	2.403		ED
	∠HGeH	106.4			∠GeGeH	112.5		
HBr	H—Br (r_e)	1.4145						MW
HCN	C—H (r_e)	1.0655			C—N (r_e)	1.1532		MW, IR
					linear			
HCNO	H—C	1.027	C—N	1.161	N—O	1.207		MW
					linear			
HCl	H—Cl (r_e)	1.2746						MW
HF	H—F (r_e)	0.9169						MW
HI	H—I (r_e)	1.6090						MW
HNCO	N—H	0.986	N—C	1.209	C—O	1.166		MW
	∠HNC	128.0						
HNCS	N—H	0.989	N—C	1.216	C—S	1.561		MW
	∠HNC	135.0	∠NCS	180				
HNO	N—H	1.063	N—O	1.212	∠HNO	108.6		UV
HNO_2	O_a		s-trans conformer		s-cis			MW
	$N—O_bH$				conformer			
			O_b—H	0.958	0.98			
			N—O_b	1.432	1.39			
			N—O_a	1.170	1.19			
			∠O_aNO_b	110.7	114			
			∠NO_bH	102.1	104			
HNO_3	H		O_c—H	0.96	N—O_c	1.41		MW
	O_c		N—O_a	1.20	N—O_b	1.21		
	N		∠HO_cN	102.2	∠O_cNO_a	113.9		
	O_a O_b		∠O_cNO_b	115.9	planar			
HNSO	N—H	1.029	N—S	1.512	S—O	1.451		MW
	∠HNS	115.8	∠NSO	120.4				
			planar					
H_2	H—H (r_e)	0.7414						UV
H_2O	O—H (r_e)	0.9575			∠HOH (θ_e)	104.51		MW, IR
H_2O_2	O—O	1.475			∠OOH	94.8		IR
	dihedral angle of internal rotation				119.8	(C_2)		
H_2S	H—S (r_e)	1.3356			∠HSH (θ_e)	92.12		MW, IR

Compound	Structure					Method
H_2SO_4		O—H	0.97	S—O$_a$	1.574	MW
		S—O$_c$	1.422	∠H$_a$O$_a$S	108.5	
		∠O$_a$SO$_b$	101.3	∠O$_c$SO$_d$	123.3	
		∠O$_a$SO$_c$	108.6	∠O$_a$SO$_d$	106.4	
	dihedral angle between the H$_a$O$_a$S and O$_a$SO$_c$ planes			20.8		
	dihedral angle between the H$_a$O$_a$S and O$_a$SO$_b$ planes			90.9		
	dihedral angle between the H$_a$SO$_b$ and O$_c$SO$_d$ planes			88.4	(C$_2$)	
H_2S_2	S—S	2.055	S—H	1.327	∠SSH	91.3	ED, MW
	dihedral angle of internal rotation	90.6	(C$_2$)			
HfCl$_4$	Hf—Cl	2.33	(T$_d$)			ED
HgCl$_2$	Hg—Cl	2.252	linear			ED
HgH	Hg—H (r_e)	1.7404				UV
HgI$_2$	Hg—I	2.553	linear			ED
IBr	I—Br (r_e)	2.4691				MW
ICN	C—I	1.995	C—N	1.159		MW
ICl	I—Cl (r_e)	2.3210				MW
IF$_5$	I—F (average)	1.860	(I—F)$_{eq}$ – (I—F)$_{ax}$	0.03		ED, MW
	∠F$_{ax}$IF$_{eq}$	82.1	(C$_{4v}$)			
IO	I—O (r_e)	1.8676				MW
I$_2$	I—I (r_e)	2.6663				R
InBr	In—Br (r_e)	2.5432				MW
InCl	In—Cl (r_e)	2.4012				MW
InF	In—F (r_e)	1.9854				MW
InH	In—H (r_e)	1.8376				UV
InI	In—I (r_e)	2.7537				MW
IrF$_6$	Ir—F	1.830	(O$_h$)			ED
KBr	K—Br (r_e)	2.8208				MW
KCl	K—Cl (r_e)	2.6667				MW
KF	K—F (r_e)	2.1716				MW
KH	K—H (r_e)	2.244				UV
KI	K—I (r_e)	3.0478				MW
KOH	O—H	0.91	K—O	2.212 linear		MW
K$_2$	K—K (r_e)	3.9051				UV
KrF$_2$	Kr—F	1.89	linear			ED
LiBr	Li—Br (r_e)	2.1704				MW
LiCl	Li—Cl (r_e)	2.0207				MW
LiF	Li—F (r_e)	1.5639				MW
LiH	Li—H (r_e)	1.5949				MW
LiI	Li—I (r_e)	2.3919				MW
Li$_2$	Li—Li (r_e)	2.6729				UV
Li$_2$Cl$_2$				Li—Cl	2.23	ED
				Cl—Cl	3.61	
				∠ClLiCl	108	
LuCl$_3$	Lu—Cl	2.417	∠ClLuCl	112	(C$_{3v}$)	ED
MgF	Mg—F (r_e)	1.7500				UV
MgH	Mg—H (r_e)	1.7297				UV
MgO	Mg—O (r_e)	1.749				UV
MnH	Mn—H (r_e)	1.7308				UV
Mo(CO)$_6$	Mo—C	2.063	C—O	1.145	(O$_h$)	ED
MoCl$_4$O	Mo—Cl	2.279	Mo—O	1.658		ED
	∠ClMoCl	87.2	(C$_{4v}$)			
MoF$_6$	Mo—F	1.820	(O$_h$)			ED
NClH$_2$	N—H	1.017	N—Cl	1.748		MW, IR
	∠HNCl	103.7	∠HNH	107		
NCl$_3$	N—Cl	1.759	∠ClNCl	107.1		ED
NF$_2$	N—F	1.3528	∠FNF	103.18		MW
NH$_2$	N—H	1.024	∠HNH	103.3		UV
NH$_2$CN	N—H	1.00	N$_a$—C	1.35		MW

Compound	Structure						Method
	H N_a—C≡N_b H	C—N_b	1.160			∠HNH	114	
		angle between the NH_2 plane and the N—C bond					142	
NH_2NO_2	N—N	1.427			N—H		1.005	MW
	∠HNH	115.2			∠ONO		130.1	
	dihedral angle between the NH_2 and NNO_2 planes						128.2	
NH_3	N—H (r_e)	1.012			∠HNH (θ_e)		106.7	IR
NH_4Cl	N—H	1.22		N—Cl	2.54	(C_{3v})		ED
NF_2CN	F_2N_b—C≡N_a		C—N_a	1.158		C—N_b	1.386	MW
	N_b—F	1.399	∠N_aCN_b	174				
	∠CN_bF	105.4	∠FN_bF	102.8				
NH	N—H (r_e)	1.0362						LMR
NH_2OH	N—H	1.02		N—O	1.453		O—H	0.962	MW
	∠HNH	107	∠HNO	103.3		∠NOH	101.4	
	The bisector of H—N—H angle is *trans* to the O—H bond							
NO	N—O (r_e)	1.1506						IR
NOCl	N—Cl	1.975	N—O	1.14		∠ONCl	113	MW
NOF	O—N	1.136	N—F	1.512		∠FNO	110.1	MW
NO_2	N—O	1.193			∠ONO		134.1	MW
NO_2Cl	N—Cl	1.840			N—O		1.202	MW
	∠ONO	130.6			(C_{2v})			
NO_2F	N—O	1.1798			N—F		1.467	MW
	∠ONO	136			(C_{2v})			
NS	N—S (r_e)	1.4940						IR
N_2	N—N (r_e)	1.0977						UV
N_2H_4	N—H	1.021			N—N		1.449	ED, MW
	∠HNH	106.6 (assumed)			∠NNH_a		112	
	∠NNH_b	106	dihedral angle of internal rotation			91	
	H_a: the H atom closer to the C_2 axis, H_b: the H atom farther from the C_2 axis							
N_2O	N—N (r_e)	1.1284			N—O (r_e)		1.1841	MW, IR
N_2O_3	O_a O_b N_a—N_b O_c	N_a—N_b	1.864		N_a—O_a	1.142	MW
		N_b—O_b	1.202		N_b—O_c	1.217	
		∠$O_a N_a N_b$	105.05				
		∠$N_a N_b O_b$	112.72				
		∠$N_a N_b O_c$	117.47				
N_2O_4	O O N—N O O	N—N	1.782		N—O	1.190	ED
		∠ONO	135.4		(D_{2h})		
NaBr	Na—Br (r_e)	2.5020						MW
NaCl	Na—Cl (r_e)	2.3609						MW
NaF	Na—F (r_e)	1.9260						MW
NaH	Na—H (r_e)	1.8873						UV
NaI	Na—I (r_e)	2.7115						MW
Na_2	Na—Na (r_e)	3.0789						UV
$NbCl_5$	Nb—Cl_{eq}	2.241		Nb—Cl_{ax}	2.338 (D_{3h})	ED		
NbO	Nb—O (r_e)	1.691						UV
$Ni(CO)_4$	Ni—C	1.838		C—O	1.141	(T_d)		ED
NiH	Ni—H (r_e)	1.476						UV
NpF_6	Np—F	1.981		(O_h)				ED
OCS	C—O (r_e)	1.1578		C—S (r_e)		1.5601		MW
OCSe	C—O	1.159		C—Se		1.709		MW
OF	O—F (r_e)	1.3579						LMR
OF_2	O—F (r_e)	1.4053			∠FOF (θ_e)		103.07 (C_{2v})	MW
$O(SiH_3)_2$	Si—H	1.486		Si—O		1.634		ED
	∠SiOSi	144.1						
O_2	O—O (r_e)	1.2074						MW
O_2F_2	O—O	1.217		F—O		1.575		MW
	∠OOF	109.5	dihedral angle of internal rotation			87.5	(C_2)	

Compound			Structure						Method
O_3	O—O (r_e)	1.2716			$\angle OOO\ (\theta_e)$	117.47	(C_{2v})		MW
OsF_6	Os—F	1.831			(O_h)				ED
OsO_4	Os—O	1.712			(T_d)				ED
PBr_3	P—Br	2.220			$\angle BrPBr$	101.0			ED
PCl_3	P—Cl	2.039			$\angle ClPCl$	100.27			ED
PCl_5	Cl$_a$Cl$_b$ / Cl$_b$—P / Cl$_b$ / Cl$_a$		P—Cl$_a$	2.124			P—Cl$_b$	2.020 (D_{3h})	ED
PF	P—F (r_e)	1.5896							UV
PF_3	P—F	1.570			$\angle FPF$	97.8			ED, MW
PF_5	P—F$_{ax}$	1.577	P—F$_{eq}$	1.534	(D_{3h})				ED
PH	P—H (r_e)	1.4223							LMR
PH_2	P—H	1.418			$\angle HPH$	91.70			UV
PH_3	P—H	1.4200			$\angle HPH$	93.345			MW
PN	N—P (r_e)	1.4909							MW
PO	O—P (r_e)	1.4759							UV
$POCl_3$	P—O	1.449			P—Cl	1.993			ED
	$\angle ClPCl$	103.3							
POF_3	P—O	1.436	P—F	1.524	$\angle FPF$	101.3			ED, MW
P_2	P—P (r_e)	1.8931							UV
P_2F_4	P—F	1.587			P—P	2.281			ED
	$\angle PPF$	95.4			$\angle FPF$	99.1			
	The two PF$_2$ planes are *trans* to each other (the *gauche* conformer is less than 10%)								
P_4	P—P	2.21			(T_d)				ED
P_4O_6	P—O	1.638	$\angle POP$	126.4	(T_d)				ED
PbH	Pb—H (r_e)	1.839							UV
PbO	Pb—O (r_e)	1.9218							MW
PbS	Pb—S (r_e)	2.2869							MW
PbSe	Pb—Se (r_e)	2.4022							MW
PbTe	Pb—Te (r_e)	2.5950							MW
PrI_3	Pr—I	2.904	$\angle IPrI$	113	(C_{3v})				ED
PtO	Pt—O (r_e)	1.7273							UV
PuF_6	Pu—F	1.971			(O_h)				ED
RbBr	Rb—Br (r_e)	2.9447							MW
RbCl	Rb—Cl (r_e)	2.7869							MW
RbF	Rb—F (r_e)	2.2703							MW
RbH	Rb—H (r_e)	2.367							UV
RbI	Rb—I (r_e)	3.1768							MW
RbOH	Rb—O	2.301	O—H	0.957	linear				MW
$ReClO_3$	Re—O	1.702			Re—Cl	2.229			MW
	$\angle ClReO$	109.4			(C_{3v})				
ReF_6	Re—F	1.832			(O_h)				ED
RuO_4	Ru—O	1.706			(T_d)				ED
SCSe	C—Se	1.693			C—S	1.553			MW
SCTe	C—S	1.557			C—Te	1.904			MW
SCl_2	S—Cl	2.006	$\angle ClSCl$	103.0	(C_{2v})				ED
SF	S—F (r_e)	1.6006							MW
SF_2	S—F	1.5921			$\angle FSF$	98.20			MW
SF_6	S—F	1.561			(O_h)				ED
SO	S—O (r_e)	1.4811							MW
$SOCl_2$	S—O	1.44			S—Cl	2.072			MW
	$\angle ClSCl$	97.2			$\angle OSCl$	108.0			
SOF_2	S—O	1.420			S—F	1.583			ED
	$\angle OSF$	106.2			$\angle FSF$	92.2			
SOF_4	F$_b$ F$_b$ / F$_a$—S—F$_a$ / O		S—O	1.403			S—F$_a$	1.575	ED
			S—F$_b$	1.552			$\angle OSF_a$	90.7	
			$\angle OSF_b$	124.9			$\angle F_aSF_b$	89.6	
			$\angle F_bSF_b$	110.2			(C_{2v})		
SO_2	S—O (r_e)	1.4308			$\angle OSO\ (\theta_e)$	119.329			MW

Compound				Structure					Method
SO_2Cl_2	S—O	1.404		S—Cl	2.011		∠OSO	123.5	ED
	∠ClSCl	100.0		(C_{2v})					
SO_2F_2	S—O	1.397		S—F	1.530		∠OSO	123	ED
	∠FSF	97		(C_{2v})					
SO_3	S—O		1.4198			(D_{3h})			IR
$S(SiH_3)_2$	Si—H	1.494		Si—S	2.136		∠SiSSi	97.4	ED
S_2	S—S (r_e)		1.8892						R
S_2Br_2	S—Br	2.24				S—S	1.98		ED
	∠SSBr	105		dihedral angle of internal rotation			83.5		
S_2Cl_2	S—Cl	2.057				S—S	1.931		ED
	∠SSCl	108.2		dihedral angle of internal rotation			84.1	(C_2)	
S_2O_2	S—O	1.458		S—S	2.025		∠OSS	112.8	MW
						planar *cis* form			
S_8				S—S	2.07				ED
				∠SSS	105				
				(D_{4d})					
$SbCl_3$	Sb—Cl	2.333				∠ClSbCl	97.2		ED
SbH_3	Sb—H	1.704				∠HSbH	91.6		MW
SeF	Se—F	1.742							MW
SeF_6	Se—F	1.69				(O_h)			ED
SeO	Se—O (r_e)	1.6393							MW
$SeOF_2$	Se—O	1.576		Se—F	1.730				MW
	∠OSeF	104.82		∠FSeF	92.22				
SeO_2	Se—O (r_e)	1.6076				∠OSeO (θ_e)	113.83		MW
SeO_3	Se—O	1.69				(D_{3h})			ED
Se_2	Se—Se (r_e)	2.1660							UV
Se_6	Se—Se	2.34				∠SeSeSe	102		ED
	six-membered ring with chair conformation								
$SiBrF_3$	Si—F	1.560		Si—Br	2.153				MW
	∠FSiBr	108.5		(C_{3v})					
$SiBrH_3$	Si—H	1.485		Si—Br	2.210				MW
	∠HSiBr	107.8		(C_{3v})					
$SiClH_3$	Si—H	1.482		Si—Cl	2.048				MW
	∠HSiCl	107.9		(C_{3v})					
$SiCl_4$	Si—Cl	2.019		(T_d)					ED
SiF	Si—F	1.6008							UV
$SiFH_3$	Si—H	1.484		Si—F	1.593				MW, IR
	∠HSiH	110.63		(C_{3v})					
SiF_2	Si—F (r_e)	1.590		∠FSiF (θ_e)	100.8				MW
SiF_3H	Si—H (r_e)	1.4468		Si—F (r_e)	1.5624				MW
	∠HSiF (θ_e)	110.64							
SiF_4	Si—F	1.553		(T_d)					ED
SiH	Si—H (r_e)	1.5201							UV
SiH_3I	Si—H	1.485		Si—I	2.437				MW
	∠HSH	107.8							
SiH_4	Si—H	1.4798		(T_d)					IR
SiN	N—Si (r_e)	1.572							UV
SiO	Si—O (r_e)	1.5097							MW
SiS	Si—S (r_e)	1.9293							MW
SiSe	Se—Si (r_e)	2.0583							MW
Si_2	Si—Si (r_e)	2.246							UV
Si_2Cl_6	Si—Si	2.32		Si—Cl	2.009				ED
	∠ClSiCl	109.7							
Si_2F_6	Si—Si	2.317		Si—F	1.564				ED
	∠FSiF	108.6							
Si_2H_6	Si—H	1.492		Si—Si	2.331				ED
	∠SiSiH	110.3		∠HSiH	108.6				
				staggered form (assumed)					

Compound		Structure					Method
$SnCl_4$	Sn—Cl	2.280		(T_d)			ED
SnH	Sn—H (r_e)	1.7815					UV
SnH_4	Sn—H	1.711		(T_d)			R, IR
SnO	Sn—O	1.8325					MW
SnS	S—Sn (r_e)	2.2090					MW
SnSe	Se—Sn (r_e)	2.3256					MW
SnTe	Sn—Te (r_e)	2.5228					MW
SrH	Sr—H (r_e)	2.1455					UV
SrO	Sr—O (r_e)	1.9198					MW
SrS	S—Sr (r_e)	2.4405					UV
$TaCl_5$	Ta—Cl$_{eq}$	2.227		Ta—Cl$_{ax}$	2.369	(D_{3h})	ED
TaO	Ta—O (r_e)	1.6875					UV
TeF_6	Te—F	1.815		(O_h)			ED
Te_2	Te—Te (r_e)	2.5574					UV
$ThCl_4$	Th—Cl	2.58		(T_d)			ED
ThF_4	Th—F	2.14		(T_d)			ED
TlBr	Tl—Br (r_e)	2.6182					MW
TlCl	Tl—Cl (r_e)	2.4848					MW
TlF	Tl—F (r_e)	2.0844					MW
TlH	Tl—H (r_e)	1.870					UV
TlI	Tl—I (r_e)	2.8137					MW
$TiBr_4$	Ti—Br	2.339		(T_d)			ED
$TiCl_4$	Ti—Cl	2.170		(T_d)			ED
TiO	Ti—O (r_e)	1.620					UV
TiS	Ti—S (r_e)	2.0825					UV
UF_6	U—F	1.996		(O_h)			ED
$V(CO)_6$	V—C	2.015		C—O	1.138		ED
			$(O_h$, involving dynamic Jahn-Teller effect)				
VCl_3O	V—O	1.570		V—Cl	2.142		ED, MW
	∠ClVCl	111.3					
VCl_4	V—Cl	2.138		$(T_d$, involving dynamic Jahn-Teller effect)			ED
VF_5	V—F (average)	1.71					ED
VO	V—O (r_e)	1.5893					UV
$W(CO)_6$	W—C	2.059	C—O	1.149		(O_h)	ED
$WClF_5$				W—Cl	2.251		MW
				W—F (average)	1.836		
				∠F$_a$WF$_b$	88.7		
WF_4O	W—O	1.666		W—F	1.847		ED
	∠FWF	86.2		(C_{4v})			
WF_6	W—F	1.832		(O_h)			ED
XeF_2	Xe—F	1.977		linear			IR
XeF_4	Xe—F	1.94		(D_{4h})			ED
XeF_6	Xe—F	1.890		(large-amplitude bending vibration around the O_h structure)			ED
XeO_4	Xe—O	1.736		(T_d)			ED
ZnH	Zn—H (r_e)	1.5949					UV
$ZrCl_4$	Zr—Cl	2.32		(T_d)			ED
ZrF_4	Zr—F	1.902		(T_d)			ED
ZrO	Zr—O (r_e)	1.7116					UV

For $WClF_5$ structure:
Cl, F$_b$; F$_b$—W—F$_b$; F$_b$, F$_a$

STRUCTURES OF ORGANIC MOLECULES

Compounds are arranged in alphabetical order by chemical name; cross references are given for common synonyms
(lengths in Å and angles in degrees)

Compound	Structure			Method

Acetaldehyde

$C_bH_3 - C_a \overset{O}{\underset{H}{}}$ (with C=O double bond, and H below)

C_a—O	1.210		ED, MW
C_b—H	1.107		
C_a—H	1.128		
C_a—C_b	1.515	$\angle C_bC_aO$	124.1
$\angle HC_bH$	109.8	$\angle C_bC_aH$	115.3

Acetamide
CH_3CONH_2

C—O	1.220	C—N	1.380	ED
C—C	1.519	N—H	1.022	
C—H	1.124	$\angle NCO$	122.0	
$\angle CCN$	115.1			

Acetic acid

$CH_3 - C \overset{O_a}{\underset{O_b-H}{}}$

		C—C	1.520	ED
		C—O_a	1.214	
		C—O_b	1.364	
C—H	1.10	$\angle CCO_a$	126.6	
$\angle CCO_b$	110.6			

Acetone
$(CH_3)_2CO$

C—C	1.520	C—O	1.213	ED, MW
C—H	1.103	$\angle CCC$	116.0	
$\angle HCH$	108.5	symmetry axis of each methyl group is tilted 2° from the C—C bond		

Acetonitrile
CH_3CN

C—H	1.107	C—C	1.468	ED, MW
C—N	1.159	$\angle CCH$	109.7	

Acetonitrile oxide
CH_3CNO

C—C	1.442	C—N	1.169	MW
N—O	1.217	(C_{3v})		

Acetyl chloride
CH_3COCl

C—H	1.105	C—O	1.187	ED, MW
C—C	1.506	C—Cl	1.798	
$\angle HCH$	108.6	$\angle OCCl$	121.2	
$\angle CCCl$	111.6			

Acetyl cyanide → Pyruvonitrile

Acetylene
$HC\equiv CH$

C—H (r_e)	1.060	C—C (r_e)	1.203	IR

Acrolein → Acrylaldehyde

Acrylaldehyde

(structure: $H_2C_a=C_bH-C_cH=O$, planar s-trans form)

		C_b—C_c	1.484	ED, MW
		C_a—C_b	1.345	
		C_c—O	1.217	
		C_a—H	1.10	
		C_c—H	1.13	
$\angle C_aC_bC_c$	120.3	$\angle C_bC_cO$	123.3	
$\angle HC_cC_b$	114	other CCH angles (average)	122	
planar s-trans form				

Acrylonitrile

(structure: $H_2C_a=C_bH-C_c\equiv N$)

		C_a—C_b	1.343	ED, MW
		C_b—C_c	1.438	
		C_c—N	1.167	
		C_a—H	1.114	
		$\angle C_aC_bC_c$	121.7	
$\angle C_bC_cN$	178	$\angle HCC$	120	

Acryloyl chloride

(structure: $H_2C_a=C_bH-C_c(=O)Cl$)

C—H	1.086 (assumed)	C_b—C_c	1.48	MW
C_c—Cl	1.82	C_a—C_b	1.35	
		C_c—O	1.19	
		$\angle C_aC_bH$	120 (assumed)	
		$\angle C_bC_aH$	121.5 (assumed)	
		$\angle C_aC_bC_c$	123	
		$\angle C_bC_cCl$	116	
		$\angle C_bC_cO$	127	

Compound	Structure					Method
Allene	C—C	1.3084	C—H		1.087	IR
$CH_2{=}C{=}CH_2$	∠HCH	118.2				
Allyl chloride			*cis* conformer	C—Cl	1.811	MW
	Cl			∠CCCl	115.2	
	$H_2C_a{=}C_bH{-}C_cH_2$		*skew* conformer	C—Cl	1.809	
	∠CCCl	109.6	CCCCl	dihedral angle of internal rotation	122.4	
Aniline	C—C	1.392	C—N		1.431	MW
$C_6H_5NH_2$	N—H	0.998	∠HNH		113.9	
	dihedral angle between the NH_2 plane and the N—C bond				140.6	
Azetidine	$CH_2{-}CH_2$		C—N		1.482	ED
			C—C		1.553	
	$CH_2{-}NH$		C—H		1.107	
	N—H	1.03	∠CNC		92.2	
	∠CCC	86.9	∠CCN		85.8	
	dihedral angle between the CCC and CNC planes				147	
Aziridine			N—H		1.016	MW
			N—C		1.475	
			C—C		1.481	
			C—H		1.084	
			∠CNC		60.3	
			∠H$_a$NC		109.3	
	∠H$_b$CH$_c$	115.7	∠H$_b$CC		117.8	
	∠H$_b$CN	118.3	∠H$_c$CC		119.3	
	∠H$_c$CN	114.3				
Azomethane	C—N	1.482	N—N		1.247	ED
$CH_3N{=}NCH_3$	∠CNN	112.3	*trans* conformer			
Benzene	C—C	1.399	C—H		1.101	ED, IR
C_6H_6						
p-Benzoquinone			C_a—O		1.225	ED
			C_b—C_b		1.344	
			C_a—C_b		1.481	
			∠$C_b$$C_a$$C_b$		118.1	
Biacetyl	C—O	1.215	C—C (average)		1.524	ED
$CH_3COCOCH_3$	C—H	1.108	∠CCO		119.5	
	∠CCC	116.2	*trans* conformer			
Bicyclo[1.1.0]butane			C_a—C_a		1.497	MW
			C_a—C_b		1.498	
			C_a—H_a		1.071	
			C_b—H_b, C_b—H_c		1.093	
			∠$H_b$$C_b$$H_c$		115.6	
	∠$C_b$$C_a$$H_a$	130.4	∠$C_a$$C_b$$C_a$		60.0	
	∠$C_a$$C_a$$H_a$	128.4	dihedral angle between the two $C_a$$C_a$$C_b$ planes		121.7	
Bicyclo[2.2.1]hepta-2,5-diene			C_a—C_b		1.535	ED
			C_b—C_b		1.343	
			C_a—C_c		1.573	
			C—H		1.12	
			∠$C_a$$C_c$$C_a$		94	
	dihedral angle between the two $C_a$$C_b$$C_b$$C_a$ planes				115.6	
			(C_{2v})			
Bicyclo[2.2.1]heptane	See the preceding molecule for the labels of the C atoms					ED
C_7H_{12}	C_a—C_b	1.54	C_b—C_b		1.56	
	C_a—C_c	1.56	C—C (average)		1.549	
	—$C_a$$C_c$$C_a$	93.1	dihedral angle between the two $C_a$$C_b$$C_b$$C_a$ planes		113.1	

Compound	Structure				Method
Bicyclo[2.2.0]hexa-2,5-diene		C_b—C_b	1.345		ED
		C_a—C_a	1.574		
		C_a—C_b	1.524		
		dihedral angle between the two $C_aC_bC_bC_a$ planes		117.3	
Bicyclo[2.2.2]octane	$HC_a(C_bH_2C_bH_2)_3C_aH$	C_a—C_b	1.54		ED
	C_b—C_b 1.55	$\angle C_aC_bC_b$	109.7		
	C—C (average) 1.542				
	large-amplitude torsional motion about the D_{3h} symmetry axis				
Bicyclo[1.1.1]pentane C_5H_8	C—C 1.557	$\angle$CCC	74.2		ED
Bicyclo[2.1.0]pentane		C_a—C_a	1.536		MW
		C_b—C_b	1.565		
		C_a—C_b	1.528		
		C_a—C_c	1.507		
	Dihedral angle between the $C_aC_aC_bC_b$ and $C_cC_aC_c$ planes		112.7		
Biphenyl		C—C (intra-ring)	1.396		ED
		(inter-ring)	1.49		
		torsional dihedral angle between the two rings		~40	
4,4'-Bipyridyl	C—C, C—N (intra-ring)	1.375			ED
		C—C (inter-ring)	1.465		
		torsional dihedral angle between the two rings		~37	
Bis (cyclopentadienyl) beryllium $(C_5H_5)_2Be$	Be—(cyclopentadienyl plane) 1.470, 1.92				ED
	C—C 1.423 (C_{5v}) (The Be atom has two equilibrium positions)				
Bis (cyclopentadienyl) iron → Ferrocene					
Bis (cyclopentadienyl) lead $(C_5H_5)_2Pb$	C—C 1.430	Pb—C	2.79		ED
	dihedral angle between the two C_5H_5 planes 40~50 (The two rings are not parallel.)				
Bis (cyclopentadienyl) manganese $(C_5H_5)_2Mn$	Mn—C 2.383 C—C 1.429 (D_{5h})				ED
Bis (cyclopentadienyl) nickel $(C_5H_5)_2Ni$	Ni—C 2.196	C—C	1.430		ED
			(D_{5h})		
Bis (cyclopentadienyl) ruthenium $(C_5H_5)_2Ru$	C—C 1.439	Ru—C	2.196		ED
Bis (cyclopentadienyl) tin $(C_5H_5)_2Sn$	C—C 1.431	Sn—C	2.71		ED
	C—H 1.14	(D_{5h})			
Bis (trifluoromethyl) peroxide CF_3OOCF_3	O—O 1.42	C—O	1.399		ED
	C—F 1.320	$\angle$COO	107		
	$\angle$FCF 109.0	COOC dihedral angle of internal rotation	123		
Borine carbonyl BH_3CO	B—H 1.194	B—C	1.540		MW
	C—O 1.131	$\angle$HBH	113.9		
	$\angle$BCO 180	(C_{3v})			
Bromobenzene		C—H	1.072		MW
		C_c—C_d	1.401		
		C_b—C_c	1.375		
		C—Br	1.85		
		C_a—C_b	1.42		
		$\angle C_bC_aC_b$	117.4		
Bromoform $CHBr_3$	C—Br 1.924	C—H	1.11		ED, MW
	$\angle$BrCBr 111.7	(C_{3v})			
Bromoiodoacetylene $IC≡CBr$	C—I 1.972	C—C	1.206		ED
	C—Br 1.795				

Compound	Structure				Method
1,3-Butadiene	C_aH_2		C_b-C_b	1.467	ED
			C_a-C_b	1.349	
	C_bH-C_bH		C—H (average)	1.108	
	C_aH_2		∠CCC	124.4	
	∠C_bC_aH	120.9	anti conformer	(C_{2h})	
1,3-Butadiyne	$HC_a{\equiv}C_b-C_b{\equiv}C_aH$		C—H	1.09	ED
	C_a-C_b	1.218	C_b-C_b	1.384	
			linear		
Butane	C—C	1.531	C—H	1.117	ED
$CH_3CH_2CH_2CH_3$	∠CCC	113.8	∠CCH	111.0	
	trans conformer 54%		dihedral angle for the gauche conformer		65
2-Butanone → Ethyl methyl ketone					
Butatriene	$H_2C_a{=}C_b{=}C_b{=}C_aH_2$		C—H	1.08	ED
	C_a-C_b	1.32	C_b-C_b	1.28 (D_{2h})	
2-Butene	$C_aH_3-C_bH{=}C_bH-C_aH_3$				ED
	C_a-C_b	cis conformer	1.506	trans conformer 1.508	
	C_b-C_b	1.346	1.347		
	∠$C_aC_bC_b$	125.4	123.8		
3-Buten-1-yne → Vinylacetylene					
tert-Butyl chloride	C—H	1.102	C—C	1.528	ED, MW
$(CH_3)_3CCl$	C—Cl	1.828	∠CCCl	107.3	
	∠CCH	110.8	∠CCC	111.6	
tert-Butyl cyanide → Pivalonitrile					
2-Butyne	$C_aH_3-C_b{\equiv}C_b-C_aH_3$	C—H	1.116		ED
	C_b-C_b	1.214	C_a-C_b	1.468	
	∠C_bC_aH	110.7			
Carbon C_2	C—C (r_e)	1.3119			UV
Carbon C_3	C—C	1.277	linear		UV
Carbon suboxide → Tricarbon dioxide					
Carbon tetrabromide	C—Br	1.935	(T_d)		ED
CBr_4					
Carbon tetrachloride	C—Cl	1.767	(T_d)		ED
CCl_4					
Carbon tetrafluoride	C—F	1.323	(T_d)		ED
CF_4					
Carbon tetraiodide	C—I	2.15	(T_d)		ED
CI_4					
Carbonyl cyanide	C—O	1.209	C—C	1.466	ED, MW
$CO(CN)_2$	C—N	1.153	∠CCC	115	
	∠CCN	180			
Chloroacetylene	C—H	1.0550	C—C	1.2033	MW
$HC{\equiv}CCl$	C—Cl	1.6368			
Chlorobenzene	C—C	1.400	C—Cl	1.737	ED
C_6H_5Cl	C—H	1.083	∠CC(Cl)C	121.7	
	∠CC(H)C	120			
Chlorobromoacetylene	Cl—C	1.636	C—C	1.206	ED
$ClC{\equiv}CBr$	C—Br	1.784			
Chlorocyanoacetylene	C—Cl	1.624	C—C	1.205	ED
$ClC{\equiv}CCN$	C—CN	1.362	C—N	1.160	
Chloroethane → Ethyl chloride					
2-Chloroethanol	C—O	1.413	C—C	1.519	ED
$ClCH_2CH_2OH$	C—Cl	1.801	C—H	1.093	
	O—H 1.033 ∠CCCl 110.7			∠CCO 113.8	
	fraction of the gauche conformer at 37°C is 92 ∼ 94%,				
	dihedral angle of internal rotation 62.4				
Chloroethylene → Vinyl chloride					
Chloroform	C—H	1.100	C—Cl	1.758	MW
$CHCl_3$	∠ClCCl	111.3	(C_{3v})		
Chloroiodoacetylene	C—Cl	1.63	C—I	1.99	MW
$ClC{\equiv}CI$	C—C	1.209 (assumed)			

Compound	Structure				Method
Chloromethane → Methyl chloride					
3-Chloropropene → Allyl chloride					
Cyanamide $H_2N_aCN_b$	N_a-C	1.346	$C-N_b$	1.160	MW
	$N-H$	1.00	$\angle HNH$	114	
	dihedral angle between the NH_2 plane and the $N-C$ bond			142	
Cyanoacetylene $H-C_b\equiv C_a-C_c\equiv N$	C_b-H	1.058	C_a-C_b	1.205	MW
	C_a-C_c	1.378	C_c-N	1.159	
Cyanocyclopropane $C_3H_5C_aN$	$C-C$ (ring)	1.513	$C-C_a$	1.472	MW
	$C-H$	1.107	C_a-N	1.157	
	$\angle HCH$	114.6	$\angle C_aCH$	119.6	
Cyanogen $(CN)_2$	$C-N$	1.163	$C-C$	1.393	ED
				linear	
Cyclobutane $(CH_2)_4$	$C-H$	1.113	$C-C$	1.555	ED
	dihedral angle between the two CCC planes 145				
Cyclobutanone			C_a-C_b	1.527	MW
			C_b-C_c	1.556	
			$\angle C_bC_aC_b$	93.1	
			$\angle C_aC_bC_c$	88.0	

Cyclobutanone structure:

$$C_cH_2 \quad \overset{C_bH_2}{\underset{C_bH_2}{\diagup\diagdown}} \quad C_a{=}O$$

Compound	Structure				Method
Cyclobutene	C_b-C_b	1.342	C_a-C_a	1.566	MW
	C_a-C_b	1.517	C_a-H	1.094	
	C_b-H	1.083			
	$\angle C_aC_bC_b$	94.2	$\angle C_bC_bH$	133.5	
	$\angle C_aC_aH$	114.5	$\angle C_aC_aC_b$	85.8	
	$\angle HC_aH$	109.2	dihedral angle between the CH_2 plane and the C_a-C_a bond	135.8	

Cyclobutene structure:
$$\begin{array}{c} H_2C_a-C_aH_2 \\ | \quad\quad | \\ HC_b{=}C_bH \end{array}$$

Compound	Structure				Method
Cyclohexane C_6H_{12}	$C-C$	1.536	$C-H$	1.119	ED
	$\angle CCC$	111.3	chair form		
Cyclohexene			C_a-C_a	1.334	ED
			C_a-C_b	1.50	
			C_b-C_c	1.52	
			C_c-C_c	1.54	
	$\angle C_aC_aC_b$	123.4	$\angle C_aC_bC_c$	112.0	
	$\angle C_bC_cC_c$	110.9	(C_2) half-chair form		

Cyclohexene structure:
$$\begin{array}{c} HC_a{=}C_aH \\ H_2C_b \diagdown \quad\quad \diagup C_bH_2 \\ C_cH_2-C_cH_2 \end{array}$$

Compound	Structure				Method
Cyclooctatetraene			C_a-C_b	1.476	ED
			$C-H$	1.100	
			C_a-C_a, C_b-C_b	1.340	
			$\angle C_bC_aC_a, \angle C_aC_bC_b$	126.1	
	dihedral angle between the $C_aC_aC_aC_a$ and $C_aC_bC_bC_a$ planes			136.9	
	tub form (D_{2d})				

Compound	Structure				Method
1,3-Cyclopentadiene			C_a-C_b	1.509	MW
			C_b-C_c	1.342	
			C_c-C_c	1.469	
			$\angle C_aC_bC_c$	109.3	
	$\angle C_bC_cC_c$	109.4	$\angle C_bC_aC_b$	102.8	

1,3-Cyclopentadiene structure:
$$\begin{array}{c} C_aH_2 \\ HC_b \diagup \quad\quad \diagdown C_bH \\ \| \quad\quad\quad \| \\ HC_c-C_cH \end{array}$$

Compound	Structure				Method
Cyclopentadienylindium	In		$In-C$	2.621	ED
			$C-C$	1.426	
			(C_{5v})		

Cyclopentadienylindium structure:
$$\begin{array}{c} In \\ | \\ HC - CH \\ HC \quad\quad CH \\ CH \end{array}$$

Compound	Structure				Method
Cyclopentane $(CH_2)_5$	$C-H$	1.114	$C-C$	1.546	ED
	$\angle CCH$	111.7			
	(The out-of-plane vibration of the C atoms is essentially free pseudorotation; average value of the displacements of the C atoms from the molecular plane 0.43)				
Cyclopentene			C_a-C_b	1.546	ED
			C_b-C_c	1.519	
			C_c-C_c	1.342	
			$\angle C_bC_cC_b$	104.0	
	$\angle C_bC_cC_c$	110.0	$\angle C_aC_bC_c$	103.0	
	dihedral angle between the $C_bC_aC_b$ and $C_bC_cC_cC_b$ planes			151.2	

Cyclopentene structure:
$$\begin{array}{c} C_aH_2 \\ H_2C_b \diagup \quad\quad \diagdown C_bH_2 \\ C_cH{=}C_cH \end{array}$$

Compound	Structure				Method
Cyclopropane	C—C	1.512	C—H	1.083	R
$(CH_2)_3$	∠HCH	114.0			
Cyclopropanone		C—H	1.086	C_a—C_b	1.475	MW
	H_2C_b $C_a{=}O$	C_b—C_b	1.575	C_a—O	1.191	
	H_2C_b	∠$C_aC_bC_b$	57.7			
	∠HC_bH	114	dihedral angle between the CH_2 plane and the C_b—C_b bond	151	
Cyclopropene		C_b—C_b	1.304		ED
	C_aH_2	C_a—C_b	1.519		
	$HC_b{=\!=\!=}C_bH$	C_b—H	1.077		
	C_a—H	1.112	∠C_bC_bH	133	
	∠HC_aH	118			
Decalin	C—C (average)	1.530	C—H (average)	1.113	ED
$C_{10}H_{18}$	∠CCC (average)	111.4			
Dewer benzene → Bicyclo[2.2.0] hexa-2,5-diene					
Diacetylene → 1,3-Butadiyne					
1,4-Diazabicyclo[2.2.2]octane	CH_2CH_2	C—N	1.472		ED
	N—CH_2CH_2—N	C—C	1.562		
	CH_2CH_2	∠NCC	110.2		
		∠CNC	108.7		
	large-amplitude torsional motion about the D_{3h} symmetry axis				
2,3-Diaza-1,3-butadiene → Formaldehyde azine					
Diazirine		C—H	1.09		MW
	CH_2 $\|\|$ N / N	C—N	1.482		
		N—N	1.228		
		∠HCH	117		
Diazoacetonitrile	H	C_b—N_b	1.280		MW
	C_b	N_b—N_c	1.132		
	C_a N_b	C_a—N_a	1.165		
	N_a N_c	C—H	1.082		
		C_a—C_b	1.424		
	∠C_aC_bH	117	∠$C_aC_bN_b$	119.5	
Diazomethane	C—H	1.075	C—N	1.32	MW, IR
CH_2N_2	N—N	1.12	∠HCH	126.0 linear	
1,2-Dibromoethane	C—C	1.506	C—Br	1.950	ED
CH_2BrCH_2Br	C—H	1.108	∠CCBr	109.5	
	∠CCH	110	fraction of the *trans* conformer at 25°C	95%	
Dibromomethane	C—H	1.08	C—Br	1.924	ED
CH_2Br_2	∠HCBr	109	∠BrCBr	113.2	
2,2′-Dichlorobiphenyl	C—C	1.398	C—C inter-ring	1.495	ED
C_6H_4Cl—C_6H_4Cl	C—Cl	1.732	C—H	1.10	
	∠CCCl	121.4	∠CCH	126	
	dihedral angle between the two aromatic rings 74 (defined to be 0 for that of the *cis* conformer)				
trans-1,4-Dichlorocyclohexane	C—H	1.102	C—Cl	1.810	ED
$C_6H_{10}Cl_2$	C—C	1.530	∠CCC	111.5	
	∠CCCl (*ee*)	108.6	∠CCCl (*aa*)	110.6	
	∠HCCl (*ee*)	111.5	∠HCCl (*aa*)	107.6	
	ee 49%		aa 51%	e: equatorial, a: axial	
1,1-Dichloroethane	C—Cl	1.766	C—C	1.540	MW
$CHCl_2CH_3$	∠ClCCl	112.0	∠CCCl	111.0	
1,2-Dichloroethane	C—C	1.531	C—Cl	1.790	ED
CH_2ClCH_2Cl	C—H	1.11	∠CCCl	109.0	
	∠CCH	113			
	fraction of the *trans* conformer at room temperature 73%, that of the *gauche* conformer 27%				

Compound	Structure				Method
1,1-Dichloroethylene	C—C	1.32 (assumed)	C—Cl	1.73	MW
CH_2=CCl_2	∠ClCC	123	(C_{2v})		
cis-1,2-Dichloroethylene	C—Cl	1.718	C—C	1.354	ED
CHCl=CHCl	∠ClCC	123.8			
Dichloromethane	C—H (r_e)	1.087	C—Cl (r_e)	1.765	MW, IR
CH_2Cl_2	∠HCH (θ_e)	111.5	∠ClCCl (θ_e)	112.0	
1,1-Difluoroethane	C—C	1.498	C—H (average)	1.081	ED
CH_3CHF_2	C—F	1.364	∠CCH (average)	111.0	
	∠CCF 110.7	dihedral angle between the two CCF planes		118.9	
1,2-Difluoroethane	C—F	1.389	C—C	1.503	ED
CH_2FCH_2F	C—H	1.103	∠CCF	110.3	
	∠CCH 111	dihedral angle of internal rotation		109	
	fraction of the *gauche* conformer at 22°C 94%				
1,1-Difluoroethane	C—C	1.340	C—F	1.315	ED, MW
CH_2=CF_2	C—H	1.091	∠CCF	124.7	
	∠CCH	119.0			
cis-1,2-Difluoroethylene	C—C	1.33	C—F	1.342	ED, MW
CHF=CHF	C—H	1.099	∠CCF	122.0	
	∠CCH	124.1			
Difluoromethane	C—H	1.093	C—F	1.357	MW
CH_2F_2	∠HCH	113.7	∠FCF	108.3	
Dimethoxymethane			C_a—O	1.432	ED
			C_b—O	1.382	
			C—H (average)	1.108	
	∠COC	114.6	∠OCO	114.3	
	∠OCH	110.3			
Dimethylacetylene → 2-Butyne					
Dimethylamine	C—H	1.106	N—H	1.00	ED
$(CH)_2NH$	C—N	1.455	∠CNC	111.8	
	∠CNH	107	∠NCH	112	
	∠HCH	107			
Dimethylberyllium	Be—C	1.698	C—H	1.127	ED
$(CH_3)_2Be$	∠BeCH	113.9	CBeC linear		
Dimethylcadmium	C—Cd	2.112	∠HCH	108.4	R
$(CH_3)_2Cd$					
Dimethyl carbonate	$C_aH_3O_a$		C_b—O_b	1.209	ED
		C_b=O_b	C_b—O_a	1.34	
	$C_aH_3O_a$		C_a—O_a	1.42	
	∠$O_aC_bO_a$	107	∠$C_bO_aC_a$	114.5	
Dimethylcyanamide	C_b—N_b	1.161	C_b—N_a	1.338	ED
$(C_aH_3)_2N_a$—C_b≡N_b	C_a—N_a	1.463	∠$C_aN_aC_a$	115.5	
	∠$C_aN_aC_b$	116.0			
1,2-Dimethyldiborane			B—B	1.799	ED
			B—C	1.580	
			B—H_b	1.358 (*cis*), 1.365 (*trans*)	
	B—H_t	1.24			
	∠BBC	122.6 (*cis*), 121.8 (*trans*)			
Dimethyl diselenide	C—H	1.13	C—Se	1.95	ED
$(CH_3)_2Se_2$	Se—Se	2.326	∠CSeSe	98.9	
	∠HCSe 108	dihedral angle between the CSeSe and SeSeC planes		88	
Dimethyl disulfide	C—S	1.816	S—S	2.029	ED
$(CH_3)_2S_2$	C—H	1.105	∠SSC	103.2	
	∠SCH 111.3	CSSC dihedral angle of internal rotation		85	
S,S'-Dimethyl dithiocarbonate	$C_aH_3SC_bSC_aH_3$		C_b—O	1.206	ED
	‖		C_b—S	1.777	
	O		C_a—S	1.802	
	∠OCS	124.9	∠CSC	99.3	
			syn-syn conformer		

Compound	Structure				Method
Dimethyl ether	C—O	1.416	C—H	1.121	ED
$(CH_3)_2O$	∠COC	112	∠HCH	108	
Dimethylglyoxal → Biacetyl					
N,N'-Dimethylhydrazine	N—N	1.42	C—N	1.46	ED
$CH_3NH—NHCH_3$	N—H	1.03	C—H	1.12	
	∠NNC 112		CNNC dihedral angle of internal rotation	90	
Dimethylmercury	C—Hg	2.083	C—H	1.160 (assumed)	ED
$(CH_3)_2Hg$	Hg···H	2.71			
Dimethylphosphine	C—P	1.848	P—H	1.419	MW
$(CH_3)_2PH$	∠CPC	99.7	∠CPH	97.0	
Dimethyl selenide	C—H	1.093	Se—C	1.943	MW
$(CH_3)_2Se$	∠CSeC	96.2	∠SeCH	108.7	
	∠HCH	110.3			
Dimethyl sulfide	C—S	1.807	C—H	1.116	ED, MW
$(CH_3)_2S$	∠CSC	99.05	∠HCH	109.3	
Dimethyl sulfone	C—H	1.114	S—O	1.435	ED
$(CH_3)_2SO_2$	S—C	1.771	∠CSC	102	
	∠OSO	121			
Dimethyl sulfoxide	C—H	1.081	C—S	1.799	MW
$(CH_3)_2SO$	S—O	1.485	∠CSC	96.6	
	∠CSO	106.7	∠HCH	110.3	
	dihedral angle between the SCC plane and the S—O bond			115.5	
Dimethylzinc	Zn—C	1.929	∠HCH	107.7	R
$(CH_3)_2Zn$					
1,4-Dioxane	C—C	1.523	C—O	1.423	ED
	C—H	1.112	∠COC	112.45	
$\begin{array}{c} CH_2CH_2 \\ O \quad\quad O \\ CH_2CH_2 \end{array}$	∠CCO	109.2	chair form		
Ethanal → Acetaldehyde					
Ethane	C—C	1.5351	C—H	1.0940	MW
C_2H_6	∠CCH	111.17	staggered conformation		
Ethanethiol	$C_bH_3—C_aH_2—SH$		$C_a—H$	1.090	MW
	$C_b—H$	1.093	$C_a—C_b$	1.530	
	$C_a—S$	1.829	S—H	1.350	
	$∠C_bC_aH$	109.6	$∠C_aC_bH$	109.7	
	$∠C_bC_aS$	108.3	$∠C_aSH$	96.4	
Ethanol	$C_bH_3C_aH_2OH$		C—C	1.512	MW
	C—O	1.431	O—H	0.971	
	$C_a—H$	1.10	$C_b—H$	1.09	
	∠CCO	107.8	∠COH	105	
	$∠C_bC_aH$	111	$∠C_aC_bH$	110	
	staggered conformation				
Ethyl chloride	$\begin{array}{c} H_b \quad\quad Cl \\ H_b—C_b—C_a—H_a \\ H_b \quad\quad H_a \end{array}$		C—C	1.528	ED, MW
			C—Cl	1.802	
			C—H	1.103	
			$C_a—H_a=C_b—H_b$ (assumed)		
			∠CCCl	110.7	
	$∠H_bC_bH_b$	109.8	$∠H_aC_aH_a$	109.2	
	$∠C_bC_aH_a$	110.6			
Ethylene	C—H	1.087	C—C	1.339	MW
$CH_2=CH_2$	∠CCH	121.3			
Ethylenediamine	C—N	1.469	C—C	1.545	ED
$H_2NCH_2CH_2NH_2$	C—H	1.11	∠CCN	110.2	
	gauche conformer dihedral angle between the NCC and CCN planes			64	
Ethylene dibromide → 1,2-Dibromoethane					
Ethylene dichloride → 1,2-Dichloroethane					
Ethyleneimine → Aziridine					

Compound	Structure					Method

Ethylene oxide

CH_2
|
O
|
CH_2

C—C	1.466	C—H	1.085	MW
C—O	1.431	∠HCH	116.6	
dihedral angle between the NH_2 plane and the N—C bond			158.0	

Ethylene sulfide → Thiirane

Ethyl methyl ether
$C_2H_5OCH_3$

C—O (average)	1.418	C—C	1.520	ED
C—H (average)	1.118	∠COC	111.9	
∠OCC	109.4	∠HCH	109.0	
fraction of the *trans* conformer at 20°C		80%		

Ethyl methyl ketone

C_aH_3 C_bH_2 C_c(=O) C_dH_3

		C—C (average)	1.518	ED
		C_c—O	1.219	
		C—H (average)	1.102	
		∠$C_aC_bC_c$	113.5	
∠C_bC_cO, ∠C_dC_cO	121.9	trans conformer	95%	

Ethyl methyl sulfide
$C_2H_5SCH_3$

C—S (average)	1.813	C—C	1.536	ED
C—H	1.111	∠CSC	97	
∠SCC	114.0	∠HCH	110	
fraction of the *gauche* conformer at 20°C	75%			

Ferrocene
$(C_5H_5)_2Fe$

C—C	1.440	C—H	1.104	ED
Fe—C	2.064	(D_{5h})		

Fluoroform
CHF_3

C—H	1.098	C—F	1.332	MW
∠FCF	108.8	(C_{3v})		

Formaldehyde
H_2CO

C—H	1.116	C—O	1.208	MW
∠HCH	116.5			

Formaldehyde azine

$H_2C=N—N=CH_2$ N—N	1.418			ED
C—N	1.277	C—H	1.094	
∠CNN	111.4	∠HCN	120.7	
fraction of the *trans* conformer at −30°C	91%			

Formaldehyde dimethylacetal →
Dimethoxymethane

Formaldoxime

H_a OH_c
 \ /
  C=N
 /
H_b

		C—H_a	1.085	MW
		C—H_b	1.086	
		C—N	1.276	
N—O	1.408	O—H_c	0.956	
∠H_bCN	115.6	∠CNO	110.2	
∠H_aCN	121.8	∠NOH_c	102.7	

Formamide

H_c O
 \ //
  N—C
 / \
H_b H_a

		C—H_a	1.125	ED, MW
		N—H	1.027	
		C—N	1.368	
C—O	1.212	∠NCO	125.0	
∠CNH (average)	119.2			

Formic acid

 O_a
 //
H—C
 \
 O_b—H

		C—O_a	1.202	MW
		C—O_b	1.343	
O_b—H	0.972	C—H	1.097	
∠HCO_a	124.1	∠O_aCO_b	124.9	
∠CO_bH	106.3	planar		

Formic acid dimer

O_a····H—O_b
HC   CH
O_b—H····O_a

		O_a···O_b	2.703	ED
		C—O_a	1.220	
		C—O_b	1.323	
		∠O_aCO_b	126.2	
		∠CO_aO_b	108.5	

Formyl radical

H
 \
  C=O
 •

C—H	1.110	C—O	1.1712	MW
∠HCO	127.43			

Compound	Structure		Method

Fulvene

C_dH_2
C_a
HC_b C_bH
HC_c—C_cH

C_a—C_d	1.349	MW
C_a—C_b	1.470	
C_b—C_c	1.355	
C_c—C_c	1.476	
C_b—H	1.078	
C_c—H	1.080	

C_d—H	1.13	$\angle C_bC_aC_b$	106.6	
$\angle C_aC_bC_c$	107.7	$\angle C_bC_cC_c$	109	
$\angle C_aC_bH$	124.7	$\angle C_bC_cH$	126.4	
$\angle HC_dH$	117			

2-Furaldehyde

H   H
C_c—C_b O_b
H—C_d C_a—C_e
O_a H

C_a—C_e	1.458	MW
C_e—O_b	1.250	
C_e—H	1.088	
$\angle C_eC_aC_b$	133.9	

$\angle C_aC_eH$	116.9	$\angle C_aC_eO$	121.6

trans conformer (with respect to the O_a and O_b atoms)

Furan

H_a O H_a
C_a C_a
C_b—C_b
H_b H_b

C_b—C_b	1.431	MW
C_a—C_b	1.361	
C_a—O	1.362	
C_a—H_a	1.075	
C_b—H_b	1.077	

$\angle C_aC_bC_b$	106.1	$\angle C_aOC_a$	106.6
$\angle C_bC_aO$	110.7	$\angle OC_aH_a$	115.9
$\angle C_bC_bH_b$	128.0		

Furfural → 2-Furaldehyde

Glycolaldehyde

H_c O_b
C_b
H_b C_a H_a
H_b O_a

C_b—O_b	1.209	MW
C_a—O_a	1.437	
C_a—C_b	1.499	
O_a—H_a	1.051	
C_b—H_c	1.102	
C_a—H_b	1.093	

$\angle C_aC_bO_b$	122.7	$\angle C_aC_bH_c$	115.3
$\angle C_bC_aO_a$	111.5	$\angle C_aO_aH_a$	101.6
$\angle C_bC_aH_b$	109.2	$\angle H_bC_aH_b$	107.6
$\angle H_bC_aO_a$	109.7		

Glyoxal
CHOCHO

C—C	1.526	C—O	1.212	ED, UV
C—H	1.132	$\angle$CCO	121.2	
$\angle$HCO	112	*trans* conformer	(C_{2h} (assumed))	

Hexachloroethane
Cl_3CCCl_3

C—C	1.56	C—Cl	1.769	ED
$\angle$CCCl	110.0			

2,4-Hexadiyne

C_aH_3—C_b≡C_c—C_c≡C_b—C_aH_3				ED
C_a—C_b	1.450	C_b—C_c	1.208	
C_c—C_c	1.377	C_a—H	1.09	

Hexafluoroethane
F_3CCF_3

C—C	1.545	C—F	1.326	ED
$\angle$CCF	109.8	staggered conformation		

Hexafluoropropene
CF_2=$CFCF_3$

average value of the C=C and C—F distances			1.329	ED
C—C	1.513	$\angle$CCC	127.8	
$\angle$FCC (CF_2)	124	$\angle$FCC (CF)	120	
$\angle$FCC (CF_3)	110			

1,3,5-Hexatriene

H_2C_a=C_bH—C_cH=C_cH—C_bH=C_aH_2				ED
C_a—C_b	1.337	C_b—C_c	1.458	
C_c—C_c	1.368	$\angle C_aC_bC_c$	121.7	
$\angle C_bC_cC_c$	124.4			

Iminocyanide radical
HNCN

N—H	1.034	N⋯N	2.470	UV
$\angle$HNC	116.5	$\angle$NCN	~180	

Iodocyanoacetylene
I—C_a≡C_b—C_c≡N

I—C_a	1.985	C_a—C_b	1.207	MW
C_b—C_c	1.370	C_c—N	1.160	

Compound	Structure				Method
Isobutane	C_a—H	1.122	C_b—H	1.113	ED, MW
$(C_bH_3)_3C_aH$	C_a—C_b	1.535	$\angle C_bC_aC_b$	110.8	
	$\angle C_aC_bH$	111.4			
Isobutylene → 2-Methylpropene					
Ketene	C—C	1.317	C—O	1.161	MW
CH_2=C=O	C—H	1.080	$\angle$HCH	123.0	
Malononitrile	C—H	1.091	C—C	1.480	MW
$C_aH_2(C_bN)_2$	C—N	1.147	$\angle$CCC	110.4	
	$\angle$HCH	108.4	$\angle$CCN	176.6	
	(The two N atoms are bent away from each other in the plane of C_b—C_a—C_b)				
Methane	C—H (r_e)	1.0870	(T_d)		MW
CH_4					
Methanethiol	C—H	1.09	C—S	1.819	MW
CH_3SH	S—H	1.34	$\angle$HSC	96.5	
	$\angle$HCH	109.8			
	angle between the CH_3 symmetry axis and the C—S bond 2.2. (The axis of the CH_3 group is tilted away from the H atom with respect to the C—S bond.)				
Methanol	C—H	1.0936	C—O	1.4246	MW
CH_3OH	O—H	0.9451	$\angle$HCH	108.63	
	$\angle$COH	108.53			
	angle between the CH_3 symmetry axis and the C—O bond (The axis of the CH_3 group is tilted away from the H atom with respect to the C—O bond.)			3.27	
Methyl radical	C—H	1.08	planar		UV
·CH_3					
N-Methylacetamide	H_3C_a C_b—N O C_cH_3 H		C_a—C_b	1.520	ED
			N—C_c	1.469	
			C—H	1.107	
	C_b—N	1.386	C_b—O	1.225	
	$\angle C_bNC_c$	119.7			
	$\angle NC_bO$	121.8			
	$\angle C_aC_bN$	114.1			
Methylacetylene → Propyne					
Methylal → Dimethoxymethane					
Methylamine	N—H	1.010	C—N	1.471	MW
CH_3NH_2	C—H	1.099	$\angle$HNH	107.1	
	$\angle$HNC	110.3	$\angle$HCH	108.0	
	dihedral angle between the CH_3 symmetry axis and the C—N bond (The axis of the CH_3 group is tilted away from the NH_2 group with respect to the C—N bond.)			2.9	
Methyl azide	CH_3 N_a—N_b—N_c		C—H	1.09	ED
			C—N_a	1.468	
			N_a—N_b	1.216	
	N_b—N_c	1.113	$\angle CN_aN_b$	116.8	
	NNN linear				
Methyl bromide	C—H (r_e)	1.086	C—Br (r_e)	1.933	MW, IR
CH_3Br	$\angle$HCH (θ_e)	111.2	(C_{3v})		
Methyl chloride	C—H	1.090	C—Cl	1.785	MW, IR
CH_3Cl	$\angle$HCH	110.8			
Methyldiazirine	CH_3CH <img N || N />	C—N	1.481	C—C	1.501	MW
	N—N	1.235	$\angle$NCN	49.3	
	dihedral angle between the CNN plane and the C—C bond		122.3		
Methylene	C—H	1.078	$\angle$HCH	130	LMR
:CH_2					
Methylenecyclopropane	C_cH_2 | C_b=C_aH_2 C_cH_2		C_a—C_b	1.332	MW
			C_b—C_c	1.457	
			C_c—C_c	1.542	
	C_c—H	1.09	$\angle C_cC_bC_c$	63.9	

Compound	Structure					Method
	$\angle HC_aH$	114.3	$\angle HC_cH$	113.5		
	dihedral angle between the C_cH_2 plane and the C_c—C_c bond			150.8		
3-Methyleneoxetane	C_cH_2		C_b—C_c	1.52		MW
	O C_b=C_aH_2		C_c—O	1.45		
	C_cH_2		C_a—C_b	1.33		
	C—H	1.09 (assumed)	$\angle C_cC_bC_c$	87		
	$\angle HC_cH$	114 (assumed)	$\angle HC_aH$	120 (assumed		
Methyl fluoride	C—H (r_e)	1.095				MW, IR
CH_3F	C—F (r_e)	1.382	$\angle HCH$ (θ_e)	110.45 (C_{3v})		
Methyl formate	C_aH_3		C_a—H	1.08		ED
	O_b		C_b—O_b	1.206		
	O_a—C_b		C—O (average)	1.393		
	H_b		C_b—H	1.101 (assumed)		
	$\angle O_aC_bO_b$	127	$\angle COC$	114		
	$\angle O_aC_aH$	110				
Methylgermane	C—H	1.083	Ge—H	1.529		MW
CH_3GeH_3	C—Ge	1.945	$\angle HCH$	108.4		
	$\angle HGeH$	109.3				
Methyl hypochlorite	C—H	1.103	O—Cl	1.674		MW
CH_3OCl	O—C	1.389	$\angle HCH$	109.6		
	$\angle COCl$	112.8				
Methylidyne radical	C—H (r_e)	1.1198				UV
:CH						
Methylidyne phosphide	H—C (r_e)	1.0692	C—P (r_e)	1.5398		MW
HCP						
Methyl iodide	C—H (r_e)	1.084	C—I (r_e)	2.132		MW, IR
CH_3I	$\angle HCH$ (θ_e)	111.2	(C_{3v})			
Methyl isocyanide	C_aH_3—N≡C_b C_a—H	1.102	C_a—N	1.424		MW
	N—C_b	1.166	$\angle NC_aH$	109.12		
Methylketene	C_cH_3		O—C_a	1.171		MW
	C_b=C_a=O		C_b—C_c	1.518		
	H		C_c—H	1.10		
	C_a—C_b	1.306	C_b—H	1.083		
	$\angle OC_aC_b$	180.5	$\angle C_aC_bC_c$	122.6		
	$\angle C_aC_bH$	113.7	$\angle C_cC_bH$	123.7		
	$\angle HCH$	109.2				
Methylmercury chloride	Hg—Cl	2.282	C—H	1.15		MW,
CH_3HgCl	Hg—C	1.99	(C_{3v})			NMR
Methyl nitrate	H_a H_a O_a		C—H_a	1.10		MW
	C N		C—H_b	1.09		
	H_b O O_b		C—O	1.437		
			O—N	1.402		
	N—O_a	1.205	N—O_b			
	$\angle OCH_a$	110	$\angle OCH_b$	103		
	$\angle CON$	112.7	$\angle ONO_a$	118.1		
	$\angle ONO_b$	112.4				
Methylphosphine	C—P	1.858	C—H	1.094		ED
CH_3PH_2						
2-Methylpropane → Isobutane						
2-Methylpropene	C_aH_3 H_c		C_a—H		1.119	ED, MW
	C_b=C_c		C_c—H_c		1.10	
	C_aH_3 H		C_a—C_b		1.508	
			C_b—C_c		1.342	
	$\angle HC_aC_b$	111.4	$\angle H_cC_cH_c$		118.5	
	(average)					
	$\angle C_aC_bC_a$	115.6	$\angle C_aC_bC_c$		122.2	
	$\angle HC_aH$	107.9	$\angle C_bC_cH$		121	

Compound	Structure						Method
Methylsilane CH_3SiH_3	C—H	1.093		C—Si		1.867	MW
	Si—H	1.485		∠HCH		107.7	
	∠HSiH	108.3		(C_{3v})			
Methylstannane CH_3SnH_3	C—Sn	2.143		Sn—H		1.700	MW
				(C_{3v})			
Methyl thiocyanate	C_aH_3	$S—C_b$	1.684		$S—C_a$	1.824	MW
	$S—C_b—N$	$C_b—N$	1.170		C—H	1.081	
	∠C_aSC_b	99.0	∠HCH	110.6	∠HCS	108.3	
Naphthalene				$C_a—C_b$		1.37	ED
				$C_b—C_b$		1.41	
				$C_a—C_c$		1.42	
				$C_c—C_c$		1.42	
				C—C (average)		1.40	
				∠$C_aC_cC_c$		119.4	
Neopentane $C(CH_3)_4$	C—C	1.537		C—H		1.114	ED
	∠CCH	112					
Nickelocene → Bis (cyclopentadienyl) nickel							
Nitromethane CH_3NO_2	C—H	1.088 (assumed)		C—N		1.489	MW
	N—O	1.224		∠NCH		107	
	∠ONO	125.3					
N-Nitrosodimethylamine $(CH_3)_2NNO$	N—O	1.235		N—N		1.344	ED
	C—N	1.461		∠ONN		113.6	
	∠CNC	123.2		∠CNN		116.4	
Nitrosomethane CH_3NO	C—N	1.49		N—O		1.22	MW
	C—H	1.084		∠CNO		112.6	
	∠NCH	109.0					
Norbornane → Bicyclo[2.2.1]heptane							
Norbornadiene → Bicyclo[2.2.1]hepta-2,5-diene							
1,2,5-Oxadiazole		O—N	1.380		∠NON	110.4	MW
		C—N	1.300		∠ONC	105.8	
		C—C	1.421		∠CCN	109.0	
		C—H	1.076		∠CCH	130.2	
		∠NCH	120.9		planar		
1,3,4-Oxadiazole		O—C	1.348		∠COC	102.0	MW
		C—N	1.297		∠OCN	113.4	
		N—N	1.399		∠CNN	105.6	
		C—H	1.075		∠OCH	118.1	
		∠NCH	128.5		planar		
Oxalic acid				C—C	1.544		ED
				$C—O_a$	1.205		
				$C—O_b$	1.336		
				$O_b—H$	1.05		
				∠CCO_a	123.1		
				∠O_aCO_b	125.0		
				∠CO_bH	104		
Oxalyl dichloride				C—O		1.182	ED
				C—C		1.534	
				C—Cl		1.744	
	∠CCO	124.2		∠CCCl		111.7	
	fraction of the *trans* conformer at 0°C 68%, that of the *gauche* conformer 32%						
Oxetane				C—O		1.448	MW
	H_2C CH_2 CH_2			C—C		1.546	
				C—H (average)		1.090	
				∠COC		92	

Compound	Structure				Method
	∠CCC	85	∠OCC	92	
	∠HCH (average)	109.9			

Oxirane → Ethylene oxide

Phenol

C—C (average)	1.397	MW
C_b—H	1.084	
C_c—H	1.076	
C_d—H	1.082	
C_a—O	1.364	
O—H	0.956	
∠COH	109.0	

Phosphirane

C—P	1.867	P—H	1.43	MW
C—C	1.502	C—H	1.09	
∠CPC	47.4	∠HCH	114.4	

∠HPC 95.2   ∠CCH 118

dihedral angle between the PCC plane and the PH bond   95.7

Piperazine

C—C	1.540	ED
C—N	1.467	
C—H	1.110	

∠CNC	109.0	∠CCN	110.4
		(C_{2h})	

Pivalonitrile
$(C_cH_3)_3C_b$—C_a≡N

C_a—C_b	1.495	C_a—N	1.159	MW
C_b—C_c	1.536	∠$C_cC_bC_c$	110.5	

Propadiene → Allene

Propane
C_3H_8

C—C	1.532	C—H	1.107	ED
∠CCC	112	∠HCH	107	

Propenal → Acrylaldehyde

Propene

C_a—H_a	1.104	ED, MW
C_a—C_b	1.341	
C_c—H_d	1.117	
C_b—C_c	1.506	
∠$C_bC_aH_{a,b,c}$	121.3	
∠$C_aC_bC_c$	124.3	

∠$C_bC_cH_d$ 110.7

1-Propenyl chloride
CH_3—C_bH=C_aH—Cl C_a—Cl   1.728   MW

∠C_bC_aCl   121.9   *trans* conformer

Propiolaldehyde
H_aC_a≡C_b—C_cH_cO

C_a—C_b	1.211	C_a—H_a	1.085	ED, MW
C_c—H_c	1.130	C_b—C_c	1.453	
∠C_bC_cO	124.2	C_c—O	1.214	
∠$C_aC_bC_c$	178.6	∠$C_bC_cH_c$	113.7	
		planar		

Propylene → Propene

Propylene oxide
$C_aH_3C_bH$——C_cH_2
          \\ O /

C_a—C_b	1.51	MW
∠$C_aC_bC_c$	121.0	

dihedral angle between the C_bC_cO plane and the C_aC_b bond   123.8

Propynal → Propiolaldehyde

Propyne
H_3C_c—C_b≡C_aH

C_c—C_b	1.459	C_c—H	1.105	MW
C_a—H	1.056	C_b—C_a	1.206	
		∠HC_cC_b	110.2	

Pyrazine

C—C	1.339	C—N	1.403	ED
C—H	1.115	∠CCN	115.6	
∠CCH	123.9			

Pyridazine

N—C_a	1.341	ED, MW
C_a—C_b	1.393	
N—N	1.330	
C_b—C_b	1.375	

∠NCC   123.7   ∠NNC   119.3

Compound	Structure				Method

Pyridine

N—C$_a$	1.340		MW
C$_b$—C$_c$	1.394		
C$_b$—H$_b$	1.081		
C$_a$—C$_b$	1.395		
C$_a$—H$_a$	1.084		
C$_c$—H$_c$	1.077		

∠C$_a$NC$_a$	116.8	∠NC$_a$C$_b$	123.9
∠C$_a$C$_b$C$_c$	118.5	∠C$_b$C$_c$C$_b$	118.3
∠NC$_a$H$_a$	115.9	∠C$_c$C$_b$H$_b$	121.3

Pyrimidine

N—C	1.340	C—C	1.393	ED
∠NCN	127.6	∠CNC	115.5	

(C$_{2v}$ assumed)

Pyrrole

N—Ca	1.370	MW
C$_b$—C$_b$	1.417
C$_a$—C$_b$	1.382
N—H	0.996
C$_a$—H$_a$	1.076

C$_b$—H$_b$	1.077	∠C$_a$NC$_a$	109.8
∠NC$_a$C$_b$	107.7	∠C$_a$C$_b$C$_b$	107.4
∠NC$_a$H$_a$	121.5	∠C$_b$C$_b$H	127.1

Pyruvonitrile

C—H	1.12		ED, MW
C—N	1.17		
C—O	1.208		
C$_b$—C$_c$	1.477		

C$_a$—C$_b$	1.518	∠HCH	109.2
∠C$_a$C$_b$O	124.5	∠C$_a$C$_b$C$_c$	114.2
∠CCN	179		

Ruthenocene → Bis (cyclopentadienyl) ruthenium

Silacyclobutane

CH$_2$—CH$_2$
CH$_2$—SiH$_2$

Si—C	1.892		ED
C—C	1.600		
Si—H	1.47		

C—H	1.14	∠CSiC	80.7
∠SiCC	84.8	∠CCC	99.8
		dihedral angle between the CCC and CSiC planes	146

Spiropentane

C$_b$—C$_b$	1.52		ED
C$_a$—C$_b$	1.47		
C—H	1.09		
∠C$_b$C$_a$C$_b$	62		

∠HCH	118	(D$_{2d}$)

Succinonitrile

CH$_2$CN
CH$_2$CN

C—C	1.561	C—C(N)	1.465	ED
C—N	1.161	C—H	1.09	
∠CCC	110.4			

fraction of the *anti* conformer at 170°C 74%,
dihedral angle of CCCC for the *gauche* conformer   75

Tetrachloroethylene	C—Cl	1.718	C—C	1.354	ED
CCl$_2$=CCl$_2$	∠ClCCl	115.7			
Tetracyanoethylene	C—N	1.162	C—C	1.435	ED
(CN)$_2$C=C(CN)$_2$	C=C	1.357	∠CC=C	121.1	

Tetrafluoro-1,3-dithietane

C—S	1.785	ED
C—F	1.314	
∠CSC	83.2	

∠FCS	113.7	(D$_{2h}$ assumed)

Tetrafluoroethylene	C—C	1.31	C—F	1.319	ED
CF$_2$=CF$_2$	∠CCF	123.8	(D$_{2h}$ assumed)		

Compound	Structure				Method
Tetrahydrofuran	C—H	1.115	C—O	1.428	ED
CH₂CH₂ | CH₂CH₂ O	C—C	1.536			
	The skeletal bending vibration of the molecular plane is essentially free pseudorotation				
Tetrahydropyran			C—O	1.420	ED
			C—C	1.531	
			C—H	1.116	
			∠COC	111.5	
			∠OCC	111.8	
	∠CCC (C)	108	∠CCC (O)	111	
	chair form				
Tetrahydrothiophene	C—S	1.839	C—H	1.120	
CH₂CH₂ | CH₂CH₂ S	C—C	1.536	∠CSC	93.4	
	∠SCC	106.1	∠CCC	105.0	
Tetramethylgermane (CH₃)₄Ge	Ge—C	1.945	C—H	1.12	ED
	∠GeCH	108	(T_d excluding the H atoms)		
Tetramethyllead (CH₃)₄Pb	Pb—C	2.238	(T_d excluding the H atoms)		ED
Tetramethylsilane (CH₃)₄Si	C—H	1.115	C—Si	1.875	ED
	∠HCH	109.8	(T_d excluding the H atoms)		
Tetramethylstannane (CH₃)₄Sn	C—Sn	2.144			ED
	C—H	1.12	(T_d excluding the H atoms)		
1,2,5-Thiadiazole	S—N	1.631	∠NSN	99.6	MW
	C—N	1.328	∠CCN	113.8	
	C—C	1.420	∠CCH	126.2	
	C—H	1.079	planar		
1,3,4-Thiadiazole	S—C	1.721	∠CSC	86.4	MW
	N—N	1.371	∠SCN	114.6	
	C—N	1.302	∠CCN	112.2	
	C—H	1.08	∠SCH	121.9	
	∠NCH	123.5	planar		
Thietane	C—S	1.847			ED, MW
CH₂—CH₂ | CH₂—S	C—C	1.549			
	C—H (average)	1.100			
	∠CSC	76.8	∠HCH (average)	112	
	dihedral angle between the CCC and CSC planes		154		
Thiirane	C—C	1.484	∠HCH	116	MW
H₂C | S H₂C	C—H	1.083	∠CSC	48.3	
	C—S	1.815	∠CCS	65.9	
	dihedral angle between the CH₂ plane and the C—C bond	152			
Thioformaldehyde CH₂S	C—S	1.611	C—H	1.093	MW
	∠HCH	116.9			
Thioformamide			N—H_a	1.002	MW
			N—H_b	1.007	
			C—N	1.358	
	C—S	1.626	C—H_c	1.10	
	∠H_aNH_b	121.7	∠H_aNC	117.9	
	∠H_bNC	120.4	∠NCS	125.3	
	∠NCH_c	108	∠SCH_c	127	

Thiolane → Tetrahydrothiophene

Compound	Structure					Method
Thiophene			C_a—H_a	1.078		MW
			C_b—H_b	1.081		
			C_a—S	1.714		
			C_a—C_b	1.370		
			C_b—C_b	1.423		
			∠C_aSC_a	92.2		
	∠S$C_a$$C_b$	115.5	∠$C_a$$C_b$$C_b$	112.5		
	∠S$C_a$$H_a$	119.9	∠$C_b$$C_b$$H_b$	124.3		
Toluene	CH_3	C—C (ring)	1.399		C—CH_3 1.524	ED
		C—H (average)	1.11			
		the difference between the C—H(CH_3) and C—H(ring): about 0.01				
1,1,1-Tribromoethane	C—Br	1.93	C—H	1.095 (assumed)		MW
CH_3CBr_3	C—C	1.51 (assumed)	∠CCBr	108		
	∠BrCBr	111	∠CCH	109.0 (assumed)		
Tribromomethane → Bromoform						
Tri-*tert*-butyl methane	C_a—C_b	1.611	C—H	1.111		ED
$HC_a[C_b(C_cH_3)_3]_3$	C_b—C_c	1.548	∠$C_a$$C_b$$C_c$	113.0		
Tricarbon dioxide	C—O	1.163	C—C	1.289		ED
OCCCO	linear (with a large-amplitude bending vibration)					
Trichloroacetonitrile	C—N	1.165	C—C	1.460		ED
CCl_3CN	C—Cl	1.763	∠ClCCl	110.0		
1,1,1-Trichloroethane	C—H	1.090	C—C	1.541		MW
CH_3CCl_3	C—Cl	1.771	∠HCH	110.0		
	∠CCH	108.9	∠ClCCl	109.4		
	∠CCCl	109.6				
Trichloro(methyl)germane	Ge—Cl	2.132	Ge—C	1.89		ED, MW
CH_3GeCl_3	C—H	1.103 (assumed)	∠ClGeCl	106.4		
	∠GeCH	110.5 (assumed)				
Trichloro(methyl)silane	C—Si	1.876	Si—Cl	2.021		MW
CH_3SiCl_3			(C_{3v})			
Trichloro(methyl)stannane	Sn—Cl	2.304	Sn—C	2.10		ED
CH_3SnCl_3	C—H	1.100	∠CSnCl	113.9		
	∠ClSnCl	104.7	∠SnCH	108		
Triethylenediamine → 1,4-Diazabicyclo [2.2.2]octane						
Trifluoroacetic acid			C—F	1.325		ED
			C—C	1.546		
			C—O_a	1.192		
	C—O_b	1.35	O—H	0.96 (assumed)		
	∠CCO_a	126.8	∠CCO_b	111.1		
	∠CCF	109.5				
1,1,1-Trifluoroethane	C—C	1.494	C—F	1.340		ED
CH_3CF_3	C—H	1.081	∠CCF	119.2		
	∠CCH	112				
Trifluoromethane → Fluoroform						
1,1,1-Trifluoro-2,2,2-trichloroethane	C—C	1.54	C—F	1.33		MW
CF_3CCl_3	C—Cl	1.77	∠CCF	110		
	∠CCCl	109.6	staggered conformation			
Trimethylaluminium	C—H	1.113	Al—C	1.957		ED
$(CH_3)_3Al$	∠AlCH	111.7	∠CAlC	120		
Trimethylamine	C—N	1.458	C—H	1.100		ED
$(CH_3)_3N$	∠CNC	110.9	∠HCH	110		
Trimethylarsine	C—As	1.979	∠CAsC	98.8		ED
$(CH_3)_3As$	∠AsCH	111.4				
Trimethylbismuth	Bi—C	2.263	C—H	1.07		ED
$(CH_3)_3Bi$	∠CBiC	97.1				
Trimethylborane	C—B	1.578	C—H	1.114		ED
$(CH_3)_3B$	∠CBC	120.0	∠BCH	112.5		

Compound	Structure				Method
Trimethyleneimine → Azetidine					
Trimethylphosphine	C—P	1.847	C—H	1.091	ED
$(CH_3)_3P$	∠CPC	98.6	∠PCH	110.7	
1,3,5-Trioxane			C—O	1.422	MW
			∠OCO	112.2	
			∠COC	110.3	

1,3,5-Trioxane structure:

H_2C O CH_2
O   O
C
H_2

Triphenylamine	C—C	1.392	C—N	1.42	ED
$(C_6H_5)_3N$	∠CNC	116	(C_3)		

torsional dihedral angle of the two phenyl rings 47° (defined to be 0 when the symmetry axis is contained in the phenyl planes)

Tropone			C_a—O	1.23	ED
			C_a—C_b	1.45	
			C_b—C_c	1.36	
			C_c—C_d	1.46	
			C_d—C_d	1.34	
			∠$C_bC_aC_b$	122	
			∠$C_aC_bC_c$	133	
	∠$C_bC_cC_d$	126	∠$C_cC_dC_d$	130	
			(C_{2v})		

Tropone structure:

O
‖
C_a
HC_b C_bH
‖   ‖
HC_c C_cH
C_d=C_d
H   H

Vinylacetylene			C_b—C_c	1.434	ED, MW
			C_a—C_b	1.344	
			C_c—C_d	1.215	
			C_a—H_a	1.11	
			C_d—H_d	1.09	
			∠$C_aC_bC_c$	123.1	
	∠$C_bC_cC_d$	178	∠$H_aC_aC_b$	119	
	∠$H_bC_aC_b$	122	∠$H_cC_bC_a$	122	
	∠$C_cC_dH_d$	182			

Vinylacetylene structure:

H_a H_c
C_a=C_b
H_b C_c≡
C_d
H_d

Vinyl chloride			C—C	1.342	ED, MW
			C—Cl	1.730	
			C—H	1.09	
	∠CCCl	122.5	∠CCH$_a$	124	
	∠CCH$_b$	120	∠CCH$_c$	121.1	

Vinyl chloride structure:

H_c Cl
C=C
H_b H_a

CHARACTERISTIC BOND LENGTHS IN FREE MOLECULES

This is a summary of typical bond lengths in gas-phase molecules. The value given for each bond is near the mid-range of values found in simple molecules. Bond lengths usually vary by 1 or 2%, and often by more, depending on the nature of the other bonds attached to the two atoms in question. References 1 and 2 give bond lengths in individual gas-phase molecules, as determined by spectroscopic and electron diffraction methods.

All bond distances are given in Å ($1 \text{ Å} = 10^{-10}$ m).

References

1. "Bond Lengths and Angles in Gas-Phase Molecules", *CRC Handbook of Chemistry and Physics*, 86th Edition, 2005, p. **9**-19.
2. Harmony, M. D., Laurie, V. W., Kuczkowski, R. L., Schwendeman, R. H., Ramsay, D. A., Lovas, F. J., Lafferty, W. J., and Maki, A. G., Molecular structure of gas-phase polyatomic molecules determined by spectroscopic methods, *J. Phys. Chem. Ref. Data* 8, 619, 1979.
3. Lide, D. R., "A survey of carbon-carbon bond lengths", *Tetrahedron* 17, 125, 1962.

A. Characteristic lengths of single bonds.

	As	Br	C	Cl	F	Ge	H	I	N	O	P	S	Sb	Se	Si
As	2.10														
Br	2.32	2.28													
C	1.96	1.94	1.53												
Cl	2.17	2.14	1.79	1.99											
F	1.71	1.76	1.39	1.63	1.41										
Ge		2.30	1.95	2.15	1.73	2.40									
H	1.51	1.41	1.09	1.28	0.92	1.53	0.74								
I		2.47	2.13	2.32	1.91	2.51	1.61	2.67							
N			1.46	1.90	1.37		1.02		1.45						
O			1.42	1.70	1.42		0.96		1.43	1.48					
P		2.22	1.85	2.04	1.57		1.42		1.65		2.25				
S		2.24	1.82	2.05	1.56		1.34					2.00			
Sb			2.33				1.70								
Se			1.95		1.71		1.47							2.33	
Si		2.21	1.87	2.05	1.58		1.48	2.44		1.63		2.14			2.33
Sn			2.14	2.28			1.71	2.67							
Te					1.82		1.66								

B. Lengths of multiple bonds (non-ring molecules).

C=C	1.34
C≡C	1.20
C=N	1.21
C≡N	1.16
C=O	1.21
C=S	1.61
N=N	1.24
N≡N	1.13
N=O	1.18
O=O	1.21

C. Effect of environment on carbon-carbon single bonds (other single bonds not shown). From Reference 3.

Configuration	C–C length	Examples of molecules
C–C	1.526	$H_3C–CH_3$
C–C=	1.501	$H_3C–CH=CH_2$
C–C≡	1.459	$H_3C–C≡CH$
=C–C=	1.467	$H_2C=CH–CH=CH_2$
≡C–C=	1.445	$HC≡C–CH=CH_2$
≡C–C≡	1.378	$HC≡C–C≡CH$

D. Some metal-carbon bond lengths in gas-phase molecules.

Al–C	1.96	Bi–C	2.26	Pb–C	2.24
B–C	1.58	Cd–C	2.11	Sn–C	2.14
Be–C	1.70	Hg–C	2.08	Zn–C	1.93

DIPOLE MOMENTS

This table gives values of the electric dipole moment for about 800 molecules. When available, values determined by microwave spectroscopy, molecular beam electric resonance, and other high-resolution spectroscopic techniques were selected. Otherwise, the values come from measurements of the dielectric constant in the gas phase or, if these do not exist, in the liquid phase. Compounds are listed by molecular formula in Hill order; compounds not containing carbon are listed first, followed by compounds containing carbon.

The dipole moment μ is given in Debye units (D). The conversion factor to SI units is 1 D = 3.33564×10^{-30} C m.

Dipole moments of individual conformers (rotational isomers) are given when they have been measured. The conformers are designated as *gauche, trans, axial*, etc. The meaning of these terms can be found in the references. In some cases an average value, obtained from measurements on the bulk gas, is also given. Other information on molecules that have been studied by spectroscopy, such as the components of the dipole moment in the molecular framework and the variation with vibrational state and isotopic species, is given in References 1 and 2.

When the accuracy of a value is explicitly stated (i.e., 1.234 ± 0.005), the stated uncertainty generally indicates two or three standard deviations. When no uncertainty is given, the value may be assumed to be precise to a few units in the last decimal place. However, if more than three decimal places are given, the exact interpretation of the final digits may require analysis of the vibrational averaging.

Values measured in the gas phase that are questionable because of undetermined error sources are indicated as approximate (≈). Values obtained by liquid phase measurements, which sometimes have large errors because of association effects, are enclosed in brackets, e.g., [1.8].

References

1. Nelson, R. D., Lide, D. R., and Maryott, A. A., *Selected Values of Electric Dipole Moments for Molecules in the Gas Phase*, Natl. Stand. Ref. Data Ser. - Nat. Bur. Stnds. 10, 1967.
2. *Landolt-Börnstein, Numerical Data and Functional Relationships in Science and Technology, New Series*, II/6 (1974), II/14a (1982), II/14b (1983), II/19c (1992), Springer-Verlag, Heidelberg.
3. Riddick, J. A., Bunger, W. B., and Sakano, T. K., *Organic Solvents, Fourth Edition*, John Wiley & Sons, New York, 1986.

Mol. form.	Name	μ/D
Compounds not containing carbon		
AgBr	Silver(I) bromide	5.62 ± 0.03
AgCl	Silver(I) chloride	6.08 ± 0.06
AgF	Silver(I) fluoride	6.22 ± 0.30
AgI	Silver(I) iodide	4.55 ± 0.05
AlF	Aluminum monofluoride	1.53 ± 0.15
AsCl$_3$	Arsenic(III) chloride	1.59 ± 0.08
AsF$_3$	Arsenic(III) fluoride	2.59 ± 0.05
AsH$_3$	Arsine	0.217 ± 0.003
BClH$_2$	Chloroborane	0.75 ± 0.05
BF	Fluoroborane(1)	≈0.5
BF$_2$H	Difluoroborane	0.971 ± 0.010
B$_4$H$_{10}$	Tetraborane	0.486 ± 0.002
B$_5$H$_9$	Pentaborane(9)	2.13 ± 0.04
B$_6$H$_{10}$	Hexaborane	2.50 ± 0.05
BaO	Barium oxide	7.954 ± 0.003
BaS	Barium sulfide	10.86 ± 0.02
BrCl	Bromine chloride	0.519 ± 0.004
BrF	Bromine fluoride	1.422 ± 0.016
BrF$_3$Si	Bromotrifluorosilane	0.83 ± 0.01
BrF$_5$	Bromine pentafluoride	1.51 ± 0.15
BrH	Hydrogen bromide	0.8272 ± 0.0003
BrH$_3$Si	Bromosilane	1.319
BrI	Iodine bromide	0.726 ± 0.003
BrK	Potassium bromide	10.628 ± 0.001
BrLi	Lithium bromide	7.268 ± 0.001
BrNO	Nitrosyl bromide	≈1.8
BrNa	Sodium bromide	9.1183 ± 0.0006
BrO	Bromine monoxide	1.76 ± 0.04
BrO$_2$	Bromine dioxide	2.8 ± 0.3
BrRb	Rubidium bromide	≈10.9
BrTl	Thallium(I) bromide	4.49 ± 0.05
CaCl	Calcium monochloride	≈3.6
ClCs	Cesium chloride	10.387 ± 0.004
ClF	Chlorine fluoride	0.888061
ClFO$_3$	Perchloryl fluoride	0.023 ± 0.001
ClF$_3$	Chlorine trifluoride	0.6 ± 0.10
ClF$_3$Si	Chlorotrifluorosilane	0.636 ± 0.004
ClGeH$_3$	Chlorogermane	2.13 ± 0.02
ClH	Hydrogen chloride	1.1086 ± 0.0003
ClHO	Hypochlorous acid	≈1.3
ClH$_3$Si	Chlorosilane	1.31 ± 0.01
ClI	Iodine chloride	1.24 ± 0.02
ClIn	Indium(I) chloride	3.79 ± 0.19
ClK	Potassium chloride	10.269 ± 0.001
ClLi	Lithium chloride	7.12887
ClNO$_2$	Nitryl chloride	0.53
ClNS	Thionitrosyl chloride	1.87 ± 0.02
ClNa	Sodium chloride	9.00117
ClO	Chlorine oxide	1.297 ± 0.001
ClRb	Rubidium chloride	10.510 ± 0.005
ClTl	Thallium(I) chloride	4.54299
Cl$_2$H$_2$Si	Dichlorosilane	1.17 ± 0.02
Cl$_2$OS	Thionyl chloride	1.45 ± 0.03
Cl$_2$O$_2$S	Sulfuryl chloride	1.81 ± 0.04
Cl$_2$S	Sulfur dichloride	0.36 ± 0.01
Cl$_3$FSi	Trichlorofluorosilane	0.49 ± 0.01
Cl$_3$HSi	Trichlorosilane	0.86 ± 0.01
Cl$_3$N	Nitrogen trichloride	0.39 ± 0.01
Cl$_3$OP	Phosphorus(V) oxychloride	2.54 ± 0.05
Cl$_3$P	Phosphorus(III) chloride	0.56 ± 0.02
CrO	Chromium monoxide	3.88 ± 0.13
CsF	Cesium fluoride	7.884 ± 0.001
CsNa	Cesium sodium	4.75 ± 0.20
CuF	Copper(I) fluoride	5.77 ± 0.29
CuO	Copper(II) oxide	4.5 ± 0.5
FGa	Gallium monofluoride	2.45 ± 0.05
FGeH$_3$	Fluorogermane	2.33 ± 0.12
FH	Hydrogen fluoride	1.826178
FHO	Hypofluorous acid	2.23 ± 0.11

Mol. form.	Name	μ/D
FH$_2$N	Fluoramide	2.27 ± 0.18
FH$_3$Si	Fluorosilane	1.2969 ± 0.0006
FI	Iodine fluoride	1.948 ± 0.020
FIn	Indium(I) fluoride	3.40 ± 0.07
FK	Potassium fluoride	8.585 ± 0.003
FLi	Lithium fluoride	6.3274 ± 0.0002
FNO	Nitrosyl fluoride	1.730 ± 0.003
FNO$_2$	Nitryl fluoride	0.466 ± 0.005
FNS	Thionitrosyl fluoride (NSF)	1.902 ± 0.012
FN$_3$	Fluorine azide	≈1.3
FNa	Sodium fluoride	8.156 ± 0.001
FO	Fluorine oxide	0.0043 ± 0.0004
FRb	Rubidium fluoride	8.5465 ± 0.0005
FS	Sulfur monofluoride	0.794 ± 0.02
FTl	Thallium(I) fluoride	4.2282 ± 0.0008
F$_2$Ge	Germanium(II) fluoride	2.61 ± 0.02
F$_2$HN	Difluoramine	1.92 ± 0.02
F$_2$H$_2$Si	Difluorosilane	1.55 ± 0.02
F$_2$N$_2$	cis-Difluorodiazine	0.16 ± 0.01
F$_2$O	Fluorine monoxide	0.308180
F$_2$OS	Thionyl fluoride	1.63 ± 0.01
F$_2$O$_2$	Fluorine dioxide	1.44 ± 0.07
F$_2$O$_2$S	Sulfuryl fluoride	1.12 ± 0.02
F$_2$S	Sulfur difluoride	1.05 ± 0.05
F$_2$Si	Difluorosilylene	1.23 ± 0.02
F$_3$HSi	Trifluorosilane	1.27 ± 0.03
F$_3$H$_3$Si$_2$	1,1,1-Trifluorodisilane	2.03 ± 0.10
F$_3$ISi	Trifluoroiodosilane	1.11 ± 0.03
F$_3$N	Nitrogen trifluoride	0.235 ± 0.004
F$_3$NO	Trifluoramine oxide	0.0390 ± 0.0004
F$_3$OP	Phosphorus(V) oxyfluoride	1.8685 ± 0.0001
F$_3$P	Phosphorus(III) fluoride	1.03 ± 0.01
F$_3$PS	Phosphorus(V) sulfide trifluoride	0.64 ± 0.02
F$_4$N$_2$	Tetrafluorohydrazine (gauche)	0.257 ± 0.002
F$_4$S	Sulfur tetrafluoride	0.632 ± 0.003
F$_4$Se	Selenium tetrafluoride	1.78 ± 0.09
F$_5$I	Iodine pentafluoride	2.18 ± 0.11
GeH$_3$N$_3$	Germylazide	2.579 ± 0.003
GeO	Germanium(II) oxide	3.2823 ± 0.0001
GeS	Germanium(II) sulfide	2.00 ± 0.06
GeSe	Germanium(II) selenide	1.65 ± 0.05
GeTe	Germanium(II) telluride	1.06 ± 0.07
HI	Hydrogen iodide	0.448 ± 0.001
HKO	Potassium hydroxide	7.415 ± 0.002
HLi	Lithium hydride	5.884 ± 0.001
HLiO	Lithium hydroxide	4.754 ± 0.002
HN	Imidogen	1.39 ± 0.07
HNO	Nitrosyl hydride	1.62 ± 0.03
HNO$_2$	Nitrous acid (cis)	1.423 ± 0.005
HNO$_2$	Nitrous acid (trans)	1.855 ± 0.016
HNO$_3$	Nitric acid	2.17 ± 0.02
HN$_3$	Hydrazoic acid	1.70 ± 0.09
HO	Hydroxyl	1.655 ± 0.001
HS	Mercapto	0.7580 ± 0.0001
H$_2$O	Water	1.8546 ± 0.0040
H$_2$O$_2$	Hydrogen peroxide	1.573 ± 0.001
H$_2$S	Hydrogen sulfide	0.97833
H$_3$N	Ammonia	1.4718 ± 0.0002
H$_3$NO	Hydroxylamine	0.59 ± 0.05
H$_3$P	Phosphine	0.5740 ± 0.0003
H$_3$Sb	Stibine	0.12 ± 0.05

Mol. form.	Name	μ/D
H$_4$N$_2$	Hydrazine	1.75 ± 0.09
H$_6$OSi$_2$	Disiloxane	0.24 ± 0.02
IK	Potassium iodide	≈10.8
ILi	Lithium iodide	7.428 ± 0.001
INa	Sodium iodide	9.236 ± 0.003
IO	Iodine monoxide	2.45 ± 0.05
IRb	Rubidium iodide	≈11.5
ITl	Thallium(I) iodide	4.61 ± 0.07
KLi	Lithium potassium	3.45 ± 0.20
KNa	Potassium sodium	2.693 ± 0.014
LaO	Lanthanum monoxide	3.207 ± 0.011
LiNa	Lithium sodium	0.463 ± 0.002
LiO	Lithium monoxide	6.84 ± 0.03
LiRb	Lithium rubidium	4.0 ± 0.1
MgO	Magnesium oxide	6.2 ± 0.6
NO	Nitric oxide	0.15872
NO$_2$	Nitrogen dioxide	0.316 ± 0.010
NP	Phosphorus nitride	2.7470 ± 0.0001
NS	Nitrogen sulfide	1.81 ± 0.02
N$_2$O	Nitrous oxide	0.16083
N$_2$O$_3$	Nitrogen trioxide	2.122 ± 0.010
NaRb	Rubidium sodium	3.1 ± 0.3
OP	Phosphorus monoxide	1.88 ± 0.07
OPb	Lead(II) oxide	4.64 ± 0.50
OS	Sulfur monoxide	1.55 ± 0.02
OS$_2$	Sulfur oxide (SSO)	1.47 ± 0.03
OSi	Silicon monoxide	3.0982
OSn	Tin(II) oxide	4.32 ± 0.22
OSr	Strontium oxide	8.900 ± 0.003
OTi	Titanium(II) oxide	2.96 ± 0.05
OY	Yttrium monoxide	4.524 ± 0.007
OZr	Zirconium(II) oxide	2.55 ± 0.01
O$_2$S	Sulfur dioxide	1.63305
O$_2$Se	Selenium dioxide	2.62 ± 0.05
O$_2$Zr	Zirconium(IV) oxide	7.80 ± 0.02
O$_3$	Ozone	0.53373
PbS	Lead(II) sulfide	3.59 ± 0.18
SSi	Silicon monosulfide	1.73 ± 0.09
SSn	Tin(II) sulfide	3.18 ± 0.16

Compounds containing carbon

Mol. form.	Name	μ/D
CBrF$_3$	Bromotrifluoromethane	0.65 ± 0.05
CBr$_2$F$_2$	Dibromodifluoromethane	0.66 ± 0.05
CClF$_3$	Chlorotrifluoromethane	0.50 ± 0.01
CClN	Cyanogen chloride	2.8331 ± 0.0002
CCl$_2$F$_2$	Dichlorodifluoromethane	0.51 ± 0.05
CCl$_2$O	Carbonyl chloride	1.17 ± 0.01
CCl$_3$F	Trichlorofluoromethane	0.46 ± 0.02
CF	Fluoromethylidyne	0.645 ± 0.005
CFN	Cyanogen fluoride	2.120 ± 0.001
CF$_2$	Difluoromethylene	0.47 ± 0.02
CF$_2$O	Carbonyl fluoride	0.95 ± 0.01
CF$_3$I	Trifluoroiodomethane	1.048 ± 0.003
CH	Methylidyne	≈1.46
CHBrClF	Bromochlorofluoromethane	1.5 ± 0.3
CHBr$_3$	Tribromomethane	0.99 ± 0.02
CHClF$_2$	Chlorodifluoromethane	1.42 ± 0.03
CHCl$_2$F	Dichlorofluoromethane	1.29 ± 0.03
CHCl$_3$	Trichloromethane	1.04 ± 0.02
CHFO	Formyl fluoride	2.081 ± 0.001
CHF$_2$N	Carboimidic difluoride	1.393 ± 0.001
CHF$_3$	Trifluoromethane	1.65150

Mol. form.	Name	μ/D	Mol. form.	Name	μ/D
CHN	Hydrogen cyanide	2.985188	$C_2HF_3O_2$	Trifluoroacetic acid	2.28 ± 0.25
CHN	Hydrogen isocyanide	3.05 ± 0.15	C_2HI	Iodoacetylene	0.02525
CHNO	Isocyanic acid (HNCO)	≈1.6	$C_2H_2Br_4$	1,1,2,2-Tetrabromoethane	[1.38]
CHNO	Fulminic acid	3.09934	$C_2H_2Cl_2$	1,1-Dichloroethene	1.34 ± 0.01
CH_2BrCl	Bromochloromethane	[1.66]	$C_2H_2Cl_2$	cis-1,2-Dichloroethene	1.90 ± 0.04
CH_2Br_2	Dibromomethane	1.43 ± 0.03	$C_2H_2Cl_2O$	Chloroacetyl chloride	2.23 ± 0.11
CH_2ClF	Chlorofluoromethane	1.82 ± 0.04	$C_2H_2Cl_4$	1,1,2,2-Tetrachloroethane	1.32 ± 0.07
CH_2Cl_2	Dichloromethane	1.60 ± 0.03	$C_2H_2F_2$	1,1-Difluoroethene	1.3893 ± 0.0002
CH_2F_2	Difluoromethane	1.9785 ± 0.02	$C_2H_2F_2$	cis-1,2-Difluoroethene	2.42 ± 0.02
CH_2I_2	Diiodomethane	[1.08]	$C_2H_2F_4$	1,1,1,2-Tetrafluoroethane	1.80 ± 0.22
CH_2N_2	Diazomethane	1.50 ± 0.01	$C_2H_2N_2S$	1,2,5-Thiadiazole	1.579 ± 0.007
CH_2N_2	Cyanamide	4.28 ± 0.10	C_2H_2O	Ketene	1.42215
CH_2N_4	1H-Tetrazole	2.19 ± 0.05	$C_2H_2O_2$	Glyoxal (cis)	4.8 ± 0.2
CH_2O	Formaldehyde	2.332 ± 0.002	C_2H_3Br	Bromoethene	1.42 ± 0.03
CH_2O_2	Formic acid	1.425 ± 0.002	C_2H_3Cl	Chloroethene	1.45 ± 0.03
CH_2S	Thioformaldehyde	1.6491 ± 0.0004	$C_2H_3ClF_2$	1-Chloro-1,1-difluoroethane	2.14 ± 0.04
CH_2Se	Selenoformaldehyde	1.41 ± 0.01	C_2H_3ClO	Acetyl chloride	2.72 ± 0.14
CH_3BCl_2	Dichloromethylborane	1.419 ± 0.013	$C_2H_3Cl_3$	1,1,1-Trichloroethane	1.755 ± 0.015
CH_3BF_2	Difluoromethylborane	1.668 ± 0.003	$C_2H_3Cl_3$	1,1,2-Trichloroethane	[1.4]
CH_3BO	Borane carbonyl	1.698 ± 0.020	C_2H_3F	Fluoroethene	1.468 ± 0.003
CH_3Br	Bromomethane	1.8203 ± 0.0004	C_2H_3FO	Acetyl fluoride	2.96 ± 0.03
CH_3Cl	Chloromethane	1.8963 ± 0.0002	$C_2H_3F_3$	1,1,1-Trifluoroethane	2.3470 ± 0.005
CH_3Cl_3Si	Methyltrichlorosilane	1.91 ± 0.01	C_2H_3HgN	Cyanomethylmercury	4.7 ± 0.1
CH_3F	Fluoromethane	1.858 ± 0.002	C_2H_3I	Iodoethene	1.311 ± 0.005
CH_3F_2OP	Methylphosphonic difluoride	3.69 ± 0.26	C_2H_3N	Acetonitrile	3.92519
CH_3F_2P	Methyldifluorophosphine	2.056 ± 0.006	C_2H_3NO	Methyl cyanate	4.26 ± 0.18
CH_3F_3Si	Trifluoromethylsilane	2.3394 ± 0.0002	C_2H_3NO	Methyl isocyanate	≈2.8
CH_3F_3Si	(Trifluoromethyl)silane	2.32 ± 0.02	C_2H_3NS	Methyl isothiocyanate	3.453 ± 0.003
CH_3I	Iodomethane	1.6406 ± 0.0004	$C_2H_3N_3$	1H-1,2,4-Triazole	2.7 ± 0.1
CH_3NO	Formamide	3.73 ± 0.07	C_2H_4BrCl	1-Bromo-2-chloroethane	[1.2]
CH_3NO_2	Nitromethane	3.46 ± 0.02	$C_2H_4Br_2$	1,2-Dibromoethane	[1.19]
CH_3N_3	Methyl azide	2.17 ± 0.04	C_2H_4ClF	1-Chloro-1-fluoroethane	2.068 ± 0.014
CH_4O	Methanol	1.70 ± 0.02	$C_2H_4Cl_2$	1,1-Dichloroethane	2.06 ± 0.04
CH_4O_2	Methylhydroperoxide	≈0.65	$C_2H_4Cl_2$	1,2-Dichloroethane	[1.83]
CH_4S	Methanethiol	1.52 ± 0.08	$C_2H_4F_2$	1,1-Difluoroethane	2.27 ± 0.05
CH_5FSi	Fluoromethylsilane	1.700 ± 0.008	$C_2H_4F_2$	1,2-Difluoroethane (gauche)	2.67 ± 0.13
CH_5ISi	Iodomethylsilane	1.862 ± 0.005	C_2H_4O	Acetaldehyde	2.750 ± 0.006
CH_5N	Methylamine	1.31 ± 0.03	C_2H_4O	Ethylene oxide	1.89 ± 0.01
CH_6OSi	Methyl silyl ether	1.15 ± 0.02	$C_2H_4O_2$	Acetic acid	1.70 ± 0.03
CH_6Si	Methylsilane	0.73456	$C_2H_4O_2$	Methyl formate	1.77 ± 0.04
CH_8B_2	Methyldiborane(6)	0.566 ± 0.006	$C_2H_4O_2$	Glycolaldehyde	2.73 ± 0.05
CIN	Cyanogen iodide	3.67 ± 0.02	C_2H_5Br	Bromoethane	2.04 ± 0.02
CO	Carbon monoxide	0.10980	C_2H_5Cl	Chloroethane	2.05 ± 0.02
COS	Carbon oxysulfide	0.715189	C_2H_5ClO	2-Chloroethanol	1.78 ± 0.09
COSe	Carbon oxyselenide	0.73 ± 0.02	$C_2H_5Cl_3Si$	Trichloroethylsilane	[2.04]
CS	Carbon monosulfide	1.958 ± 0.005	C_2H_5F	Fluoroethane	1.937 ± 0.007
CSe	Carbon monoselenide	1.99 ± 0.04	C_2H_5I	Iodoethane	1.976 ± 0.002
C_2BrF	Bromofluoroacetylene	0.448 ± 0.002	C_2H_5N	Ethyleneimine	1.90 ± 0.01
C_2ClF_3	Chlorotrifluoroethene	0.40 ± 0.10	C_2H_5NO	Acetamide	3.68 ± 0.03
C_2ClF_5	Chloropentafluoroethane	0.52 ± 0.05	C_2H_5NO	N-Methylformamide	3.83 ± 0.08
$C_2Cl_2F_2$	1,1-Dichloro-2,2-difluoroethene	0.50	$C_2H_5NO_2$	Nitroethane	3.23 ± 0.03
$C_2Cl_2F_4$	1,2-Dichloro-1,1,2,2-tetrafluoroethane	≈0.5	C_2H_6O	Ethanol (gauche)	1.68 ± 0.03
			C_2H_6O	Ethanol (trans)	1.44 ± 0.03
C_2F_3N	Trifluoroacetonitrile	1.262 ± 0.010	C_2H_6O	Ethanol (average)	1.69 ± 0.03
C_2F_3N	Trifluoroisocyanomethane	1.153 ± 0.010	C_2H_6O	Dimethyl ether	1.30 ± 0.01
C_2HBr	Bromoacetylene	0.22962	C_2H_6OS	Dimethyl sulfoxide	3.96 ± 0.04
C_2HCl	Chloroacetylene	0.44408	$C_2H_6O_2$	Ethylene glycol (average)	2.36 ± 0.10
C_2HCl_3	Trichloroethene	[0.8]	C_2H_6S	Ethanethiol (gauche)	1.61 ± 0.08
C_2HCl_5	Pentachloroethane	0.92 ± 0.05	C_2H_6S	Ethanethiol (trans)	1.58 ± 0.08
C_2HF	Fluoroacetylene	0.7207 ± 0.0003	C_2H_6S	Dimethyl sulfide	1.554 ± 0.004
C_2HF_3	Trifluoroethene	1.32 ± 0.03	$C_2H_6S_2$	1,2-Ethanedithiol	2.03 ± 0.08

Mol. form.	Name	μ/D	Mol. form.	Name	μ/D
$C_2H_6S_2$	Dimethyl disulfide	[1.85]	$C_3H_6O_2$	Ethyl formate (*gauche*)	1.81 ± 0.02
C_2H_6Si	Vinylsilane	0.657 ± 0.002	$C_3H_6O_2$	Ethyl formate (*trans*)	1.98 ± 0.02
C_2H_7N	Ethylamine (*gauche*)	1.210 ± 0.015	$C_3H_6O_2$	Ethyl formate (*average*)	1.93
C_2H_7N	Ethylamine (*trans*)	1.304 ± 0.011	$C_3H_6O_2$	Methyl acetate	1.72 ± 0.09
C_2H_7N	Ethylamine (*average*)	1.22 ± 0.10	$C_3H_6O_2$	1,3-Dioxolane	1.19 ± 0.06
C_2H_7N	Dimethylamine	1.01 ± 0.02	$C_3H_6O_2S$	Thietane 1,1-dioxide	4.8 ± 0.1
C_2H_7NO	Ethanolamine	[2.27]	$C_3H_6O_3$	1,3,5-Trioxane	2.08 ± 0.02
$C_2H_8N_2$	1,2-Ethanediamine	1.99 ± 0.10	C_3H_6S	Thietane	1.85 ± 0.09
C_3HF_3	3,3,3-Trifluoro-1-propyne	2.317 ± 0.013	C_3H_7Br	1-Bromopropane	2.18 ± 0.11
C_3HN	Cyanoacetylene	3.73172	C_3H_7Br	2-Bromopropane	2.21 ± 0.11
$C_3H_2F_2$	3,3-Difluorocyclopropene	2.98 ± 0.02	C_3H_7Cl	1-Chloropropane (*gauche*)	2.02 ± 0.03
C_3H_2O	2-Propynal	2.78 ± 0.02	C_3H_7Cl	1-Chloropropane (*trans*)	1.95 ± 0.02
$C_3H_3Cl_2F$	1,1-Dichloro-2-fluoropropene	2.43 ± 0.02	C_3H_7Cl	1-Chloropropane (*average*)	2.05 ± 0.04
C_3H_3F	3-Fluoropropyne	1.73 ± 0.02	C_3H_7Cl	2-Chloropropane	2.17 ± 0.11
$C_3H_3F_3$	3,3,3-Trifluoropropene	2.45 ± 0.05	C_3H_7F	1-Fluoropropane (*gauche*)	1.90 ± 0.10
C_3H_3N	Acrylonitrile	3.92 ± 0.07	C_3H_7F	1-Fluoropropane (*trans*)	2.05 ± 0.04
C_3H_3NO	Oxazole	1.503 ± 0.030	C_3H_7F	2-Fluoropropane	1.958 ± 0.001
C_3H_3NO	Isoxazole	2.95 ± 0.04	C_3H_7I	1-Iodopropane	2.04 ± 0.10
C_3H_4	Propyne	0.784 ± 0.001	C_3H_7I	2-Iodopropane	[1.95]
C_3H_4	Cyclopropene	0.454 ± 0.010	C_3H_7N	Allylamine	≈1.2
$C_3H_4F_2$	1,1-Difluoro-1-propene	0.889 ± 0.007	C_3H_7N	Cyclopropylamine	1.19 ± 0.01
$C_3H_4N_2$	1*H*-Pyrazole	2.20 ± 0.01	C_3H_7N	Propyleneimine (*cis*)	1.77 ± 0.09
$C_3H_4N_2$	Imidazole	3.8 ± 0.4	C_3H_7N	Propyleneimine (*trans*)	1.57 ± 0.03
C_3H_4O	Propargyl alcohol	1.13 ± 0.06	C_3H_7NO	*N,N*-Dimethylformamide	3.82 ± 0.08
C_3H_4O	Acrolein (*trans*)	3.117 ± 0.004	C_3H_7NO	*N*-Methylacetamide	[4.3]
C_3H_4O	Acrolein (*cis*)	2.552 ± 0.003	$C_3H_7NO_2$	1-Nitropropane	3.66 ± 0.07
C_3H_4O	Cyclopropanone	2.67 ± 0.13	$C_3H_7NO_2$	2-Nitropropane	3.73 ± 0.07
$C_3H_4O_2$	Vinyl formate	1.49 ± 0.01	C_3H_8	Propane	0.084 ± 0.001
$C_3H_4O_2$	2-Oxetanone	4.18 ± 0.03	C_3H_8O	1-Propanol (*gauche*)	1.58 ± 0.03
$C_3H_4O_2$	3-Oxetanone	0.887 ± 0.005	C_3H_8O	1-Propanol (*trans*)	1.55 ± 0.03
$C_3H_4O_3$	Ethylene carbonate	[4.9]	C_3H_8O	2-Propanol (*trans*)	1.58 ± 0.03
C_3H_5Br	2-Bromopropene	[1.51]	C_3H_8O	Ethyl methyl ether (*trans*)	1.17 ± 0.02
C_3H_5Br	3-Bromopropene	≈1.9	$C_3H_8O_2$	1,2-Propylene glycol	[2.25]
C_3H_5Cl	*cis*-1-Chloropropene	1.67 ± 0.08	$C_3H_8O_2$	1,3-Propylene glycol	[2.55]
C_3H_5Cl	*trans*-1-Chloropropene	1.97 ± 0.10	$C_3H_8O_2$	Ethylene glycol monomethyl ether (*gauche*)	2.36 ± 0.05
C_3H_5Cl	2-Chloropropene	1.647 ± 0.010			
C_3H_5Cl	3-Chloropropene	1.94 ± 0.10	$C_3H_8O_2$	Dimethoxymethane	[0.74]
C_3H_5ClO	Epichlorohydrin	[1.8]	$C_3H_8O_3$	Glycerol	[2.56]
C_3H_5F	*cis*-1-Fluoropropene	1.46 ± 0.03	C_3H_8S	1-Propanethiol (*gauche*)	1.683 ± 0.010
C_3H_5F	*trans*-1-Fluoropropene	≈1.9	C_3H_8S	1-Propanethiol (*trans*)	1.60 ± 0.08
C_3H_5F	2-Fluoropropene	1.61 ± 0.03	C_3H_8S	2-Propanethiol (*gauche*)	1.53 ± 0.03
C_3H_5F	3-Fluoropropene (*gauche*)	1.939 ± 0.015	C_3H_8S	2-Propanethiol (*trans*)	1.61 ± 0.03
C_3H_5F	3-Fluoropropene (*cis*)	1.765 ± 0.014	C_3H_8S	Ethyl methyl sulfide (*gauche*)	1.593 ± 0.004
C_3H_5N	Propanenitrile	4.05 ± 0.03	C_3H_8S	Ethyl methyl sulfide (*trans*)	1.56 ± 0.03
C_3H_5NO	Ethyl cyanate	4.72 ± 0.09	C_3H_9N	Propylamine	1.17 ± 0.06
C_3H_5NO	3-Hydroxypropanenitrile (*gauche*)	3.17 ± 0.02	C_3H_9N	Isopropylamine	1.19 ± 0.06
C_3H_6	Propene	0.366 ± 0.001	C_3H_9N	Trimethylamine	0.612 ± 0.003
$C_3H_6Br_2$	1,2-Dibromopropane	[1.2]	$C_3H_9O_4P$	Trimethyl phosphate	[3.18]
$C_3H_6Cl_2$	1,2-Dichloropropane	[1.85]	C_4H_4	1-Buten-3-yne	0.22 ± 0.02
$C_3H_6Cl_2$	1,3-Dichloropropane	2.08 ± 0.04	C_4H_4	Methylenecyclopropene	1.90 ± 0.01
C_3H_6O	Acetone	2.88 ± 0.03	$C_4H_4N_2$	Succinonitrile	[3.7]
C_3H_6O	Propanal (*gauche*)	2.86 ± 0.01	$C_4H_4N_2$	Pyrimidine	2.334 ± 0.010
C_3H_6O	Propanal (*cis*)	2.52 ± 0.05	$C_4H_4N_2$	Pyridazine	4.22 ± 0.02
C_3H_6O	Propanal (*average*)	2.72	C_4H_4O	Furan	0.66 ± 0.01
C_3H_6O	Allyl alcohol (*gauche*)	1.55 ± 0.08	$C_4H_4O_2$	Diketene	3.53 ± 0.07
C_3H_6O	Allyl alcohol (*average*)	1.60 ± 0.08	C_4H_4S	Thiophene	0.55 ± 0.01
C_3H_6O	Methyl vinyl ether	0.965 ± 0.002	C_4H_5N	2-Methylacrylonitrile	3.69 ± 0.18
C_3H_6O	Methyloxirane	2.01 ± 0.02	C_4H_5N	Pyrrole	1.767 ± 0.001
C_3H_6O	Oxetane	1.94 ± 0.01	C_4H_5N	Isocyanocyclopropane	4.03 ± 0.10
$C_3H_6O_2$	Propanoic acid (*cis*)	1.46 ± 0.07	C_4H_5NO	2-Methyloxazole	1.37 ± 0.07
$C_3H_6O_2$	Propanoic acid (*average*)	1.75 ± 0.09	C_4H_5NO	4-Methyloxazole	1.08 ± 0.05

Mol. form.	Name	μ/D	Mol. form.	Name	μ/D
C_4H_5NO	5-Methyloxazole	2.16 ± 0.04	C_4H_9Cl	2-Chlorobutane	2.04 ± 0.10
C_4H_5NO	4-Methylisoxazole	3.583 ± 0.005	C_4H_9Cl	1-Chloro-2-methylpropane	2.00 ± 0.10
C_4H_6	1,2-Butadiene	0.403 ± 0.002	C_4H_9Cl	2-Chloro-2-methylpropane	2.13 ± 0.04
C_4H_6	1-Butyne	0.782 ± 0.004	C_4H_9I	1-Iodobutane	[1.93]
C_4H_6	Cyclobutene	0.132 ± 0.001	C_4H_9I	2-Iodobutane	2.12 ± 0.11
C_4H_6O	Divinyl ether	0.78 ± 0.05	C_4H_9I	1-Iodo-2-methylpropane	[1.87]
C_4H_6O	3-Methoxy-1,2-propadiene	0.963 ± 0.020	C_4H_9N	Pyrrolidine	[1.57]
C_4H_6O	trans-2-Butenal	3.67 ± 0.07	C_4H_9NO	N-Methylpropanamide	3.61
C_4H_6O	2-Methylpropenal	2.68 ± 0.13	C_4H_9NO	N,N-Dimethylacetamide	[3.7]
C_4H_6O	Cyclobutanone	2.89 ± 0.03	C_4H_9NO	Morpholine	1.55 ± 0.03
C_4H_6O	2,3-Dihydrofuran	1.32 ± 0.03	C_4H_{10}	Isobutane	0.132 ± 0.002
C_4H_6O	2,5-Dihydrofuran	1.63 ± 0.01	$C_4H_{10}O$	1-Butanol	1.66 ± 0.03
$C_4H_6O_2$	trans-Crotonic acid	[2.13]	$C_4H_{10}O$	2-Butanol	[1.8]
$C_4H_6O_2$	Methacrylic acid	[1.65]	$C_4H_{10}O$	2-Methyl-1-propanol	1.64 ± 0.08
$C_4H_6O_2$	Vinyl acetate	[1.79]	$C_4H_{10}O$	2-Methyl-2-propanol	[1.66]
$C_4H_6O_2$	Methyl acrylate	[1.77]	$C_4H_{10}O$	Diethyl ether	1.15 ± 0.02
$C_4H_6O_2$	γ-Butyrolactone	4.27 ± 0.03	$C_4H_{10}O$	Methyl propyl ether (trans-trans)	1.107 ± 0.013
$C_4H_6O_2$	2,3-Dihydro-1,4-dioxin	0.939 ± 0.008	$C_4H_{10}O$	Isopropyl methyl ether	1.247 ± 0.003
$C_4H_6O_2$	3,6-Dihydro-1,2-dioxin	2.329 ± 0.001	$C_4H_{10}O_2$	1,4-Butanediol	[2.58]
$C_4H_6O_3$	Acetic anhydride	≈2.8	$C_4H_{10}O_2$	Ethylene glycol monoethyl ether	[2.08]
$C_4H_6O_3$	Propylene carbonate	[4.9]	$C_4H_{10}O_3$	Diethylene glycol	[2.31]
C_4H_6S	2,3-Dihydrothiophene	1.61 ± 0.20	$C_4H_{10}S$	1-Butanethiol	[1.53]
C_4H_6S	2,5-Dihydrothiophene	1.75 ± 0.01	$C_4H_{10}S$	2-Methyl-2-propanethiol	1.66 ± 0.03
C_4H_7N	Butanenitrile (gauche)	3.91 ± 0.04	$C_4H_{10}S$	Diethyl sulfide	1.54 ± 0.08
C_4H_7N	Butanenitrile (anti)	3.73 ± 0.06	$C_4H_{11}N$	Butylamine	≈1.0
C_4H_7N	2-Methylpropanenitrile	4.29 ± 0.09	$C_4H_{11}N$	sec-Butylamine	[1.28]
C_4H_7N	2-Isocyanopropane	4.055 ± 0.001	$C_4H_{11}N$	tert-Butylamine	[1.29]
C_4H_7NO	2-Pyrrolidone	[3.5]	$C_4H_{11}N$	Isobutylamine	[1.27]
C_4H_8	1-Butene (cis)	0.438 ± 0.007	$C_4H_{11}N$	Diethylamine	0.92 ± 0.05
C_4H_8	1-Butene (skew)	0.359 ± 0.011	$C_4H_{11}NO_2$	Diethanolamine	[2.8]
C_4H_8	cis-2-Butene	0.253 ± 0.005	$C_4H_{13}N_3$	Diethylenetriamine	[1.89]
C_4H_8	Isobutene	0.503 ± 0.010	C_5F_5N	Perfluoropyridine	0.98 ± 0.08
C_4H_8	Methylcyclopropane	0.139 ± 0.004	C_5H_3NS	2-Thiophenecarbonitrile	4.59 ± 0.02
$C_4H_8Cl_2$	1,4-Dichlorobutane	2.22 ± 0.11	C_5H_3NS	3-Thiophenecarbonitrile	4.13 ± 0.02
$C_4H_8Cl_2O$	Bis(2-chloroethyl) ether	[2.58]	C_5H_4	1,3-Pentadiyne	1.207 ± 0.001
C_4H_8O	cis-2-Buten-1-ol	1.96 ± 0.03	C_5H_4ClN	4-Chloropyridine	0.756 ± 0.005
C_4H_8O	trans-2-Buten-1-ol	1.90 ± 0.02	C_5H_4FN	3-Fluoropyridine	2.09 ± 0.26
C_4H_8O	2-Methyl-2-propenol (skew)	1.295 ± 0.022	C_5H_4O	2,4-Cyclopentadien-1-one	3.132 ± 0.007
C_4H_8O	Ethyl vinyl ether	[1.26]	C_5H_4OS	4H-Pyran-4-thione	3.95 ± 0.05
C_4H_8O	1,2-Epoxybutane	1.891 ± 0.011	$C_5H_4O_2$	Furfural	[3.54]
C_4H_8O	Butanal	2.72 ± 0.05	$C_5H_4O_2$	4H-Pyran-4-one	3.79 ± 0.02
C_4H_8O	Isobutanal (gauche)	2.69 ± 0.01	$C_5H_4S_2$	4H-Thiopyran-4-thione	3.9 ± 0.2
C_4H_8O	Isobutanal (trans)	2.86 ± 0.01	C_5H_5N	Pyridine	2.215 ± 0.010
C_4H_8O	2-Butanone	2.779 ± 0.015	C_5H_6	1,2,3-Pentatriene	0.51 ± 0.05
C_4H_8O	Tetrahydrofuran	1.75 ± 0.04	C_5H_6	1-Penten-3-yne	0.66 ± 0.02
C_4H_8OS	1,4-Oxathiane	0.295 ± 0.003	C_5H_6	cis-3-Penten-1-yne	0.78 ± 0.02
$C_4H_8O_2$	Butanoic acid	[1.65]	C_5H_6	trans-3-Penten-1-yne	1.06 ± 0.05
$C_4H_8O_2$	2-Methylpropanoic acid	[1.08]	C_5H_6	2-Methyl-1-buten-3-yne	0.513 ± 0.02
$C_4H_8O_2$	Propyl formate	[1.89]	C_5H_6	1,3-Cyclopentadiene	0.419 ± 0.004
$C_4H_8O_2$	Ethyl acetate	1.78 ± 0.09	$C_5H_6N_2$	2-Methylpyrimidine	1.676 ± 0.010
$C_4H_8O_2$	cis-2-Butene-1,4-diol	[2.48]	$C_5H_6N_2$	5-Methylpyrimidine	2.881 ± 0.006
$C_4H_8O_2$	trans-2-Butene-1,4-diol	[2.45]	C_5H_6O	2-Methylfuran	0.65 ± 0.05
$C_4H_8O_2$	1,3-Dioxane	2.06 ± 0.04	C_5H_6O	3-Methylfuran	1.03 ± 0.02
$C_4H_8O_2S$	Sulfolane	[4.8]	C_5H_6O	3-Cyclopenten-1-one	2.79 ± 0.03
C_4H_8S	3-Methylthietane	2.046 ± 0.009	$C_5H_6O_2$	5-Methyl-2(3H)-furanone	4.08 ± 0.02
C_4H_8S	Tetrahydrothiophene	[1.90]	$C_5H_6O_2$	Furfuryl alcohol	[1.92]
$C_4H_8S_2$	1,3-Dithiane	2.14 ± 0.04	C_5H_6S	2-Methylthiophene	0.674 ± 0.005
C_4H_9Br	1-Bromobutane	2.08 ± 0.10	C_5H_6S	3-Methylthiophene	0.914 ± 0.015
C_4H_9Br	2-Bromobutane	2.23 ± 0.11	C_5H_7N	3-Methyl-2-butenenitrile	4.61 ± 0.13
C_4H_9Br	2-Bromo-2-methylpropane	[2.17]	C_5H_7N	Cyclobutanecarbonitrile	4.04 ± 0.04
C_4H_9Cl	1-Chlorobutane	2.05 ± 0.04	$C_5H_7NO_2$	Ethyl cyanoacetate	[2.17]

Mol. form.	Name	μ/D
C_5H_8	cis-1,3-Pentadiene	0.500 ± 0.015
C_5H_8	trans-1,3-Pentadiene	0.585 ± 0.010
C_5H_8	2-Methyl-1,3-butadiene	0.25 ± 0.01
C_5H_8	1-Pentyne (gauche)	0.769 ± 0.028
C_5H_8	1-Pentyne (trans)	0.842 ± 0.010
C_5H_8	Cyclopentene	0.20 ± 0.02
C_5H_8	3,3-Dimethylcyclopropene	0.287 ± 0.003
C_5H_8O	Cyclopropyl methyl ketone	2.62 ± 0.25
C_5H_8O	Cyclopentanone	≈3.3
C_5H_8O	3,4-Dihydro-2H-pyran	1.400 ± 0.008
C_5H_8O	3,6-Dihydro-2H-pyran	1.283 ± 0.005
$C_5H_8O_2$	Ethyl acrylate	[1.96]
$C_5H_8O_2$	Methyl methacrylate	[1.67]
$C_5H_8O_2$	2,4-Pentanedione	[2.78]
$C_5H_8O_2$	Dihydro-3-methyl-2(3H)-furanone	4.56 ± 0.02
$C_5H_8O_2$	Dihydro-5-methyl-2(3H)-furanone	4.71 ± 0.05
$C_5H_8O_2$	Tetrahydro-4H-pyran-4-one	1.720 ± 0.003
C_5H_9N	Pentanenitrile	4.12 ± 0.08
C_5H_9N	2,2-Dimethylpropanenitrile	3.95 ± 0.04
C_5H_9N	1,2,5,6-Tetrahydropyridine	1.007 ± 0.003
C_5H_9NO	N-Methyl-2-pyrrolidone	[4.1]
C_5H_{10}	1-Pentene	≈0.5
C_5H_{10}	3-Methyl-1-butene (gauche)	0.398 ± 0.004
C_5H_{10}	3-Methyl-1-butene (trans)	0.320 ± 0.010
C_5H_{10}	1,1-Dimethylcyclopropane	0.142 ± 0.001
$C_5H_{10}O$	2,2-Dimethylpropanal	2.66 ± 0.05
$C_5H_{10}O$	2-Pentanone	[2.70]
$C_5H_{10}O$	3-Pentanone	[2.82]
$C_5H_{10}O$	Tetrahydropyran (chair)	1.58 ± 0.03
$C_5H_{10}O_2$	Pentanoic acid	[1.61]
$C_5H_{10}O_2$	3-Methylbutanoic acid	[0.63]
$C_5H_{10}O_2$	Butyl formate	[2.03]
$C_5H_{10}O_2$	Isobutyl formate	[1.88]
$C_5H_{10}O_2$	Propyl acetate	[1.78]
$C_5H_{10}O_2$	Ethyl propanoate	[1.74]
$C_5H_{10}O_2$	Tetrahydrofurfuryl alcohol	[2.1]
$C_5H_{10}O_3$	Diethyl carbonate	1.10 ± 0.06
$C_5H_{10}O_3$	Ethylene glycol monomethyl ether acetate	[2.13]
$C_5H_{10}O_3$	Ethyl lactate	[2.4]
$C_5H_{10}S$	Thiacyclohexane	1.781 ± 0.010
$C_5H_{11}Br$	1-Bromopentane	2.20 ± 0.11
$C_5H_{11}Cl$	1-Chloropentane	2.16 ± 0.11
$C_5H_{11}Cl$	1-Chloro-3-methylbutane	[1.92]
$C_5H_{11}N$	Piperidine (equitorial)	0.82 ± 0.02
$C_5H_{11}N$	Piperidine (axial)	1.19 ± 0.02
$C_5H_{11}N$	Piperidine (average)	[1.19]
$C_5H_{11}N$	N-Methylpyrrolidine	0.572 ± 0.003
C_5H_{12}	Isopentane	0.13 ± 0.05
$C_5H_{12}N_2O$	Tetramethylurea	[3.5]
$C_5H_{12}O$	1-Pentanol	[1.7]
$C_5H_{12}O$	2-Pentanol	[1.66]
$C_5H_{12}O$	3-Pentanol	[1.64]
$C_5H_{12}O$	2-Methyl-1-butanol	[1.88]
$C_5H_{12}O$	2-Methyl-2-butanol	[1.82]
$C_5H_{12}O_2$	1,5-Pentanediol	[2.5]
$C_5H_{12}O_3$	Diethylene glycol monomethyl ether	[1.6]
$C_6H_2F_4$	1,2,3,4-Tetrafluorobenzene	2.42 ± 0.05
$C_6H_2F_4$	1,2,3,5-Tetrafluorobenzene	1.46 ± 0.06
$C_6H_3F_3$	1,2,4-Trifluorobenzene	1.402 ± 0.009
$C_6H_4ClNO_2$	1-Chloro-2-nitrobenzene	4.64 ± 0.09
$C_6H_4ClNO_2$	1-Chloro-3-nitrobenzene	3.73 ± 0.07
$C_6H_4ClNO_2$	1-Chloro-4-nitrobenzene	2.83 ± 0.06
$C_6H_4Cl_2$	o-Dichlorobenzene	2.50 ± 0.05
$C_6H_4Cl_2$	m-Dichlorobenzene	1.72 ± 0.09
$C_6H_4FNO_2$	1-Fluoro-4-nitrobenzene	2.87 ± 0.06
$C_6H_4F_2$	o-Difluorobenzene	2.46 ± 0.05
$C_6H_4F_2$	m-Difluorobenzene	1.51 ± 0.02
$C_6H_4N_2$	2-Pyridinecarbonitrile	5.78 ± 0.11
$C_6H_4N_2$	3-Pyridinecarbonitrile	3.66 ± 0.11
$C_6H_4N_2$	4-Pyridinecarbonitrile	1.96 ± 0.03
$C_6H_4O_2$	3,5-Cyclohexadiene-1,2-dione	4.23 ± 0.02
C_6H_5Br	Bromobenzene	1.70 ± 0.03
C_6H_5Cl	Chlorobenzene	1.69 ± 0.03
C_6H_5ClO	p-Chlorophenol	2.11 ± 0.11
C_6H_5F	Fluorobenzene	1.60 ± 0.08
C_6H_5I	Iodobenzene	1.70 ± 0.09
C_6H_5NO	2-Pyridinecarboxaldehyde	3.56 ± 0.07
C_6H_5NO	3-Pyridinecarboxaldehyde	1.44
C_6H_5NO	4-Pyridinecarboxaldehyde	1.66
$C_6H_5NO_2$	Nitrobenzene	4.22 ± 0.08
C_6H_6	Fulvene	0.4236 ± 0.013
C_6H_6ClN	o-Chloroaniline	[1.77]
C_6H_6O	Phenol	1.224 ± 0.008
C_6H_6O	2-Vinylfuran	0.69 ± 0.07
$C_6H_6O_2$	p-Hydroquinone	2.38 ± 0.05
C_6H_6S	Benzenethiol	[1.23]
C_6H_7N	Aniline	1.13 ± 0.02
C_6H_7N	2-Methylpyridine	1.85 ± 0.04
C_6H_7N	3-Methylpyridine	[2.40]
C_6H_7N	4-Methylpyridine	2.70 ± 0.02
C_6H_8O	3-Methyl-2-cyclopenten-1-one	4.33 ± 0.002
$C_6H_8O_4$	Dimethyl maleate	[2.48]
C_6H_8Si	Phenylsilane	0.845 ± 0.012
C_6H_9F	1-Fluorocyclohexane	1.942 ± 0.010
C_6H_{10}	1-Hexyne	0.83 ± 0.05
C_6H_{10}	3,3-Dimethyl-1-butyne	0.661 ± 0.004
C_6H_{10}	Cyclohexene (half-chair)	0.332 ± 0.012
$C_6H_{10}F_2$	1,1-Difluorocyclohexane	2.556 ± 0.010
$C_6H_{10}O$	3-Methylcyclopentanone	3.14 ± 0.03
$C_6H_{10}O$	Cyclohexanone	3.246 ± 0.006
$C_6H_{10}O$	Mesityl oxide	[2.79]
$C_6H_{10}O_4$	Diethyl oxalate	[2.49]
$C_6H_{10}O_4$	Ethylene glycol diacetate	[2.34]
$C_6H_{11}Cl$	Chlorocyclohexane (equitorial)	2.44 ± 0.07
$C_6H_{11}Cl$	Chlorocyclohexane (axial)	1.91 ± 0.02
$C_6H_{11}F$	Fluorocyclohexane (equitorial)	2.11 ± 0.04
$C_6H_{11}F$	Fluorocyclohexane (axial)	1.81 ± 0.04
$C_6H_{11}N$	4-Methylpentanenitrile	[3.5]
$C_6H_{11}NO$	Caprolactam	[3.9]
$C_6H_{12}O$	Butyl vinyl ether	[1.25]
$C_6H_{12}O$	2-Hexanone	[2.66]
$C_6H_{12}O_2$	Hexanoic acid	[1.13]
$C_6H_{12}O_2$	Pentyl formate	1.90 ± 0.10
$C_6H_{12}O_2$	Butyl acetate	[1.87]
$C_6H_{12}O_2$	sec-Butyl acetate	[1.87]
$C_6H_{12}O_2$	Isobutyl acetate	[1.86]
$C_6H_{12}O_2$	Ethyl butanoate	[1.74]
$C_6H_{12}O_2$	Diacetone alcohol	[3.24]
$C_6H_{12}O_3$	Ethylene glycol monoethyl ether acetate	[2.25]

Mol. form.	Name	μ/D
C₆H₁₂O₃	Paraldehyde	1.43 ± 0.07
C₆H₁₃N	Cyclohexylamine	[1.26]
C₆H₁₄O	Dipropyl ether	1.21 ± 0.06
C₆H₁₄O	Diisopropyl ether	1.13 ± 0.10
C₆H₁₄O	Butyl ethyl ether	[1.24]
C₆H₁₄O₂	2-Methyl-2,4-pentanediol	[2.9]
C₆H₁₄O₂	Ethylene glycol monobutyl ether	[2.08]
C₆H₁₄O₂	1,1-Diethoxyethane	[1.38]
C₆H₁₄O₃	Diethylene glycol monoethyl ether	[1.6]
C₆H₁₄O₃	Diethylene glycol dimethyl ether	[1.97]
C₆H₁₅N	Dipropylamine	[1.03]
C₆H₁₅N	Diisopropylamine	[1.15]
C₆H₁₅N	Triethylamine	0.66 ± 0.05
C₆H₁₅NO₃	Triethanolamine	[3.57]
C₆H₁₅O₄P	Triethyl phosphate	[3.12]
C₆H₁₈N₃OP	Hexamethylphosphoric triamide	[5.5]
C₇H₅Cl₃	(Trichloromethyl)benzene	[2.03]
C₇H₅F₃	(Trifluoromethyl)benzene	2.86 ± 0.06
C₇H₅N	Benzonitrile	4.18 ± 0.08
C₇H₅N	Isocyanobenzene	4.018 ± 0.003
C₇H₆Cl₂	2,4-Dichlorotoluene	[1.70]
C₇H₆Cl₂	3,4-Dichlorotoluene	[2.95]
C₇H₆Cl₂	(Dichloromethyl)benzene	[2.07]
C₇H₆O	2,4,6-Cycloheptatrien-1-one	4.1 ± 0.3
C₇H₆O	Benzaldehyde	[3.0]
C₇H₆O₂	Salicylaldehyde	[2.86]
C₇H₇Cl	o-Chlorotoluene	1.56 ± 0.08
C₇H₇Cl	m-Chlorotoluene	[1.82]
C₇H₇Cl	p-Chlorotoluene	2.21 ± 0.04
C₇H₇Cl	(Chloromethyl)benzene	[1.82]
C₇H₇F	o-Fluorotoluene	1.37 ± 0.07
C₇H₇F	m-Fluorotoluene	1.82 ± 0.04
C₇H₇F	p-Fluorotoluene	2.00 ± 0.10
C₇H₇NO₃	2-Nitroanisole	[5.0]
C₇H₈	Toluene	0.375 ± 0.010
C₇H₈	2,5-Norbornadiene	0.0587 ± 0.0001
C₇H₈O	o-Cresol	[1.45]
C₇H₈O	m-Cresol	[1.48]
C₇H₈O	p-Cresol	[1.48]
C₇H₈O	Benzyl alcohol	1.71 ± 0.09
C₇H₈O	Anisole	1.38 ± 0.07
C₇H₉N	o-Methylaniline	[1.60]
C₇H₉N	m-Methylaniline	[1.45]
C₇H₉N	p-Methylaniline	[1.52]
C₇H₉N	2,4-Dimethylpyridine	[2.30]
C₇H₉N	2,6-Dimethylpyridine	[1.66]
C₇H₁₀	1,3-Cycloheptadiene	0.740
C₇H₁₂	Methylenecyclohexane	0.62 ± 0.01
C₇H₁₂O₄	Diethyl malonate	[2.54]
C₇H₁₄O	2-Heptanone	[2.59]
C₇H₁₄O	3-Heptanone	[2.78]
C₇H₁₄O	2,4-Dimethyl-3-pentanone	[2.74]
C₇H₁₄O	cis-3-Methylcyclohexanol	[1.91]
C₇H₁₄O	trans-3-Methylcyclohexanol	[1.75]
C₇H₁₄O₂	Pentyl acetate	1.75 ± 0.10
C₇H₁₄O₂	Isopentyl acetate	[1.86]
C₇H₁₅Br	1-Bromoheptane	2.16 ± 0.11
C₇H₁₆O	2-Heptanol	[1.71]
C₇H₁₆O	3-Heptanol	[1.71]
C₈H₆	Phenylacetylene	0.656 ± 0.005
C₈H₇N	Benzeneacetonitrile	[3.5]

Mol. form.	Name	μ/D
C₈H₈	Styrene	0.123 ± 0.003
C₈H₈O	Acetophenone	3.02 ± 0.06
C₈H₈O₂	Methyl benzoate	[1.94]
C₈H₈O₃	Methyl salicylate	[2.47]
C₈H₁₀	Ethylbenzene	0.59 ± 0.05
C₈H₁₀	o-Xylene	0.640 ± 0.005
C₈H₁₀O	2,4-Xylenol	[1.4]
C₈H₁₀O	2,5-Xylenol	[1.45]
C₈H₁₀O	2,6-Xylenol	[1.40]
C₈H₁₀O	3,4-Xylenol	[1.56]
C₈H₁₀O	3,5-Xylenol	[1.55]
C₈H₁₀O	Phenetole	1.45 ± 0.15
C₈H₁₀O₂	1,2-Dimethoxybenzene	[1.29]
C₈H₁₁N	N,N-Dimethylaniline	1.68 ± 0.17
C₈H₁₁N	2,4-Dimethylaniline	[1.40]
C₈H₁₁N	2,6-Dimethylaniline	[1.63]
C₈H₁₁N	2,4,6-Trimethylpyridine	[2.05]
C₈H₁₆	2-Octanone	[2.70]
C₈H₁₆O₂	Octanoic acid	[1.15]
C₈H₁₆O₂	sec-Hexyl acetate	[1.9]
C₈H₁₆O₂	Isobutyl isobutanoate	[1.9]
C₈H₁₆O₄	Diethylene glycol monoethyl ether acetate	[1.8]
C₈H₁₇Cl	1-Chlorooctane	[2.00]
C₈H₁₈O	1-Octanol	[1.76]
C₈H₁₈O	2-Octanol	[1.71]
C₈H₁₈O	2-Ethyl-1-hexanol	[1.74]
C₈H₁₈O	Dibutyl ether	1.17 ± 0.06
C₈H₁₈S	Dibutyl sulfide	[1.61]
C₈H₁₉N	Dibutylamine	[0.98]
C₉H₇N	Quinoline	2.29 ± 0.11
C₉H₇N	Isoquinoline	2.73 ± 0.14
C₉H₁₀O₂	Ethyl benzoate	2.00 ± 0.10
C₉H₁₀O₂	Benzyl acetate	[1.22]
C₉H₁₂	Isopropylbenzene	≈0.79
C₉H₁₈O	2,6-Dimethyl-4-heptanone	[2.66]
C₉H₁₈O₂	Nonanoic acid	[0.79]
C₁₀H₇Br	1-Bromonaphthalene	[1.55]
C₁₀H₇Cl	1-Chloronaphthalene	[1.57]
C₁₀H₈	Azulene	0.80 ± 0.02
C₁₀H₁₄	tert-Butylbenzene	≈0.83
C₁₀H₁₆O	Camphor, (+)	[3.1]
C₁₀H₂₀O₂	2-Ethylhexyl acetate	[1.8]
C₁₀H₂₁Br	1-Bromodecane	[1.93]
C₁₀H₂₂O	Dipentyl ether	[1.20]
C₁₀H₂₂O	Diisopentyl ether	[1.23]
C₁₁H₁₂O₂	Ethyl trans-cinnamate	[1.84]
C₁₂H₁₀	Acenaphthene	≈0.85
C₁₂H₁₀O	Diphenyl ether	≈1.3
C₁₂H₂₇BO₃	Tributyl borate	[0.77]
C₁₂H₂₇N	Tributylamine	[0.78]
C₁₂H₂₇O₄P	Tributyl phosphate	[3.07]
C₁₄H₁₂O₂	Benzyl benzoate	[2.06]
C₁₆H₂₂O₄	Dibutyl phthalate	[2.82]
C₁₈H₃₄O₂	Oleic acid	[1.18]
C₁₈H₃₄O₄	Dibutyl sebacate	[2.48]
C₂₁H₂₁O₄P	Tri-o-cresyl phosphate	[2.87]
C₂₁H₂₁O₄P	Tri-m-cresyl phosphate	[3.05]
C₂₁H₂₁O₄P	Tri-p-cresyl phosphate	[3.18]
C₂₂H₄₄O₂	Butyl stearate	[1.88]
C₂₄H₃₈O₄	Bis(2-ethylhexyl) phthalate	[2.84]

BOND DISSOCIATION ENERGIES

Yu-Ran Luo and J. Alistair Kerr

The bond dissociation energy (enthalpy) is also referred to as bond disruption energy, bond energy, bond strength, or binding energy (abbreviation: BDE, BE, or D). It is defined as the standard enthalpy change of the following fission: R–X → R + X. The BDE, denoted by $D°(R–X)$, is usually derived by the thermochemical equation, $D°(R–X) = \Delta_f H°(R) + \Delta_f H°(X) − \Delta_f H°(RX)$. The enthalpy of formation $\Delta_f H°$ of a large number of atoms, free radicals, ions, clusters and compounds is available from the website of NIST, NASA, CODATA, and IUPAC. Most authors prefer to use the BDE values at 298.15 K.

The following seven tables provide essential information of experimental BDE values of R–X and R⁺–X bonds.

(1) Table 1: Bond Dissociation Energies in Diatomic Molecules
(2) Table 2: Enthalpy of Formation of Gaseous Atoms
(3) Table 3: Bond Dissociation Energies in Polyatomic Molecules
(4) Table 4: Enthalpies of Formation of Free Radicals and Other Transient Species
(5) Table 5: Bond Dissociation Energies of Common Organic Molecules
(6) Table 6: Bond Dissociation Energies in Diatomic Cations
(7) Table 7: Bond Dissociation Energies in Polyatomic Cations

The data in these tables have been revised through September 2005.

TABLE 1. Bond Dissociation Energies in Diatomic Molecules

The BDEs in diatomic species have usually been measured by spectroscopy or mass spectrometry. In the absence of data on enthalpy function, the values at 0 K, $D°(A–B)$, are converted to $D°_{298}$ by the approximate equation:

$$D°_{298}(A–B) \approx D°(A–B) + (3/2)RT = D°(A–B) + 3.7181 \text{ kJ mol}^{-1}$$

This table has been arranged in alphabetical order of the atoms A in the diatomics A–B and all data are taken from the *Comprehensive Handbook of Chemical Bond Energies*, by Yu-Ran Luo, Taylor & Francis, 2006, which contains detailed references.

A–B	$D°_{298}$/kJ mol^{-1}	A–B	$D°_{298}$/kJ mol^{-1}	A–B	$D°_{298}$/kJ mol^{-1}	A–B	$D°_{298}$/kJ mol^{-1}
Ac–O	794	Al–Al	133±5.8	Al–Ti	263.4	As–Cl	448
Ag–Ag	162.9±2.9	Al–Ar	5.69	Al–U	326±29	As–D	270.3
Ag–Al	183.7±9.2	Al–As	202.7±7.1	Al–V	147.4±1.0	As–F	410
Ag–Au	202.5±9.6	Al–Au	325.9±6.3	Al–Xe	7.39	As–Ga	202.5±4.8
Ag–Bi	192±42	Al–Br	429.2±5.8	Am–O	553±36	As–H	274.0±2.9
Ag–Br	281.5±1.3	Al–C	267.7	Ar–Ar	4.91	As–I	296.6±24
Ag–Cl	279.1±8.4	Al–Ca	52.7	Ar–B	4.62	As–In	201±10
Ag–Cu	171.5±9.6	Al–Cl	502	Ar–Br	~5.0	As–N	489±2.1
Ag–D	226.8	Al–Co	181.6±0.2	Ar–C	5.158	As–O	484±8
Ag–Dy	130±19	Al–Cr	222.9±0.9	Ar–Ca	4.44±0.60	As–P	433.5±12.6
Ag–Eu	127±13	Al–Cu	227.1±1.2	Ar–Cd	5.57±0.05	As–S	379.5±6.3
Ag–F	356.9±5.8	Al–D	290.4	Ar–Ga	3.96	As–Sb	330.5±5.4
Ag–Ga	159±17	Al–F	675	Ar–Ge	<5.4	As–Se	96
Ag–Ge	174.5±21	Al–H	288±13	Ar–He	3.96	As–Tl	198.3±14.6
Ag–H	202.4±9.6	Al–I	369.9±2.1	Ar–Hg	5.32	Au–Au	226.2±0.5
Ag–Ho	124±19	Al–Kr	6.05	Ar–I	~5.3	Au–B	367.8±10.5
Ag–I	234±29	Al–Li	76.1	Ar–In	4.18	Au–Ba	254.8±10.0
Ag–In	166.5±4.9	Al–N	≤368±15	Ar–Kr	5.11	Au–Be	237.7±4.0
Ag–Li	186.1	Al–Ne	3.9	Ar–Li	~7.82	Au–Bi	293±8.4
Ag–Mn	99.2±21	Al–Ni	224.7±4.8	Ar–Mg	~3.7	Au–Br	213±21
Ag–Na	133.1±12.6	Al–O	501.9±10.6	Ar–Na	~4.2	Au–Ca	250.4±4.0
Ag–Nd	<213	Al–P	216.7±12.6	Ar–Ne	4.27	Au–Ce	322±18
Ag–O	221±21	Al–Pd	254.4±12.1	Ar–Si	5.86	Au–Cl	280±13
Ag–S	216.7±14.6	Al–S	332±10	Ar–Sn	<5.1	Au–Co	218.0±16.4
Ag–Se	210.0±14.6	Al–Sb	216.3±6	Ar–Tl	4.09	Au–Cr	223.7±28.9
Ag–Si	185.1±9.6	Al–Se	318±13	Ar–Xe	5.28	Au–Cs	253±3.5
Ag–Sn	136±21	Al–Si	246.9±12.6	Ar–Zn	5.0	Au–Cu	227.1±1.2
Ag–Te	195.8±14.6	Al–Te	268±13	As–As	385.8±10.5	Au–D	322.2

Bond Dissociation Energies

A-B	D^o_{298}/kJ mol^{-1}	A-B	D^o_{298}/kJ mol^{-1}	A-B	D^o_{298}/kJ mol^{-1}	A-B	D^o_{298}/kJ mol^{-1}
Au–Dy	259±24	B–Pt	477.8±16.7	Br–Br	193.859±0.120	Br–Tl	331±21
Au–Eu	245±12	B–Rh	475.8±21	Br–C	318.0±8.4	Br–Tm	299.1
Au–F	294.1	B–Ru	446.9±21	Br–Ca	339	Br–U	377±15
Au–Fe	187.0±19.3	B–S	577±9.2	Br–Cd	159±96	Br–V	439±42
Au–Ga	290±15	B–Sc	272±63	Br–Ce	373.2	Br–W	329.3
Au–Ge	273.2±14.6	B–Se	462±14.6	Br–Cl	219.32±0.05	Br–Xe	5.94±0.02
Au–H	328	B–Si	317±12	Br–Co	326±42	Br–Y	481±84
Au–Ho	267±35	B–Te	354±20	Br–Cr	328.0±24.3	Br–Yb	295.4
Au–I	276	B–Th	297±33	Br–Cs	389.1±4.2	Br–Zn	138±29
Au–In	286.0±5.7	B–Ti	272±63	Br–Cu	331±25	Br–Zr	420
Au–La	457±28	B–U	322±33	Br–D	370.74	C–C	618.3±15.4
Au–Li	284.5±6.7	B–Y	289±63	Br–Dy	339.3±10.5	C–Ce	443±30
Au–Lu	332±19	Ba–Br	402	Br–Er	361.3	C–Cl	394.9±13.4
Au–Mg	179.1±2.7	Ba–Cl	443	Br–Eu	548	C–D	341.4
Au–Mn	197.7±21	Ba–D	≤193.7	Br–F	280±12	C–F	513.8±10.0
Au–Na	215.1±12.6	Ba–F	580.6	Br–Fe	243±84	C–Fe	376.3±28.9
Au–Nd	294±29	Ba–H	192.0	Br–Ga	402±13	C–Ge	455.7±11
Au–Ni	247±16.4	Ba–I	322.6±6.3	Br–Gd	372.0	C–H	338.4±1.2
Au–O	223±21	Ba–O	562±13.4	Br–Ge	347±8	C–Hf	540±25
Au–Pb	133±42	Ba–Pd	221.8±5.0	Br–H	366.16±0.20	C–I	253.1±35.6
Au–Pd	142.7±21	Ba–Rh	259.4±25	Br–Hg	74.9	C–Ir	631±5
Au–Pr	311±25	Ba–S	418±21	Br–Ho	321.8	C–La	463±20
Au–Rb	243±3.5	Be–Be	59	Br–I	179.1±0.4	C–Mo	482±16
Au–Rh	232.6±29	Be–Br	316	Br–In	409±10	C–N	750.0±2.9
Au–S	253.6±14.6	Be–Cl	434	Br–K	379.1±4.2	C–Nb	523.8±14.5
Au–Sc	280±40	Be–D	203.1	Br–La	446.2	C–Ni	337.0
Au–Se	251.0±14.6	Be–F	573	Br–Li	418.8±4.2	C–O	1076.38±0.67
Au–Si	304.6±6.0	Be–H	221	Br–Lu	301.5	C–Os	608±25
Au–Sn	256.5±7.2	Be–I	261	Br–Mg	317.96	C–P	507.5±8.8
Au–Sr	264±42	Be–O	437	Br–Mn	314.2±9.6	C–Pd	436±20
Au–Tb	285±33	Be–S	372±59	Br–Mo	313.4	C–Pt	610±5
Au–Te	237.2±14.6	Be–T	204.4	Br–N	280.8±21	C–Rh	580±4
Au–U	318±29	Bi–Bi	204.4	Br–Na	363.1±4.2	C–Ru	648±13
Au–V	246.0±8.7	Bi–Br	240.2	Br–Nd	339.7	C–S	713.3±1.2
Au–Y	310±12	Bi–Cl	300.4±4.2	Br–Ni	360±13	C–Sc	444±21
B–B	290	Bi–D	283.7	Br–O	235.4±1.9	C–Se	590.4±5.9
B–Br	390.9±0.5	Bi–F	366.5±12.5	Br–P	≤329	C–Si	447
B–C	448±29	Bi–Ga	158.6±16.7	Br–Pb	248.5±14.6	C–Tc	564±29
B–Cd	301.0	Bi–H	≤283.3	Br–Pr	344.5	C–Th	453±17
B–Ce	305±21	Bi–I	186.1±5.8	Br–Rb	380.7±4.2	C–Ti	423±30
B–Cl	427	Bi–In	153.6±1.7	Br–S	218±17	C–U	455±15
B–D	341.0±6.3	Bi–Li	149.4	Br–Sb	314±59	C–V	423±24
B–F	732	Bi–O	337.2±12.6	Br–Sc	444±63	C–Y	418±14
B–H	345.2±2.5	Bi–P	281.7±13	Br–Se	297±84	C–Zr	495.8±38.6
B–I	361	Bi–Pb	142.4±3.0	Br–Si	358.2±8.4	Ca–Ca	16.52±0.11
B–Ir	512.2±17	Bi–S	315.5±4.6	Br–Sm	331.4	Ca–Cl	409±8.7
B–La	335±63	Bi–Sb	252.7±3.9	Br–Sn	337±13	Ca–D	≤169.9
B–N	377.9±8.7	Bi–Se	280.3±5.9	Br–Sr	365	Ca–F	529
B–Ne	3.97	Bi–Sn	193±13	Br–T	372.77	Ca–H	223.8
B–O	809	Bi–Te	232.2±11.3	Br–Tb	382.8	Ca–I	284.7±8.4
B–P	347±16.7	Bi–Tl	120.9±12.6	Br–Th	364	Ca–Kr	5.15±0.72
B–Pd	351.5±16.7	Bk–O	598	Br–Ti	373	Ca–Li	84.9±8.4

A-B	D^o_{298}/kJ mol^{-1}	A-B	D^o_{298}/kJ mol^{-1}	A-B	D^o_{298}/kJ mol^{-1}	A-B	D^o_{298}/kJ mol^{-1}
Ca–O	383.3±5.0	Cl–I	211.3±0.4	Co–S	331	D–I	302.33
Ca–Pd	347 - 360	Cl–In	436±8	Co–Si	274.4±17	D–In	246
Ca–S	335±21	Cl–K	433.0±8.4	Co–Ti	235.37±0.10	D–K	182.4
Ca–Xe	7.31±0.96	Cl–La	521.6	Co–Y	253.71±0.10	D–Li	240.24
Cd–Cd	~8.62	Cl–Li	469±13	Co–Zr	306.39±0.10	D–Lu	302
Cd–Cl	208.4	Cl–Lu	325.7±2	Cr–Cr	152.0±6	D–Mg	161.33±0.32
Cd–F	305±21	Cl–Mg	312	Cr–Cu	154.4±14.5	D–Mn	312±6
Cd–H	69.0±0.4	Cl–Mn	338.5±6.7	Cr–F	523±19	D–N	341.6
Cd–I	97.2±2.1	Cl–N	333.9±9.6	Cr–Fe	~75	D–Ni	≤302.9
Cd–In	134	Cl–Na	412.1±8.4	Cr–Ge	154±7	D–O	429.64
Cd–K	7.3	Cl–Nd	418.7	Cr–H	189.9±6.7	D–P	299.0
Cd–Kr	5.17	Cl–Ni	377.0±6.7	Cr–I	287.0±24.3	D–Pt	≤350.2
Cd–Na	10.2	Cl–O	267.47±0.08	Cr–N	377.8±18.8	D–S	350.62±1.20
Cd–Ne	3.97	Cl–P	≤376	Cr–Nb	295.72±0.06	D–Si	302.5
Cd–O	236±84	Cl–Pb	301±50	Cr–O	461±8.7	D–Sr	167.7
Cd–S	208.5±20.9	Cl–Pr	423.5	Cr–Pb	105±2	D–T	444.91
Cd–Se	127.6±25.1	Cl–Ra	343±75	Cr–S	331	D–Tl	193.0
Cd–Te	100.0±15.1	Cl–Rb	427.6±8.4	Cr–Sn	141±3	D–Zn	88.7
Cd–Xe	6.54	Cl–S	241.8	Cs–Cs	43.919±0.010	Dy–Dy	70.3
Ce–Ce	251.7	Cl–Sb	360±50	Cs–F	517.1±7.7	Dy–F	531
Ce–Cl	457.0	Cl–Sc	331	Cs–H	175.364	Dy–I	269.0±8.4
Ce–F	582±42	Cl–Se	322	Cs–Hg	8	Dy–O	615
Ce–I	333.8	Cl–Si	416.7±6.3	Cs–I	338.5±2.1	Dy–S	414±42
Ce–Ir	575±9	Cl–Sm	418.7	Cs–Na	63.2±1.3	Dy–Se	322±20
Ce–N	519±21	Cl–Sn	350±8	Cs–O	293±25	Dy–Te	234±20
Ce–O	790	Cl–Sr	409	Cs–Rb	49.57±0.01	Er–Er	75±29
Ce–Os	524±20	Cl–T	438.64	Cu–Cu	201	Er–F	565±17
Ce–Pd	319±21	Cl–Ta	544	Cu–D	270.3	Er–I	315.8
Ce–Pt	550±5	Cl–Tb	470.1	Cu–Dy	144±18	Er–O	606
Ce–Rh	545±7	Cl–Th	489	Cu–F	414	Er–S	418±21
Ce–Ru	494±12	Cl–Ti	405.4±10.5	Cu–Ga	215.9±15	Er–Se	326±20
Ce–S	569	Cl–Tl	372.8±2.1	Cu–Ge	208.8±21	Er–Te	238±20
Ce–Se	494.5±14.6	Cl–Tm	378.0	Cu–H	254.8±6	Es–O	460
Ce–Te	189.4±12.6	Cl–U	439	Cu–Ho	144±19	Eu–Eu	45.2
Cf–O	498	Cl–V	477±63	Cu–I	289±63	Eu–F	544
Cl–Cl	242.417±1.930	Cl–W	419	Cu–In	187.4±7.9	Eu–I	288.3
Cl–Co	337.6±6.7	Cl–Xe	7.08	Cu–Li	191.9	Eu–Li	268.1±12.6
Cl–Cr	377.8±6.7	Cl–Y	523±84	Cu–Na	176.1±16.7	Eu–O	473
Cl–Cs	445.7±7.7	Cl–Yb	374.5	Cu–Ni	201.7±9.6	Eu–Rh	238±34
Cl–Cu	377.8±7.5	Cl–Zn	229±8	Cu–O	287.4±11.6	Eu–S	365.7±13.4
Cl–D	436.303±0.011	Cl–Zr	530	Cu–S	274.5±14.6	Eu–Se	302.9±14.6
Cl–Dy	392.4	Cm–O	732	Cu–Se	255.2±14.6	Eu–Te	251.0±14.6
Cl–Er	448.6	Co–Co	<127	Cu–Si	221.3±6.3	F–F	158.670±0.096
Cl–Eu	405.5	Co–Cu	161.1±16.4	Cu–Sn	170±10	F–Fe	447
Cl–F	260.83	Co–D	270.2±5.8	Cu–Tb	191±18	F–Ga	584±13
Cl–Fe	329.7±6.7	Co–F	431±63	Cu–Te	230.5±14.6	F–Gd	590±17
Cl–Ga	463±13	Co–Ge	230±21	D–D	443.3197±0.0003	F–Ge	523±13
Cl–Gd	451.0	Co–H	244.9±4.8	D–F	576.174±0.096	F–H	569.658±0.012
Cl–Ge	390.8±9.6	Co–I	280±21	D–Ga	<276.5	F–Hf	650±15
Cl–H	431.361±0.013	Co–Mn	50±8	D–Ge	≤322	F–Hg	~180
Cl–Hg	92.0±9.2	Co–Nb	267.02±0.10	D–H	439.2223±0.0002	F–Ho	540
Cl–Ho	409.1	Co–O	384.5±13.4	D–Hg	42.05	F–I	≤271.5

Bond Dissociation Energies

A-B	D^o_{298}/kJ mol⁻¹	A-B	D^o_{298}/kJ mol⁻¹	A-B	D^o_{298}/kJ mol⁻¹	A-B	D^o_{298}/kJ mol⁻¹
F–In	516±13	Ga–Ga	<106.4	H–Sb	239.7±4.2	I–Nd	301.5
F–K	489.2	Ga–H	276	H–Sc	205±17	I–Ni	293±21
F–Kr	6.6	Ga–I	334±13	H–Se	312.5	I–O	240±5
F–La	659.0±17.2	Ga–In	94.0±3	H–Si	293.3±1.9	I–Pb	194±38
F–Li	577±21	Ga–Kr	4.08	H–Sn	264±17	I–Pr	306.2
F–Lu	405±19	Ga–Li	133.1±14.6	H–Sr	164±8	I–Rb	318.8±2.1
F–Mg	463	Ga–O	374±21	H–T	440.49	I–Si	243.1±8.4
F–Mn	445.2±7.5	Ga–P	229.7±12.6	H–Te	270.7±1.7	I–Sm	293.1
F–Mo	464	Ga–Sb	192.0±12.6	H–Ti	204.6±8.8	I–Sn	235±3
F–N	≤349	Ga–Te	265±21	H–Tl	195.4±4	I–Sr	301
F–Na	477.3	Ga–Xe	5.27	H–V	209.3±6.8	I–Tb	336.2
F–Nd	545.2±12.6	Gd–Gd	206.3±67.5	H–Yb	183.1±2.0	I–Te	192±42
F–Ni	428.4±19.3	Gd–I	333.8	H–Zn	85.8±2	I–Th	361±25
F–Np	430±50	Gd–O	715	He–He	3.809	I–Ti	306
F–O	220	Gd–S	526.8±10.5	He–Hg	3.8	I–Tl	285±21
F–P	≤405	Gd–Se	430±15	He–Xe	3.8	I–Tm	260.8
F–Pb	355±13	Gd–Te	341±15	Hf–Hf	328±58	I–U	299±27
F–Pr	582±46	Ge–Ge	264.4±6.8	Hf–N	535±30	I–Xe	~6.9
F–Pu	538±29	Ge–H	263.2±4.8	Hf–O	801±13	I–Y	422.6±12.5
F–Rb	494±21	Ge–I	268±25	Hg–Hg	7.79±0.24	I–Yb	257.3
F–Ru	402	Ge–Ni	290.3±10.9	Hg–I	34.69±0.96	I–Zn	153.1±6.3
F–S	343.5±6.7	Ge–O	660.3	Hg–K	8.8	I–Zr	127
F–Sb	439±96	Ge–Pd	254.7±10.5	Hg–Kr	5.75	In–In	82.0±5.7
F–Sc	599.1±13.4	Ge–S	534±3	Hg–Li	13.16±0.38	In–Kr	4.85
F–Se	339±42	Ge–Sc	270±11	Hg–Na	10.8	In–Li	92.5±14.6
F–Si	576.4±17	Ge–Se	484.7±1.7	Hg–Ne	4.14	In–O	346±30
F–Sm	565	Ge–Si	297	Hg–O	269	In–P	197.9±8.4
F–Sn	476±8	Ge–Sn	230.1±13	Hg–Rb	8.4	In–S	287.9±14.6
F–Sr	538	Ge–Te	396.7±3.3	Hg–S	217.3±22.2	In–Sb	151.9±10.5
F–T	579.009±0.108	Ge–Y	279±11	Hg–Se	144.3±30.1	In–Se	245.2±14.6
F–Ta	573±13	H–H	435.7799±0.0001	Hg–T	43.14	In–Te	215.5±14.6
F–Tb	561±42	H–Hg	39.844	Hg–Te	<142	In–Xe	6.48
F–Th	652	H–I	298.26±0.10	Hg–Tl	2.9	In–Zn	32.2
F–Ti	569±33	H–In	243.1	Hg–Xe	6.65	Ir–Ir	361±68
F–Tl	439±21	H–K	174.576	Ho–Ho	70.3	Ir–La	577±12
F–Tm	510	H–Li	238.039±0.006	Ho–I	275.1	Ir–Nb	465±25
F–U	648	H–Mg	126.4±2.9	Ho–O	606	Ir–O	414±42
F–V	590±63	H–Mn	251±5	Ho–S	428.4±14.6	Ir–Si	462.8±21
F–W	≤544	H–Mo	211±19	Ho–Se	333±15	Ir–Th	574±42
F–Xe	14.18	H–N	≤338.9	Ho–Te	≤259±15	Ir–Ti	422±13
F–Y	685.3±13.4	H–Na	185.69±0.29	I–I	152.25±0.57	Ir–Y	457±15
F–Yb	≥517.6±9.6	H–Nb	>221.9±9.6	I–In	306.9±1.1	K–K	56.96
F–Zn	364±63	H–Ni	240±8	I–K	322.5±2.1	K–Kr	4.6
F–Zr	627.2±10.5	H–O	429.91±0.29	I–Kr	5.67	K–Li	82.0±4.2
Fe–Fe	118	H–P	297.0±2.1	I–La	411.7	K–Na	65.994±0.008
Fe–Ge	210.9±29	H–Pb	≤157	I–Li	345.2±4.2	K–Zn	6.5
Fe–H	148±3	H–Pd	234±25	I–Lu	263.2	K–O	271.5±12.6
Fe–I	123	H–Pt	330	I–Mg	229	K–Rb	53.723±0.005
Fe–O	407.0±1.0	H–Rb	172.6	I–Mn	282.8±9.6	K–Xe	5.0
Fe–S	328.9±14.6	H–Rh	241.0±5.9	I–Mo	266.9	Kr–Kr	5.39
Fe–Si	297±25	H–Ru	223±15	I–N	159±17	Kr–Li	~12.1
Fm–O	443	H–S	353.57±0.30	I–Na	304.2±2.1	Kr–Mg	6.71±0.96

A-B	D^o_{298}/kJ mol^{-1}	A-B	D^o_{298}/kJ mol^{-1}	A-B	D^o_{298}/kJ mol^{-1}	A-B	D^o_{298}/kJ mol^{-1}
Kr–Na	~4.53	N–S	467±24	O–Rh	405±42	Pr–Te	326±20
Kr–Ne	4.31	N–Sb	460±84	O–Ru	528±42	Pt–Pt	306.7±1.9
Kr–O	<8	N–Sc	464±84	O–S	517.90±0.05	Pt–Si	501±18
Kr–Tl	4.14	N–Si	437.1±9.9	O–Sb	434±42	Pt–Th	551±42
Kr–Xe	5.66	N–Ta	607±84	O–Sc	671.4±1.0	Pt–Ti	397.5±10.6
Kr–Zn	5.0	N–Th	577±33	O–Se	429.7±6.3	Pt–Y	474±12
La–La	244.9	N–Ti	476±33	O–Si	799.6±13.4	Rb–Rb	48.898±0.005
La–N	519±42	N–U	531±21	O–Sm	573	Re–Re	432±30
La–O	798	N–V	523±38	O–Sn	528	Rh–Rh	235.85±0.05
La–Pt	505±12	N–Xe	26.9	O–Sr	426.3±6.3	Rh–Sc	444±11
La–Rh	550±12	N–Y	477±63	O–Ta	839	Rh–Si	395.0±18.0
La–S	573.4±1.7	N–Zr	565±25	O–Tb	694	Rh–Th	513±21
La–Se	485.7±14.6	Na–Na	74.805±0.586	O–Tc	548	Rh–Ti	390.8±14.6
La–Te	385.6±15	Na–Ne	~3.8	O–Te	377±21	Rh–U	519±17
La–Y	197±21	Na–O	270±4	O–Th	877	Rh–V	364±29
Li–Li	105.0	Na–Rb	63.887±0.024	O–Ti	668	Rh–Y	446±11
Li–Mg	67.4±6.3	Na–Xe	~5.12	O–Tl	213±84	Ru–Ru	193.0±19.3
Li–Na	87.181±0.001	Nb–Nb	513	O–Tm	514	Ru–Si	397.1±21
Li–O	340.5±6.3	Nb–Ni	271.9±0.1	O–U	755	Ru–Th	592±42
Li–Pb	78.7±8	Nb–O	726.5±10.6	O–V	637	Ru–V	414±29
Li–S	312.5±7.5	Nb–Ti	302.0±0.1	O–W	720±71	S–S	425.30
Li–Sb	169.0±10.0	Nb–V	369.3±0.1	O–Xe	36.4	S–Sb	378.7
Li–Si	149	Nd–Nd	82.8	O–Y	714.1±10.2	S–Sc	478.2±12.6
Li–Sm	193.3±18.8	Nd–O	703	O–Yb	387.7±10	S–Se	371.1±6.7
Li–Tm	276.1±14.6	Nd–S	471.5±14.6	O–Zn	≤250	S–Si	617±5
Li–Xe	~12.1	Nd–Se	393.9	O–Zr	766.1±10.6	S–Sm	389
Li–Yb	143.5±12.6	Nd–Te	305±15	Os–Os	415±77	S–Sn	467
Lr–O	665	Ne–Ne	4.070	P–P	489.1	S–Sr	338.5±16.7
Lu–Lu	142±33	Ne–Xe	4.31	P–Pt	≤416.7±16.7	S–Ta	669.5±13.5
Lu–O	669	Ne–Zn	3.92	P–Rh	353.1±16.7	S–Tb	515±42
Lu–Pt	402±34	Ni–Ni	204	P–S	442±10	S–Te	335±42
Lu–S	508.4±14.4	Ni–O	366±30	P–Sb	356.9±4.2	S–Ti	418±3
Lu–Se	418±15	Ni–Pd	140.9	P–Se	363.7±10.0	S–Tm	368±21
Lu–Te	325±15	Ni–Pt	273.7±0.3	P–Si	363.6	S–U	528.4±10.5
Md–O	418	Ni–S	356±21	P–Te	297.9±10.0	S–V	449.4±14.6
Mg–Mg	11.3	Ni–Si	318±17	P–Th	372±29	S–Y	528.4±10.5
Mg–Ne	~4.1	Ni–V	206.3±0.2	P–Tl	209±13	S–Yb	167
Mg–O	358.2±7.2	Ni–Y	283.92±0.10	P–U	293±21	S–Zn	224.8±12.6
Mg–S	234	Ni–Zr	279.8±0.1	P–W	305±4	S–Zr	572.0±11.6
Mg–Xe	9.70±1.79	No–O	268	Pb–Pb	86.6±0.8	Sb–Sb	301.7±6.3
Mn–Mn	61.6±9.6	Np–O	731	Pb–S	398	Sb–Te	277.4±3.8
Mn–O	362±25	O–O	498.36±0.17	Pb–Sb	161.5±10.5	Sb–Tl	126.7±10.5
Mn–S	301±17	O–Os	575	Pb–Se	302.9±4.2	Sc–Sc	163±21
Mn–Se	239.3±9.2	O–P	589	Pb–Te	249.8±10.5	Sc–Se	385±17
Mo–Mo	435.5±1.0	O–Pa	792	Pd–Pd	>136	Sc–Si	227.2±14
Mo–Nb	452±25	O–Pb	374	Pd–Pt	191.0	Sc–Te	289±17
Mo–O	502	O–Pd	238.1±12.6	Pd–Si	261±12	Se–Se	330.5
N–N	944.84±0.10	O–Pr	740	Pd–Y	241±15	Se–Si	538±13
N–O	631.62±0.18	O–Pt	391±42	Po–Po	187	Se–Sm	331.0±14.6
N–P	617.1±20.9	O–Pu	656.1	Pr–Pr	129.1	Se–Sn	401.2±5.9
N–Pt	374.2±9.6	O–Rb	276±12.6	Pr–S	492.5±4.6	Se–Sr	251.0±12.6
N–Pu	469±63	O–Re	627±84	Pr–Se	446.4±23.0	Se–Tb	423±20

A-B	D^o_{298}/kJ mol^{-1}	A-B	D^o_{298}/kJ mol^{-1}	A-B	D^o_{298}/kJ mol^{-1}	A-B	D^o_{298}/kJ mol^{-1}
Se–Te	293.3	Sm–Te	272.4±14.6	Te–Ti	289±17	Tm–Tm	54±17
Se–Ti	381±42	Sn–Sn	187.1±0.3	Te–Tm	182±40	U–U	222±21
Se–Tm	274±40	Sn–Te	338.1±6.3	Te–Y	339±13	V–V	269.3±0.1
Se–V	347±21	Sr–Sr	16.64±1.12	Te–Zn	117.6±18.0	V–Zr	260.6±0.3
Se–Y	435±13	T–T	446.67	Th–Th	≤289±33	W–W	666
Se–Zn	170.7±25.9	Ta–Ta	390±96	Ti–Ti	117.6	Xe–Xe	6.023
Si–Si	310	Tb–Tb	138.8	Ti–V	203.2±0.1	Y–Y	~270±39
Si–Te	448±8	Tb–Te	339±42	Ti–Zr	214.3±0.1	Yb–Yb	16.3
Si–Y	258±17	Tc–Tc	330	Tl–Tl	59.4	Zn–Zn	22.2±6.3
Sm–Sm	54±21	Te–Te	257.6±4.1	Tl–Xe	4.18	Zr–Zr	298.2±0.1

TABLE 2. Enthalpy of Formation of Gaseous Atoms

Atom	$\Delta_f H^o_{298}$/kJ mol^{-1}	Ref.	Atom	$\Delta_f H^o_{298}$/kJ mol^{-1}	Ref.	Atom	$\Delta_f H^o_{298}$/kJ mol^{-1}	Ref.
Ac	406	5	Gd	397.5±2.1	4	Pu	345	6
Ag	284.9±0.8	2	Ge	372±3	2	Ra	159	5
Al	330.9±4.0	2	H	217.998±0.006	2	Rb	80.9±0.8	2
Am	284	6	Hf	618.4±6.3	3	Re	774±6.3	1
As	302.5±13	1	Hg	61.38±0.04	2	Rh	556±4	1
Au	368.2±2.1	1	Ho	300.6±2.1	4	Ru	650.6±6.3	1
B	565±5	2	I	106.76±0.04	2	S	277.17±0.15	2
Ba	179.1±5.0	3	In	243±4	1	Sb	264.4 ±2.5	1
Be	324±5	2	Ir	669±4	1	Sc	377.8±4	1
Bi	209.6±2.1	1	K	89.0±0.8	2	Se	227.2±4	1
Bk	310	6	La	431.0±2.1	4	Si	450.0±8	2
Br	111.87±0.12	3	Li	159.3±1.0	2	Sm	206.7±2.1	4
C	716.68±0.45	2	Lu	427.6±2.1	4	Sn	301.2±1.5	2
Ca	177.8±0.8	2	Mg	147.1±0.8	2	Sr	164.0±1.7	3
Cd	111.80±0.20	2	Mn	283.3±4.2	3	Ta	782.0±2.5	1
Ce	420.1±2.1	4	Mo	658.98±3.8	3	Tb	388.7±2.1	4
Cf	196	6	N	472.68±0.40	2	Tc	678	5
Cl	121.301±0.008	2	Na	107.5±0.7	3	Te	196.6±2.1	1
Cm	386	6	Nb	733.0±8	3	Th	602±6	2
Co	426.7	3	Nd	326.9±2.1	4	Ti	473±3	2
Cr	397.48±4.2	3	Ni	430.1±8.4	3	Tl	182.2±0.4	1
Cs	76.5±1.0	2	Np	464.8	6	Tm	232.2±2.1	4
Cu	337.4±1.2	2	O	249.18±0.10	2	U	533±8	2
Dy	290.4±2.1	4	Os	787±6.3	1	V	515.5±8	3
Er	316.4±2.1	4	P	316.5±1.0	2	W	851.0±6.3	3
Es	133	6	Pa	563	5	Y	424.7±2.1	4
Eu	177.4±2.1	4	Pb	195.2±0.8	2	Yb	155.6±2.1	4
F	79.38±0.30	2	Pd	376.6±2.1	1	Zn	130.40±0.40	2
Fe	415.5±1.3	3	Pr	356.9±2.1	4	Zr	610.0±8.4	3
Ga	271.96±2.1	3	Pt	565.7±1.3	1			

References

1. Brewer, L. and Rosenblatt, G.M., *Adv. High Tem. Chem.*, 2, 1-83, 1969.
2. Cox, J.D., Wagman, D.D., and Medvedev, V.A., Eds., *CODATA Key Values for Thermodynamics*, Hemisphere Publishing Corporation, New York, 1989; updated e-version: http://www.codata.org/codata.
3. NIST Chemistry WebBook, http://webbook.nist.gov. *NIST-JANAF Thermochemical Table* 4th Edition, Ed. Chase, M.W.Jr., ACS, AIP, New York, 1998.
4. Chandrasekharaiah, M.S. and Gingerich, K.A., Thermodynamic properties of gaseous species, in *Handbook on the Chemistry and Physics of Rare Earths*; Gschneidner, K.A., Jr, Ering, L., Eds. Elsevier, Amsterdam, 1989, Vol. 12, Chapter 86, p. 409-431.
5. Lias, S.G. et al., Gas-phase ion and neutral thermochemistry, *J. Phys. Chem. Ref. Data*, 17, Supl. 1, 1988.
6. Klenschmidt, P.D., et al., *High Tem. Sci.*, 19, 267-274, 1985.

TABLE 3. Bond Dissociation Energies in Polyatomic Molecules

The D°_{298} values in polyatomic molecules are notoriously difficult to measure accurately since the mechanism of the kinetic systems involved in many of the measurements are seldom straightforward. Thus, much lively controversy has taken place in the literature and is likely to continue for some time to come. We will keep on updating and presenting our assessment of the most reliable BDE data every year.

The references relating to each of the D°_{298} values listed in Table 3 are contained in the *Comprehensive Handbook of Chemical Bond Energies*, by Yu-Ran Luo, Taylor & Francis, 2006. Many D°_{298} in Table 3 are derived from the equation

$$D^\circ_{298}(R\text{-}X) = \Delta_f H^\circ(R) + \Delta_f H^\circ(X) - \Delta_f H^\circ(RX)$$

Here, the enthalpies of formation of the atoms and radicals are taken from Tables 2 and 4, respectively, and the enthalpies of formation of the molecules are from reference sources listed in the above Comprehensive Handbook of Chemical bond Energie.

Table 3 presents H-C, C-C, C-halogen, O-, N-, S-, Si-, Ge-, Sn-, Pb-, P-, As-, Sb-, Bi-, Se-, Te-, and metal-X BDEs. The boldface in the species indicates the dissociated fragment. The metal-X BDEs are arranged on the basis of the Periodic Table with the new IUPAC notation for Groups 1 to 18, see inside front cover of this Handbook.

Bond	$D^\circ_{298}/$ kJ mol^{-1}	Bond	$D^\circ_{298}/$ kJ mol^{-1}	Bond	$D^\circ_{298}/$ kJ mol^{-1}
(1) C–H BDEs		$CH_2=CHCH_2CH_2CH_3$	348.8	$H–CH_2I$	431.6±2.8
CH_3–**H**	439.3 ±0.4	$CH_2=CHCH(CH_3)_2$	332.6±7.1	CF_3CF_2–**H**	429.7±2.1
CH_3CH_2–**H**	420.5±1.3	$CH_2=C(CH_3CH_2)CH_2$–**H**	356.1±8.4	CHF_2CF_2–**H**	431.0±18.8
$CH_3CH_2CH_2$–**H**	422.2±2.1	$(CH_2=CH)_2C(CH_3)$–**H**	322.2	CH_2FCF_2–**H**	433.0±14.6
$CH_3CH_2CH_3$	410.5±2.9	**H**-cyclo-C_3H_5	444.8±1.0	CHF_2CFH–**H**	426.8±14.6
$CH_3CH_2CH_2CH_2$–**H**	421.3	**H**-CH_2-cyclo-C_3H_5	407.5±6.7	CF_3CH_2–**H**	446.4±4.5
$CH_3CH_2CH_2CH_3$	411.1±2.2	**H**-cyclo-C_4H_7	409.2±1.3	CH_3CF_2–**H**	416.3±4.2
$(CH_3)_2CHCH_2$–**H**	419.2±4.2	**H**-cyclo-C_5H_9	400.0±4.2	CH_2FCHF–**H**	413.4±12.6
$(CH_3)_3C$–**H**	400.4±2.9	**H**-cyclo-C_6H_{11}	416.3	CHF_2CH_2–**H**	433.0±14.6
$(CH_3)_3CCH_2$–**H**	419.7±4.2	**H**-C_6H_5	472.2±2.2	CH_2FCH_2–**H**	433.5±8.4
$(CH_3CH_2)CH(CH_3)_2$	400.8	**H**-$CH_2C_6H_5$	375.5±5.0	CH_3CHF–**H**	410.9±8.4
$CH_3CH_2(CH_2)_2CH_3$	415.1	**H**-$CH(CH_3)C_6H_5$	357.3±6.3	CF_3CHCl–**H**	425.9±6.3
$(C_3H_7)CH(CH_3)_2$	396.2±8.4	**H**-$CH(C_6H_5)_2$	353.5±2.1	CF_3CClBr–**H**	404.2±6.3
$CH_3CH(CH_3)CH(CH_3)_2$	399.2±13.0	**H**-$CH(C_6H_4$-p-OH$)_2$	375.8±4.7	$CClF_2CHF$–**H**	412.1±2.1
$CH_3CH_2(CH_2)_3CH_3$	410	**H**-$C(CH_3)_2C_6H_5$	348.1±4.2	CCl_3CCl_2–**H**	397.5±8.4
$CH_3CH_2(CH_2)_4CH_3$	410	**H**-$C(C_6H_5)_3$	338.9±8.4	$CHCl_2CCl_2$–**H**	393.3±8.4
HCC–**H**	557.81±0.30	1-**H**-C_8H_7	469.4±5.4	CH_3CCl_2–**H**	397.9±5.0
HCCCC–**H**	539±12	2-**H**-C_8H_7	468.2±5.9	CH_3CHCl–**H**	406.6±1.5
$CHCCH_2$–**H**	372.0±4.2	**H**-CF_3	445.2±2.9	CH_2ClCH_2–**H**	423.1±2.4
CH_3CCCH_2–**H**	379.5	**H**-CHF_2	431.8±4.2	CH_3CBr_2–**H**	397.1±5.0
$HCCCH_2CH_3$	373.0	**H**-CH_2F	423.8±4.2	CH_2BrCH_2–**H**	415.1±8.4
$CH_2=CHCCCH_2$–**H**	363.3	**H**-$CClF_2$	421.3±8.4	CH_3CHBr–**H**	406.7±4.2
$CH_3CCCH_2CH_3$	365.3±9.6	**H**-CCl_2F	410.9±8.4	$CF_2=CF$–**H**	464.4±8.4
$HCCCH_2CH_2CH_3$	349.8±8.4	**H**-$CBrF_2$	415.5±12.6	$CF_3CF_2CF_2$–**H**	432.2
$HCCCH(CH_3)_2$	345.2±8.4	**H**-$CHClF$	421.7±10.0	CH_3CH_2CHCl–**H**	407.0±3.5
$CH_3CCCH(CH_3)_2$	344.3±11.3	**H**-CCl_3	392.5±2.5	$CH_2=CH$-CHF–**H**	370.7±4.6
HCCCCCC–**H**	~543±13	**H**-$CHCl_2$	407.1±4.2	$CH_2=CHCHCl$–**H**	370.7±4.6
$H_2C=CH$–**H**	464.2±2.5	**H**-CH_2Cl	419.0±2.3	$CH_2=CHCHBr$–**H**	374.0±4.6
$CH_2=C=CH$–**H**	371.1±12.6	**H**-$CFClBr$	413±21	**H**-C_6F_5	487.4
$CH_3CH=CH$–**H**	464.8	**H**-$CHClBr$	406.0±2.4	**H**-CH_2OH	401.92±0.63
$CH_2=CHCH_2$–**H**	369±3	**H**-CCl_2Br	387±21	CH_2CHOH	467±11
$CH_2=CH$-CH_2CH_2–**H**	410.5	**H**-$CClBr_2$	371±21	CH_3CH_2OH	401.2±4.2
$CH_2=CHCH_2CH_3$	350.6	**H**-CBr_3	399.2±8.4	CH_3CH_2OH	421.7±8
$CH_2=C(CH_3)CH_2$–**H**	372.8	**H**-$CHBr_2$	417.1±7.5	$CH_3CH_2CH_2OH$	392
$CH_2=CHCH=CHCH_2$–**H**	347.3±12.6	**H**-CH_2Br	427.2±2.4	$CH_3CH_2CH_2OH$	394.6±8.4
$(CH_2=CH)_2CH$–**H**	320.5±4.2	**H**-CI_3	423±29	$CH_3CH_2CH_2OH$	406.3±8.4
		H-CHI_2	431.0±8.4		

Bond	$D^{\circ}_{298}/$ kJ mol^{-1}	Bond	$D^{\circ}_{298}/$ kJ mol^{-1}	Bond	$D^{\circ}_{298}/$ kJ mol^{-1}
(CH$_3$)$_2$CHOH	383.7±8.4	CH$_3$C(O)OH	398.7±12.1	CH$_2$(CN)$_2$	366.5
(CH$_3$)$_2$CHOH	394.6±8.4	ClCH$_2$C(O)OH	398.9	CH$_2$(CN)(NH$_2$)	355.2
CH$_2$=CHCH$_2$OH	341.4±7.5	H–C(O)OCH$_3$	399.2±8.4	(CH$_3$)$_2$CHCN	384.5
(CH$_3$)$_3$COH	418.4±8.4	CH$_3$C(O)OCH$_3$	406.3±10.5	CH$_3$NC	389.1±12.6
(CH$_2$=CH)$_2$CHOH	288.7	CH$_3$C(O)OCH$_3$	404.6	H–HCNN	405.8±8.4
Ph$_2$CHOH	326	CH$_3$C(O)OCH$_2$CH$_3$	401.7	H–CNN	331±17
CH$_3$CH(OH)$_2$	~ 385	CH$_3$C(O)OPh	419.2±5.4	CH$_3$NO$_2$	415.4
(CH$_2$OH)$_2$	385.3	CH$_3$CH$_2$C(O)OEt	400	CH$_3$CH$_2$NO$_2$	410.5
HOCH$_2$(CH$_2$)$_2$(OH)CH–H	399.2	PhCH$_2$C(O)OEt	370.7	C$_2$H$_5$CH$_2$NO$_2$	410.5
CH$_3$OCH$_3$	402.1	Me$_2$CHC(O)OEt	387.4	Me$_2$CHNO$_2$	394.9
CHF$_2$OCF$_3$	443.5±4.2	PhCHMe(C(O)OEt)	358.2	C$_6$H$_5$C(NO$_2$)CHCH$_3$	357.3
CHF$_2$OCHF$_2$	435.1±4.2	H-furaylmethyl	361.9±8.4	H–C(S)H	399.6±5.0
CH$_3$OCF$_3$	426.8±4.2	CH$_3$NH$_2$	392.9±8.4	CH$_3$SH	392.9±8.4
CH$_3$OCH$_2$CH$_3$	389.1	CH$_3$N=CH$_2$	407.9±14.6	CH$_3$SCH$_3$	392.0±5.9
(CH$_3$)$_3$COC(CH$_3$)$_3$	402.1	CH$_3$CH$_2$NH$_2$	377.0±8.4	PhSCH$_3$	389.1
CH$_3$CH$_2$OCH$_2$CH$_3$	389.1	C$_2$H$_5$CH$_2$NH$_2$	380.7±8.4	PhCH$_2$SPh	352.3
CH$_3$CH$_2$Ot-C(CH$_3$)$_3$	405.4	C$_3$H$_7$CH$_2$NH$_2$	393.3±8.4	(PhS)$_2$CHPh	341.0
CH$_3$OPh	385.0	C$_4$H$_9$CH$_2$NH$_2$	387.7±8.4	PhSCHPh$_2$	344.8
H-2-oxiran-2-yl	420.5±6.5	HOCH$_2$CH$_2$NH$_2$	379.5±8.4	CH$_3$SOCH$_3$	393.3
H-tetrahydrofuran-2-yl	385.3±6.7	(CH$_3$CH$_2$)$_2$NH	370.7±8.4	CH$_3$SO$_2$CH$_3$	414.2
HC(O)–H	368.40±0.67	(C$_3$H$_7$CH$_2$)$_2$NH	379.9±8.4	CH$_3$SO$_2$CF$_3$	431.0
FC(O)–H	423.0	(C$_4$H$_9$CH$_2$)$_2$NH	384.5±8.4	CH$_3$SO$_2$Ph	414.2
CH$_3$C(O)–H	374.0±1.3	(C$_2$H$_5$)$_2$NCH$_2$CH$_3$	379.5±1.7	PhCH$_2$SO$_2$Me	380.7
CF$_3$C(O)–H	390.4	(C$_2$H$_5$CH$_2$)$_3$N	376.6±8.4	PhCH$_2$SO$_2$CF$_3$	372.4
C$_2$H$_5$C(O)–H	374.5	((CH$_3$)$_2$CCH$_2$)$_3$N	388.3±8.4	PhCH$_2$SO$_2$tBu	376.6
CH$_2$=CHC(O)–H	372.8	(Bu)$_2$NCH$_2$(nPr)	381±10.0	Ph$_2$CHSO$_2$Ph	365.3
C$_3$H$_7$C(O)–H	371.2	((CH$_3$)$_2$CH)$_3$N	387.0±8.4	CH$_2$(SPh)$_2$	372.4
iso-C$_3$H$_7$C(O)–H	364.5	(CH$_3$)$_2$CHNH$_2$	372.0±8.4	H–CH$_2$SiMe$_3$	418±6.3
C$_4$H$_9$C(O)–H	372.0	CH$_3$NHCH$_3$	364.0±8.4	H–CH$_2$C(CH$_3$)$_2$SiMe$_3$	409±5
(CH$_3$)$_2$CHCH$_2$C(O)–H	362.5	(CH$_3$)$_3$N	380.7±8.4	H–CH$_2$SiMe$_2$Ph	410.1
C$_2$H$_5$CH(CH$_3$)C(O)–H	360.8	tert-BuN(CH$_3$)$_2$	376.6±8.4	H–CH((CH$_3$)$_3$Si)$_2$	397±13
tert-BuC(O)–H	375.1	((HOCH$_2$CH$_2$)$_2$(CH$_3$))N	364.4±8.4	H–CH$_2$B(RO)$_2$	412.5
Et$_2$CHC(O)–H	367.2	(HOCH$_2$CH$_2$)$_3$N	379.9±8.4	H–CH((CH$_3$)$_2$P)$_2$	385±13
CH$_3$(CH$_2$)$_8$C(O)–H	373.3	((HOCH$_2$)CH(CH$_3$))$_3$N	379.9±8.4		
C$_6$H$_5$C(O)–H	371.1±10.9	PhCH$_2$NH$_2$	368.2	**(2) C–C BDEs**	
PhCH$_2$C(O)–H	362.0	PhN(CH$_2$CH$_3$)$_2$	383.3±4.2	CH$_3$–CH$_3$	377.4±0.8
PhC(CH$_3$)$_2$C(O)–H	362.9	Ph$_2$NCH$_3$	379.5±1.7	CH$_3$–C$_2$H$_5$	370.3±2.1
H–CH=C=O	448.1	PhN(CH$_2$Ph)$_2$	357.3±8.8	CH$_3$–C$_3$H$_7$	372.0±2.9
CH$_3$C(O)H	394.5±9.2	N(CH$_2$Ph)$_3$	372.8±2.5	CH$_3$–iso-C$_3$H$_7$	369.0±3.8
CH$_3$C(O)Cl	≤423.4	PhN(CH$_2$CH=CH$_2$)$_2$	339.3±2.9	CH$_3$–C$_4$H$_9$	371.5±2.9
CH$_3$CH$_2$C(O)H	383.7	N(CH$_2$CH=CH$_2$)$_3$	345.6±3.3	CH$_3$–iso-C$_4$H$_9$	370.3±4.6
CH$_3$COCH$_3$	401.2±2.9	H$_2$NNH(CH$_3$)	410	CH$_3$–sec-C$_4$H$_9$	368.2±2.9
CF$_3$C(O)CH$_3$	465.6	HNN(CH$_3$)$_2$	410	CH$_3$–tert-C$_4$H$_9$	363.6±2.9
CH$_3$COCH$_2$CH$_3$	403.8	(CH$_3$)$_2$NC$_6$H$_5$	383.7±5.4	CH$_3$–C$_5$H$_{11}$	368.4±6.3
MeCOCH$_2$Me	386.2±7.1	H–CN	528.5±0.8	CH$_3$–CH(C$_2$H$_5$)$_2$	365.7±4.2
EtCOCH$_2$Me	396.5±2.8	CH$_3$CN	405.8±4.2	CH$_3$–C(CH$_3$)$_2$(CH$_2$CH$_3$)	360.9±6.3
CH$_3$CH$_2$COC$_6$H$_5$	402.8±3.6	CH$_3$CH$_2$CN	393.3±12.6	CH$_3$–C$_6$H$_{13}$	368.2±6.3
MeCH$_2$COPh	388.7	PhCH$_2$CN	344.3	C$_2$H$_5$–C$_2$H$_5$	363.2±2.5
H–C(O)OH	404.2	C$_6$F$_5$CH$_2$CN	350.6	C$_3$H$_7$–C$_3$H$_7$	366.1±3.3

Bond	$D^\circ_{298}/$ kJ mol^{-1}	Bond	$D^\circ_{298}/$ kJ mol^{-1}	Bond	$D^\circ_{298}/$ kJ mol^{-1}
iso-C$_3$H$_7$–*iso*-C$_3$H$_7$	353.5±4.6	CHF$_2$–CHF$_2$	382.4±15.5	CF$_2$CF–CFCF$_2$	558.1±12.6
C$_4$H$_9$–C$_4$H$_9$	364.0±3.8	CClF$_2$–CClF$_2$	378.7±12.6	CH$_2$FCH$_2$–CPh$_3$	274.9±16.7
iso-C$_4$H$_9$–*iso*-C$_4$H$_9$	362.3±6.3	CF$_2$Cl–CFCl$_2$	358.6±12.6	CHF$_2$CH$_2$–CPh$_3$	264.0±16.7
sec-C$_4$H$_9$–*sec*-C$_4$H$_9$	348.5±3.3	CHF$_2$–CH$_2$F	394.1±16.7	CH$_3$–CH$_2$OH	364.8±4.2
tert-C$_4$H$_9$–*tert*-C$_4$H$_9$	322.6±4.2	CH$_2$F–CH$_2$F	368.2±8.4	CF$_3$–CH$_2$OH	405.4±6.3
CH$_3$–*cyclo*-C$_5$H$_9$	358.2±5.0	CHF$_2$–CH$_3$	405.0±8.4	C$_2$H$_5$–CH$_2$OH	356.9±5.0
CH$_3$–*cyclo*-C$_6$H$_{11}$	377.0±7.5	CH$_2$F–CH$_3$	388.3±8.4	C$_3$H$_7$–CH$_2$OH	357.3±3.3
cyclo-C$_6$H$_{11}$–*cyclo*-C$_6$H$_{11}$	369.0±8.4	CHClF–CH$_3$	399.6±12.6	*iso*-C$_3$H$_7$–CH$_2$OH	354.8±4.2
CH$_3$–CH$_2$C≡CH	320.5±5.0	CF$_2$Br–CHClF	369.4	C$_4$H$_9$–CH$_2$OH	355.6±4.2
CH$_3$–CH$_2$C≡CCH$_3$	308.4±6.3	CF$_2$Br–CH$_3$	396.6±15.1	*sec*-C$_4$H$_9$–CH$_2$OH	352.7±4.2
CH$_3$–CH(CH$_3$)C≡CH	305.4±8.4	CCl$_3$–CCl$_3$	285.8±6.3	*iso*-C$_4$H$_9$–CH$_2$OH	354.0±5.4
CH$_3$–CH(CH$_3$)C≡CCH$_3$	320.9±6.3	CCl$_3$–CClF$_2$	282.0±12.6	C$_6$H$_5$–CH$_2$OH	413.4±5.4
CH$_3$–C(CH$_3$)$_2$C≡CH	295.8±6.3	CCl$_3$–CHCl$_2$	306.3±8.4	HOH$_2$C–CH$_2$OH	358.2±6.3
CH$_3$–C(CH$_3$)$_2$C≡CCH$_3$	303.3±6.3	CCl$_3$–CH$_2$Cl	323.8±8.4	NH$_2$CH$_2$–CH$_2$OH	335.6±10.5
CH$_3$–CHCH$_2$	426.3±6.3	CCl$_3$–CH$_3$	362.3±6.3	CH$_3$–CH$_2$OCH$_3$	363.2±5.0
CH$_3$–CH=CCH$_2$	359.8±5.9	CHCl$_2$–CHCl$_2$	342.3±8.4	CH$_3$OCH$_2$–CH$_2$OCH$_3$	338.9±10.5
CH$_3$–cyclopro-en-1-yl	340.6±20.9	CHCl$_2$–CH$_2$Cl	358.6±8.4	CH$_3$–C(O)H	354.8±1.7
CH$_3$–CH$_2$CH=CH$_2$	317.6±3.8	CHCl$_2$–CH$_3$	365.1±3.3	CCl$_3$–C(O)H	309.2±5.0
CH$_3$–CH$_2$C(CH$_3$)=CH$_2$	310.0±4.2	CHBrCl–CH$_3$	384.5	CH$_3$–C(O)F	417.6±6.3
CH$_3$–CH(CH$_3$)CH=CH$_2$	302.5±6.3	CHClBr–CHClBr	323.0±12.6	CH$_3$–C(O)Cl	367.8±6.3
CH$_3$–C(CH$_3$)$_2$CH=CH$_2$	282.4±6.3	CH$_2$Cl–CH$_2$Cl	360.7±8.4	CCl$_3$–C(O)Cl	289.1±6.3
CH$_3$–*cyclo*-C$_5$H$_7$	299.2±8.4	CH$_2$Cl–CH$_3$	375.7±9.2	CHCl$_2$–C(O)Cl	312.5±8.4
CH$_3$–C$_6$H$_5$	426.8±4.2	Br$_3$C–CH$_3$	356.9±12.6	CClH$_2$–C(O)Cl	340.2±8.4
HCC–C$_6$H$_5$	590.8±5.9	Br$_3$C–CBr$_3$	278.7±16.7	C$_6$H$_5$–C(O)H	408.4±4.2
C$_2$H$_3$–C$_6$H$_5$	482.0±5.4	CHBr$_2$–CH$_3$	372.8	C$_6$H$_5$–C(O)Cl	417.6±6.3
CH$_3$–CH$_2$C$_6$H$_5$	325.1±4.2	CH$_2$Br–CH$_2$Cl	378.2	CH$_3$–C(O)CH$_3$	351.9±2.1
CH$_3$–CH(CH$_3$)C$_6$H$_5$	318.8±8.4	CH$_2$Br–CH$_2$Br	379.9±8.4	C$_2$H$_5$–C(O)CH$_3$	347.3±2.9
CH$_3$–C(CH$_3$)$_2$C$_6$H$_5$	303.3±8.4	CH$_2$I–CH$_2$I	387.0±10.5	C$_3$H$_7$–C(O)CH$_3$	348.5±2.9
CH$_3$–CH$_2$CHCHPh	295.4	CH$_3$–CH$_2$Br	381.6±8.4	*iso*-C$_3$H$_7$–C(O)CH$_3$	340.2±3.8
CH$_3$–CH(C$_6$H$_5$)$_2$	315.9±6.3	CH$_3$–CH$_2$I	384.5±8.4	C$_4$H$_7$–C(O)CH$_3$	346.9±5.4
CH$_3$–C(CH$_3$)(C$_6$H$_5$)$_2$	290.8±8.4	CF$_3$–CF$_2$CF$_3$	424.3±13.6	*tert*-C$_4$H$_9$–C(O)CH$_3$	329.3±4.2
C$_6$H$_5$–C$_6$H$_5$	478.6±6.3	CF$_3$–CF=CF$_2$	420.5	C$_6$H$_5$–C(O)CH$_3$	406.7±4.6
C$_6$H$_5$–CH$_2$C$_6$H$_5$	383.7±8.4	CH$_3$–CH$_2$CH$_2$Cl	371.4±2.8	C$_6$H$_5$CH$_2$–C(O)CH$_3$	299.7±8.4
C$_6$H$_5$CH$_2$–CH$_2$C$_6$H$_5$	272.8±9.2	CH$_3$–CHClCH$_3$	367.5±2.0	HC(O)–C(O)H	295.8±6.3
C$_6$H$_5$–CH(C$_6$H$_5$)$_2$	361.1±8.4	CH$_2$Cl–CHClCH$_3$	356.5±8.4	ClC(O)–C(O)Cl	292.5±8.4
C$_6$H$_5$–C(C$_6$H$_5$)$_3$	324.3±12.6	CH$_2$Cl–CH$_2$CClH$_2$	369.0±8.4	CH$_3$C(O)–C(O)H	302.5±8.4
Ph$_2$CH–CHPh$_2$	247.3±8.4	CH$_3$–CCl$_2$CH$_3$	362.8±8.4	CH$_3$C(O)–C(O)CH$_3$	307.1±4.2
PhCH$_2$–CPh$_3$	234.7±14.6	CH$_2$Br–CHBrCH$_3$	369.4±8.4	C$_6$H$_5$C(O)–C(O)C$_6$H$_5$	288.3±16.7
R-R, R = phenalenyl	42	CH$_2$ClCH$_2$–CHClCH$_3$	364.4±8.4	CH$_3$–C(O)OH	384.9±8.4
R-R, R = 9-phenylfluorenyl	63.6	CH$_2$ClCH$_2$–CH$_2$CClH$_2$	369.0±8.4	CF$_3$–C(O)OH	370.7±8.4
CF$_3$–CF$_3$	413.0±5.0	CH$_3$CHBr–CHBrCH$_3$	355.6±8.4	CCl$_3$–C(O)OH	310.5±12.6
CF$_3$–CHF$_2$	399.6±8.4	CF$_3$–C$_6$H$_5$	463.2±12.6	CClH$_2$–C(O)OH	357.7±8.4
CF$_3$–CClF$_2$	373.6±12.5	CCl$_3$–C$_6$H$_5$	388.7±8.4	CH$_2$Br–C(O)OH	358.2±8.4
CF$_3$–CH$_2$F	397.5±8.4	CH$_3$–C$_6$F$_5$	439.3	NH$_2$CH$_2$–C(O)OH	349.4±8.4
CF$_3$–CCl$_3$	332.2±5.4	CF$_3$–C$_6$F$_5$	435.1	CH$_3$NHCH$_2$–C(O)OH	300.4±8.4
CF$_3$–CHBrCl	377.0±10.5	CF$_3$–CH$_2$C$_6$H$_5$	365.7±12.6	C$_6$H$_5$–C(O)OH	429.7±8.4
CF$_3$–CH$_2$Br	399.6±8.4	C$_6$F$_5$–C$_6$F$_5$	488.3	C$_6$F$_5$–C(O)OH	470.0±10.5
CF$_3$–CH$_2$I	408.4±10.5	CF$_3$–CHPh$_2$	352.3±16.7	HOCH$_2$–C(O)OH	371.5±5.4
CF$_3$–CH$_3$	429.3±5.0	CF$_3$–CPh$_3$	290.8±16.7	HOC(O)–C(O)OH	334.7±6.3

Bond	$D^\circ_{298}/$ kJ mol^{-1}	Bond	$D^\circ_{298}/$ kJ mol^{-1}	Bond	$D^\circ_{298}/$ kJ mol^{-1}
CH$_3$NHCH$_2$–C(O)OH	301.2±16.7	Cl–CHF$_2$	364±8	Cl–CH$_2$C(O)OH	310.9±2.2
CH$_3$CH(NH$_2$)–C(O)OH	331.4±16.7	Cl–CH$_2$F	354.4±11.7	Cl–C(O)OC$_6$H$_5$	364
NH$_2$CH$_2$–CH$_2$C(O)OH	325.5±16.7	Cl–CF$_2$Cl	333.9±10.5	Cl–C(NO$_2$)$_3$	302.1
CN–CN	571.9±6.7	Cl–CFCl$_2$	320.9±8.4	Br–CN	364.8±4.2
HC(O)–CN	455.2±8.4	Cl–CHFCl	346.0±13.4	Br–CF$_3$	296.2±1.3
HC(S)–CN	530.1±8.4	Cl–CCl$_3$	296.6	Br–CHF$_2$	288.7±8.4
CF$_3$–CN	469.0±4.2	Cl–CHCl$_2$	320.5±6.3	Br–CF$_2$Cl	269.9±6.3
CH$_3$–CN	521.7±9.2	Cl–CH$_2$Cl	338.0±3.3	Br–CCl$_3$	231.4±4.2
NCC–CN	462.3	Cl–CBrCl$_2$	287±10.5	Br–CH$_2$Cl	277.3±3.6
C$_2$H$_5$–CN	506.7±7.5	Cl–CH$_2$Br	332.8±4.6	Br–CBr$_3$	242.3±8.4
CH$_3$–CH$_2$CN	348.1±12.6	Cl–CH$_2$I	328.2±6.9	Br–CHBr$_2$	274.9±13.0
C$_6$H$_5$–CH$_2$CN	386.6±8.4	Cl–CH$_3$	350.2±1.7	Br–CH$_2$Br	276.1±5.3
CH$_3$–CH(CH$_3$)CN	332.6±8.4	Cl–C≡CCl	443±50	Br–CH$_2$I	274.5±7.5
CH$_3$–C(CH$_3$)$_2$CN	340.6±16.7	Cl–C≡CH	435.6±8.4	Br–CH$_3$	294.1±2.1
CH$_3$–C(CH$_3$)(CN)C$_6$H$_5$	250.6	Cl–CH$_2$CN	267.4	Br–C≡CH	410.5
(Ph)$_2$(CN)C–C(CN)(Ph)$_2$	109.6	Cl–CCl=CCl$_2$	383.7	Br–CH=CH$_2$	338.3±3.1
(NO$_2$)$_3$C–C(NO$_2$)$_3$	308.8	Cl–CH=CH$_2$	396.5±4.8	Br–CF$_2$CF$_3$	283.3±6.3
C$_{58}$–C$_2$	955.2±14.5	Cl–CF=CF$_2$	434.7±8.4	Br–CClBrCF$_3$	251.0±6.3
		Cl–CF$_2$CF$_3$	346.0±7.1	Br–CF$_2$CF$_2$Br	282.8±6.7
(3) C–halogen BDEs		Cl–CF$_2$CF$_2$Cl	331.4±20.9	Br–CHClCF$_3$	274.9±6.3
F–CN	482.8	Cl–CCl$_2$CF$_3$	307.9	Br–CF$_2$CH$_3$	287.0±5.4
F–CF$_3$	546.8±2.1	Cl–CCl$_2$CCl$_3$	303.8	Br–CH$_2$CH$_2$Cl	292.5±8.4
F–CHF$_2$	533.9±5.9	Cl–CHClCCl$_3$	330.5±4.2	Br–CHClCH$_3$	272.0±8.4
F–CH$_2$F	496.2±8.8	Cl–CCl$_2$CHCl$_2$	311.7	Br–C$_2$H$_5$	292.9±4.2
F–CF$_2$Cl	511.7	Cl–CHClCH$_3$	327.9±1.8	Br–CH$_2$CH=CH$_2$	237.2±5.0
F–CFCl$_2$	482.0±10.5	Cl–CH$_2$CH$_2$Cl	345.1±5.0	Br–C$_3$H$_7$	298.3±4.2
F–CHFCl	462.3±10.0	Cl–CHBrCH$_3$	331.8±8.4	Br–iso-C$_3$H$_7$	299.2±6.3
F–CCl$_3$	439.3±4	Cl–CH$_2$CH$_3$	352.3±3.3	Br–CH$_2$CH$_2$CH$_2$Br	324.7
F–CH$_2$Cl	465.3±9.6	Cl–CH$_2$CH=CH$_2$	298.3±5.0	Br–CF$_2$CF$_2$CF$_3$	278.2±10.5
F–CH$_3$	460.2±8.4	Cl–C$_3$H$_7$	352.7±4.2	CF$_3$CFBrCF$_3$	274.2±4.6
F–C≡CH	521.3	Cl–CH$_2$CH$_2$CH$_2$Cl	348.9	Br–C$_4$H$_9$	296.6±4.2
F–C≡CF	519±21	Cl–iso-C$_3$H$_7$	354.0±6.3	Br–sec-C$_4$H$_9$	300.0±4.2
F–CF=CF$_2$	546.4±12.6	Cl–CH$_2$CHCH=CH$_2$	342.7	Br–tert-C$_4$H$_9$	292.9±6.3
F–CF$_2$CF$_3$	532.2±6.3	Cl–C$_4$H$_9$	350.6±6.3	Br–C$_6$H$_5$	336.4±6.3
F–CH$_2$CF$_3$	457.7	Cl–sec-C$_4$H$_9$	350.2±6.3	Br–C$_6$F$_5$	~328
F–CF$_2$CH$_3$	522.2±8.4	Cl–tert-C$_4$H$_9$	351.9±6.3	Br–CH$_2$C$_6$H$_5$	239.3±6.3
F–C$_2$H$_3$	517.6±12.6	CH$_2$CHCHCl(CH$_3$)	300.0±6.3	Br–CH$_2$C$_6$F$_5$	225.1±6.3
F–C$_2$H$_5$	467.4±8.4	Cl–C$_5$H$_{11}$	350.6±6.3	Br–1–C$_{10}$H$_7$	339.7
F–C$_3$H$_7$	474.9±8.4	Cl–C(CH$_3$)$_2$(C$_2$H$_5$)	352.7±6.3	Br–2–C$_{10}$H$_7$	341.8
F–iso-C$_3$H$_7$	483.8±8.4	Cl–cyclo-C$_6$H$_{11}$	360.2±6.5	Br–anthracenyl	322.6
F–tert-C$_4$H$_9$	495.8±8.4	Cl–C$_6$H$_5$	399.6±6.3	Br–C(O)CH$_3$	292.0±8.4
F–C$_6$H$_5$	525.5±8.4	Cl–C$_6$F$_5$	383.3±8.4	Br–C(O)C$_6$H$_5$	276.6±8.4
F–C$_6$F$_5$	485±25	Cl–CH$_2$C$_6$H$_5$	299.9±4.3	Br–CH$_2$C(O)CH$_3$	257.9±10.5
F–CH$_2$C$_6$H$_5$	412.8±4.2	Cl–C(O)Cl	318.8±8.4	Br–CH$_2$C(O)C$_6$H$_5$	271
F–COH	497.9±10.5	Cl–COF	376.6	Br–CH$_2$C(O)OH	257.4±3.7
F–COF	510.3	Cl–C(O)CH$_3$	354.0±8.4	Br–C(NO$_2$)$_3$	218.4
F–COCl	484.5	Cl–C(O)CH$_2$CH$_3$	353.3±6.3	I–CN	320.1
F–C(O)CH$_3$	511.7±12.6	Cl–C(O)C$_6$H$_5$	341.0±8.4	I–CF$_3$	227.2±1.3
Cl–CN	422.6±8.4	Cl–CH$_2$C(O)C$_6$H$_5$	309	I–CCl$_3$	168±42
Cl–CF$_3$	365.3±3.8				

Bond	D°_{298}/kJ mol^{-1}	Bond	D°_{298}/kJ mol^{-1}	Bond	D°_{298}/kJ mol^{-1}
I–CH$_2$Cl	221.8±4.2	C$_6$H$_5$CH$_2$O–H	442.7±8.8	HO–OCl	~146
I–CH$_2$Br	219.2±5.4	CH$_3$C(OH)O–H	446.9±6.3	HO–OBr	138.5±8.4
I–CH$_2$I	216.9±7.9	(CH$_3$)$_2$C(OH)O–H	450.6±6.3	FO–OF	199.6
I–CH$_3$	238.9±2.1	HC(O)O–H	468.6±12.6	ClO–OCl	72.4±2.8
I–CH$_2$CN	187.0±6.3	CH$_3$C(O)O–H	468.6±12.6	IO–OI	74.9±17
I–CF$_2$CF$_3$	219.2±2.1	C$_2$H$_5$C(O)O–H	472.8	trans-perp-HO–ONO	≤67.8±0.4
I–CF$_2$CF$_2$I	217.6±6.7	iso-C$_3$H$_7$C(O)O–H	472.8	cis-cis-HO–ONO	83.3±2.1
I–CH$_2$CF$_3$	235.6±4.2	C$_6$H$_5$C(O)O–H	464.4±16.7	HO–ONO$_2$	163.2±8.4
I–CHFCClF$_2$	202±2	HOO–H	367.4±2.1	HO–OCH$_3$	189.1±4.2
I–CF$_2$CH$_3$	217.6±4.2	CH$_3$OO–H	370.3±2.1	HO–OCF$_3$	201.3±20.9
I–CFICH3	218.0±4.2	CF$_3$OO–H	383	HO–OC$_2$H$_5$	178.7±6.3
CF$_3$CFICF$_3$	215.1	CH$_2$FOO–H	379	HO–O-iso-C$_3$H$_7$	185.8±6.3
I–CH=CH$_2$	259.0±4.2	CCl$_3$OO–H	386	HO–O-tert-C$_4$H$_9$	186.2±4.2
I–C$_2$H$_5$	233.5±6.3	CHCl$_2$OO–H	383	HO–OC(O)CH$_3$	169.9±2.1
I–CH$_2$CH=CH$_2$	185.8±6.3	CH$_2$ClOO–H	379	HO–OC(O)C$_2$H$_5$	169.9±2.1
I–C$_3$H$_7$	236.8±4.2	CBr$_3$OO–H	383	CH$_3$O–OCH$_3$	167.4±6.3
I–iso-C$_3$H$_7$	234.7±6.3	CH$_2$BrOO–H	379	CF$_3$O–OCF$_3$	198.7±2.1
I–C$_4$F$_9$	205.8	C$_2$H$_5$OO–H	354.8±9.2	C$_2$H$_5$O–OC$_2$H$_5$	166.1
I–tert-C$_4$H$_9$	227.2±6.3	CH$_3$CHClOO–H	377	C$_3$H$_7$O–OC$_3$H$_7$	155.2±4.2
I–C$_6$H$_5$	272.0±4.2	CH$_3$CCl$_2$OO–H	383	iso-C$_3$H$_7$O–O-iso-C$_3$H$_7$	157.7
I–C$_6$F$_5$	<301.7	CF$_3$CHClOO–H	384	sec-C$_4$H$_9$O–O-sec-C$_4$H$_9$	152.3±4.2
I–CH$_2$C$_6$H$_5$	187.8±4.8	C$_2$Cl$_5$OO–H	383	tert-BuO–O-tert-Bu	162.8±2.1
I–1-naphthyl	274.5±10.5	iso-C$_3$H$_7$OO–H	356	tert-BuCH$_2$O–OCH$_2$-tert-Bu	152.3
I–2-naphthyl	272.0±10.5	CH$_2$=CHCH$_2$OO–H	372.4	EtC(Me)$_2$O–OC(Me)$_2$Et	164.4±4.2
I–CH$_2$CN	187.0±8.4	tert-C$_4$H$_9$OO–H	352.3±8.8	(CF$_3$)$_3$CO–OC(CF$_3$)$_3$	148.5±4.6
I–CH$_2$OCH$_3$	229.4±8.4	C$_6$H$_5$OO–H	384	Ph$_3$CO–OCPh$_3$	131.4
I–CH$_2$SCH$_3$	216.8±6.3	C$_6$H$_5$CH$_2$OO–H	363	SF$_5$O–OSF$_5$	155.6
I–C(O)CH$_3$	223.0±8.4	(C$_6$H$_5$)$_2$CHOO–H	370	SF$_5$O–OOSF$_5$	126.8
I–C(O)C$_6$H$_5$	212.1±8.4	CH$_3$C(O)OO–H	386	(CH$_3$)$_3$CO–OSi(CH$_3$)$_3$	196.6
I–CH$_2$C(O)OH	197.5±2.7	CCl$_2$(CN)OO–H	384	tert-BuO–OGeEt$_3$	192.5
I–C(NO$_2$)$_3$	144.8	OHCH$_2$OO–H	368	tert-BuO–OSnEt$_3$	192.5
(4) O–X BDEs		H–ONO	330.7	CF$_3$OO–OCF$_3$	126.8±8.4
HO–H	497.10±0.29	H–OONO	299.2	HC(O)O–OH	199.2±8.4
FO–H	425.1	H–ONO$_2$	426.8	FC(O)O–OC(O)F	96.2
ClO–H	393.7	H–ONNOH	189	CH$_3$C(O)O–ONO$_2$	131.4±8.4
BrO–H	405	H–OPO$_2$	465.7±12.6	CH$_3$C(O)O–OC(O)CH$_3$	140.2±21
IO–H	403.3	H–OSO$_2$OH	441.4±14.6	CF$_3$C(O)O–OC(O)CF$_3$	125.5
CH$_3$O–H	440.2±3	H–OSiMe3	495	CF$_3$OC(O)O–OC(O)F	121.3±4.2
CF$_3$O–H	497.1	(CH$_3$)CHNO-H	354.4	CF$_3$OC(O)O–OCF$_3$	142.3±2.9
HC≡CO–H	443.1	(CH$_3$)$_2$CNO-H	354.0	CF$_3$OC(O)O–OC(O)OCF$_3$	119.2
C$_2$H$_5$O–H	441.0±5.9	(C$_6$H$_5$)CHNO-H	368.6	C$_2$H$_5$C(O)O–OC(O)C$_2$H$_5$	150.6
CH$_2$=CHO–H	355.6	PhO–H	362.8±2.9	C$_3$H$_7$C(O)O–OC(O)C$_3$H$_7$	150.6
CF$_3$CH$_2$O–H	447.7±10.5	α-tocopherol RO-H	323.4	FS(O)$_2$O–OS(O)$_2$F	92–100
C$_3$H$_7$O–H	≤433±2	β-tocopherol RO-H	335.6	HO–CF$_3$	≤482.0±1.3
iso-C$_3$H$_7$O–H	442.3±2.8	γ-tocopherol RO-H	335.1	FO–CF$_3$	408±17
C$_4$H$_9$O–H	432.3	δ-tocopherol RO-H	342.8	HO–CH$_3$	384.93±0.71
sec-C$_4$H$_9$O–H	441.4±4.2	p-C$_6$H$_5$CH$_2$-C$_6$H$_4$O-H	356.2	HO–C$_2$H$_5$	391.2±2.9
tert-C$_4$H$_9$O–H	444.9±2.8	O–O$_2$	106.6	HO–CH$_2$CF$_3$	408.4±8.4
tert-BuCH$_2$O–H	436.1	HO–OH	210.66±0.42	HO–CH$_2$CH=CH$_2$	332.6±4.2
		HO–OF	199.7±8.4		

Bond	D°_{298}/ kJ mol^{-1}	Bond	D°_{298}/ kJ mol^{-1}	Bond	D°_{298}/ kJ mol^{-1}
HO–C$_3$H$_7$	392.0±2.9	CF$_3$–OOCF$_3$	361.5±8.4	BrO–NO$_2$	118.0±6.3
HO–iso-C$_3$H$_7$	397.9±4.2	CH$_3$–OO	137.0±3.8	CH$_3$O–NO$_2$	176.1±4.2
HO–C$_4$H$_9$	389.9±4.2	CF$_3$–OO	169.0	C$_2$H$_5$O–NO$_2$	174.5±4.2
HO–sec-C$_4$H$_9$	396.1±4.2	CClF$_2$–OO	127.6	C$_3$H$_7$O–NO$_2$	177.0±4.2
HO–iso-C$_4$H$_9$	394.1±4.2	CCl$_2$F–OO	124.7	iso-C$_3$H$_7$O–NO$_2$	175.7±4.2
HO–tert-C$_4$H$_9$	398.3±4.2	CH$_2$Cl–OO	122.4±10.5	HOO–NO$_2$	99.2±4.6
HO–CH(CH$_3$)(nC$_3$H$_7$)	398.3±4.2	CHCl$_2$–OO	108.2±8.2	CH$_3$OO–NO$_2$	86.6±8.4
HO–CH(C$_2$H$_5$)$_2$	399.2±4.2	CCl$_3$–OO	92.0±6.4	CF$_3$OO–NO$_2$	105
HO–C(CH$_3$)$_2$(C$_2$H$_5$)	395.8±6.3	HC(O)–OOH	290.0	CF$_2$ClOO–NO$_2$	106.7
HO–C$_6$H$_5$	463.6±4.2	CH$_3$C(O)–OOC(O)CH$_3$	315.1	CFCl$_2$OO–NO$_2$	106.7
HO–C$_6$F$_5$	446.9±9.2	ClO–CF$_3$	≤369.9±1.3	CCl$_3$OO–NO$_2$	95.8
HO–CH$_2$C$_6$H$_5$	334.1±2.6	CH$_3$–ONO	245.2	CH$_3$N(O)-O	305.3±4.4
HO–C(CH$_3$)$_2$C$_6$H$_5$	339.3±6.3	C$_2$H$_5$–ONO	260.2	C$_6$H$_5$N(O)-O	392±8
cyclo-C$_5$H$_9$–OH	385.8±6.3	C$_3$H$_7$–ONO	249.4±6.3	C$_5$H$_5$N-O	264.9±2.0
1-C$_{10}$H$_7$–OH	468.6±6.3	iso-C$_3$H$_7$–ONO	254.4±6.3	C$_6$H$_5$N=N(O)(C$_6$H$_5$)-O	309.4±3.5
2-C$_{10}$H$_7$–OH	467.8±6.3	C$_4$H$_9$–ONO	256.5±6.3	C$_6$H$_5$(O)N=N(O)(C$_6$H$_5$)-O	309.4±3.6
(CH$_3$)$_2$(NH$_2$)C–OH	310.4±6.3	iso-C$_4$H$_9$–ONO	254.0±6.3	O–SO	551.1
CH$_3$C(O)-OH	459.4±4.2	sec-C$_4$H$_9$–ONO	253.6±6.3	O–SOF$_2$	513.3
HOCH$_2$-OH	411.3	tert-C$_4$H$_9$–ONO	252.7±6.3	O–SOCl$_2$	398.5
CH$_3$–OCH$_3$	351.9±4.2	(C$_2$H$_5$)(CH$_3$)$_2$C–ONO	254.0±8.4	O–S(OH)$_2$	493.7±25
ICH$_2$–OCH$_3$	373.2±12.6	CH$_3$–ONO$_2$	340.2	HO–SH	293.3±16.7
CH$_3$O–C$_2$H$_5$	355.2±5.4	C$_2$H$_5$–ONO$_2$	344.8	HO–SOH	313.4±12.6
CH$_3$O–CHClCH$_3$	370.3±8.4	CH$_3$O–CH$_2$CN	393.3	HO–S(OH)O$_2$	384.9±8.4
CH$_3$O–C$_3$H$_7$	358.6±6.3	O–N$_2$	167.4±0.4	HO–SCH$_3$	303.8±12.6
CH$_3$O–iso-C$_3$H$_7$	360.7±4.2	O–NO	306.24±0.41	HO–SO$_2$CH$_3$	360.2±12.6
CH$_3$O–C$_4$H$_9$	346.0±6.3	O–NO$_2$	208.6±1.4	F–OH	215.1
CH$_3$O-tert-C$_4$H$_9$	353.1±6.3	NO–NO	40.6±2.1	F–OF	164.1
C$_6$H$_5$–OCH$_3$	418.8±5.9	O$_2$N–ONO$_2$	95.4±1.5	F–OCF$_3$	200.8±4.2
C$_6$H$_5$CH(CH$_3$)–OCH$_3$	313.4±9.6	cis-HO–NO	207.0	F–OCH$_3$	>196.6
C$_6$H$_5$–OC$_6$H$_5$	326.8±4.2	trans-HO–NO	200.64±0.19	F–ONO$_2$	140.6
CH$_3$–OC(O)H	383.7±12.6	FO–NO	132.5±17	Cl–OH	233.5
HC(O)–OH	457.7±2.1	cis-ClO–NO	127.6±8.4	Cl–OCl	142
CH$_3$C(O)–OH	459.4±4.2	trans-ClO–NO	116.6±8.4	Cl–OCF$_3$	≤220.9±8.4
C$_6$H$_5$C(O)–OH	447.7±10.5	cis-BrO–NO	138.1±8.4	Cl–OCH$_3$	200.8
HO–CH$_2$C(O)OH	368.2±10.5	trans-BrO–NO	121.6±8.4	Cl–O-tert-C$_4$H$_9$	198.3
CH$_3$–OC(O)CH$_3$	380.3±12.6	trans-perp-HOO–NO	114.2±4	Cl–OOCl	91.2
HC(O)–OCH$_3$	423.8±4.2	CH$_3$O–NO	176.6±3.3	Cl–ONO$_2$	169.5
CH$_3$C(O)–OCH$_3$	424.3±6.3	C$_2$H$_5$O–NO	185.4±4.2	Br–OH	209.6±4.2
C$_6$H$_5$C(O)–OCH$_3$	421.3±12.6	C$_3$H$_7$O–NO	179.1±6.3	Br–OBr	125
C$_6$H$_5$C(O)–OC$_6$H$_5$	307.5±8.4	iso-C$_3$H$_7$O–NO	175.3±4.2	Br–O-tert-C$_4$H$_9$	183.3
CH$_3$OCH$_2$–OCH$_3$	367.5±8.4	C$_4$H$_9$O–NO	177.8±6.5	Br–ONO$_2$	143.1±6.3
CH$_3$C(O)–OC(O)CH$_3$	382.4±12.6	iso-C$_4$H$_9$O–NO	175.7±6.5	I–OH	213.4
C$_6$H$_5$C(O)–OC(O)C$_6$H$_5$	384.9±16.7	sec-C$_4$H$_9$O–NO	173.6±3.3	I–OI	130.1
CH$_3$–OOH	300.4±12.6	tert-C$_4$H$_9$O–NO	176.1±5.9	I–ONO$_2$	>140.6
C$_2$H$_5$–OOH	332.2±20.9	tert-AmO–NO	171.1±0.4		
C$_3$H$_7$–OOH	364.4	C$_6$H$_5$O–NO	87.0	**(5) N–X BDEs**	
iso-C$_3$H$_7$–OOH	298.3	HO–NO$_2$	205.4	H–NH$_2$	450.08±0.24
tert-C$_4$H$_9$–OOH	309.2±4.2	FO–NO$_2$	131.8±12.6	H–NF$_2$	316.7±10.5
CH$_3$–OOCH$_3$	292.5±8.4	ClO–NO$_2$	111.8±2.1	H–NNH	254.4
				H–N$_3$	≤389

Bond	$D^{\circ}_{298}/$ kJ mol^{-1}	Bond	$D^{\circ}_{298}/$ kJ mol^{-1}	Bond	$D^{\circ}_{298}/$ kJ mol^{-1}
H–N=CH$_2$	364±25	CH$_2$CHCH$_2$–NO	110	Br–NO$_2$	82.0±7.1
H–NO	199.5	iso-C$_3$H$_7$–NO	152.7±12.6	Br–NF$_2$	<227.2
H–NCO	460.7±2.1	tert-C$_4$H$_9$–NO	167	I–NO	75.6±4
H–NCS	≤396.6±4.6	C$_6$H$_5$–NO	226.8±2.1	I–NO$_2$	79.6±4
H–NCS	347.3±8.4	C$_6$F$_5$–NO	211.3±4.2		
CH$_3$NH$_2$	425.1±8.4	C$_6$H$_5$CH$_2$–NO	123	**(6) S–X BDEs**	
tert-BuNH$_2$	397.5±8.4	CH$_3$–NO$_2$	260.7±2.1	H–SH	381.18±0.05
C$_6$H$_5$CH$_2$NH$_2$	418.4	C$_2$H$_5$–NO$_2$	254.4	H–SCH$_3$	365.7±2.1
(CH$_3$)$_2$NH	395.8±8.4	C$_3$H$_7$–NO$_2$	256.5	H–SCHCH$_2$	351.5±8.4
H–NHNH(CH$_3$)	276±21	iso-C$_3$H$_7$–NO$_2$	259.8	H–SC$_2$H$_5$	365.3
H–NHN(CH$_3$)$_2$	356±21	C$_4$H$_9$–NO$_2$	254.8	H–SC$_3$H$_7$	365.7
NH$_2$CN	414.2	sec-C$_4$H$_9$–NO$_2$	263.2	H–S-iso-C$_3$H$_7$	369.9±8.4
(NH$_2$)$_2$C=O	464.4	tert-C$_4$H$_9$–NO$_2$	258.6	H–S-tert-C$_4$H$_9$	362.3±9.2
(NH$_2$)$_2$C=S	389.1	C$_6$H$_5$–NO$_2$	295.8±4.2	H–SOH	330.5±14.6
CH$_3$CSNH$_2$	380.7	C$_6$H$_5$CH$_2$–NO$_2$	210.3±6.3	H–SCOCH$_3$	370.7
PhCSNH$_2$	380.7	(NO$_2$)CH$_2$–NO$_2$	207.1	H–SCOPh	364
(PhNH)$_2$C=S	364.0	(NO$_2$)$_3$C–NO$_2$	176.1	H–SO$_2$CH$_3$	≤397
(NH$_2$)$_2$C=NH	435.1	CF$_3$–NF$_2$	280.7	H–SSCH$_3$	330.5±14.6
Ph$_2$C=NH	489.5	C$_6$H$_5$CH$_2$–NF$_2$	237.2±14.6	H–SPh	349.4±4.5
H–N(SiMe$_3$)$_2$	464	CH$_3$–NH$_2$	356.1±2.1	H–SSH	318.0±14.6
H–NHPh	375.3	C$_2$H$_5$–NH$_2$	352.3±6.3	H–SSSH	292.9±6.5
C$_6$H$_5$NHOH	292	C$_3$H$_7$–NH$_2$	356.1±2.9	HS–SH	270.7±8.4
C$_6$H$_5$NH(CONMe2)	387.9	iso-C$_3$H$_7$–NH$_2$	357.7±3.8	FS–SF	362.3
H–NPh$_2$	364.8	C$_4$H$_9$–NH$_2$	356.1±2.9	ClS–SCl	329.7
HN–N$_2$	63	sec-C$_4$H$_9$–NH$_2$	359.0±2.9	HS–SCH$_3$	272.0
ON–N	480.7±0.4	iso-C$_4$H$_9$–NH$_2$	254.8±5.0	HS–SPh	255.2±6.3
ON–NO	8.49±0.12	tert-C$_4$H$_9$–NH$_2$	355.6±6.3	CH$_3$S–SCH$_3$	272.8±3.8
ON–NO$_2$	42.5	pyridin-2-yl–NH$_2$	431	C$_2$H$_5$S–SC$_2$H$_5$	276.6
O$_2$N–NO$_2$	57.3±1	C$_6$H$_5$–NH$_2$	429.3±4.2	MeS–SPh	272.0±6.3
H$_2$N–NH$_2$	277.0±1.3	C$_6$H$_5$CH$_2$–NH$_2$	306.7±6.3	C$_6$H$_5$S–SC$_6$H$_5$	214.2±12.6
F$_2$N–NF$_2$	92.9±12.6	C$_6$H$_5$CH(CH)$_3$–NH$_2$	307.5±9.6	F$_5$S–SF$_5$	305±21
H$_2$N–NHCH$_3$	275.8±8.4	HC(O)–NH$_2$	421.7±8.4	HS–CH$_3$	312.5±4.2
H$_2$N–N(CH$_3$)$_2$	259.8±8.4	CH$_3$C(O)–NH$_2$	414.6±8.4	HS–C$_2$H$_5$	307.9±2.1
H$_2$N–NHC$_6$H$_5$	227.6±8.4	HS–NO	138.9	HS–C$_3$H$_7$	310.5±2.9
H$_2$N–NO$_2$	230	CH$_3$S–NO	104.6±4.2	HS–iso-C$_3$H$_7$	307.1±3.8
H$_2$NN(CH$_3$)–NO	179.6	tert-BuS–NO	115.1	HS–C$_4$H$_9$	309.2±2.9
(C$_6$H$_5$)$_2$N-NO	94.6	PhCH$_2$S–NO	120.5	HS–sec-C$_4$H$_9$	307.5±2.9
N$_3$–CH$_3$	335.1±20.5	C$_6$H$_5$S-NO	81.2±5.4	HS–iso-C$_4$H$_9$	310.0±4.6
N$_3$–C$_6$H$_5$	375.7±20.9	SCN–SCN	255.6	HS–tert-C$_4$H$_9$	301.2±3.8
N$_3$–CH$_2$C$_6$H$_5$	211.3±14.2	FSO$_2$–NF$_2$	163	HS–C$_6$H$_5$	360.7±6.3
CH$_3$–NC	413.0±3.3	F–NO	235.26	HS–CH$_2$C$_6$H$_5$	258.2±6.3
C$_2$H$_5$–NC	413.4±8.4	F–NO$_2$	221.3	HS–C(O)H	309.6±8.4
iso-C$_3$H$_7$–NC	423.0±8.4	F–NF$_2$	254.0	HS–C(O)CH$_3$	307.9±6.3
tert-C$_4$H$_9$–NC	399.6±5.4	F–NH$_2$	286.6	CH$_3$S–CH$_3$	307.9±3.3
NC–NO	204.4	Cl–NO	158.8±0.8	HOS–CH$_3$	284.9±12.6
CH$_3$–NO	172	Cl–NO$_2$	141.8±1.3	CH$_3$SO–CH$_3$	221.8±8.4
CF$_3$–NO	167	Cl–NF$_2$	~134	HOSO$_2$–CH$_3$	324.3±12.6
CCl$_3$–NO	125	Cl–NH$_2$	253.1	CH$_3$SO$_2$–CH$_3$	279.5
C$_2$H$_5$–NO	171.5	Br–NO	120.1±0.8	F$_5$S–CF$_3$	392±43
				F–SF$_5$	391.6

Bond	D°_{298} / kJ mol^{-1}
F–SO$_2$(F)	379
Cl–SF$_5$	<272
Cl–SO$_2$CH$_3$	293
Cl–SO$_2$Ph	297
Br–SBr	259±17
Br–SF$_5$	<230
I-SH	206.7±8.4
I-SCH$_3$	206.3±7.1

(7) Si-, Ge-, Sn-, and Pb–X BDEs

Bond	D°_{298} / kJ mol^{-1}
SiH$_3$–H	383.7±2.1
Me$_3$Si–H	396±7
H$_5$Si$_2$–H	373±8
(C$_2$H$_5$)$_3$Si–H	396±4
C$_6$H$_5$SiH$_2$–H	382±5
(CH$_3$S)$_3$Si–H	364.0
(iPrS)$_3$Si–H	376.6
PhMe$_2$Si–H	377±7
Ph$_2$SiH–H	379±7
Ph$_2$MeSi–H	361±10
SiF$_3$–H	432±5
SiCl$_3$–H	395±5
SiBr$_3$–H	334±8
SiH$_3$–SiH$_3$	321±4
SiH$_3$–Si$_2$H$_5$	313±8
Ph$_3$Si–SiPh$_3$	368.2
F$_3$Si–SiF$_3$	453.1±25
SiH$_3$–CH$_3$	375±5
SiF$_3$–CH$_3$	355.6
H$_3$Si–NO	158.2±5.7
H$_3$Si–PH$_2$	331.4
SiH$_3$–F	638±5
SiH$_3$–Cl	458±7
SiH$_3$–Br	376±9
SiH$_3$–I	299±8
GeH$_3$–H	348.9±8.4
Me$_3$Ge–H	364.0
Ph$_3$Ge–H	359.8
(CH$_3$)$_3$Ge-Ge(CH$_3$)$_3$	280.3
(CH$_3$)$_3$Ge-CH$_3$	288.7
Me$_3$Sn–H	326.4
Ph$_3$Sn–H	294.6
(CH$_3$)$_3$Sn-Sn(CH$_3$)$_3$	257.7
(CH$_3$)$_3$Sn-Cl	425±17
(CH$_3$)$_3$Pb-Pb(CH$_3$)$_3$	228.4
Cl$_3$Pb-Cl	271±84
(CH$_3$)$_3$Pb-CH$_3$	238±21

(8) P-, As-, Sb-, Bi–X BDEs

Bond	D°_{298} / kJ mol^{-1}
H$_2$P–H	351.0±2.1
CH$_3$PH–H	322.2±12.6
H$_2$P–PH$_2$	256.1
(C$_2$H$_5$)$_2$P–P(C$_2$H$_5$)$_2$	359.8
F$_2$P–F	549
Cl$_2$P–Cl	356±8
Br$_2$P–Br	<259
I$_2$P–I	217
H$_2$P–SiH$_3$	331.4
H$_2$As–H	319.2±0.8
H$_2$Sb–H	288.3±2.1
F$_2$Bi–F	435±19
Br$_2$Bi–Br	>297.1

(9) Se- and Te–X BDEs

Bond	D°_{298} / kJ mol^{-1}
H–SeH	334.93±0.75
H–SeC$_6$H$_5$	326.4±16.7
PhSe–SePh	280±19
H–TeH	277.0±5.0
H–TeC$_6$H$_5$	≤264
PhTe–TePh	138.1±12.6

(10) Metal-Centered BDEs
Arranged by the Periodic Table

(10.1) Group 1

Bond	D°_{298} / kJ mol^{-1}
Li–OH	431.0
Li–C$_2$H$_5$	214.6±8.4
Li–nC$_4$H$_9$	197.9±16.3
Na–OH	342.3
Na–O$_2$	<200
K–OH	359
Rb–OH	356.2±4.2
Cs–OH	373

(10.2) Group 2

Bond	D°_{298} / kJ mol^{-1}
BeO–H	469
Be(OH)–OH	476
MgO–H	441
Mg(OH)–OH	349
BrMg–CH$_3$	253
BrMg–CH$_2$CH$_3$	205
BrMg–i-C$_3$H$_7$	184
BrMg–t-C$_4$H$_9$	174
BrMg–C$_6$H$_5$	289
BrMg–CH$_2$C$_6$H$_5$	201
BrMg–C(C$_6$H$_5$)$_3$	180
Ca(OH)–OH	409
Sr(OH)–OH	407
Ba(OH)–OH	443

(10.3) Group 3

Bond	D°_{298} / kJ mol^{-1}
Sc–CH$_3$	116±29
Sc–C$_6$H$_6$	60.8
La(η^5-C$_5$Me$_5$)$_2$–CH(SiMe$_3$)$_2$	278.7±10.5
Nd(η^5-C$_5$Me$_5$)$_2$–CH(SiMe$_3$)$_2$	236.8±10.5
(η^5-C$_5$Me$_5$)$_2$Sm–H	226.8±12.6
(η^5-C$_5$Me$_5$)$_2$Sm–OCH$_3$	343.1
(η^5-C$_5$Me$_5$)$_2$Sm–(η^3-C$_3$H$_5$)	188.3±6.3
(η^5-C$_5$Me$_5$)$_2$Sm–S-nC$_3$H$_7$	295.4±10.0
(η^5-C$_5$Me$_5$)$_2$Sm–N(CH$_3$)$_2$	201.7±7.5
(η^5-C$_5$Me$_5$)$_2$Sm–SiH(SiMe$_3$)$_2$	179.9±21
(η^5-C$_5$Me$_5$)$_2$Sm–P(Et)$_2$	136.4±8.4
(η^5-C$_5$Me$_5$)$_2$Eu–I	238.9±8.4
(η^5-C$_5$Me$_5$)$_2$Yb–I	256.1±6.3
Lu(η^5-C$_5$Me$_5$)$_2$–CH(SiMe$_3$)$_2$	279.1±10.5
(η^5-C$_5$H$_4$SiMe$_3$)$_3$Th–H	277±6
(η^5-C$_5$H$_4$SiMe$_3$)$_3$Th–O	371±24
(η^5-C$_5$H$_5$)$_3$Th–CH$_3$	375±9
(η^5-C$_5$H$_5$)$_3$Th–CH$_2$Si(CH$_3$)$_3$	369±12
(C$_9$H$_7$)$_3$Th–CH$_2$C$_6$H$_5$	342±9
(η^5-C$_5$H$_4$tBu)$_3$U–H	249.7±5.7
(η^5-C$_5$H$_4$SiMe$_3$)$_3$U–H	253.7±5.1
[HB(3,5-Me$_2$Pz)$_3$]U(Cl)$_2$–Cl	422.6
(η^5-C$_5$H$_4$SiMe$_3$)$_3$U–I	265.6±4.3
(η^5-C$_5$H$_4$tBu)$_3$U–O	307±9
(η^5-C$_5$H$_4$SiMe$_3$)$_3$U–CO	43.1±0.8
(C$_9$H$_7$)$_3$U–CH$_3$	196.3±6.6
(η^5-C$_5$Me$_5$)$_2$U(Cl)–C$_6$H$_5$	358±11
(η^5-C$_5$H$_4$SiMe$_3$)$_3$U–THF	41.0±0.8

(10.4) Group 4

Bond	D°_{298} / kJ mol^{-1}
Ti(η^5-C$_5$H$_5$)$_2$–Cl	471
Ti(Cl)(η^5-C$_5$H$_5$)$_2$–Cl	390
Ti(η^5-C$_5$Me$_5$)$_2$–I	219
Ti(η^5-C$_5$H$_5$)$_2$–CO	174
Ti(CO)(η^5-C$_5$H$_5$)$_2$–CO	170
Ti–CH$_3$	174±29
Ti(Cl)(η^5-C$_5$H$_5$)$_2$–CH$_3$	276
Ti(Cl)((η^5-C$_5$H$_5$)$_2$–C$_6$H$_5$	292
Ti(C$_6$H$_6$)–C$_6$H$_5$	308.7
Zr(η^5-C$_5$Me$_5$)$_2$–H	351.0±7.5
Zr(H)(η^5-C$_5$Me$_5$)$_2$–H	326.4±4
Zr(η^5-C$_5$Me$_5$)$_2$–Cl	481.2
Zr(η^5-C$_5$Me$_5$)$_2$–Br	410.0
Zr(I)(η^5-C$_5$Me$_5$)$_2$–I	336.4±2.1
Zr(η^5-C$_5$Me$_5$)$_2$(Ph)–OH	482.4±6.3
Zr(η^5-C$_5$Me$_5$)$_2$(Ph)(OH)–OH	482.8±10.5
Zr(η^5-C$_5$Me$_5$)$_2$(NH$_2$)H–NH$_2$	421.3±15.1
Zr(η^5-C$_5$Me$_5$)$_3$–CH$_3$	276±10
Zr(η^5-C$_5$H$_5$)$_2$(C$_6$H$_5$)–C$_6$H$_5$	300±10
Zr(η^5-C$_5$H$_5$)$_2$(Si(SiMe$_3$)$_3$)–SiMe$_3$	188±30
Hf(H)(η^5-C$_5$Me$_5$)$_2$–H	346.0±7.9
Hf(η^5-C$_5$Me$_5$)(C$_4$H$_9$)–C$_4$H$_9$	274±10

(10.5) Group 5

Bond	D°_{298} / kJ mol^{-1}
(η^5-C$_5$H$_5$)(CO)$_3$V–η^2H$_2$	90±20
(η^5-C$_5$H$_5$)(CO)$_3$V–CO	146±21

Bond	$D^o_{298}/$ kJ mol^{-1}
V–CH$_3$	169±18
V–C$_6$H$_6$	76.2
V(C$_6$H$_6$)–C$_6$H$_6$	307.8
Nb(η^5-C$_5$H$_5$)$_2$H$_3$–TFE	18.8±1.3
Ta(CH$_3$)$_5$–CH$_3$	261±5
(Me$_3$SiCH$_2$)$_4$Ta–(CH$_2$SiMe$_3$)	184.1±8.4

(10.6) Group 6

Bond	$D^o_{298}/$ kJ mol^{-1}
[Cr(CO)$_3$(η^5-C$_5$Me$_5$)]$_2$–Hg	61.5
[Cr(CO)$_3$(η^5-C$_5$Me$_5$)]–Hg	111.3
Cr(CO)$_5$–Xe	37.7±3.8
(CO)$_2$(PPh$_3$)(η^5-C$_5$H$_5$)Cr–H	250.2±4.2
(η^5-C$_5$H$_5$)Cr(CO)$_3$–H,	257
Cr(CO)$_5$–H$_2$	78±4
(P(C$_6$H$_{11}$)$_3$)$_2$(CO)$_3$Cr–H$_2$	30.5±0.4
(η^6-C$_6$H$_6$)(CO)$_3$Cr–H$_2$	251±17
Cr(CO)$_5$–N$_2$	81±4
(P(C$_6$H$_{11}$)$_3$)$_2$(CO)$_3$Cr–N$_2$	38.9±0.8
(η^5-C$_5$Me$_5$)(CO)$_3$Cr–SH	193
Cr(CO)$_5$–CO	154.0±8.4
Cr(CO)$_5$–CH$_4$	~33.5±8
Cr–C$_6$H$_6$	9.6±5.8
Cr(C$_6$H$_6$)–C$_6$H$_6$	268.2±15.4
Cr(CO)$_5$–C$_6$H$_6$	57.3±3.3
(P(C$_6$H$_{11}$)$_3$)$_2$(CO)$_3$Cr–P (OMe)$_3$	68.6±2.5
(η^5-C$_5$H$_5$))Mo(CO)$_3$–H	290
Mo(η^5-C$_5$H$_5$)$_2$–H	246
Mo(H)(η^5-C$_5$H$_5$)$_2$–H	256.9±8.4
Mo(CO)$_3$(η^5-C$_5$H$_5$)–I	216.7±4.2
(η^5-C$_5$Me$_5$)$_2$Mo–O	272
(P(C$_6$H$_{11}$)$_3$)$_2$(CO)$_3$Mo–H$_2$	27.2±0.8
(P(C$_6$H$_{11}$)$_3$)$_2$(CO)$_3$Mo–N$_2$	37.7±2.5
Mo(CO)$_5$–CO	169.5±8.4
Mo(CO)$_3$ (η^5-C$_5$H$_5$)–CH$_3$	203±8
W(CO)$_5$–Xe	35.1±0.8
W(CO)$_3$(η^5-C$_5$H$_5$)–H	303
W(H)(η^5-C$_5$H$_5$)$_2$–H	310.9±4.2
W(I)(η^5-C$_5$H$_5$)$_2$–H	273±14
(CO)$_5$W–H$_2$	≥67
(P(C$_6$H$_{11}$)$_3$)(CO)$_3$W–(η^2-H$_2$)	28.5±2.1
W(CO)$_5$–CO	192.5±8.48.4
W(CH$_3$)(η^5-C$_5$H$_5$)$_2$–CH$_3$	220.9±4

(10.7) Group 7

Bond	$D^o_{298}/$ kJ mol^{-1}
F$_3$Mn–MnF$_3$	210.9±2.5
(CO)$_5$Mn–Mn(CO)$_5$	185±8
(CO)$_5$Mn–H	284.5
(PPh$_3$)Mn(CO)$_4$–H	286.2
MnBr(CO)$_4$–CO	184
(η^5-C$_5$H$_5$)(CO)$_2$Mn–CO	195.8±9.2
Mn–CH$_3$	>35±12
Mn(CO)$_5$–CH$_3$	187.0±3.8

Bond	$D^o_{298}/$ kJ mol^{-1}
Mn(CO)$_5$–C$_6$H$_5$	207±11
(CO)$_5$Mn–Re(CO)$_5$	149±11
(η^5-C$_5$H$_5$)Mn(CO)$_2$–PhMe	59.4±3.3
(CO)$_5$Tc–Tc(CO)$_5$	177.5±1.9
(CO)$_5$Re–Re(CO)$_5$	187±4.8
(CO)$_5$Re–H	313
(CO)$_5$Re–CH$_3$	220±8

(10.8) Group 8

Bond	$D^o_{298}/$ kJ mol^{-1}
(CO)$_4$Fe–Fe(CO)$_5$	171.5
(CO)$_4$Fe(H)$_x$–H	259.4±8.4
(η^5-C$_5$H$_5$)(CO)$_2$Fe–H	239
Fe(CO)$_3$(N$_2$)–N$_2$	37.7±19.2
Fe(C$_2$H$_2$)(CO)$_4$–CO	88±2.3
Fe(CO)$_2$(PMe$_3$)–CO	>125
Fe(CO)$_3$(PPh$_3$)–CO	<177.8±5
Fe–NH$_3$	31.4±4.2
Fe–CH$_2$	364±29
Fe–CH$_3$	135±29
Fe(C$_2$H$_4$)(CO)$_3$–C$_2$H$_4$	89.1±8
Fe–C$_3$H$_5$	218
Fe–C$_3$H$_6$	79
Fe(CO)$_5$–Ni(CO)$_4$	37.7
Fe(CO)$_5$–(η^3-C$_3$H$_5$)	176
Fe(C$_3$H$_6$)(CO)$_3$–C$_3$H$_6$	~79.5
(CO$_2$)(η^5-C$_5$H$_5$)Ru–H	272
(PMe$_3$)$_2$(η^5-C$_5$Me$_5$)Ru–H	167.4
(CO)$_2$(η^5-C$_5$Me$_5$)Ru–Cl	337.6
(η^5-C$_5$Me$_5$)(PMe$_3$)$_2$Ru–Cl	<138
(η^5-C$_5$Me$_5$)(PMe$_3$)$_2$Ru–OH	204.6
(CO)$_4$Ru–CO	115±1.7
(η^5-C$_5$Me$_5$)(PMe$_3$)$_2$Ru–CH$_3$	142.3
Os(H)(CO)$_4$–H	326.4
(CO)$_4$Os–CO	133±2.6
Os(C$_2$H$_2$)(CO)$_4$–CO	99.5±0.8

(10.9) Group 9

Bond	$D^o_{298}/$ kJ mol^{-1}
(CO)$_4$Co–Co(CO)$_4$	83±29
(CO)$_4$Co–Mn(CO)$_5$	96±12
(CO)$_4$Co–Re(CO)$_5$	113±15
Co(CO)$_4$–H	278
Co(CO)$_3$(PPh$_3$)–H	272
(CO)$_3$HCo–CO	~54
(η^5-C$_5$H$_5$)Co(CO)–CO	184.3±4.8
Co–CH$_2$	331±38
Co–CH$_3$	178±8
cobalamin–CH$_3$	150.6
cobinamide–iC$_4$H$_9$	104
Co–C bonds in B$_{12}$	123.8±6.3
Cl(CO)$_2$Rh–Rh(CO)$_2$Cl	94.6
HRh(m-xylyl)Rh–H	255.6±1.7
(PiPr$_3$)$_2$(Cl)Rh–H$_2$	136.0
(PiPr$_3$)$_2$(Cl)Rh–N$_2$	69.0

Bond	$D^o_{298}/$ kJ mol^{-1}
(PiPr$_3$)$_2$(Cl)Rh–CO	201.7
HRh(m-xylyl)Rh–CH$_2$OH	195.4±7.5
Ir(Cl)(CO)(PMe$_3$)$_2$–H	251
Ir(H)(η^5-C$_5$Me$_5$)(PMe$_3$)–H	310.5±21
Ir(Cl)(H)(CO)(PEt$_3$)$_2$–H	243.1
Ir(Cl)(H)(CO)(PPh$_3$)$_2$–H	246.9
(Cl)(CO)(PPh$_3$)$_2$Ir–H$_2$	62.8
(Cl)(CO)(PPh$_3$)$_2$Ir–CO	45.2
Ir(H)(η^5-C$_5$Me$_5$)(PMe$_3$)–C$_6$H$_5$	321

(10.10) Group 10

Bond	$D^o_{298}/$ kJ mol^{-1}
Ni–H$_2$O	~29
Ni(CO)$_3$–N$_2$	~42
Ni(CO)$_3$–CO	104.6±8.4
Ni–CH$_3$	208±8
Ni–C$_2$H$_2$	193±25
Ni–C$_2$H$_4$	147.3±17.6
Ni–propyne	155±21
Ni–2-butyne	121±21
Pd–OH	213
trans-Pt(PPh$_3$)$_2$(Cl)–H	307±37
[Ph$_2$PCH$_2$]$_2$MePt–H	104.6
[Ph$_2$PCH$_2$]$_2$MePt –OH	167.4
[Ph$_2$PCH$_2$]$_2$MePt –SH	90.0
Pt(η^5-C$_5$H$_5$)(CH$_3$)$_2$–CH$_3$	163±21
cis-Pt(PEt$_3$)$_2$(CH$_3$)–CH$_3$	269±13

(10.11) Group 11

Bond	$D^o_{298}/$ kJ mol^{-1}
Cu–OH	>406
Cu–CO	25±5
Cu–CH$_3$	223±5
Cu–NH$_3$	47±15
Cu(NH$_3$)–NH$_3$	83.7±4.2
Cu–C$_6$H$_6$	16.4±12.5
Cu(C$_6$H$_6$)–C$_6$H$_6$	27.0±19.3
Ag–CH$_3$	134.1±6.8
Ag–NH$_3$	8±13
Ag(NH$_3$)–NH$_3$	62.8±4.2
Au–OH	>262
Au–NH$_3$	76±6
Au–CH$_3$	≥191.6
Au–C$_6$H$_6$	8.4

(10.12) Group 12

Bond	$D^o_{298}/$ kJ mol^{-1}
Zn–CH$_3$	70±10
Zn(CH$_3$)–CH$_3$	266.5±6.3
Zn–C$_2$H$_5$	92.0±17.6
Zn(C$_2$H$_5$)–C$_2$H$_5$	219.2±8.4
Cd–CH$_3$	63.6±10.0
Cd(CH$_3$)–CH$_3$	234.3±6.3
Hg–CH$_3$	22.6±12.6
Hg(CH$_3$)–CH$_3$	239.3±6.3
ClHg–CH$_3$	280.0±12.6

Bond	$D^o_{298}/$ kJ mol^{-1}	Bond	$D^o_{298}/$ kJ mol^{-1}	Bond	$D^o_{298}/$ kJ mol^{-1}
BrHg–CH$_3$	270±38	Cl$_3$B–N(CH$_3$)$_3$	127.6	Cl$_3$Ga-S(C$_2$H$_5$)$_2$	235.1
IHg–CH$_3$	258.6±12.6	F$_2$B–CH$_3$	397 - 418	In–CH$_3$	216.3
		Al–OH	547±13	In(CH$_3$)$_1$–CH$_3$	318.8
(10.13) Group 13		Al–C$_2$H$_2$	>54	In(CH$_3$)$_2$–CH$_3$	587.4
H$_3$B–BH$_3$	172	Cl$_3$Al–N(CH$_3$)$_3$	198.7±8.4	(CH$_3$)$_3$In–N(CH$_3$)$_3$	83.3±2.1
H$_3$B–NH$_3$	130.1±4.2	(CH$_3$)$_3$Al–N(CH$_3$)$_3$	130	Tl–OH	330±30
(CH$_3$)$_3$B–NH$_3$	57.7±1.3	(CH$_3$)$_3$Al–O(CH$_3$)$_2$	92		
F$_3$B–N(CH$_3$)$_3$	130±4.6	(CH$_3$)$_3$Ga–O(C$_2$H$_5$)$_2$	50.6±0.8		

TABLE 4. Enthalpies of Formation of Free Radicals and Other Transient Species

Reference

Yu-Ran Luo, *Comprehensive Handbook of Chemical Bond Energies*, Taylor & Francis, 2006.

Radical	$\Delta_f H^o_{298}/$ kJ mol-1	Radical	$\Delta_f H^o_{298}/$ kJ mol-1	Radical	$\Delta_f H^o_{298}/$ kJ mol-1
(1) Carbon-Centered Species		•C$_4$H$_7$, cyclobutyl	219.2±4.2	•C$_5$H$_{11}$, (C$_2$H$_5$)(CH$_3$)$_2$C•	29
CH	595.8±0.6	n-C$_4$H$_9$•, n-butyl, CH$_3$CH$_2$CH$_2$C•H$_2$	77.8±2.1	•C$_6$H$_5$, phenyl	330.1±3.3
CH$_2$ (triplet)	391.2±1.6	i-C$_4$H$_9$•, i-butyl, (CH$_3$)$_2$CHC•H$_2$	70±4	•C$_6$H$_7$, cyclohexa-1,3-dien-5-yl	199.2
CH$_2$ (singlet)	428.8±1.6	s-C$_4$H$_9$•, s-butyl, CH$_3$C•HCH$_2$CH$_3$	67.8±2.1	•C$_6$H$_7$, cyclohexa-1,4-dien-3-yl	201.7±5.0
•CH$_3$, methyl	146.7±0.3	t-C$_4$H$_9$•, t-butyl, (CH$_3$)$_3$C•	48±3	•C$_6$H$_9$, CH$_3$C≡CC•(CH$_3$)$_2$	221.8±9.2
•C$_2$H, acetenyl, CH≡C•	567.4±2.1	•C$_5$H$_3$, CH≡C-C≡CC•H$_2$	579.1	•C$_6$H$_9$, (CH$_2$=CH)$_2$C•(CH$_3$)	193.7
•C$_2$H$_2$, vinylidene CH$_2$=C••	419.7±16.7	•C$_5$H$_3$, (CH≡C)$_2$C•H	573.2	•C$_6$H$_9$, cyclohexa-1-en-3-yl	119.7
•C$_2$H$_3$, vinyl, CH$_2$=C•H	299.6±3.3	•C$_5$H$_5$, CH$_2$=CHC≡CC•H$_2$	351.5	•C$_6$H$_{11}$, CH$_2$=CH(CH$_2$)$_3$C•H$_2$	158.6
•C$_2$H$_5$, ethyl, CH$_3$C•H$_2$	118.8±1.3	•C$_5$H$_5$, CH$_2$=CH-C•H-C≡CH	372.4	•C$_6$H$_{11}$, CH$_2$=CHC•H(CH$_2$)$_2$CH$_3$	89.0
•C$_3$H$_3$, propargyl, CH≡CC•H$_2$	339.0±4.2	•C$_5$H$_5$, cyclopenta-1,3-dien-5-yl	261.5±4.2	•C$_6$H$_{11}$, CH$_2$=C(CH$_3$)C•(CH$_3$)$_2$	37.7±6.3
•C$_3$H$_3$, CH$_3$C≡C•	515±13	•C$_5$H$_7$, CH$_3$C≡CC•HCH$_3$	272.8±9.2	•C$_6$H$_{11}$, (CH$_3$)$_2$C=C(CH$_3$)C•H$_2$	39.7±6.3
•C$_3$H$_3$, CH$_2$=C=CH• ↔ CH≡CC•H$_2$	339.0±4.2	•C$_5$H$_7$, CH≡CC•HC$_2$H$_5$	277.0±8.4	•C$_6$H$_{11}$, (CH$_3$)$_2$C=CHC•H(CH$_3$)	47.3
•C$_3$H$_3$, cyclopro-2-en-1-yl	439.7±17.2	•C$_5$H$_7$, CH≡CC•(CH$_3$)$_2$	257.3±9.2	•C$_6$H$_{11}$, (Z)-CH$_3$CH=CHC•(CH$_3$)$_2$	54.4
•C$_3$H$_5$, allyl, CH$_2$=CHC•H$_2$	171.0±3.0	•C$_5$H$_7$, CH$_2$=CHCH=CHC•H$_2$	205.0±12.6	•C$_6$H$_{11}$, cyclohexyl	75.3±6.3
•C$_3$H$_5$, CH$_3$CH=C•H	267±6	•C$_5$H$_7$, (CH$_2$=CH)$_2$C•H	208.0±4.2	nC$_6$H$_{13}$•, CH$_3$CH$_2$CH$_2$CH$_2$CH$_2$C•H$_2$	33.5
•C$_3$H$_5$, CH$_3$C•=CH$_2$	231.4	•C$_5$H$_7$, CH$_3$CH=C=CHC•H$_2$	278.0	•C$_6$H$_{13}$, (nC$_4$H$_9$)(CH$_3$)C•H	29.3
•C$_3$H$_5$, cyclopropyl	279.9±10.5	•C$_5$H$_7$, spiropentyl	380.7±4.2	•C$_6$H$_{13}$, 2-methyl-2-pentyl	3.3±8.4
n-C$_3$H$_7$•, n-propyl, CH$_3$CH$_2$C•H$_2$	100±2	•C$_5$H$_7$, cyclopent-1-en-3-yl	160.7±4.2	•C$_6$H$_{13}$, 3-methyl-3-pentyl	14.2
i-C$_3$H$_7$•, i-propyl, CH$_3$C•HCH$_3$	88±3	•C$_5$H$_9$, cyclopentyl	105.9±4.2	•C$_6$H$_{13}$, 2,3-dimethyl-2-butyl	3.1±10
•n-C$_4$H$_3$, CH≡CCH=C•H	547.3	•C$_5$H$_9$, CH$_2$=CHC•HCH$_2$CH$_3$	109.6±8.4	•C$_7$H$_3$, (CH≡C)$_3$C•	784.5
•i-C$_4$H$_3$, CH$_2$=C•C≡CH	499.2	•C$_5$H$_9$, CH$_3$CH=CHC•H(CH$_3$)	92	•C$_7$H$_7$, benzyl, C$_6$H$_5$C•H$_2$	208.0±1.7
•C$_4$H$_5$, CH$_3$C≡CC•H$_2$	304.5	•C$_5$H$_9$, CH$_3$CH=C(CH$_3$)C•H$_2$	92.0	•C$_7$H$_7$, quadricyclolan-5-yl	578.6±5.4
•C$_4$H$_5$, CH≡CC•HCH$_3$	316.5	•C$_5$H$_9$, CH$_2$=CHC•(CH$_3$)$_2$	87.0±8.4	•C$_7$H$_7$, quadricyclolan-4-yl	587.4±5.4
•C$_4$H$_5$, •CH=CHCHCH$_2$	364.4	•C$_5$H$_9$, CH$_2$=C(CH$_3$)C•H(CH$_3$)	93.7	•C$_7$H$_7$, norborna-2,5-dien-7-yl	511.7±7.9
•C$_4$H$_5$, CH$_2$=CHC•H$_2$	313.3	•C$_5$H$_9$, CH$_2$=C(C•H$_2$)CH$_2$CH$_3$	114.2	•C$_7$H$_7$, cyclohepta-1,3,5-trien-7-yl	285.3±12.6
•C$_4$H$_7$, CH$_3$CH=CHC•H$_2$	146±8	•C$_5$H$_9$, CH$_2$=CH(CH$_2$)$_2$C•H$_2$	179.5	•C$_7$H$_9$, CH$_2$=CH(CH=CH)$_2$CC•H$_2$	251.0
•C$_4$H$_7$, CH$_2$=CHCH$_2$C•H$_2$	192.5	nC$_5$H$_{11}$•, CH$_3$CH$_2$CH$_2$CH$_2$C•H$_2$	54.4	•C$_7$H$_9$, (CH$_2$=CH)$_3$C•	274.0
•C$_4$H$_7$, CH$_2$=C(CH$_3$)C•H$_2$	137.9	•C$_5$H$_{11}$, (C$_2$H$_5$)$_2$C•H	47.0	•C$_7$H$_{11}$, norborn-1-yl	136.4±10.5
•C$_4$H$_7$, CH$_2$=CHC•HCH$_3$	136.2	•C$_5$H$_{11}$, (nC$_3$H$_7$)(CH$_3$)C•H	50.2	•C$_7$H$_{11}$, cycloheptenyl	119.2
•C$_4$H$_7$, cyclopropylmethyl	213.8±6.7	•C$_5$H$_{11}$, (CH$_3$)$_3$C•CH$_2$	36.4±8.4	•C$_7$H$_{13}$, cycloheptyl	50.6±4.2

Radical	$\Delta_f H^o_{298}$/kJ mol-1	Radical	$\Delta_f H^o_{298}$/kJ mol-1	Radical	$\Delta_f H^o_{298}$/kJ mol-1
•C_7H_{13}, cyclo-[C•$(CH_3)(CH_2)_5$]	22.6	•$C_{11}H_9$, 1-naphthylmethyl	252.7	•CH_2Br	171.1±2.7
•C_7H_{13}, cyclo-[C•$(CH_2CH_3)(CH_2)_4$]	47.0	•$C_{11}H_{21}$, cycloundecanyl	7.5	•CI_3	424.9±2.8
•C_7H_{15}, $(nC_5H_{11})(CH_3)CH$•	8.4	•$C_{12}H_{23}$, cyclododecanyl	-38.5	•CHI_2	314.4±3.3
•C_7H_{15}, $(CH_3)_2CHCHC$•$(CH_3)_2$	-21.8±5.2	•$C_{13}H_9$, 9-fluorenyl	297.5	•CH_2I	229.7±8.4
•C_8H_7, cubyl	831.0±16.7	•$C_{13}H_{11}$, $(C_6H_5)_2C$•H	302.1±4.2	•C_2F, FC≡C•	460.0±21.0
•C_8H_7, C_6H_5C•$=CH_2$	309.6	•$C_{13}H_{11}$, 9-methyl-9-fluorenyl	268.2	•C_2Cl, ClC≡C•	568±26
•C_8H_7, $C_6H_5CH=CH$•	387.0	•$C_{14}H_{11}$, 9,10-dihydroanthracen-9-yl	261.0	•C_2F_3, $CF_2=C$•F	-192.0±8.4
•C_8H_9, C_6H_5C•$H(CH_3)$	175.7±7.5	•$C_{15}H_{11}$, 9-anthracenylmethyl	337.6	•C_2F_2H, $CF_2=C$•H	-92.9±8.4
•C_8H_9, $C_6H_5CH_2C$•H_2	236.0±7.5	•$C_{15}H_{11}$, 9-phenanthrenylmethyl	311.3	•C_2F_2H, CHF=C•F	-50.6±8.4
•C_8H_9, p-$CH_3C_6H_4C$•H_2	167.4	•$C_{16}H_{31}$, $CH_2=CHC$•$H(CH_2)_{12}CH_3$	-118.8	•CCl_2H, CHCl=C•Cl	234.7±8.4
•C_8H_9, m-$CH_3C_6H_4C$•H_2	167.4	•$C_{19}H_{15}$, trityl, $(C_6H_5)_3C$•	392.0±8.4	•$CClH_2$, $CH_2=C$•Cl	>251
•C_8H_9, o-$CH_3C_6H_4C$•H_2	167.4	•$C_{35}H_{25}$, pentamethylcyclopentadienyl	67.4	•C_2F_5, CF_3C•F_2	-892.9±4.2
•C_8H_9, 1-vinyl-cyclohexa-2,4-dienyl	247.7±14.2	CF	255.2±8	•C_2HF_4, CF_3C•HF	-680.8±9.6
•C_8H_9, 2-vinyl-cyclohexa-2,4-dienyl	249.8±14.2	CF_2	-182.0±6.3	•C_2HF_4, CHF_2C•F_2	-664.8
•C_8H_9, 3-vinyl-cyclohexa-2,4-dienyl	269.4±14.2	FC•(O)	-161.2±8.4	•$C_2H_2F_3$, CF_3C•H_2	-517.1±8.4
•C_8H_9, 6-vinyl-cyclohexa-2,4-dienyl	284.5±14.2	CHF	143.0±12.6	•$C_2H_2F_3$, CHF_2C•HF	-456.0
•C_8H_{13}, $CH_2=CHCH=CHC$•$H(CH_2)_2CH_3$	130.5	CClF	31.0±13.4	•$C_2H_2F_3$, CH_2FC•F_2	-449.8
•C_8H_{13}, $CH_2=CHC$•$H(CH_2)_3CH=CH_2$	130.5	CCl	443.1±13.0	•$C_2H_2F_2Cl$, CF_2ClC•H_2	-310.9±7.0
•C_8H_{13}, bicyclooct-1-yl	92.0	CCl_2	226	•$C_2H_3F_2$, CH_3C•F_2	-302.5±8.4
•C_8H_{15}, $CH_2=CHC$•$H(CH_2)_4CH_3$	49.8	ClC•(O)	-21.8±2.5	•$C_2H_3F_2$, CHF_2C•H_2	-285.8
•C_8H_{15}, (E)-$CH_3CH=C$•$(CH_2)_4CH_3$	29.7	CHCl	326.4±8.4	•$C_2H_3F_2$, CH_2FC•HF	-238.5
•C_8H_{15}, (Z)-$(CH_3)_2C$•$CH=CHCH(CH_3)_2$	9.2	CClBr	267	•C_2H_4F, CH_3C•HF	-70.3±8.4
•C_8H_{15}, cyclooctanyl	59.4	CBr	510±63	•C_2H_4F, CH_2FC•H_2	-59.4±8.4
•C_8H_{15}, cyclo-[C•$(CH_2CH_3)(CH_2)_5$]	10.0	CHBr	373±18	•$C_2H_2F_2Cl$, CF_2ClC•H_2	-315.2±6
•C_9H_7, indenyl	297.1	CBr_2	343.5	•C_2F_4Cl, CF_2ClC•F_2	-686.0
•C_9H_9, indanyl-1	204.2±8.4	CI	570±35	•C_2HF_3Cl, $CClF_2C$•HF	-450.6±12.6
•C_9H_{11}, 2,6-dimethylbenzyl	124.7	CI_2	468±60	•C_2F_4Cl, CF_3C•FCl	-728.0
•C_9H_{11}, 3,6-dimethylbenzyl	124.7	•CF_3	-465.7±2.1	•$C_2F_3Cl_2$, CF_3C•Cl_2	-564.0
•C_9H_{11}, 3,5-dimethylbenzyl	124.7	•CHF_2	-238.9±4.2	•C_2F_3ClBr, CF_3C•ClBr	-504.2±8.4
•C_9H_{11}, C_6H_5C•$(CH_3)_2$	133.9±4.2	•CH_2F	-31.8±4.2	•C_2Cl, ClC≡C•	534±50
•C_9H_{11}, o-•$C_6H_4C_2H_5$	279.5±7.5	•$CClF_2$	-279.0±8.4	•C_2Cl_3, $CCl_2=C$•Cl	190±50
•C_9H_{17}, cyclononanyl	52.3	•CCl_2F	-89.0±8.4	•C_2Cl_5, CCl_3C•Cl_2	35.1±5.4
•$C_{10}H_7$, naphth-1-yl	401.7±5.4	•CBrClF	-35.5±6.3	•C_2HCl_4, $CHCl_2C$•Cl_2	23.4±8.4
•$C_{10}H_7$, naphth-2-yl	400.4±5.9	•CHClF	-60.7±10.0	•C_2HCl_4, CCl_3C•HCl	51.0
•$C_{10}H_{11}$, tetralin-1-yl	154.8±5.0	•$CBrF_2$	-224.7±12.6	•$C_2H_2Cl_3$, CH_2ClC•Cl_2	26.4
•$C_{10}H_{13}$, 1-phenyl-but-4-yl	192.0	•CCl_3	71.1±2.5	•$C_2H_2Cl_3$, $CHCl_2C$•HCl	46.4
•$C_{10}H_{13}$, $(C_6H_5CH_2)(C_2H_5)C$•H	184.5	•$CHCl_2$	93.3±4.2	•$C_2H_2Cl_3$, CCl_3C•H_2	71.5±8
•$C_{10}H_{13}$, $(C_6H_5CH_2CH_2)(CH_3)C$•H	184.5	•CH_2Cl	117.2±2.9	•$C_2H_3Cl_2$, CH_3C•Cl_2	42.5±1.7
•$C_{10}H_{13}$, $(C_6H_5C$•$HCH_2CH_2CH_3$	134.7	•CHBrCl	143±6	•$C_2H_3Cl_2$, CH_2ClC•ClH	65.3
•$C_{10}H_{15}$, 1-adamantyl	51.5	•$CHBr_2$	188.3±9.2	•$C_2H_3Cl_2$, $CHCl_2C$•H_2	90.1±0.8
•$C_{10}H_{15}$, 2-adamantyl	61.9	•CBr_2Cl	163±8	•C_2H_4Cl, CH_3C•HCl	76.5±1.6
•$C_{10}H_{19}$, cyclodecanyl	32.2	•$CBrCl_2$	124±8	•C_2H_4Cl, CH_2ClC•H_2	93.0±2.4
		•CBr_3	214.8	•$C_2H_3Br_2$, CH_3C•Br_2	140.2±5.4
				•C_2H_4Br, $BrCH_2C$•H_2	135.1

Radical	$\Delta_f H^o_{298}$/ kJ mol-1	Radical	$\Delta_f H^o_{298}$/ kJ mol-1	Radical	$\Delta_f H^o_{298}$/ kJ mol-1
$^\bullet C_2H_4Br$, $CH_3C^\bullet HBr$	126.8	$C^\bullet CO$	381.2±2.1	$^\bullet C_7H_5O_2$, 3-C(O)OH-$^\bullet C_6H_4$	-35.0
$^\bullet C_2Br$, $CBrC^\bullet$	623.8	$CH_3C^\bullet(O)$	-10.3±1.8	$^\bullet C_7H_5O_2$, 4-C(O)OH-$^\bullet C_6H_4$	-36.0
$^\bullet C_2Br_3$, $CBr_2C^\bullet Br$	385.3	$CF_3C^\bullet(O)$	-608.7	$^\bullet CH_3O_2$, $C^\bullet H_2OOH$	66.1
$^\bullet C_2Br_5$, $CBr_3C^\bullet Br_2$	283.3	$CH_2ClC^\bullet(O)$	-21±12.6	$^\bullet C_2H_5O_2$, $C^\bullet H_2CH_2OOH$	46.0±4.6
$^\bullet C_3H_6Cl$, $CH_3CH_2C^\bullet HCl$	56.6	$CHCl_2C^\bullet(O)$	-17.6±23	$^\bullet C_2H_5O_2$, $CH_3CH^\bullet OOH$	26.9
$^\bullet C_3H_6Cl$, $CH_3C^\bullet ClCH_3$	29.9±0.6	$CCl_3C^\bullet(O)$	-19.7	$^\bullet C_3H_7O_2$, $CH_3CH^\bullet CH_2OOH$	10.9±5.4
$^\bullet C_3H_6Br$, $C^\bullet H_2CH_2CH_2Br$	120.1±1.3	$CH_3CH_2C^\bullet(O)$	-31.7±3.4	$^\bullet C_3H_7O_2$, $C^\bullet H_2CH(OOH)CH_3$	2.9±6.3
$^\bullet C_3H_6Br$, $CH_3C^\bullet HCH_2Br$	96.7±5.9	$CH_2CHC^\bullet(O)$	88.5	$^\bullet C_4H_9O_2$, $(CH_3)_2C^\bullet CH_2OOH$	-30.1±5.4
$^\bullet C_3H_6Br$, $CH_3CH_2C^\bullet HBr$	107.5±2.5	$CH_2C(CH_3)C^\bullet(O)$	58.6±16.7	$^\bullet C_4H_9O_2$, $C^\bullet H_2C(CH_3)_2OOH$	-26.8±5.4
$^\bullet C_6F_5$	-547.7±8.4	$CH_3CH_2CH_2C^\bullet(O)$	54.4±4.2	$^\bullet C_2H_3O_3$, $C^\bullet H_2C(O)OOH$	-137.9
$^\bullet CH_3O$, $HOC^\bullet H_2$	-17.0±0.7	$(CH_3)_2CHC^\bullet(O)$	-64.0±3.8	$^\bullet CHN_2$	494.5
$^\bullet CH_2ClO$, $HOC^\bullet ClH$	-60.7±7.5	$(CH_3)_3CC^\bullet(O)$	-102.9±6.3	$^\bullet CH_2N=CH_2$	263.6±12.6
$^\bullet CHCl_2O$, $HOC^\bullet Cl_2$	-94.1±7.5	$C_6H_5C^\bullet(O)$	116.3±10.9	$^\bullet CH_2NH_2$	151.9±8.4
$^\bullet CH_2ClO$, $ClOC^\bullet H_2$	135.6±9.2	$HC(O)CH_2^\bullet$	10.5±9.2	$CH_3C^\bullet HNH_2$	111.7±8.4
$^\bullet CH_2BrO$, $BrOC^\bullet H_2$	151±16	$ClC(O)CH_2^\bullet$	-52.7±13	$(CH_3)_2C^\bullet NH_2$	69.9±8.4
$^\bullet C_2H_3O$, $C^\bullet H=CHOH$	121±11	$E-C^\bullet HClC(O)H$	-27.2±10.5	$^\bullet CH_2NHCH_3$	156.6
$^\bullet C_2H_3O$, $C^\bullet H_2CHO$	13.0±2	$Z-C^\bullet HClC(O)H$	-23.4±10.5	$^\bullet CH_2N(CH_3)_2$	148.0
$^\bullet C_2H_5O$, $CH_3C^\bullet HOH$	-54.0	$C^\bullet Cl_2C(O)H$	-55.6±14.2	$(C_2H_5)_2NC^\bullet HCH_3$	68.6±2.1
$^\bullet C_2H_4ClO$, $CH_3C^\bullet ClOH$	-108.4±8.8	$E-C^\bullet HClC(O)Cl$	-88.7±15.1	$^\bullet CH_2N(CH_3)Ph$	266.0±12.6
$^\bullet C_2H_4ClO$, $C^\bullet H_2CHClOH$	-73.2±8.8	$C^\bullet H_2C(O)F$	-273.0±5.8	$^\bullet CN$	439.3±2.9
$^\bullet C_2H_3Cl_2O$, $C^\bullet H_2CCl_2OH$	-99.6±8.8	$Z-C^\bullet HClC(O)Cl$	-84.9±13.8	$^\bullet CH_2CN$	252.6±4
$^\bullet C_2H_5O$, $C^\bullet H_2CH_2OH$	-31±7	$C^\bullet Cl_2C(O)Cl$	-101.7±15.5	$CH_3C^\bullet HCN$	226.7±12.6
$^\bullet C_2H_3O$, oxiran-2-yl	149.8±6.3	$CH_3C(O)CH_2^\bullet$	-34±3	$^\bullet CH_2CH_2CN$	245.4±12.6
$^\bullet C_3H_5O$, $CH_2=CHC^\bullet HOH$	0±8.4	$CH_3C(O)C^\bullet HCH_3$	-70.3±7.1	$(CH_3)_2C^\bullet CN$	190.4±12.6
$^\bullet C_3H_7O$, $CH_3CH_2C^\bullet HOH$	-81±4	$CH_3C(O)C^\bullet=CH_2$	113.4	$Ph(CH_3)C^\bullet CN$	248.5±8.4
$^\bullet C_3H_7O$, $(CH_3)C^\bullet HCH_2OH$	-78.7±8.4	$C_2H_5C(O)C^\bullet HCH_3$	-107.5±20.9	$NCC^\bullet HCH_2CN$	381.8±12.6
$^\bullet C_3H_7O$, $HOCH_2CH_2C^\bullet H_2$	-66.9±8.4	$iPrC(O)C^\bullet(CH_3)_2$	-173.6±20.9	$^\bullet CH_2NC$	334.7±16.7
$^\bullet C_3H_7O$, $(CH_3)_2C^\bullet OH$	-96.4	$tC_4H_9C(O)C^\bullet H_2$	-115.5±12.6	$^\bullet C(O)NC$	210.0±10
$^\bullet C_3H_7O$, $^\bullet CH_2CH(OH)CH_3$	-62.8±11.7	$PhC(O)C^\bullet H_2$	84.5±12.6	$^\bullet C(O)NH_2$	-15.1±4
$^\bullet C_4H_9O$, $^\bullet CH_2C(OH)(CH_3)_2$	-147.3±8.4	$PhC(O)C^\bullet HCH_3$	41.4±20.9	$C^\bullet NN$	569±21
$^\bullet C_2H_5O_3$, $C^\bullet H_2OCH_2OOH$	109.6±4.2	$PhC^\bullet HC(O)CH_2Ph$	134.3±20.9	$HC^\bullet NN$	460±8
$PhCH^\bullet OH$	29.3±8.4	$PhC(O)OC^\bullet H_2$	-69.9	$H_2C^\bullet NN$	292.5±2.1
$Ph_2C^\bullet OH$	152.3±6.3	$^\bullet C(O)OH$-trans	≥-194.6±2.9	$^\bullet CH_2NO$	157±4
$^\bullet C_2H_5O$, $CH_3OC^\bullet H_2$	0±4.2	$^\bullet C(O)OH$-cis	-219.7	$^\bullet CH_2NO_2$	115.1±12.6
$^\bullet C_3H_7O$, $CH_3OC^\bullet HCH_3$	-57.7±8.4	$^\bullet C(O)OCH_3$	-161.5	$CH_3C^\bullet HNO_2$	61.9±12.6
$^\bullet C_3H_7O$, $CH_3CH_2OC^\bullet H_2$	-45.2±8.4	$C^\bullet H_2C(O)OH$	-248.9±12.0	$(CH_3)_2C^\bullet NO_2$	6.3±12.6
$^\bullet C_3H_7O$, $C^\bullet H_2CH_2OCH_3$	-7.1±4.2	$C^\bullet H(CH_3)C(O)OH$	-293±3	$PhC^\bullet HNO_2$	169.0±12.6
$^\bullet C_4H_9O$, $(CH_3)_2CHOC^\bullet H_2$	-70.3±7.1	$C^\bullet H_2C(O)OCH_3$	-236.8±8.4	$^\bullet C_6H_6N$, 3-NH_2-C_6H_4	320.1
$^\bullet C_4H_9O$, $CH_3CH_2OC^\bullet HCH_3$	-81.2±4.2	$C^\bullet H_2C(O)OCH_2CH_3$	-260.2±12.6	$^\bullet C_6H_6N$, 4-NH_2-C_6H_4	327.8
$^\bullet C_4H_9O$, $C^\bullet H_2CH(CH_3)OCH_3$	-42.3±3.8	$C^\bullet H_2C(O)OPh$	-28.0	$^\bullet C_6H_4NO_2$, 3-NO_2-C_6H_4	340.6±10.0
$^\bullet C_4H_9O$, $(CH_3)_2C^\bullet OCH_3$	-72.4±10	$^\bullet C_4H_7O$, tetrahydrofuran-2-yl	-18.0±6.3	$^\bullet C_6H_4NO_2$, 4-NO_2-C_6H_4	302.7
$^\bullet C_5H_{11}O$, $(CH_3)_3COC^\bullet H_2$	-102.5±8.4	$^\bullet C_4H_8O$, cyclopentanon-2-yl	-41.8±12.6	$^\bullet C_6H_4CH_3$, 2-Me-C_6H_4	315.1±10.5
$^\bullet C_2H_5O_2$, $HOCH_2C^\bullet HOH$	-220.1±8.4	$^\bullet C_4H_7O_2$, 1,4-dioxan-2-yl	-131.8±12.6	$^\bullet C_6H_4CH_3$, 4-Me-C_6H_4	296.6±9.6
$C^\bullet H=C=O$, ketenyl	177.5±8.8	$^\bullet C_7H_5O_2$, 2-C(O)OH-$^\bullet C_6H_4$	-33.0	$^\bullet C_6H_3N_2O_4$, 3,5-$(NO_2)_2$-C_6H_3	305.4
$HC^\bullet(O)$	42.5±0.5			$^\bullet C_7H_6NO_2$, 2-Me-4-NO_2-C_6H_3	295.4±8.4
				$^\bullet C_4H_3N$, pyrrol-2-yl	385.8

Radical	$\Delta_f H^o_{298}/$ kJ mol^{-1}	Radical	$\Delta_f H^o_{298}/$ kJ mol^{-1}	Radical	$\Delta_f H^o_{298}/$ kJ mol^{-1}
$^\bullet C_4H_3N$, pyrrol-3-yl	385.8	$^\bullet CH_2C(CH_3)_2SiMe_3$	-125	$C_6H_5O^\bullet$	48.5±2.9
$^\bullet C_4H_8N$, pyrrolidin-2-yl	142.7±12.6	$^\bullet CP$	450±9	o-Cl-$C_6H_4O^\bullet$	30.6
$^\bullet C_5H_4N$, pyrid-2-yl	362.0			$C_6Cl_5O^\bullet$	~63
$^\bullet C_5H_4N$, pyrid-3-yl	391.0	*(2) Oxygen-Centered Species*		p-Cl-$C_6H_4O^\bullet$	~9
$^\bullet C_5H_4N$, pyrid-4-yl	391.0	$HO^\bullet$	37.3±0.3	o-OH-$C_6H_4O^\bullet$	-186.3
$^\bullet C_4H_7N_2$, piperad-2-yl	119.7	$FO^\bullet$	109±10	p-OH-$C_6H_4O^\bullet$	-143.6
$^\bullet C_4H_3N_2$, pyrazin-2-yl	409.2±12.6	$ClO^\bullet$	101.63±0.1	o-CH_3O-$C_6H_4O^\bullet$	-125.5
$^\bullet C_4H_3N_2$, pyrimid-2-yl	388.0±12.6	$BrO^\bullet$	126.2±1.7	p-CH_3O-$C_6H_4O^\bullet$	-81.1
$^\bullet C_4H_3N_2$, pyrimid-4-yl	409.0±12.6	$IO^\bullet$	115.9±5.0	$C_6H_5CH_2O^\bullet$	136.0±12.6
$^\bullet C_4H_3N_2$, pyrimid-5-yl	446.4±12.6	$HOO^\bullet$	13.4±2.1	$C_{10}H_7O^\bullet$, naphthoxy-1	165.3
$^\bullet CH(NO_2)_2$	139.1	$FOO^\bullet$	25.4±2	$C_{10}H_7O^\bullet$, naphthoxy-2	174.1
$^\bullet C(NO_2)_3$	201.2	$ClOO^\bullet$	98.0±4	$HC(O)O^\bullet$	-129.7±12.6
$^\bullet CH_2C(NO_2)_3$	150.6	$BrOO^\bullet$	108±40	$FC(O)O^\bullet$	368.0
$^\bullet CH_2CH(NO_2)_2$	103.3	$IOO^\bullet$	96.6±15	$CH_3C(O)O^\bullet$	-179.9±12.6
$^\bullet CH_2CH_2C(NO_2)_3$	133.9	$OFO^\bullet$	378.6±20	$CF_3C(O)O^\bullet$	-797.0
$^\bullet CH_2N(NO_2)CH_2C(NO_2)_3$	173.6	$OClO^\bullet$	95.4	$CF_3OC(O)O^\bullet$	-958.1±16.7
$^\bullet CH_2N(NO_2)CH_2CH(NO_2)_2$	126.4	$ClOOClO^\bullet$	142±12	$C_6H_5C(O)O^\bullet$	-50.2±16.7
$^\bullet CH_2CH_2N(NO_2)CH_2C(NO_2)_3$	168.6	$ClClO^\bullet$	90±30	$CH_3OO^\bullet$	20.1±5.1
$^\bullet CH_2CH_2ONO_2$	37.7	$NCO^\bullet$	184.1	$C_2H_3OO^\bullet$, $CH_2=CHOO^\bullet$	101.7±1.7
$^\bullet CH_2(ONO_2)CHCH_2ONO_2$	-25.5	$CNO^\bullet$	386.6	$C_2H_5OO^\bullet$	-28.5±9.6
$^\bullet CH(CH_2ONO_2)_2$	-57.3	$HONNO^\bullet$	172	$C_3H_5OO^\bullet$, $CH_2=CHCH_2OO^\bullet$	88.7
$^\bullet CH_2C(CH_2ONO_2)_3$	-158.2	sym-ClO_3	217.2±21	$iC_3H_7OO^\bullet$	-65.4±11.3
$^\bullet CH_2NHNO_2$	164.8	$HSO^\bullet$	-21.8±2.1	$C_4H_7OO^\bullet$, $CH_3CH=CHCH_2OO^\bullet$	82.6±5.3
$^\bullet CH_2N(NO_2)CH_3$	149.4	$HSOO^\bullet$	112	$tC_4H_9OO^\bullet$	-101.5±9.2
$^\bullet CH_2N(NO_2)_2$	210.5	$CH_3SOO^\bullet$	76	neo-$C_5H_{11}OO^\bullet$	-115.5
$^\bullet CH_2CH_2N(NO_2)CH_3$	144.3	$CF_3SO_2O^\bullet$	-912	$HOCH_2OO^\bullet$	-162.1
$^\bullet CH_2N(NO_2)CH_2N(NO_2)CH_3$	202.1	$NCO^\bullet$	184.0	$HOOCH_2CH_2OO^\bullet$	100
$^\bullet CH_2N(NO_2)(CH_2)N(NO_2)CH_3$	173.2	$O_2NO^\bullet$	71.1	$C_6H_5CH_2OO^\bullet$	114.6±4.2
$C^\bullet(S)H$	300.4±8.4	$ONOO^\bullet$	82.8	c-$C_6H_{11}OO^\bullet$	-25.0±10.5
$^\bullet CH_2SH$	151.9±8.4	$HOS(O)_2O^\bullet$	-511.7	$(C_2H_5)N(CH3)CHOO^\bullet$	-36.0±12.6
$^\bullet CH_2SCH_3$	136.8±5.9	$CH_3O^\bullet$	21.0±2.1	$CF_3OO^\bullet$	-635.0
$^\bullet CH_2SPh$	268.6±12.6	$CF_3O^\bullet$	-635.1±7.1	$CF_2ClOO^\bullet$	-406.7±14.6
$^\bullet CH_2SOCH_3$	23.8±12.6	$CCl_3O^\bullet$	-38.1±9.2	$CFCl_2OO^\bullet$	-213.7
$HOC^\bullet(S)S$	110.5	$CH_2ClO^\bullet$	-21.3±9.2	$CH_2ClOO^\bullet$	-5.1±13.6
$^\bullet CH_2SO_2CH_3$	-177.0±12.6	$CHCl_2O^\bullet$	-32.2±9.2	$CHCl_2OO^\bullet$	-19.2±11.2
$^\bullet CH_2SO_2Ph$	-57.3±12.6	$CH_2=CH-O^\bullet$	18.4±1.3	$CCl_3OO^\bullet$	-20.9±8.9
$PhC^\bullet HSO_2CH_3$	-109.2±12.6	$CF_3CHFO^\bullet$	-851.0	$CH_3CHClOO^\bullet$	-54.7±3.4
$PhC^\bullet HSO_2Ph$	7±12.6	$C_2H_5O^\bullet$	-13.6±3.3	$CH_3CCl_2OO^\bullet$	-63.8±9.8
$Ph_2C^\bullet SO_2Ph$	102±12.6	$CH_3CHClO^\bullet$	-61.9±12.1	$CH_3OCH_2OO^\bullet$	-142.2±4.2
$Ph_2C^\bullet SPh$	435.6±12.6	$CH_3CCl_2O^\bullet$	-91.6±11.7	$CH_3C(O)CH_2OO^\bullet$	-142.1±4
$NC^\bullet(O)$	127.2	$nC_3H_7O^\bullet$	-30.1±8.4	$CH_3C(O)OO^\bullet$	-154.4±5.8
$^\bullet CNH$	207.9±12.1	$iC_3H_7O^\bullet$	-48.5±3.3	$HOOO^\bullet$	29.7±8.4
$^\bullet CNO$	323±30	$(CH_3)_2CClO^\bullet$	-108.4±8.4	$CH_3OOO^\bullet$	33.4±12.6
$^\bullet CH_2SiMe_3$	-32±6	$nC_4H_9O^\bullet$	-62.8	$C_2H_5OOO^\bullet$	5.4±12.6
		$sC_4H_9O^\bullet$	-69.5		
		$tC_4H_9O^\bullet$	-85.8±3.8		
		$CH_2=CHCH_2O^\bullet$	87.0		

Radical	$\Delta_f H^o_{298}/$ kJ mol-1	Radical	$\Delta_f H^o_{298}/$ kJ mol-1	Radical	$\Delta_f H^o_{298}/$ kJ mol-1
		iC$_3$H$_7$N$_2$•	146.0±8.4	SiCl	198.3±6.7
(3) Nitrogen-Centered Species		nC$_4$H$_9$N$_2$•	140.6±8.4	SiCl$_2$	-169±3
ON	90.29±0.17	tC$_4$H$_9$N$_2$•	97.5±4.2	•SiCl$_3$	322±8
NO$_2$	33.10±0.8	(NO$_2$)HN•	162.3	SiBr	235±46
N$_2$O	82.05±0.4	(CH$_3$)(NO$_2$)N•	139.0	SiBr$_2$	46±8
NH	357±1	(NO$_2$)$_2$N•	200.0	•SiBr$_3$	-201.7±63
•NH$_2$	186.2±1.0	CH$_3$N•CH$_2$N(NO$_2$)CH$_3$	185.4	SiI	313.8±42
•NNH	249.5			SiI$_2$	92.5±8.4
•NCO	131.8	**(4) Sulfur-Centered Species**		•SiI$_3$	35.3±63
•N$_3$	414.2±20.9	HOS•	-6.7±2.1	SiH	376.6±8.4
•N$_2$H$_3$	243.5	HC(O)S•	56.5	SiH$_2$($^1A_1$)	273±2
(Z)-N$_2$H$_2$	213.0±10.9	HS•O$_2$	-221.8	SiH$_2$($^3B_1$)	360.7
NF	209.2	HOS•O$_2$	-384.9	•SiH$_3$	200.4±2.5
•NF$_2$	42.3±8	NCS•	300±8	MeSi•H$_2$	141±6
•NHF	112±15	HS•	143.0±0.8	Me$_2$Si•H	78±6
NBr	301±21	CH$_3$S•	124.7±1.7	Me$_3$Si•	15±7
HNO	107.1±2.5	C$_2$H$_5$S•	101	•Si$_2$H$_3$	~402
FNO	-65.7±1.7	nC$_3$H$_7$S•	80	H$_3$SiSi•H$_2$	234±6
ClNO	51.71±0.42	iC$_3$H$_7$S•	74.9±8.4	C$_6$H$_5$Si•H$_2$	274
BrNO	82.13±0.8	tC$_4$H$_9$S•	43.9±8.4	H$_3$SiSi•H	312±8
INO	112.1±20.9	C$_6$H$_5$S•	242.7±4.6	MeSi•	302.2
NCO	120.9	C$_6$Cl$_5$S•	~184	MeSi•H	202±6
NCN	464.8±2.9	C$_6$H$_5$CH$_2$S•	246	Me$_2$Si••	135±8
NSi	372±63	CH$_3$S•O	-67±10	SiN	313.8±42
NH$_2$C(O)N•H	0.8±12.6	CH$_3$S•O$_2$	-239.3	•GeH$_3$	221.8±8.4
CH$_3$C(O)N•H	-6.7±12.6	HSS•	115.5±14.6	GeF	-71±10
NH$_2$C(S)N•H	194±12.6	CH$_3$SS•	68.6±8.4	GeF$_2$	-574±20
CH$_3$C(S)N•H	173±12.6	C$_2$H$_5$SS•	43.5±8.4	•GeF$_3$	-807±50
PhC(S)N•H	307±12.6	iC$_3$H$_7$SS•	13.8±8.4	GeCl	69±18
HCON•H	49.8±12.6	tC$_4$H$_9$SS•	-19.2±8.4	GeCl$_2$	-171±5
NH$_2$C(NH)N•H	250.6±12.6	HOC(S)S•	110.5±4.6	•GeCl$_3$	-268±50
•NHCN	319.2±2.9	HC(O)S•	56.5	GeBr	137±5
CH$_2$N•H	104.6±12.6	SF	13.0±6.3	GeBr$_2$	-61±5
CH$_3$N•H	184.1±8.4	SF$_2$	-296.7±16.7	•GeBr$_3$	-119±50
tBuN•H	95.4±12.6	SF$_3$	-503.0±33.5	GeI	211±25
C$_6$H$_5$CH$_2$N•H	288.3±12.6	SF$_4$	-763.2±20.9	GeI$_2$	50.2±4
C$_6$H$_5$N•H	244.3±4.2	SF$_5$	-879.9±15.1	•GeI$_3$	42±50
(CH$_3$)$_2$N•	158.2±4.2	ClS•	156.5±16.7	SnF	-95±7.2
(C$_6$H$_5$)(CH$_3$)N•	241.0±6.3	SN	263.6±105	SnF$_2$	-511±9.2
(C$_6$H$_5$)$_2$N•	366.0±6.3	SCl	156.5±16.7	•SnF$_3$	-647±50
1-pyrrolyl	269.2±12.6			SnCl	35±12
carbazol-9-yl	383.3±8.4	**(5) Si-, Ge-, Sn-, Pb-Centered Species**		SnCl$_2$	-202.6±7.1
CH$_3$N$_2$•	215.5±7.5	SiF	-20.1±12.6	•SnCl$_3$	-292±50
C$_2$H$_5$N$_2$•	187.4±10.5	SiF$_2$	-638±6	SnBr	76±12
		•SiF$_3$	-987±20		

Radical	$\Delta_f H^o_{298}/$ kJ mol-1	Radical	$\Delta_f H^o_{298}/$ kJ mol-1	Radical	$\Delta_f H^o_{298}/$ kJ mol-1
$SnBr_2$	-119±2.8	PbH	236.2±19.2	PbBr	70.9±42
$\cdot SnBr_3$	-159±50	PbF	-80.3±10.5	$PbBr_2$	-104.4±6.3
SnI	173±12	PbF_2	-435.1±8.4	$\cdot PbBr_3$	-104±80
SnI_2	-8.1±4.2	$\cdot PbF_3$	-490±60	PbI	107.4±37.7
$\cdot SnI_3$	-8±50	PbCl	15.1±50	PbI_2	-3.2±4.2
$\cdot Sn(CH_3)_3$	132.2	$PbCl_2$	-174.1±1.3	$\cdot PbI_3$	22±80
$\cdot Sn(C_6H_5)_3$	518.8±21	$\cdot PbCl_3$	-178±80		

TABLE 5. Bond Dissociation Energies of Some Organic Molecules

D^0_{298}(R-X)/ kJ mol^{-1} of some organic compounds are listed below. All data are from Tables 1 and 3.

	X = H	F	Cl	Br	I	OH	OCH_3	NH_2	NO	CH_3	$COCH_3$	CF_3	CCl_3
R = H	435.7799	569.658	431.361	366.16	298.26	497.10	440.2	450.08	199.5	439.3	374.0	445.2	392.5
CH_3	439.3	460.2	350.2	294.1	238.9	384.93	351.9	356.1	172.0	377.4	351.9	429.3	362.3
C_2H_5	420.5	447.4	352.3	292.9	233.5	391.2	355.2	352.3	171.5	370.3	347.3	–	–
$i-C_3H_7$	410.5	483.8	354.0	299.2	234.7	397.9	360.7	357.7	152.7	369.0	340.2	–	–
$t-C_4H_9$	400.4	495.8	351.9	292.9	227.2	398.3	353.1	355.6	167	363.6	329.3	–	–
C_6H_5	472.2	525.5	399.6	336.4	272.0	463.6	418.8	429.3	226.8	426.8	406.7	463.2	388.7
$C_6H_5CH_2$	375.5	412.8	299.9	239.3	187.8	334.1	–	306.7	123	325.1	299.7	365.7	–
CCl_3	392.5	439.3	296.6	231.4	168	–	–	–	125	362.3	–	332.2	285.8
CF_3	445.2	546.8	365.3	296.2	227.2	≤482.0	–	–	167	429.3	–	413.0	332.2
C_2F_5	429.7	532.2	346.0	283.3	219.2	–	–	–	–	–	–	424.3	–
CH_3CO	374.0	511.7	354.0	292.0	223.0	459.4	424.3	414.6	–	351.9	307.1	–	–
CN	528.5	482.8	422.6	364.8	320.1	–	–	–	204.4	521.7	–	469.0	–
C_6F_5	487.4	485	383.3	~328	<301.7	446.9	–	–	211.3	439.3	–	435.1	–

TABLE 6. Bond Dissociation Energies in Diatomic Cations

From thermochemistry, we have

$$D^o_{298}(A^+-B) \equiv \Delta_f H^o(A^+) + \Delta_f H^o(B) - \Delta_f H^o(AB^+) = D^o_{298}(A-B) + IP(A) - IP(AB)$$

Thus, $D^o_{298}(A^+-B)$ may be derived using the Table 1 and the ionization potentials of species A and AB. The following table has been arranged in an alphabetical order of the atoms. The **boldface** in the species indicates the dissociated fragment. The details are described in the *Comprehensive Handbook of Chemical Bond Energies*, by Yu-Ran Luo, Taylor & Francis, 2006.

A^+-B	D^o_{298} kJ/ mol^{-1}	A^+-B	D^o_{298} kJ/ mol^{-1}	A^+-B	D^o_{298} kJ/ mol^{-1}	A^+-B	D^o_{298} kJ/ mol^{-1}
Ag$^+$–Ag	167.9±8.7	**Ar$^+$**–Ne	7.5±0.8	**B$^+$**–B	187	**Ba$^+$**–O	441.4±15
Ag$^+$–Cl	32±30	**As$^+$**–As	364±22	**B$^+$**–Br	164±21	**Be$^+$**–Ar	49.0±2.4
Ag$^+$–F	24±27	**As$^+$**–H	290.8±3.0	**B$^+$**–C	284±58	**Be$^+$**–Au	410±29
Ag$^+$–H	43.5±5.9	**As$^+$**–O	495	**B$^+$**–Cl	308±21	**Be$^+$**–Be	187.0
Ag$^+$–O	123±5	**As$^+$**–P	367±59	**B$^+$**–F	460±10	**Be$^+$**–Cl	417±50
Ag$^+$–S	123±13	**As$^+$**–S	433.2±12.5	**B$^+$**–H	198±5	**Be$^+$**–F	575±98
Al$^+$–Al	121	**Au$^+$**–Al	170±30	**B$^+$**–O	326±48	**Be$^+$**–H	307.3±5.0
Al$^+$–Ar	15.47	**Au$^+$**–Au	234.5	**B$^+$**–Pt	314±98	**Be$^+$**–O	362.0±6.2
Al$^+$–Ca	148.5	**Au$^+$**–B	329±50	**B$^+$**–Se	298±98	**Bi$^+$**–Bi	199±10
Al$^+$–Cl	173±42	**Au$^+$**–Be	401±29	**B$^+$**–Si	365±15	**Bi$^+$**–O	174
Al$^+$–F	314±21	**Au$^+$**–F	79	**Ba$^+$**–Ar	11.85	**Bi$^+$**–S	179±50
Al$^+$–Kr	5.54	**Au$^+$**–Ge	292±24	**Ba$^+$**–Br	418±10	**Bi$^+$**–Se	184±29
Al$^+$–O	166.7±12.0	**Au$^+$**–H	143.5	**Ba$^+$**–Cl	468.2±10	**Bi$^+$**–Te	125±50
Al$^+$–Se	114±49	**Au$^+$**–I	230~280	**Ba$^+$**–D	245.2±9.6	**Bi$^+$**–Tl	100±42
Ar$^+$–Ar	130.323±0.087	**Au$^+$**–Xe	130±13	**Ba$^+$**–F	640±29	**Bk$^+$**–O	610
Ar$^+$–He	2.9±0.8	**B$^+$**–Ar	32.7	**Ba$^+$**–I	335±10	**Br$^+$**–Br	318.858±0.024

A^+–B	D°_{298} kJ/mol^{-1}	A^+–B	D°_{298} kJ/mol^{-1}	A^+–B	D°_{298} kJ/mol^{-1}	A^+–B	D°_{298} kJ/mol^{-1}
Br$^+$–C	451.5±8.6	**Co$^+$**–Ar	52.89±0.06	**D$^+$**–D	263.4405±0.0003	**Ga$^+$**–Cl	86±21
Br$^+$–Cl	303.000±0.048	**Co$^+$**–Br	>289	**Dy$^+$**–Br	324.2	**Ga$^+$**–F	136±15
Br$^+$–F	251.5±12.6	**Co$^+$**–C	351±29	**Dy$^+$**–Cl	407.9	**Ga$^+$**–Ga	126.3
Br$^+$–H	379.26±2.89	**Co$^+$**–Cl	285±12	**Dy$^+$**–Cu	196±42	**Ga$^+$**–I	41.6±15
Br$^+$–O	365.7±3.1	**Co$^+$**–Co	269	**Dy$^+$**–F	535±24	**Ga$^+$**–O	46±50
C$^+$–Ar	72.3	**Co$^+$**–D	199.6±5.8	**Dy$^+$**–I	279.9	**Ga$^+$**–Sb	38±96
C$^+$–Br	398±8.6	**Co$^+$**–H	195±6	**Dy$^+$**–O	597±15	**Ga$^+$**–Te	19±29
C$^+$–C	601.9±19.3	**Co$^+$**–He	16.4±0.4	**Er$^+$**–Br	315.8	**Gd$^+$**–Cd	122.5±10
C$^+$–Cl	614	**Co$^+$**–I	211.7±8.4	**Er$^+$**–Cl	406.7	**Gd$^+$**–H	179.5
C$^+$–F	721±40	**Co$^+$**–Kr	68.37±0.18	**Er$^+$**–F	546±34	**Ge$^+$**–Br	398±42
C$^+$–H	397.848±0.013	**Co$^+$**–Ne	12.8±0.4	**Er$^+$**–I	271.6	**Ge$^+$**–C	223±31
C$^+$–N	524.5±4.2	**Co$^+$**–O	317.3±4.8	**Er$^+$**–O	583±15	**Ge$^+$**–Cl	473±50
C$^+$–O	810.7±0.8	**Co$^+$**–S	288.3±8.7	**Es$^+$**–O	470±60	**Ge$^+$**–F	565±21
C$^+$–P	587±50	**Co$^+$**–Si	317.1±6.7	**Eu$^+$**–Ag	85±50	**Ge$^+$**–Ge	274±10
C$^+$–S	706.6±2.1	**Co$^+$**–Xe	85.7±6.8	**Eu$^+$**–Au	252±97	**Ge$^+$**–H	377±84
C$^+$–Se	587±50	**Cr$^+$**–Ar	31.7±3.9	**Eu$^+$**–Br	333.8	**Ge$^+$**–O	344±21
Ca$^+$–Al	144.7	**Cr$^+$**–C	277±24	**Eu$^+$**–Cl	430.7	**Ge$^+$**–S	283±21
Ca$^+$–Ar	12.99±0.60	**Cr$^+$**–Cl	>211	**Eu$^+$**–F	543±29	**Ge$^+$**–Se	234±10
Ca$^+$–Au	306±29	**Cr$^+$**–Cr	129	**Eu$^+$**–I	290.7	**Ge$^+$**–Si	268±21
Ca$^+$–Br	417.6±10	**Cr$^+$**–D	135±9	**Eu$^+$**–O	393±15	**Ge$^+$**–Te	233±19
Ca$^+$–Ca	104.1	**Cr$^+$**–F	279±42	**Eu$^+$**–S	257±32	**H$^+$**–D	261.1021±0.0002
Ca$^+$–Cl	433.4±12	**Cr$^+$**–H	136±9	**F$^+$**–Ar	161.1	**H$^+$**–H	259.4659±0.0002
Ca$^+$–F	556.5±8.4	**Cr$^+$**–He	7.8±0.4	**F$^+$**–F	325.393±0.096	**He$^+$**–H	123.9
Ca$^+$–H	284.2±10	**Cr$^+$**–Ne	9.5±0.4	**F$^+$**–He	181.62±0.08	**He$^+$**–He	229.687±0.019
Ca$^+$–I	293.7±10.8	**Cr$^+$**–O	276	**F$^+$**–Kr	152.4	**Hf$^+$**–O	724±21
Ca$^+$–Kr	18.60±0.72	**Cr$^+$**–S	258.6±16.4	**F$^+$**–Xe	188	**Hg$^+$**–Ar	22.2±1.2
Ca$^+$–Ne	4.95±0.06	**Cr$^+$**–Si	203±15	**Fe$^+$**–Ar	14.2±7.7	**Hg$^+$**–H	207
Ca$^+$–O	348±5	**Cr$^+$**–Xe	71.9±10.0	**Fe$^+$**–Br	>293	**Hg$^+$**–Hg	134
Ca$^+$–Xe	25.38±0.96	**Cs$^+$**–Ar	8.2	**Fe$^+$**–C	356.1±17.2	**Hg$^+$**–Kr	37.9±1.3
Cd$^+$–Cd	122.5±10	**Cs$^+$**–Br	60.5±10	**Fe$^+$**–Cl	>343	**Hg$^+$**–Xe	72.2±1.3
Cd$^+$–H	179.5	**Cs$^+$**–Cl	107.4±10	**Fe$^+$**–Co	259±21	**Ho$^+$**–Ag	155±61
Ce$^+$–Au	278±34	**Cs$^+$**–Cs	62.6±9.6	**Fe$^+$**–Cr	209±29	**Ho$^+$**–Au	250±60
Ce$^+$–Br	341.0	**Cs$^+$**–F	43.7±10	**Fe$^+$**–Cu	222±29	**Ho$^+$**–Br	320.6
Ce$^+$–C	254±96	**Cs$^+$**–He	5.1	**Fe$^+$**–D	227	**Ho$^+$**–Cl	410.3
Ce$^+$–Ce	207±42	**Cs$^+$**–I	29.3±10	**Fe$^+$**–F	360 - 423	**Ho$^+$**–Cu	214±35
Ce$^+$–Cl	429.5	**Cs$^+$**–Kr	15.1	**Fe$^+$**–Fe	272	**Ho$^+$**–F	542±50
Ce$^+$–F	586±63	**Cs$^+$**–Na	48.1±4.2	**Fe$^+$**–H	211.2±9.6	**Ho$^+$**–Ho	88±96
Ce$^+$–I	295.5	**Cs$^+$**–Ne	6.11	**Fe$^+$**–I	>239	**Ho$^+$**–I	270.4
Ce$^+$–Ir	530±96	**Cs$^+$**–O	59	**Fe$^+$**–Kr	33.5±6.7	**Ho$^+$**–O	551±25
Ce$^+$–N	494±63	**Cs$^+$**–Rb	68.3±10	**Fe$^+$**–N	485	**I$^+$**–Br	184.90±0.02
Ce$^+$–O	852±15	**Cs$^+$**–Xe	14.7	**Fe$^+$**–Nb	285±21	**I$^+$**–Cl	247.5±0.4
Ce$^+$–Pd	255±53	**Cu$^+$**–Ar	51.9±6.8	**Fe$^+$**–Ni	268±21	**I$^+$**–F	262.9±2.1
Ce$^+$–Pt	467±96	**Cu$^+$**–Cl	91±10	**Fe$^+$**–O	343.3±1.9	**I$^+$**–H	304.70±0.10
Ce$^+$–Rh	423±96	**Cu$^+$**–Cu	155.2±7.7	**Fe$^+$**–S	295.2±5.8	**I$^+$**–I	262.90±0.04
Ce$^+$–S	524±59	**Cu$^+$**–F	117±21	**Fe$^+$**–Sc	200±21	**I$^+$**–O	316.3±10.5
Cl$^+$–Ar	169	**Cu$^+$**–Ge	231±23	**Fe$^+$**–Si	277±9	**In$^+$**–Br	65.2±12.6
Cl$^+$–Cl	386.02±0.30	**Cu$^+$**–H	93±13	**Fe$^+$**–Ta	301±21	**In$^+$**–Cl	193±21
Cl$^+$–D	457.284±0.017	**Cu$^+$**–Kr	24.3±0.8	**Fe$^+$**–Ti	251±25	**In$^+$**–F	148±50
Cl$^+$–F	291±10	**Cu$^+$**–O	133.9±11.6	**Fe$^+$**–V	314±21	**In$^+$**–I	51.5±21
Cl$^+$–H	452.714±0.018	**Cu$^+$**–S	203.3±14.5	**Fe$^+$**–Xe	46.0±5.8	**In$^+$**–In	81±30
Cl$^+$–N	650±10	**Cu$^+$**–Si	260±8	**Ga$^+$**–Bi	62±98	**In$^+$**–S	171±50
Cl$^+$–O	468.0±2.1	**Cu$^+$**–Xe	102.1±5.8	**Ga$^+$**–Br	56.5±16	**In$^+$**–Sb	73±50

A^+–B	D^o_{298} kJ/mol^{-1}	A^+–B	D^o_{298} kJ/mol^{-1}	A^+–B	D^o_{298} kJ/mol^{-1}	A^+–B	D^o_{298} kJ/mol^{-1}
In$^+$–Se	118±50	Lu$^+$–F	376.8	Nb$^+$–Nb	576.8±9.6	Pb$^+$–Se	169.4±6.3
In$^+$–Te	41±50	Lu$^+$–H	204±15	Nb$^+$–O	688±11	Pb$^+$–Te	163±63
Ir$^+$–C	544±96	Lu$^+$–I	40.7	Nb$^+$–S	532±10	Pd$^+$–C	528±5
Ir$^+$–D	302.8±5.8	Lu$^+$–O	524±15	Nb$^+$–V	404.7±0.2	Pd$^+$–H	208.4±8.7
Ir$^+$–H	305.7±5.8	Lu$^+$–Si	107±13	Nb$^+$–Xe	73.28±0.12	Pd$^+$–O	145±11
Ir$^+$–O	247	Mg$^+$–Ar	19.20	Nd$^+$–Au	267±84	Pd$^+$–Pd	197±29
K$^+$–Ar	14±7	Mg$^+$–Au	267±29	Nd$^+$–Br	352.9	Pd$^+$–S	197±6
K$^+$–Br	35.7±10.5	Mg$^+$–Cl	327±6.5	Nd$^+$–Cl	441.4	Pd$^+$–Si	289±50
K$^+$–Cl	51±19	Mg$^+$–D	203.6±0.8	Nd$^+$–F	309.6	Pr$^+$–Au	317±81
K$^+$–He	6.00	Mg$^+$–F	477±50	Nd$^+$–I	596±32	Pr$^+$–Br	357.7
K$^+$–I	18±45	Mg$^+$–H	190.8±5.8	Nd$^+$–O	753±15	Pr$^+$–Cl	445.0
K$^+$–K	83.86±0.15	Mg$^+$–Kr	25.39	Ne$^+$–H	1239	Pr$^+$–F	557±63
K$^+$–Kr	15.8	Mg$^+$–Mg	125	Ne$^+$–He	13.0±0.8	Pr$^+$–I	317.0
K$^+$–Li	59.9±5.9	Mg$^+$–Ne	4.9±0.6	Ne$^+$–Ne	125.29±1.93	Pr$^+$–O	796±15
K$^+$–Na	58.69±0.08	Mg$^+$–O	245.2±10	Ni$^+$–Ar	53.9	Pt$^+$–Ar	36.4±8.7
K$^+$–Ne	7.79	Mg$^+$–Xe	53.74	Ni$^+$–Br	>289	Pt$^+$–B	398±105
K$^+$–O	13	Mn$^+$–Cl	>211	Ni$^+$–C	418	Pt$^+$–C	530.5±4.8
K$^+$–Xe	19.5	Mn$^+$–F	321±24	Ni$^+$–Cl	192±4	Pt$^+$–Cl	249.8±14.5
Kr$^+$–Ar	55.31±0.14	Mn$^+$–H	202.5±5.9	Ni$^+$–D	166.0±7.7	Pt$^+$–H	275±5
Kr$^+$–H	464	Mn$^+$–I	>211	Ni$^+$–F	≥456	Pt$^+$–N	326.9±9.6
Kr$^+$–He	2.1±0.8	Mn$^+$–Mn	129	Ni$^+$–H	158.1±7.7	Pt$^+$–O	318.4±6.7
Kr$^+$–Kr	110.967±0.033	Mn$^+$–O	285±13	Ni$^+$–He	12.4±0.4	Pt$^+$–Pt	318±23
Kr$^+$–N	133±13	Mn$^+$–S	247±23	Ni$^+$–I	>297	Pt$^+$–Si	515±50
Kr$^+$–Ne	3.8±0.8	Mn$^+$–Se	165±50	Ni$^+$–Ne	9.9±0.4	Pt$^+$–Xe	86.6±28.9
La$^+$–Au	436±97	Mo$^+$–C	488.2±19	Ni$^+$–Ni	208	Pu$^+$–F	562±50
La$^+$–Br	425.9	Mo$^+$–F	376±29	Ni$^+$–O	275.9±7.7	Pu$^+$–O	655
La$^+$–C	427±33	Mo$^+$–H	170±6	Ni$^+$–S	241.0±3.9	Rb$^+$–Ar	12.0
La$^+$–Cl	503.6	Mo$^+$–Mo	449.4±1.0	Ni$^+$–Si	326±6.7	Rb$^+$–Br	17.6±5.1
La$^+$–F	589±34	Mo$^+$–O	488.2±1.9	Np$^+$–F	730±100	Rb$^+$–Cl	10.5±10.5
La$^+$–H	243±9	Mo$^+$–S	355.1±5.8	Np$^+$–O	≥752	Rb$^+$–I	27±42
La$^+$–I	392.4	Mo$^+$–Xe	>53.1±6.8	O$^+$–Ar	33.8	Rb$^+$–Kr	14.9
La$^+$–Ir	356±97	N$^+$–Ar	208.4±9.6	O$^+$–F	301.8±8.4	Rb$^+$–Na	50.1±3.9
La$^+$–O	875±25	N$^+$–F	584±42	O$^+$–H	487.9±0.34	Rb$^+$–Ne	6.95
La$^+$–Pt	522±78	N$^+$–H	≥435.67±0.77	O$^+$–N	1050.64±0.13	Rb$^+$–O	29
La$^+$–Rh	345±97	N$^+$–N	843.85±0.10	O$^+$–O	647.75±0.17	Rb$^+$–Rb	75.6±9.6
La$^+$–S	629±96	N$^+$–O	115	Os$^+$–H	238.9	Rb$^+$–Xe	21.5
La$^+$–Si	277.0±9.6	Na$^+$–Ar	19±8	Os$^+$–O	418±50	Re$^+$–C	497.7±3.9
Li$^+$–Ar	33±14	Na$^+$–Br	58.2±10.6	P$^+$–C	512±42	Re$^+$–H	224.7±6.7
Li$^+$–Bi	91±50	Na$^+$–Cl	20.3±10	P$^+$–Cl	289	Re$^+$–O	435±59
Li$^+$–Br	41.8±10.6	Na$^+$–He	7.55	P$^+$–F	490.6±8.4	Rh$^+$–C	414±17
Li$^+$–Cl	66±15	Na$^+$–I	64.9±3.0	P$^+$–H	329.6±2.1	Rh$^+$–H	164.8±3.8
Li$^+$–F	7±21	Na$^+$–Kr	~24.9	P$^+$–N	483±21	Rh$^+$–O	295.0±5.8
Li$^+$–He	10.66	Na$^+$–Li	95.8±3.9	P$^+$–O	791.3±8.4	Rh$^+$–S	226±13
Li$^+$–I	51.1±6.3	Na$^+$–Na	98.64±0.29	P$^+$–P	481±50	Ru$^+$–C	453.5±10.6
Li$^+$–Kr	48.1	Na$^+$–Na	6.4	P$^+$–S	606±34	Ru$^+$–H	160.2±5.0
Li$^+$–Li	137.3±6.3	Na$^+$–Ne	~9.04	Pa$^+$–O	~800	Ru$^+$–O	372±5
Li$^+$–Ne	15.32	Na$^+$–O	37±19	Pb$^+$–Br	260±63	Ru$^+$–S	288±6
Li$^+$–O	38.9±9.6	Na$^+$–Xe	~28.6	Pb$^+$–Cl	285±63	S$^+$–C	620.8±1.3
Li$^+$–Sb	129.6±13.9	Nb$^+$–Ar	40.87±0.13	Pb$^+$–F	347±32	S$^+$–F	343.5±4.8
Li$^+$–Xe	56.4	Nb$^+$–C	509±15	Pb$^+$–O	247±8.4	S$^+$–H	348.2±1.7
Lu$^+$–Br	86.1	Nb$^+$–Fe	>251	Pb$^+$–Pb	214±29	S$^+$–N	516±34
Lu$^+$–Cl	180.6	Nb$^+$–H	220±7	Pb$^+$–S	293±50	S$^+$–O	524.3±0.4

A^+-B	D^o_{298} kJ/ mol^{-1}	A^+-B	D^o_{298} kJ/ mol^{-1}	A^+-B	D^o_{298} kJ/ mol^{-1}	A^+-B	D^o_{298} kJ/ mol^{-1}
S^+-P	573±21	Sn^+-O	281±10	Ti^+-Pt	82±96	W^+-F	444±96
S^+-S	522.4±0.5	Sn^+-S	240±19	Ti^+-S	461.1±6.8	W^+-H	222.5±5
Sc^+-C	326±6	Sn^+-Se	174±6.3	Ti^+-Si	249±16	W^+-O	695±42
Sc^+-Cl	410±42	Sn^+-Sn	193	Ti^+-Ti	229	Xe^+-Ar	13.4
Sc^+-F	605±32	Sn^+-Te	168.7±8.4	Tl^+-Br	52±50	Xe^+-H	355
Sc^+-Fe	201±21	Sr^+-Ar	13.32±2.92	Tl^+-Cl	26±4	Xe^+-Kr	41.65±0.08
Sc^+-H	235±8	Sr^+-Br	378.1±8.4	Tl^+-F	13±21	Xe^+-N	62.7±9.6
Sc^+-O	689±5	Sr^+-Cl	427±8.4	Tl^+-I	133±21	Xe^+-Ne	2.1±0.8
Sc^+-S	479.5±4.8	Sr^+-F	615±50	Tl^+-Tl	22±50	Xe^+-Xe	99.6
Sc^+-Se	475.8±8.4	Sr^+-H	209±5	Tm^+-Br	312.2	Y^+-C	281±12
Sc^+-Si	242.3±10.5	Sr^+-I	308.2	Tm^+-Cl	407.9	Y^+-F	677±21
Se^+-F	364±42	Sr^+-Kr	18.13±6.94	Tm^+-F	537±16	Y^+-H	260.5±5.8
Se^+-H	304	Sr^+-Ne	4.52±9.6	Tm^+-I	266.8	Y^+-O	718±25
Se^+-P	514±25	Sr^+-O	298.7	Tm^+-O	482±15	Y^+-Pt	466±192
Se^+-S	392±19	Sr^+-Sr	105.2±1.6	U^+-Br	345±29	Y^+-S	533.9±8
Se^+-Se	413±19	Ta^+-H	230±6	U^+-C	300±96	Y^+-Si	243±13
Si^+-Au	175±50	Ta^+-O	787±63	U^+-Cl	431±34	Y^+-Te	360±96
Si^+-B	351±15	Ta^+-Ta	666	U^+-D	283.4±9.6	Y^+-Y	281±21
Si^+-Br	276±96	Tb^+-Cu	245±34	U^+-F	668±29	Yb^+-Br	307.4
Si^+-C	365±50	Tb^+-O	722±15	U^+-H	284±8	Yb^+-Cl	399.6
Si^+-Cl	591.0±0.6	Tc^+-H	197.5	U^+-N	~485	Yb^+-F	557.5±14.4
Si^+-F	684.1±5.4	Tc^+-O	>167	U^+-O	757±42	Yb^+-I	262.0
Si^+-H	316.6±2.1	Te^+-H	305±12	U^+-P	186	Yb^+-O	376±15
Si^+-O	478±13.4	Te^+-O	339±50	U^+-S	518±29	Yb^+-Yb	238±96
Si^+-P	272±50	Te^+-P	415±97	V^+-Ar	39.39±0.12	Zn^+-Ar	28.7±1.2
Si^+-Pd	237±50	Te^+-Se	342±19	V^+-C	373±13.5	Zn^+-H	216±15
Si^+-Pt	525±50	Te^+-Te	278±29	V^+-D	202±6	Zn^+-O	161.1±4.8
Si^+-S	387.5±6.0	Th^+-Cl	499±29	V^+-Fe	314±21	Zn^+-S	198±12
Si^+-Si	334±19	Th^+-F	682±29	V^+-H	202±6	Zn^+-Si	274.1±9.6
Si^+-Te	347±50	Th^+-O	875±16	V^+-Kr	49.46±0.18	Zn^+-Zn	60±19
Sm^+-Br	343.3	Th^+-Pt	388±193	V^+-N	448.6±5.8	Zr^+-Ar	36.09±0.24
Sm^+-Cl	435.4	Th^+-Rh	504±67	V^+-Nb	403.5±0.2	Zr^+-C	445.8±15.4
Sm^+-F	620.9	Ti^+-C	395±23	V^+-O	581.6±9.6	Zr^+-H	218.8±9.6
Sm^+-I	299.1	Ti^+-Cl	426.8	V^+-S	358.9±8.7	Zr^+-N	443±46
Sm^+-O	569±15	Ti^+-F	≥456	V^+-Si	229±15	Zr^+-O	753±11
Sn^+-Br	335±50	Ti^+-H	226.6±10.6	V^+-V	302	Zr^+-S	533±21
Sn^+-Cu	184±96	Ti^+-N	501±13	V^+-Xe	66.4±0.6	Zr^+-Zr	407.0±9.6
Sn^+-F	364±29	Ti^+-O	667±7	W^+-C	483±21		

TABLE 7. Bond Dissociation Energies in Polyatomic Cations

This Table has been arranged on the basis of the Periodic Table with the IUPAC notation for Groups 1 to 18, see inside front cover of this *Handbook*. The **boldface** in the species indicates the dissociated fragment.

Reference

Luo, Y.-R. *Comprehensive Handbook of Chemical Bond Energies*, Taylor & Francis, 2006.

Bond	$D^\circ_{298}/$ kJ mol^{-1}	Bond	$D^\circ_{298}/$ kJ mol^{-1}	Bond	$D^\circ_{298}/$ kJ mol^{-1}
(1) Group 1		$K^+(H_2O)_6-H_2O$	41.8	*(3) Group 3*	
Li$^+$–H_2	27.2	**K**$^+$–NH_3	79±7	**Sc**$^+$–H_2	23.0±1.3
Li$^+$–CO	57±13	**K**$^+$–C_6H_6	80.3	**Sc**$^+$–CH_2	412±22
Li$^+$–H_2O	139±8	**K**$^+$–adenine	95.1±3.2	**Sc**$^+$–CH_3	233±10
Li$^+$–NH_3	156±8	**K**$^+$–indole	104.6±12.6	**Sc**$^+$–C_2H_2	240±20
Li$^+$–CH_4	130	**K**$^+$–Phe (phenylalanine)	150.5±5.8	**Sc**$^+$–C_2H_4	≥131
Li$^+$–CH_3OH	156±8	**K**$^+$–Tyr (tyrosine)	165.0±5.8	**Sc**$^+$–C_6H_6	222±21
Li$^+$–CH_3OCH_3	167±10	**Rb**$^+$–H_2O	66.9±12.6	**Sc**$^+$–H_2O	131
Li$^+$–pyridine	183.0±14.5	**Rb**$^+$–NH_3	78.2	**Sc**$^+$–NH	483±10
Li$^+$–Gly (glycine)	220±9	**Rb**$^+$–CH_3CN	86.6±1.3	**Sc**$^+$–NH_2	347±5
Na$^+$–H_2	10.4±0.8	**Rb**$^+$–C_6H_5OH	70.2±3.7	**Sc**$^+$–pyridine	231.5±10.3
Na$^+$–N_2	33.5	**Cs**$^+$–H_2O	57.3	**Y**$^+$–CH_2	398±13
Na$^+$–CO	31±8	**Cs**$^+$–$C_6H_5NH_2$	70.8±4.5	**Y**$^+$–CH_3	249±5.0
Na$^+$–CO_2	66.5			**Y**$^+$–C_2H_2	218±13
Na$^+$–SO_2	79.1	*(2) Group 2*		**Y**$^+$–C_2H_4	>138
Na$^+$–O_3	52.3	**CH_3Be**$^+$–CH_3	192.9±13.4	**Y**$^+$–CO	29.9±10.6
Na$^+$–H_2O	91.2±6.3	*tert*-**$C(CH_3)_3Be$**$^+$–*tert*-$C(CH_3)_3$	121.8±13.4	**Y**$^+$(O)–CO_2	86±5
Na$^+(H_2O)$–H_2O	82.0±5.8	**Mg**$^+$–OH	314±33	**La**$^+$–CH	523±33
Na$^+(H_2O)_2$–H_2O	66.1	**Mg**$^+$–CO	43.1±5.8	**La**$^+$–CH_2	401±7
Na$^+(H_2O)_3$–H_2O	52.7±0.8	**Mg**$^+$–CO_2	58.4±5.8	**La**$^+$–CH_3	217±15
Na$^+$(glycine)–H_2O	75.1±5.3	**Mg**$^+$–H_2O	122.5±12.5	**La**$^+$–C_2H_2	262±30
Na$^+$(glutamine)–H_2O	52±1	**Mg**$^+$–NH_3	158.9±11.6	**La**$^+$–C_2H_4	192.5
Na$^+$–NH_3	106.2±5.4	**Mg**$^+$–CH_4	29.8±6.8	**Lu**$^+$–CH_2	>230±6
Na$^+$–HNO_3	86.2	**Mg**$^+$–MeOH	147.6±6.8	**Lu**$^+$–CH_3	176±20
Na$^+$–CH_4	30.1	**Mg**$^+$–C_6H_6	155.2	**U**$^+$(F)–F	552±44
Na$^+$–CH_3OH	98.8±5.7	**Mg**$^+$–pyridine	200.0±6.4	**U**$^+$(F)$_2$–F	523±38
Na$^+$–CH_3CN	125.5±9.6	**Mg**$^+$–imidazole	243.9±10.4	**U**$^+$(F)$_3$–F	381±19
Na$^+$–C_2H_4	44.6±4.4	**Mg**$^{2+}(H_2O)_5$–H_2O	101.3	**U**$^+$(F)$_4$–F	243±17
Na$^+$–CH_3OCH_3	101.4±5.7	**Mg**$^{2+}(Me_2CO)_5$–Me_2CO	93.3	**U**$^+$(F)$_5$–F	26±11
Na$^+$–$CH_3C(O)H$	114.4±3.4	**Ca**$^+$–OH	435.1±14.5		
Na$^+$– MeCOMe	131.3±4.1	**Ca**$^+$–H_2O	117.2	*(4) Group 4*	
Na$^+$–C_6H_6	97.0±5.9	**Ca**$^+$–C_6H_6	134	**Ti**$^+$–CH	478±5
Na$^+$–pyrrole	103.7±4.8	**Ca**$^+$–imidazole	186.3±3.9	**Ti**$^+$–CH_2	391±15
Na$^+$ –Gly (glycine)	166.7±5.1	**Ca**$^{2+}(H_2O)_4$–H_2O	110.0±5.9	**Ti**$^+$–CH_3	213.8±3
Na$^+$ –Ala (alanine)	167±4	**Ca**$^{2+}(Me_2CO)_5$–Me_2CO	101.3	**Ti**$^+$–CH_4	70.3±2.5
Na$^+$–GlyGly (glycylglycine)	203±8	**Sr**$^+$–CO	20.3	**Ti**$^+$–C_2H_2	213±13
K$^+$–H_2	6.1±0.8	**Sr**$^+$–CO_2	41.9	**Ti**$^+$–C_2H_4	146±11
K$^+$–CO_2	35.6	**Sr**$^+$–H_2O	144.3	**Ti**$^+$–C_6H_6	259±9
K$^+$–H_2O	74.9	**Sr**$^+$–C_6H_6	117	**Ti**$^+$–CO	117.7±5.8
K$^+(H_2O)_2$–H_2O	67.4	**Sr**$^{2+}(H_2O)_5$–H_2O	87.4	**Ti**$^+$–H_2O	157.7±5.9
K$^+(H_2O)_3$–H_2O	55.2	**Ba**$^+$–OH	530.7±19.3	**Ti**$^+$–NH	466±12
K$^+(H_2O)_4$–H_2O	11.8	**Ba**$^{2+}(H_2O)_4$–H_2O	90.8	**Ti**$^+$–NH_2	356±13
K$^+(H_2O)_5$–H_2O	44.8			**Ti**$^+$–NH_3	197±7
				Ti$^+$–pyridine	217.2±9.3

Bond	$D^{\circ}_{298}/$ kJ mol^{-1}	Bond	$D^{\circ}_{298}/$ kJ mol^{-1}	Bond	$D^{\circ}_{298}/$ kJ mol^{-1}
Ti$^+$–imidazole	≤232.4±8.2	Cr$^+$–C$_6$H$_6$	170±10	Fe$^+$(η^5-C$_5$H$_5$)–H	193±21
Zr$^+$–CH	568±13	Cr$^+$–indole	196.6±16.7	(CO)$_2$)$_5$Fe$^+$–H	299±15
Zr$^+$–CH$_2$	444.8±5	Cr$^+$–CO	89.7±5.8	Fe$^+$–H$_2$	45.2±2.5
Zr$^+$–CH$_3$	227.7±9.6	Cr$^+$–OH	298±14	Fe$^+$–CH	423±29
Zr$^+$–C$_2$H$_2$	273±14	Cr$^+$–H$_2$O	132.6±8.8	Fe$^+$–CH$_2$	≤342±2
Zr$^+$–CO	77±10	Cr$^+$–N$_2$	59±4	Fe$^+$–CH$_3$	229±5
Zr$^+$–CS	242±11	Cr$^+$–NH$_3$	183±10	Fe$^+$–CH$_4$	73.2
Hf$^+$–CH$_2$	427±38	(CO)$_6$Mo$^+$–H	260±9	Fe$^+$–C$_2$H$_2$	159.0±2.1
Hf$^+$–C$_2$H$_2$	150.6	Mo$^+$–CH	509±10	Fe$^+$–C$_2$H$_3$	238±10
		Mo$^+$–CH$_2$	329±12	Fe$^+$–C$_2$H$_4$	145±11
(5) Group 5		Mo$^+$–CH$_3$	157±12	Fe$^+$–C$_2$H$_5$	233±9
(CO)$_6$V$^+$–H	220±14	Mo$^+$–CO	193.9±9.6	Fe$^+$–C$_2$H$_6$	64±6
V$^+$–H$_2$	42.7±2.1	Mo$^+$–CO$_2$	49.2±7	Fe$^+$–OH	366±12
V$^+$–CH	470±5	Mo$^+$–CS	162±18	Fe$^+$–CO	129.3±3.9
V$^+$–CH$_2$	326±6	Mo$^+$–CS$_2$	67.5±12.5	Fe$^+$D–CO	53±13
V$^+$–CH$_3$	193±7	Mo$^+$–NH	<385	Fe$^+$–CO$_2$	74.3±7.7
V$^+$–C$_2$H$_2$	172±8	Mo$^+$–pyrrole	>289	Fe$^+$–H$_2$O	128.9±0.8
V$^+$–C$_2$H$_4$	124±8	(CO)$_6$W$^+$–H	257±9	Fe$^+$–N$_2$	53±4
V$^+$–(η^5-C$_5$H$_5$)	530.7	W$^+$–CH	580±27	Fe$^+$–NH$_3$	184±12
V$^+$–C$_6$H$_6$	234±10	W$^+$–CH$_2$	456.4±5.8	Fe$^+$–CS$_2$	166.1±4.6
V$^+$–CO	114.8±2.9	W$^+$–CH$_3$	~222.9±9.6	Fe$^+$–imidazole	246.1±13.8
V$^+$–CO$_2$	72.4±3.8	(PMe$_3$)$_3$(CO)$_3$W$^+$–H	259.4	Fe$^+$–SiH	254±13
V$^+$–H$_2$O	149.8±5.0	W$^+$–pyrrole	>209	Fe$^+$–SiH$_2$	181±9
V$^+$–NH	423±29			Fe$^+$–SiH$_3$	183±9
V$^+$–NH$_2$	293±6	*(7) Group 7*		Ru$^+$(η^5-C$_5$H$_5$)$_2$–H	292±16
V$^+$–NH$_3$	192±11	(CO)$_5$Mn$^+$–H	172±10	(η^5-C$_5$Me$_5$)$_2$Ru$^+$–H	284.5
V$^+$–pyridine	218.7±13.5	Mn$^+$–H$_2$	7.9±1.7	Ru$^+$–CH	501.7±11.6
V$^+$–imidazole	≤243.4±8.0	Mn$^+$–CH$_2$	295±13	Ru$^+$–CH$_2$	344.4±4.8
Nb$^+$–H$_2$	61.9	Mn$^+$–CH$_3$	215±10	Ru$^+$–CH$_3$	160.2±5.8
Nb$^+$–CH	581±19	Mn$^+$(CO)$_5$–CH$_3$	132±15	Ru$^+$–CS	253±20
Nb$^+$–CH$_2$	428.4±8.7	Mn$^+$(CO)$_5$–CH$_4$	>30	OsO$_4$$^+$–H	552±13
Nb$^+$–CH$_3$	198.8±10.6	Mn$^+$–(η^5-C$_5$H$_5$)	326.1±9.6		
Nb$^+$–CH$_3$NH$_2$	134	Mn$^+$–C$_6$H$_6$	145±10	*(9) Group 9*	
Nb$^+$–C$_3$H$_6$	117.7	Mn$^+$–OH	332±24	(η^5-C$_5$H$_5$)(CO)$_2$Co$^+$–H	245±12
(NbFe)$^+$–C$_3$H$_4$	>163	Mn$^+$–CO	25±10	(CH$_3$OD)Co$^+$–H	147.6±7.7
Nb$^+$–CO	95.5±4.8	Mn$^+$–H$_2$O	121.8±5.9	Co$^+$–H$_2$	76.1±4.2
Nb$^+$–CS	242±11	Mn$^+$–CH$_3$OH	134±29	(η^5-C$_5$H$_5$)Co$^+$–H$_2$	67.8
Nb$_7$$^+$–N$_2$	<215	Mn$^+$–OC(CH$_3$)$_2$	159±14	Co$^+$–CH	420±37
Ta$^+$–CH	575±9	Mn$^+$–CS	80.0±21	Co$^+$–CH$_2$	317±5
Ta$^+$–CH$_2$	485±5	Mn$^+$–NH$_2$	254±20	Co$^+$–CH$_3$	203±4
Ta$^+$–CH$_3$	196±3	Mn$^+$–NH$_3$	147±8	Co$^+$–CH$_4$	96.7
Ta$^+$–C$_6$H$_6$	251~301	Tc$^+$–CH$_2$	<464	Co$^+$–C$_{60}$	243±67
		Tc$^+$–C$_2$H$_2$	<320	Co$^+$–CO	173.7±6.7
(6) Group 6		Re$^+$(CH$_3$)(CO)$_5$–H	294±13	Co$^+$–H$_2$O	164.4±5.9
(CO)$_6$Cr$^+$–H	230±10	(PMe$_3$)(CO)$_2$Re$^+$–H	300.4	Co$^+$–CS	259±33
(η^5-C$_5$H$_5$)(NO)(CO)$_2$Cr$^+$–H	207.1±14			Co$^+$–N$_2$	96.2±7.1
Cr$^+$–H$_2$	31.8±2.1	*(8) Group 8*		Co$^+$–NH$_2$	247±7
Cr$^+$–CH	294±29	Fe$^+$(O)–H	444±17	Co$^+$–NH$_3$	219±16
Cr$^+$–CH$_2$	216±4	Fe$^+$(CO)–H	120±23	Co$^+$–CH$_3$CN	>255±17
Cr$^+$–CH$_3$	110±4	Fe$^+$(H$_2$O)–H	215±14		

Bond	$D^o_{298}/$ kJ mol^{-1}	Bond	$D^o_{298}/$ kJ mol^{-1}	Bond	$D^o_{298}/$ kJ mol^{-1}
$Co^+-P(CH_3)_3$	278±11	**(11) Group 11**		**(13) Group 13**	
$Co^+-P(C_2H_5)_3$	339±16	Cu^+-H_2	51.9±0.4	B^+-H_2	15.9±0.8
$(CH)Rh^+-H$	372±21	Cu^+-CH_2	267.3±6.8	HB^+-H_2	61.5±2.1
$(\eta^5-C_5H_5)(CO)_2Rh^+-H$	287±12	Cu^+-CH_3	111±7	$(CH_3)_2B^+-CH_3$	32.6±4.2
Rh^+-CH	444±12	$Cu^+-C_2H_2$	>21.2±9.6	Al^+-H_2	5.6±0.6
Rh^+-CH_2	356±8	$Cu^+-C_2H_4$	176±14	Al^+-N_2	5.6
Rh^+-CH_3	142±6	$Cu^+-C_6H_6$	218.0±9.6	Al^+-CO_2	≥29.3
Rh^+-NO	167±21	Cu^+-CO	149±7	Al^+-H_2O	104±15
Rh^+-CS	234±19	Cu^+-N_2	89±30	Al^+-MeOH	139.7
$(CO)(\eta^5-C_5H_5)(PPh_3)Ir^+-H$	313.4	Cu^+-NO	109.0±4.8	$Al^+-EtC(O)Et$	191.2
$(CO)_2(\eta^5-C_5Me_5)Ir^+-H$	298.3	Cu^+-H_2O	160.7±7.5	$Al^+-C_6H_6$	147.3±8.4
Ir^+-CH_2	>464	Cu^+-NH_2	192±13	$Al^+-pyridine$	190.3±10.3
$Ir^+-C_2H_4$	234.3	Cu^+-NH_3	237±15	$Al^+-phenol$	154.8±16.7
(10) Group 10		Cu^+-CS	238.3±11.6	$Al^+-imidazole$	232.4±8.2
$(CO)_4Ni^+-H$	248±9	Cu^+-SiH	246±27	Ga^+-NH_3	122.5
$(\eta^5-C_5H_5)(NO)Ni^+-H$	315±14	Cu^+-SiH_2	≥231±7	In^+-NH_3	111.0
$(\eta^5-C_5H_5)(\eta^5-C_5H_5)Ni^+-H$	215±13	Cu^+-SiH_3	97±25	**(14) Group 14**	
Ni^+-H_2	72.4±1.3	Ag^+-CH_2	≥107±4	$C_{58}^+-C_2$	955±15
Ni^+-CH	301.0±11.6	Ag^+-CH_3	66.6±4.8	$C_{60}^+-C_2$	822.0±12.5
Ni^+-CH_2	306±4	$Ag^+-C_2H_5$	65.7±7.5	$C_{62}^+-C_2$	846.2±10.6
Ni^+-CH_3	169.8±6.8	$Ag^+-C_6H_6$	167±19	$C_{78}^+-C_2$	938.8±10.6
Ni^+-CH_4	96.5±4	Ag^+-O_2	29.7±0.8	HC_2^+-H	574.749
Ni^+-OH	235±19	Ag^+-CO	89±5	$C_6H_5^+-H$	376.3±4.8
Ni^+-CO	175±11	Ag^+-H_2O	134±8	$C_2H_3^+-Cl$	247±4.8
Ni^+-CO_2	104±1	Ag^+-CS	152±20	$C_2H_5^+-Br$	126.4
Ni^+-H_2O	183.7±3.3	Ag^+-NH_3	170±13	$C_6H_5^+-Br$	266.3
Ni^+-CS	234.5±9.6	Au^+-CH_2	≤372±3	$CH_3^+-H_2$	186
Ni^+-N_2	110.9±10.5	$Au^+-C_2H_4$	344.5	$CH_5^+-H_2$	7.9±0.4
Ni^+-NO	227.6±7.5	$Au^+-C_6H_6$	289±29	$C_2H_5^+-H_2$	17
Ni^+-NH_2	232.5±7.7	Au^+-CO	201±8	CO^+-N_2	67.5±19.3
Ni^+-NH_3	238±19	Au^+-H_2O	164.0±9.6	$H_2CH^+-N_2$	31.8
Pd^+-CH	536±10	Au^+-H_2S	230±25	CO^+-CO	173.7±14.6
Pd^+-CH_2	463±3	Au^+-NH_3	297±29	$CO^+(CO)-CO$	52.3
Pd^+-CH_3	258±8	Au^+-PH_3	402±33	$CO^+(CO)_2-CO$	30.2
Pd^+-CH_4	170.8±7.7	**(12) Group 12**		$CO^+(CO)_3-CO$	18.4
Pd^+-CS	200±14	Zn^+-H_2	15.7±1.7	$(CO_2)^+-CO_2$	70.3
$Pd^+-C_2H_2$	>28.9±4.8	Zn^+-CH_3	280±7	$(CO_2)^+(CO_2)-CO_2$	34.7
Pt^+-H_2	146.7±11.6	Zn^+-OH	127.2	$(CO_2)^+(CO_2)_2-CO_2$	21.3
Pt^+-CH	536.4±9.6	Zn^+-H_2O	163	$(CO_2)^+(CO_2)_3-CO_2$	20.1±1.3
Pt^+-CH_2	471	Zn^+-NO	76.2±9.6	$CH_3^+-N_2O$	221.3
Pt^+-CH_3	257.6±7.7	$Zn^+-pyrimidine$	209.6±7.7	$CH_3^+-SO_2$	253.6
Pt^+-CH_4	170.8±7.7	Zn^+-CS	149±23	CH_3^+-OCS	239.3
Pt^+-O_2	64.6±4.8	Cd^+-CH_3	228±3	$CH_3^+-CS_2$	251.9
Pt^+-CO	218.1±8.7	$Cd^+(CH_3)-CH_3$	109±3	$CH_3^+-H_2O$	279
Pt^+-CO_2	59.8±4.8	$Cd^+-C_6H_6$	136±19	$CH_3^+(H_2O)-H_2O$	106.3
Pt^+-NH_3	274±12	Hg^+-CH_3	285±3	$CH_3^+(H_2O)_2-H_2O$	87.9
$Pt^+-C_2H_4$	229.7	$Hg^+(CH_3)-CH_3$	96±3	$CH_3^+(H_2O)_3-H_2O$	61.9
				$CH_3^+(H_2O)_4-H_2O$	48.5

Bond	$D^{\circ}_{298}/$ kJ mol^{-1}	Bond	$D^{\circ}_{298}/$ kJ mol^{-1}	Bond	$D^{\circ}_{298}/$ kJ mol^{-1}
$CH_3{}^+-H_2S$	344.8	Pb^+-NH_3	118.4±0.8	(valine H)$^+$–valine	86.6±8.4
$CH_2{}^+-CH_2O$	303.0±2.9	Pb^+-CH_3OH	97.5±0.8	(betainH)$^+$–betaine	139.9±4.8
$CH_3{}^+-NH_3$	431.4	$Pb^+-CH_3NH_2$	148.1 ±1.3	$H_4P^+-H_2O$	54.4
$(CH_3)^+-CH_3$	209.2±4.2	$Pb^+-C_6H_6$	110±2	$(H_4P)^+-PH_3$	48.1
$CH_3{}^+-CH_4$	166.5	**(15) Group 15**		$AsH_2{}^+-H$	257
$CF_3{}^+-CH_4$	19.0	H_2N^+-H	544.43±0.10	I_2As^+–acetone	106±17
$(CH_5)^+-CH_4$	28.7±1.3	H_3N^+-H	515.1	I_2As^+–benzene	77±17
$C_6H_6{}^+-CH_4$	12.0	Me_3N^+-H	376	Bi^+-H_2O	95.4
$CH_3{}^+-CH_3F$	230	Et_3N^+-H	362	Bi^+-NH_3	149
$CH_3{}^+-CF_3Cl$	221	(imidazole)$^+$ –Zn	216.1±3.9	$Bi^+-C_6H_6$	≤149
$CH_3{}^+-CH_3Cl$	259	$N_2H^+-H_2$	24.7±0.8	**(16) Group 16**	
tert-$C_4H_9{}^+-CH_3OH$	63	ON^+-O_2	14.2	$(H_3O)^+-H_2$	14.6±2.1
tert-$C_4H_9{}^+-CH_3CN$	85	N^+-N_2	303.8	O^+-O_2	179.5
tert-$C_4H_9{}^+-SO_2F_2$	43.5	ON^+-N_2	21.3	$O^+(O_2)_1-O_2$	28.9
$CH_3{}^+-C_2H_3O$	338.7±2.9	$N_2{}^+-N_2$	102.3±14.6	$O^+(O_2)_2-O_2$	3.9
$CH_3{}^+-CF_3ClOCl$	252	$HN_2{}^+-N_2$	60.7	$O_2{}^+-O_2$	38.3±2.1
tert--$C_4H_9{}^+-(CH_3)_2S$	185	$N_3{}^+-N_2$	18.8±1.3	$O_2{}^+(O_2)-O_2$	24.6±1.3
tert-$C_4H_9{}^+-C_2H_5OH$	85	$O_2N^+-N_2$	19.2±1.3	$O_2{}^+(O_2)_2-O_2$	10.4±0.8
tert-$C_4H_9{}^+-C_3H_8$	27.6	$H_4N^+-N_2$	54±21	$O_2{}^+(O_2)_3-O_2$	9.0±0.8
tert-$C_4H_9{}^+-t-C_4H_9Cl$	339	ON^+-NO	59.4±0.8	$O_2{}^+(O_2)_4-O_2$	8.0±0.8
tert-$C_4H_9{}^+-(CH_3)_3CH$	30.1	ON^+-CO	27.2±1.3	$O_2{}^+(O_2)_5-O_2$	7.9±1.3
tert-$C_4H_9{}^+-C_6H_6$	92	ON^+-O_3	<58	O^+-N_2	231.4
$(C_6H_6)^+-C_6H_6$	73.6	ON^+-CO_2	32.2	$O_2{}^+-N_2$	22.6
$(C_6H_6)^+$–indole	54.8	$N_2O^+-ON_2$	72.8±6.3	$(H_3O)^+-N_2$	22.2±2.1
$C_6F_6{}^+-C_6F_6$	30.1±4	NO^+-ON_2	36.4±0.8	$O_4{}^+-N_2$	12.3
$C_{60}{}^+-C_{60}$	35.89±7.72	$(HON_2)^+-ON_2$	69.9±4	$O_2{}^+-CO$	31.8
$PhSiH_2{}^+-H$	159	ON^+-H_2O	95	$O_2{}^+-CO_2$	41.0±2.1
$Si^+(CH_3)_3-Cl$	178.5±1.9	$ON^+(H_2O)-H_2O$	67.4	$CO_2{}^+-CO_2$	65.3±4
$SiH_3{}^+-CO$	≥151	$ON^+(H_2O)_2-H_2O$	56.5	$(H_3O)^+-CO_2$	64.0
$SiF_3{}^+-CO$	174.1±1.3	$H_4N^+-H_2O$	86.2±4.2	$(H_3O)^+(CO_2)-CO_2$	51.9
$(CH_3)_3Si^+-H_2O$	125.9±7.9	$H_4N^+(H_2O)-H_2O$	72.8±4.2	$(H_3O)^+(CO_2)_2-CO_2$	43.9
$(CH_3)_3Si^+-NH_3$	194.6	$H_4N^+(H_2O)_2-H_2O$	57.3±4.2	$(H_3O)^+(CO_2)_3-CO_2$	18.0
$Si^+(CH_3)(Cl)_2-CH_3$	60.8±2.9	$H_4N^+(H_2O)_3-H_2O$	51.0	$O_2{}^+-ON_2$	56.1±4
$Si^+(CH_3)_2(Cl)-CH_3$	41.5±1.9	$H_4N^+(H_2O)_4-H_2O$	44.4	$(H_3O)^+-ON_2$	70.7±6.5
Si^+-CH_3	413.9±5.8	(glycine)H$^+-H_2O$	77.2±11.0	$(H_3O)^+(H_2O)-ON_2$	50.6±2.1
$Si^+(CH_3)-CH_3$	123±48	(tryptophan)H$^+-H_2O$	31.2±2.5	$(H_3O)^+(H_2O)_2-ON_2$	42.7±2.1
$Si^+(CH_3)_2-CH_3$	513±27	(tryptophanylglicine)H$^+-H_2O$	56.0±5.3	$O_3{}^+-O_3$	67.5±39
$Si^+(CH_3)_3-CH_3$	66.6±5.8	$H_4N^+-H_2S$	47.7	$OClO^+-OClO$	246±48
$(CH_3)_3Si^+-CH_3OH$	164.0	$H^+(NH_3)-NH_3$	108.8	$O_2{}^+-H_2O$	>67
$(CH_3)_3Si^+-(C_2H_5)_2O$	184.9	$H^+(NH_3)_2-NH_3$	69.5	$(OH)^+(H_2O)_2-H_2O$	87.4
$(CH_3)_3Si^+-C_6H_6$	100.0	$H^+(NH_3)_3-NH_3$	57.3	$(OH)^+(H_2SO_4)(H_2O)_4-H_2O$	56.9
$(CH_3)_3Si^+-CH_3NH_2$	231.8	$H^+(NH_3)_4-NH_3$	49.0	$(OH)^+(H_2SO_4)(H_2O)_5-H_2O$	49.8
$(CH_3)_3Ge^+-H_2O$	119.7±2.1	$H^+(NH_3)_5-NH_3$	29.3	$(OH)^+(H_2SO_4)(H_2O)_6-H_2O$	44.8
$(C_2H_5)_3Ge^+-H_2O$	104.2±2.1	$H^+(NH_3)_6-NH_3$	27.2	$(H_2O)^+-H_2O$	164.0
$(CH_3)_3Sn^+-NH_3$	154	$NH_4{}^+-CH_4$	15.0	$(H_3O)^+-H_2O$	140.2
$(CH_3)_3Sn^+-H_2O$	108	ON^+-CH_3OH	97.6	$(H_3O)^+(H_2O)-H_2O$	93.3
$(CH_3)_3Sn^+-(CH_3)_2CO$	157	$O_2N^+-CH_3OH$	80.3±9.6	$(H_3O)^+(H_2O)_2-H_2O$	71.1
$(CH_3)_3Sn^+-C_3H_7SH$	143	$(CH_3CNH)^+-CH_3CN$	130.1±9.6	$(H_3O)^+(H_2O)_3-H_2O$	64.0
Pb^+-H_2O	93.7	(pyridineH)$^+$–pyridine	105.4±4	$(H_3O)^+(H_2O)_4-H_2O$	54.4

Bond	$D^\circ_{298}/$ kJ mol^{-1}	Bond	$D^\circ_{298}/$ kJ mol^{-1}	Bond	$D^\circ_{298}/$ kJ mol^{-1}
$(H_3O)^+(H_2O)_5$–H_2O	49.0	S^+–CS_2	166	$I^+(CH_3I)_2$–CH_3	112.9
$(H_3O)^+(H_2O)_6$–H_2O	43.1	CS^+–CS_2	150.6	*(18) Group 18*	
$(HCOOH)H^+$–H_2O	100.8	CS_2^+–CS_2	104.2	$He^+(He)_1$–He	17.6
$CH_3OH_2^+$–H_2O	115.6	HCS_2^+–CS_2	46.4	$He^+(He)_2$–He	2.7±0.6
CH_3CHOH^+–H_2O	104.6	OS^+–SO_2	57.7	$Ne^+(Ne)$–Ne	10.3±0.6
$(CH_3)_2OH^+$–H_2O	100.4	O_2S^+–SO_2	63.6	$Ne^+(Ne)_2$–Ne	3.3±0.6
$(tetrahydrofuranH)^+$–H_2O	82.8	OCS^+–OCS	100.0	$Ar^+(Ar)$–Ar	20.4±0.6
$(furanH)^+$–H_2O	43.5	OCS^+–CO_2	72.0	$Ar^+(Ar)_2$–Ar	7.0±0.6
$furane^+$–H_2O	41.0	SO_2^+–CO_2	42.7	$Ar^+(N_2)$–Ar	25.1
$(phenol)^+$–H_2O	78.0	H_3S^+–H_2O	91.6	$Ar^+(N_2)(Ar)$–Ar	7.1
$(1\text{-naphthol})^+$–H_2O	66.4	$thiopheneH^+$–H_2O	42.7	$Ar^+(N_2)(Ar)_2$–Ar	7.1
H_3O^+–$HC(O)H$	137.7	H_3S^+–H_2S	53.6±6.3	$Kr^+(Kr)$–Kr	23.3±0.6
H_3O^+–NH_3	229.3	H_3S^+–CH_4	16.3	$Kr^+(Kr)_2$–Kr	9.0±0.6
$H_3O^+(NH_3)$–NH_3	77.0	$(CH_3)_2Se^{\bullet+}$–$Se(CH_3)_2$	~95±3	$Xe^+(Xe)$–Xe	25.2±0.6
$H_3O^+(NH_3)_2$–NH_3	71.5	$(CH_3)_2Te^{\bullet+}$–$Te(CH_3)_2$	97±2	$Xe^+(Xe)_2$–Xe	11.0±0.6
$H_3O^+(NH_3)_3$–NH_3	62.8			Ar^+–H_2	93.7
H_3O^+–PH_3	144	*(17) Group 17*		Ar^+–N_2	127.6
H_3O^+–SO_3	74	HF^+–HF	≥138	$Ar^+(N_2)$–N_2	31.0
$(HCOOH)^+$–$HCOOH$	96.5±9.6	$(H_2Cl)^+$–Cl	39.6	$Ar^+(N_2)_2$–N_2	10.9
H_3O^+–CH_4	33.5	HCl^+–HCl	83.9	Ar^+–CO	75±17
$(CH_3OH)^+$–CH_3OH	115.8±19.3	Cl^+–CCl_3	446.7±9.6	$Ar^+(CO)$–CO	13
$CH_3OH_2^+$–CH_3OH	136.4	Cl^+–C_2H_3	685.0±4.8	Kr^+–CO	103.3±7.5
H_3O^+–CH_3CN	195.4	HBr^+–HBr	96	Kr^+–CO_2	79.1±2.9
$furan^+$–$furan$	94.1	I^+–CH_3	330.0		
BH^+–B, B = tetrahydofuran	125.1	$I^+(CH_3I)$–CH_3	51.1		

ELECTRONEGATIVITY

Electronegativity is a parameter originally introduced by Pauling which describes, on a relative basis, the tendency of an atom in a molecule to attract bonding electrons. While electronegativity is not a precisely defined molecular property, the electronegativity difference between two atoms provides a useful measure of the polarity and ionic character of the bond between them. This table gives the electronegativity X, on the Pauling scale, for the most common oxidation state. Other scales are described in the references.

References

1. Pauling, L., *The Nature of the Chemical Bond, Third Edition*, Cornell University Press, Ithaca, NY, 1960.
2. Allen, L. C., *J. Am. Chem. Soc.*, 111, 9003, 1989.
3. Allred, A. L., *J. Inorg. Nucl. Chem.*, 17, 215, 1961.

Z	Symbol	X	Z	Symbol	X	Z	Symbol	X
1	H	2.20	33	As	2.18	65	Tb	—
2	He	—	34	Se	2.55	66	Dy	1.22
3	Li	0.98	35	Br	2.96	67	Ho	1.23
4	Be	1.57	36	Kr	—	68	Er	1.24
5	B	2.04	37	Rb	0.82	69	Tm	1.25
6	C	2.55	38	Sr	0.95	70	Yb	—
7	N	3.04	39	Y	1.22	71	Lu	1.0
8	O	3.44	40	Zr	1.33	72	Hf	1.3
9	F	3.98	41	Nb	1.6	73	Ta	1.5
10	Ne	—	42	Mo	2.16	74	W	1.7
11	Na	0.93	43	Tc	2.10	75	Re	1.9
12	Mg	1.31	44	Ru	2.2	76	Os	2.2
13	Al	1.61	45	Rh	2.28	77	Ir	2.2
14	Si	1.90	46	Pd	2.20	78	Pt	2.2
15	P	2.19	47	Ag	1.93	79	Au	2.4
16	S	2.58	48	Cd	1.69	80	Hg	1.9
17	Cl	3.16	49	In	1.78	81	Tl	1.8
18	Ar	—	50	Sn	1.96	82	Pb	1.8
19	K	0.82	51	Sb	2.05	83	Bi	1.9
20	Ca	1.00	52	Te	2.1	84	Po	2.0
21	Sc	1.36	53	I	2.66	85	At	2.2
22	Ti	1.54	54	Xe	2.60	86	Rn	—
23	V	1.63	55	Cs	0.79	87	Fr	0.7
24	Cr	1.66	56	Ba	0.89	88	Ra	0.9
25	Mn	1.55	57	La	1.10	89	Ac	1.1
26	Fe	1.83	58	Ce	1.12	90	Th	1.3
27	Co	1.88	59	Pr	1.13	91	Pa	1.5
28	Ni	1.91	60	Nd	1.14	92	U	1.7
29	Cu	1.90	61	Pm	—	93	Np	1.3
30	Zn	1.65	62	Sm	1.17	94	Pu	1.3
31	Ga	1.81	63	Eu	—			
32	Ge	2.01	64	Gd	1.20			

FORCE CONSTANTS FOR BOND STRETCHING

Representative force constants (f) for stretching of chemical bonds are listed in this table. Except where noted, all force constants are derived from values of the harmonic vibrational frequencies ω_e. Values derived from the observed vibrational fundamentals ν, which are noted by a, are lower than the harmonic force constants, typically by 2 to 3% in the case of heavy atoms (often by 5 to 10% if one of the atoms is hydrogen). Values are given in the SI unit newton per centimeter (N/cm), which is identical to the commonly used cgs unit mdyn/Å.

References

1. Huber, K. P., and Herzberg, G., *Molecular Spectra and Molecular Structure. IV. Constants of Diatomic Molecules*, Van Nostrand Reinhold, New York, 1979.
2. Shimanouchi, T., The Molecular Force Field, in Eyring, H., Henderson, D., and Yost, W., Eds., *Physical Chemistry: An Advanced Treatise*, Vol. IV, Academic Press, New York, 1970.
3. Tasumi, M., and Nakata, M., *Pure and Appl. Chem.*, 57, 121—147, 1985.

Bond	Molecule	f/(N/cm)	Note
H-H	H_2	5.75	
Be-H	BeH	2.27	
B-H	BH	3.05	
C-H	CH	4.48	
	CH_4	5.44	b
	C_2H_6	4.83	a,b,c
	CH_3CN	5.33	b
	CH_3Cl	5.02	a,b,c
	$CCl_2{=}CH_2$	5.57	b
	HCN	6.22	
N-H	NH	5.97	
O-H	OH	7.80	
	H_2O	8.45	
P-H	PH	3.22	
S-H	SH	4.23	
	H_2S	4.28	
F-H	HF	9.66	
Cl-H	HCl	5.16	
Br-H	HBr	4.12	
I-H	HI	3.14	
Li-H	LiH	1.03	
Na-H	NaH	0.78	
K-H	KH	0.56	
Rb-H	RbH	0.52	
Cs-H	CsH	0.47	
C-C	C_2	12.16	
	$CCl_2{=}CH_2$	8.43	
	C_2H_6	4.50	a,c
	CH_3CN	5.16	
C-F	CF	7.42	
	CH_3F	5.71	a,c
C-Cl	CCl	3.95	
	CH_3Cl	3.44	a,c
	$CCl_2{=}CH_2$	4.02	b
C-Br	CH_3Br	2.89	a,c
C-I	CH_3I	2.34	a,c
C-O	CO	19.02	
	CO_2	16.00	
	OCS	16.14	
	CH_3OH	5.42	a,c
C-S	CS	8.49	
	CS_2	7.88	

Bond	Molecule	f/(N/cm)	Note
	OCS	7.44	
C-N	CN	16.29	
	HCN	18.78	
	CH_3CN	18.33	
	CH_3NH_2	5.12	a,c
C-P	CP	7.83	
Si-Si	Si_2	2.15	
Si-O	SiO	9.24	
Si-F	SiF	4.90	
Si-Cl	SiCl	2.63	
N-N	N_2	22.95	
	N_2O	18.72	
N-O	NO	15.95	
	N_2O	11.70	
P-P	P_2	5.56	
P-O	PO	9.45	
O-O	O_2	11.77	
	O_3	5.74	a
S-O	SO	8.30	
	SO_2	10.33	a
S-S	S_2	4.96	
F-F	F_2	4.70	
Cl-F	ClF	4.48	
Br-F	BrF	4.06	
Cl-Cl	Cl_2	3.23	
Br-Cl	BrCl	2.82	
Br-Br	Br_2	2.46	
I-I	I_2	1.72	
Li-Li	Li_2	0.26	
Li-Na	LiNa	0.21	
Na-Na	Na_2	0.17	
Li-F	LiF	2.50	
Li-Cl	LiCl	1.43	
Li-Br	LiBr	1.20	
Li-I	LiI	0.97	
Na-F	NaF	1.76	
Na-Cl	NaCl	1.09	
Na-Br	NaBr	0.94	
Na-I	NaI	0.76	
Be-O	BeO	7.51	
Mg-O	MgO	3.48	
Ca-O	CaO	3.61	

[a] Derived from fundamental frequency, without anharmonicity correction.
[b] Average of symmetric and antisymmetric (or degenerate) modes.
[c] Calculated from Local Symmetry Force Field (see Reference 2).

FUNDAMENTAL VIBRATIONAL FREQUENCIES OF SMALL MOLECULES

This table lists the fundamental vibrational frequencies of selected three-, four-, and five-atom molecules. Both stable molecules and transient free radicals are included. The data have been taken from evaluated sources. In general, the selected values are based on gas-phase infrared, Raman, or ultraviolet spectra; when these were not available, liquid-phase or matrix-isolation spectra were used.

Molecules are grouped by structural type. Within each group, related molecules appear together for convenient comparison.

The vibrational modes are described by their approximate character in terms of stretching, bending, deformation, etc. However, it should be emphasized that most such descriptions are only approximate, and that the true normal mode usually involves a mixture of motions. Abbreviations are:

sym.	symmetric
antisym.	antisymmetric
str.	stretch
deform.	deformation
scis.	scissors
rock.	rocking
deg.	degenerate

In the case of free radicals, strong interactions may exist between the electronic and bending vibrational motions. Details can be found in References 3 and 4. The references should be consulted for information on the accuracy of the data and for data on other molecules not listed here.

All fundamental frequencies (more precisely, wavenumbers) are given in units of cm^{-1}.

XY_2 Molecules

Point groups $D_{\infty h}$ (linear) and C_{2v} (bent)

Molecule	Structure	Sym. str.	Bend	Antisym. str.
CO_2	Linear	1333	667	2349
CS_2	Linear	658	397	1535
C_3	Linear	1224	63	2040
CNC	Linear		321	1453
NCN	Linear	1197	423	1476
BO_2	Linear	1056	447	1278
BS_2	Linear	510	120	1015
KrF_2	Linear	449	233	590
XeF_2	Linear	515	213	555
$XeCl_2$	Linear	316		481
H_2O	Bent	3657	1595	3756
D_2O	Bent	2671	1178	2788
F_2O	Bent	928	461	831
Cl_2O	Bent	639	296	686
O_3	Bent	1103	701	1042
H_2S	Bent	2615	1183	2626
D_2S	Bent	1896	855	1999
SF_2	Bent	838	357	813
SCl_2	Bent	525	208	535
SO_2	Bent	1151	518	1362
H_2Se	Bent	2345	1034	2358
D_2Se	Bent	1630	745	1696

XY_2 Molecules

Point groups $D_{\infty h}$ (linear) and C_{2v} (bent)

Molecule	Structure	Sym. str.	Bend	Antisym. str.
NH_2	Bent	3219	1497	3301
NO_2	Bent	1318	750	1618
NF_2	Bent	1075	573	942
ClO_2	Bent	945	445	1111
CH_2	Bent		963	
CD_2	Bent		752	
CF_2	Bent	1225	667	1114
CCl_2	Bent	721	333	748
CBr_2	Bent	595	196	641
SiH_2	Bent	2032	990	2022
SiD_2	Bent	1472	729	1468
SiF_2	Bent	855	345	870
$SiCl_2$	Bent	515		505
$SiBr_2$	Bent	403		400
GeH_2	Bent	1887	920	1864
$GeCl_2$	Bent	399	159	374
SnF_2	Bent	593	197	571
$SnCl_2$	Bent	352	120	334
$SnBr_2$	Bent	244	80	231
PbF_2	Bent	531	165	507
$PbCl_2$	Bent	314	99	299
ClF_2	Bent	500		576

XYZ Molecules

Point Groups $C_{\infty v}$ (linear) and C_s (bent)

Molecule	Structure	XY str.	Bend	YZ str.
HCN	Linear	3311	712	2097
DCN	Linear	2630	569	1925
FCN	Linear	1077	451	2323
ClCN	Linear	744	378	2216
BrCN	Linear	575	342	2198
ICN	Linear	486	305	2188
CCN	Linear	1060	230	1917
CCO	Linear	1063	379	1967
HCO	Bent	2485	1081	1868
HCC	Linear	3612		1848
OCS	Linear	2062	520	859
NCO	Linear	1270	535	1921

XYZ Molecules

Point Groups $C_{\infty v}$ (linear) and C_s (bent)

Molecule	Structure	XY str.	Bend	YZ str.
NNO	Linear	2224	589	1285
HNB	Linear	3675		2035
HNC	Linear	3653		2032
HNSi	Linear	3583	523	1198
HBO	Linear		754	1817
FBO	Linear		500	2075
ClBO	Linear	676	404	1958
BrBO	Linear	535	374	1937
FNO	Bent	766	520	1844
ClNO	Bent	596	332	1800
BrNO	Bent	542	266	1799
HNF	Bent		1419	1000

XYZ Molecules

Point Groups $C_{\infty v}$ (linear) and C_s (bent)

Molecule	Structure	XY str.	Bend	YZ str.
HNO	Bent	2684	1501	1565
HPO	Bent	2095	983	1179
HOF	Bent	3537	886	1393
HOCl	Bent	3609	1242	725
HOO	Bent	3436	1392	1098
FOO	Bent	579	376	1490
ClOO	Bent	407	373	1443
BrOO	Bent			1487
HSO	Bent		1063	1009

XYZ Molecules

Point Groups $C_{\infty v}$ (linear) and C_s (bent)

Molecule	Structure	XY str.	Bend	YZ str.
NSF	Bent	1372	366	640
NSCl	Bent	1325	273	414
HCF	Bent		1407	1181
HCCl	Bent		1201	815
HSiF	Bent	1913	860	834
HSiCl	Bent		808	522
HSiBr	Bent	1548	774	408

Symmetric XY₃ Molecules

Point Groups D_{3h} (planar) and C_{3v} (pyramidal)

Molecule	Structure	Sym. str.	Sym. deform.	Deg. str.	Deg. deform.
NH₃	Pyram.	3337	950	3444	1627
ND₃	Pyram.	2420	748	2564	1191
PH₃	Pyram.	2323	992	2328	1118
AsH₃	Pyram.	2116	906	2123	1003
SbH₃	Pyram.	1891	782	1894	831
NF₃	Pyram.	1032	647	907	492
PF₃	Pyram.	892	487	860	344
AsF₃	Pyram.	741	337	702	262
PCl₃	Pyram.	504	252	482	198
PI₃	Pyram.	303	111	325	79
AsI₃	Pyram.	219	94	224	71
AlCl₃	Pyram.	375	183	595	150
SO₃	Planar	1065	498	1391	530
BF₃	Planar	888	691	1449	480
BH₃	Planar		1125	2808	1640
CH₃	Planar		606	3161	1396
CD₃	Planar		453	2369	1029
CF₃	Pyram.	1090	701	1260	510
SiF₃	Pyram.	830	427	937	290

Linear XYYX Molecules

Point Group $D_{\infty h}$

Molecule	Sym. XY str.	Antisym. XY str.	YY str.	Bend	Bend
C₂H₂	3374	3289	1974	612	730
C₂D₂	2701	2439	1762	505	537
C₂N₂	2330	2158	851	507	233

Planar X₂YZ Molecules

Point Group C_{2v}

Molecule	Sym. XY str.	YZ str.	YX₂ scis.	Antisym. XY str.	YX₂ rock	YX₂ wag
H₂CO	2783	1746	1500	2843	1249	1167
D₂CO	2056	1700	1106	2160	990	938
F₂CO	965	1928	584	1249	626	774
Cl₂CO	567	1827	285	849	440	580
O₂NF	1310	822	568	1792	560	742
O₂NCl	1286	793	370	1685	408	652

Tetrahedral XY₄ Molecules
Point Group T_d

Molecule	Sym. str.	Deg. deform.(e)	Deg. str.(f)	Deg. deform.(f)
CH₄	2917	1534	3019	1306
CD₄	2109	1092	2259	996
CF₄	909	435	1281	632
CCl₄	459	217	776	314
CBr₄	267	122	672	182
CI₄	178	90	555	125
SiH₄	2187	975	2191	914
SiD₄	1558	700	1597	681
SiF₄	800	268	1032	389
SiCl₄	424	150	621	221
GeH₄	2106	931	2114	819
GeD₄	1504	665	1522	596
GeCl₄	396	134	453	172
SnCl₄	366	104	403	134
TiCl₄	389	114	498	136
ZrCl₄	377	98	418	113
HfCl₄	382	102	390	112
RuO₄	885	322	921	336
OsO₄	965	333	960	329

References

1. T. Shimanouchi, Tables of Molecular Vibrational Frequencies, Consolidated Volume I, Natl. Stand. Ref. Data Ser. Natl. Bur. Stand. (U.S.), 39, 1972.
2. T. Shimanouchi, Tables of Molecular Vibrational Frequencies, Consolidated Volume II, *J. Phys. Chem. Ref. Data*, 6, 993, 1977.
3. G. Herzberg, *Electronic Spectra and Electronic Structure of Polyatomic Molecules*, D. Van Nostrand Co., Princeton, NJ, 1966.
4. M. E. Jacox, Ground state vibrational energy levels of polyatomic transient molecules, *J. Phys. Chem. Ref. Data*, 13, 945, 1984.

SPECTROSCOPIC CONSTANTS OF DIATOMIC MOLECULES

This table lists the leading spectroscopic constants and equilibrium internuclear distance r_e in the ground electronic state for selected diatomic molecules. The constants are those describing the vibrational and rotational energy through the expressions:

$$E_{vib}/hc = \omega_e(v + 1/2) - \omega_e x_e(v + 1/2)^2 + \cdots$$

$$E_{rot}/hc = B_v J(J + 1) - D_v[J(J + 1)]^2 + \cdots$$

where

$$B_v = B_e - \alpha_e(v+1/2) + \cdots$$

$$D_v = D_e + \cdots$$

Here v and J are the vibrational and rotational quantum numbers, respectively, h is Planck's constant, and c is the speed of light. In this customary formulation the constants ω_e, B_e, etc. have dimensions of inverse length; in this table they are given in units of cm^{-1}.

Users should note that higher order terms in the above energy expressions are required for very precise calculations; constants for many of these terms can be found in the references. Also, if the ground electronic state is not $^1\Sigma$, additional terms are needed to account for the interaction between electronic and pure rotational angular momentum. For some molecules in the table the data have been analyzed in terms of the Dunham series expansion:

$$E/hc = \Sigma_{lm} Y_{lm}(v+1/2)^l J^m(J+1)^m$$

In such cases it has been assumed that $Y_{10} = \omega_e$, $Y_{01} = B_e$, etc., although in the highest approximations these identities are not precisely correct. Some of the values of r_e in the table have been corrected for breakdown of the Born-Oppenheimer approximation, which can affect the last decimal place. Because of differences in the method of data analysis and limitations in the model, care should be taken in comparing r_e values for different molecules to a precision beyond 0.001 Å.

Molecules are listed in alphabetical order by formula as written in the most common form. In most cases this form places the more electropositive element first, but there are exceptions such as OH, NH, CH, etc.

* Indicates a value for the interval between $v = 0$ and $v = 1$ states instead of a value of ω_e.

References

1. Huber, K. P., and Herzberg, G., *Molecular Spectra and Molecular Structure IV. Constants of Diatomic Molecules*, Van Nostrand Reinhold, New York, 1979.
2. Lovas, F. J., and Tiemann, E., *J. Phys. Chem. Ref. Data*, 3, 609, 1974.
3. *Landolt-Börnstein, Numerical Data and Functional Relationships in Science and Technology, New Series*, II/6 (1974), II/14a (1982), II/14b (1983), II/19a (1992), II/19d-1 (1995), *Molecular Constants*, Springer-Verlag, Heidelberg.

Molecule	State	ω_e cm^{-1}	$\omega_e x_e$ cm^{-1}	B_e cm^{-1}	α_e cm^{-1}	D_e 10^{-6}cm^{-1}	r_e Å
^{107}Ag^{79}Br	$^1\Sigma^+$	249.57	0.63	0.064833	0.0002361	0.0175	2.39311
^{107}Ag^{35}Cl	$^1\Sigma^+$	343.49	1.17	0.12298388	0.00059541	0.06305	2.28079
^{107}Ag^{19}F	$^1\Sigma^+$	513.45	2.59	0.2657020	0.0019206	0.284	1.98318
^{107}Ag^{1}H	$^1\Sigma^+$	1759.9	34.06	6.449	0.201	344	1.618
^{107}Ag^{2}H	$^1\Sigma^+$	1250.70	17.17	3.2572	0.0722	85.9	1.6180
^{107}Ag^{127}I	$^1\Sigma^+$	206.50	0.46	0.04486821	0.0001414	0.00847	2.54463
^{107}Ag^{16}O	$^2\Pi_{1/2}$	490.2	3.1	0.3020	0.0025	0.45	2.003
^{27}Al$_2$	$^3\Sigma_g^-$	350.01	2.02	0.2054	0.0012	0.31	2.466
^{27}Al^{79}Br	$^1\Sigma^+$	378.0	1.28	0.15919713	0.00086045	0.11285	2.29481
^{27}Al^{35}Cl	$^1\Sigma^+$	481.30	1.95	0.24393012	0.00161113	0.2503	2.13011
^{27}Al^{19}F	$^1\Sigma^+$	802.3	4.77	0.5524798	0.0049841	1.0464	1.65437
^{27}Al^{1}H	$^1\Sigma^+$	1682.56	29.09	6.3907	0.1858	356.5	1.6478
^{27}Al^{2}H	$^1\Sigma^+$	1211.95	15.14	3.3186	0.0697	97	1.6463
^{27}Al^{127}I	$^1\Sigma^+$	316.1	1.0	0.11769985	0.00055859		2.53710
^{27}Al^{16}O	$^2\Sigma^+$	979.23	6.97	0.6414	0.0058	1.08	1.6179
^{27}Al^{32}S	$^2\Sigma^+$	617.1	3.33	0.2799	0.0018	0.22	2.029
^{75}As$_2$	$^1\Sigma_g^+$	429.55	1.12	0.10179	0.000333		2.1026
^{75}As^{1}H	$^3\Sigma^-$	2130*		7.3067	0.2117	327	1.52315
^{75}As^{2}H	$^3\Sigma^-$	1484*		3.6688		90	1.5306
^{75}As^{14}N	$^1\Sigma^+$	1068.54	5.41	0.54551	0.003366	0.53	1.6184
^{75}As^{16}O	$^2\Pi_{1/2}$	967.08	4.85	0.48482	0.003299	0.49	1.6236
^{197}Au$_2$	$^1\Sigma_g^+$	190.9	0.42	0.028013	0.0000723	0.00250	2.4719
^{197}Au^{1}H	$^1\Sigma^+$	2305.01	43.12	7.2401	0.2136	279	1.5239
^{197}Au^{2}H	$^1\Sigma^+$	1634.98	21.65	3.6415	0.07614	70.9	1.5238
^{11}B$_2$	$^3\Sigma_g^-$	1051.3	9.35	1.212	0.014		1.590
^{11}B^{79}Br	$^1\Sigma^+$	684.31	3.52	0.4894	0.0035	1.00	1.888
^{11}B^{35}Cl	$^1\Sigma^+$	840.29	5.49	0.684282	0.006812	1.84	1.71528
^{11}B^{19}F	$^1\Sigma^+$	1402.1	11.8	1.516950	0.019056	7.105	1.26267
^{11}B^{1}H	$^1\Sigma^+$	2366.9	49.40	12.021	0.412	1242	1.2324
^{11}B^{2}H	$^1\Sigma^+$	1703.3	28	6.54	0.17	400	1.2324
^{11}B^{14}N	$^3\Pi$	1514.6	12.3	1.666	0.025	8.1	1.281
^{11}B^{16}O	$^2\Sigma^+$	1885.69	11.81	1.7820	0.0166	6.32	1.2045

Molecule	State	ω_e cm^{-1}	$\omega_e x_e$ cm^{-1}	B_e cm^{-1}	α_e cm^{-1}	D_e 10^{-6}cm^{-1}	r_e Å
^{11}B^{32}S	$^2\Sigma^+$	1180.17	6.31	0.7949	0.0061	1.40	1.6092
^{138}Ba^{79}Br	$^2\Sigma^+$	193.77	0.41	0.0415082	0.0001219	0.00762	2.84449
^{138}Ba^{35}Cl	$^2\Sigma^+$	279.92	0.82	0.08396717	0.00033429	0.03022	2.68276
^{138}Ba^{19}F	$^2\Sigma^+$	468.9	1.79	0.2159	0.0012	0.175	2.163
^{138}Ba^{1}H	$^2\Sigma^+$	1168.31	14.50	3.38285	0.06599	112.67	2.23175
^{138}Ba^{2}H	$^2\Sigma^+$	829.77	7.32	1.7071	0.02363	28.77	2.2304
^{138}Ba^{127}I	$^2\Sigma^+$	152.14	0.27	0.02680587	0.00006634	0.00333	3.08476
^{138}Ba^{16}O	$^1\Sigma^+$	669.76	2.03	0.3126140	0.0013921	0.2724	1.93969
^{138}Ba^{32}S	$^1\Sigma^+$	379.42	0.88	0.10331	0.0003188	0.0306	2.5074
^{9}Be^{19}F	$^2\Sigma^+$	1247.36	9.12	1.4889	0.0176	8.28	1.3610
^{9}Be^{1}H	$^2\Sigma^+$	2060.78	36.31	10.3164	0.3030	1022.1	1.3426
^{9}Be^{2}H	$^2\Sigma^+$	1530.32	20.71	5.6872	0.1225	313.8	1.3419
^{9}Be^{16}O	$^1\Sigma^+$	1487.32	11.83	1.6510	0.0190	8.20	1.3309
^{9}Be^{32}S	$^1\Sigma^+$	997.94	6.14	0.79059	0.00664	2.00	1.7415
^{209}Bi$_2$	$^1\Sigma_g^+$	172.71	0.34	0.022781	0.000055	0.00150	2.6596
^{209}Bi^{1}H	$^3\Sigma^-$	1635.73	31.6	5.137	0.148	183	1.805
^{209}Bi^{2}H	$^3\Sigma^-$	1173.32	16.1	2.592	0.054	50.6	1.804
^{79}Br$_2$	$^1\Sigma_g^+$	325.32	1.08	0.082107	0.0003187	0.02092	2.2811
^{79}Br^{35}Cl	$^1\Sigma^+$	444.28	1.84	0.152470	0.000770	0.07183	2.13607
^{79}Br^{19}F	$^1\Sigma^+$	670.75	4.05	0.35584	0.00261	0.401	1.75894
^{79}Br^{16}O	$^2\Pi_{3/2}$	779	6.8	0.429598	0.003639	0.523	1.717
^{12}C$_2$	$^1\Sigma_g^+$	1854.71	13.34	1.8198	0.0177	6.92	1.2425
^{12}C^{35}Cl	$^2\Pi_{1/2}$	866.72*	6.2	0.6936	0.00672	1.9	1.6450
^{12}C^{19}F	$^2\Pi_{1/2}$	1308.1	11.10	1.4172	0.0184	6.5	1.2718
^{12}C^{1}H	$^2\Pi_{1/2}$	2858.5	63.0	14.457	0.534	1450	1.1199
^{12}C^{2}H	$^2\Pi_{1/2}$	2099.8	34.02	7.806	0.208	420	1.1190
^{12}C^{14}N	$^2\Sigma^+$	2068.59	13.09	1.8997830	0.0173717	6.4034	1.17181
^{12}C^{16}O	$^1\Sigma^+$	2169.81	13.29	1.93128075	0.01750390	6.1216	1.12823
^{12}C^{31}P	$^2\Sigma^+$	1239.67	6.86	0.7986	0.00597	1.33	1.562
^{12}C^{32}S	$^1\Sigma^+$	1285.15	6.50	0.8200434	0.0059182	1.336	1.53482
^{12}C^{80}Se	$^1\Sigma^+$	1035.36	4.86	0.5750	0.00379	0.71	1.67609
^{40}Ca^{35}Cl	$^2\Sigma^+$	367.53	1.31	0.1522302	0.0007990	0.1029	2.43676
^{40}Ca^{19}F	$^2\Sigma^+$	581.1	2.74	0.339	0.0026	0.45	1.967
^{40}Ca^{1}H	$^2\Sigma^+$	1298.34	19.10	4.2766	0.0970	183.7	2.0025
^{40}Ca^{2}H	$^2\Sigma^+$	910*		2.1769	0.035	47.9	2.002
^{40}Ca^{127}I	$^2\Sigma^+$	238.70	0.63	0.0693263	0.0002634	0.0234	2.82859
^{40}Ca^{16}O	$^1\Sigma^+$	732.03	4.83	0.444441	0.003282	0.6541	1.8221
^{40}Ca^{32}S	$^1\Sigma^+$	462.23	1.78	0.1766757	0.0008270	0.1032	2.31775
^{114}Cd^{1}H	$^2\Sigma^+$	1337.1*		5.323		314	1.781
^{114}Cd^{2}H	$^2\Sigma^+$			2.704		76	1.775
^{35}Cl$_2$	$^1\Sigma_g^+$	559.7	2.68	0.2440	0.0015	0.186	1.988
^{35}Cl^{19}F	$^1\Sigma^+$	786.15	6.16	0.516479	0.004358	0.88	1.62831
^{35}Cl^{16}O	$^2\Pi_{3/2}$	853.8	5.5	0.62345	0.0058	1.33	1.56963
^{52}Cr^{1}H	$^6\Sigma^+$	1581*	32	6.220	0.179	347	1.656
^{52}Cr^{2}H	$^6\Sigma^+$	1182*		3.14		88.8	1.664
^{52}Cr^{16}O	$^5\Pi$	898.4	6.8	0.5231	0.0070		1.615
^{133}Cs$_2$	$^1\Sigma_g^+$	42.02	0.08	0.0127	0.0000264	0.00464	4.47
^{133}Cs^{79}Br	$^1\Sigma^+$	149.66	0.37	0.03606925	0.00012401	0.00838	3.07225
^{133}Cs^{35}Cl	$^1\Sigma^+$	214.17	0.73	0.07209149	0.00033756	0.03268	2.90627
^{133}Cs^{19}F	$^1\Sigma^+$	352.56	1.62	0.18436969	0.0011756	0.20168	1.34535
^{133}Cs^{1}H	$^1\Sigma^+$	891.0	12.9	2.7099	0.0579	113	2.4938
^{133}Cs^{2}H	$^1\Sigma^+$	619.1*		1.354		20	2.505
^{133}Cs^{127}I	$^1\Sigma^+$	119.18	0.25	0.02362736	0.00006826	0.00371	3.31519
^{133}Cs^{16}O	$^2\Sigma^+$	357.5*		0.223073	0.001303	0.348	2.3007
^{63}Cu$_2$	$^1\Sigma_g^+$	264.55	1.02	0.10874	0.000614	0.0716	2.2197
^{63}Cu^{79}Br	$^1\Sigma^+$	314.8	0.96	0.10192625	0.00045214	0.04274	2.17344
^{65}Cu^{35}Cl	$^1\Sigma^+$	415.29	1.58	0.17628802	0.00099647	0.12706	2.05118
^{63}Cu^{19}F	$^1\Sigma^+$	622.7	3.95	0.3794029	0.0032298	0.563	1.74493
^{63}Cu^{1}H	$^1\Sigma^+$	1941.26	37.51	7.9441	0.2563	520	1.46263
^{63}Cu^{2}H	$^1\Sigma^+$	1384.14	18.97	4.0381	0.0917	136.2	1.4626

Molecule	State	ω_e cm^{-1}	$\omega_e x_e$ cm^{-1}	B_e cm^{-1}	α_e cm^{-1}	D_e 10^{-6}cm^{-1}	r_e Å
^{63}Cu^{127}I	$^1\Sigma^+$	264.5	0.60	0.07328742	0.00028390	0.02244	2.33832
^{63}Cu^{16}O	$^2\Pi_{3/2}$	640.17	4.43	0.44454	0.00456	0.85	1.7244
^{63}Cu^{32}S	$^2\Pi_{3/2}$	415.0	1.75	0.1891		0.18	2.051
^{19}F$_2$	$^1\Sigma_g^+$	916.64	11.24	0.89019	0.013847	3.3	1.41193
^{56}Fe^{16}O	$^5\Delta$	965*		0.650		0.72	1.444
^{69}Ga^{81}Br	$^1\Sigma^+$	263.0	0.81	0.081839	0.0003207	0.032	2.35248
^{69}Ga^{35}Cl	$^1\Sigma^+$	365.67	1.25	0.1499046	0.0007936	0.1008	2.20169
^{69}Ga^{19}F	$^1\Sigma^+$	622.2	3.2	0.3595161	0.0028642	0.50	1.77437
^{69}Ga^{1}H	$^1\Sigma^+$	1604.52	28.77	6.137	0.181	342	1.663
^{69}Ga^{2}H	$^1\Sigma^+$			3.083	0.06	84	1.663
^{69}Ga^{127}I	$^1\Sigma^+$	216.38	0.47	0.0569359	0.0001897	0.015770	2.57464
^{69}Ga^{16}O	$^2\Sigma$	767.5	6.24	0.4271		0.37	1.744
^{74}Ge^{79}Br	$^2\Pi_{1/2}$	295	0.7				
^{74}Ge^{35}Cl	$^2\Pi_{1/2}$	407.6	1.36				
^{72}Ge^{1}H	$^2\Pi_{1/2}$	1833.77	37	6.726	0.192	326	1.5880
^{72}Ge^{2}H	$^2\Pi_{1/2}$	1320.09	19	3.415	0.070	83.2	1.5874
^{74}Ge^{16}O	$^1\Sigma^+$	986.49	4.47	0.4856981	0.0030787	0.4709	1.62464
^{74}Ge^{32}S	$^1\Sigma^+$	575.8	1.80	0.18656576	0.00074910	0.07883	2.01209
^{74}Ge^{80}Se	$^1\Sigma^+$	408.7	1.36	0.09634051	0.00028904	0.02207	2.13463
^{74}Ge^{130}Te	$^1\Sigma^+$	323.9	0.75	0.06533821	0.00017246	0.012	2.34017
^{1}H$_2$	$^1\Sigma_g^+$	4401.21	121.34	60.853	3.062	47100	0.74144
^{2}H$_2$	$^1\Sigma_g^+$	3115.50	61.82	30.444	1.0786	11410	0.74152
^{3}H$_2$	$^1\Sigma_g^+$	2546.5	41.23	20.335	0.5887		0.74142
^{1}H^{81}Br	$^1\Sigma^+$	2648.97	45.22	8.46488	0.23328	345.8	1.41444
^{2}H^{81}Br	$^1\Sigma^+$	1884.75	22.72	4.245596	0.084	88.32	1.4145
^{1}H^{35}Cl	$^1\Sigma^+$	2990.95	52.82	10.59342	0.30718	531.94	1.27455
^{2}H^{35}Cl	$^1\Sigma^+$	2145.16	27.18	5.448796	0.113292	140	1.27458
^{1}H^{19}F	$^1\Sigma^+$	4138.32	89.88	20.9557	0.798	2151	0.91681
^{2}H^{19}F	$^1\Sigma^+$	2998.19	45.76	11.0102	0.3017	594	0.91694
^{1}H^{127}I	$^1\Sigma^+$	2309.01	39.64	6.4263650	0.1689	206.9	1.60916
^{202}Hg^{1}H	$^2\Sigma^+$	1203.24*		5.3888		395.3	1.7662
^{202}Hg^{2}H	$^2\Sigma^+$	896.12*		2.739		91	1.757
^{127}I$_2$	$^1\Sigma_g^+$	214.50	0.61	0.03737	0.000114	0.0043	2.666
^{127}I^{79}Br	$^1\Sigma^+$	268.64	0.81	0.0568325	0.0001969	0.0102	2.46899
^{127}I^{35}Cl	$^1\Sigma^+$	384.29	1.50	0.1141587	0.0005354	0.0403	2.32088
^{127}I^{19}F	$^1\Sigma^+$	610.24	3.12	0.2797111	0.0018738	0.2356	1.90976
^{127}I^{16}O	$^2\Pi_{3/2}$	681.5	4.3	0.34026	0.00270	0.36	1.8676
^{115}In^{81}Br	$^1\Sigma^+$	221.0	0.65	0.05489468	0.00018672	0.01350	2.54315
^{115}In^{35}Cl	$^1\Sigma^+$	317.39	1.03	0.1090583	0.0005177	0.0515	2.40117
^{115}In^{19}F	$^1\Sigma^+$	535.4	2.6	0.2623241	0.0018798	0.252	1.98540
^{115}In^{1}H	$^1\Sigma^+$	1476.0	25.61	4.995	0.143	223	1.8380
^{115}In^{2}H	$^1\Sigma^+$	1048.2	12.4	2.523	0.051	58	1.837
^{115}In^{127}I	$^1\Sigma^+$	177.08	0.34	0.03686702	0.00010411	0.00639	2.75364
^{39}K$_2$	$^1\Sigma_g^+$	92.02	0.28	0.056743	0.000165	0.0863	3.9051
^{39}K^{79}Br	$^1\Sigma^+$	213	0.80	0.08122109	0.00040481	0.04462	2.82078
^{39}K^{35}Cl	$^1\Sigma^+$	281	1.30	0.1286348	0.0007899	0.1087	2.66665
^{39}K^{19}F	$^1\Sigma^+$	426.26	2.45	0.27993741	0.00233492	0.4829	2.17146
^{39}K^{1}H	$^1\Sigma^+$	983.6	14.3	3.416400	0.085313	163.55	2.243
^{39}K^{2}H	$^1\Sigma^+$	707	7.7	1.754	0.0318	50	2.240
^{39}K^{127}I	$^1\Sigma^+$	186.53	0.57	0.06087473	0.00026776	0.02593	3.04784
^{139}La^{16}O	$^2\Sigma^+$	812.8	2.22	0.35252001	0.00142365	0.2626	1.82591
^{7}Li$_2$	$^1\Sigma_g^+$	351.43	2.61	0.67264	0.00704	9.87	2.6729
^{7}Li^{79}Br	$^1\Sigma^+$	563.2	3.5	0.555399	0.005644	2.159	2.17043
^{7}Li^{35}Cl	$^1\Sigma^+$	642.95	4.47	0.7065225	0.0080102	3.409	2.02067
^{7}Li^{19}F	$^1\Sigma^+$	910.57	8.21	1.3452583	0.0202887	11.745	1.56386
^{7}Li^{1}H	$^1\Sigma^+$	1405.65	23.20	7.51373	0.21665	862	1.59490
^{7}Li^{2}H	$^1\Sigma^+$	1054.80	12.94	4.23310	0.09155	276	1.5941
^{7}Li^{127}I	$^1\Sigma^+$	496.85	2.85	0.4431766	0.0040862	1.4104	2.39192
^{7}Li^{16}O	$^2\Pi$	814.62	7.78	1.212830	0.017899	0.1079	1.68822
^{24}Mg$_2$	$^1\Sigma_g^+$	51.12	1.64	0.09287	0.00378	1.22	3.891

Molecule	State	ω_e cm^{-1}	$\omega_e x_e$ cm^{-1}	B_e cm^{-1}	α_e cm^{-1}	D_e 10^{-6}cm^{-1}	r_e Å
^{24}Mg^{35}Cl	$^2\Sigma^+$	462.12*	2.1	0.2456154	0.0016204	0.2723	2.19639
^{24}Mg^{19}F	$^2\Sigma^+$	711.69*	4.9	0.51922	0.00470	1.080	1.7500
^{24}Mg^{1}H	$^2\Sigma^+$	1495.20	31.89	5.8257	0.1859	344	1.7297
^{24}Mg^{2}H	$^2\Sigma^+$	1077.9	16.1	3.0306	0.06289	92	1.7302
^{24}Mg^{16}O	$^1\Sigma^+$	784.78	5.26	0.57470436	0.00532377	1.2328	1.74838
^{55}Mn^{1}H	$^7\Sigma$	1548.0	28.8	5.6841	0.1570	303.9	1.7311
^{55}Mn^{2}H	$^7\Sigma$	1103	13.9	2.8957	0.051	79.5	1.7310
^{14}N$_2$	$^1\Sigma_g^+$	2358.57	14.32	1.99824	0.017318	5.76	1.09769
^{14}N^{79}Br	$^3\Sigma^-$	691.75	4.72	0.444	0.0040		1.79
^{14}N^{35}Cl	$^3\Sigma^-$	827.96	5.30	0.649770	0.006414	1.598	1.61071
^{14}N^{19}F	$^3\Sigma^-$	1141.37	8.99	1.2057	0.01492	5.39	1.3170
^{14}N^{1}H	$^3\Sigma^-$	3282.3	78.4	16.6993	0.6490	1709.7	1.0362
^{14}N^{2}H	$^3\Sigma^-$	2398	42	8.7913	0.2531	490.4	1.0361
^{14}N^{16}O	$^2\Pi_{1/2}$	1904.20	14.07	1.67195	0.0171	0.5	1.15077
^{14}N^{32}S	$^2\Pi_{1/2}$	1218.7	7.28	0.769602	0.0064	1.2	1.4940
^{23}Na$_2$	$^1\Sigma_g^+$	159.13	0.72	0.154707	0.008736	0.581	3.0789
^{23}Na^{79}Br	$^1\Sigma^+$	302	1.5	0.1512533	0.0009410	0.1554	2.50204
^{23}Na^{35}Cl	$^1\Sigma^+$	366	2.05	0.2180631	0.0016248	0.3120	2.36080
^{23}Na^{19}F	$^1\Sigma^+$	535.66	3.57	0.4369011	0.0045580	1.163	1.92595
^{23}Na^{1}H	$^1\Sigma^+$	1172.2	19.72	4.9033634	0.1370919	343.40	1.88654
^{23}Na^{2}H	$^1\Sigma^+$	826.1*		2.557089	0.051600	93.46	1.88654
^{23}Na^{127}I	$^1\Sigma^+$	258	1.1	0.1178056	0.0006478	0.0973	2.71145
^{23}Na^{16}O	$^2\Pi$	492.3		0.424630	0.004506	1.2638	2.05155
^{93}Nb^{16}O	$^4\Sigma^-$	989.0	3.8	0.4321	0.0021	0.22	1.691
^{58}Ni^{1}H	$^2\Delta_{5/2}$	1926.6	38	7.700	0.23	481	1.476
^{58}Ni^{2}H	$^2\Delta_{5/2}$	1390.1	19	3.992	0.092	130	1.465
^{16}O$_2$	$^3\Sigma_g^-$	1580.19	11.98	1.44563	0.0159	4.839	1.20752
^{16}O^{1}H	$^2\Pi_{3/2}$	3737.76	84.88	18.911	0.7242	1938	0.96966
^{16}O^{2}H	$^2\Pi_{3/2}$	2720.24	44.05	10.021	0.276	537.4	0.9698
^{31}P$_2$	$^1\Sigma_g^+$	780.77	2.84	0.30362	0.00149	0.188	1.8934
^{31}P^{35}Cl	$^3\Sigma^-$	551.38	2.23	0.2528748	0.0015119	0.2124	2.01461
^{31}P^{19}F	$^3\Sigma^-$	846.75	4.49	0.5665	0.00456		1.58938
^{31}P^{1}H	$^3\Sigma^-$	2365.2	44.5	8.5371	0.2514	436	1.42140
^{31}P^{2}H	$^3\Sigma^-$	1699.2	23.0	4.4081	0.0928	116	1.4220
^{31}P^{14}N	$^1\Sigma^+$	1337.24	6.98	0.7864854	0.0055364	1.091	1.49087
^{31}P^{16}O	$^2\Pi_{1/2}$	1233.34	6.56	0.7337	0.0055	1.3	1.4759
^{208}Pb$_2$		110.5	0.35				
^{208}Pb^{79}Br	$^2\Pi_{1/2}$	207.5	0.50				
^{208}Pb^{35}Cl	$^2\Pi_{1/2}$	303.9	0.88				
^{208}Pb^{19}F	$^2\Pi_{1/2}$	502.73	2.28	0.22875	0.001473	0.183	2.0575
^{208}Pb^{1}H	$^2\Pi_{1/2}$	1564.1	29.75	4.971	0.144	201	1.839
^{208}Pb^{16}O	$^1\Sigma^+$	720.96	3.52	0.30730373	0.00190977	0.2138	1.92181
^{208}Pb^{32}S	$^1\Sigma^+$	429.17	1.26	0.11632307	0.00043510	0.03418	2.28678
^{208}Pb^{80}Se	$^1\Sigma^+$	277.6	0.51	0.05059953	0.00012993	0.0070	2.40218
^{208}Pb^{130}Te	$^1\Sigma^+$	212.0	0.43	0.03130774	0.00006743	0.0027	2.59492
^{195}Pt^{12}C	$^1\Sigma^+$	1051.13	4.86	0.53044	0.003273	0.546	1.6767
^{195}Pt^{1}H	$^2\Delta_{5/2}$	2294.68*	46	7.1963	0.1996	261	1.52852
^{195}Pt^{2}H	$^2\Delta_{5/2}$	1644.3*	23	3.640	0.071	66	1.524
^{85}Rb^{79}Br	$^1\Sigma^+$	169.46	0.46	0.04752798	0.00018596	0.01496	2.94474
^{85}Rb^{35}Cl	$^1\Sigma^+$	228	0.92	0.0876404	0.0004537	0.04947	2.78673
^{85}Rb^{19}F	$^1\Sigma^+$	376	1.9	0.2106640	0.0015228	0.2684	2.27033
^{85}Rb^{1}H	$^1\Sigma^+$	936.9	14.21	3.020	0.072	123	2.367
^{85}Rb^{127}I	$^1\Sigma^+$	138.51	0.33	0.03283293	0.00010946	0.00738	3.17688
^{85}Rb^{16}O	$^2\Sigma^+$	388.4*		0.246481	0.002174	0.397	2.25420
^{32}S$_2$	$^3\Sigma_g^-$	725.65	2.84	0.2955	0.001570	0.19	1.8892
^{32}S^{19}F	$^2\Pi_{3/2}$			0.552174			1.60058
^{32}S^{1}H	$^2\Pi_{3/2}$	2711.6	59.9	9.5995	0.2785	480.6	1.34066
^{32}S^{2}H	$^2\Pi_{3/2}$	1885	31	4.95130	0.10308	130	1.34049
^{32}S^{16}O	$^3\Sigma^-$	1149.2	5.6	0.7208171	0.005737	1.134	1.48109
^{121}Sb^{35}Cl	$^3\Sigma^-$	374.7	0.6				

Molecule	State	ω_e cm^{-1}	$\omega_e x_e$ cm^{-1}	B_e cm^{-1}	α_e cm^{-1}	D_e 10^{-6}cm^{-1}	r_e Å
^{121}Sb^{19}F	$^3\Sigma^-$	605.0	2.6	0.2792	0.0020	0.23	1.918
^{121}Sb^{1}H	$^3\Sigma^-$			5.684		240	1.723
^{121}Sb^{2}H	$^3\Sigma^-$			2.8782		45	1.7194
^{121}Sb^{14}N	$^1\Sigma^+$	942.0	5.6				
^{121}Sb^{16}O	$^2\Pi_{1/2}$	816	4.2	0.3580	0.0022	0.270	1.826
^{45}Sc^{19}F	$^1\Sigma^+$	735.6	3.8	0.3950	0.00266		1.788
^{80}Se$_2$	$^3\Sigma_g^-$	385.30	0.96	0.08992	0.000288	0.024	2.166
^{80}Se^{1}H	$^2\Pi_{3/2}$	2400*		8.02	0.23	330	1.48
^{80}Se^{2}H	$^2\Pi_{3/2}$	1708*		3.94			1.48
^{80}Se^{16}O	$^3\Sigma^-$	914.69	4.52	0.4655	0.00323	0.5	1.648
^{28}Si$_2$	$^3\Sigma_g^-$	510.98	2.02	0.2390	0.0014	0.21	2.246
^{28}Si^{35}Cl	$^2\Pi_{1/2}$	535.60	2.17	0.2561	0.0016	0.25	2.058
^{28}Si^{19}F	$^2\Pi_{1/2}$	857.19	4.73	0.5812	0.00494	1.07	1.6011
^{28}Si^{1}H	$^2\Pi_{1/2}$	2041.80	35.51	7.4996	0.2190	397	1.5201
^{28}Si^{2}H	$^2\Pi_{1/2}$	1469.32	18.23	3.8840	0.0781	105.4	1.5199
^{28}Si^{14}N	$^2\Sigma^+$	1151.4	6.47	0.7311	0.00565	1.2	1.572
^{28}Si^{16}O	$^1\Sigma^+$	1241.54	5.97	0.7267521	0.0050379	0.9923	1.50975
^{28}Si^{32}S	$^1\Sigma^+$	749.64	2.58	0.30352788	0.00147308	0.201	1.92926
^{28}Si^{80}Se	$^1\Sigma^+$	580.0	1.78	0.1920117	0.0007767	0.0842	2.05832
^{120}Sn^{79}Br	$^2\Pi_{1/2}$	247.2	0.6				
^{120}Sn^{35}Cl	$^2\Pi_{1/2}$	351.1	1.06	0.1117	0.0004		2.361
^{118}Sn^{19}F	$^2\Pi_{1/2}$	577.6	2.69	0.2727	0.0014	0.26	1.944
^{120}Sn^{1}H	$^2\Pi_{1/2}$			5.31488		207.5	1.78146
^{120}Sn^{2}H	$^2\Pi_{1/2}$	1188.0*		2.6950	0.049	53.4	1.7770
^{120}Sn^{127}I	$^2\Pi_{1/2}$	199.0	0.6				
^{120}Sn^{16}O	$^1\Sigma^+$	822.13	3.72	0.35571998	0.00214432	0.26638	1.83251
^{120}Sn^{32}S	$^1\Sigma^+$	487.26	1.36	0.13686139	0.00050563	0.0424	2.20898
^{120}Sn^{80}Se	$^1\Sigma^+$	331.2	0.74	0.0649978	0.0001705	0.011	2.32557
^{120}Sn^{130}Te	$^1\Sigma^+$	259.5	0.50	0.04247917	0.00009543	0.0055	2.52280
^{88}Sr^{79}Br	$^2\Sigma^+$	216.60	0.52	0.0541847	0.0001827	0.01356	2.73522
^{88}Sr^{35}Cl	$^2\Sigma^+$	302.3	0.95				
^{88}Sr^{19}F	$^2\Sigma^+$	502.4	2.3	0.2505346	0.0015513	0.2498	2.07537
^{88}Sr^{1}H	$^2\Sigma^+$	1206.2	17.0	3.6751	0.0814	135	2.1456
^{88}Sr^{2}H	$^2\Sigma^+$	841	8.6	1.8609	0.0292	34.7	2.1449
^{88}Sr^{127}I	$^2\Sigma^+$	173.77	0.35	0.0367097	0.0001060	0.00655	2.94364
^{88}Sr^{16}O	$^1\Sigma^+$	653.5	3.96	0.33798	0.00219	0.36	1.91983
^{181}Ta^{16}O	$^2\Delta_{3/2}$	1028.69	3.51	0.40284	0.00182	0.2450	1.68746
^{130}Te$_2$	$^3\Sigma_g^-$	247.07	0.51	0.039681	0.000106	0.0044	2.5574
^{130}Te^{1}H	$^2\Pi_{3/2}$			5.56			1.74
^{130}Te^{16}O	0^+	797.11	4.00	0.3554	0.00237	0.27	1.825
^{232}Th^{16}O	$^1\Sigma^+$	895.77	2.39	0.332644	0.001302	0.1833	1.84032
^{48}Ti^{16}O	$^3\Delta_1$	1009.02	4.50	0.53541	0.00301	0.603	1.6202
^{205}Tl^{81}Br	$^1\Sigma^+$	192.10	0.39	0.0423899	0.0001276	0.0083	1.61817
^{205}Tl^{35}Cl	$^1\Sigma^+$	284.71	0.86	0.09139702	0.00039784	0.0377	2.48483
^{205}Tl^{19}F	$^1\Sigma^+$	476.86	2.24	0.22315014	0.00150380	0.1955	2.08439
^{205}Tl^{1}H	$^1\Sigma^+$	1390.7	22.7	4.806	0.154	254	1.870
^{205}Tl^{2}H	$^1\Sigma^+$	987.7	12.04	2.419	0.057	60	1.869
^{205}Tl^{127}I	$^1\Sigma^+$	150*		0.0271676	0.0000664	0.0036	2.81361
^{51}V^{16}O	$^4\Sigma^-$	1011.3	4.86	0.54825	0.00352	0.6	1.5893
^{89}Y^{35}Cl	$^1\Sigma$	380.7	1.3	0.1160	0.0003	0.09	2.41
^{89}Y^{19}F	$^1\Sigma^+$	631.29	2.50	0.29042	0.00163	0.237	1.9257
^{89}Y^{16}O	$^2\Sigma^+$	861.0	2.9	0.3881	0.0018	0.32	1.790
^{174}Yb^{1}H	$^2\Sigma^+$	1249.54	21.06	3.9931	0.0957	161.8	2.0526
^{174}Yb^{2}H	$^2\Sigma^+$	886.6	10.57	2.01162	0.03425	41.60	2.0516
^{64}Zn^{35}Cl	$^2\Sigma$	390.5	1.6				
^{64}Zn^{19}F	$^2\Sigma$	628	3.5				
^{64}Zn^{1}H	$^2\Sigma^+$	1607.6	55.14	6.6794	0.2500	466	1.5949
^{64}Zn^{2}H	$^2\Sigma^+$	1072	28	3.350		124	1.6054
^{64}Zn^{127}I	$^2\Sigma$	223.4	0.6				
^{90}Zr^{16}O	$^1\Sigma^+$	969.8	4.9	0.42263	0.0023	0.319	1.7116

INFRARED CORRELATION CHARTS

Far Infrared Region

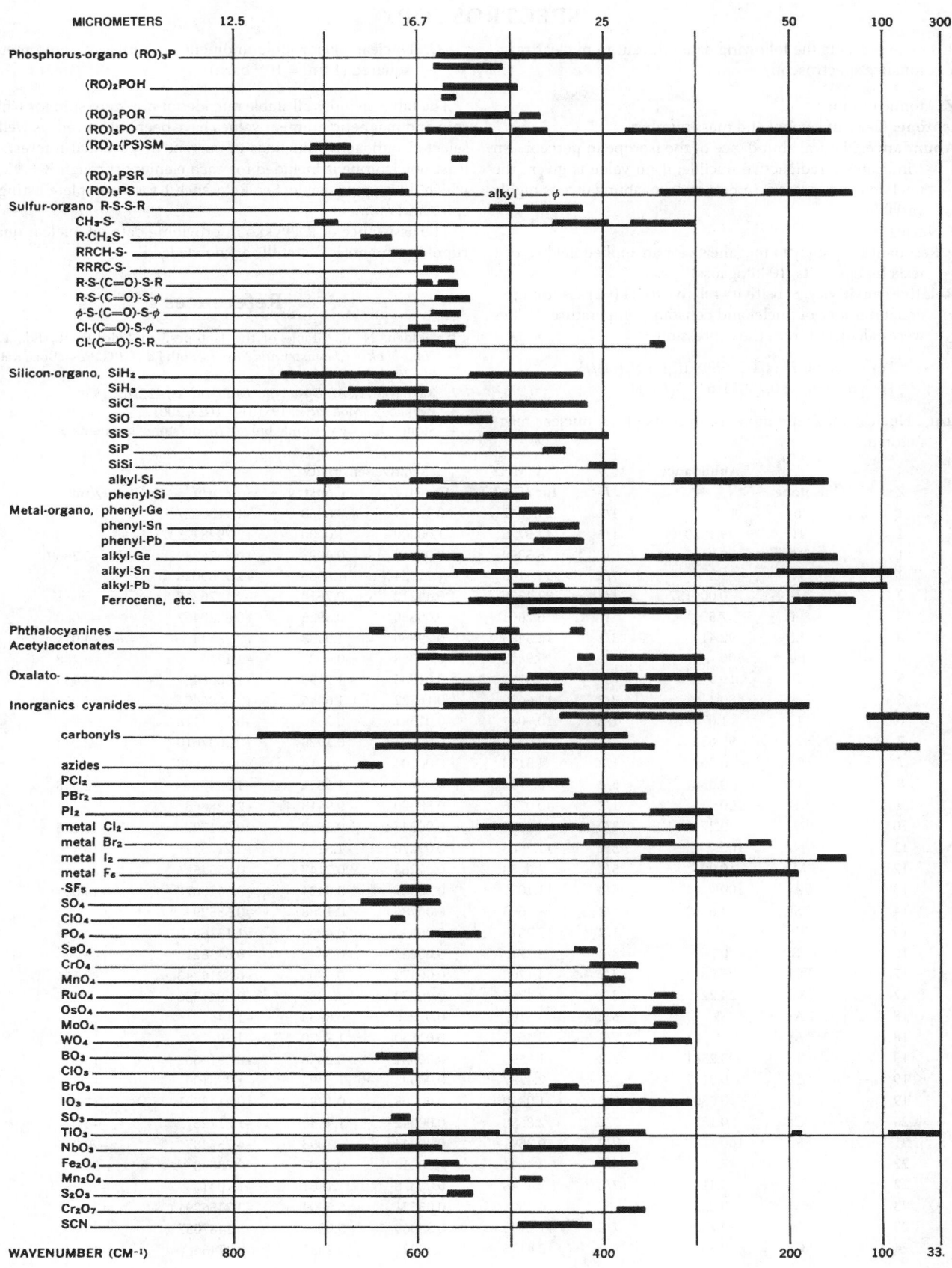

NUCLEAR SPINS, MOMENTS, AND OTHER DATA RELATED TO NMR SPECTROSCOPY

This table presents the following data relevant to nuclear magnetic resonance spectroscopy:

Z: Atomic number

Isotope: Element symbol and mass number

Abundance: Natural abundance of the isotope in percent. An * indicates a radioactive nuclide; if no value is given, the nuclide is not present in nature or its abundance is highly variable.

I: Nuclear spin

ν: Resonant frequency in megahertz for an applied field H_0 of 1 tesla (in cgs units, 10 kilogauss)

Relative sensitivity: Sensitivity relative to ^{1}H (=1) assuming an equal number of nuclei and constant temperature. Values were calculated from the expressions:

For constant H_0: $0.0076508(\mu/\mu_N)^3(I + 1)/I^2$
For constant ν: $0.23871(\mu/\mu_N)(I + 1)$

μ/μ_N: Nuclear magnetic moment in units of the nuclear magneton μ_N

Q: Nuclear quadrupole moment in units of femtometers squared (1 fm^2 = 10^{-2} barn)

The table includes all stable nuclides of non-zero spin for which spin and magnetic moment values have been measured, as well as selected radioactive nuclides of current or potential interest. At least one isotope is included for each element through $Z = 95$ for which data are available. See Reference 1 for a complete listing of spins and moments.

The assistance of P. Pyykko in providing data on nuclear quadrupole moments is gratefully acknowledged.

References

1. Holden, N. E., "Table of the Isotopes," in Lide, D. R., Ed., *CRC Handbook of Chemistry and Physics*, 86th Ed., CRC Press, Boca Raton, FL, 2005.
2. Raghavan, P., *At. Data Nuc. Data Tables*, 42, 189, 1989.
3. Pyykko, P., *Mol. Phys.*, 19, 1617–1629, 2001.
4. Stone, N. J., <www.nndc.bnl.gov/nndc/stone_moments/>

Z	Isotope	Abundance %	I	ν/MHz for H_0 = 1 T	Relative sensitivity Const. H_0	Const. ν	μ/μ_N	Q/fm^2
1	^{1}n	*	1/2	29.1647	0.32139	0.6850	−1.91304272	
1	^{1}H	99.9850	1/2	42.5775	1.00000	1.0000	+2.792847337	
1	^{2}H	0.0115	1	6.5359	0.00965	0.4093	+0.857438228	+0.2860
1	^{3}H	*	1/2	45.4148	1.21354	1.0666	+2.9789625	
2	^{3}He	0.000137	1/2	32.4360	0.44212	0.7618	−2.1276248	
3	^{6}Li	7.59	1	6.2661	0.00850	0.3925	+0.8220467	−0.0808
3	^{7}Li	92.41	3/2	16.5483	0.29356	1.9433	+3.25644	−4.01
4	^{9}Be	100	3/2	5.9842	0.01388	0.7027	−1.1776	+5.288
5	^{10}B	19.9	3	4.5752	0.01985	1.7193	+1.800645	+8.459
5	^{11}B	80.1	3/2	13.6630	0.16522	1.6045	+2.688649	+4.059
6	^{13}C	1.07	1/2	10.7084	0.01591	0.2515	+0.7024118	
7	^{14}N	99.632	1	3.0777	0.00101	0.1928	+0.4037610	+2.044
7	^{15}N	0.368	1/2	4.3173	0.00104	0.1014	−0.2831888	
8	^{17}O	0.038	5/2	5.7742	0.02910	1.5822	−1.89379	−2.558
9	^{19}F	100	1/2	40.0776	0.83400	0.9413	+2.628868	
10	^{21}Ne	0.27	3/2	3.3631	0.00246	0.3949	−0.661797	+10.155
11	^{23}Na	100	3/2	11.2688	0.09270	1.3233	+2.217522	+10.4
12	^{25}Mg	10.00	5/2	2.6083	0.00268	0.7147	−0.85545	+19.94
13	^{27}Al	100	5/2	11.1031	0.20689	3.0424	+3.641507	+14.66
14	^{29}Si	4.6832	1/2	8.4655	0.00786	0.1988	−0.55529	
15	^{31}P	100	1/2	17.2515	0.06652	0.4052	+1.13160	
16	^{33}S	0.76	3/2	3.2717	0.00227	0.3842	+0.6438212	−6.78
17	^{35}Cl	75.78	3/2	4.1765	0.00472	0.4905	+0.8218743	−8.165
17	^{37}Cl	24.22	3/2	3.4765	0.00272	0.4083	+0.6841236	−6.435
18	^{37}Ar	*	3/2	5.819	0.01276	0.6833	+1.145	
18	^{39}Ar	*	7/2	3.46	0.01130	1.7079	−1.59	
19	^{39}K	93.2581	3/2	1.9893	0.00051	0.2336	+0.3914662	+5.85
19	^{40}K	0.0117	4	2.4737	0.00523	1.5493	−1.298100	−7.3
19	^{41}K	6.7302	3/2	1.0919	0.00008	0.1282	+0.2148701	+7.11
20	^{43}Ca	0.135	7/2	2.8688	0.00642	1.4150	−1.31726	−4.08
21	^{45}Sc	100	7/2	10.3591	0.30244	5.1093	+4.756487	−22.0
22	^{47}Ti	7.44	5/2	2.4041	0.00210	0.6587	−0.78848	+30.2
22	^{49}Ti	5.41	7/2	2.4048	0.00378	1.1861	−1.10417	+24.7
23	^{50}V	0.250	6	4.2505	0.05571	5.5904	+3.345689	+21
23	^{51}V	99.750	7/2	11.2133	0.38360	5.5306	+5.1487057	−5.2
24	^{53}Cr	9.501	3/2	2.4115	0.00091	0.2832	−0.47454	−15

Z	Isotope	Abundance %	I	v/MHz for $H_0 = 1$ T	Relative sensitivity Const. H_0	Relative sensitivity Const. v	μ/μ_N	Q/fm^2
25	^{55}Mn	100	5/2	10.5763	0.17881	2.8980	+3.46872	+33
26	^{57}Fe	2.119	1/2	1.3816	0.00003	0.0324	+0.0906230	
27	^{59}Co	100	7/2	10.077	0.27841	4.9702	+4.627	+42
28	^{61}Ni	1.1399	3/2	3.8114	0.00359	0.4476	−0.75002	+16.2
29	^{63}Cu	69.17	3/2	11.2982	0.09342	1.3268	+2.22329	−22.0
29	^{65}Cu	30.83	3/2	12.1030	0.11484	1.4213	+2.38167	−20.4
30	^{67}Zn	4.10	5/2	2.6694	0.00287	0.7314	+0.875479	+15.0
31	^{69}Ga	60.108	3/2	10.2478	0.06971	1.2034	+2.01659	+17.1
31	^{71}Ga	39.892	3/2	13.0208	0.14300	1.5291	+2.56227	+10.7
32	^{73}Ge	7.73	9/2	1.4897	0.00141	1.1546	−0.8794677	−19.6
33	^{75}As	100	3/2	7.3150	0.02536	0.8590	+1.439475	+31.4
34	^{77}Se	7.63	1/2	8.1571	0.00703	0.1916	+0.53506	
35	^{79}Br	50.69	3/2	10.7042	0.07945	1.2570	+2.106400	+31.3
35	^{81}Br	49.31	3/2	11.5384	0.09951	1.3550	+2.270562	+26.2
36	^{83}Kr	11.49	9/2	1.6442	0.00190	1.2744	−0.970669	+25.9
37	^{85}Rb	72.17	5/2	4.1254	0.01061	1.1304	+1.35303	+27.6
37	^{87}Rb	27.83	3/2	13.9811	0.17703	1.6418	+2.75124	+13.35
38	^{87}Sr	7.00	9/2	1.8525	0.00272	1.4358	−1.093603	+33.5
39	^{89}Y	100	1/2	2.0949	0.00012	0.0492	−0.1374154	
40	^{91}Zr	11.22	5/2	3.9748	0.00949	1.0891	−1.30362	−17.6
41	^{93}Nb	100	9/2	10.4523	0.48821	8.1011	+6.1705	−32
42	^{95}Mo	15.92	5/2	2.7874	0.00327	0.7638	−0.9142	−2.2
42	^{97}Mo	9.55	5/2	2.8463	0.00349	0.7799	−0.9335	+25.5
43	^{99}Tc	*	9/2	9.6294	0.38174	7.4633	+5.6847	−12.9
44	^{99}Ru	12.76	5/2	1.9553	0.00113	0.5358	−0.6413	+7.9
44	^{101}Ru	17.06	5/2	2.1916	0.00159	0.6005	−0.7188	+45.7
45	^{103}Rh	100	1/2	1.3477	0.00003	0.0317	−0.08840	
46	^{105}Pd	22.33	5/2	1.957	0.00113	0.5364	−0.642	+66.0
47	^{107}Ag	51.839	1/2	1.7331	0.00007	0.0407	−0.1136796	
47	^{109}Ag	48.161	1/2	1.9924	0.00010	0.0468	−0.1306906	
48	^{111}Cd	12.80	1/2	9.0692	0.00966	0.2130	−0.5948861	
48	^{113}Cd	12.22	1/2	9.4871	0.01106	0.2228	−0.6223009	
49	^{113}In	4.29	9/2	9.3655	0.35121	7.2588	+5.5289	+79.9
49	^{115}In	95.71	9/2	9.3856	0.35348	7.2744	+5.5408	+81
50	^{115}Sn	0.34	1/2	14.0077	0.03561	0.3290	−0.91883	
50	^{117}Sn	7.68	1/2	15.2610	0.04605	0.3584	−1.00104	
50	^{119}Sn	8.59	1/2	15.9660	0.05273	0.3750	−1.04728	
51	^{121}Sb	57.21	5/2	10.2551	0.16302	2.8100	+3.3634	−36
51	^{123}Sb	42.79	7/2	5.5532	0.04659	2.7389	+2.5498	−49
52	^{123}Te	0.89	1/2	11.2349	0.01837	0.2639	−0.7369478	
52	^{125}Te	7.07	1/2	13.5454	0.03220	0.3181	−0.8885051	
53	^{127}I	100	5/2	8.5778	0.09540	2.3504	+2.813273	−71.0
54	^{129}Xe	26.44	1/2	11.8604	0.02162	0.2786	−0.7779763	
54	^{131}Xe	21.18	3/2	3.5159	0.00282	0.4129	+0.6918619	−11.4
55	^{133}Cs	100	7/2	5.6234	0.04838	2.7735	+2.582025	−0.343
56	^{135}Ba	6.592	3/2	4.2582	0.00500	0.5001	+0.837943	+16.0
56	^{137}Ba	11.232	3/2	4.7634	0.00700	0.5594	+0.937365	+24.5
57	^{138}La	0.090	5	5.6615	0.09404	5.3188	+3.713646	+45
57	^{139}La	99.910	7/2	6.0612	0.06058	2.9895	+2.7830455	+20
58	^{137}Ce	*	3/2	4.88	0.00752	0.5729	0.96	
58	^{139}Ce	*	3/2	5.39	0.01012	0.6326	1.06	
58	^{141}Ce	*	7/2	2.37	0.00364	1.1708	1.09	
59	^{141}Pr	100	5/2	13.0359	0.33483	3.5720	+4.2754	−5.9
60	^{143}Nd	12.2	7/2	2.319	0.00339	1.1440	−1.065	−63
60	^{145}Nd	8.3	7/2	1.429	0.00079	0.7047	−0.656	−33
61	^{143}Pm	*	5/2	11.59	0.23510	3.1748	+3.80	
61	^{147}Pm	*	7/2	5.62	0.04827	2.7714	+2.58	+74
62	^{147}Sm	14.99	7/2	1.7748	0.00152	0.8753	−0.8149	−26
62	^{149}Sm	13.82	7/2	1.4631	0.00085	0.7216	−0.6718	+7.4
63	^{151}Eu	47.81	5/2	10.5856	0.17929	2.9006	+3.4718	+90.3

Z	Isotope	Abundance %	I	v/MHz for $H_0 = 1$ T	Relative sensitivity Const. H_0	Const. v	μ/μ_N	Q/fm^2
63	^{153}Eu	52.19	5/2	4.6745	0.01544	1.2809	+1.5331	+241
64	^{155}Gd	14.80	3/2	1.312	0.00015	0.1541	−0.2582	+127
64	^{157}Gd	15.65	3/2	1.720	0.00033	0.2020	−0.3385	+135
65	^{159}Tb	100	3/2	10.23	0.06945	1.2019	+2.014	+143.2
66	^{161}Dy	18.91	5/2	1.4654	0.00048	0.4015	−0.4806	+250.7
66	^{163}Dy	24.90	5/2	2.0508	0.00130	0.5619	+0.6726	+265
67	^{165}Ho	100	7/2	9.0883	0.20423	4.4825	+4.173	+358
68	^{167}Er	22.93	7/2	1.2281	0.00050	0.6057	−0.5639	+356.5
69	^{169}Tm	100	1/2	3.531	0.00057	0.0829	−0.2316	
70	^{171}Yb	14.28	1/2	7.5261	0.00552	0.1768	+0.49367	
70	^{173}Yb	16.13	5/2	2.0730	0.00135	0.5680	−0.67989	+280
71	^{175}Lu	97.41	7/2	4.8626	0.03128	2.3983	+2.2327	+349
71	^{176}Lu	2.59	7	3.451	0.03975	6.0516	+3.169	+497
72	^{177}Hf	18.60	7/2	1.7282	0.00140	0.8524	+0.7935	+336.5
72	^{179}Hf	13.62	9/2	1.0856	0.00055	0.8414	−0.6409	+379.3
73	^{180}Ta	0.012	9	4.087	0.10610	11.5175	+4.825	
73	^{181}Ta	99.988	7/2	5.1627	0.03744	2.5463	+2.3705	+317
74	^{183}W	14.31	1/2	1.7957	0.00008	0.0422	+0.1177848	
75	^{185}Re	37.40	5/2	9.7176	0.13870	2.6627	+3.1871	+218
75	^{187}Re	62.60	5/2	9.8170	0.14300	2.6900	+3.2197	+207
76	^{187}Os	1.96	1/2	0.9856	0.00001	0.0231	+0.06465189	
76	^{189}Os	16.15	3/2	3.3536	0.00244	0.3938	+0.659933	+85.6
77	^{191}Ir	37.3	3/2	0.7658	0.00003	0.0899	+0.1507	+81.6
77	^{193}Ir	62.7	3/2	0.8319	0.00004	0.0977	+0.1637	+75.1
78	^{195}Pt	33.832	1/2	9.2922	0.01039	0.2182	+0.60952	
79	^{197}Au	100	3/2	0.7406	0.00003	0.0870	+0.145746	+54.7
80	^{199}Hg	16.87	1/2	7.7123	0.00594	0.1811	+0.5058855	
80	^{201}Hg	13.18	3/2	2.8469	0.00149	0.3343	−0.5602257	+38.6
81	^{203}Tl	29.524	1/2	24.7316	0.19598	0.5809	+1.6222579	
81	^{205}Tl	70.476	1/2	24.9749	0.20182	0.5866	+1.6382146	
82	^{207}Pb	22.1	1/2	9.0340	0.00955	0.2122	+0.59258	
83	^{209}Bi	100	9/2	6.9630	0.14433	5.3967	+4.1106	−51.6
84	^{209}Po	*	1/2	11.7	0.02096	0.2757	+0.77	
86	^{211}Rn	*	1/2	9.16	0.00997	0.2152	+0.601	
87	^{223}Fr	*	3/2	5.95	0.01362	0.6982	+1.17	+117
88	^{223}Ra	*	3/2	1.3746	0.00017	0.1614	+0.2705	+125
88	^{225}Ra	*	1/2	11.187	0.01814	0.2627	−0.7338	
89	^{227}Ac	*	3/2	5.6	0.01131	0.6564	+1.1	+170
90	^{229}Th	*	5/2	1.40	0.00042	0.3843	+0.46	+430
91	^{231}Pa	*100	3/2	10.2	0.06903	1.1995	2.01	−172
92	^{235}U	*0.7200	7/2	0.83	0.00015	0.4082	−0.38	+493.6
93	^{237}Np	*	5/2	9.57	0.13264	2.6234	+3.14	+388.6
94	^{239}Pu	*	1/2	3.09	0.00038	0.0727	+0.203	
95	^{243}Am	*	5/2	4.6	0.01446	1.2532	+1.5	+421

PROTON NMR CHEMICAL SHIFTS FOR CHARACTERISTIC ORGANIC STRUCTURES

The chart below summarizes the range of chemical shifts for protons in several classes or organic compounds and substituent groups. The chemical shifts δ are given in parts per million relative to tetramethylsilane.

Reference

Mohacsi, E., *J. Chem. Edu.*, 41, 38, 1964 (with permission)

¹³C-NMR ABSORPTIONS OF MAJOR FUNCTIONAL GROUPS

The table below lists the range of ^{13}C chemical shifts δ in parts per million relative to tetramethylsilane, in descending order, for various functional groups. Examples of simple compounds for each family are given to illustrate the correlations. The shifts for the carbons of interest, which are italicized, are given in parentheses; when two or more values appear, they refer to the sequence of italicized carbon atoms from left to right in the formula.

References

1. Yoder, C. H. and Schaeffer, C. D., Jr., *Introduction to Multinuclear NMR: Theory and Application*, Benjamin/Cummings, Menlo Park, CA, 1987.
2. Silverstein, R. M., Bassler, G. C., and Morrill, T. C., *Spectrometric Identification of Organic Compounds*, John Wiley & Sons, New York, 1981.
3. Brown, D. W., A Short Set of ^{13}C NMR Correlation Tables, *J. Chem. Educ.*, 62, 209, 1985.

δ (ppm)	Group	Family	Example	(δ of italicized carbon)
220-165	>C=O	Ketones	$(CH_3)_2CO$	(206.0)
			$(CH_3)_2CHCOCH_3$	(212.1)
		Aldehydes	CH_3CHO	(199.7)
		α,β-Unsaturated carbonyls	$CH_3CH{=}CHCHO$	(192.4)
			$CH_2{=}CHCOCH_3$	(169.9)
		Carboxylic acids	HCO_2H	(166.0)
			CH_3CO_2H	(178.1)
		Amides	$HCONH_2$	(165.0)
			CH_3CONH_2	(172.7)
		Esters	$CH_3CO_2CH_2CH_3$	(170.3)
			$CH_2{=}CHCO_2CH_3$	(165.5)
140-120	>C=C<	Aromatic	C_6H_6	(128.5)
		Alkenes	$CH_2{=}CH_2$	(123.2)
			$CH_2{=}CHCH_3$	(115.9, 136.2)
			$CH_2{=}CHCH_2Cl$	(117.5, 133.7)
			$CH_3CH{=}CHCH_2CH_3$	(132.7)
125-115	-CN	Nitriles	$CH_3\text{-}CN$	(117.7)
80-70	-CC-	Alkynes	$HCCH$	(71.9)
			CH_3CCH	(73.9)
70-45	-C-O	Esters	$CH_3OOCH_2CH_3$	(57.6, 67.9)
		Alcohols	$HOCH_3$	(49.0)
			$HOCH_2CH_3$	(57.0)
40-20	-C-NH₂	Amines	CH_3NH_2	(26.9)
			$CH_3CH_2NH_2$	(35.9)
30-15	-S-CH₃	Sulfides (thioethers)	$C_6H_5\text{-}S\text{-}CH_3$	15.6
30-(-2.3)	-C-H	Alkanes, cycloalkanes	CH_4	(-2.3)
			CH_3CH_3	(5.7)
			$CH_3CH_2CH_3$	(15.8, 16.3)
			$CH_3CH_2CH_2CH_3$	(13.4, 25.2)
			$CH_3CH_2CH_2CH_2CH_3$	(13.9, 22.8, 34.7)
		Cyclohexane		(26.9)

Section 10
Atomic, Molecular, and Optical Physics

LINE SPECTRA OF THE ELEMENTS

Joseph Reader and Charles H. Corliss

The original tables from which this table was derived were prepared under the auspices of the Committee on Line Spectra of the Elements of the National Academy of Sciences-National Research Council. The table contains the outstanding spectral lines of neutral (I) and singly ionized (II) atoms of the elements from hydrogen through plutonium (Z = 1–94); selected strong lines from doubly ionized (III), triply ionized (IV), and quadruply ionized (V) atoms are also included. Listed are lines that appear in emission from the vacuum ultraviolet to the far infrared. These lines were selected from much larger lists in such a way as to include the stronger observed lines in each spectral region. A more extensive list may be found in Reference 1.

The data were compiled by the following contributors.

J. G. Conway — Lawrence Berkeley Laboratory
C. H. Corliss — National Bureau of Standards
R. D. Cowan — Los Alamos Scientific Laboratory
C. R. Cowley — University of Michigan
Henry M. and Hannah Crosswhite — Argonne National Laboratory
S. P. Davis — University of California, Berkeley
V. Kaufman — National Bureau of Standards
R. L. Kelly — Naval Postgraduate School
J. F. Kielkopf — University of Louisville
W. C. Martin — National Bureau of Standards
T. K. McCubbin — Pennsylvania State University
L. J. Radziemski — Los Alamos Scientific Laboratory
J. Reader — National Bureau of Standards
C. J. Sansonetti — National Bureau of Standards
G. V. Shalimoff — Lawrence Berkeley Laboratory
R. W. Stanley — Purdue University
J. O. Stoner, Jr. — University of Arizona
H. H. Stroke — New York University
D. R. Wood — Wright State University
E. F. Worden — Lawrence Livermore Laboratory
J. J. Wynne — International Business Machines Corporation
R. Zalubas — National Bureau of Standards

All wavelengths are given in Ångstrom units (10^{-10} m). Below 2000 Å the wavelengths are in vacuum (except for the Cu II line at 1999.698 Å, which is in air); above 2000 Å the wavelengths are in air. Wavelengths given to three decimal places have an uncertainty of less than 0.001 Å and are therefore suitable for calibration purposes. In the air region, the elements used most commonly for calibration are Ne, Ar, Kr, Fe, Th, and Hg; in the vacuum region, the most common are C, N, O, Si, Cu.

All data refer to natural isotopic abundance of the elements except that Kr I and Kr II lines below 11,000 Å given to three decimal places are for ^{86}Kr. A separate table for ^{198}Hg contains accurately known wavelengths that are frequently used for calibration.

A large number of the lines for neutral and singly ionized atoms were extracted from the National Bureau of Standards (NBS) *Tables of Spectral Line Intensities* (Reference 2). The intensities of these lines represent quantitative estimates of relative line strengths that take account of varying detection sensitivity at different wavelengths. They are on a linear scale. For nearly all of the other lines the intensities represent qualitative estimates of the relative strengths of lines not greatly separated in wavelength. Because different observers frequently use different scales for their intensity estimates, these intensities are useful only as a rough indication of the appearance of a spectrum. In some cases the intensity scale is not intended to be linear. In the first and second spectra the intensities of the lines of the singly ionized atom (II) relative to those of the neutral atom (I) should be used with caution, inasmuch as the concentration of ions in a light source depends greatly on the excitation conditions.

Descriptive symbols that follow the wavelength have the following meanings:

c — complex
d — line consists of two unresolved lines
h — hazy
l — shaded to longer wavelengths
s — shaded to shorter wavelengths
p — perturbed by a close line
r — easily reversed
w — wide

The table is arranged alphabetically by element name (not symbol); for each element the lines are listed by wavelength. References to the sources of data for each element are given at the end of the table, starting on page 10-89.

General References

1. Reader, J., Corliss, C. H., Wiese, W. L., and Martin, G. A., *Tables of Line Spectra of the Elements, Part 1. Wavelengths and Intensities*, Nat. Stand. Ref. Data Sys.- Nat. Bur. Standards (U.S.), No. 68, 1980.
2. Meggers, W. F., Corliss, C. H., and Scribner, B. F., *Tables of Spectral Line Intensities, Part 1. Arranged by Elements*, Nat. Bur. Stand. (U.S.), Monograph 145, 1975.
3. Fuhr, J. R., Martin, W. C., Musgrove, A., Sugar, J., and Wiese, W. L., "NIST Atomic Spectroscopic Database" ver. 1.1, January 1996. *NIST Physical Reference Data*, National Institute of Standards and Technology, Gaithersburg, MD. Available at the WWW address: http://physics.nist.gov/PhysRefData/contents.html

Actinium Ac Z = 89

Intensity	Wavelength/Å	
2000 h	2952.55	III
2000 h	3392.78	III
3000	3487.59	III
2000 s	3863.12	II
3000 s	4088.44	II
3000 s	4168.40	II
100	4179.98	I
20	4183.12	I
20	4194.40	I
20 l	4384.53	I
20	4396.71	I
2000 h	4413.09	III
20	4462.73	I
3000 h	4569.87	III
1000	5910.85	II
20	6359.86	I
20 l	6691.27	I

Aluminum Al Z = 13

Intensity	Wavelength/Å	
900	125.53	V
800	126.07	V
800	130.41	V
1000	130.85	V
900	131.00	V
900	131.44	V
800	160.07	IV
1000	278.69	V
900	281.39	V
70	486.884	III
30	486.912	III
250	511.138	III
150	511.191	III
500	560.317	III
200	560.433	III
100	670.068	III
200	671.118	III
500	695.829	III
400	696.217	III
200	725.683	III
300	726.915	III
400	855.034	III
500	856.746	III
400	892.024	III
50	893.887	III
450	893.897	III
800	1042.17	IV
50	1191.812	II
900	1237.19	IV
900	1257.62	IV
800	1264.18	IV
1000	1272.76	IV
150	1350.18	II
800	1384.13	III
800	1447.51	IV
800	1494.79	IV
1000	1526.14	V
800	1537.54	IV
800	1539.830	II
1000	1557.25	IV
100	1569.385	II
900	1582.04	IV
800	1584.46	IV
125	1596.059	II
700	1605.766	III
100	1611.814	III
800	1611.874	III
150	1625.627	II
800	1639.06	IV
100	1644.235	II
100	1644.809	II
1000	1670.787	II
100	1686.250	II
800	1719.440	II
500	1721.244	II
900	1721.271	II
500	1724.952	II
900	1724.984	II
350	1760.104	II
300	1761.975	II
290	1763.00	I
500	1763.869	II
700	1763.952	II
450	1765.64	I
300	1765.815	II
450	1766.38	I
400	1767.731	II
450	1769.14	I
1000	1818.56	IV
600	1828.588	II
400	1832.837	II
250	1834.808	II
1000	1854.716	III
300	1855.929	II
700	1858.026	II
120	1859.980	II
1000	1862.311	II
600	1862.790	III
200	1929.978	II
150	1931.048	II
200	1932.377	II
400	1934.503	II
150	1934.713	II
300	1935.840	III
200	1935.949	III
150	1936.907	II
220	1939.261	II
700	1990.531	II
150	2016.052	II
150	2016.234	II
100	2016.368	II
200	2074.008	II
700	2094.264	II
150	2094.744	II
300	2094.791	II
100	2095.104	II
200	2095.141	II
400	2269.10	I
120	2269.22	I
140	2321.56	I
460	2367.05	I
110	2367.61	I
110	2368.11	I
180	2369.30	I
140	2370.22	I
160	2372.07	I
850	2373.12	I
170	2373.35	I
110	2373.57	I
240	2567.98	I
480	2575.10	I
110	2637.70	II
150	2652.48	I
200	2660.39	II
160	2669.17	II
650	2816.19	II
150	3041.28	II
360	3050.07	I
450	3057.14	I
150	3074.64	II
4500 r	3082.153	I
7200 r	3092.710	I
1800 r	3092.839	I
150	3428.92	II
150	3443.64	I
900	3492.23	IV
800	3508.46	IV
450	3586.56	II
360	3587.07	II
290	3587.45	II
870	3601.63	III
220	3651.06	II
110	3651.10	II
150	3654.98	II
290	3655.00	II
450	3900.68	II
4500 r	3944.006	I
9000 r	3961.520	I
110	3995.86	II
290	4226.81	II
870	4529.19	III
150	4585.82	II
110	4588.19	II
550	4666.80	II
110	4898.76	II
110	4902.77	II
150	5280.21	II
290	5283.77	II
150	5285.85	II
110	5312.32	II
220	5316.07	II
150	5371.84	II
180	5557.06	I
110	5557.95	I
450	5593.23	II
1200	5696.60	III
1000	5722.73	III
110	5853.62	II
220	5971.94	II
290	6001.76	II
220	6001.88	II
450	6006.42	II
150	6061.11	II
290	6068.43	II
110	6068.53	II
450	6073.23	II
110	6181.57	II
150	6181.68	II
290	6182.28	II
220	6182.45	II
450 h	6183.42	II
450	6201.52	II
360	6201.70	II
290	6226.18	II
360	6231.78	II
450	6243.36	II
450	6335.74	II
360	6696.02	I
230	6698.67	I
110	7361.57	I
140	7362.30	I
230	7835.31	I
290	7836.13	I
110	8075.35	I
290	8640.70	II
360	8772.87	I
450	8773.90	I
110	8828.91	I
180	8841.28	I
140	8923.56	I
150	9290.65	II
110	9290.75	II
150	10076.29	II
110	10768.36	I
140	10782.04	I
110	10872.98	I
230	10891.73	I
450	11253.19	I
570	11254.88	I
570	13123.41	I
450	13150.76	I
230	16718.96	I
300	16750.56	I
140	16763.36	I
300	21093.04	I
360	21163.75	I

Antimony Sb Z = 51

Intensity	Wavelength/Å	
15	722.86	III
15	732.33	III
	861.5	IV
4	876.84	II
4	921.07	II
6	983.57	II
15	999.62	III
6	1001.13	II
6	1009.43	II
40	1011.94	III
6	1052.21	II
8	1056.27	II
8	1057.32	II
40	1065.90	III
6	1073.81	II
30	1075.82	III
	1087.6	IV

Intensity	Wavelength/Å		Intensity	Wavelength/Å		Intensity	Wavelength/Å		Intensity	Wavelength/Å	
8	1104.32	V	100 r	2139.69	I	15	4514.50	II	4	447.53	V
30	1151.49	III	10	2141.80	II	30	4596.90	II	18	449.06	V
40	1157.74	III	50 r	2141.83	I	20	4599.09	II	4	449.49	V
	1199.1	IV	100 r	2144.86	I	15	4604.77	II	3	458.12	V
50	1205.20	III	1500 r	2175.81	I	30	4647.32	II	2	458.98	V
50	1210.64	III	250 r	2179.19	I	20	4675.74	II	6 p	461.23	V
12	1226.00	V	200 r	2201.32	I	40	4711.26	II	3	462.42	V
6	1230.30	II	300 r	2208.45	I	20	4757.81	II	7	463.94	V
8	1274.98	II	150 r	2220.73	I	20	4765.36	II	30	487.227	II
20	1306.69	III	100	2221.98	I	30	4784.03	II	50	490.650	II
8	1327.40	II	120 r	2224.93	I	20	4802.01	II	30	490.701	II
6	1358.04	II	300 r	2262.51	I	20	4832.82	II	30	519.327	II
8	1384.70	II	120	2288.98	I	20	4877.24	II	3	522.09	V
20	1404.18	III	150 r	2293.44	I	15	4947.40	II	5	524.19	V
6	1407.83	II	300 r	2306.46	I	15	5044.56	II	6	527.69	V
8	1436.49	II	2500 r	2311.47	I	20	5238.94	II	30	542.912	II
20 r	1486.57	I	150	2315.89	I	20	5354.24	II	200	543.203	II
40 h	1491.36	I	400 h	2373.67	I	40 h	5556.10	I	70	547.461	II
	1499.2	IV	300 h	2383.64	I	100 l	5632.02	I	2	554.50	V
12	1505.70	V	100	2395.22	I	30	5639.75	II	70	556.817	II
50 r	1512.57	I	150	2422.13	I	60 h	5830.34	I	5	558.48	V
12	1524.47	V	250	2426.35	I	100	6005.21	II	70	573.362	II
120 r	1532.74	I	400 r	2445.51	I	20	6053.41	II	30	576.736	II
80 r	1535.06	I	400	2478.32	I	30	6079.80	II	70	580.263	II
6	1565.51	II	150	2480.44	I	50	6130.04	II	30	583.437	II
8	1576.11	II	100	2510.54	I	20	6154.94	II	70	597.700	II
7	1581.36	II	2000 r	2528.52	I	20	6611.49	I	30	602.858	II
80 r	1599.96	I	15	2528.54	II	30	6647.44	II	30	612.372	II
10	1606.98	II	10	2567.75	II	30 h	7648.28	I	6	623.77	IV
200 w	1612.8	I	150	2574.06	I	80	7844.44	I	3	635.12	V
100 w	1623.3	I	15	2590.13	III	200	7924.65	I	500	661.867	II
20	1657.04	II	1500 r	2598.05	I	60	8411.69	I	30	664.562	II
100 w	1662.6	I	500 r	2598.09	I	150	8572.64	I	200	666.011	II
15	1673.89	III	300 r	2612.31	I	100	8619.55	I	1000	670.946	II
15	1711.84	III	12	2617.17	III	400	9518.68	I	3000	671.851	II
80 r	1716.93	I	200 r	2652.60	I	400	9949.14	I	70	676.242	II
150 r	1717.45	I	20	2669.39	III	200	10078.49	I	30	677.952	II
150 r	1723.43	I	300 r	2670.64	I	300	10261.01	I	30	679.218	II
15	1725.33	III	200 r	2682.76	I	200	10585.60	I	200	679.401	II
100 r	1736.19	I	120	2692.25	I	1000	10677.41	I	10	683.28	IV
100 h	1765.76	I	150 r	2718.90	I	800	10741.94	I	7	688.39	IV
100 r	1780.87	I	400 r	2769.95	I	80	10794.11	I	12 p	689.01	IV
100 r	1788.24	I	1000 r	2877.92	I	600	10839.73	I	6	699.41	IV
150	1800.18	I	15	2980.96	II	200	10868.58	I	8	700.28	IV
50 r	1810.50	I	500 r	3029.83	I	400	10879.55	I	3	705.35	V
80 r	1814.20	I	600 r	3232.52	I	300	11012.79	I	5	709.20	V
100	1829.50	I	20	3241.28	II	150	11266.23	I	4	715.60	V
50 r	1868.17	I	700 r	3267.51	I	5	12116.06	I	3	715.65	V
300 r	1871.15	I	15	3498.46	II				200	718.090	II
150 r	1882.56	I	25	3637.80	II	*Argon Ar Z = 18*			3000	723.361	II
100	1927.08	I	250	3637.83	I	3	336.56	V	2	725.11	V
200 r	1950.39	I	20	3722.78	II	3	337.56	V	500	725.548	II
60 r	2029.49	I	200 r	3722.79	I	6	338.00	V	70	730.930	II
70 r	2039.77	I	20	3850.22	II	2	338.43	V	200	740.269	II
150 r	2049.57	I	200	4033.55	I	2	339.01	V	200	744.925	II
1000 r	2068.33	I	20	4033.56	II	3	339.89	V	70	745.322	II
100	2079.56	I	20	4133.63	II	3	350.88	V	4	754.20	IV
50 r	2098.41	I	15	4140.54	II	4	396.87	IV	5	761.47	IV
80 r	2118.48	I	15	4195.17	II	4	398.55	IV	12	769.15	III
100 r	2127.39	I	20	4219.07	II	2	436.67	V	5	800.57	IV
50 r	2137.05	I	20	4314.32	II	5	446.00	V	10	801.09	IV
						8	446.95	V			

Intensity	Wavelength/Å		Intensity	Wavelength/Å		Intensity	Wavelength/Å		Intensity	Wavelength/Å	
10	801.41	IV	12	2293.03	III	12	2913.00	IV	35	3718.206	II
5	801.91	IV	4	2299.72	IV	11	2926.33	IV	70	3729.309	II
20	802.859	I	10	2300.85	III	200	2942.893	II	50	3737.889	II
100	806.471	I	15	2302.17	III	100	2979.050	II	150	3765.270	II
60	806.869	I	9	2317.00	III	10	3010.02	III	50	3766.119	II
30	807.218	I	15	2317.47	III	12	3024.05	III	20	3770.369	I
40	807.653	I	12	2318.04	III	50	3033.508	II	20	3770.520	II
50	809.927	I	10	2319.13	III	6	3037.98	IV	25	3780.840	II
120	816.232	I	10	2319.37	III	12	3054.82	III	20	3795.37	III
70	816.464	I	9	2345.17	III	10	3064.77	III	25	3803.172	II
80	820.124	I	7	2351.67	III	8	3077.40	IV	50	3809.456	II
4	822.16	V	9	2360.26	III	10	3078.15	III	7	3834.679	I
120	825.346	I	10	2395.63	III	50	3093.402	II	70	3850.581	II
120	826.365	I	12	2399.15	III	7	3110.41	III	10	3858.32	III
5	827.05	V	10	2413.20	III	7	3127.90	III	35	3868.528	II
3	827.35	V	7	2415.61	III	8	3200.37	I	7	3907.84	III
150	834.392	I	10	2418.82	III	20	3243.689	II	35	3925.719	II
4 p	834.88	V	5	2420.456	III	25	3285.85	III	50	3928.623	II
100	835.002	I	12	2423.52	III	25	3293.640	II	25	3932.547	II
2	836.13	V	12	2423.93	III	20	3301.88	III	70	3946.097	II
15	840.03	IV	7	2443.69	III	20	3307.228	II	7	3947.505	I
100	842.805	I	8	2447.71	IV	15	3311.25	III	35	3948.979	I
20	843.77	IV	8	2472.95	III	7	3319.34	I	8	3960.53	III
25	850.60	IV	7	2476.10	III	7	3323.59	III	20	3979.356	II
180	866.800	I	12	2488.86	III	25	3336.13	III	35	3994.792	II
150	869.754	I	12	2513.28	IV	20	3344.72	III	50	4013.857	II
10	871.10	III	10	2516.789	II	25	3350.924	II	6	4023.60	III
9	875.53	III	6	2518.40	IV	15	3358.49	III	50	4033.809	II
180 r	876.058	I	9	2525.69	IV	7	3361.28	III	20	4035.460	II
12	878.73	III	10	2534.709	II	7	3373.47	I	150	4042.894	II
8	879.62	III	15	2562.087	II	25	3376.436	II	50	4044.418	I
180 r	879.947	I	12	2562.17	IV	25	3388.531	II	100	4052.921	II
9	883.18	III	10	2568.07	IV	15	3391.85	III	200	4072.005	II
10	887.40	III	7	2569.53	IV	7	3393.73	I	70	4072.385	II
150	894.310	I	12	2599.47	IV	7	3417.49	III	25	4076.628	II
5	900.36	IV	10	2608.06	IV	9	3424.25	III	35	4079.574	II
9	901.17	IV	7	2608.44	IV	8	3438.04	III	25	4082.387	II
1000	919.781	II	12	2615.68	IV	7	3461.07	I	150	4103.912	II
1000	932.054	II	6	2619.98	IV	9	3471.32	III	300	4131.724	II
1000 r	1048.220	I	12	2621.36	IV	70	3476.747	II	5	4146.70	III
500 r	1066.660	I	12	2624.92	IV	20	3478.232	II	35	4156.086	II
7	1669.67	III	7	2631.90	III	20	3480.55	III	400	4158.590	I
7	1673.42	III	15	2640.34	IV	50	3491.244	II	50	4164.180	I
7	1675.48	III	10	2654.63	III	100	3491.536	II	35	4179.297	II
9	1914.40	III	8	2674.02	III	12	3499.67	III	50	4181.884	I
7	1915.56	III	9	2678.38	III	15	3503.58	III	100	4190.713	I
10	2125.16	III	9	2682.63	IV	70	3509.778	II	50	4191.029	I
15	2133.87	III	10	2724.84	III	8	3511.12	III	200	4198.317	I
10	2138.59	III	14	2757.92	IV	70	3514.388	II	400	4200.674	I
10	2148.73	III	7	2762.23	III	70	3545.596	II	25	4218.665	II
15	2166.19	III	10	2776.26	IV	70	3545.845	II	25	4222.637	II
10	2168.26	III	12	2784.47	IV	7	3554.306	I	25	4226.988	II
20	2170.23	III	14	2788.96	IV	100	3559.508	II	100	4228.158	II
25	2177.22	III	7	2797.11	IV	100	3561.030	II	100	4237.220	II
8	2184.06	III	16	2809.44	IV	70	3576.616	II	25	4251.185	I
10	2188.22	III	10	2830.25	IV	25	3581.608	II	200	4259.362	I
15	2192.06	III	7	2842.88	III	50	3582.355	II	100	4266.286	I
7	2248.73	III	8	2855.29	III	70	3588.441	II	70	4266.527	II
10	2279.10	III	6	2874.40	IV	7	3606.522	I	150	4272.169	I
7	2281.22	III	9	2884.12	III	25	3622.138	II	550	4277.528	II
7	2282.21	III	25	2891.612	II	20	3639.833	II	20	4282.898	II

Intensity	Wavelength/Å		Intensity	Wavelength/Å		Intensity	Wavelength/Å		Intensity	Wavelength/Å	
100	4300.101	I	25	5165.773	II	7	6951.478	I	200	10673.565	I
25	4300.650	II	20	5187.746	I	7	6960.250	I	11	10681.773	I
70	4309.239	II	20	5216.814	II	10000	6965.431	I	7	10683.034	II
200	4331.200	II	7	5221.271	I	150	7030.251	I	30	10733.87	I
50	4332.030	II	5	5421.352	I	10000	7067.218	I	30	10759.16	I
100	4333.561	I	10	5451.652	I	100	7068.736	I	7	10812.896	II
50	4335.338	I	25	5495.874	I	25	7107.478	I	11	11078.869	I
25	4345.168	I	5	5506.113	I	25	7125.820	I	30	11106.46	I
800	4348.064	II	25	5558.702	I	1000	7147.042	I	12	11441.832	I
50	4352.205	II	10	5572.541	I	15	7158.839	I	400	11488.109	I
25	4362.066	II	35	5606.733	I	70	7206.980	I	200	11668.710	I
50	4367.832	II	20	5650.704	I	15	7265.172	I	12	11719.488	I
200	4370.753	II	10	5739.520	I	7	7270.664	I	200	12112.326	I
70	4371.329	II	5	5834.263	I	2000	7272.936	I	50	12139.738	I
50	4375.954	II	10	5860.310	I	35	7311.716	I	50	12343.393	I
150	4379.667	II	15	5882.624	I	25	7316.005	I	200	12402.827	I
50	4385.057	II	25	5888.584	I	5	7350.814	I	200	12439.321	I
70	4400.097	II	50	5912.085	I	70	7353.293	I	100	12456.12	I
200	4400.986	II	15	5928.813	I	200	7372.118	I	200	12487.663	I
400	4426.001	II	5	5942.669	I	20	7380.426	II	150	12702.281	I
150	4430.189	II	7	5987.302	I	10000	7383.980	I	30	12733.418	I
50	4430.996	II	5	5998.999	I	20	7392.980	I	12	12746.232	I
50	4433.838	II	5	6025.150	I	15	7412.337	I	200	12802.739	I
20	4439.461	II	70	6032.127	I	10	7425.294	I	50	12933.195	I
35	4448.879	II	35	6043.223	I	25	7435.368	I	500	12956.659	I
100	4474.759	II	10	6052.723	I	10	7436.297	I	200	13008.264	I
200	4481.811	II	20	6059.372	I	20000	7503.869	I	200	13213.99	I
100	4510.733	I	7	6098.803	I	15000	7514.652	I	200	13228.107	I
20	4522.323	I	10	6105.635	I	25000	7635.106	I	100	13230.90	I
20	4530.552	II	100	6114.923	II	15000	7723.761	I	500	13272.64	I
400	4545.052	II	10	6145.441	I	10000	7724.207	I	1000	13313.210	I
20	4564.405	II	7	6170.174	I	10	7891.075	I	1000	13367.111	I
400	4579.350	II	150	6172.278	II	20000	7948.176	I	30	13499.41	I
400	4589.898	II	10	6173.096	I	20000	8006.157	I	1000	13504.191	I
15	4596.097	I	10	6212.503	I	25000	8014.786	I	11	13573.617	I
550	4609.567	II	5	6215.938	I	7	8053.308	I	30	13599.333	I
7	4628.441	I	25	6243.120	II	20000	8103.693	I	400	13622.659	I
35	4637.233	II	7	6296.872	I	35000	8115.311	I	200	13678.550	I
400	4657.901	II	15	6307.657	I	10000	8264.522	I	1000	13718.577	I
15	4702.316	I	7	6369.575	I	20	8392.27	I	10	13825.715	I
20	4721.591	II	20	6384.717	I	15000	8408.210	I	10	13907.478	I
550	4726.868	II	70	6416.307	I	20000	8424.648	I	200	14093.640	I
50	4732.053	II	25	6483.082	II	15000	8521.442	I	100	15046.50	I
300	4735.906	II	15	6538.112	I	7	8605.776	I	25	15172.69	I
800	4764.865	II	15	6604.853	I	4500	8667.944	I	10	15329.34	I
550	4806.020	II	25	6638.221	II	20	8771.860	II	30	15989.49	I
150	4847.810	II	20	6639.740	II	180	8849.91	I	30	16519.86	I
50	4865.910	II	50	6643.698	II	20	9075.394	I	500	16940.58	I
800	4879.864	II	5	6660.676	I	35000	9122.967	I	12	18427.76	I
70	4889.042	II	5	6664.051	I	550	9194.638	I	50	20616.23	I
20	4904.752	II	25	6666.359	II	15000	9224.499	I	30	20986.11	I
35	4933.209	II	100	6677.282	I	400	9291.531	I	20	23133.20	I
200	4965.080	II	35	6684.293	II	1600	9354.220	I	20	23966.52	I
50	5009.334	II	150	6752.834	I	25000	9657.786	I			
70	5017.163	II	5	6756.163	I	4500	9784.503	I			
70	5062.037	II	15	6766.612	I	180	10052.06	I			
20	5090.495	II	20	6861.269	II	30	10332.72	I			
100	5141.783	II	150	6871.289	I	100	10467.177	II			
70	5145.308	II	5	6879.582	I	1600	10470.054	I			
5	5151.391	I	10	6888.174	I	13	10478.034	I			
15	5162.285	I	50	6937.664	I	180	10506.50	I			

Arsenic As Z = 33

Intensity	Wavelength/Å	
510	871.7	III
325	889.0	III
325	927.5	III
325	937.2	III
325	953.6	III
325	963.8	III
250	987.7	V

Intensity	Wavelength/Å		Intensity	Wavelength/Å		Intensity	Wavelength/Å		Intensity	Wavelength/Å	
340	1021.96	II	300	2831.164	II	14	587.57	III	10	3193.91	I
250	1029.5	V	100 r	2860.44	I	18	647.27	III	25 h	3203.70	I
340	1082.35	II	300	2884.406	II	300	719.86	V	30	3221.63	I
500	1139.40	II	80	2926.3	III	150	721.85	V	40	3222.19	I
615	1149.31	II	615	2959.572	II	1000	766.87	V	50	3261.96	I
555	1181.51	II	300	3003.819	II	40000	794.89	IV	60 r	3262.34	I
555	1189.87	II	300	3116.516	II	300	877.41	V	40	3281.50	I
615	1196.38	II	340	3842.60	II	50000	923.74	IV	15	3281.77	I
615	1196.56	II	325	3922.6	III	200	946.26	V	50	3322.80	I
340	1207.44	II	715	4190.082	II	200	1486.72	II	80 h	3356.80	I
800	1211.17	II	615	4197.40	II	400	1504.01	II	50	3368.18	III
800	1218.10	II	615	4242.982	II	300	1554.38	II	60 r	3377.08	I
340	1223.15	II	500	4315.657	II	200	1572.73	II	20	3377.39	I
760	1241.31	II	500	4323.867	II		1573.92	II	70 r	3420.32	I
965	1243.08	II	500	4336.64	II		1630.40	II	25	3421.01	I
870	1245.67	II	500	4352.145	II	100	1674.51	II	30 h	3421.48	I
800	1258.58	II	425	4352.864	II	400	1694.37	II	40	3463.74	I
965	1263.77	II	375	4371.17	II		1697.16	II	200 r	3501.11	I
800	1266.34	II	615	4427.106	II		1761.75	II	80 h	3524.97	I
800	1267.59	II	615	4431.562	II		1771.03	II	30 h	3531.35	I
715	1280.99	II	715	4458.469	II		1786.93	II	80 h	3544.66	I
715	1287.54	II	340	4461.075	II	100	1904.15	II	20 h	3547.68	I
715	1305.70	II	715	4466.348	II	500	1924.70	II	100	3552.45	II
340	1307.74	II	500	4474.46	II		1985.60	II	200	3567.73	II
760	1333.15	II	800	4494.230	II	300	1999.54	II	100	3576.28	II
965	1341.55	II	850	4507.659	II	10	2001.30	III	30	3577.62	I
760	1355.93	II	615	4539.74	II		2009.20	II	80 h	3579.67	I
965	1369.77	II	715	4543.483	II	400	2023.95	II	200	3596.57	II
800	1373.65	II	615	4602.427	II		2052.68	II	40	3630.64	I
1000	1375.07	II	340	4629.787	II		2054.57	II	40 h	3636.83	I
760	1375.78	II	340	4707.586	II	500	2214.7	II	20 h	3688.47	I
800	1394.64	II	340	4730.67	II	800	2245.61	II	400	3735.75	II
800	1400.31	II	340	4888.557	II	1000	2254.73	II	200	3816.69	II
500	1448.59	II	340	5105.58	II	1400	2304.24	II	200	3842.80	II
500	1558.88	II	500	5107.55	II	60	2331.10	III	100	3854.76	II
500	1570.99	II	425	5231.38	II	2000	2335.27	II	20	3889.33	I
100 r	1593.60	I	500	5331.23	II	190	2347.58	II	1400 l	3891.78	II
500	1660.55	II	340	5497.727	II	40	2512.28	III	20	3892.65	I
340	1860.34	II	425	5558.09	II	40	2523.83	III	40	3909.91	I
1000 r	1890.42	I	425	5651.32	II	60	2528.51	II	500	3914.73	II
500	1912.94	II	425	6110.07	II	50	2559.54	III	25	3926.85	III
800 r	1937.59	I	500	6170.27	II	8 h	2596.64	I	50	3935.72	I
585 r	1972.62	I	300	6511.74	II	100	2634.78	II	20	3937.87	I
170 r	1990.35	I	300	7092.27	II	40	2681.89	III	200	3939.67	II
100 r	1991.13	I	300	7102.72	III	8	2702.63	I	500	3949.51	II
100 r	1995.43	I	340	7990.53	II	18	2771.36	II	25	3993.06	III
230 r	2003.34	I	300	8174.51	II	15	2785.28	I	80	3993.40	I
100 r	2009.19	I	200	9300.61	I	100 r	3071.58	I	30	3995.66	I
200	2263.2	IV	230	9597.95	I	40	3079.14	III	300	4036.26	II
350 r	2288.12	I	290	9626.70	I	10 h	3108.21	I	200	4083.77	II
200	2301.0	IV	230	9833.76	I	8	3132.60	I	30 h	4084.86	I
350 r	2349.84	I	170	9915.71	I	8 h	3135.72	I	1500 h	4130.66	II
100 r	2370.77	I	290	9923.05	I	10	3137.70	I	20	4132.43	I
135 r	2381.18	I	290	10024.04	I	10	3155.34	I	200	4166.00	II
250	2417.5	IV	170	10614.07	I	10	3155.67	I	500	4216.04	II
250	2454.0	IV				12	3158.05	I	800	4267.95	II
170 r	2456.53	I	**Astatine At** Z = 85			12 h	3158.54	I	100	4283.10	I
200	2461.4	IV	8	2162.25	I	25	3165.60	I	300	4287.80	II
340	2602.00	II	10	2244.01	I	15 h	3173.69	I	200	4297.60	II
170 r	2780.22	I				30	3183.16	I	800	4309.32	II
300	2830.359	II	**Barium Ba** Z = 56			15	3183.96	I	20 h	4323.00	I
			14	555.48	III						

Intensity	Wavelength/Å		Intensity	Wavelength/Å		Intensity	Wavelength/Å		Intensity	Wavelength/Å	
600	4325.73	II	300	6110.78	I	120 h	11697.45	I		107.38	I
200	4326.74	II	400	6135.83	II	120	13207.30	I	3	509.99	III
300	4329.62	II	20000	6141.72	II	120	13810.50	I	2	549.31	III
80	4350.33	I	150	6341.68	I	120	14077.90	I	6	582.08	III
60	4402.54	I	500	6378.91	II	120	15000.40	I	4	661.32	III
400	4405.23	II	10	6383.76	III	120	20712.00	I	8	675.59	III
40	4431.89	I	90	6450.85	I	150	25515.70	I		714.0	II
60 h	4488.98	I	150	6482.91	I	150	29223.90	I	4	725.59	III
50 h	4493.64	I	12000	6496.90	II				5	725.71	II
40	4505.92	I	300	6498.76	I	**Beryllium Be Z = 4**			5	743.58	II
200	4509.63	II	150	6527.31	I		58.13	IV	7	746.23	III
60 h	4523.17	I	3000	6595.33	I		58.57	IV	2	767.75	III
130	4524.93	II	150	6654.10	I		59.32	IV	8	775.37	II
65000	4554.03	II	1500	6675.27	I		60.74	IV	20	842.06	II
40	4573.85	I	1800	6693.84	I		64.06	IV		865.3	II
80	4579.64	I	1000	6769.62	II		75.93	IV	2	925.25	II
30	4599.75	I	600	6865.69	I	1 h	76.10	III	10	943.56	II
20 h	4619.92	I	300 h	6867.85	I	2	76.48	III	10	973.27	II
25 h	4628.33	I	1000	6874.09	II	3	78.53	III		981.4	II
300	4644.10	II	6000	7059.94	I	4	78.66	III		1020.1	II
30	4673.62	I	2400 hs	7120.33	I	1 h	78.92	III	8	1026.93	II
35	4691.62	I	600	7195.24	I	5	81.89	III	5	1036.32	II
20	4700.43	I	600 hl	7228.84	I	10	82.38	III	15	1048.23	II
800	4708.94	II	3000	7280.30	I		82.58	II	1	1114.69	III
40	4726.44	I	1200	7392.41	I	20	83.20	III	20	1143.03	II
800	4843.46	II	300	7417.53	I		83.66	II		1155.9	II
300	4847.14	II	900 hl	7459.78	I	30	84.76	III	60	1197.19	II
200	4850.84	II	600	7488.08	I	50	88.31	III	2	1213.12	III
30 h	4877.65	I	450 hl	7636.90	I		89.16	I	1	1214.32	III
400	4899.97	II	600 hl	7642.91	I		89.80	II	2	1362.25	III
15	4902.90	I	1800	7672.09	I		90.04	II	1	1401.52	III
20000	4934.09	II	1200	7780.48	I		90.21	I	10	1421.26	III
8	4947.35	I	180 h	7839.57	I		90.67	I	5	1422.86	III
1000	4957.15	II	1500	7905.75	I		91.06	II		1426.12	I
300	4997.81	II	600	7911.34	I		91.36	II	1	1435.17	III
1000	5013.00	II	900 h	8210.24	I		91.74	II	2	1440.77	III
20 h	5159.94	I	8	8308.69	III		92.19	I		1491.76	I
20	5267.03	I	1800 h	8559.97	I		92.61	II	20	1512.30	II
800	5361.35	II	100	8710.74	II		93.14	II	60	1512.43	II
1000	5391.60	II	100	8737.71	II		93.42	II	100	1661.49	I
200	5421.05	II	300 h	8799.76	I		93.93	II	2 h	1754.69	III
100	5424.55	I	300	8860.98	I		94.78	II	15	1776.12	II
200	5428.79	II	450	8914.99	I		95.76	II	20	1776.34	III
300	5480.30	II	300	9219.69	I		96.29	I		1907.	I
200	5519.05	I	300	9308.08	I		97.24	I		1909.0	II
1000 r	5535.48	I	300 h	9324.58	I		97.44	I		1912.	I
20 h	5620.40	I	1500	9370.06	I		97.86	I	3	1917.03	III
10	5680.18	I	300	9455.92	I		97.97	I		1919.	I
400	5777.62	I	8	9521.76	III		98.12	I	5	1929.67	I
800	5784.18	II	450	9589.37	I		98.37	I	10	1943.68	I
100	5800.23	I	900	9608.88	I		98.66	I	60 h	1954.97	III
20	5805.69	I	300 h	9645.72	I		98.94	I		1956.	I
150	5826.28	I	1500 hl	9830.37	I		99.19	I	50	1964.59	I
2800	5853.68	II	900	10001.08	I	100	100.25	III	5	1985.13	I
15	5907.64	I	600	10032.10	I		100.86	I		1997.95	I
100	5971.70	I	1200 h	10233.23	I		101.20	I		1997.98	I
800	5981.25	II	300	10471.26	I		102.13	I	60	1998.01	I
100	5997.09	I	120 hl	10791.25	I		102.49	II		2033.25	I
300	5999.85	II	180 hl	11012.69	I		104.40	II		2033.28	I
100	6019.47	I	150 h	11114.42	I		104.67	I		2033.38	I
200	6063.12	I	240	11303.04	I		105.80	I	50	2055.90	I
							107.26	I			

Intensity	Wavelength/Å		Intensity	Wavelength/Å		Intensity	Wavelength/Å		Intensity	Wavelength/Å	
100	2056.01	I		3163.	I	64	5270.28	II	60	14644.75	I
75 h	2076.94	III		3168.	I	500	5270.81	II	200	16157.72	I
60 h	2080.38	III		3180.7	II	20	5403.04	II	80	17855.38	I
25	2118.56	III		3187.	I	20	5410.21	II	120	17856.63	I
15 h	2122.27	III	20	3193.81	I		5558.	I	100	18143.54	I
10	2125.57	I	20	3197.10	II	140 h	6142.01	III	160	31775.05	I
20	2125.68	I	30	3197.15	II	10	6229.11	I	200	31778.70	I
15 h	2127.20	III	20	3208.60	I	16	6279.43	II			
5	2137.25	III		3220.	I	30	6279.73	II	**Bismuth Bi** *Z = 83*		
25	2145.	I	60	3229.63	I	30	6473.54	I	6	420.7	IV
55	2174.99	I	2	3233.52	II	60	6547.89	II	6	431.2	IV
55	2175.10	I	10	3241.62	II	60	6558.36	II	2	488.39	V
5	2191.57	III	30	3241.83	II	30	6564.52	I	3	563.62	V
	2273.5	II	15	3269.02	I	2 h	6636.44	II	5	670.76	III
	2324.6	II	100	3274.58	I	1	6756.72	II	6	686.88	V
	2337.0	I	30	3274.67	I	2	6757.13	II	5	730.71	V
950	2348.61	I	30	3282.91	I	30	6786.56	I	10	738.17	V
20	2350.66	I	30	3321.01	I	1 h	6884.22	I	4	775.16	III
60	2350.71	I	30	3321.09	I	6 h	6884.44	I	6	790.5	IV
200	2350.83	I	220	3321.34	I	100	6982.75	I	6	790.6	IV
2	2413.34	II	20	3345.43	I	6 h	7154.40	I	8	792.5	IV
16	2413.46	II	60	3367.63	I	40 h	7154.65	I	10	820.3	IV
20	2453.84	II		3405.6	II	100	7209.13	I	9	822.9	IV
	2480.6	I	5	3451.37	I	3	7401.20	II	12	824.9	IV
35	2494.54	I	300	3455.18	I	2	7401.43	II	15 d	864.45	V
35	2494.58	I	20	3476.56	I	10	7551.90	I	15	872.6	IV
100	2494.73	I	300	3515.54	I	10 h	7618.68	I	12	923.9	IV
16	2507.43	II	10	3555.	I	20 h	7618.88	I	15	943.3	IV
5	2617.99	II	100	3720.36	III	60	8090.06	I	25	1039.99	III
20	2618.13	II		3720.92	III	5 h	8158.99	I	50 h	1045.76	III
100	2650.45	I		3722.98	III	10 h	8159.24	I	30	1051.81	III
60	2650.55	I	100	3736.30	I	4	8254.07	I	15	1058.88	II
200	2650.62	I	700	3813.45	I	10 h	8287.07	I	20	1085.47	II
60	2650.69	I	40	3865.13	I	30	8547.36	I	10	1099.20	II
100	2650.76	I	80	3865.42	I	60	8547.67	I	24	1103.4	IV
5	2697.46	II	1	3865.51	I	300	8801.37	I	20	1139.01	III
20	2697.58	II	6	3865.72	I	6	8882.18	I	50	1224.64	III
20	2728.88	II	100	3866.03	I	40	9190.45	I	10	1225.43	II
30	2738.05	I	90 h	4249.14	III	20 h	9243.92	I	15	1232.78	II
	2764.2	II	100	4253.05	I	1 h	9343.89	II	10	1241.05	II
20	2898.13	I	60	4253.76	I	40	9392.74	I	10	1265.35	II
10	2898.19	I	300	4360.66	II	2	9476.43	II	15	1283.73	II
20	2898.25	I	500	4360.99	II	16	9477.03	II	10	1306.18	II
30	2986.06	I	400	4407.94	I	20	9847.32	I	60	1317.0	IV
10	2986.42	I	2	4485.52	III	10 h	9895.63	I	20	1325.46	II
60	3019.33	I	100 h	4487.30	III	20 h	9895.96	I	40	1326.84	III
30	3019.49	I	1	4495.09	III	80	9939.78	I	20	1329.47	II
30	3019.53	I	140 h	4497.8	III	16	10095.52	II	60	1346.12	III
20	3019.60	I		4526.6	I	20	10095.73	II	20	1350.07	II
10	3046.52	II		4548.	I	60	10119.92	I	25	1372.61	II
30	3046.69	II	12	4572.66	I	80	10331.03	I	15	1376.02	II
	3090.3	I	700	4673.33	II	30	11066.46	I	20	1393.92	II
10	3110.81	I	1000	4673.42	II		11173.	II	35	1423.33	III
10	3110.92	I	6	4709.37	I	1	11173.73	II	35	1423.52	III
20	3110.99	I	200	4828.16	II	120	11496.39	I	45	1436.83	II
	3120.	I	40	4849.16	I	2 h	11625.16	II	25	1447.94	II
480	3130.42	II	2 h	4858.22	II		11659.	II	50	1455.11	II
320	3131.07	II	80	5087.75	I	2	11660.25	II	60 h	1461.00	III
	3136.	I	8	5218.12	II	100	12095.36	II	25	1462.14	II
	3150.	I	20	5218.33	II	30	12098.18	II	35	1486.93	II
	3160.6	I	3	5255.86	II	100	14643.92	I	20	1502.50	II
									40	1520.57	II

Intensity	Wavelength/Å		Intensity	Wavelength/Å		Intensity	Wavelength/Å		Intensity	Wavelength/Å	
40	1533.17	II	100	2924.	IV	60 h	5144.3	II			
30	1536.77	II	100	2933.	IV	20	5201.5	II	*Boron B Z = 5*		
35	1538.06	II	100	2936.	IV	75 h	5209.2	II		41.00	V
20	1563.67	II	15	2936.7	II	40 h	5270.3	II	30	48.59	V
40	1573.70	II	3200	2938.30	I	10	5397.8	II	10	52.68	IV
60	1591.79	II	20	2950.4	II	10 c	5552.35	I	30	60.31	IV
25	1601.58	II	12	2963.4	II	3	5599.41	I		194.37	V
60 h	1606.40	III	2800	2989.03	I	20	5655.2	II		262.37	V
40	1609.70	II	700	2993.34	I	40 h	5719.2	II	160	344.0	IV
40	1611.38	II	100	3012.	IV	6	5742.55	I	450	385.0	IV
20	1652.81	II	2400	3024.64	I	12	5818.3	II	40	411.80	III
20	1749.29	II	60	3034.87	I	20	5860.2	II	285	418.7	IV
80	1777.11	II	100	3042.	IV	20	5973.0	II	20	510.77	III
60	1787.47	II	9000 c	3067.72	I	15	6059.1	II	40	510.85	III
70	1791.93	II	140	3076.66	I	15	6128.0	II		512.53	V
70	1823.80	II	35	3115.0	III	6	6134.82	I	150	518.24	III
100	1902.41	II	100	3239.	IV	3	6475.73	I	75	518.27	III
9000	1954.53	I	550 c	3397.21	I	3	6476.24	I	110	677.00	III
7000	1960.13	I	10	3430.83	II	15	6497.7	II	160	677.14	III
25	1989.35	II	12	3431.23	II	10	6577.2	II	40	693.95	II
7000	2021.21	I	40 h	3451.0	III	40 h	6600.2	II	40	731.36	II
9000	2061.70	I	40	3473.8	III	50 h	6808.6	II	40	731.44	II
45 h	2068.9	II	35	3485.5	III	4 h	6991.12	I		749.74	V
4600	2110.26	I	500 c	3510.85	I	12	7033.	II	40	758.48	III
2500	2133.63	I	380 c	3596.11	I	2	7036.15	I	70	758.67	III
15	2143.40	II	45	3613.4	III	10 h	7381.	II	110	882.54	II
15	2143.46	II	100	3643.	IV	2	7502.33	I	110	882.68	II
60	2186.9	II	12	3654.2	II	10 h	7637.	II	40	984.67	II
40 h	2214.0	II	100	3682.	IV	10	7750.	II	110	1081.88	II
360	2228.25	I	50	3695.32	III	3	7838.70	I	110	1082.07	II
1700	2230.61	I	50	3695.68	III	2	7840.33	I	70	1112.2	IV
340	2276.58	I	100	3734.	IV	20	7965.	II	450	1168.9	IV
100	2311.	IV	70 h	3792.5	II	40	8008.	III	70	1170.9	IV
100	2326.	IV	12	3811.1	II	12 h	8050.	II	110	1230.16	II
16	2368.12	II	20	3815.8	II	50	8070.	III	220	1362.46	II
12	2368.25	II	10	3845.8	II	15	8328.	II	70	1600.46	I
100	2376.	IV	30	3863.9	II	15	8388.	II	120	1600.73	I
190	2400.88	I	100	3868.	IV	30	8532.	II	160	1623.58	II
75 h	2414.6	III	40 h	4079.1	II	2	8544.54	I	110	1623.77	II
10	2501.0	II	10	4097.2	II	1	8579.74	I	220	1624.02	II
25	2515.69	I	140	4121.53	I	25	8653.	II	70	1624.16	II
70	2524.49	I	140	4121.86	I	2	8754.88	I	160	1624.34	II
20 h	2544.5	II	75 h	4259.4	II	3	8761.54	I	100	1663.04	I
700	2627.91	I	25	4272.0	II	25	8863.	II	150	1666.87	I
100	2629.	IV	70 h	4301.7	II	2	8907.81	I	200	1667.29	I
100	2677.	IV	12 h	4339.8	II	2000 d	9657.04	I	150	1817.86	I
12	2693.0	II	25 h	4340.5	II	40	9827.78	I	200	1818.37	I
280 c	2696.76	I	12 h	4379.4	II	20	10104.5	I	300	1825.91	I
20	2713.3	II	25 h	4476.8	II	15	10138.8	I	300	1826.41	I
140 d	2730.50	I	60 h	4705.3	II	20	10300.6	I	110	1842.81	II
100	2767.	IV	600 c	4722.52	I	20	10536.19	I	20	1953.83	III
100	2772.	IV	30	4730.3	II	50	11072.44	I	550	2065.78	III
360	2780.52	I	20	4749.7	II	1500 d	11710.37	I	250	2066.38	I
100	2786.	IV	40 h	4797.4	III	40	11999.49	I	250	2066.65	I
15	2803.42	II	12	4908.2	II	200	12165.08	I	100	2066.93	I
11	2803.70	II	10	4916.6	II	200	12690.04	I	300	2067.19	I
12	2805.3	II	12	4969.7	II	100	12817.8	I	450	2067.23	III
140 c	2809.62	I	20	4993.6	II	200	14330.5	I	160	2077.09	III
100	2842.	IV	45 h	5079.3	III	50	16001.5	I	500	2088.91	I
80 h	2855.6	III	10	5091.6	II	60	22551.6	I	500	2089.57	I
4000	2897.98	I	50 h	5124.3	II				70	2220.30	II
									40	2234.09	III

Intensity	Wavelength/Å		Intensity	Wavelength/Å		Intensity	Wavelength/Å		Intensity	Wavelength/Å	
70	2234.59	III	1000	619.87	IV	500	3041.18	IV	1200	7827.23	I
40	2323.03	II	1000	630.14	IV	500	3074.42	III	2500 s	7881.45	I
40	2328.67	II	1000	642.23	IV	500	3349.64	III	2500	7881.57	I
40	2393.20	II	1000	661.53	IV	500	3380.56	IV	2500	7925.81	I
220	2395.05	II	1000	683.51	IV	500	3540.16	III	30000 c	7938.68	I
40	2459.69	II	1000	697.72	IV	500	3562.43	III	3000	7947.94	I
40	2459.90	II	1000	715.39	IV	1200	3815.65	I	3000	7950.18	I
1000	2496.77	I	1000	731.00	IV	1500	3992.36	I	8000	7978.44	I
1000	2497.73	I	1000	800.12	IV	1000	4223.89	II	10000	7978.57	I
70	2524.7	IV	700	812.95	V	2000	4365.14	I	30000	7989.94	I
160	2530.3	IV	1000	813.66	IV	1000	4365.60	II	2000	8026.35	I
450	2821.68	IV	1000	850.81	V	1500	4425.14	I	2500	8026.54	I
70	2824.57	IV	1000	889.23	II	10000	4441.74	I	30000	8131.52	I
285	2825.85	IV	1000	948.97	II	10000	4472.61	I	1000 c	8152.65	I
160	2918.08	II	1000	1015.54	II	20000	4477.72	I	10000	8153.75	I
110	3032.26	II	1000	1049.00	II	1000	4490.42	I	25000	8154.00	I
70	3179.33	II	1000	1069.15	V	3000	4513.44	I	5000	8246.86	I
110	3323.18	II	900	1112.13	V	15000	4525.59	I	15000	8264.96	I
110	3323.60	II	1000	1143.56	V	3000	4575.74	I	75000 c	8272.44	I
450	3451.29	II	1000	1189.28	I	2500	4614.58	I	20000	8334.70	I
285	4121.93	II	1000	1189.50	I	2500	4752.28	I	10000	8343.70	I
110	4194.79	II	1000	1210.73	I	4000	4780.31	I	1200	8384.04	I
40	4242.98	III	1000	1221.13	I	1600	4785.19	I	40000	8446.55	I
70	4243.61	III	1000	1223.24	I	4000	4979.76	I	4000	8477.45	I
110	4472.10	II	1200	1224.41	I	1200	5395.48	I	1500	8513.38	I
110	4472.85	II	1200	1226.90	I	1200	5466.22	I	1000	8557.73	I
220	4487.05	III	7500	1232.43	I	1800	5852.08	I	1000	8566.28	I
360	4497.73	III	1200	1243.90	I	1600	5940.48	I	20000	8638.66	I
70	4784.21	II	1500	1251.66	I	2400	6122.14	I	4000	8698.53	I
110	4940.38	II	1000	1255.80	I	40000	6148.60	I	10000 c	8793.47	I
110	6080.44	II	1500	1259.20	I	2000	6177.39	I	15000	8819.96	I
70	6285.47	II	1200	1261.66	I	1500	6335.48	I	25000	8825.22	I
70	7030.20	II	1200	1266.20	I	60000	6350.73	I	4000	8888.98	I
40	7031.90	II	1000	1279.48	I	2500	6410.32	I	30000	8897.62	I
110	7835.25	III	1000	1286.26	I	1800	6483.56	I	6000	8932.40	I
70	7841.41	III	3000	1309.91	I	1000	6514.62	I	1800	8949.39	I
20	8667.22	I	3000	1316.74	I	20000	6544.57	I	9000	8964.00	I
70	8668.57	I	1000	1317.37	I	1500	6548.09	I	30000	9166.06	I
800	11660.04	I	2000	1317.70	I	50000 c	6559.80	I	15000	9173.63	I
570	11662.47	I	12000	1384.60	I	1000	6571.31	I	20000	9178.16	I
125	15629.08	I	3000	1449.90	I	1800	6579.14	I	40000	9265.42	I
200	16240.38	I	50000	1488.45	I	20000	6582.17	I	15000	9320.86	I
250	16244.67	I	30000	1531.74	I	1500	6620.47	I	6000	9793.48	I
235	18994.33	I	25000	1540.65	I	50000 c	6631.62	I	10000	9896.40	I
			30000	1574.84	I	20000	6682.28	I	3000	10140.08	I
Bromine Br	**Z = 35**		20000	1576.39	I	10000	6692.13	I	6000	10237.74	I
700	379.73	IV	25000	1582.31	I	8000	6728.28	I	1000	10299.62	I
700	400.37	IV	75000	1633.40	I	2000	6760.06	I	1500	10377.65	I
800	482.11	V	1000	2133.79	IV	2000	6779.48	I	30000	10457.96	I
900	531.97	V	1000	2145.02	IV	2200	6786.74	I	1000	10742.14	I
1000	545.43	IV	1000	2257.21	IV	6500	6790.04	I	3000	10755.92	I
1000	547.90	V	1000	2272.73	IV	1600 c	6791.48	I	1700	13217.17	I
1000	559.76	IV	1000	2307.40	IV	1800	6861.15	I	1800	14354.57	I
1000	569.19	IV	1000	2408.16	IV	10000	7005.19	I	1250	14888.70	I
1000	576.59	IV	1000	2411.58	IV	2000	7260.45	I	1800	16731.19	I
1000	585.10	IV	700	2491.14	IV	10000	7348.51	I	1200	18568.31	I
1000	586.71	IV	1000	2581.19	IV	40000	7512.96	I	3500	19733.62	I
1000	597.51	IV	600	2661.40	IV	1600	7591.61	I	1000	20281.73	I
1000	600.09	IV	1000	2842.88	IV	1800	7595.07	I	1000	20624.67	I
1000	601.27	IV	1100 h	2907.71	IV	2000	7616.41	I	1200	21787.24	I
1000	607.03	IV	500 h	2972.26	II	30000	7803.02	I	4000	22865.65	I
1000	617.85	IV									

Intensity	Wavelength/Å		Intensity	Wavelength/Å		Intensity	Wavelength/Å		Intensity	Wavelength/Å	
1000	23513.15	I	75	1793.40	III	30	2823.19	II	1000	5378.13	II
500	28346.50	I	40	1823.41	III	200	2836.900	I	200	5381.89	II
500	30380.85	I	100	1827.70	II	25	2856.46	II	40	5843.30	II
600	31630.13	I	50	1844.66	III	100	2868.180	I	50	5880.22	II
150	38345.75	I	40	1851.13	III	200 r	2880.767	I	300	6099.142	I
120	39964.36	I	40	1855.85	III	50 r	2881.224	I	100	6111.49	I
			200	1856.67	III	200	2914.67	II	100	6325.166	I
Cadmium Cd Z = 48			150	1874.08	III	50	2927.87	II	30	6330.013	I
50	427.01	IV	300	1922.23	II	200	2929.27	II	400	6354.72	II
50	447.85	IV	100	1943.54	II	1000 r	2980.620	I	500	6359.98	II
60	480.90	IV	40	1965.54	II	200 r	2981.362	I	2000	6438.470	I
70	493.00	IV	30	1986.89	II	50	2981.845	I	400	6464.94	II
70	495.13	IV	200	1995.43	II	50	3030.60	II	25	6567.65	II
70	498.14	IV	100	2007.49	II	150	3080.822	I	500	6725.78	II
70	498.53	IV	50	2032.45	II	25	3081.48	II	100	6759.19	II
80	504.09	IV	75	2036.23	II	30	3082.593	I	30	6778.116	I
70	504.20	IV	40	2039.83	III	100	3092.34	II	50	7237.01	II
70	504.50	IV	50	2045.61	III	200	3133.167	I	100	7284.38	II
80	506.31	IV	75	2087.91	III	50	3146.79	II	1000	7345.670	I
60	508.01	IV	150	2096.00	II	150	3250.33	II	50	8066.99	II
50	508.95	IV	50	2111.60	III	300	3252.524	I	5	8200.309	I
70	509.55	IV	1000 r	2144.41	II	300	3261.055	I	20	9289.	I
70	511.40	IV	50	2155.06	II	50	3343.21	II	15	11652.	I
80	513.00	IV	100	2187.79	II	50	3385.49	II	35	14487.	I
70	514.50	IV	1000	2194.56	II	30	3388.88	II	80	15708.	I
60	519.42	IV	1000	2265.02	II	800	3403.652	I	55 d	19120.	I
80	524.41	IV	1500 r	2288.022	I	50	3417.49	II	25	24371.	I
70	524.47	IV	1000	2312.77	II	50	3442.42	II	35	25448.	I
70	525.10	IV	200	2321.07	II	100	3464.43	II			
60	525.19	IV	40	2376.82	II	1000	3466.200	I	**Calcium Ca Z = 20**		
70	527.07	IV	50	2418.69	II	800	3467.655	I	250	190.46	V
80	531.09	IV	50	2469.73	II	25	3483.08	II	250	196.97	V
80	531.51	IV	40	2487.93	II	150	3495.44	II	300	199.55	V
70	534.29	IV	40	2495.58	II	25	3499.952	I	250	200.51	V
70	536.77	IV	50	2509.11	II	100	3524.11	II	265	257.98	V
60	540.90	IV	30	2516.22	II	100	3535.69	II	400	267.77	V
70	541.74	IV	25 h	2525.196	I	1000	3610.508	I	300	270.31	V
80	542.60	IV	50	2544.613	I	800	3612.873	I	400	280.99	V
80	546.55	IV	50	2551.98	II	60	3614.453	I	300	284.98	V
60	553.06	IV	25	2553.465	I	20	3649.558	I	450 c	286.96	V
80	554.05	IV	3	2565.789	I	10	3981.926	I	500	322.17	V
60	567.01	IV	500	2572.93	II	100	4029.12	II	300	323.22	V
150	1118.16	IV	50	2580.106	I	200	4134.77	II	300	330.94	V
100	1164.65	IV	30	2592.026	I	50	4141.49	II	300	334.55	V
100	1183.40	IV	25 h	2602.048	I	100	4285.08	II	250 c	342.45	IV
100	1256.00	II	50	2628.979	I	8	4306.672	I	250	343.93	IV
150	1296.43	II	40	2632.190	I	100	4412.41	II	450	352.92	V
100	1326.50	II	75	2639.420	I	3	4412.989	I	250	377.18	V
60	1370.48	IV	40	2659.23	II	1000	4415.63	II	200	387.08	V
150	1370.91	II	50 h	2660.325	I	30	4440.45	II	750	425.00	V
60	1418.89	IV	25	2668.20	II	8	4662.352	I	600	434.57	IV
200	1514.26	II	50	2672.62	II	200	4678.149	I	250	437.77	IV
50	1545.17	III	100	2677.540	I	30	4744.69	II	750	443.82	IV
200	1571.58	II	25	2677.748	I	300	4799.912	I	500	450.57	IV
100	1668.60	II	50	2707.00	II	50	4881.72	II	500	558.60	V
50	1702.47	II	75	2712.505	I	50	5025.50	II	400	637.93	V
40	1707.16	III	50	2733.820	I	1000 h	5085.822	I	300	643.12	V
40	1722.95	III	1000	2748.54	II	6	5154.660	I	400	646.57	V
50	1724.41	II	100 h	2763.894	I	100	5268.01	II	750	656.00	IV
40	1747.67	III	50 h	2764.230	I	100	5271.60	II	300	656.76	V
40	1773.06	III	50	2774.958	I	1000	5337.48	II	500	669.70	IV
100	1785.84	II							24	1341.89	II

Intensity	Wavelength/Å		Intensity	Wavelength/Å		Intensity	Wavelength/Å		Intensity	Wavelength/Å	
12	1342.54	II	20	4489.18	II	40	7820.78	II	250	371.75	III
20	1433.75	II	19	4499.88	III	60	7843.38	II	150	371.78	III
20	1545.29	III	23	4526.94	I	20	8017.50	II	650	384.03	IV
60	1649.86	II	22	4578.55	I	20	8020.50	II	700	384.18	IV
20	1807.34	II	23	4581.40	I	70	8133.05	II	500	386.203	III
40	1814.50	II	23	4581.47	I	100	8201.72	II	400	419.52	IV
40	1838.01	II	24	4585.87	I	110	8248.80	II	500	419.71	IV
60	1840.06	II	24	4585.96	I	70	8254.73	II	200	450.734	III
20	1843.09	II	20	4685.27	I	130	8498.02	II	400	459.46	III
40	1850.69	II	30	4716.74	II	170	8542.09	II	500	459.52	III
17	2123.03	III	40	4721.03	II	160	8662.14	II	570	459.63	III
16	2152.43	III	40	4799.97	II	100	8912.07	II	250	511.522	III
16	2687.76	III	25	4878.13	I	110	8927.36	II	250	535.288	III
19	2881.78	III	70	5001.48	II	110	9213.90	II	300	538.080	III
21	2899.79	III	80	5019.97	II	90	9312.00	II	350	538.149	III
19	2924.33	III	40	5021.14	II	100	9319.56	II	400	538.312	III
20	2988.63	III	23	5041.62	I	110	9320.65	II	350	574.281	III
10	3006.86	I	25	5188.85	I	25	9416.97	I	9	595.022	II
15	3028.59	III	22	5261.71	I	100	9567.97	II	30	687.053	II
3	3055.32	I	23	5262.24	I	110	9599.24	II	50	687.345	II
19	3119.67	III	22	5264.24	I	80	9601.82	II	10	858.092	II
170	3158.87	II	24	5265.56	I	80	9854.74	II	20	858.559	II
180	3179.33	II	25	5270.27	I	110	9890.63	II	30	903.624	II
150	3181.28	II	60	5285.27	II	90	9931.39	II	60	903.962	II
20	3316.51	II	70	5307.22	II	100	10223.04	II	150	904.142	II
12	3361.92	I	50	5339.19	II	20	10343.81	I	30	904.480	II
19	3372.67	III	27	5349.47	I	20	11838.99	II	800	977.03	III
20	3461.87	II	23	5512.98	I	25	12816.04	I	9	1009.86	II
13	3487.60	I	25	5581.97	I	24	12823.86	I	10	1010.08	II
18	3537.77	III	27	5588.76	I	25	12909.10	I	10	1010.37	II
20	3644.41	I	24	5590.12	I	30	13033.57	I	80	1036.337	II
30	3683.70	II	26	5594.47	I	21	13086.44	I	150	1037.018	II
40	3694.11	II	25	5598.49	I	24	13134.95	I	150	1157.910	I
170	3706.03	II	24	5601.29	I	20	16150.77	I	150	1158.019	I
180	3736.90	II	24	5602.85	I	22	16157.36	I	150	1158.035	I
20	3755.67	II	30	5857.45	I	21	16197.04	I	370	1174.93	III
30	3758.39	II	27	6102.72	I	20	18925.47	I	350	1175.26	III
230	3933.66	II	29	6122.22	I	24	18970.14	I	330	1175.59	III
220	3968.47	II	22	6161.29	I	30	19046.14	I	500	1175.71	III
50	4097.10	II	30	6162.17	I	48	19309.20	I	350	1175.99	III
60	4109.82	II	22	6163.76	I	49	19452.99	I	370	1176.37	III
30	4110.28	II	24	6166.44	I	47	19505.72	I	150	1188.992	I
40	4206.18	II	26	6169.06	I	50	19776.79	I	150	1189.447	I
50	4220.07	II	28	6169.56	I	35	19853.10	I	200	1189.631	I
50	4226.73	I	35	6439.07	I	34	19862.22	I	300	1193.009	I
24	4283.01	I	30	6449.81	I	23	19917.19	I	300	1193.031	I
22	4289.36	I	22	6455.60	I	24	19933.70	I	300	1193.240	I
22	4298.99	I	80	6456.87	II	25	22624.93	I	300	1193.264	I
25	4302.53	I	34	6462.57	I	30	22651.23	I	100	1193.393	I
20	4302.81	III	29	6471.66	I				150	1193.649	I
23	4307.74	I	32	6493.78	I	**Carbon C Z = 6**			150	1193.679	I
22	4318.65	I	28	6499.65	I	110	34.973	V	100	1194.064	I
20	4355.08	I	23	6572.78	I	450	40.268	V	100	1194.488	I
19	4399.59	III	30	6717.69	I	110	227.19	V	100	1261.552	I
25	4425.44	I	33	7148.15	I	250	244.91	IV	250	1277.245	I
26	4434.96	I	31	7202.19	I	160	248.66	V	250	1277.282	I
25	4435.69	I	33	7326.15	I	160	248.74	V	300	1277.513	I
30	4454.78	I	30	7575.81	II	200	289.14	IV	300	1277.550	I
28	4455.89	I	60	7581.11	II	250	289.23	IV	200	1280.333	I
20	4456.61	I	80	7601.30	II	570	312.42	IV	100	1311.363	I
20	4472.04	II	20	7602.32	II	500	312.46	IV	9	1323.951	II
						250	371.69	III			

Intensity	Wavelength/Å		Intensity	Wavelength/Å		Intensity	Wavelength/Å		Intensity	Wavelength/Å	
120	1329.578	I	1000	4267.26	II	250	9620.80	I	40000	3085.10	III
120	1329.600	I	200	4325.56	III	300	9658.44	I	20000	3106.98	III
150	1334.532	II	600	4647.42	III	200	10683.08	I	30000	3110.53	III
300	1335.708	II	520	4650.25	III	300	10691.25	I	30000	3121.56	III
100	1354.288	I	375	4651.47	III	12	11619.29	I	20000	3141.29	III
150	1355.84	I	200 w	4658.30	IV	23	11628.83	I	20000	3143.96	III
120	1364.164	I	200	4665.86	III	13	11658.85	I	20000	3147.06	III
100	1459.032	I	200	4771.75	I	47	11659.68	I	710	3194.83	II
200	1463.336	I	200	4932.05	I	24	11669.63	I	990	3201.71	II
120	1467.402	I	5	4943.88	V	85	11748.22	I	710	3218.94	II
150	1481.764	I	5	4944.56	V	142	11753.32	I	880	3221.17	II
1000	1548.202	IV	200	5052.17	I	114	11754.76	I	710	3227.11	II
900	1550.774	IV	350	5132.94	II	11	11777.54	I	20000	3228.57	III
150	1560.310	I	350	5133.28	II	17	11892.91	I	710	3234.16	II
400	1560.683	I	350	5143.49	II	30	11895.75	I	990	3272.25	II
400	1560.708	I	570	5145.16	II	26	12614.10	I	20000	3353.29	III
100	1561.341	I	400	5151.09	II	20	13502.27	I	10000	3395.77	III
400	1561.438	I	300	5380.34	I	38	14399.65	I	30000	3427.36	III
150	1656.266	I	250	5648.07	II	16	14403.25	I	40000	3443.63	III
120	1656.928	I	350	5662.47	II	61	14420.12	I	30000	3454.39	III
300	1657.008	I	450	5695.92	III	12	14429.03	I	40000	3459.39	III
120	1657.380	I	250	5801.33	IV	13	14442.24	I	60000	3470.92	III
120	1657.907	I	200	5811.98	IV	12	16559.66	I	710	3485.05	II
150	1658.122	I	150	5826.42	III	50	16890.38	I	50000	3497.81	III
500	1751.823	I	570	5889.77	II	10	17338.56	I	60000	3504.64	III
1000	1930.905	I	350	5891.59	II	11	17448.60	I	770	3539.08	II
250	2162.94	III	200	6001.13	I	13	18139.80	I	50000	3544.07	III
40	2270.91	V	250	6006.03	I	23	19721.99	I	1200	3560.80	II
5	2277.25	V	110	6007.18	I				1000	3577.45	II
20	2277.92	V	150	6010.68	I	*Cerium Ce Z = 58*			1800	3655.85	II
800	2296.87	III	300	6013.22	I	300	399.36	V	880	3660.64	II
800	2478.56	I	250	6014.84	I	200	482.96	V	880	3667.98	II
250	2509.12	II	800	6578.05	II	40	741.79	IV	1000	3709.29	II
350	2512.06	II	570	6582.88	II	30	754.60	IV	1000	3709.93	II
200 l	2524.41	IV	200	6587.61	I	75	1332.16	IV	1400	3716.37	II
300 s	2529.98	IV	150	6744.38	III	75	1372.72	IV	800	3728.42	II
250 h	2574.83	II	250	6783.90	II	100	2000.42	IV	860	3786.63	II
150	2697.75	III	150 h	7037.25	III	100	2009.94	IV	2500	3801.52	II
110 l	2724.85	III	250	7113.18	I	10000	2318.64	III	800	3803.09	II
150 l	2725.30	III	250	7115.19	I	10000	2372.34	III	1000	3808.11	II
150 l	2725.90	III	250	7115.63	II	10000	2380.12	III	1100	3838.54	II
350 l	2741.28	II	200	7116.99	I	10000	2431.45	III	860	3848.59	II
250	2746.49	II	350	7119.90	II	15000	2439.80	III	860	3853.15	II
1000	2836.71	II	800	7231.32	II	10000	2454.32	III	1200	3854.18	II
800	2837.60	II	1000	7236.42	II	10000	2469.95	III	1200	3854.31	II
200	2982.11	III	150	7612.65	III	10000	2483.82	III	1100	3878.36	II
800 h	2992.62	II	90 w	7726.2	IV	10000	2497.50	III	1500	3882.45	II
350	3876.19	II	200	7860.89	I	20000	2531.99	III	1000	3889.98	II
350	3876.41	II	200	8058.62	I	10000	2603.59	III	770	3907.29	II
350	3876.66	II	300 h	8196.48	III	340	2651.01	II	980	3912.44	II
570	3918.98	II	150	8332.99	III	270	2830.90	II	770	3918.28	II
800	3920.69	II	520	8335.15	I	250	2874.14	II	770	3931.09	II
150	4056.06	III	300	8500.32	III	10000	2923.81	III	770	3940.34	II
200	4067.94	III	250	9061.43	I	10000	2931.54	III	2000	3942.15	II
250	4068.91	III	200	9062.47	I	400	2976.91	II	2700	3942.75	II
250	4070.26	III	200	9078.28	I	10000	3022.75	III	770	3943.89	II
250	4074.52	II	250	9088.51	I	50000	3031.58	III	3100	3952.54	II
350 l	4075.85	II	450	9094.83	I	95000	3055.59	III	980	3956.28	II
150	4162.86	III	300	9111.80	I	20000	3056.56	III	770	3960.91	II
250 h	4186.90	III	800	9405.73	I	40000	3057.23	III	770	3967.05	II
800	4267.00	II	150	9603.03	I	20000	3057.58	III	770	3978.65	II
						680	3063.01	II			

Intensity	Wavelength/Å		Intensity	Wavelength/Å		Intensity	Wavelength/Å		Intensity	Wavelength/Å	
770	3984.68	II	280	5159.69	I	25	7329.91	I	630	2845.70	III
700	3992.39	II	280	5161.48	I	25	7397.77	I	3100	2859.32	III
910	3993.82	II	370	5187.46	II	25	7616.11	II	200	2893.85	III
2800	3999.24	II	260	5223.46	I	25	7689.17	II	180	2921.13	III
910	4003.77	II	260	5245.92	I	22	7844.94	II	3200	2976.86	III
2700	4012.39	II	340	5274.23	II	22	7857.54	II	210	3001.28	III
910	4014.90	II	450	5353.53	II	30	8025.56	II	1700	3066.59	III
840	4024.49	II	300	5393.40	II	25	8772.14	II	1100 c	3149.36	III
840	4028.41	II	280	5409.23	II	30	8891.20	II	1400	3152.36	III
840	4031.34	II	260	5512.08	II				8400	3268.32	III
2100	4040.76	II	300	5696.99	I	**Cesium Ce Z = 55**			1300	3315.51	III
910	4042.58	II	370	5699.23	I	10000	614.01	III	550	3340.60	III
700	4053.51	II	240	5719.03	I	2000	638.17	III	430	3344.02	III
1100	4071.81	II	230	5940.86	I	2500	666.25	III	1200	3349.46	III
1800	4073.48	II	55	6001.90	I	5000	691.60	III	400	3463.45	III
1500	4075.71	II	55	6005.86	I	3500	703.89	III	580	3476.83	III
1500	4075.85	II	55	6006.82	I	15000	718.14	II	480	3559.82	III
910	4083.23	II	75	6013.42	I	20000	721.79	III	7200	3597.45	III
770	4118.14	II	110	6024.20	I	20000	722.20	III	1300	3608.31	III
980	4123.87	II	10000	6032.54	III	5000	731.56	III	2300	3618.19	III
980	4127.37	II	110	6043.39	II	12000	740.29	III	300 c	3641.34	III
2700	4133.80	II	55	6047.40	I	15000	808.76	II	520	3651.08	III
2000	4137.65	II	10000	6060.91	III	15000	813.84	II	4800	3661.40	III
770	4142.40	II	45	6098.34	II	7500	830.39	III	640	3699.50	III
980	4149.94	II	45	6123.67	I	35000	901.27	II	430	3837.46	III
1400	4151.97	II	35	6143.36	II	15000	920.35	III	2100 c	3876.15	I
1300	4165.61	II	35	6186.17	I	40000	926.66	II	2900	3888.37	III
3500	4186.60	II	35	6208.98	I	25000 c	1054.79	III	600 c	3888.61	I
840	4198.72	II	35	6228.94	I	17 c	1673.99	III	2700	3925.60	III
910	4202.94	II	23	6232.45	II	12	1705.25	III	680 c	4001.70	III
1500	4222.60	II	28	6237.45	I	10	1801.83	III	3100	4006.55	III
770	4227.75	II	45	6272.05	II	20 c	1822.40	III	420	4006.78	III
980	4239.92	II	35	6295.58	I	11	1823.93	III	520	4043.42	III
1100	4248.68	II	28	6299.51	II	12	1824.70	III	14000	4264.70	II
2000	4289.94	II	23	6300.21	I	12	1841.80	III	18000 w	4277.13	II
1500	4296.67	II	35	6310.01	I	25	1915.50	III	370	4403.86	III
770	4300.33	II	35	6343.95	II	25 c	1923.29	III	1200	4410.22	III
770	4306.72	II	35	6371.11	II	12	1961.33	III	940	4425.68	III
980	4337.77	II	28	6386.84	I	17	1996.56	III	530	4471.48	III
700	4349.79	II	23	6393.02	II	710	2035.11	III	12000	4501.55	III
910	4364.66	II	35	6430.07	I	120	2056.43	III	1200	4506.72	III
910	4382.17	II	23	6436.40	I	330	2076.43	III	590	4522.86	III
700	4386.84	II	35	6458.03	I	540	2077.30	III	20000	4526.74	II
1700	4391.66	II	28	6467.39	I	410	2088.68	III	1000 c	4555.28	I
980	4418.78	II	35	6473.72	I	210	2101.63	III	460 c	4593.17	I
770	4449.34	II	23	6513.59	II	200	2141.47	III	99900	4603.79	II
2400	4460.21	II	45	6555.65	I	1000	2316.88	III	420 h	4620.61	III
1400	4471.24	II	23	6579.10	I	230	2325.95	III	210	4665.52	III
700	4479.36	II	22	6612.06	I	390	2340.49	III	25000	4830.19	II
700	4483.90	II	30	6628.93	I	1600	2455.81	III	140	4851.59	III
840	4486.91	II	22	6652.72	II	1600	2477.57	III	19000	4870.04	II
770	4523.08	II	26	6700.66	I	890	2485.45	III	37000	4952.85	II
840	4527.35	II	35	6704.27	I	410	2495.07	III	370	5035.72	III
840	4528.47	II	30	6774.28	II	1400	2525.67	III	27000	5043.80	II
840	4539.75	II	35	6775.59	I	430	2573.05	III	75000	5227.04	II
2100	4562.36	II	30	6924.81	I	16000	2596.86	III	29000	5249.38	II
1100	4572.28	II	30	6986.02	I	390	2610.12	III	11000	5274.05	II
840	4593.93	II	35	7061.75	II	6200	2630.51	III	10000 c	5349.13	II
1700	4628.16	II	35	7086.35	II	370	2700.32	III	22000	5370.99	II
310	4737.28	II	22	7238.36	II	710	2701.20	III	230	5380.79	III
470	5079.68	II	25	7252.75	I	390	2776.44	III	60 c	5465.94	I
						270	2810.87	III			

Intensity	Wavelength/Å		Intensity	Wavelength/Å		Intensity	Wavelength/Å		Intensity	Wavelength/Å	
37	5502.88	I	1500	17012.32	I	2500	728.951	II	12000	1396.527	I
39000	5563.02	II	760	20138.47	I	2000	777.562	II	500	1441.470	II
100	5635.21	I	880	22811.86	I	5000	787.580	II	500	1528.569	II
210 c	5664.02	I	1100	23037.98	I	5000	788.740	II	500	1542.942	II
27	5745.72	I	3900	23344.47	I	5000	793.342	II	500	1558.144	II
24000	5831.14	II	4400	24251.21	I	500	834.84	IV	500	1565.050	II
59 c	5838.83	I	850	24374.96	I	500	834.97	IV	600	1822.50	III
300	5845.14	I	890 d	25763.51	I	6000	839.297	II	500	1828.40	III
51000	5925.63	II	500	25764.73	I	8000	839.599	II	500	1857.488	II
140	5950.14	III	680 c	29310.06	I	600	840.93	IV	500	1901.61	III
110	5979.97	III	2800	30103.27	I	5000	851.691	II	500	1983.61	III
640 c	6010.49	I	610 c	30953.06	I	2000	888.026	II	450 h	1997.370	II
86	6034.09	I	1100	34900.13	I	2000	893.549	II	450	2032.116	II
150	6043.99	III	190	36131.00	I	2000	961.499	II	350 h	2088.583	II
870	6079.86	III	2 c	39177.28	I	500	973.21	IV	350 h	2091.458	II
9800	6128.61	II	2 d	39421.25	I	600	977.56	IV	700	2253.07	III
330	6150.42	III	1	39424.11	I	40	978.284	I	500	2268.95	III
1000	6213.10	I				700	984.95	IV	500	2278.34	III
170	6217.60	I	**Chlorine Cl Z = 17**			25	998.372	I	700	2283.93	III
450	6242.96	III	500	392.43	V	25	998.432	I	600	2323.50	III
320 c	6354.55	I	800	486.17	IV	75	1002.346	I	500	2336.45	III
510	6456.33	II	800	534.73	IV	500	1005.28	III	600	2340.64	III
8300	6495.53	II	700	535.67	IV	600	1008.78	III	600	2359.67	III
10000 w	6536.44	II	600	536.15	IV	150	1013.664	I	600	2370.37	III
490	6586.51	I	900	537.61	IV	700	1015.02	III	700	2416.42	III
97	6628.66	I	500	538.03	V	90	1025.553	I	600	2447.14	III
8800	6646.57	II	600	538.12	IV	6000	1063.831	II	600	2448.58	III
3300 c	6723.28	I	800	542.23	V	3000	1067.945	II	500	2486.91	III
9600	6724.47	II	600	542.30	V	9000	1071.036	II	500	2532.48	III
400	6753.12	III	1000	545.11	V	6000	1071.767	II	600	2580.67	III
200	6824.65	I	600	546.33	V	5000	1075.230	II	500	2603.59	III
300	6870.45	I	1000	547.63	V	5000	1079.080	II	500	2632.67	III
37000	6955.50	II	500	549.22	IV	200	1084.667	I	500	2633.18	III
4800	6973.30	I	700	552.02	IV	200	1085.171	I	600	2665.54	III
16000	6979.67	I	600	553.30	IV	250	1085.304	I	700	2710.37	III
980	6983.49	I	700	554.62	IV	400	1088.06	I	500	2724.03	IV
13000 w	7149.54	II	600	556.23	III	350	1090.271	I	500	2751.23	IV
1900 c	7219.60	III	700	556.61	III	250	1090.982	I	700	2782.47	IV
790	7228.53	I	700	557.12	III	250	1092.437	I	600	2965.56	III
130	7279.90	I	350	559.305	II	400	1094.769	I	500	3063.13	IV
1100	7279.96	I	700	561.53	III	350	1095.148	I	600	3076.68	IV
2600 c	7608.90	I	700	561.68	III	350	1095.662	I	600	3104.46	III
3300	7943.88	I	700	561.74	III	400	1095.797	I	800	3139.34	III
22000	7997.44	II	400	571.904	II	250	1096.810	I	900	3191.45	III
3500	8015.73	I	800	574.406	II	300	1097.369	I	700	3289.80	III
510	8078.94	I	500	601.50	IV	200	1098.068	I	700	3320.57	III
4500	8079.04	I	500	604.59	IV	200	1099.523	I	800	3329.06	III
59000 c	8521.13	I	500	606.35	III	500	1107.528	I	900	3340.42	III
15000 c	8761.41	I	700	618.057	II	800	1139.214	II	800	3392.89	III
61000 c	8943.47	I	600	619.982	II	800	1167.148	I	800	3393.45	III
18000	9172.32	I	800	620.298	II	3000	1179.293	I	900	3530.03	III
5200	9208.53	I	700	626.735	II	1200	1188.774	I	800	3560.68	III
19000	10024.36	I	800	635.881	II	900	1201.353	I	900	3602.10	III
4800	10123.41	I	1000	636.626	II	3000	1335.726	I	800	3612.85	III
26000	10123.60	I	1000	650.894	II	10000	1347.240	I	700	3622.69	III
2900	13424.31	I	1000	659.811	II	5000	1351.657	I	700	3656.95	III
38000 c	13588.29	I	1300	661.841	II	12000	1363.447	I	700	3670.28	III
8400	13602.56	I	2000	663.074	II	2500	1373.116	I	700	3682.05	III
5700	13758.81	I	1500	682.053	II	20000	1379.528	I	600	3705.45	III
55000 c	14694.91	I	1500	687.656	II	25000	1389.693	I	600	3707.34	III
820	16535.63	I	1500	693.594	II	20000	1389.957	I	800	3720.45	III
			2000	725.271	II						

Intensity	Wavelength/Å		Intensity	Wavelength/Å		Intensity	Wavelength/Å		Intensity	Wavelength/Å	
800	3748.81	III	2300	7899.31	I	310	13296.0	I	130	2383.33	I
500	3779.35	III	1800	7915.08	I	550	13346.8	I	140	2408.62	I
10000	3850.99	II	3000	7924.645	I	525	13821.7	I	170	2496.31	I
25000	3860.83	II	2100	7933.89	I	294	14931.7	I	110	2502.53	I
500	3925.87	III	1700	7935.012	I	269	15108.0	I	190	2504.31	I
700	3991.50	III	650	7952.52	I	381	15465.1	I	110	2516.92	I
600	4018.50	III	1500	7974.72	I	1094	15520.3	I	390	2519.52	I
600	4059.07	III	1300	7976.97	I	1487	15730.1	I	190	2527.12	I
500	4104.23	III	600	7980.60	I	2780	15869.7	I	160	2549.54	I
500	4106.83	III	2900	7997.85	I	277	15883.3	I	130	2560.69	I
10000 h	4132.50	II	2200	8015.61	I	342	15928.9	I	150	2571.74	I
500	4608.21	III	1100	8023.33	I	735	15960.0	I	100	2577.65	I
40	4623.938	I	400	8051.07	I	283	15970.5	I	380	2591.85	I
50	4654.040	I	1700	8084.51	I	259	16198.5	I	250	2653.59	II
80	4661.208	I	2200	8085.56	I	717	19755.3	I	250	2658.59	II
45	4691.523	I	3000	8086.67	I	100	24470.0	I	320	2663.42	II
40	4721.255	I	1300	8087.73	I		39716.0	I	440	2666.02	II
45	4740.729	I	2500	8194.42	I		40085.5	I	280	2668.71	II
13000	4781.32	II	2200	8199.13	I		40089.5	I	350	2671.81	II
99000	4794.55	II	2200	8200.21	I		40532.2	I	280	2672.83	II
29000	4810.06	II	800	8203.78	I				1800	2677.16	II
16000	4819.47	II	18000	8212.04	I	*Chromium Cr Z = 24*			320	2678.79	II
81000	4896.77	II	3000	8220.45	I	100	438.62	V	230	2687.09	II
47000	4904.78	II	20000	8221.74	I	100	464.02	V	280	2691.04	II
26000	4917.73	II	18000	8333.31	I	100	620.66	IV	180	2698.41	II
10000	4995.48	II	99900	8375.94	I	100	629.26	IV	180	2698.69	II
26000	5078.26	II	400	8406.199	I	80	630.30	IV	110	2701.99	I
30	5099.789	I	15000	8428.25	I	100	666.55	IV	140	2712.31	II
56000	5217.94	II	2200	8467.34	I	100	693.92	IV	170	2722.75	II
23000	5221.36	II	2200	8550.44	I	60	1030.47	III	420 h	2726.51	I
15000	5392.12	II	20000	8575.24	I	100	1033.69	III	280 h	2731.91	I
99000	5423.23	II	750	8578.02	I	100	1036.03	III	170 h	2736.47	I
10000	5423.51	II	75000	8585.97	I	80	1055.89	IV	250	2743.64	II
19000	5443.37	II	450	8628.54	I	80	1068.41	III	110 h	2748.29	I
10000	5444.21	II	300	8641.71	I	100	1116.48	V	330	2748.98	II
40	5532.162	I	3500	8686.26	I	150	1121.07	V	390	2750.73	II
50 d	5796.305	I	2200	8912.92	I	150	1127.63	V	280	2751.87	II
45	5799.914	I	3000	8948.06	I	100	1263.50	V	110 h	2752.88	I
30	5856.742	I	2000	9038.982	I	100	1417.42	IV	150	2757.10	I
50	6019.812	I	2500	9045.43	I	150	1465.86	V	350	2757.72	II
200	6140.245	I	1000	9069.656	I	150	1497.97	V	750	2762.59	II
160	6194.757	I	2000	9073.17	I	170	1519.03	V	750	2766.54	II
150	6434.833	I	7500	9121.15	I	220	1579.70	V	250 h	2769.92	I
300	6932.903	I	3000	9191.731	I	170	1591.72	V	610	2780.70	I
300	6981.886	I	500	9197.596	I	150	1603.19	V	180	2822.37	II
600	7086.814	I	4000	9288.86	I	120	1672.66	IV	180	2830.47	II
7500	7256.62	I	1500	9393.862	I	120	1758.51	IV	2500	2835.63	II
5000	7414.11	I	3500	9452.10	I	140	1802.72	IV	110	2840.02	II
550	7462.370	I	500	9486.964	I	130	1812.41	IV	1700	2843.25	II
550	7489.47	I	1000	9584.801	I	200	1837.44	V	1200	2849.84	II
700	7492.118	I	3500	9592.22	I	140	1873.89	IV	120	2851.36	II
11000	7547.072	I	250	9632.509	I	140	1967.18	IV	880	2855.68	II
2300	7672.42	I	1000	9702.439	I	120	1972.07	IV	610	2858.91	II
450	7702.828	I	250	9744.426	I	19000	2055.52	II	440	2860.93	II
7000	7717.581	I	200	9807.057	I	14000	2061.49	II	790	2862.57	II
10000	7744.97	I	400	9875.970	I	8900	2065.42	II	750	2865.11	II
2200	7769.16	I	331	10392.549	I	200	2226.72	III	610	2866.74	II
650	7771.09	I	300	11123.05	I	200	2235.91	III	480	2867.65	II
2200	7821.36	I	269	11409.69	I	150	2237.59	III	210	2870.44	II
1700	7830.75	I	1000	11436.33	I	150	2244.10	III	110	2871.63	I
3000	7878.22	I	350	13243.8	I	150	2284.44	III	160	2873.48	II
						150	2324.88	III			

Intensity	Wavelength/Å		Intensity	Wavelength/Å		Intensity	Wavelength/Å		Intensity	Wavelength/Å	
320	2875.99	II	220	3197.08	II	110	3807.93	I	110	4374.16	I
230	2876.24	II	170	3209.18	II	180	3815.43	I	530	4384.98	I
180	2877.98	II	140	3217.40	II	180	3819.56	I	110	4458.54	I
120	2879.27	I	120	3245.54	I	130	3826.42	I	660	4496.86	I
170	2887.00	I	130	3251.84	I	130	3830.03	I	380	4526.47	I
700	2889.29	I	130	3257.82	I	380	3841.28	I	380	4530.74	I
370	2893.25	I	130	3339.80	II	190	3848.98	I	240	4535.72	I
190	2894.17	I	110	3342.59	II	140	3849.36	I	240	4540.50	I
210	2896.75	I	170	3358.50	II	290	3850.04	I	240	4540.72	I
180	2905.49	I	160	3360.30	II	140	3852.22	I	140	4544.62	I
260	2909.05	I	430	3368.05	II	190	3854.22	I	600	4545.96	I
260	2910.90	I	140	3382.68	II	110	3855.29	I	120	4565.51	I
250	2911.14	I	170	3403.32	II	140	3855.57	I	120	4571.68	I
480	2967.64	I	360	3408.76	II	260	3857.63	I	360	4580.06	I
480	2971.11	I	210	3421.21	II	660	3883.29	I	360	4591.39	I
210	2971.91	II	270	3422.74	II	570	3885.22	I	480	4600.75	I
480	2975.48	I	140	3433.31	II	380	3886.79	I	240	4613.37	I
190	2979.74	II	270	3433.60	I	260	3894.04	I	600	4616.14	I
350	2980.79	I	160	3436.19	I	360	3902.92	I	550	4626.19	I
110	2985.32	II	140	3441.44	I	960	3908.76	I	1600	4646.17	I
480	2985.85	I	170	3445.62	I	120 hd	3911.82	I	570	4651.28	I
1500	2986.00	I	170	3447.43	I	120	3915.84	I	840	4652.16	I
2100	2986.47	I	190	3453.33	I	190	3916.24	I	240 d	4698.46	I
660	2988.65	I	130	3455.60	I	1900	3919.16	I	190	4708.04	I
160	2989.19	II	100	3460.43	I	600	3921.02	I	240	4718.43	I
480	2991.89	I	120	3550.64	I	600	3928.64	I	120	4730.71	I
230	2994.07	I	130	3566.16	I	410	3941.49	I	140	4737.35	I
300	2995.10	I	130	3573.64	I	1900	3963.69	I	340	4756.11	I
700	2996.58	I	330 h	3574.80	I	120	3969.06	I	190	4789.32	I
210	2998.79	I	19000	3578.69	I	1600	3969.75	I	120	4801.03	I
1100	3000.89	I	160 h	3584.33	I	1600	3976.66	I	110	4829.38	I
750	3005.06	I	130	3585.30	II	960	3983.91	I	140	4870.80	I
140	3013.03	I	17000	3593.49	I	190	3984.34	I	130	4887.01	I
710	3013.71	I	350	3601.67	I	160	3989.99	I	260	4922.27	I
710	3014.76	I	13000	3605.33	I	960	3991.12	I	110	4936.33	I
1400	3014.92	I	130	3632.84	I	160	3991.67	I	70	4942.50	I
710	3015.19	I	350	3636.59	I	190	3992.84	I	110	4954.81	I
2800	3017.57	I	630	3639.80	I	160	4001.44	I	60	5013.32	I
430	3018.50	I	220	3641.83	I	120	4012.47	II	70	5166.23	I
240	3018.82	I	220	3649.00	I	120	4026.17	I	70	5184.59	I
430	3020.67	I	170	3653.91	I	190	4039.10	I	70	5192.00	I
2800	3021.56	I	220	3656.26	I	160	4048.78	I	85	5196.44	I
1100	3024.35	I	130	3663.21	I	120	4058.77	I	5300	5204.52	I
170	3029.16	I	120	3685.55	I	140	4126.52	I	8400	5206.04	I
710	3030.24	I	130	3686.80	I	120	4153.82	I	11000	5208.44	I
140	3031.35	I	130	3687.25	I	140	4163.62	I	85	5224.94	I
390	3034.19	I	130	3730.81	I	170	4174.80	I	290	5247.56	I
550	3037.04	I	150	3732.03	I	170	4179.26	I	530	5264.15	I
550	3040.85	I	480	3743.58	I	110	4209.37	I	180	5265.72	I
110	3050.14	II	570	3743.88	I	20000	4254.35	I	95 h	5275.17	I
710	3053.88	I	340	3749.00	I	110	4263.14	I	70 h	5276.03	I
240	3118.65	II	230	3757.66	I	16000	4274.80	I	340	5296.69	I
430	3120.37	II	260	3768.24	I	10000	4289.72	I	70 h	5297.36	I
470	3124.94	II	130	3791.38	I	780	4337.57	I	660	5298.27	I
120	3128.70	II	130	3792.14	I	1100	4339.45	I	85	5300.75	I
590	3132.06	II	120	3793.29	I	380	4339.72	I	340 h	5328.34	I
140	3136.68	II	130	3793.88	I	1900	4344.51	I	70 h	5329.17	I
140	3147.23	II	140	3797.13	I	380	4351.05	I	780	5345.81	I
100	3155.15	I	200	3797.72	I	2300	4351.77	I	380	5348.32	I
100	3163.76	I	530	3804.80	I	570	4359.63	I	40	5400.61	I
240	3180.70	II	110	3806.83	I	530	4371.28	I	1400	5409.79	I

Intensity	Wavelength/Å		Intensity	Wavelength/Å		Intensity	Wavelength/Å		Intensity	Wavelength/Å	
24	5628.64	I	1800	1955.17	I	200	2347.39	II	800	2580.32	II
7	5642.36	I	1500	1958.55	I	1600	2352.85	I	300 d	2582.22	II
24	5664.04	I	1500	1961.59	I	200 d	2353.41	II	500	2587.22	II
24	5694.73	I	1500 h	1968.69	I	2000	2353.42	I	500	2587.52	II
40	5698.33	I	1500 h	1968.93	I	500	2363.80	II	200	2588.91	II
24	5702.31	I	3000	1970.71	I	400	2378.62	II	100 p	2605.71	II
24	5712.78	I	1800 h	1971.16	I	1400	2380.48	I	100	2612.50	II
24 h	5783.11	I	1800 h	1972.52	I	200	2381.76	II	100	2614.36	II
30 h	5783.93	I	1500	1973.85	I	300 p	2383.45	II	100 p	2628.77	II
24 h	5785.00	I	1800	1976.97	I	1400	2384.86	I	100	2632.26	II
19 h	5785.82	I	2400 h	1980.89	I	200	2386.36	II	100	2636.07	II
60 h	5787.99	I	1500	1989.80	I	500	2388.92	II	310	2646.42	I
180 h	5791.00	I	1800	1990.34	I	200	2397.38	II	770	2648.64	I
35	6330.10	I	1500 l	1998.49	I	1100 d	2402.06	I	100	2653.72	II
22	6362.87	I	1500	2002.32	I	200 p	2404.16	II	100	2663.53	II
19	6661.08	I	900	2008.04	I	5300	2407.25	I	200	2666.73	II
21 h	6883.03	I	50	2011.51	II	5300	2411.62	I	100	2675.85	II
27 h	6924.13	I	1200 h	2014.58	I	1600	2412.76	I	100	2684.42	II
30 h	6978.48	I	900	2016.17	I	4800	2414.46	I	100	2702.02	II
85	7355.90	I	50	2022.35	II	4800	2415.30	I	200	2706.62	II
130	7400.21	I	50	2027.04	II	300	2417.65	II	200	2707.35	II
150	7462.31	I	900	2031.96	I	4100	2424.93	I	190	2715.99	I
40	8947.15	I	1500	2039.95	I	3300	2432.21	I	100	2727.78	II
19	8976.83	I	1200	2041.11	I	2900	2436.66	I	80	2734.54	II
			50	2065.54	II	2400	2439.05	I	190	2745.10	I
Cobalt Co *Z = 27*			1500 h	2077.76	I	200	2442.63	II	100	2753.22	II
20	355.52	V	900	2085.67	I	200 d	2446.03	II	190	2764.19	I
18	355.88	V	900	2087.55	I	200 p	2447.69	II	100	2766.70	II
12	356.06	V	900	2089.35	I	200	2450.00	II	100	2774.97	II
66	609.16	IV	900	2093.40	I	200	2464.20	II	100	2791.00	II
70	609.21	IV	900	2094.86	I	200	2486.44	II	100	2793.73	II
64	609.28	IV	900	2095.77	I	200	2498.82	II	150	2815.56	I
10	1018.36	V	1200	2097.51	I	570	2504.52	I	80	2835.63	II
10	1021.14	V	1500	2104.73	I	500	2506.46	II	80	2847.35	II
15	1231.73	V	1500	2106.80	I	360	2506.88	I	80	2871.22	II
50	1277.01	V	900	2108.98	I	200	2511.16	II	190	2886.44	I
80	1299.58	II	900 s	2117.68	I	860	2517.87	I	100	2918.38	II
80	1306.95	II	900	2137.78	I	500	2519.82	II	100	2930.24	II
50	1345.67	V	900	2138.97	I	4300	2521.36	I	100	2954.73	II
1000	1696.01	III	900	2163.03	I	200 h	2524.65	II	690	2987.16	I
800	1697.99	III	1100	2174.60	I	300	2524.97	II	690	2989.59	I
1000	1707.35	III	200	2193.60	II	500	2528.62	II	60	3022.59	II
5000	1760.35	III	200	2256.73	II	2900	2528.97	I	3100	3044.00	I
5000	1773.57	III	150	2260.00	II	200 p	2530.09	II	1700	3061.82	I
2000	1780.05	III	200	2283.52	II	720	2530.13	I	80	3387.70	II
3000	1782.97	III	1000	2286.15	II	860	2532.18	I	1100	3388.17	I
1000	1787.08	III	200	2291.98	II	200 d	2533.82	II	2200	3395.38	I
1000	1789.07	III	300 d	2293.38	II	2900	2535.96	I	11000	3405.12	I
1000	1823.08	III	300	2301.40	II	860	2536.49	I	4500	3409.18	I
2000	1830.09	III	800 d	2307.85	II	300	2541.94	II	6700	3412.34	I
2000	1831.44	III	2600	2309.02	I	1700	2544.25	I	2200	3412.63	I
5000	1835.00	III	500	2311.60	II	200	2546.74	II	2700	3417.16	I
1500	1842.34	I	500	2314.05	II	340	2548.34	I	50	3423.84	II
1800	1847.89	I	300	2314.96	II	310	2553.37	I	2500	3431.58	I
1800	1852.71	I	200 p	2317.06	II	310	2555.07	I	4500	3433.04	I
2400	1855.05	I	2400	2323.14	I	300	2559.41	II	1600	3442.93	I
2000	1863.83	III	300 p	2324.31	II	200	2560.03	II	8800	3443.64	I
1500	1878.28	I	200 d	2326.11	II	960	2562.15	I	50	3446.39	II
1800	1936.58	I	500	2326.47	II	500	2564.04	II	4100	3449.17	I
1500	1946.79	I	1400	2335.99	I	1100	2567.35	I	2100	3449.44	I
1500	1951.90	I	1600	2338.67	I	960	2574.35	I	21000	3453.50	I
1800	1954.22	I									

Intensity	Wavelength/Å		Intensity	Wavelength/Å		Intensity	Wavelength/Å		Intensity	Wavelength/Å	
1000	3455.23	I	90	4549.66	I	200	974.759	II	250	1418.426	II
5100	3462.80	I	140	4565.59	I	250	977.567	II	250	1421.759	II
5100	3465.80	I	190	4581.60	I	100	987.657	II	200	1427.829	II
8000	3474.02	I	120	4629.38	I	250	992.953	II	400	1430.243	II
1900	3483.41	I	85	4663.41	I	300	1004.055	II	250	1434.904	II
4800	3489.40	I	110	4792.86	I	300	1008.569	II	150	1436.236	II
2400	3495.69	I	100	4840.27	I	300	1008.728	II	150	1442.139	II
50	3501.72	II	150	4867.88	I	300	1010.269	II	200	1445.984	II
9600	3502.28	I	80 h	4964.18	II	250	1012.597	II	200	1449.058	II
7000	3506.32	I	50	5212.71	I	500	1018.707	II	250	1450.304	II
50	3507.77	II	50	5230.22	I	500	1027.831	II	200	1452.294	II
2900	3509.84	I	50	5247.93	I	250	1028.328	II	300	1458.002	II
1400	3510.43	I	50	5342.71	I	200	1030.263	II	250	1459.412	II
4800	3512.64	I	50	5352.05	I	600	1036.470	II	200	1463.752	II
3800	3513.48	I				600	1039.348	II	400	1463.838	II
4800	3518.35	I	**Copper Cu Z = 29**			600	1039.582	II	200	1466.070	II
1300	3520.08	I	80	685.141	II	800	1044.519	II	400	1470.697	II
2700	3521.57	I	100	709.313	II	800	1044.744	II	200	1472.395	II
3800	3523.43	I	100	718.179	II	500	1049.755	II	250	1473.978	II
60	3523.51	II	150	724.489	II	600	1054.690	II	200	1474.935	II
6400	3526.85	I	200	735.520	II	400	1055.797	II	150	1476.059	II
2700	3529.03	I	250	736.032	II	600	1056.955	II	300 r	1481.23	III
7300	3529.81	I	80	779.295	II	400	1058.799	II	200	1481.544	II
1900	3533.36	I	100	797.455	II	600	1059.096	II	200	1485.328	II
50	3545.03	II	150	810.998	II	600	1060.634	II	750	1488.831	II
1100	3560.89	I	200	813.883	II	600	1063.005	II	300	1492.834	II
80	3561.07	II	300	826.996	II	200	1065.782	II	250	1493.366	II
8800	3569.38	I	150	848.808	II	200	1066.134	II	250	1495.430	II
50	3574.95	II	250	851.303	II	500	1069.195	II	350	1496.687	II
1600	3574.96	I	250	858.487	II	300	1073.745	II	150	1503.368	II
60	3575.32	I	400	861.994	II	200	1088.395	II	250	1504.757	II
2500	3575.36	I	400	865.390	II	300	1094.402	II	200	1505.388	II
60	3577.96	II	250	869.336	II	250	1097.053	II	300	1508.632	II
1000	3585.16	I	150	873.263	II	150	1119.947	II	350	1510.506	II
6700	3587.19	I	200	876.723	II	200	1142.640	II	200	1512.465	II
1900	3594.87	I	250	877.012	II	300	1144.856	II	200	1513.366	II
1600	3602.08	I	200	877.555	II	100	1250.048	II	500	1514.492	II
100	3621.21	II	500	878.699	II	150	1265.506	II	200	1517.631	II
1000	3627.81	I	100	884.133	II	300	1275.572	II	500	1519.492	II
80	3643.61	II	250	885.847	II	150	1282.455	II	600	1519.837	II
60	3681.35	II	600	886.943	II	150	1287.468	II	200	1520.540	II
1100	3745.50	I	600	890.567	II	150	1298.395	II	200	1524.860	II
1400	3842.05	I	500	892.414	II	300	1308.297	II	150	1525.764	II
6900	3845.47	I	800	893.678	II	300	1314.337	II	500	1531.856	II
5500	3873.12	I	400	894.227	II	100	1320.686	II	300	1532.131	II
2800	3873.96	I	600	896.759	II	100	1326.395	II	250	1533.986	II
7900	3894.08	I	400	896.976	II	150	1350.594	II	250	1535.002	II
1500	3935.97	I	600	901.073	II	250	1351.837	II	500	1537.559	II
80 h	3963.10	II	400	906.113	II	150	1355.305	II	200	1540.239	II
6000	3995.31	I	800	914.213	II	300	1358.773	II	300	1540.389	II
970	3997.91	I	600	922.019	II	200	1359.009	II	300	1540.588	II
350	4020.90	I	500	924.239	II	200	1362.600	II	750	1541.703	II
370	4045.39	I	400	935.232	II	250	1367.951	II	400	1544.677	II
350	4066.37	I	600	935.898	II	200	1371.840	II	100	1547.958	II
830	4092.39	I	600	943.335	II	300 r	1376.79	III	300	1550.653	II
550	4110.54	I	600	945.525	II	200 r	1377.49	III	300	1551.389	II
2800	4118.77	I	500	945.965	II	100	1393.128	II	500	1552.646	II
4400	4121.32	I	200	954.383	II	100	1398.642	II	250	1553.896	II
90	4190.71	I	250	956.290	II	150	1402.777	II	400	1555.134	II
90	4469.56	I	400	958.154	II	150	1407.169	II	500	1555.703	II
690	4530.96	I	200	960.414	II	100	1414.898	II	300	1558.345	II
			250	968.042	II						

Intensity	Wavelength/Å		Intensity	Wavelength/Å		Intensity	Wavelength/Å		Intensity	Wavelength/Å	
400	1565.924	II	900	2135.981	II	800	2769.669	II	500	4043.751	II
400	1566.415	II	400	2148.984	II	200	2791.795	II	2000	4062.64	I
100	1569.416	II	150	2161.320	II	170	2799.528	II	120	4068.106	II
300	1579.492	II	1300 r	2165.09	I	100	2810.804	II	500	4131.363	II
300	1580.626	II	250	2174.982	II	1250 r	2824.37	I	200	4143.017	II
400	1581.995	II	1600 r	2178.94	I	350	2837.368	II	300	4153.623	II
500	1583.682	II	700	2179.410	II	100	2857.748	II	500	4161.140	II
400	1590.165	II	1700 r	2181.72	I	600	2877.100	II	370	4164.284	II
600	1593.556	II	700	2189.630	II	270	2884.196	II	400	4171.851	II
500 r	1593.75	III	900	2192.268	II	2500 r	2961.16	I	500	4179.512	II
400	1598.402	II	400	2195.683	II	100	2986.335	II	500	4211.866	II
400	1602.388	II	1700 r	2199.58	I	2000	2997.36	I	320	4230.449	II
200	1604.848	II	1300 r	2199.75	I	2000	3010.84	I	200	4255.635	II
300	1605.281	II	100	2200.509	II	2500	3036.10	I	950	4275.11	I
400	1606.834	II	200	2209.806	II	2500	3063.41	I	300	4279.962	II
250	1608.639	II	750	2210.268	II	1400	3073.80	I	500	4292.470	II
150	1610.296	II	1600 r	2214.58	I	1500	3093.99	I	400	4365.370	II
200	1617.915	II	250	2215.106	II	1250	3099.93	I	100	4444.831	II
600	1621.426	II	1000 r	2215.65	I	2000	3108.60	I	400	4506.002	II
400	1622.428	II	750	2218.108	II	1400 h	3126.11	I	150	4516.049	II
250	1630.268	II	2100 r	2225.70	I	1500	3194.10	I	150	4541.032	II
100	1636.605	II	150	2226.780	II	1400	3208.23	I	500	4555.920	II
1000 r	1642.21	III	1600 r	2227.78	I	1500 h	3243.16	I	100	4596.906	II
250	1649.458	II	350	2228.868	II	10000 r	3247.54	I	120	4649.271	II
30 r	1655.32	I	2500 r	2230.08	I	10000 r	3273.96	I	2000	4651.12	I
200	1656.322	II	1100 r	2238.45	I	1400 h	3282.72	I	120	4661.363	II
200	1660.001	II	900	2242.618	II	400	3290.418	II	320	4671.702	II
300	1663.002	II	2300 r	2244.26	I	1500 h	3290.54	I	300	4673.577	II
100	1672.776	II	1000	2247.002	II	110	3300.881	II	450	4681.994	II
30	1688.09	I	1300 r	2260.53	I	250	3301.229	II	100	4758.433	II
30	1691.08	I	2200 r	2263.08	I	2500 h	3307.95	I	400	4812.948	II
30 r	1703.84	I	150	2263.786	II	200	3316.276	II	120	4851.262	II
50 r	1713.36	I	200	2276.258	II	1500	3337.84	I	300	4854.988	II
150	1717.721	II	100	2286.645	II	150	3338.648	II	100	4873.304	II
50 r	1725.66	I	2500 r	2293.84	I	200	3365.648	II	150	4901.427	II
100	1736.551	II	170	2294.368	II	450	3370.454	II	1000	4909.734	II
50 r	1741.57	I	1000	2303.12	I	300	3374.952	II	500	4918.376	II
150	1753.281	II	150	2369.890	II	200	3380.712	II	200	4926.424	II
200 r	1774.82	I	2500 r	2392.63	I	100	3384.945	II	900	4931.698	II
100 r	1825.35	I	120	2403.337	II	1250 h	3483.76	I	120	4943.026	II
250	1929.751	II	1500	2406.66	I	1250	3524.23	I	700	4953.724	II
250	1944.597	II	1000 r	2441.64	I	2000	3530.38	I	500	4985.506	II
100	1946.493	II	100	2485.792	II	1400	3599.13	I	400	5006.801	II
200	1957.518	II	2000 r	2492.15	I	1400	3602.03	I	350	5009.851	II
150	1970.495	II	150	2506.273	II	1000	3686.555	II	400	5012.620	II
150	1977.027	II	120	2526.593	II	150	3786.270	II	350	5021.279	II
500	1979.956	II	300	2544.805	II	170	3797.849	II	200	5039.016	II
300	1989.855	II	100	2571.756	II	100	3818.879	II	300	5047.348	II
250	1999.698	II	150	2590.529	II	140	3826.921	II	900	5051.793	II
270	2035.854	II	200	2600.270	II	160	3864.137	II	400	5058.910	II
250	2037.127	II	2500 r	2618.37	I	280	3884.131	II	500	5065.459	II
350	2043.802	II	200	2666.291	II	150	3892.924	II	450	5067.094	II
300	2054.980	II	750	2689.300	II	170	3903.177	II	350	5072.302	II
100	2078.663	II	700	2700.962	II	140	3920.654	II	450	5088.277	II
110	2098.398	II	650	2703.184	II	120	3933.268	II	420	5093.816	II
320	2104.797	II	700	2713.508	II	120	3987.024	II	350	5100.067	II
300	2112.100	II	650	2718.778	II	150	3993.302	II	1500	5105.54	I
320	2117.310	II	300	2721.677	II	140	4003.476	II	250	5124.476	II
350	2122.980	II	120	2737.342	II	1250	4022.63	I	2000	5153.24	I
350	2126.044	II	270	2745.271	II	100	4032.647	II	100	5158.093	II
420	2134.341	II	2500 r	2766.37	I	600	4043.484	II	100	5183.367	II

Intensity	Wavelength/Å		Intensity	Wavelength/Å		Intensity	Wavelength/Å		Intensity	Wavelength/Å	
2500	5218.20	I	270	6879.404	II	1200	3156.52	II	830	3577.98	II
100	5269.991	II	220	6937.553	II	670	3162.83	II	440	3580.04	II
100	5276.525	II	150	6952.871	II	1000	3169.99	II	3300	3585.06	II
1650	5292.52	I	150	6977.572	II	470	3215.19	II	1400	3585.78	II
100	5368.383	II	200	7022.860	II	830	3216.63	II	560	3586.11	II
1500	5700.24	I	300	7194.896	II	490	3235.89	II	1100	3591.41	II
1500	5782.13	I	400	7326.008	II	490	3245.12	II	560	3591.81	II
150	5805.989	II	300	7331.694	II	1200	3251.27	II	560	3592.11	II
100	5833.515	II	250	7382.277	II	890	3280.09	II	1800	3595.04	II
200	5897.971	II	1000	7404.354	II	490	3282.77	II	560	3600.38	II
120	5937.577	II	270	7434.156	II	1100	3308.88	II	1800	3606.12	II
400	5941.196	II	500	7562.015	II	780	3316.32	II	440	3618.51	II
100	5993.260	II	700	7652.333	II	1000	3319.88	II	560	3620.16	II
650	6000.120	II	1000	7664.648	II	780	3341.00	II	470	3624.27	II
100	6023.264	II	150	7681.788	II	510	3353.58	II	1100	3629.42	II
250	6072.218	II	450	7744.097	II	510	3368.11	II	4000	3630.24	II
150	6080.343	II	800	7778.738	II	5300	3385.02	II	440	3632.78	II
150	6099.990	II	750	7805.184	II	610	3388.85	II	1100	3640.25	II
160	6107.412	II	1500	7807.659	II	3800	3393.57	II	11000	3645.40	II
300	6114.493	II	1000	7825.654	II	1300	3396.16	II	1000	3648.78	II
600	6150.384	II	350	7860.577	II	5300	3407.80	II	700	3664.62	II
750	6154.222	II	300	7890.567	II	1300	3413.78	II	990	3672.30	II
500	6172.037	II	700	7902.553	II	530	3414.82	II	420	3672.70	II
550	6186.884	II	1500	7933.13	I	780	3419.63	II	1400	3674.08	II
400	6188.676	II	400	7944.438	II	530	3425.06	II	2200	3676.59	II
300	6198.092	II	400	7972.033	II	1900	3434.37	II	640	3678.51	I
470	6204.261	II	1200	7988.163	II	560	3440.93	II	820	3684.85	I
450	6208.457	II	2000	8092.63	I	1300	3441.45	II	1300	3685.78	I
750	6216.939	II	500	8277.560	II	3800	3445.57	II	4700	3694.81	II
700	6219.844	II	800	8283.160	II	830	3446.99	II	990	3698.21	II
500	6261.848	II	250	8503.396	II	2700	3454.32	II	540	3701.63	II
1000	6273.349	II	750	8511.061	II	1300	3456.56	II	440	3707.57	II
350	6288.696	II	200	8609.134	II	4400	3460.97	II	440	3708.22	II
900	6301.009	II	500	9813.213	II	720	3468.43	II	420	3710.07	II
550	6305.972	II	250	9827.978	II	560	3471.14	II	1600	3724.45	II
400	6312.492	II	200	9830.798	II	560 d	3471.53	II	930	3739.34	I
120	6326.466	II	600	9861.280	II	1300	3477.07	II	1200	3747.82	II
400	6373.268	II	600	9864.137	II	4400	3494.49	II	1400	3753.51	II
750	6377.840	II	200	9883.969	II	560	3496.34	II	1400	3753.75	II
400	6403.384	II	550	9916.419	II	830	3498.71	II	1200	3757.05	I
850	6423.884	II	500	9917.954	II	830	3504.53	II	4700	3757.37	II
200	6442.965	II	550	9925.594	II	830	3505.45	II	640	3767.63	I
750	6448.559	II	450	9938.998	II	1300	3506.81	II	640	3773.05	I
170	6466.246	II	500	9960.354	II	560	3517.26	II	420	3781.47	I
950	6470.168	II	450	10006.588	II	4400	3523.98	II	3300	3786.18	II
750	6481.437	II	550	10022.969	II	22000	3531.70	II	1600	3788.44	II
400	6484.421	II	550	10038.093	II	4400	3534.96	II	700	3791.87	II
220	6517.317	II	650	10054.938	II	5500	3536.02	II	510	3804.14	II
400	6530.083	II	450	10080.354	II	4400	3538.52	II	580	3806.27	II
120	6551.286	II				1700	3542.33	II	470	3812.27	I
200	6577.080	II				1400	3546.83	II	470	3813.67	II
750	6624.292	II				4400	3550.22	II	1400	3816.76	II
800	6641.396	II				2200	3551.62	II	700	3825.68	II
450	6660.962	II				440 h	3558.23	II	2300	3836.50	II
100	6770.362	II				440	3559.30	II	1400	3841.31	II
300	6806.216	II				2200	3563.15	II	420	3846.34	II
400	6809.647	II				560	3563.69	II	420	3847.02	I
320	6823.202	II				780	3573.83	II	1200	3853.03	II
250	6844.157	II				1400	3574.15	II	420	3858.40	I
320	6868.791	II				4400	3576.24	II	560	3868.45	II
270	6872.231	II				1700	3576.87	II	1600	3868.81	I

Dysprosium Dy Z = 66

Intensity	Wavelength/Å	
260	2356.91	II
240	2410.01	II
260	2439.84	II
220	2585.30	I
440	2634.80	II
220	2755.75	II
300	2816.39	II
390	2913.95	II
610	3038.28	II
830	3135.38	II
500	3141.14	II

Intensity	Wavelength/Å		Intensity	Wavelength/Å		Intensity	Wavelength/Å		Intensity	Wavelength/Å	
820	3869.86	II	990	4612.26	I	1000	3070.40	III	3200	3944.42	I
7000	3872.11	II	170	4731.84	II	610	3073.34	II	2700	3973.04	I
1200	3873.99	II	120 h	4775.79	I	720	3082.08	II	3200	3973.58	I
470	3879.11	II	480	4957.34	II	610	3084.02	II	1400	3974.72	II
5800	3898.53	II	70	5022.12	I	770	3122.72	II	810	3977.02	I
540	3914.87	II	160	5042.63	I	1500	3166.25	III	1100	3982.33	I
540	3915.59	II	95	5070.68	I	870	3181.92	II	810	3987.66	I
540 d	3917.29	I	120	5077.67	I	870	3220.73	II	14000	4007.96	I
420	3927.86	I	80	5090.38	II	610	3223.31	II	1100	4012.58	I
540	3930.14	I	80	5110.32	I	2300	3230.58	II	3000	4020.51	I
2100	3931.52	II	130 h	5120.04	I	2700	3264.78	II	1000	4046.96	I
10000	3944.68	II	190	5139.60	II	720	3279.33	II	940	4055.47	II
800	3957.79	II	110	5169.69	II	720	3280.22	II	690	4059.78	I
14000	3968.39	II	80	5185.30	I	2000	3301.23	III	3500	4087.63	I
2700	3978.57	II	290	5192.86	II	2300	3312.42	II	1100	4098.10	I
1400	3981.92	II	95	5197.66	II	770	3323.19	II	6900	4151.11	I
1600	3983.65	II	70	5259.88	I	770	3332.70	II	1000	4190.70	I
800	3984.21	II	130	5260.56	I	1300	3346.04	II	1400	4218.43	I
540	3991.32	II	65	5267.11	I	1400	3364.08	II	690	4286.56	I
1600	3996.69	II	55	5282.07	I	1400 d	3368.02	II	40000	4290.06	III
8000	4000.45	II	160	5301.58	I	7700	3372.71	II	20000	4386.86	III
420	4005.84	I	65	5340.30	I	970	3374.17	II	810	4409.34	I
540	4011.29	II	85	5389.58	II	1700	3385.08	II	1000	4606.61	I
540	4013.82	I	80	5419.13	I	2300	3392.00	II	570	4675.62	II
540	4014.70	II	70	5423.32	I	770	3441.13	II	15000	4735.56	III
420	4027.78	II	95	5451.11	I	970	3471.71	II	2000	4783.12	III
520 d	4028.32	II	65	5547.27	I	610	3479.41	II	250	5007.25	I
520	4032.47	II	100	5639.50	I	970	3485.85	II	200	5035.94	I
420	4033.65	II	55 h	5645.99	I	6700	3499.10	II	210	5042.05	II
420	4036.32	II	80	5652.01	I	610	3502.78	I	120	5124.56	I
12000	4045.97	I	70 h	5718.46	I	610	3524.91	II	130	5127.41	I
1600	4050.56	II	55	5745.53	I	820	3549.84	II	120	5131.53	I
520	4055.14	II	55 h	5868.11	II	1500	3558.02	I	130	5133.83	II
2500	4073.12	II	70	5945.80	I	1000	3559.90	II	170	5164.77	I
7400	4077.96	II	120	5974.49	I	920	3570.75	II	130	5172.78	I
3900	4103.30	II	140	5988.56	I	1000	3580.52	II	160	5188.90	II
860	4103.87	I	140	6088.26	I	610	3590.76	I	150	5206.52	I
1500	4111.34	II	100	6168.43	I	610	3599.50	II	140	5255.93	II
490	4124.63	II	270	6259.09	I	1000	3599.83	II	80	5272.91	I
990	4129.42	II	160	6579.37	I	3100	3616.56	II	90	5348.06	I
1200	4143.10	II	75	6667.86	I	720	3628.04	I	60	5414.63	II
990	4146.06	I	180	6835.42	I	1000	3633.54	II	180	5456.62	I
5700	4167.97	I	80	6852.96	I	1600	3638.68	I	90	5468.32	I
930	4183.72	I	65	6899.32	II	900	3645.94	II	80	5485.97	II
12000	4186.82	I	55	7426.86	II	7900	3692.65	II	80	5593.46	I
2200	4191.64	I	55	7543.73	I	1300	3729.52	II	60	5611.82	I
6800	4194.84	I	80	7662.36	I	900	3742.64	II	70	5622.01	I
800	4198.02	I	100	8201.57	II	900	3747.43	I	80	5626.53	II
680	4201.30	I	45	8791.39	II	1800	3786.84	II	90	5640.36	I
680	4202.24	I				1600	3810.33	I	70	5664.95	I
16000	4211.72	I	*Erbium Er Z = 68*			4000	3816.78	III	70	5719.55	I
1800	4213.18	I	600	2277.65	III	3600	3830.48	II	100	5739.19	I
3700	4215.16	I	290	2586.73	II	680	3855.90	I	290	5762.80	I
4400	4218.09	I	490	2670.26	II	7500	3862.85	I	70	5784.66	I
4400	4221.11	I	500	2739.27	III	1500	3880.61	II	70	5800.79	I
2700	4225.16	I	610	2755.63	II	1200	3882.89	II	430	5826.79	I
1000	4308.63	II	1000	2904.47	II	4200	3892.68	I	100	5850.07	I
540	4409.38	II	1500	2910.36	II	5200	3896.23	II	120	5855.31	I
740	4449.70	II	1500	2964.52	II	11000	3906.31	II	140	5872.35	I
420	4577.78	I	1200	3002.41	II	3200	3937.01	I	120	5881.14	I
2100	4589.36	I	1000	3055.10	III	2100	3938.63	II	8000	5903.30	III

Intensity	Wavelength/Å	
70	6022.56	I
70	6061.25	I
60	6076.45	II
360	6221.02	I
55	6262.56	I
60	6268.87	I
130	6308.77	I
55	6326.13	I
55	6492.35	I
60	6583.48	I
70	6601.11	I
70	6759.87	I
35	6790.92	I
70	6848.10	I
55	6865.13	I
55	7459.55	I
120	7469.51	I
35	7680.01	I
35	7797.47	I
35	7921.85	I
30	7937.84	I
35	8312.82	I
55	8409.90	I
9	8866.84	II

Europium Eu Z = 63

Intensity	Wavelength/Å	
30	2124.69	III
200	2350.51	III
4000	2375.46	III
100 d	2435.14	III
1000	2444.38	III
4000	2445.99	III
2000	2513.76	III
200	2522.14	III
160	2564.17	II
110	2568.17	II
230	2577.14	II
1000	2638.77	II
380	2641.27	II
640	2668.34	II
110	2673.42	II
250	2678.29	II
250	2685.66	II
550	2692.03	II
700	2701.14	II
800	2701.90	II
240	2705.28	II
180	2709.99	I
700	2716.98	II
4200	2727.78	II
160	2740.62	II
120	2744.26	II
480	2781.89	II
1900	2802.84	II
220	2811.75	II
3400	2813.94	II
550	2816.18	II
2000	2820.78	II
400 cw	2828.72	II
260	2859.67	II
280	2862.57	II
200	2892.54	I

Intensity	Wavelength/Å	
140	2893.03	I
360	2893.83	I
3200	2906.68	II
160	2908.99	I
850	2925.04	II
200 cw	2952.68	II
260	2960.21	II
300	2991.33	II
100 c	3023.93	III
200 c	3026.79	III
320 cw	3054.94	II
120	3058.98	I
220	3077.36	II
120	3097.45	II
320	3106.18	I
950	3111.43	I
120	3130.73	II
50 c	3171.00	III
50 c	3183.78	III
420	3210.57	I
1000	3212.81	I
420	3213.75	I
150	3272.77	II
210	3277.78	II
150	3301.95	II
140	3308.02	II
140	3313.33	II
950	3334.33	I
110	3350.40	I
140	3369.06	II
190	3391.99	II
280	3396.58	II
150	3425.02	II
150	3441.00	II
130	3461.38	II
470 cw	3521.09	II
150	3542.15	II
180	3552.52	II
150	3603.20	II
6400	3688.42	II
20000 cw	3724.94	II
350	3741.31	II
260	3761.12	II
39000 cw	3819.67	II
140	3844.23	II
190	3865.57	I
150	3884.75	I
28000 cw	3907.10	II
32000 cw	3930.48	II
30000 cw	3971.96	II
180	4011.69	II
150	4017.58	II
120	4039.19	I
120	4085.38	II
33000 cw	4129.70	II
60000 cw	4205.05	II
150	4298.73	I
240	4355.09	II
14000 cw	4435.56	II
3000	4522.57	II
11000	4594.03	I

Intensity	Wavelength/Å	
9800	4627.22	I
8300	4661.88	I
110	4867.62	I
150	4907.18	I
180	4911.40	I
180	5013.17	I
170	5022.91	I
110	5029.54	I
170	5114.37	I
170	5129.10	I
210	5133.52	I
270	5160.07	I
210	5166.70	I
200	5199.85	I
110	5200.96	I
120	5206.44	I
750	5215.10	I
300	5223.49	I
120	5239.24	I
200	5266.40	I
390	5271.96	I
110	5272.48	I
150	5282.82	I
120	5291.26	I
120	5294.64	I
540	5357.61	I
120	5361.61	I
110	5376.94	I
120	5392.94	I
450	5402.77	I
380	5451.51	I
260	5452.94	I
120	5488.65	I
120	5510.52	I
200	5547.44	I
150	5570.33	I
200	5577.14	I
120	5580.03	I
210	5645.80	I
330	5765.20	I
180	5783.69	I
170	5818.74	II
600 cw	5830.98	I
330	5966.07	II
480 cw	5967.10	I
170	5972.75	I
240	5992.83	I
110	6012.56	I
420	6018.15	I
170	6029.00	I
420	6049.51	II
140	6057.36	I
240	6083.84	I
240	6099.35	I
120	6118.78	I
330	6173.05	II
110	6178.76	I
260 cw	6188.13	I
140	6195.07	I
240	6262.25	I
170	6299.77	I

Intensity	Wavelength/Å	
230	6303.41	II
120 cw	6350.04	I
120 cw	6400.93	I
180	6410.04	I
140	6411.32	I
830	6437.64	II
120	6457.96	I
1400	6645.11	II
50	6666.35	III
140	6802.72	I
360	6864.54	I
120	7040.20	I
330	7077.10	II
570	7194.81	II
570	7217.55	II
540	7301.17	II
720	7370.22	II
300	7426.57	II
160	7583.91	I
60 cw	7742.57	I
70	7746.19	I
35	7887.99	I
24 cw	8209.80	I
21 cw	8642.67	I
18	8870.30	I

Fluorine F Z = 9

Intensity	Wavelength/Å	
50	148.00	V
50	163.56	V
90	165.98	V
100	166.18	V
50	186.84	V
60	190.57	V
70	190.84	V
50	196.39	IV
60	196.45	IV
70	200.09	IV
80	201.16	IV
90	208.25	IV
90	240.08	IV
100	251.03	IV
140	419.65	IV
150	420.05	IV
160	420.73	IV
100	429.51	III
110	430.15	III
150	430.76	IV
90	464.29	III
120	465.98	V
130	490.57	IV
160	491.00	IV
50	497.38	IV
60	497.83	IV
70	498.80	IV
90	506.16	V
100	508.08	V
120	508.39	III
60	514.08	V
90	525.29	V
100	526.30	V
120	567.69	III
110	567.75	III

Intensity	Wavelength/Å		Intensity	Wavelength/Å		Intensity	Wavelength/Å		Intensity	Wavelength/Å	
140	570.64	IV	110	1839.30	III	270	3847.09	II	8000	6870.22	I
140	571.30	IV	120	1839.97	III	260	3849.99	II	15000	6902.48	I
150	571.39	IV	110	1840.14	III	250	3851.67	II	6000	6909.82	I
160	572.66	IV	100	2027.44	III	5	3898.48	I	4000	6966.35	I
90	605.67	II	120	2030.32	III	8	3930.69	I	45000	7037.47	I
100	606.80	II	120	2217.17	III	5	3934.26	I	30000	7127.89	I
90	630.20	III	50	2298.29	IV	5	3948.56	I	15000	7202.36	I
100	647.77	V	40	2451.58	IV	240	4024.73	II	1000	7309.03	I
110	647.87	V	120	2452.07	III	220	4025.01	II	15000	7311.02	I
130	654.03	V	50	2456.92	IV	230	4025.49	II	700	7314.30	I
120	656.12	III	130	2464.85	III	200	4103.51	II	5000	7331.96	I
130	656.87	III	130	2470.29	III	200	4246.23	II	120	7336.77	III
110	657.23	V	120	2478.73	III	200	4299.17	II	130	7354.94	III
140	657.33	V	150	2484.37	III	140 h	4420.30	III	10000	7398.69	I
140	658.33	III	120	2542.77	III	120 h	4427.35	III	4000	7425.65	I
140	676.12	IV	120	2580.04	III	120 h	4432.32	III	2200	7482.72	I
130	677.15	IV	130	2583.81	III	140 h	4479.99	III	2500	7489.16	I
150	677.22	IV	120	2593.23	III	6	4960.65	I	900	7514.92	I
130	678.99	IV	130	2595.53	III	150	5012.54	III	5000	7552.24	I
160	679.21	IV	140	2599.28	III	160	5110.99	III	5000	7573.38	I
60	757.04	V	130	2625.01	III	15	5230.41	I	7000	7607.17	I
150	806.96	I	140	2629.70	III	12	5279.01	I	18000	7754.70	I
125	809.60	I	120	2656.44	III	18	5540.52	I	15000	7800.21	I
500	951.87	I	130	2755.55	III	12	5552.43	I	300	7879.18	I
1000	954.83	I	160	2759.63	III	10	5577.33	I	500	7898.59	I
750	955.55	I	120	2788.15	III	20	5624.06	I	350	7936.31	I
500	958.52	I	160	2811.45	III	12	5626.93	I	300	7956.32	I
20	972.40	I	40	2820.74	IV	15	5659.15	I	80	8016.01	II
350	973.90	I	50	2826.13	IV	40	5667.53	I	1000	8040.93	I
100	976.22	I	140	2833.99	III	90	5671.67	I	900	8075.52	I
40	976.51	I	150	2835.63	III	18	5689.14	I	350	8077.52	I
100	977.75	I	150	2860.33	III	25	5700.82	I	350	8126.56	I
60	1082.31	V	120	2862.86	III	25	5707.31	I	600	8129.26	I
70	1088.39	V	140	2887.58	III	140	5753.17	III	300	8159.51	I
80	1219.03	III	150	2889.45	III	120	5761.20	III	600	8179.34	I
80	1266.87	III	120	2905.30	III	12	5950.15	I	300	8191.24	I
90	1267.71	III	140	2913.29	III	25	5959.19	I	350	8208.63	I
70	1297.54	III	160	2916.34	III	70	5965.28	I	2500	8214.73	I
70	1359.92	III	140	2932.49	III	50	5994.43	I	3000	8230.77	I
110	1498.93	III	140	2994.28	III	150	6015.83	I	500	8232.19	I
120	1502.01	III	120 h	2997.21	III	80	6038.04	I	1500	8274.62	I
110	1504.18	III	130	2997.53	III	900	6047.54	I	2000	8298.58	I
140	1504.79	III	120	2999.47	III	100	6080.11	I	600	8302.40	I
130	1506.30	III	130	3039.25	III	150	6091.82	III	900	8807.58	I
110	1506.77	III	120	3039.75	III	140	6125.50	III	1000	8900.92	I
100	1553.02	III	160	3042.80	III	800	6149.76	I	300	8912.78	I
110	1557.59	III	150	3049.14	III	400	6210.87	I	350	9025.49	I
100	1563.73	III	140	3113.62	III	130	6233.57	III	400	9042.10	I
100	1565.54	III	160	3115.70	III	13000	6239.65	I	350	9178.68	I
100	1623.40	III	180	3121.54	III	10000	6348.51	I	200	9433.67	I
100	1650.76	III	140	3124.79	III	140	6363.05	III	25	9505.30	I
130	1670.39	III	140	3134.23	III	8000	6413.65	I	12	9662.04	I
140	1677.40	III	140	3146.99	III	450	6569.69	I	25	9734.34	I
100	1716.99	III	180	3174.17	III	300	6580.39	I	15	9822.11	I
120	1770.09	III	170	3174.76	III	400	6650.41	I	12	9902.65	I
150	1770.67	III	120	3214.00	III	1800	6690.48	I	80 h	10047.98	II
110	1772.93	III	200	3501.45	II	400	6708.28	I	15	10285.45	I
140	1773.36	III	200	3501.57	II	7000	6773.98	I	20	10862.31	I
160	1791.65	III	200	3502.96	II	1500	6795.53	I			
110	1803.03	III	6	3594.10	I	9000	6834.26	I			
170	1805.90	III	12	3668.17	I	50000	6856.03	I			

Francium Fr Z = 87

	7177.	I

Intensity	Wavelength/Å		Intensity	Wavelength/Å		Intensity	Wavelength/Å		Intensity	Wavelength/Å	
Gadolinium Gd Z = 64			2000	3684.13	I	700	4694.33	I	50	6752.67	II
1200	1007.24	IV	3100	3687.74	II	410	4743.65	I	26	6786.33	II
1200	1063.84	IV	2000	3697.73	II	470	4767.24	I	100	6828.25	I
1600	1476.98	IV	1300	3699.73	II	300	4784.62	I	100	6916.57	I
1500	1705.03	IV	2700	3712.70	II	320	4821.69	I	50	6985.89	II
1600	1706.01	IV	2000	3713.57	I	280	4934.12	I	75	6991.92	I
2000	1736.24	IV	1400	3716.36	II	750	5015.04	I	60	6996.76	II
1500	1815.32	IV	2000	3717.48	I	75	5039.09	I	45	7006.16	II
2200	1975.24	III	1800 d	3719.45	II	5000	5091.70	III	35	7122.57	I
3400	2018.07	III	1500	3730.84	II	130	5098.38	II	170	7168.37	I
2800	2359.31	III	4500	3743.47	II	910	5103.45	I	28	7189.57	II
1400	2397.87	IV	1400	3758.31	II	180	5108.91	II	28	7262.66	I
2800	2697.39	III	8700	3768.39	II	120	5125.56	II	35	7441.85	I
2800	2703.28	III	1400	3770.69	II	860	5155.84	I	40	7464.36	I
2700	2727.89	III	2900	3783.05	I	190	5176.28	II	55	7562.97	I
9000	2904.73	III	5100	3796.37	II	410	5197.77	I	80	7733.50	I
9500	2955.53	III	3700	3813.97	II	280	5219.40	I	35	7846.35	II
1200	2999.04	II	3300	3850.69	II	130	5233.93	I	35	7856.93	I
2100	3010.13	II	5100	3850.97	II	320	5251.18	I	25	7930.25	II
1900	3027.60	II	4300	3852.45	II	120	5252.14	II	18	8146.15	I
2100	3032.84	II	1600	3866.99	I	140	5255.80	I	21	8668.63	I
1600	3034.05	II	1500	3894.70	II	280	5283.08	I	21 h	8832.06	II
2100	3081.99	II	2200	3916.51	II	280	5301.67	I	14 h	8849.14	I
3500	3100.50	II	1200	3934.79	I	220	5302.76	I	18 h	8867.31	I
930	3145.00	II	1400	3945.54	I	280	5307.30	I	5000	14332.88	III
980	3156.53	II	1200	3957.67	II	130	5321.50	I			
980	3161.37	II	1100	4023.14	I	280	5321.78	I	*Gallium Ga Z = 31*		
4000	3176.66	III	1100	4028.15	I	110	5327.32	I	14	294.53	IV
1400	3331.38	II	1400	4037.33	II	170	5333.30	I	61	295.67	IV
1100	3336.18	II	1600	4045.01	I	300	5343.00	I	30	298.44	V
5400	3350.47	II	1300	4049.43	II	200	5348.67	I	30	300.01	V
4300	3358.62	II	2200	4049.86	II	300	5350.38	I	30	302.86	V
5400	3362.23	II	2600	4053.64	I	240	5353.26	I	41	304.99	IV
1100	3392.53	II	2600	4058.22	I	3000	5365.96	III	30	307.03	V
1100 d	3407.56	II	1900	4063.39	II	150	5370.63	I	30	308.26	V
6900	3422.47	II	1300	4078.44	II	4000	5553.30	III	30	311.79	V
1700	3439.21	II	2800	4078.70	I	3000	5587.88	III	30	313.68	V
2700	3439.99	II	1500	4085.56	II	190	5617.91	I	40	319.41	V
1400	3450.38	II	1100	4092.71	I	110	5632.25	I	40	322.31	V
1100	3451.23	II	2600	4098.61	II	260	5643.24	I	50	322.99	V
2700	3463.98	II	2200	4130.37	II	3000	5658.98	III	30	323.10	V
1700	3467.27	II	1100	4132.28	II	390	5696.22	I	40	324.25	V
1700	3468.99	II	2400	4175.54	I	120	5733.86	II	40	324.95	V
1400	3473.22	II	2400	4184.25	II	240	5791.38	I	40	326.14	V
2200	3481.28	II	2200	4190.78	I	220	5851.63	I	30	326.77	V
1700	3481.80	II	1300	4212.00	II	280	5856.22	I	30	328.65	V
1700	3494.40	II	4800	4225.85	I	110	5904.56	I	25	423.18	IV
1400	3505.51	II	1700	4251.73	II	170	5911.45	II	16	439.92	IV
1100	3512.50	II	1600	4262.09	I	85	5930.29	I	50	620.00	III
4300	3545.80	II	1100	4306.34	I	85	5936.84	I	40	622.01	III
3900	3549.36	II	1800	4313.84	I	65	5937.71	I	90	806.51	III
1400	3557.05	II	2600 d	4325.57	II	110 h	5988.02	I	90	817.30	III
5400	3584.96	II	1900	4327.12	I	430	6114.07	I	50	828.70	III
1100	3592.71	II	1000	4344.30	II	75	6305.15	II	20	878.17	V
1100	3604.87	I	2200	4346.46	I	40	6331.35	I	40	973.21	V
6100	3646.19	II	1400	4401.86	I	40	6380.95	II	40	989.75	V
3900	3654.62	II	1400	4422.41	I	40 h	6538.15	I	90	1014.47	V
3100	3656.15	II	1100	4430.63	I	55	6564.78	I	90	1019.71	V
1400	3662.26	II	1100	4519.66	I	50	6634.36	II	120	1050.48	V
2700	3664.60	II	910	4537.81	I	35	6681.23	II	80	1054.56	V
2000	3671.20	II	520	4614.50	I	85	6730.73	I	90	1058.12	V
									80	1066.69	V

Intensity	Wavelength/Å		Intensity	Wavelength/Å		Intensity	Wavelength/Å		Intensity	Wavelength/Å	
60	1069.60	V	85	1338.09	IV				100	1742.195	I
80	1073.77	V	77	1347.03	IV				50	1746.065	I
90	1078.83	V	76	1351.06	IV	\multicolumn{3}{Germanium Ge Z = 32}		200	1750.043	I
110	1079.60	V	70	1353.92	III	700	294.51	V	100	1758.279	I
60	1080.99	V	74	1364.63	IV	1000	295.64	V	100 h	1764.185	I
80	1085.00	III	60	1395.54	IV	200	304.98	V	100 h	1765.284	I
250	1085.01	V	77	1402.55	IV	20	621.52	V	50 h	1766.433	I
80	1087.37	V	70	1405.32	IV	50	724.21	V	200	1774.176	I
90	1091.71	V	73	1465.87	IV	60	746.88	V	200	1785.046	I
100	1094.36	V	90	1495.07	III	60	760.05	V	100 h	1793.071	I
80	1095.10	V	50	1534.46	III	10	862.234	II	75 h	1801.432	I
160	1102.83	V	10	1813.98	II	15	875.493	II	200 h	1841.328	I
140	1103.03	V	15	1845.30	II	15	905.977	II	200 h	1842.410	I
60	1105.61	III	20	2091.34	II	20	920.554	II	100 h	1844.410	I
75	1105.62	V	90	2417.70	III	300	971.35	V	100 h	1845.872	I
70	1106.17	V	90	2423.98	III	300	990.66	V	100 h	1846.958	I
80	1118.34	V	15	2424.36	III	50	999.101	V	200	1853.134	I
120	1126.40	V	10	2632.66	I	300	1004.38	V	500 r	1860.086	I
130	1128.10	V	10	2665.05	I	100	1016.638	II	100	1865.052	I
120	1128.53	V	20	2700.47	II	900	1045.71	V	300 r	1874.256	I
100	1129.94	V	15	2780.15	II	700	1072.66	V	100	1895.197	I
130	1136.07	V	50	3521.77	III	100	1075.072	II	500 r	1904.702	I
67	1137.06	IV	80	3581.19	III	300	1085.51	II	50 h	1908.434	I
130	1150.23	V	100	3589.34	III	40	1088.45	III	30	1912.409	I
90	1150.27	III	10	3731.10	III	800	1089.49	V	300 r	1917.592	I
70	1156.10	IV	10	3806.60	III	200	1098.71	II	100 h	1923.467	I
120	1156.51	V	10	4032.99	I	500	1106.74	II	500 r	1929.826	I
35	1157.74	V	10	4172.04	I	1000	1116.94	V	10 h	1934.048	I
70	1163.60	IV	15	4251.16	II	500	1120.46	II	100 r	1937.483	I
25	1169.40	V	10 h	4254.04	II	700	1163.39	V	500	1938.008	II
75	1170.58	IV	10	4255.77	II	200	1164.27	II	100 r	1938.300	I
48	1171.71	IV	40	4262.00	II	500	1181.19	II	500	1938.891	II
40	1178.95	V	100	4380.69	III	500	1181.65	II	30 s	1944.116	I
68	1185.23	IV	150	4381.76	III	200	1188.73	II	200	1944.731	I
40	1186.06	IV	100	4863.00	III	20	1188.99	IV	200	1955.115	I
73	1190.89	IV	150	4993.78	III	300	1191.26	II	500	1962.013	I
73	1193.02	IV	10	5808.28	III	700	1222.30	V	30 h	1963.373	I
75	1195.02	IV	20	5848.25	III	20	1229.81	IV	30	1965.383	I
69	1201.54	IV	15	5993.51	III	500	1237.059	II	200	1970.880	I
72	1206.89	IV	10	6334.2	II	500	1261.905	II	200	1979.274	II
80	1213.17	V	2000	6396.56	I	100	1264.710	II	300 h	1987.849	I
63	1216.15	IV	1000	6413.44	I	200	1401.24	II	300	1988.267	I
50	1228.03	IV	10 h	7251.4	I	200	1538.091	II	500 r	1998.887	I
60	1236.38	IV	20 h	7403.0	I	500	1576.855	II	200	2011.29	I
60	1238.59	IV	30 h	7464.0	I	75	1581.070	II	1700	2019.068	I
75	1245.53	IV	10 h	7620.5	I	100	1602.486	II	2400 r	2041.712	I
83	1258.77	IV	50 h	7734.77	I	3 r	1615.57	I	1600 r	2043.770	I
81	1264.66	IV	100 h	7800.01	I	2 r	1624.130	I	420	2054.461	I
82	1267.15	IV	15 h	8002.55	I	2 r	1630.173	I	220 h	2057.238	I
90	1267.16	III	20 h	8074.25	I	3 r	1636.31	I	750 r	2065.215	I
81	1279.24	IV	100 h	8311.86	I	4 r	1639.730	I	2600 r	2068.656	I
15	1283.64	V	200 h	8386.49	I	2	1647.531	I	420	2086.021	I
80	1285.33	IV	200 h	9492.92	I	200	1649.194	II	2000 r	2094.258	I
80	1293.46	III	200 h	9493.12	I	2	1651.528	I	25	2104.45	III
60	1295.36	III	300 h	9589.36	I	4 r	1651.955	I	240	2105.824	I
82	1295.86	IV	100 h	10905.95	I	3	1661.345	I	95 h	2124.744	I
83	1299.46	IV	400	11949.12	I	4 r	1663.539	I	50 h	2186.451	I
82	1303.53	IV	200	12109.78	I	10 h	1665.275	I	340 r	2198.714	I
80	1309.68	IV	60	14996.64	I	4	1667.802	I	15	2220.375	I
80	1314.82	IV	60	22016.81	I	3 r	1670.608	I	18	2256.001	I
60	1323.15	III	70	22568.71	I	100 r	1691.090	I	18	2314.201	I
						200 r	1716.784	I			
						100 h	1739.102	I			

Germanium Ge Z = 32 (heading appears in column 3 between rows)

Intensity	Wavelength/Å		Intensity	Wavelength/Å		Intensity	Wavelength/Å		Intensity	Wavelength/Å	
24	2327.918	I	1000	5893.389	II	60	1368.62	I	100	1624.34	I
15	2359.233	I	500	6021.041	II	70	1374.82	I	300 d	1629.13	III
20	2379.144	I	75	6283.452	II	50	1375.76	I	250	1638.88	III
10	2389.472	I	100	6336.377	II	180	1377.73	III	50	1639.90	I
15	2397.885	I	100	6484.181	II	150	1378.69	III	100	1644.17	III
130	2417.367	I	6	6557.488	I	150	1379.98	III	150	1646.67	I
30	2436.412	I	50	7049.369	II	125	1380.53	III	250	1652.74	III
30	2488.25	IV	30	7145.390	II	200	1381.36	III	250	1664.77	III
90	2497.962	I	5	7353.334	I	80	1382.75	I	100	1665.76	I
500	2500.54	II	7	7384.208	I	50	1385.33	I	100	1668.11	III
70	2533.230	I	10	7833.575	I	300	1385.79	III	125	1673.93	III
20	2542.44	IV	10	8031.039	I	100	1389.41	III	1000	1693.94	III
3	2556.298	I	6	8044.165	I	180	1391.46	III	150	1697.09	III
28	2589.188	I	10	8256.013	I	60	1392.27	I	200	1698.98	III
500	2592.534	I	10	8482.21	I	180	1396.00	III	200	1699.34	I
8	2644.184	I	9	8700.60	I	50	1402.12	I	200	1700.00	III
1200	2651.172	I	5	9068.785	I	100	1402.91	III	200	1702.25	III
500	2651.568	I	5	9095.957	I	70	1407.38	I	100	1707.53	III
500	2691.341	I	6	9398.868	I	100	1408.45	I	250	1710.16	III
850	2709.624	I	20	9474.993	II	225	1409.50	III	200	1715.69	III
400	2729.78	II	20	9475.645	II	250	1413.80	III	100	1716.71	III
40	2740.426	I	4	9492.559	I	100	1414.27	III	300	1717.83	III
650	2754.588	I	7	9625.664	I	100	1417.09	III	500	1727.31	III
70	2793.925	I	5	10039.436	I	125	1417.39	III	100 d	1733.17	III
80	2829.008	I	4	10200.952	I	150	1427.42	III	300	1738.48	III
1000	2831.843	II	10	10382.427	I	300	1428.93	III	150	1744.39	III
1000	2845.527	II	10	10404.913	I	80	1429.19	I	500	1746.10	III
750	3039.067	I	8	10734.068	I	250	1430.06	III	500	1756.92	III
600	3067.021	I	8	10947.416	I	275	1433.37	III	500	1761.95	III
20	3124.816	I	10	11125.130	I	50	1435.79	I	300	1767.42	III
35	3211.86	III	230	11252.83	I	250	1435.81	III	100	1774.42	III
40	3255.05	III	600	11714.76	I	300	1439.12	III	800	1775.17	III
110	3269.489	I	1300	12069.20	I	200	1441.21	III	200	1776.40	III
40	3434.03	III	1050	12391.58	I	150	1446.37	III	100	1780.57	III
300	3499.21	II	235	13107.61	I	250	1448.42	III	60	1783.22	II
60	3554.19	IV	470	14822.38	I	250	1454.95	III	300	1786.11	III
50	3676.65	IV	150	16759.79	I	100	1464.72	III	150	1792.65	III
200	4178.96	III	135	17214.34	I	150	1471.28	III	500	1793.76	III
70	4226.562	I	70	18811.86	I	100	1474.73	III	200	1801.98	III
200	4260.85	III	62	19279.24	I	150	1481.10	III	400	1805.24	III
150	4291.71	III	28	20673.64	I	100	1481.76	I	100	1809.81	III
10	4685.829	I				300	1487.15	III	400	1821.17	III
1000	4741.806	II	**Gold Au Z = 79**			250	1487.91	III	400	1844.89	III
1000	4814.608	II	100	843.44	III	200	1489.47	III	150	1848.83	III
50	4824.097	II	100	845.14	III	250	1500.37	III	500	1861.80	III
100	5131.752	II	200	945.10	III	200	1502.47	III	150	1871.92	III
200	5178.648	II	100	1040.63	III	200	1503.74	III	100	1879.83	I
3	5194.583	I	80	1044.49	III	100	1542.00	III	150	1918.28	III
6	5265.892	I	80	1046.81	III	100	1548.50	III	100	1932.04	III
6	5513.263	I	100 h	1239.96	III	200	1567.54	III	100	1935.42	III
8	5564.741	I	100	1278.51	III	200	1574.85	III	200	1948.79	III
8	5607.010	I	100	1314.84	III	200	1579.44	III	100	1958.47	III
6	5616.135	I	100 h	1328.37	I	150	1584.10	III	400	1989.63	III
7	5621.426	I	200	1336.72	III	200	1587.16	I	150	1996.85	III
6	5664.226	I	180	1341.68	III	200	1589.56	III	11000	2012.00	I
5	5664.842	I	100	1348.89	III	150	1593.41	III	2600	2021.38	I
9	5691.954	I	150	1350.32	III	70	1598.24	I	150	2082.09	II
6	5701.776	I	150	1355.61	III	200	1600.51	III	300	2083.09	III
5	5717.877	I	150	1356.13	III	250	1617.16	III	60	2110.68	II
6	5801.029	I	50	1363.98	I	100	1617.78	III	100	2159.08	III
9	5802.093	I	500	1365.40	III	500	1621.93	III	80	2167.33	III
			200	1367.17	III						

Intensity	Wavelength/Å		Intensity	Wavelength/Å		Intensity	Wavelength/Å		Intensity	Wavelength/Å	
200	2172.20	III	100	3927.69	I	370	1437.27	V	200	2537.33	II
100	2184.11	III	400	4040.93	I	500	1437.73	V	320	2551.40	II
500	2188.97	III	700	4065.07	I	370	1445.40	V	400 h	2560.74	III
70	2248.56	II	100	4084.10	I	270	1457.91	V	250	2563.61	II
80	2263.62	II	100	4241.80	I	100	1717.21	IV	300 h	2567.46	III
300	2322.27	III	200	4315.11	I	270	1719.32	V	890	2571.67	II
180	2352.65	I	120 h	4437.27	I	550	1729.08	V	320	2573.90	II
100	2382.40	III	250	4488.25	I	750	1731.83	V	320	2576.82	II
120	2387.75	I	900 h	4607.51	I	750	1733.96	V	300	2578.14	II
150	2402.71	III	100 h	4620.56	I	440	1741.74	V	320	2582.54	II
150	2405.12	III	500	4792.58	I	1000	1749.11	V	390	2606.37	II
2600	2427.95	I	100	4811.60	I	1000	1750.19	V	450	2607.03	II
60	2533.52	II	100	5147.44	I	500	1760.89	V	230	2613.60	II
250	2641.48	I	300	5230.26	I	370	1765.62	V	450	2622.74	II
3400	2675.95	I	100 h	5261.76	I	270	1774.02	V	160	2637.00	I
1100	2748.25	I	100	5655.77	I	6200	2012.78	II	1100	2638.71	II
100	2780.82	I	100 h	5721.36	I	8500	2028.18	II	1100	2641.41	II
1000	2802.04	II	300	5837.37	I	300	2070.94	III	160	2642.75	I
300	2819.79	II	100 h	5862.93	I	1200	2096.18	II	670	2647.29	II
100	2822.55	II	300 h	5956.96	I	200 h	2099.30	III	160	2657.84	II
100 h	2825.44	II	600	6278.17	I	200 h	2110.31	III	210	2661.88	II
300	2837.85	II	100	6562.68	I	200	2155.66	III	290	2683.35	II
100	2846.92	II	600	7510.73	I	200	2183.50	III	200	2687.22	III
100	2856.74	II	10	8145.06	I	200	2195.44	III	670	2705.61	I
300	2883.45	I	10	9254.28	I	540	2210.82	II	210	2712.42	II
300	2891.96	I				200	2234.59	III	250	2718.59	I
100	2893.25	II	*Hafnium Hf* Z = 72			320	2254.01	II	710	2738.76	II
100	2907.04	II	220	545.41	V	160	2255.15	II	200	2743.64	I
300	2913.52	II	180	600.00	V	250	2266.83	II	360	2751.81	II
300	2918.24	II	200	618.27	IV	620	2277.16	II	500	2753.60	III
100	2954.22	II	200	644.54	IV	200 h	2313.44	III	450	2761.63	I
100 h	2990.27	II	400	647.39	IV	230	2321.14	II	160	2766.96	I
300	2994.80	II	600	665.65	IV	580	2322.47	II	170	2773.02	I
320	3029.20	I	200	673.49	IV	300	2323.25	II	980	2773.36	II
300	3065.42	I	270	867.25	V	300	2324.89	II	180	2774.02	II
100	3122.50	II	180	875.88	V	200	2332.97	II	390	2779.37	I
1600	3122.78	I	180	885.58	V	300	2336.47	III	230	2808.00	II
100	3194.72	I	180	896.14	V	200	2337.33	II	230	2813.86	II
100	3227.99	III	180	901.54	V	230	2343.32	II	170	2814.48	II
300	3230.63	I	180	919.10	V	320	2347.44	II	230	2817.68	I
300	3308.30	I	270	921.67	V	540	2351.22	II	200	2819.74	I
300	3309.64	I	245	951.62	V	250	2380.30	II	1200	2820.22	II
100	3309.86	III	180	960.12	V	250	2383.540	III	490	2822.68	II
100	3320.12	I	180	964.74	V	170	2393.18	II	180	2833.28	I
100	3355.15	I	160	971.51	V	450	2393.36	II	410	2845.83	I
100 h	3391.31	I	160	1092.76	V	670	2393.83	II	270	2849.21	II
100	3395.40	I	160	1201.76	V	540	2405.42	II	270	2850.96	I
100	3467.21	I	270	1232.03	V	370	2410.14	II	180	2851.21	II
300	3557.36	I	200	1233.59	V	320	2417.69	II	180	2860.56	I
300	3586.73	I	160	1237.42	V	390	2447.25	II	760	2861.01	II
100	3611.57	I	160	1239.53	V	450	2460.49	II	760	2861.70	II
100 h	3631.31	I	160	1244.46	V	400	2461.74	III	2100	2866.37	I
300	3637.90	I	440	1396.66	V	430	2464.19	II	210	2887.14	I
100 h	3645.02	I	270	1400.09	V	210	2469.18	II	800	2889.62	I
100	3650.74	I	160	1401.70	V	2000	2495.16	III	1800	2898.26	I
100	3709.62	I	370	1407.17	V	290	2496.99	II	1200	2904.41	I
100	3796.01	I	370	1408.38	V	580	2512.69	II	890	2904.75	I
100	3874.73	I	270	1412.28	V	580	2513.03	II	2000	2916.48	I
100 h	3892.26	I	270	1413.51	V	1000	2515.16	III	580	2918.58	I
400	3897.86	I	160	1421.96	V	890	2516.88	II	320	2919.59	II
300	3909.38	I	220	1422.53	V	340	2531.19	II	180	2924.62	I
			370	1433.43	V						

Intensity	Wavelength/Å		Intensity	Wavelength/Å		Intensity	Wavelength/Å		Intensity	Wavelength/Å	
490	2929.63	II	670	3312.86	I	650	3793.37	II	160	5719.18	I
450	2929.90	I	180	3317.99	II	850 d	3800.38	I	95	6098.67	I
710	2937.80	II	890	3332.73	I	320	3811.78	I	95	6185.13	I
2000	2940.77	I	370	3352.06	II	1300	3820.73	I	85	6789.27	I
160	2944.71	I	230	3358.91	I	280	3830.02	I	160	6818.94	I
1200	2950.68	I	180	3360.06	I	800	3849.18	I	160	7063.83	I
1100	2954.20	I	180	3378.93	I	600	3858.31	I	570	7131.81	I
540	2958.02	I	230	3384.70	II	230	3860.91	I	650	7237.10	I
1400	2964.88	I	170	3386.21	I	200	3872.55	II	410	7240.87	I
620	2966.93	I	800	3389.83	II	160	3877.10	II	360	7624.40	I
710	2968.81	II	230	3392.81	I	380	3880.82	II	110	7740.17	I
890	2975.88	II	230	3394.59	II	200	3882.52	I	310	7845.35	I
1100	2980.81	I	230	3397.26	I	200	3889.23	I	130	7920.71	I
210	2982.72	I	230	3397.60	I	200	3889.33	I	250	7994.73	I
170	3000.10	II	2300	3399.80	II	620	3899.94	I	130	8204.58	I
800	3005.56	I	170	3400.21	I	620	3918.09	II	150	8546.48	I
1100	3012.90	II	180	3402.51	I	200	3923.90	II	160	8640.06	I
540	3016.78	I	230	3410.17	II	320	3931.38	I	40	8711.24	I
1100	3016.94	II	230	3417.34	I	410	3951.83	I	65	9004.73	I
980	3018.31	I	410	3419.18	I	160	3968.01	I			
1200	3020.53	I	200	3428.37	II	200	3973.48	I	*Helium He Z = 2*		
410	3031.16	II	250	3438.24	II	180	4032.27	I	15	231.454	II
710	3050.76	I	710	3472.40	I	230	4062.84	I	20	232.584	II
1100	3057.02	I	200	3478.99	II	180	4083.35	I	30	234.347	II
850	3067.41	I	480	3479.28	II	540	4093.16	II	50	237.331	II
2100	3072.88	I	250	3495.75	II	1100	4174.34	I	100	243.027	II
170	3074.10	I	250	3497.16	I	160	4206.58	II	300	256.317	II
250	3074.79	I	980	3497.49	I	190	4209.70	I	1000	303.780	II
430	3080.84	I	1200	3505.23	II	170	4228.08	I	500	303.786	II
200	3096.76	I	980	3523.02	I	170	4232.44	II	10	320.293	I
340	3101.40	II	980	3535.54	II	170	4260.98	I	2	505.500	I
710	3109.12	II	760	3536.62	I	200	4263.39	I	3	505.684	I
710	3131.81	I	180	3548.81	I	170	4272.85	II	4	505.912	I
850	3134.72	II	540	3552.70	II	320	4294.79	I	5	506.200	I
170	3139.65	II	1300	3561.66	II	160	4330.27	I	7	506.570	I
220	3145.32	II	270	3567.36	I	180	4336.66	II	10	507.058	I
220	3148.41	I	1100	3569.04	II	250	4356.33	I	15	507.718	I
450	3156.63	I	210	3597.42	II	180	4370.97	II	20	508.643	I
270	3159.82	I	540	3599.87	I	160	4417.91	I	25	509.998	I
710	3162.61	II	800	3616.89	I	200	4438.04	I	35	512.098	I
450	3168.39	I	320	3630.87	II	250	4565.94	I	50	515.616	I
890	3172.94	I	800	3644.36	II	500 d	4598.80	I	100	522.213	I
450	3176.86	II	320	3649.10	I	230	4620.86	I	400	537.030	I
220	3181.01	I	200	3651.84	I	210	4655.19	I	1000	584.334	I
360	3193.53	II	220	3665.35	II	120	4699.01	I	50	591.412	I
670	3194.19	II	200	3672.27	I	160	4782.74	I	5	958.70	II
200	3196.93	I	480	3675.74	I	310	4800.50	I	6	972.11	II
310	3206.11	I	2200	3682.24	I	130	4859.24	I	8	992.36	II
180	3210.98	I	280	3696.51	I	120	4975.25	I	15	1025.27	II
180	3217.30	II	240	3699.72	II	95	5018.20	I	30	1084.94	II
180	3220.61	II	340	3701.15	II	95	5047.45	I	35	1215.09	II
360	3247.66	I	1000	3717.80	I	75	5170.18	I	50	1215.17	II
220	3249.53	I	650	3719.28	II	230	5181.86	I	120	1640.34	II
890	3253.70	II	160	3729.10	I	110	5243.99	I	180	1640.47	II
270	3255.28	II	460	3733.79	I	120	5294.87	I	7	2385.40	II
180	3273.66	II	160	3737.88	II	110	5354.73	I	9	2511.20	II
200 h	3279.67	III	400	3746.80	I	110	5373.86	I	50	2577.6	I
270	3279.98	II	170	3766.92	II	75	5452.92	I	1	2723.19	I
160	3291.05	I	200	3768.25	I	230	5550.60	I	12	2733.30	II
210	3306.12	I	1400	3777.64	I	230	5552.12	I	2	2763.80	I
340	3310.27	I	1400	3785.46	I	95	5613.27	I	10	2818.2	I
									4	2829.08	I

Intensity	Wavelength/Å		Intensity	Wavelength/Å		Intensity	Wavelength/Å		Intensity	Wavelength/Å	
10	2945.11	I	2000	10830.34	I	1600	3461.97	II	1300	4125.65	I
40	3013.7	I	9	10913.05	I	810 c	3473.91	II	4300	4127.16	I
20	3187.74	I	3	10917.10	I	5400 c	3474.26	II	1500	4136.22	I
3	3202.96	II	4	11626.4	II	6300	3484.84	II	980 cw	4152.61	II
15	3203.10	II	30	11969.12	I	2500 c	3494.76	II	8100	4163.03	I
1	3354.55	I	20	12527.52	I	810 c	3498.88	II	2500	4173.23	I
2	3447.59	I	50	12784.99	I	810	3510.73	I	540	4194.35	I
1	3587.27	I	20	12790.57	I	4100 c	3515.59	II	2000	4227.04	I
3	3613.64	I	7	12845.96	I	410 c	3519.94	II	1300 cw	4254.43	I
2	3634.23	I	10	12968.45	I	630	3540.76	II	490	4264.05	I
3	3705.00	I	2	12984.89	I	1600	3546.05	II	1300	4350.73	I
1	3732.86	I	12	15083.64	I	1100 c	3556.78	II	300	4477.64	II
10	3819.607	I	200	17002.47	I	410	3560.15	II	290	4629.10	II
1	3819.76	I	1	18555.55	I	410 c	3573.24	II	80	4701.69	II
500	3888.65	I	6	18636.8	II	630 c	3574.80	II	130	4709.84	II
20	3964.729	I	500	18685.34	I	810	3579.12	I	130 c	4717.52	I
1	4009.27	I	200	18697.23	I	410	3580.75	II	290	4742.04	II
50	4026.191	I	100	19089.38	I	410	3581.83	II	100 c	4757.01	I
5	4026.36	I	20	19543.08	I	630 c	3592.23	II	65	4782.92	I
12	4120.82	I	1000	20581.30	I	1100 cw	3598.77	II	290	4939.01	I
2	4120.99	I	80	21120.07	I	540 c	3600.95	II	250 c	4967.21	II
3	4143.76	I	10	21121.43	I	410	3618.43	I	220	4979.97	I
10	4387.929	I	20	21132.03	I	430 c	3626.69	II	90	4995.05	I
3	4437.55	I	3	30908.5	II	490	3627.25	II	130	5042.37	I
200	4471.479	I	4	40478.90	I	430 c	3631.76	II	80	5093.07	I
25	4471.68	I				430 c	3638.30	II	140	5127.81	I
6	4685.4	II	*Holmium Ho* Z = 67			1600 c	3662.29	I	130	5142.59	II
30	4685.7	II	170	2502.91	II	1400	3667.97	I	110	5143.22	II
30	4713.146	I	170	2533.80	I	720	3679.19	I	160	5149.59	II
4	4713.38	I	190	2605.86	II	670	3679.70	I	90 c	5167.88	I
20	4921.931	I	270	2750.35	II	720	3682.65	I	130 c	5182.11	I
100	5015.678	I	270	2769.89	II	580	3690.65	I	90	5190.11	II
10	5047.74	I	300	2824.20	II	410	3700.04	I	65	5251.82	I
5	5411.52	II	270 c	2831.69	II	490 c	3702.35	II	90	5301.25	I
500	5875.62	I	270	2849.10	II	430	3712.88	I	80	5330.11	I
100	5875.97	I	360	2880.26	II	450	3720.72	I	90	5359.99	I
8	6560.10	II	460	2880.98	II	1100	3731.40	I	100	5407.08	I
100	6678.15	I	570 c	2909.41	II	810	3736.35	I	70	5566.52	I
3	6867.48	I	410 c	2979.63	II	3200 cw	3748.17	II	65	5627.60	I
200	7065.19	I	410	2987.64	II	8900 c	3796.75	II	140	5659.58	I
30	7065.71	I	480 c	3049.38	II	8900 c	3810.73	II	140 c	5691.47	I
50	7281.35	I	410 c	3054.00	II	410 c	3835.35	II	140 c	5696.57	I
1	7816.15	I	500 c	3057.45	II	1300 cw	3837.51	II	140 c	5860.28	I
2	8361.69	I	500 c	3082.34	II	410 c	3842.05	II	70 c	5882.99	I
2	9063.27	I	910	3084.36	II	1100	3843.86	II	70	5921.76	I
2	9210.34	I	430 c	3086.54	II	490 c	3846.73	II	70 cw	5948.03	I
10	9463.61	I	760	3118.50	II	1800 c	3854.07	II	70	5972.76	I
4	9516.60	I	580 c	3166.62	II	2700 c	3861.68	II	90	5973.52	I
3	9526.17	I	810	3173.78	II	3000 c	3888.96	II	230 c	5982.90	I
1	9529.27	I	810 c	3181.50	II	13000 c	3891.02	II	120	6081.79	I
1	9603.42	I	980 c	3281.97	II	1300 cw	3905.68	II	70	6208.65	I
3	9702.60	I	630 c	3337.23	II	580	3955.73	I	70 c	6305.36	I
6	10027.73	I	980 c	3343.58	II	490	3959.68	I	70	6550.97	I
2	10031.16	I	8100 c	3398.98	II	2700	4040.81	I	260	6604.94	I
15	10123.6	II	810 c	3410.26	II	5400 c	4045.44	II	120	6628.99	I
1	10138.50	I	1400 c	3414.90	II	8100	4053.93	I	55 cw	6694.32	I
10	10311.23	I	5400	3416.46	II	1700	4065.09	II	55 c	6785.43	I
2	10311.54	I	1200	3421.63	II	720	4068.05	I	40 cw	6939.49	I
3	10667.65	I	2000 c	3425.34	II	8900	4103.84	I	45 cw	6950.39	I
300	10829.09	I	2000 c	3428.13	II	2900	4108.62	I	140	7555.09	I
1000	10830.25	I	3200	3453.14	II	1500	4120.20	I	40 c	7628.42	I
			16000 c	3456.00	II						

Intensity	Wavelength/Å	
50 c	7693.15	I
60 cw	7815.48	I
60	7894.64	I
50	8512.94	I
40	8670.19	I
90	8915.98	II

Hydrogen H Z = 1

Intensity	Wavelength/Å	
15	926.226	I
20	930.748	I
30	937.803	I
50	949.743	I
100	972.537	I
300	1025.722	I
1000	1215.668	I
500	1215.674	I
5	3835.384	I
6	3889.049	I
8	3970.072	I
15	4101.74	I
30	4340.47	I
80	4861.33	I
120	6562.72	I
180	6562.852	I
5	9545.97	I
7	10049.4	I
12	10938.1	I
20	12818.1	I
40	18751.0	I
5	21655.3	I
8	26251.5	I
15	40511.6	I
4	46525.1	I
6	74578.	I

Indium In Z = 49

Intensity	Wavelength/Å	
17	378.61	V
17	386.21	V
14	388.91	V
25	393.89	V
25	400.57	V
25	402.39	V
622	472.71	IV
689	479.39	IV
709	498.62	IV
10	882.24	III
10	890.84	III
10	915.87	III
85	954.67	IV
87	973.50	IV
86	991.60	IV
89	1024.68	IV
85	1024.79	IV
88	1031.45	IV
82	1031.98	IV
80	1054.43	IV
84	1063.03	IV
83	1068.25	IV
82	1069.82	IV
86	1077.64	IV
90	1082.10	IV
83	1086.33	IV

Intensity	Wavelength/Å	
82	1096.81	IV
84	1097.18	IV
85	1116.10	IV
80	1124.06	IV
90	1131.46	IV
85	1144.43	IV
80	1145.41	IV
89	1146.62	IV
83	1154.11	IV
84	1154.60	IV
90	1157.71	IV
90	1157.82	IV
85	1159.78	IV
88	1176.50	IV
85	1191.58	IV
83	1204.87	IV
90	1206.55	IV
88	1221.50	IV
85	1221.90	IV
85	1233.58	IV
87	1235.84	IV
90	1373.20	IV
88	1398.77	IV
81	1412.09	IV
100	1625.42	III
100	1748.83	III
30	1842.41	III
40	1850.30	III
30	2154.08	III
50	2205.28	II
40	2281.64	II
100 c	2306.05	II
90 d	2313.21	II
70 d	2327.95	II
80 h	2334.57	II
110 d	2382.63	II
70 h	2427.20	II
100	2447.90	II
110 d	2488.62	II
90	2488.95	II
80	2498.59	II
100	2499.60	II
90 d	2500.99	II
110 d	2512.31	II
100	2521.37	I
100	2527.41	III
160 d	2554.44	II
1100	2560.15	I
200	2601.76	I
90 d	2654.70	II
100 d	2662.63	II
140 d	2668.65	II
140 d	2674.56	II
80	2683.12	II
1600	2710.26	I
300	2713.94	I
80	2726.15	III
130 d	2749.75	II
700	2753.88	I
90 d	2818.97	II
180 c	2836.92	I

Intensity	Wavelength/Å	
80	2865.68	II
120 d	2890.18	II
1100	2932.63	I
100	2941.05	II
100	2982.80	III
110 c	2999.40	II
100	3008.08	III
8000	3039.36	I
110 d	3099.80	II
180 c	3101.8	II
130 c	3138.60	II
80 c	3142.75	II
130 d	3146.70	II
150	3155.77	II
100 c	3158.40	II
90 c	3176.30	II
90 d	3198.11	II
13000	3256.09	I
3000	3258.56	I
90 c	3338.50	II
100 c	3404.28	II
110 d	3438.40	II
180 c	3693.91	II
95 c	3708.13	II
380 w	3716.14	II
120 c	3718.30	II
160 c	3718.72	II
160 c	3723.40	II
170 w	3795.21	II
230 c	3799.21	II
250 c	3834.65	II
200 c	3842.18	II
100	3852.82	III
100	3889.78	II
100 c	3902.07	II
250 w	3962.35	II
120 c	4004.66	II
140 d	4013.92	II
100	4023.77	III
150	4032.32	III
410 w	4056.94	II
100	4071.57	III
100	4072.93	III
17000	4101.76	I
140 c	4205.14	II
100 d	4213.04	II
110 d	4219.66	II
100	4252.68	III
150 d	4372.87	II
150 c	4500.78	II
18000	4511.31	I
110 c	4549.01	II
140 c	4570.85	II
180 w	4578.02	II
180 w	4578.40	II
140 c	4616.08	II
170 c	4617.17	II
250 c	4620.14	II
150 w	4620.70	II
170 c	4627.30	II
140 c	4637.04	II

Intensity	Wavelength/Å	
380 c	4638.16	II
220 c	4644.58	II
360 c	4655.62	II
320 w	4656.74	II
190 c	4681.11	II
450 w	4684.8	II
90 d	4907.06	II
150 c	4973.77	II
80 h	5109.36	II
100 w	5115.14	II
140 c	5117.40	II
270 c	5120.80	II
200 w	5121.75	II
80 d	5129.85	II
240 c	5175.42	II
140 c	5184.44	II
200	5248.77	III
150 c	5309.45	II
80	5411.41	II
140 c	5418.45	II
220 w	5436.70	II
130 c	5497.50	II
140 c	5507.08	II
320 c	5513.00	II
250 w	5523.28	II
130 c	5536.50	II
190 w	5555.45	II
240 c	5576.90	II
200 w	5636.70	II
100	5645.15	III
160 c	5708.50	II
100 c	5721.80	II
100	5819.50	III
210 c	5853.15	II
490 w	5903.4	II
260 w	5915.4	II
120 c	5918.78	II
130 c	6062.9	II
250 c	6095.95	II
210 c	6108.66	II
180 w	6115.9	II
230 w	6128.7	II
240 w	6129.4	II
320 w	6132.1	II
150 c	6140.0	II
90	6143.23	II
140 c	6148.10	II
190 w	6149.5	II
80	6161.15	II
180 w	6162.45	II
200	6197.72	III
100 c	6224.28	II
280 w	6228.3	II
140 w	6231.1	II
270 w	6304.8	II
290 w	6362.3	II
300 w	6469.0	II
210 c	6541.20	II
190 c	6751.88	II
180 c	6765.9	II
100 c	6783.72	II

Intensity	Wavelength/Å		Intensity	Wavelength/Å		Intensity	Wavelength/Å		Intensity	Wavelength/Å	
320 w	6891.5	II	600	1267.60	I	10000	5119.29	I	800 c	8169.38	I
380 w	7182.9	II	1500	1275.26	I	3000 c	5161.20	II	500 d	8222.57	I
180 c	7255.0	II	3000	1289.40	I	1000	5234.57	I	4000	8240.05	I
210 c	7276.5	II	10000	1300.34	I	3000 c	5245.71	II	10000 c	8393.30	I
180 c	7303.4	II	3000	1302.98	I	10000	5338.22	II	1000	8486.11	I
320 c	7350.6	II	3000	1313.95	I	5000 c	5345.15	II	1500 c	8664.95	I
100 c	7632.7	II	3000	1317.54	I	600 c	5427.06	I	500 c	8700.80	I
100 c	7682.9	II	2000	1330.19	I	3000	5435.83	II	250 d	8748.22	I
210 c	7740.7	II	20000	1336.52	II	2000 c	5464.62	II	1000	8853.24	I
100 c	7776.96	II	5000	1355.10	I	10000	5625.69	II	2000	8853.80	I
180 c	7789.0	II	3000	1357.97	I	2000 c	5690.91	II	3000	8857.50	I
90 c	7840.9	II	5000	1360.97	I	4000 c	5710.53	II	1000 d	8898.50	I
240 c	8227.0	II	3000	1361.11	I	1000 d	5764.33	I	400	8964.69	I
100 w	8813.5	II	2500	1367.71	I	2000	5894.03	I	400	8993.13	I
80 c	8832.6	II	2500	1368.22	I	5000	5950.25	II	5000	9022.40	I
120 c	9197.7	II	4000	1383.23	I	300	5984.86	I	15000	9058.33	I
120 c	9202.0	II	3000	1390.75	I	2000 d	6024.08	I	1000	9098.86	I
220 w	9213.0	II	2000	1392.90	I	2000 c	6074.98	II	12000	9113.91	I
160 d	9241.1	II	2000	1400.01	I	1000	6082.43	I	600	9128.03	I
100	9977.86	I	8000	1425.49	I	2000 c	6127.49	II	600	9227.74	I
200	10257.03	I	5000	1446.26	I	800	6191.88	I	1000	9335.05	I
100 h	10744.31	I	5000	1453.18	I	500	6213.10	I	4000	9426.71	I
20	11334.72	I	5000	1457.39	I	800	6244.48	I	3000	9427.15	I
20	11731.48	I	5000	1457.47	I	1000	6293.98	I	2000	9598.22	I
10	12912.59	I	10000	1457.98	I	500	6313.13	I	2000	9649.61	I
9	13429.96	I	2500	1458.79	I	800	6330.37	I	3000 d	9653.06	I
7	14719.08	I	4000	1459.15	I	400	6333.50	I	5000	9731.73	I
6	22291.06	I	2500	1465.83	I	2000	6337.85	I	500	10003.05	I
7	23879.13	I	1000	1485.92	I	1000	6339.44	I	750	10131.16	I
			5000	1492.89	I	500	6359.16	I	1000	10238.82	I
Iodine I Z = 53			5000	1507.04	I	1000	6566.49	I	400	10375.20	I
30	363.78	V	5000	1514.68	I	2000	6583.75	I	400	10391.74	I
36	380.74	V	15000	1518.05	I	1000	6585.27	I	5000	10466.54	I
45	565.53	V	2500	1526.45	I	5000	6619.66	I	400	11236.56	I
50	607.57	V	5000	1593.58	I	500	6661.11	I	350	11558.46	I
6	612.46	IV	5000	1617.60	I	500 c	6697.29	I	320	11778.34	I
6	666.81	III	2500	1640.78	I	400	6732.03	I	450	11996.86	I
8	705.11	III	15000	1702.07	I	4000	6812.57	II	300	12033.69	I
7	784.64	III	12000	1782.76	I	500	6989.78	I	150	12304.58	I
7	784.80	III	5000	1799.09	I	500	7120.05	I	60	13149.16	I
8	795.52	III	75000	1830.38	I	1200	7122.05	I	140	13958.27	I
7	919.28	IV	15000	1844.45	I	2000	7142.06	I	200	14287.02	I
10000	1034.66	II	2000	2061.63	I	1000	7164.79	I	100	14460.00	I
8	1094.20	III	7	2361.13	IV	400 d	7191.66	I	225	15032.57	I
10000	1139.80	II	6	2372.45	IV	700	7227.30	I	105	15528.65	I
10000	1160.56	II	7	2376.46	IV	1000	7236.78	I	150	16037.33	I
20000	1166.48	II	8	2387.11	IV	500	7237.84	I	15	18275.71	I
10000	1178.65	II	9	2426.10	IV	5000	7402.06	I	20	18348.52	I
15000	1187.34	II	8	2475.35	IV	1000	7410.50	I	15	18982.41	I
10000	1190.85	II	8	2519.74	IV	500	7416.48	I	35	19070.17	I
200	1218.41	I	8	2545.67	IV	5000	7468.99	I	110	19105.12	I
20000	1220.89	II	7	2545.71	III	500 c	7490.52	I	50	19370.02	I
600	1224.05	I	8	2652.23	IV	2000	7554.18	I	10	20648.69	I
600	1224.08	I	5000	3078.75	II	500 d	7556.65	I	220	22183.03	I
500	1228.89	I	200	4102.23	I	2000 c	7700.20	I	150	22226.53	I
20000	1234.06	II	200	4129.21	I	600	7897.98	I	30	22309.21	I
600	1251.34	I	100 d	4134.15	I	500	7969.48	I	32	24420.82	I
8	1252.35	III	500 d	4321.84	I	1000	8003.63	I	12	27365.42	I
2500	1259.15	I	250	4763.31	I	99000	8043.74	I	9	27573.05	I
3000	1259.51	I	1000	4862.32	I	300 d	8065.70	I	10	30361.93	I
800	1261.27	I	200	4916.94	I	1000	8090.76	I	8	30383.88	I
600	1267.57	I									

Intensity	Wavelength/Å		Intensity	Wavelength/Å		Intensity	Wavelength/Å		Intensity	Wavelength/Å	
10	34295.73	I	540	2333.30	I	230	2572.70	I	330	3039.26	I
9	34513.11	I	740	2333.84	I	740	2577.26	I	300	3047.16	I
3	40228.54	I	580	2334.50	I	740	2592.06	I	300	3049.44	I
2	41633.80	I	1600	2343.18	I	740	2599.04	I	300	3057.28	I
			740	2343.61	I	700	2608.25	I	1600	3068.89	I
Iridium Ir Z = 77			580	2355.00	I	1800	2611.30	I	320	3083.22	I
9900	2010.65	I	230	2357.53	II	210	2614.98	I	240	3086.44	I
8700	2022.35	I	410	2358.16	I	330	2617.78	I	390	3088.04	I
15000	2033.57	I	500	2360.73	I	210	2619.88	I	510	3100.29	I
6200	2052.22	I	2500	2363.04	I	250	2625.32	I	510	3100.45	I
5000	2060.64	I	370	2368.04	II	700	2634.17	I	340	3120.76	I
3700	2083.22	I	3500	2372.77	I	250	2639.42	I	200	3121.78	I
3100	2085.74	I	290	2375.09	II	3500	2639.71	I	3400	3133.32	I
17000	2088.82	I	250	2377.28	I	210	2644.19	I	490	3168.88	I
14000	2092.63	I	250	2377.98	I	1800	2661.98	I	370	3177.58	I
2700	2112.68	I	500	2379.38	I	350	2662.63	I	370	3198.92	I
1800	2119.54	I	540	2381.62	I	2700	2664.79	I	610	3212.12	I
2000	2125.44	I	210	2383.17	I	520	2669.91	I	370	3219.51	I
4500	2126.81	II	1300	2386.89	I	520	2671.84	I	5100	3220.78	I
2000	2127.52	I	2500	2390.62	I	330	2673.61	I	300	3229.28	I
4500	2127.94	I	2700	2391.18	I	270	2692.34	I	470	3241.52	I
3700	2148.22	I	230	2407.59	I	3000	2694.23	I	200	3262.01	I
2500	2150.54	I	290	2409.37	I	330	2772.46	I	390	3266.44	I
3500	2152.68	II	290	2410.17	I	250	2775.55	I	200	3322.60	I
2900	2155.81	I	290	2410.73	I	520	2781.29	I	560	3368.48	I
7900	2158.05	I	540	2413.31	I	330	2785.22	I	660	3437.02	I
2100	2162.88	I	370	2415.86	I	540	2797.35	I	410	3448.97	I
5800	2169.42	II	620	2418.11	I	1600	2797.70	I	3200	3513.64	I
4500	2175.24	I	210	2424.89	I	380	2798.18	I	220	3515.95	I
2700	2178.17	I	370	2424.99	I	410	2800.82	I	410	3522.03	I
1600	2187.43	II	290	2425.66	I	680	2823.18	I	320	3558.99	I
1100	2190.38	II	540	2427.61	I	1200	2824.45	I	1200	3573.72	I
740	2191.64	I	540	2431.24	I	820	2836.40	I	320	3594.39	I
910	2208.09	II	1300	2431.94	I	1100	2839.16	I	220	3609.77	I
1300	2220.37	I	270	2435.14	I	820	2840.22	I	660	3628.67	I
790	2221.07	II	250	2445.34	I	3800	2849.72	I	220	3636.20	I
2500	2242.68	II	250	2447.76	I	380	2875.60	I	300	3661.71	I
620	2245.76	II	910	2452.81	I	380	2875.98	I	300	3664.62	I
2100	2253.38	I	1300	2455.61	I	270	2877.68	I	320	3674.98	I
2100	2255.10	I	230	2455.87	I	820	2882.64	I	200	3687.08	I
1400	2255.81	I	210	2457.03	I	650	2897.15	I	200	3731.36	II
350	2258.51	I	210	2457.23	I	260	2901.95	I	530	3747.20	I
1400	2258.86	I	870	2467.30	I	260	2904.80	I	3100	3800.12	I
830	2264.61	I	3300	2475.12	I	200	2907.24	I	230	3817.24	I
1100	2266.33	I	210	2478.11	I	440	2916.36	I	480	3902.51	I
1000	2268.90	I	2100	2481.18	I	230	2918.57	I	480	3915.38	I
660	2280.00	I	620	2493.08	I	4400	2924.79	I	400	3934.84	I
950	2281.02	II	210	2496.27	I	1200	2934.64	I	590	3976.31	I
660	2281.91	I	250	2502.63	I	880	2936.68	I	460	3992.12	I
330	2284.60	I	4100	2502.98	I	250	2938.47	I	350	4033.76	I
330	2295.08	I	210	2513.71	I	2700	2943.15	I	370	4069.92	I
790	2298.05	I	990	2533.13	I	230	2946.97	I	150	4070.68	I
460	2299.53	I	1100	2534.46	I	200	2949.76	I	100	4092.61	I
910	2300.50	I	580	2537.22	I	1200	2951.22	I	140	4115.78	I
2700	2304.22	I	580	2542.02	I	200	2974.95	I	90	4172.56	I
410	2305.47	I	7900	2543.97	I	440	2980.65	I	260	4268.10	I
210	2307.27	I	790	2546.03	I	300	2996.08	I	220	4311.50	I
910	2308.93	I	210	2551.40	I	220	3002.25	I	160	4399.47	I
460	2315.38	I	210	2555.35	I	600	3003.63	I	65	4403.78	I
410	2321.45	I	910	2564.18	I	270	3017.31	I	110	4426.27	I
410	2321.58	I	210	2569.88	I	380	3029.36	I	55	4478.48	I
210	2327.98	I									

Intensity	Wavelength/Å		Intensity	Wavelength/Å		Intensity	Wavelength/Å		Intensity	Wavelength/Å	
55	4545.68	I	10	844.28	III	500	1479.47	V	15	1651.58	IV
30	4548.48	I	10 p	861.83	III	10 h	1505.17	III	15	1652.90	IV
35	4568.09	I	10	891.17	III	13	1526.60	IV	13	1653.41	IV
75	4616.39	I	10	950.33	III	13	1530.26	IV	13	1656.11	IV
26	4656.18	I	10	981.37	III	14	1532.63	IV	15	1656.65	IV
50	4728.86	I	10 w	983.88	III	13	1532.91	IV	14	1660.10	IV
26	4756.46	I	12	1055.27	II	15	1533.86	IV	13	1662.32	IV
65	4778.16	I	15	1068.36	II	13	1533.95	IV	13	1662.52	IV
30	4795.67	I	15	1071.60	II	14	1536.58	IV	13	1663.54	IV
50	4938.09	I	15	1096.89	II	10 h	1538.63	III	13	1668.09	IV
26	4970.48	I	12	1099.12	II	13	1542.16	IV	12	1670.74	II
25	4999.74	I	18	1112.09	II	14	1542.70	IV	14	1671.04	IV
25	5002.74	I	12	1121.99	II	12 h	1550.20	III	13	1673.68	IV
30	5014.98	I	12	1122.86	II	13	1566.26	IV	14	1675.66	IV
30	5123.66	I	12	1128.07	II	14	1568.27	IV	13	1681.36	IV
35	5364.32	I	12	1130.43	II	13	1591.51	IV	15	1687.69	IV
75	5449.50	I	15	1133.41	II	13	1592.05	IV	15	1698.88	IV
45	5625.55	I	12	1133.68	II	13	1598.01	IV	12	1702.04	II
35	5894.06	I	12	1138.64	II	13	1600.58	IV	13	1709.81	IV
20	6110.67	I	12	1142.33	II	10 h	1601.21	III	15	1711.41	IV
12	6288.28	I	12	1143.23	II	13	1601.67	IV	14	1712.76	IV
10	6686.08	I	18	1144.95	II	13	1603.18	IV	14	1717.90	IV
6	7834.32	I	12	1147.41	II	13	1603.73	IV	14	1718.16	IV
			15	1148.29	II	13	1604.88	IV	14	1719.46	IV
Iron Fe Z = 26			12	1151.16	II	13	1605.68	IV	14	1722.71	IV
350	386.16	V	12	1267.44	II	15	1605.97	IV	14	1724.06	IV
350	386.88	V	12	1272.00	II	13	1606.98	IV	16	1725.63	IV
400	387.20	V	400	1317.86	V	17	1609.10	IV	13	1761.08	IV
400	387.50	V	400	1323.27	V	14	1609.83	IV	12	1761.38	II
400	387.76	V	400	1330.40	V	13	1610.47	IV	20	1785.26	II
400	387.78	V	400	1359.01	V	13	1611.20	IV	20	1786.74	II
400	395.90	V	600	1361.82	V	13	1613.64	IV	18	1788.07	II
400	404.62	V	700	1373.59	V	15	1614.02	IV	13	1792.10	IV
400	405.50	V	600	1373.67	V	13	1614.64	IV	13	1796.93	IV
800	407.42	V	500	1376.34	V	13	1615.00	IV	13	1827.98	IV
600	407.44	V	500	1378.56	V	16	1616.68	IV	10	1869.83	III
400	407.49	V	800	1387.94	V	14	1617.68	IV	12	1877.99	III
500	407.75	V	400	1397.97	V	12	1618.47	II	10	1882.05	III
400	409.71	V	600	1400.24	V	14	1619.02	IV	12	1886.76	III
400	410.20	V	800	1402.39	V	13	1621.16	IV	13	1890.67	III
600	411.55	V	400	1406.67	V	14	1621.57	IV	11	1893.98	III
700	417.39	V	500	1406.82	V	13	1623.38	IV	20	1895.46	III
700	418.04	V	400	1407.25	V	13	1623.53	IV	10 s	1907.58	III
500	418.47	V	600	1409.45	V	15	1626.47	IV	19	1914.06	III
700	421.06	V	400	1415.20	V	14	1626.90	IV	15	1915.08	III
500	421.78	V	600	1420.46	V	13	1628.54	IV	15	1922.79	III
500	422.31	V	800	1430.57	V	13	1630.18	IV	10 p	1926.01	III
500	426.06	V	13	1431.43	IV	17	1631.08	IV	18	1926.30	III
500	426.11	V	800	1440.53	V	15	1631.12	II	15	1930.39	III
350	426.97	V	400	1442.22	V	14	1632.40	IV	14	1931.51	III
17	525.69	IV	800	1446.62	V	13	1634.01	IV	30	1934.538	I
15	526.29	IV	700	1448.85	V	18	1635.40	II	25	1937.269	I
13	526.63	IV	400	1449.93	V	15	1636.32	II	14	1937.34	III
14	536.61	IV	700	1456.16	V	15	1639.40	II	10 l	1938.90	III
15	537.10	IV	500	1459.83	V	14	1639.40	IV	14 s	1943.48	III
13	537.26	IV	400	1460.73	V	16	1640.04	IV	12	1945.34	III
14	537.79	IV	500	1462.63	V	14	1640.16	IV	50	1946.988	I
13	537.94	IV	700	1464.68	V	12	1641.76	II	10	1950.33	III
13	552.14	IV	500	1465.38	V	15	1641.87	IV	12	1951.01	III
14	607.53	IV	400	1466.65	V	15	1647.09	IV	25	1951.571	I
13	608.80	IV	500	1469.00	V	12	1647.16	II	30	1952.59	I
10	813.38	III									

Intensity	Wavelength/Å		Intensity	Wavelength/Å		Intensity	Wavelength/Å		Intensity	Wavelength/Å	
11	1952.65	III	60	2264.389	I	80	2406.97	II	500	2493.26	II
30	1953.005	I	80	2267.085	I	300	2410.52	II	60	2494.000	I
13	1953.32	III	10	2267.42	III	200	2411.07	II	50	2494.251	I
10	1953.49	III	80	2267.469	I	150	2413.31	II	100	2495.87	I
10 w	1954.22	III	50	2270.862	I	80	2417.87	II	600	2496.533	I
60	1957.823	I	150	2272.070	I	60	2420.396	I	150	2498.90	I
11	1958.58	III	150	2276.026	I	60	2423.089	I	1000	2501.132	I
60	1960.144	I	80	2279.937	I	150	2424.14	II	50	2501.693	I
13	1960.32	III	150	2284.086	I	120	2428.36	II	80	2506.09	II
30	1961.25	I	150	2287.250	I	120	2430.08	II	500	2507.900	I
50	1962.111	I	300	2292.524	I	80	2432.26	II	50	2508.753	I
12	1963.11	II	10	2293.06	III	60	2438.182	I	1000	2510.835	I
15	1987.50	III	15	2295.86	III	150	2439.30	II	120	2511.76	II
14	1991.61	III	200	2297.787	I	150	2439.74	I	80	2512.275	I
13	1994.07	III	600	2298.169	I	100	2442.57	I	400	2512.365	I
12	1995.56	III	80	2299.220	I	250	2443.872	I	80	2516.570	I
12	1996.42	III	300	2300.142	I	100	2444.51	II	300	2517.661	I
10	2061.55	III	50	2301.684	I	50	2445.212	I	800	2518.102	I
12	2068.24	III	100	2303.424	I	100	2445.57	II	150	2519.629	I
14	2078.99	III	150	2303.581	I	60	2447.709	I	50	2522.480	I
100	2084.122	I	120	2308.999	I	100	2453.476	I	4000	2522.849	I
10	2084.35	III	150	2313.104	I	1500	2457.598	I	200	2523.66	I
12	2090.14	III	10 p	2317.70	III	150	2458.78	II	500	2524.293	I
15	2097.48	III	10	2319.22	III	80	2461.28	II	100	2525.02	I
12	2097.69	III	200	2320.358	I	100	2461.86	II	200	2525.39	II
12	2103.80	III	10 p	2321.71	III	100	2462.181	I	300	2526.29	II
10	2107.32	III	10	2326.95	III	1500	2462.647	I	2000	2527.435	I
15	2151.78	III	100	2327.40	II	50	2463.730	I	800	2529.135	I
12	2157.71	III	100	2331.31	II	800	2465.149	I	250	2529.55	II
50	2157.794	I	300	2332.80	II	60	2467.732	I	150	2529.836	I
12	2158.47	III	10 p	2336.77	III	600	2468.879	I	200	2530.687	I
10	2161.27	III	200	2338.01	II	80	2470.67	II	120	2533.63	II
40	2166.773	I	10	2338.96	III	80	2470.965	I	100	2534.42	II
12	2166.95	III	600	2343.49	II	800	2472.336	I	120	2535.49	II
12	2171.04	III	80	2343.96	II	1000	2472.895	I	400	2535.607	I
15	2174.66	III	150	2344.28	II	200	2473.16	I	200	2536.792	I
300	2178.118	I	200	2348.11	II	600	2474.814	I	200	2536.80	II
12	2180.41	III	250	2348.30	II	60	2476.657	I	100	2538.80	II
250	2186.486	I	200	2359.12	II	120	2479.480	I	100	2538.91	II
60	2186.892	I	150	2360.00	II	1200	2479.776	I	150	2538.99	II
120	2187.195	I	120	2360.29	II	100	2480.16	II	50	2539.357	I
250	2191.839	I	200	2364.83	II	80	2482.12	II	200	2540.66	II
150	2196.043	I	80	2365.76	II	100	2482.66	II	600	2540.972	I
80	2200.390	I	80	2368.59	II	10000	2483.271	I	80	2541.10	II
80	2200.724	I	80	2369.456	I	300	2483.533	I	300	2542.10	I
15	2208.41	II	80	2369.95	II	1000	2484.185	I	250	2543.92	I
10 p	2208.85	III	120	2371.430	I	50	2485.990	I	150	2544.70	I
20	2213.65	II	300	2373.624	I	800	2486.373	I	800	2545.978	I
12	2218.26	II	150	2373.74	II	100	2486.691	I	80	2546.67	II
20	2220.38	II	120	2374.518	I	100	2487.066	I	100	2548.74	II
10	2221.83	III	120	2376.43	II	120	2487.370	I	80	2549.08	II
10	2229.27	III	80	2379.27	II	4000	2488.143	I	80	2549.39	II
10	2232.43	III	120	2380.76	II	100	2488.945	I	600	2549.613	I
10	2232.69	III	150	2381.835	I	80	2489.48	II	400	2562.53	II
10	2235.91	III	1000	2382.04	II	1000	2489.750	I	200	2563.48	II
10	2238.16	III	300	2388.63	II	50	2489.913	I	150	2574.36	II
12 p	2241.54	III	200	2389.973	I	3000	2490.644	I	300	2576.691	I
50	2250.790	I	1000	2395.62	II	100	2490.71	II	100	2582.58	II
60	2251.874	I	300	2399.24	II	2000	2491.155	I	1500	2584.54	I
300	2259.511	I	800	2404.88	II	100	2491.40	II	650	2585.88	II
12	2261.59	III	250	2406.66	II	100	2493.18	II	90	2591.54	II

Intensity	Wavelength/Å		Intensity	Wavelength/Å		Intensity	Wavelength/Å		Intensity	Wavelength/Å	
90	2593.73	II	80	2744.527	I	500	2999.512	I	80	3246.005	I
650	2598.37	II	300	2746.48	II	120	3000.451	I	80	3265.046	I
2000	2599.40	II	100	2749.32	II	800	3000.948	I	50	3265.617	I
300	2599.57	I	500	2749.48	II	12	3001.62	III	13	3266.88	III
60	2605.657	I	1200	2750.140	I	60	3001.655	I	50	3271.000	I
300	2606.51	II	80	2753.29	II	12 h	3007.28	III	11	3276.08	III
800	2606.827	I	150	2754.032	I	200	3007.282	I	150	3286.75	I
650	2607.09	II	100	2754.426	I	500	3008.14	I	10	3288.81	III
600	2611.87	II	800	2755.73	II	120	3009.569	I	120	3305.97	I
320	2613.82	II	250	2756.328	I	15	3013.17	III	200	3306.343	I
320	2617.62	II	100	2757.316	I	60	3017.627	I	400	3355.227	I
250	2618.018	I	120	2761.780	I	60	3018.983	I	80	3355.517	I
90	2620.41	II	150	2761.81	II	500	3020.491	I	60	3369.546	I
400	2623.53	I	150	2762.026	I	1500	3020.639	I	120	3370.783	I
200	2625.67	I	120	2762.772	I	600	3021.073	I	50	3378.678	I
150	2628.29	II	120	2763.109	I	500	3024.032	I	50	3380.110	I
250	2631.05	II	80	2766.910	I	150	3025.638	I	60	3383.978	I
250	2631.32	II	250	2767.522	I	500	3025.842	I	50	3392.304	I
100	2632.237	I	300	2772.07	I	80	3030.148	I	150	3392.651	I
300	2635.809	I	600	2778.220	I	60	3031.214	I	150	3399.333	I
50	2641.646	I	3000	2788.10	I	60	3034.484	I	80	3404.353	I
200	2643.998	I	200	2797.78	I	800	3037.389	I	500	3407.458	I
300	2666.812	I	400	2804.521	I	80	3041.637	I	250	3413.131	I
60	2666.965	I	1500	2806.98	I	800	3047.604	I	60	3424.284	I
600	2679.062	I	10 p	2813.24	III	600	3057.446	I	500	3427.119	I
500	2684.75	II	2500	2813.287	I	1000	3059.086	I	60	3428.748	I
400	2689.212	I	300	2823.276	I	250	3067.244	I	6000	3440.606	I
10 h	2695.13	III	600	2825.56	I	120	3075.719	I	2500	3440.989	I
200	2699.106	I	50	2825.687	I	120	3091.577	I	1000	3443.876	I
80	2706.012	I	120	2828.808	I	80	3098.189	I	200	3445.149	I
400	2706.582	I	1500	2832.436	I	100	3099.895	I	1200	3465.860	I
60	2708.571	I	120	2835.950	I	100	3099.968	I	2000	3475.450	I
200	2711.655	I	200	2838.119	I	60	3100.303	I	500	3476.702	I
80	2714.41	II	200	2843.631	I	100	3100.665	I	2500	3490.574	I
50	2716.257	I	1000	2843.977	I	10 p	3136.43	III	500	3497.840	I
50	2717.786	I	100	2845.594	I	10	3174.09	III	10	3501.76	III
250	2718.436	I	800	2851.797	I	80	3175.445	I	250	3513.817	I
4000	2719.027	I	50	2869.307	I	10	3175.99	III	300	3521.261	I
100	2719.420	I	50	2872.334	I	10	3178.01	III	400	3526.040	I
50	2720.197	I	80	2874.172	I	150	3184.895	I	100	3526.166	I
1500	2720.903	I	50	2894.504	I	250	3191.659	I	60	3526.237	I
400	2723.578	I	12	2904.43	III	500	3193.226	I	60	3526.381	I
150	2724.953	I	10	2907.50	III	800	3193.299	I	60	3526.467	I
50	2726.235	I	12	2907.70	III	200	3196.928	I	100	3533.199	I
80	2727.54	II	120	2912.157	I	80	3199.500	I	200	3536.556	I
200	2728.020	I	120	2929.007	I	50	3205.398	I	300	3541.083	I
50	2728.820	I	1200	2936.903	I	100	3211.88	I	250	3542.075	I
80	2728.90	II	60	2941.343	I	200	3214.011	I	80	3553.739	I
1000	2733.581	I	1000	2947.876	I	200	3214.396	I	400	3554.925	I
60	2734.005	I	600	2953.940	I	60	3215.938	I	200	3556.878	I
50	2734.268	I	250	2957.364	I	50	3217.377	I	400	3558.515	I
500	2735.475	I	150	2965.254	I	80	3219.583	I	1000	3565.379	I
50	2735.612	I	1500	2966.898	I	60	3219.766	I	1200	3570.097	I
500	2737.310	I	120	2969.36	I	300	3222.045	I	800	3570.25	I
120	2737.83	I	800	2970.099	I	600	3225.78	I	120	3571.996	I
400	2739.55	II	1200	2973.132	I	80	3227.796	I	100	3573.393	I
250	2742.254	I	500	2973.235	I	50	3233.967	I	60	3573.829	I
800	2742.405	I	600	2981.445	I	120	3234.613	I	60	3573.888	I
200	2743.20	II	1000	2983.570	I	300	3236.222	I	4000	3581.19	I
150	2743.565	I	1000	2994.427	I	100	3239.433	I	150	3582.199	I
200	2744.068	I	250	2994.502	I	80	3244.187	I	150	3584.660	I

Intensity	Wavelength/Å		Intensity	Wavelength/Å		Intensity	Wavelength/Å		Intensity	Wavelength/Å	
120	3584.929	I	6000	3737.131	I	1200	3927.920	I	40	4195.329	I
300	3585.319	I	100	3738.306	I	2000	3930.296	I	150	4198.304	I
150	3585.705	I	400	3743.362	I	60	3948.774	I	40	4199.095	I
10	3586.04	III	6000	3745.561	I	60	3949.953	I	300	4202.029	I
200	3586.103	I	1200	3745.899	I	50	3951.164	I	40	4203.984	I
400	3586.984	I	3000	3748.262	I	50	3952.601	I	80	4206.696	I
100	3594.633	I	80	3748.964	I	16	3954.33	III	80	4210.343	I
11	3600.94	III	3000	3749.485	I	60	3956.454	I	400	4216.183	I
150	3603.204	I	1500	3758.232	I	250	3956.68	I	100	4219.360	I
11	3603.88	III	400	3760.05	I	60	3966.614	I	50	4222.212	I
200	3605.454	I	1500	3763.788	I	11	3968.72	III	11	4222.27	III
500	3606.680	I	400	3765.54	I	100	3969.257	I	50	4225.956	I
1500	3608.859	I	600	3767.191	I	80	3977.741	I	200	4227.423	I
250	3610.16	I	60	3776.452	I	10 w	3979.42	III	100	4233.602	I
60	3612.068	I	250	3785.95	I	40	3981.771	I	13	4235.56	III
150	3617.788	I	100	3786.68	I	50	3983.956	I	250	4235.936	I
1500	3618.768	I	250	3787.880	I	60	3994.114	I	50	4238.809	I
200	3621.462	I	250	3790.092	I	200	3997.392	I	12	4243.75	III
150	3622.004	I	150	3794.34	I	40	3998.053	I	50	4247.425	I
150	3623.19	I	400	3795.002	I	400	4005.241	I	200	4250.118	I
100	3631.096	I	120	3797.518	I	60	4009.713	I	300	4250.787	I
1200	3631.463	I	250	3798.511	I	100	4021.867	I	40	4258.315	I
60	3632.041	I	400	3799.547	I	10	4035.42	III	800	4260.473	I
100	3638.298	I	200	3805.345	I	50	4040.638	I	250	4271.153	I
200	3640.389	I	80	3806.696	I	4000	4045.813	I	1200	4271.759	I
80	3643.717	I	600	3812.964	I	11	4053.11	III	12 h	4273.40	III
1500	3647.842	I	60	3813.059	I	1500	4063.594	I	12	4279.72	III
250	3649.506	I	1500	3815.840	I	50	4066.975	I	1200	4282.402	I
80	3650.279	I	2500	3820.425	I	50	4067.977	I	14 h	4286.16	III
200	3651.467	I	150	3821.179	I	1200	4071.737	I	80	4291.462	I
120	3670.024	I	80	3824.306	I	40	4076.629	I	16 h	4296.85	III
150	3670.089	I	2500	3824.444	I	12	4081.00	III	250	4299.234	I
100	3676.311	I	1500	3825.880	I	40	4100.737	I	18 h	4304.78	III
150	3677.629	I	1200	3827.823	I	40	4107.489	I	1200	4307.901	I
1500	3679.913	I	1000	3834.222	I	150	4118.544	I	20 h	4310.36	III
200	3682.242	I	120	3839.257	I	10	4120.90	III	150	4315.084	I
120	3683.054	I	500	3840.437	I	11	4122.02	III	1500	4325.761	I
150	3684.107	I	800	3841.047	I	11	4122.78	III	80	4352.734	I
120	3685.998	I	120	3843.256	I	40	4127.608	I	80	4369.771	I
500	3687.456	I	80	3846.800	I	400	4132.058	I	11 h	4372.31	III
120	3689.477	I	200	3849.96	I	80	4134.676	I	14 h	4372.53	III
150	3694.008	I	120	3850.817	I	40	4136.997	I	18 h	4372.81	III
120	3695.051	I	2500	3856.372	I	15	4137.76	III	800	4375.929	I
150	3701.086	I	150	3859.212	I	13	4139.35	III	3000	4383.544	I
80	3704.462	I	10000	3859.911	I	200	4143.415	I	1200	4404.750	I
1200	3705.566	I	150	3865.523	I	800	4143.869	I	300	4415.122	I
60	3707.041	I	60	3867.215	I	40	4153.898	I	12	4419.60	III
150	3707.821	I	250	3872.501	I	50	4154.500	I	600	4427.299	I
300	3707.919	I	150	3873.761	I	60	4156.799	I	400	4461.652	I
600	3709.246	I	250	3878.018	I	18	4164.73	III	120	4466.551	I
120	3716.442	I	2000	3878.573	I	13	4166.84	III	80	4476.017	I
8000	3719.935	I	4000	3886.282	I	50	4172.744	I	80	4482.169	I
1500	3722.563	I	200	3887.048	I	13	4174.26	III	200	4482.252	I
120	3724.377	I	300	3888.513	I	60	4174.912	I	50	4489.739	I
60	3725.491	I	800	3895.656	I	50	4175.635	I	50	4528.613	I
60	3727.093	I	1200	3899.707	I	50	4177.593	I	30	4647.433	I
500	3727.619	I	400	3902.945	I	120	4181.754	I	30	4736.771	I
150	3732.396	I	250	3906.479	I	50	4184.891	I	50	4859.741	I
1200	3733.317	I	80	3916.731	I	120	4187.038	I	120	4871.317	I
5000	3734.864	I	600	3920.258	I	120	4187.795	I	60	4872.136	I
120	3735.324	I	1200	3922.911	I	80	4191.430	I	30	4878.208	I

Intensity	Wavelength/Å		Intensity	Wavelength/Å		Intensity	Wavelength/Å		Intensity	Wavelength/Å	
100	4890.754	I	10	5368.06	III	80	7187.313	I	50	625.02	III
250	4891.492	I	400	5371.489	I	30	7207.381	I	30	625.76	III
30	4903.309	I	11 l	5375.47	III	30	7445.746	I	45	628.59	III
150	4918.992	I	40	5393.167	I	40	7495.059	I	50	630.04	III
500	4920.502	I	300	5397.127	I	60	7511.045	I	35	633.09	III
1500	4957.597	I	250	5405.774	I	80	7937.131	I	120	637.87	V
80	5001.862	I	250	5429.695	I	60	7945.984	I	50	639.98	III
30	5005.711	I	100	5434.523	I	80	7998.939	I	60	646.41	III
100	5006.117	I	200	5446.871	I	60	8046.047	I	50	651.20	III
60	5012.067	I	120	5455.609	I	50	8085.176	I	50	659.72	III
30	5014.941	I	25	5497.516	I	150	8220.41	I	30	664.86	III
150	5041.755	I	20	5501.464	I	120	8327.053	I	40	672.34	III
30	5049.819	I	30	5506.778	I	20	8331.908	I	35	672.85	III
30	5051.634	I	30	5569.618	I	120	8387.770	I	35	676.57	III
25	5074.748	I	60	5572.841	I	30	8468.404	I	35	680.13	III
150	5110.357	I	120	5586.755	I	15	8514.069	I	35	683.68	III
40	5139.251	I	200	5615.644	I	60	8661.898	I	35	686.25	III
100	5139.462	I	20	5624.541	I	150	8688.621	I	45	687.98	III
25	5151.910	I	50	5662.515	I	52	11422.32	I		690.86	V
12	5156.12	III	11	5719.88	III	87	11439.12	I		691.75	V
80	5166.281	I	10	5756.38	III	91	11593.59	I	45	691.93	III
2500	5167.487	I	20	5762.990	I	255	11607.57	I	50	695.61	III
80	5168.897	I	18	5833.93	III	160	11638.26	I	30	698.05	III
500	5171.595	I	10	5854.62	III	230	11689.98	I	50	708.36	III
50	5191.454	I	30	5862.353	I	160	11783.26	I	600	708.85	V
80	5192.343	I	15	5891.91	III	580	11882.84	I	50	714.00	III
200	5194.941	I	30	5914.114	I	225	11884.08	I	100 p	722.04	III
10	5199.08	III	10 p	5920.13	III	1030	11973.05	I	60	729.40	II
30	5204.582	I	18 p	5929.69	III	96	14400.56	I	30	746.70	III
25	5215.179	I	10	5952.31	III	72	14512.23	I	200	761.18	II
150	5216.274	I	14	5953.62	III	50	14555.06	I	100	763.98	II
60	5226.862	I	12	5979.32	III	40	14826.43	I	60	766.20	II
1000	5227.150	I	30	5986.956	I	94	15294.58	I	200	771.03	II
250	5232.939	I	12 h	5989.08	III	41	15769.42	I	60 p	773.69	II
10	5235.66	III	18	5999.54	III	105	18856.65	I	200	782.10	II
18	5243.31	III	16	6032.59	III				100	783.72	II
13 l	5260.34	III	13	6036.56	III	***Krypton Kr*** $Z = 36$			60	785.97	III
100	5266.555	I	11	6048.72	III	30	467.35	III		793.44	IV
1200	5269.537	I	11	6054.18	III	150	472.16	V		794.11	IV
800	5270.357	I	40	6065.482	I	100	484.39	V	7	805.76	IV
14	5272.98	III	30	6102.159	I	250	496.25	V		810.70	V
15	5276.48	III	40	6136.614	I	120	500.77	V	18	816.82	IV
30	5281.789	I	40	6137.694	I	200	507.20	V	60	818.15	II
16	5282.30	III	40	6191.558	I	30	540.86	III	60	830.38	II
60	5283.621	I	30	6213.429	I	60	548.04	V	50	837.66	III
12	5284.83	III	30	6219.279	I	30	565.64	III	22	842.04	IV
11	5298.12	III	40	6230.726	I	30	569.16	III	100	844.06	II
12	5299.93	III	20	6246.317	I	30	571.98	III	50	854.73	III
25	5302.299	I	80	6247.56	II	30	579.83	III	60	862.58	III
14 w	5302.60	III	30	6252.554	I	30	585.14	III	60	864.82	II
10	5306.76	III	20	6393.602	I	30	585.96	III	60	868.87	II
10	5322.74	III	30	6399.999	I	30	593.70	III	40	870.84	III
150	5324.178	I	20	6411.647	I	30	594.10	III	50	876.08	III
800	5328.038	I	20	6421.349	I	30	596.41	III	200	884.14	II
300	5328.531	I	30	6430.844	I	40	600.17	III	1000	886.30	II
100	5332.899	I	200	6456.38	II	30	603.67	III	400	891.01	II
80	5339.928	I	60	6494.981	I	50	605.86	III	75	897.81	III
500	5341.023	I	20	6546.239	I	35	606.47	III	200	911.39	II
11	5346.88	III	20	6592.913	I	50	611.12	III	2000	917.43	II
12	5353.77	III	40	6677.989	I	35	616.72	III	50	945.44	I
12	5363.76	III	25	7164.443	I	40	621.45	III	50	946.54	I
						45	622.80	III			

Intensity	Wavelength/Å		Intensity	Wavelength/Å		Intensity	Wavelength/Å		Intensity	Wavelength/Å	
20	951.06	I	60	2992.22	III	300	4057.037	II	200	5468.17	II
50	953.40	I	50	3022.30	III	300	4065.128	II	10	5501.43	III
50	963.37	I	80	3024.45	III	50	4067.37	III	500	5562.224	I
2000	964.97	II	50	3046.93	III	500	4088.337	II	2000	5570.288	I
50	987.29	III	30	3056.72	III	250	4098.729	II	80	5580.386	I
100	1001.06	I	60	3063.13	III	100	4109.248	II	100	5649.561	I
100	1003.55	I	40	3097.16	III	40	4131.33	III	400	5681.89	II
100	1030.02	I	60	3112.25	III	250	4145.122	II	200 h	5690.35	II
30	1158.74	III	30	3120.61	III	40	4154.46	III	100	5832.855	I
200	1164.87	I	100	3124.39	III	150	4250.580	II	3000	5870.914	I
650	1235.84	I	60	3141.35	III	1000	4273.969	I	200	5992.22	II
6	1638.82	III	3	3142.01	IV	100	4282.967	I	60	5993.849	I
6	1914.09	III	100	3189.11	III	600	4292.923	II	10 h	6037.17	III
3	2237.34	IV	80	3191.21	III	200	4300.49	II	60	6056.125	I
6	2291.26	IV	6	3224.99	IV	500 h	4317.81	II	10 h	6078.38	III
3	2329.3	IV	40	3239.52	III	400	4318.551	I	10	6310.22	III
4	2336.75	IV	40	3240.44	III	1000	4319.579	I	300	6420.18	II
4	2348.27	IV	300	3245.69	III	150 h	4322.98	II	100	6421.026	I
3	2358.5	IV	3	3261.70	IV	100	4351.359	I	200	6456.288	I
3	2388.05	IV	150	3264.81	III	3000	4355.477	II	150	6570.07	II
40	2393.94	III	100	3268.48	III	500	4362.641	I	60	6699.228	I
4	2416.9	IV	30	3271.65	III	200	4369.69	II	100	6904.678	I
3	2428.04	IV	30	3285.89	III	800	4376.121	I	250	7213.13	II
5	2442.68	IV	30	3304.75	III	300 h	4386.54	II	100	7224.104	I
4	2451.7	IV	50	3311.47	III	200	4399.965	I	80	7287.258	I
6	2459.74	IV	200	3325.75	III	100	4425.189	I	400	7289.78	II
100 h	2464.77	II	60	3330.76	III	500	4431.685	II	400	7407.02	II
5	2474.06	IV	50	3342.48	III	600	4436.812	II	60	7425.541	I
60	2492.48	II	100	3351.93	III	600	4453.917	I	200	7435.78	II
40	2494.01	III	40	3374.96	III	800	4463.689	I	100	7486.862	I
4	2517.0	IV	100	3439.46	III	800	4475.014	II	300	7524.46	II
5	2518.02	IV	70	3474.65	III	400 h	4489.88	II	1000	7587.411	I
6	2519.38	IV	100	3488.59	III	600	4502.353	I	2000	7601.544	I
5	2524.5	IV	200	3507.42	III	400 h	4523.14	II	150	7641.16	II
5	2546.0	IV	100	3564.23	III	200 h	4556.61	II	1000	7685.244	I
6	2547.0	IV	100 h	3607.88	II	800	4577.209	II	1200	7694.538	I
4	2558.08	IV	200	3631.889	II	300	4582.978	II	250	7735.69	II
30	2563.25	III	30	3641.34	III	150 h	4592.80	II	150	7746.827	I
3	2586.9	IV	250	3653.928	II	500	4615.292	II	800	7854.821	I
5	2606.17	IV	80	3665.324	I	1000	4619.166	II	200	7913.423	I
10	2609.5	IV	150	3669.01	II	800	4633.885	II	180	7928.597	I
8	2615.3	IV	100	3679.559	I	2000	4658.876	II	200	7933.22	II
7	2621.11	IV	80	3686.182	II	500	4680.406	II	120	7973.62	II
60	2639.76	III	30	3690.65	III	100	4691.301	II	100	7982.401	I
30	2680.32	III	300 h	3718.02	II	200	4694.360	II	1500	8059.503	I
40	2681.19	III	200	3718.595	II	3000	4739.002	II	4000	8104.364	I
80 h	2712.40	II	150	3721.350	II	300	4762.435	II	6000	8112.899	I
3	2730.55	IV	200	3741.638	II	1000	4765.744	II	60	8132.967	I
8	2748.18	IV	150	3744.80	II	300	4811.76	II	3000	8190.054	I
6	2774.70	IV	80	3754.245	II	300	4825.18	II	200	8202.72	II
3	2829.60	IV	500	3778.089	II	800	4832.077	II	80	8218.365	I
100	2833.00	II	500	3783.095	II	700	4846.612	II	3000	8263.240	I
3	2836.08	IV	3	3809.30	IV	150	4857.20	II	100	8272.353	I
30	2841.00	III	5	3860.58	IV	300	4945.59	II	1500	8281.050	I
30	2851.16	III	40 h	3868.70	III	20 h	5016.45	III	5000	8298.107	I
5	2853.0	IV	150 h	3875.44	II	200	5022.40	II	100	8412.430	I
3	2859.3	IV	150	3906.177	II	250	5086.52	II	3000	8508.870	I
50	2870.61	III	200	3920.081	II	400 h	5125.73	II	150	8764.110	I
100	2892.18	III	5	3934.29	IV	500	5208.32	II	6000	8776.748	I
30	2909.17	III	100	3994.840	II	200	5308.66	II	2000	8928.692	I
50	2952.56	III	100 h	3997.793	II	500	5333.41	II	500	9238.48	II

Intensity	Wavelength/Å	
500 hl	9293.82	II
200 h	9320.99	II
300	9361.95	II
100	9362.082	I
200 h	9402.82	II
200 h	9470.93	II
500	9577.52	II
500 h	9605.80	II
400 h	9619.61	II
200	9663.34	II
200 h	9711.60	II
2000	9751.758	I
500	9803.14	II
500	9856.314	I
1000	10221.46	II
100	11187.108	I
200	11257.711	I
150	11259.126	I
500	11457.481	I
150	11792.425	I
1500	11819.377	I
600	11997.105	I
160	12077.224	I
100	12861.892	I
1100	13177.412	I
1000	13622.415	I
2400	13634.220	I
800	13658.394	I
200	13711.036	I
600	13738.851	I
150	13974.027	I
550	14045.657	I
140	14104.298	I
180	14402.22	I
2000	14426.793	I
100	14517.84	I
1600	14734.436	I
550	14762.672	I
450	14765.472	I
400	14961.894	I
120	15005.307	I
140	15209.526	I
1700	15239.615	I
130	15326.480	I
1500	15334.958	I
700	15372.037	I
200	15474.026	I
180	15681.02	I
120	15820.09	I
200	16726.513	I
2000	16785.128	I
1000	16853.488	I
2400	16890.441	I
1600	16896.753	I
1800	16935.806	I
600	17098.771	I
700	17367.606	I
120	17404.443	I
150	17616.854	I
650	17842.737	I
700	18002.229	I

Intensity	Wavelength/Å	
2600	18167.315	I
100	18399.786	I
150	18580.896	I
300	18696.294	I
170	18785.460	I
200	18797.703	I
140	20209.878	I
300	20423.964	I
140	20446.971	I
600	21165.471	I
1800	21902.513	I
120	22485.775	I
180	23340.416	I
120	24260.506	I
180	24292.221	I
600	25233.820	I
180	28610.55	I
1000	28655.72	I
150	28769.71	I
140	28822.49	I
300	29236.69	I
300	30663.54	I
300	30979.16	I
500	39300.6	I
1100	39486.52	I
220	39557.25	I
100	39572.60	I
1400	39588.4	I
1100	39589.6	I
500	39954.8	I
300	39966.6	I
1300	40306.1	I
250	40685.16	I

Lanthanum La Z = 57

Intensity	Wavelength/Å	
100	344.12	IV
400	390.72	V
1000	432.11	V
2500	435.28	V
10000	463.14	IV
5000	482.16	V
7000	498.08	V
15000	499.54	IV
10000	503.58	V
12000	526.76	V
10000	531.07	V
15000	533.23	V
8000	547.44	V
40000	552.02	IV
5000	600.24	V
30000	631.26	IV
400	796.99	III
2000	870.40	III
1000	882.34	III
400	942.86	III
50000	1081.61	III
95000	1099.73	III
2000	1255.63	III
10000	1349.18	III
25000	1368.04	IV
20000	1463.47	IV
15000	1507.87	IV

Intensity	Wavelength/Å	
10000	1523.79	III
4000	1808.66	IV
5000	1902.97	IV
4000 c	2197.45	IV
770	2256.76	II
25000 w	2417.58	IV
50000	2532.75	IV
45000	2582.05	IV
95000 w	2597.50	IV
70000 w	2662.75	IV
420	2808.39	II
50000 w	2848.30	IV
30000 c	2962.58	IV
70000 w	3009.51	IV
90000 c	3056.68	IV
1000	3171.63	III
1500	3171.74	III
510	3245.13	II
550	3265.67	II
800	3303.11	II
1500	3337.49	II
870	3344.56	II
1500	3380.91	II
320	3628.83	II
1000	3645.42	II
550	3713.54	II
2400	3759.08	II
3700	3790.83	II
3900	3794.78	II
600	3840.72	II
1600	3849.02	II
3400	3871.64	II
1700	3886.37	II
1300	3916.05	II
1100	3921.54	II
2200	3929.22	II
9000	3949.10	II
4400	3988.52	II
3600	3995.75	II
2800	4031.69	II
3000	4042.91	II
850	4067.39	II
2800	4077.35	II
5500	4086.72	II
4400	4123.23	II
550	4141.74	II
1100	4151.97	II
1500	4196.55	II
1600	4238.38	II
480	4269.50	II
600	4286.97	II
600	4296.05	II
440	4322.51	II
4600	4333.74	II
550	4354.40	II
2000	4429.90	II
850	4522.37	II
420	4526.12	II
400	4558.46	II
400	4574.88	II
410	4613.39	II

Intensity	Wavelength/Å	
410	4619.88	II
540	4655.50	II
360	4662.51	II
230	4692.50	II
230	4728.42	II
500	4740.28	II
390	4743.09	II
320	4748.73	II
320	4860.91	II
850	4899.92	II
1000	4920.98	II
1000	4921.79	II
370	4949.77	I
340	4970.39	II
370	4986.83	II
720	4999.47	II
210	5050.57	I
470	5114.56	II
470	5122.99	II
450	5145.42	I
290	5158.69	I
580	5177.31	I
850	5183.42	II
260	5188.22	II
720	5211.86	I
520	5234.27	I
340	5253.46	I
370	5271.19	I
370	5301.98	II
180	5303.55	II
500	5455.15	I
470	5501.34	I
240	5648.25	I
180	5740.66	I
370	5769.34	I
320	5789.24	I
450	5791.34	I
140	5821.99	I
320	5930.62	I
720	6249.93	I
260 d	6262.30	II
450	6394.23	I
250	6455.99	I
180	6709.50	I
110	7045.96	I
160	7066.23	II
50	7161.25	I
110 w	7282.34	II
110 w	7334.18	I
75 cw	7483.50	II
50	7498.83	I
85	7539.23	I
40	7964.83	I
75	8086.05	I
85	8324.69	I
95	8346.53	I
65	8545.44	I
300	8583.45	III
40	8674.43	I
35	8825.82	I
120	9184.38	III

Intensity	Wavelength/Å	
100	9212.63	III
140	10284.79	III

Lead Pb Z = 82

Intensity	Wavelength/Å	
10	496.38	IV
12	499.94	IV
14	529.78	IV
20	570.16	IV
10	648.50	IV
20	703.73	V
12	749.46	V
10	752.52	V
10	761.09	IV
18	767.45	V
18	769.49	V
14	771.42	V
14	782.79	V
15	797.02	V
18	802.07	IV
12	802.82	IV
18	809.63	V
10	812.59	IV
10	827.41	IV
12	832.60	IV
12	845.94	IV
18	857.64	IV
16	862.33	IV
20	863.97	V
14	870.44	IV
6	873.71	II
12	879.96	IV
18	883.90	V
14	884.96	IV
14	884.99	IV
14	888.37	V
8	889.68	II
16	890.72	IV
14	894.40	V
12	896.08	V
12	908.51	IV
14	915.71	V
12	917.90	IV
12	918.09	V
12	920.28	V
12	920.66	V
10	922.12	IV
12	922.49	IV
10	927.64	IV
14	932.20	IV
12	954.35	V
10	967.23	II
10	986.71	II
10	995.89	II
10	1016.61	II
14	1028.61	IV
20	1032.05	IV
16	1041.24	IV
18	1044.14	IV
12	1048.9	III
10	1049.82	II
10	1050.77	II
10	1051.26	V

Intensity	Wavelength/Å	
15	1056.53	IV
10	1060.66	II
12	1072.09	IV
18	1080.81	IV
20	1084.17	IV
10	1088.86	V
10	1103.94	II
10	1108.43	II
10	1109.84	II
20	1116.08	IV
10	1119.57	II
10	1121.36	II
10	1133.14	II
18	1137.84	IV
14	1144.93	IV
12	1157.88	V
14	1185.43	V
20	1189.95	IV
10	1203.63	II
10	1231.20	II
11	1233.50	V
10	1291.10	IV
20	1313.05	IV
10	1331.65	II
10	1335.20	II
12	1343.06	IV
10	1348.37	II
16	1388.94	IV
18	1400.26	IV
10	1404.34	IV
10	1433.96	II
10	1512.42	II
14	1535.71	IV
20	1553.1	III
10	1671.53	II
10	1682.15	II
20	1726.75	II
10	1796.670	II
10	1822.050	II
10	1904.77	I
7	1921.471	II
12	1959.34	IV
16	1973.16	IV
10	1998.83	V
5 r	2022.02	I
10	2042.58	IV
12	2049.34	IV
8 r	2053.28	I
12	2079.22	IV
6	2111.758	I
10	2115.066	I
15	2154.01	IV
500 r	2170.00	I
7	2175.580	I
12	2177.46	IV
7	2187.888	I
8	2189.603	I
10	2203.534	II
20	2237.425	I
20	2246.86	I
25	2246.89	I

Intensity	Wavelength/Å	
20	2259.01	V
150	2332.418	I
16	2359.53	IV
180	2388.797	I
550 r	2393.792	I
140	2399.597	I
320 r	2401.940	I
320 r	2411.734	I
16	2417.61	IV
15	2424.81	V
150 r	2443.829	I
160 r	2446.181	I
130 r	2476.378	I
80 r	2577.260	I
500 r	2613.655	I
900 r	2614.175	I
160	2628.262	I
4	2634.256	II
10	2657.094	I
700	2663.154	I
10	2697.541	I
25000 r	2801.995	I
100	2822.58	I
14000 r	2823.189	I
35000 r	2833.053	I
6	2840.557	II
14000 r	2873.311	I
3	2914.442	II
15	2966.460	I
15	2972.991	I
15	2980.157	I
4	2986.876	II
10	3043.85	III
150	3118.894	I
10	3137.81	III
10	3176.50	III
600	3220.528	I
100	3229.613	I
400	3240.186	I
200	3262.355	I
35000	3572.729	I
50000 r	3639.568	I
20000	3671.491	I
70000 r	3683.462	I
10	3713.982	II
25000	3739.935	I
12	3854.08	III
15000	4019.632	I
95000	4057.807	I
14000	4062.136	I
10	4157.814	I
10000	4168.033	I
8	4272.66	III
200	4340.413	I
10	4496.15	IV
6	4499.34	III
16	4534.60	IV
7	4571.21	III
10	4579.051	II
6	4761.12	III
1000	5005.416	I

Intensity	Wavelength/Å	
100	5006.572	I
50	5089.484	I
10	5107.242	I
2000	5201.437	I
10	5372.099	II
40	5692.346	I
200	5895.624	I
2000	6001.862	I
500	6011.667	I
500	6059.356	I
40	6081.409	II
50	6110.520	I
100	6235.266	I
50 c	6660.20	II
20000	7228.965	I
10	7346.676	I
20	7809.259	I
5	7896.737	I
10	8168.001	I
6	8191.886	I
5	8217.711	I
40	8272.690	I
20	8409.384	I
10	8478.492	I
5	8722.810	I
10	8857.457	I
8	9293.476	I
15	9438.05	I
15	9604.297	I
15	9674.351	I
200	10290.458	I
100	10498.965	I
50	10649.249	I
15	10886.688	I
40	10969.53	I
	13512.6	I
	14743.0	I
	15349.6	I
	39039.4	I

Lithium Li Z = 3

Intensity	Wavelength/Å	
	102.9	III
	103.4	III
	104.1	III
	105.5	III
	108.0	III
	113.9	III
	125.5	II
	135.0	III
	136.5	II
	140.5	II
	167.21	II
	168.74	II
	171.58	II
	178.02	II
	199.28	II
	207.5	II
	456.	II
	483.	II
	540.	II
	540.0	III
	729.	II

Intensity	Wavelength/Å		Intensity	Wavelength/Å		Intensity	Wavelength/Å		Intensity	Wavelength/Å	
	729.1	III		2313.49	I	5	3199.33	II	48	8126.23	I
	800.	II		2315.08	I	2	3199.43	II	48	8126.45	I
	820.	II		2316.95	I	17	3232.66	I		8517.37	II
	861.	II		2319.18	I		3249.87	II		9581.42	II
	905.5	II		2321.88	I		3306.28	II		10120.	II
	917.5	II		2325.11	I		3488.	I		12232.	I
	936.	II		2329.02	I		3579.8	I		12782.	I
	945.	II		2329.84	II		3618.	I		13566.	I
	965.	II		2333.94	I		3662.	I		17552.	I
	972.	II	3	2336.88	II		3684.32	II		18697.	I
	988.	II	5	2336.91	II	1	3714.00	II		19290.	I
	1018.	II	2	2337.00	II	5	3714.16	II		24467.	I
	1032.	II		2340.15	I	6 d	3714.27	II		40475.	I
	1036.	II		2348.22	I	8	3714.29	II			
	1093.	II		2358.93	I	7 d	3714.40	II	*Lutetium Lu Z = 71*		
	1103.	II		2373.54	I	10	3714.41	II	100	563.72	V
	1109.	II		2381.54	II	1	3714.51	II	500	810.73	III
	1116.	II		2383.20	II		3714.58	II	2000	832.28	III
	1132.1	II	1	2394.39	I	3	3718.7	I	100	861.92	V
	1141.	II		2402.33	II	6	3794.72	I	400	876.80	IV
	1166.4	II		2410.84	II	20	3915.30	I	100	880.32	V
	1198.09	II	3	2425.43	I	20	3915.35	I	100	891.81	V
	1215.	II		2429.81	II	10	3985.48	I	100	914.72	V
	1238.	II		2460.2	I	10	3985.54	I	400	1001.18	III
	1253.8	II	10	2475.06	I	40	4132.56	I	100	1272.42	IV
	1420.89	II		2506.94	II	40	4132.62	I	800	1333.79	IV
	1424.	II		2508.78	II		4196.	I	400	1429.38	IV
3	1492.93	II		2518.	I	20	4273.07	I	200	1441.76	V
5	1492.97	II		2539.49	II	20	4273.13	I	200	1453.35	V
1	1493.04	II	24	2551.7	II	5	4325.42	II	200	1468.99	V
	1555.	II		2559.	II	5	4325.47	II	400	1472.12	V
3	1653.08	II	15	2562.31	I	1	4325.54	II	200	1473.71	V
5	1653.13	II		2605.08	II		4516.45	II	200	1485.58	V
1	1653.21	II	2	2657.29	II	13	4602.83	I	400	1511.26	IV
	1681.66	II	3	2657.30	II	13	4602.89	I	600	1772.57	IV
	1755.33	II		2674.46	II		4607.34	II	100 c	1786.25	V
	2009.	II		2728.24	II		4671.51	II	1000	1854.57	III
	2039.	I	5	2728.29	II	6	4671.65	II	1500	2065.35	III
	2068.	II	2	2728.32	II	2	4671.70	II	1500 c	2070.56	III
	2131.	II	3	2730.47	II	3	4678.06	II	600 c	2086.47	IV
	2164.	II	1	2730.55	II	1	4678.29	II	1000 c	2104.41	IV
	2173.4	I	5	2741.20	I		4760.	I	1000 c	2108.31	IV
	2183.	II		2766.99	II		4763.	II	1700 h	2195.54	II
	2214.	II		2790.31	II		4788.36	II	1000	2236.14	III
	2222.	II		2801.	I		4843.0	II	2000	2236.22	III
	2237.	II		2846.	I	4	4881.32	II	95	2276.94	II
h	2249.21	II		2868.	I	4	4881.39	II	190	2297.41	II
	2286.82	II		2895.	I	1	4881.49	II	1300	2392.19	II
	2302.57	II	2	2934.02	II	8	4971.66	I	120	2399.14	II
	2303.33	II	2	2934.07	II	8	4971.75	I	80	2419.21	II
	2304.59	I	5	2934.12	II		5037.92	II	130	2459.64	II
	2304.92	I	1	2934.25	II		5271.	I	370	2536.95	II
	2305.36	I		2968.	I		5315.	I	930	2571.23	II
	2305.83	I	3	3029.12	II		5395.	I	1700	2578.79	II
	2306.29	I	3	3029.14	II		5440.	I	4500 c	2603.35	III
	2306.82	I		3144.	I	600 c	5483.55	II	1800	2613.40	II
	2307.44	I	3	3155.31	II	600 c	5485.65	II	18000	2615.42	II
	2308.97	I	4	3155.33	II	320	6103.54	I	1800	2619.26	II
	2309.88	I	1	3196.26	II	320	6103.65	I	2700	2657.80	II
	2310.94	I	9	3196.33	II	3600	6707.76	I	570 h	2685.08	I
	2312.11	I	4	3196.36	II	3600	6707.91	I	4200	2701.71	II
									180 d	2719.09	I

Intensity	Wavelength/Å		Intensity	Wavelength/Å		Intensity	Wavelength/Å		Intensity	Wavelength/Å	
480 h	2728.95	I	250	4281.03	I	3000	323.31	IV	1	2613.36	I
3600	2754.17	II	330 d	4295.97	I	30	353.09	V	2	2614.73	I
750 h	2765.74	I	150	4309.57	I	150	857.29	IV	3	2617.51	I
2000	2772.55	III	190 c	4430.48	I	50	919.03	IV	3	2628.66	I
2700	2796.63	II	190	4450.81	I	250	1037.41	IV	6	2630.05	I
270 c	2834.35	II	3300	4518.57	I	300	1210.99	IV	8	2632.87	I
330 h	2845.13	I	100 h	4648.21	I	300	1342.19	IV	2	2644.80	I
3000	2847.51	II	1000	4658.02	I	800	1346.57	IV	3	2646.21	I
570 h	2885.14	I	85 h	4659.03	I	300	1346.68	IV	4	2649.06	I
6300	2894.84	II	150	4785.42	II	600	1352.05	IV	8	2660.76	II
4500	2900.30	II	85	4815.05	I	900	1384.46	IV	8	2660.82	II
300	2903.05	I	460	4904.88	I	500	1385.77	IV	6	2668.12	I
9000	2911.39	II	180	4942.34	I	800	1387.53	IV	8	2669.55	I
270 h	2949.73	I	800	4994.13	II	300	1404.68	IV	10	2672.46	I
1200	2951.69	II	800	5001.14	I	1000	1409.36	IV	3	2693.72	I
4200	2963.32	II	140	5134.05	I	500	1437.53	IV	5	2695.18	I
2400	2969.82	II	2700	5135.09	I	1000	1437.64	IV	6	2698.14	I
1800	2989.27	I	170	5196.61	I	300	1447.42	IV	8	2731.99	I
3000	3020.54	II	500	5402.57	I	300	1459.54	IV	10	2733.49	I
2100	3056.72	II	140 c	5421.90	I	400	1459.62	IV	12	2736.53	I
1000	3057.86	III	100	5437.88	I	400	1481.51	IV	5	2765.22	I
7500	3077.60	II	2100	5476.69	II	350	1490.45	IV	7	2768.34	I
390	3080.11	I	550	5736.55	I	300	1495.50	IV	38	2776.69	I
5100 h	3081.47	I	80	5800.59	I	300	1607.11	IV	32	2778.27	I
3000	3118.43	I	690 cw	5983.9	II	5	1668.43	I	90	2779.83	I
2400	3171.36	I	140	5997.13	I	500	1683.02	IV	8	2781.29	I
260	3191.80	II	1400	6004.52	I	10	1683.41	I	32	2781.42	I
1400	3198.12	II	440	6055.03	I	400	1698.81	IV	36	2782.97	I
4800	3254.31	II	150	6159.94	II	15	1707.06	I	1000	2795.53	II
3800	3278.97	I	600	6198.13	III	40	1734.84	II	600	2802.70	II
7600	3281.74	I	160	6199.66	II	50	1737.62	II	3	2809.76	I
6200	3312.11	I	2100	6221.87	II	20	1747.80	I	2	2811.11	I
7600	3359.56	I	80	6235.36	II	40	1750.65	II	1	2811.78	I
6200	3376.50	I	160	6242.34	II	50	1753.46	II	12	2846.72	I
950	3385.50	I	70 h	6345.35	I	30	1827.93	I	12	2846.75	I
160 h	3391.55	I	1100	6463.12	II	300	1844.17	IV	14	2848.34	I
1400	3396.82	I	29	6477.67	I	9	2025.82	I	14	2848.42	I
4100	3397.07	II	55 c	6523.18	I	25	2064.90	III	16	2851.65	I
4800	3472.48	II	35 cw	6611.28	II	20	2091.96	III	16	2851.66	I
8300 c	3507.39	II	23 c	6677.14	I	20	2177.70	III	6000	2852.13	I
1600	3508.42	I	30 c	6793.77	I	3	2329.58	II	2	2902.92	I
4800	3554.43	II	45	6917.31	I	20	2395.15	III	4	2906.36	I
4800	3567.84	I	23	7031.24	I	6	2449.57	II	3	2915.45	I
340	3596.34	I	45	7125.84	II	1	2557.23	I	10	2936.74	I
800	3623.99	II	14 ch	7237.98	I	1	2560.94	I	12	2938.47	I
680	3636.25	I	11 c	7441.52	I	1	2562.26	I	2	2942.00	I
2600	3647.77	I	9 c	8178.16	I	1	2564.94	I	13	2942.00	I
110	3756.70	I	17	8382.08	I	1	2570.91	I	20	3091.08	I
110	3756.79	I	35	8459.19	II	1	2572.25	I	22	3092.99	I
150	3800.67	I	10 d	8478.50	I	2	2574.94	I	14	3096.90	I
2700	3841.18	I	29 c	8508.08	I	1	2577.89	I	9	3104.71	II
530	3876.65	II	35 c	8610.98	I	1	2580.59	I	8	3104.81	II
50	3918.86	I				1	2584.22	I	6	3168.98	II
480	3968.46	I	*Magnesium Mg Z = 12*			2	2585.56	I	6	3172.71	II
670	4054.45	I	400	146.95	IV	3	2588.28	I	7	3175.78	II
310	4122.49	I	20	186.51	III	1	2591.89	I	2	3197.62	I
3100	4124.73	I	20	187.20	III	1	2593.23	I	17	3329.93	I
150 c	4131.79	I	10	188.53	III	2	2595.97	I	6	3332.15	I
460	4154.08	I	100	231.73	III	2	2602.50	I	9	3336.68	I
1600	4184.25	II	80	234.26	III	4	2603.85	I	7	3535.04	II
150	4277.50	I	35	276.58	V	5	2606.62	I	8	3538.86	II
			4000	320.99	IV						

Intensity	Wavelength/Å		Intensity	Wavelength/Å		Intensity	Wavelength/Å		Intensity	Wavelength/Å	
7	3549.52	II	12	8712.69	I	90	1264.41	IV	500	2066.38	III
8	3553.37	II	13	8717.83	I	500	1283.58	III	1000	2069.02	III
140	3829.30	I	10	8734.99	II	400	1287.59	III	30	2076.21	II
300	3832.30	I	17	8736.02	I	300	1291.62	III	900	2077.38	III
500	3838.29	I	11	8745.66	II	1000	1360.72	III	800	2084.23	III
8	3848.24	II	14	8806.76	I	800	1365.20	III	600	2090.05	III
7	3850.40	II	10	8824.32	II	500 h	1609.17	III	1500	2092.16	I
3	3878.31	I	11	8835.08	II	1000	1614.14	III	500	2094.78	III
3	3895.57	I	20	8923.57	I	2000	1620.60	III	20	2097.46	II
4	3903.86	I	10	8997.16	I	500	1633.80	III	500	2097.93	III
6	3938.40	I	14	9218.25	II	80	1667.00	IV	500	2099.97	III
8	3986.75	I	13	9244.27	II	80	1698.30	IV	20	2102.50	II
10	4057.50	I	12	9246.50	I	20	1726.47	II	1700	2109.58	I
15	4167.27	I	30	9255.78	I	30	1732.70	II	30	2113.96	II
20	4351.91	I	10	9327.54	II	50	1733.55	II	1000	2169.78	III
9	4384.64	II	10	9340.54	II	40	1734.49	II	700	2174.15	III
10	4390.59	II	25	9414.96	I	30	1737.93	II	900	2176.87	III
8	4428.00	II	17	9429.81	I	20	1740.16	II	800	2181.86	III
9	4433.99	II	19	9432.76	I	20	1742.00	II	800	2184.87	III
14	4481.16	II	20	9438.78	I	85	1742.10	IV	290	2208.81	I
13	4481.33	II	12	9631.89	II	85	1766.27	IV	540	2213.85	I
28	4571.10	I	11	9632.43	II	80	1795.65	IV	900	2220.55	III
10	4730.03	I	15	9953.20	I	80	1795.79	IV	770	2221.84	I
7	4851.10	II	15	9983.20	I	30	1853.27	II	1000	2227.42	III
75	5167.33	I	17	9986.47	I	20	1857.92	II	20	2373.36	II
220	5172.68	I	18	9993.21	I	50	1902.95	II	20	2427.38	II
400	5183.61	I	14	10092.16	II	20	1907.84	II	50	2427.72	II
8	5264.21	II	35	10811.08	I	75	1910.25	IV	30	2427.94	II
7	5264.37	II	11	10914.23	II	30	1911.41	II	30	2437.37	II
9	5401.54	II	10	10951.78	II	20 d	1914.68	II	20	2437.84	II
6	5528.41	I	25	10953.32	I	100	1915.10	II	30	2452.49	II
30	5711.09	I	27	10957.30	I	20	1918.64	II	50	2499.00	II
10	6318.72	I	28	10965.45	I	30	1919.64	II	30	2507.60	II
9	6319.24	I	15	11032.10	I	80	1921.25	II	20	2516.60	II
7	6319.49	I	14	11033.66	I	20	1923.07	II	30	2516.74	II
10	6346.74	II	45	11828.18	I	20	1923.34	II	20	2521.66	II
9	6346.96	II	30	12083.66	I	30	1925.52	II	20	2530.72	II
11	6545.97	II	28	14877.62	I	50	1926.59	II	20	2531.80	II
7	6781.45	II	35	15024.99	I	30	1931.40	II	50	2532.78	II
8	6787.85	II	30	15040.24	I	500	1941.28	III	50	2533.33	II
7	6812.86	II	25	15047.70	I	800	1943.21	III	30	2534.10	II
8	6819.27	II	10	15765.84	I	20	1945.15	II	80	2534.22	III
10	7193.17	I	30	17108.66	I	20	1947.93	II	100	2535.66	II
10	7291.06	I	5	26392.90	I	20	1950.14	II	30	2535.98	II
12	7387.69	I				500	1952.36	III	100	2537.92	II
20	7657.60	I	**Manganese Mn Z = 25**			1000	1952.52	III	50	2541.11	II
19	7659.15	I	600	410.30	V	30	1953.23	II	80	2542.92	II
17	7659.90	I	600	410.60	V	20 d	1954.81	II	50	2543.45	II
15	7691.55	I	600	415.62	V	30	1959.25	II	100	2548.75	II
12	7877.05	II	650	415.98	V	20	1969.24	II	50	2551.85	II
13	7896.37	II	600	428.59	V	500	1978.95	III	30	2553.27	II
10	8098.72	I	600	435.67	V	30	1994.23	II	75	2556.57	II
9	8115.22	II	1000	441.72	V	9700	1996.06	I	30	2556.89	II
8	8120.43	II	850	442.49	V	14000	1999.51	I	95	2558.59	II
10	8209.84	I	60	579.79	IV	18000	2003.85	I	30	2559.41	II
20	8213.03	I	60	581.44	IV	1000 w	2027.83	III	150	2563.65	II
10	8213.99	II	60	581.65	IV	500 w	2028.14	III	30	2565.22	II
11	8234.64	II	60	585.21	IV	50	2037.31	II	580	2572.76	I
10	8310.26	I	90	1242.25	IV	40	2037.64	II	480	2575.51	I
15	8346.12	I	90	1244.50	IV	40	2039.97	II	12000	2576.10	II
10	8710.18	I	95	1251.93	IV	500	2049.68	III	550	2584.31	I
			95	1257.28	IV						

Intensity	Wavelength/Å		Intensity	Wavelength/Å		Intensity	Wavelength/Å		Intensity	Wavelength/Å	
30	2588.97	II	70	2886.68	II	290	3619.28	I	800	4451.59	I
45	2589.71	II	160	2889.58	II	220	3623.79	I	160	4453.00	I
250	2592.94	I	55	2892.39	II	140	3629.74	I	130	4455.01	I
6200	2593.73	II	50	2898.70	II	100	3660.40	I	160	4455.32	I
250	2595.76	I	80	2900.16	II	280	3693.67	I	110	4455.82	I
95	2598.90	II	140 h	2914.60	I	180	3696.57	I	210	4457.55	I
30	2602.72	II	190 h	2925.57	I	210	3706.08	I	270	4458.26	I
45	2603.72	II	1100	2933.06	II	130	3718.93	I	150	4461.08	I
4300	2605.69	II	1500	2939.30	II	130	3731.93	I	510	4462.02	I
190	2610.20	II	250 h	2940.39	I	260	3790.22	I	290	4464.68	I
500	2618.14	II	1900	2949.20	II	110	3800.55	I	200	4470.14	I
140	2622.90	I	30	3019.92	II	3200	3806.72	I	130	4472.79	I
150	2624.04	I	55	3031.06	II	700	3809.59	I	170	4490.08	I
40	2624.80	II	30	3035.35	II	2100	3823.51	I	240	4498.90	I
200	2625.58	II	330	3044.57	I	390	3823.89	I	240	4502.22	I
190	2632.35	II	120	3045.59	I	200	3829.68	I	160	4709.72	I
130	2638.17	II	200	3047.04	I	480	3833.86	I	180	4727.48	I
80	2639.84	II	30	3050.65	II	1300	3834.36	I	130	4739.11	I
27	2650.99	II	250	3054.36	I	350	3839.78	I	1000	4754.04	I
60	2655.91	II	140	3062.12	I	670	3841.08	I	180	4761.53	I
30	2666.77	II	170	3066.02	I	350	3843.98	I	750	4762.38	I
30	2667.03	II	170	3070.27	I	120	3926.47	I	300	4765.86	I
110	2672.59	II	160	3073.13	I	130	3982.58	I	500	4766.43	I
55	2673.37	II	140 h	3178.50	I	150	3985.24	I	940	4783.42	I
55	2674.43	II	220	3212.88	I	190	3986.83	I	1000	4823.52	I
45	2680.34	II	1000	3228.09	I	150	3987.10	I	19	5004.91	I
30	2680.68	II	300	3230.72	I	1500	4018.10	I	30	5074.79	I
30	2681.25	II	850	3236.78	I	150	4026.44	I	200	5079.20	III
55	2684.55	II	330	3243.78	I	27000	4030.76	I	150	5100.03	III
55	2685.94	II	650	3248.52	I	19000	4033.07	I	60	5117.94	I
110	2688.25	II	100	3251.14	I	11000	4034.49	I	50	5150.89	I
27	2693.19	II	310	3252.95	I	1500	4035.73	I	50	5196.59	I
55	2695.36	II	310	3256.14	I	5600	4041.36	I	85	5255.32	I
27	2698.97	II	220	3258.41	I	210 d	4045.13	I	160	5341.06	I
85	2701.00	II	180	3260.23	I	1100	4048.76	I	19	5349.88	I
50	2701.17	II	180	3264.71	I	150	4055.21	I	95	5377.63	I
160	2701.70	II	200	3330.78	II	1900	4055.54	I	95	5394.67	I
100	2703.98	II	720	3441.99	II	210	4057.95	I	50	5399.49	I
130	2705.74	II	50	3460.03	II	1100	4058.93	I	95	5407.42	I
80	2707.53	II	360	3460.33	II	150	4059.39	I	35	5413.69	I
110	2708.45	II	360 h	3474.04	II	730	4061.74	I	85	5420.36	I
45	2709.96	II		3474.13	II	730	4063.53	I	35	5432.55	I
80	2710.33	II	290	3482.91	II	290	4070.28	I	150	5454.07	III
110	2711.58	II	180	3488.68	II	730	4079.24	I	12	5457.47	I
30	2716.80	II	140	3495.84	II	730	4079.42	I	60	5470.64	I
30	2717.53	II	50	3496.81	II	1100	4082.94	I	200	5474.68	III
30	2719.01	II	100	3497.54	II	1100	4083.63	I	40	5481.40	I
50	2719.74	II	360	3531.85	I	200	4110.90	I	30	5505.87	I
30	2722.10	II	1100	3532.12	I	150	4131.12	I	50	5516.77	I
30	2724.46	II	1300	3547.80	I	120	4135.04	I	40	5537.76	I
55	2728.61	II	1100	3548.03	I	150	4176.60	I	21	5551.98	I
6200	2794.82	I	390	3548.20	I	120	4189.99	I	200	5946.65	III
5100	2798.27	I	2200	3569.49	I	370	4235.14	I	140	6013.50	I
220	2799.84	I	720	3569.80	I	510	4235.29	I	200	6016.64	I
3700	2801.06	I	1400	3577.88	I	190	4239.72	I	290	6021.80	I
110	2809.11	I	720	3586.54	I	290	4257.66	I	200	6231.21	III
60	2815.02	II	290	3595.12	I	290	4265.92	I	17	6440.97	I
30	2816.33	II	150	3601.72	III	270	4281.10	I	24	6491.71	I
60	2870.08	II	420	3607.54	I	50	4323.63	II	14 h	6942.52	I
30	2872.94	II	420	3608.49	I	350	4414.88	I	12	6989.96	I
80	2879.49	II	360	3610.30	I	210	4436.35	I	14	7069.84	I

Intensity	Wavelength/Å		Intensity	Wavelength/Å		Intensity	Wavelength/Å		Intensity	Wavelength/Å	
12	7184.25	I	30	3801.660	I	40	1783.70	II	50	2698.83	I
24 h	7283.82	I	20	3901.867	I	30	1796.22	II	50	2699.38	I
35 h	7302.89	I	60	3906.372	I	200	1796.90	II	80	2705.36	II
50	7326.51	I	200	3983.839	II	60	1798.74	II	70	2724.43	III
12	7680.20	I	1800	4046.572	I	30	1803.89	II	80	2752.78	I
12 h	8672.06	I	150	4077.838	I	40	1808.29	II	20	2759.71	I
12 h	8701.05	I	40	4108.057	I	400	1820.34	II	6	2769.22	III
17 h	8703.76	I	250	4339.224	I	5	1832.74	I	40	2803.46	I
30 h	8740.93	I	400	4347.496	I	1000	1849.50	I	30	2804.43	I
			4000	4358.337	I	160	1869.23	II	2	2805.34	I
Mercury 198 Hg Z = 80			80	4916.068	I	300	1870.55	II	2	2806.77	I
80	1250.564	I	1100	5460.753	I	200	1875.54	II	150	2814.93	II
8	1259.242	I	160	5675.922	I	1	1894.77	III	3	2844.76	III
100	1268.825	I	240	5769.598	I	20	1900.28	II	750	2847.68	II
5	1307.751	I	280	5790.663	I	30	1927.60	II	50	2856.94	I
20	1402.619	I	20	6072.713	I	300	1942.27	II	150	2893.60	I
10	1435.503	I	30	6234.402	I	100	1972.94	II	150	2916.27	II
1000	1849.492	I	160	6716.429	I	200	1973.89	II	60	2925.41	I
60	2262.210	II	250	6907.461	I	150	1987.98	II	400	2947.08	II
20	2302.065	I	240	11287.407	I	90	2026.97	II	1200	2967.28	I
20	2345.440	I				90	2052.93	II	300	3021.50	I
100	2378.325	I	*Mercury Hg Z = 80*			70	2148.00	II	120	3023.47	I
20	2380.004	I	3	621.44	III	5	2247.55	I	30	3025.61	I
40	2399.349	I	2	679.68	III	60	2262.23	II	50	3027.49	I
20	2399.729	I	2	878.59	III	20	2302.06	I	15	3090.05	III
20	2446.900	I	1	886.48	III	7	2314.15	III	400	3125.67	I
15	2464.064	I	400	893.08	II	15	2323.20	I	320	3131.55	I
40	2481.999	I	300	915.83	II	5	2340.57	I	320	3131.84	I
30	2482.713	I	150	923.39	II	20	2345.43	I	400	3208.20	II
40	2483.821	I	200	940.80	II	20	2352.48	I	400	3264.06	II
90	2534.769	I	100	962.74	II	100	2378.32	I	5	3283.02	II
15000	2536.506	I	50	969.13	II	20	2380.00	I	12	3312.28	III
25	2563.861	I	1	988.89	III	4	2380.55	III	80	3341.48	I
25	2576.290	I	2	1009.29	III	40	2399.38	I	100	3385.25	II
250	2652.043	I	5	1068.03	III	20	2399.73	I	8	3389.01	III
400	2653.683	I	800	1099.26	II	10	2400.49	I	5	3450.77	III
100	2655.130	I	2	1161.95	III	60	2407.35	II	400	3451.69	II
50	2698.831	I	80	1250.58	I	50	2414.13	II	3	3500.35	III
80	2752.783	I	8	1259.24	I	8	2431.65	III	4	3538.88	III
20	2759.710	I	100	1268.82	I	5	2441.06	I	200	3549.42	II
40	2803.471	I	5	1307.75	I	20	2446.90	I	5	3557.24	III
30	2804.438	I	300	1307.93	II	15	2464.06	I	2800	3650.15	I
750	2847.675	II	400	1321.71	II	5	2480.56	III	300	3654.84	I
50	2856.939	I	400	1331.74	II	40	2482.00	I	80	3662.88	I
150	2893.598	I	80	1350.07	II	30	2482.72	I	240	3663.28	I
150	2916.227	II	200	1361.27	II	40	2483.82	I	30	3701.44	I
60	2925.413	I	20	1402.62	I	7	2484.50	III	35	3704.17	I
1200	2967.283	I	200	1414.43	II	90	2534.77	I	30	3801.66	I
300	3021.500	I	10	1435.51	I	15000	2536.52	I	15	3803.51	III
120	3023.476	I	15	1619.46	II	25	2563.86	I	100	3806.38	II
30	3025.608	I	120	1623.95	II	25	2576.29	I	20	3901.87	I
50	3027.490	I	20	1628.25	II	5	2578.91	I	60	3906.37	I
400	3125.670	I	150	1649.94	II	2	2612.92	III	100	3918.92	II
320	3131.551	I	50	1653.64	II	4	2617.97	III	200	3983.96	II
320	3131.842	I	200	1672.41	II	15	2625.19	I	1800	4046.56	I
80	3341.481	I	9	1681.40	III	5	2639.78	I	150	4077.83	I
2800	3650.157	I	100	1702.73	II	250	2652.04	I	40	4108.05	I
300	3654.839	I	100	1707.40	II	400	2653.69	I	70	4122.07	III
80	3662.883	I	120	1727.18	II	100	2655.13	I	10	4140.34	III
240	3663.281	I	250	1732.14	II	3	2670.49	III	100	4216.74	III
30	3701.432	I	15	1759.75	III	5	2674.91	I			
35	3704.170	I	20	1775.68	I						

Intensity	Wavelength/Å		Intensity	Wavelength/Å		Intensity	Wavelength/Å		Intensity	Wavelength/Å	
250	4339.22	I	120	13209.95	I	150	2269.71	III	80	2834.39	II
400	4347.49	I	140	13426.57	I	200	2294.97	III	80	2835.33	II
4000	4358.33	I	60	13468.38	I	160	2304.25	II	160	2842.15	II
100	4398.62	II	80	13505.58	I	160	2306.97	II	1700	2848.23	II
15	4470.58	III	500	13570.21	I	150	2330.93	III	370	2853.23	II
12	4552.84	III	450	13673.51	I	110	2332.12	II	370	2863.81	II
90	4660.28	II	200	13950.55	I	190	2341.59	II	220	2866.69	II
50	4797.01	III	500	15295.82	I	100	2359.76	III	1700	2871.51	II
80	4855.72	II	100	16881.48	I	110	2389.20	II	85	2872.88	II
10	4869.85	III	400	16920.16	I	140	2403.61	II	220	2879.05	II
5	4883.00	I	300	16942.00	I	120	2413.01	II	65	2888.15	II
5	4889.91	I	500	17072.79	I	85	2498.28	II	1300	2890.99	II
80	4916.07	I	400	17109.93	I	200	2506.19	III	95	2891.28	II
5	4970.37	I	20	17116.75	I	440	2538.46	II	190	2892.81	II
80	4973.57	III	20	17198.67	I	330	2542.67	II	950	2894.45	II
5	4980.64	I	20	17213.20	I	80	2558.88	II	140	2897.63	II
20	5102.70	I	70	17329.41	I	85	2564.34	II	70	2900.80	II
40	5120.64	I	30	17436.18	I	250	2593.70	II	290	2903.07	II
100	5128.45	II	50	18130.38	I	250	2602.80	II	80	2907.12	II
20	5137.94	I	40	19700.17	I	400	2616.78	I	600	2909.12	II
30	5210.82	III		22493.28	I	440	2629.85	I	1100	2911.92	II
20	5290.74	I	250	23253.07	I	330	2636.67	II	120	2918.83	II
5	5316.78	I		32148.06	I	720	2638.76	II	1300	2923.39	II
60	5354.05	I		36303.03	I	410	2640.99	I	140	2924.32	II
30	5384.63	I				600	2644.35	II	1100	2930.50	II
1100	5460.74	I	*Molybdenum Mo Z = 42*			370	2646.49	II	800	2934.30	II
30	5549.63	I	50	867.92	IV	640	2649.46	I	95	2940.10	II
160	5675.86	I	100	884.19	IV	480	2653.35	II	110	2941.22	II
6	5695.71	III	60	886.05	IV	560 h	2655.03	I	150	2944.82	II
240	5769.60	I	50	891.74	IV	640	2660.58	II	140	2946.69	II
100	5789.66	I	100	1169.33	III	720	2672.84	II	95	2947.28	II
280	5790.66	I	100	1254.93	III	250	2673.27	II	125	2947.32	III
140	5803.78	I	100	1258.52	III	1000	2679.85	I	95	2955.84	II
60	5859.25	I	100	1262.21	III	95	2681.36	II	240	2956.06	II
60	5871.73	II	100	1263.74	III	640	2683.23	II	70	2956.90	II
20	5871.98	I	100	1274.37	III	880	2684.14	II	95	2960.24	II
20	6072.72	I	100	1276.40	III	560	2687.99	II	250	2963.79	II
1000	6149.50	II	200	1277.40	III	480	2701.42	II	210	2965.27	II
25	6220.35	III	200	1277.58	III	190	2713.51	II	70	2971.91	II
30	6234.40	I	200	1278.40	III	290	2717.35	II	250	2972.61	II
35	6418.98	III	150	1281.90	III	85	2726.97	II	80	2975.40	II
40	6501.38	III	150	1283.60	III	140	2729.68	III	95	2992.84	II
80	6521.13	II	100	1854.73	III	80	2730.20	II	95	3027.77	II
10	6584.26	III	80	1926.26	IV	330	2732.88	II	100	3060.78	II
6	6610.12	III	100	1929.24	IV	160	2736.96	II	800	3064.28	I
30	6709.29	III	80	1971.06	IV	80 h	2737.88	II	250	3065.04	II
160	6716.43	I	70	2010.92	IV	290	2746.30	II	800	3074.37	I
250	6907.52	I	19000	2015.11	II	110	2756.07	II	85	3077.66	II
250	7081.90	I	40000	2020.30	II	220	2763.62	II	800	3085.62	I
200	7091.86	I	21000	2038.44	II	240	2769.76	II	270	3087.62	II
40	7346.37	II	17000	2045.98	II	160	2773.78	II	190	3092.07	II
100	7485.87	II	50	2060.38	IV	190	2774.39	II	560	3094.66	I
12	7517.46	III	4800	2081.68	II	1700	2775.40	II	560	3101.34	I
20	7728.82	I	2400	2089.52	II	65	2777.86	II	1400	3112.12	I
7	7808.10	III	2200	2092.50	II	880	2780.04	II	290	3122.00	II
100	7944.66	II	4000	2093.11	II	400	2784.99	II	14000	3132.59	I
25	7946.75	III	2700	2100.84	II	100	2807.74	III	110	3138.72	II
50	7984.51	III	1500	2104.29	II	400	2807.76	II	220	3152.82	II
5	8151.64	III	1400	2108.02	II	1700	2816.15	II	55	3155.64	II
2000	10139.75	I	100	2184.37	III	220	2817.44	II	6000	3158.16	I
240	11287.40	I	100	2211.02	III	80	2827.74	II	8700	3170.35	I
			400	2269.69	II						

Intensity	Wavelength/Å		Intensity	Wavelength/Å		Intensity	Wavelength/Å		Intensity	Wavelength/Å	
95	3172.03	II	580	3886.82	I	65	5279.65	I	95	3007.97	II
160	3172.74	II	19000	3902.96	I	210	5280.86	I	95	3014.19	II
120 d	3187.59	II	65	3941.48	II	55	5292.08	I	95	3018.35	II
7600	3193.97	I	1400	4062.08	I	55	5295.47	I	140	3056.71	II
880	3205.88	I	2300	4069.88	I	55	5313.89	I	130	3069.73	II
3000	3208.83	I	1300	4081.44	I	80	5354.88	I	160	3075.38	II
560	3215.07	I	940	4084.38	I	65	5356.48	I	240	3092.92	II
880	3228.22	I	730	4107.47	I	560 hl	5360.56	I	260	3115.18	II
600	3229.79	I	630	4120.10	I	110 hl	5364.28	I	290	3133.60	II
1100	3233.14	I	2900	4143.55	I	65	5394.52	I	220	3134.90	II
950	3237.08	I	480	4185.82	I	50	5400.47	I	170	3141.46	II
65	3240.71	II	2500	4188.32	I	55	5435.68	I	170	3142.44	II
950	3256.21	I	1500	4232.59	I	65	5437.75	I	150	3203.47	II
480	3264.40	I	890	4276.91	I	50	5501.54	I	220	3259.24	II
800	3270.90	I	1200	4277.24	I	7800	5506.49	I	220	3265.12	II
200	3271.69	III	1400	4288.64	I	5200	5533.05	I	320	3275.22	II
1100	3289.02	I	680	4292.13	I	50	5543.12	I	290	3285.10	II
950	3290.82	I	890	4293.21	I	55	5556.28	I	410	3328.28	II
190	3292.31	II	840	4326.14	I	2500	5570.45	I	320	3353.59	II
100	3313.62	II	1900	4381.64	I	100	5610.93	I	410	3560.75	II
190	3320.90	II	2500	4411.57	I	330	5632.47	I	470	3587.51	II
640	3323.95	I	990	4434.95	I	50	5634.86	I	370	3615.82	II
1300	3344.75	I	480	4457.36	I	230	5650.13	I	410	3653.15	II
95	3346.40	II	630	4474.56	I	55	5674.47	I	470	3662.26	II
1600	3358.12	I	400	4536.80	I	460	5689.14	I	540	3665.18	II
950	3363.78	I	460	4626.47	I	80	5705.72	I	540	3672.36	II
950	3379.97	I	640	4707.26	I	210	5722.74	I	580	3673.54	II
1900 d	3384.62	I	700	4731.44	I	620	5751.40	I	1200	3685.80	II
130	3395.36	II	770	4760.19	I	520	5791.85	I	440	3687.30	II
640	3404.34	I	410	4819.25	I	55 h	5849.73	I	410	3689.69	II
1300	3405.94	I	410	4830.51	I	50 h	5851.52	I	410	3697.56	II
640	3437.22	I	180	5014.60	I	520	5858.27	I	470	3713.70	II
130	3446.08	II	80	5029.00	I	50	5869.33	I	640 d	3714.73	II
3200	3447.12	I	65	5030.78	I	820	5888.33	I	470	3715.68	II
640	3449.07	I	100	5047.71	I	50 h	5893.38	I	410	3718.54	II
950	3456.39	I	50	5055.00	I	160 h	5928.88	I	410	3721.35	II
640	3460.78	I	200	5059.88	I	35	6025.49	I	780	3723.50	II
800	3504.41	I	100	5080.02	I	1300	6030.66	I	410	3724.87	II
560	3508.12	I	100	5096.65	I	40	6101.87	I	710	3728.13	II
480	3521.41	I	130	5097.52	I	40	6357.22	I	470	3730.58	II
640	3537.28	I	130	5109.71	I	35	6401.07	I	1000 d	3735.54	II
520	3558.10	I	80	5114.97	I	100	6424.37	I	440	3737.10	II
400	3563.14	I	150	5145.38	I	230	6619.13	I	1000	3738.06	II
1400	3581.89	I	110	5147.39	I	50	6650.38	I	580	3752.49	II
1400	3624.46	I	80	5163.19	I	110	6733.98	I	510	3757.82	II
1000	3635.43	I	100	5167.76	I	50	6746.27	I	930	3758.95	II
400	3657.35	I	160 d	5171.08	I	35	6753.97	I	930	3763.47	II
540	3664.81	I	230 h	5172.94	I	40	6838.88	I	510	3769.65	II
590	3672.82	I	160 h	5174.18	I	35	6914.01	I	1400	3775.50	II
1300	3680.60	I	110	5200.17	I	110	7109.87	I	710	3779.47	II
65	3688.31	II	50	5200.74	I	150	7242.50	I	580	3780.40	II
180	3692.64	II	50	5211.86	I	40	7245.85	I	510	3781.32	II
1400	3694.94	I	80	5219.40	I	40	7391.36	I	2400	3784.25	II
500	3727.69	I	65	5231.06	I	140	7485.74	I	370	3801.12	II
80	3744.37	II	100	5234.26	I	27	7720.77	I	1200	3803.47	II
29000	3798.25	I	460 h	5238.20	I	40 h	8328.44	I	2500	3805.36	II
520	3826.70	I	230 h	5240.88	I	45 h	8389.32	I	470	3807.23	II
940	3828.87	I	110 h	5242.81	I	45 h	8483.39	I	540	3808.77	II
1700	3833.75	I	100	5245.51	I				440	3809.06	II
29000	3864.11	I	150	5259.04	I				580	3810.49	II
580	3869.08	I	65	5261.14	I				710	3814.73	II

Neodymium Nd Z = 60

Intensity	Wavelength/Å	
75	2764.98	I
80	2993.20	II

Intensity	Wavelength/Å		Intensity	Wavelength/Å		Intensity	Wavelength/Å		Intensity	Wavelength/Å	
410	3822.47	II	1000	4021.78	II	240	4719.02	I	20	7192.01	II
1200	3826.42	II	1200	4023.00	II	240	4811.34	II	15	7236.54	II
540	3828.85	II	410	4030.47	II	350	4825.48	II	12	7316.81	II
440	3829.16	II	1200	4031.82	II	280	4859.02	II	10	7406.62	II
510	3830.47	II	3000	4040.80	II	350	4883.81	I	10	7418.18	II
740	3836.54	II	410	4043.59	II	220	4890.70	II	12	7511.16	II
1700	3838.98	II	410	4048.81	II	240	4891.07	I	17	7513.73	II
410 d	3841.82	II	850	4051.15	II	280	4896.93	I	12	7528.99	II
1700 d	3848.24	II	850	4059.96	II	210	4901.84	I	10	7538.26	II
1500	3848.52	II	4700	4061.09	II	330	4920.68	II	12	7696.56	II
470	3850.22	II	1100	4069.28	II	470	4924.53	I	10	7750.95	II
2400 d	3851.66	II	710	4075.12	II	260	4944.83	I	10	7808.47	II
3700 d	3863.33	II	470	4075.28	II	290	4954.78	I	12	7863.04	II
850	3869.07	II	470	4080.23	II	290	4959.13	II	12	7917.01	II
470	3875.87	II	1400	4109.08	II	250	4989.94	II	12	7958.95	I
1100	3878.58	II	2500	4109.46	II	360	5076.59	II	12	7965.73	II
1000	3879.55	II	510	4110.48	II	360	5092.80	II	15	7982.09	II
780	3880.38	II	410	4123.88	II	360	5107.59	II	12	7982.68	II
1200	3880.78	II	470	4133.36	II	340	5123.79	II	12	8000.76	II
540	3887.87	II	510	4135.33	II	680	5130.60	II	10	8120.93	II
1300	3889.93	II	3000	4156.08	II	500	5191.45	II	12	8122.07	II
1300	3890.58	II	510	4156.26	II	630	5192.62	II	12	8141.75	II
1300	3890.94	II	410	4168.00	II	330	5200.12	II	12	8143.27	II
580	3891.51	II	810	4175.61	II	310	5212.37	II	10	8231.52	II
470	3892.06	II	2400	4177.32	II	450	5234.20	II	10	8307.72	II
810	3894.63	II	640	4179.59	II	250	5239.79	II	12	8346.36	II
440	3897.63	II	470	4205.60	II	720	5249.59	II	17	8839.10	II
2000	3900.21	II	470	4211.29	II	360	5255.51	II			
1300	3901.84	II	440	4227.73	II	590	5273.43	II	*Neon Ne Z = 10*		
1700	3905.89	II	1300	4232.38	II	680	5293.17	II	66	119.01	V
510	3907.84	II	2000	4247.38	II	220	5311.46	II	200	122.52	V
2000	3911.16	II	850	4252.44	II	500	5319.82	II	66	125.12	V
850	3912.23	II	410	4261.84	II	290	5361.47	II	45	131.99	V
440	3915.13	II	470	4282.44	II	160	5431.53	II	50	132.04	V
610	3915.95	II	710	4284.52	II	240	5594.43	II	150	140.76	V
1100	3920.96	II	5400	4303.58	II	220	5620.54	I	150	140.79	V
510	3927.10	II	470	4314.52	II	140 d	5675.97	I	100	142.44	V
610	3934.82	II	1100	4325.76	II	220	5688.53	II	100	142.50	V
410	3936.11	II	510	4327.93	II	130	5702.24	II	150	142.72	V
510	3938.86	II	540	4338.70	II	160	5708.28	II	100	143.27	V
2000	3941.51	II	680	4351.29	II	100	5729.29	I	150	143.34	V
2000	3951.16	II	850	4358.17	II	160	5804.02	II	150	147.13	V
810	3952.20	II	470 d	4374.93	II	80	5811.57	II	66	151.23	V
590	3958.00	II	710	4385.66	II	70	5825.87	II	120	151.42	V
510	3962.21	II	540	4400.83	II	80	5842.39	II	15	151.82	IV
1400	3963.12	II	510	4411.06	II	55	5858.91	I	15	152.23	IV
1100	3973.30	II	580	4446.39	II	45	6007.67	I	45	154.50	V
740	3973.69	II	1400	4451.57	II	45	6034.24	II	15	158.65	IV
740	3976.85	II	740	4462.99	II	55	6066.03	I	15	158.82	IV
740	3979.49	II	410	4501.82	II	45	6178.59	I	100	164.02	V
470	3986.25	II	250	4516.36	II	45	6223.39	I	100	164.14	V
1400	3990.10	II	340	4541.27	II	55	6310.49	I	80	172.62	IV
1000	3991.74	II	340	4542.61	II	65	6385.20	II	500	173.93	V
1100	3994.68	II	340	4563.22	II	45	6630.14	I	80	177.16	IV
410	4000.50	II	300	4621.94	I	45	6650.57	II	150	186.58	IV
540	4004.02	II	510	4634.24	I	40	6740.11	II	100	194.28	IV
410	4007.43	II	340	4641.10	I	40	6900.43	II	100	208.48	IV
3700	4012.25	II	250	4645.77	II	35	7037.30	II	100	208.73	IV
540	4012.70	II	300	4649.67	I	40	7066.89	II	80	208.90	IV
1000	4020.87	II	310	4683.45	I	29	7129.35	II	150	212.56	IV
1000	4021.34	II	470	4706.54	II	24	7189.42	II	140	223.24	IV
									120	223.60	IV

Intensity	Wavelength/Å		Intensity	Wavelength/Å		Intensity	Wavelength/Å		Intensity	Wavelength/Å	
140	234.32	IV	100	542.07	IV	300	2216.07	III	150	2963.24	II
120	234.70	IV	150	543.89	IV	10	2220.81	IV	150	2967.18	II
20	251.14	III	400	568.42	V	75	2227.42	V	100	2973.10	II
20	251.56	III	250	569.76	V	110	2232.41	V	15	2974.72	I
20	251.73	III	500	569.83	V	65	2245.48	V	100	2979.46	II
40	267.06	III	250	572.11	V	250	2258.02	IV	12	2982.67	I
40	267.52	III	800	572.34	V	65	2259.57	V	150	3001.67	II
20	267.71	III	35	587.213	I	175	2262.08	IV	120 p	3017.31	II
40	283.18	III	35	589.179	I	240	2263.21	III	300	3027.02	II
160	283.21	III	35	589.911	I	65	2263.39	V	300	3028.86	II
110	283.69	III	70	591.830	I	110	2264.54	IV	100	3030.79	II
40	283.89	III	100	595.920	I	200	2264.91	III	120	3034.46	II
220	301.12	III	75	598.706	I	250	2265.71	V	100	3035.92	II
220	313.05	III	35	598.891	I	550	2285.79	IV	100	3037.72	II
220	313.68	III	70	600.036	I	30	2293.14	IV	100	3039.59	II
40	313.95	III	170	602.726	I	250	2293.49	IV	100	3044.09	II
90	352.956	I	170	615.628	I	250	2350.84	IV	100	3045.56	II
60	354.962	I	170	618.672	I	450	2352.52	IV	120	3047.56	II
50	357.83	IV	120	619.102	I	700	2357.96	IV	100	3054.34	II
400	357.96	V	200	626.823	I	250	2362.68	IV	100	3054.68	II
500	358.47	V	200	629.739	I	250	2363.28	IV	100	3059.11	II
200	358.72	IV	1000	735.896	I	110	2365.49	IV	100	3062.49	II
500	359.38	V	400	743.720	I	350	2372.16	IV	100	3063.30	II
90	361.433	II	60	993.88	I	65	2384.20	IV	100	3070.89	II
60	362.455	II	70	1068.65	I	350	2384.95	IV	100	3071.53	II
1000	365.59	V	90	1131.72	I	300	2412.73	III	100	3075.73	II
220	379.31	III	100	1131.85	II	240	2412.94	III	120	3088.17	II
125	387.14	IV	90	1229.83	I	200	2413.78	III	100	3092.09	II
100	388.22	IV	20	1255.03	III	200	2473.40	III	120	3092.90	II
150	405.854	II	110	1255.68	III	80 p	2562.12	II	100	3094.01	II
120	407.138	II	160	1257.19	III	90 w	2567.12	II	100	3095.10	II
800	416.20	V	90	1418.38	I	800	2590.04	III	100	3097.13	II
150	421.61	IV	90	1428.58	I	600	2593.60	III	100	3117.98	II
200	445.040	II	90	1436.09	I	400	2595.68	III	120	3118.16	II
300	446.256	II	120	1681.68	II	300	2610.03	III	10	3126.199	I
250	446.590	II	180	1688.36	II	240	2613.41	III	300	3141.33	II
180	447.815	II	100	1888.11	II	200	2615.87	III	100	3143.72	II
150	454.654	II	100	1889.71	II	80	2623.11	II	100 p	3148.68	II
200	455.274	II	200	1907.49	II	80	2629.89	II	100	3164.43	II
10	456.275	II	500	1916.08	II	90 w	2636.07	II	100	3165.65	II
120	456.348	II	300	1930.03	II	80	2638.29	II	100	3188.74	II
90	456.896	II	200	1938.83	II	200	2638.70	III	120	3194.58	II
1000	460.728	II	100 c	1945.46	II	200	2641.07	III	500	3198.59	II
500	462.391	II	80	2007.01	II	80	2644.10	II	60	3208.96	II
140	469.77	IV	65	2018.44	IV	600	2677.90	III	120	3209.36	II
200	469.82	IV	110	2022.19	IV	500	2678.64	III	120	3213.74	II
180	469.87	IV	80	2025.56	II	80	2762.92	II	150	3214.33	II
140	469.92	IV	150	2085.47	II	90	2792.02	II	150	3218.19	II
250	480.41	V	200	2086.96	III	80	2794.22	II	120	3224.82	II
150	481.28	V	300	2089.43	III	100	2809.48	II	120	3229.57	II
250	481.36	V	240	2092.44	III	80	2906.59	II	200	3230.07	II
500	482.99	V	400	2095.54	III	80	2906.82	II	120	3230.42	II
285	488.10	III	180	2096.11	II	90	2910.06	II	120	3232.02	II
220	488.87	III	120	2096.25	II	90	2910.41	II	150	3232.37	II
450	489.50	III	200	2161.22	III	80	2911.14	II	100	3243.40	II
70	489.64	III	300	2163.77	III	80	2915.12	II	100	3244.10	II
220	490.31	III	200	2180.89	III	80	2925.62	II	100	3248.34	II
360	491.05	III	30	2203.88	IV	80 w	2932.10	II	100	3250.36	II
120	521.74	IV	200	2209.35	III	80	2940.65	II	150	3297.73	II
140	521.82	IV	200	2211.85	III	90	2946.04	II	150	3309.74	II
80	541.13	IV	240	2213.76	III	150	2955.72	II	300	3319.72	II

Intensity	Wavelength/Å		Intensity	Wavelength/Å		Intensity	Wavelength/Å		Intensity	Wavelength/Å	
1000	3323.74	II	250	3727.11	II	100	6074.338	I	1000	8571.352	I
150	3327.15	II	800	3766.26	II	80	6096.163	I	4000	8591.259	I
100	3329.16	II	1000	3777.13	II	60	6128.450	I	6000	8634.647	I
200	3334.84	II	100	3818.43	II	100	6143.063	I	3000	8647.041	I
150	3344.40	II	120	3829.75	II	120	6163.594	I	15000	8654.383	I
300	3345.45	II	150	4219.74	II	250	6182.146	I	4000	8655.522	I
150	3345.83	II	100	4233.85	II	150	6217.281	I	100	8668.26	II
200	3355.02	II	120	4250.65	II	150	6266.495	I	5000	8679.492	I
120	3357.82	II	120	4369.86	II	60	6304.789	I	5000	8681.921	I
200	3360.60	II	70	4379.40	II	100	6334.428	I	2000	8704.112	I
120	3362.16	II	150	4379.55	II	120	6382.992	I	4000	8771.656	I
100	3362.71	II	100	4385.06	II	200	6402.246	I	12000	8780.621	I
120	3367.22	II	200	4391.99	II	150	6506.528	I	10000	8783.753	I
12	3369.808	I	150	4397.99	II	60	6532.882	I	500	8830.907	I
40	3369.908	I	150	4409.30	II	150	6598.953	I	7000	8853.867	I
100	3371.80	II	100	4413.22	II	70	6652.093	I	1000	8865.306	I
500	3378.22	II	100	4421.39	II	90	6678.276	I	1000	8865.755	I
150	3388.42	II	100 p	4428.52	II	20	6717.043	I	3000	8919.501	I
120	3388.94	II	100 p	4428.63	II	100	6929.467	I	2000	8988.57	I
300	3392.80	II	150 p	4430.90	II	90	7024.050	I	100	9079.46	II
100	3404.82	II	150 p	4430.94	II	100	7032.413	I	6000	9148.67	I
120	3406.95	II	120	4457.05	II	50	7051.292	I	6000	9201.76	I
100	3413.15	II	100	4522.72	II	80	7059.107	I	4000	9220.06	I
120	3416.91	II	10	4537.754	I	100	7173.938	I	2000	9221.58	I
120	3417.69	II	10	4540.380	I	150	7213.20	II	2000	9226.69	I
50	3417.904	I	100	4569.06	II	150	7235.19	II	1000	9275.52	I
15	3418.006	I	15	4704.395	I	100	7245.167	I	200	9287.56	II
120	3428.69	II	12	4708.862	I	150	7343.94	II	6000	9300.85	I
60	3447.703	I	10	4710.067	I	40	7472.439	I	1500	9310.58	I
50	3454.195	I	10	4712.066	I	90	7488.871	I	3000	9313.97	I
100	3456.61	II	15	4715.347	I	100	7492.10	II	6000	9326.51	I
100	3459.32	II	10	4752.732	I	150	7522.82	II	2000	9373.31	I
25	3460.524	I	12	4788.927	I	80	7535.774	I	5000	9425.38	I
30	3464.339	I	10	4790.22	I	60	7544.044	I	3000	9459.21	I
30	3466.579	I	10	4827.344	I	100	7724.628	I	5000	9486.68	I
60	3472.571	I	10	4884.917	I	120	7740.74	II	5000	9534.16	I
150	3479.52	II	4	5005.159	I	300	7839.055	I	3000	9547.40	I
200	3480.72	II	10	5037.751	I	120	7926.20	II	120	9577.01	II
200	3481.93	II	10	5144.938	I	400	7927.118	I	1000	9665.42	I
25	3498.064	I	25	5330.778	I	700	7936.996	I	100	9808.86	II
30	3501.216	I	20	5341.094	I	2000	7943.181	I	800	10295.42	I
25	3515.191	I	8	5343.283	II	2000	8082.458	I	2000	10562.41	I
150	3520.472	I	60	5400.562	I	100	8084.34	II	1500	10798.07	I
120	3542.85	II	5	5562.766	I	1000	8118.549	I	2000	10844.48	I
120	3557.80	II	10	5656.659	I	600	8128.911	I	3000	11143.020	I
100	3561.20	II	5	5719.225	I	3000	8136.406	I	3500	11177.528	I
250	3568.50	II	12	5748.298	I	2500	8259.379	I	1600	11390.434	I
100	3574.18	II	80	5764.419	I	100	8264.81	II	1100	11409.134	I
200	3574.61	II	12	5804.450	I	2500	8266.077	I	3000	11522.746	I
50	3593.526	I	40	5820.156	I	800	8267.117	I	1500	11525.020	I
30	3593.640	I	500	5852.488	I	6000	8300.326	I	950	11536.344	I
15	3600.169	I	100	5872.828	I	100	8315.00	II	500	11601.537	I
20	3633.665	I	100	5881.895	I	1500	8365.749	I	1200	11614.081	I
150	3643.93	II	60	5902.462	I	100	8372.11	II	300	11688.002	I
200	3664.07	II	60	5906.429	I	8000	8377.606	I	2000	11766.792	I
20	3682.243	I	100	5944.834	I	1000	8417.159	I	1500	11789.044	I
12	3685.736	I	100	5965.471	I	4000	8418.427	I	500	11789.889	I
200	3694.21	II	120	5975.534	I	1500	8463.358	I	1000	11984.912	I
10	3701.225	I	80	5987.907	I	800	8484.444	I	3000	12066.334	I
150	3709.62	II	100	6029.997	I	5000	8495.360	I	800	12459.389	I
250	3713.08	II				600	8544.696	I	1000	12689.201	I

Intensity	Wavelength/Å		Intensity	Wavelength/Å		Intensity	Wavelength/Å		Intensity	Wavelength/Å	
1100	12912.014	I	300 l	3986.89	I	300	8155.11	I	300	863.22	III
700	13219.241	I	300 s	5044.66	I	300 l	8167.42	I	300	867.51	III
800	15230.714	I	300 l	5601.70	I	300 l	8183.06	I	300	973.79	III
400	17161.930	I	300 l	5652.75	I	300 l	8188.61	I	400	979.59	III
400	18035.80	I	300 l	5784.39	I	300 l	8247.82	I	500	1317.22	II
1000	18083.21	I	300 l	5878.04	I	300 l	8287.11	I	76	1398.19	IV
350	18221.11	I	300 s	6011.22	I	300 s	8287.75	I	74	1411.45	IV
250	18227.02	I	300	6056.09	I	300 l	8306.22	I	70	1438.82	IV
2500	18276.68	I	300 s	6073.90	I	300 s	8313.66	I	73	1449.01	IV
2000	18282.62	I	300 s	6080.05	I	1000 l	8339.12	I	76	1452.22	IV
1200	18303.97	I	300 l	6120.49	I	300	8356.79	I	73	1482.25	IV
250	18359.12	I	300	6188.59	I	300 l	8367.11	I	72	1489.83	IV
1200	18384.85	I	300 l	6200.00	I	3000	8372.88	I	75	1525.31	IV
2000	18389.95	I	300 s	6215.90	I	3000	8529.96	I	74	1527.68	IV
1000	18402.84	I	300 s	6317.84	I	1000 s	8696.23	I	74	1527.80	IV
1200	18422.39	I	300 l	6341.38	I	1000 s	8906.02	I	76	1534.71	IV
300	18458.65	I	300 l	6566.11	I	1000	8942.70	I	73	1537.25	IV
400	18475.79	I	300 l	6720.68	I	1000 s	9004.75	I	75	1543.41	IV
900	18591.55	I	300 s	6751.32	I	1000 l	9006.31	I	74	1546.23	IV
1600	18597.70	I	300 s	6795.21	I	10000 l	9016.18	I	300	1604.54	III
350	18618.96	I	300 l	6802.62	I	3000 l	9141.30	I	300	1652.87	III
550	18625.16	I	300 l	6805.81	I	3000 s	9379.33	I	400	1687.90	III
1200	21041.295	I	300 s	6816.44	I	3000 l	9468.66	I	1000	1692.51	III
750	21708.145	I	300 l	6865.45	I	3000 s	9679.13	I	800	1709.90	III
300	22247.35	I	300 s	6907.13	I	3000 l	9930.55	I	650	1715.30	III
350	22428.13	I	300 h	6912.91	I	10000 l	10091.99	I	500	1719.46	III
2250	22530.40	I	1000 s	6930.31	I	10000 s	10817.45	I	400	1722.28	III
400	22661.81	I	300 l	6963.63	I	10000 l	11695.15	I	500	1738.25	III
600	23100.51	I	3000 s	6972.09	I	10000 l	11776.64	I	300	1739.78	III
1000	23260.30	I	300	7014.02	I	10000 s	12148.18	I	1000	1741.55	II
1050	23373.00	I	300 l	7018.91	I	10000 s	12377.42	I	300	1741.96	III
850	23565.36	I	300 s	7039.14	I	10000 l	12407.99	I	550	1747.01	III
3500	23636.52	I	300 s	7080.01	I	10000 l	13834.33	I	300	1752.43	III
300	23701.64	I	300 l	7174.83	I				400	1753.01	III
1100	23709.2	I	300 l	7184.93	I	*Nickel Ni Z = 28*			800	1764.69	III
1800	23951.42	I	300 l	7284.28	I	55	315.24	V	500	1767.94	III
600	23956.46	I	300 l	7292.29	I	56	315.71	V	2000	1769.64	III
1000	23978.12	I	300 l	7332.52	I	72	354.18	V	400	1776.07	III
200	24098.54	I	300 s	7370.60	I	76	354.42	V	300	1807.24	III
500	24161.42	I	300 l	7381.03	I	68	354.49	V	300	1819.28	III
600	24249.64	I	300 l	7381.65	I	500	630.71	III	800	1823.06	III
1500	24365.05	I	300 l	7402.70	I	500	676.94	III	400	1830.01	III
800	24371.60	I	300 s	7512.22	I	300	713.33	III	650	1847.28	III
400	24447.85	I	300 l	7515.15	I	300	713.38	III	800	1854.15	III
700	24459.4	I	300 l	7546.05	I	500	718.48	III	300	1858.75	III
300	24776.46	I	300 l	7624.83	I	300	722.09	III	1000	2165.55	II
550	24928.88	I	300	7626.85	I	500	729.82	III	2000	2169.10	II
250	25161.69	I	300 s	7681.01	I	400	731.70	III	2000	2174.67	II
650	25524.37	I	300 s	7685.25	I	300	732.16	III	1500	2175.15	II
125	28386.21	I	1000 l	7735.14	I	300	747.99	III	2500	2185.50	II
150	30200.	I	300 l	7761.61	I	300	750.05	III	3000	2192.09	II
250	33173.09	I	1000 l	7765.75	I	300	757.80	III	5000	2205.55	II
450	33352.35	I	300 s	7776.07	I	400	770.22	III	4000	2206.72	II
1300	33901.	I	300	7787.46	I	500	778.81	III	6000	2216.48	II
2200	33912.10	I	1000 l	7791.38	I	300	788.04	III	1000	2264.46	II
600	34131.31	I	300 l	7851.44	I	500	811.57	III	2000	2270.21	II
100	34471.44	I	300 l	7887.88	I	500	826.14	III	1600	2289.98	I
120	35834.78	I	300 l	7901.71	I	500	842.14	III	630	2300.78	I
			300 l	7975.98	I	400	845.24	III	1000	2303.00	II
Neptunium Np Z = 93			300 h	8080.32	I	300	847.43	III	2000	2310.96	I
300	3481.93	I	300 s	8124.59	I	300	860.64	III	1700	2312.34	I
300 h	3501.50	I				300	862.88	III			

Intensity	Wavelength/Å		Intensity	Wavelength/Å		Intensity	Wavelength/Å		Intensity	Wavelength/Å	
1400	2313.66	I	530	3612.74	I	500 w	9900.92	II	100	2362.06	III
1400	2313.98	I	6600	3619.39	I				80	2362.50	III
1000	2316.04	II	200	3664.10	I	*Niobium Nb Z = 41*			80	2365.70	III
1400	2317.16	I	130	3669.24	I	80	464.55	V	100	2372.73	III
2600	2320.03	I	180	3670.43	I	80	468.32	V	170	2376.40	II
1900	2321.38	I	260	3674.15	I	80	763.77	V	110	2387.09	II
1400	2325.79	I	160	3688.42	I	80	774.02	V	100	2387.41	III
940	2329.96	I	80	3693.93	I	60	993.54	IV	140	2387.52	II
1200	2345.54	I	120	3722.48	I	400	1005.72	IV	80	2388.23	III
400	2347.52	I	150	3736.81	I	500	1007.05	IV	45	2388.27	II
1000	2375.42	II	60	3739.23	I	500	1010.19	IV	160	2398.48	II
240	2386.58	I	600	3775.57	I	100	1116.08	IV	80	2404.89	III
1000	2394.52	II	700	3783.53	I	150	1120.02	IV	55	2405.34	II
2000	2416.13	II	700	3807.14	I	100	1258.87	V	55	2405.85	II
240	2419.31	I	110	3831.69	I	60	1314.56	III	140	2412.46	II
160	2472.06	I	1200	3858.30	I	80	1445.43	III	100	2413.94	III
150	2798.65	I	110	3973.56	I	80	1445.98	III	160	2416.99	II
250	2821.29	I	110	4401.55	I	80	1447.09	III	140	2418.69	II
500	2943.91	I	85	4459.04	I	100	1456.68	III	100	2421.91	III
570	2981.65	I	55	4470.48	I	80	1484.73	III	75	2433.80	II
500	2992.60	I	65	4605.00	I	100	1495.94	III	40	2435.95	II
1000	2994.46	I	75	4648.66	I	80	1498.02	III	45	2437.42	II
4000	3002.49	I	110	4714.42	I	80	1499.45	III	40	2442.14	II
2200	3003.63	I	45	4786.54	I	100	1501.99	III	28	2442.68	II
3700	3012.00	I	45	4855.41	I	60	1502.30	IV	65	2451.87	II
1700	3037.94	I	40	4904.41	I	80	1513.81	III	65	2453.95	II
3500	3050.82	I	45	4980.16	I	60	1524.36	IV	100	2456.99	III
1500	3054.32	I	45	4984.13	I	100	1524.91	III	55	2458.09	II
1900	3057.64	I	50	5017.59	I	100	1590.21	III	65	2462.89	I
500	3064.62	I	100	5035.37	I	80	1598.86	III	80	2468.72	III
2600	3101.55	I	100	5080.52	I	80	1604.72	III	80	2475.87	III
1300	3101.88	I	65	5081.11	I	80	1639.51	III	110	2477.38	II
2900	3134.11	I	40 h	5146.48	I	100	1682.77	III	65	2478.29	II
1100	3232.96	I	40 h	5155.76	I	100	1705.44	III	65	2479.94	II
600	3243.06	I	180	5476.91	I	100	1707.14	III	35	2483.88	II
660	3315.66	I	23	5709.56	I	100	1758.33	V	100	2499.73	III
2000	3331.88	II	16	5754.68	I	100	1877.34	V	110	2511.00	II
2900	3369.57	I	10	5857.76	I	100	1892.92	III	110	2521.40	II
3300	3380.57	I	10	5892.88	I	60	1922.41	IV	390	2544.80	II
1300	3391.05	I	10	6108.12	I	100	1938.84	III	100	2545.64	III
3300	3392.99	I	10	6176.81	I	60	1978.22	IV	110	2551.38	II
8200	3414.76	I	10	6191.18	I	3300	2029.32	II	130	2556.94	II
1600	3423.71	I	13	6256.36	I	65	2032.53	IV	80	2557.94	III
2600	3433.56	I	16	6643.64	I	3000	2032.99	II	130	2562.41	II
990	3437.28	I	22	6767.77	I	2000	2109.42	II	110	2571.33	II
4800	3446.26	I	10	6914.56	I	1700	2125.21	II	390	2583.99	II
1300	3452.89	I	26	7122.20	I	1100	2126.54	II	390	2590.94	II
5000	3458.47	I	16	7393.60	I	80 h	2130.24	III	80	2598.86	III
5000	3461.65	I	16	7409.35	I	1500	2131.18	II	80	2633.17	III
1600	3472.54	I	23	7422.28	I	80	2273.92	III	200	2642.24	II
550	3483.77	I	13	7522.76	I	100	2275.23	III	320	2646.26	II
5500	3492.96	I	19	7555.60	I	80	2279.36	III	330	2647.50	I
660	3500.85	I	23	7617.00	I	100	2281.51	III	330	2654.45	I
2600	3510.34	I	16	7714.32	I	80	2284.40	III	310	2656.08	II
6600	3515.05	I	19	7727.61	I	100	2290.36	III	80	2657.99	III
660	3519.77	I	19	7748.89	I	370	2295.68	II	110	2665.25	II
8200	3524.54	I	10	7788.94	I	280	2302.08	II	110	2666.59	II
5000	3566.37	I	13	7797.59	I	100	2313.30	III	110	2667.30	II
990	3571.87	I	1000	8096.75	II	100	2338.09	III	400	2671.93	II
1300	3597.70	I	700	8121.48	II	80	2344.12	III	200	2673.57	II
1300	3610.46	I	9	8862.55	I	90	2349.21	III	200	2675.94	II
						80	2355.54	III			

Intensity	Wavelength/Å		Intensity	Wavelength/Å		Intensity	Wavelength/Å		Intensity	Wavelength/Å	
160	2691.77	II	80	3142.26	III	500	3584.97	I	350	4143.21	I
1000	2697.06	II	390	3145.40	II	750	3589.11	I	870	4150.12	I
320	2698.86	II	1200	3163.40	II	500	3589.36	I	4400	4152.58	I
320	2702.20	II	150	3175.78	II	500	3593.97	I	870	4163.47	I
150	2702.52	II	390	3180.29	II	500	3602.56	I	4400	4163.66	I
470	2716.62	II	300	3191.10	II	300	3619.51	II	4000	4164.66	I
470	2721.98	II	150	3191.43	II	420	3649.85	I	3500	4168.13	I
310	2733.26	II	1000	3194.98	II	400	3651.19	II	310	4184.44	I
110	2737.09	II	120	3203.35	II	200	3659.61	II	1200	4190.88	I
240	2768.13	II	300	3206.34	II	630	3660.37	I	870	4192.07	I
310	2773.20	I	390	3215.60	II	900	3664.70	I	870	4195.09	I
270	2780.24	II	800	3225.48	II	1500	3697.85	I	1300	4195.66	I
110	2793.05	II	140	3229.56	II	330	3711.34	I	310	4198.51	I
190	2827.08	II	400	3236.40	II	3300	3713.01	I	350	4201.52	I
250	2841.15	II	200	3247.47	II	480	3716.99	I	870	4205.31	I
280	2842.65	II	120	3248.94	II	2700	3726.24	I	350	4214.73	I
160	2846.28	II	320	3254.07	II	2700	3739.80	I	420	4217.94	I
240	2861.09	II	230	3260.56	II	670	3740.73	II	420	4229.15	I
100	2865.61	II	160	3263.37	II	1700	3742.39	I	770	4262.05	I
500	2868.52	II	200	3283.46	II	530	3763.49	I	420	4266.02	I
800	2875.39	II	160	3292.02	II	350	3765.08	I	400	4286.99	I
270	2876.95	II	320	3296.01	I	530	3771.85	I	580	4299.60	I
530	2877.03	II	400	3312.60	I	870	3781.01	I	580	4300.99	I
100	2880.72	II	120	3319.58	II	1700	3787.06	I	390	4311.27	I
570	2883.18	II	130	3341.60	II	1300	3790.15	I	350	4326.33	I
280	2888.83	II	1300	3341.97	I	3500	3791.21	I	390	4331.37	I
470	2897.81	II	1300	3343.71	I	2700	3798.12	I	330	4410.21	I
400	2899.24	II	1700	3349.06	I	2700	3802.92	I	150	4503.04	I
470	2908.24	II	420	3349.52	I	670	3803.88	I	530	4523.41	I
670	2910.59	II	340	3354.74	I	530	3804.74	I	480	4546.82	I
470	2911.74	II	1700	3358.42	I	670	3810.49	I	370	4564.53	I
1100	2927.81	II	130	3365.58	II	530	3811.03	I	720	4573.08	I
110	2931.47	II	340	3366.96	I	530	3815.51	I	480	4581.62	I
870	2941.54	II	130	3369.16	II	210	3818.86	II	1200	4606.77	I
110 h	2945.88	II	350	3374.92	I	670	3824.88	I	170	4616.17	I
110	2946.12	II	170	3386.24	II	350	3835.18	I	450	4630.11	I
110	2946.90	II	350	3392.34	I	350	3863.38	I	450	4648.95	I
1100	2950.88	II	230	3408.68	II	530	3877.56	I	450	4663.83	I
400	2972.57	II	180	3409.19	II	870	3878.82	I	340	4666.24	I
320	2974.10	II	230	3412.94	II	670	3883.14	I	240	4667.22	I
210	2977.68	II	230	3425.42	II	1100	3885.44	I	580	4672.09	I
200	2982.11	II	230	3426.57	III	670	3885.68	I	530	4675.37	I
330	2990.26	II	180	3432.70	II	580	3891.30	I	320	4685.14	I
470	2994.73	II	180	3440.59	II	670	3914.70	I	130 c	4706.14	I
80	3001.84	III	200	3479.56	II	530	3920.20	I	260	4708.29	I
140	3024.74	II	100	3484.05	II	670	3937.44	I	150	4713.50	I
350	3028.44	II	500	3498.63	I	520	3943.67	I	220 c	4749.70	I
300	3032.77	II	460	3507.96	I	910 d	3966.09	I	130 c	4967.78	I
100	3044.76	II	200	3510.26	II	1100	4032.52	I	190	4988.97	I
100	3055.52	II	200	3515.42	II	16000 c	4058.94	I	230	5017.75	I
220	3064.53	II	200	3517.67	II	350	4060.79	I	150	5026.36	I
110	3069.68	II	2000	3535.30	I	12000	4079.73	I	210	5039.04	I
100	3070.90	II	1300	3537.48	I	440	4100.40	I	170	5058.01	I
110	3071.56	II	250	3540.96	II	6700	4100.92	I	130	5065.25	I
100	3073.24	II	500	3544.02	I	310	4116.90	I	750	5078.96	I
400	3076.87	II	300	3550.45	I	5300	4123.81	I	420	5095.30	I
110	3080.35	II	1000	3554.66	I	670	4129.43	I	170	5100.16	I
1800	3094.18	II	630	3563.50	I	770	4129.93	I	170	5120.30	I
140	3099.19	II	630	3563.62	I	2300	4137.10	I	210	5134.75	I
270	3127.53	II	1500	3575.85	I	440	4139.44	I	250	5160.33	I
1500	3130.79	II	5000	3580.27	I	2700	4139.71	I	250	5164.38	I

Intensity	Wavelength/Å	
230	5180.31	I
190	5189.20	I
170	5193.08	I
150	5195.84	I
150	5232.81	I
150 d	5251.62	I
270	5271.53	I
130 c	5276.20	I
250	5318.60	I
460	5344.17	I
340	5350.74	I
110	5437.27	I
85	5551.35	I
170	5642.11	I
130	5664.71	I
170	5665.63	I
130	5729.19	I
110	5760.34	I
110	5819.43	I
130 d	5838.64	I
190 cw	5900.62	I
150	5983.22	I
75	6221.96	I
85 c	6430.46	I
65	6544.61	I
210 cw	6660.84	I
150 cw	6677.33	I
130 c	6723.62	I
85	6828.11	I
85	6990.32	I
190 c	7046.81	I
130	7159.43	I
190 cw	7372.50	I
65	7515.93	I
170 c	7574.58	I
75 c	7726.68	I
35	7885.31	I
40	8135.20	I
29 cw	8320.93	I
29	8346.08	I
35	8905.78	I

Nitrogen N Z = 7

Intensity	Wavelength/Å	
400	181.75	IV
52	186.069	V
62	186.153	V
400	191.7	IV
400	192.9	IV
500	196.87	IV
500	197.23	IV
500	202.60	IV
500	205.94	IV
500	205.97	IV
500	206.03	IV
90	209.303	V
500	217.20	IV
500 d	217.90	IV
500 d	223.4	IV
800 w	225.12	IV
800	225.21	IV
600 w	234.12	IV
600 w	234.20	IV

Intensity	Wavelength/Å	
600 w	234.25	IV
550	236.07	IV
500	237.99	IV
500 w	238.7	IV
600	238.80	IV
500 w	239.62	IV
900	247.20	IV
90	247.561	V
120	247.706	V
500 w	248.43	IV
500 w	248.46	IV
500 w	248.48	IV
500	257.95	III
650	258.50	III
700	259.19	III
800	260.09	III
600	260.45	IV
800	261.28	III
500	262.91	III
500	265.23	III
500	265.27	III
150	266.196	V
200	266.379	V
500	268.70	III
650	270.99	IV
250	283.42	IV
300	283.48	IV
350	283.58	IV
600	285.56	IV
600 w	297.7	IV
700	297.82	IV
650	300.32	IV
90	303.123	IV
500	303.28	IV
150	314.715	III
200	314.850	III
90	314.877	III
150	315.053	IV
120	322.503	IV
150	322.570	IV
200	322.724	IV
120	323.175	IV
600	323.26	III
300	335.050	IV
500	338.35	III
500	340.20	III
500 w	351.93	IV
500	351.98	III
700	353.06	IV
120	362.833	III
150	362.881	III
150	362.946	III
90	362.985	III
300	374.204	III
350	374.441	III
500	387.48	III
500	420.77	IV
250	451.869	III
300	452.226	III
650	463.74	IV
285	644.634	II

Intensity	Wavelength/Å	
360	644.837	II
450	645.178	II
140	647.50	I
360	660.286	II
170	671.016	II
285	671.386	II
150	671.630	II
160	671.773	II
170	672.001	II
500	684.996	III
570	685.513	III
650	685.816	III
500	686.335	III
350	692.70	I
90	713.518	V
150	713.860	V
285	746.984	II
150	748.195	V
200	748.291	V
500	763.336	III
570	764.359	III
570	765.148	IV
250	771.544	III
300	771.901	III
350	772.385	III
200	772.891	III
150	772.975	III
650	775.965	II
90	885.67	I
90	909.697	I
80	910.278	I
40	910.645	I
450	915.612	II
450	915.962	II
550	916.012	II
650	916.701	II
520	921.992	IV
500	922.519	IV
480	923.057	IV
520	924.283	IV
90	953.415	I
100	953.655	I
130	953.970	I
1000	955.335	IV
130	963.990	I
115	964.626	I
70	965.041	I
650	979.842	III
700	979.919	III
900	989.790	III
700	991.514	III
1000	991.579	III
150 w	1036.16	IV
90	1067.614	I
60	1068.612	I
90	1078.71	IV
450	1083.990	II
600	1084.580	II
430	1085.546	II
650	1085.701	II
175	1097.237	I

Intensity	Wavelength/Å	
115	1098.095	I
115	1098.260	I
105	1100.360	I
40	1100.465	I
90	1101.291	I
360	1134.165	I
385	1134.415	I
410	1134.980	I
105	1143.65	I
130	1163.884	I
60	1164.206	I
105	1164.325	I
270	1167.448	I
105	1168.334	I
60	1168.417	I
195	1168.536	I
230	1176.510	I
105	1176.630	I
195	1177.695	I
500	1183.031	III
570	1184.550	III
90	1188.01	IV
410	1199.550	I
385	1200.223	I
360	1200.710	I
175	1225.026	I
160	1225.37	I
130	1228.41	I
160	1228.79	I
1000	1238.821	V
900	1242.804	V
360	1243.179	I
315	1243.306	I
290	1310.540	I
250	1310.95	I
230	1319.00	I
315	1319.68	I
115	1326.57	I
115	1327.92	I
150	1387.371	III
360	1411.94	I
700	1492.625	I
490	1492.820	I
640	1494.675	I
90	1549.336	V
200 l	1616.33	V
350 l	1619.69	V
1000	1718.55	IV
250	1729.945	III
775	1742.729	I
700	1745.252	I
570	1747.848	III
350	1751.218	III
650	1751.657	III
150	1804.486	III
200	1805.669	III
150	1846.42	III
90 w	1860.37	V
350	1885.06	III
400	1885.22	III
200	1907.99	III

Intensity	Wavelength/Å		Intensity	Wavelength/Å		Intensity	Wavelength/Å		Intensity	Wavelength/Å	
150	1919.55	III	360	3919.00	II	185	5281.20	I	450	7762.24	II
150	1919.77	III	90	3938.52	III	140	5292.68	I	400	8184.87	I
300	1920.65	III	450	3955.85	II	90	5314.35	III	400	8188.02	I
150	1920.84	III	1000	3995.00	II	200	5320.82	III	250	8200.36	I
200	1921.30	III	150	3998.63	III	150	5327.18	III	300	8210.72	I
200	2064.01	III	200	4003.58	III	450	5495.67	II	570	8216.34	I
250	2064.42	III	360	4035.08	II	285	5535.36	II	400	8223.14	I
120	2068.68	III	550	4041.31	II	650	5666.63	II	400	8242.39	I
90	2071.09	III	360	4043.53	II	550	5676.02	II	550	8438.74	II
90	2080.34	IV	150	4057.76	IV	870	5679.56	II	500	8567.74	I
160	2095.53	II	250	4097.33	III	450	5686.21	II	570	8594.00	I
70	2096.20	II	140	4099.94	I	450	5710.77	II	650	8629.24	I
110	2096.86	II	200	4103.43	III	285	5747.30	II	500	8655.89	I
90	2117.59	III	185	4109.95	I	700	5752.50	I	220	8676.08	II
90	2121.50	III	285	4176.16	II	240	5764.75	I	700	8680.28	I
110	2130.18	II	120	4195.76	III	265	5829.54	I	650	8683.40	I
160	2142.78	II	150	4200.10	III	235	5854.04	I	500	8686.15	I
90	2147.31	III	285	4227.74	II	360	5927.81	II	110	8687.43	II
200	2188.20	III	285	4236.91	II	550	5931.78	II	110 h	8699.00	II
150	2188.38	III	220	4237.05	II	285	5940.24	II	500	8703.25	I
160	2206.09	II	450	4241.78	II	650	5941.65	II	160 h	8710.54	II
160	2286.69	II	90	4332.91	III	285	5952.39	II	570	8711.70	I
110	2288.44	II	120	4345.68	III	160	5999.43	I	500	8718.83	I
220	2316.49	II	300	4379.11	III	210	6008.47	I	250	8728.89	I
160	2316.69	II	285	4432.74	II	285	6167.76	II	200	8747.36	I
285	2317.05	II	650	4447.03	II	360	6379.62	II	500	9386.80	I
90 w	2318.09	IV	90	4510.91	III	150	6380.77	IV	570	9392.79	I
160	2461.27	II	120	4514.86	III	185	6411.65	I	250	9460.68	I
150	2477.69	IV	360	4530.41	II	210	6420.64	I	200	9863.33	I
110	2496.83	II	550	4601.48	II	210	6423.02	I	160 h	9865.41	II
70	2496.97	II	350	4603.73	V	210	6428.32	I	110 h	9868.21	II
110	2520.22	II	90	4606.33	IV	185	6437.68	I	160 h	9887.39	II
160	2520.79	II	450	4607.16	II	235	6440.94	I	220 h	9891.09	II
220	2522.23	II	360	4613.87	II	90	6454.11	III	160 h	9961.86	II
110	2590.94	II	250	4619.98	V	185	6457.90	I	220 h	9969.34	II
250	2645.65	IV	450	4621.39	II	120	6467.02	III	285 h	10023.27	II
300	2646.18	IV	870	4630.54	II	300	6468.44	I	220 h	10035.45	II
350	2646.96	IV	90	4634.14	III	265	6481.71	I	220 h	10065.15	II
250 w	2682.18	III	120	4640.64	III	750	6482.05	II	160 h	10070.12	II
90	2689.20	III	550	4643.08	II	360	6482.70	I	250	10105.13	I
160	2709.84	II	285	4788.13	II	300	6483.75	I	300	10108.89	I
110	2799.22	II	450	4803.29	II	325	6484.80	I	350	10112.48	I
110	2823.64	II	180	4847.38	I	160	6491.22	I	400	10114.64	I
60 l	2859.16	V	90	4858.82	III	210	6499.54	I	110 h	10126.27	II
160	2885.27	II	150	4867.15	III	185	6506.31	I	250	10539.57	I
90 l	2974.52	V	285	4895.11	II	750	6610.56	II	200	12074.51	I
150 w	2980.78	V	160	4914.94	I	185	6622.54	I	380	12186.82	I
250 w	2981.31	V	210	4935.12	I	185	6636.94	I	225	12288.97	I
60 w	2998.43	V	200 w	4944.56	V	235	6644.96	I	290	12328.76	I
220	3006.83	II	160	4950.23	I	185	6646.50	I	310	12381.65	I
90	3078.25	IV	350	4963.98	I	235	6653.46	I	180	12438.40	I
120	3367.34	III	285	4987.37	II	210	6656.51	I	510	12461.25	I
360	3437.15	II	450	4994.36	II	185	6722.62	I	920	12469.62	I
90	3463.37	IV	650	5001.48	II	210	7398.64	I	500	13429.61	I
570	3478.71	IV	360	5002.70	II	160	7406.12	I	840	13581.33	I
500	3482.99	IV	870	5005.15	II	265	7406.24	I	180	13587.73	I
400	3484.96	IV	550	5007.32	II	685	7423.64	I	180	13602.27	I
90	3747.54	IV	450	5010.62	II	785	7442.29	I	290	13624.18	I
90	3754.67	III	360	5016.39	II	900	7468.31	I	250	14757.07	I
120	3771.05	III	360	5025.66	II	185	7608.80	I	100	14868.87	I
285	3838.37	II	550	5045.10	II	60 w	7618.46	V	160	14966.60	I

Intensity	Wavelength/Å		Intensity	Wavelength/Å		Intensity	Wavelength/Å		Intensity	Wavelength/Å	
180	15582.27	I	960	2770.71	I	35	5203.23	I	150	202.393	V
120 s	17516.58	I	2800	2806.91	I	45	5255.82	I	110	203.78	V
100 l	17584.86	I	5100	2838.63	I	55	5265.15	I	150	203.82	V
100	17878.26	I	2300	2844.40	I	40	5298.78	I	100	203.85	V
			1500	2850.76	I	110	5376.79	I	200	203.89	V
Osmium Os *Z = 76*			1500	2860.96	I	120	5416.34	I	100	203.94	V
9600	2001.45	I	9600	2909.06	I	45	5416.69	I	110	207.18	IV
13000	2003.73	I	2100	2912.33	I	28	5417.51	I	150	207.24	IV
17000	2010.15	I	2100	2919.79	I	55	5443.31	I	300	207.794	V
29000	2018.14	I	1100 h	2948.23	I	22	5446.93	I	150	215.040	V
14000	2022.76	I	1400	2949.53	I	22	5457.30	I	200	215.103	V
14000	2028.23	I	4400	3018.04	I	28	5470.00	I	250	215.245	V
18000	2034.44	I	1100	3030.70	I	22	5509.33	I	250	216.018	V
26000	2045.36	I	2900	3040.90	I	270	5523.53	I	520	220.352	V
8600	2058.69	I	120	3042.74	II	22	5546.82	I	80	227.372	V
13000	2061.69	I	8600	3058.66	I	80	5584.44	I	80	227.469	V
7800	2067.21	II	1100	3077.72	I	35	5620.08	I	150	227.511	V
4200	2070.67	II	3100	3156.25	I	22	5642.56	I	80	227.549	V
7200	2076.95	I	180	3173.93	II	28	5645.25	I	80	227.634	V
14000	2079.97	I	150	3213.31	II	28	5680.88	I	80	227.689	V
2900	2082.54	I	1900	3232.06	I	170	5721.93	I	150	231.823	V
2900	2089.03	I	3100	3262.29	I	22	5765.05	I	140	233.46	IV
2900	2089.21	I	3100	3267.94	I	170	5780.82	I	150	233.50	IV
6000	2097.60	I	1200	3290.26	I	40	5800.60	I	110	233.52	IV
5300	2100.63	I	7600	3301.56	I	110	5857.76	I	200	233.56	IV
2100	2117.66	I	960	3336.15	I	28	5860.64	I	110	233.60	IV
4800	2117.96	I	960	3370.59	I	65	5996.00	I	90	238.36	IV
5300	2137.11	I	620	3387.84	I	35	6227.70	I	180	238.57	IV
2600	2154.59	I	620	3401.86	I	22	6269.41	I	110	248.459	V
1300	2157.84	I	620	3504.66	I	22	6403.15	I	110	252.56	IV
1200	2158.53	I	1200	3528.60	I	27	6729.56	I	110	252.95	IV
3100	2166.90	I	1200	3560.86	I	22	7145.54	I	150	253.08	IV
1100	2167.75	I	620	3598.11	I	26	7602.95	I	300	260.39	IV
2100	2171.65	I	95	3604.48	II	7	8041.29	I	250	260.56	IV
1100	2234.61	I	480	3670.89	I				80 d	264.34	III
1300	2252.15	I	3700	3752.52	I	*Oxygen O* *Z = 8*			110	264.48	III
2000	2255.85	II	2100	3782.20	I	80	124.616	V	110	266.97	III
1400	2264.60	I	730	3876.77	I	110	135.523	V	150	266.98	III
1400	2282.26	II	1000	3963.63	I	80	138.109	V	150	267.03	III
500	2367.35	II	730	3977.23	I	110	139.029	V	150	277.38	III
2600	2377.03	I	960	4066.69	I	80	151.447	V	300	279.63	IV
1700	2387.29	I	1200	4112.02	I	110	151.477	V	375	279.94	IV
1100	2395.88	I	2500	4135.78	I	150	151.546	V	110	285.71	IV
200	2423.07	II	1200	4173.23	I	80	164.574	V	150	285.84	IV
1400	2424.97	I	1200	4211.86	I	110	164.657	V	110	286.448	V
110	2454.91	II	4900	4260.85	I	80	164.709	V	80	295.62	III
1800	2461.42	I	560	4293.95	I	80	166.235	V	110	295.66	III
110	2468.90	II	560	4311.40	I	150	167.99	V	120	295.72	III
530	2486.24	II	4900	4420.47	I	110	170.219	V	150	303.41	III
4500	2488.55	I	540	4550.41	I	450	172.169	V	150	303.46	III
2600	2498.41	I	670	4793.99	I	250	185.745	V	140	303.52	III
2400	2513.25	I	55	5031.83	I	375	192.751	V	160	303.62	III
780	2538.00	II	45	5039.12	I	450	192.799	V	160	303.69	III
1000	2542.51	I	35	5072.88	I	520	192.906	V	250	303.80	III
1000	2590.76	I	35	5074.77	I	80	193.003	V	200	305.60	III
1800	2613.06	I	35	5079.09	I	200	194.593	V	250	305.66	III
3800	2637.13	I	90	5103.50	I	150	195.86	IV	190	305.70	III
1900	2644.11	I	55	5110.81	I	200	196.01	IV	300	305.77	III
1900	2658.60	I	140	5149.74	I	80	202.161	V	190	305.84	III
2100	2689.82	I	40	5193.52	I	80	202.224	V	200	306.62	IV
3000	2714.64	I	270	5202.63	I	80	202.283	V	150	306.88	IV
1300	2720.04	I				80	202.334	V			

Intensity	Wavelength/Å		Intensity	Wavelength/Å		Intensity	Wavelength/Å		Intensity	Wavelength/Å	
450	320.979	III	775	760.445	V	220	1763.22	III	775	2789.85	V
300	328.45	III	640	761.128	V	220	1764.48	III	160	2836.26	IV
250	328.74	III	700	762.003	V	750	1767.78	III	160	2921.45	IV
300	345.31	III	70	770.793	I	550	1768.24	III	200	2941.33	V
110	355.14	III	90	771.056	I	360	1771.67	III	210	2941.65	V
90	355.33	III	520	774.518	V	110	1773.00	III	80	2959.68	III
80	355.47	III	70	775.321	I	110	1773.85	III	265	2972.29	I
200	359.02	III	200	779.734	IV	220	1779.16	III	250	2983.78	III
190	359.22	III	315	779.821	IV	160	1781.03	III	80	3017.63	III
150	359.38	III	360	779.912	IV	160	1784.85	III	80	3023.45	III
210	373.80	III	200	779.997	IV	220	1789.66	III	80	3043.02	III
200	374.00	III	640	787.711	IV	110	1848.26	III	200	3047.13	III
300	374.08	III	520	790.109	IV	110	1856.62	III	110	3059.30	III
190	374.16	III	700	790.199	IV	285	1872.78	III	460	3063.42	IV
200	374.33	III	70	791.973	I	285	1872.87	III	410	3071.61	IV
210	374.44	III	300	796.66	II	285	1874.94	III	80	3121.71	III
450	395.558	III	200	802.200	IV	160	1920.04	III	160	3122.62	II
300	434.98	III	160	802.255	IV	110	1920.75	III	220	3129.44	II
800	507.391	III	90	804.267	I	110	1921.52	III	110	3132.86	III
900	507.683	III	70	804.848	I	220	1923.49	III	450	3134.82	II
1000	508.182	III	70	805.295	I	110	1923.82	III	285	3138.44	II
1000	525.795	III	80	805.810	I	110	1926.94	III	160	3144.66	V
250	537.83	II	240	832.762	II	360	2013.27	III	160	3209.66	IV
300	538.26	II	600	832.927	III	160	2026.96	III	80	3238.57	III
220	539.09	II	450	833.332	II	220	2045.67	III	200	3260.98	III
200	539.55	II	780	833.742	III	160	2052.74	III	300	3265.46	III
150	539.85	II	600	834.467	II	30 d	2283.42	II	80	3267.31	III
700	553.330	IV	600	835.096	III	30 d	2284.89	II	220	3270.98	II
775	554.075	IV	800	835.292	III	110	2293.32	II	220	3273.52	II
850	554.514	IV	40	877.879	I	200	2300.35	II	220	3277.69	II
700	555.261	IV	130	921.296	IV	30 d	2313.05	II	360	3287.59	II
700	597.818	III	160	921.366	IV	30 d	2316.12	II	160	3305.15	II
1000	599.598	III	80	922.008	I	30 d	2316.79	II	160	3306.60	II
580	608.398	IV	200	923.367	IV	50 d	2319.68	II	80	3312.30	III
110	609.70	III	130	923.433	IV	30 d	2322.15	II	110	3340.74	III
640	609.829	IV	90	935.193	I	30 d	2339.31	II	230	3348.08	IV
160	610.04	III	40	948.686	I	200 d	2390.44	III	270	3349.11	IV
200	610.75	III	90	971.738	I	80	2394.33	III	160	3354.27	IV
100	610.85	III	40	976.448	I	110	2411.60	II	200	3375.40	IV
270	616.952	IV	160	988.773	I	80	2422.84	III	220	3377.20	II
150	617.005	IV	40	990.204	I	80	2425.55	II	130	3378.06	IV
200	617.036	IV	250	1025.762	I	250	2433.56	II	360	3381.20	IV
520	624.617	IV	90	1027.431	I	80 d	2436.06	II	360	3385.52	IV
580	625.130	IV	160	1039.230	I	80 d	2438.83	III	285	3390.25	II
640	625.852	IV	60	1040.942	I	80	2444.26	II	270	3396.79	IV
1000	629.730	V	40	1152.152	I	300	2445.55	II	360	3403.52	IV
150	644.148	II	900	1302.168	I	200	2449.372	IV	220	3407.38	II
200	672.95	II	600	1304.858	I	200	2450.040	IV	230	3409.66	IV
150	673.77	II	300	1306.029	I	200	2454.99	III	160	3409.84	II
230	681.272	V	200	1338.612	IV	200	2493.44	IV	410	3411.69	IV
70	685.544	I	130	1342.992	IV	200	2493.77	IV	230	3413.64	IV
800	702.332	III	230	1343.512	IV	200	2507.73	IV	80	3444.10	III
800	702.822	III	640	1371.292	V	230	2509.19	IV	80	3455.12	III
900	702.899	III	160	1476.89	III	200	2517.2	IV	285	3470.81	II
1000	703.850	III	160 w	1506.72	V	200	2558.06	III	200	3489.83	IV
900	718.484	II	285	1590.01	III	80	2687.53	III	160	3492.24	IV
600	718.562	II	160	1591.33	III	110	2695.49	III	230	3560.39	IV
70	744.794	I	315 w	1643.68	V	300	2733.34	II	270	3563.33	IV
700	758.678	V	160	1707.996	V	110	2747.46	II	80	3698.70	III
640	759.441	V	220	1760.12	III	1000	2781.01	V	80	3702.75	III
580	760.228	V	110	1760.42	III	920	2786.99	V	80	3703.37	III

Intensity	Wavelength/Å		Intensity	Wavelength/Å		Intensity	Wavelength/Å		Intensity	Wavelength/Å	
110	3707.24	III	50	4469.41	II	235	7947.55	I	540	11297.68	I
220	3712.75	II	360	4590.97	II	210	7950.80	I	590	11302.38	I
110	3715.08	III	285	4596.17	II	185	7952.16	I	265	11358.69	I
315 w	3725.93	IV	80 d	4609.39	II	110	7981.94	I	490	12464.02	I
285	3727.33	II	160	4638.85	II	135	7982.40	I	450	12570.04	I
360	3729.03	IV	360	4641.81	II	190	7986.98	I	120	12990.77	I
410	3736.85	IV	450	4649.14	II	135	7987.33	I	160	13076.91	I
160	3739.92	II	160	4650.84	II	250	7995.07	I	700	13163.89	I
110	3744.00	III	360	4661.64	II	400	8221.82	I	750	13164.85	I
230	3744.89	IV	285	4676.23	II	265	8227.65	I	640	13165.11	I
360	3749.49	II	220	4699.21	II	265	8230.02	I	160	16212.06	I
150	3754.67	III	285	4705.36	II	325	8233.00	I	120	17966.70	I
80	3757.21	III	160	4924.60	II	120	8235.35	I	590	18021.21	I
250	3759.87	III	230 w	4930.27	V	120	8426.16	I	120	18041.48	I
110	3791.26	III	220	4943.06	II	810	8446.25	I	120	18042.19	I
160	3803.14	II	135	5329.10	I	1000	8446.36	I	120	18046.23	I
120	3823.41	I	160	5329.68	I	935	8446.76	I	140	18229.23	I
450	3911.96	II	190	5330.74	I	325	8820.43	I	540	18243.63	I
160	3919.29	II	90	5435.18	I	160 d	9057.01	I	140	26173.56	I
185	3947.29	I	110	5435.78	I	120	9118.29	I			
160	3947.48	I	135	5436.86	I	80	9134.71	I	*Palladium Pd Z = 46*		
140	3947.59	I	120	5577.34	I	80	9150.14	I	200	705.49	III
220	3954.37	II	110	5592.37	III	80	9151.48	I	200	727.72	III
100	3954.61	I	130	5597.91	V	235	9156.01	I	500	763.06	III
200	3961.59	III	160	5958.39	I	450	9260.81	I	500	766.42	III
450	3973.26	II	190	5958.58	I	490	9260.84	I	2000	781.02	III
220	3982.20	II	80	5995.28	I	450	9260.94	I	500	794.08	III
160	4069.90	II	160	6046.23	I	400	9262.58	I	500	797.52	III
285	4072.16	II	190	6046.44	I	540	9262.67	I	500	800.03	III
450	4075.87	II	110	6046.49	I	590	9262.77	I	500	800.10	III
80 d	4083.91	II	100	6106.27	I	490	9265.94	I	500	803.67	III
50 d	4087.14	II	400	6155.98	I	640	9266.01	I	500	825.35	III
150 d	4089.27	II	450	6156.77	I	185	9399.19	I	500	840.58	III
110	4097.24	II	490	6158.18	I	120	9481.16	I	500	856.47	III
220	4105.00	II	80	6256.83	I	120 d	9482.88	I	500	864.04	III
285	4119.22	II	100	6261.55	I	235	9487.43	I	500	880.59	III
100	4123.99	V	100	6366.34	I	140	9492.71	I	500	888.84	III
160	4132.81	II	100	6374.32	I	265	9497.97	I	1000	889.29	III
50	4146.06	II	320	6453.60	I	160	9499.30	I	300	1596.89	III
220	4153.30	II	360	6454.44	I	235	9505.59	I	500	1741.62	III
285	4185.46	II	400	6455.98	I	210	9521.96	I	4000	1782.55	III
450	4189.79	II	130	6500.24	V	120	9523.36	I	400	1843.49	III
80	4233.27	I	80	6604.91	I	120	9523.96	I	1500	1851.59	III
50 d	4253.74	II	100	6653.83	I	100	9528.28	I	2000	1852.27	III
50 d	4253.98	II	360	7001.92	I	100	9622.13	I	1000	1859.21	III
50 d	4275.47	II	450	7002.23	I	120	9625.29	I	1500	1874.63	III
50 d	4303.78	II	210	7156.70	I	160	9677.38	I	2000	1885.83	III
285	4317.14	II	400	7254.15	I	80	9694.66	I	1000	1887.40	III
160	4336.86	II	450	7254.45	I	65	9694.91	I	1500	1891.34	III
220	4345.56	II	320	7254.53	I	235	9741.50	I	4000	1914.62	III
285	4349.43	II	210	7476.44	I	235	9760.65	I	1000	1930.33	III
220	4366.90	II	100	7477.24	I	120	9909.05	I	2000	1941.64	III
100	4368.25	I	120	7479.08	I	140	9936.98	I	800	2002.16	III
220	4395.95	II	120	7480.67	I	120	9940.41	I	1000	2004.47	III
450	4414.91	II	100	7706.75	I	160	9995.31	I	500	2055.11	III
285	4416.98	II	870	7771.94	I	120 d	10421.18	I	500	2149.82	III
160	4448.21	II	810	7774.17	I	590	11286.34	I	500	2177.55	III
160	4452.38	II	750	7775.39	I	640	11286.91	I	500	2177.63	III
50	4465.45	II	80	7886.27	I	490	11287.02	I	100 r	2231.59	II
50 d	4466.28	II	100	7943.15	I	490	11287.32	I	200 r	2296.53	II
50	4467.83	II	100	7947.17	I	490	11295.10	I	100	2426.87	II
									100	2430.94	II

Intensity	Wavelength/Å		Intensity	Wavelength/Å		Intensity	Wavelength/Å		Intensity	Wavelength/Å	
100	2433.11	II	45	7915.80	I	25	1379.429	I	700	3175.09	V
100	2435.32	II	55	8132.82	I	25	1381.469	I	520	3204.04	V
150	2446.17	II	45	8300.83	I	15	1381.637	I	300	3219.307	III
1100	2447.91	I	65	8761.35	I	500	1484.507	IV	400	3233.602	III
100	2457.29	II				400	1487.788	IV	650	3347.736	IV
150	2469.29	II	*Phosphorus P Z = 15*			350	1502.228	III	570	3364.467	IV
100	2471.18	II	250	328.78	V	80	1532.51	II	400	3371.122	IV
1700	2476.42	I	150	359.899	IV	120	1535.90	II	300	3957.641	III
250	2486.52	II	500	388.318	IV	450	1610.50	V	350	3978.307	III
300	2488.92	II	250	389.50	V	150	1618.632	III	400	4059.312	III
200	2498.81	II	300	390.70	V	200	1618.907	III	300	4080.084	III
150	2505.73	II	300	445.158	IV	140	1671.070	I	500	4222.195	III
150	2551.84	II	375	475.60	V	100	1671.510	I	350	4246.720	III
150	2565.51	II	120	498.180	III	180	1671.680	I	400	4420.71	II
100	2569.56	II	520	542.57	V	140	1672.035	I	250	4479.776	III
150	2658.75	II	600	544.92	V	140	1672.474	I	250	4540.288	IV
1900	2763.09	I	200	569.853	III	600	1674.591	I	250	4541.112	IV
150 h	2776.85	II	200	581.831	III	600	1679.695	I	500	4588.04	II
100 h	2787.92	II	350	629.008	IV	140	1685.976	I	500	4589.86	II
200	2854.59	II	400	629.914	IV	100	1694.028	I	600	4602.08	II
100 h	2871.37	II	500	631.779	IV	100	1694.486	I	300	4626.70	II
100 h	2878.01	II	450	673.90	V	100	1706.376	I	300	4658.31	II
520	2922.49	I	10	810.24	II	100	1707.553	I	500	4943.53	II
650	3002.65	I	650	823.179	IV	600	1774.951	I	300	4954.39	II
1500	3027.91	I	700	824.730	IV	500	1782.838	I	300	4969.71	II
1100	3065.31	I	800	827.932	IV	400	1787.656	I	100	5079.381	I
2600	3114.04	I	300	847.669	III	140	1834.801	I	100	5098.221	I
11000	3242.70	I	350	855.624	III	140	1847.165	I	100	5100.974	I
2700	3251.64	I	500	859.652	III	100	1849.820	I	140	5109.628	I
3500	3258.78	I	10	865.44	II	140	1851.194	I	140	5154.844	I
3600	3302.13	I	450	865.45	V	100	1852.069	I	180	5162.290	I
5000	3373.00	I	600	871.39	V	500	1858.886	I	300	5253.52	II
24000	3404.58	I	700	877.476	IV	400	1859.393	I	140	5293.539	I
13000	3421.24	I	300	913.971	III	140	1864.348	I	400	5296.13	II
5000	3433.45	I	300	917.120	III	650	1888.523	IV	250	5316.07	II
6400	3441.40	I	350	918.665	III	180	1905.481	I	300	5344.75	II
7700	3460.77	I	1000	950.655	IV	140	1906.403	I	180	5345.851	I
10000	3481.15	I	250	1003.598	III	280	1907.665	I	100	5364.631	I
2000	3489.77	I	570	1025.563	IV	280	2023.489	I	250	5378.20	II
12000	3516.94	I	500	1028.096	IV	180	2024.516	I	300	5386.88	II
12000	3553.08	I	570	1030.517	IV	400	2032.432	I	400	5425.91	II
4500	3571.16	I	500	1033.111	IV	400	2033.477	I	100	5428.094	I
20000	3609.55	I	500	1035.517	IV	400	2135.465	I	400	5450.74	II
20000	3634.70	I	900	1117.98	V	400	2136.182	I	140	5458.305	I
5500	3690.34	I	570	1118.551	IV	400	2149.145	I	180	5477.672	I
1400	3718.91	I	700	1128.01	V	280	2152.940	I	140	5477.860	I
1500	3799.19	I	20	1249.82	II	500	2154.080	I	140	5478.267	I
1500	3832.29	I	20	1301.87	II	180	2235.732	I	100	5514.774	I
2200	3894.20	I	20	1304.47	II	450	2440.93	V	100	5516.997	I
1500	3958.64	I	15	1304.68	II	250	2478.256	IV	250	5588.34	II
290	4087.34	I	35	1305.48	II	750	2533.976	I	500	6024.18	II
2500	4212.95	I	60	1310.70	II	950	2535.603	I	400	6034.04	II
180	4473.59	I	500	1334.808	III	750	2553.262	I	500	6043.12	II
160	5163.84	I	650	1344.327	III	500	2554.915	I	250	6055.50	II
120	5295.63	I	300	1344.845	III	250	2605.506	IV	150	6083.409	III
55	5542.80	I	500	1366.695	IV	300	2632.713	III	350	6087.82	II
75	5670.07	I	15	1372.033	I	400	2644.295	IV	180	6097.690	I
55 h	5695.09	I	400	1372.674	IV	400	2728.770	IV	350	6165.59	II
65	6784.52	I	15	1373.500	I	500	2739.309	IV	500	6199.024	I
75	7368.12	I	10	1374.732	I	250	2739.872	IV	180	6210.499	I
120	7764.03	I	15	1377.080	I	450	2978.55	V	140	6375.681	I
			15	1377.937	I						

Intensity	Wavelength/Å		Intensity	Wavelength/Å		Intensity	Wavelength/Å		Intensity	Wavelength/Å	
100	6388.579	I	228	15962.53	I	240	2308.04	I	70	2738.48	I
250	6435.32	II	296	16254.77	I	50	2310.96	II	70	2747.61	I
600	6459.99	II	203	16292.97	I	90	2315.50	I	80	2753.86	I
600	6503.46	II	1627	16482.92	I	220	2318.29	I	200	2754.92	I
600	6507.97	II	588	16590.07	I	100	2326.10	I	30	2769.84	I
100	6717.411	I	225	16613.05	I	170	2340.18	I	500	2771.67	I
150	6992.690	III	221	16738.68	I	280	2357.10	I	40	2773.24	I
100	7102.200	I	419	16803.39	I	180	2368.28	I	20	2774.00	I
100	7158.367	I	471	17112.48	I	50	2377.28	II	50	2774.77	II
180	7165.465	I	289	17286.91	I	130	2383.64	I	50	2793.27	I
180	7175.102	I	299	17423.67	I	40	2386.81	I	100	2794.21	II
180	7176.660	I	287	23844.97	I	120	2389.53	I	40 h	2799.98	II
200	7443.657	IV	311	29097.16	I	35	2396.17	I	140	2803.24	I
250	7845.63	II				70	2401.87	I	10	2808.51	I
100	8046.801	I	**Platinum Pt Z = 78**			200	2403.09	I	50	2818.25	I
150	8113.528	III	30	1621.66	II	100	2418.06	I	30 h	2822.27	II
140	8278.058	I	30	1723.13	II	50	2424.87	II	1400	2830.30	I
100	8367.856	I	30	1751.70	II	80	2428.04	I	70	2834.71	I
140	8531.475	I	50 r	1777.09	II	50	2428.20	I	16	2853.11	I
140	8613.835	I	30	1781.86	II	25	2429.10	I	80 h	2860.68	II
180	8637.578	I	30	1879.09	II	180	2436.69	I	40 h	2865.05	II
400	8741.529	I	40	1883.05	II	650	2440.06	I	40 h	2875.85	II
100	8872.174	I	50	1889.52	II	60	2450.97	I	100 h	2877.52	II
180	9175.819	I	50	1911.70	II	440	2467.44	I	25	2888.20	I
950	9193.85	I	30	1929.25	II	35	2471.01	I	25	2893.22	I
600	9278.88	I	30	1929.68	II	1000	2487.17	I	600	2893.86	I
1250	9304.94	I	30	1939.80	II	25	2488.74	II	300	2897.87	I
500	9323.50	I	30	1949.90	II	200	2490.12	I	60	2905.90	I
950	9435.069	I	30	1983.74	II	160	2495.82	I	120	2912.26	I
950	9441.86	I	40	2014.93	II	240	2498.50	I	120	2913.54	I
600	9452.83	I	3200	2030.63	I	50	2505.93	I	70	2919.34	I
1250	9493.56	I	4400	2032.41	I	120	2508.50	I	30	2921.38	I
1700	9525.73	I	100	2036.46	II	50	2514.07	I	1700	2929.79	I
1500	9545.18	I	40	2041.57	II	60	2515.03	I	30	2942.76	I
280	9556.81	I	5500	2049.37	I	240	2515.58	I	30	2944.75	I
1700	9563.439	I	1500	2067.50	I	140	2524.30	I	25	2959.10	I
280	9593.50	I	3000	2084.59	I	40	2529.41	I	60	2960.75	I
750	9609.04	I	1000	2103.33	I	50	2536.49	I	1800	2997.97	I
400	9638.939	I	30	2115.57	II	160	2539.20	I	35	3001.17	II
500	9676.24	I	950	2128.61	I	18	2549.46	I	220	3002.27	I
180	9706.533	I	30	2130.69	II	50	2552.25	I	30	3017.88	I
1500	9734.750	I	1900	2144.23	I	50	2596.00	I	30 h	3031.22	II
280	9736.680	I	100	2144.24	II	70	2603.14	I	130	3036.45	I
1500	9750.77	I	600	2165.17	I	30	2616.76	II	800	3042.64	I
600	9790.21	I	1500	2174.67	I	50	2619.57	I	3200	3064.71	I
1700	9796.85	I	30	2190.32	II	30	2625.34	II	30	3071.94	I
280	9834.80	I	400	2202.22	I	1100	2628.03	I	130	3100.04	I
400	9903.68	I	50 h	2202.58	II	130	2639.35	I	320	3139.39	I
280	9976.67	I	320	2222.61	I	1000	2646.89	I	140	3156.56	I
229	10084.27	I	50 h	2233.11	II	500	2650.86	I	120	3200.71	I
458	10511.58	I	30 h	2240.99	II	20	2658.17	I	320	3204.04	I
962	10529.52	I	100	2245.52	II	2800	2659.45	I	30	3230.29	I
1235	10581.57	I	150	2249.30	I	40	2674.57	I	20	3233.42	I
415	10596.90	I	30	2251.52	II	440	2677.15	I	20	3250.36	I
435	10681.40	I	30 h	2251.92	II	200	2698.43	I	40	3251.98	I
265	10813.13	I	190	2268.84	I	2000	2702.40	I	160	3255.92	I
764	11183.23	I	30 h	2271.72	II	1600	2705.89	I	25	3268.42	I
402	11186.75	I	280	2274.38	I	60	2713.13	I	25	3281.97	I
479	14241.64	I	50 h	2287.50	II	1300	2719.04	I	120	3290.22	I
256	14307.83	I	30	2288.20	II	130	2729.92	I	500	3301.86	I
714	15711.52	I	150	2289.27	I	1800	2733.96	I	60	3315.05	I
			150	2292.40	I						

Intensity	Wavelength/Å		Intensity	Wavelength/Å		Intensity	Wavelength/Å		Intensity	Wavelength/Å	
35	3323.80	I	10000	7068.90	I	250	388.92	IV			
340	3408.13	I	10000	8691.94	I	250	389.07	IV	10	514.94	III
35	3427.93	I	3000	9533.07	I	250	389.07	V	50	520.61	III
60	3483.43	I	3000	12144.46	I	250	390.11	V	250	523.00	IV
160	3485.27	I	3000	16897.38	I	250	390.42	IV	25	523.79	III
120	3628.11	I				300	390.57	IV	200	526.45	IV
70	3638.79	I	*Polonium Po Z = 84*			200	391.46	IV	150	527.62	IV
70	3643.17	I	1500 w	2450.08	I	200	392.47	IV	40	529.80	III
50	3663.10	I	1500 w	2558.01	I	500	393.14	IV	15	539.71	III
80	3671.99	I	2500 w	3003.21	I	250	395.40	V	15	546.12	III
80	3674.04	I	1200	4170.52	I	200	398.36	V	750	580.32	V
35	3699.91	I	800	4493.21	I	15	398.63	III	250	585.51	V
18	3706.53	I	500	8618.26	I	200	398.88	V	500	586.32	V
80	3818.69	I				200	399.75	V	30	600.77	II
40	3900.73	I	*Potassium K Z = 19*			400	400.21	IV	250	602.27	V
110	3922.96	I	100	214.35	V	20	402.10	III	400	603.43	V
35	3948.40	I	150	271.82	IV	300	402.91	IV	25	607.93	II
100	3966.36	I	100	273.06	IV	250	403.97	IV	30	612.62	II
20	3996.57	I	150	282.35	V	150	404.41	IV	250	638.67	V
110	4118.69	I	150	293.33	V	30	406.48	III	750	646.19	IV
80	4164.56	I	300	294.84	V	250	408.08	IV	300	687.50	V
40	4192.43	I	200	296.17	V	40	408.96	III	20	708.84	III
18	4327.06	I	200	297.06	V	50	413.79	III	300	720.43	V
18	4391.83	I	200	300.25	V	30	414.87	III	400	724.42	V
80	4442.55	I	200	300.50	V	250	415.05	V	600	731.86	V
14	4445.55	I	200	311.24	V	200	415.79	V	500	737.14	IV
25	4498.76	I	250	312.77	V	30	416.00	III	500	741.95	IV
12	4520.90	I	200	315.18	V	150	417.28	IV	500	745.26	IV
35	4552.42	I	250	327.38	V	30	417.54	III	400	746.35	IV
12	4879.53	I	25	330.68	III	30	418.62	III	300	749.99	IV
14	5044.04	I	300	340.46	IV	400	422.18	V	150	754.19	IV
30	5059.48	I	150	340.74	IV	300	425.16	V	400	754.67	IV
35	5227.66	I	30	341.92	III	500	425.59	V	20	765.31	III
40	5301.02	I	15	348.00	III	75	434.72	III	30	765.64	III
12	5368.99	I	200	349.50	V	50	435.68	III	150	770.29	V
12	5390.79	I	300	354.93	IV	250	438.02	V	150	771.46	V
14	5475.77	I	150	356.26	IV	25	441.81	II	35	778.53	III
14	5478.50	I	300	359.73	IV	200	442.30	IV	20	872.31	III
6	5763.57	I	200	359.91	IV	300	443.57	IV	10	873.86	III
20	5840.12	I	250	362.08	IV	75	444.34	III	15	874.04	III
8	5844.84	I	150	362.15	IV	200	445.61	V	6	2550.02	III
6	6026.04	I	150	363.02	IV	250	446.83	IV	5	2635.11	III
7	6318.37	I	500	372.15	V	75	448.60	III	5	2689.90	III
8	6326.58	I	200	372.46	V	750	448.60	IV	5	2938.45	III
9	6523.45	I	200	372.77	V	200	449.71	V	5	2986.20	III
10	6710.42	I	300	375.96	IV	200	452.90	V	6	2992.42	III
20	6760.02	I	300	375.96	V	250	455.67	V	6	3052.07	III
60	6842.60	I	250	377.76	V	400	456.33	IV	5	3056.84	III
20	7113.73	I	30	379.12	III	400	456.33	V	5	3062.18	II
10	8224.74	I	300	379.12	V	75	466.79	III	4	3101.79	I
			300	379.88	IV	100	470.09	III	3	3102.04	I
Plutonium Pu Z = 94			25	380.48	III	75	471.57	III	7	3217.16	I
10000	2806.11	II	250	380.48	IV	45	474.92	III	6	3217.62	I
10000	2950.06	II	200	381.70	IV	10	476.03	II	11	3446.37	I
10000	3000.31	II	30	382.23	III	40	479.18	III	10	3447.38	I
10000	3200.23	II	300	382.23	IV	10	482.11	III	3	3648.84	I
10000	3418.88	II	150	382.49	IV	10	482.41	III	4	3648.98	I
10000	3805.93	I	200	382.65	IV	200	482.71	V	18	4044.14	I
10000	4097.12	I	300	382.91	IV	200	483.75	V	17	4047.21	I
10000	4170.95	I	250	384.10	IV	30	495.14	II	10	4641.88	I
10000	4367.41	I	200	386.61	IV	75	497.10	III	11	4642.37	I
10000	5590.54	I	300	387.80	V						

Intensity	Wavelength/Å		Intensity	Wavelength/Å		Intensity	Wavelength/Å		Intensity	Wavelength/Å	
4	4740.91	I	8	10487.11	I	680	3880.47	II	620	4171.82	II
6	4744.35	I	17	11019.87	I	440 c	3885.19	II	730	4172.25	II
5	4753.93	I	16	11022.67	I	440 c	3889.34	II	5200	4179.39	II
7	4757.39	I	17	11690.21	I	770 c	3908.05	II	2500	4189.48	II
5	4786.49	I	16	11769.62	I	630	3912.90	II	560 c	4191.60	II
7	4791.05	I	17	11772.83	I	310	3913.55	II	2500 c	4206.72	II
6	4799.75	I		12522.11	I	1300 c	3918.85	II	500	4208.32	II
8	4804.35	I		13377.86	I	420	3919.63	II	320	4211.86	II
7	4849.86	I		13397.09	I	960	3925.47	II	320	4217.81	II
8	4856.09	I		15163.08	I	480	3927.46	II	3800	4222.93	II
8	4863.48	I		15168.40	I	370	3929.29	II	3800	4225.35	II
9	4869.76	I		40158.37	I	370	3935.82	II	320	4233.11	II
8	4942.02	I				730 c	3947.63	II	320 c	4236.15	II
9	4950.82	I	**Praseodymium Pr Z = 59**			900 c	3949.43	II	960	4241.01	II
9	4956.15	I	7000	865.90	V	900 c	3953.51	II	340	4243.51	II
10	4965.03	I	5000	869.17	V	380	3956.75	II	840 c	4247.63	II
10	5084.23	I	2000	1228.59	IV	470	3962.45	II	500	4254.40	II
11	5097.17	I	5000	1293.22	IV	560	3964.26	II	320	4269.09	II
11	5099.20	I	5000	1295.28	IV	1600 c	3964.81	II	790 c	4272.27	II
12	5112.25	I	5000	1321.36	IV	560 c	3966.57	II	470 c	4280.07	II
12	5323.28	I	5000	1333.57	IV	500	3971.16	II	790 c	4282.42	II
13	5339.69	I	5000	1354.66	IV	320	3971.67	II	450 c	4298.98	II
12	5342.97	I	2000	1360.64	IV	620 c	3972.14	II	1500	4305.76	II
14	5359.57	I	2000	1365.77	IV	320	3974.85	II	1300	4333.97	II
16	5782.38	I	5000	1374.41	IV	1300 c	3989.68	II	360	4338.70	II
17	5801.75	I	5000	1435.56	IV	340	3992.16	II	620 cw	4344.30	II
15	5812.15	I	2000	1520.98	IV	1600	3994.79	II	470 c	4347.49	II
17	5831.89	I	5000	1574.55	IV	560 c	3997.04	II	340	4350.40	II
8	6120.27	II	5000	1575.10	IV	320	3999.12	II	450	4354.91	II
7	6307.29	II	3000	1578.38	IV	620 c	4000.17	II	410 c	4359.79	II
19	6911.08	I	2000	1622.30	IV	730	4004.70	II	1200	4368.33	II
12	6936.28	I	10000	1884.87	IV	1900	4008.69	II	320	4371.62	II
20	6938.77	I	2000	2083.23	IV	620	4010.60	II	430	4405.83	II
7	6964.18	I	3300	2246.20	V	730	4015.39	II	1700	4408.82	II
12	6964.67	I	2000 c	2378.98	IV	620	4020.96	II	410	4413.77	II
25	7664.90	I	40 h	2598.04	II	470	4022.71	II	1200 c	4429.13	II
24	7698.96	I	100 h	2707.37	II	360	4025.54	II	730	4449.83	II
5	7955.37	I	60	2760.35	II	360 c	4029.72	II	960	4468.66	II
4	7956.83	I	270	3168.24	II	730 c	4031.75	II	1100	4496.46	II
7	8078.11	I	200 d	3195.99	II	960	4033.83	II	790	4510.15	II
6	8079.62	I	190	3219.48	II	730	4038.45	II	340 c	4534.15	II
9	8250.18	I	200	3584.21	II	470	4039.34	II	340	4535.92	II
8	8251.74	I	250	3645.66	II	1300	4044.81	II	270 c	4628.74	II
3	8390.22	I	250	3646.30	II	340	4047.08	II	270 c	4672.09	II
11	8503.45	I	370	3668.83	II	450	4051.13	II	290	4695.77	I
10	8505.11	I	290	3714.05	II	2200	4054.88	II	250	4736.69	I
4	8763.96	I	410	3739.18	II	2200	4056.54	II	200	4924.60	I
3	8767.05	I	680	3761.87	II	450	4058.80	II	320	4939.74	I
13	8902.19	I	680	3800.30	II	3400	4062.81	II	380	4951.37	I
12	8904.02	I	390	3811.84	II	500 c	4079.77	II	270	5034.41	II
5	8923.31	I	1300 h	3816.02	II	500 c	4080.98	II	320	5045.52	I
4	8925.44	I	680	3818.28	II	790	4081.85	II	360	5110.38	II
7	9347.24	I	310	3821.80	II	500	4083.34	II	560	5110.76	II
3	9349.25	I	960	3830.72	II	560	4096.82	II	410	5129.52	II
6	9351.59	I	480	3840.99	II	380	4098.40	II	620	5173.90	II
15	9595.70	I	580	3846.59	II	2900 c	4100.72	II	360	5206.55	II
14	9597.83	I	1200	3850.79	II	1700 c	4118.46	II	360	5219.05	II
6	9949.67	I	720 c	3851.55	II	340	4130.77	II	560	5220.11	II
5	9954.14	I	960	3852.80	II	1500 c	4141.22	II	680	5259.73	II
9	10479.63	I	480 c	3865.45	II	2700	4143.11	II	340 c	5292.02	II
5	10482.15	I	480	3876.19	II	1700 c	4164.16	II	340	5292.62	II
			1700 c	3877.18	II						

Intensity	Wavelength/Å	
430	5322.76	II
65	5509.15	II
150	5535.17	II
110	5623.05	II
90	5624.45	II
90	5756.17	II
90	5779.28	I
160 d	5815.17	II
90	5823.72	II
90	5859.68	II
160	5939.90	II
7000 w	5956.05	III
90	5956.60	II
110	5967.82	II
90	6006.33	II
150	6017.80	II
150	6025.72	II
140	6055.13	II
65	6087.52	II
9000 w	6090.02	III
65	6114.38	II
65	6148.23	I
5000	6160.24	III
190	6161.18	II
270	6165.94	II
45	6244.35	II
110	6281.28	II
55 c	6359.03	I
55	6411.23	I
45	6429.63	II
45	6431.84	II
45	6486.55	I
45	6566.77	II
55	6616.67	I
75	6656.83	II
55	6673.41	II
75	6673.78	II
35 c	6747.09	I
55 cw	6798.60	I
35 cw	6827.60	II
7000	6910.14	III
40	7021.51	II
5000	7030.39	III
4500	7076.62	III
20	7114.55	I
24	7227.70	II
16	7407.56	II
20 c	7451.74	II
14	7541.02	II
20	7645.66	II
16	7721.84	I
14	7871.67	I
14	8067.44	I
10 cw	8122.78	II
11	8141.10	I
5000 w	8602.74	III
10	8714.59	II

Promethium Pm Z = 61

Intensity	Wavelength/Å	
1000	3892.15	II
1000	3910.26	II
1000	3919.10	II
1000	3957.74	II
1000 r	3998.96	II
1000	4417.96	II
900 r	4728.36	I
900	6100.21	I
1000 d	6520.45	I

Protactinium Pa Z = 91

Intensity	Wavelength/Å	
3000	2599.16	II
3000	2699.22	II
3000	2822.79	II
3000 h	2871.42	II
3000 h	2891.14	II
3000 l	3011.10	II
3000 s	3033.59	II
3000 l	3071.24	II
3000 l	3093.23	II
3000 l	3126.23	II
3000 l	3146.28	II
3000 l	3170.89	II
3000 l	3171.54	II
3000 l	3240.58	II
3000	3274.46	II
3000 l	3332.69	II
3000 s	3346.66	II
3000 l	3452.82	II
3000	3504.97	I
3000 s	3530.65	II
3000	3570.56	I
3000	3571.82	I
3000	3618.07	I
10000	3636.52	I
3000	3702.74	I
3000	3752.67	I
3000	3873.35	I
3000	3931.83	I
3000 s	3952.62	II
10000 l	3957.85	II
3000 s	3970.07	II
3000	3981.82	I
10000	3982.23	I
3000 l	4012.96	II
3000 s	4018.21	II
3000	4030.16	II
3000 s	4046.93	II
10000 s	4056.20	II
10000 s	4070.40	II
3000 l	4176.18	II
10000 l	4217.23	II
10000 s	4248.08	II
3000 s	4291.34	II
3000 s	4601.43	II
3000 l	6035.78	I
3000	6162.56	I
3000 l	6358.61	I
3000	6379.25	I
3000 l	6438.97	I
3000 h	6792.75	I
10000	6945.72	I
3000	6960.09	I
3000 h	6961.78	I
3000 s	6992.73	I
3000	7076.27	I
3000 h	7100.94	I
10000 s	7114.89	I
3000 h	7171.55	I
3000	7227.13	I
3000	7318.79	I
10000 l	7368.25	I
3000 h	7471.89	I
10000 h	7493.15	I
3000 h	7558.26	I
10000 h	7608.20	I
10000	7626.79	I
10000 s	7635.18	I
10000	7669.34	I
3000	7679.20	I
10000 h	7749.19	I
3000	7872.95	I
3000 l	7945.56	I
10000	8039.34	I
10000 h	8099.84	I
10000	8199.04	I
10000	8271.87	I
3000 s	8358.98	I
3000 s	8369.60	I
3000 h	8441.04	I
10000 h	8532.66	I
10000 s	8572.96	I
3000 h	8639.91	I
3000 h	8653.51	I
10000	8735.27	I
3000	10923.32	I
10000	11791.73	I
10000	14344.76	I
3000	18478.61	I

Radium Ra Z = 88

Intensity	Wavelength/Å	
100	3649.55	II
200	3814.42	II
100	4340.64	II
100	4682.28	II
100	4825.91	I
50	5660.81	II
50	7141.21	I
50	8019.70	II

Radon Rn Z = 86

Intensity	Wavelength/Å	
100	4349.60	I
200	7055.42	I
100	7268.11	I
300	7450.00	I
100	7809.82	I
100	8099.51	I
100	8270.96	I
100	8600.07	I

Rhenium Re Z = 75

Intensity	Wavelength/Å	
25000	2003.53	I
16000	2017.87	I
27000	2049.08	I
10000	2085.59	I
9800	2097.12	I
3400	2139.04	II
3700	2156.67	I
4900	2167.94	I
3400	2176.21	I
4200 c	2214.26	II
5200 c	2275.25	II
2900	2287.51	I
2700	2294.49	I
390	2298.09	II
610	2302.99	I
680	2306.54	I
800	2322.49	I
300	2328.66	I
860	2344.78	I
230	2349.39	I
680	2352.07	I
250	2356.50	I
1200	2365.90	I
570	2367.68	I
520	2369.27	I
220	2370.76	II
320	2375.07	I
370	2379.77	I
340	2388.57	I
230	2393.65	I
320	2394.37	I
320	2396.79	I
210 d	2400.72	I
210	2401.68	I
1500	2405.06	I
740	2405.60	I
320	2406.70	I
270	2410.37	I
1200	2419.81	I
300	2421.73	I
300	2421.88	I
2500	2428.58	I
490	2431.54	I
420	2432.18	I
340 c	2441.47	I
230	2442.51	I
250	2444.94	I
610	2446.98	I
610	2449.71	I
390	2461.20	I
800 c	2461.84	II
1200	2483.92	I
390	2485.81	I
980	2487.33	I
370	2496.04	I
370	2501.72	I
570	2502.35	II
230	2504.60	II
270	2505.94	I
1800 c	2508.99	I
570	2520.01	I
540	2521.50	I
370	2534.80	I
570	2540.51	I
740 d	2544.74	I
370	2545.48	I
300	2552.02	I
370	2554.63	II

Intensity	Wavelength/Å		Intensity	Wavelength/Å		Intensity	Wavelength/Å		Intensity	Wavelength/Å	
1000	2556.51	I	320	3069.94	I	240	3453.50	I	110 c	5752.93	I
250	2559.08	I	260	3071.16	I	55000 c	3460.46	I	110 cw	5776.83	I
340	2564.19	I	550	3082.43	I	40000 c	3464.73	I	550	5834.31	I
540	2568.64	II	340	3088.76	I	400	3467.96	I	200	6307.70	I
370	2571.81	II	700	3100.67	I	240	3476.44	I	200	6321.90	I
380	2586.79	I	700	3108.81	I	400	3480.38	I	100 cw	6605.19	I
290	2599.86	I	340	3110.86	I	320	3480.85	I	180 c	6813.41	I
290	2603.89	I	340 c	3118.19	I	240	3482.23	I	260	6829.90	I
660	2608.50	II	340	3121.36	I	560	3503.06	I	50 cw	7640.94	I
610 d	2611.54	I	420	3128.94	I	320	3516.65	I	65 cw	7912.94	I
310	2635.83	II	260	3134.02	I	320	3517.33	I			
550	2636.64	I	250	3141.38	I	320	3537.46	I	*Rhodium Rh* $Z = 45$		
270	2642.75	I	440	3151.64	I	240	3549.89	I	50	813.44	III
270	2649.05	I	330	3153.79	I	240	3570.26	I	80	882.51	III
660	2651.90	I	360 c	3158.31	I	360	3579.12	I	100	925.75	III
400	2654.12	I	220	3164.52	I	810 c	3580.15	II	150	937.28	III
220	2663.63	I	700	3168.37	I	650	3580.97	I	500	991.62	III
940	2674.34	I	220	3174.61	I	810	3583.02	I	400	992.48	III
220	2688.53	I	440	3177.71	I	320	3617.08	I	500 d	1009.60	III
1300	2715.47	I	260	3178.61	I	810	3637.84	I	200	1012.22	III
220	2732.21	I	600	3182.87	I	440	3651.97	I	200	1015.17	III
610	2733.04	II	1100	3184.76	I	320	3670.53	I	200	1073.87	III
220	2758.00	I	1100	3185.57	I	860 c	3689.50	I	150	1784.24	III
210	2763.79	I	260	3190.78	I	1500 c	3691.48	I	200	1784.94	III
310	2767.74	I	260	3192.36	I	520	3703.24	I	150	1796.50	III
220	2768.85	I	220	3198.58	I	240	3709.93	I	200	1816.03	III
220	2769.32	I	1100 c	3204.25	I	360 c	3717.28	I	1000	1832.05	III
350	2770.42	I	380	3235.94	I	4000	3725.76	I	500	1859.85	III
550	2783.57	I	600	3258.85	I	240 c	3735.01	I	800	1880.66	III
220	2791.29	I	600	3259.55	I	810	3735.31	I	500	1884.91	III
220	2814.68	I	300	3268.89	I	910	3740.10	I	500	1887.36	III
880	2819.95	I	280	3296.70	I	300 cw	3745.44	I	700	1888.62	III
310	2834.08	I	280	3296.99	I	700	3787.52	I	800	1901.32	III
220	2843.00	I	280	3301.60	I	240	3869.94	I	500	1910.16	III
270	2850.98	I	240	3302.23	I	240	3875.26	I	600	1919.37	III
240	2867.19	I	320	3303.21	II	240	3876.86	I	500	1927.07	III
2900	2887.68	I	280	3303.75	I	380 c	3917.27	I	700	1931.79	III
490	2896.01	I	240	3313.95	I	550	3929.85	I	500	1954.25	III
830 c	2902.48	I	600	3322.48	I	280	3961.04	I	500	1994.26	III
210	2905.58	I	2000	3338.18	I	350 c	3962.48	I	800	2013.71	III
550	2909.82	I	1600	3342.24	I	220	4033.31	I	500	2017.47	III
830 c	2927.42	I	810	3344.32	I	240	4081.43	I	500	2028.53	III
270	2930.61	I	320	3346.20	I	240 c	4110.89	I	800	2036.72	III
440	2943.14	I	240 d	3356.33	I	240 cw	4133.42	I	600	2037.61	III
270	2962.27	I	240	3377.74	I	1800	4136.45	I	1000	2040.18	III
720	2965.11	I	320	3379.06	II	700	4144.36	I	3000	2048.67	III
1500	2965.76	I	320	3379.70	I	220	4182.90	I	2000	2064.11	III
310	2976.29	I	240	3389.43	I	220	4183.06	I	800	2076.84	III
210	2978.15	I	4000	3399.30	I	650	4221.08	I	1000	2118.53	III
220	2980.82	I	650	3404.72	I	3600 c	4227.46	I	1000	2118.63	III
220	2982.19	I	650	3405.89	I	260 c	4257.60	I	1000	2139.44	III
220	2988.47	I	240	3408.67	I	380	4358.69	I	1000	2152.23	III
1800	2992.36	I	320	3409.83	I	360 cw	4394.38	I	3000	2158.17	III
5500	2999.60	I	320	3417.77	I	2600	4513.31	I	3000	2163.19	III
350	3001.14	I	810	3419.41	I	260	4516.64	I	3000	2167.33	III
220	3004.14	I	8000	3424.62	I	500	4522.73	I	150	2276.21	II
500	3016.02	I	400	3426.19	I	2200 cw	4889.14	I	140	2288.57	I
300	3016.49	I	300	3427.61	I	220	4923.90	I	110	2309.82	I
380	3030.45	I	320	3437.71	I	1300	5270.95	I	350	2322.58	I
240	3047.25	I	400	3449.37	I	1600 cw	5275.56	I	140	2326.47	I
1600	3067.40	I	16000 c	3451.88	I	100	5667.88	I	190	2334.77	II
									300	2361.92	I

Intensity	Wavelength/Å		Intensity	Wavelength/Å		Intensity	Wavelength/Å		Intensity	Wavelength/Å	
110	2368.34	I	2300	3280.55	I	240	3754.12	I	16	6253.72	I
270	2382.89	I	2300	3283.57	I	380	3754.27	I	29	6319.53	I
230	2383.40	I	280	3289.14	I	490	3755.58	I	40	6752.35	I
270	2386.14	II	210	3294.28	I	1000	3760.40	I	13	6827.33	I
80	2415.84	II	260	3300.46	I	2300	3765.08	I	11	6857.68	I
130	2427.68	I	50	3310.69	III	490	3769.97	I	20	6879.94	I
230	2429.52	I	4200	3323.09	I	380	3778.13	I	65	6965.67	I
110	2437.90	I	330	3338.54	I	1000	3788.47	I	16	6979.15	I
330	2440.34	I	280	3360.80	I	1300	3792.18	I	16	7001.58	I
90	2461.04	II	420	3368.38	I	3800	3793.22	I	18	7101.64	I
130	2473.09	I	1100	3372.25	I	4900	3799.31	I	15	7104.45	I
150	2487.47	I	110	3377.14	I	760	3805.92	I	18	7268.18	I
100	2490.77	II	110	3385.78	I	1300	3806.76	I	35	7270.82	I
130	2502.46	I	5600	3396.82	I	470	3815.01	I	18 h	7442.39	I
300	2504.29	II	820	3399.70	I	760	3816.47	I	12	7475.74	I
150	2505.67	I	160	3406.55	I	1300	3818.19	I	12	7495.24	I
350	2509.70	I	820	3412.27	I	3800	3822.26	I	11	7557.67	I
300	2511.03	II	330	3421.22	I	2300	3828.48	I	29	7791.61	I
200	2515.75	I	120 d	3424.38	I	2000	3833.89	I	55	7824.91	I
130	2520.53	II	8200	3434.89	I	5900	3856.52	I	21	8029.91	I
110	2537.04	II	1400	3440.53	I	490	3870.01	I	29	8045.36	I
350	2545.70	I	120	3447.74	I	380	3877.34	I	15	8136.20	I
550	2555.36	I	120	3450.29	I	120	3913.51	I	8	8425.59	I
150	2622.58	I	400	3455.22	I	240	3922.19	I			
230	2625.88	I	180	3457.07	I	2000	3934.23	I	*Rubidium Rb Z = 37*		
100	2630.42	I	220	3457.93	I	590	3942.72	I	30	465.85	III
110	2647.28	I	5900	3462.04	I	3800	3958.86	I	40	481.118	II
400	2652.66	I	180	3469.62	I	45	3964.54	II	500	482.83	III
100	2680.63	I	4700	3470.66	I	380	3975.31	I	500	489.66	III
400	2703.73	I	120	3472.25	I	240	3984.40	I	600	493.48	III
100	2715.31	II	4700	3474.78	I	240	3995.61	I	90	497.430	II
180	2718.54	I	2100	3478.91	I	380	3996.15	I	20	508.434	II
160	2728.94	I	110	3494.44	I	120	4023.14	I	150	513.266	II
100	2771.51	I	1200	3498.73	I	560	4082.78	I	300	530.173	II
130	2783.03	I	5900	3502.52	I	140	4097.52	I	75	533.801	II
150	2826.43	I	2800	3507.32	I	120	4119.68	I	1200	535.86	III
180	2826.68	I	8800	3528.02	I	1100	4121.68	I	40	542.887	II
280	2862.94	I	880 d	3538.14	I	1500	4128.87	I	200	555.036	II
110	2878.66	I	280	3541.91	I	2100	4135.27	I	1200	556.19	III
140	2882.37	I	1200	3543.95	I	240	4154.37	I	1500	566.71	III
160	2907.21	I	1800	3549.54	I	330	4196.50	I	1000	572.82	III
65	2910.17	II	1200	3570.18	I	3300	4211.14	I	1500	576.65	III
180	2924.02	I	4700	3583.10	I	820	4288.71	I	2500	579.63	III
130	2929.11	I	4700	3596.19	I	4200	4374.80	I	1500	581.26	III
130	2931.94	I	5900	3597.15	I	130	4569.00	I	2500	589.419	II
230	2968.66	I	3100	3612.47	I	150	4675.03	I	1000	594.94	III
160	2977.68	I	1800	3626.59	I	70	4745.11	I	1300	595.88	III
450	2986.20	I	8200	3657.99	I	70	5090.63	I	1200	598.49	III
110	3004.46	I	1300	3666.22	I	60	5155.54	I	1500	643.878	II
50	3006.43	III	560	3681.04	I	60	5175.97	I	25	663.76	IV
130	3023.91	I	1900	3690.70	I	95	5193.14	I	3000	697.049	II
50	3052.44	III	9400	3692.36	I	130	5354.40	I	6000	711.187	II
180	3083.96	I	940	3695.52	I	95	5390.44	I	25	716.24	IV
140	3121.76	I	280	3698.26	I	160	5599.42	I	50	740.85	IV
240	3123.70	I	380	3698.60	I	40	5686.38	I	10000	741.456	II
130	3155.78	I	7600	3700.91	I	29	5792.66	I	5000	769.04	III
140	3189.05	I	940	3713.02	I	40	5806.91	I	25	776.89	IV
470	3191.19	I	650	3735.28	I	35	5831.58	I	2500	815.28	III
190	3197.13	I	420	3737.27	I	130	5983.60	I	15	850.18	IV
520	3263.14	I	420	3744.17	I	35	6102.72	I	1000	1604.12	II
520	3271.61	I	1200	3748.22	I	14	6199.99	I	5000	1760.50	II
									2000	2068.92	II

Intensity	Wavelength/Å		Intensity	Wavelength/Å		Intensity	Wavelength/Å		Intensity	Wavelength/Å	
10000	2075.95	II	40 l	8271.41	I	110	2517.32	II	650	3669.49	I
30000	2143.83	II	30	8271.71	I	150	2535.59	II	550	3726.10	I
10000	2217.08	II	40 l	8868.512	I	550	2549.58	I	8700	3726.93	I
5000	2291.71	II	30	8868.852	I	370	2609.06	I	11000	3728.03	I
50000	2472.20	II	30 l	9522.65	I	830	2612.07	I	7100	3730.43	I
1000	2631.75	III	20 l	9540.18	I	460	2642.96	I	3500	3742.28	I
2000	2956.07	III	2000 c	9689.05	II	330	2651.84	I	870	3742.78	I
500	3086.84	III	35 l	10075.282	I	400	2659.62	I	2800	3745.59	I
500	3111.36	III	30 l	10075.708	I	330	2661.61	II	760	3753.54	I
5000 c	3148.90	II	100	13235.17	I	690	2678.76	II	870	3755.93	I
25	3157.54	I	20	13442.81	I	330	2692.06	II	1200	3759.84	I
50	3227.98	I	30	13443.57	I	200	2712.41	II	600	3761.51	I
500	3286.41	III	75	13665.01	I	690	2719.52	I	600	3767.35	I
60	3348.72	I	1000	14752.41	I	140	2725.47	II	1500	3777.59	I
75	3350.82	I	800	15288.43	I	310	2734.35	II	600	3782.74	I
100	3587.05	I	150	15289.48	I	1800	2735.72	I	3900	3786.06	I
40	3591.57	I	20	22529.65	I	100	2778.38	II	6000	3790.51	I
5000	3600.60	II	4	27314.31	I	110	2787.83	II	760	3798.05	I
10000	3600.64	II				350	2810.03	I	7600	3798.90	I
25000	3940.51	II	*Ruthenium Ru Z = 44*			1700	2810.55	I	7600	3799.35	I
1000	4201.80	I	250	850.09	III	350	2818.36	I	600	3812.72	I
500	4215.53	I	200	850.30	III	400	2829.16	I	760	3817.27	I
90000	4244.40	II	250	919.74	III	640	2854.07	I	760	3819.03	I
15000	4273.14	II	500	940.09	III	420	2861.41	I	650	3822.09	I
20000	4571.77	II	500	966.54	III	550	2866.64	I	550	3824.93	I
10000	4648.57	II	750	974.14	III	1800	2874.98	I	760	3831.80	I
30000	4775.95	II	900	979.43	III	740	2886.54	I	930	3839.70	I
2	5087.987	I	500	981.35	III	370	2908.88	I	760	3850.43	I
2	5132.471	I	900	986.84	III	1100	2916.26	I	1300	3857.55	I
10	5150.134	I	900	994.56	III	180	2945.67	II	650	3862.69	I
10000	5152.08	II	300	1001.65	III	370	2949.50	I	1300	3867.84	I
1	5165.023	I	500	1009.13	III	550	2965.16	I	650	3892.21	I
2	5165.142	I	900	1009.87	III	170	2965.55	II	760	3909.08	I
15	5195.278	I	500	1014.68	III	140	2976.59	II	1500	3923.47	I
2	5233.968	I	800	1190.51	III	550	2976.92	I	3300	3925.92	I
20	5260.034	I	500	1200.07	III	1400	2988.95	I	600	3931.76	I
1	5260.228	I	500	1207.17	III	460	2994.96	I	760	3945.57	I
3	5322.380	I	500	1209.77	III	440	3006.59	I	600	3978.44	I
40	5362.601	I	300	1211.31	III	330	3017.24	I	600	3979.42	I
4	5390.568	I	500	1941.35	III	310	3020.88	I	870	3984.86	I
75	5431.532	I	500	2009.28	III	390	3064.84	I	1500	4022.16	I
3	5431.830	I	2400	2076.43	I	330	3096.57	I	600	4023.83	I
6	5578.788	I	2600	2083.77	I	830	3099.28	I	1400	4051.40	I
40	5647.774	I	2400	2090.89	I	740	3100.84	I	710	4054.05	I
20	5653.750	I	690	2255.52	I	490	3294.11	I	760	4068.37	I
60	5724.121	I	780	2272.09	I	370	3301.59	I	980	4076.73	I
3	5724.614	I	780	2279.57	I	930	3339.55	I	6000	4080.60	I
75	6070.755	I	480	2317.80	I	3100	3417.35	I	930	4097.79	I
30 c	6159.626	I	120	2334.96	II	4900	3428.31	I	1900	4112.74	I
75 c	6206.309	I	190 h	2342.85	II	6400	3436.74	I	2000	4144.16	I
120 c	6298.325	I	310	2351.33	I	8300	3498.94	I	650	4145.74	I
5	6299.224	I	170	2357.91	II	640	3514.49	I	870	4167.51	I
10000	6458.33	II	780	2402.72	II	790	3539.37	I	550	4197.58	I
5000	6560.81	II	150	2407.92	II	690	3570.59	I	550	4198.88	I
100 l	7279.997	I	180	2455.53	II	6400	3589.22	I	7600	4199.90	I
150	7408.173	I	150	2456.44	II	6900	3593.02	I	1500	4206.02	I
200 l	7618.933	I	370	2456.57	II	6400	3596.18	I	5400	4212.06	I
300	7757.651	I	280	2478.93	II	1300	3599.76	I	760	4214.44	I
60	7759.436	I	140	2498.42	II	3100	3634.93	I	930	4217.27	I
90000 c	7800.27	I	140	2498.57	II	6200	3661.35	I	550	4230.31	I
45000 c	7947.60	I	260	2507.01	II	830	3663.37	I	760	4241.05	I
			110	2513.32	II						

Intensity	Wavelength/Å		Intensity	Wavelength/Å		Intensity	Wavelength/Å		Intensity	Wavelength/Å	
760	4243.06	I	26	6982.01	I	480	3854.56	I	730	4716.10	I
760	4284.33	I	26	7027.98	I	400	3858.74	I	770	4728.42	I
550	4295.93	I	35	7238.92	I	3700	3885.29	II	470	4745.68	II
3700	4297.71	I	16	7393.93	I	1600	3896.98	II	730	4760.27	I
930	4307.60	I	18	7468.91	I	1300	3903.42	II	580	4783.10	I
550	4319.87	I	26	7485.79	I	2500	3922.40	II	350	4785.86	I
550	4342.07	I	70	7499.75	I	1900	3928.28	II	970	4841.70	I
710	4354.13	I	26	7559.61	I	1300	3941.87	II	730	4883.97	I
870	4361.21	I	18	7621.50	I	470	3951.89	I	630	4910.40	I
2400	4372.21	I	18	7722.87	I	1500	3963.00	II	350	4913.25	II
870	4385.39	I	22	7791.86	I	1500	3971.40	II	430	4918.99	I
1300	4385.65	I	30	7847.80	I	620	3974.66	I	400	5044.28	I
1700	4390.44	I	80	7881.49	I	1000	3976.43	II	540	5071.20	I
1600	4410.03	I	18	8264.96	I	1500	3990.00	II	510	5117.16	I
1100	4460.04	I	22	8710.84	I	1400	4064.58	II	350	5122.14	I
5400	4554.51	I				1000	4092.27	II	360	5155.03	II
1700	4584.44	I	*Samarium Sm Z = 62*			1900	4118.55	II	470	5175.42	I
720	4647.61	I	150	2789.38	II	1200	4152.21	II	250	5200.59	I
1400	4709.48	I	410	3152.52	II	1000	4188.13	II	260	5251.92	I
500	4757.84	I	720	3183.92	II	1100	4203.05	II	400	5271.40	I
550	4869.15	I	600	3211.73	II	1000	4225.33	II	250	5282.91	I
160	5011.23	I	530	3216.85	II	1200	4236.74	II	220	5453.00	I
450	5057.33	I	600	3218.61	II	2100	4256.39	II	230	5493.72	I
120	5076.32	I	720	3230.56	II	1300	4262.68	II	230	5516.09	I
200	5093.83	I	720	3236.64	II	1200	4279.68	II	140	5550.40	I
530	5136.55	I	720	3239.66	II	2200	4280.79	II	140	5659.86	I
170	5142.76	I	720	3250.37	II	710	4282.21	I	120	5696.73	I
250	5147.24	I	850	3254.38	II	470	4282.83	I	85	5706.20	I
110	5151.07	I	1700	3306.39	II	1600	4296.74	I	70	5773.77	I
500	5155.14	I	1200	3321.18	II	1900	4318.94	II	60	5778.33	I
920	5171.03	I	1200	3365.86	II	470	4319.53	I	70 d	5786.98	II
180	5195.02	I	1200	3382.40	II	1800	4329.02	II	60	5788.38	I
130	5284.08	I	4200	3568.27	II	440	4330.02	I	60	5800.52	I
260	5309.27	I	4200	3592.60	II	1300	4334.15	II	65	5802.84	I
110	5335.93	I	1700	3604.28	II	880	4336.14	I	65	5867.79	I
130	5361.77	I	3400	3609.49	II	1100	4347.80	II	50	5874.21	I
110 h	5401.04	I	1700	3621.23	II	440	4362.91	I	50	5898.96	I
80	5484.32	I	3400	3634.29	II	530	4380.42	I	65	5965.71	II
130	5510.71	I	2200	3661.36	II	1600	4390.86	II	50	6045.00	I
90	5559.75	I	2200	3670.84	II	410	4401.17	I	50	6070.06	I
290	5636.24	I	1100	3693.99	II	470	4419.33	I	45	6084.12	I
180	5699.05	I	1600	3728.47	II	1500	4420.53	II	45 h	6159.56	I
65	5814.98	I	2100	3731.26	II	2900	4424.34	II	45	6256.54	I
55	5919.34	I	1600	3735.98	II	470	4429.66	I	100	6267.28	II
80	5921.45	I	2900	3739.12	II	1600	4433.88	II	140	6569.31	II
21 h	5973.38	I	1200	3743.87	II	1800	4434.32	II	110	6589.72	II
16	5988.67	I	930	3745.46	I	530	4441.81	I	50	6671.51	I
35	5993.65	I	800	3756.41	I	440	4442.28	I	120 d	6731.84	II
18	6116.77	I	1200	3757.53	II	710	4445.15	I	95	6794.20	II
26	6199.42	I	1900	3760.69	II	1300	4452.73	II	120	6860.93	I
26	6225.20	I	1100	3764.37	II	1200	4454.63	II	120	6955.29	II
18	6295.22	I	370 d	3773.33	I	1000	4458.52	II	90	7020.44	II
16	6390.23	I	1100	3778.14	II	2200	4467.34	II	90	7039.22	II
26 h	6444.84	I	1500	3788.12	II	810	4470.89	I	90	7042.24	II
21	6663.14	I	1600	3793.97	II	370	4499.11	I	90	7051.52	II
55	6690.00	I	1600	3797.73	II	440	4581.73	I	90	7082.37	II
21	6766.95	I	1600	3826.20	II	380	4649.49	I	26	7088.30	I
30	6775.02	I	1100	3831.50	II	470 d	4670.75	I	30	7095.50	I
21	6824.17	I	560	3834.48	I	1100	4674.60	II	30	7104.54	I
26	6911.48	I	1600	3843.50	II	370	4688.73	I	26	7115.96	I
110	6923.23	I	530	3853.30	I	530	4704.40	II	85 d	7149.60	II
			2700	3854.21	II						

Intensity	Wavelength/Å	
23	7213.82	I
60	7240.90	II
26	7347.30	I
30	7444.56	I
26	7445.41	I
13	7470.76	I
45	7645.09	II
12	7645.82	I
40 w	7835.08	II
16	7895.96	I
90	7928.14	II
40	8048.70	II
16	8065.16	I
45	8068.46	II
40 w	8305.79	II
19	8383.71	I
45 w	8485.99	II
45 w	8708.43	II
95	8913.66	II

Scandium Sc Z = 21

Intensity	Wavelength/Å	
350	180.14	V
500	243.87	V
500	252.85	V
500	253.73	V
900	283.91	V
800	284.45	V
600	288.29	V
900	289.59	V
15	289.85	IV
1000 d	291.93	V
800	293.25	V
15	296.31	IV
15	299.04	IV
700	300.00	V
1000	573.36	V
600	587.94	V
10	785.12	IV
25	1168.61	III
15	1550.80	IV
180	1603.06	III
150	1610.19	III
160	2010.42	III
12	2118.97	IV
11	2185.43	IV
11	2205.46	IV
14	2222.22	IV
11	2271.33	IV
110	2438.62	I
560	2545.22	II
2900	2552.37	II
560	2555.82	II
2300	2560.25	II
1100	2563.21	II
11	2586.93	IV
120	2692.78	I
350	2699.07	III
360	2706.77	I
210	2707.95	I
580	2711.35	I
230	2734.05	III
340	2965.86	I
1200	2974.01	I
1400	2980.75	I
340	2988.95	I
2200	3015.36	I
2700	3019.34	I
360	3030.76	I
120 h	3056.31	I
130	3065.11	II
990	3251.32	II
1500	3255.69	I
4400	3269.91	I
5500	3273.63	I
110 d	3343.28	II
270	3352.05	II
9900	3353.73	II
2000	3359.68	II
1700	3361.27	II
1700	3361.94	II
4000	3368.95	II
6600	3372.15	II
130	3418.51	I
200	3429.21	I
200	3429.48	I
270	3431.36	I
530	3435.56	I
270	3457.45	I
180	3462.19	I
130 d	3469.65	I
110	3471.13	I
200	3498.91	I
2700	3535.73	II
6600	3558.55	II
6100	3567.70	II
13000	3572.53	II
9900	3576.35	II
7700	3580.94	II
4000	3589.64	II
4000	3590.48	II
28000	3613.84	II
110	3617.43	I
20000	3630.75	II
13000	3642.79	II
6600	3645.31	II
110	3646.90	I
5300	3651.80	II
110	3664.25	II
290	3666.54	II
75 h	3717.10	I
270	3833.07	II
610	3843.03	II
20000	3907.49	I
23000	3911.81	I
4400	3933.38	I
5500	3996.61	I
530	4014.49	II
20000	4020.40	I
20000	4023.69	I
220	4030.67	I
140	4031.39	I
220	4043.80	I
200	4046.48	I
2700	4047.79	I
120	4049.95	I
5500	4054.55	I
220	4056.59	I
100	4068.66	III
160 h	4074.97	I
160	4078.57	I
6100	4082.40	I
200	4086.67	I
400	4087.16	I
440 h	4133.00	I
530 h	4140.30	I
720	4152.36	I
1100 h	4165.19	I
110 h	4218.26	I
110 h	4219.73	I
180	4231.93	I
200	4233.61	I
400	4238.05	I
15000	4246.83	II
290	4294.77	II
350	4305.71	II
4200	4314.09	II
3300	4320.74	II
2400	4325.01	II
180	4354.61	II
110	4358.64	I
2000	4374.46	II
130	4384.81	II
1100	4400.37	II
880	4415.56	II
120 h	4557.24	I
160 h	4573.99	I
350	4670.40	II
120	4706.97	I
120	4709.34	I
200	4728.77	I
490	4729.23	I
590	4734.10	I
690	4737.65	I
790	4741.02	I
1200	4743.81	I
200	4753.16	I
220	4779.35	I
170	4839.44	I
90	4909.76	I
90	4922.84	I
90	4934.25	I
170	4954.06	I
120	4973.66	I
150	4980.37	I
140	4991.92	I
530	5031.02	II
250	5064.32	I
530	5070.23	I
250	5075.81	I
2100	5081.56	I
1200	5083.72	I
1100	5085.55	I
750	5086.95	I
390	5087.14	I
270	5089.89	I
390	5096.73	I
620	5099.23	I
370	5101.12	I
180	5109.06	I
150	5112.86	I
320	5116.69	I
390	5210.52	I
280	5219.67	I
350	5239.82	II
280	5258.33	I
210	5285.76	I
120	5341.05	I
350	5349.30	I
120	5349.71	I
210	5355.75	I
530	5356.10	I
270	5375.35	I
370	5392.08	I
270	5446.20	I
120	5451.34	I
750	5481.99	I
530	5484.62	I
570	5514.22	I
660	5520.50	I
660	5526.82	II
70	5564.86	I
110	5591.33	I
80	5640.98	II
250	5657.88	II
1500	5671.81	I
1200	5686.84	I
1100	5700.21	I
10	5706.82	IV
190	5708.61	I
880	5711.75	I
230	5717.28	I
180	5724.08	I
14	5771.63	IV
620	6210.68	I
320	6239.78	I
120	6245.63	II
110	6249.96	I
80	6256.01	III
250	6258.96	I
750	6305.67	I
60	6378.82	I
90	6413.35	I
60	6604.60	II
65	6737.87	I
50	6819.52	I
50	6835.03	I
90	7449.16	III
55 h	7741.17	I
30	7800.44	I
19 h	8761.40	I
50	8829.78	III
30 h	8834.35	I
400	22051.86	I
150	22065.05	I

Intensity	Wavelength/Å		Intensity	Wavelength/Å		Intensity	Wavelength/Å		Intensity	Wavelength/Å	
Selenium Se	*Z = 34*		285	2057.5	III	150	8440.47	I	100	1190.42	II
360	613.0	V	500	2074.8	I	150	8450.38	I	200	1193.28	II
360	652.7	IV	285	2136.6	IV	150	8742.33	I	250	1194.50	II
450	670.1	IV	500	2164.2	I	300	8918.86	I	100	1197.39	II
360	724.3	III	600	2413.5	I	200	9001.97	I	30	1206.51	III
450	746.4	IV	300	2548.0	I	200	9038.61	I	30	1206.53	III
450	759.1	V	360	2665.5	IV	100	9432.50	I	9	1207.52	III
360	808.7	V	285	2724.3	IV	200	10217.25	I	10	1210.46	III
360	830.3	V	285	2767.2	III	377	10307.45	I	50	1226.81	II
360	832.7	II	220	2773.8	III	900	10327.26	I	100	1227.60	II
450	839.5	V	160	2951.6	IV	640	10386.36	I	150	1228.75	II
360	843.0	III	450	3387.2	III	275	11946.87	I	200	1229.39	II
360	845.8	V	450	3413.9	III	170	11952.64	I	100	1246.74	II
360	912.9	II	450	3457.8	III	205	11972.93	I	150	1248.43	II
360	959.6	IV	450	3637.6	III	315	14817.93	I	100	1250.09	II
360	974.8	III	450	3738.7	III	410	14917.47	I	150	1250.43	II
450	996.7	IV	450	3800.9	III	500	15151.44	I	200	1251.16	II
360	1013.4	II	450	4169.1	III	320	15471.00	I	40	1256.49	I
360	1014.0	II	360	4175.3	II	265	15520.97	I	50	1258.80	I
450	1033.6	II	450	4180.9	II	395	15618.40	I	1000	1260.42	II
450	1049.6	II	285	4382.9	II	360	16659.44	I	2000	1264.73	II
360	1057.4	II	285	4446.0	II	505	16813.78	I	200	1265.02	II
360	1094.7	V	220	4449.2	II	205	16866.54	I	17	1294.54	III
360	1099.1	III	285	4467.6	II	235	21374.24	I	14	1296.73	III
450	1119.2	III	500	4730.8	I	680	21442.56	I	15	1298.89	III
360	1141.9	II	400	4739.0	I	415	21473.48	I	18	1298.96	III
450	1192.3	II	300	4742.2	I	270	21716.36	I	14	1301.15	III
450	1227.6	V	285	4840.6	II	240	21730.60	I	16	1303.32	III
285	1291.0	II	360	4845.0	II	150	23388.85	I	100	1304.37	II
285	1308.9	II	450	5227.5	II	265	24148.18	I	50 h	1305.59	II
285	1314.4	IV	360	5305.4	II	375	24385.99	I	200	1309.27	II
120	1435.3	I	100	5365.5	I	255	25017.51	I	13	1312.59	III
120	1435.8	I	120	5369.9	I	510	25127.43	I	100	1346.87	II
150	1449.2	I	110	5374.1	I				100	1348.54	II
150	1500.9	I	285	5522.4	II	*Silicon Si*	*Z = 14*		150	1350.06	II
250	1530.4	I	285	5566.9	II	10	85.18	V	100	1352.64	II
150	1531.3	I	285	5866.3	II	15	96.44	V	100	1353.72	II
200	1531.8	I	450	6056.0	II	10	97.14	V	15	1393.76	IV
150	1575.3	I	285	6303.8	III	20	117.86	V	12	1402.77	IV
150	1577.6	I	200	6325.6	I	20	118.97	V	13	1417.24	III
150	1577.9	I	360	6444.2	II	4	457.82	IV	90 h	1485.02	II
150	1579.5	I	285	6490.5	II	8	566.61	III	100 h	1485.51	II
200	1580.0	I	285	6535.0	II	8	653.33	III	12	1500.24	III
150	1587.5	I	150	6831.3	I	7	815.05	IV	10	1501.19	III
150	1593.2	I	120	6990.690	I	8	818.13	IV	9	1501.87	III
250	1606.5	I	100	6991.792	I	9	823.41	III	100 h	1509.10	II
200	1617.4	I	200	7010.809	I	40 h	845.77	II	50 h	1512.07	II
150	1621.2	I	150	7013.875	I	100	889.72	II	60 p	1516.91	II
150	1643.4	I	300	7062.065	I	200	892.00	II	500	1526.72	II
250	1671.2	I	200	7575.1	I	9	967.95	III	1000	1533.45	II
250	1675.3	I	250	7583.4	I	100	989.87	II	150	1594.55	I
250	1690.7	I	150	7592.2	I	200	992.68	II	100	1622.87	I
250	1793.3	I	300	8001.0	I	10	993.52	III	300	1629.43	I
300	1795.3	I	200	8036.4	I	13	994.79	III	200	1629.92	I
300	1855.2	I	150	8093.2	I	16	997.39	III	100	1667.62	I
250	1858.8	I	150	8094.7	I	50	1023.69	II	100	1668.52	I
400	1898.6	I	180	8149.3	I	8	1066.63	IV	100	1672.59	I
350	1913.8	I	150	8152.0	I	14	1108.37	III	200	1675.20	I
300	1919.2	I	200	8157.7	I	16	1109.97	III	200	1696.20	I
500	1960.9	I	180	8163.1	I	18	1113.23	III	200	1697.94	I
500	2039.8	I	150	8182.9	I	8	1122.49	IV	100 h	1770.92	I
						10	1128.34	IV			

Intensity	Wavelength/Å		Intensity	Wavelength/Å		Intensity	Wavelength/Å		Intensity	Wavelength/Å	
100 h	1776.83	I	300	2904.28	II	16	4716.65	III	90	6131.850	I
100 h	1799.12	I	500	2905.69	II	50	4782.991	I	100	6142.487	I
150	1808.00	II	55	2970.355	I	35	4792.212	I	100	6145.015	I
500 h	1814.07	I	150	2987.645	I	80	4792.324	I	160	6155.134	I
200	1816.92	II	50	3006.739	I	15	4813.33	III	160	6237.320	I
200	1836.51	I	75	3020.004	I	16	4819.72	III	40	6238.287	I
200	1841.44	I	100 h	3030.00	II	18	4828.97	III	125	6243.813	I
9	1842.55	III	9	3040.93	III	30	4947.607	I	125	6244.468	I
200	1843.77	I	100 h	3043.69	II	40	5006.061	I	180	6254.188	I
300	1845.51	I	50 h	3048.30	II	1000	5041.03	II	45	6331.954	I
400	1847.47	I	150 h	3053.18	II	1000	5055.98	II	1000	6347.10	II
200	1848.14	I	25	3086.24	III	10 h	5091.42	III	1000	6371.36	II
500	1850.67	I	20	3093.42	III	100	5181.90	II	45	6526.609	I
200	1852.46	I	16	3096.83	III	100 h	5185.25	II	45	6527.199	I
500 h	1874.84	I	9	3165.71	IV	200 h	5192.86	II	45	6555.462	I
200	1881.85	I	16	3185.13	III	500 h	5202.41	II	50 h	6660.52	II
200	1887.70	I	13	3186.02	III	100 h	5405.34	II	100	6671.88	II
200 h	1893.25	I	150	3188.97	II	100 h	5438.62	II	7	6701.21	IV
1000 h	1901.33	I	150	3193.09	II	100 h	5456.45	II	50 h	6717.04	II
100 h	1902.46	II	100	3195.41	II	500 h	5466.43	II	100	6721.853	I
50 h	1910.62	II	14	3196.50	III	500 h	5466.87	II	50	6829.82	II
50	1941.67	II	200	3199.51	II	100 h	5469.21	II	30	6848.568	I
100	1949.56	II	100 h	3203.87	II	200 h	5496.45	II	80	6976.523	I
100	1954.97	I	200 h	3210.03	II	35	5517.535	I	180	7003.567	I
50	2058.65	II	15	3210.55	III	100 h	5540.74	II	180	7005.883	I
50	2059.01	II	75	3214.66	II	150 h	5576.66	II	90	7017.646	I
200	2072.02	II	12	3230.50	III	30	5622.221	I	250	7034.903	I
200	2072.70	II	14	3233.95	III	100 h	5632.97	II	6 h	7047.94	IV
100	2124.12	I	15	3241.62	III	200 h	5639.48	II	200	7165.545	I
50 h	2136.56	II	12	3258.66	III	90	5645.611	I	100	7226.206	I
110	2207.98	I	10	3276.26	III	150 h	5660.66	II	100	7235.326	I
115	2210.89	I	300	3333.14	II	80	5665.554	I	180	7250.625	I
110	2211.74	I	500	3339.82	II	1000 h	5669.56	II	160	7275.294	I
120	2216.67	I	15	3486.91	III	120	5684.484	I	400	7289.173	I
120	2218.06	I	9	3525.94	III	300 h	5688.81	II	375	7405.774	I
10	2296.87	III	20	3590.47	III	100	5690.425	I	200	7409.082	I
10	2308.19	III	8	3762.44	IV	90	5701.105	I	275	7415.946	I
100 h	2356.30	II	20 c	3791.41	III	200 h	5701.37	II	425	7423.497	I
30 h	2357.18	II	25	3796.11	III	100 h	5706.37	II	9 h	7466.32	III
50 h	2357.97	II	30	3806.54	III	160	5708.397	I	12 h	7612.36	III
300	2435.15	I	100 h	3853.66	II	20	5739.73	III	100	7680.267	I
11	2449.48	III	500 h	3856.02	II	45	5747.667	I	6 h	7723.82	IV
425	2506.90	I	200 h	3862.60	II	45	5753.625	I	30	7800.008	I
375	2514.32	I	300	3905.523	I	45	5754.220	I	400	7848.80	II
500	2516.113	I	20	3924.47	III	45	5762.977	I	500	7849.72	II
7	2517.51	IV	10	4088.85	IV	70	5772.145	I	30	7849.967	I
350	2519.202	I	70	4102.936	I	70	5780.384	I	90	7918.386	I
425	2524.108	I	9	4116.10	IV	90	5793.071	I	120	7932.349	I
450	2528.509	I	300 h	4128.07	II	100	5797.859	I	140	7944.001	I
110	2532.381	I	500 h	4130.89	II	150 h	5800.47	II	35	7970.306	I
25	2541.82	III	100 h	4190.72	II	200	5806.74	II	35	8035.619	I
10	2546.09	III	50	4198.13	II	50	5846.13	II	70	8093.241	I
14	2559.21	III	9	4338.50	III	300 h	5868.40	II	9 h	8102.86	III
30	2563.679	I	30	4552.62	III	40	5873.764	I	11 h	8103.45	III
85	2568.641	I	25	4567.82	III	10 h	5898.79	III	35	8230.642	I
45	2577.151	I	20	4574.76	III	150	5915.22	II	9 h	8262.57	III
190	2631.282	I	100	4621.42	II	200	5948.545	I	40	8443.982	I
11	2640.79	III	150	4621.72	II	500	5957.56	II	40	8501.547	I
14	2655.51	III	9 h	4631.24	IV	500	5978.93	II	60	8502.221	I
9	2817.11	III	10 h	4654.32	IV	90	6125.021	I	40	8536.165	I
1000	2881.579	I	9	4683.02	III	85	6131.574	I	120	8556.780	I

Intensity	Wavelength/Å		Intensity	Wavelength/Å		Intensity	Wavelength/Å		Intensity	Wavelength/Å	
50	8648.462	I	95	1432.60	II	28000 r	3382.89	I	12	162.448	IV
40	8728.011	I	100	1464.72	II	30	3469.16	I	10	163.190	IV
75	8742.451	I	50	1466.23	II	70	3475.82	II	12	168.411	IV
100	8752.009	I	50 r	1507.37	I	80	3495.28	II	10	168.546	IV
35	8790.389	I	100 r	1515.63	I	50	3542.61	I	5	183.95	III
100	9412.72	II	50 r	1548.58	I	50 h	3624.68	I	10	190.445	IV
100	9413.506	I	100	1555.16	II	75	3682.46	II	10	199.772	IV
30	10371.269	I	100	1644.50	II	30	3682.50	I	8	202.49	III
120	10585.141	I	60	1651.52	I	80	3683.34	II	8	202.76	III
120	10603.431	I	50	1652.10	I	50 h	3709.20	I	8 p	203.06	III
120	10660.975	I	700	1656.18	III	200	3810.94	I	8	203.28	III
30	10694.251	I	120	1682.82	II	50	3811.78	I	8	203.33	III
30	10727.408	I	500	1693.51	III	100 h	3840.74	I	15	229.87	III
60	10749.384	I	10	1708.11	I	50 h	3907.41	I	50 c	250.52	III
30	10784.550	I	50	1709.27	I	50	3909.31	II	30	251.37	III
80	10786.856	I	125	1736.44	II	50 h	3914.40	I	25	266.90	III
140	10827.091	I	750	1751.03	III	70	3920.10	II	70	267.65	III
60	10843.854	I	10 h	1766.14	I	60	3949.43	II	50	267.87	III
130	10869.541	I	75	1790.37	II	100 h	3981.58	I	50	268.63	III
30	10882.802	I	600	1917.08	III	70	3985.19	II	20 p	272.08	III
30	10885.336	I	700	1957.62	III	100 h	4055.48	I	20	272.45	III
80	10979.308	I	100	1967.38	II	80	4085.91	II	10	319.644	IV
30	10982.061	I	600	1975.92	III	100	4185.48	II	300	372.08	II
80	11017.965	I	500	1977.03	III	90 h	4210.96	I	350	376.38	II
370	11984.19	I	600	2000.24	III	100	4212.82	I	100	378.14	III
220	11991.57	I	150	2015.96	II	50	4311.07	I	70	380.10	III
440	12031.51	I	150	2033.98	II	50 h	4476.04	I	12	408.684	IV
190	15888.39	I	200	2061.17	I	30 h	4615.69	I	10	409.614	IV
95	16060.03	I	100	2069.85	I	80	4620.04	II	15	410.372	IV
110	19722.50	I	80 r	2113.82	II	50	4620.46	II	10	411.334	IV
			60	2145.60	II	60 h	4668.48	I	13	412.242	IV
Silver Ag Z = 47			600	2161.89	III	30 h	4677.60	I	11	1582.18	IV
25	730.83	II	50	2186.76	II	100	4788.40	II	11 d	1583.98	IV
30	752.80	II	60	2229.53	II	30 h	4847.82	I	12	1584.14	IV
400	799.41	III	100 r	2246.43	II	100	4874.10	I	12 d	1587.05	IV
15	1005.32	II	500	2246.51	III	80	5027.35	II	11	1615.92	IV
10	1065.49	II	75 r	2248.74	II	1000	5209.08	I	12	1618.57	IV
12	1072.23	II	75	2280.03	II	1000	5465.50	I	11	1655.47	IV
250	1074.22	II	30 h	2309.56	I	100	5471.55	I	15 c	1701.97	IV
150	1107.03	II	700	2310.04	III	100	5667.34	I	20 d	1887.47	III
150	1112.46	II	70 r	2317.05	II	10 h	6268.50	I	12	1960.76	IV
60	1195.83	II	80 r	2320.29	II	320	7687.78	I	11	1965.08	IV
50	1223.33	II	70 r	2324.68	II	25	8005.4	II	12 d	2106.33	IV
50	1240.80	II	80 r	2331.40	II	500	8273.52	I	30	2230.33	III
50	1246.87	II	70	2357.92	II	25	8403.8	II	16	2232.19	III
55	1256.81	II	50 h	2375.02	I	30 h	8645.70	I	20 h	2246.70	III
55	1257.55	II	75	2411.41	II	10 h	8704.85	I	300	2315.65	II
50	1266.63	II	90 r	2413.23	II	12	8747.6	II	18	2386.99	III
70	1273.67	II	100 r	2437.81	II	15	9000.9	II	17	2394.03	III
65	1297.51	II	80	2447.93	II	10	12551.0	I	300	2420.99	II
85	1311.20	II	80	2473.84	II	60	16819.5	I	300	2424.73	II
55	1313.81	II	60	2506.63	II	20	17416.7	I	25	2459.31	III
50	1314.61	II	50 h	2575.63	I	15	18307.9	I	18	2468.85	III
60	1323.84	II	60	2660.49	II	15	18382.3	I	20	2474.73	III
60	1342.09	II	60	2721.77	I				1000	2493.15	II
50	1342.57	II	75	2767.54	II	**Sodium Na Z = 11**			25	2497.03	III
70	1346.62	II	100 h	2824.39	I	7	142.232	IV	17	2510.26	III
50	1353.54	II	30 h	3130.02	I	8	146.064	IV	20	2543.84	I
150	1364.50	II	90	3180.70	II	9	150.298	IV	10	2543.87	I
100	1396.00	II	100	3267.35	II	8	150.687	IV	70	2593.87	I
100	1410.93	II	55000 r	3280.68	I	8	155.510	IV	35	2593.92	I
90	1419.72	II				8	156.537	IV			

Intensity	Wavelength/Å	
850	2611.81	II
850	2661.00	II
1000	2671.83	II
200	2680.34	I
100	2680.43	I
1000	2841.72	II
400	2852.81	I
200	2853.01	I
2	2893.62	I
1100	2904.92	II
1100	2917.52	II
1100	2919.05	II
1200	2919.85	II
1300	2920.95	II
1000	2923.49	II
1200	2951.24	II
1100	2952.40	II
1000	2977.13	II
1100	2979.66	II
1100	2980.63	II
1300	2984.19	II
1700	3124.42	II
2500	3135.48	II
1700	3137.86	II
2000	3149.28	II
2000	3163.74	II
1000	3179.06	II
1700	3189.79	II
1600	3212.19	II
1500	3257.96	II
1700	3285.60	II
1700	3301.35	II
1200	3302.37	I
600	3302.98	I
1500	3304.96	II
1000	3318.04	II
50	3426.86	I
1500	3533.05	I
1200	3631.27	II
6	4238.99	I
10	4242.08	I
1	4249.41	I
2	4252.52	I
15	4273.64	I
20	4276.79	I
2	4287.84	I
3	4291.01	I
30	4321.40	I
40	4324.62	I
3	4341.49	I
5	4344.74	I
40	4390.03	I
60	4393.34	I
5	4419.88	I
8	4423.25	I
60	4494.18	I
100	4497.66	I
10	4541.63	I
15	4545.19	I
120	4664.811	I
200	4668.560	I

Intensity	Wavelength/Å	
20	4747.941	I
30	4751.822	I
200	4978.541	I
400	4982.813	I
40	5148.838	I
80	5153.402	I
280	5682.633	I
70	5688.193	I
560	5688.205	I
80000	5889.950	I
40000	5895.924	I
120	6154.225	I
240	6160.747	I
130	6530.70	II
130	6544.04	II
130	6545.75	II
20	7373.23	I
10	7373.49	I
50	7809.78	I
25	7810.24	I
4400	8183.256	I
800	8194.790	I
8800	8194.824	I
100	8649.92	I
60	8650.89	I
25	8942.96	I
40	9153.88	I
60	9465.94	I
80	9961.28	I
20	10566.00	I
60	10572.28	I
200	10746.44	I
80	10749.29	I
120	10834.87	I
35	11190.19	I
50	11197.21	I
400	11381.45	I
1000	11403.78	I
400	12679.17	I
60	14767.48	I
100	14779.73	I
60	16373.85	I
100	16388.85	I
400	18465.25	I
50	22056.44	I
25	22083.67	I
60	23348.41	I
100	23379.13	I

Strontium Sr Z = 38

Intensity	Wavelength/Å	
15	298.12	IV
15	300.12	IV
125	330.67	III
500	351.62	III
75	358.80	III
250	363.49	III
150	371.21	III
20	378.53	IV
75	392.44	IV
50	393.00	IV
50	396.22	IV
1000	437.24	III

Intensity	Wavelength/Å	
1875	491.79	III
1250	507.04	III
3750	514.38	III
10	517.28	V
2500	562.75	III
25	578.01	V
30	624.93	V
25	642.23	V
50	649.21	V
25	660.94	V
200	664.43	IV
35	686.23	V
100	710.35	IV
12	747.82	V
50	1025.23	III
35	1125.49	III
50	1236.23	III
1400	2152.84	II
1400	2165.96	II
100	2273.71	III
100	2340.13	III
50	2346.97	IV
160	2428.10	I
100	2486.52	III
40	2555.60	IV
40	2571.04	IV
100	3002.61	III
200	3012.32	III
10	3019.29	IV
100	3021.73	III
50	3061.43	III
50	3182.61	III
100	3235.39	III
400	3351.25	I
650	3380.71	II
50	3430.76	III
950	3464.46	II
600	3969.26	I
1300	4030.38	I
46000	4077.71	II
32000	4215.52	II
9	4298.57	IV
340	4305.45	II
65000	4607.33	I
9	4685.08	IV
3200	4722.28	I
2200	4741.92	I
1400	4784.32	I
4800	4811.88	I
3600	4832.08	I
3000	4872.49	I
2000	4876.32	I
1000	4891.98	I
8000	4962.26	I
1300	4967.94	I
800 h	5156.07	I
1400	5222.20	I
2000	5225.11	I
2000	5229.27	I
2800	5238.55	I
4800	5256.90	I

Intensity	Wavelength/Å	
40	5257.71	III
40	5443.48	III
1500	5450.84	I
7000	5480.84	I
3500	5504.17	I
2600	5521.83	I
2000	5534.81	I
2000	5540.05	I
1000	6380.75	I
900 h	6386.50	I
600 h	6388.24	I
9000	6408.47	I
5500	6504.00	I
1000	6546.79	I
1700	6550.26	I
3000	6617.26	I
1800	6791.05	I
4800	6878.38	I
1200	6892.59	I
5500	7070.10	I
2500	7309.41	I
500	7621.50	I
400 h	7673.06	I
200 h	8422.80	I
120	8505.69	II
200	8688.91	II
100	9294.10	I
400 h	9448.95	I
600	9596.00	I
300	9624.70	I
100	9638.10	I
100 h	9647.70	II
300	10036.66	II
1000	10327.31	II
200	10914.88	II
700	11241.25	I
100	12014.76	II
60	12445.90	II
40	12495.00	I
75	12974.70	II
100	13123.80	II
50	17447.40	I
230	20261.40	I
120	20700.70	I
30	26023.60	I

Sulfur S Z = 16

Intensity	Wavelength/Å	
5	437.4	V
5	438.2	V
5	439.6	V
20	519.3	IV
20	520.1	IV
40	520.8	IV
20	522.0	IV
20	522.5	IV
20	551.2	IV
40	652.5	IV
40	653.0	IV
70	653.6	IV
40	654.0	IV
70	655.6	IV
20	655.9	IV

Intensity	Wavelength/Å		Intensity	Wavelength/Å		Intensity	Wavelength/Å		Intensity	Wavelength/Å	
110	657.3	IV	355	1316.542	I	285	3902.0	I	160	9035.9	I
40	658.3	V	290	1316.618	I	160	3928.6	III	450	9212.9	I
70	659.8	V	375	1323.515	I	360	3933.3	II	450	9228.1	I
40	660.9	IV	355	1326.643	I	450	4120.8	I	450	9237.5	I
160	661.4	IV	775	1381.552	I	280	4142.3	II	285	9413.5	I
110	663.2	V	710	1385.510	I	360	4145.1	II	285	9421.9	I
40	663.7	IV	960	1388.435	I	450	4153.1	II	285	9437.1	I
40	664.8	IV	640	1389.154	I	450	4162.7	II	650	9649.9	I
70	666.1	IV	775	1392.588	I	360	4253.6	III	450	9672.3	I
20	678.1	V	1000	1396.112	I	450	4694.1	I	450	9680.8	I
40	680.3	V	300	1409.337	I	285	4695.4	I	450	9693.7	I
110	680.9	V	510	1425.030	I	160	4696.2	I	285	9697.3	I
40	681.6	V	425	1433.280	I	280	4716.2	II	285	9739.7	I
20	693.5	V	300	1436.968	I	450	4815.5	II	285	9932.3	I
70	729.5	III	300	1448.229	I	360	4924.1	II	285	9949.8	I
110	732.42	III	425	1472.972	I	450	4925.3	II	285	9958.9	I
70	735.2	III	550	1473.995	I	285	4993.5	I	285	10455.5	I
70	738.5	III	300	1474.380	I	360	5428.6	II	285	10459.5	I
110	744.9	IV	355	1481.665	I	650	5432.8	II			
110	748.4	IV	485	1483.039	I	1000	5453.8	II	*Tantalum Ta Z = 73*		
110	750.2	IV	300	1483.233	I	1000	5473.6	II	60	493.07	V
110	753.8	IV	330	1485.622	I	1000	5509.7	II	1000	890.87	V
285	786.5	V	390	1487.150	I	280	5564.9	II	500	947.30	V
70	789.0	III	20	1624.0	IV	1000	5606.1	II	67	999.34	IV
70	796.7	III	20	1629.2	IV	450	5640.0	II	79	1116.10	IV
70	800.5	IV	680	1666.688	I	450	5640.3	II	78	1136.17	IV
70	804.0	IV	640	1687.530	I	280	5647.0	II	85	1175.51	IV
70	809.7	IV	710	1807.311	I	650	5659.9	II	80	1189.28	IV
110	816.0	IV	680	1820.343	I	450	5664.7	II	80	1192.67	IV
70	824.9	III	640	1826.245	I	160	5706.1	I	85	1213.09	IV
70	836.3	III	710	1900.286	I	450	5819.2	II	500	1213.42	V
160	849.2	V	550	1914.698	I	450	6052.7	I	85	1215.53	IV
110	852.2	V	20	2387.0	IV	280	6286.4	II	90	1223.73	IV
220	854.8	V	40	2398.9	IV	450	6287.1	II	88	1238.12	IV
110	857.9	V	110	2460.5	III	450	6305.5	II	95	1240.06	IV
110	860.5	V	110	2489.6	III	450	6312.7	II	87	1258.34	IV
40	906.9	II	160	2496.2	III	280	6384.9	II	94	1264.91	IV
40	910.5	II	160	2499.1	III	280	6397.3	II	98	1272.42	IV
40	912.7	II	220	2508.2	III	280	6398.0	II	94	1275.48	IV
40	937.4	II	70	2636.9	III	360	6413.7	II	86	1275.94	IV
40	937.7	II	220	2665.4	III	160	6743.6	I	92	1308.51	IV
160	1062.7	IV	110	2691.8	III	285	6748.8	I	87	1315.58	IV
160	1073.0	IV	110	2702.8	III	450	6757.2	I	92	1332.38	IV
70	1073.5	IV	220	2718.9	III	450	7579.0	I	86	1343.30	IV
285	1077.1	III	110	2721.4	III	450	7629.8	I	92	1365.88	IV
40	1102.3	II	220	2726.8	III	285	7686.1	I	5000	1392.56	V
70	1194.0	III	220	2731.1	III	450	7696.7	I	91	1398.78	IV
70	1201.0	III	110	2741.0	III	1000	7924.0	I	93	1413.40	IV
40	1234.1	II	285	2756.9	III	160	7928.8	I	91	1454.32	IV
40	1250.5	II	110	2775.2	III	285	7930.3	I	92	1464.41	IV
110	1253.8	II	160	2785.5	III	450	7931.7	I	93	1469.82	IV
110	1259.5	II	110	2863.5	III	450	7967.4	I	90	1495.25	IV
275	1270.782	I	160	2904.3	III	450	7967.4	II	95	1514.19	IV
250	1277.216	I	160	2986.0	III	450	8314.7	I	85	1607.70	IV
280	1295.653	I	110	3097.5	IV	450	8314.7	II	7000	1709.10	V
275	1302.337	I	110	3497.3	III	450	8585.6	I	85	1712.16	IV
235	1302.863	I	160	3632.0	III	285	8680.5	I	85	1716.13	IV
235	1303.110	I	110	3709.4	III	450	8694.7	I	85	2055.75	IV
245	1303.430	I	160	3717.8	III	360	8874.5	I	1100	2140.13	II
260	1305.883	I	160	3838.3	III	110	8882.5	I	1500	2146.87	II
265	1310.194	I	285	3867.6	I	220	8884.2	I	1200	2182.71	II
									1100	2193.88	II

Intensity	Wavelength/Å		Intensity	Wavelength/Å		Intensity	Wavelength/Å		Intensity	Wavelength/Å	
1500	2196.03	II	770	2775.88	I	200	5037.37	I	75	7006.96	I
90	2199.58	IV	680	2796.34	I	100	5067.87	I	150	7148.63	I
1500	2199.67	II	680	2797.76	II	110	5115.84	I	110	7172.90	I
90	2207.64	IV	510	2806.58	I	100	5141.62	I	140	7301.74	I
1400 d	2210.03	II	640	2844.25	I	100	5143.69	I	160	7346.41	I
1400	2239.48	II	560	2848.52	I	330	5156.56	I	140 c	7352.86	I
1200	2250.76	II	1500	2850.49	I	110	5212.74	I	100	7356.96	I
840	2261.42	II	1900	2850.98	I	110 d	5218.45	I	90 cw	7369.09	I
990	2262.30	II	360	2861.98	I	140	5341.05	I	160	7407.89	I
990	2272.59	II	470	2871.42	I	200	5402.51	I	100	7882.37	I
790	2285.25	II	380	2880.02	I	130	5419.19	I	75	8026.50	I
600	2286.59	II	770	2891.84	I	90	5518.91	I	75	8281.62	I
990	2289.16	II	560	2902.05	I	150	5645.91	I			
440	2302.24	II	310	2915.49	I	130	5664.90	I	*Technetium Tc Z = 43*		
440	2302.93	II	410	2925.19	I	130	5776.77	I	10000 c	3636.07	I
440	2312.60	II	310	2932.70	I	90	5780.71	I	20000 c	4031.63	I
420	2315.46	II	1700	2933.55	I	130	5811.10	I	15000	4095.67	I
690	2331.98	II	470	2940.06	I	240	5877.36	I	20000	4262.27	I
550	2332.19	II	1200	2940.22	I	130	5882.30	I	30000	4297.06	I
250	2357.30	I	510	2951.92	I	90	5901.91	I	20000	4853.59	I
260	2361.09	I	340	2953.56	I	90	5918.95	I			
600	2364.24	II	1500	2963.32	I	130	5939.76	I	*Tellurium Te Z = 52*		
320	2371.58	I	770	2965.13	II	240	5944.02	I	8	802.28	II
1400	2387.06	II	770	2965.54	I	190 c	5997.23	I	8	1059.51	II
2400	2400.63	II	340	2969.47	I	100	6020.72	I	8	1077.66	II
320	2416.89	II	430	2975.56	I	250	6045.39	I	10	1161.42	II
360	2427.64	I	1800	3012.54	II	100	6047.25	I	10	1174.34	II
360	2429.71	II	290 d	3027.48	I	100	6101.58	I	12	1175.79	II
480	2432.70	II	530	3049.56	I	65	6144.56	I	9	1208.54	II
380	2470.90	II	530	3069.24	I	130	6154.50	I	9	1220.98	II
600	2474.62	I	360	3077.24	I	150	6256.68	I	9	1253.62	II
500	2484.95	I	560	3103.25	I	150	6268.70	I	9	1270.52	II
600	2488.70	II	380	3124.97	I	150	6309.58	I	10	1324.92	II
500	2490.46	I	380	3130.58	I	75	6325.08	I	9	1363.24	II
600	2504.45	I	270	3132.64	I	65	6341.17	I	8	1366.73	II
600	2507.45	I	320	3170.29	I	75	6356.16	I	10	1374.80	II
1200 d	2526.35	I	270	3173.59	I	65	6360.84	I	10	1608.41	II
600	2532.12	II	600	3180.95	I	90	6389.45	I	10	1613.15	II
1200	2559.43	I	300	3223.83	I	65	6428.60	I	5	1655.4	I
460	2562.10	I	1100	3311.16	I	250	6430.79	I	5	1688.5	I
600	2577.37	II	680	3318.84	I	200	6450.36	I	6	1700.0	I
600	2603.49	II	330 d	3330.99	II	380	6485.37	I	5	1708.0	I
1400	2608.63	I	640	3371.54	I	65	6505.52	I	10	1822.4	I
1200	2635.58	II	360	3385.05	I	100	6514.39	I	26000	2002.02	I
860	2636.90	I	450	3406.94	I	100	6516.10	I	6500	2081.16	I
2400	2647.47	I	490	3480.52	I	100	6574.84	I	18000	2142.81	I
2600	2653.27	I	380	3497.85	I	110	6611.95	I	3200	2147.25	I
1900	2656.61	I	490	3511.04	I	75	6621.30	I	500	2259.02	I
1500	2661.34	I	750	3607.41	I	100	6673.73	I	1200	2383.26	I
770	2675.90	II	980	3626.62	I	180	6675.53	I	1500	2385.78	I
1500	2685.17	II	500	3642.06	I	75 c	6740.73	I	50	2438.69	II
470	2694.52	II	210	3918.51	I	75	6771.74	I	120	2530.72	I
1000	2698.30	I	210	3970.10	I	160 c	6813.25	I	100	2649.66	II
1200	2710.13	I	210	3996.17	I	210	6866.23	I	80	2661.10	II
2600	2714.67	I	410	4061.40	I	180	6875.27	I	110	2677.13	I
470	2727.44	II	310	4067.91	I	150	6902.10	I	100	2858.29	II
1200	2748.78	I	300	4205.88	I	140	6927.38	I	150	2895.41	II
860	2749.83	I	360 c	4510.98	I	140	6928.54	I	70	2967.29	II
410	2752.49	II	340	4574.31	I	65	6951.26	I	70	3047.00	II
1000	2758.31	I	260	4619.51	I	180	6966.13	I	100	3175.14	I
430	2761.68	II	450	4681.88	I	110 d	6995.22	I	60	3256.80	II
									60	3329.22	II

Intensity	Wavelength/Å		Intensity	Wavelength/Å		Intensity	Wavelength/Å		Intensity	Wavelength/Å	
150	3406.79	II	689	9956.30	I	810 d	3472.79	II	390	4196.74	I
50	3442.25	II	325	9977.13	I	810	3500.84	II	650	4203.74	I
50	3521.11	II	5950	10051.41	I	5700	3509.17	II	600	4206.49	I
50	3552.19	II	4097	10091.01	I	1300	3523.66	II	480	4215.09	I
100	3611.78	II	381	10118.08	I	1100	3540.24	II	480	4232.82	I
50	3617.57	II	397	10300.56	I	810	3543.89	II	650	4266.34	I
50	4006.52	II	745	10493.57	I	3200	3561.74	II	760 cw	4278.52	II
70	4127.32	II	1880	10918.34	I	810	3567.35	II	450	4310.42	I
100	4169.77	II	10200	11089.56	I	4200	3568.52	II	2200	4318.83	I
80	4225.73	II	508	11163.74	I	1600	3568.98	II	600	4322.23	I
100	4261.11	II	6620	11487.23	I	1100	3579.20	II	600	4325.83	II
60	4273.43	II	1580	13247.75	I	710	3585.03	II	3000	4326.43	I
80	4285.85	II	1050	14513.51	I	810	3596.38	II	600	4332.12	I
150	4364.00	II	1480	15452.45	I	1600	3600.44	II	870	4336.43	I
75	4385.10	II	2430	15546.23	I	810	3625.54	II	600	4337.64	I
170	4478.63	II	3760	16403.90	I	2300	3650.40	II	1700	4338.41	I
80	4537.07	II	1960	17303.54	I	810	3654.88	II	700	4340.62	I
100	4557.78	II	2780	18291.59	I	2000	3658.88	II	870	4356.81	I
70	4630.62	II	1020	21043.73	I	3800	3676.35	II	330	4382.45	I
100	4641.12	II	464	21602.50	I	810	3682.26	II	300	4388.23	I
180	4654.37	II	74	22555.29	I	450	3693.58	I	260	4390.91	I
200	4686.91	II	38	26539.17	I	450	3700.12	I	350	4423.10	I
100	4696.38	II				4700	3702.86	II	240	4436.12	I
100	4706.53	II	**Terbium Tb**	*Z* = 65		2400	3703.92	II	240	4448.04	I
100	4766.05	II	1000	1259.40	IV	1000 d	3711.76	II	430	4493.07	I
100	4784.87	II	1000	1327.67	IV	650	3745.04	I	75	4514.31	II
100	4827.14	II	1000	1373.86	IV	870	3747.17	II	110	4549.07	I
150	4831.28	II	5000	1595.39	IV	870	3747.34	II	110	4550.45	I
150	4842.90	II	2000	1633.19	IV	1100	3755.24	II	110	4556.46	I
130	4865.12	II	2000	2027.79	IV	650	3759.35	I	110	4563.69	II
200	4866.24	II	1000	2089.98	IV	1700	3765.14	I	210	4578.69	II
8	5083.0	I	1000	2332.54	IV	2100	3776.49	II	65	4584.84	II
50	5449.84	II	110	2584.61	II	600	3783.53	I	65	4591.56	II
50	5487.95	II	110	2608.57	II	410	3789.92	I	75 d	4626.32	II
150	5576.35	II	130	2628.69	II	760 d	3806.85	II	95	4626.94	II
150	5649.26	II	140	2669.29	II	1500	3830.26	I	65	4632.07	I
100	5666.20	II	190	2704.07	II	540	3833.42	I	65 h	4636.59	I
200	5708.12	II	270	2769.53	II	920 d	3842.50	II	85	4641.00	II
150	5755.85	II	320	2897.44	II	3700	3848.73	II	210	4641.98	II
100	5974.68	II	250	2956.21	II	3500 w	3874.17	II	260 cw	4645.31	II
50	6367.13	II	230	3010.59	II	480	3888.22	I	80	4647.23	I
10 h	6790.0	I	230	3016.18	II	490	3894.64	I	80	4662.79	I
20 h	6837.6	I	460	3053.55	II	2400	3899.20	II	80	4676.90	I
20 h	6854.7	I	460	3070.05	II	1600	3901.33	I	70 c	4681.87	I
15 h	7191.1	I	670	3078.86	II	480	3908.06	I	80	4688.63	II
20 h	7263.5	I	480	3082.36	II	650	3915.43	I	80	4693.11	II
12	7460.98	II	480	3089.58	II	760	3925.45	II	200	4702.41	II
15	7468.75	II	480	3102.96	II	810 d	3939.52	II	110	4707.94	II
15	7921.69	II	440	3139.64	II	2200 d	3976.84	II	80	4739.93	I
15	7943.14	II	480	3187.26	II	1800	3981.87	II	70	4747.80	I
10	7950.34	II	480	3199.56	II	970	4002.59	II	410 cw	4752.53	II
30 h	8061.4	I	1100	3218.93	II	1900	4005.47	II	180	4786.78	I
10	8122.44	II	1200	3219.98	II	760	4012.75	II	100	4813.77	I
20	8186.44	II	480	3252.32	II	870	4032.28	I	80	4875.57	II
15	8273.53	II	760	3280.31	II	2100	4033.03	II	80	4881.15	II
15	8672.95	II	760	3281.40	II	430	4054.12	I	95	4915.90	I
10	8733.81	II	1000	3285.04	II	410	4060.37	I	65	4931.79	I
205	8758.18	I	1500	3293.07	II	1300	4061.58	I	85	4993.82	II
81	9004.37	I	3800	3324.40	II	650	4105.37	I	110	5078.25	I
5660	9722.74	I	760	3349.42	II	1100	4144.41	II	75	5089.12	II
532	9868.92	I	760	3364.93	II	350	4158.53	I	85	5186.13	I
			810	3454.06	II						

Intensity	Wavelength/Å	
120	5228.12	I
75	5248.71	I
75 w	5262.11	II
75	5281.05	I
65	5304.72	I
110	5319.23	I
65 w	5337.90	I
160	5354.88	I
75	5369.72	I
75	5375.98	I
50	5424.10	II
55	5459.81	I
55	5509.61	I
50	5514.54	I
65	5524.12	I
85 c	5747.58	I
75	5795.64	I
75	5803.13	II
65	5815.36	I
65	5851.07	I
65	5870.62	I
65 c	5920.78	I
75	5967.34	II
35	6331.68	II
35 cw	6518.68	I
35	6581.82	I
90	6677.94	II
40 cw	6702.61	I
130	6794.58	II
55	6896.37	II
45 h	6899.95	I
40	6901.98	I
65	7204.28	I
40	7257.73	I
45	7348.88	II
45	7496.12	I
27 h	7582.03	II
45	7590.24	I
65	7596.44	I
30	7627.81	I
30	7737.63	I
30	7855.79	II
27	7927.90	II
30	8025.42	II
30	8085.06	II
65	8194.82	II
95	8212.57	I
40	8450.06	II
30 h	8511.80	I
45	8583.45	II
30	8603.40	I
65	8765.74	II

Thallium Tl Z = 81

Intensity	Wavelength/Å	
10	570.49	IV
5 r	670.87	II
15 r	696.30	II
5 r	709.23	II
10 r	817.18	II
5 r	836.34	II
8 r	1018.85	II
30	1028.69	IV

Intensity	Wavelength/Å	
20	1034.73	IV
20	1036.61	IV
10 r	1049.73	II
8 r	1050.30	II
5 r	1074.97	II
30	1079.68	IV
10 r	1130.17	II
15 r	1162.55	II
10 r	1167.43	II
10 r	1183.41	II
12 r	1194.84	II
5 r	1246.00	II
10	1266.33	III
15 r	1307.50	II
8 r	1310.20	II
25 r	1321.71	II
8 r	1330.40	II
10 r	1373.52	II
10	1477.14	III
8 r	1489.65	I
10 r	1499.30	II
10 r	1507.82	II
15 r	1561.58	II
10 r	1568.57	II
7 r	1593.26	II
5 h	1616.	I
5	1685.40	I
10 r	1792.76	II
12 r	1814.85	II
25 r	1908.64	II
100 r	2007.56	I
100 r	2210.71	I
30	2298.04	II
140	2315.98	I
900 h	2379.69	I
20	2530.86	II
700	2580.14	I
420	2709.23	I
4400 d	2767.87	I
10	2849.80	II
2800	2918.32	I
20	3091.56	II
15	3185.51	II
15	3186.56	II
15	3187.74	II
1200	3229.75	I
15	3291.01	II
15	3369.15	II
9	3456.34	III
20000	3519.24	I
5000	3529.43	I
8	3540.08	II
9	3560.68	II
12000 w	3775.72	I
10	3832.30	II
10	3887.15	II
7	4109.85	III
6	4269.81	III
20	4274.98	II
40	4306.80	II
20	4737.05	II

Intensity	Wavelength/Å	
15	4981.35	II
25	5078.54	II
25	5152.14	II
18000	5350.46	I
15 d	5384.85	II
10	5410.97	II
25	5949.48	II
10	6179.98	II
10	6378.32	II
16 h	6549.84	I
10	6966.5	II
10	7815.80	I
20	8373.6	I
10	8474.27	I
10	8664.1	II
20	9130.	II
20	9130.5	I
40	9509.4	I
20	9930.4	I
30	10011.9	I
40	10488.80	I
1000	11512.82	I
150	12736.4	I
700	13013.2	I

Thorium Th Z = 90

Intensity	Wavelength/Å	
150	1707.37	IV
200	1959.02	IV
200	2002.34	IV
200	2413.50	III
200	2427.94	III
200	2431.68	III
200	2441.24	III
500	2565.593	II
480	2692.415	II
520	2747.156	II
410	2752.166	II
800	2832.315	II
1200	2837.295	II
100	2848.084	I
550	2870.406	II
100	2936.086	I
100	2943.729	I
420	3049.092	II
450	3067.729	II
670	3078.828	II
480	3080.217	II
510	3108.296	II
100	3116.263	I
510	3119.526	II
510	3122.963	II
480	3125.507	II
100	3136.216	I
420	3139.306	II
420	3142.835	II
420	3175.726	II
1100	3180.193	II
770	3188.233	II
560	3221.292	II
560	3229.009	II
480	3235.84	II
590	3238.116	II

Intensity	Wavelength/Å	
910	3256.274	II
180	3257.366	I
910	3262.668	II
620	3287.789	II
910	3291.739	II
620	3292.520	II
240	3301.650	I
480	3304.238	I
510	3321.450	II
840	3325.120	II
250	3330.476	I
620	3334.604	II
620	3337.870	II
310	3348.768	I
980	3351.228	II
620	3358.602	II
250	3374.974	I
1300	3392.035	II
200	3396.727	I
250	3398.544	I
200	3405.558	I
250	3413.012	I
390	3421.210	I
270	3423.989	I
980	3433.998	II
770	3435.976	II
1300	3469.920	II
170	3471.218	I
200	3486.552	I
670	3539.587	II
180	3544.018	I
170	3549.595	I
200	3555.013	I
530	3559.451	II
200	3576.557	I
270	3592.780	I
270	3598.120	I
980	3609.445	II
200	3612.427	I
480	3615.133	II
270	3635.943	I
210	3642.248	I
170	3649.735	I
220	3663.202	I
280	3669.968	I
700	3675.567	II
150	3682.486	I
170	3692.566	I
180	3698.105	I
340	3706.767	I
590	3719.435	I
770	3721.825	II
1300	3741.183	II
310	3747.539	I
650	3752.569	II
180	3770.056	I
590	3803.075	I
450	3828.384	I
840	3839.746	II
450	3863.405	II
210	3875.374	I

Intensity	Wavelength/Å	
340	3895.419	I
590	3929.669	II
200	3932.911	I
390	3967.392	I
200	3972.155	I
150	3980.089	I
530	3994.549	II
220	4008.210	I
220	4009.056	I
280	4012.495	I
4200	4019.129	II
250	4030.842	I
250	4036.047	I
250	4063.407	I
700	4086.520	II
700	4094.747	II
150	4100.341	I
840	4108.421	II
240	4112.754	I
280	4115.758	I
1100	4116.713	II
200	4127.411	I
200	4134.067	I
450	4149.986	II
620	4178.060	II
620	4208.890	II
110	4253.538	II
110	4260.333	I
480	4277.313	II
700	4282.042	II
130	4337.277	I
1300	4381.860	II
1100	4391.110	II
110	4498.940	I
280	4510.527	II
90	4723.438	I
50	4840.843	I
280	4863.163	II
260	5017.255	II
110	5067.974	I
120	5148.211	II
95	5216.596	II
110	5231.160	I
95	5247.654	II
60	5343.581	I
60	5587.026	I
95	5707.103	II
70	5760.551	I
85	5989.044	II
60	6169.822	I
50	6182.622	I
50	6274.116	II
50	6274.117	II
50	6355.911	II
60	6457.283	I
50	6462.614	I
50 h	6531.342	I
55	6989.656	I
30	7045.795	II
30	7084.171	I
30	7168.896	I
40	7191.132	II
35	7208.006	I
50	7525.508	II
30	7647.380	I
30	8330.451	I
40	8967.641	I
20	9833.42	I
20	10726.93	I
20	10942.24	II
30	11230.259	I
20	11984.67	II
20	17208.22	II
15	18811.88	I
10	22264.35	II

Thulium Tm Z = 69

Intensity	Wavelength/Å	
5000	2185.94	III
360	2284.79	II
20000	2296.21	III
5000	2305.03	III
20000	2311.16	III
5000	2312.72	III
5000	2326.19	III
6000	2328.50	III
6000	2329.29	III
3000	2331.80	III
3000	2357.05	III
4000	2406.63	III
450	2409.02	II
450	2426.17	II
770	2480.13	II
30000	2489.44	III
2000	2504.71	III
1300	2509.08	II
3000	2519.78	III
130	2527.02	I
10000	2552.46	III
360	2552.76	I
540	2561.65	II
430	2588.27	II
170 h	2596.49	I
810	2607.06	II
730	2624.33	II
5000	2682.32	III
2000	2707.03	III
3000	2719.47	III
540	2721.19	II
3000	2724.44	III
4000	2727.56	III
680	2794.60	II
730	2797.27	II
2000	2806.77	III
580	2827.92	II
200	2854.17	I
1600	2869.23	II
1000	2947.72	III
490	2973.22	I
1000	2998.28	III
1500	3015.30	II
360	3081.12	I
7400	3131.26	II
2300	3133.89	II
1900	3151.04	II
1500	3157.34	II
450	3172.65	I
2300	3172.83	II
1200	3236.81	II
1600	3240.23	II
2300	3241.54	II
320	3246.96	I
1900	3258.05	II
1600	3266.64	II
1200	3267.40	II
1100	3276.81	II
1200	3283.40	II
1200	3285.61	II
2300	3291.00	II
2000	3302.46	II
1200	3309.80	II
230	3349.99	I
4000	3362.61	II
1700	3397.50	II
850	3410.05	I
340	3412.59	I
340	3416.59	I
6400	3425.08	II
340	3429.33	I
4900	3441.50	II
4900	3453.66	II
8500	3462.20	II
210	3467.51	I
340	3476.69	I
340	3480.98	I
420	3487.38	I
340	3499.95	I
250	3517.60	I
1700	3535.52	II
420	3537.91	I
210	3555.82	I
340	3560.92	I
420	3563.88	I
1300	3566.47	II
420	3567.36	I
280	3586.07	I
2100	3608.77	II
1000	3629.09	III
380	3638.41	I
1100	3668.09	II
4800	3700.26	II
3800	3701.36	II
7700	3717.91	I
2400	3734.12	II
5000	3744.06	I
1700	3751.81	I
6000	3761.33	II
4800	3761.91	II
7100	3795.75	II
770	3798.54	I
600	3807.72	I
290	3826.39	I
1300	3838.20	II
290	3840.87	I
8900	3848.02	II
6800	3883.13	I
1800	3883.44	II
5400	3887.35	I
440	3896.62	I
3500	3916.48	I
1500	3949.27	I
1500	3958.10	II
1800	3996.52	II
220	4024.23	I
380	4044.47	I
10000	4094.19	I
9500	4105.84	I
1100	4138.33	I
8800	4187.62	I
6000	4203.73	I
380	4222.67	I
3000	4242.15	II
270	4271.71	I
150	4298.36	I
2700	4359.93	I
1400	4386.43	I
200	4394.42	I
140	4396.50	I
120	4454.03	I
540	4481.26	II
150	4519.60	I
260	4522.57	II
110	4548.60	I
270	4599.02	I
300	4615.94	II
80	4626.33	II
95	4626.56	II
110	4634.26	II
120	4655.09	I
160	4681.92	I
120	4691.11	I
110	4724.26	I
680	4733.34	I
70	4759.90	I
80	4831.20	II
140	4957.18	I
160	5009.77	II
160	5034.22	II
150	5060.90	I
95	5113.97	I
80	5213.38	I
650	5307.12	I
80	5346.49	II
270	5631.41	I
520	5675.84	I
40	5684.76	II
35	5709.97	II
190	5764.29	I
35	5838.76	II
240	5895.63	I
140	5971.26	I
200	6460.26	I
95	6604.96	I
110	6779.77	I
120	6844.26	I
80	6845.76	I

Intensity	Wavelength/Å		Intensity	Wavelength/Å		Intensity	Wavelength/Å		Intensity	Wavelength/Å	
10	6937.37	I	120 r	1823.00	I	110	3330.62	I	15	498.26	V
10	7017.90	I	9	1831.89	II	60	3351.97	II	14	502.08	V
12	7034.34	I	50 r	1848.75	I	10	3472.46	II	13	526.57	V
10	7106.14	I	200 r	1860.32	I	11	3575.45	II	18	779.07	IV
17	7272.62	I	80	1886.05	I	280 r	3801.02	I	20	1298.66	III
14	7310.51	I	100	1891.40	I	10	5332.36	II	20	1298.97	III
14	7432.18	I	12	1899.91	II	20	5561.95	II	23	1455.19	III
75	7481.08	I	50	1909.30	I	25	5588.92	II	20	1467.34	IV
75	7490.20	I	80	1925.31	I	500	5631.71	I	11	1717.40	V
140	7558.33	I	500	1941.86	III	15	5799.18	II	10	1841.49	V
80	7731.53	I	150	1952.15	I	50	5925.44	I	20	2067.56	IV
40	7856.08	I	50 h	1977.6	I	100	5970.30	I	18	2103.16	IV
55	7927.51	I	80	1984.20	I	150	6037.70	I	180	2273.28	I
110	7930.84	I	50	2040.66	I	250	6069.00	I	190	2279.96	I
95	8017.90	I	50	2054.03	I	100	6073.46	I	190	2305.67	I
27	8472.01	II	70	2058.31	I	400	6149.71	I	22	2413.99	III
			80	2068.58	I	200	6154.60	I	25	2516.05	III
Tin Sn Z = 50			100	2072.89	I	150	6171.50	I	360	2525.60	II
7	169.47	II	100	2073.08	I	100	6310.78	I	24	2527.84	III
150	361.01	V	200	2096.39	I	70	6453.50	II	210	2529.85	I
100	753.01	III	100	2100.93	I	25	6844.05	II	190	2531.25	II
200	910.92	III	100 r	2113.93	I	20	7191.40	II	190	2534.62	II
500	956.25	IV	50	2121.26	I	10	7387.79	II	130	2535.87	II
7	985.13	II	40 r	2148.73	I	13	7741.80	II	23	2540.06	III
500	1019.72	IV	20 r	2151.43	I	100	7754.97	I	24	2563.44	III
1000	1044.49	IV	30	2151.54	II	100 h	8030.5	I	23	2565.42	III
1000	1073.41	IV	80	2171.32	I	200	8114.09	I	22	2567.56	III
200	1089.35	V	150 r	2194.49	I	80	8357.04	I	270	2599.92	I
8	1108.19	II	300 r	2199.34	I	300	8422.72	I	340	2605.15	I
1000	1119.34	IV	400 r	2209.65	I	400	8552.60	I	510	2611.28	I
1000	1139.29	III	80 r	2231.72	I	50 h	8681.7	I	300	2619.94	I
1000	1158.33	III	400 r	2246.05	I	50 h	9410.86	I	640	2641.10	I
200	1160.74	V	60	2251.17	I	80 h	9415.37	I	800	2644.26	I
10	1161.43	II	400 r	2268.91	I	150	9616.40	I	950	2646.64	I
1000	1184.25	III	200 r	2286.68	I	50	9741.1	I	250	2742.32	I
2000	1210.52	III	600 r	2317.23	I	100 h	9742.8	I	250	2802.50	I
9	1219.07	II	300 r	2334.80	I	300 h	9805.38	I	190	2841.94	II
13	1223.70	II	1000 r	2354.84	I	500	9850.52	I	180	2877.44	I
11	1243.00	II	22	2368.33	II	54	10894.00	I	280	2884.11	II
2000	1251.38	V	100	2408.15	I	70	11191.85	I	450	2912.08	I
1000	1259.92	III	800 r	2421.70	I	56	11277.66	I	340	2928.34	I
20	1290.86	II	1000 r	2429.49	I	200	11454.59	I	1100	2942.00	I
200	1294.36	V	15	2448.98	II	200	11616.26	I	1300	2948.26	I
1000	1305.97	III	300	2483.39	I	258	11739.78	I	1600	2956.13	I
1000	1314.55	IV	13	2483.48	II	96	11825.18	I	22	2984.75	III
20	1316.59	II	10	2486.99	II	106	11835.82	I	1300 d	3066.22	II
1000	1327.34	III	200	2495.70	I	254	11932.99	I	1100	3072.97	II
1000	1347.65	III	400	2546.55	I	48	12009.50	I	1600	3075.22	II
1000	1386.74	III	500 r	2571.58	I	111	12313.24	I	2300	3078.64	II
25	1400.52	II	200	2594.42	I	42	12530.87	I	3600	3088.02	II
1000	1437.52	IV	200 r	2661.24	I	42	12536.5	I	720	3119.72	I
20	1475.15	II	700 r	2706.51	I	89	12888.5	I	500	3161.20	II
9	1489.22	II	150	2779.81	I	187	12981.7	I	780	3161.77	II
1000	1570.36	III	1400 r	2839.99	I	187	13018.5	I	1000	3162.57	II
10 r	1737.21	I	1000 r	2863.32	I	68	13081.5	I	1600	3168.52	II
15 r	1751.46	I	700 r	3009.14	I	378	13460.2	I	2400	3186.45	I
20 r	1764.98	I	850 r	3034.12	I	144	13608.2	I	1000	3190.87	II
30 r	1790.75	I	12	3047.50	II	40	20861.7	I	3100	3191.99	I
80 r	1804.60	I	550 r	3175.05	I	4	24738.2	I	3800	3199.92	I
15	1811.34	II	550 r	3262.34	I				780	3202.54	II
500	1811.71	III	50	3283.21	II	**Titanium Ti Z = 22**			1100	3217.06	II
40 r	1815.74	I				17	252.96	V			

Intensity	Wavelength/Å		Intensity	Wavelength/Å		Intensity	Wavelength/Å		Intensity	Wavelength/Å	
1300	3222.84	II	950	3962.85	I	840	5020.03	I	20	8466.87	III
6600	3234.52	II	950	3964.27	I	840	5022.87	I	90	8675.39	I
5200	3236.57	II	4800	3981.76	I	1200	5035.91	I			
4100	3239.04	II	570	3982.48	I	840	5036.47	I	*Tungsten W Z = 74*		
2600	3241.99	II	5700	3989.76	I	740	5038.40	I	5800	2001.71	II
1200	3248.60	II	7800	3998.64	I	1200	5039.95	I	13000	2008.07	II
1200	3252.91	II	950	4008.93	I	1400	5064.66	I	5100	2009.98	II
1200	3254.25	II	1200	4024.57	I	1100	5173.75	I	4100	2010.23	II
1200	3261.60	II	840	4078.47	I	1300	5192.98	I	4100	2014.23	II
840	3314.42	I	890	4286.01	I	1400	5210.39	I	7300	2026.08	II
2900	3322.94	II	840	4287.40	I	17	5278.12	III	15000	2029.98	II
2100	3329.46	II	950	4289.07	I	20	5398.93	IV	5300	2049.63	II
1800	3335.20	II	840	4290.94	I	340	5512.53	I	9700	2079.11	II
1100	3340.34	II	840	4295.76	I	270	5514.35	I	6100	2094.75	II
5700	3341.88	I	2000	4298.66	I	320	5514.54	I	2100	2118.87	II
4300	3349.04	II	200	4300.05	II	250	5644.14	I	2400	2121.59	II
12000	3349.41	II	2900	4300.56	I	130	5675.44	I	1500	2166.32	II
4100	3354.64	I	4100	4301.09	I	95	5689.47	I	1300	2204.48	II
7200	3361.21	II	6000	4305.92	I	95	5715.13	I	460	2249.80	I
1100	3370.44	I	1200	4314.80	I	85	5739.51	I	510	2277.58	I
4300	3371.45	I	330	4395.04	II	400	5866.46	I	530 d	2294.49	I
5700	3372.80	II	890	4427.10	I	230	5899.32	I	340	2309.02	I
2900 d	3377.48	I	230	4443.80	II	120	5918.55	I	440	2313.17	I
1400	3380.28	II	840	4449.15	I	150	5922.12	I	460	2321.63	I
5700	3383.76	II	550	4450.90	I	120	5941.76	I	390 d	2326.56	I
1400	3385.95	I	840	4453.32	I	300	5953.17	I	320	2354.61	I
1400	3387.84	II	950	4455.33	I	200	5965.84	I	580	2360.44	I
1100	3394.58	II	1100	4457.43	I	270	5978.56	I	850	2363.07	I
890	3444.31	II	240	4468.50	II	340	5999.04	I	510	2374.47	I
600	3461.50	II	530	4481.26	I	110	6064.63	I	670	2384.82	I
600	3477.18	II	780	4512.74	I	120	6085.23	I	730	2397.09	II
480	3491.05	II	1000	4518.03	I	120	6091.17	I	560	2397.73	I
890	3504.89	II	1000	4522.80	I	120	6126.22	I	560	2397.98	I
600	3510.84	II	780	4527.31	I	17	6246.65	IV	1700 d	2405.58	I
17	3576.44	IV	6000	4533.24	I	380	6258.10	I	610	2415.68	I
600	3610.16	I	240	4533.97	II	380	6258.70	I	870	2424.21	I
4800	3635.46	I	3600	4534.78	I	300	6261.10	I	1800	2435.96	I
6600	3642.68	I	2400	4535.58	I	55	6546.28	I	580	2444.06	I
7200	3653.50	I	1200	4535.92	I	65	6554.23	I	780	2451.48	II
600	3671.67	I	1200	4536.05	I	75	6556.07	I	870	2452.00	I
3100	3685.20	II	720	4544.69	I	18	6621.58	III	630	2454.98	I
600	3689.91	I	950	4548.77	I	18	6667.99	III	780	2455.51	I
2900	3729.82	I	240	4549.63	II	80	6743.12	I	780	2456.53	I
3300	3741.06	I	15	4549.84	III	20	7072.64	III	1100	2459.30	I
330	3741.64	II	950	4552.46	I	18	7084.57	III	1400	2466.85	I
5200	3752.86	I	720	4555.49	I	260	7209.44	I	480	2472.51	I
3300	3759.30	II	240	4571.98	II	130	7244.86	I	1200	2474.15	I
2900	3761.32	II	15 d	4572.20	III	130	7251.72	I	870	2480.13	I
840	3786.04	I	950	4617.27	I	120	7344.72	I	1500	2481.44	I
500	3882.89	I	480	4623.09	I	90	7357.74	I	480 d	2482.10	I
530	3900.54	II	720	4656.47	I	60	7364.11	I	580	2484.74	I
2600	3904.78	I	840	4667.59	I	60	7978.88	I	390	2487.50	I
500	3913.46	II	950	4681.92	I	55	8024.84	I	390	2489.23	II
500	3914.34	I	470	4840.87	I	75	8364.24	I	630	2495.26	I
15	3915.47	III	400	4885.08	I	100	8377.85	I	680	2504.70	I
1100	3924.53	I	380	4899.91	I	100	8382.54	I	75	2510.47	II
890	3929.88	I	5800	4981.73	I	75	8396.87	I	310	2520.46	I
1100	3947.78	I	4600	4991.07	I	120	8412.36	I	780	2521.32	I
4500	3948.67	I	4000	4999.51	I	170	8426.52	I	270	2522.04	II
4500	3956.34	I	3600	5007.21	I	490	8434.94	I	780	2523.41	I
5200	3958.21	I	3200 d	5014.19	I	240	8435.70	I	430	2527.76	I
									780	2533.64	I

Intensity	Wavelength/Å		Intensity	Wavelength/Å		Intensity	Wavelength/Å		Intensity	Wavelength/Å	
1200	2547.14	I	440	3046.44	I	2200	4302.11	I	580	2931.41	II
780	2550.38	I	810	3049.69	I	200	4378.48	I	530 p	2940.37	II
2700	2551.35	I	180	3073.28	I	180	4384.85	I	1300	2941.92	II
730	2561.97	I	180 d	3084.83	I	200	4408.28	I	830	2943.90	II
870	2580.49	I	370	3093.50	I	640	4484.19	I	580	2956.06	II
390	2584.39	I	240	3107.23	I	170	4588.73	I	580	2967.94	II
390	2589.17	II	240	3108.02	I	640	4659.87	I	580	2971.06	II
370	2601.96	I	230	3117.57	I	640	4680.51	I	530	2984.61	II
680	2606.39	I	260	3120.18	I	790	4843.81	I	630	3022.21	II
370	2608.32	I	290	3163.42	I	380	4886.90	I	630	3031.99	II
970	2613.08	I	320	3176.60	I	220	4982.59	I	580	3050.20	II
480	2613.82	I	190	3181.82	I	820	5053.28	I	630	3057.91	II
400	2620.25	I	390	3191.57	I	770	5224.66	I	630	3062.54	II
400	2622.21	I	390	3198.84	I	220	5514.68	I	580	3072.78	II
400	2625.22	I	520	3207.25	I	65	5648.37	I	580	3093.01	II
400	2632.48	I	1000	3215.56	I	55	5735.09	I	580	3102.39	II
400	2632.70	I	190	3232.49	I	45	5804.85	I	970	3111.62	II
810	2633.13	I	210	3254.36	I	40	5902.64	I	530	3119.35	II
400 d	2638.62	I	210	3259.66	I	55	5947.57	I	680	3124.95	II
650	2646.18	I	210 d	3266.62	I	55	5965.86	I	530	3139.61	II
400	2646.73	I	730	3300.82	I	55	6012.78	I	680	3149.24	II
1600	2656.54	I	440	3311.38	I	40	6021.52	I	530	3153.11	II
810	2662.84	I	440	3326.20	I	45	6292.02	I	730	3229.50	II
810	2671.47	I	440	3331.69	I	35	6404.21	I	680	3232.16	II
650	2677.28	I	390	3373.75	I	40	6445.12	I	730	3291.33	II
2100	2681.42	I	230	3429.59	I	17	6611.62	I	1100	3305.89	II
650	2695.67	I	240	3443.00	I	13	6678.42	I	730	3390.38	I
650	2699.59	I	400	3495.24	I	15	6693.08	I	580	3424.56	II
400	2700.01	I	650	3545.22	I	13	6984.27	I	580	3435.49	I
400	2706.58	I	240	3570.65	I	15	7140.52	I	630	3466.30	I
400	2708.59	I	1900	3617.52	I	9	7162.64	I	680	3482.49	II
400 d	2708.80	I	650	3682.08	I	11	7200.16	I	1600	3489.37	I
400	2715.50	I	400	3683.30	I	10	7278.24	I	530	3496.41	II
2100	2718.91	I	570	3688.06	I	15	7285.81	I	630	3500.08	I
2600	2724.35	I	810	3707.92	I	15	7296.55	I	780	3507.34	I
400	2725.03	I	510	3757.92	I	10	7509.00	I	1600	3514.61	I
650	2748.84	I	680	3760.13	I	17	7569.92	I	630	3533.57	II
400	2762.34	I	1000	3768.45	I	17	7614.15	I	530	3540.47	II
400	2764.27	II	340	3773.71	I	13	7688.97	I	1200	3550.82	II
400	2769.74	I	1000	3780.77	I	11	7784.15	I	680	3555.32	I
810	2770.88	I	290	3809.22	I	22	8017.19	I	1200	3561.80	I
810	2774.00	I	190	3810.38	I	22	8055.64	I	2300	3566.59	I
810	2774.48	I	260	3810.79	I	13	8123.82	I	530	3569.08	I
810	2792.70	I	1400	3817.48	I	10	8338.08	I	630	3578.72	II
400	2799.93	I	1100	3835.06	I	27	8585.11	I	3200	3584.88	I
810	2818.06	I	730	3846.22	I	10	8594.42	I	840	3638.20	I
1600	2831.38	I	1800	3867.99	I	13	8865.53	I	2800	3670.07	II
810	2833.63	I	730	3881.41	I				1100	3701.52	II
810	2848.02	I	8600	4008.75	I	**Uranium U Z = 92**			600	3738.04	II
1500	2896.44	I	540	4015.22	I	440	2565.41	II	680	3746.42	II
690	2935.00	I	910	4045.59	I	610	2635.53	II	950	3748.68	II
2400	2944.40	I	730	4069.95	I	830	2793.94	II	600	3751.17	I
2400	2946.99	I	5000	4074.36	I	870	2802.56	II	1900	3782.84	II
730 d	2979.71	I	1000	4102.70	I	630	2807.05	II	570	3793.10	II
360	3013.79	I	540	4137.46	I	630	2817.96	II	1900	3811.99	I
520	3016.47	I	450	4171.17	I	870	2821.12	II	750	3826.51	II
770	3017.44	I	220	4207.05	I	680	2828.90	II	2000	3831.46	II
210	3024.93	I	250	4219.37	I	920	2832.06	II	1200	3839.63	I
310 d	3026.67	I	540	4244.36	I	970	2865.68	II	2400	3854.64	II
440 d	3041.73	I	1400	4269.38	I	1200	2889.62	II	4900	3859.57	II
270	3043.80	I	4100	4294.61	I	780	2906.80	II	1900	3865.92	II
						780	2908.28	II			

Intensity	Wavelength/Å		Intensity	Wavelength/Å		Intensity	Wavelength/Å		Intensity	Wavelength/Å	
1500	3871.03	I	500	1006.46	III	750	3190.68	II	510	4332.82	I
1000	3881.45	II	500	1149.94	III	1100	3267.70	II	760	4341.01	I
2200	3890.36	II	100	1426.65	IV	900	3271.12	II	1000	4352.87	I
2000	3932.02	II	1000	1643.03	III	750	3276.12	II	12000	4379.24	I
1200	3943.82	I	1000	1650.14	III	80 h	3514.25	IV	7000	4384.72	I
1200	3985.79	II	100	1680.20	V	560	3517.30	II	4800	4389.97	I
1000	4042.75	I	1000	1694.78	III	560	3533.68	I	3600	4395.23	I
1600	4050.04	II	1000	1760.07	III	560	3545.20	II	1400	4400.58	I
880	4062.54	II	1000	1788.26	III	560	3556.80	II	2300	4406.64	I
2200	4090.13	II	1000	1794.60	III	560	3589.76	II	2800	4407.64	I
810	4116.10	II	1000	1812.19	III	490	3592.02	II	3600	4408.20	I
880	4153.97	I	300	1861.56	IV	560	3592.53	I	4600	4408.51	I
1400	4171.59	II	500	1939.06	IV	100	3679.86	III	640	4416.47	I
1000	4241.67	II	400	1951.43	IV	1300	3688.07	I	640	4421.57	I
600	4472.33	II	500	1997.72	IV	1000	3690.28	I	640	4437.84	I
620	4543.63	II	2100	2092.44	I	1500	3692.22	I	830	4441.68	I
170	4689.07	II	500	2268.30	IV	1000	3695.86	I	640	4444.21	I
150	4756.81	I	1000	2292.86	III	3800	3703.58	I	610	4452.01	I
110	5008.21	II	2500	2330.42	III	1800	3704.70	I	1000	4459.76	I
170	5027.38	I	2500	2371.06	III	320	3715.47	II	2000	4460.29	I
80	5160.32	II	1000	2382.46	III	250	3727.34	II	610	4462.36	I
70	5280.38	I	240	2507.78	I	280	3732.76	II	510	4577.17	I
80	5475.70	II	410	2526.22	I	520	3790.32	I	640	4580.40	I
70	5480.26	II	210	2527.90	II	1100	3794.96	I	830	4586.36	I
70	5481.20	II	80 h	2570.72	IV	570	3799.91	I	1300	4594.11	I
160	5492.95	II	230	2574.02	I	570	3803.47	I	230	4619.77	I
70	5780.59	I	250	2593.05	III	1000	3813.49	I	100	4635.18	I
70	5798.53	II	250	2595.10	III	1300	3818.24	I	130	4646.40	I
230	5915.39	I	80 h	2645.54	IV	1700	3828.56	I	160	4670.49	I
100	5976.32	I	180	2661.42	I	2600	3840.75	I	130	4776.36	I
90	6077.29	I	1100	2687.96	II	1200	3855.37	I	110	4786.51	I
55	6372.46	I	680	2700.94	II	3000	3855.84	I	130	4796.92	I
90	6395.42	I	530	2706.17	II	1300	3864.86	I	130	4807.53	I
110	6449.16	I	640	2715.69	II	1500	3875.08	I	130	4827.45	I
90	6826.92	I	180	2731.35	I	700	3890.18	I	150	4831.64	I
45	7533.93	I	240	2864.36	I	2400	3902.25	I	120	4832.43	I
50	7881.94	I	900	2891.64	II	700	3909.89	I	320	4851.48	I
35	8445.39	I	900	2892.66	II	540	3990.57	I	480	4864.74	I
75	8607.95	I	1400	2893.32	II	430	3998.73	I	620	4875.48	I
30	8757.76	I	900	2906.46	II	170	4005.71	II	740	4881.56	I
100	10554.93	I	2400	2908.82	II	1100	4090.58	I	110	5128.53	I
75	11167.84	I	710	2923.62	I	1800	4092.69	I	110	5138.42	I
100	11384.13	I	2400	2924.02	II	890	4095.49	I	110	5192.99	I
100	11859.42	I	1700	2924.64	II	2800	4099.80	I	110	5194.83	I
100	11908.83	I	900	2941.37	II	590	4102.16	I	110	5234.07	I
100	12250.46	I	1100	2944.57	II	2800	4105.17	I	110	5240.87	I
100	13185.16	I	410	2962.77	I	2300	4109.79	I	100	5401.93	I
75	13306.23	I	600	2968.38	II	8900	4111.78	I	140	5415.26	I
100	13961.58	I	1200	3056.33	I	4300	4115.18	I	140	5584.50	I
75	18634.43	I	1400	3060.46	I	1800	4116.47	I	100	5592.42	I
75	21910.22	I	2400	3066.38	I	2000	4123.57	I	200	5624.60	I
			3800	3093.11	II	3100	4128.07	I	400	5627.64	I
Vanadium V Z = 23			3000	3102.30	II	3100	4132.02	I	110	5657.44	I
20	225.46	V	2600	3110.71	II	2300	4134.49	I	110	5668.36	I
20	251.66	V	2000	3118.38	II	20	4200.32	V	310	5670.85	I
20	286.84	V	1500	3125.28	II	360	4232.46	I	1200	5698.52	I
35	483.01	V	3200	3183.41	I	560	4268.64	I	920	5703.56	I
50	633.94	III	5300	3183.98	I	460	4271.55	I	570	5706.98	I
200	677.34	IV	3800	3185.40	I	460	4276.96	I	850	5727.03	I
500	684.37	IV	410	3187.71	II	430	4284.06	I	230	5731.25	I
400	737.85	IV	530	3188.51	II	460	4330.02	I	230	5737.06	I
100	864.27	III									

Intensity	Wavelength/Å		Intensity	Wavelength/Å		Intensity	Wavelength/Å		Intensity	Wavelength/Å	
450	6039.73	I	500	1032.44	II	2	3400.07	I	500 h	4310.51	II
480	6081.44	I	700	1037.68	II	2	3418.37	I	1000 l	4330.52	II
1300	6090.22	I	1100	1041.31	II	2	3420.00	I	200 h	4369.20	II
600	6119.52	I	10	1047.8	III	3	3442.66	I	100 l	4373.78	II
450	6199.19	I	1000	1048.27	II	60	3444.2	III	500 h	4393.20	II
450	6216.37	I	1200	1051.92	II	70	3454.2	III	500 l	4395.77	II
430	6230.74	I	12	1066.4	III	100 w	3458.7	III	200 l	4406.88	II
710	6243.10	I	2000	1074.48	II	100 h	3461.26	II	150 l	4416.07	II
280	6251.82	I	600	1083.86	II	40	3468.22	III	50	4434.2	III
130	6268.82	I	1200	1100.43	II	4	3469.81	I	500 h	4448.13	II
170	6274.65	I	30	1130.3	III	4	3472.36	I	100 w	4462.1	III
200	6285.16	I	600	1158.47	II	5	3506.74	I	1000 h	4462.19	II
200	6292.83	I	250	1169.63	II	80	3522.83	III	500 l	4480.86	II
170	6296.49	I	800 p	1183.05	II	50	3542.3	III	100 l	4521.86	II
110	6531.43	I	250	1192.04	I	10	3549.86	I	100 w	4569.1	III
65 c	6753.00	I	25	1232.1	III	50	3552.1	III	100 w	4570.1	III
50 c	6766.49	I	600	1244.76	II	10	3554.04	I	100 w	4641.4	III
40	6784.98	I	250	1250.20	I	40	3561.4	III	30	4673.7	III
40	7338.92	I	1000	1295.59	I	100	3579.7	III	60	4683.57	III
35	7356.54	I	600	1469.61	I	80	3583.6	III	30	4723.60	III
29 c	8027.39	I	80	2668.98	III	100 w	3595.4	III	600	4734.152	I
120 w	8116.80	I	100	2717.33	III	100	3606.06	III	100 w	4757.3	III
70 c	8161.07	I	30	2814.45	III	40	3607.0	III	150	4792.619	I
60 c	8919.85	I	40	2815.91	III	15	3610.32	I	500	4807.02	I
			30	2827.45	III	8	3613.06	I	400	4829.71	I
Xenon Xe Z = 54			40	2847.65	III	100 w	3615.9	III	300	4843.29	I
8	657.8	III	30	2862.40	III	40	3623.1	III	40	4869.5	III
8	660.1	III	200	2864.73	II	600	3624.08	III	500	4916.51	I
9	673.8	III	80 w	2871.10	III	6	3633.06	I	500	4923.152	I
9	674.0	III	60 w	2871.24	III	10	3669.91	I	200 l	4971.71	II
9	676.6	III	30	2871.7	III	50	3676.67	III	400	4972.71	II
10	694.0	III	150 h	2895.22	II	40	3685.90	I	300	4988.77	II
20	698.5	III	30	2896.62	III	40	3693.49	I	100 l	4991.17	II
12	705.1	III	50	2906.6	III	40	3776.3	III	200	5028.280	I
10	721.2	III	40	2911.89	III	300	3781.02	III	200	5044.92	II
15	731.0	III	80 w	2912.36	III	100	3841.5	III	1000	5080.62	II
10	733.3	III	40	2940.2	III	200	3877.8	III	300	5122.42	II
350	740.41	II	60	2945.2	III	60	3880.5	III	100	5125.70	II
15	742.6	III	40	2947.5	III	100 l	3907.91	II	100	5178.82	II
10	756.0	III	40	2948.1	III	500	3922.55	III	300	5188.04	II
10	761.5	III	80 w	2970.47	III	300	3950.59	III	400	5191.37	II
10	769.1	III	400	2979.32	II	100	4037.59	II	100	5192.10	II
25	779.1	III	40	2992.87	III	200	4050.07	III	60	5239.0	III
15	792.9	III	30	3004.25	III	200 l	4057.46	II	500	5260.44	II
12	796.1	III	100 h	3017.43	II	60	4060.4	III	500	5261.95	II
15	802.0	III	100	3023.81	III	100 h	4098.89	II	2000	5292.22	II
350	803.07	II	40	3083.5	III	100	4109.1	III	300	5309.27	II
25	823.2	III	50	3091.1	III	100	4145.7	III	1000	5313.87	II
30	824.9	III	30	3106.46	III	200 l	4158.04	II	2000	5339.33	II
25	853.0	III	300	3128.87	II	1000 h	4180.10	II	200	5363.20	II
600	880.80	II	100 w	3138.3	III	500 h	4193.15	II	30	5367.1	III
350	885.54	II	80 c	3150.82	III	300 h	4208.48	II	200	5368.07	II
15	889.3	III	40	3185.2	III	100 h	4209.47	II	500	5372.39	II
20	894.0	III	100	3242.86	III	300 h	4213.72	II	100	5392.80	I
20	896.0	III	80	3268.98	III	100	4215.60	II	50	5401.0	III
600	925.87	II	30	3287.82	III	300 h	4223.00	II	3000	5419.15	II
250	935.40	II	80 w	3301.55	III	400 h	4238.25	II	800	5438.96	II
10	965.5	III	40	3331.6	III	500 h	4245.38	II	300	5445.45	II
800	972.77	II	30	3358.0	III	100 l	4251.57	II	200	5450.45	II
700	976.68	II	200 h	3366.72	II	30	4285.9	III	400	5460.39	II
35	1003.4	III	80	3384.12	III	500 h	4296.40	II	1000	5472.61	II
35	1017.7	III									

Intensity	Wavelength/Å		Intensity	Wavelength/Å		Intensity	Wavelength/Å		Intensity	Wavelength/Å	
100 l	5494.86	II	200	6528.65	II	300	8739.39	I	500	30794.18	I
40	5524.4	III	100	6533.16	I	100	8758.20	I	6000	31069.23	I
200	5525.53	II	1000	6595.01	II	5000	8819.41	I	125	31336.01	I
600	5531.07	II	100	6595.56	I	300	8862.32	I	550	31607.91	I
100	5566.62	I	400	6597.25	II	200	8908.73	I	100	32293.08	I
300	5616.67	II	100	6598.84	II	200	8930.83	I	1800	32739.26	I
300	5659.38	II	150	6668.92	I	1000	8952.25	I	3500	33666.69	I
600	5667.56	II	300	6694.32	II	100	8981.05	I	150	34014.67	I
150	5670.91	II	200	6728.01	I	200	8987.57	I	450	34335.27	I
100	5695.75	I	150	6788.71	II	400	9045.45	I	170	34744.00	I
200	5699.61	II	100	6790.37	II	500	9162.65	I	5000	35070.25	I
200	5716.10	II	1000	6805.74	II	100	9167.52	I	110	35246.92	I
500	5726.91	II	200	6827.32	I	100	9374.76	I	250	36209.21	I
500	5751.03	II	100	6872.11	I	200	9513.38	I	150	36231.74	I
300	5758.65	II	300	6882.16	I	50 h	9591.35	II	450	36508.36	I
300	5776.39	II	80	6910.22	II	150	9685.32	I	850	36788.83	I
100	5815.96	II	100	6925.53	I	50 l	9698.68	II	140	38685.98	I
300	5823.89	I	800 h	6942.11	II	100	9718.16	I	175	38737.82	I
150	5824.80	I	100	6976.18	I	2000	9799.70	I	270	38939.60	I
100	5875.02	I	2000	6990.88	II	3000	9923.19	I	120	39955.14	I
300	5893.29	II	150	7082.15	II	100	10838.37	I			
100	5894.99	I	500	7119.60	I	90	11742.01	I	*Ytterbium Yb*	*Z = 70*	
200	5905.13	II	50 s	7147.50	II	375	12235.24	I	1000	1050.24	IV
100	5934.17	I	200	7149.03	II	100	12257.76	I	1000	1054.46	IV
500	5945.53	II	500	7164.83	II	300	12590.20	I	5000	1134.43	IV
300	5971.13	II	100	7284.34	II	2500	12623.391	I	900	1316.04	IV
2000	5976.46	II	200	7301.80	II	250	13544.15	I	800	1326.36	IV
200	6008.92	II	200	7339.30	II	2000	13657.055	I	900	1350.26	IV
1000	6036.20	II	100	7386.00	I	1250	14142.444	I	80	1561.42	III
2000	6051.15	II	150	7393.79	I	800	14240.96	I	80 h	1765.21	III
600	6093.50	II	300	7548.45	II	375	14364.99	I	800	1791.06	IV
1500	6097.59	II	200	7584.68	I	140	14660.81	I	100	1863.32	III
400	6101.43	II	80	7618.57	II	3000	14732.806	I	800	1873.91	III
100	6115.08	II	500	7642.02	I	100	15099.72	I	500	1898.25	III
100	6146.45	II	100	7643.91	I	2500	15418.394	I	500	1998.82	III
150	6178.30	I	200	7670.66	II	150	15557.13	I	900	2116.65	IV
120	6179.66	I	60	7787.04	II	250	15979.54	I	2500	2116.67	II
300	6182.42	I	100	7802.65	I	100	16039.90	I	800	2123.32	IV
500	6194.07	II	100	7881.32	I	1000	16053.28	I	3000	2126.74	II
100	6198.26	I	300	7887.40	I	125	16554.49	I	800	2139.99	IV
60	6205.97	III	500	7967.34	I	1500	16728.15	I	20000	2144.77	IV
100	6220.02	II	100	8029.67	I	1500	17325.77	I	15000	2154.18	IV
25	6221.7	III	200	8057.26	I	350	18788.13	I	370	2161.60	II
60	6238.2	III	150	8061.34	I	150	20187.19	I	850	2185.71	II
60	6259.05	III	100	8101.98	I	3000	20262.242	I	640	2224.46	II
500	6270.82	II	150 h	8151.80	II	250	21470.09	I	300	2240.11	III
400	6277.54	II	100	8171.02	I	1250	23193.33	I	300	2305.32	III
100	6284.41	II	700	8206.34	I	110	23279.54	I	140	2320.81	II
100	6286.01	I	10000	8231.635	I	1800	24824.71	I	170	2390.74	II
250	6300.86	II	500	8266.52	I	175	25145.84	I	460	2464.50	I
500	6318.06	I	7000	8280.116	I	2000	26269.08	I	140	2512.06	II
400	6343.96	II	2000	8346.82	I	2500	26510.86	I	270	2538.67	II
600	6356.35	II	100	8347.24	II	250	28381.54	I	2000	2567.61	III
200	6375.28	II	2000	8409.19	I	750	28582.25	I	1000	2579.57	III
100	6397.99	II	50 h	8515.19	II	300	29384.41	I	800	2599.14	III
300	6469.70	I	200	8576.01	I	150	29448.06	I	600	2621.11	III
150	6472.84	I	50 h	8604.23	II	100	29649.58	I	1000	2642.56	III
120	6487.76	I	250	8648.54	I	100	29813.62	I	1000	2651.74	III
100	6498.72	I	100	8692.20	I	600	30253.14	I	700	2652.25	III
200 h	6504.18	I	200	8696.86	I	1500	30475.46	I	990	2653.75	II
300	6512.83	II	50 h	8716.19	II	100	30504.12	I	200	2665.04	II
									2000	2666.13	III

Intensity	Wavelength/Å		Intensity	Wavelength/Å		Intensity	Wavelength/Å		Intensity	Wavelength/Å	
2000	2666.99	III	140	3418.39	I	170	4786.61	II	500	344.59	V
390	2671.96	I	360	3426.04	I	35	4816.43	I	900	355.86	IV
390	2672.66	II	240	3431.11	I	40	4837.46	I	300	370.42	IV
170	2718.35	II	85	3452.40	I	40 h	4894.60	I	300	372.05	V
230	2748.66	II	500	3454.08	II	27	4912.36	I	400	379.96	V
1300	2750.48	II	190 d	3458.29	II	710	4935.50	I	500	386.82	IV
170	2776.28	II	360	3460.27	I	140	4966.90	I	300	403.45	V
600	2795.60	III	2400	3464.37	I	30	5067.80	I	300	420.74	V
1000	2803.43	III	500	3476.30	II	70	5069.14	I	600	425.03	IV
600	2816.92	III	500	3478.84	II	220	5074.34	I	300	473.10	IV
1000	2818.72	III	50	3517.00	I	50	5076.74	I	4000	584.98	V
140	2821.15	II	230	3520.29	II	60	5196.08	I	2000	630.97	V
190	2830.99	II	35	3559.03	I	85	5211.60	I	5000	805.20	III
230 h	2847.18	II	200	3560.33	II	100	5244.11	I	7000	809.92	III
360	2851.13	II	170	3560.70	II	150 h	5277.04	I	15000	989.21	III
430	2859.80	II	360	3585.47	II	170	5335.15	II	25000	996.37	III
140	2861.21	II	200	3619.80	II	30 h	5351.29	I	5000	1314.51	III
200	2867.06	II	240	3637.76	II	150	5352.95	II	4000	1334.04	III
45	2873.49	I	70	3648.15	I	30	5363.66	I	4000	2068.98	III
200	2888.04	II	90	3655.73	I	40	5449.27	II	10000	2127.98	III
3600	2891.38	II	240	3669.69	II	60	5481.92	I	16000	2191.16	III
600	2898.30	III	140	3675.08	II	40	5505.49	I	350	2243.06	II
1000	2906.31	III	32000	3694.19	II	17	5524.54	I	10000	2284.34	III
170	2914.21	II	70	3700.58	I	85 h	5539.05	I	10000	2327.31	III
140	2915.28	II	400	3711.91	III	2400	5556.47	I	50	2354.20	I
280	2919.35	II	180	3734.69	I	60	5651.98	II	50000	2367.23	III
35	2934.36	I	550	3770.10	I	220	5719.99	I	40000	2414.64	III
140	2945.91	II	80	3774.32	I	27	5771.66	II	560	2422.20	II
2000	2970.56	II	60 h	3791.74	I	35	5833.99	II	60	2694.21	I
200	2983.99	II	170	3839.91	I	35	5837.14	II	95	2723.00	I
170	2994.80	II	340	3872.85	I	27	5854.51	I	70	2742.53	I
800	2998.00	III	340	3900.85	I	17	5989.33	I	140	2760.10	I
310	3005.77	II	140	3911.27	I	40	5991.51	II	90000	2817.04	III
160	3017.56	II	500	3931.23	III	60	6152.57	II	45	2822.56	I
160	3026.67	II	32000	3987.99	I	60	6274.78	II	70	2854.43	II
2000	3029.49	III	930	3990.88	I	200	6328.52	III	95	2886.48	I
920	3031.11	II	50	4007.36	I	35 h	6400.35	I	160	2919.05	I
3000	3092.50	III	2000	4028.14	III	35 h	6417.91	I	99000	2946.01	III
28	3100.74	I	70	4052.28	I	340	6489.06	I	390	2948.40	I
170	3107.90	II	440	4089.68	I	180	6667.82	I	350	2964.96	I
190	3117.81	II	470	4149.07	I	25	6727.61	II	480	2974.59	I
4000	3126.01	III	120	4174.56	I	690	6799.60	I	750	2984.26	I
1000	3138.58	III	340	4180.81	II	9 h	7244.41	I	140	2996.94	I
230	3140.94	II	300	4213.64	III	8 h	7305.22	I	130	3021.73	I
28	3162.29	I	150 d	4218.56	II	10 h	7313.05	I	190	3045.37	I
800	3191.35	III	120	4231.97	I	16 h	7350.04	I	95	3095.88	II
390	3192.88	II	70	4277.74	I	25	7448.28	I	110	3173.06	II
240	3201.16	II	120	4305.97	I	30 h	7527.46	I	220	3179.41	II
2000	3228.58	III	60 h	4393.69	I	750	7699.48	I	70	3191.31	I
35	3239.58	I	60 h	4430.21	I	100	7971.46	III	2300	3195.62	II
18000	3289.37	II	440	4439.19	I	70 h	8922.56	II	2200	3200.27	II
130	3305.25	I	85 h	4482.42	I	200	10110.60	III	2200	3203.32	II
140	3305.73	II	100	4517.58	III	100	10830.36	III	3900	3216.69	II
80	3319.41	I	85 h	4563.95	I				6200	3242.28	II
2000	3325.51	III	640	4576.21	I	*Yttrium Y Z = 39*			4700	3327.89	II
240	3337.17	II	200	4582.36	I	150	264.64	IV	85	3388.59	I
280 d	3342.93	II	70	4589.21	I	150	273.03	IV	85	3412.47	I
240	3375.48	II	140	4590.83	I	900	333.09	V	170	3485.73	I
2000	3384.01	III	40	4684.27	I	500	333.80	V	1700	3496.09	II
140	3387.50	I	190	4726.08	II	400	335.14	V	3900	3549.01	II
50	3412.45	I	170 h	4781.87	I	500	336.62	V	130	3551.80	I
						500	339.02	V			

Intensity	Wavelength/Å		Intensity	Wavelength/Å		Intensity	Wavelength/Å		Intensity	Wavelength/Å	
540	3552.69	I	300	4487.47	I	160	5706.73	I	100	1767.69	III
170	3558.76	I	500	4505.95	I	90	5743.85	I	100	1797.64	II
190	3571.43	I	890	4527.25	I	75	5765.64	I	100 d	1811.05	II
260	3576.05	I	440	4527.80	I	100	5781.69	II	100 d	1833.57	II
3300	3584.52	II	100	4544.32	I	120	6009.19	I	100	1864.12	II
300	3587.75	I	100	4559.37	I	120	6023.41	I	100	1866.08	II
100	3589.69	I	130	4596.55	I	120	6135.04	I	100	1872.13	II
2800	3592.92	I	95	4604.80	I	150	6138.43	I	100 d	1918.96	II
10000	3600.73	II	2000	4643.70	I	1200	6191.73	I	100 d	1929.67	II
6200	3601.92	II	200 h	4658.32	I	300	6222.59	I	100	1969.40	II
7800	3611.05	II	2000	4674.84	I	1000	6435.00	I	100	1982.11	II
4300	3620.94	I	180	4696.81	I	90	6538.60	I	100	1986.99	II
1900	3628.71	II	170	4728.53	I	70	6557.39	I	500	2025.48	II
7800	3633.12	II	160	4752.79	I	95	6613.75	II	500	2062.00	II
3000	3664.61	II	410	4760.98	I	40	6650.61	I	200	2064.23	II
170	3692.53	I	120	4781.04	I	150	6687.58	I	120	2079.08	I
13000	3710.30	II	170	4786.89	I	70	6700.71	I	300	2099.94	II
1200	3747.55	II	180	4799.30	I	190	6793.71	I	200	2102.18	II
10000	3774.33	II	140	4819.64	I	21	6815.16	I	800 r	2138.56	I
1400	3776.56	II	120	4822.13	I	45	6845.24	I	1000	2501.99	II
7400	3788.70	II	770	4839.87	I	29	6887.22	I	150	2515.81	I
1300	3818.35	II	550	4845.68	I	24 h	6950.31	I	1000	2557.95	II
4000	3832.88	II	410	4852.69	I	24	6979.88	I	300	2582.49	I
80	3876.82	I	120	4854.25	I	29	7052.94	I	200	2608.56	I
480	3878.28	II	890	4854.87	II	35	7191.66	I	300	2608.64	I
4400	3950.36	II	330	4859.84	I	35	7264.17	II	200	2670.53	I
3600	3982.60	II	1900	4883.69	II	50	7346.46	I	300	2684.16	I
940	4039.83	I	95	4893.44	I	29	7450.30	II	300	2712.49	I
2400	4047.64	I	1100	4900.12	II	9000	7558.71	III	200	2756.45	I
9400	4077.38	I	100	4906.11	I	35	7563.13	I	300	2770.86	I
2000	4083.71	I	150	4921.87	I	29	7855.52	I	300	2770.98	I
9900	4102.38	I	120	4974.30	I	110	7881.90	II	400	2800.87	I
8900	4128.31	I	100	5006.97	I	10000	7991.43	III	100	2801.06	I
7500	4142.85	I	75	5070.21	I	24	8344.43	I	200	3035.78	I
100 h	4157.63	I	75	5072.19	I	10000	8796.21	III	200	3072.06	I
2400	4167.52	I	1100	5087.42	II	95	8800.62	I	300	3196.31	II
2000	4174.14	I	180	5135.20	I	19 h	8835.85	II	500 r	3282.33	I
8000	4177.54	II	960	5200.41	II				800	3302.58	I
160	4217.80	I	1500	5205.72	II	**Zinc Zn Z = 30**			700 r	3302.94	I
280 h	4220.63	I	10000	5238.10	III	200	425.90	IV	800	3345.02	I
600	4235.73	II	180	5240.81	I	200	428.54	IV	500	3345.57	I
2200	4235.94	I	75	5380.62	I	200	430.59	IV	50	3883.34	I
300	4251.20	I	220	5402.78	II	1000	677.63	III	300	4680.14	I
360 h	4302.30	I	90	5424.37	I	750	677.96	III	400	4722.15	I
2800	4309.63	II	190	5438.24	I	200	713.90	III	400	4810.53	I
110	4330.78	I	710	5466.46	I	60	1193.23	II	800	4911.62	II
440 h	4348.79	I	100	5468.47	I	50	1239.12	IV	500	4924.03	II
120	4357.73	I	240	5497.41	II	50	1249.69	IV	200	5181.98	I
800	4358.73	II	300	5503.45	I	500	1265.74	IV	500	5894.33	II
120	4366.03	I	250	5509.90	II	500	1306.66	IV	500	6021.18	II
12000	4374.94	II	120	5521.63	I	200	1456.72	III	500	6102.49	II
150 h	4375.61	I	740	5527.54	I	200	1459.98	IV	500	6214.61	II
100	4387.74	I	120	5544.50	I	300	1499.42	III	1000 h	6362.34	I
1800	4398.02	II	180	5577.42	I	300	1500.42	III	300	7588.5	II
890	4422.59	II	620	5581.87	I	300	1505.92	III	300	7732.5	II
100	4443.66	I	120	5606.33	I	300	1515.85	III	100	11054.25	I
130	4446.63	I	560	5630.13	I	300	1552.30	III	100	13053.63	I
170	4475.72	I	120	5644.69	I	90	1572.99	II	100	13150.59	I
180	4476.96	I	120	5648.47	I	200	1629.19	III	100	14038.70	I
160	4477.45	I	740	5662.94	II	200	1639.33	III	20	16483.45	I
110	4487.28	I	90	5675.27	I	200	1673.05	III	20	16491.98	I
						80 d	1735.61	II			

Intensity	Wavelength/Å		Intensity	Wavelength/Å		Intensity	Wavelength/Å		Intensity	Wavelength/Å	
20	16505.23	I	620	2814.90	I	1000	3430.53	II	770	3998.97	II
10	24375.02	I	390	2818.74	II	4700	3438.23	II	400	4023.98	I
			530	2825.56	II	600	3447.36	I	770	4024.92	I
Zirconium Zr Z = 40			710	2837.23	I	410	3457.56	II	990	4027.20	I
500	304.01	V	660	2844.58	II	820	3463.02	II	400	4029.68	II
60	480.66	IV	350	2848.52	I	600	3471.19	I	490	4030.04	I
60	497.23	IV	350	2851.97	II	1200	3479.39	II	400	4035.89	I
60	500.22	IV	340	2869.81	II	1300	3481.15	II	610	4043.58	I
600	628.66	IV	490	2875.98	II	4100	3496.21	II	490	4044.56	I
500	633.56	IV	300	2915.99	II	820	3505.67	II	400	4045.61	II
50	690.39	III	270	2918.24	II	1000	3509.32	I	610	4048.67	II
2000	740.61	V	320	2926.99	II	2000	3519.60	I	770	4055.03	I
10000	800.00	V	320	2948.94	II	440	3525.81	II	600	4055.71	I
10000	806.89	V	320	2955.78	II	440	3533.22	I	1500	4064.16	I
10000	812.05	V	320	2960.87	I	630	3542.62	II	2000	4072.70	I
3000	841.40	V	320	2962.68	II	1800	3547.68	I	240	4078.31	I
300	863.65	IV	320	2968.96	II	630	3550.46	I	2000	4081.22	I
500	864.59	IV	320	2978.05	II	1800	3551.95	II	400	4121.46	I
9000	1183.97	IV	820	2985.39	I	2100	3556.60	II	1200	4149.20	II
9000	1201.77	IV	320	3003.74	II	1100	3566.10	I	400	4161.21	II
10000	1219.86	IV	820	3011.75	I	2100	3572.47	II	400	4166.36	I
500	1303.93	V	350	3020.47	II	1100	3575.79	I	660	4187.56	I
500 p	1323.81	V	500	3028.04	II	1300	3576.85	II	400	4194.76	I
1000	1469.47	IV	880	3029.52	I	880	3586.29	I	610	4199.09	I
10000	1546.17	IV	350 d	3036.39	II	3500	3601.19	I	610	4201.46	I
10000	1598.95	IV	690	3054.84	II	690	3611.89	II	610	4208.98	II
5000	1607.95	IV	690	3106.58	II	1100	3613.10	II	400	4213.86	I
100	1612.38	III	350	3120.74	I	1100	3614.77	II	2000	4227.76	I
700	1725.02	V	500	3129.18	II	1100	3623.86	I	2000	4239.31	I
200	1790.19	III	500	3129.76	II	1100	3663.65	I	770	4240.34	I
150	1793.56	III	350	3132.07	I	390	3671.27	II	770	4241.20	I
125	1798.13	III	690	3138.68	II	800	3674.72	II	1200	4241.69	I
600	1860.86	V	540	3164.31	II	390	3697.46	II	550	4282.20	I
200	1940.25	III	880	3165.97	II	960	3698.17	II	550	4294.79	I
600	2028.54	V	880	3182.86	II	720	3709.26	II	550	4341.13	I
125	2070.43	III	540	3191.21	I	560	3745.98	II	1000	4347.89	I
200	2086.78	III	540	3212.01	I	880	3751.60	II	290	4359.74	II
10000	2091.49	IV	760	3214.19	II	480	3764.39	I	310	4360.81	I
10000	2092.36	IV	630	3231.69	II	480	3766.72	I	350	4366.45	I
600	2132.42	V	630	3234.12	I	340	3766.82	II	550	4507.12	I
10000	2163.68	IV	760	3241.05	II	720	3780.54	I	610	4535.75	I
100	2175.80	III	1000	3273.05	II	560	3791.40	I	490	4542.22	I
100	2191.15	III	1300	3279.26	II	560	3822.41	I	490	4575.52	I
10000	2286.67	IV	880	3284.71	II	2200	3835.96	I	350	4602.57	I
100	2301.60	III	540	3305.15	II	1300	3836.76	II	700	4633.98	I
90	2539.65	I	880	3306.28	II	550	3843.02	II	2300	4687.80	I
570	2567.64	II	380	3322.99	II	550	3847.01	I	510	4688.45	I
1600	2568.87	II	380	3326.80	II	550	3849.25	I	1900	4710.08	I
2100	2571.39	II	380	3334.25	II	2900	3863.87	I	1400	4739.48	I
250	2620.56	III	760	3340.56	II	770	3864.34	I	870	4772.31	I
200	2643.79	III	380	3344.79	II	990	3877.60	I	700	4815.63	I
150	2664.26	III	760	3356.09	II	1500	3885.42	I	250	5046.58	I
1800	2678.63	II	540	3357.26	II	2900	3890.32	I	360	5064.91	I
90	2687.75	I	380	3374.73	II	2000	3891.38	I	470	5078.25	I
750	2700.13	II	570	3387.87	II	610	3921.79	I	300	5155.45	I
1300	2722.61	II	760	3388.30	II	1200	3929.53	I	200	5158.00	I
800	2726.49	II	5700	3391.98	II	940	3958.22	II	100	5191.60	II
1400	2734.86	II	570	3393.12	II	490	3966.66	I	270	5385.14	I
1100	2742.56	II	570	3404.83	II	990	3968.26	I	160	5664.51	I
660	2745.86	II	760	3410.25	II	660	3973.50	I	160	5797.74	I
660	2752.21	II	380	3414.66	I	770	3991.13	II	340	5879.80	I
530	2758.81	II									

Intensity	Wavelength/Å		Intensity	Wavelength/Å		Intensity	Wavelength/Å		Intensity	Wavelength/Å	
170	6045.85	I	300	6313.02	I	170	7103.72	I	790	8070.08	I
170	6121.91	I	150	6953.84	I	590	7169.09	I	390	8132.99	I
680	6127.44	I	150	6990.84	I	160	7944.61	I	280	8212.53	I
340	6134.55	I	540	7097.70	I	160	8005.27	I			
440	6143.20	I	280	7102.91	I	150	8063.09	I			

Sources of Data for Each Element

Numbers following the element name refer to the references on the following pages.

Actinium: 193
Aluminum: 6,8,81,89,127,144,146,227,228,282
Americium: 92
Antimony: 164,167,194,386,406
Argon: 190,203,204,219,367,368,372,373,374,375,414,421
Arsenic: 163,168,197,244,280
Astatine: 188
Barium: 1,78,111,252,259,277,279
Berkelium: 53,339
Beryllium: 15,44,73,102,115,134,135,171,175,198,335
Bismuth: 1,357,358,359,360,361
Boron: 66,69,74,94,104,171,221,222
Bromine: 42,122,124,139,142,240,243,246,248,249,250,316
Cadmium: 44,285,296,353,399
Calcium: 16,25,70,150,270
Californium: 52,331
Carbon: 22,66,211
Cerium: 1,136,166,261,305
Cesium: 78,82,154,155,200,201,259,263,325
Chlorine: 11,28,30,31,85,233,238,239
Chromium: 1,379,380,412
Cobalt: 1,100,125,159,236,276,291
Copper: 199,273,290,295,324
Curium: 51,332
Dysprosium: 1
Einsteinium: 333
Erbium: 1,301
Europium: 1,312
Fluorine: 68,169,224,225,226
Francium: 408
Gadolinium: 1,46,137,151,152
Gallium: 2,19,62,132,140,141,143,195,281
Germanium: 5,119,293,340,341,342
Gold: 38,72,234,393,395
Hafnium: 1,369,404,410,425
Helium: 16,94,173,183,317
Holmium: 1
Hydrogen: 214
Indium: 1,132,348,349,350,351,352,353,435,436
Iodine: 20,21,58,84,124,153,161,176,184
Iridium: 1
Iron: 56,63,71,101,105,138,174,278,381,382
Krypton: 61,121,123,147,208,232,366,390,409,417,421
Lanthanum: 1,78,79,220,309
Lead: 54,64,106,256,274,297,283,329,330
Lithium: 3,15,17,18,37,44,112,284,321,335
Lutetium: 1,148,310,401
Magnesium: 4,7,49,83,103,128,129,177,217,269,315,335
Manganese: 1,126,385,405,433
Mercury (198): 43,50,69,145,229,242

Mercury (Natural): 34,45,90,117,133,189,235,304,327,328,343
Molybdenum: 1,383,420
Neodymium: 1
Neon: 56,58,69,118,150,230,364,365,371,388,389,400,402,413,430
Neptunium: 93
Nickel: 1,294,415,416,422
Niobium: 1,392,407,431
Nitrogen: 66,107,108,212,213,318
Osmium: 1
Oxygen: 23,24,36,66,69,209,210,215
Palladium: 1,287,424
Phosphorus: 179,180,182,336
Platinum: 1,288
Plutonium: 91
Polonium: 47,48
Potassium: 32,59,60,75,76,86,150,160,172,268,314,322
Praseodymium: 1,149,306,308,337,338
Promethium: 196,260
Protactinium: 96
Radium: 253,254
Radon: 251
Rhenium: 1
Rhodium: 1,396
Rubidium: 12,109,130,241,257,258,262,264
Ruthenium: 1,423
Samarium: 1
Scandium: 1,88,150,298,323
Selenium: 9,80,181,216,245,247,275
Silicon: 87,170,237,292,319,320
Silver: 13,99,255,286,289,363,387,398
Sodium: 178,205,206,207,268,299,334
Strontium: 1,109,110,218,231,265,279,313
Sulfur: 29,144,202,209,210,266
Tantalum: 1,411,426
Technetium: 35
Tellurium: 1,344,345,346,347
Terbium: 1,302
Thallium: 1,195,348,354,355,356
Thorium: 1,97,98,156,157,165,434
Thulium: 1,307
Tin: 187,191,399,423
Titanium: 1,378,427,428
Tungsten: 1
Uranium: 1,303
Vanadium: 1,394,397,432
Xenon: 33,116,118,120,232,384,391,429
Ytterbium: 1,40,192,311
Yttrium: 1,77,265,419
Zinc: 39,55,113,131,185,186,370,376,377
Zirconium: 1,362,403,418

References

1. Meggers, W. F., Corliss, C. H., and Scribner, B. F., *Natl. Bur. Stand. (U.S.) Monogr.*, 145, Washington, D.C., 1975.
2. Aksenov, V. P. and Ryabtsev, A. N., *Opt. Spectrosc.*, 37, 860, 1970.
3. Andersen, N., Bickel, W. S., Carriveau, G. W., Jensen, K., and Veje, E., *Phys. Scr.*, 4, 113, 1971.
4. Andersson, E. and Johannesson, G. A., *Phys. Scr.*, 3, 203, 1971.
5. Andrew, K. L. and Meissner, K. W., *J. Opt. Soc. Am.*, 49, 146, 1959.
6. Artru, M. C. and Brillet, W. U. L., *J. Opt. Soc. Am.*, 64, 1063, 1974.
7. Artru, M. C. and Kaufman, V., *J. Opt. Soc. Am.*, 62, 949, 1972.
8. Artru, M. C. and Kaufman, V., *J. Opt. Soc. Am.*, 65, 594, 1975.
9. Badami, J. S. and Rao, K. R., *Proc. R. Soc. London*, 140(A), 387, 1933.
10. Baird, K. M. and Smith, D. S., *J. Opt. Soc. Am.*, 48, 300, 1958.
11. Bashkin, S. and Martinson, I., *J. Opt. Soc. Am.*, 61, 1686, 1971.
12. Beacham, J. R., Ph.D. thesis, Purdue University, 1970.
13. Benschop, H., Joshi, Y. N., and van Kleef, T. A. M., *Can. J. Phys.*, 53, 700, 1975.
14. Berry, H. G., Bromander, J., and Buchta, R., *Phys. Scr.*, 1, 181, 1970.
15. Berry, H. G., Bromander, J., Martinson, I., and Buchta, R., *Phys. Scr.*, 3, 63, 1971.
16. Berry, H. G., Desesquelles, J., and Dufay, M., *Phys. Rev. Sect A.*, 6, 600, 1972.
17. Berry, H. G., Desesquelles, J., and Dufay, M., *Nucl. Instrum. Methods*, 110, 43, 1973.
18. Berry, H. G., Pinnington, E. H., and Subtil, J. L., *J. Opt. Soc. Am.*, 62, 767, 1972.
19. Bidelman, W. P. and Corliss, C. H., *Astrophys. J.*, 135, 968, 1962.
20. Bloch, L. and Bloch, E., *Ann. Phys.* (Paris), 10(11), 141, 1929.
21. Bloch, L., Bloch, E., and Felici, N., *J. Phys. Radium*, 8, 355, 1937.
22. Bockasten, K., *Ark. Fys.*, 9, 457, 1955.
23. Bockasten, K., Hallin, R., Johansson, K. B., and Tsui, P., *Phys. Lett.* (Netherlands), 8, 181, 1964.
24. Bockasten, K. and Johansson, K. B., *Ark. Fys.*, 38, 563, 1969.
25. Borgstrom, A., *Ark. Fys.*, 38, 243, 1968.
26. Borgstrom, A., *Phys. Scr.*, 3, 157, 1971.
27. Bowen, I. S., *Phys. Rev.*, 29, 231, 1927.
28. Bowen, I. S., *Phys. Rev.*, 31, 34, 1928.
29. Bowen, I. S., *Phys. Rev.*, 39, 8, 1932.
30. Bowen, I. S., *Phys. Rev.*, 45, 401, 1934.
31. Bowen, I. S., *Phys. Rev.*, 46, 377, 1934.
32. Bowen, I. S., *Phys. Rev.*, 46, 791, 1934.
33. Boyce, J. C., *Phys. Rev.*, 49, 730, 1936.
34. Boyce, J. C. and Robinson, H. A., *J. Opt. Soc. Am.*, 26, 133, 1936.
35. Bozman, W. R., Meggers, W. F., and Corliss, C. H., *J. Res. Natl. Bur. Stand. Sect. A*, 71, 547, 1967.
36. Bromander, J., *Ark. Fys.*, 40, 257, 1969.
37. Bromander, J. and Buchta, R., *Phys. Scr.*, 1, 184, 1970.
38. Brown, C. M. and Ginter, M. L., *J. Opt. Soc. Am.*, 68, 243, 1978.
39. Brown, C. M., Tilford, S. G., and Ginter, M. L., *J. Opt. Soc. Am.*, 65, 1404, 1975.
40. Bryant, B. W., *Johns Hopkins Spectroscopic Report* No. 21, 1961.
41. Buchet, J. P., Buchet-Poulizac, M. C., Berry, H. G., and Drake, G. W. F., *Phys. Rev. Sect. A*, 7, 922, 1973.
42. Budhiraja, C. J. and Joshi, Y. N., *Can. J. Phys.*, 49, 391, 1971.
43. Burns, K. and Adams, K. B., *J. Opt. Soc. Am.*, 42, 56, 1952.
44. Burns, K. and Adams, K. B., *J. Opt. Soc. Am.*, 46, 94, 1956.
45. Burns, K., Adams, K. B., and Longwell, J., *J. Opt. Soc. Am.*, 40, 339, 1950.
46. Callahan, W. R., Ph.D. thesis, Johns Hopkins University, 1962.
47. Charles, G. W., *J. Opt. Soc. Am.*, 56, 1292, 1966.
48. Charles, G. W., Hunt, D. J., Pish, G., and Timma, D. L., *J. Opt. Soc. Am.*, 45, 869, 1955.
49. Codling, K., *Proc. Phys. Soc.*, 77, 797, 1961.
50. Comite Consultatif Pour La Definition du Metre, *J. Phys. Chem. Ref. Data*, 3, 852, 1974.
51. Conway, J. G., Blaise, J., and Verges, J., *Spectrochim. Acta Part B*, 31, 31, 1976.
52. Conway, J. G., Worden, E. F., Blaise, J., and Verges, J., *Spectrochim. Acta Part B*, 32, 97, 1977.
53. Conway, J. G., Worden, E. F., Blaise, J., Camus, P., and Verges, J., *Spectrochim. Acta Part B*, 32, 101, 1977.
54. Crooker, A. M., *Can. J. Res. Sect.* A, 14, 115, 1936.
55. Crooker, A. M. and Dick, K. A., *Can. J. Phys.*, 46, 1241, 1968.
56. Crosswhite, H. M., *J. Res. Natl. Bur. Stand. Sect. A*, 79, 17, 1975.
58. Crosswhite, H. M. and Dieke, G. H., *American Institute of Physics Handbook*, Section 7, 1972.
59. de Bruin, T. L., *Z. Phys.*, 38, 94, 1926.
60. de Bruin, T. L., *Z. Phys.*, 53, 658, 1929.
61. de Bruin, T. L., Humphreys, C. J., and Meggers, W. F., *J. Res. Natl. Bur. Stand.*, 11, 409, 1933.
62. Dick, K. A., *J. Opt. Soc. Am.*, 64, 702, 1973.
63. Dobbie, J. C., *Ann. Solar Phys. Observ.* (Cambridge), 5, 1, 1938.
64. Earls, L. T. and Sawyer, R. A., *Phys. Rev.*, 47, 115, 1935.
65. Edlen, B., *Z. Phys.*, 85, 85, 1933.
66. Edlen, B., *Nova Acta Reglae Soc. Sci. Ups.*, (IV) 9, No. 6, 1934.
67. Edlen, B., *Z. Phys.*, 93, 726, 1935.
68. Edlen, B., *Z. Phys.*, 94, 47, 1935.
69. Edlen, B., *Rep. Prog. Phys.*, 26, 181, 1963.
70. Edlen, B. and Risberg, P., *Ark. Fys.*, 10, 553, 1956.
71. Edlen, B. and Swings, P., *Astrophys. J.*, 95, 532, 1942.
72. Ehrhardt, J. C. and Davis, S. P., *J. Opt. Soc. Am.*, 61, 1342, 1971.
73. Eidelsberg, M., *J. Phys. B*, 5, 1031, 1972.
74. Eidelsberg, M., *J. Phys. B*, 7, 1476, 1974.
75. Ekberg, J. O. and Svensson, L. A., *Phys. Scr.*, 2, 283, 1970.
76. Ekefors, E., *Z. Phys.*, 71, 53, 1931.
77. Epstein, G. L. and Reader, J., *J. Opt. Soc. Am.*, 65, 310, 1975.
78. Epstein, G. L. and Reader, J., *J. Opt. Soc. Am.*, 66, 590, 1976.
79. Epstein, G. L. and Reader, J., unpublished.
80. Eriksson, K. B. S., *Phys. Lett. A.*, 41, 97, 1972.
81. Eriksson, K. B. S. and Isberg, H. B. S., *Ark. Fys.*, 23, 527, 1963.
82. Eriksson, K. B. S. and Wenaker, I., *Phys. Scr.*, 1, 21, 1970.
83. Esteva, J. M. and Mehlman, G., *Astrophys. J.*, 193, 747, 1974.
84. Even-Zohar, M. and Fraenkel, B. S., *J. Phys. B*, 5, 1596, 1972.
85. Fawcett, B. C., *J. Phys. B*, 3, 1732, 1970.
86. Fawcett, B. C., Culham Laboratory Report ARU-R4, 1971.
87. Ferner, E., *Ark. Mat. Astron. Fys.*, 28(A), 4, 1941.
88. Fischer, R. A., Knopf, W. C., and Kinney, F. E., *Astrophys. J.*, 130, 683, 1959.
89. Fowler, A., *Report on Series in Line Spectra*, Fleetway Press, London, 1922.
90. Fowles, G. R., *J. Opt. Soc. Am.*, 44, 760, 1954.
91. Fred, M., *Argonne Natl. Lab.*, unpublished, 1977.
92. Fred, M. and Tomkins, F. S., *J. Opt. Soc. Am.*, 47, 1076, 1957.
93. Fred, M., Tomkins, F. S., Blaise, J. E., Camus, P., and Verges, J., Argonne National Laboratory Report No. 76–68, 1976.
94. Garcia, J. D. and Mack, J. E., *J. Opt. Soc. Am.*, 55, 654, 1965.
96. Giacchetti, A., *Argonne Natl. Lab.*, unpublished, 1975.
97. Giacchetti, A., Blaise, J., Corliss, C. H., and Zalubas, R., *J. Res. Natl. Bur. Stand. Sect. A*, 78, 247, 1974.
98. Giacchetti, A., Stanley, R. W., and Zalubas, R., *J. Opt. Soc. Am.*, 69, 474, 1970.
99. Gilbert, W. P., *Phys. Rev.*, 47, 847, 1935.
100. Gilroy, H. T., *Phys. Rev.*, 38, 2217, 1931.
101. Glad, S., *Ark. Fys.*, 10, 291, 1956.
102. Goldsmith, S., *J. Phys. B*, 2, 1075, 1969.
103. Goorvitch, D., Mehlmam-Balloffet, G., and Valero, F. P. J., *J. Opt. Soc. Am.*, 60, 1458, 1970.
104. Goorvitch, D. and Valero, F. P. J., *Astrophys. J.*, 171, 643, 1972.
105. Green, L. C., *Phys. Rev.*, 55, 1209, 1939.
106. Gutman, F., *Diss. Abstr. Int.* B, 31, 363, 1970.
107. Hallin, R., *Ark. Fys.*, 31, 511, 1966.
108. Hallin, R., *Ark. Fys.*, 32, 201, 1966.
109. Hansen, J. E. and Persson, W., *J. Opt. Soc. Am.*, 64, 696, 1974.
110. Hansen, J. E. and Persson, W., *Phys. Scr.*, 13, 166, 1976.
111. Hellintin, P., *Phys. Scr.*, 13, 155, 1976.
112. Herzberg, G. and Moore, H. R., *Can. J. Phys.*, 37, 1293, 1959.
113. Hetzler, C. W., Boreman, R. W., and Burns, K., *Phys. Rev.*, 48, 656, 1935.
114. Holmstrom, J. E. and Johansson, L., *Ark. Fys.*, 40, 133, 1969.

115. Hontzeas, S., Martinson, I., Erman, P., and Buchta, R., *Nucl. Instrum. Methods*, 110, 51, 1973.
116. Humphreys, C. J., *J. Res. Natl. Bur. Stand.*, 22, 19, 1939.
117. Humphreys, C. J., *J. Opt. Soc. Am.*, 43, 1027, 1953.
118. Humphreys, C. J., *J. Phys. Chem. Ref. Data*, 2, 519, 1973.
119. Humphreys, C. J. and Andrew, K. L., *J. Opt. Soc. Am.*, 54, 1134, 1964.
120. Humphreys, C. J. and Meggers, W. F., *J. Res. Natl. Bur. Stand.*, 10, 139, 1933.
121. Humphreys, C. J. and Paul, E., Jr., *J. Opt. Soc. Am.*, 60, 200, 1970.
122. Humphreys, C. J. and Paul, E., Jr., *J. Opt. Soc. Am.*, 62, 432, 1972.
123. Humphreys, C. J., Paul, E., Jr., Cowan, R. D., and Andrew, K. L., *J. Opt. Soc. Am.*, 57, 855, 1967.
124. Humphreys, C. J., Paul, E., Jr., and Minnhagen, L., *J. Opt. Soc. Am.*, 61, 110, 1971.
125. Iglesias, L., Inst. of Optics, Madrid, unpublished, 1977.
126. Iglesias, L. and Velasco, R., *Publ. Inst. Opt. Madrid*, No. 23, 1964.
127. Isberg, B., *Ark. Fys.*, 35, 551, 1967.
128. Johannesson, G. A., Lundstrom, T., and Minnhagen, L., *Phys. Scr.*, 6, 129, 1972.
129. Johannesson, G. A. and Lundstrom, T., *Phys. Scr.*, 8, 53, 1973.
130. Johansson, I. *Ark. Fys.*, 20, 135, 1961.
131. Johansson, I. and Contreras, R., *Ark. Fys.*, 37, 513, 1968.
132. Johansson, I. and Litzen, U., *Ark. Fys.*, 34, 573, 1967.
133. Johansson, I. and Svensson, K. F., *Ark. Fys.*, 16, 353, 1960.
134. Johansson, L., *Ark. Fys.*, 20, 489, 1961.
135. Johansson, L., *Ark. Fys.*, 23, 119, 1963.
136. Johansson, S. and Litzen, U., *Phys. Scr.*, 6, 139, 1972.
137. Johansson, S. and Litzen, U., *Phys. Scr.*, 8, 43, 1973.
138. Johansson, S. and Litzen, U., *Phys. Scr.*, 10, 121, 1974.
139. Joshi, Y. N., St. Francis Xavier Univ., Nova Scotia, unpublished.
140. Joshi, Y. N., Bhatia, K. S., and Jones, W. E., *Sci. Light Tokyo*, 21, 113, 1972.
141. Joshi, Y. N., Bhatia, K. S., and Jones, W. E., *Spectrochim. Acta Part B*, 28, 149, 1973.
142. Joshi, Y. N. and Budhiraja, C. J., *Can. J. Phys.*, 49, 670, 1971.
143. Joshi, Y. N. and van Kleef, T. A. M., *Can. J. Phys.*, 52, 1891, 1974.
144. Kaufman, V., *Natl. Bur. Stand.*, unpublished.
145. Kaufman, V., *J. Opt. Soc. Am.*, 52, 866, 1962.
146. Kaufman, V., Artru, M. C., and Brillet, W. U. L., *J. Opt. Soc. Am.*, 64, 197, 1974.
147. Kaufman, V. and Humphreys, C. J., *J. Opt. Soc. Am.*, 59, 1614, 1969.
148. Kaufman, V. and Sugar, J., *J. Opt. Soc. Am.*, 61, 1693, 1971.
149. Kaufman, V. and Sugar, J., *J. Res. Natl. Bur. Stand. Sect. A*, 71, 583, 1967.
150. Kelly, R. L. and Palumbo, L. J., *Naval Research Laboratory Report 7599*, Washington,   DC., 1973.
151. Kielkopf, J. F., *Univ. of Louisville*, unpublished. 1975.
152. Kielkopf, J. F., *Univ. of Louisville*, unpublished, 1976.
153. Kiess, C. C. and Corliss, C. H., *J. Res. Natl. Bur. Stand. Sect. A*, 63, 1, 1959.
154. Kleiman, H., *J. Opt. Soc. Am.*, 52, 441, 1962.
155. Eriksson, K. B., Johansson, I., and Norlen, G., *Ark. Fys.*, 28, 233, 1964.
156. Klinkenberg, P. F. A., *Physica*, 15, 774, 1949.
157. Klinkenberg, P. F. A., *Physica*, 16, 618, 1950.
158. Krishnamurty, S. G., *Proc. Phys. Soc. London*, 48, 277, 1936.
159. Kruger, P. G. and Gilroy, H. T., *Phys. Rev.*, 48, 720, 1935.
160. Kruger, P. G. and Pattin, H. S., *Phys. Rev.*, 52, 621, 1937.
161. Lacroute, P., *Ann. Phys.* (Paris), 3, 5, 1935.
162. Lang, R. J., *Phys. Rev.*, 30, 762, 1927.
163. Lang, R. J., *Phys. Rev.*, 32, 737, 1928.
164. Lang, R. J., *Phys. Rev.*, 35, 445, 1930.
165. Lang, R. J., *Can. J. Res. Sect. A*, 14, 43, 1936.
166. Lang, R. J., *Can. J. Res. Sect. A*, 14, 127, 1936.
167. Lang, R. J. and Vestine, E. H., *Phys. Rev.*, 42, 233, 1932.
168. Li, H. and Andrew, K. L., *J. Opt. Soc. Am.*, 61, 96, 1971.
169. Liden, K., *Ark. Fys.*, 1, 229, 1949.
170. Litzen, U., *Ark. Fys.*, 28, 239, 1965.
171. Litzen, U., *Phys. Scr.*, 1, 251, 1970.
172. Litzen, U., *Phys. Scr.*, 1, 253, 1970.
173. Litzen, U., *Phys. Scr.*, 2, 103, 1970.
174. Litzen, U. and Verges, I., *Phys. Scr.*, 13, 240, 1976.
175. Lofstrand, B., *Phys. Scr.*, 8, 57, 1973.
176. Luc-Koenig, E., Morillon, C., and Verges, J., *Phys. Scr.*, 12, 199, 1975.
177. Lundstrom, T., *Phys. Scr.*, 7, 62, 1973.
178. Lundstrom, T. and Minnhagen, L., *Phys. Scr.*, 5, 243, 1972.
179. Magnusson, C. E. and Zetterberg, P. O., *Phys. Scr.*, 10, 177, 1974.
180. Magnusson, C. E., and Zetterberg, P. O., *Phys. Scr.*, 15, 237, 1977.
181. Martin, D. C., *Phys. Rev.*, 48. 938, 1935.
182. Svendenius, N., *Phys. Scr.*, 22, 240, 1980.
183. Martin, W. C., *J. Res. Natl. Bur. Stand. Sect. A*, 64, 19, 1960.
184. Martin, W. C. and Corliss, C. H., *J. Res. Natl. Bur. Stand. Sect. A*, 64, 443, 1960.
185. Martin, W. C. and Kaufman, V., *J. Res. Natl. Bur. Stand. Sect. A*, 74, 11, 1970.
186. Martin, W. C. and Kaufman, V., *J. Opt. Soc. Am.*, 60, 1096, 1970.
187. McCormick, W. W. and Sawyer, R. A., *Phys. Rev.*, 54, 71, 1938.
188. McLaughlin, R., *J. Opt. Soc. Am.*, 54, 965, 1964.
189. McLennan, J. C., McLay, A. B., and Crawford, M. F., *Proc. R. Soc. London Ser. A*, 134, 41, 1931.
190. Meissner, K. W., *Z. Phys.*, 39, 172, 1926.
191. Meggers, W. F., *J. Res. Natl. Bur. Stand.*, 24, 153, 1940.
192. Meggers, W. F. and Corliss, C. H., *J. Res. Natl. Bur. Stand. Sect. A*, 70, 63, 1966.
193. Meggers, W. F., Fred, M., and Tomkins, F. S., *J. Res. Natl. Bur. Stand.*, 58, 297, 1957.
194. Meggers, W. F. and Humphreys, C. J., *J. Res. Natl. Bur. Stand.*, 28, 463, 1942.
195. Meggers, W. F. and Murphy, R. J., *J. Res. Natl. Bur. Stand.*, 48, 334, 1952.
196. Meggers, W. F., Scribner, B. F., and Bozman, W. R., *J. Res. Natl. Bur. Stand.*, 46, 85, 1951.
197. Meggers, W. F., Shenstone, A. G., and Moore, C. E., *J. Res. Natl. Bur. Stand.*, 45, 346, 1950.
198. Mehlman, G. and Esteva, J. M., *Astrophys. J.*, 188, 191, 1974.
199. Meinders, E., *Physica*, 84(C), 117, 1976.
200. Sansonetti, C. J., Dissertation, Purdue University, 1981.
201. Sansonetti, C. J., *Natl. Bur. Stand. (U.S.)*, unpublished.
202. Millikan, R. A. and Bowen, I. S., *Phys. Rev.*, 25, 600, 1925.
203. Minnhagen, L., *J. Opt. Soc. Am.*, 61, 1257, 1925.
204. Minnhagen, L., *J. Opt. Soc. Am.*, 63, 1185, 1973.
205. Minnhagen, L., *Phys. Scr.*, 11, 38, 1975.
206. Minnhagen, L., *J. Opt. Soc. Am.*, 66, 659, 1976.
207. Minnhagen, L. and Nietsche, H., *Phys. Scr.*, 5, 237, 1972.
208. Minnhagen. L., Strihed, H., and Petersson, B., *Ark. Fys.*, 39, 471, 1969.
209. Moore, C. E., *Natl. Bur. Stand. (U.S.) Circ.*, 488, 1950.
210. Moore, C. E., *Revised Multiplet Table*, Princeton University Observatory No. 20, 1945.
211. Moore, C. E., National Standard Reference Data Series - National Bureau of Standards 3, Sect. 3, 1970.
212. Moore, C. E., National Standard Reference Data Series - National Bureau of Standards 3, Sect. 4, 1971.
213. Moore, C. E., National Standard Reference Data Series - National Bureau of Standards 3, Sect. 5, 1975.
214. Moore, C. E., National Standard Reference Data Series - National Bureau of Standards 3, Sect. 6, 1972.
215. Moore, C. E., National Standard Reference Data Series - National Bureau of Standards 3, Sect. 7, 1975.
216. Morillon, C. and Verges, J., *Phys. Scr.*, 10, 227, 1974.
217. Newsom, G. H., *Astrophys. J.*, 166, 243, 1971.
218. Newsom, G. H., O'Connor, S., and Learner, R. C. M., *J. Phys. B*, 6, 2162, 1973.
219. Norlen, G., *Phys. Scr.*, 8, 249, 1973.
220. Odabasi, H., *J. Opt. Soc. Am.*, 57, 1459, 1967.
221. Olme, A., *Ark. Fys.*, 40, 35, 1969.
222. Olme, A., *Phys. Scr.*, 1, 256, 1970.
223. Johansson, S., and Litzen, U., *J. Opt. Soc. Am.*, 61, 1427, 1971.
224. Palenius, H. P., *Ark. Fys.*, 39, 15, 1969.
225. Palenius, H. P., *Phys. Scr.*, 1, 113, 1970.
226. Palenius, H. P., *Univ. of Lund, Sweden*, unpublished.
227. Paschen, F., *Ann. Phys.*, Series 5, 12, 509, 1932.
228. Paschen, F. and Ritschl, R., *Ann. Phys.*, Series 5, 18, 867, 1933.

229. Peck, E. R., Khanna, B. N., and Anderholm, N. C., *J. Opt. Soc. Am.*, 52, 53, 1962.
230. Persson, W., *Phys. Scr.*, 3, 133, 1971.
231. Persson, W. and Valind, S., *Phys. Scr.*, 5, 187, 1972.
232. Petersson, B., *Ark. Fys.*, 27, 317, 1964.
233. Phillips, L. W. and Parker, W. L., *Phys. Rev.*, 60, 301, 1941.
234. Platt, J. R. and Sawyer, R. A., *Phys. Rev.*, 60, 866, 1941.
235. Plyer, E. K., Blaine, L. R., and Tidwell, E., *J. Res. Natl. Bur. Stand.*, 55, 279, 1955.
236. Poppe, R., van Kleef, T. A. M., and Raassen, A. J. J., *Physica*, 77, 165, 1974.
237. Radziemski, L. J., Jr. and Andrew, K. L., *J. Opt. Soc. Am.*, 55, 474, 1965.
238. Radziemski, L. J., Jr. and Kaufman, V., *J. Opt. Soc. Am.*, 59, 424, 1969.
239. Radziemski, L. J., Jr. and Kaufman, V., *J. Opt. Soc. Am.*, 64, 366, 1974.
240. Ramanadham, R. and Rao, K. R., *Indian J. Phys.*, 18, 317, 1944.
241. Ramb, R., *Ann. Phys.*, 10, 311, 1931.
242. Rank, D. H., Bennett, J. M., and Bennett, H. E., *J. Opt. Soc. Am.*, 40, 477, 1950.
243. Rao, A. S. and Krishnamurty, S. G., *Proc. Phys. Soc. London*, 46, 531, 1943.
244. Rao, K. R., *Proc. R. Soc. London, Ser. A*, 134, 604, 1932.
245. Rao, K. R. and Badami, J. S., *Proc. R. Soc. London Ser. A*, 131, 154, 1931.
246. Rao, K. R. and Krishnamurty, S. G., *Proc. R. Soc. London Ser. A*, 161, 38, 1937.
247. Rao, K. R. and Murti, S. G. K., *Proc. R. Soc. London* Ser. A, 145, 681, 1934.
248. Rao, Y. B., *Indian J. Phys.*, 32, 497, 1958.
249. Rao, Y. B., *Indian J. Phys.*, 33, 546, 1959.
250. Rao, Y. B., *Indian J. Phys.*, 35, 386, 1961.
251. Rasmussen, E., *Z. Phys.*, 80, 726, 1933.
252. Rasmussen, E., *Z. Phys.*, 83, 404, 1933.
253. Rasmussen, E., *Z. Phys.*, 86, 24, 1934.
254. Rasmussen, E., *Z. Phys.*, 87, 607, 1934.
255. Rasmussen, E., *Phys. Rev.*, 57, 840, 1940.
256. Rau, A. S. and Narayan, A. L., *Z. Phys.*, 59, 687, 1930.
257. Reader, J., *J. Opt. Soc. Am.*, 65, 286, 1975.
258. Reader, J., *J. Opt. Soc. Am.*, 65, 988, 1975.
259. Reader, J., *J. Opt Soc. Am.*, 73, 349, 1983.
260. Reader, J. and Davis, S., *J. Res. Natl. Bur. Stand. Sect. A*, 71, 587, 1967, and unpublished.
261. Reader, J. and Ekberg, J. O., *J. Opt. Soc. Am.*, 62, 464, 1972.
262. Reader, J. and Epstein, G. L., *J. Opt. Soc. Am.*, 62, 1467, 1972.
263. Reader, J. and Epstein, G. L., *J. Opt. Soc. Am.*, 65, 638, 1975.
264. Reader, J. and Epstein, G. L., *Natl. Bur. Stand.*, unpublished.
265. Reader, J., Epstein, G. L., and Ekberg, J. O., *J. Opt. Soc. Am.*, 62, 273, 1972.
266. Kaufman, V., *Phys. Scr.*, 26, 439, 1982.
267. Ricard, R., Givord, M., and George, F., *C. R. Acad. Sci. Paris*, 205, 1229, 1937.
268. Risberg, P., *Ark. Fys.*, 10, 583, 1956.
269. Risberg, G., *Ark. Fys.*, 28, 381, 1965.
270. Risberg, G., *Ark. Fys.*, 37, 231, 1968.
271. Robinson, H. A., *Phys. Rev.*, 49, 297, 1936.
272. Robinson, H. A., *Phys. Rev.*, 50, 99, 1936.
273. Ross, C. B., Jr., Doctoral dissertation, Purdue University, 1969.
274. Ross, C. B., Wood, D. R., and Scholl, P. S., *J. Opt. Soc. Am.*, 66, 36, 1976.
275. Ruedy, J. E. and Gibbs, R. C., *Phys. Rev.*, 46, 880, 1934.
276. Russell, H. N., King, R. B., and Moore, C. E., *Phys. Rev.*, 58, 407, 1940.
277. Russell, H. N. and Moore, C. E., *J. Res. Natl. Bur. Stand.*, 55, 299, 1955.
278. Russell, H. N., Moore, C. E., and Weeks, D. W., *Trans. Am. Philos. Soc.*, 34(2), 111, 1944.
279. Saunders, F., Schneider, E., and Buckingham, E., *Proc. Natl. Acad. Sci.*, 20, 291, 1934.
280. Sawyer, R. A. and Humphreys, C. J., *Phys. Rev.*, 32, 583, 1928.
281. Sawyer, R. A. and Lang, R. J., *Phys. Rev.*, 34, 712, 1929.
282. Sawyer, R. A. and Paschen, F., *Ann. Phys.*, 84(4),1, 1927.
283. Scholl, P. S., M.S. thesis, Wright State Univ., 1975.
284. Schurmann, D., *Z. Phys.*, 17, 4, 1975.
285. Seguier, J., *C. R. Acad. Sci. Paris*, 256, 1703, 1963.
286. Shenstone, A. G., *Phys. Rev.*, 31, 317, 1928.
287. Shenstone, A. G., *Phys. Rev.*, 32, 30, 1928.
288. Shenstone, A. G., *Trans. R. Soc. London*, 237(A), 57, 1938.
289. Shenstone, A. G., *Phys. Rev.*, 57, 894, 1940.
290. Shenstone, A. G. *Philos. Trans. R. Soc. London Ser. A*, 241, 297, 1948.
291. Shenstone, A. G., *Can. J. Phys.*, 38, 677, 1960.
292. Shenstone, A. G., *Proc. R. Soc. London*, 261(A), 153, 1961.
293. Shenstone, A. G., *Proc. R. Soc. London*, 276(A), 293, 1963.
294. Shenstone, A. G., *J. Res. Natl. Bur. Stand. Sect. A*, 74, 801, 1970.
295. Shenstone, A. G., *J. Res. Natl. Bur. Stand. Sect. A*, 79, 497, 1975.
296. Shenstone, A. G. and Pittenger, J. T., *J. Opt. Soc. Am.*, 39, 219, 1949.
297. Smith, S., *Phys. Rev.*, 36, 1, 1930.
298. Smitt, R., *Phys. Scr.*, 8, 292, 1973.
299. Soderqvist, J., *Ark. Mat. Astronom. Fys.*, 32(A), 1, 1946.
300. Sommer, L. A., *Ann. Phys.*, 75, 163, 1924.
301. Spector, N., *J. Opt. Soc. Am.*, 63, 358, 1973.
302. Spector, N. and Sugar, J., *J. Opt. Soc. Am.*, 66, 436, 1976.
303. Steinhaus, D. W., Radziemski, L. J., Jr., and Blaise, J., *Los Alamos Sci. Lab.*, unpublished, 1975.
304. Subbaraya, T. S., *Z. Phys.*, 78, 541, 1932.
305. Sugar, J., *J. Opt. Soc. Am.*, 55, 33, 1965.
306. Sugar, J., *J. Res. Natl. Bur. Stand. Sect. A*, 73, 333, 1969.
307. Sugar, J., *J. Opt. Soc. Am.*, 60, 454, 1970.
308. Sugar, J., *J. Res. Natl. Bur. Stand. Sect. A*, 78, 555, 1974.
309. Sugar, J. and Kaufman, V., *J. Opt. Soc. Am.*, 55, 1283, 1965.
310. Sugar, J. and Kaufman, V., *J. Opt. Soc. Am.*, 62, 562, 1972.
311. Sugar, J., Kaufman, V., and Spector, N., *J. Res. Natl. Bur. Stand., Sect. A*, 83, 233, 1978.
312. Sugar, J. and Spector, N., *J. Opt. Soc. Am.*, 64, 1484, 1974.
313. Sullivan, F. J. *Univ. Pittsburgh Bull.*, 35, 1, 1938.
314. Svensson, L. A. and Ekberg, J. O., *Ark. Fys.*, 37, 65, 1968.
315. Swensson, J. W. and Risberg, G., *Ark. Fys.*, 31, 237, 1966.
316. Tech, J. L., *J. Res. Natl. Bur. Stand. Sect. A*, 67, 505, 1963.
317. Tech, J. L. and Ward, J. F., *Phys. Rev. Lett.*, 27, 367, 1971.
318. Tilford, S. G., *J. Opt. Soc. Am.*, 53, 1051, 1963.
319. Toresson, Y. G., *Ark. Fys.*, 17, 179, 1960.
320. Toresson, Y. G., *Ark. Fys.*, 18, 389, 1960.
321. Toresson, Y. G. and Edlen, B., *Ark. Fys.*, 23, 117, 1963.
322. Tsien, W. Z., *Chin. J. Phys.*, Peiping, 3, 117, 1939.
323. van Deurzen, C. H. H., Conway, J., and Davis, S. P., *J. Opt. Soc. Am.*, 63, 158, 1973.
324. van Kleef, T. A. M., Raassen, A. J. J., and Joshi, Y. N., *Physica*, 84(C), 401, 1976.
325. Sansonetti, C. J., Andrew, K. L., and Verges, J., *J. Opt. Soc. Am.*, 71, 423, 1981.
326. Wheatley, M. A. and Sawyer, R. A., *Phys. Rev.*, 61, 591, 1942.
327. Wilkinson, P. G., *J. Opt. Soc. Am.*, 45, 862, 1955.
328. Wilkinson, P. G. and Andrew, K. L., *J. Opt. Soc. Am.*, 53, 710, 1963.
329. Wood, D. and Andrew, K. L., *J. Opt. Soc. Am.*, 58, 818, 1968.
330. Wood, D. R., Ron, C. B., Scholl, P. S., and Hoke, M., *J. Opt. Soc. Am.*, 64, 1159, 1974.
331. Worden, E. F. and Conway, J. G., *Lawrence Livermore Lab.*, unpublished, 1977.
332. Worden, E. F., Hulet, E. K., Gutmacher, R. G., Conway, J. G., *At. Data Nucl. Data Tables*, 18, 459, 1976.
333. Worden, E. F., Lougheed, R. W., Gutmacher, R. G., and Conway, J. G., *J. Opt. Soc. Am.*, 64, 77, 1974.
334. Wu, C. M., Ph.D. thesis, University of British Columbia, 1971.
335. Zaidel, A. N., Prokofev, V. K., Raiskii, S. M., Slavnyi, V. A., and Schreider, E. Y., *Tables of Spectral Lines*, 3rd ed., Plenum, New York, 1970.
336. Zetterberg, P. O. and Magnusson, C. E., *Phys. Scr.*, 15, 189, 1977.
337. Sugar, J., *J. Opt. Soc. Am.*, 55, 1058, 1965.
338. Sugar, J., *J. Opt. Soc. Am.*, 61, 727, 1971.
339. Worden, E. F., and Conway, J. G., *At. Data Nucl. Data Tables*, 22, 329, 1978.
340. Kaufman, V. and Edlen, B., *J. Phys. Chem. Ref. Data*, 3, 825, 1974.
341. Lang, R. J., *Phys. Rev.*, 34, 697, 1929.
342. Ryabtsev, A. N., *Opt. Spectros.*, 39, 455, 1975.

343. Foster, E. W., *Proc. R. Soc. London*, 200(A), 429, 1950.
344. Morillon, C. and Verges, J., *Phys. Scr.*, 12, 129, 1975.
345. Ruedy, J. E., *Phys. Rev.*, 41, 588, 1932.
346. McLennan, J. C., McLay, A. B., and McLeod, J. H., *Philos. Mag.*, 4, 486, 1927.
347. Handrup, M. B. and Mack, J. E., *Physica*, 30, 1245, 1964.
348. Clearman, H. E., *J. Opt. Soc. Am.*, 42, 373, 1952.
349. Paschen, F., *Ann. Physik*, 424, 148, 1938.
350. Paschen, F. and Campbell, J. S., *Ann. Phys.*, 31(5), 29, 1938.
351. Nodwell, R., *Univ. of British Columbia, Vancouver,* unpublished, 1955.
352. Gibbs, R. C. and White, H. E., *Phys. Rev.*, 31, 776, 1928.
353. Green, M., *Phys. Rev.*, 60, 117, 1941.
354. Ellis, C. B. and Sawyer, R. A., *Phys. Rev.*, 49, 145, 1936.
355. McLennan, J. C., McLay, A. B., and Crawford, M. F., *Proc. R. Soc. London Ser. A*, 125, 50, 1929.
356. Mack, J. E. and Fromer, M., *Phys. Rev.*, 48, 346, 1935.
357. Humphreys, C. J. and Paul, E., U.S. Nav. Ord. Lab., Navord Rep. 4589, 25, 1956.
358. Walters, F. M., *Sci. Pap. Bur. Stand.*, 17, 161, 1921.
359. Crawford, M. F. and McLay, A. B., *Proc. R. Soc. London Ser. A*, 143, 540, 1934.
360. McLay, A. D. and Crawford, M. F., *Phys. Rev.*, 44, 986, 1933.
361. Schoepfle, G. K., *Phys. Rev.*, 47, 232, 1935.
362. Acquista, N., and Reader, J., *J. Opt. Soc. Am.*, 70, 789, 1980.
363. Benschop, H., Joshi, Y. N., and van Kleef, T. A. M., *Can. J. Phys.*, 53, 498, 1975.
364. Bockasten, K., Hallin, R., and Hughes, T. P., *Proc. Phys. Soc.*, 81, 522, 1963.
365. Boyce, J. C., *Phys. Rev.*, 46, 378, 1934.
366. Boyce, J. C., *Phys. Rev.*, 47, 718, 1935.
367. Boyce, J. C., *Phys. Rev.*, 48, 396, 1935.
368. Boyce, J. C., *Phys. Rev.*, 49, 351, 1936.
369. Corliss, C. H. and Meggers, W. F., *J. Res. Natl. Bur. Stand.*, 61, 269, 1958.
370. Crooker, A. M. and Dick, K. A., *Can. J. Phys.*, 42, 766, 1964.
371. De Bruin, T. L., *Z. Physik*, 77, 505, 1932.
372. De Bruin, T. L., *Proc. Roy. Acad. Amsterdam*, 36, 727, 1933.
373. De Bruin, T. L., *Zeeman Verhandelingen*, (The Hague), 1935, p. 415.
374. De Bruin, T. L., *Physica*, 3, 809, 1936.
375. De Bruin, T. L., Proc. Roy. Acad. Amsterdam, 40, 339, 1937.
376. Dick, K. A., *Can. J. Phys.*, 46, 1291, 1968.
377. Dick, K. A., unpublished, 1978.
378. Edlen, B. and Swensson, J. W., *Phys. Scr.*, 12, 21, 1975.
379. Ekberg, J. O., *Phys. Scr.*, 7, 55, 1973.
380. Ekberg, J. O., *Phys. Scr.*, 7, 59, 1973.
381. Ekberg, J. O., *Phys. Scr.*, 12, 42, 1975.
382. Ekberg, J. O. and Edlen, B., *Phys. Scr.*, 18, 107, 1978.
383. Eliason, A. Y., *Phys. Rev.*, 43, 745, 1933.
384. Gallardo, M., Massone, C. A., Tagliaferri, A. A., Garavaglia, M., and Persson, W., *Phys. Scr.*, 19, 538, 1979.
385. Garcia-Riquelme, O., *Optica Pura Y Aplicada*, 1, 53, 1968.
386. Gibbs, R. C., Vieweg, A. M., and Gartlein, C. W., *Phys. Rev.*, 34, 406, 1929.
387. Gilbert, W. P., *Phys. Rev.*, 48, 338, 1935.
388. Goldsmith, S. and Kaufman, A. S., *Proc. Phys. Soc.*, 81, 544, 1963.

389. Hermansdorfer, H., *J. Opt. Soc. Am.*, 62, 1149, 1972.
390. Humphreys, C. J., *Phys. Rev.*, 47, 712, 1935.
391. Humphreys, C. J., *J. Res. Natl. Bur. Stand.*, 16, 639, 1936.
392. Iglesias, L., *J. Opt. Soc. Am.*, 45, 856, 1955.
393. Iglesias, L., *J. Res. Natl. Bur. Stand.*, 64A, 481, 1960.
394. Iglesias, L., Anales Fisica Y Quimica, 58A, 191, 1962.
395. Iglesias, L., *J. Res. Natl. Bur. Stand.*, 70A, 465, 1966.
396. Iglesias, L., *Can. J. Phys.*, 44, 895, 1966.
397. Iglesias, L., *J. Res. Natl. Bur. Stand.*, 72A, 295, 1968.
398. Joshi, Y. N., *Can. Spectrosc.*, 15, 96, 1970.
399. Joshi, Y. N. and van Kleef, T. A. M., *Can. J. Phys.*, 55, 714, 1977.
400. Kaufman, A. S., Hughes, T. P., and Williams, R. V., *Proc. Phys. Soc.*, 76, 17, 1960.
401. Kaufman, V. and Sugar, J., *J. Opt. Soc. Am.*, 68, 1529, 1978.
402. Keussler, V., *Z. Physik*, 85, 1, 1933.
403. Kiess, C. C., *J. Res. Natl. Bur. Stand.*, 56, 167, 1956.
404. Klinkenberg, P. F. A., van Kleef, T. A. M., and Noorman, P. E., *Physica*, 27, 1177, 1961.
405. Kovalev, V. I., Romanos, A. A., and Ryabtsev, A. N., *Opt. Spectrosc.*, 43, 10, 1977.
406. Lang, R. J., *Proc. Natl. Acad. Sci.*, 13, 341, 1927.
407. Lang, R. J., *Zeeman Verhandelingen*, (The Hague), 44, 1935.
408. Liberman, S., et al., *C. R. Acad. Sci.* (Paris), 286, 253, 1978.
409. Livingston, A. E., *J. Phys.*, B9, L215, 1976.
410. Meijer, F. G., *Physica*, 72, 431, 1974.
411. Meijer, F. G. and Metsch, B. C., *Physica*, 94C, 259, 1978.
412. Moore, F. L., thesis, Princeton, 1949.
413. Paul, F. W. and Polster, H. D., *Phys. Rev.*, 59, 424, 1941.
414. Phillips, L. W. and Parker, W. L., *Phys. Rev.*, 60, 301, 1941.
415. Poppe, R., *Physica*, 81C, 351, 1976.
416. Raassen, A. J. J., van Kleef, T. A. M., and Metsch, B. C., *Physica*, 84C, 133, 1976.
417. Rao, A. B. and Krishnamurty, S. G., *Proc. Phys. Soc. (London)*, 51, 772, 1939.
418. Reader, J. and Acquista, N., *J. Opt. Soc. Am.*, 69, 239, 1979.
419. Reader, J. and Epstein, G. L., *J. Opt. Soc. Am.*, 62, 619, 1972.
420. Rico, F. R., *Anales, Real Soc. Esp. Fis. Quim.*, 61, 103, 1965.
421. Schonheit, E., *Optik*, 23, 409, 1966.
422. Shenstone, A. G., *J. Opt. Soc. Am.*, 44, 749, 1954.
423. Shenstone, A. G., unpublished, 1958.
424. Shenstone, A. G., *J. Res. Natl. Bur. Stand.*, 67A, 87, 1963.
425. Sugar, J. and Kaufman, V., *J. Opt. Soc. Am.*, 64, 1656, 1974.
426. Sugar, J. and Kaufman, V., *Phys. Rev.*, C12, 1336, 1975.
427. Svensson, L. A., *Phys. Scr.*, 13, 235, 1976.
428. Swensson, J. W. and Edlen, B., *Phys. Scr.*, 9, 335, 1974.
429. Tagliaferri, A. A., Gallego Lluesma, E., Garavaglia, M., Gallardo, M., and Massone, C. A., *Optica Pura Y Aplica*, 7, 89.
430. Tilford, S. G. and Giddings, L. E., *Astrophys.* J., 141, 1222, 1965.
431. Trawick, M. W., *Phys. Rev.*, 46, 63, 1934.
432. Van Deurzen, C. H. H., *J. Opt. Soc. Am.*, 67, 476, 1977.
433. Yarosewick, S. L. and Moore, F. L., *J. Opt. Soc. Am.*, 57, 1381, 1967.
434. Zalubas, R., unpublished, 1979.
435. Bhatia, K. S., Jones, W. E., and Crooker, A. M., *Can. J. Phys.*, 50, 2421, 1972.
436. van Kleef, T. A. M. and Joshi, Y. N., *Phys. Scr.*, 24, 557, 1981.

NIST ATOMIC TRANSITION PROBABILITIES

J. R. Fuhr and W. L. Wiese

For the current edition of this *Handbook*, we include more accurate data for Fe I and Fe II,[1] and Ba I and Ba II.[2] The new tables contain critically evaluated atomic transition probabilities for over 10500 selected lines of all elements for which reliable data are available on an absolute scale. The material is largely for neutral and singly ionized spectra, but also includes a number of prominent lines of more highly charged ions of important elements.

Many of the data are obtained from comprehensive compilations of the Data Center on Atomic Transition Probabilities at the National Institute of Standards and Technology. Specifically, data have been taken from three recent comprehensive critical compilations on C, N and O,[3] on Sc through Mn[4] and Fe through Ni,[5] and special compilations on neutral and singly ionized iron and barium. Material from earlier compilations for the elements H through Ne[6] and Na through Ca[7] was supplemented by more recent material taken directly from the original literature. For the highly charged ions, some of the data were derived from studies of the systematic behavior of transition probabilities.[8-10] Most of the original literature is cited in the above tables and in recent bibliographies;[11,12] for lack of space, individual literature references are not cited here.

The wavelength range for the neutral species is normally the visible spectrum or shorter wavelengths; only the very prominent near infrared lines are included. For the higher ions, most of the strong lines are located in the far UV. The tabulation is limited to electric dipole — including intercombination — lines and comprises essentially the fairly strong transitions with estimated uncertainties in the 10% to 50% range. With the exception of hydrogen, helium, and the alkali metals, most transitions are between states with low principal quantum numbers.

The transition probability, A, is given in units of 10^8 s^{-1} and is listed to as many digits as is consistent with the indicated accuracy. The power of 10 is indicated by the E notation (i.e., E-02 means 10^{-2}). Generally, the estimated uncertainties of the A-values are ±25 to 50% for two-digit numbers, ±10 to 25% for three-digit numbers and ±1% or better for four- and five-digit numbers.

Each transition is identified by the wavelength λ in ångströms (1 Å = 10^{-10} m); and the statistical weights, g_i and g_k, of the lower (i) and upper (k) states [the product $g_k A$ (or $g_i f$) is needed for many applications]. Whenever the wavelengths of individual lines within a multiplet are extremely close, only an average wavelength for the multiplet as well as the multiplet A-value are given, and this is indicated by an asterisk (*) to the left of the wavelength. This also has been done when the transition probability for an entire multiplet has been taken from the literature and values for individual lines cannot be determined because of insufficient knowledge of the coupling of electrons. The wavelength data have been taken either from recent compilations or from the original literature cited in bibliographies published by the Atomic Energy Levels Data Center[13,14] at the National Institute of Standards and Technology. Wavelength values are consistent with those given in the table "Line Spectra of the Elements," which appears elsewhere in this *Handbook*.

In addition to the transition probability A, the atomic oscillator strength f and the line strength S are often used in the literature. The conversion factors between these quantities are (for electric-dipole transitions):

$$g_i f = 1.499 \cdot 10^{-8} \lambda^2 g_k A = 303.8 \lambda^{-1} S$$

where λ is in ångströms, A is in 10^8 s^{-1}, and S is in atomic units, which are

$$a_0^2 e^2 = 7.188 \cdot 10^{-59} \text{ m}^2 \text{C}^2.$$

The table for hydrogen is presented first, followed by the tables for other elements in alphabetical sequence by element name (not symbol). Within each element, the tables are ordered by increasing ionization stage (e.g., Al I, Al II, etc.).

The transition probabilities for hydrogen and hydrogen-like ions are known precisely. Because of the hydrogen degeneracy, a "transition" is actually the sum of all fine-structure transitions between the principal quantum numbers; therefore, the hydrogen table gives weighted average A-values. For hydrogen-like ions of nuclear charge Z, the following scaling laws hold:

$$A_Z = Z^4 A_{\text{Hydrogen}}$$

$$f_Z = f_{\text{Hydrogen}}$$

$$S_Z = Z^{-2} S_{\text{Hydrogen}}$$

$$\lambda_Z = Z^{-2} \lambda_{\text{Hydrogen}}$$

For very highly charged hydrogen-like ions, starting at about Z>25, relativistic corrections[15] must be applied.

References

1. J. R. Fuhr and W. L. Wiese, A Critical Compilation of Atomic Transition Probabilities for Neutral and Singly-Ionized Iron, *J. Phys. Chem. Ref. Data*, to be published, 2006.
2. J. Z. Klose, J. R. Fuhr, and W. L. Wiese, Critically Evaluated Atomic Transition Probabilities for Ba I and Ba II, *J. Phys. Chem. Ref. Data*, 31, 217 (2002).
3. W. L. Wiese, J. R. Fuhr and T. M. Deters, Atomic Transition Probabilities for Carbon, Nitrogen and Oxygen, *J. Phys. Chem. Ref. Data*, Monograph 7, 1996.
4. G. A. Martin, J. R. Fuhr, and W. L. Wiese, Atomic Transition Probabilities — Scandium through Manganese, *J. Phys. Chem. Ref. Data, 17, Suppl.* 3, 1988.
5. J. R. Fuhr, G. A. Martin, and W. L. Wiese, Atomic Transition Probabilities — Iron through Nickel, *J. Phys. Chem. Ref. Data, 17, Suppl.* 4, 1988.
6. W. L. Wiese, M. W. Smith, and B. M. Glennon, Atomic Transition Probabilities (H through Ne — A Critical Data Compilation), National Standard Reference Data Series, National Bureau of Standards 4, Vol. I, U.S. Government Printing Office, Washington, D.C., 1966.
7. W. L. Wiese, M. W. Smith, and B. M. Miles, Atomic Transition Probabilities (Na through Ca — A Critical Data Compilation), National Standard Reference Data Series, National Bureau of Standards 22, Vol. II, U. S. Government Printing Office, Washington, D.C., 1969.
8. W. L. Wiese and A. W. Weiss, *Phys. Rev.*, 175, 50, 1968.
9. M. W. Smith and W. L Wiese, *Astrophys. J., Suppl. Ser.*, 23, No. 196, 103, 1971.
10. G. A. Martin and W. L. Wiese, *J. Phys. Chem. Ref. Data*, 5, 537, 1976.
11. J. R. Fuhr, H. R. Felrice, K. Olsen, J. Hwang, and S. Kotochigova, Atomic Transition Probability Bibliographic Database, <http://physics.nist.gov/fvalbib>.
12. W. L. Wiese, Reports on Astronomy, *Trans. Int. Astron. Union*, 18A, 116-123, 1982; 19A, 122-138, 1985; 20A, 117-123, 1988; 21A, 111-

116, 1991; 22A, 109-114, 1994; 23A, 13-17, 1997; 24A, 2-9, 1999; 25A, 383-391, 2003, D. Reidel, Kluwer, Dordrecht, Holland.

13. C. E. Moore, Bibliography on the Analyses of Optical Atomic Spectra, National Bureau of Standards Special Publication 306 — Section 1, 1968; Sections 2-4, 1969.

14. L. Hagan and W. C. Martin, Bibliography on Atomic Energy Levels and Spectra (July 1968 through June 1971), National Bureau of Standards Special Publication 363, 1972; L. Hagan, Bibliography on Atomic Energy Levels and Spectra (July 1971 through June 1975),

National Bureau of Standards Special Publication 363, Supplement 1, 1977; R. Zalubas and A. Albright, Bibliography on Atomic Energy Levels and Spectra (July 1975 through June 1979), National Bureau of Standards Special Publication 363, Supplement 2, 1980; A. Musgrove and R. Zalubas, Bibliography on Atomic Energy Levels and Spectra (July 1979 through December 1983), National Bureau of Standards Special Publication 363, Supplement 3, 1985.

15. S. M. Younger and A. Weiss, *J. Res. Natl. Bur. Stand.*, 79A, 629, 1975.

λ	Weights		A	λ	Weights		A	λ	Weights		A
Å	g_i	g_k	$10^8\ s^{-1}$	Å	g_i	g_k	$10^8\ s^{-1}$	Å	g_i	g_k	$10^8\ s^{-1}$
Hydrogen				3721.94	8	392	1.303E-04	75004	72	128	1.561E-03
				3734.37	8	338	1.893E-04	123680	72	98	4.561E-03
H I				3750.15	8	288	2.834E-04				
912.768	2	1800	5.167E-06	3770.63	8	242	4.397E-04	*Aluminum*			
912.839	2	1682	6.122E-06	3797.90	8	200	7.122E-04				
912.918	2	1568	7.297E-06	3835.38	8	162	1.216E-03	*Al I*			
913.006	2	1458	8.753E-06	3889.05	8	128	2.215E-03	2263.5	2	4	6.6E-01
913.104	2	1352	1.057E-05	3970.07	8	98	4.389E-03	2269.1	4	6	7.9E-01
913.215	2	1250	1.286E-05	4101.73	8	72	9.732E-03	2269.2	4	4	1.3E-01
913.339	2	1152	1.578E-05	4340.46	8	50	2.530E-02	2367.1	2	4	7.2E-01
913.480	2	1058	1.952E-05	4861.32	8	32	8.419E-02	2373.1	4	6	8.6E-01
913.641	2	968	2.438E-05	6562.80	8	18	4.410E-01	2373.4	4	4	1.4E-01
913.826	2	882	3.077E-05	8392.40	18	800	1.517E-05	2568.0	2	4	2.3E-01
914.039	2	800	3.928E-05	8413.32	18	722	1.964E-05	2575.1	4	6	2.8E-01
914.286	2	722	5.077E-05	8437.96	18	648	2.580E-05	2575.4	4	4	4.4E-02
914.576	2	648	6.654E-05	8467.26	18	578	3.444E-05	2652.5	2	2	1.33E-01
914.919	2	578	8.858E-05	8502.49	18	512	4.680E-05	2660.4	4	2	2.64E-01
915.329	2	512	1.200E-04	8545.39	18	450	6.490E-05	3082.2	2	4	6.3E-01
915.824	2	450	1.657E-04	8598.40	18	392	9.211E-05	3092.7	4	6	7.4E-01
916.429	2	392	2.341E-04	8665.02	18	338	1.343E-04	3092.8	4	4	1.2E-01
917.181	2	338	3.393E-04	8750.48	18	288	2.021E-04	3944.0	2	2	4.93E-01
918.129	2	288	5.066E-04	8862.79	18	242	3.156E-04	3961.5	4	2	9.8E-01
919.351	2	242	7.834E-04	9014.91	18	200	5.156E-04	6696.0	2	4	1.69E-02
920.963	2	200	1.263E-03	9229.02	18	162	8.905E-04	6698.7	2	2	1.69E-02
923.150	2	162	2.143E-03	9545.97	18	128	1.651E-03	7835.3	4	6	5.7E-02
926.226	2	128	3.869E-03	10049.4	18	98	3.358E-03	7836.1	6	8	6.2E-02
930.748	2	98	7.568E-03	10938.1	18	72	7.783E-03				
937.803	2	72	1.644E-02	12818.1	18	50	2.201E-02	*Al II*			
949.743	2	50	4.125E-02	16407.2	32	288	1.620E-04	1047.9	1	3	3.6E-01
972.537	2	32	1.278E-01	16806.5	32	242	2.556E-04	1048.6	3	5	4.8E-01
1025.72	2	18	5.575E-01	17362.1	32	200	4.235E-04	1539.8	3	5	8.8E+00
1215.67	2	8	4.699E+00	18174.1	32	162	7.459E-04	1670.8	1	3	1.46E+01
3662.26	8	1800	2.847E-06	18751.0	18	32	8.986E-02	1719.4	1	3	6.79E+00
3663.40	8	1682	3.374E-06	19445.6	32	128	1.424E-03	1764.0	5	5	9.8E+00
3664.68	8	1568	4.022E-06	21655.3	32	98	3.041E-03	1772.8	1	3	9.5E+00
3666.10	8	1458	4.826E-06	26251.5	32	72	7.711E-03	1777.0	5	7	1.7E+01
3667.68	8	1352	5.830E-06	27575	50	288	1.402E-04	*1819.0	15	15	5.6E+00
3669.46	8	1250	7.096E-06	28722	50	242	2.246E-04	1855.9	1	3	8.32E-01
3671.48	8	1152	8.707E-06	30384	50	200	3.800E-04	1858.0	3	3	2.48E+00
3673.76	8	1058	1.078E-05	32961	50	162	6.908E-04	1862.3	5	3	4.12E+00
3676.36	8	968	1.347E-05	37395	50	128	1.388E-03	1931.0	3	1	1.08E+01
3679.35	8	882	1.700E-05	40511.5	32	50	2.699E-02	1990.5	3	5	1.47E+01
3682.81	8	800	2.172E-05	43753	72	288	1.288E-04	2816.2	3	1	3.83E+00
3686.83	8	722	2.809E-05	46525	50	98	3.253E-03	4663.1	5	3	5.3E-01
3691.55	8	648	3.685E-05	46712	72	242	2.110E-04	6226.2	1	3	6.2E-01
3697.15	8	578	4.910E-05	51273	72	200	3.688E-04	6231.8	3	5	8.4E-01
3703.85	8	512	6.658E-05	59066	72	162	7.065E-04	6243.4	5	7	1.1E+00
3711.97	8	450	9.210E-05	74578	50	72	1.025E-02	6335.7	5	3	1.4E-01
								6823.4	3	3	3.4E-01
								6837.1	5	3	5.7E-01

λ Å	g_i	g_k	A $10^8\ s^{-1}$	λ Å	g_i	g_k	A $10^8\ s^{-1}$	λ Å	g_i	g_k	A $10^8\ s^{-1}$
6920.3	3	1	9.6E-01	*308.6	2	6	9.9E+01	4587.21	3	1	4.9E-03
7042.1	3	5	5.9E-01	*341.3	6	2	1.3E+02	4589.29	3	5	6.2E-05
7056.7	3	3	5.8E-01	550.05	2	4	8.55E+00	4596.10	3	3	9.47E-04
7471.4	5	7	9.4E-01	568.12	2	2	7.73E+00	4628.44	3	5	3.83E-04
Al III				1997	2	4	1.07E+00	4642.15	3	5	9.6E-04
*560.36	2	6	4.0E-01	2069	2	2	9.7E-01	4647.49	3	3	1.2E-03
695.83	2	4	7.4E-01	*4761	2	6	2.55E-01	4702.32	3	3	1.09E-03
696.22	2	2	7.2E-01	5172	2	4	3.95E-02	4746.82	3	1	3.6E-03
*1352.8	10	14	4.40E+00	5551	4	6	3.85E-02	4752.94	3	3	4.5E-03
1379.7	2	2	4.59E+00	5687	4	4	6.0E-03	4768.68	3	5	8.6E-03
1384.1	4	2	9.1E+00					4798.74	7	9	8.8E-04
1605.8	2	4	1.22E+01	*Argon*				4835.97	7	9	9.3E-04
1611.8	4	4	2.42E+00					4836.70	3	5	1.02E-03
1611.9	4	6	1.45E+01	*Ar I*				4876.26	3	5	7.8E-03
1854.7	2	4	5.40E+00	1048.22	1	3	5.36E+00	4886.29	7	9	1.2E-03
1862.8	2	2	5.33E+00	1066.66	1	3	1.29E+00	4887.95	3	3	1.3E-02
*1935.9	10	14	1.22E+01	3406.18	3	1	3.9E-03	4894.69	3	1	1.8E-02
3601.6	6	4	1.34E+00	3461.08	3	5	6.7E-04	4921.04	5	7	5.9E-04
3601.9	4	4	1.49E-01	3554.30	5	5	2.7E-03	4937.72	7	5	3.6E-04
3612.4	4	2	1.5E+00	3563.29	1	3	1.2E-03	4956.75	7	9	1.8E-03
				3567.66	5	7	1.1E-03	4989.95	5	7	1.1E-03
Al X				3572.30	3	1	5.1E-03	5032.03	7	5	8.2E-04
39.925	1	3	2.22E+03	3606.52	3	1	7.6E-03	5048.81	3	5	4.6E-03
51.979	1	3	4.8E+03	3632.68	3	5	6.6E-04	5054.18	3	3	4.5E-03
55.227	1	3	5.2E+03	3634.46	3	3	1.3E-03	5056.53	3	1	5.7E-03
55.272	3	5	7.2E+03	3643.12	3	5	2.4E-04	5060.08	7	9	3.7E-03
55.376	5	7	9.5E+03	3649.83	3	1	8.0E-03	5070.99	5	3	2.6E-03
59.107	3	5	4.6E+03	3659.53	3	3	4.4E-04	5073.08	3	5	5.9E-04
332.78	1	3	5.6E+01	3670.67	3	5	3.1E-04	5078.03	7	7	4.7E-04
394.83	3	1	8.3E+01	3675.23	3	3	4.9E-04	5087.09	5	7	1.6E-03
395.36	3	5	1.2E+01	3770.37	1	3	7.0E-04	5104.74	3	5	8.7E-04
397.76	1	3	1.7E+01	3834.68	3	1	7.5E-03	5118.21	5	7	2.7E-03
400.43	3	3	1.3E+01	3894.66	3	3	5.7E-04	5127.80	5	5	3.3E-04
401.12	5	5	3.6E+01	3947.50	5	5	5.6E-04	5151.39	3	1	2.39E-02
403.55	3	1	4.9E+01	3948.98	5	3	4.55E-03	5152.30	3	5	1.1E-03
406.31	5	3	1.9E+01	4044.42	3	5	3.33E-03	5162.29	3	3	1.90E-02
670.06	3	5	9.8E+00	4045.96	3	3	4.1E-04	5177.54	7	5	2.4E-03
2535	1	3	3.8E-01	4054.53	3	3	2.7E-04	5192.72	7	7	1.2E-04
				4158.59	5	5	1.40E-02	5194.02	3	1	7.8E-03
Al XI				4164.18	5	3	2.88E-03	5210.49	7	7	1.1E-03
*36.675	2	6	1.5E+03	4181.88	1	3	5.61E-03	5214.77	5	3	2.1E-03
39.091	2	4	2.6E+03	4190.71	5	5	2.80E-03	5216.28	5	3	1.3E-03
39.180	4	6	3.1E+03	4191.03	1	3	5.39E-03	5221.27	7	9	8.8E-03
39.530	2	2	1.8E+02	4198.32	3	1	2.57E-02	5241.09	5	5	1.3E-03
39.623	4	2	3.7E+02	4200.67	5	7	9.67E-03	5246.24	5	7	1.2E-03
48.298	2	4	3.09E+03	4251.18	5	3	1.11E-03	5249.20	5	5	7.9E-04
48.338	2	2	3.08E+03	4259.36	3	1	3.98E-02	5252.79	5	7	5.4E-03
52.299	2	4	8.1E+03	4266.29	3	5	3.12E-03	5254.47	3	5	3.6E-03
52.446	4	6	9.6E+03	4272.17	3	3	7.97E-03	5286.07	5	7	9.6E-04
52.458	4	4	1.6E+03	4300.10	3	5	3.77E-03	5290.00	5	3	9.0E-04
54.217	2	2	4.8E+02	4333.56	3	5	5.68E-03	5309.52	5	5	1.2E-03
54.388	4	2	9.6E+02	4335.34	3	3	3.87E-03	5317.73	5	7	2.6E-03
*99.083	2	6	2.2E+02	4345.17	3	3	2.97E-03	5373.50	3	5	2.7E-03
103.6	2	4	4.2E+02	4363.79	3	3	1.2E-04	5393.27	5	5	9.6E-04
103.8	4	6	5.0E+02	4424.00	1	3	7.3E-05	5410.48	5	7	2.0E-03
*141.6	2	6	4.07E+02	4510.73	3	1	1.18E-02	5421.35	7	5	6.0E-03
150.31	2	4	8.5E+02	4522.32	1	3	8.98E-04	5439.99	3	3	1.9E-03
150.61	4	6	9.9E+02	4544.75	3	3	8.3E-04	5442.24	7	7	9.3E-04
157.0	2	2	1.3E+02	4554.32	3	5	3.8E-04	5451.65	3	5	4.7E-03
157.4	4	2	2.6E+02	4584.96	3	5	1.6E-03	5457.42	5	3	3.6E-03
*205.0	2	6	6.3E+01	4586.61	3	3	2.3E-03				

λ Å	Weights g_i	g_k	A $10^8 s^{-1}$	λ Å	Weights g_i	g_k	A $10^8 s^{-1}$	λ Å	Weights g_i	g_k	A $10^8 s^{-1}$
5459.65	7	7	3.8E-04	6013.68	7	5	1.4E-03	6779.93	1	3	1.21E-03
5467.16	5	5	7.6E-04	6025.15	5	3	9.0E-03	6818.29	3	1	2.0E-03
5473.46	5	3	2.0E-03	6043.22	5	7	1.47E-02	6827.25	5	3	2.4E-03
5490.12	5	5	8.5E-04	6052.73	3	5	1.9E-03	6851.88	3	5	6.7E-04
5492.09	3	1	5.6E-03	6064.76	5	7	5.8E-04	6871.29	3	3	2.78E-02
5495.87	7	9	1.69E-02	6081.25	3	3	7.5E-04	6879.59	3	3	1.8E-03
5506.11	5	7	3.6E-03	6085.86	3	3	9.0E-05	6887.10	5	7	1.3E-03
5524.96	7	7	1.7E-03	6090.79	1	3	3.0E-03	6888.17	3	5	2.5E-03
5528.97	1	3	1.2E-03	6098.81	3	3	5.2E-03	6925.01	3	3	1.2E-03
5534.49	5	3	2.7E-03	6101.16	3	3	3.3E-03	6937.67	3	1	3.08E-02
5540.87	7	5	4.1E-04	6104.58	3	1	3.4E-03	6951.46	5	5	2.2E-03
5552.77	3	3	7.9E-04	6105.64	3	5	1.21E-02	6960.23	5	5	2.4E-03
5558.70	3	5	1.42E-02	6113.46	3	5	4.7E-04	6965.43	5	3	6.39E-02
5559.66	3	5	2.2E-03	6119.66	3	3	5.1E-04	6992.17	3	1	7.5E-03
5572.54	5	7	6.6E-03	6121.86	3	5	1.3E-04	7030.25	7	5	2.67E-02
5574.22	3	5	4.6E-04	6127.42	5	3	1.1E-03	7067.22	5	5	3.80E-02
5581.87	7	5	5.6E-04	6128.73	3	5	8.6E-04	7068.73	5	3	2.0E-02
5588.72	5	5	1.5E-03	6145.44	5	7	7.6E-03	7086.70	1	3	1.5E-03
5597.48	5	7	4.2E-03	6155.24	5	3	5.1E-03	7107.48	5	5	4.5E-03
5606.73	3	3	2.20E-02	6165.12	5	5	9.89E-04	7125.83	3	3	6.0E-03
5618.01	3	3	2.1E-03	6170.17	5	5	5.0E-03	7147.04	5	3	6.25E-03
5620.92	3	1	3.6E-03	6173.10	3	5	6.7E-03	7158.83	3	1	2.1E-02
5623.78	5	5	1.4E-03	6179.41	5	3	6.6E-04	7162.57	1	3	5.8E-04
5635.58	3	5	9.6E-04	6212.50	5	7	3.9E-03	7206.98	5	3	2.48E-02
5637.33	1	3	9.1E-04	6215.94	5	5	5.7E-03	7229.93	5	5	6.6E-04
5639.12	1	3	2.1E-03	6230.93	5	5	1.2E-04	7265.17	3	3	1.7E-03
5641.39	3	5	8.7E-04	6243.40	3	1	1.3E-03	7270.66	7	7	1.1E-03
5648.69	5	3	1.2E-03	6244.73	3	5	2.0E-04	7272.93	3	3	1.83E-02
5650.70	3	1	3.20E-02	6248.41	3	5	6.8E-04	7285.44	5	3	1.2E-03
5659.13	5	5	2.6E-03	6278.65	5	7	2.0E-04	7311.72	3	3	1.7E-02
5681.90	5	7	2.0E-03	6296.87	3	5	9.0E-03	7316.01	3	3	9.6E-03
5683.73	5	5	2.0E-03	6307.66	5	5	6.0E-03	7350.78	3	1	1.2E-02
5700.87	5	7	5.9E-03	6309.14	3	3	7.6E-04	7353.32	5	7	9.6E-03
5712.51	1	3	8.7E-04	6364.89	3	1	5.6E-03	7372.12	7	9	1.9E-02
5739.52	3	5	8.7E-03	6369.58	5	3	4.2E-03	7383.98	3	5	8.47E-02
5772.11	5	7	2.0E-03	6384.72	3	3	4.21E-03	7392.97	5	3	7.2E-03
5773.99	5	5	1.1E-03	6416.31	3	5	1.16E-02	7412.33	3	3	3.9E-03
5783.54	3	5	8.1E-04	6431.56	5	3	5.1E-04	7422.26	3	5	6.6E-04
5789.48	5	5	4.6E-04	6466.55	1	3	1.5E-03	7425.29	5	7	3.1E-03
5790.40	5	3	3.4E-04	6481.14	1	3	9.4E-04	7435.33	5	5	9.0E-03
5802.08	5	3	4.2E-03	6513.85	3	3	5.4E-04	7436.25	7	5	2.7E-03
5843.77	3	5	3.3E-04	6538.11	7	7	1.1E-03	7471.17	3	3	2.2E-04
5882.62	3	1	1.23E-02	6596.12	7	5	2.3E-04	7484.24	3	5	3.4E-03
5888.58	7	5	1.29E-02	6598.68	5	5	3.6E-04	7503.84	3	1	4.45E-01
5916.58	5	3	5.9E-04	6604.02	7	5	2.8E-03	7510.42	5	5	4.5E-03
5927.11	7	7	3.7E-04	6604.85	5	7	1.3E-04	7514.65	3	1	4.02E-01
5928.81	5	3	1.1E-02	6632.09	3	3	5.3E-04	7618.33	3	5	2.9E-03
5940.86	1	3	1.2E-03	6656.88	3	3	3.1E-04	7628.86	3	5	2.9E-03
5942.67	5	5	1.8E-03	6660.68	3	1	7.8E-03	7635.11	5	5	2.45E-01
5943.89	7	5	3.6E-04	6664.05	5	5	1.5E-03	7670.04	5	3	2.8E-03
5949.26	3	3	1.5E-03	6677.28	3	1	2.36E-03	7704.81	5	7	6.3E-04
5964.48	1	3	7.7E-04	6684.73	3	5	3.9E-04	7723.76	5	5	5.18E-02
5968.32	3	3	1.8E-03	6698.47	3	3	2.5E-04	7724.21	1	3	1.17E-01
5971.60	3	1	1.1E-02	6698.88	5	3	1.6E-03	7798.55	3	5	8.7E-04
5981.90	5	7	1.2E-04	6719.22	1	3	2.4E-03	7868.20	1	3	3.50E-03
5987.30	7	7	1.2E-03	6722.88	5	7	3.2E-04	7891.08	5	5	9.5E-03
5988.13	3	5	6.1E-04	6752.84	3	5	1.93E-02	7916.45	3	3	1.2E-03
5994.66	3	5	2.6E-04	6754.37	3	3	2.1E-03	7948.18	1	3	1.86E-01
5999.00	5	5	1.4E-03	6756.10	5	5	3.6E-03	8006.16	3	5	4.90E-02
6005.73	5	3	1.4E-03	6766.61	5	3	4.0E-03	8014.79	5	5	9.28E-02

λ Å	g_i	g_k	A 10^8 s^{-1}
8037.23	1	3	3.59E-03
8046.13	3	1	1.12E-02
8053.31	5	3	8.6E-03
8066.60	5	5	1.4E-03
8103.69	3	3	2.5E-01
8115.31	5	7	3.31E-01
8264.52	3	3	1.53E-01
8384.73	5	7	2.4E-03
8408.21	3	5	2.23E-01
8424.65	3	5	2.15E-01
8490.30	3	5	9.6E-04
8521.44	3	3	1.39E-01
8605.78	5	5	1.04E-02
8620.46	1	3	9.2E-03
8667.94	1	3	2.43E-02
8761.69	3	5	9.5E-03
8784.61	3	1	2.4E-03
8799.08	5	3	4.6E-03
8962.19	3	3	1.6E-03
9075.42	3	1	1.2E-02
9122.97	5	3	1.89E-01
9194.64	3	3	1.76E-02
9224.50	3	5	5.03E-02
9291.53	3	1	3.26E-02
9354.22	3	3	1.06E-02
9657.78	3	3	5.43E-02
9784.50	3	5	1.47E-02
10470.05	1	3	9.8E-03
10478.0	3	3	2.44E-02
10950.7	5	3	3.96E-03
11078.9	5	5	8.3E-03
11393.7	3	1	2.22E-02
11441.8	5	3	1.39E-02
11467.5	3	5	3.69E-03
11488.11	3	3	1.9E-03
11668.7	5	5	3.76E-02
11719.5	5	3	9.52E-03
12026.6	1	3	4.2E-03
12112.2	7	7	3.1E-02
12139.8	3	3	4.5E-02
12343.7	5	7	2.0E-02
12402.9	3	3	1.1E-01
12439.2	3	5	4.9E-02
12456.1	5	3	8.9E-02
12487.6	7	5	1.1E-01
12554.4	7	5	1.2E-03
12702.4	3	3	7.1E-02
12733.6	5	5	1.1E-01
12746.3	3	3	2.0E-02
12802.7	5	5	5.7E-02
12933.3	3	1	1.0E-01
12956.6	3	3	7.4E-02
13008.5	5	3	8.9E-02
13214.7	3	1	8.1E-02
13273.1	5	7	1.5E-01
13313.4	3	5	1.3E-01
13504.0	5	7	1.1E-01
13599.2	5	5	2.2E-02
13622.4	3	5	7.3E-02
13678.5	3	5	6.2E-02

λ Å	g_i	g_k	A 10^8 s^{-1}
14093.6	1	3	4.3E-02
14739.1	5	7	8.8E-04
15046.4	1	3	5.2E-02
15172.3	1	3	1.3E-02
15329.6	5	5	1.2E-03
15555.5	5	7	9.8E-05
15734.9	5	3	2.9E-04
15816.8	5	3	8.7E-04
15989.3	1	3	1.9E-02
16122.7	5	3	3.9E-04
16180.0	5	5	1.2E-03
16264.1	3	3	3.0E-04
16520.1	3	5	2.6E-03
16739.8	3	5	3.1E-03
16940.4	5	5	2.5E-02
20317.0	1	3	1.6E-03
20616.5	5	5	3.9E-03
20812.0	5	7	7.6E-04
21332.2	3	3	3.2E-04
21534.9	3	5	1.1E-03
22039.2	3	1	1.2E-03
22077.4	5	3	1.4E-03
23133.4	3	3	1.7E-03
23844.8	9	7	1.1E-02
23967.5	3	1	3.6E-03

Ar II

λ Å	g_i	g_k	A 10^8 s^{-1}
2317.7	6	4	1.4E-01
2891.6	4	2	1.82E-01
2942.9	4	4	5.3E-01
2979.1	2	2	4.16E-01
3033.5	2	4	9.9E-02
3139.0	6	6	5.2E-01
3169.7	4	6	4.9E-01
3181.0	6	4	3.7E-01
3212.5	4	4	5.2E-02
3221.6	6	6	1.8E-02
3226.0	4	4	2.1E-02
3243.7	4	2	1.06E+00
3249.8	2	4	6.3E-01
3263.6	2	4	1.55E-01
3281.7	2	2	4.2E-01
3430.4	6	8	6.2E-02
3454.1	6	4	3.14E-01
3466.3	8	6	3.0E-02
3476.7	6	6	1.25E+00
3491.2	4	4	1.79E+00
3491.5	6	8	2.31E+00
3509.8	2	2	2.55E+00
3514.4	4	6	1.36E+00
3520.0	6	6	5.2E-01
3521.3	8	8	2.27E-01
3535.3	2	4	5.7E-01
3548.5	4	4	8.7E-01
3550.0	6	6	2.6E-02
3556.9	2	2	5.0E-02
3559.5	6	8	2.88E+00
3565.0	2	4	5.5E-01
3576.6	6	8	2.75E+00
3581.6	2	4	1.76E+00

λ Å	g_i	g_k	A 10^8 s^{-1}
3582.4	4	6	2.53E+00
3588.4	8	10	3.03E+00
3605.9	4	6	4.4E-02
3656.0	6	6	7.6E-02
3682.5	4	2	1.7E-02
3709.9	4	4	4.7E-02
3717.2	6	8	5.2E-02
3729.3	6	4	4.80E-01
3746.9	4	6	2.1E-02
3763.5	8	6	1.78E-01
3766.1	4	4	7.4E-02
3777.5	2	2	1.1E-02
3780.8	8	8	7.7E-01
3786.4	8	6	1.5E-02
3799.4	6	4	1.7E-01
3808.6	6	6	1.0E-02
3826.8	6	6	2.81E-01
3841.5	4	2	2.69E-01
3844.7	6	8	4.8E-02
3845.4	6	4	1.6E-02
3850.6	4	4	3.87E-01
3868.5	4	6	1.4E+00
3872.1	4	4	1.5E-01
3875.3	4	2	8.2E-02
3880.3	2	2	2.32E-01
3891.4	2	2	4.3E-02
3892.0	6	4	6.3E-02
3900.6	4	6	7.2E-02
3911.6	2	4	7.7E-02
3914.8	4	4	3.7E-02
3928.6	2	4	2.44E-01
3931.2	2	4	2.0E-02
3932.5	4	4	9.3E-01
3944.3	8	6	4.1E-02
3952.7	4	4	2.08E-01
3958.4	6	4	3.8E-02
3968.4	6	6	4.8E-02
3979.4	4	2	9.8E-01
3988.2	6	6	4.1E-02
3992.1	4	6	1.6E-02
4013.9	8	8	1.05E-01
4031.4	4	2	7.5E-02
4035.5	4	6	4.4E-02
4038.8	6	8	1.2E-02
4042.9	4	4	4.06E-01
4045.7	4	4	1.6E-02
4052.9	2	4	6.7E-01
4065.1	4	4	1.1E-02
4072.0	6	6	5.8E-01
4079.6	6	4	1.19E-01
4082.4	6	6	2.9E-02
4112.8	4	4	1.1E-02
4128.6	8	6	1.4E-02
4131.7	4	2	8.5E-01
4178.4	6	4	1.2E-02
4202.0	2	4	2.1E-02
4228.2	4	6	1.31E-01
4237.2	4	4	1.12E-01
4266.5	6	6	1.64E-01
4277.5	6	4	8.0E-01

λ	Weights		A	λ	Weights		A	λ	Weights		A
Å	g_i	g_k	10^8 s^{-1}	Å	g_i	g_k	10^8 s^{-1}	Å	g_i	g_k	10^8 s^{-1}
4282.9	4	2	1.32E-01	6684.3	8	6	1.07E-01	713.81	2	2	2.4E+01
4300.6	6	6	5.7E-02	6756.6	4	4	2.0E-02	*Ar IX*			
4331.2	4	4	5.74E-01	6863.5	6	6	2.5E-02	48.739	1	3	1.69E+03
4332.0	4	2	1.92E-01	7233.5	2	4	3.7E-02				
4348.1	6	8	1.17E+00	7380.4	4	4	5.6E-02	*Ar XIII*			
4352.2	2	2	2.12E-01	7589.3	6	4	1.07E-01	162.96	5	3	3.4E+02
4362.1	4	6	5.5E-02					*163.08	9	3	5.3E+02
4370.8	4	4	6.6E-01	*Ar III*				184.90	5	5	1.66E+02
4371.3	6	4	2.21E-01	769.15	5	3	6.0E+00	186.38	1	3	8.8E+01
4376.0	4	2	2.05E-01	871.10	5	3	1.59E+00	*207.89	9	9	9.5E+01
4379.7	2	2	1.00E+00	875.53	3	1	3.74E+00	*245.10	9	15	3.7E+01
4383.8	4	4	1.1E-02	878.73	5	5	2.79E+00				
4400.1	4	4	1.60E-01	879.62	3	3	9.2E-01	*Ar XIV*			
4401.0	8	6	3.04E-01	883.18	1	3	1.22E+00	180.29	2	4	4.5E+01
4412.9	6	8	6.1E-02	887.40	3	5	9.0E-01	183.41	2	2	1.69E+02
4420.9	2	4	3.1E-02	3024.1	5	7	2.6E+00	187.95	4	4	1.97E+02
4426.0	4	6	8.17E-01	3027.2	5	5	6.4E-01	191.35	4	2	7.5E+01
4430.2	2	4	5.69E-01	3054.8	3	5	1.9E+00	194.39	2	2	4.6E+01
4431.0	6	6	1.09E-01	3064.8	3	3	1.0E+00	203.35	4	2	7.8E+01
4460.6	4	6	1.5E-02	3078.2	1	3	1.4E+00				
4474.8	4	2	2.90E-01	3285.9	5	7	2.0E+00	*Ar XV*			
4481.8	6	6	4.55E-01	3301.9	5	5	2.0E+00	25.05	1	3	1.7E+04
4491.0	6	4	4.6E-02	3311.3	5	3	2.0E+00	221.10	1	3	9.55E+01
4530.5	6	4	2.1E-02	3336.1	7	9	2.0E+00	*265.3	9	9	8.1E+01
4545.1	4	4	4.71E-01	3344.7	5	7	1.8E+00				
4579.4	2	2	8.0E-01	3352.1	7	7	2.2E-01	*Ar XVI*			
4589.9	4	6	6.64E-01	3358.5	3	5	1.6E+00	*23.52	2	6	1.43E+04
4598.8	4	4	6.7E-02	3361.3	5	5	3.0E-01	*24.96	6	10	4.4E+04
4609.6	6	8	7.89E-01	3472.6	5	7	2.0E-01	353.88	2	4	1.5E+01
4637.2	6	6	7.1E-02	3480.6	7	7	1.6E+00	389.11	2	2	1.1E+01
4657.9	4	2	8.92E-01	3499.7	3	3	1.3E+00	1268	2	4	1.9E+00
4726.9	4	4	5.88E-01	3500.6	3	5	2.6E-01	1401	2	2	1.4E+00
4732.1	6	4	6.7E-02	3502.7	5	3	4.3E-01	2975	2	4	9.0E-02
4735.9	6	4	5.80E-01	3503.6	5	5	1.2E+00	3514	4	6	6.5E-02
4764.9	2	4	6.4E-01	3511.7	7	5	2.6E-01				
4806.0	6	6	7.80E-01					*Arsenic*			
4847.8	4	2	8.49E-01	*Ar IV*							
4879.9	4	6	8.23E-01	840.03	4	2	2.73E+00	*As I*			
4889.0	2	2	1.9E-01	843.77	4	4	2.70E+00	1890.4	4	6	2.0E+00
4904.8	6	8	3.7E-02	850.60	4	6	2.63E+00	1937.6	4	4	2.0E+00
4933.2	4	4	1.44E-01	*Ar VI*				1972.6	4	2	2.0E+00
4965.1	2	4	3.94E-01	292.15	2	2	6.9E+01	2288.1	6	4	2.8E+00
4972.2	2	2	9.7E-02	294.05	4	2	1.36E+02	2344.0	2	4	3.5E-01
5009.3	4	6	1.51E-01	*Ar VII*				2349.8	4	2	3.1E+00
5017.2	4	6	2.07E-01	*250.41	9	3	2.78E+02	2369.7	4	4	6.0E-01
5017.6	4	4	1.1E-02	*477.54	9	15	9.92E+01	2370.8	4	6	4.2E-01
5062.0	2	4	2.23E-01	585.75	1	3	7.83E+01	2456.5	6	4	7.2E-02
5141.8	6	8	8.1E-02	*637.30	9	9	6.7E+01	2492.9	4	2	1.2E-01
5145.3	4	6	1.06E-01					2745.0	2	4	2.6E-01
5176.2	6	6	1.7E-02	*Ar VIII*				2780.2	4	4	7.8E-01
6103.5	2	2	1.7E-02	158.92	2	4	1.1E+02	2860.4	2	2	5.5E-01
6114.9	10	8	2.00E-01	159.18	2	2	1.11E+02	2898.7	4	2	9.9E-02
6138.7	6	4	1.2E-02	229.44	2	2	1.12E+02				
6172.3	8	6	2.00E-01	230.88	4	2	2.21E+02	*Barium*			
6243.1	8	6	3.0E-02	337.09	4	4	1.2E+01				
6483.1	4	2	1.06E-01	337.26	6	4	1.0E+02	*Ba I*			
6638.2	6	4	1.37E-01	338.22	4	2	1.1E+02	2380.66	1	3	2.62E-04
6639.7	4	2	1.69E-01	519.43	2	4	6.3E+01	2380.75	1	3	2.44E-04
6643.7	10	8	1.47E-01	526.46	4	6	7.2E+01	2380.86	1	3	2.55E-04
6666.4	2	2	8.8E-02	526.87	4	4	1.2E+01	2380.97	1	3	2.57E-04
				700.24	2	4	2.55E+01	2381.08	1	3	2.50E-04
								2381.21	1	3	2.91E-04
								2381.34	1	3	3.15E-04
								2381.48	1	3	3.41E-04

λ	Weights		A	λ	Weights		A	λ	Weights		A
Å	g_i	g_k	10^8 s^{-1}	Å	g_i	g_k	10^8 s^{-1}	Å	g_i	g_k	10^8 s^{-1}
2381.63	1	3	3.22E-04	4189.44	3	5	1.13E-03	9704.31	3	1	1.60E-01
2381.79	1	3	3.44E-04	4190.76	3	5	1.28E-03	9821.48	3	1	5.50E-02
2381.97	1	3	3.39E-04	4192.20	3	5	1.36E-03	10370.30	3	5	1.30E-02
2382.15	1	3	3.63E-04	4193.81	3	5	1.58E-03	10540.10	5	3	1.80E-02
2382.36	1	3	3.64E-04	4195.59	3	5	1.78E-03	10649.10	5	5	2.70E-02
2382.57	1	3	3.14E-04	4323.00	3	5	8.80E-02	11075.70	3	3	3.10E-05
2382.80	1	3	3.69E-04	4402.54	3	5	2.70E-01	11303.00	5	3	1.10E-03
2383.06	1	3	3.57E-04	4488.98	5	7	2.80E-01	11373.70	3	1	1.30E-01
2383.34	1	3	3.37E-04	4493.64	5	5	1.95E-01	12342.30	3	3	9.00E-04
2383.63	1	3	3.70E-04	4573.85	3	1	1.21E+00	14723.10	3	5	8.60E-03
2383.96	1	3	3.20E-04	4579.64	5	5	7.00E-01	14999.90	5	3	2.50E-03
2384.32	1	3	2.97E-04	4599.72	3	1	4.07E-01	17186.90	3	1	2.70E-02
2384.71	1	3	3.00E-04	4619.92	1	3	2.70E-02	18202.80	5	3	1.20E-02
2385.15	1	3	2.50E-04	4700.42	3	3	6.10E-02	21567.70	5	3	2.60E-03
2385.62	1	3	2.30E-04	4726.43	5	3	3.30E-01	30685.30	5	3	6.50E-03
2386.15	1	3	2.03E-04	4801.30	9	3	1.39E-01				
2386.74	1	3	1.87E-04	4902.85	5	3	5.40E-02	***Ba II***			
2387.40	1	3	1.37E-04	5169.53	5	3	9.00E-04	1622.43	2	4	2.46E-02
2388.13	1	3	9.66E-05	5519.04	3	5	5.70E-01	1630.36	2	2	2.42E-02
2388.96	1	3	8.37E-05	5535.48	1	3	1.19E+00	1761.74	4	4	1.00E-02
2399.39	1	3	1.10E-04	5777.62	5	7	8.00E-01	1771.10	4	2	1.00E-01
2402.07	1	3	4.60E-04	5784.04	3	5	2.10E-01	1786.95	6	4	9.00E-02
2405.30	1	3	4.90E-04	5800.23	5	5	2.39E-01	1892.49	2	4	1.30E-01
2409.23	1	3	8.60E-04	5826.27	5	3	4.50E-01	1954.28	4	6	1.40E-01
2414.08	1	3	1.50E-03	5971.70	5	5	1.62E-01	1955.05	4	4	2.40E-02
2420.11	1	3	2.30E-03	5997.09	3	3	2.80E-01	1985.75	2	4	2.00E-01
2427.41	1	3	5.60E-03	6019.47	3	1	8.10E-01	1999.55	2	4	7.12E-02
2438.81	1	3	1.40E-03	6063.11	5	3	5.60E-01	2009.28	2	2	6.51E-02
2452.33	1	3	8.10E-04	6083.39	3	1	1.10E-01	2024.06	2	2	6.86E-02
2472.74	1	3	4.60E-03	6129.23	3	1	6.00E-02	2052.75	4	6	2.20E-01
3071.58	1	3	4.20E-01	6309.36	3	3	2.00E-04	2054.09	4	4	3.70E-02
3501.11	1	3	3.50E-01	6341.68	5	7	1.16E-01	2079.98	4	2	1.17E-01
3889.33	1	3	1.10E-02	6450.85	3	5	1.10E-01	2153.93	2	4	3.43E-01
4132.43	1	3	1.50E-02	6498.76	7	7	5.40E-01	2200.89	2	2	1.13E-01
4175.69	3	5	1.97E-04	6527.31	5	5	3.30E-01	2214.76	4	4	1.60E-02
4175.91	3	5	1.98E-04	6527.40	15	15	6.15E-01	2232.79	4	6	3.69E-01
4176.12	3	5	2.08E-04	6595.33	3	3	3.80E-01	2235.38	4	4	6.13E-02
4176.36	3	5	2.19E-04	6675.27	5	3	1.89E-01	2245.69	4	2	1.60E-01
4176.60	3	5	2.26E-04	6693.84	7	5	1.46E-01	2254.78	6	4	1.40E-01
4176.86	3	5	2.48E-04	6986.80	5	3	5.20E-03	2285.99	4	2	2.03E-01
4177.15	3	5	2.77E-04	7059.94	7	9	5.00E-01	2528.41	2	4	6.91E-01
4177.44	3	5	3.03E-04	7120.33	3	5	1.10E-01	2634.78	4	6	7.33E-01
4177.74	3	5	3.14E-04	7195.23	1	3	5.60E-02	2641.37	4	4	1.21E-01
4178.07	3	5	3.07E-04	7213.60	5	5	6.50E-04	2647.26	2	2	2.26E-01
4178.43	3	5	3.64E-04	7280.30	5	7	3.20E-01	2771.35	4	2	3.95E-01
4178.80	3	5	4.01E-04	7392.41	3	3	1.81E-01	3390.18	4	6	4.54E-03
4179.20	3	5	4.31E-04	7417.54	7	5	7.70E-03	3412.44	6	8	4.77E-03
4179.64	3	5	4.46E-04	7488.08	7	7	7.30E-02	3413.95	6	6	3.18E-04
4180.09	3	5	4.53E-04	7528.18	5	5	2.70E-02	3552.45	4	6	3.87E-03
4180.57	3	5	4.55E-04	7610.48	5	5	1.10E-02	3576.28	6	8	4.07E-03
4181.09	3	5	4.99E-04	7644.90	9	3	5.03E-01	3578.57	6	6	2.71E-04
4181.66	3	5	5.42E-04	7672.09	3	5	1.52E-01	3891.78	2	4	2.17E+00
4182.27	3	5	6.11E-04	7780.48	5	5	7.60E-02	4130.65	4	6	2.18E+00
4182.94	3	5	6.65E-04	7877.80	3	5	1.60E-02	4166.00	4	4	3.54E-01
4183.64	3	5	6.70E-04	7905.75	5	3	2.65E-01	4216.07	2	4	5.09E-02
4184.40	3	5	7.93E-04	8147.70	5	5	6.30E-02	4267.92	6	8	3.10E-01
4185.25	3	5	8.43E-04	8560.00	5	5	2.00E-01	4309.26	8	10	3.10E-01
4186.16	3	5	9.24E-04	8654.08	5	7	3.10E-03	4325.75	4	6	5.65E-02
4187.15	3	5	9.90E-04	9370.12	5	5	7.60E-02	4329.56	4	4	9.39E-03
4188.25	3	5	1.03E-03	9645.60	7	5	1.10E-01	4524.93	2	2	6.63E-01

λ Å	g_i	g_k	A 10^8 s^{-1}	λ Å	g_i	g_k	A 10^8 s^{-1}	λ Å	g_i	g_k	A 10^8 s^{-1}
4554.03	2	4	1.11E+00	24612.50	4	4	4.75E-03	2798.7	6	6	3.6E-02
4708.90	2	4	8.47E-02	24699.00	4	2	9.98E-02	2898.0	4	2	1.53E+00
4843.48	4	6	9.34E-02	25923.20	6	4	3.66E-02	2938.3	6	4	1.23E+00
4847.19	2	2	3.49E-02	27687.20	2	4	6.10E-02	2989.0	4	4	5.5E-01
4850.92	4	4	1.55E-02	29058.90	4	2	2.89E-02	2993.3	4	6	1.6E-01
4899.93	4	2	1.04E+00	30196.00	2	2	4.70E-02	3024.6	6	6	8.8E-01
4934.08	2	2	9.53E-01	42934.70	6	8	4.82E-03	3067.7	4	2	2.07E+00
4957.09	6	8	5.13E-01	43294.30	4	6	4.39E-03	3076.7	4	4	3.5E-02
4997.79	4	2	6.37E-02	47520.80	6	6	2.37E-04	3397.2	6	4	1.81E-01
5012.95	8	10	5.15E-01					3402.9	6	6	1.6E-02
5185.06	2	4	1.10E-02	*Beryllium*				3510.9	6	4	6.8E-02
5267.01	2	2	1.00E-02					3596.1	2	4	1.98E-01
5361.35	4	6	4.01E-02	*Be I*				3888.2	2	2	6.9E-02
5391.59	6	8	4.22E-02	1491.8	1	3	1.3E-02	4121.5	2	2	1.64E-01
5413.57	6	6	9.21E-05	1661.5	1	3	2.0E-01	4308.5	2	4	1.6E-02
5421.06	6	6	2.77E-03	2348.6	1	3	5.55E+00	4493.0	2	4	1.5E-02
5428.84	6	4	1.92E-03	*2494.7	9	15	1.6E+00	4722.5	4	2	1.17E-01
5480.25	8	6	1.78E-03	*2650.6	9	9	4.24E+00	6134.8	4	4	1.8E-02
5784.15	2	4	1.59E-01	4572.7	3	5	7.9E-01				
5853.67	4	4	6.00E-02	*Be II*				*Boron*			
5981.26	4	6	1.73E-01	1197.1	2	2	4.7E-01				
5999.91	4	4	2.86E-02	1197.2	4	2	9.4E-01	*B I*			
6135.60	2	2	6.64E-02	1512.3	2	4	9.2E+00	1378.6	2	4	3.50E+00
6141.71	6	4	4.12E-01	1512.4	4	6	1.1E+01	1378.9	2	2	1.40E+01
6378.92	4	2	1.18E-01	1776.1	2	2	1.4E+00	1378.9	4	4	1.75E+01
6496.90	4	2	3.10E-01	1776.3	4	2	2.9E+00	1379.2	4	2	7.0E+00
6769.48	6	8	9.35E-01	*2453.8	2	6	1.42E-01	1465.5	2	4	3.34E+00
6874.08	8	10	9.26E-01	3046.5	2	4	4.8E-01	1465.7	4	4	6.7E+00
6995.14	6	8	8.90E-03	3046.7	4	6	5.9E-01	1465.8	6	4	1.00E+01
7115.03	8	10	8.80E-03	3130.4	2	4	1.14E+00	1825.9	2	4	1.76E+00
8496.80	2	4	3.31E-02	3131.1	2	2	1.15E+00	1826.4	4	6	2.11E+00
8591.43	6	6	6.19E-04	3241.6	2	2	1.41E-01	2088.9	2	4	2.8E-01
8661.90	6	4	1.27E-02	3241.8	4	2	2.8E-01	2089.6	4	6	3.3E-01
8703.69	4	6	3.69E-02	3274.6	2	4	1.9E-01	2496.8	2	2	8.64E-01
8710.77	6	8	7.88E-01	3274.7	2	2	1.9E-01	2497.7	4	2	1.73E+00
8719.12	4	4	6.12E-03	4360.7	2	4	9.2E-01				
8737.75	4	6	7.29E-01	4361.0	4	6	1.1E+00	*Bromine*			
8760.61	8	6	1.17E-02	*5255.9	2	6	2.56E-02				
8897.46	6	6	4.93E-02	5270.3	2	2	3.30E-01	*Br I*			
9603.12	2	4	4.16E-01	5270.8	4	2	6.6E-01	1488.5	4	4	1.2E+00
10115.00	4	6	4.27E-01	6279.4	2	4	1.2E-01	1540.7	4	4	1.4E+00
10212.80	4	4	6.92E-02	6279.7	4	6	1.43E-01	1574.8	2	4	2.0E-01
10709.80	6	6	9.90E-05	6756.7	2	2	5.1E-02	1576.4	4	6	2.1E-02
10768.00	2	4	5.56E-02	6757.1	4	2	1.02E-01	1633.4	2	4	8.1E-02
10769.70	6	4	2.04E-03	7401.2	2	4	3.0E-02	4365.1	2	4	7.5E-03
10993.40	8	6	1.83E-03	7401.4	2	2	3.0E-02	4425.1	4	2	4.2E-03
11088.50	4	6	6.11E-02					4441.7	6	4	7.5E-03
11127.50	4	4	1.01E-02	*Bismuth*				4472.6	4	4	9.3E-03
11519.50	2	2	2.47E-02					4477.7	6	8	1.3E-02
11577.10	2	2	1.75E-01	*Bi I*				4513.4	6	4	2.8E-03
11931.90	4	2	4.44E-02	1954.5	4	6	1.2E+00	4525.6	6	6	7.2E-03
12475.00	4	2	2.80E-01	2021.2	4	4	6.0E-02	4575.7	4	4	1.6E-02
13057.80	2	4	2.14E-01	2061.7	4	6	9.9E-01	4614.6	4	6	5.4E-03
14211.50	2	2	1.66E-01	2110.3	4	2	9.1E-01	4979.8	4	2	2.6E-03
17738.90	6	8	2.16E-01	2177.3	4	2	2.6E-02	5245.1	2	4	3.1E-03
18530.70	8	10	1.96E-01	2228.3	4	4	8.9E-01	5345.4	2	4	7.6E-03
18729.70	2	4	1.23E-01	2230.6	4	6	2.6E+00	7348.5	4	6	1.2E-01
19642.60	4	6	1.28E-01	2276.6	4	4	2.5E-01	7513.0	6	4	1.2E-01
19845.10	4	4	2.07E-02	2515.7	4	6	4.3E-02	7803.0	2	4	5.3E-02
22994.70	2	2	6.18E-02	2627.9	4	4	4.7E-01	7938.7	6	6	1.9E-01
				2696.8	4	6	6.4E-02	8131.5	2	4	3.8E-02
				2780.5	4	2	3.09E-01	8343.7	2	2	2.2E-01

λ Å	Weights g_i	g_k	A 10^8 s^{-1}	λ Å	Weights g_i	g_k	A 10^8 s^{-1}	λ Å	Weights g_i	g_k	A 10^8 s^{-1}
8446.6	4	4	1.2E-01	4289.4	1	3	6.0E-01	2103.2	2	4	8.2E-01
8638.7	6	4	9.7E-02	4299.0	3	3	4.66E-01	2112.8	4	6	9.7E-01
Br II				4302.5	5	5	1.36E+00	2113.2	4	4	1.6E-01
4704.9	5	7	1.1E+00	4307.7	3	1	1.99E+00	2197.8	2	2	3.1E-01
4785.5	5	5	9.4E-01	4318.7	5	3	7.4E-01	2208.6	4	2	6.2E-01
4816.7	5	3	1.1E+00	4355.1	5	7	1.9E-01	3158.9	2	4	3.1E+00
				4425.4	1	3	4.98E-01	3179.3	4	6	3.6E+00
Cadmium				4435.0	3	5	6.7E-01	3181.3	4	4	5.8E-01
				4435.7	3	3	3.42E-01	3706.0	2	2	8.8E-01
Cd I				4454.8	5	7	8.7E-01	3736.9	4	2	1.7E+00
2288.0	1	3	5.3E+00	4455.9	5	5	2.0E-01	3933.7	2	4	1.47E+00
2836.9	1	3	2.8E-01	4526.9	5	3	4.1E-01	3968.5	2	2	1.4E+00
2880.8	3	5	4.2E-01	4578.6	3	5	1.76E-01				
2881.2	3	3	2.4E-01	4581.4	5	7	2.09E-01	*Ca III*			
2980.6	5	7	5.9E-01	4585.9	7	9	2.29E-01	357.97	1	3	8.8E+02
2981.4	5	5	1.5E-01	4685.3	3	5	8.0E-02	439.69	1	3	1.9E-01
3261.1	1	3	4.06E-03	4878.1	5	7	1.88E-01	490.55	1	3	1.6E-02
3403.7	1	3	7.7E-01	5041.6	5	3	3.3E-01				
3466.2	3	5	1.2E+00	5188.9	3	5	4.0E-01	*Ca V*			
3467.7	3	3	6.7E-01	5261.7	3	3	1.5E-01	558.60	5	3	2.2E+01
3610.5	5	7	1.3E+00	5262.2	3	1	6.0E-01	637.93	5	3	3.9E+00
3612.9	5	5	3.5E-01	5264.2	5	5	9.1E-02	643.12	3	1	9.1E+00
4140.5	3	5	4.7E-02	5265.6	5	3	4.4E-01	646.57	5	5	6.9E+00
4662.4	3	5	5.5E-02	5270.3	7	5	5.0E-01	647.88	3	3	2.3E+00
4678.1	1	3	1.3E-01	5582.0	5	7	6.0E-02	651.55	1	3	2.9E+00
4799.9	3	3	4.1E-01	5588.8	7	7	4.9E-01	656.76	3	5	2.1E+00
5085.8	5	3	5.6E-01	5590.1	3	5	8.3E-02				
6438.5	3	5	5.9E-01	5594.5	5	5	3.8E-01	**Ca VII**			
				5598.5	3	3	4.3E-01	550.20	5	5	1.8E+01
Cd II				5601.3	7	5	8.6E-02	624.39	1	3	3.3E+00
2144.4	2	4	2.8E+00	5602.9	5	3	1.4E-01	630.54	3	5	4.5E+00
2265.0	2	2	3.0E+00	5857.5	3	5	6.6E-01	630.79	3	3	2.2E+00
2572.9	2	2	1.7E+00	6102.7	1	3	9.6E-02	639.15	5	7	5.7E+00
2748.5	4	2	2.8E+00	6122.2	3	3	2.87E-01	640.41	5	5	1.3E+00
4415.6	4	6	1.4E-02	6161.3	5	5	3.3E-02				
				6162.2	5	3	3.54E-01	*Ca VIII*			
Calcium				6163.8	3	3	5.6E-02	182.71	2	2	1.6E+02
				6166.4	3	1	2.2E-01	184.16	4	2	3.2E+02
Ca I				6169.1	5	3	1.7E-01				
2275.5	1	3	3.01E-01	6169.6	7	5	1.9E-01	*Ca IX*			
2995.0	1	3	3.67E-01	6439.1	7	9	5.3E-01	163.23	5	3	3.76E+02
2997.3	3	5	2.41E-01	6449.8	3	5	9.0E-02	371.89	1	3	8.8E+01
2999.6	3	3	2.79E-01	6462.6	5	7	4.7E-01	373.81	3	5	1.16E+02
3000.9	3	1	1.58E+00	6471.7	7	7	5.9E-02	378.08	5	7	1.5E+02
3006.9	5	5	7.5E-01	6493.8	3	5	4.4E-01	395.03	3	5	2.2E+02
3009.2	5	3	4.30E-01	6499.7	5	5	8.1E-02	466.24	1	3	1.12E+02
3344.5	1	3	1.51E-01					498.01	3	5	2.49E+01
3350.2	3	5	1.78E-01	*Ca II*				506.18	5	5	7.2E+01
3361.9	5	7	2.23E-01	1341.9	2	4	1.5E-02	515.57	5	3	3.75E+01
3624.1	1	3	2.12E-01	1342.5	2	2	1.5E-02				
3630.8	3	5	2.97E-01	1649.9	2	4	3.2E-03	*Ca X*			
3631.0	3	3	1.53E-01	1652.0	2	2	3.1E-03	110.96	2	4	2.9E+02
3644.4	5	7	3.55E-01	1673.9	2	4	2.24E-01	111.20	2	2	2.92E+02
3644.8	5	5	9.4E-02	1680.1	4	6	2.65E-01	151.84	2	2	2.3E+02
3870.5	3	5	7.2E-02	1680.1	4	4	4.41E-02	153.02	4	2	4.5E+02
3957.1	3	3	9.8E-02	1807.3	2	4	3.54E-01	206.57	4	4	2.9E+01
3973.7	5	3	1.75E-01	1814.5	4	6	4.2E-01	206.75	6	4	2.6E+02
4092.6	3	5	1.1E-01	1814.7	4	4	7.0E-02	207.39	4	2	2.8E+02
4094.9	5	7	1.2E-01	1843.1	2	2	1.6E-01	411.70	2	4	8.3E+01
4098.5	7	9	1.3E-01	1850.7	4	2	3.08E-01	419.75	4	6	9.5E+01
4108.5	5	7	9.0E-01					420.47	4	4	1.6E+01
4226.7	1	3	2.18E+00					557.76	2	4	3.50E+01
4283.0	3	5	4.34E-01					574.01	2	2	3.2E+01

λ Å	g_i	g_k	A 10^8 s^{-1}
Ca XI			
30.448	1	3	6.2E+03
30.867	1	3	4.9E+04
35.212	1	3	2.0E+03
Ca XII			
140.05	4	2	3.7E+02
147.27	2	2	1.6E+02
Ca XV			
141.69	5	3	4.08E+02
*142.23	9	3	6.3E+02
161.00	5	5	1.9E+02
Ca XVII			
19.558	1	3	3.8E+04
21.198	3	5	4.9E+04
192.82	1	3	1.21E+02
218.82	3	5	2.76E+01
223.02	1	3	3.44E+01
228.72	3	3	2.37E+01
232.83	5	5	6.5E+01
244.06	5	3	3.28E+01
Ca XVIII			
*18.71	2	6	2.31E+04
*19.74	6	10	7.0E+04
302.19	2	4	2.0E+01
344.76	2	2	1.3E+01
Carbon			
C I			
945.191	1	3	3.79E+00
945.338	3	3	1.14E+01
945.579	5	3	1.89E+01
1193.24	5	7	1.22E+00
1260.74	1	3	5.32E-01
1260.93	3	1	1.70E+00
1261.00	3	3	4.42E-01
1261.12	3	5	3.71E-01
1261.43	5	3	7.06E-01
1261.55	5	5	1.27E+00
1274.11	5	7	1.03E-02
1277.25	1	3	1.27E+00
1277.28	3	5	1.73E+00
1277.51	3	3	9.12E-01
1277.55	5	7	2.31E+00
1277.72	5	5	6.35E-01
1277.95	5	3	5.56E-02
1279.23	5	7	1.10E-01
1279.89	3	5	3.08E-01
1280.14	1	3	3.11E-01
1280.33	5	5	5.77E-01
1280.40	3	3	1.73E-01
1280.60	3	1	8.22E-01
1280.85	5	3	3.33E-01
1328.83	1	3	7.95E-01
1329.09	3	1	2.41E+00
1329.58	5	5	1.79E+00
1329.60	5	3	1.00E+00
1355.84	5	7	1.04E+00
1364.16	5	5	1.57E-01

λ Å	g_i	g_k	A 10^8 s^{-1}
1431.60	5	7	2.11E+00
1432.10	5	5	2.01E+00
1432.53	5	3	2.11E+00
1459.03	5	3	4.76E-01
1463.34	5	7	1.88E+00
1467.40	5	3	5.49E-01
1468.41	5	3	3.90E-02
1470.09	5	7	1.37E-02
1472.23	5	3	8.01E-03
1481.76	5	5	3.92E-01
1560.31	1	3	6.57E-01
1561.34	5	5	2.94E-01
1561.44	5	7	1.18E+00
1656.27	3	5	8.58E-01
1656.93	1	3	1.13E+00
1657.01	5	5	2.52E+00
1657.38	3	3	8.64E-01
1657.91	3	1	3.43E+00
1658.12	5	3	1.44E+00
1751.83	1	3	9.07E-01
1763.91	1	3	3.59E-02
1765.37	1	3	1.04E-02
1930.90	5	3	3.51E+00
2478.56	1	3	3.40E-01
2902.23	1	3	4.32E-03
2903.27	3	3	1.29E-02
2905.00	5	3	2.15E-02
4371.37	3	3	1.27E-02
4762.31	1	3	3.37E-03
4762.53	3	5	2.72E-03
4766.67	3	3	2.36E-03
4770.03	3	1	1.07E-02
4771.74	5	5	7.97E-03
4775.90	5	3	4.84E-03
4812.92	1	3	4.03E-04
4817.37	3	3	8.76E-04
4826.80	5	3	6.28E-04
4932.05	3	1	6.02E-02
5023.84	7	9	1.81E-03
5039.06	7	9	4.73E-03
5041.48	3	5	5.25E-03
5041.79	5	7	3.28E-03
5052.17	3	5	2.60E-02
5380.34	3	3	1.86E-02
5545.05	3	3	3.04E-03
5668.94	3	3	2.35E-02
5793.12	7	5	3.44E-03
5794.47	5	5	6.44E-04
5800.23	3	3	1.04E-03
5800.60	5	3	3.04E-03
5805.20	3	1	4.12E-03
6001.12	5	5	3.22E-03
6006.02	7	5	1.79E-02
6007.18	3	3	5.34E-03
6010.68	3	1	2.13E-02
6013.17	7	5	1.79E-02
6013.21	7	9	4.35E-03
6014.83	5	3	1.60E-02
6016.45	5	7	3.86E-03
6587.61	3	3	5.09E-02

λ Å	g_i	g_k	A 10^8 s^{-1}
6655.52	3	3	5.03E-03
6828.12	3	5	9.89E-03
7111.47	3	5	2.17E-02
7113.18	7	9	2.47E-02
7115.17	5	7	2.19E-02
7115.18	3	1	4.43E-02
7116.99	7	5	3.26E-02
7119.66	5	3	3.12E-02
7860.88	5	5	1.53E-02
8058.62	5	5	1.09E-02
8335.15	3	1	3.51E-01
9061.44	3	5	7.31E-02
9062.49	1	3	9.48E-02
9078.29	3	3	7.07E-02
9088.51	3	1	3.00E-01
9094.83	5	5	2.28E-01
9111.81	5	3	1.35E-01
9405.73	3	5	2.91E-01
9603.03	1	3	3.06E-02
9620.78	3	3	8.62E-02
9658.43	5	3	1.25E-01
C II			
687.345	4	6	2.84E+01
858.092	2	2	1.18E+00
858.559	4	2	2.35E+00
903.623	2	4	6.85E+00
903.962	2	2	2.74E+01
904.142	4	4	3.42E+01
904.480	4	2	1.37E+01
1009.86	2	4	5.71E+00
1010.08	4	4	1.14E+01
1010.37	6	4	1.71E+01
1036.34	2	2	7.61E+00
1037.02	4	2	1.52E+01
1323.91	4	4	4.33E+00
1323.95	6	4	4.49E+00
1334.53	2	4	2.37E+00
1335.71	4	6	2.84E+00
2091.14	2	4	1.00E-01
2091.19	4	6	1.69E-01
2091.65	6	8	2.41E-01
2093.16	6	6	7.20E-02
2173.85	2	4	2.31E-01
2174.17	2	2	2.31E-01
2509.13	2	4	4.53E-01
2511.74	4	2	9.04E-02
2512.06	4	6	5.42E-01
2727.31	2	4	6.63E-02
2728.72	4	4	3.31E-01
2729.21	2	2	2.65E-01
2730.63	4	2	1.32E-01
5132.95	2	4	3.89E-01
5133.28	4	6	2.80E-01
5137.26	2	2	1.55E-01
5139.17	4	4	1.24E-01
5143.49	4	2	7.73E-01
5145.16	6	6	6.49E-01
5151.08	6	4	4.16E-01
5640.55	2	4	9.89E-02

λ Å	g_i	g_k	A 10^8 s^{-1}	λ Å	g_i	g_k	A 10^8 s^{-1}	λ Å	g_i	g_k	A 10^8 s^{-1}
5648.07	4	4	1.97E-01	386.203	1	3	3.46E+01	9705.41	3	5	5.93E-02
5662.46	6	4	2.93E-01	459.466	1	3	5.91E+01	9706.44	3	3	3.29E-02
5818.31	2	2	3.38E-02	459.514	3	5	7.97E+01	9715.09	5	7	7.88E-02
5822.98	2	4	3.38E-03	459.627	5	7	1.06E+02	9717.75	5	5	1.97E-02
5823.18	4	2	3.38E-02	574.281	3	5	6.24E+01	9718.79	5	3	2.19E-03
5827.85	4	4	2.16E-02	977.020	1	3	1.767E+01				
5836.37	6	4	4.22E-02	1174.93	3	5	3.293E+00	*C IV*			
5843.62	6	6	1.20E-02	1175.26	1	3	4.385E+00	*312.43	2	6	4.63E+01
5856.06	8	6	5.31E-02	1175.59	3	3	3.287E+00	*384.13	6	10	1.76E+02
6095.29	2	4	4.20E-01	1175.71	5	5	9.856E+00	1548.19	2	4	2.65E+00
6098.51	4	6	5.03E-01	1175.99	3	1	1.313E+01	1550.77	2	2	2.64E+00
6102.56	4	4	8.37E-02	1176.37	5	3	5.468E+00	5801.31	2	4	3.17E-01
6578.05	2	4	3.63E-01	1247.38	3	1	2.082E+01	5811.97	2	2	3.16E-01
6582.88	2	2	3.62E-01	2296.87	3	5	1.376E+00				
6724.56	2	4	3.17E-02	2849.05	3	1	1.95E-01	*C V*			
6727.07	2	2	6.34E-02	3703.70	3	3	5.90E-01	34.9728	1	3	2.554E+03
6727.26	4	6	2.96E-02	4325.56	3	5	1.24E-01	40.2678	1	3	8.873E+03
6731.07	4	4	5.06E-02	4647.42	3	5	7.26E-01	*227.19	3	9	1.363E+02
6733.58	4	2	6.32E-02	4650.25	3	3	7.25E-01	247.315	1	3	1.278E+02
6734.00	6	8	1.80E-02	4651.02	3	5	2.28E-01	*248.71	9	15	4.247E+02
6738.61	6	6	7.23E-02	4651.47	3	1	7.24E-01	*260.19	9	3	6.680E+01
6742.43	6	4	4.41E-02	4652.05	1	3	3.04E-01	267.267	3	5	3.947E+02
6750.54	8	8	1.08E-01	4659.06	3	3	2.27E-01	*2273.9	3	9	5.646E-01
6755.16	8	6	2.38E-02	4663.64	3	1	9.05E-01	3526.66	1	3	1.663E-01
6779.94	4	6	2.56E-01	4665.86	5	5	6.78E-01	8420.72	3	5	6.898E-02
6780.59	2	4	1.52E-01	4673.95	5	3	3.75E-01	*8433.2	3	9	6.868E-02
6783.91	6	8	3.65E-01	5244.66	1	3	5.30E-02	8448.12	3	1	6.832E-02
6787.21	2	2	3.04E-01	5253.58	3	3	1.58E-01	8449.19	3	3	6.829E-02
6791.47	4	4	1.94E-01	5272.52	5	3	2.61E-01				
6798.10	4	2	6.04E-02	5695.92	3	5	4.27E-01	*Cesium*			
6800.69	6	6	1.09E-01	5858.34	3	1	1.34E-01				
6812.28	6	4	1.80E-02	5863.25	3	3	3.35E-02	*Cs I*			
7046.25	4	2	3.20E-01	5871.68	5	3	1.00E-01	3203.5	2	4	7.6E-06
7053.09	4	4	3.19E-01	5880.56	5	5	1.99E-02	3205.3	2	4	7.9E-06
7063.68	4	6	3.17E-01	5894.07	7	5	1.11E-01	3207.5	2	4	8.5E-06
7112.48	2	4	2.94E-01	6727.48	1	3	1.12E-01	3210.0	2	4	9.4E-06
7113.04	4	6	3.15E-01	6731.04	3	5	1.50E-01	3212.8	2	4	1.19E-05
7115.63	6	8	3.60E-01	6742.15	3	3	8.32E-02	3216.2	2	4	1.49E-05
7119.76	4	4	1.17E-01	6744.39	5	7	1.99E-01	3220.1	2	4	1.7E-05
7119.91	8	10	4.19E-01	6762.17	5	5	4.95E-02	3220.2	2	2	1.07E-07
7125.72	6	6	1.02E-01	6773.39	5	3	5.47E-03	3224.8	2	4	2.0E-05
7132.47	6	4	8.33E-03	6851.18	3	5	7.60E-03	3225.0	2	2	1.43E-07
7134.10	8	8	5.93E-02	6853.68	5	7	5.64E-03	3230.5	2	4	2.5E-05
7231.33	2	4	3.52E-01	6857.24	3	3	3.79E-02	3230.7	2	2	1.97E-07
7236.42	4	6	4.22E-01	6862.69	5	5	3.51E-02	3237.4	2	4	2.8E-05
7237.17	4	4	7.03E-02	6868.78	5	3	1.26E-02	3237.6	2	2	2.63E-07
8028.85	2	2	1.71E-02	6872.04	7	7	4.46E-02	3245.9	2	4	3.45E-05
8037.73	2	4	4.26E-02	6881.10	7	5	7.80E-03	3246.2	2	2	3.7E-07
8039.40	4	2	8.51E-02	7353.88	5	3	3.09E-02	3256.7	2	4	4.25E-05
8048.31	4	4	1.36E-01	7707.43	3	5	1.30E-01	3257.1	2	2	7.0E-07
8062.10	4	6	3.04E-02	7771.76	3	1	1.77E-01	3270.5	2	4	5.6E-05
8062.80	6	4	4.56E-02	7780.41	3	3	1.76E-01	3271.0	2	2	9.8E-07
8076.64	6	6	7.05E-02	7796.00	3	5	1.75E-01	3288.6	2	4	1.0E-04
9238.30	4	6	3.34E-02	8500.32	1	3	1.01E-01	3289.3	2	2	2.7E-06
9251.01	2	4	2.77E-02	9593.32	3	3	5.32E-03	3313.1	2	4	1.6E-04
9863.06	2	4	5.56E-02	9651.47	5	5	1.57E-02	3314.0	2	2	5.2E-06
9870.78	4	6	9.31E-02	9696.48	5	7	7.53E-03	3347.5	2	4	2.2E-04
9882.68	6	8	1.33E-01	9696.54	3	5	7.12E-03	3348.8	2	2	1.1E-05
				9699.57	7	9	8.47E-03	3397.9	2	4	4.0E-04
C III				9701.10	1	3	4.40E-02	3400.0	2	2	2.4E-05
310.170	1	3	6.56E+00					3476.8	2	4	6.6E-04
								3480.0	2	2	6.6E-05

λ	Weights		A
Å	g_i	g_k	10^8 s^{-1}
3611.4	2	4	1.5E-03
3617.3	2	2	2.5E-04
3876.1	2	4	3.8E-03
3888.6	2	2	9.7E-04
4555.3	2	4	1.88E-02
4593.2	2	2	8.0E-03
8521.1	2	4	3.276E-01
8943.5	2	2	2.87E-01

Chlorine

Cl I

λ	Weights		A
1188.8	4	6	2.33E+00
1188.8	4	4	2.71E-01
1201.4	2	4	2.39E+00
1335.7	4	2	1.74E+00
1347.2	4	4	4.19E+00
1351.7	2	2	3.23E+00
1363.4	2	4	7.5E-01
4323.3	4	4	1.1E-02
4363.3	4	6	6.8E-03
4379.9	4	4	1.4E-02
4389.8	6	8	1.4E-02
4526.2	4	4	5.1E-02
4601.0	2	2	4.2E-02
4661.2	2	4	1.2E-02
7256.6	6	4	1.5E-01
7414.1	6	4	4.7E-02
7547.1	4	4	1.2E-01
7717.6	4	4	3.0E-02
7745.0	2	4	6.3E-02
7769.2	6	6	6.0E-02
7821.4	6	8	9.8E-02
7830.8	4	4	9.7E-02
7878.2	6	6	1.8E-02
7899.3	4	6	5.1E-02
7924.6	2	4	2.1E-02
7935.0	6	8	3.9E-02
7997.9	4	4	2.1E-02

Cl II

λ	Weights		A
3329.1	5	7	1.5E+00
3522.1	7	7	1.4E+00
3798.8	5	7	1.6E+00
3805.2	7	9	1.8E+00
3809.5	3	5	1.5E+00
3851.0	5	7	1.8E+00
3851.4	5	5	1.6E+00
3854.7	3	5	2.2E+00
3861.9	5	7	2.4E+00
3868.6	7	9	2.7E+00
3913.9	9	9	8.2E-01
3990.2	5	7	8.4E-01
4132.5	5	5	1.6E+00
4276.5	9	7	7.6E-01
4768.7	3	5	7.7E-01
4781.3	5	7	1.0E+00
4794.6	5	7	1.04E+00
4810.1	5	5	9.9E-01
4819.5	5	3	1.00E+00
4904.8	5	7	8.1E-01

λ	Weights		A
4917.7	3	5	7.5E-01
5078.3	7	7	7.7E-01
5219.1	3	9	8.6E-01
5392.1	5	7	1.0E+00

Cl III

λ	Weights		A
2298.5	4	4	4.2E+00
2340.6	6	6	4.2E+00
2370.4	8	6	2.8E+00
2531.8	2	4	4.4E+00
2532.5	4	6	5.3E+00
2577.1	4	6	4.3E+00
2580.7	6	8	4.7E+00
2601.2	2	4	4.6E+00
2603.6	4	6	5.0E+00
2609.5	6	8	5.7E+00
2617.0	8	10	6.6E+00
2661.6	4	6	3.4E+00
2665.5	6	8	4.8E+00
2691.5	4	4	3.5E+00
2710.4	4	6	3.5E+00
3340.4	6	6	1.5E+00
3392.9	4	4	1.9E+00
3393.5	6	6	1.9E+00
3530.0	6	8	1.8E+00
3560.7	4	6	1.7E+00
3602.1	6	8	1.7E+00
3612.9	4	6	1.2E+00
3720.5	4	6	1.7E+00

Chromium

Cr I

λ	Weights		A
1999.95	9	9	1.4E+00
2383.30	9	11	4.1E-01
2389.21	3	5	2.3E-01
2408.60	9	7	6.7E-01
2408.72	7	5	2.9E-01
2492.57	3	5	4.5E-01
2495.08	3	3	2.7E-01
2496.30	5	7	5.6E-01
2502.55	7	9	2.2E-01
2504.31	7	9	4.5E-01
2508.11	5	5	2.1E-01
2508.97	5	3	3.8E-01
2527.11	9	9	5.3E-01
2549.55	3	3	4.8E-01
2560.70	5	5	4.3E-01
2571.74	7	5	6.4E-01
2577.66	7	7	2.6E-01
2591.84	9	7	6.5E-01
2620.48	5	3	1.9E-01
2673.64	3	3	1.8E-01
2701.99	9	11	2.1E-01
2726.50	5	7	7.5E-01
2731.90	5	5	7.8E-01
2736.46	5	3	7.5E-01
2752.85	3	3	8.7E-01
2757.09	5	5	6.8E-01
2761.74	5	3	6.8E-01
2764.36	7	7	3.7E-01

λ	Weights		A
2769.90	7	5	1.1E+00
2780.70	9	7	1.4E+00
2879.27	5	7	2.1E-01
2887.00	3	5	2.7E-01
2889.22	9	9	6.6E-01
2893.25	7	7	5.2E-01
2894.17	1	3	3.3E-01
2896.76	5	5	3.0E-01
2905.48	3	1	1.3E+00
2909.05	5	3	6.8E-01
2910.89	7	5	3.4E-01
2911.15	9	7	2.6E-01
2967.64	7	9	3.9E-01
2971.10	5	7	7.1E-01
2975.48	3	5	8.9E-01
2980.78	1	3	5.10E-01
2988.64	5	7	5.2E-01
2991.88	3	1	3.0E+00
2994.06	5	5	2.5E-01
2995.09	5	5	4.3E-01
2996.57	5	3	2.0E+00
2998.78	5	3	4.07E-01
3000.88	7	5	1.6E+00
3005.06	9	7	9.2E-01
3013.72	3	5	8.3E-01
3015.20	1	3	1.63E+00
3020.67	3	3	1.5E+00
3021.58	9	11	2.91E+00
3024.36	5	5	1.27E+00
3029.17	5	3	3.8E-01
3030.25	7	7	1.1E+00
3031.35	5	3	3.1E-01
3034.19	7	7	3.5E-01
3037.05	9	9	5.4E-01
3040.84	7	5	7.4E-01
3053.87	9	7	7.97E-01
3148.44	9	11	5.6E-01
3155.16	11	13	5.7E-01
3163.76	13	15	6.0E-01
3237.73	9	9	1.3E+00
3238.09	11	11	2.0E-01
3578.68	7	9	1.48E+00
3593.48	7	7	1.50E+00
3605.32	7	5	1.62E+00
3639.80	13	11	1.8E+00
3743.89	13	13	7.61E-01
3757.66	7	7	4.13E-01
3768.24	5	5	5.10E-01
3804.80	9	9	6.9E-01
3963.69	13	15	1.3E+00
3969.75	11	13	1.2E+00
3983.90	7	9	1.05E+00
3991.12	5	7	1.07E+00
4001.44	9	11	6.8E-01
4039.10	15	15	6.7E-01
4048.78	13	13	6.4E-01
4058.78	11	11	6.7E-01
4065.71	9	11	3.5E-01
4165.52	11	13	7.5E-01
4204.48	13	11	3.1E-01

λ Å	Weights g_i	g_k	A $10^8\,\text{s}^{-1}$	λ Å	Weights g_i	g_k	A $10^8\,\text{s}^{-1}$	λ Å	Weights g_i	g_k	A $10^8\,\text{s}^{-1}$
4254.33	7	9	3.15E-01	5783.89	5	5	2.02E-01	1127.63	9	11	3.5E+01
4263.15	15	17	6.4E-01	5787.97	5	7	2.35E-01	1465.86	5	3	1.1E+01
4274.81	7	7	3.07E-01					1481.65	3	1	1.0E+01
4275.98	11	11	2.2E-01	*Cr II*				1519.03	5	7	9.5E+00
4280.42	13	15	4.7E-01	2653.57	4	6	3.5E-01	1579.70	7	9	8.6E+00
4289.73	7	5	3.16E-01	2658.59	2	4	5.8E-01				
4291.97	7	5	2.4E-01	2666.02	6	8	5.9E-01	*Cr VI*			
4297.75	11	13	4.9E-01	2668.71	4	2	1.4E+00	161.687	6	6	1.7E+02
4298.05	9	9	2.6E-01	2671.80	6	4	1.0E+00	168.088	4	6	2.0E+02
4300.52	9	7	1.9E-01	2672.83	8	6	5.5E-01	201.007	4	4	2.5E+03
4301.19	11	9	2.6E-01	2744.97	4	6	8.5E-01	201.224	4	6	1.8E+02
4302.78	11	11	2.5E-01	2787.61	6	6	1.5E+00	201.388	6	4	2.7E+02
4319.66	5	3	1.8E-01	2822.38	14	16	2.3E+00	201.606	6	6	2.6E+03
4337.25	5	7	2.0E-01	2835.63	10	12	2.0E+00	202.442	6	4	1.0E+03
4373.65	9	9	2.8E-01	2840.01	10	12	2.7E+00	202.739	4	2	1.2E+03
4376.80	13	13	3.2E-01	2843.24	8	10	6.4E-01	226.241	6	8	7.2E+02
4413.86	7	5	2.7E-01	2849.83	6	8	9.2E-01	227.202	4	6	6.6E+02
4422.70	5	5	2.7E-01	2851.35	8	10	2.2E+00				
4424.29	9	7	2.1E-01	2856.77	4	6	4.3E-01	*Cr X*			
4429.93	3	3	2.4E-01	2857.40	6	8	2.8E-01	216.72	6	8	9.0E+02
4432.16	1	3	1.8E-01	2860.92	2	4	6.9E-01	223.86	4	2	7.7E+02
4432.77	15	15	4.9E-01	2862.57	8	8	6.3E-01	224.74	4	4	7.6E+02
4443.72	3	1	4.5E-01	2866.72	4	4	1.2E+00	226.24	4	6	7.3E+02
4482.88	3	3	3.0E-01	2867.09	4	4	1.1E+00	227.42	4	4	5.2E+02
4490.55	9	7	3.9E-01	2867.65	2	2	1.1E+00	227.50	4	6	1.8E+01
4492.31	5	3	4.47E-01	2870.43	6	6	1.3E+00	228.63	6	4	8.1E+01
4495.28	9	7	2.0E-01	2873.81	4	2	8.8E-01	228.71	6	6	4.5E+02
4500.29	7	7	2.1E-01	2880.86	6	4	7.9E-01	231.21	2	4	1.2E+02
4506.84	13	11	2.7E-01	2898.53	10	12	1.2E+00	232.96	4	4	4.4E+02
4540.72	11	11	3.14E-01	2921.81	8	10	9.0E-01	242.20	2	4	5.0E+01
4564.17	11	13	5.1E-01	2930.83	2	4	1.1E+00	244.19	4	6	5.8E+01
4595.60	13	13	4.7E-01	2935.12	6	8	1.8E+00	395.984	4	4	2.4E+01
4622.47	7	7	4.1E-01	2953.34	2	2	1.8E+00	398.150	6	6	2.1E+01
4663.33	3	3	2.0E-01	2966.03	10	8	5.4E-01				
4665.90	3	3	3.0E-01	2971.90	14	14	2.0E+00	*Cr XI*			
4689.38	7	5	2.3E-01	2979.73	12	12	1.8E+00	214.31	5	7	1.4E+01
4698.46	9	7	2.2E-01	2985.32	10	10	2.2E+00	226.45	5	7	6.0E+02
4708.02	11	9	4.31E-01	2989.18	8	8	2.2E+00	232	3	1	4.1E+02
4718.43	13	11	3.4E-01	3118.64	2	4	1.7E+00	235.53	5	7	5.5E+02
4730.69	7	5	3.83E-01	3120.36	4	6	1.5E+00	240.76	1	3	4.8E+02
4737.33	9	7	3.38E-01	3122.59	12	12	4.4E-01	250.28	5	7	1.0E+01
4741.09	3	5	2.2E-01	3128.69	4	4	8.1E-01	366.491	3	3	1.2E+01
4752.07	13	13	6.2E-01	3136.68	6	6	6.4E-01	366.942	3	1	3.0E+01
4756.09	11	9	4.0E-01	4588.22	8	6	1.2E-01	374.927	5	5	2.3E+01
4792.49	7	5	2.6E-01					422.083	3	5	1.0E+01
4801.02	9	7	3.06E-01	*Cr V*							
4816.13	9	9	1.8E-01	434.306	9	9	1.5E+01	*Cr XII*			
4870.79	7	9	3.5E-01	436.351	9	7	2.4E+01	216	4	6	2.4E+02
4887.01	9	11	3.2E-01	436.601	7	5	2.1E+01	218	6	8	2.4E+02
4922.28	11	13	4.0E-01	437.420	7	7	1.4E+01	239	2	2	1.6E+02
4966.80	3	1	3.0E-01	437.655	5	5	1.3E+01	244.70	2	4	3.0E+02
5204.51	5	3	5.09E-01	441.056	5	3	2.3E+01	247	4	2	2.4E+02
5206.02	5	5	5.14E-01	456.357	1	3	9.5E+00	247	2	2	3.3E+02
5208.42	5	7	5.06E-01	456.637	3	1	3.3E+01	248	6	8	1.4E+02
5243.38	5	3	2.19E-01	456.743	3	3	9.1E+00	250	6	8	3.5E+02
5297.37	7	9	3.88E-01	457.028	5	5	2.7E+01	250	6	6	2.2E+02
5297.99	7	7	3.0E-01	457.504	5	3	1.2E+01	251.52	4	6	3.4E+02
5328.36	9	11	6.2E-01	464.015	9	7	3.6E+01	252	4	6	2.0E+02
5329.17	9	9	2.25E-01	469.634	5	5	2.3E+01	256	2	2	1.5E+02
5783.11	3	3	2.1E-01	1106.25	7	9	1.2E+01	259	2	4	3.2E+02
				1121.07	7	9	2.1E+01	269	2	2	2.1E+02
								300.32	2	2	1.4E+02

λ	Weights		A	λ	Weights		A	λ	Weights		A
Å	g_i	g_k	$10^8\ s^{-1}$	Å	g_i	g_k	$10^8\ s^{-1}$	Å	g_i	g_k	$10^8\ s^{-1}$
305.81	4	4	2.76E+02	125.3	6	8	5.4E+02	123.87	6	4	3.9E+02
309	4	2	2.7E+02	148.5	2	4	2.18E+02	125.51	4	4	3.4E+02
309	6	6	1.6E+02	149.1	2	2	2.1E+02	128.10	6	6	2.8E+02
311.55	4	2	1.6E+02	157.1	2	4	3.3E+02	136.52	4	2	1.66E+02
324	4	6	2.2E+02	158.4	4	6	3.7E+02	139.87	4	4	1.49E+02
327	6	8	2.2E+02	187.02	4	6	9.3E+02	140.82	4	2	2.66E+02
332.06	6	4	1.4E+02	187.30	6	8	9.6E+02	155.46	2	2	2.84E+02
				189.1	2	2	2.13E+02	157.40	4	4	2.83E+02
Cr XIII				191.0	4	2	4.11E+02				
49.59	1	3	9.9E+02	222.9	4	2	2.2E+02	*Cr XIX*			
67.01	1	3	1.67E+03	346.3	4	6	2.4E+02	14.73	3	3	7.1E+04
228	5	7	1.8E+02	346.5	6	8	2.5E+02	14.80	1	3	1.3E+05
267.73	5	7	1.9E+02					14.81	5	3	3.4E+04
270	3	1	1.7E+02	*Cr XV*				14.84	5	7	1.3E+05
276.4	5	7	2.2E+02	18.497	1	3	1.62E+05	109.64	3	3	2.46E+02
277	1	3	2.1E+02	18.782	1	3	2.8E+04	110.37	5	3	6.0E+02
279.32	3	5	3.5E+02	19.015	1	3	6.3E+02	113.97	5	3	5.5E+02
286	3	1	4.6E+02	20.863	1	3	6.0E+03	118.31	3	1	3.29E+02
328.29	1	3	1.86E+02	21.153	1	3	5.6E+03	118.67	5	3	2.1E+02
345	7	9	1.74E+02	102	3	3	1.6E+02	118.83	3	3	1.35E+02
				102.18	5	3	7.0E+02	126.30	1	3	1.56E+02
Cr XIV				103	3	1	3.8E+02	126.33	5	5	4.35E+02
*38.036	2	6	2.47E+02	105	7	5	5.3E+02	130.99	7	5	2.9E+02
39.796	2	4	3.05E+02	111.27	3	3	1.7E+02	134.89	3	1	1.98E+02
40.018	4	6	3.6E+02					138.15	3	1	1.75E+02
40.782	2	4	3.9E+02	*Cr XVI*				138.45	5	5	1.71E+02
40.800	2	2	3.9E+02	17.073	4	6	1.2E+04	140.92	5	3	1.38E+02
41.556	2	4	4.5E+02	17.242	2	4	8.6E+04	143.57	3	1	7.2E+02
41.788	4	6	5.3E+02	17.299	4	4	2.5E+04	163.94	5	5	3.1E+02
44.597	2	4	7.1E+02	17.372	4	4	1.4E+05	179.18	3	1	1.45E+02
44.869	4	6	8.3E+02	17.438	4	2	1.1E+05				
46.125	4	2	3.1E+02	17.514	2	4	1.1E+05	*Cr XX*			
46.468	2	4	6.6E+02	17.587	2	4	2.0E+04	14.13	2	4	1.1E+05
46.527	2	2	6.7E+02	17.656	2	2	2.0E+04	14.26	4	6	1.3E+05
48.300	4	6	5.9E+02	19.442	4	2	9.9E+03	128.42	4	4	3.8E+02
48.338	6	8	6.3E+02	19.714	2	2	1.1E+04	131.31	6	4	1.27E+02
50.821	2	4	1.2E+03					133.82	2	4	8.3E+01
51.172	4	6	1.4E+03	*Cr XVII*				135.26	4	2	2.41E+02
51.180	4	4	2.3E+02	16.31	5	3	9.6E+03	140.75	4	4	1.35E+02
52.321	4	6	1.0E+03	16.32	5	7	3.2E+04	148.99	6	4	1.75E+02
52.363	6	8	1.1E+03	16.37	3	1	9.7E+04	156.00	2	4	8.4E+01
53.760	2	2	3.0E+02	16.44	5	7	1.3E+05	167.97	6	6	1.12E+02
54.164	4	2	5.9E+02	16.59	3	1	5.7E+04	180.85	4	4	1.6E+02
60.699	4	6	2.05E+03	16.65	5	5	1.1E+04				
60.756	6	8	2.19E+03	16.66	1	3	1.8E+05	*Cr XXI*			
63.324	2	4	1.07E+03	16.68	5	7	6.8E+04	12.97	3	1	4.8E+04
63.539	2	2	1.13E+03	16.80	5	7	4.4E+04	12.98	5	5	3.9E+04
68.594	2	4	1.98E+03	16.97	1	3	2.63E+04	13.02	3	5	3.8E+04
69.213	4	6	2.31E+03	16.97	3	3	1.5E+04	13.02	5	7	3.9E+04
69.247	4	4	3.8E+02	17.968	5	3	8.6E+03	13.08	1	3	5.2E+04
86.060	4	6	5.3E+03	18.336	5	3	1.7E+04	13.22	3	1	4.6E+04
86.169	6	8	5.9E+03	18.336	5	5	1.6E+04	13.34	3	5	5.2E+04
86.185	6	6	3.9E+02	18.389	1	3	9.2E+03	13.49	1	3	9.0E+04
101.05	6	4	4.4E+02					13.53	3	3	6.6E+04
101.42	4	2	4.83E+02	*Cr XVIII*				13.55	3	5	1.2E+05
104.4	4	6	3.0E+02	95.77	4	2	3.08E+02	13.65	5	7	1.5E+05
104.5	6	8	3.1E+02	102.32	4	4	1.54E+02	13.66	3	1	1.2E+05
109.8	2	4	2.3E+02	104.98	6	4	8.7E+02	13.67	5	3	3.9E+04
110.4	4	6	2.8E+02	106.84	4	2	3.4E+02	13.68	3	3	8.2E+04
118.3	4	2	2.1E+02	110.41	4	2	7.9E+02	13.75	5	3	4.5E+04
125.2	4	6	5.0E+02	112.27	4	2	4.24E+02	13.75	5	5	9.5E+04
				119.62	2	2	3.2E+02				

λ Å	g_i	g_k	A 10^8 s⁻¹	λ Å	g_i	g_k	A 10^8 s⁻¹	λ Å	g_i	g_k	A 10^8 s⁻¹
13.76	1	3	1.51E+05	2424.93	10	10	3.2E+00	3523.42	4	2	9.8E-01
13.78	5	7	1.7E+05	2432.21	8	8	2.6E+00	3526.85	10	10	1.3E-01
13.84	5	7	2.59E+05	2436.66	6	6	2.6E+00	3529.03	6	8	8.8E-02
13.87	3	5	8.5E+04	2439.04	4	4	2.7E+00	3529.82	8	10	4.6E-01
13.92	3	5	8.5E+04	2460.80	4	6	1.2E-01	3533.36	4	6	9.1E-02
13.93	5	7	4.2E+04	2467.69	6	8	7.0E-02	3560.89	4	4	2.3E-01
13.95	5	5	3.8E+04	2470.27	10	12	1.5E-01	3564.95	6	8	7.0E-02
14.04	3	5	1.2E+05	2476.64	10	8	2.2E-01	3569.37	8	8	1.6E+00
14.24	1	3	1.41E+05	2504.52	10	8	1.8E-01	3574.97	6	6	1.5E-01
				2511.02	10	10	9.2E-01	3575.36	8	8	9.6E-02
Cr XXII				2521.36	10	8	3.0E+00	3585.15	8	8	7.1E-02
2.190	4	2	1.7E+06	2528.97	8	6	2.8E+00	3587.19	6	6	1.4E+00
2.191	2	2	2.5E+06	2530.13	6	6	7.1E-02	3594.87	6	6	9.2E-02
2.198	4	4	4.5E+06	2535.96	6	4	1.9E+00	3602.08	4	4	1.0E-01
2.199	2	4	2.3E+06	2536.50	8	8	3.0E-01	3704.06	6	8	1.2E-01
2.202	4	6	1.6E+06	2544.25	4	2	3.0E+00	3745.49	8	8	7.5E-02
2.203	4	2	1.3E+06	2562.12	4	4	3.9E-01	3842.05	8	6	1.3E-01
13.149	2	4	1.29E+05	2567.34	6	6	3.0E-01	3845.47	8	10	4.6E-01
13.292	4	6	1.54E+05	2574.35	8	8	1.7E-01	3861.16	6	4	1.4E-01
				2685.34	6	8	5.5E-02	3873.12	10	8	1.2E-01
Cr XXIII				3017.55	8	6	6.9E-02	3873.95	8	6	1.0E-01
1.7632	1	3	3.68E+05	3044.00	10	10	1.9E-01	3881.87	6	4	8.2E-02
1.8557	1	3	8.97E+05	3048.89	6	4	7.5E-02	3894.07	6	8	6.9E-01
2.095	3	1	3.5E+06	3061.82	8	8	1.6E-01	3894.98	4	2	8.8E-02
2.101	1	3	2.0E+06	3072.34	6	6	1.5E-01	3935.96	8	10	6.2E-02
2.101	5	5	7.9E+05	3086.78	4	4	1.9E-01	3995.31	8	10	2.5E-01
2.102	3	5	2.1E+06	3354.37	8	6	1.1E-01	3997.90	6	8	7.0E-02
2.103	3	5	1.2E+06	3367.11	10	8	6.0E-02	4092.39	8	8	5.7E-02
2.104	1	3	1.4E+06	3385.22	8	6	1.1E-01	4110.53	6	6	5.5E-02
2.105	3	3	9.6E+05	3388.16	6	4	2.4E-01	4118.77	6	8	1.6E-01
2.106	3	3	2.0E+06	3395.37	6	8	2.9E-01	4121.32	8	10	1.9E-01
2.107	5	5	2.3E+06	3405.12	10	10	1.0E+00	5146.75	8	8	1.5E-01
2.107	3	5	3.3E+06	3409.17	8	8	4.2E-01	5212.70	10	10	1.9E-01
2.109	5	3	1.7E+06	3412.34	8	10	6.1E-01	5265.79	6	8	5.0E-02
2.113	3	5	5.9E+05	3412.63	10	8	1.2E-01	5280.63	10	8	2.8E-01
2.119	3	1	2.7E+05	3414.74	4	4	8.8E-02	5352.05	12	10	2.7E-01
2.129	3	1	5.1E+05	3417.15	6	6	3.2E-01	5477.09	6	8	6.8E-02
2.1818	1	3	3.37E+06	3431.58	8	6	1.1E-01	5483.96	8	10	7.3E-02
2.1923	1	3	2.34E+05	3433.05	4	4	1.0E+00	6082.43	10	10	5.4E-02
				3442.92	6	4	1.2E-01	6455.00	8	10	9.0E-02
Cobalt				3443.64	8	8	6.9E-01	7838.12	8	10	5.4E-02
				3449.17	6	6	7.6E-01	8093.93	12	10	2.0E-01
Co I				3449.44	10	10	1.8E-01	8372.79	10	10	8.7E-02
2287.80	8	8	8.6E-01	3453.51	10	12	1.1E+00				
2295.22	10	8	2.2E-01	3455.24	4	2	1.9E-01	*Co II*			
2309.03	10	10	5.6E-01	3462.80	4	6	7.9E-01	2286.15	11	13	3.3E+00
2323.13	8	8	5.0E-01	3465.79	10	12	9.2E-02	2307.85	9	11	2.6E+00
2325.53	6	8	1.1E-01	3474.02	6	8	5.6E-01	2311.61	7	9	2.8E+00
2335.98	6	6	5.1E-01	3483.41	8	10	5.5E-02	2314.05	5	7	2.8E+00
2338.66	4	4	7.7E-01	3489.40	8	6	1.3E+00	2314.97	3	5	2.7E+00
2353.36	8	10	1.5E-01	3491.32	4	4	5.0E-02	2330.36	5	3	1.32E+00
2355.48	6	8	1.3E-01	3495.68	4	6	4.9E-01	2344.28	3	3	1.5E+00
2358.18	4	6	1.4E-01	3502.28	10	8	8.0E-01	2353.41	7	7	1.9E+00
2365.06	10	10	1.3E-01	3502.63	6	6	5.2E-02	2363.80	9	9	2.1E+00
2371.85	6	8	7.3E-02	3506.32	8	6	8.2E-01	2378.62	11	9	1.9E+00
2384.86	10	8	2.4E-01	3509.84	6	8	3.2E-01	2383.45	9	7	1.8E+00
2392.03	6	6	4.0E-01	3512.64	6	4	1.0E+00	2388.92	11	11	2.8E+00
2402.06	8	6	5.1E-01	3513.48	8	10	7.8E-02	2389.54	5	3	1.5E+00
2407.25	10	12	3.6E+00	3518.34	6	4	1.6E-01	2404.17	3	3	1.5E+00
2412.76	4	6	6.5E-01	3521.58	10	8	1.8E-01	2417.66	9	9	8.5E-01
2414.46	6	8	3.4E+00								
2415.29	4	6	3.6E+00								

λ Å	Weights g_i	g_k	A $10^8\ \text{s}^{-1}$	λ Å	Weights g_i	g_k	A $10^8\ \text{s}^{-1}$	λ Å	Weights g_i	g_k	A $10^8\ \text{s}^{-1}$
Copper				4194.8	17	17	7.2E-01	3212.8	8	8	2.9E-01
				4211.7	17	19	2.08E+00	3213.8	8	6	1.8E-01
Cu I				4218.1	15	15	1.85E+00	3235.1	8	10	1.0E-02
*2024.3	2	6	9.8E-02	4221.1	15	17	1.52E+00	3241.4	8	8	2.3E-02
2165.1	2	4	5.1E-01	4225.2	13	15	4.5E+00	3246.0	8	8	1.4E-02
2178.9	2	4	9.13E-01	4268.3	15	15	3.6E-02	3247.6	8	8	2.3E-02
2181.7	2	2	1.0E+00	4276.7	13	13	7.3E-01	3322.3	8	6	3.5E-02
2225.7	2	2	4.6E-01	4292.0	15	15	5.8E-02	3334.3	8	6	3.4E-01
2244.3	2	4	1.19E-02	4577.8	17	19	2.2E-02	3350.4	8	10	1.5E-02
2441.6	2	2	2.0E-02	4589.4	17	15	1.3E-01	3353.7	8	8	5.8E-03
2492.2	2	4	3.11E-02	4612.3	17	15	8.2E-02	3457.1	8	8	8.4E-03
2618.4	6	4	3.07E-01	5077.7	17	17	5.7E-03	3467.9	8	8	1.0E-02
2766.4	4	4	9.6E-02	5301.6	17	15	1.1E-02	3589.3	8	6	6.9E-03
2824.4	6	6	7.8E-02	5547.3	17	17	2.7E-03	4594.0	8	10	1.4E+00
2961.2	6	8	3.76E-02	5639.5	17	19	4.7E-03	4627.2	8	8	1.3E+00
3063.4	4	4	1.55E-02	5974.5	17	17	4.0E-03	4661.9	8	8	1.3E+00
3194.1	4	4	1.55E-02	5988.6	17	15	5.3E-03	5645.8	8	6	5.4E-03
3247.5	2	4	1.39E+00	6010.8	15	15	2.6E-02	5765.2	8	8	1.1E-02
3274.0	2	2	1.37E+00	6088.3	15	13	3.5E-02	6018.2	8	10	8.5E-03
3337.8	6	8	3.8E-03	6168.4	15	17	2.5E-02	6291.3	8	6	1.8E-03
4022.6	2	4	1.90E-01	6259.1	17	19	8.5E-03	6864.5	8	10	5.8E-03
4062.6	4	6	2.10E-01	6579.4	17	15	7.5E-03	7106.5	8	8	2.6E-03
4249.0	2	2	1.95E-01								
4275.1	6	8	3.45E-01	*Erbium*				*Fluorine*			
4480.4	2	2	3.0E-02								
4509.4	4	2	2.75E-01	*Er I*				*F I*			
4530.8	4	2	8.4E-02	3862.9	13	13	2.5E+00	806.96	4	6	3.3E+00
4539.7	6	4	2.12E-01	4008.0	13	15	2.6E+00	809.60	2	4	2.8E+00
4587.0	8	6	3.20E-01	4151.1	13	11	1.8E+00	951.87	4	2	2.6E+00
4651.1	10	8	3.80E-01					954.83	4	4	5.77E+00
4704.6	8	8	5.5E-02	*Europium*				955.55	2	2	5.1E+00
5105.5	6	4	2.0E-02					958.52	2	4	1.3E+00
5153.2	2	4	6.0E-01	*Eu I*				6239.7	6	4	2.5E-01
5218.2	4	6	7.5E-01	2372.9	8	6	1.9E-01	6348.5	4	4	1.8E-01
5220.1	4	4	1.50E-01	2375.3	8	8	2.0E-01	6413.7	2	4	1.1E-01
5292.5	8	8	1.09E-01	2379.7	8	10	2.0E-01	6708.3	6	4	1.4E-02
5700.2	4	4	2.4E-03	2619.3	8	10	7.0E-03	6774.0	6	6	1.0E-01
5782.1	4	2	1.65E-02	2643.8	8	8	6.6E-03	6795.5	4	2	5.2E-02
				2659.4	8	10	1.2E-02	6834.3	4	4	2.1E-01
Cu II				2682.6	8	6	1.2E-02	6856.0	6	8	4.94E-01
2489.7	5	5	1.5E-02	2710.0	8	10	1.4E-01	6870.2	2	2	3.8E-01
2544.8	9	7	1.1E+00	2724.0	8	8	1.2E-01	6902.5	4	6	3.2E-01
2689.3	7	7	4.1E-01	2731.4	8	8	3.1E-02	6909.8	2	4	2.2E-01
2701.0	5	5	6.7E-01	2732.6	8	6	3.7E-02	6966.4	4	2	1.1E-01
2703.2	3	3	1.2E+00	2735.3	8	10	4.7E-02	7037.5	4	4	3.0E-01
2713.5	5	5	6.8E-01	2738.6	8	10	1.3E-02	7127.9	2	2	3.8E-01
				2743.3	8	6	1.1E-01	7309.0	6	8	4.7E-01
Dysprosium				2745.6	8	6	5.0E-02	7311.0	4	2	3.9E-01
				2747.8	8	8	5.2E-02	7314.3	4	6	4.8E-01
Dy I				2772.9	8	6	1.0E-02	7332.0	6	4	3.1E-01
2862.7	17	15	6.5E-02	2878.9	8	10	2.8E-02	7398.7	6	6	2.85E-01
2964.6	17	17	6.5E-02	2892.5	8	8	1.0E-01	7425.7	4	2	3.4E-01
3147.7	15	17	1.1E-01	2893.0	8	6	1.0E-01	7482.7	4	4	5.6E-02
3263.2	15	13	1.4E-01	2909.0	8	10	6.9E-02	7489.2	2	2	1.1E-01
3511.0	15	13	3.1E-01	2958.9	8	6	1.6E-02	7514.9	2	2	5.2E-02
3571.4	15	13	2.0E-01	3059.0	8	8	3.8E-02	7552.2	4	6	7.8E-02
3757.1	17	19	3.0E+00	3067.0	8	10	9.1E-03	7573.4	2	4	1.0E-01
3868.8	17	17	3.1E+00	3106.2	8	10	5.5E-02	7607.2	4	4	7.0E-02
3967.5	17	19	8.7E-01	3111.4	8	10	3.0E-01	7754.7	4	6	3.82E-01
4046.0	17	15	1.5E+00	3168.3	8	10	6.9E-02	7800.2	2	4	2.1E-01
4103.9	13	11	1.7E+00	3185.5	8	10	5.8E-03				
4186.8	17	17	1.32E+00	3210.6	8	8	1.1E-01				

λ Å	g_i	g_k	A 10^8 s^{-1}	λ Å	g_i	g_k	A 10^8 s^{-1}	λ Å	g_i	g_k	A 10^8 s^{-1}
Gallium				1075.1	4	2	1.3E+00	4024.0	3	1	1.128E-02
Ga I				1237.1	2	4	1.9E+01	*4026.2	9	15	1.160E-01
2195.4	2	2	1.9E-02	1261.9	4	6	2.2E+01	*4120.8	9	3	4.453E-02
2199.7	4	2	3.3E-02	1264.7	4	4	3.5E+00	4143.8	3	5	4.881E-02
2214.4	4	6	1.2E-02	1602.5	2	2	3.4E+00	4169.0	3	1	1.830E-02
2235.9	4	2	4.3E-02	1649.2	4	2	6.5E+00	4387.9	3	5	8.989E-02
2255.0	2	2	3.1E-02	4741.8	2	4	4.6E-01	4437.6	3	1	3.269E-02
2259.2	4	6	3.1E-02	4814.6	4	6	5.1E-01	*4471.5	9	15	2.458E-01
2294.2	2	4	7.0E-02	4824.1	4	4	8.6E-02	*4713.2	9	3	9.521E-02
2297.9	4	2	5.8E-02	5131.8	4	6	1.9E+00	4921.9	3	5	1.986E-01
2338.2	4	6	9.8E-02	5178.5	6	6	1.3E-01	5015.7	1	3	1.337E-01
2371.3	2	2	5.7E-02	5178.6	6	8	2.0E+00	5047.7	3	1	6.771E-02
2418.7	4	2	1.0E-01	5893.4	2	4	9.2E-01	*5875.7	9	15	7.070E-01
2450.1	2	4	2.8E-01	6021.0	2	2	8.4E-01	6678.2	3	5	6.371E-01
2500.2	4	6	3.4E-01	6336.4	2	2	4.4E-01	*7065.2	9	3	2.785E-01
2659.9	2	2	1.2E-01	6484.2	4	2	8.5E-01	7281.4	3	1	1.830E-01
2719.7	4	2	2.3E-01					*8361.7	3	9	3.813E-03
2874.2	2	4	1.2E+00	*Gold*				*9463.6	3	9	5.687E-03
2943.6	4	6	1.4E+00					9603.4	1	3	5.829E-03
2944.2	4	4	2.7E-01	*Au I*				*9702.6	9	3	8.651E-03
4033.0	2	2	4.9E-01	2427.95	2	4	1.99E+00	*10311	9	15	1.995E-02
4172.0	4	2	9.2E-01	2675.95	2	2	1.64E+00	*10668	9	3	1.447E-02
				3122.78	6	4	1.90E-01	*10830	3	9	1.022E-01
Ga II				6278.30	4	2	3.4E-02	*10913	15	21	1.980E-02
829.60	1	3	2.2E-01					10917	5	7	1.608E-02
1414.4	1	3	1.88E+01	*Helium*				*10997	15	9	1.425E-03
								11013	1	3	9.250E-03
Germanium				*He I*				11045	3	5	1.846E-02
				510.00	1	3	4.622E-01	11226	3	1	1.117E-02
Ge I				512.10	1	3	7.317E-01	*11969	9	15	3.478E-02
1944.7	3	1	7.0E-01	515.62	1	3	1.258E+00	*12528	3	9	7.093E-03
1955.1	3	3	2.8E-01	522.21	1	3	2.436E+00	12756	5	3	1.275E-03
1988.3	5	3	2.5E-01	537.03	1	3	5.663E+00	*12785	15	21	4.134E-02
1998.9	5	5	5.5E-01	584.33	1	3	1.799E+01	12791	5	7	3.248E-02
2041.7	1	3	1.1E+00	*2677.1	3	9	4.417E-03	*12846	9	3	2.732E-02
2065.2	3	3	8.5E-01	*2696.1	3	9	6.023E-03	12968	3	5	3.362E-02
2068.7	3	5	1.2E+00	*2723.2	3	9	8.500E-03	*12985	15	9	2.729E-03
2086.0	3	5	4.0E-01	*2763.8	3	9	1.251E-02				
2094.3	5	7	9.7E-01	*2829.1	3	9	1.939E-02	*Indium*			
2105.8	5	5	1.7E-01	*2945.1	3	9	3.201E-02				
2256.0	5	5	3.2E-02	*3187.7	3	9	5.636E-02	*In I*			
2417.4	5	5	9.6E-01	3231.3	1	3	5.102E-03	2560.2	2	4	4.0E-01
2498.0	1	3	1.3E-01	3258.3	1	3	6.963E-03	2710.3	4	6	4.0E-01
2533.2	3	3	1.0E-01	3296.8	1	3	9.843E-03	3039.4	2	4	1.3E+00
2589.2	5	3	5.1E-02	3354.6	1	3	1.454E-02	3256.1	4	6	1.3E+00
2592.5	3	5	7.1E-01	3447.6	1	3	2.269E-02	4101.8	2	2	5.6E-01
2651.2	5	5	2.0E+00	*3554.4	9	15	7.597E-03	4511.3	4	2	1.02E+00
2651.6	1	3	8.5E-01	*3563.0	9	3	4.836E-03				
2691.3	3	3	6.1E-01	*3587.3	9	15	1.811E-02	*In II*			
2709.6	3	1	2.8E+00	3613.6	1	3	3.802E-02	2941.1	3	1	1.4E+00
2754.6	5	3	1.1E+00	*3634.2	9	15	2.606E-02				
3039.1	5	3	2.8E+00	*3652.0	9	3	9.744E-03	*Iodine*			
3124.8	5	5	3.1E-02	*3705.0	9	15	3.953E-02				
3269.5	5	3	2.9E-01	*3819.6	9	15	6.435E-02	*I I*			
4226.6	1	3	2.1E-01	3833.6	3	5	9.647E-03	1782.8	4	4	2.71E+00
4685.8	1	3	9.5E-02	*3867.5	9	3	2.447E-02	1830.4	4	6	1.6E-01
				3871.8	3	5	1.339E-02				
Ge II				*3888.7	3	9	9.475E-02	*Iridium*			
999.10	2	4	1.9E+00	3926.5	3	5	1.937E-02				
1016.6	4	6	2.1E+00	3935.9	3	1	7.448E-03	*Ir I*			
1017.1	4	4	3.5E-01	3964.7	1	3	6.951E-02	2475.12	10	10	2.1E-01
1055.0	2	2	6.9E-01	4009.3	3	5	2.961E-02	2502.98	10	12	3.2E-01
								2639.71	10	10	4.7E-01
								2661.98	10	10	2.5E-01

λ Å	Weights g_i	g_k	A 10^8 s^{-1}	λ Å	Weights g_i	g_k	A 10^8 s^{-1}	λ Å	Weights g_i	g_k	A 10^8 s^{-1}
2664.79	10	8	4.0E-01	2443.87	11	11	5.89E-02	2606.83	9	11	2.43E-01
2694.23	10	12	4.8E-01	2453.48	9	7	1.89E-01	2609.22	7	7	4.60E-01
2849.72	10	10	2.2E-01	2453.57	11	13	1.23E-01	2618.02	7	7	1.50E-01
2853.31	10	10	2.0E-03	2457.60	11	11	4.81E-01	2623.53	7	9	2.13E-01
2882.64	10	8	7.2E-02	2462.18	7	5	1.10E-01	2632.24	5	5	1.21E-01
2924.79	10	12	1.42E-01	2462.65	9	9	5.85E-02	2635.72	11	9	4.29E-02
2934.64	8	10	2.0E-01	2463.73	7	5	1.64E-01	2635.81	5	7	2.11E-01
2951.22	10	8	2.8E-02	2465.15	9	9	4.35E-01	2641.03	9	7	7.71E-02
3003.63	8	10	5.9E-02	2468.88	11	11	2.40E-01	2641.64	9	7	6.47E-02
3168.88	8	10	5.47E-02	2470.97	9	11	2.36E-02	2644.00	3	5	2.34E-01
3220.78	10	8	2.4E-01	2472.34	11	13	7.21E-02	2656.15	13	15	1.63E-01
3558.99	6	8	1.5E-02	2472.87	5	3	2.10E-01	2662.06	7	5	4.64E-02
3573.72	8	10	5.4E-02	2472.89	7	7	1.30E+00	2666.81	11	9	8.91E-02
3617.21	6	8	2.0E-02	2473.16	9	9	2.75E-02	2666.97	9	11	5.16E-02
3628.67	8	8	2.8E-02	2474.81	7	7	6.13E-01	2669.49	11	13	1.34E-01
3661.71	8	8	4.0E-02	2476.66	5	3	3.05E-01	2679.02	9	11	1.10E-01
3734.77	8	8	2.7E-02	2479.48	5	5	2.10E-01	2679.06	11	11	1.50E-01
4033.76	8	10	2.7E-02	2479.78	5	5	1.74E+00	2689.21	9	7	1.68E-01
4069.92	6	8	3.6E-02	2483.27	9	11	4.80E+00	2689.83	7	9	3.04E-02
4913.35	12	12	3.3E-02	2483.53	5	5	2.09E-01	2697.02	7	9	3.51E-02
4939.24	10	12	2.5E-03	2484.19	3	3	2.26E+00	2699.11	9	9	5.59E-02
				2485.99	9	9	2.94E-02	2701.91	9	7	1.05E-01
Iron				2486.69	7	9	1.47E-01	2702.45	13	11	4.23E-02
				2487.07	3	3	6.40E-01	2706.01	13	13	2.28E-01
Fe I				2488.14	7	9	4.20E+00	2706.58	7	5	2.69E-01
1934.54	9	7	2.5E-01	2489.75	1	3	2.31E+00	2708.57	9	9	6.49E-01
1937.27	9	7	2.2E-01	2489.91	3	5	8.72E-02	2710.54	5	7	5.99E-02
1940.66	7	5	2.6E-01	2490.64	5	7	3.44E+00	2711.66	9	11	4.99E-02
2132.02	9	9	7.6E-02	2491.16	3	5	2.91E+00	2716.26	9	9	3.70E-02
2145.19	7	7	5.7E-02	2491.99	9	9	3.25E-01	2716.42	11	9	4.96E-02
2153.01	5	5	6.9E-02	2494.00	3	5	8.89E-02	2718.44	5	3	3.79E-01
2161.58	3	5	5.0E-02	2496.53	9	11	2.15E-01	2719.03	9	7	1.42E+00
2166.77	9	7	2.7E+00	2501.13	9	7	6.75E-01	2719.06	7	7	7.40E-01
2171.30	5	7	5.1E-02	2501.69	11	9	3.69E-02	2719.42	11	11	3.20E-01
2173.21	3	5	8.3E-02	2505.01	9	11	2.56E-01	2720.90	7	5	1.04E+00
2176.84	1	3	1.0E-01	2506.57	7	9	2.04E-01	2723.58	5	3	5.69E-01
2191.84	5	5	1.2E+00	2507.90	7	9	1.93E-01	2724.95	7	9	4.76E-02
2196.04	3	3	1.2E+00	2510.83	7	5	1.29E+00	2726.06	3	1	5.52E-01
2200.39	1	3	8.9E-01	2517.66	5	7	1.58E-01	2728.02	9	9	3.45E-02
2200.72	3	5	2.8E-01	2518.10	5	3	1.93E+00	2728.82	9	9	2.98E-01
2259.51	9	11	5.66E-02	2519.63	3	5	1.34E-01	2731.28	5	7	6.84E-02
2272.07	7	9	2.92E-02	2522.85	9	9	2.13E+00	2733.58	11	9	7.10E-01
2276.03	9	7	1.25E-01	2524.29	3	1	3.23E+00	2735.48	9	7	5.03E-01
2284.09	7	5	1.29E-01	2527.27	13	13	3.46E-01	2737.31	3	3	7.25E-01
2287.25	5	3	2.23E-01	2527.44	7	7	1.93E+00	2737.64	13	11	1.14E-01
2292.52	7	9	2.96E-02	2529.14	5	5	9.91E-01	2742.25	7	5	3.41E-01
2294.41	3	1	3.61E-01	2529.31	5	7	4.86E+00	2742.41	5	5	4.70E-01
2297.79	7	7	1.44E-01	2529.84	3	3	3.83E-01	2743.57	7	7	4.84E-02
2298.17	9	9	3.09E-01	2533.14	11	11	2.07E-01	2744.07	1	3	3.09E-01
2299.22	5	5	7.03E-02	2535.61	1	3	9.59E-01	2744.53	5	3	2.53E-01
2300.14	5	7	4.99E-02	2537.17	13	15	3.70E+00	2750.14	7	7	2.74E-01
2301.68	1	3	8.68E-02	2537.46	9	11	3.19E-02	2753.69	3	1	4.00E-01
2303.58	3	5	4.83E-02	2540.97	3	5	9.59E-01	2754.03	5	5	7.29E-02
2309.00	3	5	1.02E-01	2542.10	11	13	4.47E+00	2755.18	11	9	5.13E-02
2313.10	5	7	1.18E-01	2543.92	9	11	4.70E+00	2756.33	3	5	1.41E-01
2320.36	7	9	1.41E-01	2545.98	5	7	7.16E-01	2757.32	3	3	2.85E-01
2373.62	7	7	6.53E-02	2549.61	7	9	2.31E-01	2761.78	5	5	1.94E-01
2389.97	5	7	4.47E-02	2576.69	11	11	1.13E-01	2762.03	7	7	1.76E-01
2438.18	11	9	7.09E-02	2584.54	11	13	3.15E-01	2767.52	9	9	1.48E-01
2439.74	13	13	3.46E+00	2599.57	9	9	1.47E-01	2769.30	13	13	1.80E-01
2442.57	11	11	3.12E+00								

λ	Weights		A	λ	Weights		A	λ	Weights		A
Å	g_i	g_k	10^8 s^{-1}	Å	g_i	g_k	10^8 s^{-1}	Å	g_i	g_k	10^8 s^{-1}
2772.07	11	11	2.34E-02	2973.13	5	7	1.35E-01	3099.89	3	3	1.93E-01
2772.11	5	7	4.12E-02	2973.24	7	9	1.83E-01	3099.97	9	9	8.23E-02
2778.22	11	11	9.08E-02	2976.13	5	7	9.70E-02	3100.30	5	5	1.87E-01
2780.70	9	9	9.01E-02	2980.53	7	7	1.66E-01	3100.67	7	7	1.35E-01
2784.34	11	11	2.30E-02	2981.45	7	5	6.53E-02	3100.84	13	11	2.73E-02
2787.93	9	11	2.27E-02	2981.85	7	9	1.86E-01	3101.00	9	9	5.53E-02
2788.10	11	13	5.92E-01	2982.23	9	7	3.47E-02	3112.08	11	11	5.24E-02
2789.80	11	9	2.36E-01	2983.57	9	7	2.79E-01	3119.49	11	9	8.28E-02
2797.78	9	9	4.52E-02	2987.29	9	7	5.25E-02	3120.44	9	7	7.26E-02
2803.61	9	9	1.04E-01	2990.39	9	11	3.5E-01	3125.68	13	11	8.46E-02
2804.52	9	9	1.05E-01	2994.43	7	5	4.39E-01	3132.52	9	7	3.39E-01
2804.86	9	7	2.40E-01	2996.39	3	5	1.70E-01	3142.45	7	7	3.93E-02
2806.98	9	11	1.15E-01	2999.51	11	11	1.70E-01	3142.89	5	5	5.65E-02
2812.04	9	9	5.00E-02	3000.45	9	11	5.41E-02	3143.99	9	9	6.10E-01
2813.29	9	11	3.42E-01	3000.95	5	3	6.42E-01	3145.06	9	9	4.65E-02
2823.28	7	7	1.51E-01	3003.03	7	5	7.50E-02	3147.79	7	7	7.59E-02
2825.56	7	9	1.32E-01	3004.11	11	11	2.79E-02	3153.20	7	9	7.91E-02
2832.44	7	9	2.38E-01	3005.30	13	15	2.94E-02	3154.50	5	7	4.64E-02
2834.75	9	11	5.41E-02	3007.15	9	7	7.34E-02	3156.27	7	7	6.36E-01
2838.12	5	5	1.28E-01	3008.14	3	1	1.07E+00	3157.04	9	11	1.26E-01
2843.63	9	7	6.96E-02	3009.09	13	11	7.77E-02	3157.89	5	7	1.61E-01
2843.98	5	7	3.17E-01	3009.57	9	9	1.43E-01	3160.66	9	9	1.93E-01
2845.59	7	5	7.86E-02	3011.48	7	9	3.79E-01	3161.95	11	13	4.65E-02
2851.80	3	5	3.37E-01	3015.92	11	9	6.3E-02	3165.86	7	9	5.35E-02
2853.77	7	9	5.91E-02	3016.18	5	3	8.85E-02	3166.44	9	7	1.14E-01
2863.43	9	9	4.13E-02	3018.98	7	7	1.03E-01	3171.35	9	7	1.85E-01
2868.45	5	3	1.45E-01	3019.29	9	11	2.33E-02	3175.44	11	11	1.44E-01
2877.30	9	9	4.61E-02	3020.49	5	5	1.94E-01	3178.01	11	9	1.28E-01
2883.75	11	11	2.91E-02	3020.64	9	9	7.59E-01	3180.22	7	9	4.42E-01
2887.81	11	13	7.98E-02	3021.07	7	7	4.55E-01	3181.52	7	5	1.84E-01
2892.48	9	9	8.78E-02	3024.03	3	5	4.87E-02	3182.06	9	9	3.23E-02
2894.50	5	5	4.83E-01	3025.64	13	13	5.86E-01	3182.97	5	7	1.42E-01
2895.03	7	7	4.24E-02	3025.84	1	3	3.48E-01	3188.57	11	11	5.00E-02
2899.41	5	3	4.68E-01	3026.46	5	5	1.10E-01	3188.82	3	5	2.53E-01
2901.91	11	11	1.78E-01	3030.15	11	11	5.04E-01	3190.65	9	11	5.75E-02
2907.52	9	11	1.61E-01	3031.64	3	3	1.38E-01	3190.82	9	9	5.55E-02
2908.86	7	9	8.98E-02	3037.39	3	5	2.91E-01	3192.80	3	5	5.01E-01
2918.02	13	13	1.18E+00	3040.43	9	11	2.45E-02	3193.30	5	7	3.07E-01
2919.84	9	11	7.44E-02	3041.64	7	9	4.24E-02	3194.42	5	3	1.08E-01
2920.69	5	5	6.38E-02	3041.74	7	9	5.20E-02	3196.12	11	9	1.40E-01
2923.29	11	11	1.39E+00	3042.02	3	5	4.70E-02	3196.93	9	11	5.97E-01
2923.85	11	11	2.97E-01	3042.66	5	7	5.20E-02	3199.53	9	9	2.23E-01
2925.36	7	9	1.69E-01	3047.60	5	7	2.84E-01	3202.56	9	7	6.18E-02
2929.01	7	5	5.10E-02	3053.07	3	5	1.53E-01	3205.40	3	3	9.77E-01
2929.12	9	9	1.53E+00	3055.26	7	5	9.48E-02	3210.23	9	11	1.15E-01
2936.90	9	9	1.40E-01	3057.45	11	9	3.13E-01	3210.83	5	3	9.24E-01
2947.36	5	5	9.30E-02	3059.09	7	9	1.63E-01	3211.61	9	9	3.07E-02
2947.88	7	7	1.83E-01	3060.54	9	7	6.75E-02	3211.99	11	9	4.64E-01
2948.43	9	9	3.32E-01	3066.48	9	7	9.11E-02	3214.01	7	7	8.38E-01
2953.49	7	7	3.64E-01	3067.00	11	13	1.71E-01	3214.06	7	5	1.18E+00
2953.94	5	5	1.89E-01	3067.12	5	7	3.89E-02	3215.94	5	5	6.19E-01
2954.65	5	7	1.06E-01	3067.24	9	7	3.12E-01	3217.38	11	9	1.50E-01
2957.36	3	3	1.77E-01	3068.17	5	3	1.11E-01	3219.58	7	9	4.64E-01
2957.48	5	3	1.31E-01	3073.98	11	9	3.83E-02	3219.80	9	7	3.61E-01
2959.99	11	13	5.02E-01	3075.72	7	5	3.14E-01	3221.92	3	3	1.22E-01
2960.66	11	9	8.48E-02	3078.43	1	3	1.52E-01	3222.07	11	11	8.65E-01
2965.25	1	3	1.16E-01	3079.99	9	11	8.35E-02	3225.79	11	13	1.18E+00
2966.90	9	11	2.72E-01	3083.74	5	3	3.08E-01	3227.80	9	7	4.96E-01
2968.48	3	3	8.26E-02	3091.58	3	1	5.53E-01	3228.25	5	3	3.72E-01
2970.10	3	5	1.08E-01	3098.19	11	11	7.52E-02	3229.99	9	11	1.06E-01

λ Å	g_i	g_k	A 10^8 s^{-1}	λ Å	g_i	g_k	A 10^8 s^{-1}	λ Å	g_i	g_k	A 10^8 s^{-1}
3230.21	5	5	2.06E-01	3406.80	3	3	2.08E-01	3534.53	11	11	2.20E-02
3230.96	7	5	3.7E-01	3407.46	7	9	6.09E-01	3536.56	5	7	9.95E-01
3233.05	13	15	4.19E-01	3410.17	3	5	5.07E-01	3537.73	5	3	1.33E-01
3233.97	9	9	2.08E-01	3411.35	9	9	6.0E-02	3537.89	11	11	8.0E-02
3239.43	9	9	2.95E-01	3413.13	5	7	3.23E-01	3540.12	7	9	9.48E-02
3244.19	9	11	3.06E-01	3415.53	3	5	4.64E-02	3541.08	9	11	8.65E-01
3246.96	5	3	1.09E-01	3417.84	3	3	4.01E-01	3542.08	7	9	9.51E-01
3248.20	7	7	1.92E-01	3418.51	3	1	9.88E-01	3543.67	3	5	1.6E-01
3250.76	11	11	2.85E-02	3422.66	3	5	1.38E-01	3545.64	9	9	2.05E-01
3252.91	9	11	2.20E-02	3424.28	7	7	1.61E-01	3547.19	9	9	7.13E-02
3253.60	7	9	1.62E-01	3425.01	9	7	2.57E-01	3552.11	3	5	4.8E-02
3254.36	11	13	4.24E-01	3426.63	5	3	1.94E-01	3552.83	5	5	1.74E-01
3257.23	9	9	4.76E-02	3426.67	11	11	1.07E-01	3553.74	11	9	1.09E+00
3257.59	7	5	8.94E-02	3427.12	7	9	5.04E-01	3554.50	3	5	9.87E-02
3259.99	7	9	2.99E-02	3428.19	5	5	1.71E-01	3554.92	11	13	1.40E+00
3264.51	5	3	1.01E-01	3431.81	5	7	5.53E-02	3556.88	9	11	4.1E-01
3265.62	7	5	3.06E-01	3440.61	9	7	1.71E-01	3558.52	5	7	1.77E-01
3271.00	5	3	6.4E-01	3440.99	7	5	1.24E-01	3559.50	3	3	2.2E-01
3271.48	7	7	8.47E-02	3443.88	5	3	7.92E-02	3560.70	7	9	7.4E-02
3280.26	9	11	4.21E-01	3445.15	5	7	2.34E-01	3565.38	7	9	4.29E-01
3282.89	3	5	3.42E-01	3447.28	5	5	1.07E-01	3567.03	5	7	8.34E-02
3284.59	5	5	5.64E-02	3450.33	3	3	2.34E-01	3568.82	7	9	6.72E-02
3286.75	7	7	5.99E-01	3451.91	3	5	1.13E-01	3568.97	11	9	4.64E-02
3290.99	3	5	7.58E-02	3458.30	3	1	2.92E-01	3570.10	9	11	6.76E-01
3292.02	7	9	5.77E-01	3459.91	5	3	2.17E-01	3572.00	11	11	2.89E-01
3292.59	3	3	3.0E-01	3465.86	3	3	1.19E-01	3572.59	9	9	3.31E-02
3298.13	3	5	9.01E-02	3468.84	9	11	2.61E-02	3573.39	5	7	1.05E-01
3305.97	5	7	4.05E-01	3469.01	9	9	8.58E-02	3573.83	13	13	2.41E-02
3306.34	9	9	5.74E-01	3475.45	5	5	9.75E-02	3573.89	9	7	5.73E-01
3306.35	3	5	4.84E-01	3475.65	7	5	8.61E-02	3575.11	3	3	1.60E-01
3307.23	13	13	1.97E-01	3476.34	7	7	2.70E-01	3575.25	11	9	7.43E-02
3310.34	11	11	3.78E-02	3476.85	7	9	3.21E-02	3575.37	5	5	3.06E-01
3310.49	7	9	6.17E-02	3485.34	5	3	1.30E-01	3576.76	11	9	8.8E-02
3314.74	5	7	7.25E-01	3489.67	11	13	7.47E-02	3578.38	1	3	7.82E-02
3319.25	9	9	3.73E-02	3490.57	7	7	6.14E-02	3581.19	11	13	1.02E+00
3322.47	9	11	8.21E-02	3495.29	9	7	9.46E-02	3581.65	11	9	3.21E-02
3323.74	5	5	2.8E-01	3497.10	7	7	9.02E-02	3581.81	3	5	8.68E-02
3328.87	11	11	2.21E-01	3500.56	7	5	5.28E-02	3582.20	13	11	2.35E-01
3335.77	3	5	7.48E-02	3505.06	5	3	1.77E-01	3584.66	11	11	3.29E-01
3336.26	9	9	4.91E-02	3506.50	5	5	7.35E-02	3584.79	7	5	1.56E-01
3337.67	11	9	6.06E-02	3508.47	9	11	6.46E-02	3584.96	11	9	6.74E-01
3340.56	5	5	4.95E-02	3513.82	11	11	3.40E-02	3585.19	11	9	3.19E-02
3341.91	11	11	3.02E-02	3516.41	7	9	3.6E-02	3585.32	7	7	1.17E-01
3342.29	3	3	9.42E-02	3516.56	7	5	6.82E-02	3585.71	9	9	3.75E-02
3347.93	5	5	4.91E-02	3521.26	9	9	6.14E-02	3586.11	13	11	7.02E-01
3354.06	1	3	1.34E-01	3522.27	11	11	5.03E-02	3586.74	13	13	3.62E-02
3355.23	9	9	2.59E-01	3522.90	5	7	3.51E-02	3586.98	5	5	1.66E-01
3369.55	9	9	2.15E-01	3523.31	5	3	1.06E-01	3587.24	7	9	7.73E-02
3370.78	11	11	2.89E-01	3524.07	7	5	9.9E-02	3588.53	9	7	7.21E-02
3380.11	7	7	1.66E-01	3524.24	5	7	5.04E-02	3588.61	11	11	1.19E-01
3383.69	5	3	8.33E-02	3526.17	7	7	4.14E-02	3588.92	5	3	2.15E-01
3383.98	7	7	6.52E-02	3526.24	7	9	1.70E-01	3589.45	9	7	1.05E-01
3392.30	5	5	9.93E-02	3526.38	7	7	4.13E-01	3594.63	9	9	3.14E-01
3392.65	7	7	1.88E-01	3526.47	5	5	1.29E-01	3595.30	5	5	8.21E-02
3394.58	5	3	8.70E-02	3526.67	5	5	5.26E-01	3597.02	5	3	1.8E-01
3399.33	5	5	2.76E-01	3527.79	9	9	2.17E-01	3599.63	11	9	2.33E-01
3402.26	13	13	2.19E-01	3529.82	3	3	7.75E-01	3602.46	7	7	1.02E-01
3403.29	5	7	3.98E-02	3530.39	13	13	4.65E-02	3602.53	7	5	2.12E-01
3404.35	5	7	1.09E-01	3533.01	1	3	8.52E-01	3603.20	11	11	2.59E-01
3406.44	3	5	2.7E-01	3533.20	3	5	8.25E-01	3603.82	3	3	1.70E-01

λ Å	Weights g_i	g_k	A 10^8 s^{-1}	λ Å	Weights g_i	g_k	A 10^8 s^{-1}	λ Å	Weights g_i	g_k	A 10^8 s^{-1}
3605.45	9	9	4.66E-01	3689.46	9	9	3.70E-01	3781.94	5	7	3.8E-02
3605.50	13	11	2.12E-01	3690.45	1	3	1.22E-01	3785.95	11	13	4.14E-02
3606.68	11	13	8.29E-01	3690.73	11	11	2.99E-01	3786.19	5	5	1.3E-01
3608.14	9	11	6.22E-02	3694.01	5	7	8.35E-01	3787.16	5	5	9.9E-02
3608.86	3	5	8.13E-01	3695.05	7	9	2.01E-01	3787.88	3	5	1.29E-01
3610.16	13	13	5.90E-01	3697.43	7	7	1.94E-01	3789.18	9	11	2.16E-02
3610.69	5	3	1.05E-01	3698.60	5	7	3.6E-02	3789.82	9	7	4.1E-02
3612.07	11	13	1.11E-01	3699.14	5	7	4.9E-02	3793.48	7	7	7.92E-02
3613.44	7	7	7.0E-02	3701.09	7	9	6.35E-01	3794.34	9	11	4.15E-02
3617.79	5	7	7.09E-01	3702.03	3	1	3.7E-01	3795.00	5	7	1.15E-01
3618.30	11	9	4.89E-02	3703.55	7	7	3.84E-02	3797.51	13	13	4.57E-01
3618.39	9	9	8.88E-02	3703.69	9	11	6.31E-02	3798.51	9	11	3.23E-02
3618.77	5	7	7.22E-01	3703.82	1	3	1.02E-01	3799.55	7	9	7.31E-02
3621.46	9	11	4.45E-01	3704.46	11	9	1.42E-01	3801.68	5	7	6.26E-02
3621.72	11	9	1.07E-01	3707.92	7	5	3.32E-01	3801.98	11	13	3.7E-02
3622.00	7	7	5.14E-01	3709.25	9	7	1.56E-01	3802.28	5	5	5.63E-02
3623.19	13	13	6.68E-02	3711.22	7	9	3.62E-02	3804.01	11	9	4.6E-02
3625.14	11	9	8.15E-02	3711.41	3	5	1.28E-01	3805.34	9	11	8.60E-01
3630.35	9	7	1.04E-01	3716.44	9	7	3.49E-01	3806.22	3	3	2.5E-01
3631.10	11	11	2.15E-01	3718.41	7	7	5.17E-02	3806.70	11	11	4.35E-01
3631.46	7	9	5.17E-01	3719.93	9	11	1.62E-01	3807.54	3	5	9.37E-02
3632.04	3	5	6.74E-01	3721.50	5	5	1.94E-01	3808.73	9	9	3.54E-02
3632.56	11	9	5.69E-02	3722.56	5	5	4.97E-02	3810.76	5	3	1.94E-01
3633.07	9	11	3.54E-02	3724.38	5	7	1.04E-01	3812.96	7	5	7.91E-02
3634.33	9	7	1.05E-01	3726.93	5	5	4.57E-01	3813.06	7	7	5.52E-02
3636.22	5	7	2.20E-01	3727.09	9	7	1.71E-01	3813.88	13	11	6.62E-02
3637.87	9	9	5.9E-02	3727.62	7	5	2.24E-01	3815.84	9	7	1.12E+00
3638.30	7	9	2.36E-01	3727.81	7	5	1.91E-01	3816.34	5	7	4.16E-02
3640.39	9	11	3.57E-01	3730.39	9	11	9.73E-02	3817.64	11	11	7.7E-02
3644.80	7	5	8.3E-02	3730.46	7	9	3.09E-02	3819.49	7	5	4.9E-02
3645.07	9	9	2.91E-02	3730.95	5	7	3.50E-02	3820.43	11	9	6.67E-01
3645.82	1	3	4.87E-01	3732.40	5	5	2.69E-01	3821.18	11	13	5.54E-01
3647.42	3	3	3.38E-01	3734.86	11	11	9.01E-01	3821.83	5	5	7.30E-02
3647.84	9	11	2.91E-01	3735.32	9	9	2.70E-01	3825.88	9	7	5.97E-01
3649.51	11	9	3.94E-01	3737.13	7	9	1.41E-01	3827.82	7	5	1.05E+00
3650.03	7	7	2.26E-01	3738.31	11	13	3.44E-01	3829.45	3	3	1.32E-01
3650.28	11	11	6.15E-02	3740.24	7	9	1.3E-01	3833.31	9	9	4.68E-02
3651.47	7	9	5.83E-01	3742.62	9	9	6.75E-02	3834.22	7	5	4.52E-01
3655.46	5	5	1.18E-01	3743.36	5	3	2.60E-01	3836.33	5	5	3.29E-01
3659.52	9	9	6.31E-02	3743.47	11	11	6.05E-01	3839.26	9	9	2.35E-01
3664.54	7	9	4.68E-02	3744.10	5	3	3.17E-01	3840.44	5	3	4.70E-01
3666.24	11	9	3.87E-02	3745.56	5	7	1.15E-01	3841.05	5	3	1.36E+00
3667.25	9	7	1.3E-01	3746.93	7	7	2.33E-01	3843.26	9	7	3.70E-01
3668.21	7	9	3.2E-02	3748.26	3	5	9.15E-02	3845.70	5	7	5.89E-02
3669.15	9	7	8.03E-02	3748.96	9	11	1.48E-01	3845.99	9	7	4.5E-02
3669.52	9	7	2.34E-01	3749.49	9	9	7.63E-01	3846.41	11	9	1.68E-01
3670.02	3	5	8.60E-02	3753.61	7	5	1.22E-01	3846.80	7	7	6.20E-01
3670.09	11	13	7.20E-02	3756.94	11	11	2.2E-01	3849.97	3	1	6.05E-01
3674.76	5	3	7.91E-02	3757.45	5	3	8.26E-02	3852.57	7	9	3.26E-02
3676.31	9	11	4.63E-02	3758.23	7	7	6.34E-01	3854.37	9	7	5.07E-02
3677.31	5	7	2.28E-01	3759.15	13	11	4.55E-02	3856.37	7	5	4.64E-02
3677.63	7	5	6.08E-01	3760.05	13	15	4.47E-02	3859.21	13	11	7.25E-02
3682.17	7	5	1.04E-01	3760.53	3	5	5.50E-02	3859.91	9	9	9.69E-02
3682.24	5	5	1.5E+00	3762.20	9	11	2.4E-02	3865.52	3	3	1.55E-01
3684.11	9	7	2.97E-01	3763.79	5	5	5.44E-01	3867.22	5	5	3.16E-01
3684.14	9	7	9.29E-02	3765.54	13	15	9.51E-01	3871.75	11	11	5.83E-02
3686.00	9	11	3.34E-01	3765.70	11	11	2.36E-02	3872.50	5	5	1.05E-01
3687.46	11	9	8.00E-02	3767.19	3	3	6.39E-01	3873.76	11	9	6.57E-02
3687.66	9	9	7.38E-02	3778.51	7	5	1.17E-01	3878.02	7	7	7.72E-02
3688.46	7	9	7.3E-02	3779.45	3	3	1.05E-01	3878.67	9	7	7.02E-02

λ Å	g_i	g_k	A 10^8 s^{-1}	λ Å	g_i	g_k	A 10^8 s^{-1}	λ Å	g_i	g_k	A 10^8 s^{-1}
3878.73	3	3	5.34E-01	3997.39	9	11	1.26E-01	4175.64	3	5	1.14E-01
3883.28	7	7	1.28E-01	3998.05	11	9	5.70E-02	4181.75	5	7	2.32E-01
3884.36	11	9	3.99E-02	4005.24	7	5	2.04E-01	4182.38	5	5	5.04E-02
3885.51	3	5	7.26E-02	4006.31	11	9	5.1E-02	4184.89	5	5	1.03E-01
3886.28	7	7	5.29E-02	4009.71	3	5	4.64E-02	4187.04	7	5	2.15E-01
3887.05	9	9	3.52E-02	4014.53	9	7	1.53E-01	4187.80	9	7	1.52E-01
3888.51	5	5	2.50E-01	4017.15	9	11	3.25E-02	4191.43	5	3	2.73E-01
3888.82	5	3	1.95E-01	4018.27	5	7	3.44E-02	4195.33	11	11	1.11E-01
3891.93	3	3	2.71E-01	4021.87	7	9	8.55E-02	4196.21	7	7	1.09E-01
3893.39	11	11	1.00E-01	4024.73	7	9	8.09E-02	4198.25	9	9	1.47E-01
3894.01	5	5	1.03E-01	4030.49	9	11	1.04E-01	4198.30	11	9	8.03E-02
3897.89	11	13	6.20E-02	4030.89	9	7	5.02E-02	4198.63	5	5	1.25E-01
3900.52	7	7	7.9E-02	4031.96	3	5	7.6E-02	4199.10	9	11	4.92E-01
3902.95	7	7	2.14E-01	4040.64	5	5	4.8E-02	4200.92	7	9	6.25E-02
3903.90	9	9	7.61E-02	4043.90	7	7	8.69E-02	4202.03	9	9	8.22E-02
3906.75	5	7	7.05E-02	4044.61	5	3	8.17E-02	4203.94	13	13	2.97E-02
3907.93	7	5	6.67E-02	4045.59	9	7	7.39E-02	4203.98	3	5	7.37E-02
3909.66	3	5	5.7E-02	4045.81	9	9	8.62E-01	4210.34	3	3	1.48E-01
3911.00	9	9	2.68E-02	4054.87	5	3	9.61E-02	4217.55	3	5	2.46E-01
3916.73	13	11	9.83E-02	4058.22	9	7	4.47E-02	4219.36	11	13	2.88E-01
3918.42	3	1	4.22E-01	4062.44	3	3	1.85E-01	4222.21	7	7	5.76E-02
3918.64	7	7	1.17E-01	4063.59	7	7	6.65E-01	4224.17	9	11	1.06E-01
3919.07	9	9	3.72E-02	4067.98	9	9	1.51E-01	4224.51	3	5	6.81E-02
3925.64	5	5	8.04E-02	4070.77	7	5	1.1E-01	4225.45	5	7	1.65E-01
3925.94	1	3	1.67E-01	4071.74	5	5	7.64E-01	4227.43	11	13	5.29E-01
3926.01	7	7	7.26E-02	4073.76	5	3	1.68E-01	4233.60	3	5	1.85E-01
3928.08	9	9	5.64E-02	4074.79	9	9	3.43E-02	4235.94	9	9	1.88E-01
3931.12	5	7	4.8E-02	4076.63	9	9	1.32E-01	4238.81	7	9	2.41E-01
3932.63	9	11	2.70E-02	4076.80	5	7	3.81E-02	4245.26	1	3	9.0E-02
3933.60	3	5	5.92E-02	4079.17	5	5	5.4E-02	4246.09	7	5	5.85E-02
3935.81	5	5	1.14E-01	4080.21	3	1	2.3E-01	4247.43	9	11	1.94E-01
3941.28	5	5	9.1E-02	4084.49	11	9	8.66E-02	4250.12	5	7	2.07E-01
3942.44	3	5	9.62E-02	4085.00	3	5	4.7E-02	4250.79	7	7	1.02E-01
3946.99	9	11	3.91E-02	4085.30	7	7	8.92E-02	4260.47	11	11	3.99E-01
3947.53	5	5	5.12E-02	4085.98	7	5	5.1E-02	4267.83	1	3	8.17E-02
3948.10	7	9	1.31E-01	4098.18	7	7	7.49E-02	4271.15	7	9	1.82E-01
3948.77	11	9	2.08E-01	4107.49	5	3	1.74E-01	4271.76	9	11	2.28E-01
3949.95	7	5	4.79E-02	4109.80	3	3	1.51E-01	4276.68	9	9	2.6E-02
3951.16	3	5	4.29E-01	4112.96	11	13	1.1E-01	4282.40	7	5	1.21E-01
3952.60	11	11	2.97E-02	4118.55	11	13	4.96E-01	4294.12	9	9	3.12E-02
3953.15	7	9	2.97E-02	4125.62	9	11	9.9E-02	4299.23	9	11	1.29E-01
3955.34	3	3	1.5E-01	4126.18	11	11	4.2E-02	4307.90	7	9	3.38E-01
3956.46	13	11	1.76E-01	4127.61	1	3	1.43E-01	4309.03	13	13	1.96E-02
3956.68	11	13	1.22E-01	4132.06	5	7	1.18E-01	4315.08	5	5	7.76E-02
3957.02	5	7	1.67E-01	4132.90	3	5	7.70E-02	4325.76	5	7	5.16E-01
3960.28	5	7	4.10E-02	4134.68	5	7	1.25E-01	4327.10	5	5	1.12E-01
3963.10	3	5	1.5E-01	4137.00	3	5	2.75E-01	4369.77	9	9	6.09E-02
3967.42	9	7	1.52E-01	4142.59	3	5	7.5E-02	4383.55	9	11	5.00E-01
3967.96	7	9	6.09E-02	4143.41	9	9	2.70E-01	4388.41	7	7	1.03E-01
3969.26	9	7	2.26E-01	4143.87	7	9	1.33E-01	4401.29	7	7	6.4E-02
3970.39	3	1	3.9E-01	4149.37	11	13	4.23E-02	4404.75	7	9	2.75E-01
3971.32	11	9	4.97E-02	4153.90	7	9	2.05E-01	4415.12	5	7	1.19E-01
3973.65	5	7	5.81E-02	4154.50	5	3	2.64E-01	4422.57	3	3	8.72E-02
3976.61	3	5	1.20E-01	4154.81	9	11	1.40E-01	4433.22	5	3	2.1E-01
3977.74	5	5	6.41E-02	4156.80	5	5	1.20E-01	4443.19	1	3	1.02E-01
3981.77	9	9	3.57E-02	4157.78	5	7	2.18E-01	4455.03	9	7	4.1E-02
3983.96	9	7	5.72E-02	4158.79	3	5	1.6E-01	4466.55	5	7	1.20E-01
3985.39	5	5	8.53E-02	4170.90	5	5	6.29E-02	4469.38	5	7	1.59E-01
3989.86	5	7	5.3E-02	4172.12	7	5	9.80E-02	4476.02	3	5	1.01E-01
3996.96	9	9	7.95E-02	4172.64	11	11	2.24E-02	4484.22	7	9	5.04E-02

λ Å	g_i	g_k	A 10^8 s^{-1}	λ Å	g_i	g_k	A 10^8 s^{-1}	λ Å	g_i	g_k	A 10^8 s^{-1}
4485.68	3	3	1.1E-01	5364.87	5	7	5.59E-01	22473.28	11	11	3.32E-02
4494.56	5	7	3.45E-02	5367.47	7	9	7.13E-01	23566.67	9	11	2.21E-02
4528.61	7	9	5.44E-02	5369.96	9	11	7.22E-01	24547.95	11	9	3.72E-02
4547.85	5	7	4.48E-02	5373.71	7	9	3.7E-02	24729.10	13	11	5.08E-02
4556.13	7	5	1.05E-01	5383.37	11	13	7.81E-01	*Fe II*			
4619.29	7	5	5.2E-02	5393.17	7	9	4.91E-02	1055.26	10	8	4.6E-01
4638.01	7	7	3.37E-02	5398.28	5	5	9.0E-02	1063.97	10	8	3.5E-01
4654.61	7	7	3.68E-02	5404.15	9	11	6.92E-01	1068.35	8	8	1.59E+00
4667.45	7	9	6.03E-02	5410.91	7	9	6.33E-01	1071.58	6	8	1.14E+00
4673.16	5	7	3.81E-02	5415.20	11	13	7.67E-01	1096.88	10	8	2.26E+00
4678.85	7	9	4.97E-02	5463.28	9	9	2.9E-01	1112.05	10	12	2.0E-01
4736.77	9	11	4.78E-02	5473.90	7	7	5.2E-02	1121.97	10	8	1.92E+00
4789.65	5	7	4.57E-02	5476.29	7	9	2.87E-02	1122.84	8	6	1.81E+00
4800.65	7	9	3.01E-02	5476.56	9	9	8.70E-02	1125.45	10	8	1.03E+00
4859.74	5	3	1.62E-01	5563.60	5	7	3.4E-02	1127.10	10	10	5.9E-02
4871.32	7	5	2.44E-01	5569.62	5	3	2.34E-01	1128.05	2	4	1.40E+00
4872.14	3	3	2.54E-01	5572.84	7	5	2.28E-01	1130.44	6	8	3.1E-01
4878.21	1	3	1.21E-01	5576.09	3	1	2.5E-01	1133.40	8	10	2.6E-01
4890.76	5	5	2.25E-01	5586.76	9	7	2.19E-01	1133.67	10	8	3.1E-01
4891.49	9	7	3.08E-01	5594.66	9	9	5.20E-02	1138.63	8	8	5.5E-01
4903.31	3	5	6.58E-02	5602.95	3	3	1.00E-01	1142.37	10	8	2.6E-01
4918.99	7	7	1.79E-01	5615.64	11	9	2.64E-01	1143.23	10	10	9.8E-01
4920.50	11	9	3.58E-01	5624.54	5	5	7.41E-02	1144.94	10	12	3.52E+00
4957.30	9	9	1.18E-01	5633.95	11	13	7.7E-02	1147.41	8	8	1.24E+00
4957.60	13	11	4.22E-01	5638.26	9	7	4.4E-02	1148.28	8	10	3.35E+00
4966.09	11	11	3.31E-02	5649.99	3	5	5.1E-02	1151.15	6	8	2.23E+00
4973.10	3	3	1.1E-01	5655.18	7	9	4.7E-02	1267.42	8	6	9.3E-01
4978.60	5	3	1.19E-01	5658.82	7	7	4.34E-02	1272.61	6	4	3.3E-01
4985.25	5	5	1.48E-01	5662.52	11	9	6.18E-02	1272.65	6	4	2.2E-01
4988.95	7	7	5.2E-02	5679.02	5	7	3.7E-02	1371.02	14	12	1.74E+00
5001.86	9	7	3.7E-01	5686.53	9	11	6.71E-02	1563.79	8	8	1.33E+00
5006.12	11	11	5.87E-02	5705.99	7	9	6.1E-02	1580.63	8	10	5.8E-01
5014.94	7	5	2.64E-01	5753.12	3	5	8.26E-02	1588.69	10	8	4.9E-03
5021.59	7	9	6.18E-02	5762.99	5	7	9.6E-02	1608.45	10	8	1.91E+00
5022.24	5	3	2.4E-01	5816.37	9	11	4.49E-02	1608.54	10	8	2.1E-02
5048.44	3	5	4.88E-02	5905.67	5	3	1.1E-01	1610.92	10	10	1.94E-01
5068.77	9	7	3.37E-02	6301.50	5	5	6.43E-02	1611.20	10	8	4.40E-02
5074.75	9	11	1.4E-01	6400.00	7	9	9.27E-02	1618.47	8	8	5.53E-01
5090.77	7	5	1.9E-01	6411.65	5	7	4.43E-02	1621.25	8	8	1.3E-02
5121.64	5	5	7.9E-02	6419.95	7	7	1.2E-01	1621.69	8	6	1.32E+00
5137.38	11	9	1.0E-01	6469.19	3	3	8.3E-02	1623.09	8	8	1.99E-01
5139.25	7	5	9.16E-02	6496.47	5	5	7.8E-02	1625.52	8	10	4.04E-01
5139.46	9	9	8.69E-02	6569.22	7	9	6.0E-02	1625.91	6	8	1.02E-01
5184.27	5	7	3.8E-02	6633.75	7	7	3.44E-02	1629.16	6	6	8.66E-01
5191.46	5	3	2.32E-01	6841.34	5	7	3.4E-02	1631.13	6	4	6.93E-01
5192.34	7	7	1.34E-01	6855.16	7	9	2.86E-02	1633.91	6	8	3.85E-01
5208.59	7	5	6.23E-02	7187.32	9	11	8.36E-02	1634.35	4	6	3.21E-01
5215.18	5	3	1.10E-01	7511.02	11	11	1.35E-01	1635.40	8	6	2.28E+00
5226.86	5	5	1.36E-01	8220.38	13	11	1.69E-01	1636.33	4	4	9.63E-01
5232.94	9	11	1.94E-01	8699.45	7	9	4.08E-02	1637.40	10	8	3.57E-01
5235.39	9	7	3.75E-02	9012.07	11	9	4.46E-02	1639.40	2	4	6.85E-01
5242.49	13	11	2.38E-02	9401.11	9	11	2.64E-02	1641.76	6	4	1.76E+00
5263.31	5	5	6.36E-02	9414.04	7	9	3.98E-02	1647.16	6	6	4.98E-01
5266.56	7	9	1.10E-01	9443.80	5	7	6.39E-02	1661.32	10	8	1.2E-02
5273.16	1	3	8.12E-02	9569.91	11	11	2.50E-02	1663.70	6	8	9.9E-03
5281.79	5	7	5.00E-02	9626.50	9	9	4.51E-02	1670.75	10	8	1.06E+00
5283.62	7	7	1.02E-01	9738.57	11	13	7.64E-02	1676.86	8	8	6.75E-02
5302.30	3	5	9.04E-02	9763.38	3	5	5.42E-02	1688.40	6	8	2.53E-02
5324.18	9	9	2.06E-01	9861.74	7	9	5.49E-02	1702.04	10	12	1.02E+00
5339.93	5	7	6.36E-02	9889.04	9	11	2.22E-02				

λ Å	g_i	g_k	A 10^8 s^{-1}	λ Å	g_i	g_k	A 10^8 s^{-1}	λ Å	g_i	g_k	A 10^8 s^{-1}
1761.37	8	8	1.42E+00	2352.31	2	4	4.38E+00	2412.01	6	8	1.66E-01
1785.27	6	8	1.2E+01	2353.47	12	14	4.98E+00	2413.31	2	4	1.02E+00
1786.75	6	6	1.2E+01	2353.68	8	8	1.30E+00	2414.10	14	12	1.05E-02
1788.08	6	4	4.6E+00	2354.48	10	8	8.13E-01	2414.51	10	8	4.2E-03
1796.98	6	8	3.0E-03	2354.89	6	4	2.67E-01	2416.45	8	10	2.38E+00
1818.52	8	8	5.70E-02	2356.21	6	8	7.1E-03	2417.87	12	12	9.5E-01
1833.08	6	8	2.2E-02	2359.11	4	6	5.0E-01	2418.17	6	8	2.0E-02
1863.11	6	8	2.4E-03	2359.60	10	10	2.25E-01	2418.44	6	8	2.28E+00
2020.75	6	8	1.83E-01	2360.00	10	10	3.59E-01	2419.89	10	10	2.2E-02
2057.33	6	8	2.80E-02	2360.29	8	6	6.23E-01	2422.69	6	8	1.46E+00
2074.19	10	8	2.30E-02	2360.53	6	8	2.22E-01	2422.93	10	8	2.94E-02
2078.16	10	10	2.84E-02	2361.73	8	8	2.40E-01	2423.21	4	6	1.40E+00
2097.02	8	8	1.07E-02	2362.02	8	8	1.41E-01	2424.15	10	12	2.21E+00
2122.45	10	8	4.8E-03	2363.86	8	10	5.3E+00	2424.39	6	8	1.61E-01
2146.37	8	8	7.1E-03	2364.83	8	8	5.90E-01	2424.50	8	8	2.9E-03
2162.02	10	10	2.54E-01	2365.76	6	6	2.16E+00	2424.59	6	6	1.24E+00
2182.36	10	8	8.6E-02	2366.59	6	6	1.01E-01	2424.65	8	8	6.55E-02
2187.68	8	8	2.87E-02	2366.88	8	10	3.51E-02	2428.08	12	10	7.0E-03
2189.03	10	10	1.97E-02	2368.60	6	4	6.06E-01	2428.36	8	10	2.68E+00
2191.98	8	8	7.54E-01	2369.95	10	12	5.9E+00	2428.80	4	4	1.38E+00
2201.59	6	8	7.77E-01	2370.50	4	4	1.73E-01	2429.04	2	4	1.23E+00
2208.41	10	10	1.59E+00	2372.36	10	8	6.6E-03	2429.39	4	4	6.9E-01
2209.03	10	8	1.27E+00	2373.74	10	10	4.25E-01	2429.86	8	8	1.51E+00
2213.66	14	14	3.26E-01	2375.19	4	2	9.81E-01	2430.08	8	10	1.91E+00
2218.26	8	10	1.57E+00	2376.43	12	14	6.4E+00	2432.26	6	8	1.57E+00
2220.38	12	12	4.19E-01	2378.55	8	8	1.70E-01	2432.87	14	14	2.86E+00
2228.73	6	8	1.59E+00	2378.70	8	8	1.49E-01	2433.50	10	12	1.30E-01
2249.18	10	8	3.00E-02	2379.28	8	8	2.73E-01	2434.06	8	6	7.2E-01
2250.18	4	4	1.67E-02	2379.42	10	10	3.68E-01	2434.24	8	10	2.01E+00
2250.94	6	6	3.19E-02	2380.76	6	8	3.10E-01	2434.73	12	12	2.79E+00
2251.56	8	6	9.8E-03	2382.04	10	12	3.13E+00	2434.95	4	6	1.39E+00
2253.13	8	8	4.41E-02	2382.36	4	6	3.19E-02	2435.00	8	8	2.02E+00
2254.41	4	2	5.5E-03	2382.90	12	14	1.62E-01	2436.62	6	8	2.70E+00
2255.77	6	4	4.75E-01	2383.06	8	6	1.0E-01	2439.30	12	14	2.25E+00
2260.08	10	10	3.18E-02	2383.24	6	6	3.59E-01	2440.42	6	8	1.18E+00
2260.24	2	2	3.4E-02	2384.39	4	4	3.22E-01	2441.13	10	10	8.95E-01
2260.86	4	6	2.16E-02	2385.01	6	8	3.60E-02	2442.38	10	12	2.75E+00
2262.69	4	4	1.98E-02	2388.39	10	12	2.02E-01	2443.71	8	10	1.44E+00
2266.00	6	6	1.0E-02	2388.63	8	8	1.05E+00	2444.52	6	8	2.78E+00
2267.59	6	8	3.69E-02	2390.10	14	16	5.5E+00	2445.11	12	12	2.03E+00
2268.56	2	4	6.0E-03	2390.76	6	6	1.17E+00	2445.57	4	6	2.07E+00
2268.82	8	8	3.97E-03	2391.48	8	10	3.77E-02	2445.80	4	6	1.23E+00
2279.92	8	10	4.49E-02	2394.00	12	10	9.4E-02	2446.11	8	8	1.06E+00
2292.42	12	10	8.42E-03	2395.42	6	4	2.67E-01	2446.47	12	14	2.99E-01
2296.88	6	8	1.82E-02	2395.63	8	10	2.59E+00	2447.21	6	6	1.15E+00
2312.22	10	8	9.3E-03	2396.72	10	12	2.15E-01	2447.33	4	2	2.56E+00
2327.40	6	4	6.55E-01	2399.24	6	6	1.39E+00	2447.76	12	10	1.97E+00
2327.88	10	12	1.08E+00	2400.05	12	14	4.57E+00	2449.96	4	4	1.24E+00
2331.31	10	8	3.17E-01	2401.29	6	8	1.89E+00	2450.21	2	4	1.26E+00
2332.80	8	6	1.31E+00	2402.45	10	10	5.8E-01	2453.98	8	10	1.31E+00
2338.01	4	4	1.13E+00	2402.60	6	8	2.17E-02	2454.58	14	12	1.16E+00
2338.54	10	12	5.6E-02	2402.63	8	8	8.19E-01	2455.71	8	8	1.01E+00
2343.50	10	8	1.73E+00	2404.43	4	2	6.44E-01	2455.90	4	6	1.73E+00
2343.96	8	6	3.13E-01	2404.89	6	8	1.96E+00	2457.10	6	4	4.71E-01
2344.28	2	4	9.27E-01	2406.09	6	8	2.05E-02	2458.78	10	12	2.31E+00
2345.34	14	12	7.3E-01	2406.66	4	4	1.61E+00	2458.97	6	4	2.51E+00
2348.12	10	8	6.50E-01	2410.27	8	8	7.65E-01	2460.44	10	12	5.39E+00
2348.30	6	6	1.15E+00	2410.52	4	6	1.55E+00	2461.28	6	8	2.34E+00
2351.20	12	10	7.19E-01	2411.07	2	2	2.37E-01	2461.86	8	10	2.43E+00
2351.67	6	6	1.80E+00	2411.81	10	12	4.33E+00	2463.28	12	10	7.1E-01

λ (Å)	g_i	g_k	A (10^8 s^{-1})	λ (Å)	g_i	g_k	A (10^8 s^{-1})	λ (Å)	g_i	g_k	A (10^8 s^{-1})
2464.01	10	8	1.32E+00	2525.92	8	8	7.4E-01	2559.93	6	8	2.47E-01
2464.91	6	4	2.22E+00	2526.08	10	8	3.52E-01	2560.28	4	4	1.77E+00
2465.91	8	6	1.62E+00	2526.29	6	6	2.47E+00	2561.59	10	10	1.1E-02
2466.50	2	4	2.40E+00	2527.10	12	10	3.67E-01	2562.09	4	2	1.62E+00
2466.67	4	2	2.64E+00	2527.71	10	8	9.1E-01	2562.54	8	6	1.79E+00
2466.82	6	4	1.77E+00	2528.68	10	8	2.3E-02	2563.48	6	4	1.51E+00
2468.30	10	10	9.8E-02	2529.08	4	6	1.80E+00	2566.22	8	10	2.61E-01
2469.37	10	8	2.23E-02	2529.23	12	10	3.27E-01	2566.40	8	6	2.29E+00
2469.52	8	6	2.58E+00	2529.55	10	10	2.20E+00	2566.62	10	12	7.1E-02
2470.41	6	6	6.0E-01	2530.10	4	6	6.6E-01	2566.91	4	2	1.15E+00
2470.67	8	6	1.54E+00	2533.63	12	12	1.92E+00	2568.41	2	4	4.77E-01
2470.85	8	8	5.4E-03	2534.42	8	8	1.83E+00	2568.89	8	8	2.8E-02
2471.28	10	8	4.15E-01	2535.36	6	4	2.46E+00	2569.78	2	4	1.11E+00
2472.61	8	10	3.22E+00	2535.49	10	8	7.47E-01	2570.55	6	8	1.1E-03
2473.32	2	2	2.74E+00	2536.67	12	12	5.7E-01	2570.85	8	6	1.84E+00
2475.12	4	6	3.72E+00	2536.81	10	10	1.69E+00	2571.55	10	10	2.89E-02
2475.54	6	8	3.18E+00	2536.84	12	14	6.8E-01	2572.97	6	8	7.89E-02
2476.27	8	10	9.7E-02	2537.14	10	10	1.44E+00	2573.21	8	10	1.42E-01
2477.35	8	8	1.70E-01	2538.21	14	12	1.26E+00	2573.76	8	8	2.3E-02
2478.57	6	6	9.1E-01	2538.40	6	8	3.7E-02	2574.37	6	4	2.43E+00
2480.16	10	8	1.55E+00	2538.50	8	6	5.9E-01	2576.86	10	12	1.32E+00
2481.05	12	12	1.46E-01	2538.68	6	8	7.4E-01	2577.43	6	8	7.8E-03
2482.12	14	14	6.5E-01	2538.91	10	8	1.28E+00	2577.92	2	2	1.24E+00
2482.33	4	4	2.23E+00	2538.99	14	12	1.93E+00	2580.72	8	6	2.2E-02
2482.66	12	10	1.25E+00	2539.81	8	8	5.6E-02	2581.11	6	6	7.61E-02
2482.87	6	4	1.69E+00	2540.52	2	2	1.26E+00	2582.41	6	8	2.22E-01
2483.72	8	10	5.4E-01	2540.66	6	8	1.70E+00	2582.58	4	4	8.80E-01
2484.24	4	6	8.3E-02	2541.10	8	6	9.6E-01	2583.05	8	10	2.16E-02
2484.44	8	8	2.16E+00	2541.84	8	6	8.2E-01	2583.35	8	10	9.0E-03
2489.11	4	4	2.6E-02	2542.32	4	2	3.9E-03	2585.62	10	10	3.09E-01
2489.48	12	12	5.1E-01	2542.74	2	2	1.61E+00	2585.88	10	8	8.94E-01
2489.83	12	12	1.94E+00	2543.38	10	12	6.7E-01	2586.06	6	4	5.8E-02
2490.71	10	12	1.44E+00	2543.43	6	4	8.3E-01	2587.95	8	10	1.69E+00
2490.86	8	8	8.8E-01	2544.97	4	6	3.93E-01	2588.19	2	2	1.5E-01
2491.40	10	8	1.01E+00	2545.22	8	10	5.3E-01	2588.80	8	8	8.4E-02
2492.34	10	12	2.30E-01	2545.44	8	10	1.52E-01	2590.55	4	6	7.9E-02
2493.26	14	16	3.04E+00	2545.53	8	10	1.2E-02	2591.54	6	6	5.72E-01
2493.88	6	6	1.74E+00	2546.67	8	8	7.98E-01	2592.79	14	16	2.74E+00
2494.12	12	10	2.97E-02	2547.34	8	8	2.28E-01	2593.73	2	4	1.63E-01
2497.68	10	12	8.4E-03	2548.16	6	8	8.0E-03	2594.96	8	8	1.0E-01
2497.82	6	6	1.68E+00	2548.32	4	6	2.69E-01	2595.28	12	10	1.67E-03
2500.92	6	8	2.41E+00	2548.59	10	10	2.67E-01	2595.30	10	8	1.2E-02
2501.35	2	2	1.48E+00	2548.74	4	2	2.43E+00	2598.37	8	6	1.43E+00
2502.39	8	8	1.43E+00	2548.92	12	10	6.0E-01	2599.40	10	10	2.35E+00
2503.33	12	12	7.3E-01	2549.08	10	8	1.89E+00	2604.05	8	8	1.49E-01
2503.54	8	8	3.32E-01	2549.40	4	4	1.65E+00	2604.67	8	10	1.2E-02
2503.57	10	10	2.53E-01	2549.46	6	6	1.12E+00	2605.04	6	8	2.34E+00
2503.88	10	10	2.23E+00	2549.77	8	6	2.35E-01	2605.31	4	4	1.99E+00
2506.09	10	10	9.9E-01	2550.03	10	10	1.74E+00	2605.43	6	6	3.40E-01
2506.80	8	10	1.98E+00	2550.15	8	10	3.91E-01	2605.90	4	2	1.27E+00
2508.34	8	10	3.79E-01	2550.57	12	12	1.6E-02	2606.52	6	6	2.31E+00
2510.57	8	8	1.54E-01	2550.68	12	12	1.07E+00	2607.09	6	4	1.73E+00
2511.76	8	10	2.30E+00	2551.20	10	8	2.48E-01	2608.85	10	8	5.0E-02
2513.15	10	8	2.49E-01	2554.94	8	8	2.6E-02	2609.13	8	10	2.77E-01
2514.38	8	8	2.11E+00	2555.07	6	8	1.96E-01	2609.44	6	8	6.0E-02
2517.14	2	4	9.2E-01	2555.45	4	6	2.49E-01	2609.87	8	8	1.34E-01
2519.05	8	6	2.10E+00	2557.08	8	10	2.8E-02	2611.07	4	6	7.28E-02
2521.09	6	4	2.05E+00	2557.51	10	8	1.53E-01	2611.34	8	6	1.4E-02
2521.82	8	8	2.36E+00	2559.24	8	8	6.4E-02	2611.87	8	8	1.20E+00
2525.39	14	14	1.91E+00	2559.77	6	8	2.42E-01	2613.57	10	12	2.0E-02

λ	Weights		A	λ	Weights		A	λ	Weights		A
Å	g_i	g_k	10^8 s^{-1}	Å	g_i	g_k	10^8 s^{-1}	Å	g_i	g_k	10^8 s^{-1}
2613.82	4	2	2.12E+00	2699.20	4	4	6.2E-01	2770.50	12	10	4.08E-02
2614.19	8	10	3.3E-02	2703.99	8	8	1.38E+00	2771.19	10	12	4.3E-02
2614.59	10	8	3.37E-02	2704.58	8	8	1.66E-02	2771.56	6	4	1.9E-02
2614.87	8	6	3.5E-02	2707.13	4	6	8.3E-01	2772.73	6	8	1.1E-03
2617.62	6	6	4.88E-01	2709.06	4	6	3.88E-01	2774.69	2	4	2.73E-01
2619.08	10	10	2.48E-01	2709.38	6	4	2.78E-03	2775.34	6	6	1.5E-04
2620.17	6	6	1.1E-01	2709.99	6	8	8.7E-03	2776.18	6	8	2.66E-02
2620.41	4	4	4.30E-02	2711.84	12	14	4.36E-01	2776.91	8	8	4.08E-01
2620.70	8	8	3.43E-01	2712.39	10	12	1.29E-01	2779.30	10	8	1.00E+00
2621.67	2	2	5.60E-01	2714.41	8	6	5.70E-01	2779.91	2	4	2.56E-01
2623.13	14	14	8.8E-02	2716.22	6	6	1.15E+00	2780.05	2	2	3.3E-01
2623.72	6	6	1.92E-01	2716.44	6	6	2.8E-02	2783.69	12	10	1.06E+00
2625.49	12	14	2.55E+00	2716.57	14	12	1.35E+00	2784.28	2	4	3.4E-02
2625.67	8	10	3.52E-01	2716.70	8	8	1.02E-03	2785.19	12	10	1.53E+00
2626.50	4	6	3.48E-01	2717.88	16	14	1.51E+00	2787.24	8	6	1.83E-01
2626.70	8	8	1.94E-02	2718.64	10	8	1.18E+00	2790.56	8	10	2.1E-02
2628.29	2	4	8.74E-01	2719.30	6	8	4.44E-01	2793.89	10	12	1.26E-01
2628.58	6	6	3.4E-02	2721.81	12	10	5.1E-02	2796.63	10	10	2.0E-01
2629.59	6	8	7.2E-01	2722.06	8	8	1.42E-01	2797.92	10	10	3.2E-02
2630.07	4	6	5.1E-01	2722.74	6	8	8.2E-01	2799.29	10	8	1.55E-01
2631.05	4	6	8.16E-01	2724.88	6	6	9.58E-02	2799.72	10	10	5.0E-03
2631.32	6	8	6.29E-01	2726.52	6	8	5.0E-02	2804.02	6	6	1.6E-02
2631.61	10	12	6.6E-01	2727.38	12	10	3.12E-01	2805.32	4	6	2.5E-02
2633.20	6	4	1.21E+00	2727.54	6	4	9.38E-01	2805.79	8	8	3.22E-02
2636.70	4	4	8.8E-02	2728.91	8	10	1.25E-01	2809.78	8	8	3.10E-01
2637.50	6	6	6.2E-01	2730.73	4	4	2.79E-01	2811.27	12	10	1.2E-02
2637.64	2	4	6.6E-01	2732.01	10	8	7.05E-02	2812.49	4	4	2.9E-02
2639.57	2	2	8.0E-01	2732.45	10	10	9.8E-04	2813.61	8	10	3.40E-02
2641.12	4	4	3.7E-02	2732.94	8	6	9.5E-01	2817.09	6	4	3.37E-01
2642.01	6	6	2.29E-01	2736.49	2	4	1.5E-02	2819.34	12	12	9.7E-03
2646.21	12	10	1.44E-02	2736.97	4	2	1.22E+00	2826.03	8	6	4.5E-02
2649.47	6	8	1.98E+00	2739.55	8	8	2.21E+00	2827.43	12	14	2.4E-02
2650.48	6	8	1.60E+00	2741.39	6	6	2.03E-01	2828.63	12	10	6.9E-02
2651.30	12	12	4.0E-03	2743.20	2	4	1.97E+00	2831.56	4	6	7.6E-01
2652.57	10	8	4.45E-02	2744.90	6	8	3.62E-02	2833.09	6	6	4.55E-01
2654.63	4	4	8.1E-01	2746.48	4	6	2.05E+00	2835.71	4	6	5.1E-01
2657.92	10	10	3.2E-02	2746.98	6	6	1.69E+00	2836.19	4	4	5.4E-02
2658.25	8	8	2.12E-01	2749.18	4	4	1.21E+00	2836.51	2	4	9.8E-02
2659.06	10	10	2.5E-03	2749.32	6	8	2.16E+00	2837.30	10	12	1.9E-02
2662.56	2	2	1.33E+00	2749.49	2	2	1.16E+00	2838.22	4	2	8.6E-01
2664.66	8	10	1.91E+00	2750.01	10	10	1.8E-02	2839.51	10	8	1.47E+00
2666.64	6	8	1.87E+00	2751.13	4	4	2.92E-01	2839.80	8	10	5.8E-01
2667.22	4	6	1.02E+00	2752.15	4	4	7.7E-01	2840.34	12	12	7.7E-02
2669.93	2	4	5.2E-01	2753.29	10	12	1.89E+00	2840.65	2	4	7.6E-01
2670.38	6	8	6.0E-02	2754.89	8	6	1.21E+00	2840.76	10	12	1.49E-01
2671.39	2	4	6.5E-01	2755.74	8	10	2.15E+00	2841.36	10	10	4.3E-03
2680.23	6	8	1.10E-01	2756.51	6	8	7.3E-02	2842.08	8	8	1.5E-02
2682.51	8	10	9.2E-01	2757.03	10	8	8.07E-02	2843.32	10	10	1.40E-02
2683.00	4	6	7.3E-01	2759.33	8	8	2.7E-04	2843.48	4	6	9.6E-02
2684.75	8	10	1.57E+00	2761.81	2	4	1.38E-01	2844.96	2	2	5.5E-01
2684.96	12	12	6.4E-03	2762.33	6	6	6.0E-01	2845.60	8	6	1.57E+00
2686.11	6	6	9.4E-03	2762.45	6	4	3.3E-02	2847.21	8	6	1.7E-04
2686.39	6	4	1.6E-02	2763.66	14	12	1.34E+00	2847.77	4	4	5.1E-01
2691.74	10	8	5.04E-02	2763.91	8	6	2.9E-02	2848.11	6	6	9.9E-01
2692.60	10	12	1.40E+00	2764.79	12	12	1.1E-02	2848.32	6	4	1.59E+00
2692.83	8	6	1.64E-02	2765.13	10	8	1.47E+00	2848.91	12	10	5.3E-02
2693.86	8	6	4.2E-02	2767.50	12	14	1.58E+00	2849.61	10	12	4.6E-02
2697.33	4	4	2.48E-01	2768.93	4	6	4.75E-02	2852.87	2	4	1.65E-02
2697.46	4	2	1.65E+00	2769.15	8	10	6.6E-02	2853.21	6	6	2.3E-02
2697.73	10	8	2.6E-02	2769.35	12	14	2.07E-01	2855.67	8	10	9.2E-02

λ Å	g_i	g_k	A 10^8 s^{-1}	λ Å	g_i	g_k	A 10^8 s^{-1}	λ Å	g_i	g_k	A 10^8 s^{-1}
2856.15	10	10	5.0E-02	2975.94	2	2	9.1E-03	3210.44	2	4	3.63E-02
2856.38	6	8	4.42E-01	2978.85	10	10	7.2E-03	3211.08	6	8	2.1E-03
2856.91	8	8	1.32E+00	2979.35	2	4	1.61E-02	3213.31	4	6	6.12E-02
2857.17	6	8	1.22E-01	2980.96	6	8	1.1E-02	3227.74	6	8	8.9E-02
2857.42	6	6	2.0E-02	2982.06	4	6	2.41E-01	3231.71	6	8	1.4E-02
2858.34	10	12	4.85E-01	2984.82	6	6	4.29E-01	3232.78	8	6	5.0E-02
2861.17	4	4	1.7E-03	2985.54	2	4	2.39E-01	3237.40	4	4	1.8E-02
2864.97	8	8	4.3E-02	2997.30	6	8	8.6E-02	3237.82	2	4	6.8E-02
2868.87	6	6	7.3E-03	2998.85	6	8	4.2E-03	3241.69	8	8	1.9E-03
2869.16	8	10	1.4E-02	3000.06	8	6	3.0E-02	3243.72	10	8	5.1E-02
2869.31	4	6	4.04E-01	3002.32	6	8	2.0E-02	3247.17	4	6	7.1E-02
2869.70	8	6	1.1E-02	3002.65	4	6	1.79E-01	3247.39	8	8	6.0E-03
2870.61	8	10	7.5E-03	3004.26	8	8	8.6E-03	3255.89	8	8	2.78E-03
2871.06	10	12	2.2E-02	3020.01	12	10	6.4E-04	3257.36	8	6	1.5E-03
2871.13	12	10	3.0E-02	3021.42	8	6	3.8E-03	3258.77	6	8	9.39E-02
2872.38	10	8	1.70E-01	3036.96	6	6	2.22E-01	3259.05	8	10	6.7E-02
2873.40	8	10	4.56E-01	3044.84	8	10	1.2E-02	3266.94	10	10	4.5E-03
2875.35	8	10	1.35E-01	3048.99	4	4	3.84E-01	3267.04	8	10	2.0E-04
2876.80	8	8	9.56E-02	3056.80	14	12	1.7E-02	3268.51	8	6	6.8E-03
2879.25	10	8	3.6E-02	3062.24	12	10	1.36E-01	3269.77	10	8	5.2E-03
2880.76	8	8	2.21E-02	3065.32	6	6	2.9E-02	3273.49	8	8	8.5E-03
2883.71	12	14	1.48E-01	3070.69	10	8	1.28E-02	3276.60	6	8	1.0E-02
2884.76	6	8	2.46E-01	3071.12	2	4	2.59E-01	3277.35	8	10	3.31E-03
2885.93	14	12	3.8E-02	3076.44	4	6	3.75E-01	3279.64	10	10	5.8E-03
2886.24	12	10	6.9E-03	3077.17	14	12	1.35E-01	3281.29	6	6	2.31E-03
2887.31	6	4	1.9E-02	3078.68	6	8	5.5E-01	3285.41	4	2	4.5E-04
2888.10	4	6	6.1E-02	3089.38	6	8	2.2E-02	3289.35	8	8	2.1E-02
2892.83	4	6	1.8E-03	3096.29	8	8	1.9E-02	3295.23	6	8	3.6E-03
2894.78	10	12	5.7E-02	3101.89	6	8	9.1E-03	3295.82	4	4	2.04E-03
2895.22	8	10	1.09E-01	3105.17	4	2	7.5E-02	3297.88	6	4	1.0E-02
2897.27	6	4	1.8E-01	3105.55	2	2	7.0E-02	3302.86	6	8	2.78E-04
2902.32	6	8	8.81E-03	3106.57	8	8	1.88E-02	3303.46	2	2	6.5E-04
2902.46	10	10	3.2E-02	3114.30	4	4	6.4E-02	3304.43	6	8	2.0E-03
2906.12	2	4	4.4E-02	3114.69	2	4	2.5E-02	3313.99	2	4	1.4E-04
2907.86	8	6	1.3E-03	3116.58	6	4	5.5E-02	3323.06	8	10	1.4E-02
2910.76	8	8	1.5E-02	3129.01	8	10	2.3E-03	3325.01	8	8	3.35E-03
2916.15	8	8	4.8E-04	3131.72	12	10	6.6E-03	3360.11	12	12	2.1E-03
2917.08	6	8	2.7E-02	3133.05	4	6	1.5E-02	3366.97	8	6	2.2E-02
2917.47	6	8	1.4E-03	3135.36	6	6	8.8E-02	3381.01	6	4	3.0E-02
2922.02	8	10	3.8E-02	3144.75	8	6	2.7E-02	3388.14	8	10	3.8E-04
2926.59	8	10	5.1E-02	3146.75	10	10	4.9E-06	3395.33	8	8	3.66E-03
2934.49	8	10	5.6E-03	3154.20	10	10	2.06E-01	3398.36	14	14	2.5E-03
2939.51	6	4	4.0E-03	3155.95	10	8	4.17E-03	3416.02	4	2	2.6E-03
2944.40	4	2	3.5E-01	3162.80	8	8	5.5E-02	3425.57	6	8	2.1E-04
2945.26	6	6	5.6E-04	3163.09	6	6	1.92E-03	3436.11	8	6	5.7E-03
2947.65	6	4	2.01E-01	3166.67	6	4	1.4E-03	3442.22	8	6	3.2E-03
2949.18	10	8	2.45E-01	3167.86	8	8	1.59E-01	3453.62	8	10	8.5E-03
2953.77	6	8	5.2E-02	3170.34	4	2	8.2E-03	3456.92	8	6	7.1E-03
2954.05	8	8	1.2E-02	3177.53	8	8	1.74E-01	3463.96	6	6	7.6E-05
2959.60	8	6	9.7E-02	3179.50	6	8	1.11E-01	3464.50	10	8	2.2E-03
2959.84	8	6	1.36E-01	3180.15	4	6	7.7E-02	3468.68	8	8	2.0E-02
2961.28	4	2	8.9E-03	3183.11	4	6	9.80E-03	3475.74	6	8	1.7E-04
2964.13	8	6	4.6E-02	3185.32	2	4	3.00E-03	3487.99	4	6	1.7E-04
2964.62	2	2	6.5E-02	3186.74	4	4	3.85E-02	3493.47	10	10	3.2E-02
2965.03	4	4	9.43E-02	3187.30	10	10	5.0E-02	3494.67	4	6	7.1E-04
2965.41	6	4	1.1E-02	3192.07	8	10	5.2E-03	3495.62	10	8	2.62E-03
2968.74	8	10	2.4E-03	3192.91	6	6	1.27E-02	3499.88	8	8	4.29E-03
2969.94	8	6	2.28E-01	3193.80	2	2	5.4E-02	3503.47	2	2	2.6E-04
2970.52	4	6	2.70E-02	3193.86	8	8	3.86E-02	3507.40	2	4	4.1E-04
2970.69	10	8	4.15E-02	3196.07	6	8	1.61E-02	3508.20	2	4	1.2E-04

λ Å	Weights g_i	g_k	A 10^8 s^{-1}	λ Å	Weights g_i	g_k	A 10^8 s^{-1}	λ Å	Weights g_i	g_k	A 10^8 s^{-1}
3614.88	8	8	4.2E-03	4833.20	12	10	4.6E-06	5965.62	10	10	2.19E-01
3621.27	2	4	2.2E-02	4840.00	10	8	4.0E-06	5991.38	12	10	4.2E-05
3624.89	2	2	2.4E-02	4893.82	8	6	2.5E-05	6084.11	10	8	3.0E-05
3632.29	10	10	1.2E-03	4923.93	6	4	4.28E-02	6113.32	8	6	1.7E-05
3711.98	4	6	1.5E-03	4990.51	6	8	5.2E-01	6129.70	10	10	3.2E-06
3748.48	6	4	3.4E-02	4993.36	10	8	6.9E-05	6147.74	4	2	1.3E-03
3759.46	4	2	3.2E-02	5000.74	2	4	1.8E-05	6149.26	2	2	1.3E-03
3814.12	4	6	4.9E-03	5001.96	12	14	1.57E+00	6175.15	8	8	1.8E-03
3824.93	6	6	3.2E-05	5018.44	6	6	2.0E-02	6179.38	8	6	4.6E-04
3827.08	6	8	2.5E-03	5030.63	10	10	7.1E-01	6238.39	4	4	7.5E-04
3906.04	6	8	1.1E-02	5035.71	10	12	9.4E-01	6239.95	2	4	1.1E-04
3914.50	6	4	4.6E-05	5100.66	10	8	2.0E-05	6247.56	6	4	1.6E-03
3935.96	8	10	8.3E-03	5132.67	10	10	2.0E-05	6305.30	10	10	1.4E-03
3938.29	6	6	6.1E-05	5136.80	6	4	2.8E-05	6331.95	6	8	1.8E-03
3938.97	4	6	8.4E-03	5144.35	4	6	8.5E-01	6369.46	6	4	1.40E-04
3945.21	4	4	3.9E-05	5149.47	8	10	9.0E-01	6383.72	6	6	1.1E-03
3974.17	4	6	6.3E-05	5169.03	6	8	4.22E-02	6416.92	6	6	3.6E-04
4024.55	6	6	2.5E-03	5197.58	6	4	5.4E-03	6432.68	6	6	8.5E-05
4075.95	6	4	1.6E-05	5227.48	12	14	1.22E+00	6446.41	8	10	1.3E-03
4087.28	6	4	3.0E-05	5234.62	8	6	2.5E-03	6456.38	8	6	1.7E-03
4122.67	6	6	3.3E-04	5247.95	4	6	1.43E+00	6516.08	6	8	8.3E-05
4124.79	6	8	3.4E-05	5251.23	6	8	8.0E-01	7222.39	4	2	2.5E-04
4128.75	6	4	2.6E-04	5262.48	4	6	8.0E-07	7224.49	2	2	2.8E-04
4173.46	6	6	4.43E-03	5264.18	8	10	4.76E-01	7301.56	6	6	2.1E-05
4178.86	6	8	1.72E-03	5264.81	6	4	3.52E-04	7320.65	6	4	1.4E-04
4180.98	4	4	2.2E-04	5272.40	6	6	3.9E-03	7449.33	4	6	1.68E-04
4233.17	6	8	7.22E-03	5276.00	10	8	3.76E-03	7462.41	6	6	2.7E-04
4258.15	4	4	3.1E-04	5284.11	6	8	1.9E-04	7479.69	6	8	3.5E-05
4273.33	4	2	9.1E-04	5306.18	6	8	3.28E-01	7515.83	8	6	8.1E-05
4296.57	4	6	7.0E-04	5316.22	14	14	3.69E-01	7711.72	8	8	4.94E-04
4303.18	4	4	2.20E-03	5316.62	12	10	3.89E-03				
4351.77	4	6	4.86E-03	5316.78	8	6	6.5E-04	**Fe III**			
4369.41	2	4	2.3E-04	5325.55	8	8	8.0E-04	1843.4	9	7	4.8E+00
4384.32	12	10	7.2E-05	5387.06	12	14	5.2E-01	1844.3	7	5	4.9E+00
4385.39	2	2	4.5E-03	5395.86	6	8	5.5E-01	1846.9	5	3	5.5E+00
4413.60	10	10	2.2E-05	5402.06	10	12	5.6E-01	1854.38	3	1	5.7E+00
4416.83	2	4	2.1E-03	5414.07	8	8	9.4E-05	1865.20	7	7	6.1E+00
4472.93	6	4	2.5E-04	5425.26	10	10	9.2E-05	1893.98	11	9	5.5E+00
4489.18	8	6	5.9E-04	5427.83	12	10	5.9E-03	1896.80	13	11	5.0E+00
4491.41	4	4	1.89E-03	5429.99	8	10	6.0E-01	1904.3	5	5	5.7E+00
4508.29	4	2	7.3E-03	5465.93	6	8	6.2E-01	1907.58	15	13	5.3E+00
4515.34	6	6	2.37E-03	5482.31	10	12	4.78E-01	1915.08	13	15	6.0E+00
4520.22	10	8	9.8E-04	5493.83	8	10	4.01E-01	1922.79	11	13	5.5E+00
4522.63	6	4	8.4E-03	5506.19	12	14	1.14E+00	1930.39	9	11	5.1E+00
4534.17	4	6	2.3E-04	5510.78	10	12	2.28E-01	1931.51	9	11	5.3E+00
4541.52	4	4	8.6E-04	5525.12	10	8	3.17E-05	1937.35	7	9	5.1E+00
4549.19	4	6	9.2E-03	5529.05	6	6	2.01E-01	1943.48	5	7	5.0E+00
4549.47	8	6	1.00E-02	5534.85	12	10	3.0E-04	1950.33	13	15	5.5E+00
4555.89	8	8	2.26E-03	5544.76	12	12	2.49E-01	1951.01	11	11	5.3E+00
4576.34	6	6	6.4E-04	5607.14	6	8	4.63E-05	1952.65	9	9	4.9E+00
4582.84	6	8	3.44E-04	5627.50	8	6	2.93E-05	1953.32	7	7	5.1E+00
4583.84	10	8	7.22E-03	5725.96	6	6	5.1E-06	1987.50	13	13	4.9E+00
4620.52	8	8	2.53E-04	5783.63	8	10	4.62E-01				
4629.34	10	10	1.72E-03	5813.68	6	4	8.8E-04	**Fe VII**			
4635.32	6	8	1.0E-02	5823.15	8	10	2.0E-04	150.807	5	7	1.3E+03
4656.98	6	6	1.37E-04	5824.41	6	6	8.3E-07	150.852	7	9	1.3E+03
4666.76	8	10	1.3E-04	5885.01	4	6	6.4E-01	151.023	9	11	1.6E+03
4670.18	6	8	3.2E-05	5902.83	8	10	4.98E-01	151.046	7	7	2.2E+02
4720.15	4	6	7.5E-06	5955.70	6	8	4.19E-01	151.145	9	9	2.1E+02
4731.45	6	8	2.8E-04	5961.71	10	12	7.4E-01	151.432	5	7	2.2E+02
								151.512	5	5	5.3E+02

λ Å	Weights g_i	g_k	A 10^8 s^{-1}	λ Å	Weights g_i	g_k	A 10^8 s^{-1}	λ Å	Weights g_i	g_k	A 10^8 s^{-1}
151.675	7	7	3.9E+02	92.81	9	11	3.7E+03	217	6	8	4.0E+02
151.782	9	9	2.4E+02	92.87	11	13	3.9E+03	217	6	6	2.6E+02
154.307	3	1	8.9E+02	93.433	9	11	3.2E+03	219	2	4	4.8E+02
154.335	5	7	1.2E+03	179.762	5	7	1.67E+03	219	4	6	2.4E+02
154.363	3	3	4.2E+02					219.123	4	6	3.9E+02
154.565	5	3	3.5E+02	*Fe XII*				220	4	4	3.2E+02
154.650	5	5	8.8E+02	65.905	4	4	2.0E+03	221	4	6	5.9E+02
154.848	1	3	7.7E+02	66.526	6	8	1.7E+03	226	2	4	3.9E+02
154.921	3	5	9.7E+02	66.960	4	6	1.6E+03	234	2	2	2.8E+02
154.941	3	3	2.4E+02	67.164	4	2	1.1E+03	264.787	4	4	3.38E+02
154.949	5	7	1.0E+03	67.821	4	6	1.4E+03	265	4	4	1.5E+02
155.994	9	11	1.8E+03	68.382	2	4	1.7E+03	266	6	4	1.7E+02
158.481	9	9	2.3E+03	80.541	6	6	8.7E+02	268	6	6	2.1E+02
165.087	1	3	6.9E+02	81.943	6	4	1.4E+03	268	4	2	3.3E+02
165.919	7	5	2.8E+03	82.226	4	2	1.9E+03	270.524	4	2	2.1E+02
166.365	9	7	2.9E+03	84.48	4	6	4.5E+03	274.203	2	2	1.8E+02
173.441	9	9	3.6E+03	84.48	8	10	4.9E+03	280	4	6	2.8E+02
176.744	9	9	2.7E+03	84.52	10	12	5.2E+03	283	6	8	2.7E+02
176.928	7	7	2.4E+03	84.52	6	8	4.0E+03	288.45	6	4	1.6E+02
177.172	5	5	1.5E+03	84.85	6	8	2.3E+03				
235.221	5	3	1.7E+02	85.14	8	10	3.4E+03	*Fe XV*			
240.053	3	1	1.3E+02	85.477	10	12	4.6E+03	38.95	1	3	1.69E+03
243.379	9	7	2.1E+02	186.880	6	8	1.0E+03	52.911	1	3	2.94E+03
				192.394	4	2	9.0E+02	59.404	3	5	3.4E+03
Fe VIII				193.509	4	4	9.1E+02	63.959	5	7	1.6E+03
112.472	4	4	3.6E+02	195.119	4	6	8.6E+02	65.370	1	3	3.2E+03
112.486	6	6	4.3E+02					65.612	3	3	9.8E+02
116.196	4	6	4.5E+02	*Fe XIII*				66.238	5	3	1.6E+03
117.197	6	8	3.8E+02	62.353	1	3	2.0E+03	68.860	9	11	9.2E+02
167.486	4	4	3.0E+03	62.46	5	7	1.2E+03	69.7	3	1	1.9E+03
168.172	6	6	3.1E+03	62.699	3	5	2.3E+03	69.942	3	5	7.4E+03
168.545	6	4	2.0E+03	63.188	5	7	3.9E+03	69.989	5	7	7.9E+03
168.929	4	2	2.1E+03	64.139	1	3	2.1E+03	70.052	7	9	8.8E+03
185.213	6	8	1.0E+03	74.845	5	5	1.0E+03	70.224	1	3	4.13E+03
186.601	4	6	9.4E+02	75.892	5	3	7.7E+02	70.53	7	5	2.6E+02
				76.117	5	3	2.1E+03	70.59	7	7	1.7E+03
Fe X				78.452	9	11	6.3E+03	73.199	7	9	8.8E+03
76.822	2	2	1.8E+03	84.270	7	9	5.5E+03	73.473	5	7	6.2E+03
77.865	4	6	1.6E+03	107.384	7	5	1.8E+03	233.857	5	7	2.2E+02
100.026	8	10	2.6E+03					235	1	3	2.5E+02
101.733	6	8	1.8E+03	*Fe XIV*				243	1	3	2.4E+02
101.846	4	6	1.7E+03	58.963	2	4	2.7E+03	243	5	7	2.3E+02
102.095	10	12	2.9E+03	59.579	4	6	3.1E+03	243.790	3	5	4.2E+02
102.192	10	12	2.9E+03	69.176	4	6	5.6E+02	248	3	1	5.4E+02
102.829	4	6	2.1E+03	69.386	2	4	7.6E+02	284.160	1	3	2.28E+02
103.319	6	8	2.6E+03	69.66	2	2	8.9E+02				
103.724	6	8	1.7E+03	69.66	6	6	1.3E+03	*Fe XVI*			
104.638	8	10	2.1E+03	70.251	6	4	8.1E+02	31.041	2	4	5.2E+02
174.534	4	6	1.8E+03	70.613	4	2	1.7E+03	31.242	4	6	6.1E+02
175.266	2	4	1.72E+03	72.80	10	12	7.9E+03	32.166	2	4	6.8E+02
				76.022	4	6	6.6E+03	32.192	2	2	6.7E+02
Fe XI				76.152	6	8	7.0E+03	32.433	2	4	7.7E+02
72.166	5	7	2.9E+03	91.009	6	4	5.1E+02	32.652	4	6	9.1E+02
72.310	5	5	1.5E+03	91.273	4	2	5.6E+02	34.857	2	4	1.23E+03
72.635	5	7	1.6E+03	188	4	6	2.7E+02	35.106	4	6	1.44E+03
91.394	5	7	2.6E+03	190	6	8	2.8E+02	35.333	4	6	6.4E+02
91.472	7	9	2.5E+03	207	2	2	2.1E+02	35.368	6	8	6.8E+02
91.63	3	5	2.3E+03	211.316	2	4	3.6E+02	36.01	4	2	5.0E+02
91.63	7	9	3.4E+03	213	4	2	2.8E+02	36.749	2	4	1.1E+03
91.63	5	7	2.8E+03	214	2	2	4.0E+02	36.803	2	2	1.2E+03
91.733	9	11	4.1E+03	216	6	8	1.7E+02	37.096	4	6	1.0E+03

λ Å	Weights g_i	g_k	A 10^8 s^{-1}	λ Å	Weights g_i	g_k	A 10^8 s^{-1}	λ Å	Weights g_i	g_k	A 10^8 s^{-1}
37.138	6	8	1.07E+03	13.68	3	1	8.0E+04	8.65	5	7	3.9E+04
39.827	2	4	2.1E+03	13.69	5	7	2.3E+04	8.66	5	5	4.4E+03
40.153	4	6	2.5E+03	13.700	1	3	2.7E+05	8.74	1	3	2.5E+04
40.161	4	4	4.1E+02	13.71	5	5	2.2E+04	9.42	3	1	4.3E+04
40.199	4	6	1.7E+03	13.738	5	7	1.0E+04	9.42	3	3	3.3E+04
40.245	6	8	1.8E+03	13.796	5	7	7.0E+04	9.44	3	5	1.7E+04
41.91	2	2	4.72E+02	13.83	5	5	1.4E+04	9.45	1	3	5.2E+04
42.30	4	2	9.2E+02	13.934	1	3	4.51E+04	9.46	5	3	1.5E+04
46.661	4	6	3.46E+03	13.961	3	3	2.0E+04	9.47	5	7	4.9E+04
46.718	6	8	3.7E+03	14.668	5	7	1.1E+04	9.47	5	5	6.1E+03
50.350	2	4	1.86E+03	14.671	5	3	1.1E+04	9.52	3	3	8.1E+03
50.555	2	2	1.98E+03	14.929	3	3	1.2E+04	9.58	5	5	5.2E+03
54.142	2	4	3.41E+03	14.966	5	3	2.5E+04	9.59	5	5	1.0E+04
54.728	4	6	4.16E+03	14.995	5	5	2.2E+04	9.67	1	3	5.7E+04
54.769	4	4	6.97E+02	15.015	1	3	1.4E+04	9.68	5	7	4.0E+03
62.879	2	2	1.05E+03	16.668	3	1	1.1E+04	9.74	5	3	5.3E+03
63.719	4	2	2.18E+03					12.02	1	3	1.3E+04
66.263	4	6	9.39E+03	*Fe XX*				12.13	3	3	1.8E+04
66.368	6	8	1.00E+04	12.67	6	6	1.0E+04	12.18	5	7	2.2E+04
66.392	6	6	6.69E+02	12.69	4	6	1.2E+04	12.19	5	3	6.4E+03
76.502	6	4	6.7E+02	12.73	4	2	4.0E+04	12.21	3	1	1.5E+05
76.796	4	2	7.72E+02	12.77	4	4	2.1E+05	12.21	3	3	1.2E+05
80.192	4	6	5.2E+02	12.78	4	2	6.9E+04	12.25	1	3	2.1E+05
80.270	6	8	5.4E+02	12.78	2	4	1.4E+05	12.28	5	3	5.2E+04
85.587	2	4	4.0E+02	12.79	6	4	1.7E+04	12.30	5	7	2.1E+05
86.133	4	6	4.8E+02	12.82	4	4	1.1E+05	12.36	3	3	3.6E+04
96.256	4	6	8.7E+02	12.88	6	4	2.7E+04	12.37	5	7	3.1E+05
96.348	6	8	9.3E+02	12.89	4	4	4.4E+04	12.38	5	3	6.9E+03
117.2	2	4	3.93E+02	12.90	4	2	6.2E+03	12.47	5	7	5.8E+04
117.7	2	2	3.9E+02	12.90	4	6	1.4E+05	12.47	5	3	1.3E+04
123.4	2	4	5.9E+02	12.92	2	4	1.7E+04	12.49	5	7	1.3E+04
124.5	4	6	7.0E+02	12.93	4	6	1.6E+05	12.53	5	5	1.5E+04
144.06	4	6	1.6E+03	12.93	2	2	1.2E+04	12.57	1	3	7.2E+04
144.25	6	8	1.6E+03	12.98	2	2	6.7E+04	12.73	5	5	8.2E+03
148	4	2	6.5E+02	12.99	6	6	5.1E+04	12.95	3	5	6.2E+03
266.7	4	6	3.9E+02	13.00	6	4	1.1E+04	13.00	1	3	7.2E+03
267.0	6	8	4.3E+02	13.01	2	4	3.0E+04	13.03	5	5	1.3E+04
				13.03	4	2	8.6E+04	13.14	3	1	2.0E+04
Fe XVII				13.07	6	4	8.2E+03	13.41	1	3	7.3E+03
11.023	1	3	2.1E+04	13.13	2	4	8.9E+04				
12.123	1	3	8.0E+04	13.24	4	4	1.2E+04	*Fe XXII*			
12.264	1	3	5.9E+04	13.28	4	4	6.1E+03	9.002	4	6	5.5E+04
12.526	1	3	3.0E+03	13.70	4	6	1.1E+04	9.006	6	8	5.7E+04
12.681	1	3	3.5E+03	13.71	2	2	9.9E+03	9.006	6	6	5.3E+04
13.823	1	3	3.3E+04	13.78	4	4	1.0E+04	9.163	4	6	6.9E+04
13.891	1	3	3.4E+03	13.79	6	6	1.2E+04	9.183	6	8	8.3E+04
15.015	1	3	2.28E+05	13.83	4	2	9.8E+03	9.241	4	6	5.1E+04
15.262	1	3	6.0E+04	13.90	4	2	1.2E+04	11.748	4	4	1.2E+05
16.777	1	3	8.29E+03	13.98	6	4	1.6E+04	11.748	4	6	1.6E+05
17.054	1	3	9.33E+03	13.99	4	2	2.2E+04	11.748	4	2	1.8E+05
41.37	9	11	4.8E+03	14.05	4	4	1.7E+04	11.763	2	4	1.6E+05
49.427	3	3	4.0E+03	14.23	2	2	6.3E+03	11.789	2	2	2.6E+05
50.26	7	9	6.0E+03					11.789	6	8	1.2E+05
58.76	9	11	1.2E+04	*Fe XXI*				11.797	2	4	1.7E+05
				8.53	3	1	1.8E+04	11.823	6	4	7.9E+04
Fe XIX				8.53	3	5	6.1E+03	11.837	6	8	2.3E+05
13.413	5	3	1.3E+04	8.53	3	3	1.5E+04	11.837	6	6	1.7E+05
13.426	5	7	4.8E+04	8.56	5	7	2.0E+04	11.886	4	6	1.3E+05
13.47	3	1	1.5E+05	8.56	1	3	2.1E+04	11.898	2	4	8.2E+04
13.520	5	7	2.0E+05	8.56	5	3	6.5E+03	11.922	4	6	1.8E+05
13.56	3	5	1.0E+04	8.64	5	7	1.5E+04				

λ Å	Weights g_i	g_k	A 10^8 s^{-1}	λ Å	Weights g_i	g_k	A 10^8 s^{-1}	λ Å	Weights g_i	g_k	A 10^8 s^{-1}
11.976	6	8	5.9E+04	1.8614	4	4	6.24E+06	4463.7	3	3	2.3E-02
12.027	2	4	6.9E+04	1.8626	2	4	3.16E+06	4502.4	3	5	9.2E-03
12.045	6	8	2.4E+05	1.8627	2	2	5.47E+06	5562.2	5	5	2.8E-03
12.045	4	4	9.7E+04	1.8637	2	2	1.91E+06	5570.3	5	3	2.1E-02
12.053	4	6	6.1E+04	1.8655	4	6	2.14E+06	5649.6	1	3	3.7E-03
12.077	2	4	1.0E+05	1.8672	4	2	1.63E+06	5870.9	3	5	1.8E-02
12.077	4	6	2.4E+05	1.8678	4	4	3.5E+05	6904.7	3	5	1.3E-02
12.095	6	6	7.8E+04	1.8721	4	6	3.2E+05	7224.1	3	5	1.4E-02
12.193	2	4	7.2E+04	1.8721	2	2	2.0E+05	7587.4	3	1	5.1E-01
12.193	4	6	9.9E+04	1.8730	2	4	1.5E+05	7601.5	5	5	3.1E-01
12.325	2	2	1.5E+05	1.8739	4	4	8.3E+04	7685.2	3	1	4.9E-01
				1.891	2	2	9.7E+04	7694.5	5	3	5.6E-02
Fe XXIII				1.897	4	2	9.8E+04	7854.8	1	3	2.3E-01
7.733	5	7	3.0E+04	8.231	2	4	6.10E+04	8059.5	1	3	1.9E-01
7.849	5	7	4.9E+04	8.316	4	6	7.07E+04	8104.4	5	5	1.3E-01
8.307	1	3	4.8E+04	10.619	2	4	7.28E+04	8112.9	5	7	3.6E-01
8.529	1	3	4.3E+04	10.663	2	2	7.51E+04	8190.1	3	5	1.1E-01
8.550	3	5	6.0E+04	11.030	2	4	1.84E+05	8263.2	3	5	3.5E-01
8.552	3	3	3.2E+04	11.171	4	6	2.18E+05	8281.1	3	3	1.9E-01
8.614	5	7	7.7E+04					8298.1	3	3	3.2E-01
8.664	3	3	4.4E+04	*Fe XXV*				8508.9	3	3	2.4E-01
8.669	5	7	6.1E+04	1.4607	1	3	2.54E+05	8776.7	3	5	2.7E-01
8.672	1	3	6.8E+04	1.4945	1	3	5.05E+05	8928.7	5	3	3.7E-01
8.752	5	7	1.2E+05	1.5730	1	3	1.24E+06				
8.764	5	7	4.6E+04	1.5749	1	3	1.5E+05	*Kr II*			
8.814	3	5	6.2E+04	1.778	3	3	8.7E+04	4250.6	4	4	1.2E-01
10.902	5	5	5.3E+04	1.782	3	1	4.69E+06	4292.9	4	4	9.6E-01
10.910	3	1	6.7E+04	1.787	1	3	2.57E+06	4355.5	6	8	1.0E+00
10.927	5	7	6.0E+04	1.787	5	5	1.19E+06	4431.7	2	2	1.8E+00
10.934	3	5	5.4E+04	1.788	3	5	2.68E+06	4436.8	2	4	6.6E-01
10.979	1	3	7.9E+04	1.788	3	5	1.63E+06	4577.2	6	8	9.6E-01
11.018	1	3	4.9E+04	1.789	1	3	1.78E+06	4583.0	6	4	7.6E-01
11.086	3	1	6.5E+04	1.790	3	3	1.23E+06	4615.3	4	4	5.4E-01
11.165	3	5	6.7E+04	1.791	3	5	4.10E+06	4619.2	4	6	8.1E-01
11.255	3	3	3.7E+04	1.791	3	3	2.59E+06	4633.9	4	6	7.1E-01
11.298	1	3	1.3E+05	1.792	3	1	4.92E+06	4658.9	6	4	6.5E-01
11.325	3	5	1.7E+05	1.792	5	5	2.81E+06	4739.0	6	6	7.6E-01
11.338	3	3	9.3E+04	1.793	3	1	2.67E+06	4762.4	2	4	4.2E-01
11.429	3	1	1.7E+05	1.794	5	3	2.22E+06	4765.7	4	6	6.7E-01
11.433	3	3	1.2E+05	1.797	3	5	8.8E+05	4811.8	2	4	1.7E-01
11.441	5	7	2.2E+05	1.798	3	3	1.0E+05	4825.2	2	4	1.9E-01
11.445	5	5	5.6E+04	1.800	1	3	8.6E+04	4832.1	4	2	7.3E-01
11.485	3	5	1.40E+05	1.802	3	1	4.1E+05	5208.3	4	4	1.4E-01
11.491	5	3	5.9E+04	1.810	3	1	5.9E+05	5308.7	4	6	2.4E-02
11.519	5	5	1.16E+05	1.8502	1	3	4.57E+06	7407.0	6	6	7.0E-02
11.520	1	3	2.16E+05	1.8593	1	3	4.42E+05				
11.524	5	7	2.3E+05	10.038	3	3	8.08E+04	*Lead*			
11.593	5	7	3.58E+05								
11.613	3	5	1.0E+05	*Krypton*				*Pb I*			
11.615	3	3	4.4E+04					2022.0	1	3	5.2E-02
11.691	5	7	7.7E+04	*Kr I*				2053.3	1	3	1.2E-01
11.698	5	5	7.3E+04	1164.9	1	3	3.16E+00	2170.0	1	3	1.5E+00
11.737	3	5	1.8E+05	1235.8	1	3	3.12E+00	2401.9	3	3	1.9E-01
11.898	1	3	2.03E+05	4274.0	5	5	2.6E-02	2446.2	3	3	2.5E-01
				4351.4	3	1	3.2E-02	2476.4	3	5	2.8E-01
Fe XXIV				4362.6	5	3	8.4E-03	2577.3	5	3	5.0E-01
1.8523	2	2	1.0E+05	4376.1	3	1	5.6E-02	2613.7	3	3	2.7E-01
1.8552	2	4	4.82E+06	4400.0	3	5	2.0E-02	2614.2	3	5	1.9E+00
1.8563	4	2	2.43E+06	4410.4	3	3	4.4E-03	2628.3	5	3	3.1E-02
1.8572	2	2	3.06E+06	4425.2	3	3	9.7E-03	2657.1	3	5	9.8E-04
1.858	2	4	1.2E+05	4453.9	3	5	7.8E-03	2663.2	5	5	7.1E-01

λ Å	Weights g_i	g_k	A 10^8 s^{-1}
2802.0	5	7	1.6E+00
2823.2	5	5	2.6E-01
2833.1	1	3	5.8E-01
2873.3	5	5	3.7E-01
3572.7	5	3	9.9E-01
3639.6	3	3	3.4E-01
3671.5	5	3	4.4E-01
3683.5	3	1	1.5E+00
3739.9	5	5	7.3E-01
4019.6	5	7	3.5E-02
4057.8	5	3	8.9E-01
4062.1	5	3	9.2E-01
4168.0	5	5	1.2E-02
5005.4	1	3	2.7E-01
5201.4	1	3	1.9E-01
7229.0	5	3	8.9E-03

Lithium

Li I

λ Å	g_i	g_k	A 10^8 s^{-1}
*2741.2	2	6	1.3E-02
*3232.7	2	6	1.17E-02
*4602.9	6	10	2.23E-01
*6103.6	6	10	6.860E-01
*6707.8	2	6	3.691E-01

Lutetium

Lu I

λ Å	g_i	g_k	A 10^8 s^{-1}
3376.5	4	4	2.23E+00
3567.8	4	6	5.9E-01
3620.3	6	4	1.1E-02
3841.2	6	6	2.5E-01
4518.6	4	4	2.1E-01

Magnesium

Mg I

λ Å	g_i	g_k	A 10^8 s^{-1}
2025.8	1	3	8.4E-01
*2779.8	9	9	5.2E+00
*2850.0	9	15	2.3E-01
2852.1	1	3	4.95E+00
*3094.9	9	15	5.2E-01
3329.9	1	3	3.3E-02
3332.2	3	3	9.7E-02
3336.7	5	3	1.6E-01
*3835.3	9	15	1.68E+00
4703.0	3	5	2.55E-01
5167.3	1	3	1.16E-01
5172.7	3	3	3.46E-01
5183.6	5	3	5.75E-01
5528.4	3	5	1.99E-01

Mg II

λ Å	g_i	g_k	A 10^8 s^{-1}
1239.9	2	4	1.4E-02
1240.4	2	2	1.4E-02
*2660.8	10	14	3.8E-01
2790.8	2	4	4.0E+00
2795.5	2	4	2.6E+00
2797.9	4	4	7.9E-01
2798.1	4	6	4.8E+00
2802.7	2	2	2.6E+00
2928.8	2	2	1.2E+00
2936.5	4	2	2.3E+00

λ Å	g_i	g_k	A 10^8 s^{-1}
*3104.8	10	14	8.1E-01
3848.2	6	4	2.8E-02
3848.3	4	4	3.0E-03
3850.4	4	2	3.0E-02
*4481.2	10	14	2.23E+00
9218.3	2	4	3.6E-01
9244.3	2	2	3.6E-01

Mg IV

λ Å	g_i	g_k	A 10^8 s^{-1}
320.99	4	2	1.2E+02
323.31	2	2	5.9E+01
1219.0	6	6	5.9E+00
1375.5	4	4	4.5E+00
1459.6	6	4	4.6E+00
1495.5	4	6	6.4E+00
1510.7	4	4	6.7E+00
1683.0	6	8	5.8E+00
1698.8	4	6	3.9E+00
1893.9	6	6	2.8E+00

Mg VI

λ Å	g_i	g_k	A 10^8 s^{-1}
*269.92	10	6	3.1E+02
*292.53	6	6	9.0E+01
*314.64	6	2	1.8E+02
*349.15	10	10	6.1E+01
*387.94	6	10	1.3E+01
399.29	4	2	2.8E+01
400.68	4	4	2.8E+01
403.32	4	6	2.7E+01

Mg VII

λ Å	g_i	g_k	A 10^8 s^{-1}
277.01	3	3	9.5E+01
278.41	5	3	1.5E+02
280.74	5	3	2.0E+02
319.02	5	5	8.9E+01
*366.42	9	9	4.4E+01
*433.04	9	15	1.6E+01
1334.3	5	5	5.3E+00
1410.0	5	5	2.57E+00
1487.0	3	5	3.02E+00
1487.9	5	7	3.66E+00

Mg VIII

λ Å	g_i	g_k	A 10^8 s^{-1}
*74.976	6	10	4.3E+03
315.02	4	4	1.2E+02
*342.29	10	6	6.3E+01
353.86	4	4	3.89E+01
356.00	6	4	5.7E+01
*428.52	10	10	3.24E+01
*434.62	6	10	1.6E+01
*489.33	6	6	3.9E+01
*686.92	6	10	9.4E+00

Mg IX

λ Å	g_i	g_k	A 10^8 s^{-1}
62.751	1	3	2.87E+03
*67.189	9	15	6.20E+03
*71.965	9	3	1.22E+03
72.312	3	5	4.43E+03
77.737	3	1	3.92E+02
368.07	1	3	5.27E+01
438.69	3	1	7.9E+01
*443.74	9	9	4.19E+01
749.55	3	5	8.2E+00

λ Å	g_i	g_k	A 10^8 s^{-1}
1639.8	3	5	2.1E+00
2814.2	1	3	3.35E-01

Mg X

λ Å	g_i	g_k	A 10^8 s^{-1}
57.876	2	4	2.09E+03
57.920	2	2	2.09E+03
63.152	2	4	5.6E+03
63.295	4	6	6.7E+03
609.79	2	4	7.53E+00
624.94	2	2	7.01E+00
2212.5	2	4	9.64E-01
2278.7	2	2	8.82E-01
5918.7	2	4	3.20E-02
6229.6	4	6	3.30E-02

Mg XI

λ Å	g_i	g_k	A 10^8 s^{-1}
7.310	1	3	1.15E+04
7.473	1	3	2.27E+04
7.850	1	3	5.50E+04
9.169	1	3	1.97E+05

Manganese

Mn I

λ Å	g_i	g_k	A 10^8 s^{-1}
2794.82	6	8	3.7E+00
2798.27	6	6	3.6E+00
2801.08	6	4	3.7E+00
3007.65	6	8	1.8E-01
3011.38	8	10	3.1E-01
3016.45	10	12	2.9E-01
3043.36	8	8	5.9E-01
3044.57	10	8	5.7E-01
3045.59	10	10	6.7E-01
3045.80	8	10	1.7E-01
3047.03	12	12	6.1E-01
3054.36	8	6	4.6E-01
3070.27	6	6	1.9E-01
3073.18	4	4	3.7E-01
3082.71	14	14	2.9E-01
3110.68	6	8	2.7E-01
3113.80	12	10	2.6E-01
3118.10	4	6	1.7E-01
3122.88	10	10	1.9E-01
3126.85	8	6	2.3E-01
3132.28	10	10	2.1E-01
3132.79	8	8	2.7E-01
3175.58	8	10	1.8E-01
3201.11	4	6	2.2E-01
3228.09	10	12	6.4E-01
3230.23	10	12	1.9E-01
3230.72	8	8	3.5E-01
3240.88	6	4	2.2E-01
3243.78	6	6	5.3E-01
3251.13	4	2	2.3E-01
3252.95	4	4	1.8E-01
3256.14	4	6	5.0E-01
3258.41	2	2	9.7E-01
3260.24	2	4	3.8E-01
3267.79	14	14	3.5E-01
3268.72	6	8	3.3E-01
3270.35	12	12	2.6E-01
3273.02	10	10	2.7E-01

λ (Å)	g_i	g_k	A (10^8 s^{-1})	λ (Å)	g_i	g_k	A (10^8 s^{-1})	λ (Å)	g_i	g_k	A (10^8 s^{-1})
3298.23	6	4	2.8E-01	4070.28	2	2	2.3E-01	3482.90	5	5	2.0E-01
3303.28	4	4	1.9E-01	4079.42	2	4	3.8E-01	3488.68	3	3	2.5E-01
3463.66	8	8	3.2E-01	4082.95	4	6	2.95E-01				
3470.01	6	8	2.4E-01	4083.63	6	8	2.8E-01	*Mn VI*			
3511.83	12	12	2.7E-01	4089.94	8	10	1.7E-01	307.999	9	9	3.7E+01
3535.30	10	10	1.7E-01	4105.37	10	8	1.7E-01	309.440	9	7	5.7E+01
3559.81	6	6	2.1E-01	4135.03	12	12	3.0E-01	309.579	7	5	4.4E+01
3577.87	10	8	9.4E-01	4141.06	10	10	2.6E-01	310.058	7	7	3.4E+01
3595.11	6	4	1.8E-01	4148.80	8	8	2.3E-01	310.182	5	5	2.8E+01
3601.27	12	10	2.3E-01	4176.61	14	12	2.4E-01	311.748	5	3	5.7E+01
3607.53	8	8	2.3E-01	4189.99	12	10	2.0E-01	320.598	3	5	1.5E+01
3608.49	6	6	3.6E-01	4201.78	10	8	2.3E-01	320.681	1	3	2.2E+01
3610.30	4	4	4.2E-01	4235.30	8	6	9.17E-01	320.874	3	1	7.8E+01
3635.70	10	8	2.1E-01	4239.74	4	2	3.9E-01	320.979	3	3	2.2E+01
3660.40	12	14	9.1E-01	4257.67	2	2	3.7E-01	321.176	5	5	6.0E+01
3675.67	6	8	2.2E-01	4265.93	4	4	4.92E-01	321.541	5	3	2.7E+01
3676.96	10	12	7.3E-01	4281.10	6	6	2.3E-01	325.146	9	7	1.3E+02
3680.15	12	10	1.9E-01	4411.87	12	10	2.6E-01	328.431	5	5	4.4E+01
3682.09	8	10	7.6E-01	4414.89	8	6	2.93E-01	328.558	3	5	1.2E+01
3684.87	6	8	2.6E-01	4419.77	10	8	2.1E-01	329.043	1	3	1.1E+01
3706.08	12	14	1.4E+00	4436.36	6	4	4.37E-01	1236.23	5	3	1.3E+01
3718.92	10	12	9.6E-01	4451.58	8	8	7.98E-01	1255.77	3	1	1.2E+01
3731.94	8	10	1.0E+00	4453.01	4	2	5.44E-01	1285.10	5	7	1.1E+01
3771.44	14	14	1.9E-01	4455.82	4	6	1.7E-01	1333.87	7	9	1.0E+01
3773.86	12	12	2.5E-01	4457.04	6	4	2.34E-01				
3800.55	6	8	2.7E-01	4457.55	6	6	4.27E-01	*Mercury*			
3806.72	10	12	5.9E-01	4458.26	6	8	4.62E-01				
3823.51	8	10	5.21E-01	4461.09	8	8	1.7E-01	*Hg I*			
3823.89	6	6	2.31E-01	4462.03	8	10	7.00E-01	2536.52	1	3	8.00E-02
3833.87	4	4	3.14E-01	4464.68	6	6	4.39E-01	2652.04	3	5	3.88E-01
3834.37	6	8	4.29E-01	4470.14	4	4	3.00E-01	2655.13	3	5	1.1E-01
3839.78	2	2	4.64E-01	4472.79	2	2	4.35E-01	2752.78	1	3	6.10E-02
3841.07	4	6	3.3E-01	4479.40	8	10	3.4E-01	2856.94	3	1	1.1E-02
3843.99	2	4	2.11E-01	4490.08	2	4	2.49E-01	2893.60	3	3	1.6E-01
3889.46	12	14	3.1E-01	4498.90	4	6	2.49E-01	2925.4	5	3	7.7E-02
3898.37	6	8	1.7E-01	4502.22	6	8	1.86E-01	2967.3	1	3	4.5E-01
3899.34	4	6	2.4E-01	4605.37	10	12	3.6E-01	3021.50	5	7	5.09E-01
3924.08	2	4	9.4E-01	4626.54	12	14	3.6E-01	3023.48	5	5	9.4E-02
3926.48	6	8	5.4E-01	4709.71	8	8	1.72E-01	3027.49	5	5	2.0E-02
3951.98	2	2	3.1E-01	4727.46	6	6	1.7E-01	3125.66	3	5	6.56E-01
3952.84	6	6	4.1E-01	4739.11	4	4	2.40E-01	3341.48	5	3	1.68E-01
3975.88	2	4	1.8E-01	4754.05	6	8	3.03E-01	3650.15	5	7	1.3E+00
3982.16	4	2	3.5E-01	4761.53	2	4	5.35E-01	3654.83	5	5	1.8E-01
3982.58	6	4	2.3E-01	4762.38	8	10	7.83E-01	4046.56	1	3	2.1E-01
3982.90	6	4	5.5E-01	4765.86	4	6	4.1E-01	4077.81	3	1	4.0E-02
3991.60	2	2	2.1E-01	4766.43	6	8	4.6E-01	4108.1	3	1	3.0E-02
4011.91	8	8	2.3E-01	4783.43	8	8	4.01E-01	4339.22	3	5	2.88E-02
4018.11	10	8	2.54E-01	4823.53	10	8	4.99E-01	4347.50	3	5	8.4E-02
4030.76	6	8	1.7E-01	6013.48	4	6	1.72E-01	4358.34	3	3	5.57E-01
4033.07	6	6	1.65E-01	6021.79	8	6	3.32E-01	4916.07	3	1	5.8E-02
4034.49	6	4	1.58E-01					5025.64	3	3	2.7E-04
4041.36	10	10	7.87E-01	*Mn II*				5460.75	5	3	4.87E-01
4048.75	6	4	7.5E-01	2593.72	7	7	2.6E+00	5769.59	3	5	2.36E-01
4052.48	6	8	3.8E-01	2605.68	7	5	2.7E+00	6234.4	1	3	5.3E-03
4055.55	8	8	4.31E-01	2933.05	5	3	2.0E+00	6716.4	1	3	4.3E-03
4058.94	4	2	7.25E-01	2939.31	5	5	1.9E+00	6907.5	3	5	2.8E-02
4061.74	8	6	1.9E-01	2949.20	5	7	1.9E+00	7728.8	1	3	9.7E-03
4063.53	6	6	1.69E-01	3441.99	9	7	4.3E-01	10139.79	3	1	2.71E-01
4065.08	12	14	2.5E-01	3460.32	7	5	3.2E-01				
4066.24	10	8	2.2E-01	3474.13	5	3	1.5E-01	*Molybdenum*			
								Mo I			
								2616.79	3	5	7.34E-01

λ Å	Weights g_i	g_k	A 10^8 s^{-1}	λ Å	Weights g_i	g_k	A 10^8 s^{-1}	λ Å	Weights g_i	g_k	A 10^8 s^{-1}
2621.06	7	7	1.16E-01	3036.31	3	5	5.81E-01	3228.21	5	7	3.85E-01
2628.96	3	3	2.81E-01	3041.70	13	11	5.94E-01	3229.79	9	11	1.44E-01
2629.85	5	7	7.75E-01	3046.80	13	11	1.63E-01	3233.14	13	13	6.33E-01
2631.50	1	3	2.54E-01	3047.31	11	9	5.01E-01	3237.06	7	9	2.95E-01
2638.30	5	5	7.57E-01	3055.32	9	7	4.29E-01	3244.47	5	3	2.80E-01
2640.98	7	5	1.20E+00	3057.56	7	5	2.64E-01	3247.61	5	5	1.71E-01
2644.36	5	7	1.96E-01	3061.59	7	5	4.41E-01	3249.93	5	3	1.87E-01
2649.46	7	9	9.84E-01	3064.27	13	13	8.46E-01	3251.65	3	5	3.05E-01
2655.02	9	7	4.08E-01	3065.04	13	13	3.08E-01	3256.21	5	3	6.89E-01
2658.11	7	7	6.43E-01	3069.51	5	5	1.52E-01	3256.72	3	3	1.31E-01
2665.09	7	9	1.32E-01	3069.96	11	11	2.72E-01	3259.16	11	13	1.62E-01
2679.85	9	11	1.31E+00	3070.89	9	11	1.87E-01	3262.63	7	9	3.62E-01
2684.16	9	9	4.18E-01	3074.37	11	11	1.42E+00	3264.40	11	9	5.42E-01
2706.11	3	5	2.03E-01	3079.88	9	11	9.55E-01	3265.14	5	7	2.60E-01
2710.74	3	3	1.57E-01	3080.40	7	9	3.61E-01	3266.16	9	11	1.95E-01
2725.15	3	5	2.79E-01	3081.16	3	5	2.35E-01	3270.90	7	7	3.59E-01
2728.71	3	3	1.26E-01	3085.62	9	9	1.63E+00	3276.07	11	9	1.18E-01
2733.39	5	7	2.95E-01	3089.13	11	9	1.53E-01	3285.03	1	3	1.41E-01
2743.71	1	3	2.47E-01	3089.71	5	7	2.34E-01	3285.35	9	7	4.49E-01
2745.38	13	11	1.29E-01	3094.66	7	7	1.63E+00	3287.38	5	5	1.38E-01
2751.47	7	9	2.54E-01	3099.92	9	7	1.45E-01	3289.01	9	9	5.08E-01
2756.26	5	3	1.18E-01	3100.88	7	9	1.20E+00	3290.82	7	5	5.44E-01
2761.53	9	11	2.06E-01	3101.34	5	5	1.92E+00	3305.56	5	3	1.74E-01
2763.02	3	1	4.44E-01	3106.34	7	5	2.21E-01	3305.91	7	9	3.06E-01
2766.25	3	5	1.17E-01	3117.54	13	13	1.89E-01	3307.13	7	9	1.25E-01
2787.83	9	7	2.85E-01	3123.03	3	3	2.81E-01	3312.33	7	5	1.62E-01
2792.96	5	3	1.53E-01	3125.03	5	3	1.98E-01	3323.95	9	7	2.82E-01
2798.02	7	5	1.22E-01	3132.59	7	9	1.79E+00	3325.13	5	3	2.26E-01
2801.47	5	7	1.24E-01	3135.90	9	11	3.68E-01	3325.67	5	5	1.72E-01
2825.68	5	7	2.53E-01	3136.75	9	11	1.57E-01	3327.30	1	3	2.88E-01
2826.75	7	7	4.23E-01	3142.75	3	5	4.10E-01	3336.56	9	9	1.64E-01
2876.54	9	9	2.84E-01	3147.35	13	11	2.41E-01	3340.16	5	3	1.20E-01
2886.60	11	11	4.74E-01	3155.19	7	7	2.75E-01	3344.73	3	5	6.04E-01
2906.06	3	3	8.04E-01	3158.17	7	7	4.63E-01	3346.83	11	11	1.13E-01
2913.52	5	3	1.38E-01	3170.34	7	7	1.37E+00	3347.00	3	3	2.72E-01
2915.38	5	3	7.31E-01	3171.38	5	7	2.03E-01	3358.12	5	7	7.59E-01
2918.84	5	3	3.79E-01	3175.59	13	11	8.40E-01	3361.37	9	9	1.38E-01
2930.39	1	3	1.91E-01	3179.78	11	13	2.33E-01	3363.78	5	7	2.74E-01
2936.50	11	11	2.33E-01	3183.03	11	9	3.98E-01	3363.87	5	7	1.39E-01
2945.43	7	7	3.66E-01	3184.58	7	5	2.77E-01	3373.81	3	3	2.03E-01
2945.66	3	3	4.08E-01	3185.10	7	7	2.54E-01	3375.22	7	7	1.38E-01
2946.01	5	5	1.68E-01	3185.71	5	3	6.10E-01	3375.65	7	9	1.56E-01
2951.45	9	9	1.43E-01	3188.10	7	9	3.45E-01	3378.19	3	1	1.88E-01
2959.48	9	11	1.75E-01	3188.41	5	7	4.40E-01	3378.46	13	13	3.75E-01
2972.96	5	3	2.69E-01	3192.79	9	11	1.88E-01	3379.96	5	5	4.11E-01
2977.27	9	7	3.28E-01	3193.98	7	5	1.53E+00	3382.48	3	3	2.66E-01
2978.28	7	5	1.50E-01	3194.88	9	11	1.75E-01	3384.61	7	9	7.32E-01
2983.04	1	3	2.82E-01	3195.96	9	7	4.10E-01	3385.87	9	11	3.30E-01
2987.92	3	5	8.43E-01	3197.18	1	3	1.47E-01	3389.79	5	7	1.85E-01
2988.23	5	7	4.28E-01	3198.85	15	13	7.22E-01	3392.17	9	9	1.97E-01
2988.68	7	9	1.61E-01	3200.89	3	5	1.82E-01	3393.65	11	11	2.08E-01
2989.80	9	7	9.27E-01	3205.22	1	3	4.27E-01	3404.33	7	7	2.10E-01
3000.24	9	9	1.40E-01	3205.43	9	11	2.55E-01	3413.37	11	11	1.25E-01
3000.44	5	5	1.25E-01	3205.89	9	9	5.35E-01	3415.27	9	9	1.83E-01
3000.85	5	7	2.58E-01	3208.84	7	5	2.77E-01	3415.61	7	9	1.29E-01
3001.43	5	5	2.31E-01	3210.97	7	5	6.94E-01	3416.14	9	11	2.45E-01
3007.71	7	5	1.90E-01	3214.44	9	7	2.01E-01	3418.52	5	3	1.41E-01
3013.39	7	5	6.06E-01	3215.07	3	5	4.20E-01	3419.69	7	7	1.15E-01
3016.78	9	9	2.75E-01	3216.78	15	13	2.10E-01	3420.04	5	5	3.28E-01
3025.00	5	5	8.49E-01	3221.73	3	1	1.41E+00	3422.31	9	9	2.52E-01

λ Å	Weights g_i	g_k	A 10^8 s^{-1}	λ Å	Weights g_i	g_k	A 10^8 s^{-1}	λ Å	Weights g_i	g_k	A 10^8 s^{-1}
3425.13	11	11	2.29E-01	3598.88	13	11	5.67E-01	3777.72	13	11	1.66E-01
3427.90	11	13	4.09E-01	3600.73	9	9	2.07E-01	3788.25	7	9	2.87E-01
3434.79	7	7	1.75E-01	3601.88	7	9	1.15E-01	3794.43	9	9	1.22E-01
3435.45	15	15	1.50E+00	3602.94	5	7	2.96E-01	3797.47	7	5	1.48E-01
3437.21	11	9	8.06E-01	3604.07	9	7	3.25E-01	3798.25	7	9	6.90E-01
3438.87	1	3	2.34E-01	3610.61	5	3	1.78E-01	3801.84	9	7	3.16E-01
3441.87	5	3	1.34E-01	3611.99	7	7	1.16E-01	3805.99	5	5	2.44E-01
3442.66	3	3	2.94E-01	3615.16	7	9	1.96E-01	3819.78	9	11	1.47E-01
3445.03	7	9	1.53E-01	3623.22	11	9	5.58E-01	3824.78	5	7	1.40E-01
3445.26	7	5	2.96E-01	3624.46	9	11	5.27E-01	3827.15	7	7	1.94E-01
3445.80	9	9	1.14E-01	3624.62	5	7	1.37E-01	3828.88	7	7	1.35E-01
3447.12	9	11	8.75E-01	3638.20	5	3	3.51E-01	3830.81	5	5	1.83E-01
3447.29	5	3	1.79E-01	3638.21	5	3	3.33E-01	3831.07	7	9	1.20E-01
3449.07	7	9	1.52E-01	3640.62	7	5	1.94E-01	3832.11	9	9	3.05E-01
3449.85	5	7	1.65E-01	3647.84	7	7	2.11E-01	3833.75	9	9	1.70E-01
3452.60	7	7	2.48E-01	3648.70	7	5	1.15E-01	3834.64	3	5	1.20E-01
3456.15	5	5	3.60E-01	3654.58	3	3	1.80E-01	3846.18	7	7	1.26E-01
3456.52	3	3	2.96E-01	3657.36	5	7	2.03E-01	3847.25	3	1	2.41E-01
3460.22	5	3	2.77E-01	3658.13	9	9	1.86E-01	3848.30	9	9	1.26E-01
3460.78	9	7	6.03E-01	3659.36	7	9	6.70E-01	3851.99	11	9	1.78E-01
3465.84	3	1	9.99E-01	3660.92	3	5	1.34E-01	3864.10	7	7	6.24E-01
3466.19	9	7	2.11E-01	3662.15	7	9	1.45E-01	3866.69	3	5	1.74E-01
3466.96	7	7	1.52E-01	3662.99	11	11	3.48E-01	3867.67	5	3	2.22E-01
3467.85	5	7	2.63E-01	3663.27	7	5	2.30E-01	3869.08	5	3	1.35E-01
3469.22	5	3	6.96E-01	3664.81	11	13	9.54E-01	3874.15	7	5	1.67E-01
3469.63	13	15	1.51E-01	3664.88	1	3	1.92E-01	3902.95	7	5	6.17E-01
3470.92	3	5	2.91E-01	3669.34	9	7	2.16E-01	3909.54	9	7	1.13E-01
3475.03	3	3	4.68E-01	3672.81	9	11	1.95E-01	3911.94	5	5	1.15E-01
3479.42	7	5	2.26E-01	3672.82	9	9	1.13E-01	3915.43	5	5	1.40E-01
3483.67	7	7	1.13E-01	3676.23	3	1	5.22E-01	3916.43	5	3	1.78E-01
3483.83	7	5	1.41E-01	3680.68	11	11	2.96E-01	3919.55	11	13	2.24E-01
3489.43	7	7	3.27E-01	3681.72	9	7	1.68E-01	3955.48	13	11	1.71E-01
3504.41	7	9	8.06E-01	3683.01	3	5	1.20E-01	3973.76	11	13	4.39E-01
3505.31	7	9	2.25E-01	3687.96	5	7	2.12E-01	3977.90	9	7	1.35E-01
3508.11	9	9	1.59E-01	3688.97	11	9	3.26E-01	3980.20	5	3	2.70E-01
3510.77	13	13	4.75E-01	3690.59	11	9	2.07E-01	3991.85	11	9	1.29E-01
3517.55	11	11	5.41E-01	3694.94	5	7	6.36E-01	4010.13	5	3	4.38E-01
3518.21	3	3	3.64E-01	3696.04	11	11	3.59E-01	4021.01	9	11	2.65E-01
3521.38	9	9	1.39E-01	3698.07	7	5	1.48E-01	4051.18	13	11	1.36E-01
3521.41	9	11	6.06E-01	3708.55	7	9	1.28E-01	4062.08	11	9	1.96E-01
3524.65	5	3	3.10E-01	3715.75	9	7	2.38E-01	4069.88	13	11	3.25E-01
3524.98	7	9	2.25E-01	3718.48	5	7	1.34E-01	4076.19	9	9	1.16E-01
3538.92	11	11	2.24E-01	3720.25	7	9	2.86E-01	4084.37	9	7	1.94E-01
3540.57	5	3	4.46E-01	3725.55	7	7	1.60E-01	4102.15	5	3	1.22E-01
3542.17	7	5	4.93E-01	3727.68	9	11	1.51E-01	4107.46	7	5	2.02E-01
3552.71	9	7	3.64E-01	3728.30	7	5	1.55E-01	4120.09	13	15	6.05E-01
3555.64	3	3	3.46E-01	3728.50	7	9	2.20E-01	4131.92	9	11	1.56E-01
3558.09	5	7	5.43E-01	3733.02	7	7	1.45E-01	4148.98	9	11	1.56E-01
3563.75	1	3	1.53E-01	3733.41	13	13	2.80E-01	4157.40	13	11	2.17E-01
3566.05	9	9	2.67E-01	3735.62	11	11	1.66E-01	4157.90	9	11	1.60E-01
3566.74	7	7	1.43E-01	3742.28	7	7	1.56E-01	4185.82	11	13	3.82E-01
3570.64	15	15	7.18E-01	3747.19	5	7	3.07E-01	4188.32	11	13	3.32E-01
3573.88	3	5	3.58E-01	3748.48	9	11	3.95E-01	4194.56	11	11	2.70E-01
3580.54	13	11	5.49E-01	3755.10	3	5	1.41E-01	4232.59	9	11	3.17E-01
3581.88	11	13	3.81E-01	3755.16	9	9	2.48E-01	4240.83	5	5	1.68E-01
3584.25	3	3	1.73E-01	3758.52	9	9	1.22E-01	4246.02	11	13	2.00E-01
3585.57	7	5	3.95E-01	3759.60	9	7	1.82E-01	4251.88	13	11	1.76E-01
3588.95	7	7	1.18E-01	3760.88	9	9	2.16E-01	4254.95	7	9	2.01E-01
3590.74	7	9	2.23E-01	3768.73	9	9	2.88E-01	4269.28	11	11	1.36E-01
3595.55	5	5	2.32E-01	3769.99	7	9	2.46E-01	4276.91	7	9	2.85E-01

λ Å	Weights g_i	g_k	A 10^8 s^{-1}	λ Å	Weights g_i	g_k	A 10^8 s^{-1}	λ Å	Weights g_i	g_k	A 10^8 s^{-1}
4277.24	9	11	1.35E-01	5926.37	7	7	1.63E-01	735.90	1	3	6.11E+00
4317.92	15	15	1.28E-01	5928.88	7	9	5.32E-01	743.72	1	3	4.86E-01
4325.80	3	3	1.84E-01	7154.11	9	9	3.45E-01	3369.8	5	5	1.0E-03
4326.14	5	7	2.56E-01					3369.9	5	3	7.6E-03
4340.74	5	7	1.23E-01	*Neodymium*				3375.6	5	3	2.2E-03
4381.63	13	13	2.93E-01					3417.9	3	5	9.2E-03
4382.41	11	13	3.83E-01	*Nd II*				3418.0	3	3	2.2E-03
4409.94	13	13	1.38E-01	3780.4	16	18	1.4E-01	3423.9	3	3	1.0E-03
4411.69	11	11	2.63E-01	3805.4	14	16	6.9E-01	3447.7	5	5	2.1E-02
4434.95	9	9	2.51E-01	3807.2	10	12	4.9E-02	3450.8	5	3	4.9E-03
4446.42	11	11	1.90E-01	3863.3	8	10	1.5E-01	3454.2	3	1	3.7E-02
4457.35	7	7	1.28E-01	3941.5	10	10	6.1E-01	3460.5	1	3	7.0E-03
4474.57	5	5	2.10E-01	3951.2	12	12	6.0E-01	3464.3	5	5	6.7E-03
4491.65	11	11	2.09E-01	3973.3	18	18	6.3E-01	3466.6	1	3	1.3E-02
4536.80	13	15	5.03E-01	3979.5	10	12	2.7E-01	3472.6	5	7	1.7E-02
4598.23	1	3	1.47E-01	3990.1	16	16	5.2E-01	3498.1	3	5	5.1E-03
4624.23	9	9	1.32E-01	4012.3	18	20	5.5E-01	3501.2	3	3	1.2E-02
4633.08	3	5	2.35E-01	4061.1	16	18	4.4E-01	3510.7	5	3	2.2E-03
4649.06	3	1	1.25E-01	4106.6	14	16	6.8E-02	3515.2	3	5	6.9E-03
4652.24	5	7	1.55E-01	4109.5	14	16	3.7E-01	3520.5	3	1	9.3E-02
4686.08	3	3	1.72E-01	4133.4	14	12	1.5E-01	3593.5	3	5	9.9E-03
4688.21	13	15	1.54E-01	4156.1	12	14	3.4E-01	3593.6	3	3	6.6E-03
4707.25	7	9	3.63E-01	4205.6	18	16	1.8E-01	3600.2	3	3	4.3E-03
4718.86	5	5	2.17E-01	4284.5	18	18	8.5E-02	3633.7	3	1	1.1E-02
4723.05	9	9	1.23E-01	4303.6	8	10	4.7E-01	3682.2	3	5	1.6E-02
4731.44	9	11	4.49E-01	4325.8	16	16	1.6E-01	3685.7	3	3	3.9E-02
4758.50	11	9	3.01E-01	4358.2	14	14	1.5E-01	3701.2	3	5	2.2E-03
4760.18	11	13	4.67E-01	4382.7	12	10	4.0E-02	4536.3	3	3	5.0E-03
4764.11	9	7	2.16E-01	4400.8	10	10	6.8E-02	4702.5	3	3	2.1E-02
4811.05	13	11	4.36E-01	4451.6	12	14	2.5E-01	4708.9	3	3	4.2E-02
4819.25	11	9	2.71E-01	4456.4	16	18	6.4E-02	4955.4	3	3	3.3E-03
4830.51	9	7	4.07E-01	4463.0	14	16	1.8E-01	5113.7	3	3	1.0E-02
4858.39	13	11	1.24E-01	4958.1	12	10	1.2E-02	5120.5	3	3	5.6E-03
4868.02	7	5	3.11E-01	5130.6	22	20	1.6E-01	5154.4	3	3	1.9E-02
5037.18	9	7	1.14E-01	5192.6	20	18	1.7E-01	5191.3	3	3	1.3E-02
5044.36	7	5	1.31E-01	5249.6	18	16	1.8E-01	5326.4	3	3	6.8E-03
5047.70	3	1	2.61E-01	5276.9	12	10	1.2E-01	5333.3	3	3	5.3E-03
5163.18	9	11	2.03E-01	5293.2	16	14	1.2E-01	5341.1	3	3	1.1E-01
5171.06	5	7	1.84E-01	5302.3	20	18	1.1E-01	5400.6	3	1	9.0E-03
5172.94	5	5	4.11E-01	5311.5	14	12	1.1E-01	5418.6	3	3	5.2E-03
5174.18	5	3	5.83E-01	5319.8	12	10	1.6E-01	5433.7	3	3	2.83E-03
5191.45	7	9	1.62E-01	5357.0	18	16	1.8E-01	5652.6	3	3	8.9E-03
5238.21	7	9	3.74E-01	5371.9	20	20	5.1E-02	5662.5	3	3	6.9E-03
5240.87	7	7	3.89E-01	5485.7	18	18	5.7E-02	5852.5	3	1	6.82E-01
5242.80	7	5	2.01E-01	5594.4	16	16	7.0E-02	5868.4	3	3	1.4E-02
5261.53	5	7	1.13E-01	5620.6	18	18	1.3E-01	5881.9	5	3	1.15E-01
5280.85	5	5	1.28E-01	5688.5	14	14	5.9E-02	5913.6	3	3	4.8E-02
5355.52	9	9	1.21E-01	5718.1	16	16	8.7E-02	5939.3	5	3	2.00E-03
5356.46	11	11	2.11E-01	5726.8	10	10	5.6E-02	5944.8	5	5	1.13E-01
5360.51	9	11	6.19E-01	5740.9	12	12	7.2E-02	5961.6	3	3	3.3E-02
5364.28	9	9	2.26E-01	5804.0	10	10	4.6E-02	5975.5	5	3	3.51E-02
5460.50	5	3	3.46E-01	5865.1	16	18	1.3E-02	6030.0	3	3	5.61E-02
5493.76	7	5	2.13E-01	6051.9	12	10	1.1E-02	6046.1	3	3	2.26E-03
5506.49	5	7	3.61E-01					6074.3	3	1	6.03E-01
5533.03	5	5	3.72E-01	*Neon*				6096.2	3	5	1.81E-01
5570.44	5	3	3.30E-01					6118.0	5	3	6.09E-03
5849.71	3	3	3.02E-01	*Ne I*				6128.5	3	3	6.7E-03
5851.50	3	5	1.55E-01	615.63	1	3	3.8E-01	6143.1	5	5	2.82E-01
5893.36	5	5	2.60E-01	618.67	1	3	9.3E-01	6150.3	3	3	1.5E-02
5895.93	5	7	3.12E-01	619.10	1	3	3.3E-01	6163.6	1	3	1.46E-01
				626.82	1	3	7.4E-01				
				629.74	1	3	4.8E-01				

λ Å	Weights g_i	g_k	A 10^8 s^{-1}	λ Å	Weights g_i	g_k	A 10^8 s^{-1}	λ Å	Weights g_i	g_k	A 10^8 s^{-1}
6217.3	5	3	6.37E-02	462.39	2	2	2.3E+01	3243.4	6	6	2.3E-01
6266.5	1	3	2.49E-01	1907.5	4	2	2.8E-01	3244.1	6	8	1.5E+00
6273.0	3	3	9.7E-03	1916.1	4	4	6.9E-01	3248.1	4	4	2.4E-01
6293.7	3	3	6.39E-03	1930.0	2	2	5.7E-01	3255.4	6	4	3.8E-02
6304.8	3	5	4.16E-02	1938.8	2	4	1.3E-01	3263.4	2	4	3.9E-01
6328.2	5	3	3.39E-02	2858.0	6	6	7.9E-01	3269.9	4	6	5.1E-01
6330.9	3	3	2.3E-02	2870.0	6	6	1.7E-01	3270.8	6	4	5.7E-02
6334.4	5	5	1.61E-01	2873.0	6	4	3.8E-01	3297.7	6	6	4.3E-01
6351.9	1	3	3.45E-03	2876.3	4	6	7.8E-01	3309.7	4	2	3.1E-01
6383.0	3	3	3.21E-01	2876.5	6	4	3.3E-01	3310.5	4	4	6.9E-02
6401.1	3	3	1.39E-02	2878.1	2	2	6.9E-02	3311.3	4	2	2.6E-01
6402.2	5	7	5.14E-01	2888.4	4	6	7.0E-02	3314.7	6	6	4.4E-02
6506.5	3	5	3.00E-01	2891.5	4	4	6.1E-02	3319.7	4	2	1.6E+00
6532.9	1	3	1.08E-01	2897.0	6	8	5.2E-02	3320.2	8	6	2.1E-01
6599.0	3	3	2.32E-01	2906.8	2	4	5.5E-01	3323.7	4	4	1.6E+00
6602.9	3	3	5.9E-03	2910.1	4	2	1.7E+00	3327.2	4	4	9.1E-01
6652.1	3	1	2.9E-03	2910.4	2	4	5.9E-01	3329.2	8	8	8.8E-01
6678.3	3	5	2.33E-01	2916.2	6	4	9.6E-02	3330.7	6	6	3.9E-02
6717.0	3	3	2.17E-01	2925.6	2	2	5.6E-01	3334.8	6	8	1.8E+00
6721.1	3	3	4.9E-04	2933.7	6	6	6.9E-02	3336.1	4	6	1.1E+00
6929.5	3	5	1.74E-01	2955.7	6	4	1.2E+00	3344.4	2	2	1.5E+00
7024.1	3	3	1.89E-02	3001.7	4	4	8.7E-01	3345.5	6	4	1.4E+00
7032.4	5	3	2.53E-01	3017.3	6	6	3.5E-01	3345.8	4	4	2.2E-01
7051.3	3	3	3.0E-02	3027.0	6	6	1.4E+00	3353.6	4	2	1.2E-01
7059.1	3	5	6.8E-02	3028.7	4	2	8.5E-01	3355.0	4	6	1.3E+00
7173.9	3	5	2.87E-02	3028.9	2	4	4.7E-01	3356.3	6	6	2.0E-01
7245.2	3	3	9.35E-02	3034.5	6	8	3.1E+00	3357.8	6	6	5.0E-01
7304.8	1	3	2.55E-03	3037.7	4	4	2.1E+00	3360.3	2	4	8.6E-01
7438.9	1	3	2.31E-02	3045.6	2	2	2.5E+00	3360.6	2	4	8.2E-01
7472.4	3	3	4.0E-02	3047.6	4	6	1.8E+00	3362.9	4	2	3.5E-01
7535.8	3	3	4.3E-01	3054.7	2	4	9.4E-01	3371.8	4	6	2.2E-01
7937.0	5	5	7.8E-03	3092.9	6	6	1.3E+00	3374.1	4	4	3.0E-01
8082.5	3	3	1.2E-03	3097.1	8	8	1.3E+00	3378.2	2	2	1.7E+00
8118.5	3	3	4.9E-02	3118.0	8	6	4.2E-02	3379.3	2	2	3.0E-01
8128.9	3	5	7.2E-03	3134.1	6	4	2.6E-01	3386.2	4	6	5.5E-02
8259.4	5	5	2.03E-02	3140.4	8	6	2.4E-01	3388.4	4	6	2.2E+00
8571.4	3	3	5.5E-02	3151.1	6	6	4.8E-02	3390.6	2	4	7.7E-02
8582.9	3	5	1.00E-02	3154.8	8	6	1.8E-02	3392.8	2	4	4.4E-01
8647.0	5	5	3.91E-02	3164.4	8	8	1.6E-01	3404.8	4	6	1.9E+00
8681.9	3	3	2.1E-01	3165.7	6	6	1.2E-01	3407.0	6	8	2.3E+00
8767.5	3	3	1.1E-03	3173.6	6	4	4.5E-02	3411.4	4	2	6.1E-01
8771.7	3	3	1.6E-01	3176.1	4	6	6.0E-02	3413.2	4	4	1.8E+00
8783.8	3	5	3.13E-01	3187.6	4	6	1.4E-02	3414.9	4	6	1.8E-02
8865.3	3	3	9.4E-03	3188.7	6	6	3.9E-01	3416.9	6	6	6.4E-01
9201.8	3	3	9.1E-02	3190.9	4	6	1.5E-01	3417.7	6	8	1.6E+00
9433.0	3	3	1.1E-03	3194.6	4	4	5.2E-01	3438.9	2	2	1.4E+00
9486.7	3	3	2.5E-02	3198.6	6	8	1.7E+00	3440.7	2	4	3.5E-01
9534.2	3	3	6.3E-02	3198.9	4	4	2.3E-01	3453.1	4	4	4.6E-01
10621	3	3	2.4E-03	3209.0	8	8	1.6E-01	3454.8	4	4	1.6E+00
11409	3	3	4.2E-02	3209.4	2	4	6.0E-01	3456.6	2	4	9.6E-01
11525	3	3	8.4E-02	3213.7	2	4	1.7E+00	3457.1	4	6	9.9E-02
11767	3	3	6.9E-02	3214.3	4	6	2.2E+00	3459.3	6	6	1.6E+00
12459	3	3	1.5E-02	3218.2	8	10	3.6E+00	3475.2	4	4	1.2E-02
				3224.8	6	8	3.5E-01	3477.6	4	6	4.3E-01
Ne II				3229.5	8	8	1.3E-01	3481.9	4	2	1.4E+00
*357.03	6	10	3.8E+01	3229.6	8	10	3.6E+00	3503.6	2	2	2.0E+00
*361.77	6	2	1.6E+01	3230.1	6	6	1.8E+00	3522.7	4	2	2.3E-02
*406.28	6	10	1.8E+01	3230.4	4	6	1.4E-01	3538.0	4	2	7.6E-01
*446.37	6	6	4.07E+01	3232.0	6	4	2.7E-01	3539.9	4	4	3.6E-02
460.73	4	2	4.7E+01	3232.4	4	4	1.6E+00	3542.2	6	4	6.0E-01

λ Å	Weights g_i	g_k	A 10^8 s^{-1}
3542.9	4	6	1.2E+00
3546.2	2	4	6.3E-02
3551.6	2	4	3.7E-02
3557.8	2	2	1.9E-01
3561.2	4	6	2.1E-01
3565.8	4	4	6.2E-01
3568.5	6	8	1.4E+00
3571.2	4	4	6.3E-01
3574.2	6	6	1.0E+00
3574.6	4	6	1.3E+00
3590.4	4	6	3.6E-02
3594.2	4	2	1.3E+00
3612.3	2	4	2.6E-01
3628.0	4	4	6.0E-01
3632.7	4	4	1.3E-01
3643.9	4	4	3.2E-01
3644.9	2	4	9.9E-01
3659.9	4	6	6.7E-02
3664.1	6	4	7.0E-01
3679.8	4	2	3.2E-01
3694.2	6	6	1.0E+00
3697.1	2	2	2.8E-01
3701.8	4	6	2.7E-01
3709.6	4	2	1.1E+00
3713.1	4	6	1.3E+00
3721.8	4	6	2.0E-01
3726.9	4	4	1.2E-01
3727.1	2	4	9.8E-01
3734.9	4	4	1.9E-01
3744.6	2	4	2.6E-01
3751.2	2	2	1.8E-01
3753.8	4	6	4.5E-01
3766.3	4	6	2.9E-01
3777.1	2	4	4.2E-01
3800.0	4	4	3.7E-01
3818.4	2	4	6.1E-01
3829.8	4	6	8.4E-01
3942.3	4	6	1.0E-02

Ne V

λ Å	Weights g_i	g_k	A 10^8 s^{-1}
*142.61	9	9	6.7E+02
*143.32	9	15	1.2E+03
147.13	5	7	1.5E+03
151.23	5	5	3.38E+02
154.50	1	3	7.0E+02
*167.69	9	9	1.5E+02
*358.93	9	3	2.1E+02
365.59	5	3	1.35E+02
*482.15	9	9	3.01E+01
*571.04	9	15	1.0E+01
2259.6	3	5	1.9E+00
2265.7	5	7	2.4E+00

Ne VII

λ Å	Weights g_i	g_k	A 10^8 s^{-1}
97.502	1	3	1.07E+03
*115.46	9	3	4.8E+02
116.69	3	5	1.6E+03
127.66	3	1	1.9E+02
465.22	1	3	4.09E+01
558.61	3	5	8.11E+00
559.95	1	3	1.07E+01
561.38	3	3	7.99E+00
561.73	5	5	2.39E+01
562.99	3	1	3.17E+01
564.53	5	3	1.31E+01

Ne VIII

λ Å	Weights g_i	g_k	A 10^8 s^{-1}
*88.09	2	6	8.4E+02
*98.208	6	10	2.77E+03
770.41	2	4	5.90E+00
780.32	2	2	5.69E+00
2820.7	2	4	7.20E-01
2860.1	2	2	6.88E-01

Nickel

Ni I

λ Å	Weights g_i	g_k	A 10^8 s^{-1}
1963.85	7	7	1.1E-01
1976.87	7	9	1.1E+00
1981.61	5	5	1.3E-01
1990.25	5	7	8.3E-01
2007.01	5	5	1.7E-01
2007.69	7	7	9.0E-02
2014.25	3	5	9.3E-01
2025.40	7	5	2.3E-01
2026.62	9	7	2.4E-01
2047.35	7	5	1.8E-01
2052.04	9	9	9.7E-02
2055.50	5	3	3.3E-01
2059.92	7	5	2.1E-01
2060.20	5	3	2.3E-01
2064.39	3	1	4.0E-01
2069.52	5	5	1.1E-01
2085.57	5	5	2.6E+00
2089.09	7	5	9.7E-02
2095.13	5	7	1.1E-01
2114.43	5	5	9.7E-02
2121.40	7	5	2.8E-01
2124.80	5	3	3.8E-01
2147.80	5	3	4.7E-01
2157.83	5	3	4.1E-01
2158.31	7	5	6.9E-01
2161.04	5	5	1.3E-01
2173.54	5	3	1.5E-01
2174.48	3	1	8.9E-01
2182.38	7	5	1.3E-01
2183.91	5	5	1.2E-01
2190.22	5	5	3.0E-01
2197.35	3	3	7.8E-01
2201.59	5	3	7.3E-01
2221.94	5	3	2.2E-01
2244.46	5	5	3.8E-01
2253.57	7	7	1.9E-01
2254.81	9	9	9.6E-02
2258.15	7	5	1.7E-01
2259.56	5	3	2.0E-01
2261.42	9	7	9.1E-02
2287.32	3	5	1.8E-01
2289.98	9	7	2.1E+00
2293.11	5	5	3.8E-01
2300.77	7	7	7.5E-01
2302.97	3	3	4.5E-01
2307.35	5	7	1.6E-01
2312.34	7	7	5.5E+00
2313.98	5	5	5.0E+00
2317.16	7	5	3.8E+00
2320.03	9	11	6.9E+00
2321.38	5	7	5.6E+00
2324.65	7	9	1.8E-01
2325.79	7	9	3.5E+00
2329.96	5	3	5.3E+00
2345.54	9	7	2.2E+00
2346.63	7	5	5.5E-01
2347.51	9	9	2.2E-01
2348.73	7	7	2.2E-01
2419.31	7	5	2.0E-01
2943.91	7	5	1.1E-01
2981.65	5	3	2.8E-01
3002.48	7	7	8.0E-01
3003.62	5	5	6.9E-01
3012.00	5	5	1.3E+00
3037.93	7	7	2.8E-01
3050.82	7	9	6.0E-01
3054.31	5	5	4.0E-01
3057.64	3	3	1.0E+00
3064.62	5	7	1.1E-01
3101.56	5	7	6.3E-01
3101.88	5	7	4.9E-01
3134.11	3	5	7.3E-01
3225.02	5	3	9.3E-02
3369.56	9	7	1.8E-01
3380.57	5	3	1.3E+00
3392.98	7	7	2.4E-01
3414.76	7	9	5.5E-01
3423.71	3	3	3.3E-01
3433.56	7	7	1.7E-01
3446.26	5	5	4.4E-01
3452.88	5	5	9.8E-02
3458.46	3	5	6.1E-01
3461.66	7	9	2.7E-01
3472.55	5	7	1.2E-01
3483.77	5	3	1.4E-01
3492.96	5	3	9.8E-01
3510.33	3	1	1.2E+00
3515.05	5	7	4.2E-01
3524.54	7	5	1.0E+00
3566.37	5	5	5.6E-01
3597.70	3	3	1.4E-01
3619.39	5	7	6.6E-01
4027.67	5	7	1.3E-01
4295.88	9	7	1.7E-01
4401.54	9	11	3.8E-01
4462.46	3	5	1.7E-01
4470.48	5	7	1.9E-01
4600.37	5	3	2.6E-01
4604.99	9	7	2.3E-01
4606.23	5	3	1.0E-01
4648.66	11	9	2.4E-01
4686.22	5	5	1.4E-01
4701.54	9	9	1.4E-01
4714.42	13	11	4.6E-01
4715.78	7	7	2.0E-01

λ Å	g_i	g_k	A 10^8 s^-1
4732.47	7	9	9.3E-02
4752.43	3	3	2.0E-01
4756.52	9	9	1.5E-01
4786.54	11	11	1.8E-01
4812.00	3	1	9.5E-02
4829.03	5	7	1.9E-01
4831.18	9	7	1.6E-01
4838.64	9	7	2.2E-01
4855.41	5	5	5.7E-01
4904.41	5	3	6.2E-02
4912.03	3	3	1.5E-01
4913.97	1	3	2.2E-01
4918.36	9	7	2.3E-01
4935.83	7	5	2.4E-01
4937.34	9	9	1.2E-01
4953.20	5	5	1.2E-01
4980.17	9	11	1.9E-01
5000.34	7	7	1.4E-01
5012.46	7	7	1.1E-01
5017.58	11	11	2.0E-01
5035.37	7	9	5.7E-01
5042.20	3	5	1.4E-01
5048.85	7	7	1.6E-01
5080.53	9	11	3.2E-01
5081.11	7	9	5.7E-01
5082.35	3	3	2.5E-01
5084.08	7	9	3.1E-01
5099.95	7	7	2.9E-01
5115.40	11	9	2.2E-01
5129.37	7	5	1.2E-01
5155.14	5	5	1.1E-01
5155.76	5	7	2.9E-01
5176.57	5	5	1.8E-01
5371.33	7	7	1.6E-01
5476.91	1	3	9.5E-02
5637.12	3	3	1.1E-01
5664.02	5	7	1.1E-01
5695.00	3	3	1.7E-01
6086.29	3	5	1.1E-01
6175.42	3	3	1.7E-01
7122.24	5	7	2.1E-01
7381.94	9	11	9.7E-02
7422.30	7	5	1.8E-01
7727.66	7	7	1.1E-01

Ni II

λ Å	g_i	g_k	A 10^8 s^-1
2165.55	10	10	2.4E+00
2169.10	8	8	1.58E+00
2174.67	8	10	1.43E+00
2175.15	6	6	1.77E+00
2184.61	4	4	2.90E+00
2201.41	4	6	1.3E+00
2206.72	6	8	1.66E+00
2216.48	10	12	3.4E+00
2220.40	6	8	2.3E+00
2222.96	10	10	9.8E-01
2224.86	8	8	1.55E+00
2226.33	6	6	1.3E+00
2253.85	4	6	1.98E+00
2264.46	6	8	1.43E+00
2270.21	8	10	1.56E+00
2278.77	8	6	2.8E+00
2287.09	6	4	2.8E+00
2296.55	8	8	1.98E+00
2297.14	6	4	2.70E+00
2297.49	4	2	3.0E+00
2298.27	6	6	2.8E+00
2303.00	8	6	2.9E+00
2316.04	10	8	2.88E+00
2334.58	8	8	8.0E-01
2375.42	6	8	6.6E-01
2394.52	8	10	1.70E+00
2416.13	6	8	2.1E+00
2437.89	8	10	5.4E-01
2510.87	8	10	5.8E-01

Ni III

λ Å	g_i	g_k	A 10^8 s^-1
1692.51	11	13	7.9E+00
1709.90	9	11	6.3E+00
1719.46	5	7	6.0E+00
1722.28	3	5	5.9E+00
1724.52	3	1	6.7E+00
1741.96	9	7	5.7E+00
1752.43	7	5	5.5E+00
1760.56	5	3	6.5E+00
1769.64	11	11	6.2E+00
1823.06	9	9	5.6E+00

Ni XIV

λ Å	g_i	g_k	A 10^8 s^-1
164.13	6	8	1.2E+03
168	2	4	2.4E+02
168.12	4	2	8.5E+02
169.69	4	4	9.8E+02
170.50	4	4	7.1E+02
171.37	4	6	9.4E+02
172.16	6	6	4.7E+02
172.80	6	4	1.4E+02
177.28	4	4	5.6E+02
178	2	4	8.9E+01
181	4	6	7.4E+01
182.14	4	2	1.5E+02
196	4	2	3.8E+01
288.894	4	4	4.6E+01
292.399	6	6	3.6E+01

Ni XV

λ Å	g_i	g_k	A 10^8 s^-1
50.249	5	7	6.8E+03
60.890	9	11	1.0E+04
64.635	7	9	9.6E+03
163.64	5	7	5.6E+01
173.73	5	7	7.6E+02
175	3	1	5.7E+02
179.28	5	7	7.5E+02
181	1	3	6.8E+02
269	3	1	5.3E+01
278.386	5	5	4.3E+01

Ni XVI

λ Å	g_i	g_k	A 10^8 s^-1
166	4	6	3.1E+02
168	6	8	3.2E+02
182	2	2	2.5E+02
185.23	2	4	4.2E+02
187	4	6	1.2E+02
187	4	2	3.3E+02
188	2	2	4.7E+02
190	6	8	2.0E+02
192	6	8	4.54E+02
192	6	6	3.1E+02
194	4	6	2.8E+02
194	2	4	5.5E+02
194	2	2	1.1E+02
194	4	4	3.5E+02
194.04	4	6	4.6E+02
195.27	4	4	9.5E+01
196	4	6	6.7E+02
197	4	6	1.5E+02
197	4	2	1.2E+02
199	2	4	4.9E+02
206	2	2	3.7E+02
217	4	4	1.1E+02
218.391	2	4	9.5E+01
223.119	2	2	1.3E+02
231	4	4	1.6E+02
232.475	4	4	4.07E+02
233	6	4	2.4E+02
235	4	2	3.8E+02
235	6	6	2.5E+02
236	4	4	1.2E+02
237.875	4	2	2.6E+02
238	6	4	1.3E+02
239.550	2	2	2.6E+02
245	4	4	1.4E+02
245	4	6	3.2E+02
249	6	8	3.3E+02
249	6	4	1.2E+02
250	4	2	1.6E+02
254	6	4	1.8E+02

Ni XVII

λ Å	g_i	g_k	A 10^8 s^-1
30.919	1	3	2.77E+03
42.855	1	3	4.75E+03
54.451	9	11	1.5E+04
55.361	1	3	6.7E+03
57.348	7	9	1.4E+04
197.39	1	3	1.6E+02
199.87	3	5	2.1E+02
204	3	3	1.8E+02
205	3	1	2.4E+02
206	1	3	3.0E+02
207.50	5	7	2.5E+02
215.89	3	5	4.8E+02
216	1	3	2.7E+02
217	5	7	2.4E+02
227	5	5	1.6E+02
249.180	1	3	2.75E+02
281.50	3	1	2.1E+02
282	3	1	2.4E+02
284	5	3	1.5E+02
292	5	7	2.2E+02

Ni XVIII

λ Å	g_i	g_k	A 10^8 s^-1
24.881	2	4	8.6E+02
25.070	4	6	9.9E+02

λ Å	Weights g_i	g_k	A 10^8 s^{-1}	λ Å	Weights g_i	g_k	A 10^8 s^{-1}	λ Å	Weights g_i	g_k	A 10^8 s^{-1}
26.02	2	4	1.26E+03	11.48	3	1	1.1E+05	137.01	4	4	2.6E+02
26.020	2	4	1.1E+03	11.48	1	3	4.0E+05	138.80	4	6	7.2E+01
26.046	2	2	1.1E+03	11.517	5	7	1.4E+05	153.47	2	2	1.27E+02
26.218	4	6	1.5E+03	11.539	5	7	1.2E+05	159.69	2	4	8.9E+01
27.98	4	6	1.0E+03	11.67	1	3	8.0E+04				
27.982	2	4	2.0E+03	11.72	3	3	2.3E+04	*Ni XXV*			
28.018	6	8	1.1E+03	12.454	5	3	3.3E+04	9.30	3	1	9.3E+04
28.220	4	6	2.33E+03	12.472	3	3	1.8E+04	9.31	5	7	8.2E+04
29.383	4	6	1.58E+03	12.502	5	5	2.8E+04	9.32	3	5	7.8E+04
29.422	6	8	1.69E+03					9.34	1	3	1.1E+05
29.779	2	4	1.9E+03	*Ni XXII*				9.42	3	1	9.0E+04
29.829	2	2	1.9E+03	72.52	4	2	2.84E+02	9.49	3	5	8.9E+04
31.845	4	6	2.7E+03	84.06	6	4	1.2E+03	9.60	1	3	1.8E+05
31.890	6	8	3.0E+03	84.24	4	2	5.6E+02	9.63	3	5	2.4E+05
32.034	2	4	3.4E+03	85.86	4	2	4.9E+02	9.64	3	3	1.3E+05
32.340	4	6	4.0E+03	88.00	4	2	1.2E+03	9.71	3	1	2.3E+05
36.990	4	6	5.5E+03	95.95	2	2	4.4E+02	9.71	3	3	1.8E+05
37.049	6	8	5.9E+03	98.16	4	4	5.2E+02	9.74	5	7	3.0E+05
41.015	2	4	2.97E+03	98.58	4	4	2.45E+02	9.75	3	5	1.3E+05
41.218	2	2	3.2E+03	100.60	6	6	3.9E+02	9.76	1	3	3.03E+05
43.814	2	4	5.5E+03	101.31	6	4	4.83E+02	9.76	5	3	7.5E+04
44.365	4	6	6.8E+03	103.31	4	2	2.66E+02	9.78	5	7	2.9E+05
44.405	4	4	1.14E+03	106.04	4	4	2.36E+02	9.86	5	7	4.8E+05
52.615	4	6	1.5E+04	106.16	4	2	5.1E+02	9.87	3	5	2.03E+05
52.720	6	8	1.6E+04	124.31	2	2	3.7E+02	9.92	5	5	1.3E+05
52.745	6	6	1.06E+03	126.32	4	4	3.3E+02	9.94	5	7	1.29E+05
59.950	6	4	9.6E+02					9.97	3	5	2.5E+05
60.212	4	2	1.1E+03	*Ni XXIII*				10.08	1	3	2.80E+05
63.512	4	6	7.9E+02	87.66	3	3	2.8E+02				
63.589	6	8	8.5E+02	88.11	5	3	8.3E+02	*Ni XXVI*			
69.075	4	6	8.0E+02	90.49	3	3	1.77E+02	1.5930	4	2	3.4E+06
76.254	4	6	1.38E+03	90.96	5	3	2.5E+02	1.5935	2	2	4.0E+06
76.359	6	8	1.47E+03	91.83	5	3	7.5E+02	1.5973	4	4	8.1E+06
99.275	2	4	1.0E+03	92.32	3	1	4.39E+02	1.5977	2	4	4.4E+06
100.4	4	6	1.2E+03	100.42	1	3	2.1E+02	1.5982	2	2	7.3E+06
114.46	4	6	2.5E+03	102.08	5	5	5.3E+02	1.5996	2	2	2.7E+06
114.74	6	8	2.7E+03	103.23	3	3	2.4E+02	1.6005	4	6	2.7E+06
				103.67	5	5	1.78E+02	1.6036	4	2	2.1E+06
Ni XIX				104.70	3	1	2.94E+02	9.390	2	4	2.59E+05
9.140	1	3	3.1E+04	106.02	5	5	2.87E+02	9.535	4	6	2.96E+05
9.153	1	3	5.2E+03	108.27	7	5	3.32E+02				
9.977	1	3	1.1E+05	111.23	3	1	2.26E+02	*Ni XXVII*			
10.110	1	3	9.4E+04	111.78	5	3	2.19E+02	1.2534	1	3	3.35E+05
10.283	1	3	4.7E+03	111.86	1	3	1.7E+02	1.2824	1	3	6.38E+05
10.433	1	3	5.1E+03	112.55	3	1	1.0E+03	1.3500	1	3	1.63E+06
11.539	1	3	4.8E+04	128.87	5	5	4.02E+02	1.3516	1	3	2.4E+05
11.599	1	3	6.3E+03	133.54	3	3	1.86E+02	1.531	3	3	2.0E+05
12.435	1	3	3.66E+05	137.55	3	1	2.53E+02	1.534	3	1	6.9E+06
12.656	1	3	1.0E+05					1.537	5	5	2.3E+06
13.779	1	3	1.23E+04	*Ni XXIV*				1.537	1	3	3.7E+06
14.043	1	3	1.31E+04	101.13	6	4	1.63E+02	1.538	3	5	3.9E+06
40.7	3	3	6.4E+03	102.11	4	4	5.4E+02	1.539	1	3	2.6E+06
40.7	3	1	8.4E+03	103.43	2	4	1.3E+02	1.539	3	5	2.6E+06
41.132	7	9	9.4E+03	103.53	4	2	4.17E+02	1.540	3	3	1.7E+06
				104.64	2	2	4.7E+02	1.541	3	5	5.5E+06
Ni XXI				106.68	4	2	3.67E+02	1.542	3	3	3.6E+06
11.13	3	3	1.7E+04	113.14	4	4	1.65E+02	1.542	5	5	3.5E+06
11.23	5	3	1.7E+04	118.52	2	4	1.5E+02	1.544	5	3	3.2E+06
11.239	5	7	5.7E+04	122.72	6	4	2.17E+02	1.546	3	5	1.6E+06
11.28	3	1	2.2E+05	134.73	6	6	1.44E+02	1.547	3	3	2.1E+05
11.318	5	7	2.8E+05	135.47	4	4	8.0E+01	1.549	1	3	2.0E+05

λ Å	Weights g_i	Weights g_k	A 10^8 s^{-1}	λ Å	Weights g_i	Weights g_k	A 10^8 s^{-1}	λ Å	Weights g_i	Weights g_k	A 10^8 s^{-1}
1.551	3	1	8.2E+05	6926.67	4	6	7.75E-03	9968.51	6	4	4.50E-03
1.558	3	1	6.5E+05	6945.18	6	6	1.83E-02	9980.42	4	6	8.10E-03
1.5883	1	3	6.02E+06	6951.60	2	4	1.03E-02	9997.73	8	8	9.20E-03
1.5963	1	3	7.70E+05	6960.50	4	4	4.67E-03				
				6973.07	2	2	3.83E-03	*N II*			
Nitrogen				6979.18	6	4	9.83E-03	474.891	5	5	9.66E+00
				6982.03	4	2	2.04E-02	475.647	1	3	1.17E+01
N I				7423.64	2	4	5.95E-02	475.698	3	5	1.58E+01
1163.88	6	6	7.52E-01	7442.30	4	4	1.24E-01	475.757	3	3	8.75E+00
1164.00	4	6	1.27E-02	7468.31	6	4	1.93E-01	475.803	5	7	2.10E+01
1164.21	6	4	5.17E-02	7898.98	6	4	2.82E-01	475.884	5	5	5.25E+00
1164.32	4	4	6.94E-01	7899.28	4	4	3.28E-02	508.697	5	5	1.91E+00
1167.45	6	8	1.29E+00	7915.42	4	2	3.13E-01	510.758	5	7	1.87E+01
1168.42	6	6	4.24E-02	8184.86	4	6	8.58E-02	513.849	5	5	1.24E+01
1168.54	4	6	1.24E+00	8188.01	2	4	1.27E-01	529.355	1	3	7.23E+00
1176.51	6	4	9.22E-01	8200.36	2	2	4.95E-02	529.413	3	1	2.43E+01
1176.63	4	4	1.02E-01	8210.72	4	4	4.84E-02	529.491	3	3	6.75E+00
1177.69	4	2	1.02E+00	8216.34	6	6	2.23E-01	529.637	3	5	4.92E+00
1199.55	4	6	4.01E+00	8223.13	4	2	2.64E-01	529.722	5	3	1.03E+01
1200.22	4	4	3.99E+00	8242.39	6	4	1.36E-01	529.867	5	5	1.94E+01
1200.71	4	2	3.98E+00	8567.74	2	4	4.92E-02	533.511	1	3	2.39E+01
1310.54	4	6	8.42E-01	8594.00	2	2	2.09E-01	533.581	3	5	3.20E+01
1316.29	4	6	1.42E-02	8629.24	4	4	2.66E-01	533.650	3	3	1.66E+01
1492.63	6	4	3.13E+00	8655.88	4	2	1.05E-01	533.729	5	7	4.13E+01
1492.82	4	4	3.51E-01	8680.28	6	8	2.46E-01	533.815	5	5	9.19E+00
1494.68	4	2	3.72E+00	8683.40	4	6	1.80E-01	547.818	5	3	2.16E+00
3822.03	2	2	3.70E-02	8686.15	2	4	1.09E-01	559.762	1	3	1.14E+01
3830.43	4	4	4.67E-02	8703.25	2	2	2.10E-01	574.650	5	7	3.60E+01
3834.22	4	2	1.89E-02	8711.70	4	4	1.28E-01	582.156	5	5	2.85E+01
4099.94	2	4	3.48E-02	8718.84	6	6	6.75E-02	635.197	1	3	2.33E+01
4109.95	4	6	3.90E-02	8728.90	4	2	3.76E-02	644.634	1	3	1.21E+01
4113.97	4	4	6.62E-03	8747.37	6	4	1.04E-02	644.837	3	3	3.64E+01
4137.64	2	4	2.80E-03	9028.92	2	2	3.02E-01	645.178	5	3	6.07E+01
4143.43	4	4	6.09E-03	9045.88	6	8	2.80E-01	660.286	5	3	3.69E+01
4151.48	6	4	1.01E-02	9049.49	6	6	1.88E-02	671.016	3	5	2.47E+00
4249.87	4	2	2.59E-02	9049.89	4	6	2.60E-01	671.386	5	5	7.40E+00
4264.00	6	4	2.26E-02	9060.48	2	4	2.95E-01	671.411	1	3	3.04E+00
4356.29	6	8	5.10E-02	9187.45	6	6	2.44E-01	671.630	3	3	2.27E+00
4385.54	2	2	8.84E-03	9187.86	4	4	1.76E-02	671.773	3	1	9.85E+00
4392.41	4	2	1.76E-02	9207.59	6	4	2.70E-02	672.001	5	3	3.87E+00
4435.43	2	4	7.51E-03	9208.00	4	4	2.33E-01	745.841	1	3	1.25E+01
4442.45	4	4	3.81E-02	9386.81	2	4	2.24E-01	746.984	5	3	3.85E+01
4669.89	4	4	7.49E-03	9392.79	4	6	2.63E-01	748.369	5	3	3.83E+00
4914.94	2	2	8.08E-03	9460.68	4	4	3.98E-02	775.965	5	5	3.08E+01
4935.12	4	2	1.76E-02	9776.90	2	4	1.18E-02	915.612	1	3	4.38E+00
5199.84	2	2	1.87E-02	9786.78	4	6	1.13E-02	915.962	3	1	1.32E+01
5201.61	2	4	1.87E-02	9788.29	2	2	2.99E-02	1083.99	1	3	2.18E+00
5281.20	6	6	2.45E-03	9798.56	4	4	2.75E-02	1085.55	5	5	9.47E-01
5344.05	6	6	6.10E-04	9810.01	4	2	5.30E-02	1085.70	5	7	3.87E+00
5356.62	4	6	1.41E-03	9814.02	6	8	6.56E-03	3408.13	3	1	2.19E-01
5367.01	4	4	1.07E-03	9822.75	6	6	4.95E-02	3437.14	3	1	2.07E+00
5372.61	2	4	8.34E-04	9834.61	6	4	4.50E-02	3593.60	3	5	1.21E-01
5378.27	2	2	1.66E-03	9863.33	8	8	9.62E-02	3609.10	3	3	1.41E-01
6606.18	4	6	8.87E-04	9872.15	8	6	2.97E-02	3615.86	3	1	1.53E-01
6622.54	6	6	7.93E-03	9883.38	2	2	2.93E-02	3829.80	3	5	2.42E-01
6626.99	2	4	2.20E-03	9905.52	4	2	3.11E-03	3838.37	5	5	6.98E-01
6636.94	4	4	1.40E-02	9909.22	2	4	7.58E-03	3842.19	1	3	3.06E-01
6644.96	8	6	3.49E-02	9931.47	4	4	3.64E-02	3847.40	3	3	2.22E-01
6646.50	2	2	2.18E-02	9947.07	6	8	1.08E-02	3855.10	3	1	8.82E-01
6653.46	6	4	2.74E-02	9965.75	4	6	7.60E-03	3856.06	5	3	3.71E-01
6656.51	4	2	2.17E-02								

λ Å	Weights g_i	g_k	A 10^8 s^{-1}	λ Å	Weights g_i	g_k	A 10^8 s^{-1}	λ Å	Weights g_i	g_k	A 10^8 s^{-1}
3919.00	3	3	6.76E-01	5025.66	7	7	1.07E-01	5931.78	3	5	4.27E-01
3955.85	3	5	1.31E-01	5040.71	7	5	3.78E-03	5940.24	3	3	2.26E-01
3995.00	3	5	1.35E+00	5045.10	5	3	3.42E-01	5941.65	5	7	5.54E-01
4114.33	3	3	1.42E-03	5073.59	3	3	2.59E-02	5952.39	5	5	1.27E-01
4124.08	3	5	3.20E-01	5168.05	3	5	3.06E-01	5960.91	5	5	1.34E-02
4133.67	5	5	5.30E-01	5170.16	3	3	6.54E-01	6065.00	3	5	2.21E-03
4145.77	7	5	7.36E-01	5171.27	3	1	8.71E-01	6284.32	5	3	7.74E-02
4374.99	3	5	5.55E-03	5171.47	5	7	5.81E-01	6379.62	3	3	6.11E-02
4447.03	3	5	1.14E+00	5172.34	3	5	6.01E-01	6482.05	3	3	3.01E-01
4459.94	3	1	1.12E-01	5172.97	1	3	5.01E-01	6610.56	5	7	6.34E-01
4465.53	3	3	2.36E-02	5173.39	5	7	7.36E-01	6857.03	5	5	2.53E-01
4477.68	5	3	8.85E-02	5174.46	5	5	5.07E-01	6869.58	5	5	2.51E-01
4488.09	5	5	1.30E-02	5175.89	7	9	8.93E-01	6887.83	5	7	2.49E-01
4507.56	7	5	1.00E-01	5176.57	5	3	2.17E-01	7762.24	5	5	8.74E-02
4564.76	3	5	1.41E-02	5177.06	3	3	5.00E-01	8438.74	1	3	2.24E-01
4601.48	3	5	2.35E-01	5179.34	7	9	8.67E-01	8831.75	1	3	8.42E-03
4607.15	1	3	3.26E-01	5179.52	9	11	1.07E+00	8855.30	3	3	2.51E-02
4613.87	3	3	2.26E-01	5180.36	5	5	4.28E-01	8893.29	5	3	4.12E-02
4621.39	3	1	9.55E-01	5183.20	7	7	2.88E-01				
4630.54	5	5	7.72E-01	5184.96	7	7	3.20E-01	*N III*			
4643.09	5	3	4.51E-01	5185.09	5	3	7.11E-02	374.198	2	4	9.89E+01
4654.53	3	5	2.43E-02	5186.21	7	5	5.76E-02	451.871	2	2	1.03E+01
4667.21	3	3	2.99E-02	5190.38	9	9	1.77E-01	452.227	4	2	2.05E+01
4674.91	3	1	1.05E-01	5191.96	7	5	4.25E-02	684.998	2	4	9.63E+00
4694.27	1	3	1.23E-01	5199.50	9	7	1.51E-02	685.515	2	2	3.83E+01
4695.90	3	5	1.29E-01	5313.42	3	3	1.41E-01	685.817	4	4	4.54E+01
4697.64	3	3	3.06E-02	5320.20	5	3	4.20E-01	686.336	4	2	1.95E+01
4698.55	3	1	3.67E-01	5320.96	3	5	2.52E-01	763.334	2	2	9.58E+00
4700.03	5	7	1.05E-01	5327.76	5	5	4.65E-02	764.351	4	2	1.85E+01
4702.50	5	5	9.15E-02	5338.73	5	7	1.85E-01	771.545	2	4	8.19E+00
4704.25	5	3	2.13E-01	5340.21	7	5	2.59E-01	771.901	4	4	1.64E+01
4706.40	7	9	6.09E-02	5351.23	7	7	3.67E-01	772.384	6	4	2.45E+01
4709.58	7	7	1.82E-01	5383.72	3	5	3.31E-03	772.889	6	4	2.09E+01
4712.07	7	5	1.46E-01	5452.07	1	3	8.89E-02	772.955	4	2	2.34E+01
4718.38	9	9	3.02E-01	5454.22	3	1	3.34E-01	979.832	4	4	8.84E+00
4721.58	9	7	7.75E-02	5462.58	3	3	1.00E-01	979.905	6	6	9.21E+00
4774.24	3	5	3.24E-02	5478.09	3	5	4.75E-02	989.799	2	4	4.18E+00
4779.72	3	3	2.52E-01	5480.05	5	3	1.30E-01	991.511	4	4	8.17E-01
4781.19	5	7	2.05E-02	5495.65	5	5	2.40E-01	991.577	4	6	4.97E+00
4788.14	5	5	2.52E-01	5526.23	3	5	2.13E-01	1747.85	2	4	1.28E+00
4793.65	5	3	7.77E-02	5530.24	5	7	4.04E-01	1751.22	4	4	2.48E-01
4803.29	7	7	3.18E-01	5535.35	7	9	6.04E-01	1751.66	4	6	1.51E+00
4810.30	7	5	4.75E-02	5535.38	3	3	4.53E-01	2972.55	2	2	6.67E-01
4860.17	3	5	1.61E-02	5540.06	3	1	6.03E-01	2977.33	4	2	3.32E-01
4987.38	3	1	7.48E-01	5543.47	5	5	3.51E-01	2978.84	2	4	1.66E-01
4991.24	3	5	3.54E-01	5551.92	7	7	2.00E-01	2983.64	4	4	8.24E-01
4994.36	5	7	2.62E-01	5552.68	5	3	1.50E-01	3342.76	2	2	3.80E-01
4994.37	3	3	7.60E-01	5565.26	7	5	3.97E-02	3353.98	2	4	7.66E-01
4997.22	3	3	1.96E-01	5666.63	3	5	3.74E-01	3354.32	4	6	5.51E-01
5001.13	3	5	9.76E-01	5676.02	1	3	2.96E-01	3355.46	4	2	7.51E-01
5001.47	5	7	1.05E+00	5679.56	5	7	5.25E-01	3358.78	2	2	3.05E-01
5002.70	1	3	8.45E-02	5686.21	3	3	1.94E-01	3360.98	4	4	2.44E-01
5005.15	7	9	1.16E+00	5710.77	5	5	1.24E-01	3365.80	4	2	1.52E+00
5005.30	5	5	6.51E-02	5730.66	5	3	1.34E-02	3367.36	6	6	1.27E+00
5007.33	3	5	7.89E-01	5747.30	3	5	3.40E-02	3374.07	6	4	8.13E-01
5010.62	3	3	2.19E-01	5767.45	3	3	2.44E-02	3745.95	2	4	1.90E-01
5011.31	5	3	5.84E-01	5893.15	5	7	2.88E-01	3752.63	2	2	6.67E-02
5012.04	7	7	5.19E-01	5897.25	3	5	2.16E-01	3754.69	4	4	3.78E-01
5016.38	5	5	1.62E-01	5899.83	1	3	1.60E-01	3762.60	4	4	4.24E-02
5023.05	7	5	3.61E-01	5927.81	1	3	3.22E-01	3771.03	6	4	5.59E-01

λ Å	Weights g_i	g_k	A 10^8 s^{-1}	λ Å	Weights g_i	g_k	A 10^8 s^{-1}	λ Å	Weights g_i	g_k	A 10^8 s^{-1}
3771.36	6	4	8.28E-02	7371.51	4	4	3.53E-02	7103.24	1	3	6.28E-02
3792.97	8	6	1.03E-01	7404.54	6	6	3.61E-02	7109.35	3	5	8.46E-02
3934.50	2	4	7.49E-01	8307.51	2	4	1.65E-02	7111.28	3	3	4.70E-02
3938.51	4	6	8.96E-01	8344.95	2	2	6.52E-02	*7116.8	9	15	1.12E-01
3942.88	4	4	1.49E-01	8386.39	4	4	8.03E-02	7122.98	5	7	1.12E-01
4097.36	2	4	8.70E-01	8424.56	4	2	3.17E-02	7127.25	5	5	2.80E-02
4103.39	2	2	8.67E-01					7127.25	5	3	3.11E-03
4195.74	2	4	9.37E-01	*N IV*				9165.07	3	5	4.23E-02
4200.07	4	6	1.12E+00	247.205	1	3	1.19E+02	9182.16	5	7	4.45E-02
4215.77	4	4	1.85E-01	*283.52	9	15	3.05E+02	9222.99	7	9	4.95E-02
4318.78	2	4	5.40E-02	*322.64	9	3	8.99E+01	9247.04	5	5	7.66E-03
4321.22	2	2	1.08E-01	335.047	3	5	1.845E+02	9311.55	7	7	5.36E-03
4321.39	4	6	5.03E-02	387.356	3	1	2.55E+01				
4325.43	4	4	8.60E-02	765.147	1	3	2.320E+01	*N V*			
4327.69	6	8	3.06E-02	*923.16	9	9	1.759E+01	*209.29	2	6	1.21E+02
4327.88	4	2	1.07E-01	955.334	3	1	2.919E+01	*247.66	6	10	4.26E+02
4332.95	6	6	1.23E-01	1718.55	3	5	2.321E+00	1238.82	2	4	3.40E+00
4337.01	6	4	7.47E-02	2649.88	3	3	1.07E+00	1242.80	2	2	3.37E+00
4345.81	8	8	1.82E-01	3052.20	1	3	1.33E-01	4603.74	2	4	4.14E-01
4351.11	8	6	4.01E-02	3059.60	3	3	3.95E-01	4619.97	2	2	4.10E-01
4510.88	2	4	2.84E-01	3075.19	5	3	6.48E-01				
4510.96	4	6	4.77E-01	3443.61	3	5	3.46E-01	*N VI*			
4514.85	6	8	6.80E-01	3445.22	1	3	4.60E-01	24.8980	1	3	5.158E+03
4518.14	2	2	5.65E-01	3454.65	3	3	3.42E-01	28.7870	1	3	1.809E+04
4523.56	4	4	3.61E-01	3461.36	3	1	1.36E+00	*161.220	3	9	2.859E+02
4530.86	4	2	1.12E-01	3463.36	5	5	1.02E+00	173.275	1	3	2.697E+02
4534.58	6	6	2.01E-01	3474.53	5	3	5.61E-01	*173.93	9	15	8.756E+02
4547.30	6	4	3.33E-02	3478.72	3	5	1.06E+00	185.192	3	5	8.205E+02
4634.13	2	4	6.36E-01	*3480.8	3	9	1.06E+00	*1901	3	9	6.780E-01
4640.64	4	6	7.60E-01	3483.00	3	3	1.06E+00	2896.4	1	3	2.079E-01
4641.85	4	4	1.26E-01	3484.93	3	1	1.06E+00	*6991.1	3	9	8.384E-02
4858.70	2	4	4.35E-01	3689.94	3	1	9.10E-02	9622.0	1	3	3.276E-02
4858.98	4	6	4.66E-01	3694.14	3	3	2.27E-02				
4861.27	6	8	5.32E-01	3707.39	5	3	6.73E-02	*Oxygen*			
4867.12	4	4	1.73E-01	3714.43	5	5	1.34E-02				
4867.17	8	10	6.18E-01	3735.43	7	5	7.37E-02	*O I*			
4873.60	6	6	1.50E-01	3747.54	3	5	9.92E-01	791.973	5	5	4.94E+00
4881.78	6	4	1.22E-02	4057.76	3	5	6.62E-01	792.938	1	3	2.19E+00
4884.14	8	8	8.71E-02	4740.26	3	5	1.53E-02	792.967	3	5	1.64E+00
4896.58	8	6	5.86E-03	4747.96	3	3	7.60E-02	877.798	5	3	2.85E+00
5260.86	2	2	2.80E-02	4752.49	5	7	1.13E-02	877.879	5	5	5.12E+00
5270.57	2	4	6.95E-02	4762.09	5	5	6.99E-02	922.008	5	7	1.23E+00
5272.68	4	2	1.39E-01	4769.86	5	3	2.50E-02	935.193	5	5	1.33E+00
5282.43	4	4	2.21E-02	4786.92	7	7	8.79E-02	1028.16	1	3	4.22E-01
5297.75	4	6	4.93E-02	4796.66	7	5	1.53E-02	1152.15	5	5	5.28E+00
5298.95	6	4	7.38E-02	5200.41	3	5	2.67E-01	1217.65	1	3	2.06E+00
5314.36	6	6	1.14E-01	5204.28	5	7	3.55E-01	1302.17	5	3	3.41E+00
5320.87	6	8	5.68E-01	5205.15	1	3	1.97E-01	1304.86	3	3	2.03E+00
5327.19	4	6	5.29E-01	5226.70	3	3	1.46E-01	1306.03	1	3	6.76E-01
5352.46	6	6	3.72E-02	5245.60	5	5	8.66E-02	3823.41	7	7	6.63E-03
6365.84	2	2	2.18E-01	5272.35	5	3	9.48E-03	3823.87	5	3	1.87E-03
6394.75	2	4	2.15E-01	5288.25	5	3	3.22E-02	3824.35	5	5	5.19E-03
6445.34	2	4	8.89E-02	5736.93	3	5	1.84E-01	3825.02	3	3	5.59E-03
6450.79	2	2	1.77E-01	5776.31	1	3	1.85E-02	3825.19	5	7	8.31E-04
6454.08	4	6	1.49E-01	5784.76	3	1	5.51E-02	3855.01	5	5	1.63E-02
6463.09	4	4	1.13E-01	5795.09	3	3	1.37E-02	3947.29	5	7	4.91E-03
6467.02	6	8	2.11E-01	5812.31	3	5	1.36E-02	3947.48	5	5	4.88E-03
6468.57	4	2	3.52E-02	5826.43	5	3	2.25E-02	3947.59	5	3	4.87E-03
6478.76	6	6	6.31E-02	5843.84	5	5	4.01E-02	3951.93	3	1	3.10E-03
6487.84	6	4	1.05E-02	6380.75	1	3	1.42E-01	3952.98	5	3	1.29E-03
								3953.00	1	3	1.03E-03
								3954.52	3	5	7.73E-04

λ (Å)	g_i	g_k	A (10^8 s^{-1})	λ (Å)	g_i	g_k	A (10^8 s^{-1})	λ (Å)	g_i	g_k	A (10^8 s^{-1})
3954.61	5	5	2.32E-03	9260.85	3	3	3.34E-01	3843.58	4	6	3.55E-02
3997.95	5	3	2.41E-02	9260.94	3	5	1.56E-01	3847.89	2	2	1.95E-01
4217.09	3	1	5.44E-03	9262.58	5	3	1.11E-01	3850.80	4	6	6.00E-03
4222.77	5	3	2.26E-03	9262.67	5	5	2.60E-01	3851.03	4	4	1.59E-01
4222.82	1	3	1.81E-03	9262.78	5	7	2.97E-01	3851.47	8	8	2.72E-02
4233.27	5	5	4.04E-03	9265.83	7	5	2.97E-02	3856.13	4	2	2.28E-01
4368.19	3	1	7.56E-03	9265.93	7	7	1.48E-01	3857.16	6	6	6.59E-02
4368.24	3	5	7.59E-03	9266.01	7	9	4.45E-01	3863.50	6	8	6.49E-02
4967.38	3	5	4.43E-03	9482.89	5	3	2.34E-01	3864.13	2	2	9.12E-02
4967.88	5	7	8.44E-03	9622.11	5	3	5.22E-04	3864.43	6	6	2.15E-01
4968.79	7	9	1.27E-02	9622.16	3	3	1.57E-03	3864.67	6	4	1.80E-01
5019.29	5	5	7.13E-03	9625.26	7	5	3.25E-04	3874.09	2	4	3.26E-02
5020.22	7	5	9.98E-03	9625.30	7	7	1.85E-03	3875.80	8	6	3.38E-02
5329.11	3	5	9.48E-03	9694.66	5	7	4.54E-04	3882.19	8	8	5.50E-01
5329.69	5	7	1.81E-02	9694.91	5	5	4.54E-04	3882.45	4	4	8.94E-02
5330.74	7	9	2.71E-02	9695.06	5	3	4.54E-04	3883.14	8	6	1.13E-01
5435.18	3	5	7.74E-03					3893.52	4	6	1.89E-02
5435.77	5	5	1.29E-02	*O II*				3907.45	6	6	8.64E-02
5436.86	7	5	1.80E-02	429.918	4	2	4.25E+01	3911.96	6	4	1.09E+00
5512.60	3	5	2.69E-03	430.041	4	4	4.13E+01	3912.12	4	4	1.41E-01
5512.77	5	7	3.58E-03	430.176	4	6	4.36E+01	3919.27	4	2	1.22E+00
5554.83	3	3	5.83E-03	483.760	4	2	2.05E+01	3945.04	2	4	2.05E-01
5555.00	5	5	9.71E-03	483.980	6	4	1.80E+01	3954.36	2	2	8.57E-01
5958.39	3	5	6.80E-03	484.027	4	4	3.22E+00	3973.26	4	4	1.04E+00
5958.58	5	7	9.06E-03	485.087	6	8	2.60E+01	3982.71	4	2	4.27E-01
6046.23	3	3	1.05E-02	485.470	6	6	1.20E+00	4069.62	2	4	1.52E+00
6046.44	5	3	1.75E-02	485.518	4	6	1.93E+01	4069.88	4	6	1.53E+00
6046.49	1	3	3.50E-03	2290.85	2	4	7.41E-02	4072.15	6	8	1.98E+00
6155.99	3	5	2.67E-02	2293.30	2	2	3.25E-01	4075.86	8	10	2.11E+00
6156.78	5	7	5.08E-02	2300.33	4	4	4.17E-01	4078.84	4	4	5.52E-01
6158.19	7	9	7.62E-02	2302.81	4	2	1.67E-01	4084.65	6	6	7.28E-02
6324.84	7	5	3.76E-05	2365.14	4	2	1.52E-01	4085.11	6	6	4.55E-01
6453.60	3	5	1.65E-02	2375.72	6	4	1.35E-01	4092.93	8	8	2.65E-01
6454.44	5	5	2.75E-02	2406.38	6	4	1.85E-01	4094.14	6	4	4.70E-02
6455.98	7	5	3.85E-02	2407.48	4	4	2.25E-01	4096.53	4	6	1.73E-01
6726.28	5	5	1.18E-05	2411.60	4	2	2.05E-01	4097.22	2	4	3.62E-01
6726.54	5	3	6.44E-06	2411.64	2	2	1.10E-01	4103.00	2	2	5.09E-01
7001.92	3	5	2.65E-02	2415.13	4	2	2.20E-01	4104.72	4	6	3.14E-01
7002.23	5	7	3.53E-02	2418.46	6	6	2.30E-01	4104.99	4	4	9.14E-01
7254.15	3	3	2.24E-02	2425.57	6	6	1.77E-01	4106.02	8	6	1.70E-02
7254.45	5	3	3.73E-02	2433.54	2	4	4.21E-01	4109.84	6	6	1.21E-02
7254.53	1	3	7.45E-03	2436.06	4	4	1.69E-01	4110.19	6	4	2.54E-01
7771.94	5	7	3.69E-01	2444.25	4	4	7.56E-02	4110.79	4	2	7.70E-01
7774.17	5	5	3.69E-01	2445.53	4	6	4.98E-01	4112.02	6	6	1.81E-01
7775.39	5	3	3.69E-01	2517.96	4	6	7.72E-02	4113.83	8	6	2.41E-01
7981.94	3	3	2.33E-04	2523.21	2	2	9.63E-02	4119.22	6	8	1.33E+00
7982.40	1	3	3.09E-04	2526.87	4	4	1.20E-01	4120.28	6	6	2.15E-01
7986.98	3	5	4.19E-04	2530.28	6	8	8.16E-02	4120.55	6	4	2.60E-01
7987.33	5	5	1.41E-04	2571.46	2	4	1.15E-01	4121.46	2	2	5.60E-01
7995.07	5	7	5.63E-04	2575.28	4	6	1.37E-01	4129.32	4	2	1.79E-01
8221.82	7	7	2.89E-01	3134.73	8	6	1.23E+00	4132.80	2	4	9.13E-01
8227.65	5	3	8.13E-02	3273.43	8	6	9.99E-01	4140.70	4	4	4.09E-02
8230.00	5	5	2.26E-01	3377.15	2	2	1.27E+00	4153.30	4	6	7.91E-01
8233.00	3	3	2.43E-01	3390.21	2	4	1.22E+00	4156.53	6	4	2.11E-01
8235.35	3	5	4.86E-02	3407.28	6	6	1.02E+00	4169.22	6	6	2.71E-01
8446.25	3	1	3.22E-01	3712.74	2	4	2.84E-01	4185.44	6	8	1.91E+00
8446.36	3	5	3.22E-01	3727.32	4	4	5.81E-01	4189.58	8	8	7.06E-02
8446.76	3	3	3.22E-01	3749.48	6	4	9.31E-01	4189.79	8	10	1.98E+00
8820.42	5	7	2.93E-01	3833.07	6	8	1.02E-02	4192.51	6	4	3.21E-01
9260.81	3	1	4.46E-01	3842.81	2	4	7.45E-02	4196.27	4	4	3.56E-02

λ (Å)	g_i	g_k	A (10^8 s^{-1})	λ (Å)	g_i	g_k	A (10^8 s^{-1})	λ (Å)	g_i	g_k	A (10^8 s^{-1})
4196.70	4	2	3.56E-01	4864.88	4	2	8.07E-02	597.814	1	3	1.49E+01
4317.14	2	4	3.70E-01	4871.52	4	6	5.60E-01	599.590	5	5	5.41E+01
4319.63	4	6	2.55E-01	4872.02	4	4	9.34E-02	702.337	1	3	6.06E+00
4319.87	2	2	5.62E-01	4890.86	4	2	4.80E-01	702.838	3	1	1.83E+01
4325.76	2	2	1.47E-01	4906.83	4	4	4.54E-01	832.929	1	3	3.41E+00
4327.46	6	6	6.76E-01	4924.53	4	6	5.43E-01	835.092	5	5	1.44E+00
4327.85	6	4	7.24E-02	4941.07	2	4	5.87E-01	835.289	5	7	5.99E+00
4328.59	4	2	1.12E+00	4943.01	4	6	7.78E-01	1679.03	3	5	6.57E-01
4331.47	4	6	4.82E-02	4955.71	4	4	1.82E-01	1686.73	3	3	6.48E-01
4331.86	4	4	6.50E-01	5159.94	2	2	3.29E-01	1760.41	3	5	8.38E-01
4336.86	4	4	1.57E-01	5175.90	4	2	1.49E-01	1764.46	5	5	2.50E+00
4345.56	4	2	8.31E-01	5190.50	2	4	1.26E-01	1766.63	1	3	1.11E+00
4347.22	6	4	1.19E-01	5206.65	4	4	3.58E-01	1772.28	3	1	3.29E+00
4347.41	4	4	9.32E-01	5583.22	2	4	2.17E-02	1772.97	5	3	1.37E+00
4349.43	6	6	6.91E-01	5611.07	2	2	2.14E-02	2390.43	3	3	1.62E+00
4351.26	6	6	9.89E-01	6627.37	4	4	1.73E-01	2454.97	3	1	3.43E+00
4351.46	4	6	5.82E-02	6641.03	2	2	9.88E-02	2665.68	3	5	6.75E-01
4359.40	4	6	1.44E-02	6666.66	4	2	6.78E-02	2674.58	5	5	1.11E+00
4366.89	6	4	3.98E-01	6677.87	2	4	3.37E-02	2683.66	3	1	1.85E+00
4369.27	4	4	3.57E-01	6717.75	2	2	1.33E-01	2686.15	7	5	1.54E+00
4395.93	6	6	3.91E-01	6721.39	4	2	1.81E-01	2687.55	3	3	1.84E+00
4405.98	6	4	4.30E-02	6810.48	6	8	1.64E-03	2695.48	3	5	1.82E+00
4414.90	4	6	8.34E-01	6844.10	4	6	2.97E-03	2794.14	3	1	1.82E-01
4416.97	2	4	7.13E-01	6846.80	8	8	3.17E-02	2798.93	3	3	4.52E-02
4443.01	6	6	5.05E-01	6869.48	6	6	5.35E-02	2809.66	5	3	1.34E-01
4443.52	6	8	1.89E-02	6884.88	4	4	6.12E-02	2818.70	5	5	2.66E-02
4447.68	8	6	2.52E-02	6895.10	10	8	2.72E-01	2836.31	7	5	1.46E-01
4448.19	8	8	5.10E-01	6906.44	8	6	2.48E-01	2959.69	3	5	1.83E+00
4452.38	4	4	1.37E-01	6907.87	4	2	3.03E-01	2983.78	3	5	2.15E+00
4466.24	2	4	9.00E-01	6910.56	6	4	2.43E-01	2992.08	3	5	9.32E-02
4467.46	2	2	9.00E-01					2996.48	3	3	4.64E-01
4563.18	4	4	7.18E-03	*O III*				2997.69	5	7	6.88E-02
4590.97	6	8	8.85E-01	263.694	1	3	3.32E+01	3004.34	5	5	4.27E-01
4595.96	6	6	4.87E-02	263.727	3	5	4.48E+01	3008.78	5	3	1.53E-01
4596.18	4	6	8.34E-01	263.773	3	3	2.49E+01	3017.62	7	7	5.38E-01
4638.86	2	4	3.71E-01	263.817	5	7	5.97E+01	3023.43	3	5	4.79E-01
4641.81	4	6	5.96E-01	263.861	5	5	1.49E+01	3024.36	7	5	9.39E-02
4649.13	6	8	7.81E-01	277.386	5	7	9.43E+01	3024.54	1	3	6.16E-01
4650.84	2	2	6.86E-01	279.788	5	5	4.25E+01	3035.41	3	3	4.59E-01
4661.63	4	4	4.10E-01	295.942	1	3	5.56E+01	3042.07	3	1	1.94E+00
4673.73	4	2	1.35E-01	303.413	1	3	4.29E+01	3047.10	5	5	1.49E+00
4676.23	6	6	2.05E-01	303.461	3	1	1.29E+02	3059.28	5	3	8.72E-01
4690.89	2	4	1.86E-01	303.517	3	3	3.21E+01	3064.98	1	3	2.17E-01
4691.42	2	2	7.43E-01	303.622	3	5	3.21E+01	3068.13	3	1	6.49E-01
4696.35	6	4	3.25E-02	303.695	5	3	5.34E+01	3068.26	3	3	5.41E-02
4698.44	6	6	6.59E-02	303.800	5	5	9.61E+01	3068.67	3	5	2.27E-01
4699.01	6	8	9.88E-01	305.596	1	3	1.20E+02	3074.14	5	7	1.84E-01
4699.22	4	6	9.36E-01	305.656	3	5	1.62E+02	3074.72	5	3	3.76E-01
4701.18	4	6	9.23E-01	305.702	3	3	9.01E+01	3075.13	5	5	1.61E-01
4701.71	4	2	3.69E-01	305.767	5	7	2.16E+02	3075.95	7	9	1.07E-01
4703.16	4	6	9.20E-01	305.836	5	5	5.40E+01	3083.65	7	7	3.20E-01
4705.35	6	8	1.10E+00	320.978	5	7	2.17E+02	3084.64	7	5	2.55E-01
4710.01	4	6	2.98E-01	328.448	5	5	1.04E+02	3088.04	9	9	5.30E-01
4741.70	6	6	4.71E-02	345.312	1	3	1.35E+02	3095.79	9	7	1.35E-01
4751.28	6	8	6.39E-02	374.073	5	5	2.85E+01	3115.67	3	1	1.39E+00
4752.69	6	6	1.45E-02	395.557	5	3	2.80E+01	3121.63	3	3	1.38E+00
4844.92	4	6	1.02E-02	507.388	1	3	1.61E+01	3132.79	3	5	1.37E+00
4856.39	4	6	5.58E-02	507.680	3	3	4.82E+01	3198.18	3	5	9.57E-02
4856.76	4	4	1.00E-01	508.178	5	3	8.04E+01	3201.14	3	3	4.77E-01
4860.97	2	4	4.70E-01	525.794	5	3	9.60E+01	3202.51	5	7	7.08E-02

λ Å	g_i	g_k	A 10^8 s^{-1}	λ Å	g_i	g_k	A 10^8 s^{-1}	λ Å	g_i	g_k	A 10^8 s^{-1}
3207.61	5	5	4.40E-01	3698.72	5	7	7.62E-01	625.127	4	4	2.13E+01
3210.58	5	3	1.58E-01	3703.36	7	9	1.14E+00	625.853	6	4	3.19E+01
3216.07	7	7	5.58E-01	3704.75	3	3	8.53E-01	779.736	6	4	1.46E+00
3221.21	7	5	9.75E-02	3707.27	3	5	7.34E-01	779.820	4	4	1.31E+01
3260.86	5	7	1.68E+00	3709.54	3	1	1.13E+00	779.912	6	6	1.36E+01
3265.33	7	9	1.88E+00	3712.49	5	5	6.59E-01	779.997	4	6	9.70E-01
3267.20	3	5	1.58E+00	3714.03	3	3	4.06E-01	787.710	2	4	5.95E+00
3281.83	5	5	2.89E-01	3715.09	5	7	9.73E-01	790.112	4	4	1.18E+00
3284.45	7	7	2.06E-01	3720.89	7	7	3.74E-01	790.199	4	6	7.08E+00
3299.39	1	3	1.64E-01	3721.95	5	3	2.80E-01	921.296	2	4	2.21E+00
3312.33	3	3	4.60E-01	3725.31	5	5	2.41E-01	921.365	2	2	8.83E+00
3326.06	3	3	2.65E-01	3728.51	5	7	1.29E+00	923.367	4	4	1.10E+01
3330.30	3	5	6.81E-01	3728.84	7	9	1.45E+00	923.436	4	2	4.39E+00
3330.32	3	5	4.76E-01	3729.80	3	5	1.22E+00	1338.61	2	4	2.17E+00
3332.41	5	3	7.92E-01	3732.13	5	3	2.67E-02	1342.99	4	4	4.29E-01
3332.93	5	7	5.04E-01	3734.83	7	5	7.40E-02	1343.51	4	6	2.57E+00
3336.67	3	3	3.76E-01	3742.63	5	5	2.24E-01	2120.58	2	2	1.05E+00
3336.69	5	5	8.77E-02	3746.90	7	7	1.59E-01	2132.64	4	4	1.29E+00
3340.76	5	3	6.57E-01	3754.70	3	5	7.53E-01	2493.39	2	4	1.18E+00
3344.20	5	5	1.25E-01	3757.23	1	3	5.56E-01	2493.75	4	6	8.48E-01
3344.51	5	7	3.48E-01	3759.88	5	7	9.79E-01	2493.99	2	2	6.09E-01
3347.98	7	5	4.86E-01	3774.03	3	3	3.91E-01	2499.27	2	2	4.68E-01
3350.62	5	3	1.12E+00	3791.28	5	5	2.24E-01	2501.81	4	4	3.73E-01
3350.92	7	7	9.91E-01	3810.98	5	3	2.37E-02	2507.73	4	2	2.32E+00
3355.86	7	7	6.89E-01	3816.75	5	3	9.63E-02	2509.22	6	6	1.94E+00
3362.31	7	5	6.87E-01	3961.57	5	7	1.25E+00	2510.58	4	2	1.19E+00
3376.61	3	1	1.49E+00	4072.64	1	3	3.37E-01	2517.37	6	4	1.24E+00
3376.76	3	3	1.12E+00	4073.98	3	5	4.54E-01	2781.22	2	2	1.03E-01
3377.26	3	5	5.20E-01	4081.02	5	7	6.02E-01	2803.57	6	4	1.26E-01
3382.61	5	7	9.86E-01	4089.30	3	3	2.49E-01	2805.87	2	4	2.90E-01
3383.31	5	3	3.70E-01	4103.07	5	5	1.48E-01	2812.50	6	6	3.58E-02
3383.81	5	5	8.62E-01	4118.60	5	3	1.63E-02	2816.53	4	4	5.74E-01
3384.90	7	9	1.48E+00	4440.09	5	3	4.42E-01	2829.17	8	6	1.56E-01
3394.22	7	7	4.88E-01	4447.69	5	5	4.40E-01	2836.27	6	4	8.43E-01
3395.43	7	5	9.75E-02	4461.61	5	7	4.36E-01	2916.31	2	4	1.06E+00
3406.88	1	3	1.93E-01	4524.22	3	1	3.38E-01	2921.46	4	6	1.27E+00
3408.13	3	1	5.79E-01	4532.78	5	3	1.40E-01	2926.18	4	4	2.11E-01
3415.26	3	3	1.44E-01	4535.29	3	3	8.40E-02	3063.43	2	4	1.30E+00
3428.63	3	5	1.42E-01	4555.39	5	5	2.49E-01	3071.60	2	2	1.29E+00
3430.57	5	3	2.37E-01	4557.91	3	5	8.27E-02	3177.89	2	4	7.59E-02
3444.05	5	5	4.21E-01	5268.30	1	3	3.50E-01	3180.77	2	2	1.51E-01
3446.68	3	5	9.71E-01	5508.24	5	5	1.06E-01	3180.99	4	6	7.06E-02
3447.15	1	3	8.09E-01	5592.25	3	3	3.27E-01	3185.74	4	4	1.21E-01
3447.97	5	7	1.19E+00					3188.22	6	8	4.28E-02
3450.91	7	9	1.44E+00	*O IV*				3188.64	4	2	1.50E-01
3451.30	3	3	8.06E-01	238.360	2	4	2.96E+02	3194.78	6	6	1.71E-01
3454.84	5	5	6.89E-01	238.570	4	6	3.54E+02	3199.58	6	4	1.04E-01
3454.99	9	11	1.72E+00	238.579	4	4	5.90E+01	3209.65	8	8	2.53E-01
3459.48	5	3	1.14E-01	279.631	2	2	2.68E+01	3216.31	8	6	5.56E-02
3459.94	7	7	5.14E-01	279.933	4	2	5.34E+01	3348.06	2	4	8.51E-01
3466.13	9	9	2.84E-01	553.329	2	4	1.22E+01	3349.11	4	6	1.02E+00
3466.85	7	5	6.82E-02	554.076	2	2	4.86E+01	3354.27	4	2	7.71E-01
3475.24	9	7	2.42E-02	554.513	4	4	6.06E+01	3362.55	4	4	7.65E-01
3520.94	1	3	1.50E-01	555.263	4	2	2.41E+01	3375.40	4	6	7.56E-01
3531.22	3	1	4.45E-01	608.397	2	2	1.21E+01	3378.02	4	4	1.66E-01
3533.38	3	3	1.11E-01	609.829	4	2	2.40E+01	3381.21	4	6	7.19E-01
3534.90	3	5	1.11E-01	616.952	6	4	2.60E+01	3381.30	2	4	4.28E-01
3555.24	5	3	1.82E-01	617.005	4	4	2.89E+00	3385.52	6	8	1.02E+00
3556.78	5	5	3.26E-01	617.036	4	2	2.89E+01	3390.19	2	2	8.49E-01
3695.38	3	5	4.01E-01	624.619	2	4	1.07E+01	3396.80	4	4	5.40E-01

λ (Å)	g_i	g_k	A (10^8 s^{-1})	λ (Å)	g_i	g_k	A (10^8 s^{-1})	λ (Å)	g_i	g_k	A (10^8 s^{-1})
3405.77	4	2	1.67E-01	2781.01	3	5	1.40E+00	*173.03	6	10	8.78E+02
3409.70	6	6	3.00E-01	*2784.0	3	9	1.40E+00	1031.91	2	4	4.16E+00
3411.30	4	4	1.69E-01	2786.99	3	3	1.39E+00	1037.61	2	2	4.09E+00
3411.69	4	6	1.02E+00	2789.85	3	1	1.38E+00	3811.35	2	4	5.14E-01
3425.55	6	4	4.94E-02	3058.68	3	5	1.39E+00	3834.24	2	2	5.05E-01
3489.89	4	6	7.29E-01	3144.66	3	5	8.86E-01				
3492.21	2	4	6.06E-01	3219.24	3	1	1.54E-01	*O VII*			
3493.43	4	4	1.21E-01	3222.29	1	3	1.16E-01	18.6270	1	3	9.365E+03
3560.39	4	6	1.03E+00	3227.54	3	3	3.38E-02	21.6020	1	3	3.309E+04
3563.33	6	8	1.10E+00	3239.21	3	3	3.28E-01	*120.33	3	9	5.334E+02
3593.08	6	6	7.15E-02	3248.28	5	3	1.18E-01	128.411	1	3	8.982E+02
3725.89	2	4	5.61E-01	3263.54	5	5	1.86E-02	*128.46	9	15	1.615E+03
3725.94	4	6	6.01E-01	3275.64	5	3	4.76E-01	135.820	3	5	1.523E+03
3729.03	6	8	6.86E-01	3297.62	7	5	1.30E-01	*1630.3	3	9	7.935E-01
3736.68	4	4	2.23E-01	3690.17	3	5	1.97E-02	2448.98	1	3	2.514E-01
3736.85	8	10	7.95E-01	3698.36	3	3	1.03E-01	*5933.1	3	9	1.002E-01
3744.89	6	6	1.92E-01	3702.72	5	7	1.41E-02	8241.76	1	3	3.864E-02
3758.39	8	8	1.11E-01	3717.31	5	5	9.63E-02				
3930.68	2	2	3.80E-02	3725.63	5	3	2.91E-02	*Phosphorus*			
3942.06	2	4	9.42E-02	3746.64	7	7	1.18E-01				
3945.31	4	2	1.88E-01	3761.58	7	5	1.61E-02	*P I*			
3956.77	4	4	2.98E-02	4119.37	3	5	3.66E-01	1671.7	4	2	3.9E-01
3974.58	4	6	6.62E-02	4120.49	3	1	3.33E-01	1674.6	4	4	4.0E-01
3977.09	6	4	9.91E-02	4123.96	5	7	4.81E-01	1679.7	4	6	3.9E-01
3995.08	6	6	1.52E-01	4125.49	1	3	2.70E-01	1775.0	4	6	2.17E+00
4687.03	2	4	2.79E-01	4134.11	3	3	3.34E-01	1782.9	4	4	2.14E+00
4772.60	2	4	1.23E-01	4153.27	3	3	1.92E-01	1787.7	4	2	2.13E+00
4779.10	2	2	2.45E-01	4158.86	3	5	3.39E-01	2135.5	4	4	2.11E-01
4783.42	4	6	2.06E-01	4178.46	5	5	1.12E-01	2136.2	6	4	2.83E+00
4794.18	4	4	1.56E-01	4213.35	5	3	1.19E-02	2149.1	4	2	3.18E+00
4798.27	6	8	2.91E-01	4522.66	5	3	1.02E-02	2152.9	2	4	4.85E-01
4813.15	6	6	8.65E-02	4554.53	3	5	2.41E-01	2154.1	4	4	1.73E-01
5305.51	4	4	6.10E-02	5114.06	1	3	1.80E-01	2154.1	4	6	5.8E-01
5362.51	6	6	6.12E-02	5339.94	1	3	1.85E-02	2534.0	2	4	2.00E-01
6876.49	2	4	1.88E-02	5349.74	3	1	7.04E-02	2535.6	4	4	9.5E-01
6931.60	2	2	7.35E-02	5372.71	3	3	1.42E-02	2553.3	2	2	7.1E-01
7004.11	4	4	8.90E-02	5414.59	3	5	9.29E-03	2554.9	4	2	3.00E-01
7061.30	4	2	3.48E-02	5428.38	5	3	2.68E-02				
				5471.12	5	5	4.86E-02	*P II*			
O V				5571.81	1	3	8.33E-02	1301.9	1	3	5.0E-01
172.169	1	3	2.94E+02	5580.12	3	5	1.11E-01	1304.5	3	1	1.5E+00
*192.85	9	15	6.90E+02	5583.23	3	3	6.20E-02	1304.7	3	3	3.7E-01
*215.17	9	3	1.83E+02	*5589.9	9	15	1.49E-01	1305.5	3	5	3.8E-01
220.353	3	5	4.292E+02	5597.89	5	7	1.48E-01	1309.9	5	3	6.2E-01
248.460	3	1	5.59E+01	5604.27	5	5	3.68E-02	1310.7	5	5	1.1E+00
629.732	1	3	2.872E+01	5607.41	5	3	4.08E-03	4475.3	5	7	1.3E+00
758.677	3	5	5.547E+00	6330.05	5	7	1.21E-01	4499.2	5	7	1.4E+00
759.442	1	3	7.373E+00	6460.12	3	5	9.37E-02	4530.8	3	5	1.0E+00
760.227	3	3	5.514E+00	6466.14	5	7	1.01E-01	4554.8	3	5	9.6E-01
760.446	5	5	1.652E+01	6500.24	7	9	1.11E-01	4588.0	5	7	1.7E+00
761.128	3	1	2.197E+01	6543.77	5	5	1.64E-02	4589.9	3	5	1.6E+00
762.004	5	3	9.125E+00	6601.28	7	7	1.14E-02	4602.1	7	9	1.9E+00
774.518	3	1	3.804E+01	6764.72	1	3	4.37E-02	4943.5	7	5	6.3E-01
1371.30	3	5	3.336E+00	6789.62	3	5	5.79E-02	5253.5	3	5	1.0E+00
2729.31	3	5	4.52E-01	6817.40	3	3	3.00E-02	5425.9	5	5	6.9E-01
2731.45	1	3	5.90E-01	6828.95	5	7	7.35E-02	6024.2	3	5	5.1E-01
2743.61	3	3	4.38E-01	6878.76	5	5	1.65E-02	6043.1	5	7	6.8E-01
2752.23	3	1	1.82E+00								
2755.13	5	5	1.37E+00	*O VI*				*P III*			
2769.69	5	3	7.88E-01	*150.10	2	6	2.62E+02	1334.8	2	4	5.5E-01
								1344.3	4	6	6.4E-01
								1344.8	4	4	1.1E-01

λ Å	g_i	g_k	A 10^8 s^{-1}	λ Å	g_i	g_k	A 10^8 s^{-1}	λ Å	g_i	g_k	A 10^8 s^{-1}
4057.4	4	4	1.0E-01	4734.2	15	13	2.5E-02	3692.36	10	8	9.1E-01
4059.3	6	4	9.0E-01	4879.1	15	15	1.8E-02	3700.91	8	10	3.9E-01
4080.1	4	2	9.9E-01	4886.0	15	15	1.3E-02	3713.02	4	4	8.3E-02
Potassium				4912.6	17	15	5.7E-02	3788.47	4	6	1.4E-01
				5034.4	19	19	1.1E-01	3793.22	8	6	4.2E-01
K I				5110.8	21	19	2.78E-01	3799.31	8	8	5.5E-01
4044.1	2	4	1.24E-02	5135.1	17	17	1.25E-01	3806.76	6	6	6.2E-02
4047.2	2	2	1.24E-02	5173.9	19	17	3.18E-01	3818.19	6	4	5.8E-01
5084.2	2	2	3.50E-03	5219.1	15	15	9.5E-02	3822.26	6	6	8.5E-01
5099.2	4	2	7.0E-03	5220.1	17	15	2.35E-01	3828.48	6	6	6.2E-01
5323.3	2	2	6.3E-03	5251.7	15	13	1.1E-02	3833.89	6	4	5.8E-01
5339.7	4	2	1.26E-02	5259.7	15	13	2.24E-01	3856.52	8	10	5.9E-01
5343.0	2	4	4.0E-03	5292.6	13	13	9.3E-02	3872.39	4	6	6.7E-03
5359.6	4	6	4.6E-03	5810.6	17	19	2.3E-02	3877.34	8	6	3.7E-02
5782.4	2	2	1.23E-02	5879.3	15	15	7.6E-02	3913.51	8	8	2.5E-03
5801.8	4	2	2.46E-02	6200.8	15	17	1.8E-02	3922.19	4	2	6.25E-02
5812.2	2	4	2.8E-03	6278.7	13	15	2.6E-02	3934.23	8	8	1.58E-01
5831.9	4	6	3.2E-03	6398.0	11	13	1.9E-02	3942.72	4	2	7.15E-01
6911.1	2	2	2.72E-02	*Rhodium*				3958.86	6	8	5.5E-01
6938.8	4	2	5.4E-02					3984.40	4	4	1.1E-01
7664.9	2	4	3.87E-01	*Rh I*				3995.61	4	6	4.7E-02
7699.0	2	2	3.82E-01	3083.96	8	6	4.8E-02	4053.44	2	2	2.8E-02
K II				3114.91	6	4	4.45E-02	4056.34	6	4	9.5E-03
607.93	1	3	1.3E-02	3121.76	6	6	1.1E-01	4082.78	6	4	1.4E-01
K III				3123.70	10	8	4.6E-02	4097.52	2	4	7.0E-02
2550.0	6	4	2.0E+00	3137.71	4	6	3.3E-02	4121.68	6	6	9.8E-02
2635.1	4	4	1.2E+00	3189.05	6	6	3.03E-01	4128.87	6	8	1.73E-01
2992.4	6	8	2.5E+00	3197.13	6	4	4.35E-02	4135.27	8	8	1.0E-01
3052.1	4	6	1.7E+00	3263.14	6	6	1.3E-01	4196.50	6	8	3.9E-02
3202.0	4	4	1.8E+00	3271.61	6	4	2.0E-01	4211.14	8	10	1.62E-01
3289.1	4	6	2.0E+00	3280.55	8	8	2.36E-01	4244.44	4	4	6.5E-03
3322.4	6	6	1.3E+00	3283.57	6	8	4.4E-01	4278.60	4	6	9.2E-03
3421.8	2	4	1.5E+00	3289.14	4	4	1.0E-01	4288.71	6	8	6.1E-02
K XVI				3323.09	8	10	6.3E-01	4373.04	2	4	1.8E-02
206.27	1	3	9.4E+01	3331.09	4	2	5.40E-02	4374.80	8	10	1.64E-01
K XVII				3338.54	8	6	3.5E-02	4379.92	6	6	2.48E-02
22.020	2	4	4.7E+04	3360.80	4	4	1.2E-01	4492.47	6	6	4.5E-03
22.163	4	6	5.6E+04	3368.38	6	4	1.1E-01	4528.72	6	8	1.35E-02
22.18	4	4	9.3E+03	3396.82	10	10	6.5E-01	4548.73	4	6	5.5E-03
22.60	2	2	2.5E+03	3399.70	6	8	1.2E-01	4551.64	4	4	4.00E-02
22.76	4	2	4.7E+03	3462.04	6	6	6.2E-01	4565.19	4	4	1.1E-02
Praseodymium				3470.66	4	4	8.5E-01	4569.00	6	8	1.0E-02
				3478.91	6	6	3.32E-01	4608.12	2	2	2.1E-02
Pr II				3484.04	6	8	9.3E-03	4675.03	8	8	6.4E-03
3997.0	15	15	1.87E-01	3498.73	4	6	2.12E-01	4721.00	6	4	3.43E-03
4062.8	13	15	1.00E+00	3502.52	10	10	4.3E-01	4745.11	6	6	5.2E-03
4100.7	17	19	8.4E-01	3507.32	6	8	3.4E-01	4755.58	4	4	6.0E-03
4143.1	15	17	5.8E-01	3528.02	8	8	8.5E-01	4842.43	6	8	1.6E-03
4179.4	13	15	5.2E-01	3543.95	4	4	4.65E-01	4963.71	2	2	3.0E-02
4222.9	11	13	3.91E-01	3549.54	6	6	2.22E-01	4977.75	4	4	9.8E-03
4241.0	17	15	2.30E-01	3570.18	4	6	1.82E-01	4979.18	4	6	1.0E-02
4359.8	15	15	1.1E-01	3583.10	8	10	2.6E-01	5090.63	6	6	5.0E-03
4405.8	17	17	9.0E-02	3596.19	6	4	5.5E-01	5120.69	6	8	3.1E-03
4429.3	15	15	2.28E-01	3597.15	6	8	5.9E-01	5130.76	4	4	4.35E-03
4449.8	13	13	1.24E-01	3612.47	4	2	8.90E-01	5155.54	2	4	9.8E-03
4468.7	11	13	1.54E-01	3620.46	6	4	8.5E-02	5184.19	6	8	1.6E-03
4510.2	13	15	1.16E-01	3654.87	8	8	6.0E-02	5212.73	4	2	5.95E-03
4534.2	15	17	4.9E-02	3657.99	8	6	8.8E-01	5292.14	10	10	3.7E-03
				3666.22	6	8	8.4E-02	5390.44	4	6	9.5E-03
				3690.70	6	4	3.23E-01	5424.72	4	4	5.0E-03

λ Å	Weights g_i	g_k	A 10^8 s^{-1}	λ Å	Weights g_i	g_k	A 10^8 s^{-1}	λ Å	Weights g_i	g_k	A 10^8 s^{-1}
5599.42	6	8	1.3E-02	3911.81	6	8	1.79E+00	4983.43	4	4	2.58E-01
5983.60	10	10	2.1E-02	3933.38	6	6	1.62E-01	4991.91	6	6	3.8E-01
				3996.60	4	6	1.65E-01	4995.00	4	6	5.9E-02
Rubidium				4020.39	4	4	1.63E+00	5018.41	6	4	2.09E-01
				4023.22	4	4	3.0E-01	5021.52	4	4	2.30E-01
Rb I				4023.68	6	6	1.65E+00	5064.31	8	10	7.3E-02
3022.5	2	4	4.13E-05	4031.38	6	6	2.9E-01	5066.38	6	6	3.6E-02
3032.0	2	4	4.93E-05	4036.86	6	4	7.9E-02	5070.17	6	8	1.16E-01
3044.2	2	4	8.2E-05	4043.80	8	8	3.11E-01	5072.71	2	4	2.0E-02
3060.2	2	4	1.05E-04	4047.80	6	4	1.54E-01	5075.82	4	6	1.15E-01
3082.0	2	4	1.49E-04	4051.83	8	6	7.7E-02	5080.22	4	4	4.1E-02
3112.6	2	4	2.5E-04	4054.54	4	2	1.67E-01	5081.56	10	10	7.6E-01
3113.1	2	2	1.3E-04	4067.00	6	8	1.91E-01	5083.72	8	8	6.2E-01
3157.5	2	4	3.38E-04	4067.63	10	8	4.1E-02	5085.55	6	6	5.7E-01
3158.3	2	2	2.0E-04	4074.96	4	6	3.7E-01	5086.94	4	4	6.6E-01
3228.0	2	4	6.4E-04	4078.56	2	4	4.3E-01	5096.72	6	4	1.69E-01
3229.2	2	2	3.8E-04	4080.57	4	4	6.6E-02	5099.27	4	6	1.50E-01
3348.7	2	4	1.37E-03	4082.39	6	4	2.73E-01	5101.12	10	8	8.8E-02
3350.8	2	2	8.9E-04	4086.66	6	8	3.7E-01	5331.79	4	4	1.11E-01
3587.1	2	4	3.97E-03	4087.47	4	6	1.12E-01	5339.43	6	6	1.06E-01
3591.6	2	2	2.9E-03	4093.12	4	4	1.23E-01	5341.07	4	2	3.8E-01
4201.8	2	4	1.8E-02	4094.86	6	6	1.44E-01	5349.34	6	4	5.9E-01
4215.5	2	2	1.5E-02	4098.36	8	8	8.7E-02	5350.28	8	8	6.8E-02
7800.3	2	4	3.70E-01	4132.98	4	6	1.19E+00	5355.79	6	4	3.0E-01
7947.6	2	2	3.40E-01	4140.27	6	8	1.17E+00	5356.10	8	6	5.7E-01
				4147.38	6	6	1.74E-01	5375.37	8	6	3.4E-01
Scandium				4161.85	8	8	1.77E-01	5392.06	10	8	4.2E-01
				4171.53	6	4	1.36E-01	5416.16	4	6	4.4E-02
Sc I				4186.42	6	8	8.4E-02	5416.41	6	6	2.0E-02
2116.7	4	4	2.0E-01	4187.61	8	6	1.28E-01	5425.55	6	8	4.5E-02
2120.4	6	6	2.0E-01	4193.53	4	6	6.1E-02	5429.42	2	4	9.0E-02
2262.3	4	4	5.8E-02	4204.52	6	8	3.5E-02	5432.98	4	4	5.4E-02
2266.6	4	2	4.8E-01	4205.20	10	8	1.12E-01	5433.25	6	4	9.7E-02
2270.9	6	4	4.6E-01	4212.32	4	6	1.58E-01	5438.28	4	6	3.4E-02
2280.8	4	6	2.8E-01	4212.48	6	6	8.6E-02	5439.04	2	2	1.74E-01
2289.6	6	6	4.1E-02	4216.08	2	4	2.36E-01	5442.62	4	2	2.15E-01
2311.29	4	6	4.1E-02	4218.23	4	4	2.26E-01	5446.20	8	8	2.8E-01
2315.69	4	4	2.5E-01	4225.54	6	8	9.5E-02	5451.37	6	6	1.50E-01
2320.32	6	6	2.4E-01	4225.69	4	6	7.6E-02	5455.24	4	4	6.6E-02
2324.75	6	4	4.1E-02	4231.64	4	4	1.31E-01	5464.95	4	2	3.2E-02
2328.19	4	6	4.6E-02	4233.59	6	6	4.0E-01	5468.40	6	4	9.7E-02
2334.67	4	2	1.7E-01	4238.05	8	8	7.1E-01	5472.19	8	6	9.7E-02
2346.03	6	4	1.3E-01	4239.55	6	4	2.27E-01	5482.01	8	8	5.2E-01
2429.19	4	4	2.8E-01	4246.14	8	6	1.15E-01	5484.63	6	6	5.2E-01
2438.63	6	6	2.1E-01	4542.55	6	4	1.28E-01	5514.23	6	8	4.1E-01
2468.40	4	2	4.9E-02	4544.67	8	6	1.33E-01	5520.52	8	10	4.3E-01
2692.78	4	2	1.61E-01	4706.94	4	6	2.81E-01	5526.10	4	4	7.1E-02
2699.02	4	6	2.4E-02	4709.31	6	8	4.0E-01	5541.07	6	6	5.5E-02
2706.74	4	4	3.1E-01	4711.72	2	4	1.81E-01	5631.04	2	4	3.0E-02
2707.93	6	4	1.49E-01	4714.30	4	4	2.14E-01	5671.83	10	12	5.4E-01
2711.34	6	6	3.2E-01	4719.31	6	6	1.04E-01	5686.86	8	10	4.9E-01
2965.88	4	6	7.5E-02	4728.77	8	8	1.16E-01	5700.19	6	8	4.6E-01
2974.01	4	4	5.5E-01	4729.20	4	4	2.20E-01	5708.64	10	10	4.7E-02
2980.76	6	6	5.4E-01	4729.24	6	6	1.93E-01	5711.79	4	6	4.5E-01
2988.97	6	4	6.9E-02	4734.11	4	2	1.10E+00	5717.31	8	8	7.5E-02
3015.37	4	6	7.8E-01	4737.65	6	4	8.8E-01	5724.13	6	6	7.4E-02
3019.35	6	8	8.7E-01	4741.02	8	6	9.1E-01	5988.43	6	6	6.6E-02
3030.76	6	6	1.00E-01	4743.82	10	8	9.8E-01	6026.16	4	4	7.2E-02
3255.68	4	4	3.2E-01	4973.67	4	2	8.4E-01	6146.20	6	8	4.2E-02
3269.90	4	2	3.13E+00	4980.36	6	4	5.6E-01	6198.43	4	6	3.5E-02
3273.63	6	4	2.81E+00								
3907.48	4	6	1.66E+00								

λ Å	Weights g_i	Weights g_k	A 10^8 s^{-1}	λ Å	Weights g_i	Weights g_k	A 10^8 s^{-1}	λ Å	Weights g_i	Weights g_k	A 10^8 s^{-1}
6249.96	6	8	3.2E-02	3590.47	7	5	2.9E-01	7680.3	3	5	4.6E-02
6262.22	4	6	8.4E-02	3613.83	7	9	1.48E+00	7918.4	3	5	5.2E-02
6280.16	2	4	4.0E-02	3630.74	5	7	1.20E+00	7932.3	5	7	5.1E-02
6284.16	6	6	3.9E-02	3642.78	3	5	1.13E+00	7944.0	7	9	5.8E-02
6284.73	4	4	7.1E-02	3645.31	7	7	2.74E-01	7970.3	5	5	7.1E-03
6293.02	2	2	1.04E-01	3651.80	5	5	3.0E-01				
7741.16	10	10	3.8E-02	3859.59	7	5	1.1E+00	*Si II*			
7800.42	8	8	5.1E-02	4246.82	5	5	1.29E+00	989.87	2	4	6.7E+00
				4314.08	9	7	4.1E-01	992.68	4	6	8.0E+00
Sc II				4320.75	7	5	4.0E-01	1020.7	2	2	1.3E+00
1880.6	5	3	5.0E+00	4325.00	5	3	4.3E-01	1190.4	2	4	6.9E+00
2064.3	7	5	2.2E+00	4374.46	9	9	1.48E-01	1193.3	2	2	2.8E+01
2068.0	5	3	2.0E+00	4400.39	7	7	1.43E-01	1194.5	4	4	3.6E+01
2273.1	1	3	7.7E+00	4415.54	5	5	1.47E-01	1197.4	4	2	1.4E+01
2545.20	5	5	4.0E-01	4670.41	5	7	1.16E-01	1248.4	4	4	1.3E+01
2552.35	7	5	2.21E+00	5031.01	5	3	3.5E-01	1251.2	6	4	1.9E+01
2555.79	3	3	6.9E-01	5239.81	1	3	1.39E-01	1260.4	2	4	2.0E+01
2560.23	5	3	2.01E+00	5526.79	9	7	3.3E-01	1264.7	4	6	2.3E+01
2563.19	3	1	2.70E+00	5657.91	5	5	1.04E-01	1304.4	2	2	3.6E+00
2611.19	5	5	2.2E+00	5669.06	3	1	1.31E-01	1309.3	4	2	7.0E+00
2667.70	3	5	1.5E+00					1526.7	2	2	3.73E+00
2746.36	3	1	3.9E+00	*Silicon*				1533.5	4	2	7.4E+00
2782.31	5	5	1.3E+00					1808.0	2	4	3.7E-02
2789.15	7	7	1.3E+00	*Si I*				2904.3	4	6	6.7E-01
2801.31	9	9	1.3E+00	1977.6	1	3	1.8E-01	2905.7	6	8	7.1E-01
2819.49	3	5	2.3E+00	1979.2	3	1	5.1E-01	3210.0	4	6	4.6E-01
2822.12	5	7	2.5E+00	1980.6	3	3	1.3E-01	4128.1	4	6	1.32E+00
2826.64	7	9	2.8E+00	1983.2	3	5	1.4E-01	4130.9	6	8	1.42E+00
2870.85	5	3	1.1E+00	1986.4	5	3	2.1E-01	5041.0	2	4	9.8E-01
2912.98	5	3	1.1E+00	1989.0	5	5	4.1E-01	5056.0	4	6	1.2E+00
2979.68	3	5	1.2E+00	2208.0	1	3	3.11E-01	5957.6	2	2	4.2E-01
2988.92	5	7	2.9E+00	2210.9	3	5	4.16E-01	5978.9	4	2	8.1E-01
3039.92	7	9	3.5E+00	2211.7	3	3	2.32E-01	6347.1	2	4	7.0E-01
3045.73	5	7	3.68E+00	2216.7	5	7	5.5E-01	6371.4	2	2	6.9E-01
3052.92	7	9	3.92E+00	2218.1	5	5	1.38E-01	7848.8	4	6	3.9E-01
3060.54	7	7	3.0E-01	2506.9	3	5	4.66E-01	7849.7	6	8	4.2E-01
3065.12	9	11	4.00E+00	2514.3	1	3	6.1E-01				
3075.36	9	9	2.5E-01	2516.1	5	5	1.21E+00	*Si III*			
3128.27	3	3	1.9E+00	2519.2	3	3	4.56E-01	883.40	5	7	6.3E+01
3133.07	5	5	1.8E+00	2524.1	3	1	1.81E+00	994.79	3	3	7.89E+00
3139.72	7	7	2.1E+00	2528.5	5	3	7.7E-01	997.39	5	3	1.31E+01
3190.98	3	3	1.1E+00	2532.4	1	3	2.6E-01	1141.6	3	5	3.0E+01
3199.33	5	3	1.9E+00	2631.3	1	3	9.7E-01	1144.3	5	7	3.9E+01
3312.72	5	7	1.2E+00	2881.6	5	3	1.89E+00	1161.6	5	5	1.6E+01
3320.40	5	3	1.2E+00	3905.5	1	3	1.18E-01	1206.5	1	3	2.59E+01
3343.23	9	7	1.1E+00	4738.8	3	3	1.0E-02	1206.5	3	3	4.89E+01
3353.72	5	7	1.51E+00	4783.0	5	3	1.7E-02	1207.5	5	5	1.9E+01
3359.67	5	5	2.16E-01	4792.3	5	5	1.7E-02	1294.5	3	5	5.42E+00
3361.26	3	3	3.4E-01	4818.1	5	7	1.1E-02	1296.7	1	3	7.19E+00
3361.93	3	1	1.17E+00	4821.2	3	5	8.0E-03	1298.9	3	3	5.36E+00
3368.94	5	3	8.3E-01	4947.6	3	1	4.2E-02	1299.0	5	5	1.61E+01
3372.15	7	5	9.9E-01	5006.1	3	5	2.8E-02	1301.2	3	1	2.13E+01
3379.16	3	3	2.5E+00	5622.2	3	3	1.6E-02	1303.3	5	3	8.85E+00
3535.71	5	3	6.1E-01	5690.4	3	3	1.2E-02	1328.8	1	3	2.7E+01
3558.53	5	7	3.0E-01	5708.4	5	5	1.4E-02	1417.2	3	1	2.60E+01
3567.70	3	5	3.5E-01	5754.2	5	3	1.5E-02	1435.8	5	7	2.1E+01
3572.53	7	7	1.38E+00	5772.1	3	1	3.6E-02	1589.0	5	5	1.1E+01
3576.34	5	5	1.06E+00	5948.5	3	5	2.2E-02	1778.7	7	9	4.4E+00
3580.93	3	3	1.23E+00	7226.2	3	5	7.9E-03	1783.1	5	7	3.8E+00
3589.63	5	3	4.6E-01	7405.8	3	5	3.7E-02	3241.6	5	3	2.3E+00
				7409.1	5	7	2.3E-02	*3486.9	15	21	1.8E+00

λ	Weights		A	λ	Weights		A	λ	Weights		A
Å	g_i	g_k	10^8 s^{-1}	Å	g_i	g_k	10^8 s^{-1}	Å	g_i	g_k	10^8 s^{-1}
3590.5	3	5	3.9E+00	225.03	3	3	1.2E+02	4393.3	4	4	1.6E-03
4552.6	3	5	1.26E+00	227.01	5	3	2.0E+02	4393.3	4	6	9.2E-03
4554.0	5	3	7.6E-01	227.30	5	3	2.3E+02	4494.2	2	4	1.2E-02
4567.8	3	3	1.25E+00	258.10	5	5	1.04E+02	4497.7	4	6	1.4E-02
4683.0	5	5	9.5E-01	*294.37	9	9	5.9E+01	4497.7	4	4	2.4E-03
4716.7	5	7	2.8E+00	*347.36	9	15	2.2E+01	4664.8	2	4	2.33E-02
5451.5	3	5	6.0E-01					4668.6	4	4	4.1E-03
5473.1	5	7	7.9E-01	*Si X*				4668.6	4	6	2.5E-02
5716.3	9	7	1.9E-01	253.77	2	4	2.9E+01	4747.9	2	2	6.3E-03
5739.7	1	3	4.7E-01	256.57	2	2	1.1E+02	4751.8	4	2	1.27E-02
7462.6	5	3	4.9E-01	258.35	4	4	1.4E+02	4978.5	2	4	4.1E-02
7466.3	7	5	5.4E-01	261.05	4	2	5.4E+01	4982.8	4	4	8.2E-03
7612.4	3	5	1.1E+00	272.00	2	2	3.0E+01	4982.8	4	6	4.89E-02
				277.26	4	2	5.7E+01	5148.8	2	2	1.17E-02
Si IV				287.08	2	4	2.6E+01	5153.4	4	2	2.33E-02
457.82	2	4	3.6E+00	289.19	4	4	5.0E+01	5682.6	2	4	1.03E-01
458.16	2	2	3.6E+00	292.22	6	4	7.3E+01	5688.2	4	6	1.2E-01
515.12	2	2	4.1E+00	*347.73	10	10	4.3E+01	5688.2	4	4	2.1E-02
516.35	4	2	8.2E+00	*353.09	6	10	2.1E+01	5890.0	2	4	6.11E-01
*560.50	6	10	1.0E+00					5895.9	2	2	6.10E-01
*749.94	10	14	1.45E+01	*Si XI*				6154.2	2	2	2.6E-02
815.05	2	2	1.23E+01	43.763	1	3	6.11E+03	6160.8	4	2	5.2E-02
818.13	4	2	2.44E+01	*49.116	9	3	2.45E+03	8183.3	2	4	4.53E-01
*860.74	10	6	1.8E+00	49.222	3	5	8.9E+03	8194.8	4	6	5.4E-01
*1066.6	10	14	3.91E+01	52.296	3	1	7.6E+02	8194.8	4	4	9.0E-02
1122.5	2	4	2.05E+01	303.30	1	3	6.42E+01	11381	2	4	8.9E-02
1128.3	4	4	4.03E+00	358.29	3	1	1.03E+02	11404	4	2	1.76E-01
1128.3	4	6	2.42E+01	358.63	3	5	1.38E+01				
1393.8	2	4	7.73E+00	361.41	1	3	1.80E+01	*Na II*			
1402.8	2	2	7.58E+00	364.50	3	3	1.32E+01	300.15	1	3	3.0E+01
*1724.1	10	6	5.5E+00	365.42	5	5	3.90E+01	301.44	1	3	4.9E+01
				368.28	3	1	5.1E+01	372.08	1	3	3.4E+01
Si V				371.48	5	3	2.07E+01				
96.439	1	3	4.8E+02	604.14	3	5	1.12E+01	*Na III*			
97.143	1	3	2.0E+03	2300.8	1	3	4.34E-01	378.14	4	2	7.7E+01
117.86	1	3	3.0E+02					380.10	2	2	3.7E+01
				Si XII				1991.0	4	6	8.3E+00
Si VI				*40.924	2	6	4.42E+03	2004.2	2	4	4.6E+00
246.00	4	2	1.7E+02	*44.118	6	10	1.4E+04	2011.9	6	8	8.4E+00
249.12	2	2	8.5E+01	499.43	2	4	9.56E+00	2151.5	2	4	4.4E+00
				520.72	2	2	8.47E+00	2174.5	4	6	5.3E+00
Si VII				1862	2	4	1.15E+00	2230.3	6	8	3.7E+00
217.83	5	3	4.3E+02	1949	2	2	1.0E+00	2232.2	4	6	3.3E+00
272.64	5	3	5.1E+01	4620	2	4	4.6E-02	2246.7	4	6	2.4E+00
274.18	3	1	1.2E+02	4942	4	6	4.5E-02	2459.3	4	6	3.0E+00
275.35	5	5	8.9E+01					2468.9	2	4	2.4E+00
275.67	3	3	3.0E+01	*Silver*				2497.0	6	6	1.7E+00
276.84	1	3	3.9E+01								
278.45	3	5	2.9E+01	*Ag I*				*Na V*			
				2061.2	2	4	3.1E-02	*307.89	10	6	2.0E+02
Si VIII				2069.9	2	2	1.5E-02	*333.46	6	6	5.6E+01
214.76	4	2	4.1E+02	3280.7	2	4	1.4E+00	*369.01	10	6	1.2E+02
216.92	6	4	3.6E+02	3382.9	2	2	1.3E+00	*400.72	10	10	5.0E+01
232.86	2	2	8.0E+01	5209.1	2	4	7.5E-01	*445.14	6	10	7.1E+00
235.56	4	4	9.7E+01	5465.5	4	6	8.6E-01	459.90	4	2	2.3E+01
250.45	2	2	7.7E+01	5471.6	4	4	1.4E-01	461.05	4	4	2.3E+01
250.79	4	2	1.6E+02					463.26	4	6	2.2E+01
314.31	4	2	5.2E+01	*Sodium*				510.10	2	2	5.6E+01
316.20	4	4	5.0E+01					511.19	4	4	6.8E+01
319.83	4	6	4.9E+01	*Na I*							
				3302.4	2	4	2.81E-02	*Na VI*			
Si IX				3303.0	2	2	2.81E-02	313.75	5	3	1.3E+02
223.73	1	3	4.2E+01	4390.0	2	4	7.7E-03				

λ Å	g_i	g_k	A 10⁸ s⁻¹	λ Å	g_i	g_k	A 10⁸ s⁻¹	λ Å	g_i	g_k	A 10⁸ s⁻¹
361.25	5	5	7.7E+01	2428.1	1	3	1.7E-01	1820.3	3	3	2.2E+00
*416.53	9	9	3.7E+01	2569.5	1	3	5.3E-02	1826.2	1	3	7.2E-01
*492.80	9	15	1.3E+01	2931.8	1	3	1.9E-02	4694.1	5	7	6.7E-03
1550.6	5	5	4.35E+00	4607.3	1	3	2.01E+00	4695.4	5	5	6.7E-03
1567.8	5	3	2.68E+00					4696.2	5	5	6.5E-03
1608.5	3	1	2.6E+00	**Sr II**				6403.6	3	3	5.7E-03
1649.4	5	5	2.05E+00	2018.7	2	2	1.2E-01	6408.1	5	5	9.5E-03
1741.5	3	5	2.59E+00	2051.9	4	2	2.4E-01	6415.5	7	5	1.3E-02
1747.5	5	7	3.1E+00	2282.0	2	4	8.3E-01	*6751.2	15	25	7.9E-02
				2322.4	4	6	9.1E-01	7679.6	3	5	1.2E-02
Na VII				2324.5	4	4	1.5E-01	7686.1	5	5	2.0E-02
*94.409	6	10	2.7E+03	2423.5	2	2	2.4E-01	7696.7	7	5	2.8E-02
*105.27	6	2	4.5E+02	2471.6	4	2	4.8E-01				
353.29	4	4	1.0E+02	3464.5	4	6	3.1E+00	**S II**			
381.30	4	2	4.0E+01	3474.9	4	4	5.1E-01	1124.4	2	4	1.0E+00
397.49	4	4	3.5E+01	4077.7	2	4	1.42E+00	1125.0	4	4	4.6E+00
399.18	6	4	5.2E+01	4161.8	2	2	6.5E-01	1131.0	2	2	3.5E+00
*483.28	10	10	2.9E+01	4215.5	2	2	1.27E+00	1131.6	4	2	1.4E+00
486.74	2	4	1.1E+01	4305.5	4	2	1.4E+00	1250.5	4	2	4.6E-01
491.95	4	6	1.3E+01	4414.8	4	6	1.1E-01	1253.8	4	4	4.2E-01
555.80	4	4	2.3E+01	4417.5	4	4	1.8E-02	1259.5	4	6	3.4E-01
777.83	4	6	6.8E+00	4585.9	4	2	7.0E-02	4463.6	8	6	5.3E-01
				5303.1	2	4	1.9E-01	4483.4	6	4	3.1E-01
Na VIII				5379.1	4	6	2.2E-01	4486.7	4	2	6.6E-01
*83.34	9	15	3.94E+03	5385.5	4	4	3.7E-02	4524.7	4	4	9.3E-02
*89.88	9	3	8.09E+02	5723.7	2	2	7.1E-02	4525.0	6	4	1.2E+00
90.536	3	5	2.86E+03	5819.0	4	2	1.4E-01	4552.4	4	2	1.2E+00
411.15	1	3	4.42E+01	8688.9	4	6	5.5E-01	4656.7	2	4	9.0E-02
1239.4	3	3	3.02E+00	8719.6	4	4	9.7E-02	4716.2	4	4	2.9E-01
1802.7	3	1	2.70E+00					4815.5	6	4	8.8E-01
1867.7	3	5	2.01E+00	**Sulfur**				4885.6	2	4	1.7E-01
2059.1	3	5	1.80E+00					4917.2	2	2	6.6E-01
2558.2	5	3	2.26E-02	**S I**				4924.1	4	6	2.2E-01
2772.0	3	5	4.19E-01	1295.7	5	5	4.9E+00	4925.3	2	4	2.4E-01
3021.0	5	7	4.90E-01	1296.2	5	3	2.7E+00	4942.5	2	2	1.5E-01
3108.9	1	3	2.58E-01	1302.3	3	5	1.8E+00	4991.9	4	4	1.5E-01
3182.3	1	3	2.92E-01	1302.9	3	3	1.6E+00	5009.5	4	2	7.0E-01
				1303.1	3	1	6.6E+00	5014.0	4	4	8.4E-01
Na IX				1303.4	5	3	1.9E+00	5027.2	4	2	2.6E-01
70.615	2	4	1.35E+03	1305.9	1	3	2.4E+00	5032.4	6	6	8.1E-01
70.653	2	2	1.35E+03	1401.5	5	3	9.1E-01	5047.3	4	2	3.6E-01
77.764	2	4	3.6E+03	1409.3	3	3	5.0E-01	5103.3	6	4	5.0E-01
77.911	4	6	4.3E+03	1412.9	1	3	1.6E-01	5142.3	2	2	1.9E-01
681.72	2	4	6.63E+00	1425.0	5	7	4.5E+00	5201.0	4	4	7.5E-01
694.17	2	2	6.30E+00	1425.2	5	5	1.2E+00	5201.3	6	4	6.5E-02
2487.7	2	4	8.32E-01	1433.3	3	5	3.3E+00	5212.6	4	6	9.8E-02
2535.8	2	2	7.89E-01	1433.3	3	3	1.9E+00	5212.6	6	6	8.5E-01
6841.8	2	4	2.59E-02	1437.0	1	3	2.4E+00	5320.7	6	8	9.2E-01
7103.4	4	6	2.78E-02	1448.2	5	3	7.3E+00	5345.7	4	6	8.8E-01
				1473.0	5	7	4.2E-01	5345.7	6	6	1.1E-01
Strontium				1474.0	5	7	1.6E+00	5428.6	2	4	4.2E-01
				1474.4	5	5	5.0E-01	5432.8	4	6	6.8E-01
Sr I				1474.6	5	3	6.2E-02	5453.8	6	8	8.5E-01
2206.2	1	3	6.6E-03	1481.7	3	5	1.7E-01	5473.6	2	2	7.3E-01
2211.3	1	3	8.5E-03	1483.0	3	5	1.2E+00	5509.7	4	4	4.0E-01
2217.8	1	3	1.2E-02	1483.2	3	3	7.5E-01	5526.2	8	8	8.1E-02
2226.3	1	3	1.6E-02	1487.2	1	3	8.7E-01	5536.8	4	6	6.6E-02
2237.7	1	3	2.3E-02	1666.7	5	5	6.3E+00	5556.0	4	2	1.1E-01
2253.3	1	3	3.7E-02	1687.5	1	3	9.4E-01	5564.9	6	6	1.7E-01
2275.3	1	3	6.7E-02	1782.3	1	3	1.9E+00	5578.8	6	6	1.1E-01
2307.3	1	3	1.2E-01	1807.3	5	3	3.8E+00				
2354.3	1	3	1.8E-01								

λ Å	Weights g_i	g_k	A $10^8\,s^{-1}$	λ Å	Weights g_i	g_k	A $10^8\,s^{-1}$	λ Å	Weights g_i	g_k	A $10^8\,s^{-1}$
5606.1	10	8	5.4E-01	706.48	2	4	4.17E+01	3260.2	4	4	5.8E-03
5616.6	4	4	1.2E-01	712.68	4	6	4.85E+01	3337.8	6	6	1.3E-02
5640.0	4	6	6.6E-01	712.84	4	4	8.1E+00	3383.9	6	4	5.3E-03
5645.6	6	4	1.8E-02	933.38	2	4	1.7E+01	3406.9	4	6	6.8E-02
5647.0	2	4	5.7E-01	944.52	2	2	1.6E+01	3419.7	8	8	1.91E-02
5659.9	6	4	4.6E-01					3463.8	4	6	2.62E-02
5664.7	4	2	5.8E-01	**S VII**				3484.6	4	4	8.5E-03
5819.2	4	4	8.5E-02	60.161	1	3	9.46E+03	3488.8	6	4	7.3E-03
6305.5	8	6	1.8E-01	60.804	1	3	5.1E+02	3497.9	6	8	4.9E-02
6312.7	6	4	3.0E-01	72.029	1	3	8.61E+02	3505.0	8	6	2.72E-02
								3553.4	4	6	3.3E-03
S III				**S VIII**				3607.4	6	8	4.6E-02
2496.2	7	5	2.5E+00	198.55	4	2	2.5E+02	3625.2	10	8	1.0E-02
2508.2	5	3	2.3E+00	202.61	2	2	1.2E+02	3626.6	8	10	7.1E-02
2636.9	3	5	4.5E-01					3642.1	10	12	5.5E-02
2665.4	5	5	1.4E+00	**S XI**				3657.5	6	6	4.3E-03
2680.5	1	3	6.2E-01	*189.90	9	3	4.3E+02	3731.0	4	6	5.3E-03
2691.8	3	3	4.6E-01	190.37	5	3	2.8E+02	3754.5	8	8	6.5E-03
2702.8	3	1	1.9E+00	215.95	5	5	1.4E+02	3784.3	4	6	4.3E-02
2718.9	3	3	1.2E+00	217.63	1	3	7.2E+01	3792.1	4	4	9.0E-03
2721.4	5	3	7.7E-01	239.81	1	3	2.6E+01	3826.9	6	6	5.2E-03
2726.8	3	5	6.0E-01	242.57	3	5	1.9E+01	3836.6	8	10	4.0E-03
2731.1	5	5	1.1E+00	242.82	3	3	1.9E+01	3848.1	10	8	1.30E-02
2756.9	7	7	1.4E+00	246.90	5	5	5.4E+01	3858.6	10	10	2.5E-03
2785.5	3	3	6.1E-01	247.12	5	3	3.0E+01	3918.5	4	2	2.5E-02
2856.0	5	7	5.1E+00	*288.49	9	15	2.9E+01	3922.8	4	4	3.98E-02
2863.5	7	9	5.7E+00					3996.2	2	4	3.35E-02
2872.0	3	5	4.7E+00	**S XII**				3999.3	4	4	1.8E-02
2950.2	3	5	3.0E+00	212.14	2	4	3.7E+01	4003.7	10	8	3.1E-03
2964.8	5	7	4.0E+00	215.18	2	2	1.4E+02	4006.8	6	8	7.6E-03
3662.0	3	3	6.4E-01	218.20	4	4	1.7E+02	4026.9	4	4	3.60E-02
3717.8	5	3	1.0E+00	221.44	4	2	6.4E+01	4029.9	10	10	2.8E-02
3778.9	3	5	4.4E-01	227.50	2	2	3.7E+01	4030.7	8	10	2.3E-03
3831.8	1	3	5.6E-01	234.48	4	2	6.8E+01	4040.9	10	12	7.3E-03
3837.8	3	3	4.2E-01					4061.4	2	4	6.5E-02
3838.3	5	5	1.3E+00	**S XIII**				4064.6	4	4	3.83E-02
3860.6	3	1	1.6E+00	32.236	1	3	1.09E+04	4067.2	6	4	6.8E-02
3899.1	5	3	6.7E-01	37.600	3	1	1.3E+03	4067.9	6	8	8.4E-03
4253.6	5	7	1.2E+00	256.66	1	3	8.7E+01	4097.2	10	8	2.1E-03
4285.0	3	5	9.0E-01	299.89	3	5	1.78E+01	4105.0	6	4	1.1E-02
				303.37	1	3	2.28E+01	4136.2	8	6	1.82E-02
S IV				307.36	3	3	1.64E+01	4147.9	10	8	1.79E-02
551.17	2	2	2.06E+01	308.91	5	5	4.82E+01	4175.2	6	8	2.8E-02
554.07	4	2	4.08E+01	312.68	3	1	6.3E+01	4205.9	8	10	8.9E-03
3097.5	2	4	2.6E+00	316.84	5	3	2.50E+01	4303.0	6	6	2.08E-02
3117.7	2	2	2.5E+00	500.42	3	5	1.43E+01	4378.8	8	6	4.8E-03
								4386.1	4	6	1.0E-02
S V				**S XIV**				4402.5	6	6	2.28E-02
437.37	1	3	1.12E+01	*30.434	2	6	8.28E+03	4415.7	2	4	2.53E-02
438.19	3	3	3.33E+01	*32.517	6	10	2.6E+04	4441.0	4	6	7.5E-03
439.65	5	3	5.5E+01	417.67	2	4	1.2E+01	4441.7	10	8	9.0E-03
*661.52	9	15	6.44E+01	445.71	2	2	1.0E+01	4473.5	6	8	1.36E-02
*679.01	9	15	8.6E+01	1550	2	4	1.4E+00	4511.0	10	12	1.56E-02
*690.75	9	9	5.0E+01	1663	2	2	1.2E+00	4511.5	10	8	3.6E-03
786.48	1	3	5.25E+01	3967	2	4	5.4E-02	4514.2	10	10	3.1E-03
*854.85	9	9	4.18E+01	4153	4	6	5.7E-02	4521.1	10	10	2.3E-03
								4530.9	4	6	2.42E-02
S VI				**Tantalum**				4547.2	4	6	5.3E-03
248.99	2	4	3.1E+01					4553.7	6	8	9.5E-03
249.27	2	2	3.1E+01	**Ta I**				4565.9	8	8	2.5E-02
388.94	2	2	4.5E+01	3127.9	4	6	5.7E-03				
390.86	4	2	8.8E+01	3168.3	4	4	6.0E-03				
				3170.3	8	10	8.5E-02				
				3205.5	6	8	5.6E-03				

λ (Å)	Weights g_i	g_k	A (10^8 s^{-1})	λ (Å)	Weights g_i	g_k	A (10^8 s^{-1})	λ (Å)	Weights g_i	g_k	A (10^8 s^{-1})
4574.3	4	4	1.2E-02	5944.0	4	6	2.13E-02	2601.1	8	6	1.7E-01
4580.7	8	10	2.1E-03	5997.2	10	10	2.4E-02	2622.5	8	10	6.1E-02
4619.5	6	4	5.3E-02	6020.7	2	4	1.0E-02	2841.1	6	6	2.0E-01
4633.1	4	4	1.2E-02	6045.4	6	8	2.6E-02	2854.2	8	6	2.7E-01
4669.1	6	4	2.85E-02	6047.3	8	10	9.0E-03	2914.8	8	8	7.7E-02
4681.9	6	6	1.5E-02	6249.8	6	6	3.5E-03	2933.0	8	8	1.0E-01
4684.9	10	8	2.8E-03	6258.7	6	8	3.3E-03	2973.2	8	8	2.3E-01
4685.3	6	8	3.4E-03	6309.6	4	6	1.83E-02	3046.9	8	8	1.8E-01
4691.9	2	4	4.08E-02	6360.8	6	8	4.6E-03	3081.1	8	8	1.9E-01
4706.1	6	6	1.4E-02	6428.6	6	6	6.0E-03	3122.5	6	6	5.2E-01
4740.2	4	4	5.0E-02	6430.8	8	8	2.9E-02	3142.4	6	6	8.8E-02
4758.0	4	6	7.5E-03	6450.4	8	10	2.2E-02	3172.7	8	8	1.8E-01
4769.0	8	8	2.8E-02	6485.4	10	10	5.8E-02	3233.7	8	10	5.1E-02
4780.9	10	8	2.16E-02	6514.4	6	4	2.2E-02	3247.0	6	8	3.0E-01
4812.8	4	4	1.2E-02	6516.1	6	8	1.25E-02	3251.8	6	4	5.2E-01
4825.4	6	6	2.63E-02	6612.0	6	4	1.9E-02	3380.7	6	8	2.0E-01
4832.2	4	4	1.7E-02	6673.7	2	4	9.0E-03	3406.0	6	8	1.5E-01
4852.2	4	4	1.7E-02	6771.7	4	4	5.8E-03	3410.1	8	10	1.0E-01
4884.0	6	8	1.1E-02	6866.2	8	6	2.58E-02	3416.6	8	8	5.7E-02
4904.6	12	10	1.95E-02	6927.4	10	12	1.01E-02	3418.6	6	6	1.1E-01
4920.9	8	10	2.1E-03	6928.5	10	8	1.69E-02	3563.9	8	8	9.8E-02
4921.3	2	4	1.2E-02	6951.3	10	10	3.7E-03	3567.4	8	10	4.2E-02
4926.0	4	4	1.5E-02	6953.9	6	8	8.3E-03	3744.1	8	8	9.5E-01
4936.4	8	6	4.5E-02	6966.1	8	8	1.2E-02	3751.8	8	10	1.9E-01
4969.7	4	4	1.0E-02	6969.5	10	10	2.9E-03	3798.5	6	4	1.2E+00
5012.5	4	4	1.9E-02	7407.9	6	4	2.0E-02	3807.7	6	6	3.9E-01
5037.4	10	8	4.4E-02					3883.1	8	6	1.0E+00
5043.3	6	4	2.73E-02	**Thallium**				3887.4	8	8	3.8E-01
5067.9	8	6	2.92E-02					3916.5	8	6	1.5E+00
5069.9	10	12	1.7E-03	*Tl I*				3949.3	6	6	1.0E+00
5082.3	10	12	1.9E-03	2104.6	2	4	4.0E-02	4022.6	6	8	4.0E-02
5087.4	6	4	1.5E-02	2118.9	2	2	2.0E-02	4044.5	6	4	2.9E-01
5090.7	8	6	9.5E-03	2129.3	2	4	5.8E-02	4094.2	8	6	9.0E-01
5095.3	6	6	5.0E-03	2151.9	2	2	3.1E-02	4105.8	8	10	6.0E-01
5136.5	2	2	4.5E-02	2168.6	2	4	9.8E-02	4138.3	6	4	7.0E-01
5141.6	4	2	1.2E-02	2237.8	2	4	1.9E-01	4158.6	8	8	5.5E-02
5143.7	6	4	1.7E-02	2316.0	2	2	7.8E-02	4187.6	8	8	6.1E-01
5147.6	6	4	9.0E-03	2379.7	2	4	4.4E-01	4203.7	8	10	2.5E-01
5161.8	4	6	6.3E-03	2507.9	4	2	1.1E-02	4222.7	6	8	1.5E-01
5218.7	8	6	8.2E-03	2538.2	4	2	1.6E-02	4271.7	6	6	1.1E-01
5235.4	6	6	4.7E-03	2580.1	2	2	1.8E-01	4359.9	8	6	1.3E-01
5295.0	6	6	7.5E-03	2609.0	4	6	1.0E-01	4386.4	8	8	4.2E-02
5336.1	6	8	5.5E-03	2609.8	4	4	1.9E-02	4394.4	6	4	1.1E-01
5349.6	6	4	2.2E-02	2665.6	4	2	5.7E-02	4643.1	6	6	3.4E-02
5354.7	4	4	6.5E-03	2709.2	4	6	1.7E-01	4681.9	6	8	3.9E-02
5396.0	6	8	2.5E-03	2710.7	4	4	3.7E-02	4691.1	6	6	3.9E-02
5402.5	4	2	1.4E-02	2767.9	2	4	1.26E+00	5307.1	8	10	2.3E-02
5435.3	4	6	1.1E-02	2826.2	4	2	8.0E-02	5658.3	6	8	1.0E-02
5499.4	10	10	6.1E-03	2918.3	4	6	4.2E-01	5675.8	8	10	1.3E-02
5518.9	8	10	3.8E-02	2921.5	4	4	7.6E-02	5760.2	6	6	1.3E-02
5620.7	8	10	6.0E-03	3229.8	4	2	1.73E-01				
5640.2	6	8	4.9E-03	3519.2	4	6	1.24E+00	**Tin**			
5645.9	6	8	1.43E-02	3529.4	4	4	2.20E-01				
5699.2	6	6	4.2E-03	3775.7	2	2	6.25E-01	*Sn I*			
5767.9	6	8	2.6E-03	5350.5	4	2	7.05E-01	2073.1	1	3	3.6E-02
5780.7	4	6	3.3E-03					2199.3	3	5	2.9E-01
5811.1	8	6	5.7E-03	**Thulium**				2209.7	5	5	5.6E-01
5849.7	10	8	2.8E-03					2246.1	1	3	1.6E+00
5877.4	10	12	2.3E-02	*Tm I*				2268.9	5	7	1.2E+00
5939.8	2	4	1.6E-02	2513.8	8	10	6.9E-02	2286.7	5	5	3.1E-01
				2527.0	8	8	1.7E-01	2317.2	5	7	2.0E+00
				2596.5	8	10	1.6E-01				

λ Å	Weights g_i	g_k	A 10^8 s^{-1}	λ Å	Weights g_i	g_k	A 10^8 s^{-1}	λ Å	Weights g_i	g_k	A 10^8 s^{-1}
2334.8	3	3	6.6E-01	5596.2	4	4	1.5E-01	4055.01	1	3	2.8E-01
2354.8	3	5	1.7E+00	5797.2	6	6	2.8E-01	4060.26	3	5	2.4E-01
2380.7	3	5	3.1E-02	5799.2	6	8	8.1E-01	4064.20	3	3	2.4E-01
2408.2	5	3	1.8E-01	6453.5	2	4	1.2E+00	4065.09	3	1	7.0E-01
2421.7	5	7	2.5E+00	6761.5	2	2	3.2E-01	4186.12	9	9	2.10E-01
2429.5	5	7	1.5E+00	6844.1	2	2	6.6E-01	4266.23	5	5	3.1E-01
2433.5	5	3	8.0E-03					4284.99	5	5	3.2E-01
2455.2	5	5	1.1E-02	*Titanium*				4289.07	5	5	3.0E-01
2476.4	5	3	1.1E-02					4290.93	3	3	4.5E-01
2483.4	5	5	2.1E-01	*Ti I*				4295.75	3	1	1.3E+00
2491.8	1	3	1.7E-01	2276.75	7	5	1.3E+00	4393.93	9	11	3.3E-01
2495.7	5	5	6.2E-01	2280.00	9	7	9.4E-01	4417.27	11	9	3.6E-01
2523.9	5	3	7.4E-02	2299.86	5	5	6.9E-01	4449.14	11	11	9.7E-01
2546.6	1	3	2.1E-01	2302.75	7	7	5.7E-01	4450.90	9	9	9.6E-01
2558.0	1	3	3.4E-01	2305.69	9	9	5.2E-01	4453.31	5	5	5.98E-01
2571.6	5	7	4.5E-01	2424.26	9	9	1.7E-01	4453.71	7	7	4.7E-01
2594.4	5	5	3.0E-01	2520.54	5	3	3.8E-01	4455.32	7	7	4.8E-01
2636.9	1	3	1.1E-01	2529.87	7	5	3.8E-01	4457.43	9	9	5.6E-01
2661.2	3	3	1.1E-01	2541.92	9	7	4.3E-01	4465.81	5	7	3.28E-01
2706.5	3	5	6.6E-01	2599.91	5	5	6.7E-01	4481.26	7	7	5.7E-01
2761.8	5	5	3.7E-03	2605.16	7	7	6.4E-01	4496.15	7	5	4.4E-01
2779.8	5	7	1.8E-01	2611.29	9	9	6.4E-01	4518.02	7	9	1.72E-01
2785.0	5	3	1.4E-01	2611.47	7	5	3.3E-01	4522.80	5	7	1.9E-01
2788.0	1	3	1.4E-01	2619.94	9	7	2.1E-01	4527.31	3	5	2.2E-01
2812.6	1	3	2.3E-01	2631.55	7	7	1.7E-01	4533.24	11	11	8.83E-01
2813.6	5	5	1.2E-01	2632.42	5	5	2.7E-01	4534.78	9	9	6.87E-01
2840.0	5	5	1.7E+00	2641.12	5	3	1.8E+00	4544.69	5	3	3.3E-01
2850.6	5	5	3.3E-01	2644.28	7	5	1.4E+00	4548.76	7	5	2.85E-01
2863.3	1	3	5.4E-01	2646.65	9	7	1.5E+00	4552.45	9	7	2.1E-01
2913.5	1	3	8.3E-01	2733.27	5	5	1.9E+00	4563.43	9	11	2.1E-01
3009.1	3	3	3.8E-01	2735.30	3	1	4.1E+00	4617.27	7	9	8.51E-01
3032.8	1	3	6.2E-01	2912.07	5	7	1.3E+00	4623.10	5	7	5.74E-01
3034.1	3	1	2.0E+00	2942.00	5	5	1.0E+00	4639.94	3	3	6.64E-01
3141.8	1	3	1.9E-01	2948.26	7	7	9.3E-01	4640.43	3	1	5.0E-01
3175.1	5	3	1.0E+00	2956.13	9	9	9.7E-01	4645.19	3	1	8.57E-01
3218.7	1	3	4.7E-02	2956.80	7	5	1.8E+00	4650.02	5	3	2.6E-01
3223.6	5	5	1.2E-03	3186.45	5	7	8.0E-01	4742.79	9	9	5.3E-01
3262.3	5	3	2.7E+00	3191.99	7	9	8.5E-01	4758.12	11	11	7.13E-01
3330.6	5	5	2.0E-01	3199.92	9	11	9.4E-01	4759.27	13	13	7.40E-01
3655.8	1	3	4.1E-02	3341.88	5	7	6.5E-01	4778.26	9	9	2.0E-01
3801.0	5	3	2.8E-01	3354.63	7	9	6.9E-01	4805.42	5	7	5.8E-01
4524.7	1	3	2.6E-01	3370.44	5	3	7.6E-01	4840.87	5	5	1.76E-01
5631.7	1	3	2.4E-02	3371.45	9	11	7.2E-01	4856.01	13	15	5.2E-01
5970.3	5	3	9.6E-02	3377.58	7	5	6.9E-01	4885.08	11	13	4.90E-01
6037.7	5	5	5.0E-02	3385.94	9	7	5.0E-01	4913.62	7	9	4.44E-01
6069.0	1	3	4.6E-02	3635.46	5	7	8.04E-01	4928.34	3	5	6.2E-01
6073.5	3	1	6.3E-02	3642.68	7	9	7.74E-01	4981.73	11	13	6.60E-01
6171.5	3	3	4.9E-02	3653.50	9	11	7.54E-01	4989.14	7	5	3.25E-01
				3724.57	9	9	9.1E-01	4991.07	9	11	5.84E-01
Sn II				3725.16	5	3	7.3E-01	4999.50	7	9	5.27E-01
2368.3	4	2	4.4E-03	3729.81	5	5	4.27E-01	5000.99	9	7	3.52E-01
2449.0	4	6	3.7E-01	3741.06	7	7	4.17E-01	5007.21	5	7	4.92E-01
2487.0	6	8	5.5E-01	3752.86	9	9	5.04E-01	5014.28	3	5	6.8E-01
3283.2	4	6	1.0E+00	3786.04	5	3	1.4E+00	5036.47	7	9	3.94E-01
3352.0	6	8	1.0E+00	3948.67	5	3	4.85E-01	5038.40	5	7	3.87E-01
3472.5	2	4	1.6E-01	3956.34	7	5	3.00E-01	5062.11	5	3	2.98E-01
3575.5	4	6	1.3E-01	3958.21	9	7	4.05E-01	5210.39	9	9	3.57E-02
5332.4	2	4	8.6E-01	3981.76	5	5	3.76E-01	5222.69	3	3	1.95E-01
5562.0	4	6	1.2E+00	3989.76	7	7	3.79E-01	5224.30	11	11	3.6E-01
5588.9	4	6	8.5E-01	3998.64	9	9	4.08E-01	5259.98	5	7	2.3E-01
				4013.24	7	5	2.0E-01				

λ Å	Weights g_i	g_k	A 10^8 s^{-1}	λ Å	Weights g_i	g_k	A 10^8 s^{-1}	λ Å	Weights g_i	g_k	A 10^8 s^{-1}
5351.07	7	7	3.4E-01	3023.67	8	8	1.0E+00	3510.86	8	8	9.3E-01
5503.90	11	9	2.6E-01	3029.76	10	10	3.5E-01	3520.27	2	4	4.8E-01
5774.04	9	11	5.5E-01	3056.75	2	4	3.2E-01	3535.41	4	6	5.5E-01
5785.98	11	13	6.1E-01	3058.08	6	6	5.0E-01	3641.33	4	2	4.9E-01
5804.27	13	15	6.8E-01	3066.34	4	4	3.3E-01	3706.23	4	4	3.1E-01
6098.66	9	7	2.5E-01	3071.25	6	4	3.6E-01	3741.64	6	6	6.2E-01
6220.46	9	7	1.8E-01	3072.99	4	2	1.6E+00	3757.70	4	4	4.1E-01
Ti II				3075.23	6	4	1.13E+00	3759.30	8	8	9.4E-01
2440.91	4	4	5.1E-01	3078.65	8	6	1.09E+00	3761.33	6	6	9.9E-01
2451.18	6	6	4.5E-01	3081.52	10	8	1.1E+00	4911.18	6	4	3.2E-01
2525.59	10	8	5.6E-01	3088.04	10	8	1.25E+00	*Ti III*			
2531.28	8	6	4.9E-01	3089.44	8	6	1.3E+00	865.79	5	3	6.6E+01
2534.63	6	4	5.4E-01	3097.20	4	6	4.4E-01	1002.37	5	5	7.6E+00
2535.89	4	2	6.8E-01	3103.81	10	8	1.1E+00	1004.67	7	5	4.3E+01
2555.99	6	8	3.2E-01	3105.10	2	4	6.3E-01	1005.80	3	3	1.3E+01
2635.44	4	4	1.9E+00	3106.26	6	6	7.8E-01	1007.16	5	3	3.8E+01
2638.56	6	6	1.7E+00	3117.67	4	2	1.1E+00	1008.12	3	1	5.1E+01
2642.02	8	8	1.9E+00	3119.83	6	4	5.9E-01	1286.37	9	9	2.0E+00
2645.86	10	10	2.7E+00	3127.86	6	6	1.6E+00	1289.30	7	7	2.2E+00
2746.54	6	8	2.6E+00	3128.50	8	8	1.1E+00	1291.62	5	5	2.4E+00
2751.59	8	10	3.7E+00	3161.23	4	2	5.9E-01	1293.23	9	7	1.0E+00
2752.68	8	10	1.1E+00	3161.80	6	4	4.6E-01	1298.97	7	5	4.9E+00
2757.62	6	8	7.2E-01	3162.59	8	6	3.9E-01	1327.59	5	3	3.2E+00
2758.35	4	6	9.9E-01	3168.55	10	8	4.1E-01	1420.44	1	3	1.2E+00
2758.79	2	4	4.4E-01	3181.73	6	8	4.6E-01	1421.63	3	1	4.0E+00
2764.28	4	4	7.4E-01	3182.54	4	6	4.3E-01	1422.41	5	5	3.0E+00
2804.82	6	8	4.6E+00	3189.49	4	4	9.2E-01	1424.14	5	3	1.6E+00
2810.30	8	10	5.1E+00	3190.91	6	8	1.3E+00	1455.19	9	7	6.4E+00
2817.83	10	12	3.8E+00	3202.56	4	6	1.1E+00	1498.70	5	5	2.8E+00
2819.87	8	8	6.5E-01	3224.25	12	10	7.0E-01	2007.36	3	3	3.4E+00
2821.26	6	8	7.9E-01	3228.62	4	2	2.0E+00	2007.60	1	3	1.2E+00
2827.12	8	10	1.0E+00	3232.29	8	6	6.0E-01	2010.80	5	3	5.4E+00
2828.06	12	14	4.4E+00	3234.51	10	10	1.38E+00	2097.30	5	7	3.3E+00
2828.64	6	6	1.2E+00	3236.13	4	4	7.0E-01	2099.86	3	5	2.5E+00
2828.83	10	10	9.1E-01	3236.58	8	8	1.11E+00	2104.86	3	3	1.1E+00
2834.02	10	12	7.9E-01	3239.04	6	6	9.87E-01	2105.09	1	3	1.7E+00
2836.47	8	8	1.2E+00	3239.66	6	4	9.4E-01	2199.22	3	3	5.7E+00
2839.64	12	12	8.3E-01	3241.99	4	4	1.16E+00	2237.77	7	7	2.4E+00
2845.93	10	10	1.2E+00	3251.91	6	4	3.38E-01	2331.35	3	1	4.3E+00
2851.11	2	4	4.1E-01	3252.92	8	6	3.9E-01	2331.66	3	3	1.2E+00
2856.10	12	12	1.5E+00	3272.07	2	4	3.2E-01	2339.00	5	3	3.0E+00
2862.33	4	6	4.0E-01	3278.28	4	4	9.6E-01	2346.79	7	5	3.3E+00
2877.47	8	8	5.7E-01	3278.91	6	4	1.0E+00	2374.99	5	3	4.0E+00
2884.13	10	10	5.2E-01	3282.32	2	2	1.6E+00	2413.99	5	7	3.8E+00
2910.65	8	8	4.6E-01	3287.66	8	10	1.4E+00	2516.05	7	9	3.4E+00
2926.64	10	8	8.9E-01	3315.32	2	4	3.8E-01	2567.56	3	3	2.3E+00
2931.10	6	6	3.2E+00	3321.70	4	4	7.2E-01	2984.75	5	5	1.9E+00
2936.02	4	6	2.7E+00	3322.94	10	10	3.96E-01	3066.51	3	3	2.5E+00
2938.57	6	8	2.4E+00	3329.46	8	8	3.25E-01	3228.89	3	3	1.5E+00
2941.90	8	10	1.8E+00	3332.11	6	4	1.1E+00	3278.31	7	9	3.4E+00
2942.97	8	8	1.1E+00	3340.34	4	4	3.6E-01	3320.94	3	5	2.8E+00
2945.30	10	12	2.7E+00	3361.23	8	10	1.1E+00	3340.20	7	9	3.7E+00
2952.00	8	8	3.0E-01	3372.80	6	8	1.11E+00	3346.18	9	11	3.7E+00
2954.59	10	12	4.0E+00	3383.77	4	6	1.09E+00	3354.71	11	13	4.4E+00
2958.80	8	10	4.0E+00	3452.49	2	2	7.7E-01	3397.24	3	1	1.8E+00
2979.06	4	6	1.2E+00	3456.40	4	4	8.2E-01	3404.46	3	3	1.8E+00
2990.06	6	8	5.6E-01	3465.56	4	2	4.1E-01	3417.62	3	5	1.9E+00
3017.17	12	12	3.6E-01	3483.63	10	8	9.7E-01	3915.47	9	11	2.1E+00
3022.64	10	10	1.2E+00	3492.37	8	6	9.8E-01	4119.14	5	5	9.9E-01
				3504.90	10	10	8.2E-01				

λ Å	Weights g_i	g_k	A 10^8 s^{-1}	λ Å	Weights g_i	g_k	A 10^8 s^{-1}	λ Å	Weights g_i	g_k	A 10^8 s^{-1}
4213.26	9	11	2.2E+00	439.513	3	3	7.5E+00	60.762	2	2	3.5E+02
4215.53	9	11	2.2E+00	439.745	3	1	2.1E+01	61.286	4	2	1.8E+02
4247.62	11	13	1.1E+00	447.484	5	5	1.6E+01	62.433	4	6	2.08E+02
4248.54	5	7	2.3E+00	447.701	5	3	6.5E+00	62.470	6	8	2.22E+02
4250.09	3	5	9.5E-01	507.174	3	5	6.5E+00	65.540	4	6	3.2E+02
4259.01	11	13	9.4E-01	516.215	5	7	6.9E+00	65.577	6	8	3.5E+02
4269.84	9	11	1.7E+00					67.171	2	4	6.2E+02
4285.61	13	15	3.0E+00	*Ti X*				67.555	4	6	7.2E+02
4288.66	11	13	1.1E+00	253	4	6	2.1E+02	70.986	4	6	5.7E+02
4296.70	11	13	1.6E+00	254	6	8	2.3E+02	71.031	6	8	6.1E+02
4319.56	9	11	1.1E+00	281	2	2	1.1E+02	71.545	2	2	1.8E+02
4343.25	3	1	1.0E+00	289.579	2	4	2.5E+02	71.987	4	2	3.48E+02
4378.94	3	5	1.6E+00	290.294	4	6	1.1E+02	82.121	2	4	5.9E+02
4433.91	11	13	1.8E+00	291	4	2	1.8E+02	82.307	4	6	1.13E+03
4440.66	1	3	1.2E+00	291	2	2	2.3E+02	82.344	2	2	5.8E+02
4533.26	3	5	1.5E+00	292	6	8	1.1E+02	82.368	6	8	1.2E+03
4576.53	9	7	1.3E+00	293.684	6	8	2.97E+02	89.844	2	4	9.9E+02
4628.07	3	1	1.5E+00	293.798	6	6	1.7E+02	90.512	4	6	1.16E+03
4652.86	7	9	2.6E+00	295.584	4	6	2.9E+02	90.547	4	4	1.9E+02
4874.00	5	7	1.5E+00	296	4	6	1.4E+02	116.497	4	6	3.0E+03
4914.32	3	3	1.1E+00	297	4	6	9.9E+01	116.597	6	8	3.2E+03
4971.19	9	11	2.1E+00	298	4	6	4.3E+02	116.62	6	6	2.1E+02
5083.80	5	3	9.7E-01	302	2	2	1.6E+02	139.884	6	4	2.6E+02
5278.33	3	3	9.4E-01	305	2	4	2.5E+02	140.361	4	2	2.9E+02
7506.87	11	13	1.1E+00	317	2	2	1.5E+02	141.6	4	6	1.7E+02
				355.815	2	2	1.3E+02	141.7	6	8	1.7E+02
Ti IV				360.133	4	4	2.19E+02	169.7	4	6	2.8E+02
423.49	4	6	4.9E+01	363	4	2	2.1E+02	169.8	6	8	2.9E+02
424.16	6	8	5.3E+01	363	6	6	1.3E+02	207.2	2	4	1.5E+02
433.63	4	2	5.5E+00	365.628	4	2	1.2E+02	208.5	4	6	1.8E+02
433.76	6	4	5.0E+00	382	4	6	1.8E+02	252.8	4	6	4.8E+02
729.36	4	2	5.7E+00	385	6	8	1.8E+02	253.1	6	8	5.2E+02
1183.64	2	2	6.9E+00	389.99	6	4	1.1E+02	257.5	4	2	2.4E+02
1195.21	4	2	1.4E+01								
1451.74	2	4	1.8E+01	*Ti XI*				*Ti XIII*			
1467.34	4	6	2.1E+01	65.403	1	3	5.1E+02	23.356	1	3	1.02E+05
2067.56	2	4	5.1E+00	87.725	1	3	8.5E+02	23.698	1	3	1.2E+04
2103.16	2	2	5.0E+00	266	5	7	1.8E+02	23.991	1	3	3.4E+02
2541.79	4	6	6.9E+00	308.250	3	5	1.3E+02	26.641	1	3	4.06E+03
2546.88	6	8	7.4E+00	313.229	5	7	1.6E+02	26.960	1	3	3.06E+03
2862.60	4	2	4.1E+00	318	3	1	1.4E+02	117.1	3	3	1.3E+02
3576.44	4	6	4.6E+00	322.75	5	7	1.99E+02	117.3	3	1	2.8E+02
				323	1	3	1.8E+02	120.2	5	3	5.4E+02
Ti VIII				327.192	3	5	2.9E+02	120.2	7	5	4.4E+02
249	6	4	1.0E+01	332	3	1	3.25E+02	128.7	3	3	1.2E+02
258.610	6	8	7.5E+02	386.140	1	3	1.48E+02				
269.533	4	6	6.0E+02	408	7	9	1.37E+02	*Ti XIV*			
272.037	4	4	4.3E+02	425.74	3	1	1.2E+02	21.341	4	6	9.8E+03
272.843	6	4	6.2E+01	446.69	3	1	1.2E+02	21.522	2	4	4.5E+04
276.701	2	4	9.3E+01	453	5	7	1.3E+02	21.657	4	4	1.3E+04
277.813	4	4	3.8E+02					21.733	4	4	8.8E+04
289.375	2	4	3.6E+01	*Ti XII*				21.82	4	2	6.4E+04
478.971	4	4	1.7E+01	52.896	2	4	1.61E+02	21.883	2	4	7.0E+04
480.376	6	6	1.5E+01	53.140	4	6	1.9E+02	21.958	2	4	1.2E+04
				53.433	2	4	2.1E+02	22.05	2	2	1.4E+04
Ti IX				53.457	2	2	2.1E+02	24.592	4	2	6.1E+03
267.941	5	7	5.1E+02	55.181	2	4	2.4E+02	24.891	2	2	7.5E+03
278.713	5	7	4.7E+02	55.443	4	6	2.81E+02				
281.446	3	1	3.2E+02	59.133	2	4	3.72E+02	*Ti XV*			
285.128	1	3	4.1E+02	59.435	4	6	4.41E+02	20.19	5	7	6.9E+03
433.567	1	3	6.9E+00	60.701	2	4	3.4E+02	20.234	5	7	1.9E+04

λ	Weights		A	λ	Weights		A	λ	Weights		A
Å	g_i	g_k	$10^8\,\text{s}^{-1}$	Å	g_i	g_k	$10^8\,\text{s}^{-1}$	Å	g_i	g_k	$10^8\,\text{s}^{-1}$
20.234	3	3	4.9E+04	163.049	3	1	6.2E+02	10.690	4	6	1.58E+04
20.246	1	3	4.2E+04	186.863	5	5	2.66E+02	*11.452	2	6	1.7E+04
20.250	5	3	6.5E+03	207.73	3	1	1.07E+02	11.872	2	4	2.8E+04
20.29	3	3	1.1E+04					11.958	4	6	3.4E+04
20.30	1	3	3.4E+04	*Ti XVIII*				11.958	4	4	5.6E+03
20.30	1	1	5.8E+04	17.22	2	4	7.3E+04	15.211	2	4	3.50E+04
20.313	5	3	7.5E+04	17.365	4	6	8.6E+04	15.253	2	2	3.58E+04
20.418	5	7	8.0E+04	17.39	4	4	1.4E+04	15.907	2	4	8.84E+04
20.538	3	3	3.8E+04	133.852	2	4	5.2E+01	16.049	4	6	1.05E+05
20.54	3	1	4.1E+04	144.759	4	4	3.2E+02	16.067	4	4	1.8E+04
20.551	1	3	1.3E+04	150.15	6	4	1.15E+02	31.586	4	6	5.49E+03
20.689	5	7	4.3E+04	153.15	4	2	1.97E+02	45.650	2	4	9.6E+03
20.698	1	3	1.1E+05	153.23	2	4	6.7E+01	45.996	4	6	1.1E+04
20.771	5	3	1.1E+04	159.00	4	4	1.16E+02				
20.897	5	7	2.85E+04	166.225	6	4	1.54E+02	*Ti XXI*			
20.928	5	5	8.4E+03	179.902	2	4	6.3E+01	2.0633	1	3	1.32E+05
21.065	3	3	1.1E+04	189.663	6	6	9.6E+01	2.1108	1	3	2.60E+05
21.079	1	3	1.58E+04	191.23	4	4	6.6E+01	2.2211	1	3	6.35E+05
21.102	3	5	1.3E+04	197.838	4	6	4.56E+01	2.497	3	1	2.4E+06
22.482	5	3	6.4E+03	208.07	4	4	1.2E+02	2.505	5	5	3.5E+05
22.936	5	5	1.1E+04					2.505	1	3	1.4E+06
22.966	5	3	1.1E+04	*Ti XIX*				2.507	3	5	1.4E+06
23.034	1	3	6.3E+03	15.67	3	1	3.3E+04	2.508	3	5	7.9E+05
				15.68	5	5	2.7E+04	2.510	3	3	6.9E+05
Ti XVI				15.74	5	7	2.7E+04	2.510	1	3	9.6E+05
110.561	4	2	3.36E+02	15.75	3	5	2.4E+04	2.511	3	3	1.4E+06
116.198	4	4	1.45E+02	15.83	1	3	3.2E+04	2.512	5	5	1.8E+06
118.215	6	4	7.4E+02	15.86	1	3	2.9E+04	2.512	3	1	1.4E+06
121.382	4	2	2.4E+02	16.02	3	1	3.1E+04	2.513	3	1	2.7E+06
124.805	4	2	6.1E+02	16.18	3	5	3.8E+04	2.513	3	5	2.4E+06
129.075	4	2	3.81E+02	16.41	1	3	6.1E+04	2.514	5	3	1.2E+06
134.724	2	2	2.6E+02	16.43	3	5	8.2E+04	2.520	3	5	2.6E+05
138.800	6	4	3.5E+02	16.46	3	3	4.4E+04	2.527	3	1	1.2E+05
143.459	4	4	2.8E+02	16.51	5	7	1.0E+05	2.539	3	1	4.1E+05
145.665	6	6	2.3E+02	16.55	5	5	2.7E+04	2.6102	1	3	2.40E+06
157.812	4	2	1.32E+02	16.61	3	1	8.0E+04	2.6227	1	3	1.12E+05
161.168	4	4	1.2E+02	16.64	3	3	5.3E+04				
163.610	4	2	1.92E+02	16.69	1	3	1.02E+05	*Tungsten*			
169.740	4	6	1.0E+02	16.71	3	5	7.3E+04				
176.267	2	2	2.45E+02	16.72	5	3	3.3E+04	*W I*			
178.240	4	4	2.52E+02	16.72	5	5	7.3E+04	2879.4	1	3	2.4E-01
				16.74	5	7	1.2E+05	2911.0	1	3	7.7E-02
Ti XVII				16.77	3	3	2.6E+04	2923.5	7	9	1.54E-02
18.05	3	3	4.5E+04	16.80	5	7	1.81E+05	2935.0	3	5	1.5E-01
18.13	5	3	2.4E+04	16.85	3	5	4.4E+04	3013.8	7	9	6.4E-02
18.13	1	3	8.1E+04	17.08	3	5	8.3E+04	3016.5	9	11	9.27E-02
18.176	5	7	9.2E+04	17.36	1	3	9.5E+04	3017.4	7	9	1.21E-01
123.654	3	3	2.3E+02					3024.9	3	3	1.4E-01
124.553	5	3	5.2E+02	*Ti XX*				3046.4	3	5	5.8E-02
127.782	5	3	4.6E+02	2.629	2	4	4.9E+04	3049.7	7	5	1.7E-01
135.202	3	1	2.93E+02	2.6295	4	4	3.2E+06	3064.9	5	7	1.1E-02
136.160	5	3	1.95E+02	2.631	2	2	6.1E+05	3084.9	5	5	1.3E-02
136.393	3	3	1.14E+02	2.6319	2	4	1.5E+06	3093.5	7	9	4.4E-02
141.948	5	5	3.87E+02	2.632	2	2	2.7E+06	3107.2	5	7	2.33E-02
142.589	1	3	1.35E+02	2.6355	4	6	1.2E+06	3108.0	7	9	1.58E-02
144.405	5	5	9.4E+01	8.621	4	2	1.1E+06	3145.5	9	9	4.8E-03
146.067	7	5	2.6E+02	9.788	4	6	5.26E+03	3170.2	7	5	6.0E-03
154.133	3	1	1.63E+02	10.046	2	4	7.29E+03	3176.6	3	5	2.12E-02
156.54	3	1	1.44E+02	10.109	4	6	8.6E+03	3183.5	7	7	2.64E-03
158.469	5	5	1.4E+02	*10.278	2	6	8.4E+03	3184.4	5	3	2.3E-02
159.62	5	3	1.03E+02	10.620	2	4	1.34E+04	3191.6	1	3	3.2E-02

λ Å	Weights g_i	g_k	A 10^8 s^{-1}	λ Å	Weights g_i	g_k	A 10^8 s^{-1}	λ Å	Weights g_i	g_k	A 10^8 s^{-1}
3198.8	7	9	4.6E-02	4294.6	7	5	1.2E-01	6285.9	7	5	6.6E-03
3207.3	7	9	3.0E-02	4302.1	7	7	3.6E-02	6292.0	3	5	2.26E-03
3208.3	5	5	4.4E-02	4355.2	9	9	5.1E-03	6303.2	9	9	1.84E-03
3215.6	9	11	2.1E-01	4361.8	9	7	1.64E-03	6404.2	5	7	1.50E-03
3221.9	5	7	1.61E-02	4378.5	7	5	3.48E-03	6439.7	9	9	1.29E-03
3223.1	5	3	3.53E-03	4458.1	3	5	4.2E-03	6445.1	7	5	6.4E-03
3232.5	9	9	2.4E-02	4466.3	7	5	1.5E-02	6532.4	3	5	4.6E-03
3235.1	7	5	2.68E-03	4472.5	13	11	1.55E-03	6538.1	11	9	2.7E-03
3259.7	7	7	1.3E-02	4484.2	3	5	5.6E-03	6563.2	5	5	2.04E-03
3300.8	7	9	8.1E-02	4492.3	9	11	3.6E-03	6814.9	9	9	1.46E-03
3311.4	7	5	5.6E-02	4495.3	11	11	3.3E-03	7285.8	13	11	1.47E-03
3363.3	9	7	6.6E-03	4504.8	9	7	7.0E-03	7569.9	5	3	3.73E-03
3371.0	7	5	1.0E-02	4552.5	9	9	1.42E-03	7664.9	5	3	3.80E-03
3371.4	3	3	6.7E-03	4586.8	1	3	4.20E-03	8017.2	5	7	1.6E-03
3386.1	7	7	2.64E-03	4592.6	7	9	3.4E-03	8358.7	5	7	1.89E-03
3413.0	7	9	9.7E-03	4609.9	7	9	1.42E-02	9381.4	9	7	1.53E-03
3459.5	9	9	2.04E-03	4613.3	9	9	2.9E-03				
3510.0	7	9	5.2E-03	4634.8	9	9	8.8E-03	*Uranium*			
3545.2	1	3	3.2E-02	4659.9	1	3	1.0E-02				
3570.6	5	3	6.7E-03	4680.5	7	7	1.4E-02	*U I*			
3606.1	3	5	9.6E-03	4720.4	3	5	3.22E-03	3553.0	13	13	2.0E-02
3617.5	7	7	1.1E-01	4729.6	7	5	7.8E-03	3553.0	9	7	1.4E-02
3631.9	3	5	1.3E-02	4752.6	3	3	5.20E-03	3553.4	15	13	2.2E-02
3675.6	9	11	1.20E-02	4757.5	7	5	2.72E-03	3554.5	11	9	8.4E-03
3682.1	9	11	2.0E-02	4757.8	11	9	4.1E-03	3554.9	15	17	7.9E-03
3707.9	7	7	2.9E-02	4788.4	9	11	2.6E-02	3555.3	13	15	2.7E-02
3757.9	7	9	1.38E-02	4843.8	5	5	1.9E-02	3555.8	13	11	4.1E-03
3760.1	5	7	1.99E-02	4886.9	9	11	8.1E-03	3556.9	13	11	7.5E-03
3768.5	3	3	3.47E-02	4924.6	13	11	1.75E-03	3557.8	13	13	2.9E-02
3780.8	7	5	4.2E-02	4931.6	7	5	1.0E-02	3558.0	11	13	1.6E-02
3809.2	7	5	9.0E-03	4948.6	9	11	1.36E-03	3558.6	9	7	3.9E-02
3817.5	7	7	3.1E-02	4972.6	9	11	3.9E-03	3559.4	7	9	1.5E-02
3829.1	3	3	3.83E-03	4982.6	1	3	4.17E-03	3560.3	9	7	6.4E-02
3835.1	5	5	5.2E-02	4986.9	11	9	6.3E-03	3561.4	15	13	5.5E-02
3846.3	3	5	2.14E-02	5006.2	9	7	1.2E-02	3561.5	9	9	2.5E-02
3847.5	1	3	8.3E-03	5015.3	7	9	5.4E-03	3561.8	13	11	5.7E-02
3864.3	5	5	5.6E-03	5040.4	3	5	5.2E-03	3563.7	13	13	2.9E-02
3868.0	7	9	4.6E-02	5053.3	3	3	1.9E-02	3563.8	7	7	1.1E-02
3881.4	7	7	3.6E-02	5071.5	13	11	3.4E-03	3565.0	13	11	2.9E-02
3968.5	1	3	5.07E-03	5117.6	11	11	1.61E-03	3566.0	13	15	1.7E-02
3975.5	9	11	4.1E-03	5124.2	5	5	4.0E-03	3566.6	11	11	2.4E-01
4001.4	9	9	5.6E-03	5141.2	7	9	1.12E-03	3568.8	13	13	3.8E-02
4008.8	7	9	1.63E-01	5224.7	7	5	1.2E-02	3569.1	17	15	1.1E-01
4019.3	5	3	6.7E-03	5243.0	9	7	1.1E-02	3569.4	9	9	1.5E-02
4028.8	1	3	2.0E-02	5254.5	7	5	3.86E-03	3570.1	13	11	1.3E-02
4045.6	7	5	2.88E-02	5268.6	9	9	1.4E-03	3570.2	11	9	5.3E-03
4055.2	7	9	1.79E-03	5500.5	11	9	6.9E-03	3570.6	13	15	2.7E-02
4070.0	7	5	3.60E-02	5514.7	5	3	7.3E-03	3570.7	15	15	1.2E-02
4070.6	3	5	5.6E-03	5537.7	9	11	2.2E-03	3571.2	11	11	6.3E-03
4074.4	7	7	1.0E-01	5617.1	7	7	1.47E-03	3571.6	17	15	1.3E-01
4088.3	5	3	4.13E-03	5631.9	9	7	1.43E-03	3572.9	13	15	1.5E-02
4102.7	9	7	4.9E-02	5660.7	13	11	6.8E-03	3573.9	13	11	4.0E-02
4115.6	11	11	4.8E-03	5675.4	5	5	2.20E-03	3574.1	13	15	3.5E-02
4137.5	5	7	8.4E-03	5796.5	9	7	2.21E-03	3574.8	13	15	1.9E-02
4171.2	7	9	8.6E-03	5891.6	7	7	1.47E-03	3577.1	17	15	4.3E-02
4203.8	9	7	4.9E-03	5947.6	5	7	2.40E-03	3577.5	15	13	7.8E-03
4219.4	9	7	6.1E-03	5965.9	7	5	1.0E-02	3577.8	11	11	8.3E-03
4244.4	9	11	1.38E-02	6021.5	5	3	8.7E-03	3577.9	13	13	2.3E-02
4269.4	7	5	3.04E-02	6081.4	5	3	4.7E-03	3578.3	13	11	2.0E-02
4283.8	9	7	1.69E-03	6203.5	7	7	3.0E-03	3580.0	9	9	1.2E-02
								3580.2	11	9	2.9E-02

λ Å	Weights g_i	g_k	A 10^8 s^{-1}
3580.4	11	13	7.5E-03
3580.9	13	13	2.1E-02
3582.6	13	13	2.9E-02
3584.6	7	5	2.4E-02
3584.9	13	15	1.8E-01
3585.4	11	11	1.9E-02
3585.8	11	9	2.8E-02
3587.8	9	11	1.3E-02
3588.3	7	9	1.8E-02
3589.7	11	13	2.1E-02
3589.8	15	13	5.9E-02
3590.7	9	7	2.2E-02
3591.7	11	9	5.3E-02
3593.0	11	11	1.4E-02
3593.2	13	15	4.2E-02
3593.7	11	11	7.2E-02

Vanadium

V I

λ Å	Weights g_i	g_k	A 10^8 s^{-1}
3043.12	6	8	2.3E-01
3050.39	10	8	5.3E-01
3053.65	4	4	1.3E+00
3056.33	6	6	1.3E+00
3060.46	8	8	1.4E+00
3066.37	10	10	2.1E+00
3066.53	6	4	3.2E-01
3075.93	4	6	2.8E-01
3080.33	2	4	2.7E-01
3083.54	6	8	2.5E-01
3087.06	2	2	9.2E-01
3088.11	4	6	4.9E-01
3089.13	4	4	5.3E-01
3093.79	6	6	4.1E-01
3094.69	2	4	4.3E-01
3112.92	4	2	5.0E-01
3183.41	6	8	2.4E+00
3183.96	8	10	2.5E+00
3183.98	4	6	2.4E+00
3185.38	10	12	2.7E+00
3198.01	6	6	3.9E-01
3202.39	8	8	4.0E-01
3205.58	8	10	1.3E+00
3207.41	10	10	2.6E-01
3212.43	10	12	1.4E+00
3218.87	8	6	3.5E-01
3233.19	10	8	3.2E-01
3273.03	8	8	2.7E-01
3284.36	10	10	2.8E-01
3309.18	4	4	3.2E-01
3329.85	6	4	7.7E-01
3356.35	4	6	3.1E-01
3365.55	2	4	4.8E-01
3376.05	4	4	3.2E-01
3377.39	4	2	9.0E-01
3377.62	6	6	6.0E-01
3397.58	6	4	2.3E-01
3400.39	8	8	2.5E-01
3529.73	4	6	4.1E-01
3533.68	6	8	5.2E-01
3533.76	2	4	3.7E-01

λ Å	Weights g_i	g_k	A 10^8 s^{-1}
3543.49	2	2	6.7E-01
3545.33	4	4	3.7E-01
3553.27	6	6	2.2E-01
3555.14	4	2	2.6E-01
3663.60	4	6	3.1E+00
3667.74	6	8	2.7E+00
3672.41	12	12	9.2E-01
3673.41	8	10	2.7E+00
3676.70	14	14	1.3E+00
3680.12	10	12	2.2E+00
3686.26	10	12	2.3E-01
3687.50	12	14	2.9E+00
3688.07	8	8	3.5E-01
3690.28	2	4	4.5E-01
3692.22	6	6	5.4E-01
3695.34	14	16	2.8E+00
3695.86	4	4	6.6E-01
3703.57	10	8	9.2E-01
3704.70	8	6	6.6E-01
3705.04	6	4	3.6E-01
3706.03	10	10	5.2E-01
3708.71	12	12	4.4E-01
3790.46	10	8	2.3E-01
3794.96	10	10	2.3E-01
3806.79	10	10	2.5E-01
3818.24	4	2	6.73E-01
3828.56	6	4	5.33E-01
3840.75	8	6	5.48E-01
3855.36	4	4	3.30E-01
3855.85	10	8	5.78E-01
3863.86	8	6	3.1E-01
3864.86	6	6	2.70E-01
3871.07	10	8	2.8E-01
3875.07	8	8	2.36E-01
3902.26	10	10	2.68E-01
3921.86	4	2	2.7E-01
3922.43	6	6	2.6E-01
3930.02	10	10	3.3E-01
3934.01	8	8	6.2E-01
3992.80	12	10	1.2E+00
3998.73	14	12	1.0E+00
4050.96	10	10	1.4E+00
4051.35	12	12	1.3E+00
4090.57	8	10	8.5E-01
4092.68	8	10	2.30E-01
4095.48	6	8	7.2E-01
4099.78	6	8	4.10E-01
4102.15	4	6	7.1E-01
4104.77	10	8	2.1E+00
4105.16	4	6	4.9E-01
4109.78	2	4	5.00E-01
4111.78	10	10	1.01E+00
4115.18	8	8	5.80E-01
4116.47	6	6	3.2E-01
4116.59	2	2	2.90E-01
4123.50	4	2	1.00E+00
4128.06	6	4	7.70E-01
4131.99	8	6	5.5E-01
4134.49	10	8	2.90E-01
4232.46	10	10	9.8E-01

λ Å	Weights g_i	g_k	A 10^8 s^{-1}
4232.95	8	8	7.7E-01
4268.64	14	14	1.2E+00
4271.55	12	12	9.6E-01
4276.95	10	10	9.4E-01
4284.05	8	8	1.2E+00
4291.82	12	14	8.8E-01
4296.10	10	12	7.7E-01
4297.67	8	10	7.0E-01
4298.03	6	8	7.8E-01
4379.23	10	12	1.1E+00
4384.71	8	10	1.1E+00
4389.98	6	8	6.9E-01
4395.22	4	6	5.5E-01
4400.57	2	4	3.4E-01
4406.64	10	10	2.2E-01
4407.63	8	8	4.4E-01
4408.20	6	6	6.0E-01
4416.47	4	2	2.6E-01
4452.01	14	16	9.2E-01
4457.75	10	12	2.7E-01
4460.33	10	8	3.0E-01
4462.36	12	14	7.6E-01
4468.00	8	10	2.3E-01
4469.71	10	12	6.2E-01
4474.04	10	8	4.7E-01
4496.06	8	6	4.0E-01
4514.18	6	4	3.3E-01
4524.21	12	10	3.0E-01
4525.17	4	2	4.1E-01
4529.58	10	8	2.4E-01
4545.40	10	12	7.6E-01
4560.72	8	10	7.0E-01
4571.79	6	8	6.0E-01
4578.73	4	6	6.8E-01
4706.16	6	4	2.4E-01
4757.47	2	2	7.6E-01
4766.62	6	4	5.6E-01
4776.36	8	6	5.1E-01
4786.50	10	8	4.7E-01
4796.92	12	10	4.8E-01
4807.52	14	12	5.8E-01
5193.00	12	12	4.0E-01
5195.39	8	8	2.3E-01
5234.08	10	10	4.9E-01
5240.87	12	12	4.3E-01
5415.25	12	14	3.1E-01
5487.91	12	10	2.9E-01
5507.75	10	8	3.5E-01
6090.21	8	6	2.60E-01

V II

λ Å	Weights g_i	g_k	A 10^8 s^{-1}
2527.90	13	13	6.1E-01
2528.47	9	9	5.2E-01
2528.83	11	11	5.3E-01
2554.04	9	9	5.4E-01
2589.10	9	9	7.7E-01
2640.86	5	7	1.2E+00
2677.80	3	5	3.4E-01
2679.33	7	7	3.4E-01
2683.09	1	3	3.4E-01

λ	Weights		A	λ	Weights		A	λ	Weights		A
Å	g_i	g_k	10^8 s^{-1}	Å	g_i	g_k	10^8 s^{-1}	Å	g_i	g_k	10^8 s^{-1}
2687.96	9	9	7.6E-01	3271.12	7	9	6.9E-01	1308.06	7	9	7.9E+00
2689.88	3	1	9.2E-01	3276.12	9	11	5.2E-01	1309.50	5	5	8.7E+00
2690.25	7	5	3.4E-01	3279.84	9	11	5.8E-01	1312.72	7	7	8.6E+00
2690.79	5	3	5.2E-01	3287.71	5	7	7.5E-01	1317.57	5	7	8.7E+00
2700.94	9	11	3.5E-01	3337.85	5	7	5.3E-01	1321.92	7	9	9.9E+00
2706.17	7	9	3.4E-01	3517.30	9	7	3.8E-01	1326.81	3	5	4.0E+00
2734.22	9	7	6.2E-01	3530.77	5	3	4.5E-01	1329.29	5	5	1.5E+01
2753.41	13	11	4.2E-01	3545.19	7	5	4.3E-01	1329.97	3	3	4.8E+00
2784.20	9	9	1.3E+00	3556.80	9	7	5.1E-01	1330.36	1	3	6.0E+00
2787.91	7	9	5.0E-01	3592.01	7	5	4.4E-01	1331.67	3	1	1.7E+01
2825.86	9	7	1.2E+00	3618.92	3	5	3.3E-01	1332.46	5	3	7.5E+00
2843.82	7	5	9.9E-01					1334.49	9	9	8.3E+00
2847.57	9	7	4.6E-01	*V III*				1355.13	7	9	2.5E+01
2854.34	11	9	5.0E-01	2318.06	8	10	4.6E+00	1356.53	5	3	4.9E+00
2862.31	11	11	3.6E-01	2323.82	6	8	3.8E+00	1395.00	5	7	1.4E+01
2868.11	5	3	2.1E+00	2330.42	10	10	3.2E+00	1400.42	5	7	7.5E+00
2869.13	13	11	4.8E-01	2331.75	8	8	2.5E+00	1403.62	7	9	8.4E+00
2882.49	5	5	4.2E-01	2334.21	6	6	2.2E+00	1412.69	3	3	1.1E+01
2884.78	3	3	5.6E-01	2337.13	4	4	2.7E+00	1414.41	5	7	1.2E+01
2889.61	3	1	1.9E+00	2343.10	6	8	3.6E+00	1414.84	5	5	4.6E+00
2891.64	5	3	1.4E+00	2358.73	6	8	4.2E+00	1418.53	7	7	5.2E+00
2892.43	9	9	3.6E-01	2366.31	8	10	4.2E+00	1419.58	7	9	1.3E+01
2892.65	7	5	1.3E+00	2371.06	10	12	5.2E+00	1423.72	3	5	7.1E+00
2893.31	9	7	1.2E+00	2373.06	4	6	2.9E+00	1426.65	9	11	2.2E+01
2903.07	3	5	3.4E-01	2382.46	8	10	5.0E+00	1429.11	5	5	5.0E+00
2906.45	7	7	7.8E-01	2393.58	6	8	4.3E+00	1434.84	7	7	5.4E+00
2908.81	11	9	1.6E+00	2404.18	4	6	2.5E+00	1451.04	3	3	7.0E+00
2910.01	5	5	1.1E+00	2516.14	10	10	3.7E+00	1454.00	5	3	1.1E+01
2910.38	3	3	1.2E+00	2521.55	8	8	3.5E+00	1520.14	5	7	7.2E+00
2911.05	7	9	3.7E-01	2548.21	6	4	2.0E+00	1522.49	3	5	5.5E+00
2912.46	11	9	5.0E-01	2554.22	8	6	1.2E+00	1601.92	3	3	1.2E+01
2915.88	9	7	4.9E-01	2593.05	6	6	2.8E+00	1611.88	7	7	5.2E+00
2924.02	11	11	1.7E+00	2595.10	8	8	2.8E+00	1806.18	5	3	7.3E+00
2924.63	9	9	1.2E+00					1809.85	3	1	7.2E+00
2930.80	7	7	5.8E-01	*V IV*				1817.68	5	3	4.8E+00
2941.37	11	9	3.5E-01	677.345	9	9	6.7E+00	1825.84	7	5	5.3E+00
2944.57	9	7	7.6E-01	680.632	9	7	1.2E+01	1861.56	5	7	6.6E+00
2948.08	9	11	4.0E-01	681.145	7	5	1.1E+01	1939.07	7	9	5.8E+00
2952.07	7	5	7.2E-01	682.455	7	7	6.5E+00	1951.43	5	5	5.0E+00
2955.58	7	9	3.3E-01	682.923	5	5	6.9E+00	1963.10	3	5	4.8E+00
2968.37	7	9	7.0E-01	684.450	7	5	7.7E+00	1997.72	7	7	4.7E+00
2972.26	5	7	5.2E-01	691.530	5	3	1.1E+01	2084.43	5	5	4.0E+00
2973.98	9	11	3.5E-01	723.537	3	1	1.5E+01	2120.05	7	9	8.1E+00
2985.18	7	9	4.4E-01	724.068	5	5	1.1E+01	2141.20	3	5	7.0E+00
3001.20	7	7	7.5E-01	724.809	5	3	5.6E+00	2146.83	7	9	6.6E+00
3014.82	5	3	8.9E-01	737.854	9	7	2.4E+01	2149.85	5	7	5.1E+00
3016.78	7	5	5.0E-01	750.110	5	5	1.0E+01	2151.09	7	9	4.3E+00
3020.21	9	7	5.0E-01	884.146	1	3	4.7E+00	2155.34	11	13	1.2E+01
3048.21	11	13	7.0E-01	1071.05	5	5	6.1E+00	2446.80	9	11	5.3E+00
3063.25	9	11	1.0E+00	1110.72	3	3	5.0E+00	2570.72	9	11	7.6E+00
3100.94	7	7	5.8E-01	1112.20	7	7	6.3E+00	3284.56	7	9	5.3E+00
3113.56	11	11	5.0E-01	1112.44	5	5	5.0E+00	3496.42	7	9	4.4E+00
3122.89	11	13	7.6E-01	1127.84	7	5	8.9E+00	3514.25	9	11	4.7E+00
3134.93	13	13	5.9E-01	1131.26	9	7	9.4E+00				
3136.50	11	11	5.3E-01	1194.46	7	5	1.0E+01	*Xenon*			
3139.73	9	9	5.2E-01	1226.52	5	5	1.5E+01				
3151.32	3	5	4.4E-01	1243.72	3	1	9.4E+00	*Xe I*			
3190.69	9	9	3.3E-01	1247.07	5	3	4.7E+00	1043.8	1	3	5.9E-01
3250.78	11	9	5.2E-01	1272.97	3	1	2.7E+01	1047.1	1	3	1.3E+00
3251.87	5	7	3.5E-01	1304.17	3	5	1.5E+01	1050.1	1	3	8.5E-02
				1305.42	5	7	7.0E+00	1056.1	1	3	2.45E+00

λ Å	Weights g_i	g_k	A 10^8 s^{-1}
1061.2	1	3	1.9E-01
1068.2	1	3	3.99E+00
1085.4	1	3	4.10E-01
1099.7	1	3	4.34E-01
1110.7	1	3	1.5E+00
1129.3	1	3	4.4E-02
1170.4	1	3	1.6E+00
1192.0	1	3	6.2E+00
1250.2	1	3	1.4E-01
1295.6	1	3	2.46E+00
1469.6	1	3	2.81E+00
4501.0	5	3	6.2E-03
4524.7	5	5	2.1E-03
4624.3	5	5	7.2E-03
4671.2	5	7	1.0E-02
4807.0	3	1	2.4E-02
7119.6	7	9	6.6E-02
7967.3	1	3	3.0E-03
8409.2	5	3	1.0E-02

Xe II

λ Å	g_i	g_k	A 10^8 s^{-1}
4180.1	4	4	2.2E+00
4330.5	6	8	1.4E+00
4414.8	6	6	1.0E+00
4603.0	4	4	8.2E-01
4844.3	6	8	1.1E+00
4876.5	6	8	6.3E-01
5260.4	2	4	2.2E-01
5262.0	4	4	8.5E-01
5292.2	6	6	8.9E-01
5372.4	4	2	7.1E-01
5419.2	4	6	6.2E-01
5439.0	4	2	7.4E-01
5472.6	8	8	9.9E-02
5531.1	8	6	8.8E-02
5719.6	4	6	6.1E-02
5976.5	4	4	2.8E-01
6036.2	6	6	7.5E-02
6051.2	8	6	1.7E-01
6097.6	6	4	2.6E-01
6270.8	4	6	1.8E-01
6277.5	4	6	3.6E-02
6805.7	8	6	6.1E-02
6990.9	10	8	2.7E-01

Ytterbium

Yb I

λ Å	g_i	g_k	A 10^8 s^{-1}
2464.5	1	3	9.1E-01
2672.0	1	3	1.18E-01
3464.4	1	3	6.2E-01
3988.0	1	3	1.76E+00
5556.5	1	3	1.14E-02

Yb II

λ Å	g_i	g_k	A 10^8 s^{-1}
3289.4	2	4	1.8E+00
3694.2	2	2	1.4E+00

Yttrium

Y I

λ Å	g_i	g_k	A 10^8 s^{-1}
2948.41	4	4	3.5E-01
2974.59	4	6	3.5E-01
2984.25	6	8	4.8E-01
2995.26	6	4	5.1E-02
2996.94	4	6	8.4E-02
3005.26	4	4	4.8E-02
3022.28	6	6	6.6E-02
3045.36	6	6	1.07E-01
3053.95	6	4	1.9E-03
3155.65	4	6	2.7E-03
3172.84	4	4	9.9E-03
3185.96	6	8	1.2E-03
3209.38	6	6	3.0E-03
3227.16	6	4	1.10E-03
3484.05	4	6	1.2E-02
3549.66	6	6	1.0E-03
3552.69	4	4	2.3E-01
4077.36	4	6	1.1E+00
4083.71	4	4	2.5E-01
4102.36	6	8	1.3E+00
4128.30	6	6	1.6E+00
4142.84	4	4	1.6E+00
4167.51	6	6	2.38E-01
4235.93	6	4	3.0E-01
4352.40	4	4	6.7E-03
4379.33	6	4	7.83E-01
4385.47	4	4	6.9E-02
4394.01	8	8	1.9E-02
4409.70	4	6	2.7E-03
4417.43	10	8	3.2E-02
4437.34	6	6	8.64E-02
4443.65	10	8	1.1E-01
4459.01	4	6	1.8E-02
4476.95	8	6	2.8E-01
4491.74	10	10	2.3E-02
4514.01	4	6	3.34E-01
4527.78	8	6	8.33E-01
4534.09	6	8	4.4E-02
4544.31	6	6	4.10E-01
4559.36	2	4	4.0E-01
4581.33	6	4	1.5E-01
4613.00	6	4	1.8E-01
4643.70	4	6	1.8E-01
4653.78	4	6	1.6E-01
4674.85	6	8	1.3E-01
4725.84	4	4	1.5E-01
4762.96	6	4	4.2E-02
4780.16	2	4	8.9E-02
4781.03	8	10	1.0E-01
4799.30	6	8	1.6E-01
4804.31	6	4	2.6E-01
4804.80	4	4	3.84E-01
4821.63	6	6	1.0E-01
4845.67	8	8	6.8E-01
4852.68	6	6	6.2E-01
4856.71	6	6	2.0E-01
4859.84	4	4	7.26E-01
4893.44	6	4	2.2E-01
4900.08	8	6	2.0E-01
4906.11	10	8	1.2E-01
4950.01	8	6	2.0E-02
4963.49	4	4	1.4E-02
4981.97	4	6	4.7E-03
5004.44	6	4	1.2E-02
5205.01	4	4	8.4E-03
5258.47	6	6	2.9E-03
5271.82	8	6	1.1E-02
5380.63	6	4	3.2E-01
5381.24	4	4	9.9E-03
5388.39	6	8	1.1E-02
5390.81	8	6	2.9E-02
5401.88	6	8	6.0E-03
5424.36	6	4	3.47E-01
5466.24	4	4	1.0E-01
5466.47	10	12	6.3E-01
5469.10	4	6	3.6E-03
5513.65	6	6	2.39E-01
5519.88	4	6	1.2E-02
5526.43	6	4	3.9E-03
5527.56	8	10	5.4E-01
5541.63	8	8	5.2E-02
5551.00	4	4	6.9E-02
5573.03	6	4	1.8E-02
5594.12	6	8	5.0E-02
5606.34	10	10	5.84E-02
5619.96	6	4	2.0E-02
5630.14	4	6	4.9E-01
5641.78	2	4	1.9E-02
5675.27	6	6	9.3E-02
5675.64	4	6	4.3E-02
5693.63	4	4	1.1E-01
5714.94	8	6	2.0E-02
5729.25	6	6	2.2E-03
5732.09	6	6	7.5E-02
5740.22	8	6	4.0E-02
5757.59	4	6	7.6E-03
5788.36	4	4	9.4E-03
5844.13	6	4	5.6E-03
5879.93	4	2	8.5E-02
5902.91	6	8	4.0E-02
6087.94	6	4	1.1E-01
6191.72	4	4	4.7E-02
6222.58	4	6	5.9E-03
6402.01	6	4	2.7E-03
6435.02	6	6	4.0E-02
6437.17	10	8	4.8E-02
6538.57	10	10	1.5E-01
6622.48	8	6	4.5E-03
6815.15	2	4	7.18E-02
7009.89	2	4	4.4E-02
7035.15	4	4	6.3E-02

Y II

λ Å	g_i	g_k	A 10^8 s^{-1}
3112.03	1	3	1.3E-02
3179.42	3	5	3.8E-02
3195.62	3	3	8.23E-01
3200.27	5	5	4.8E-01
3203.32	3	1	2.77E+00
3216.69	5	3	2.0E+00
3242.28	7	5	2.0E+00
3448.81	5	5	4.1E-02
3467.88	5	3	2.7E-02

λ	Weights		A	λ	Weights		A	λ	Weights		A
Å	g_i	g_k	10^8 s^{-1}	Å	g_i	g_k	10^8 s^{-1}	Å	g_i	g_k	10^8 s^{-1}
3496.08	1	3	3.49E-01	4374.95	5	5	9.97E-01	*Zinc*			
3549.01	5	7	3.97E-01	4398.01	5	3	1.16E-01	*Zn I*			
3584.51	3	5	4.02E-01	4422.59	3	1	1.83E-01	748.29	1	3	6.0E-02
3600.74	7	7	1.4E+00	4682.33	5	5	1.9E-02	765.60	1	3	7.6E-02
3601.91	3	3	1.13E+00	4786.58	7	7	2.1E-02	792.05	1	3	5.7E-02
3611.04	5	5	1.04E+00	4823.31	5	5	4.3E-02	793.85	1	3	1.8E-01
3628.70	5	3	3.3E-01	4854.87	5	3	3.9E-01	809.92	1	3	2.6E-01
3664.62	7	5	3.7E-01	4881.44	5	3	1.5E-03	1109.1	1	3	3.05E-01
3710.29	7	9	1.5E+00	4883.69	9	7	4.7E-01	2138.6	1	3	7.09E+00
3747.55	3	3	1.9E-01	4900.11	7	5	4.51E-01	3075.9	1	3	3.29E-04
3774.34	5	7	1.1E+00	4982.13	7	9	1.5E-02	3282.3	1	3	9.0E-01
3776.56	5	3	2.42E-01	5087.42	9	9	2.0E-01	3302.6	3	5	1.2E+00
3788.70	3	5	8.1E-01	5119.11	5	7	1.6E-02	3302.9	3	3	6.7E-01
3818.34	5	5	9.70E-02	5200.41	5	5	1.3E-01	3345.0	5	7	1.7E+00
3832.90	7	7	3.0E-01	5205.73	7	7	1.6E-01	3345.6	5	5	4.0E-01
3878.29	7	5	2.9E-02	5289.82	7	5	6.7E-03	3345.9	5	3	4.5E-02
3930.66	5	5	2.1E-02	5320.78	9	7	3.9E-03	6362.3	3	5	4.74E-01
3950.36	3	5	2.80E-01	5473.39	3	5	4.3E-02	11054	3	1	2.43E-01
3951.59	5	3	1.5E-02	5480.73	1	3	7.62E-02				
3982.60	5	5	2.7E-01	5497.41	5	5	1.2E-01	*Zn II*			
4124.91	5	7	1.8E-02	5509.90	5	5	4.24E-02	2025.5	2	4	3.3E+00
4177.54	5	5	5.27E-01	5544.61	3	1	1.8E-01	2064.2	2	4	4.6E+00
4199.27	3	5	5.36E-03	5546.01	5	3	5.8E-02	2099.9	4	6	5.6E+00
4204.69	1	3	2.20E-02	5728.89	5	5	3.0E-02	2102.2	4	4	9.3E-01
4235.73	5	5	2.3E-02	6613.74	5	7	1.7E-02	4911.6	4	6	1.6E+00
4309.62	7	5	1.29E-01	6832.48	5	5	3.3E-03				
4358.73	3	3	5.55E-02	7264.16	5	3	1.3E-02				

ELECTRON AFFINITIES

Thomas M. Miller

Electron affinity is defined as the energy difference between the lowest (ground) state of the neutral and the lowest state of the corresponding negative ion. The accuracy of electron affinity measurements has been greatly improved since the advent of laser photodetachment experiments with negative ions. Electron affinities can be determined with optical precision, though a detailed understanding of atomic and molecular states and splittings is required to specify the photodetachment threshold corresponding to the electron affinity.

Atomic and molecular electron affinities are discussed in two excellent articles reviewing photodetachment studies which appear in *Gas Phase Ion Chemistry*, Vol. 3, Bowers, M. T., Ed., Academic Press, Orlando, 1984: Chapter 21 by Drzaic, P. S., Marks, J., and Brauman, J. I., "Electron Photodetachment from Gas Phase Negative Ions," p. 167, and Chapter 22 by Mead, R. D., Stevens, A. E., and Lineberger, W. C., "Photodetachment in Negative Ion Beams," p. 213. Persons interested in photodetachment details should consult these articles and the critical reviews of Andersen, T., Haugen, H. K., and Hotop, H., *J. Phys. Chem. Ref. Data*, 28, 1511, 1999, Hotop, H. and Lineberger, W. C., *J. Phys. Chem. Ref. Data*, 14, 731, 1985, and Andersen, T., Haugen, H. K., and Hotop, H., *J. Phys. Chem. Ref. Data* 28, 1511, 1999. For simplicity in the tables below, any electron affinity which was discussed in the articles by Drzaic et al. or Hotop and Lineberger is referenced to these sources, where original references are given. The develop-

ment of cluster-ion photodetachment apparatuses has brought an explosion of electron affinity estimates for atomic and molecular clusters. The policy in this tabulation is to list the electron affinities for the atoms, diatoms, and triatoms, if adiabatic electron affinities have been determined, but to refer the reader to original sources for higher-order clusters. Additional data on molecular electron affinities may be found in Lias, S. G., Bartmess, J. E., Liebman, J. F., Holmes, J. L., Levin, R. D., and Mallard, W. G., *Gas Phase Ion and Neutral Thermochemistry*, *J. Phys. Chem. Ref. Data*, 17, (Supplement No. 1), 1988 and on the NIST WebBook at the Internet address http://webbook.nist.gov/.

For the present tabulation the 2002 CODATA value e/hc = 8065.54445 ± 0.00069 cm^{-1} eV^{-1} (http://physics.nist.gov) has been used to convert electron affinities from the units used in spectroscopic work, cm^{-1}, into eV for these tables. The 86 ppb uncertainty in e/hc is insignificant compared to uncertainties in the electron affinity measurements.

Abbreviations used in the tables: calc = calculated value; PT = photodetachment threshold using a lamp as a light source; LPT = laser photodetachment threshold; LPES = laser photoelectron spectroscopy; DA = dissociative attachment; attach = electron attachment/detachment equilibrium; e-scat = electron scattering; kinetic = dissociation kinetics; Knud=Knudsen cell; CT = charge transfer; CD = collisional detachment; and ZEKE = zero electron kinetic energy spectroscopy.

TABLE 1. Atomic Electron Affinities

Atomic number	Atom	Electron affinity in eV	Uncertainty in eV	Method	Ref.	
1	H	0.754195	0.000019	LPT	89	
		0.75420812	-	calc	205	
	D	0.754593	0.000074	LPT	89	deuterium
	D	0.75465624	-	calc	205	deuterium
	T	0.75480540	-	calc	205	tritium
2	He	not stable	-	calc	1	
3	Li	0.618049	0.000020	LPT	185	
4	Be	not stable	-	calc	1	
5	B	0.279723	0.000025	LPES	191	
6	C	1.262119	0.000020	LPT	28	
7	N	not stable	-	DA	1	
8	O	1.4611096	0.0000007	LPT	4	
9	F	3.4011895	0.0000025	LPT	227	
10	Ne	not stable	-	calc	1	
11	Na	0.547926	0.000025	LPT	1	
12	Mg	not stable	-	e-scat	1	
13	Al	0.43283	0.00005	LPES	208	
14	Si	1.3895220	0.0000024	LPES	227	
15	P	0.7465	0.0003	LPT	1	
16	S	2.077103	0.000001	LPT	1	
17	Cl	3.612724	0.000027	LPT	52	
18	Ar	not stable	-	calc	1	
19	K	0.50147	0.00010	LPT	1	
20	Ca	0.02455	0.00010	LPT	44	
21	Sc	0.188	0.020	LPES	1	
22	Ti	0.079	0.014	LPES	1	
23	V	0.525	0.012	LPES	1	
24	Cr	0.666	0.012	LPES	1	

Atomic number	Atom	Electron affinity in eV	Uncertainty in eV	Method	Ref.
25	Mn	not stable	-	calc	1
26	Fe	0.151	0.003	LPES	27
27	Co	0.662	0.003	LPES	27
28	Ni	1.156	0.010	LPES	1
29	Cu	1.235	0.005	LPES	37
30	Zn	not stable	-	e-scat	1
31	Ga	0.43	0.03	LPES	183
32	Ge	1.232712	0.000015	LPES	28
33	As	0.814	0.008	LPES	200
34	Se	2.020670	0.000025	LPT	1
35	Br	3.363588	0.000002	LPT	74
36	Kr	not stable	-	calc	1
37	Rb	0.48592	0.00002	LPT	1
38	Sr	0.048	0.006	LPT	122
39	Y	0.307	0.012	LPES	1
40	Zr	0.426	0.014	LPES	1
41	Nb	0.893	0.025	LPES	1
42	Mo	0.748	0.002	LPES	127
43	Tc	0.55	0.20	calc	1
44	Ru	1.05	0.15	calc	1
45	Rh	1.137	0.008	LPES	1
46	Pd	0.562	0.005	LPES	116
47	Ag	1.302	0.007	LPES	1
48	Cd	not stable	-	e-scat	1
49	In	0.3	0.2	PT	1
50	Sn	1.112067	0.000015	LPES	28
51	Sb	1.046	0.005	LPES	108
52	Te	1.970876	0.000007	LPT	261
53	I	3.059037	0.000010	LPT	92
54	Xe	not stable	-	calc	1
55	Cs	0.471626	0.000025	LPT	1
56	Ba	0.14462	0.00006	LPT	195
57	La	0.47	0.02	LPT	184
58	Ce	0.955	0.026	LPES	269
59	Pr	0.962	0.024	LPES	225
63	Eu	0.864	0.024	LPES	268
69	Tm	1.029	0.022	LPES	264
70	Yb	-0.020	-	calc	196
71	Lu	0.34	0.01	LPT	223
72	Hf	»0	-	calc	1
73	Ta	0.322	0.012	LPES	1
74	W	0.815	0.002	LPES	37
75	Re	0.15	0.15	calc	1
76	Os	1.1	0.2	calc	1
77	Ir	1.5638	0.0005	LPT	141
78	Pt	2.128	0.002	LPT	1
79	Au	2.30863	0.00003	LPT	1
80	Hg	not stable	-	e-scat	1
81	Tl	0.2	0.2	PT	1
82	Pb	0.364	0.008	LPES	1
83	Bi	0.942362	0.000013	LPT	262
84	Po	1.9	0.3	calc	1
85	At	2.8	0.2	calc	1
86	Rn	not stable	-	calc	1
87	Fr	0.46	-	calc	82
88	Ra	0.10	-	calc	273
89	Ac	0.35	-	calc	207
118	ekaradon	0.056	0.01	calc	140
121	ekaactinium	0.57	-	calc	207

TABLE 2. Electron Affinities for Diatomic Molecules

Molecule	Electron affinity in eV	Uncertainty in eV	Method	Ref.	Molecule	Electron affinity in eV	Uncertainty in eV	Method	Ref.
Ag_2	1.023	0.007	LPES	37	MnH	0.869	0.010	LPES	9
AgO	1.654	0.002	LPES	233	MnO	1.375	0.010	LPES	158
Al_2	1.10	0.15	LPES	68	MoO	1.290	0.006	LPES	127
AlO	2.60	0.02	LPES	143	NH	0.370	0.004	LPT	32
AlP	2.043	0.020	LPES	218	NO	0.026	0.005	LPES	73
AlS	2.60	0.03	LPES	129	NRh	1.51	0.02	LPES	206
As_2	0.739	0.008	LPES	200	NS	1.194	0.011	LPES	2
AsH	1.0	0.1	PT	2	Na_2	0.430	0.015	LPES	104
AsO	1.286	0.008	LPES	198	NaBr	0.788	0.010	LPES	30
Au_2	1.938	0.007	LPES	37	NaCl	0.727	0.010	LPES	30
AuO	2.374	0.007	LPES	282	NaF	0.520	0.010	LPES	30
AuPd	1.88	—	LPES	220	NaI	0.865	0.010	LPES	30
AuS	2.469	0.006	LPES	282	NaK	0.465	0.030	LPES	104
BN	3.160	0.005	LPES	189	NbO	1.29	0.02	LPES	174
BO	2.508	0.008	LPES	6	Ni_2	0.926	0.010	LPES	112
BeH	0.7	0.1	PT	2	NiCu	0.889	0.010	LPES	128
Bi_2	1.271	0.008	LPES	119	NiAg	0.979	0.010	LPES	128
Br_2	2.55	0.10	CT	2	NiD	0.477	0.007	LPES	29
BrO	2.353	0.006	LPES	88	NiH	0.481	0.007	LPES	29
C_2	3.269	0.006	LPES	87	NiO	1.470	0.003	LPES	146
CH	1.238	0.008	LPES	2	O_2	0.450	0.002	LPES	222
CN	3.862	0.004	LPES	111	OD	1.825533	0.000037	LPT	142
CRh	1.46	0.02	LPES	206	OH	1.8276487	0.000011	LPT	226
CS	0.205	0.021	LPES	2	ORh	1.58	0.02	LPES	206
CaH	0.93	0.05	PT	2	P_2	0.589	0.025	LPES	42
Cl_2	2.38	0.10	CT	2	PH	1.027	0.006	LPES	281
ClO	2.275	0.006	LPES	88	PO	1.092	0.010	LPES	2
Co_2	1.110	0.008	LPES	27	Pb_2	1.366	0.010	LPES	117
CoD	0.680	0.010	LPES	29	PbO	0.722	0.006	LPES	105
CoH	0.671	0.010	LPES	29	PbS	1.049	0.010	LPES	228
Cr_2	0.505	0.005	LPES	114	Pd_2	1.685	0.008	LPES	112
CrD	0.568	0.010	LPES	29	PdCO	0.604	0.010	LPES	160
CrH	0.563	0.010	LPES	29	PdO	1.570	0.006	LPES	290
CrO	1.221	0.006	LPES	5	Pt_2	1.898	0.008	LPES	112
Cs_2	0.469	0.015	LPES	104	PtN	1.240	0.010	LPES	46
CsCl	0.455	0.010	LPES	30	Rb_2	0.498	0.015	LPES	104
CsO	0.273	0.012	LPES	133	RbCl	0.544	0.010	LPES	30
Cu_2	0.836	0.006	LPES	37	RbCs	0.478	0.020	LPES	104
CuO	1.777	0.006	LPES	118	Re_2	1.571	0.008	LPES	33
F_2	3.08	0.10	CT	2	S_2	1.670	0.015	LPES	53
FO	2.272	0.006	LPES	88	SD	2.315	0.002	LPES	10
Fe_2	0.902	0.008	LPES	27	SF	2.285	0.006	LPES	93
FeD	0.932	0.015	LPES	9	SH	2.314343	0.000004	LPT	47
FeH	0.934	0.011	LPES	9	SO	1.125	0.005	LPES	84
FeO	1.493	0.005	LPES	45	Sb_2	1.282	0.008	LPES	108
GaAs	1.949	0.020	LPES	218	ScO	1.35	0.02	LPES	171
GaO	2.612	0.008	LPES	279	Se_2	1.94	0.07	LPES	38
GaP	1.988	0.020	LPES	218	SeH	2.212519	0.000025	LPT	48
Ge_2	2.035	0.001	LPES	123	SeO	1.456	0.020	LPES	41
I_2	2.55	0.05	CT	2	Si_2	2.201	0.010	LPES	100
IBr	2.55	0.10	CT	2	SiF	0.81	0.02	LPES	278
IO	2.378	0.006	LPES	88	SiH	1.277	0.009	LPES	2
InP	1.845	0.020	LPES	218	SiN	2.949	0.008	LPES	274
K_2	0.497	0.012	LPES	104	Sn_2	1.962	0.010	LPES	117
KBr	0.642	0.010	LPES	30	SnO	0.598	0.006	LPES	168
KCl	0.582	0.010	LPES	30	SnPb	1.569	0.008	LPES	117
KCs	0.471	0.020	LPES	104	Te_2	1.92	0.07	LPES	38
KI	0.728	0.010	LPES	30	TeH	2.102	0.015	LPES	39
KRb	0.486	0.020	LPES	104	TeO	1.697	0.022	LPES	40
LiCl	0.593	0.010	LPES	30	TiO	1.30	0.03	LPES	172
LiD	0.337	0.012	LPES	102	VO	1.229	0.008	LPES	170
LiH	0.342	0.012	LPES	102	YO	1.35	0.02	LPES	171
MgCl	1.589	0.011	LPES	31	ZnF	1.974	0.008	LPES	179
MgH	1.05	0.06	PT	2	ZnH	<0.95	—	PT	2
MgI	1.899	0.018	LPES	31	ZnO	2.087	0.008	LPES	179
MgO	1.630	0.025	LPES	178	ZrO	1.3	0.3	LPES	173
MnD	0.866	0.010	LPES	9					

TABLE 3. Electron Affinities for Triatomic Molecules

Molecule	Electron affinity in eV	Uncertainty in eV	Method	Ref.		Molecule	Electron affinity in eV	Uncertainty in eV	Method	Ref.	
Ag_3	2.32	0.05	LPES	37		Ge_3	2.23	0.01	LPES	123	
$AgCN$	1.588	0.010	LPES	163		GeH_2	1.097	0.015	LPES	28	
Al_3	1.4	0.15	LPES	68		HCO	0.313	0.005	LPES	35	
AlO_2	4.23	0.02	LPES	143		HCl_2	4.896	0.005	LPES	69	
AlP_2	1.933	0.007	LPES	217		HNO	0.338	0.015	LPES	14	
Al_2P	2.513	0.020	LPES	217		HO_2	1.078	0.006	LPES	15	
Al_2S	0.80	0.12	LPES	129		HS_2	1.907	0.015	LPES	53	
As_3	1.45	0.03	LPES	200		I_3	4.226	0.013	LPES	162	
AsH_2	1.27	0.03	PT	2		InP_2	1.61	0.05	LPES	137	
Au_3	3.7	0.3	LPES	37		In_2P	2.36	0.05	LPES	137	
$AuBr_2$	4.46	0.07	LPES	294		K_3	0.956	0.050	LPES	18	
$AuCl_2$	4.60	0.07	LPES	294		MnD_2	0.465	0.014	LPES	34	
AuI_2	4.18	0.07	LPES	294		MnH_2	0.444	0.016	LPES	34	
Au_2H	3.55	0.03	LPES	276		MnO_2	2.06	0.03	LPES	158	
Au_2Pd	3.80	—	LPES	220		N_3	2.70	0.12	PT	2	
BO_2	4.3	0.2	CT	98		N_3	2.68	0.01	LPT	255	
B_2N	3.098	0.005	LPES	193		NCN	2.484	0.006	LPES	154	
B_3	2.82	0.02	LPES	221		NCO	3.609	0.005	LPES	111	
Bi_3	1.60	0.03	LPES	119		NCS	3.537	0.005	LPES	111	
C_3	1.981	0.020	LPES	11		NH_2	0.771	0.005	LPES	58	
CBr_2	1.88	0.07	LPES	235		N_2O	-0.03	0.10	calc	59	
CCl_2	1.59	0.07	LPES	235		NO_2	2.273	0.005	LPES	63	
CD_2	0.645	0.006	LPES	12		$(NO)R$	R=Ar,Kr,Xe	—	LPES	90	
CDF	0.535	0.005	LPES	95		Na_3	1.019	0.060	LPES	18	
CF_2	0.180	0.020	LPES	235		$NaCS_2$	0.80	0.05	LPES	278	
CH_2	0.652	0.006	LPES	12		Na_2CS_2	0.25	0.05	LPES	278	
$CHBr$	1.454	0.005	LPES	95		Nb_3	1.032	0.010	LPES	175	
$CHCl$	1.210	0.005	LPES	95		Ni_3	1.41	0.05	LPES	55	
CHF	0.542	0.005	LPES	95		$NiCN$	1.771	0.010	LPES	287	
CHI	1.42	0.17	LPES	95		$NiCO$	0.804	0.012	LPES	2	
CI_2	2.09	0.07	LPES	235		NiD_2	1.926	0.007	LPES	34	
C_2Cr	2.30	1.617	0.015	271		NiH_2	1.934	0.008	LPES	34	
C_2H	2.969	0.006	LPES	87		NiO_2	3.05	0.01	LPES	214	ONiO
C_2Nb	1.380	0.025	LPES	243		NiO_2	0.82	0.03	LPES	214	Ni(O₂)
C_2O	2.289	0.018	LPES	180		O_3	2.1028	0.0025	LPT	2	
COS	-0.04	—	LPES	272		O_2Ar	0.52	0.02	LPES	75	
CS_2	0.58	0.05	LPES	278		$OClO$	2.140	0.008	LPES	88	
C_2Ti	1.542	0.020	LPES	147		OIO	2.577	0.008	LPES	88	
CoD_2	1.465	0.013	LPES	34		PH_2	1.263	0.006	LPES	281	
CoH_2	1.450	0.014	LPES	34		P_2H	1.514	0.010	LPES	281	
CrH_2	>2.5	—	LPES	34		PO_2	3.42	0.01	LPES	124	
Cr_2D	1.464	0.005	LPES	107		Pd_3	<1.5	0.1	LPES	55	
Cr_2H	1.474	0.005	LPES	107		$PdCN$	2.543	0.007	LPES	287	
CrO_2	2.413	0.008	LPES	144	OCrO	$PdCO$	0.606	0.010	LPES	293	
CrO_2	1.5	0.06	LPES	241	Cr(O₂)	Pt_3	1.87	0.02	LPES	55	
Cs_3	0.864	0.030	LPES	18		$PtCN$	3.191	0.003	LPES	287	
Cu_3	2.11	0.05	LPES	37		$PtCO$	1.212	0.010	LPES	293	
$CuCN$	1.466	0.010	LPES	163		Rb_3	0.920	0.030	LPES	18	
$CuCl_2$	4.35	0.05	LPES	177		ReO_2	2.5	0.1	LPES	216	
$CuBr_2$	4.35	0.05	LPES	177		S_3	2.093	0.025	LPES	16	
DCO	0.301	0.005	LPES	35		SO_2	1.107	0.008	LPES	16	
DNO	0.330	0.015	LPES	14		S_2O	1.877	0.008	LPES	16	
DO_2	1.077	0.005	LPES	15		Sb_3	1.85	0.03	LPES	108	
DS_2	1.912	0.015	LPES	53		SeO_2	1.823	0.050	LPES	38	
Fe_3	1.43	0.06	LPES	149		SiF_2	0.10	0.10	LPES	278	
FeC_2	1.9782	0.0006	LPES	254		Si_2F	1.99	0.28	LPES	17	
$FeCO$	1.157	0.005	LPES	103		SiH_2	1.124	0.020	LPES	2	
FeD_2	1.038	0.013	LPES	34		Si_2H	2.31	0.01	LPES	182	
FeH_2	1.049	0.014	LPES	34		Si_3	2.29	0.02	LPES	110	
FeO_2	2.358	0.030	LPES	130		Sn_3	2.24	0.01	LPES	289	
Fe_2H	0.564	0.019	LPES	254		$SnCN$	1.922	0.006	LPES	292	
Fe_2O	1.60	0.02	LPES	152		Ta_3	1.36	0.03	LPES	169	
$GaAs_2$	1.894	0.033	LPES	192		TiO_2	1.59	0.03	LPES	172	
GaP_2	1.666	0.041	LPES	192		V_3	1.107	0.010	LPES	176	
Ga_2As	2.428	0.020	LPES	192		VO_2	2.3	0.2	CT	101	
Ga_2P	2.481	0.015	LPES	192		WO_2	1.958	0.050	LPES	233	

TABLE 4. Electron Affinities for Larger Polyatomic Molecules

Molecule	Electron affinity in eV	Uncertainty in eV	Method	Ref.	
Ag_n	$n=1\text{-}60$	—	LPES	37	
Al_n	$n=3\text{-}32$	—	LPES	68	
Al_5	2.23	0.05	LPES	238	
Al_2C_2	0.64	0.05	LPES	239	acetylide
Al_3C	2.56	0.06	LPES	161	
Al_3C_2	2.19	0.03	LPES	244	
Al_3Ge_2	2.43	0.03	LPES	244	
Al_3Si_2	2.36	0.03	LPES	244	
Al_3O	1.00	0.15	LPES	68	
$Al_5H_2O_5$	3.10	0.10	LPES	283	
Al_5O_4	3.50	0.05	LPES	283	
Al_nO_m	$n=1,2$	$m=1\text{-}5$	LPES	143	
Al_nO_m	$n=3\text{-}7$	$m=2\text{-}5$	LPES	267	
Al_nP_m	$n=1\text{-}4$	$m=1\text{-}4$	LPES	217	
Al_nS_m	$n=1\text{-}5$	$m=1\text{-}3$	LPES	129	
$Ar(H_2O)_n$	$n=2,6,7$	—	LPES	77	
Ar_nBr	$n=2\text{-}9$	—	ZEKE	212	
Ar_nI	$n=2\text{-}19$	—	ZEKE	212	
As_4	<0.8	—	LPES	200	
As_5	≈1.7	—	LPES	200	
As_5	≈3.5	—	LPES	253	
Au_n	$n=1\text{-}233$	—	LPES	37	
AuF_6	7.5	estimate	CT	98	
Au_3Pd	2.51	—	LPES	220	
Au_4Pd	2.69	—	LPES	220	
Au_6	2.06	0.02	LPES	288	
$Au_6(CO)$	2.04	0.05	LPES	288	
$Au_6(CO)_2$	2.03	0.05	LPES	288	
$Au_6(CO)_3$	1.95	0.05	LPES	288	
$Au_{12}Nb$	3.70	0.03	LPES	275	
$Au_{12}Ta$	3.77	0.03	LPES	275	
$Au_{12}V$	3.76	0.03	LPES	275	
B_5	2.33	0.02	LPES	245	
BD_3	0.027	0.014	LPES	62	
BH_3	0.038	0.015	LPES	62	
B_6Li	2.3	0.1	LPES	298	
B_3N	2.098	0.035	LPES	193	
Bi_n	$n=2\text{-}9$	—	LPES	213	
Bi_4	1.05	0.010	LPES	119	
Bi_5	2.87	0.02	LPES	253	
$Br(CO_2)$	3.582	0.017	LPES	131	
$Br(H_2O)_n$	$n=1\text{-}4$	—	LPES	250	
C_n	$n=2\text{-}84$	—	LPES	70	
C_nCr	$n=2\text{-}8$	—	LPES	271	
C_nNb	$n=2\text{-}7$	—	LPES	243	
$(CO_2)_n$	$n=1,2$	—	LPES	75	
$(CS)_n$	$n=2$	—	LPES	75	
$(CS_2)_n$	$n=1,2$	—	LPES	75	
CAl_3Ge	2.70	0.06	LPES	224	
CAl_3Si	2.77	0.06	LPES	224	
CCl_4	≤1.14	—	CT	266	
$CCoNO_3$	1.73	0.03	LPES	199	$Co(CO_2)NO$
CDO_2	3.510	0.015	LPES	109	
CF_3	1.82	0.05	LPES	187	
CF_3Br	0.91	0.2	CD	2	
CF_3I	1.57	0.2	CD	2	
CFO_2	4.277	0.030	LPES	131	
$CHCl_3$	≤0.78	—	CT	266	
CHO_2	3.498	0.015	LPES	109	

Molecule	Electron affinity in eV	Uncertainty in eV	Method	Ref.	
CH_2O_4	2.1	0.2	PT	2	$CO_3(H_2O)$
CH_2S	0.465	0.023	LPES	53	
CD_3NO_2	0.24	0.08	LPES	211	
CD_3O	1.559	0.004	LPES	194	
CD_3O_2	1.154	0.004	LPES	188	d_3-methyl peroxyl radical
CD_3S	1.856	0.006	LPT	2	
CD_3S_2	1.748	0.022	LPES	53	
CH_3	0.08	0.03	LPES	2	
CH_3I	0.11	0.02	LPES	277	
CH_3NO_2	0.26	0.08	LPES	211	
CH_3O	1.572	0.004	LPES	194	
CH_3O_2	1.161	0.005	LPES	188	methyl peroxyl radical
CH_3S	1.867	0.004	LPES	166	
CH_3S_2	1.757	0.022	LPES	53	
CH_3Si	0.852	0.010	LPES	97	CH_3-Si
CH_3Si	2.010	0.010	LPES	97	$CH_2=SiH$
CH_4N	0.432	0.015	LPES	215	
CH_5Si	1.19	0.04	LPT	65	CH_3SiH_2
CO_3	2.69	0.14	LPES	2	
C_2F_2	2.255	0.006	LPES	106	difluorovinylidene
C_2DN	2.009	0.020	LPES	219	DCCN
C_2DN	1.877	0.010	LPES	219	DCNC
C_2DO	2.350	0.020	LPES	13	
C_2HF	1.718	0.006	LPES	106	monofluorovinylidene
C_2HN	2.003	0.014	LPES	219	HCCN
C_2HN	1.883	0.013	LPES	219	HCNC
C_2HO	2.338	0.008	LPES	190	
C_2HNPd	2.17	0.03	LPES	291	
C_2HPd	1.98	0.03	LPES	287	
C_2HPt	2.650	0.010	LPES	287	
C_2D_2	0.492	0.006	LPES	83	vinylidene-d_2
C_2HD	0.489	0.006	LPES	83	vinylidene-d_1
C_2HFe	1.4512	0.0025	LPES	254	
C_2HNi	1.063	0.019	LPES	254	
C_2H_2	0.490	0.006	LPES	83	vinylidene
C_2H_2FO	2.22	0.09	PT	2	acetyl fluoride enolate
C_2D_2N	1.538	0.012	LPES	21	cyanomethyl-d_2 radical
C_2D_2N	1.070	0.024	LPES	21	isocyanomethyl-d_2 radical
C_2H_2Fe	1.328	0.019	LPES	254	
C_2H_2N	1.543	0.014	LPES	21	cyanomethyl radical
C_2H_2N	1.059	0.024	LPES	21	isocyanomethyl radical
C_2H_2Ni	2.531	0.005	LPES	287	$HNiC_2H$
C_2H_3	0.667	0.024	LPES	90	vinyl
C_2H_3Fe	1.587	0.019	LPES	254	
C_2H_3Ni	1.103	0.019	LPES	254	
C_2D_3O	1.81897	0.00012	LPT	22	acetaldehyde-d_3 enolate
C_2H_3O	1.82476	0.00012	LPT	22	acetaldehyde enolate
C_2D_5O	1.699	0.004	LPES	194	ethoxide-d_3
C_2H_5N	0.56	0.01	PT	2	ethyl nitrine
C_2H_5O	1.712	0.004	LPES	194	ethoxide
$C_2H_5O_2$	1.186	0.004	LPES	188	ethyl peroxyl radical
C_2H_5S	1.953	0.006	LPT	2	ethyl sulfide
C_2H_5S	0.868	0.051	LPES	53	CH_3SCH_2
$C_2H_7O_2$	2.26	0.08	PT	50	MeOHOMe
C_3Fe	1.69	0.08	LPES	132	
C_3H	1.858	0.023	LPES	11	
C_3HFe	1.58	0.06	LPES	132	
C_3H_2	1.794	0.008	LPES	153	
$C_3H_2F_3O$	2.625	0.010	LPT	113	1,1,1-trifluoroacetone enolate
C_3H_3	0.893	0.025	LPES	24	propargyl radical

Molecule	Electron affinity in eV	Uncertainty in eV	Method	Ref.	
C_3H_2D	0.88	0.15	LPES	24	propargyl-d_1 radical
C_3D_2H	0.907	0.023	LPES	24	propargyl-d_2 radical
C_3H_3N	1.247	0.012	LPES	21	CH_3CH-CN
C_3D_5	0.464	0.006	LPES	138	allyl-d_5
C_3H_5	0.481	0.008	LPES	138	allyl
C_3H_5	0.397	0.069	kinetic	155	cyclopropyl
C_3H_4D	0.373	0.019	LPES	25	allyl-d_1
C_3H_5O	1.758	0.019	LPT	113	acetone enolate
C_3H_5O	1.621	0.006	LPT	113	propionaldehyde enolate
$C_3H_5O_2$	1.80	0.06	PT	2	methyl acetate enolate
C_3H_7O	1.789	0.033	LPES	23	propyl oxide
C_3H_7O	1.847	0.004	LPES	194	isopropyl oxide
C_3H_7S	2.00	0.02	PT	2	propyl sulfide
C_3H_7S	2.02	0.02	PT	2	isopropyl sulfide
C_3O	1.34	0.15	LPES	11	
C_3O_2	0.85	0.15	LPES	11	
C_3Ti	1.561	0.015	LPES	147	
$C_4F_4Cl_2$	0.87	0.08	attach	258	1,2-dichlorotetrafluoro-cyclobutene
$C_4F_4O_3$	0.5	0.2	CD	2	tetrafluorosuccinic anhydride
C_4F_8	0.63	0.05	attach	256	octafluorocyclobutane
C_4Fe	<2.2	0.2	LPES	132	
C_4HFe	1.67	0.06	LPES	132	
C_4H_2Fe	1.633	0.019	LPES	254	
$C_4H_2O_3$	1.44	0.10	CT	61	maleic anhydride
C_4H_3Fe	1.182	0.019	LPES	254	
C_4H_3Ni	0.824	0.019	LPES	254	
C_4D_4	0.909	0.015	LPES	125	vinylvinylidene-d_4
C_4H_4	0.914	0.015	LPES	125	vinylvinylidene
C_4H_4N	2.145	0.010	LPES	265	pyrrolyl
$C_4H_4N_3O$	0.75	—	LPES	285	NO (pyrimidine)
C_4H_5O	1.801	0.008	LPT	113	cyclobutanone enolate
C_4H_6	0.431	0.006	LPES	135	trimethylenemethane
$C_4H_6O_2$	0.69	0.10	CT	61	2,3-butanedione
C_4H_6D	0.493	0.008	LPES	138	2-methylallyl-d_7
C_4H_7	0.505	0.006	LPES	138	2-methylallyl
C_4H_7O	1.67	0.05	PT	2	butyraldehyde enolate
C_4H_5DO	1.67	0.05	PT	2	2-butanone-3-d_1 enolate
$C_4H_5D_2O$	1.75	0.06	PT	2	2-butanone-3,3-d_2 enolate
C_4H_9O	1.909	0.004	LPES	194	t-butoxyl
C_4H_9S	2.03	0.02	PT	2	n-butyl sulfide
C_4H_9S	2.07	0.02	PT	2	t-butyl sulfide
C_4O	2.05	0.15	LPES	11	
C_4O_2	2.0	0.2	LPES	11	
C_4Ti	1.494	0.020	LPES	147	
C_5	2.853	0.001	LPT	99	
C_5F_5N	0.70	0.05	attach	259	pentafluoropyridine
$C_5F_6O_3$	1.5	0.2	CD	2	hexafluoroglutaric anhydride
C_5HF_4N	0.40	0.08	attach	259	tetrafluoropyridine
C_5D_5	1.790	0.008	LPES	11	cyclopentadienyl-d_5
C_5H_5	1.804	0.007	LPES	11	cyclopentadienyl
$C_5H_5NO_2$	1.39	—	LPES	285	O_2 (pyridine)
$C_5H_5N_2O$	0.62	—	LPES	285	NO (pyridine)
C_5H_7	0.91	0.03	PT	2	pentadienyl
$C_5H_7NO_3$	1.87	—	LPES	285	O_2 (pyridine · H_2O)
C_5H_7O	1.598	0.007	LPT	113	cyclopentanone enolate
C_5H_9O	1.69	0.05	PT	2	3-penanone enolate
$C_5H_{11}O$	1.93	0.05	LPT	2	neopentoxyl
$C_5H_{11}S$	2.09	0.02	PT	2	pentyl sulfide
C_5O_2	1.2	0.2	LPES	11	
C_5Ti	1.748	0.050	LPES	147	

Molecule	Electron affinity in eV	Uncertainty in eV	Method	Ref.	
C_6	4.180	0.001	LPT	8	
$C_6Br_4O_2$	2.44	0.20	CT	2	tetrabromo-BQ
$C_6Cl_4O_2$	2.78	0.10	CT	61	tetrachloro-BQ
$C_6F_4O_2$	2.70	0.10	CT	61	tetrafluoro-BQ
C_6F_5Br	1.15	0.11	CT	67	pentafluorobromobenzene
C_6F_5Cl	0.75	0.05	attach	260	pentafluorochlorobenzene
C_6F_5I	1.41	0.11	CT	67	pentafluoroiodobenzene
$C_6F_5NO_2$	1.52	0.11	CT	67	pentafluoro-NB
C_6F_6	0.53	0.05	attach	257	hexafluorobenzene
C_6F_{10}	>1.4	0.3	CT	2	perfluorocyclohexane
$C_6H_2Cl_2O_2$	2.48	0.10	CT	61	2,6-dichloro-BQ
$C_6H_3F_2NO_2$	1.17	0.10	CT	61	2,4-difluoro-NB
C_6D_4	0.551	0.010	LPES	36	o-benzyne-d_4
C_6H_4	0.560	0.010	LPES	36	o-benzyne
$C_6H_4BrNO_2$	1.16	0.10	CT	61	o-bromo-NB
$C_6H_4BrNO_2$	1.32	0.10	CT	61	m-bromo-NB
$C_6H_4BrNO_2$	1.29	0.10	CT	61	p-bromo-NB
$C_6H_4ClNO_2$	1.14	0.10	CT	61	o-chloro-NB
$C_6H_4ClNO_2$	1.28	0.10	CT	61	m-chloro-NB
$C_6H_4ClNO_2$	1.26	0.10	CT	61	p-chloro-NB
C_6H_4ClO	≤2.58	0.08	PT	2	o-chlorophenoxide
$C_6H_4FNO_2$	1.07	0.10	CT	61	o-fluoro-NB
$C_6H_4FNO_2$	1.23	0.10	CT	61	m-fluoro-NB
$C_6H_4FNO_2$	1.12	0.10	CT	61	p-fluoro-NB
$C_6H_4N_2O_4$	1.65	0.10	CT	61	o-diNB
$C_6H_4N_2O_4$	1.65	0.10	CT	61	m-diNB
$C_6H_4N_2O_4$	2.00	0.10	CT	61	p-diNB
$C_6H_4O_2$	1.860	0.005	LPES	284	1,4-benzoquinone (BQ)
C_6D_5	1.092	0.020	LPES	26	phenyl-d_5
C_6D_5N	1.44	0.02	LPES	96	phenylnitrene-d_5
$C_6H_2O_2$	1.859	0.005	LPES	232	dehydrobenzoquinone
$C_6H_3O_2$	<2.18	—	LPES	232	benzoquinonide
C_6H_5	1.096	0.006	LPES	26	phenyl
C_6H_5N	1.429	0.011	LPT	115	phenylnitrene
$C_6H_5NO_2$	1.00	0.01	LPES	164	nitrobenzene (NB)
C_6H_5O	2.253	0.006	LPES	26	phenoxyl
C_6H_5S	<2.47	0.06	PT	2	thiophenoxide
C_6H_5NH	1.70	0.03	PT	2	anilide
C_6H_5NO	0.44	—	LPES	285	NO (benzene)
$C_6H_6O_2$	1.06	—	LPES	285	O_2 (benzene)
C_6H_7	<1.67	0.04	PT	2	methylchylopentadienyl
C_6H_8	0.855	0.010	LPES	203	$(CH_2)_2C-C(CH_2)_2$
C_6H_8Si	1.435	0.004	LPT	65	$C_6H_5SiH_3$
C_6H_9	0.654	0.010	LPES	203	$CH_2=C(CH_3)-C(CH_2)_2$
C_6H_9O	1.526	0.010	LPT	113	cyclohexanone enolate
C_6H_{10}	0.645	0.015	LPES	126	t-butyl vinylidene
$C_6H_{11}O$	1.755	+0.05/-0.005	LPT	113	pinacolone enolate
$C_6H_{11}O$	1.82	0.06	PT	2	3,3-dimethylbutananl enolate
C_6N_4	2.3	0.3	PT	2	TCNE
C_7F_5N	1.11	0.11	CT	67	pentafluorobenzonitrile
C_7F_8	0.86	0.11	CT	67	octafluorotoluene
C_7F_{14}	1.08	0.10	CT	61	perfluoromethylcyclohexane
C_7HF_5O	1.10	0.11	CT	67	pentafluorobenzaldehyde
$C_7H_3N_3O_4$	2.16	0.10	CT	61	3,5-$(NO_2)_2$-benzonitrile
$C_7H_4F_3NO_2$	1.41	0.10	CT	61	m-trifluoromethyl-NB
$C_7H_4N_2O_2$	1.61	0.10	CT	61	o-cyano-NB
$C_7H_4N_2O_2$	1.56	0.10	CT	61	m-cyno-NB
$C_7H_4N_2O_2$	1.72	0.10	CT	61	p-cyano-NB
C_7H_6Br	1.308	0.008	LPES	167	o-bromobenzyl
C_7H_6Br	1.307	0.008	LPES	167	m-bromobenzyl·

Molecule	Electron affinity in eV	Uncertainty in eV	Method	Ref.	
C_7H_6Br	1.229	0.008	LPES	167	p-bromobenzyl
C_7H_6Cl	1.257	0.008	LPES	167	o-chlorobenzyl
C_7H_6Cl	1.272	0.008	LPES	167	m-chlorobenzyl
C_7H_6Cl	1.174	0.008	LPES	167	p-chlorobenzyl
C_7H_6F	1.091	0.008	LPES	167	o-fluorobenzyl
C_7H_6F	1.173	0.008	LPES	167	m-fluorobenzyl
C_7H_6F	0.937	0.008	LPES	167	p-fluorobenzyl
C_7H_6FO	2.218	0.010	LPT	2	m-fluoroacetophenone enolate
C_7H_6FO	2.176	0.010	LPT	2	p-fluoroacetophenone enolate
$C_7H_6FeO_3$	0.990	0.10	CT	120	η_4-1,3-butadiene-Fe(CO)$_3$
$C_7H_6N_2O_4$	1.77	0.05	PT	60	3,4-dintrotoluene
$C_7H_6N_2O_4$	1.77	0.05	PT	60	2,3-dinitrotoluene
$C_7H_6N_2O_4$	1.60	0.05	PT	60	2,4-dinitrotoluene
$C_7H_6N_2O_4$	1.55	0.05	PT	60	2,6-dinitrotoluene
$C_7H_6O_2$	1.85	0.10	CT	61	o-CH$_3$-BQ
C_7H_7	0.912	0.006	LPES	26	benzyl
C_7H_7	0.868	0.006	LPES	136	1-quadricyclanide
C_7H_7	0.962	0.006	LPES	136	2-quadricyclanide
C_7H_7	1.286	0.006	LPES	136	norbornadienide
C_7H_7	0.39	0.04	LPES	136	cycloheptatrienide
C_7H_7	3.046	0.006	LPES	136	1-(1,6-heptadiynide)
C_7H_7	>1.140	0.006	LPES	136	3-(1,6-heptadiynide)
$C_7H_7NO_2$	0.92	0.10	CT	61	o-methyl-NB
$C_7H_7NO_2$	0.99	0.10	CT	61	m-methyl-NB
$C_7H_7NO_2$	0.95	0.10	CT	61	p-methyl-NB
$C_7H_7NO_3$	1.04	0.10	CT	61	m-OCH$_3$-NB
$C_7H_7NO_3$	0.91	0.10	CT	61	p-OCH$_3$-NB
C_7H_7O	<2.36	0.06	PT	2	o-methyl phenoxide
C_7H_7O	2.14	0.02	PT	50	benzyloxide
C_7H_8FO	<3.05	0.06	PT	50	PhCH$_2$OHF
C_7H_9	1.27	0.03	PT	2	heptatrienyl
C_7H_9O	1.61	0.05	PT	2	2-norbornanone enolate
C_7H_9Si	1.33	0.04	LPT	65	C$_6$H$_5$(CH$_3$)SiH
$C_7H_{11}O$	1.598	0.007	LPT	113	cycloheptanone enolate
$C_7H_{11}O$	1.49	0.04	PT	2	2,5-dimethyl-cyclopentanone enolate
$C_7H_{13}O$	1.72	0.06	PT	2	4-heptanone enolate
$C_7H_{13}O$	1.46	0.04	PT	2	diisopropyl ketone enolate
$C_8F_{14}N_2$	1.89	0.10	CT	51	1,4-(CN)$_2$C$_6$F$_4$
$C_8H_3F_5O$	0.88	0.11	CT	67	pentafluoroacetophenone
$C_8H_3F_6NO_2$	1.79	0.10	CT	61	3,5-(CF$_3$)$_2$-NB
$C_8H_4F_3N$	0.70	0.05	attach	263	o-trifluoromethylbenzonitrile
$C_8H_4F_3N$	0.67	0.05	attach	263	m-trifluoromethylbenzonitrile
$C_8H_4F_3N$	0.83	0.05	attach	263	p-trifluoromethylbenzonitrile
$C_8H_4O_3$	1.21	0.10	CT	61	phthalic anhydride
C_8H_6	1.044	0.008	LPES	148	
C_8H_7	1.091	0.008	LPES	134	
C_8H_7O	2.057	0.010	PT	2	acetophenone enolate
C_8H_7O	2.10	0.08	LPT	2	phenylacetaldehyde enolate
C_8H_8	0.55	0.02	CT	134	cyclooctatetraene
C_8H_8	0.919	0.008	LPES	139	m-xylylene
$C_8H_9NO_2$	1.21	0.05	PT	60	3,5-dimethyl-NB
$C_8H_9NO_2$	2.61	0.05	PT	60	2,6-dimethyl-NB
$C_8H_9NO_2$	0.86	0.10	CT	61	2,3-dimethyl-NB
$C_8H_{13}O$	1.63	0.06	PT	2	cyclooctanone enolate
C_8S_2	0.049	0.005	LPES	230	bithiophene
$C_9H_8FeO_3$	0.76	0.10	CT	120	η_4-1,3-cyclohexadiene-Fe(CO)$_3$
C_9H_9O	2.030	0.010	LPT	2	m-methylacetophenone enolate
C_9H_9SiN	1.43	0.10	PT	2	trimethylsilylnitrene
$C_9H_{11}NO_2$	0.70	0.10	CT	61	2,4,6-trimethyl-NB

Molecule	Electron affinity in eV	Uncertainty in eV	Method	Ref.	
$C_9H_{15}O$	1.69	0.06	PT	2	cyclononanone enolate
$C_{10}H_4C_{12}O_2$	2.19	0.10	CT	61	2,3-dichloro-1,4-naphthoquinone
$C_{10}H_6N_2O_4$	1.78	0.10	CT	61	1,3-dinitronaphthalene
$C_{10}H_6N_2O_4$	1.77	0.10	CT	61	1,5-dinitronaphthalene
$C_{10}H_6O_2$	1.81	0.10	CT	61	1,4-naphthoquinone
$C_{10}H_7$	1.403	0.015	LPES	197	1-naphthyl radical
$C_{10}H_7NO_2$	1.23	0.10	CT	61	1-nitronaphthalene
$C_{10}H_7NO_2$	1.18	0.10	CT	61	2-nitronaphthalene
$C_{10}H_8$	0.790	0.008	LPES	230	azulene
$C_{10}H_8CrO_3$	0.93	0.10	CT	120	η_4-1,3,5-cycloheptatriene Cr(CO)$_3$
$C_{10}H_8FeO_3$	0.98	0.10	CT	120	η_4-1,3,5-cycloheptatriene-Fe(CO)$_3$
$C_{10}H_8NO$	0.66	—	LPES	285	NO$^-$(naphthlene)
$C_{10}H_8O_2$	1.41	—	LPES	285	O$_2$(naphthlene)
$C_{10}H_{10}O_3$	2.09	—	LPES	285	O$_2$(naphthlene · H$_2$O)
$C_{10}H_{12}O_4$	2.72	—	LPES	285	O$_2$(naphthlene · (H$_2$O)$_2$)
$C_{10}H_{17}O$	1.83	0.06	PT	2	cyclodecanone enolate
$C_{11}H_8FeO_3$	1.29	0.10	CT	120	η_4-1,3-butadiene-Fe(CO)$_3$
$C_{12}F_{10}$	0.82	0.11	CT	67	decafluorobiphenyl
$C_{12}H_4N_4$	2.8	0.3	CD	2	TCNQ
$C_{12}H_9$	1.07	0.10	PT	2	perinaphthenyl
$C_{12}H_{12}NO$	0.79	—	LPES	285	NO (benzene)$_2$
$C_{12}H_{15}O$	2.032	0.010	LPT	2	*t*-butylacetophenone enolate
$C_{12}H_{21}O$	1.90	0.07	PT	2	cyclododecanone enolate
$C_{13}F_{10}O$	1.52	0.11	CT	67	decafluorobenzophenone
$C_{13}H_9FO$	0.64	0.10	CT	61	4-fluorobenzophenone
$C_{13}H_{10}O$	0.62	0.10	CT	61	benzophenone
$C_{14}H_9NO_2$	1.43	0.10	CT	61	9-nitroanthracene
$C_{14}H_{10}$	0.530	0.005	LPES	286	anthracene
$C_{14}H_{12}O$	0.770	0.005	LPES	286	anthracene · H$_2$O
$(C_{14}H_{10})_n$	n=1-16	—	LPES	231	anthracene clusters
$C_{16}H_{10}$	0.406	0.010	LPES	270	pyrene
$C_{18}H_{12}$	1.04	0.10	CT	66	tetracene
$C_{20}H_{12}$	0.79	0.10	CT	66	benz[a]pyrene
$C_{20}H_{12}$	0.973	0.005	LPES	236	perylene
$C_{20}H_{16}NO$	1.06	—	LPES	285	NO (naphthalene)$_2$
$C_{22}H_{14}$	1.35	0.10	CT	66	pentacene
$C_{44}Cl_{28}FeN_4$	2.59	0.11	CT	186	FeTPPCl$_{28}$
$C_{44}Cl_8F_{20}FeN_4$	3.21	0.03	CT	186	FeTPPβCl$_8$
$C_{44}Cl_9F_{20}FeN_4$	3.35	0.03	CT	186	FeTPPF$_{20}$βCl$_8$Cl
$C_{44}H_8F_{20}FeN_4$	2.15	0.15	CT	186	FeTPPF$_{20}$
$C_{44}H_8ClF_{20}FeN_4$	3.14	0.03	CT	186	FeTPPF$_{20}$Cl
$C_{44}H_8Cl_{21}FeN_4$	2.93	0.23	CT	186	FeTPPoCl$_{20}$Cl
$C_{44}H_{12}Cl_{17}FeN_4$	3.14	0.03	CT	186	FeTPPoCl$_8$βCl$_8$Cl
$C_{44}H_{20}Cl_8FeN_4$	1.86	0.03	CT	186	FeTPPoCl$_8$
$C_{44}H_{20}Cl_9FeN_4$	2.10	0.19	CT	186	FeTPPoCl$_8$Cl
$C_{44}H_{28}FeN_4$	1.87	0.03	CT	186	iron tetraphenylporphyrin (FeTPP)
$C_{44}H_{28}NiN_4$	1.51	0.01	CT	186	nickel tetraphenylporphyrin (NiTPP)
$C_{44}H_{28}ClFeN_4$	2.15	0.15	CT	186	FeTPPCl
$C_{44}H_{30}N_4$	1.69	0.01	CT	186	H$_2$ tetraphenylporphyrin
$C_{45}H_{29}NiN_4O$	1.74	0.01	CT	186	NiTPPCHO
$C_{52}H_{39}FeN_7O$	1.97	0.03	CT	186	FeTPP-val
C_{60}	2.65	0.05	LPT	201	
$C_{60}F_2$	2.74	0.07	Knud	202	
$C_{64}H_{64}FeN_8O_4$	2.07	0.03	CT	186	FeTPP-piv
$C_{70}F_2$	2.80	0.07	Knud	202	
(benzene)$_n$	n=53-124	—	LPES	248	
(toluene)$_n$	n=33-139	—	LPES	248	
CeF_4	3.8	0.4	CT	98	
$Cl(CO_2)$	3.907	0.010	LPES	131	
$Cl(H_2O)$	n=1-4	—	LPES	250	

Molecule	Electron affinity in eV	Uncertainty in eV	Method	Ref.	
Co_n	n=1-108	—	LPES	251	
$CoBr_3$	4.6	0.1	LPES	249	
$CoCl_3$	4.7	0.1	LPES	249	
CoF_4	6.4	0.3	CT	98	
$Cr(CO)_3$	1.349	0.006	LPES	94	
CrO_3	3.66	0.02	LPES	241	
CrO_4	4.98	0.09	LPES	241	
CrO_5	4.4	0.1	LPES	241	
CsO_4	2.5	0.2	LPES	252	
Cu_n	n=1-411	—	LPES	37	
$CuBr_2$	4.35	0.05	LPES	237	
$Cu_n(CN)_m$	n=1-6	m=1-6	LPES	159	
$CuCl_2$	4.35	0.05	LPES	237	
$F(H_2O)_n$	n=1-4	—	LPES	242	
$F(H_2O)_n$	n=1-4	—	LPES	250	
Fe_n	n=3-34	—	LPES	149	
$Fe(CO)_2$	1.22	0.02	LPES	2	
$Fe(CO)_3$	1.8	0.2	LPES	2	
$Fe(CO)_4$	2.4	0.3	LPES	2	
$FeBr_3$	4.26	0.06	LPES	249	
$FeBr_4$	5.50	0.08	LPES	249	
$FeCl_3$	4.22	0.06	LPES	249	
$FeCl_4$	6.00	0.08	LPES	249	
FeF_3	3.6	0.1	CT	98	
FeF_4	6.0	estimate	CT	98	
Fe_2H_2	0.942	0.019	LPES	254	
Fe_nO_m	n=1-4	m=1-6	LPES	152	
Ga_2As_3	2.783	0.024	LPES	192	
Ga_xAs_y	n=2-50	n=x+y	LPES	229	
Ga_2P_3	2.991	0.026	LPES	192	
Ge_n	n=3-15	—	LPES	71	
Ge_xAs_y	n=5-30	n=x+y	LPES	72	
GeH_3	<1.74	0.04	PT	2	
$H(NH_3)_n$	n=1,2	—	LPES	76	
HNO_3	0.57	0.15	CD	2	
$(H_2O)_n$	n=2-19	—	LPES	77	
$I(CO_2)$	3.225	0.001	LPES	131	
$I(H_2O)_n$	n=1-4	—	LPES	250	
In_xP_y	n=2-8	n=x+y	LPES	137	
IrF_4	4.7	0.3	CT	98	
IrF_6	6.5	0.4	CT	98	
K_n	n=2-7	—	LPES	18	
KO_4	2.8	0.2	LPES	252	
LiO_4	3.3	0.2	LPES	252	
$MnBr_3$	5.03	0.06	LPES	249	
$MnCl_3$	5.07	0.06	LPES	249	
MnF_4	5.5	0.2	CT	98	
MnO_3	3.335	0.010	LPES	158	
$Mo(CO)_3$	1.337	0.006	LPES	94	
MoF_5	3.5	0.2	CT	98	
MoF_6	3.8	0.2	CT	98	
MoO_3	3.17	0.02	LPES	280	
Mo_2O_2	2.24	0.02	LPES	280	
Mo_2O_3	2.33	0.07	LPES	280	
Mo_2O_4	2.13	0.04	LPES	280	
N_2CD	2.622	0.005	LPES	154	NCND
N_2CH	2.622	0.005	LPES	154	NCNH
$(NH_3)_n$	n=41-1100	—	LPES	77	
$NH_2(NH_3)_n$	n=1,2	—	LPES	78	
$NO(H_2O)_n$	n=1,2	—	LPES	75	

Molecule	Electron affinity in eV	Uncertainty in eV	Method	Ref.	
NO_3	3.937	0.014	LPES	85	
$NO_3(H_2O)_n$	$n=0-6$	—	LPES	240	
$NO(N_2O)_n$	$n=1,2$	—	LPES	79	
$(NO)_2$	>2.1	—	LPES	75	
$(N_2O)_n$	$n=1,2$	—	LPES	81	
Na_n	$n=2-5$	—	LPES	18	
$(NaF)_n$	$n=1-7,12$	—	LPES	64	
$Na(NaF)_n$	$n=5,7-12$	—	LPES	64	
NaO_4	3.1	0.2	LPES	252	
NaO_5	3.2	0.2	LPES	252	
$NaSO_3$	2.3	0.2	LPES	252	
Nb_n	$n=6-17$	—	LPES	181	
Nb_8	1.513	0.008	LPES	157	
Nb_3O	1.393	0.006	LPES	169	
Ni_n	$n=1-100$	—	LPES	247	
$Ni_n(benzene)_m$	$n=1-3$	$m=1,2$	LPES	295	
$NiBr_3$	4.94	0.08	LPES	249	
$NiCl_3$	5.20	0.08	LPES	249	
$Ni(CO)_2$	0.643	0.014	LPES	2	
$Ni(CO)_3$	1.077	0.013	LPES	2	
$Ni(CO)H$	1.126	0.010	LPES	293	HNiCO
$OH(H_2O)$	<2.95	0.15	PT	2	
$OH(NH_3)$	2.35	0.07	LPES	234	
$OH(N_2O)$	2.14	0.02	LPES	209	
$OH(N_2O)_n$	$n=1-5$	—	LPES	209	
OsF_4	3.9	0.3	CT	98	
OsF_6	6.0	0.3	CT	98	
P_5	3.88	0.03	LPES	253	
PBr_3	1.59	0.15	CD	2	
PBr_2Cl	1.63	0.20	CD	2	
PCl_2Br	1.52	0.20	CD	2	
PCl_3	0.82	0.10	CD	2	
PF_5	0.75	0.15	CT	121	
PO_3	4.95	0.06	LPES	156	
$POCl_2$	3.83	0.25	CD	2	
$POCl_3$	1.41	0.20	CD	2	
PtF_4	5.5	0.3	CT	98	
PtF_6	7.0	0.4	CT	98	
ReF_6	4.7	estimate	CT	98	
ReO_3	3.6	0.1	LPES	216	
RhF_4	5.4	0.3	CT	98	
RuF_4	4.8	0.3	CT	98	
RuF_5	5.2	0.4	CT	98	
RuF_6	7.5	0.3	CT	98	
SF_4	1.5	0.2	CT	91	
SF_5	4.23	0.12	e-scat	204	
SF_6	1.05	0.10	CT	56	
SO_3	1.97	0.10	LPES	165	
$(SO_2)_2$	0.6	0.2	LPES	80	
Sb_n	$n=2-9$	—	LPES	213	
Sb_5	3.46	0.03	LPES	253	
$ScBr_4$	6.13	0.08	LPES	249	
$ScCl_4$	6.89	0.08	LPES	249	
SeF_6	2.9	0.2	CD	2	
Si_4	2.13	0.01	LPES	110	
Si_5	2.59	0.02	LPES	110	
Si_7	1.85	0.02	LPES	110	
Si_n	$n=3-20$	—	LPES	71	
Si_2C_3	1.766	0.012	LPES	296	linear Si-C$_3$-Si

Molecule	Electron affinity in eV	Uncertainty in eV	Method	Ref.
SiD_3	1.386	0.022	LPES	43
SiF_3	2.41	0.22	LPES	17
SiF_4	≤ 0	—	LPES	17
SiF_5	≥ 4.66	—	LPES	17
Si_nF	n=2-11	—	LPES	17
SiH_3	1.406	0.014	LPES	43
Si_3H	2.53	0.01	LPES	182
Si_4H	2.68	0.01	LPES	182
Si_nNa_m	n=4-11	m=1-3	LPES	210
Sn_n	n=1-12	—	LPES	289
$SnCH_2CN$	1.57	0.02	LPES	292
$Sn(CN)_2$	2.622	0.004	LPES	292
$Sn(CN)(CH_2CN)$	2.29	0.05	LPES	292
Ta_3O	1.583	0.010	LPES	169
TeF_6	3.34	0.17	CD	2
Ti_n	n=1-130	—	LPES	151
TiO_3	4.2	—	LPES	172
UF_5	3.7	0.2	CT	98
UF_6	5.1	0.2	CT	98
UO_3	<2.1	—	CT	98
V_n	n=3-65	—	LPES	150
VF_4	3.5	0.2	CT	98
V_2O_n	n=3-7	—	LPES	246
V_3O	1.218	0.008	LPES	169
V_4O_{10}	4.2	0.6	CT	101
$W(CO)_3$	1.859	0.006	LPES	94
WF_5	1.25	0.3	CD	18
WF_6	>3.5	—	CT	19
WO_3	3.33	+0.04/-0.15	LPT	86
WO_3	3.9	0.2	CT	98

References

1. Hotop, H., and Lineberger, W. C., *J. Phys. Chem. Ref. Data*, 14, 731, 1985.
2. Drzaic, P. S., Marks, J., and Brauman, J. I., in *Gas Phase Ion Chemistry*, Vol. 3, Bowers, M. T., Ed., Academic Press, Orlando, 1984, p. 167. The reference for C_6H_4ClO should read "Richardson, et al., 1975c".
3. Schulz, P. A., Mead, R. D., Jones, P. L., and Lineberger, W. C., *J. Chem. Phys.*, 77, 1153, 1982.
4. Neumark, D. M., Lykke, K. R., Anderson, T., and Lineberger, W. C., *Phys. Rev. A*, 32, 1890, 1985. EA(O) = 11,784.645 ± 0.008 cm⁻¹.
5. Wenthold, P. G., Gunion, R. F., and Lineberger, W. C., *Chem. Phys. Lett.*, 258, 101, 1996.
6. Wenthold, P. G., Kim, J. B., Jonas, K. L., and Lineberger, W. C., *J. Phys. Chem. A* 101, 4472, 1997.
7. Klein, R., McGinnis, R. P., and Leone, S. R., *Chem. Phys. Lett.*, 100, 475, 1983.
8. Arnold, D. W., Bradforth, S. E., Kitsopoulos, T. N., and Neumark, D. M., *J. Chem. Phys.*, 95, 8753, 1991; linear C_n.
9. Stevens, A. E., Fiegerle, C. S., and Lineberger, W. C., *J. Chem. Phys.*, 78, 5420, 1983.
10. Breyer, F., Frey, P., and Hotop, H., *Z. Phys. A* 300, 7, 1981.
11. Oakes, J. M., and Ellison, G. B., *Tetrahedron*, 42, 6263, 1986.
12. Leopold, D. G., Murray, K. K., Miller, A. E. S., and Lineberger, W. C., *J. Chem. Phys.*, 83, 4849, 1985.
13. Oakes, J. M., Jones, M.E., Bierbaum, V. M., and Ellison, G. B., *J. Phys. Chem.*, 87, 4810, 1983.
14. Ellis, H. B., Jr. and Ellison, G. B., *J. Chem. Phys.*, 78, 6541, 1983.
15. Ramond, T. M., Blanksby, S. J., Kato, S., Bierbaum, V. M., Davico, G. E., Schwartz, R. L., Lineberger, W. C., and Ellison, G. B., *J. Phys. Chem. A* 106, 9641, 2002.
16. Nimlos, M. E., and Ellison, G. B., *J. Chem. Phys.*, 90, 2574, 1986.
17. Kawamata, H., Negishi, Y., Kishi, R., Iwata, S., Nakajima, A., and Kaya, K., *J. Chem. Phys.* 105, 5369, 1996.
18. McHugh, K. M., Eaton, J. G., Lee, G. H., Sarkas, H. W., Kidder, L. H., Snodgrass, J. T., Manaa, M. R., and Bowen, K. H., *J. Chem. Phys.*, 91, 3792, 1989. See also Ref. 104.
19. Viggiano, A. A., Paulson, J. F., Dale, F., Henchman, M., Adams, N. G., and Smith, D., *J. Phys. Chem.*, 89, 2264, 1985. The upper limit given in this paper (≤3.4 eV) was later found to be incorrect when rapid charge transfer from HCO_2^- to WF_6 was observed (unpublished).
20. Burnett, S. M., Stevens, A. E., Fiegerle, C. S., and Lineberger, W. C., *Chem. Phys. Lett.*, 100, 124, 1983.
21. Moran, S., Ellis, H. B., DeFrees, D. J., McLean, A. D., and Ellison, G. B., *J. Am. Chem. Soc.*, 109, 5996, 1987; Moran, S., Ellis, H. B., DeFrees, D. J., McLean, A. D., Paulson, S. E., and Ellison, G. B., *J. Am. Chem. Soc.*, 109, 6004, 1987; see also Lykke, K. R., Neumark, D. M., Andersen, T., Trapa, V. J., and Lineberger, W. C., *J. Chem. Phys.*, 87, 6842, 1987.
22. Mead, R. D., Lykke, K. R., Lineberger, W. C., Marks, J., and Brauman, J. I., *J. Chem. Phys.*, 81, 4883, 1984; Lykke, K. R., Mead, R. D., and Lineberger, W. C., *Phys. Rev. Lett.*, 52, 2221, 1984. The EAs are 14,717.7 ± 1.0 cm⁻¹ for acetaldehyde enolate and 14,671.0 ± 1.0 cm⁻¹ for acetaldehyde-d_3 enolate.
23. Ellison, G. B., Engelking, P. C., and Lineberger, W. C., *J. Chem. Phys.*, 86, 4873, 1982.
24. Oakes, J. M., and Ellison, G. B., *J. Am. Chem. Soc.*, 105, 2969, 1983.
25. Ellison, G. B., and Oakes, J. M., *J. Am. Chem. Soc.*, 106, 7734, 1984. EA(allyl) and EA(allyl-d_5) are 0.119 and 0.083 eV too low, respectively,

in this work, according to Ref. 138. Therefore, EA(allyl-d$_1$) is likely too low by a similar amount.

26. Gunion, R. F., Gilles, M. K., Polak, M. L., and Lineberger, W. C., *Int. J. Mass Spectrom. Ion Processes*, 117, 601, 1992; see also Ref. 136.

27. Leopold, D. G., and Lineberger, W. C., *J. Chem. Phys.*, 85, 51, 1986.

28. Scheer, M., Bilodeau, R. C., Brodie, C. A., and Haugen, H. K., *Phys. Rev. A*, 58, 2844, 1998.

29. Miller, A. E. S., Fiegerle, C. S., and Lineberger, W. C., *J. Chem. Phys.*, 87, 1549, 1987.

30. Miller, T. M., Leopold, D. G., Murray, K. K., and Lineberger, W. C., *J. Chem. Phys.*, 85, 2368, 1986.

31. Miller, T. M., and Lineberger, W. C., *Chem. Phys. Lett.*, 146, 364, 1988.

32. Neumark, D. M., Lykke, K. R., Andersen, T., and Lineberger, W. C., *J. Chem. Phys.*, 83, 4364, 1985.

33. Leopold, D. G., Miller, T. M., and Lineberger, W. C., *J. Am. Chem. Soc.*, 108, 178, 1986.

34. Miller, A. E. S., Fiegerle, C. S., and Lineberger, W. C., *J. Chem. Phys.*, 84, 4127, 1986.

35. Murray, K. K., Miller, T. M., Leopold, D. G., and Lineberger, W. C., *J. Chem. Phys.*, 84, 2520, 1986.

36. Leopold, D. G., Miller, A. E. S., and Lineberger, W. C., *J. Am. Chem. Soc.*, 108, 1379, 1986.

37. Li, J., Li, X., Zhai, H. J., and Wang, L.-S., Science 299, 864, 2003; Hakkinen, H., Yoon, B., Landman, U., Li, X., Zhai, H. J., and Wang, L.-S., *J. Phys. Chem. A* 107, 6168, 2003; Taylor, K. J., Pettiette-Hall, C. L., Cheshnovsky, O., and Smalley, R. E., *J. Chem. Phys.* 96, 3319, 1992; Handschuh, H., Cha, C.-Y., Bechthold, P. S., Ganteför, G., and Eberhardt, W., *J. Chem. Phys.*, 102, 6406, 1995; Cha, C.-Y., Ganteför, G., and Eberhardt, W., *J. Chem. Phys.*, 99, 6308, 1993; Ho, J., Ervin, K. M., and Lineberger, W. C., *J. Chem. Phys.*, 93, 6987, 1990; Leopold, D. G., Ho, J., and Lineberger, W. C., *J. Chem. Phys.*, 86, 1715, 1987; Pettiette, C. L., Yang, S. H., Craycraft, M. J., Conceicao, J., Laaksonen, R. T., Cheshnovsky, O., and Smalley, R. E., *J. Chem. Phys.*, 88, 5377, 1988.

38. Snodgrass, J. T., Coe, J. V., McHugh, K. M., Friedhoff, C. B., and Bowen, K. H., *J. Phys. Chem.*, 93, 1249, 1989.

39. Friedhoff, C. B., Snodgrass, J. T., Coe, J. V., McHugh, K. M., and Bowen, K. H., *J. Chem. Phys.*, 84, 1051, 1986.

40. Friedhoff, C. B., Coe, J. V., Snodgrass, J. T., McHugh, K. M., and Bowen, K. H., *Chem. Phys. Lett.*, 124, 268, 1986.

41. Coe, J. V., Snodgrass, J. T., Friedhoff, C. B., McHugh, K. M., and Bowen, K. H., *J. Chem. Phys.*, 84, 619, 1986.

42. Snodgrass, J. T., Coe, J. V., Friedhoff, C. B., McHugh, K. M., and Bowen, K. H., *Chem. Phys. Lett.*, 122, 352, 1985.

43. Nimlos, M. R., and Ellison, G. B., *J. Am. Chem. Soc.*, 108, 6522, 1986.

44. Petrunin, V., Andersen, H., Balling, P., and Andersen, T., *Phys. Rev. Lett.*, 76, 744, 1996.

45. Andersen, T., Lykke, K. R., Neumark, D. M., and Lineberger, W. C., *J. Chem. Phys.*, 86, 1858, 1987.

46. Murray, K. K., Lykke, K. R., and Lineberger, W. C., *Phys. Rev. A*, 36, 699, 1987.

47. Mansour, N. B., and Larson, D. J., *Abstracts of the XV Int. Conf. on the Phys. of Electronic and Atomic Collisions*, p. 70, 1987. EA(SH) = 18666.44 ± 0.03 cm^{-1}.

48. Stonemann, R. C., and Larson, D. J., *Phys. Rev. A*, 35, 2928, 1987. EA(SeH) = 17,845.17 ± 0.20 cm^{-1}.

49. Nimlos, M. R., Harding, L. B., and Ellison, G. B., *J. Chem. Phys.*, 87, 5116, 1987.

50. Moylan, C. R., Dodd, J. A., Han, C.-C., and Braumann, J. I., *J. Chem. Phys.*, 86, 5350, 1987.

51. Chowdhury, S., Grimsrud, E. P., Heinis, T., and Kebarle, P., *J. Am. Chem. Soc.*, 108, 3630, 1986.

52. Berzinsh, U., Gustafsson, M., Hanstorp, D., Klinkmueller, A. E., Ljungblad, U., Maartensson-Pendrill, A.-M., *Phys. Rev. A* 51, 231, 1995. EA(Cl) = 29138.59 ± 0.22 cm^{-1}.

53. Moran, S., and Ellison, G. B., *J. Phys. Chem.*, 92, 1794, 1988.

54. Murray, K. K., Leopold, D. G., Miller, T. M., and Lineberger, W. C., *J. Chem. Phys.*, 89, 5442, 1988.

55. Ervin, K. M., Ho, J., and Lineberger, W. C., *J. Chem. Phys.*, 89, 4514, 1988.

56. Grimsrud, E. P., Chowdhury, S., and Kebarle, P., *J. Chem. Phys.*, 85, 4989, 1985.

57. Fischer, C. F., *Phys. Rev. A*, 39, 963, 1989.

58. Wickham-Jones, C. T., Ervin, K. M., Ellision, G. B., and Lineberger, W. C., *J. Chem. Phys.*, 91, 2762, 1989.

59. Kryachko, E. S., Vinckier, C., and Nguyen, M. T., *J. Chem. Phys.*, 114, 7911, 2001.

60. Mock, R. S., and Grimsrud, E. P., *J. Am. Chem. Soc.*, 111, 2861, 1989.

61. Chowdhury, S., Heinis, T., Grimsrud, E. P., and Kebarle, P., *J. Phys. Chem.*, 90, 2747, 1986. The uncertainty and other results are quoted in Ref. 60.

62. Wickham-Jones, C. T., Moran, S., and Ellison, G. B., *J. Chem. Phys.*, 90, 795, 1989.

63. Ervin, K. M., Ho, J., and Lineberger, W. C., *J. Phys. Chem.*, 92, 5405, 1988.

64. Miller, T. M., and Lineberger, W. C., Int. *J. Mass Spectrom. Ion Processes*, 102, 239, 1990.

65. Wetzel, D. M., Salomon, K. E., Berger, S., and Brauman, J. I., *J. Am. Chem. Soc.*, 111, 3835, 1989.

66. Crocker, L., Wang, T., and Kebarle, P., *J. Am. Chem. Soc.*, 115, 7818, 1993.

67. Dillow, G. W., and Kebarle, P., *J. Am. Chem. Soc.*, 111, 5592, 1989.

68. Gantefor, G., Gausa, M., Meiwes-Broer, K. H., and Lutz, H. O., Z. Phys. D, 9, 253, 1988; Taylor, K. J., Petteitte, C. L., Craycraft, M. J., Chesnovsky, O., and Smalley, R. E., *Chem. Phys. Lett.*, 152, 347, 1988.

69. Metz, R. B., Kitsopoulos, T., Weaver, A., and Neumark, D. M., *J. Chem. Phys.*, 88, 1463, 1988.

70. Yang, S., Pettiette, C. L., Conceicao, J., Cheshnovsky, O., and Smalley, R. E., *Chem. Phys. Lett.*, 139, 233, 1987; Yang, S., Taylor, K. J., Craycraft, M. J., Conceicao, J., Pettiette, C. L., Cheshnovsky, O., and Smalley, R. E., *Chem. Phys. Lett.*, 144, 431, 1988; Arnold, D. W., Bradforth, S. E., Kitsopoulos, T. N., and Neumark, D. M., *J. Chem. Phys.*, 95, 5479, 1991.

71. Cheshnovsky, O., Yang, S., Pettiette, C. L., Craycraft, M. J., Liu, Y., and Smalley, R. E., *Chem. Phys. Lett.*, 138, 119, 1987.

73. Travers, M. J., Cowles, D. C., and Ellison, G. B., *Chem. Phys. Lett.*, 164, 449, 1989.

74. Blondel, C., Cacciani, P., Delsart, C., and Trainham, R., *Phys. Rev. A*, 40, 3698, 1989. EA(Br) = 27,129.170 ± 0.015 cm^{-1} and EA(F) = 27,432.440 ± 0.025 cm^{-1}.

75. Bowen, K. H., and Eaton, J. G., in *The Structure of Small Molecules and Ions*, Naaman, R., and Vager, Z., Eds., Plenum, New York, 1988, pp. 147-169; Arnold, S. T., Eaton, J. G., Patel-Mistra, D., Sarkas, H. W., and Bowen, K. H., in *Ion and Cluster Ion Spectroscopy and Structure*, Maier, J. P., Ed., Elsevier Science, New York, 1989, p. 417.

76. Snodgrass, J. T., Coe, J. V., Friedhoff, C. B., McHugh, K. M., and Bowen, K. H., *Faraday Disc. Chem. Soc.*, 88, 1988.

77. Lee, G. H., Arnold, S. T., Eaton, J. G., Sarkas, H. W., Bowen, K. H., Ludewigt, C., and Haberland, H., *Z. Phys. D - Atoms, Mol. and Clusters*, 20, 9, 1991; Coe, J. V., Lee, G. H., Eaton, J. G., Arnold, S. T., Sarkas, H. W., Bowen, K. H., Ludewigt, C., Haberland, H., and Worsnop, D. R., *J. Chem. Phys.*, 92, 3980, 1990.

78. Snodgrass, J. T., Coe, J. V., Freidhoff, C. B., McHugh, K. M., and Bowen, K. H., to be published, quoted in Ref. 75.

79. Coe, J. V., Snodgrass, J. T., Friedhoff, C. B., McHugh, K. M., and Bowen, K. H., *J. Chem. Phys.*, 87, 4302, 1987.

80. Friedhoff, C. B., Snodgrass, J. T., and Bowen, K. H., to be published, quoted in Ref. 75.

81. Coe, J. V., Snodgrass, J. T., Friedhoff, C. B., McHugh, K. M., and Bowen, K. H., *Chem. Phys. Lett.*, 124, 274, 1986.

82. Eliav, E., Vilkas, M. J., Ishikawa, Y., and Kaldor, U., *J. Chem. Phys.* 123, 224113(5), 2005.

83. Ervin, K. M., Ho, J., and Lineberger, W. C., *J. Chem. Phys.*, 91, 5974, 1989.

84. Polak, M. L., Fiala, B. L., Ervin, K. M., and Lineberger, W. C., *J. Chem. Phys.*, 94, 6924, 1991.

85. Weaver, A., Arnold, D. W., Bradforth, S. E., Neumark, D. M., *J. Chem. Phys.*, 94, 1740, 1991.

86. Walter, C. W., Devynck, P., Hertzler, C. F., Bae, Y. K., Smith, G. P., and Peterson, J. R., *Bull. Am. Phys. Soc.*, 35, 1163, 1990.

87. Ervin, K. M., and Lineberger, W. C., *J. Phys. Chem.*, 95, 1167, 1991.

88. Gilles, M. K., Polak, M. L., and Lineberger, W. C., *J. Chem. Phys.*, 96, 8012, 1992.

89. Lykke, K. R., Murray, K. K., and Lineberger, W. C., *Phys. Rev. A*, 43, 6104, 1991. EA(H) = 6082.99 ± 0.15 cm^{-1} and EA(D) = 6086.2 ± 0.6 cm^{-1}.

90. Ervin, K. M., Gronert, S., Barlow, S. E., Gilles, M. K., Harrison, A. G., Bierbaum, V. M., DePuy, C. H., Lineberger, W. C., and Ellison, G. B., *J. Am. Chem. Soc.*, 112, 5750, 1990.

91. Viggiano, A. A., Miller, T. M., Miller, A. E. S., Morris, R. A., Van Doren, J. M., and Paulson, J. F., *Int. J. Mass Spectrom. Ion Processes*, 109, 327, 1991.

92. Hanstorp, D., and Gustafsson, M., *J. Phys. B: At. Mol. Opt. Phys.*, 25, 1773, 1992. EA(I) = 24,672.7956 ± 0.079 cm^{-1}.

93. Polak, M. L., Gilles, M. K., and Lineberger, W. C., *J. Chem. Phys.*, 96, 7191, 1992.

94. Bengali, A. A., Casey, S. M., Cheng, C.-L., Dick, J. P., Fenn, P. T., Villalta, P. W., and Leopold, D. G., *J. Am. Chem. Soc.*, 114, 5257, 1992.

95. Gilles, M. K., Ervin, K. M., Ho, J., and Lineberger, W. C., *J. Phys. Chem.*, 96, 1130, 1992.

96. Travers, M. J., Cowles, D. C., Clifford, E. P., and Ellison, G. B., *J. Am. Chem. Soc.*, 114, 8699, 1992.

97. Bengali, A. A., and Leopold, D. G., *J. Am. Chem. Soc.*, 114, 9192, 1992.

98. Rudnyi, E. B., Kaibicheva, E. A., and Sidorov, L. N., *Rapid Comm. in Mass Spectrom.*, 6, 356, 1992; Sidorov, L. N., High Temp. Sci., 29, 153, 1990. See also Srivastava, R. D., Uy, O. M., and Farber, M., *Trans. Faraday Soc.*, 67, 2941, 1971.

99. Kitsopoulos, T. N., Chick, C. J., Zhao, Y., and Neumark, D. M., *J. Chem. Phys.*, 95, 5479, 1991.

100. Arnold, C. C., Kitsopoulos, T. N., and Neumark, D. M., *J. Chem. Phys.*, 99, 766, 1993.

101. Rudnyi, E. B., Kaibicheva, E. A., and Sidorov, L. N., *J. Chem. Thermodynamics*, 25, 929, 1993.

102. Sarkas, H. W., Hendricks, J. H., Arnold, S. T., and Bowen, K. H., *J. Chem. Phys.* 100, 1884, 1994.

103. Villalta, P. W., and Leopold, D. G., *J. Chem. Phys.* 98, 7730, 1993.

104. Eaton, J. G., Sarkas, H. W., Arnold, S. T., McHugh, K. M., and Bowen, K. H., *Chem. Phys. Lett.*, 193, 141, 1992. See also Ref. 18.

105. Polak, M. L., Gilles, M. K., Gunion, R. F., and Lineberger, W. C., *Chem. Phys. Lett.*, 210, 55, 1993.

106. Gilles, M. K., Lineberger, W. C., and Ervin, K. M., *J. Am. Chem. Soc.*, 115, 1031, 1993.

107. Casey, S. M., and Leopold, D. G., *Chem. Phys. Lett.*, 201, 205, 1993.

108. Polak, M. L., Gerber, G., Ho, J., and Lineberger, W. C., *J. Chem. Phys.*, 97, 8990, 1992.

109. Kim, E. H., Bradforth, S. E., Arnold, D.W., Metz, R. B., and Neumark, D. M., *J. Chem. Phys.*, 103, 7801, 1995.

110. Xu, C., Taylor, T. R., Burton, G. R., and Neumark, D. M., *J. Chem. Phys.*, 108, 1395, 1998.

111. Bradforth, S. E., Kim, E. H., Arnold, D. W., and Neumark, D. M., *J. Chem. Phys.*, 98, 800, 1993.

112. Ho, J., Polak, M. L., Ervin, K. M., and Lineberger, W. C., *J. Chem. Phys.*, 99, 8542, 1993.

113. Brinkman, E. A., Berger, S., Marks, J., and Brauman, J. I., *J. Chem. Phys.*, 99, 7586, 1993.

114. Casey, S. M., and Leopold, D. G., *J. Phys. Chem.*, 97, 816, 1993.

115. McDonald, R. N., and Davidson, S. J., *J. Am. Chem. Soc.*, 115, 10857, 1993.

116. Ho, J., Ervin, K. M., Polak, M. L., Gilles, M. K., and Lineberger, W. C., *J. Chem. Phys.*, 95, 4845, 1991.

117. Ho, J., Polak, M. L., and Lineberger, W. C., *J. Chem. Phys.*, 96, 144, 1992. See also Reference 289.

118. Polak, M. L., Gilles, M. K., Ho, J., and Lineberger, W. C., *J. Phys. Chem.*, 95, 3460, 1991.

119. Polak, M. L., Ho, J., Gerber, G., and Lineberger, W. C., *J. Chem. Phys.*, 95, 3053, 1991.

120. Sharpe, P., and Kebarle, P., *J. Am. Chem. Soc.*, 115, 782, 1993.

121. Miller, T. M., Miller, A. E. S., Viggiano, A. A., Morris, R. A., and Paulson, J. F., *J. Chem. Phys.*, 100, 7200, 1994. Accurate calculations have yielded a higher result (0.90 eV); see Lau, J. K.-C. and Li, W.-K., *J. Mol. Struct. (Theochem)* 578, 221, 2002.

122. Berkovits, D., Boaretto, E., Gehlberg, S., Heber, O., and Paul, M., *Phys. Rev. Lett.*, 75, 414, 1995.

123. Arnold, C. C., Xu, C., Burton, G. R., and Neumark, D. M., *J. Chem. Phys.*, 102, 6982, 1995. Burton, G. R., Xu, C., Arnold, C. C., and Neumark, D. M., *J. Chem. Phys.* 104, 2757 1996.

124. Xu, C., de Beer, E., and Neumark, D. M., *J. Chem. Phys.*, 104, 2749, 1996.

125. Gunion, R. F., Koppel, H., Leach, G. W., and Lineberger, W. C., *J. Chem. Phys.*, 103, 1250, 1995.

126. Gunion, R. F., and Lineberger, W. C., *J. Phys. Chem.*, 100, 4395, 1996.

127. Gunion, R. F., Dixon-Warren, St. J., and Lineberger, W. C., *J. Chem. Phys.*, 104, 1765, 1996.

128. Dixon-Warren, St. J., Gunion, R. F., and Lineberger, W. C., *J. Chem. Phys.*, 104, 4902, 1996.

129. Nakajima, A., Zhang, N., Kawamata, H., Hayase, T., Nakao, K., and Kaya, K., *Chem. Phys. Lett.*, 241, 295, 1995; Nakajima, A., Taguwa, T., Nakao, K., Hoshino, K., Iwata, S., and Kaya, K., *J. Chem. Phys.*, 102, 660, 1995.

130. Fan, J., and Wang, L.-S., *J. Chem. Phys.*, 102, 8714, 1995.

131. Arnold, D. W., Bradforth, S. E., Kim, E. H., and Neumark, D. M., *J. Chem. Phys.*, 102, 3493, 1995; Zhao, Y., Arnold, C. C., and Neumark, D. M., *J. Chem. Soc. Faraday Trans.* 2, 89, 1449, 1992.

132. Fan, J., Lou, L., and Wang, L.-S., *J. Chem. Phys.*, 102, 2701, 1995.

133. Sarkas, H. W., Hendricks, J. H., Arnold, S. T., Slager, V. L., and Bowen, K. H., *J. Chem. Phys.*, 100, 3358, 1994.

134. Kato, S., Lee, H. S., Gareyev, R., Wenthold, P. G., Lineberger, W. C., DePuy, C. H., and Bierbaum, V. M., *J. Am. Chem. Soc.*, 119, 7863, 1997. See also Miller, T. M., Viggiano, A. A., and Miller, A. E. S., *J. Phys. Chem. A* 106, 10200, 2002.

135. Wenthold, P. G., Hu, J., Squires, R. R., and Lineberger, W. C., *J. Am. Chem. Soc.*, 118, 475, 1996.

136. Gunion, R. F., Karney, W., Wenthold, P. G., Borden, W. T., and Lineberger, W. C., *J. Am. Chem. Soc.*, 118, 5074, 1996. The numbers in the abstract for 1,6-heptadiyne were misprinted. EA(cycloheptatrienide) quoted here derives from the LPES data combined with other thermochemical data in Ref. 136.

137. Xu, C., de Beer, E., Arnold, D. W., Arnold, C. C., and Neumark, D. M., *J. Chem. Phys.*, 101, 5406, 1996.

138. Wenthold, P. G., Polak, M. L., and Lineberger, W. C., *J. Phys. Chem.*, 100, 6920, 1996.

139. Wenthold, P. G., Kim, J. B., and Lineberger, W. C., *J. Am. Chem. Soc.*, 119, 1354, 1997.

140. Eliav, E., Kaldor, U., Ishikawa, Y., and Pyykko, P., *Phys. Rev. Lett.*, 77, 5350, 1996.

141. Davies, B. J., Ingram, C. W., Larson, D. J., and Ljungblad, U., *J. Chem. Phys.* 106, 5783, 1997. EA(Ir) = 12,613 ± 4 cm^{-1}.

142. Smith, J. R., Kim, J. B., and Lineberger, W. C., *Phys. Rev. A.*, 55, 2036, 1997. EA(OH) = 14,741.02 ± 0.03 cm^{-1}. Schulz, P. A., Mead, R. D., Jones, P. L., and Lineberger, W. C., *J. Chem. Phys.* 77, 1153, 1982. EA(OD) = 14,723.92 ± 0.30 cm^{-1}. See also Rudmin, J. D., Ratliff, L. P., Yukich, J. N., and Larson, D. J., *J. Phys. B: At. Mol. Opt. Phys.*, 29 L881, 1996.

143. Desai, S. R., Wu, H., Rohlfing, C. M., and Wang, L.-S., Int. *J. Chem. Phys.*, 106, 1309, 1997.

144. Wenthold, P. G., Jonas, K.-L., and Lineberger, W. C., *J. Chem. Phys.*, 106, 9961, 1997.

145. Wu, H., and Wang, L.-S., *J. Chem. Phys.*, 107, 16, 1997.

146. Moravec, V. D., and Jarrold, C. C., *J. Chem. Phys.* 108, 1804, 1998.

147. Wang, X.-B., Ding, C.-F., and Wang, L.-S., *J. Phys. Chem. A*, 101, 7699, 1997.

148. Wenthold, P. G., and Lineberger, W. C., *J. Am. Chem. Soc.*, 19, 7772, 1997.

149. Wang, L.-S., Li, X., and Zhang, H.-F., Chem. Phys., 262, 53, 2000; Wang, L.-S., Cheng, H.-S., and Fan, J., *J. Chem. Phys.*, 102, 9480, 1995.

150. Wu, H., Desai, S. R., and Wang, L.-S., *Phys. Rev. Lett.*, 77, 2436, 1996.

151. Liu, S.-R., Zhai, H.-J., Castro, M., and Wang, L.-S., *J. Chem. Phys.* 118, 2108, 2003.

152. Gutsev, G. L., Bauschlicher, C. W., Zhai, H.-J., and Wang, L.-S., *J. Chem. Phys.* 119, 11135, 2003; Wang, L.-S., Wu, H., and Desai, S. R., *Phys. Rev. Lett.*, 76, 4853, 1996.

153. Robinson, M. S., Polak, M. L, Bierbaum, V. M., DePuy, C. H., and Lineberger, W. C., *J. Am. Chem. Soc.*, 117, 6766, 1995.

154. Clifford, E. P., Wenthold, P. G., Lineberger, W. C., Petersson, G. A., and Ellison, G. B., *J. Phys. Chem.*, 101, 4338, 1997.

155. Seburg, R. A., and Squires, R. R., *Int. J. Mass Spectrom. Ion Processes*, 167/168, 541, 1997.

156. Wang, X.-B., and Wang, L.-S., *Chem. Phys. Lett.*, 313, 179, 1999.

157. Marcy, T. P., and Leopold, D. G., *Int. J. Mass Spectrom.*, 195/196, 653, 2000.

158. Gutsev, G. L., Rao, B. K., Jena, P., Li, X., and Wang, L.-S., *J. Chem. Phys.*, 113, 1473, 2000.

159. Negishi, Y., Yasuike, T., Hayakawa, F., Kizawa, M., Yabushita, S., and Nakajima, A., *J. Chem. Phys.*, 113, 1725, 2000.

160. Klopcic, S. A., Moravec, V. D., and Jarrold, C. C., *J. Chem. Phys.*, 110, 8986, 1999.

161. Boldyrev, A. I., Simons, J., Li, X., Chen, W., and Wang, L.-S., *J. Chem. Phys.* 110, 8980, 1999.

162. Taylor, T. R., Asmis, K. R., Zanni, M. T., and Neumark, D. M., *J. Chem. Phys.*, 110, 7607, 1999.

163. Boldyrev, A., Li, X., and Wang, L.-S., *J. Chem. Phys.*, 112, 3627, 2000.

164. Defrançois, C., Périquet, V., Lyapustina, S. A., Lippa, T. P., Robinson, D. W., Bowen, K. H., Nonaka, H., and Compton, R. N., *J. Chem. Phys.*, 111, 4569, 1999.

165. Dobrin, S., Boo, B. H., Alconcel, L. S., and Continetti, R. E., *J. Phys. Chem. A* 104, 10695, 2000.

166. Schwartz, R. L., Davico, G. E., and Lineberger, W. C., *J. Electron Spectros. and Related Phenomena*, 108, 163, 2000.

167. Kim, J. B., Wenthold, P. G., and Lineberger, W. C., *J. Phys. Chem.* 103, 10833, 1999.

168. Davico, G. E., Ramond, T. M., and Lineberger, W. C., *J. Chem. Phys.*, 113, 8852, 2000.

169. Green, S. M. E., Alex, S., Fleischer, N. L., Millam, E. L., Marcy, T. P., and Leopold, D. G., *J. Chem. Phys.* 114, 2653, 2001.

170. Wu, H., and Wang, L.-S., *J. Chem. Phys.* 108, 5310, 1998.

171. Wu, H., and Wang, L.-S., *J. Phys. Chem. A* 102, 9129, 1998.

172. Wu, H., and Wang, L.-S., *J. Chem. Phys.* 107, 8221, 1997.

173. Thomas, O. C., Xu, S. J., Lippa, T. P., and Bowen, K. H., *J. Cluster Science* 10, 525, 1999.

174. Wang, L.-S., private communication quoted in Ref. 169.

175. Marcy, T. P., PhD dissertation, quoted in Ref. 169.

176. Alex, S., Green, M. E., and Leopold, D. G., unpublished, quoted in Ref. 169.

177. Wang, X.-B., Wang, L.-S., Brown, R., Schwerdtfeger, P., Schröder, D., and Schwarz, H., *J. Chem. Phys.* 114, 7388, 2001.

178. Kim, J. H., Li, X., Wang, L.-S., de Clercq, H. L., Fancher, C. A., Thomas, O. C., and Bowen, K. H., *J. Phys. Chem. A* 105, 5709, 2001.

179. Moravec, V. D., Klopcic, S. A., Chatterjee, B., and Jarrold, C. C., *Chem. Phys. Lett.* 341, 313, 2001.

180. Zengin, V., Persson, B. J., Strong, K. M., and Continetti, R. E., *J. Chem. Phys.* 105, 9740, 1996.

181. Kietzmann, H., Morenzin, J., Bechthold, P. S., Ganteför, G., and Eberhardt, W., *J. Chem. Phys.* 109, 2275, 1998.

182. Xu, C., Taylor, T. R., Burton, G. R., and Neumark, D. M., *J. Chem. Phys.* 108, 7645, 1998.

183. Williams, W. W., Carpenter, D. L., Covington, A. M., Koepnick, M. C., Calabrese, D., and Thompson, J. S., *J. Phys. B: At. Mol. Opt. Phys.* 31, L341, 1998.

184. Covington, A. M., Calabrese, D., Thompson, J. S., and Kvale, T. J., *J. Phys. B: At. Mol. Opt. Phys.* 31, L855, 1998.

185. Haeffler, G., Hanstrorp, D., Kiyan, I., Klinkmueller, A. E., Ljungblad, U., Pegg, D. J., *Phys. Rev. A* 53, 4127, 1996.

186. Chen, H. L., Ellis, Jr., P. E., Wijesekera, T., Hagan, T. E., Groh, S. E., Lyons, J. E., and Ridge, D. P., *J. Am. Chem. Soc.* 116, 1086, 1994.

187. Deyerl, H.-J., Alconcel, L. S., and Continetti, R. E., *J. Phys. Chem. A* 105, 552, 2001.

188. Blanksby, S. J., Ramond, T. M., Davico, G. E., Nimlos, M. R., Kato, S., Bierbaum, V. M., Lineberger, W. C., Ellison, G. B., Okumura, M., *J. Am. Chem. Soc.* 123, 9585, 2001.

189. Asmis, K. R., Taylor, T. R., Xu, C., and Neumark, D. M., *Chem. Phys. Lett.* 295, 75, 1998.

190. Schäfer-Bung, B., Engels, B., Taylor, T. R., Neumark, D. M., Botschwina, P., and Peric, M., *J. Chem. Phys.* 115, 1777, 2001.

191. Scheer, M., Bilodeau, R. C., and Haugen, H. K., Phys. Rev. Lett. 80, 2562, 1998.

192. Taylor, T. R., Gómez, H., Asmis, K. R., and Neumark, D. M., *J. Chem. Phys.* 115, 4620, 2001.

193. Asmis, K. R., Taylor, T. R., and Neumark, D. M., *J. Chem. Phys.* 111, 8838, 1999 and 111, 10491, 1999.

194. Ramond, T. M., Davico, G. E., Schwartz, R. L., and Lineberger, W. C., *J. Chem. Phys.* 112, 1158, 2000.

195. Petrunin, V. V., Voldstad, J. D., Balling, P., Kristensen, P., Andersen, T., and Haugen, H. K., *Phys. Rev. Lett.* 75, 1911, 1995.

196. Dzuba, V. A., and Gribakin, G. F., *J. Phys. B: At. Mol. Opt. Phys.* 31, L483, 1998.

197. Ervin, K. M., Ramond, T. M., Davico, G. E., G. E., Schwartz, R. L., Casey, S. M., and Lineberger, W. C., *J. Phys. Chem.* 105, 10822, 2001.

198. Lippa, T. P., Xu, S.-J., Lyapustina, S. A., and Bowen, K. H., *J. Chem. Phys.* 109, 9263, 1998.

199. Turner, N. J., Martel, A. A., and Waller, I. M., *J. Phys. Chem.* 98, 474, 1994.

200. Lippa, T. P., Xu, S.-J., Lyapustina, S. A., Nilles, J. M., and Bowen, K. H., *J. Chem. Phys.* 109, 10727, 1998.

201. Wang, L.-S., Concericao, J., Changming, C., and Smalley, R. E., *Chem. Phys. Lett.* 182, 5, 1991.

202. Boltalina, O. V., Sidorov, L. N., Sukhanova, E. V., and Sorokin, I. D., *Chem. Phys. Lett.* 230, 567, 1994.

203. Clifford, E. P., Wenthold, P. G., Lineberger, W. C., Ellison, G. B., Wang, C. X., Grabowski, J. J., Vila, F., and Jordan, K. D., *J. Chem. Soc. Perkin Trans. 2*, 1015, 1998.

204. Spanel, P., Matejcik, S., and Smith, D., J. Phys. B: At. Mol. Phys. **28** 2941 (1995). See Miller, A. E. S., Miller, T. M., Viggiano, A. A., Morris, R. A., Van Doren, J. M., Arnold, S. T., and Paulson, J. F., *J. Chem. Phys.* 102, 8865, 1995 for interpretation in terms of EA(SF$_5$).

205. Kinghom, D. B., and Adamowicz, L., *J. Chem. Phys.* 106, 4589, 1997. EA(H) = 6083.0994 cm^{-1}, EA(D) = 6086.7137 cm^{-1}, and EA(T) = 6087.9168 cm^{-1}.

206. Xi, L., and Wang, L.-S., *J. Chem. Phys.* 109, 5264, 1998.

207. Ephraim, E., Shmulyian, S., Kaldor, U., and Isikawa, Y., *J. Chem. Phys.* 109, 3954, 1998. Also EA(La) = 0.35 eV.

208. Scheer, M., Bilodeau, R. C., Thogersen, J., and Haugen, H. K., *Phys. Rev. A* 57, R1493, 1998.

209. Kim, J. B., Wenthold, P. G., and Lineberger, W. C., *J. Chem. Phys.* 108, 830, 1998.

210. Kishi, R., Kawamata, H., Negishi, Y., Iwata, S., Nakajima, A., and Kaya, K., *J. Chem. Phys.* 107, 10029, 1997.

211. Compton, R. N., Carman, Jr., H. S., Desfrançois, C., Abdoul-Carmine, J., Schermann, J. P., Hendricks, J. H., Lyapustina, S. A., and Bowen, K. H., *J. Chem. Phys.* 105, 3472, 1996.

212. Yourshaw, I., Zhao, Y., and Neumark, D. M., *J. Chem. Phys.* 105, 351, 1996.

213. Gausa, M., Kaschner, R., Seifert, G., Faehrmann, J. H., Lutz, H. O., and Meiwes-Broer, K., *J. Chem. Phys.* 104, 9719, 1996.

214. Wu, H., and Wang, L.-S., *J. Chem. Phys.* 107, 16, 1997; Moravec, V. D., and Jarrold, C. C., *J. Chem. Phys.* 108, 1804, 1998.

215. Radisic, D., Xu, S., and Bowen, K. H., *Chem. Phys. Lett.* 354, 9, 2002.

216. Pramann, A., and Rademann, K., *Chem. Phys. Lett.* 343, 99, 2001.

217. Gómez, H., Taylor, T. R., and Neumark, D. M., *J. Phys. Chem. A* 105, 6886, 2001.

218. Gómez, H., Taylor, T. R., Zhao, Y., and Neumark, D. M., *J. Chem. Phys.* 117, 8644, 2002.

219. Nimlos, M. R., Davico, G., Geise, C. M., Wenthold, P. G., Lineberger, W. C., Blanksby, S. J., Hadad, C. M., Petersson, G. A., Ellison, G. B., *J. Chem. Phys.* 117, 4323, 2002.

220. Koyasu, K., Mitsui, M., Nakajima, A., and Kaya, K., *Chem. Phys. Lett.* 358, 224, 2002.

221. Zhai, H.-J., Wang, L.-S., Alexandrova, A. N., Boldyrev, A. I., and Zakrzewski, V. G., *J. Phys. Chem. A* 107, 9319, 2003.

222. Schiedt, J., and Weinkauf, R., *Z. Naturforsch. A* 50, 1041, 1995. See also Ervin, K. M., Anusiewicz, I., Skurski, P., Simons, J., and Lineberger, W. C., *J. Phys. Chem. A* 107, 8521, 2003 [EA(O$_2$) = 0.448 ± 0.006 eV].

223. Davis, V. T., and Thompson, J. S., *J. Phys. B: At. Mol. Opt. Phys.* 34, L433, 2001.

224. Li, X., Zhai, H.-J., and Wang, L.-S., *Chem. Phys. Lett.* 357, 415, 2002.

225. Davis, V. T., and Thompson, J. S., *J. Phys. B: At. Mol. Opt. Phys.* 35, L11, 2002.

226. Goldfarb, F., Drag, C., Chaibi, W., Kröger, S., Blondel, C., and Delsart, C., *J. Chem. Phys.* 122, 014308, 2005. EA(OH) = 14740.982(7) cm^{-1}.

227. Blondel, C., Delsart, C., and Goldfarb, F., *J. Phys. B: At. Mol. Opt. Phys.* 34, L281, 2001. EA(F) = 11207.252(18) cm^{-1} and EA(Si) = 27432.446(19) cm^{-1}.

228. Fancher, C. A., de Clercq, H. L., and Bowen, K. H., *Chem. Phys. Lett.* 366, 197, 2002.

229. Jin, C., Taylor, K. J., Conceicao, J., and Smalley, R. E., *Chem. Phys. Lett.* 175, 17, 1990.

230. Schiedt, J., Knott, W. J., Le Barbu, K., Schlag, E. W., and Weinkauf, R., *J. Chem. Phys.* 113, 9470, 2000.

231. Song, J. K., Lee, N. K., Kim, J. H., Han, S. Y., and Kim, S. K., *J. Chem. Phys.* 119, 3071, 2003.

232. Davico, G. E., Schwartz, R. L., Ramond, T. M., and Lineberger, W. C., *J. Amer. Chem. Soc.* 121, 6047, 1999.

233. Andrews, D. H., Gianola, A. J., and Lineberger, W. C., *J. Chem. Phys.* 117, 4074, 2002.

234. Schwartz, R. L., Davico, G. E., Kim, J. B., and Lineberger, W. C., *J. Chem. Phys.* 112, 4966, 2000.

235. Schwartz, R. L., Davico, G. E., Ramond, T. M., and Lineberger, W. C., *J. Phys. Chem. A* 103, 8213, 1999.

236. Schiedt, J., and Weinkauf, R., *Chem. Phys. Lett.* 274, 18, 1997.

237. Wang, X.-B., Wang, L.-S., Brown, R., Schwerdtfeger, P., Schröder, D., and Schwartz, H., *J. Chem. Phys.* 114, 7388, 2003.

238. Geske, G. D., Boldyrev, A. I., Li, X., and Wang, L.-S., *J. Chem. Phys.* 113, 5130, 2000.

239. Cannon, N. A., Boldyrev, A. I., Li, X., and Wang, L.-S., *J. Chem. Phys.* 113, 2671, 2000.

240. Wang, X.-B., Yang, X., and Wang, L.-S., *J. Chem. Phys.* 116, 561, 2002.

241. Gutsev, G. L., Jena, P., Zhai, H.-J., and Wang, L.-S., *J. Chem. Phys.* 115, 7935, 2001.

242. Yang, X., Wang, X.-B., and Wang, L.-S., *J. Chem. Phys.* 115, 2889, 2001.

243. Zhai, H.-J., Liu, S.-R., Li, X., and Wang, L.-S., *J. Chem. Phys.* 115, 5170, 2001.

244. Li, X., Wang, L.-S., Cannon, N. A., and Boldyrev, A. I., *J. Chem. Phys.* 116, 1330, 2002.

245. Zhai, H.-J., Wang, L.-S., Alexandrova, A. N., and Boldyrev, A. I., *J. Chem. Phys.* 117, 7917, 2002.

246. Zhai, H.-J., and Wang, L.-S., *J. Chem. Phys.* 117, 7882, 2002.

247. Liu, S.-R., Zhai, H.-J., and Wang, L.-S., *J. Chem. Phys.* 117, 9758, 2002.

248. Mitsui, M., Nakajima, A., and Kaya, K., *J. Chem. Phys.* 117, 9740, 2002.

249. Yang, X., Wang, X.-B., Wang, L.-S., Niu, S., and Ichiye, T., *J. Chem. Phys.* 119, 8311, 2003.

250. Kim, J., Lee, H. M., Suh, S. B., Majumdar, D., and Kim, K. S., *J. Chem. Phys.* 113, 5259, 2000.

251. Liu, S. R., Zhai, H. J., and Wang, L.-S., *Phys. Rev. B* 64, 153402, 2001.

252. Zhai, H. J., Yang, X., Wang, X. B., Wang, L.-S., Elliott, B., and Boldyrev, A. I., *J. Am. Chem. Soc.* 124, 6742, 2002.

253. Zhai, H. J., Wang, L.-S., Kuznetsov, A. E., and Boldyrev, A. I., *J. Phys. Chem. A* 106, 5600, 2002.

254. Drechsler, G., and Boesl, U., *Int. J. Mass Spectrom.* 228, 1067, 2003.

255. Illenberger, E, Comita, P. B., Brauman, J. I., Fenzlaff, H. P., Heni, M., Heinrich, N., Koch, and W., Fenking, G., *Ber. Bunsen-Ges. Phys. Chem.* 89, 1026, 1985; Jackson, R. L., Pellerite, M. J., and Brauman, J. I., *J. Am. Chem. Soc.* 103, 1802, 1981.

256. Miller, T. M., Friedman, J. F., and Viggiano, A. A., *J. Chem. Phys.* 120, 7024, 2004.

257. Miller, T. M., Van Doren, J. M., and Viggiano, A. A., *Int. J. Mass Spectrom.* 233, 67, 2004.

258. Van Doren, J. M., McSweeney, S. A., Hargus, M. D., Kerr, D. M., Miller, T. M., Arnold, S. T., and Viggiano, A. A., *Int. J. Mass Spectrom.* 228, 541, 2003.

259. Van Doren, J. M., Miller, T. M., and Viggiano, A. A., *J. Chem. Phys.* 123, 114303, 2005.

260. Miller, T. M., and Viggiano, A. A., *Phys. Rev. A* 71, 012702, 2005.

261. Haeffler, G., Klinkmueller, A. E., Rangell, J., Berzinsh, U., and Hanstorp, D., *Z. Phys. D* 38, 211, 1996.

262. Bilodeau, R. C., and Haugen, H. K., *Phys. Rev. A* 64, 024501, 2001.

263. Miller, T. M., Viggiano, A. A., Friedman, J. F., and Van Doren, J. M., *J. Chem. Phys.* 121, 9993, 2004.

264. Davis, V. T., and Thompson, J. S., *Phys. Rev. A* 65, 010501, 2001. Theoretical work implies that the measured EA(Tm) was actually for a long-lived excited anion state, and that Tm does not form a stable anion. See O'Malley, S. M., and Beck, D. R., *Phys. Rev. A* 70, 022502, 2004.

265. Gianola, A. J., Ichino, T., Hoenigman, R. L., Kato, S., Bierbaum, V. M., and Lineberger, W. C., *J. Phys. Chem. A* 108, 10326, 2004.

266. Staneke, P. O., Groothuis, G., Ingemann, S., and Nibbering, N. M. M., *Int. J. Mass Spectrom. Ion Processes* 142, 83, 1995. A Gaussian-3 calculation yields EA(CCl$_4$) = 0.994 eV [Ed.].

267. Meloni, G., Ferguson, M. J., and Neumark, D. M., *Phys. Chem. Chem. Phys.* 5, 4073, 2003.

268. Davis, V. T., and Thompson, J. S., *J. Phys. B: At. Mol. Opt. Phys.* 37, 1961, 2004.

269. Davis, V. T., and Thompson, J. S., *Phys. Rev. Lett.* 88, 073003, 2002.

270. Ando, N., Kokubo, S., Mitsui, M., and Nakajima, A., *Chem. Phys. Lett.* 389, 279, 2004.

271. Zhai, H.-J., and Wang, L.-S., *J. Chem. Phys.* 120, 8996, 2004.

272. Surber, E., and Sanov, A., *J. Chem. Phys.* 116, 5921, 2002.

273. Dzuba, V. A., and Gribakin, G. F., *Phys. Rev. A* 55, 2443, 1997.

274. Meloni, G., Sheehan, S. M., Ferguson, M. J., and Neumark, D. M., *J. Phys. Chem. A* 108, 9750, 2004.

275. Zhai, H.-J., Li, J., and Wang, L.-S., *J. Chem. Phys.* 121, 8369, 2004.

276. Zhai, H.-J., Kiran, B., and Wang, L.-S., *J. Chem. Phys.* 121, 8231, 2004.

277. Kim, J., Kelley, J. A., Ayotte, P., Nielsen, S. B., Weddle, G. H., and Johnson, M. A., *J. Am. Soc. Mass Spectrom.* 10, 810, 1999.

278. Misaizu, F., Tsunoyama, H., Yasumura, Y., Ohshimo, K., and Ohno, K., *Chem. Phys. Lett.* 389, 241, 2004. See also Ref. 236 for comments on EA(CS).

279. Meloni, G., Sheehan, S. M., and Neumark, D. M., *J. Chem. Phys.* 122, 074317, 2005

280. Yoder, B. L., Maze, J. T., Raghavachari, K., and Jarrold, C. C., *J. Chem. Phys.* 122, 094313, 2005.

281. Ervin, K. M., and Lineberger, W. C., *J. Chem. Phys.* 122, 194303, 2005.

282. Ichino, T., Gianola, A. J., Andrews, D. H., and Lineberger, W. C., *J. Phys. Chem. A* 108, 11307, 2004.

283. Das, U., Raghavachari, K., and Jarrold, C. C., *J. Chem. Phys.* 122, 014313, 2005.

284. Schiedt, J., and Weinkauf, R., *J. Chem. Phys.* 110, 304, 1999.

285. Le Barbu, K., Schiedt, J., Weinkauf, R., Schlag, E. W., Nilles, J. M., Xu, S.-J., Thomas, O. C., and Bowen, J. H., *J. Chem. Phys.* 116, 9663, 2002. Uncertainties not stated.

286. Schiedt, J., and Weinkauf, R., *Chem. Phys. Lett.* 266, 201, 1997. The uncertainty for EA(anthracene) quoted as ±0.008 eV in a later paper (Ref. 230).

287. Chatterjee, B., Akin, F. A., Jarrold, C. C., and Raghavachari, K., *J. Phys. Chem. A* 109, 6880, 2005.

288. Zhai, H.-J., Kiran, B., Dai, B., Li, J., and Wang, L.-S., *J. Amer. Chem. Soc.* 127, 12098, 2005.

289. Moravec, V. D., Klopcic, S. A., and Jarrold, C. C., *J. Chem. Phys.* 110, 5079, 1999.

290. Klopcic, S. A., Moravec, V. D., and Jarrold, C. C., *J. Chem. Phys.* 110, 10216, 1999.

291. Moravec, V. D., and Jarrold, C. C., *J. Chem. Phys.*, 112, 792, 2000.

292. Moravec, V. D., and Jarrold, C. C., *J. Chem. Phys.* 113, 1035, 2000.

293. Chatterjee, B., Akin, F. A., Jarrold, C. C., and Raghavachari, K., *J. Chem. Phys.* 119, 10591, 2003.

294. Schröder, D., Brown, R., Schwerdtfeger, P., Wang, X. B., Yang, X., Wang, L. S., and Schwarz, H., *Angew. Chem. Int. Ed.* 42, 311, 2003.
295. Zheng, W., Nilles, J. M., Thomas, O. C., and Bowen, K. H., *J. Chem. Phys.* 122, 044306, 2005.
296. Duan, X., Burggraf, L. W., Weeks, D. E., Davico, G. E., Schwartz, R. L., and Lineberger, W. C., *J. Chem. Phys.* 116, 3601, 2002.
297. Andrews, D. H., Gianola, A. J., and Lineberger, W. C., *J. Chem. Phys.* 117, 4074, 2002.
298. Alexandrova, A. N., Boldyrev, A. I., Zhai, H.-J., and Wang, L.-S., *J. Chem. Phys.* 122, 054313, 2005.

PROTON AFFINITIES

Proton affinity is a useful parameter for describing gas phase ion-molecule reactions in fields such as atmospheric chemistry, plasma chemistry, mass spectrometry, and astrophysics. The proton affinity E_{pa} (often designated in the literature as PA) of a molecular species M is defined as the negative of the enthalpy change for the gas phase reaction

$$M + H^+ \rightarrow MH^+.$$

A closely related quantity is the gas phase basicity $\Delta_{base}G°$ (often designated as GB), which is the negative of the Gibbs energy change for the same reaction. Thus the two are related by

$$\Delta_{base}G° = E_{pa} + T\Delta S,$$

where T is the temperature and ΔS is the entropy change in the reaction (which can be calculated if the molecular structure of M and M^+ is known).

Direct measurement of the proton affinity is possible for only a few molecules, mainly olefins and carbonyl compounds. However, these measurements have been used to establish a scale of E_{pa} values that permits proton affinities to be determined for many other molecules, including unstable species and reaction intermediates. The basis for this scale is described by Hunter and Lias in Reference 1.

The E_{pa} and $\Delta_{base}G°$ values at a temperature of 298 K are tabulated below for selected molecules. Many values are given to one decimal place, but the majority are not accurate to better than one or two kilojoules per mole. The methods of measurement are described in Reference 1, which contains a much more extensive and detailed tabulation.

Compounds are listed by molecular formula in the Hill order, but with all compounds that do not contain carbon appearing before those that do contain carbon.

References

1. Hunter, E. P. L., and Lias, S. G., *J. Phys. Chem. Ref. Data* 27, 413, 1998.
2. Hunter, E. P., and Lias, S. G., "Proton Affinity Evaluation", in *NIST Chemistry WebBook*, NIST Standard Reference Database No. 69, Linstrom, P. J., and Mallard, W. G., Eds., March 2003, National Institute of Standards and Technology, Gaithersburg, MD 20899, <http://webbook.nist.gov>.
3. Do, K., Klein, T. P., Pommerening, C. A., Bachrach, S. M., and Sunderlin, L. S., *J. Am. Chem. Soc.* 120, 6093, 1998.
4. Kim, H.-T., Green, R. J., Qian, J., and Anderson, S. L., *J. Chem. Phys.* 112, 5717, 2000.
5. Park, S. T., Kim, S. K., and Kim, M. S., *J. Chem. Phys.* 114, 5568, 2001.
6. Hiraoka, K., Mizuno, T., Eguchi, D., Takao, T., and Ino, S., *J. Chem. Phys.* 116, 7574, 2002.
7. Oresmaa, L. O., Haukka, M., Vainiotalo, P., and Pakkanen, T. A., *J. Org. Chem.* 67, 8216, 2002.
8. Wang, F., Ma, S., Zhang, D., and Cooks, R. G., *J. Phys. Chem. A* 102, 2988, 1998.
9. Bouchoux, G., Gal, J.-F., Szulejko, J. E., McMahon, T. B., Tortajada, J., Luna, A., Yanez, M., and Mo, O., *J. Phys. Chem. A* 102, 9183, 1998.
10. van Beelen, E., Koblenz, T. A., Ingemann, S. and Hammerum, S., *J. Phys. Chem. A* 108, 2728, 2004.

Molecular formula	Name	E_{pa} kJ/mol	$\Delta_{base}G°$ kJ/mol	Notes
Ar	Argon	369.2	346.3	
AsF$_3$	Arsenic(III) fluoride	636.7	604.2	
AsH$_3$	Arsine	747.9	712.0	
BHO$_2$	Metaboric acid	763.0	730.5	
BH$_3$O$_3$	Boric acid	728.1	698.4	
B$_2$H$_6$	Diborane	615	586.0	
B$_3$H$_6$N$_3$	Borazine	802.5	772.8	
B$_4$H$_{10}$	Tetraborane(10)	605	572.5	
B$_5$H$_9$	Pentaborane(9)	699.4	666.9	
BaO	Barium oxide	1215.4	1187.6	
Br	Bromine (atomic)	554.4	531.2	
BrH	Hydrogen bromide	584.2	557.7	
BrLi	Lithium bromide	819	792.5	
CaO	Calcium oxide	1190.6	1162.3	
Cl	Chlorine (atomic)	513.6	490.1	
ClH	Hydrogen chloride	556.9	530.1	
ClLi	Lithium chloride	827	800.5	
Co	Cobalt	742.7	719.8	
Cr	Chromium	791.3	768.4	
CsHO	Cesium hydroxide	1117.9	1092.2	
Cs$_2$O	Cesium oxide	1442.9	1412.2	
Cu	Copper	655.3	632.4	
F	Fluorine (atomic)	340.1	315.1	
FH	Hydrogen fluoride	484	456.7	
FO	Fluorine oxide	508.7	482.2	
F$_2$	Fluorine	332	305.5	
F$_2$O$_2$S	Sulfuryl fluoride	605.5	580.5	

Molecular formula	Name	E_{pa} kJ/mol	$\Delta_{base}G°$ kJ/mol	Notes
F_3N	Nitrogen trifluoride	568.4	538.6	
F_3OP	Phosphoryl fluoride	694.0	664.2	
F_3P	Phosphorus(III) fluoride	695.3	662.8	
F_4Si	Tetrafluorosilane	502.9	476.6	
F_6S	Sulfur hexafluoride	575.3	550.7	
Fe	Iron	754	731.1	
FeO	Iron(II) oxide	907	880.5	
GeH_4	Germane	713.4	687.1	
HI	Hydrogen iodide	627.5	601.3	
HKO	Potassium hydroxide	1101.8	1075.4	
HLi	Lithium hydride	1021.7	996.4	
HLiO	Lithium hydroxide	1000.1	972.1	
HNO_3	Nitric acid	751.4	731.5	
HN_3	Hydrazoic acid	756.0	723.5	
HNa	Sodium hydride	1095	1070.6	
HNaO	Sodium hydroxide	1071.8	1044.8	
HO	Hydroxyl	593.2	564.0	
HO_2	Hydroperoxy	660	627.5	
HP	Phosphorus monohydride	670.3	639.6	
H_2	Hydrogen	422.3	394.7	
$H_2N_2O_2$	Nitramide	757.4	725.0	
H_2O	Water	691	660.0	
H_2O_2	Hydrogen peroxide	674.5	643.8	
H_2O_4S	Sulfuric acid	717	681	Ref. 3
H_2P	Phosphino	709.2	675.7	
H_2S	Hydrogen sulfide	705	673.8	
H_2Se	Hydrogen selenide	707.8	676.4	
H_2Si	Silylene	839.2	804.1	
H_2Te	Hydrogen telluride	735.9	704.5	
H_3N	Ammonia	853.6	819.0	
H_3P	Phosphine	785	750.9	
H_4N_2	Hydrazine	853.2	822.4	
H_4Si	Silane	639.7	613.4	
H_6OSi_2	Disiloxane	749	718.3	
He	Helium	177.8	148.5	
I	Iodine (atomic)	608.2	583.5	
K_2O	Potassium oxide	1342.5	1311.8	
Kr	Krypton	424.6	402.4	
La	Lanthanum	1013	991.9	
Li_2	Dilithium	1162	1133.1	
Li_2O	Lithium oxide	1206	1175.3	
Lu	Lutetium	992	970.6	
Mg	Magnesium	819.6	797.3	
MgO	Magnesium oxide	988	959.4	
Mg_2	Dimagnesium	919	886.5	
Mn	Manganese	797.3	774.4	
N	Nitrogen (atomic)	342.2	318.7	
NO	Nitric oxide	531.8	505.3	
NO_2	Nitrogen dioxide	591.0	560.3	
NP	Phosphorus nitride	789.4	757.0	
N_2	Nitrogen	493.8	464.5	
N_2O	Nitrous oxide	549.8	523.3	Protonation at N
N_2O	Nitrous oxide	575.2	548.7	Protonation at O
Na_2	Disodium	1146.8	1118.2	
Na_2O	Sodium oxide	1375.9	1345.2	
Ne	Neon	198.8	174.4	
Ni	Nickel	737	714.1	
O	Oxygen (atomic)	485.2	459.6	
OP	Phosphorus monoxide	682	649.5	
OSi	Silicon monoxide	777.8	750.4	Protonation at O

Molecular formula	Name	E_{pa} kJ/mol	$\Delta_{base}G°$ kJ/mol	Notes
OSi	Silicon monoxide	533	500.5	Protonation at Si
OSr	Strontium oxide	1209	1180.7	
O_2	Oxygen	421	396.3	
O_2S	Sulfur dioxide	672.3	643.3	
O_3	Ozone	625.5	595.9	
O_3S	Sulfur trioxide	588.3	560.3	
O_4Os	Osmium(VIII) oxide	676.9	650.6	
P	Phosphorus	626.8	604.8	
Pd	Palladium	696	673.4	
Rh	Rhodium	768	745.4	
Ru	Ruthenium	774	751.4	
S	Sulfur	664.3	640.2	
SSi	Silicon monosulfide	627	596.6	Protonation at Si
SSi	Silicon monosulfide	683	660.2	Protonation at S
Sc	Scandium	914	892.0	
Si	Silicon	837	814.1	
Ti	Titanium	876	853.7	
U	Uranium	995.2	973.2	
V	Vanadium	859.4	836.8	
Xe	Xenon	499.6	478.1	
Y	Yttrium	967	945.9	
Zn	Zinc	608.6	586.0	
$CBrF_3$	Bromotrifluoromethane	580.0	550.3	
CBrN	Cyanogen bromide	749.8	719.2	
$CClF_3$	Chlorotrifluoromethane	571.3	541.5	
CClN	Cyanogen chloride	722.1	691.5	
CCl_2	Dichloromethylene	861	828.5	
CCl_2S	Carbonothioic dichloride	752.5	721.8	
CFN	Cyanogen fluoride	632	601.3	
CF_2	Difluoromethylene	765	732.5	
CF_2O	Carbonyl fluoride	666.7	637.0	
CF_3I	Trifluoroiodomethane	628.0	598.2	
CF_3NO	Trifluoronitrosomethane	703.3	670.8	
CF_4	Tetrafluoromethane	529.3	503.7	
CHCl	Chloromethylene	874.1	839.9	
CHF	Fluoromethylene	797.9	763.8	
CHF_3	Trifluoromethane	619.5	589.7	
CHF_3O_3S	Trifluoromethanesulfonic acid	699.4	666.9	
CHN	Hydrogen cyanide	712.9	681.6	
CHN	Hydrogen isocyanide	772.3	739.8	
CHNO	Isocyanic acid (HNCO)	753	718.8	
CHNO	Fulminic acid	758	725.5	
CHO	Oxomethyl (HCO)	636	601.8	
CHO_2	Formyloxyl	623.4	590.9	
CH_2F_2	Difluoromethane	620.5	589.7	
CH_2N_2	Diazomethane	858.9	826.7	
CH_2N_2	Cyanamide	805.6	774.9	
CH_2O	Formaldehyde	712.9	683.3	
CH_2O_2	Formic acid	742.0	710.3	
CH_2S	Thioformaldehyde	759.7	730.5	
CH_2Se	Selenoformaldehyde	764.0	734.9	
CH_3Br	Bromomethane	664.2	638.0	
CH_3Cl	Chloromethane	647.3	621.1	
CH_3F	Fluoromethane	598.9	571.5	
CH_3I	Iodomethane	691.7	665.5	
CH_3NO	Formamide	822.2	791.2	
CH_3NO_2	Nitromethane	754.6	721.6	
CH_3NO_2	Methyl nitrite	798.9	766.4	
CH_3NO_3	Methyl nitrate	733.6	714.8	
CH_3N_3	Methyl azide	833	800.5	

Molecular formula	Name	E_{pa} kJ/mol	$\Delta_{base}G°$ kJ/mol	Notes
CH_4	Methane	543.5	520.6	
CH_4N	Methylamidogen	832.8	801.6	
CH_4N_2O	Urea	873.5	841.6	Protonation at O; Ref. 8
CH_4N_2S	Thiourea	893.7	863.9	
CH_4O	Methanol	754.3	724.5	
CH_4O_3S	Methanesulfonic acid	761.3	728.9	
CH_4S	Methanethiol	773.4	742	
CH_5N	Methylamine	899.0	864.5	
CH_5NO	O-Methylhydroxylamine	844.8	812.3	
CH_5N_3	Guanidine	986.3	949.4	
CH_5P	Methylphosphine	851.5	817.6	
CH_6N_2	Methylhydrazine	898.8	866.4	
CN	Cyanide	>595	>564	Protonation at N
CNS	Thiocyanate	751	718.5	
CO	Carbon monoxide	594	562.8	Protonation at C
CO	Carbon monoxide	426.3	402.2	Protonation at O
COS	Carbon oxysulfide	628.5	602.6	Protonation at S
$COSe$	Carbon oxyselenide	670	644.1	Protonation at Se
CO_2	Carbon dioxide	540.5	515.8	
CS	Carbon monosulfide	791.5	760	
CS_2	Carbon disulfide	681.9	657.7	
CSe	Carbon monoselenide	831.8	800.2	Protonation at C
CSe_2	Carbon diselenide	725	700.9	
C_2ClF_3O	Trifluoroacetyl chloride	681.6	649.8	
C_2Cl_3N	Trichloroacetonitrile	723.2	692.6	
C_2F_3N	Trifluoroacetonitrile	688.4	657.7	
C_2H	Ethynyl	753	720.8	
C_2HCl_3O	Trichloroacetaldehyde	722.3	690.5	
$C_2HCl_3O_2$	Trichloroacetic acid	770.0	739.1	
C_2HF	Fluoroacetylene	686	661.3	
C_2HF_3	Trifluoroethene	699.4	666.9	
$C_2HF_3O_2$	Trifluoroacetic acid	711.7	680.7	
C_2H_2	Acetylene	641.4	616.7	
C_2H_2ClN	Chloroacetonitrile	745.7	715.1	
$C_2H_2F_2$	1,1-Difluoroethene	734	705.1	
$C_2H_2F_2$	trans-1,2-Difluoroethene	688.6	657.9	
C_2H_2O	Ketene	825.3	793.6	
$C_2H_3ClO_2$	Chloroacetic acid	765.4	734.5	
$C_2H_3Cl_3O$	2,2,2-Trichloroethanol	729.3	698.9	
C_2H_3F	Fluoroethene	729	700.1	
$C_2H_3FO_2$	Fluoroacetic acid	765.4	734.5	
$C_2H_3F_3O$	2,2,2-Trifluoroethanol	700.2	669.9	
$C_2H_3F_3O$	Methyl trifluoromethyl ether	719.2	690.0	
C_2H_3N	Acetonitrile	779.2	748	
C_2H_3N	Isocyanomethane	839.1	806.6	
C_2H_3NO	Methyl isocyanate	764.4	732.0	
C_2H_3NS	Methyl thiocyanate	796.7	766.1	
C_2H_3NS	Methyl isothiocyanate	799.2	766.7	
$C_2H_3N_3$	1H-1,2,3-Triazole	879.3	847.4	
$C_2H_3N_3$	1H-1,2,4-Triazole	886.0	855.9	
C_2H_4	Ethylene	680.5	651.5	
$C_2H_4F_2O$	2,2-Difluoroethanol	727.4	697.0	
$C_2H_4F_3N$	2,2,2-Trifluoroethylamine	846.8	812.9	
$C_2H_4N_2$	Aminoacetonitrile	824.9	791.0	
C_2H_4O	Acetaldehyde	768.5	736.5	
C_2H_4O	Oxirane	774.2	745.3	
$C_2H_4O_2$	Acetic acid	783.7	752.8	
$C_2H_4O_2$	Methyl formate	782.5	751.5	
C_2H_4S	Thiirane	807.4	777.6	
C_2H_5Br	Bromoethane	696.2	669.7	

Molecular formula	Name	E_{pa} kJ/mol	$\Delta_{base}G°$ kJ/mol	Notes
C_2H_5BrO	2-Bromoethanol	766.1	735.7	
C_2H_5Cl	Chloroethane	693.4	666.9	
C_2H_5ClO	2-Chloroethanol	766.1	735.7	
C_2H_5F	Fluoroethane	683.4	655.8	
C_2H_5FO	2-Fluoroethanol	715.6	685.2	
C_2H_5I	Iodoethane	724.8	698.3	
C_2H_5N	Ethenamine	898.9	866.5	
C_2H_5N	Ethyleneimine	905.5	872.5	
C_2H_5NO	Acetamide	863.6	832.6	
C_2H_5NO	N-Methylformamide	851.3	820.3	
$C_2H_5NO_2$	Nitroethane	765.7	733.2	
$C_2H_5NO_2$	Ethyl nitrite	818.9	786.4	
$C_2H_5NO_2$	Glycine	886.5	852.2	
$C_2H_5NO_2$	Acetohydroxamic acid	854.0	823.0	
C_2H_5NS	Thioacetamide	884.6	852.8	
C_2H_6	Ethane	596.3	569.9	
C_2H_6Hg	Dimethyl mercury	771.6	740.8	
$C_2H_6N_2$	Ethanimidamide	970.7	938.2	
$C_2H_6N_2$	trans-Dimethyldiazene	865.1	834.4	
$C_2H_6N_2O$	2-Aminoacetamide		882.3	
$C_2H_6N_2O_2$	N-Methyl-N-nitromethanamine	828.3	795.8	
C_2H_6O	Ethanol	776.4	746	
C_2H_6O	Dimethyl ether	792	764.5	
C_2H_6OS	Dimethyl sulfoxide	884.4	853.7	
$C_2H_6O_2$	1,2-Ethanediol	815.9	773.6	
C_2H_6S	Ethanethiol	789.6	758.4	
C_2H_6S	Dimethyl sulfide	830.9	801.2	
$C_2H_6S_2$	Dimethyl disulfide	815.3	782.8	
C_2H_7N	Ethylamine	912.0	878	
C_2H_7N	Dimethylamine	929.5	896.5	
C_2H_7NO	Ethanolamine	930.3	896.8	
$C_2H_7O_3P$	Dimethyl hydrogen phosphite	894.8	862.4	
C_2H_7P	Dimethylphosphine	912.0	877.9	
$C_2H_8N_2$	1,2-Ethanediamine	951.6	912.5	
$C_2H_8N_2$	1,1-Dimethylhydrazine	927.1	894.7	
C_2N_2	Cyanogen	674.7	645.8	
C_2O	Dicarbon monoxide	774.7	747.0	
C_3	Carbon trimer	767.0	736.3	
C_3F_6O	Perfluoroacetone	670.4	639.7	
C_3HN	Cyanoacetylene	751.2	720.5	
$C_3H_2F_6O$	1,1,1,3,3,3-Hexafluoro-2-propanol	686.6	656.2	
$C_3H_2N_2$	Malononitrile	723.0	694.1	
C_3H_3	2-Propynyl	741	708.5	
$C_3H_3Cl_3O$	1,1,1-Trichloro-2-propanone	768.3	736.3	
$C_3H_3F_3O$	1,1,1-Trifluoroacetone	723.9	692.0	
$C_3H_3F_3O_2$	Methyl trifluoroacetate	740.5	709.6	
C_3H_3N	Acrylonitrile	784.7	753.7	
C_3H_3NO	Oxazole	876.4	844.5	
C_3H_3NO	Isoxazole	848.6	816.8	
C_3H_3NO	2-Oxopropanenitrile	746.9	716.2	
C_3H_3NS	Thiazole	904	872.1	
$C_3H_3N_3$	1,3,5-Triazine	848.8	819.6	
C_3H_4	Allene	775.3	745.8	
C_3H_4	Propyne	748.2	723.0	
C_3H_4	Cyclopropene	818.5	787.8	
C_3H_4ClN	3-Chloropropanenitrile	773.1	742.4	
$C_3H_4N_2$	1H-Pyrazole	894.1	860.5	
$C_3H_4N_2$	Imidazole	942.8	909.2	
$C_3H_4N_2S$	2-Thiazolamine	930.6	898.7	
C_3H_4O	Acrolein	797.0	765.1	

Molecular formula	Name	E_{pa} kJ/mol	$\Delta_{base}G°$ kJ/mol	Notes
C_3H_4O	1-Propen-1-one	834.1	803.4	
$C_3H_4O_3$	Ethylene carbonate	814.2	784.4	
C_3H_5	Allyl	736	707.4	
C_3H_5	Cyclopropyl	738.9	702.0	
$C_3H_5ClO_2$	Ethyl chloroformate	764.8	733.8	
C_3H_5FO	1-Fluoro-2-propanone	795.4	763.5	
$C_3H_5F_3O$	2,2,2-Trifluoroethyl methyl ether	747.6	718.4	
C_3H_5N	Propanenitrile	794.1	763.0	
C_3H_5N	2-Propyn-1-amine	887.4	853.5	
C_3H_5N	Ethyl isocyanide	851.3	818.9	
C_3H_5NO	Acrylamide	870.7	839.8	
C_3H_5NO	Methoxyacetonitrile	758.1	727.4	
C_3H_5NO	2-Azetidinone	852.6	821.7	
C_3H_5NS	(Methylthio)acetonitrile	784.8	754.1	
$C_3H_5N_3$	1H-Pyrazol-3-amine	921.5	889.6	
$C_3H_5N_3$	1H-Pyrazol-4-amine	907.6	874.0	
C_3H_6	Propene	741.6		Ref. 5
C_3H_6	Cyclopropane	750.3	722.2	
$C_3H_6N_2$	3-Aminopropanenitrile	866.4	832.5	
$C_3H_6N_2$	Dimethylcyanamide	852.1	821.4	
$C_3H_6N_2S$	2-Imidazolidinethione	921.9	891.2	
C_3H_6O	Methyl vinyl ether	859.2	830.3	
C_3H_6O	Propanal	786.0	754.0	
C_3H_6O	Acetone	812	782.1	
C_3H_6O	Oxetane	801.3	773.9	
$C_3H_6O_2$	Propanoic acid	797.2	766.2	
$C_3H_6O_2$	Ethyl formate	799.4	768.4	
$C_3H_6O_2$	Methyl acetate	821.6	790.7	
$C_3H_6O_3$	Dimethyl carbonate	830.2	799.2	
C_3H_6S	(Methylthio)ethene	858.2	829.3	
C_3H_6S	Thietane	834.8	805.0	
C_3H_6S	Methylthiirane	833.3	801.5	
C_3H_7N	Allylamine	909.5	875.5	
C_3H_7N	Cyclopropylamine	904.7	869.9	
C_3H_7N	Azetidine	943.4	908.6	
C_3H_7N	1-Methylaziridine	934.8	904.1	
C_3H_7N	Propyleneimine	925.1	892.1	
C_3H_7NO	N,N-Dimethylformamide	887.5	856.6	
C_3H_7NO	N-Methylacetamide	888.5	857.6	
C_3H_7NO	Propanamide	876.2	845.3	
$C_3H_7NO_2$	Isopropyl nitrite	845.5	813.0	
$C_3H_7NO_2$	L-Alanine	901.6	867.7	
$C_3H_7NO_2$	Sarcosine	921.2	888.7	
$C_3H_7NO_2S$	L-Cysteine	903.2	869.3	
$C_3H_7NO_3$	L-Serine	914.6	880.7	
C_3H_8	Propane	625.7	607.8	
$C_3H_8N_2O$	N,N'-Dimethylurea	903.3	873.5	
$C_3H_8N_2S$	N,N'-Dimethylthiourea	926.0	895.1	
C_3H_8O	1-Propanol	786.5	756.1	
C_3H_8O	2-Propanol	793.0	762.6	
C_3H_8O	Ethyl methyl ether	808.6	781.2	
$C_3H_8O_2$	1,3-Propanediol	876.2	825.9	
$C_3H_8O_2$	2-Methoxyethanol	768.8	729.8	
$C_3H_8O_3$	Glycerol	874.8	820	
C_3H_8S	1-Propanethiol	794.9	763.6	
C_3H_8S	2-Propanethiol	803.6	772.3	
C_3H_8S	Ethyl methyl sulfide	846.5	815.3	
C_3H_9As	Trimethylarsine	897.3	864.9	
$C_3H_9BO_3$	Trimethyl borate	815.8	783.4	
C_3H_9N	Propylamine	917.8	883.9	

Molecular formula	Name	E_{pa} kJ/mol	$\Delta_{base}G°$ kJ/mol	Notes
C_3H_9N	Isopropylamine	923.8	889.0	
C_3H_9N	Ethylmethylamine	942.2	909.2	
C_3H_9N	Trimethylamine	948.9	918.1	
C_3H_9NO	2-Methoxyethylamine	928.6	894.6	
C_3H_9NO	Trimethylamine oxide	983.2	953.5	
C_3H_9NO	3-Amino-1-propanol	962.5	917.3	
$C_3H_9O_3P$	Trimethyl phosphite	929.7	899.9	
$C_3H_9O_4P$	Trimethyl phosphate	890.6	860.8	
C_3H_9P	Trimethylphosphine	958.8	926.3	
$C_3H_{10}N_2$	1,3-Propanediamine	987.0	940.0	
$C_3H_{10}OSi$	Trimethylsilanol	814.0	781.5	
C_4F_8	Perfluorocyclobutane	>544		Ref. 6
C_4H_2	1,3-Butadiyne	737.2	712.8	
$C_4H_4F_6O$	Bis(2,2,2-trifluoroethyl) ether	702.3	674.9	
$C_4H_4N_2$	Pyrazine	877.1	847.0	
$C_4H_4N_2$	Pyrimidine	885.8	855.7	
$C_4H_4N_2$	Pyridazine	907.2	877.1	
$C_4H_4N_2O_2$	Uracil	872.7	841.7	
$C_4H_4N_2S_2$	2,4(1H,3H)-Pyrimidinedithione	911.4	880.5	
C_4H_4O	Furan	812	781	Ref. 10
$C_4H_4O_3$	Succinic anhydride	797		Ref. 9
C_4H_4S	Thiophene	815.0	784.3	
$C_4H_5Cl_3O_2$	Ethyl trichloroacetate	790.4	759.4	
$C_4H_5F_3O_2$	Ethyl trifluoroacetate	758.8	727.9	
C_4H_5N	Pyrrole	875.4	843.8	
C_4H_5N	Cyclopropanecarbonitrile	808.2	777.5	
$C_4H_5NO_2$	Ethyl cyanoformate	745.7	714.7	
C_4H_5NS	2-Methylthiazole	930.6	898.7	
$C_4H_5N_3O$	Cytosine	949.9	918	
C_4H_6	1,2-Butadiene	778.9	749.8	
C_4H_6	1,3-Butadiene	783.4	757.6	
C_4H_6	2-Butyne	775.8	745.1	
C_4H_6	Cyclobutene	784.4	753.6	
$C_4H_6F_3NO$	2,2,2-Trifluoro-N,N-dimethylacetamide	849.0	818.0	
$C_4H_6N_2$	1-Methylimidazol	959.6	927.7	
$C_4H_6N_2$	2-Methyl-1H-imidazole	963.4	929.6	
$C_4H_6N_2$	4-Methyl-1H-imidazole	952.8	920.9	
$C_4H_6N_2$	1-Methyl-1H-pyrazole	912.0	880.1	
$C_4H_6N_2$	3-Methyl-1H-pyrazole	906.0	874.2	
$C_4H_6N_2$	4-Methyl-1H-pyrazole	906.8	873.4	
C_4H_6O	2-Methylpropenal	808.7	776.8	
C_4H_6O	3-Buten-2-one	834.7	802.8	
C_4H_6O	Cyclobutanone	802.5	772.7	
C_4H_6O	2,3-Dihydrofuran	866.9	834.4	
C_4H_6O	2,5-Dihydrofuran	823.4	796	
$C_4H_6O_2$	trans-2-Butenoic acid	824.0	793	
$C_4H_6O_2$	Methacrylic acid	816.7	785.7	
$C_4H_6O_2$	Cyclopropanecarboxylic acid	821.4	790.4	
$C_4H_6O_2$	Vinyl acetate	813.9	782.9	
$C_4H_6O_2$	Methyl acrylate	825.8	794.8	
$C_4H_6O_2$	2,3-Butanedione	801.9	770.1	
$C_4H_6O_2$	γ-Butyrolactone	840.0	808.1	
$C_4H_6O_2$	2,3-Dihydro-1,4-dioxin	823.5	792.8	
$C_4H_6O_3$	Acetic anhydride	844		Ref. 9
C_4H_7	2-Methylallyl	778	747.3	
C_4H_7N	Butanenitrile	798.4	767.7	
C_4H_7N	2-Methylpropanenitrile	803.6	772.8	
C_4H_7N	1-Isocyanopropane	856.8	824.3	
C_4H_7NO	2-Butenamide	887.1	856.1	

Molecular formula	Name	E_{pa} kJ/mol	$\Delta_{base}G°$ kJ/mol	Notes
C_4H_7NO	2-Methyl-2-propenamide	880.4	849.4	
$C_4H_7NO_4$	L-Aspartic acid	908.9	875	
C_4H_8	trans-2-Butene	747	719.9	
C_4H_8	Isobutene	802.1	775.6	
$C_4H_8N_2$	(Dimethylamino)acetonitrile	884.5	853.7	
$C_4H_8N_2O_3$	L-Asparagine	929	891.5	
$C_4H_8N_2O_3$	N-Glycylglycine		882	
C_4H_8O	Ethyl vinyl ether	870.1	840.4	
C_4H_8O	2-Methoxy-1-propene	894.9	866.1	
C_4H_8O	Butanal	792.7	760.8	
C_4H_8O	Isobutanal	797.3	765.5	
C_4H_8O	2-Butanone	827.3	795.5	
C_4H_8O	Tetrahydrofuran	822.1	794.7	
$C_4H_8O_2$	Propyl formate	804.9	773.9	
$C_4H_8O_2$	Isopropyl formate	811.3	780.3	
$C_4H_8O_2$	Ethyl acetate	835.7	804.7	
$C_4H_8O_2$	Methyl propanoate	830.2	799.2	
$C_4H_8O_2$	1,3-Dioxane	825.4	796.2	
$C_4H_8O_2$	1,4-Dioxane	797.4	770.0	
$C_4H_8O_3$	Ethyl methyl carbonate	842.7	810.8	
C_4H_8S	Tetrahydrothiophene	849.1	819.3	
C_4H_9N	Pyrrolidine	948.3	915.3	
C_4H_9NO	N-Methylpropanamide	920.4	889.4	
C_4H_9NO	2-Methylpropanamide	878.6	846.7	
C_4H_9NO	N-Ethylacetamide	898.0	867.0	
C_4H_9NO	N,N-Dimethylacetamide	908.0	877.0	
C_4H_9NO	Morpholine	924.3	891.2	
$C_4H_9NO_2$	tert-Butyl nitrite	863.9	831.4	
$C_4H_9NO_2$	Ethyl N-methylcarbamate	888.8	857.8	
$C_4H_9NO_3$	L-Threonine	922.5	888.5	
C_4H_9NS	N,N-Dimethylthioacetamide	925.3	894.4	
C_4H_{10}	Isobutane	677.8	671.3	
$C_4H_{10}N_2$	Piperazine	943.7	914.7	
$C_4H_{10}N_2$	3-Ethyl-3-methyldiaziridine	903.8	871.3	
$C_4H_{10}O$	1-Butanol	789.2	758.9	
$C_4H_{10}O$	2-Butanol	815.7	784.6	
$C_4H_{10}O$	2-Methyl-1-propanol	793.7	762.2	
$C_4H_{10}O$	2-Methyl-2-propanol	802.6	772.2	
$C_4H_{10}O$	Diethyl ether	828.4	801	
$C_4H_{10}O$	Methyl propyl ether	814.9	785.7	
$C_4H_{10}O$	Isopropyl methyl ether	826.3	797.1	
$C_4H_{10}O_2$	1,4-Butanediol	915.6	854.9	
$C_4H_{10}O_2$	1,2-Dimethoxyethane	858.0	820.2	
$C_4H_{10}O_3$	1,2,4-Butanetriol	905.9	841	
$C_4H_{10}S$	1-Butanethiol	801.7	770.5	
$C_4H_{10}S$	2-Methyl-1-propanethiol	802.6	771.4	
$C_4H_{10}S$	2-Methyl-2-propanethiol	816.4	785.1	
$C_4H_{10}S$	Diethyl sulfide	856.7	827.0	
$C_4H_{11}N$	Butylamine	921.5	886.6	
$C_4H_{11}N$	tert-Butylamine	934.1	899.9	
$C_4H_{11}N$	Isobutylamine	924.8	890.8	
$C_4H_{11}N$	Diethylamine	952.4	919.4	
$C_4H_{11}N$	Isopropylmethylamine	952.4	919.4	
$C_4H_{11}N$	Ethyldimethylamine	960.1	929.1	
$C_4H_{11}NO$	N-Ethyl-N-hydroxyethanamine	914.7	882.2	
$C_4H_{11}NO$	4-Amino-1-butanol	984.5	932.1	
$C_4H_{11}NO_2$	Diethanolamine	953	920	
$C_4H_{12}N_2$	1,4-Butanediamine	1005.6	954.3	
$C_4H_{12}N_2$	N,N'-Dimethyl-1,2-ethanediamine	989.2	946.9	
$C_4H_{12}Sn$	Tetramethylstannane	823.7	797.4	

Molecular formula	Name	E_{pa} kJ/mol	$\Delta_{base}G°$ kJ/mol	Notes
$C_4H_{14}OSi_2$	1,1,3,3-Tetramethyldisiloxane	845.3	814.6	
C_4NiO_4	Nickel carbonyl	742.3	716.0	
C_5F_5N	Perfluoropyridine	764.9	733.0	
C_5FeO_5	Iron pentacarbonyl	833.0	798.5	
$C_5H_3ClN_4$	6-Chloro-1H-purine	873.6	841.7	
C_5H_4BrN	2-Bromopyridine	904.8	873.0	
C_5H_4BrN	3-Bromopyridine	910.0	878.2	
C_5H_4BrN	4-Bromopyridine	917.8	886.0	
C_5H_4ClN	2-Chloropyridine	900.9	869	
C_5H_4ClN	3-Chloropyridine	903.4	871.5	
C_5H_4ClN	4-Chloropyridine	916.1	884.2	
C_5H_4FN	3-Fluoropyridine	902.0	870.1	
C_5H_4FN	2-Fluoropyridine	884.6	852.7	
$C_5H_4N_2O_2$	4-Nitropyridine	874.3	842.5	
$C_5H_4N_2O_3$	4-Nitropyridine 1-oxide	868.0	837.3	
$C_5H_4N_4$	1H-Purine	920.1	888.2	
$C_5H_4N_4O$	Hypoxanthine	912.3	880.5	
C_5H_5	Cyclopentadienyl	831.5	799.1	
C_5H_5N	Pyridine	930	898.1	
C_5H_5NO	3-Pyridinol	929.5	897.7	
C_5H_5NO	Pyridine-1-oxide	923.6	892.9	
$C_5H_5N_5$	Adenine	942.8	912.5	
$C_5H_5N_5O$	Guanine	959.5	927.6	
C_5H_6	1,3-Cyclopentadiene	821.6	798.4	
$C_5H_6N_2$	2-Pyridinamine	947.2	915.3	
$C_5H_6N_2$	3-Pyridinamine	954.4	922.6	
$C_5H_6N_2$	4-Pyridinamine	979.7	947.8	
$C_5H_6N_2O_2$	Thymine	880.9	850.0	
C_5H_6O	2-Methylfuran	865.9	833.5	
C_5H_6O	3-Methylfuran	854.0	821.5	
$C_5H_6O_3$	Glutaric anhydride	816		Ref. 9
$C_5H_6O_3$	3-Methylsuccinic anhydride	807		Ref. 9
C_5H_6S	2-Methylthiophene	859.0	826.5	
$C_5H_7F_3O_2$	Propyl trifluoroacetate	763.9	732.9	
C_5H_8	$trans$-1,3-Pentadiene	834.1	804.4	
C_5H_8	2-Methyl-1,3-butadiene	826.4	797.6	
C_5H_8	2-Pentyne	810.2	778.0	
C_5H_8	3-Methyl-1-butyne	814.9	787.8	
C_5H_8	Cyclopentene	766.3	733.8	
C_5H_8	1-Methylcyclobutene	841.5	807.3	
C_5H_8	Vinylcyclopropane	816.3	787.5	
C_5H_8	3,3-Dimethylcyclopropene	847.8	817.1	
$C_5H_8N_2$	1,3-Dimethyl-1H-pyrazole	933.9	902.3	
$C_5H_8N_2$	1,4-Dimethyl-1H-imidazole	976.7	944.9	
$C_5H_8N_2$	1,5-Dimethyl-1H-pyrazole	934.3	902.8	
$C_5H_8N_2$	3,4-Dimethyl-1H-pyrazole	927.3	895.4	
$C_5H_8N_2$	3,5-Dimethyl-1H-pyrazole	933.5	900.1	
$C_5H_8N_2$	1,2-Dimethyl-1H-imidazole	984.7	952.6	
$C_5H_8N_2$	1,5-Dimethyl-1H-imidazole	977.6	945.8	
C_5H_8O	$trans$-2-Pentenal	839.0	807.2	
C_5H_8O	3-Methyl-2-butenal	856.9	825.0	
C_5H_8O	3-Methyl-3-buten-2-one	843.1	811.3	
C_5H_8O	Cyclopropyl methyl ketone	854.9	823	
C_5H_8O	Cyclopentanone	823.7	794.0	
C_5H_8O	3,4-Dihydro-2H-pyran	865.8	833.4	
$C_5H_8O_2$	3-Methyl-2-butenoic acid	822.9	791.9	
$C_5H_8O_2$	cis-2-Methyl-2-butenoic acid	822.5	791.5	
$C_5H_8O_2$	Cyclobutanecarboxylic acid	817.4	786.4	
$C_5H_8O_2$	Methyl trans-2-butenoate	851.3	820.4	
$C_5H_8O_2$	Methyl methacrylate	831.4	800.5	

Molecular formula	Name	E_{pa} kJ/mol	$\Delta_{base}G°$ kJ/mol	Notes
$C_5H_8O_2$	Methyl cyclopropanecarboxylate	842.1	811.2	
$C_5H_8O_2$	2,4-Pentanedione	873.5	836.8	
C_5H_9N	2-Isocyano-2-methylpropane	870.7	838.3	
C_5H_9N	3-(Dimethylamino)-1-propyne	940.3	909.5	
C_5H_9N	Pentanenitrile	802.4	771.7	
C_5H_9N	2,2-Dimethylpropanenitrile	810.9	780.2	
C_5H_9NO	3-Ethoxypropanenitrile	807.2	776.5	
C_5H_9NO	N,N-Dimethyl-2-propenamide	904.3	873.4	
C_5H_9NO	N-Methyl-2-pyrrolidone	923.5	891.6	
$C_5H_9NO_2$	L-Proline	920.5	886.0	
$C_5H_9NO_4$	L-Glutamic acid	913.0	879.1	
$C_5H_9N_3$	Histamine	999.8	961.9	
C_5H_{10}	2-Methyl-2-butene	808.8	779.9	
$C_5H_{10}N_2O$	1,3-Dimethyl-2-imidazolidinone	918.4	886.0	
$C_5H_{10}N_2O_3$	L-Glutamine	937.8	900	
$C_5H_{10}O$	Allyl ethyl ether	833.7	804.5	
$C_5H_{10}O$	Pentanal	796.6	764.8	
$C_5H_{10}O$	2-Pentanone	832.7	800.9	
$C_5H_{10}O$	3-Pentanone	836.8	807	
$C_5H_{10}O$	3-Methyl-2-butanone	836.3	804.4	
$C_5H_{10}O$	Tetrahydropyran	822.8	795.4	
$C_5H_{10}O$	2-Methyltetrahydrofuran	840.8	811.6	
$C_5H_{10}O_2$	Butyl formate	806.0	775	
$C_5H_{10}O_2$	Propyl acetate	836.6	805.6	
$C_5H_{10}O_2$	Isopropyl acetate	836.6	805.6	
$C_5H_{10}O_2$	Methyl butanoate	836.4	805.4	
$C_5H_{10}O_2$	Methyl isobutanoate	836.6	805.7	
$C_5H_{10}O_2$	cis-1,2-Cyclopentanediol	885.6	853.1	
$C_5H_{10}S$	Thiacyclohexane	855.8	826.0	
$C_5H_{11}N$	Allyldimethylamine	957.8	926.8	
$C_5H_{11}N$	Piperidine	954.0	921	
$C_5H_{11}N$	N-Methylpyrrolidine	965.6	934.8	
$C_5H_{11}NO$	2,2-Dimethylpropanamide	889.0	857.2	
$C_5H_{11}NO_2$	L-Valine	910.6	876.7	
$C_5H_{11}NO_2S$	L-Methionine	935.4	901.5	
$C_5H_{12}N_2O$	Tetramethylurea	930.6	899.6	
$C_5H_{12}N_2S$	Tetramethylthiourea	947.6	916.6	
$C_5H_{12}O$	2,2-Dimethyl-1-propanol	795.5	765.2	
$C_5H_{12}O$	Butyl methyl ether	820.3	791.2	
$C_5H_{12}O$	Methyl tert-butyl ether	841.6	812.4	
$C_5H_{12}O$	Ethyl isopropyl ether	842.7	813.5	
$C_5H_{12}S$	2,2-Dimethyl-1-propanethiol	809.5	778.2	
$C_5H_{12}Si$	Vinyltrimethylsilane	833	804.1	
$C_5H_{13}N$	Pentylamine	923.5	889.5	
$C_5H_{13}N$	2-Methyl-2-butanamine	937.8	903.6	
$C_5H_{13}N$	2,2-Dimethylpropylamine	928.3	894.0	
$C_5H_{13}N$	Ethylisopropylamine	960.0	926.7	
$C_5H_{13}N$	N,N-Dimethyl-1-propanamine	962.8	931.9	
$C_5H_{13}N$	Diethylmethylamine	971.0	940.0	
$C_5H_{13}N_3$	1,1,3,3-Tetramethylguanidine	1031.6	997.4	
$C_5H_{14}N_2$	N,N,N',N'-Tetramethylmethanediamine	952.2	919.8	
$C_5H_{14}N_2$	N,N-Dimethyl-1,3-propanediamine	1025.0	975.3	
$C_5H_{14}N_2$	1,5-Pentanediamine	999.6	946.2	
C_6CrO_6	Chromium carbonyl	739.2	714.6	
C_6F_6	Hexafluorobenzene	648.0	624.4	
C_6HF_5	Pentafluorobenzene	690.4	662.7	
$C_6H_2F_4$	1,2,3,4-Tetrafluorobenzene	700.4	672.7	
$C_6H_2F_4$	1,2,3,5-Tetrafluorobenzene	747.3	719.6	
$C_6H_2F_4$	1,2,4,5-Tetrafluorobenzene	746.5	718.8	
$C_6H_3F_3$	1,2,3-Trifluorobenzene	724.3	696.6	

Molecular formula	Name	E_{pa} kJ/mol	$\Delta_{base}G°$ kJ/mol	Notes
$C_6H_3F_3$	1,2,4-Trifluorobenzene	729.5	699.4	
$C_6H_3F_3$	1,3,5-Trifluorobenzene	741.9	715.4	
C_6H_4	Benzyne	841	808.5	
$C_6H_4F_2$	o-Difluorobenzene	731.2	703.5	
$C_6H_4F_2$	m-Difluorobenzene	749.7	722	
$C_6H_4F_2$	p-Difluorobenzene	718.7	692.8	
$C_6H_4N_2$	2-Pyridinecarbonitrile	872.9	841	
$C_6H_4N_2$	3-Pyridinecarbonitrile	877.0	845.1	
$C_6H_4N_2$	4-Pyridinecarbonitrile	880.6	848.8	
$C_6H_4O_2$	p-Benzoquinone	799.1	769.3	
C_6H_5	Phenyl	884	851.5	
C_6H_5Br	Bromobenzene	754.1	725.8	
C_6H_5Cl	Chlorobenzene	753.1	724.6	
C_6H_5F	Fluorobenzene	755.9	726.6	
C_6H_5NO	Nitrosobenzene	854.3	823.6	
C_6H_5NO	4-Pyridinecarboxaldehyde	904.6	872.8	
$C_6H_5NO_2$	Nitrobenzene	800.3	769.5	
$C_6H_5N_3$	Azidobenzene	820	787.5	
C_6H_5O	Phenoxy	873.2		Ref. 4
C_6H_6	Benzene	750.4	725.4	
C_6H_6BrN	3-Bromoaniline	873.2	841.4	
C_6H_6ClN	3-Chloroaniline	868.1	836.3	
C_6H_6ClN	4-Chloroaniline	873.8	842.0	
C_6H_6ClN	2-Chloro-4-methylpyridine	921.2	889.4	
C_6H_6ClN	2-Chloro-6-methylpyridine	908.0	876.2	
C_6H_6ClNO	2-Chloro-6-methoxypyridine	909.9	878.0	
C_6H_6FN	3-Fluoroaniline	867.3	835.5	
C_6H_6FN	4-Fluoroaniline	871.5	839.7	
C_6H_6IN	3-Iodoaniline	878.7	846.8	
C_6H_6N	Anilino	949.8	917.4	
$C_6H_6N_2O$	3-Pyridinecarboxamide	918.3	886.4	
$C_6H_6N_2O_2$	4-Nitroaniline	866.0	834.2	
$C_6H_6N_4$	6-Methyl-1H-purine	939.2	907.3	
C_6H_6O	Bis(2-propynyl) ether	783.9	756.5	
C_6H_6O	Phenol	817.3	786.3	
C_6H_7N	Bis(2-propynyl)amine	910.0	876.9	
C_6H_7N	Aniline	882.5	850.6	
C_6H_7N	2-Methylpyridine	949.1	917.3	
C_6H_7N	3-Methylpyridine	943.4	911.6	
C_6H_7N	4-Methylpyridine	947.2	915.3	
C_6H_7NO	1-Methyl-2(1H)-pyridinone	925.8	894.8	
C_6H_7NO	2-Aminophenol	898.8	866.9	
C_6H_7NO	3-Aminophenol	898.8	866.9	
C_6H_7NO	2-Methoxypyridine	934.7	902.8	
C_6H_7NO	3-Methoxypyridine	942.7	910.9	
C_6H_7NO	4-Methoxypyridine	961.7	929.8	
C_6H_7NO	3-Methylpyridine-1-oxide	935.2	902.8	
C_6H_8	1,3-Cyclohexadiene	837	804.5	
C_6H_8	1,4-Cyclohexadiene	837	808.0	
$C_6H_8N_2$	1,2-Benzenediamine	896.5	865.8	
$C_6H_8N_2$	1,3-Benzenediamine	929.9	899.2	
$C_6H_8N_2$	1,4-Benzenediamine	905.9	874.0	
$C_6H_8N_2O$	Bis(2-cyanoethyl) ether	813.8	786.4	
C_6H_8O	2,4-Dimethylfuran	894.7	862.3	
C_6H_8O	2,5-Dimethylfuran	865.9	835.2	
C_6H_8O	3,4-Dimethylfuran	869.0	838.3	
$C_6H_8O_2$	1,3-Cyclohexanedione	881.2	849.4	
$C_6H_8O_2$	1,4-Cyclohexanedione	812.5	782.7	
$C_6H_8O_2$	1,2-Cyclohexanedione	849.6	818.9	
$C_6H_8O_3$	4-Methylglutaric anhydride	820		Ref. 9

Molecular formula	Name	E_{pa} kJ/mol	$\Delta_{base}G°$ kJ/mol	Notes
$C_6H_9F_3O_2$	Butyl trifluoroacetate	764.8	733.8	
C_6H_9N	2,5-Dimethylpyrrole	918.7	887.1	
$C_6H_9N_3O_2$	L-Histidine	988	950.2	
C_6H_{10}	Methylenecyclopentane	832.4	803.5	
C_6H_{10}	(1-Methylvinyl)cyclopropane	871.6	842.7	
C_6H_{10}	2-Methyl-1,3-pentadiene	864.9	836	
C_6H_{10}	3-Methyl-1,3-pentadiene	852.3	823.4	
C_6H_{10}	2,3-Dimethyl-1,3-butadiene	835.0	807.8	
C_6H_{10}	1-Hexyne	799.8	774.8	
C_6H_{10}	2-Hexyne	806.1	781.1	
C_6H_{10}	Cyclohexene	784.5	752.0	
C_6H_{10}	1-Methylcyclopentene	816.5	787.1	
$C_6H_{10}N_2$	1,3,5-Trimethyl-1H-pyrazole	949.3	917.4	
$C_6H_{10}N_2$	3,4,5-Trimethyl-1H-pyrazole	949.3	916.0	
$C_6H_{10}O$	7-Oxabicyclo[2.2.1]heptane	844.2	816.8	
$C_6H_{10}O$	7-Oxabicyclo[4.1.0]heptane	848.1	815.6	
$C_6H_{10}O$	trans-3-Hexen-2-one	865.6	833.8	
$C_6H_{10}O$	Diallyl ether	827.4	800.0	
$C_6H_{10}O$	Cyclohexanone	841.0	811.2	
$C_6H_{10}O$	Mesityl oxide	878.7	846.9	
$C_6H_{10}O_2$	Cyclopentanecarboxylic acid	817.4	786.4	
$C_6H_{10}O_2$	2,5-Hexanedione	892.0	851.8	
$C_6H_{11}N$	N-Allyl-2-propen-1-amine	949.3	916.3	
$C_6H_{11}NO$	1-Methyl-2-piperidinone	924.4	892.6	
$C_6H_{11}N_3O_4$	N-(N-Glycylglycyl)glycine	966.8	916.8	
C_6H_{12}	1-Hexene	805.2	776.3	
C_6H_{12}	2-Methyl-2-pentene	812	783.1	
C_6H_{12}	2,3-Dimethyl-2-butene	813.9	785.9	
C_6H_{12}	Cyclohexane	686.9	666.9	
$C_6H_{12}N_2$	Triethylenediamine	963.4	934.6	
$C_6H_{12}N_2O_3$	N-L-Alanyl-L-alanine		905.6	
$C_6H_{12}O$	Oxepane	834.2	806.8	
$C_6H_{12}O$	3-Hexanone	843.2	811.3	
$C_6H_{12}O$	3,3-Dimethyl-2-butanone	840.1	808.2	
$C_6H_{12}O_2$	cis-1,3-Cyclohexanediol	882.2	849.7	
$C_6H_{12}O_2$	trans-1,3-Cyclohexanediol	828.6	797.9	
$C_6H_{12}O_2$	Methyl 2,2-dimethylpropanoate	845.2	814.2	
$C_6H_{12}O_2$	Diacetone alcohol	822.9	791.1	
$C_6H_{12}O_6$	α-D-Glucose		778.9	
$C_6H_{12}O_6$	β-D-Glucose		778.9	
$C_6H_{13}N$	N,N,2-Trimethylpropenylamine	967.0	934.5	
$C_6H_{13}N$	Cyclohexylamine	934.4	899.6	
$C_6H_{13}N$	1-Methylpiperidine	971.1	940.1	
$C_6H_{13}N$	Hexahydro-1H-azepine	956.7	923.5	
$C_6H_{13}NO$	N,N-Dimethylbutanamide	921.7	890.8	
$C_6H_{13}NO$	N,N-Diethylacetamide	925.4	894.4	
$C_6H_{13}NO_2$	L-Leucine	914.6	880.6	
$C_6H_{13}NO_2$	L-Isoleucine	917.4	883.5	
$C_6H_{14}N_2O_2$	L-Lysine	996	951.0	
$C_6H_{14}N_4O_2$	L-Arginine	1051.0	1006.6	
$C_6H_{14}O$	Dipropyl ether	837.9	810.5	
$C_6H_{14}O$	Diisopropyl ether	855.5	828.1	
$C_6H_{14}O$	tert-Butyl ethyl ether	856.0	826.9	
$C_6H_{14}O_3$	Diethylene glycol dimethyl ether	918.8	870.9	
$C_6H_{14}S$	Dipropyl sulfide	864.7	834.9	
$C_6H_{14}S$	Diisopropyl sulfide	876.4	846.6	
$C_6H_{15}N$	Butyldimethylamine	969.2	938.2	
$C_6H_{15}N$	Isobutyldimethylamine	968.7	937.8	
$C_6H_{15}N$	Hexylamine	927.5	893.5	
$C_6H_{15}N$	Dipropylamine	962.3	929.3	

Molecular formula	Name	E_{pa} kJ/mol	$\Delta_{base}G°$ kJ/mol	Notes
$C_6H_{15}N$	Diisopropylamine	971.9	938.6	
$C_6H_{15}N$	Triethylamine	981.8	951	
$C_6H_{15}NO$	6-Amino-1-hexanol	969.0	915.7	
$C_6H_{15}OP$	Triethylphosphine oxide	936.6	906.8	
$C_6H_{15}O_4P$	Triethyl phosphate	909.3	879.6	
$C_6H_{15}P$	Triethylphosphine	984.5	952.0	
$C_6H_{16}N_2$	1,6-Hexanediamine	999.5	946.2	
$C_6H_{16}N_2$	N,N,N',N'-Tetramethyl-1,2-ethanediamine	1012.8	970.6	
$C_6H_{16}OSi$	Triethylsilanol	822.1	794.8	
$C_6H_{18}N_3OP$	Hexamethylphosphoric triamide	958.6	928.7	
$C_6H_{18}N_3P$	Hexamethylphosphorous triamide	930.1	897.7	
$C_6H_{18}OSi_2$	Hexamethyldisiloxane	846.4	816.2	
C_6MoO_6	Molybdenum hexacarbonyl	762.6	738.1	
C_6O_6W	Tungsten carbonyl	758.0	733.4	
$C_7H_4N_2O_2$	3-Nitrobenzonitrile	781.4	750.7	
$C_7H_4N_2O_2$	4-Nitrobenzonitrile	775.7	745.1	
C_7H_5ClO	3-Chlorobenzaldehyde	813.0	781.1	
C_7H_5ClO	4-Chlorobenzaldehyde	831.3	799.4	
C_7H_5FO	3-Fluorobenzaldehyde	814.3	782.5	
C_7H_5FO	4-Fluorobenzaldehyde	827.1	795.3	
C_7H_5N	Benzonitrile	811.5	780.9	
C_7H_5N	Isocyanobenzene	868.4	836.0	
C_7H_5NO	Benzoxazole	891.6	859.8	
$C_7H_5NO_3$	4-Nitrobenzaldehyde	795.1	763.2	
C_7H_6ClNO	3-Chlorobenzamide	877.2	846.3	
C_7H_6ClNO	4-Chlorobenzamide	877.2	846.3	
C_7H_6F	m-Fluorobenzyl	836.5	804	
C_7H_6FNO	3-Fluorobenzamide	877.2	846.3	
C_7H_6FNO	4-Fluorobenzamide	877.2	846.3	
$C_7H_6F_3N$	3-(Trifluoromethyl)aniline	856.9	825.1	
$C_7H_6N_2$	1H-Benzimidazole	953.8	920.5	
$C_7H_6N_2$	1H-Indazole	900.8	868.9	
$C_7H_6N_2$	3-Aminobenzonitrile	842.3	810.4	
$C_7H_6N_2$	1H-Pyrrolo[2,3-b]pyridine	940.2	908.3	
$C_7H_6N_2O_3$	4-Nitrobenzamide	845.3	814.4	
$C_7H_6N_2O_3$	3-Nitrobenzamide	854.2	823.2	
C_7H_6O	Benzaldehyde	834.0	802.1	
C_7H_6O	2,4,6-Cycloheptatrien-1-one	920.8	891.0	
$C_7H_6O_2$	Benzoic acid	821.1	790.1	
C_7H_7	Benzyl	831.4	800.7	
C_7H_7Br	2-Bromotoluene	775.3	745.8	
C_7H_7Br	3-Bromotoluene	782.0	752.5	
C_7H_7Br	4-Bromotoluene	775.3	745.8	
C_7H_7Cl	2-Chlorotoluene	790.5	761.1	
C_7H_7Cl	3-Chlorotoluene	783.9	754.5	
C_7H_7Cl	4-Chlorotoluene	762.9	735.2	
C_7H_7F	2-Fluorotoluene	773.3	743.8	
C_7H_7F	3-Fluorotoluene	785.4	756.0	
C_7H_7F	4-Fluorotoluene	763.8	736.1	
C_7H_7I	1-Iodo-2-methylbenzene	780.3	750.8	
C_7H_7N	4-Vinylpyridine	944.1	912.3	
C_7H_7NO	1-(3-Pyridinyl)ethanone	916.2	884.3	
C_7H_7NO	1-(4-Pyridinyl)ethanone	914.7	882.9	
C_7H_7NO	4-Aminobenzaldehyde	910.4	878.6	
C_7H_7NO	Benzamide	892.1	861.2	
$C_7H_7NO_2$	Methyl 3-pyridinecarboxylate	925.6	893.8	
$C_7H_7NO_2$	Methyl 4-pyridinecarboxylate	926.6	894.7	
$C_7H_7NO_2$	Aniline-2-carboxylic acid	901.5	869.0	
$C_7H_7NO_2$	Aniline-3-carboxylic acid	864.7	832.3	

Molecular formula	Name	E_{pa} kJ/mol	$\Delta_{base}G°$ kJ/mol	Notes
$C_7H_7NO_2$	Aniline-4-carboxylic acid	864.7	832.3	
$C_7H_7NO_2$	4-Nitrotoluene	815.2	782.7	
$C_7H_7NO_3$	4-Nitrobenzenemethanol	810.5	778.0	
$C_7H_7N_3$	1-Methyl-1H-benzotriazole	931.2	898.7	
C_7H_7O	2-Methylphenoxy	874.5	842	
C_7H_8	Toluene	784.0	756.3	
C_7H_8	2,5-Norbornadiene	849.3	820.3	
$C_7H_8N_2O$	4-Aminobenzamide	927.9	896.9	
$C_7H_8N_2O$	3-Aminobenzamide	900.9	869.9	
$C_7H_8N_2O_2$	N-Methyl-4-nitroaniline	891.6	865.1	
C_7H_8O	o-Cresol	832	800	Ref. 10
C_7H_8O	m-Cresol	841	809	Ref. 10
C_7H_8O	p-Cresol	814	782	Ref. 10
C_7H_8O	Benzyl alcohol	778.3	748.0	
C_7H_8O	Anisole	839.6	807.2	
$C_7H_8O_2$	2,6-Dimethyl-4H-pyran-4-one	941.5	907.3	
$C_7H_8O_2S$	Methyl phenyl sulfone	812.7	780.3	
C_7H_8S	(Methylthio)benzene	872.6	843.7	
C_7H_9N	Benzylamine	913.3	879.4	
C_7H_9N	2-Methylaniline	890.9	859.1	
C_7H_9N	3-Methylaniline	895.8	864.0	
C_7H_9N	4-Methylaniline	896.7	864.8	
C_7H_9N	N-Methylaniline	916.6	890.1	
C_7H_9N	2-Ethylpyridine	952.4	920.6	
C_7H_9N	3-Ethylpyridine	947.4	915.5	
C_7H_9N	4-Ethylpyridine	951.1	919.2	
C_7H_9N	2,3-Dimethylpyridine	958.9	927.0	
C_7H_9N	2,4-Dimethylpyridine	962.9	930.8	
C_7H_9N	2,5-Dimethylpyridine	958.8	926.9	
C_7H_9N	2,6-Dimethylpyridine	963.0	931.1	
C_7H_9N	3,4-Dimethylpyridine	957.3	925.5	
C_7H_9N	3,5-Dimethylpyridine	955.4	923.5	
C_7H_9NO	2-Methoxyaniline	905.2	873.3	
C_7H_9NO	3-Methoxyaniline	913.0	881.1	
C_7H_9NO	4-Methoxyaniline	900.3	868.5	
C_7H_{10}	Bicyclo[2.2.1]hept-2-ene	836.5	804.0	
$C_7H_{10}N_2$	N,N-Dimethyl-2-pyridinamine	968.2	941.6	
$C_7H_{10}N_2$	N,N-Dimethyl-4-pyridinamine	997.6	971.1	
$C_7H_{10}O$	Dicyclopropyl ketone	880.4	850.6	
$C_7H_{10}O$	Bicyclo[2.2.1]heptan-2-one	847.4	815.5	
$C_7H_{11}N$	Cyclohexanecarbonitrile	815.0	784.4	
C_7H_{12}	2,4-Dimethyl-1,3-pentadiene	886.5	857.6	
C_7H_{12}	1-Methylcyclohexene	825.1	792.6	
C_7H_{12}	1,2-Dimethylcyclopentene	822.6	791.9	
$C_7H_{12}N_2$	2,3,4,6,7,8-Hexahydropyrrolo[1,2-a]pyrimidine	1038.3	1005.9	
$C_7H_{12}O$	Cycloheptanone	845.6	815.9	
$C_7H_{12}O$	4-Methylcyclohexanone	844.9	813.0	
$C_7H_{12}O_2$	Cyclohexanecarboxylic acid	823.8	792.8	
$C_7H_{13}N$	1-Azabicyclo[2.2.2]octane	983.3	952.5	
C_7H_{14}	2,4-Dimethyl-2-pentene	812	783.1	
$C_7H_{14}O$	Methoxycyclohexane	840.5	811.3	
$C_7H_{14}O$	4-Heptanone	845.0	815.3	
$C_7H_{14}O$	2,4-Dimethyl-3-pentanone	850.3	820.5	
$C_7H_{14}O$	Cyclohexanemethanol	802.1	771.7	
$C_7H_{15}N$	Cyclohexanemethanamine	926.6	895.8	
$C_7H_{16}O$	tert-Butyl isopropyl ether	870.7	841.5	
$C_7H_{17}N$	Heptylamine	923.2	889.3	
$C_7H_{17}N$	Methyldipropylamine	983.5	950.9	
$C_7H_{17}N$	Diethylpropylamine	978.8	947.9	

Molecular formula	Name	E_{pa} kJ/mol	$\Delta_{base}G°$ kJ/mol	Notes
$C_7H_{18}N_2$	1,7-Heptanediamine	998.5	944.9	
$C_7H_{18}N_2$	N,N,N',N'-Tetramethyl-1,3-propanediamine	1035.2	985.4	
$C_8H_4F_3N$	3-(Trifluoromethyl)benzonitrile	791.4	760.8	
$C_8H_4F_3N$	4-(Trifluoromethyl)benzonitrile	787.2	758.3	
$C_8H_4N_2$	m-Dicyanobenzene	779.3	750.4	
$C_8H_4N_2$	p-Dicyanobenzene	779.0	751.8	
C_8H_5Cl	1-Chloro-4-ethynylbenzene	832.4	801.7	
$C_8H_5Cl_3O$	2,2,2-Trichloro-1-phenylethanone	818.9	787.0	
$C_8H_5F_3O$	2,2,2-Trifluoro-1-phenylethanone	799.2	767.4	
$C_8H_5F_3O$	4-(Trifluoromethyl)benzaldehyde	805.6	773.8	
C_8H_5NO	4-Formylbenzonitrile	796.9	766.3	
C_8H_6	Phenylacetylene	832.0	801.3	
C_8H_6ClN	4-(Chloromethyl)benzonitrile	812.8	782.1	
C_8H_6ClN	3-(Chloromethyl)benzonitrile	811.2	780.6	
$C_8H_6N_2$	Quinoxaline	903.8	873.7	
$C_8H_6N_2$	Cinnoline	936.3	904.4	
C_8H_7Br	1-Bromo-4-vinylbenzene	838.7	809.8	
C_8H_7Br	1-Bromo-3-vinylbenzene	822.4	793.5	
C_8H_7ClO	1-(3-Chlorophenyl)ethanone	846.9	815.1	
C_8H_7ClO	1-(4-Chlorophenyl)ethanone	856.6	824.8	
$C_8H_7ClO_2$	Methyl 4-chlorobenzoate	842.1	811.1	
$C_8H_7ClO_2$	Methyl 3-chlorobenzoate	835.4	804.4	
C_8H_7FO	1-(4-Fluorophenyl)ethanone	858.6	826.8	
C_8H_7N	Benzeneacetonitrile	805.5	774.8	
C_8H_7N	1H-Indole	933.4	901.9	
$C_8H_7NO_3$	1-(4-Nitrophenyl)ethanone	824.3	792.5	
$C_8H_7NO_3$	1-(3-Nitrophenyl)ethanone	826.0	794.1	
$C_8H_7NO_4$	Methyl 3-nitrobenzoate	815.7	784.7	
$C_8H_7NO_4$	Methyl 4-nitrobenzoate	813.2	782.3	
C_8H_8	Styrene	839.5	809.2	
$C_8H_8N_2$	1-Methyl-1H-benzimidazole	967.0	935.2	
$C_8H_8N_2$	2-Methyl-2H-indazole	941.4	909.6	
$C_8H_8N_2$	1-Methyl-1H-indazole	922.4	890.5	
C_8H_8O	3-Methylbenzaldehyde	840.0	808.1	
C_8H_8O	4-Methylbenzaldehyde	851.8	820.0	
C_8H_8O	Acetophenone	861.1	829.3	
$C_8H_8O_2$	o-Toluic acid	838.8	807.8	
$C_8H_8O_2$	m-Toluic acid	829.8	798.8	
$C_8H_8O_2$	p-Toluic acid	836.7	805.7	
$C_8H_8O_2$	Methyl benzoate	850.5	819.5	
$C_8H_8O_2$	3-Methoxybenzaldehyde	844.1	812.2	
$C_8H_8O_2$	4-Methoxybenzaldehyde	881.1	849.3	
$C_8H_8O_2$	1-(3-Hydroxyphenyl)ethanone	863.6	831.8	
$C_8H_8O_2$	1-(4-Hydroxyphenyl)ethanone	883.7	851.9	
$C_8H_8O_3$	Methyl 4-hydroxybenzoate	863.4	832.5	
$C_8H_8O_3$	Methyl 3-hydroxybenzoate	850.0	819.1	
C_8H_9N	2,3-Dihydro-1H-indole	957.1	926.3	
C_8H_9NO	3-Methylbenzamide	900.9	869.9	
C_8H_9NO	4-Methylbenzamide	900.9	869.9	
C_8H_9NO	1-(4-Aminophenyl)ethanone	908.8	877.0	
$C_8H_9NO_2$	3-Methoxybenzamide	900.9	869.9	
$C_8H_9NO_2$	2,4-Dimethyl-1-nitrobenzene	831.0	798.5	
$C_8H_9NO_2$	Methyl 4-aminobenzoate	883.9	853.0	
$C_8H_9NO_2$	4-Methoxybenzamide	900.3	869.4	
C_8H_{10}	Ethylbenzene	788.0	760.3	
C_8H_{10}	o-Xylene	796.0	768.3	
C_8H_{10}	m-Xylene	812.1	786.2	
C_8H_{10}	p-Xylene	794.4	766.8	
$C_8H_{10}ClN$	4-Chloro-N,N-dimethylaniline	922.9	896.4	

Molecular formula	Name	E_{pa} kJ/mol	$\Delta_{base}G°$ kJ/mol	Notes
$C_8H_{10}N_2O_2$	N,N-Dimethyl-4-nitroaniline	896.7	870.2	
$C_8H_{10}N_2O_2$	N,N-Dimethyl-3-nitroaniline	894.1	867.6	
$C_8H_{10}O$	Benzyl methyl ether	816.7	787.5	
$C_8H_{10}O$	2-Methylanisole	850	818	Ref. 10
$C_8H_{10}O$	3-Methylanisole	860	828	Ref. 10
$C_8H_{10}O$	4-Methylanisole	841	809	Ref. 10
$C_8H_{11}N$	4-Isopropylpyridine	955.7	923.8	
$C_8H_{11}N$	3-Ethylaniline	897.9	866.1	
$C_8H_{11}N$	N-Ethylaniline	924.8	892.9	
$C_8H_{11}N$	N,N-Dimethylaniline	941.1	909.2	
$C_8H_{11}N$	2,6-Dimethylaniline	901.7	869.8	
$C_8H_{11}N$	Benzeneethanamine	936.2	902.3	
$C_8H_{11}N$	2-Propylpyridine	955.7	923.8	
C_8H_{12}	2-Methyl-2-norbornene	845	812.5	
$C_8H_{12}N_2$	N,N-Dimethyl-1,4-benzenediamine	955.0	928.4	
$C_8H_{12}N_2O_2$	Ethyl 1,5-dimethyl-1H-pyrazole-3-carboxylate	933.4	901.5	
$C_8H_{14}O$	Cyclooctanone	849.4	819.6	
$C_8H_{14}O$	1-Cyclohexylethanone	841.4	809.5	
$C_8H_{14}O_2$	Methyl cyclohexanecarboxylate	846.2	815.3	
$C_8H_{16}O$	2,2,4-Trimethyl-3-pentanone	856.9	825.0	
$C_8H_{17}N$	Cyclohexyldimethylamine	983.6	952.6	
$C_8H_{18}O$	Dibutyl ether	845.7	818.3	
$C_8H_{18}O$	Di-sec-butyl ether	865.9	838.5	
$C_8H_{18}O$	Di-tert-butyl ether	887.4	860.0	
$C_8H_{18}O_4$	Triethylene glycol dimethyl ether	946.6	892.4	
$C_8H_{18}O_5$	Tetraethylene glycol		>910	
$C_8H_{18}S$	Dibutyl sulfide	871.8	842.1	
$C_8H_{18}S$	Di-tert-butyl sulfide	893.8	864.0	
$C_8H_{19}N$	N-Ethyl-N-isopropyl-2-propanamine	994.3	963.5	
$C_8H_{19}N$	Octylamine	928.9	895.0	
$C_8H_{19}N$	Dibutylamine	968.5	935.3	
$C_8H_{19}N$	Di-sec-butylamine	980.7	947.5	
$C_8H_{19}N$	Diisobutylamine	958.1	925.1	
$C_8H_{20}N_2$	N,N,N',N'-Tetramethyl-1,4-butanediamine	1046.3	992.7	
$C_8H_{20}N_2$	Tetraethylhydrazine	964.3	935.3	
$C_9H_7MnO_3$	Manganese 2-methylcyclopentadienyl tricarbonyl	833.8	801.3	
C_9H_7N	Quinoline	953.2	921.4	
C_9H_7N	Isoquinoline	951.7	919.9	
C_9H_7NO	4-Acetylbenzonitrile	826.8	795.0	
C_9H_8	Indene	848.8	819.6	
C_9H_8O	2-Methylbenzofuran	859.6	827.2	
$C_9H_8O_3$	Methyl 4-formylbenzoate	832.9	801.9	
C_9H_9Cl	1-Chloro-4-isopropenylbenzene	854.3	825.4	
$C_9H_9ClO_2$	3-Chloro-4-methoxyacetophenone	883.7	851.9	
C_9H_{10}	2-Methylstyrene	855.2	826.3	
C_9H_{10}	3-Methylstyrene	849.4	820.5	
C_9H_{10}	4-Methylstyrene	861.7	832.8	
C_9H_{10}	cis-1-Propenylbenzene	836.4	807.5	
C_9H_{10}	trans-1-Propenylbenzene	834.2	805.3	
C_9H_{10}	Isopropenylbenzene	864.2	835.3	
C_9H_{10}	Cyclopropylbenzene	834.9	802.4	
$C_9H_{10}N_2$	4-(Dimethylamino)benzonitrile	889.1	862.6	
$C_9H_{10}O$	1-(3-Methylphenyl)ethanone	868.2	836.4	
$C_9H_{10}O$	1-Phenyl-1-propanone	867.4	835.6	
$C_9H_{10}O$	1-Phenyl-2-propanone	842.6	810.8	
$C_9H_{10}O$	4-Methylacetophenone	875.5	843.6	
$C_9H_{10}OS$	4-Acetylthioanisole	888.2	856.3	
$C_9H_{10}O_2$	Methyl 2-methylbenzoate	858.3	827.3	

Molecular formula	Name	E_{pa} kJ/mol	$\Delta_{base}G°$ kJ/mol	Notes
$C_9H_{10}O_2$	Methyl 3-methylbenzoate	857.7	826.8	
$C_9H_{10}O_2$	1-(3-Methoxyphenyl)ethanone	871.2	839.3	
$C_9H_{10}O_2$	Methyl 4-methylbenzoate	861.5	830.6	
$C_9H_{10}O_2$	4-Acetylanisole	895.6	863.7	
$C_9H_{10}O_3$	Methyl 4-methoxybenzoate	870.6	839.6	
$C_9H_{10}O_3$	Methyl 3-methoxybenzoate	856.7	825.8	
$C_9H_{11}N$	5,6,7,8-Tetrahydroquinoline	966.0	934.1	
$C_9H_{11}N$	5,6,7,8-Tetrahydroisoquinoline	966.6	934.7	
$C_9H_{11}NO$	4-(Dimethylamino)benzaldehyde	924.8	898.3	
$C_9H_{11}NO$	N,N-Dimethylbenzamide	932.7	901.8	
$C_9H_{11}NO_2$	1,3,5-Trimethyl-2-nitrobenzene	823.8	793.1	
$C_9H_{11}NO_2$	L-Phenylalanine	922.9	888.9	
$C_9H_{11}NO_3$	L-Tyrosine	926	892.1	
C_9H_{12}	Propylbenzene	790.1	762.4	
C_9H_{12}	Isopropylbenzene	791.6	763.9	
C_9H_{12}	1,3,5-Trimethylbenzene	836.2	808.6	
$C_9H_{12}N_2$	3-(2-Pyrrolidinyl)pyridine, (S)-	964.0	931.0	
$C_9H_{12}N_2O_6$	Uridine	947.6	916.6	
$C_9H_{12}O_3$	1,3,5-Trimethoxybenzene	926.7	898.2	
$C_9H_{13}N$	N-Ethyl-N-methylaniline	939.0	912.4	
$C_9H_{13}N$	2,6-Diethylpyridine	972.3	940.4	
$C_9H_{13}N$	4-tert-Butylpyridine	957.7	925.8	
$C_9H_{13}N$	2-tert-Butylpyridine	961.7	929.8	
$C_9H_{13}N$	2-Methyl-N,N-dimethylaniline	951.8	925.3	
$C_9H_{13}N$	3-Methyl-N,N-dimethylaniline	942.1	915.7	
$C_9H_{13}N$	4-Methyl-N,N-dimethylaniline	950.0	918.1	
$C_9H_{13}N$	N,N-Dimethylbenzylamine	968.4	937.4	
$C_9H_{13}NO$	4-Methoxy-N,N-dimethylaniline	949.1	922.4	
$C_9H_{13}N_3O_5$	Cytidine	982.5	950.0	
$C_9H_{14}O$	Isophorone	893.5	861.6	
$C_9H_{15}N$	N,N-Diallyl-2-propen-1-amine	972.3	941.3	
$C_9H_{15}N$	N-(1-Cyclopenten-1-yl)pyrrolidine	1019.2	988.4	
$C_9H_{16}O$	Cyclononanone	852.6	822.8	
$C_9H_{17}N_3O_4$	N-(N-L-Alanyl-L-alanyl)-L-alanine		924.1	
$C_9H_{18}O$	5-Nonanone	853.7	821.9	
$C_9H_{18}O$	Di-tert-butyl ketone	861.3	831.5	
$C_9H_{19}N$	1-Isobutylpiperidine	974.5	943.5	
$C_9H_{19}N$	2,2,6,6-Tetramethylpiperidine	987.0	953.9	
$C_9H_{21}N$	Tripropylamine	991.0	960.1	
$C_{10}H_8$	Naphthalene	802.9	779.4	
$C_{10}H_8$	Azulene	925.2	896	
$C_{10}H_8N_2$	2,2'-Bipyridine	965		Ref. 7
$C_{10}H_9N$	1-Naphthylamine	907.0	875.1	
$C_{10}H_{10}Fe$	Ferrocene	863.6	841.3	
$C_{10}H_{10}N_2$	1,8-Naphthalenediamine	944.5	912.1	
$C_{10}H_{10}N_2$	1-Methyl-3-phenyl-1H-pyrazole	932.6	900.8	
$C_{10}H_{10}N_2$	1-Methyl-5-phenyl-1H-pyrazole	932.4	900.5	
$C_{10}H_{10}Ni$	Nickelocene	935.7	907.3	
$C_{10}H_{10}O_2$	1,4-Diacetylbenzene	850.8	821.0	
$C_{10}H_{10}O_2$	1,3-Diacetylbenzene	852.0	822.3	
$C_{10}H_{10}O_3$	4-Acetylphenyl acetate	853.2	821.3	
$C_{10}H_{10}O_4$	Dimethyl isophthalate	843.5	814.3	
$C_{10}H_{10}O_4$	Dimethyl terephthalate	843.2	812.3	
$C_{10}H_{10}Ru$	Ruthenocene	899.1	876.8	
$C_{10}H_{12}$	1-Methyl-3-(1-methylvinyl)benzene	867.6	838.7	
$C_{10}H_{12}$	1-Methyl-4-(1-methylvinyl)benzene	881.8	852.9	
$C_{10}H_{12}$	1-Methyl-2-(1-methylvinyl)benzene	857.8	828.9	
$C_{10}H_{12}$	1,2,3,4-Tetrahydronaphthalene	809.7	782.1	
$C_{10}H_{12}O$	1-(2,4-Dimethylphenyl)ethanone	882.6	850.8	
$C_{10}H_{12}O$	1-(2,5-Dimethylphenyl)ethanone	873.5	841.6	

Molecular formula	Name	E_{pa} kJ/mol	$\Delta_{base}G°$ kJ/mol	Notes
$C_{10}H_{12}O$	1-(3,4-Dimethylphenyl)ethanone	882.8	851.0	
$C_{10}H_{12}O_2$	Methyl 2,5-dimethylbenzoate	864.7	833.7	
$C_{10}H_{12}O_2$	Methyl 2,4-dimethylbenzoate	868.2	837.2	
$C_{10}H_{12}O_2$	Methyl 3,5-dimethylbenzoate	864.3	833.4	
$C_{10}H_{13}N$	1-Phenylpyrrolidine	941.6	915.1	
$C_{10}H_{13}NO$	4'-(Dimethylamino)acetophenone	932.8	906.3	
$C_{10}H_{13}NO$	N,N,3-Trimethylbenzamide	927.0	896.0	
$C_{10}H_{13}NO$	N,N,4-Trimethylbenzamide	927.0	896.0	
$C_{10}H_{13}NO$	1-[3-(Dimethylamino)phenyl]ethanone	928.0	901.5	
$C_{10}H_{13}N_5O_3$	2'-Deoxyadenosine	991.5	959.1	
$C_{10}H_{13}N_5O_4$	Adenosine	989.3	956.8	
$C_{10}H_{13}N_5O_5$	Guanosine	993.4	960.9	
$C_{10}H_{14}$	Butylbenzene	791.9	764.2	
$C_{10}H_{14}$	1,2,3,5-Tetramethylbenzene	845.6	816.5	
$C_{10}H_{14}ClN$	4-Chloro-N,N-diethylaniline	931.0	899.2	
$C_{10}H_{14}N_2$	L-Nicotine	963.4	932.6	
$C_{10}H_{14}N_2O$	N,N-Diethyl-3-pyridinecarboxamide	940.9	909.0	
$C_{10}H_{14}N_2O_5$	Thymidine	948.3	915.9	
$C_{10}H_{15}N$	N,N,2,6-Tetramethylaniline	954.1	923.2	
$C_{10}H_{15}N$	N,N,3,5-Tetramethylaniline	956.1	924.3	
$C_{10}H_{15}N$	N,N-Diethylaniline	959.8	927.9	
$C_{10}H_{16}N_2$	N,N,N',N'-Tetramethyl-1,2-benzenediamine	982.6	950.2	
$C_{10}H_{17}N$	Tricyclo[3.3.1.1³,⁷]decan-1-amine	948.8	916.3	
$C_{10}H_{22}O$	Dipentyl ether	852.7	825.3	
$C_{10}H_{22}O_5$	Tetraethylene glycol dimethyl ether	953.8	897.8	
$C_{10}H_{23}N$	Decylamine	930.4	896.5	
$C_{10}H_{24}N_2$	N,N,N',N'-Tetramethyl-1,6-hexanediamine	1035.8	982.2	
$C_{11}H_9N$	4-Phenylpyridine	939.7	907.8	
$C_{11}H_{10}$	1-Methylnaphthalene	834.8	805.3	
$C_{11}H_{10}$	2-Methylnaphthalene	831.9	802.4	
$C_{11}H_{12}N_2O_2$	L-Tryptophan	948.9	915	
$C_{11}H_{14}O_2$	Methyl 2,4,6-trimethylbenzoate	866.3	835.3	
$C_{11}H_{15}N$	1-Phenylpiperidine	952.9	926.4	
$C_{11}H_{15}N$	Tricyclo[3.3.1.1³,⁷]decane-1-carbonitrile	834.4	803.8	
$C_{11}H_{16}$	Pentamethylbenzene	850.7	823.5	
$C_{11}H_{17}N$	N,N-Diethyl-4-methylaniline	962.8	931.0	
$C_{11}H_{17}N$	2-Hexylpyridine	963.6	931.7	
$C_{11}H_{18}O$	1,4,7,7-Tetramethylbicyclo[2.2.1]heptan-2-one	863.3	831.4	
$C_{11}H_{24}O_4$	2,6,10,14-Tetraoxapentadecane		895.1	
$C_{12}H_8N_2$	Phenazine	938.4	908.3	
$C_{12}H_9NO$	Phenyl-3-pyridinylmethanone	934.1	902.3	
$C_{12}H_{10}$	Acenaphthene	851.7	821.0	
$C_{12}H_{10}$	Biphenyl	813.6	782.9	
$C_{12}H_{16}O$	1-(4-tert-Butylphenyl)ethanone	882.5	850.6	
$C_{12}H_{18}$	Hexamethylbenzene	860.6	836.0	
$C_{12}H_{18}O$	1-Tricyclo[3.3.1.1³,⁷]dec-1-ylethanone	864.9	833.1	
$C_{12}H_{19}N$	N,N-Dipropylaniline	963.0	931.1	
$C_{12}H_{20}O$	2,5-Di-tert-butylfuran	894.7	863.9	
$C_{12}H_{27}N$	Tributylamine	998.5	967.6	
$C_{13}H_9N$	Acridine	972.6	940.7	
$C_{13}H_{10}$	9H-Fluorene	831.5	803.8	
$C_{13}H_{10}O$	Benzophenone	882.3	852.5	
$C_{13}H_{12}$	2-Methylbiphenyl	815.9	783.4	
$C_{13}H_{12}$	3-Methylbiphenyl	828.0	795.5	
$C_{13}H_{12}$	4-Methylbiphenyl	817.9	785.4	
$C_{13}H_{12}$	Diphenylmethane	802.0	769.5	
$C_{13}H_{13}P$	Methyldiphenylphosphine	972.1	939.7	

Molecular formula	Name	E_{pa} kJ/mol	$\Delta_{base}G°$ kJ/mol	Notes
$C_{13}H_{21}N$	2,6-Di-*tert*-butylpyridine	982.9	951	
$C_{13}H_{21}N$	2,4-Di-*tert*-butylpyridine	983.8	952.0	
$C_{13}H_{21}NO$	*N,N*-Dimethyltricyclo[3.3.1.1^{3,7}]decane-1-carboxamide	949.4	917.6	
$C_{14}H_{10}$	Anthracene	877.3	846.6	
$C_{14}H_{10}$	Phenanthrene	825.7	795.0	
$C_{14}H_{12}$	1,1-Diphenylethene	885.7	856.9	
$C_{14}H_{14}$	1,2-Diphenylethane	801.8	774.1	
$C_{14}H_{18}$	1,2,3,4,5,6,7,8-Octahydroanthracene	845.4	814.7	
$C_{14}H_{18}$	1,2,3,4,5,6,7,8-Octahydrophenanthrene	846.2	815.5	
$C_{14}H_{23}N$	4-Octylaniline	894.5	862	
$C_{15}H_{12}$	2-Methylanthracene	887.5	855.1	
$C_{15}H_{12}$	9-Methylanthracene	896.5	865.8	
$C_{15}H_{12}N_2$	3,5-Diphenyl-1*H*-pyrazole	946.3	912.7	
$C_{15}H_{16}$	1,3-Diphenylpropane	820.1	787.6	
$C_{15}H_{18}$	1,4-Dimethyl-7-isopropylazulene	983.1	950.6	
$C_{15}H_{24}$	1,3-Di-*tert*-butyl-5-methylbenzene	853.7	826.0	
$C_{16}H_{10}$	Fluoranthene	828.6	800.9	
$C_{16}H_{10}$	Pyrene	869.2	840.1	
$C_{16}H_{18}$	1,4-Diphenylbutane	822.0	779.8	
$C_{17}H_{20}$	1,5-Diphenylpentane	824.7	782.4	
$C_{18}H_{12}$	Chrysene	840.9	810.1	
$C_{18}H_{12}$	Naphthacene	905.5	876.5	
$C_{18}H_{12}$	Triphenylene	819.2	791.2	
$C_{18}H_{15}As$	Triphenylarsine	908.9	876.4	
$C_{18}H_{15}AsO$	Triphenylarsine oxide	906.2	876.4	
$C_{18}H_{15}N$	Triphenylamine	908.9	876.4	
$C_{18}H_{15}OP$	Triphenylphosphine oxide	906.2	876.4	
$C_{18}H_{15}P$	Triphenylphosphine	972.8	940.4	
$C_{18}H_{15}PS$	Triphenylphosphine sulfide	906.2	876.4	
$C_{18}H_{15}Sb$	Triphenylstibine	845.5	813.1	
$C_{18}H_{22}$	1,6-Diphenylhexane	826.1	783.8	
$C_{18}H_{30}$	1,3,5-Tri-*tert*-butylbenzene	848.8	822.3	
$C_{20}H_{12}$	Perylene	888.6	859.6	
$C_{22}H_{12}$	Benzo[ghi]perylene	876.0	845.2	
$C_{22}H_{14}$	Picene	851.3	820.6	
$C_{24}H_{12}$	Coronene	861.3	835.0	
C_{60}	Carbon (fullerene-C_{60})		827.5	
C_{70}	Carbon (fullerene-C_{70})		827.5	

ATOMIC AND MOLECULAR POLARIZABILITIES

Thomas M. Miller

The *polarizability* of an atom or molecule describes the response of the electron cloud to an external field. The atomic or molecular energy shift ΔW due to an external electric field E is proportional to E^2 for external fields which are weak compared to the internal electric fields between the nucleus and electron cloud. The *electric dipole polarizability* α is the constant of proportionality defined by $\Delta W = -\alpha E^2/2$. The induced electric dipole moment is αE. *Hyperpolarizabilities*, coefficients of higher powers of E, are less often required. Technically, the polarizability is a tensor quantity but for spherically symmetric charge distributions reduces to a single number. In any case, an *average polarizability* is usually adequate in calculations. Frequency-dependent or *dynamic polarizabilities* are needed for electric fields which vary in time, except for frequencies which are much lower than electron orbital frequencies, where *static polarizabilities* suffice.

Polarizabilities for atoms and molecules in excited states are found to be larger than for ground states and may be positive or negative. Molecular polarizabilities are very slightly temperature dependent since the size of the molecule depends on its rovibrational state. Only in the case of dihydrogen has this effect been studied enough to warrant consideration in Table 3.

Polarizabilities are normally expressed in cgs units of cm^3. Ground state polarizabilities are in the range of 10^{-24} cm^3 = 1 Å^3 and hence are often given in Å^3 units. Theorists tend to use atomic units of a_o^3 where a_o is the Bohr radius. The conversion is $\alpha(cm^3)$ = $0.148184 \times 10^{-24} \times \alpha(a_o^3)$. Polarizabilities are only recently encountered in SI units, $C\,m^2/V = J/(V/m)^2$. The conversion from cgs units to SI units is $\alpha(C\,m^2/V) = 4\pi\varepsilon_o \times 10^{-6}\,\alpha(cm^3)$, where ε_o is the permittivity of free space in SI units and the factor 10^{-6} simply converts cm^3 into m^3. Thus, $\alpha(C\,m^2/V) = 1.11265 \times 10^{-16} \times \alpha(cm^3)$. Persons measuring excited state polarizabilities by optical methods tend to use units of $MHz/(V/cm)^2$, where the energy shift, ΔW, is expressed in frequency units with a factor of h understood. The polarizability is $-2\,\Delta W/E^2$. The conversion into cgs units is $\alpha(cm^3) = 5.95531 \times 10^{-16} \times \alpha[MHz/(V/cm)^2]$.

The polarizability appears in many formulas for low-energy processes involving the valence electrons of atoms or molecules. These formulas are given below in cgs units: the polarizability α is in cm^3; masses m or μ are in grams; energies are in ergs; and electric charges are in esu, where $e = 4.8032 \times 10^{-10}$ esu. The symbol $\alpha(\nu)$ denotes a frequency (ν) dependent polarizability, where $\alpha(\nu)$ reduces to the static polarizability a for $\nu = 0$. For further information, see Bonin, K. D., and Kresin, V. V., *Electric Dipole Polarizabilities of Atoms, Molecules, and Clusters*, World Scientific, Sinapore, 1997; Bonin, K. D., and Kadar-Kallen, *Int. J. Mod. Phys. B*, 24, 3313, 1994; and Miller, T. M., and Bederson, B., *Advances in Atomic and Molecular Physics*, 13, 1, 1977, and Gould, H., and Miller, T. M., *Advances in Atomic, Molecular, and Optical Physics*, 51, 243, 2005. Details on polarizability-related interactions, especially in regard to hyperpolarizabilities and nonlinear optical phenomena, are given by Bogaard, M. P., and Orr, B. J., in P*hysical Chemistry, Series Two, Vol.2, Molecular Structure and Properties*, Buckingham, A. D., Ed., Butterworths, London, 1975, pp. 149-194. A tabulation of tensor and hyperpolarizabilities is included. The gas number density, n, in Table 1 is usually taken to be that of 1 atm at 0°C in reporting experimental data.

TABLE 1. Formulas Involving Polarizability

Description	Formula	Remarks
Lorentz-Lorenz relation	$\alpha(\nu) = \dfrac{3}{4\pi n}\left[\dfrac{\eta^2(\nu)-1}{\eta^2(\nu)+2}\right]$	For a gas of atoms or nonpolar molecules; the index of refraction is $\eta(\nu)$
Refraction by polar molecules	$\alpha(\nu) + \dfrac{d^2}{3kT} = \dfrac{3}{4\pi n}\left[\dfrac{\eta^2(\nu)-1}{\eta^2(\nu)+2}\right]$	The dipole moment is d, in esu·cm (= 10^{-18} D)
Dielectric constant (dimensionless)	$\kappa(\nu) = 1 + 4\pi n\ \alpha(\nu)$	From the Lorentz-Lorenz relation for the usual case of $\kappa(\nu) \approx 1$
Index of refraction (dimensionless)	$\eta(\nu) = 1 + 2\pi n\ \alpha(\nu)$	From $\eta^3(\nu) = \kappa(\nu)$
Diamagnetic susceptibility	$\chi_m = e^2\left(a_o N\alpha\right)^{1/2} / 4m_e c^2$	From the approximation that the static polarizability is given by the variational formula $\alpha = (4/9a_o)\Sigma(N_i r_i^2)^2$; N is the number of electrons, m_e is the electron mass; a crude approximation is $\chi_m = (E_i/4m_e c^2)\alpha$, where E_i is the ionization energy
Long-range electron- or ion-molecule interaction energy	$V(r) = -e^2\alpha / 2r^4$	The target molecule polarizability is α
Ion mobility in a gas	$\kappa = -13.87 / (\alpha\mu)^{1/2}\ cm^2 / V\cdot s$	This one formula is not in cgs units. Enter α in Å^3 or $10^{-24}\,cm^3$ units and the reduced mass μ of the ion-molecule pair in amu. Classical limit; pure polarization potential
Langevin capture cross section	$\sigma(\nu_o) = (2\pi e / \nu_o)(\alpha / \mu)^{1/2}$	The relative velocity of approach for an ion-molecule pair is ν_o; the target molecular polarizability is α and the reduced mass of the ion-molecule pair is μ
Langevin reaction rate coefficient	$k = 2\pi e(\alpha / \mu)^{1/2}$	Collisional rate coefficient for an ion-molecule reaction
Rate coefficient for polar molecules	$k_d = 2\pi e\left[(\alpha / \mu)^{1/2} + cd(2 / \mu\pi kT)^{1/2}\right]$	The dipole moment of the neutral is d in esu·cm; the number c is a "locking factor" that depends on α and d, and is between 0 and 1

Description	Formula	Remarks
Modified effective range cross section for electron-neutral scattering	$\sigma(k) = 4\pi A^2$ $+ 32\pi^4 \, e^2 \alpha A k \, / \, 3h^2$ $+ ...$	Here, k is the electron momentum divided by $h/2\pi$, where h is Planck's constant; A is called the "scattering length"; the reduced mass is μ
van der Waals constant between two systems A, B	$C_6 = \dfrac{3}{2}\left[\dfrac{\alpha^A \alpha^B E^A E^B}{E^A + E^B}\right]$	For the interaction potential term $V_6(r) = -C_6 r^{-6}$; $E^{A,B}$ represents average dipole transition energies and $\alpha^{A,B}$ the respective polarizabilities of A, B
Dipole-quadrupole constant between two systems A, B	$C_8 = \dfrac{15}{4}\left[\dfrac{\alpha^A \alpha_q^B E^A E_q^B}{E^A + E_q^B}\right]$ $+ \dfrac{15}{4}\left[\dfrac{\alpha_q^A \alpha^B E_q^A E^B}{E_q^A + E^B}\right]$	For the interaction potential term $V_8(r) = -C_8 r^{-8}$; $E_q^{A,B}$ represents average quadrupole transition energies and $\alpha_q^{A,B}$ are the respective quadrupole polarizabilities of A, B
van der Waals constant between an atom and a surface	$C_3 = \dfrac{\alpha g E^A E^S}{8(E^A + E^S)}$	For an interaction potential $V_3(r) = -C_3 r^{-3}$; $E^{A,S}$ are characteristic energies of the atom and surface; $g = 1$ for a free-electron metal and $g = (\varepsilon_\infty - 1)/(\varepsilon_\infty + 1)$ for an ionic crystal
Relationship between $\alpha(\nu)$ and oscillator strengths	$\alpha(\nu) = \dfrac{e^2 h^2}{4\pi^2 m_e} \Sigma \dfrac{f_k}{E_k^2 - (h\nu)^2}$	Here, f_k is the oscillator strength from the ground state to an excited state k, with excitation energy E_k. This formula is often used to estimate static polarizabilities ($\nu = 0$)
Dynamic polarizability	$\alpha(\nu) = \dfrac{\alpha E_r^2}{E_r^2 - (h\nu)^2}$	Approximate variation of the frequency-dependent polarizability $\alpha(\nu)$ from $\nu = 0$ up to the first dipole-allowed electronic transition, of energy E_r; the static dipole polarizability is $\alpha(0)$; infrared contributions ignored
Rayleigh scattering cross section	$\alpha(\nu) = \dfrac{8\pi}{9c^4}(2\pi\nu)^4$ $\times [3\alpha^2(\nu) + 2\gamma^2(\nu)/3]$	The photon frequency is ν; the polarizability anisotropy (the difference between polarizabilities parallel and perpendicular to the molecular axis) is $\gamma(\nu)$
Verdet constant	$V(\nu) = \dfrac{\nu n}{2m_e c^2}\left[\dfrac{d\alpha(\nu)}{d\nu}\right]$	Defined from $\theta = V(\nu)B$, where θ is the angle of rotation of linearly polarized light through a medium of number density n, per unit length, for a longitudinal magnetic field strength B (Faraday effect)

TABLE 2. Static Average Electric Dipole Polarizabilities for Ground State Atoms (in Units of 10^{-24} cm^3)

Atomic number	Atom	Polarizability	Estimated accuracy (%)	Method	Ref.	Atomic number	Atom	Polarizability	Estimated accuracy (%)	Method	Ref.
1	H	0.666793	"exact"	calc	MB77					index/	
2	He	0.2050522	"exact"	calc	LJS04	18	Ar	1.6411	0.05	diel	NB65/OC67
		0.2050	0.1	index/ diel	NB65/OC67	19	K	43.4	2	beam	MB77
3	Li	24.33	0.7	beam	MJBTV06	20	Ca	22.8	2	calc	MB77
4	Be	5.60	2	calc	MB77			29.4	6	calc	BM02
5	B	3.03	2	calc	MB77			25.0	8	beam	MB77
6	C	1.76	2	calc	MB77	21	Sc	17.8	25	calc	D84
				calc/		22	Ti	14.6	25	calc	D84
7	N	1.10	2	index	MB77	23	V	12.4	25	calc	D84
				calc/		24	Cr	11.6	25	calc	D84
8	O	0.802	2	index	MB77	25	Mn	9.4	25	calc	D84
9	F	0.557	2	calc	MB77	26	Fe	8.4	25	calc	D84
10	Ne	0.3956	0.1	diel	OC67	27	Co	7.5	25	calc	D84
				inter-		28	Ni	6.8	25	calc	D84
11	Na	24.11	0.12	ferom	ESCHP94	29	Cu	6.2	6	calc	BM02
12	Mg	10.6	2	calc	MB77			6.1	25	calc	D84
		11.1	5	calc	S71	30	Zn	5.75	2	index	GHM96
		10.6	5	calc	BM02			6.1	6	calc	BM02
13	Al	6.8	4.4	beam	MMD90			5.6	25	calc	D84
14	Si	5.38	2	calc	MB77	31	Ga	8.12	2	calc	MB77
15	P	3.63	2	calc	MB77	32	Ge	6.07	2	calc	MB77
16	S	2.90	2	calc	MB77	33	As	4.31	2	calc	MB77
17	Cl	2.18	2	calc	MB77	34	Se	3.77	2	calc	MB77
						35	Br	3.05	2	calc	MB77

Atomic number	Atom	Polariz-ability	Estimated accuracy (%)	Method	Ref.	Atomic number	Atom	Polariz-ability	Estimated accuracy (%)	Method	Ref.
36	Kr	2.4844	0.05	diel	OC67	69	Tm	21.8	25	calc	D84
37	Rb	47.3	2	beam	MB77	70	Yb	21.0	25	calc	D84
38	Sr	27.6	8	beam	MB77	71	Lu	21.9	25	calc	D84
		23.5	6	calc	BM02	72	Hf	16.2	25	calc	D84
39	Y	22.7	25	calc	D84	73	Ta	13.1	25	calc	D84
40	Zr	17.9	25	calc	D84	74	W	11.1	25	calc	D84
41	Nb	15.7	25	calc	D84	75	Re	9.7	25	calc	D84
42	Mo	12.8	25	calc	D84	76	Os	8.5	25	calc	D84
43	Tc	11.4	25	calc	D84	77	Ir	7.6	25	calc	D84
44	Ru	9.6	25	calc	D84	78	Pt	6.5	25	calc	D84
45	Rh	8.6	25	calc	D84	79	Au	5.8	25	calc	D84
46	Pd	4.8	25	calc	D84	80	Hg	5.02	1	index	GH96
47	Ag	7.2	25	calc	D84			5.7	25	calc	D84
48	Cd	7.36	3	index	GH95	81	Tl	7.6	15	beam	NYU84
		7.4	6	calc	BM02			7.5	25	calc	D84
		7.2	25	calc	D84	82	Pb	6.8	25	calc	D84
49	In	10.2	12	beam	GMBSJ84	83	Bi	7.4	25	calc	D84
		9.1	25	calc	D84	84	Po	6.8	25	calc	D84
50	Sn	7.7	25	calc	D84	85	At	6.0	25	calc	D84
51	Sb	6.6	25	calc	D84	86	Rn	5.3	25	calc	D84
52	Te	5.5	25	calc	D84	87	Fr	48.60	2	calc	LSMS05
53	I	5.35	25	index	A56			47.1	5	calc	DJSB99
		4.7	25	calc	D84	88	Ra	38.3	25	calc	D84
54	Xe	4.044	0.5	diel	MB77	89	Ac	32.1	25	calc	D84
55	Cs	59.42	0.13	beam	AG03	90	Th	32.1	25	calc	D84
56	Ba	39.7	8	beam	MB77	91	Pa	25.4	25	calc	D84
57	La	31.1	25	calc	D84	92	U	24.9	6	beam	KB94
58	Ce	29.6	25	calc	D84	93	Np	24.8	25	calc	D84
59	Pr	28.2	25	calc	D84	94	Pu	24.5	25	calc	D84
60	Nd	31.4	25	calc	D84	95	Am	23.3	25	calc	D84
61	Pm	30.1	25	calc	D84	96	Cm	23.0	25	calc	D84
62	Sm	28.8	25	calc	D84	97	Bk	22.7	25	calc	D84
63	Eu	27.7	25	calc	D84	98	Cf	20.5	25	calc	D84
64	Gd	23.5	25	calc	D84	99	Es	19.7	25	calc	D84
65	Tb	25.5	25	calc	D84	100	Fm	23.8	25	calc	D84
66	Dy	24.5	25	calc	D84	101	Md	18.2	25	calc	D84
67	Ho	23.6	25	calc	D84	102	No	17.5	25	calc	D84
68	Er	22.7	25	calc	D84	119	ekafrancium	24.26	2	cal	LSMS05

[a] Methods: calc = calculated value; beam = atomic beam deflection technique; interferom = atomic beam interference; index = determination based on the measured index of refraction; diel = determination based on the measured dielectric constant.

References

A56. Atoji, M., *J. Chem. Phys.* 25, 174, 1956. Semiempirical method based on molecular polarizabilities and atomic radii.

AG03. Amini, J. M., and Gould, H., *Phys. Rev. Lett.* 91, 153001, 2003.

BM02. Bromley, M. W. J., and Mitroy, J., *Phys. Rev. A* 65, 062505, 2002; 062506, 2002.

D84. Doolen, G. D., Los Alamos National Laboratory, unpublished. A relativistic linear response method was used. The method is that described by Zangwill, A., and Soven, P., *Phys. Rev. A* 21, 1561, 1980. Adjustments of less than 10% across the periodic table have been made to these results to bring them into agreement with accurate experimental values where available, for the purpose of presenting "recommended" polarizabilities in Table 2.

DJSB99. Derevianko, A., Johnson, W. R., Safronova, M. S., and Babb, J. F., *Phys. Rev. Lett.* 82, 3589, 1999.

ESCHP94. Ekstrom, C. R., Schmiedmayer, J., Chapman, M. S., Hammond, T. D., and Pritchard, D. E., *Phys. Rev. A* 51, 3883, 1995. See theoretical work by Thakkar, A. J., and Lupinetti, C., *Chem. Phys. Lett.* 402, 270, 2005.

GH95. Goebel, D., and Holm, U., *Phys. Rev. A* 52, 3691, 1995.

GH96. Goebel, D., and Holm, U., *J. Chem. Phys.* 100, 7710, 1996.

GHM96. Goebel, D., Holm, U., and Maroulis, G., *Phys. Rev. A* 54, 1973, 1996.

GMBSJ84. Guella, T. P., Miller, T. M., Bederson, B., Stockdale, J. A. D., and Jaduszliwer, B., *Phys. Rev. A* 29, 2977, 1984.

KB94. Kadar-Kallen, M. A., and Bonin, K. D., *Phys. Rev. Lett.* 72, 828, 1994.

LJS04. Łach, G., Jeziorski, B., and Szalewicz, K., *Phys. Rev. Lett.* 92, 233001, 2004.

LS04. Lim, I. S., and Schwerdtfeger, P., *Phys. Rev. A* 70, 062501, 2004.

LSMS05. Lim, L. S., Schwerdfeger, P., Metz, B., Stoll, H., *J. Chem. Phys.* 122, 104103, 2005.

MB77. Miller, T. M., and Bederson, B., *Adv. At. Mol. Phys.* 13, 1, 1977. For simplicity, any value in Table 2 which has not changed since this 1977 review is referenced as MB77. Persons interested in original references and further details should consult MB77.

MJBTV05. Miffre, A., Jacquey, M., Büchner, M., Trénec, G., and Vigué, J., *Phys. Rev. A* 73, 011603 (R), 2006.

MMD90. Milani, P., Moullet, I., and de Heer, W. A., *Phys. Rev. A* 42, 5150, 1990. See theoretical comments on this result, in Fuentealba, P., *Chem. Phys. Lett.* 397, 459, 2004, and in Lupinetti, C., and Thakkar, A. J., *J. Chem. Phys.* 122, 044301, 2005.

NB65. Newell, A. C., and Baird, R. D., *J. Appl. Phys.* 36, 3751, 1965.

NYU84. Preliminary value from the New York University group. See GMBSJ84.

OC67. Orcutt, R. H., and Cole, R. H., *J. Chem. Phys.* 46, 697, 1967; see also the later references from this group, given following the tables.

S71. Stwalley, W. C., *J. Chem. Phys.* 54, 4517, 1971.

TABLE 3. Average Electric Dipole Polarizabilities for Ground State Diatomic Molecules (in Units of 10^{-24} cm^3)

Molecule	Polarizability	Ref.	Molecule	Polarizability	Ref.
Al_2	19	23		5.35	2
BH	3.32*	1	HgCl	7.4*	9
Br_2	7.02	2	ICl	12.3	2
CO	1.95	3	K_2	77	22
Cl_2	4.61	3		72	21
Cs_2	104	22	Li_2	32.8	29
CsK	89	22		34	22
D_2 ($v=0,J=0$)	0.7921*	5	LiCl	3.46*	10
D_2 (293 K)	0.7954	6	LiF	10.8*	11
DCl	2.84	2	LiH	3.84*	12
F_2	1.38*	7		3.68*	13
H_2 ($v=0,J=0$)	0.8023*	5		3.88*	14
H_2 (293 K)	0.8045*	5	N_2	1.7403	6,8
H_2 (293 K)	0.8042	6	NO	1.70	2
H_2 (322 K)	0.8059	8	Na_2	40	22
HBr	3.61	3		38	21
HCl	2.63	3	NaK	51	22
	2.77	2	NaLi	40	4
HD ($v=0,J=0$)	0.7976*	5	O_2	1.5812	6
HF	0.80	27	Rb_2	79	22
HI	5.44	3			

TABLE 4. Average Electric Dipole Polarizabilities for Ground State Triatomic Molecules (in Units of 10^{-24} cm^3)

Molecule	Polarizability	Ref.	Molecule	Polarizability	Ref.
BeH_2	4.34*	14	HgI_2	19.1	2
CO_2	2.911	8	Li_3	34.5	29
CS_2	8.74	3	$LiNa_2$	61.2	30
	8.86	2	Li_2Na	35.4	30
D_2O	1.26	2	N_2O	3.03	8
H_2O	1.45	2	NO_2	3.02	2†
H_2S	3.782	3	Na_3	70	21
	3.95	2	O_3	3.21	2
HCN	2.59	3	OCS	5.71	2
	2.46	2		5.2	15
$HgBr_2$	14.5	2	SO_2	3.72	3
$HgCl_2$	11.6	2		4.28	2

TABLE 5. Average Electric Dipole Polarizabilities for Ground State Inorganic Polyatomic Molecules (Larger than Triatomic) (in Units of 10^{-24} cm^3)

Molecule	Polarizability	Ref.	Molecule	Polarizability	Ref.
$AsCl_3$	14.9	2	$(HgCl)_2$	14.7	9
AsN_3	5.75	2	K_n	n=2,5,7-9,11,20	21
BCl_3	9.38	20	$(KBr)_2$	42.0	16
BF_3	3.31	2	$(KCl)_2$	32.1	16
$(BN_3)_2$	5.73	2	$(KF)_2$	21.0	16
$(BH_2N)_3$	8.0	2†	$(KI)_2$	36.3	16
ClF_3	6.32	2	Li_n	n=2-22	29
$(CsBr)_2$	54.5	16	$(LiBr)_2$	18.9	16
$(CsCl)_2$	42.4	16	$(LiCl)_2$	13.1	16
$(CsF)_2$	28.4	16	$(LiF)_2$	6.9	16
$(CsI)_2$	51.8	16	$(LiI)_2$	23.4	16
Ga_nAs_m	$n+m$=4-30	28	$LiNa_3$	75.6	30
$GeCl_4$	15.1	2	Li_2Na_2	60.0	30
GeH_3Cl	6.7	2†	Li_3Na	54.8	30

Molecule	Polarizability	Ref.
ND_3	1.70	2
NF_3	3.62	2
NH_3	2.81	20
	2.10	2
	2.26	3
	2.22*	33
$(NO_2)_2$	6.69	2
Na_n	n=1-40	21
$(NaBr)_2$	26.8	16
$(NaCl)_2$	23.4	16
$(NaF)_2$	20.7	16
$(NaI)_2$	26.9	16
OsO_4	8.17	2
PCl_3	12.8	2
PF_5	6.10	2
PH_3	4.84	2
$(RbBr)_2$	48.2	16
$(RbCl)_2$	43.2	16
$(RbF)_2$	40.7	16

Molecule	Polarizability	Ref.
$(RbI)_2$	46.3	16
SF_6	6.54	8
$(SF_5)_2$	13.2	2
SO_3	4.84	2
SO_2Cl_2	10.5	2
SeF_6	7.33	2
SiF_4	5.45	2
SiH_4	5.44	2
$(SiH_3)_2$	11.1	2
$SiHCl_3$	10.7	2
SiH_2Cl_2	8.92	2
SiH_3Cl	7.02	2
$SnBr_4$	22.0	2
$SnCl_4$	18.0	2
	13.8	15
SnI_4	32.3	2
TeF_6	9.00	2
$TiCl_4$	16.4	2
UF_6	12.5	2

TABLE 6. Average Electric Dipole Polarizabilities for Ground State Hydrocarbon Molecules (in Units of 10^{-24} cm^3)

Molecule	Name	Polarizability	Ref.
CH_4	methane	2.593	8
C_2H_2	acetylene	3.33	3
		3.93	2
C_2H_4	ethylene	4.252	8
C_2H_6	ethane	4.47	3
		4.43	2
C_3H_4	propyne	6.18	2
C_3H_6	propene	6.26	2
	cyclopropane	5.66	2
C_3H_8	propane	6.29	3
		6.37	2
C_4H_6	1-butyne	7.41	2†
	1,3-butadiene	8.64	2
C_4H_8	1-butene	7.97	2
		8.52	2
	trans-2-butene	8.49	2
	2-methylpropene	8.29	2
C_4H_{10}	butane	8.20	2
	isobutane	8.14	27
C_5H_6	1,3-cyclopentadiene	8.64	2
C_5H_8	1-pentyne	9.12	2
	trans-1,3-pentadiene	10.0	2
	isoprene	9.99	2
C_5H_{10}	cyclopentane	9.15	18
	1-pentene	9.65	27
	2-pentene	9.84	27
C_5H_{12}	pentane	9.99	2
	neopentane	10.20	18
C_6H_6	benzene	10.0	25
		10.32	3
		10.74	2
C_6H_{10}	1-hexyne	10.9	2†
	2-ethyl-1,3-butadiene	11.8	2†
	3-methyl-1,3-pentadiene	11.8	2†
	2-methyl-1,3-pentadiene	12.1	2†
	2,3-dimethyl-1,3-butadiene	11.8	2†
	cyclohexene	10.7	2†
C_6H_{12}	cyclohexane	11.0	18

Molecule	Name	Polarizability	Ref.
		10.87	15
	1-hexene	11.65	27
C_6H_{14}	hexane	11.9	2
C_7H_8	toluene	11.8	25
		12.26	15
		12.3	2
C_7H_{12}	1-heptyne	12.8	2†
C_7H_{14}	methylcyclohexane	13.1	2
	1-heptene	13.51	27
C_7H_{16}	heptane	13.61	2
C_8H_8	styrene	15.0	2
		14.41	27
C_8H_{10}	ethylbenzene	14.2	2
	o-xylene	14.9	2
		14.1	15
	p-xylene	13.7	25
		14.2	15
		14.9	2
	m-xylene	14.2	15
C_8H_{16}	ethylcyclohexane	15.9	2
C_8H_{18}	n-octane	15.9	2
	3-methylheptane	15.44	27
	2,2,4-trimethylpentane	15.44	27
C_9H_{10}	α- methylstyrene	16.05	27
C_9H_{12}	isopropylbenzene	16.0	2†
	1,3,5-trimethylbenzene	15.5	25
		16.14	27
C_9H_{18}	isopropylcyclohexane	17.2	2
C_9H_{20}	nonane	17.36	27
$C_{10}H_8$	naphthalene	16.5	17
		17.48	27
$C_{10}H_{14}$	durene	17.3	25
	tert- butylbenzene	17.2	25
		17.8	2†
$C_{10}H_{20}$	tert-butylcyclohexane	19.8	2
$C_{10}H_{22}$	decane	19.10	27
$C_{11}H_{10}$	α-methylnaphthalene	19.35	27
	β-methylnaphthalene	19.52	27

Molecule	Name	Polarizability	Ref.	Molecule	Name	Polarizability	Ref.
$C_{11}H_{14}$	α,β,β-trimethylstyrene	19.64	27		phenanthrene	36.8*	17
$C_{11}H_{16}$	pentamethylbenzene	19.1	25			24.70	27
$C_{11}H_{24}$	undecane	21.03	27	$C_{14}H_{22}$	p-di-tert-butylbenzene	24.5	25
$C_{12}H_{10}$	acenaphthene	20.61	27	$C_{16}H_{10}$	pyrene	28.22	27
$C_{12}H_{12}$	α-ethylnaphthalene	21.19	27	$C_{17}H_{12}$	2,3- benzfluorene	30.21	27
	β-ethylnaphthalene	21.36	27	$C_{18}H_{12}$	naphthacene	32.27	27
$C_{12}H_{18}$	hexamethylbenzene	20.9	25		1,2- benzanthracene	32.86	27
$C_{12}H_{26}$	dodecane	22.75	27		chrysene	33.06	27
$C_{13}H_{10}$	fluorene	21.68	27		triphenylene	31.07	27
$C_{14}H_{10}$	anthracene	25.4	17	$C_{18}H_{30}$	1,3,5-tri-tert-butylbenzene	31.8	25
		25.93	27	$C_{24}H_{12}$	coronene	42.50	27

TABLE 7. Average Electric Dipole Polarizabilities for Ground State Organic Halides (in Units of 10^{-24} cm^3)

Molecule	Name	Polarizability	Ref.	Molecule	Name	Polarizability	Ref.
CBr_2F_2	dibromodifluoromethane	9.0	2[†]		trans-dichloroethylene	8.15	27
$CClF_3$	chlorotrifluoromethane	5.72	20		cis-dichloroethylene	8.03	27
		5.59	2	$C_2H_2Cl_2F_2$	1,1-dichloro-2,2-difluoroethane	8.4	2[†]
CCl_2F_2	dichlorodifluoromethane	7.93	20	$C_2H_2Cl_2O$	chloroacetyl chloride	8.92	2
		7.81	2	$C_2H_2Cl_3F$	1,2,2-trichloro-1-fluoroethane	10.2	2[†]
CCl_2O	phosgene	7.29	2	$C_2H_2Cl_4$	1,1,2,2-tetrachloroethane	12.1	2[†]
CCl_2S	thiophosgene	10.2	2	C_2H_2ClN	chloroacetonitrile	6.10	18
CCl_3F	trichlorofluoromethane	9.47	2	$C_2H_2F_2$	1,1-difluoroethylene	5.01	20
CCl_3NO_2	trichloronitromethane	10.8	2[†]	C_2H_3Br	bromoethylene	7.59	2
CCl_4	carbon tetrachloride	11.2	2	C_2H_3Cl	chloroethylene	6.41	2
		10.5	3	$C_2H_3ClF_2$	1-chloro-1,1-difluoroethane	8.05	2
CF_4	carbon tetrafluoride	3.838	8	C_2H_3ClO	acetyl chloride	6.62	2
CF_2O	carbonylfluoride	1.88*	17	$C_2H_3ClO_2$	methyl chloroformate	7.1	2[†]
$CHBr_3$	bromoform	11.8	27	$C_2H_3Cl_3$	1,1,1-trichloroethane	10.7	2
$CHBrF_2$	bromodifluoromethane	5.7	2[†]	$C_2H_3F_3$	1,1,1-trifluoroethane	4.4	2[†]
$CHClF_2$	chlorodifluoromethane	6.38	20	C_2H_3I	iodoethylene	9.3	2[†]
		5.91	2	C_2H_4BrCl	1-bromo-2-chloroethane	9.5	2[†]
$CHCl_2F$	dichlorofluoromethane	6.82	2	$C_2H_4Br_2$	1,2-dibromoethane	10.7	2[†]
$CHCl_3$	chloroform	9.5	8	C_2H_4ClF	1-chloro-2-fluoroethane	6.5	2[†]
		8.23	27	$C_2H_4ClNO_2$	1-chloro-1-nitroethane	10.9	2
CHF_3	fluoroform	3.52	20	$C_2H_4Cl_2$	1,1-dichloroethane	8.64	2
		3.57	8		1,2-dichloroethane	8.0	2[†]
$CHFO$	fluoroformaldehyde	1.76*	17	C_2H_5Br	bromoethane	8.05	2
CHI_3	iodoform	18.0	27			7.28	27
CH_2Br_2	dibromomethane	9.32	2	C_2H_5Cl	chloroethane	7.27	20
		8.68	27			8.29	2
CH_2ClNO_2	chloronitromethane	6.9	2[†]			6.4	15
CH_2Cl_2	dichloromethane	6.48	3	C_2H_5ClO	2-chloroethanol	7.1	2[†]
		7.93	2			6.88	27
CH_2I_2	diiodomethane	12.90	27		chloromethyl methyl ether	7.1	2[†]
CH_3Br	bromomethane	5.87	20	C_2H_5F	fluoroethane	4.96	2
		6.03	2	C_2H_5I	iodoethane	10.0	2
		5.55	15	$C_3H_4Cl_2$	dichloropropene	10.1	2[†]
CH_3Cl	chloromethane	5.35	20	C_3H_5Cl	chloropropene	8.3	2
		4.72	8	C_3H_5ClO	chloroacetone	8.4	2[†]
CH_3F	fluoromethane	2.97	8	$C_3H_5ClO_2$	ethyl chloroformate	9.0	2[†]
CH_3I	iodomethane	7.97	2	$C_3H_6ClNO_2$	1-chloro-1-nitropropane	10.4	2[†]
C_2ClF_5	chloropentafluoroethane	6.3	2[†]	$C_3H_6Cl_2$	dichloropropane	10.9	2[†]
$C_2Cl_2F_4$	1,2-dichlorotetrafluoroethane	8.5	2[†]	C_3H_7Br	1-bromopropane	9.4	2[†]
C_2Cl_3N	trichloroacetonitrile	10.42	18			9.07	27
C_2F_6	hexafluoroethane	6.82	2		2-bromopropane	9.6	2[†]
C_2HBr	bromoacetylene	7.39	2	C_3H_7Cl	chloropropane	10.0	2
C_2HCl	chloroacetylene	6.07	2	C_3H_7ClO	β-chloroethyl methyl ether	8.71	27
C_2HCl_3	trichloroethlyene	10.03	27		2-chloro-1-propanol	8.89	27
C_2HCl_5	pentachloroethane	14.0	2		3-chloro-1-propanol	8.84	27
$C_2H_2Cl_2$	1,1-dichloroethylene	7.83	27				

Molecule	Name	Polarizability	Ref.
C_3H_7I	1-iodopropane	11.5	2[†]
C_4H_5Cl	4-chloro-1,2-butadiene	10.0	2[†]
C_4H_7Cl	1-chloro-2-methylpropene	10.8	2
$C_4H_7ClO_2$	2-chlorobutyric acid	10.87	27
	3-chlorobutyric acid	10.80	27
	4-chlorobutyric acid	10.69	27
$C_4H_8Cl_2$	1,4-dichlorobutane	12.0	2[†]
C_4H_9Br	bromobutane	13.9	2
		10.86	27
C_4H_9Cl	1-chlorobutane	11.3	2
	1-chloro-2-methylpropane	11.1	2
	2-chloro-2-methylpropane	12.5	2[†]
	2-chlorobutane	12.4	2
C_4H_9ClO	β-chloroethyl ethyl ether	10.56	27
	2-chloro-1-butanol	10.70	27
	3-chloro-1-butanol	10.38	27
C_4H_9I	1-iodobutane	13.3	2[†]
		12.65	27
$C_5H_9ClO_2$	methyl 2-chlorobutanoate	12.33	27
	methyl 3-chlorobutanoate	12.31	27
	methyl 4-chlorobutanoate	12.27	27
	2-chloropentanoic acid	12.69	27
	3-chloropentanoic acid	12.57	27
	4-chloropentanoic acid	12.53	27
$C_5H_{11}Br$	1-bromopentane	13.1	2[†]
$C_5H_{11}Cl$	1-chloropropane	12.0	2[†]
$C_5H_{11}F$	fluoropentane	9.95	27
C_6F_6	hexafluorobenzene	9.58	27
C_6HF_5	pentafluorobenzene	9.63	27
$C_6H_2Cl_2O_2$	2,5-dichloro-1,4-benzoquinone	18.4	2
$C_6H_2F_4$	1,2,3,4-tetrafluorobenzene	9.69	27
	1,2,4,5-tetrafluorobenzene	9.69	27
$C_6H_3F_3$	1,3,5-trifluorobenzene	9.74	27
C_6H_4BrF	p-bromofluorobenzene	13.4	2[†]
$C_6H_4ClNO_2$	chloronitrobenzene	14.6	2[†]
$C_6H_4Cl_2$	o-dichlorobenzene	14.17	27
	m-dichlorobenzene	14.23	27
	p-dichlorobenzene	14.20	27
C_6H_4FI	p-fluoroiodobenzene	15.5	2[†]
$C_6H_4FNO_2$	p-fluoronitrobenzene	12.8	2[†]
$C_6H_4F_2$	o-difluorobenzene	9.80	27
	m-difluorobenzene	10.3	2[†]
	p-difluorobenzene	9.80	27
C_6H_5Br	bromobenzene	14.7	2
		13.62	27
C_6H_5Cl	chlorobenzene	14.1	2
		12.3	15
C_6H_5ClO	chlorophenol	13.0	2[†]
C_6H_5F	fluorobenzene	10.3	2
C_6H_5I	iodobenzene	15.5	2[†]
$C_6H_{11}ClO_2$	ethyl 2-chlorobutanoate	14.16	27
	ethyl 3-chlorobutanoate	14.13	27
	ethyl 4-chlorobutanoate	14.11	27
$C_6H_{13}Br$	bromohexane	14.44	27
$C_6H_{13}F$	fluorohexane	11.80	27
C_7H_7Br	p-bromotoluene	14.80	27
C_7H_7Cl	p-chlorotoluene	13.70	27
C_7H_7F	p-fluorotoluene	11.70	27
C_7H_7I	p-iodotoluene	17.10	27
$C_7H_{15}Br$	1-bromoheptane	16.8	2[†]
		16.23	27
$C_7H_{15}F$	fluoroheptane	13.66	27
$C_8H_{17}Br$	bromooctane	18.02	27
$C_8H_{17}F$	fluorooctane	15.46	27
$C_9H_{19}Br$	bromononane	19.81	27
$C_9H_{19}F$	fluorononane	17.34	27
$C_{10}F_8$	octafluoronaphthalene	17.64	27
$C_{10}H_7Br$	α-bromonaphthalene	20.34	27
$C_{10}H_7Cl$	α-chloronaphthalene	19.30	27
	β-chloronaphthalene	19.58	27
$C_{10}H_7I$	α-iodonaphthalene	22.41	27
	β-iodonaphthalene	22.95	27
$C_{10}H_{21}Br$	bromodecane	21.60	27
$C_{10}H_{21}F$	fluorodecane	19.18	27
$C_{11}H_{23}F$	fluoroundecane	21.00	27
$C_{12}H_{25}Br$	bromododecane	25.18	27
$C_{12}H_{25}F$	fluorododecane	22.83	27
$C_{12}H_8Br_2O$	4,4'-dibromodiphenyl ether	27.8	2[†]
$C_{12}H_9BrO$	4-bromodiphenyl ether	24.2	2[†]
$C_{13}H_{11}BrO$	p-bromophenyl-p-tolyl ether	26.6	2[†]
$C_{14}H_9Br$	9-bromoanthracene	28.32	27
$C_{14}H_9Cl$	9-chloroanthracene	27.35	27
$C_{14}H_9F$	fluoranthracene	28.34	27
$C_{14}H_{29}F$	fluorotetradecane	26.57	27
$C_{16}H_{33}Br$	bromohexadecane	32.34	27
$C_{18}H_{37}Br$	bromooctadecane	35.92	27

TABLE 8. Static Average Electric Dipole Polarizabilities for Other Ground State Organic Molecules (in Units of 10^{-24} cm³)

Molecule	Name	Polarizability	Ref.
CN_4O_8	tetranitromethane	15.3	2
CH_2O	formaldehyde	2.8	2[†]
		2.45	18
CH_2O_2	formic acid	3.4	2[†]
CH_3NO	formamide	4.2	2[†]
		4.08	18
CH_3NO_2	nitromethane	7.37	2
CH_4O	methanol	3.29	2
		3.23	15
		3.32	18
CH_5N	methyl amine	4.7	2
		4.01	19
		4.01*	33
C_2N_2	cyanogen	7.99	2
C_2H_2O	ketene	4.4	2[†]
C_2H_3N	acetonitrile	4.40	2[†]
		4.48	18
C_2H_4O	acetaldehyde	4.6	2[†]
		4.59	18
	ethylene oxide	4.43	18
$C_2H_4O_2$	acetic acid	5.1	2[†]
	methyl formate	5.05	27
$C_2H_4O_4$	formic acid dimer	12.7	2
C_2H_5NO	acetamide	5.67	18
	N-methyl formamide	5.91	18
$C_2H_5NO_2$	nitroethane	9.63	2
	ethyl nitrite	7.0	15
C_2H_6O	ethanol	5.41	2
		5.11	18
	methyl ether	5.29	20

Molecule	Name	Polarizability	Ref.
		5.84	2
		5.16	15
$C_2H_6O_2$	ethylene glycol	5.7	2[†]
		5.61	27
$C_2H_6O_2S$	dimethyl sulfone	7.3	2[†]
C_2H_6S	ethanethiol	7.41	2
C_2H_7N	ethyl amine	7.10	2
	dimethyl amine	6.37	2
		5.90*	33
$C_2H_8N_2$	ethylene diamine	7.2	2[†]
$C_3H_2N_2$	malononitrile	5.79	18
C_3H_3N	acrylonitrile	8.05	2
$C_3H_4N_2$	pyrazole	7.23	27
C_3H_4O	propenal	6.38	2[†]
C_3H_5N	propionitrile	6.70	2
		6.24	18
		6.27*	32
C_3H_6O	acetone	6.33	15
		6.4	2[†]
		6.39	18
	allyl alcohol	7.65	2
	propionaldehyde	6.50	2
$C_3H_6O_2$	propionic acid	6.9	2[†]
	ethyl formate	8.01	2
		6.88	27
	methyl acetate	6.94	2
		6.81	27
$C_3H_6O_3$	dimethyl carbonate	7.7	2[†]
C_3H_7NO	N-methyl acetamide	7.82	18
	N,N-dimethyl formamide	7.81	18
$C_3H_7NO_2$	nitropropane	8.5	2[†]
C_3H_8O	2-propanol	7.61	2
		6.97	18
	1-propanol	6.74	2
	ethyl methyl ether	7.93	2
$C_3H_8O_2$	dimethoxymethane	7.7	2[†]
	ethylene glycol monomethyl ether	7.44	27
C_3H_9N	propylamine	7.70	27
		9.20	2
	isopropylamine	7.77	27
	trimethylamine	8.15	2
		7.78*	33
$C_4H_2N_2$	fumaronitrile	11.8	2
$C_4H_4N_2$	succinonitrile	8.1	2[†]
	pyrimidine	8.53*	17
	pyridazine	9.27*	17
$C_4H_4O_2$	diketene	8.0	2[†]
C_4H_4S	thiophene	9.67	2
C_4H_5N	methacrylonitrile	8.0	2[†]
	trans-crotononitrile	8.2	2[†]
$C_4H_6N_2$	N-methylpyrazole	8.99	27
C_4H_6O	crotonaldehyde	8.5	2[†]
	methacrylaldehyde	8.3	2[†]
$C_4H_6O_2$	biacetyl	8.2	2[†]
$C_4H_6O_3$	acetic anhydride	8.9	2[†]
C_4H_6S	divinyl sulfide	10.9	2[†]
C_4H_7N	butyronitrile	8.4	2[†]
	isobutyronitrile	8.05	18
		8.05*	32
C_4H_8O	butanal	8.2	2[†]
	methyl ethyl ketone	8.13	15
	trans-2,3-epoxy butane	8.22*	17
$C_4H_8O_2$	ethyl acetate	9.7	2
		8.62	27

Molecule	Name	Polarizability	Ref.
	1,4-dioxane	10.0	2
	p-dioxane	8.60	18
	2-methyl-1,3-dioxolane	9.44	15
	butyric acid	8.58	27
	methyl propionate	8.97	27
$C_4H_9NO_2$	1-nitrobutane	10.4	2[†]
	2-methyl-2-nitropropane	10.3	2[†]
$C_4H_{10}O$	ethyl ether	10.2	2
		8.73	15
	1-butanol	8.88	2
	2-methylpropanol	8.92	2
	methyl propyl ether	8.86	27
	ethylene glycol monoethyl ether	9.28	27
$C_4H_{10}S$	ethyl sulfide	10.8	2
$C_4H_{11}N$	butylamine	13.5	2
	diethylamine	10.2	2
		9.61	27
C_5H_5N	pyridine	9.5	15
		9.18	27
	4-cyano-1,3-butadiene	10.5	2[†]
$C_5H_8N_2$	1,5-dimethylpyrazole	10.72	27
$C_5H_8O_2$	acetyl acetone	10.5	2[†]
C_5H_9N	valeronitrile	10.4	2
	22-DMPN	9.59	18
$C_5H_{10}O$	diethyl ketone	9.93	15
	methyl propyl ketone	9.93	15
$C_5H_{10}O_2$	ethyl propionate	10.41	27
	methyl butanoate	10.41	27
$C_5H_{10}O_3$	diethyl carbonate	11.3	2
$C_5H_{12}O$	ethyl propyl ether	10.68	27
$C_5H_{12}O_4$	tetramethyl orthocarbonate	13.0	2[†]
$C_6H_4N_2O_4$	p-dinitrobenzene	18.4	2
$C_6H_4O_2$	p-benzoquinone	14.5	2
$C_6H_5NO_2$	nitrobenzene	14.7	2
		12.92	15
C_6H_6O	phenol	11.1	2[†]
		9.94*	17
C_6H_7N	aniline	12.1	2[†]
$C_6H_8N_2$	phenylenediamine	13.8	2[†]
	phenylhydrazine	12.91	27
$C_6H_{10}N_2$	1-ethyl-5-methylpyrazole	12.50	27
$C_6H_{10}O_3$	ethyl acetoacetate	12.9	2[†]
$C_6H_{12}N_2$	dimethylketazine	15.6	2
$C_6H_{12}O$	cyclohexanol	11.56	18
$C_6H_{12}O_2$	amyl formate	14.2	2
$C_6H_{12}O_3$	paraldehyde	17.9	2
$C_6H_{14}O$	propyl ether	12.8	2
		12.5	15
$C_6H_{14}O_2$	1,1-diethoxyethane	13.2	2[†]
	1,2-diethoxyethane	11.3	2[†]
$C_6H_{15}N$	triethylamine	13.1	2
		13.38	27
	dipropylamine	13.29	27
$C_7H_4N_2O_2$	p-cyanonitrobenzene	19.0	2
C_7H_5N	benzonitrile	12.5	2[†]
$C_7H_7NO_3$	nitroanisole	15.7	2[†]
C_7H_8O	anisole	13.1	2[†]
C_7H_9NO	o-anisidine	14.2	2[†]
$C_7H_{10}N_2$	1,1-methylphenylhydrazine	14.81	27
$C_7H_{14}O$	cyclohexyl methyl ether	13.4	2[†]
	2,4-dimethyl-3-pentanone	13.5	15
$C_7H_{14}O_2$	pentyl acetate	14.9	2
$C_8H_4N_2$	p-dicyanobenzene	19.2	2

Molecule	Name	Polarizability	Ref.		Molecule	Name	Polarizability	Ref.
$C_8H_6N_2$	quinoxaline	15.13	27			2-methylquinoline	18.65	27
C_8H_8O	acetophenone	15.0	2			1-methylisoquinoline	18.28	27
$C_8H_8O_2$	2,5-dimethyl-1,4-benzoquinone	18.8	2		$C_{10}H_{10}Fe$	ferrocene	17.1	26
$C_8H_{10}O$	phenetole	14.9	2		$C_{10}H_{10}N_2$	2,3-dimethylquinoxaline	18.70	27
$C_8H_{11}N$	N-dimethylaniline	16.2	2†		$C_{10}H_{14}BeO_4$	beryllium acetylacetonate	34.1	2
$C_8H_{12}N_2$	1,1-ethylphenylhydrazine	16.62	27		$C_{11}H_8O$	1-naphthaldehyde	19.75	27
$C_8H_{12}O_2$	ethyl sorbate	17.2	2†			2-naphthaldehyde	20.06	27
	tetramethylcyclobutane-1,3-dione	18.6	2		$C_{12}H_8N_2$	phenazine	23.43	27
$C_8H_{14}O_4$	diethyl succinate	16.8	2†		$C_{12}H_9NO_3$	4-nitrodiphenyl ether	24.7	2†
$C_8H_{18}O$	butyl ether	17.2	2		$C_{14}H_8O_2$	anthraquinone	24.46	27
C_9H_7N	quinoline	15.70	27		$C_{14}H_{14}O$	di-p-tolyl ether	24.9	2†
	isoquinoline	16.43	27		$C_{15}H_{21}AlO_6$	aluminum acetylacetonate	51.9	2
$C_9H_{10}O_2$	ethyl benzoate	16.9	2†		$C_{15}H_{21}CrO_6$	chromium acetylacetonate	53.7	2
$C_9H_{21}N$	tripropylamine	18.87	27		$C_{15}H_{21}FeO_6$	ferric acetylacetonate	58.1	2
$C_{10}H_9N$	α-naphthylamine	19.50	27		$C_{20}H_{28}O_8Th$	thorium acetylacetonate	79.0	2
	β-naphthylamine	19.73	27		C_{60}	buckminsterfullerene	76.5	24
							79	31

Note: All polarizabilities in the tables are experimental values except those values marked by an asterisk (*), which indicates a calculated result. The experimental polarizabilities are mostly determined by measurements of a dielectric constant or refractive index which are quite accurate (0.5% or better). However, one should treat many of the results with several percent of caution because of the age of the data and because some of the results refer to optical frequencies rather than static. Comments given with the references are intended to allow one to judge the degree of caution required. Interested persons should consult these references. In many cases, the reference given is to a theoretical paper in which the experimental results are quoted. These papers, noted in the References, contain valuable information on polarizability calculations and experimental data which often includes the tensor components of the polarizability.

An emperical additive formula for molecular polarizabilities at 589 nm frequency has been given in Bosque, R., and Sales, J., *J. Chem. Inf. Comput. Sci.* 42, 1154, 2002: $\alpha = 0.32 + 1.51\#C + 0.17\#H + 0.57\#O + 1.05\#N + 2.99\#S + 2.48\#P + 2.16\#Cl + 3.29\#Br + 5.45\#I$, where "#C" denotes the number of carbon atoms in the molecule, etc.

References

Kagawa, H., Ichimura, A., Kamka, N. A., and Mori, K., J. Mol. Structure (Theochem) 546, 127, 2001. Parameters were developed for rapid estimation of molecular polarizabilities; this paper contains references for the measured polarizabilities of 371 molecules.

1. McCullough, E. A., Jr., *J. Chem. Phys.*, 63, 5050, 1975. This calculation is for the parallel component, not the average polarizability.

2. Maryott, A. A., and Buckley, F., *U. S. National Bureau of Standards Circular No. 537*, 1953. A tabulation of dipole moments, dielectric constants, and molar refractions measured between 1910 and 1952, and used here to determine polarizabilities if no more recent result exists. The polarizability is $3/(4\pi N_A)$ times the molar polarization or molar refraction, where N_A is Avogadro's number. The value $3/(4\pi N_A)$ = 0.3964308 x 10^{-24} cm³ was used for this conversion. A dagger (†) following the reference number in the tables indicates that the polarizability was derived from the molar refraction and hence may not include some low-frequency contributions to the static polarizability; these "static" polarizabilities are therefore low by 1 to 30%.

3. Hirschfelder, J. O., Curtis, C. F., and Bird, R. B., *Molecular Theory of Gases and Liquids*, Wiley, New York, 1954, p. 950. Fundamental information on molecular polarizabilities.

4. Miller, T. M., and Bederson, B., *Adv. At. Mol. Phys.*, 13, 1, 1977. Review emphasizing atomic polarizabilities and measurement techniques. The data quoted in Table 3 are accurate to 8 to 12%.

5. Kolos, W., and Wolniewicz, L., *J. Chem. Phys.*, 46, 1426, 1967. Highly accurate molecular hydrogen calculations. See also recent work by Machado, A.M., and Masilli, M., *J. Chem. Phys.* 120, 7505, 2004.

6. Newell, A. C., and Baird, R. C., *J. Appl. Phys.*, 36, 3751, 1965. Highly accurate refractive index measurements at 47.7 GHz (essentially static).

7. Jao, T. C., Beebe, N. H. F., Person, W. B., and Sabin, J. R., *Chem. Phys. Lett.*, 26, 474, 1974. Tensor polarizabilities, derivatives, and other results are reported.

8. Orcutt, R. H., and Cole, R. H., *J. Chem. Phys.*, 46, 697, 1967 (He, Ne, Ar, Kr, H_2, N_2); Sutter, H., and Cole, R. H., *J. Chem. Phys.*, 52, 132, 1970 (CF_3H, CFH_3, $CClF_3$, $CClH_3$); Bose, T. K., and Cole, R. H., *J. Chem. Phys.*, 52, 140, 1970 (CO_2), and 54, 3829, 1971 (C_2H_4); Nelson, R. D., and Cole, R. H., *J. Chem. Phys.*, 54, 4033, 1971 (SF_6, $CClF_3$); Bose, T. K., Sochanski, J. S., and Cole, R. H., *J. Chem. Phys.*, 57, 3592, 1972 (CH_4, CF_4); Kirouac, S., and Bose, T. K., *J. Chem. Phys.*, 59, 3043, 1973 (N_2O), and 64, 1580, 1976 (He). Highly accurate dielectric constant measurements. These modern data give the most accurate polarizabilities available. A criticism of the interpretation of these data in the case of polar molecules is given in Ref. 20, p. 2905.

9. Huestis, D. L., Technical Report #MP 78-25, SRI International (project PYU 6158), Menlo Park, CA 94025. Molar refractions for mercury-chlorine compounds are analyzed.

10. Bounds, D. G., Clarke, J. H. R., and Hinchliffe, A., *Chem. Phys. Lett.*, 45, 367, 1977. Theoretical tensor polarizability for LiCl.

11. Kolker, H. J., and Karplus, M., *J. Chem. Phys.*, 39, 2011, 1963. Theoretical.

12. Cutschick, V. P., and McKoy, V., *J. Chem. Phys.*, 58, 2397, 1973. Theoretical tensor polarizabilities.

13. Gready, J. E., Bacskay, G. B., and Hush, N. S., *Chem. Phys.*, 22, 141, 1977, and 23, 9, 1977. Theoretical.

14. Amos, A. T., and Yoffe, J. A., *J. Chem. Phys.*, 63, 4723, 1975. Theoretical.

15. Stuart, H. A., *Landolt-Börnstein Zahlenwerte and Funktionen*, Vol. 1, Part 3, Eucken, A., and Hellwege, K. H., Eds., Springer-Verlag, Berlin, 1951, p. 511. Tabulation of molecular polarizabilities. Two misprints in the chemical symbols have been corrected.

16. Guella, T., Miller, T. M., Stockdale, J. A. D., Bederson, B., and Vuskovic, L., *J. Chem. Phys.*, 94, 6857, 1991. Beam measurements with accuracies between 12-24%.

17. Marchese, F. T., and Jaff, H. H., *Theoret. Chim. Acta* (Berlin), 45, 241, 1977. Theoretical and experimental tensor polarizabilities are tabulated in this paper.

18. Applequist, J., Carl, J. R., and Fung, K.-K., *J. Am. Chem. Soc.*, 94, 2952, 1972. Excellent reference on the calculation of molecular polarizabilities, including extensive tables of tensor polarizabilities, both theoretical and experimental, at 589.3 nm wavelength.

19. Bridge, N. J., and Buckingham, A. D., *Proc. Roy. Soc.* (London), A295, 334, 1966. Measured tensor polarizabilities at 633 nm wavelength.

20. Barnes, A. N. M., Turner, D. J., and Sutton, L. E., *Trans. Faraday Soc.*, 67, 2902, 1971. Dielectric constants yielding polarizabilities accurate from 0.3-8%.

21. Rayane, D., Allouche, A. R., Benichou, E., Antoine, R., Aubert-Frecon, M., Dugourd, Ph., Broyer, M., Ristori, C., Chandezon, F., Huber, B. A., and Guet, C., Eur. Phys. J. D 9, 243, 1999. See also Knight, W. D., Clemenger, K., de Heer, W. A., and Saunders, W. A., Phys. Rev. B 31, 2539, 1985. These data probably correspond to a very low internal temperature.

22. Tarnovsky, V., Bunimovicz, M., Vuskovic, L., Stumpf, B., and Bederson, B., *J. Chem. Phys.*, 98, 3894, 1993. These data correspond to internal temperatures 480-948 K.

23. Milani, P., Moullet, I., and de Heer, W. A., *Phys. Rev. A*, 42, 5150, 1990. Beam measurements accurate to 11%.

24. Antoine, R., Dugourd, P., Rayane, D., Benichou, E., Broyer, M., Chandezon, F., and Guet, C., *J. Chem. Phys.*, 110, 9771, 1999.

25. Aroney, M. J., and Pratten, S. J., *J. Chem. Soc., Faraday Trans.*, 1, 80, 1201, 1984. Uncertainties in the range 1-3%.

26. Le Fevre, R. J. W., Murthy, D. S. N., and Saxby, J. D., *Aust. J. Chem.*, 24, 1057, 1971. Kerr effect.

27. No, K. T., Cho, K. H., Jhon, M. S., and Scheraga, H. A., J. Am. Chem. Soc., 115, 2005, 1993. Theoretical; these results are quoted in numerous valuable papers on calculated polarizabilities, e.g., Miller, K. J., and Savchik, J. A., *J. Am. Chem. Soc.*, 101, 7206, 1979.

28. Schlecht, S., Schäfer, R., Woenckhaus, J., Becker, J. A., *Chem. Phys. Lett.*, 246, 315, 1995.

29. Benichou, E., Antoine, R., Rayane, D., Vezin, B., Dalby, F. W., Dugourd, P., Ristori, C., Chandezon, F., Huber, B. A., Rocco, J. C., Blundell, S. A., and Guet, C., *Phys. Rev. A* 59, R1, 1999. See also Rayane, D., Allouche, A. R., Benichou, E., Antoine, R., Aubert-Frecon, M., Dugourd, Ph., Broyer, M., Ristori, C., Chandezon, F., Huber, B. A., and Guet, C., *Eur. Phys. J.* D 9, 243, 1999.

30. Antoine, R., Rayane, D., Allouche, A. R., Aubert-Frécon, M., Benichou, E., Dalby, F. W., Dugourd, P., Broyer, M., and Guet, C., *J. Chem. Phys.*, 110, 5568, 1999.

31. Ballard, A., Bonin, K., and Louderback, J., *J. Chem. Phys.* 113, 5732, 2000.

32. Ritchie, G.L.D., and Watson, J.N., *J. Phys. Chem. A* 108, 4515, 2004. These measurements are at 632.8 nm frequency, and are stated accurate to 0.4%.

33. Ritchie, G.L.D., and Blanch, E.W., *J. Phys. Chem. A* 107, 2093, 2003. These measurements are at 632.8 nm frequncy, and are stated accurate to better than 1%.

IONIZATION ENERGIES OF ATOMS AND ATOMIC IONS

The ionization energies (often called ionization potentials) of neutral and partially ionized atoms are listed in this table. Data were obtained from the compilations cited below, supplemented by results from the recent research literature. Values for the first and second ionization energies come from Reference 6. All values are given in electron volts (eV).

Following the traditional spectroscopic notation, columns are headed I, II, III, etc. up to XXX, where I indicates the neutral atom, II the singly ionized atom, III the doubly ionized atom, etc. The first section of the table includes spectra I to VIII of all the elements through rutherfordium; subsequent sections cover higher spectra (ionization stages) for those elements for which data are available.

References

1. Moore, C. E., *Ionization Potentials and Ionization Limits Derived from the Analysis of Optical Spectra*, Natl. Stand. Ref. Data Ser. — Natl. Bur. Stand. (U.S.) No. 34, 1970.
2. Martin, W. C., Zalubas, R., and Hagan, L., *Atomic Energy Levels — The Rare Earth Elements*, Natl. Stand. Ref. Data Ser. — Natl. Bur. Stand. (U.S.), No. 60, 1978.
3. Sugar, J. and Corliss, C., *Atomic Energy Levels of the Iron Period Elements: Potassium through Nickel, J. Phys. Chem. Ref. Data*, Vol.14, Suppl. 2, 1985.
4. References to papers in *J. Phys. Chem. Ref. Data*, in the period 1973—91 covering other elements may be found in the cumulative index to that journal.
5. Martin, W.C., and Wiese, W.L., in *Atomic, Molecular, and Optical Physics Handbook*, Drake, G.W.F., Ed., AIP Press, New York, 1996.
6. Sansonetti, J. E., Martin, W. C., and Young, S. L., Handbook of Basic Atomic Spectroscopic Data (version 1.1), NIST Physical Data web site <http://physics.nist.gov/PhysRefData/Handbook> (October 2004); *J. Phys. Chem. Ref. Data* (to be published).

Neutral Atoms to +7 Ions

Z	Element	I	II	III	IV	V	VI	VII	VIII
1	H	13.598443							
2	He	24.587387	54.417760						
3	Li	5.391719	75.6400	122.45429					
4	Be	9.32270	18.21114	153.89661	217.71865				
5	B	8.29802	25.1548	37.93064	259.37521	340.22580			
6	C	11.26030	24.3833	47.8878	64.4939	392.087	489.99334		
7	N	14.5341	29.6013	47.44924	77.4735	97.8902	552.0718	667.046	
8	O	13.61805	35.1211	54.9355	77.41353	113.8990	138.1197	739.29	871.4101
9	F	17.4228	34.9708	62.7084	87.1398	114.2428	157.1651	185.186	953.9112
10	Ne	21.56454	40.96296	63.45	97.12	126.21	157.93	207.2759	239.0989
11	Na	5.139076	47.2864	71.6200	98.91	138.40	172.18	208.50	264.25
12	Mg	7.646235	15.03527	80.1437	109.2655	141.27	186.76	225.02	265.96
13	Al	5.985768	18.82855	28.44765	119.992	153.825	190.49	241.76	284.66
14	Si	8.15168	16.34584	33.49302	45.14181	166.767	205.27	246.5	303.54
15	P	10.48669	19.7695	30.2027	51.4439	65.0251	220.421	263.57	309.60
16	S	10.36001	23.33788	34.79	47.222	72.5945	88.0530	280.948	328.75
17	Cl	12.96763	23.8136	39.61	53.4652	67.8	97.03	114.1958	348.28
18	Ar	15.759610	27.62966	40.74	59.81	75.02	91.009	124.323	143.460
19	K	4.3406633	31.63	45.806	60.91	82.66	99.4	117.56	154.88
20	Ca	6.11316	11.87172	50.9131	67.27	84.50	108.78	127.2	147.24
21	Sc	6.56149	12.79977	24.75666	73.4894	91.65	110.68	138.0	158.1
22	Ti	6.82812	13.5755	27.4917	43.2672	99.30	119.53	140.8	170.4
23	V	6.74619	14.618	29.311	46.709	65.2817	128.13	150.6	173.4
24	Cr	6.76651	16.4857	30.96	49.16	69.46	90.6349	160.18	184.7
25	Mn	7.43402	15.6400	33.668	51.2	72.4	95.6	119.203	194.5
26	Fe	7.9024	16.1877	30.652	54.8	75.0	99.1	124.98	151.06
27	Co	7.88101	17.084	33.50	51.3	79.5	102.0	128.9	157.8
28	Ni	7.6398	18.16884	35.19	54.9	76.06	108	133	162
29	Cu	7.72638	20.2924	36.841	57.38	79.8	103	139	166
30	Zn	9.394199	17.96439	39.723	59.4	82.6	108	134	174
31	Ga	5.999301	20.51514	30.71	64				
32	Ge	7.89943	15.93461	34.2241	45.7131	93.5			
33	As	9.7886	18.5892	28.351	50.13	62.63	127.6		
34	Se	9.75239	21.19	30.8204	42.9450	68.3	81.7	155.4	
35	Br	11.8138	21.591	36	47.3	59.7	88.6	103.0	192.8
36	Kr	13.99961	24.35984	36.950	52.5	64.7	78.5	111.0	125.802
37	Rb	4.177128	27.2895	40	52.6	71.0	84.4	99.2	136
38	Sr	5.69485	11.0301	42.89	57	71.6	90.8	106	122.3
39	Y	6.2173	12.224	20.52	60.597	77.0	93.0	116	129
40	Zr	6.63390	13.1	22.99	34.34	80.348			
41	Nb	6.75885	14.0	25.04	38.3	50.55	102.057	125	

Neutral Atoms to +7 Ions

Z	Element	I	II	III	IV	V	VI	VII	VIII
42	Mo	7.09243	16.16	27.13	46.4	54.49	68.8276	125.664	143.6
43	Tc	7.28	15.26	29.54					
44	Ru	7.36050	16.76	28.47					
45	Rh	7.45890	18.08	31.06					
46	Pd	8.3369	19.43	32.93					
47	Ag	7.57623	21.47746	34.83					
48	Cd	8.99382	16.90831	37.48					
49	In	5.78636	18.8703	28.03	54				
50	Sn	7.34392	14.6322	30.50260	40.73502	72.28			
51	Sb	8.60839	16.63	25.3	44.2	56	108		
52	Te	9.0096	18.6	27.96	37.41	58.75	70.7	137	
53	I	10.45126	19.1313	33					
54	Xe	12.12984	20.9750	32.1230					
55	Cs	3.893905	23.15744						
56	Ba	5.211664	10.00383						
57	La	5.5769	11.059	19.1773	49.95	61.6			
58	Ce	5.5387	10.85	20.198	36.758	65.55	77.6		
59	Pr	5.473	10.55	21.624	38.98	57.53			
60	Nd	5.5250	10.72	22.1	40.4				
61	Pm	5.582	10.90	22.3	41.1				
62	Sm	5.6437	11.07	23.4	41.4				
63	Eu	5.67038	11.25	24.92	42.7				
64	Gd	6.14980	12.09	20.63	44.0				
65	Tb	5.8638	11.52	21.91	39.79				
66	Dy	5.9389	11.67	22.8	41.47				
67	Ho	6.0215	11.80	22.84	42.5				
68	Er	6.1077	11.93	22.74	42.7				
69	Tm	6.18431	12.05	23.68	42.7				
70	Yb	6.25416	12.176	25.05	43.56				
71	Lu	5.42586	13.9	20.9594	45.25	66.8			
72	Hf	6.82507	15	23.3	33.33				
73	Ta	7.54957							
74	W	7.86403	16.1						
75	Re	7.83352							
76	Os	8.43823							
77	Ir	8.96702							
78	Pt	8.9588	18.563						
79	Au	9.22553	20.20						
80	Hg	10.4375	18.7568	34.2					
81	Tl	6.108194	20.4283	29.83					
82	Pb	7.41663	15.03248	31.9373	42.32	68.8			
83	Bi	7.2855	16.703	25.56	45.3	56.0	88.3		
84	Po	8.414							
85	At								
86	Rn	10.7485							
87	Fr	4.072741							
88	Ra	5.278423	10.14715						
89	Ac	5.17	11.75						
90	Th	6.3067	11.9	20.0	28.8				
91	Pa	5.89							
92	U	6.1941	10.6						
93	Np	6.2657							
94	Pu	6.0260	11.2						
95	Am	5.9738							
96	Cm	5.9914							
97	Bk	6.1979							
98	Cf	6.2817	11.8						
99	Es	6.42	12.0						
100	Fm	6.50							
101	Md	6.58							
102	No	6.65							
103	Lr	4.9							
104	Rf	6.0							

+8 Ions to +15 Ions

Z	Element	IX	X	XI	XII	XIII	XIV	XV	XVI
9	F	1103.1176							
10	Ne	1195.8286	1362.1995						
11	Na	299.864	1465.121	1648.702					
12	Mg	328.06	367.50	1761.805	1962.6650				
13	Al	330.13	398.75	442.00	2085.98	2304.1410			
14	Si	351.12	401.37	476.36	523.42	2437.63	2673.182		
15	P	372.13	424.4	479.46	560.8	611.74	2816.91	3069.842	
16	S	379.55	447.5	504.8	564.44	652.2	707.01	3223.78	3494.1892
17	Cl	400.06	455.63	529.28	591.99	656.71	749.76	809.40	3658.521
18	Ar	422.45	478.69	538.96	618.26	686.10	755.74	854.77	918.03
19	K	175.8174	503.8	564.7	629.4	714.6	786.6	861.1	968
20	Ca	188.54	211.275	591.9	657.2	726.6	817.6	894.5	974
21	Sc	180.03	225.18	249.798	687.36	756.7	830.8	927.5	1009
22	Ti	192.1	215.92	265.07	291.500	787.84	863.1	941.9	1044
23	V	205.8	230.5	255.7	308.1	336.277	896.0	976	1060
24	Cr	209.3	244.4	270.8	298.0	354.8	384.168	1010.6	1097
25	Mn	221.8	248.3	286.0	314.4	343.6	403.0	435.163	1134.7
26	Fe	233.6	262.1	290.2	330.8	361.0	392.2	457	489.256
27	Co	186.13	275.4	305	336	379	411	444	511.96
28	Ni	193	224.6	321.0	352	384	430	464	499
29	Cu	199	232	265.3	369	401	435	484	520
30	Zn	203	238	274	310.8	419.7	454	490	542
36	Kr	230.85	268.2	308	350	391	447	492	541
37	Rb	150	277.1						
38	Sr	162	177	324.1					
39	Y	146.2	191	206	374.0				
42	Mo	164.12	186.4	209.3	230.28	279.1	302.60	544.0	570

+16 Ions to +23 Ions

Z	Element	XVII	XVIII	XIX	XX	XXI	XXII	XXIII	XXIV
17	Cl	3946.2960							
18	Ar	4120.8857	4426.2296						
19	K	1033.4	4610.8	4934.046					
20	Ca	1087	1157.8	5128.8	5469.864				
21	Sc	1094	1213	1287.97	5674.8	6033.712			
22	Ti	1131	1221	1346	1425.4	6249.0	6625.82		
23	V	1168	1260	1355	1486	1569.6	6851.3	7246.12	
24	Cr	1185	1299	1396	1496	1634	1721.4	7481.7	7894.81
25	Mn	1224	1317	1437	1539	1644	1788	1879.9	8140.6
26	Fe	1266	1358	1456	1582	1689	1799	1950	2023
27	Co	546.58	1397.2	1504.6	1603	1735	1846	1962	2119
28	Ni	571.08	607.06	1541	1648	1756	1894	2011	2131
29	Cu	557	633	670.588	1697	1804	1916	2060	2182
30	Zn	579	619	698	738	1856			
36	Kr	592	641	786	833	884	937	998	1051
42	Mo	636	702	767	833	902	968	1020	1082

+24 Ions to +29 Ions

Z	Element	XXV	XXVI	XXVII	XXVIII	XXIX	XXX
25	Mn	8571.94					
26	Fe	8828	9277.69				
27	Co	2219.0	9544.1	10012.12			
28	Ni	2295	2399.2	10288.8	10775.40		
29	Cu	2308	2478	2587.5	11062.38	11567.617	
36	Kr	1151	1205.3	2928	3070	3227	3381
42	Mo	1263	1323	1387	1449	1535	1601

IONIZATION ENERGIES OF GAS-PHASE MOLECULES

Sharon G. Lias

This table presents values for the first ionization energies (IP) of approximately 1000 molecules and atoms. Substances are listed by molecular formula in the modified Hill order (see Preface). Values enclosed in parentheses are considered not to be well established. Data appearing in the 1988 reference were updated in 1996 for inclusion in the database of ionization energies available at the Internet site of the Standard Reference Data program of the National Institute of Standards and Technology (http://webbook.nist.gov). The list appearing here includes these updates.

The list also includes values for enthalpies of formation of the ions at 298 K, $\Delta_f H_{ion}$, given according to the ion convention used by mass spectrometrists; to convert these values to the electron convention used by thermodynamicists, add 6 kJ/mol. Details on the calculation of $\Delta_f H_{ion}$, as well as data for a much larger number of molecules, may be found in the reference and on the Internet site.

Reference

Lias, S. G., Bartmess, J. E., Liebman, J. F., Holmes, J. L., Levin, R. D., and Mallard, W.G., *Gas-Phase Ion and Neutral Thermochemistry*, *J. Phys. Chem. Ref. Data*, Vol. 17, Suppl. No. 1, 1988.

Mol. form.	Name	IP/eV	$\Delta_f H_{ion}$ kJ/mol
Substances not containing carbon			
Ac	Actinium	5.17	905
Ag	Silver	7.57624	1016
AgCl	Silver(I) chloride	($\leq$ 10.08)	$\leq$ 1065
AgF	Silver(I) fluoride	(11.0 ± 0.3)	1071
Al	Aluminum	5.98577	905
AlBr	Aluminum monobromide	(9.3)	913
$AlBr_3$	Aluminum tribromide	(10.4)	593
AlCl	Aluminum monochloride	9.4	855
$AlCl_3$	Aluminum trichloride	(12.01)	573
AlF	Aluminum monofluoride	9.73 ± 0.01	673
AlF_3	Aluminum trifluoride	$\leq$ 15.45	$\leq$ 282
AlI	Aluminum monoiodide	9.3 ± 0.3	965
AlI_3	Aluminum triiodide	(9.1)	673
Am	Americium	5.9738 ± 0.0002	860
Ar	Argon	15.75962	1521
As	Arsenic	9.8152	1250
$AsCl_3$	Arsenic(III) chloride	(10.55 ± 0.025)	754
AsF_3	Arsenic(III) fluoride	(12.84 ± 0.05)	452
AsH_3	Arsine	(9.89)	1021
Au	Gold	9.22567	1254
B	Boron	8.29803	1363
BBr_3	Boron tribromide	(10.51)	809
BCl_3	Boron trichloride	11.60 ± 0.02	718
BF	Fluoroborane	11.12 ± 0.01	957
	Difluoroborane	(9.4)	317
BF_3	Boron trifluoride	15.7 ± 0.3	365
BH	Boron monohydride	(9.77)	1385
BH_3	Borane	12.026 ± 0.024	1261
BI_3	Boron triiodide	(9.25 ± 0.03)	964
BO_2	Boron dioxide	(13.5 ± 0.3)	1001
B_2H_6	Diborane	11.38 ± 0.05	1134
B_2O_3	Boron oxide	13.5 ± 0.15	460
B_4H_{10}	Tetraborane	10.76 ± 0.04	1105
B_5H_9	Pentaborane(9)	9.90 ± 0.04	1028
B_6H_{10}	Hexaborane	(9.0)	965
Ba	Barium	5.21170	683
BaO	Barium oxide	6.91 ± 0.06	543
Be	Beryllium	9.32263	1224
BeO	Beryllium oxide	(10.1 ± 0.4)	1111
Bi	Bismuth	7.2855	908
$BiCl_3$	Bismuth trichloride	(10.4)	736

Mol. form.	Name	IP/eV	$\Delta_f H_{ion}$ kJ/mol
Bk	Berkelium	6.23	911
Br	Bromine (atomic)	11.81381	1252
BrCl	Bromine chloride	11.01	1079
BrF	Bromine fluoride	11.86	1086
BrF_5	Bromine pentafluoride	13.172 ± 0.002	840
BrH	Hydrogen bromide	11.66 ± 0.03	1087
BrH_3Si	Bromosilane	10.6	943
BrI	Iodine bromide	9.790 ± 0.004	986
BrK	Potassium bromide	7.85 ± 0.1	578
BrLi	Lithium bromide	(8.7)	685
BrNO	Nitrosyl bromide	10.17 ± 0.03	1065
BrNa	Sodium bromide	8.31 ± 0.1	660
BrO	Bromine monoxide	10.46 ± 0.02	1135
BrRb	Rubidium bromide	7.94 ± 0.03	583
BrTl	Thallium(I) bromide	9.14 ± 0.02	844
Br_2	Bromine	10.516 ± 0.005	1046
Br_2Hg	Mercury(II) bromide	10.560 ± 0.003	935
Br_2Sn	Tin(II) bromide	9.0	839
Br_3Ga	Gallium(III) bromide	10.40	711
Br_3P	Phosphorus(III) bromide	9.7	798
Br_4Hf	Hafnium(IV) bromide	(10.9)	366
Br_4Sn	Tin(IV) bromide	10.6	709
Br_4Ti	Titanium(IV) bromide	10.3	375
Br_4Zr	Zirconium(IV) bromide	(10.7)	388
Ca	Calcium	6.11316	768
CaCl	Calcium monochloride	5.86 ± 0.07	462
CaO	Calcium oxide	6.66 ± 0.18	668
Cd	Cadmium	8.99367	980
Ce	Cerium	5.5387	957
Cf	Californium	6.30	805
Cl	Chlorine (atomic)	12.96764	1373
ClCs	Cesium chloride	(7.84 ± 0.05)	510
ClF	Chlorine fluoride	12.66 ± 0.01	1171
$ClFO_3$	Perchloryl fluoride	(12.945 ± 0.005)	1224
ClF_2	Chlorine difluoride	(12.77 ± 0.05)	1128
ClF_3	Chlorine trifluoride	(12.65 ± 0.05)	1057
ClF_5S	Sulfur chloride pentafluoride	(12.335 ± 0.005)	144
ClH	Hydrogen chloride	12.749 ± 0.009	1137
ClHO	Hypochlorous acid	(11.12 ± 0.01)	993
ClH_3Si	Chlorosilane	11.4	899
ClI	Iodine chloride	10.088 ± 0.01	991
ClIn	Indium(I) chloride	(9.51)	842
ClK	Potassium chloride	(8.0 ± 0.4)	557
ClLi	Lithium chloride	9.57	727
ClNO	Nitrosyl chloride	10.87 ± 0.01	1099
$ClNO_2$	Nitryl chloride	(11.84)	1155
ClNa	Sodium chloride	8.92 ± 0.06	681
ClO	Chlorine monoxide	10.95	1159
ClO_2	Chlorine dioxide	10.33 ± 0.02	1093
ClRb	Rubidium chloride	(8.50 ± 0.03)	590
ClTl	Thallium(I) chloride	9.70 ± 0.03	869
Cl_2	Chlorine	11.480 ± 0.005	1108
Cl_2CrO_2	Chromyl chloride	11.6	580
Cl_2Ge	Germanium(II) chloride	(10.20 ± 0.05)	813
Cl_2H_2Si	Dichlorosilane	11.4	765
Cl_2Hg	Mercury(II) chloride	11.380 ± 0.003	952
Cl_2O	Chlorine oxide	10.94	1135
Cl_2OS	Thionyl chloride	10.96	844
Cl_2O_2S	Sulfuryl chloride	12.05	807
Cl_2Pb	Lead(II) chloride	(10.2)	791

Mol. form.	Name	IP/eV	$\Delta_f H_{ion}$ kJ/mol
Cl_2S	Sulfur dichloride	9.45 ± 0.03	895
Cl_2Si	Dichlorosilylene	(10.93 ± 0.10)	887
Cl_2Sn	Tin(II) chloride	(10.0)	760
Cl_3Ga	Gallium(III) chloride	11.52	664
Cl_3HSi	Trichlorosilane	(11.7)	648
Cl_3N	Nitrogen trichloride	(10.12 ± 0.1)	1244
Cl_3OP	Phosphorus(V) oxychloride	11.36 ± 0.02	540
Cl_3OV	Vanadyl trichloride	(11.6)	425
Cl_3P	Phosphorus(III) chloride	9.91	668
Cl_3PS	Phosphorus(V) sulfide trichloride	9.71 ± 0.03	573
Cl_3Sb	Antimony(III) chloride	(≤ 10.7)	s719
Cl_4Ge	Germanium(IV) chloride	11.68 ± 0.05	629
Cl_4Hf	Hafnium(IV) chloride	(11.7)	246
Cl_4Si	Tetrachlorosilane	11.79 ± 0.01	527
Cl_4Sn	Tin(IV) chloride	11.7 ± 0.2	656
Cl_4Ti	Titanium(IV) chloride	(11.5)	349
Cl_4V	Vanadium(IV) chloride	(9.2)	361
Cl_4Zr	Zirconium(IV) chloride	(11.2)	210
Cl_5Mo	Molybdenum(V) chloride	(8.7)	392
Cl_5Nb	Niobium(V) chloride	(10.97)	356
Cl_5P	Phosphorus(V) chloride	(10.2)	608
Cl_5Ta	Tantalum(V) chloride	(11.08)	303
Cl_6W	Tungsten(VI) chloride	(9.5)	348
Cm	Curium	6.02	966
Co	Cobalt	7.8810	1187
Cr	Chromium	6.76664	1050
Cs	Cesium	3.89390	452
CsF	Cesium fluoride	(8.80 ± 0.10)	489
CsNa	Cesium sodium	(4.05 ± 0.04)	535
Cu	Copper	7.72638	1084
CuF	Copper(I) fluoride	10.15 ± 0.02	984
Dy	Dysprosium	5.9389	862
Er	Erbium	6.1078	907
Es	Einsteinium	6.42	753
Eu	Europium	5.6704	723
F	Fluorine (atomic)	17.42282	1761
FGa	Gallium monofluoride	(9.6 ± 0.5)	700
FH	Hydrogen fluoride	16.044 ± 0.003	1276
FHO	Hypofluorous acid	12.71 ± 0.01	1130
FH_3Si	Fluorosilane	11.7	752
FI	Iodine fluoride	10.54 ± 0.01	922
FIn	Indium monofluoride	(9.6 ± 0.5)	740
FNO	Nitrosyl fluoride	12.63 ± 0.03	1152
FNO_2	Nitryl fluoride	(13.09)	1154
FNS	Thionitrosyl fluoride (NSF)	11.51 ± 0.04	1090
FO	Fluorine monoxide	12.78 ± 0.03	1342
FO_2	Fluorine superoxide (FOO)	(12.6 ± 0.2)	1228
FS	Sulfur fluoride	10.09	986
FTl	Thallium(I) fluoride	10.52	835
F_2	Fluorine	15.697 ± 0.003	1515
F_2Ge	Germanium(II) fluoride	(≤ 11.65)	551
F_2HN	Difluoramine	(11.53 ± 0.08)	1046
F_2H_2Si	Difluorosilane	(12.2)	386
F_2Mg	Magnesium fluoride	(13.6 ± 0.3)	588
F_2N	Difluoroamidogen	11.628 ± 0.01	1155
F_2N_2	trans-Difluorodiazine	(12.8)	1315
F_2O	Fluorine monoxide	13.11 ± 0.01	1290
F_2OS	Thionyl fluoride	12.25	688
F_2O_2S	Sulfuryl fluoride	13.04 ± 0.01	501
F_2Pb	Lead(II) fluoride	(11.5)	679

Mol. form.	Name	IP/eV	$\Delta_f H_{ion}$ kJ/mol
F_2S	Sulfur difluoride	(10.08)	676
F_2Si	Difluorosilylene	10.78 ± 0.05	450
F_2Sn	Tin(II) fluoride	(11.1)	586
F_2Xe	Xenon difluoride	12.35 ± 0.01	1083
F_3HSi	Trifluorosilane	(14.0)	150
F_3N	Nitrogen trifluoride	13.00 ± 0.02	1125
F_3NO	Trifluoramine oxide	13.31 ± 0.06	1121
F_3OP	Phosphorus(V) oxyfluoride	12.76 ± 0.01	−24
F_3P	Phosphorus(III) fluoride	11.60 ± 0.05	161
F_3PS	Phosphorus(V) sulfide trifluoride	≤ 11.05 ± 0.035	≤ 58
F_3Si	Trifluorosilyl	(9.99)	− 32
F_4Ge	Germanium(IV) fluoride	(15.5)	307
F_4N_2	Tetrafluorohydrazine	11.94 ± 0.03	1119
F_4S	Sulfur tetrafluoride	12.0 ± 0.3	399
F_4Si	Tetrafluorosilane	15.24 ± 0.14	−144
F_4Xe	Xenon tetrafluoride	12.65 ± 0.1	1016
F_5I	Iodine pentafluoride	12.943 ± 0.005	408
F_5P	Phosphorus(V) fluoride	(15.1)	−137
F_5S	Sulfur pentafluoride	9.60 ± 0.05	10
F_6Mo	Molybdenum(VI) fluoride	(14.5 ± 0.1)	−159
F_6S	Sulfur hexafluoride	15.32 ± 0.02	258
F_6U	Uranium(VI) fluoride	14.00 ± 0.10	−796
Fe	Iron	7.9024	1177
Fm	Fermium	6.50	627
Ga	Gallium	5.99930	851
GaI_3	Gallium(III) iodide	9.40	765
Gd	Gadolinium	6.1500	991
Ge	Germanium	7.900	1139
GeH_4	Germane	≤ 10.53	≤ 1108
GeI_4	Germanium(IV) iodide	(9.42)	850
GeO	Germanium(II) oxide	11.25 ± 0.01	1044
GeS	Germanium(II) sulfide	(9.98)	1055
H	Hydrogen (atomic)	13.59844	1530
HI	Hydrogen iodide	10.386 ± 0.001	1028
HLi	Lithium hydride	7.7	882
HN	Imidogen	≤ 13.49 ± 0.01	1678
HNO	Nitrosyl hydride	(10.1)	1075
HNO_2	Nitrous acid	≤ 11.3	≤ 1011
HNO_3	Nitric acid	11.95 ± 0.01	1019
HN_3	Hydrazoic acid	10.72 ± 0.025	1328
HO	Hydroxyl	13.0170 ± 0.0002	1294
HO_2	Hydroperoxy	11.35 ± 0.01	1106
HS	Mercapto	10.4219 ± 0.0004	1145
H_2	Hydrogen	15.42593 ± 0.00005	1488
H_2N	Amidogen	11.14 ± 0.01	1264
H_2O	Water	12.6206 ± 0.0020	976
H_2O_2	Hydrogen peroxide	10.58 ± 0.04	885
H_2S	Hydrogen sulfide	10.457 ± 0.012	989
H_2Se	Hydrogen selenide	9.892 ± 0.005	984
H_2Si	Silylene	8.244 ± 0.025	1084
H_3N	Ammonia	10.070 ± 0.020	925
H_3NO	Hydroxylamine	(10.00)	923
H_3P	Phosphine	9.869 ± 0.002	958
H_3Sb	Stibine	9.54 ± 0.03	1067
H_4N_2	Hydrazine	8.1 ± 0.15	877
H_4Si	Silane	11.00 ± 0.02	1095
H_4Sn	Stannane	(10.75)	1200
H_6Si_2	Disilane	9.74 ± 0.02	1019
H_8Si_3	Trisilane	(9.2)	1009
He	Helium	24.58741	2372

Mol. form.	Name	IP/eV	$\Delta_f H_{ion}$ kJ/mol
Hf	Hafnium	6.82507 ± 0.00004	1278
Hg	Mercury	10.43750	1069
HgI$_2$	Mercury(II) iodide	9.5088 ± 0.0022	900
Ho	Holmium	6.0216	882
I	Iodine (atomic)	10.45126	1115
IK	Potassium iodide	(7.21 ± 0.3)	570
ILi	Lithium iodide	(7.5)	633
INa	Sodium iodide	7.64 ± 0.02	659
ITl	Thallium(I) iodide	8.47 ± 0.02	826
I$_2$	Iodine	9.3074 ± 0.0002	960
I$_4$Ti	Titanium(IV) iodide	(9.1)	602
I$_4$Zr	Zirconium(IV) iodide	(9.3)	500
In	Indium	5.78636	802
Ir	Iridium	9.1	1543
K	Potassium	4.34066	508
KLi	Lithium potassium	4.57 ± 0.04	512
KNa	Potassium sodium	4.41636 ± 0.00017	561
K$_2$	Dipotassium	4.0637 ± 0.0002	519
Kr	Krypton	13.99961	1351
La	Lanthanum	5.5770	969
Li	Lithium	5.39172	680
LiNa	Lithium sodium	5.05 ± 0.04	571
LiO	Lithium monoxide	(8.44)	894
LiRb	Lithium rubidium	4.3 ± 0.1	486
Li$_2$	Dilithium	5.1127 ± 0.0003	709
Lu	Lutetium	5.42585	950
Md	Mendelevium	6.58	635
Mg	Magnesium	7.64624	885
MgO	Magnesium oxide	(8.76 ± 0.22)	901
Mn	Manganese	7.43402	998
Mo	Molybdenum	7.09243	1343
N	Nitrogen (atomic)	14.53414	1875
NO	Nitric oxide	9.26438 ± 0.00005	985
NO$_2$	Nitrogen dioxide	9.586 ± 0.002	958
NP	Phosphorus nitride	11.84 ± 0.04	1247
NS	Nitrogen sulfide	8.87 ± 0.01	1119
N$_2$	Nitrogen	15.5808	1503
N$_2$O	Nitrous oxide	12.886	1325
N$_2$O$_4$	Nitrogen tetroxide	(10.8)	1050
N$_2$O$_5$	Nitrogen pentoxide	(11.9)	1161
Na	Sodium	5.13908	603
NaRb	Rubidium sodium	4.32 ± 0.04	480
Na$_2$	Disodium	4.894 ± 0.003	614
Nb	Niobium	6.75885	1384
Nd	Neodymium	5.5250	859
Ne	Neon	21.56454	2081
Ni	Nickel	7.6398	1167
No	Nobelium	6.65	642
Np	Neptunium	6.2657 ± 0.0003	1069
O	Oxygen (atomic)	13.61806	1563
OPb	Lead(II) oxide	9.08 ± 0.10	939
OS	Sulfur monoxide	10.294 ± 0.004	998
OS$_2$	Sulfur oxide (SSO)	10.584 ± 0.005	971
OSi	Silicon monoxide	11.49 ± 0.20	1008
OSn	Tin(II) oxide	9.60 ± 0.02	944
OSr	Strontium oxide	6.6 ± 0.2	623
O$_2$	Oxygen	12.0697 ± 0.0002	1165
O$_2$S	Sulfur dioxide	12.349 ± 0.001	894
O$_2$Th	Thorium(IV) oxide	(8.7 ± 0.15)	342
O$_2$Ti	Titanium(IV) oxide	(9.54 ± 0.1)	623

Mol. form.	Name	IP/eV	$\Delta_f H_{ion}$ kJ/mol
O_2U	Uranium(IV) oxide	(5.4 ± 0.1)	57
O_3	Ozone	12.43	1342
O_3S	Sulfur trioxide	12.82 ± 0.03	841
O_3U	Uranium(VI) oxide	(10.5 ± 0.5)	214
O_4Os	Osmium(VIII) oxide	(12.32)	850
O_4Ru	Ruthenium(VIII) oxide	12.15 ± 0.03	988
O_7Re_2	Rhenium(VII) oxide	(12.7 ± 0.2)	125
Os	Osmium	8.7	1630
P	Phosphorus	10.48669	1328
P_2	Diphosphorus	10.53	1160
Pa	Protactinium	5.89	1133
Pb	Lead	7.41666	911
PbS	Lead(II) sulfide	(8.5 ± 0.5)	954
Pd	Palladium	8.3367	1181
Pm	Promethium	5.55	536
Pr	Praseodymium	5.464	883
Pt	Platinum	9.0	1433
Pu	Plutonium	6.025	926
Ra	Radium	5.27892	668
Rb	Rubidium	4.17713	484
Re	Rhenium	7.88	1530
Rh	Rhodium	7.45890	1276
Rn	Radon	10.74850	1037
Ru	Ruthenium	7.36050	1355
S	Sulfur	10.36001	1277
SSn	Tin(II) sulfide	(8.8)	966
S_2	Disulfur	9.356 ± 0.002	1031
Sb	Antimony	8.64	1096
Sc	Scandium	6.56144	1010
Se	Selenium	9.75238	1168
Si	Silicon	8.15169	1238
Sm	Samarium	5.6437	751
Sn	Tin	7.34381	1011
Sr	Strontium	5.69484	713
Ta	Tantalum	7.89	1544
Tb	Terbium	5.8639	955
Tc	Technetium	7.28	1380
Te	Tellurium	9.0096	1066
Th	Thorium	6.308 ± 0.003	1207
Ti	Titanium	6.8282	1127
Tl	Thallium	6.10829	771
Tm	Thulium	6.18431	827
U	Uranium	6.19405	1129
V	Vanadium	6.746 ± 0.002	1166
W	Tungsten	7.98	1621
Xe	Xenon	12.12987	1170
Y	Yttrium	6.217	1022
Yb	Ytterbium	6.25416	754
Zn	Zinc	9.39405	1037
Zr	Zirconium	6.63390	1251

Substances containing carbon

C	Carbon	11.26030	1803
$CBrClF_2$	Bromochlorodifluoromethane	(11.21)	642
$CBrCl_3$	Bromotrichloromethane	(10.6)	980
$CBrF_3$	Bromotrifluoromethane	(11.40)	451
CBr_2F_2	Dibromodifluoromethane	11.03 ± 0.04	683
CBr_4	Tetrabromomethane	(10.31 ± 0.02)	1079
CCl	Chloromethylidyne	(8.9 ± 0.2)	1244
$CClF_3$	Chlorotrifluoromethane	12.6 ± 0.2	505

Mol. form.	Name	IP/eV	$\Delta_f H_{ion}$ kJ/mol
CClN	Cyanogen chloride	12.34 ± 0.01	1329
CCl$_2$	Dichloromethylene	(9.27)	1058
CCl$_2$F$_2$	Dichlorodifluoromethane	12.05 ± 0.24	685
CCl$_2$O	Carbonyl chloride	(11.5)	888
CCl$_3$F	Trichlorofluoromethane	11.77 ± 0.02	868
CCl$_4$	Tetrachloromethane	11.47 ± 0.01	1010
CF	Fluoromethylidyne	9.11 ± 0.01	1134
CFN	Cyanogen fluoride	13.34 ± 0.02	1325
CF$_2$	Difluoromethylene	11.44 ± 0.03	899
CF$_2$O	Carbonyl fluoride	13.035 ± 0.030	617
CF$_3$	Trifluoromethyl	8.7 ± 0.2	379
CF$_3$I	Trifluoroiodomethane	10.23	397
CH	Methylidyne	10.64 ± 0.01	1622
CHBrCl$_2$	Bromodichloromethane	10.6	973
CHBr$_2$Cl	Chlorodibromomethane	10.59 ± 0.01	1030
CHBr$_3$	Tribromomethane	10.48 ± 0.02	1035
CHCl	Chloromethylene	9.84	1247
CHClF$_2$	Chlorodifluoromethane	(12.2)	693
CHCl$_2$F	Dichlorofluoromethane	(11.5)	829
CHCl$_3$	Trichloromethane	11.37 ± 0.02	992
CHF	Fluoromethylene	10.06 ± 0.05	1121
CHF$_3$	Trifluoromethane	(13.86)	643
CHI$_3$	Triiodomethane	9.25 ± 0.02	1010
CHN	Hydrogen cyanide	13.60 ± 0.01	1447
CHN	Hydrogen isocyanide	(12.5 ± 0.1)	1407
CHNO	Isocyanic acid	11.595 ± 0.005	1016
CHNO	Fulminic acid	(10.83)	1263
CHO	Oxomethyl (HCO)	(8.55)	826
CH$_2$	Methylene	10.396 ± 0.003	1392
CH$_2$BrCl	Bromochloromethane	10.77 ± 0.01	1085
CH$_2$Br$_2$	Dibromomethane	(10.50 ± 0.02)	1013
CH$_2$ClF	Chlorofluoromethane	11.71 ± 0.01	870
CH$_2$Cl$_2$	Dichloromethane	11.32 ± .01	996
CH$_2$F$_2$	Difluoromethane	12.71	774
CH$_2$I$_2$	Diiodomethane	9.46 ± 0.02	1030
CH$_2$N$_2$	Diazomethane	8.999 ± 0.001	1098
CH$_2$N$_2$	Cyanamide	(10.4)	1137
CH$_2$O	Formaldehyde	10.88 ± 0.01	941
CH$_2$O$_2$	Formic acid	11.33 ± 0.01	715
CH$_3$	Methyl	9.843 ± 0.002	1095
CH$_3$BO	Borane carbonyl	11.14 ± 0.02	962
CH$_3$Br	Bromomethane	10.541 ± 0.003	979
CH$_3$Cl	Chloromethane	11.22 ± 0.01	1001
CH$_3$Cl$_3$Si	Methyltrichlorosilane	(11.36 ± 0.03)	548
CH$_3$F	Fluoromethane	12.47 ± 0.02	956
CH$_3$I	Iodomethane	9.538	936
CH$_3$NO	Formamide	10.16 ± 0.06	796
CH$_3$NO$_2$	Nitromethane	11.08 ± 0.07	994
CH$_3$N$_3$	Methyl azide	9.81 ± 0.02	1227
CH$_3$O	Methoxy	(10.72)	1050
CH$_4$	Methane	12.61 ± 0.01	1143
CH$_4$N$_2$O	Urea	9.7	690
CH$_4$O	Methanol	10.85 ± 0.01	845
CH$_4$S	Methanethiol	9.44 ± 0.005	888
CH$_5$N	Methylamine	(8.80)	826
CH$_6$N$_2$	Methylhydrazine	7.7 ± 0.15	835
CH$_6$Si	Methylsilane	(10.7)	1003
CN	Cyanide	13.5984	1748
CNO	Cyanate	11.76 ± 0.01	1290
CO	Carbon monoxide	14.014 ± 0.0003	1242

Mol. form.	Name	IP/eV	$\Delta_f H_{ion}$ kJ/mol
COS	Carbon oxysulfide	11.18 ± 0.01	936
COSe	Carbon oxyselenide	10.36 ± 0.01	929
CO_2	Carbon dioxide	13.773 ± 0.002	935
CS	Carbon sulfide	11.33 ± 0.01	1361
CS_2	Carbon disulfide	10.0685 ± 0.0020	1089
C_2	Dicarbon	(11.4 ± 0.3)	2000
$C_2Br_2F_4$	1,2-Dibromotetrafluoroethane	(11.1)	280
C_2ClF_3	Chlorotrifluoroethylene	9.81 ± 0.03	373
C_2ClF_5	Chloropentafluoroethane	(12.6)	99
C_2Cl_2	Dichloroacetylene	9.9	1165
$C_2Cl_2F_4$	1,2-Dichlorotetrafluoroethane	12.2	252
$C_2Cl_3F_3$	1,1,1-Trichlorotrifluoroethane	11.5	386
$C_2Cl_3F_3$	1,1,2-Trichlorotrifluoroethane	11.99 ± 0.02	429
C_2Cl_4	Tetrachloroethylene	9.326 ± 0.001	887
$C_2Cl_4F_2$	1,1,2,2-Tetrachloro-1,2-difluoroethane	(11.3)	563
C_2Cl_4O	Trichloroacetyl chloride	(11.0)	827
C_2Cl_6	Hexachloroethane	(11.1)	920
C_2F_3N	Trifluoroacetonitrile	13.93 ± 0.07	845
C_2F_4	Tetrafluoroethylene	10.12 ± 0.02	315
C_2F_6	Hexafluoroethane	(13.6)	-30
C_2H	Ethynyl	(11.61 ± 0.07)	1685
C_2HBr	Bromoacetylene	10.31 ± 0.02	1242
$C_2HBrClF_3$	2-Bromo-2-chloro-1,1,1-trifluoroethane	(11.0)	363
C_2HCl	Chloroacetylene	10.58 ± 0.02	1276
C_2HClF_2	1-Chloro-2,2-difluoroethylene	9.80 ± 0.04	628
C_2HCl_3	Trichloroethylene	9.46 ± 0.02	894
C_2HCl_3O	Dichloroacetyl chloride	(10.9)	809
C_2HCl_5	Pentachloroethane	(11.0)	919
C_2HF	Fluoroacetylene	11.26	1195
C_2HF_3	Trifluoroethylene	10.14	489
$C_2HF_3O_2$	Trifluoroacetic acid	11.46	75
C_2H_2	Acetylene	11.400 ± 0.002	1328
$C_2H_2Cl_2$	1,1-Dichloroethylene	9.81 ± 0.04	949
$C_2H_2Cl_2$	cis-1,2-Dichloroethylene	9.66 ± 0.01	936
$C_2H_2Cl_2$	trans-1,2-Dichloroethylene	9.64 ± 0.02	934
$C_2H_2Cl_2O$	Chloroacetyl chloride	(≤ 10.3)	815
$C_2H_2Cl_4$	1,1,1,2-Tetrachloroethane	(11.1)	920
$C_2H_2Cl_4$	1,1,2,2-Tetrachloroethane	(≤ 11.62)	≤ 971
$C_2H_2F_2$	1,1-Difluoroethylene	10.29 ± 0.01	650
$C_2H_2F_2$	cis-1,2-Difluoroethylene	10.23 ± 0.02	690
C_2H_2O	Ketene	9.617 ± 0.003	880
$C_2H_2O_2$	Glyoxal	10.2	773
$C_2H_2S_2$	Thiirene	8.61	892
C_2H_3Br	Bromoethylene	9.83 ± 0.02	1028
C_2H_3Cl	Chloroethylene	9.99 ± 0.02	985
$C_2H_3ClF_2$	1-Chloro-1,1-difluoroethane	(11.98)	626
C_2H_3ClO	Acetyl chloride	10.82 ± 0.04	801
C_2H_3ClO	Chloroacetaldehyde	(10.48)	815
$C_2H_3ClO_2$	Chloroacetic acid	(10.7)	597
$C_2H_3Cl_3$	1,1,1-Trichloroethane	(11.0)	917
$C_2H_3Cl_3$	1,1,2-Trichloroethane	(11.0)	911
C_2H_3F	Fluoroethylene	10.36 ± 0.01	861
C_2H_3FO	Acetyl fluoride	(11.5)	667
$C_2H_3F_3$	1,1,1-Trifluoroethane	13.3 ± 0.5	536
C_2H_3N	Acetonitrile	12.20 ± 0.01	1253
C_2H_3NO	Methylisocyanate	(10.67)	900
C_2H_4	Ethylene	10.5138 ± 0.0006	1067
$C_2H_4Br_2$	1,2-Dibromoethane	10.35 ± 0.04	961
$C_2H_4Cl_2$	1,1-Dichloroethane	11.04 ± 0.02	935
$C_2H_4Cl_2$	1,2-Dichloroethane	11.04 ± 0.02	931

Mol. form.	Name	IP/eV	$\Delta_f H_{ion}$ kJ/mol
$C_2H_4F_2$	1,1-Difluoroethane	(11.87)	643
C_2H_4O	Acetaldehyde	10.229 ± 0.0007	821
C_2H_4O	Ethylene oxide	10.56 ± 0.01	966
$C_2H_4O_2$	Acetic acid	10.65 ± 0.02	595
$C_2H_4O_2$	Methyl formate	10.835 ± 0.005	690
C_2H_5Br	Bromoethane	10.29 ± 0.01	931
C_2H_5Cl	Chloroethane	10.98 ± 0.02	947
C_2H_5ClO	2-Chloroethanol	(10.5)	756
C_2H_5F	Fluoroethane	(11.78)	873
C_2H_5I	Iodoethane	9.3492 ± 0.0006	893
C_2H_5N	Ethyleneimine	(9.5 ± 0.3)	1044
C_2H_5NO	Acetamide	9.65 ± 0.03	693
C_2H_5NO	N-Methylformamide	9.83 ± 0.04	760
$C_2H_5NO_2$	Nitroethane	10.88 ± 0.05	948
C_2H_6	Ethane	11.56 ± 0.02	1031
$C_2H_6Cl_2Si$	Dichlorodimethylsilane	(10.7)	576
C_2H_6O	Ethanol	10.43 ± 0.05	772
C_2H_6O	Dimethyl ether	10.025 ± 0.025	783
C_2H_6OS	Dimethyl sulfoxide	9.10 ± 0.03	727
$C_2H_6O_2$	Ethylene glycol	10.16	593
C_2H_6S	Ethanethiol	9.31 ± 0.03	851
C_2H_6S	Dimethyl sulfide	8.69 ± 0.02	801
$C_2H_6S_2$	Dimethyl disulfide	(7.4 ± 0.3)	690
C_2H_7N	Ethylamine	8.86 ± 0.02	808
C_2H_7N	Dimethylamine	8.24 ± 0.08	777
C_2H_7NO	Ethanolamine	8.96	664
$C_2H_8N_2$	1,2-Ethanediamine	(8.6)	812
$C_2H_8N_2$	1,1-Dimethylhydrazine	7.29 ± 0.05	787
C_2N_2	Cyanogen	13.37 ± 0.01	1597
C_3F_6	Perfluoropropene	10.60 ± 0.03	−103
C_3F_6O	Perfluoroacetone	(11.57 ± 0.13)	−282
C_3F_8	Perfluoropropane	(13.38)	−491
C_3HN	Cyanoacetylene	11.64 ± 0.01	1475
C_3H_2O	2-Propynal	(10.7 ± 0.1)	1145
$C_3H_3F_3$	3,3,3-Trifluoropropene	(10.9)	437
C_3H_3N	2-Propenenitrile	10.91 ± 0.01	1237
C_3H_3NO	Oxazole	(9.9)	940
C_3H_3NO	Isoxazole	(9.93)	1038
C_3H_4	Allene	9.692 ± 0.004	1126
C_3H_4	Propyne	10.37 ± 0.01	1187
C_3H_4	Cyclopropene	9.67 ± 0.01	1209
$C_3H_4N_2$	Imidazole	(8.81)	997
C_3H_4O	Propargyl alcohol	10.49 ± 0.02	1060
C_3H_4O	Acrolein	10.103 ± 0.006	900
C_3H_4O	Cyclopropanone	(9.1 ± 0.1)	895
$C_3H_4O_2$	Propenoic acid	10.60	701
$C_3H_4O_2$	2-Oxetanone	(9.70 ± 0.01)	653
C_3H_5Br	3-Bromopropene	(9.96)	1008
C_3H_5Cl	3-Chloropropene	10.04 ± 0.01	965
C_3H_5ClO	Epichlorohydrin	(10.64)	919
$C_3H_5ClO_2$	Methyl chloroacetate	(10.3)	575
C_3H_5F	3-Fluoropropene	(10.11)	821
C_3H_5N	Propanenitrile	11.84 ± 0.02	1194
C_3H_5NO	Acrylamide	(9.5)	720
C_3H_6	Propene	9.73 ± 0.02	959
C_3H_6	Cyclopropane	9.86	1005
$C_3H_6Br_2$	1,2-Dibromopropane	(10.1)	903
$C_3H_6Br_2$	1,3-Dibromopropane	(≤ 10.2)	≤ 919
$C_3H_6Cl_2$	1,2-Dichloropropane	10.8 ± 0.1	886
$C_3H_6Cl_2$	1,3-Dichloropropane	10.89 ± 0.04	892

Mol. form.	Name	IP/eV	$\Delta_f H_{ion}$ kJ/mol
C_3H_6O	Allyl alcohol	9.67 ± 0.05	808
C_3H_6O	Methyl vinyl ether	8.95 ± 0.01	763
C_3H_6O	Propanal	9.96 ± 0.01	772
C_3H_6O	Acetone	9.703 ± 0.006	719
C_3H_6O	Methyloxirane	(10.22)	892
C_3H_6O	Oxetane	9.65 ± 0.01	851
$C_3H_6O_2$	Propanoic acid	10.525 ± 0.003	568
$C_3H_6O_2$	Ethyl formate	10.61 ± 0.01	639
$C_3H_6O_2$	Methyl acetate	10.25 ± 0.02	579
$C_3H_6O_2$	1,3-Dioxolane	(9.9)	658
$C_3H_6O_3$	1,3,5-Trioxane	(10.3)	528
C_3H_7Br	1-Bromopropane	10.18 ± 0.01	898
C_3H_7Br	2-Bromopropane	10.10 ± 0.03	877
C_3H_7Cl	1-Chloropropane	10.81 ± 0.01	911
C_3H_7Cl	2-Chloropropane	10.79 ± 0.02	896
C_3H_7F	1-Fluoropropane	(11.3)	806
C_3H_7F	2-Fluoropropane	(11.08)	776
C_3H_7I	1-Iodopropane	9.25 ± 0.01	860
C_3H_7I	2-Iodopropane	9.19 ± 0.02	845
C_3H_7N	Allylamine	(8.76)	891
C_3H_7N	Cyclopropylamine	(8.8)	926
C_3H_7N	Propyleneimine	(9.0)	960
C_3H_7NO	N,N-Dimethylformamide	(9.12)	688
$C_3H_7NO_2$	1-Nitropropane	(10.81)	919
$C_3H_7NO_2$	2-Nitropropane	(10.71)	894
C_3H_8	Propane	10.95 ± 0.05	952
C_3H_8O	1-Propanol	10.18 ± 0.06	727
C_3H_8O	2-Propanol	10.17 ± 0.02	709
C_3H_8O	Ethyl methyl ether	9.72 ± 0.07	722
$C_3H_8O_2$	Dimethoxymethane	9.7	588
C_3H_8S	1-Propanethiol	9.20 ± 0.01	819
C_3H_8S	2-Propanethiol	9.145 ± 0.005	806
C_3H_8S	Ethyl methyl sulfide	(8.55)	765
$C_3H_9BO_3$	Trimethyl borate	(10.0)	65
C_3H_9ClSi	Trimethylchlorosilane	(10.15)	624
C_3H_9N	Propylamine	(8.78)	777
C_3H_9N	Isopropylamine	(8.72)	758
C_3H_9N	Trimethylamine	7.82 ± 0.06	731
C_3H_9NO	3-Amino-1-propanol	(9.0)	651
$C_4H_2O_3$	Maleic anhydride	(10.8)	645
C_4H_4	1-Buten-3-yne	9.58 ± 0.02	1230
$C_4H_4N_2$	Succinonitrile	(12.1 ± 0.25)	1377
$C_4H_4N_2$	Pyrimidine	9.23	1087
$C_4H_4N_2$	Pyridazine	8.67 ± 0.03	1112
C_4H_4O	Furan	8.883 ± 0.003	822
$C_4H_4O_2$	Diketene	(9.6 ± 0.02)	736
$C_4H_4O_3$	Succinic anhydride	(10.6)	500
$C_4H_4O_4$	Fumaric acid	(10.7)	355
C_4H_4S	Thiophene	8.86 ± 0.02	970
C_4H_5N	Methylacrylonitrile	10.34	1127
C_4H_5N	Pyrrole	8.207 ± 0.005	900
C_4H_5N	Cyclopropanecarbonitrile	(10.25)	1173
C_4H_6	1,2-Butadiene	(9.03)	1034
C_4H_6	1,3-Butadiene	9.082 ± 0.004	986
C_4H_6	1-Butyne	10.19 ± 0.02	1148
C_4H_6	2-Butyne	9.59 ± 0.03	1071
C_4H_6	Cyclobutene	9.43 ± 0.02	1067
C_4H_6O	Divinyl ether	(8.7)	827
C_4H_6O	trans-2-Butenal	9.73 ± 0.01	835
C_4H_6O	2-Methylpropenal	(9.92)	834

Mol. form.	Name	IP/eV	$\Delta_f H_{ion}$ kJ/mol
C_4H_6O	Cyclobutanone	(9.35)	815
$C_4H_6O_2$	cis-Crotonic acid	(10.08)	625
$C_4H_6O_2$	trans-Crotonic acid	(9.9)	604
$C_4H_6O_2$	Methacrylic acid	(10.15)	611
$C_4H_6O_2$	Vinyl acetate	9.19 ± 0.05	572
$C_4H_6O_2$	Methyl acrylate	(9.9)	641
$C_4H_6O_3$	Acetic anhydride	(10.0)	398
$C_4H_6O_4$	Dimethyl oxalate	(10.0)	287
C_4H_6S	2,5-Dihydrothiophene	(8.4)	898
C_4H_7N	Butanenitrile	(11.2)	1110
C_4H_7N	2-Methylpropanenitrile	(11.3)	1115
C_4H_7NO	2-Pyrrolidone	(9.2)	674
C_4H_8	1-Butene	9.55 ± 0.06	921
C_4H_8	cis-2-Butene	9.11 ± 0.01	871
C_4H_8	trans-2-Butene	9.10 ± 0.01	866
C_4H_8	Isobutene	9.239 ± 0.003	875
C_4H_8	Cyclobutane	(9.82 ± 0.05)	976
C_4H_8	Methylcyclopropane	(9.46)	936
$C_4H_8Br_2$	1,4-Dibromobutane	(10.15)	879
C_4H_8O	Ethyl vinyl ether	(8.98)	709
C_4H_8O	1,2-Epoxybutane	(≤ 10.15)	862
C_4H_8O	Butanal	9.84 ± 0.02	742
C_4H_8O	Isobutanal	9.71 ± 0.01	721
C_4H_8O	2-Butanone	9.52 ± 0.04	678
C_4H_8O	Tetrahydrofuran	9.38 ± 0.05	721
$C_4H_8O_2$	Butanoic acid	10.17 ± 0.05	509
$C_4H_8O_2$	2-Methylpropanoic acid	10.33 ± 0.03	516
$C_4H_8O_2$	Propyl formate	10.52 ± 0.02	555
$C_4H_8O_2$	Ethyl acetate	10.01 ± 0.05	522
$C_4H_8O_2$	Methyl propanoate	10.15 ± 0.03	548
$C_4H_8O_2$	1,3-Dioxane	9.8	607
$C_4H_8O_2$	1,4-Dioxane	9.19 ± 0.01	571
$C_4H_8O_2S$	Sulfolane	(9.8)	577
C_4H_8S	Tetrahydrothiophene	8.38	774
C_4H_9Br	1-Bromobutane	(10.12)	869
C_4H_9Br	2-Bromobutane	10.01 ± 0.02	845
C_4H_9Br	1-Bromo-2-methylpropane	10.09 ± 0.02	861
C_4H_9Br	2-Bromo-2-methylpropane	9.92 ± 0.03	823
C_4H_9Cl	1-Chlorobutane	10.67 ± 0.03	875
C_4H_9Cl	2-Chlorobutane	10.53	857
C_4H_9Cl	1-Chloro-2-methylpropane	10.73 ± 0.07	877
C_4H_9Cl	2-Chloro-2-methylpropane	(10.61)	842
C_4H_9I	1-Iodobutane	9.23 ± 0.01	840
C_4H_9I	2-Iodobutane	9.10 ± 0.02	815
C_4H_9I	1-Iodo-2-methylpropane	9.19 ± 0.01	824
C_4H_9I	2-Iodo-2-methylpropane	(9.02)	798
C_4H_9N	Pyrrolidine	(8.0)	769
C_4H_9NO	N,N-Dimethylacetamide	8.81 ± 0.03	616
C_4H_9NO	Morpholine	(8.2)	841
C_4H_{10}	Butane	10.53 ± 0.10	890
C_4H_{10}	Isobutane	(10.57)	886
$C_4H_{10}O$	1-Butanol	9.99 ± 0.05	689
$C_4H_{10}O$	2-Butanol	9.88 ± 0.03	658
$C_4H_{10}O$	2-Methyl-1-propanol	10.02 ± 0.04	683
$C_4H_{10}O$	2-Methyl-2-propanol	9.90 ± 0.02	642
$C_4H_{10}O$	Diethyl ether	9.51 ± 0.03	666
$C_4H_{10}O$	Methyl propyl ether	9.41 ± 0.07	670
$C_4H_{10}O$	Isopropyl methyl ether	9.45 ± 0.04	661
$C_4H_{10}O_2$	Ethylene glycol monoethyl ether	(9.6)	529
$C_4H_{10}O_2$	Ethylene glycol dimethyl ether	(9.3)	558

Mol. form.	Name	IP/eV	$\Delta_f H_{ion}$ kJ/mol
$C_4H_{10}S$	1-Butanethiol	9.14 ± 0.01	794
$C_4H_{10}S$	2-Butanethiol	(9.10)	781
$C_4H_{10}S$	2-Methyl-1-propanethiol	(9.12)	783
$C_4H_{10}S$	2-Methyl-2-propanethiol	(9.03)	762
$C_4H_{10}S$	Diethyl sulfide	(8.43)	730
$C_4H_{10}S$	Methyl propyl sulfide	(8.8)	767
$C_4H_{10}S$	Isopropyl methyl sulfide	(8.7)	749
$C_4H_{10}S_2$	Diethyl disulfide	(8.27)	724
$C_4H_{11}N$	Butylamine	8.7 ± 0.1	748
$C_4H_{11}N$	sec-Butylamine	8.46 ± 0.1	711
$C_4H_{11}N$	tert-Butylamine	8.46 ± 0.1	695
$C_4H_{11}N$	Isobutylamine	8.50 ± 0.1	721
$C_4H_{11}N$	Diethylamine	7.85 ± 0.1	684
$C_4H_{12}Si$	Tetramethylsilane	9.80 ± 0.04	713
$C_4H_{12}Sn$	Tetramethylstannane	8.89 ± 0.05	837
C_4NiO_4	Nickel carbonyl	8.27 ± 0.04	200
$C_5H_4O_2$	Furfural	9.22 ± 0.01	739
C_5H_5N	Pyridine	9.25	1031
C_5H_6	1-Penten-3-yne	9.00 ± 0.01	1119
C_5H_6	cis-3-Penten-1-yne	9.14 ± 0.04	1137
C_5H_6	trans-3-Penten-1-yne	9.05 ± 0.01	1128
C_5H_6	2-Methyl-1-buten-3-yne	9.25 ± 0.02	1152
C_5H_6	1,3-Cyclopentadiene	8.55 ± 0.02	955
C_5H_6O	2-Methylfuran	8.38 ± 0.02	729
C_5H_6O	3-Methylfuran	(8.64)	763
C_5H_6S	2-Methylthiophene	(8.14)	867
C_5H_6S	3-Methylthiophene	(8.40)	893
C_5H_8	cis-1,3-Pentadiene	8.63 ± 0.03	914
C_5H_8	trans-1,3-Pentadiene	8.59 ± 0.02	905
C_5H_8	1,4-Pentadiene	9.60 ± 0.02	1032
C_5H_8	2-Methyl-1,3-butadiene	8.84 ± 0.01	928
C_5H_8	1-Pentyne	10.10 ± 0.01	1119
C_5H_8	Cyclopentene	9.01 ± 0.01	905
C_5H_8	Spiropentane	(9.26)	1078
C_5H_8O	Cyclopropyl methyl ketone	(≤ 9.46)	796
C_5H_8O	Cyclopentanone	9.26 ± 0.01	701
C_5H_8O	3,4-Dihydro-2H-pyran	8.35 ± 0.01	681
$C_5H_8O_2$	Ethyl acrylate	(≤ 10.3)	617
$C_5H_8O_2$	Methyl methacrylate	(9.7)	589
$C_5H_8O_2$	2,4-Pentanedione	8.85 ± 0.01	469
C_5H_9NO	N-Methyl-2-pyrrolidone	(≤ 9.17)	≤ 676
C_5H_{10}	1-Pentene	9.51 ± 0.01	896
C_5H_{10}	cis-2-Pentene	9.01 ± 0.03	843
C_5H_{10}	trans-2-Pentene	9.04 ± 0.01	841
C_5H_{10}	2-Methyl-1-butene	9.12 ± 0.01	844
C_5H_{10}	3-Methyl-1-butene	9.52 ± 0.01	891
C_5H_{10}	2-Methyl-2-butene	8.69 ± 0.01	796
C_5H_{10}	Cyclopentane	(10.33 ± 0.15)	918
$C_5H_{10}O$	2,2-Dimethylpropanal	9.51 ± 0.01	675
$C_5H_{10}O$	Cyclopentanol	(9.72)	695
$C_5H_{10}O$	Pentanal	9.74 ± 0.04	709
$C_5H_{10}O$	2-Pentanone	9.38 ± 0.01	646
$C_5H_{10}O$	3-Pentanone	9.31 ± 0.01	640
$C_5H_{10}O$	3-Methyl-2-butanone	9.30 ± 0.01	635
$C_5H_{10}O$	Tetrahydropyran	9.25 ± 0.01	670
$C_5H_{10}O_2$	Pentanoic acid	(≤ 10.53)	≤ 527
$C_5H_{10}O_2$	3-Methylbutanoic acid	(≤ 10.51)	≤ 499
$C_5H_{10}O_2$	Butyl formate	10.52 ± 0.02	584
$C_5H_{10}O_2$	Propyl acetate	(≤ 9.92)	501
$C_5H_{10}O_2$	Isopropyl acetate	9.99 ± 0.03	482

Mol. form.	Name	IP/eV	$\Delta_f H_{ion}$ kJ/mol
$C_5H_{10}O_2$	Ethyl propanoate	(10.00)	500
$C_5H_{10}O_2$	Methyl butanoate	(10.07)	520
$C_5H_{10}S$	Thiacyclohexane	(8.2)	728
$C_5H_{11}Br$	1-Bromopentane	10.10 ± 0.01	846
$C_5H_{11}I$	1-Iodopentane	9.20 ± 0.01	817
$C_5H_{11}N$	Piperidine	8.03 ± 0.11	726
$C_5H_{11}N$	N-Methylpyrrolidine	≤ 8.41 ± 0.02	≤ 809
C_5H_{12}	Pentane	10.28 ± 0.10	845
C_5H_{12}	Isopentane	10.32 ± 0.05	843
C_5H_{12}	Neopentane	(≤ 10.2)	≤ 818
$C_5H_{12}O$	1-Pentanol	(10.00)	668
$C_5H_{12}O$	2-Pentanol	(9.78)	630
$C_5H_{12}O$	3-Pentanol	9.78	628
$C_5H_{12}O$	2-Methyl-1-butanol	(9.86)	649
$C_5H_{12}O$	2-Methyl-2-butanol	(9.8)	615
$C_5H_{12}O$	3-Methyl-2-butanol	(9.88 ± 0.13)	637
$C_5H_{12}O$	Butyl methyl ether	(9.4 ± 0.1)	648
$C_5H_{12}O$	Methyl tert-butyl ether	(9.24)	608
$C_5H_{12}O$	Ethyl propyl ether	(9.45)	640
$C_5H_{12}S$	tert-Butyl methyl sulfide	(8.38)	687
$C_5H_{12}S$	Ethyl propyl sulfide	(8.50)	716
$C_5H_{12}S$	Ethyl isopropyl sulfide	(8.35)	689
C_6BrF_5	Bromopentafluorobenzene	(9.67)	222
C_6ClF_5	Chloropentafluorobenzene	(9.72)	126
C_6Cl_6	Hexachlorobenzene	(8.98)	822
C_6F_6	Hexafluorobenzene	9.89 ± 0.04	8
C_6F_{12}	Perfluorocyclohexane	(13.2)	−1095
C_6HF_5	Pentafluorobenzene	(9.63)	122
C_6HF_5O	Pentafluorophenol	(9.20)	−71
$C_6H_2F_4$	1,2,3,4-Tetrafluorobenzene	(9.53)	284
$C_6H_2F_4$	1,2,3,5-Tetrafluorobenzene	(9.53)	263
$C_6H_2F_4$	1,2,4,5-Tetrafluorobenzene	(9.35)	254
$C_6H_3Cl_3$	1,2,4-Trichlorobenzene	(9.04)	880
$C_6H_3Cl_3$	1,3,5-Trichlorobenzene	9.32 ± 0.02	899
$C_6H_4ClNO_2$	1-Chloro-3-nitrobenzene	(9.92 ± 0.1)	995
$C_6H_4ClNO_2$	1-Chloro-4-nitrobenzene	(9.96 ± 0.1)	999
$C_6H_4Cl_2$	o-Dichlorobenzene	9.06 ± 0.02	907
$C_6H_4Cl_2$	m-Dichlorobenzene	9.10 ± 0.02	906
$C_6H_4Cl_2$	p-Dichlorobenzene	8.92 ± 0.02	885
$C_6H_4FNO_2$	1-Fluoro-4-nitrobenzene	(9.90)	826
$C_6H_4F_2$	o-Difluorobenzene	9.29 ± 0.01	602
$C_6H_4F_2$	m-Difluorobenzene	9.33 ± 0.01	591
$C_6H_4F_2$	p-Difluorobenzene	9.1589 ± 0.0003	577
$C_6H_4O_2$	p-Benzoquinone	10.01 ± 0.06	844
C_6H_5Br	Bromobenzene	9.00 ± 0.02	971
C_6H_5Cl	Chlorobenzene	9.07 ± 0.02	930
C_6H_5ClO	m-Chlorophenol	8.655 ± 0.001	680
C_6H_5ClO	p-Chlorophenol	(≤ 8.69)	≤ 692
C_6H_5F	Fluorobenzene	9.20 ± 0.01	772
C_6H_5I	Iodobenzene	8.685	1003
$C_6H_5NO_2$	Nitrobenzene	9.86 ± 0.02	1019
$C_6H_5NO_3$	o-Nitrophenol	(9.1)	782
$C_6H_5NO_3$	m-Nitrophenol	(9.0)	755
$C_6H_5NO_3$	p-Nitrophenol	(9.1)	761
C_6H_6	Benzene	9.24378 ± 0.00007	975
C_6H_6	Fulvene	(8.36)	1031
C_6H_6ClN	o-Chloroaniline	(8.50)	883
C_6H_6ClN	m-Chloroaniline	(8.09)	835
C_6H_6ClN	p-Chloroaniline	(≤ 8.18)	≤ 844
$C_6H_6N_2O_2$	o-Nitroaniline	(8.27)	861

Mol. form.	Name	IP/eV	$\Delta_f H_{ion}$ kJ/mol
$C_6H_6N_2O_2$	*m*-Nitroaniline	(8.31)	865
$C_6H_6N_2O_2$	*p*-Nitroaniline	(8.34)	859
C_6H_6O	Phenol	8.49 ± 0.02	723
$C_6H_6O_2$	*p*-Hydroquinone	7.94 ± 0.01	503
C_6H_6S	Benzenethiol	(8.32)	915
C_6H_7N	Aniline	7.720 ± 0.002	832
C_6H_7N	2-Methylpyridine	(9.02)	970
C_6H_7N	3-Methylpyridine	(9.04)	979
C_6H_7N	4-Methylpyridine	(9.04)	976
$C_6H_8N_2$	*o*-Phenylenediamine	(7.2)	787
$C_6H_8N_2$	*m*-Phenylenediamine	(7.14)	777
$C_6H_8N_2$	*p*-Phenylenediamine	(6.87 ± 0.05)	759
C_6H_{10}	1,5-Hexadiene	9.27 ± 0.05	978
C_6H_{10}	1-Hexyne	10.03 ± 0.05	1089
C_6H_{10}	3,3-Dimethyl-1-butyne	9.90 ± 0.04	1060
C_6H_{10}	Cyclohexene	8.945 ± 0.01	859
$C_6H_{10}O$	Cyclohexanone	9.14 ± 0.01	656
$C_6H_{10}O$	Mesityl oxide	9.10 ± 0.01	694
$C_6H_{10}O_4$	Diethyl oxalate	(9.8)	205
$C_6H_{11}NO$	Caprolactam	(9.07 ± 0.02)	629
C_6H_{12}	1-Hexene	9.44 ± 0.04	869
C_6H_{12}	*cis*-2-Hexene	(8.97 ± 0.01)	818
C_6H_{12}	*trans*-2-Hexene	(8.97 ± 0.01)	814
C_6H_{12}	2-Methyl-1-pentene	(9.08 ± 0.01)	817
C_6H_{12}	4-Methyl-1-pentene	9.45 ± 0.01	862
C_6H_{12}	2-Methyl-2-pentene	(8.58)	761
C_6H_{12}	4-Methyl-*cis*-2-pentene	8.98 ± 0.01	809
C_6H_{12}	4-Methyl-*trans*-2-pentene	(8.97 ± 0.01)	804
C_6H_{12}	2-Ethyl-1-butene	(9.06 ± 0.02)	818
C_6H_{12}	2,3-Dimethyl-1-butene	(9.07 ± 0.01)	812
C_6H_{12}	2,3-Dimethyl-2-butene	8.27 ± 0.01	729
C_6H_{12}	Cyclohexane	9.86 ± 0.03	828
C_6H_{12}	Methylcyclopentane	(9.85)	845
$C_6H_{12}O$	Hexanal	9.72 ± 0.05	691
$C_6H_{12}O$	2-Hexanone	9.3 ± 0.1	626
$C_6H_{12}O$	3-Hexanone	9.12 ± 0.02	600
$C_6H_{12}O$	3-Methyl-2-pentanone	9.21 ± 0.01	600
$C_6H_{12}O$	4-Methyl-2-pentanone	9.30 ± 0.01	609
$C_6H_{12}O$	2-Methyl-3-pentanone	9.10 ± 0.01	592
$C_6H_{12}O$	3,3-Dimethyl-2-butanone	9.12 ± 0.02	589
$C_6H_{12}O$	Cyclohexanol	(9.75)	651
$C_6H_{12}O_2$	Hexanoic acid	≤ 10.12	≤ 463
$C_6H_{12}O_2$	Butyl acetate	(9.92 ± .05)	471
$C_6H_{12}O_2$	*sec*-Butyl acetate	9.90	453
$C_6H_{12}O_2$	Methyl 2,2-dimethylpropanoate	(9.90 ± 0.04)	466
$C_6H_{13}I$	1-Iodohexane	9.179	794
$C_6H_{13}N$	Cyclohexylamine	(8.86)	750
C_6H_{14}	Hexane	10.13	810
C_6H_{14}	2-Methylpentane	(10.12)	802
C_6H_{14}	3-Methylpentane	(10.08)	801
C_6H_{14}	2,2-Dimethylbutane	(10.06)	787
C_6H_{14}	2,3-Dimethylbutane	(10.02)	791
$C_6H_{14}O$	1-Hexanol	(9.89)	639
$C_6H_{14}O$	2-Hexanol	(9.80 ± 0.03)	611
$C_6H_{14}O$	3-Hexanol	(9.63 ± 0.03)	599
$C_6H_{14}O$	Dipropyl ether	(9.27)	602
$C_6H_{14}O$	Diisopropyl ether	9.20 ± 0.05	569
$C_6H_{14}O$	Butyl ethyl ether	(9.36)	610
$C_6H_{14}O$	Methyl pentyl ether	(≤ 9.67)	≤ 657
$C_6H_{14}O_2$	1,1-Diethoxyethane	(9.2)	434

Mol. form.	Name	IP/eV	$\Delta_f H_{ion}$ kJ/mol
$C_6H_{14}O_3$	Diethylene glycol dimethyl ether	≤ 9.8	≤ 448
$C_6H_{14}S$	Dipropyl sulfide	8.30 ± 0.02	676
$C_6H_{14}S$	Diisopropyl sulfide	(8.2 ± 0.2)	649
$C_6H_{15}N$	Hexylamine	(8.63 ± 0.05)	699
$C_6H_{15}N$	Dipropylamine	(7.84 ± 0.02)	641
$C_6H_{15}N$	Diisopropylamine	(7.73 ± 0.03)	602
$C_6H_{15}N$	Triethylamine	(7.50 ± 0.02)	631
$C_6H_{15}NO_3$	Triethanolamine	(7.9)	206
$C_7H_3F_5$	2,3,4,5,6-Pentafluorotoluene	(9.4)	64
C_7H_5ClO	Benzoyl chloride	(9.53)	815
$C_7H_5Cl_3$	(Trichloromethyl)benzene	(≤ 9.60)	≤ 914
$C_7H_5F_3$	(Trifluoromethyl)benzene	9.685 ± 0.005	335
C_7H_5N	Benzonitrile	9.70 ± 0.01	1154
C_7H_6O	Benzaldehyde	9.49 ± 0.02	878
$C_7H_6O_2$	Benzoic acid	(9.3)	604
C_7H_7Br	p-Bromotoluene	8.67 ± 0.02	908
C_7H_7Cl	o-Chlorotoluene	(8.7 ± 0.1)	856
C_7H_7Cl	m-Chlorotoluene	(8.83)	869
C_7H_7Cl	p-Chlorotoluene	(8.69)	855
C_7H_7Cl	(Chloromethyl)benzene	9.10 ± 0.02	897
C_7H_7F	o-Fluorotoluene	8.91 ± 0.01	709
C_7H_7F	m-Fluorotoluene	8.91 ± 0.01	709
C_7H_7F	p-Fluorotoluene	8.79 ± 0.01	701
C_7H_7NO	Benzamide	(9.25)	792
$C_7H_7NO_2$	o-Nitrotoluene	9.24	946
$C_7H_7NO_2$	m-Nitrotoluene	9.45 ± 0.1	941
$C_7H_7NO_2$	p-Nitrotoluene	9.46 ± 0.05	942
C_7H_8	Toluene	8.8276 ± 0.0006	901
C_7H_8O	o-Cresol	(8.24)	670
C_7H_8O	m-Cresol	8.29 ± 0.07	668
C_7H_8O	p-Cresol	(8.3)	675
C_7H_8O	Benzyl alcohol	(8.3)	701
C_7H_8O	Anisole	8.22 ± 0.03	725
C_7H_9N	Benzylamine	(8.64)	917
C_7H_9N	o-Methylaniline	(7.44 ± 0.02)	772
C_7H_9N	m-Methylaniline	(7.50 ± 0.02)	778
C_7H_9N	p-Methylaniline	(7.24 ± 0.02)	753
C_7H_9N	N-Methylaniline	7.34 ± 0.04	792
C_7H_9N	2,3-Dimethylpyridine	(8.85 ± 0.02)	922
C_7H_9N	2,4-Dimethylpyridine	(8.85 ± 0.03)	918
C_7H_9N	2,5-Dimethylpyridine	(≤ 8.80 ± 0.05)	≤ 916
C_7H_9N	2,6-Dimethylpyridine	8.86 ± 0.03	913
C_7H_9N	3,4-Dimethylpyridine	(≤ 9.15)	≤ 953
C_7H_9N	3,5-Dimethylpyridine	(≤ 9.25)	≤ 965
$C_7H_{10}O$	Dicyclopropyl ketone	(9.1)	1041
C_7H_{14}	1-Heptene	9.34 ± 0.10	839
C_7H_{14}	trans-3-Heptene	(8.92)	790
C_7H_{14}	Cycloheptane	9.97	844
C_7H_{14}	Methylcyclohexane	9.64	775
C_7H_{14}	cis-1,2-Dimethylcyclopentane	(9.92 ± 0.05)	828
C_7H_{14}	trans-1,2-Dimethylcyclopentane	9.7 ± 0.2	799
$C_7H_{14}O$	1-Heptanal	(9.65)	668
$C_7H_{14}O$	2-Heptanone	9.28 ± 0.10	594
$C_7H_{14}O$	3-Heptanone	9.18 ± 0.08	589
$C_7H_{14}O$	4-Heptanone	9.10 ± 0.06	577
$C_7H_{14}O$	5-Methyl-2-hexanone	(9.28)	586
$C_7H_{14}O$	2,4-Dimethyl-3-pentanone	8.95 ± 0.01	552
$C_7H_{14}O$	1-Methylcyclohexanol	(9.8 ± 0.2)	586
C_7H_{16}	Heptane	9.93 ± 0.10	771
$C_7H_{16}O$	1-Heptanol	(9.84)	614

Mol. form.	Name	IP/eV	$\Delta_f H_{ion}$ kJ/mol
$C_7H_{16}O$	2-Heptanol	(9.70)	580
$C_7H_{16}O$	3-Heptanol	(9.68)	578
$C_7H_{16}O$	4-Heptanol	(9.61)	572
$C_7H_{16}O$	Ethyl pentyl ether	(≤ 9.49)	≤ 602
$C_8H_4O_3$	Phthalic anhydride	(10.1)	603
$C_8H_6O_4$	Isophthalic acid	(9.98)	268
$C_8H_6O_4$	Terephthalic acid	(9.86)	232
C_8H_7N	2-Methylbenzonitrile	(≤ 9.38)	1085
C_8H_7N	3-Methylbenzonitrile	(≤ 9.34)	1085
C_8H_7N	4-Methylbenzonitrile	9.32 ± 0.02	1083
C_8H_7N	Indole	7.7602 ± 0.0006	908
C_8H_8	Styrene	8.464 ± 0.001	964
C_8H_8O	p-Tolualdehyde	(9.33)	825
C_8H_8O	Acetophenone	9.29 ± 0.03	810
$C_8H_8O_2$	o-Toluic acid	(9.1)	558
$C_8H_8O_2$	m-Toluic acid	(9.43)	579
$C_8H_8O_2$	p-Toluic acid	(9.23)	560
$C_8H_8O_2$	Benzeneacetic acid	(8.26)	479
$C_8H_8O_2$	Methyl benzoate	9.32 ± 0.03	611
C_8H_{10}	Ethylbenzene	8.77 ± 0.01	876
C_8H_{10}	o-Xylene	8.56 ± 0.01	844
C_8H_{10}	m-Xylene	8.56 ± 0.01	843
C_8H_{10}	p-Xylene	8.44 ± 0.01	832
$C_8H_{10}O$	p-Ethylphenol	(7.84)	613
$C_8H_{10}O$	2,3-Xylenol	(8.26)	640
$C_8H_{10}O$	2,4-Xylenol	(8.0)	609
$C_8H_{10}O$	2,6-Xylenol	(8.05)	615
$C_8H_{10}O$	3,4-Xylenol	(8.09)	624
$C_8H_{10}O$	Phenetole	(8.13)	683
$C_8H_{11}N$	2,4,6-Trimethylpyridine	(≤ 8.9)	≤ 880
$C_8H_{11}N$	N-Ethylaniline	(≤ 7.67)	≤ 794
$C_8H_{11}N$	N,N-Dimethylaniline	7.12 ± 0.02	787
C_8H_{14}	1-Octyne	(9.95 ± 0.02)	1040
C_8H_{14}	2-Octyne	9.31 ± 0.01	961
C_8H_{14}	3-Octyne	9.22 ± 0.01	952
C_8H_{14}	4-Octyne	9.20 ± 0.01	946
C_8H_{16}	1-Octene	9.43 ± 0.01	829
C_8H_{16}	Cyclooctane	9.75 ± 0.05	816
C_8H_{16}	Ethylcyclohexane	(9.54)	748
C_8H_{16}	1,1-Dimethylcyclohexane	(9.42)	728
C_8H_{16}	cis-1,2-Dimethylcyclohexane	(<9.78)	772
C_8H_{16}	trans-1,2-Dimethylcyclohexane	9.41	728
C_8H_{16}	cis-1,3-Dimethylcyclohexane	(<9.98)	778
C_8H_{16}	trans-1,3-Dimethylcyclohexane	9.53	743
C_8H_{16}	cis-1,4-Dimethylcyclohexane	(<9.93)	782
C_8H_{16}	trans-1,4-Dimethylcyclohexane	(9.56)	738
C_8H_{16}	Propylcyclopentane	(9.34)	753
$C_8H_{16}O$	2,2,4-Trimethyl-3-pentanone	(8.80)	511
C_8H_{18}	Octane	9.80 ± 0.10	737
C_8H_{18}	2-Methylheptane	(9.84)	734
C_8H_{18}	2,2,4-Trimethylpentane	(9.86)	713
C_8H_{18}	2,2,3,3-Tetramethylbutane	9.8	720
$C_8H_{18}O$	Dibutyl ether	(9.28)	s 560
$C_8H_{18}O$	Di-sec-butyl ether	(9.11)	511
$C_8H_{18}O$	Di-tert-butyl ether	8.88 ± 0.07	493
$C_8H_{18}S$	Dibutyl sulfide	(8.2)	624
$C_8H_{18}S$	Di-tert-butyl sulfide	(8.0)	583
$C_8H_{18}S$	Diisobutyl sulfide	(8.34)	625
$C_8H_{19}N$	Dibutylamine	(7.69)	586
$C_8H_{19}N$	Diisobutylamine	(7.8)	574

Mol. form.	Name	IP/eV	$\Delta_f H_{ion}$ kJ/mol
$C_8H_{20}Si$	Tetraethylsilane	(8.9)	595
C_9H_7N	Quinoline	8.62 ± 0.01	1041
C_9H_7N	Isoquinoline	8.53 ± 0.03	1032
C_9H_8	Indene	8.14 ± 0.01	949
C_9H_{10}	o-Methylstyrene	(8.20)	908
C_9H_{10}	m-Methylstyrene	(8.15)	899
C_9H_{10}	p-Methylstyrene	(8.1)	895
C_9H_{10}	Cyclopropylbenzene	(8.35)	956
C_9H_{10}	Indan	(8.3)	864
$C_9H_{10}O_2$	Ethyl benzoate	(8.9)	537
C_9H_{12}	Propylbenzene	8.713 ± 0.010	848
C_9H_{12}	Isopropylbenzene	8.73 ± 0.01	847
C_9H_{12}	1,2,3-Trimethylbenzene	8.42 ± 0.02	803
C_9H_{12}	1,2,4-Trimethylbenzene	8.27 ± 0.01	784
C_9H_{12}	1,3,5-Trimethylbenzene	8.41 ± 0.01	796
$C_9H_{13}N$	N,N-Dimethyl-o-toluidine	7.40 ± 0.02	814
$C_9H_{14}O$	Isophorone	(≤ 9.07)	≤ 670
C_9H_{18}	Butylcyclopentane	(9.95)	793
C_9H_{18}	Propylcyclohexane	(9.46)	720
C_9H_{18}	Isopropylcyclohexane	(9.33)	704
$C_9H_{18}O$	2-Nonanone	(9.16)	545
$C_9H_{18}O$	5-Nonanone	(9.07)	530
$C_9H_{18}O$	2,6-Dimethyl-4-heptanone	9.01 ± 0.06	512
C_9H_{20}	Nonane	9.71 ± 0.10	709
$C_{10}F_8$	Perfluoronaphthalene	8.85	−368
$C_{10}H_7Br$	1-Bromonaphthalene	8.08 ± 0.03	955
$C_{10}H_7Cl$	1-Chloronaphthalene	(8.13)	906
$C_{10}H_8$	Naphthalene	8.1442 ± 0.0009	936
$C_{10}H_8$	Azulene	7.38 ± 0.05	1001
$C_{10}H_8O$	1-Naphthol	7.76 ± 0.03	719
$C_{10}H_8O$	2-Naphthol	7.87 ± 0.06	729
$C_{10}H_{10}O_4$	Dimethyl phthalate	(9.64 ± 0.07)	277
$C_{10}H_{12}$	1,2,3,4-Tetrahydronaphthalene	8.46 ± 0.02	841
$C_{10}H_{14}$	Butylbenzene	8.69 ± 0.02	826
$C_{10}H_{14}$	sec-Butylbenzene	8.68 ± 0.02	820
$C_{10}H_{14}$	tert-Butylbenzene	8.68 ± 0.05	816
$C_{10}H_{14}$	Isobutylbenzene	8.69 ± 0.02	817
$C_{10}H_{14}$	p-Cymene	(8.29)	771
$C_{10}H_{14}$	o-Diethylbenzene	(≤ 8.51)	≤ 804
$C_{10}H_{14}$	m-Diethylbenzene	(8.49)	798
$C_{10}H_{14}$	p-Diethylbenzene	(8.40)	790
$C_{10}H_{14}$	1,2,4,5-Tetramethylbenzene	8.04 ± 0.02	730
$C_{10}H_{14}O$	p-tert-Butylphenol	(7.8)	552
$C_{10}H_{16}$	α-Pinene	(8.07)	808
$C_{10}H_{16}O$	Camphor	(8.76)	577
$C_{10}H_{18}$	cis-Decahydronaphthalene	9.36 ± 0.04	734
$C_{10}H_{18}$	trans-Decahydronaphthalene	9.34 ± 0.04	720
$C_{10}H_{20}$	1-Decene	9.42 ± 0.05	786
$C_{10}H_{20}$	Butylcyclohexane	(9.41)	695
$C_{10}H_{22}$	Decane	(9.65)	682
$C_{11}H_{10}$	1-Methylnaphthalene	7.97 ± 0.03	882
$C_{11}H_{10}$	2-Methylnaphthalene	7.91 ± 0.08	877
$C_{11}H_{16}$	p-tert-Butyltoluene	(8.12)	730
$C_{11}H_{24}$	Undecane	(9.56)	650
$C_{11}H_{24}$	2-Methyldecane	(9.7)	658
$C_{12}H_8$	Acenaphthylene	(8.22)	1053
$C_{12}H_9N$	Carbazole	(7.57)	961
$C_{12}H_{10}$	Acenaphthene	7.75 ± 0.07	903
$C_{12}H_{10}$	Biphenyl	8.23 ± 0.10	977
$C_{12}H_{10}N_2O$	trans-Azoxybenzene	(8.1)	1123

Mol. form.	Name	IP/eV	$\Delta_f H_{ion}$ kJ/mol
$C_{12}H_{10}O$	Diphenyl ether	(8.09)	766
$C_{12}H_{11}N$	Diphenylamine	7.16 ± 0.04	908
$C_{12}H_{18}$	5,7-Dodecadiyne	(8.67)	1079
$C_{12}H_{18}$	Hexamethylbenzene	7.85 ± 0.01	670
$C_{12}H_{22}$	Cyclohexylcyclohexane	(9.41)	690
$C_{12}H_{27}N$	Tributylamine	(7.4)	492
$C_{13}H_{10}$	9H-Fluorene	7.91 ± 0.02	952
$C_{13}H_{10}O$	Benzophenone	9.08 ± 0.05	926
$C_{13}H_{12}$	Diphenylmethane	(8.55)	963
$C_{14}H_{10}$	Anthracene	7.439 ± 0.006	948
$C_{14}H_{10}$	Phenanthrene	7.8914 ± 0.0006	966
$C_{14}H_{10}$	Diphenylacetylene	7.94 ± 0.03	1168
$C_{14}H_{12}$	cis-Stilbene	(7.80)	1005
$C_{14}H_{12}$	trans-Stilbene	7.656 ± 0.001	973
$C_{14}H_{14}$	1,2-Diphenylethane	8.9 ± 0.1	1002
$C_{16}H_{10}$	Fluoranthene	7.9 ± 0.1	1052
$C_{16}H_{10}$	Pyrene	7.4256 ± 0.0006	935
$C_{18}H_{12}$	Chrysene	7.60 ± 0.01	1017
$C_{18}H_{14}$	o-Terphenyl	(7.99)	1056
$C_{18}H_{14}$	m-Terphenyl	(8.01)	1057
$C_{18}H_{14}$	p-Terphenyl	7.80 ± 0.03	1037
$C_{20}H_{12}$	Perylene	6.960 ± 0.001	981
$C_{24}H_{12}$	Coronene	7.29 ± 0.01	1026

X-RAY ATOMIC ENERGY LEVELS

The energy levels in this table are the values recommended by Bearden and Burr on the basis of a thorough review of the literature on x-ray wavelengths and related data. All values are in electron volts (eV). Values in parentheses are interpolated, and an asterisk * indicates a level which is not resolved from the level above it. See Reference 1 for uncertainties in the levels and a complete description of how the recommended values were obtained.

References

1. Bearden, J. A., and Burr, A. F., *Rev. Mod. Phys.*, 39, 125, 1967; also published as *X-Ray Wavelengths and X-Ray Atomic Energy Levels*, Natl. Stand. Ref. Data Sys. — Natl. Bur. Standards (U.S.), No. 14, 1967.
2. Gray, D. E., Ed., *American Institute of Physics Handbook, Third Edition*, pp. 7-158 to 7-167, McGraw-Hill, New York, 1972.

Level	¹H	²He	³Li	⁴Be	⁵B	⁶C	⁷N	⁸O
K	13.59811	24.58678	54.75	111.0	188.0	283.8	401.6	532.0
L_I								23.7
$L_{II,III}$					4.7	6.4	9.2	7.1

Level	⁹F	¹⁰Ne	¹¹Na	¹²Mg	¹³Al	¹⁴Si	¹⁵P	¹⁶S
K	685.4	866.9	1072.1	1305.0	1559.6	1838.9	2145.5	2472.0
L_I	(31)	(45)	63.3	89.4	117.7	148.7	189.3	229.2
$L_{II,III}$	8.6	18.3	31.1	51.4	73.1	99.2	132.2	164.8

Level	¹⁷Cl	¹⁸Ar	¹⁹K	²⁰Ca	²¹Sc	²²Ti	²³V	²⁴Cr
K	2822.4	3202.9	3607.4	4038.1	4492.8	4966.4	5465.1	5989.2
L_I	270.2	320	377.1	437.8	500.4	563.7	628.2	694.6
L_{II}	201.6	247.3	296.3	350.0	406.7	461.5	520.5	583.7
L_{III}	200.0	245.2	293.6	346.4	402.2	455.5	512.9	574.5
M_I	17.5	25.3	33.9	43.7	53.8	60.3	66.5	74.1
$M_{II,III}$	6.8	12.4	17.8	25.4	32.3	34.6	37.8	42.5
$M_{IV,V}$					6.6	3.7	2.2	2.3

Level	²⁵Mn	²⁶Fe	²⁷Co	²⁸Ni	²⁹Cu	³⁰Zn	³¹Ga	³²Ge
K	6539.0	7112.0	7708.9	8332.8	8978.9	9658.6	10367.1	11103.1
L_I	769.0	846.1	925.6	1008.1	1096.1	1193.6	1297.7	1414.3
L_{II}	651.4	721.1	793.6	871.9	951.0	1042.8	1142.3	1247.8
L_{III}	640.3	708.1	778.6	854.7	931.1	1019.7	1115.4	1216.7
M_I	83.9	92.9	100.7	111.8	119.8	135.9	158.1	180.0
M_{II}	48.6	54.0	59.5	68.1	73.6	86.6	106.8	127.9
M_{III}	48.6*	54.0*	59.5*	68.1*	73.6*	86.6*	102.9	120.8
$M_{IV,V}$	3.3	3.6	2.9	3.6	1.6	8.1	17.4	28.7

Level	³³As	³⁴Se	³⁵Br	³⁶Kr	³⁷Rb	³⁸Sr	³⁹Y	⁴⁰Zr
K	11866.7	12657.8	13473.7	14325.6	15199.7	16104.6	17038.4	17997.6
L_I	1526.5	1653.9	1782.0	1921.0	2065.1	2216.3	2372.5	2531.6
L_{II}	1358.6	1476.2	1596.0	1727.2	1863.9	2006.8	2155.5	2306.7
L_{III}	1323.1	1435.8	1549.9	1674.9	1804.4	1939.6	2080.0	2222.3
M_I	203.5	231.5	256.5		322.1	357.5	393.6	430.3
M_{II}	146.4	168.2	189.3	222.7	247.4	279.8	312.4	344.2
M_{III}	140.5	161.9	181.5	213.8	238.5	269.1	300.3	330.5
M_{IV}	41.2	56.7	70.1	88.9	111.8	135.0	159.6	182.4
M_V	41.2*	56.7*	69.0	88.9*	110.3	133.1	157.4	180.0
N_I			27.3	24.0	29.3	37.7	45.4	51.3
N_{II}	2.5	5.6	5.2	10.6	14.8	19.9	25.6	28.7
N_{III}	2.5*	5.6*	4.6	10.6*	14.0	19.9*	25.6*	28.7*

Level	⁴¹Nb	⁴²Mo	⁴³Tc	⁴⁴Ru	⁴⁵Rh	⁴⁶Pd	⁴⁷Ag	⁴⁸Cd
K	18985.6	19999.5	21044.0	22117.2	23219.9	24350.3	25514.0	26711.2
L_I	2697.7	2865.5	3042.5	3224.0	3411.9	3604.3	3805.8	4018.0
L_{II}	2464.7	2625.1	2793.2	2966.9	3146.1	3330.3	3523.7	3727.0
L_{III}	2370.5	2520.2	2676.9	2837.9	3003.8	3173.3	3351.1	3537.5
M_I	468.4	504.6		585.0	627.1	669.9	717.5	770.2
M_{II}	378.4	409.7	444.9	482.8	521.0	559.1	602.4	650.7
M_{III}	363.0	392.3	425.0	460.6	496.2	531.5	571.4	616.5
M_{IV}	207.4	230.3	256.4	283.6	311.7	340.0	372.8	410.5

Level	^{41}Nb	^{42}Mo	^{43}Tc	^{44}Ru	^{45}Rh	^{46}Pd	^{47}Ag	^{48}Cd
M$_V$	204.6	227.0	252.9	279.4	307.0	334.7	366.7	403.7
N$_I$	58.1	61.8		74.9	81.0	86.4	95.2	107.6
N$_{II}$	33.9	34.8	38.9	43.1	47.9	51.1	62.6	66.9
N$_{III}$	33.9*	34.8*	38.9*	43.1*	47.9*	51.1*	55.9	66.9*
N$_{IV,V}$	3.2	1.8		2.0	2.5	1.5	3.3	9.3

Level	^{49}In	^{50}Sn	^{51}Sb	^{52}Te	^{53}I	^{54}Xe	^{55}Cs	^{56}Ba
K	27939.9	29200.1	30491.2	31813.8	33169.4	34561.4	35984.6	37440.6
L$_I$	4237.5	4464.7	4698.3	4939.2	5188.1	5452.8	5714.3	5988.8
L$_{II}$	3938.0	4156.1	4380.4	4612.0	4852.1	5103.7	5359.4	5623.6
L$_{III}$	3730.1	3928.8	4132.2	4341.4	4557.1	4782.2	5011.9	5247.0
M$_I$	825.6	883.8	943.7	1006.0	1072.1		1217.1	1292.8
M$_{II}$	702.2	756.4	811.9	869.7	930.5	999.0	1065.0	1136.7
M$_{III}$	664.3	714.4	765.6	818.7	874.6	937.0	997.6	1062.2
M$_{IV}$	450.8	493.3	536.9	582.5	631.3		739.5	796.1
M$_V$	443.1	484.8	527.5	572.1	619.4	672.3	725.5	780.7
N$_I$	121.9	136.5	152.0	168.3	186.4		230.8	253.0
N$_{II}$	77.4	88.6	98.4	110.2	122.7	146.7	172.3	191.8
N$_{III}$	77.4*	88.6*	98.4*	110.2*	122.7*	146.7*	161.6	179.7
N$_{IV}$	16.2	23.9	31.4	39.8	49.6		78.8	92.5
N$_V$	16.2*	23.9*	31.4*	39.8*	49.6*		76.5	89.9
O$_I$	0.1	0.9	6.7	11.6	13.6		22.7	39.1
O$_{II}$	0.8	1.1	2.1	2.3	3.3		13.1	16.6
O$_{III}$	0.8*	1.1*	2.1*	2.3*	3.3*		11.4	14.6

Level	^{57}La	^{58}Ce	^{59}Pr	^{60}Nd	^{61}Pm	^{62}Sm	^{63}Eu	^{64}Gd
K	38924.6	40443.0	41990.6	43568.9	45184.0	46834.2	48519.0	50239.1
L$_I$	6266.3	6548.8	6834.8	7126.0	7427.9	7736.8	8052.0	8375.6
L$_{II}$	5890.6	6164.2	6440.4	6721.5	7012.8	7311.8	7617.1	7930.3
L$_{III}$	5482.7	5723.4	5964.3	6207.9	6459.3	6716.2	6976.9	7242.8
M$_I$	1361.3	1434.6	1511.0	1575.3		1722.8	1800.0	1880.8
M$_{II}$	1204.4	1272.8	1337.4	1402.8	1471.4	1540.7	1613.9	1688.3
M$_{III}$	1123.4	1185.4	1242.2	1297.4	1356.9	1419.8	1480.6	1544.0
M$_{IV}$	848.5	901.3	951.1	999.9	1051.5	1106.0	1160.6	1217.2
M$_V$	831.7	883.3	931.0	977.7	1026.9	1080.2	1130.9	1185.2
N$_I$	270.4	289.6	304.5	315.2		345.7	360.2	375.8
N$_{II}$	205.8	223.3	236.3	243.3	242	265.6	283.9	288.5
N$_{III}$	191.4	207.2	217.6	224.6	242*	247.4	256.6	270.9
N$_{IV,V}$	98.9	110.0	113.2	117.5	120.4	129.0	133.2	140.5
N$_{VI,VII}$		0.1	2.0	1.5		5.5	0.0	0.1
O$_I$	32.3	37.8	37.4	37.5		37.4	31.8	36.1
O$_{II,III}$	14.4	19.8	22.3	21.1		21.3	22.0	20.3

Level	^{65}Tb	^{66}Dy	^{67}Ho	^{68}Er	^{69}Tm	^{70}Yb	^{71}Lu	^{72}Hf
K	51995.7	53788.5	55617.7	57485.5	59389.6	61332.3	63313.8	65350.8
L$_I$	8708.0	9045.8	9394.2	9751.3	10115.7	10486.4	10870.4	11270.7
L$_{II}$	8251.6	8580.6	8917.8	9264.3	9616.9	9978.2	10348.6	10739.4
L$_{III}$	7514.0	7790.1	8071.1	8357.9	8648.0	8943.6	9244.1	9560.7
M$_I$	1967.5	2046.8	2128.3	2206.5	2306.8	2398.1	2491.2	2600.9
M$_{II}$	1767.7	1841.8	1922.8	2005.8	2089.8	2173.0	2263.5	2365.4
M$_{III}$	1611.3	1675.6	1741.2	1811.8	1884.5	1949.8	2023.6	2107.6
M$_{IV}$	1275.0	1332.5	1391.5	1453.3	1514.6	1576.3	1639.4	1716.4
M$_V$	1241.2	1294.9	1351.4	1409.3	1467.7	1527.8	1588.5	1661.7
N$_I$	397.9	416.3	435.7	449.1	471.7	487.2	506.2	538.1
N$_{II}$	310.2	331.8	343.5	366.2	385.9	396.7	410.1	437.0
N$_{III}$	385.0	292.9	306.6	320.0	336.6	343.5	359.3	380.4
N$_{IV}$	147.0	154.2	161.0	176.7	179.6	198.1	204.8	223.8
N$_V$	147.0*	154.2*	161.0*	167.6	179.6*	184.9	195.0	213.7
N$_{VI,VII}$	2.6	4.2	3.7	4.3	5.3	6.3	6.9	17.1
O$_I$	39.0	62.9	51.2	59.8	53.2	54.1	56.8	64.9
O$_{II}$	25.4	26.3	20.3	29.4	32.3	23.4	28.0	38.1
O$_{III}$	25.4*	26.3*	20.3*	29.4*	32.3*	23.4*	28.0*	30.6

Level	^{73}Ta	^{74}W	^{75}Re	^{76}Os	^{77}Ir	^{78}Pt	^{79}Au	^{80}Hg
K	67416.4	69525.0	71676.4	73870.8	76111.0	78394.8	80724.9	83102.3
L$_I$	11681.5	12099.8	12526.7	12968.0	13418.5	13879.9	14352.8	14839.3
L$_{II}$	11136.1	11544.0	11958.7	12385.0	12824.1	13272.6	13733.6	14208.7
L$_{III}$	9881.1	10206.8	10535.3	10870.9	11215.2	11563.7	11918.7	12283.9
M$_I$	2708.0	2819.6	2931.7	3048.5	3173.7	3296.0	3424.9	3561.6
M$_{II}$	2468.7	2574.9	2681.6	2792.2	2908.7	3026.5	3147.8	3278.5
M$_{III}$	2194.0	2281.0	2367.3	2457.2	2550.7	2645.4	2743.0	2847.1
M$_{IV}$	1793.2	1871.6	1948.9	2030.8	2116.1	2201.9	2291.1	2384.9
M$_V$	1735.1	1809.2	1882.9	1960.1	2040.4	2121.6	2205.7	2294.9
N$_I$	565.5	595.0	625.0	654.3	690.1	722.0	758.8	800.3
N$_{II}$	464.8	491.6	517.9	546.5	577.1	609.2	643.7	676.9
N$_{III}$	404.5	425.3	444.4	468.2	494.3	519.0	545.4	571.0
N$_{IV}$	241.3	258.8	273.7	289.4	311.4	330.8	352.0	378.3
N$_V$	229.3	245.4	260.2	272.8	294.9	313.3	333.9	359.8
N$_{VI}$	25.0	36.5	40.6	46.3	63.4	74.3	86.4	102.2
N$_{VII}$	25.0*	33.6	40.6*	46.3*	60.5	71.1	82.8	98.5
O$_I$	71.1	77.1	82.8	83.7	95.2	101.7	107.8	120.3
O$_{II}$	44.9	46.8	45.6	58.0	63.0	65.3	71.7	80.5
O$_{III}$	36.4	35.6	34.6	45.4	50.5	51.7	53.7	57.6
O$_{IV,V}$	5.7	6.1	3.5		3.8	2.2	2.5	6.4

Level	^{81}Tl	^{82}Pb	^{83}Bi	^{84}Po	^{85}At	^{86}Rn	^{87}Fr	^{88}Ra
K	85530.4	88004.5	90525.9	93105.0	95729.9	98404	101137	103921.9
L$_I$	15346.7	15860.8	16387.5	16939.3	17493	18049	18639	19236.7
L$_{II}$	14697.9	15200.0	15711.1	16244.3	16784.7	17337.1	17906.5	18484.3
L$_{III}$	12657.5	13035.2	13418.6	13813.8	14213.5	14619.4	15031.2	15444.4
M$_I$	3704.1	3850.7	3999.1	4149.4	(4317)	(4482)	(4652)	4822.0
M$_{II}$	3415.7	3554.2	3696.3	3854.1	4008	4159	4327	4489.5
M$_{III}$	2956.6	3066.4	3176.9	3301.9	3426	3538	3663	3791.8
M$_{IV}$	2485.1	2585.6	2687.6	2798.0	2908.7	3021.5	3136.2	3248.4
M$_V$	2389.3	2484.0	2579.6	2683.0	2786.7	2892.4	2999.9	3104.9
N$_I$	845.5	893.6	938.2	995.3	(1042)	(1097)	(1153)	1208.4
N$_{II}$	721.3	763.9	805.3	851	886	929	980	1057.6
N$_{III}$	609.0	644.5	678.9	705	740	768	810	879.1
N$_{IV}$	406.6	435.2	463.6	500.2	533.2	566.6	603.3	635.9
N$_V$	386.2	412.9	440.0	473.4			577	602.7
N$_{VI}$	122.8	142.9	161.9					298.9
N$_{VII}$	118.5	138.1	157.4					298.9*
O$_I$	136.3	147.3	159.3					254.4
O$_{II}$	99.6	104.8	116.8					200.4
O$_{III}$	75.4	86.0	92.8					152.8
O$_{IV}$	15.3	21.8	26.5	31.4				67.2
O$_V$	13.1	19.2	24.4	31.4*				67.2*
P$_I$		3.1						43.5
P$_{II,III}$		0.7	2.7					18.8

Level	^{89}Ac	^{90}Th	^{91}Pa	^{92}U	^{93}Np	^{94}Pu	^{95}Am	^{96}Cm
K	106755.3	109650.9	112601.4	115606.1	118678	121818	125027	128220
L$_I$	19840	20472.1	21104.6	21757.4	22426.8	23097.2	23772.9	24460
L$_{II}$	19083.2	19693.2	20313.7	20947.6	21600.5	22266.2	22944.0	23779
L$_{III}$	15871.0	16300.3	16733.1	17166.3	17610.0	18056.8	18504.1	18930
M$_I$	(5002)	5182.3	5366.9	5548.0	5723.2	5932.9	6120.5	6288
M$_{II}$	4656	4830.4	5000.9	5182.2	5366.2	5541.2	5710.2	5895
M$_{III}$	3909	4046.1	4173.8	4303.4	4434.7	4556.6	4667.0	4797
M$_{IV}$	3370.2	3490.8	3611.2	3727.6	3850.3	3972.6	4092.1	4227
M$_V$	3219.0	3332.0	3441.8	3551.7	3665.8	3778.1	3886.9	3971
N$_I$	(1269)	1329.5	1387.1	1440.8	1500.7	1558.6	1617.1	1643
N$_{II}$	1080	1168.2	1224.3	1272.6	1327.7	1372.1	1411.8	1440
N$_{III}$	890	967.3	1006.7	1044.9	1086.8	1114.8	(1135.7)	1154
N$_{IV}$	674.9	714.1	743.4	780.4	815.9	848.9	878.7	

Level	89Ac	90Th	91Pa	92U	93Np	94Pu	95Am	96Cm
N_V		676.4	708.2	737.7	770.3	801.4	827.6	
N_{VI}		344.4	371.2	391.3	415.0	445.8		
N_{VII}		335.2	359.5	380.9	404.4	432.4		
O_I		290.2	309.6	323.7		351.9		385
O_{II}		229.4	222.9	259.3	283.4	274.1		
O_{III}		181.8	222.9*	195.1	206.1	206.5		
O_{IV}		94.3	94.1	105.0	109.3	116.0	115.8	
O_V		87.9	94.1*	96.3	101.3	105.4	103.3	
P_I		59.5		70.7				
P_{II}		49.0		42.3				
P_{III}		43.0		32.3				

Level	97Bk	98Cf	99Es	100Fm	101Md	102No	103Lr
K	131590	135960	139490	143090	146780	150540	154380
L_I	25275	26110	26900	27700	28530	29380	30240
L_{II}	24385	25250	26020	26810	27610	28440	29280
L_{III}	19452	19930	20410	20900	21390	21880	22360
M_I	6556	6754	6977	7205	7441	7675	7900
M_{II}	6147	6359	6574	6793	7019	7245	7460
M_{III}	4977	5109	5252	5397	5546	5688	5710
M_{IV}	4366	4497	4630	4766	4903	5037	5150
M_V	4132	4253	4374	4498	4622	4741	4860
N_I	1755	1799	1868	1937	2010	2078	2140
N_{II}	1554	1616	1680	1747	1814	1876	1930
N_{III}	1235	1279	1321	1366	1410	1448	1480
O_I	398	419	435	454	472	484	490

ELECTRON BINDING ENERGIES OF THE ELEMENTS

Gwyn P. Williams

This table gives the binding energies in electron volts (eV) for selected electronic levels of the elements. For metallic elements the binding energy is referred to the Fermi level; for semiconductors, to the valence band maximum; and for gases and insulators, to the vacuum level. The atomic number is listed after the element name.

References

1. Fluggle and Martensson, *J. Elect. Spect.*, 21, 275, 1980.
2. Cardona, M. and Ley, L., *Photoemission from Solids*, Springer Verlag, Heidelberg, 1978.
3. Bearden, J. A. and Burr, A. F., *Rev. Mod. Phys.*, 39, 125, 1967.

Actinium (89)

K	1s	106755
L I	2s	19840
L II	$2p_{1/2}$	19083
L III	$2p_{3/2}$	15871
M I	3s	5002
M II	$3p_{1/2}$	4656
M III	$3p_{3/2}$	3909
M IV	$3d_{3/2}$	3370
M V	$3d_{5/2}$	3219
N I	4s	1269[a]
N II	$4p_{1/2}$	1080[a]
N III	$4p_{3/2}$	890[a]
N IV	$4d_{3/2}$	675[a]
N V	$4d_{5/2}$	639[a]
N VI	$4f_{5/2}$	319[a]
N VII	$4f_{7/2}$	319[a]
O I	5s	272[a]
O II	$5p_{1/2}$	215[a]
O III	$5p_{3/2}$	167[a]
O IV	$5d_{3/2}$	80[a]
O V	$5d_{5/2}$	80[a]
P I	6s	—
P II	$6p_{1/2}$	—
P III	$6p_{3/2}$	—

Aluminum (13)

K	1s	1559.0
L I	2s	117.8[a]
L II	$2p_{1/2}$	72.9[a]
L III	$2p_{3/2}$	72.5[a]

Antimony (51)

K	1s	30419
L I	2s	4698
L II	$2p_{1/2}$	4380
L III	$2p_{3/2}$	4132
M I	3s	946[b]
M II	$3p_{1/2}$	812.7[b]
M III	$3p_{3/2}$	766.4[b]
M IV	$3d_{3/2}$	537.5[b]
M V	$3d_{5/2}$	528.2[b]
N I	4s	153.2[b]
N II	$4p_{1/2}$	95.6[b,c]
N III	$4p_{3/2}$	95.6[b]
N IV	$4d_{3/2}$	33.3[b]
N V	$4d_{5/2}$	32.1[b]

Argon (18)

K	1s	3205.9[a]
L I	2s	326.3[a]
L II	$2p_{1/2}$	250.6[a]
L III	$2p_{3/2}$	248.4[a]
M I	3s	29.3[a]
M II	$3p_{1/2}$	15.9[a]
M III	$3p_{3/2}$	15.7[a]

Arsenic (33)

K	1s	11867
L I	2s	1527.0[a,d]
L II	$2p_{1/2}$	1359.1[a,d]
L III	$2p_{3/2}$	1323.6[a,d]
M I	3s	204.7[a]
M II	$3p_{1/2}$	146.2[a]
M III	$3p_{3/2}$	141.2[a]
M IV	$3d_{3/2}$	41.7[a]
M V	$3d_{5/2}$	41.7[a]

Astatine (85)

K	1s	95730
L I	2s	17493
L II	$2p_{1/2}$	16785
L III	$2p_{3/2}$	14214
M I	3s	4317
M II	$3p_{1/2}$	4008
M III	$3p_{3/2}$	3426
M IV	$3d_{3/2}$	2909
M V	$3d_{5/2}$	2787
N I	4s	1042[a]
N II	$4p_{1/2}$	886[a]
N III	$4p_{3/2}$	740[a]
N IV	$4d_{3/2}$	533[a]
N V	$4d_{5/2}$	507[a]
N VI	$4f_{5/2}$	210[a]
N VII	$4f_{7/2}$	210[a]
O I	5s	195[a]
O II	$5p_{1/2}$	148[a]
O III	$5p_{3/2}$	115[a]
O IV	$5d_{3/2}$	40[a]
O V	$5d_{5/2}$	40[a]

Barium (56)

K	1s	37441
L I	2s	5989
L II	$2p_{1/2}$	5624
L III	$2p_{3/2}$	5247
M I	3s	1293[a,d]
M II	$3p_{1/2}$	1137[a,d]
M III	$3p_{3/2}$	1063[a,d]
M IV	$3d_{3/2}$	795.7[a]
M V	$3d_{5/2}$	780.5[a]
N I	4s	253.5[b]
N II	$4p_{1/2}$	192
N III	$4p_{3/2}$	178.6[b]
N IV	$4d_{3/2}$	92.6[b]
N V	$4d_{5/2}$	89.9[b]
N VI	$4f_{5/2}$	—
N VII	$4f_{7/2}$	—
O I	5s	30.3[b]
O II	$5p_{1/2}$	17.0[b]
O III	$5p_{3/2}$	14.8[b]

Beryllium (4)

K	1s	111.5[a]

Bismuth (83)

K	1s	90526
L I	2s	16388
L II	$2p_{1/2}$	15711
L III	$2p_{3/2}$	13419
M I	3s	3999
M II	$3p_{1/2}$	3696
M III	$3p_{3/2}$	3177
M IV	$3d_{3/2}$	2688
M V	$3d_{5/2}$	2580
N I	4s	939[b]
N II	$4p_{1/2}$	805.2[b]
N III	$4p_{3/2}$	678.8[b]
N IV	$4d_{3/2}$	464.0[b]
N V	$4d_{5/2}$	440.1[b]
N VI	$4f_{5/2}$	162.3[b]
N VII	$4f_{7/2}$	157.0[b]
O I	5s	159.3[a,d]
O II	$5p_{1/2}$	119.0[b]
O III	$5p_{3/2}$	92.6[b]
O IV	$5d_{3/2}$	26.9[b]
O V	$5d_{5/2}$	23.8[b]

Boron (5)

K	1s	188[a]

Bromine (35)

K	1s	13474
L I	2s	1782[a]
L II	$2p_{1/2}$	1596[a]
L III	$2p_{3/2}$	1550[a]
M I	3s	257[a]
M II	$3p_{1/2}$	189[a]
M III	$3p_{3/2}$	182[a]
M IV	$3d_{3/2}$	70[a]
M V	$3d_{5/2}$	69[a]

Cadmium (48)

K	1s	26711
L I	2s	4018
L II	$2p_{1/2}$	3727

Calcium (20)

K	1s	4038.5[a]
L I	2s	438.4[b]
L II	$2p_{1/2}$	349.7[b]
L III	$2p_{3/2}$	346.2[b]
M I	3s	44.3[b]
M II	$3p_{1/2}$	25.4[b]
M III	$3p_{3/2}$	25.4[b]

Carbon (6)

K	1s	284.2[a]

Cerium (58)

K	1s	40443
L I	2s	6548
L II	$2p_{1/2}$	6164
L III	$2p_{3/2}$	5723
M I	3s	1436[a,d]
M II	$3p_{1/2}$	1274[a,d]
M III	$3p_{3/2}$	1187[a,d]
M IV	$3d_{3/2}$	902.4[a]
M V	$3d_{5/2}$	883.8[a]
N I	4s	291.0[a]
N II	$4p_{1/2}$	223.3
N III	$4p_{3/2}$	206.5[a]
N IV	$4d_{3/2}$	109[a]
N V	$4d_{5/2}$	—
N VI	$4f_{5/2}$	0.1
N VII	$4f_{7/2}$	0.1
O I	5s	37.8
O II	$5p_{1/2}$	19.8[a]
O III	$5p_{3/2}$	17.0[a]

Cesium (55)

K	1s	35985
L I	2s	5714
L II	$2p_{1/2}$	5359
L III	$2p_{3/2}$	5012
M I	3s	1211[a,d]
M II	$3p_{1/2}$	1071[a]

L-column overlap for Antimony/Cadmium section continued:

L III	$2p_{3/2}$	3538
M I	3s	772.0[b]
M II	$3p_{1/2}$	652.6[b]
M III	$3p_{3/2}$	618.4[b]
M IV	$3d_{3/2}$	411.9[b]
M V	$3d_{5/2}$	405.2[b]
N I	4s	109.8[b]
N II	$4p_{1/2}$	63.9[b,c]
N III	$4p_{3/2}$	63.9[b,c]
N IV	$4d_{3/2}$	11.7[b]
N V	$4d_{5/2}$	10.7[b]

M III	$3p_{3/2}$	1003[a]
M IV	$3d_{3/2}$	740.5[a]
M V	$3d_{5/2}$	726.6[a]
N I	4s	232.3[a]
N II	$4p_{1/2}$	172.4[a]
N III	$4p_{3/2}$	161.3[a]
N IV	$4d_{3/2}$	79.8[a]
N V	$4d_{5/2}$	77.5[a]
N VI	$4f_{5/2}$	—
N VII	$4f_{7/2}$	—
O I	5s	22.7
O II	$5p_{1/2}$	14.2[a]
O III	$5p_{3/2}$	12.1[a]

Chlorine (17)

K	1s	2822.0
L I	2s	270[a]
L II	$2p_{1/2}$	202[a]
L III	$2p_{3/2}$	200[a]

Chromium(24)

K	1s	5989
L I	2s	696.0[b]
L II	$2p_{1/2}$	583.8[b]
L III	$2p_{3/2}$	574.1[b]
M I	3s	74.1[b]
M II	$3p_{1/2}$	42.2[b]
M III	$3p_{3/2}$	42.2[b]

Cobalt (27)

K	1s	7709
L I	2s	925.1[b]
L II	$2p_{1/2}$	793.2[b]
L III	$2p_{3/2}$	778.1[b]
M I	3s	101.0[b]
M II	$3p_{1/2}$	58.9[b]
M III	$3p_{3/2}$	58.9[b]

Copper (29)

K	1s	8979
L I	2s	1096.7[b]
L II	$2p_{1/2}$	952.3[b]
L III	$2p_{3/2}$	932.5[b]
M I	3s	122.5[b]
M II	$3p_{1/2}$	77.3[b]
M III	$3p_{3/2}$	75.1[b]

Dysprosium (66)

K	1s	53789
L I	2s	9046
L II	$2p_{1/2}$	8581
L III	$2p_{3/2}$	7790
M I	3s	2047
M II	$3p_{1/2}$	1842
M III	$3p_{3/2}$	1676
M IV	$3d_{3/2}$	1333
M V	$3d_{5/2}$	1292[a]
N I	4s	414.2[a]
N II	$4p_{1/2}$	333.5[a]
N III	$4p_{3/2}$	293.2[a]
N IV	$4d_{3/2}$	153.6[a]
N V	$4d_{5/2}$	153.6[a]
N VI	$4f_{5/2}$	8.0[a]
N VII	$4f_{7/2}$	4.3[a]
O I	5s	49.9[a]

O II	$5p_{1/2}$	26.3
O III	$5p_{3/2}$	26.3

Erbium (68)

K	1s	57486
L I	2s	9751
L II	$2p_{1/2}$	9264
L III	$2p_{3/2}$	8358
M I	3s	2206
M II	$3p_{1/2}$	2006
M III	$3p_{3/2}$	1812
M IV	$3d_{3/2}$	1453
M V	$3d_{5/2}$	1409
N I	4s	449.8[a]
N II	$4p_{1/2}$	366.2
N III	$4p_{3/2}$	320.2[a]
N IV	$4d_{3/2}$	167.6[a]
N V	$4d_{5/2}$	167.6[a]
N VI	$4f_{5/2}$	—
N VII	$4f_{7/2}$	4.7[a]
O I	5s	50.6[a]
O II	$5p_{1/2}$	31.4[a]
O III	$5p_{3/2}$	24.7[a]

Europium (63)

K	1s	48519
L I	2s	8052
L II	$2p_{1/2}$	7617
L III	$2p_{3/2}$	6977
M I	3s	1800
M II	$3p_{1/2}$	1614
M III	$3p_{3/2}$	1481
M IV	$3d_{3/2}$	1158.6[a]
M V	$3d_{5/2}$	1127.5[a]
N I	4s	360
N II	$4p_{1/2}$	284
N III	$4p_{3/2}$	257
N IV	$4d_{3/2}$	133
N V	$4d_{5/2}$	1227[a]
N VI	$4f_{5/2}$	0
N VII	$4f_{7/2}$	0
O I	5s	32
O II	$5p_{1/2}$	22
O III	$5p_{3/2}$	22

Fluorine (9)

K	1s	696.7[a]

Francium (87)

K	1s	101137
L I	2s	18639
L II	$2p_{1/2}$	17907
L III	$2p_{3/2}$	15031
M I	3s	4652
M II	$3p_{1/2}$	4327
M III	$3p_{3/2}$	3663
M IV	$3d_{3/2}$	3136
M V	$3d_{5/2}$	3000
N I	4s	1153[a]
N II	$4p_{1/2}$	980[a]
N III	$4p_{3/2}$	810[a]
N IV	$4d_{3/2}$	603[a]
N V	$4d_{5/2}$	577[a]
N VI	$4f_{5/2}$	268[a]
N VII	$4f_{7/2}$	268[a]

O I	5s	234[a]
O II	$5p_{1/2}$	182[a]
O III	$5p_{3/2}$	140[a]
O IV	$5d_{3/2}$	58[a]
O V	$5d_{5/2}$	58[a]
P I	6s	34
P II	$6p_{1/2}$	15
P III	$6p_{3/2}$	15

Gadolinium (64)

K	1s	50239
L I	2s	8376
L II	$2p_{1/2}$	7930
L III	$2p_{3/2}$	7243
M I	3s	1881
M II	$3p_{1/2}$	1688
M III	$3p_{3/2}$	1544
M IV	$3d_{3/2}$	1221.9[a]
M V	$3d_{5/2}$	1189.6[a]
N I	4s	378.6[a]
N II	$4p_{1/2}$	286
N III	$4p_{3/2}$	271
N IV	$4d_{3/2}$	—
N V	$4d_{5/2}$	142.6[a]
N VI	$4f_{5/2}$	8.6[a]
N VII	$4f_{7/2}$	8.6[a]
O I	5s	36
O II	$5p_{1/2}$	20
O III	$5p_{3/2}$	20

Gallium (31)

K	1s	10367
L I	2s	1299.0[a,d]
L II	$2p_{1/2}$	1143.2[b]
L III	$2p_{3/2}$	1116.4[b]
M I	3s	159.5[b]
M II	$3p_{1/2}$	103.5[b]
M III	$3p_{3/2}$	100.0[b]
M IV	$3d_{3/2}$	18.7[b]
M V	$3d_{5/2}$	18.7[b]

Germanium (32)

K	1s	11103
L I	2s	1414.6[a,d]
L II	$2p_{1/2}$	1248.1[a,d]
L III	$2p_{3/2}$	1217.0[a,d]
M I	3s	180.1[a]
M II	$3p_{1/2}$	124.9[a]
M III	$3p_{3/2}$	120.8[a]
M IV	$3d_{3/2}$	29.8[a]
M V	$3d_{5/2}$	29.2[a]

Gold (79)

K	1s	80725
L I	2s	14353
L II	$2p_{1/2}$	13734
L III	$2p_{3/2}$	11919
M I	3s	3425
M II	$3p_{1/2}$	3148
M III	$3p_{3/2}$	2743
M IV	$3d_{3/2}$	2291
M V	$3d_{5/2}$	2206
N I	4s	762.1[b]
N II	$4p_{1/2}$	642.7[b]
N III	$4p_{3/2}$	546.3[b]

N IV	$4d_{3/2}$	353.2[b]
N V	$4d_{5/2}$	335.1[b]
N VI	$4f_{5/2}$	87.6[b]
N VII	$4f_{7/2}$	83.9[b]
O I	5s	107.2[a,d]
O II	$5p_{1/2}$	74.2[b]
O III	$5p_{3/2}$	57.2[b]

Hafnium (72)

K	1s	65351
L I	2s	11271
L II	$2p_{1/2}$	10739
L III	$2p_{3/2}$	9561
M I	3s	2601
M II	$3p_{1/2}$	2365
M III	$3p_{3/2}$	2107
M IV	$3d_{3/2}$	1176
M V	$3d_{5/2}$	1662
N I	4s	538[a]
N II	$4p_{1/2}$	438.2[b]
N III	$4p_{3/2}$	380.7[b]
N IV	$4d_{3/2}$	220.0[b]
N V	$4d_{5/2}$	211.5[b]
N VI	$4f_{5/2}$	15.9[b]
N VII	$4f_{7/2}$	14.2[b]
O I	5s	64.2[b]
O II	$5p_{1/2}$	38[a]
O III	$5p_{3/2}$	29.9[b]

Helium (2)

K	1s	24.6[a]

Holmium (67)

K	1s	55618
L I	2s	9394
L II	$2p_{1/2}$	8918
L III	$2p_{3/2}$	8071
M I	3s	2128
M II	$3p_{1/2}$	1923
M III	$3p_{3/2}$	1741
M IV	$3d_{3/2}$	1392
M V	$3d_{5/2}$	1351
N I	4s	432.4[a]
N II	$4p_{1/2}$	343.5
N III	$4p_{3/2}$	308.2[a]
N IV	$4d_{3/2}$	160[a]
N V	$4d_{5/2}$	160[a]
N VI	$4f_{5/2}$	8.6[a]
N VII	$4f_{7/2}$	5.2[a]
O I	5s	49.3[a]
O II	$5p_{1/2}$	30.8[a]
O III	$5p_{3/2}$	24.1[a]

Hydrogen (1)

K	1s	13.6

Indium (49)

K	1s	27940
L I	2s	4238
L II	$2p_{1/2}$	3938
L III	$2p_{3/2}$	3730
M I	3s	827.2[b]
M II	$3p_{1/2}$	703.2[b]
M III	$3p_{3/2}$	665.3[b]
M IV	$3d_{3/2}$	451.4[b]

Level	Subshell	Energy
M V	$3d_{5/2}$	443.9[b]
N I	4s	122.9[b]
N II	$4p_{1/2}$	73.5[b,c]
N III	$4p_{3/2}$	73.5[b,c]
N IV	$4d_{3/2}$	17.7[b]
N V	$4d_{5/2}$	16.9[b]

Iodine (53)

Level	Subshell	Energy
K	1s	33169
L I	2s	5188
L II	$2p_{1/2}$	4852
L III	$2p_{3/2}$	4557
M I	3s	1072[a]
M II	$3p_{1/2}$	931[a]
M III	$3p_{3/2}$	875[a]
M IV	$3d_{3/2}$	631[a]
M V	$3d_{5/2}$	620[a]
N I	4s	186[a]
N II	$4p_{1/2}$	123[a]
N III	$4p_{3/2}$	123[a]
N IV	$4d_{3/2}$	50[a]
N V	$4d_{5/2}$	50[a]

Iridium (77)

Level	Subshell	Energy
K	1s	76111
L I	2s	13419
L II	$2p_{1/2}$	12824
L III	$2p_{3/2}$	11215
M I	3s	3174
M II	$3p_{1/2}$	2909
M III	$3p_{3/2}$	2551
M IV	$3d_{3/2}$	2116
M V	$3d_{5/2}$	2040
N I	4s	691.1[b]
N II	$4p_{1/2}$	577.8[b]
N III	$4p_{3/2}$	495.8[b]
N IV	$4d_{3/2}$	311.9[b]
N V	$4d_{5/2}$	296.3[b]
N VI	$4f_{5/2}$	63.8[b]
N VII	$4f_{7/2}$	60.8[b]
O I	52	95.2[a,d]
O II	$5p_{1/2}$	63.0[a,d]
O III	$5p_{3/2}$	48.0[b]

Iron (26)

Level	Subshell	Energy
K	1s	7112
L I	2s	844.6[b]
L II	$2p_{1/2}$	719.9[b]
L III	$2p_{3/2}$	706.8[b]
M I	3s	91.3[b]
M II	$3p_{1/2}$	52.7[b]
M III	$3p_{3/2}$	52.7[b]

Krypton (36)

Level	Subshell	Energy
K	1s	14326
L I	2s	1921
L II	$2p_{1/2}$	1730.9[a]
L III	$2p_{3/2}$	1678.4[a]
M I	3s	292.8[a]
M II	$3p_{1/2}$	222.2[a]
M III	$3p_{3/2}$	214.4[a]
M IV	$3d_{3/2}$	95.0[a]
M V	$3d_{5/2}$	93.8[a]
N I	4s	27.5[a]
N II	$4p_{1/2}$	14.1[a]
N III	$4p_{3/2}$	14.1[a]

Lanthanum (57)

Level	Subshell	Energy
K	1s	38925
L I	2s	6266
L II	$2p_{1/2}$	5891
L III	$2p_{3/2}$	5483
M I	3s	1362[a,d]
M II	$3p_{1/2}$	1209[a,d]
M III	$3p_{3/2}$	1128[a,d]
M IV	$3d_{3/2}$	853[a]
M V	$3d_{5/2}$	836[a]
N I	4s	247.7[a]
N II	$4p_{1/2}$	205.8
N III	$4p_{3/2}$	196.0[a]
N IV	$4d_{3/2}$	105.3[a]
N V	$4d_{5/2}$	102.5[a]
N VI	$4f_{5/2}$	—
N VII	$4f_{7/2}$	—
O I	5s	34.3[a]
O II	$5p_{1/2}$	19.3[a]
O III	$5p_{3/2}$	16.8[a]

Lead (82)

Level	Subshell	Energy
K	1s	88005
L I	2s	15861
L II	$2p_{1/2}$	15200
L III	$2p_{3/2}$	13055
M I	3s	3851
M II	$3p_{1/2}$	3554
M III	$3p_{3/2}$	3066
M IV	$3d_{3/2}$	2586
M V	$3d_{5/2}$	2484
N I	4s	891.8[b]
N II	$4p_{1/2}$	761.9[b]
N III	$4p_{3/2}$	643.5[b]
N IV	$4d_{3/2}$	434.3[b]
N V	$4d_{5/2}$	412.2[b]
N VI	$4f_{5/2}$	141.7[b]
N VII	$4f_{7/2}$	136.9[b]
O I	5s	147[a,d]
O II	$5p_{1/2}$	106.4[b]
O III	$5p_{3/2}$	83.3[b]
O IV	$5d_{3/2}$	20.7[b]
O V	$5d_{5/2}$	18.1[b]

Lithium (3)

Level	Subshell	Energy
K	1s	54.7[a]

Lutetium

Level	Subshell	Energy
K	1s	63314
L I	2s	10870
L II	$2p_{1/2}$	10349
L III	$2p_{3/2}$	9244
M I	3s	2491
M II	$3p_{1/2}$	2264
M III	$3p_{3/2}$	2024
M IV	$3d_{3/2}$	1639
M V	$3d_{5/2}$	1589
N I	4s	506.8[a]
N II	$4p_{1/2}$	412.4[a]
N III	$4p_{3/2}$	359.2[a]
N IV	$4d_{3/2}$	206.1[a]
N V	$4d_{5/2}$	196.3[a]
N VI	$4f_{5/2}$	8.9[a]
N VII	$4f_{7/2}$	7.5[a]
O I	5s	57.3[a]
O II	$5p_{1/2}$	33.6[a]
O III	$5p_{3/2}$	26.7[a]

Magnesium (12)

Level	Subshell	Energy
K	1s	1303.0[b]
L I	2s	88.6[a]
L II	$2p_{1/2}$	49.6[b]
L III	$2p_{3/2}$	49.2[a]

Manganese (25)

Level	Subshell	Energy
K	1s	6539
L I	2s	769.1[b]
L II	$2p_{1/2}$	649.9[b]
L III	$2p_{3/2}$	638.7[b]
M I	3s	82.3[b]
M II	$3p_{1/2}$	47.2[b]
M III	$3p_{3/2}$	47.2[b]

Mercury (80)

Level	Subshell	Energy
K	1s	83102
L I	2s	14839
L II	$2p_{1/2}$	14209
L III	$2p_{3/2}$	12284
M I	3s	3562
M II	$3p_{1/2}$	3279
M III	$3p_{3/2}$	2847
M IV	$3d_{3/2}$	2385
M V	$3d_{5/2}$	2295
N I	4s	802.2[b]
N II	$4p_{1/2}$	680.2[b]
N III	$4p_{3/2}$	576.6[b]
N IV	$4d_{3/2}$	378.2[b]
N V	$4d_{5/2}$	358.8[b]
N VI	$4f_{5/2}$	104.0[b]
N VII	$4f_{7/2}$	99.9[b]
O I	5s	127[b]
O II	$5p_{1/2}$	83.1[b]
O III	$5p_{3/2}$	64.5[b]
O IV	$5d_{3/2}$	9.6[b]
O V	$5d_{5/2}$	7.8[b]

Molybdenum (42)

Level	Subshell	Energy
K	1s	20000
L I	2s	2866
L II	$2p_{1/2}$	2625
L III	$2p_{3/2}$	2520
M I	3s	506.3[b]
M II	$3p_{1/2}$	411.6[b]
M III	$3p_{3/2}$	394.0[b]
M IV	$3d_{3/2}$	231.1[b]
M V	$3d_{5/2}$	227.9[b]
N I	4s	63.2[b]
N II	$4p_{1/2}$	37.6[b]
N III	$4p_{3/2}$	35.5[b]

Neodymium (60)

Level	Subshell	Energy
K	1s	43569
L I	2s	7126
L II	$2p_{1/2}$	6722
L III	$2p_{3/2}$	6208
M I	3s	1575
M II	$3p_{1/2}$	1403
M III	$3p_{3/2}$	1297
M IV	$3d_{3/2}$	1003.3[a]
M V	$3d_{5/2}$	980.4[a]
N I	4s	319.2[a]
N II	$4p_{1/2}$	243.3
N III	$4p_{3/2}$	224.6
N IV	$4d_{3/2}$	120.5[a]
N V	$4d_{5/2}$	120.5[a]
N VI	$4f_{5/2}$	1.5
N VII	$4f_{7/2}$	1.5
O I	5s	37.5
O II	$5p_{1/2}$	21.1
O III	$5p_{3/2}$	21.1

Neon (10)

Level	Subshell	Energy
K	1s	870.2[a]
L I	2s	48.5[a]
L II	$2p_{1/2}$	21.7[a]
L III	$2p_{3/2}$	21.6[a]

Nickel (28)

Level	Subshell	Energy
K	1s	8333
L I	2s	1008.6[b]
L II	$2p_{1/2}$	870.0[b]
L III	$2p_{3/2}$	852.7[b]
M I	3s	110.8[b]
M II	$3p_{1/2}$	68.0[b]
M III	$3p_{3/2}$	66.2[b]

Niobium (41)

Level	Subshell	Energy
K	1s	18986
L I	2s	2698
L II	$2p_{1/2}$	2465
L III	$2p_{3/2}$	2371
M I	3s	466.6[b]
M II	$3p_{1/2}$	376.1[b]
M III	$3p_{3/2}$	360.6[b]
M IV	$3d_{3/2}$	205.0[b]
M V	$3d_{5/2}$	202.3[b]
N I	4s	56.4[b]
N II	$4p_{1/2}$	32.6[b]
N III	$4p_{3/2}$	30.8[b]

Nitrogen (7)

Level	Subshell	Energy
K	1s	409.9[a]
L I	2s	37.3[a]

Osmium (76)

Level	Subshell	Energy
K	1s	73871
L I	2s	12968
L II	$2p_{1/2}$	12385
L III	$2p_{3/2}$	10871
M I	3s	3049
M II	$3p_{1/2}$	2792
M III	$3p_{3/2}$	2457
M IV	$3d_{3/2}$	2031
M V	$3d_{5/2}$	1960
N I	4s	658.2[b]
N II	$4p_{1/2}$	549.1[b]
N III	$4p_{3/2}$	470.7[b]
N IV	$4d_{3/2}$	293.1[b]
N V	$4d_{5/2}$	278.5[b]
N VI	$4f_{5/2}$	53.4[b]
N VII	$4f_{7/2}$	50.7[b]
O I	5s	84[a]
O II	$5p_{1/2}$	58[a]

O III	$5p_{3/2}$	44.5[b]

Oxygen (8)

K	1s	543.1[a]
L I	2s	41.6[a]

Palladium (46)

K	1s	24350
L I	2s	3604
L II	$2p_{1/2}$	3330
L III	$2p_{3/2}$	3173
M I	3s	671.6[b]
M II	$3p_{1/2}$	559.9[b]
M III	$3p_{3/2}$	532.3[b]
M IV	$3d_{3/2}$	340.5[b]
M V	$3d_{5/2}$	335.2[b]
N I	4s	87.1[a,d]
N II	$4p_{1/2}$	55.7[b,c]
N III	$4p_{3/2}$	50.9[b,c]

Phosphorus (15)

K	1s	2145.5
L I	2s	189[a]
L II	$2p_{1/2}$	136[a]
L III	$2p_{3/2}$	135[a]

Platinum (78)

K	1s	78395
L I	2s	13880
L II	$2p_{1/2}$	13273
L III	$2p_{3/2}$	11564
M I	3s	3296
M II	$3p_{1/2}$	3027
M III	$3p_{3/2}$	2645
M IV	$3d_{3/2}$	2202
M V	$3d_{5/2}$	2122
N I	4s	725.4[b]
N II	$4p_{1/2}$	609.1[b]
N III	$4p_{3/2}$	519.4[b]
N IV	$4d_{3/2}$	331.6[b]
N V	$4d_{5/2}$	314.6[b]
N VI	$4f_{5/2}$	74.5[b]
N VII	$4f_{7/2}$	71.2[b]
O I	5s	101.7[a,d]
O II	$5p_{1/2}$	65.3[a,b]
O III	$5p_{3/2}$	51.7[b]

Polonium (84)

K	1s	93105
L I	2s	16939
L II	$2p_{1/2}$	16244
L III	$2p_{3/2}$	13814
M I	3s	4149
M II	$3p_{1/2}$	3854
M III	$3p_{3/2}$	3302
M IV	$3d_{3/2}$	2798
M V	$3d_{5/2}$	2683
N I	4s	995[a]
N II	$4p_{1/2}$	851[a]
N III	$4p_{3/2}$	705[a]
N IV	$4d_{3/2}$	500[a]
N V	$4d_{5/2}$	473[a]
N VI	$4f_{5/2}$	184[a]
N VII	$4f_{7/2}$	184[a]
O I	5s	177[a]

O II	$5p_{1/2}$	132[a]
O III	$5p_{3/2}$	104[a]
O IV	$5d_{3/2}$	31[a]
O V	$5d_{5/2}$	31[a]

Potassium (19)

K	1s	3608.4[a]
L I	2s	378.6[a]
L II	$2p_{1/2}$	297.3[a]
L III	$2p_{3/2}$	294.6[a]
M I	3s	34.8[a]
M II	$3p_{1/2}$	18.3[a]
M III	$3p_{3/2}$	18.3[a]

Praseodymium (59)

K	1s	41991
L I	2s	6835
L II	$2p_{1/2}$	6440
L III	$2p_{3/2}$	5964
M I	3s	1511
M II	$3p_{1/2}$	1337
M III	$3p_{3/2}$	1242
M IV	$3d_{3/2}$	948.3[a]
M V	$3d_{5/2}$	928.8[a]
N I	4s	304.5
N II	$4p_{1/2}$	236.3
N III	$4p_{3/2}$	217.6
N IV	$4d_{3/2}$	115.1[a]
N V	$4d_{5/2}$	115.1[a]
N VI	$4f_{5/2}$	2.0
N VII	$4f_{7/2}$	2.0
O I	5s	37.4
O II	$5p_{1/2}$	22.3
O III	$5p_{3/2}$	22.3

Promethium (61)

K	1s	45184
L I	2s	7428
L II	$2p_{1/2}$	7013
L III	$2p_{3/2}$	6459
M I	3s	—
M II	$3p_{1/2}$	1471.4
M III	$3p_{3/2}$	1357
M IV	$3d_{3/2}$	1052
M V	$3d_{5/2}$	1027
N I	4s	—
N II	$4p_{1/2}$	242
N III	$4p_{3/2}$	242
N IV	$4d_{3/2}$	120
N V	$4d_{5/2}$	120

Protactinium (91)

K	1s	112601
L I	2s	21105
L II	$2p_{1/2}$	20314
L III	$2p_{3/2}$	16733
M I	3s	5367
M II	$3p_{1/2}$	5001
M III	$3p_{3/2}$	4174
M IV	$3d_{3/2}$	3611
M V	$3d_{5/2}$	3442
N I	4s	1387[a]
N II	$4p_{1/2}$	1224[a]
N III	$4p_{3/2}$	1007[a]
N IV	$4d_{3/2}$	743[a]

N V	$4d_{5/2}$	708[a]
N VI	$4f_{5/2}$	371[a]
N VII	$4f_{7/2}$	360[a]
O I	5s	310[a]
O II	$5p_{1/2}$	232[a]
O III	$5p_{3/2}$	232[a]
O IV	$5d_{3/2}$	94[a]
O V	$5d_{5/2}$	94[a]
P I	6s	—
P II	$6p_{1/2}$	—
P III	$6p_{3/2}$	—

Radium (88)

K	1s	103922
L I	2s	19237
L II	$2p_{1/2}$	18484
L III	$2p_{3/2}$	15444
M I	3s	4822
M II	$3p_{1/2}$	4490
M III	$3p_{3/2}$	3792
M IV	$3d_{3/2}$	3248
M V	$3d_{5/2}$	3105
N I	4s	1208[a]
N II	$4p_{1/2}$	1058
N III	$4p_{3/2}$	879[a]
N IV	$4d_{3/2}$	636[a]
N V	$4d_{5/2}$	603[a]
N VI	$4f_{5/2}$	299[a]
N VII	$4f_{7/2}$	299[a]
O I	5s	254[a]
O II	$5p_{1/2}$	200[a]
O III	$5p_{3/2}$	153[a]
O IV	$5d_{3/2}$	68[a]
O V	$5d_{5/2}$	68[a]
P I	6s	44
P II	$6p_{1/2}$	19
P III	$6p_{3/2}$	19

Radon (86)

K	1s	98404
L I	2s	18049
L II	$2p_{1/2}$	17337
L III	$2p_{3/2}$	14619
M I	3s	4482
M II	$3p_{1/2}$	4159
M III	$3p_{3/2}$	3538
M IV	$3d_{3/2}$	3022
M V	$3d_{5/2}$	2892
N I	4s	1097[a]
N II	$4p_{1/2}$	929[a]
N III	$4p_{3/2}$	768[a]
N IV	$4d_{3/2}$	567[a]
N V	$4d_{5/2}$	541[a]
N VI	$4f_{5/2}$	238[a]
N VII	$4f_{7/2}$	238[a]
O I	5s	214[a]
O II	$5p_{1/2}$	164[a]
O III	$5p_{3/2}$	127[a]
O IV	$5d_{3/2}$	48[a]
O V	$5d_{5/2}$	48[a]
P I	6s	26

Rhenium (75)

K	1s	71676

L I	2s	12527
L II	$2p_{1/2}$	11959
L III	$2p_{3/2}$	10535
M I	3s	2932
M II	$3p_{1/2}$	2682
M III	$3p_{3/2}$	2367
M IV	$3d_{3/2}$	1949
M V	$3d_{5/2}$	1883
N I	4s	625.4[b]
N II	$4p_{1/2}$	518.7[b]
N III	$4p_{3/2}$	446.8[b]
N IV	$4d_{3/2}$	273.9[b]
N V	$4d_{5/2}$	260.5[b]
N VI	$4f_{5/2}$	42.9[a]
N VII	$4f_{7/2}$	40.5[a]
O I	5s	83[b]
O II	$5p_{1/2}$	45.6[b]
O III	$5p_{3/2}$	34.6[a,d]

Rhodium (45)

K	1s	23220
L I	2s	3412
L II	$2p_{1/2}$	3146
L III	$2p_{3/2}$	3004
M I	3s	628.1[b]
M II	$3p_{1/2}$	521.3[b]
M III	$3p_{3/2}$	496.5[b]
M IV	$3d_{3/2}$	311.9[b]
M V	$3d_{5/2}$	307.2[b]
N I	4s	81.4[a,d]
N II	$4p_{1/2}$	50.5[b]
N III	$4p_{3/2}$	47.3[b]

Rubidium (37)

K	1s	15200
L I	2s	2065
L II	$2p_{1/2}$	1864
L III	$2p_{3/2}$	1804
M I	3s	326.7[a]
M II	$3p_{1/2}$	248.7[a]
M III	$3p_{3/2}$	239.1[a]
M IV	$3d_{3/2}$	113.0[a]
M V	$3d_{5/2}$	112[a]
N I	4s	30.5[a]
N II	$4p_{1/2}$	16.3[a]
N III	$4p_{3/2}$	15.3[a]

Ruthenium (44)

K	1s	22117
L I	2s	3224
L II	$2p_{1/2}$	2967
L III	$2p_{3/2}$	2838
M I	3s	586.2[b]
M II	$3p_{1/2}$	483.3[b]
M III	$3p_{3/2}$	461.5[b]
M IV	$3d_{3/2}$	284.2[b]
M V	$3d_{5/2}$	280.0[b]
N I	4s	75.0[b]
N II	$4p_{1/2}$	46.5[b]
N III	$4p_{3/2}$	43.2[b]

Samarium (62)

K	1s	46834
L I	2s	7737
L II	$2p_{1/2}$	7312

Level	Orbital	Energy
L III	$2p_{3/2}$	6716
M I	$3s$	1723
M II	$3p_{1/2}$	1541
M III	$3p_{3/2}$	1419.8
M IV	$3d_{3/2}$	1110.9[a]
M V	$3d_{5/2}$	1083.4[a]
N I	$4s$	347.2[a]
N II	$4p_{1/2}$	265.6
N III	$4p_{3/2}$	247.4
N IV	$4d_{3/2}$	129.0
N V	$4d_{5/2}$	129.0
N VI	$4f_{5/2}$	5.2
N VII	$4f_{7/2}$	5.2
O I	$5s$	37.4
O II	$5p_{1/2}$	21.3
O III	$5p_{3/2}$	21.3

Scandium (21)

Level	Orbital	Energy
K	$1s$	4492
L I	$2s$	498.0[a]
L II	$2p_{1/2}$	403.6[a]
L III	$2p_{3/2}$	389.7[a]
M I	$3s$	51.1[a]
M II	$3p_{1/2}$	28.3[a]
M III	$3p_{3/2}$	28.3[a]

Selenium (34)

Level	Orbital	Energy
K	$1s$	12658
L I	$2s$	1652.0[a,d]
L II	$2p_{1/2}$	1474.3[a,d]
L III	$2p_{3/2}$	1433.9[a,d]
M I	$3s$	229.6[a]
M II	$3p_{1/2}$	166.5[a]
M III	$3p_{3/2}$	160.7[a]
M IV	$3d_{3/2}$	55.5[a]
M V	$3d_{5/2}$	54.6[a]

Silicon (14)

Level	Orbital	Energy
K	$1s$	1839
L I	$2s$	149.7[a,d]
L II	$2p_{1/2}$	99.8[a]
L III	$2p_{3/2}$	99.2[a]

Silver (47)

Level	Orbital	Energy
K	$1s$	25514
L I	$2s$	3806
L II	$2p_{1/2}$	3524
L III	$2p_{3/2}$	3351
M I	$3s$	719.0[b]
M II	$3p_{1/2}$	603.8[b]
M III	$3p_{3/2}$	573.0[b]
M IV	$3d_{3/2}$	374.0[b]
M V	$3d_{5/2}$	368.0[b]
N I	$4s$	97.0[b]
N II	$4p_{1/2}$	63.7[b]
N III	$4p_{3/2}$	58.3[b]

Sodium (11)

Level	Orbital	Energy
K	$1s$	1070.8[b]
L I	$2s$	63.5[b]
L II	$2p_{1/2}$	30.4[b]
L III	$2p_{3/2}$	30.5[a]

Strontium (38)

Level	Orbital	Energy
K	$1s$	16105
L I	$2s$	2216
L II	$2p_{1/2}$	2007
L III	$2p_{3/2}$	1940
M I	$3s$	358.7[b]
M II	$3p_{1/2}$	280.3[b]
M III	$3p_{3/2}$	270.0[b]
M IV	$3d_{3/2}$	136.0[b]
M V	$3d_{5/2}$	134.2[b]
N I	$4s$	38.9[b]
N II	$4p_{1/2}$	21.6[b]
N III	$4p_{3/2}$	20.1[b]

Sulfur (16)

Level	Orbital	Energy
K	$1s$	2472
L I	$2s$	230.9[a,d]
L II	$2p_{1/2}$	163.6[a]
L III	$2p_{3/2}$	162.5[a]

Tantalum (73)

Level	Orbital	Energy
K	$1s$	67416
L I	$2s$	11682
L II	$2p_{1/2}$	11136
L III	$2p_{3/2}$	9881
M I	$3s$	2708
M II	$3p_{1/2}$	2469
M III	$3p_{3/2}$	2194
M IV	$3d_{3/2}$	1793
M V	$3d_{5/2}$	1735
N I	$4s$	563.4[b]
N II	$4p_{1/2}$	463.4[b]
N III	$4p_{3/2}$	400.9[b]
N IV	$4d_{3/2}$	237.9[b]
N V	$4d_{5/2}$	226.4[b]
N VI	$4f_{5/2}$	23.5[b]
N VII	$4f_{7/2}$	21.6[b]
O I	$5s$	69.7[b]
O II	$5p_{1/2}$	42.2[a]
O III	$5p_{3/2}$	32.7[b]

Technetium (43)

Level	Orbital	Energy
K	$1s$	21044
L I	$2s$	3043
L II	$2p_{1/2}$	2793
L III	$2p_{3/2}$	2677
M I	$3s$	586.1[a]
M II	$3p_{1/2}$	447.6[a]
M III	$3p_{3/2}$	417.7[a]
M IV	$3d_{3/2}$	257.6[a]
M V	$3d_{5/2}$	253.9[a]
N I	$4s$	69.5[a]
N II	$4p_{1/2}$	42.3[a]
N III	$4p_{3/2}$	39.9[a]

Tellurium (52)

Level	Orbital	Energy
K	$1s$	31814
L I	$2s$	4939
L II	$2p_{1/2}$	4612
L III	$2p_{3/2}$	4341
M I	$3s$	1006[b]
M II	$3p_{1/2}$	870.8[b]
M III	$3p_{3/2}$	820.0[b]
M IV	$3d_{3/2}$	583.4[b]
M V	$3d_{5/2}$	573.0[b]
N I	$4s$	169.4[b]
N II	$4p_{1/2}$	103.3[b,c]
N III	$4p_{3/2}$	103.3[b,c]
N IV	$4d_{3/2}$	41.9[b]
N V	$4d_{5/2}$	40.4[b]

Terbium (65)

Level	Orbital	Energy
K	$1s$	51996
L I	$2s$	8708
L II	$2p_{1/2}$	8252
L III	$2p_{3/2}$	7514
M I	$3s$	1968
M II	$3p_{1/2}$	1768
M III	$3p_{3/2}$	1611
M IV	$3d_{3/2}$	1267.9[a]
M V	$3d_{5/2}$	1241.1[a]
N I	$4s$	396.0[a]
N II	$4p_{1/2}$	322.4[a]
N III	$4p_{3/2}$	284.1[a]
N IV	$4d_{3/2}$	150.5[a]
N V	$4d_{5/2}$	150.5[a]
N VI	$4f_{5/2}$	7.7[a]
N VII	$4f_{7/2}$	2.4[a]
O I	$5s$	45.6[a]
O II	$5p_{1/2}$	28.7[a]
O III	$5p_{3/2}$	22.6[a]

Thallium (81)

Level	Orbital	Energy
K	$1s$	85530
L I	$2s$	15347
L II	$2p_{1/2}$	14698
L III	$2p_{3/2}$	12658
M I	$3s$	3704
M II	$3p_{1/2}$	3416
M III	$3p_{3/2}$	2957
M IV	$3d_{3/2}$	2485
M V	$3d_{5/2}$	2389
N I	$4s$	846.2[b]
N II	$4p_{1/2}$	720.5[b]
N III	$4p_{3/2}$	609.5[b]
N IV	$4d_{3/2}$	405.7[b]
N V	$4d_{5/2}$	385.0[b]
N VI	$4f_{5/2}$	122.2[b]
N VII	$4f_{7/2}$	117.8[b]
O I	$5s$	136[a,d]
O II	$5p_{1/2}$	94.6[b]
O III	$5p_{3/2}$	73.5[b]
O IV	$5d_{3/2}$	14.7[b]
O V	$5d_{5/2}$	12.5[b]

Thorium (90)

Level	Orbital	Energy
K	$1s$	109651
L I	$2s$	20472
L II	$2p_{1/2}$	19693
L III	$2p_{3/2}$	16300
M I	$3s$	5182
M II	$3p_{1/2}$	4830
M III	$3p_{3/2}$	4046
M IV	$3d_{3/2}$	3491
M V	$3d_{5/2}$	3332
N I	$4s$	1330[a]
N II	$4p_{1/2}$	1168[a]
N III	$4p_{3/2}$	966.4[b]
N IV	$4d_{3/2}$	712.1[b]
N V	$4d_{5/2}$	675.2[b]
N VI	$4f_{5/2}$	342.4[b]
N VII	$4f_{7/2}$	333.1[b]
O I	$5s$	290[a,c]
O II	$5p_{1/2}$	229[a,c]
O III	$5p_{3/2}$	182[a,c]
O IV	$5d_{3/2}$	92.5[b]
O V	$5d_{5/2}$	85.4[b]
P I	$6s$	41.4[b]
P II	$6p_{1/2}$	24.5[b]
P III	$6p_{3/2}$	16.6[b]

Thulium (69)

Level	Orbital	Energy
K	$1s$	59390
L I	$2s$	10116
L II	$2p_{1/2}$	9617
L III	$2p_{3/2}$	8648
M I	$3s$	2307
M II	$3p_{1/2}$	2090
M III	$3p_{3/2}$	1885
M IV	$3d_{3/2}$	1515
M V	$3d_{5/2}$	1468
N I	$4s$	470.9[a]
N II	$4p_{1/2}$	385.9[a]
N III	$4p_{3/2}$	332.6[a]
N IV	$4d_{3/2}$	175.5[a]
N V	$4d_{5/2}$	175.5[a]
N VI	$4f_{5/2}$	—
N VII	$4f_{7/2}$	4.6
O I	$5s$	54.7[a]
O II	$5p_{1/2}$	31.8[a]
O III	$5p_{3/2}$	25.0[a]

Tin (50)

Level	Orbital	Energy
K	$1s$	29200
L I	$2s$	4465
L II	$2p_{1/2}$	4156
L III	$2p_{3/2}$	3929
M I	$3s$	884.7[b]
M II	$3p_{1/2}$	756.5[b]
M III	$3p_{3/2}$	714.6[b]
M IV	$3d_{3/2}$	493.2[b]
M V	$3d_{5/2}$	484.9[b]
N I	$4s$	137.1[b]
N II	$4p_{1/2}$	83.6[b,c]
N III	$4p_{3/2}$	83.6[b,c]
N IV	$4d_{3/2}$	24.9[b]
N V	$4d_{5/2}$	23.9[b]

Titanium (22)

Level	Orbital	Energy
K	$1s$	4966
L I	$2s$	560.9[b]
L II	$2p_{1/2}$	460.2[b]
L III	$2p_{3/2}$	453.8[b]
M I	$3s$	58.7[b]
M II	$3p_{1/2}$	32.6[b]
M III	$3p_{3/2}$	32.6[b]

Tungsten (74)

Level	Orbital	Energy
K	$1s$	69525
L I	$2s$	12100
L II	$2p_{1/2}$	11544
L III	$2p_{3/2}$	10207
M I	$3s$	2820
M II	$3p_{1/2}$	2575
M III	$3p_{3/2}$	2281
M IV	$3d_{3/2}$	1949

M V	$3d_{5/2}$	1809
N I	4s	594.1[b]
N II	$4p_{1/2}$	490.4[b]
N III	$4p_{3/2}$	423.6[b]
N IV	$4d_{3/2}$	255.9[b]
N V	$4d_{5/2}$	243.5[b]
N VI	$4f_{5/2}$	33.6[a]
N VII	$4f_{7/2}$	31.4[b]
O I	5s	75.6[b]
O II	$5p_{1/2}$	453[a,d]
O III	$5p_{3/2}$	36.8[b]

Uranium (92)

K	1s	115606
L I	2s	21757
L II	$2p_{1/2}$	20948
L III	$2p_{3/2}$	17166
M I	3s	5548
M II	$3p_{1/2}$	5182
M III	$3p_{3/2}$	4303
M IV	$3d_{3/2}$	3728
M V	$3d_{5/2}$	3552
N I	4s	1439[a,d]
N II	$4p_{1/2}$	1271[a,d]
N III	$4p_{3/2}$	1043[b]
N IV	$4d_{3/2}$	778.3[b]
N V	$4d_{5/2}$	736.2[b]
N VI	$4f_{5/2}$	388.2[a]
N VII	$4f_{7/2}$	377.4[b]
O I	5s	321[a,c,d]

O II	$5p_{1/2}$	257[a,c,d]
O III	$5p_{3/2}$	192[a,c,d]
O IV	$5d_{3/2}$	102.8[b]
O V	$5d_{5/2}$	94.2[b]
P I	6s	43.9[b]
P II	$6p_{1/2}$	26.8[b]
P III	$6p_{3/2}$	16.8[b]

Vanadium (23)

K	1s	5465
L I	2s	626.7[b]
L II	$2p_{1/2}$	519.8[b]
L III	$2p_{3/2}$	521.1[b]
M I	3s	66.3[b]
M II	$3p_{1/2}$	37.2[b]
M III	$3p_{3/2}$	37.2[b]

Xenon (54)

K	1s	34561
L I	2s	5453
L II	$2p_{1/2}$	5107
L III	$2p_{3/2}$	4786
M I	3s	1148.7[a]
M II	$3p_{1/2}$	1002.1[a]
M III	$3p_{3/2}$	940.6[a]
M IV	$3d_{3/2}$	689.0[a]
M V	$3d_{5/2}$	676.4[a]
N I	4s	213.2[a]
N II	$4p_{1/2}$	146.7
N III	$4p_{3/2}$	145.5[a]
N IV	$4d_{3/2}$	69.5[a]
N V	$4d_{5/2}$	67.5[a]
N VI	$4f_{5/2}$	—

N VII	$4f_{7/2}$	—
O I	5s	23.3[a]
O II	$5p_{1/2}$	13.4[a]
O III	$5p_{3/2}$	12.1[a]

Ytterbium (70)

K	1s	61332
L I	2s	10486
L II	$2p_{1/2}$	9978
L III	$2p_{3/2}$	8944
M I	3s	2398
M II	$3p_{1/2}$	2173
M III	$3p_{3/2}$	1950
M IV	$3d_{3/2}$	1576
M V	$3d_{5/2}$	1528
N I	4s	480.5[a]
N II	$4p_{1/2}$	388.7[a]
N III	$4p_{3/2}$	339.7[a]
N IV	$4d_{3/2}$	191.2[a]
N V	$4d_{5/2}$	182.4[a]
N VI	$4f_{5/2}$	2.5[a]
N VII	$4f_{7/2}$	1.3[a]
O I	5s	52.0[a]
O II	$5p_{1/2}$	30.3[a]
O III	$5p_{3/2}$	24.1[a]

Yttrium (39)

K	1s	17038
L I	2s	2373
L II	$2p_{1/2}$	2156
L III	$2p_{3/2}$	2080
M I	3s	392.0[a,d]
M II	$3p_{1/2}$	310.6[a]

M III	$3p_{3/2}$	298.8[a]
M IV	$3d_{3/2}$	157.7[a]
M V	$3d_{5/2}$	155.8[b]
N I	4s	43.8[a]
N II	$4p_{1/2}$	24.4[a]
N III	$4p_{3/2}$	23.1[a]

Zinc (30)

K	1s	9659
L I	2s	1196.2[a]
L II	$2p_{1/2}$	1044.9[a]
L III	$2p_{3/2}$	1021.8[a]
M I	3s	139.8[a]
M II	$3p_{1/2}$	91.4[a]
M III	$3p_{3/2}$	88.6[a]
M IV	$3d_{3/2}$	10.2[a]
M V	$3d_{5/2}$	10.1[a]

Zirconium (40)

K	1s	17998
L I	2s	2532
L II	$2p_{1/2}$	2307
L III	$2p_{3/2}$	2223
M I	3s	430.3[b]
M II	$3p_{1/2}$	343.5[b]
M III	$3p_{3/2}$	329.8[b]
M IV	$3d_{3/2}$	181.1[b]
M V	$3d_{5/2}$	178.8[b]
N I	4s	50.6[a]
N II	$4p_{1/2}$	28.5[b]
N III	$4p_{3/2}$	27.1[b]

[a] Reference 1.
[b] Reference 2 (remaining values from Reference 3).
[c] One-particle approximation not valid.
[d] Derived using energy differences from Reference 3.

NATURAL WIDTH OF X-RAY LINES

Natural widths of K X-ray lines in eV:

Element	Kα_1	Kα_2	Kβ_1	Kβ_3	Element	Kα_1	Kα_2	Kβ_1	Kβ_3
Ca	1.00	0.98			Ce	18.60	19.50	20.60	18.60
Ti	1.45	2.13			Nd	21.50	21.50	23.25	21.33
Cr	2.05	2.64			Sm	26.00	24.70	25.65	24.65
Fe	2.45	3.20			Gd	29.50	28.00	29.37	28.00
Ni	3.00	3.70			Dy	33.90	32.20	32.73	32.00
Zn	3.40	3.96			Er	35.00	35.50	36.20	35.70
Ge	3.75	4.18			Yb	38.80	40.60	41.43	41.15
Se	4.10	4.43			Hf	42.70	44.30	46.00	46.10
Kr	4.23	4.62			W	46.80	48.00	51.83	51.50
Sr	5.17	4.97			Os	49.00	49.40	55.90	55.95
Zr	5.70	5.25			Pt	54.10	54.30	59.98	62.13
Mo	6.82	6.80			Hg	64.75	68.20	65.75	68.95
Ru	7.41	7.96			Pb	67.10	72.30	72.20	73.80
Pd	8.80	9.20			Po	73.20	75.10	78.60	80.10
Cd	9.80	10.40			Rn	80.00	81.50	85.50	86.50
Sn	11.20	12.40	11.80	11.00	Ra	87.00	88.20	94.20	95.50
Te	12.80	14.20	13.30	13.10	Th	94.70	95.00	99.70	101.00
Xe	14.20	15.10	15.30	14.50	U	103.00	104.30	105.00	107.30
Ba	16.10	16.80	18.15	16.70					

From Salem, S. I. and Lee, P. L., *At. Data Nucl. Data Tables*, 18, 233, 1976.

Natural widths of L X-ray lines in eV:

Element	Lα_1	Lα_2	Lβ_1	Lβ_2	Lβ_3	Lβ_4	Lγ_1
Zr	1.68	1.52	1.87	5.13	5.50	5.60	3.34
Mo	1.86	1.80	2.03	5.30	5.90	5.78	3.76
Ru	2.03	1.98	2.18	5.45	6.35	5.96	4.15
Pd	2.21	2.16	2.36	5.63	6.80	6.18	4.50
Gd	2.43	2.40	2.54	5.82	7.23	6.28	4.83
Sn	2.62	2.62	2.75	6.10	7.70	6.60	5.23
Tc	2.88	2.88	2.96	6.25	8.22	6.82	5.60
Xe	3.15	3.15	3.20	6.43	8.70	7.15	5.95
Ba	3.39	3.45	3.45	6.70	9.20	7.42	6.35
Ce	3.70	3.78	3.73	6.86	9.70	7.82	6.75
Nd	3.93	4.08	4.00	7.18	10.30	8.15	7.16
Sm	4.13	4.50	4.33	7.42	10.80	8.60	7.50
Cd	4.46	4.90	4.63	7.70	11.20	9.08	7.83
Dy	4.81	5.35	5.03	7.90	11.50	9.60	8.30
Er	5.17	5.73	5.45	8.28	11.85	10.03	8.75
Yb	5.40	6.22	5.90	8.58	12.20	11.00	9.20
Hf	5.83	6.70	6.36	8.92	12.40	12.80	9.63
W	6.50	7.20	6.90	9.06	13.10	14.60	10.20
Os	7.04	7.70	7.42	9.60	14.60	16.50	10.65
Pt	7.60	8.28	8.00	9.95	16.10	18.00	11.20
Hg	8.10	8.80	8.70	10.40	17.40	19.70	11.80
Pb	8.82	9.35	9.35	10.75	18.65	21.30	12.30
Po	9.50	9.95	10.10	11.25	19.90	22.70	13.05
Rn	10.03	10.50	10.65	11.65	21.00	24.00	13.55
Ra	11.00	11.20	11.60	12.20	22.00	25.20	14.30
Th	11.90	11.80	12.40	12.80	22.85	26.35	15.00
U	12.40	12.40	13.50	13.30	23.70	27.50	15.70
Pu	13.20	13.00	14.10	13.90	24.10	28.30	16.40
Cm	14.80	13.60	15.70	14.60	25.00	29.40	17.10

PHOTON ATTENUATION COEFFICIENTS

Martin J. Berger and John H. Hubbell

This table gives mass attenuation coefficients for photons for all elements at energies between 1 keV (soft x-rays) and 1 GeV (hard gamma rays). The mass attenuation coefficient μ describes the attenuation of radiation as it passes through matter by the relation

$$I(x)/I_o = e^{-\mu\rho x}$$

where I_o is the initial intensity, $I(x)$ the intensity after path length x, and ρ is the mass density of the element in question. To a high approximation the mass attenuation coefficient is additive for the elements present, independent of the way in which they are bound in chemical compounds.

The power of ten is indicated beside each number in the table; i.e., 7.41 + 03 means 7.41×10^3. A vertical line between two columns indicates that an absorption edge lies between those energy values. The various edges are labeled at the bottom of the table.

The attenuation coefficients were calculated with the computer program XCOM (Reference 1), which uses a cross-section database compiled at the Photon and Charged Particle Data Center at the National Institute of Standards and Technology. Their accuracy has been confirmed at all energies by extensive comparisons with experimental attenuation coefficients. Such comparisons for X-ray energies up to 100 keV can be found in Reference 2.

References

1. Berger, M. J. and Hubbell, J. H., *National Bureau of Standards Report* NBSIR-87-3597, 1987.
2. Saloman, E. B., Hubbell, J. H., and Scofield, J. H., *Atomic Data and Nuclear Data Tables*, 38, 1, 1988.

Mass attenuation coefficient, cm²/g

Photon energy, MeV

	Atomic no.	0.001	0.002	0.005	0.01	0.02	0.05	0.1	0.2	0.5
H	1	7.21 + 00	1.06 + 00	4.19-01	3.85-01	3.69-01	3.36-01	2.94-01	2.43-01	1.73-01
He	2	6.08 + 01	6.86 + 00	5.77-01	2.48-01	1.96-01	1.70-01	1.49-01	1.22-01	8.71-02
Li	3	2.34 + 02	2.71 + 01	1.62 + 00	3.40-01	1.86-01	1.49-01	1.29-01	1.06-01	7.53-02
Be	4	6.04 + 02	7.47 + 01	4.37 + 00	6.47-01	2.25-01	1.55-01	1.33-01	1.09-01	7.74-02
B	5	1.23 + 03	1.60 + 02	9.68 + 00	1.25 + 00	3.01-01	1.66-01	1.39-01	1.14-01	8.07-02
C	6	2.21 + 03	3.03 + 02	1.91 + 01	2.37 + 00	4.42-01	1.87-01	1.51-01	1.23-01	8.72-02
N	7	3.31 + 03	4.77 + 02	3.14 + 01	3.88 + 00	6.18-01	1.98-01	1.53-01	1.23-01	8.72-02
O	8	4.59 + 03	6.95 + 02	4.79 + 01	5.95 + 00	8.65-01	2.13-01	1.55-01	1.24-01	8.73-02
F	9	5.65 + 03	9.05 + 02	6.51 + 01	8.21 + 00	1.13 + 00	2.21-01	1.50-01	1.18-01	8.27-02
Ne	10	7.41 + 03	1.24 + 03	9.34 + 01	1.20 + 01	1.61 + 00	2.58-01	1.60-01	1.24-01	8.66-02
Na	11	6.54 + 02	1.52 + 03	1.19 + 02	1.56 + 01	2.06 + 00	2.80-01	1.59-01	1.20-01	8.37-02
Mg	12	9.22 + 02	1.93 + 03	1.58 + 02	2.11 + 01	2.76 + 00	3.29-01	1.69-01	1.24-01	8.65-02
Al	13	1.19 + 03	2.26 + 03	1.93 + 02	2.62 + 01	3.44 + 00	3.68-01	1.70-01	1.22-01	8.44-02
Si	14	1.57 + 03	2.78 + 03	2.45 + 02	3.39 + 01	4.46 + 00	4.38-01	1.84-01	1.28-01	8.75-02
P	15	1.91 + 03	3.02 + 02	2.86 + 02	4.04 + 01	5.35 + 00	4.92-01	1.87-01	1.25-01	8.51-02
S	16	2.43 + 03	3.85 + 02	3.49 + 02	5.01 + 01	6.71 + 00	5.85-01	2.02-01	1.30-01	8.78-02
Cl	17	2.83 + 03	4.52 + 02	3.90 + 02	5.73 + 01	7.74 + 00	6.48-01	2.05-01	1.27-01	8.45-02
Ar	18	3.18 + 03	5.12 + 02	4.23 + 02	6.32 + 01	8.63 + 00	7.01-01	2.04-01	1.20-01	7.96-02
K	19	4.06 + 03	6.59 + 02	5.19 + 02	7.91 + 01	1.09 + 01	8.68-01	2.34-01	1.32-01	8.60-02
Ca	20	4.87 + 03	8.00 + 02	6.03 + 02	9.34 + 01	1.31 + 01	1.02 + 00	2.57-01	1.38-01	8.85-02
Sc	21	5.24 + 03	8.70 + 02	6.31 + 02	9.95 + 01	1.41 + 01	1.09 + 00	2.58-01	1.31-01	8.31-02
Ti	22	5.87 + 03	9.86 + 02	6.84 + 02	1.11 + 02	1.59 + 01	1.21 + 00	2.72-01	1.31-01	8.19-02
V	23	6.50 + 03	1.11 + 03	9.29 + 01	1.22 + 02	1.77 + 01	1.35 + 00	2.88-01	1.32-01	8.07-02
Cr	24	7.40 + 03	1.28 + 03	1.08 + 02	1.39 + 02	2.04 + 01	1.55 + 00	3.17-01	1.38-01	8.28-02
Mn	25	8.09 + 03	1.42 + 03	1.21 + 02	1.51 + 02	2.25 + 01	1.71 + 00	3.37-01	1.39-01	8.19-02
Fe	26	9.09 + 03	1.63 + 03	1.40 + 02	1.71 + 02	2.57 + 01	1.96 + 00	3.72-01	1.46-01	8.41-02
Co	27	9.80 + 03	1.78 + 03	1.54 + 02	1.84 + 02	2.80 + 01	2.14 + 00	3.95-01	1.48-01	8.32-02
Ni	28	9.86 + 03	2.05 + 03	1.79 + 02	2.09 + 02	3.22 + 01	2.47 + 00	4.44-01	1.58-01	8.70-02
Cu	29	1.06 + 04	2.15 + 03	1.90 + 02	2.16 + 02	3.38 + 01	2.61 + 00	4.58-01	1.56-01	8.36-02
Zn	30	1.55 + 03	2.37 + 03	2.12 + 02	2.33 + 02	3.72 + 01	2.89 + 00	4.97-01	1.62-01	8.45-02
Ga	31	1.70 + 03	2.52 + 03	2.27 + 02	3.42 + 01	3.93 + 01	3.08 + 00	5.20-01	1.62-01	8.24-02
Ge	32	1.89 + 03	2.71 + 03	2.47 + 02	3.74 + 01	4.22 + 01	3.34 + 00	5.55-01	1.66-01	8.21-02
As	33	2.12 + 03	2.93 + 03	2.71 + 02	4.12 + 01	4.56 + 01	3.63 + 00	5.97-01	1.72-01	8.26-02
Se	34	2.32 + 03	3.10 + 03	2.90 + 02	4.41 + 01	4.82 + 01	3.86 + 00	6.28-01	1.74-01	8.13-02
Br	35	2.62 + 03	3.41 + 03	3.21 + 02	4.91 + 01	5.27 + 01	4.26 + 00	6.86-01	1.84-01	8.33-02
Kr	36	2.85 + 03	3.60 + 03	3.43 + 02	5.26 + 01	5.55 + 01	4.52 + 00	7.22-01	1.87-01	8.23-02
Rb	37	3.17 + 03	3.41 + 03	3.74 + 02	5.77 + 01	5.98 + 01	4.92 + 00	7.80-01	1.96-01	8.36-02
Sr	38	3.49 + 03	2.59 + 03	4.06 + 02	6.27 + 01	6.39 + 01	5.31 + 00	8.37-01	2.04-01	8.44-02
Y	39	3.86 + 03	7.42 + 02	4.42 + 02	6.87 + 01	6.86 + 01	5.76 + 00	9.05-01	2.15-01	8.61-02
Zr	40	4.21 + 03	8.12 + 02	4.76 + 02	7.42 + 01	7.24 + 01	6.17 + 00	9.66-01	2.24-01	8.69-02
Nb	41	4.60 + 03	8.89 + 02	5.13 + 02	8.04 + 01	7.71 + 01	6.64 + 00	1.04 + 00	2.34-01	8.83-02
Mo	42	4.94 + 03	9.60 + 02	5.45 + 02	8.58 + 01	1.31 + 01	7.04 + 00	1.10 + 00	2.42-01	8.85-02
Tc	43	5.36 + 03	1.04 + 03	5.84 + 02	9.23 + 01	1.41 + 01	7.52 + 00	1.17 + 00	2.53-01	8.97-02
Ru	44	5.72 + 03	1.12 + 03	6.17 + 02	9.80 + 01	1.50 + 01	7.92 + 00	1.23 + 00	2.62-01	8.99-02
Rh	45	6.17 + 03	1.21 + 03	6.59 + 02	1.05 + 02	1.61 + 01	8.45 + 00	1.31 + 00	2.74-01	9.13-02
Pd	46	6.54 + 03	1.29 + 03	6.91 + 02	1.11 + 02	1.70 + 01	8.85 + 00	1.38 + 00	2.83-01	9.13-02
Ag	47	7.04 + 03	1.40 + 03	7.39 + 02	1.19 + 02	1.84 + 01	9.45 + 00	1.47 + 00	2.97-01	9.32-02
Cd	48	7.35 + 03	1.47 + 03	7.69 + 02	1.24 + 02	1.92 + 01	9.78 + 00	1.52 + 00	3.04-01	9.25-02
In	49	7.81 + 03	1.58 + 03	8.13 + 02	1.32 + 02	2.04 + 01	1.03 + 01	1.61 + 00	3.17-01	9.37-02
Sn	50	8.16 + 03	1.66 + 03	8.47 + 02	1.38 + 02	2.15 + 01	1.07 + 01	1.68 + 00	3.26-01	9.37-02

L_3 | L_1

L_2

K EDGE

Mass attenuation coefficient, cm²/g

Photon energy, MeV

	Atomic no.	1.0	2.0	5.0	10.0	20.0	50.0	100.0	500.0	1000.0
H	1	1.26-01	8.77-02	5.05-02	3.25-02	2.15-02	1.42-02	1.19-02	1.14-02	1.16-02
He	2	6.36-02	4.42-02	2.58-02	1.70-02	1.18-02	8.61-03	7.78-03	7.79-03	7.95-03
Li	3	5.50-02	3.83-02	2.26-02	1.53-02	1.11-02	8.68-03	8.21-03	8.61-03	8.87-03
Be	4	5.65-02	3.94-02	2.35-02	1.63-02	1.23-02	1.02-02	9.94-03	1.08-02	1.12-02
B	5	5.89-02	4.11-02	2.48-02	1.76-02	1.37-02	1.19-02	1.19-02	1.32-02	1.37-02
C	6	6.36-02	4.44-02	2.71-02	1.96-02	1.58-02	1.43-02	1.46-02	1.64-02	1.70-02
N	7	6.36-02	4.45-02	2.74-02	2.02-02	1.67-02	1.57-02	1.63-02	1.85-02	1.92-02
O	8	6.37-02	4.46-02	2.78-02	2.09-02	1.77-02	1.71-02	1.79-02	2.06-02	2.13-02
F	9	6.04-02	4.23-02	2.66-02	2.04-02	1.77-02	1.75-02	1.86-02	2.14-02	2.21-02
Ne	10	6.32-02	4.43-02	2.82-02	2.20-02	1.95-02	1.96-02	2.11-02	2.43-02	2.51-02
Na	11	6.10-02	4.28-02	2.75-02	2.18-02	1.97-02	2.03-02	2.19-02	2.53-02	2.62-02
Mg	12	6.30-02	4.43-02	2.87-02	2.31-02	2.13-02	2.23-02	2.42-02	2.81-02	2.90-02
Al	13	6.15-02	4.32-02	2.84-02	2.32-02	2.17-02	2.31-02	2.52-02	2.93-02	3.03-02
Si	14	6.36-02	4.48-02	2.97-02	2.46-02	2.34-02	2.52-02	2.76-02	3.23-02	3.34-02
P	15	6.18-02	4.36-02	2.91-02	2.45-02	2.36-02	2.58-02	2.84-02	3.33-02	3.45-02
S	16	6.37-02	4.50-02	3.04-02	2.59-02	2.53-02	2.79-02	3.08-02	3.62-02	3.75-02
Cl	17	6.13-02	4.33-02	2.95-02	2.55-02	2.52-02	2.81-02	3.11-02	3.67-02	3.80-02
Ar	18	5.76-02	4.07-02	2.80-02	2.45-02	2.45-02	2.76-02	3.07-02	3.62-02	3.75-02
K	19	6.22-02	4.40-02	3.05-02	2.70-02	2.74-02	3.11-02	3.46-02	4.09-02	4.24-02
Ca	20	6.39-02	4.52-02	3.17-02	2.84-02	2.90-02	3.32-02	3.71-02	4.40-02	4.56-02
Sc	21	5.98-02	4.24-02	3.00-02	2.72-02	2.80-02	3.23-02	3.62-02	4.30-02	4.45-02
Ti	22	5.89-02	4.18-02	2.98-02	2.73-02	2.84-02	3.30-02	3.71-02	4.40-02	4.56-02
V	23	5.79-02	4.11-02	2.96-02	2.74-02	2.88-02	3.36-02	3.78-02	4.49-02	4.65-02
Cr	24	5.93-02	4.21-02	3.06-02	2.86-02	3.03-02	3.56-02	4.01-02	4.76-02	4.93-02
Mn	25	5.85-02	4.16-02	3.04-02	2.87-02	3.07-02	3.63-02	4.09-02	4.86-02	5.04-02
Fe	26	5.99-02	4.26-02	3.15-02	2.99-02	3.22-02	3.83-02	4.33-02	5.15-02	5.33-02
Co	27	5.91-02	4.20-02	3.13-02	3.00-02	3.26-02	3.88-02	4.40-02	5.23-02	5.41-02
Ni	28	6.16-02	4.39-02	3.29-02	3.18-02	3.48-02	4.17-02	4.73-02	5.61-02	5.81-02
Cu	29	5.90-02	4.20-02	3.18-02	3.10-02	3.41-02	4.10-02	4.66-02	5.53-02	5.72-02
Zn	30	5.94-02	4.24-02	3.22-02	3.18-02	3.51-02	4.24-02	4.82-02	5.72-02	5.91-02
Ga	31	5.77-02	4.11-02	3.16-02	3.13-02	3.48-02	4.22-02	4.80-02	5.70-02	5.89-02
Ge	32	5.73-02	4.09-02	3.16-02	3.16-02	3.53-02	4.30-02	4.89-02	5.80-02	6.00-02
As	33	5.73-02	4.09-02	3.19-02	3.21-02	3.60-02	4.40-02	5.01-02	5.95-02	6.15-02
Se	34	5.62-02	4.01-02	3.14-02	3.19-02	3.60-02	4.41-02	5.03-02	5.97-02	6.17-02
Br	35	5.73-02	4.09-02	3.23-02	3.29-02	3.74-02	4.60-02	5.24-02	6.22-02	6.43-02
Kr	36	5.63-02	4.02-02	3.20-02	3.28-02	3.74-02	4.61-02	5.26-02	6.25-02	6.46-02
Rb	37	5.69-02	4.06-02	3.25-02	3.36-02	3.85-02	4.75-02	5.43-02	6.45-02	6.67-02
Sr	38	5.71-02	4.08-02	3.29-02	3.41-02	3.93-02	4.87-02	5.56-02	6.61-02	6.83-02
Y	39	5.80-02	4.14-02	3.35-02	3.50-02	4.05-02	5.03-02	5.75-02	6.83-02	7.06-02
Zr	40	5.81-02	4.15-02	3.38-02	3.55-02	4.12-02	5.13-02	5.87-02	6.98-02	7.22-02
Nb	41	5.87-02	4.18-02	3.44-02	3.63-02	4.22-02	5.27-02	6.03-02	7.17-02	7.42-02
Mo	42	5.84-02	4.16-02	3.44-02	3.65-02	4.26-02	5.33-02	6.10-02	7.26-02	7.51-02
Tc	43	5.88-02	4.19-02	3.48-02	3.71-02	4.35-02	5.45-02	6.24-02	7.43-02	7.68-02
Ru	44	5.85-02	4.16-02	3.48-02	3.73-02	4.39-02	5.50-02	6.30-02	7.51-02	7.77-02
Rh	45	5.89-02	4.20-02	3.53-02	3.80-02	4.48-02	5.63-02	6.45-02	7.69-02	7.94-02
Pd	46	5.85-02	4.16-02	3.52-02	3.80-02	4.50-02	5.66-02	6.49-02	7.73-02	8.00-02
Ag	47	5.92-02	4.21-02	3.58-02	3.88-02	4.61-02	5.81-02	6.67-02	7.93-02	8.20-02
Cd	48	5.83-02	4.14-02	3.54-02	3.85-02	4.59-02	5.79-02	6.64-02	7.91-02	8.18-02
In	49	5.85-02	4.15-02	3.56-02	3.90-02	4.65-02	5.88-02	6.75-02	8.04-02	8.32-02
Sn	50	5.80-02	4.11-02	3.55-02	3.90-02	4.66-02	5.90-02	6.78-02	8.07-02	8.35-02

Mass attenuation coefficient, cm²/g

Photon energy, MeV

	Atomic no.	0.001	0.002	0.005	0.01	0.02	0.05	0.1	0.2	0.5
Sb	51	8.58 + 03	1.77 + 03	8.85 + 02	1.46 + 02	2.27 + 01	1.12 + 01	1.76 + 00	3.38-01	9.45-02
Te	52	8.43 + 03	1.83 + 03	9.01 + 02	1.50 + 02	2.34 + 01	1.14 + 01	1.80 + 00	3.43-01	9.33-02
I	53	9.10 + 03	2.00 + 03	8.43 + 02	1.63 + 02	2.54 + 01	1.23 + 01	1.94 + 00	3.66-01	9.70-02
Xe	54	9.41 + 03	2.09 + 03	6.39 + 02	1.69 + 02	2.65 + 01	1.27 + 01	2.01 + 00	3.76-01	9.70-02
Cs	55	9.37 + 03	2.23 + 03	2.30 + 02	1.79 + 02	2.82 + 01	1.34 + 01	2.12 + 00	3.94-01	9.91-02
ZB	56	8.54 + 03	2.32 + 03	2.41 + 02	1.86 + 02	2.94 + 01	1.38 + 01	2.20 + 00	4.05-01	9.92-02
La	57	9.09 + 03	2.46 + 03	2.58 + 02	1.97 + 02	3.12 + 01	1.45 + 01	2.32 + 00	4.24-01	1.01-01
Ce	58	9.71 + 03	2.61 + 03	2.74 + 02	2.08 + 02	3.31 + 01	1.52 + 01	2.45 + 00	4.45-01	1.04-01
Pr	59	1.06 + 04	2.77 + 03	2.92 + 02	2.21 + 02	3.53 + 01	1.60 + 01	2.59 + 00	4.69-01	1.07-01
Nd	60	6.63 + 03	2.88 + 03	3.06 + 02	2.30 + 02	3.68 + 01	1.65 + 01	2.69 + 00	4.84-01	1.08-01
Pm	61	2.06 + 03	3.05 + 03	3.26 + 02	2.44 + 02	3.92 + 01	1.73 + 01	2.84 + 00	5.10-01	1.12-01
Sm	62	2.11 + 03	3.12 + 03	3.36 + 02	2.50 + 02	4.03 + 01	1.77 + 01	2.90 + 00	5.19-01	1.11-01
Eu	63	2.22 + 03	3.28 + 03	3.54 + 02	2.63 + 02	4.24 + 01	1.85 + 01	3.04 + 00	5.43-01	1.14-01
Gd	64	2.29 + 03	3.36 + 03	3.65 + 02	2.69 + 02	4.36 + 01	3.86 + 00	3.11 + 00	5.54-01	1.14-01
Tb	65	2.40 + 03	3.51 + 03	3.84 + 02	2.82 + 02	4.59 + 01	4.06 + 00	3.25 + 00	5.77-01	1.17-01
Dy	66	2.49 + 03	3.47 + 03	3.99 + 02	2.90 + 02	4.76 + 01	4.23 + 00	3.36 + 00	5.95-01	1.18-01
Ho	67	2.62 + 03	3.59 + 03	4.17 + 02	3.01 + 02	4.98 + 01	4.43 + 00	3.49 + 00	6.18-01	1.20-01
Er	68	2.75 + 03	3.52 + 03	4.36 + 02	3.13 + 02	5.20 + 01	4.63 + 00	3.63 + 00	6.41-01	1.23-01
Tm	69	2.90 + 03	3.69 + 03	4.57 + 02	2.83 + 02	5.45 + 01	4.87 + 00	3.78 + 00	6.68-01	1.26-01
Yb	70	3.02 + 03	3.80 + 03	4.72 + 02	2.94 + 02	5.63 + 01	5.04 + 00	3.88 + 00	6.86-01	1.27-01
Lu	71	3.19 + 03	3.45 + 03	4.94 + 02	2.21 + 02	5.88 + 01	5.28 + 00	4.03 + 00	7.13-01	1.30-01
Hf	72	3.34 + 03	3.60 + 03	5.11 + 02	2.30 + 02	6.09 + 01	5.48 + 00	4.15 + 00	7.34-01	1.32-01
Ta	73	3.51 + 03	3.77 + 03	5.33 + 02	2.38 + 02	6.33 + 01	5.72 + 00	4.30 + 00	7.60-01	1.35-01
W	74	3.68 + 03	3.92 + 03	5.53 + 02	9.69 + 01	6.57 + 01	5.95 + 00	4.44 + 00	7.84-01	1.38-01
Re	75	3.87 + 03	3.77 + 03	5.76 + 02	1.01 + 02	6.84 + 01	6.21 + 00	4.59 + 00	8.12-01	1.41-01
Os	76	4.03 + 03	2.22 + 03	5.93 + 02	1.04 + 02	7.04 + 01	6.41 + 00	4.70 + 00	8.33-01	1.43-01
Ir	77	4.24 + 03	1.03 + 03	6.18 + 02	1.09 + 02	7.32 + 01	6.69 + 00	4.86 + 00	8.63-01	1.46-01
Pt	78	4.43 + 03	1.08 + 03	6.40 + 02	1.13 + 02	7.57 + 01	6.95 + 00	4.99 + 00	8.90-01	1.49-01
Au	79	4.65 + 03	1.14 + 03	6.66 + 02	1.18 + 02	7.88 + 01	7.26 + 00	5.16 + 00	9.22-01	1.53-01
Hg	80	4.83 + 03	1.18 + 03	6.87 + 02	1.22 + 02	8.12 + 01	7.50 + 00	5.28 + 00	9.46-01	1.56-01
Tl	81	5.01 + 03	1.23 + 03	7.07 + 02	1.26 + 02	8.36 + 01	7.75 + 00	5.40 + 00	9.69-01	1.58-01
Pb	82	5.21 + 03	1.29 + 03	7.30 + 02	1.31 + 02	8.64 + 01	8.04 + 00	5.55 + 00	9.99-01	1.61-01
Bi	83	5.44 + 03	1.35 + 03	7.58 + 02	1.36 + 02	8.95 + 01	8.38 + 00	5.74 + 00	1.03 + 00	1.66-01
Po	84	5.72 + 03	1.42 + 03	7.93 + 02	1.43 + 02	9.35 + 01	8.80 + 00	5.99 + 00	1.08 + 00	1.71-01
At	85	5.87 + 03	1.49 + 03	8.25 + 02	1.49 + 02	9.70 + 01	9.19 + 00	6.17 + 00	1.12 + 00	1.77-01
Rn	86	5.83 + 03	1.49 + 03	8.16 + 02	1.48 + 02	9.56 + 01	9.12 + 00	6.09 + 00	1.10 + 00	1.73-01
Fr	87	6.08 + 03	1.56 + 03	8.49 + 02	1.54 + 02	9.93 + 01	9.52 + 00	1.66 + 00	1.14 + 00	1.78-01
Ra	88	6.20 + 03	1.62 + 03	8.74 + 02	1.59 + 02	1.02 + 02	9.85 + 00	1.71 + 00	1.17 + 00	1.82-01
Ac	89	6.47 + 03	1.70 + 03	8.69 + 02	1.65 + 02	1.06 + 02	1.03 + 01	1.79 + 00	1.21 + 00	1.87-01
Th	90	6.61 + 03	1.74 + 03	8.88 + 02	1.69 + 02	9.37 + 01	1.05 + 01	1.83 + 00	1.23 + 00	1.90-01
Pa	91	6.53 + 03	1.83 + 03	8.76 + 02	1.77 + 02	7.03 + 01	1.10 + 01	1.92 + 00	1.29 + 00	1.97-01
U	92	6.63 + 03	1.86 + 03	8.89 + 02	1.79 + 02	7.11 + 01	1.12 + 01	1.95 + 00	1.30 + 00	1.98-01
Np	93	6.95 + 03	1.96 + 03	9.32 + 02	1.87 + 02	7.45 + 01	1.18 + 01	2.05 + 00	1.35 + 00	2.05-01
Pu	94	7.19 + 03	2.04 + 03	9.65 + 02	1.94 + 02	7.71 + 01	1.22 + 01	2.13 + 00	1.39 + 00	2.10-01
Am	95	7.37 + 03	2.10 + 03	9.90 + 02	1.98 + 02	7.93 + 01	1.25 + 01	2.19 + 00	1.42 + 00	2.14-01
Cm	96	7.54 + 03	2.15 + 03	1.02 + 03	2.03 + 02	8.14 + 01	1.28 + 01	2.25 + 00	1.44 + 00	2.18-01
Bk	97	7.84 + 03	2.25 + 03	1.06 + 03	2.10 + 02	8.39 + 01	1.34 + 01	2.35 + 00	1.50 + 00	2.25-01
Cf	98	7.89 + 03	2.31 + 03	9.27 + 02	2.15 + 02	8.58 + 01	1.37 + 01	2.41 + 00	1.52 + 00	2.29-01
Es	99	7.79 + 03	2.40 + 03	9.59 + 02	2.22 + 02	4.01 + 01	1.42 + 01	2.51 + 00	1.57 + 00	2.36-01
Fm	100	7.13 + 03	2.46 + 03	9.77 + 02	2.26 + 02	4.09 + 01	1.45 + 01	2.57 + 00	1.59 + 00	2.39-01

N_5 M_5 M_3 L_3 K EDGE

N_4 M_4 M_2 L_2

N_3 M_1 L_1

N_2

N_1

	Atomic no.	Mass attenuation coefficient, cm²/g Photon energy, MeV								
		1.0	2.0	5.0	10.0	20.0	50.0	100.0	500.0	1000.0
Sb	51	5.80-02	4.10-02	3.56-02	3.92-02	4.70-02	5.96-02	6.85-02	8.16-02	8.44-02
Te	52	5.67-02	4.01-02	3.49-02	3.86-02	4.64-02	5.89-02	6.77-02	8.07-02	8.35-02
I	53	5.84-02	4.12-02	3.61-02	4.00-02	4.82-02	6.13-02	7.04-02	8.40-02	8.69-02
Xe	54	5.78-02	4.08-02	3.58-02	3.99-02	4.82-02	6.12-02	7.04-02	8.40-02	8.69-02
Cs	55	5.85-02	4.12-02	3.64-02	4.06-02	4.91-02	6.25-02	7.19-02	8.58-02	8.88-02
ZB	56	5.80-02	4.08-02	3.61-02	4.04-02	4.90-02	6.25-02	7.19-02	8.58-02	8.88-02
La	57	5.88-02	4.12-02	3.66-02	4.11-02	5.00-02	6.37-02	7.34-02	8.76-02	9.06-02
Ce	58	5.96-02	4.18-02	3.73-02	4.19-02	5.10-02	6.52-02	7.50-02	8.96-02	9.27-02
Pr	59	6.07-02	4.24-02	3.80-02	4.29-02	5.23-02	6.68-02	7.69-02	9.19-02	9.50-02
Nd	60	6.07-02	4.24-02	3.81-02	4.30-02	5.26-02	6.72-02	7.74-02	9.25-02	9.56-02
Pm	61	6.19-02	4.31-02	3.88-02	4.40-02	5.38-02	6.89-02	7.94-02	9.48-02	9.81-02
Sm	62	6.11-02	4.24-02	3.83-02	4.35-02	5.34-02	6.84-02	7.88-02	9.41-02	9.73-02
Eu	63	6.19-02	4.28-02	3.88-02	4.42-02	5.42-02	6.96-02	8.02-02	9.57-02	9.90-02
Gd	64	6.12-02	4.23-02	3.84-02	4.38-02	5.38-02	6.91-02	7.97-02	9.51-02	9.83-02
Tb	65	6.20-02	4.27-02	3.89-02	4.45-02	5.47-02	7.03-02	8.11-02	9.67-02	1.00-01
Dy	66	6.20-02	4.26-02	3.90-02	4.46-02	5.49-02	7.06-02	8.15-02	9.72-02	1.00-01
Ho	67	6.26-02	4.29-02	3.93-02	4.50-02	5.55-02	7.14-02	8.24-02	9.83-02	1.02-01
Er	68	6.32-02	4.32-02	3.96-02	4.55-02	5.61-02	7.23-02	8.34-02	9.95-02	1.03-01
Tm	69	6.40-02	4.36-02	4.01-02	4.61-02	5.70-02	7.35-02	8.48-02	1.01-01	1.04-01
Yb	70	6.40-02	4.35-02	4.00-02	4.61-02	5.70-02	7.35-02	8.49-02	1.01-01	1.04-01
Lu	71	6.48-02	4.39-02	4.05-02	4.66-02	5.77-02	7.45-02	8.60-02	1.02-01	1.06-01
Hf	72	6.50-02	4.39-02	4.05-02	4.68-02	5.80-02	7.48-02	8.64-02	1.03-01	1.06-01
Ta	73	6.57-02	4.41-02	4.08-02	4.72-02	5.85-02	7.56-02	8.73-02	1.04-01	1.07-01
W	74	6.62-02	4.43-02	4.10-02	4.75-02	5.89-02	7.62-02	8.80-02	1.05-01	1.08-01
Re	75	6.69-02	4.46-02	4.14-02	4.79-02	5.95-02	7.70-02	8.89-02	1.06-01	1.09-01
Os	76	6.71-02	4.46-02	4.13-02	4.79-02	5.96-02	7.71-02	8.90-02	1.06-01	1.10-01
Ir	77	6.79-02	4.50-02	4.17-02	4.84-02	6.02-02	7.80-02	9.01-02	1.07-01	1.11-01
Pt	78	6.86-02	4.52-02	4.20-02	4.87-02	6.06-02	7.86-02	9.08-02	1.08-01	1.12-01
Au	79	6.95-02	4.57-02	4.24-02	4.93-02	6.14-02	7.95-02	9.19-02	1.09-01	1.13-01
Hg	80	6.99-02	4.57-02	4.25-02	4.94-02	6.15-02	7.98-02	9.22-02	1.10-01	1.13-01
Tl	81	7.03-02	4.58-02	4.25-02	4.94-02	6.16-02	8.00-02	9.24-02	1.10-01	1.14-01
Pb	82	7.10-02	4.61-02	4.27-02	4.97-02	6.21-02	8.06-02	9.31-02	1.11-01	1.15-01
Bi	83	7.21-02	4.66-02	4.32-02	5.03-02	6.28-02	8.15-02	9.42-02	1.12-01	1.16-01
Po	84	7.39-02	4.75-02	4.40-02	5.12-02	6.40-02	8.32-02	9.61-02	1.15-01	1.18-01
At	85	7.54-02	4.82-02	4.46-02	5.20-02	6.49-02	8.44-02	9.76-02	1.16-01	1.20-01
Rn	86	7.30-02	4.65-02	4.30-02	5.01-02	6.26-02	8.14-02	9.42-02	1.12-01	1.16-01
Fr	87	7.45-02	4.72-02	4.36-02	5.08-02	6.35-02	8.26-02	9.56-02	1.14-01	1.18-01
Ra	88	7.53-02	4.75-02	4.38-02	5.10-02	6.38-02	8.31-02	9.61-02	1.15-01	1.19-01
Ac	89	7.69-02	4.82-02	4.44-02	5.17-02	6.47-02	8.43-02	9.75-02	1.16-01	1.20-01
Th	90	7.71-02	4.81-02	4.42-02	5.15-02	6.45-02	8.40-02	9.72-02	1.16-01	1.20-01
Pa	91	7.94-02	4.93-02	4.52-02	5.26-02	6.59-02	8.60-02	9.95-02	1.19-01	1.23-01
U	92	7.90-02	4.88-02	4.46-02	5.19-02	6.51-02	8.49-02	9.83-02	1.17-01	1.21-01
Np	93	8.13-02	4.99-02	4.56-02	5.30-02	6.65-02	8.68-02	1.01-01	1.20-01	1.24-01
Pu	94	8.26-02	5.05-02	4.60-02	5.34-02	6.71-02	8.76-02	1.01-01	1.21-01	1.25-01
Am	95	8.33-02	5.06-02	4.60-02	5.34-02	6.70-02	8.77-02	1.02-01	1.21-01	1.25-01
Cm	96	8.41-02	5.08-02	4.60-02	5.34-02	6.70-02	8.77-02	1.02-01	1.21-01	1.26-01
Bk	97	8.62-02	5.18-02	4.68-02	5.42-02	6.81-02	8.92-02	1.03-01	1.24-01	1.28-01
Cf	98	8.70-02	5.20-02	4.68-02	5.42-02	6.81-02	8.92-02	1.04-01	1.24-01	1.28-01
Es	99	8.89-02	5.28-02	4.74-02	5.48-02	6.89-02	9.04-02	1.05-01	1.25-01	1.29-01
Fm	100	8.94-02	5.28-02	4.72-02	5.45-02	6.86-02	9.00-02	1.05-01	1.25-01	1.29-01

CLASSIFICATION OF ELECTROMAGNETIC RADIATION

Hans Dolezalek

Basic Conversions:

$$c = \lambda \nu = \nu/k$$
$$\nu = c/\lambda = ck$$

$$\lambda = c/\nu = 1/k$$
$$k = \nu/c = 1/\lambda$$
$$c = \text{speed of light} = 2.99792458 \times 10^8 \text{ m/s}$$

Frequency (ν)	Wavelength (λ)	Wave number (k)	Names of bands	Approximate photon energies
$3 \times 10^0 - 3 \times 10^1$ Hz 3 — 30 Hz	$10^8 - 10^7$ m 100 — 10 Mm	$10^{-8} - 10^{-7}$ m^{-1} 10 — 100 Gm^{-1}	ELF-(ELF 1), ITU band no. 1	
$3 \times 10^1 - 3 \times 10^2$ Hz 30 — 300 Hz	$10^7 - 10^6$ m 10 — 1 Mm	$10^{-7} - 10^{-6}$ m^{-1} 100 Gm^{-1} — 1Mm^{-1}	SLF-(ELF 2), ITU band no. 2, megameter waves	
$3 \times 10^2 - 3 \times 10^3$ Hz 300 Hz — 3 kHz	$10^6 - 10^5$ m 1 Mm — 100 km	$10^{-6} - 10^{-5}$ m^{-1} 1 — 10 Mm^{-1}	ULF-(ELF 3), ITU band no. 3	
$3 \times 10^3 - 3 \times 10^4$ Hz 3 — 30 kHz	$10^5 - 10^4$ m 100 — 10 km	$10^{-5} - 10^{-4}$ m^{-1} 10 — 100 Mm^{-1}	VLF, ITU band no. 4, myriameter waves	
$3 \times 10^4 - 3 \times 10^5$ Hz 30 — 300 kHz	$10^4 - 10^3$ m 10 — 1 km	$10^{-4} - 10^{-3}$ m^{-1} 100 Mm^{-1} — 1 km^{-1}	LF, ITU band no. 5, kilometer waves	
$3 \times 10^5 - 3 \times 10^6$ Hz 300 kHz — 3 MHz	$10^3 - 10^2$ m 1 km — 100 m	$10^{-3} - 10^{-2}$ m^{-1} 1 — 10 km^{-1}	MF, ITU band no. 6, hectometer waves	
$3 \times 10^6 - 3 \times 10^7$ Hz 3 — 30 MHz	$10^2 - 10^1$ m 100 — 10 m	$10^{-2} - 10^{-1}$ m^{-1} 10 — 100 km^{-1}	HF, ITU band no. 7, decameter waves	
$3 \times 10^7 - 3 \times 10^8$ Hz 30 — 300 MHz	$10^1 - 10^0$ m 10 — 1 m	$10^{-1} - 10^0$ m^{-1} 100 km^{-1} — 1 mm^{-1}	VHF, ITU band no. 8, meter waves	
$3 \times 10^8 - 3 \times 10^9$ Hz 300 MHz — 3 GHz	$10^0 - 10^{-1}$ 1 m — 100 mm	$10^0 - 10^1$ m^{-1} 1 — 10 m^{-1}	UHF, ITU band no. 9, decimeter waves[a]	
$3 \times 10^9 - 3 \times 10^{10}$ Hz 3 — 30 GHz	$10^{-1} - 10^{-2}$ 100 — 10 mm	$10^1 - 10^2$ m^{-1} 10 — 100 m^{-1}	SHF, ITU band no. 10, centimeter waves[a]	
$3 \times 10^{10} - 3 \times 10^{11}$ Hz 30 — 300 GHz	$10^{-2} - 10^{-3}$ m 10 — 1 mm	$10^2 - 10^3$ m^{-1} 100 m^{-1} — 1 mm^{-1} (1 — 10 cm^{-1})	EHF, ITU band no. 11, millimeter waves	
$3 \times 10^{11} - 3 \times 10^{12}$ Hz 300 GHz — 3 THz	$10^{-3} - 10^{-4}$ 1 mm — 100 µm	$10^3 - 10^4$ m^{-1} (1 — 10 mm^{-1}) (10 — 100 cm^{-1})	Part of micrometer waves, includes part of far or thermal infrared; ITU band no. 12	
$3 \times 10^{12} - 3 \times 10^{13}$ Hz 3 — 30 THz	$10^{-4} - 10^{-5}$ 100 — 10 µm	$10^4 - 10^5$ m^{-1} 10 — 100 mm^{-1} (100 — 1000 cm^{-1})	Part of micrometer waves includes part of far (thermal) infrared	
$3 \times 10^{13} - 3 \times 10^{14}$ Hz 30 — 300 THz	$10^{-5} - 10^{-6}$ m 10 — 1 µm (100,000 — 10,000 Å)	$10^5 - 10^6$ m^{-1} 100 mm^{-1} — 1µm^{-1}	Part of µm waves, part of infrared	$(1.6 - 16) \times 10^{-20}$ joule {0.1 — 1 eV}
$3 \times 10^{14} - 3 \times 10^{15}$ Hz 300 THz — 3 PHz	$10^{-6} - 10^{-7}$ m 1 µm — 100 m (10,000 — 1000 Å)	$10^6 - 10^7$ m^{-1} 1 — 10 µm^{-1}	Near infrared, visible, near ultraviolet	$(1.6 - 16) \times 10^{-19}$ joule {1 — 10 eV}
$3 \times 10^{15} - 3 \times 10^{16}$ Hz 3 — 30 PHz	$10^{-7} - 10^{-8}$ m 100 — 10 nm (1000 — 100 Å)	$10^7 - 10^8$ m^{-1} 10 — 100 µm^{-1}	Part of vacuum ultraviolet	$(1.6 - 16) \times 10^{-18}$ joule {10 — 100 eV}
$3 \times 10^{16} - 3 \times 10^{17}$ Hz 30 — 300 PHz	$10^{-8} - 10^{-9}$ m 10 — 1 nm (100 — 10 Å)	$10^8 - 10^9$ m^{-1} 100 µm^{-1} — 1 nm^{-1}	Part of soft X-rays	$(1.6 - 16) \times 10^{-17}$ joule {100 — 1000 eV}
$3 \times 10^{17} - 3 \times 10^{18}$ Hz 300 PHz — 3 Ehz	$10^{-9} - 10^{-10}$ m 1 nm — 100 pm (10 — 1 Å)	$10^9 - 10^{10}$ m^{-1} 1 — 10 nm^{-1}	Part of soft X-rays	$(1.6 - 16) \times 10^{-16}$ joule {1 — 10 keV}
$3 \times 10^{18} - 3 \times 10^{19}$ Hz 3 — 30 Ehz	$10^{-10} - 10^{-11}$ m 100 — 10 pm (1 — 0.1 Å)	$10^{10} - 10^{11}$ m^{-1} 10 — 100 nm^{-1}	Hard X-rays and part of soft γ-rays	$(1.6 - 16) \times 10^{-15}$ joule {10 — 100 keV}
$3 \times 10^{19} - 3 \times 10^{20}$ Hz 30 — 300 Ehz	$10^{-11} - 10^{-12}$ m 10 — 1 pm (0.1 — 0.01 Å)	$10^{11} - 10^{12}$ m^{-1} 100 nm^{-1} — 1 pm^{-1}	Part of soft and part of hard γ-rays (limit at 510 keV)	$(1.6 - 16) \times 10^{-14}$ joule {100 keV — 1 MeV}
$3 \times 10^{20} - 3 \times 10^{21}$ Hz 300 — 3000 EHz	$10^{-12} - 10^{-13}$ m 1 pm — 100 fm (0.01 — 0.001 Å)	$10^{12} - 10^{13}$ m^{-1} 1 — 10 pm^{-1}	Part of hard γ-rays and part of "cosmic" γ-rays	$(1.6 - 16) \times 10^{-13}$ joule {1 — 10 MeV}
$3 \times 10^{21} - 3 \times 10^{22}$ Hz 3000 — 30,000 EHz	$10^{-13} - 10^{-14}$ m 100 — 10 fm (0.001 — 0.0001 Å)	$10^{13} - 10^{14}$ m^{-1} 10 — 100 pm^{-1}	γ-rays produced by cosmic rays	$(1.6 - 16) \times 10^{-12}$ joule {10 — 100 MeV}

Note: Abbreviations used in this table: Å— ångstrom (1 Å=10^{-10} m); EHz—exahertz (10^{18} hertz); EHF—extremely high frequency; ELF— extremely low frequency; eV—electron volt (1 eV = 1.60218×10^{-19} joule); fm—femtometer (10^{-15} m); GHz—gigahertz (10^9 hertz); Gm—gigameter (10^9 m); HF—high frequency; Hz—hertz (s^{-1}); ITU—International Telecommunications Union; keV— kiloelectron volt (10^3 eV); km—kilometer (10^3 m); LF—low frequency; m—meter; MeV— megaelectron volt (10^6 eV); MF—medium frequency; MHz—megahertz (10^6 hertz); Mm—megameter (10^6 meter); mm—millimeter (10^{-3} meter); μm—micrometer (10^{-6} meter); nm—nanometer (10^{-9} meter); PHz—petahertz (10^{15} hertz); pm—picometer (10^{-12} meter); SHF—super high frequency; SLF—super low frequency; THz—terahertz; UHF— ultra high frequency; ULF—ultra low frequency; VHF—very high frequency; VLF—very low frequency.

[a] Also called "microwaves"; not to be confused with "micrometer waves".

LETTER DESIGNATIONS OF MICROWAVE BANDS

Frequency (GHz)	Wavelength (cm)	Wavenumber (cm⁻¹)	Band
1—2	30—15	0.033—0.067	L-Band
1—4	15—7.5	0.067—0.133	S-Band
4—8	7.5—3.7	0.133—0.267	C-Band
8—12	3.7—2.5	0.267—0.4	X-Band
12—18	2.5—1.7	0.4—0.6	Ku-Band
18—27	1.7—1.1	0.6—0.9	K-Band
27—40	1.1—0.75	0.9—1.33	Ka-Band

SENSITIVITY OF THE HUMAN EYE TO LIGHT OF DIFFERENT WAVELENGTHS

The human eye responds to electromagnetic radiation in the wavelength range from about 360 nm (violet) to 820 nm (red), with a peak sensitivity near 555 nm. While the detailed shape of this response curve depends on the individual person, studies on representative samples of human subjects have led to adoption of a standard function relating the perceived brightness (luminous flux) to the actual power of the spectral radiation. This function is referred to as $V(\lambda)$, the photopic spectral luminous efficiency function, and it plays an important role in photometry.

The function $V(\lambda)$, as adopted by the International Commission on Illumination (CIE), is tabulated and plotted below.

References

1. *The Basis for Physical Photometry*, CIE Publication #18.2, 1983.
2. *CIE Standard Colorimetric Observers*, ISO/CIE #10527, 1991.
3. *Kaye and Laby Tables of Physical and Chemical Constants, Sixteenth Edition*, Longman Group Ltd., Harlow, Essex, 1995.

λ/nm	$V(\lambda)$	λ/nm	$V(\lambda)$	λ/nm	$V(\lambda)$
360	0.000004	520	0.710000	670	0.032000
370	0.000012	530	0.862000	680	0.017000
380	0.000039	540	0.954000	690	0.008210
390	0.000120	550	0.994950	700	0.004102
400	0.000396	555	1.000000	710	0.002091
410	0.001210	560	0.995000	720	0.001047
420	0.004000	570	0.952000	730	0.000520
430	0.011600	580	0.870000	740	0.000249
440	0.023000	590	0.757000	750	0.000120
450	0.038000	600	0.631000	760	0.000060
460	0.060000	610	0.503000	770	0.000030
470	0.090980	620	0.381000	780	0.000015
480	0.139020	630	0.265000	790	0.000007
490	0.208020	640	0.175000	800	0.000004
500	0.323000	650	0.107000	810	0.000002
510	0.503000	660	0.061000	820	0.000001

Spectral Luminous Efficiency Function

BLACK BODY RADIATION

The total power radiated from an ideal black body and the wavelength corresponding to maximum power are given here as a function of absolute temperature. Constants used in the calculation are taken from the table "Fundamental Physical Constants" in

Section 1. The radiated power in a band $\Delta\lambda$ at λ_{max} may be calculated from:

$$P_{max} = 0.657548\ (\Delta\lambda/\lambda_{max})\ P_{tot}$$

T/K	P_{tot}	$\lambda_{max}/\mu m$	T/K	P_{tot}	$\lambda_{max}/\mu m$	T/K	P_{tot}	$\lambda_{max}/\mu m$
50	0.354 W/m²	57.955	740	17.004	3.916	1520	302.689	1.906
100	5.671	28.978	750	17.942	3.864	1540	318.937	1.882
150	28.707	19.318	760	18.918	3.813	1560	335.831	1.858
200	90.728	14.489	770	19.934	3.763	1580	353.387	1.834
250	221.504	11.591	780	20.989	3.715	1600	371.623	1.811
273	314.973	10.614	790	22.087	3.668	1620	390.555	1.789
280	348.541	10.349	800	23.226	3.622	1640	410.202	1.767
290	401.064	9.992	810	24.410	3.577	1660	430.581	1.746
300	459.311	9.659	820	25.638	3.534	1680	451.710	1.725
310	523.684	9.348	830	26.911	3.491	1700	473.607	1.705
320	594.596	9.055	840	28.232	3.450	1720	496.290	1.685
330	672.478	8.781	850	29.600	3.409	1740	519.779	1.665
340	757.771	8.523	860	31.018	3.369	1760	544.093	1.646
350	850.931	8.279	870	32.486	3.331	1780	569.249	1.628
360	952.428	8.049	880	34.006	3.293	1800	595.267	1.610
370	1.063 kW/m²	7.832	890	35.578	3.256	1820	622.168	1.592
380	1.182	7.626	900	37.204	3.220	1840	649.970	1.575
390	1.312	7.430	910	38.886	3.184	1860	678.694	1.558
400	1.452	7.244	920	40.623	3.150	1880	708.359	1.541
410	1.602	7.068	930	42.418	3.116	1900	738.987	1.525
420	1.764	6.899	940	44.272	3.083	1920	770.597	1.509
430	1.939	6.739	950	46.187	3.050	1940	803.210	1.494
440	2.125	6.586	960	48.162	3.018	1960	836.848	1.478
450	2.325	6.439	970	50.201	2.987	1980	871.531	1.464
460	2.539	6.299	980	52.303	2.957	2000	907.282	1.449
470	2.767	6.165	990	54.471	2.927	2020	944.121	1.435
480	3.010	6.037	1000	56.705	2.898	2040	982.071	1.420
490	3.269	5.914	1020	61.379	2.841	2060	1.021 MW/m²	1.407
500	3.544	5.796	1040	66.337	2.786	2080	1.061	1.393
510	3.836	5.682	1060	71.589	2.734	2100	1.103	1.380
520	4.146	5.573	1080	77.147	2.683	2120	1.145	1.367
530	4.474	5.467	1100	83.022	2.634	2140	1.189	1.354
540	4.822	5.366	1120	89.227	2.587	2160	1.234	1.342
550	5.189	5.269	1140	95.773	2.542	2180	1.281	1.329
560	5.577	5.175	1160	102.672	2.498	2200	1.328	1.317
570	5.986	5.084	1180	109.939	2.456	2220	1.377	1.305
580	6.417	4.996	1200	117.584	2.415	2240	1.428	1.294
590	6.871	4.911	1220	125.621	2.375	2260	1.479	1.282
600	7.349	4.830	1240	134.063	2.337	2280	1.532	1.271
610	7.851	4.750	1260	142.924	2.300	2300	1.587	1.260
620	8.379	4.674	1280	152.217	2.264	2320	1.643	1.249
630	8.933	4.600	1300	161.955	2.229	2340	1.700	1.238
640	9.514	4.528	1320	172.154	2.195	2360	1.759	1.228
650	10.122	4.458	1340	182.827	2.163	2380	1.819	1.218
660	10.760	4.391	1360	193.989	2.131	2400	1.881	1.207
670	11.427	4.325	1380	205.655	2.100	2420	1.945	1.197
680	12.124	4.261	1400	217.838	2.070	2440	2.010	1.188
690	12.853	4.200	1420	230.556	2.041	2460	2.077	1.178
700	13.615	4.140	1440	243.822	2.012	2480	2.145	1.168
710	14.410	4.081	1460	257.652	1.985	2500	2.215	1.159
720	15.239	4.025	1480	272.063	1.958	2550	2.398	1.136
730	16.103	3.970	1500	287.070	1.932	2600	2.591	1.115

T/K	P_{tot}	$\lambda_{max}/\mu m$	T/K	P_{tot}	$\lambda_{max}/\mu m$	T/K	P_{tot}	$\lambda_{max}/\mu m$
2650	2.796	1.093	3600	9.524	0.805	5100	38.362	0.568
2700	3.014	1.073	3650	10.065	0.794	5200	41.461	0.557
2750	3.243	1.054	3700	10.627	0.783	5300	44.743	0.547
2800	3.485	1.035	3750	11.214	0.773	5400	48.217	0.537
2850	3.741	1.017	3800	11.824	0.763	5500	51.889	0.527
2900	4.011	0.999	3850	12.458	0.753	5600	55.767	0.517
2950	4.294	0.982	3900	13.118	0.743	5700	59.858	0.508
3000	4.593	0.966	3950	13.804	0.734	5800	64.170	0.500
3050	4.907	0.950	4000	14.517	0.724	5900	68.712	0.491
3100	5.237	0.935	4100	16.024	0.707	6000	73.490	0.483
3150	5.583	0.920	4200	17.645	0.690	6500	101.222	0.446
3200	5.946	0.906	4300	19.386	0.674	7000	136.149	0.414
3250	6.326	0.892	4400	21.254	0.659	7500	179.418	0.386
3300	6.725	0.878	4500	23.253	0.644	8000	232.264	0.362
3350	7.142	0.865	4600	25.389	0.630	8500	296.004	0.341
3400	7.578	0.852	4700	27.670	0.617	9000	372.042	0.322
3450	8.033	0.840	4800	30.101	0.604	9500	461.867	0.305
3500	8.509	0.828	4900	32.689	0.591	10000	567.051	0.290
3550	9.006	0.816	5000	35.441	0.580			

The curves below show, for various temperatures, the fraction of radiant power as a function of wavelength. The function plotted is $P_\lambda/\Delta\lambda\, P_{tot}$, where P_λ is the power at wavelength λ in a small interval $\Delta\lambda$ (in μm), and P_{tot} is the total power.

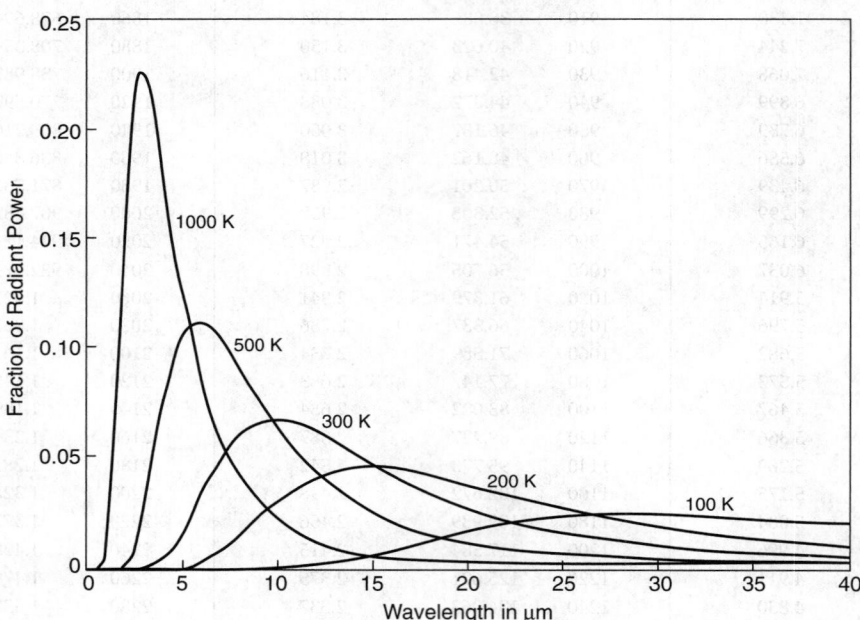

CHARACTERISTICS OF INFRARED DETECTORS

This graph summarizes the wavelength response of some semi-conductors used as detectors for infrared radiation. The quantity $D^*(\lambda)$ is the signal to noise ratio for an incident radiant power density of 1 W/cm² and a bandwidth of 1 Hz (60° field of view). The Ge, InAs, and InSb detectors are photovoltaics, while the HgCdTe series are photoconductive devices. The cutoff wavelength of the latter can be varied by adjusting the relative amounts of Hg, Cd, and Te (three examples are shown at 77 K). The graph also shows the theoretical background limited sensitivity for ideal detectors which introduce no intrinsic noise.

Reference

Infrared Detectors 1995, EG&G Judson, Montgomeryville, PA.

INDEX OF REFRACTION OF INORGANIC CRYSTALS

This table lists the index of refraction of selected crystalline inorganic compounds. When, available, values are give as a function of wavelength in the range from the ultraviolet to the far infrared region. For each compound a value at 589 nm, the wavelength of the principal sodium line, is given. The data have been taken from the references indicated; in many cases, data from a reference have been fitted to generate the index of refraction at the wavelengths used in this table. All values refer to ambient temperature. Entries marked · * are based on extrapolation beyond the range of available experimental data.

Compounds belonging to the cubic crystal system have only a single refractive index value, but other systems are anisotropic, so that the crystal is characterized by two or three unique indexes. Hexagonal, rhombohedral, and tetragonal crystals have two unique indexes which are traditionally labeled n_o and n_e for "ordinary ray" and "extraordinary ray". Orthorhombic, monoclinic, and triclinic crystals are characterized by three indexes which are here called n_x, n_y, and n_z. The table indicates the crystal system for each entry in order to identify the material uniquely.

The refractive index and other optical properties for metals, semiconductors, and certain other compounds can be found in the tables "Optical Properties of Selected Elements" and "Optical Properties of Selected Inorganic and Organic Solids" in Section 12 of this Handbook.

References

1. Li, H. H., "Refractive Index of Alkali Halides and its Wavelength and Temperature Derivatives", *J. Phys. Chem. Ref. Data* 5, 329, 1976.
2. Li, H. H., "Refractive Index of Alkaline Earth Halides and its Wavelength and Temperature Derivatives", *J. Phys. Chem. Ref. Data* 9, 161, 1980.
3. Li, H. H., "Refractive Index of ZnS, ZnSe, and ZnTe and its Wavelength and Temperature Derivatives", *J. Phys. Chem. Ref. Data* 13, 103, 1984.
4. Shannon, R. D., Shannon, R. C., Medenbach, O., and Fischer, R. X., "Refractive Index and Dispersion of Fluorides and Oxides", *J. Phys. Chem. Ref. Data* 31, 931, 2002.
5. Gray, D. E., ed., *American Institute of Physics Handbook*, Sec. 6b, p. **6**-12, McGraw-Hill, New York, 1972.
6. Landolt-Börnstein Numerical Data and Functional Relationships in Science and *Technology*, III/11, Elastic, Piezoelectric, Pyroelectric, Piezooptic, Electrooptic Constants, and Nonlinear Dielectric Susceptibilities of Crystals, Springer-Verlag, Berlin, 1979.
7. Landolt-Börnstein Numerical Data and Functional Relationships in Science and *Technology*, III/30A, High Frequency Properties of Dielectric Crystals. Piezooptic and Electrooptic Constants, Springer-Verlag, Berlin, 1996.
8. Weber, M. J., *CRC Handbook of Laser Science and Technology*, Vol. IV. Optical Materials. Part 2: Properties, CRC Press, Boca Raton, FL, 1986.

Compound	Crystal system	Ray	300 nm	589 nm	750 nm	1 μm	2 μm	5 μm	10 μm	20 μm	Ref.
AgCl	cub	n		2.0668	2.0401	2.0224	2.0062	1.9975	1.9803	1.9069	5
AlPO$_4$	rhomb	n_o		1.5247	1.5203	1.5161	1.5034				6
	rhomb	n_e		1.5338	1.5290	1.5245	1.5116				6
Al$_2$O$_3$	hex	n_o		1.7673							4
	hex	n_e		1.7598							4
As$_2$O$_3$[a]	cub	n		1.7537							4
BaF$_2$	cub	n	1.5010	1.4744	1.4712	1.4686	1.4647	1.4511	1.4014		2
BaO	cub	n		1.9841							4
BaSO$_4$	orth	n_x		1.6362							4
	orth	n_y		1.6374							4
	orth	n_z		1.6480							4
BaTiO$_3$	tetr	n_o		2.4405							4
	tetr	n_e		2.3831							4
BaWO$_4$	tetr	n_o		1.8426							4
	tetr	n_e		1.8405							4
BeO	hex	n_o		1.7184							4
	hex	n_e		1.7342							4
BeSO$_4$·4H$_2$O	tetr	n_o		1.4713							4
	tetr	n_e		1.4328							4
CaCO$_3$[b]	hex	n_o	1.7216	1.6584	1.6503	1.6436	1.6249				5
	hex	n_e	1.5145	1.4864	1.4828	1.4801	1.4753				5
CaF$_2$	cub	n	1.4540	1.4338	1.4311	1.4289	1.4239	1.3990	1.299		2
CaO	cub	n		1.8396							4
CaSO$_4$	orth	n_x		1.5698							4
	orth	n_y		1.5755							4
	orth	n_z		1.6137							4
CaSO$_4$·2H$_2$O	monocl	n_x		1.5207							4
	monocl	n_y		1.5227							4
	monocl	n_z		1.5304							4
CaWO$_4$	tetr	n_o		1.9195							4
	tetr	n_e		1.9355							4
CdS	hex	n_o		2.507	2.390	2.334					5

Compound	Crystal system	Ray	Index of Refraction at the Indicated Wavelength								Ref.
			300 nm	589 nm	750 nm	1 μm	2 μm	5 μm	10 μm	20 μm	
	hex	n_e		2.525	2.409	2.352					5
CdSe	hex	n_o			2.68*	2.5502	2.4682	2.4483	2.4331		7
	hex	n_e			2.69*	2.5696	2.4873	2.4676	2.4514		7
CdTe	cub	n							2.6724	2.6302	7
CeF_3	hex	n_o		1.6183							4
	hex	n_e		1.6113							4
CsBr	cub	n	1.8047	1.6974	1.6861	1.6784	1.6711	1.6678	1.6630	1.6439	1
CsCl	cub	n	1.712	1.640	1.631	1.626	1.620	1.616	1.606	1.563	1
$CsClO_4$	orth	n_x		1.4752							4
	orth	n_y		1.4788							4
	orth	n_z		1.4804							4
CsF	cub	n	1.506	1.477	1.474	1.472	1.469*	1.461*	1.436*	1.32*	1
CsI	cub	n	1.9790	1.7873	1.7694	1.7576	1.7465	1.7428	1.7396	1.7280	1
Cs_2SO_4	orth	n_x		1.5598							4
	orth	n_y		1.5644							4
	orth	n_z		1.5662							4
CuBr	cub	n		2.117							7
CuCl	cub	n		1.9727	1.9391				1.9245		7
$CuSO_4 \cdot 5H_2O$	tricl	n_x		1.5140							4
	tricl	n_y		1.5367							4
	tricl	n_z		1.5436							4
Dy_2O_3	cub	n		1.9757							4
FeF_2	tetr	n_o		1.514							4
	tetr	n_e		1.524							4
Gd_2O_3	cub	n		1.96							4
HgS	rhomb	n_o		2.9413	2.7770	2.7120	2.6305		2.6018		6
	rhomb	n_e		3.3072	3.0896	3.0050	2.8776		2.8522		6
KBr	cub	n	1.6482	1.5598	1.5498	1.5444	1.5383	1.5345	1.5264	1.4924	1
KCl	cub	n	1.5455	1.4902	1.4840	1.4798	1.4753	1.4704	1.4564	1.3946	1
$KClO_4$	orth	n_x		1.4730							4
	orth	n_y		1.4736							4
	orth	n_z		1.4768							4
KF	cub	n	1.380	1.362	1.360	1.358	1.355	1.344	1.304*	1.09*	1
KH_2AsO_4	tetr	n_o		1.5674							7
	tetr	n_e		1.5179							7
KH_2PO_4	tetr	n_o	1.5450	1.5093	1.5030	1.4957					5
	tetr	n_e	1.4977	1.4682	1.4641	1.4606					5
KI	cub	n	1.834*	1.665	1.650	1.640	1.631	1.627	1.620	1.593	1
KIO_3	tricl	n_x		1.6959							7
	tricl	n_y		1.8317							7
	tricl	n_z		1.8343							7
KIO_4	tetr	n_o		1.6205							4
	tetr	n_e		1.6476							4
$KNbO_3$	orth	n_x		2.2480	2.3395	2.2612					7
	orth	n_y		2.3464	2.2959	2.2622					7
	orth	n_x		2.1803	2.1457	2.1288					7
K_2SO_4	orth	n_x		1.4934							4
	orth	n_y		1.4947							4
	orth	n_z		1.4973							4
LaF_3	hex	n_o		1.605							4
	hex	n_e		1.599							4
LiBr	cub	n	1.810	1.783	1.781	1.778	1.774*	1.756*	1.68*	1.33*	1
LiCl	cub	n	1.677	1.662	1.660	1.658	1.654*	1.62*	1.53*		1
$LiClO_4 \cdot 3H_2O$	hex	n_o		1.4832							4
	hex	n_e		1.4384							4
LiF	cub	n	1.4087	1.3921	1.3895	1.3871	1.3786	1.3266	1.1005		1
LiI	cub	n	1.979	1.955	1.952	1.950	1.948*	1.940*	1.91*	1.77*	1
$LiIO_3$	hex	n_o		1.8875	1.8713	1.8589	1.8410				6
	hex	n_e		1.7400	1.7268	1.7179	1.7062				6
$LiNbO_3$	rhomb	n_o		2.3007	2.2632	2.2370					7

Compound	Crystal system	Ray	Index of Refraction at the Indicated Wavelength								Ref.
			300 nm	589 nm	750 nm	1 μm	2 μm	5 μm	10 μm	20 μm	
	rhomb	n_e		2.2116	2.1804	2.1567					7
LiTaO$_3$	rhomb	n_o		2.1864	2.1590	2.1391	2.1066				7
	rhomb	n_e		2.1908	2.1634	2.1432	2.1115				7
Li$_2$SO$_4$·H$_2$O	monocl	n_x		1.4615							4
	monocl	n_y		1.4765							4
	monocl	n_z		1.4863							4
Lu$_2$O$_3$	cub	n		1.9349							4
MgF$_2$	tetr	n_o	1.3930	1.3776	1.375	1.373	1.368	1.34	1.21		2
	tetr	n_e	1.4055	1.3894	1.387	1.385	1.379	1.34	1.21		2
MgO	cub	n		1.7355	1.7283	1.7228	1.7084	1.6361			5
MgSO$_4$·7H$_2$O	orth	n_x		1.4326							4
	orth	n_y		1.4555							4
	orth	n_z		1.4607							4
MnF$_2$	tetr	n_o		1.472							4
	tetr	n_e		1.501							4
NH$_4$H$_2$AsO$_4$	tetr	n_o	1.6401	1.5777	1.5704	1.5583					7
	tetr	n_e	1.5754	1.5232	1.5179	1.5101					7
NH$_4$H$_2$PO$_4$	tetr	n_o	1.5668	1.5247	1.5187	1.5084					7
	tetr	n_e	1.5137	1.4797	1.4754	1.4694					7
NaBr	cub	n	1.748	1.642	1.631	1.623	1.616	1.609	1.593*	1.520*	1
NaBrO$_3$	cub	n		1.6168							4
NaCl	cub	n	1.6066	1.5441	1.5369	1.5320	1.5265	1.5188	1.4947	1.382*	1
NaClO$_3$	cub	n		1.5151							7
NaF	cub	n	1.3424	1.3252	1.3231	1.3214	1.3179	1.3017	1.2400		1
NaH$_2$PO$_4$·2H$_2$O	orth	n_x		1.4400							7
	orth	n_y		1.4628							7
	orth	n_z		1.4814							7
NaI	cub	n	1.93*	1.774	1.758	1.74	1.73*	1.73*	1.71*	1.66*	1
NaNO$_2$	orth	n_x		1.6547							7
	orth	n_y		1.3455							7
	orth	n_z		1.4125							7
NaNO$_3$	rhomb	n_o		1.5840							5
	rhomb	n_e		1.3340							5
Na$_2$HPO$_4$·7H$_2$O	monocl	n_x		1.4411							4
	monocl	n_y		1.4423							4
	monocl	n_z		1.4525							4
Na$_2$SO$_4$	orth	n_x		1.4669							4
	orth	n_y		1.4730							4
	orth	n_z		1.4809							4
NdF$_3$	hex	n_o		1.6191							4
	hex	n_e		1.6132							4
Nd$_2$O$_3$	cub	n		1.92							4
NiF$_2$	tetr	n_o		1.526							4
	tetr	n_e		1.561							4
NiSO$_4$·6H$_2$O	tetr	n_o		1.5107							4
	tetr	n_e		1.4870							4
PbF$_2$	cub	n	1.94*	1.767	1.754	1.745	1.73	1.70	1.66	1.32	5
PbSO$_4$	orth	n_x		1.8780							4
	orth	n_y		1.8834							4
	orth	n_z		1.8945							4
PrF$_3$	hex	n_o		1.6207							4
	hex	n_e		1.6146							4
RbBr	cub	n	1.639	1.553	1.544	1.538	1.532	1.530	1.525	1.505*	1
RbCl	cub	n	1.549	1.493	1.487	1.483	1.479	1.475	1.465	1.424*	1
RbClO$_4$	orth	n_x		1.4691							4
	orth	n_y		1.4701							4
	orth	n_z		1.4732							4
RbF	cub	n	1.428*	1.397	1.394	1.391	1.388	1.379	1.346	1.19*	1
RbH$_2$AsO$_4$	tetr	n_o	1.6183	1.5603	1.5538	1.5432					7
	tetr	n_e	1.5718	1.5232	1.5184	1.5121					7

Index of Refraction at the Indicated Wavelength

Compound	Crystal system	Ray	300 nm	589 nm	750 nm	1 μm	2 μm	5 μm	10 μm	20 μm	Ref.
RbH_2PO_4	tetr	n_o	1.5434	1.5078	1.5021	1.4941					7
	tetr	n_e	1.5106	1.4791	1.4754	1.4704					7
RbI	cub	n	1.808	1.647	1.633	1.623	1.615	1.612	1.608	1.595	1
Rb_2SO_4	orth	n_x		1.5131							4
	orth	n_y		1.5133							4
	orth	n_z		1.5144							4
$Sb_2O_3{}^c$	cub	n		2.8017							4
Sc_2O_3	cub	n		1.9943							4
$SiO_2{}^d$	hex	n_o	1.5733	1.5442	1.5394	1.5350	1.5209				5
	hex	n_e	1.5882	1.5534	1.5484	1.5438	1.5291				5
SnO_2	tetr	n_o		1.993							4
	tetr	n_e		2.088							4
SrF_2	cub	n	1.459	1.4380	1.435	1.433	1.429	1.412	1.35		2
SrO	cub	n		1.8710							4
$SrSO_4$	orth	n_x		1.6214							4
	orth	n_y		1.6231							4
	orth	n_z		1.6303							4
$SrTiO_3$	cub	n		2.4082	2.3525	2.3160	2.2676	2.1205			5
$SrWO_4$	tetr	n_o		1.8618							4
	tetr	n_e		1.8719							4
TbF_3	hex	n_o		1.6034							4
	hex	n_e		1.5603							4
TeO_2	tetr	n_o		2.2738		2.2080					7
	tetr	n_e		2.4295		2.3520					7
ThO_2	cub	n		2.1113							4
$TiO_2{}^e$	tetr	n_o		2.612	2.533	2.485	2.399	2.220			5
	tetr	n_e		2.910	2.805	2.748					5
	tetr	n_o		2.562							4
	tetr	n_e		2.489							4
TlBr	cub	n		2.418	2.350	2.289	2.103	1.984	2.339	2.322	5
TlCl	cub	n		2.247	2.198	2.145	1.986	1.891	2.193		5
$TlClO_4$	orth	n_x		1.6427							4
	orth	n_y		1.6446							4
	orth	n_z		1.6542							4
Tl_2SO_4	orth	n_x		1.8604							4
	orth	n_y		1.8676							4
	orth	n_z		1.8857							4
Y_2O_3	cub	n		1.930							4
Yb_2O_3	cub	n		1.9468							4
ZnF_2	tetr	n_o		1.495							4
	tetr	n_e		1.525							4
ZnO	hex	n_o		2.0036	1.9662	1.9435	1.9197				7
	hex	n_e		2.0199	1.9821	1.9589	1.9330				7
ZnS^f	cub	n		2.3691	2.3232	2.2932	2.2633				7
ZnS^g	hex	n_o		2.372	2.331	2.303	2.26	2.25	2.20		3,5
	hex	n_e		2.368	2.327	2.301					5
ZnSe	cub	n		2.6222	2.5384	2.4888	2.4462	2.4296	2.4065		3
ZnTe	cub	n		3.060	2.880	2.789	2.719	2.698	2.684		3
$ZrSiO_4{}^h$	tetr	n_o		1.9255							4
	tetr	n_e		1.9843							4

* Provisional value based on extrapolation beyond the range of experimental data.

a Arsenolite
b Calcite
c Senarmontite
d α-Quartz
e Rutile
f Sphalerite
g Wurtzite
h Zircon

REFRACTIVE INDEX AND TRANSMITTANCE OF REPRESENTATIVE GLASSES

Typical values of the index of refraction and internal transmittance (fraction of light transmitted through a one centimeter thickness) are tabulated here for selected types of glasses, as well as for synthetic fused (vitreous) silica. Nominal compositions are given in the first part of the table. The second part gives the index of refraction, relative to air, and the internal transmittance for representative samples of each glass at wavelengths in the infrared, visible, and near-ultraviolet regions. It should be emphasized that wide variation of these parameters may be found among subtypes of each glass. More detailed data may be found in Reference 3.

Assuming that the Lambert-Beer Law is followed, the transmittance of a glass plate of thickness d (in centimeters) can be obtained by raising the transmittance value in the table to the power d.

References

1. Weber, M. J., *CRC Handbook of Laser Science and Technology*, Vol. IV, Part 2, CRC Press, Boca Raton, FL ,1988.
2. Gray, D. E., Ed., *American Institute of Physics Handbook, Third Edition*, McGraw Hill, New York, 1972.
3. *Schott Optical Glass*, Schott Glass Technologies, Inc., 400 York Ave., Duryea, PA 18642.
4. Kaye, G. W. C., and Laby, T. H., *Tables of Physical and Chemical Constants, 15th Edition*, Longman, London, 1986.

Type	Name	Composition in percent by mass									
		SiO_2	B_2O_3	Al_2O_3	Na_2O	K_2O	CaO	BaO	ZnO	PbO	P_2O_5
PK	Phosphate crown		3	10		12	5				70
PSK	Dense phosphate crown		3	5			4	28			60
BK	Borosilicate crown	70	10		8	8	1	3			
K	Crown	74			9	11	6				
ZK	Zinc crown	71			17				12		
BaK	Barium crown	60	3		3	10		19	5		
SK	Dense crown	39	15	5				41			
KF	Crown flint	67		2	16				3	12	
BaLF	Barium light flint	51			6	5		20	14	4	
SSK	Extra dense crown	35	10	5				42	8		
LLF	Extra light flint	63			5	8				24	
BaF	Barium flint	46				8		16	8	22	
LF	Light flint	53			5	8				34	
F	Flint	47			2	7				44	
BaSF	Dense barium flint	43			1	7		11	5	33	
SF	Dense flint	33				5				62	
KzFS	Short flint										
SiO_2	Fused silica	100									

Type	Index of refraction				Transmittance of 1 cm plate			
	1.060 µm	546.1 nm	365.0 nm	312.6 nm	1.060 µm	546.1 nm	365.0 nm	310 nm
PK	1.51519	1.52736	1.54503	1.5574	0.997	0.998	0.987	0.46
PSK	1.54154	1.55440	1.57342	1.5868	0.996	0.998	0.984	0.46
BK	1.50669	1.51872	1.53627	1.5486	0.999	0.998	0.987	0.35
K	1.50091	1.51314	1.53189	1.5454	0.998	0.998	0.988	0.40
ZK	1.52220	1.53534	1.55588	1.5708	0.996	0.998	0.976	0.27
BaK	1.55695	1.57124	1.59407	1.6108	0.998	0.997	0.986	0.28
SK	1.59490	1.60994	1.63398		0.998	0.998	0.959	0.28
KF	1.50586	1.51978	1.54251	1.5600	0.998	0.996	0.989	0.49
BaLF	1.57579	1.59166	1.61804		0.996	0.998	0.933	0.010
SSK	1.60402	1.61993	1.64595		0.999	0.998	0.915	0.010
LaK	1.69710	1.71616	1.74573		0.999	0.998	0.882	0.17
LLF	1.52775	1.54344	1.57038		0.998	0.997	0.990	0.32
BaF	1.56873	1.58565	1.61524		0.999	0.997	0.992	0.004
LF	1.56594	1.58482	1.61926		0.999	0.998	0.981	0.008
F	1.58636	1.60718	1.64606		0.997	0.998	0.959	
BaSF	1.60889	1.62987	1.66926		0.999	0.998	0.857	
SF	1.71350	1.74620	1.8145		0.998	0.997	0.650	
KzFS	1.59680	1.61639	1.64849	1.6739		0.998	0.672	0.012
SiO_2	1.44968	1.46008	1.47435[a]	1.53430[b]				

[a] At 366.3 nm.

[b] At 213.9 nm.

INDEX OF REFRACTION OF WATER

This table gives the index of refraction of liquid water at atmospheric pressure, relative to a vacuum, at several temperatures and wavelengths. It is generated from the formulation in Reference 1, which covers a wide range of temperature, pressure, and wavelength. The wavelengths listed here correspond to prominent lines of cadmium (226.50 and 361.05 nm), potassium (404.41 nm), sodium (589.00 nm), Ne (632.80 nm, from a helium - neon laser), and mercury (1.01398 μm).

References

1. Schiebener, P., Straub, J., Levelt Sengers, J. M. H., and Gallagher, J. S., *J. Phys. Chem. Ref. Data*, 19, 677, 1990; 19, 1617, 1990.
2. Marsh, K. N., Editor, *Recommended Reference Materials for the Realization of Physicochemical Properties*, Blackwell Scientific Publications, Oxford, 1987.

$t/°C$	226.50 nm	361.05 nm	404.41 nm	589.00 nm	632.80 nm	1.01398 μm
0	1.39450	1.34896	1.34415	1.33432	1.33306	1.32612
10	1.39422	1.34870	1.34389	1.33408	1.33282	1.32591
20	1.39336	1.34795	1.34315	1.33336	1.33211	1.32524
30	1.39208	1.34682	1.34205	1.33230	1.33105	1.32424
40	1.39046	1.34540	1.34065	1.33095	1.32972	1.32296
50	1.38854	1.34373	1.33901	1.32937	1.32814	1.32145
60	1.38636	1.34184	1.33714	1.32757	1.32636	1.31974
70	1.38395	1.33974	1.33508	1.32559	1.32438	1.31784
80	1.38132	1.33746	1.33284	1.32342	1.32223	1.31576
90	1.37849	1.33501	1.33042	1.32109	1.31991	1.31353
100	1.37547	1.33239	1.32784	1.31861	1.31744	1.31114

INDEX OF REFRACTION OF LIQUIDS FOR CALIBRATION PURPOSES

This table gives the index of refraction of six liquids which are available in highly pure form and whose index of refraction has been accurately measured as a function of wavelength and temperature. They are therefore useful for calibration of refractometers. The estimated uncertainty in the values is:

2,2,4-Trimethylpentane	±0.00003
Hexadecane	±0.00008
trans-Bicyclo[4.0.0]decane	±0.00008
1-Methylnaphthalene	±0.00008
Toluene	±0.00003
Methylcyclohexane	±0.00003

Full details are given in the references. This table is reprinted from Reference 1 by permission of the International Union of Pure and Applied Chemistry.

References

1. Marsh, K. N., Editor, *Recommended Reference Materials for the Realization of Physicochemical Properties*, Blackwell Scientific Publications, Oxford, 1987.
2. Tilton, L. W., *J. Opt. Soc. Am.*, 32, 71, 1941.

λ	2,2,4-Trimethylpentane			Hexadecane		
nm	20°C	25°C	30°C	20°C	25°C	30°C
667.81	1.38916	1.38670	1.38424	1.43204	1.43001	1.42798
656.28	1.38945	1.38698	1.38452	1.43235	1.43032	1.42829
589.26	1.39145	1.38898	1.38650	1.43453	1.43250	1.43047
546.07	1.39316	1.39068	1.38820	1.43640	1.43436	1.43232
501.57	1.39544	1.39294	1.39044	1.43888	1.43684	1.43480
486.13	1.39639	1.39389	1.39138	1.43993	1.43788	1.43583
435.83	1.40029	1.39776	1.39523	1.44419	1.44213	1.44007

λ	trans-Bicyclo[4.4.0]decane			1-Methylnaphthalene		
nm	20°C	25°C	30°C	20°C	25°C	30°C
667.81	1.46654	1.46438	1.46222	1.60828	1.60592	1.60360
656.28	1.46688	1.46472	1.46256	1.60940	1.60703	1.60471
589.26	1.46932	1.46715	1.46498	1.61755	1.61512	1.61278
546.07	1.47141	1.46923	1.46705	1.62488	1.62240	1.62005
501.57	1.47420	1.47200	1.46980	1.63513	1.63259	1.63022
486.13	1.47535	1.47315	1.47095	1.63958	1.63701	1.63463
435.83	1.48011	1.47789	1.47567		1.65627	1.65386

λ	Toluene			Methylcyclohexane		
nm	20°C	25°C	30°C	20°C	25°C	30°C
667.81	1.49180	1.48903	1.48619	1.42064	1.41812	1.41560
656.28	1.49243	1.48966	1.48682	1.42094	1.41842	1.41591
589.26	1.49693	1.49413	1.49126	1.42312	1.42058	1.41806
546.07	1.50086	1.49803	1.49514	1.42497	1.42243	1.41989
501.57	1.50620	1.50334	1.50041	1.42744	1.42488	1.42233
486.13	1.50847	1.50559	1.50265	1.42847	1.42590	1.42334
435.83	1.51800	1.51506	1.51206	1.43269	1.43010	1.42752

INDEX OF REFRACTION OF AIR

This is a table of the index of refraction n of dry air at 15°C and a pressure of 101.325 kPa and containing 0.045% by volume of carbon dioxide ("standard air"). The index of refraction is defined by $n = \lambda_{vac}/\lambda_{air}$ where λ is the wavelength of the radiation. The index is calculated from the expression

$$(n-1) \times 10^8 = 8342.54 + 2406147(130 - \sigma^2)^{-1} + 15998(38.9 - \sigma^2)^{-1}$$

where $\sigma = 1/\lambda_{vac}$ and λ_{vac} has units of μm. The equation is valid for λ_{vac} from 200 nm to 2 μm. The table also gives the correction $(n-1)\lambda_{air}$ which must be added to the wavelength in air to obtain λ_{vac}.

If the air is at a temperature t in °C (ITS-90) and a pressure p in pascals, a value of $(n-1)$ from this table should be multiplied by

$$p[1 + p(60.1 - 0.972t) \times 10^{-10}]/96095.43(1 + 0.003661t)$$

References

1. Birch, K. P., and Downs, M. J., *Metrologia*, 31, 315, 1994.
2. Edlen, B., *Metrologia* 2, 71, 1966.

λ_{vac}	$(n-1) \times 10^8$	$\lambda_{vac} - \lambda_{air}$	λ_{vac}	$(n-1) \times 10^8$	$\lambda_{vac} - \lambda_{air}$	λ_{vac}	$(n-1) \times 10^8$	$\lambda_{vac} - \lambda_{air}$
200 nm	32409	0.06480 nm	540	27804	0.15010	880	27462	0.24160
210	31748	0.06665	550	27784	0.15277	890	27458	0.24431
220	31226	0.06868	560	27765	0.15544	900	27454	0.24701
230	30801	0.07082	570	27747	0.15811	910	27449	0.24972
240	30447	0.07305	580	27730	0.16079	920	27445	0.25243
250	30148	0.07535	590	27714	0.16347	930	27441	0.25513
260	29892	0.07769	600	27698	0.16614	940	27437	0.25784
270	29670	0.08009	610	27684	0.16882	950	27434	0.26055
280	29477	0.08251	620	27670	0.17151	960	27430	0.26326
290	29307	0.08497	630	27657	0.17419	970	27427	0.26597
300	29157	0.08745	640	27644	0.17688	980	27423	0.26868
310	29023	0.08995	650	27632	0.17956	990	27420	0.27138
320	28904	0.09247	660	27621	0.18225			
330	28796	0.09500	670	27610	0.18494	1.00 μm	27417	0.0002741
340	28700	0.09755	680	27600	0.18763	1.05	27402	0.0002876
350	28612	0.10011	690	27590	0.19032	1.10	27390	0.0003012
360	28532	0.10269	700	27581	0.19301	1.15	27379	0.0003148
370	28460	0.10527	710	27572	0.19570	1.20	27370	0.0003283
380	28393	0.10786	720	27563	0.19840	1.25	27361	0.0003419
390	28332	0.11046	730	27555	0.20109	1.30	27354	0.0003555
400	28276	0.11307	740	27547	0.20379	1.35	27347	0.0003691
410	28224	0.11569	750	27539	0.20649	1.40	27341	0.0003827
420	28177	0.11831	760	27532	0.20918	1.45	27336	0.0003963
430	28132	0.12094	770	27525	0.21188	1.50	27331	0.0004099
440	28091	0.12357	780	27518	0.21458	1.55	27327	0.0004234
450	28053	0.12620	790	27511	0.21728	1.60	27323	0.0004370
460	28018	0.12885	800	27505	0.21998	1.65	27319	0.0004506
470	27985	0.13149	810	27499	0.22268	1.70	27316	0.0004642
480	27954	0.13414	820	27493	0.22538	1.75	27313	0.0004778
490	27925	0.13679	830	27488	0.22808	1.80	27310	0.0004914
500	27897	0.13945	840	27482	0.23079	1.85	27307	0.0005050
510	27872	0.14211	850	27477	0.23349	1.90	27305	0.0005187
520	27848	0.14477	860	27472	0.23619	1.95	27303	0.0005323
530	27825	0.14743	870	27467	0.23890	2.00	27301	0.0005459

CHARACTERISTICS OF LASER SOURCES

William F. Krupke

Light amplification by stimulated emission of radiation was first demonstrated by Maiman in 1960, the result of a population inversion produced between energy levels of chromium ions in a ruby crystal when irradiated with a xenon flashlamp. Since then population inversions and coherent emission have been generated in literally thousands of substances (neutral and ionized gases, liquids, and solids) using a variety of incoherent excitation techniques (optical pumping, electrical discharges, gas-dynamic flow, electron-beams, chemical reactions, nuclear decay).

The extrema of laser output parameters which have been demonstrated to date and the laser media used are summarized in Table 1. Note that the extreme power and energy parameters listed in this table were attained with laser *systems* rather than with simple laser oscillators.

Laser sources are commonly classified in terms of the state-of-matter of the active medium: gas, liquid, and solid. Each of these classes is further subdivided into one or more types as shown in Table 2. A well-known representative example of each type of laser is also given in Table 2 together with its nominal operation wavelength and the methods by which it is pumped.

The various lasers together cover a wide spectral range from the far ultraviolet to the far infrared. The particular wavelength of emission (usually a narrow line) is presented for some six dozen lasers in Figures 1A and 1B.

By suitably designing the excitation source and/or by controlling the laser resonator structure, laser systems can provide continuous or pulsed radiation as shown in Table 3.

Besides the method of excitation and the temporal behavior of a laser, there are many other parameters that characterize its operation and efficiency, as shown in Tables 4 and 5.

Although many lasers only emit in one or more narrow spectral "lines", an increasing number of lasers can be tuned by changing the composition or the pressure of the medium, or by varying the wavelength of the pump bands. The spectral regions in which these tunable lasers operate are presented in Figure 2.

Reference

Krupke, W. F., in *Handbook of Laser Science and Technology*, Vol. I, Weber, M. J., Ed., CRC Press, Boca Raton, FL, 1986.

TABLE 1. Extrema of Output Parameters of Laser Devices or Systems

Parameter	Value	Laser medium
Peak power	1×10^{14} W (collimated)	Nd:glass
Peak power density	10^{18} W/cm^2 (focused)	Nd:glass
Pulse energy	$>10^5$ J	CO_2, Nd:glass
Average power	10^5 W	CO_2
Pulse duration	3×10^{-15} s continuous wave (cw)	Rh6G dye; various gases, liquids, solids
Wavelength	60 nm $\leftrightarrow$ 385 µm	Many required
Efficiency (nonlaser pumped)	70%	CO
Beam quality	Diffraction limited	Various gases, liquids, solids
Spectral linewidth	20 Hz (for 10^{-1} s)	Neon-helium
Spatial coherence	10 m	Ruby

TABLE 2. Classes, Types, and Representative Examples of Laser Sources

Class	Type (characteristic)	Representative example	Nominal operating wavelength (nm)	Method(s) of excitation
Gas	Atom, neutral (electronic transition)	Neon-Helium (Ne-He)	633	Glow discharge
	Atom, ionic (electronic transition)	Argon (Ar$^+$)	488	Arc discharge
	Molecule, neutral (electronic transition)	Krypton fluoride (KrF)	248	Glow discharge; e-beam
	Molecule, neutral (vibrational transition)	Carbon dioxide (CO_2)	10600	Glow discharge; gasdynamic flow
	Molecule, neutral (rotational transition)	Methyl fluoride (CH_3F)	496000	Laser pumping
	Molecule, ionic (electronic transition)	Nitrogen ion (N$_2^+$)	420	E-beam
Liquid	Organic solvent (dye-chromophore)	Rhodamine dye (Rh6G)	580–610	Flashlamp; laser pumping
	Organic solvent (rare earth chelate)	Europium:TTF	612	Flashlamp
	Inorganic solvent (trivalent rare earth ion)	Neodymium:POCl$_4$	1060	Flashlamp
Solid	Insulator, crystal (impurity)	Neodymium:YAG	1064	Flashlamp, arc lamp
	Insulator, crystal (stoichiometric)	Neodymium:UP(NdP$_5$O$_{14}$)	1052	Flashlamp
	Insulator, crystal (color center)	F$_2^-$:LiF	1120	Laser pumping
	Insulator, amorphous (impurity)	Neodymium:glass	1061	Flashlamp
	Semiconductor (p-n junction)	GaAs	820	Injection current
	Semiconductor (electron-hole plasma)	GaAs	890	E- beam, laser pumping

TABLE 3. Temporal Characteristics of Lasers and Laser Systems

Form	Technique	Pulse width range (s)
Continuous wave	Excitation is continuous; resonator Q is held constant at some moderate value	∞
Pulsed	Excitation is pulsed; resonator Q is held constant at some moderate value	$10^{-8} - 10^{-3}$
Q-Switched	Excitation is continuous or pulsed; resonator Q is switched from a very low value to a moderate value	$10^{-8} - 10^{-6}$
Cavity dumped	Excitation is continuous or pulsed; resonator Q is switched from a very high value to a low value	$10^{-7} - 10^{-5}$
Mode locked	Excitation is continuous or pulsed; phase or loss of the resonator modes is modulated at a rate related to the resonator transit time	$10^{-12} - 10^{-9}$

TABLE 4. Properties and Performance of Some Continuous Wave (CW) Lasers

Parameter	Unit	Gas			Liquid	Solid	
		Neon helium	Argon ion	Carbon dioxide	Rhodamine 6G dye	Nd:YAG	GaAs
Excitation method		DC discharge	DC discharge	DC discharge	Ar$^+$ laser pump	Krypton arc lamp	DC injection
Gain medium composition		Neon:helium	Argon	CO_2:N_2:He	Rh 6G:H_2O	Nd:YAG	p:n:GaAs
Gain medium density	Torr	0.1:1.0	0.4	0.4:0.8:5.0			
	ions/cm^3				2(18):2(22)	1.5(20):2(22)	2(19):3(18):3(22)
Wavelength	nm	633	488	10600	590	1064	810
Laser cross-section	cm^{-2}	3(-13)	1.6(-12)	1.5(-16)	1.8(-16)	7(-19)	~6(-15)
Radiative lifetime (upper level)	s	~1(-7)	7.5(-9)	4(-3)	6.5(-9)	2.6(-4)	~1(-9)
Decay lifetime (upper level)	s	~1(-7)	~5.0(-9)	~4(-3)	6.0(-9)	2.3(-4)	~1(-9)
Gain bandwidth	nm	2(-3)	5(-3)	1.6(-2)	80	0.5	10
Type, gain saturation		Inhomogeneous	Inhomogeneous	Homogeneous	Homogeneous	Homogeneous	Homogeneous
Homogeneous saturation flux	W cm^{-2}			~20	3(5)	2.3(3)	~2(4)
Decay lifetime (lower level)	s	~1(-8)	~4(-10)	~5(-6)	<1(-12)	<1(-7)	<1(-12)
Inversion density	cm^{-3}	~1(9)	2(10)	2(15)	2(16)	6(16)	1(16)
Small signal gain coefficient	cm^{-1}	~1(-3)	~3(-2)	1(-2)	4	5(-2)	40
Pump power density	W cm^{-3}	3	900	0.15	1(6)	150	7(7)
Output power density	W cm^{-3}	2.6(-3)	~1	2(-2)	3(5)	95	5(6)
Laser size (diameter: length)	cm:cm	0.5:100	0.3:100	5.0:600	1(-3):0.3	0.6:10	5(-4):7(-3);2(-2)[a]
Excitation current/ voltage	A/V	3(-2):2(3)	30:300	0.1:1.5(4)		90:125	1.0/1.7
Excitation current density	A cm^{-2}	0.15	600	6(-3)		140	4.5(3)
Excitation power	W	60	9(3)	1.5(3)	4	1.1(4)	1.7
Output power	W	0.06	10	240	0.3	300	0.12
Efficiency[b]	%	0.1	0.1	13	7	2.6	7

[a] Junction thickness:width:length.
[b] Pressure dependent.

TABLE 5. Properties and Performance of Some Pulsed Lasers

| Parameter | Unit | Gas | | | | Liquid | Solid | |
		Carbon dixoxide		Krypton fluoride		Rhodamine 6G	Nd:YAG	Nd:glass
Excitation method		TEA-discharge	E- beam/sust.	Glow discharge	E-beam	Xenon flashlamp	Xenon flashlamp	Xenon flashlamp
Gain medium composition		$CO_2:N_2:He$	$CO_2:N_2:He$	$He:Kr:F_2$	$Ar:Kr:F_2$	Rh6G:alcohol	Nd:YAG	Nd:Glass
Gain medium density	torr	100:50:600	240:240:320	1070:70:3	1235:52:3			–
	ions/cm^3					1(18):1.5(22)	1.5(20):1(22)	3(20):2(22)
Wavelength	nm	10600	10600	249	249	590	1064	1061
Laser cross-section	cm^{-2}	2(-18)	2(-18)	2(-16)	2(-16)	1.8(-16)	7(-19)	2.8(-20)
Radiative lifetime (upper level)	s	4(-3)	4(-3)	7(-9)	7(-9)	6.5(-9)	2.6(-4)	4.1(-4)
Decay lifetime (upper level)	s	~1(-4)	5(-5)	2(-9)	3(-9)	6.0(-9)	2.3(-4)	3.7(-4)
Gain bandwidth	nm	1	1	2	2	80	0.5	26
Homogeneous saturation fluence	J/cm^2	0.2	0.2	4(-3)	4(-3)	2(-3)	0.6	~5
Decay lifetime (lower level)	s	5(-8)[a]	1(-8)[a]	< 1(-12)	<1(-12)	<1(-12)	<1(-7)	<1(-8)
Inversion density	cm^{-3}	3(17)	6(17)	4(14)	2(14)	2(16)	4(17)	3(18)
Small signal gain coefficient	cm^{-1}	2(-2)	4(-2)	8–92)	4(-2)	4	0.3	8(-2)
Medium excitation energy density	J/cm^3	0.1	0.36	0.15	0.13	2.8	0.15	0.6
Output energy density	J/cm^3	2(-2)	1.8(-2)	1.5(-3)	1.2(-2)	0.85	5(-2)	2(-2)
Laser dimensions	cm: cm: cm	4.5:4.5:87	10:10:100	1.5:4.5:100	8.5:10:100	1.2:25	0.6:7.5	0.6:8.3
Excitation current/voltage	A/V	6(4)/3.3(3)	2.4(4)/4(4)	2.5(4)/1.5(5)	1.2(4)/2.5(5)	2(5)/2.5(4)		
Excitation current density	A cm^2	8.5	22	170	11.5	2.6(3)		
Excitation peak power	W	2(8)	9(8)	4(9)	3(9)	5.4(9)	4(4)	9(4)
Output pulse energy	J	35	180	1	102	32	0.1	1.0
Output pulse length	s	1(-6)	4(-6)	2.5(-8)	6(-7)	3.2(-6)	2(-8)	1(-4)
Output pulse power	W	3.5(7)	4(7)	4(7)	2(8)	1(7)	5(6)	1(4)
Efficiency	%	17	5	1	10[b]	0.2	1.5	3.7

[a] Pressure dependent.
[b] Intrinsic efficiency ≡ energy output/energy deposited in gas.

FIGURE 1A. Wavelengths of lasers operating in the 120 to 1200 nm spectral region.

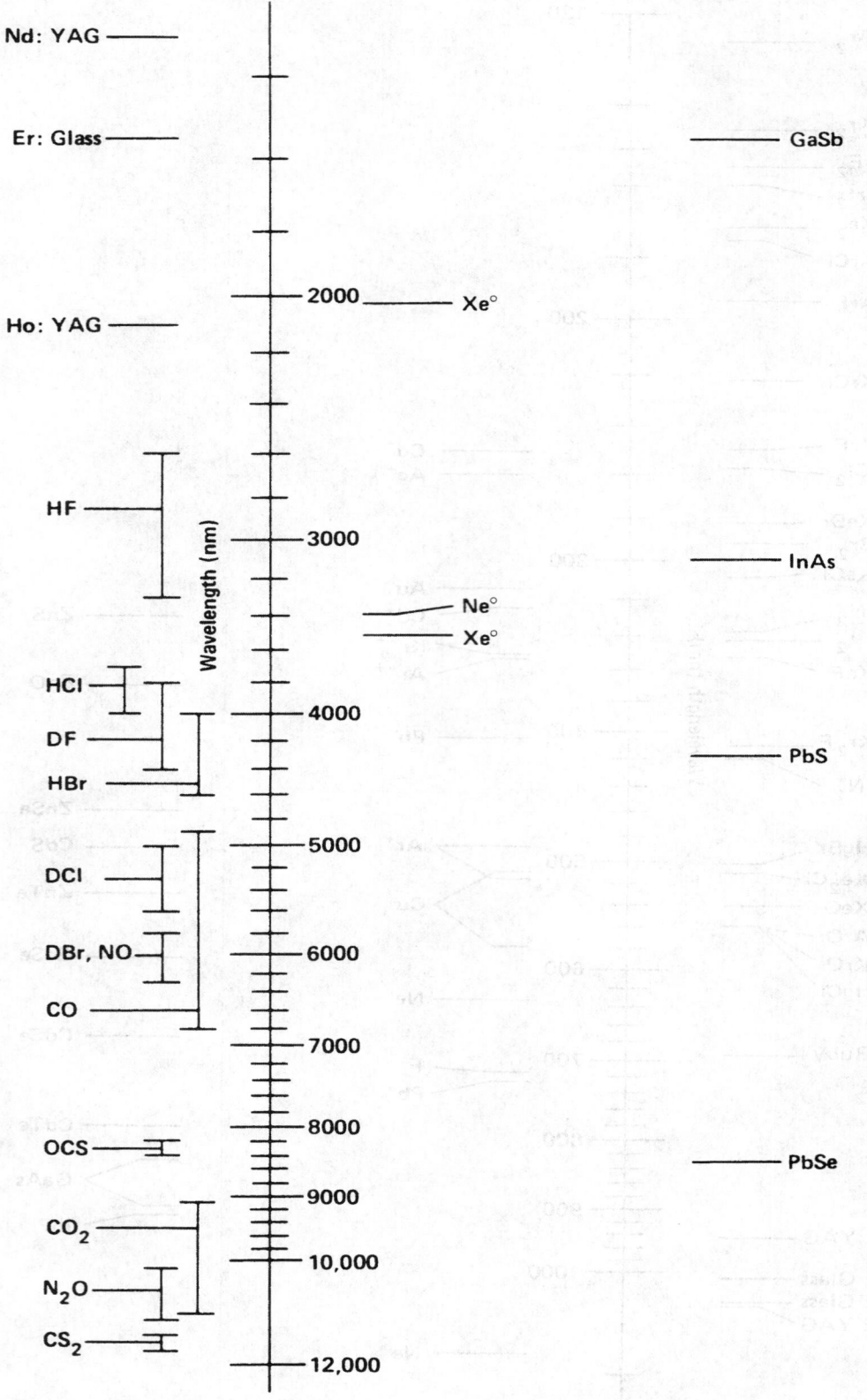

FIGURE 1B. Wavelength of lasers operating in the 1300 to 12,000 nm spectral region.

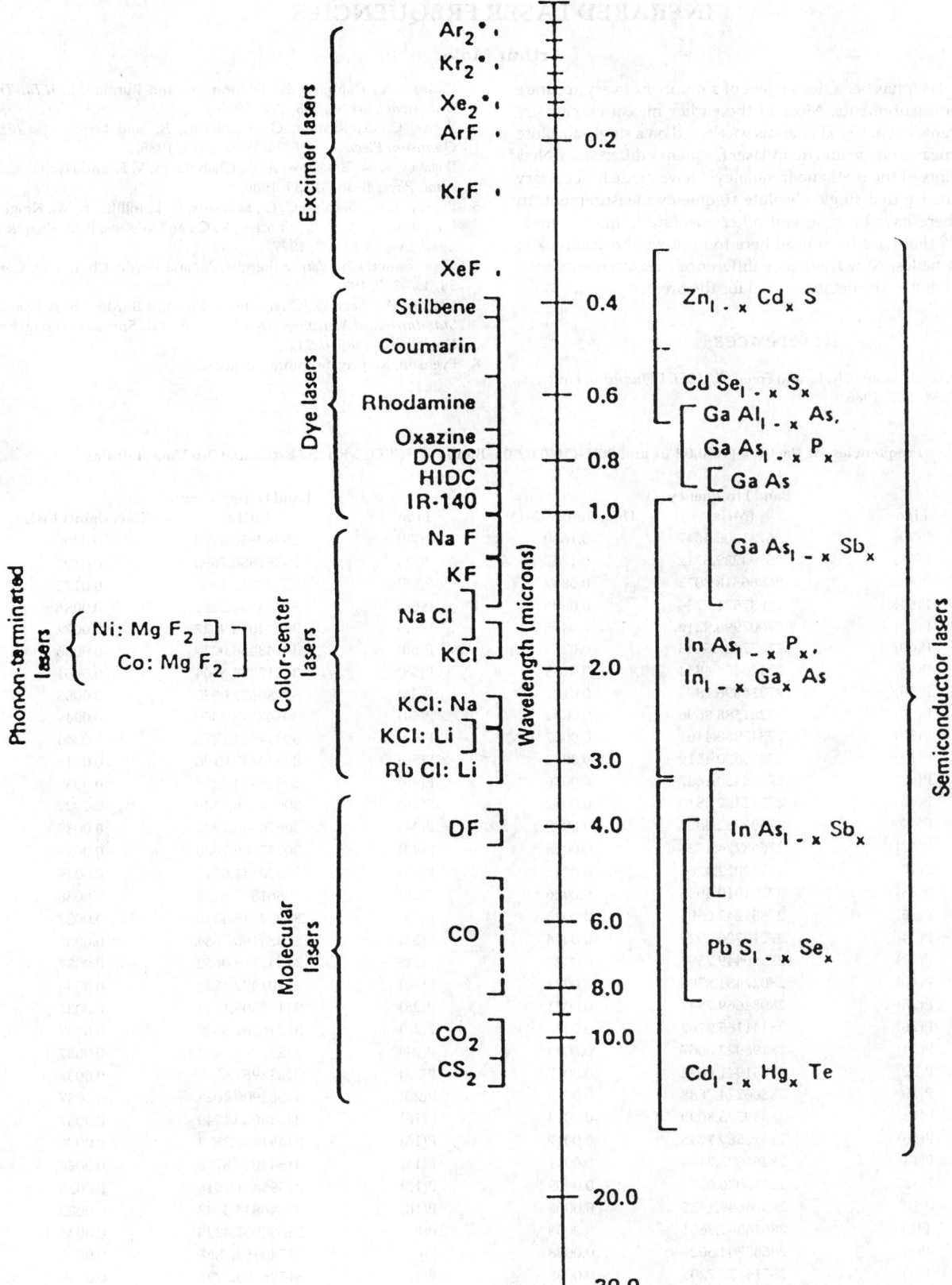

FIGURE 2. Spectral tuning ranges of various types of tunable lasers.

INFRARED LASER FREQUENCIES

Arthur Maki

The CO_2 laser has been the subject of a number of very accurate frequency measurements. Most of the earlier measurements are given by Bradley et al.[1] That analysis was based on a single absolute frequency measurement and many laser frequency differences. New measurements of the methane frequency[2-4] have made it necessary to slightly revise that single absolute frequency measurement. In addition, there have been several other absolute frequency measurements[5-7] that have been used here to improve the accuracy of the present tables. New frequency difference measurements have also been added to the database used for the present tables.[8]

References

1. Bradley, L. C., Soohoo, K. L., and Freed, C., *IEEE J. Quantum Electron.*, QE-22, 234–267, 1986.
2. Clairon, A., Dahmani, B., Filimon, A., and Rutman, J., *IEEE Trans. Inst. Meas.*, IM-34, 265–268, 1985.
3. Weiss, C. O., Kramer, G., Lipphardt, B., and Garcia, E., *IEEE J. Quantum Electron.*, QE-24, 1970–1972, 1988.
4. Bagayev, S. N., Baklanov, A. E., Chebotayev, V. P., and Dychkov, A. S., *Appl. Phys.*, B-48, 31–35, 1989.
5. Blaney, T. G., Bradley, C. C., Edwards, G. J., Jolliffe, B. W., Knight, D. J. E., Rowley, W. R. C., Shotten, K. C., and Woods, P. T., *Proc. R. Soc. Lond.*, A-355, 61–88, 1977.
6. Chardonnet, Ch., Van Lerberghe, A., and Bordé, Ch. J., *Opt. Comm.*, 58, 333–337, 1986.
7. Clairon, A., Acef, O., Chardonnet, Ch., and Bordé, Ch. J., *Frequency Standards and Metrology*, De Marchi, A., Ed., Springer-Verlag, Berlin, Heidelberg, 1989, p. 212.
8. Evenson, K., private communication.

Frequencies for the $00°1-(10°0,02°0)_I$ and $00°1-(10°0,02°0)_{II}$ Bands of $^{12}C^{16}O_2$ with the Estimated 2-σ Uncertainties

Line	Band I frequency (MHz)	Uncertainty (MHz)	Line	Band II frequency (MHz)	Uncertainty (MHz)
P(70)	26721305.4647	0.1680	P(70)	29789856.3783	0.0308
P(68)	26794232.6712	0.1217	P(68)	29861850.7690	0.0192
P(66)	26866318.8073	0.0867	P(66)	29933216.1760	0.0122
P(64)	26937571.7234	0.0606	P(64)	30003944.2861	0.0086
P(62)	27007998.9216	0.0415	P(62)	30074026.9127	0.0072
P(60)	27077607.5643	0.0279	P(60)	30143456.0039	0.0066
P(58)	27146404.4834	0.0185	P(58)	30212223.6504	0.0061
P(56)	27214396.1873	0.0121	P(56)	30280322.0930	0.0055
P(54)	27281588.8696	0.0081	P(54)	30347743.7306	0.0049
P(52)	27347988.4161	0.0057	P(52)	30414481.1273	0.0044
P(50)	27413600.4119	0.0043	P(50)	30480527.0196	0.0041
P(48)	27478430.1487	0.0036	P(48)	30545874.3239	0.0039
P(46)	27542482.6310	0.0032	P(46)	30610516.1429	0.0039
P(44)	27605762.5826	0.0030	P(44)	30674445.7724	0.0039
P(42)	27668274.4525	0.0028	P(42)	30737656.7080	0.0039
P(40)	27730022.4206	0.0027	P(40)	30800142.6511	0.0039
P(38)	27791010.4036	0.0026	P(38)	30861897.5150	0.0038
P(36)	27851242.0594	0.0025	P(36)	30922915.4310	0.0037
P(34)	27910720.7927	0.0024	P(34)	30983190.7534	0.0037
P(32)	27969449.7593	0.0023	P(32)	31042718.0652	0.0037
P(30)	28027431.8708	0.0022	P(30)	31101492.1833	0.0036
P(28)	28084669.7981	0.0021	P(28)	31159508.1631	0.0037
P(26)	28141165.9762	0.0020	P(26)	31216761.3029	0.0037
P(24)	28196922.6067	0.0019	P(24)	31273247.1487	0.0037
P(22)	28251941.6622	0.0017	P(22)	31328961.4978	0.0037
P(20)	28306224.8888	0.0016	P(20)	31383900.4028	0.0037
P(18)	28359773.8090	0.0014	P(18)	31438060.1749	0.0037
P(16)	28412589.7245	0.0012	P(16)	31491437.3872	0.0036
P(14)	28464673.7184	0.0011	P(14)	31544028.8776	0.0036
P(12)	28516026.6574	0.0009	P(12)	31595831.7516	0.0036
P(10)	28566649.1935	0.0008	P(10)	31646843.3843	0.0035
P(8)	28616541.7661	0.0008	P(8)	31697061.4225	0.0035
P(6)	28665704.6027	0.0008	P(6)	31746483.7868	0.0035
P(4)	28714137.7205	0.0008	P(4)	31795108.6724	0.0035
P(2)	28761840.9272	0.0008	P(2)	31842934.5511	0.0035
R(0)	28832026.2198	0.0008	R(0)	31913172.5691	0.0035
R(2)	28877902.4382	0.0007	R(2)	31958996.0621	0.0034
R(4)	28923046.4303	0.0006	R(4)	32004017.3822	0.0034

Line	Band I frequency (MHz)	Uncertainty (MHz)	Line	Band II frequency (MHz)	Uncertainty (MHz)
R(6)	28967457.0657	0.0005	R(6)	32048236.2498	0.0034
R(8)	29011133.0054	0.0003	R(8)	32091652.6619	0.0034
R(10)	29054072.7010	0.0001	R(10)	32134266.8917	0.0034
R(12)	29096274.3935	0.0003	R(12)	32176079.4878	0.0034
R(14)	29137736.1129	0.0005	R(14)	32217091.2721	0.0035
R(16)	29178455.6759	0.0007	R(16)	32257303.3386	0.0036
R(18)	29218430.6852	0.0009	R(18)	32296717.0510	0.0037
R(20)	29257658.5269	0.0010	R(20)	32335334.0408	0.0038
R(22)	29296136.3689	0.0011	R(22)	32373156.2044	0.0039
R(24)	29333861.1583	0.0012	R(24)	32410185.7003	0.0041
R(26)	29370829.6191	0.0011	R(26)	32446424.9459	0.0042
R(28)	29407038.2491	0.0011	R(28)	32481876.6140	0.0042
R(30)	29442483.3168	0.0011	R(30)	32516543.6293	0.0042
R(32)	29477160.8582	0.0012	R(32)	32550429.1641	0.0042
R(34)	29511066.6733	0.0013	R(34)	32583536.6340	0.0042
R(36)	29544196.3221	0.0015	R(36)	32615869.6937	0.0041
R(38)	29576545.1205	0.0017	R(38)	32647432.2320	0.0040
R(40)	29608108.1360	0.0019	R(40)	32678228.3665	0.0039
R(42)	29638880.1831	0.0022	R(42)	32708262.4386	0.0038
R(44)	29668855.8183	0.0024	R(44)	32737539.0081	0.0039
R(46)	29698029.3350	0.0027	R(46)	32766062.8469	0.0041
R(48)	29726394.7582	0.0032	R(48)	32793838.9334	0.0045
R(50)	29753945.8385	0.0037	R(50)	32820872.4463	0.0055
R(52)	29780676.0464	0.0042	R(52)	32847168.7576	0.0071
R(54)	29806578.5659	0.0047	R(54)	32872733.4269	0.0099
R(56)	29831646.2878	0.0052	R(56)	32897572.1935	0.0141
R(58)	29855871.8032	0.0058	R(58)	32921690.9701	0.0202
R(60)	29879247.3960	0.0074	R(60)	32945095.8355	0.0288
R(62)	29901765.0357	0.0113	R(62)	32967793.0268	0.0407
R(64)	29923416.3695	0.0186	R(64)	32989788.9322	0.0567
R(66)	29944192.7145	0.0302	R(66)	33011090.0831	0.0780
R(68)	29964085.0488	0.0475	R(68)	33031703.1467	0.1060
R(70)	29983084.0036	0.0720	R(70)	33051634.9172	0.1423

Frequencies for the $00°1\text{-}(10°0,02°0)_I$ and $00°1\text{-}(10°0,02°0)_{II}$ Bands of $^{13}C^{16}O_2$ with the Estimated 2-σ Uncertainties

Line	Band I frequency (MHz)	Uncertainty (MHz)	Line	Band II frequency (MHz)	Uncertainty (MHz)
P(66)	25523832.1808	0.7836	P(66)	28512082.5283	1.2894
P(64)	25590013.4703	0.5415	P(64)	28585121.9396	0.9194
P(62)	25655543.6502	0.3629	P(62)	28657449.4180	0.6420
P(60)	25720428.2487	0.2339	P(60)	28729056.6374	0.4375
P(58)	25784672.4840	0.1430	P(58)	28799935.4147	0.2897
P(56)	25848281.2771	0.0810	P(56)	28870077.7187	0.1853
P(54)	25911259.2627	0.0405	P(54)	28939475.6771	0.1135
P(52)	25973610.8005	0.0157	P(52)	29008121.5846	0.0659
P(50)	26035339.9857	0.0045	P(50)	29076007.9109	0.0357
P(48)	26096450.6582	0.0079	P(48)	29143127.3077	0.0180
P(46)	26156946.4123	0.0101	P(46)	29209472.6164	0.0090
P(44)	26216830.6053	0.0101	P(44)	29275036.8754	0.0058
P(42)	26276106.3655	0.0090	P(42)	29339813.3270	0.0050
P(40)	26334776.6003	0.0077	P(40)	29403795.4243	0.0044
P(38)	26392844.0030	0.0068	P(38)	29466976.8383	0.0037
P(36)	26450311.0599	0.0063	P(36)	29529351.4635	0.0032
P(34)	26507180.0565	0.0061	P(34)	29590913.4252	0.0029
P(32)	26563453.0836	0.0060	P(32)	29651657.0844	0.0028
P(30)	26619132.0428	0.0058	P(30)	29711577.0447	0.0028
P(28)	26674218.6515	0.0055	P(28)	29770668.1566	0.0031
P(26)	26728714.4479	0.0054	P(26)	29828925.5239	0.0035
P(24)	26782620.7952	0.0054	P(24)	29886344.5074	0.0041
P(22)	26835938.8858	0.0054	P(22)	29942920.7308	0.0046
P(20)	26888669.7451	0.0055	P(20)	29998650.0838	0.0051
P(18)	26940814.2347	0.0055	P(18)	30053528.7271	0.0054

Line	Band I frequency (MHz)	Uncertainty (MHz)	Line	Band II frequency (MHz)	Uncertainty (MHz)
P(16)	26992373.0555	0.0055	P(16)	30107553.0955	0.0055
P(14)	27043346.7508	0.0054	P(14)	30160719.9016	0.0055
P(12)	27093735.7083	0.0052	P(12)	30213026.1388	0.0054
P(10)	27143540.1624	0.0051	P(10)	30264469.0839	0.0054
P(8)	27192760.1962	0.0049	P(8)	30315046.2994	0.0054
P(6)	27241395.7431	0.0048	P(6)	30364755.6359	0.0055
P(4)	27289446.5880	0.0047	P(4)	30413595.2335	0.0056
P(2)	27336912.3682	0.0046	P(2)	30461563.5231	0.0057
R(0)	27407012.8882	0.0045	P(0)	30531879.5415	0.0057
R(2)	27453013.4589	0.0043	P(2)	30577664.6138	0.0056
R(4)	27498426.5430	0.0040	P(4)	30622575.1885	0.0054
R(6)	27543251.1200	0.0037	P(6)	30666611.0128	0.0051
R(8)	27587486.0225	0.0034	P(8)	30709772.1257	0.0047
R(10)	27631129.9356	0.0031	P(10)	30752058.8571	0.0045
R(12)	27674181.3963	0.0029	P(12)	30793471.8269	0.0044
R(14)	27716638.7917	0.0029	P(14)	30834011.9425	0.0043
R(16)	27758500.3577	0.0029	P(16)	30873680.3976	0.0044
R(18)	27799764.1770	0.0029	P(18)	30912478.6694	0.0044
R(20)	27840428.1773	0.0030	P(20)	30950408.5159	0.0044
R(22)	27880490.1283	0.0029	P(22)	30987471.9732	0.0043
R(24)	27919947.6395	0.0029	P(24)	31023671.3517	0.0042
R(26)	27958798.1567	0.0028	P(26)	31059009.2327	0.0042
R(28)	27997038.9591	0.0028	P(28)	31093488.4642	0.0042
R(30)	28034667.1551	0.0027	P(30)	31127112.1569	0.0043
R(32)	28071679.6785	0.0027	P(32)	31159883.6793	0.0045
R(34)	28108073.2842	0.0026	P(34)	31191806.6529	0.0046
R(36)	28143844.5432	0.0026	P(36)	31222884.9469	0.0048
R(38)	28178989.8377	0.0026	P(38)	31253122.6730	0.0053
R(40)	28213505.3554	0.0028	P(40)	31282524.1795	0.0061
R(42)	28247387.0838	0.0033	P(42)	31311094.0452	0.0077
R(44)	28280630.8035	0.0046	P(44)	31338837.0736	0.0108
R(46)	28313232.0818	0.0083	P(46)	31365758.2858	0.0173
R(48)	28345186.2652	0.0161	P(48)	31391862.9147	0.0295
R(50)	28376488.4720	0.0301	P(50)	31417156.3972	0.0505
R(52)	28407133.5839	0.0531	P(52)	31441644.3679	0.0845
R(54)	28437116.2372	0.0887	P(54)	31465332.6516	0.1366
R(56)	28466430.8141	0.1419	P(56)	31488227.2557	0.2138
R(58)	28495071.4324	0.2188	P(58)	31510334.3631	0.3247
R(60)	28523031.9357	0.3271	P(60)	31531660.3243	0.4800
R(62)	28550305.8819	0.4763	P(62)	31552211.6497	0.6932
R(64)	28576886.5323	0.6781	P(64)	31571995.0017	0.9805
R(66)	28602766.8393	0.9467	P(66)	31591017.1868	1.3619

Frequencies for the 00°1-(10°0,02°0)$_I$ and 00°1-(10°0,02°0)$_{II}$ Bands of $^{12}C^{18}O_2$ with the Estimated 2-σ Uncertainties

Line	Band I frequency (MHz)	Uncertainty (MHz)	Line	Band II frequency (MHz)	Uncertainty (MHz)
P(70)	27045326.3119	0.4540	P(70)	30695237.5856	0.0858
P(68)	27114914.0922	0.3324	P(68)	30755520.2231	0.0570
P(66)	27183635.7945	0.2392	P(66)	30815311.4928	0.0364
P(64)	27251496.4118	0.1688	P(64)	30874607.2084	0.0223
P(62)	27318500.7361	0.1165	P(62)	30933403.2309	0.0131
P(60)	27384653.3618	0.0783	P(60)	30991695.4724	0.0075
P(58)	27449958.6881	0.0510	P(58)	31049479.9009	0.0049
P(56)	27514420.9224	0.0319	P(56)	31106752.5446	0.0041
P(54)	27578044.0828	0.0191	P(54)	31163509.4964	0.0040
P(52)	27640832.0010	0.0108	P(52)	31219746.9183	0.0040
P(50)	27702788.3248	0.0059	P(50)	31275461.0455	0.0039
P(48)	27763916.5206	0.0035	P(48)	31330648.1908	0.0039
P(46)	27824219.8762	0.0028	P(46)	31385304.7490	0.0039
P(44)	27883701.5029	0.0026	P(44)	31439427.2006	0.0039
P(42)	27942364.3379	0.0025	P(42)	31493012.1163	0.0038
P(40)	28000211.1464	0.0024	P(40)	31546056.1605	0.0038

Line	Band I frequency (MHz)	Uncertainty (MHz)	Line	Band II frequency (MHz)	Uncertainty (MHz)
P(38)	28057244.5242	0.0022	P(38)	31598556.0954	0.0037
P(36)	28113466.8992	0.0021	P(36)	31650508.7847	0.0037
P(34)	28168880.5335	0.0020	P(34)	31701911.1970	0.0037
P(32)	28223487.5256	0.0019	P(32)	31752760.4093	0.0037
P(30)	28277289.8118	0.0017	P(30)	31803053.6105	0.0037
P(28)	28330289.1679	0.0016	P(28)	31852788.1043	0.0038
P(26)	28382487.2111	0.0015	P(26)	31901961.3125	0.0038
P(24)	28433885.4012	0.0013	P(24)	31950570.7773	0.0038
P(22)	28484485.0420	0.0012	P(22)	31998614.1649	0.0038
P(20)	28534287.2828	0.0011	P(20)	32046089.2669	0.0037
P(18)	28583293.1193	0.0010	P(18)	32092994.0036	0.0037
P(16)	28631503.3952	0.0010	P(16)	32139326.4254	0.0036
P(14)	28678918.8025	0.0009	P(14)	32185084.7154	0.0036
P(12)	28725539.8830	0.0010	P(12)	32230267.1907	0.0036
P(10)	28771367.0288	0.0010	P(10)	32274872.3041	0.0037
P(8)	28816400.4829	0.0010	P(8)	32318898.6455	0.0038
P(6)	28860640.3403	0.0011	P(6)	32362344.9434	0.0039
P(4)	28904086.5477	0.0011	P(4)	32405210.0652	0.0041
P(2)	28946738.9048	0.0011	P(2)	32447493.0185	0.0041
R(0)	29009228.1702	0.0010	P(0)	32509824.0580	0.0042
R(2)	29049894.0586	0.0010	P(2)	32550648.1723	0.0042
R(4)	29089764.2368	0.0009	P(4)	32590887.7542	0.0042
R(6)	29128837.8426	0.0008	P(6)	32630542.4457	0.0041
R(8)	29167113.8668	0.0008	P(8)	32669612.0295	0.0041
R(10)	29204591.1529	0.0009	P(10)	32708096.4282	0.0040
R(12)	29241268.3964	0.0010	P(12)	32745995.7040	0.0040
R(14)	29277144.1444	0.0011	P(14)	32783310.0573	0.0040
R(16)	29312216.7955	0.0012	P(16)	32820039.8258	0.0040
R(18)	29346484.5984	0.0012	P(18)	32856185.4827	0.0040
R(20)	29379945.6517	0.0013	P(20)	32891747.6358	0.0040
R(22)	29412597.9024	0.0013	P(22)	32926727.0254	0.0040
R(24)	29444439.1458	0.0013	P(24)	32961124.5220	0.0040
R(26)	29475467.0236	0.0014	P(26)	32994941.1249	0.0040
R(28)	29505679.0230	0.0015	P(28)	33028177.9594	0.0040
R(30)	29535072.4755	0.0016	P(30)	33060836.2743	0.0040
R(32)	29563644.5557	0.0018	P(32)	33092917.4394	0.0041
R(34)	29591392.2794	0.0020	P(34)	33124422.9429	0.0043
R(36)	29618312.5023	0.0023	P(36)	33155354.3878	0.0046
R(38)	29644401.9182	0.0028	P(38)	33185713.4894	0.0049
R(40)	29669657.0575	0.0036	P(40)	33215502.0716	0.0056
R(42)	29694074.2853	0.0053	P(42)	33244722.0637	0.0068
R(44)	29717649.7992	0.0082	P(44)	33273375.4969	0.0092
R(46)	29740379.6276	0.0128	P(46)	33301464.5003	0.0134
R(48)	29762259.6274	0.0200	P(48)	33328991.2976	0.0199
R(50)	29783285.4820	0.0307	P(50)	33355958.2027	0.0294
R(52)	29803452.6988	0.0461	P(52)	33382367.6161	0.0427
R(54)	29822756.6072	0.0681	P(54)	33408222.0209	0.0607
R(56)	29841192.3558	0.0985	P(56)	33433523.9780	0.0848
R(58)	29858754.9100	0.1401	P(58)	33458276.1228	0.1165
R(60)	29875439.0495	0.1960	P(60)	33482481.1601	0.1576
R(62)	29891239.3658	0.2702	P(62)	33506141.8605	0.2104
R(64)	29906150.2589	0.3673	P(64)	33529261.0556	0.2775
R(66)	29920165.9352	0.4930	P(66)	33551841.6335	0.3621
R(68)	29933280.4042	0.6540	P(68)	33573886.5352	0.4679
R(70)	29945487.4756	0.8581	P(70)	33595398.7493	0.5992

Frequencies for the $00°1$-$(10°0,02°0)_I$ and $00°1$-$(10°0,02°0)_{II}$ Bands of $^{13}C^{18}O_2$ with the Estimated 2-σ Uncertainties

Line	frequency (MHz)	Uncertainty (MHz)	Line	frequency (MHz)	Uncertainty (MHz)
P(70)	25967863.7652	1.1146	P(70)	28960476.2278	0.4069
P(68)	26033448.2798	0.8152	P(68)	29022326.9578	0.2861
P(66)	26098273.9159	0.5860	P(66)	29083661.3546	0.1961

Line	Band I frequency (MHz)	Uncertainty (MHz)	Line	Band II frequency (MHz)	Uncertainty (MHz)
P(64)	26162346.4813	0.4129	P(64)	29144473.5795	0.1303
P(62)	26225671.5466	0.2844	P(62)	29204757.8761	0.0833
P(60)	26288254.4494	0.1906	P(60)	29264508.5768	0.0507
P(58)	26350100.2984	0.1237	P(58)	29323720.1086	0.0290
P(56)	26411213.9778	0.0772	P(56)	29382386.9988	0.0152
P(54)	26471600.1504	0.0459	P(54)	29440503.8809	0.0073
P(52)	26531263.2618	0.0258	P(52)	29498065.4997	0.0038
P(50)	26590207.5442	0.0138	P(50)	29555066.7172	0.0032
P(48)	26648437.0195	0.0077	P(48)	29611502.5178	0.0031
P(46)	26705955.5026	0.0057	P(46)	29667368.0132	0.0031
P(44)	26762766.6051	0.0055	P(44)	29722658.4475	0.0034
P(42)	26818873.7378	0.0055	P(42)	29777369.2022	0.0039
P(40)	26874280.1143	0.0056	P(40)	29831495.8006	0.0044
P(38)	26928988.7531	0.0056	P(38)	29885033.9125	0.0049
P(36)	26983002.4809	0.0056	P(36)	29937979.3584	0.0053
P(34)	27036323.9351	0.0055	P(34)	29990328.1139	0.0054
P(32)	27088955.5657	0.0054	P(32)	30042076.3132	0.0055
P(30)	27140899.6384	0.0051	P(30)	30093220.2534	0.0055
P(28)	27192158.2363	0.0049	P(28)	30143756.3978	0.0054
P(26)	27242733.2620	0.0047	P(26)	30193681.3793	0.0053
P(24)	27292626.4396	0.0044	P(24)	30242992.0038	0.0052
P(22)	27341839.3165	0.0042	P(22)	30291685.2529	0.0051
P(20)	27390373.2651	0.0040	P(20)	30339758.2870	0.0049
P(18)	27438229.4843	0.0037	P(18)	30387208.4477	0.0048
P(16)	27485409.0008	0.0035	P(16)	30434033.2603	0.0046
P(14)	27531912.6704	0.0033	P(14)	30480230.4356	0.0045
P(12)	27577741.1795	0.0031	P(12)	30525797.8725	0.0044
P(10)	27622895.0455	0.0031	P(10)	30570733.6593	0.0043
P(8)	27667374.6182	0.0031	P(8)	30615036.0750	0.0043
P(6)	27711180.0803	0.0033	P(6)	30658703.5912	0.0044
P(4)	27754311.4480	0.0034	P(4)	30701734.8727	0.0045
P(2)	27796768.5718	0.0036	P(2)	30744128.7785	0.0045
R(0)	27859189.3155	0.0036	P(0)	30806522.5414	0.0045
R(2)	27899959.0889	0.0035	P(2)	30847319.2956	0.0044
R(4)	27940052.7921	0.0033	P(4)	30887476.2168	0.0043
R(6)	27979469.5315	0.0031	P(6)	30926993.0424	0.0042
R(8)	28018208.2478	0.0028	P(8)	30965869.7046	0.0041
R(10)	28056267.7161	0.0026	P(10)	31004106.3298	0.0040
R(12)	28093646.5448	0.0025	P(12)	31041703.2379	0.0040
R(14)	28130343.1757	0.0025	P(14)	31078660.9408	0.0040
R(16)	28166355.8825	0.0025	P(16)	31114980.1420	0.0040
R(18)	28201682.7706	0.0025	P(18)	31150661.7340	0.0041
R(20)	28236321.7757	0.0025	P(20)	31185706.7976	0.0042
R(22)	28270270.6628	0.0024	P(22)	31220116.5992	0.0043
R(24)	28303527.0249	0.0024	P(24)	31253892.5891	0.0043
R(26)	28336088.2817	0.0023	P(26)	31287036.3991	0.0044
R(28)	28367951.6781	0.0024	P(28)	31319549.8396	0.0043
R(30)	28399114.2823	0.0025	P(30)	31351434.8973	0.0043
R(32)	28429572.9843	0.0026	P(32)	31382693.7318	0.0042
R(34)	28459324.4940	0.0028	P(34)	31413328.6728	0.0042
R(36)	28488365.3390	0.0029	P(36)	31443342.2165	0.0041
R(38)	28516691.8625	0.0029	P(38)	31472737.0219	0.0040
R(40)	28544300.2211	0.0031	P(40)	31501515.9074	0.0039
R(42)	28571186.3823	0.0032	P(42)	31529681.8467	0.0040
R(44)	28597346.1222	0.0032	P(44)	31557237.9646	0.0042
R(46)	28622775.0223	0.0038	P(46)	31584187.5329	0.0046
R(48)	28647468.4672	0.0071	P(48)	31610533.9656	0.0057
R(50)	28671421.6417	0.0148	P(50)	31636280.8146	0.0088
R(52)	28694629.5272	0.0286	P(52)	31661431.7650	0.0151
R(54)	28717086.8993	0.0510	P(54)	31685990.6298	0.0261
R(56)	28738788.3239	0.0852	P(56)	31709961.3449	0.0434

Line	Band I frequency (MHz)	Uncertainty (MHz)	Line	Band II frequency (MHz)	Uncertainty (MHz)
R(58)	28759728.1540	0.1355	P(58)	31733347.9642	0.0693
R(60)	28779900.5263	0.2075	P(60)	31756154.6537	0.1068
R(62)	28799299.3572	0.3078	P(62)	31778385.6867	0.1594
R(64)	28817918.3393	0.4447	P(64)	31800045.4375	0.2317
R(66)	28835750.9374	0.6283	P(66)	31821138.3761	0.3291
R(68)	28852790.3843	0.8707	P(68)	31841669.0622	0.4581
R(70)	28869029.6768	1.1863	P(70)	31861642.1394	0.6268

Frequencies for the $01^{1e}1\text{-}(11^{1e}0,03^{1e}0)_I$ and $01^{1e}1\text{-}(11^{1e}0,03^{1e}0)_{II}$ Bands of $^{12}C^{16}O_2$ with the Estimated 2-σ Uncertainties

Line	Band I frequency (MHz)	Uncertainty (MHz)	Line	Band II frequency (MHz)	Uncertainty (MHz)
P(59)	26125213.2723	1.6633	P(59)	30427055.2899	0.1962
P(57)	26191576.6703	1.0880	P(57)	30494640.3229	0.1332
P(55)	26257240.7898	0.6844	P(55)	30561557.5929	0.0865
P(53)	26322208.2302	0.4094	P(53)	30627802.0344	0.0530
P(51)	26386481.4313	0.2286	P(51)	30693368.7014	0.0306
P(49)	26450062.6783	0.1155	P(49)	30758252.7710	0.0175
P(47)	26512954.1076	0.0498	P(47)	30822449.5469	0.0123
P(45)	26575157.7109	0.0191	P(45)	30885954.4624	0.0114
P(43)	26636675.3402	0.0160	P(43)	30948763.0834	0.0109
P(41)	26697508.7115	0.0182	P(41)	31010871.1119	0.0100
P(39)	26757659.4084	0.0177	P(39)	31072274.3882	0.0091
P(37)	26817128.8857	0.0160	P(37)	31132968.8940	0.0091
P(35)	26875918.4726	0.0144	P(35)	31192950.7549	0.0102
P(33)	26934029.3751	0.0131	P(33)	31252216.2430	0.0118
P(31)	26991462.6787	0.0119	P(31)	31310761.7788	0.0134
P(29)	27048219.3509	0.0106	P(29)	31368583.9339	0.0147
P(27)	27104300.2431	0.0096	P(27)	31425679.4328	0.0155
P(25)	27159706.0925	0.0093	P(25)	31482045.1550	0.0157
P(23)	27214437.5237	0.0097	P(23)	31537678.1367	0.0154
P(21)	27268495.0505	0.0104	P(21)	31592575.5725	0.0147
P(19)	27321879.0769	0.0108	P(19)	31646734.8172	0.0137
P(17)	27374589.8987	0.0108	P(17)	31700153.3868	0.0127
P(15)	27426627.7040	0.0104	P(15)	31752828.9602	0.0119
P(13)	27477992.5747	0.0098	P(13)	31804759.3803	0.0113
P(11)	27528684.4867	0.0096	P(11)	31855942.6551	0.0113
P(9)	27578703.3113	0.0101	P(9)	31906376.9582	0.0116
P(7)	27628048.8151	0.0113	P(7)	31956060.6304	0.0122
P(5)	27676720.6609	0.0127	P(5)	32004992.1796	0.0129
P(3)	27724718.4080	0.0141	P(3)	32053170.2819	0.0136
R(1)	27841759.7696	0.0152	P(1)	32170312.0391	0.0149
R(3)	27887393.2105	0.0146	P(3)	32215845.0845	0.0151
R(5)	27932349.2934	0.0135	P(5)	32260620.8121	0.0152
R(7)	27976627.0108	0.0124	P(7)	32304638.8261	0.0152
R(9)	28020225.2521	0.0115	P(9)	32347898.8990	0.0150
R(11)	28063142.8031	0.0110	P(11)	32390400.9714	0.0148
R(13)	28105378.3457	0.0109	P(13)	32432145.1513	0.0145
R(15)	28146930.4576	0.0109	P(15)	32473131.7137	0.0142
R(17)	28187797.6116	0.0107	P(17)	32513361.0997	0.0140
R(19)	28227978.1750	0.0103	P(19)	32552833.9153	0.0140
R(21)	28267470.4088	0.0099	P(21)	32591550.9309	0.0141
R(23)	28306272.4666	0.0099	P(23)	32629513.0796	0.0143
R(25)	28344382.3939	0.0107	P(25)	32666721.4564	0.0144
R(27)	28381798.1267	0.0122	P(27)	32703177.3164	0.0142
R(29)	28418517.4902	0.0141	P(29)	32738882.0732	0.0136
R(31)	28454538.1976	0.0165	P(31)	32773837.2976	0.0136
R(33)	28489857.8477	0.0213	P(33)	32808044.7156	0.0174
R(35)	28524473.9240	0.0312	P(35)	32841506.2063	0.0279
R(37)	28558383.7917	0.0486	P(37)	32874223.8000	0.0462
R(39)	28591584.6963	0.0754	P(39)	32906199.6761	0.0735
R(41)	28624073.7602	0.1131	P(41)	32937436.1606	0.1114
R(43)	28655847.9806	0.1644	P(43)	32967935.7238	0.1624
R(45)	28686904.2261	0.2328	P(45)	32997700.9775	0.2292

Line	Band I frequency (MHz)	Uncertainty (MHz)	Line	Band II frequency (MHz)	Uncertainty (MHz)
R(47)	28717239.2334	0.3239	P(47)	33026734.6728	0.3151
R(49)	28746849.6038	0.4465	P(49)	33055039.6965	0.4238
R(51)	28775731.7988	0.6142	P(51)	33082619.0689	0.5595
R(53)	28803882.1361	0.8465	P(53)	33109475.9403	0.7272

Frequencies for the $01^{1f}1\text{-}(11^{1f}0,03^{1f}0)_{I}$ and $01^{1f}1\text{-}(11^{1f}0,03^{1f}0)_{II}$ Bands of $^{12}C^{16}O_2$ with the Estimated 2-σ Uncertainties

Line	Band I frequency (MHz)	Uncertainty (MHz)	Line	Band II frequency (MHz)	Uncertainty (MHz)
P(60)	26051570.0104	4.4521	P(60)	30355115.0204	0.2752
P(58)	26120964.4932	3.0629	P(58)	30425283.5969	0.1926
P(56)	26189552.8496	2.0516	P(56)	30494732.8293	0.1301
P(54)	26257339.6006	1.3305	P(54)	30563455.6325	0.0840
P(52)	26324329.0344	0.8289	P(52)	30631445.1076	0.0512
P(50)	26390525.2136	0.4901	P(50)	30698694.5456	0.0292
P(48)	26455931.9824	0.2698	P(48)	30765197.4310	0.0163
P(46)	26520552.9722	0.1334	P(46)	30830947.4444	0.0111
P(44)	26584391.6075	0.0551	P(44)	30895938.4662	0.0104
P(42)	26647451.1105	0.0181	P(42)	30960164.5794	0.0105
P(40)	26709734.5057	0.0151	P(40)	31023620.0723	0.0105
P(38)	26771244.6242	0.0174	P(38)	31086299.4415	0.0107
P(36)	26831984.1067	0.0157	P(36)	31148197.3941	0.0114
P(34)	26891955.4069	0.0126	P(34)	31209308.8510	0.0126
P(32)	26951160.7945	0.0105	P(32)	31269628.9481	0.0138
P(30)	27009602.3576	0.0096	P(30)	31329153.0395	0.0147
P(28)	27067282.0045	0.0092	P(28)	31387876.6994	0.0151
P(26)	27124201.4662	0.0090	P(26)	31445795.7236	0.0149
P(24)	27180362.2977	0.0089	P(24)	31502906.1318	0.0141
P(22)	27235765.8792	0.0090	P(22)	31559204.1695	0.0128
P(20)	27290413.4182	0.0093	P(20)	31614686.3091	0.0113
P(18)	27344305.9494	0.0096	P(18)	31669349.2515	0.0098
P(16)	27397444.3368	0.0097	P(16)	31723189.9280	0.0086
P(14)	27449829.2733	0.0096	P(14)	31776205.5007	0.0081
P(12)	27501461.2824	0.0096	P(12)	31828393.3642	0.0085
P(10)	27552340.7179	0.0101	P(10)	31879751.1463	0.0095
P(8)	27602467.7649	0.0111	P(8)	31930276.7092	0.0107
P(6)	27651842.4399	0.0125	P(6)	31979968.1497	0.0120
P(4)	27700464.5912	0.0139	P(4)	32028823.8002	0.0131
P(2)	27748333.8988	0.0148	P(2)	32076842.2290	0.0139
R(2)	27864709.8633	0.0146	P(2)	32193218.1935	0.0150
R(4)	27909939.2762	0.0135	P(4)	32238298.4853	0.0151
R(6)	27954412.3294	0.0122	P(6)	32282538.0393	0.0153
R(8)	27998127.7801	0.0112	P(8)	32325936.7244	0.0153
R(10)	28041084.2173	0.0107	P(10)	32368494.6458	0.0154
R(12)	28083280.0620	0.0108	P(12)	32410212.1438	0.0155
R(14)	28124713.5668	0.0110	P(14)	32451089.7941	0.0155
R(16)	28165382.8151	0.0112	P(16)	32491128.4063	0.0154
R(18)	28205285.7213	0.0111	P(18)	32530329.0234	0.0152
R(20)	28244420.0302	0.0110	P(20)	32568692.9211	0.0149
R(22)	28282783.3158	0.0114	P(22)	32606221.6061	0.0147
R(24)	28320372.9812	0.0129	P(24)	32642916.8154	0.0144
R(26)	28357186.2574	0.0149	P(26)	32678780.5147	0.0140
R(28)	28393220.2023	0.0168	P(28)	32713814.8971	0.0133
R(30)	28428471.6994	0.0175	P(30)	32748022.3813	0.0120
R(32)	28462937.4565	0.0165	P(32)	32781405.6101	0.0106
R(34)	28496614.0042	0.0142	P(34)	32813967.4482	0.0122
R(36)	28529497.6934	0.0163	P(36)	32845710.9809	0.0212
R(38)	28561584.6939	0.0309	P(38)	32876639.5111	0.0385
R(40)	28592870.9914	0.0593	P(40)	32906756.5580	0.0646
R(42)	28623352.3850	0.1054	P(42)	32936065.8540	0.1015
R(44)	28653024.4839	0.1779	P(44)	32964571.3426	0.1518
R(46)	28681882.7038	0.2907	P(46)	32992277.1760	0.2184
R(48)	28709922.2632	0.4635	P(48)	33019187.7118	0.3048
R(50)	28737138.1785	0.7231	P(50)	33045307.5105	0.4152

INFRARED AND FAR-INFRARED ABSORPTION FREQUENCY STANDARDS

Arthur Maki

Aside from the CO_2 laser transitions, the absorption spectrum of CO has been more accurately and thoroughly measured than any other spectrum. A bibliography of earlier measurements on CO is given by Maki and Wells,[1] and the present tables were calculated from the measurements referred to in that work. In addition, some new and very accurate frequency measurements[2,3] have been made and were incorporated in the present tables. The frequencies of the rotational transitions of HF and HCl were calculated from constants obtained from fitting the measurements of Evenson et al.[4,5] and Jennings and Wells.[6]

A new report on infrared wavenumber standards from the International Union of Pure and Applied Chemistry, Commission on Molecular Structure and Spectroscopy, may be found in Reference 7.

References

1. Maki, A. G. and Wells, J. S., *Wavenumber Calibration Tables from Heterodyne Frequency Measurements*, NIST Special Publication 821, U.S. Dept. of Commerce, Washington, D.C., 1991.
2. Evenson, K. and Stroh, F., private communication.
3. George, T., Urban, W., and co-workers, private communication.
4. Jennings, D. A., Evenson, K. M., Zink, L. R., Demuynck, C., Destombes, J. L., Lemoine, B., and Johns, J. W. C., *J. Mol. Spectrosc.*, 122, 477–480, 1987.
5. Nolt, I. G., Radostitz, J. V., DiLonardo, G., Evenson, K. M., Jennings, D. A., Leopold, K. R., Vanek, M. D., Zink, L. R., Hinz, A., and Chance, K. V., *J. Mol. Spectrosc.*, 125, 274–287, 1987.
6. Jennings, D. A. and Wells, J. S., *J. Mol. Spectrosc.*, 130, 267–268, 1988.
7. High Resolution Wavenumber Standards for the Infrared, *Pure Appl. Chemistry*, 68, 193, 1996.

Wavenumbers for the $v = 1 - 0$ Band of CO

Wavenumber (unc)*/cm^{-1}	Transition	Wavenumber (unc)/cm^{-1}	Transition
		2147.081132(01)	R(0)
2139.426071(01)	P(1)	2150.856006(01)	R(1)
2135.546178(01)	P(2)	2154.595581(01)	R(2)
2131.631574(01)	P(3)	2158.299710(01)	R(3)
2127.682404(01)	P(4)	2161.968245(01)	R(4)
2123.698816(01)	P(5)	2165.601041(01)	R(5)
2119.680957(01)	P(6)	2169.197949(01)	R(6)
2115.628973(01)	P(7)	2172.758824(01)	R(7)
2111.543012(01)	P(8)	2176.283519(01)	R(8)
2107.423221(01)	P(9)	2179.771887(01)	R(9)
2103.269746(01)	P(10)	2183.223782(01)	R(10)
2099.082734(01)	P(11)	2186.639057(01)	R(11)
2094.862333(01)	P(12)	2190.017565(01)	R(12)
2090.608688(01)	P(13)	2193.359161(01)	R(13)
2086.321947(01)	P(14)	2196.663698(01)	R(14)
2082.002256(01)	P(15)	2199.931030(01)	R(15)
2077.649762(01)	P(16)	2203.161010(01)	R(16)
2073.264612(01)	P(17)	2206.353492(01)	R(17)
2068.846952(01)	P(18)	2209.508331(02)	R(18)
2064.396929(01)	P(19)	2212.625379(02)	R(19)
2059.914688(02)	P(20)	2215.704492(02)	R(20)
2055.400377(02)	P(21)	2218.745522(02)	R(21)
2050.854140(02)	P(22)	2221.748326(03)	R(22)
2046.276126(03)	P(23)	2224.712755(03)	R(23)
2041.666479(03)	P(24)	2227.638666(03)	R(24)
2037.025345(03)	P(25)	2230.525912(04)	R(25)
2032.352870(04)	P(26)	2233.374349(04)	R(26)
2027.649200(04)	P(27)	2236.183829(04)	R(27)
2022.914480(04)	P(28)	2238.954210(05)	R(28)
2018.148857(05)	P(29)	2241.685344(05)	R(29)
2013.352474(05)	P(30)	2244.377088(06)	R(30)
2008.525477(06)	P(31)	2247.029296(07)	R(31)
2003.668012(06)	P(32)	2249.641824(08)	R(32)
1998.780224(07)	P(33)	2252.214527(10)	R(33)
1993.862257(09)	P(34)	2254.747262(14)	R(34)
1988.914257(11)	P(35)	2257.239883(18)	R(35)
1983.936367(14)	P(36)	2259.692248(24)	R(36)
1978.928733(18)	P(37)	2262.104213(33)	R(37)
1973.891500(25)	P(38)	2264.475634(45)	R(38)

Wavenumbers for the $v = 1 - 0$ Band of CO

Wavenumber (unc)*/cm⁻¹	Transition	Wavenumber (unc)/cm⁻¹	Transition
1968.824811(34)	P(39)	2266.806368(61)	R(39)
1963.728813(46)	P(40)	2269.096273(81)	R(40)
1958.603648(61)	P(41)	2271.345206(106)	R(41)
1953.449462(82)	P(42)	2273.553027(139)	R(42)

* The uncertainty in the last digits (twice the standard error) is given in parentheses.

Wavenumbers for the $v = 2 - 0$ Band of CO

Wavenumber (unc)*/cm⁻¹	Transition	Wavenumber (unc)/cm⁻¹	Transition
		4263.837198(02)	R(0)
4256.217140(02)	P(1)	4267.542066(02)	R(1)
4252.302244(02)	P(2)	4271.176630(02)	R(2)
4248.317633(02)	P(3)	4274.740746(02)	R(3)
4244.263453(02)	P(4)	4278.234264(02)	R(4)
4240.139852(02)	P(5)	4281.657039(02)	R(5)
4235.946975(02)	P(6)	4285.008924(02)	R(6)
4231.684972(02)	P(7)	4288.289772(02)	R(7)
4227.353987(02)	P(8)	4291.499437(02)	R(8)
4222.954169(02)	P(9)	4294.637773(02)	R(9)
4218.485665(02)	P(10)	4297.704631(02)	R(10)
4213.948620(02)	P(11)	4300.699868(02)	R(11)
4209.343182(02)	P(12)	4303.623334(02)	R(12)
4204.669499(02)	P(13)	4306.474886(02)	R(13)
4199.927716(02)	P(14)	4309.254375(02)	R(14)
4195.117980(02)	P(15)	4311.961657(02)	R(15)
4190.240439(02)	P(16)	4314.596584(02)	R(16)
4185.295239(02)	P(17)	4317.159011(02)	R(17)
4180.282526(02)	P(18)	4319.648791(02)	R(18)
4175.202447(02)	P(19)	4322.065779(03)	R(19)
4170.055149(03)	P(20)	4324.409829(03)	R(20)
4164.840777(03)	P(21)	4326.680794(03)	R(21)
4159.559478(03)	P(22)	4328.878530(03)	R(22)
4154.211398(03)	P(23)	4331.002889(04)	R(23)
4148.796683(04)	P(24)	4333.053728(04)	R(24)
4143.315479(04)	P(25)	4335.030899(05)	R(25)
4137.767932(04)	P(26)	4336.934259(06)	R(26)
4132.154187(05)	P(27)	4338.763661(07)	R(27)
4126.474391(06)	P(28)	4340.518961(09)	R(28)
4120.728689(07)	P(29)	4342.200014(11)	R(29)
4114.917226(09)	P(30)	4343.806675(16)	R(30)
4109.040148(12)	P(31)	4345.338799(21)	R(31)
4103.097600(16)	P(32)	4346.796243(29)	R(32)
4097.089728(21)	P(33)	4348.178862(40)	R(33)
4091.016676(29)	P(34)	4349.486513(54)	R(34)
4084.878591(40)	P(35)	4350.719052(73)	R(35)
4078.675618(54)	P(36)	4351.876336(96)	R(36)
4072.407901(73)	P(37)	4352.958224(127)	R(37)
4066.075588(97)	P(38)	4353.964572(166)	R(38)
4059.678822(127)	P(39)	4354.895240(214)	R(39)

* The uncertainty in the last digits (twice the standard error) is given in parentheses.

Wavenumbers for the $v = 3 - 0$ Band of CO

Wavenumber (unc)*/cm⁻¹	Transition	Wavenumber (unc)/cm⁻¹	Transition
		6354.179057(13)	R(0)
6346.594000(13)	P(1)	6357.813923(13)	R(1)
6342.644103(13)	P(2)	6361.343487(13)	R(2)
6338.589491(13)	P(3)	6364.767599(13)	R(3)
6334.430309(13)	P(4)	6368.086115(13)	R(4)

Wavenumbers for the $\nu = 3 - 0$ Band of CO

Wavenumber (unc)*/cm^{-1}	Transition	Wavenumber (unc)/cm^{-1}	Transition
6330.166705(13)	P(5)	6371.298887(13)	R(5)
6325.798826(13)	P(6)	6374.405768(12)	R(6)
6321.326819(13)	P(7)	6377.406611(12)	R(7)
6316.750831(12)	P(8)	6380.301271(12)	R(8)
6312.071008(12)	P(9)	6383.089600(12)	R(9)
6307.287498(12)	P(10)	6385.771452(12)	R(10)
6302.400447(12)	P(11)	6388.346680(13)	R(11)
6297.410003(12)	P(12)	6390.815139(13)	R(12)
6292.316311(13)	P(13)	6393.176681(13)	R(13)
6287.119520(13)	P(14)	6395.431160(13)	R(14)
6281.819775(13)	P(15)	6397.578430(13)	R(15)
6276.417224(13)	P(16)	6399.618344(13)	R(16)
6270.912012(13)	P(17)	6401.550757(13)	R(17)
6265.304287(13)	P(18)	6403.375523(13)	R(18)
6259.594194(13)	P(19)	6405.092495(14)	R(19)
6253.781880(13)	P(20)	6406.701527(14)	R(20)
6247.867492(14)	P(21)	6408.202474(14)	R(21)
6241.851176(14)	P(22)	6409.595189(15)	R(22)
6235.733077(14)	P(23)	6410.879527(15)	R(23)
6229.513342(15)	P(24)	6412.055343(16)	R(24)
6223.192117(15)	P(25)	6413.122491(17)	R(25)
6216.769547(16)	P(26)	6414.080825(19)	R(26)
6210.245778(17)	P(27)	6414.930201(23)	R(27)
6203.620957(19)	P(28)	6415.670474(28)	R(28)
6196.895229(23)	P(29)	6416.301500(37)	R(29)
6190.068739(28)	P(30)	6416.823133(50)	R(30)
6183.141633(37)	P(31)	6417.235231(67)	R(31)
6176.114058(50)	P(32)	6417.537649(90)	R(32)
6168.986159(67)	P(33)		
6161.758082(90)	P(34)		

* The uncertainty in the last digits (twice the standard error) is given in parentheses.

Frequencies and Wavenumbers for the Rotational Lines of CO

Frequency/MHz	Uncertainty*/MHz	J'	J''	Wavenumber/cm^{-1}	Uncertainty*/cm^{-1}
115271.2029	0.0004	1	0	3.84503345	0.00000001
230538.0016	0.0008	2	1	7.68991999	0.00000003
345795.9923	0.0012	3	2	11.53451273	0.00000004
461040.7712	0.0016	4	3	15.37866477	0.00000005
576267.9350	0.0019	5	4	19.22222923	0.00000006
691473.0809	0.0021	6	5	23.06505926	0.00000007
806651.8065	0.0023	7	6	26.90700800	0.00000008
921799.7104	0.0025	8	7	30.74792863	0.00000008
1036912.3919	0.0027	9	8	34.58767438	0.00000009
1151985.4515	0.0029	10	9	38.42609848	0.00000010
1267014.4906	0.0031	11	10	42.26305422	0.00000010
1381995.1119	0.0034	12	11	46.09839491	0.00000011
1496922.9195	0.0038	13	12	49.93197392	0.00000013
1611793.5189	0.0042	14	13	53.76364468	0.00000014
1726602.5173	0.0047	15	14	57.59326065	0.00000016
1841345.5237	0.0052	16	15	61.42067535	0.00000017
1956018.1486	0.0057	17	16	65.24574239	0.00000019
2070616.0050	0.0061	18	17	69.06831542	0.00000020
2185134.7075	0.0065	19	18	72.88824816	0.00000022
2299569.8733	0.0069	20	19	76.70539441	0.00000023
2413917.1217	0.0071	21	20	80.51960806	0.00000024
2528172.0747	0.0073	22	21	84.33074306	0.00000024
2642330.3567	0.0074	23	22	88.13865346	0.00000025
2756387.5949	0.0075	24	23	91.94319341	0.00000025

Frequencies and Wavenumbers for the Rotational Lines of CO

Frequency/MHz	Uncertainty*/MHz	J'	J''	Wavenumber/cm^{-1}	Uncertainty*/cm^{-1}
2870339.4194	0.0077	25	24	95.74421713	0.00000026
2984181.4631	0.0080	26	25	99.54157896	0.00000027
3097909.3621	0.0085	27	26	103.33513334	0.00000028
3211518.7558	0.0090	28	27	107.12473480	0.00000030
3325005.2869	0.0096	29	28	110.91023800	0.00000032
3438364.6013	0.0102	30	29	114.69149772	0.00000034
3551592.3489	0.0107	31	30	118.46836884	0.00000036
3664684.1829	0.0111	32	31	122.24070637	0.00000037
3777635.7608	0.0118	33	32	126.00836545	0.00000039
3890442.7435	0.0137	34	33	129.77120137	0.00000046
4003100.7965	0.0179	35	34	133.52906952	0.00000060
4115605.5892	0.0254	36	35	137.28182546	0.00000085
4227952.7954	0.0370	37	36	141.02932487	0.00000123
4340138.0932	0.0531	38	37	144.77142361	0.00000177
4452157.1657	0.0746	39	38	148.50797766	0.00000249
4564005.7001	0.1025	40	39	152.23884318	0.00000342

* The uncertainty given is twice the standard error.

Frequencies and Wavenumbers for the Rotational Lines of HF

Frequency/MHz	Uncertainty*/MHz	J'	J''	Wavenumber/cm^{-1}	Uncertainty*/cm^{-1}
1232476.21	0.12	1	0	41.110981	0.000004
2463428.09	0.19	2	1	82.171116	0.000006
3691334.81	0.25	3	2	123.129676	0.000008
4914682.58	0.51	4	3	163.936165	0.000017
6131968.11	1.10	5	4	204.540439	0.000037
7341702.00	2.00	6	5	244.892818	0.000067
8542412.1	3.21	7	6	284.944197	0.000107
9732646.8	4.72	8	7	324.646153	0.000157
10910978.2	6.51	9	8	363.951056	0.000217
12076004.8	8.55	10	9	402.81216	0.000285
13226355.2	10.81	11	10	441.18372	0.000361
14360689.8	13.25	12	11	479.02105	0.00044
15477704.4	15.86	13	12	516.28065	0.00053
16576131.8	18.61	14	13	552.92024	0.00062
17654744.4	21.48	15	14	588.89888	0.00072
18712356.5	24.44	16	15	624.17703	0.00082
19747825.6	27.43	17	16	658.71656	0.00092
20760054.3	30.32	18	17	692.4809	0.00101
21747991.7	32.91	19	18	725.4349	0.00110
22710634.7	34.94	20	19	757.5452	0.00117
23647028.7	36.08	21	20	788.7800	0.00120
24556268.8	35.93	22	21	819.1090	0.00120
25437499.9	34.12	23	22	848.5037	0.00114
26289917.4	30.32	24	23	876.9373	0.00101
27112767.2	24.41	25	24	904.38457	0.00081
27905345.6	16.88	26	25	930.82214	0.00056
28666999.3	10.80	27	26	956.22817	0.00036
29397124.8	14.65	28	27	980.58253	0.00049
30095168.2	24.62	29	28	1003.86676	0.00082
30760624.2	33.36	30	29	1026.0640	0.00111
31393035.7	36.17	31	30	1047.1590	0.00121

* The uncertainty given is twice the standard error.

Frequencies and Wavenumbers for the Rotational Lines of H^{35}Cl

Frequency/MHz	Uncertainty*/MHz	J′	J″	Wavenumber/cm^{-1}	Uncertainty*/cm^{-1}
1876226.517	0.065	3	2	62.584180	0.000002
2499864.439	0.066	4	3	83.386502	0.000002
3121986.563	0.064	5	4	104.138262	0.000002
3742216.601	0.076	6	5	124.826909	0.000003
4360180.042	0.098	7	6	145.439951	0.000003
4975504.51	0.11	8	7	165.964966	0.000004
5587820.10	0.12	9	8	186.389615	0.000004
6196759.76	0.22	10	9	206.701656	0.000007
6801959.63	0.50	11	10	226.888951	0.000017
7403059.41	1.02	12	11	246.939481	0.000034
7999702.7	1.8	13	12	266.841359	0.000062
8591537.3	3.1	14	13	286.582837	0.000103
9178215.8	4.8	15	14	306.152324	0.000161

* The uncertainty given is twice the standard error.

Frequencies and Wavenumbers for the Rotational Lines of H^{37}Cl

Frequency/MHz	Uncertainty*/MHz	J′	J″	Wavenumber/cm^{-1}	Uncertainty*/cm^{-1}
1873410.72	0.05	3	2	62.490255	0.000002
2496115.33	0.05	4	3	83.261445	0.000002
3117308.69	0.05	5	4	103.982225	0.000002
3736615.64	0.06	6	5	124.640082	0.000002
4353662.84	0.08	7	6	145.222561	0.000003
4968079.04	0.09	8	7	165.717279	0.000003
5579495.53	0.10	9	8	186.111938	0.000003
6187546.42	0.19	10	9	206.394332	0.000006
6791869.04	0.45	11	10	226.552365	0.000015
7392104.3	0.9	12	11	246.574057	0.000030
7987896.9	1.6	13	12	266.447561	0.000054
8578896.1	2.7	14	13	286.161170	0.000089

* The uncertainty given is twice the standard error.

Section 11
Nuclear and Particle Physics

SUMMARY TABLES OF PARTICLE PROPERTIES

SUMMARY TABLES OF PARTICLE PROPERTIES

Extracted from the Particle Listings of the
Review of Particle Physics
Published in Eur. Jour. Phys. **C3**, 1 (1998)
Available at http://pdg.lbl.gov

Particle Data Group Authors:

C. Caso, G. Conforto, A. Gurtu, M. Aguilar-Benitez, C. Amsler,
R.M. Barnett, P.R. Burchat, C.D. Carone, O. Dahl, M. Doser,
S. Eidelman, J.L. Feng, M. Goodman, C. Grab, D.E. Groom,
K. Hagiwara, K.G. Hayes, J.J. Hernández, K. Hikasa, K. Honscheid,
F. James, M.L. Mangano, A.V. Manohar, K. Mönig, H. Murayama,
K. Nakamura, K.A. Olive, A. Piepke, M. Roos, R.H. Schindler,
R.E. Shrock, M. Tanabashi, N.A. Törnqvist, T.G. Trippe, P. Vogel,
C.G. Wohl, R.L. Workman, W.-M. Yao

Technical Associates: B. Armstrong, J.L. Casas Serradilla,
B.B. Filimonov, P.S. Gee, S.B. Lugovsky, S. Mankov, F. Nicholson

**Other Authors who have made substantial contributions
to reviews since the 1994 edition:**

K.S. Babu, D. Besson, O. Biebel, R.N. Cahn, R.L. Crawford, R.H. Dalitz,
T. Damour, K. Desler, R.J. Donahue, D.A. Edwards, J. Erler,
V.V. Ezhela, A. Fassò, W. Fetscher, D. Froidevaux, T.K. Gaisser,
L. Garren, S. Geer, H.-J. Gerber, F.J. Gilman, H.E. Haber, C. Hagmann,
I. Hinchliffe, C.J. Hogan, G. Höhler, J.D. Jackson, K.F. Johnson,
D. Karlen, B. Kayser, K. Kleinknecht, I.G. Knowles, C. Kolda, P. Kreitz,
P. Langacker, R. Landua, L. Littenberg, D.M. Manley, J. March-Russell,
T. Nakada, H. Quinn, G. Raffelt, B. Renk, M.T. Ronan, L.J. Rosenberg,
M. Schmitt, D.N. Schramm, D. Scott, T. Sjöstrand, G.F. Smoot,
S. Spanier, M. Srednicki, T. Stanev, M. Suzuki, N.P. Tkachenko,
G. Valencia, K. van Bibber, R. Voss, L. Wolfenstein, S. Youssef

(Approximate closing date for data: January 1, 1998)

GAUGE AND HIGGS BOSONS

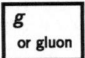

$I(J^{PC}) = 0\Gamma1(1^{--})$

Mass $m < 2 \times 10^{-16}$ eV
Charge $q < 5 \times 10^{-30}\, e$
Mean life $\tau =$ Stable

---

g
or gluon

$I(J^P) = 0(1^-)$

Mass $m = 0$ [a]
SU(3) color octet

W

$J = 1$

Charge $= \pm 1\, e$
Mass $m = 80.41 \pm 0.10$ GeV
$m_Z - m_W = 10.78 \pm 0.10$ GeV
$m_{W^+} - m_{W^-} = -0.2 \pm 0.6$ GeV
Full width $\Gamma = 2.06 \pm 0.06$ GeV

W^- modes are charge conjugates of the modes below.

W^+ DECAY MODES		Fraction (Γ_i/Γ)	Confidence level	p (MeV/c)
$\ell^+ \nu$	[b]	(10.74 ± 0.33) %		–
$e^+ \nu$		(10.9 ± 0.4) %		40205
$\mu^+ \nu$		(10.2 ± 0.5) %		40205
$\tau^+ \nu$		(11.3 ± 0.8) %		40185
hadrons		(67.8 ± 1.0) %		–
$\pi^+ \gamma$		$< 2.2 \quad \times 10^{-4}$	95%	40205

Z

$J = 1$

Charge $= 0$
Mass $m = 91.187 \pm 0.007$ GeV [c]
Full width $\Gamma = 2.490 \pm 0.007$ GeV
$\Gamma(\ell^+\ell^-) = 83.83 \pm 0.27$ MeV [b]
Γ(invisible) $= 498.3 \pm 4.2$ MeV [d]
Γ(hadrons) $= 1740.7 \pm 5.9$ MeV
$\Gamma(\mu^+\mu^-)/\Gamma(e^+e^-) = 1.000 \pm 0.005$
$\Gamma(\tau^+\tau^-)/\Gamma(e^+e^-) = 0.998 \pm 0.005$ [e]

Average charged multiplicity

$\langle N_{charged} \rangle = 21.00 \pm 0.13$

Couplings to leptons

$g_V^\ell = -0.0377 \pm 0.0007$
$g_A^\ell = -0.5008 \pm 0.0008$
$g^{\nu_e} = 0.53 \pm 0.09$
$g^{\nu_\mu} = 0.502 \pm 0.017$

Asymmetry parameters [f]

$A_e = 0.1519 \pm 0.0034$
$A_\mu = 0.102 \pm 0.034$
$A_\tau = 0.143 \pm 0.008$
$A_c = 0.59 \pm 0.19$
$A_b = 0.89 \pm 0.11$

Charge asymmetry (%) at Z pole

$A_{FB}^{(0\ell)} = 1.59 \pm 0.18$
$A_{FB}^{(0u)} = 4.0 \pm 7.3$
$A_{FB}^{(0s)} = 9.9 \pm 3.1 \quad (S = 1.2)$
$A_{FB}^{(0c)} = 7.32 \pm 0.58$
$A_{FB}^{(0b)} = 10.02 \pm 0.28$

Z DECAY MODES		Fraction (Γ_i/Γ)	Confidence level	p (MeV/c)
$e^+ e^-$		(3.366 ± 0.008) %		45594
$\mu^+ \mu^-$		(3.367 ± 0.013) %		45593
$\tau^+ \tau^-$		(3.360 ± 0.015) %		45559
$\ell^+ \ell^-$	[b]	(3.366 ± 0.006) %		–
invisible		(20.01 ± 0.16) %		–
hadrons		(69.90 ± 0.15) %		–
$(u\bar{u} + c\bar{c})/2$		(10.1 ± 1.1) %		–
$(d\bar{d} + s\bar{s} + b\bar{b})/3$		(16.6 ± 0.6) %		–
$c\bar{c}$		(12.4 ± 0.6) %		–
$b\bar{b}$		(15.16 ± 0.09) %		–
$g g g$		$< 1.1 \quad$ %	95%	–
$\pi^0 \gamma$		$< 5.2 \quad \times 10^{-5}$	95%	45593
$\eta \gamma$		$< 5.1 \quad \times 10^{-5}$	95%	45592
$\omega \gamma$		$< 6.5 \quad \times 10^{-4}$	95%	45590
$\eta'(958) \gamma$		$< 4.2 \quad \times 10^{-5}$	95%	45588
$\gamma \gamma$		$< 5.2 \quad \times 10^{-5}$	95%	45594
$\gamma \gamma \gamma$		$< 1.0 \quad \times 10^{-5}$	95%	45594
$\pi^\pm W^\mp$	[g]	$< 7 \quad \times 10^{-5}$	95%	10139
$\rho^\pm W^\mp$	[g]	$< 8.3 \quad \times 10^{-5}$	95%	10114
$J/\psi(1S) X$		$(3.66 \pm 0.23) \times 10^{-3}$		–
$\psi(2S) X$		$(1.60 \pm 0.29) \times 10^{-3}$		–
$\chi_{c1}(1P) X$		$(2.9 \pm 0.7) \times 10^{-3}$		–
$\chi_{c2}(1P) X$		$< 3.2 \quad \times 10^{-3}$	90%	–
$\Upsilon(1S)\, X + \Upsilon(2S)\, X$ $+ \Upsilon(3S)\, X$		$(1.0 \pm 0.5) \times 10^{-4}$		–
$\Upsilon(1S) X$		$< 5.5 \quad \times 10^{-5}$	95%	–
$\Upsilon(2S) X$		$< 1.39 \quad \times 10^{-4}$	95%	–
$\Upsilon(3S) X$		$< 9.4 \quad \times 10^{-5}$	95%	–
$(D^0/\overline{D^0}) X$		(20.7 ± 2.0) %		–
$D^\pm X$		(12.2 ± 1.7) %		–
$D^*(2010)^\pm X$	[g]	(11.4 ± 1.3) %		–
$B_s^0 X$		seen		–
anomalous $\gamma +$ hadrons	[h]	$< 3.2 \quad \times 10^{-3}$	95%	–
$e^+ e^- \gamma$	[h]	$< 5.2 \quad \times 10^{-4}$	95%	45594
$\mu^+ \mu^- \gamma$	[h]	$< 5.6 \quad \times 10^{-4}$	95%	45593
$\tau^+ \tau^- \gamma$	[h]	$< 7.3 \quad \times 10^{-4}$	95%	45559
$\ell^+ \ell^- \gamma \gamma$	[i]	$< 6.8 \quad \times 10^{-6}$	95%	–
$q \bar{q} \gamma \gamma$	[i]	$< 5.5 \quad \times 10^{-6}$	95%	–

$\nu\bar{\nu}\gamma\gamma$		[i] $<$ 3.1	$\times 10^{-6}$	95%	45594
$e^{\pm}\mu^{\mp}$	LF	[g] $<$ 1.7	$\times 10^{-6}$	95%	45593
$e^{\pm}\tau^{\mp}$	LF	[g] $<$ 9.8	$\times 10^{-6}$	95%	45576
$\mu^{\pm}\tau^{\mp}$	LF	[g] $<$ 1.2	$\times 10^{-5}$	95%	45576

Higgs Bosons — H^0 and $H^{\pm}$, Searches for

H^0 Mass $m >$ 77.5 GeV, CL = 95%

H_1^0 in Supersymmetric Models ($m_{H_1^0} < m_{H_2^0}$)

Mass $m >$ 62.5 GeV, CL = 95%

A^0 Pseudoscalar Higgs Boson in Supersymmetric Models [j]

Mass $m >$ 62.5 GeV, CL = 95%　$\tan\beta >1$

$H^{\pm}$ Mass $m >$ 54.5 GeV, CL = 95%

See the Particle Listings for a Note giving details of Higgs Bosons.

Heavy Bosons Other Than Higgs Bosons, Searches for

Additional W Bosons

W_R — right-handed W

Mass $m >$ 549 GeV

(assuming light right-handed neutrino)

W' with standard couplings decaying to $e\nu, \mu\nu$

Mass $m >$ 720 GeV, CL = 95%

Additional Z Bosons

Z'_{SM} with standard couplings

Mass $m >$ 690 GeV, CL = 95%　　($p\bar{p}$ direct search)

Mass $m >$ 779 GeV, CL = 95%　　(electroweak fit)

Z_{LR} of $SU(2)_L \times SU(2)_R \times U(1)$

(with $g_L = g_R$)

Mass $m >$ 630 GeV, CL = 95%　　($p\bar{p}$ direct search)

Mass $m >$ 389 GeV, CL = 95%　　(electroweak fit)

Z_{χ} of $SO(10) \to SU(5) \times U(1)_{\chi}$

(coupling constant derived from G.U.T.)

Mass $m >$ 595 GeV, CL = 95%　　($p\bar{p}$ direct search)

Mass $m >$ 321 GeV, CL = 95%　　(electroweak fit)

Z_{ψ} of $E_6 \to SO(10) \times U(1)_{\psi}$

(coupling constant derived from G.U.T.)

Mass $m >$ 590 GeV, CL = 95%　　($p\bar{p}$ direct search)

Mass $m >$ 160 GeV, CL = 95%　　(electroweak fit)

Z_{η} of $E_6 \to SU(3) \times SU(2) \times U(1) \times U(1)_{\eta}$

(coupling constant derived from G.U.T.);

charges are $Q_{\eta} = \sqrt{3/8}Q_{\chi} - \sqrt{5/8}Q_{\psi}$)

Mass $m >$ 620 GeV, CL = 95%　　($p\bar{p}$ direct search)

Mass $m >$ 182 GeV, CL = 95%　　(electroweak fit)

Scalar Leptoquarks

Mass $m >$ 225 GeV, CL = 95%　(1st generation, pair prod.)

Mass $m >$ 237 GeV, CL = 95%　(1st gener., single prod.)

Mass $m >$ 119 GeV, CL = 95%　(2nd gener., pair prod.)

Mass $m >$ 73 GeV, CL = 95%　(2nd gener., single prod.)

Mass $m >$ 99 GeV, CL = 95%　(3rd gener., pair prod.)

(See the Particle Listings for assumptions on leptoquark quantum numbers and branching fractions.)

Axions (A^0) and Other Very Light Bosons, Searches for

The standard Peccei-Quinn axion is ruled out. Variants with reduced couplings or much smaller masses are constrained by various data. The Particle Listings in the full *Review* contain a Note discussing axion searches.

The best limit for the half-life of neutrinoless double beta decay with Majoron emission is $> 7.2 \times 10^{24}$ years (CL = 90%).

NOTES

In this Summary Table:

When a quantity has "(S = ...)" to its right, the error on the quantity has been enlarged by the "scale factor" S, defined as $S = \sqrt{\chi^2/(N-1)}$, where N is the number of measurements used in calculating the quantity. We do this when $S > 1$, which often indicates that the measurements are inconsistent. When $S > 1.25$ we also show in the Particle Listings an ideogram of the measurements. For more about S, see the Introduction.

A decay momentum p is given for each decay mode. For a 2-body decay, p is the momentum of each decay product in the rest frame of the decaying particle. For a 3-or-more-body decay, p is the largest momentum any of the products can have in this frame.

[a] Theoretical value. A mass as large as a few MeV may not be precluded.

[b] ℓ indicates each type of lepton (e, μ, and τ), not sum over them.

[c] The Z-boson mass listed here corresponds to a Breit-Wigner resonance parameter. It lies approximately 34 MeV above the real part of the position of the pole (in the energy-squared plane) in the Z-boson propagator.

[d] This partial width takes into account Z decays into $\nu\bar{\nu}$ and any other possible undetected modes.

[e] This ratio has not been corrected for the τ mass.

[f] Here $A \equiv 2g_V g_A/(g_V^2 + g_A^2)$.

[g] The value is for the sum of the charge states of particle/antiparticle states indicated.

[h] See the Z Particle Listings for the γ energy range used in this measurement.

[i] For $m_{\gamma\gamma} = (60 \pm 5)$ GeV.

[j] The limits assume no invisible decays.

LEPTON SUMMARY TABLE

LEPTONS

e
$J = \frac{1}{2}$

Mass $m = 0.51099907 \pm 0.00000015$ MeV [a]
$\quad\quad = (5.485799111 \pm 0.000000012) \times 10^{-4}$ u
$(m_{e^+} - m_{e^-})/m < 4 \times 10^{-8}$, CL = 90%
$|q_{e^+} + q_{e^-}|/e < 4 \times 10^{-8}$
Magnetic moment $\mu = 1.001159652193 \pm 0.000000000010 \; \mu_B$
$(g_{e^+} - g_{e^-})/g_{\text{average}} = (-0.5 \pm 2.1) \times 10^{-12}$
Electric dipole moment $d = (0.18 \pm 0.16) \times 10^{-26} \; e\,\text{cm}$
Mean life $\tau > 4.3 \times 10^{23}$ yr, CL = 68% [b]

μ
$J = \frac{1}{2}$

Mass $m = 105.658389 \pm 0.000034$ MeV [c]
$\quad\quad = 0.113428913 \pm 0.000000017$ u
Mean life $\tau = (2.19703 \pm 0.00004) \times 10^{-6}$ s
$\tau_{\mu^+}/\tau_{\mu^-} = 1.00002 \pm 0.00008$
$\quad c\tau = 658.654$ m
Magnetic moment $\mu = 1.0011659230 \pm 0.0000000084 \; e\hbar/2m_\mu$
$(g_{\mu^+} - g_{\mu^-})/g_{\text{average}} = (-2.6 \pm 1.6) \times 10^{-8}$
Electric dipole moment $d = (3.7 \pm 3.4) \times 10^{-19} \; e\,\text{cm}$

Decay parameters [d]

$\rho = 0.7518 \pm 0.0026$
$\eta = -0.007 \pm 0.013$
$\delta = 0.749 \pm 0.004$
$\xi P_\mu = 1.003 \pm 0.008$ [e]
$\xi P_\mu \delta/\rho > 0.99682$, CL = 90% [e]
$\xi' = 1.00 \pm 0.04$
$\xi'' = 0.7 \pm 0.4$
$\alpha/A = (0 \pm 4) \times 10^{-3}$
$\alpha'/A = (0 \pm 4) \times 10^{-3}$
$\beta/A = (4 \pm 6) \times 10^{-3}$
$\beta'/A = (2 \pm 6) \times 10^{-3}$
$\overline{\eta} = 0.02 \pm 0.08$

μ^+ modes are charge conjugates of the modes below.

μ^- DECAY MODES		Fraction (Γ_i/Γ)	Confidence level	p (MeV/c)
$e^- \overline{\nu}_e \nu_\mu$		$\approx 100\%$		53
$e^- \overline{\nu}_e \nu_\mu \gamma$	[f]	$(1.4 \pm 0.4)\%$		53
$e^- \overline{\nu}_e \nu_\mu e^+ e^-$	[g]	$(3.4 \pm 0.4) \times 10^{-5}$		53

Lepton Family number (LF) violating modes

			Fraction (Γ_i/Γ)		Confidence level	p (MeV/c)
$e^- \nu_e \overline{\nu}_\mu$	LF	[h]	< 1.2	%	90%	53
$e^- \gamma$	LF		< 4.9	$\times 10^{-11}$	90%	53
$e^- e^+ e^-$	LF		< 1.0	$\times 10^{-12}$	90%	53
$e^- 2\gamma$	LF		< 7.2	$\times 10^{-11}$	90%	53

τ
$J = \frac{1}{2}$

Mass $m = 1777.05^{+0.29}_{-0.26}$ MeV
Mean life $\tau = (290.0 \pm 1.2) \times 10^{-15}$ s
$\quad c\tau = 86.93 \; \mu$m
Magnetic moment anomaly > -0.052 and < 0.058, CL = 95%
Electric dipole moment $d > -3.1$ and $< 3.1 \times 10^{-16} \; e\,\text{cm}$, CL = 95%

Weak dipole moment

$\text{Re}(d^w_\tau) < 0.56 \times 10^{-17} \; e\,\text{cm}$, CL = 95%
$\text{Im}(d^w_\tau) < 1.5 \times 10^{-17} \; e\,\text{cm}$, CL = 95%

Weak anomalous magnetic dipole moment

$\text{Re}(\alpha^w_\tau) < 4.5 \times 10^{-3}$, CL = 90%
$\text{Im}(\alpha^w_\tau) < 9.9 \times 10^{-3}$, CL = 90%

Decay parameters

See the τ Particle Listings for a note concerning τ-decay parameters.

$\rho^\tau(e \text{ or } \mu) = 0.748 \pm 0.010$
$\rho^\tau(e) = 0.745 \pm 0.012$
$\rho^\tau(\mu) = 0.741 \pm 0.030$
$\xi^\tau(e \text{ or } \mu) = 1.01 \pm 0.04$
$\xi^\tau(e) = 0.98 \pm 0.05$
$\xi^\tau(\mu) = 1.07 \pm 0.08$
$\eta^\tau(e \text{ or } \mu) = 0.01 \pm 0.07$
$\eta^\tau(\mu) = -0.10 \pm 0.18$
$(\delta\xi)^\tau(e \text{ or } \mu) = 0.749 \pm 0.026$
$(\delta\xi)^\tau(e) = 0.733 \pm 0.033$
$(\delta\xi)^\tau(\mu) = 0.78 \pm 0.05$
$\xi^\tau(\pi) = 0.99 \pm 0.05$
$\xi^\tau(\rho) = 0.996 \pm 0.010$
$\xi^\tau(a_1) = 1.02 \pm 0.04$
$\xi^\tau(\text{all hadronic modes}) = 0.997 \pm 0.009$

τ^+ modes are charge conjugates of the modes below. "$h^\pm$" stands for $\pi^\pm$ or $K^\pm$. "ℓ" stands for e or μ. "Neutral" means neutral hadron whose decay products include γ's and/or π^0's.

τ^- DECAY MODES		Fraction (Γ_i/Γ)	Scale factor/ Confidence level	p (MeV/c)
Modes with one charged particle				
$\text{particle}^- \geq 0 \text{ neutrals} \geq 0 K^0_L \nu_\tau$ ("1-prong")		$(84.71 \pm 0.13)\%$	S=1.2	–
$\text{particle}^- \geq 0 \text{ neutrals} \geq 0 K^0 \nu_\tau$		$(85.30 \pm 0.13)\%$	S=1.2	–
$\mu^- \overline{\nu}_\mu \nu_\tau$	[i]	$(17.37 \pm 0.09)\%$		885
$\mu^- \overline{\nu}_\mu \nu_\tau \gamma$	[g]	$(3.0 \pm 0.6) \times 10^{-3}$		–
$e^- \overline{\nu}_e \nu_\tau$	[i]	$(17.81 \pm 0.07)\%$		889
$h^- \geq 0 \text{ neutrals} \geq 0 K^0_L \nu_\tau$		$(49.52 \pm 0.16)\%$	S=1.2	–
$h^- \geq 0 K^0_L \nu_\tau$		$(12.32 \pm 0.12)\%$	S=1.5	–
$h^- \nu_\tau$		$(11.79 \pm 0.12)\%$	S=1.5	–
$\pi^- \nu_\tau$	[i]	$(11.08 \pm 0.13)\%$	S=1.4	883
$K^- \nu_\tau$	[i]	$(7.1 \pm 0.5) \times 10^{-3}$		820
$h^- \geq 1 \text{ neutrals} \nu_\tau$		$(36.91 \pm 0.17)\%$	S=1.2	–
$h^- \pi^0 \nu_\tau$		$(25.84 \pm 0.14)\%$	S=1.1	–
$\pi^- \pi^0 \nu_\tau$	[i]	$(25.32 \pm 0.15)\%$	S=1.1	878
$\pi^- \pi^0 \text{non-}\rho(770) \nu_\tau$		$(3.0 \pm 3.2) \times 10^{-3}$		878
$K^- \pi^0 \nu_\tau$	[i]	$(5.2 \pm 0.5) \times 10^{-3}$		814
$h^- \geq 2\pi^0 \nu_\tau$		$(10.79 \pm 0.16)\%$	S=1.2	–
$h^- 2\pi^0 \nu_\tau$		$(9.39 \pm 0.14)\%$	S=1.2	–
$h^- 2\pi^0 \nu_\tau (\text{ex.} K^0)$		$(9.23 \pm 0.14)\%$	S=1.2	–
$\pi^- 2\pi^0 \nu_\tau (\text{ex.} K^0)$	[i]	$(9.15 \pm 0.15)\%$	S=1.2	862
$K^- 2\pi^0 \nu_\tau (\text{ex.} K^0)$	[i]	$(8.0 \pm 2.7) \times 10^{-4}$		796
$h^- \geq 3\pi^0 \nu_\tau$		$(1.40 \pm 0.11)\%$	S=1.1	–
$h^- 3\pi^0 \nu_\tau$		$(1.23 \pm 0.10)\%$	S=1.1	–
$\pi^- 3\pi^0 \nu_\tau (\text{ex.} K^0)$	[i]	$(1.11 \pm 0.14)\%$		836
$K^- 3\pi^0 \nu_\tau (\text{ex.} K^0)$		$(4.3^{+10.0}_{-2.9}) \times 10^{-4}$		766
$h^- 4\pi^0 \nu_\tau (\text{ex.} K^0)$		$(1.7 \pm 0.6) \times 10^{-3}$		–
$h^- 4\pi^0 \nu_\tau (\text{ex.} K^0, \eta)$	[i]	$(1.1 \pm 0.6) \times 10^{-3}$		–
$K^- \geq 0\pi^0 \geq 0 K^0 \nu_\tau$		$(1.66 \pm 0.10)\%$		–
$K^- \geq 1 (\pi^0 \text{ or } K^0) \nu_\tau$		$(9.5 \pm 1.0) \times 10^{-3}$		–
Modes with K^0's				
$K^0 (\text{particles})^- \nu_\tau$		$(1.66 \pm 0.09)\%$	S=1.4	–
$h^- \overline{K}^0 \geq 0 \text{ neutrals} \geq 0 K^0_L \nu_\tau$		$(1.62 \pm 0.09)\%$	S=1.4	–
$h^- \overline{K}^0 \nu_\tau$		$(9.9 \pm 0.8) \times 10^{-3}$	S=1.5	–
$\pi^- \overline{K}^0 \nu_\tau$	[i]	$(8.3 \pm 0.8) \times 10^{-3}$	S=1.4	812
$\pi^- \overline{K}^0 (\text{non-}K^*(892)^-) \nu_\tau$		$< 1.7 \times 10^{-3}$	CL=95%	812
$K^- K^0 \nu_\tau$	[i]	$(1.59 \pm 0.24) \times 10^{-3}$		737
$h^- \overline{K}^0 \pi^0 \nu_\tau$		$(5.5 \pm 0.5) \times 10^{-3}$		–
$\pi^- \overline{K}^0 \pi^0 \nu_\tau$	[i]	$(3.9 \pm 0.5) \times 10^{-3}$		794
$\overline{K}^0 \rho^- \nu_\tau$		$(1.9 \pm 0.7) \times 10^{-3}$		–
$K^- K^0 \pi^0 \nu_\tau$	[i]	$(1.51 \pm 0.29) \times 10^{-3}$		685
$\pi^- \overline{K}^0 \pi^0 \pi^0 \nu_\tau$		$(6 \pm 4) \times 10^{-4}$		–
$K^- K^0 \pi^0 \pi^0 \nu_\tau$		$< 3.9 \times 10^{-4}$	CL=95%	–
$\pi^- K^0 \overline{K}^0 \nu_\tau$	[i]	$(1.21 \pm 0.21) \times 10^{-3}$	S=1.2	682

11-3

Mode	Value		
$\pi^- K_S^0 K_S^0 \nu_\tau$	$(3.0 \pm 0.5) \times 10^{-4}$	S=1.2	–
$\pi^- K_S^0 K_L^0 \nu_\tau$	$(6.0 \pm 1.0) \times 10^{-4}$	S=1.2	–
$\pi^- K_S^0 K_S^0 \pi^0 \nu_\tau$	$< 2.0 \times 10^{-4}$	CL=95%	–
$\pi^- K_S^0 K_L^0 \pi^0 \nu_\tau$	$(3.1 \pm 1.2) \times 10^{-4}$		–
$K^- K^0 \geq 0$ neutrals ν_τ	$(3.1 \pm 0.4) \times 10^{-3}$		–
$K^0 h^+ h^- h^- \geq 0$ neutrals ν_τ	$< 1.7 \times 10^{-3}$	CL=95%	–
$K^0 h^+ h^- h^- \nu_\tau$	$(2.3 \pm 2.0) \times 10^{-4}$		–

Modes with three charged particles

Mode	Value		
$h^- h^- h^+ \geq 0$ neut. ν_τ ("3-prong")	(15.18 ± 0.13) %	S=1.2	–
$h^- h^- h^+ \geq 0$ neutrals ν_τ (ex. $K_S^0 \to \pi^+\pi^-$)	(14.60 ± 0.13) %	S=1.2	–
$\pi^- \pi^+ \pi^- \geq 0$ neutrals ν_τ	(14.60 ± 0.14) %		–
$h^- h^- h^+ \nu_\tau$	(9.96 ± 0.10) %	S=1.1	–
$h^- h^- h^+ \nu_\tau$ (ex.K^0)	(9.62 ± 0.10) %	S=1.1	–
$h^- h^- h^+ \nu_\tau$ (ex.K^0,ω)	(9.57 ± 0.10) %	S=1.1	–
$\pi^- \pi^+ \pi^- \nu_\tau$	(9.56 ± 0.11) %	S=1.1	–
$\pi^- \pi^+ \pi^- \nu_\tau$ (ex.K^0)	(9.52 ± 0.11) %	S=1.1	–
$\pi^- \pi^+ \pi^- \nu_\tau$ (ex.K^0,ω)	[l] (9.23 ± 0.11) %	S=1.1	–
$h^- h^- h^+ \geq 1$ neutrals ν_τ	(5.18 ± 0.11) %	S=1.2	–
$h^- h^- h^+ \geq 1$ neutrals ν_τ (ex. $K_S^0 \to \pi^+\pi^-$)	(4.98 ± 0.11) %	S=1.2	–
$h^- h^- h^+ \pi^0 \nu_\tau$	(4.50 ± 0.09) %	S=1.1	–
$h^- h^- h^+ \pi^0 \nu_\tau$ (ex.K^0)	(4.31 ± 0.09) %	S=1.1	–
$h^- h^- h^+ \pi^0 \nu_\tau$ (ex. K^0, ω)	(2.59 ± 0.09) %		–
$\pi^- \pi^+ \pi^- \pi^0 \nu_\tau$	(4.35 ± 0.10) %		–
$\pi^- \pi^+ \pi^- \pi^0 \nu_\tau$ (ex.K^0)	(4.22 ± 0.10) %		–
$\pi^- \pi^+ \pi^- \pi^0 \nu_\tau$ (ex.K^0,ω)	[l] (2.49 ± 0.10) %		–
$h^- (\rho\pi)^0 \nu_\tau$	(2.88 ± 0.35) %		–
$(a_1(1260) h)^- \nu_\tau$	< 2.0 %	CL=95%	–
$h^- \rho^0 \pi^0 \nu_\tau$	(1.35 ± 0.20) %		–
$h^- \rho^+ h^- \nu_\tau$	$(4.5 \pm 2.2) \times 10^{-3}$		–
$h^- \rho^- h^+ \nu_\tau$	(1.17 ± 0.23) %		–
$h^- h^- h^+ 2\pi^0 \nu_\tau$	$(5.4 \pm 0.4) \times 10^{-3}$		–
$h^- h^- h^+ 2\pi^0 \nu_\tau$ (ex.K^0)	$(5.3 \pm 0.4) \times 10^{-3}$		–
$h^- h^- h^+ 2\pi^0 \nu_\tau$ (ex.K^0,ω,η)	[l] $(1.1 \pm 0.4) \times 10^{-3}$		–
$h^- h^- h^+ \geq 3\pi^0 \nu_\tau$	[l] $(1.4 ^{+0.9}_{-0.7}) \times 10^{-3}$	S=1.5	–
$h^- h^- h^+ 3\pi^0 \nu_\tau$	$(2.9 \pm 0.8) \times 10^{-4}$		–
$K^- h^+ h^- \geq 0$ neutrals ν_τ	$(5.4 \pm 0.7) \times 10^{-4}$	S=1.1	–
$K^- \pi^+ \pi^- \geq 0$ neutrals ν_τ	$(3.1 \pm 0.6) \times 10^{-3}$	S=1.1	–
$K^- \pi^+ \pi^- \nu_\tau$	$(2.3 \pm 0.4) \times 10^{-3}$		–
$K^- \pi^+ \pi^- \nu_\tau$ (ex.K^0)	[l] $(1.8 \pm 0.5) \times 10^{-3}$		–
$K^- \pi^+ \pi^- \pi^0 \nu_\tau$	$(8 \pm 4) \times 10^{-4}$		–
$K^- \pi^+ \pi^- \pi^0 \nu_\tau$ (ex.K^0)	[l] $(2.4 ^{+4.3}_{-1.6}) \times 10^{-4}$		–
$K^- \pi^+ K^- \geq 0$ neut. ν_τ	$< 9 \times 10^{-4}$	CL=95%	–
$K^- K^+ \pi^- \geq 0$ neut. ν_τ	$(2.3 \pm 0.4) \times 10^{-3}$		–
$K^- K^+ \pi^- \nu_\tau$	[l] $(1.61 \pm 0.26) \times 10^{-3}$		685
$K^- K^+ \pi^- \pi^0 \nu_\tau$	[l] $(6.9 \pm 3.0) \times 10^{-4}$		–
$K^- K^+ K^- \geq 0$ neut. ν_τ	$< 2.1 \times 10^{-3}$	CL=95%	–
$K^- K^+ K^- \nu_\tau$	1.9×10^{-4}	CL=90%	–
$\pi^- K^+ \pi^- \geq 0$ neut. ν_τ	$< 2.5 \times 10^{-3}$	CL=95%	–
$e^- e^- e^+ \bar\nu_e \nu_\tau$	$(2.8 \pm 1.5) \times 10^{-5}$		889
$\mu^- e^- e^+ \bar\nu_\mu \nu_\tau$	$< 3.6 \times 10^{-5}$	CL=90%	885

Modes with five charged particles

Mode	Value		
$3h^- 2h^+ \geq 0$ neutrals ν_τ (ex. $K_S^0 \to \pi^-\pi^+$) ("5-prong")	$(9.7 \pm 0.7) \times 10^{-4}$		–
$3h^- 2h^+ \nu_\tau$ (ex.K^0)	[l] $(7.5 \pm 0.7) \times 10^{-4}$		–
$3h^- 2h^+ \pi^0 \nu_\tau$ (ex.K^0)	[l] $(2.2 \pm 0.5) \times 10^{-4}$		–
$3h^- 2h^+ 2\pi^0 \nu_\tau$	$< 1.1 \times 10^{-4}$	CL=90%	–

Miscellaneous other allowed modes

Mode	Value		
$(5\pi)^- \nu_\tau$	$(7.4 \pm 0.7) \times 10^{-3}$		–
$4h^- 3h^+ \geq 0$ neutrals ν_τ ("7-prong")	$< 2.4 \times 10^{-6}$	CL=90%	–
$K^*(892)^- \geq 0(h^0 \neq K_S^0)\nu_\tau$	(1.94 ± 0.31) %		–
$K^*(892)^- \geq 0$ neutrals ν_τ	(1.33 ± 0.13) %		–
$K^*(892)^- \nu_\tau$	(1.28 ± 0.08) %		665
$K^*(892)^0 K^- \geq 0$ neutrals ν_τ	$(3.2 \pm 1.4) \times 10^{-3}$		–
$K^*(892)^0 K^- \nu_\tau$	$(2.1 \pm 0.4) \times 10^{-3}$		539
$\overline{K}^*(892)^0 \pi^- \geq 0$ neutrals ν_τ	$(3.8 \pm 1.7) \times 10^{-3}$		–
$\overline{K}^*(892)^0 \pi^- \nu_\tau$	$(2.2 \pm 0.5) \times 10^{-3}$		653
$(\overline{K}^*(892)\pi)^- \nu_\tau \to \pi^- \overline{K}^0 \pi^0 \nu_\tau$	$(1.1 \pm 0.5) \times 10^{-3}$		–
$K_1(1270)^- \nu_\tau$	$(4 \pm 4) \times 10^{-3}$		433
$K_1(1400)^- \nu_\tau$	$(8 \pm 4) \times 10^{-3}$		335

Mode	Value		
$K_2^*(1430)^- \nu_\tau$	$< 3 \times 10^{-3}$	CL=95%	317
$\eta \pi^- \nu_\tau$	$< 1.4 \times 10^{-4}$	CL=95%	798
$\eta \pi^- \pi^0 \nu_\tau$	[l] $(1.74 \pm 0.24) \times 10^{-3}$		778
$\eta \pi^- \pi^0 \pi^0 \nu_\tau$	$(1.4 \pm 0.7) \times 10^{-4}$		746
$\eta K^- \nu_\tau$	$(2.7 \pm 0.6) \times 10^{-4}$		720
$\eta \pi^+ \pi^- \pi^- \geq 0$ neutrals ν_τ	$< 3 \times 10^{-3}$	CL=90%	–
$\eta \pi^- \pi^+ \pi^- \nu_\tau$	$(3.4 \pm 0.8) \times 10^{-4}$		–
$\eta a_1(1260)^- \nu_\tau \to \eta \pi^- \rho^0 \nu_\tau$	$< 3.9 \times 10^{-4}$	CL=90%	–
$\eta \eta \pi^- \nu_\tau$	$< 1.1 \times 10^{-4}$	CL=95%	637
$\eta \eta \pi^- \pi^0 \nu_\tau$	$< 2.0 \times 10^{-4}$	CL=95%	559
$\eta'(958) \pi^- \nu_\tau$	$< 7.4 \times 10^{-5}$	CL=90%	–
$\eta'(958) \pi^- \pi^0 \nu_\tau$	$< 8.0 \times 10^{-5}$	CL=90%	–
$\phi \pi^- \nu_\tau$	$< 2.0 \times 10^{-4}$	CL=90%	585
$\phi K^- \nu_\tau$	$< 6.7 \times 10^{-5}$	CL=90%	–
$f_1(1285) \pi^- \nu_\tau$	$(5.8 \pm 2.3) \times 10^{-4}$		–
$f_1(1285) \pi^- \nu_\tau \to \eta \pi^- \pi^+ \pi^- \nu_\tau$	$(1.9 \pm 0.7) \times 10^{-4}$		–
$h^- \omega \geq 0$ neutrals ν_τ	(2.36 ± 0.08) %		–
$h^- \omega \nu_\tau$	[l] (1.93 ± 0.06) %		–
$h^- \omega \pi^0 \nu_\tau$	[l] $(4.3 \pm 0.5) \times 10^{-3}$		–
$h^- \omega 2\pi^0 \nu_\tau$	$(1.9 \pm 0.8) \times 10^{-4}$		–

Lepton Family number (LF), Lepton number (L), or Baryon number (B) violating modes
(In the modes below, ℓ means a sum over e and μ modes)

L means lepton number violation (e.g. $\tau^- \to e^+ \pi^- \pi^-$). Following common usage Γ LF means lepton family violation *and not* lepton number violation (e.g. $\tau^- \to e^- \pi^+ \pi^-$). B means baryon number violation.

Mode	Type	Value		
$e^- \gamma$	LF	$< 2.7 \times 10^{-6}$	CL=90%	888
$\mu^- \gamma$	LF	$< 3.0 \times 10^{-6}$	CL=90%	885
$e^- \pi^0$	LF	$< 3.7 \times 10^{-6}$	CL=90%	883
$\mu^- \pi^0$	LF	$< 4.0 \times 10^{-6}$	CL=90%	880
$e^- K^0$	LF	$< 1.3 \times 10^{-3}$	CL=90%	819
$\mu^- K^0$	LF	$< 1.0 \times 10^{-3}$	CL=90%	815
$e^- \eta$	LF	$< 8.2 \times 10^{-6}$	CL=90%	804
$\mu^- \eta$	LF	$< 9.6 \times 10^{-6}$	CL=90%	800
$e^- \rho^0$	LF	$< 2.0 \times 10^{-6}$	CL=90%	722
$\mu^- \rho^0$	LF	$< 6.3 \times 10^{-6}$	CL=90%	718
$e^- K^*(892)^0$	LF	$< 5.1 \times 10^{-6}$	CL=90%	663
$\mu^- K^*(892)^0$	LF	$< 7.5 \times 10^{-6}$	CL=90%	657
$e^- \overline{K}^*(892)^0$	LF	$< 7.4 \times 10^{-6}$	CL=90%	663
$\mu^- \overline{K}^*(892)^0$	LF	$< 7.5 \times 10^{-6}$	CL=90%	657
$e^- \phi$	LF	$< 6.9 \times 10^{-6}$	CL=90%	596
$\mu^- \phi$	LF	$< 7.0 \times 10^{-6}$	CL=90%	590
$\pi^- \gamma$	L	$< 2.8 \times 10^{-4}$	CL=90%	883
$\pi^- \pi^0$	L	$< 3.7 \times 10^{-4}$	CL=90%	878
$e^- e^+ e^-$	LF	$< 2.9 \times 10^{-6}$	CL=90%	888
$e^- \mu^+ \mu^-$	LF	$< 1.8 \times 10^{-6}$	CL=90%	882
$e^+ \mu^- \mu^-$	LF	$< 1.5 \times 10^{-6}$	CL=90%	882
$\mu^- e^+ e^-$	LF	$< 1.7 \times 10^{-6}$	CL=90%	885
$\mu^+ e^- e^-$	LF	$< 1.5 \times 10^{-6}$	CL=90%	885
$\mu^- \mu^+ \mu^-$	LF	$< 1.9 \times 10^{-6}$	CL=90%	873
$e^- \pi^+ \pi^-$	LF	$< 2.2 \times 10^{-6}$	CL=90%	877
$e^+ \pi^- \pi^-$	L	$< 1.9 \times 10^{-6}$	CL=90%	877
$\mu^- \pi^+ \pi^-$	LF	$< 8.2 \times 10^{-6}$	CL=90%	866
$\mu^+ \pi^- \pi^-$	L	$< 3.4 \times 10^{-6}$	CL=90%	866
$e^- \pi^+ K^-$	LF	$< 6.4 \times 10^{-6}$	CL=90%	814
$e^- \pi^- K^+$	LF	$< 3.8 \times 10^{-6}$	CL=90%	814
$e^+ \pi^- K^-$	L	$< 2.1 \times 10^{-6}$	CL=90%	814
$e^- K^+ K^-$	LF	$< 6.0 \times 10^{-6}$	CL=90%	739
$e^+ K^- K^-$	L	$< 3.8 \times 10^{-6}$	CL=90%	739
$\mu^- \pi^+ K^-$	LF	$< 7.5 \times 10^{-6}$	CL=90%	800
$\mu^- \pi^- K^+$	LF	$< 7.4 \times 10^{-6}$	CL=90%	800
$\mu^+ \pi^- K^-$	L	$< 7.0 \times 10^{-6}$	CL=90%	800
$\mu^- K^+ K^-$	LF	$< 1.5 \times 10^{-5}$	CL=90%	699
$\mu^+ K^- K^-$	L	$< 6.0 \times 10^{-6}$	CL=90%	699
$e^- \pi^0 \pi^0$	LF	$< 6.5 \times 10^{-6}$	CL=90%	878
$\mu^- \pi^0 \pi^0$	LF	$< 1.4 \times 10^{-5}$	CL=90%	867
$e^- \eta \eta$	LF	$< 3.5 \times 10^{-5}$	CL=90%	700
$\mu^- \eta \eta$	LF	$< 6.0 \times 10^{-5}$	CL=90%	654
$e^- \pi^0 \eta$	LF	$< 2.4 \times 10^{-5}$	CL=90%	798
$\mu^- \pi^0 \eta$	LF	$< 2.2 \times 10^{-5}$	CL=90%	784
$\overline{p} \gamma$	LB	$< 2.9 \times 10^{-4}$	CL=90%	641
$\overline{p} \pi^0$	LB	$< 6.6 \times 10^{-4}$	CL=90%	632
$\overline{p} \eta$	LB	$< 1.30 \times 10^{-4}$	CL=90%	476
e^- light boson	LF	$< 2.7 \times 10^{-3}$	CL=95%	–
μ^- light boson	LF	$< 5 \times 10^{-3}$	CL=95%	–

Heavy Charged Lepton Searches

$L^{\pm}$ – charged lepton

Mass $m > 80.2$ GeV, CL = 95% $m_{\nu} \approx 0$

$L^{\pm}$ – stable charged heavy lepton

Mass $m > 84.2$ GeV, CL = 95%

Neutrinos

See the Particle Listings for a Note "Neutrino Mass" giving details of neutrinos, masses, mixing, and the status of experimental searches.

  ν_e $J = \frac{1}{2}$

Mass m: Unexplained effects have resulted in significantly negative m^2 in the new, precise tritium beta decay experiments. It is felt that a real neutrino mass as large as 10–15 eV would cause observable spectral distortions even in the presence of the end-point count excesses.

Mean life/mass, $\tau/m_{\nu_e} > 7 \times 10^9$ s/eV   (solar)

Mean life/mass, $\tau/m_{\nu_e} > 300$ s/eV, CL = 90%   (reactor)

Magnetic moment $\mu < 1.8 \times 10^{-10} \mu_B$, CL = 90%

  ν_μ $J = \frac{1}{2}$

Mass $m < 0.17$ MeV, CL = 90%

Mean life/mass, $\tau/m_{\nu_\mu} > 15.4$ s/eV, CL = 90%

Magnetic moment $\mu < 7.4 \times 10^{-10} \mu_B$, CL = 90%

  ν_τ $J = \frac{1}{2}$

Mass $m < 18.2$ MeV, CL = 95%

Magnetic moment $\mu < 5.4 \times 10^{-7} \mu_B$, CL = 90%

Electric dipole moment $d < 5.2 \times 10^{-17}$ e cm, CL = 95%

Number of Light Neutrino Types

(including ν_e, ν_μ, and ν_τ)

Number $N = 2.994 \pm 0.012$ (Standard Model fits to LEP data)

Number $N = 3.07 \pm 0.12$ (Direct measurement of invisible Z width)

Massive Neutrinos and Lepton Mixing, Searches for

For excited leptons, see Compositeness Limits below.

See the Particle Listings for a Note "Neutrino Mass" giving details of neutrinos, masses, mixing, and the status of experimental searches.

While no direct, uncontested evidence for massive neutrinos or lepton mixing has been obtained, suggestive evidence has come from solar neutrino observations, from anomalies in the relative fractions of ν_e and ν_μ observed in energetic cosmic-ray air showers, and possibly from a $\overline{\nu}_e$ appearance experiment at Los Alamos. Sample limits are:

Stable Neutral Heavy Lepton Mass Limits

Mass $m > 45.0$ GeV, CL = 95%   (Dirac)

Mass $m > 39.5$ GeV, CL = 95%   (Majorana)

Neutral Heavy Lepton Mass Limits

Mass $m > 69.0$ GeV, CL = 95%   (Dirac ν_L coupling to e, μ, τ with $|U_{\ell j}|^2 > 10^{-12}$)

Mass $m > 58.2$ GeV, CL = 95%   (Majorana ν_L coupling to e, μ, τ with $|U_{\ell j}|^2 > 10^{-12}$)

Solar Neutrinos

Detectors using gallium ($E_\nu \gtrsim 0.2$ MeV), chlorine ($E_\nu \gtrsim 0.8$ MeV), and Čerenkov effect in water ($E_\nu \gtrsim 7$ MeV) measure significantly lower neutrino rates than are predicted from solar models. The deficit in the solar neutrino flux compared with solar model calculations could be explained by oscillations with $\Delta m^2 \leq 10^{-5}$ eV^2 causing the disappearance of ν_e.

Atmospheric Neutrinos

Underground detectors observing neutrinos produced by cosmic rays in the atmosphere have measured a ν_μ/ν_e ratio much less than expected and also a deficiency of upward going ν_μ compared to downward. This could be explained by oscillations leading to the disappearance of ν_μ with $\Delta m^2 \approx 10^{-3}$ to 10^{-2} eV^2.

ν oscillation: $\overline{\nu}_e \not\rightarrow \overline{\nu}_e$ (θ = mixing angle)

$\Delta m^2 < 9 \times 10^{-4}$ eV^2, CL = 90%   (if $\sin^2 2\theta = 1$)

$\sin^2 2\theta < 0.02$, CL = 90%   (if $\Delta(m^2)$ is large)

ν oscillation: ν_μ ($\overline{\nu}_\mu$) $\rightarrow$ ν_e ($\overline{\nu}_e$) (any combination)

$\Delta m^2 < 0.075$ eV^2, CL = 90%   (if $\sin^2 2\theta = 1$)

$\sin^2 2\theta < 1.8 \times 10^{-3}$, CL = 90%   (if $\Delta(m^2)$ is large)

NOTES

In this Summary Table:

When a quantity has "(S = ...)" to its right, the error on the quantity has been enlarged by the "scale factor" S, defined as S = $\sqrt{\chi^2/(N-1)}$, where N is the number of measurements used in calculating the quantity. We do this when S > 1, which often indicates that the measurements are inconsistent. When S > 1.25, we also show in the Particle Listings an ideogram of the measurements. For more about S, see the Introduction.

A decay momentum p is given for each decay mode. For a 2-body decay, p is the momentum of each decay product in the rest frame of the decaying particle. For a 3-or-more-body decay, p is the largest momentum any of the products can have in this frame.

[a] The uncertainty in the electron mass in unified atomic mass units (u) is ten times smaller than that given by the 1986 CODATA adjustment, quoted in the Table of Physical Constants (Section 1). The conversion to MeV via the factor 931.49432(28) MeV/u is more uncertain because of the electron charge uncertainty. Our value in MeV differs slightly from the 1986 CODATA result.

[b] This is the best "electron disappearance" limit. The best limit for the mode $e^- \rightarrow \nu\gamma$ is > 2.35×10^{25} yr (CL=68%).

[c] The muon mass is most precisely known in u (unified atomic mass units). The conversion factor to MeV via the factor 931.49432(28) MeV/u is more uncertain because of the electron charge uncertainty.

[d] See the "Note on Muon Decay Parameters" in the μ Particle Listings for definitions and details.

[e] P_μ is the longitudinal polarization of the muon from pion decay. In standard $V-A$ theory, $P_\mu = 1$ and $\rho = \delta = 3/4$.

[f] This only includes events with the γ energy > 10 MeV. Since the $e^- \overline{\nu}_e \nu_\mu$ and $e^- \overline{\nu}_e \nu_\mu \gamma$ modes cannot be clearly separated, we regard the latter mode as a subset of the former.

[g] See the μ Particle Listings for the energy limits used in this measurement.

[h] A test of additive vs. multiplicative lepton family number conservation.

[i] Basis mode for the τ.

QUARKS

The u-, d-, and s-quark masses are estimates of so-called "current-quark masses," in a mass-independent subtraction scheme such as $\overline{\text{MS}}$ at a scale $\mu \approx 2$ GeV. The c- and b-quark masses are estimated from charmonium, bottomonium, D, and B masses. They are the "running" masses in the $\overline{\text{MS}}$ scheme. These can be different from the heavy quark masses obtained in potential models.

u

$$I(J^P) = \tfrac{1}{2}(\tfrac{1}{2}^+)$$

Mass $m = 1.5$ to 5 MeV [a]   Charge $= \tfrac{2}{3}\,e$ $I_z = +\tfrac{1}{2}$

$m_u/m_d = 0.20$ to 0.70

d

$$I(J^P) = \tfrac{1}{2}(\tfrac{1}{2}^+)$$

Mass $m = 3$ to 9 MeV [a]   Charge $= -\tfrac{1}{3}\,e$ $I_z = -\tfrac{1}{2}$

$m_s/m_d = 17$ to 25

$\overline{m} = (m_u + m_d)/2 = 2$ to 6 MeV

s

$$I(J^P) = 0(\tfrac{1}{2}^+)$$

Mass $m = 60$ to 170 MeV [a]   Charge $= -\tfrac{1}{3}\,e$ Strangeness $= -1$

$(m_s - (m_u + m_d)/2)/(m_d - m_u) = 34$ to 51

c

$$I(J^P) = 0(\tfrac{1}{2}^+)$$

Mass $m = 1.1$ to 1.4 GeV   Charge $= \tfrac{2}{3}\,e$ Charm $= +1$

b

$$I(J^P) = 0(\tfrac{1}{2}^+)$$

Mass $m = 4.1$ to 4.4 GeV   Charge $= -\tfrac{1}{3}\,e$ Bottom $= -1$

t

$$I(J^P) = 0(\tfrac{1}{2}^+)$$

Charge $= \tfrac{2}{3}\,e$ Top $= +1$

Mass $m = 173.8 \pm 5.2$ GeV   (direct observation of top events)

Mass $m = 170 \pm 7\ (+14)$ GeV   (Standard Model electroweak fit, assuming $M_H = M_Z$. Number in parentheses is shift from changing M_H to 300 GeV.

b' (4th Generation) Quark, Searches for

Mass $m > 128$ GeV, CL $= 95\%$ ($p\bar{p}$, charged current decays)

Mass $m > 46.0$ GeV, CL $= 95\%$ ($e^+ e^-$, all decays)

Free Quark Searches

All searches since 1977 have had negative results.

NOTES

[a] The ratios m_u/m_d and m_s/m_d are extracted from pion and kaon masses using chiral symmetry. The estimates of u and d masses are not without controversy and remain under active investigation. Within the literature there are even suggestions that the u quark could be essentially massless. The s-quark mass is estimated from SU(3) splittings in hadron masses.

LIGHT UNFLAVORED MESONS ($S = C = B = 0$)

For $I = 1$ ($\pi\Gamma b\Gamma\rho\Gamma a$): $u\overline{d}\Gamma(u\overline{u}-d\overline{d})/\sqrt{2}\Gamma\,d\overline{u}$;
for $I = 0$ ($\eta\Gamma\eta'\Gamma h\Gamma h'\Gamma\omega\Gamma\phi\Gamma f\Gamma f'$): $c_1(u\overline{u} + d\overline{d}) + c_2(s\overline{s})$

$\pi^\pm$ $\qquad I^G(J^P) = 1^-(0^-)$

Mass $m = 139.56995 \pm 0.00035$ MeV
Mean life $\tau = (2.6033 \pm 0.0005) \times 10^{-8}$ s (S = 1.2)
$c\tau = 7.8045$ m

$\pi^\pm \to \ell^\pm\nu\gamma$ form factors [a]

$F_V = 0.017 \pm 0.008$
$F_A = 0.0116 \pm 0.0016$ (S = 1.3)
$R = 0.059^{+0.009}_{-0.008}$

π^- modes are charge conjugates of the modes below.

π^+ DECAY MODES		Fraction (Γ_i/Γ)	Confidence level	p (MeV/c)
$\mu^+\nu_\mu$	[b]	(99.98770±0.00004) %		30
$\mu^+\nu_\mu\gamma$	[c]	(1.24 ±0.25) × 10^{-4}		30
$e^+\nu_e$	[b]	(1.230 ±0.004) × 10^{-4}		70
$e^+\nu_e\gamma$	[c]	(1.61 ±0.23) × 10^{-7}		70
$e^+\nu_e\pi^0$		(1.025 ±0.034) × 10^{-8}		4
$e^+\nu_e e^+ e^-$		(3.2 ±0.5) × 10^{-9}		70
$e^+\nu_e\nu\overline{\nu}$		< 5 × 10^{-6}	90%	70

Lepton Family number (LF) or Lepton number (L) violating modes

			Fraction	Confidence level	p (MeV/c)
$\mu^+\overline{\nu}_e$	L	[d]	< 1.5 × 10^{-3}	90%	30
$\mu^+\nu_e$	LF	[d]	< 8.0 × 10^{-3}	90%	30
$\mu^- e^+ e^+ \nu$	LF		< 1.6 × 10^{-6}	90%	30

π^0 $\qquad I^G(J^{PC}) = 1^-(0^{-+})$

Mass $m = 134.9764 \pm 0.0006$ MeV
$m_{\pi^\pm} - m_{\pi^0} = 4.5936 \pm 0.0005$ MeV
Mean life $\tau = (8.4 \pm 0.6) \times 10^{-17}$ s (S = 3.0)
$c\tau = 25.1$ nm

π^0 DECAY MODES		Fraction (Γ_i/Γ)	Scale factor/ Confidence level	p (MeV/c)
2γ		(98.798±0.032) %	S=1.1	67
$e^+ e^-\gamma$		(1.198±0.032) %	S=1.1	67
γ positronium		(1.82 ±0.29) × 10^{-9}		67
$e^+ e^+ e^- e^-$		(3.14 ±0.30) × 10^{-5}		67
$e^+ e^-$		(7.5 ±2.0) × 10^{-8}		67
4γ		< 2 × 10^{-8}	CL=90%	67
$\nu\overline{\nu}$	[e]	< 8.3 × 10^{-7}	CL=90%	67
$\nu_e\overline{\nu}_e$		< 1.7 × 10^{-6}	CL=90%	67
$\nu_\mu\overline{\nu}_\mu$		< 3.1 × 10^{-6}	CL=90%	67
$\nu_\tau\overline{\nu}_\tau$		< 2.1 × 10^{-6}	CL=90%	67

Charge conjugation (C) or Lepton Family number (LF) violating modes

3γ	C	< 3.1 × 10^{-8}	CL=90%	67
$\mu^+ e^- + e^- \mu^+$	LF	< 1.72 × 10^{-8}	CL=90%	26

η $\qquad I^G(J^{PC}) = 0^+(0^{-+})$

Mass $m = 547.30 \pm 0.12$ MeV
Full width $\Gamma = 1.18 \pm 0.11$ keV [f] (S = 1.8)

C-nonconserving decay parameters

$\pi^+\pi^-\pi^0$	Left-right asymmetry	= $(0.09 \pm 0.17) \times 10^{-2}$
$\pi^+\pi^-\pi^0$	Sextant asymmetry	= $(0.18 \pm 0.16) \times 10^{-2}$
$\pi^+\pi^-\pi^0$	Quadrant asymmetry	= $(-0.17 \pm 0.17) \times 10^{-2}$
$\pi^+\pi^-\gamma$	Left-right asymmetry	= $(0.9 \pm 0.4) \times 10^{-2}$
$\pi^+\pi^-\gamma$	β (D-wave)	= 0.05 ± 0.06 (S = 1.5)

Dalitz plot parameter

$\pi^0\pi^0\pi^0$ $\quad \alpha = -0.039 \pm 0.015$

η DECAY MODES		Fraction (Γ_i/Γ)	Scale factor/ Confidence level	p (MeV/c)
Neutral modes				
neutral modes		(71.5 ±0.6) %	S=1.4	–
2γ	[f]	(39.21±0.34) %	S=1.4	274
$3\pi^0$		(32.2 ±0.4) %	S=1.3	178
$\pi^0 2\gamma$		(7.1 ±1.4) × 10^{-4}		257
other neutral modes		< 2.8 %	CL=90%	
Charged modes				
charged modes		(28.5 ±0.6) %	S=1.4	–
$\pi^+\pi^-\pi^0$		(23.1 ±0.5) %	S=1.4	173
$\pi^+\pi^-\gamma$		(4.77±0.13) %	S=1.3	235
$e^+ e^-\gamma$		(4.9 ±1.1) × 10^{-3}		274
$\mu^+\mu^-\gamma$		(3.1 ±0.4) × 10^{-4}		252
$e^+ e^-$		< 7.7 × 10^{-5}	CL=90%	274
$\mu^+\mu^-$		(5.8 ±0.8) × 10^{-6}		252
$\pi^+\pi^- e^+ e^-$		(1.3 $^{+1.2}_{-0.8}$) × 10^{-3}		235
$\pi^+\pi^- 2\gamma$		< 2.1 × 10^{-3}		235
$\pi^+\pi^-\pi^0\gamma$		< 6 × 10^{-4}	CL=90%	173
$\pi^0\mu^+\mu^-\gamma$		< 3 × 10^{-6}	CL=90%	210

**Charge conjugation (C)ΓParity (P)Γ
Charge conjugation × Parity (CP)Γor
Lepton Family number (LF) violating modes**

$\pi^+\pi^-$	$P\Gamma CP$	< 9 × 10^{-4}	CL=90%	235
3γ	C	< 5 × 10^{-4}	CL=95%	274
$\pi^0 e^+ e^-$	C [g]	< 4 × 10^{-5}	CL=90%	257
$\pi^0\mu^+\mu^-$	C [g]	< 5 × 10^{-6}	CL=90%	210
$\mu^+ e^- + \mu^- e^+$	LF	< 6 × 10^{-6}	CL=90%	263

$f_0(400\text{–}1200)$ [h] or σ $\qquad I^G(J^{PC}) = 0^+(0^{++})$

Mass $m = (400\text{–}1200)$ MeV
Full width $\Gamma = (600\text{–}1000)$ MeV

$f_0(400\text{–}1200)$ DECAY MODES	Fraction (Γ_i/Γ)	p (MeV/c)
$\pi\pi$	dominant	–
$\gamma\gamma$	seen	–

$\rho(770)$ [j]

$$I^G(J^{PC}) = 1^+(1^{--})$$

Mass $m = 770.0 \pm 0.8$ MeV (S = 1.8)
Full width $\Gamma = 150.7 \pm 1.1$ MeV
$\Gamma_{ee} = 6.77 \pm 0.32$ keV

$\rho(770)$ DECAY MODES	Fraction (Γ_i/Γ)	Scale factor/ Confidence level	p (MeV/c)
$\pi\pi$	~ 100 %		358

$\rho(770)^\pm$ decays

$\pi^\pm\gamma$	$(4.5 \pm 0.5) \times 10^{-4}$	S=2.2	372
$\pi^\pm\eta$	$< 6 \times 10^{-3}$	CL=84%	146
$\pi^\pm\pi^+\pi^-\pi^0$	$< 2.0 \times 10^{-3}$	CL=84%	249

$\rho(770)^0$ decays

$\pi^+\pi^-\gamma$	$(9.9 \pm 1.6) \times 10^{-3}$		358
$\pi^0\gamma$	$(6.8 \pm 1.7) \times 10^{-4}$		372
$\eta\gamma$	$(2.4 ^{+0.8}_{-0.9}) \times 10^{-4}$	S=1.6	189
$\mu^+\mu^-$	[j] $(4.60 \pm 0.28) \times 10^{-5}$		369
e^+e^-	[j] $(4.49 \pm 0.22) \times 10^{-5}$		384
$\pi^+\pi^-\pi^0$	$< 1.2 \times 10^{-4}$	CL=90%	319
$\pi^+\pi^-\pi^+\pi^-$	$< 2 \times 10^{-4}$	CL=90%	246
$\pi^+\pi^-\pi^0\pi^0$	$< 4 \times 10^{-5}$	CL=90%	252

$\omega(782)$

$$I^G(J^{PC}) = 0^-(1^{--})$$

Mass $m = 781.94 \pm 0.12$ MeV (S = 1.5)
Full width $\Gamma = 8.41 \pm 0.09$ MeV
$\Gamma_{ee} = 0.60 \pm 0.02$ keV

$\omega(782)$ DECAY MODES	Fraction (Γ_i/Γ)	Scale factor/ Confidence level	p (MeV/c)
$\pi^+\pi^-\pi^0$	$(88.8 \pm 0.7$) %		327
$\pi^0\gamma$	$(8.5 \pm 0.5$) %		379
$\pi^+\pi^-$	(2.21 ± 0.30) %		365
neutrals (excluding $\pi^0\gamma$)	$(5.3 ^{+8.7}_{-3.5}) \times 10^{-3}$		–
$\eta\gamma$	$(6.5 \pm 1.0) \times 10^{-4}$		199
$\pi^0 e^+ e^-$	$(5.9 \pm 1.9) \times 10^{-4}$		379
$\pi^0\mu^+\mu^-$	$(9.6 \pm 2.3) \times 10^{-5}$		349
e^+e^-	$(7.07 \pm 0.19) \times 10^{-5}$	S=1.1	391
$\pi^+\pi^-\pi^0\pi^0$	< 2 %	CL=90%	261
$\pi^+\pi^-\gamma$	$< 3.6 \times 10^{-3}$	CL=95%	365
$\pi^+\pi^-\pi^+\pi^-$	$< 1 \times 10^{-3}$	CL=90%	256
$\pi^0\pi^0\gamma$	$(7.2 \pm 2.5) \times 10^{-5}$		367
$\mu^+\mu^-$	$< 1.8 \times 10^{-4}$	CL=90%	376
3γ	$< 1.9 \times 10^{-4}$	CL=95%	391

Charge conjugation (C) violating modes

$\eta\pi^0$	C $< 1 \times 10^{-3}$	CL=90%	162
$3\pi^0$	C $< 3 \times 10^{-4}$	CL=90%	329

$\eta'(958)$

$$I^G(J^{PC}) = 0^+(0^{-+})$$

Mass $m = 957.78 \pm 0.14$ MeV
Full width $\Gamma = 0.203 \pm 0.016$ MeV (S = 1.3)

$\eta'(958)$ DECAY MODES	Fraction (Γ_i/Γ)	Scale factor/ Confidence level	p (MeV/c)
$\pi^+\pi^-\eta$	$(43.8 \pm 1.5$) %	S=1.1	232
$\rho^0\gamma$ (including non-resonant $\pi^+\pi^-\gamma$)	$(30.2 \pm 1.3$) %	S=1.1	169
$\pi^0\pi^0\eta$	$(20.7 \pm 1.3$) %	S=1.2	239
$\omega\gamma$	(3.01 ± 0.30) %		160
$\gamma\gamma$	(2.11 ± 0.13) %	S=1.2	479
$3\pi^0$	$(1.54 \pm 0.26) \times 10^{-3}$		430
$\mu^+\mu^-\gamma$	$(1.03 \pm 0.26) \times 10^{-4}$		467
$\pi^+\pi^-\pi^0$	< 5 %	CL=90%	427
$\pi^0\pi^0$	< 4 %	CL=90%	118
$\pi^+\pi^+\pi^-\pi^-$	< 1 %	CL=90%	372
$\pi^+\pi^+\pi^-\pi^-$ neutrals	< 1 %	CL=95%	–
$\pi^+\pi^+\pi^-\pi^-\pi^0$	< 1 %	CL=90%	298
6π	< 1 %	CL=90%	189
$\pi^+\pi^- e^+ e^-$	$< 6 \times 10^{-3}$	CL=90%	458
$\pi^0\gamma\gamma$	$< 8 \times 10^{-4}$	CL=90%	469
$4\pi^0$	$< 5 \times 10^{-4}$	CL=90%	379
e^+e^-	$< 2.1 \times 10^{-7}$	CL=90%	479

Charge conjugation (C) or Parity (P) violating modes

$\pi^+\pi^-$	P,CP	< 2 %	CL=90%	458
$\pi^0\pi^0$	P,CP	$< 9 \times 10^{-4}$	CL=90%	459
$\pi^0 e^+ e^-$	C [g] < 1.3 %		CL=90%	469
$\eta e^+ e^-$	C [g] < 1.1 %		CL=90%	322
3γ	C $< 1.0 \times 10^{-4}$		CL=90%	479
$\mu^+\mu^-\pi^0$	C [g] $< 6.0 \times 10^{-5}$		CL=90%	445
$\mu^+\mu^-\eta$	C [g] $< 1.5 \times 10^{-5}$		CL=90%	274

$f_0(980)$ [k]

$$I^G(J^{PC}) = 0^+(0^{++})$$

Mass $m = 980 \pm 10$ MeV
Full width $\Gamma = 40$ to 100 MeV

$f_0(980)$ DECAY MODES	Fraction (Γ_i/Γ)	Confidence level	p (MeV/c)
$\pi\pi$	dominant		470
$K\overline{K}$	seen		–
$\gamma\gamma$	$(1.19 \pm 0.33) \times 10^{-5}$		490
e^+e^-	$< 3 \times 10^{-7}$	90%	490

$a_0(980)$ [k]

$$I^G(J^{PC}) = 1^-(0^{++})$$

Mass $m = 983.4 \pm 0.9$ MeV
Full width $\Gamma = 50$ to 100 MeV

$a_0(980)$ DECAY MODES	Fraction (Γ_i/Γ)	p (MeV/c)
$\eta\pi$	dominant	321
$K\overline{K}$	seen	–
$\gamma\gamma$	seen	492

$\phi(1020)$

$$I^G(J^{PC}) = 0^-(1^{--})$$

Mass $m = 1019.413 \pm 0.008$ MeV
Full width $\Gamma = 4.43 \pm 0.05$ MeV

$\phi(1020)$ DECAY MODES	Fraction (Γ_i/Γ)	Scale factor/ Confidence level	p (MeV/c)
K^+K^-	$(49.1 \pm 0.8$) %	S=1.3	127
$K^0_L K^0_S$	$(34.1 \pm 0.6$) %	S=1.2	110
$\rho\pi + \pi^+\pi^-\pi^0$	$(15.5 \pm 0.7$) %	S=1.5	–
$\eta\gamma$	(1.26 ± 0.06) %	S=1.1	363
$\pi^0\gamma$	$(1.31 \pm 0.13) \times 10^{-3}$		501
e^+e^-	$(2.99 \pm 0.08) \times 10^{-4}$	S=1.2	510
$\mu^+\mu^-$	$(2.5 \pm 0.4) \times 10^{-4}$		499
$\eta e^+ e^-$	$(1.3 ^{+0.8}_{-0.6}) \times 10^{-4}$		363
$\pi^+\pi^-$	$(8 ^{+5}_{-4}) \times 10^{-5}$	S=1.5	490
$\omega\gamma$	< 5 %	CL=84%	210
$\rho\gamma$	$< 7 \times 10^{-4}$	CL=90%	219
$\pi^+\pi^-\gamma$	$< 3 \times 10^{-5}$	CL=90%	490
$f_0(980)\gamma$	$< 1 \times 10^{-4}$	CL=90%	39
$\pi^0\pi^0\gamma$	$< 1 \times 10^{-3}$	CL=90%	492
$\pi^+\pi^-\pi^+\pi^-$	$< 8.7 \times 10^{-4}$	CL=90%	410
$\pi^+\pi^+\pi^-\pi^-\pi^0$	$< 1.5 \times 10^{-4}$	CL=95%	341
$\pi^0 e^+ e^-$	$< 1.2 \times 10^{-4}$	CL=90%	501
$\pi^0\eta\gamma$	$< 2.5 \times 10^{-3}$	CL=90%	346
$a_0(980)\gamma$	$< 5 \times 10^{-3}$	CL=90%	36
$\eta'(958)\gamma$	$(1.2 ^{+0.7}_{-0.5}) \times 10^{-4}$		–
$\mu^+\mu^-\gamma$	$(2.3 \pm 1.0) \times 10^{-5}$		–

$h_1(1170)$

$$I^G(J^{PC}) = 0^-(1^{+-})$$

Mass $m = 1170 \pm 20$ MeV
Full width $\Gamma = 360 \pm 40$ MeV

$h_1(1170)$ DECAY MODES	Fraction (Γ_i/Γ)	p (MeV/c)
$\rho\pi$	seen	310

$b_1(1235)$ $\qquad I^G(J^{PC}) = 1^+(1^{+-})$

Mass $m = 1229.5 \pm 3.2$ MeV (S = 1.6)
Full width $\Gamma = 142 \pm 9$ MeV (S = 1.2)

$b_1(1235)$ DECAY MODES	Fraction (Γ_i/Γ)	Confidence level	p (MeV/c)
$\omega\pi$	dominant		348
[D/S amplitude ratio $= 0.29 \pm 0.04$]			
$\pi^\pm\gamma$	$(1.6\pm0.4)\times10^{-3}$		608
$\eta\rho$	seen		–
$\pi^+\pi^+\pi^-\pi^0$	< 50 %	84%	536
$(K\overline{K})^\pm\pi^0$	< 8 %	90%	248
$K^0_S K^0_L \pi^\pm$	< 6 %	90%	238
$K^0_S K^0_S \pi^\pm$	< 2 %	90%	238
$\phi\pi$	< 1.5 %	84%	146

$a_1(1260)$ [l] $\qquad I^G(J^{PC}) = 1^-(1^{++})$

Mass $m = 1230 \pm 40$ MeV [m]
Full width $\Gamma = 250$ to 600 MeV

$a_1(1260)$ DECAY MODES	Fraction (Γ_i/Γ)	p (MeV/c)
$\rho\pi$	dominant	356
[D/S amplitude ratio $= -0.100 \pm 0.028$]		
$\pi\gamma$	seen	607
$\pi(\pi\pi)_{S\text{-wave}}$	possibly seen	575

$f_2(1270)$ $\qquad I^G(J^{PC}) = 0^+(2^{++})$

Mass $m = 1275.0 \pm 1.2$ MeV
Full width $\Gamma = 185.5^{+3.8}_{-2.7}$ MeV (S = 1.5)

$f_2(1270)$ DECAY MODES	Fraction (Γ_i/Γ)	Scale factor/ Confidence level	p (MeV/c)
$\pi\pi$	$(84.6^{+2.5}_{-1.3})$ %	S=1.3	622
$\pi^+\pi^-2\pi^0$	$(7.2^{+1.5}_{-2.7})$ %	S=1.3	562
$K\overline{K}$	(4.6 ± 0.4) %	S=2.8	403
$2\pi^+2\pi^-$	(2.8 ± 0.4) %	S=1.2	559
$\eta\eta$	$(4.5\pm1.0)\times10^{-3}$	S=2.4	327
$4\pi^0$	$(3.0\pm1.0)\times10^{-3}$		564
$\gamma\gamma$	$(1.32^{+0.17}_{-0.16})\times10^{-5}$		637
$\eta\pi\pi$	$< 8 \times10^{-3}$	CL=95%	475
$K^0 K^-\pi^+ +$ c.c.	$< 3.4 \times10^{-3}$	CL=95%	293
e^+e^-	$< 9 \times10^{-9}$	CL=90%	637

$f_1(1285)$ $\qquad I^G(J^{PC}) = 0^+(1^{++})$

Mass $m = 1281.9 \pm 0.6$ MeV (S = 1.7)
Full width $\Gamma = 24.0 \pm 1.2$ MeV (S = 1.4)

$(4\pi = \rho(\pi\pi)_{P\text{wave}})$

$f_1(1285)$ DECAY MODES	Fraction (Γ_i/Γ)	Scale factor/ Confidence level	p (MeV/c)
4π	(35 ± 4) %	S=1.6	563
$\pi^0\pi^0\pi^+\pi^-$	(23.5 ± 3.0) %	S=1.6	566
$2\pi^+2\pi^-$	(11.7 ± 1.5) %	S=1.6	563
$\rho^0\pi^+\pi^-$	(11.7 ± 1.5) %	S=1.6	340
$4\pi^0$	$< 7 \times10^{-4}$	CL=90%	568
$\eta\pi\pi$	(50 ± 18) %		479
$a_0(980)\pi$ [ignoring $a_0(980)\to K\overline{K}$]	(34 ± 8) %	S=1.2	234
$\eta\pi\pi$ [excluding $a_0(980)\pi$]	(15 ± 7) %	S=1.1	–
$K\overline{K}\pi$	(9.6 ± 1.2) %	S=1.5	308
$K\overline{K}^*(892)$	not seen		–
$\gamma\rho^0$	(5.4 ± 1.2) %	S=2.3	410
$\phi\gamma$	$(7.9\pm3.0)\times10^{-4}$		236

$\eta(1295)$ $\qquad I^G(J^{PC}) = 0^+(0^{-+})$

Mass $m = 1297.0 \pm 2.8$ MeV
Full width $\Gamma = 53 \pm 6$ MeV

$\eta(1295)$ DECAY MODES	Fraction (Γ_i/Γ)	p (MeV/c)
$\eta\pi^+\pi^-$	seen	488
$a_0(980)\pi$	seen	245
$\eta\pi^0\pi^0$	seen	–
$\eta(\pi\pi)_{S\text{-wave}}$	seen	–

$\pi(1300)$ $\qquad I^G(J^{PC}) = 1^-(0^{-+})$

Mass $m = 1300 \pm 100$ MeV [m]
Full width $\Gamma = 200$ to 600 MeV

$\pi(1300)$ DECAY MODES	Fraction (Γ_i/Γ)	p (MeV/c)
$\rho\pi$	seen	406
$\pi(\pi\pi)_{S\text{-wave}}$	seen	–

$a_2(1320)$ $\qquad I^G(J^{PC}) = 1^-(2^{++})$

Mass $m = 1318.1 \pm 0.6$ MeV (S = 1.1)
Full width $\Gamma = 107 \pm 5$ MeV [m] ($K^\pm K^0_S$ and $\eta\pi$ modes)

$a_2(1320)$ DECAY MODES	Fraction (Γ_i/Γ)	Scale factor/ Confidence level	p (MeV/c)
$\rho\pi$	(70.1 ± 2.7) %	S=1.2	419
$\eta\pi$	(14.5 ± 1.2) %		535
$\omega\pi\pi$	(10.6 ± 3.2) %	S=1.3	362
$K\overline{K}$	(4.9 ± 0.8) %		437
$\eta'(958)\pi$	$(5.3\pm0.9)\times10^{-3}$		287
$\pi^\pm\gamma$	$(2.8\pm0.6)\times10^{-3}$		652
$\gamma\gamma$	$(9.4\pm0.7)\times10^{-6}$		659
$\pi^+\pi^-\pi^-$	< 8 %	CL=90%	621
e^+e^-	$< 2.3 \times10^{-7}$	CL=90%	659

$f_0(1370)$ [k] $\qquad I^G(J^{PC}) = 0^+(0^{++})$

Mass $m = 1200$ to 1500 MeV
Full width $\Gamma = 200$ to 500 MeV

$f_0(1370)$ DECAY MODES	Fraction (Γ_i/Γ)	p (MeV/c)
$\pi\pi$	seen	–
4π	seen	–
$4\pi^0$	seen	–
$2\pi^+2\pi^-$	seen	–
$\pi^+\pi^-2\pi^0$	seen	–
$2(\pi\pi)_{S\text{-wave}}$	seen	–
$\eta\eta$	seen	–
$K\overline{K}$	seen	–
$\gamma\gamma$	seen	–
e^+e^-	not seen	–

$f_1(1420)$ [n] $\qquad I^G(J^{PC}) = 0^+(1^{++})$

Mass $m = 1426.2 \pm 1.2$ MeV (S = 1.3)
Full width $\Gamma = 55.0 \pm 3.0$ MeV

$f_1(1420)$ DECAY MODES	Fraction (Γ_i/Γ)	p (MeV/c)
$K\overline{K}\pi$	dominant	439
$K\overline{K}^*(892) +$ c.c.	dominant	155
$\eta\pi\pi$	possibly seen	571

$\omega(1420)$ [o] $\qquad I^G(J^{PC}) = 0^-(1^{--})$

Mass $m = 1419 \pm 31$ MeV
Full width $\Gamma = 174 \pm 60$ MeV

$\omega(1420)$ DECAY MODES	Fraction (Γ_i/Γ)	p (MeV/c)
$\rho\pi$	dominant	488

$\eta(1440)$ [p] $I^G(J^{PC}) = 0^+(0^{-+})$

Mass $m = 1400 - 1470$ MeV [m]
Full width $\Gamma = 50 - 80$ MeV [m]

$\eta(1440)$ DECAY MODES	Fraction (Γ_i/Γ)	p (MeV/c)
$K\overline{K}\pi$	seen	–
$K\overline{K}^*(892)+$ c.c.	seen	–
$\eta\pi\pi$	seen	–
$\quad a_0(980)\pi$	seen	–
$\quad \eta(\pi\pi)_{S\text{-wave}}$	seen	–
4π	seen	–

$a_0(1450)$ $I^G(J^{PC}) = 1^-(0^{++})$

Mass $m = 1474 \pm 19$ MeV
Full width $\Gamma = 265 \pm 13$ MeV

$a_0(1450)$ DECAY MODES	Fraction (Γ_i/Γ)	p (MeV/c)
$\pi\eta$	seen	613
$\pi\eta'(958)$	seen	392
$K\overline{K}$	seen	530

$\rho(1450)$ [q] $I^G(J^{PC}) = 1^+(1^{--})$

Mass $m = 1465 \pm 25$ MeV [m]
Full width $\Gamma = 310 \pm 60$ MeV [m]

$\rho(1450)$ DECAY MODES	Fraction (Γ_i/Γ)	Confidence level	p (MeV/c)
$\pi\pi$	seen		719
4π	seen		665
$\omega\pi$	<2.0 %	95%	512
e^+e^-	seen		732
$\eta\rho$	<4 %		317
$\phi\pi$	<1 %		358
$K\overline{K}$	$<1.6 \times 10^{-3}$	95%	541

$f_0(1500)$ [r] $I^G(J^{PC}) = 0^+(0^{++})$

Mass $m = 1500 \pm 10$ MeV   (S = 1.3)
Full width $\Gamma = 112 \pm 10$ MeV

$f_0(1500)$ DECAY MODES	Fraction (Γ_i/Γ)	p (MeV/c)
$\eta\eta'(958)$	seen	–
$\eta\eta$	seen	513
4π	seen	–
$\quad 4\pi^0$	seen	690
$\quad 2\pi^+2\pi^-$	seen	686
2π	seen	–
$\quad \pi^+\pi^-$	seen	737
$\quad 2\pi^0$	seen	738
$K\overline{K}$	seen	563

$f_2'(1525)$ $I^G(J^{PC}) = 0^+(2^{++})$

Mass $m = 1525 \pm 5$ MeV [m]
Full width $\Gamma = 76 \pm 10$ MeV [m]

$f_2'(1525)$ DECAY MODES	Fraction (Γ_i/Γ)	p (MeV/c)
$K\overline{K}$	(88.8 ± 3.1) %	581
$\eta\eta$	(10.3 ± 3.1) %	531
$\pi\pi$	$(8.2 \pm 1.5) \times 10^{-3}$	750
$\gamma\gamma$	$(1.32 \pm 0.21) \times 10^{-6}$	763

$\omega(1600)$ [s] $I^G(J^{PC}) = 0^-(1^{--})$

Mass $m = 1649 \pm 24$ MeV   (S = 2.3)
Full width $\Gamma = 220 \pm 35$ MeV   (S = 1.6)

$\omega(1600)$ DECAY MODES	Fraction (Γ_i/Γ)	p (MeV/c)
$\rho\pi$	seen	637
$\omega\pi\pi$	seen	601
e^+e^-	seen	824

$\omega_3(1670)$ $I^G(J^{PC}) = 0^-(3^{--})$

Mass $m = 1667 \pm 4$ MeV
Full width $\Gamma = 168 \pm 10$ MeV [m]

$\omega_3(1670)$ DECAY MODES	Fraction (Γ_i/Γ)	p (MeV/c)
$\rho\pi$	seen	647
$\omega\pi\pi$	seen	614
$b_1(1235)\pi$	possibly seen	359

$\pi_2(1670)$ $I^G(J^{PC}) = 1^-(2^{-+})$

Mass $m = 1670 \pm 20$ MeV [m]
Full width $\Gamma = 258 \pm 18$ MeV [m]   (S = 1.7)

$\pi_2(1670)$ DECAY MODES	Fraction (Γ_i/Γ)	p (MeV/c)
3π	(95.8 ± 1.4) %	806
$f_2(1270)\pi$	(56.2 ± 3.2) %	325
$\rho\pi$	(31 ± 4) %	649
$f_0(1370)\pi$	(8.7 ± 3.4) %	–
$K\overline{K}^*(892)+$ c.c.	(4.2 ± 1.4) %	453

$\phi(1680)$ $I^G(J^{PC}) = 0^-(1^{--})$

Mass $m = 1680 \pm 20$ MeV [m]
Full width $\Gamma = 150 \pm 50$ MeV [m]

$\phi(1680)$ DECAY MODES	Fraction (Γ_i/Γ)	p (MeV/c)
$K\overline{K}^*(892)+$ c.c.	dominant	463
$K_S^0 K\pi$	seen	620
$K\overline{K}$	seen	681
e^+e^-	seen	840
$\omega\pi\pi$	not seen	622

$\rho_3(1690)$ $I^G(J^{PC}) = 1^+(3^{--})$

J^P from the 2π and $K\overline{K}$ modes.
Mass $m = 1691 \pm 5$ MeV [m]
Full width $\Gamma = 160 \pm 10$ MeV [m]   (S = 1.5)

$\rho_3(1690)$ DECAY MODES	Fraction (Γ_i/Γ)	Scale factor	p (MeV/c)
4π	(71.1 ± 1.9) %		788
$\quad \pi^\pm\pi^+\pi^-\pi^0$	(67 ± 22) %		788
$\quad \omega\pi$	(16 ± 6) %		656
$\pi\pi$	(23.6 ± 1.3) %		834
$K\overline{K}\pi$	(3.8 ± 1.2) %		628
$K\overline{K}$	(1.58 ± 0.26) %	1.2	686
$\eta\pi^+\pi^-$	seen		728

$\rho(1700)$ [q] $I^G(J^{PC}) = 1^+(1^{--})$

Mass $m = 1700 \pm 20$ MeV [m]   ($\eta\rho^0$ and $\pi^+\pi^-$ modes)
Full width $\Gamma = 240 \pm 60$ MeV [m]   ($\eta\rho^0$ and $\pi^+\pi^-$ modes)

$\rho(1700)$ DECAY MODES	Fraction (Γ_i/Γ)	p (MeV/c)
$\rho\pi\pi$	dominant	640
$2(\pi^+\pi^-)$	large	792
$\rho^0\pi^+\pi^-$	large	640
$\rho^\pm\pi^\mp\pi^0$	large	642
$\pi^+\pi^-$	seen	838
$\pi^-\pi^0$	seen	839
$K\overline{K}^*(892)+$ c.c.	seen	479
$\eta\rho$	seen	533
$K\overline{K}$	seen	692
e^+e^-	seen	850
$\pi^0\omega$	seen	662

$f_J(1710)$ [t] $I^G(J^{PC}) = 0^+(\text{even}^{++})$

Mass $m = 1712 \pm 5$ MeV   (S = 1.1)
Full width $\Gamma = 133 \pm 14$ MeV   (S = 1.2)

$f_J(1710)$ DECAY MODES	Fraction (Γ_i/Γ)	p (MeV/c)
$K\overline{K}$	seen	690
$\eta\eta$	seen	648
$\pi\pi$	seen	837

$\pi(1800)$ $I^G(J^{PC}) = 1^-(0^{-+})$

Mass $m = 1801 \pm 13$ MeV   (S = 1.9)
Full width $\Gamma = 210 \pm 15$ MeV

$\pi(1800)$ DECAY MODES	Fraction (Γ_i/Γ)	p (MeV/c)
$\pi^+\pi^-\pi^-$	seen	–
$f_0(980)\pi^-$	seen	623
$f_0(1370)\pi^-$	seen	–
$\rho\pi^-$	not seen	728
$\eta\eta\pi^-$	seen	–
$a_0(980)\eta$	seen	459
$f_0(1500)\pi^-$	seen	240
$\eta\eta'(958)\pi^-$	seen	–
$K_0^*(1430)K^-$	seen	–
$K^*(892)K^-$	not seen	560

$\phi_3(1850)$ $I^G(J^{PC}) = 0^-(3^{--})$

Mass $m = 1854 \pm 7$ MeV
Full width $\Gamma = 87^{+28}_{-23}$ MeV   (S = 1.2)

$\phi_3(1850)$ DECAY MODES	Fraction (Γ_i/Γ)	p (MeV/c)
$K\overline{K}$	seen	785
$K\overline{K}^*(892)+$ c.c.	seen	602

$f_2(2010)$ $I^G(J^{PC}) = 0^+(2^{++})$

Seen by one group only.
Mass $m = 2011^{+60}_{-80}$ MeV
Full width $\Gamma = 202 \pm 60$ MeV

$f_2(2010)$ DECAY MODES	Fraction (Γ_i/Γ)	p (MeV/c)
$\phi\phi$	seen	–

$a_4(2040)$ $I^G(J^{PC}) = 1^-(4^{++})$

Mass $m = 2020 \pm 16$ MeV
Full width $\Gamma = 387 \pm 70$ MeV

$a_4(2040)$ DECAY MODES	Fraction (Γ_i/Γ)	p (MeV/c)
$K\overline{K}$	seen	892
$\pi^+\pi^-\pi^0$	seen	–
$\eta\pi^0$	seen	941

$f_4(2050)$ $I^G(J^{PC}) = 0^+(4^{++})$

Mass $m = 2044 \pm 11$ MeV   (S = 1.4)
Full width $\Gamma = 208 \pm 13$ MeV   (S = 1.2)

$f_4(2050)$ DECAY MODES	Fraction (Γ_i/Γ)	p (MeV/c)
$\omega\omega$	(26 ± 6) %	658
$\pi\pi$	(17.0 ± 1.5) %	1012
$K\overline{K}$	$(6.8^{+3.4}_{-1.8}) \times 10^{-3}$	895
$\eta\eta$	$(2.1 \pm 0.8) \times 10^{-3}$	863
$4\pi^0$	< 1.2 %	977

$f_2(2300)$ $I^G(J^{PC}) = 0^+(2^{++})$

Mass $m = 2297 \pm 28$ MeV
Full width $\Gamma = 149 \pm 40$ MeV

$f_2(2300)$ DECAY MODES	Fraction (Γ_i/Γ)	p (MeV/c)
$\phi\phi$	seen	529

$f_2(2340)$ $I^G(J^{PC}) = 0^+(2^{++})$

Mass $m = 2339 \pm 60$ MeV
Full width $\Gamma = 319^{+80}_{-70}$ MeV

$f_2(2340)$ DECAY MODES	Fraction (Γ_i/Γ)	p (MeV/c)
$\phi\phi$	seen	573

STRANGE MESONS
$(S = \pm 1, C = B = 0)$

$K^+ = u\bar{s}$, $K^0 = d\bar{s}$, $\overline{K^0} = \bar{d}s$, $K^- = \bar{u}s$, similarly for K^*'s

$K^\pm$　　　　　　　　　　　　$I(J^P) = \frac{1}{2}(0^-)$

Mass $m = 493.677 \pm 0.016$ MeV [u]　(S = 2.8)
Mean life $\tau = (1.2386 \pm 0.0024) \times 10^{-8}$ s　(S = 2.0)
　　$c\tau = 3.713$ m

Slope parameter g [v]

(See Particle Listings for quadratic coefficients)
　$K^+ \to \pi^+\pi^+\pi^- = -0.2154 \pm 0.0035$　(S = 1.4)
　$K^- \to \pi^-\pi^-\pi^+ = -0.217 \pm 0.007$　(S = 2.5)
　$K^\pm \to \pi^\pm\pi^0\pi^0 = 0.594 \pm 0.019$　(S = 1.3)

$K^\pm$ decay form factors [a,w]

K^+_{e3}　$\lambda_+ = 0.0286 \pm 0.0022$
$K^+_{\mu3}$　$\lambda_+ = 0.032 \pm 0.008$　(S = 1.6)
$K^+_{\mu3}$　$\lambda_0 = 0.006 \pm 0.007$　(S = 1.6)
K^+_{e3}　$|f_S/f_+| = 0.084 \pm 0.023$　(S = 1.2)
K^+_{e3}　$|f_T/f_+| = 0.38 \pm 0.11$　(S = 1.1)
$K^+_{\mu3}$　$|f_T/f_+| = 0.02 \pm 0.12$
$K^+ \to e^+\nu_e\gamma$　$|F_A + F_V| = 0.148 \pm 0.010$
$K^+ \to \mu^+\nu_\mu\gamma$　$|F_A + F_V| < 0.23$ CL = 90%
$K^+ \to e^+\nu_e\gamma$　$|F_A - F_V| < 0.49$
$K^+ \to \mu^+\nu_\mu\gamma$　$|F_A - F_V| = -2.2$ to 0.3

K^- modes are charge conjugates of the modes below.

K^+ DECAY MODES		Fraction (Γ_i/Γ)	Scale factor/ Confidence level	p (MeV/c)
$\mu^+\nu_\mu$		(63.51 ± 0.18) %	S=1.3	236
$e^+\nu_e$		$(1.55\pm0.07)\times10^{-5}$		247
$\pi^+\pi^0$		(21.16 ± 0.14) %	S=1.1	205
$\pi^+\pi^+\pi^-$		(5.59 ± 0.05) %	S=1.8	125
$\pi^+\pi^0\pi^0$		(1.73 ± 0.04) %	S=1.2	133
$\pi^0\mu^+\nu_\mu$		(3.18 ± 0.08) %	S=1.5	215
Called $K^+_{\mu3}$.				
$\pi^0 e^+\nu_e$		(4.82 ± 0.06) %	S=1.3	228
Called K^+_{e3}.				
$\pi^0\pi^0 e^+\nu_e$		$(2.1\pm0.4)\times10^{-5}$		206
$\pi^+\pi^- e^+\nu_e$		$(3.91\pm0.17)\times10^{-5}$		203
$\pi^+\pi^-\mu^+\nu_\mu$		$(1.4\pm0.9)\times10^{-5}$		151
$\pi^0\pi^0\pi^0 e^+\nu_e$		$< 3.5 \times10^{-6}$	CL=90%	135
$\pi^+\gamma\gamma$	[x]	$(1.10\pm0.32)\times10^{-6}$		227
$\pi^+3\gamma$	[x]	$< 1.0 \times10^{-4}$	CL=90%	227
$\mu^+\nu_\mu\nu\bar{\nu}$		$< 6.0 \times10^{-6}$	CL=90%	236
$e^+\nu_e\nu\bar{\nu}$		$< 6 \times10^{-5}$	CL=90%	247
$\mu^+\nu_\mu e^+ e^-$		$(1.3\pm0.4)\times10^{-7}$		236
$e^+\nu_e e^+ e^-$		$(3.0^{+3.0}_{-1.5})\times10^{-8}$		247
$\mu^+\nu_\mu\mu^+\mu^-$		$< 4.1 \times10^{-7}$	CL=90%	185
$\mu^+\nu_\mu\gamma$	[x,y]	$(5.50\pm0.28)\times10^{-3}$		236
$\pi^+\pi^0\gamma$	[x,y]	$(2.75\pm0.15)\times10^{-4}$		205
$\pi^+\pi^0\gamma$(DE)	[x,z]	$(1.8\pm0.4)\times10^{-5}$		205
$\pi^+\pi^+\pi^-\gamma$	[x,y]	$(1.04\pm0.31)\times10^{-4}$		125
$\pi^+\pi^0\pi^0\gamma$	[x,y]	$(7.5^{+5.5}_{-3.0})\times10^{-6}$		133
$\pi^0\mu^+\nu_\mu\gamma$	[x,y]	$< 6.1 \times10^{-5}$	CL=90%	215
$\pi^0 e^+\nu_e\gamma$	[x,y]	$(2.62\pm0.20)\times10^{-4}$		228
$\pi^0 e^+\nu_e\gamma$(SD)	[aa]	$< 5.3 \times10^{-5}$	CL=90%	228
$\pi^0\pi^0 e^+\nu_e\gamma$		$< 5 \times10^{-6}$	CL=90%	206

Lepton Family number (LF), Lepton number (L), $\Delta S = \Delta Q$ (SQ) violating modes, or $\Delta S = 1$ weak neutral current ($S1$) modes

$\pi^+\pi^+ e^-\bar{\nu}_e$	SQ	$< 1.2 \times10^{-8}$	CL=90%	203
$\pi^+\pi^+\mu^-\bar{\nu}_\mu$	SQ	$< 3.0 \times10^{-6}$	CL=95%	151
$\pi^+ e^+ e^-$	S1	$(2.74\pm0.23)\times10^{-7}$		227
$\pi^+\mu^+\mu^-$	S1	$(5.0\pm1.0)\times10^{-8}$		172
$\pi^+\nu\bar{\nu}$	S1	$(4.2^{+9.7}_{-3.5})\times10^{-10}$		227
$\mu^-\nu e^+ e^+$	LF	$< 2.0 \times10^{-8}$	CL=90%	236
$\mu^+\nu_e$	LF	[d] $< 4 \times10^{-3}$	CL=90%	236
$\pi^+\mu^+ e^-$	LF	$< 2.1 \times10^{-10}$	CL=90%	214
$\pi^+\mu^- e^+$	LF	$< 7 \times10^{-9}$	CL=90%	214
$\pi^-\mu^+ e^+$	L	$< 7 \times10^{-9}$	CL=90%	214
$\pi^- e^+ e^+$	L	$< 1.0 \times10^{-8}$	CL=90%	227
$\pi^-\mu^+\mu^+$	L	[d] $< 1.5 \times10^{-4}$	CL=90%	172
$\mu^+\bar{\nu}_e$	L	[d] $< 3.3 \times10^{-3}$	CL=90%	236
$\pi^0 e^+\bar{\nu}_e$	L	$< 3 \times10^{-3}$	CL=90%	228

K^0　　　　　　　　　　　　$I(J^P) = \frac{1}{2}(0^-)$

50% K_S, 50% K_L
　Mass $m = 497.672 \pm 0.031$ MeV
　$m_{K^0} - m_{K^\pm} = 3.995 \pm 0.034$ MeV　(S = 1.1)
　$|m_{K^0} - m_{\overline{K^0}}| / m_{average} < 10^{-18}$ [bb]

K^0_S　　　　　　　　　　　　$I(J^P) = \frac{1}{2}(0^-)$

Mean life $\tau = (0.8934 \pm 0.0008) \times 10^{-10}$ s
　$c\tau = 2.6762$ cm

CP-violation parameters [cc]
　$\mathrm{Im}(\eta_{+-0}) = -0.002 \pm 0.008$
　$\mathrm{Im}(\eta_{000})^2 < 0.1$ CL = 90%

K^0_S DECAY MODES		Fraction (Γ_i/Γ)	Scale factor/ Confidence level	p (MeV/c)
$\pi^+\pi^-$		(68.61 ± 0.28) %	S=1.2	206
$\pi^0\pi^0$		(31.39 ± 0.28) %	S=1.2	209
$\pi^+\pi^-\gamma$	[y,dd]	$(1.78\pm0.05)\times10^{-3}$		206
$\gamma\gamma$		$(2.4\pm0.9)\times10^{-6}$		249
$\pi^+\pi^-\pi^0$		$(3.4^{+1.1}_{-0.9})\times10^{-7}$		133
$3\pi^0$		$< 3.7 \times10^{-5}$	CL=90%	139
$\pi^\pm e^\mp\nu$	[ee]	$(6.70\pm0.07)\times10^{-4}$	S=1.1	229
$\pi^\pm\mu^\mp\nu$	[ee]	$(4.69\pm0.06)\times10^{-4}$	S=1.1	216

$\Delta S = 1$ weak neutral current ($S1$) modes

$\mu^+\mu^-$	S1	$< 3.2 \times10^{-7}$	CL=90%	225
$e^+ e^-$	S1	$< 1.4 \times10^{-7}$	CL=90%	249
$\pi^0 e^+ e^-$	S1	$< 1.1 \times10^{-6}$	CL=90%	231

K^0_L　　　　　　　　　　　　$I(J^P) = \frac{1}{2}(0^-)$

$m_{K_L} - m_{K_S} = (0.5301 \pm 0.0014) \times 10^{10}\ \hbar\ \mathrm{s}^{-1}$
　　　　　　　$= (3.489 \pm 0.009) \times 10^{-12}$ MeV
Mean life $\tau = (5.17 \pm 0.04) \times 10^{-8}$ s　(S = 1.1)
　$c\tau = 15.51$ m

Slope parameter g [v]

(See Particle Listings for quadratic coefficients)
　$K^0_L \to \pi^+\pi^-\pi^0 = 0.670 \pm 0.014$　(S = 1.6)

K_L decay form factors [w]

K^0_{e3}　$\lambda_+ = 0.0300 \pm 0.0016$　(S = 1.2)
$K^0_{\mu3}$　$\lambda_+ = 0.034 \pm 0.005$　(S = 2.3)
$K^0_{\mu3}$　$\lambda_0 = 0.025 \pm 0.006$　(S = 2.3)
K^0_{e3}　$|f_S/f_+| < 0.04$ CL = 68%
K^0_{e3}　$|f_T/f_+| < 0.23$ CL = 68%
$K^0_{\mu3}$　$|f_T/f_+| = 0.12 \pm 0.12$
$K_L \to e^+ e^-\gamma$:　$\alpha_{K^*} = -0.28 \pm 0.08$

CP-violation parameters [cc]

$\delta = (0.327 \pm 0.012)\%$

$|\eta_{00}| = (2.275 \pm 0.019) \times 10^{-3}$ (S = 1.1)

$|\eta_{+-}| = (2.285 \pm 0.019) \times 10^{-3}$

$|\eta_{00}/\eta_{+-}| = 0.9956 \pm 0.0023$ [ff]   (S = 1.8)

$\epsilon'/\epsilon = (1.5 \pm 0.8) \times 10^{-3}$ [ff]   (S = 1.8)

$\phi_{+-} = (43.5 \pm 0.6)^{\circ}$

$\phi_{00} = (43.4 \pm 1.0)^{\circ}$

$\phi_{00} - \phi_{+-} = (-0.1 \pm 0.8)^{\circ}$

j for $K_L^0 \to \pi^+\pi^-\pi^0 = 0.0011 \pm 0.0008$

$|\eta_{+-\gamma}| = (2.35 \pm 0.07) \times 10^{-3}$

$\phi_{+-\gamma} = (44 \pm 4)^{\circ}$

$|\epsilon'_{+-\gamma}|/\epsilon < 0.3$ CL = 90%

$\Delta S = -\Delta Q$ in $K_{\ell 3}^0$ decay

Re $x = 0.006 \pm 0.018$ (S = 1.3)

Im $x = -0.003 \pm 0.026$ (S = 1.2)

CPT-violation parameters

Re $\Delta = 0.018 \pm 0.020$

Im $\Delta = 0.02 \pm 0.04$

K_L^0 DECAY MODES		Fraction (Γ_i/Γ)	Scale factor/ Confidence level	p (MeV/c)
$3\pi^0$		$(21.12 \pm 0.27)\%$	S=1.1	139
$\pi^+\pi^-\pi^0$		$(12.56 \pm 0.20)\%$	S=1.7	133
$\pi^{\pm}\mu^{\mp}\nu$	[gg]	$(27.17 \pm 0.25)\%$	S=1.1	216
Called $K_{\mu 3}^0$.				
$\pi^{\pm}e^{\mp}\nu_e$	[gg]	$(38.78 \pm 0.27)\%$	S=1.1	229
Called K_{e3}^0.				
2γ		$(5.92 \pm 0.15) \times 10^{-4}$		249
3γ		$< 2.4 \times 10^{-7}$	CL=90%	249
$\pi^0 2\gamma$	[hh]	$(1.70 \pm 0.28) \times 10^{-6}$		231
$\pi^0\pi^{\pm}e^{\mp}\nu$	[gg]	$(5.18 \pm 0.29) \times 10^{-5}$		207
$(\pi\mu\,\text{atom})\nu$		$(1.06 \pm 0.11) \times 10^{-7}$		–
$\pi^{\pm}e^{\mp}\nu_e\gamma$	[y,gg,hh]	$(3.62^{+0.26}_{-0.21}) \times 10^{-3}$		229
$\pi^+\pi^-\gamma$	[y,hh]	$(4.61 \pm 0.14) \times 10^{-5}$		206
$\pi^0\pi^0\gamma$		$< 5.6 \times 10^{-6}$		209

Charge conjugation × Parity (CP, CPV) or Lepton Family number (LF) violating modes or $\Delta S = 1$ weak neutral current ($S1$) modes

		Fraction		p (MeV/c)
$\pi^+\pi^-$	CPV	$(2.067 \pm 0.035) \times 10^{-3}$	S=1.1	206
$\pi^0\pi^0$	CPV	$(9.36 \pm 0.20) \times 10^{-4}$		209
$\mu^+\mu^-$	$S1$	$(7.2 \pm 0.5) \times 10^{-9}$	S=1.4	225
$\mu^+\mu^-\gamma$	$S1$	$(3.25 \pm 0.28) \times 10^{-7}$		225
e^+e^-	$S1$	$< 4.1 \times 10^{-11}$	CL=90%	249
$e^+e^-\gamma$	$S1$	$(9.1 \pm 0.5) \times 10^{-6}$		249
$e^+e^-\gamma\gamma$	$S1$ [hh]	$(6.5 \pm 1.2) \times 10^{-7}$		249
$\pi^+\pi^-e^+e^-$	$S1$ [hh]	$< 4.6 \times 10^{-7}$	CL=90%	206
$\mu^+\mu^-e^+e^-$	$S1$	$(2.9^{+6.7}_{-2.4}) \times 10^{-9}$		225
$e^+e^-e^+e^-$	$S1$	$(4.1 \pm 0.8) \times 10^{-8}$	S=1.2	249
$\pi^0\mu^+\mu^-$	$CPS1$ [ii]	$< 5.1 \times 10^{-9}$	CL=90%	177
$\pi^0 e^+e^-$	$CPS1$ [ii]	$< 4.3 \times 10^{-9}$	CL=90%	231
$\pi^0\nu\bar{\nu}$	$CPS1$ [jj]	$< 5.8 \times 10^{-5}$	CL=90%	231
$e^{\pm}\mu^{\mp}$	LF [gg]	$< 3.3 \times 10^{-11}$	CL=90%	238
$e^{\pm}e^{\pm}\mu^{\mp}\mu^{\mp}$	LF [gg]	$< 6.1 \times 10^{-9}$	CL=90%	–

$K^*(892)$ $I(J^P) = \frac{1}{2}(1^-)$

$K^*(892)^{\pm}$ mass $m = 891.66 \pm 0.26$ MeV

$K^*(892)^0$ mass $m = 896.10 \pm 0.28$ MeV   (S = 1.4)

$K^*(892)^{\pm}$ full width $\Gamma = 50.8 \pm 0.9$ MeV

$K^*(892)^0$ full width $\Gamma = 50.5 \pm 0.6$ MeV   (S = 1.1)

$K^*(892)$ DECAY MODES	Fraction (Γ_i/Γ)	Confidence level	p (MeV/c)
$K\pi$	$\sim 100\%$		291
$K^0\gamma$	$(2.30 \pm 0.20) \times 10^{-3}$		310
$K^{\pm}\gamma$	$(9.9 \pm 0.9) \times 10^{-4}$		309
$K\pi\pi$	$< 7 \times 10^{-4}$	95%	224

$K_1(1270)$ $I(J^P) = \frac{1}{2}(1^+)$

Mass $m = 1273 \pm 7$ MeV [m]

Full width $\Gamma = 90 \pm 20$ MeV [m]

$K_1(1270)$ DECAY MODES	Fraction (Γ_i/Γ)	p (MeV/c)
$K\rho$	$(42 \pm 6)\%$	76
$K_0^*(1430)\pi$	$(28 \pm 4)\%$	–
$K^*(892)\pi$	$(16 \pm 5)\%$	301
$K\omega$	$(11.0 \pm 2.0)\%$	–
$K f_0(1370)$	$(3.0 \pm 2.0)\%$	–

$K_1(1400)$ $I(J^P) = \frac{1}{2}(1^+)$

Mass $m = 1402 \pm 7$ MeV

Full width $\Gamma = 174 \pm 13$ MeV   (S = 1.6)

$K_1(1400)$ DECAY MODES	Fraction (Γ_i/Γ)	p (MeV/c)
$K^*(892)\pi$	$(94 \pm 6)\%$	401
$K\rho$	$(3.0 \pm 3.0)\%$	298
$K f_0(1370)$	$(2.0 \pm 2.0)\%$	–
$K\omega$	$(1.0 \pm 1.0)\%$	285
$K_0^*(1430)\pi$	not seen	–

$K^*(1410)$ $I(J^P) = \frac{1}{2}(1^-)$

Mass $m = 1414 \pm 15$ MeV   (S = 1.3)

Full width $\Gamma = 232 \pm 21$ MeV   (S = 1.1)

$K^*(1410)$ DECAY MODES	Fraction (Γ_i/Γ)	Confidence level	p (MeV/c)
$K^*(892)\pi$	$> 40\%$	95%	408
$K\pi$	$(6.6 \pm 1.3)\%$		611
$K\rho$	$< 7\%$	95%	309

$K_0^*(1430)$ [kk] $I(J^P) = \frac{1}{2}(0^+)$

Mass $m = 1429 \pm 6$ MeV

Full width $\Gamma = 287 \pm 23$ MeV

$K_0^*(1430)$ DECAY MODES	Fraction (Γ_i/Γ)	p (MeV/c)
$K\pi$	$(93 \pm 10)\%$	621

$K_2^*(1430)$ $I(J^P) = \frac{1}{2}(2^+)$

$K_2^*(1430)^{\pm}$ mass $m = 1425.6 \pm 1.5$ MeV   (S = 1.1)

$K_2^*(1430)^0$ mass $m = 1432.4 \pm 1.3$ MeV

$K_2^*(1430)^{\pm}$ full width $\Gamma = 98.5 \pm 2.7$ MeV   (S = 1.1)

$K_2^*(1430)^0$ full width $\Gamma = 109 \pm 5$ MeV   (S = 1.9)

$K_2^*(1430)$ DECAY MODES	Fraction (Γ_i/Γ)	Scale factor/ Confidence level	p (MeV/c)
$K\pi$	$(49.9 \pm 1.2)\%$		622
$K^*(892)\pi$	$(24.7 \pm 1.5)\%$		423
$K^*(892)\pi\pi$	$(13.4 \pm 2.2)\%$		375
$K\rho$	$(8.7 \pm 0.8)\%$	S=1.2	331
$K\omega$	$(2.9 \pm 0.8)\%$		319
$K^+\gamma$	$(2.4 \pm 0.5) \times 10^{-3}$	S=1.1	627
$K\eta$	$(1.5^{+3.4}_{-1.0}) \times 10^{-3}$	S=1.3	492
$K\omega\pi$	$< 7.2 \times 10^{-4}$	CL=95%	110
$K^0\gamma$	$< 9 \times 10^{-4}$	CL=90%	631

$K^*(1680)$ $I(J^P) = \frac{1}{2}(1^-)$

Mass $m = 1717 \pm 27$ MeV   (S = 1.4)
Full width $\Gamma = 322 \pm 110$ MeV   (S = 4.2)

$K^*(1680)$ DECAY MODES	Fraction (Γ_i/Γ)	p (MeV/c)
$K\pi$	(38.7 ± 2.5) %	779
$K\rho$	$(31.4^{+4.7}_{-2.1})$ %	571
$K^*(892)\pi$	$(29.9^{+2.2}_{-4.7})$ %	615

$K_2(1770)$ [ll] $I(J^P) = \frac{1}{2}(2^-)$

Mass $m = 1773 \pm 8$ MeV
Full width $\Gamma = 186 \pm 14$ MeV

$K_2(1770)$ DECAY MODES	Fraction (Γ_i/Γ)	p (MeV/c)
$K\pi\pi$		–
$K_2^*(1430)\pi$	dominant	287
$K^*(892)\pi$	seen	653
$K f_2(1270)$	seen	–
$K\phi$	seen	441
$K\omega$	seen	608

$K_3^*(1780)$ $I(J^P) = \frac{1}{2}(3^-)$

Mass $m = 1776 \pm 7$ MeV   (S = 1.1)
Full width $\Gamma = 159 \pm 21$ MeV   (S = 1.3)

$K_3^*(1780)$ DECAY MODES	Fraction (Γ_i/Γ)	Confidence level	p (MeV/c)
$K\rho$	(31 ± 9) %		612
$K^*(892)\pi$	(20 ± 5) %		651
$K\pi$	(18.8 ± 1.0) %		810
$K\eta$	(30 ± 13) %		715
$K_2^*(1430)\pi$	< 16 %	95%	284

$K_2(1820)$ [mm] $I(J^P) = \frac{1}{2}(2^-)$

Mass $m = 1816 \pm 13$ MeV
Full width $\Gamma = 276 \pm 35$ MeV

$K_2(1820)$ DECAY MODES	Fraction (Γ_i/Γ)	p (MeV/c)
$K_2^*(1430)\pi$	seen	325
$K^*(892)\pi$	seen	680
$K f_2(1270)$	seen	186
$K\omega$	seen	638

$K_4^*(2045)$ $I(J^P) = \frac{1}{2}(4^+)$

Mass $m = 2045 \pm 9$ MeV   (S = 1.1)
Full width $\Gamma = 198 \pm 30$ MeV

$K_4^*(2045)$ DECAY MODES	Fraction (Γ_i/Γ)	p (MeV/c)
$K\pi$	(9.9 ± 1.2) %	958
$K^*(892)\pi\pi$	(9 ± 5) %	800
$K^*(892)\pi\pi\pi$	(7 ± 5) %	764
$\rho K\pi$	(5.7 ± 3.2) %	742
$\omega K\pi$	(5.0 ± 3.0) %	736
$\phi K\pi$	(2.8 ± 1.4) %	591
$\phi K^*(892)$	(1.4 ± 0.7) %	363

CHARMED MESONS
$(C = \pm 1)$

$D^+ = c\overline{d}$, $D^0 = c\overline{u}$, $\overline{D}^0 = \overline{c}u$, $D^- = \overline{c}d$,   similarly for D^*'s

$D^\pm$ $I(J^P) = \frac{1}{2}(0^-)$

Mass $m = 1869.3 \pm 0.5$ MeV   (S = 1.1)
Mean life $\tau = (1.057 \pm 0.015) \times 10^{-12}$ s
$c\tau = 317$ μm

CP-violation decay-rate asymmetries

$A_{CP}(K^+K^-\pi^\pm) = -0.017 \pm 0.027$
$A_{CP}(K^\pm K^{*0}) = -0.02 \pm 0.05$
$A_{CP}(\phi\pi^\pm) = -0.014 \pm 0.033$
$A_{CP}(\pi^+\pi^-\pi^\pm) = -0.02 \pm 0.04$

$D^+ \to \overline{K}^*(892)^0 \ell^+ \nu_\ell$ form factors

$r_2 = 0.72 \pm 0.09$
$r_v = 1.85 \pm 0.12$
$\Gamma_L/\Gamma_T = 1.23 \pm 0.13$
$\Gamma_+/\Gamma_- = 0.16 \pm 0.04$

D^- modes are charge conjugates of the modes below.

D^+ DECAY MODES		Fraction (Γ_i/Γ)	Scale factor/ Confidence level	p (MeV/c)
Inclusive modes				
e^+ anything		(17.2 ± 1.9) %		–
K^- anything		(24.2 ± 2.8) %	S=1.4	–
$\overline{K}^0$ anything + K^0 anything		(59 ± 7) %		–
K^+ anything		(5.8 ± 1.4) %		–
η anything	[nn]	< 13 %	CL=90%	–
Leptonic and semileptonic modes				
$\mu^+ \nu_\mu$		$< 7.2 \times 10^{-4}$	CL=90%	932
$\overline{K}^0 \ell^+ \nu_\ell$	[oo]	(6.8 ± 0.8) %		868
$\overline{K}^0 e^+ \nu_e$		(6.7 ± 0.9) %		868
$\overline{K}^0 \mu^+ \nu_\mu$		$(7.0^{+3.0}_{-2.0})$ %		865
$K^-\pi^+ e^+ \nu_e$		$(4.1^{+0.9}_{-0.7})$ %		863
$\overline{K}^*(892)^0 e^+ \nu_e$		(3.2 ± 0.33) %		720
$\times$ B$(\overline{K}^{*0} \to K^-\pi^+)$				
$K^-\pi^+ e^+ \nu_e$ nonresonant		$< 7 \times 10^{-3}$	CL=90%	863
$K^-\pi^+ \mu^+ \nu_\mu$		(3.2 ± 0.4) %	S=1.1	851
$\overline{K}^*(892)^0 \mu^+ \nu_\mu$		(2.9 ± 0.4) %		715
$\times$ B$(\overline{K}^{*0} \to K^-\pi^+)$				
$K^-\pi^+ \mu^+ \nu_\mu$ nonresonant		$(2.7 \pm 1.1) \times 10^{-3}$		851
$(\overline{K}^*(892)\pi)^0 e^+ \nu_e$		< 1.2 %	CL=90%	714
$(\overline{K}\pi\pi)^0 e^+ \nu_e$ non-$\overline{K}^*(892)$		$< 9 \times 10^{-3}$	CL=90%	846
$K^-\pi^+\pi^0 \mu^+ \nu_\mu$		$< 1.4 \times 10^{-3}$	CL=90%	825
$\pi^0 \ell^+ \nu_\ell$	[pp]	$(3.1 \pm 1.5) \times 10^{-3}$		930

Fractions of some of the following modes with resonances have already appeared above as submodes of particular charged-particle modes.

$\overline{K}^*(892)^0 \ell^+ \nu_\ell$	[oo]	(4.7 ± 0.4) %		720
$\overline{K}^*(892)^0 e^+ \nu_e$		(4.8 ± 0.5) %		720
$\overline{K}^*(892)^0 \mu^+ \nu_\mu$		(4.4 ± 0.6) %	S=1.1	715
$\rho^0 e^+ \nu_e$		$(2.2 \pm 0.8) \times 10^{-3}$		776
$\rho^0 \mu^+ \nu_\mu$		$(2.7 \pm 0.7) \times 10^{-3}$		772
$\phi e^+ \nu_e$		< 2.09 %	CL=90%	657
$\phi \mu^+ \nu_\mu$		< 3.72 %	CL=90%	651
$\eta \ell^+ \nu_\ell$		$< 5 \times 10^{-3}$	CL=90%	–
$\eta'(958) \mu^+ \nu_\mu$		$< 9 \times 10^{-3}$	CL=90%	684
Hadronic modes with a $\overline{K}$ or $\overline{K}K\overline{K}$				
$\overline{K}^0 \pi^+$		(2.89 ± 0.26) %	S=1.1	862
$K^-\pi^+\pi^+$	[qq]	(9.0 ± 0.6) %		845
$\overline{K}^*(892)^0 \pi^+$		(1.27 ± 0.13) %		712
$\times$ B$(\overline{K}^{*0} \to K^-\pi^+)$				
$\overline{K}_0^*(1430)^0 \pi^+$		(2.3 ± 0.3) %		368
$\times$ B$(\overline{K}_0^*(1430)^0 \to K^-\pi^+)$				
$\overline{K}^*(1680)^0 \pi^+$		$(3.7 \pm 0.8) \times 10^{-3}$		65
$\times$ B$(\overline{K}^*(1680)^0 \to K^-\pi^+)$				
$K^-\pi^+\pi^+$ nonresonant		(8.5 ± 0.8) %		845
$\overline{K}^0 \pi^+\pi^0$	[qq]	(9.7 ± 3.0) %	S=1.1	845

$\overline{K}^0 \rho^+$		(6.6 ±2.5) %		680
$\overline{K}^*(892)^0 \pi^+$		(6.3 ±0.4) × 10^{-3}		712
× B($\overline{K}^{*0} \to \overline{K}^0 \pi^0$)				
$\overline{K}^0 \pi^+ \pi^0$ nonresonant		(1.3 ±1.1) %		845
$K^- \pi^+ \pi^+ \pi^0$	[qq]	(6.4 ±1.1) %		816
$\overline{K}^*(892)^0 \rho^+$ total		(1.4 ±0.9) %		423
× B($\overline{K}^{*0} \to K^- \pi^+$)				
$\overline{K}_1(1400)^0 \pi^+$		(2.2 ±0.6) %		390
× B($\overline{K}_1(1400)^0 \to K^- \pi^+ \pi^0$)				
$K^- \rho^+ \pi^+$ total		(3.1 ±1.1) %		616
$K^- \rho^+ \pi^+$ 3-body		(1.1 ±0.4) %		616
$\overline{K}^*(892)^0 \pi^+ \pi^0$ total		(4.5 ±0.9) %		687
× B($\overline{K}^{*0} \to K^- \pi^+$)				
$\overline{K}^*(892)^0 \pi^+ \pi^0$ 3-body		(2.8 ±0.9) %		687
× B($\overline{K}^{*0} \to K^- \pi^+$)				
$K^*(892)^- \pi^+ \pi^+$ 3-body		(7 ±3) × 10^{-3}		688
× B($K^{*-} \to K^- \pi^0$)				
$K^- \pi^+ \pi^+ \pi^0$ nonresonant	[rr]	(1.2 ±0.6) %		816
$\overline{K}^0 \pi^+ \pi^+ \pi^-$	[qq]	(7.0 ±0.9) %		814
$\overline{K}^0 a_1(1260)^+$		(4.0 ±0.9) %		328
× B($a_1(1260)^+ \to \pi^+ \pi^+ \pi^-$)				
$\overline{K}_1(1400)^0 \pi^+$		(2.2 ±0.6) %		390
× B($\overline{K}_1(1400)^0 \to \overline{K}^0 \pi^+ \pi^-$)				
$K^*(892)^- \pi^+ \pi^+$ 3-body		(1.4 ±0.6) %		688
× B($K^{*-} \to \overline{K}^0 \pi^-$)				
$\overline{K}^0 \rho^0 \pi^+$ total		(4.2 ±0.9) %		614
$\overline{K}^0 \rho^0 \pi^+$ 3-body		(5 ±5) × 10^{-3}		614
$\overline{K}^0 \pi^+ \pi^+ \pi^-$ nonresonant		(8 ±4) × 10^{-3}		814
$K^- \pi^+ \pi^+ \pi^+ \pi^-$	[qq]	(7.2 ±1.0) × 10^{-3}		772
$\overline{K}^*(892)^0 \pi^+ \pi^+ \pi^-$		(5.4 ±2.3) × 10^{-3}		642
× B($\overline{K}^{*0} \to K^- \pi^+$)				
$\overline{K}^*(892)^0 \rho^0 \pi^+$		(1.9 $^{+1.1}_{-1.0}$) × 10^{-3}		242
× B($\overline{K}^{*0} \to K^- \pi^+$)				
$\overline{K}^*(892)^0 \pi^+ \pi^+ \pi^-$ no-ρ		(2.9 ±1.1) × 10^{-3}		642
× B($\overline{K}^{*0} \to K^- \pi^+$)				
$K^- \rho^0 \pi^+ \pi^+$		(3.1 ±0.9) × 10^{-3}		529
$K^- \pi^+ \pi^+ \pi^+ \pi^-$ nonresonant		< 2.3 × 10^{-3}	CL=90%	772
$K^- \pi^+ \pi^+ \pi^0 \pi^0$		(2.2 $^{+5.0}_{-0.9}$) %		775
$\overline{K}^0 \pi^+ \pi^+ \pi^- \pi^0$		(5.4 $^{+3.0}_{-1.4}$) %		773
$\overline{K}^0 \pi^+ \pi^+ \pi^+ \pi^- \pi^-$		(8 ±7) × 10^{-4}		714
$K^- \pi^+ \pi^+ \pi^+ \pi^- \pi^0$		(2.0 ±1.8) × 10^{-3}		718
$\overline{K}^0 \overline{K}^0 K^+$		(1.8 ±0.8) %		545

Fractions of some of the following modes with resonances have already appeared above as submodes of particular charged-particle modes.

$\overline{K}^0 \rho^+$		(6.6 ±2.5) %		680
$\overline{K}^0 a_1(1260)^+$		(8.0 ±1.7) %		328
$\overline{K}^0 a_2(1320)^+$		< 3 × 10^{-3}	CL=90%	199
$\overline{K}^*(892)^0 \pi^+$		(1.90±0.19) %		712
$\overline{K}^*(892)^0 \rho^+$ total	[rr]	(2.1 ±1.3) %		423
$\overline{K}^*(892)^0 \rho^+$ S-wave	[rr]	(1.6 ±1.6) %		423
$\overline{K}^*(892)^0 \rho^+$ P-wave		< 1 × 10^{-3}	CL=90%	423
$\overline{K}^*(892)^0 \rho^+$ D-wave		(10 ±7) × 10^{-3}		423
$\overline{K}^*(892)^0 \rho^+$ D-wave longitudinal		< 7 × 10^{-3}	CL=90%	423
$\overline{K}_1(1270)^0 \pi^+$		< 7 × 10^{-3}	CL=90%	487
$\overline{K}_1(1400)^0 \pi^+$		(4.9 ±1.2) %		390
$\overline{K}^*_1(1410)^0 \pi^+$		< 7 × 10^{-3}	CL=90%	382
$\overline{K}^*_0(1430)^0 \pi^+$		(3.7 ±0.4) %		368
$\overline{K}^*(1680)^0 \pi^+$		(1.43±0.30) %		65
$\overline{K}^*(892)^0 \pi^+ \pi^0$ total		(6.7 ±1.4) %		687
$\overline{K}^*(892)^0 \pi^+ \pi^0$ 3-body	[rr]	(4.2 ±1.4) %		687
$K^*(892)^- \pi^+ \pi^+$ 3-body		(2.0 ±0.9) %		688
$K^- \rho^+ \pi^+$ total		(3.1 ±1.1) %		616
$K^- \rho^+ \pi^+$ 3-body		(1.1 ±0.4) %		616
$\overline{K}^0 \rho^0 \pi^+$ total		(4.2 ±0.9) %	CL=90%	614
$\overline{K}^0 \rho^0 \pi^+$ 3-body		(5 ±5) × 10^{-3}		614
$\overline{K}^0 f_0(980) \pi^+$		< 5 × 10^{-3}	CL=90%	461
$\overline{K}^*(892)^0 \pi^+ \pi^+ \pi^-$		(8.1 ±3.4) × 10^{-3}	S=1.7	642
$\overline{K}^*(892)^0 \rho^0 \pi^+$		(2.9 $^{+1.7}_{-1.5}$) × 10^{-3}	S=1.8	242
$\overline{K}^*(892)^0 \pi^+ \pi^+ \pi^-$ no-ρ		(4.3 ±1.7) × 10^{-3}		642
$K^- \rho^0 \pi^+ \pi^+$		(3.1 ±0.9) × 10^{-3}		529

Pionic modes

$\pi^+ \pi^0$		(2.5 ±0.7) × 10^{-3}		925
$\pi^+ \pi^+ \pi^-$		(3.6 ±0.4) × 10^{-3}		908
$\rho^0 \pi^+$		(1.05±0.31) × 10^{-3}		769
$\pi^+ \pi^+ \pi^-$ nonresonant		(2.2 ±0.4) × 10^{-3}		908
$\pi^+ \pi^+ \pi^- \pi^0$		(1.9 $^{+1.5}_{-1.2}$) %		882
$\eta \pi^+$ × B($\eta \to \pi^+ \pi^- \pi^0$)		(1.7 ±0.6) × 10^{-3}		848
$\omega \pi^+$ × B($\omega \to \pi^+ \pi^- \pi^0$)		< 6 × 10^{-3}	CL=90%	764
$\pi^+ \pi^+ \pi^+ \pi^- \pi^-$		(2.1 ±0.4) × 10^{-3}		845
$\pi^+ \pi^+ \pi^+ \pi^- \pi^- \pi^0$		(2.9 $^{+2.9}_{-2.0}$) × 10^{-3}		799

Fractions of some of the following modes with resonances have already appeared above as submodes of particular charged-particle modes.

$\eta \pi^+$		(7.5 ±2.5) × 10^{-3}		848
$\rho^0 \pi^+$		(1.05±0.31) × 10^{-3}		769
$\omega \pi^+$		< 7 × 10^{-3}	CL=90%	764
$\eta \rho^+$		< 1.2 %	CL=90%	658
$\eta'(958) \pi^+$		< 9 × 10^{-3}	CL=90%	680
$\eta'(958) \rho^+$		< 1.5 %	CL=90%	355

Hadronic modes with a $K\overline{K}$ pair

$K^+ \overline{K}^0$		(7.4 ±1.0) × 10^{-3}		792
$K^+ K^- \pi^+$	[qq]	(8.8 ±0.8) × 10^{-3}		744
$\phi \pi^+$ × B($\phi \to K^+ K^-$)		(3.0 ±0.3) × 10^{-3}		647
$K^+ \overline{K}^*(892)^0$		(2.8 ±0.4) × 10^{-3}		610
× B($\overline{K}^{*0} \to K^- \pi^+$)				
$K^+ K^- \pi^+$ nonresonant		(4.5 ±0.9) × 10^{-3}		744
$K^0 \overline{K}^0 \pi^+$		—		741
$K^*(892)^+ \overline{K}^0$		(2.1 ±1.0) %		611
× B($K^{*+} \to K^0 \pi^+$)				
$K^+ K^- \pi^+ \pi^0$		—		682
$\phi \pi^+ \pi^0$ × B($\phi \to K^+ K^-$)		(1.1 ±0.5) %		619
$\phi \rho^+$ × B($\phi \to K^+ K^-$)		< 7 × 10^{-3}	CL=90%	268
$K^+ K^- \pi^+ \pi^0$ non-ϕ		(1.5 $^{+0.7}_{-0.6}$) %		682
$K^+ \overline{K}^0 \pi^+ \pi^-$		< 2 %	CL=90%	678
$K^0 K^- \pi^+ \pi^+$		(1.0 ±0.6) %		678
$K^*(892)^+ \overline{K}^*(892)^0$		(1.2 ±0.5) %		273
× $B^2(K^{*+} \to K^0 \pi^+)$				
$K^0 K^- \pi^+ \pi^+$ non-$K^{*+} \overline{K}^{*0}$		< 7.9 × 10^{-3}	CL=90%	678
$K^+ K^- \pi^+ \pi^+ \pi^-$		—		600
$\phi \pi^+ \pi^+ \pi^-$		< 1 × 10^{-3}	CL=90%	565
× B($\phi \to K^+ K^-$)				
$K^+ K^- \pi^+ \pi^+ \pi^-$ nonresonant		< 3 %	CL=90%	600

Fractions of the following modes with resonances have already appeared above as submodes of particular charged-particle modes.

$\phi \pi^+$		(6.1 ±0.6) × 10^{-3}		647
$\phi \pi^+ \pi^0$		(2.3 ±1.0) %		619
$\phi \rho^+$		< 1.4 %	CL=90%	268
$\phi \pi^+ \pi^+ \pi^-$		< 2 × 10^{-3}	CL=90%	565
$K^+ \overline{K}^*(892)^0$		(4.2 ±0.5) × 10^{-3}		610
$K^*(892)^+ \overline{K}^0$		(3.2 ±1.5) %		611
$K^*(892)^+ \overline{K}^*(892)^0$		(2.6 ±1.1) %		273

Doubly Cabibbo suppressed (DC) modes Γ ΔC = 1 weak neutral current (C1) modes Γ or Lepton Family number (LF) or Lepton number (L) violating modes

$K^+ \pi^+ \pi^-$	DC	(6.8 ±1.5) × 10^{-4}		845
$K^+ \rho^0$	DC	(2.5 ±1.2) × 10^{-4}		681
$K^*(892)^0 \pi^+$	DC	(3.6 ±1.6) × 10^{-4}		712
$K^+ \pi^+ \pi^-$ nonresonant	DC	(2.4 ±1.2) × 10^{-4}		845
$K^+ K^+ K^-$	DC	< 1.4 × 10^{-4}	CL=90%	550
ϕK^+	DC	< 1.3 × 10^{-4}	CL=90%	527
$\pi^+ e^+ e^-$	C1	< 6.6 × 10^{-5}	CL=90%	929
$\pi^+ \mu^+ \mu^-$	C1	< 1.8 × 10^{-5}	CL=90%	917
$\rho^+ \mu^+ \mu^-$	C1	< 5.6 × 10^{-4}	CL=90%	759
$K^+ e^+ e^-$	[ss]	< 2.0 × 10^{-4}	CL=90%	869
$K^+ \mu^+ \mu^-$	[ss]	< 9.7 × 10^{-5}	CL=90%	856
$\pi^+ e^+ \mu^-$	LF	< 1.1 × 10^{-4}	CL=90%	926
$\pi^+ e^- \mu^+$	LF	< 1.3 × 10^{-4}	CL=90%	926
$K^+ e^+ \mu^-$	LF	< 1.3 × 10^{-4}	CL=90%	866

$K^+ e^- \mu^+$	LF	$< 1.2 \times 10^{-4}$	CL=90%	866
$\pi^- e^+ e^+$	L	$< 1.1 \times 10^{-4}$	CL=90%	929
$\pi^- \mu^+ \mu^+$	L	$< 8.7 \times 10^{-5}$	CL=90%	917
$\pi^- e^+ \mu^+$	L	$< 1.1 \times 10^{-4}$	CL=90%	926
$\rho^- \mu^+ \mu^+$	L	$< 5.6 \times 10^{-4}$	CL=90%	759
$K^- e^+ e^+$	L	$< 1.2 \times 10^{-4}$	CL=90%	869
$K^- \mu^+ \mu^+$	L	$< 1.2 \times 10^{-4}$	CL=90%	856
$K^- e^+ \mu^+$	L	$< 1.3 \times 10^{-4}$	CL=90%	866
$K^*(892)^- \mu^+ \mu^+$	L	$< 8.5 \times 10^{-4}$	CL=90%	703

$$\boxed{D^0} \qquad\qquad I(J^P) = \tfrac{1}{2}(0^-)$$

Mass $m = 1864.6 \pm 0.5$ MeV (S = 1.1)

$m_{D^\pm} - m_{D^0} = 4.76 \pm 0.10$ MeV (S = 1.1)

Mean life $\tau = (0.415 \pm 0.004) \times 10^{-12}$ s

$c\tau = 124.4\ \mu$m

$|m_{D_1^0} - m_{D_2^0}| < 24 \times 10^{10}\ \hbar\ \mathrm{s}^{-1}\Gamma$ CL = 90% [tt]

$|\Gamma_{D_1^0} - \Gamma_{D_2^0}|/\Gamma_{D^0} < 0.20\Gamma$ CL = 90% [tt]

$\Gamma(K^+ \ell^- \bar\nu_\ell\ (\text{via } \overline{D}{}^0))/\Gamma(K^- \ell^+ \nu_\ell) < 0.005\Gamma$ CL = 90%

$\frac{\Gamma(K^+\pi^-\ \text{or}\ K^+\pi^-\pi^+\pi^-\ (\text{via}\overline{D}{}^0))}{\Gamma(K^-\pi^+\ \text{or}\ K^-\pi^+\pi^+\pi^-)} < 0.0085\ (\text{or} < 0.0037)\Gamma$ CL = 90% [uu]

CP-violation decay-rate asymmetries

$A_{CP}(K^+K^-) = 0.026 \pm 0.035$

$A_{CP}(\pi^+\pi^-) = -0.05 \pm 0.08$

$A_{CP}(K_S^0 \phi) = -0.03 \pm 0.09$

$A_{CP}(K_S^0 \pi^0) = -0.018 \pm 0.030$

$\overline{D}{}^0$ modes are charge conjugates of the modes below.

D^0 DECAY MODES		Fraction (Γ_i/Γ)	Scale factor/ Confidence level	p (MeV/c)
Inclusive modes				
e^+ anything		(6.75 ± 0.29) %		—
μ^+ anything		(6.6 ± 0.8) %		—
K^- anything		(53 ± 4) %	S=1.3	—
$\overline{K}{}^0$ anything $+ K^0$ anything		(42 ± 5) %		—
K^+ anything		$(3.4 ^{+0.6}_{-0.4})$ %		—
η anything	[nn]	< 13 %	CL=90%	
Semileptonic modes				
$K^- \ell^+ \nu_\ell$	[oo]	(3.50 ± 0.17) %	S=1.3	867
$K^- e^+ \nu_e$		(3.66 ± 0.18) %		867
$K^- \mu^+ \nu_\mu$		(3.23 ± 0.17) %		863
$K^- \pi^0 e^+ \nu_e$		$(1.6 ^{+1.3}_{-0.5})$ %		861
$\overline{K}{}^0 \pi^- e^+ \nu_e$		$(2.8 ^{+1.7}_{-0.9})$ %		860
$\overline{K}{}^*(892)^- e^+ \nu_e$		(1.35 ± 0.22) %		719
$\quad \times B(K^{*-} \to \overline{K}{}^0\pi^-)$				
$K^*(892)^- \ell^+ \nu_\ell$		—		
$\overline{K}{}^*(892)^0 \pi^- e^+ \nu_e$		—		708
$K^- \pi^+ \pi^- \mu^+ \nu_\mu$		$< 1.2 \times 10^{-3}$	CL=90%	821
$(\overline{K}{}^*(892)\pi)^- \mu^+ \nu_\mu$		$< 1.4 \times 10^{-3}$	CL=90%	693
$\pi^- e^+ \nu_e$		$(3.7 \pm0.6) \times 10^{-3}$		927

A fraction of the following resonance mode has already appeared above as a submode of a charged-particle mode.

$K^*(892)^- e^+ \nu_e$		(2.02 ± 0.33) %		719

Hadronic modes with a $\overline{K}$ or $\overline{K}K\overline{K}$

$K^- \pi^+$		(3.85 ± 0.09) %		861
$\overline{K}{}^0 \pi^0$		(2.12 ± 0.21) %	S=1.1	860
$\overline{K}{}^0 \pi^+ \pi^-$	[qq]	(5.4 ± 0.4) %	S=1.2	842
$\overline{K}{}^0 \rho^0$		(1.21 ± 0.17) %		676
$\overline{K}{}^0 f_0(980)$		$(3.0 \pm0.8) \times 10^{-3}$		549
$\quad \times B(f_0 \to \pi^+\pi^-)$				
$\overline{K}{}^0 f_2(1270)$		$(2.4 \pm0.9) \times 10^{-3}$		263
$\quad \times B(f_2 \to \pi^+\pi^-)$				
$\overline{K}{}^0 f_0(1370)$		$(4.3 \pm1.3) \times 10^{-3}$		—
$\quad \times B(f_0 \to \pi^+\pi^-)$				
$K^*(892)^- \pi^+$		(3.4 ± 0.3) %		711
$\quad \times B(K^{*-} \to \overline{K}{}^0\pi^-)$				
$K_0^*(1430)^- \pi^+$		$(6.4 \pm1.6) \times 10^{-3}$		364
$\quad \times B(K_0^*(1430)^- \to \overline{K}{}^0\pi^-)$				
$\overline{K}{}^0 \pi^+ \pi^-$ nonresonant		(1.47 ± 0.24) %		842

$K^- \pi^+ \pi^0$	[qq]	(13.9 ± 0.9) %	S=1.3	844
$K^- \rho^+$		(10.8 ± 1.0) %		678
$K^*(892)^- \pi^+$		(1.7 ± 0.2) %		711
$\quad \times B(K^{*-} \to K^-\pi^0)$				
$\overline{K}{}^*(892)^0 \pi^0$		(2.1 ± 0.3) %		709
$\quad \times B(\overline{K}{}^{*0} \to K^-\pi^+)$				
$K^- \pi^+ \pi^0$ nonresonant		$(6.9 \pm2.5) \times 10^{-3}$		844
$\overline{K}{}^0 \pi^0 \pi^0$		—		843
$\overline{K}{}^*(892)^0 \pi^0$		(1.1 ± 0.2) %		709
$\quad \times B(\overline{K}{}^{*0} \to \overline{K}{}^0\pi^0)$				
$\overline{K}{}^0 \pi^0 \pi^0$ nonresonant		$(7.9 \pm2.1) \times 10^{-3}$		843
$K^- \pi^+ \pi^+ \pi^-$	[qq]	(7.6 ± 0.4) %	S=1.1	812
$K^- \pi^+ \rho^0$ total		(6.3 ± 0.4) %		612
$K^- \pi^+ \rho^0$ 3-body		$(4.8 \pm2.1) \times 10^{-3}$		612
$\overline{K}{}^*(892)^0 \rho^0$		$(9.8 \pm2.2) \times 10^{-3}$		418
$\quad \times B(\overline{K}{}^{*0} \to K^-\pi^+)$				
$K^- a_1(1260)^+$		(3.6 ± 0.6) %		327
$\quad \times B(a_1(1260)^+ \to \pi^+\pi^+\pi^-)$				
$\overline{K}{}^*(892)^0 \pi^+ \pi^-$ total		(1.5 ± 0.4) %		683
$\quad \times B(\overline{K}{}^{*0} \to K^-\pi^+)$				
$\overline{K}{}^*(892)^0 \pi^+ \pi^-$ 3-body		$(9.5 \pm2.1) \times 10^{-3}$		683
$\quad \times B(\overline{K}{}^{*0} \to K^-\pi^+)$				
$K_1(1270)^- \pi^+$	[rr]	$(3.6 \pm1.0) \times 10^{-3}$		483
$\quad \times B(K_1(1270)^- \to K^-\pi^+\pi^-)$				
$K^- \pi^+ \pi^+ \pi^-$ nonresonant		(1.76 ± 0.25) %		812
$\overline{K}{}^0 \pi^+ \pi^- \pi^0$	[qq]	(10.0 ± 1.2) %		812
$\overline{K}{}^0 \eta \times B(\eta \to \pi^+\pi^-\pi^0)$		$(1.6 \pm0.3) \times 10^{-3}$		772
$\overline{K}{}^0 \omega \times B(\omega \to \pi^+\pi^-\pi^0)$		(1.9 ± 0.4) %		670
$K^*(892)^- \rho^+$		(4.1 ± 1.6) %		422
$\quad \times B(K^{*-} \to \overline{K}{}^0\pi^-)$				
$\overline{K}{}^*(892)^0 \rho^0$		$(4.9 \pm1.1) \times 10^{-3}$		418
$\quad \times B(\overline{K}{}^{*0} \to \overline{K}{}^0\pi^0)$				
$K_1(1270)^- \pi^+$	[rr]	$(5.1 \pm1.4) \times 10^{-3}$		483
$\quad \times B(K_1(1270)^- \to \overline{K}{}^0\pi^-\pi^0)$				
$\overline{K}{}^*(892)^0 \pi^+ \pi^-$ 3-body		$(4.8 \pm1.1) \times 10^{-3}$		683
$\quad \times B(\overline{K}{}^{*0} \to \overline{K}{}^0\pi^0)$				
$\overline{K}{}^0 \pi^+ \pi^- \pi^0$ nonresonant		(2.1 ± 2.1) %		812
$K^- \pi^+ \pi^0 \pi^0$		(15 ± 5) %		815
$K^- \pi^+ \pi^+ \pi^- \pi^0$		(4.1 ± 0.4) %		771
$\overline{K}{}^*(892)^0 \pi^+ \pi^- \pi^0$		(1.2 ± 0.6) %		641
$\quad \times B(\overline{K}{}^{*0} \to K^-\pi^+)$				
$\overline{K}{}^*(892)^0 \eta$		$(2.9 \pm0.8) \times 10^{-3}$		580
$\quad \times B(\overline{K}{}^{*0} \to K^-\pi^+)$				
$\quad \times B(\eta \to \pi^+\pi^-\pi^0)$				
$K^- \pi^+ \omega \times B(\omega \to \pi^+\pi^-\pi^0)$		(2.7 ± 0.5) %		605
$\overline{K}{}^*(892)^0 \omega$		$(7 \pm3) \times 10^{-3}$		406
$\quad \times B(\overline{K}{}^{*0} \to K^-\pi^+)$				
$\quad \times B(\omega \to \pi^+\pi^-\pi^0)$				
$\overline{K}{}^0 \pi^+ \pi^+ \pi^- \pi^-$		$(5.8 \pm1.6) \times 10^{-3}$		768
$\overline{K}{}^0 \pi^+ \pi^- \pi^0 \pi^0 (\pi^0)$		$(10.6 ^{+7.3}_{-3.0})$ %		771
$\overline{K}{}^0 K^+ K^-$		$(9.4 \pm1.0) \times 10^{-3}$		544
$\overline{K}{}^0 \phi \times B(\phi \to K^+K^-)$		$(4.3 \pm0.5) \times 10^{-3}$		520
$\overline{K}{}^0 K^+ K^-$ non-ϕ		$(5.1 \pm0.8) \times 10^{-3}$		544
$K_S^0 K_S^0 K_S^0$		$(8.4 \pm1.5) \times 10^{-4}$		538
$K^+ K^- K^- \pi^+$		$(2.1 \pm0.5) \times 10^{-4}$		434
$K^+ K^- \overline{K}{}^0 \pi^0$		$(7.2 ^{+4.8}_{-3.5}) \times 10^{-3}$		435

Fractions of many of the following modes with resonances have already appeared above as submodes of particular charged-particle modes. (Modes for which there are only upper limits and $\overline{K}{}^*(892)\rho$ submodes only appear below.)

$\overline{K}{}^0 \eta$		$(7.1 \pm1.0) \times 10^{-3}$		772
$\overline{K}{}^0 \rho^0$		(1.21 ± 0.17) %		676
$K^- \rho^+$		(10.8 ± 1.0) %	S=1.2	678
$\overline{K}{}^0 \omega$		(2.1 ± 0.4) %		670
$\overline{K}{}^0 \eta'(958)$		(1.72 ± 0.26) %		565
$\overline{K}{}^0 f_0(980)$		$(5.7 \pm1.6) \times 10^{-3}$		549
$\overline{K}{}^0 \phi$		$(8.6 \pm1.0) \times 10^{-3}$		520
$K^- a_1(1260)^+$		(7.3 ± 1.1) %		327
$\overline{K}{}^0 a_1(1260)^0$		< 1.9 %	CL=90%	322
$\overline{K}{}^0 f_2(1270)$		$(4.2 \pm1.5) \times 10^{-3}$		263
$K^- a_2(1320)^+$		$< 2 \times 10^{-3}$	CL=90%	197
$\overline{K}{}^0 f_0(1370)$		$(7.0 \pm2.1) \times 10^{-3}$		—
$K^*(892)^- \pi^+$		(5.1 ± 0.4) %	S=1.2	711
$\overline{K}{}^*(892)^0 \pi^0$		(3.2 ± 0.4) %		709
$\overline{K}{}^*(892)^0 \pi^+ \pi^-$ total		(2.3 ± 0.5) %		683
$\overline{K}{}^*(892)^0 \pi^+ \pi^-$ 3-body		(1.43 ± 0.32) %		683

$K^- \pi^+ \rho^0$ total		(6.3 ±0.4) %	612
$K^- \pi^+ \rho^0$ 3-body		(4.8 ±2.1) × 10^{-3}	612
$\overline{K}^*(892)^0 \rho^0$		(1.47±0.33) %	418
$\overline{K}^*(892)^0 \rho^0$ transverse		(1.5 ±0.5) %	418
$\overline{K}^*(892)^0 \rho^0$ S-wave		(2.8 ±0.6) %	418
$\overline{K}^*(892)^0 \rho^0$ S-wave long.	< 3	× 10^{-3} CL=90%	418
$\overline{K}^*(892)^0 \rho^0$ P-wave	< 3	× 10^{-3} CL=90%	418
$\overline{K}^*(892)^0 \rho^0$ D-wave		(1.9 ±0.6) %	418
$K^*(892)^- \rho^+$		(6.1 ±2.4) %	422
$K^*(892)^- \rho^+$ longitudinal		(2.9 ±1.2) %	422
$K^*(892)^- \rho^+$ transverse		(3.2 ±1.8) %	422
$K^*(892)^- \rho^+$ P-wave	< 1.5	% CL=90%	422
$K^- \pi^+ f_0(980)$	< 1.1	% CL=90%	459
$\overline{K}^*(892)^0 f_0(980)$	< 7	× 10^{-3} CL=90%	–
$K_1(1270)^- \pi^+$	[rr]	(1.06±0.29) %	483
$K_1(1400)^- \pi^+$	< 1.2	% CL=90%	386
$\overline{K}_1(1400)^0 \pi^0$	< 3.7	% CL=90%	387
$K^*(1410)^- \pi^+$	< 1.2	% CL=90%	378
$K^*_0(1430)^- \pi^+$		(1.04±0.26) %	364
$K^*_2(1430)^- \pi^+$	< 8	× 10^{-3} CL=90%	367
$\overline{K}^*_2(1430)^0 \pi^0$	< 4	× 10^{-3} CL=90%	363
$\overline{K}^*(892)^0 \pi^+ \pi^- \pi^0$		(1.8 ±0.9) %	641
$\overline{K}^*(892)^0 \eta$		(1.9 ±0.5) %	580
$K^- \pi^+ \omega$		(3.0 ±0.6) %	605
$\overline{K}^*(892)^0 \omega$		(1.1 ±0.5) %	406
$K^- \pi^+ \eta'(958)$		(7.0 ±1.8) × 10^{-3}	479
$\overline{K}^*(892)^0 \eta'(958)$	< 1.1	× 10^{-3} CL=90%	99

Pionic modes

$\pi^+ \pi^-$		(1.53±0.09) × 10^{-3}	922
$\pi^0 \pi^0$		(8.5 ±2.2) × 10^{-4}	922
$\pi^+ \pi^- \pi^0$		(1.6 ±1.1) %	S=2.7	907
$\pi^+ \pi^+ \pi^- \pi^-$		(7.4 ±0.6) × 10^{-3}	879
$\pi^+ \pi^+ \pi^- \pi^- \pi^0$		(1.9 ±0.4) %	844
$\pi^+ \pi^+ \pi^+ \pi^- \pi^- \pi^-$		(4.0 ±3.0) × 10^{-4}	795

Hadronic modes with a $K\overline{K}$ pair

$K^+ K^-$		(4.27±0.16) × 10^{-3}	791
$K^0 \overline{K}^0$		(6.5 ±1.8) × 10^{-3}	S=1.2	788
$K^0 K^- \pi^+$		(6.4 ±1.0) × 10^{-3}	S=1.1	739
$\overline{K}^*(892)^0 K^0$	< 1.1	× 10^{-3} CL=90%	605
× B($\overline{K}^{*0} \to K^- \pi^+$)			
$K^*(892)^+ K^-$		(2.3 ±0.5) × 10^{-3}	610
× B($K^{*+} \to K^0 \pi^+$)			
$K^0 K^- \pi^+$ nonresonant		(2.3 ±2.3) × 10^{-3}	739
$\overline{K}^0 K^+ \pi^-$		(5.0 ±1.0) × 10^{-3}	739
$K^*(892)^0 \overline{K}^0$	< 5	× 10^{-4} CL=90%	605
× B($K^{*0} \to K^+ \pi^-$)			
$K^*(892)^- K^+$		(1.2 ±0.7) × 10^{-3}	610
× B($K^{*-} \to \overline{K}^0 \pi^-$)			
$\overline{K}^0 K^+ \pi^-$ nonresonant		(3.9 $^{+2.3}_{-1.9}$) × 10^{-3}	739
$K^+ K^- \pi^0$		(1.3 ±0.4) × 10^{-3}	742
$K^0_S K^0_S \pi^0$	< 5.9	× 10^{-4}	739
$K^+ K^- \pi^+ \pi^-$	[vv]	(2.52±0.24) × 10^{-3}	676
$\phi \pi^+ \pi^-$ × B($\phi \to K^+ K^-$)		(5.3 ±1.4) × 10^{-4}	614
$\phi \rho^0$ × B($\phi \to K^+ K^-$)		(3.0 ±1.6) × 10^{-4}	260
$K^+ K^- \rho^0$ 3-body		(9.1 ±2.3) × 10^{-4}	309
$K^*(892)^0 K^- \pi^+$ +c.c.	[ww] < 5	× 10^{-4}	528
× B($K^{*0} \to K^+ \pi^-$)			
$K^*(892)^0 \overline{K}^*(892)^0$		(6 ±2) × 10^{-4}	257
× B^2($K^{*0} \to K^+ \pi^-$)			
$K^+ K^- \pi^+ \pi^-$ non-ϕ		—	676
$K^+ K^- \pi^+ \pi^-$ nonresonant	< 8	× 10^{-4} CL=90%	676
$K^0 \overline{K}^0 \pi^+ \pi^-$		(6.9 ±2.7) × 10^{-3}	673
$K^+ K^- \pi^+ \pi^- \pi^0$		(3.1 ±2.0) × 10^{-3}	600

Fractions of most of the following modes with resonances have already appeared above as submodes of particular charged-particle modes.

$\overline{K}^*(892)^0 K^0$	< 1.6	× 10^{-3} CL=90%	605
$K^*(892)^+ K^-$		(3.5 ±0.8) × 10^{-3}	610
$K^*(892)^0 \overline{K}^0$	< 8	× 10^{-4} CL=90%	605
$K^*(892)^- K^+$		(1.8 ±1.0) × 10^{-3}	610
$\phi \pi^0$	< 1.4	× 10^{-3} CL=90%	644
$\phi \eta$	< 2.8	× 10^{-3} CL=90%	489

$\phi \omega$	< 2.1	× 10^{-3} CL=90%	239
$\phi \pi^+ \pi^-$		(1.08±0.29) × 10^{-3}	614
$\phi \rho^0$		(6 ±3) × 10^{-4}	260
$\phi \pi^+ \pi^-$ 3-body		(7 ±5) × 10^{-4}	614
$K^*(892)^0 K^- \pi^+$ + c.c.	[ww] < 8	× 10^{-4} CL=90%	–
$K^*(892)^0 \overline{K}^*(892)^0$		(1.4 ±0.5) × 10^{-3}	257

Doubly Cabibbo suppressed (DC) modesΓ
ΔC = 2 forbidden via mixing (C2M) modesΓ
ΔC = 1 weak neutral current (C1) modesΓor
Lepton Family number (LF) violating modes

$K^+ \ell^- \overline{\nu}_\ell$ (via $\overline{D}^0$)	C2M	< 1.7	× 10^{-4} CL=90%	–
$K^+ \pi^-$ or	C2M	< 1.0	× 10^{-3} CL=90%	–
$\quad K^+ \pi^- \pi^+ \pi^-$ (via $\overline{D}^0$)				
$K^+ \pi^-$	DC	(2.8 ±0.9) × 10^{-4}		861
$K^+ \pi^-$ (via $\overline{D}^0$)		< 1.9	× 10^{-4} CL=90%	861
$K^+ \pi^- \pi^+ \pi^-$	DC	(1.9 ±2.7) × 10^{-4}		812
$K^+ \pi^- \pi^+ \pi^-$ (via $\overline{D}^0$)		< 4	× 10^{-4} CL=90%	812
μ^- anything (via $\overline{D}^0$)		< 4	× 10^{-4} CL=90%	–
$e^+ e^-$	C1	< 1.3	× 10^{-5} CL=90%	932
$\mu^+ \mu^-$	C1	< 4.1	× 10^{-6} CL=90%	926
$\pi^0 e^+ e^-$	C1	< 4.5	× 10^{-5} CL=90%	927
$\pi^0 \mu^+ \mu^-$	C1	< 1.8	× 10^{-4} CL=90%	915
$\eta e^+ e^-$	C1	< 1.1	× 10^{-4} CL=90%	852
$\eta \mu^+ \mu^-$	C1	< 5.3	× 10^{-4} CL=90%	838
$\rho^0 e^+ e^-$	C1	< 1.0	× 10^{-4} CL=90%	773
$\rho^0 \mu^+ \mu^-$	C1	< 2.3	× 10^{-4} CL=90%	756
$\omega e^+ e^-$	C1	< 1.8	× 10^{-4} CL=90%	768
$\omega \mu^+ \mu^-$	C1	< 8.3	× 10^{-4} CL=90%	751
$\phi e^+ e^-$	C1	< 5.2	× 10^{-5} CL=90%	654
$\phi \mu^+ \mu^-$	C1	< 4.1	× 10^{-4} CL=90%	631
$\overline{K}^0 e^+ e^-$	[ss] < 1.1	× 10^{-4} CL=90%	866
$\overline{K}^0 \mu^+ \mu^-$	[ss] < 2.6	× 10^{-4} CL=90%	852
$\overline{K}^*(892)^0 e^+ e^-$	[ss] < 1.4	× 10^{-4} CL=90%	717
$\overline{K}^*(892)^0 \mu^+ \mu^-$	[ss] < 1.18	× 10^{-3} CL=90%	698
$\pi^+ \pi^- \pi^0 \mu^+ \mu^-$	C1	< 8.1	× 10^{-4} CL=90%	863
$\mu^\pm e^\mp$	LF [gg] < 1.9	× 10^{-5} CL=90%	929
$\pi^0 e^\pm \mu^\mp$	LF [gg] < 8.6	× 10^{-5} CL=90%	924
$\eta e^\pm \mu^\mp$	LF [gg] < 1.0	× 10^{-4} CL=90%	848
$\rho^0 e^\pm \mu^\mp$	LF [gg] < 4.9	× 10^{-5} CL=90%	769
$\omega e^\pm \mu^\mp$	LF [gg] < 1.2	× 10^{-4} CL=90%	764
$\phi e^\pm \mu^\mp$	LF [gg] < 3.4	× 10^{-5} CL=90%	648
$\overline{K}^0 e^\pm \mu^\mp$	LF [gg] < 1.0	× 10^{-4} CL=90%	862
$\overline{K}^*(892)^0 e^\pm \mu^\mp$	LF [gg] < 1.0	× 10^{-4} CL=90%	712

$D^*(2007)^0$	$I(J^P) = \frac{1}{2}(1^-)$
	Γ JΓ P need confirmation.

Mass $m = 2006.7 \pm 0.5$ MeV　(S = 1.1)
$m_{D^{*0}} - m_{D^0} = 142.12 \pm 0.07$ MeV
Full width Γ < 2.1 MeVΓCL = 90%

$\overline{D}^*(2007)^0$ modes are charge conjugates of modes below.

$D^*(2007)^0$ DECAY MODES	Fraction (Γ_i/Γ)	p (MeV/c)
$D^0 \pi^0$	(61.9±2.9) %	43
$D^0 \gamma$	(38.1±2.9) %	137

$D^*(2010)^\pm$	$I(J^P) = \frac{1}{2}(1^-)$
	Γ JΓ P need confirmation.

Mass $m = 2010.0 \pm 0.5$ MeV　(S = 1.1)
$m_{D^*(2010)^+} - m_{D^+} = 140.64 \pm 0.10$ MeV　(S = 1.1)
$m_{D^*(2010)^+} - m_{D^0} = 145.397 \pm 0.030$ MeV
Full width Γ < 0.131 MeVΓCL = 90%

$D^*(2010)^-$ modes are charge conjugates of the modes below.

$D^*(2010)^\pm$ DECAY MODES	Fraction (Γ_i/Γ)	p (MeV/c)
$D^0 \pi^+$	(68.3±1.4) %	39
$D^+ \pi^0$	(30.6±2.5) %	38
$D^+ \gamma$	(1.1 $^{+2.1}_{-0.7}$) %	136

$I(J^P) = \frac{1}{2}(1^+)$
Γ J P need confirmation.

Mass $m = 2422.2 \pm 1.8$ MeV   (S = 1.2)
Full width $\Gamma = 18.9^{+4.6}_{-3.5}$ MeV

$\overline{D}_1(2420)^0$ modes are charge conjugates of modes below.

$D_1(2420)^0$ DECAY MODES	Fraction (Γ_i/Γ)	p (MeV/c)
$D^*(2010)^+\pi^-$	seen	355
$D^+\pi^-$	not seen	474

$I(J^P) = \frac{1}{2}(2^+)$

$J^P = 2^+$ assignment strongly favored (ALBRECHT 89B).

Mass $m = 2458.9 \pm 2.0$ MeV   (S = 1.2)
Full width $\Gamma = 23 \pm 5$ MeV

$\overline{D}_2^*(2460)^0$ modes are charge conjugates of modes below.

$D_2^*(2460)^0$ DECAY MODES	Fraction (Γ_i/Γ)	p (MeV/c)
$D^+\pi^-$	seen	503
$D^*(2010)^+\pi^-$	seen	387

$I(J^P) = \frac{1}{2}(2^+)$

$J^P = 2^+$ assignment strongly favored (ALBRECHT 89B).
Mass $m = 2459 \pm 4$ MeV   (S = 1.7)
$m_{D_2^*(2460)^\pm} - m_{D_2^*(2460)^0} = 0.9 \pm 3.3$ MeV   (S = 1.1)
Full width $\Gamma = 25^{+8}_{-7}$ MeV

$D_2^*(2460)^-$ modes are charge conjugates of modes below.

$D_2^*(2460)^\pm$ DECAY MODES	Fraction (Γ_i/Γ)	p (MeV/c)
$D^0\pi^+$	seen	508
$D^{*0}\pi^+$	seen	390

CHARMED, STRANGE MESONS ($C = S = \pm1$)

$D_s^+ = c\bar{s}$, $D_s^- = \bar{c}s$ similarly for D_s^*'s

$D_s^\pm$
was $F^\pm$

$I(J^P) = 0(0^-)$

Mass $m = 1968.5 \pm 0.6$ MeV   (S = 1.1)
$m_{D_s^\pm} - m_{D^\pm} = 99.2 \pm 0.5$ MeV   (S = 1.1)
Mean life $\tau = (0.467 \pm 0.017) \times 10^{-12}$ s
     $c\tau = 140$ μm

D_s^+ form factors
    $r_2 = 1.6 \pm 0.4$
    $r_V = 1.5 \pm 0.5$
    $\Gamma_L/\Gamma_T = 0.72 \pm 0.18$

Branching fractions for modes with a resonance in the final state include all the decay modes of the resonance. D_s^- modes are charge conjugates of the modes below.

D_s^+ DECAY MODES	Fraction (Γ_i/Γ)	Scale factor/ Confidence level	p (MeV/c)
Inclusive modes			
K^- anything	(13 $^{+14}_{-12}$) %		–
$\overline{K}^0$ anything + K^0 anything	(39 ±28) %		–
K^+ anything	(20 $^{+18}_{-14}$) %		–
non-$K\overline{K}$ anything	(64 ±17) %		–
e^+ anything	(8 $^{+6}_{-5}$) %		–
ϕ anything	(18 $^{+15}_{-10}$) %		–
Leptonic and semileptonic modes			
$\mu^+\nu_\mu$	(4.0 $^{+2.2}_{-2.0}$) $\times 10^{-3}$	S=1.4	981
$\tau^+\nu_\tau$	(7 ± 4) %		182
$\phi\ell^+\nu_\ell$	[xx] (2.0 $\pm$ 0.5) %		–
$\eta\ell^+\nu_\ell + \eta'(958)\ell^+\nu_\ell$	[xx] (3.4 $\pm$ 1.0) %		–
$\eta\ell^+\nu_\ell$	(2.5 $\pm$ 0.7) %		–
$\eta'(958)\ell^+\nu_\ell$	(8.8 $\pm$ 3.4) $\times 10^{-3}$		–
Hadronic modes with a $K\overline{K}$ pair (including from a ϕ)			
$K^+\overline{K}^0$	(3.6 $\pm$ 1.1) %		850
$K^+K^-\pi^+$	[qq] (4.4 $\pm$ 1.2) %	S=1.1	805
$\phi\pi^+$	[yy] (3.6 $\pm$ 0.9) %		712
$K^+\overline{K}^*(892)^0$	[yy] (3.3 $\pm$ 0.9) %		682
$f_0(980)\pi^+$	[yy] (1.8 $\pm$ 0.8) %	S=1.3	732
$K^+\overline{K}_0^*(1430)^0$	[yy] (7 $\pm$ 4) $\times 10^{-3}$		186
$f_J(1710)\pi^+ \to K^+K^-\pi^+$	[zz] (1.5 $\pm$ 1.9) $\times 10^{-3}$		204
$K^+K^-\pi^+$ nonresonant	(9 $\pm$ 4) $\times 10^{-3}$		805
$K^0\overline{K}^0\pi^+$	–		802
$K^*(892)^+\overline{K}^0$	[yy] (4.3 $\pm$ 1.4) %		683
$K^+K^-\pi^+\pi^0$	–		748
$\phi\pi^+\pi^0$	[yy] (9 $\pm$ 5) %		687
$\phi\rho^+$	[yy] (6.7 $\pm$ 2.3) %		407
$\phi\pi^+\pi^0$ 3-body	[yy] < 2.6 %	CL=90%	687
$K^+K^-\pi^+\pi^0$ non-ϕ	< 9 %	CL=90%	748
$K^+\overline{K}^0\pi^+\pi^-$	< 2.8 %	CL=90%	744
$K^0K^-\pi^+\pi^+$	(4.3 $\pm$ 1.5) %		744
$K^*(892)^+\overline{K}^*(892)^0$	[yy] (5.8 $\pm$ 2.5) %		412
$K^0K^-\pi^+\pi^+$ non-$K^{*+}\overline{K}^{*0}$	< 2.9 %	CL=90%	744
$K^+K^-\pi^+\pi^+\pi^-$	(8.3 $\pm$ 3.3) $\times 10^{-3}$		673
$\phi\pi^+\pi^+\pi^-$	[yy] (1.18 $\pm$ 0.35) %		640
$K^+K^-\pi^+\pi^+\pi^-$ non-ϕ	(3.0 $^{+3.0}_{-2.0}$) $\times 10^{-3}$		673
Hadronic modes without K's			
$\pi^+\pi^+\pi^-$	(1.0 $\pm$ 0.4) %	S=1.2	959
$\rho^0\pi^+$	< 8 $\times 10^{-4}$	CL=90%	827
$f_0(980)\pi^+$	[yy] (1.8 $\pm$ 0.8) %	S=1.7	732
$f_2(1270)\pi^+$	[yy] (2.3 $\pm$ 1.3) $\times 10^{-3}$		559
$f_0(1500)\pi^+ \to \pi^+\pi^-\pi^+$	[aaa] (2.8 $\pm$ 1.6) $\times 10^{-3}$		391
$\pi^+\pi^+\pi^-$ nonresonant	< 2.8 $\times 10^{-3}$	CL=90%	959
$\pi^+\pi^+\pi^-\pi^0$	< 12 %	CL=90%	935
$\eta\pi^+$	[yy] (2.0 $\pm$ 0.6) %		902

$\omega\pi^+$	[yy]	$(3.1 \pm 1.4) \times 10^{-3}$		822
$\pi^+\pi^+\pi^+\pi^-\pi^-$		$(6.9 \pm 3.0) \times 10^{-3}$		899
$\pi^+\pi^+\pi^-\pi^0\pi^0$		—		902
$\eta\rho^+$	[yy]	$(10.3 \pm 3.2) \%$		727
$\eta\pi^+\pi^0$ 3-body	[yy]	$< 3.0 \quad \%$	CL=90%	886
$\pi^+\pi^+\pi^+\pi^-\pi^-\pi^0$		$(4.9 \pm 3.2) \%$		856
$\eta'(958)\pi^+$	[yy]	$(4.9 \pm 1.8) \%$		743
$\pi^+\pi^+\pi^+\pi^-\pi^-\pi^0\pi^0$		—		803
$\eta'(958)\rho^+$	[yy]	$(12 \quad \pm 4 \quad) \%$		470
$\eta'(958)\pi^+\pi^0$ 3-body	[yy]	$< 3.1 \quad \%$	CL=90%	720

Modes with one or three K's

$K^0\pi^+$		$< 8 \quad \times 10^{-3}$	CL=90%	916
$K^+\pi^+\pi^-$		$(1.0 \pm 0.4) \%$		900
$K^+\rho^0$		$< 2.9 \quad \times 10^{-3}$	CL=90%	747
$K^*(892)^0\pi^+$	[yy]	$(6.5 \pm 2.8) \times 10^{-3}$		773
$K^+K^+K^-$		$< 6 \quad \times 10^{-4}$	CL=90%	628
ϕK^+	[yy]	$< 5 \quad \times 10^{-4}$	CL=90%	607

$\Delta C = 1$ weak neutral current ($C1$) modes, or Lepton number (L) violating modes

$\pi^+\mu^+\mu^-$	[ss]	$< 4.3 \quad \times 10^{-4}$	CL=90%	968
$K^+\mu^+\mu^-$	C1	$< 5.9 \quad \times 10^{-4}$	CL=90%	909
$K^*(892)^+\mu^+\mu^-$	C1	$< 1.4 \quad \times 10^{-3}$	CL=90%	765
$\pi^-\mu^+\mu^+$	L	$< 4.3 \quad \times 10^{-4}$	CL=90%	968
$K^-\mu^+\mu^+$	L	$< 5.9 \quad \times 10^{-4}$	CL=90%	909
$K^*(892)^-\mu^+\mu^+$	L	$< 1.4 \quad \times 10^{-3}$	CL=90%	765

$$I(J^P) = 0(?^?)$$

J^P is natural, width and decay modes consistent with 1^-.

Mass $m = 2112.4 \pm 0.7$ MeV $\quad$ (S = 1.1)

$m_{D_s^{*\pm}} - m_{D_s^\pm} = 143.8 \pm 0.4$ MeV

Full width $\Gamma < 1.9$ MeV, CL = 90%

D_s^{*-} modes are charge conjugates of the modes below.

D_s^{*+} DECAY MODES	Fraction (Γ_i/Γ)	p (MeV/c)
$D_s^+\gamma$	$(94.2\pm2.5) \%$	139
$D_s^+\pi^0$	$(5.8\pm2.5) \%$	48

$D_{s1}(2536)^\pm$

$$I(J^P) = 0(1^+)$$
J, P need confirmation.

Mass $m = 2535.35 \pm 0.34 \pm 0.5$ MeV

Full width $\Gamma < 2.3$ MeV, CL = 90%

$D_{s1}(2536)^-$ modes are charge conjugates of the modes below.

$D_{s1}(2536)^+$ DECAY MODES	Fraction (Γ_i/Γ)	p (MeV/c)
$D^*(2010)^+ K^0$	seen	150
$D^*(2007)^0 K^+$	seen	169
$D^+ K^0$	not seen	382
$D^0 K^+$	not seen	392
$D_s^{*+}\gamma$	possibly seen	389

$D_{sJ}(2573)^\pm$

$$I(J^P) = 0(?^?)$$

J^P is natural, width and decay modes consistent with 2^+.

Mass $m = 2573.5 \pm 1.7$ MeV

Full width $\Gamma = 15^{+5}_{-4}$ MeV

$D_{sJ}(2573)^-$ modes are charge conjugates of the modes below.

$D_{sJ}(2573)^+$ DECAY MODES	Fraction (Γ_i/Γ)	p (MeV/c)
$D^0 K^+$	seen	436
$D^*(2007)^0 K^+$	not seen	245

BOTTOM MESONS
($B = \pm 1$)

$B^+ = u\bar{b}, B^0 = d\bar{b}, \overline{B}^0 = \bar{d}\,b, B^- = \bar{u}\,b \quad$ similarly for B^*'s

B-particle organization

Many measurements of B decays involve admixtures of B hadrons. Previously we arbitrarily included such admixtures in the $B^\pm$ section, but because of their importance we have created two new sections: "$B^\pm/B^0$ Admixture" for $\Upsilon(4S)$ results and "$B^\pm/B^0/B_s^0/b$-baryon Admixture" for results at higher energies. Most inclusive decay branching fractions are found in the Admixture sections. B^0-$\overline{B}^0$ mixing data are found in the B^0 section, while B_s^0-$\overline{B}_s^0$ mixing data and B-$\overline{B}$ mixing data for a B^0/B_s^0 admixture are found in the B_s^0 section. CP-violation data are found in the B^0 section. b-baryons are found near the end of the Baryon section.

The organization of the B sections is now as follows, where bullets indicate particle sections and brackets indicate reviews.

[Production and Decay of b-flavored Hadrons]
[Semileptonic Decays of B Mesons]
- $B^\pm$
    mass, mean life
    branching fractions
- B^0
    mass, mean life
    branching fractions
    polarization in B^0 decay
    B^0-$\overline{B}^0$ mixing
    [B^0-$\overline{B}^0$ Mixing and CP Violation in B Decay]
    CP violation
- $B^\pm$ B^0 Admixtures
    branching fractions
- $B^\pm/B^0/B_s^0/b$-baryon Admixtures
    mean life
    production fractions
    branching fractions
- B^*
    mass
- B_s^0
    mass, mean life
    branching fractions
    polarization in B_s^0 decay
    B_s^0-$\overline{B}_s^0$ mixing
    B-$\overline{B}$ mixing (admixture of B^0, B_s^0)

At end of Baryon Listings:
- Λ_b
    mass, mean life
    branching fractions
- b-baryon Admixture
    mean life
    branching fractions

$B^{\pm}$ $I(J^P) = \frac{1}{2}(0^-)$

I, J, P need confirmation. Quantum numbers shown are quark-model predictions.

Mass $m_{B^{\pm}} = 5278.9 \pm 1.8$ MeV
Mean life $\tau_{B^{\pm}} = (1.65 \pm 0.04) \times 10^{-12}$ s
$c\tau = 495\ \mu$m

B^- modes are charge conjugates of the modes below. Modes which do not identify the charge state of the B are listed in the $B^{\pm}/B^0$ ADMIXTURE section.

The branching fractions listed below assume 50% $B^0\overline{B}^0$ and 50% B^+B^- production at the $\Upsilon(4S)$. We have attempted to bring older measurements up to date by rescaling their assumed $\Upsilon(4S)$ production ratio to 50:50 and their assumed D, D_s, D^* and ψ branching ratios to current values whenever this would affect our averages and best limits significantly.

Indentation is used to indicate a subchannel of a previous reaction. All resonant subchannels have been corrected for resonance branching fractions to the final state so the sum of the subchannel branching fractions can exceed that of the final state.

B^+ DECAY MODES	Fraction (Γ_i/Γ)	Scale factor/ Confidence level	p (MeV/c)
Semileptonic and leptonic modes			
$\ell^+ \nu_\ell$ anything	[pp] (10.3 $\pm$0.9) %		–
$\overline{D}^0 \ell^+ \nu_\ell$	[pp] (1.86$\pm$0.33) %		–
$\overline{D}^*(2007)^0 \ell^+ \nu_\ell$	[pp] (5.3 $\pm$0.8) %		–
$\pi^0 e^+ \nu_e$	< 2.2 $\times 10^{-3}$	CL=90%	2638
$\omega \ell^+ \nu_\ell$	[pp] < 2.1 $\times 10^{-4}$	CL=90%	–
$\rho^0 \ell^+ \nu_\ell$	[pp] < 2.1 $\times 10^{-4}$	CL=90%	–
$e^+ \nu_e$	< 1.5 $\times 10^{-5}$	CL=90%	2639
$\mu^+ \nu_\mu$	< 2.1 $\times 10^{-5}$	CL=90%	2638
$\tau^+ \nu_\tau$	< 5.7 $\times 10^{-4}$	CL=90%	2340
$e^+ \nu_e \gamma$	< 2.0 $\times 10^{-4}$	CL=90%	–
$\mu^+ \nu_\mu \gamma$	< 5.2 $\times 10^{-5}$	CL=90%	–
D, D^*, or D_s modes			
$\overline{D}^0 \pi^+$	(5.3 $\pm$0.5) $\times 10^{-3}$		2308
$\overline{D}^0 \rho^+$	(1.34 $\pm$0.18) %		2238
$\overline{D}^0 \pi^+ \pi^+ \pi^-$	(1.1 $\pm$0.4) %		2289
$\overline{D}^0 \pi^+ \pi^+ \pi^-$ nonresonant	(5 $\pm$4) $\times 10^{-3}$		2289
$\overline{D}^0 \pi^+ \rho^0$	(4.2 $\pm$3.0) $\times 10^{-3}$		2209
$\overline{D}^0 a_1(1260)^+$	(5 $\pm$4) $\times 10^{-3}$		2123
$D^*(2010)^- \pi^+ \pi^+$	(2.1 $\pm$0.6) $\times 10^{-3}$		2247
$D^- \pi^+ \pi^+$	< 1.4 $\times 10^{-3}$	CL=90%	2299
$\overline{D}^*(2007)^0 \pi^+$	(4.6 $\pm$0.4) $\times 10^{-3}$		2256
$D^*(2010)^+ \pi^0$	< 1.7 $\times 10^{-4}$	CL=90%	2254
$\overline{D}^*(2007)^0 \rho^+$	(1.55 $\pm$0.31) %		2183
$\overline{D}^*(2007)^0 \pi^+ \pi^+ \pi^-$	(9.4 $\pm$2.6) $\times 10^{-3}$		2236
$\overline{D}^*(2007)^0 a_1(1260)^+$	(1.9 $\pm$0.5) %		2062
$D^*(2010)^- \pi^+ \pi^+ \pi^0$	(1.5 $\pm$0.7) %		2235
$D^*(2010)^- \pi^+ \pi^+ \pi^+ \pi^-$	< 1 %	CL=90%	2217
$\overline{D}_1^*(2420)^0 \pi^+$	(1.5 $\pm$0.6) $\times 10^{-3}$	S=1.3	2081
$\overline{D}_1^*(2420)^0 \rho^+$	< 1.4 $\times 10^{-3}$	CL=90%	1997
$\overline{D}_2^*(2460)^0 \pi^+$	< 1.3 $\times 10^{-3}$	CL=90%	2064
$\overline{D}_2^*(2460)^0 \rho^+$	< 4.7 $\times 10^{-3}$	CL=90%	1979
$\overline{D}^0 D_s^+$	(1.3 $\pm$0.4) %		1815
$\overline{D}^0 D_s^{*+}$	(9 $\pm$4) $\times 10^{-3}$		1734
$\overline{D}^*(2007)^0 D_s^+$	(1.2 $\pm$0.5) %		1737
$\overline{D}^*(2007)^0 D_s^{*+}$	(2.7 $\pm$1.0) %		1650
$D_s^+ \pi^0$	< 2.0 $\times 10^{-4}$	CL=90%	2270
$D_s^{*+} \pi^0$	< 3.3 $\times 10^{-4}$	CL=90%	2214
$D_s^+ \eta$	< 5 $\times 10^{-4}$	CL=90%	2235
$D_s^{*+} \eta$	< 8 $\times 10^{-4}$	CL=90%	2177
$D_s^+ \rho^0$	< 4 $\times 10^{-4}$	CL=90%	2198
$D_s^{*+} \rho^0$	< 5 $\times 10^{-4}$	CL=90%	2139
$D_s^+ \omega$	< 5 $\times 10^{-4}$	CL=90%	2195
$D_s^{*+} \omega$	< 7 $\times 10^{-4}$	CL=90%	2136
$D_s^+ a_1(1260)^0$	< 2.2 $\times 10^{-3}$	CL=90%	2079
$D_s^{*+} a_1(1260)^0$	< 1.6 $\times 10^{-3}$	CL=90%	2014
$D_s^+ \phi$	< 3.2 $\times 10^{-4}$	CL=90%	2141
$D_s^{*+} \phi$	< 4 $\times 10^{-4}$	CL=90%	2079

B^+ DECAY MODES	Fraction (Γ_i/Γ)		p (MeV/c)
$D_s^+ \overline{K}^0$	< 1.1 $\times 10^{-3}$	CL=90%	2241
$D_s^{*+} \overline{K}^0$	< 1.1 $\times 10^{-3}$	CL=90%	2184
$D_s^+ \overline{K}^*(892)^0$	< 5 $\times 10^{-4}$	CL=90%	2171
$D_s^{*+} \overline{K}^*(892)^0$	< 4 $\times 10^{-4}$	CL=90%	2110
$D_s^- \pi^+ K^+$	< 8 $\times 10^{-4}$	CL=90%	2222
$D_s^{*-} \pi^+ K^+$	< 1.2 $\times 10^{-3}$	CL=90%	2164
$D_s^- \pi^+ K^*(892)^+$	< 6 $\times 10^{-3}$	CL=90%	2137
$D_s^{*-} \pi^+ K^*(892)^+$	< 8 $\times 10^{-3}$	CL=90%	2075
Charmonium modes			
$J/\psi(1S) K^+$	(9.9 $\pm$1.0) $\times 10^{-4}$		1683
$J/\psi(1S) K^+ \pi^+ \pi^-$	(1.4 $\pm$0.6) $\times 10^{-3}$		1612
$J/\psi(1S) K^*(892)^+$	(1.47 $\pm$0.27) $\times 10^{-3}$		1571
$J/\psi(1S) \pi^+$	(5.0 $\pm$1.5) $\times 10^{-5}$		1727
$J/\psi(1S) \rho^+$	< 7.7 $\times 10^{-4}$	CL=90%	1613
$J/\psi(1S) a_1(1260)^+$	< 1.2 $\times 10^{-3}$	CL=90%	1414
$\psi(2S) K^+$	(6.9 $\pm$3.1) $\times 10^{-4}$	S=1.3	1284
$\psi(2S) K^*(892)^+$	< 3.0 $\times 10^{-3}$	CL=90%	1115
$\psi(2S) K^+ \pi^+ \pi^-$	(1.9 $\pm$1.2) $\times 10^{-3}$		909
$\chi_{c1}(1P) K^+$	(1.0 $\pm$0.4) $\times 10^{-3}$		1411
$\chi_{c1}(1P) K^*(892)^+$	< 2.1 $\times 10^{-3}$	CL=90%	1265
K or K^* modes			
$K^0 \pi^+$	(2.3 $\pm$1.1) $\times 10^{-5}$		2614
$K^+ \pi^0$	< 1.6 $\times 10^{-5}$	CL=90%	2615
$\eta' K^+$	(6.5 $\pm$1.7) $\times 10^{-5}$		2528
$\eta' K^*(892)^+$	< 1.3 $\times 10^{-4}$	CL=90%	2472
ηK^+	< 1.4 $\times 10^{-5}$	CL=90%	2587
$\eta K^*(892)^+$	< 3.0 $\times 10^{-5}$	CL=90%	2534
$K^*(892)^0 \pi^+$	< 4.1 $\times 10^{-5}$	CL=90%	2561
$K^*(892)^+ \pi^0$	< 9.9 $\times 10^{-5}$	CL=90%	2562
$K^+ \pi^- \pi^+$ nonresonant	< 2.8 $\times 10^{-5}$	CL=90%	2609
$K^- \pi^+ \pi^+$ nonresonant	< 5.6 $\times 10^{-5}$	CL=90%	–
$K_1(1400)^0 \pi^+$	< 2.6 $\times 10^{-3}$	CL=90%	2451
$K_2^*(1430)^0 \pi^+$	< 6.8 $\times 10^{-4}$	CL=90%	2443
$K^+ \rho^0$	< 1.9 $\times 10^{-5}$	CL=90%	2559
$K^0 \rho^+$	< 4.8 $\times 10^{-5}$	CL=90%	2559
$K^*(892)^+ \pi^+ \pi^-$	< 1.1 $\times 10^{-3}$	CL=90%	2556
$K^*(892)^+ \rho^0$	< 9.0 $\times 10^{-4}$	CL=90%	2505
$K_1(1400)^+ \rho^0$	< 7.8 $\times 10^{-4}$	CL=90%	2389
$K_2^*(1430)^+ \rho^0$	< 1.5 $\times 10^{-3}$	CL=90%	2382
$K^+ \overline{K}^0$	< 2.1 $\times 10^{-5}$	CL=90%	2592
$K^+ K^- \pi^+$ nonresonant	< 7.5 $\times 10^{-5}$	CL=90%	–
$K^+ K^- K^+$	< 2.0 $\times 10^{-5}$	CL=90%	2522
$K^+ \phi$	< 1.2 $\times 10^{-5}$	CL=90%	2516
$K^+ K^- K^+$ nonresonant	< 3.8 $\times 10^{-5}$	CL=90%	2516
$K^*(892)^+ K^+ K^-$	< 1.6 $\times 10^{-3}$	CL=90%	2466
$K^*(892)^+ \phi$	< 7.0 $\times 10^{-3}$	CL=90%	2460
$K_1(1400)^+ \phi$	< 1.1 $\times 10^{-3}$	CL=90%	2339
$K_2^*(1430)^+ \phi$	< 3.4 $\times 10^{-3}$	CL=90%	2332
$K^+ f_0(980)$	< 8 $\times 10^{-5}$	CL=90%	2524
$K^*(892)^+ \gamma$	(5.7 $\pm$3.3) $\times 10^{-5}$		2564
$K_1(1270)^+ \gamma$	< 7.3 $\times 10^{-3}$	CL=90%	2486
$K_1(1400)^+ \gamma$	< 2.2 $\times 10^{-3}$	CL=90%	2453
$K_2^*(1430)^+ \gamma$	< 1.4 $\times 10^{-3}$	CL=90%	2447
$K^*(1680)^+ \gamma$	< 1.9 $\times 10^{-3}$	CL=90%	2361
$K_3^*(1780)^+ \gamma$	< 5.5 $\times 10^{-3}$	CL=90%	2343
$K_4^*(2045)^+ \gamma$	< 9.9 $\times 10^{-3}$	CL=90%	2243
Light unflavored meson modes			
$\pi^+ \pi^0$	< 2.0 $\times 10^{-5}$	CL=90%	2636
$\pi^+ \pi^+ \pi^-$	< 1.3 $\times 10^{-4}$	CL=90%	2630
$\rho^0 \pi^+$	< 4.3 $\times 10^{-5}$	CL=90%	2582
$\pi^+ f_0(980)$	< 1.4 $\times 10^{-4}$	CL=90%	2547
$\pi^+ f_2(1270)$	< 2.4 $\times 10^{-4}$	CL=90%	2483
$\pi^+ \pi^- \pi^+$ nonresonant	< 4.1 $\times 10^{-5}$	CL=90%	–
$\pi^+ \pi^0 \pi^0$	< 8.9 $\times 10^{-4}$	CL=90%	2631
$\rho^+ \pi^0$	< 7.7 $\times 10^{-5}$	CL=90%	2582
$\pi^+ \pi^- \pi^+ \pi^0$	< 4.0 $\times 10^{-3}$	CL=90%	2621
$\rho^+ \rho^0$	< 1.0 $\times 10^{-3}$	CL=90%	2525
$a_1(1260)^+ \pi^0$	< 1.7 $\times 10^{-3}$	CL=90%	2494
$a_1(1260)^0 \pi^+$	< 9.0 $\times 10^{-4}$	CL=90%	2494
$\omega \pi^+$	< 4.0 $\times 10^{-4}$	CL=90%	2580

$\eta\pi^+$	< 1.5	$\times 10^{-5}$	CL=90%	2609
$\eta'\pi^+$	< 3.1	$\times 10^{-5}$	CL=90%	2550
$\eta'\rho^+$	< 4.7	$\times 10^{-5}$	CL=90%	2493
$\eta\rho^+$	< 3.2	$\times 10^{-5}$	CL=90%	2554
$\pi^+\pi^+\pi^+\pi^-\pi^-$	< 8.6	$\times 10^{-4}$	CL=90%	2608
$\rho^0 a_1(1260)^+$	< 6.2	$\times 10^{-4}$	CL=90%	2434
$\rho^0 a_2(1320)^+$	< 7.2	$\times 10^{-4}$	CL=90%	2411
$\pi^+\pi^+\pi^+\pi^-\pi^-\pi^0$	< 6.3	$\times 10^{-3}$	CL=90%	2592
$a_1(1260)^+ a_1(1260)^0$	< 1.3	%	CL=90%	2335

Baryon modes

$p\bar{p}\pi^+$	< 1.6	$\times 10^{-4}$	CL=90%	2439
$p\bar{p}\pi^+$ nonresonant	< 5.3	$\times 10^{-5}$	CL=90%	–
$p\bar{p}\pi^+\pi^+\pi^-$	< 5.2	$\times 10^{-4}$	CL=90%	2369
$p\bar{p}K^+$ nonresonant	< 8.9	$\times 10^{-5}$	CL=90%	–
$p\bar{\Lambda}$	< 6	$\times 10^{-5}$	CL=90%	2430
$p\bar{\Lambda}\pi^+\pi^-$	< 2.0	$\times 10^{-4}$	CL=90%	2367
$\bar{\Delta}^0 p$	< 3.8	$\times 10^{-4}$	CL=90%	2402
$\Delta^{++}\bar{p}$	< 1.5	$\times 10^{-4}$	CL=90%	2402
$\Lambda_c^- p\pi^+$	(6.2 ±2.7) $\times 10^{-4}$			–
$\Lambda_c^- p\pi^+\pi^0$	< 3.12	$\times 10^{-3}$	CL=90%	–
$\Lambda_c^- p\pi^+\pi^+\pi^-$	< 1.46	$\times 10^{-3}$	CL=90%	–
$\Lambda_c^- p\pi^+\pi^+\pi^-\pi^0$	< 1.34	%	CL=90%	–

Lepton Family number (LF) or Lepton number (L) violating modesΓor ΔB = 1 weak neutral current (B1) modes

$\pi^+ e^+ e^-$	B1	< 3.9	$\times 10^{-3}$	CL=90%	2638
$\pi^+ \mu^+ \mu^-$	B1	< 9.1	$\times 10^{-3}$	CL=90%	2633
$K^+ e^+ e^-$	B1	< 6	$\times 10^{-5}$	CL=90%	2616
$K^+ \mu^+ \mu^-$	B1	< 1.0	$\times 10^{-5}$	CL=90%	2612
$K^*(892)^+ e^+ e^-$	B1	< 6.9	$\times 10^{-4}$	CL=90%	2564
$K^*(892)^+ \mu^+ \mu^-$	B1	< 1.2	$\times 10^{-3}$	CL=90%	2560
$\pi^+ e^+ \mu^-$	LF	< 6.4	$\times 10^{-3}$	CL=90%	2637
$\pi^+ e^- \mu^+$	LF	< 6.4	$\times 10^{-3}$	CL=90%	2637
$K^+ e^+ \mu^-$	LF	< 6.4	$\times 10^{-3}$	CL=90%	2615
$K^+ e^- \mu^+$	LF	< 6.4	$\times 10^{-3}$	CL=90%	2615
$\pi^- e^+ e^+$	L	< 3.9	$\times 10^{-3}$	CL=90%	2638
$\pi^- \mu^+ \mu^+$	L	< 9.1	$\times 10^{-3}$	CL=90%	2633
$\pi^- e^+ \mu^+$	LF	< 6.4	$\times 10^{-3}$	CL=90%	2637
$K^- e^+ e^+$	L	< 3.9	$\times 10^{-3}$	CL=90%	2616
$K^- \mu^+ \mu^+$	L	< 9.1	$\times 10^{-3}$	CL=90%	2612
$K^- e^+ \mu^+$	LF	< 6.4	$\times 10^{-3}$	CL=90%	2615

 B^0

$I(J^P) = \frac{1}{2}(0^-)$

IΓ JΓ P need confirmation. Quantum numbers shown are quark-model predictions.

> Mass $m_{B^0} = 5279.2 \pm 1.8$ MeV
> $m_{B^0} - m_{B^\pm} = 0.35 \pm 0.29$ MeV (S = 1.1)
> Mean life $\tau_{B^0} = (1.56 \pm 0.04) \times 10^{-12}$ s
> $c\tau = 468\ \mu m$
> $\tau_{B^+}/\tau_{B^0} = 1.02 \pm 0.04$ (average of direct and inferred)
> $\tau_{B^+}/\tau_{B^0} = 1.04 \pm 0.04$ (direct measurements)
> $\tau_{B^+}/\tau_{B^0} = 0.95^{+0.15}_{-0.12}$ (inferred from branching fractions)

B^0-$\bar{B}^0$ mixing parameters

> $\chi_d = 0.172 \pm 0.010$
> $\Delta m_{B^0} = m_{B^0_H} - m_{B^0_L} = (0.464 \pm 0.018) \times 10^{12}\ \hbar\ s^{-1}$
> $x_d = \Delta m_{B^0}/\Gamma_{B^0} = 0.723 \pm 0.032$

CP violation parameters

> $|\mathrm{Re}(\epsilon_{B^0})| = 0.002 \pm 0.008$

$\bar{B}^0$ modes are charge conjugates of the modes below. Reactions indicate the weak decay vertex and do not include mixing. Modes which do not identify the charge state of the B are listed in the $B^\pm/B^0$ ADMIXTURE section.

The branching fractions listed below assume 50% $B^0\bar{B}^0$ and 50% B^+B^- production at the $\Upsilon(4S)$. We have attempted to bring older measurements up to date by rescaling their assumed $\Upsilon(4S)$ production ratio to 50:50 and their assumed $D\Gamma D_s\Gamma D^*\Gamma$ and ψ branching ratios to current values whenever this would affect our averages and best limits significantly.

Indentation is used to indicate a subchannel of a previous reaction. All resonant subchannels have been corrected for resonance branching fractions to the final state so the sum of the subchannel branching fractions can exceed that of the final state.

B^0 DECAY MODES	Fraction (Γ_i/Γ)	Scale factor/ Confidence level	p (MeV/c)
$\ell^+\nu_\ell$ anything	[pp] (10.5 ± 0.8) %		–
$D^-\ell^+\nu_\ell$	[pp] (2.00± 0.25) %		–
$D^*(2010)^-\ell^+\nu_\ell$	[pp] (4.60± 0.27) %		–
$\rho^-\ell^+\nu_\ell$	[pp] (2.5 $^{+0.8}_{-1.0}$) $\times 10^{-4}$		–
$\pi^-\ell^+\nu_\ell$	(1.8 ± 0.6) $\times 10^{-4}$		–

Inclusive modes

K^+ anything	(78 ±80) %		–

DΓ D*Γor D_s modes

$D^-\pi^+$	(3.0 ± 0.4) $\times 10^{-3}$			2306
$D^-\rho^+$	(7.9 ± 1.4) $\times 10^{-3}$			2236
$\overline{D}^0\pi^+\pi^-$	< 1.6	$\times 10^{-3}$	CL=90%	2301
$D^*(2010)^-\pi^+$	(2.76± 0.21) $\times 10^{-3}$			2254
$D^-\pi^+\pi^+\pi^-$	(8.0 ± 2.5) $\times 10^{-3}$			2287
$(D^-\pi^+\pi^+\pi^-)$ nonresonant	(3.9 ± 1.9) $\times 10^{-3}$			2287
$D^-\pi^+\rho^0$	(1.1 ± 1.0) $\times 10^{-3}$			2207
$D^-a_1(1260)^+$	(6.0 ± 3.3) $\times 10^{-3}$			2121
$D^*(2010)^-\pi^+\pi^0$	(1.5 ± 0.5) %			2247
$D^*(2010)^-\rho^+$	(6.7 ± 3.3) $\times 10^{-3}$			2181
$D^*(2010)^-\pi^+\pi^+\pi^-$	(7.6 ± 1.7) $\times 10^{-3}$	S=1.3	2235
$(D^*(2010)^-\pi^+\pi^+\pi^-)$ non-resonant	(0.0 ± 2.5) $\times 10^{-3}$			2235
$D^*(2010)^-\pi^+\rho^0$	(5.7 ± 3.1) $\times 10^{-3}$			2151
$D^*(2010)^-a_1(1260)^+$	(1.30± 0.27) %			2061
$D^*(2010)^-\pi^+\pi^+\pi^-\pi^0$	(3.4 ± 1.8) %			2218
$\overline{D}_2^*(2460)^-\pi^+$	< 2.2	$\times 10^{-3}$	CL=90%	2064
$\overline{D}_2^*(2460)^-\rho^+$	< 4.9	$\times 10^{-3}$	CL=90%	1979
$D^-D_s^+$	(8.0 ± 3.0) $\times 10^{-3}$			1812
$D^*(2010)^-D_s^+$	(9.6 ± 3.4) $\times 10^{-3}$			1735
$D^-D_s^{*+}$	(1.0 ± 0.5) %			1731
$D^*(2010)^-D_s^{*+}$	(2.0 ± 0.7) %			1649
$D_s^+\pi^-$	< 2.8	$\times 10^{-4}$	CL=90%	2270
$D_s^{*+}\pi^-$	< 5	$\times 10^{-4}$	CL=90%	2214
$D_s^+\rho^-$	< 7	$\times 10^{-4}$	CL=90%	2198
$D_s^{*+}\rho^-$	< 8	$\times 10^{-4}$	CL=90%	2139
$D_s^+a_1(1260)^-$	< 2.6	$\times 10^{-3}$	CL=90%	2079
$D_s^{*+}a_1(1260)^-$	< 2.2	$\times 10^{-3}$	CL=90%	2014
$D_s^- K^+$	< 2.4	$\times 10^{-4}$	CL=90%	2242
$D_s^{*-}K^+$	< 1.7	$\times 10^{-4}$	CL=90%	2185
$D_s^- K^*(892)^+$	< 9.9	$\times 10^{-4}$	CL=90%	2172
$D_s^{*-}K^*(892)^+$	< 1.1	$\times 10^{-3}$	CL=90%	2112
$D_s^-\pi^+K^0$	< 5	$\times 10^{-3}$	CL=90%	2221
$D_s^{*-}\pi^+K^0$	< 3.1	$\times 10^{-3}$	CL=90%	2164
$D_s^-\pi^+K^*(892)^0$	< 4	$\times 10^{-3}$	CL=90%	2136
$D_s^{*-}\pi^+K^*(892)^0$	< 2.0	$\times 10^{-3}$	CL=90%	2074
$\overline{D}^0\pi^0$	< 1.2	$\times 10^{-4}$	CL=90%	2308
$\overline{D}^0\rho^0$	< 3.9	$\times 10^{-4}$	CL=90%	2238
$\overline{D}^0\eta$	< 1.3	$\times 10^{-4}$	CL=90%	2274
$\overline{D}^0\eta'$	< 9.4	$\times 10^{-4}$	CL=90%	2198
$\overline{D}^0\omega$	< 5.1	$\times 10^{-4}$	CL=90%	2235
$\overline{D}^*(2007)^0\pi^0$	< 4.4	$\times 10^{-4}$	CL=90%	2256
$\overline{D}^*(2007)^0\rho^0$	< 5.6	$\times 10^{-4}$	CL=90%	2183
$\overline{D}^*(2007)^0\eta$	< 2.6	$\times 10^{-4}$	CL=90%	2220
$\overline{D}^*(2007)^0\eta'$	< 1.4	$\times 10^{-3}$	CL=90%	2141
$\overline{D}^*(2007)^0\omega$	< 7.4	$\times 10^{-4}$	CL=90%	2180
$D^*(2010)^+D^*(2010)^-$	< 2.2	$\times 10^{-3}$	CL=90%	1711
$D^*(2010)^+D^-$	< 1.8	$\times 10^{-3}$	CL=90%	1790
$D^+D^*(2010)^-$	< 1.2	$\times 10^{-3}$	CL=90%	1790

Charmonium modes

$J/\psi(1S)K^0$	(8.9 ± 1.2) $\times 10^{-4}$			1683
$J/\psi(1S)K^+\pi^-$	(1.1 ± 0.6) $\times 10^{-3}$			1652
$J/\psi(1S)K^*(892)^0$	(1.35± 0.18) $\times 10^{-3}$			1570
$J/\psi(1S)\pi^0$	< 5.8	$\times 10^{-5}$	CL=90%	1728
$J/\psi(1S)\eta$	< 1.2	$\times 10^{-3}$	CL=90%	1672
$J/\psi(1S)\rho^0$	< 2.5	$\times 10^{-4}$	CL=90%	1614

$J/\psi(1S)\omega$	< 2.7	$\times 10^{-4}$	CL=90%	1609
$\psi(2S)K^0$	< 8	$\times 10^{-4}$	CL=90%	1283
$\psi(2S)K^+\pi^-$	< 1	$\times 10^{-3}$	CL=90%	1238
$\psi(2S)K^*(892)^0$	(1.4 ± 0.9) $\times 10^{-3}$			1113
$\chi_{c1}(1P)K^0$	< 2.7	$\times 10^{-3}$	CL=90%	1411
$\chi_{c1}(1P)K^*(892)^0$	< 2.1	$\times 10^{-3}$	CL=90%	1263

K or K* modes

$K^+\pi^-$	(1.5 $^{+0.5}_{-0.4}$) $\times 10^{-5}$			2615
$K^0\pi^0$	< 4.1	$\times 10^{-5}$	CL=90%	2614
$\eta'K^0$	(4.7 $^{+2.8}_{-2.2}$) $\times 10^{-5}$			2528
$\eta'K^*(892)^0$	< 3.9	$\times 10^{-5}$	CL=90%	2472
$\eta K^*(892)^0$	< 3.0	$\times 10^{-5}$	CL=90%	2534
ηK^0	< 3.3	$\times 10^{-5}$	CL=90%	2593
K^+K^-	< 4.3	$\times 10^{-6}$	CL=90%	2593
$K^0\overline{K}^0$	< 1.7	$\times 10^{-5}$	CL=90%	2592
$K^+\rho^-$	< 3.5	$\times 10^{-5}$	CL=90%	2559
$K^0\rho^0$	< 3.9	$\times 10^{-5}$	CL=90%	2559
$K^0 f_0(980)$	< 3.6	$\times 10^{-4}$	CL=90%	2523
$K^*(892)^+\pi^-$	< 7.2	$\times 10^{-5}$	CL=90%	2562
$K^*(892)^0\pi^0$	< 2.8	$\times 10^{-5}$	CL=90%	2562
$K_2^*(1430)^+\pi^-$	< 2.6	$\times 10^{-3}$	CL=90%	2445
$K^0 K^+K^-$	< 1.3	$\times 10^{-3}$	CL=90%	2522
$K^0\phi$	< 8.8	$\times 10^{-5}$	CL=90%	2516
$K^-\pi^+\pi^+\pi^-$	[bbb] < 2.3	$\times 10^{-4}$	CL=90%	2600
$K^*(892)^0\pi^+\pi^-$	< 1.4	$\times 10^{-3}$	CL=90%	2556
$K^*(892)^0\rho^0$	< 4.6	$\times 10^{-4}$	CL=90%	2504
$K^*(892)^0 f_0(980)$	< 1.7	$\times 10^{-4}$	CL=90%	2467
$K_1(1400)^+\pi^-$	< 1.1	$\times 10^{-3}$	CL=90%	2451
$K^-a_1(1260)^+$	[bbb] < 2.3	$\times 10^{-4}$	CL=90%	2471
$K^*(892)^0 K^+K^-$	< 6.1	$\times 10^{-4}$	CL=90%	2466
$K^*(892)^0\phi$	< 4.3	$\times 10^{-5}$	CL=90%	2459
$K_1(1400)^0\rho^0$	< 3.0	$\times 10^{-3}$	CL=90%	2389
$K_1(1400)^0\phi$	< 5.0	$\times 10^{-3}$	CL=90%	2339
$K_2^*(1430)^0\rho^0$	< 1.1	$\times 10^{-3}$	CL=90%	2380
$K_2^*(1430)^0\phi$	< 1.4	$\times 10^{-3}$	CL=90%	2330
$K^*(892)^0\gamma$	(4.0 ± 1.9) $\times 10^{-5}$			2564
$K_1(1270)^0\gamma$	< 7.0	$\times 10^{-3}$	CL=90%	2486
$K_1(1400)^0\gamma$	< 4.3	$\times 10^{-3}$	CL=90%	2453
$K_2^*(1430)^0\gamma$	< 4.0	$\times 10^{-4}$	CL=90%	2445
$K^*(1680)^0\gamma$	< 2.0	$\times 10^{-3}$	CL=90%	2361
$K_3^*(1780)^0\gamma$	< 1.0	%	CL=90%	2343
$K_4^*(2045)^0\gamma$	< 4.3	$\times 10^{-3}$	CL=90%	2244
$\phi\phi$	< 3.9	$\times 10^{-5}$	CL=90%	2435

Light unflavored meson modes

$\pi^+\pi^-$	< 1.5	$\times 10^{-5}$	CL=90%	2636
$\pi^0\pi^0$	< 9.3	$\times 10^{-6}$	CL=90%	2636
$\eta\pi^0$	< 8	$\times 10^{-6}$	CL=90%	2609
$\eta\eta$	< 1.8	$\times 10^{-5}$	CL=90%	2582
$\eta'\pi^0$	< 1.1	$\times 10^{-5}$	CL=90%	2551
$\eta'\eta'$	< 4.7	$\times 10^{-5}$	CL=90%	2460
$\eta'\eta$	< 2.7	$\times 10^{-5}$	CL=90%	2522
$\eta'\rho^0$	< 2.3	$\times 10^{-5}$	CL=90%	2493
$\eta\rho^0$	< 1.3	$\times 10^{-5}$	CL=90%	2554
$\pi^+\pi^-\pi^0$	< 7.2	$\times 10^{-4}$	CL=90%	2631
$\rho^0\pi^0$	< 2.4	$\times 10^{-5}$	CL=90%	2582
$\rho^\mp\pi^\pm$	[gg] < 8.8	$\times 10^{-5}$	CL=90%	2582
$\pi^+\pi^-\pi^+\pi^-$	< 2.3	$\times 10^{-4}$	CL=90%	2621
$\rho^0\rho^0$	< 2.8	$\times 10^{-4}$	CL=90%	2525
$a_1(1260)^\mp\pi^\pm$	[gg] < 4.9	$\times 10^{-4}$	CL=90%	2494
$a_2(1320)^\mp\pi^\pm$	[gg] < 3.0	$\times 10^{-4}$	CL=90%	2473
$\pi^+\pi^-\pi^0\pi^0$	< 3.1	$\times 10^{-3}$	CL=90%	2622
$\rho^+\rho^-$	< 2.2	$\times 10^{-3}$	CL=90%	2525
$a_1(1260)^0\pi^0$	< 1.1	$\times 10^{-3}$	CL=90%	2494
$\omega\pi^0$	< 4.6	$\times 10^{-4}$	CL=90%	2580
$\pi^+\pi^+\pi^-\pi^-\pi^0$	< 9.0	$\times 10^{-3}$	CL=90%	2609
$a_1(1260)^+\rho^-$	< 3.4	$\times 10^{-3}$	CL=90%	2434
$a_1(1260)^0\rho^0$	< 2.4	$\times 10^{-3}$	CL=90%	2434
$\pi^+\pi^+\pi^+\pi^-\pi^-\pi^-$	< 3.0	$\times 10^{-3}$	CL=90%	2592
$a_1(1260)^+a_1(1260)^-$	< 2.8	$\times 10^{-3}$	CL=90%	2336
$\pi^+\pi^+\pi^+\pi^-\pi^-\pi^-\pi^0$	< 1.1	%	CL=90%	2572

Baryon modes

$p\overline{p}$	< 1.8	$\times 10^{-5}$	CL=90%	2467
$p\overline{p}\pi^+\pi^-$	< 2.5	$\times 10^{-4}$	CL=90%	2406
$p\overline{\Lambda}\pi^-$	< 1.8	$\times 10^{-4}$	CL=90%	2401
$\Delta^0\overline{\Delta}^0$	< 1.5	$\times 10^{-3}$	CL=90%	2334
$\Delta^{++}\Delta^{--}$	< 1.1	$\times 10^{-4}$	CL=90%	2334
$\overline{\Sigma}_c^{--}\Delta^{++}$	< 1.0	$\times 10^{-3}$	CL=90%	1839
$\Lambda_c^- p\pi^+\pi^-$	(1.3 ± 0.6) $\times 10^{-3}$			–
$\Lambda_c^- p$	< 2.1	$\times 10^{-4}$	CL=90%	2021
$\Lambda_c^- p\pi^0$	< 5.9	$\times 10^{-4}$	CL=90%	–
$\Lambda_c^- p\pi^+\pi^-\pi^0$	< 5.07	$\times 10^{-3}$	CL=90%	–
$\Lambda_c^- p\pi^+\pi^-\pi^+\pi^-$	< 2.74	$\times 10^{-3}$	CL=90%	–

Lepton Family number (LF) violating modes or $\Delta B = 1$ weak neutral current (B1) modes

$\gamma\gamma$	B1	< 3.9	$\times 10^{-5}$	CL=90%	2640
e^+e^-	B1	< 5.9	$\times 10^{-6}$	CL=90%	2640
$\mu^+\mu^-$	B1	< 6.8	$\times 10^{-7}$	CL=90%	2637
$K^0 e^+e^-$	B1	< 3.0	$\times 10^{-4}$	CL=90%	2616
$K^0\mu^+\mu^-$	B1	< 3.6	$\times 10^{-4}$	CL=90%	2612
$K^*(892)^0 e^+e^-$	B1	< 2.9	$\times 10^{-4}$	CL=90%	2564
$K^*(892)^0\mu^+\mu^-$	B1	< 2.3	$\times 10^{-5}$	CL=90%	2559
$K^*(892)^0\nu\overline{\nu}$	B1	< 1.0	$\times 10^{-3}$	CL=90%	2244
$e^\pm\mu^\mp$	LF	[gg] < 5.9	$\times 10^{-6}$	CL=90%	2639
$e^\pm\tau^\mp$	LF	[gg] < 5.3	$\times 10^{-4}$	CL=90%	2341
$\mu^\pm\tau^\mp$	LF	[gg] < 8.3	$\times 10^{-4}$	CL=90%	2339

$B^\pm/B^0$ ADMIXTURE

The branching fraction measurements are for an admixture of B mesons at the $\Upsilon(4S)$. The values quoted assume that B($\Upsilon(4S) \to B\overline{B}$) = 100%.

For inclusive branching fractions, e.g., $B \to D^\pm$ anything, the treatment of multiple D's in the final state must be defined. One possiblity would be to count the number of events with one-or-more D's and divide by the total number of B's. Another possibility would be to count the total number of D's and divide by the total number of B's, which is the definition of average multiplicity. The two definitions are identical when only one of the specified particles is allowed in the final state. Even though the "one-or-more" definition seems sensible, for practical reasons inclusive branching fractions are almost always measured using the multiplicity definition. For heavy final state particles, authors call their results inclusive branching fractions while for light particles some authors call their results multiplicities. In the B sections, we list all results as inclusive branching fractions, adopting a multiplicity definition. This means that inclusive branching fractions can exceed 100% and that inclusive partial widths can exceed total widths, just as inclusive cross sections can exceed total cross sections.

$\overline{B}$ modes are charge conjugates of the modes below. Reactions indicate the weak decay vertex and do not include mixing.

B DECAY MODES		Fraction (Γ_i/Γ)	Scale factor/ Confidence level	p (MeV/c)
Semileptonic and leptonic modes				
$B \to e^+\nu_e$ anything	[ccc]	(10.41±0.29) %	S=1.2	–
$B \to \overline{p}e^+\nu_e$ anything		< 1.6	$\times 10^{-3}$ CL=90%	–
$B \to \mu^+\nu_\mu$ anything	[ccc]	(10.3 ±0.5) %		–
$B \to \ell^+\nu_\ell$ anything	[pp,ccc]	(10.45±0.21) %		–
$B \to D^-\ell^+\nu_\ell$ anything	[pp]	(2.7 ±0.8) %		–
$B \to \overline{D}^0\ell^+\nu_\ell$ anything	[pp]	(7.0 ±1.4) %		–
$B \to \overline{D}^{**}\ell^+\nu_\ell$	[pp,ddd]	(2.7 ±0.7) %		–
$B \to \overline{D}_1(2420)\ell^+\nu_\ell$ anything		(7.4 ±1.6) $\times 10^{-3}$		–
$B \to D\pi\ell^+\nu_\ell$ anything + $D^*\pi\ell^+\nu_\ell$ anything		(2.3 ±0.4) %		–
$B \to \overline{D}_2^*(2460)\ell^+\nu_\ell$ anything		< 6.5	$\times 10^{-3}$ CL=95%	–
$B \to D^{*-}\pi^+\ell^+\nu_\ell$ anything		(1.00±0.34) %		–
$B \to D_s^-\ell^+\nu_\ell$ anything	[pp] <	9	$\times 10^{-3}$ CL=90%	–
$B \to D_s^-\ell^+\nu_\ell K^+$ anything	[pp] <	6	$\times 10^{-3}$ CL=90%	–
$B \to D_s^-\ell^+\nu_\ell K^0$ anything	[pp] <	9	$\times 10^{-3}$ CL=90%	–
$B \to K^+\ell^+\nu_\ell$ anything	[pp]	(6.0 ±0.5) %		–
$B \to K^-\ell^+\nu_\ell$ anything	[pp]	(10 ±4) $\times 10^{-3}$		–
$B \to K^0/\overline{K}^0\ell^+\nu_\ell$ anything	[pp]	(4.4 ±0.5) %		–

D, D^*, or D_s modes

$B \to D^{\pm}$ anything		$(24.1 \pm 1.9)\%$	—
$B \to D^0/\overline{D}^0$ anything		$(63.1 \pm 2.9)\%$	S=1.1 —
$B \to D^*(2010)^{\pm}$ anything		$(22.7 \pm 1.6)\%$	—
$B \to D^*(2007)^0$ anything		$(26.0 \pm 2.7)\%$	—
$B \to D_s^{\pm}$ anything	[gg]	$(10.0 \pm 2.5)\%$	—
$b \to c\overline{c}s$		$(22 \pm 4)\%$	—
$B \to D_s D, D_s^* D, D_s D^*$ or $D_s^* D^*$	[gg]	$(4.9 \pm 1.3)\%$	—
$B \to D^*(2010)\gamma$		$< 1.1 \times 10^{-3}$	CL=90% —
$B \to D_s^+ \pi^-, D_s^{*+}\pi^-,$ $D_s^+\rho^-, D_s^{*+}\rho^-, D_s^+\pi^0,$ $D_s^{*+}\pi^0, D_s^+\eta, D_s^{*+}\eta,$ $D_s^+\rho^0, D_s^{*+}\rho^0, D_s^+\omega,$ $D_s^{*+}\omega$	[gg]	$< 5 \times 10^{-4}$	CL=90% —
$B \to D_{s1}(2536)^+$ anything		$< 9.5 \times 10^{-3}$	CL=90% —

Charmonium modes

$B \to J/\psi(1S)$ anything		$(1.13 \pm 0.06)\%$	—
$B \to J/\psi(1S)$ (direct) anything		$(8.0 \pm 0.8) \times 10^{-3}$	—
$B \to \psi(2S)$ anything		$(3.5 \pm 0.5) \times 10^{-3}$	—
$B \to \chi_{c1}(1P)$ anything		$(4.2 \pm 0.7) \times 10^{-3}$	—
$B \to \chi_{c1}(1P)$ (direct) anything		$(3.7 \pm 0.7) \times 10^{-3}$	—
$B \to \chi_{c2}(1P)$ anything		$< 3.8 \times 10^{-3}$	CL=90% —
$B \to \eta_c(1S)$ anything		$< 9 \times 10^{-3}$	CL=90% —

K or K^* modes

$B \to K^{\pm}$ anything	[gg]	$(78.9 \pm 2.5)\%$	—
$B \to K^+$ anything		$(66 \pm 5)\%$	—
$B \to K^-$ anything		$(13 \pm 4)\%$	—
$B \to K^0/\overline{K}^0$ anything	[gg]	$(64 \pm 4)\%$	—
$B \to K^*(892)^{\pm}$ anything		$(18 \pm 6)\%$	—
$B \to K^*(892)^0/\overline{K}^*(892)^0$ anything	[gg]	$(14.6 \pm 2.6)\%$	—
$B \to K_1(1400)\gamma$		$< 4.1 \times 10^{-4}$	CL=90% —
$B \to K_2^*(1430)\gamma$		$< 8.3 \times 10^{-4}$	CL=90% —
$B \to K_2(1770)\gamma$		$< 1.2 \times 10^{-3}$	CL=90% —
$B \to K_3^*(1780)\gamma$		$< 3.0 \times 10^{-3}$	CL=90% —
$B \to K_4^*(2045)\gamma$		$< 1.0 \times 10^{-3}$	CL=90% —
$B \to \overline{b} \to \overline{s}\gamma$		$(2.3 \pm 0.7) \times 10^{-4}$	—
$B \to \overline{b} \to \overline{s}$ gluon		$< 6.8\%$	CL=90% —

Light unflavored meson modes

$B \to \pi^{\pm}$ anything	[gg,eee]	$(359 \pm 7)\%$	—
$B \to \eta$ anything		$(17.6 \pm 1.6)\%$	—
$B \to \rho^0$ anything		$(21 \pm 5)\%$	—
$B \to \omega$ anything		$< 81\%$	CL=90% —
$B \to \phi$ anything		$(3.5 \pm 0.7)\%$	S=1.8 —

Baryon modes

$B \to \Lambda_c^{\pm}$ anything		$(6.4 \pm 1.1)\%$	—
$B \to \Lambda_c^- e^+$ anything		$< 3.2 \times 10^{-3}$	CL=90% —
$B \to \Lambda_c^-$ p anything		$(3.6 \pm 0.7)\%$	—
$B \to \Lambda_c^- p e^+ \nu_e$		$< 1.5 \times 10^{-3}$	CL=90% —
$B \to \overline{\Sigma}_c^{--}$ anything		$(4.2 \pm 2.4) \times 10^{-3}$	—
$B \to \overline{\Sigma}_c^-$ anything		$< 9.6 \times 10^{-3}$	CL=90% —
$B \to \overline{\Sigma}_c^0$ anything		$(4.6 \pm 2.4) \times 10^{-3}$	—
$B \to \overline{\Sigma}_c^0 N (N = p$ or $n)$		$< 1.5 \times 10^{-3}$	CL=90% —
$B \to \Xi_c^0$ anything $\times \mathcal{B}(\Xi_c^0 \to \Xi^- \pi^+)$		$(1.4 \pm 0.5) \times 10^{-4}$	—
$B \to \Xi_c^+$ anything $\times \mathcal{B}(\Xi_c^+ \to \Xi^- \pi^+ \pi^+)$		$(4.5 ^{+1.3}_{-1.2}) \times 10^{-4}$	—
$B \to p/\overline{p}$ anything	[gg]	$(8.0 \pm 0.4)\%$	—
$B \to p/\overline{p}$ (direct) anything	[gg]	$(5.5 \pm 0.5)\%$	—
$B \to \Lambda/\overline{\Lambda}$ anything	[gg]	$(4.0 \pm 0.5)\%$	—
$B \to \Xi^-/\overline{\Xi}^+$ anything	[gg]	$(2.7 \pm 0.6) \times 10^{-3}$	—
$B \to$ baryons anything		$(6.8 \pm 0.6)\%$	—
$B \to p\overline{p}$ anything		$(2.47 \pm 0.23)\%$	—
$B \to \Lambda\overline{p}/\overline{\Lambda}p$ anything	[gg]	$(2.5 \pm 0.4)\%$	—
$B \to \Lambda\overline{\Lambda}$ anything		$< 5 \times 10^{-3}$	CL=90% —

Lepton Family number (LF) violating modes or $\Delta B = 1$ weak neutral current (B1) modes

$B \to e^+ e^- s$	B1	$< 5.7 \times 10^{-5}$	CL=90%	—
$B \to \mu^+ \mu^- s$	B1	$< 5.8 \times 10^{-5}$	CL=90%	—
$B \to e^{\pm} \mu^{\mp} s$	LF	$< 2.2 \times 10^{-5}$	CL=90%	—

$B^{\pm}/B^0/B_s^0/b$-baryon ADMIXTURE

These measurements are for an admixture of bottom particles at high energy (LEP, Tevatron, $Sp\overline{p}S$).

Mean life $\tau = (1.564 \pm 0.014) \times 10^{-12}$ s

Mean life $\tau = (1.72 \pm 0.10) \times 10^{-12}$ s   Charged b-hadron admixture

Mean life $\tau = (1.58 \pm 0.14) \times 10^{-12}$ s   Neutral b-hadron admixture

$\tau_{\text{charged } b-\text{hadron}}/\tau_{\text{neutral } b-\text{hadron}} = 1.09 \pm 0.13$

The branching fraction measurements are for an admixture of B mesons and baryons at energies above the $\Upsilon(4S)$. Only the highest energy results (LEP, Tevatron, $Sp\overline{p}S$) are used in the branching fraction averages. The production fractions give our best current estimate of the admixture at LEP.

For inclusive branching fractions, e.g., $B \to D^{\pm}$ anything, the treatment of multiple D's in the final state must be defined. One possibity would be to count the number of events with one-or-more D's and divide by the total number of B's. Another possibility would be to count the total number of D's and divide by the total number of B's, which is the definition of average multiplicity. The two definitions are identical when only one of the specified particles is allowed in the final state. Even though the "one-or-more" definition seems sensible, for practical reasons inclusive branching fractions are almost always measured using the multiplicity definition. For heavy final state particles, authors call their results inclusive branching fractions while for light particles some authors call their results multiplicities. In the B sections, we list all results as inclusive branching fractions, adopting a multiplicity definition. This means that inclusive branching fractions can exceed 100% and that inclusive partial widths can exceed total widths, just as inclusive cross sections can exceed total cross sections.

The modes below are listed for a $\overline{b}$ initial state. b modes are their charge conjugates. Reactions indicate the weak decay vertex and do not include mixing.

$\overline{b}$ DECAY MODES		Fraction (Γ_i/Γ)	Confidence level	p (MeV/c)

PRODUCTION FRACTIONS

The production fractions for weakly decaying b-hadrons at the Z have been calculated from the best values of mean lives, mixing parameters, and branching fractions in this edition by the LEP B Oscillation Working Group as described in the note "Production and Decay of b-Flavored Hadrons" in the $B^{\pm}$ Particle Listings. Values assume

$$B(\overline{b} \to B^+) = B(\overline{b} \to B^0)$$
$$B(\overline{b} \to B^+) + B(\overline{b} \to B^0) + B(\overline{b} \to B_s^0) + B(b \to \Lambda_b) = 100\%.$$

The notation for production fractions varies in the literature (f_{B^0}, $f(b \to \overline{B}^0)$, $Br(b \to \overline{B}^0)$). We use our own branching fraction notation here, $B(\overline{b} \to B^0)$.

B^+	$(39.7 ^{+1.8}_{-2.2})\%$	—
B^0	$(39.7 ^{+1.8}_{-2.2})\%$	—
B_s^0	$(10.5 ^{+1.8}_{-1.7})\%$	—
Λ_b	$(10.1 ^{+3.9}_{-3.1})\%$	—

DECAY MODES

Semileptonic and leptonic modes

ν anything		$(23.1 \pm 1.5)\%$	—
$\ell^+ \nu_\ell$ anything	[pp,ccc]	$(10.99 \pm 0.23)\%$	—
$e^+ \nu_e$ anything	[ccc]	$(10.9 \pm 0.5)\%$	—
$\mu^+ \nu_\mu$ anything	[ccc]	$(10.8 \pm 0.5)\%$	—
$D^- \ell^+ \nu_\ell$ anything	[pp]	$(2.02 \pm 0.29)\%$	—
$\overline{D}^0 \ell^+ \nu_\ell$ anything	[pp]	$(6.5 \pm 0.6)\%$	—
$D^{*-} \ell^+ \nu_\ell$ anything	[pp]	$(2.76 \pm 0.29)\%$	—
$\overline{D}_j^0 \ell^+ \nu_\ell$ anything	[pp,fff]	seen	—
$D_j^- \ell^+ \nu_\ell$ anything	[pp,fff]	seen	—
$\overline{D}_2^*(2460)^0 \ell^+ \nu_\ell$ anything		seen	—
$D_2^*(2460)^- \ell^+ \nu_\ell$ anything		seen	—
$\tau^+ \nu_\tau$ anything		$(2.6 \pm 0.4)\%$	—
$\overline{c} \to \ell^- \overline{\nu}_\ell$ anything	[pp]	$(7.8 \pm 0.6)\%$	—

Charmed meson and baryon modes

$\overline{D}^0$ anything	(60.1 ± 3.2) %		–
D^- anything	(23.7 ± 2.3) %		–
$\overline{D}_s$ anything	(18 ± 5) %		–
Λ_c anything	(9.7 ± 2.9) %		–
$\overline{c}/c$ anything	[eee] (117 ± 4) %		–

Charmonium modes

$J/\psi(1S)$ anything	(1.16 ± 0.10) %		–
$\psi(2S)$ anything	(4.8 ± 2.4) × 10^{-3}		–
$\chi_{c1}(1P)$ anything	(1.8 ± 0.5) %		–

K or K* modes

$\overline{s}\gamma$	< 5.4 × 10^{-4}	90%	–
$K^\pm$ anything	(88 ±19) %		–
K^0_S anything	(29.0 ± 2.9) %		–

Pion modes

π^0 anything	[eee] (278 ±60) %		–

Baryon modes

$p/\overline{p}$ anything	(14 ± 6) %		–

Other modes

charged anything	[eee] (497 ± 7) %		–
hadron$^+$ hadron$^-$	(1.7 $^{+1.0}_{-0.7}$) × 10^{-5}		–
charmless	(7 ±21) × 10^{-3}		–

Baryon modes

$\Lambda/\overline{\Lambda}$ anything	(5.9 ± 0.6) %		–

$\Delta B = 1$ weak neutral current (B1) modes

$\mu^+\mu^-$ anything	B1	< 3.2 × 10^{-4}	90%	–

B^* $\qquad\qquad I(J^P) = \frac{1}{2}(1^-)$

I, J, P need confirmation. Quantum numbers shown are quark-model predictions.

Mass $m_{B^*} = 5324.9 \pm 1.8$ MeV

$m_{B^*} - m_B = 45.78 \pm 0.35$ MeV

B^* DECAY MODES	Fraction (Γ_i/Γ)	p (MeV/c)
$B\gamma$	dominant	46

BOTTOM, STRANGE MESONS
$(B = \pm1, S = \mp1)$

$B^0_s = s\overline{b}$, $\overline{B}^0_s = \overline{s}b$ similarly for B^*_s's

---

B^0_s $\qquad\qquad I(J^P) = 0(0^-)$

I, J, P need confirmation. Quantum numbers shown are quark-model predictions.

Mass $m_{B^0_s} = 5369.3 \pm 2.0$ MeV

Mean life $\tau = (1.54 \pm 0.07) \times 10^{-12}$ s

$c\tau = 462~\mu$m

B^0_s-$\overline{B}^0_s$ mixing parameters

χ_B at high energy $= f_d\chi_d + f_s\chi_s = 0.118 \pm 0.006$

$\Delta m_{B^0_s} = m_{B^0_{sH}} - m_{B^0_{sL}} > 9.1 \times 10^{12}~\hbar$ s^{-1}, CL = 95%

$x_s = \Delta m_{B^0_s}/\Gamma_{B^0_s} > 14.0$, CL = 95%

$\chi_s > 0.4975$, CL = 95%

These branching fractions all scale with B($\overline{b} \to B^0_s$), the LEP B^0_s production fraction. The first four were evaluated using B($\overline{b} \to B^0_s$) = $(10.5 ^{+1.8}_{-1.7})$% and the rest assume B($\overline{b} \to B^0_s$) = 12%.

The branching fraction B($B^0_s \to D^-_s \ell^+ \nu_\ell$ anything) is not a pure measurement since the measured product branching fraction B($\overline{b} \to B^0_s$) × B($B^0_s \to D^-_s \ell^+ \nu_\ell$ anything) was used to determine B($\overline{b} \to B^0_s$) as described in the note on "Production and Decay of b-Flavored Hadrons."

B^0_s DECAY MODES		Fraction (Γ_i/Γ)		Confidence level	p (MeV/c)
D^-_s anything		(92 ±33) %			–
$D^-_s \ell^+ \nu_\ell$ anything	[ggg]	(8.1 ± 2.5) %			–
$D^-_s \pi^+$		< 13	%		2321
$J/\psi(1S)\phi$		(9.3 ± 3.3) × 10^{-4}			1590
$J/\psi(1S)\pi^0$		< 1.2 × 10^{-3}		90%	1788
$J/\psi(1S)\eta$		< 3.8 × 10^{-3}		90%	1735
$\psi(2S)\phi$		seen			1122
$\pi^+\pi^-$		< 1.7 × 10^{-4}		90%	1122
$\pi^0\pi^0$		< 2.1 × 10^{-4}		90%	2861
$\eta\pi^0$		< 1.0 × 10^{-3}		90%	2655
$\eta\eta$		< 1.5 × 10^{-3}		90%	2628
π^+K^-		< 2.1 × 10^{-4}		90%	2660
K^+K^-		< 5.9 × 10^{-5}		90%	2639
$p\overline{p}$		< 5.9 × 10^{-5}		90%	2515
$\gamma\gamma$		< 1.48 × 10^{-4}		90%	2685
$\phi\gamma$		< 7 × 10^{-4}		90%	2588

Lepton Family number (LF) violating modes or
$\Delta B = 1$ weak neutral current (B1) modes

$\mu^+\mu^-$	B1	< 2.0 × 10^{-6}		90%	2682
e^+e^-	B1	< 5.4 × 10^{-5}		90%	2864
$e^\pm\mu^\mp$	LF	[gg] < 4.1 × 10^{-5}		90%	2864
$\phi\nu\overline{\nu}$	B1	< 5.4 × 10^{-3}		90%	–

$c\overline{c}$ MESONS

$\eta_c(1S)$

$$I^G(J^{PC}) = 0^+(0^{-+})$$

Mass $m = 2979.8 \pm 2.1$ MeV $\quad (S = 2.1)$
Full width $\Gamma = 13.2^{+3.8}_{-3.2}$ MeV

$\eta_c(1S)$ DECAY MODES	Fraction (Γ_i/Γ)	Confidence level	p (MeV/c)
Decays involving hadronic resonances			
$\eta'(958)\pi\pi$	(4.1 ± 1.7) %		1319
$\rho\rho$	(2.6 ± 0.9) %		1275
$K^*(892)^0 K^-\pi^+ +$ c.c.	(2.0 ± 0.7) %		1273
$K^*(892)\overline{K}^*(892)$	$(8.5 \pm 3.1)\times 10^{-3}$		1193
$\phi\phi$	$(7.1 \pm 2.8)\times 10^{-3}$		1086
$a_0(980)\pi$	< 2 %	90%	1323
$a_2(1320)\pi$	< 2 %	90%	1193
$K^*(892)\overline{K} +$ c.c.	< 1.28 %	90%	1307
$f_2(1270)\eta$	< 1.1 %	90%	1142
$\omega\omega$	$< 3.1 \times 10^{-3}$	90%	1268
Decays into stable hadrons			
$K\overline{K}\pi$	(5.5 ± 1.7) %		1378
$\eta\pi\pi$	(4.9 ± 1.8) %		1425
$\pi^+\pi^- K^+ K^-$	$(2.0 ^{+0.7}_{-0.6})$ %		1342
$2(K^+ K^-)$	(2.1 ± 1.2) %		1053
$2(\pi^+\pi^-)$	(1.2 ± 0.4) %		1457
$p\overline{p}$	$(1.2 \pm 0.4)\times 10^{-3}$		1157
$K\overline{K}\eta$	< 3.1 %	90%	1262
$\pi^+\pi^- p\overline{p}$	< 1.2 %	90%	1023
$\Lambda\overline{\Lambda}$	$< 2 \times 10^{-3}$	90%	987
Radiative decays			
$\gamma\gamma$	$(3.0 \pm 1.2)\times 10^{-4}$		1489

$J/\psi(1S)$

$$I^G(J^{PC}) = 0^-(1^{--})$$

Mass $m = 3096.88 \pm 0.04$ MeV
Full width $\Gamma = 87 \pm 5$ keV
$\Gamma_{ee} = 5.26 \pm 0.37$ keV $\quad$ (Assuming $\Gamma_{ee} = \Gamma_{\mu\mu}$)

$J/\psi(1S)$ DECAY MODES	Fraction (Γ_i/Γ)	Scale factor/ Confidence level	p (MeV/c)
hadrons	(87.7 ± 0.5) %		—
$\quad$ virtual $\gamma \to$ hadrons	(17.0 ± 2.0) %		—
$e^+ e^-$	(6.02 ± 0.19) %		1548
$\mu^+\mu^-$	(6.01 ± 0.19) %		1545
Decays involving hadronic resonances			
$\rho\pi$	(1.27 ± 0.09) %		1449
$\quad \rho^0\pi^0$	$(4.2 \pm 0.5)\times 10^{-3}$		1449
$a_2(1320)\rho$	(1.09 ± 0.22) %		1125
$\omega\pi^+\pi^+\pi^-\pi^-$	$(8.5 \pm 3.4)\times 10^{-3}$		1392
$\omega\pi^+\pi^-$	$(7.2 \pm 1.0)\times 10^{-3}$		1435
$\quad \omega f_2(1270)$	$(4.3 \pm 0.6)\times 10^{-3}$		1143
$K^*(892)^0\overline{K}_2^*(1430)^0 +$ c.c.	$(6.7 \pm 2.6)\times 10^{-3}$		1005
$\omega K^*(892)\overline{K} +$ c.c.	$(5.3 \pm 2.0)\times 10^{-3}$		1098
$K^+\overline{K}^*(892)^- +$ c.c.	$(5.0 \pm 0.4)\times 10^{-3}$		1373
$K^0\overline{K}^*(892)^0 +$ c.c.	$(4.2 \pm 0.4)\times 10^{-3}$		1371
$\omega\pi^0\pi^0$	$(3.4 \pm 0.8)\times 10^{-3}$		1436
$b_1(1235)^\pm\pi^\mp$	$[gg]$ $(3.0 \pm 0.5)\times 10^{-3}$		1299
$\omega K^\pm K_S^0\pi^\mp$	$[gg]$ $(3.0 \pm 0.7)\times 10^{-3}$		1210
$b_1(1235)^0\pi^0$	$(2.3 \pm 0.6)\times 10^{-3}$		1299
$\phi K^*(892)\overline{K} +$ c.c.	$(2.04 \pm 0.28)\times 10^{-3}$		969
$\omega K\overline{K}$	$(1.9 \pm 0.4)\times 10^{-3}$		1268
$\quad \omega f_J(1710) \to \omega K\overline{K}$	$(4.8 \pm 1.1)\times 10^{-4}$		878
$\phi 2(\pi^+\pi^-)$	$(1.60 \pm 0.32)\times 10^{-3}$		1318
$\Delta(1232)^{++}\overline{p}\pi^-$	$(1.6 \pm 0.5)\times 10^{-3}$		1030
$\omega\eta$	$(1.58 \pm 0.16)\times 10^{-3}$		1394
$\phi K\overline{K}$	$(1.48 \pm 0.22)\times 10^{-3}$		1179
$\quad \phi f_J(1710) \to \phi K\overline{K}$	$(3.6 \pm 0.6)\times 10^{-4}$		875
$p\overline{p}\omega$	$(1.30 \pm 0.25)\times 10^{-3}$	$S=1.3$	769
$\Delta(1232)^{++}\overline{\Delta}(1232)^{--}$	$(1.10 \pm 0.29)\times 10^{-3}$		938
$\Sigma(1385)^-\overline{\Sigma}(1385)^+$ (or c.c.)	$[gg]$ $(1.03 \pm 0.13)\times 10^{-3}$		692
$p\overline{p}\eta'(958)$	$(9 \pm 4)\times 10^{-4}$	$S=1.7$	596
$\phi f_2'(1525)$	$(8 \pm 4)\times 10^{-4}$	$S=2.7$	871
$\phi\pi^+\pi^-$	$(8.0 \pm 1.2)\times 10^{-4}$		1365
$\phi K^\pm K_S^0\pi^\mp$	$[gg]$ $(7.2 \pm 0.9)\times 10^{-4}$		1114
$\omega f_1(1420)$	$(6.8 \pm 2.4)\times 10^{-4}$		1062
$\phi\eta$	$(6.5 \pm 0.7)\times 10^{-4}$		1320
$\Xi(1530)^-\overline{\Xi}^+$	$(5.9 \pm 1.5)\times 10^{-4}$		597
$p K^-\overline{\Sigma}(1385)^0$	$(5.1 \pm 3.2)\times 10^{-4}$		645
$\omega\pi^0$	$(4.2 \pm 0.6)\times 10^{-4}$	$S=1.4$	1447
$\phi\eta'(958)$	$(3.3 \pm 0.4)\times 10^{-4}$		1192
$\phi f_0(980)$	$(3.2 \pm 0.9)\times 10^{-4}$	$S=1.9$	1182
$\Xi(1530)^0\overline{\Xi}^0$	$(3.2 \pm 1.4)\times 10^{-4}$		608
$\Sigma(1385)^-\overline{\Sigma}^+$ (or c.c.)	$[gg]$ $(3.1 \pm 0.5)\times 10^{-4}$		857
$\phi f_1(1285)$	$(2.6 \pm 0.5)\times 10^{-4}$	$S=1.1$	1032
$\rho\eta$	$(1.93 \pm 0.23)\times 10^{-4}$		1398
$\omega\eta'(958)$	$(1.67 \pm 0.25)\times 10^{-4}$		1279
$\omega f_0(980)$	$(1.4 \pm 0.5)\times 10^{-4}$		1271
$\rho\eta'(958)$	$(1.05 \pm 0.18)\times 10^{-4}$		1283
$p\overline{p}\phi$	$(4.5 \pm 1.5)\times 10^{-5}$		527
$a_2(1320)^\pm\pi^\mp$	$[gg]$ $< 4.3 \times 10^{-3}$	CL=90%	1263
$K\overline{K}_2^*(1430) +$ c.c.	$< 4.0 \times 10^{-3}$	CL=90%	1159
$K_2^*(1430)^0\overline{K}_2^*(1430)^0$	$< 2.9 \times 10^{-3}$	CL=90%	588
$K^*(892)^0\overline{K}^*(892)^0$	$< 5 \times 10^{-4}$	CL=90%	1263
$\phi f_2(1270)$	$< 3.7 \times 10^{-4}$	CL=90%	1036
$p\overline{p}\rho$	$< 3.1 \times 10^{-4}$	CL=90%	779
$\phi\eta(1440) \to \phi\eta\pi\pi$	$< 2.5 \times 10^{-4}$	CL=90%	946
$\omega f_2'(1525)$	$< 2.2 \times 10^{-4}$	CL=90%	1003
$\Sigma(1385)^0\overline{\Lambda}$	$< 2 \times 10^{-4}$	CL=90%	911
$\Delta(1232)^+\overline{p}$	$< 1 \times 10^{-4}$	CL=90%	1100
$\Sigma^0\overline{\Lambda}$	$< 9 \times 10^{-5}$	CL=90%	1032
$\phi\pi^0$	$< 6.8 \times 10^{-6}$	CL=90%	1377
Decays into stable hadrons			
$2(\pi^+\pi^-)\pi^0$	(3.37 ± 0.26) %		1496
$3(\pi^+\pi^-)\pi^0$	(2.9 ± 0.6) %		1433
$\pi^+\pi^-\pi^0$	(1.50 ± 0.20) %		1533
$\pi^+\pi^-\pi^0 K^+ K^-$	(1.20 ± 0.30) %		1368
$4(\pi^+\pi^-)\pi^0$	$(9.0 \pm 3.0)\times 10^{-3}$		1345
$\pi^+\pi^- K^+ K^-$	$(7.2 \pm 2.3)\times 10^{-3}$		1407
$K\overline{K}\pi$	$(6.1 \pm 1.0)\times 10^{-3}$		1440
$p\overline{p}\pi^+\pi^-$	$(6.0 \pm 0.5)\times 10^{-3}$	$S=1.3$	1107
$2(\pi^+\pi^-)$	$(4.0 \pm 1.0)\times 10^{-3}$		1517
$3(\pi^+\pi^-)$	$(4.0 \pm 2.0)\times 10^{-3}$		1466
$n\overline{n}\pi^+\pi^-$	$(4 \pm 4)\times 10^{-3}$		1106
$\Sigma^0\overline{\Sigma}^0$	$(1.27 \pm 0.17)\times 10^{-3}$		992
$2(\pi^+\pi^-)K^+ K^-$	$(3.1 \pm 1.3)\times 10^{-3}$		1320
$p\overline{p}\pi^+\pi^-\pi^0$	$[hhh]$ $(2.3 \pm 0.9)\times 10^{-3}$	$S=1.9$	1033
$p\overline{p}$	$(2.14 \pm 0.10)\times 10^{-3}$		1232
$p\overline{p}\eta$	$(2.09 \pm 0.18)\times 10^{-3}$		948
$p\overline{n}\pi^-$	$(2.00 \pm 0.10)\times 10^{-3}$		1174
$n\overline{n}$	$(1.9 \pm 0.5)\times 10^{-3}$		1231
$\Xi^-\overline{\Xi}^+$	$(1.8 \pm 0.4)\times 10^{-3}$	$S=1.8$	818
$\Lambda\overline{\Lambda}$	$(1.35 \pm 0.14)\times 10^{-3}$	$S=1.2$	1074
$p\overline{p}\pi^0$	$(1.09 \pm 0.09)\times 10^{-3}$		1176
$\Lambda\overline{\Sigma}^-\pi^+$ (or c.c.)	$[gg]$ $(1.06 \pm 0.12)\times 10^{-3}$		945
$p K^-\overline{\Lambda}$	$(8.9 \pm 1.6)\times 10^{-4}$		876
$2(K^+ K^-)$	$(7.0 \pm 3.0)\times 10^{-4}$		1131
$p K^-\overline{\Sigma}^0$	$(2.9 \pm 0.8)\times 10^{-4}$		820
$K^+ K^-$	$(2.37 \pm 0.31)\times 10^{-4}$		1468
$\Lambda\overline{\Lambda}\pi^0$	$(2.2 \pm 0.7)\times 10^{-4}$		998
$\pi^+\pi^-$	$(1.47 \pm 0.23)\times 10^{-4}$		1542
$K_S^0 K_L^0$	$(1.08 \pm 0.14)\times 10^{-4}$		1466
$\Lambda\overline{\Sigma}^+ +$ c.c.	$< 1.5 \times 10^{-4}$	CL=90%	1032
$K_S^0 K_S^0$	$< 5.2 \times 10^{-6}$	CL=90%	1466
Radiative decays			
$\gamma\eta_c(1S)$	(1.3 ± 0.4) %		116
$\gamma\pi^+\pi^- 2\pi^0$	$(8.3 \pm 3.1)\times 10^{-3}$		1518
$\gamma\eta\pi\pi$	$(6.1 \pm 1.0)\times 10^{-3}$		1487
$\gamma\eta(1440) \to \gamma K\overline{K}\pi$	$[p]$ $(9.1 \pm 1.8)\times 10^{-4}$		1223
$\gamma\eta(1440) \to \gamma\gamma\rho^0$	$(6.4 \pm 1.4)\times 10^{-5}$		1223
$\gamma\eta(1440) \to \gamma\eta\pi^+\pi^-$	$(3.4 \pm 0.7)\times 10^{-4}$		—
$\gamma\rho\rho$	$(4.5 \pm 0.8)\times 10^{-3}$		1343
$\gamma\eta'(958)$	$(4.31 \pm 0.30)\times 10^{-3}$		1400
$\gamma 2\pi^+ 2\pi^-$	$(2.8 \pm 0.5)\times 10^{-3}$	$S=1.9$	1517
$\gamma f_4(2050)$	$(2.7 \pm 0.7)\times 10^{-3}$		874
$\gamma\omega\omega$	$(1.59 \pm 0.33)\times 10^{-3}$		1337
$\gamma\eta(1440) \to \gamma\rho^0\rho^0$	$(1.7 \pm 0.4)\times 10^{-3}$	$S=1.3$	1223
$\gamma f_2(1270)$	$(1.38 \pm 0.14)\times 10^{-3}$		1286
$\gamma f_J(1710) \to \gamma K\overline{K}$	$(8.5 ^{+1.2}_{-0.9})\times 10^{-4}$	$S=1.2$	1075

$\gamma\eta$	$(8.6\ \pm 0.8\ )\times 10^{-4}$	1500
$\gamma f_1(1420)\to\gamma K\overline{K}\pi$	$(8.3\ \pm 1.5\ )\times 10^{-4}$	1220
$\gamma f_1(1285)$	$(6.5\ \pm 1.0\ )\times 10^{-4}$	1283
$\gamma f_2'(1525)$	$(4.7\ ^{+0.7}_{-0.5}\ )\times 10^{-4}$	1173
$\gamma\phi\phi$	$(4.0\ \pm 1.2\ )\times 10^{-4}$ S=2.1	1166
$\gamma p\overline{p}$	$(3.8\ \pm 1.0\ )\times 10^{-4}$	1232
$\gamma\eta(2225)$	$(2.9\ \pm 0.6\ )\times 10^{-4}$	834
$\gamma\eta(1760)\to\gamma\rho^0\rho^0$	$(1.3\ \pm 0.9\ )\times 10^{-4}$	1048
$\gamma\pi^0$	$(3.9\ \pm 1.3\ )\times 10^{-5}$	1546
$\gamma p\overline{p}\pi^+\pi^-$	$<\ 7.9\ \times 10^{-4}$ CL=90%	1107
$\gamma\gamma$	$<\ 5\ \times 10^{-4}$ CL=90%	1548
$\gamma\Lambda\overline{\Lambda}$	$<\ 1.3\ \times 10^{-4}$ CL=90%	1074
3γ	$<\ 5.5\ \times 10^{-5}$ CL=90%	1548
$\gamma f_J(2220)$	$>\ 2.50\ \times 10^{-3}$ CL=99.9%	–
$\gamma f_0(1500)$	$(5.7\ \pm 0.8\ )\times 10^{-4}$	1184
$\gamma e^+ e^-$	$(8.8\ \pm 1.4\ )\times 10^{-3}$	–

$\chi_{c0}(1P)$ $I^G(J^{PC})=0^+(0^{++})$

Mass $m=3417.3\pm 2.8$ MeV
Full width $\Gamma=14\pm 5$ MeV

$\chi_{c0}(1P)$ DECAY MODES	Fraction (Γ_i/Γ)	Confidence level	p (MeV/c)
Hadronic decays			
$2(\pi^+\pi^-)$	(3.7 ± 0.7) %		1679
$\pi^+\pi^- K^+ K^-$	(3.0 ± 0.7) %		1580
$\rho^0\pi^+\pi^-$	(1.6 ± 0.5) %		1608
$3(\pi^+\pi^-)$	(1.5 ± 0.5) %		1633
$K^+\overline{K}^*(892)^0\pi^- +$ c.c.	(1.2 ± 0.4) %		1522
$\pi^+\pi^-$	$(7.5\pm 2.1)\times 10^{-3}$		1702
$K^+ K^-$	$(7.1\pm 2.4)\times 10^{-3}$		1635
$\pi^+\pi^- p\overline{p}$	$(5.0\pm 2.0)\times 10^{-3}$		1320
$p\overline{p}$	$<\ 9.0\ \times 10^{-4}$	90%	1427
Radiative decays			
$\gamma J/\psi(1S)$	$(6.6\pm 1.8)\times 10^{-3}$		303
$\gamma\gamma$	$<\ 5\ \times 10^{-4}$	95%	1708

$\chi_{c1}(1P)$ $I^G(J^{PC})=0^+(1^{++})$

Mass $m=3510.53\pm 0.12$ MeV
Full width $\Gamma=0.88\pm 0.14$ MeV

$\chi_{c1}(1P)$ DECAY MODES	Fraction (Γ_i/Γ)	p (MeV/c)
Hadronic decays		
$3(\pi^+\pi^-)$	(2.2 ± 0.8) %	1683
$2(\pi^+\pi^-)$	(1.6 ± 0.5) %	1727
$\pi^+\pi^- K^+ K^-$	$(9\ \pm 4\ )\times 10^{-3}$	1632
$\rho^0\pi^+\pi^-$	$(3.9\pm 3.5)\times 10^{-3}$	1659
$K^+\overline{K}^*(892)^0\pi^- +$ c.c.	$(3.2\pm 2.1)\times 10^{-3}$	1576
$\pi^+\pi^- p\overline{p}$	$(1.4\pm 0.9)\times 10^{-3}$	1381
$p\overline{p}$	$(8.6\pm 1.2)\times 10^{-5}$	1483
$\pi^+\pi^- +\ K^+ K^-$	$<\ 2.1\ \times 10^{-3}$	–
Radiative decays		
$\gamma J/\psi(1S)$	(27.3 ± 1.6) %	389

$\chi_{c2}(1P)$ $I^G(J^{PC})=0^+(2^{++})$

Mass $m=3556.17\pm 0.13$ MeV
Full width $\Gamma=2.00\pm 0.18$ MeV

$\chi_{c2}(1P)$ DECAY MODES	Fraction (Γ_i/Γ)	Confidence level	p (MeV/c)
Hadronic decays			
$2(\pi^+\pi^-)$	(2.2 ± 0.5) %		1751
$\pi^+\pi^- K^+ K^-$	(1.9 ± 0.5) %		1656
$3(\pi^+\pi^-)$	(1.2 ± 0.8) %		1707
$\rho^0\pi^+\pi^-$	$(7\ \pm 4\ )\times 10^{-3}$		1683
$K^+\overline{K}^*(892)^0\pi^- +$ c.c.	$(4.8\pm 2.8)\times 10^{-3}$		1601
$\pi^+\pi^- p\overline{p}$	$(3.3\pm 1.3)\times 10^{-3}$		1410
$\pi^+\pi^-$	$(1.9\pm 1.0)\times 10^{-3}$		1773
$K^+ K^-$	$(1.5\pm 1.1)\times 10^{-3}$		1708
$p\overline{p}$	$(10.0\pm 1.0)\times 10^{-5}$		1510
$J/\psi(1S)\pi^+\pi^-\pi^0$	$<\ 1.5\ $ %	90%	185

Radiative decays		
$\gamma J/\psi(1S)$	(13.5 ± 1.1) %	430
$\gamma\gamma$	$(1.6\pm 0.5)\times 10^{-4}$	1778

$\psi(2S)$ $I^G(J^{PC})=0^-(1^{--})$

Mass $m=3686.00\pm 0.09$ MeV
Full width $\Gamma=277\pm 31$ keV   (S = 1.1)
$\Gamma_{ee}=2.14\pm 0.21$ keV   (Assuming $\Gamma_{ee}=\Gamma_{\mu\mu}$)

$\psi(2S)$ DECAY MODES	Fraction (Γ_i/Γ)	Scale factor/ Confidence level	p (MeV/c)
hadrons	(98.10 ± 0.30) %		–
virtual $\gamma\to$ hadrons	$(2.9\ \pm 0.4\ )$ %		–
$e^+ e^-$	$(8.5\ \pm 0.7\ )\times 10^{-3}$		1843
$\mu^+\mu^-$	$(7.7\ \pm 1.7\ )\times 10^{-3}$		1840
Decays into $J/\psi(1S)$ and anything			
$J/\psi(1S)$ anything	$(54.2\ \pm 3.0\ )$ %		–
$J/\psi(1S)$ neutrals	$(22.8\ \pm 1.7\ )$ %		–
$J/\psi(1S)\pi^+\pi^-$	$(30.2\ \pm 1.9\ )$ %		477
$J/\psi(1S)\pi^0\pi^0$	$(17.9\ \pm 1.8\ )$ %		481
$J/\psi(1S)\eta$	$(2.7\ \pm 0.4\ )$ %	S=1.7	200
$J/\psi(1S)\pi^0$	$(9.7\ \pm 2.1\ )\times 10^{-4}$		527
$J/\psi(1S)\mu^+\mu^-$	$(10.0\ \pm 3.3\ )\times 10^{-3}$		–
Hadronic decays			
$3(\pi^+\pi^-)\pi^0$	$(3.5\ \pm 1.6\ )\times 10^{-3}$		1746
$2(\pi^+\pi^-)\pi^0$	$(3.0\ \pm 0.8\ )\times 10^{-3}$		1799
$\pi^+\pi^- K^+ K^-$	$(1.6\ \pm 0.4\ )\times 10^{-3}$		1726
$\pi^+\pi^- p\overline{p}$	$(8.0\ \pm 2.0\ )\times 10^{-4}$		1491
$K^+\overline{K}^*(892)^0\pi^- +$ c.c.	$(6.7\ \pm 2.5\ )\times 10^{-4}$		1673
$2(\pi^+\pi^-)$	$(4.5\ \pm 1.0\ )\times 10^{-4}$		1817
$\rho^0\pi^+\pi^-$	$(4.2\ \pm 1.5\ )\times 10^{-4}$		1751
$\overline{p}p$	$(1.9\ \pm 0.5\ )\times 10^{-4}$		1586
$3(\pi^+\pi^-)$	$(1.5\ \pm 1.0\ )\times 10^{-4}$		1774
$\overline{p}p\pi^0$	$(1.4\ \pm 0.5\ )\times 10^{-4}$		1543
$K^+ K^-$	$(1.0\ \pm 0.7\ )\times 10^{-4}$		1776
$\pi^+\pi^-\pi^0$	$(9\ \pm 5\ )\times 10^{-5}$		1830
$\rho\pi$	$<\ 8.3\ \times 10^{-5}$ CL=90%		1760
$\pi^+\pi^-$	$<\ 8\ \pm 5\ \times 10^{-5}$		1838
$\Lambda\overline{\Lambda}$	$<\ 4\ \times 10^{-4}$ CL=90%		1467
$\Xi^-\overline{\Xi}^+$	$<\ 2\ \times 10^{-4}$ CL=90%		1285
$K^+ K^-\pi^0$	$<\ 2.96\ \times 10^{-5}$ CL=90%		1754
$K^+\overline{K}^*(892)^- +$ c.c.	$<\ 5.4\ \times 10^{-5}$ CL=90%		1698
Radiative decays			
$\gamma\chi_{c0}(1P)$	$(9.3\ \pm 0.9\ )$ %		261
$\gamma\chi_{c1}(1P)$	$(8.7\ \pm 0.8\ )$ %		171
$\gamma\chi_{c2}(1P)$	$(7.8\ \pm 0.8\ )$ %		127
$\gamma\eta_c(1S)$	$(2.8\ \pm 0.6\ )\times 10^{-3}$		639
$\gamma\eta'(958)$	$<\ 1.1\ \times 10^{-3}$ CL=90%		1719
$\gamma\gamma$	$<\ 1.6\ \times 10^{-4}$ CL=90%		1843
$\gamma\eta(1440)\to\gamma K\overline{K}\pi$	$<\ 1.2\ \times 10^{-4}$ CL=90%		1569

$\psi(3770)$ $I^G(J^{PC})=?^?(1^{--})$

Mass $m=3769.9\pm 2.5$ MeV   (S = 1.8)
Full width $\Gamma=23.6\pm 2.7$ MeV   (S = 1.1)
$\Gamma_{ee}=0.26\pm 0.04$ keV   (S = 1.2)

$\psi(3770)$ DECAY MODES	Fraction (Γ_i/Γ)	Scale factor	p (MeV/c)
$D\overline{D}$	dominant		242
$e^+ e^-$	$(1.12\pm 0.17)\times 10^{-5}$	1.2	1885

$\psi(4040)$ [iii] $I^G(J^{PC})=?^?(1^{--})$

Mass $m=4040\pm 10$ MeV
Full width $\Gamma=52\pm 10$ MeV
$\Gamma_{ee}=0.75\pm 0.15$ keV

$\psi(4040)$ DECAY MODES	Fraction (Γ_i/Γ)	p (MeV/c)
$e^+ e^-$	$(1.4\pm 0.4)\times 10^{-5}$	2020
$D^0\overline{D}^0$	seen	777
$D^*(2007)^0\overline{D}^0 +$ c.c.	seen	578
$D^*(2007)^0\overline{D}^*(2007)^0$	seen	232

$\psi(4160)$ [iii]	$I^G(J^{PC}) = ?^?(1^{--})$

Mass $m = 4159 \pm 20$ MeV
Full width $\Gamma = 78 \pm 20$ MeV
$\Gamma_{ee} = 0.77 \pm 0.23$ keV

$\psi(4160)$ DECAY MODES	Fraction (Γ_i/Γ)	p (MeV/c)
$e^+ e^-$	$(10\pm4) \times 10^{-6}$	2079

$\psi(4415)$ [iii]	$I^G(J^{PC}) = ?^?(1^{--})$

Mass $m = 4415 \pm 6$ MeV
Full width $\Gamma = 43 \pm 15$ MeV   (S = 1.8)
$\Gamma_{ee} = 0.47 \pm 0.10$ keV

$\psi(4415)$ DECAY MODES	Fraction (Γ_i/Γ)	p (MeV/c)
hadrons	dominant	–
$e^+ e^-$	$(1.1\pm0.4) \times 10^{-5}$	2207

$b\overline{b}$ MESONS

$\Upsilon(1S)$	$I^G(J^{PC}) = 0^-(1^{--})$

Mass $m = 9460.37 \pm 0.21$ MeV   (S = 2.7)
Full width $\Gamma = 52.5 \pm 1.8$ keV
$\Gamma_{ee} = 1.32 \pm 0.05$ keV

$\Upsilon(1S)$ DECAY MODES	Fraction (Γ_i/Γ)	Scale factor/ Confidence level	p (MeV/c)
$\tau^+ \tau^-$	$(2.67^{+0.14}_{-0.16})$ %		4384
$e^+ e^-$	(2.52 ± 0.17) %		4730
$\mu^+ \mu^-$	(2.48 ± 0.07) %	S=1.1	4729
Hadronic decays			
$J/\psi(1S)$ anything	$(1.1 \pm0.4) \times 10^{-3}$		4223
$\rho\pi$	< 2 $\times 10^{-4}$	CL=90%	4698
$\pi^+ \pi^-$	< 5 $\times 10^{-4}$	CL=90%	4728
$K^+ K^-$	< 5 $\times 10^{-4}$	CL=90%	4704
$p\overline{p}$	< 5 $\times 10^{-4}$	CL=90%	4636
Radiative decays			
$\gamma 2h^+ 2h^-$	$(7.0 \pm1.5) \times 10^{-4}$		4720
$\gamma 3h^+ 3h^-$	$(5.4 \pm2.0) \times 10^{-4}$		4703
$\gamma 4h^+ 4h^-$	$(7.4 \pm3.5) \times 10^{-4}$		4679
$\gamma \pi^+ \pi^- K^+ K^-$	$(2.9 \pm0.9) \times 10^{-4}$		4686
$\gamma 2\pi^+ 2\pi^-$	$(2.5 \pm0.9) \times 10^{-4}$		4720
$\gamma 3\pi^+ 3\pi^-$	$(2.5 \pm1.2) \times 10^{-4}$		4703
$\gamma 2\pi^+ 2\pi^- K^+ K^-$	$(2.4 \pm1.2) \times 10^{-4}$		4658
$\gamma \pi^+ \pi^- p\overline{p}$	$(1.5 \pm0.6) \times 10^{-4}$		4604
$\gamma 2\pi^+ 2\pi^- p\overline{p}$	$(4 \pm6) \times 10^{-5}$		4563
$\gamma 2K^+ 2K^-$	$(2.0 \pm2.0) \times 10^{-5}$		4601
$\gamma \eta'(958)$	< 1.3 $\times 10^{-3}$	CL=90%	4682
$\gamma \eta$	< 3.5 $\times 10^{-4}$	CL=90%	4714
$\gamma f_2'(1525)$	< 1.4 $\times 10^{-4}$	CL=90%	4607
$\gamma f_2(1270)$	< 1.3 $\times 10^{-4}$	CL=90%	4644
$\gamma \eta(1440)$	< 8.2 $\times 10^{-5}$	CL=90%	4624
$\gamma f_J(1710) \to \gamma K \overline{K}$	< 2.6 $\times 10^{-4}$	CL=90%	4576
$\gamma f_0(2200) \to \gamma K^+ K^-$	< 2 $\times 10^{-4}$	CL=90%	4475
$\gamma f_J(2220) \to \gamma K^+ K^-$	< 1.5 $\times 10^{-5}$	CL=90%	4469
$\gamma \eta(2225) \to \gamma \phi\phi$	< 3 $\times 10^{-3}$	CL=90%	4469
γX	< 3 $\times 10^{-5}$	CL=90%	–
X = pseudoscalar with $m < 7.2$ GeV)			
$\gamma X \overline{X}$	< 1 $\times 10^{-3}$	CL=90%	–
$X \overline{X}$ = vectors with $m < 3.1$ GeV)			

$\chi_{b0}(1P)$ [iii]	$I^G(J^{PC}) = 0^+(0^{++})$

J needs confirmation.

Mass $m = 9859.8 \pm 1.3$ MeV

$\chi_{b0}(1P)$ DECAY MODES	Fraction (Γ_i/Γ)	Confidence level	p (MeV/c)
$\gamma \Upsilon(1S)$	<6 %	90%	391

$\chi_{b1}(1P)$ [iii]	$I^G(J^{PC}) = 0^+(1^{++})$

J needs confirmation.

Mass $m = 9891.9 \pm 0.7$ MeV

$\chi_{b1}(1P)$ DECAY MODES	Fraction (Γ_i/Γ)	p (MeV/c)
$\gamma \Upsilon(1S)$	(35 ± 8) %	422

$\chi_{b2}(1P)$ [iii]	$I^G(J^{PC}) = 0^+(2^{++})$

J needs confirmation.

Mass $m = 9913.2 \pm 0.6$ MeV

$\chi_{b2}(1P)$ DECAY MODES	Fraction (Γ_i/Γ)	p (MeV/c)
$\gamma \Upsilon(1S)$	(22 ± 4) %	443

$\Upsilon(2S)$	$I^G(J^{PC}) = 0^-(1^{--})$

Mass $m = 10.02330 \pm 0.00031$ GeV
Full width $\Gamma = 44 \pm 7$ keV
$\Gamma_{ee} = 0.520 \pm 0.032$ keV

$\Upsilon(2S)$ DECAY MODES	Fraction (Γ_i/Γ)	Confidence level	p (MeV/c)
$\Upsilon(1S) \pi^+ \pi^-$	(18.5 ± 0.8) %		475
$\Upsilon(1S) \pi^0 \pi^0$	(8.8 ± 1.1) %		480
$\tau^+ \tau^-$	(1.7 ± 1.6) %		4686
$\mu^+ \mu^-$	(1.31 ± 0.21) %		5011
$e^+ e^-$	(1.18 ± 0.20) %		5012
$\Upsilon(1S) \pi^0$	< 8 $\times 10^{-3}$	90%	531
$\Upsilon(1S) \eta$	< 2 $\times 10^{-3}$	90%	127
$J/\psi(1S)$ anything	< 6 $\times 10^{-3}$	90%	4533
Radiative decays			
$\gamma \chi_{b1}(1P)$	(6.7 ± 0.9) %		131
$\gamma \chi_{b2}(1P)$	(6.6 ± 0.9) %		110
$\gamma \chi_{b0}(1P)$	(4.3 ± 1.0) %		162
$\gamma f_J(1710)$	< 5.9 $\times 10^{-4}$	90%	4866
$\gamma f_2'(1525)$	< 5.3 $\times 10^{-4}$	90%	4896
$\gamma f_2(1270)$	< 2.41 $\times 10^{-4}$	90%	4931

$\chi_{b0}(2P)$ [iii]	$I^G(J^{PC}) = 0^+(0^{++})$

J needs confirmation.

Mass $m = 10.2321 \pm 0.0006$ GeV

$\chi_{b0}(2P)$ DECAY MODES	Fraction (Γ_i/Γ)	p (MeV/c)
$\gamma \Upsilon(2S)$	(4.6 ± 2.1) %	210
$\gamma \Upsilon(1S)$	$(9 \pm6) \times 10^{-3}$	746

$\chi_{b1}(2P)$ [iii]	$I^G(J^{PC}) = 0^+(1^{++})$

J needs confirmation.

Mass $m = 10.2552 \pm 0.0005$ GeV
$m_{\chi_{b1}(2P)} - m_{\chi_{b0}(2P)} = 23.5 \pm 1.0$ MeV

$\chi_{b1}(2P)$ DECAY MODES	Fraction (Γ_i/Γ)	Scale factor	p (MeV/c)
$\gamma \Upsilon(2S)$	(21 ± 4) %	1.5	229
$\gamma \Upsilon(1S)$	(8.5 ± 1.3) %	1.3	764

$\chi_{b2}(2P)$ [iii]	$I^G(J^{PC}) = 0^+(2^{++})$

J needs confirmation.

Mass $m = 10.2685 \pm 0.0004$ GeV
$m_{\chi_{b2}(2P)} - m_{\chi_{b1}(2P)} = 13.5 \pm 0.6$ MeV

$\chi_{b2}(2P)$ DECAY MODES	Fraction (Γ_i/Γ)	p (MeV/c)
$\gamma \Upsilon(2S)$	(16.2 ± 2.4) %	242
$\gamma \Upsilon(1S)$	(7.1 ± 1.0) %	776

$\Upsilon(3S)$ $I^G(J^{PC}) = 0^-(1^{--})$

Mass $m = 10.3553 \pm 0.0005$ GeV
Full width $\Gamma = 26.3 \pm 3.5$ keV

$\Upsilon(3S)$ DECAY MODES	Fraction (Γ_i/Γ)	Scale factor/ Confidence level	p (MeV/c)
$\Upsilon(2S)$ anything	$(10.6\ \pm 0.8\ )\ \%$		296
$\Upsilon(2S)\pi^+\pi^-$	$(\ 2.8\ \pm 0.6\ )\ \%$	S=2.2	177
$\Upsilon(2S)\pi^0\pi^0$	$(\ 2.00\pm 0.32)\ \%$		190
$\Upsilon(2S)\gamma\gamma$	$(\ 5.0\ \pm 0.7\ )\ \%$		327
$\Upsilon(1S)\pi^+\pi^-$	$(\ 4.48\pm 0.21)\ \%$		814
$\Upsilon(1S)\pi^0\pi^0$	$(\ 2.06\pm 0.28)\ \%$		816
$\Upsilon(1S)\eta$	$< 2.2\ \ \ \times 10^{-3}$	CL=90%	–
$\mu^+\mu^-$	$(\ 1.81\pm 0.17)\ \%$		5177
e^+e^-	seen		5177

Radiative decays

$\gamma\chi_{b2}(2P)$	$(11.4\ \pm 0.8\ )\ \%$	S=1.3	87
$\gamma\chi_{b1}(2P)$	$(11.3\ \pm 0.6\ )\ \%$		100
$\gamma\chi_{b0}(2P)$	$(\ 5.4\ \pm 0.6\ )\ \%$	S=1.1	123

$\Upsilon(4S)$ or $\Upsilon(10580)$ $I^G(J^{PC}) = ?^?(1^{--})$

Mass $m = 10.5800 \pm 0.0035$ GeV
Full width $\Gamma = 10 \pm 4$ MeV
$\Gamma_{ee} = 0.248 \pm 0.031$ keV   (S = 1.3)

$\Upsilon(4S)$ DECAY MODES	Fraction (Γ_i/Γ)	Confidence level	p (MeV/c)
$B\bar{B}$	$> 96\ \ \ \%$	95%	–
non-$B\bar{B}$	$< 4\ \ \ \%$	95%	–
e^+e^-	$(\ 2.8\pm 0.7) \times 10^{-5}$		5290
$J/\psi(3097)$ anything	$(\ 2.2\pm 0.7) \times 10^{-3}$		–
D^{*+} anything $+$ c.c.	$< 7.4\ \ \ \%$	90%	5099
ϕ anything	$< 2.3\ \ \ \times 10^{-3}$	90%	5240
$\Upsilon(1S)$ anything	$< 4\ \ \ \times 10^{-3}$	90%	1053

$\Upsilon(10860)$ $I^G(J^{PC}) = ?^?(1^{--})$

Mass $m = 10.865 \pm 0.008$ GeV   (S = 1.1)
Full width $\Gamma = 110 \pm 13$ MeV
$\Gamma_{ee} = 0.31 \pm 0.07$ keV   (S = 1.3)

$\Upsilon(10860)$ DECAY MODES	Fraction (Γ_i/Γ)	p (MeV/c)
e^+e^-	$(2.8\pm 0.7) \times 10^{-6}$	5432

$\Upsilon(11020)$ $I^G(J^{PC}) = ?^?(1^{--})$

Mass $m = 11.019 \pm 0.008$ GeV
Full width $\Gamma = 79 \pm 16$ MeV
$\Gamma_{ee} = 0.130 \pm 0.030$ keV

$\Upsilon(11020)$ DECAY MODES	Fraction (Γ_i/Γ)	p (MeV/c)
e^+e^-	$(1.6\pm 0.5) \times 10^{-6}$	5509

NOTES

In this Summary Table:

When a quantity has "(S = ...)" to its right, the error on the quantity has been enlarged by the "scale factor" S, defined as $S = \sqrt{\chi^2/(N-1)}$, where N is the number of measurements used in calculating the quantity. We do this when $S > 1$, which often indicates that the measurements are inconsistent. When $S > 1.25$, we also show in the Particle Listings an ideogram of the measurements. For more about S, see the Introduction.

A decay momentum p is given for each decay mode. For a 2-body decay, p is the momentum of each decay product in the rest frame of the decaying particle. For a 3-or-more-body decay, p is the largest momentum any of the products can have in this frame.

[a] See the "Note on $\pi^\pm \to \ell^\pm \nu \gamma$ and $K^\pm \to \ell^\pm \nu \gamma$ Form Factors" in the $\pi^\pm$ Particle Listings for definitions and details.

[b] Measurements of $\Gamma(e^+\nu_e)/\Gamma(\mu^+\nu_\mu)$ always include decays with γ's, and measurements of $\Gamma(e^+\nu_e\gamma)$ and $\Gamma(\mu^+\nu_\mu\gamma)$ never include low-energy γ's. Therefore, since no clean separation is possible, we consider the modes with γ's to be subreactions of the modes without them, and let $[\Gamma(e^+\nu_e) + \Gamma(\mu^+\nu_\mu)]/\Gamma_{total} = 100\%$.

[c] See the $\pi^\pm$ Particle Listings for the energy limits used in this measurement; low-energy γ's are not included.

[d] Derived from an analysis of neutrino-oscillation experiments.

[e] Astrophysical and cosmological arguments give limits of order 10^{-13}; see the π^0 Particle Listings.

[f] See the "Note on the Decay Width $\Gamma(\eta \to \gamma\gamma)$" in our 1994 edition, Phys. Rev. **D50**, 1 August 1994, Part I, p. 1451.

[g] C parity forbids this to occur as a single-photon process.

[h] See the "Note on scalar mesons" in the $f_0(1370)$ Particle Listings. The interpretation of this entry as a particle is controversial.

[i] See the "Note on $\rho(770)$" in the $\rho(770)$ Particle Listings.

[j] The e^+e^- branching fraction is from $e^+e^- \to \pi^+\pi^-$ experiments only. The $\omega\rho$ interference is then due to $\omega\rho$ mixing only, and is expected to be small. If $e\mu$ universality holds, $\Gamma(\rho^0 \to \mu^+\mu^-) = \Gamma(\rho^0 \to e^+e^-) \times 0.99785$.

[k] See the "Note on scalar mesons" in the $f_0(1370)$ Particle Listings.

[l] See the "Note on $a_1(1260)$" in the $a_1(1260)$ Particle Listings.

[m] This is only an educated guess; the error given is larger than the error on the average of the published values. See the Particle Listings for details.

[n] See the "Note on the $f_1(1420)$" in the $\eta(1440)$ Particle Listings.

[o] See also the $\omega(1600)$ Particle Listings.

[p] See the "Note on the $\eta(1440)$" in the $\eta(1440)$ Particle Listings.

[q] See the "Note on the $\rho(1450)$ and the $\rho(1700)$" in the $\rho(1700)$ Particle Listings.

[r] See the "Note on non-$q\bar{q}$ mesons" in the Particle Listings (see the index for the page number).

[s] See also the $\omega(1420)$ Particle Listings.

[t] See the "Note on $f_J(1710)$" in the $f_J(1710)$ Particle Listings.

[u] See the note in the $K^\pm$ Particle Listings.

[v] The definition of the slope parameter g of the $K \to 3\pi$ Dalitz plot is as follows (see also "Note on Dalitz Plot Parameters for $K \to 3\pi$ Decays" in the $K^\pm$ Particle Listings):

$$|M|^2 = 1 + g(s_3 - s_0)/m_{\pi^+}^2 + \cdots.$$

[w] For more details and definitions of parameters see the Particle Listings.

[x] See the $K^\pm$ Particle Listings for the energy limits used in this measurement.

[y] Most of this radiative mode, the low-momentum γ part, is also included in the parent mode listed without γ's.

[z] Direct-emission branching fraction.

[aa] Structure-dependent part.

[bb] Derived from measured values of ϕ_{+-}, ϕ_{00}, $|\eta|$, $|m_{K_L^0} - m_{K_S^0}|$ and $\tau_{K_S^0}$, as described in the introduction to "Tests of Conservation Laws."

[cc] The CP-violation parameters are defined as follows (see also "Note on CP Violation in $K_S \to 3\pi$" and "Note on CP Violation in K_L^0 Decay" in the Particle Listings):

$$\eta_{+-} = |\eta_{+-}|e^{i\phi_{+-}} = \frac{A(K_L^0 \to \pi^+\pi^-)}{A(K_S^0 \to \pi^+\pi^-)} = \epsilon + \epsilon'$$

$$\eta_{00} = |\eta_{00}|e^{i\phi_{00}} = \frac{A(K_L^0 \to \pi^0\pi^0)}{A(K_S^0 \to \pi^0\pi^0)} = \epsilon - 2\epsilon'$$

$$\delta = \frac{\Gamma(K_L^0 \to \pi^-\ell^+\nu) - \Gamma(K_L^0 \to \pi^+\ell^-\nu)}{\Gamma(K_L^0 \to \pi^-\ell^+\nu) + \Gamma(K_L^0 \to \pi^+\ell^-\nu)}$$

$$\text{Im}(\eta_{+-0})^2 = \frac{\Gamma(K_S^0 \to \pi^+\pi^-\pi^0)^{CP \text{ viol.}}}{\Gamma(K_L^0 \to \pi^+\pi^-\pi^0)}$$

$$\text{Im}(\eta_{000})^2 = \frac{\Gamma(K_S^0 \to \pi^0\pi^0\pi^0)}{\Gamma(K_L^0 \to \pi^0\pi^0\pi^0)} .$$

where for the last two relations CPT is assumed valid, i.e., $\text{Re}(\eta_{+-0}) \simeq 0$ and $\text{Re}(\eta_{000}) \simeq 0$.

[dd] See the K_S^0 Particle Listings for the energy limits used in this measurement.

[ee] Calculated from K_L^0 semileptonic rates and the K_S^0 lifetime assuming $\Delta S = \Delta Q$.

[ff] ϵ'/ϵ is derived from $|\eta_{00}/\eta_{+-}|$ measurements using theoretical input on phases.

[gg] The value is for the sum of the charge states of particle/antiparticle states indicated.

[hh] See the K_L^0 Particle Listings for the energy limits used in this measurement.

[ii] Allowed by higher-order electroweak interactions.

[jj] Violates CP in leading order. Test of direct CP violation since the indirect CP-violating and CP-conserving contributions are expected to be suppressed.

[kk] See the "Note on $f_0(1370)$" in the $f_0(1370)$ Particle Listings and in the 1994 edition.

[ll] See the note in the $L(1770)$ Particle Listings in Reviews of Modern Physics **56** No. 2 Pt. II (1984), p. S200. See also the "Note on $K_2(1770)$ and the $K_2(1820)$" in the $K_2(1770)$ Particle Listings .

[mm] See the "Note on $K_2(1770)$ and the $K_2(1820)$" in the $K_2(1770)$ Particle Listings .

[nn] This is a weighted average of $D^\pm$ (44%) and D^0 (56%) branching fractions. See "D^+ and $D^0 \to$ (η anything) / (total D^+ and D^0)" under "D^+ Branching Ratios" in the Particle Listings.

[oo] This value averages the e^+ and μ^+ branching fractions, after making a small phase-space adjustment to the μ^+ fraction to be able to use it as an e^+ fraction; hence our ℓ^+ here is really an e^+.

[pp] An ℓ indicates an e or a μ mode, not a sum over these modes.

[qq] The branching fraction for this mode may differ from the sum of the submodes that contribute to it, due to interference effects. See the relevant papers in the Particle Listings.

[rr] The two experiments measuring this fraction are in serious disagreement. See the Particle Listings.

[ss] This mode is not a useful test for a $\Delta C=1$ weak neutral current because both quarks must change flavor in this decay.

[tt] The D_1^0-D_2^0 limits are inferred from the D^0-$\overline{D}^0$ mixing ratio $\Gamma(K^+\ell^-\overline{\nu}_\ell(\text{via } \overline{D}^0)) / \Gamma(K^-\ell^+\nu_\ell)$.

[uu] The larger limit (from E791) allows interference between the doubly Cabibbo-suppressed and mixing amplitudes; the smaller limit (from E691) doesn't. See the papers for details.

[vv] The experiments on the division of this charge mode amongst its submodes disagree, and the submode branching fractions here add up to considerably more than the charged-mode fraction.

[ww] However, these upper limits are in serious disagreement with values obtained in another experiment.

[xx] For now, we average together measurements of the $X e^+\nu_e$ and $X\mu^+\nu_\mu$ branching fractions. This is the average, not the sum.

[yy] This branching fraction includes all the decay modes of the final-state resonance.

[zz] This value includes only K^+K^- decays of the $f_J(1710)$, because branching fractions of this resonance are not known.

[aaa] This value includes only $\pi^+\pi^-$ decays of the $f_0(1500)$, because branching fractions of this resonance are not known.

[bbb] B^0 and B_s^0 contributions not separated. Limit is on weighted average of the two decay rates.

[ccc] These values are model dependent. See 'Note on Semileptonic Decays' in the B^+ Particle Listings.

[ddd] D^{**} stands for the sum of the $D(1\,^1P_1)$, $D(1\,^3P_0)$, $D(1\,^3P_1)$, $D(1\,^3P_2)$, $D(2\,^1S_0)$, and $D(2\,^1S_1)$ resonances.

[eee] Inclusive branching fractions have a multiplicity definition and can be greater than 100%.

[fff] D_J represents an unresolved mixture of pseudoscalar and tensor D^{**} (P-wave) states.

[ggg] Not a pure measurement. See note at head of B_s^0 Decay Modes.

[hhh] Includes $p\overline{p}\pi^+\pi^-\gamma$ and excludes $p\overline{p}\eta$, $p\overline{p}\omega$, $p\overline{p}\eta'$.

[iii] J^{PC} known by production in e^+e^- via single photon annihilation. I^G is not known; interpretation of this state as a single resonance is unclear because of the expectation of substantial threshold effects in this energy region.

[jjj] Spectroscopic labeling for these states is theoretical, pending experimental information.

See also the table of suggested $q\bar{q}$ quark-model assignments in the Quark Model section.

- Indicates particles that appear in the preceding Meson Summary Table. We do not regard the other entries as being established.
- † Indicates that the value of J given is preferred, but needs confirmation.

LIGHT UNFLAVORED ($S = C = B = 0$)

	$I^G(J^{PC})$		$I^G(J^{PC})$
• $\pi^{\pm}$	$1^-(0^-)$	• $X(1650)$	$0^+(?^{?-})$
• π^0	$1^-(0^{-+})$	• $\omega_3(1670)$	$0^-(3^{--})$
• η	$0^+(0^{-+})$	• $\pi_2(1670)$	$1^-(2^{-+})$
• $f_0(400\text{--}1200)$	$0^+(0^{++})$	• $\phi(1680)$	$0^-(1^{--})$
• $\rho(770)$	$1^+(1^{--})$	• $\rho_3(1690)$	$1^+(3^{--})$
• $\omega(782)$	$0^-(1^{--})$	• $\rho(1700)$	$1^+(1^{--})$
• $\eta'(958)$	$0^+(0^{-+})$	• $f_J(1710)$	$0^+(\text{even}^{++})$
• $f_0(980)$	$0^+(0^{++})$	$\eta(1760)$	$0^+(0^{-+})$
• $a_0(980)$	$1^-(0^{++})$	$X(1775)$	$1^-(?^{-+})$
• $\phi(1020)$	$0^-(1^{--})$	• $\pi(1800)$	$1^-(0^{-+})$
• $h_1(1170)$	$0^-(1^{+-})$	$f_2(1810)$	$0^+(2^{++})$
• $b_1(1235)$	$1^+(1^{+-})$	• $\phi_3(1850)$	$0^-(3^{--})$
• $a_1(1260)$	$1^-(1^{++})$	$\eta_2(1870)$	$0^+(2^{-+})$
• $f_2(1270)$	$0^+(2^{++})$	$X(1910)$	$0^+(?^{?+})$
• $f_1(1285)$	$0^+(1^{++})$	$f_2(1950)$	$0^+(2^{++})$
• $\eta(1295)$	$0^+(0^{-+})$	$X(2000)$	$1^-(?^{?+})$
• $\pi(1300)$	$1^-(0^{-+})$	• $f_2(2010)$	$0^+(2^{++})$
• $a_2(1320)$	$1^-(2^{++})$	$f_0(2020)$	$0^+(0^{++})$
• $f_0(1370)$	$0^+(0^{++})$	• $a_4(2040)$	$1^-(4^{++})$
$h_1(1380)$	$?^-(1^{+-})$	• $f_4(2050)$	$0^+(4^{++})$
$\hat{\rho}(1405)$	$1^-(1^{-+})$	$f_0(2060)$	$0^+(0^{++})$
• $f_1(1420)$	$0^+(1^{++})$	$\pi_2(2100)$	$1^-(2^{-+})$
• $\omega(1420)$	$0^-(1^{--})$	$f_2(2150)$	$0^+(2^{++})$
$f_2(1430)$	$0^+(2^{++})$	$\rho(2150)$	$1^+(1^{--})$
• $\eta(1440)$	$0^+(0^{-+})$	$f_0(2200)$	$0^+(0^{++})$
• $a_0(1450)$	$1^-(0^{++})$	$f_J(2220)$	$0^+(2^{++}$ or $4^{++})$
• $\rho(1450)$	$1^+(1^{--})$	$\eta(2225)$	$0^+(0^{-+})$
• $f_0(1500)$	$0^+(0^{++})$	$\rho_3(2250)$	$1^+(3^{--})$
$f_1(1510)$	$0^+(1^{++})$	• $f_2(2300)$	$0^+(2^{++})$
• $f'_2(1525)$	$0^+(2^{++})$	$f_4(2300)$	$0^+(4^{++})$
$f_2(1565)$	$0^+(2^{++})$	• $f_2(2340)$	$0^+(2^{++})$
• $\omega(1600)$	$0^-(1^{--})$	$\rho_5(2350)$	$1^+(5^{--})$
$X(1600)$	$2^+(2^{++})$	$a_6(2450)$	$1^-(6^{++})$
$f_2(1640)$	$0^+(2^{++})$	$f_6(2510)$	$0^+(6^{++})$
$\eta_2(1645)$	$0^+(2^{-+})$	$X(3250)$	$?^?(?^{??})$

OTHER LIGHT UNFLAVORED ($S = C = B = 0$)

$e^+e^-(1100\text{--}2200)$	$?^?(1^{--})$
$\overline{N}N(1100\text{--}3600)$	
$X(1900\text{--}3600)$	

STRANGE ($S = \pm 1$, $C = B = 0$)

	$I(J^P)$
• $K^{\pm}$	$1/2(0^-)$
• K^0	$1/2(0^-)$
• K^0_S	$1/2(0^-)$
• K^0_L	$1/2(0^-)$
• $K^*(892)$	$1/2(1^-)$
• $K_1(1270)$	$1/2(1^+)$
• $K_1(1400)$	$1/2(1^+)$
• $K^*(1410)$	$1/2(1^-)$
• $K^*_0(1430)$	$1/2(0^+)$
• $K^*_2(1430)$	$1/2(2^+)$
$K(1460)$	$1/2(0^-)$
$K_2(1580)$	$1/2(2^-)$
$K_1(1650)$	$1/2(1^+)$
• $K^*(1680)$	$1/2(1^-)$
• $K_2(1770)$	$1/2(2^-)$
• $K^*_3(1780)$	$1/2(3^-)$
• $K_2(1820)$	$1/2(2^-)$
$K(1830)$	$1/2(0^-)$
$K^*_0(1950)$	$1/2(0^+)$
$K^*_2(1980)$	$1/2(2^+)$
• $K^*_4(2045)$	$1/2(4^+)$
$K_2(2250)$	$1/2(2^-)$
$K_3(2320)$	$1/2(3^+)$
$K^*_5(2380)$	$1/2(5^-)$
$K_4(2500)$	$1/2(4^-)$
$K(3100)$	$?^?(?^{??})$

CHARMED ($C = \pm 1$)

	$I(J^P)$
• $D^{\pm}$	$1/2(0^-)$
• D^0	$1/2(0^-)$
• $D^*(2007)^0$	$1/2(1^-)$
• $D^*(2010)^{\pm}$	$1/2(1^-)$
• $D_1(2420)^0$	$1/2(1^+)$
$D_1(2420)^{\pm}$	$1/2(?)$
• $D^*_2(2460)^0$	$1/2(2^+)$
• $D^*_2(2460)^+$	$1/2(2^+)$

CHARMED, STRANGE ($C = S = \pm 1$)

	$I(J^P)$
• $D^{\pm}_s$	$0(0^-)$
• $D^{*\pm}_s$	$0(?^?)$
• $D_{s1}(2536)^{\pm}$	$0(1^+)$
• $D_{sJ}(2573)^{\pm}$	$0(?^?)$

BOTTOM ($B = \pm 1$)

	$I(J^P)$
• $B^{\pm}$	$1/2(0^-)$
• B^0	$1/2(0^-)$
• B^*	$1/2(1^-)$
$B^*_J(5732)$	$?^?(?^?)$

BOTTOM, STRANGE ($B = \pm 1$, $S = \mp 1$)

	$I^G(J^{PC})$
• B^0_s	$0(0^-)$
B^*_s	$0(1^-)$
$B^*_{sJ}(5850)$	$?(?^?)$

BOTTOM, CHARMED ($B = C = \pm 1$)

$B^{\pm}_c$	$0(0^-)$

$c\bar{c}$

	$I^G(J^{PC})$
• $\eta_c(1S)$	$0^+(0^{-+})$
• $J/\psi(1S)$	$0^-(1^{--})$
• $\chi_{c0}(1P)$	$0^+(0^{++})$
• $\chi_{c1}(1P)$	$0^+(1^{++})$
$h_c(1P)$	$?^?(?^{??})$
• $\chi_{c2}(1P)$	$0^+(2^{++})$
$\eta_c(2S)$	$?^?(?^{?+})$
• $\psi(2S)$	$0^-(1^{--})$
• $\psi(3770)$	$?^?(1^{--})$
• $\psi(4040)$	$?^?(1^{--})$
• $\psi(4160)$	$?^?(1^{--})$
• $\psi(4415)$	$?^?(1^{--})$

$b\bar{b}$

	$I^G(J^{PC})$
• $\Upsilon(1S)$	$0^-(1^{--})$
• $\chi_{b0}(1P)$	$0^+(0^{++})$
• $\chi_{b1}(1P)$	$0^+(1^{++})$
• $\chi_{b2}(1P)$	$0^+(2^{++})$
• $\Upsilon(2S)$	$0^-(1^{--})$
• $\chi_{b0}(2P)$	$0^+(0^{++})$
• $\chi_{b1}(2P)$	$0^+(1^{++})$
• $\chi_{b2}(2P)$	$0^+(2^{++})$
• $\Upsilon(3S)$	$0^-(1^{--})$
• $\Upsilon(4S)$	$?^?(1^{--})$
• $\Upsilon(10860)$	$?^?(1^{--})$
• $\Upsilon(11020)$	$?^?(1^{--})$

NON-$q\bar{q}$ CANDIDATES

Non-$q\bar{q}$ Candidates

BARYON SUMMARY TABLE

This short table gives the name, the quantum numbers (where known), and the status of baryons in the Review. Only the baryons with 3- or 4-star status are included in the main Baryon Summary Table. Due to insufficient data or uncertain interpretation, the other entries in the short table are not established as baryons. The names with masses are of baryons that decay strongly. See our 1986 edition (Physics Letters **170B**) for listings of evidence for Z baryons (KN resonances).

Name	J^P	Status	Name	J^P	Status	Name	J^P	Status	Name	J^P	Status	Name	J^P	Status
p	P_{11}	****	$\Delta(1232)$	P_{33}	****	Λ	P_{01}	****	Σ^+	P_{11}	****	Ξ^0	P_{11}	****
n	P_{11}	****	$\Delta(1600)$	P_{33}	***	$\Lambda(1405)$	S_{01}	****	Σ^0	P_{11}	****	Ξ^-	P_{11}	****
$N(1440)$	P_{11}	****	$\Delta(1620)$	S_{31}	****	$\Lambda(1520)$	D_{03}	****	Σ^-	P_{11}	****	$\Xi(1530)$	P_{13}	****
$N(1520)$	D_{13}	****	$\Delta(1700)$	D_{33}	****	$\Lambda(1600)$	P_{01}	***	$\Sigma(1385)$	P_{13}	****	$\Xi(1620)$		*
$N(1535)$	S_{11}	****	$\Delta(1750)$	P_{31}	*	$\Lambda(1670)$	S_{01}	****	$\Sigma(1480)$		*	$\Xi(1690)$		***
$N(1650)$	S_{11}	****	$\Delta(1900)$	S_{31}	**	$\Lambda(1690)$	D_{03}	****	$\Sigma(1560)$		**	$\Xi(1820)$	D_{13}	***
$N(1675)$	D_{15}	****	$\Delta(1905)$	F_{35}	****	$\Lambda(1800)$	S_{01}	***	$\Sigma(1580)$	D_{13}	**	$\Xi(1950)$		***
$N(1680)$	F_{15}	****	$\Delta(1910)$	P_{31}	****	$\Lambda(1810)$	P_{01}	***	$\Sigma(1620)$	S_{11}	**	$\Xi(2030)$		***
$N(1700)$	D_{13}	***	$\Delta(1920)$	P_{33}	***	$\Lambda(1820)$	F_{05}	****	$\Sigma(1660)$	P_{11}	***	$\Xi(2120)$		*
$N(1710)$	P_{11}	***	$\Delta(1930)$	D_{35}	***	$\Lambda(1830)$	D_{05}	****	$\Sigma(1670)$	D_{13}	****	$\Xi(2250)$		**
$N(1720)$	P_{13}	****	$\Delta(1940)$	D_{33}	*	$\Lambda(1890)$	P_{03}	****	$\Sigma(1690)$		**	$\Xi(2370)$		**
$N(1900)$	P_{13}	**	$\Delta(1950)$	F_{37}	****	$\Lambda(2000)$		*	$\Sigma(1750)$	S_{11}	***	$\Xi(2500)$		*
$N(1990)$	F_{17}	**	$\Delta(2000)$	F_{35}	**	$\Lambda(2020)$	F_{07}	*	$\Sigma(1770)$	P_{11}	*			
$N(2000)$	F_{15}	**	$\Delta(2150)$	S_{31}	*	$\Lambda(2100)$	G_{07}	****	$\Sigma(1775)$	D_{15}	****	Ω^-		****
$N(2080)$	D_{13}	**	$\Delta(2200)$	G_{37}	*	$\Lambda(2110)$	F_{05}	***	$\Sigma(1840)$	P_{13}	*	$\Omega(2250)^-$		***
$N(2090)$	S_{11}	*	$\Delta(2300)$	H_{39}	**	$\Lambda(2325)$	D_{03}	*	$\Sigma(1880)$	P_{11}	**	$\Omega(2380)^-$		**
$N(2100)$	P_{11}	*	$\Delta(2350)$	D_{35}	*	$\Lambda(2350)$	H_{09}	***	$\Sigma(1915)$	F_{15}	****	$\Omega(2470)^-$		**
$N(2190)$	G_{17}	****	$\Delta(2390)$	F_{37}	*	$\Lambda(2585)$		**	$\Sigma(1940)$	D_{13}	***			
$N(2200)$	D_{15}	**	$\Delta(2400)$	G_{39}	**				$\Sigma(2000)$	S_{11}	*	Λ_c^+		****
$N(2220)$	H_{19}	****	$\Delta(2420)$	$H_{3,11}$	****				$\Sigma(2030)$	F_{17}	****	$\Lambda_c(2593)^+$		***
$N(2250)$	G_{19}	****	$\Delta(2750)$	$I_{3,13}$	**				$\Sigma(2070)$	F_{15}	*	$\Lambda_c(2625)^+$		***
$N(2600)$	$I_{1,11}$	***	$\Delta(2950)$	$K_{3,15}$	**				$\Sigma(2080)$	P_{13}	**	$\Sigma_c(2455)$		****
$N(2700)$	$K_{1,13}$	**							$\Sigma(2100)$	G_{17}	*	$\Sigma_c(2520)$		***
									$\Sigma(2250)$		***	Ξ_c^+		***
									$\Sigma(2455)$		**	Ξ_c^0		***
									$\Sigma(2620)$		**	$\Xi_c(2645)$		***
									$\Sigma(3000)$		*	Ω_c^0		***
									$\Sigma(3170)$		*			
												Λ_b^0		***
												Ξ_b^0, Ξ_b^-		*

****   Existence is certain, and properties are at least fairly well explored.

***   Existence ranges from very likely to certain, but further confirmation is desirable and/or quantum numbers, branching fractions, etc. are not well determined.

**   Evidence of existence is only fair.

*   Evidence of existence is poor.

N BARYONS
$(S = 0, I = 1/2)$
$p\Gamma N^+ = uud; \quad n\Gamma N^0 = udd$

p $I(J^P) = \frac{1}{2}(\frac{1}{2}^+)$

Mass $m = 938.27231 \pm 0.00028$ MeV [a]
$\quad = 1.007276470 \pm 0.000000012$ u
$\left|\frac{q_{\bar{p}}}{m_{\bar{p}}}\right|/\left(\frac{q_p}{m_p}\right) = 1.0000000015 \pm 0.0000000011$
$|q_p + q_{\bar{p}}|/e < 2 \times 10^{-5}$
$|q_p + q_e|/e < 1.0 \times 10^{-21}$ [b]
Magnetic moment $\mu = 2.79284739 \pm 0.00000006 \ \mu_N$
Electric dipole moment $d = (-4 \pm 6) \times 10^{-23} \ e\,cm$
Electric polarizability $\bar{\alpha} = (12.1 \pm 0.9) \times 10^{-4} \ fm^3$
Magnetic polarizability $\bar{\beta} = (2.1 \pm 0.9) \times 10^{-4} \ fm^3$
Mean life $\tau > 1.6 \times 10^{25}$ years   (independent of mode)
$\quad\quad > 10^{31}$ to 5×10^{32} years [c]   (mode dependent)

Below Γ for N decays Γ p and n distinguish proton and neutron partial life-times. See also the "Note on Nucleon Decay" in our 1994 edition (Phys. Rev. **D50** Γ 1673) for a short review.

The "partial mean life" limits tabulated here are the limits on $\tau/B_i \Gamma$ where τ is the total mean life and B_i is the branching fraction for the mode in question.

p DECAY MODES	Partial mean life (10^{30} years)	Confidence level	p (MeV/c)
Antilepton + meson			
$N \to e^+ \pi$	$> 130 \ (n)\Gamma > 550 \ (p)$	90%	459
$N \to \mu^+ \pi$	$> 100 \ (n)\Gamma > 270 \ (p)$	90%	453
$N \to \nu\pi$	$> 100 \ (n)\Gamma > 25 \ (p)$	90%	459
$p \to e^+ \eta$	> 140	90%	309
$p \to \mu^+ \eta$	> 69	90%	296
$n \to \nu\eta$	> 54	90%	310
$N \to e^+ \rho$	$> 58 \ (n)\Gamma > 75 \ (p)$	90%	153
$N \to \mu^+ \rho$	$> 23 \ (n)\Gamma > 110 \ (p)$	90%	119
$N \to \nu\rho$	$> 19 \ (n)\Gamma > 27 \ (p)$	90%	153
$p \to e^+ \omega$	> 45	90%	142
$p \to \mu^+ \omega$	> 57	90%	104
$n \to \nu\omega$	> 43	90%	144
$N \to e^+ K$	$> 1.3 \ (n)\Gamma > 150 \ (p)$	90%	337
$p \to e^+ K^0_S$	> 76	90%	337
$p \to e^+ K^0_L$	> 44	90%	337
$N \to \mu^+ K$	$> 1.1 \ (n)\Gamma > 120 \ (p)$	90%	326
$p \to \mu^+ K^0_S$	> 64	90%	326
$p \to \mu^+ K^0_L$	> 44	90%	326
$N \to \nu K$	$> 86 \ (n)\Gamma > 100 \ (p)$	90%	339
$p \to e^+ K^*(892)^0$	> 52	90%	45
$N \to \nu K^*(892)$	$> 22 \ (n)\Gamma > 20 \ (p)$	90%	45
Antilepton + mesons			
$p \to e^+ \pi^+ \pi^-$	> 21	90%	448
$p \to e^+ \pi^0 \pi^0$	> 38	90%	449
$n \to e^+ \pi^- \pi^0$	> 32	90%	449
$p \to \mu^+ \pi^+ \pi^-$	> 17	90%	425
$p \to \mu^+ \pi^0 \pi^0$	> 33	90%	427
$n \to \mu^+ \pi^- \pi^0$	> 33	90%	427
$n \to e^+ K^0 \pi^-$	> 18	90%	319
Lepton + meson			
$n \to e^- \pi^+$	> 65	90%	459
$n \to \mu^- \pi^+$	> 49	90%	453
$n \to e^- \rho^+$	> 62	90%	154
$n \to \mu^- \rho^+$	> 7	90%	120
$n \to e^- K^+$	> 32	90%	340
$n \to \mu^- K^+$	> 57	90%	330
Lepton + mesons			
$p \to e^- \pi^+ \pi^+$	> 30	90%	448
$n \to e^- \pi^+ \pi^0$	> 29	90%	449
$p \to \mu^- \pi^+ \pi^+$	> 17	90%	425
$n \to \mu^- \pi^+ \pi^0$	> 34	90%	427
$p \to e^- \pi^+ K^+$	> 20	90%	320
$p \to \mu^- \pi^+ K^+$	> 5	90%	279

	Partial mean life (10^{30} years)	Confidence level	p (MeV/c)
Antilepton + photon(s)			
$p \to e^+ \gamma$	> 460	90%	469
$p \to \mu^+ \gamma$	> 380	90%	463
$n \to \nu\gamma$	> 24	90%	470
$p \to e^+ \gamma\gamma$	> 100	90%	469
Three (or more) leptons			
$p \to e^+ e^+ e^-$	> 510	90%	469
$p \to e^+ \mu^+ \mu^-$	> 81	90%	457
$p \to e^+ \nu\nu$	> 11	90%	469
$n \to e^+ e^- \nu$	> 74	90%	470
$n \to \mu^+ e^- \nu$	> 47	90%	464
$n \to \mu^+ \mu^- \nu$	> 42	90%	458
$p \to \mu^+ e^+ e^-$	> 91	90%	464
$p \to \mu^+ \mu^+ \mu^-$	> 190	90%	439
$p \to \mu^+ \nu\nu$	> 21	90%	463
$p \to e^- \mu^+ \mu^+$	> 6	90%	457
$n \to 3\nu$	> 0.0005	90%	470
Inclusive modes			
$N \to e^+$ anything	$> 0.6 \ (n\Gamma p)$	90%	–
$N \to \mu^+$ anything	$> 12 \ (n\Gamma p)$	90%	–
$N \to e^+ \pi^0$ anything	$> 0.6 \ (n\Gamma p)$	90%	–
$\Delta B = 2$ dinucleon modes			

The following are lifetime limits per iron nucleus.

$pp \to \pi^+ \pi^+$	> 0.7	90%	–
$pn \to \pi^+ \pi^0$	> 2	90%	–
$nn \to \pi^+ \pi^-$	> 0.7	90%	–
$nn \to \pi^0 \pi^0$	> 3.4	90%	–
$pp \to e^+ e^+$	> 5.8	90%	–
$pp \to e^+ \mu^+$	> 3.6	90%	–
$pp \to \mu^+ \mu^+$	> 1.7	90%	–
$pn \to e^+ \bar{\nu}$	> 2.8	90%	–
$pn \to \mu^+ \bar{\nu}$	> 1.6	90%	–
$nn \to \nu_e \bar{\nu}_e$	> 0.000012	90%	–
$nn \to \nu_\mu \bar{\nu}_\mu$	> 0.000006	90%	–

$\bar{p}$ DECAY MODES

$\bar{p}$ DECAY MODES	Partial mean life (years)	Confidence level	p (MeV/c)
$\bar{p} \to e^- \gamma$	> 1848	95%	469
$\bar{p} \to e^- \pi^0$	> 554	95%	459
$\bar{p} \to e^- \eta$	> 171	95%	309
$\bar{p} \to e^- K^0_S$	> 29	95%	337
$\bar{p} \to e^- K^0_L$	> 9	95%	337

n $I(J^P) = \frac{1}{2}(\frac{1}{2}^+)$

Mass $m = 939.56563 \pm 0.00028$ MeV [a]
$\quad = 1.008664904 \pm 0.000000014$ u
$m_n - m_p = 1.293318 \pm 0.000009$ MeV
$\quad = 0.001388434 \pm 0.000000009$ u
Mean life $\tau = 886.7 \pm 1.9$ s $(S = 1.2)$
$\quad c\tau = 2.658 \times 10^8$ km
Magnetic moment $\mu = -1.9130428 \pm 0.0000005 \ \mu_N$
Electric dipole moment $d < 0.97 \times 10^{-25} \ e\,cm \Gamma CL = 90\%$
Electric polarizability $\alpha = (0.98^{+0.19}_{-0.23}) \times 10^{-3} \ fm^3$ $(S = 1.1)$
Charge $q = (-0.4 \pm 1.1) \times 10^{-21} \ e$
Mean $n\bar{n}$-oscillation time $> 1.2 \times 10^8$ s $\Gamma CL = 90\%$ [d] (bound n)
$\quad\quad > 0.86 \times 10^8$ s $\Gamma CL = 90\%$ (free n)

Decay parameters [e]

$pe^- \bar{\nu}_e$	$g_A/g_V = -1.2670 \pm 0.0035$	$(S = 1.9)$
"	$A = -0.1162 \pm 0.0013$	$(S = 1.8)$
"	$B = 0.990 \pm 0.008$	
"	$a = -0.102 \pm 0.005$	
"	$\phi_{AV} = (180.07 \pm 0.18)^\circ$ [f]	
"	$D = (-0.5 \pm 1.4) \times 10^{-3}$	

n DECAY MODES	Fraction (Γ_i/Γ)	Confidence level	p (MeV/c)
$pe^- \bar{\nu}_e$	100 %		1.19
Charge conservation (Q) violating mode			
$p\nu_e \bar{\nu}_e$	$Q < 8 \times 10^{-27}$	68%	1.29

$N(1440) P_{11}$ $I(J^P) = \frac{1}{2}(\frac{1}{2}^+)$

Breit-Wigner mass = 1430 to 1470 ($\approx$ 1440) MeV
Breit-Wigner full width = 250 to 450 ($\approx$ 350) MeV
p_{beam} = 0.61 GeV/c $4\pi\lambda^2$ = 31.0 mb
Re(pole position) = 1345 to 1385 ($\approx$ 1365) MeV
-2Im(pole position) = 160 to 260 ($\approx$ 210) MeV

$N(1440)$ DECAY MODES	Fraction (Γ_i/Γ)	p (MeV/c)
$N\pi$	60–70 %	397
$N\pi\pi$	30–40 %	342
$\quad \Delta\pi$	20–30 %	143
$\quad N\rho$	<8 %	†
$\quad N(\pi\pi)_{S\text{-wave}}^{I=0}$	5–10 %	–
$p\gamma$	0.035–0.048 %	414
$\quad p\gamma\,\Gamma\text{helicity}=1/2$	0.035–0.048 %	414
$n\gamma$	0.009–0.032 %	413
$\quad n\gamma\,\Gamma\text{helicity}=1/2$	0.009–0.032 %	413

$N(1520) D_{13}$ $I(J^P) = \frac{1}{2}(\frac{3}{2}^-)$

Breit-Wigner mass = 1515 to 1530 ($\approx$ 1520) MeV
Breit-Wigner full width = 110 to 135 ($\approx$ 120) MeV
p_{beam} = 0.74 GeV/c $4\pi\lambda^2$ = 23.5 mb
Re(pole position) = 1505 to 1515 ($\approx$ 1510) MeV
-2Im(pole position) = 110 to 120 ($\approx$ 115) MeV

$N(1520)$ DECAY MODES	Fraction (Γ_i/Γ)	p (MeV/c)
$N\pi$	50–60 %	456
$N\pi\pi$	40–50 %	410
$\quad \Delta\pi$	15–25 %	228
$\quad N\rho$	15–25 %	†
$\quad N(\pi\pi)_{S\text{-wave}}^{I=0}$	<8 %	–
$p\gamma$	0.46–0.56 %	470
$\quad p\gamma\,\Gamma\text{helicity}=1/2$	0.001–0.034 %	470
$\quad p\gamma\,\Gamma\text{helicity}=3/2$	0.44–0.53 %	470
$n\gamma$	0.30–0.53 %	470
$\quad n\gamma\,\Gamma\text{helicity}=1/2$	0.04–0.10 %	470
$\quad n\gamma\,\Gamma\text{helicity}=3/2$	0.25–0.45 %	470

$N(1535) S_{11}$ $I(J^P) = \frac{1}{2}(\frac{1}{2}^-)$

Breit-Wigner mass = 1520 to 1555 ($\approx$ 1535) MeV
Breit-Wigner full width = 100 to 250 ($\approx$ 150) MeV
p_{beam} = 0.76 GeV/c $4\pi\lambda^2$ = 22.5 mb
Re(pole position) = 1495 to 1515 ($\approx$ 1505) MeV
-2Im(pole position) = 90 to 250 ($\approx$ 170) MeV

$N(1535)$ DECAY MODES	Fraction (Γ_i/Γ)	p (MeV/c)
$N\pi$	35–55 %	467
$N\eta$	30–55 %	182
$N\pi\pi$	1–10 %	422
$\quad \Delta\pi$	<1 %	242
$\quad N\rho$	<4 %	†
$\quad N(\pi\pi)_{S\text{-wave}}^{I=0}$	<3 %	–
$\quad N(1440)\pi$	<7 %	†
$p\gamma$	0.15–0.35 %	481
$\quad p\gamma\,\Gamma\text{helicity}=1/2$	0.15–0.35 %	481
$n\gamma$	0.004–0.29 %	480
$\quad n\gamma\,\Gamma\text{helicity}=1/2$	0.004–0.29 %	480

$N(1650) S_{11}$ $I(J^P) = \frac{1}{2}(\frac{1}{2}^-)$

Breit-Wigner mass = 1640 to 1680 ($\approx$ 1650) MeV
Breit-Wigner full width = 145 to 190 ($\approx$ 150) MeV
p_{beam} = 0.96 GeV/c $4\pi\lambda^2$ = 16.4 mb
Re(pole position) = 1640 to 1680 ($\approx$ 1660) MeV
-2Im(pole position) = 150 to 170 ($\approx$ 160) MeV

$N(1650)$ DECAY MODES	Fraction (Γ_i/Γ)	p (MeV/c)
$N\pi$	55–90 %	547
$N\eta$	3–10 %	346
ΛK	3–11 %	161
$N\pi\pi$	10–20 %	511
$\quad \Delta\pi$	1–7 %	344
$\quad N\rho$	4–12 %	†
$\quad N(\pi\pi)_{S\text{-wave}}^{I=0}$	<4 %	–
$\quad N(1440)\pi$	<5 %	147
$p\gamma$	0.04–0.18 %	558
$\quad p\gamma\,\Gamma\text{helicity}=1/2$	0.04–0.18 %	558
$n\gamma$	0.003–0.17 %	557
$\quad n\gamma\,\Gamma\text{helicity}=1/2$	0.003–0.17 %	557

$N(1675) D_{15}$ $I(J^P) = \frac{1}{2}(\frac{5}{2}^-)$

Breit-Wigner mass = 1670 to 1685 ($\approx$ 1675) MeV
Breit-Wigner full width = 140 to 180 ($\approx$ 150) MeV
p_{beam} = 1.01 GeV/c $4\pi\lambda^2$ = 15.4 mb
Re(pole position) = 1655 to 1665 ($\approx$ 1660) MeV
-2Im(pole position) = 125 to 155 ($\approx$ 140) MeV

$N(1675)$ DECAY MODES	Fraction (Γ_i/Γ)	p (MeV/c)
$N\pi$	40–50 %	563
ΛK	<1 %	209
$N\pi\pi$	50–60 %	529
$\quad \Delta\pi$	50–60 %	364
$\quad N\rho$	< 1–3 %	†
$p\gamma$	0.004–0.023 %	575
$\quad p\gamma\,\Gamma\text{helicity}=1/2$	0.0–0.015 %	575
$\quad p\gamma\,\Gamma\text{helicity}=3/2$	0.0–0.011 %	575
$n\gamma$	0.02–0.12 %	574
$\quad n\gamma\,\Gamma\text{helicity}=1/2$	0.006–0.046 %	574
$\quad n\gamma\,\Gamma\text{helicity}=3/2$	0.01–0.08 %	574

$N(1680) F_{15}$ $I(J^P) = \frac{1}{2}(\frac{5}{2}^+)$

Breit-Wigner mass = 1675 to 1690 ($\approx$ 1680) MeV
Breit-Wigner full width = 120 to 140 ($\approx$ 130) MeV
p_{beam} = 1.01 GeV/c $4\pi\lambda^2$ = 15.2 mb
Re(pole position) = 1665 to 1675 ($\approx$ 1670) MeV
-2Im(pole position) = 105 to 135 ($\approx$ 120) MeV

$N(1680)$ DECAY MODES	Fraction (Γ_i/Γ)	p (MeV/c)
$N\pi$	60–70 %	567
$N\pi\pi$	30–40 %	532
$\quad \Delta\pi$	5–15 %	369
$\quad N\rho$	3–15 %	†
$\quad N(\pi\pi)_{S\text{-wave}}^{I=0}$	5–20 %	–
$p\gamma$	0.21–0.32 %	578
$\quad p\gamma\,\Gamma\text{helicity}=1/2$	0.001–0.011 %	578
$\quad p\gamma\,\Gamma\text{helicity}=3/2$	0.20–0.32 %	578
$n\gamma$	0.021–0.046 %	577
$\quad n\gamma\,\Gamma\text{helicity}=1/2$	0.004–0.029 %	577
$\quad n\gamma\,\Gamma\text{helicity}=3/2$	0.01–0.024 %	577

$N(1700)\ D_{13}$ $I(J^P) = \frac{1}{2}(\frac{3}{2}^-)$

Breit-Wigner mass = 1650 to 1750 ($\approx$ 1700) MeV
Breit-Wigner full width = 50 to 150 ($\approx$ 100) MeV
p_{beam} = 1.05 GeV/c $4\pi\lambda^2$ = 14.5 mb
Re(pole position) = 1630 to 1730 ($\approx$ 1680) MeV
-2Im(pole position) = 50 to 150 ($\approx$ 100) MeV

$N(1700)$ DECAY MODES	Fraction (Γ_i/Γ)	p (MeV/c)
$N\pi$	5–15 %	580
ΛK	<3 %	250
$N\pi\pi$	85–95 %	547
$N\rho$	<35 %	†
$p\gamma$	0.01–0.05 %	591
$p\gamma\ \Gamma$ helicity=1/2	0.0–0.024 %	591
$p\gamma\ \Gamma$ helicity=3/2	0.002–0.026 %	591
$n\gamma$	0.01–0.13 %	590
$n\gamma\ \Gamma$ helicity=1/2	0.0–0.09 %	590
$n\gamma\ \Gamma$ helicity=3/2	0.01–0.05 %	590

$N(1710)\ P_{11}$ $I(J^P) = \frac{1}{2}(\frac{1}{2}^+)$

Breit-Wigner mass = 1680 to 1740 ($\approx$ 1710) MeV
Breit-Wigner full width = 50 to 250 ($\approx$ 100) MeV
p_{beam} = 1.07 GeV/c $4\pi\lambda^2$ = 14.2 mb
Re(pole position) = 1670 to 1770 ($\approx$ 1720) MeV
-2Im(pole position) = 80 to 380 ($\approx$ 230) MeV

$N(1710)$ DECAY MODES	Fraction (Γ_i/Γ)	p (MeV/c)
$N\pi$	10–20 %	587
ΛK	5–25 %	264
$N\pi\pi$	40–90 %	554
$\Delta\pi$	15–40 %	393
$N\rho$	5–25 %	48
$N(\pi\pi)^{I=0}_{S\text{-wave}}$	10–40 %	–
$p\gamma$	0.002–0.05%	598
$p\gamma\ \Gamma$ helicity=1/2	0.002–0.05%	598
$n\gamma$	0.0–0.02%	597
$n\gamma\ \Gamma$ helicity=1/2	0.0–0.02%	597

$N(1720)\ P_{13}$ $I(J^P) = \frac{1}{2}(\frac{3}{2}^+)$

Breit-Wigner mass = 1650 to 1750 ($\approx$ 1720) MeV
Breit-Wigner full width = 100 to 200 ($\approx$ 150) MeV
p_{beam} = 1.09 GeV/c $4\pi\lambda^2$ = 13.9 mb
Re(pole position) = 1650 to 1750 ($\approx$ 1700) MeV
-2Im(pole position) = 110 to 390 ($\approx$ 250) MeV

$N(1720)$ DECAY MODES	Fraction (Γ_i/Γ)	p (MeV/c)
$N\pi$	10–20 %	594
ΛK	1–15 %	278
$N\pi\pi$	>70 %	561
$N\rho$	70–85 %	104
$p\gamma$	0.003–0.10 %	604
$p\gamma\ \Gamma$ helicity=1/2	0.003–0.08 %	604
$p\gamma\ \Gamma$ helicity=3/2	0.001–0.03 %	604
$n\gamma$	0.002–0.39 %	603
$n\gamma\ \Gamma$ helicity=1/2	0.0–0.002 %	603
$n\gamma\ \Gamma$ helicity=3/2	0.001–0.39 %	603

$N(2190)\ G_{17}$ $I(J^P) = \frac{1}{2}(\frac{7}{2}^-)$

Breit-Wigner mass = 2100 to 2200 ($\approx$ 2190) MeV
Breit-Wigner full width = 350 to 550 ($\approx$ 450) MeV
p_{beam} = 2.07 GeV/c $4\pi\lambda^2$ = 6.21 mb
Re(pole position) = 1950 to 2150 ($\approx$ 2050) MeV
-2Im(pole position) = 350 to 550 ($\approx$ 450) MeV

$N(2190)$ DECAY MODES	Fraction (Γ_i/Γ)	p (MeV/c)
$N\pi$	10–20 %	888

$N(2220)\ H_{19}$ $I(J^P) = \frac{1}{2}(\frac{9}{2}^+)$

Breit-Wigner mass = 2180 to 2310 ($\approx$ 2220) MeV
Breit-Wigner full width = 320 to 550 ($\approx$ 400) MeV
p_{beam} = 2.14 GeV/c $4\pi\lambda^2$ = 5.97 mb
Re(pole position) = 2100 to 2240 ($\approx$ 2170) MeV
-2Im(pole position) = 370 to 570 ($\approx$ 470) MeV

$N(2220)$ DECAY MODES	Fraction (Γ_i/Γ)	p (MeV/c)
$N\pi$	10–20 %	905

$N(2250)\ G_{19}$ $I(J^P) = \frac{1}{2}(\frac{9}{2}^-)$

Breit-Wigner mass = 2170 to 2310 ($\approx$ 2250) MeV
Breit-Wigner full width = 290 to 470 ($\approx$ 400) MeV
p_{beam} = 2.21 GeV/c $4\pi\lambda^2$ = 5.74 mb
Re(pole position) = 2080 to 2200 ($\approx$ 2140) MeV
-2Im(pole position) = 280 to 680 ($\approx$ 480) MeV

$N(2250)$ DECAY MODES	Fraction (Γ_i/Γ)	p (MeV/c)
$N\pi$	5–15 %	923

$N(2600)\ I_{1,11}$ $I(J^P) = \frac{1}{2}(\frac{11}{2}^-)$

Breit-Wigner mass = 2550 to 2750 ($\approx$ 2600) MeV
Breit-Wigner full width = 500 to 800 ($\approx$ 650) MeV
p_{beam} = 3.12 GeV/c $4\pi\lambda^2$ = 3.86 mb

$N(2600)$ DECAY MODES	Fraction (Γ_i/Γ)	p (MeV/c)
$N\pi$	5–10 %	1126

Δ BARYONS
($S = 0,\ I = 3/2$)
$\Delta^{++} = uuu\Gamma$ $\Delta^+ = uud\Gamma$ $\Delta^0 = udd\Gamma$ $\Delta^- = ddd$

$\Delta(1232)\ P_{33}$ $I(J^P) = \frac{3}{2}(\frac{3}{2}^+)$

Breit-Wigner mass (mixed charges) = 1230 to 1234 ($\approx$ 1232) MeV
Breit-Wigner full width (mixed charges) = 115 to 125 ($\approx$ 120) MeV
p_{beam} = 0.30 GeV/c $4\pi\lambda^2$ = 94.8 mb
Re(pole position) = 1209 to 1211 ($\approx$ 1210) MeV
-2Im(pole position) = 98 to 102 ($\approx$ 100) MeV

$\Delta(1232)$ DECAY MODES	Fraction (Γ_i/Γ)	p (MeV/c)
$N\pi$	>99 %	227
$N\gamma$	0.52–0.60 %	259
$N\gamma\ \Gamma$ helicity=1/2	0.11–0.13 %	259
$N\gamma\ \Gamma$ helicity=3/2	0.41–0.47 %	259

$\Delta(1600)\ P_{33}$ $I(J^P) = \frac{3}{2}(\frac{3}{2}^+)$

Breit-Wigner mass = 1550 to 1700 ($\approx$ 1600) MeV
Breit-Wigner full width = 250 to 450 ($\approx$ 350) MeV
p_{beam} = 0.87 GeV/c $4\pi\lambda^2$ = 18.6 mb
Re(pole position) = 1500 to 1700 ($\approx$ 1600) MeV
-2Im(pole position) = 200 to 400 ($\approx$ 300) MeV

$\Delta(1600)$ DECAY MODES	Fraction (Γ_i/Γ)	p (MeV/c)
$N\pi$	10–25 %	512
$N\pi\pi$	75–90 %	473
$\Delta\pi$	40–70 %	301
$N\rho$	<25 %	†
$N(1440)\pi$	10–35 %	74
$N\gamma$	0.001–0.02 %	525
$N\gamma\ \Gamma$ helicity=1/2	0.0–0.02 %	525
$N\gamma\ \Gamma$ helicity=3/2	0.001–0.005 %	525

$\Delta(1620)\ S_{31}$ $I(J^P) = \frac{3}{2}(\frac{1}{2}^-)$

Breit-Wigner mass = 1615 to 1675 ($\approx$ 1620) MeV
Breit-Wigner full width = 120 to 180 ($\approx$ 150) MeV
p_{beam} = 0.91 GeV/c $4\pi\lambda^2$ = 17.7 mb
Re(pole position) = 1580 to 1620 ($\approx$ 1600) MeV
-2Im(pole position) = 100 to 130 ($\approx$ 115) MeV

$\Delta(1620)$ DECAY MODES	Fraction (Γ_i/Γ)	p (MeV/c)
$N\pi$	20–30 %	526
$N\pi\pi$	70–80 %	488
$\Delta\pi$	30–60 %	318
$N\rho$	7–25 %	†
$N\gamma$	0.004–0.044 %	538
$N\gamma\,\Gamma$ helicity=1/2	0.004–0.044 %	538

$\Delta(1700)\ D_{33}$ $I(J^P) = \frac{3}{2}(\frac{3}{2}^-)$

Breit-Wigner mass = 1670 to 1770 ($\approx$ 1700) MeV
Breit-Wigner full width = 200 to 400 ($\approx$ 300) MeV
p_{beam} = 1.05 GeV/c $4\pi\lambda^2$ = 14.5 mb
Re(pole position) = 1620 to 1700 ($\approx$ 1660) MeV
-2Im(pole position) = 150 to 250 ($\approx$ 200) MeV

$\Delta(1700)$ DECAY MODES	Fraction (Γ_i/Γ)	p (MeV/c)
$N\pi$	10–20 %	580
$N\pi\pi$	80–90 %	547
$\Delta\pi$	30–60 %	385
$N\rho$	30–55 %	†
$N\gamma$	0.12–0.26 %	591
$N\gamma\,\Gamma$ helicity=1/2	0.08–0.16 %	591
$N\gamma\,\Gamma$ helicity=3/2	0.025–0.12 %	591

$\Delta(1905)\ F_{35}$ $I(J^P) = \frac{3}{2}(\frac{5}{2}^+)$

Breit-Wigner mass = 1870 to 1920 ($\approx$ 1905) MeV
Breit-Wigner full width = 280 to 440 ($\approx$ 350) MeV
p_{beam} = 1.45 GeV/c $4\pi\lambda^2$ = 9.62 mb
Re(pole position) = 1800 to 1860 ($\approx$ 1830) MeV
-2Im(pole position) = 230 to 330 ($\approx$ 280) MeV

$\Delta(1905)$ DECAY MODES	Fraction (Γ_i/Γ)	p (MeV/c)
$N\pi$	5–15 %	713
$N\pi\pi$	85–95 %	687
$\Delta\pi$	<25 %	542
$N\rho$	>60 %	421
$N\gamma$	0.01–0.03 %	721
$N\gamma\,\Gamma$ helicity=1/2	0.0–0.1 %	721
$N\gamma\,\Gamma$ helicity=3/2	0.004–0.03 %	721

$\Delta(1910)\ P_{31}$ $I(J^P) = \frac{3}{2}(\frac{1}{2}^+)$

Breit-Wigner mass = 1870 to 1920 ($\approx$ 1910) MeV
Breit-Wigner full width = 190 to 270 ($\approx$ 250) MeV
p_{beam} = 1.46 GeV/c $4\pi\lambda^2$ = 9.54 mb
Re(pole position) = 1830 to 1880 ($\approx$ 1855) MeV
-2Im(pole position) = 200 to 500 ($\approx$ 350) MeV

$\Delta(1910)$ DECAY MODES	Fraction (Γ_i/Γ)	p (MeV/c)
$N\pi$	15–30 %	716
$N\gamma$	0.0–0.2 %	725
$N\gamma\,\Gamma$ helicity=1/2	0.0–0.2 %	725

$\Delta(1920)\ P_{33}$ $I(J^P) = \frac{3}{2}(\frac{3}{2}^+)$

Breit-Wigner mass = 1900 to 1970 ($\approx$ 1920) MeV
Breit-Wigner full width = 150 to 300 ($\approx$ 200) MeV
p_{beam} = 1.48 GeV/c $4\pi\lambda^2$ = 9.37 mb
Re(pole position) = 1850 to 1950 ($\approx$ 1900) MeV
-2Im(pole position) = 200 to 400 ($\approx$ 300) MeV

$\Delta(1920)$ DECAY MODES	Fraction (Γ_i/Γ)	p (MeV/c)
$N\pi$	5–20 %	722

$\Delta(1930)\ D_{35}$ $I(J^P) = \frac{3}{2}(\frac{5}{2}^-)$

Breit-Wigner mass = 1920 to 1970 ($\approx$ 1930) MeV
Breit-Wigner full width = 250 to 450 ($\approx$ 350) MeV
p_{beam} = 1.50 GeV/c $4\pi\lambda^2$ = 9.21 mb
Re(pole position) = 1840 to 1940 ($\approx$ 1890) MeV
-2Im(pole position) = 200 to 300 ($\approx$ 250) MeV

$\Delta(1930)$ DECAY MODES	Fraction (Γ_i/Γ)	p (MeV/c)
$N\pi$	10–20 %	729
$N\gamma$	0.0–0.02 %	737
$N\gamma\,\Gamma$ helicity=1/2	0.0–0.01 %	737
$N\gamma\,\Gamma$ helicity=3/2	0.0–0.01 %	737

$\Delta(1950)\ F_{37}$ $I(J^P) = \frac{3}{2}(\frac{7}{2}^+)$

Breit-Wigner mass = 1940 to 1960 ($\approx$ 1950) MeV
Breit-Wigner full width = 290 to 350 ($\approx$ 300) MeV
p_{beam} = 1.54 GeV/c $4\pi\lambda^2$ = 8.91 mb
Re(pole position) = 1880 to 1890 ($\approx$ 1885) MeV
-2Im(pole position) = 210 to 270 ($\approx$ 240) MeV

$\Delta(1950)$ DECAY MODES	Fraction (Γ_i/Γ)	p (MeV/c)
$N\pi$	35–40 %	741
$N\pi\pi$		716
$\Delta\pi$	20–30 %	574
$N\rho$	<10 %	469
$N\gamma$	0.08–0.13 %	749
$N\gamma\,\Gamma$ helicity=1/2	0.03–0.055 %	749
$N\gamma\,\Gamma$ helicity=3/2	0.05–0.075 %	749

$\Delta(2420)\ H_{3,11}$ $I(J^P) = \frac{3}{2}(\frac{11}{2}^+)$

Breit-Wigner mass = 2300 to 2500 ($\approx$ 2420) MeV
Breit-Wigner full width = 300 to 500 ($\approx$ 400) MeV
p_{beam} = 2.64 GeV/c $4\pi\lambda^2$ = 4.68 mb
Re(pole position) = 2260 to 2400 ($\approx$ 2330) MeV
-2Im(pole position) = 350 to 750 ($\approx$ 550) MeV

$\Delta(2420)$ DECAY MODES	Fraction (Γ_i/Γ)	p (MeV/c)
$N\pi$	5–15 %	1023

Λ BARYONS
(S = −1, I = 0)

$\Lambda^0 = uds$

Λ

$I(J^P) = 0(\frac{1}{2}^+)$

Mass $m = 1115.683 \pm 0.006$ MeV
Mean life $\tau = (2.632 \pm 0.020) \times 10^{-10}$ s (S = 1.6)
  $c\tau = 7.89$ cm
Magnetic moment $\mu = -0.613 \pm 0.004 \ \mu_N$
Electric dipole moment $d < 1.5 \times 10^{-16}$ e cm ΓCL = 95%

Decay parameters

$p\pi^-$	$\alpha_- = 0.642 \pm 0.013$	
"	$\phi_- = (-6.5 \pm 3.5)^\circ$	
"	$\gamma_- = 0.76$ [g]	
"	$\Delta_- = (8 \pm 4)^\circ$ [g]	
$n\pi^0$	$\alpha_0 = +0.65 \pm 0.05$	
$pe^-\bar{\nu}_e$	$g_A/g_V = -0.718 \pm 0.015$ [e]	

Λ DECAY MODES	Fraction (Γ_i/Γ)	p (MeV/c)
$p\pi^-$	(63.9 ± 0.5) %	101
$n\pi^0$	(35.8 ± 0.5) %	104
$n\gamma$	$(1.75 \pm 0.15) \times 10^{-3}$	162
$p\pi^-\gamma$	[h] $8.4 \pm 1.4) \times 10^{-4}$	101
$pe^-\bar{\nu}_e$	$(8.32 \pm 0.14) \times 10^{-4}$	163
$p\mu^-\bar{\nu}_\mu$	$(1.57 \pm 0.35) \times 10^{-4}$	131

Λ(1405) S_{01}

$I(J^P) = 0(\frac{1}{2}^-)$

Mass $m = 1407 \pm 4$ MeV
Full width $\Gamma = 50.0 \pm 2.0$ MeV
Below $\overline{K}N$ threshold

Λ(1405) DECAY MODES	Fraction (Γ_i/Γ)	p (MeV/c)
$\Sigma\pi$	100 %	152

Λ(1520) D_{03}

$I(J^P) = 0(\frac{3}{2}^-)$

Mass $m = 1519.5 \pm 1.0$ MeV [i]
Full width $\Gamma = 15.6 \pm 1.0$ MeV [i]
  $p_{beam} = 0.39$ GeV/c $4\pi\lambda^2 = 82.8$ mb

Λ(1520) DECAY MODES	Fraction (Γ_i/Γ)	p (MeV/c)
$N\overline{K}$	45 ± 1%	244
$\Sigma\pi$	42 ± 1%	267
$\Lambda\pi\pi$	10 ± 1%	252
$\Sigma\pi\pi$	0.9 ± 0.1%	152
$\Lambda\gamma$	0.8 ± 0.2%	351

Λ(1600) P_{01}

$I(J^P) = 0(\frac{1}{2}^+)$

Mass $m = 1560$ to $1700 (\approx 1600)$ MeV
Full width $\Gamma = 50$ to $250 (\approx 150)$ MeV
  $p_{beam} = 0.58$ GeV/c $4\pi\lambda^2 = 41.6$ mb

Λ(1600) DECAY MODES	Fraction (Γ_i/Γ)	p (MeV/c)
$N\overline{K}$	15–30 %	343
$\Sigma\pi$	10–60 %	336

Λ(1670) S_{01}

$I(J^P) = 0(\frac{1}{2}^-)$

Mass $m = 1660$ to $1680 (\approx 1670)$ MeV
Full width $\Gamma = 25$ to $50 (\approx 35)$ MeV
  $p_{beam} = 0.74$ GeV/c $4\pi\lambda^2 = 28.5$ mb

Λ(1670) DECAY MODES	Fraction (Γ_i/Γ)	p (MeV/c)
$N\overline{K}$	15–25 %	414
$\Sigma\pi$	20–60 %	393
$\Lambda\eta$	15–35 %	64

Λ(1690) D_{03}

$I(J^P) = 0(\frac{3}{2}^-)$

Mass $m = 1685$ to $1695 (\approx 1690)$ MeV
Full width $\Gamma = 50$ to $70 (\approx 60)$ MeV
  $p_{beam} = 0.78$ GeV/c $4\pi\lambda^2 = 26.1$ mb

Λ(1690) DECAY MODES	Fraction (Γ_i/Γ)	p (MeV/c)
$N\overline{K}$	20–30 %	433
$\Sigma\pi$	20–40 %	409
$\Lambda\pi\pi$	~ 25 %	415
$\Sigma\pi\pi$	~ 20 %	350

Λ(1800) S_{01}

$I(J^P) = 0(\frac{1}{2}^-)$

Mass $m = 1720$ to $1850 (\approx 1800)$ MeV
Full width $\Gamma = 200$ to $400 (\approx 300)$ MeV
  $p_{beam} = 1.01$ GeV/c $4\pi\lambda^2 = 17.5$ mb

Λ(1800) DECAY MODES	Fraction (Γ_i/Γ)	p (MeV/c)
$N\overline{K}$	25–40 %	528
$\Sigma\pi$	seen	493
$\Sigma(1385)\pi$	seen	345
$N\overline{K}^*(892)$	seen	†

Λ(1810) P_{01}

$I(J^P) = 0(\frac{1}{2}^+)$

Mass $m = 1750$ to $1850 (\approx 1810)$ MeV
Full width $\Gamma = 50$ to $250 (\approx 150)$ MeV
  $p_{beam} = 1.04$ GeV/c $4\pi\lambda^2 = 17.0$ mb

Λ(1810) DECAY MODES	Fraction (Γ_i/Γ)	p (MeV/c)
$N\overline{K}$	20–50 %	537
$\Sigma\pi$	10–40 %	501
$\Sigma(1385)\pi$	seen	356
$N\overline{K}^*(892)$	30–60 %	†

Λ(1820) F_{05}

$I(J^P) = 0(\frac{5}{2}^+)$

Mass $m = 1815$ to $1825 (\approx 1820)$ MeV
Full width $\Gamma = 70$ to $90 (\approx 80)$ MeV
  $p_{beam} = 1.06$ GeV/c $4\pi\lambda^2 = 16.5$ mb

Λ(1820) DECAY MODES	Fraction (Γ_i/Γ)	p (MeV/c)
$N\overline{K}$	55–65 %	545
$\Sigma\pi$	8–14 %	508
$\Sigma(1385)\pi$	5–10 %	362

Λ(1830) D_{05}

$I(J^P) = 0(\frac{5}{2}^-)$

Mass $m = 1810$ to $1830 (\approx 1830)$ MeV
Full width $\Gamma = 60$ to $110 (\approx 95)$ MeV
  $p_{beam} = 1.08$ GeV/c $4\pi\lambda^2 = 16.0$ mb

Λ(1830) DECAY MODES	Fraction (Γ_i/Γ)	p (MeV/c)
$N\overline{K}$	3–10 %	553
$\Sigma\pi$	35–75 %	515
$\Sigma(1385)\pi$	>15 %	371

$\boxed{\Lambda(1890)\ P_{03}}$ $\qquad\qquad I(J^P) = 0(\frac{3}{2}^+)$

Mass $m = 1850$ to 1910 (≈ 1890) MeV
Full width $\Gamma = 60$ to 200 (≈ 100) MeV
$p_{beam} = 1.21$ GeV/c $\qquad 4\pi\lambdabar^2 = 13.6$ mb

$\Lambda(1890)$ DECAY MODES	Fraction (Γ_i/Γ)	p (MeV/c)
$N\overline{K}$	20–35 %	599
$\Sigma\pi$	3–10 %	559
$\Sigma(1385)\pi$	seen	420
$N\overline{K}^*(892)$	seen	233

$\boxed{\Lambda(2100)\ G_{07}}$ $\qquad\qquad I(J^P) = 0(\frac{7}{2}^-)$

Mass $m = 2090$ to 2110 (≈ 2100) MeV
Full width $\Gamma = 100$ to 250 (≈ 200) MeV
$p_{beam} = 1.68$ GeV/c $\qquad 4\pi\lambdabar^2 = 8.68$ mb

$\Lambda(2100)$ DECAY MODES	Fraction (Γ_i/Γ)	p (MeV/c)
$N\overline{K}$	25–35 %	751
$\Sigma\pi$	~ 5 %	704
$\Lambda\eta$	<3 %	617
ΞK	<3 %	483
$\Lambda\omega$	<8 %	443
$N\overline{K}^*(892)$	10–20 %	514

$\boxed{\Lambda(2110)\ F_{05}}$ $\qquad\qquad I(J^P) = 0(\frac{5}{2}^+)$

Mass $m = 2090$ to 2140 (≈ 2110) MeV
Full width $\Gamma = 150$ to 250 (≈ 200) MeV
$p_{beam} = 1.70$ GeV/c $\qquad 4\pi\lambdabar^2 = 8.53$ mb

$\Lambda(2110)$ DECAY MODES	Fraction (Γ_i/Γ)	p (MeV/c)
$N\overline{K}$	5–25 %	757
$\Sigma\pi$	10–40 %	711
$\Lambda\omega$	seen	455
$\Sigma(1385)\pi$	seen	589
$N\overline{K}^*(892)$	10–60 %	524

$\boxed{\Lambda(2350)\ H_{09}}$ $\qquad\qquad I(J^P) = 0(\frac{9}{2}^+)$

Mass $m = 2340$ to 2370 (≈ 2350) MeV
Full width $\Gamma = 100$ to 250 (≈ 150) MeV
$p_{beam} = 2.29$ GeV/c $\qquad 4\pi\lambdabar^2 = 5.85$ mb

$\Lambda(2350)$ DECAY MODES	Fraction (Γ_i/Γ)	p (MeV/c)
$N\overline{K}$	~ 12 %	915
$\Sigma\pi$	~ 10 %	867

$\boxed{\begin{array}{c} \boldsymbol{\Sigma}\ \textbf{BARYONS} \\ (\boldsymbol{S = -1,\ I = 1}) \\ \Sigma^+ = uus,\ \Sigma^0 = uds,\ \Sigma^- = dds \end{array}}$

$\boxed{\Sigma^+}$ $\qquad\qquad I(J^P) = 1(\frac{1}{2}^+)$

Mass $m = 1189.37 \pm 0.07$ MeV $\quad$ (S = 2.2)
Mean life $\tau = (0.799 \pm 0.004) \times 10^{-10}$ s
$\quad c\tau = 2.396$ cm
Magnetic moment $\mu = 2.458 \pm 0.010\ \mu_N$ $\quad$ (S = 2.1)
$\Gamma(\Sigma^+ \to n\ell^+\nu)/\Gamma(\Sigma^- \to n\ell^-\overline{\nu}) < 0.043$

Decay parameters

$p\pi^0$	$\alpha_0 = -0.980^{+0.017}_{-0.015}$
"	$\phi_0 = (36 \pm 34)^\circ$
"	$\gamma_0 = 0.16$ [g]
"	$\Delta_0 = (187 \pm 6)^\circ$ [g]
$n\pi^+$	$\alpha_+ = 0.068 \pm 0.013$
"	$\phi_+ = (167 \pm 20)^\circ$ $\quad$ (S = 1.1)
"	$\gamma_+ = -0.97$ [g]
"	$\Delta_+ = (-73^{+133}_{-10})^\circ$ [g]
$p\gamma$	$\alpha_\gamma = -0.76 \pm 0.08$

Σ^+ DECAY MODES	Fraction (Γ_i/Γ)	Confidence level	p (MeV/c)
$p\pi^0$	(51.57 ± 0.30) %		189
$n\pi^+$	(48.31 ± 0.30) %		185
$p\gamma$	$(1.23 \pm 0.05) \times 10^{-3}$		225
$n\pi^+\gamma$	[h] $(4.5 \pm 0.5) \times 10^{-4}$		185
$\Lambda e^+\nu_e$	$(2.0 \pm 0.5) \times 10^{-5}$		71

$\Delta S = \Delta Q$ (SQ) violating modes or
$\Delta S = 1$ weak neutral current ($S1$) modes

					p (MeV/c)
$ne^+\nu_e$	SQ	< 5	$\times 10^{-6}$	90%	224
$n\mu^+\nu_\mu$	SQ	< 3.0	$\times 10^{-5}$	90%	202
pe^+e^-	$S1$	< 7	$\times 10^{-6}$		225

$\boxed{\Sigma^0}$ $\qquad\qquad I(J^P) = 1(\frac{1}{2}^+)$

Mass $m = 1192.642 \pm 0.024$ MeV
$m_{\Sigma^-} - m_{\Sigma^0} = 4.807 \pm 0.035$ MeV $\quad$ (S = 1.1)
$m_{\Sigma^0} - m_\Lambda = 76.959 \pm 0.023$ MeV
Mean life $\tau = (7.4 \pm 0.7) \times 10^{-20}$ s
$\quad c\tau = 2.22 \times 10^{-11}$ m
Transition magnetic moment $|\mu_{\Sigma\Lambda}| = 1.61 \pm 0.08\ \mu_N$

Σ^0 DECAY MODES	Fraction (Γ_i/Γ)	Confidence level	p (MeV/c)
$\Lambda\gamma$	100 %		74
$\Lambda\gamma\gamma$	< 3 %	90%	74
Λe^+e^-	[J] 5×10^{-3}		74

Σ^-		$I(J^P) = 1(\frac{1}{2}^+)$

Mass $m = 1197.449 \pm 0.030$ MeV   (S = 1.2)
$m_{\Sigma^-} - m_{\Sigma^+} = 8.08 \pm 0.08$ MeV   (S = 1.9)
$m_{\Sigma^-} - m_\Lambda = 81.766 \pm 0.030$ MeV   (S = 1.2)
Mean life $\tau = (1.479 \pm 0.011) \times 10^{-10}$ s   (S = 1.3)
$\quad c\tau = 4.434$ cm
Magnetic moment $\mu = -1.160 \pm 0.025\ \mu_N$ (S = 1.7)

Decay parameters

$n\pi^-$	$\alpha_- = -0.068 \pm 0.008$	
"	$\phi_- = (10 \pm 15)°$	
"	$\gamma_- = 0.98$ [g]	
"	$\Delta_- = (249^{+12}_{-120})°$ [g]	
$ne^-\bar{\nu}_e$	$g_A/g_V = 0.340 \pm 0.017$ [e]	
"	$f_2(0)/f_1(0) = 0.97 \pm 0.14$	
"	$D = 0.11 \pm 0.10$	
$\Lambda e^-\bar{\nu}_e$	$g_V/g_A = 0.01 \pm 0.10$ [e] (S = 1.5)	
"	$g_{WM}/g_A = 2.4 \pm 1.7$ [e]	

Σ^- DECAY MODES	Fraction (Γ_i/Γ)	p (MeV/c)
$n\pi^-$	(99.848 ± 0.005) %	193
$n\pi^-\gamma$	[h] $(4.6 \pm 0.6) \times 10^{-4}$	193
$ne^-\bar{\nu}_e$	$(1.017 \pm 0.034) \times 10^{-3}$	230
$n\mu^-\bar{\nu}_\mu$	$(4.5 \pm 0.4) \times 10^{-4}$	210
$\Lambda e^-\bar{\nu}_e$	$(5.73 \pm 0.27) \times 10^{-5}$	79

$\Sigma(1385)\ P_{13}$		$I(J^P) = 1(\frac{3}{2}^+)$

$\Sigma(1385)^+$ mass $m = 1382.8 \pm 0.4$ MeV   (S = 2.0)
$\Sigma(1385)^0$ mass $m = 1383.7 \pm 1.0$ MeV   (S = 1.4)
$\Sigma(1385)^-$ mass $m = 1387.2 \pm 0.5$ MeV   (S = 2.2)
$\Sigma(1385)^+$ full width $\Gamma = 35.8 \pm 0.8$ MeV
$\Sigma(1385)^0$ full width $\Gamma = 36 \pm 5$ MeV
$\Sigma(1385)^-$ full width $\Gamma = 39.4 \pm 2.1$ MeV   (S = 1.7)
$\quad$ Below $\overline{K}N$ threshold

$\Sigma(1385)$ DECAY MODES	Fraction (Γ_i/Γ)	p (MeV/c)
$\Lambda\pi$	88 ± 2 %	208
$\Sigma\pi$	12 ± 2 %	127

$\Sigma(1660)\ P_{11}$		$I(J^P) = 1(\frac{1}{2}^+)$

Mass $m = 1630$ to 1690 (≈ 1660) MeV
Full width $\Gamma = 40$ to 200 (≈ 100) MeV
$\quad p_{beam} = 0.72$ GeV/c $4\pi\lambdabar^2 = 29.9$ mb

$\Sigma(1660)$ DECAY MODES	Fraction (Γ_i/Γ)	p (MeV/c)
$N\overline{K}$	10–30 %	405
$\Lambda\pi$	seen	439
$\Sigma\pi$	seen	385

$\Sigma(1670)\ D_{13}$		$I(J^P) = 1(\frac{3}{2}^-)$

Mass $m = 1665$ to 1685 (≈ 1670) MeV
Full width $\Gamma = 40$ to 80 (≈ 60) MeV
$\quad p_{beam} = 0.74$ GeV/c $4\pi\lambdabar^2 = 28.5$ mb

$\Sigma(1670)$ DECAY MODES	Fraction (Γ_i/Γ)	p (MeV/c)
$N\overline{K}$	7–13 %	414
$\Lambda\pi$	5–15 %	447
$\Sigma\pi$	30–60 %	393

$\Sigma(1750)\ S_{11}$		$I(J^P) = 1(\frac{1}{2}^-)$

Mass $m = 1730$ to 1800 (≈ 1750) MeV
Full width $\Gamma = 60$ to 160 (≈ 90) MeV
$\quad p_{beam} = 0.91$ GeV/c $4\pi\lambdabar^2 = 20.7$ mb

$\Sigma(1750)$ DECAY MODES	Fraction (Γ_i/Γ)	p (MeV/c)
$N\overline{K}$	10–40 %	486
$\Lambda\pi$	seen	507
$\Sigma\pi$	<8 %	455
$\Sigma\eta$	15–55 %	81

$\Sigma(1775)\ D_{15}$		$I(J^P) = 1(\frac{5}{2}^-)$

Mass $m = 1770$ to 1780 (≈ 1775) MeV
Full width $\Gamma = 105$ to 135 (≈ 120) MeV
$\quad p_{beam} = 0.96$ GeV/c $4\pi\lambdabar^2 = 19.0$ mb

$\Sigma(1775)$ DECAY MODES	Fraction (Γ_i/Γ)	p (MeV/c)
$N\overline{K}$	37–43%	508
$\Lambda\pi$	14–20%	525
$\Sigma\pi$	2–5%	474
$\Sigma(1385)\pi$	8–12%	324
$\Lambda(1520)\pi$	17–23%	198

$\Sigma(1915)\ F_{15}$		$I(J^P) = 1(\frac{5}{2}^+)$

Mass $m = 1900$ to 1935 (≈ 1915) MeV
Full width $\Gamma = 80$ to 160 (≈ 120) MeV
$\quad p_{beam} = 1.26$ GeV/c $4\pi\lambdabar^2 = 12.8$ mb

$\Sigma(1915)$ DECAY MODES	Fraction (Γ_i/Γ)	p (MeV/c)
$N\overline{K}$	5–15 %	618
$\Lambda\pi$	seen	622
$\Sigma\pi$	seen	577
$\Sigma(1385)\pi$	<5 %	440

$\Sigma(1940)\ D_{13}$		$I(J^P) = 1(\frac{3}{2}^-)$

Mass $m = 1900$ to 1950 (≈ 1940) MeV
Full width $\Gamma = 150$ to 300 (≈ 220) MeV
$\quad p_{beam} = 1.32$ GeV/c $4\pi\lambdabar^2 = 12.1$ mb

$\Sigma(1940)$ DECAY MODES	Fraction (Γ_i/Γ)	p (MeV/c)
$N\overline{K}$	<20 %	637
$\Lambda\pi$	seen	639
$\Sigma\pi$	seen	594
$\Sigma(1385)\pi$	seen	460
$\Lambda(1520)\pi$	seen	354
$\Delta(1232)\overline{K}$	seen	410
$N\overline{K}^*(892)$	seen	320

$\Sigma(2030)\ F_{17}$		$I(J^P) = 1(\frac{7}{2}^+)$

Mass $m = 2025$ to 2040 (≈ 2030) MeV
Full width $\Gamma = 150$ to 200 (≈ 180) MeV
$\quad p_{beam} = 1.52$ GeV/c $4\pi\lambdabar^2 = 9.93$ mb

$\Sigma(2030)$ DECAY MODES	Fraction (Γ_i/Γ)	p (MeV/c)
$N\overline{K}$	17–23 %	702
$\Lambda\pi$	17–23 %	700
$\Sigma\pi$	5–10 %	657
ΞK	<2 %	412
$\Sigma(1385)\pi$	5–15 %	529
$\Lambda(1520)\pi$	10–20 %	430
$\Delta(1232)\overline{K}$	10–20 %	498
$N\overline{K}^*(892)$	<5 %	438

Σ(2250)

$I(J^P) = 1(?^?)$

Mass $m = 2210$ to $2280 \ (\approx 2250)$ MeV
Full width $\Gamma = 60$ to $150 \ (\approx 100)$ MeV
$p_{beam} = 2.04$ GeV/c $\qquad 4\pi\lambda^2 = 6.76$ mb

Σ(2250) DECAY MODES	Fraction (Γ_i/Γ)	p (MeV/c)
$N\overline{K}$	<10 %	851
$\Lambda\pi$	seen	842
$\Sigma\pi$	seen	803

Ξ BARYONS
$(S = -2, I = 1/2)$
$\Xi^0 = uss \qquad \Xi^- = dss$

Ξ⁰

$I(J^P) = \frac{1}{2}(\frac{1}{2}^+)$

P is not yet measured; + is the quark model prediction.

Mass $m = 1314.9 \pm 0.6$ MeV
$m_{\Xi^-} - m_{\Xi^0} = 6.4 \pm 0.6$ MeV
Mean life $\tau = (2.90 \pm 0.09) \times 10^{-10}$ s
$\quad c\tau = 8.71$ cm
Magnetic moment $\mu = -1.250 \pm 0.014 \ \mu_N$

Decay parameters

$\Lambda\pi^0$	$\alpha = -0.411 \pm 0.022$ (S = 2.1)
"	$\phi = (21 \pm 12)^\circ$
"	$\gamma = 0.85$ [g]
"	$\Delta = (218^{+12}_{-19})^\circ$ [g]
$\Lambda\gamma$	$\alpha = 0.4 \pm 0.4$
$\Sigma^0\gamma$	$\alpha = 0.20 \pm 0.32$

Ξ⁰ DECAY MODES	Fraction (Γ_i/Γ)	Confidence level	p (MeV/c)
$\Lambda\pi^0$	(99.54 ± 0.05) %		135
$\Lambda\gamma$	$(1.06 \pm 0.16) \times 10^{-3}$		184
$\Sigma^0\gamma$	$(3.5 \pm 0.4) \times 10^{-3}$		117
$\Sigma^+ e^- \overline{\nu}_e$	$< 1.1 \times 10^{-3}$	90%	120
$\Sigma^+ \mu^- \overline{\nu}_\mu$	$< 1.1 \times 10^{-3}$	90%	64

$\Delta S = \Delta Q$ (SQ) violating modes or
$\Delta S = 2$ forbidden ($S2$) modes

$\Sigma^- e^+ \nu_e$	SQ	<	9	$\times 10^{-4}$	90%	112
$\Sigma^- \mu^+ \nu_\mu$	SQ	<	9	$\times 10^{-4}$	90%	49
$p\pi^-$	S2	<	4	$\times 10^{-5}$	90%	299
$p e^- \overline{\nu}_e$	S2	<	1.3	$\times 10^{-3}$		323
$p \mu^- \overline{\nu}_\mu$	S2	<	1.3	$\times 10^{-3}$		309

Ξ⁻

$I(J^P) = \frac{1}{2}(\frac{1}{2}^+)$

P is not yet measured; + is the quark model prediction.

Mass $m = 1321.32 \pm 0.13$ MeV
Mean life $\tau = (1.639 \pm 0.015) \times 10^{-10}$ s
$\quad c\tau = 4.91$ cm
Magnetic moment $\mu = -0.6507 \pm 0.0025 \ \mu_N$

Decay parameters

$\Lambda\pi^-$	$\alpha = -0.456 \pm 0.014$ (S = 1.8)
"	$\phi = (4 \pm 4)^\circ$
"	$\gamma = 0.89$ [g]
"	$\Delta = (188 \pm 8)^\circ$ [g]
$\Lambda e^- \overline{\nu}_e$	$g_A/g_V = -0.25 \pm 0.05$ [e]

Ξ⁻ DECAY MODES	Fraction (Γ_i/Γ)	Confidence level	p (MeV/c)
$\Lambda\pi^-$	(99.887 ± 0.035) %		139
$\Sigma^- \gamma$	$(1.27 \pm 0.23) \times 10^{-4}$		118
$\Lambda e^- \overline{\nu}_e$	$(5.63 \pm 0.31) \times 10^{-4}$		190
$\Lambda \mu^- \overline{\nu}_\mu$	$(3.5^{+3.5}_{-2.2}) \times 10^{-4}$		163
$\Sigma^0 e^- \overline{\nu}_e$	$(8.7 \pm 1.7) \times 10^{-5}$		122
$\Sigma^0 \mu^- \overline{\nu}_\mu$	$< 8 \times 10^{-4}$	90%	70
$\Xi^0 e^- \overline{\nu}_e$	$< 2.3 \times 10^{-3}$	90%	6

$\Delta S = 2$ forbidden ($S2$) modes

$n\pi^-$	S2	<	1.9	$\times 10^{-5}$	90%	303
$n e^- \overline{\nu}_e$	S2	<	3.2	$\times 10^{-3}$	90%	327
$n \mu^- \overline{\nu}_\mu$	S2	<	1.5	%	90%	314
$p\pi^-\pi^-$	S2	<	4	$\times 10^{-4}$	90%	223
$p\pi^- e^- \overline{\nu}_e$	S2	<	4	$\times 10^{-4}$	90%	304
$p\pi^- \mu^- \overline{\nu}_\mu$	S2	<	4	$\times 10^{-4}$	90%	250
$p \mu^- \mu^-$	L	<	4	$\times 10^{-4}$	90%	272

Ξ(1530) P_{13}

$I(J^P) = \frac{1}{2}(\frac{3}{2}^+)$

$\Xi(1530)^0$ mass $m = 1531.80 \pm 0.32$ MeV (S = 1.3)
$\Xi(1530)^-$ mass $m = 1535.0 \pm 0.6$ MeV
$\Xi(1530)^0$ full width $\Gamma = 9.1 \pm 0.5$ MeV
$\Xi(1530)^-$ full width $\Gamma = 9.9^{+1.7}_{-1.9}$ MeV

Ξ(1530) DECAY MODES	Fraction (Γ_i/Γ)	Confidence level	p (MeV/c)
$\Xi\pi$	100 %		152
$\Xi\gamma$	<4 %	90%	200

Ξ(1690)

$I(J^P) = \frac{1}{2}(?^?)$

Mass $m = 1690 \pm 10$ MeV [i]
Full width $\Gamma < 50$ MeV

Ξ(1690) DECAY MODES	Fraction (Γ_i/Γ)	p (MeV/c)
$\Lambda\overline{K}$	seen	240
$\Sigma\overline{K}$	seen	51
$\Xi^-\pi^+\pi^-$	possibly seen	214

Ξ(1820) D_{13}

$I(J^P) = \frac{1}{2}(\frac{3}{2}^-)$

Mass $m = 1823 \pm 5$ MeV [i]
Full width $\Gamma = 24^{+15}_{-10}$ MeV [i]

Ξ(1820) DECAY MODES	Fraction (Γ_i/Γ)	p (MeV/c)
$\Lambda\overline{K}$	large	400
$\Sigma\overline{K}$	small	320
$\Xi\pi$	small	413
$\Xi(1530)\pi$	small	234

Ξ(1950)

$I(J^P) = \frac{1}{2}(?^?)$

Mass $m = 1950 \pm 15$ MeV [i]
Full width $\Gamma = 60 \pm 20$ MeV [i]

Ξ(1950) DECAY MODES	Fraction (Γ_i/Γ)	p (MeV/c)
$\Lambda\overline{K}$	seen	522
$\Sigma\overline{K}$	possibly seen	460
$\Xi\pi$	seen	518

Ξ(2030)

$I(J^P) = \frac{1}{2}(\geq \frac{5}{2}^?)$

Mass $m = 2025 \pm 5$ MeV [i]
Full width $\Gamma = 20^{+15}_{-5}$ MeV [i]

Ξ(2030) DECAY MODES	Fraction (Γ_i/Γ)	p (MeV/c)
$\Lambda\overline{K}$	~ 20 %	589
$\Sigma\overline{K}$	~ 80 %	533
$\Xi\pi$	small	573
$\Xi(1530)\pi$	small	421
$\Lambda\overline{K}\pi$	small	501
$\Sigma\overline{K}\pi$	small	430

Ω BARYONS
$(S=-3, I=0)$

$\Omega^- = sss$

$\boxed{\Omega^-}$ $\qquad\qquad I(J^P) = 0(\tfrac{3}{2}^+)$

J^P is not yet measured; $\tfrac{3}{2}^+$ is the quark model prediction.

Mass $m = 1672.45 \pm 0.29$ MeV
Mean life $\tau = (0.822 \pm 0.012) \times 10^{-10}$ s
$\quad c\tau = 2.46$ cm
Magnetic moment $\mu = -2.02 \pm 0.05\ \mu_N$

Decay parameters

ΛK^-	$\alpha = -0.026 \pm 0.026$
$\Xi^0 \pi^-$	$\alpha = 0.09 \pm 0.14$
$\Xi^- \pi^0$	$\alpha = 0.05 \pm 0.21$

Ω^- DECAY MODES	Fraction (Γ_i/Γ)	Confidence level	p (MeV/c)
ΛK^-	(67.8 ± 0.7) %		211
$\Xi^0 \pi^-$	(23.6 ± 0.7) %		294
$\Xi^- \pi^0$	(8.6 ± 0.4) %		290
$\Xi^- \pi^+ \pi^-$	$(4.3^{+3.4}_{-1.3}) \times 10^{-4}$		190
$\Xi(1530)^0 \pi^-$	$(6.4^{+5.1}_{-2.0}) \times 10^{-4}$		17
$\Xi^0 e^- \overline{\nu}_e$	$(5.6 \pm 2.8) \times 10^{-3}$		319
$\Xi^- \gamma$	$< 4.6 \times 10^{-4}$	90%	314

$\Delta S = 2$ forbidden (S2) modes

$\Lambda \pi^-$	S2	$< 1.9 \times 10^{-4}$	90%	449

$\boxed{\Omega(2250)^-}$ $\qquad\qquad I(J^P) = 0(?^?)$

Mass $m = 2252 \pm 9$ MeV
Full width $\Gamma = 55 \pm 18$ MeV

$\Omega(2250)^-$ DECAY MODES	Fraction (Γ_i/Γ)	p (MeV/c)
$\Xi^- \pi^+ K^-$	seen	531
$\Xi(1530)^0 K^-$	seen	437

CHARMED BARYONS
$(C=+1)$

$\Lambda_c^+ = udc\Gamma \quad \Sigma_c^{++} = uuc\Gamma \quad \Sigma_c^+ = udc\Gamma \quad \Sigma_c^0 = ddc\Gamma$
$\Xi_c^+ = usc\Gamma \quad \Xi_c^0 = dsc\Gamma \quad \Omega_c^0 = ssc$

 $\boxed{\Lambda_c^+}$ $\qquad\qquad I(J^P) = 0(\tfrac{1}{2}^+)$

J not confirmed; $\tfrac{1}{2}$ is the quark model prediction.

Mass $m = 2284.9 \pm 0.6$ MeV
Mean life $\tau = (0.206 \pm 0.012) \times 10^{-12}$ s
$\quad c\tau = 61.8\ \mu$m

Decay asymmetry parameters

$\Lambda \pi^+$	$\alpha = -0.98 \pm 0.19$
$\Sigma^+ \pi^0$	$\alpha = -0.45 \pm 0.32$
$\Lambda \ell^+ \nu_\ell$	$\alpha = -0.82^{+0.11}_{-0.07}$

Nearly all branching fractions of the Λ_c^+ are measured relative to the $pK^-\pi^+$ model butthere are no model-independent measurements of this branching fraction. We explain how we arrive at our value of $B(\Lambda_c^+ \to pK^-\pi^+)$ in a Note at the beginning of the branching-ratio measurements in the Listings. When this branching fraction is eventually well determined all the other branching fractions will slide up or down proportionally as the true value differs from the value we use here.

Λ_c^+ DECAY MODES		Fraction (Γ_i/Γ)	Scale factor/ Confidence level	p (MeV/c)
Hadronic modes with a p and one $\overline{K}$				
$p\overline{K}^0$		(2.5 ± 0.7) %		872
$pK^-\pi^+$	[k]	(5.0 ± 1.3) %		822
$p\overline{K}^*(892)^0$	[l]	(1.8 ± 0.6) %		681
$\Delta(1232)^{++}K^-$		$(8 \pm 5) \times 10^{-3}$		709
$\Lambda(1520)\pi^+$	[l]	$(4.5^{+2.5}_{-2.1}) \times 10^{-3}$		626
$pK^-\pi^+$ nonresonant		(2.8 ± 0.9) %		822
$p\overline{K}^0\eta$		(1.3 ± 0.4) %		567
$p\overline{K}^0\pi^+\pi^-$		(2.4 ± 1.1) %		753
$pK^-\pi^+\pi^0$		seen		758
$pK^*(892)^-\pi^+$	[l]	(1.1 ± 0.6) %		579
$p(K^-\pi^+)_{\text{nonresonant}}\pi^0$		(3.6 ± 1.2) %		758
$\Delta(1232)\overline{K}^*(892)$		seen		416
$pK^-\pi^+\pi^+\pi^-$		$(1.1 \pm 0.8) \times 10^{-3}$		670
$pK^-\pi^+\pi^0\pi^0$		$(8 \pm 4) \times 10^{-3}$		676
$pK^-\pi^+\pi^0\pi^0\pi^0$		$(5.0 \pm 3.4) \times 10^{-3}$		573
Hadronic modes with a p and zero or two K's				
$p\pi^+\pi^-$		$(3.5 \pm 2.0) \times 10^{-3}$		926
$pf_0(980)$	[l]	$(2.8 \pm 1.9) \times 10^{-3}$		621
$p\pi^+\pi^+\pi^-\pi^-$		$(1.8 \pm 1.2) \times 10^{-3}$		851
pK^+K^-		$(2.3 \pm 0.9) \times 10^{-3}$		615
$p\phi$	[l]	$(1.2 \pm 0.5) \times 10^{-3}$		589
Hadronic modes with a hyperon				
$\Lambda\pi^+$		$(9.0 \pm 2.8) \times 10^{-3}$		863
$\Lambda\pi^+\pi^0$		(3.6 ± 1.3) %		843
$\Lambda\rho^+$		< 5 %	CL=95%	638
$\Lambda\pi^+\pi^+\pi^-$		(3.3 ± 1.0) %		806
$\Lambda\pi^+\eta$		(1.7 ± 0.6) %		690
$\Sigma(1385)^+\eta$	[l]	$(8.5 \pm 3.3) \times 10^{-3}$		569
$\Lambda K^+\overline{K}^0$		$(6.0 \pm 2.1) \times 10^{-3}$		441
$\Sigma^0\pi^+$		$(9.9 \pm 3.2) \times 10^{-3}$		824
$\Sigma^+\pi^0$		(1.00 ± 0.34) %		826
$\Sigma^+\eta$		$(5.5 \pm 2.3) \times 10^{-3}$		712
$\Sigma^+\pi^+\pi^-$		(3.4 ± 1.0) %		803
$\Sigma^+\rho^0$		< 1.4 %	CL=95%	578
$\Sigma^-\pi^+\pi^+$		(1.8 ± 0.8) %		798
$\Sigma^0\pi^+\pi^0$		(1.8 ± 0.8) %		802
$\Sigma^0\pi^+\pi^+\pi^-$		(1.1 ± 0.4) %		762
$\Sigma^+\pi^+\pi^-\pi^0$		$-$		766
$\Sigma^+\omega$	[l]	(2.7 ± 1.0) %		568
$\Sigma^+\pi^+\pi^+\pi^-\pi^-$		$(3.0^{+4.1}_{-2.1}) \times 10^{-3}$		707
$\Sigma^+K^+K^-$		$(3.5 \pm 1.2) \times 10^{-3}$		346
$\Sigma^+\phi$	[l]	$(3.5 \pm 1.7) \times 10^{-3}$		292
$\Sigma^+K^+\pi^-$		$(7^{+6}_{-4}) \times 10^{-3}$		668
Ξ^0K^+		$(3.9 \pm 1.4) \times 10^{-3}$		652
$\Xi^-K^+\pi^+$		$(4.9 \pm 1.7) \times 10^{-3}$		564
$\Xi(1530)^0K^+$	[l]	$(2.6 \pm 1.0) \times 10^{-3}$		471
Semileptonic modes				
$\Lambda\ell^+\nu_\ell$	[m]	(2.0 ± 0.6) %		$-$
$\Lambda e^+\nu_e$		(2.1 ± 0.6) %		$-$
$\Lambda\mu^+\nu_\mu$		(2.0 ± 0.7) %		$-$
e^+ anything		(4.5 ± 1.7) %		$-$
pe^+ anything		(1.8 ± 0.9) %		$-$
Λe^+ anything		$-$		$-$
$\Lambda\mu^+$ anything		$-$		$-$
$\Lambda\ell^+\nu_\ell$ anything		$-$		$-$
Inclusive modes				
p anything		(50 ± 16) %		$-$
p anything (no Λ)		(12 ± 19) %		$-$
p hadrons		$-$		$-$
n anything		(50 ± 16) %		$-$
n anything (no Λ)		(29 ± 17) %		$-$
Λ anything		(35 ± 11) %	S=1.4	$-$
$\Sigma^\pm$ anything	[n]	(10 ± 5) %		$-$

$\Delta C = 1$ weak neutral current ($C1$) modes or Lepton number (L) violating modes

$p\mu^+\mu^-$	$C1$	$<$ 3.4	$\times 10^{-4}$	CL=90%	936
$\Sigma^-\mu^+\mu^+$	L	$<$ 7.0	$\times 10^{-4}$	CL=90%	811

$\Lambda_c(2593)^+$ $I(J^P) = 0(\frac{1}{2}^-)$

The spin-parity follows from the fact that $\Sigma_c(2455)\pi$ decays with little available phase space are dominant.

Mass $m = 2593.9 \pm 0.8$ MeV
$m - m_{\Lambda_c^+} = 308.9 \pm 0.6$ MeV   (S = 1.1)
Full width $\Gamma = 3.6^{+2.0}_{-1.3}$ MeV

$\Lambda_c^+\pi\pi$ and its submode $\Sigma_c(2455)\pi$ — the latter just barely — are the only strong decays allowed to an excited Λ_c^+ having this mass; and the $\Lambda_c^+\pi^+\pi^-$ mode seems to be largely via $\Sigma_c^{++}\pi^-$ or $\Sigma_c^0\pi^+$.

$\Lambda_c(2593)^+$ DECAY MODES	Fraction (Γ_i/Γ)	p (MeV/c)
$\Lambda_c^+\pi^+\pi^-$	[o] $\approx$ 67 %	124
$\Sigma_c(2455)^{++}\pi^-$	24 $\pm$ 7 %	17
$\Sigma_c(2455)^0\pi^+$	24 $\pm$ 7 %	23
$\Lambda_c^+\pi^+\pi^-$ 3-body	18 $\pm$ 10 %	124
$\Lambda_c^+\pi^0$	not seen	261
$\Lambda_c^+\gamma$	not seen	290

$\Lambda_c(2625)^+$ $I(J^P) = 0(?^?)$

J^P is expected to be $3/2^-$.

Mass $m = 2626.6 \pm 0.8$ MeV   (S = 1.2)
$m - m_{\Lambda_c^+} = 341.7 \pm 0.6$ MeV   (S = 1.6)
Full width $\Gamma < 1.9$ MeV CL = 90%

$\Lambda_c^+\pi\pi$ and its submode $\Sigma(2455)\pi$ are the only strong decays allowed to an excited Λ_c^+ having this mass.

$\Lambda_c(2625)^+$ DECAY MODES	Fraction (Γ_i/Γ)	p (MeV/c)
$\Lambda_c^+\pi^+\pi^-$	seen	184
$\Sigma_c(2455)^{++}\pi^-$	small	100
$\Sigma_c(2455)^0\pi^+$	small	101
$\Lambda_c^+\pi^+\pi^-$ 3-body	large	184
$\Lambda_c^+\pi^0$	not seen	293
$\Lambda_c^+\gamma$	not seen	319

$\Sigma_c(2455)$ $I(J^P) = 1(\frac{1}{2}^+)$

J^P not confirmed; $\frac{1}{2}^+$ is the quark model prediction.

$\Sigma_c(2455)^{++}$ mass $m = 2452.8 \pm 0.6$ MeV
$\Sigma_c(2455)^+$ mass $m = 2453.6 \pm 0.9$ MeV
$\Sigma_c(2455)^0$ mass $m = 2452.2 \pm 0.6$ MeV
$m_{\Sigma_c^{++}} - m_{\Lambda_c^+} = 167.87 \pm 0.19$ MeV
$m_{\Sigma_c^+} - m_{\Lambda_c^+} = 168.7 \pm 0.6$ MeV
$m_{\Sigma_c^0} - m_{\Lambda_c^+} = 167.30 \pm 0.20$ MeV
$m_{\Sigma_c^{++}} - m_{\Sigma_c^0} = 0.57 \pm 0.23$ MeV
$m_{\Sigma_c^+} - m_{\Sigma_c^0} = 1.4 \pm 0.6$ MeV

$\Lambda_c^+\pi$ is the only strong decay allowed to a Σ_c having this mass.

$\Sigma_c(2455)$ DECAY MODES	Fraction (Γ_i/Γ)	p (MeV/c)
$\Lambda_c^+\pi$	$\approx$ 100 %	90

$\Sigma_c(2520)$ $I(J^P) = 1(?^?)$

$\Sigma_c(2520)^{++}$ mass $m = 2519.4 \pm 1.5$ MeV
$\Sigma_c(2520)^0$ mass $m = 2517.5 \pm 1.4$ MeV
$m_{\Sigma_c(2520)^{++}} - m_{\Lambda_c^+} = 234.5 \pm 1.4$ MeV
$m_{\Sigma_c(2520)^0} - m_{\Lambda_c^+} = 232.6 \pm 1.3$ MeV
$m_{\Sigma_c(2520)^{++}} - m_{\Sigma_c(2520)^0} = 1.9 \pm 1.9$ MeV
$\Sigma_c(2520)^{++}$ full width $\Gamma = 18 \pm 5$ MeV
$\Sigma_c(2520)^0$ full width $\Gamma = 13 \pm 5$ MeV

Ξ_c^+ $I(J^P) = \frac{1}{2}(\frac{1}{2}^+)$

$I(J^P)$ not confirmed; $\frac{1}{2}(\frac{1}{2}^+)$ is the quark model prediction.

Mass $m = 2465.6 \pm 1.4$ MeV
Mean life $\tau = (0.35^{+0.07}_{-0.04}) \times 10^{-12}$ s
$c\tau = 106$ μm

Ξ_c^+ DECAY MODES	Fraction (Γ_i/Γ)	p (MeV/c)
$\Lambda K^-\pi^+\pi^+$	seen	784
$\Lambda \overline{K}^*(892)^0\pi^+$	not seen	601
$\Sigma(1385)^+K^-\pi^+$	not seen	676
$\Sigma^+K^-\pi^+$	seen	808
$\Sigma^+\overline{K}^*(892)^0$	seen	653
$\Sigma^0K^-\pi^+\pi^+$	seen	733
$\Xi^0\pi^+$	seen	875
$\Xi^-\pi^+\pi^+$	seen	850
$\Xi(1530)^0\pi^+$	not seen	748
$\Xi^0\pi^+\pi^0$	seen	854
$\Xi^0\pi^+\pi^+\pi^-$	seen	817
$\Xi^0 e^+\nu_e$	seen	882

Ξ_c^0 $I(J^P) = \frac{1}{2}(\frac{1}{2}^+)$

$I(J^P)$ not confirmed; $\frac{1}{2}(\frac{1}{2}^+)$ is the quark model prediction.

Mass $m = 2470.3 \pm 1.8$ MeV   (S = 1.3)
$m_{\Xi_c^0} - m_{\Xi_c^+} = 4.7 \pm 2.1$ MeV   (S = 1.2)
Mean life $\tau = (0.098^{+0.023}_{-0.015}) \times 10^{-12}$ s
$c\tau = 29$ μm

Ξ_c^0 DECAY MODES	Fraction (Γ_i/Γ)	p (MeV/c)
$\Lambda \overline{K}^0$	seen	864
$\Xi^-\pi^+$	seen	875
$\Xi^-\pi^+\pi^+\pi^-$	seen	816
$pK^-\overline{K}^*(892)^0$	seen	406
Ω^-K^+	seen	522
$\Xi^-e^+\nu_e$	seen	882
$\Xi^-\ell^+$ anything	seen	–

$\Xi_c(2645)$ $I(J^P) = ?(?^?)$

$\Xi_c(2645)^+$ mass $m = 2644.6 \pm 2.1$ MeV   (S = 1.2)
$\Xi_c(2645)^0$ mass $m = 2643.8 \pm 1.8$ MeV
$m_{\Xi_c(2645)^+} - m_{\Xi_c^0} = 174.3 \pm 1.1$ MeV
$m_{\Xi_c(2645)^0} - m_{\Xi_c^+} = 178.2 \pm 1.1$ MeV
$\Xi_c(2645)^+$ full width $\Gamma < 3.1$ MeV CL = 90%
$\Xi_c(2645)^0$ full width $\Gamma < 5.5$ MeV CL = 90%

$\Xi_c\pi$ is the only strong decay allowed to a Ξ_c resonance having this mass.

$\Xi_c(2645)$ DECAY MODES	Fraction (Γ_i/Γ)	p (MeV/c)
$\Xi_c^0\pi^+$	seen	101
$\Xi_c^+\pi^-$	seen	107

Ω_c^0 $I(J^P) = 0(\frac{1}{2}^+)$

$I(J^P)$ not confirmed; $0(\frac{1}{2}^+)$ is the quark model prediction.

Mass $m = 2704 \pm 4$ MeV $(S = 1.8)$
Mean life $\tau = (0.064 \pm 0.020) \times 10^{-12}$ s
$c\tau = 19$ μm

Ω_c^0 DECAY MODES	Fraction (Γ_i/Γ)	p (MeV/c)
$\Sigma^+ K^- K^- \pi^+$	seen	697
$\Xi^- K^- \pi^+ \pi^+$	seen	838
$\Omega^- \pi^+$	seen	827
$\Omega^- \pi^- \pi^+ \pi^+$	seen	759

BOTTOM BARYONS
$(B = -1)$

$\Lambda_b^0 = u d b, \Xi_b^0 = u s b, \Xi_b^- = d s b$

Λ_b^0 $I(J^P) = 0(\frac{1}{2}^+)$

$I(J^P)$ not yet measured; $0(\frac{1}{2}^+)$ is the quark model prediction.

Mass $m = 5624 \pm 9$ MeV $(S = 1.8)$
Mean life $\tau = (1.24 \pm 0.08) \times 10^{-12}$ s
$c\tau = 372$ μm

These branching fractions are actually an average over weakly decaying b-baryons weighted by their production rates in Z decay (or high-energy $p\bar{p}$), branching ratios, and detection efficiencies. They scale with the LEP Λ_b production fraction B($b \to \Lambda_b$) and are evaluated for our value B($b \to \Lambda_b$) = $(10.1^{+3.9}_{-3.1})$%.

The branching fractions B(b-baryon $\to \Lambda \ell^- \bar{\nu}_\ell$ anything) and B($\Lambda_b^0 \to \Lambda_c^+ \ell^- \bar{\nu}_\ell$ anything) are not pure measurements because the underlying measured products of these with B($b \to \Lambda_b$) were used to determine B($b \to \Lambda_b$), as described in the note "Production and Decay of b-Flavored Hadrons."

Λ_b^0 DECAY MODES	Fraction (Γ_i/Γ)	Confidence level	p (MeV/c)
$J/\psi(1S)\Lambda$	$(4.7 \pm 2.8) \times 10^{-4}$		1744
$\Lambda_c^+ \pi^-$	seen		2345
$\Lambda_c^+ a_1(1260)^-$	seen		2156
$\Lambda_c^+ \ell^- \bar{\nu}_\ell$ anything	[p] $(9.0^{+3.1}_{-3.8})$ %		–
$p\pi^-$	$< 5.0 \times 10^{-5}$	90%	2732
pK^-	$< 5.0 \times 10^{-5}$	90%	2711

b-baryon ADMIXTURE (Λ_b, Ξ_b, Σ_b, Ω_b)

Mean life $\tau = (1.20 \pm 0.07) \times 10^{-12}$ s

These branching fractions are actually an average over weakly decaying b-baryons weighted by their production rates in Z decay (or high-energy $p\bar{p}$), branching ratios, and detection efficiencies. They scale with the LEP Λ_b production fraction B($b \to \Lambda_b$) and are evaluated for our value B($b \to \Lambda_b$) = $(10.1^{+3.9}_{-3.1})$%.

The branching fractions B(b-baryon $\to \Lambda \ell^- \bar{\nu}_\ell$ anything) and B($\Lambda_b^0 \to \Lambda_c^+ \ell^- \bar{\nu}_\ell$ anything) are not pure measurements because the underlying measured products of these with B($b \to \Lambda_b$) were used to determine B($b \to \Lambda_b$), as described in the note "Production and Decay of b-Flavored Hadrons."

b-baryon ADMIXTURE ($\Lambda_b, \Xi_b, \Sigma_b, \Omega_b$)	Fraction (Γ_i/Γ)	p (MeV/c)
$p\mu^- \bar{\nu}$ anything	(4.9 ± 2.4) %	–
$\Lambda \ell^- \bar{\nu}_\ell$ anything	$(3.1^{+1.0}_{-1.2})$ %	–
$\Lambda/\bar{\Lambda}$ anything	(35^{+12}_{-14}) %	–
$\Xi^- \ell^- \bar{\nu}_\ell$ anything	$(5.5^{+2.0}_{-2.4}) \times 10^{-3}$	–

NOTES

This Summary Table only includes established baryons. The Particle Listings include evidence for other baryons. The masses, widths, and branching fractions for the resonances in this Table are Breit-Wigner parameters, but pole positions are also given for most of the N and Δ resonances.

For most of the resonances, the parameters come from various partial-wave analyses of more or less the same sets of data, and it is not appropriate to treat the results of the analyses as independent or to average them together. Furthermore, the systematic errors on the results are not well understood. Thus, we usually only give ranges for the parameters. We then also give a best guess for the mass (as part of the name of the resonance) and for the width. The *Note on N and Δ Resonances* and the *Note on Λ and Σ Resonances* in the Particle Listings review the partial-wave analyses.

When a quantity has "(S = ...)" to its right, the error on the quantity has been enlarged by the "scale factor" S, defined as $S = \sqrt{\chi^2/(N-1)}$, where N is the number of measurements used in calculating the quantity. We do this when $S > 1$, which often indicates that the measurements are inconsistent. When $S > 1.25$ we also show in the Particle Listings an ideogram of the measurements. For more about S, see the Introduction.

A decay momentum p is given for each decay mode. For a 2-body decay, p is the momentum of each decay product in the rest frame of the decaying particle. For a 3-or-more-body decay, p is the largest momentum any of the products can have in this frame. For any resonance, the *nominal* mass is used in calculating p. A dagger ("†") in this column indicates that the mode is forbidden when the nominal masses of resonances are used, but is in fact allowed due to the nonzero widths of the resonances.

[a] The masses of the p and n are most precisely known in u (unified atomic mass units). The conversion factor to MeV, 1 u = 931.49432 ± 0.00028 MeV, is less well known than are the masses in u.

[b] The limit is from neutrality-of-matter experiments; it assumes $q_n = q_p + q_e$. See also the charge of the neutron.

[c] The first limit is geochemical and independent of decay mode. The second entry, a range of limits, assumes the dominant decay modes are among those investigated. For antiprotons the best limit, inferred from the observation of cosmic ray $\bar{p}$'s is $\tau_{\bar{p}} > 10^7$ yr, the cosmic-ray storage time, but this limit depends on a number of assumptions. The best direct observation of stored antiprotons gives $\tau_{\bar{p}}/B(\bar{p} \to e^- \gamma) > 1848$ yr.

[d] There is some controversy about whether nuclear physics and model dependence complicate the analysis for bound neutrons (from which the best limit comes). The second limit here is from reactor experiments with free neutrons.

[e] The parameters g_A, g_V, and g_{WM} for semileptonic modes are defined by $\bar{B}_f[\gamma_\lambda(g_V + g_A\gamma_5) + i(g_{WM}/m_{B_i}) \sigma_{\lambda\nu} q^\nu]B_i$, and ϕ_{AV} is defined by $g_A/g_V = |g_A/g_V|e^{i\phi_{AV}}$. See the "Note on Baryon Decay Parameters" in the neutron Particle Listings.

[f] Time-reversal invariance requires this to be 0° or 180°.

[g] The decay parameters γ and Δ are calculated from α and ϕ using

$$\gamma = \sqrt{1-\alpha^2} \cos\phi, \qquad \tan\Delta = -\frac{1}{\alpha} \sqrt{1-\alpha^2} \sin\phi.$$

See the "Note on Baryon Decay Parameters" in the neutron Particle Listings.

[h] See the Particle Listings for the pion momentum range used in this measurement.

[i] The error given here is only an educated guess. It is larger than the error on the weighted average of the published values.

[j] A theoretical value using QED.

[k] See the "Note on Λ_c^+ Branching Fractions" in the Branching Fractions of the Λ_c^+ Particle Listings.

[l] This branching fraction includes all the decay modes of the final-state resonance.

[m] An ℓ indicates an e or a μ mode, not a sum over these modes.

[n] The value is for the sum of the charge states of particle/antiparticle states indicated.

[o] Assuming isospin conservation, so that the other third is $\Lambda_c^+ \pi^0 \pi^0$.

[p] Not a pure measurement. See note at head of Λ_b^0 Decay Modes.

MONOPOLES, SUPERSYMMETRY, COMPOSITENESS, etc., SEARCHES FOR

Magnetic Monopole Searches

Isolated supermassive monopole candidate events have not been confirmed. The most sensitive experiments obtain negative results.
Best cosmic-ray supermassive monopole flux limit:
$$< 1.0 \times 10^{-15} \text{ cm}^{-2}\text{sr}^{-1}\text{s}^{-1} \quad \text{for } 1.1 \times 10^{-4} < \beta < 0.1$$

Supersymmetric Particle Searches

Limits are based on the Minimal Supersymmetric Standard Model.
Assumptions include: 1) $\tilde{\chi}_1^0$ (or $\tilde{\gamma}$) is lightest supersymmetric particle;
2) R-parity is conserved; 3) All scalar quarks (except $\tilde{t}_L$ and $\tilde{t}_R$) are degenerate in mass, and $m_{\tilde{q}_R} = m_{\tilde{q}_L}$. 4) Limits for selectrons and smuons
refer to the $\tilde{\ell}_R$ states.
See the Particle Listings for a Note giving details of supersymmetry.

$\tilde{\chi}_i^0$ — neutralinos (mixtures of $\tilde{\gamma}$, $\tilde{Z}^0$, and $\tilde{H}_i^0$)
  Mass $m_{\tilde{\chi}_1^0} > 10.9$ GeV, CL = 95%
  Mass $m_{\tilde{\chi}_2^0} > 45.3$ GeV, CL = 95%   [$\tan\beta > 1$]
  Mass $m_{\tilde{\chi}_3^0} > 75.8$ GeV, CL = 95%   [$\tan\beta > 1$]
  Mass $m_{\tilde{\chi}_4^0} > 127$ GeV, CL = 95%   [$\tan\beta > 3$]

$\tilde{\chi}_i^\pm$ — charginos (mixtures of $\tilde{W}^\pm$ and $\tilde{H}_i^\pm$)
  Mass $m_{\tilde{\chi}_1^\pm} > 65.7$ GeV, CL = 95%   [$m_{\tilde{\chi}_1^\pm} - m_{\tilde{\chi}_1^0} \geq 2$ GeV]
  Mass $m_{\tilde{\chi}_2^\pm} > 99$ GeV, CL = 95%   [GUT relations assumed]

$\tilde{\nu}$ — scalar neutrino (sneutrino)
  Mass $m > 37.1$ GeV, CL = 95%   [one flavor]
  Mass $m > 43.1$ GeV, CL = 95%   [three degenerate flavors]

$\tilde{e}$ — scalar electron (selectron)
  Mass $m > 58$ GeV, CL = 95%   [$m_{\tilde{e}_R} - m_{\tilde{\chi}_1^0} \geq 4$ GeV]

$\tilde{\mu}$ — scalar muon (smuon)
  Mass $m > 55.6$ GeV, CL = 95%   [$m_{\tilde{\mu}_R} - m_{\tilde{\chi}_1^0} \geq 4$ GeV]

$\tilde{\tau}$ — scalar tau (stau)
  Mass $m > 45$ GeV, CL = 95%   [if $m_{\tilde{\chi}_1^0} < 38$ GeV]

$\tilde{q}$ — scalar quark (squark)
  These limits include the effects of cascade decays, evaluated assuming a fixed value of the parameters μ and $\tan\beta$. The limits are weakly sensitive to these parameters over much of parameter space. Limits assume GUT relations between gaugino masses and the gauge coupling; in particular that for $|\mu|$ not small, $m_{\tilde{\chi}_1^0} \approx m_{\tilde{g}}/6$.
  Mass $m > 176$ GeV, CL = 95%   [any $m_{\tilde{g}} < 300$ GeV, $\mu = -250$ GeV, $\tan\beta = 2$]
  Mass $m > 224$ GeV, CL = 95%   [$m_{\tilde{g}} \leq m_{\tilde{q}}$, $\mu = -400$ GeV, $\tan\beta = 4$]

$\tilde{g}$ — gluino
  There is some controversy on whether gluinos in a low-mass window ($1 \lesssim m_{\tilde{g}} \lesssim 5$ GeV) are excluded or not. See the Supersymmetry Listings for details.
  The limits summarised here refer to the high-mass region ($m_{\tilde{g}} \gtrsim 5$ GeV), and include the effects of cascade decays, evaluated assuming a fixed value of the parameters μ and $\tan\beta$. The limits are weakly sensitive to these parameters over much of parameter space. Limits assume GUT relations between gaugino masses and the gauge coupling; in particular that for $|\mu|$ not small, $m_{\tilde{\chi}_1^0} \approx m_{\tilde{g}}/6$.
  Mass $m > 173$ GeV, CL = 95%   [any $m_{\tilde{q}}$, $\mu = -200$ GeV, $\tan\beta = 2$]
  Mass $m > 212$ GeV, CL = 95%   [$m_{\tilde{g}} \geq m_{\tilde{q}}$, $\mu = -250$ GeV, $\tan\beta = 2$]

Quark and Lepton Compositeness, Searches for

Scale Limits Λ for Contact Interactions (the lowest dimensional interactions with four fermions)

If the Lagrangian has the form
$$\pm \frac{g^2}{2\Lambda^2}\, \overline{\psi}_L \gamma_\mu \psi_L \overline{\psi}_L \gamma^\mu \psi_L$$
(with $g^2/4\pi$ set equal to 1), then we define $\Lambda \equiv \Lambda_{LL}^\pm$. For the full definitions and for other forms, see the Note in the Listings on Searches for Quark and Lepton Compositeness in the full *Review* and the original literature.

$\Lambda_{LL}^+(eeee) > 2.4$ TeV, CL = 95%
$\Lambda_{LL}^-(eeee) > 3.6$ TeV, CL = 95%
$\Lambda_{LL}^+(ee\mu\mu) > 2.6$ TeV, CL = 95%
$\Lambda_{LL}^-(ee\mu\mu) > 2.9$ TeV, CL = 95%
$\Lambda_{LL}^+(ee\tau\tau) > 1.9$ TeV, CL = 95%
$\Lambda_{LL}^-(ee\tau\tau) > 3.0$ TeV, CL = 95%
$\Lambda_{LL}^+(\ell\ell\ell\ell) > 3.5$ TeV, CL = 95%
$\Lambda_{LL}^-(\ell\ell\ell\ell) > 3.8$ TeV, CL = 95%
$\Lambda_{LL}^+(eeqq) > 2.5$ TeV, CL = 95%
$\Lambda_{LL}^-(eeqq) > 3.7$ TeV, CL = 95%
$\Lambda_{LL}^+(eebb) > 3.1$ TeV, CL = 95%
$\Lambda_{LL}^-(eebb) > 2.9$ TeV, CL = 95%
$\Lambda_{LL}^+(\mu\mu qq) > 2.9$ TeV, CL = 95%
$\Lambda_{LL}^-(\mu\mu qq) > 4.2$ TeV, CL = 95%
$\Lambda_{LR}^\pm(\nu_\mu \nu_e \mu e) > 3.1$ TeV, CL = 90%
$\Lambda_{LL}^\pm(qqqq) > 1.6$ TeV, CL = 95%

Excited Leptons

The limits from $\ell^{*+}\ell^{*-}$ do not depend on λ (where λ is the $\ell\ell^*$ transition coupling). The λ-dependent limits assume chiral coupling, except for the third limit for e^* which is for nonchiral coupling. For chiral coupling, this limit corresponds to $\lambda_\gamma = \sqrt{2}$.

$e^{*\pm}$ — excited electron
  Mass $m > 85.0$ GeV, CL = 95%   (from $e^{*+}e^{*-}$)
  Mass $m > 91$ GeV, CL = 95%   (if $\lambda_Z > 1$)
  Mass $m > 194$ GeV, CL = 95%   (if $\lambda_\gamma = 1$)

$\mu^{*\pm}$ — excited muon
  Mass $m > 85.3$ GeV, CL = 95%   (from $\mu^{*+}\mu^{*-}$)
  Mass $m > 91$ GeV, CL = 95%   (if $\lambda_Z > 1$)

$\tau^{*\pm}$ — excited tau
  Mass $m > 84.6$ GeV, CL = 95%   (from $\tau^{*+}\tau^{*-}$)
  Mass $m > 90$ GeV, CL = 95%   (if $\lambda_Z > 0.18$)

ν^* — excited neutrino
  Mass $m > 84.9$ GeV, CL = 95%   (from $\nu^*\bar{\nu}^*$)
  Mass $m > 91$ GeV, CL = 95%   (if $\lambda_Z > 1$)
  Mass $m =$ none 40–96 GeV, CL = 95%   (from $ep \rightarrow \nu^* X$)

q^* — excited quark
  Mass $m > 45.6$ GeV, CL = 95%   (from $q^*\bar{q}^*$)
  Mass $m > 88$ GeV, CL = 95%   (if $\lambda_Z > 1$)
  Mass $m > 570$ GeV, CL = 95%   ($p\bar{p} \rightarrow q^* X$)

Color Sextet and Octet Particles

Color Sextet Quarks (q_6)
  Mass $m > 84$ GeV, CL = 95%   (Stable q_6)

Color Octet Charged Leptons (ℓ_8)
  Mass $m > 86$ GeV, CL = 95%   (Stable ℓ_8)

Color Octet Neutrinos (ν_8)
  Mass $m > 110$ GeV, CL = 90%   ($\nu_8 \rightarrow \nu g$)

TESTS OF CONSERVATION LAWS

TESTS OF CONSERVATION LAWS

Revised by L. Wolfenstein and T.G. Trippe⎡May 1998.

In keeping with the current interest in tests of conservation laws⎡ we collect together a Table of experimental limits on all weak and electromagnetic decays⎡mass differences⎡and moments⎡and on a few reactions⎡whose observation would violate conservation laws. The Table is given only in the full *Review of Particle Physics*⎡not in the Particle Physics Booklet. For the benefit of Booklet readers⎡we include the best limits from the Table in the following text. Limits in this text are for CL=90% unless otherwise specified. The Table is in two parts: "Discrete Space-Time Symmetries"⎡*i.e.*⎡CT⎡T⎡TCP⎡and CPT; and "Number Conservation Laws"⎡*i.e.*⎡lepton⎡baryon⎡hadronic flavor⎡and charge conservation. The references for these data can be found in the the Particle Listings in the *Review*. A discussion of these tests follows.

CPT INVARIANCE

General principles of relativistic field theory require invariance under the combined transformation CPT. The simplest tests of CPT invariance are the equality of the masses and lifetimes of a particle and its antiparticle. The best test comes from the limit on the mass difference between K^0 and $\overline{K}^0$. Any such difference contributes to the CP-violating parameter ϵ. Assuming CPT invariance⎡ϕ_ϵ⎡the phase of ϵ should be very close to 44°. (See the "Note on CP Violation in K_L^0 Decay" in the Particle Listings.) In contrast⎡if the entire source of CP violation in K^0 decays were a $K^0 - \overline{K}^0$ mass difference⎡ϕ_ϵ would be $44° + 90°$. Assuming that there is no other source of CPT violation than this mass difference⎡it is possible to deduce that [1]

$$m_{\overline{K}^0} - m_{K^0} \approx \frac{2(m_{K_L^0} - m_{K_S^0})\,|\eta|\,(\frac{2}{3}\phi_{+-} + \frac{1}{3}\phi_{00} - \phi_0)}{\sin\phi_0},$$

where $\phi_0 = 43.5°$ with an uncertainty of less than 0.1°. Using our best values of the CP-violation parameters⎡we get $|(m_{\overline{K}^0} - m_{K^0})/m_{K^0}| \leq 10^{-18}$. Limits can also be placed on specific CPT-violating decay amplitudes. Given the small value of $(1 - |\eta_{00}/\eta_{+-}|)$⎡the value of $\phi_{00} - \phi_{+-}$ provides a measure of CPT violation in $K_L^0 \to 2\pi$ decay. Results from CERN [1] and Fermilab [2] indicate no CPT-violating effect.

CP AND *T* INVARIANCE

Given CPT invariance⎡CP violation and T violation are equivalent. So far the only evidence for CP or T violation comes from the measurements of η_{+-}⎡η_{00}⎡and the semileptonic decay charge asymmetry for K_L⎡*e.g.*⎡$|\eta_{+-}| = |A(K_L^0 \to \pi^+\pi^-)/A(K_S^0 \to \pi^+\pi^-)| = (2.285 \pm 0.019) \times 10^{-3}$ and $[\Gamma(K_L^0 \to \pi^- e^+ \nu) - \Gamma(K_L^0 \to \pi^+ e^- \bar{\nu})]/[\text{sum}] = (0.333 \pm 0.014)\%$. Other searches for CP or T violation divide into (a) those that involve weak interactions or parity violation⎡and (b) those that involve processes otherwise allowed by the strong or electromagnetic interactions. In class (a) the most sensitive are probably the searches for an electric dipole moment of the neutron⎡measured to be $< 1.0 \times 10^{-25}$ e cm⎡and the electron $(-0.18 \pm 0.16) \times 10^{-26}$ e cm. A nonzero value requires both P and T violation. Class (b) includes the search for C violation in η decay⎡believed to be an electromagnetic process⎡*e.g.*⎡as measured by $\Gamma(\eta \to \mu^+\mu^-\pi^0)/\Gamma(\eta \to \text{all}) < 5 \times 10^{-6}$⎡and searches for T violation in a number of nuclear and electromagnetic reactions.

CONSERVATION OF LEPTON NUMBERS

Present experimental evidence and the standard electroweak theory are consistent with the absolute conservation of three separate lepton numbers: electron number L_e⎡muon number L_μ⎡and tau number L_τ. Searches for violations are of the following types:

a) $\Delta L = 2$ for one type of lepton. The best limit comes from the search for neutrinoless double beta decay $(Z, A) \to (Z + 2, A) + e^- + e^-$. The best laboratory limit is $t_{1/2} > 1.1 \times 10^{25}$ yr (CL=90%) for ^{76}Ge.

b) Conversion of one lepton type to another. For purely leptonic processes⎡the best limits are on $\mu \to e\gamma$ and $\mu \to 3e$⎡measured as $\Gamma(\mu \to e\gamma)/\Gamma(\mu \to \text{all}) < 5 \times 10^{-11}$ and $\Gamma(\mu \to 3e)/\Gamma(\mu \to \text{all}) < 1.0 \times 10^{-12}$. For semileptonic processes⎡the best limit comes from the coherent conversion process in a muonic atom⎡$\mu^- + (Z, A) \to e^- + (Z, A)$⎡measured as $\Gamma(\mu^-\text{Ti} \to e^-\text{Ti})/\Gamma(\mu^-\text{Ti} \to \text{all}) < 4 \times 10^{-12}$. Of special interest is the case in which the hadronic flavor also changes⎡as in $K_L \to e\mu$ and $K^+ \to \pi^+e^-\mu^+$⎡measured as $\Gamma(K_L \to e\mu)/\Gamma(K_L \to \text{all}) < 3.3 \times 10^{-11}$ and $\Gamma(K^+ \to \pi^+e^-\mu^+)/\Gamma(K^+ \to \text{all}) < 2.1 \times 10^{-10}$. Limits on the conversion of τ into e or μ are found in τ decay and are much less stringent than those for $\mu \to e$ conversion⎡*e.g.*⎡$\Gamma(\tau \to \mu\gamma)/\Gamma(\tau \to \text{all}) < 3.0 \times 10^{-6}$ and $\Gamma(\tau \to e\gamma)/\Gamma(\tau \to \text{all}) < 2.7 \times 10^{-6}$.

c) Conversion of one type of lepton into another type of antilepton. The case most studied is $\mu^- + (Z, A) \to e^+ + (Z - 2, A)$⎡the strongest limit being $\Gamma(\mu^-\text{Ti} \to e^+\text{Ca})/\Gamma(\mu^-\text{Ti} \to \text{all}) < 9 \times 10^{-11}$.

d) Relation to neutrino mass. If neutrinos have mass⎡then it is expected even in the standard electroweak theory that the lepton numbers are not separately conserved⎡as a consequence of lepton mixing analogous to Cabibbo quark mixing. However⎡in this case lepton-number-violating processes such as $\mu \to e\gamma$ are expected to have extremely small probability. For small neutrino masses⎡the lepton-number violation would be observed first in neutrino oscillations⎡which have been the subject of extensive experimental searches. For example⎡searches for $\bar{\nu}_e$ disappearance⎡which we label as $\bar{\nu}_e \nrightarrow \bar{\nu}_e$⎡give measured limits $\Delta(m^2) < 9 \times 10^{-4}$ eV^2 for $\sin^2(2\theta) = 1$⎡and $\sin^2(2\theta) < 0.02$ for large $\Delta(m^2)$⎡where θ is the neutrino mixing angle. Possible evidence for mixing has come from two sources. The deficit in the solar neutrino flux compared with solar model calculations could be explained by oscillations with $\Delta(m^2) \leq 10^{-5}$ eV^2 causing the disappearance of ν_e. In addition underground detectors observing neutrinos produced by cosmic rays in the atmosphere have measured a ν_μ/ν_e ratio much less than expected and also a deficiency of upward going ν_μ compared to downward. This could be explained by oscillations leading to the disappearance of ν_μ with $\Delta(m^2)$ of the order 10^{-2}–10^{-3} eV^2.

CONSERVATION OF HADRONIC FLAVORS

In strong and electromagnetic interactions⎡hadronic flavor is conserved⎡*i.e.* the conversion of a quark of one flavor (d, u, s, c, b, t) into a quark of another flavor is forbidden. In the Standard Model⎡the weak interactions violate these conservation laws in a manner described by the Cabibbo-Kobayashi-Maskawa mixing (see the section "Cabibbo-Kobayashi-Maskawa Mixing Matrix"). The way in which these conservation laws are violated is tested as follows:

a) $\Delta S = \Delta Q$ rule. In the semileptonic decay of strange particles⎡the strangeness change equals the change in charge of the hadrons. Tests come from limits on decay rates such as

$\Gamma(\Sigma^+ \to ne^+\nu)/\Gamma(\Sigma^+ \to \text{all}) < 5 \times 10^{-6}$ and from a detailed analysis of $K_L \to \pi e\nu$, which yields the parameter x, measured to be $(\text{Re}\,x, \text{Im}\,x) = (0.006 \pm 0.018, -0.003 \pm 0.026)$. Corresponding rules are $\Delta C = \Delta Q$ and $\Delta B = \Delta Q$.

b) Change of flavor by two units. In the Standard Model this occurs only in second-order weak interactions. The classic example is $\Delta S = 2$ via $K^0 - \overline{K}^0$ mixing, which is directly measured by $m(K_S) - m(K_L) = (3.489 \pm 0.009) \times 10^{-12}$ MeV. There is now evidence for $B^0 - \overline{B}^0$ mixing ($\Delta B = 2$), with the corresponding mass difference between the eigenstates $(m_{B^0_H} - m_{B^0_L}) = (0.723 \pm 0.032)\,\Gamma_{B^0} = (3.05 \pm 0.12) \times 10^{-10}$ MeV, and for $B^0_s - \overline{B}^0_s$ mixing, with $(m_{B^0_{sH}} - m_{B^0_{sL}}) > 14\,\Gamma_{B^0_s}$ or $> 6 \times 10^{-9}$ MeV (CL=95%). No evidence exists for $D^0 - \overline{D}^0$ mixing, which is expected to be much smaller in the Standard Model.

c) Flavor-changing neutral currents. In the Standard Model the neutral-current interactions do not change flavor. The low rate $\Gamma(K_L \to \mu^+\mu^-)/\Gamma(K_L \to \text{all}) = (7.2 \pm 0.5) \times 10^{-9}$ puts limits on such interactions; the nonzero value for this rate is attributed to a combination of the weak and electromagnetic interactions. The best test should come from $K^+ \to \pi^+\nu\overline{\nu}$, which occurs in the Standard Model only as a second-order weak process with a branching fraction of $(1 \text{ to } 8) \times 10^{-10}$. Observation of one event has been reported [4] yielding $\Gamma(K^+ \to \pi^+\nu\overline{\nu})/\Gamma(K^+ \to \text{all}) = (4.2^{+9.7}_{-3.5}) \times 10^{-10}$. Limits for charm-changing or bottom-changing neutral currents are much less stringent: $\Gamma(D^0 \to \mu^+\mu^-)/\Gamma(D^0 \to \text{all}) < 4 \times 10^{-6}$ and $\Gamma(B^0 \to \mu^+\mu^-)/\Gamma(B^0 \to \text{all}) < 7 \times 10^{-7}$. One cannot isolate flavor-changing neutral current (FCNC) effects in non leptonic decays. For example, the FCNC transition $s \to d + (\overline{u} + u)$ is equivalent to the charged-current transition $s \to u + (\overline{u} + d)$. Tests for FCNC are therefore limited to hadron decays into lepton pairs. Such decays are expected only in second-order in the electroweak coupling in the Standard Model.

References

1. R. Carosi *et al.*, Phys. Lett. **B237**, 303 (1990).
2. M. Karlsson *et al.*, Phys. Rev. Lett. **64**, 2976 (1990); L.K. Gibbons *et al.*, Phys. Rev. Lett. **70**, 1199 (1993).
3. B. Schwingenheuer *et al.*, Phys. Rev. Lett. **74**, 4376 (1995).
4. S. Adler *et al.*, Phys. Rev. Lett. **79**, 2204 (1997).

TESTS OF DISCRETE SPACE-TIME SYMMETRIES

CHARGE CONJUGATION (C) INVARIANCE

$\Gamma(\pi^0 \to 3\gamma)/\Gamma_{\text{total}}$	$<3.1 \times 10^{-8}$, CL = 90%
η C-nonconserving decay parameters	
$\pi^+\pi^-\pi^0$ left-right asymmetry parameter	$(0.09 \pm 0.17) \times 10^{-2}$
$\pi^+\pi^-\pi^0$ sextant asymmetry parameter	$(0.18 \pm 0.16) \times 10^{-2}$
$\pi^+\pi^-\pi^0$ quadrant asymmetry parameter	$(-0.17 \pm 0.17) \times 10^{-2}$
$\pi^+\pi^-\gamma$ left-right asymmetry parameter	$(0.9 \pm 0.4) \times 10^{-2}$
$\pi^+\pi^-\gamma$ parameter β (D-wave)	0.05 ± 0.06 (S = 1.5)
$\Gamma(\eta \to 3\gamma)/\Gamma_{\text{total}}$	$<5 \times 10^{-4}$, CL = 95%
$\Gamma(\eta \to \pi^0 e^+ e^-)/\Gamma_{\text{total}}$	[a] $<4 \times 10^{-5}$, CL = 90%
$\Gamma(\eta \to \pi^0 \mu^+ \mu^-)/\Gamma_{\text{total}}$	[a] $<5 \times 10^{-6}$, CL = 90%
$\Gamma(\omega(782) \to \eta\pi^0)/\Gamma_{\text{total}}$	$<1 \times 10^{-3}$, CL = 90%
$\Gamma(\omega(782) \to 3\pi^0)/\Gamma_{\text{total}}$	$<3 \times 10^{-4}$, CL = 90%
$\Gamma(\eta'(958) \to \pi^0 e^+ e^-)/\Gamma_{\text{total}}$	[a] $<1.3 \times 10^{-2}$, CL = 90%
$\Gamma(\eta'(958) \to \eta e^+ e^-)/\Gamma_{\text{total}}$	[a] $<1.1 \times 10^{-2}$, CL = 90%
$\Gamma(\eta'(958) \to 3\gamma)/\Gamma_{\text{total}}$	$<1.0 \times 10^{-4}$, CL = 90%
$\Gamma(\eta'(958) \to \mu^+ \mu^- \pi^0)/\Gamma_{\text{total}}$	[a] $<6.0 \times 10^{-5}$, CL = 90%
$\Gamma(\eta'(958) \to \mu^+ \mu^- \eta)/\Gamma_{\text{total}}$	[a] $<1.5 \times 10^{-5}$, CL = 90%

PARITY (P) INVARIANCE

e electric dipole moment	$(0.18 \pm 0.16) \times 10^{-26}$ e cm
μ electric dipole moment	$(3.7 \pm 3.4) \times 10^{-19}$ e cm
τ electric dipole moment (d_τ)	> -3.1 and $< 3.1 \times 10^{-16}$ e cm, CL = 95%
$\Gamma(\eta \to \pi^+\pi^-)/\Gamma_{\text{total}}$	$<9 \times 10^{-4}$ CL = 90%
$\Gamma(\eta'(958) \to \pi^+\pi^-)/\Gamma_{\text{total}}$	$<2 \times 10^{-2}$ CL = 90%
$\Gamma(\eta'(958) \to \pi^0\pi^0)/\Gamma_{\text{total}}$	$<9 \times 10^{-4}$ CL = 90%
p electric dipole moment	$(-4 \pm 6) \times 10^{-23}$ e cm
n electric dipole moment	$<0.97 \times 10^{-25}$ e cm, CL = 90%
Λ electric dipole moment	$<1.5 \times 10^{-16}$ e cm, CL = 95%

TIME REVERSAL (T) INVARIANCE

Limits on e, μ, τ, p, n, and Λ electric dipole moments under Parity Invariance above are also tests of Time Reversal Invariance.

μ decay parameters	
transverse e^+ polarization normal to plane of μ spin, e^+ momentum	0.007 ± 0.023
α'/A	$(0 \pm 4) \times 10^{-3}$
β'/A	$(2 \pm 6) \times 10^{-3}$
τ electric dipole moment (d_τ)	> -3.1 and $< 3.1 \times 10^{-16}$ e cm, CL = 95%
Im(ξ) in $K^\pm_{\mu3}$ decay (from transverse μ pol.)	-0.017 ± 0.025
Im(ξ) in $K^0_{\mu3}$ decay (from transverse μ pol.)	-0.007 ± 0.026
$n \to pe^-\nu$ decay parameters	
ϕ_{AV}, phase of g_A relative to g_V	[b] $(180.07 \pm 0.18)^\circ$
triple correlation coefficient D	$(-0.5 \pm 1.4) \times 10^{-3}$
triple correlation coefficient D for $\Sigma^- \to ne^-\overline{\nu}_e$	0.11 ± 0.10

CP INVARIANCE

$\text{Re}(d^W_\tau)$	$<0.56 \times 10^{-17}$ e cm, CL = 95%
$\text{Im}(d^W_\tau)$	$<1.5 \times 10^{-17}$ e cm, CL = 95%
$\Gamma(\eta \to \pi^+\pi^-)/\Gamma_{\text{total}}$	$<9 \times 10^{-4}$, CL = 90%
$\Gamma(\eta'(958) \to \pi^+\pi^-)/\Gamma_{\text{total}}$	$<2 \times 10^{-2}$, CL = 90%
$\Gamma(\eta'(958) \to \pi^0\pi^0)/\Gamma_{\text{total}}$	$<9 \times 10^{-4}$, CL = 90%
$K^\pm \to \pi^\pm\pi^+\pi^-$ rate difference/average	$(0.07 \pm 0.12)\%$
$K^\pm \to \pi^\pm\pi^0$ rate difference/average	$(0.0 \pm 0.6)\%$
$K^\pm \to \pi^\pm\pi^0\gamma$ rate difference/average	$(0.9 \pm 3.3)\%$
$(g_{\tau^+} - g_{\tau^-}) / (g_{\tau^+} + g_{\tau^-})$ for $K^\pm \to \pi^\pm\pi^+\pi^-$	$(-0.7 \pm 0.5)\%$
CP-violation parameters in K^0_S decay	
$\text{Im}(\eta_{+-0}) = \text{Im}(A(K^0_S \to \pi^+\pi^-\pi^0, \text{CP-violating}) / A(K^0_L \to \pi^+\pi^-\pi^0))$	-0.002 ± 0.008
$\text{Im}(\eta_{000})^2 = \Gamma(K^0_S \to 3\pi^0) / \Gamma(K^0_L \to 3\pi^0)$	<0.1, CL = 90%
charge asymmetry j for $K^0_L \to \pi^+\pi^-\pi^0$	0.0011 ± 0.0008
$|\epsilon'_{+-\gamma}|/\epsilon$ for $K^0_L \to \pi^+\pi^-\gamma$	<0.3, CL = 90%
$\Gamma(K^0_L \to \pi^0\mu^+\mu^-)/\Gamma_{\text{total}}$	[c] $<5.1 \times 10^{-9}$, CL = 90%
$\Gamma(K^0_L \to \pi^0 e^+ e^-)/\Gamma_{\text{total}}$	[c] $<4.3 \times 10^{-9}$, CL = 90%
$\Gamma(K^0_L \to \pi^0\nu\overline{\nu})/\Gamma_{\text{total}}$	[d] $<5.8 \times 10^{-5}$, CL = 90%
$A_{CP}(K^+K^-\pi^\pm)$ in $D^\pm \to K^+K^-\pi^\pm$	-0.017 ± 0.027
$A_{CP}(K^\pm K^{*0})$ in $D^+ \to K^+\overline{K}^{*0}$ and $D^- \to K^-K^{*0}$	-0.02 ± 0.05
$A_{CP}(\phi\pi^\pm)$ in $D^\pm \to \phi\pi^\pm$	-0.014 ± 0.033
$A_{CP}(\pi^+\pi^-\pi^\pm)$ in $D^\pm \to \pi^+\pi^-\pi^\pm$	-0.02 ± 0.04
$A_{CP}(K^+K^-)$ in $D^0\overline{D}^0 \to K^+K^-$	0.026 ± 0.035
$A_{CP}(\pi^+\pi^-)$ in $D^0\overline{D}^0 \to \pi^+\pi^-$	-0.05 ± 0.08
$A_{CP}(K^0_S\phi)$ in $D^0\overline{D}^0 \to K^0_S\phi$	-0.03 ± 0.09
$A_{CP}(K^0_S\pi^0)$ in $D^0\overline{D}^0 \to K^0_S\pi^0$	-0.018 ± 0.030
$|\text{Re}(\epsilon_{B^0})|$	0.002 ± 0.008
$[\alpha_-(\Lambda) + \alpha_+(\overline{\Lambda})] / [\alpha_-(\Lambda) - \alpha_+(\overline{\Lambda})]$	-0.03 ± 0.06

Limits are given at the 90% confidence level, while errors are given as ± 1 standard deviation.

CP VIOLATION OBSERVED

K_L^0 branching ratios

charge asymmetry in $K_{\ell 3}^0$ decays

$\delta(\mu) = [\Gamma(\pi^- \mu^+ \nu_\mu)$ $(0.304 \pm 0.025)\%$
    $- \Gamma(\pi^+ \mu^- \bar{\nu}_\mu)]/\text{sum}$

$\delta(e) = [\Gamma(\pi^- e^+ \nu_e)$ $(0.333 \pm 0.014)\%$
    $- \Gamma(\pi^+ e^- \bar{\nu}_e)]/\text{sum}$

parameters for $K_L^0 \to 2\pi$ decay

$|\eta_{00}| = |A(K_L^0 \to 2\pi^0) /$ $(2.275 \pm 0.019) \times 10^{-3}$ (S = 1.1)
    $A(K_S^0 \to 2\pi^0)|$

$|\eta_{+-}| = |A(K_L^0 \to \pi^+ \pi^-) /$ $(2.285 \pm 0.019) \times 10^{-3}$
    $A(K_S^0 \to \pi^+ \pi^-)|$

$\epsilon'/\epsilon \approx \text{Re}(\epsilon'/\epsilon) = (1 - |\eta_{00}/\eta_{+-}|)/3$ [e] $(1.5 \pm 0.8) \times 10^{-3}$ (S = 1.8)

ϕ_{+-} phase of η_{+-} $(43.5 \pm 0.6)°$

ϕ_{00} phase of η_{00} $(43.4 \pm 1.0)°$

parameters for $K_L^0 \to \pi^+ \pi^- \gamma$ decay

$|\eta_{+-\gamma}| = |A(K_L^0 \to \pi^+ \pi^- \gamma \, CP$ $(2.35 \pm 0.07) \times 10^{-3}$
    violating$)/A(K_S^0 \to \pi^+ \pi^- \gamma)|$

$\phi_{+-\gamma} = $ phase of $\eta_{+-\gamma}$ $(44 \pm 4)°$

$\Gamma(K_L^0 \to \pi^+ \pi^-)/\Gamma_{\text{total}}$ $(2.067 \pm 0.035) \times 10^{-3}$ (S = 1.1)

$\Gamma(K_L^0 \to \pi^0 \pi^0)/\Gamma_{\text{total}}$ $(9.36 \pm 0.20) \times 10^{-4}$

CPT INVARIANCE

$(m_{W^+} - m_{W^-}) / m_{\text{average}}$ -0.002 ± 0.007

$(m_{e^+} - m_{e^-}) / m_{\text{average}}$ $<4 \times 10^{-8}$ CL = 90%

$|q_{e^+} + q_{e^-}|/e$ $<2 \times 10^{-18}$

$(g_{e^+} - g_{e^-}) / g_{\text{average}}$ $(-0.5 \pm 2.1) \times 10^{-12}$

$(\tau_{\mu^+} - \tau_{\mu^-}) / \tau_{\text{average}}$ $(2 \pm 8) \times 10^{-5}$

$(g_{\mu^+} - g_{\mu^-}) / g_{\text{average}}$ $(-2.6 \pm 1.6) \times 10^{-8}$

$(m_{\pi^+} - m_{\pi^-}) / m_{\text{average}}$ $(2 \pm 5) \times 10^{-4}$

$(\tau_{\pi^+} - \tau_{\pi^-}) / \tau_{\text{average}}$ $(6 \pm 7) \times 10^{-4}$

$(m_{K^+} - m_{K^-}) / m_{\text{average}}$ $(-0.6 \pm 1.8) \times 10^{-4}$

$(\tau_{K^+} - \tau_{K^-}) / \tau_{\text{average}}$ $(0.11 \pm 0.09)\%$ (S = 1.2)

$K^{\pm} \to \mu^{\pm} \nu_\mu$ rate difference/average $(-0.5 \pm 0.4)\%$

$K^{\pm} \to \pi^{\pm} \pi^0$ rate difference/average   [f] $(0.8 \pm 1.2)\%$

$|m_{K^0} - m_{\overline{K^0}}| / m_{\text{average}}$ [g] $<10^{-18}$

phase difference $\phi_{00} - \phi_{+-}$ $(-0.1 \pm 0.8)°$

CPT-violation parameters in K^0 decay

real part of Δ 0.018 ± 0.020

imaginary part of Δ 0.02 ± 0.04

$(|\frac{q_{\bar{p}}}{m_{\bar{p}}}| - |\frac{q_p}{m_p}|)/|\frac{q}{m}|_{\text{average}}$ $(1.5 \pm 1.1) \times 10^{-9}$

$|q_p + q_{\bar{p}}|/e$ $<2 \times 10^{-5}$

$(\mu_p + \mu_{\bar{p}}) / |\mu|_{\text{average}}$ $(-2.6 \pm 2.9) \times 10^{-3}$

$(m_n - m_{\bar{n}}) / m_{\text{average}}$ $(9 \pm 5) \times 10^{-5}$

$(m_\Lambda - m_{\bar{\Lambda}}) / m_\Lambda$ $(-1.0 \pm 0.9) \times 10^{-5}$

$(\tau_\Lambda - \tau_{\bar{\Lambda}}) / \tau_{\text{average}}$ 0.04 ± 0.09

$(\mu_{\Sigma^+} + \mu_{\overline{\Sigma}^-}) / |\mu|_{\text{average}}$ 0.014 ± 0.015

$(m_{\Xi^-} - m_{\overline{\Xi}^+}) / m_{\text{average}}$ $(1.1 \pm 2.7) \times 10^{-4}$

$(\tau_{\Xi^-} - \tau_{\overline{\Xi}^+}) / \tau_{\text{average}}$ 0.02 ± 0.18

$(m_{\Omega^-} - m_{\overline{\Omega}^+}) / m_{\text{average}}$ $(0 \pm 5) \times 10^{-4}$

Limits are given at the 90% confidence level, while errors are given as ± 1 standard deviation.

TESTS OF NUMBER CONSERVATION LAWS

LEPTON FAMILY NUMBER

Lepton family number conservation means separate conservation of each of L_e, L_μ, L_τ.

$\Gamma(Z \to e^\pm \mu^\mp)/\Gamma_{\text{total}}$ [h] $<1.7 \times 10^{-6}$ CL = 95%

$\Gamma(Z \to e^\pm \tau^\mp)/\Gamma_{\text{total}}$ [h] $<9.8 \times 10^{-6}$ CL = 95%

$\Gamma(Z \to \mu^\pm \tau^\mp)/\Gamma_{\text{total}}$ [h] $<1.2 \times 10^{-5}$ CL = 95%

limit on $\mu^- \to e^-$ conversion

$\sigma(\mu^- {}^{32}\text{S} \to e^- {}^{32}\text{S}) /$ $<7 \times 10^{-11}$ CL = 90%
    $\sigma(\mu^- {}^{32}\text{S} \to \nu_\mu {}^{32}\text{P}^*)$

$\sigma(\mu^- \text{Ti} \to e^- \text{Ti}) /$ $<4.3 \times 10^{-12}$ CL = 90%
    $\sigma(\mu^- \text{Ti} \to \text{capture})$

$\sigma(\mu^- \text{Pb} \to e^- \text{Pb}) /$ $<4.6 \times 10^{-11}$ CL = 90%
    $\sigma(\mu^- \text{Pb} \to \text{capture})$

limit on muonium $\to$ antimuonium
    conversion $R_g = G_C / G_F$ <0.018 CL = 90%

$\Gamma(\mu^- \to e^- \nu_e \bar{\nu}_\mu)/\Gamma_{\text{total}}$ [i] $<1.2 \times 10^{-2}$ CL = 90%

$\Gamma(\mu^- \to e^- \gamma)/\Gamma_{\text{total}}$ $<4.9 \times 10^{-11}$ CL = 90%

$\Gamma(\mu^- \to e^- e^+ e^-)/\Gamma_{\text{total}}$ $<1.0 \times 10^{-12}$ CL = 90%

$\Gamma(\mu^- \to e^- 2\gamma)/\Gamma_{\text{total}}$ $<7.2 \times 10^{-11}$ CL = 90%

$\Gamma(\tau^- \to e^- \gamma)/\Gamma_{\text{total}}$ $<2.7 \times 10^{-6}$ CL = 90%

$\Gamma(\tau^- \to \mu^- \gamma)/\Gamma_{\text{total}}$ $<3.0 \times 10^{-6}$ CL = 90%

$\Gamma(\tau^- \to e^- \pi^0)/\Gamma_{\text{total}}$ $<3.7 \times 10^{-6}$ CL = 90%

$\Gamma(\tau^- \to \mu^- \pi^0)/\Gamma_{\text{total}}$ $<4.0 \times 10^{-6}$ CL = 90%

$\Gamma(\tau^- \to e^- K^0)/\Gamma_{\text{total}}$ $<1.3 \times 10^{-3}$ CL = 90%

$\Gamma(\tau^- \to \mu^- K^0)/\Gamma_{\text{total}}$ $<1.0 \times 10^{-3}$ CL = 90%

$\Gamma(\tau^- \to e^- \eta)/\Gamma_{\text{total}}$ $<8.2 \times 10^{-6}$ CL = 90%

$\Gamma(\tau^- \to \mu^- \eta)/\Gamma_{\text{total}}$ $<9.6 \times 10^{-6}$ CL = 90%

$\Gamma(\tau^- \to e^- \rho^0)/\Gamma_{\text{total}}$ $<2.0 \times 10^{-6}$ CL = 90%

$\Gamma(\tau^- \to \mu^- \rho^0)/\Gamma_{\text{total}}$ $<6.3 \times 10^{-6}$ CL = 90%

$\Gamma(\tau^- \to e^- K^*(892)^0)/\Gamma_{\text{total}}$ $<5.1 \times 10^{-6}$ CL = 90%

$\Gamma(\tau^- \to \mu^- K^*(892)^0)/\Gamma_{\text{total}}$ $<7.5 \times 10^{-6}$ CL = 90%

$\Gamma(\tau^- \to e^- \overline{K}^*(892)^0)/\Gamma_{\text{total}}$ $<7.4 \times 10^{-6}$ CL = 90%

$\Gamma(\tau^- \to \mu^- \overline{K}^*(892)^0)/\Gamma_{\text{total}}$ $<7.5 \times 10^{-6}$ CL = 90%

$\Gamma(\tau^- \to e^- \phi)/\Gamma_{\text{total}}$ $<6.9 \times 10^{-6}$ CL = 90%

$\Gamma(\tau^- \to \mu^- \phi)/\Gamma_{\text{total}}$ $<7.0 \times 10^{-6}$ CL = 90%

$\Gamma(\tau^- \to e^- e^+ e^-)/\Gamma_{\text{total}}$ $<2.9 \times 10^{-6}$ CL = 90%

$\Gamma(\tau^- \to e^- \mu^+ \mu^-)/\Gamma_{\text{total}}$ $<1.8 \times 10^{-6}$ CL = 90%

$\Gamma(\tau^- \to e^+ \mu^- \mu^-)/\Gamma_{\text{total}}$ $<1.5 \times 10^{-6}$ CL = 90%

$\Gamma(\tau^- \to \mu^- e^+ e^-)/\Gamma_{\text{total}}$ $<1.7 \times 10^{-6}$ CL = 90%

$\Gamma(\tau^- \to \mu^+ e^- e^-)/\Gamma_{\text{total}}$ $<1.5 \times 10^{-6}$ CL = 90%

$\Gamma(\tau^- \to \mu^- \mu^+ \mu^-)/\Gamma_{\text{total}}$ $<1.9 \times 10^{-6}$ CL = 90%

$\Gamma(\tau^- \to e^- \pi^+ \pi^-)/\Gamma_{\text{total}}$ $<2.2 \times 10^{-6}$ CL = 90%

$\Gamma(\tau^- \to \mu^- \pi^+ \pi^-)/\Gamma_{\text{total}}$ $<8.2 \times 10^{-6}$ CL = 90%

$\Gamma(\tau^- \to e^- \pi^+ K^-)/\Gamma_{\text{total}}$ $<6.4 \times 10^{-6}$ CL = 90%

$\Gamma(\tau^- \to e^- \pi^- K^+)/\Gamma_{\text{total}}$ $<3.8 \times 10^{-6}$ CL = 90%

$\Gamma(\tau^- \to e^- K^+ K^-)/\Gamma_{\text{total}}$ $<6.0 \times 10^{-6}$ CL = 90%

$\Gamma(\tau^- \to \mu^- \pi^+ K^-)/\Gamma_{\text{total}}$ $<7.5 \times 10^{-6}$ CL = 90%

$\Gamma(\tau^- \to \mu^- \pi^- K^+)/\Gamma_{\text{total}}$ $<7.4 \times 10^{-6}$ CL = 90%

$\Gamma(\tau^- \to \mu^- K^+ K^-)/\Gamma_{\text{total}}$ $<1.5 \times 10^{-5}$ CL = 90%

$\Gamma(\tau^- \to e^- \pi^0 \pi^0)/\Gamma_{\text{total}}$ $<6.5 \times 10^{-6}$ CL = 90%

$\Gamma(\tau^- \to \mu^- \pi^0 \pi^0)/\Gamma_{\text{total}}$ $<1.4 \times 10^{-5}$ CL = 90%

$\Gamma(\tau^- \to e^- \eta \eta)/\Gamma_{\text{total}}$ $<3.5 \times 10^{-5}$ CL = 90%

$\Gamma(\tau^- \to \mu^- \eta \eta)/\Gamma_{\text{total}}$ $<6.0 \times 10^{-5}$ CL = 90%

$\Gamma(\tau^- \to e^- \pi^0 \eta)/\Gamma_{\text{total}}$ $<2.4 \times 10^{-5}$ CL = 90%

$\Gamma(\tau^- \to \mu^- \pi^0 \eta)/\Gamma_{\text{total}}$ $<2.2 \times 10^{-5}$ CL = 90%

$\Gamma(\tau^- \to e^- \text{light boson})/\Gamma_{\text{total}}$ $<2.7 \times 10^{-3}$ CL = 95%

$\Gamma(\tau^- \to \mu^- \text{light boson})/\Gamma_{\text{total}}$ $<5 \times 10^{-3}$ CL = 95%

ν oscillations. (For other lepton mixing effects in particle decays, see the Particle Listings.)

$\bar{\nu}_e \not\to \bar{\nu}_e$

    $\Delta(m^2)$ for $\sin^2(2\theta) = 1$ $<9 \times 10^{-4}$ eV^2, CL = 90%

    $\sin^2(2\theta)$ for "Large" $\Delta(m^2)$ <0.02, CL = 90%

$\nu_e \to \nu_\tau$

    $\Delta(m^2)$ for $\sin^2(2\theta) = 1$ <9 eV^2, CL = 90%

    $\sin^2(2\theta)$ for "Large" $\Delta(m^2)$ <0.25, CL = 90%

$\bar{\nu}_e \to \bar{\nu}_\tau$

    $\sin^2(2\theta)$ for "Large" $\Delta(m^2)$ <0.7, CL = 90%

$\nu_\mu \to \nu_e$

$\Delta(m^2)$ for $\sin^2(2\theta) = 1$ <0.09 eV^2 CL = 90%

$\sin^2(2\theta)$ for "Large" $\Delta(m^2)$ $<3.0 \times 10^{-3}$ CL = 90%

$\overline{\nu}_\mu \rightarrow \overline{\nu}_e$

$\Delta(m^2)$ for $\sin^2(2\theta) = 1$ <0.14 eV^2 CL = 90%

$\sin^2(2\theta)$ for "Large" $\Delta(m^2)$ <0.004 CL = 95%

$\nu_\mu(\overline{\nu}_\mu) \rightarrow \nu_e(\overline{\nu}_e)$

$\Delta(m^2)$ for $\sin^2(2\theta) = 1$ <0.075 eV^2 CL = 90%

$\sin^2(2\theta)$ for "Large" $\Delta(m^2)$ $<1.8 \times 10^{-3}$ CL = 90%

$\nu_\mu \rightarrow \nu_\tau$

$\Delta(m^2)$ for $\sin^2(2\theta) = 1$ <0.9 eV^2 CL = 90%

$\sin^2(2\theta)$ for "Large" $\Delta(m^2)$ <0.004 CL = 90%

$\overline{\nu}_\mu \rightarrow \overline{\nu}_\tau$

$\Delta(m^2)$ for $\sin^2(2\theta) = 1$ <2.2 eV^2 CL = 90%

$\sin^2(2\theta)$ for "Large" $\Delta(m^2)$ $<4.4 \times 10^{-2}$ CL = 90%

$\nu_\mu(\overline{\nu}_\mu) \rightarrow \nu_\tau(\overline{\nu}_\tau)$

$\Delta(m^2)$ for $\sin^2(2\theta) = 1$ <1.5 eV^2 CL = 90%

$\sin^2(2\theta)$ for "Large" $\Delta(m^2)$ $<8 \times 10^{-3}$ CL = 90%

$\nu_e \not\rightarrow \nu_e$

$\Delta(m^2)$ for $\sin^2(2\theta) = 1$ <0.17 eV^2 CL = 90%

$\sin^2(2\theta)$ for "Large" $\Delta(m^2)$ $<7 \times 10^{-2}$ CL = 90%

$\nu_\mu \not\rightarrow \nu_\mu$

$\Delta(m^2)$ for $\sin^2(2\theta) = 1$ <0.23 or >1500 eV^2

$\sin^2(2\theta)$ for $\Delta(m^2) = 100$eV^2 [j] <0.02 CL = 90%

$\overline{\nu}_\mu \not\rightarrow \overline{\nu}_\mu$

$\Delta(m^2)$ for $\sin^2(2\theta) = 1$ <7 or >1200 eV^2

$\sin^2(2\theta)$ for 190 eV$^2 < \Delta(m^2) <$ [k] <0.02 CL = 90%
320 eV^2

$\Gamma(\pi^+ \rightarrow \mu^+ \nu_e)/\Gamma_{\text{total}}$ [l] $<8.0 \times 10^{-3}$ CL = 90%

$\Gamma(\pi^+ \rightarrow \mu^- e^+ e^+ \nu)/\Gamma_{\text{total}}$ $<1.6 \times 10^{-6}$ CL = 90%

$\Gamma(\pi^0 \rightarrow \mu^+ e^- + \mu^- e^+)/\Gamma_{\text{total}}$ $<1.72 \times 10^{-8}$ CL = 90%

$\Gamma(\eta \rightarrow \mu^+ e^- + \mu^- e^+)/\Gamma_{\text{total}}$ $<6 \times 10^{-6}$ CL = 90%

$\Gamma(K^+ \rightarrow \mu^- \nu e^+ e^+)/\Gamma_{\text{total}}$ $<2.0 \times 10^{-8}$ CL = 90%

$\Gamma(K^+ \rightarrow \mu^+ \nu_e)/\Gamma_{\text{total}}$ [l] $<4 \times 10^{-3}$ CL = 90%

$\Gamma(K^+ \rightarrow \pi^+ \mu^+ e^-)/\Gamma_{\text{total}}$ $<2.1 \times 10^{-10}$ CL = 90%

$\Gamma(K^+ \rightarrow \pi^+ \mu^- e^+)/\Gamma_{\text{total}}$ $<7 \times 10^{-9}$ CL = 90%

$\Gamma(K^0_L \rightarrow e^\pm \mu^\mp)/\Gamma_{\text{total}}$ [h] $<3.3 \times 10^{-11}$ CL = 90%

$\Gamma(K^0_L \rightarrow e^\pm e^\pm \mu^\mp \mu^\mp)/\Gamma_{\text{total}}$ [h] $<6.1 \times 10^{-9}$ CL = 90%

$\Gamma(D^+ \rightarrow \pi^+ e^+ \mu^-)/\Gamma_{\text{total}}$ $<1.1 \times 10^{-4}$ CL = 90%

$\Gamma(D^+ \rightarrow \pi^+ e^- \mu^+)/\Gamma_{\text{total}}$ $<1.3 \times 10^{-4}$ CL = 90%

$\Gamma(D^+ \rightarrow K^+ e^+ \mu^-)/\Gamma_{\text{total}}$ $<1.3 \times 10^{-4}$ CL = 90%

$\Gamma(D^+ \rightarrow K^+ e^- \mu^+)/\Gamma_{\text{total}}$ $<1.2 \times 10^{-4}$ CL = 90%

$\Gamma(D^0 \rightarrow \mu^\pm e^\mp)/\Gamma_{\text{total}}$ [h] $<1.9 \times 10^{-5}$ CL = 90%

$\Gamma(D^0 \rightarrow \pi^0 e^\pm \mu^\mp)/\Gamma_{\text{total}}$ [h] $<8.6 \times 10^{-5}$ CL = 90%

$\Gamma(D^0 \rightarrow \eta e^\pm \mu^\mp)/\Gamma_{\text{total}}$ [h] $<1.0 \times 10^{-4}$ CL = 90%

$\Gamma(D^0 \rightarrow \rho^0 e^\pm \mu^\mp)/\Gamma_{\text{total}}$ [h] $<4.9 \times 10^{-5}$ CL = 90%

$\Gamma(D^0 \rightarrow \omega e^\pm \mu^\mp)/\Gamma_{\text{total}}$ [h] $<1.2 \times 10^{-4}$ CL = 90%

$\Gamma(D^0 \rightarrow \phi e^\pm \mu^\mp)/\Gamma_{\text{total}}$ [h] $<3.4 \times 10^{-5}$ CL = 90%

$\Gamma(D^0 \rightarrow \overline{K}^0 e^\pm \mu^\mp)/\Gamma_{\text{total}}$ [h] $<1.0 \times 10^{-4}$ CL = 90%

$\Gamma(D^0 \rightarrow \overline{K}^*(892)^0 e^\pm \mu^\mp)/\Gamma_{\text{total}}$ [h] $<1.0 \times 10^{-4}$ CL = 90%

$\Gamma(B^+ \rightarrow \pi^+ e^+ \mu^-)/\Gamma_{\text{total}}$ $<6.4 \times 10^{-3}$ CL = 90%

$\Gamma(B^+ \rightarrow \pi^+ e^- \mu^+)/\Gamma_{\text{total}}$ $<6.4 \times 10^{-3}$ CL = 90%

$\Gamma(B^+ \rightarrow K^+ e^+ \mu^-)/\Gamma_{\text{total}}$ $<6.4 \times 10^{-3}$ CL = 90%

$\Gamma(B^+ \rightarrow K^+ e^- \mu^+)/\Gamma_{\text{total}}$ $<6.4 \times 10^{-3}$ CL = 90%

$\Gamma(B^+ \rightarrow \pi^- e^+ \mu^+)/\Gamma_{\text{total}}$ $<6.4 \times 10^{-3}$ CL = 90%

$\Gamma(B^+ \rightarrow K^- e^+ \mu^+)/\Gamma_{\text{total}}$ $<6.4 \times 10^{-3}$ CL = 90%

$\Gamma(B^0 \rightarrow e^\pm \mu^\mp)/\Gamma_{\text{total}}$ [h] $<5.9 \times 10^{-6}$ CL = 90%

$\Gamma(B^0 \rightarrow e^\pm \tau^\mp)/\Gamma_{\text{total}}$ [h] $<5.3 \times 10^{-4}$ CL = 90%

$\Gamma(B^0 \rightarrow \mu^\pm \tau^\mp)/\Gamma_{\text{total}}$ [h] $<8.3 \times 10^{-4}$ CL = 90%

$\Gamma(B \rightarrow e^\pm \mu^\mp s)/\Gamma_{\text{total}}$ $<2.2 \times 10^{-5}$ CL = 90%

$\Gamma(B^0_s \rightarrow e^\pm \mu^\mp)/\Gamma_{\text{total}}$ [h] $<4.1 \times 10^{-5}$ CL = 90%

TOTAL LEPTON NUMBER

Violation of total lepton number conservation also implies violation of lepton family number conservation.

limit on $\mu^- \rightarrow e^+$ conversion

$\sigma(\mu^- {}^{32}\text{S} \rightarrow e^+ {}^{32}\text{Si}^*) /$
$\sigma(\mu^- {}^{32}\text{S} \rightarrow \nu_\mu {}^{32}\text{P}^*)$ $<9 \times 10^{-10}$ CL = 90%

$\sigma(\mu^- {}^{127}\text{I} \rightarrow e^+ {}^{127}\text{Sb}^*) /$
$\sigma(\mu^- {}^{127}\text{I} \rightarrow \text{anything})$ $<3 \times 10^{-10}$ CL = 90%

$\sigma(\mu^- \text{Ti} \rightarrow e^+ \text{Ca}) /$
$\sigma(\mu^- \text{Ti} \rightarrow \text{capture})$ $<8.9 \times 10^{-11}$ CL = 90%

$\Gamma(\tau^- \rightarrow \pi^- \gamma)/\Gamma_{\text{total}}$ $<2.8 \times 10^{-4}$ CL = 90%

$\Gamma(\tau^- \rightarrow \pi^- \pi^0)/\Gamma_{\text{total}}$ $<3.7 \times 10^{-4}$ CL = 90%

$\Gamma(\tau^- \rightarrow e^+ \pi^- \pi^-)/\Gamma_{\text{total}}$ $<1.9 \times 10^{-6}$ CL = 90%

$\Gamma(\tau^- \rightarrow \mu^+ \pi^- \pi^-)/\Gamma_{\text{total}}$ $<3.4 \times 10^{-6}$ CL = 90%

$\Gamma(\tau^- \rightarrow e^+ \pi^- K^-)/\Gamma_{\text{total}}$ $<2.1 \times 10^{-6}$ CL = 90%

$\Gamma(\tau^- \rightarrow e^+ K^- K^-)/\Gamma_{\text{total}}$ $<3.8 \times 10^{-6}$ CL = 90%

$\Gamma(\tau^- \rightarrow \mu^+ \pi^- K^-)/\Gamma_{\text{total}}$ $<7.0 \times 10^{-6}$ CL = 90%

$\Gamma(\tau^- \rightarrow \mu^+ K^- K^-)/\Gamma_{\text{total}}$ $<6.0 \times 10^{-6}$ CL = 90%

$\Gamma(\tau^- \rightarrow \overline{p}\gamma)/\Gamma_{\text{total}}$ $<2.9 \times 10^{-4}$ CL = 90%

$\Gamma(\tau^- \rightarrow \overline{p}\pi^0)/\Gamma_{\text{total}}$ $<6.6 \times 10^{-4}$ CL = 90%

$\Gamma(\tau^- \rightarrow \overline{p}\eta)/\Gamma_{\text{total}}$ $<1.30 \times 10^{-3}$ CL = 90%

$\nu_e \rightarrow (\overline{\nu}_e)_L$

$\alpha\Delta(m^2)$ for $\sin^2(2\theta) = 1$ <0.14 eV^2 CL = 90%

$\alpha^2\sin^2(2\theta)$ for "Large" $\Delta(m^2)$ <0.032 CL = 90%

$\nu_\mu \rightarrow (\overline{\nu}_e)_L$

$\alpha\Delta(m^2)$ for $\sin^2(2\theta) = 1$ <0.16 eV^2 CL = 90%

$\alpha^2\sin^2(2\theta)$ for "Large" $\Delta(m^2)$ <0.001 CL = 90%

$\Gamma(\pi^+ \rightarrow \mu^+ \overline{\nu}_e)/\Gamma_{\text{total}}$ [l] $<1.5 \times 10^{-3}$ CL = 90%

$\Gamma(K^+ \rightarrow \pi^- \mu^+ e^+)/\Gamma_{\text{total}}$ $<7 \times 10^{-9}$ CL = 90%

$\Gamma(K^+ \rightarrow \pi^- e^+ e^+)/\Gamma_{\text{total}}$ $<1.0 \times 10^{-8}$ CL = 90%

$\Gamma(K^+ \rightarrow \pi^- \mu^+ \mu^+)/\Gamma_{\text{total}}$ [l] $<1.5 \times 10^{-4}$ CL = 90%

$\Gamma(K^+ \rightarrow \mu^+ \overline{\nu}_e)/\Gamma_{\text{total}}$ [l] $<3.3 \times 10^{-3}$ CL = 90%

$\Gamma(K^+ \rightarrow \pi^0 e^+ \overline{\nu}_e)/\Gamma_{\text{total}}$ $<3 \times 10^{-3}$ CL = 90%

$\Gamma(D^+ \rightarrow \pi^- e^+ e^+)/\Gamma_{\text{total}}$ $<1.1 \times 10^{-4}$ CL = 90%

$\Gamma(D^+ \rightarrow \pi^- \mu^+ \mu^+)/\Gamma_{\text{total}}$ $<8.7 \times 10^{-5}$ CL = 90%

$\Gamma(D^+ \rightarrow \pi^- e^+ \mu^+)/\Gamma_{\text{total}}$ $<1.1 \times 10^{-4}$ CL = 90%

$\Gamma(D^+ \rightarrow \rho^- \mu^+ \mu^+)/\Gamma_{\text{total}}$ $<5.6 \times 10^{-4}$ CL = 90%

$\Gamma(D^+ \rightarrow K^- e^+ e^+)/\Gamma_{\text{total}}$ $<1.2 \times 10^{-4}$ CL = 90%

$\Gamma(D^+ \rightarrow K^- \mu^+ \mu^+)/\Gamma_{\text{total}}$ $<1.2 \times 10^{-4}$ CL = 90%

$\Gamma(D^+ \rightarrow K^- e^+ \mu^+)/\Gamma_{\text{total}}$ $<1.3 \times 10^{-4}$ CL = 90%

$\Gamma(D^+ \rightarrow K^*(892)^- \mu^+ \mu^+)/\Gamma_{\text{total}}$ $<8.5 \times 10^{-4}$ CL = 90%

$\Gamma(D_s^+ \rightarrow \pi^- \mu^+ \mu^+)/\Gamma_{\text{total}}$ $<4.3 \times 10^{-4}$ CL = 90%

$\Gamma(D_s^+ \rightarrow K^- \mu^+ \mu^+)/\Gamma_{\text{total}}$ $<5.9 \times 10^{-4}$ CL = 90%

$\Gamma(D_s^+ \rightarrow K^*(892)^- \mu^+ \mu^+)/\Gamma_{\text{total}}$ $<1.4 \times 10^{-3}$ CL = 90%

$\Gamma(B^+ \rightarrow \pi^- e^+ e^+)/\Gamma_{\text{total}}$ $<3.9 \times 10^{-3}$ CL = 90%

$\Gamma(B^+ \rightarrow \pi^- \mu^+ \mu^+)/\Gamma_{\text{total}}$ $<9.1 \times 10^{-3}$ CL = 90%

$\Gamma(B^+ \rightarrow K^- e^+ e^+)/\Gamma_{\text{total}}$ $<3.9 \times 10^{-3}$ CL = 90%

$\Gamma(B^+ \rightarrow K^- \mu^+ \mu^+)/\Gamma_{\text{total}}$ $<9.1 \times 10^{-3}$ CL = 90%

$\Gamma(\Xi^- \rightarrow p \mu^- \mu^-)/\Gamma_{\text{total}}$ $<4 \times 10^{-4}$ CL = 90%

$\Gamma(\Lambda_c^+ \rightarrow \Sigma^- \mu^+ \mu^+)/\Gamma_{\text{total}}$ $<7.0 \times 10^{-4}$ CL = 90%

BARYON NUMBER

$\Gamma(\tau^- \rightarrow \overline{p}\gamma)/\Gamma_{\text{total}}$ $<2.9 \times 10^{-4}$ CL = 90%

$\Gamma(\tau^- \rightarrow \overline{p}\pi^0)/\Gamma_{\text{total}}$ $<6.6 \times 10^{-4}$ CL = 90%

$\Gamma(\tau^- \rightarrow \overline{p}\eta)/\Gamma_{\text{total}}$ $<1.30 \times 10^{-3}$ CL = 90%

p mean life $>1.6 \times 10^{25}$ years

A few examples of proton or bound neutron decay follow. For limits on many other nucleon decay channels see the Baryon Summary Table.

$\tau(N \rightarrow e^+ \pi)$ $> 130\ (n)$, $> 550\ (p) \times 10^{30}$ years, CL = 90%

$\tau(N \rightarrow \mu^+ \pi)$ $> 100\ (n)$, $> 270\ (p) \times 10^{30}$ years, CL = 90%

$\tau(N \rightarrow e^+ K)$ $> 1.3\ (n)$, $> 150\ (p) \times 10^{30}$ years, CL = 90%

$\tau(N \rightarrow \mu^+ K)$ $> 1.1\ (n)$, $> 120\ (p) \times 10^{30}$ years, CL = 90%

limit on $n\overline{n}$ oscillations (bound n)   [m] $>1.2 \times 10^8$ s, CL = 90%

limit on $n\overline{n}$ oscillations (free n) $>0.86 \times 10^8$ s, CL = 90%

Limits are given at the 90% confidence level, while errors are given as ± 1 standard deviation.

ELECTRIC CHARGE (Q)

e mean life / branching fraction	[n]	$>4.3 \times 10^{23}$ yr ΓCL = 68%
$\Gamma(n \to p\nu_e\overline{\nu}_e)/\Gamma_{\text{total}}$		$<8 \times 10^{-27}$ ΓCL = 68%

$\Delta S = \Delta Q$ RULE

Allowed in second-order weak interactions.

$\Gamma(K^+ \to \pi^+\pi^+ e^-\overline{\nu}_e)/\Gamma_{\text{total}}$	$<1.2 \times 10^{-8}$ ΓCL = 90%
$\Gamma(K^+ \to \pi^+\pi^+\mu^-\overline{\nu}_\mu)/\Gamma_{\text{total}}$	$<3.0 \times 10^{-6}$ ΓCL = 95%
$x = \text{A}(\overline{K}^0 \to \pi^-\ell^+\nu) = \text{A}(\Delta S = -\Delta Q)/\text{A}(\Delta S = \Delta Q)$	
real part of x	0.006 ± 0.018 (S = 1.3)
imaginary part of x	-0.003 ± 0.026 (S = 1.2)
$\Gamma(\Sigma^+ \to n\ell^+\nu)/\Gamma(\Sigma^- \to n\ell^-\overline{\nu})$	<0.043
$\Gamma(\Sigma^+ \to ne^+\nu_e)/\Gamma_{\text{total}}$	$<5 \times 10^{-6}$ ΓCL = 90%
$\Gamma(\Sigma^+ \to n\mu^+\nu_\mu)/\Gamma_{\text{total}}$	$<3.0 \times 10^{-5}$ ΓCL = 90%
$\Gamma(\Xi^0 \to \Sigma^- e^+\nu_e)/\Gamma_{\text{total}}$	$<9 \times 10^{-4}$ ΓCL = 90%
$\Gamma(\Xi^0 \to \Sigma^-\mu^+\nu_\mu)/\Gamma_{\text{total}}$	$<9 \times 10^{-4}$ ΓCL = 90%

$\Delta S = 2$ FORBIDDEN

Allowed in second-order weak interactions.

$\Gamma(\Xi^0 \to p\pi^-)/\Gamma_{\text{total}}$	$<4 \times 10^{-5}$ ΓCL = 90%
$\Gamma(\Xi^0 \to pe^-\overline{\nu}_e)/\Gamma_{\text{total}}$	$<1.3 \times 10^{-3}$
$\Gamma(\Xi^0 \to p\mu^-\overline{\nu}_\mu)/\Gamma_{\text{total}}$	$<1.3 \times 10^{-3}$
$\Gamma(\Xi^- \to n\pi^-)/\Gamma_{\text{total}}$	$<1.9 \times 10^{-5}$ ΓCL = 90%
$\Gamma(\Xi^- \to ne^-\overline{\nu}_e)/\Gamma_{\text{total}}$	$<3.2 \times 10^{-3}$ ΓCL = 90%
$\Gamma(\Xi^- \to n\mu^-\overline{\nu}_\mu)/\Gamma_{\text{total}}$	$<1.5 \times 10^{-2}$ ΓCL = 90%
$\Gamma(\Xi^- \to p\pi^-\pi^-)/\Gamma_{\text{total}}$	$<4 \times 10^{-4}$ ΓCL = 90%
$\Gamma(\Xi^- \to p\pi^- e^-\overline{\nu}_e)/\Gamma_{\text{total}}$	$<4 \times 10^{-4}$ ΓCL = 90%
$\Gamma(\Xi^- \to p\pi^-\mu^-\overline{\nu}_\mu)/\Gamma_{\text{total}}$	$<4 \times 10^{-4}$ ΓCL = 90%
$\Gamma(\Omega^- \to \Lambda\pi^-)/\Gamma_{\text{total}}$	$<1.9 \times 10^{-4}$ ΓCL = 90%

$\Delta S = 2$ VIA MIXING

Allowed in second-order weak interactions, e.g. mixing.

$m_{K^0_L} - m_{K^0_S}$	$(0.5301 \pm 0.0014) \times 10^{10} \, \hbar \, \text{s}^{-1}$
$m_{K^0_L} - m_{K^0_S}$	$(3.489 \pm 0.009) \times 10^{-12}$ MeV

$\Delta C = 2$ VIA MIXING

Allowed in second-order weak interactions, e.g. mixing.

$\left|m_{D^0_1} - m_{D^0_2}\right|$	[o]	$<24 \times 10^{10} \, \hbar \, \text{s}^{-1}$ ΓCL = 90%
$\left|\Gamma_{D^0_1} - \Gamma_{D^0_2}\right|/\Gamma_{D^0}$ mean life	[o]	<0.20 ΓCL = 90%
difference/average		
$\Gamma(K^+\ell^-\overline{\nu}_\ell(\text{via } \overline{D}^0))/\Gamma(K^-\ell^+\nu_\ell)$		<0.005 ΓCL = 90%
$\Gamma(K^+\pi^-$ or $K^+\pi^-\pi^+\pi^- (\text{via } \overline{D}^0))/$	[p]	< 0.0085 (or < 0.0037) ΓCL = 90%
$\Gamma(K^-\pi^+$ or $K^-\pi^+\pi^+\pi^-)$		
$\Gamma(D^0 \to K^+\ell^-\overline{\nu}_\ell(\text{via } \overline{D}^0))/\Gamma_{\text{total}}$		$<1.7 \times 10^{-4}$ ΓCL = 90%
$\Gamma(D^0 \to K^+\pi^-$ or $K^+\pi^-\pi^+\pi^-$ (via $\overline{D}^0))/\Gamma_{\text{total}}$		$<1.0 \times 10^{-3}$ ΓCL = 90%

$\Delta B = 2$ VIA MIXING

Allowed in second-order weak interactions, e.g. mixing.

χ_d	0.172 ± 0.010
$\Delta m_{B^0} = m_{B^0_H} - m_{B^0_L}$	$(0.464 \pm 0.018) \times 10^{12} \, \hbar \, \text{s}^{-1}$
$x_d = \Delta m_{B^0}/\Gamma_{B^0}$	0.723 ± 0.032
χ_B at high energy	0.118 ± 0.006
$\Delta m_{B^0_s} = m_{B^0_{sH}} - m_{B^0_{sL}}$	$>9.1 \times 10^{12} \, \hbar \, \text{s}^{-1}$ ΓCL = 95%
$x_s = \Delta m_{B^0_s}/\Gamma_{B^0_s}$	>14.0 ΓCL = 95%
χ_s	>0.4975 ΓCL = 95%

$\Delta S = 1$ WEAK NEUTRAL CURRENT FORBIDDEN

Allowed by higher-order electroweak interactions.

$\Gamma(K^+ \to \pi^+ e^+ e^-)/\Gamma_{\text{total}}$		$(2.74 \pm 0.23) \times 10^{-7}$
$\Gamma(K^+ \to \pi^+\mu^+\mu^-)/\Gamma_{\text{total}}$		$(5.0 \pm 1.0) \times 10^{-8}$
$\Gamma(K^+ \to \pi^+\nu\overline{\nu})/\Gamma_{\text{total}}$		$(4.2^{+9.7}_{-3.5}) \times 10^{-10}$
$\Gamma(K^0_S \to \mu^+\mu^-)/\Gamma_{\text{total}}$		$<3.2 \times 10^{-7}$ ΓCL = 90%
$\Gamma(K^0_S \to e^+ e^-)/\Gamma_{\text{total}}$		$<1.4 \times 10^{-7}$ ΓCL = 90%
$\Gamma(K^0_S \to \pi^0 e^+ e^-)/\Gamma_{\text{total}}$		$<1.1 \times 10^{-6}$ ΓCL = 90%
$\Gamma(K^0_L \to \mu^+\mu^-)/\Gamma_{\text{total}}$		$(7.2 \pm 0.5) \times 10^{-9}$ (S = 1.4)
$\Gamma(K^0_L \to \mu^+\mu^-\gamma)/\Gamma_{\text{total}}$		$(3.25 \pm 0.28) \times 10^{-7}$
$\Gamma(K^0_L \to e^+ e^-)/\Gamma_{\text{total}}$		$<4.1 \times 10^{-11}$ ΓCL = 90%
$\Gamma(K^0_L \to e^+ e^-\gamma)/\Gamma_{\text{total}}$		$(9.1 \pm 0.5) \times 10^{-6}$
$\Gamma(K^0_L \to e^+ e^-\gamma\gamma)/\Gamma_{\text{total}}$	[q]	$(6.5 \pm 1.2) \times 10^{-7}$
$\Gamma(K^0_L \to \pi^+\pi^- e^+ e^-)/\Gamma_{\text{total}}$	[q]	$<4.6 \times 10^{-7}$ ΓCL = 90%
$\Gamma(K^0_L \to \mu^+\mu^- e^+ e^-)/\Gamma_{\text{total}}$		$(2.9^{+6.7}_{-2.4}) \times 10^{-9}$
$\Gamma(K^0_L \to e^+ e^- e^+ e^-)/\Gamma_{\text{total}}$		$(4.1 \pm 0.8) \times 10^{-8}$ (S = 1.2)
$\Gamma(K^0_L \to \pi^0\mu^+\mu^-)/\Gamma_{\text{total}}$		$<5.1 \times 10^{-9}$ ΓCL = 90%
$\Gamma(K^0_L \to \pi^0 e^+ e^-)/\Gamma_{\text{total}}$		$<4.3 \times 10^{-9}$ ΓCL = 90%
$\Gamma(K^0_L \to \pi^0\nu\overline{\nu})/\Gamma_{\text{total}}$		$<5.8 \times 10^{-5}$ ΓCL = 90%
$\Gamma(\Sigma^+ \to pe^+ e^-)/\Gamma_{\text{total}}$		$<7 \times 10^{-6}$

$\Delta C = 1$ WEAK NEUTRAL CURRENT FORBIDDEN

Allowed by higher-order electroweak interactions.

$\Gamma(D^+ \to \pi^+ e^+ e^-)/\Gamma_{\text{total}}$	$<6.6 \times 10^{-5}$ ΓCL = 90%
$\Gamma(D^+ \to \pi^+\mu^+\mu^-)/\Gamma_{\text{total}}$	$<1.8 \times 10^{-5}$ ΓCL = 90%
$\Gamma(D^+ \to \rho^+\mu^+\mu^-)/\Gamma_{\text{total}}$	$<5.6 \times 10^{-4}$ ΓCL = 90%
$\Gamma(D^0 \to e^+ e^-)/\Gamma_{\text{total}}$	$<1.3 \times 10^{-5}$ ΓCL = 90%
$\Gamma(D^0 \to \mu^+\mu^-)/\Gamma_{\text{total}}$	$<4.1 \times 10^{-6}$ ΓCL = 90%
$\Gamma(D^0 \to \pi^0 e^+ e^-)/\Gamma_{\text{total}}$	$<4.5 \times 10^{-5}$ ΓCL = 90%
$\Gamma(D^0 \to \pi^0\mu^+\mu^-)/\Gamma_{\text{total}}$	$<1.8 \times 10^{-4}$ ΓCL = 90%
$\Gamma(D^0 \to \eta e^+ e^-)/\Gamma_{\text{total}}$	$<1.1 \times 10^{-4}$ ΓCL = 90%
$\Gamma(D^0 \to \eta\mu^+\mu^-)/\Gamma_{\text{total}}$	$<5.3 \times 10^{-4}$ ΓCL = 90%
$\Gamma(D^0 \to \rho^0 e^+ e^-)/\Gamma_{\text{total}}$	$<1.0 \times 10^{-4}$ ΓCL = 90%
$\Gamma(D^0 \to \rho^0\mu^+\mu^-)/\Gamma_{\text{total}}$	$<2.3 \times 10^{-4}$ ΓCL = 90%
$\Gamma(D^0 \to \omega e^+ e^-)/\Gamma_{\text{total}}$	$<1.8 \times 10^{-4}$ ΓCL = 90%
$\Gamma(D^0 \to \omega\mu^+\mu^-)/\Gamma_{\text{total}}$	$<8.3 \times 10^{-4}$ ΓCL = 90%
$\Gamma(D^0 \to \phi e^+ e^-)/\Gamma_{\text{total}}$	$<5.2 \times 10^{-5}$ ΓCL = 90%
$\Gamma(D^0 \to \phi\mu^+\mu^-)/\Gamma_{\text{total}}$	$<4.1 \times 10^{-4}$ ΓCL = 90%
$\Gamma(D^0 \to \pi^+\pi^-\pi^0\mu^+\mu^-)/\Gamma_{\text{total}}$	$<8.1 \times 10^{-4}$ ΓCL = 90%
$\Gamma(D^+_s \to K^+\mu^+\mu^-)/\Gamma_{\text{total}}$	$<5.9 \times 10^{-4}$ ΓCL = 90%
$\Gamma(D^+_s \to K^*(892)^+\mu^+\mu^-)/\Gamma_{\text{total}}$	$<1.4 \times 10^{-3}$ ΓCL = 90%
$\Gamma(\Lambda^+_c \to p\mu^+\mu^-)/\Gamma_{\text{total}}$	$<3.4 \times 10^{-4}$ ΓCL = 90%

Limits are given at the 90% confidence level Γ while errors are given as ± 1 standard deviation.

$\Delta B = 1$ WEAK NEUTRAL CURRENT FORBIDDEN

Allowed by higher-order electroweak interactions.

$\Gamma(B^+ \to \pi^+ e^+ e^-)/\Gamma_{\text{total}}$	$<3.9 \times 10^{-3}$, CL = 90%
$\Gamma(B^+ \to \pi^+ \mu^+ \mu^-)/\Gamma_{\text{total}}$	$<9.1 \times 10^{-3}$, CL = 90%
$\Gamma(B^+ \to K^+ e^+ e^-)/\Gamma_{\text{total}}$	$<6 \times 10^{-5}$, CL = 90%
$\Gamma(B^+ \to K^+ \mu^+ \mu^-)/\Gamma_{\text{total}}$	$<1.0 \times 10^{-5}$, CL = 90%
$\Gamma(B^+ \to K^*(892)^+ e^+ e^-)/\Gamma_{\text{total}}$	$<6.9 \times 10^{-4}$, CL = 90%
$\Gamma(B^+ \to K^*(892)^+ \mu^+ \mu^-)/\Gamma_{\text{total}}$	$<1.2 \times 10^{-3}$, CL = 90%
$\Gamma(B^0 \to \gamma\gamma)/\Gamma_{\text{total}}$	$<3.9 \times 10^{-5}$, CL = 90%
$\Gamma(B^0 \to e^+ e^-)/\Gamma_{\text{total}}$	$<5.9 \times 10^{-6}$, CL = 90%
$\Gamma(B^0 \to \mu^+ \mu^-)/\Gamma_{\text{total}}$	$<6.8 \times 10^{-7}$, CL = 90%
$\Gamma(B^0 \to K^0 e^+ e^-)/\Gamma_{\text{total}}$	$<3.0 \times 10^{-4}$, CL = 90%
$\Gamma(B^0 \to K^0 \mu^+ \mu^-)/\Gamma_{\text{total}}$	$<3.6 \times 10^{-4}$, CL = 90%
$\Gamma(B^0 \to K^*(892)^0 e^+ e^-)/\Gamma_{\text{total}}$	$<2.9 \times 10^{-4}$, CL = 90%
$\Gamma(B^0 \to K^*(892)^0 \mu^+ \mu^-)/\Gamma_{\text{total}}$	$<2.3 \times 10^{-5}$, CL = 90%
$\Gamma(B^0 \to K^*(892)^0 \nu\bar{\nu})/\Gamma_{\text{total}}$	$<1.0 \times 10^{-3}$, CL = 90%
$\Gamma(B \to e^+ e^- s)/\Gamma_{\text{total}}$	$<5.7 \times 10^{-5}$, CL = 90%
$\Gamma(B \to \mu^+ \mu^- s)/\Gamma_{\text{total}}$	$<5.8 \times 10^{-5}$, CL = 90%
$\Gamma(\bar{b} \to \mu^+ \mu^- \text{anything})/\Gamma_{\text{total}}$	$<3.2 \times 10^{-4}$, CL = 90%
$\Gamma(B_s^0 \to \mu^+ \mu^-)/\Gamma_{\text{total}}$	$<2.0 \times 10^{-6}$, CL = 90%
$\Gamma(B_s^0 \to e^+ e^-)/\Gamma_{\text{total}}$	$<5.4 \times 10^{-5}$, CL = 90%
$\Gamma(B_s^0 \to \phi\nu\bar{\nu})/\Gamma_{\text{total}}$	$<5.4 \times 10^{-3}$, CL = 90%

NOTES

In this Summary Table:

When a quantity has "(S = ...)" to its right, the error on the quantity has been enlarged by the "scale factor" S, defined as $S = \sqrt{\chi^2/(N-1)}$, where N is the number of measurements used in calculating the quantity. We do this when $S > 1$, which often indicates that the measurements are inconsistent. When $S > 1.25$, we also show in the Particle Listings an ideogram of the measurements. For more about S, see the Introduction.

[a] C parity forbids this to occur as a single-photon process.

[b] Time-reversal invariance requires this to be 0° or 180°.

[c] Allowed by higher-order electroweak interactions.

[d] Violates CP in leading order. Test of direct CP violation since the indirect CP-violating and CP-conserving contributions are expected to be suppressed.

[e] ϵ'/ϵ is derived from $|\eta_{00}/\eta_{+-}|$ measurements using theoretical input on phases.

[f] Neglecting photon channels. See e.g. A. Pais and S.B. Treiman, Phys. Rev. **D12**, 2744 (1975).

[g] Derived from measured values of ϕ_{+-}, ϕ_{00}, $|\eta|$, $|m_{K_L^0} - m_{K_S^0}|$, and $\tau_{K_S^0}$, as described in the introduction to "Tests of Conservation Laws."

[h] The value is for the sum of the charge states of particle/antiparticle states indicated.

[i] A test of additive vs. multiplicative lepton family number conservation.

[j] $\Delta(m^2) = 100 \text{ eV}^2$.

[k] $190 \text{ eV}^2 < \Delta(m^2) < 320 \text{ eV}^2$.

[l] Derived from an analysis of neutrino-oscillation experiments.

[m] There is some controversy about whether nuclear physics and model dependence complicate the analysis for bound neutrons (from which the best limit comes). The second limit here is from reactor experiments with free neutrons.

[n] This is the best "electron disappearance" limit. The best limit for the mode $e^- \to \nu\gamma$ is $> 2.35 \times 10^{25}$ yr (CL=68%).

[o] The D_1^0-D_2^0 limits are inferred from the D^0-$\bar{D}^0$ mixing ratio $\Gamma(K^+ \ell^- \bar{\nu}_\ell (\text{via } \bar{D}^0)) / \Gamma(K^- \ell^+ \nu_\ell)$.

[p] The larger limit (from E791) allows interference between the doubly Cabibbo-suppressed and mixing amplitudes; the smaller limit (from E691) doesn't. See the papers for details.

[q] See the K_L^0 Particle Listings for the energy limits used in this measurement.

TABLE OF THE ISOTOPES

Norman E. Holden

This table presents an evaluated set of values for the experimental quantities that characterize the decay of radioactive nuclides. A list of the major references used in this evaluation is given below. When uncertainties are not listed, they are assumed to be five or less in the last digit quoted. If they exceed five in the last digit, the value is prefaced by an approximate sign. For quasi-stable nuclides, the measured width, Γ, of the resonance is given. To estimate the approximate half-life, the Heisenberg relationship may be used, the half-life = 4.56×10^{-22} seconds / Γ(MeV). The effective literature cutoff date for data in this edition of the Table is December, 2005.

Table Layout

Column No.	Column title	Description
1	Isotope or Element	For elements, the atomic number and chemical symbol are listed. For nuclides, the mass number and chemical symbol are listed. Isomers are indicated by the addition of m, m1, or m2.
2	Isotopic Abundance	In atom percent.
3	Atomic Mass or Atomic Weight	Atomic mass relative to ^{12}C = 12. Atomic weight of elements is given on the same scale.
4	Half-life/Resonance Width	Half-life in decimal notation. μs = microseconds; ms = milliseconds; s = seconds; m = minutes; h = hours; d = days; and y = years. For quasi-stable nuclides, the measured width at half maximum of the energy resonance is given
5	Decay Mode/Energy	Decay modes are α = alpha particle emission; β = negative beta emission; $\beta+$ = positron emission; EC = orbital electron capture; IT = isomeric transition from upper to lower isomeric state; n = neutron emission; sf = spontaneous fission; $\beta\beta$ = double beta decay. Total disintegration energy in MeV units.
6	Particle Energy/Intensity	End point energies of beta transitions and discrete energies of alpha particles in MeV and their intensities in percent.
7	Spin and Parity	Nuclear spin or angular momentum of the nuclides in units of $h/2\pi$; parity is positive or negative.
8	Magnetic Dipole Moment	Magnetic dipole moments in nuclear magneton units.
9	Electric Quadrupole Moment	Electric quadrupole moments in barn units (10^{-24} cm^2).
10	Gamma Ray Energy/Intensity	Gamma ray energies in MeV and intensities in percent. Ann. rad. refers to the 511.006 keV photons emitted in the annihilation of positrons in matter.

General Nuclear Data References

The following references represent the major sources of the nuclear data presented, along with subsequent published journal articles and reports:

1. G. Audi, O. Bersillon, J. Blachot, A.H. Wapstra, *The Nubase Evaluation of Nuclear and Decay Properties,* Nuclear Physics A729, 3 (2003).
2. G. Audi, A.H. Wapstra, C. Thibault, *The AME2003 Atomic Mass Evaluation (II)*, Nuclear Physics A729, 337 (2003).
3. International Commission on Atomic Weights, *Atomic Weights of the Elements - 1999*, Pure & Applied Chemistry 75, 667 (2001).
4. E.M. Baum, H.D. Knox, T.R. Miller, *Chart of the Nuclides, 16th Edition,* Knolls Atomic Power Lab. (2002)
5. N.E. Holden, *Total and Spontaneous Fission Half-lives for Uranium, Plutonium, Americium and Curium Nuclides,* Pure & Applied Chemistry 61, 1483 (1989).
6. N.E. Holden, *Half-lives of Selected Nuclides,* Pure & Applied Chemistry 62, 941 (1990).
7. N.E. Holden, *Review of Thermal Neutron Cross Sections and Isotopic Composition of the Elements*, BNL-NCS-42224 (March 1989).
8. P. Raghavan, *Table of Nuclear Moments*, Atomic Data Nuclear Data Tables 42, 189 (1989).
9. E. Brown, R. Firestone, *Radioactivity Handbook*, Wiley Interscience Press (1986).
10. J.K. Tuli, *Nuclear Wallet Cards*, Brookhaven National Laboratory (April 2005).
11. N.E. Holden, D.C. Hoffman, *Spontaneous Fission Half-lives for Ground State Nuclides*, Pure & Applied Chemistry 72 1525 (2000).
12. N. Stone, *Table of New Nuclear Moments*, private communication, www.nndc.bnl.gov/nndc/stone_moments/moments.html (Dec 2000)

This research was carried out under the auspices of the US Department of Energy Contract No. DE-AC02-98CH10886. The author is at Brookhaven National Laboratory, Upton, NY, and can be contacted at holden@bnl.gov.

Elem. or Isot.	Natural Abundance (Atom %)	Atomic Mass or Weight	Half-life/ Resonance Width (MeV)	Decay Mode/ Energy (/MeV)	Particle Energy/ Intensity (MeV/%)	Spin ($h/2\pi$)	Nuclear Magnetic Mom. (nm)	Elect. Quadr. Mom. (b)	γ-Energy / Intensity (MeV/%)
$_0$n		1.008664916	614. s	β- /0.78235	0.782/100.	1/2+	-1.913043		
				β-, γ	/ < 0.069				
$_1$H		**1.00794(7)**							
^{1}H	99.9885(70)	1.007825032	> 2.8 × 10²³ y			1/2+	+2.79285		
^{2}H	0.0115(70)	2.014101778				1+	+0.85744	+2.86 mb	
^{3}H		3.016049278	12.33 y	β-/0.01859	0.01860/100.	1/2+	+2.97896		
^{4}H		4.0278	Γ ~ 3	n/	/100	2-			
^{5}H		5.0353	Γ < 0.5	n/	/100	(1/2+)			
^{6}H		6.0449	Γ = 1.6(4)	n/		(2-)			
^{7}H		7.053	Γ ~ 20.						
$_2$He		**4.002602(2)**							
^{3}He	0.000134(3)	3.016029319				1/2+	-2.12762		
^{4}He	99.999866(3)	4.002603254				0+			
^{5}He		5.01222	Γ = 0.60(2)	n, α		3/2-			
^{6}He		6.018889	0.807 s	β-/3.508	3.510/100.	0+			
				β, d	/0.00076				
^{7}He		7.02802	Γ = 0.15(2)	n		(3/2)-			
^{8}He		8.03392	0.119 s	β-/10.65	/84.	0+			0.9807/84.
				n/	/16.				0.4776/5.
				β, t	/0.82				
^{9}He		9.04395	Γ = 0.10(6)	n	/100	(1/2-)			
^{10}He		10.0524	Γ = 0.3(2)	2n	/100	0+			
$_3$Li		**6.941(2)**							
^{4}Li		4.0272	Γ = 6.0	p/	/100	2-			
^{5}Li		5.01254	Γ = 1.2	p/α		3/2-			
^{6}Li	7.59(4)	6.01512280				1+	+0.82205	-0.8 mb	
^{7}Li	92.41(4)	7.0160046				3/2-	+3.25644	-0.0400	
^{8}Li		8.0224874	0.840 s	β-/16.004	12.5/100.	2+	+1.6536	+0.0314	
				α/	α(1.6)				
^{9}Li		9.026790	0.178 s	β-/13.606	13.5/75.	3/2-	3.4368	-0.0306	
				β-/	11/25.				
^{10}Li		10.03548	Γ = 0.11(5)	n	/7.	1+			
^{11}Li		11.04380	8.8 ms	β-/20.6	/8.3	3/2(-)	3.668	-0.031	3.368/33.
				β-, n	/85.7				0.320/7.
				β-, 2n	/4.1				2.590/8.
				β-, 3n	/1.9				5.958/3.
				β-, d	/>0.01				2.895/1.5
				β-, t	/0.02				2.811/1.1
^{12}Li		12.054	< 0.01 μs						
$_4$Be		**9.012182(3)**							
^{5}Be		5.041		p, ³He		(1/2+)			
^{6}Be		6.01973	Γ = 0.092(6)	2p,α		0+			
^{7}Be		7.0169298	53.28 d	EC/0.8618		3/2-	-1.4		0.4776/10.4
^{8}Be		8.00530510	Γ = 6.8(17)eV	2α/0.046		0+			
^{9}Be	100.	9.0121822				3/2-	-1.1776	+0.0529	
^{10}Be		10.0135338	1.52 × 10⁶ y	β-/0.5559	0.555/100.	0+			
^{11}Be		11.02166	13.8 s	β-, β-α/11.51	11.48/61.	1/2+			2.125/35.5
^{12}Be		12.02692	22.0 ms	β-, (n)/11.71	n//0.5	0+			(0.95 - 4.4)
^{13}Be		13.0357	Γ ~ 1.						
^{14}Be		14.0429	4.6 ms	β-/16.2		0+			3.5346/0.9
				β-, n	0.288/94.				3.6845/7.
				β-, 2n	/6.				
				β-, α	/<0.012				
				β-, t	/<0.04				
^{15}Be		15.053	< 0.2 μs	β-					

Elem. or Isot.	Natural Abundance (Atom %)	Atomic Mass or Weight	Half-life/ Resonance Width (MeV)	Decay Mode/ Energy (/MeV)	Particle Energy/ Intensity (MeV/%)	Spin ($h/2\pi$)	Nuclear Magnetic Mom. (nm)	Elect. Quadr. Mom. (b)	γ-Energy / Intensity (MeV/%)
^{16}Be		16.062	< 0.2 μs	β-		0+			
$_5$B		**10.811(7)**							
^{7}B		7.0299	Γ = 1.4(2)	p, α		(3/2-)			
^{8}B		8.024607	0.770 s	β+, 2α/17.979	13.7(β+)/93.	2+	1.0355	0.068	ann.rad.
^{9}B		9.013329	Γ = 0.5(2) keV	p, 2α/		3/2-			
^{10}B	19.9(7)	10.0129370				3+	+1.8006	+0.085	
^{11}B	80.1(7)	11.0093054				3/2-	+2.6886	+0.0406	
^{12}B		12.014352	0.0202 s	β-/13.369		1+	+1.0027	0.0132	4.438/1.3
				β- α/1.6/					3.215/0.00065
^{13}B		13.017780	0.0174 s	β- /13.437	13.4	3/2-	+3.1778	0.037	3.68/7.6
				β- n/0.25/	2.43(n)/0.09				
					3.55(n)/0.16				
^{14}B		14.02540	14. ms	β-/20.64		2-	1.185	0.0298	6.094/90.
^{15}B		15.03110	9.9 ms	β-, (n)/19.09	n//99.7	(3/2-)	2.66	0.038	
^{16}B		16.0398	Γ < 0.1	n					
^{17}B		17.0470	5.1 ms	β-, (n)/22.7			2.54	0.039	
^{18}B		18.056	< 0.026 μs			0-			
^{19}B		19.0637	2.9 ms	β-, (n)/26.5	1n//72.	(3/2-)			
					2n//16.				
					3n// < 9.				
$_6$C		**12.0107(8)**							
^{8}C		8.03768	Γ = 0.25(4)	p		0+			
^{9}C		9.031037	127. ms	β+, p, 2α/16.498		(3/2-)	-1.391		ann.rad.
^{10}C		10.0168532	19.3 s	β+/3.648	1.865	0+			ann.rad.
									0.71829/100.
^{11}C		11.011434	20.3 m	β+, EC/1.982	0.9608/99.	3/2-	-0.964	0.0333	ann.rad.
^{12}C	98.93(8)	12.000000000				0+			
^{13}C	1.07(8)	13.003354838				½-	+0.70241		
^{14}C		14.003241989	5715. y	β-/0.15648	0.1565/100.	0+			
^{15}C		15.010599	2.45 s	β-/9.772	4.51/68.	½+	1.32		5.298/68.
					9.82/32.				(7.30−9.05)
^{16}C		16.014701	~ 0.750 s	β-/8.012	β/3.3, 4.3/84, 16	0+			
				β, n	n/0.8, 1.7/84, 16				
^{17}C		17.02259	0.19 s	β-/13.17		3/2+			1.375
				β-, n	n/1.6−3.7/11.				1.849
									1.906
^{18}C		18.02676	0.092 s	β-/11.81		0+			
				β-, n	n/0.88−4.59/21.				
^{19}C		19.0348	0.05 s	n		½+			
^{20}C		20.0403	0.02 s	β,n	1n// ~ 65.	0+			
					2n// < 19.				
^{21}C		21.049	< 0.03 μs						
^{22}C		22.057	6 ms	β-, n	1n// ~ 61.	0+			
					2n// < 37.				
$_7$N		**14.0067(2)**							
^{10}N		10.0417	Γ = 2.3(16)						
^{11}N		11.02609	Γ ~ 1.			½+			
^{12}N		12.018613	11.00 ms	β+, β+α/17.338	16.38/95.	1+	+0.457	+10. mb	ann.rad.
									4.438/2.
^{13}N		13.0057386	9.97 m	β+ /2.2204	1.190/100.	½-	0.3222		
^{14}N	99.636(20)	14.003074005				1+	+0.40376	+0.0200	
^{15}N	0.364(20)	15.00010898				½-	-0.28319		
^{16}N		16.006102	7.13 s	β- /10.419	4.27/68.	2-			6.129/68.8
					10.44/26.		1.986	18 mb	7.115/4.7
				β-, α	1.85/.0012				(0.99−8.87)
^{17}N		17.00845	4.17 s	β-, β- n/8.68	3.7/100.	½-	0.352		0.871/3.

Elem. or Isot.	Natural Abundance (Atom %)	Atomic Mass or Weight	Half-life/ Resonance Width (MeV)	Decay Mode/ Energy (/MeV)	Particle Energy/ Intensity (MeV/%)	Spin ($h/2\pi$)	Nuclear Magnetic Mom. (nm)	Elect. Quadr. Mom. (b)	γ-Energy / Intensity (MeV/%)
				0.4−1.7n/95.					2.1842/0.3
				β- α/	8.0, 8.2				
^{18}N		18.01408	0.62 s	β- /13.90	9.4/100.	1-	0.328	0.012	0.822/48.
				β-, α	/12.				1.65/47.
				β-,n	/14				1.982/77.
									(0.535−7.13)
^{19}N		19.01703	0.32 s	β-/12.53			< 0.32		(0.096−3.14)
^{20}N		20.0234	0.14 s	β- /17.97					
^{21}N		21.0271	0.08 s						
^{22}N		22.0344	0.02 s	β-,n	1n// ~ 41.				
					2n// < 13.				
^{23}N		23.0412	15. ms	β-, n	n// ~ 42.				
					2n// ~ 8.				
					3n// < 3.4				
^{24}N		24.0510	< 0.052 μs						
^{25}N		25.061	< 0.26 μs						
$_8$O		**15.9994(3)**							
^{12}O		12.03441	Γ = 0.51(16)	2p		0+			
^{13}O		13.02481	8.9 ms	β+, p/17.77	1.560 (p)	(3/2-)	1.389	0.011	ann.rad.
					p/(1.00 - 13.5)				4.438/0.56
^{14}O		14.0085963	70.63 s	β+ /5.1430	1.81/99.	0+			ann.rad.
									2.312/99.4
^{15}O		15.0030656	122.2 s	β+ /2.754	1.723/100.	1/2-	0.7195		ann.rad.
^{16}O	99.757(16)	15.9949146196				0+			
^{17}O	0.038(1)	16.9991317				5/2+	-1.8938	-0.026	
^{18}O	0.205(14)	17.999161				0+			
^{19}O		19.003580	26.9 s	β- /4.820	3.25/60.	5/2+	1.5320	3.7 mb	0.197/95.9
					4.60/40.				1.3569/50.4
									(0.11−4.18)
^{20}O		20.004077	13.5 s	β- /3.814		0+			1.057/100.
^{21}O		21.00866	3.4 s	β- /8.11					(0.28−4.6)
^{22}O		22.0100	2.2 s	β- /6.5		0+			0.072/100
									0.638/98
									1.862/63
									(0.918-2.499)
^{23}O		23.0157	0.08 s						
^{24}O		24.0205	~ 65. ms	β-, n	n//18.	0+			1.83/28.
									0.52/14.
									1.31/12.
^{25}O		25.0295	< 0.05 μs						
^{26}O		26.0383	< 0.04 μs			0+			
^{27}O		27.048	< 0.026 μs						
^{28}O		28.058	< 0.10 μs			0+			
$_9$F		**18.9984032(5)**							
^{14}F		14.0351							
^{15}F		15.0180	Γ = 0.8(3)	p		(1/2+)			
^{16}F		16.01147	Γ = 0.037(14)	p		0-			
^{17}F		17.0020952	64.5 s	β+ /2.761	1.75/	5/2+	+4.721	0.058	ann.rad.
^{18}F		18.000938	1.829 h	β+, EC/1.656	0.635/97.	1+			ann.rad.
^{19}F	100.	18.9984032				½+	+2.62887	0.072	
^{20}F		19.9999813	11.00 s	β- /7.0245	5.398/100.	2+	+2.0934	0.042	1.634/100.
									3.33/0.009
^{21}F		20.999949	4.16 s	β- /5.684	3.7/8.	5/2+	3.9		0.3507/90.
					5.0/63.				1.395/15.
					5.4/29.				(1.746−4.684)
^{22}F		22.00300	4.23 s	β- /10.82	3.48/15.	4+			1.2746/100.
					4.67/7.				2.0826/82.

Elem. or Isot.	Natural Abundance (Atom %)	Atomic Mass or Weight	Half-life/ Resonance Width (MeV)	Decay Mode/ Energy (/MeV)	Particle Energy/ Intensity (MeV/%)	Spin ($h/2\pi$)	Nuclear Magnetic Mom. (nm)	Elect. Quadr. Mom. (b)	γ-Energy / Intensity (MeV/%)
					5.50/62.				(0.82–4.37)
^{23}F		23.0036	2.2 s	β- /8.5		5/2+			1.701/48.
									2.129/34.
									(0.493−3.83)
^{24}F		24.0081	0.3 s	β- /13.5					1.9816/
^{25}F		25.0121	~ 50. ms	β-, (n)	n//14.				1.70/39.
									(0.57−2.19)
^{26}F		26.0196	10. ms	β-, (n)	n//11.				2.02/67.
									1.67/19.
^{27}F		27.0268	5.0 ms	β-, (n)	n//90.				2.02/18.
^{28}F		28.036	< 0.04 μs						
^{29}F		29.043	2.5 ms	β-, (n)	n//100.				
^{30}F		30.053	< 0.26 μs						
^{31}F		31.060	> 0.26 μs						
$_{10}$**Ne**		**20.1797(6)**							
^{16}Ne		16.02576	Γ = 0.12(4)	2p		0+			
^{17}Ne		17.01767	109. ms	β+, p/14.53	1.4−10.6/6.9	1/2-	0.787		ann.rad./
				β+, α	/0.014				0.495
^{18}Ne		18.0057082	1.668 s	β+ /4.446	3.416/92.	0+			ann.rad./
									1.0413/7.8
									(0.658−1.70)
^{19}Ne		19.0018802	17.22 s	β+ /3.238	2.24/99.	1/2+	-1.885		ann.rad./
									(0.11−1.55)
^{20}Ne	90.48(3)	19.992440175				0+			
^{21}Ne	0.27(1)	20.99384668				3/2+	-0.66180	+0.103	
^{22}Ne	9.25(3)	21.99138511				0+		-0.19	
^{23}Ne		22.9944669	37.2 s	β- /4.376	3.95/32.	5/2+	-1.08	+0.15	0.440/33.
					4.39/67.				(1.64−2.98)
^{24}Ne		23.9936108	3.38 m	β- /2.47	1.10/8.	0+			0.4723/100.
					1.98/92.				0.874/7.9
^{25}Ne		24.99774	0.61 s	β- /7.30	6.3/	1/2+	-1.006		0.0895/96.
					7.3/				(0.98−3.69)
^{26}Ne		26.00046	197 ms	β-, n/7.3	n//0.13	0+			0.082/100
									1.278/6
									0.233/5
									0.151/3
									1.211/1
									2.489/1
^{27}Ne		27.0076	31. ms	β-, n/12.7	n//2.	(3/2+)			
^{28}Ne		28.0121	19. ms	β-, n/12.3	n//12.	0+			2.06/19.
					2n//3.				0.86/3.
^{29}Ne		29.0194	15. ms	β-, (n)/15.4	n//29.	(3/2+)			2.92/55.
				β-, 2n	2n//4.				(0.22−1.18)
^{30}Ne		30.025	7. ms	β-, (n)	n//9.	0+			0.151/9.
^{31}Ne		31.033	3. ms						
^{32}Ne		32.040	~ 3.5 ms			0+			
^{33}Ne		33.049	< 0.26 μs						
^{34}Ne		34.057	> 1.5 μs			0+			
$_{11}$**Na**		**22.98976928(2)**							
^{18}Na		18.02597	Γ = 0.34(9)						
^{19}Na		19.01388	0.03 s	β+, p/11.18					
^{20}Na		20.00735	0.446 s	β+ /13.89		2+	+0.3694	~ + 0.04	ann.rad./
				α	2.15/				1.634/79.
^{21}Na		20.997655	22.48 s	β+ /3.547	2.50/95.	3/2+	+2.3863	~ +0.05	ann.rad./
									0.351/5.
^{22}Na		21.9944364	2.605 y	β+ /90/2.842	0.545/90.	3+	+1.746	+0.19	ann.rad./
				EC/10/					1.2745/99.9

Elem. or Isot.	Natural Abundance (Atom %)	Atomic Mass or Weight	Half-life/ Resonance Width (MeV)	Decay Mode/ Energy (/MeV)	Particle Energy/ Intensity (MeV/%)	Spin ($h/2\pi$)	Nuclear Magnetic Mom. (nm)	Elect. Quadr. Mom. (b)	γ-Energy / Intensity (MeV/%)
^{23}Na	100.	22.989769281				3/2+	+2.21752	+0.106	
^{24m}Na			20.2 ms	I.T., β-		1+			0.4723/100.
^{24}Na		23.9909628	14.96 h	β- /5.5158	1.389/>99.	4+	+1.690		1.3686/100.
									2.754/100.
									(0.997−4.238)
^{25}Na		24.989954	59.3 s	β- /3.835	2.6/7.	5/2+	+3.683	~ −0.06	0.3897/12.7
					3.15/25.				0.5850/13.
					4.0/65.				0.9747/14.9
									(0.836−2.80)
^{26}Na		25.99263	1.071 s	β- /9.31		3+	+2.851	-5.3 mb	1.809/98.9
									(0.24-7.37)
^{27}Na		26.994077	0.290 s	β- /9.01	7.95/	5/2+	+3.90	-7.2 mb	0.9847/87.4
				β-, n/					1.698/11.9
^{28}Na		27.99894	31. ms	β- /14.0	12.3/	1+	+2.42	+0.04	1.473/37.
				β-, n/					2.389/18.6
^{29}Na		29.00286	44. ms	β-, n/13.3	11.5/	3/2+	+2.46	+86. mb	2.560/36.
									(1.04−3.99)
^{30}Na		30.00898	50. ms	β-, n/17.5	n//30.	2+	+2.07		1.483/46.
^{31}Na		31.0136	17.2 ms	β-, n/15.9	n//37.	3/2-	+2.30		1.483/14.
									(0.05−3.54)
^{32}Na		32.0205	13.5 ms	β- /19.1					0.240−3.935
^{33}Na		33.027	8.0 ms	β- /20.	/ ~ 38				0.886/16
				β-, n	0.8,1.02/47(6)				0.546/6.4
				β-, 2n	/13(3)				0.050−2.55
^{34}Na		34.035	5. ms	β- /24.					
^{35}Na		35.042	1.5 ms	β- /24					
^{36}Na		36.051	< 0.26 μs						
^{37}Na		37.059	> 1.5 μs						
$_{12}$Mg		**24.3050(6)**							
^{19}Mg		19.0355	< 0.02 μs						
^{20}Mg		20.01886	96. ms	β+ /10.73	/70	0+			
				β+, p	/30				
^{21}Mg		21.01171	122. ms	β+, p/13.10		5/2+			0.332/51.
^{22}Mg		21.999574	3.876 s	β+ /4.786	3.05/	0+			0.0729/60.
									0.5820/100.
									(1.28−1.93)
^{23}Mg		22.994124	11.32 s	β+ /4.057	3.09/92.	3/2+	0.536	1.25	0.440/8.2
^{24}Mg	78.99(4)	23.98504170				0+			
^{25}Mg	10.00(1)	24.98583692				5/2+	-0.85545	+0.200	
^{26}Mg	11.01(3)	25.98259293				0+			
^{27}Mg		26.98434059	9.45 m	β- /2.6103	1.59/41.	1/2+			0.17068/0.9
					1.75/58.				0.84376/72.
					2.65/0.3				1.01443/28.
^{28}Mg		27.983877	20.9 h	β- /1.832	0.459/95.	0+			0.0306/95.
									0.4006/36.
									0.9418/36.
									1.342/54.
^{29}Mg		28.98860	1.3 s	β- /7.55	5.4/	3/2+			0.960/15.
									1.398/16.
									2.224/36.
^{30}Mg		29.99043	0.32 s	β- /7.0		0+			0.224/85.
^{31}Mg		30.99655	0.24 s	β- /11.7	8.4/29.9	(3/2+)	-0.8836		1.613/47.
									0.947/37.
									(0.666-4.640)
				β-, n	/1.7				
^{32}Mg		31.99898	0.12 s	β- /10.3		0+			2.765/25.
^{33}Mg		33.00525	91. ms	β- /13.7	/83.				1.848/
				β-, n	/17.				

Elem. or Isot.	Natural Abundance (Atom %)	Atomic Mass or Weight	Half-life/ Resonance Width (MeV)	Decay Mode/ Energy (/MeV)	Particle Energy/ Intensity (MeV/%)	Spin ($h/2\pi$)	Nuclear Magnetic Mom. (nm)	Elect. Quadr. Mom. (b)	γ-Energy / Intensity (MeV/%)
^{34}Mg		34.0095	0.02 s	β- /11.3		0+			
^{35}Mg		35.0173	0.07 s			(7/2-)			
^{36}Mg		36.023	4. ms			0+			
^{37}Mg		37.031	> 0.26 µs			(7/2-)			
^{38}Mg		38.038	> 0.26 µs			0+			
^{39}Mg		39.047	< 0.26 µs						
^{40}Mg		40.054				0+			
$_{13}$Al		26.9815386(8)							
^{21}Al		21.0280	< 0.035 µs						
^{22}Al		22.0195	59. ms	β+ /18.6	p/1.3/18.	4+			ann.rad./
				β+, p, 2p, α/	α/3.3/0.3				
^{23m}Al			~ 0.35 s	β+, p/0.17					0.554
									0.839
^{23}Al		23.00727	0.47 s	β+ /12.24					ann.rad./
				β+, p/					
^{24m}Al			0.129 s	I.T./0.4259					
				β+	13.3	1+			1.3686/5.3
^{24}Al		23.999939	2.07 s	β+ /13.878,p	3.40/48.	4+			1.078(2)/16.
					4.42/41.				1.368(2)/96.
					6.80/3.				2.753(2)/43.
					8.74/8.				4.315(3)/15.
									5.392(3)/20.
									7.0662(2)/41.
^{25}Al		24.9904281	7.17 s	β+ /4.277	3.27/	5/2+	3.646		ann.rad./
									1.6115(2)/100.
									0.975(2)/5.
^{26m}Al			6.345 s	β+ /	3.2/	0+			ann.rad./
^{26}Al		25.9868917	7.1×10^5 y	β+ /82/4.0042	1.16/	5+	+2.804	+0.17	ann.rad./
				EC/18					1.8087/99.8
^{27}Al	100.	26.9815386				5/2+	+3.64151	+0.140	
^{28}Al		27.9819103	2.25 m	β- /4.6422	2.865/100.	3+	3.24	0.18	1.7778(6)/100.
^{29}Al		28.980445	6.5 m	β- /3.680	1.4/30.	5/2+			1.2732(8)/89.
					2.5/70.				2.0282(8)/4.
									2.4262(8)/7.
^{30}Al		29.98296	3.68 s	β- /8.56	5.05/	3+	3.01		1.26313(3)/35.
									2.23525(5)/65.
^{31}Al		30.98395	0.64 s	β- /8.00	6.25/	5/2+			0.75223(3)/18.
									1.69473(3)/59.
									2.31664(4)/73.
^{32}Al		31.9881	33. ms	β- /13.0		1+	1.96		
^{33}Al		32.9908	41.7 ms	β- /12.0	/91.5				1.940/2.5
				β-, n	/8.5				(1.01−4.34)
^{34}Al		33.9969	56. ms	β- /17.1	4.255/44	4			0.929/57
				β-, n	/26.				(0.12−4.26)
^{35}Al		34.9999	38. ms	β-/14.3	0.974/48	5/2+			0.064/45.
				β-, n	/ 38.				(0.12−5.63)
^{36}Al		36.0062	0.09 s	β-/18.3					
				β-, n	/<31.				
^{37}Al		37.0107	11. ms	β-/16.					
^{38}Al		38.017	> 7.6 ms						
^{39}Al		39.023	> 8. ms						
^{40}Al		40.031	> 0.26 µs						
^{41}Al		41.038	> 0.26 µs						
$_{14}$Si		28.0855(3)							
^{22}Si		22.0345	29. ms	β+, p	1.99/20	0+			
^{23}Si		23.0255	40.7 ms	β+, p/5.9	1.32,(0.6−11.6)				
^{24}Si		24.01155	0.14 s	β+, p/10.81	1.44,3.92,1.09	0+			ann.rad./

Elem. or Isot.	Natural Abundance (Atom %)	Atomic Mass or Weight	Half-life/ Resonance Width (MeV)	Decay Mode/ Energy (/MeV)	Particle Energy/ Intensity (MeV/%)	Spin ($h/2\pi$)	Nuclear Magnetic Mom. (nm)	Elect. Quadr. Mom. (b)	γ-Energy / Intensity (MeV/%)
					(1.66−4.47)				
25Si		25.00411	221 ms	β+, p/12.74	p/4.25/9.5	5/2+			ann.rad./
					p/0.40/4.75				
					p/0.56-6.80				
26Si		25.992330	2.23 s	β+ /5.066	3.282/	0+			ann.rad./
									0.8294(8)/22.
27Si		26.9867049	4.14 s	β+ /4.8118	3.85/100.	5/2+	-0.8554		ann.rad./
									2.211(5)/0.2
28Si	92.223(19)	27.976926533				0+			
29Si	4.685(8)	28.97649470				1/2+	-0.5553		
30Si	3.092(11)	29.97377017				0+			
31Si		30.97536323	2.62 h	β- /1.4920	1.471/99.9	3/2+			1.2662(5)/0.05
32Si		31.97414808	1.6×10^2 y	β- /0.224	0.213/100.	0+			
33Si		32.97800	6.1 s	β- /5.85	3.92	(3/2+)	1.21		1.4313(5)/13.
									1.8477/100.
									2.538(2)/10.
34Si		33.97858	2.8 s	β- /4.60	3.09/	0+			0.42907(5)/60.
									1.17852(2)/64.
									1.60756(5)/36.
35Si		34.98458	0.9 s	β- /10.50					
36Si		35.9866	0.5 s	β- /7.9		0+			
				β-, n	/~ 12.				
37Si		36.9929	~ 0.09 s	β- /12.5					
				β-, n	/~ 17.				
38Si		37.9956	> 1 μs	β- /10.7		0+			
				β-, n					
39Si		39.0021	48. ms	β- /14.8					
40Si		40.006	33. ms			0+			
41Si		41.015	20. ms						
42Si		42.020	13. ms			0+			
43Si		43.029	> 0.26 μs						
15P		30.973762(2)							
24P		24.034							
25P		25.0203	< 0.03 μs						
26P		26.0118	44. ms	β+, p/18.1	p/0.41/18.0	3+			
					p/1.98/2.4				
					p/0.78-7.49				
27P		26.99923	0.3 s	β+, p/11.63	p/0.73, 0.61/0.07	1/2+			
28P		27.992315	270. ms	β+ /14.332	3.94/13.	3+			ann.rad./
					5.25/13.				1.779(2)/98.
					6.96/16.				2.839(2)/2.8
					8.8/7.				3.040(2)/3.2
					11.49/52.				4.498(2)/12.
									7.537(2)/9.
29P		28.981801	4.14 s	β+ /4.9431	3.945/98.	1/2+	1.2349		ann.rad./
									1.273/1.32
									2.426/0.39
30P		29.9783138	2.50 m	β+ /4.2323	3.245/99.9	1+			ann.rad./
									2.230(3)/0.07
31P	100.	30.9737616				1/2+	+1.13160		
32P		31.9739073	14.28 d	β- /1.7106	1.710/100.	1+	-0.2524		
33P		32.971726	25.3 d	β- /0.249	0.249/100.	1/2+			
34P		33.973636	12.4 s	β- /5.374	3.2/15.	1+			1.78−4.1/
					5.1/85.				2.127(5)/15.
35P		34.973314	47. s	β- /3.989	2.34/100.	1/2+			1.572(1)/100.
36P		35.97826	5.7 s	β- /10.41					0.902/77.
									3.291/100.
37P		36.97961	2.3 s	β- /7.90					0.6462/

Elem. or Isot.	Natural Abundance (Atom %)	Atomic Mass or Weight	Half-life/ Resonance Width (MeV)	Decay Mode/ Energy (/MeV)	Particle Energy/ Intensity (MeV/%)	Spin ($h/2\pi$)	Nuclear Magnetic Mom. (nm)	Elect. Quadr. Mom. (b)	γ-Energy / Intensity (MeV/%)
									1.5829/
^{38}P		37.9842	0.6 s	β- /12.4					1.2923/
				β-, n	/~ 12.				2.224/
^{39}P		38.9862	0.3 s	β- /10.5					
				β-, n	/26				
^{40}P		39.9913	0.15 s	β- /14.5					
				β-,n	/~ 30.				
^{41}P		40.9943	0.10 s	β-/~ 13.8					
				β-, n	/~ 30.				
^{42}P		42.0010	49. ms	β-/17.					
				β-, n	/ ~ 50.				
^{43}P		43.006	36. ms	β-/16.					
				β-, n	/100.				
^{44}P		44.013	19. ms						
^{45}P		45.019	> 0.2 μs						
^{46}P		46.027	> 0.2 μs						
$_{16}$S		**32.065(5)**							
^{26}S		26.0279	~ 10 ms			0$^+$			
^{27}S		27.0188	16. ms	β+, 2p/18.3	p/2.26, 7.80				
^{28}S		28.0044	0.13 s			0$^+$			
^{29}S		28.99661	0.188 s	β+ /13.79		5/2+			ann.rad./
				β+, p/					
^{30}S		29.984903	1.18 s	β+ /6.138	4.42/78.	0+			ann.rad./
					5.08/20.				0.678/79.
^{31}S		30.979555	2.56 s	β+ /5.396	4.39/99.	1/2+	0.48793		ann.rad./
									1.2662(5)/1.2
^{32}S	94.99(26)	31.9720710				0+			
^{33}S	0.75(2)	32.9714588				3/2+	+0.64382	-0.068	
^{34}S	4.25(24)	33.9678669				0+			
^{35}S		34.9690322	87.2 d	β- /0.1672	0.1674/100.	3/2+	+1.00	+0.047	
^{36}S	0.01(1)	35.9670808				0+			
^{37}S		36.9711256	5.05 m	β- /4.8653	1.64/94.	7/2-			0.9083(4)/0.06
					4.75/5.6				3.1033(2)/94.2
^{38}S		37.97116	2.84 h	β- /2.94	1.00/	0+			0.1962(4)/0.2
									1.9421(3)/84.
^{39}S		38.97513	11.5 s	β- /6.64					1.301/52.
									1.697/44.
^{40}S		39.9755	9. s	β- /4.7		0+			
^{41}S		40.9796	~ 2.6 s	β- /8.7					
				β-, n					
^{42}S		41.9810	~ 0.56 s	β- /7.8		0+			
				β-, n	/ < 4.				
^{43}S		42.9872	0.26 s	β-/12.					
				β-, n	/ ~ 40				
^{44}S		43.9902	0.10 s	β-/9.		0+			
				β-, n	/18.				
^{45}S		44.997	68. ms	β-/14.					
				β-, n	/54.				
^{46}S		46.001	0.05 s			0+			
^{47}S		47.009	> 0.2 μs						
^{48}S		48.014	> 0.2 μs			0+			
^{49}S		49.024	< 0.2 μs						
$_{17}$Cl		**35.453(2)**							
^{28}Cl		28.029							
^{29}Cl		29.0141	< 0.02 μs						
^{30}Cl		30.0048	< 0.03 μs						
^{31}Cl		30.99241	0.15 s	β+, p/11.98	0.986, 1.52/0.7	3/2+			ann.rad./

Elem. or Isot.	Natural Abundance (Atom %)	Atomic Mass or Weight	Half-life/ Resonance Width (MeV)	Decay Mode/ Energy (/MeV)	Particle Energy/ Intensity (MeV/%)	Spin ($h/2\pi$)	Nuclear Magnetic Mom. (nm)	Elect. Quadr. Mom. (b)	γ-Energy / Intensity (MeV/%)
^{32}Cl		31.98569	297. ms	β+ /12.69	9.47/50.	1+	1.11		ann.rad./
				β+, α	/0.05				2.2305/92
				β+, p	/0.026				(1.55−4.77)
^{33}Cl		32.9774519	2.511 s	β+ /5.583	4.51/98.	3/2+	+0.755		ann.rad./
									0.8409/0.52
									1.966/0.45
									2.866/0.44
^{34m}Cl			32.2 m	β+/	1.35/24.	3+			ann.rad./
					2.47/28.				
				I.T./					0.1457(8)/42.
									2.1276(5)/42.
^{34}Cl		33.9737628	1.528 s	β+ /5.4922	4.50/100.	0+			ann.rad./
^{35}Cl	75.76(10)	34.96885268				3/2+	+0.82187	−0.0825	
^{36}Cl		35.9683070	3.01×10^5 y	β- /0.7086	0.7093/98.	0+	+1.28547	−0.018	
				β+, EC/1.1421	0.115/0.002				ann.rad./
^{37}Cl	24.24(10)	36.96590259				3/2+	+0.68412	−0.0649	
^{38m}Cl			0.715 s	I.T./		5-			0.6714/100
^{38}Cl		37.9680104	37.2 m	β- /4.9168	1.11/31.	2-	2.05		1.64216(1)/33.3
					2.77/11.				2.16760(2)/44.8
					4.91/58.				
^{39}Cl		38.968008	55.6 m	β- /3.442	1.91/85.	3/2+			0.25026(1)/47.
					2.18/8.				1.26720(5)/54.
					3.45/7.				0.986−1.517
^{40}Cl		39.97042	1.38 m	β- /7.48		2-			0.6431(3)/6.
									1.4608(1)/77.
									2.8402(2)/17.
^{41}Cl		40.9707	34. s	β- /5.7	3.8/				(0.167−1.359)
^{42}Cl		41.9733	6.8 s	β- /9.4					
^{43}Cl		42.9741	3.3 s	β- /8.0					
^{44}Cl		43.9783	∼ 0.43 s	β-/12.3					
				β-, n	/ < 8.				
^{45}Cl		44.9803	0.40 s	β-/11.					
				β-, n	/24.				
^{46}Cl		45.984	0.23 s	β-/14.9					
				β-, n	/ ∼ 60				
^{47}Cl		46.989	0.10 s	β-/15.					
				β-, n	/ < 3.				
^{48}Cl		47.995	> 0.2 μs						
^{49}Cl		48.000	> 0.17 s						
^{50}Cl		50.008							
^{51}Cl		51.014	> 0.2 μs						
$_{18}$Ar		39.948(1)							
^{30}Ar		30.0216	< 0.02 μs			0+			
^{31}Ar		31.0121	∼ 14.1 ms	β+ /18.4	p/2.08/100.	5/2+			
				β+, p	/55.				
				β+, 2p	/2.5				
				β+, 3p	/0.11				
^{32}Ar		31.997638	98. ms	β+ /11.2	p/3.35, 0.6−5.55	0+			ann.rad./
				β+, p	/43.				
^{33}Ar		32.9899257	174. ms	β+ /11.62	3.17,2.10	1/2+	−0.72		ann.rad./
				β+, p/	(1.32−5.72)				0.810(2)/48.
^{34}Ar		33.9802712	0.844 s	β+/6.061	5.0/95.	0+			ann.rad./
									0.6658(1)/2.5
									3.1290(1)/1.3
^{35}Ar		34.975258	1.77 s	β+/5.965	4.94/93.	3/2+	+0.633	−0.08	ann.rad./
									1.2185(5)/1.22
									1.763(1)/0.25

Elem. or Isot.	Natural Abundance (Atom %)	Atomic Mass or Weight	Half-life/ Resonance Width (MeV)	Decay Mode/ Energy (/MeV)	Particle Energy/ Intensity (MeV/%)	Spin ($h/2\pi$)	Nuclear Magnetic Mom. (nm)	Elect. Quadr. Mom. (b)	γ-Energy / Intensity (MeV/%)
									2.964(1)/0.2
^{36}Ar	0.3365(30)	35.96754511				0+			
^{37}Ar		36.9667763	35.0 d	EC/.813		3/2+	+1.15	+0.076	
^{38}Ar	0.0632(5)	37.9627324				0+			
^{39}Ar		38.964313	268. y	β-/0.565	0.565/100.	7/2-	-1.59	-0.12	
^{40}Ar	99.6003(30)	39.962383123				0+			
^{41}Ar		40.9645006	1.82 h	β-/2.492	1.198/	7/2-			1.29364(5)/99.
									1.6770(3)/0.05
^{42}Ar		41.96305	33. y	β-/0.60	0.60/100.	0+			
^{43}Ar		42.96564	5.4 m	β-/4.6					0.4791(2)/10.
									0.7380(1)/43.
									0.9752(1)/100.
									1.4400(3)/39.
^{44}Ar		43.964924	11.87 m	β-/3.55		0+			0.182–1.866
^{45}Ar		44.968040	21.5 s	β-/6.9		7/2-			0.0610/25.
									1.020/35.
									3.707/34.
^{46}Ar		45.96809	8.4 s	β-/5.70		0+			1.944/
^{47}Ar		46.9722	1.23 s	β-					0.36/100
									1.66/53
									1.74/41
									(2.02 - 4.01)
^{48}Ar		47.9745	0.48 s			0+			
^{49}Ar		48.981	0.17 s	β-,n	n// ~ 65.				
^{50}Ar		49.984	~ 0.085 s	β-,n	n// ~ 35.	0+			
^{51}Ar		50.992	> 0.2 µs	β-					
^{52}Ar		51.997	10 ms	β-		0+			
^{53}Ar		53.005		β-					
$_{19}$K		**39.0983(1)**							
^{32}K		32.022							
^{33}K		33.0073	< 0.025 µs						
^{34}K		33.9984	< 0.04 µs						
^{35}K		34.98801	0.19 s	β+ /11.88		3/2+			ann.rad./
				β+, p/	/0.37				1.751/14.
									2.5698/26.
									2.9827/51.
^{36}K		35.98129	0.342 s	β+ /12.81	5.3/42.	2+	+0.548		ann.rad./
					9.9/44.				1.97044(5)/82.
				β+, p	/0.048				2.20783(5)/30.
									2.43343(2)/32.
^{37}K		36.9733759	1.23 s	β+ /6.149	5.13/	3/2+	+0.2032		ann.rad./
									2.7944(8)/2.
									3.602(2)/0.05
^{38m}K			0.924 s	β+ /6.742	5.02/100.	0+			ann.rad./
^{38}K		37.9690812	7.63 m	β+ /5.913	2.60/99.8	3+	+1.37		ann.rad./
									2.1675(3)/99.8
									3.9356(5)/0.2
^{39}K	93.2581(44)	38.9637067				3/2+	+0.39146	+0.049	
^{40}K	0.0117(1)	39.9639985	1.248×10^9 y	β- /1.3111	1.312/89.	4-	-1.29810	-0.061	ann.rad./
				β+, EC/1.505	1.50/10.7				1.4608/10.5
^{41}K	6.7302(44)	40.9618258				3/2+	+0.21487	+0.060	
^{42}K		41.9624028	12.36 h	β- /3.525	1.97/19.	2-	-1.1425		0.31260(2)/0.3
					3.523/81.				1.5246(3)/18.1
^{43}K		42.96072	22.3 h	β- /1.82	0.465/8.	3/2+	+0.163		0.2211(2)/4.
					0.825/87.				0.3729(2)/88.
					1.24/3.5				0.3971(2)/11.
					1.814/1.3				0.6178(2)/81.
^{44}K		43.96156	22.1 m	β- /5.66	5.66/34.	2-	-0.856		0.36821/2.2

Elem. or Isot.	Natural Abundance (Atom %)	Atomic Mass or Weight	Half-life/ Resonance Width (MeV)	Decay Mode/ Energy (/MeV)	Particle Energy/ Intensity (MeV/%)	Spin ($h/2\pi$)	Nuclear Magnetic Mom. (nm)	Elect. Quadr. Mom. (b)	γ-Energy / Intensity (MeV/%)
									1.15700(1)/58.
									2.15079(2)/22.
^{45}K		44.96070	17.8 m	β- /4.20	1.1/23.	3/2+	+0.173		0.1743(5)/80.
					2.1/69.				1.2607(8)/7.
					4.0/8.				1.7056(6)/69.
									2.3542(5)/14.
^{46}K		45.96198	1.8 m	β- /7.72	6.3/	2-	-1.05		1.347(1)/91.
									3.700(5)/28.
^{47}K		46.96168	17.5 s	β- /6.64	4.1/99.	½+	+1.93		0.56474(3)/15.
					6.0/1.				0.58575(3)/85.
									2.0131/100
^{48}K		47.96551	6.8 s	β- /12.09	5.0/	(2-)			0.67122(1)/4.
									0.6723(5)/20.
									0.78016(1)/32.
									3.83153(7)/80.
^{49}K		48.9675	1.26 s	β- /11.0					2.025/
									2.252/
^{50}K		49.9728	0.472 s	β- /14.2					
^{51}K		50.976	0.365 s	β- /					
^{52}K		51.983	0.105 s	β-					
^{53}K		52.987	30. ms	β-		3/2+			
^{54}K		53.994	10. ms	β-					
$_{20}$Ca		**40.078(4)**							
^{34}Ca		34.0141	< 0.035 μs			0+			
^{35}Ca		35.0049	25.7 ms	β+, p/15.6	p/1.43/49				
					1.9−8.8				
^{36}Ca		35.99309	0.10 s	β+, (p)/10.99	2.52	0+			ann.rad./
				β+, n/					
^{37}Ca		36.98587	0.18 s	β+ /11.64	3.103	3/2+			ann.rad./
				β+, n/					1.369
^{38}Ca		37.976318	0.44 s	β+ /6.74		0+			ann.rad./
									1.5677(5)/25.
									3.210(2)/1.
^{39}Ca		38.970720	0.861 s	β+ /6.531	5.49/100.	3/2+	1.02168		ann.rad./
^{40}Ca	96.941(156)	39.9625910				0+			
^{41}Ca		40.9622781	1.02×10^5 y	EC/0.4214		7/2-	-1.5948	-0.090	
^{42}Ca	0.647(23)	41.9586180				0+			
^{43}Ca	0.135(10)	42.9587666				7/2-	-1.3173	-0.055	
^{44}Ca	2.086(110)	43.9554818				0+			
^{45}Ca		44.9561866	162.7 d	β- /0.257	0.257/100.	7/2-	-1.327	+0.05	
^{46}Ca	0.004(3)	45.953693	> 0.4×10^{16} y	β-β-		0+			
^{47}Ca		46.954546	4.536 d	β- /1.992	0.684/84.	7/2-	-1.38	+0.02	1.297/75
					1.98/16.				(0.041−1.88)
^{48}Ca	0.187(21)	47.952534	4.3×10^{19} y	β-β-		0+			
			> 7.1×10^{19} y	β-					
^{49}Ca		48.955674	8.72 m	β- /5.262	0.89/7.	3/2-			3.0844(1)/90.7
					1.95/92.				4.0719(1)/8.12
									(0.143 - 4.738)
^{50}Ca		49.95752	14. s	β- /4.97	3.12/	0+			0.2569/98.
									(0.0715−1.59)
^{51}Ca		50.9615	10. s	β- /7.3		(3/2-)			
^{52}Ca		51.965	4.6 s	β- /8.0		0+			
^{53}Ca		52.9701	0.09 s	β- /10.9					
^{54}Ca		53.974	> 0.3 μs			0+			
^{55}Ca		54.981	> 0.3 μs						
^{56}Ca		55.986	> 0.3 μs			0+			
$_{21}$Sc		**44.955912(6)**							

Elem. or Isot.	Natural Abundance (Atom %)	Atomic Mass or Weight	Half-life/ Resonance Width (MeV)	Decay Mode/ Energy (/MeV)	Particle Energy/ Intensity (MeV/%)	Spin $(h/2\pi)$	Nuclear Magnetic Mom. (nm)	Elect. Quadr. Mom. (b)	γ-Energy / Intensity (MeV/%)
^{36}Sc		36.0149	0.102 s						
^{37}Sc		37.0031	0.181 s						
^{38}Sc		37.9947	< 0.3 μs						
^{39}Sc		38.98479	< 0.3 μs	p					
^{40}Sc		39.977967	0.182 s	β+ /14.320	5.73/50.	4-			ann.rad./
					7.53/15.				0.752/41.
					8.76/15.				3.732/99.5
					9.58/20.				(1.12–3.92)
^{41}Sc		40.9692511	0.596 s	β+ /6.4953	5.61/100.	7/2-	+5.431	-0.156	ann.rad./
^{42m}Sc			61.6 s	β+ /	2.82/	7+			ann.rad./
									0.4375(5)/100.
									1.2270(5)/100.
									1.5245(5)/100.
^{42}Sc		41.9655164	0.682 s	β+ /6.4259	5.32/100.	0+			ann.rad./
^{43}Sc		42.961151	3.89 h	β+, EC/2.221	0.82/22.	7/2-	+4.62	-0.26	ann.rad./
					1.22/78.				0.3729(1)/22.
^{44m}Sc			58.2 h	I.T./0.27		6+	+3.88		0.27124(1)/87.
				EC/3.926					(1.00–1.16)
^{44}Sc		43.959403	3.93 h	β+, EC/3.653	1.47/	2+	+2.56	+0.10	ann.rad./
									1.157/100
^{45}Sc	100.	44.955912				7/2-	+4.75649	-0.220	
^{46m}Sc			18.7 s	I.T./0.14253		1-			0.14253(2)/62.
^{46}Sc		45.955172	83.81 d	β- /2.367	0.357/100.	4+	+3.03	+0.12	0.8893/100
									1.121/100
^{47}Sc		46.952408	3.349 d	β- /0.600	0.439/69.	7/2-	+5.34	-0.22	0.15938(1)/68.
					0.601/31.				
^{48}Sc		47.95223	43.7 h	β- /3.99	0.655/	6+			0.9835/100
									1.03750(1)/97.
									1.3121/100
^{49}Sc		48.950024	57.3 m	β- /2.006	2.00/99.9.	7/2-			1.7619(3)/0.05
^{50}Sc		49.95219	1.71 m	β- /6.89	3.05/76.	(5+)			0.5235(1)/88.
					3.60/24.				1.1210(1)/100.
									1.5537(2)/100.
^{51}Sc		50.95360	12.4 s	β- /6.51	4.4/	7/2-			1.4373(4)/52.
					5.0/				0.718–2.144
^{52}Sc		51.9567	8.2 s	β- /9.0		(3+)			
^{53}Sc		52.9596	> 3. ms	β- /8.1					
^{54m}Sc			~ 7 μs	I.T.		(5+)			0.110/IT
^{54}Sc		53.9633	0.27 s	β- /11.6					0.100/50
									1.70/40
									0.50/40
^{55}Sc		54.968	0.103 s	β- /13					0.593(1)/40
^{56m}Sc			0.06 s						1.161/21
									0.690/19
^{56}Sc		55.973	35. ms	β-		(1+)			1.129/48
^{57}Sc		56.978	13. ms	β-					
^{58}Sc		57.984	12. ms	β-					
$_{22}$Ti		47.867(1)							
^{38}Ti		38.0098	< 0.12 μs			0+			
^{39}Ti		39.0016	30. ms	β+ /15.4					
^{40}Ti		39.9905	54. ms	β+ /11.7	p/2.16/29	0+			
				β+, p	3.73/23				
					1.70/22				
					0.242–5.74				
^{41}Ti		40.9832	80. ms	β+, p/12.93	p/4.73/107	3/2+			ann.rad./
					3.10/67				
					3.75/39				
					0.744–6.73				

Elem. or Isot.	Natural Abundance (Atom %)	Atomic Mass or Weight	Half-life/ Resonance Width (MeV)	Decay Mode/ Energy (/MeV)	Particle Energy/ Intensity (MeV/%)	Spin ($h/2\pi$)	Nuclear Magnetic Mom. (nm)	Elect. Quadr. Mom. (b)	γ-Energy / Intensity (MeV/%)
^{42}Ti		41.97303	0.20 s	β+ /7.000	6.0/	0+			ann.rad./
									0.6107(5)/56.
^{43}Ti		42.96852	0.50 s	β+ /6.87	5.80/	7/2-	0.85		ann.rad./
^{44}Ti		43.959690	60. y	EC/0.268		0+			0.06787/91
									0.07832/97
^{45}Ti		44.958126	3.078 h	β+/86/2.062	1.04	7/2-	0.095	0.015	ann.rad./
				EC/14/					(0.36−1.66)
^{46}Ti	8.25(3)	45.952632				0+			
^{47}Ti	7.44(2)	46.951763				5/2-	-0.78848	+0.30	
^{48}Ti	73.72(3)	47.947946				0+			
^{49}Ti	5.41(2)	48.947870				7/2-	-1.10417	+0.24	
^{50}Ti	5.18(2)	49.944791				0+			
^{51}Ti		50.946615	5.76 m	β- /2.471	1.50/92.	3/2-			0.3197(2)/93.
					2.13/				0.6094−0.9291
^{52}Ti		51.94690	1.7 m	β- /1.97	1.8/100.	0+			0.0170(5)/100.
									0.1245/100
^{53}Ti		52.9497	33. s	β- /5.0	(2.2−3)/	3/2-			0.1008(1)/20.
									0.1276(1)/45.
									0.2284(1)/39.
									1.6755(5)/45.
									(1.72−2.8)/
^{54}Ti		53.9511	1.5 s	β- /4.3		0+			
^{55}Ti		54.9553	1.3 s	β- /7.4					0.672/44
									(0.32 - 1.83)
^{56}Ti		55.9582	0.20 s	β- /7.0		0+			
^{57}Ti		56.9640	98. ms	β- /11.					
^{58}Ti		57.967	53. ms	β-		0+			0.114
^{59}Ti		58.973	30. ms	β-					
^{60}Ti		59.978	22. ms	β-		0+			
^{61}Ti		60.983	> 0.3 μs						
$_{23}$V		50.9415(1)							
^{40}V		40.0111							
^{41}V		40.9998							
^{42}V		41.9912	< 0.055 μs						
^{43}V		42.9807	> 0.8 s	β+ /11.3					
^{44}V		43.9741	0.09 s	β+, α/13.7					ann.rad./
^{45}V		44.96578	0.54 s	β+ /7.13		7/2-			
^{46}V		45.960201	0.4223 s	β+ /7.051	6.03/100.	0+			ann.rad./
^{47}V		46.954909	32.6 m	β+, EC/2.928	1.90/99.+	3/2-			ann.rad./
									1.7949(8)/0.19
									(0.2−2.16)
^{48}V		47.952254	15.98 d	β+ /4.012	0.698/50.	4+	2.01		ann.rad./
									0.9835/100
									(1.3−2.4)
^{49}V		48.948516	337. d	EC/0.602		7/2-	4.47		
^{50}V	0.250(4)	49.947159	1.4×10^{17} y	EC	/82.7	6+	+3.34569	+0.21	
				β-	/17.3				
^{51}V	99.750(4)	50.943960				7/2-	+5.148706	-0.04	
^{52}V		51.944776	3.76 m	β- /3.976	2.47/	3+			1.4341(1)/100.
^{53}V		52.944338	1.56 m	β- /3.436	2.52/	7/2-			1.0060(5)/90.
									1.2891(3)/10.
^{54m}V			0.9 μs	I.T.		(5+)			0.108/IT
^{54}V		53.94644	49.8 s	β- /7.04	1.00/5.	3+			0.8348/97.
					2.00/12.				0.9887/80.
					2.95/45.				2.259/46.
					5.20/11.				(0.56−3.38)
^{55}V		54.9472	6.5 s	β- /6.0	6.0/	(7/2-)			0.5177/73.
									(0.224−1.21)

Elem. or Isot.	Natural Abundance (Atom %)	Atomic Mass or Weight	Half-life/ Resonance Width (MeV)	Decay Mode/ Energy (/MeV)	Particle Energy/ Intensity (MeV/%)	Spin ($h/2\pi$)	Nuclear Magnetic Mom. (nm)	Elect. Quadr. Mom. (b)	γ-Energy / Intensity (MeV/%)
^{56}V		55.9505	0.22 s	β- /9.1					1.01/30.
									0.688/26.
									(0.82 − 1.32)
^{57}V		56.9526	0.35 s	β- /8.1					0.268/52.
									0.692/20.
									(0.25 − 1.31)
^{58}V		57.9567	0.19 s	β- /11.6					0.880/62
									1.056/28
									2.217/13
									(1.04 − 1.57)
^{59}V		58.9602	97. ms	β- /9.9					0.90/80.
^{60m}V			0.12 s						
^{60}V		59.9650	0.07 s	β- /14.					0.102−0.208
^{61}V		60.9685	47. ms						(0.071-1.144)
^{62}V		61.9738	34. ms						
^{63}V		62.978	17. ms						
^{64}V		63.983	> 0.3 µs						
$_{24}$Cr		**51.9961(6)**							
^{42}Cr		42.0064	13. ms	β+, p	p/1.90/29	0+			
					p/1.50−3.7				
^{43}Cr		42.9977	21. ms	β+, p	p/3.83/18				
					p/4.29/15				
					p/1.01−4.59				
^{44}Cr		43.98555	53. ms	β+, (p)/10.3	p/0.95−3.1	0+			
^{45}Cr		44.9796	0.05 s	β+, p/12.5		7/2-			ann.rad./
^{46}Cr		45.96836	0.3 s	β+ /7.60		0+			ann.rad./
^{47}Cr		46.96290	0.51 s	β+ /7.45		3/2-			ann.rad./
^{48}Cr		47.95403	21.6 h	EC/1.66		0+			ann.rad./
									0.116(2)/95.
									0.305(10)/100.
^{49}Cr		48.951336	42.3 m	β+, EC/2.631	1.39/	5/2-	0.476		ann.rad./
					1.45/				0.09064(1)/51.
					1.54/				0.15293(1)/27.
									(0.062-1.6)
^{50}Cr	4.345(13)	49.946044	> 1.3 × 10^{18} y	β+EC		0+			
^{51}Cr		50.944767	27.70 d	EC/0.7527		7/2-	-0.934		0.3201/10.2
^{52}Cr	83.789(18)	51.940508				0+			
^{53}Cr	9.501(17)	52.940649				3/2-	-0.47454	-0.15	
^{54}Cr	2.365(7)	53.938880				0+			
^{55}Cr		54.940840	3.497 m	β- /2.603	2.5/	3/2-			1.5282(2)/0.04
									(0.13−2.37)
^{56}Cr		55.940653	5.9 m	β- /1.62	1.50/100.	0+			0.026(2)/100.
									0.083(3)/100.
^{57}Cr		56.943613	21. s	β- /5.1	3.3/	3/2-	0.0834		0.850/8.
					3.5/				(0.083-2.62)
^{58}Cr		57.9444	7.0 s	β- /4.0		0+			(0.131−0.683)
^{59m}Cr			0.10 ms	I.T.		(9/2+)			0.208/IT
									0.193
									0.102
^{59}Cr		58.9486	1.0 s	β- /7.7					1.236
^{60}Cr		59.9500	0.6 s	β- /6.0		0+			
^{61}Cr		60.9547	0.26 s	β- /8.8					0.354−1.860
^{62}Cr		61.9566	0.19 s	β- /7.3		0+			(0.156-1.215)
^{63}Cr		62.9619	0.129 s	β-					(0.250-3.454)
^{64}Cr		63.9644	0.043 s	β-		0+			0.188
^{65}Cr		64.9702	0.027 s	β-					0.272, 1.368
^{66}Cr		65.973	0.01 s	β-		0+			
^{67}Cr		66.980	> 0.3 µs						

Elem. or Isot.	Natural Abundance (Atom %)	Atomic Mass or Weight	Half-life/ Resonance Width (MeV)	Decay Mode/ Energy (/MeV)	Particle Energy/ Intensity (MeV/%)	Spin ($h/2\pi$)	Nuclear Magnetic Mom. (nm)	Elect. Quadr. Mom. (b)	γ-Energy / Intensity (MeV/%)
$_{25}$Mn		54.938045(5)							
^{44}Mn		44.0069	< 0.105 μs						
^{45}Mn		44.9945	< 0.07 μs						
^{46}Mn		45.9867	34. ms	β+ /17.1					
				β+, p	// ~ 58				
^{47}Mn		46.9761	~ 0.1 s	β+ /12.3					
^{48}Mn		47.9685	0.15 s	β+ /13.5	5.79/58.	4+			
					4.43/10.				
^{49}Mn		48.95962	0.38 s	β+ /7.72	6.69/	5/2-			ann.rad./
^{50m}Mn			1.74 m	β+ /7.887	3.54/	5+			ann.rad./
									1.0980/94.
									0.783/91.
									(0.66−3.11)
^{50}Mn		49.954238	0.283 s	β+ /7.6330	6.61/	0+			ann.rad./
^{51}Mn		50.948211	46.2 m	β+, EC/3.208	2.2/	5/2-	3.568	0.4	ann.rad./
									0.7491(1)/0.26
									(1.148−1.164)
^{52m}Mn			21.1 m	β+ /98/5.09	2.631/	2+	0.0076		ann.rad./
				I.T./2/0.378					0.3778 (I.T.)
									1.43406(1)/98.
									(0.7−4.8)
^{52}Mn		51.945566	5.591 d	β+ /4.712	0.575/	6+	+3.063	+0.5	ann.rad./
				EC/					0.74421(1)/90.
									1.4341/100
^{53}Mn		52.941290	3.7×10^6 y	EC/0.5970		7/2-	5.024		
^{54}Mn		53.940359	312.1 d	EC/1.377		3+	+3.282	+0.33	0.8340/100
			6.7×10^8 y	β+	//1.3 × 10^{-7}				
^{55}Mn	100.	54.938045				5/2-	+3.4687	+0.32	
^{56}Mn		55.938905	2.579 h	β- /3.6954	0.718/18.	3+	+3.2266		0.84675/98.9
					1.028/34.				1.81072(4)/26.3
									2.113/13.8
									(1.04 − 3.37)
^{57}Mn		56.938285	1.45 m	β- /2.691		5/2-			
^{58}Mn		57.93998	65 s	β- /6.25	3.8/	3+			0.45916(2)/20.
					5.1/				0.81076(1)/82.
									1.32309(5)/53.
^{59}Mn		58.94044	4.6 s	β- /5.19	4.5/	5/2-			0.726/
									0.473/
									0.287−2.35
^{60m}Mn			1.77 s	β- /IT	5.7/	3+			0.824/
^{60}Mn		59.9429	50. s	β- /8.6		0+			1.969/
^{61}Mn		60.9447	0.67 s	β- /7.4		(5/2)-			
^{62}Mn		61.9484	0.67 s	β- /10.4		(3+)			0.877/
									0.942−1.299
^{63}Mn		62.9502	0.28 s	β- /8.8					0.356,0.450
^{64m}Mn			> 0.1 ms						0.135/IT
^{64}Mn		63.9543	87 ms	β- /11.8					0.746
^{65}Mn		64.9563	0.092 s	β- /10.					0.366
^{66}Mn		65.9611	64 ms						0.471
^{67}Mn		66.9641	45 ms						
^{68}Mn		67.969	~ 28 ms						
^{69}Mn		68.973	14 ms						
$_{26}$Fe		55.845(2)							
^{45}Fe		45.0146	1.8 ms	2p /1.14	p// ~ 59.				
^{46}Fe		46.0008	11. ms	β+ /13.1	p// ~ 36.	0+			
^{47}Fe		46.9929	21.7 ms	β+ /15.6	p//87.				

Elem. or Isot.	Natural Abundance (Atom %)	Atomic Mass or Weight	Half-life/ Resonance Width (MeV)	Decay Mode/ Energy (/MeV)	Particle Energy/ Intensity (MeV/%)	Spin ($h/2\pi$)	Nuclear Magnetic Mom. (nm)	Elect. Quadr. Mom. (b)	γ-Energy / Intensity (MeV/%)
^{48}Fe		47.9805	~ 44. ms	β+ /11.2		0+			
^{49}Fe		48.9736	70. ms	β+ /13.0		(7/2-)			ann.rad./
^{50}Fe		49.9630	0.15 s	β+ /8.2		0+			0.651
^{51}Fe		50.95682	0.31 s	β+ /8.02		(5/2-)			ann.rad./
^{52m}Fe			46. s	β+ /4.4		(12+)			ann.rad./
									(0.622–2.286)/
^{52}Fe		51.94811	8.28 h	β+ /57/2.37	0.804/	0+			ann.rad./
				EC/43/					0.16868(1)/99.
				I.T./					0.377 (I.T.)/
^{53m}Fe			2.6 m	I.T./3.0407		19/2-			0.7011(1)/99.
									1.0115(1)/87.
									1.3281(1)/87.
									2.3396(1)/13.
^{53}Fe		52.945308	8.51 m	β+ /3.743	2.40/42.	7/2-			ann.rad./
					2.80/57.				0.3779(1)/42.
									(1.2–3.2)
^{54}Fe	5.845(35)	53.939611	> 3.1 × 10^{22} y	EC-EC		0+			
^{55}Fe		54.938293	2.73 y	EC/0.2314		3/2-			Mn x-ray
^{56}Fe	91.754(36)	55.934938				0+			
^{57}Fe	2.119(10)	56.935394				½-	+0.0906	0.16	
^{58}Fe	0.282(4)	57.933276				0+			
^{59}Fe		58.934876	44.51 d	β- /1.565	0.273/48.	3/2-	-0.336		1.099/57
					0.475/51.				1.292/43.
									(0.14–1.48)
^{60}Fe		59.934072	1.5 × 10^6 y	β- /0.237	0.184/100.	0+			0.0586/100
^{61m}Fe			0.25 μs	I.T.		(9/2+)			0.654/IT
									0.207
^{61}Fe		60.93675	6.0 m	β- /3.98	2.5/13.				1.205/44.
					2.63/54.				1.028/43.
					2.80/31.				(0.12–3.37)
^{62}Fe		61.93677	68. s	β- /2.53	2.5/100.	0+			0.5061(1)/100.
^{63}Fe		62.9404	6. s	β- /6.3		5/2-			0.995/
									(1.365–1.427)
^{64}Fe		63.9412	2.0 s	β- /4.9		0+			
^{65m}Fe			0.4 μs	I.T.		(5/2-)			0.364/IT
^{65}Fe		64.9454	1.3 s	β- /7.9					
^{66}Fe		65.9468	0.44 s	β- /5.7		0+			0.471–1.425
^{67m}Fe			~ 0.04 ms	I.T.		(5/2-)			0.367/IT
^{67}Fe		66.9510	0.48 s	β- /8.8					0.189
^{68}Fe		67.954	0.19 s	β- / ~ 7.6		0+			
^{69}Fe		68.959	0.11 s						
^{70}Fe		69.961	0.10 s			0+			
^{71}Fe		70.967	> 0.3 μs						
^{72}Fe		71.970	> 0.3 μs			0+			
$_{27}$Co		58.933195(5)							
^{47}Co		47.0115							
^{48}Co		48.0018							
^{49}Co		48.9897	< 0.035 μs						
^{50}Co		49.9815	44. ms	β+ /17.0	2.03–2.79				
^{51}Co		50.9707	> 0.2 μs	β+ /12.8					
^{52}Co		51.9636	0.12 s	β+ /14.0					0.849–1.942
^{53m}Co			0.25 s	β+, p/		19/2-			ann.rad./
^{53}Co		52.95422	0.24 s	β+ /8.30		7/2-			ann.rad./
^{54m}Co			1.46 m	β+ /8.44	4.25/100.	7+			ann.rad./
									0.411(1)/99.
									1.130(1)/100.
									1.408(1)/100.
^{54}Co		53.948460	0.1932 s	β+ /8.2430	7.34/100.	0+			ann.rad./

Elem. or Isot.	Natural Abundance (Atom %)	Atomic Mass or Weight	Half-life/ Resonance Width (MeV)	Decay Mode/ Energy (/MeV)	Particle Energy/ Intensity (MeV/%)	Spin ($h/2\pi$)	Nuclear Magnetic Mom. (nm)	Elect. Quadr. Mom. (b)	γ-Energy / Intensity (MeV/%)
^{55}Co		54.941999	17.53 h	β+ /3.4513	0.53/	7/2-	+4.822		ann.rad./
				EC/	1.03/				0.9312/75.
					1.50/				0.4772/20.
									(0.092−3.11)
^{56}Co		55.939839	77.3 d	β+/4.566	1.459/18.	4+	3.85	+0.25	ann.rad./
				EC/					0.8468/99.9
									1.2383/68.
									(0.26−3.61)
^{57}Co		56.936291	271.8 d	EC/0.8361		7/2-	+4.72	+0.5	0.12206/86
									(0.014−0.706)
^{58m}Co			9.1 h	I.T./		5+			0.02489/0.035
^{58}Co		57.935753	70.88 d	β+ /2.307		2+	+4.04	+0.22	ann.rad./
				EC/					0.81076/99
^{59}Co	100.	58.933195				7/2-	+4.63	+0.41	
^{60m}Co			10.47 m	I.T./99.8/0.059		2+	+4.40	+0.3	0.0586/2.0
				β- /0.2/1.56					
^{60}Co		59.933817	5.271 y	β- /2.824	0.315/99.7	5+	+3.799	+0.44	1.1732/100
									1.3325/100
^{61}Co		60.932476	1.650 h	β- /1.322	1.22/95.	7/2-			0.0674/86.
									0.842−0.909
^{62m}Co			13.9 m	β- /	0.88/25.	5+			1.1635(3)/70.
					2.88/75.				1.1730(3)/98.
									2.0039(3)/19.
^{62}Co		61.93405	1.50 m	β- /5.32	1.03/10.	2+			1.1292(3)/13.
					1.76/5.				1.1730(3)/83.
					2.9/20.				1.9851(1)/3.
					4.05/60.				2.3020(1)/19.
^{63}Co		62.93361	27.5 s	β- /3.67	3.6/	7/2-			0.08713(1)/49.
									0.9817(3)/2.6
									0.156−2.17
^{64}Co		63.93581	0.30 s	β- /7.31	7.0/	1+			
^{65}Co		64.93648	1.14 s	β- /5.96		(7/2)-			
^{66m2}Co			> 0.1 ms	I.T.		(8-)			0.252/IT
									0.214
									0.175
^{66m1}Co			1.2 μs	I.T.		(5+)			0.175/IT
^{66}Co		65.9398	0.25 s	β- /10.0					(1.245−1.425)
^{67}Co		66.9409	0.43 s	β- /8.4					0.694
^{68}Co		67.9449	0.19 s	β- /11.7					
^{69}Co		68.9463	0.20 s	β- /9.3					
^{70}Co		69.951	0.12 s	β- 13.					1.26/102
									0.97/100
									(0.45 − 0.92)
^{71}Co		70.953	97. ms	β-					0.566/100
				β-,n	// > 3				(0.25 − 0.77)
^{72}Co		71.958	60. ms	β-					1.096/100
				β-,n	// > 6				0.845
									(0.455 − 1.197)
^{73}Co		72.960	41. ms	β-					0.524/100
				β-,n	// > 9				(0.24 − 0.76)
^{74}Co		73.965	30. ms	β-					0.739
				β-,n	// > 26				1.024
^{75}Co		74.968	> 0.3 μs						
$_{28}$Ni		58.6934(2)							
^{48}Ni		48.020	~ 2.1 ms	2p	p// ~ 25	0+			
^{49}Ni		49.0097	12. ms						
^{50}Ni		49.9959	12. ms	β+, p	p//70.	0+			
^{51}Ni		50.9877	> 0.2 μs	β+ /16.0					

Elem. or Isot.	Natural Abundance (Atom %)	Atomic Mass or Weight	Half-life/ Resonance Width (MeV)	Decay Mode/ Energy (/MeV)	Particle Energy/ Intensity (MeV/%)	Spin ($h/2\pi$)	Nuclear Magnetic Mom. (nm)	Elect. Quadr. Mom. (b)	γ-Energy / Intensity (MeV/%)
^{52}Ni		51.9757	38. ms	β+ /11.7		0+			
^{53}Ni		52.9685	0.05 s	β+, p/13.3		7/2-			ann.rad./
^{54}Ni		53.95791	0.10 s	β+ /8.80		0+			0.937
^{55}Ni		54.95133	0.20 s	β+ /8.70	7.66/	7/2-			ann.rad./
^{56}Ni		55.94213	6.08 d	EC/2.14		0+			0.15838/99
				β+ /<10^{-6}					0.81185(3)/87.
									0.2695−0.7500
^{57}Ni		56.939794	35.6 h	β+ /3.264	0.712/10.	3/2-	-0.798		ann.rad./
				EC/	0.849/76.				1.3776/78.
									(0.127−3.177)
^{58}Ni	68.0769(89)	57.935343	>4 × 10^{19} y	EC-EC		0+			
^{59}Ni		58.934347	~ 7.6 × 10^4 y	EC/		3/2-			
^{60}Ni	26.2231(77)	59.930786				0+			
^{61}Ni	1.1399(6)	60.931056				3/2-	-0.75002	+0.16	
^{62}Ni	3.6345(17)	61.928345				0+			
^{63}Ni		62.929669	100. y	β- /0.066945	0.065/	½-			
^{64}Ni	0.9256(9)	63.927966				0+			
^{65}Ni		64.930084	2.517 h	β- /2.137	0.65/30.	5/2-	0.69		0.36627(3)/5.
					1.020/11.				1.11553(4)/16.
					2.140/58.				1.48184(5)/23.
^{66}Ni		65.929139	54.6 h	β- /0.23		0+			
^{67m}Ni			13.3 μs	I.T.		9/2+			0.313/IT
									0.694
^{67}Ni		66.931569	21. s	β- /3.56	3.8/	½-	+0.601		1.0722/100.
									1.6539/100.
									(0.10−1.98)
^{68m2}Ni			0.34 μs						0.511
^{68m1}Ni			0.86 ms	I.T.		(5-)			0.814/IT
									2.033
^{68}Ni		67.931869	29. s	β- /2.06		0+			
^{69m2}Ni			0.44 μs	I.T.		(17/2)			0.148/IT
									0.593
									1.959
^{69m1}Ni			3.5 s						
^{69}Ni		68.935610	11. s	β- /5.4					0.6807(3)/100.
									(0.207−1.213)
^{70m}Ni			0.21 μs	I.T.		(8+)			0.183/IT
									0.448
									0.970
									1.259
^{70}Ni		69.9365	6.0 s	β- /3.5		0+			
^{71}Ni		70.9407	2.56 s	β- /6.9					
^{72}Ni		71.9421	1.6 s	β- /5.2		0+			
^{73}Ni		72.9465	0.84 s	β- /9.					
^{74}Ni		73.9481	0.9 s	β- /7.		0+			
^{75}Ni		74.9529	0.34 s						
^{76}Ni		75.955	0.24 s			0+			
^{77}Ni		76.961	0.13 s						
^{78}Ni		77.963	~ 0.11 s			0+			
$_{29}$Cu		**63.546(3)**							
^{52}Cu		51.9972							
^{53}Cu		52.9856	< 0.3 μs						
^{54}Cu		53.9767	< 0.075 μs						
^{55}Cu		54.9661	> 0.2 μs	β+ /13.2					
^{56}Cu		55.9586	93. ms	β+ /15.3					0.511/233
									2.700/100
									0.9507−3.287
^{57}Cu		56.94921	196. ms	β+ /8.77		3/2-			0.77−3.01

Elem. or Isot.	Natural Abundance (Atom %)	Atomic Mass or Weight	Half-life/ Resonance Width (MeV)	Decay Mode/ Energy (/MeV)	Particle Energy/ Intensity (MeV/%)	Spin ($h/2\pi$)	Nuclear Magnetic Mom. (nm)	Elect. Quadr. Mom. (b)	γ-Energy/ Intensity (MeV/%)
[58]Cu		57.944539	3.21 s	β+ /8.563	4.5/15.	1+			ann.rad./
				EC/	7.439/83.				0.0403(4)/5.
									1.4483(2)/11.
									1.4546(2)/16.
[59]Cu		58.939498	1.36 m	β+ /4.800	1.9/	3/2-	+ 1.89		ann.rad./
					3.75/				0.3393(1)/8.
									0.8780(1)/12.
									1.3015(1)/15.
									(0.4−2.6)
[60]Cu		59.937365	23.7 m	β+ /6.127	2.00/69.	2+	+1.219		ann.rad./
				EC/	3.00/18.				1.3325/88.
					3.92/6.				1.7915/45.
									(0.12−5.048)
[61]Cu		60.933458	3.35 h	β+ /2.237	0.56/3.	3/2-	+2.14		ann.rad./
					0.94/5.				0.2830/13.
					1.15/2.				0.6560/11.
					1.220/51.				(0.067−2.123)
[62]Cu		61.932584	9.67 m	β+ /98/3.948	2.93/98.	1+	-0.380		ann.rad./
				EC/					1.17302(1)/0.6
									(0.87−3.37)
[63]Cu	69.15(15)	62.929598				3/2-	+2.2233	-0.211	
[64]Cu		63.929764	12.701 h	β- /38/0.579	0.578/	1+	-0.217		ann.rad./
				β+ /19/1.6751	0.65/				1.3459(3)/0.6
				EC/41/					
[65]Cu	30.85(15)	64.927790				3/2-	+2.3817	-0.195	
[66]Cu		65.928869	5.09 m	β- /2.642	1.65/6.	1+	-0.282		0.8330(1)/0.22
					2.7/94.				1.0392(2)/9.2
[67]Cu		66.927730	2.580 d	β- /0.58	0.395/56.	3/2-	+ 2.54		0.09125(1)/7.
					0.484/23.				0.09325(1)/17.
					0.577/20.				0.18453(1)/47.
[68m]Cu			3.79 m	I.T./86/		6-	+1.24		0.0843(5)/70.
				β- /14/1.8					0.1112(5)/18.
									0.5259(5)/74.
									(0.64−1.34)
[68]Cu		67.929611	31. s	β- /4.46	3.5/40.	1+	+2.48		1.0774(5)/58.
					4.6/31.				1.2613(5)/17.
									(0.15−2.34)
[69m]Cu			0.36 µs	I.T.		(13/2+)			0.075/IT
									0.190/IT
									0.680
									1.871
[69]Cu		68.929429	2.8 m	β- /2.68	2.48/80.	3/2-	+2.84		0.5307(3)/3.
									0.8340(5)/6.
									1.0065(8)/10.
[70m2]Cu			6.6 s	β/93					0.8849/100
									1.072/19
				IT/7					0.141/ IT
[70m1]Cu			33. s	β- /52	2.52/10.	5-	+1.9		0.8848(2)/100.
				IT/48					0.9017(2)/90.
									1.2517(5)/60.
									(0.39−3.06)
[70]Cu		69.932392	44.5 s	β- /6.60	5.42/54.	1+	+1.5		0.8848(2)/100.
					6.09/46.				0.9017/99.7
									(0.438 - 3.062)
[71m]Cu			0.28 µs	I.T.		(19/2)			0.133/IT
									0.494
									0.939
									1.189
[71]Cu		70.932677	20. s	β- /4.56		3/2-			0.490/
[72m]Cu			1.76 µs	I.T.		(4-)			0.051/IT

Elem. or Isot.	Natural Abundance (Atom %)	Atomic Mass or Weight	Half-life/ Resonance Width (MeV)	Decay Mode/ Energy (/MeV)	Particle Energy/ Intensity (MeV/%)	Spin ($h/2\pi$)	Nuclear Magnetic Mom. (nm)	Elect. Quadr. Mom. (b)	γ-Energy / Intensity (MeV/%)
									0.082
									0.138
^{72}Cu		71.935820	6.6 s	β- /8.2		(1+)			0.652/
^{73}Cu		72.936675	4.2 s	β- /6.3	5.8/43				0.450/100
					6.25/42				0.307−1.559
^{74}Cu		73.93988	1.59 s	β- /9.9					
^{75}Cu		74.942	1.2 s	β- /7.9					
^{76m}Cu			1.2 s						
^{76}Cu		75.94528	0.64 s	β- /11.					
^{77}Cu		76.9479	0.46 s	β- / ∼ 10.					
^{78}Cu		77.9520	0.33 s	β- /12.					
^{79}Cu		78.9546	0.19 s	β- /11.					
^{80}Cu		79.961	> 0.3 μs						
$_{30}$Zn		69.409(4)							
^{54}Zn		53.9930	∼ 3.2 ms	2p	p//87	0+			
^{55}Zn		54.9840	> 1.6 μs						
^{56}Zn		55.9724	0.04 s			0+			
^{57}Zn		56.9648	0.04 s	β+, p/14.6		(7/2-)			ann.rad./
^{58}Zn		57.95459	0.09 s	β+		0+			
^{59}Zn		58.94926	183. ms	β+, p/9.09	8.1/	3/2-			ann.rad./
									(0.491−0.914)
^{60}Zn		59.94183	2.40 m	β+ /97/4.16		0+			ann.rad./
				EC/3/					0.669/47.
									(0.062−0.947)
^{61}Zn		60.93951	1.485 m	β+ /5.64	4.38/68.	3/2-			ann.rad./
									0.4748/17.
									(0.15−3.52)
^{62}Zn		61.93433	9.22 h	β+ /3/1.63	0.66/7.	0+			ann.rad./
				EC/93/					0.0408/25
									0.5967/26.
									(0.20−1.526)/
^{63}Zn		62.933212	38.5 m	β+ /93/3.367	1.02/	3/2-	-0.28164	+0.29	ann.rad./
				EC/7/	1.40/				0.66962(5)/8.4
					1.71/				0.96206(5)/6.6
					2.36/84.				(0.24−3.1)
^{64}Zn	48.268(321)	63.929142	> 4.3 × 10^{18} y	EC-β+		0+			
^{65}Zn		64.929241	244.0 d	β+ /98/1.3514	0.325/	5/2-	+0.7690	-0.023	ann.rad./
				EC/1.5/					1.1155/49.8
^{66}Zn	27.975(77)	65.926033				0+			
^{67}Zn	4.102(21)	66.927127				5/2-	+0.8755	+0.15	
^{68}Zn	19.024(123)	67.924844				0+			
^{69m}Zn			13.76 h	I.T./99+/0.439		9/2+			0.4390(2)/95.
^{69}Zn		68.926550	56. m	β- /0.906	0.905/99.9	½-			0.318/
^{70}Zn	0.631(9)	69.925319	> 1.3 × 10^{16} y	β-β-		0+			
^{71m}Zn			3.97 h	β- /	1.45/	9/2+			0.3864/93.
									0.4874/62.
									0.6203/57.
									(0.099−2.489)
^{71}Zn		70.92772	2.4 m	β- /2.81		½-			0.5116(1)/30.
									0.9103(1)/7.5
									(0.12−2.29)
^{72}Zn		71.92686	46.5 h	β- /0.46	0.25/14.	0+			0.0164(3)/8.
					0.30/86.				0.1447(1)/83.
									0.1915(2)/9.4
^{73m}Zn			6. s	I.T./0.196		(7/2+)			0.042
^{73}Zn		72.92978	24. s	β- /4.29	4.7/	(1/2-)			0.216(1)/100.
									0.496−0.911
^{74}Zn		73.92946	1.60 m	β- /2.3	2.1/	0+			0.0565/

Elem. or Isot.	Natural Abundance (Atom %)	Atomic Mass or Weight	Half-life/ Resonance Width (MeV)	Decay Mode/ Energy (/MeV)	Particle Energy/ Intensity (MeV/%)	Spin ($h/2\,\pi$)	Nuclear Magnetic Mom. (nm)	Elect. Quadr. Mom. (b)	γ-Energy/ Intensity (MeV/%)
									0.1401/
									(0.05−0.35)
^{75}Zn		74.9329	10.2 s	β- /6.0					0.229/
^{76}Zn		75.9333	5.7 s	β- /4.2	3.6/	0+			0.119/
^{77m}Zn			1.0 s	β- /		(1/2-)			0.772
^{77}Zn		76.9370	2.1 s	β- /7.3	4.8/				0.189/
^{78m}Zn			> 0.03 ms						1.070
^{78}Zn		77.9384	1.5 s	β- /6.4		0+			0.225/
^{79}Zn		78.9427	1.0 s	β- /8.6					0.702/
^{80}Zn		79.9443	0.54 s	β- /7.3		0+			0.713/
									0.2248/
^{81}Zn		80.9505	0.29 s	β- /11.9					
^{82}Zn		81.9544	> 0.15 µs			0+			
^{83}Zn		82.9610	> 0.15 µs						
$_{31}$Ga		**69.723(1)**							
^{56}Ga		55.9949							
^{57}Ga		56.9829							
^{58}Ga		57.9743							
^{59}Ga		58.9634	< 0.043 µs						
^{60}Ga		59.9571	0.07 s	β+					1.004
				β+, p	// ~ 1.6				3.848
				β+, α	// ~ 0.02				1.555−2.559
^{61}Ga		60.9495	0.17 s	β+ /9.0		3/2-			0.088−1.362
^{62}Ga		61.94418	116.0 ms	β+ /9.17	8.3/	0+			ann.rad./
				EC/					0.954/0.0012
^{63}Ga		62.939294	32. s	β+ /5.5	4.5/				ann.rad./
				EC/					0.6271(2)/10.
									0.6370(2)/11.
									1.0652(4)/45.
^{64m}Ga			0.022 ms						0.0429
^{64}Ga		63.936839	2.63 m	β+ /7.165	2.79/	0+			ann.rad./
					6.05/				0.80785(1)/14.
									0.99152(1)/43.
									1.38727(1)/12.
									3.3659(1)/13.
^{65}Ga		64.932735	15.2 m	β+ /86/3.255	0.82/10.	3/2-			ann.rad./
				EC/	1.39/19.				0.1151(2)/55.
					2.113/56.				0.1530(2)/96.
					2.237/15.				0.2069(2)/39.
									(0.06−2.4)
^{66}Ga		65.931589	9.5 h	β+ /56/5.175	0.74/1.	0+			ann.rad./
				EC/43/	1.84/54.				1.03935(8)/38.
					4.153/51.				2.7523(1)/23.
									(0.28−5.01)
^{67}Ga		66.928202	3.261 d	EC/1.001		3/2-	+1.8507	0.20	0.09332/37.
									0.18459/20.
									0.30024/17.
									(0.091−0.89)
^{68}Ga		67.927980	1.130 h	β+ /90/2.921	1.83/	1+	0.01175	0.028	ann.rad./
				EC/10/					1.0774(1)/3.
									(0.57−2.33)/
^{69}Ga	60.108(9)	68.925574				3/2-	+2.01659	+0.17	
^{70}Ga		69.926022	21.1 m	EC/0.2/0.655		1+			0.1755(5)/0.15
				β- /99.8/1.656	1.65/99.				1.042(5)/0.48
^{71}Ga	39.892(9)	70.924701	> 2.4 × 10²⁶ y	β-		3/2-	+2.56227	+0.11	
^{72}Ga		71.926366	14.10 h	β- /4.001	0.64/40.	3-	-0.13224	+0.5	0.8340/95.53
					1.51/9.				2.202/26.9
					2.52/8.				0.630/26.2

Elem. or Isot.	Natural Abundance (Atom %)	Atomic Mass or Weight	Half-life/ Resonance Width (MeV)	Decay Mode/ Energy (/MeV)	Particle Energy/ Intensity (MeV/%)	Spin ($h/2\pi$)	Nuclear Magnetic Mom. (nm)	Elect. Quadr. Mom. (b)	γ-Energy / Intensity (MeV/%)
					3.15/11.				(0.113−3.678)
[73]Ga		72.925175	74.87 h	β- /1.59		3/2-			0.05344(5)/10.
									0.29732(5)/47.
									(0.01−1.00)/
[74m]Ga			10. s	I.T./		1+			0.0565(1)/75.
[74]Ga		73.926946	8.1 m	β- /5.4	2.6/	3-			0.5959/92.
									2.354/45.
									(0.23−3.99)
[75]Ga		74.926500	2.10 m	β- /3.39	3.3/	3/2-			0.2529/
									0.5746/
									(0.12−2.10)
[76]Ga		75.928828	29. s	β- /7.0		3-			0.5629/66.
									0.5455/26.
									(0.34−4.25)
[77]Ga		76.929154	13.0 s	β- /5.3	5.2/				0.469/
									0.459/
[78]Ga		77.931608	5.09 s	β- /8.2		3+			0.619/77.
									1.187/20.
[79]Ga		78.9329	2.85 s	β- /7.0	4.6/				0.465/
[80]Ga		79.9365	1.68 s	β- /10.4	10./				0.659/
[81]Ga		80.9378	1.22 s	β- /8.3	5.1/				0.217/
[82]Ga		81.9430	0.599 s	β- /12.6					1.348/
[83]Ga		82.9470	0.308 s	β- /~ 11.5					
[84]Ga		83.9527	~ 0.085 s	β- /14					
[85]Ga		84.9570	> 0.3 μs						
[86]Ga		85.963	> 0.3 μs						
[32]Ge		72.64(1)							
[58]Ge		57.9910				0+			
[59]Ge		58.9818							
[60]Ge		59.9702	> 0.11 μs			0+			
[61]Ge		60.9638	0.04 s	β+ /13.6					
[62]Ge		61.9547	0.13 s			0+			
[63]Ge		62.9496	0.15 s	β- /9.8					
[64]Ge		63.94165	1.06 m	β+ /4.4	3.0/	0+			ann.rad./
				EC/					0.1282(2)/11.
				β+, p					0.4270(3)/37.
									0.6671(3)/17.
[65]Ge		64.9394	31. s	β+ /6.2	0.82/10.				ann.rad./
				EC/	1.39/19.				0.0620/27.
				EC, p	2.113/56.				0.6497/33.
					2.237/15.				0.8091/21.
				β+, p	//0.011				(0.19−3.28)
[66]Ge		65.93384	2.26 h	β+ /27/2.10		0+			ann.rad./
				EC/73/					0.0438/29.
									0.3819/28.
									(0.022−1.77)
[67]Ge		66.932734	19.0 m	β+ /96/4.225	1.6/	½-			ann.rad./
				EC/4/	2.3/				0.1670/84.
					3.15/				(0.25−3.73)
[68]Ge		67.92809	270.8 d	EC/0.11		0+			Ga k x-ray/39.
[69]Ge		68.927965	1.63 d	β+ /36/2.2273	0.70/	5/2-	0.735	0.02	ann.rad./
				EC/64/	1.2/				0.574/13.
									1.1068/36.
									(0.2−2.04)
[70]Ge	20.38(18)	69.924247				0+			
[71m]Ge			20.4 ms		I.T./0.0234	9/2+			0.1749
[71]Ge		70.924951	11.2 d	EC/0.229		½-	+0.547		
[72]Ge	27.31(26)	71.922076				0+			

Elem. or Isot.	Natural Abundance (Atom %)	Atomic Mass or Weight	Half-life/ Resonance Width (MeV)	Decay Mode/ Energy (/MeV)	Particle Energy/ Intensity (MeV/%)	Spin ($h/2\pi$)	Nuclear Magnetic Mom. (nm)	Elect. Quadr. Mom. (b)	γ-Energy / Intensity (MeV/%)
^{73}Ge	7.76(8)	72.923459	> 1.8×10^{23} y	β-		9/2+	-0.879467	-0.17	
^{74}Ge	36.72(15)	73.921178				0+			
^{75m}Ge			48. s	I.T./		7/2+			0.13968(3)/39.
^{75}Ge		74.922859	1.380 h	β- /1.177	1.19/	½-	+0.510		0.26461(5)/11.
									0.41931(5)/0.2
^{76}Ge	7.83(7)	75.921403	1.6×10^{21} y	β-β-		0+			
^{77m}Ge			53. s	I.T./20/		½-			1.605/0.22
				β- /80/2.861	2.9/				1.676/0.16
									0.195–1.482
^{77}Ge		76.923549	11.25 h	β- /2.702	0.71/23.	7/2+			0.2110/29.
					1.38/35.				0.2155/27.
					2.19/42.				0.2644/51.
									(0.15–2.35)
^{78}Ge		77.922853	1.45 h	β- /0.95	0.70/	0+			0.2773(5)/96.
									0.2939(5)/4.
^{79m}Ge			39. s	β- /IT		7/2+			
^{79}Ge		78.9254	19.1 s	β- /4.2	4.0/20.	½-			0.1096/21.
					4.3/80.				(0.10–2.59)
									0.5427(4)/15.
^{80}Ge		79.92537	29.5 s	β- /2.67	2.4/	0+			0.1104(4)/6.
									0.2656(4)/25.
^{81m}Ge			~ 7.6 s	β- /	3.75/	½+			0.3362(4)/
									0.7935(4)/
^{81}Ge		80.9288	~ 7.6 s	β- /6.2	3.44/	9/2+			0.1976(4)/21.
									0.3362(4)/100.
^{82}Ge		81.9296	4.6 s	β- /4.7	1.093/80	0+			1.093/
^{83}Ge		82.9346	1.9 s	β- /8.9					
^{84}Ge		83.9375	0.98 s	β- /7.7		0+			
^{85}Ge		84.9430	0.54 s	β- /10.					
^{86}Ge		85.9465	> 0.3 μs			0+			
^{87}Ge		86.9525	> 0.3 μs						
^{88}Ge		87.957	> 0.3 μs			0+			
^{89}Ge		88.964	> 0.3 μs						
$_{33}$As		74.92160(2)							
^{60}As		59.993							
^{61}As		60.981							
^{62}As		61.9732							
^{63}As		62.9637	< 0.043 μs						
^{64}As		63.9576	0.02 s						
^{65}As		64.9496	0.13 s	β+ /9.4					
^{66m2}As			8. μs						
^{66m1}As			1.1 μs						
^{66}As		65.945	95.8 ms	β+ /9.55					
^{67}As		66.9392	42. s	β+ /6.0	5.0/	5/2-			0.121/
				EC/					0.123/
									0.244/
^{68}As		67.93677	2.53 m	β+ /8.1		3+			ann.rad./
									0.652/32.
									0.762/33.
									1.016/77.
									(0.61–3.55)
^{69}As		68.93227	15.2 m	β+ /98/4.01	2.95/	5/2-	+ 1.623		ann.rad./
				EC/2/					0.0868(5)/1.5
									0.1458(3)/2.4
^{70}As		69.93092	52.6 m	β+ /84/6.22	1.44/	4+	+2.1061	+0.09	ann.rad./
				EC/16/2.14					1.0395(7)/82.
				/2.89					(0.17–4.4)
^{71}As		70.927112	2.72 d	β+ /32/2.013		5/2-	+1.6735	-0.02	ann.rad./

Elem. or Isot.	Natural Abundance (Atom %)	Atomic Mass or Weight	Half-life/ Resonance Width (MeV)	Decay Mode/ Energy (/MeV)	Particle Energy/ Intensity (MeV/%)	Spin (h/2 π)	Nuclear Magnetic Mom. (nm)	Elect. Quadr. Mom. (b)	γ-Energy / Intensity (MeV/%)
				EC/68/					0.1749(2)/84.
									1.0957(2)/4.2
^{72}As		71.926752	26.0 h	β+ /77/4.356	0.669/5.	2-	-2.1566	-0.08	ann.rad./
					1.884/12.				0.83395(5)/80.
					2.498/62.				1.0507(1)/9.6
					3.339/19.				(0.1−4.0)
^{73}As		72.923825	80.3 d	EC/0.341		3/2-			0.0133/0.1
									0.0534/10.5
									Se k x-ray/90.
^{74}As		73.923829	17.78 d	β+ /31/2.562	0.94/26.	2-	-1.597		ann.rad./
				EC/37/	1.53/3.				0.59588(1)/60.
				β- /1.353	0.71/16.				0.6084(1)/0.6
					1.35/16.				0.6348(1)/15.
^{75m}As			0.017 s						
^{75}As	100.	74.921597				3/2-	+1.43947	+0.31	
^{76}As		75.922394	26.3 h	β- /2.962	0.54/3.	2-	-0.903		0.5591(1)/45.
					1.785/8.				0.65703(5)/6.2
					2.410/36.				1.21602(1)/3.4
					2.97/51.				(0.3−2.67)
^{77}As		76.920647	38.8 h	β- /0.683	0.70/98.	3/2-	+1.295		0.2391(2)/1.6
									0.2500(3)/0.4
									0.5208/0.43
^{78}As		77.92183	1.512 h	β- /4.21	3.00/12.	2-			0.6136(3)/54.
					3.70/17.				0.6954(3)/18.
					4.42/37.				1.3088(3)/10.
^{79m}As			1.21 μs	I.T.		9/2+			0.542/IT
									0.231
^{79}As		78.92095	9.0 m	β- /2.28	1.80/95.	3/2-			0.0955(5)/16.
									0.3645(5)/1.9
^{80}As		79.92253	16. s	β- /5.64	3.38/	1+			0.6662(2)/42.
									(2.5−3.0)
^{81}As		80.92213	33. s	β- /3.856		3/2-			0.4676(2)/20.
									0.4911(2)/8.
^{82m}As			13.7 s	β- /	3.6/	5-			0.6544(1)/77.
									0.344/65.
									(0.561 − 1.894)
^{82}As		81.9245	19. s	β- /7.4	7.2/80.	(2-)			0.6544(1)/54.
									(0.755 − 3.667)
^{83}As		82.9250	13.4 s	β- /5.5					0.7345/100.
									1.1131/34.
									2.0767/28.
^{84m}As			0.6 s	β-					
^{84}As		83.9291	4. s	β-, n/7.2		1-			0.6671(2)/21.
									1.4439(5)/49.
									(0.325−5.150)
^{85}As		84.9320	2.03 s	β-, n/8.9		3/2-			0.667(1)/42.
									1.4551(2)/100.
^{86}As		85.9365	0.95 s	β-, n/11.4					0.704/
^{87}As		86.9399	0.49 s	β-, n/10.					0.704/
^{88}As		87.9449	> 0.3 μs						
^{89}As		88.9494	> 0.3 μs						
^{90}As		89.956	> 0.3 μs						
^{91}As		90.960	> 0.3 μs						
^{92}As		91.967	> 0.3 μs						
$_{34}$Se		78.96(3)							
^{64}Se			> 0.18 μs			0+			
^{65}Se		64.965	0.011 s	β+ /60/14.					

Elem. or Isot.	Natural Abundance (Atom %)	Atomic Mass or Weight	Half-life/ Resonance Width (MeV)	Decay Mode/ Energy (/MeV)	Particle Energy/ Intensity (MeV/%)	Spin $(h/2\pi)$	Nuclear Magnetic Mom. (nm)	Elect. Quadr. Mom. (b)	γ-Energy / Intensity (MeV/%)
				β+, p	3.55/				
^{66}Se		65.9552	0.03 s			0+			
^{67}Se		66.9501	0.13 s	β+ /10.2					ann.rad./
				β+, (p)/					0.352
^{68}Se		67.94180	36. s	β+ /4.7		0+			ann.rad./
									(0.050−0.426)
^{69}Se		68.93956	27.4 s	β+ /6.78	5.006/				ann.rad./
				EC/					0.0664(4)/27.
				β+, p	// ~ 0.045				0.0982(4)/63.
^{70}Se		69.9334	41.1 m	β+ /2.4		0+			ann.rad
									0.04951(5)/35.
									0.4262(2)/29.
^{71}Se		70.93224	4.7 m	β+ /4.4	3.4/36.	5/2-			ann.rad
				EC/					0.1472(3)/47.
									0.8309(3)/13.
									1.0960(3)/10.
^{72}Se		71.92711	8.5 d	EC/0.34		0+			0.0460(2)/57.
^{73m}Se			40. m	I.T./73/0.0257	0.85	3/2-			ann.rad.
				β+ /27/2.77	1.45/				0.0257(2)/27.
					1.70/				0.2538(1)/2.5
^{73}Se		72.92677	7.1 h	β+ /65/2.74	0.80/	9/2+	0.86		ann.rad
				EC/35/	1.32/95.				0.0670(1)/72.
					1.68/1.				0.3609(1)/97.
									(0.6−1.5)
^{74}Se	0.89(4)	73.922476				0+			
^{75}Se		74.922523	119.78 d	EC/0.864		5/2+	0.67	1.0	0.13600/55
									0.26465/58
									(0.024−0.821)
^{76}Se	9.37(29)	75.919214				0+			
^{77m}Se			17.4 s	I.T./		7/2+			0.1619(2)/52.
^{77}Se	7.63(16)	76.919914				½-	+0.53506		
^{78}Se	23.77(28)	77.917309				0+			
^{79m}Se			3.92 m	I.T./					0.09573(3)/9.5
^{79}Se		78.918499	2.9×10^5 y	β- /0.151		7/2+	-1.018	+0.8	
^{80}Se	49.61(41)	79.916521				0+			
^{81m}Se			57.3 m	I.T./99/0.1031		7/2+			0.1031(3)/9.7
									0.2602(2)/0.06
									0.2760/0.06
^{81}Se		80.917993	18.5 m	β- /1.585	1.6/98.	½-			0.2759/0.85
									0.2901/0.75
									0.8283/0.32
^{82}Se	8.73(22)	81.916699	$\sim 1 \times 10^{20}$ y	β-β-		0+			
^{83m}Se			1.17 m	β- /3.96	2.88/	½-			0.35666(6)/17.
					3.92/				0.9879(1)/15.
									1.0305(1)/21.
									2.0514(2)/11.
									(0.19−3.1)
^{83}Se		82.919118	22.3 m	β- /3.668	0.93/	9/2+			0.22516(6)/33.
					1.51/				0.35666(6)/69.
									0.51004(8)/45.
									(0.21−2.42)
^{84}Se		83.91846	3.3 m	β- /1.83	1.41/100.	0+			0.4088(5)/100.
^{85}Se		84.92225	32. s	β- /6.18	5.9/	5/2+			0.3450(1)/22.
									0.6094(1)/41.
^{86}Se		85.92427	15. s	β- /5.10		0+			2.0124(1)/24.
									2.4433(8)/100.
									2.6619(1)/49.
^{87}Se		86.92852	5.4 s	β- /7.28		5/2+			0.468(1)/100.
				n/					1.4979(1)/23.
^{88}Se		87.93142	1.5 s	β-, n/6.85		0+			0.5346/

Elem. or Isot.	Natural Abundance (Atom %)	Atomic Mass or Weight	Half-life/ Resonance Width (MeV)	Decay Mode/ Energy (/MeV)	Particle Energy/ Intensity (MeV/%)	Spin ($h/2\pi$)	Nuclear Magnetic Mom. (nm)	Elect. Quadr. Mom. (b)	γ-Energy / Intensity (MeV/%)
^{89}Se		88.9365	0.41 s	β-, n/9.0					
^{90}Se		89.9400	> 0.3 μs			0+			
^{91}Se		90.9460	0.27 s	β-, n/8.					
^{92}Se		91.950	> 0.3 μs			0+			
^{93}Se		92.956	> 0.3 μs						
^{94}Se		93.960	> 0.3 μs			0+			
$_{35}$Br		79.904(1)							
^{67}Br		66.9648							
^{68}Br		67.9585	< 1.5 μs						
^{69}Br		68.9501	< 0.024 μs	β+ /9.6					
^{70m}Br			2.2 s			9+			
^{70}Br		69.9446	~ 0.08 s	β+ /10.0	/0.75				
^{71}Br		70.939	21. s	β+ /6.9					
^{72}Br		71.9366	1.31 m	β+ /8.7		3	~ 0.55		0.4547−1.3167
^{73}Br		72.93169	3.4 m	β+ /4.7	3.7/	3/2-			ann.rad
									0.065−0.700
^{74m}Br			46. m	β+ /	4.5/	4-	1.82		ann.rad
									0.6348
									0.7285
									(0.2−4.38)
^{74}Br		73.92989	25.4 m	β+ /6.91					ann.rad
									0.6341
									0.6348
									(0.2−4.7)
^{75}Br		74.92578	1.62 h	β+ /76/3.03		3/2-	+0.75		ann.rad
									0.28650
									(0.1−1.56)
^{76m}Br			1.4 s	I.T./5.05		4+			0.104548
									0.05711
^{76}Br		75.92454	16.0 h	β+ /57/4.96	1.9/	1-	0.54821	0.270	ann.rad
					3.68/				0.55911
									1.85368
									(0.4−4.6)
^{77m}Br			4.3 m	I.T./0.1059		9/2+			0.1059
^{77}Br		76.921379	2.376 d	EC/99/1.365		3/2-	0.973	+0.53	ann.rad.
									0.23898
									0.52069
									(0.08−1.2)
^{78}Br		77.921146	6.45 m	β+ /92/3.574	1.2/	1+	0.13		ann.rad.
				EC/8/	2.5/				0.61363
									(0.7−3.0)
^{79m}Br			4.86 s	I.T./0.207		9/2+			0.2072
^{79}Br	50.69(7)	78.918337				3/2-	+2.106400	+0.331	
^{80m}Br			4.42 h	I.T./0.04885		5-	+1.3177	+0.75	Br k x-ray
									0.03705/39.1
									0.04885/0.3
^{80}Br		79.918529	17.66 m	β- /92/2.004	1.38 β-/7.6	1+	0.5140	0.196	ann.rad.
				EC/5.7/1.8706	1.99 β-/82				0.6169/6.7
				β+ /2.6/	0.85 β+ /2.8				(0.64−1.45)
^{81}Br	49.31(7)	80.916291				3/2-	+2.270562	+0.276	
^{82m}Br			6.1 m	I.T./98/0.046		2-			0.046/0.24
				β- /2 /3.139					(0.62−2.66)
^{82}Br		81.916804	1.471 d	β- /3.093	0.444/	5-	+1.6270	0.751	0.5544/71
									0.61905/43
									0.77649/84
									(0.013−1.96)
^{83}Br		82.915180	2.40 h	β- /0.972	0.395/1	3/2-			0.52964
					0.925/99				(0.12−0.68)

Table of the Isotopes

Elem. or Isot.	Natural Abundance (Atom %)	Atomic Mass or Weight	Half-life/ Resonance Width (MeV)	Decay Mode/ Energy (/MeV)	Particle Energy/ Intensity (MeV/%)	Spin ($h/2\pi$)	Nuclear Magnetic Mom. (nm)	Elect. Quadr. Mom. (b)	γ-Energy / Intensity (MeV/%)
^{84m}Br			6.0 m	β- /4.97	2.2/100	(6-)			0.4240/100
									0.8817/98
									1.4637/101
^{84}Br		83.91648	31.8 m	β- /4.65	2.70/11	2-	2.		0.8816/41
					3.81/20				1.8976/13
					4.63/34				(0.23–4.12)
^{85}Br		84.91561	2.87 m	β- /2.87	2.57	3/2-			0.80241/2.56
									0.92463/1.6
									(0.09–2.4)
^{86}Br		85.91880	55.5 s	β- /7.63	3.3	(2-)			1.56460/64
					7.4				2.75106/21
									(0.5–6.8)
^{87}Br		86.92071	55.6 s	β- /6.85	6.1/	3/2-			1.41983
				n/					1.4762
									(0.2–6.1)
^{88m}Br			5.1 µs						
^{88}Br		87.92407	16.3 s	β- /8.96		1-			0.7649
				n/					0.7753
									0.8021
									(0.1–6.99)
^{89}Br		88.92640	4.35 s	β- /8.16		3/2-			0.7753
				n/					1.0978
^{90}Br		89.9306	1.91 s	β- /10.4	8.3/	2-			0.6555
				n/	9.8/				0.7071
									1.3626
^{91}Br		90.9340	0.54 s	β- /90 /9.80					0.263
				β- n/10 /					0.803
^{92}Br		91.93926	0.31 s	β- /12.20					0.740
				β- n/					
^{93}Br		92.9431	0.10 s	β- /11					0.117
				β- n	//11				(0.237–3.606)
^{94}Br		93.9487	0.07 s	β- n/					
^{95}Br		94.9529	> 0.3 µs						
^{96}Br		95.959	> 0.3 µs						
^{97}Br		96.963	> 0.3 µs						
$_{36}$Kr		83.798(2)							
^{69}Kr		68.9652	0.03 s	β+, (p)	4.07/				
^{70}Kr		69.9553	0.06 s			0+			
^{71}Kr		70.950	100. ms	β+, EC/10.1					(0.198–0.207)
^{72}Kr		71.94209	17.1 s	β+ /5.0		0+			ann.rad
				EC/					0.3099/15.3
									0.4150/12.8
									(0.305 – 3.305)
^{73}Kr		72.93929	28. s	β+ /6.7		5/2-			ann.rad.
				EC/					0.1781/66
				β+, p/	/0.25				(0.06–0.86)
^{74}Kr		73.933084	11.5 m	β+ /3.1		0+			ann.rad.
				EC/					0.08970/31
									0.2030/20
									(0.010–1.06)
^{75}Kr		74.93095	4.3 m	β+ /4.90	3.2/	5/2+	-0.531	+1.1	ann.rad.
				EC/					0.1325/68
									0.1547/21
									(0.02–1.7)
^{76}Kr		75.925910	14.8 h	EC/1.31		0+			Br k x-ray
									0.270/21
									0.3158/39
									(0.03–1.07)

Elem. or Isot.	Natural Abundance (Atom %)	Atomic Mass or Weight	Half-life/ Resonance Width (MeV)	Decay Mode/ Energy (/MeV)	Particle Energy/ Intensity (MeV/%)	Spin ($h/2\pi$)	Nuclear Magnetic Mom. (nm)	Elect. Quadr. Mom. (b)	γ-Energy / Intensity (MeV/%)
^{77}Kr		76.924670	1.24 h	β+ /80/3.06		5/2+	-0.583	+0.9	ann.rad.
				EC/20/	1.55/				0.1297/80
					1.70/				0.1465/38
					1.87/				(0.02−2.3)
^{78}Kr	0.355(3)	77.920365	> 2.3 × 10^{20} y	EC-EC		0+			
^{79m}Kr			53. s	I.T./0.1299		7/2+	-0.786	+0.40	Kr x-ray
^{79}Kr		78.920082	1.455 d	β+ /7 /1.626		½-	+0.536		ann.rad.
				EC/93 /					0.2613/13
									0.39756/19
									0.6061/8
									(0.04−1.3)
^{80}Kr	2.286(10)	79.916379				0+			
^{81m}Kr			13.1 s	I.T./0.1904		½-	+0.586		0.1904
^{81}Kr		80.916592	2.1 × 10^5 y	EC/0.2807		7/2+	-0.908	+0.63	Br k x-ray
									0.2760
^{82}Kr	11.593(31)	81.913484				0+			
^{83m}Kr			1.86 h	I.T./0.0416		½-	+0.591		Kr k x-ray
									0.00940
									0.03216
^{83}Kr	11.500(19)	82.914136				9/2+	-0.970699	+0.259	
^{84}Kr	56.987(15)	83.911507				0+			
^{85m}Kr			4.48 h	β- /79 /	0.83/79	½-	+ 0.633		0.30487
				I.T./21 /0.305					0.15118
^{85}Kr		84.912527	10.73 y	β- /0.687	0.15/0.4	9/2+	1.005	+0.43	0.51399
^{86}Kr	17.279(41)	85.9106107				0+			
^{87}Kr		86.9133549	1.27 h	β- /3.887	1.33/8	5/2+	-1.023	-0.30	0.40258/49.6
					3.49/43				2.5548/9.2
					3.89/30				(0.13−3.31)
^{88}Kr		87.91445	2.84 h	β- /2.91		0+			0.19632/26.
									2.392/34.6
									(0.03−2.8)
^{89}Kr		88.9176	3.15 m	β- /4.99	3.8/	5/2+	-0.330	+0.16	0.19746
					4.6/				0.2209/19.9
					4.9/				0.5858/16.4
									1.4728/6.8
									(0.2−4.7)
^{90}Kr		89.91952	32.3 s	β- /4.39	2.6/77	0+			0.12182/32.9
					2.8/6				0.5395/28.6
									1.1187/36.2
									(0.1−4.2)
^{91}Kr		90.9235	8.6 s	β- /6.4	4.33/	5/2+	-0.583	+0.30	0.10878/43.5
					4.59/				0.50658/19.
									(0.2−4.4)
^{92}Kr		91.92616	1.84 s	β- /5.99		0+			0.1424/66.
				n/					(0.14−3.7)
^{93}Kr		92.9313	1.29 s	β- /8.6	7.1/	½+	-0.413		0.1820
				n/					0.2534/42.
									0.32309/24.6
									(0.057−4.03)
^{94}Kr		93.9344	0.21 s	β- /7.3		0+			0.2196/67
				n	n//1.0				0.6293/100.
									(0.098−0.985)
^{95}Kr		94.9398	0.10 s	β- /9.7	n//2.9		- 0.410		
^{96}Kr		95.9431	~ 80 ms	β- ,n	n//3.7	0+			
^{97}Kr		96.9486	0.06 s	β- ,n	n//7.				
^{98}Kr		97.952	0.05 s	β- ,n	n//7.	0+			
^{99}Kr		98.958	0.04 s	β- ,n	n// ~ 11.				
^{100}Kr		99.9611	> 0.34 µs			0+			

Table of the Isotopes

Elem. or Isot.	Natural Abundance (Atom %)	Atomic Mass or Weight	Half-life/ Resonance Width (MeV)	Decay Mode/ Energy (/MeV)	Particle Energy/ Intensity (MeV/%)	Spin ($h/2\pi$)	Nuclear Magnetic Mom. (nm)	Elect. Quadr. Mom. (b)	γ-Energy / Intensity (MeV/%)
$_{37}$Rb		85.4678(3)							
^{71}Rb		70.9653							
^{72}Rb		71.9591	< 1.5 µs						
^{73}Rb		72.9506	< 0.03 µs						
^{74}Rb		73.944265	64.8 ms	β+ /10.4					0.456/0.0025
									(0.053 − 4.244)
^{75}Rb		74.93857	19. s	β+ /7.02	2.31/				ann.rad
									0.179
^{76}Rb		75.935072	39. s	β+ /8.50	4.7/	1-	-0.372623	+0.4	ann.rad
									0.4240/92.
									(0.064−1.68)
^{77}Rb		76.93041	3.8 m	β+ /5.34	3.86/	3/2-	+0.654468	+0.70	ann.rad
									0.0665/59
									(0.04−2.82)
^{78m}Rb			5.7 m	I.T./0.1034		4-	+2.549	+0.81	ann.rad
				β+ /	3.4				0.4553/81.
				EC/					(0.103−4.01)
^{78}Rb		77.92814	17.7 m	β+ /7.22		0+			ann.rad
				EC/					0.4553/63.
									(0.42−5.57)
^{79}Rb		78.92399	23. m	β+ /84/3.65		5/2+	+0.3358	-0.10	ann.rad.
				EC/16 /					0.68812/23.
									(0.017−3.02)
^{80}Rb		79.92252	34. s	β+ /5.72	4.1/22	1+	-0.0836	+0.35	ann.rad.
					4.7/74				0.6167/25.
^{81m}Rb			30.5 m	I.T./0.85	1.4	9/2+	+5.598	-0.74	ann.rad.
				β+, EC/					(0.085−1.9)
^{81}Rb		80.91900	4.57 h	β+ /27/2.24	1.05/	3/2-	+2.060	+0.40	ann.rad./
				EC/73					0.19030/64.
									(0.05−1.9)
^{82m}Rb			6.47 h	β+/26/	0.80/	5-	+1.5100	+1.0	ann.rad./
				EC/74/					0.5544/63.
									0.7765/85.
									(0.092−2.3)
^{82}Rb		81.918209	1.258 m	β+/96/4.40	3.3/	1+	+0.554508	+0.19	ann.rad./
				EC/4/					0.7665/13.
									(0.47−3.96)
^{83}Rb		82.91511	86.2 d	EC/0.91		5/2-	+1.425	+0.20	Kr x-ray
									0.5205/46.
									(0.03−0.80)
^{84m}Rb			20.3 m	I.T./0.216		6-	+0.2129	+0.6	0.2163/34.
									0.2482/63.
									0.4645/32.
^{84}Rb		83.914385	32.9 d	β+/22/2.681	0.780/11	2-	-1.32412	-0.015	ann.rad./
				EC/75 /	1.658/11				0.8817/68.
				β-/3/0.894	0.893/				(1.02−1.9)
^{85}Rb	72.17(2)	84.91178974				5/2-	+1.353	+0.23	
^{86m}Rb			1.018 m	I.T./0.5560		6-	+1.815	+0.37	0.556/98.
^{86}Rb		85.9111674	18.65 d	β-/1.775	1.774/8.8	2-	-1.6920	+0.19	1.0768/8.8
^{87}Rb	27.83(2)	86.90918053	4.88×10^{10} y	β-/0.283	0.273/100	3/2-	+2.7512	+0.13	
^{88}Rb		87.9113156	17.7 m	β-/5.316	5.31	2-	0.508		0.8980/14.4
									1.8360/22.8
									(0.34−4.85)
^{89}Rb		88.91228	15.4 m	β-/4.50	1.26/38	3/2-	+2.304	+0.14	1.032/58.
					1.9/5				1.248/42.
					2.2/34				2.1960/13
					4.49/18				(0.12−4.09)

Elem. or Isot.	Natural Abundance (Atom %)	Atomic Mass or Weight	Half-life/ Resonance Width (MeV)	Decay Mode/ Energy (/MeV)	Particle Energy/ Intensity (MeV/%)	Spin ($h/2\pi$)	Nuclear Magnetic Mom. (nm)	Elect. Quadr. Mom. (b)	γ-Energy / Intensity (MeV/%)
^{90m}Rb			4.3 m	β-/4.50	1.7/	4-	+1.616	+0.20	0.1069(IT)
					6.5/				0.8317/94
									(0.20−5.00)
^{90}Rb		89.91480	2.6 m	β-/6.59	6.6	1-			0.8317/28.
									(0.31−5.60)
^{91}Rb		90.91654	58.0 s	β-/5.861	5.9	3/2-	+2.182	+0.15	0.0936/34.
									(0.35−4.70)
^{92}Rb		91.91073	4.48 s	β-/8.11	8.1/94	1-			0.8148/8.
									(0.1−6.1)
^{93}Rb		92.92204	5.85 s	β-/7.46	7.4/	5/2-	+1.410	+0.18	0.2134/4.8
				n/1					0.4326/12.5
									0.9861/4.9
									(0.16−5.41)
^{94}Rb		93.92641	2.71 s	β-/10.31	9.5/	3	+1.498	+0.16	0.8369/87.
				n/10					1.5775/32.
									(0.12−6.35)
^{95}Rb		94.92930	0.377 s	β-/9.30	8.6/	5/2-	+1.334	+0.21	0.352/65.
				n/8					0.680/22.
									(0.20−2.27)
^{96m}Rb			1.7 μs						0.2999
									0.4612
									0.2400
									0.093−0.369
^{96}Rb		95.93427	0.199 s	β-/11.76	10.8/	2+	+1.466	+0.25	0.815/76.
				n/13/					(0.20−5.42)
^{97}Rb		96.93735	0.169 s	β-/10.42	10.0	3/2+	+1.841	+0.58	0.167/100.
				n/27/					0.585/79.
									0.599/56.
									1.258/52.
									(0.14−2.08)
^{98}Rb		97.94179	0.107 s	β-/12.34	0.144/				
				n/13					(0.07−3.68)
^{99}Rb		98.9454	59. ms	β-/11.3					
^{100}Rb		99.9499	53. ms	β- 13.5					0.129
									(0.058−4.483)
^{101}Rb		100.9532	0.03 s	β- 11.8					
^{102}Rb		101.9589	0.09 s	β-					
$_{38}$**Sr**		**87.62(1)**							
^{73}Sr		72.966	> 25 ms						
^{74}Sr		73.9563	> 1.5 μs			0+			
^{75}Sr		74.9499	88. ms	β+ ,p	p//5.				0.144/4.5
^{76}Sr		75.94177	7.9 s	β+ /6.1		0+			
^{77}Sr		76.93795	9.0 s	β+ /6.9	5.6		-0.35	+1.4	0.147
				β+, p	//0.08				
^{78}Sr		77.93218	2.7 m	β+ /3.76		0+			(0.047−0.793)
^{79}Sr		78.92971	2.1 m	β+ /5.32	4.1	3/2-	-0.474	+0.74	ann.rad./
									0.039/28.
									0.105/22.
									(0.135−0.612)
^{80}Sr		79.92452	1.77 h	β+ /1.87		0+			ann.rad./
									0.174/10.
									0.589/39.
									(0.24−0.55)
^{81}Sr		80.92321	22.3 m	β+ /87/3.93	2.43/	1/2-	+0.544		ann.rad./
				EC/13/	2.68/				0.148/31.
									0.1534/35
									(0.06−1.7)
^{82}Sr		81.91840	25.36 d	EC/0.18		0+			Rb x-ray

Elem. or Isot.	Natural Abundance (Atom %)	Atomic Mass or Weight	Half-life/ Resonance Width (MeV)	Decay Mode/ Energy (/MeV)	Particle Energy/ Intensity (MeV/%)	Spin ($h/2\pi$)	Nuclear Magnetic Mom. (nm)	Elect. Quadr. Mom. (b)	γ-Energy / Intensity (MeV/%)
^{83m}Sr			5.0 s	I.T./0.2591		½−	+0.582		0.2591/87.5
^{83}Sr		82.91756	1.350 d	β+/24/2.28	0.465/	7/2+	−0.898	+0.79	ann.rad./
				EC/76/	0.803/				0.3816/12.
					1.227/				0.3816
									0.7627/30.
									(0.094−2.15)
^{84}Sr	0.56(1)	83.913425				0+			
^{85m}Sr			1.127 h	I.T./87/0.2387		½−	+0.601		0.2318/84.
				EC/13					(0.15−0.24)
^{85}Sr		84.912933	64.85 d	EC/1.065		9/2+	−1.001	+0.30	0.51399/99.3
^{86}Sr	9.86(1)	85.909260				0+			
^{87m}Sr			2.81 h	I.T./0.3884		½−	+0.63		0.3884(IT)
^{87}Sr	7.00(1)	86.908877				9/2+	−1.093	+0.34	
^{88}Sr	82.58(1)	87.905612				0+			
^{89}Sr		88.907451	50.52 d	β−/1.497	1.492/100	5/2+	−1.149	−0.3	0.9092
^{90}Sr		89.907738	29.1 y	β−/0.546	0.546/100	0+			
^{91}Sr		90.910203	9.5 h	β−/2.70	0.61/7	5/2+	−0.887	+0.044	0.5556/61.
					1.09/33				0.7498/24.
					1.36/29				1.0243/33.
					2.66/26				(0.12−2.4)
^{92}Sr		91.911038	2.71 h	β−/1.91	0.55/96	0+			1.3831/90.
					1.5/3				(0.24−1.1)
^{93}Sr		92.91403	7.4 m	β−/4.08	2.2/10	5/2+	−0.794	+0.26	0.5903/
					2.6/25				0.7104
					3.2/65				0.87573
									0.8883/
									(0.17−3.97)
^{94}Sr		93.91536	1.25 m	β−/3.511	2.1/	0+			0.6219
					3.3/				0.7043
									0.7241
									0.8064
									1.4283
^{95}Sr		94.91936	25.1 s	β−/6.08		½+	−0.5379		0.6859
					6.1/50				0.8269
									2.7173
									2.9332
^{96}Sr		95.92170	1.06 s	β−/5.37	4.2/	0+			0.1222
									0.5305
									0.8094
									0.9318
^{97}Sr		96.92615	0.42 s	β−/7.47	5.3	(1/2+)	−0.500		0.2164
									0.3071
									0.6522
									0.9538
									1.2580
									1.9050
^{98}Sr		97.92845	0.65 s	β−/5.83	5.1	0+			0.0365
									0.1190
									0.4286
									0.4447
									0.5636
^{99}Sr		98.9332	0.27 s	β−/8.0			−0.26	0.8	
^{100}Sr		99.9354	0.201 s	β−/7.1		0+			
^{101}Sr		100.9405	0.115 s	β−/9.5					
^{102}Sr		101.9430	68. ms	β−/8.8		0+			
^{103}Sr		102.9490	> 0.3 µs						
^{104}Sr		103.952	> 0.3 µs			0+			
^{105}Sr		104.959	> 0.3 µs						

Elem. or Isot.	Natural Abundance (Atom %)	Atomic Mass or Weight	Half-life/ Resonance Width (MeV)	Decay Mode/ Energy (/MeV)	Particle Energy/ Intensity (MeV/%)	Spin ($h/2\pi$)	Nuclear Magnetic Mom. (nm)	Elect. Quadr. Mom. (b)	γ-Energy / Intensity (MeV/%)
$_{39}$Y		88.90585(2)							
^{76}Y		75.9585	> 0.2 μs						
^{77}Y		76.9497	~ 57. ms						
^{78m}Y			5.8 s			(5+)			
^{78}Y		77.9436	53 ms	β+/10.5					0.279/100
									0.504/90
									0.713/40
^{79}Y		78.9374	15. s	β+/7.1					(0.152−1.106)
^{80m}Y			4.8 s						0.2285
^{80}Y		79.9343	30. s	β+/7.0	5.5	(4−)			ann.rad./
					5.0/				0.3858/100
									0.5951/42
									0.756−1.396
^{81}Y		80.9291	1.21 m	β+/5.5	3.7/				ann.rad./
					4.2/				0.428
									0.469
^{82}Y		81.9268	9.5 s	β+/7.8	6.3/	1+			ann.rad./
									0.5736
									0.6017
									0.7375
^{83m}Y			2.85 m	β+/95/4.6	2.9	1/2−			ann.rad./
				EC/5 /					0.2591
									0.4218
									0.4945
^{83}Y		82.92235	7.1 m	β+/4.47	3.3	9/2+			ann.rad./
				EC/					0.0355
									0.4899
									0.8821
									(0.03−3.4)
^{84m}Y			4.6 s	β+/		1+			ann.rad./
				EC/					0.7930
^{84}Y		83.9204	40. m	β+/6.4	1.64/47	5−			ann.rad./
				EC/	2.24/25				0.4628
					2.64/21				0.6606
					3.15/7				0.7931
									0.9744
									1.0398
									(0.2−3.3)
^{85m}Y			4.9 h	β+/70/		9/2+	6.2		ann.rad./
				EC/30/					0.2317
									0.5356
									2.1238
									(0.1−3.1)
									0.7673
^{85}Y		84.91643	2.6 h	β+/55/3.26	1.54/	1/2−			ann.rad./
				EC/45/					0.2317
									0.5045
									0.9140
									(0.07−1.4)
^{86m}Y			48. m	I.T./99/		8+	4.8		ann.rad./
				β+/					0.0102(IT)
				EC/					0.2080
									(0.09−1.1)
^{86}Y		85.91489	14.74 h	β+/5.24		4−	<0.6		ann.rad./
				EC/					0.3070
									0.6277
									1.0766
									1.1531

Elem. or Isot.	Natural Abundance (Atom %)	Atomic Mass or Weight	Half-life/ Resonance Width (MeV)	Decay Mode/ Energy (/MeV)	Particle Energy/ Intensity (MeV/%)	Spin ($h/2\pi$)	Nuclear Magnetic Mom. (nm)	Elect. Quadr. Mom. (b)	γ-Energy / Intensity (MeV/%)
									1.9207
									(0.1−3.8)
87mY			13. h	I.T./98/		9/2+	6.1		0.3807
				β+/0.7/	1.15/0.7				
				EC/					
87Y		86.910876	3.35 d	EC/99+/1.862	0.78/	1/2-			0.3880
									0.4870
88Y		87.909501	106.6 d	EC/99+/3.623	0.76/	4-			ann.rad./
				β+/0.2/					0.89802
									1.83601
									2.73404
									3.2190
89mY			15.7 s	I.T./0.909		9/2+			0.9092(IT)
89Y	100.	88.905848				1/2-	-0.13742		
90mY			3.24 h	I.T./99+/	0.68204	7+	5.1		0.2025
				β-/0.002/					0.4794
									0.6820
90Y		89.907152	2.67 d	β-/2.282	2.28/	2-	-1.630	-0.155	
91mY			49.7 m	I.T./0.555		9/2+	5.96		0.5556(IT)
91Y		90.907305	58.5 d	β-/1.544	1.545/	1/2-	0.1641		1.208
92Y		91.90895	3.54 h	β-/3.63	3.64/	2-			0.4485
									0.5611
									0.9345
									1.4054
									(0.4−3.3)
93mY			0.82 s	I.T./0.759		9/2+			0.1686(IT)
									0.5902
93Y		92.90958	10.2 h	β-/2.87	2.88/90	1/2-			0.2669
									0.9471
									1.9178
94mY			1.4 µs						0.4322
									0.7699
									1.2024
94Y		93.91160	18.7 m	β-/4.919	4.92/	2-			0.3816
									0.9188
									1.1389
									(0.3−4.1)
95Y		94.91282	10.3 m	β-/4.42		1/2-			0.4324
									0.9542
									2.1760
									3.5770
96mY			9.6 s	β-/		(3+)			0.1467
									0.6174
									0.9150
									1.1071
									1.7507
96Y		95.91589	6.2 s	β- /7.09	7.12/	0-			1.594
97mY			1.21 s	β- /7.4	4.8/	9/2+			0.1614
					6.0/				0.9700
									1.1030
97Y		96.91813	3.76 s	β- /6.69	6.7	1/2-			0.2969
									1.9960
									3.2876
									3.4013
98mY			2.1 s	β- /9.8	5.5/	(4-)			0.2415
									0.6205
									0.6473
									1.2228
									1.8016
98Y		97.92220	0.59 s	β- /8.83	8.7/	1+			0.2131

Elem. or Isot.	Natural Abundance (Atom %)	Atomic Mass or Weight	Half-life/ Resonance Width (MeV)	Decay Mode/ Energy (/MeV)	Particle Energy/ Intensity (MeV/%)	Spin ($h/2\pi$)	Nuclear Magnetic Mom. (nm)	Elect. Quadr. Mom. (b)	γ-Energy / Intensity (MeV/%)
									1.2228
									1.5907
									2.9413
									4.4501
^{99m}Y			0.011 ms						
^{99}Y		98.92464	1.47 s	β- /7.57		1/2-			0.1218/43.8
				n	/2.5/				0.5362
									0.7242
									1.0130
^{100m}Y			0.94 s	β-, n /		3+			
^{100}Y		99.9278	0.73 s	β-, n/9.3	n/1.8/	1+			
^{101}Y		100.9303	0.43 s	β-, n/8.6	n/1.5/	(5/2+)			
^{102}Y		101.9336	0.36 s	β-, n/9.9	n/4.0/				
^{103}Y		102.9367	0.23 s	β-, n	n/8.3/				
^{104}Y		103.9411	0.18 s						
^{105}Y		104.9449	> 0.15 μs						
^{106}Y		105.950	> 0.15 μs						
^{107}Y		106.9414	> 0.15 μs						
^{108}Y		107.959	> 0.15 μs						
$_{40}$Zr		91.224(2)							
^{78}Zr		77.9552	> 0.2 μs			0+			
^{79}Zr		78.9492	0.06 s						
^{80}Zr		79.940	~ 4.5 s	β+ /8.0		0+			0.290
									0.538
^{81}Zr		80.9372	5.3 s	β+ /7.2	6.1	(3/2-)			
^{82}Zr		81.9311	32. s	β+ /4.0	3.	0+			ann.rad./
^{83m}Zr			7. s	β+ /7.0		(7/2+)			ann.rad./
^{83}Zr		82.9287	44. s	β+ /5.9	4.8	(1/2-)			ann.rad./
				EC					0.0556
									0.1050
									0.2560
									0.474
									1.525
^{84}Zr		83.9233	26. m	β+ /2.7		0+			ann.rad./
				EC/					0.0449
									0.1125
									0.3729
									0.667
^{85m}Zr			10.9 s	I.T./0.2922		½-			ann.rad./
				β+, EC/					0.2922(IT)
									0.4165
^{85}Zr		84.9215	7.9 m	β+ /4.7	3.1	7/2+			ann.rad./
				EC/					0.2663
									0.4163
									0.4543
^{86}Zr		85.91647	16.5 h	EC/1.47		0+			0.0280
									0.243
									0.612
^{87m}Zr			14.0 s	I.T./0.3362		½-	+ 0.64		0.1352(IT)
									0.2010
^{87}Zr		86.91482	1.73 h	β+ /3.67	2.26	9/2+	- 0.895	+ 0.42	ann.rad./
				EC/					0.3811
									1.228
^{88m}Zr			1.4 μs			(8+)			0.077
^{88}Zr		87.91023	83.4 d	EC/0.67		0+			0.3929
^{89m}Zr			4.18 m	I.T./94/0.5877		½-	+ 0.80		ann.rad./
				β+ /1.5/					0.5877(IT)
				EC/4.7/					1.507

Table of the Isotopes

Elem. or Isot.	Natural Abundance (Atom %)	Atomic Mass or Weight	Half-life/ Resonance Width (MeV)	Decay Mode/ Energy (/MeV)	Particle Energy/ Intensity (MeV/%)	Spin ($h/2\pi$)	Nuclear Magnetic Mom. (nm)	Elect. Quadr. Mom. (b)	γ-Energy / Intensity (MeV/%)
^{89}Zr		88.908889	3.27 d	β+ /23/2.832	0.9/	9/2+	-1.05	+ 0.28	ann.rad./
				EC/77/					0.9092
^{90m}Zr			0.809 s	I.T./		5-	6.3		0.1326
									2.1862
									2.3189(IT)
^{90}Zr	51.45(40)	89.904704				0+			
^{91}Zr	11.22(5)	90.905646				5/2+	-1.30362	-0.18	
^{92}Zr	17.15(8)	91.905041				0+			
^{93}Zr		92.906476	1.5×10^6 y	β- /0.091		5/2+			0.0304
^{94}Zr	17.38(28)	93.906315	$>10^{17}$ y	β-β-		0+			
^{95}Zr		94.908043	64.02 d	β- /1.125	0.366/55	5/2+	1.13	+0.29	0.7242
					0.400/44				0.7567
^{96}Zr	2.80(9)	95.908273	3×10^{19} y	β-β-		0+			
			$>1.7 \times 10^{18}$ y	β-					
^{97}Zr		96.910953	16.75 h	β- /2.658	1.91/	½-	- 0.937		0.7434
^{98}Zr		97.91274	30.7 s	β- /2.26	2.2/100	0+			
^{99}Zr		98.91651	2.2 s	β- /4.56	3.9/	½+	- 0.930		0.4692/55.2
					3.5/				0.5459/48
									0.028−1.321
^{100}Zr		99.91776	7.1 s	β- /3.34		0+			0.4006
									0.5043
^{101}Zr		100.92114	2.1 s	β- /5.49	6.2/	3/2-	- 0.27	+ 0.81	0.1194
									0.2057
									0.2089
^{102}Zr		101.92298	2.9 s	β- /4.61		0+			
^{103}Zr		102.9266	1.3 s	β- /7.0					
^{104}Zr		103.9288	1.2 s	β- /5.9		0+			
^{105}Zr		104.9331	~ 1. s	β- /8.5					
^{106}Zr		105.9359	> 0.24 µs			0+			
^{107}Zr		106.9408	> 0.24 µs						
^{108}Zr		107.944	> 0.15 µs			0+			
^{109}Zr		108.9492	> 0.15 µs						
^{110}Zr		109.953	> 0.15 µs			0+			
$_{41}$Nb		92.90638(2)							
^{81}Nb		80.949	<0.08 µs						
^{82}Nb		81.9431	50 ms	β+ /11.					
^{83}Nb		82.9367	4.1 s	β+ /7.5					
^{84}Nb		83.9336	10. s	β+, EC/9.6		(3+)			0.540
									(0.456-1.427)
^{85m}Nb			3. s						0.069
^{85}Nb		84.9279	21. s	β+ /6.0					
^{86m}Nb			56. s	β+					
^{86}Nb		85.9250	1.46 m	β+ /8.0					ann.rad./
									0.751
									1.003
^{87m}Nb			3.7 m	β+ /		1/2-			ann.rad./
				EC/					0.1352
									0.2010
^{87}Nb		86.92036	2.6 m	β+ /5.2/		(9/2+)			ann.rad./
				EC/					0.2010
									0.4706
									0.6165
									1.0665
									1.8842
^{88m}Nb			7.7 m	β+ /		4-			ann.rad./
				EC/					0.2625
									0.3996
									1.0569

Elem. or Isot.	Natural Abundance (Atom %)	Atomic Mass or Weight	Half-life/ Resonance Width (MeV)	Decay Mode/ Energy (/MeV)	Particle Energy/ Intensity (MeV/%)	Spin $(h/2\pi)$	Nuclear Magnetic Mom. (nm)	Elect. Quadr. Mom. (b)	γ-Energy / Intensity (MeV/%)
88Nb		87.9183	14.3 m	β+ /7.6	3.2/	8+			1.0825
				EC/					ann.rad./
									1.0570
									1.0828
89mNb			2.0 h	β+ /	3.3/	9/2+			(0.07–2.5)
				EC/					0.5880/10(D)
89Nb		88.91342	1.10 h	β+ /74/4.29	2.8/	1/2-	+6.216		(0.17–4.0)
				EC/26 /					ann.rad./
									0.5074
									0.5880
									0.7696
90mNb			18.8 s	I.T./0.1246		4-			1.2775
									0.002
90Nb		89.911265	14.6 h	β/53 /6.111	0.86/5	8+	4.961		0.1225
				EC/47 /	1.5/92				ann.rad./
									0.1412
									1.1292
									2.1862
									2.3189
91mNb			62. d	I.T./97 /		1/2-			(0.1–3.3)
				EC/3 /					0.1045(IT)
91Nb		90.906996	7×10^2 y	EC/1.253		9/2+			1.2050
92mNb			10.13 d	EC/99+/		2+	6.114		Mo k x-ray
									0.9126
									0.9345
92Nb		91.907194	3.7×10^7 y	EC/2.006		7+			1.8475
									0.5611
93mNb			16.1 y	I.T./0.0304		1/2-			0.9345
									Nb x-ray
93Nb	100.	92.906378				9/2+	+6.1705	-0.32	0.0304
94mNb			6.26 m	I.T./99+	/2.086	3+			Nb k x-ray
				β- /0.5/					0.0409
94Nb		93.907284	2.4×10^4 y	β- /2.045	0.47/	6+			0.87109
									0.70263
95mNb			3.61 d	I.T./97.5/	0.2357	1/2-			0.87109
				β- /2.5 /					0.2040
									0.2356
95Nb		94.906836	34.97 d	β- /0.926	0.160/	9/2+	6.141		0.76578
96Nb		95.908101	23.4 h	β- /3.187	0.5/10	6+	4.976		0.7782
					0.75/90				0.2191–1.498
97mNb			58.1 s	I.T./0.7434	0.734/98	1/2-			0.7434
97Nb		96.908099	1.23 h	β- /1.934	1.27/98	9/2+	6.15		0.4809
									0.6579
98mNb			51. m	β- /4.67		5+			0.7874
98Nb		97.91033	2.9 s	β- /4.59	4.6/	1+			0.1726–1.89
									0.6451
									0.7874
99mNb			2.6 m	β- /	3.2/	1/2-			1.0243
									0.0978/100
99Nb		98.91162	15.0 s	β- /3.64	3.5/100	9/2+			(0.138–3.010)
									0.0977
100m2Nb			0.013 ms						0.1378/3.1
100m1Nb			3.0 s	β- /6.74	5.8				
									Nb k x-ray
									0.159
									0.6364
100Nb		99.91418	1.5 s	β- /6.25	6.2/				1.0637
					5.3/				0.5354
									0.6001–1.566

Table of the Isotopes

Elem. or Isot.	Natural Abundance (Atom %)	Atomic Mass or Weight	Half-life/ Resonance Width (MeV)	Decay Mode/ Energy (/MeV)	Particle Energy/ Intensity (MeV/%)	Spin ($h/2\pi$)	Nuclear Magnetic Mom. (nm)	Elect. Quadr. Mom. (b)	γ-Energy / Intensity (MeV/%)
^{101}Nb		100.91525	7.1 s	β- /4.57	4.3/				0.1105−0.810
^{102m}Nb			4.3 s	β- /					
^{102}Nb		101.91804	1.3 s	β- /7.21	7.2/				0.2960−2.184
^{103}Nb		102.9191	1.5 s	β- /5.53	5.3/	5/2+			
^{104m}Nb			0.9 s	β-, n/	n/0.06				
^{104}Nb		103.9225	4.8 s	β-, n/8.1	n/0.05				
^{105}Nb		104.9239	3.0 s	β-, n/6.5	n/1.7				
^{106}Nb		105.9280	1.0 s	β-, n/9.3	n/4.5				
^{107}Nb		106.9303	0.30 s	β-, n/7.9	n/6.0				
^{108}Nb		107.9348	0.19 s	β, n/	n/6.2				(0.193−0.590)
^{109}Nb		108.9376	0.19 s	β, n/	n/31				
^{110}Nb		109.9424	0.17 s	β, n/	n/40				
^{111}Nb		110.9457	> 0.15 µs						
^{112}Nb		111.951	> 0.15 µs						
^{113}Nb		112.955	> 0.15 µs						
$_{42}$Mo		95.94(2)							
^{83}Mo		82.9487	~ 6. ms						
^{84}Mo		83.9401	~ 3.7 s	β+ /6.		0+			
^{85}Mo		84.9366	3.2 s	β+/8.1		½+			
^{86}Mo		85.9307	19. s	β+ /4.8		0+			
^{87}Mo		86.9273	14. s	EC, β+/6.5					(0.752−1.004)
^{88}Mo		87.92195	8.0 m	β+ /3.4		0+	+0.5		ann.rad./
				EC					0.0800
									0.1399
									0.1707
^{89m}Mo			0.19 s	I.T./0.118		½-			0.118(IT)
									0.268
^{89}Mo		88.91948	2.2 m	β+ /5.58		9/2+			ann.rad./
				EC/					0.659
									0.803
									1.155
									1.272
^{90m}Mo			1.2 µs						0.063
^{90}Mo		89.91394	5.7 h	β+ /25/2.489	1.085/	0+			ann.rad./
				EC/75 /					0.04274
									0.12237
									0.25734
^{91m}Mo			1.08 m	I.T./50/0.653		½-			ann.rad./
				β+, EC/50 /	2.5/				0.6529
					2.8/				1.2081
					4.0/				1.5080
									2.2407
^{91}Mo		90.91175	15.5 m	β+ /94/4.43	3.44/94	9/2-			ann.rad./
				EC/6/					1.6373
									2.6321
									3.0286
									(0.1−4.2)
^{92}Mo	14.77(31)	91.906811	> 3 × 10^{17} y	β+-EC		0+			
^{93m}Mo			6.9 h	I.T./99+ /2.425		21/2+	+9.21		0.26306(IT)
									0.68461
									1.47711
^{93}Mo		92.906813	3.5 × 10^3 y	EC/0.405		5/2+			0.0304
^{94}Mo	9.23(10)	93.905088				0+			
^{95}Mo	15.90(9)	94.905842				5/2+	-0.9142	-0.02	
^{96}Mo	16.68(1)	95.904680				0+			
^{97}Mo	9.56(5)	96.906022				5/2+	-0.9335	+0.26	
^{98}Mo	24.19(26)	97.905408				0+			
^{99}Mo		98.907712	2.7476 d	β- /1.357	0.45/14	½+	0.375		0.144048

Elem. or Isot.	Natural Abundance (Atom %)	Atomic Mass or Weight	Half-life/ Resonance Width (MeV)	Decay Mode/ Energy (/MeV)	Particle Energy/ Intensity (MeV/%)	Spin ($h/2\pi$)	Nuclear Magnetic Mom. (nm)	Elect. Quadr. Mom. (b)	γ-Energy / Intensity (MeV/%)
					0.84/2				0.18109
					1.21/84				0.36644
									0.73947
^{100}Mo	9.67(20)	99.90748	~1 × 10^{19} y	β-β-		0+			
^{101}Mo		100.91035	14.6 m	β- /2.82	2.23/	1/2+			0.0063
					0.7/				0.19193
									0.5909
									(0.0809−2.405)
^{102}Mo		101.91030	11.3 m	β- /1.01	1.2/	0+			0.1493/89.
									0.2116/100.
									0.2243/32.
^{103}Mo		102.9132	1.13 m	β- /3.8		3/2+			0.1028(2)/
									0.1440(2)
									0.2511(2)
^{104}Mo		103.9138	1.00 m	β- /2.16		0+			0.0686(1)/100.
									0.4239(4)/21.
^{105}Mo		104.9170	36. s	β- /4.95		3/2+			0.0642/
									0.0856/
									0.2495/
^{106}Mo		105.91814	8.4 s	β- /3.52		0+			0.1894(2)/22.
									0.3644(2)/6.
									0.3723(2)/12.
^{107}Mo		106.9217	3.5 s	β- /6.2					
^{108}Mo		107.9235	1.1 s	β- /5.1		0+			(0.028−0.636)
^{109}Mo		108.9278	0.5 s	β- /7.2					
^{110}Mo		109.9297	0.27 s	β- /5.7		0+			Tc k x-ray
									0.142
									(0.039−0.599)
^{111}Mo		110.9344	> 0.15 μs						
^{112}Mo		111.937	> 0.15 μs			0+			
^{113}Mo		112.942	> 0.15 μs						
^{114}Mo		113.945	> 0.15 μs			0+			
^{115}Mo		114.950	> 0.15 μs						
^{116}Mo			> 0.15 μs			0+			
^{117}Mo			> 0.15 μs						
$_{43}$Tc									
^{85}Tc		84.9488	< 0.1 ms						
^{86}Tc		85.9429	0.05 s	β+ /11.9					
^{87}Tc		86.9365	2.4 s	β+ /8.6					
^{88}Tc		87.9327	5.8 s	β+ /10.1					
^{89m}Tc			13. s						
^{89}Tc		88.9272	13. s	β+ /7.5					
^{90m}Tc			49.2 s	β$^+$	5.3/	6+			ann.rad./
									0.9479/
									1.0542/
^{90}Tc		89.9236	8.3 s	β+ /8.9	7.0/15	1+			ann.rad./
					7.9/95.				0.9479/
^{91m}Tc			3.3 m	β+		½+			ann.rad./170.
				EC					0.8110(5)/5.
									1.6052(1)/7.8
									1.6339(1)/9.1
									1.9023(1)/6.
									2.4509(1)/13.5
^{91}Tc		90.9184	3.14 m	β+ /6.2	5.2	9/2+			ann.rad./200.
^{92}Tc		91.91526	4.4 m	β+ /7.87	4.1	8+			ann.rad./200.
				EC					0.0850/
									0.1475
									0.3293

Elem. or Isot.	Natural Abundance (Atom %)	Atomic Mass or Weight	Half-life/ Resonance Width (MeV)	Decay Mode/ Energy (/MeV)	Particle Energy/ Intensity (MeV/%)	Spin ($h/2\pi$)	Nuclear Magnetic Mom. (nm)	Elect. Quadr. Mom. (b)	γ-Energy / Intensity (MeV/%)
									0.7731
									1.5096
⁹³ᵐTc			43. m	I.T./13		½-			0.3924(IT)
				EC/20					0.9437
									2.6445
⁹³Tc		92.910249	2.73 h	β+ /13/3.201	0.81	9/2+	6.26		ann.rad./
				EC/87/					1.3629
									1.4771
									1.5203
									(0.1−3.0)
⁹⁴ᵐTc			52. m	β+ /72/4.33		2+			ann.rad./
				EC/28/					0.8710
									1.8686
⁹⁴Tc		93.909657	4.88 h	β+ /11/4.256		7+	5.08		ann.rad./
				EC/89/					0.4491
									0.7026
									0.8496
									0.8710
⁹⁵ᵐTc			61. d	I.T./4/		1/2-			ann.rad./
				β+ /0.3	0.5/				0.0389(IT)
				EC/96	0.7/				0.2041
									0.5821
									0.5821
									0.8351
⁹⁵Tc		94.90766	20.0 h	EC/100/1.691		9/2+	5.89		0.7657
									1.0738
⁹⁶ᵐTc			52. m	I.T./90/		4+			0.0342(IT)
				β+, EC/2/					0.7782
									1.2002
⁹⁶Tc		95.90787	4.3 d	EC/2.973		7+	+5.04		Mo k x-ray
									0.7782
									0.8125
									0.8498
									1.12168
⁹⁷ᵐTc			91. d	I.T./0.0965		1/2-			Tc k x-ray
				EC	/3.9				0.0965
⁹⁷Tc		96.906365	4.2 × 10⁶ y	EC/100/0.320		9/2+			Mo k x-ray
⁹⁸Tc		97.907216	~ 6.6 × 10⁶ y	β- /1.80	0.40/100	6+			0.65241
									0.74535
⁹⁹ᵐTc			6.01 h	I.T./100/0.142		1/2-			Tc k x-ray
									0.14049
									0.14261
⁹⁹Tc		98.906255	2.13 × 10⁵ y	β- /0.294	0.293/100	9/2+	+5.6847	-0.129	
¹⁰⁰Tc		99.907658	15.8 s	β- /3.202	2.2/	1+			0.5396
				EC /1.8(10)⁻³/	2.9/				0.5908
					3.3				(0.3 79−2.30)
¹⁰¹Tc		100.90732	14.2 m	β- /1.61	1.32/	9/2+			0.1272
									0.1841
									0.3068
									0.5451
									(0.073−0.969)
¹⁰²ᵐTc			4.4 m	I.T./2/4.8	1.8/				0.4184
				β- /98/					0.4752
									0.6281
									0.6302
									1.0464
									1.1033
									1.6163
									2.2447
¹⁰²Tc		101.90922	5.3 s	β- /4.53	3.4/	1+			0.4686

Elem. or Isot.	Natural Abundance (Atom %)	Atomic Mass or Weight	Half-life/ Resonance Width (MeV)	Decay Mode/ Energy (/MeV)	Particle Energy/ Intensity (MeV/%)	Spin ($h/2\pi$)	Nuclear Magnetic Mom. (nm)	Elect. Quadr. Mom. (b)	γ-Energy / Intensity (MeV/%)
					4.2				0.4751
					2.2/				1.1055
^{103}Tc		102.90918	54. s	β- /2.66	2.0/	5/2+			0.1361
					2.2/				0.1743
									0.2104
									0.3464
									0.5629
									(0.13–1.0)
^{104m}Tc			0.005 ms						
^{104}Tc		103.91145	18.2 m	β- /5.60	5.3/	(3+)			0.3483
									0.3580
									0.5305
									0.5351
									0.8844
									0.8931
									1.6768
									(0.3–3.7)
^{105}Tc		104.9117	7.6 m	β- /3.6	3.4/	5/2+			0.1079
									0.1432
									0.3215
^{106}Tc		105.91436	36. s	β- /6.55		2+			0.2703
									0.5222
									1.9694
									2.2393
									2.7893
^{107}Tc		106.9151	21.2 s	β- /4.8					0.1027
									0.1063
									0.1770
									0.4587
^{108}Tc		107.9185	5.1 s	β- /7.72		(3)			0.2422
									0.4656
									0.7078
									0.7326
									1.5835
^{109}Tc		108.9200	1.4 s	β- /6.3	p/0.08				
^{110}Tc		109.9238	0.83 s	β- /8.8	p/0.04				0.2407
^{111}Tc		110.9257	0.30 s	β- .n/7.0	n/0.85				0.150/92.7
									0.063–1.435
^{112}Tc		111.9292	0.26 s	β, n	n/2.6				
^{113}Tc		112.9316	0.15 s	β-, n/8.	/2.1				0.0985/100
									0.0658–1.520
^{114}Tc		113.936	0.15 s	β-, n	/1.3				
^{115}Tc		114.939	> 0.15 μs						
^{116}Tc		115.943	> 0.15 μs						
^{117}Tc		116.946	> 0.15 μs						
^{118}Tc		117.951	> 0.15 μs						
$_{44}$Ru		101.07(2)							
^{87}Ru		86.949	> 1.5 μs						
^{88}Ru		87.9403	1.2 s			0+			
^{89}Ru		88.9361	1.4 s	β+ .p/8.					
^{90}Ru		89.9299	12. s	β+ /5.9		0+			ann.rad./
									0.155–1.551
^{91}Ru		90.9263	7.9 s	β+, EC/7.4		9/2+			ann.rad./
									(0.205-1.998)
^{92}Ru		91.9201	3.7 m	β+ /53/4.5 EC/47/		0+			ann.rad./
									0.1346
									0.2138
									0.2593

Table of the Isotopes

Elem. or Isot.	Natural Abundance (Atom %)	Atomic Mass or Weight	Half-life/ Resonance Width (MeV)	Decay Mode/ Energy (/MeV)	Particle Energy/ Intensity (MeV/%)	Spin ($h/2\pi$)	Nuclear Magnetic Mom. (nm)	Elect. Quadr. Mom. (b)	γ-Energy / Intensity (MeV/%)
^{93m}Ru			10.8 s	I.T./21/		1/2-			ann.rad./
				β+, EC/79/	5.3/				0.7344
									1.1112
									1.3962
									2.0931
^{93}Ru		92.9171	1.0 m	β+ /6.3		9/2+			ann.rad./
				EC/					0.6807
									1.4349
									(0.5−4.2)weak
^{94}Ru		93.91136	52. m	EC/100/1.59		0+			0.3672
									0.5247
									0.8922
^{95}Ru		94.91041	1.64 h	EC/85/2.57	1.20/	5/2+	0.86		ann.rad./
				β+ /15/	0.91/				0.3364
									0.6268
									0.036−2.424
^{96}Ru	5.54(14)	95.90760	> 3.1 × 10^{16} y	β+β+		0+			
^{97}Ru		96.90756	2.89 d	EC/1.12		5/2+	-0.78		Tc k x-ray
									0.2157
									0.3245
									0.4606
^{98}Ru	1.87(3)	97.90529				0+			
^{99}Ru	12.76(14)	98.905939				5/2+	-0.6413	+0.079	
^{100}Ru	12.60(7)	99.904220				0+			
^{101}Ru	17.06(2)	100.905582				5/2+	-0.7188	+0.46	
^{102}Ru	31.55(14)	101.904349				0+			
^{103}Ru		102.906324	39.27 d	β- /0.763	0.223	3/2+	0.206	+0.62	0.05329
									0.29498
									0.4438
									0.49708
									0.55704
									0.61033
									(0.04−1.6)
^{104}Ru	18.62(27)	103.905433				0+			
^{105}Ru		104.907753	4.44 h	β- /1.917	1.11/22	3/2+	-0.3		0.12968
					1.134/13				0.1491
					1.187/49				0.2629
									0.31664
									0.46943
									0.67634
									0.72420
									(0.1−1.8)
^{106}Ru		105.90733	1.020 y	β- /0.0394	0.0394/100	0+			
^{107}Ru		106.9099	3.8 m	β- /2.9	2.1/				0.1939
					3.2/				0.3741
									0.4625
									0.8488
^{108}Ru		107.9102	4.5 m	β- /1.4	1.2/	0+			0.0923
									0.1651
									0.4339
									0.4975
									0.6189
^{109}Ru		108.9132	34.5 s	β- /4.2					0.1164
									0.3584
^{110}Ru		109.9141	15. s	β- /2.81		0+			0.1121
									0.3737
									0.4397
									0.7967
^{111}Ru		110.9177	1.5 s	β- /5.5					
^{112}Ru		111.9190	4.5 s	β- /4.5		0+			

Elem. or Isot.	Natural Abundance (Atom %)	Atomic Mass or Weight	Half-life/ Resonance Width (MeV)	Decay Mode/ Energy (/MeV)	Particle Energy/ Intensity (MeV/%)	Spin $(h/2\pi)$	Nuclear Magnetic Mom. (nm)	Elect. Quadr. Mom. (b)	γ-Energy / Intensity (MeV/%)
113mRu			0.6 s						
113Ru		112.9225	0.80 s	β- /7.					0.2632
									0.048−2.418
114Ru		113.9243	0.57 s	β- /6.1		0+			0.127/24
									(0.053−0.180)
115Ru		114.9287	~ 0.74 s	β- /8.					
116Ru		115.931	> 0.15 μs			0+			
117Ru		116.936	> 0.15 μs						
118Ru		117.938	> 0.15 μs			0+			
119Ru		118.943	> 0.15 μs						
120Ru		119.945	> 0.15 μs			0+			
45Rh		102.90550(2)							
89Rh		88.9488	> 0.15 μs						
90mRh			~ 12. ms						
90Rh		89.9429	1.0 s						
91mRh			1.5 s	IT					0.387
91Rh		90.9366	1.5 s						(0.438-0.973)
92mRh			0.5 s						0.866
92Rh		91.9320	4.7 s	β+ /11.1					(0.163-0.991)
93Rh		92.9257	12. s	β+ /8.1					(0.138−1.493)
94mRh			25.8 s	β+ /		8+			ann.rad./
									0.1264
									0.3117
									0.7562
									1.0752
									1.4307
94Rh		93.9217	1.18 m	β+ /9.6	6.4/	3+			ann.rad./
									0.1461
									0.3117
									0.7562
									1.4307
95mRh			1.96 m	I.T./88/		½+			ann.rad./
				β+, EC/12/					0.5433(IT)
									0.7837
95Rh		94.9159	5.0 m	β+ /5.1	3.2	9/2+			ann.rad./
									0.2293
									0.4103
									0.6610
									0.9416
									1.3520
									(0.2−3.8)
96mRh			1.51 m	I.T./60/0.052		2+			ann.rad./
				β+, EC/40/	4.70/				Tc,Ru x-rays
									0.8326
									1.0985
									1.6921
									(0.4−3.3)
96Rh		95.91446	9.6 m	β+/6.45	3.3/	5+			ann.rad./
				EC/					0.4299
									0.6315
									0.6853
									0.7418
									0.8326
									(0.2−3.4)
97mRh			46. m	I.T./5 /	2.6/	1/2-			ann.rad./
				β+, EC/95/					0.1886
									0.4215
									2.2452

Table of the Isotopes

Elem. or Isot.	Natural Abundance (Atom %)	Atomic Mass or Weight	Half-life/ Resonance Width (MeV)	Decay Mode/ Energy (/MeV)	Particle Energy/ Intensity (MeV/%)	Spin ($h/2\pi$)	Nuclear Magnetic Mom. (nm)	Elect. Quadr. Mom. (b)	γ-Energy / Intensity (MeV/%)
^{97}Rh		96.91134	31.0 m	β+ /3.52	2.1/	9/2+			ann.rad./
									0.1886
									0.3892
									0.4515
									0.8398
									0.8788
									(0.2−3.5)
^{98m}Rh			3.5 m	β+ /		5+			ann.rad./
									0.6154
									0.6524
									0.7452
^{98}Rh		97.91071	8.7 m	β+ /90/5.06	3.4/	2+			ann.rad./
									0.6524
									0.7623
^{99m}Rh			4.7 h	β+ /8/	.74/	9/2+	5.67		ann.rad./
				EC/92/					0.2766/
									0.3408
									0.6178
									1.2612
^{99}Rh		98.90813	16. d	β+/4/2.10	0.54/	1/2-			ann.rad./
				EC/97/	0.68/				0.0894/
									0.3530
									0.5277
									(0.1−2.0)
^{100m}Rh			4.7 m	I.T./99/		5+			ann.rad./
				β+ /0.4/					0.0748/
									0.2647(IT)
^{100}Rh		99.90812	20.8 h	β+ /3.63	2.62/	1-			0.4462
				EC/	2.07/				0.5396
									0.5882
									0.8225
									1.5534
									2.3761
^{101m}Rh			4.35 d	EC/92/		9/2+	+5.51		Rh k x-ray
				I.T./8/0.1573					0.1272/
									0.3069
									0.5451
^{101}Rh		100.90616	3.3 y	EC/0.54		1/2-			Ru k x-ray
									0.1272
									0.1980
									0.3252
^{102m}Rh			3.74 y	EC/2.323		6$^+$	4.04		0.4751
				IT/0.0419					0.6313
			> 1.2×10^6 y	β$^+$	/<0.00025				0.6975
									0.7668
									1.0466
									1.1032
^{102}Rh		101.906843	207. d	EC/62			0.5		ann.rad./
				β- /19/					0.4686
				β+ /14/					0.4751
									0.5566
									0.6280
									1.1032
									(0.4−1.6)
^{103m}Rh			56.12 m	IT		7/2+	4.54		
^{103}Rh	100.	102.905504				1/2-	-0.0884		
^{104m}Rh			4.36 m	I.T./99+ /		5+			Rh k x-ray
				β-	1.3/				0.0514
									0.0971
									0.5558

Elem. or Isot.	Natural Abundance (Atom %)	Atomic Mass or Weight	Half-life/ Resonance Width (MeV)	Decay Mode/ Energy (/MeV)	Particle Energy/ Intensity (MeV/%)	Spin ($h/2\pi$)	Nuclear Magnetic Mom. (nm)	Elect. Quadr. Mom. (b)	γ-Energy / Intensity (MeV/%)
^{104}Rh		103.906656	42.3 s	β-/99+/2.441	1.88/2	1+			0.3581
				EC/0.4/1.141	2.44/98				0.5558
									1.2370
									(0.35−1.8)
^{105m}Rh			43. s	I.T./1.296		1/2-			Rh k x-ray
									0.1296
^{105}Rh		104.905694	35.4 h	β- /0.567	0.247/30	7/2+	+4.45		0.2801
					0.567/70				0.3061
									0.3189
^{106m}Rh			2.18 h	β- /	0.92/	6+			0.2217
									0.4510
									0.5119
									0.6162
									0.7173
									0.7484
									1.0458
									1.5277
^{106}Rh		105.90729	29.9 s	β- /3.54	2.4/2	1+	+2.58		0.51186/
					3.0/12				0.61612
					3.54/79				0.62187
									(0.05−3.04)
^{107}Rh		106.90675	21.7 m	β- /1.51	1.20/65	7/2+			0.2776
					1.5/17				0.3028
									0.3925
^{108m}Rh			6.0 m	β- /	1.57/				0.4339
									0.4973
									0.6189
^{108}Rh		107.9087	17. s	β- /4.5		1+			0.4046
									0.4339
									0.4973
									0.5811
									0.6146
									0.9014
									0.9471
^{109}Rh		108.90874	1.34 m	β- /2.59	2.25/	7/2+			0.1134
									0.1780
									0.2914
									0.3254
									0.3268
									0.4261
^{110m}Rh			29. s	β- /	6/				(0.1−1.6)
									0.3737
									0.4397
^{110}Rh		109.91114	3.1 s	β- /5.4	5.5/	1+			0.7967
									0.3737
									0.4400
									0.5463
									0.6877
									0.8381
									0.9045
^{111}Rh		110.91159	11. s	β- /3.7					0.275
^{112m}Rh			6.8 s	β- /					
^{112}Rh		111.9144	3.5 s	β- /6.2		1+			0.3489
^{113}Rh		112.91553	0.9 s	β- /4.9					0.1285
^{114m}Rh			1.9 s	β- /					(0.103-1.923)
^{114}Rh		113.9188	1.8 s	β- /6.5		1+			(0.276-0.783)
^{115}Rh		114.9203	0.99 s	β- /6.0					
^{116m}Rh			0.9 s	β- /					0.3405
^{116}Rh		115.9241	0.7 s	β- /8.0		1+			0.340
									0.398−1.665

Table of the Isotopes

Elem. or Isot.	Natural Abundance (Atom %)	Atomic Mass or Weight	Half-life/ Resonance Width (MeV)	Decay Mode/ Energy (/MeV)	Particle Energy/ Intensity (MeV/%)	Spin ($h/2\pi$)	Nuclear Magnetic Mom. (nm)	Elect. Quadr. Mom. (b)	γ-Energy / Intensity (MeV/%)
117Rh		116.9260	0.44 s	β- /7.					0.0346
									0.1317
118Rh		117.9301	~ 0.30 ms						0.379
									0.575
									0.370−1.037
119Rh		118.932	0.17 s						
120Rh		119.936	0.12 s						
121Rh		120.939	> 0.15 µs						
122Rh		121.943							
46Pd		106.42(1)							
91Pd		90.949	> 1.5 µs						
92Pd		91.9404	1.0 s			0+			
93Pd		92.9359	1.2 s	β+, p		9/2+			0.240/81
									0.382−0.864
94Pd		93.9288	9. s	EC, β+ /~ 6.6		0+			0.5582
									(0.0546−0.798)
95mPd		94.92684	13.4 s	EC, β+ /10.2		21/2+			
95Pd		94.9247							
96Pd		95.9182	2.03 m	EC, β+ /3.5	1.15/	0+			0.1248
									0.4995
97Pd		96.9165	3.1 m	β+, EC/4.8	3.5/	5/2+			ann.rad./
									0.2653
									0.4752
									0.7927
									(0.2−3.4)
98Pd		97.91272	17.7 m	β+ /1.87		0+			ann.rad./
				EC/					0.0677
									0.1125
									0.6630
									0.8379
99Pd		98.91177	21.4 m	β+ /49/3.37	2.18/	5/2+			ann.rad./
				EC/51/					0.1360
									0.2636
									0.6734
									(0.2−2.85)
100Pd		99.90851	3.7 d	EC/0.36		0+			0.03271
									0.0748
									0.0840
101Pd		100.90829	8.4 h	β+ /5/1.980	0.776/	5/2+	-0.66		ann.rad./
				EC/95/					0.0244
									0.2963
									0.5904
102Pd	1.02(1)	101.905609				0+			
103Pd		102.906087	16.99 d	EC/0.543		5/2+			Rh k x-ray
									0.03975
									0.3575
									0.4971
104Pd	11.14(8)	103.904036				0+			
105Pd	22.33(8)	104.905085				5/2+	-0.642	+0.66	
106Pd	27.33(3)	105.903486				0+			
107mPd			20.9 s	I.T./0.2149		11/2-			Pd k x-ray
									0.2149(IT)
107Pd		106.905133	6.5 × 10^6 y	β- /0.033	0.03/	5/2+			
108Pd	26.46(9)	107.903893				0+			
109mPd			4.75 m	I.T./0.1889		11/2-			Pd x-ray
									0.1889(IT)
109Pd		108.905950	13.5 h	β- /1.116	1.028	5/2+			0.0880
									(0.08−1.0)

Elem. or Isot.	Natural Abundance (Atom %)	Atomic Mass or Weight	Half-life/ Resonance Width (MeV)	Decay Mode/ Energy (/MeV)	Particle Energy/ Intensity (MeV/%)	Spin ($h/2\pi$)	Nuclear Magnetic Mom. (nm)	Elect. Quadr. Mom. (b)	γ-Energy / Intensity (MeV/%)
[110]Pd	11.72(9)	109.905153				0+			
[111m]Pd			5.5 h	I.T./73/0.172		11/2-			0.0704
				β- /27/	0.35				0.1722
					0.77				0.3912
									(0.1−1.97)
[111]Pd		110.90767	23.4 m	β- /2.19	2.2/95	5/2+			0.0598
									0.2454
									0.5800
									0.6504
									1.3885
									1.4590
[112]Pd		111.90731	21.04 h	β- /0.29	0.28/	0+			0.018
[113m]Pd			1.48 m	β- /		5/2+			0.0959
[113]Pd		112.91015	1.64 m	β- /3.34					0.0958
									0.4824
									0.6436
									0.7394
[114]Pd		113.91036	2.48 m	β- /1.45		0+			0.1266
									0.2320
									0.5582
									0.5760
[115m]Pd			50. s			(9/2-)			0.089
[115]Pd		114.9137	25. s	β- /4.58		(3/2+)			0.1255
									0.2554
									0.3428
[116]Pd		115.9142	12.7 s	β- /2.61		0+			0.1015
									0.1147
									0.1778
[117m]Pd			19. ms			(9/2-)			0.203
[117]Pd		116.9178	4.4 s	β- /5.7		(3/2+)			0.2473
									0.077−0.403
[118]Pd		117.9190	2.4 s	β- /4.1		0+			0.1254
									0.028−0.596
[119]Pd		118.9231	0.9 s	β- /6.5					0.2566
									0.070−0.326
[120]Pd		119.9247	0.5 s	β- /5.0		0+			0.1581
									0.053−0.595
[121]Pd		120.9289	> 0.24 μs						
[122]Pd		121.9306	> 0.24 μs			0+			
[123]Pd		122.935	> 0.15 μs						
[124]Pd		123.9369				0+			
[47]Ag		107.8682(2)							
[93]Ag		92.950							
[94m]Ag			0.41 s	β+	p/0.79/1.9				(0.096-1.092)
				β+,p/27.	p/1.01/2.2				
[94]Ag		93.9428	0.60 s	β+					(0.659-0.905)
				β+,p/20.					
[95]Ag		94.9355	2.0 s	β+, p/					(0.089−2.940)
[96m]Ag			4.4 s	β+		8+			
				β+, p	/8.				
[96]Ag		95.9307	7. s	β+ /11.6		2+			ann.rad./
				EC/					0.1248
				β+, p	/18.				0.4995
									(0.1066−1.416)
[97]Ag		96.9240	19. s	β+ /7.0					ann.rad./
				EC/					0.6862
									1.2941
									(0.352−3.294)

Table of the Isotopes

Elem. or Isot.	Natural Abundance (Atom %)	Atomic Mass or Weight	Half-life/ Resonance Width (MeV)	Decay Mode/ Energy (/MeV)	Particle Energy/ Intensity (MeV/%)	Spin ($h/2\pi$)	Nuclear Magnetic Mom. (nm)	Elect. Quadr. Mom. (b)	γ-Energy / Intensity (MeV/%)
^{98}Ag	97.9216		47.6 s	β+ /8.4		5+			ann.rad./
				EC/	/36.				0.5711
				β+, p	/0.11				0.6786
									0.8631
									(0.153–1.185)
^{99m}Ag			11. s	I.T./100/		½−			Ag k x-ray
									0.1636(IT)
									0.3426
^{99}Ag		98.9176	2.07 m	β+ /87 5.4		9/2+			ann.rad./
				EC/13/					0.2199
									0.2645
									0.8056
									0.8323
									(0.2–3.5)
^{100m}Ag			2.3 m	β+ /		2+			ann.rad./
				EC/					0.6657
									1.6941
^{100}Ag		99.9161	2.0 m	β+/7.1	4.7/	5+			ann.rad./
				EC/					0.2807
									0.4503
									0.6657
									0.7508
									0.7732
^{101m}Ag			3.1 s	I.T./0.23		½−			Ag k x-ray
									0.0981
									0.176(IT)
^{101}Ag		100.9128	11.1 m	β+ /69/4.2	2.7/	9/2+	5.7		ann.rad./
				EC/31/					0.2610
					2.18/				0.2747
					2.73/				0.3269
					3.38/				0.4392
									0.6673
									1.1739
									(0.2–3.1)
^{102m}Ag			7.8 m	β+ /38/	3.4	2+	+4.14		ann.rad./
				EC/13/					0.5567
				I.T./49/					0.9777
									1.8347
									2.0545
									2.1594
									3.2386
^{102}Ag		101.91169	13.0 m	β+ /78/5.92	2.26/	5+	4.6		ann.rad./
				EC/22/					0.5564
									0.7193
									0.163–2.242
^{103m}Ag			5.7 s	I.T./0.134		1/2−			Ag k x-ray
									0.1344
^{103}Ag		102.90897	1.10 h	β+ /28/2.69	1.7	7/2+	+4.47		ann.rad./
				EC/72/	1.3				0.1187
									0.1482
^{104m}Ag			33. m	β+ /64/	2.71/	2+	+3.7		ann.rad./
				EC/36/					0.5558
				I.T./0.07/					0.7657
									(0.5–3.4)
^{104}Ag		103.90863	69. m	β+ /16/4.28	0.99/	5+	3.92		ann.rad./
				EC/84/					0.5558
									0.9259
									0.9416
									(0.18–2.27)
^{105m}Ag			7.2 m	I.T./98/0.0255		7/2+	+4.41		Ag x-ray

Elem. or Isot.	Natural Abundance (Atom %)	Atomic Mass or Weight	Half-life/ Resonance Width (MeV)	Decay Mode/ Energy (/MeV)	Particle Energy/ Intensity (MeV/%)	Spin ($h/2\pi$)	Nuclear Magnetic Mom. (nm)	Elect. Quadr. Mom. (b)	γ-Energy / Intensity (MeV/%)
				EC/2 /					0.3063
									0.3192
									(0.1−1.0)
^{105}Ag		104.90653	41.3 d	EC/1.35		1/2-	0.1014		0.0640
									0.2804
									0.3445
									0.4434
^{106m}Ag			8.4 d	EC/		6+	3.71	+1.1	Pd k x-ray
									0.4510
									0.5118
									0.7173
									1.0458
^{106}Ag		105.90667	24.0 m	β+ /59/2.965	/1.96	1+	+2.85		ann.rad./
				EC/41 /					0.5119
^{107m}Ag			44.2 s	I.T./0.093		7/2+	+4.40	1.0	Ag x-ray
									0.0931
^{107}Ag	51.839(8)	106.905097				1/2-	-0.11357		
^{108m}Ag			418. y	EC/92/		6+	3.580	+1.3	Ag k x-ray
				I.T./8 /0.079					Pd k x-ray
									0.43392
									0.61427
									0.72290
^{108}Ag		107.905956	2.39 m	β- /97/1.65	1.02/1.7	1+	+2.6884		ann.rad./
				EC/2/	1.65/96				0.43392
				β- /1/1.92	0.88/0.3				0.61885
									0.63298
^{109m}Ag			39.8 s	I.T./0.088		7/2+	+4.40	+1.0	Ag k x-ray
									0.0880
^{109}Ag	48.161(8)	108.904752				1/2-	-0.13069		
^{110m}Ag			249.8 d	β- /99/	0.087	6+	+3.60	+1.4	0.65774
				I.T./1 /0.1164	0.530				0.76393
									0.88467
									0.93748
									1.38427
									(0.447−1.56)
^{110}Ag		109.906107	24.6 s	β- /2.892	2.22/5	1+	+2.7271	0.2	0.65774
					2.89/95				0.8154
									1.1257
^{111m}Ag			1.08 m	IT/99/0.0598		7/2+			Ag k x-ray
				β- /1/					0.0598
									0.2454
^{111}Ag		110.905294	7.47 d	β- /1.037	1.035/	1/2-	-0.146		0.2454
									0.3421
^{112}Ag		111.90701	3.13 h	β- /3.96	3.94/	2-	0.0547		0.6067
					3.4				0.6174
									1.3877
									(0.4−2.9)
^{113m}Ag			1.14 m	I.T./80 /0.043		7/2+			0.1422
				β- /20 /	1.5				0.2983
									0.3161
									0.3923
^{113}Ag		112.90657	5.3 h	β- /2.02	2.01/	1/2-	0.159		0.2588
									0.2986
^{114}Ag		113.90880	4.6 s	β- /5.08	4.9/	1+			0.5582
									0.5760
									1.9946
^{115m}Ag			18.7 s	β- /		7/2+			0.1134
									0.1315
									0.2288
									0.3887

Table of the Isotopes

Elem. or Isot.	Natural Abundance (Atom %)	Atomic Mass or Weight	Half-life/ Resonance Width (MeV)	Decay Mode/ Energy (/MeV)	Particle Energy/ Intensity (MeV/%)	Spin ($h/2\pi$)	Nuclear Magnetic Mom. (nm)	Elect. Quadr. Mom. (b)	γ-Energy / Intensity (MeV/%)
^{115}Ag		114.90876	20. m	β- /3.10		1/2-			0.1316
									0.2128
									0.2291
									0.4727
									(0.13−2.49)
^{116m2}Ag			20. s	β-,IT/7	IT/0.0479				
^{116m}Ag			9.8 s	β-/92 /	3.2/	5+			0.5134
					2.9				0.7055
				I.T./8	IT/.0809				0.255−2.838
^{116}Ag		115.91136	2.68 m	β- /6.16	5.3	2-			0.5134
									0.6993
									2.4779
^{117m}Ag			5.3 s	β- /	3.2/	7/2+			0.1354
									0.2981
									0.3868
									0.1571
^{117}Ag		116.91168	1.22 m	β- /4.18	2.3	1/2-			0.1354
									0.3377
^{118m}Ag			2.8 s	β- /59/					0.1277
				I.T./41 /0.1277					0.4878
									0.6771
									0.7709
									(0.190-2.778)
^{118}Ag		117.9146	4.0 s	β- /7.1					0.4878
									0.6771
									3.2259
^{119}Ag		118.9157	2.1 s	β- /5.35		7/2+			0.0674
									0.3662
									0.3991
									0.6264
^{120m}Ag			0.40 s	β- /63.					0.2030
				I.T./37.					0.5059
									0.6978
									0.8300
									(0.115-1.644)
^{120}Ag		119.9188	1.23 s	β- /8.2					0.5059
				β-,n	n//<0.0030%				0.6978
									0.8171
									(0.442-3.044)
^{121}Ag		120.9199	0.78 s	β- /6.4					0.1150
									0.3148
									0.3537
									0.3696
									0.5007
									1.5105
									(0.11−2.5)
^{122m}Ag			1. s	β- /					
^{122}Ag		121.9235	0.44 s	β- /9.2					
^{123}Ag		122.9249	0.31 s	β- /7.4					
^{124}Ag		123.9286	0.22 s	β- /10.1					
^{125}Ag		124.9304	0.17 s	β-					
^{126}Ag		125.9345	0.11 s	β-					
^{127}Ag		126.9368	0.11 s	β-					
^{128}Ag		127.9412	58 ms	β-					
^{129m}Ag			0.16 s						
^{129}Ag		128.9437	~ 46. ms	β-, n					
^{130}Ag		129.9505	~ 35 ms						
$_{48}$Cd		112.411(8)							

Elem. or Isot.	Natural Abundance (Atom %)	Atomic Mass or Weight	Half-life/ Resonance Width (MeV)	Decay Mode/ Energy (/MeV)	Particle Energy/ Intensity (MeV/%)	Spin ($h/2\pi$)	Nuclear Magnetic Mom. (nm)	Elect. Quadr. Mom. (b)	γ-Energy / Intensity (MeV/%)
^{95}Cd		94.950							
^{96}Cd		95.9398				0+			
^{97}Cd		96.9349	3. s	β+, (p)					
^{98}Cd		97.9274	9.2 s	β+ /5.4		0+			
				(p)	/0.025				
^{99}Cd		98.9250	16. s	β+, EC/6.9					ann.rad./
^{100}Cd		99.9203	1.1 m	β+, EC/3.9		0+			ann.rad./
									(0.090−1.043)
^{101}Cd		100.9187	1.2 m	β+ /83/5.5	4.5	5/2+			In k x-ray
				EC/17/					0.0985
									1.7225
									0.31−2.84)
^{102}Cd		101.91446	5.8 m	β+ /27/2.59		0+			ann.rad./
				EC/73					0.0974
									0.4810
									1.0366
									1.3598
^{103}Cd		102.91342	7.5 m	β+ /33/4.14		5/2+	-0.81	-0.8	ann.rad./
				EC/67/					Ag k x-ray
									1.0799
									1.4487
									1.4618
									(0.1−2.8)
^{104}Cd		103.90985	58. m	EC/1.14		0+			Ag k x-ray
									0.0835
									0.7093
^{105}Cd		104.90947	55.5 m	β+ /26/2.739	1.69/	5/2+	-0.7393	+0.43	Ag k x-ray
				EC/74/					0.3469
									0.6072
									0.9618
									1.3025
									(0.25−2.4)
^{106}Cd	1.25(6)	105.90646	> 5.8 × 10^{17} y	EC, EC		0+			
^{107}Cd		106.90662	6.52 h	EC/99+/1.417		5/2+	-0.615055	+0.68	Ag k x-ray
				β+ /					0.0931
									0.8289
^{108}Cd	0.89(3)	107.90418	>4.1 × 10^{17} y	EC EC		0+			
^{109}Cd		108.904982	462.0 d	EC/0.214		5/2+	-0.827846	+0.69	Ag k x-ray
									0.08804
^{110}Cd	12.49(18)	109.903002				0+			
^{111m}Cd			48.5 m	I.T./		11/2-			Cd k x-ray
									0.1508(IT)
									0.2454
^{111}Cd	12.80(12)	110.904178				1/2+	-0.594886		
^{112}Cd	24.13(21)	111.902758				0+			
^{113m}Cd			14.1 y	β- /99.9/0.59	0.59/99.9	11/2-	-1.087	-0.71	0.2637
^{113}Cd	12.22(12)	112.904402	8.2 × 10^{15} y	β-		1/2+	-0.622301		
^{114}Cd	28.73(42)	113.903359	>6.0 × 10^{17} y	β-β-		0+			
^{115m}Cd			44.6 d	β- /1.629	0.68/1.6	11/2-	-1.042	-0.54	0.48450
					1.62/97				0.93381
									1.29064
^{115}Cd		114.905431	2.228 d	β- /1.446	0.593/42	1/2+	-0.648426		0.23141
					1.11/58				0.26085
									0.33624
									0.49227
									0.52780
^{116}Cd	7.49(18)	115.904756	3.8 × 10^{19} y	β-β-		0+			
^{117m}Cd			3.4 h	β- /2.66	0.72/	11/2-			0.1586
									0.5529
									0.37−2.42

Elem. or Isot.	Natural Abundance (Atom %)	Atomic Mass or Weight	Half-life/ Resonance Width (MeV)	Decay Mode/ Energy (/MeV)	Particle Energy/ Intensity (MeV/%)	Spin ($h/2\pi$)	Nuclear Magnetic Mom. (nm)	Elect. Quadr. Mom. (b)	γ-Energy / Intensity (MeV/%)
^{117}Cd		116.907219	2.49 h	β- /2.52	0.67/51	1/2+			0.2209
					2.2/10				0.2733
									0.3445
									1.3033
^{118}Cd		117.90692	50.3 m	β- /0.52		0+			
^{119m}Cd			2.20 m	β- /		11/2-			0.1056
									0.7208
									1.0250
									2.0213
^{119}Cd		118.9099	2.69 m	β- /3.8	~ 3.5/	1/2+			0.1340
									0.2929
									0.3429
^{120}Cd		119.90985	50.8 s	β- /1.76	1.5/	0+			
^{121m}Cd			8. s	β- /		11/2-			0.1008
									0.9878
									1.0209
									1.1815
									2.0594
^{121}Cd		120.9130	13.5 s	β- /4.9		(3/2+)			0.2102
									0.3242
									0.3492
									1.0403
^{122}Cd		121.91333	5.3 s	β- /3.0		0+			
^{123m}Cd			1.9 s	β- /					
^{123}Cd		122.91700	2.09 s	β- /6.12		3+			
^{124}Cd		123.9177	1.24 s	β- /4.17		0+			0.0365
									0.0628
									0.1799
^{125m}Cd			0.66 s	β- /					
^{125}Cd		124.9213	0.68 s	β- /7.16		3/2+			
^{126}Cd		125.9224	0.52 s	β- /5.49		0+			0.2601
^{127}Cd		126.9264	0.4 s	β- /8.5		3/2+			
^{128}Cd		127.9278	0.28 s	β- /7.1		0+			0.247
^{129}Cd		128.9322	0.24 s	β- /5.9					0.281
^{130}Cd		129.9339	0.162 s	β- /		0+			
				β-, n	/~ 3.5				
^{131}Cd		130.9407	68 ms						
^{132}Cd		131.9456	0.10 s	β-, n/	/60	0+			
^{133}Cd			0.06 s						
$_{49}$In		114.818(3)							
^{97}In		96.950							
^{98m}In			~ 0.03 s						
^{98}In		97.9421	1. s						
^{99}In		98.9342	~ 3.8 s	β+ /8.9					
^{100}In		99.9311	5.9 s	β+, (p)/10.5					(0.297-1.365)
^{101}In		100.9263	15. s	β+ /7.3					
^{102}In		101.9241	23. s	EC/8.9		(5)			0.1566
									0.7767
									(0.397−0.923)
^{103m}In			34. s						
^{103}In		102.91991	1.1 m	β+, EC/6.05	4.2	9/2+			ann.rad./
				EC	/45				0.1879
									(0.157−3.98)
^{104m}In			16. s	IT/0.0935					
^{104}In		103.9183	1.84 m	β+, EC/7.9	4.8	5+	+4.44	+0.7	ann.rad./
									0.6580
									0.8341
									0.8781

Elem. or Isot.	Natural Abundance (Atom %)	Atomic Mass or Weight	Half-life/ Resonance Width (MeV)	Decay Mode/ Energy (/MeV)	Particle Energy/ Intensity (MeV/%)	Spin ($h/2\pi$)	Nuclear Magnetic Mom. (nm)	Elect. Quadr. Mom. (b)	γ-Energy / Intensity (MeV/%)
^{105m}In			43. s	I.T.		½−			In k x-ray
									0.6740
^{105}In		104.91467	5.1 m	β+, EC/4.85	3.7	9/2+	+5.675	+0.83	0.1310
									0.2600
									0.6038
^{106m}In			5.3 m	β+ /85/	4.90	3+			ann.rad./
				EC/15/					0.6326
									0.8611
									1.7164
^{106}In		105.91347	6.2 m	β+ /65/6.52	2.6	7+	+4.92	+0.97	ann.rad./
				EC/35/					0.2259
									0.6327
									0.8611
									0.9978
									1.0091
^{107m}In			51. s	I.T./0.6786		½−			In k x-ray
									0.6785
^{107}In		106.91030	32.4 m	β+ /35/3.43	2.20/	9/2+	+5.59	+0.81	ann.rad./
				E.C./65/					Cd k x-ray
									0.2050
									0.3209
									0.5055
									(0.2−2.99)
^{108m}In			57. m	β+ /53/	1.3	6+	+4.94	+0.47	ann.rad./
				EC/47/					Cd k x-ray
									0.6329
									1.9863
									3.4522
^{108}In		107.90970	40. m	β+ /33/5.15	3.49/	3+	+4.56	+1.01	ann.rad./
				EC/67/					Cd k x-ray
									0.2429
									0.6331
									0.8756
^{109m}In			1.3 m	I.T./0.650		½−			In k x-ray
									0.6498
^{109}In		108.90715	4.17 h	β+ /8/2.02	0.79/	9/2+	+5.54	+0.84	ann.rad./
				EC/92/					Cd k x-ray
									0.2035
									0.6235
^{110m}In			4.9 h	EC/		7+	+4.72	+1.00	Cd k x-ray
									0.6577
									0.8847
									0.9375
									(0.1−1.98)
^{110}In		109.90717	1.15 h	β+ /62/3.88	2.22/	2+	+4.37	+0.35	ann.rad./
				EC/38/					Cd k x-ray
									0.6577
									(0.6−3.6)
^{111m}In			7.7 m	I.T./0.537		½−	+5.53		In k x-ray
									0.537
^{111}In		110.905103	2.8049 d	EC/0.866		9/2+	+5.50	+0.80	Cd k x-ray
									0.1712
									0.2453
^{112m}In			20.8 m	I.T./0.155		4+			In k x-ray
									0.1555
^{112}In		111.90553	14.4 m	β+ /22/2.586		1+	+2.82	+0.09	ann.rad./
				EC/34/					Cd k x-ray
				β− /0.663					0.6171
^{113m}In			1.658 h	I.T./0.3917		½−	−0.210		In k x-ray
									0.3917

Table of the Isotopes

Elem. or Isot.	Natural Abundance (Atom %)	Atomic Mass or Weight	Half-life/ Resonance Width (MeV)	Decay Mode/ Energy (/MeV)	Particle Energy/ Intensity (MeV/%)	Spin ($h/2\pi$)	Nuclear Magnetic Mom. (nm)	Elect. Quadr. Mom. (b)	γ-Energy / Intensity (MeV/%)
^{113}In	4.29(5)	112.904058				9/2+	+5.529	+0.80	
^{114m}In			49.51 d	I.T./97/0.190		5+	+4.65	+0.74	In k x-ray
				EC/3 /					0.19027
^{114}In		113.904914	1.198 m	β- /97/1.989	1.984/	1+	+2.82		Cd k x-ray
				EC/3/1.453					0.5584
									0.5727
									1.2998
^{115m}In			4.486 h	I.T./95/0.336		½-	-0.255		In k x-ray
				β- /5 /0.83					0.3362
									0.4974
^{115}In	95.71(5)	114.903878	4.4×10^{14} y	β- /0.495		9/2+	+5.541	+0.81	
^{116m2}In			2.16 s	I.T./0.162		8-	+3.22	+0.31	In k x-ray
				EC	/0.023				0.1624
^{116m1}In			54.1 m	β- /	1.0	5+	+4.43	+0.80	0.13792
									0.41688/27
									1.09723/58.5
									1.29349/85
^{116}In		115.905260	14.1 s	β- /3.274	3.3/99	1+	2.788	0.11	0.46313
									1.2526
									1.29349
^{117m}In			1.94 h	β- /53/1.769	1.77/	½-	-0.2517		In k x-ray
				I.T./47 /					0.15855
									0.31531
									0.55294
^{117}In		116.90451	44. m	β- /1.455	0.74/	9/2+	+5.52	+0.83	0.15855
									0.3966
									0.55294
^{118m2}In			8.5 s	I.T./98/		(8-)	+3.32	+0.44	In k x-ray
				β- /2/					0.1382
^{118m1}In			4.40 m	β- /	1.3	5+	+4.23	+0.80	0.2086
					2.0				0.6833
									1.2295
^{118}In		117.90635	5.0 s	β- /4.42	4.2/	1+			0.5282
									1.1734
									1.2295
									2.0432
^{119m}In			17.9 m	β- /97/	2.7/	½-	-0.32		0.3114
				I.T./3/0.311					0.7631
^{119}In		118.90585	2.3 m	β- /2.36	1.6/	9/2+	+5.52	+0.85	0.0239
									0.6495
									0.7631
									1.2149
^{120m2}In			47 s	β- /6.1		8-	+3.692	+0.53	1.171
									1.023
^{120m1}In			46. s	β- /5.8	2.2/	5+	+4.30	+0.81	1.171
									1.023
^{120}In		119.90796	3.1 s	β- /5.37	5.6/	(1+)			0.4146
					3.1/				0.5924
									0.8637
									1.0232
									1.1714
									(0.4–2.7)
^{121m}In			3.8 m	β- /99/	3.7/	1/2-	-0.36		0.0601
				I.T./1/0.313					0.3136
									0.9256
									1.0412
									1.1022
									1.1204
^{121}In		120.90785	23. s	β- /3.36	2.5	9/2+	+5.50	+0.81	0.2620
									0.6573

Elem. or Isot.	Natural Abundance (Atom %)	Atomic Mass or Weight	Half-life/ Resonance Width (MeV)	Decay Mode/ Energy (/MeV)	Particle Energy/ Intensity (MeV/%)	Spin ($h/2\pi$)	Nuclear Magnetic Mom. (nm)	Elect. Quadr. Mom. (b)	γ-Energy / Intensity (MeV/%)
									0.9256
^{122m}In			10. s	β- /	4.4/	8-	+3.78	+0.59	1.0014
									1.1403
^{122}In		121.91028	1.5 s	β- /6.37	5.3/	(1+)			0.2391
									1.0014
									1.1403
									1.164
									1.1903
^{123m}In			47. s	β- /	4.6/	(1/2-)	-0.40		0.1258
									1.170
									3.234
^{123}In		122.91044	6.0 s	β- /4.39	3.3/	(9/2+)	+5.49	+0.76	0.6188
									1.0197
									1.1305
^{124m}In			3.4 s	β-		8-	+3.89	+0.66	0.1029
									0.9699
									1.0729
									1.1316
^{124}In		123.91318	3.18 s	β- /7.36	5/	3+	+4.04	+0.61	0.7070
									0.9978
									1.1316
									3.2142
									(0.3−4.6)
^{125m}In			12.2 s	β- /	5.5/	1/2-	-0.43		0.1876
^{125}In		124.91360	2.33 s	β- /5.42	4.1/	9/2+	+5.50	+0.71	0.4260
									1.0318
									1.3350
^{126m}In			1.53 s		4.9/	3+	+4.03	+0.49	0.9086
									0.9696
									1.1411
^{126}In		125.91646	1.63 s	β- /8.21	4.2/	8-	+4.06		0.1118
									0.9086
									1.1411
^{127m}In			3.73 s	β- /	6.4/	(1/2-)			0.2523
									3.074
^{127}In		126.91735	1.14 s	β- /6.51	4.9/	(9/2+)	+5.52	+0.59	0.4680
									0.6461
									0.8051
									1.5977
^{128m}In			0.7 s	β- /	5.4/	(8-)			1.8670
									1.9739
									(0.1205−2.12)
^{128}In		127.92017	0.80 s	β- /8.98	5.0/	3+			0.9352
									1.1688
									3.5198
									4.2970
^{129m}In			1.23 s	β- /98/ n/2/	~ 7.5/	1/2-			0.3153
									0.9067
									1.2220
^{129}In		128.9217	0.63 s	β- /7.66	5.5/	9/2+			0.2853
									0.7693
									1.8650
									2.1180
^{130m2}In			0.53 s	β- /	8.8/	5+			0.0892
									0.7744
									1.2212
^{130m1}In			0.51 s	β- /	6.1/	10-			0.0892
									0.1298
									0.7744
									1.2212

Elem. or Isot.	Natural Abundance (Atom %)	Atomic Mass or Weight	Half-life/ Resonance Width (MeV)	Decay Mode/ Energy (/MeV)	Particle Energy/ Intensity (MeV/%)	Spin $(h/2\pi)$	Nuclear Magnetic Mom. (nm)	Elect. Quadr. Mom. (b)	γ-Energy / Intensity (MeV/%)
									1.9052
^{130}In		129.92497	0.29 s	β- /10.25	10.0/	1-			
^{131m2}In			0.3 s	β- /		(21/2+)			
^{131m1}In			0.35 s	β- /		(1/2-)			
^{131}In		130.92685	0.28 s	β- /9.18	6.4/	(9/2+)			0.3328
									2.433
^{132}In		131.9330	~ 0.206 s	β- /13.6	6.0/	(7-)			0.1320
					8.8/				0.2992
									0.3747
									4.0406
^{133}In		132.9378	0.165 s	β-, (n)					
^{134}In		133.9442	0.14 s						(0.354−2.005)
^{135}In		134.9493	0.09 s						
$_{50}$**Sn**		**118.710(7)**							
^{99}Sn		98.949							
^{100}Sn		99.939	1.0 s	β+ /7.3	3.4/	0+			
^{101}Sn		100.9361	3. s	β+ /9.					
^{102}Sn		101.9303	3.8 s	β+ /5.8		0+			
^{103}Sn		102.9281	7. s	β+ /7.7					1.3558
				β+,p	p//1.2				(0.351-2.813)
				EC	/ 20.				
^{104}Sn		103.9231	21. s	β+, EC/4.5		0+			
^{105}Sn		104.9214	28. s	β+ /6.3					In-x-ray
									(0.2879−3.819)
^{106}Sn		105.91688	2.0 m	β+ /20/3.18		0+			ann.rad./
				EC/80/					In k x-ray
									0.3865
									0.4772
^{107}Sn		106.9156	2.92 m	EC/5.0	1.2/				0.4218
				β+ /					0.6105
									0.6785
									1.0013
									1.1290
									1.542
^{108}Sn		107.91193	10.3 m	β+ /1/2.09	0.36/	0+			In k x-ray
				EC/99/					0.2724
									0.3965
									(0.105−1.68)
^{109}Sn		108.91128	18.0 m	β+ /9/3.85	1.52/	7/2+	-1.08	+0.3	ann.rad./
				EC/91/					In k x-ray
									0.6498
									1.0992
^{110}Sn		109.90784	4.17 h	EC/0.64		0+			In k x-ray
									0.283
^{111}Sn		110.90773	35. m	β+ /31/2.45	1.5/	7/2+	+0.61	+0.2	In k x-ray
				EC/69/					0.7620
									1.1530
									1.9147
^{112}Sn	0.97(1)	111.904818				0+			
^{113m}Sn			21.4 m	I.T./92/0.077		7/2+			Sn k x-ray
				EC/8/					In x-ray
									0.0774
^{113}Sn		112.905171	115.1 d	EC/1.036		½+	-0.879		In k x-ray
									0.25511
									0.39169
^{114}Sn	0.66(1)	113.902779				0+			
^{115}Sn	0.34(1)	114.903342				½+	-0.9188		
^{116}Sn	14.54(9)	115.901741				0+			

Elem. or Isot.	Natural Abundance (Atom %)	Atomic Mass or Weight	Half-life/ Resonance Width (MeV)	Decay Mode/ Energy (/MeV)	Particle Energy/ Intensity (MeV/%)	Spin ($h/2\pi$)	Nuclear Magnetic Mom. (nm)	Elect. Quadr. Mom. (b)	γ-Energy / Intensity (MeV/%)
117mSn			14.0 d	I.T./0.3146		11/2-	-1.396	-0.4	Sn k x-ray
									0.15856
117Sn	7.68(7)	116.902952				½+	-1.0010		
118Sn	24.22(9)	117.901603				0+			
119mSn			293. d	I.T./0.0896		11/2-	-1.4	0.21	Sn k x-ray
									0.02387
119Sn	8.59(4)	118.903308				½+	-1.0473		
120Sn	32.58(9)	119.902195				0+			
121mSn			44. y	I.T./78/0.006		11/2-	-1.388	-0.14	Sn k x-ray
				β- /22/	0.354/				0.03715
121Sn		120.904236	1.128 d	β- /0.388	0.383/100	3/2+	0.698	-0.02	
122Sn	4.63(3)	121.903439				0+			
123mSn			40.1 m	β- /1.428	1.26/99	3/2+			0.1603
									0.3814
123Sn		122.905721	129.2 d	β- /1.404	1.42/99.4	11/2-	-1.370	+0.03	0.1603
									1.0302
									1.0886
124Sn	5.79(5)	123.905274	> 2.2 × 10^{18} y	β-β-		0+			
125mSn			9.51 m	β- /2.387	2.03/98	3/2+	+0.764	+0.8	0.3321
									1.4040
125Sn		124.907784	9.63 d	β- /2.364	2.35/82	11/2-	-1.35	+0.1	1.0671
									(0.2−2.3)
126Sn		125.90765	2.34 × 10^5 y	β- /0.38	0.25/100	0+			0.0643
									0.0876
									0.4148
									0.6663
									0.6950
127mSn			4.15 m	β- /3.21	2.72/	3/2+	+0.757	+0.60	0.4909
									1.3480
									1.5640
127Sn		126.91036	2.12 h	β- /3.20	2.42/	11/2-	-1.33	+0.3	0.8231
					3.2/				1.0956
									(0.120−2.84)
128mSn			6.5 s	IT/0.091		(7-)			
128Sn		127.91054	59.1 m	β- /1.27	0.48/	0+			0.4823
					0.63/				0.5573
									0.6805
129mSn			6.9 m	β- /		11/2-	-1.30	-0.2	1.1611
129Sn		128.91348	2.4 m	β- /4.0		3/2+	+0.754	+0.05	0.6456
130mSn			1.7 m	β- /		(7-)	-0.381	-0.4	0.1449
									0.8992
130Sn		129.91397	3.7 m	β- /2.15	1.10/	0+			0.0700
									0.1925
									0.7798
131mSn			1.02 m	β- /	3.4/	11/2-	-1.28	+0.02	0.3043
									0.4500
									0.7985
									1.2260
									(0.08−3.21)
131Sn		130.91700	39. s	β- /4.69	3.8/	3/2+	+0.747	-0.04	see 131mSn
132Sn		131.91782	40. s	β- /3.12	1.8/	0+			0.0855
									0.2467
									0.3402
									0.8985
133Sn		132.92383	1.44 s	β- /7.8	7.5/	7/2-			
134Sn		133.9283	1.04 s	β- /6.8		0+			(0.053-2.417)
135Sn		134.9347	0.53 s	β-					(0.053-0.830)
				β-,n	/21.				0.733−1.855
136Sn		135.9393	0.25 s	β-, n	/30.	0+			
137Sn		136.946	0.19 s	β-, n	/~ 58				

Elem. or Isot.	Natural Abundance (Atom %)	Atomic Mass or Weight	Half-life/ Resonance Width (MeV)	Decay Mode/ Energy (/MeV)	Particle Energy/ Intensity (MeV/%)	Spin ($h/2\pi$)	Nuclear Magnetic Mom. (nm)	Elect. Quadr. Mom. (b)	γ-Energy / Intensity (MeV/%)
^{138}Sn			0.15 s			0+			
$_{51}$Sb		121.760(1)							
^{103}Sb		102.9397	> 1.5 μs						
^{104}Sb		103.9365	0.5 s						
^{105}Sb		104.9315	1.1 s	β+,p	p//<0.1				
^{106}Sb		105.9288	0.6 s	β+ /10.5					
^{107}Sb		106.9242	4.0 s	β+ /7.9					1.280
									0.1515
									0.6666
									0.553−2.046
^{108}Sb		107.9222	7.0 s	β+ /9.5					(0.151−1.280)
^{109}Sb		108.91813	17.3 s	β+ /6.38	4.42/	5/2+			0.925
				EC/	4.67/				1.062
					4.33/				0.261−2.127
^{110}Sb		109.9168	24. s	β+ /9.0	6.8/	3+			ann.rad./
				EC/					0.6365
									0.9847
									1.2117
									1.2433
^{111}Sb		110.91316	1.25 m	β+ /87/4.47	3.3/	5/2+			ann.rad./
				EC/13 /					0.1002
									0.1545
									0.4891
									1.0326
^{112}Sb		111.91240	51.4 s	β+ /90/7.06	4.75/	3+			ann.rad./
				EC/10/					0.6700
									0.9909
									1.2571
^{113}Sb		112.90937	6.7 m	β+ /65/3.91	2.42/	5/2+			ann.rad./
									(0.3−3.6)
				EC/35/					Sn k x-ray
									0.3324
									0.4980
^{114}Sb		113.90927	3.49 m	β+ /78/5.9	3.4/	3+	1.7		ann.rad./
				EC/22/					Sn k x-ray
									0.8876
									1.2999
^{115}Sb		114.90660	32.1 m	β+ /67/3.03	1.51/	5/2+	+3.46	-0.4	ann.rad./
				EC/33/					Sn k x-ray
									0.4973
^{116m}Sb			1.00 h	β+ /78/	1.16/	8-	2.6		ann.rad./
				EC/22/					Sn k x-ray
									0.4073
									0.5429
									0.9725
									1.2935
									(0.0998−1.501)
^{116}Sb		115.90679	16. m	β+ /50/4.707	1.3/	3+	2.72		ann.rad./
				EC/50/	2.3/				Sn k x-ray
									0.93180
									1.29354
									(0.138−3.903)
^{117}Sb		116.90484	2.80 h	β+ /2/1.76	0.57/	5/2+	+3.4		Sn k x-ray
				EC/98/					0.1586
^{118m}Sb			5.00 h	EC/99/		8-	2.3		Sn k x-ray
									0.25368
									1.05069
									1.22964

Elem. or Isot.	Natural Abundance (Atom %)	Atomic Mass or Weight	Half-life/ Resonance Width (MeV)	Decay Mode/ Energy (/MeV)	Particle Energy/ Intensity (MeV/%)	Spin ($h/2\pi$)	Nuclear Magnetic Mom. (nm)	Elect. Quadr. Mom. (b)	γ-Energy / Intensity (MeV/%)
^{118}Sb		117.905529	3.6 m	β+ /74/3.657	2.65/	1+	2.5		ann.rad./
				EC/26/					Sn k x-ray
									1.22964
^{119}Sb		118.90394	38.1 h	EC/0.59		5/2+	+3.45	-0.4	Sn k x-ray
									0.0239
^{120m}Sb			5.76 d	EC/		8-	2.34		Sn k x-ray
									0.0898
									0.19730
									1.02301
									1.17121
^{120}Sb		119.90507	15.89 m	β+ /41/2.68	1.72/	1+	+2.3		ann.rad./
				EC/59/					Sn k x-ray
									0.7038
									1.17121
^{121}Sb	57.21(5)	120.903816				5/2+	+3.363	-0.4	
^{122m}Sb			4.19 m	I.T./0.162		8-			Sb x-ray
									0.0614
									0.0761
^{122}Sb		121.905174	2.72 d	β- /98/1.979	1.414/65	2-	-1.90	+0.9	0.56409
				β+ /2/1.620	1.980/26				0.69277
									1.14050
									1.2569
^{123}Sb	42.79(5)	122.904214				7/2+	+2.550	-0.5	
^{124m2}Sb			20.3 m	I.T./0.035		8-			
^{124m1}Sb			1.6 m	I.T./80/	1.2/	5+			0.4984
				β- /20/	1.7/				0.6027
									0.6458
									1.1010
^{124}Sb		123.905936	60.20 d	β- /2.905	0.61/52	3-	1.2	+1.9	0.60271/97.8
					2.301/23				0.64583/7.4
									0.72277/10.5
									1.69094/48.2
									(0.0274−2.808)
^{125}Sb		124.905254	2.758 y	β- /0.767	0.13/30	7/2+	+2.63		0.0355
					0.302/45				0.17632
					0.62/13				0.38044
									0.42786
									0.46336
									0.60060
									0.63595
^{126m2}Sb			11. s	I.T./		3-			L x-ray
									0.0227
^{126m1}Sb			19.0 m	β- /86 /	1.9	5+			0.4148
				I.T./14 /					0.6663
									0.6950
^{126}Sb		125.90725	12.4 d	β- /3.67	1.9	8-	1.3		0.2786
									0.4148/83.3
									0.6663/99.7
									0.6950/99
									0.7205
^{127}Sb		126.90692	3.84 d	β- /1.581	0.89/	7/2+	2.70		0.2524
					1.10/				0.2908
					1.50/				0.4121
									0.4370
									0.6857
									0.7837
^{128m}Sb			10.1 m	β- /96/	2.6/	5+			0.3140
				I.T./4/					0.5941
									0.7432
									0.7539

Elem. or Isot.	Natural Abundance (Atom %)	Atomic Mass or Weight	Half-life/ Resonance Width (MeV)	Decay Mode/ Energy (/MeV)	Particle Energy/ Intensity (MeV/%)	Spin ($h/2\pi$)	Nuclear Magnetic Mom. (nm)	Elect. Quadr. Mom. (b)	γ-Energy / Intensity (MeV/%)
128Sb		127.90917	9.1 h	β- /4.38	2.3/	8-	1.3		0.2148
									0.3141
									0.5265
									0.7433
									0.7540
129mSb			17.7 m	β- /					0.4338
									0.6578
									0.7598
129Sb		128.90915	4.40 h	β- /2.38	0.65/	7/2-	2.82		0.0278
									0.1808
									0.3594
									0.4596
									0.5447
									0.8128
									0.9146
									1.0301
130mSb			6.5 m	β- /2.6	2.12/				0.1023
									0.7934
									0.8394
130Sb		129.91166	38.4 m	β- /4.96	2.9/	8-			0.1823
									0.3309
									0.4680
									0.7394
									0.8394
131Sb		130.91198	23.0 m	β- /3.20	1.31/	7/2+			0.6423
					3.0/				0.6579
									0.9331
									0.9434
132mSb			2.8 m	β- /	3.9/	4+			0.1034
									0.3538
									0.6968
									0.9739
									0.9896
132Sb		131.91447	4.2 m	β- /5.49		8-			0.1034
									0.1506
									0.6968
									0.9739
133Sb		132.91525	2.5 m	β- /4.00	1.20/	7/2+	3.00		0.4235
									0.6318
									0.8165
									1.0764
134mSb			10.4 s	β- /	6.1	7-			
134Sb		133.92038	0.8 s	β- /8.4	8.4	0-			0.1152
									0.2970
									0.7063
									1.2791
135Sb		134.9252	1.71 s	β- /8.12		7/2+			1.127
									1.279
136Sb		135.9304	0.82 s	β- /9.3					
137Sb		136.9353	> 0.15 µs						
138Sb		137.9408	> 0.15 µs						
139Sb		138.9460	> 0.15 µs						
52Te		127.60(3)							
105Te		104.9436							
106Te		105.9375	0.07 ms	α/4.3	/100	0+			
107Te		106.9350	3.1 ms	α/ 70/	3.86(1)/				(0.090-0.721)
				β+, EC/10.1					
108Te		107.9294	2.1 s	α /68 /	3.314(4)/	0+			

Elem. or Isot.	Natural Abundance (Atom %)	Atomic Mass or Weight	Half-life/ Resonance Width (MeV)	Decay Mode/ Energy (/MeV)	Particle Energy/ Intensity (MeV/%)	Spin ($h/2\pi$)	Nuclear Magnetic Mom. (nm)	Elect. Quadr. Mom. (b)	γ-Energy / Intensity (MeV/%)
				β+, EC/32 /6.8					
¹⁰⁹Te		108.9274	4.6 s	β+ EC/96 /8.7					0.7523
				α/4 /	3.107(4)/				0.287−2.045
¹¹⁰Te		109.9224	19. s	β+, EC/4.5		0+			ann.rad./
									0.2191
									0.6059
¹¹¹Te		110.9211	19.3 s	β+, EC/8.0		(7/2+)			ann.rad./
									0.267
									0.322
									0.341
¹¹²Te		111.9170	2.0 m	β+, EC/4.3		0+			ann.rad./
									0.2962
									0.3727
									0.4187
¹¹³Te		112.9159	1.7 s	β+ /85/5.7	4.5/	(7/2+)			ann.rad./
				EC/15/					Sb k x-ray
									0.8144
									1.0181
									1.1812
¹¹⁴Te		113.91209	15. m	β+ /40/3.2		0+			ann.rad./
				EC/60/					Sb k x-ray
									0.0838
									0.0903
¹¹⁵ᵐTe			6.7 m	β+ /45/		(1/2+)			ann.rad./
				EC/55/					Sb k x-ray
									0.7236
									0.7704
¹¹⁵Te		114.91190	5.8 m	β+ /45/4.6	2.7/	7/2+			ann.rad./
				EC/55/					Sb k x-ray
									0.7236
									1.3268
									1.3806
									(0.22−2.7)
¹¹⁶Te		115.90846	2.49 h	EC/1.5		0+			Sb k x-ray
									0.0937
¹¹⁷Te		116.90865	1.03 h	EC/75/3.54	1.78/	½+			ann.rad./
				β+ /25/					Sb k x-ray
									0.9197
									1.7164
									2.3000
¹¹⁸Te		117.90583	6.00 d	EC/0.28		0+			Sb k x-ray
¹¹⁹ᵐTe			4.69 d	EC/		11/2-	0.89		Sb k x-ray
									0.15360
									0.2705
									1.21271
¹¹⁹Te		118.90640	16.0 h	β+ /2/2.293	0.627/	½+	0.25		ann.rad.
				EC/98/					Sb k x-ray
									0.6440
									0.6998
¹²⁰Te	0.09(1)	119.90402				0+			
¹²¹ᵐTe			~ 154. d	I.T. (89%)		11/2-	0.90		Te k x-ray
				EC (11%)					0.2122
¹²¹Te		120.90494	16.8 d	EC/1.04		½+			Sb k x-ray
									0.5076
									0.5731
¹²²Te	2.55(12)	121.903044				0+			
¹²³ᵐTe			119.7 d	I.T./0.247		11/2-	-0.93		Te k x-ray
									0.1590/84.1
¹²³Te	0.89(3)	122.904270	> 9.2 × 10¹⁶ y	EC/0.051		½+	-0.73695		
¹²⁴Te	4.74(14)	123.902818				0+			

Table of the Isotopes

Elem. or Isot.	Natural Abundance (Atom %)	Atomic Mass or Weight	Half-life/ Resonance Width (MeV)	Decay Mode/ Energy (/MeV)	Particle Energy/ Intensity (MeV/%)	Spin ($h/2\pi$)	Nuclear Magnetic Mom. (nm)	Elect. Quadr. Mom. (b)	γ-Energy / Intensity (MeV/%)
^{125m}Te			58. d	I.T./0.145		11/2-	-0.99	-0.06	Te k x-ray
									0.0355
^{125}Te	7.07(15)	124.904431				½+	-0.8885		
^{126}Te	18.84(25)	125.903312				0+			
^{127m}Te			109. d	I.T./98/0.088		11/2-	-1.04		Te k x-ray
				β- /2/0.77					0.0883
^{127}Te		126.905226	9.4 h	β- /0.698	0.696/	3/2+	0.64		0.3603
^{128}Te	31.74(8)	127.904463	2.2×10^{24} y	β-β-		0+			
^{129m}Te			33.6 d	I.T./63/0.105		11/2-	-1.09		Te k x-ray
				β- /37/	1.60/				0.45984
									0.6959
^{129}Te		128.906598	1.16 h	β- /1.498	0.99/9	3/2+	0.70	0.06	0.0278
					1.45/89				0.45984
									0.48728
^{130}Te	34.08(62)	129.906224	8×10^{20} y	β-β-		0+			
^{131m}Te			1.35 d	β- /78/2.4	0.42/	11/2-	-1.04		0.0811
				I.T./22/0.18					0.1021
									0.14973
									0.77369
									0.79375
									0.85225
^{131}Te		130.908524	25.0 m	β- /2.233	1.35/12	3/2+	0.70		0.14973
					1.69/22				0.45327
					2.14/60				0.49269
^{132}Te		131.90855	3.26 d	β- /0.51	0.215	0+			0.049725
									0.11198
									0.22830
^{133m}Te			55.4 m	β- /82/	2.4/30	11/2-			Te k x-ray
				I.T./18/0.334					0.0949
									0.1689
									0.3121
									0.3341
^{133}Te		132.91096	12.4 m	β- /2.94	2.25/25	3/2+			0.3121
					2.65				0.4079
									1.3334
^{134}Te		133.91137	42. m	β- /1.51	0.6/	0+			0.7672/29
					0.7/				0.0794-0.9255
^{135}Te		134.9165	19.0 s	β- /6.0	5.4/				0.267
					6.0				0.603
									0.870
^{136}Te		135.92010	17.5 s	β- /5.1	2.5/	0+			2.0779/25
									0.0873-3.235
^{137}Te		136.9253	2.5 s	β- /98 /6.9	6.8	7/2-			0.2436
				n/2 /					
^{138}Te		137.9292	1.4 s	β- /6.4		0+			
^{139}Te		138.9347	> 0.15 μs						
^{140}Te		139.9389	> 0.15 μs			0+			
^{141}Te		140.9447	> 0.15 μs						
^{142}Te		141.949	> 0.15 μs			0+			
$_{53}$I		126.90447(3)							
^{108}I		107.9435	0.04 s	α/91/4.	3.95				
^{109}I		108.9382	0.11 ms	p					0.593/100
									0.717/63
									0.496-1.057
^{110}I		109.9352	0.65 s	β+, EC/83/11.4					ann.rad./
				α/17/~ 3.6	3.457(10)/				
				p/11/					
^{111}I		110.9303	2.5 s	β+, E.C./8.5					ann.rad./

Elem. or Isot.	Natural Abundance (Atom %)	Atomic Mass or Weight	Half-life/ Resonance Width (MeV)	Decay Mode/ Energy (/MeV)	Particle Energy/ Intensity (MeV/%)	Spin ($h/2\pi$)	Nuclear Magnetic Mom. (nm)	Elect. Quadr. Mom. (b)	γ-Energy / Intensity (MeV/%)
									0.2665
									0.3215
									0.3412
¹¹²I		111.9280	3.4 s	β+, EC/10.2					ann.rad./
									0.6889
									0.7869
¹¹³I		112.9236	5.9 s	β+, EC/7.6					ann.rad./
									0.4625/100
									0.6224/74
									0.0550−1.422
¹¹⁴I		113.9219	2.1 s	β+, EC/8.7					ann.rad./
									0.6826
									0.7088
¹¹⁵I		114.9181	1.3 m	β+, EC/6.7		5/2+			ann.rad./
									0.275
									0.284
									0.460
									0.709
¹¹⁶I		115.9168	2.9 s	β+ /97/7.8 EC/3/	6.7/	1+			ann.rad./
									0.5402
									0.6789
¹¹⁷I		116.91365	2.22 m	β+, EC/4.7	3.2/	(5/2+)	3.1		ann.rad./
									0.2744
									0.3259
¹¹⁸ᵐI			8.5 m	β+, EC/ I.T.	4.9/	7-	4.2		ann.rad./
									0.104
									0.5998
									0.6052
									0.6138
¹¹⁸I		117.91307	14. m	β+, EC/7.0		2-	2.0		ann.rad./
									0.5448
									0.6052
									1.3384
¹¹⁹I		118.91007	19. m	β+ /54/3.5 EC/46/	2.4/	(5/2+)	+2.9		ann.rad./
									Te k x-ray
									0.2575
¹²⁰ᵐI			53. m	β+ /80/ EC/20/	3.8		4.2		ann.rad.
									Te k x-ray
									0.4257
									0.5604
									0.6147
									1.3459
¹²⁰I		119.91005	1.35 h	β+ /56/5.62 EC/	4.03 4.60	2-	1.23		ann.rad./
									Te k x-ray
									0.5604
									0.6411
									1.5230
									(0.111−3.1)
¹²¹I		120.90737	2.12 h	β+ /13/2.27 EC/87/	1.2/	5/2+	2.3		ann.rad./
									Te k x-ray
									0.2122
									(0.14−1.1)
¹²²I		121.90759	3.6 m	β+ /4.234 EC/	3.1/	1+	+0.94		ann.rad./
									Te k x-ray
									0.5641
¹²³I		122.905589	13.2 h	EC/1.242		5/2+	2.82		Te k x-ray
									0.1590
¹²⁴I		123.906210	4.18 d	β+ /23/3.160 EC/77/	1.54/ 2.14/ 0.75/	2-	1.44		ann.rad./ Te k x-ray 0.6027/62.9

Elem. or Isot.	Natural Abundance (Atom %)	Atomic Mass or Weight	Half-life/ Resonance Width (MeV)	Decay Mode/ Energy (/MeV)	Particle Energy/ Intensity (MeV/%)	Spin ($h/2\pi$)	Nuclear Magnetic Mom. (nm)	Elect. Quadr. Mom. (b)	γ-Energy / Intensity (MeV/%)
									0.7228/10.3
									1.6910/11.2
									(0.31−1.73)
^{125}I		124.904630	59.4 d	EC/0.1861		5/2+	2.82	-0.89	Te k x-ray
									0.0355
^{126}I		125.905624	13.0 d	EC/		2-	1.44		ann.rad./
				β+ /2.155	1.13/				Te k x-ray
				β- /1.258/47	0.87/				0.3887
					1.25/				0.6622
^{127}I	100.	126.904473				5/2+	+2.8133	-0.79	
^{128}I		127.905809	25.00 m	β- /2.118	2.13/	1+			Te k x-ray
				EC/1.251					0.44287
									0.52658
^{129}I		128.904988	1.7×10^7 y	β- /0.194	0.15/	7/2+	+2.621	-0.55	Xe k x-ray
									0.0396
^{130m}I			9.0 m	I.T./83/0.048		2+			I k x-ray
				β- /17/					0.5361
^{130}I		129.906674	12.36 h	β- /2.949	1.04/	5+	3.35		0.4180
					0.62				0.5361
									0.6685
									0.7395
^{131}I		130.906125	8.021 d	β- /0.971	0.606/	7/2+	+2.742	-0.40	0.08017
									0.28431
									0.36446
									0.63699
^{132m}I			1.39 h	IT		8-			
^{132}I		131.90800	2.28 h	β- /14/3.58	0.80/	4+	3.09	0.09	I k x-ray
				I.T./86/	1.03/				0.0980
					1.2/				0.5059
					1.6/				0.52264
					2.16/				0.63019
									0.6506
									0.66768
									0.77260
									0.95457
^{133m}I			9. s	I.T./1.63		19/2-			I Kx-ray
									0.0730
									0.6474
									0.9126
^{133}I		132.907797	20.8 h	β- /1.77	1.24/85	7/2+	+2.86	-0.27	0.51056
									0.52989
									0.87537
^{134m}I			3.7 m	I.T./98/0.316		8-			I k x-ray
				β- /2/					0.0444
									0.2719
^{134}I		133.90974	52.6 m	β- /4.05	1.2/	4+			0.1354
									0.84702
									0.88409
^{135}I		134.91005	6.57 h	β- /2.63	0.9/	7/2+	2.94		0.2884
					1.3/				0.41768
									0.52658
									1.13156
									1.26046
^{136m}I			47. s	β- /	4.7/	6-			0.1973
					5.2/				0.3468
									0.3701
									0.3814
									1.3130
									(0.16−2.36)
^{136}I		135.91465	1.39 m	β- /6.93	4.3/	2-			0.3447

Elem. or Isot.	Natural Abundance (Atom %)	Atomic Mass or Weight	Half-life/ Resonance Width (MeV)	Decay Mode/ Energy (/MeV)	Particle Energy/ Intensity (MeV/%)	Spin ($h/2\pi$)	Nuclear Magnetic Mom. (nm)	Elect. Quadr. Mom. (b)	γ-Energy / Intensity (MeV/%)
					5.6/				1.3130
									1.3211
									2.2896
									(0.3−6.1)
^{137}I		136.91787	24.5 s	β- /5.88	5.0/	(7/2+)			0.6010
									1.2180
									1.2201
									1.3026
									1.5343
									(0.25−4.4)
^{138}I		137.9224	6.5 s	β- /7.8	6.9/	2-			0.4836
					7.4/				0.5888
									0.8752
									(0.4−5.3)
^{139}I		138.92610	2.30 s	β- /6.81					0.192
				n/					0.198
									0.273
									0.382
									0.386
									0.468
									0.683
									1.313
^{140}I		139.9310	0.86 s	β- /8.8		(3)			0.372
				n/					0.377
									0.457
^{141}I		140.9350	0.45 s	β- /7.8					
^{142}I		141.9402	~ 0.2 s	β-					
^{143}I		142.9446	> 0.15 μs						
^{144}I		143.9500	> 0.15 μs						
$_{54}$Xe		131.293(6)							
^{110}Xe		109.9443	0.11 s	β+ /9.2		0+			
				α	/~ 64				
^{111m}Xe			0.9 s	EC, β+					
^{111}Xe		110.9416	0.7 s	EC, β+ /10.6					
				α/	3.58(1)/				
^{112}Xe		111.9356	3. s	EC, β+ /7.2	α/0.8/	0+			
^{113}Xe		112.9333	2.8 s	EC, β+ /9.1					
^{114}Xe		113.92798	10.0 s	β+, EC/5.9		0+			ann.rad./
									0.1031
									0.1616
									0.3085
									0.6826
									0.7088
^{115}Xe		114.92629	18. s	β+, EC/7.6		(5/2+)			ann.rad./
^{116}Xe		115.92158	56. s	β+, EC/4.3	3.3/	0+			ann.rad./
									0.1042
									0.1916
									0.2477
									0.3107
									0.4127
^{117}Xe		116.92036	1.02 m	β+, EC/6.5		(5/2+)	-0.594	+1.16	ann.rad./
									0.2214
									0.5190
									0.6389
									0.6613
^{118}Xe		117.91618	~ 4. m	β+, EC/3.	2.7/	0+			ann.rad./
									0.0535
									0.0600

Elem. or Isot.	Natural Abundance (Atom %)	Atomic Mass or Weight	Half-life/ Resonance Width (MeV)	Decay Mode/ Energy (/MeV)	Particle Energy/ Intensity (MeV/%)	Spin ($h/2\pi$)	Nuclear Magnetic Mom. (nm)	Elect. Quadr. Mom. (b)	γ-Energy / Intensity (MeV/%)
									0.1199
^{119}Xe		118.91541	5.8 m	β+, EC/5.0	3.5/	7/2+	-0.654	+1.31	0.0873
									0.1000
									0.2318
									0.4615
^{120}Xe		119.91178	40. m	β+, EC/97/1.96		0+			I k x-ray
				β+ /3/					0.0251
									0.0726
									0.1781
									(0.1−1.03)
^{121}Xe		120.91146	39. m	β+ /44/3.73	2.8/	5/2+	-0.701	+1.33	ann.rad./
				EC/56/					I k x-ray
									0.1328
									0.2527
									0.4452
									(0.1−3.1)
^{122}Xe		121.90837	20.1 h	EC/0.9		0+			I k x-ray
									0.3501
^{123}Xe		122.90848	2.00 h	β+ /23/2.68	1.51/	½+	-0.150		ann.rad./
				EC/77/					I k x-ray
									0.1489
									0.1781
									(0.1−2.1)
^{124}Xe	0.0952(3)	123.905893	> 10^{17} y	β-β-		0+			
^{125m}Xe			57. s	I.T./0.252		(9/2-)	-0.745	+0.42	Xe k x-ray
									0.1111
									0.141
^{125}Xe		124.906395	17.1 h	EC/1.653	0.47/	½+	-0.269		I k x-ray
									0.1884
									0.2434
^{126}Xe	0.0890(2)	125.90427				0+			
^{127m}Xe			1.15 m	I.T./0.297		(9/2-)	-0.884	+0.69	Xe k x-ray
									0.1246
									0.1725
^{127}Xe		126.905184	36.34 d	EC/0.662		½+	-0.504		I k x-ray
									0.1721
									0.2029
									0.3750
^{128}Xe	1.9102(8)	127.903531				0+			
^{129m}Xe			8.89 d	I.T./0.236		11/2-	-0.891	+0.64	Xe k x-ray
									0.0396
									0.1966
^{129}Xe	26.4006(82)	128.904779				½+	-0.7780		
^{130}Xe	4.0710(13)	129.903508				0+			
^{131m}Xe			11.9 d	I.T./0.164		11/2-	-0.9940	+0.73	Xe k x-ray
									0.16398
^{131}Xe	21.2324(30)	130.905082				3/2+	+0.69186	-0.12	
^{132}Xe	26.9086(33)	131.904153				0+			
^{133m}Xe			2.19 d	I.T./0.233		11/2-	-1.082	+0.77	Xe k x-ray
									0.23325
^{133}Xe		132.905911	5.243 d	β- /0.427	0.346/99	3/2+	+0.813	+0.14	Cs k x-ray
									0.080998
									0.1606
^{134}Xe	10.4357(21)	133.905394	> 1.1 × 10^{16} y	β-β-		0+			
^{135m}Xe			15.3 m	I.T./		11/2-	1.103	+0.62	Xe k x-ray
									0.52658
^{135}Xe		134.907227	9.10 h	β-/1.15	0.91/	3/2+	0.903	+0.21	0.24975
									0.60807
^{136}Xe	8.8573(44)	135.90722	> 2.4 × 10^{21} y	β-β-		0+			
^{137}Xe		136.91156	3.82 m	β- /4.17	4.1/	7/2-	-0.970	-0.49	0.45549

Elem. or Isot.	Natural Abundance (Atom %)	Atomic Mass or Weight	Half-life/ Resonance Width (MeV)	Decay Mode/ Energy (/MeV)	Particle Energy/ Intensity (MeV/%)	Spin (ℏ/2π)	Nuclear Magnetic Mom. (nm)	Elect. Quadr. Mom. (b)	γ-Energy / Intensity (MeV/%)
					3.6/				0.8489
									0.9822
									1.2732
									1.7834
									2.8498
^{138}Xe		137.91395	14.1 m	β- /2.77	0.8/	0+			0.1538
					2.4/				0.2426
									0.2583
									0.4345
									1.76826
									2.0158
^{139}Xe		138.91879	39.7 s	β- /5.06	4.5/		-0.304	+0.40	0.1750
					5.0/				0.2186
									0.2965
									(0.1−3.37)
^{140}Xe		139.9216	13.6 s	β- /4.1	2.6	0+			0.0801
									0.6220
									0.8055
									1.4137
									(0.04−2.3)
^{141}Xe		140.9266	1.72 s	β- /6.2	6.2/	5/2+	+0.010	-0.58	0.1187
									0.9095
									(0.05−2.55)
^{142}Xe		141.9297	1.22 s	β- /5.0	3.7/	0+			0.0338
					4.2/				0.0729
									0.2038
									0.3091
									0.4145
									0.5382
									0.5718
									0.6181
									0.6448
^{143m}Xe			0.96 s	β-					
^{143}Xe		142.9351	0.30 s	β- /7.3			-0.460	+0.93	
^{144}Xe		143.9385	1.2 s	β- /6.1		0+			
^{145}Xe		144.9441	0.9 s	β-, (n)					
^{146}Xe		145.9478	> 0.15 μs			0+			
^{147}Xe		146.9536	> 0.15 μs						
$_{55}$Cs		132.9054519(2)							
^{112}Cs		111.9503	0.5 ms	p	0.81				
^{113}Cs		112.9445	17. μs	p	0.96				
^{114}Cs		113.9414	0.58 s	β+, EC/11.8		1+			ann.rad./
									0.6826
									0.7088
^{115}Cs		114.9359	~ 1.4 s	β+, EC/8.4					ann.rad./
^{116m}Cs			0.7 s	β, EC/					ann.rad./
									0.3935
^{116}Cs		115.9334	3.8 s	β+, EC/10.8					ann.rad./
									0.3935
									0.5243
									0.6151
									0.6223
^{117m}Cs			6.5 s	β+, EC/					
^{117}Cs		116.9287	~ 8.4 s	β+, EC/7.5					ann.rad./
^{118m}Cs			17. s	β+, EC/			5.		
^{118}Cs		117.92656	14. s	β+, EC/9.		2	+3.88	+1.4	ann.rad./
									0.3372
									0.4727

Elem. or Isot.	Natural Abundance (Atom %)	Atomic Mass or Weight	Half-life/ Resonance Width (MeV)	Decay Mode/ Energy (/MeV)	Particle Energy/ Intensity (MeV/%)	Spin ($h/2\pi$)	Nuclear Magnetic Mom. (nm)	Elect. Quadr. Mom. (b)	γ-Energy / Intensity (MeV/%)
									0.5865
									0.5906
^{119m}Cs			29. s			3/2	+0.84	+0.9	
^{119}Cs		118.92238	43. s	β+, EC/6.3		9/2+	+5.5	+2.8	ann.rad./
									0.169
									0.176
									0.224
									0.257
^{120m}Cs			60. s	β+, EC/					
^{120}Cs		119.92068	64. s	β+, EC/7.92		2+	+3.87	+1.45	ann.rad./
									0.3224
									0.4735
									0.5534
									(0.3−3.28)
^{121m}Cs			2.0 m	I.T./60/		(9/2+)	+5.41	+2.7	ann.rad./
				β+ /40/	4.4				0.1794
									0.1961
^{121}Cs		120.91723	2.3 m	β+, EC/5.40	4.38/	3/2+	+0.77	+0.84	ann.rad./
									0.1537
									(0.08−0.56)
^{122m2}Cs			4.4 m	β+, EC		8-	+4.77	+3.3	ann.rad./
^{122m1}Cs			0.36 s	IT					0.3311
									0.4971
									0.6385
									(0.27−2.22)
^{122}Cs		121.91611	21. s	β+, EC/7.1	5.8/	(1+)	-0.133	-0.19	ann.rad./
									0.3311
									0.5120
									0.8179
^{123m}Cs			1.6 s	I.T./		11/2-			Cs k x-ray
									0.0946
^{123}Cs		122.91300	5.87 m	β+ /75/4.20	3.0/	1/2+	+1.38		ann.rad./
				EC/25/					Xe k x-ray
									0.0974
									0.5964
^{124m}Cs			6.3 s	IT		7+			
^{124}Cs		123.91226	30. s	β+ /9 /5.92	~ 5.	1+	+0.673	-0.74	ann.rad./
				EC/8 /					Xe k x-ray
									0.3539
									0.4925
									0.9418
^{125}Cs		124.90973	45. m	β+ /40/3.09	2.06/	1/2+	+1.41		ann.rad./
				EC/60/					Xe k x-ray
									0.112
									0.526
^{126}Cs		125.90945	1.64 m.	β+ /81/4.83	3.4	1+	+0.78	-0.68	ann.rad./
				EC/19/	3.7/				Xe k x-ray
									0.3886
									0.4912
									0.9252
^{127}Cs		126.90742	6.2 h	β+ /96/2.08	0.65/	1/2+	+1.46		Xe k x-ray
				EC/4/	1.06				0.1247
									0.4119
^{128}Cs		127.90775	3.62 m	β+ /68/3.930	2.44/	1+	+0.97	-0.57	ann.rad./
				EC/32 /	2.88/				Xe k x-ray
									0.4429
^{129}Cs		128.90606	1.336 d	EC/1.195		1/2+	+1.49		Xe k x-ray
									0.3719
									0.4115
^{130m}Cs			3.5 m	IT, β+, EC		5-	+0.629	+1.45	

Elem. or Isot.	Natural Abundance (Atom %)	Atomic Mass or Weight	Half-life/ Resonance Width (MeV)	Decay Mode/ Energy (/MeV)	Particle Energy/ Intensity (MeV/%)	Spin ($h/2\pi$)	Nuclear Magnetic Mom. (nm)	Elect. Quadr. Mom. (b)	γ-Energy / Intensity (MeV/%)
^{130}Cs		129.90671	29.21 m	β+ /55/2.98	1.98/	1+	+1.46	-0.06	ann.rad./
				EC/43/					Xe k x-ray
				β- /1.6/0.37	0.44/1.6				0.5361
^{131}Cs		130.90546	9.69 d	EC/0.352		5/2+	+3.54	-0.58	Xe k x-ray
^{132}Cs		131.906434	6.48 d	EC/98/		2-	+2.22	+0.51	Xe k x-ray
				β+ /0.3/2.120					0.4646
				β- / /1.280					0.6302
									0.66769
^{133}Cs	100.	132.90545193				7/2+	+2.582	-0.00355	
^{134m}Cs			2.91 h	I.T./0.139		8-	+1.098	+1.0	Cs k x-ray/
									0.12749
^{134}Cs		133.90671848	2.065 y	β- /2.059	0.089/27	4+	+2.994	+0.39	0.56327
					0.658/70				0.56935
				EC/1.22					0.60473
									0.79584
^{135m}Cs			53. m	I.T./1.627		19/2-	+2.18	+0.9	0.7869
									0.8402
^{135}Cs		134.905977	2.3×10^{6} y	β- /0.269	0.205/100	7/2+	+2.732	+0.05	
^{136m}Cs			19. s	I.T./		8-	+1.32	+0.7	
^{136}Cs		135.907312	13.16 d	β- /2.548	0.341/	5+	+3.71	+0.2	0.06691
									0.34057
									0.81850
									1.04807
^{137}Cs		136.907089	30.2 y	β- /1.176	0.514/95	7/2+	+2.84	+0.05	Ba k x-ray
									0.66164
^{138m}Cs			2.9 m	I.T./75	/0.080	6-	+1.71	-0.40	Cs k x-ray
				β- /25 /	3.3				0.0799
									0.1917
									0.4628
									1.43579
^{138}Cs		137.91102	32.2 m	β- /5.37	2.9/	3-	+0.700	+0.12	0.1381
									0.46269
									1.00969
									1.43579
									2.21788
^{139}Cs		138.913364	9.3 m	β- /4.213	4.21	7/2+	+2.70	-0.07	0.6272
									1.2832
									(0.4−3.66)
^{140}Cs		139.91728	1.06 m	β- /6.22	5.7/	1-	+0.13390	-0.11	0.5283
					6.21/				0.6023
									0.9084
									(0.41−3.94)
^{141}Cs		140.92005	24.9 s	β- /5.26	5.20/	7/2+	+2.44	-0.4	Ba k x-ray
									0.0485
									0.5616
									0.5887
									1.1940
									(0.05−3.33)
^{142}Cs		141.92430	1.8 s	β- /7.31	6.9/				0.3596
					7.28				0.9668
									1.1759
									1.3265
^{143}Cs		142.92735	1.78 s	β- /6.24	6.1	(3/2+)	+0.87	+0.47	0.1955
									0.2324
									0.3064
									(0.17−1.98)
^{144}Cs		143.93208	1.01 s	β- /8.47	8.46/	1	-0.546	+0.30	0.1993
					7.9/				0.5598
									0.6392
									0.7587

Table of the Isotopes

Elem. or Isot.	Natural Abundance (Atom %)	Atomic Mass or Weight	Half-life/ Resonance Width (MeV)	Decay Mode/ Energy (/MeV)	Particle Energy/ Intensity (MeV/%)	Spin $(h/2\pi)$	Nuclear Magnetic Mom. (nm)	Elect. Quadr. Mom. (b)	γ-Energy / Intensity (MeV/%)
^{145}Cs		144.93553	0.59 s	β- /7.89	7.4/	3/2+	+0.784	+0.6	0.1126
					7.9/				0.1755
									0.1990
^{146}Cs		145.9403	0.322 s	β-, (n)/9.38	~ 9.0	2-	-0.515	+0.22	
^{147}Cs		146.9442	0.227 s	β-, (n)/9.3					(0.024-2.2798)
^{148}Cs		147.9492	0.15 s	β-, (n)/10.5					
^{149}Cs		148.9529	> 50 ms						
^{150}Cs		149.9582	> 50 ms						
^{151}Cs		150.9622	> 50 ms						
$_{56}$Ba		137.327(7)							
^{114}Ba		113.9507	0.43 s	β+, (p)	p/20	0+			
				α	/0.9				
^{115}Ba		114.947	0.45 s	β+, (p)	p/<15				
^{116}Ba		115.9414	1.3 s	β+, (p)	p/3	0+			
^{117}Ba		116.9385	1.8 s	β+, (p), EC/8.4	p/16	(3/2-)			(0.0457−0.364)
^{118}Ba		117.9330	5.2 s	β+,		0+			(0.040−0.156)
^{119}Ba		118.9307	5.4 s	β+, EC/8.					
^{120}Ba		119.9260	24. s	β+, EC/5.0		0+			ann.rad./
									0.140
									(0.075−0.146)
^{121}Ba		120.9241	30. s	β+, EC/6.8		5/2+	+0.660	+1.8	ann.rad./
^{122}Ba		121.91990	2.0 m	β+, EC/3.8		0+			ann.rad./
^{123}Ba		122.91878	2.7 m	β+, EC/5.5			-0.68	+1.5	ann.rad./
									0.0306
									0.0927
									0.1161
									0.1235
^{124}Ba		123.91509	12. m	β+, EC/2.65		0+			ann.rad./
									0.1695
									0.1888
									1.2160
^{125m}Ba			8. m	β+, EC/	4.5		0.174		
^{125}Ba		124.9145	3.5 m	β+, EC/4.6	3.4	½+	+0.18		ann.rad./
									0.0550
									0.0776
									0.0854
									0.1409
^{126}Ba		125.91125	1.65 h	β+ /2/1.67		0+			Cs k x-ray
				EC/98 /					0.2179
									0.2336
									0.2576
^{127m}Ba			1.9 s	IT		7/2-	-0.723	1.6	
^{127}Ba		126.91109	12.9 m	β+ /54/3.5		1/2+	+0.083		ann.rad./
				EC/46/					Cs k x-ray
									0.1148
									0.1808
									(0.07−2.5)
^{128}Ba		127.90832	2.43 d	EC/0.52		0+			Cs k x-ray
									0.27344
^{129m}Ba			2.17 h	EC/98/		7/2+	+0.93	+1.6	Cs k x-ray
				β+ /2/					0.1769
									0.1823
									0.2023
									1.4593
^{129}Ba		128.90868	2.2 h	β+ /20/2.43	1.42/	1/2+	-0.40		ann.rad./
				EC/80/					Cs k x-ray
									0.1291
									0.2143

Elem. or Isot.	Natural Abundance (Atom %)	Atomic Mass or Weight	Half-life/ Resonance Width (MeV)	Decay Mode/ Energy (/MeV)	Particle Energy/ Intensity (MeV/%)	Spin ($h/2\pi$)	Nuclear Magnetic Mom. (nm)	Elect. Quadr. Mom. (b)	γ-Energy / Intensity (MeV/%)
									0.2208
^{130m}Ba			9.5 ms	I.T./2.475	/100.	8-	-0.04	+2.8	0.080−0.802
^{130}Ba	0.106(1)	129.906321	2.2×10^{21} y	β+β+		0+			
^{131m}Ba			14.6 m	I.T./0.187		9/2-	-0.87	+1.5	Ba k x-ray
									0.1085
^{131}Ba		130.906941	11.7 d	EC/1.37		1/2+	0.7081		Cs k x-ray
									0.12381/28.4
									0.21608/21.3
									0.49636/42.9
									(0.0549−1.171)
^{132}Ba	0.101(1)	131.905061	1.3×10^{21} y	EC EC		0+			
^{133m}Ba			1.621 d	I.T./0.288		11/2-	-0.91	+0.9	Ba k x-ray
									0.2761
^{133}Ba		132.906008	10.53 y	EC/0.517		1/2+	0.7717		Cs k x-ray
									0.08099
									0.35600
^{134}Ba	2.417(18)	133.904508				0+			
^{135m}Ba			1.20 d	I.T./0.2682		11/2-	-1.00	+1.0	Ba k x-ray
									0.2682
^{135}Ba	6.592(12)	134.9056886				3/2+	+0.838	+0.16	
^{136m}Ba			0.308 s	I.T./2.0305		7-			Ba k x-ray
									0.8185
									1.0481
^{136}Ba	7.854(24)	135.9045759				0+			
^{137m}Ba			2.552 m	I.T./0.6617		11/2-	-0.99	+0.8	Ba k x-ray
									0.66164
^{137}Ba	11.232(24)	136.9058274				3/2+	+0.9374	+0.245	
^{138}Ba	71.698(42)	137.9052472				0+			
^{139}Ba		138.9088412	1.396 h	β- /2.317	2.14/27	7/2-	-0.97	-0.57	0.16585
					2.27/72				1.2544
									1.42033
^{140}Ba		139.91060	12.75 d	β- /1.05	0.48	0+			0.16268
					1.0/				0.30485
					1.02/				0.53727
^{141}Ba		140.91441	18.3 m	β- /3.22	2.59/	3/2-	-0.34	+0.45	0.1903
					2.73/				0.2770
									0.3042
									(0.1−2.5)
^{142}Ba		141.91645	10.7 m	β- /2.212	1.0/	0+			0.23152
					1.10/				0.25512
									0.3090
									1.2040
^{143}Ba		142.92063	14.3 s	β- /4.24	4.2/	5/2+	+0.44	-0.88	0.1786
									0.21148
									0.7988
									(0.17−2.4)
^{144}Ba		143.92295	11.4 s	β- /3.1	2.4/	0+			La k x-ray
					2.9/				0.10386
									0.1566
									0.1728
									0.3882
									0.43048
^{145}Ba		144.9276	4.0 s	β- /4.9	4.9/	(5/2-)	-0.28	+1.22	La k x-ray
									0.0918
									0.09709
^{146}Ba		145.9302	2.20 s	β- /4.12	3.9/	0+			0.0644
									0.2513
									0.3270
									0.3329
									0.3622

Elem. or Isot.	Natural Abundance (Atom %)	Atomic Mass or Weight	Half-life/ Resonance Width (MeV)	Decay Mode/ Energy (/MeV)	Particle Energy/ Intensity (MeV/%)	Spin ($h/2\pi$)	Nuclear Magnetic Mom. (nm)	Elect. Quadr. Mom. (b)	γ-Energy / Intensity (MeV/%)
^{147}Ba		146.9349	0.892 s	β- /5.75	5.5/				
^{148}Ba		147.9377	0.64 s	β-, n/5.11		0+			
^{149}Ba		148.9426	0.36 s	β-, (n)/7.3					
^{150}Ba		149.9457	0.3 s			0+			
^{151}Ba		150.9508	> 0.15 μs						
^{152}Ba		151.9543				0+			
^{153}Ba		151.960							
$_{57}$**La**		**138.90547(7)**							
^{117}La		116.9501	23 ms	p	0.806/	3/2+			
^{118}La		117.9467							
^{119}La		118.9410							
^{120}La		119.9381	2.8 s	EC, β+ /11.					
^{121}La		120.9330	5.3 s						
^{122}La		121.9307	9. s	EC, β+/~ 9.7					
^{123}La		122.9262	17. s	EC/7.					
^{124}La		123.9246	30. s	EC/ ~ 8.8		(7+)			
^{125m}La			0.39 s						
^{125}La		124.92082	1.2 m	β+, EC/5.6		11/2-			ann.rad./
									0.0436
									0.0676
^{126m}La			< 50. s						
^{126}La		125.9195	54. s	β+, EC/7.6					ann.rad./
									0.256
									0.455
									0.117−3.853
^{127}La		126.91638	3.8 m	β+, EC/4.7		3/2+			ann.rad./
									0.025
									0.0562
^{128}La		127.9156	5.0 m	β+ /80/6.7		(5-)			ann.rad./
				EC/20/					Ba k x-ray
									0.2841/87
									0.4793/54
									(0.315−2.212)
^{129m}La			0.56 s	IT		(11/2-)			
^{129}La		128.91269	11.6 m	β+ /58/3.72	2.42/	3/2+			ann.rad./
				EC/42/					Ba k x-ray
									0.1105
									0.2786
									(0.1−1.8)
^{130}La		129.91237	8.7 m	β+ /78/5.6		3+			ann.rad./
				EC/22/					Ba k x-ray
									0.3573/81
									0.5506/27
									(0.1965−1.989)
^{131}La		130.91007	59. m	β+ /76/3.0	1.42/	3/2+			ann.rad./
				EC/24/	1.94/				Ba k x-ray
									0.1085
									0.3658
									0.5263
^{132m}La			24. m	I.T./76/		6-			La k x-ray
				β+, EC/24/					0.1352
									0.4645
^{132}La		131.91010	4.8 h	β+ /40/4.71	2.6/	2-			ann.rad./
				EC/60/	3.2				Ba k x-ray
					3.7/				0.4645
									0.5671
^{133}La		132.90822	3.91 h	β+ /4/2.2	1.2/	5/2+			Ba k x-ray
				EC/96/					0.2788

Elem. or Isot.	Natural Abundance (Atom %)	Atomic Mass or Weight	Half-life/ Resonance Width (MeV)	Decay Mode/ Energy (/MeV)	Particle Energy/ Intensity (MeV/%)	Spin ($h/2\pi$)	Nuclear Magnetic Mom. (nm)	Elect. Quadr. Mom. (b)	γ-Energy / Intensity (MeV/%)
									0.2901
									0.3024
^{134}La		133.90851	6.5 m	β+ /63/3.71	2.67/	1+			ann.rad./
				EC/37/					Ba k x-ray
									0.6047
									(0.5−1.9)
^{135}La		134.90698	19.5 h	EC/1.20		5/2+			Ba k x-ray
									0.4805
^{136}La		135.9076	9.87 m	β+ /36/2.9	1.8/	1+			ann.rad./
				EC/64/					Ba k x-ray
									0.8185
^{137}La		136.90649	6×10^4 y	EC/0.60		7/2+	+2.70	+0.2	0.2836
^{138}La	0.0888(6)	137.907112	1.06×10^{11} y			5+	+3.7136	+0.4	1.4358/65
									0.7887/35
^{139}La	99.9112(6)	138.906353				7/2+	+2.7830	+0.20	
^{140}La		139.909478	1.678 d	β- /3.762	1.35	3-	+0.73	+0.09	
					1.24/				
					1.67/				
^{141}La		140.910962	3.90 h	β- /2.502	2.43/	7/2+			
^{142}La		141.91408	1.54 h	β- /4.505	2.11/	2-			
					2.98/				
					4.52/				
^{143}La		142.91606	14.1 m	β- /3.43	3.3/	7/2-			
^{144}La		143.91960	40.7 s	β- /5.5	4.1/				
^{145}La		144.9216	24. s	β- /4.1	4.1/	3/2+			
^{146m}La			10.0 s	β- /6.7	5.5/	(6)			
^{146}La		145.9258	6.3 s	β- /6.6	6.2/	(2-)			
^{147}La		146.9282	4.02 s	β- /5.0	4.6/				
^{148}La		147.9322	1.1 s	β- /7.26		2-			
^{149}La		148.9347	1.10 s	β- /5.5					0.1335
									0.009−1.709
^{150}La		149.9388	0.51 s						x-ray
									(0.097−0.209)
^{151}La		150.9417	> 0.15 μs						
^{152}La		151.9462	> 0.15 μs						
^{153}La		152.950	> 0.15 μs						
^{154}La		153.955							
^{155}La		154.958							
$_{58}$Ce		**140.116(1)**							
^{119}Ce		118.953							
^{120}Ce		119.947				0+			
^{121}Ce		120.943	1.1 s	β+, p					
^{122}Ce		121.9379				0+			
^{123}Ce		122.9354	3.8 s	β+, EC/~ 8.6					ann.rad./
^{124}Ce		123.9304	6. s	EC/~ 5.6		0+			
^{125}Ce		124.9284	9.6 s	β+, EC/7.		7/2-			ann.rad./
									0.1346
									0.1666
									0.056−1.329
^{126}Ce		125.92397	50. s	EC/4.		0+			
^{127}Ce		126.9227	29. s	β+, EC/6.1					ann.rad./
									(0.058−1.961)
^{128}Ce		127.91891	4.1 m	β+, EC/3.2		0+			ann.rad./
									(0.023−0.880)
^{129}Ce		128.91810	3.5 m	β+, EC/5.6					ann.rad./
									(0.0675−1.015)
^{130}Ce		129.91474	26. m	β+, EC/2.2		0+			ann.rad./
									La k x-ray

Elem. or Isot.	Natural Abundance (Atom %)	Atomic Mass or Weight	Half-life/ Resonance Width (MeV)	Decay Mode/ Energy (/MeV)	Particle Energy/ Intensity (MeV/%)	Spin ($h/2\pi$)	Nuclear Magnetic Mom. (nm)	Elect. Quadr. Mom. (b)	γ-Energy / Intensity (MeV/%)
									0.047−1.431
^{131m}Ce			5. m	β+ EC					ann.rad./
									0.2304
									0.3955
									0.4213
^{131}Ce		130.91442	10. m	β+, EC/4.0	2.8/				ann.rad.
									0.119
									0.169
									0.414
^{132m}Ce			9.4 ms	IT/2.340					0.3255
									0.10−0.955
^{132}Ce		131.91146	3.5 h	EC/1.3		0+			La k x-ray
									0.1554
									0.1821
^{133m}Ce			1.6 h	β+, EC/		½+			ann.rad.
									0.0769
									0.0973
									0.5577
^{133}Ce		132.91152	5.4 h	β+/8/2.9	1.3/	9/2-			ann.rad.
				EC/92/					0.0584
									0.1308
									0.4722
									0.5104
^{134}Ce		133.90892	3.16 d	EC/0.5		0+			La k x-ray
									0.1304
									0.1623
									0.6047
^{135m}Ce			20. s	I.T./0.446		11/2-			Ce k x-ray
									0.0826
									0.1497
									0.2134
^{135}Ce		134.90915	17.7 h	β+/1 /2.026	0.8/	1/2+			La k x-ray
				EC/99 /					0.0345
									0.2656
									0.3001
									0.6068
^{136}Ce	0.185(2)	135.90717	> 0.7 × 10^{14} y	EC EC		0+			
^{137m}Ce			1.43 d	I.T./99 /0.254		11/2-	1.0		Ce k x-ray
				EC/0.8 /					0.1693
									0.2543
^{137}Ce		136.90781	9.0 h	β+/1.222		3/2+	0.96		La k x-ray
									0.4472
^{138}Ce	0.251(2)	137.90599	> 0.9 × 10^{14} y	EC EC		0+			
^{139m}Ce			56.4 s	I.T./0.7542		11/2-			Ce k x-ray
									0.7542
^{139}Ce		138.90665	137.6 d	EC/0.28		3/2+	1.06		La k x-ray
^{140}Ce	88.450(51)	139.905439				0+			0.16585
^{141}Ce		140.908276	32.50 d	β-/0.581	0.436/69	7/2-	1.1		Pr k x-ray
					0.581/31				0.14544/48.0
^{142}Ce	11.114(51)	141.909244	> 1.6 × 10^{17} y	β- β-		0+			
^{143}Ce		142.912386	1.38 d	β-/1.462	1.404/	3/2-	0.43		Pr k x-ray
					1.110/47				0.0574
									0.2933
^{144}Ce		143.913647	284.6 d	β-/0.319	0.185/20	0+			Pr k x-ray
					0.318/				0.0801
									0.1335
^{145}Ce		144.91723	3.00 m	β-/2.54	1.7/24	3/2-			Pr k x-ray
					1.3				0.0627
									0.7245
^{146}Ce		145.9188	13.5 m	β-/1.04	0.7/90	0+			Pr k x-ray

Elem. or Isot.	Natural Abundance (Atom %)	Atomic Mass or Weight	Half-life/ Resonance Width (MeV)	Decay Mode/ Energy (/MeV)	Particle Energy/ Intensity (MeV/%)	Spin ($h/2\pi$)	Nuclear Magnetic Mom. (nm)	Elect. Quadr. Mom. (b)	γ-Energy / Intensity (MeV/%)
									0.0986
									0.2182
									0.3167
									0.0930
^{147}Ce		146.92267	56. s	β-/3.29	3.3/				0.2687
^{148}Ce		147.92443	56. s	β-/2.1	1.66/	0+			0.0904
									0.0985
									0.1212
									0.2918
^{149}Ce		148.9284	5.2 s	β-/4.2					0.0577
									0.0864
									0.3800
									0.1099
^{150}Ce		149.93041	4.4 s	β-/3.0		0+			0.0526
^{151}Ce		150.9340	1.0 s	β-/5.3					Pr k x-ray
^{152}Ce		151.9365	1.4 s	β-/4.4		0+			0.098
									0.115
^{153}Ce		152.9406	> 0.15 µs						
^{154}Ce		153.9434	> 0.15 µs			0+			
^{155}Ce		154.948	> 0.15 µs						
^{156}Ce		155.951				0+			
^{157}Ce		156.956							
$_{59}$Pr		140.90765(2)							
^{121}Pr		120.955	0.01 s	p	p/0.882				
^{122}Pr		121.9518							
^{123}Pr		122.946							
^{124}Pr		123.943	1.2 s	β+, EC/12.					ann.rad./
^{125}Pr		124.9378	~ 3.3 s	β+					ann.rad./
									0.1358
^{126}Pr		125.9353	3.1 s	β+, EC/~ 10.4					ann.rad./
									(0.170−0.985)
^{127}Pr		126.9308	4.2 s	β+ /~ 7.5					ann.rad./
									(0.028−0.8949)
^{128}Pr		127.92879	3.0 s	β+, EC/~ 9.3					ann.rad./
									0.207/100
									0.400−1.373
^{129}Pr		128.92510	32 s	β+, EC/5.8					ann.rad./
									(0.0395−1.865)
^{130}Pr		129.9236	40. s	β+, EC/8.1					ann.rad./
^{131m}Pr			5.7 s						(0.06−0.16)
^{131}Pr		130.9203	1.7 m	β+, EC/5.3				~ 5.5	ann.rad./
									(0.059−0.980)
^{132}Pr		131.9193	1.6 m	β+, EC/7.1					ann.rad./
									0.325
									0.496
									0.533
^{133m}Pr			1.1 s	IT/0.192					0.1305
									0.0617
^{133}Pr		132.91633	6.5 m	β+, EC/4.3		5/2+			ann.rad./
									0.074
									0.1343
									0.2419
									0.3156
									0.3308
									0.4650
^{134m}Pr			~ 11. m	β+, EC/					ann.rad./
									0.294
									0.460

Table of the Isotopes

Elem. or Isot.	Natural Abundance (Atom %)	Atomic Mass or Weight	Half-life/ Resonance Width (MeV)	Decay Mode/ Energy (/MeV)	Particle Energy/ Intensity (MeV/%)	Spin ($h/2\pi$)	Nuclear Magnetic Mom. (nm)	Elect. Quadr. Mom. (b)	γ-Energy / Intensity (MeV/%)
									0.495
									0.632
^{134}Pr		133.91571	17. m	β+, EC/6.2		2+			ann.rad./
									0.294
									0.495
^{135}Pr		134.91311	24. m	β+, EC/3.7	2.5/	3/2+			ann.rad./
									0.0826
									0.2135
									0.2961
									0.5832
^{136}Pr		135.91269	13.1 m	β+ /57 /5.13 EC/43	2.98/	2+			ann.rad./ Ce k x-ray
									0.5398
									0.5522
^{137}Pr		136.91071	1.28 h	β+ /26 /2.70 EC/74 /	1.68/	5/2+			ann.rad./ Ce k x-ray
									0.4339
									0.5140
									0.8367
									(0.16−1.8)
^{138m}Pr			2.1 h	β+ /24 / EC/76 /	1.65/	7-			ann.rad./ Ce k x-ray
									0.3027
									0.7887
									1.0378
									(0.07−2.0)
^{138}Pr		137.91075	1.45 m	β+ /75 /4.44 EC/25 /	3.42/	1+			ann.rad./ Ce k x-ray
									0.7887
^{139}Pr		138.90894	4.41 h	β+ /8 /2.129 EC/92 /	1.09/	5/2+			ann.rad./ Ce k x-ray
									0.2551
									1.3473
									1.6307
^{140}Pr		139.90908	3.39 m	β+ /51 /3.39 EC/49 /	2.37/	1+			ann.rad./ Ce k x-ray
									0.3069
									1.5965
^{141}Pr	100.	140.907653				5/2+	+4.275	-0.08	
^{142m}Pr			14.6 m	I.T./0.004	c.e.	5-	2.2		
^{142}Pr		141.910045	19.12 h	β- /2.162 EC/0.744	0.58/4 2.16/96	2-	+0.234	+0.030	0.5088 1.57580
^{143}Pr		142.910817	13.57 d	β- /0.934	0.933/	7/2+	+2.70	+0.8	0.7420
^{144m}Pr			7.2 m	IT/99+/0.059 β- /		3-			Pr k x-ray
									0.0590
									0.6965
									0.8142
^{144}Pr		143.913305	17.28 m	β- /2.998	0.807/1 2.30/ 2.996/98	0-			0.69649 1.48912 2.18562
^{145}Pr		144.91451	5.98 h	β- /1.81	1.80/97	7/2+			0.0725 0.6758 0.7483
^{146}Pr		145.9176	24.2 m	β- /4.2	2.2/30 3.7/10 4.2/40	2-			0.4539/48 1.5247
^{147}Pr		146.91900	13.4 m	β- /2.69	1.5/ 2.1/	3/2+			0.3146/24. 0.5779/16 0.6413/19.
^{148m}Pr			2.0 m	β- /	4.0/	(4)			0.3016

Elem. or Isot.	Natural Abundance (Atom %)	Atomic Mass or Weight	Half-life/ Resonance Width (MeV)	Decay Mode/ Energy (/MeV)	Particle Energy/ Intensity (MeV/%)	Spin ($h/2\pi$)	Nuclear Magnetic Mom. (nm)	Elect. Quadr. Mom. (b)	γ-Energy / Intensity (MeV/%)
					3.8/				0.4506
									0.6975
^{148}Pr		147.92213	2.27 m	β- /4.9	4.8/	1⁻			0.3017
					4.5/				
^{149}Pr		148.9237	2.3 m	β- /3.40	3.0	(5/2+)			0.1085
									0.1385
									0.1651
^{150}Pr		149.92667	6.2 s	β- /5.7		1⁻			0.1302
					∼ 5.5				0.8044
									0.8527
^{151}Pr		150.92832	22.4 s	β- /4.2					
^{152}Pr		151.9315	3.2 s	β- /6.7		4⁺			0.0726
									0.164
									0.285
^{153}Pr		152.9338	4.3 s	β- /5.5					
^{154}Pr		153.9375	2.3 s	β- /7.9					
^{155}Pr		154.9401	> 0.3 μs						
^{156}Pr		155.9443	> 0.3 μs						
^{157}Pr		156.9474							
^{158}Pr		157.952							
^{159}Pr		158.956							
$_{60}$Nd		144.242(3)							
^{124}Nd		123.952				0+			
^{125}Nd		124.9489	0.6 s	β+, p					
^{126}Nd		125.9432				0+			
^{127}Nd		126.9405	1.8 s	β+, EC/9.		(5/2)			ann.rad./
^{128}Nd		127.9354	4. s	β+, EC/6.					ann.rad./
^{129}Nd		128.9332	4.9 s	β+, EC/8.		5/2(-)			ann.rad./
									(0.091−0.875)
^{130}Nd		129.92851	28. s	β+, EC/5.		0+			ann.rad./
^{131}Nd		130.92725	0.5 m	β+, EC/6.6					ann.rad./
									(0.09−0.36)
^{132}Nd		131.92332	1.5 m	β+, EC/3.7		0+			ann.rad./
									(0.099−0.567)
^{133}Nd		132.92235	1.2 m	β+, EC/5.6					ann.rad./
									(0.06−0.37)
^{134}Nd		133.91879	∼ 8.5 m	β+ /17 /2.8		0+			ann.rad./
				EC/83 /					Pr k x-ray
									0.1631/58
									(0.09−1.00)
^{135m}Nd			5.5 m	β+ /					
^{135}Nd		134.91818	12. m	β+ /65 /4.8		9/2-	-0.78	+2.0	ann.rad./
				EC/35 /					Pr k x-ray
									0.0415/23.
									0.204/51.
									(0.11−1.8)
^{136}Nd		135.91498	50.6 m	EC/94 /2.21	1.04/	0+			Pr Kx-ray
				β+ /6 /					0.0401/21.
									0.1091/35.
									(0.10−0.97)
^{137m}Nd			1.6 s	I.T./0.5196		11/2-			Nd k x-ray
									0.1084
									0.1775
									0.2337
^{137}Nd		136.91457	38. m	β+ /40 /3.69	1.7/20	1/2+	-0.63		ann.rad./
				EC/60 /	2.40/20				Pr k x-ray
									0.0755
									0.5806

Table of the Isotopes

Elem. or Isot.	Natural Abundance (Atom %)	Atomic Mass or Weight	Half-life/ Resonance Width (MeV)	Decay Mode/ Energy (/MeV)	Particle Energy/ Intensity (MeV/%)	Spin ($h/2\pi$)	Nuclear Magnetic Mom. (nm)	Elect. Quadr. Mom. (b)	γ-Energy / Intensity (MeV/%)
^{138}Nd		137.91195	5.1 h	EC/1.1		0+			Pr k x-ray
									0.1995
									0.3258
^{139m}Nd			5.5 h	I.T./12 /0.231	1.17/	11/2-			Nd k x-ray
				β+ /88 /					Pr k x-ray
									0.1139/34.
									0.7382/30.
^{139}Nd		138.91198	30. m	β+ /25 /2.79	1.77/	3/2+	0.91	+0.3	ann.rad./
				EC/75 /					Pr k x-ray
									0.4050
^{140}Nd		139.90955	3.37 d	EC /0.22		0+			Pr k x-ray
^{141m}Nd			1.04 m	IT/99+/0.756		11/2-			Nd k x-ray
									0.7565
^{141}Nd		140.909610	2.49 h	EC/98 /1.823	0.802/	3/2+	1.01	+0.3	Pr k x-ray
				β+ /2 /					(0.15−1.7)
^{142}Nd	27.153(39)	141.907723				0+			
^{143}Nd	12.173(27)	142.909814				7/2-	-1.07	-0.60	
^{144}Nd	23.798(18)	143.910087	2.1×10^{15} y	α	1.83	0+			
^{145}Nd	8.293(12)	144.912574				7/2-	-0.66	-0.31	
^{146}Nd	17.189(33)	145.913117				0+			
^{147}Nd		146.916100	10.98 d	β- /0.896	0.805/	5/2-	0.58	0.9	Pr k x-ray
									0.53102
									0.09111−0.686
^{148}Nd	5.756(21)	147.916893				0+			
^{149}Nd		148.920149	1.73 h	β- /1.691	1.03/25	5/2-	0.35	1.3	Pr k x-ray
					1.13/26				0.1143/19.
					1.42/				0.2113/27.
									(0.026−1.6)
^{150}Nd	5.638(29)	149.920891	1.4×10^{20} y	β-β-		0+			
^{151}Nd		150.923829	12.4 m	β- /2.442	1.2/	(3/2+)			Pm k x-ray
									0.1168
									0.2557
									1.1806
									(0.10−1.9)m
^{152}Nd		151.92468	11.4 m	β- /1.1		0+			0.2785/29.
									0.2501/18.
									(0.016−0.66)
^{153}Nd		152.92770	28.9 s	β- /3.6					0.418
^{154}Nd		153.9295	25.9 s	β- /2.8		0+			0.1519
									0.7998
^{155}Nd		154.9329	8.9 s	β- /5.0					0.1807
^{156}Nd		155.9350	5.5 s	β- /4.1		0+			0.0848
^{157}Nd		156.9390	> 0.3 μs						
^{158}Nd		157.9416	> 0.3 μs			0+			
^{159}Nd		158.946							
^{160}Nd		159.949				0+			
^{161}Nd		160.954							

$_{61}$Pm

Elem. or Isot.	Natural Abundance (Atom %)	Atomic Mass or Weight	Half-life/ Resonance Width (MeV)	Decay Mode/ Energy (/MeV)	Particle Energy/ Intensity (MeV/%)	Spin ($h/2\pi$)	Nuclear Magnetic Mom. (nm)	Elect. Quadr. Mom. (b)	γ-Energy / Intensity (MeV/%)
^{128}Pm		127.9484	1.0 s	β+, p					ann.rad.
^{129}Pm		128.9432	~ 2.4 s						
^{130}Pm		129.9405	2.5 s	β+, EC/11.					0.1589
									0.326−1.062
^{131}Pm		130.9359	~ 6.3 s	β+					0.185
									0.220
									0.146
^{132}Pm		131.9338	6. s	β+, EC/10.					ann.rad./
^{133}Pm		132.92978	12. s	β+, EC/~ 7.0					ann.rad./
^{134}Pm		133.9284	24. s	β+, EC/~ 8.9		(5+)			ann.rad./

Elem. or Isot.	Natural Abundance (Atom %)	Atomic Mass or Weight	Half-life/ Resonance Width (MeV)	Decay Mode/ Energy (/MeV)	Particle Energy/ Intensity (MeV/%)	Spin ($h/2\pi$)	Nuclear Magnetic Mom. (nm)	Elect. Quadr. Mom. (b)	γ-Energy / Intensity (MeV/%)
									0.294
									0.495
^{135}Pm		134.9249	0.8 m	β+, EC/6.0		11/2-			(0.13−0.47)
^{136}Pm		135.9236	1.8 m	β+ /89 /7.9		(3+)			ann.rad./
				EC/11 /					Nd k x-ray
									0.3735
									0.6027
^{137}Pm		136.92048	2.4 m	β+, EC/5.6		(11/2-)			ann.rad./
									0.1086
									0.1775
^{138m}Pm			3.2 m	β+ /50 /~ 7.0	3.9/	3+	3.		ann.rad./
				EC/50 /					Nd k x-ray
									0.5209
									0.7290
^{138}Pm		137.91955	10. s	β+ /6.9	6.1/	1+			ann.rad./
^{139m}Pm			0.18 s	IT/		(11/2-)			0.1887
^{139}Pm		138.91680	4.14 m	β+ /68 /4.52	3.52/	(5/2+)			ann.rad./
				EC/32 /					Nd k x-ray
									0.4028
									(0.27−2.4)
^{140m}Pm			5.87 m	β+ /70 /	3.2	7/2-			ann.rad./
				EC/30 /					Nd k x-ray
									0.4199
									0.7738
									1.0283
^{140}Pm		139.91604	9.2 s	β+ /89 /6.09	5.07/74	1+			ann.rad./
				EC/11 /					Nd k x-ray
									0.7738
									1.4898
^{141}Pm		140.91356	20.9 m	β+ /52 /3.72	2.71	5/2+			ann.rad./
				EC/48 /					Nd k x-ray
									0.8862
									1.2233
^{142m}Pm			67 µs						(0.208-0.882)
^{142}Pm		141.91287	40.5 s	β+ /86 /4.87	3.8/	1+			ann.rad./
				EC/20 /					Nd k x-ray
									0.6414
									1.5758
^{143}Pm		142.910933	265. d	EC/1.041		5/2+	3.8		Nd k x-ray
				β+ /< 6 × 10^{-6}/					0.7420
^{144}Pm		143.912591	360. d	EC/2.332		5-	1.7		Nd k x-ray
				β+ /7 × 10^{-6}/					0.6180
									0.6965
^{145}Pm		144.912749	17.7 y	EC/0.163		5/2+	+3.8	+0.2	Nd k x-ray
									0.0723
^{146}Pm		145.914696	5.53 y	EC/63 /1.472		3-			Nd k x-ray
				β- /37 /1.542	0.795/				0.4538
									0.7362
									0.7474
^{147}Pm		146.915139	2.623 y	β- /0.224	0.224/	7/2+	+2.6	+0.7	0.1213
									0.1974
^{148m}Pm			41.3 d	β- /95 /2.6	0.4/60	6-	1.8		0.5503/94.
				I.T./5 /0.137	0.5/17				0.6300/89.
					0.7/21				0.7257/33
^{148}Pm		147.91748	5.37 d	β- /2.47	1.02/	1-	+2.0	+0.2	0.5503
					2.47/				0.9149
									1.4651
^{149}Pm		148.918334	2.212 d	β- /1.071	0.78/9	7/2+	3.3		0.2859
					1.072/90				0.5909
									0.8594

Elem. or Isot.	Natural Abundance (Atom %)	Atomic Mass or Weight	Half-life/ Resonance Width (MeV)	Decay Mode/ Energy (/MeV)	Particle Energy/ Intensity (MeV/%)	Spin ($h/2\pi$)	Nuclear Magnetic Mom. (nm)	Elect. Quadr. Mom. (b)	γ-Energy / Intensity (MeV/%)
^{150}Pm		149.92098	2.68 h	β- /3.45	1.6/	(1-)			0.3339/69.
					2.3/				1.1658/16.
					1.8/				1.3245/17.
									(0.25−2.9)
^{151}Pm		150.92121	1.183 d	β- /1.187	0.84/	5/2+	+1.8	1.9	0.1677/8
									0.2751/7
									0.3401/22
^{152m2}Pm			15. m	β-, I.T./		(>6)			(0.14−1.4)
^{152m1}Pm			7.5 m	β- /		(4-)			0.1218
									0.2447
									0.3404
									1.0971
									1.4375
^{152}Pm		151.92350	4.1 m	β- /3.5	3.5/20	1+			0.1218
					3.50/60				(0.12−2.1)
^{153}Pm		152.92412	5.4 m	β- /1.90	1.7/	(5/2-)			0.0910
									0.1198
									0.1273
^{154m}Pm			2.7 m	β- /	2.0/				0.0820
									0.1848
									1.4403
^{154}Pm		153.92646	1.7 m	β- /4.1	1.9/				0.0820
									0.8396
									1.3940
									2.0589
									(0.08−2.8)
^{155}Pm		154.92810	48. s	β- /3.2		(5/2-)			(0.05−0.78)
^{156}Pm		155.93106	26.7 s	β- /5.16					
^{157}Pm		156.9330	10.9 s	β- /4.6					
^{158}Pm		157.9366	5. s	β- /6.3					
^{159}Pm		158.9390	1.5 s						(0.072-0.261)
^{160}Pm		159.9430							
^{161}Pm		160.9459							
^{162}Pm		161.950							
^{163}Pm		162.954							
$_{62}$Sm		150.36(2)							
^{129}Sm		128.954	~ 0.55 s	β+, p					
^{130}Sm		129.9489				0+			
^{131}Sm		130.9461	1.2 s	β+, EC/					ann.rad./
^{132}Sm		131.9407	4.0 s	β+		0+			
^{133}Sm		132.9387	2.9 s	β+, EC/~ 8.4		5/2+			ann.rad./
									0.3696
									0.0845
^{134}Sm		133.9340	11. s	β+, EC/5.		0+			ann.rad./
^{135}Sm		134.9325	10. s	β+, EC/7.		7/2+			ann.rad./
^{136}Sm		135.92828	42. s	β+, EC/4.5		0+			ann.rad./
^{137}Sm		136.92697	45. s	β+, EC/6.1					ann.rad./
^{138}Sm		137.92324	3.0 m	β+, EC/3.9		0+			ann.rad./
									0.0536
									0.0747
^{139m}Sm			10. s	I.T./94 /0.457		(11/2-)	1.1		Sm k x-ray
				β+ /6 /	4.7				0.1118
									0.1553
									0.1901
									0.2673
^{139}Sm		138.92230	2.6 m	β+ /75 /5.5	4.1/	½+	-0.53		Pm k x-ray
				EC/25 /					0.3678
									0.4028

Elem. or Isot.	Natural Abundance (Atom %)	Atomic Mass or Weight	Half-life/ Resonance Width (MeV)	Decay Mode/ Energy (/MeV)	Particle Energy/ Intensity (MeV/%)	Spin ($h/2\pi$)	Nuclear Magnetic Mom. (nm)	Elect. Quadr. Mom. (b)	γ-Energy / Intensity (MeV/%)
									(0.27−2.4)
^{140}Sm		139.91900	14.8 m	β+, EC/3.4	1.9/	0+			ann.rad./
									Pm k x-ray
									0.1396
									0.2255
^{141m}Sm			22.6 m	β+ /32 /	1.6/	11/2−	−0.83	+1.6	(0.07−1.7)
				EC/68 /	2.19/				ann.rad./
				I.T./0.3 /0.1758					Pm k x-ray
									0.1966
									0.4318
									0.7774
^{141}Sm		140.91848	10.2 m	β+ /52 /4.54	3.2/	½+	−0.74		ann.rad./
				EC/48 /					Pm k x-ray
									0.4382
^{142}Sm		141.91520	1.208 h	β+ /6 /2.10	1.0/	0+			ann.rad./
				EC/94 /					Pm k x-ray
^{143m}Sm			1.10 m	IT/99/0.7540		11/2−			Sm k x-ray
									0.7540
^{143}Sm		142.914628	8.83 m	β+ /46 /3.443	2.47/	3/2+	+1.01	+0.4	ann.rad./
				EC/54 /					Pm k x-ray
									1.0565
^{144}Sm	3.083(20)	143.911999				0+			
^{145}Sm		144.913410	340. d	EC/0.617		7/2−	−1.12	−0.60	Pm k x-ray
									0.0613
									0.4924
^{146}Sm		145.913041	1.03×10^{8} y	α/	2.50/	0+			
^{147}Sm	15.017(75)	146.914898	1.06×10^{11} y	α/	2.23/	7/2−	−0.815	−0.26	
^{148}Sm	11.254(51)	147.914823	7×10^{15} y	α/	1.96/	0+			
^{149}Sm	13.830(56)	148.917185	10^{16} y	α/		7/2−	−0.672	+0.075	
^{150}Sm	7.351(36)	149.917276				0+			
^{151}Sm		150.919932	90. y	β- /0.0768	0.076/	5/2−	−0.363	+0.7	0.02154
^{152}Sm	26.735(48)	151.919732				0+			
^{153}Sm		152.922097	1.929 d	β- /0.808	0.64/	3/2+	−0.0216	+1.3	Eu k x-ray
					0.69/				0.0697/4.7
									0.10318/29
									0.075−0.714
^{154}Sm	22.730(78)	153.922209				0+			
^{155}Sm		154.924640	22.2 m	β- /1.627	1.52	3/2−		1.1	Eu k x-ray
									0.1043/75.
^{156}Sm		155.92553	9.4 h	β- /0.72	0.43/	0+			0.0872
					0.71/				0.1657
									0.2038
^{157}Sm		156.92836	8.0 m	β- /2.7	2.4/	3/2−			Eu k x-ray
									0.1964
									0.1978
									0.3942
^{158}Sm		157.9300	5.5 m	β- /2.0		0+			0.1894/100.
									0.3636/82.
^{159}Sm		158.9332	11.3 s	β- /3.8					0.1898
^{160}Sm		159.9351	9.6 s	β- /3.6		0+			0.110
^{161}Sm		160.9388	~ 4.8 s						0.264
^{162}Sm		161.941	2.4 s			0+			(0.036-0.741)
^{163}Sm		162.945							
^{164}Sm		163.948				0+			
^{165}Sm		164.953							
$_{63}$**Eu**		**151.964(1)**							
^{130}Eu		129.964	0.9 ms	p	1.027/				
^{131}Eu		130.9578	~ 26. ms	β+, p	p/0.95				

Elem. or Isot.	Natural Abundance (Atom %)	Atomic Mass or Weight	Half-life/ Resonance Width (MeV)	Decay Mode/ Energy (/MeV)	Particle Energy/ Intensity (MeV/%)	Spin ($h/2\pi$)	Nuclear Magnetic Mom. (nm)	Elect. Quadr. Mom. (b)	γ-Energy / Intensity (MeV/%)
^{132}Eu		131.9544							
^{133}Eu		132.9492							
^{134}Eu		133.9465	0.5 s	EC, β+					ann.rad./
^{135}Eu		134.9418	1.5 s	EC, β+ /~ 8.7					ann.rad./
^{136m}Eu			~ 3.2 s			7+			0.255
^{136}Eu		135.9396	~ 3.9 s	EC, β+ /10.		1+			ann.rad./
^{137}Eu		136.9356	11. s	EC/~ 7.5		11/2-			ann.rad./
^{138}Eu		137.93371	12. s	EC, β+ /~ 9.2		7+	5		ann.rad./
^{139}Eu		138.92979	18. s	EC, β+ /6.7			6		ann.rad./
^{140m}Eu			0.125 s	EC, β+					ann.rad./
^{140}Eu		139.9281	1.51 s	EC, β+ /8.4		1-			ann.rad./
^{141m}Eu			3.0 s	β+ /58 /		11/2-			ann.rad./
				EC/9 /					Eu k x-ray
				I.T./33 /0.0964					(0.09−1.6)
^{141}Eu		140.92493	40. s	β+ /81 /5.6		5/2+	+3.49	+0.85	ann.rad./
				EC/15 /					Sm k x-ray
									0.3845
									0.3940
^{142m}Eu			1.22 m	β+ /83 /	4.8/	8-	+2.98	+1.4	ann.rad./
				EC/17 /					Sm k x-ray
									0.5566
									0.7680
									1.0233
^{142}Eu		141.92343	2.4 s	β- /94/7.4	7.0/	1+	+1.54	+0.12	ann.rad./
				EC/6 /					0.7680
^{143}Eu		142.92030	2.62 m	β+ /72/5.17	4.1/	5/2+	+3.67	+0.51	ann.rad./
				EC/28/	5.1/				Sm k x-ray
									0.1107/7
									1.5368/3.
									1.9127/2.
^{144}Eu		143.91882	10.2 s	β+ /86 /6.33	5.31/	1+	+1.89	+0.10	ann.rad./
				EC/13 /					Sm k x-ray
									1.6601
^{145}Eu		144.916265	5.93 d	β+ /2 /2.660	0.79/	5/2+	+4.00	+0.29	ann.rad./
				EC/98 /1.71					Sm k x-ray
									0.6535
									0.8937
									1.6587
^{146}Eu		145.91721	4.57 d	β+ /5 /3.88	1.47/	4-	+1.42	-0.18	ann.rad./
				EC/95 /					Sm k x-ray
									0.6336
									0.6341
									0.7470
									(0.27−2.64)
^{147}Eu		146.916746	24.4 d	EC/99. /1.722		5/2+	+3.72	+0.53	Sm k x-ray
				β+ /0.4 /					0.12113/20.6
									0.19725/24.0
									(0.601-1.077)
^{148}Eu		147.91809	54.5 d	EC/3.11	0.92	5-	+2.34	+0.35	Sm k x-ray
									0.5503/99.
									0.6299/71.
									(0.067−2.17)
^{149}Eu		148.917931	93.1 d	EC/0.692		5/2+	+3.57	+0.75	Sm k x-ray
									0.2770/4.1
									0.3275/4.8
^{150}Eu		149.91970	36. y	EC/2.26		5-	+2.71	+1.13	Sm k x-ray
									0.3340
									0.4394
									0.5843
									(0.25−1.8)

Elem. or Isot.	Natural Abundance (Atom %)	Atomic Mass or Weight	Half-life/ Resonance Width (MeV)	Decay Mode/ Energy (/MeV)	Particle Energy/ Intensity (MeV/%)	Spin ($h/2\pi$)	Nuclear Magnetic Mom. (nm)	Elect. Quadr. Mom. (b)	γ-Energy / Intensity (MeV/%)
150mEu			12.8 h	β- /92 /	1.013/	0-			Sm k x-ray
				β+ /0.4 /	1.24/				0.3339
				EC/8 /					0.4065
151Eu	47.81(6)	150.919850				5/2+	+3.472	+0.90	
152m2Eu			1.60 h	I.T./0.1478		8-			Eu k x-ray
									0.0898
152m1Eu			9.30 h	β- /72 /	1.85/	0-			Sm k x-ray
				EC/28 /	0.89/				0.12178
									0.84153
									0.96334
152Eu		151.921745	13.5 y	EC/72 /1.874	0.69/	3-	-1.941	+2.71	Sm k x-ray
				β- /28 /1.818	1.47/				Gd k x-ray
									0.12178
									0.34427
									1.40802
									(0.252−1.528)
153Eu	52.19(6)	152.921230				5/2+	+1.533	+2.41	
154mEu			46.1 m	I.T./~ 0.16		8-			Eu k x-ray
									0.0682
									0.1009
154Eu		153.922979	8.59 y	β- /99.9/1.969	0.27/29	3-	-2.01	+2.8	Gd k x-ray
				EC/0.02/0.717	0.58/38				0.12299/40.
					0.84/17				0.72331/20.
					0.98/4				1.2745/36
					1.87/11				(0.059-1.90)
155Eu		154.922893	4.76 y	β- /0.252	0.15/	5/2+	+1.52	+2.4	Gd k x-ray
									0.0865/30
									0.1053/20
156Eu		155.92475	15.2 d	β- /2.451	0.30/11	1+	≈1.1		0.08899/9.
					0.49/30				0.64623/7.
					1.2/12				0.723441/6.
					2.45/31				0.8118/10.
157Eu		156.92542	15.13 h	β- /1.36	0.98/	(5/2+)	+1.50	+2.6	Gd k x-ray
					1.30/41				0.0639/100.
									0.3705/48.
									0.4107/76.
158Eu		157.9279	45.9 m	β- /3.5	2.5/	(1-)	+1.44	+0.7	0.0795
									0.8976
									0.9442
									0.9771
159Eu		158.92909	18.1 m	β- /2.51	2.4/	(5/2+)	+1.38	+2.7	0.0678
					2.57/				0.0786
									0.0957
160Eu		159.9320	38. s	β- /4.1	2.7/	(0-)			0.0753
					4.1/				0.1735
									0.4131
									0.5155
									0.8217
									0.9110
									0.9246
161Eu		160.9337	27. s	β- /3.7					0.0719
162Eu		161.9370	11. s	β- /5.6					
163Eu		162.9392							
164Eu		163.943							
165Eu		164.946							
166Eu		165.950							
167Eu		166.953							
64Gd		157.25(3)							

Elem. or Isot.	Natural Abundance (Atom %)	Atomic Mass or Weight	Half-life/ Resonance Width (MeV)	Decay Mode/ Energy (/MeV)	Particle Energy/ Intensity (MeV/%)	Spin ($h/2\pi$)	Nuclear Magnetic Mom. (nm)	Elect. Quadr. Mom. (b)	γ-Energy / Intensity (MeV/%)
^{135}Gd		134.953	1.1 s	β+					(0.163−0.360)
^{136}Gd		135.9473				0+			
^{137}Gd		136.9450	7. s	EC, β+ /~ 8.8					ann.rad./
^{138}Gd		137.9401	~ 4.7 s	EC, β+		0+			0.0647
^{139m}Gd			~ 4.8 s						0.1216
^{139}Gd		138.9382	5. s	EC, β+ /~ 7.7					0.104−0.323
^{140}Gd		139.93367	16. s	EC/4.8		0+			0.1748
^{141m}Gd			25. s	EC, β+ /		11/2-			ann.rad./
^{141}Gd		140.93213	21. s	β+ /7.3		½+			ann.rad./
^{142}Gd		141.92812	1.17 m	EC, β+ /4.2		0+			ann.rad./
^{143m}Gd			1.84 m	β+ /67 /		11/2-			ann.rad./
				EC/33 /					Eu k x-ray
				I.T./					0.1176
									0.2719
									0.5880
									0.6681
									0.7999
^{143}Gd		142.9268	39. s	β+ /82 /6.0		1/2+			ann.rad./
				EC/18 /					Eu k x-ray
									0.2048
									0.2588
^{144}Gd		143.92296	4.5 m	β+ /45 /4.3	3.3/	0+			ann.rad./
				EC/55 /					Eu k x-ray
									0.3332
^{145m}Gd			1.44 m	I.T./95 /0.749		11/2-	-1.0		0.0273
				β+ /4 /5.7					0.3295
									0.3866
									0.7214
^{145}Gd		144.92171	23.4 m	β+ /33 /5.05	2.5/	1/2+	-0.74		ann.rad./
				EC/67 /					Eu k x-ray
									1.7579
									1.8806
									(0.32−3.69)
^{146}Gd		145.918311	48.3 d	EC/99.9 /1.03	0.35/	0+			Eu k x-ray
				β+ /0.2					0.1147
									0.1155
									0.1546
^{147}Gd		146.919094	1.588 d	EC/99.8 /2.188	0.93/	7/2-	1.0		Eu k x-ray
				EC/0.2 /					0.2293
									0.3699
									0.3960
									0.9289
									(0.1−1.8)
^{148}Gd		147.918115	71. y	α/3.27	3.1828/	0+			
^{149}Gd		148.919341	9.3 d	EC/1.32		7/2-	0.9		Eu k x-ray
									0.1496
									0.2985
									0.3465
^{150}Gd		149.91866	1.8×10^6 y	α/2.80	2.73/	0+			Eu k x-ray
^{151}Gd		150.920348	124. d	EC/0.464		7/2-	0.8		
									0.1536
									0.2432
^{152}Gd	0.20(1)	151.919791				0+			
^{153}Gd		152.921750	240. d	EC/0.485		3/2-	0.4		Eu k x-ray
									0.09743
									0.10318
^{154}Gd	2.18(3)	153.920867				0+			
^{155}Gd	14.80(12)	154.922622				3/2-	-0.259	+1.30	
^{156}Gd	20.47(9)	155.922123				0+			
^{157}Gd	15.65(2)	156.923960				3/2-	-0.340	+1.36	

Elem. or Isot.	Natural Abundance (Atom %)	Atomic Mass or Weight	Half-life/ Resonance Width (MeV)	Decay Mode/ Energy (/MeV)	Particle Energy/ Intensity (MeV/%)	Spin ($h/2\pi$)	Nuclear Magnetic Mom. (nm)	Elect. Quadr. Mom. (b)	γ-Energy / Intensity (MeV/%)
158Gd	24.84(7)	157.924104				0+			
159Gd		158.926389	18.6 h	β- 0.971	0.971/58	3/2-	-0.44		Tb k x-ray
					0.913/29				0.36351
					0.607/12				0.058-0.855
160Gd	21.86(19)	159.927054	> 1.9 × 10^19 y	β- β-		0+			
161Gd		160.929669	3.66 m	β- /1.956	1.56/85	5/2-			Tb k x-ray
									0.1023
									0.3149
									0.3609
162Gd		161.930985	8.4 m	β- /1.39	1.0/	0+			0.4030
									0.4421
163Gd		162.9340	1.13 m	β- /3.1					0.2868
									0.214
									1.685
164Gd		163.9359	45. s	β- /2.3		0+			
165Gd		164.9394	10 s	β-					
166Gd		165.942	~ 4.8 s			0+			(0.040-1.015)
167Gd		166.946							
168Gd		167.948				0+			
169Gd		168.953							
65Tb		158.92535(2)							
135Tb			0.9 ms	p	p/1.179				
138Tb		137.9532							
139Tb		138.9483	1.6 s						0.109
									0.120
140Tb		139.946	2.4 s	β+, EC/11					0.329
									0.355−0.740
141Tb		140.9415	3.5 s	β+, EC/~ 8.3					
142m2Tb			25 μs						
142mTb			0.30 s	β+, EC/		4-			
142Tb		141.9387	0.60 s	β+, EC/10.		0+			
143Tb		142.9351	12. s	β+, EC/7.4		11/2-			
144mTb			4.1 s	IT		5-			
144Tb		143.93305	< 1.5 s	β+, EC/8.4		1+			
145mTb			30. s	β+, EC/~ 6.6		11/2-			ann.rad./
									0.2577
									0.5370
									0.9876
145Tb		144.9293		β+, EC/6.5		½+			
146mTb			23. s	β+ /76 /		(5-)			ann.rad./
				EC/24 /					Gd k x-ray
									1.0789
									1.5795
146Tb		145.92725	~ 8. s	β+ /8.1		1+			
147mTb			1.8 m	β+ /35 /		11/2-			ann.rad./
				EC/65 /					Gd k x-ray
									1.3977
									1.7978
147Tb		146.92405	1.6 h	β+ /42 /4.61		5/2+	+1.70		ann.rad./
				EC/58 /					Gd k x-ray
									0.6944
									1.1522
									(0.120−3.318)
148mTb			2.3 m	β+ /25 /		9+			ann.rad./
				EC/75 /					Gd k x-ray
									0.3945
									0.6319
									0.7845

Elem. or Isot.	Natural Abundance (Atom %)	Atomic Mass or Weight	Half-life/ Resonance Width (MeV)	Decay Mode/ Energy (/MeV)	Particle Energy/ Intensity (MeV/%)	Spin ($h/2\pi$)	Nuclear Magnetic Mom. (nm)	Elect. Quadr. Mom. (b)	γ-Energy / Intensity (MeV/%)
									0.8824
^{148}Tb		147.92427	1.00 h	β+, EC/5.69		2-	-1.75	-0.3	ann.rad./
									Gd k x-ray
									0.4888
									0.7845
									(0.14−3.8)
^{149m}Tb			4.16 m	EC/88 /		11/2-			ann.rad./
				β+ /12 /					Gd k x-ray
									0.1650
									0.7960
^{149}Tb		148.923246	4.13 h	β+ /4 /3.636	1.8/	½+	+1.35		Gd k x-ray
				α/16/	3.97/				0.1650
									0.3522
									0.3886
									(0.1−3.2)
^{150m}Tb			6.0 m	β+ /17 /					ann.rad./
				EC/83 /					Gd k x-ray
									0.4384
									0.6380
									0.6504
									0.8275
^{150}Tb		149.92366	3.3 h	β+, EC/4.66		2-	-0.90		ann.rad./
									0.4963
									0.6380
									(0.3−4.29)
^{151m}Tb			25. s	I.T./95 /		11/2-			0.0229
				β+, EC/7 /					0.0495
									0.3797
									0.8305
^{151}Tb		150.923103	17.61 h	β+/1 /2.565	0.70/	1/2+	+0.92		Gd k x-ray
				EC/99 /					0.1083
									0.2517
									0.2870
									(0.1−1.8)
^{152m}Tb			4.3 m	I.T./79 /0.5018		(8+)			Tb k x-ray
				EC/21 /4.35					Gd k x-ray
									0.2833
									0.3443
									0.4111
^{152}Tb		151.92407	17.5 h	β+ /20 /3.99	2.5/	2-	-0.58	+0.3	ann.rad./
				EC/80 /	2.8/				Gd k x-ray
									0.3443
									(0.2−2.88)
^{153}Tb		152.923435	2.34 d	EC/1.570		5/2+	+3.44	+1.1	Gd k x-ray
									0.2119
									(0.05−1.1)
^{154m2}Tb			23.1 h	EC/98 /		(7-)	0.9		Gd k x-ray
				I.T./2 /					0.1231
									0.2479
									0.3467
									1.4199
^{154m1}Tb			9. h	β+ /78 /		(3-)	1.7	+3.	Gd k x-ray
				I.T./22 /					0.1231
									0.2479
									0.5401
									(0.12−2.57)
^{154}Tb		153.92468	21.5 h	EC/99 /3.56	1.86/	0-			Gd k x-ray
				β+ /1 /	2.45				0.1231
									1.2744
									2.1872

Elem. or Isot.	Natural Abundance (Atom %)	Atomic Mass or Weight	Half-life/ Resonance Width (MeV)	Decay Mode/ Energy (/MeV)	Particle Energy/ Intensity (MeV/%)	Spin ($h/2\pi$)	Nuclear Magnetic Mom. (nm)	Elect. Quadr. Mom. (b)	γ-Energy / Intensity (MeV/%)
									(0.12−3.14)
155Tb		154.92351	5.3 d	EC/0.82		3/2+	+2.01	+1.41	Gd k x-ray
									0.08654
									0.10530
156m2Tb			1.02 d	I.T./		(7-)			Tb k x-ray
									0.0496
156m1Tb			5.3 h	I.T./0.0884		(0+)			Tb k x-ray
									0.0884
156Tb		155.924747	5.3 d	EC/2.444		3-	~ 1.7	+2.	Gd k x-ray
									0.08896
									0.19921
									0.53435
									1.22245
157Tb		156.924025	1.1×10^2 y	EC/0.0601		3/2+	+2.01	+1.4	Gd k x-ray
									0.0545
158mTb			10.5 s	I.T./0.11		0-			Gd k x-ray
									0.0110
158Tb		157.925413	1.8×10^2 y	EC/80 /1.220		3-	+1.76	+2.7	Gd k x-ray
				β- /20 /0.937					0.0795
									0.9442
									0.9621
159Tb	100.	158.925347				3/2+	+2.014	+1.43	
160Tb		159.927168	72.3 d	β- /1.835	0.57/47	3-	+1.79	3.8	Dy k x-ray
					0.86/27				0.08678
									0.29857
									0.87936
									0.96615
161Tb		160.927570	6.91 d	β- /0.593	0.46/23	3/2+	2.2	+1.2	Dy k x-ray
					0.52/66				0.02565
					0.6/10				0.04892
									0.07458
162Tb		161.92949	7.6 m	β- /2.51	1.4	(1/2-)			Dy k x-ray
									0.2600
									0.8075
									0.8882
163Tb		162.930648	19.5 m	β- /1.785	0.80/	3/2+			Dy k x-ray
									0.3511
									0.3897
									0.4945
164Tb		163.9334	3.0 m	β- /3.9	1.7/	(5+)			Dy k x-ray
									0.1689
									0.2157
									0.6110
									0.6885
									0.7548
165Tb		164.9349	2.1 m	β- /3.0		3/2+			0.5389
									1.1785
									1.2920
									1.6648
166Tb		165.9380	26 s	β-/					
167Tb		166.9401	19 s						0.057
									0.070
168Tb		167.944	8 s						0.075−0.227
169Tb		168.946							
170Tb		169.950							
171Tb		170.953							
66Dy		162.500(1)							
139Dy		138.960	0.6 s	β+, p					

Table of the Isotopes

Elem. or Isot.	Natural Abundance (Atom %)	Atomic Mass or Weight	Half-life/ Resonance Width (MeV)	Decay Mode/ Energy (/MeV)	Particle Energy/ Intensity (MeV/%)	Spin ($h/2\pi$)	Nuclear Magnetic Mom. (nm)	Elect. Quadr. Mom. (b)	γ-Energy / Intensity (MeV/%)
^{140}Dy		139.954				0+			
^{141}Dy		140.9514	0.9 s	EC, β+ /9.					
^{142}Dy		141.9464	2.3 s	EC, β+ /7.1		0+			
^{143}Dy		142.9438	3.9 s	EC, β+ /~ 8.8					
^{144}Dy		143.93925	9.1 s	EC, β+ /~ 6.2		0+			
^{145m}Dy		144.9365	14. s	EC, β+		11/2-			
^{146m}Dy			0.15 s	I.T.		10+			
^{146}Dy		145.93285	30. s	EC, β+ /5.2		0+			
^{147m}Dy			56. s	I.T./40 /		(11/2-)	-0.66	+0.7	Dy k x-ray
				β+, EC/60 /					0.072
									0.6787
^{147}Dy		146.93109	75. s	EC, β+ /6.37		½+	-0.92		ann.rad./
									0.1007
									0.2534
									0.3653
^{148}Dy		147.92715	3.1 m	β+ /4 /2.68	1.2/	0+			ann.rad./
				EC/96 /					Tb k x-ray
									0.6202
^{149}Dy		148.92731	4.2 m	β+, EC/3.81		(7/2-)	-0.12	-0.62	ann.rad./
									0.1008
									0.1063
									0.2534
									0.6536
									0.7894
									1.7765
									1.8062
^{150}Dy		149.925585	7.18 m	β+, EC/67 /1.79		0+			Tb k x-ray
				α/33 /	4.233/				0.3967
^{151}Dy		150.926185	17. m	β+ /5 /2.871		7/2-	-0.95	-0.30	Tb k x-ray
				EC/89 /					0.1764
				α /6 /	4.067/				0.3030
									0.3861
									0.5463
									(0.16−2.09)
^{152}Dy		151.92472	2.37 h	EC/0.60		0+			Tb k x-ray
				α /	3.63/				0.2569
^{153}Dy		152.925765	6.3 h	β+ /1 /2.171	0.89/	(7/2-)	-0.78	~-0.15	Tb k x-ray
				EC/99 /					0.0807
				α /0.01 /	3.46/				0.0997
									0.2137
									(0.08−1.66)
^{154}Dy		153.92442	3. × 10⁶ y	α/2.95	2.87/	0+			
^{155}Dy		154.92575	9.9 h	β+ /2 /2.095	0.845/	3/2-	-0.385	+1.04	Tb k x-ray
				EC/98 /					0.0655
									0.2269
^{156}Dy	0.056(3)	155.92428				0+			
^{157}Dy		156.92547	8.1 h	EC/1.34		3/2-	-0.301	+1.30	Tb k x-ray
									(0.0609−1.319)
^{158}Dy	0.095(3)	157.924409				0+			
^{159}Dy		158.925739	144. d	EC/0.366		3/2-	-0.354	+1.37	Tb k x-ray
									0.3262
^{160}Dy	2.329(18)	159.925198				0+			
^{161}Dy	18.889(42)	160.926933				5/2+	-0.480	+2.51	
^{162}Dy	25.475(36)	161.926798				0+			
^{163}Dy	24.896(42)	162.928731				5/2-	+0.673	+2.65	
^{164}Dy	28.260(54)	163.929175				0+			
^{165m}Dy			1.26 m	I.T./98 /0.108		1/2-			Dy k x-ray
				β- /2 /					0.1082
									0.5155
^{165}Dy		164.931703	2.33 h	β- /1.286	1.29/	7/2+	-0.52	+3.5	Ho k x-ray

Elem. or Isot.	Natural Abundance (Atom %)	Atomic Mass or Weight	Half-life/ Resonance Width (MeV)	Decay Mode/ Energy (/MeV)	Particle Energy/ Intensity (MeV/%)	Spin ($h/2\pi$)	Nuclear Magnetic Mom. (nm)	Elect. Quadr. Mom. (b)	γ-Energy / Intensity (MeV/%)
									0.09468/3.8
^{166}Dy		165.932807	3.400 d	β- /0.486	0.40/	0+			Ho k x-ray
									0.0282
									0.0825
^{167}Dy		166.9357	6.2 m	β- /~ 2.35	1.78	(1/2-)			Ho k x-ray
									0.2593
									0.3103
									0.5697
									(0.06−1.4)
^{168}Dy		167.9371	8.5 m	β- /1.6		0+			Ho k x-ray
									0.1925
									0.4867
^{169}Dy		168.9403	~ 39. s	β- /3.2					
^{170}Dy		169.9424				0+			
^{171}Dy		170.9462							
^{172}Dy		171.9488				0+			
^{173}Dy		172.953							
$_{67}$Ho		164.93032(2)							
^{140}Ho		139.969	6 ms	p/	p/1.09				
^{141m}Ho			8 µs	p/	p/1.23				
^{141}Ho		140.963	4.2 ms	β+, p	p/1.71				
^{142}Ho		141.960	0.4 s	EC/β+, p					0.307
^{143}Ho		142.9546	> 0.2 µs						
^{144}Ho		143.9515	0.7 s	β+, EC/12					
^{145}Ho		144.9472	2.4 s	β+					
^{146}Ho		145.9446	3.3 s	β+, EC/10.7		(10+)			ann.rad./
^{147}Ho		146.94006	5.8 s	β+, EC/8.2		11/2-			ann.rad./
^{148m}Ho			9. s	β+, EC/		4-			ann.rad./
^{148}Ho		147.9377	2. s	β+, EC/9.4		1+			ann.rad./
									0.6615
									1.6883
^{149m}Ho			21. s	β+, EC/		11/2-			ann.rad./
									1.0733
									1.0911
^{149}Ho		148.93378	> 30. s	β+, EC/6.01		1/2+			
^{150m}Ho			25. s	β+, EC/		(9+)			ann.rad./
									0.3939
									0.5511
									0.6534
									0.8034
^{150}Ho		149.93350	1.3 m	β+, EC/6.6					ann.rad./
									0.5913
									0.6534
									0.8034
^{151m}Ho			47. s	β+, EC/87 /					ann.rad./
				α/13	4.605/				0.2102
									0.4889
									0.6948
									0.7762
^{151}Ho		150.93169	35.2 s	β+, EC/80/5.13					ann.rad./
				α/20 /	4.519/				0.3522
									0.5274
									0.9676
									1.0471
^{152m}Ho			50. s	β+, EC/90/		(9+)	+5.9	-1.	ann.rad./
				α/10/	4.453/				0.4929
									0.6138
									0.6474

Elem. or Isot.	Natural Abundance (Atom %)	Atomic Mass or Weight	Half-life/ Resonance Width (MeV)	Decay Mode/ Energy (/MeV)	Particle Energy/ Intensity (MeV/%)	Spin ($h/2\pi$)	Nuclear Magnetic Mom. (nm)	Elect. Quadr. Mom. (b)	γ-Energy / Intensity (MeV/%)
									0.6835
^{152}Ho		151.93171	2.4 m	β+, EC/88/6.47		(3+)	-1.02	+0.1	ann.rad./
				α/12/	4.387/				0.6140
									0.6476
^{153m}Ho			9.3 m	β+, EC/99+/4.12		5/2	+1.19		ann.rad./
				α/	4.01/				0.0905
									0.1089
									0.1618
									0.2302
									0.2707
									0.3659
									0.4565
^{153}Ho		152.93020	2.0 m	β+, EC/99+/4.13		11/2-	+6.8	-1.1	ann.rad./
				α/	3.91/				0.2958
									0.3346
									0.4381
									0.6383
^{154m}Ho			3.3 m	β+, EC/		(8+)	5.7	-1.0	ann.rad./
									0.3346
									0.4124
									0.4771
^{154}Ho		153.93060	12. m	β+, EC/5.75		1-	-0.64	+0.2	ann.rad./
									Dy k x-ray
									0.3346
									0.5700
									0.8734
^{155}Ho		154.92910	48. m	β+/6/3.10		(5/2+)	+3.51	+1.5	ann.rad./
				EC/94 /					Dy k x-ray
									0.0474
									0.1363
									0.3254
									(0.06−2.24)
^{156m}Ho			5.8 m	I.T./0.0352			+2.99	+2.3	ann.rad./
				β+ /25 /	1.8/				Dy k x-ray
				EC/75 /	2.9/				0.1378
									0.2666
									(0.28−2.9)
^{156}Ho		155.92984	56. m	β+, EC/4.4		(5+)			ann.rad./
									0.1378
									0.2665
^{157}Ho		156.92826	12.6 m	β+/5/2.54	1.18/	7/2-	+4.35	+3.0	ann.rad./
				EC/95/					Dy k x-ray
									0.2800
									0.3411
^{158m2}Ho			28. m	I.T./44/		2-	+2.44	+1.6	ann.rad./
				EC/56/					Dy k x-ray
									0.0989
									0.2182
^{158m1}Ho			21. m	β+, EC/		(9+)			ann.rad./
									0.0981
									0.1664
									0.2182
									0.3205
									0.4062
									0.9774
									1.0532
									0.4846
^{158}Ho		157.92894	11.3 m	β+/8/4.24	1.30/	5+	+3.77	+4.1	ann.rad./
				EC/92/					Dy k x-ray
									0.0989

Elem. or Isot.	Natural Abundance (Atom %)	Atomic Mass or Weight	Half-life/ Resonance Width (MeV)	Decay Mode/ Energy (/MeV)	Particle Energy/ Intensity (MeV/%)	Spin ($h/2\pi$)	Nuclear Magnetic Mom. (nm)	Elect. Quadr. Mom. (b)	γ-Energy / Intensity (MeV/%)
									0.2182
									0.9488
^{159m}Ho			8.3 s	IT/0.206		1/2+			Ho k x-ray
									0.1660
									0.2059
^{159}Ho		158.927712	33.0 m	EC/1.838		7/2-	+4.28	+3.2	Dy k x-ray
									0.1210
									0.1320
									0.2529
									0.3096
									(0.06−1.2)
^{160m2}Ho			3. s			1+			
^{160m}Ho			5.0 h	IT/67/0.060		2-	+2.52	+1.8	0.0868
				EC/33/3.35					0.1970
									0.6464
									0.7281
									0.8791
									0.9619
									0.9658
^{160}Ho		159.92873	25.6 m	β+, EC/3.29	0.57/	5+	+3.71	+4.0	See Ho[166m]
									0.7282
									0.8794
^{161m}Ho			6.8 s	IT/0.211					Ho k x-ray
									0.2112
^{161}Ho		160.927855	2.48 h	EC/0.859		7/2-	+4.25	+3.2	Dy k x-ray
									0.0256
									0.0592
									0.0774
									0.1031
^{162m}Ho			1.12 h	IT/61/		6-	+3.60	+4.	Dy k x-ray
				EC/39/					Ho k x-ray
									0.0807
									0.1850
									0.2828
									0.9372
									1.2200
^{162}Ho		161.929096	15. m	EC/96 /0.295		1+			Dy k x-ray
				β+ /4 /					0.0807
									1.3196
									1.3728
^{163m}Ho			1.09 s	I.T./0.298		(1/2+)			Ho k x-ray
									0.2798
^{163}Ho		162.928734	4.57 × 10³ y	EC/0.00258		7/2-	+4.23	+3.6	Dy M x-rays
^{164m}Ho			38. m	I.T./0.140		(6-)			Ho k x-ray
									0.0373
									0.0566
									0.0940
^{164}Ho		163.930234	29. m	EC/58 /0.987		1+			Dy k x-ray
				β- /42 /0.963					0.0734
									0.0914
^{165}Ho	100.	164.930322				7/2-	+4.17	+3.49	
^{166m}Ho			1.2 × 10³ y	β- /		7-	3.6	-3.	Er k x-ray
									0.18407
									0.71169
									0.81031
^{166}Ho		165.932284	1.117 d	β- /1.855	1.776/48	0-			Er k x-ray
					1.855/51				0.08057
									1.37943
^{167}Ho		166.93313	3.1 h	β- /1.007	0.31/43	(7/2-)			Er k x-ray
					0.61/21				0.0793

Elem. or Isot.	Natural Abundance (Atom %)	Atomic Mass or Weight	Half-life/ Resonance Width (MeV)	Decay Mode/ Energy (/MeV)	Particle Energy/ Intensity (MeV/%)	Spin ($h/2\pi$)	Nuclear Magnetic Mom. (nm)	Elect. Quadr. Mom. (b)	γ-Energy / Intensity (MeV/%)
					0.96/15				0.0835
					0.97/15				0.2379
									0.3213
									0.3465
^{168m}Ho			2.2 m	I.T./					
^{168}Ho		167.93552	3.0 m	β- /2.91	2.0/	3+			Er k x-ray
									0.7413
									0.8159
									0.8211
									(0.08−2.34)
^{169}Ho		168.93687	4.7 m	β- /2.12	1.2/	(7/2-)			
					2.0/				0.1496
									0.7610
									0.7784
									0.7884
									0.8529
^{170m}Ho			43. s	β- /		1+			0.0787
									0.8123
									1.8940
									1.9726
^{170}Ho		169.93962	2.8 m	β- /3.87		6+			Er k x-ray
									0.1816
									0.2582
									0.8902
									0.9321
									0.9414
									1.1387
^{171}Ho		170.941	53 s	β- /					
^{172}Ho		171.9448	25. s	β- /					Er k x-ray
									(0.077−1.186)
^{173}Ho		172.9473							
^{174}Ho		173.951							
^{175}Ho		174.954							
$_{68}$Er		167.259(3)							
^{144}Er		143.9604	> 0.2 μs			0+			
^{145}Er		144.9574	0.9 s	β+					
^{146}Er		145.9520	~ 1.7 s	β+		0+			
^{147}Er		146.9495	2.5 s	E.C, β+ /~ 9.1					
^{148}Er		147.9446	4.5 s	β+, EC/6.8		0+			
^{149m}Er			10. s	IT		11/2-			
^{149}Er		148.94231	10.7 s	ECβ+ /8.1		½+			
^{150}Er		149.93791	18. s	β+ /36 /4.11		0+			ann.rad./
				EC/64 /					Ho k x-ray
									0.4758
^{151}Er		150.93745	23. s	β+, EC/5.2		7/2-			ann.rad./
^{152}Er		151.93505	10.2 s	β+, EC/10/3.11		0+			ann.rad./
				α/90/	4.804/				
^{153}Er		152.935063	37.1 s	α/	4.674		-0.934	-0.42	0.351
				β+, EC/47/4.56	4.35/				(0.0945−1.700)
^{154}Er		153.93278	3.7 m	β+, EC/99+/2.03		0+			ann.rad./
				α/0.5/	4.166/				
^{155}Er		154.93321	5.3 m	β+, EC/47/3.84		(7/2-)	-0.669	-0.27	ann.rad./
				EC/53 /					Ho k x-ray
									0.1101
									0.2415
^{156}Er		155.93107	20. m	β+, EC/1.7		0+			ann.rad./
									0.0298
									0.0352
									0.0522
									0.1336

Elem. or Isot.	Natural Abundance (Atom %)	Atomic Mass or Weight	Half-life/ Resonance Width (MeV)	Decay Mode/ Energy (/MeV)	Particle Energy/ Intensity (MeV/%)	Spin ($h/2\pi$)	Nuclear Magnetic Mom. (nm)	Elect. Quadr. Mom. (b)	γ-Energy / Intensity (MeV/%)
[157]Er		156.93192	25. m	β+, EC/3.5		3/2-	-0.412	+0.92	ann.rad./
									0.117
									0.385
									1.320
									1.660
									1.820
									2.000
[158]Er		157.92989	2.2 h	EC/99.5 /1.78	0.74/	0+			Ho k x-ray
				β+ /0.5 /					0.0719
									0.2486
									0.3868
[159]Er		158.930684	36. m	β+ /7 /2.769		3/2-	-0.304	+1.17	ann.rad./
				EC/93 /					Ho k x-ray
									0.6245
									0.6493
									(0.07-2.5)
[160]Er		159.92908	1.191 d	EC/0.33		0+			Ho k x-ray
									(0.05-0.96)
[161]Er		160.93000	3.21 h	EC/2.00		3/2-	-0.37	+1.36	Ho k x-ray
									0.8265
									(0.07-1.74)
[162]Er	0.139(5)	161.928778				0+			
[163]Er		162.93003	1.25 h	EC/1.210		5/2-	+0.557	+2.55	Ho k x-ray
									0.4361
									0.4399
									1.1135
[164]Er	1.601(3)	163.929200				0+			
[165]Er		164.930726	10.36 h	EC/0.376		5/2-	+0.643	+2.71	Ho k x-ray
[166]Er	33.503(36)	165.930293				0+			
[167m]Er			2.27 s	I.T./0.208		½-			Er k x-ray
									0.2078
[167]Er	22.869(9)	166.932048				7/2+	-0.5639	+3.57	
[168]Er	26.978(18)	167.932370				0+			
[169]Er		168.934590	9.40 d	β- /0.351	0.35/~ 100	½-	+0.485		Tm k x-ray
									0.1098
									0.1182
[170]Er	14.910(36)	169.935464				0+			
[171]Er		170.938030	7.52 h	β- /1.491		5/2-	0.66	2.9	Tm k x-ray
									0.11160
									0.29591
									0.30832
									(0.08-1.4)
[172]Er		171.939356	2.05 d	β-/0.891	0.28/48	0+			Tm k x-ray
					0.36/46				0.0597
									0.4073
									0.6101
[173]Er		172.9424	1.4 m	β- /2.6		(7/2-)			Tm k x-ray
									0.1928
									0.1992
									0.8952
[174]Er		173.9442	3.1 m	β- /1.8		0+			Tm k x-ray
									(0.100-0.152)
[175]Er		174.9478	1.2 m	β-					(0.0765-1.168)
[176]Er		175.9501				0+			
[177]Er		176.954							
[69]Tm		168.93421(2)							
[144]Tm			~ 1.9 µs	p	1.70, 1.43				
[145]Tm		144.9701	3.1 µs	p// ~ 10	1.73/91				

Elem. or Isot.	Natural Abundance (Atom %)	Atomic Mass or Weight	Half-life/ Resonance Width (MeV)	Decay Mode/ Energy (/MeV)	Particle Energy/ Intensity (MeV/%)	Spin ($h/2\pi$)	Nuclear Magnetic Mom. (nm)	Elect. Quadr. Mom. (b)	γ-Energy / Intensity (MeV/%)
					1.4/9				
^{146m}Tm			0.198 s	β+, p	p/1.118/100				
					1.01/				
					0.89/8				
^{146}Tm		145.9664	0.08 s	β+/14.	p/1.19/100				
				p	1.01/28				
					0.94/22				
^{147m}Tm			0.4 ms	β+, p	p/1.115				
^{147}Tm		146.9610	0.56 s	EC, β+/85	~ 10.7				
				p/15/	1.052/				
^{148m}Tm		147.9578	0.7 s	β+, EC/12.					ann.rad./
^{148}Tm									
^{149}Tm		148.9527	0.9 s	β+, EC/~ 9.2		11/2-			
^{150}Tm		149.9500	2.3 s	β+, EC/~ 11.5		6-			(0.1007−2.177)
^{151}Tm		150.94548	4. s	β+, EC/7.5					ann.rad./
^{152m}Tm			8. s	β+, EC/		9+			
^{152}Tm		151.9444	5. s	β+, EC/8.8					ann.rad./
^{153}Tm		152.94201	1.6 s	β+, EC/10 /6.46					ann.rad./
				α/90 /	5.109/				
^{154m}Tm			3.3 s	β+, EC/15 /	α/5.031/100				ann.rad./
				α/	4.84/0.24				0.4605−0.7960
^{154}Tm		153.94157	8.1 s	β+, EC/56 /7.4	α/4.956/100				ann.rad./
				α/44 /	4.83/0.45				
^{155}Tm		154.93920	30. s	β+, EC/5.58					0.0315
				α/	4.46/				0.0638
									0.0881
									0.2268
									0.5320
									0.6067
^{156m}Tm			19. s	α/	4.46/				
^{156}Tm		155.93898	1.40 m	β+, EC/7.6		2-	+0.40	-0.5	ann.rad./
				α/	4.23/				0.3446
									0.4529
									0.5860
^{157}Tm		156.93697	3.6 m	β+, EC/4.5	2.6	½	+0.48		ann.rad./
				α/	3.97/				0.1104
									0.3484
									0.3855
									0.4550
									(0.1−1.58)
^{158}Tm		157.93698	4.0 m	β+, EC/74 /6.5		(2-)	+0.04	+0.7	ann.rad./
				EC/26 /					Er k x-ray
									0.1921
									0.3351
									0.6280
									1.1498
									(0.18−2.81)
^{159}Tm		158.93498	9.1 m	β+/23 /3.9		5/2+	+3.42	+1.9	ann.rad./
				EC/77 /					Er k x-ray
									0.0591
									0.0848
									0.2713
									(0.05−1.27)
^{160m}Tm			1.24 m	IT		(5)			
^{160}Tm		159.93526	9.4 m	β+/15 /5.9		1-	+0.16	+0.58	ann.rad./
				EC/85 /					Er k x-ray
									0.1264
									0.2642
									0.7285
									0.8544

Elem. or Isot.	Natural Abundance (Atom %)	Atomic Mass or Weight	Half-life/ Resonance Width (MeV)	Decay Mode/ Energy (/MeV)	Particle Energy/ Intensity (MeV/%)	Spin ($h/2\pi$)	Nuclear Magnetic Mom. (nm)	Elect. Quadr. Mom. (b)	γ-Energy / Intensity (MeV/%)
									0.8614
									1.3685
^{161}Tm		160.93355	31. m	β+, EC/3.2		7/2+	+2.40	+2.9	ann.rad./
									Er k x-ray
									0.0595
									0.0844
									1.6481
									(0.04−2.15)
^{162m}Tm			24. s	I.T./90 /		5+			Tm k x-ray
				β+, EC/10 /					Er k x-ray
									0.0669
									0.8115
									0.9003
^{162}Tm		161.93400	21.7 m	β+ /8 /4.81		1-	+0.07	+0.69	ann.rad./
				EC/92 /					Er k x-ray
									0.1020
									0.7987
									(0.1−3.75)
^{163}Tm		162.93265	1.81 h	EC/98 /2.439		½+	-0.082		Er k x-ray
				β+ /1 /					0.0692
									0.1043
									0.2414
^{164m}Tm			5.1 m	I.T./80 /		6-			0.0914
				β+, EC/20 /					0.1394
									0.2081
									0.2405
									0.3149
^{164}Tm		163.93356	2.0 m	β+ /36 /3.96	2.94/	1+	+2.38	+0.71	ann.rad./
				EC/64 /					Er k x-ray
									0.0914
^{165}Tm		164.932435	1.253 d	EC/1.593		½+	-0.139		Er k x-ray
									0.0472
									0.0544
									0.29728
									0.80636
^{166}Tm		165.93355	7.70 h	EC/98 /3.04		2+	+0.092	+2.14	Er k x-ray
				β+ /2 /					0.0806
									0.1844
									0.7789
									1.2734
									2.0524
^{167}Tm		166.932852	9.24 d	EC/0.748		½+	-0.197		Er k x-ray
									0.0571
									0.20778
^{168}Tm		167.934173	93.1 d	EC/1.679		3+	+0.23	+3.2	Er k x-ray
									0.19825
									0.4475
									0.81595
^{169}Tm	100	168.934213				½+	-0.232	-1.2	
^{170}Tm		169.935801	128.6 d	β- /99.8/0.968	0.883/24	1-	+0.247	+0.74	Yb k x-ray
				EC/0.2 /0.314	0.968/76				0.08425
^{171}Tm		170.936429	1.92 y	β- /0.096	0.03/2	½+	-0.230		0.06674
					0.096/98				
^{172}Tm		171.93840	2.65 d	β- /1.88	1.79/36	2-			Yb k x-ray
					1.88/29				0.07879
									1.38722
									1.46601
									1.52982
									1.60861
^{173}Tm		172.939604	8.2 h	β- /1.298	0.80/21	½+			Yb k x-ray

Elem. or Isot.	Natural Abundance (Atom %)	Atomic Mass or Weight	Half-life/ Resonance Width (MeV)	Decay Mode/ Energy (/MeV)	Particle Energy/ Intensity (MeV/%)	Spin ($h/2\,\pi$)	Nuclear Magnetic Mom. (nm)	Elect. Quadr. Mom. (b)	γ-Energy / Intensity (MeV/%)
					0.86/71				0.3988
									0.4613
^{174m}Tm			2.29 s						
^{174}Tm		173.94217	5.4 m	β- /3.08	0.70/14	(4-)			Yb k x-ray
					1.20/83				0.07664
									0.17669
									0.27332
									0.3666
									0.99205
									(0.08−1.6)
^{175}Tm		174.94384	15.2 m	β- /2.39	0.9/36	(1/2+)			Yb k x-ray
					1.9/23				0.36396
									0.51487
									0.94125
									0.98247
^{176}Tm		175.9470	1.9 m	β-/4.2	2.0/	(4+)			Yb k x-ray
					1.2/				0.1898
									0.3819
									1.0691
^{177}Tm		176.9490	1.4 m	β-					
^{178}Tm		177.9526							
^{179}Tm		178.955							
$_{70}$Yb		173.04(3)							
^{148}Yb		147.967				0+			
^{149}Yb		148.964	0.7 s	β+, p	p/2.5−6.4/				0.647
^{150}Yb		149.9584	> 0.2 μs			0+			
^{151}Yb		150.9554	1.6 s	β+ /8.5					
^{152}Yb		151.9503	3.2 s	β+ EC/5.5		0+			
^{153}Yb		152.9495	4. s	β+ EC/6.7					
^{154}Yb		153.94639	0.40 s	β+ EC/7 /4.49		0+			ann.rad./
				α/93 /	5.32/				
^{155}Yb		154.9458	1.7 s	β+, EC/16 /6.0			-0.8	-1.	ann.rad./
				α/84 /	5.19/				
^{156}Yb		155.94282	26. s	β+, EC/21/3.57		0+			ann.rad./
				α/79 /	4.69/				
^{157}Yb		156.94263	39. s	β+, EC/99+/5.5			-0.64		ann.rad./
				α/0.5/	4.69/				0.231
									(0.035−0.670)
^{158}Yb		157.93987	1.5 m	β+, EC/1.9		0+			ann.rad./
									0.0741
									0.2526
^{159}Yb		158.94005	1.4 m	EC, β+/5.1			-0.37	-.022	Tm k x-ray
									0.1661
									0.1772
									0.3297
									0.3903
^{160}Yb		159.93755	4.8 m	β+, EC/2.0		0+			ann.rad./
									0.1404
									0.1737
									0.2158
^{161}Yb		160.93790	4.2 m	β+, EC/3.9		3/2-	-0.33	+1.03	ann.rad./
									Tm k x-ray
									0.0782
									0.5999
									0.6315
^{162}Yb		161.93577	18.9 m	β+, EC/1.7		0+			ann.rad./
									Tm k x-ray
									0.1188

Elem. or Isot.	Natural Abundance (Atom %)	Atomic Mass or Weight	Half-life/ Resonance Width (MeV)	Decay Mode/ Energy (/MeV)	Particle Energy/ Intensity (MeV/%)	Spin ($h/2\pi$)	Nuclear Magnetic Mom. (nm)	Elect. Quadr. Mom. (b)	γ-Energy / Intensity (MeV/%)
									0.1635
^{163}Yb		162.93633	11.1 m	β+ /26 /3.4	1.4/	3/2-	-0.37	+1.24	ann.rad./
									Tm k x-ray
									0.0636
									0.8603
									(0.06 −1.9)
^{164}Yb		163.93449	1.26 h	EC/1.0		0+			Tm k x-ray
									0.0914
									0.6752
^{165}Yb		164.93528	9.9 m	β+ /10 /2.76	1.58/	(5/2-)	+0.48	+2.48	ann.rad./
				EC/90 /					Tm k x-ray
									0.0801
									1.0903
^{166}Yb		165.93388	2.363 d	EC/0.30		0+			Tm k x-ray
									0.0828
									0.1844
									0.7789
									1.2734
									2.0524
^{167}Yb		166.934950	17.5 m	β+ /0.5 /1.954	0.639/	5/2-	+0.62	+2.70	Tm k x-ray
				EC/99.5 /					0.06296
									0.10616
									0.11337
									0.17633
^{168}Yb	0.13(1)	167.933897				0+			
^{169m}Yb			46. s	I.T./0.0242		1/2-			Yb L x-ray
									0.0242
^{169}Yb		168.935190	32.02 d	EC/0.909		7/2+	-0.63	+3.5	0.1979/35.9
									0.3078/10.05
									0.0207−0.2611
^{170}Yb	3.04(15)	169.934762				0+			
^{171}Yb	14.28(57)	170.936326				1/2-	+0.49367		
^{172}Yb	21.83(67)	171.936382				0+			
^{173}Yb	16.13(27)	172.938211				5/2-	-0.67989	+2.80	
^{174}Yb	31.83(92)	173.938862				0+			
^{175}Yb		174.941277	4.19 d	β- /0.470	0.466/73	7/2-	0.77		Lu k x-ray
					0.071/21				0.3963/13
					0.353/6.2				(0.114−0.28)
^{176m}Yb			11.4 s	I.T./1.051		(8-)			Yb k x-ray
									0.0961
									0.1901
									0.2929
									0.3897
^{176}Yb	12.76(41)	175.942572	10^{26} y	β-β-		0+			
^{177m}Yb			6.41 s	I.T./0.3315		1/2-			Yb k x-ray
									0.1131
									0.2084
^{177}Yb		176.945261	1.9 h	β- /1.399	1.40	9/2+			Lu k x-ray
									0.1504
^{178}Yb		177.94665	1.23 h	β- /0.65	0.25/	0+			0.1415
									0.3246
									0.3516
									0.3815
									0.6125
^{179}Yb		178.9502	8. m	β- /2.4					
^{180}Yb		179.9523	2. m	β-		0+			0.1028−0.4423
^{181}Yb		180.9562							
$_{71}$Lu		174.967(1)							

Elem. or Isot.	Natural Abundance (Atom %)	Atomic Mass or Weight	Half-life/ Resonance Width (MeV)	Decay Mode/ Energy (/MeV)	Particle Energy/ Intensity (MeV/%)	Spin ($h/2\pi$)	Nuclear Magnetic Mom. (nm)	Elect. Quadr. Mom. (b)	γ-Energy / Intensity (MeV/%)
150mLu			0.045 ms	p/1.29					
150Lu	149.973		43. ms	p					
151mLu			16 μs	p/1.31					
151Lu	150.9676		0.08 s	p/1.231					
152Lu	151.9641		0.7 s						
153Lu	152.9588		0.9 s						
154Lu	153.9575		1.0 s	β+, EC/10.8					
155mLu			2.6 ms	α/7.41					
155Lu	154.95432		0.07 s	EC/8.0					
				α/	5.66/90				
156mLu			0.20 s	α/	5.57/				
156Lu	155.9530		~ 0.5 s	β+, EC/9.5					ann.rad./
				α/	5.45/				
157mLu			~ 9.6 s	α	4.925/				
157Lu	156.95010		4.8 s	β+, EC/94 /6.93					ann.rad./
				α/	5.00/				
158Lu	157.94931		10.4 s	β+, EC/99 /8.0					ann.rad./
				α/	4.67/				0.3682
									0.4770
159Lu	158.94663		12.3 s	β+, EC/6.0					ann.rad./
									0.1505
									0.1875
									0.3693
160Lu	159.9460		36.1 s	β+, EC/7.3					ann.rad./
									0.2434
									0.3957
									0.5773
161Lu	160.94357		1.2 m	β+, EC/5.3					ann.rad./
									0.0437
									0.0671
									0.1003
									0.1108
									0.1562
									0.2562
162mLu			~ 1.5 m	EC/		4-			
162Lu	161.9433		1.37 m	β+, EC/6.9		1-			ann.rad./
									0.1666
									0.6314
163Lu	162.94118		4.1 m	β+, EC/4.6					ann.rad./
									0.0539
									0.0581
									0.1504
									0.1631
									0.3717
164Lu	163.94134		3.14 m	β+, EC/6.3	1.6/				0.1238
					3.8/				0.2621
									0.7404
									0.8639
									0.8804
165Lu	164.93941		10.7 m	β+, EC/3.9	2.06/	1/2+			ann.rad./
									0.1206
									0.1324
									0.1742
									0.2036
									(0.04−2.0)
166m2Lu			2.1 m	β+ /35 /		(0-)			ann.rad./
				EC/65 /					Yb k x-ray
									1.0673
									1.2566
									2.0986

Elem. or Isot.	Natural Abundance (Atom %)	Atomic Mass or Weight	Half-life/ Resonance Width (MeV)	Decay Mode/ Energy (/MeV)	Particle Energy/ Intensity (MeV/%)	Spin ($h/2\pi$)	Nuclear Magnetic Mom. (nm)	Elect. Quadr. Mom. (b)	γ-Energy / Intensity (MeV/%)
166m1Lu			1.4 m	β+, EC/58 /		(3-)			ann.rad./
				I.T./42 /0.0344					0.1024
									0.2281
									0.2861
									0.8119
									0.8301
166Lu		165.93986	2.8 m	β+ /25 /5.5		(6-)			ann.rad./
				EC/75 /					Yb k x-ray
									0.1024
									0.2281
									0.3375
									0.3679
167Lu		166.93827	52. m	β+ /2 /3.1	2.1/	7/2+			Yb k x-ray
				EC/98 /					0.0297
									0.2392
									(0.03−2.0)
168mLu			6.7 m	β+ /12 /		3+			ann.rad./
				EC/88 /					Yb k x-ray
				IT/<0.8					0.1988/190
									0.8960/100
									0.9792/128
									0.018−2.65
168Lu		167.93874	5.5 m	β+ /6 /4.5	1.2/	(6-)			ann.rad./
				EC/94 /					Yb k x-ray
									0.1114
									0.1124
									0.2286
									0.3483
									1.4836
169mLu			2.7 m	I.T./0.0290		1/2-			Lu L x-ray
									0.0290
169Lu		168.93765	1.419 d	EC/2.293	1.271/	7/2+	2.30	3.5	Yb k x-ray
									0.19121
									0.9606
									(0.08−2.1)
170mLu			0.7 s	I.T./0.0929		4-			Lu L x-ray
									0.04449
									0.0484
170Lu		169.93848	2.01 d	EC/3.46	2.44/	0+			Yb k x-ray
									0.58711
									0.5908
									1.28029
									(0.1−3.38)
171mLu			1.31 m	I.T./0.0711		1/2-			Lu k x-ray
									0.07119
171Lu		170.937913	8.24 d	EC/1.479	0.362/	7/2+	2.30	3.42	Yb k x-ray
									0.01939
									0.66744
									(0.02−1.3)
172mLu			3.7 m	I.T./0.0419		1-			Lu L x-rays
									0.04186
172Lu		171.939086	6.64 d	EC/2.519		4-	2.90	3.80	Yb k x-ray
									0.18156
									1.09367
									(0.07−2.2)
173Lu		172.938931	1.37 y	EC/0.671		7/2+	2.28	3.63	Yb k x-ray
									0.07860
									0.27198
174mLu			142. d	IT/99.3/	0.17086	6-	1.50		Lu k x-ray
				EC/0.7 /					0.067055

Elem. or Isot.	Natural Abundance (Atom %)	Atomic Mass or Weight	Half-life/ Resonance Width (MeV)	Decay Mode/ Energy (/MeV)	Particle Energy/ Intensity (MeV/%)	Spin ($h/2\pi$)	Nuclear Magnetic Mom. (nm)	Elect. Quadr. Mom. (b)	γ-Energy / Intensity (MeV/%)
^{174}Lu		173.940338	3.3 y	EC/1.374		1-	1.9		Yb k x-ray
									0.07664
									1.2419
^{175}Lu	97.41(2)	174.940772				7/2+	+2.2327	+3.49	
^{176m}Lu			3.66 h	β- /1.315	1.229/	1-	+0.318	-1.47	Hf k x-ray
					1.317/				0.088372
^{176}Lu	2.59(2)	175.942686	3.73×10^{10} y	β- /1.192		7-	+3.169	+4.92	Hf k x-ray
				β+/ < 0.9					0.20187
									0.30691
^{177m2}Lu			6. m	β-		39/2-			0.089
^{177m}Lu			160.7 d	IT/22/0.9702		23/2-	2.33	5.4	Lu k x-ray
				β- /78					Hf k x-ray
									0.11295
									0.20836
									0.37850
									0.41853
^{177}Lu		176.943758	6.65 d	β- /0.498	0.497/	7/2+	+2.239	+3.39	0.11295
									0.20836
^{178m}Lu			23.1 m	β- /		(9-)			0.2166
									0.3317
^{178}Lu		177.945955	28.5 m	β-/2.099	2.03/	1+			Hf k x-ray
									0.0932
									1.3099
									1.3408
									(0.09−1.7)
^{179}Lu		178.94733	4.6 h	β- /1.405	1.35/	7/2+			0.2143
									0.3377
^{180}Lu		179.9499	5.7 m	β- /3.1	1.49/				0.40795/50.
									(0.07−1.9)
^{181}Lu		180.9520	3.5 m	β- /2.5		(7/2+)			0.0458
									0.2059
									0.5749
^{182}Lu		181.9550	2.0 m	β- /~ 4.1					0.0978
									0.7208
									0.8182
^{183}Lu		182.9576	58. s	β- /		7/2+			
^{184}Lu		183.9609	20 s	β-					
$_{72}$**Hf**		**178.49(2)**							
^{153}Hf		152.971	> 0.2 μs						
^{154}Hf		153.965	2. s	EC, β+/~ 6.7		0+			
^{155}Hf		154.9634	0.9 s	EC, β+/8.					
^{156}Hf		155.9594	25. ms	α/		0+			
^{157}Hf		156.9584	0.11 s	α/					
^{158}Hf		157.95480	2.9 s	EC/54 /5.1		0+			
				α/46 /	5.27/				
^{159}Hf		158.95400	5.6 s	β+, EC/88 /6.9					ann.rad./
				α/12 /	5.09/				
^{160}Hf		159.95068	~ 12. s	β+, EC/97 /4.9		0+			ann.rad./
				α/4.78					
^{161}Hf		160.95028	17. s	α/	4.60/				
^{162}Hf		161.94721	38. s	β+, EC/3.7		0+			ann.rad./
									0.1739
									0.1963
									0.4101
^{163}Hf		162.94709	40. s	β+, EC/5.5					ann.rad./
									0.0454
									0.0621
									0.0710

Elem. or Isot.	Natural Abundance (Atom %)	Atomic Mass or Weight	Half-life/ Resonance Width (MeV)	Decay Mode/ Energy (/MeV)	Particle Energy/ Intensity (MeV/%)	Spin ($h/2\pi$)	Nuclear Magnetic Mom. (nm)	Elect. Quadr. Mom. (b)	γ-Energy / Intensity (MeV/%)
									0.6882
^{164}Hf		163.94438	2.8 m	EC, β+/3.0		0+			
^{165}Hf		164.94457	1.32 m	EC/4.6		11/2-			
^{166}Hf		165.94218	6.8 m	EC/93 /2.3		0+			ann.rad./
				β+ /7 /					Lu k x-ray
									0.0788
^{167}Hf		166.94260	2.0 m	β+ /40 /4.0		(5/2-)			ann.rad./
				EC/60 /					Lu k x-ray
									0.1754
									0.3152
^{168}Hf		167.94057	25.9 m	β+, EC/1.8		0+			ann.rad./
									(0.0144−1.311)
^{169}Hf		168.94126	3.25 m	EC/85 /3.3		(5/2-)			ann.rad./
				β+ /15 /					Lu k x-ray
									0.3695
									0.4929
^{170}Hf		169.93961	16.0 h	EC/1.1		0+			Lu k x-ray
									0.0985
									0.1202
									0.1647
									0.5729
									0.6207
^{171m}Hf			30. s			(1/2-)	+0.53		
^{171}Hf		170.94049	12.2 h	EC, β+ /2.4		7/2+	-0.67	+3.46	ann.rad./
									Lu k x-ray
									0.1221
									0.6620
									1.0714
^{172}Hf		171.93945	1.87 y	EC/0.35		0+			Lu k x-ray
									0.02399
									0.12582
									(0.0818−0.123)
^{173}Hf		172.94051	23.6 h	EC/1.6		½-			Lu k x-ray
									0.12367
									0.13963
									0.29697
									0.31124
									(0.1−2.1)
^{174}Hf	0.16(1)	173.940046	2.0×10^{15} y			0+			
^{175}Hf		174.941509	71. d	EC/0.686		5/2-	-0.60	+2.7	Lu k x-ray
									0.08936
									0.34340
^{176}Hf	5.26(7)	175.941409				0+			
^{177m2}Hf			51.4 m	I.T./2.740		37/2-			Hf k x-ray
									0.2140
									0.2951
									0.3115
									0.3267
^{177m1}Hf			1.1 s	I.T./		23/2+			Hf k x-ray
									0.20836
									0.22847
									0.37851
^{177}Hf	18.60(9)	176.943221				7/2-	+0.7935	+0.337	
^{178m2}Hf			31. y	I.T./		16+	+8.16	+6.00	Hf k x-ray
									0.32555
									0.42635
									0.089−0.574
^{178m1}Hf			4.0 s	I.T./		8-			Hf k x-ray
									0.21342
									0.32555

Elem. or Isot.	Natural Abundance (Atom %)	Atomic Mass or Weight	Half-life/ Resonance Width (MeV)	Decay Mode/ Energy (/MeV)	Particle Energy/ Intensity (MeV/%)	Spin ($h/2\pi$)	Nuclear Magnetic Mom. (nm)	Elect. Quadr. Mom. (b)	γ-Energy / Intensity (MeV/%)
									0.42635
^{178}Hf	27.28(7)	177.943699				0+			
^{179m2}Hf			25.1 d	I.T./1.1057		25/2-	7.4		Hf k x-ray
									0.1227
									0.1461
									0.3626
									0.4537
^{179m1}Hf			18.7 s	I.T./0.375		½-			Hf k x-ray
									0.1607
									0.2141
^{179}Hf	13.62(2)	178.945816				9/2+	-0.641	+3.79	
^{180m}Hf			5.52 h	I.T./1.1416		8-	+9.	+4.6	Hf k x-ray
									0.2152
									0.3323
									0.4432
^{180}Hf	35.08(16)	179.946550				0+			
^{181m}Hf			1.5 ms	/1.738		25/2-			
^{181}Hf		180.949101	42.4 d	β- /1.027	0.408/	1/2-			Ta k x-ray
									0.13294/54
									0.48200/100
									0.3459/20
^{182m}Hf			62. m	β- /54 /1.60	0.49/43	8-			Hf k x-ray
				IT/46 /1.173	0.95/10				0.0509
									0.2244
									0.3441
									0.4558
									0.5066
									0.9428
^{182}Hf		181.95055	8.9 × 10^6 y	β- /0.37		0+			Ta k x-ray
									0.2704/79
									(0.098-0.270)
^{183}Hf		182.95353	1.07 h	β- /2.01	1.18/68	3/2-			Ta k x-ray
					1.54/25				0.0732
									0.4591
									0.7837
^{184}Hf		183.95545	4.1 h	β- /1.34	0.74/38	0+			Ta k x-ray
					0.85/16				0.0414
					1.10/46				0.1391
									0.3449
^{185}Hf		184.9588	~ 3.5 m	β- /					0.165
^{186}Hf		185.9609	~ 2.6 m			0+			0.738
^{187}Hf		186.9646	> 0.3 µs						
^{188}Hf		187.967	> 0.3 µs			0+			
$_{73}$Ta		180.94788(2)							
^{155}Ta		154.975	12 µs	p/1.77					
^{156}Ta		155.9723	0.11 s	β+ /~ 11.6					
				p/	1.02/~ 100				
^{157}Ta		156.9682	10 ms	α/	6.117				
				p/	0.927/3.4				
^{158}Ta		157.9667	37. ms	α/	6.05/100				
					5.97/100				
^{159}Ta		158.96302	0.6 s	β+, EC/20 /8.5	α/5.52/34				ann.rad./
				α/80 /	5.60/55				
^{160}Ta		159.9615	1.4 s	β+, EC/10.1					ann.rad./
				α	5.41/				
^{161}Ta		160.9584	3.16 s	β+, EC/7.5					ann.rad./
				α/	5.15				
^{162}Ta		161.9573	4. s	EC/8.6					

Elem. or Isot.	Natural Abundance (Atom %)	Atomic Mass or Weight	Half-life/ Resonance Width (MeV)	Decay Mode/ Energy (/MeV)	Particle Energy/ Intensity (MeV/%)	Spin ($h/2\pi$)	Nuclear Magnetic Mom. (nm)	Elect. Quadr. Mom. (b)	γ-Energy / Intensity (MeV/%)
^{163}Ta		162.95433	10.6 s	EC/6.8					
^{164}Ta		163.95353	14.2 s	β+ /8.5		3+			ann.rad./
				α/	4.62/				0.2110
									0.3768
^{165}Ta		164.95077	31. s	ECβ+/5.9					
^{166}Ta		165.95051	34. s	β+ /82 /7.7					ann.rad./
				EC/18 /					Hf k x-ray
									0.1587
									0.3117
									0.8101
^{167}Ta		166.94809	1.4 m	β+, EC/5.6					ann.rad./
^{168}Ta		167.94805	2.4 m	β+ /77 /6.7		3+			ann.rad./
				EC/23 /					Hf k x-ray
									0.1239
									0.2615
									0.7502
^{169}Ta		168.94601	4.9 m	β+, EC/4.4					ann.rad./
									0.0288
									0.1535
									0.1924
^{170}Ta		169.94618	6.8 m	β +/70 /6.0		(3+)			ann.rad./
				EC/35 /					Hf k x-ray
									0.1008
									0.2212
^{171}Ta		170.94448	23.3 m	β+, EC/3.7		(5/2-)			0.0496
									0.5018
									0.5064
									(0.05−1.02)
^{172}Ta		171.94490	36.8 m	β+ /25 /4.9		(3-)			ann.rad./
				EC/75 /					Hf k x-ray
									0.21396
									1.10923
									(0.09 −3.8)
^{173}Ta		172.94375	3.6 h	β+ /24 /3.7		(5/2-)	1.70	-1.9	ann.rad./
				EC/76 /					Hf k x-ray
									0.06972
									0.17219
									(0.06 −2.7)
^{174}Ta		173.94445	1.12 h	β+ /27 /3.8		(3+)			ann.rad./
				EC/73 /					Hf k x-ray
									0.09089
									0.20638
									(0.09−3.64)
^{175}Ta		174.94374	10.5 h	EC/2.0		7/2+	2.27	+3.7	Hf k x-ray
									0.2077
									0.2671
									0.3487
^{176}Ta		175.94486	8.1 h	EC/3.1		1-			Hf k x-ray
									0.08837
									1.15735
^{177}Ta		176.944472	2.356 d	EC/1.166		7/2+	2.25		Hf k x-ray
									0.11295
									(0.07−1.06)
^{178m}Ta			2.4 h	EC/		(7-)			Hf k x-ray
									0.08886
									0.21342
									0.32555
									0.42635
^{178}Ta		177.94578	9.29 m	EC/99 /1.9		1+	+2.74	+0.65	ann.rad./
				β+ /1 /					Hf k x-ray

Elem. or Isot.	Natural Abundance (Atom %)	Atomic Mass or Weight	Half-life/ Resonance Width (MeV)	Decay Mode/ Energy (/MeV)	Particle Energy/ Intensity (MeV/%)	Spin ($h/2\pi$)	Nuclear Magnetic Mom. (nm)	Elect. Quadr. Mom. (b)	γ-Energy / Intensity (MeV/%)
									0.09316
^{179}Ta		178.945930	1.8 y	EC/0.110		7/2+	2.29	3.37	Hf k x-ray
^{180m}Ta	0.0120(2)		>1.2 × 10^{15} y			(9-)	4.82		
^{180}Ta		179.947465	8.15 h	EC/87 /0.854		1+			Hf k x-ray
				β- /13 /0.708	0.61/3				W k x-ray
					0.71/10				0.09333
									0.10340
^{181}Ta	99.9880(2)	180.947996				7/2+	+2.370	+3.3	
^{182m}Ta			15.8 m	I.T./0.5198		10-			Ta k x-ray
									0.14678
									0.17157
^{182}Ta		181.950152	114.43 d	β- /1.814	0.25/30	3-	+3.02	+2.6	W k x-ray
					0.44/20				1.12127/100
					0.52/40				1.22138/79
									0.085−1.289
^{183}Ta		182.951373	5.1 d	β- /1.070	0.45/5	7/2+	+2.36		W k x-ray
					0.62/91				0.0847
									0.0991
									0.1079
									0.2461
									0.3540
^{184}Ta		183.95401	8.7 h	β- /2.87	1.11/15	(5-)			W k x-ray
					1.17/81				0.2528/44.
									0.4140/74.
									(0.09−1.4)
^{185}Ta		184.95556	49. m	β- /1.99	1.21/5	(7/2+)			W k x-ray
					1.77/81				0.0697
									0.1739
									0.1776
^{186}Ta		185.9586	10.5 m	β- /3.9	2.2/	(3-)			W k x-ray
									0.1979
									0.2149
									0.5106
									(0.09−1.5)
^{187}Ta		186.9605	> 0.3 μs						
^{188}Ta		187.9637	5 μs						0.292
^{189}Ta		188.9658	> 0.3 μs						
$_{74}$W		183.84(1)							
^{158m}W			0.14 ms	α	8.28(3)/				
^{158}W		157.975	1.3 ms	α/	6.433/96	0+			
^{159}W		158.9729	7. ms	α/					
^{160}W		159.9685	0.08 s	α/	5.92/	0+			
^{161}W		160.9674	0.41 s	β+, EC/18 /8.1					
				α/82 /	5.78/				
^{162}W		161.9635	1.39 s	β+, EC/54 /5.8		0+			
				α/46 /	5.54/				
^{163}W		162.9625	2.8 s	β+, EC/59 /7.5					
				α/41 /	5.38/				
^{164}W		163.95895	6. s	β+, EC/97 /5.0		0+			ann.rad./
				α/3 /	5.15/				
^{165}W		164.95828	5.1 s	β+, EC/99 /7.0					ann.rad./
				α/1 /	4.91/				
^{166}W		165.95503	16. s	β+, EC/99 /4.2		0+			ann.rad./
				α/1 /	4.74/				
^{167}W		166.95482	20. s	EC/5.6					
^{168}W		167.95181	53. s	EC/3.8		0+			ann.rad./
				α/10^{-5}/	4.40(1)				Ta k x-ray
									0.1755

Elem. or Isot.	Natural Abundance (Atom %)	Atomic Mass or Weight	Half-life/ Resonance Width (MeV)	Decay Mode/ Energy (/MeV)	Particle Energy/ Intensity (MeV/%)	Spin ($h/2\pi$)	Nuclear Magnetic Mom. (nm)	Elect. Quadr. Mom. (b)	γ-Energy / Intensity (MeV/%)
									(0.037−0.573)
^{169}W		168.95178	1.3 m	EC/5.4					ann.rad./
									Ta k x-ray
									0.123
									(0.097−0.699)
^{170}W		169.94923	2.4 m	EC/2.2		0+			ann.rad./
									Ta k x-ray
									0.3162
									(0.060−0.144)
^{171}W		170.94945	2.4 m	EC/4.6					ann.rad./
									Ta k x-ray
									0.1842
									(0.052−0.479)
^{172}W		171.94729	6.6 m	β+, EC/2.5		0+			ann.rad./
									Ta k x-ray
									0.0389
									(0.034−0.674)
^{173}W		172.94769	6.3 m	EC/4.0					ann.rad./
									Ta k x-ray
									0.4576
									(0.035−0.623)
^{174}W		173.94608	35. m	EC/1.9		0+			ann.rad./
									Ta k x-ray
									0.3287
									0.4288
									(0.056−0.429)
^{175}W		174.94672	35. m	EC/2.9		½-			(0.015−0.27)
^{176}W		175.94563	2.5 h	β+, EC/0.8		0+			0.03358
									0.06129
									0.09487
									0.10020
^{177}W		176.94664	2.21 h	EC/2.0		(1/2-)			Ta k x-ray
									0.15505
									0.18569
									0.42694
^{178}W		177.94588	21.6 d	EC/0.091		0+			Ta k x-ray
^{179m}W			6.4 m	IT/99.7/0.222		(1/2-)			W k x-ray
				EC/0.3/					0.2220
^{179}W		178.94707	38. m	EC/1.06		(7/2-)			Ta k x-ray
									0.0307
^{180}W	0.12(1)	179.946704	1.8×10^{18} y	α/		0+			
^{181}W		180.948197	121.1 d	EC/0.188		9/2+			Ta k x-ray
									0.13617
									0.15221
^{182}W	26.50(16)	181.948204	$> 7.7 \times 10^{21}$ y	α/		0+			
^{183m}W			5.15 s	I.T./		(11/2+)			W k x-ray
									0.0465
									0.0526
									0.0991
									0.1605
^{183}W	14.31(4)	182.950223	$> 4.1 \times 10^{21}$ y	α/		½-	+0.1177848		
^{184}W	30.64(2)	183.950931	$> 8.9 \times 10^{21}$ y	α/		0+			
^{185m}W			1.6 m	I.T./0.1974		11/2+			W k x-ray
									0.0659
									0.1315
									0.1737
^{185}W		184.953419	74.8 d	β- /0.433	0.433/99.9	3/2-			0.12536
^{186}W	28.43(19)	185.954364	$> 8.2 \times 10^{21}$ y	α/		0+			
^{187m}W			1.6 μs	IT	0.411	11/2+			(0.014-0.287)
^{187}W		186.957161	23.9 h	β- /1.311	0.624/66	3/2-	0.62		Re k x-ray

Elem. or Isot.	Natural Abundance (Atom %)	Atomic Mass or Weight	Half-life/ Resonance Width (MeV)	Decay Mode/ Energy (/MeV)	Particle Energy/ Intensity (MeV/%)	Spin ($h/2\pi$)	Nuclear Magnetic Mom. (nm)	Elect. Quadr. Mom. (b)	γ-Energy / Intensity (MeV/%)
					1.315/16				0.68572/33
					0.081−1.18				0.134−0.773
^{188}W		187.958489	69.78 d	β- /0.349	0.349/99	0+			0.0636
									0.2271
									0.2907
^{189}W		188.9619	9.7 m	β- /2.5	1.4/	(3/2-)			0.2604
					2.5/				(0.1262-1.466)
^{190m}W			~ 0.06 ms						(0.0585-0.694)
^{190}W		189.9632	30. m	β- /1.3	0.95/	0+			Re k x-ray
									0.1576
									0.1621
^{191}W		190.9666	> 0.3 μs						
^{192}W		191.968	> 0.3 μs			0+			
$_{75}$Re		186.207(1)							
^{160}Re		159.9821	0.7 ms	p/	1.261(6)/91				
				α/	6.54/				
^{161}Re		160.9776	14 ms	α/	6.24				
				p	1.35				
^{162}Re		161.9760	0.10 s	α/	6.12/94				
					6.09/94				
^{163}Re		162.97208	0.26 s	β+, EC/9.0	α/5.87/32				
				α/	5.92/66				
^{164}Re		163.9703	0.9 s	β+, EC/10.7					
				α/	5.78/				
^{165m}Re			~ 2.37 s	α/	5.502/				
^{165}Re		164.96709	2.6 s	β+, EC/87 /8.1					
				α/	5.49/ < 5.				
^{166}Re		165.9658	2.5 s	β+, EC/9.4					
				α/	5.50/				
^{167m}Re			6.2 s	α, EC/					
^{167}Re		166.9626	3.4 s	β+, EC/7.4					
				α/	5.015/				
^{168}Re		167.96157	4.4 s	β+, EC/9.1					
				α/	4.833/				0.1117
^{169m}Re			8.1 s	α	4.70/				
					4.87/				
^{169}Re		168.95879	16. s						
^{170}Re		169.95822	9.2 s	β+, EC/9.0					0.1560
									0.3055
									0.4125
^{171}Re		170.95572	15.2 s	EC/~ 5.7					
^{172m}Re			55. s	β+, EC/		(2)			ann.rad./
									0.1234
									0.2537
									0.3504
^{172}Re		171.9554	15. s	β+, EC/7.3					ann.rad./
									0.1234
									0.2537
^{173}Re		172.95324	2.0 m	EC/~ 3.9					ann.rad./
^{174}Re		173.95312	2.4 m	β+, EC/5.6					ann.rad./
									0.1119
									0.2430
^{175}Re		174.95138	5.8 m	β+, EC/4.3					ann.rad./
^{176}Re		175.95162	5.3 m	β+, EC/5.6		(3+)			ann.rad./
									0.1089
									0.2406
^{177}Re		176.95033	14. m	EC/78 /3.4		(5/2-)			ann.rad./
				β+ /22 /					W k x-ray

Elem. or Isot.	Natural Abundance (Atom %)	Atomic Mass or Weight	Half-life/ Resonance Width (MeV)	Decay Mode/ Energy (/MeV)	Particle Energy/ Intensity (MeV/%)	Spin ($h/2\pi$)	Nuclear Magnetic Mom. (nm)	Elect. Quadr. Mom. (b)	γ-Energy / Intensity (MeV/%)
									0.0797
									0.0843
									0.1968
^{178}Re		177.95099	13.2 m	β+ /11 /4.7	3.3/	(3+)			ann.rad./
				EC/89 /					W k x-ray
									0.1059
									0.2373
									0.9391
^{179m}Re			0.47 ms						
^{179}Re		178.94999	19.7 m	EC/99 /2.71	0.95/	(5/2+)	2.8		W k x-ray
				β+ /1 /					0.1199
									0.2900
									0.4154
									0.4302
									1.6803
^{180}Re		179.95079	2.45 m	EC/92 /3.80	1.76/	1-	1.6		ann.rad./
				β+ /8 /					W k x-ray
									0.1036
									0.9028
									(0.07−2.2)
^{181}Re		180.95007	20. h	EC /1.74		5/2+	3.19		W k x-ray
									0.3607
									0.3655
									0.6390
^{182m}Re			12.7 h	EC/	0.55/	2+	3.3	+1.8	W k x-ray
					1.74/				0.0677
									1.1214
									1.2215
									(0.06−2.2)
^{182}Re		181.9512	2.67 d	EC/2.8		(7+)	2.8	+4.1	W k x-ray
									0.0678
									0.2293
									1.1213
									1.2214
^{183}Re		182.95082	70. d	EC/0.56		(5/2+)	+3.17	+2.3	W k x-ray
									0.16232
^{184m}Re			165. d	I.T./75 /0.188		8+	+2.9		Re k x-ray
				EC/25 /					0.1047
									0.2165
									0.92093
									(0.10−1.1)
^{184}Re		183.952521	38. d	EC/1.48		3-	+2.53	+2.8	W k x-ray
									0.79207
									0.90328
									(0.1−1.4)
^{185}Re	37.40(2)	184.952955				5/2+	+3.1871	+2.18	
^{186m}Re			2.0 × 10⁵ y	I.T./0.150		8+			Re k x-ray
									0.0590
^{186}Re		185.954986	3.718 d	β- /92 /1.070	0.973/21	1-	+1.739	+0.62	W k x-ray
				EC/8 /0.582	1.07/71				0.1227/0.6
									0.1372/9.5
									(0.63−0.77)
^{187}Re	62.60(2)	186.955753	4.2 × 10¹⁰ y	β- /0.00266	0.0025/	5/2+	+3.2197	+2.07	
^{188m}Re			18.6 m	I.T./0.172		(6-)			Re k x-ray
									0.0925
									0.1059
^{188}Re		187.958114	17.00 h	β- /2.120	1.962/20	1-	+1.788	+0.57	Os k x-ray
					2.118/79				0.15502
									0.309−2.022
^{189}Re		188.95923	24. h	β- /1.01	1.01/	(5/2+)			0.1471

Elem. or Isot.	Natural Abundance (Atom %)	Atomic Mass or Weight	Half-life/ Resonance Width (MeV)	Decay Mode/ Energy (/MeV)	Particle Energy/ Intensity (MeV/%)	Spin ($h/2\pi$)	Nuclear Magnetic Mom. (nm)	Elect. Quadr. Mom. (b)	γ-Energy / Intensity (MeV/%)
									0.2167
									0.2194
									0.2451
^{190m}Re			3.0 h	β- /51 /		(6-)			Re k x-ray
				I.T./49 /					0.1191
									0.2238
									0.6731
									(0.1−1.79)
^{190}Re		189.9618	3.0 m	β- /3.2	1.8/	(2-)			Os k x-ray
									0.1867
									0.5580
									0.6051
^{191}Re		190.96313	9.7 m	β- /2.05	1.8/				
^{192m}Re			~ 0.12 ms						(0.0606-0.146)
^{192}Re		191.9660	16. s	β- /4.2	~ 2.5/				(0.2−0.75)
^{193m}Re			~ 0.08 ms						(0.061-0.146)
^{193}Re		192.9675	> 0.3 μs						
^{194}Re		193.9704	> 0.3 μs						
$_{76}$Os		190.23(3)							
^{162}Os		161.984	1.8 ms	α/	6.60	0+			
^{163}Os		162.9827	5.5 ms	α/	6.51				
^{164}Os		163.9780	0.04 s	α		0+			
^{165}Os		164.9768	0.07 s	α					
^{166}Os		165.97269	0.18 s	β+, EC/28 /6.3	6.27/	0+			ann. rad./
				α/72 /	5.98/				
^{167}Os		166.9716	0.7 s	β+, EC/76 /8.2					ann.rad./
				α/24 /	5.84/				
^{168}Os		167.96780	2.2 s	β+, EC/51 /5.7		0+			ann. rad./
				α/49 /					
^{169}Os		168.96702	3.3 s	β+, EC/89 /7.7	5.57/80				ann.rad./
				α/13 /	5.51/12				
					5.54/8				
^{170}Os		169.96358	7.1 s	β+, EC/5.0		0+			ann.rad./
				α/	5.40/				(0.162−0.216)
^{171}Os		170.96319	8.4 s	β+, EC/98 /7.1	α/5.24/93.5				ann.rad./
				α/19 /	5.17/6.5				0.190−0.705
^{172}Os		171.96002	19. s	β+, EC/99 /4.5		0+			ann.rad./
				α/1.1/	5.10/				(0.063−1.120)
^{173}Os		172.95981	16. s	β+, EC/6.3					ann.rad./
				α/0.4 /	4.94/				0.142−0.299
^{174}Os		173.95706	44. s	β+, EC/3.9		0+			0.118
				α/0.02 /	4.76/				0.138 / 0.001
									0.158
									0.325
^{175}Os		174.95695	1.4 m	β+, EC/5.3					0.125
									0.181
									0.248
^{176}Os		175.95481	3.6 m	β+, EC/3.2		0+			0.8155
									0.7758
									0.8573
									1.2093
									1.2909
^{177}Os		176.95497	2.8 m	β+, EC/4.5		(1/2-)			0.0848
									0.1958
									0.3002
									1.2686
^{178}Os		177.95325	5.0 m	β+, EC/2.3		0+			ann.rad./
									0.5946

Elem. or Isot.	Natural Abundance (Atom %)	Atomic Mass or Weight	Half-life/ Resonance Width (MeV)	Decay Mode/ Energy (/MeV)	Particle Energy/ Intensity (MeV/%)	Spin ($h/2\pi$)	Nuclear Magnetic Mom. (nm)	Elect. Quadr. Mom. (b)	γ-Energy / Intensity (MeV/%)
									0.6850
									0.9687
									1.3311
^{179}Os		178.95382	7. m	β+, EC/3.7					ann.rad./
									0.0654
									0.2186
									0.5938
^{180}Os		179.95238	21.5 m	β+, EC/1.5		0+			Re k x-ray
									0.0202−0.7174
^{181m}Os			1.75 h	EC/		(1/2−)			ann.rad./
									0.0489
^{181}Os		180.95324	2.7 m	EC/2.9		(7/2−)			ann.rad./
									0.11794
									0.23868
									0.8267
									(0.07−2.64)
^{182}Os		181.95211	21.5 h	EC/0.9		0+			Re k x-ray
									0.1802
									0.5100
^{183m}Os			9.9 h	EC/84 /		½−			Os k x-ray
				I.T./16 /					Re k x-ray
									1.1020
									1.1080
^{183}Os		182.95313	13. h	EC/2.1		9/2+	−0.79	+3.1	Re k x-ray
									0.1144
									0.3818
^{184}Os	0.02(1)	183.952489				0+			
^{185}Os		184.954042	93.6 d	EC/1.013		½−			Re k x-ray
									0.6461
									0.8748
									0.8805
^{186}Os	1.59(3)	185.953838	2. × 10^{15} y	α/	~ 2.75/	0+			
^{187}Os	1.96(2)	186.955750				½−	+0.0646519		
^{188}Os	13.24(8)	187.955838				0+			
^{189m}Os			5.8 h	I.T./0.0308		9/2−			Os L x-ray
									0.0308
^{189}Os	16.15(5)	188.958148				3/2+	+0.65993	+0.86	
^{190m}Os			9.9 m	I.T./1.705		10−	−0.6		Os k x-ray
									0.1867
									0.3611
									0.5026
									0.6161
^{190}Os	26.26(2)	189.958447				0+			
^{191m}Os			13.1 h	I.T./0.0744		3/2−			Os k x-ray
									0.0744
^{191}Os		190.960930	15.4 d	β− /0.314	0.140/100	9/2−		+2.5	Ir k x-ray
									0.1294
^{192m}Os			6.0 s	I.T./2.0154		(10−)			Os k x-ray
									0.2058/65.9
									0.5692/70
									(0.201−1.000)
^{192}Os	40.78(19)	191.961481				0+			
^{193}Os		192.964152	30.5 h	β− /1.141	1.04/20	3/2−	+0.730	+0.47	Ir k x-ray
									0.1389
									0.4605
^{194}Os		193.965182	6.0 y	β− /0.097	0.054/33	0+			Ir L x-ray
					0.096/67				0.0429
^{195}Os		194.968	6.5 m	β− /2.0	2.0/				
^{196}Os		195.96964	34.9 m	β− /1.16	0.84/	0+			0.1262/5
									0.4079/5.9

Elem. or Isot.	Natural Abundance (Atom %)	Atomic Mass or Weight	Half-life/ Resonance Width (MeV)	Decay Mode/ Energy (/MeV)	Particle Energy/ Intensity (MeV/%)	Spin ($h/2\pi$)	Nuclear Magnetic Mom. (nm)	Elect. Quadr. Mom. (b)	γ-Energy / Intensity (MeV/%)
^{197}Os			2.8 m	β-					0.2239
									(0.0412-0.406)
$_{77}$Ir		192.217(3)							
^{164}Ir		163.9922	0.06 ms	p	1.78				
^{165}Ir		164.9875	0.3 ms	p/87	1.71				
				α/13	6.72				
^{166m}Ir			14.3 ms	α/98.2	6.545				
				p/1.8	1.32				
^{166}Ir		165.9858	0.010 s	α/93	6.56				
				p/6.9	1.15				
^{167m}Ir			26. ms	α/48, β+	6.39/90				
				p/32	1.25/0.42				
^{167}Ir		166.98167	32. ms	α/80, β+	6.35/48				
				p/0.4	1.06/39.3				
^{168}Ir		167.9799	0.16 s	α/82					
^{169m}Ir			280. ms	α/	6.12/59				
^{169}Ir		168.97630	353. ms	α/	5.99/42				
^{170}Ir		169.9750	0.43 s	α/	6.03/				
^{171}Ir		170.97163	1.3 s	α/	5.91/				
^{172}Ir		171.9705	2.1 s	α/	5.811/				0.228
									(0.379−0.475)
^{173}Ir		172.96750	3.0 s	α/	5.665/				0.0493
									(0.092−0.296)
^{174}Ir		173.96686	4. s	α/	5.478/				0.1587
									(0.276−1.33)
^{175}Ir		174.96411	~ 4.5 s	α/	5.393/				0.1056
^{176}Ir		175.96365	8. s	EC, β+/80					0.260
				α/3.2/	5.118/				(0.135−0.415)
^{177}Ir		176.96130	30. s	EC, β+/5.7					0.184
				α/0.06/	5.011/				(0.062−0.194)
^{178}Ir		177.96108	12. s	β+, EC/6.3					
									0.1320
									0.2667
									0.3633
^{179}Ir		178.95912	4. m	EC/4.9					0.0975
									(0.045−0.220)
^{180}Ir		179.95923	1.5 m	EC/6.4					0.2765
									((0.132−1.106)
^{181}Ir		180.95763	4.9 m	β+, EC/4.1		(7/2+)			ann.rad./
									0.1076
									(0.0196−1.715)
^{182}Ir		181.95808	15. m	β+ /44 /5.6					ann.rad./
				EC/56 /					Os k x-ray
									0.1273
									0.2370
^{183}Ir		182.95685	57. m	β+, EC/3.5					ann.rad./
									0.0877
									0.2285
									0.2824
^{184}Ir		183.95748	3.0 h	β+ /12 /4.6	2.3/	5-	0.70	+2.41	ann.rad./
				EC/88 /	2.9/				Os k x-ray
									0.11968
									0.2640
									0.3904
^{185}Ir		184.95670	14. h	β+ /3 /2.4		(5/2-)	2.60	-2.1	ann.rad./
				EC/97 /					Os k x-ray
									0.2543
									1.8288

Elem. or Isot.	Natural Abundance (Atom %)	Atomic Mass or Weight	Half-life/ Resonance Width (MeV)	Decay Mode/ Energy (/MeV)	Particle Energy/ Intensity (MeV/%)	Spin ($h/2\pi$)	Nuclear Magnetic Mom. (nm)	Elect. Quadr. Mom. (b)	γ-Energy / Intensity (MeV/%)
[186m]Ir			1.7 h	EC /		(2-)	0.64	+1.46	Os k x-ray
									0.1371
									0.7675
[186]Ir		185.95795	15.7 h	EC/98 /3.83		(5+)	3.9	-2.55	Os k x-ray
				β+ /2 /					0.1372
									0.2968
									0.4348
									(0.13-3.0)
[187]Ir		186.95736	10.5 h	EC/1.50		3/2+		+0.94	Os k x-ray
									0.0743
									0.4009
									0.4271
									0.6109
									0.9128
[188]Ir		187.95885	1.72 d	β+ /2.81	1.13/	(2-)	0.30	+0.48	Os k x-ray
				EC/99+ /	1.64/				0.1550
									0.4780
									0.6330
									2.2146
[189]Ir		188.95872	13.2 d	EC/0.53		3/2+	0.13	+0.88	Os k x-ray
									0.2449
[190m2]Ir			3.09 h	β+, EC/95 /		(11-)			0.376
				I.T./5 /					
[190m1]Ir			1.12 h	I.T. /0.0263		7+			Ir L x-ray
[190]Ir		189.960546	11.8 d	EC/2.0		(4+)	0.04	+2.8	Os k x-ray
									0.1867
									0.4072
									0.5186
									0.5580
									0.6051
									(0.2-1.4)
[191m]Ir			4.93 s	I.T./0.1714		11/2-	+0.603		Ir k x-ray
									0.1294
[191]Ir	37.3(2)	190.960594				3/2+	+0.151	+0.82	
[192m2]Ir			241. y	I.T./0.161		(9+)			Ir k x-ray
[192m1]Ir			1.44 m	I.T./0.0580		(1+)			Ir L x-ray
									0.0580
									0.3165
[192]Ir		191.962605	73.83 d	β- /1.460		(4-)	+1.92	+2.15	Pt k x-ray
									0.31649/83.
									0.46806/48.
[193m]Ir			10.53 d	I.T./0.0802		11/2-			Ir L x-ray
									0.0803
[193]Ir	62.7(2)	192.962926				3/2+	+0.164	+0.75	
[194m]Ir			170. d	β- /		11			Pt k x-ray
									0.3284
									0.4829
									0.5624
[194]Ir		193.965078	19.3 h	β-/2.247	1.92/9	1-	+0.39	+0.34	0.2935
					2.25/86				0.3284
									0.6451
									(0.1-2.2)
[195m]Ir			3.9 h	β- /	0.41/	(11/2-)			Pt k x-ray
					0.97/				0.3199/9.6
									0.3649/9.5
									0.4329/9.6
									0.6849/9.6
[195]Ir		194.965980	2.8 h	β- /1.120	1.0/80	(3/2+)			Pt k x-ray
					1.11/13				0.0989/9.7
[196m]Ir			1.40 h	β-/	1.16/				Pt k x-ray

Elem. or Isot.	Natural Abundance (Atom %)	Atomic Mass or Weight	Half-life/ Resonance Width (MeV)	Decay Mode/ Energy (/MeV)	Particle Energy/ Intensity (MeV/%)	Spin ($h/2\pi$)	Nuclear Magnetic Mom. (nm)	Elect. Quadr. Mom. (b)	γ-Energy / Intensity (MeV/%)
									0.3557
									0.3935
									0.4471
									0.5214
									0.6473
^{196}Ir		195.96840	52. s	β- /3.21	2.1/15	0-			0.3329
					3.2/80				0.3557
									0.7796
^{197m}Ir			8.9 m	β- /		(11/2-)			0.3465
				I.T./					see Ir[197]
^{197}Ir		196.96965	5.8 m	β- /2.16	1.5/	(3/2+)			0.0531
					2.0/				0.1351
									0.4306
									0.4697
^{198}Ir		197.9723	8. s	β- /4.1					0.4074
									0.5070
^{199}Ir		198.97380							
$_{78}$Pt		195.084(9)							
^{166}Pt		165.995	0.3 ms	α/	7.11/	0+			
^{167}Pt		166.930	0.9 ms	α/	6.98/				
^{168}Pt		167.9882	2.1 ms	α	6.82	0+			0.582/69
									0.594/69
									0.725/62
^{169}Pt		168.9867	7.0 ms	α	6.69				
^{170}Pt		169.98250	14.0 ms	α	6.55	0+			0.509/100
									0.662/86
									0.214−0.726
^{171}Pt		170.9812	0.05 s	α	6.45				0.4450
									(0.1564-1.208)
^{172}Pt		171.97735	0.10 s	α/	6.31/94	0+			
^{173}Pt		172.9764	0.36 s	β+, EC/8.2	6.23				
				α/	6.20/				
^{174}Pt		173.97282	0.89 s	β+, EC/17 /5.6		0+			
				α/83 /	6.040/				
^{175}Pt		174.97242	2.5 s	β+, EC/65 /7.6					0.0774
				α/35 /	5.831/5				0.1354
					5.96/54				0.2128
					6.038/				
^{176}Pt		175.96895	6.3 s	β+, EC/60 /5.1		0+			ann.rad./
				α/40 /	5.528/0.6				0.2277
					5.750/41				
^{177}Pt		176.96847	11. s	EC/91 /6.8	5.53/				0.0908
				α/9 /	5.485/3				
					5.525/6				
^{178}Pt		177.96565	21. s	EC/93 /4.5		0+			
				α/7 /	5.286/0.2				
					5.442/7				
^{179}Pt		178.96536	33. s	β+, EC/5.7			+0.43		
				α/	5.16/				
^{180}Pt		179.96303	52. s	β+, EC/99.7 /3.7	0+				
				α/0.3 /	5.140/				
^{181}Pt		180.96310	51. s	β+, EC/5.2			+0.48		
^{182}Pt		181.96117	2.7 m	β+, EC/2.9		0+			ann.rad./
									0.1360
									0.1460
									0.2100
^{183m}Pt			43. s	β+, EC/		(7/2-)	+0.78	+3.4	ann.rad./
				I.T./					0.3132/26

Elem. or Isot.	Natural Abundance (Atom %)	Atomic Mass or Weight	Half-life/ Resonance Width (MeV)	Decay Mode/ Energy (/MeV)	Particle Energy/ Intensity (MeV/%)	Spin ($h/2\pi$)	Nuclear Magnetic Mom. (nm)	Elect. Quadr. Mom. (b)	γ-Energy / Intensity (MeV/%)
									0.3164/59
									0.6296/100
									0.058−1.75
¹⁸³Pt		182.96160	7. m	β+, EC/4.6			+0.50		ann.rad./
									0.119/100
									0.307/93
									0.260/90
									0.058−1.377
¹⁸⁴Pt		183.95992	17.3 m	β+, EC/2.3		0+			ann.rad./
									0.1549
									0.1919
									0.5484
¹⁸⁵ᵐPt			33. m	β+, EC/		½−	+0.5		
¹⁸⁵Pt		184.96062	1.18 h	β+, EC/3.8		(9/2+)	−0.75	+3.7	ann.rad./
									0.1353
									0.1974
									0.2296
									0.2551
¹⁸⁶Pt		185.95935	2.0 h	β+, EC/1.38		0+			ann.rad./
									0.6115
									0.6892
¹⁸⁷Pt		186.96059	2.35 h	β+, EC/3.1		3/2−	−0.41	−1.1	ann.rad./
									Ir k x-ray
									0.1064
									0.1100
									0.2015
									0.2849
									0.7092
¹⁸⁸Pt		187.95940	10.2 d	EC/0.51		0+			Ir k x-ray
									0.1876
									0.1951
¹⁸⁹Pt		188.96083	10.9 h	β+, EC/1.97		3/2−	−0.43	−1.2	Ir k x-ray
									0.0943
									0.6076
									0.7214
									(0.09−1.47)
¹⁹⁰Pt	0.014(1)	189.95993	4.5 × 10¹¹ y			0+			
¹⁹¹Pt		190.961677	2.86 d	EC/1.02		(3/2−)	−0.50	−0.9	Ir k x-ray
									0.3599
									0.4094
									0.5389
¹⁹²Pt	0.782(7)	191.961038				0+			
¹⁹³ᵐPt			4.33 d	I.T./0.1498		13/2+	−0.75		Pt k x-ray
									0.1355
¹⁹³Pt		192.962988	60. y	EC/0.0566		(1/2−)	+0.60		Ir k x-rays
¹⁹⁴Pt	32.967(99)	193.962680				0+			
¹⁹⁵ᵐPt			4.01 d	I.T./0.2952		13/2+	−0.61	+1.4	Pt k x-ray
									0.0989
¹⁹⁵Pt	33.832(10)	194.964791				1/2−	+0.6095		
¹⁹⁶Pt	25.242(41)	195.964952				0+			
¹⁹⁷ᵐPt			1.590 h	I.T./97 /		13/2+			Pt k x-ray
				β− /3 /					0.0530
									0.3465
¹⁹⁷Pt		196.967340	19.9 h	β− /0.719		1/2−	0.51		Au k x-ray
									0.1914
									0.2688
¹⁹⁸Pt	7.163(55)	197.967893				0+			
¹⁹⁹ᵐPt			13.6 s	I.T./0.424		13/2+			Pt k x-ray
									0.3919
¹⁹⁹Pt		198.970593	30.8 m	β− /1.70	0.90/18	(5/2−)			0.3170/3.88

Elem. or Isot.	Natural Abundance (Atom %)	Atomic Mass or Weight	Half-life/ Resonance Width (MeV)	Decay Mode/ Energy (/MeV)	Particle Energy/ Intensity (MeV/%)	Spin ($h/2\pi$)	Nuclear Magnetic Mom. (nm)	Elect. Quadr. Mom. (b)	γ-Energy / Intensity (MeV/%)
					1.14/14				0.49375/4.47
									0.5430/11.7
									(0.055−1.293)
^{200}Pt		199.971441	12.5 h	β- /~ 0.66		0+			Au k x-ray
									0.13590
									0.22747
									0.24371
^{201}Pt		200.97451	2.5 m	β- /2.66		(5/2-)			0.070
									0.152
									0.222
									1.760
^{202m}Pt			0.3 ms						(0.535-0.719)
^{202}Pt		201.9757	1.8 d			0+			0.440
$_{79}$Au		196.966569(4)							
^{170m}Au			0.62 ms	p/58	1.74/				
				α/42	7.11/				
^{170}Au		169.9961	0.30 ms	p/89	1.46/				
				α/11	7.00/				
^{171m}Au			1.09 ms	α/66	6.995				
				p/34	1.694				
^{171}Au		170.99188	0.022 ms	p/100	1.437				
^{172}Au		171.9900	4 ms	α/7.02	6.86				
^{173m}Au			15 ms	α/92	6.732				
^{173}Au		172.98624	0.02 s	α/94	6.672				
^{174}Au		173.9848	0.14 s	α	6.54				
^{175}Au		174.98127	0.15 s	α					
^{176}Au		175.9801	0.9 s	β+, EC/10.5					
				α/	6.260/80				
					6.290/20				
^{177}Au		176.97687	1.2 s	α/	6.115/				
					6.150/				
^{178}Au		177.9760	2.6 s	α/	5.920/				
^{179}Au		178.97321	7.5 s	α/	5.85/				
^{180}Au		179.97252	8.1 s	EC/8.6	5.65				0.1522
				α/	5.61				0.2564
					5.50				0.5242
									0.6765
									0.8084
									0.8597
^{181}Au		180.97008	11.4 s	EC/97.5/6.3	5.482/				
				α/2.7/					
^{182}Au		181.96962	21. s	β+, EC/6.9					ann.rad./
				α/0.13/					0.1549
									0.2649
									(0.13−1.4)
^{183}Au		182.96759	42. s	EC/5.5			+1.97		0.1630
				α/0.8/					0.2730
									0.3625
^{184m}Au			48 s	I.T.		(2+)	+1.44	+1.9	0.069(IT)
^{184}Au		183.96745	21. s	EC, β+/7.1		(5+)	+2.07	+4.7	
				α/0.013/					
^{185m}Au			6.8 m	β+, EC/					
				I.T./0.145					
^{185}Au		184.96579	4.3 m	β+, EC/4.71		(5/2-)	+2.17	-1.1	ann.rad./
				α/0.26/					
^{186m}Au			< 2. m	β+, EC/					0.1915
^{186}Au		185.96595	10.7 m	β+, EC/6.0		3-	-1.26	+3.1	ann.rad./

Elem. or Isot.	Natural Abundance (Atom %)	Atomic Mass or Weight	Half-life/ Resonance Width (MeV)	Decay Mode/ Energy (/MeV)	Particle Energy/ Intensity (MeV/%)	Spin ($h/2\pi$)	Nuclear Magnetic Mom. (nm)	Elect. Quadr. Mom. (b)	γ-Energy / Intensity (MeV/%)
				$\alpha/8(10)^{-4}/$					0.1915
									0.2988
187mAu			2.3 s	IT		9/2-			
187Au		186.96457	8.3 m	β+, EC /3.60		1/2+	+0.54		ann.rad./
									0.9152
									1.2668
									1.3321
									1.4081
188Au		187.96532	8.8 m	β+, EC /5.3		(1-)	-0.07		ann.rad./
									0.2660
									0.3404
									0.6061
189mAu			4.6 m	β+, EC/		11/2-	+6.19		0.1667
189Au		188.96395	28.7 m	EC/96 /3.2		1/2+	+0.49		ann.rad./
				β+ /4 /					Pt k x-ray
									0.4478
									0.7133
									0.8128
190Au		189.96470	43. m	β+ /2 /4.44		1-	-0.07		ann.rad./
				EC/98 /					Pt k x-ray
									0.2958
									0.3018
									0.5977
191mAu			0.9 s	I.T./0.2663		(11/2-)	6.6		Au k x-ray
									0.2414
									0.2526
191Au		190.96370	3.2 h	EC/1.83		3/2+	+0.137	+0.72	Pt k x-ray
									0.5864/16
									(0.088−1.30)
192Au		191.96481	4.9 h	β+ /5 /3.52	2.19/	1-	-0.011	-0.23	ann.rad./
				EC/95 /	2.49/				Pt k x-ray
									0.2959
									0.3165
193mAu			3.9 s	I.T./0.2901		11/2-	6.2	+1.98	Au k x-ray
									0.2580
193Au		192.96415	17.6 h	EC/1.07		3/2+	+0.140	+0.66	Pt k x-ray
									0.1862
									0.2556
194Au		193.96537	1.64 d	β+ /3 /2.49	1.49/	1-	+0.076	-0.24	ann.rad./
				EC/97 /					Pt k x-ray
									0.2935
									0.3284/61
195mAu			30.5 s	I.T./0.3186		11/2-	6.2	+1.9	Au k x-ray
									0.2617
195Au		194.965035	186.10 d	EC/0.227		3/2+	+0.149	+0.61	Pt k x-ray
196m2Au			9.7 h	I.T./0.5954		12-	5.7		Au k x-ray
									0.1478
									0.1883
196m1Au			8.1 s	I.T./0.0846		8+			0.0847
196Au		195.966570	6.17 d	EC/92 /1.506		2-	+0.591	0.81	Pt k x-ray
197mAu			7.8 s	I.T./0.4094		11/2-	+6.0	+1.7	Au k x-ray
				β- /8 /0.686					0.1302
									0.2790
197Au	100.	196.966569				3/2+	+0.14575	+0.55	
198mAu			2.30 d	I.T./0.812		(12-)			Au k x-ray
									0.0972
									0.1803
									0.2419
198Au		197.968242	2.695 d	β- /1.372	0.290/1	2-	+0.5934	+0.64	Hg k x-ray
					0.961/99				0.411794

Table of the Isotopes

Elem. or Isot.	Natural Abundance (Atom %)	Atomic Mass or Weight	Half-life/ Resonance Width (MeV)	Decay Mode/ Energy (/MeV)	Particle Energy/ Intensity (MeV/%)	Spin ($h/2\pi$)	Nuclear Magnetic Mom. (nm)	Elect. Quadr. Mom. (b)	γ-Energy / Intensity (MeV/%)
^{199}Au		198.968765	3.14 d	β- /0.453	0.25/22	3/2+	+0.2715	+0.51	Hg k x-ray
					0.292/72				0.15837
					0.462/6				0.20820
^{200m}Au			18.7 h	β- /84 /1.0	0.56/	12-	5.9		Au k x-ray
				I.T./16 /					0.2559/71
									0.3680/77
									0.4978/73
									0.5793/72
									(0.084−0.904)
^{200}Au		199.97073	48.4 m	β- /2.24	0.7/15	1-			0.3679/19
					2.2/77				1.2254/10.6
									(0.077−1.570)
^{201}Au		200.971657	26. m	β- /1.28	1.27/82	3/2+			(0.027−0.732)
^{202}Au		201.9738	29. s	β- /3.0		(1-)			0.4396
^{203}Au		202.975155	1.0 m	β- /2.14	~ 1.9/	3/2+			(0.04−0.37)
^{204}Au		203.9777	40. s	β- /4.5		(2-)			0.4366
									1.5113
^{205}Au		204.9799	31. s	β- /					(0.38−1.33)
$_{80}$Hg		200.59(2)							
^{171}Hg		171.0038	0.06 ms	α	7.49				
^{172}Hg		171.9988	0.3 ms	α	7.36	0+			
^{173}Hg		172.9972	0.8 ms	α	7.20				
^{174}Hg		173.99286	1.9 ms	α	7.07	0+			
^{175}Hg		174.9914	0.02 s	α					
^{176}Hg		175.98736	21 ms	α	6.74/94	0+			
^{177m}Hg			1.5 µs	IT					0.246
^{177}Hg		176.9863	0.13 s	α	6.58				
^{178}Hg		177.98248	0.26 s	EC/50 /6.1		0+			
				α/50 /	6.43/				
^{179}Hg		178.98183	1.05 s	EC/8.0					
				α/	6.29/				
^{180}Hg		179.97827	2.6 s	EC/5.5		0+			0.1250
				α/	6.12/33				0.3005
					5.69/.03				0.3812
^{181}Hg		180.97782	3.6 s	β+ EC/76 /~ 7.3		(1/2-)	+0.507		0.0663
				α/24 /					0.0811
									0.0924
									0.1474
									0.1587
									0.2142
									0.2398
^{182}Hg		181.97469	10.8 s	β+, EC/85/5.0		0+			0.129/122
				α/15/	5.87/8.6				0.2176/66
					5.45/0.03				0.0256−0.543
^{183}Hg		182.97445	9. s	β+, EC/77/6.3		½-	+0.524		0.0714
				α/	5.83/				0.0874
					5.91/				0.1538
^{184}Hg		183.97171	30.9 s	β+, EC/99/4.1		0+			0.1565/102
				α/1/	5.54/1.3				0.2367/100
					5.07/0.002				0.2384/18
									(0.018-0.4227)
^{185m}Hg			21. s	β+, EC, IT, α/	5.37/	13/2+	-1.02	+0.2	0.211
									0.292
^{185}Hg		184.97190	51. s	β+, EC/95/5.8		½-	+0.509		0.02−0.55
^{186}Hg		185.96936	1.4 m	β+, EC/3.3		0+			0.1119
				α	5.09/0.02				0.2518
^{187m}Hg			1.7 m	β+, EC/		13/2+	-1.04	+0.5	see Hg187
^{187}Hg		186.96981	2.4 m	β+, EC/4.9		3/2-	-0.594	-0.8	0.1034/32

Elem. or Isot.	Natural Abundance (Atom %)	Atomic Mass or Weight	Half-life/ Resonance Width (MeV)	Decay Mode/ Energy (/MeV)	Particle Energy/ Intensity (MeV/%)	Spin ($h/2\pi$)	Nuclear Magnetic Mom. (nm)	Elect. Quadr. Mom. (b)	γ-Energy / Intensity (MeV/%)
									0.2334/100
									0.2403/33
									0.27151/31
									0.3763/38
									0.5254/30
									0.10−2.18
^{188}Hg		187.96758	3.2 m	β+, EC/2.3		0+			0.0988
				α	4.61				0.1148
									0.1424
									0.1900
^{189m}Hg			8.6 m	EC/		13/2+	−1.06	+0.7	0.0780
									0.3210
									0.4345
									0.5655
									(0.08−2.170)
^{189}Hg		188.96819	7.6 m	EC/4.2		3/2-	−0.6086	−0.8	0.2005
									0.2038
									0.2386
									0.2485
^{190}Hg		189.66632	20.0 m	EC/1.5		0+			0.1296
									0.1426
^{191m}Hg			51. m	β+ /6 / EC/94 /		13/2+	−1.07	+0.6	ann.rad./ Au k x-ray
									0.2741
									0.4203
									0.5787
									(0.07−1.9)
^{191}Hg		190.96716	50. m	β+, EC/3.2		(3/2-)	−0.62	−0.8	0.1963
									0.2247
									0.2524
^{192}Hg		191.96563	5.0 h	EC/~ 0.5		0+			Au k x-ray
									0.1572
									0.2748
									0.3065
^{193m}Hg			11.8 h	β+, EC/91 / I.T./9 /0.2901		13/2+	−1.05843	+0.92	Hg k x-ray
									0.1866
									0.2580
									0.4076
									0.5733
									0.9324
									(0.1−1.96)
^{193}Hg		192.96667	3.8 h	EC, B+/2.34		3/2-	−0.6276	−0.7	0.1866
									0.2580
									0.8611
^{194}Hg		193.96544	520. y	EC/0.04		0+			Au L x-rays
^{195m}Hg			1.67 d	I.T./(54)/0.3186 EC/(46)/		13/2+	−1.04465	+1.1	Hg k x-ray Au k x-ray
									0.2617
									0.5603
									0.7798
^{195}Hg		194.96672	10.5 h	EC/1.51		1/2-	+0.541475		Au k x-ray
									0.0614
									0.7798
^{196}Hg	0.15(1)	195.965833	>2.5 × 10^{18} y			0+			
^{197m}Hg			23.8 h	I.T./(93)/0.2989		13/2+	−1.02768	+1.2	Hg k x-ray Au k x-ray
									0.13398
^{197}Hg		196.967213	2.69 d	EC/0.600		1/2-	+0.527374		Au k x-ray
									0.07735
^{198}Hg	9.97(20)	197.9667690				0+			

Elem. or Isot.	Natural Abundance (Atom %)	Atomic Mass or Weight	Half-life/ Resonance Width (MeV)	Decay Mode/ Energy (/MeV)	Particle Energy/ Intensity (MeV/%)	Spin ($h/2\pi$)	Nuclear Magnetic Mom. (nm)	Elect. Quadr. Mom. (b)	γ-Energy / Intensity (MeV/%)
^{199m}Hg			42.7 m	I.T./0.532		13/2+	-1.014703	+1.2	Hg k x-ray
									0.15841
^{199}Hg	16.87(22)	198.9682799				1/2-	+0.505885		
^{200}Hg	23.10(19)	199.9683260				0+			
^{201}Hg	13.18(9)	200.970302				3/2-	-0.560226	+0.39	
^{202}Hg	29.86(26)	201.970643				0+			
^{203}Hg		202.972873	46.61 d	β- /0.492	0.213/100	5/2-	+0.8489	+0.34	Tl k x-ray
									0.279188
^{204}Hg	6.87(15)	203.9734939				0+			
^{205}Hg		204.976073	5.2 m	β- /1.531	1.33/4	1/2-	+0.6010		0.20378
									(0.2−1.4)
^{206}Hg		205.97751	8.2 m	β- /1.31	0.935/34	0+			Tl k x-ray
					1.3/63				0.3052
									0.6502
^{207}Hg		206.9826	2.9 m	β- /4.8		(9/2+)			
^{208}Hg		207.9859	41. m	β-		0+			0.474
^{209}Hg		208.9910	36 s	β-					0.324
^{210}Hg		209.9945	> 0.3 µs	β-		0+			
$_{81}$Tl		204.3833(2)							
^{176}Tl		176.0006	5 ms	p	1.26/∼ 100				
^{177m}Tl			0.23 ms	p/51	1.95				
				α/49	7.48				
^{177}Tl		176.99643	0.017 s	α/73					
				p/27					
^{178}Tl		177.9949	0.25 s	α/	6.704				
					6.785				
					6.62				
					6.859				
^{179m}Tl			1.7 ms	α	/7.21/80				
				α	/7.10/20				
^{179}Tl		178.99109	0.3 s	α	6.57/				
^{180}Tl		179.9899	1.5 s	α//8	6.28/30				
					6.36/30				
					6.21/18				
					6.56/15				
					6.47/7				
^{181m}Tl			1.4 ms	α	6.58/100				
^{181}Tl		180.98626	3.2 ms	α/ < 10	6.19/100				
^{182}Tl		181.9857	3. s	β+, EC/10.9					0.351
									(0.26−0.41)
^{183m}Tl			53. ms	α	6.33/80	9/2-			0.0618
					6.38/16				(0.046-0.0894)
					6.46/4				
^{183}Tl		182.98219	5. s	β+, EC/7.7		½+			0.208
^{184}Tl		183.98187	11. s	β+, EC/(98)/9.2					0.2868
				α/(2)/	6.16/				0.3399
									0.3667
^{185m}Tl			1.8 s	I.T./0.453		(9/2-)			0.1688
				α/5.97	6.01				0.2840
^{185}Tl		184.9788	20. s	EC/β+/6.6					
^{186m}Tl			4. s	I.T./0.374					0.3738
^{186}Tl		185.9783	28. s	β+, EC/7.5					0.3567
									0.4026
									0.4053
^{187m}Tl			15.6 s	I.T./∼ 0.33		(9/2+)	+3.8	-2.4	0.2995
^{187}Tl		186.97591	50. s	β+, EC/6.0		½+	1.6		
^{188m}Tl			1.18 m	β+, EC/		(7+)			Hg k x-ray
									0.4129

Elem. or Isot.	Natural Abundance (Atom %)	Atomic Mass or Weight	Half-life/ Resonance Width (MeV)	Decay Mode/ Energy (/MeV)	Particle Energy/ Intensity (MeV/%)	Spin (h/2 π)	Nuclear Magnetic Mom. (nm)	Elect. Quadr. Mom. (b)	γ-Energy / Intensity (MeV/%)
									0.5043
									0.5921
188Tl		187.97601	1.2 m	β+, EC/7.8		(2-)	+0.48	+0.13	see Tl[188m]
									0.4129
189mTl			1.4 m	β+, EC/		(9/2-)	+3.878	-2.29	0.2156
									0.2284
									0.3175
									0.4452
189Tl		188.97359	2.3 m	β+, EC/5.2		(1/2+)			0.3337
									0.4510
									0.5223
									0.9422
190mTl			3.7 m	β+, EC/	4.2/	(7+)	+0.495	+0.29	0.1968
									0.4164
									0.7311
190Tl		189.97388	2.6 m	β+, EC/7.0	5.7/	(2-)	+0.25	-0.33	0.4164
									0.6254
									0.6838
									1.0999
191mTl			5.2 m	β+, EC/(98)/		(9/2+)	+3.903	-2.3	0.2157
									0.2647
									0.3256
									0.3359
191Tl		190.97179				(1/2)	1.59		
192mTl			10.8 m	β+, EC/		(7+)	+0.518	0.46	0.1740
									0.4228
									0.6348
									0.7863
									0.7455
192Tl		191.97223	9.6 m	β+, EC/6.4		(2-)	+0.20	-0.33	0.3975
									0.4228
									0.6908
193mTl			2.1 m	I.T./(75)/		(9/2-)	+3.948	-2.2	0.3650
193Tl		192.9707	22. m	β+, EC/3.6		(1/2+)	+1.591		0.2077
									0.3244
									0.3440
									0.6761
									1.0447
									1.5793
194mTl			32.8 m	β+ /(20)/~ 0.30 EC/(80)/		(7+)	+0.540	+0.61	ann.rad./ Hg k x-ray
									0.4282
									0.6363
									0.7490
194Tl		193.9712	33.0 m	β+, EC/5.3		2-	0.140	-0.28	0.4279/75.2
									0.6452/10.8
									(0.395-1.623)
195mTl			3.6 s	I.T./0.483		9/2-			Tl k x-ray
									0.0990
									0.3836
195Tl		194.96977	1.16 h	EC/97/2.8 β+ /(3)/		1/2+	+1.58		ann.rad./ Hg k x-ray
									0.2422
									0.5635
									0.8845
									1.3639
									(0.13−2.5)
196mTl			1.41 h	β+, EC/95/4.9		(7+)	0.55	+0.76	0.0840
									0.4261
									0.6353

Elem. or Isot.	Natural Abundance (Atom %)	Atomic Mass or Weight	Half-life/ Resonance Width (MeV)	Decay Mode/ Energy (/MeV)	Particle Energy/ Intensity (MeV/%)	Spin ($h/2\pi$)	Nuclear Magnetic Mom. (nm)	Elect. Quadr. Mom. (b)	γ-Energy / Intensity (MeV/%)
									0.6954
									(0.08−1.0)
¹⁹⁶Tl		195.97048	1.84 h	β+ /(15)/4.4		2-	+0.072	-0.18	ann.rad./
				EC/(85)/					Hg k x-ray
									0.4257
									0.6105
									(0.03−2.4)
¹⁹⁷ᵐTl			0.54 s	IT/53/0.608		9/2-			Tl k x-ray
				β+, EC/47/					0.2262
									0.4118
									0.5872
									0.6367
¹⁹⁷Tl		196.96958	2.83 h	β+ /(1)/2.18		1/2+	+1.58		Hg k x-ray
				EC/(99)/					0.1522/8.2
									0.4258
¹⁹⁸ᵐTl			1.87 h	β+, EC/(53)/		7+	+0.64		Hg k x-ray
				IT/47/0.5347					Tl k x-ray
									0.4118
									0.5872
									0.6367
¹⁹⁸Tl		197.9405	5.3 h	EC, β+ /(1)/3.5	1.4/	2-			Hg k x-ray
					2.1/				0.4118
					2.4/				0.6367
									0.6759
									(0.23−2.8)
¹⁹⁹Tl		198.96988	7.4 h	EC/1.4		1/2-	+1.60		Hg k x-ray
									0.2082
									0.2473
									0.4555
²⁰⁰Tl		199.97096	1.087 d	EC/2.46	1.07/	2-	0.04		Hg k x-ray
					1.44/				0.36799
									1.2057
									(0.11−2.3)
²⁰¹Tl		200.97082	3.038 d	EC/0.48		1/2+	+1.605		Hg k x-ray
									0.13528
									0.16740/10.0
²⁰²Tl		201.97211	12.47 d	EC/1.36		2-	0.06		Hg k x-ray
									0.43957
²⁰³Tl	29.524(14)	202.972344				1/2+	+1.622258		
²⁰⁴Tl		203.973864	3.78 y	β- /97/0.7637	0.763/97	2-	0.09		Hg k x-ray
				EC/(3)/0.347					
²⁰⁵Tl	70.476(14)	204.974428				1/2+	+1.638215		
²⁰⁶ᵐTl			3.76 m	I.T./2.644		12-			Tl k x-ray
									0.2166
									0.2661
									0.4534
									0.6866
									1.0219
²⁰⁶Tl		205.976110	4.20 m	β- /1.533	1.53/99.9	0-			Pb k x-ray
									0.80313
²⁰⁷ᵐTl			1.3 s	I.T./1.350		11/2-			Tl k x-ray
									0.3501
									1.0000
²⁰⁷Tl		206.97742	4.77 m	β- /1.423	1.43/99.8	1/2+	+1.88		0.89723
²⁰⁸Tl		207.982019	3.053 m	β- /5.001	1.28/23	(5+)	+0.29		Pb k x-ray
					1.52/22				0.27728
					1.796/51				0.51061
									0.58302
									2.61448
²⁰⁹Tl		208.98536	2.16 m	β-/3.98	1.8 /100	(1/2+)			Pb k x-ray

Elem. or Isot.	Natural Abundance (Atom %)	Atomic Mass or Weight	Half-life/ Resonance Width (MeV)	Decay Mode/ Energy (/MeV)	Particle Energy/ Intensity (MeV/%)	Spin ($h/2\pi$)	Nuclear Magnetic Mom. (nm)	Elect. Quadr. Mom. (b)	γ-Energy / Intensity (MeV/%)
									1.5670/100
									0.4651/95
									(0.12−1.33)
^{210}Tl		209.99007	1.30 m	β- /5.48	1.3/25	(5+)			Pb k x-ray
					1.9/56				0.081
									0.2981
									0.79788
^{211}Tl		210.9935	> 0.3 µs	β-					
^{212}Tl		211.9982	> 0.3 µs	β-					
$_{82}$Pb		207.2(1)							
^{178}Pb		178.00383	~ 0.2 ms			0+			
^{179}Pb		179.0022							
^{180}Pb		179.99792	5 ms	α/	7.25	0+			
^{181}Pb		180.9966	0.05 s	α/	7.07				
^{182}Pb		181.99267	55 ms	α	6.90	0+			
^{183m}Pb			0.42 s	α	6.70/82.7	13/2+			
					6.86/1.9				
^{183}Pb		182.99187	0.54 s	α/	6.57/4.3	(3/2-)			
					6.78/11.0				
^{184}Pb		183.98814	0.48 s	α/~ 80	6.63/	0+			
^{185m}Pb			4.3 s	α	6.41/100	13/2+	-1.2		
^{185}Pb		184.98761	6.3 s	α/	6.29/56	3/2-	-1.1		0.205
					6.49/44				0.269
					6.55/<1.4				
^{186}Pb		185.98424	5. s	β+, EC/95/5.5		0+			
				α/(5)/	6.32/				
					6.34/<100				
					6.01/<0.2				
^{187m}Pb			15.2 s	β+, EC/	5.99/	(1/2-)			0.0674
				α/12	6.19/				0.2080
									0.2755
									0.2995
									0.4487
									0.7477
^{187}Pb		186.98392	18.3 s	EC/7.2		13/2+			0.1930
				α/7	6.08/				0.3314
									0.3435
									0.3934
^{188}Pb		187.98087	23. s	EC/(78)/4.8		0+			0.1850
				α/(22)/	5.98/<10				0.7582
					5.61/<0.1				
^{189}Pb		188.98081	51. s	EC/6.1					
				α/	5.58/				
^{190}Pb		189.97808	1.2 m	β+ (13)/4.1		0+			ann.rad./
				EC/(86)/					Tl k x-ray
				α/(0.9)/	5.58/				0.1415
									0.1512
									0.9422
^{191m}Pb			2.2 m	β+, EC/		13/2+	-1.17	+0.085	ann.rad./
									0.3871
									0.6135
									0.7122
^{191}Pb		190.97827	1.3 m	β+, EC/5.5					ann.rad./
									0.9368
^{192}Pb		191.97579	3.5 m	β+, EC/~ 3.4		0+			ann.rad./
				α/.006/	5.11				0.1675
									0.6082
									1.1954

Elem. or Isot.	Natural Abundance (Atom %)	Atomic Mass or Weight	Half-life/ Resonance Width (MeV)	Decay Mode/ Energy (/MeV)	Particle Energy/ Intensity (MeV/%)	Spin ($h/2\pi$)	Nuclear Magnetic Mom. (nm)	Elect. Quadr. Mom. (b)	γ-Energy / Intensity (MeV/%)
^{193m}Pb			5.8 m	β+, EC/		13/2+	-1.15	+0.19	ann.rad./
									0.3650
									0.3922
^{193}Pb		192.97617	~ 2. m	EC/5.2		3/2 (-)			
^{194}Pb		193.97401	10. m	β+, EC/2.7		0+			ann.rad./
				α	4.64				0.2036
^{195m}Pb			15. m	β+ /(8)/		13/2+	-1.132	+0.30	ann.rad./
				EC/(92)/					Tl k x-ray
									0.3836
									0.3942
									0.8784
^{195}Pb		194.97454	~ 15. m	β+, EC/5.8					ann.rad./
									0.3836
									0.3937
									0.7776
^{196}Pb		195.97277	37. m	β+, EC/2.1		0+			Tl k x-ray
									0.2531
									0.5021
^{197m}Pb			43. m	EC/79/		13/2+	-1.104	+0.38	Tl k x-ray
				β+ /2/					0.3079
				IT/19/0.3193					0.3877
									0.7743
									(0.2–2.2)
^{197}Pb		196.97343	~ 8. m	EC/97/3.6		(3/2-)	-1.075	-0.08	Tl k x-ray
				β+ /3/					0.3755
									0.3858
									0.7611
^{198}Pb		197.97203	2.4 h	EC/1.4		0+			Tl k x-ray
									0.1734
									0.2903
									0.3654
^{199m}Pb			12.2 m	IT/93/0.4248		13/2+			Pb k x-ray
				β+, EC/(7)/					0.4255
^{199}Pb		198.97292	1.5 h	EC/(99)/2.9		5/2-	-1.074	+0.08	Tl k x-ray
				β+ /(1)/					0.3534
									0.7202
									1.1350
									(0.22–2.4)
^{200}Pb		199.97183	21.5 h	EC/0.81		0+			Tl k x-ray
									0.14763
^{201m}Pb			1.02 m	I.T./0.6291		13/2+			Pb k x-ray
									0.6288
^{201}Pb		200.97289	9.33 h	EC/1.90		5/2-	+0.675	-0.009	Tl k x-ray
									0.33120
									0.36131
									(0.11–1.8)
^{202m}Pb			3.53 h	IT/90/2.170		9-	-0.228	+0.58	Pb k x-ray
				β+ /10/					Tl k x-ray
									0.42219
									0.78700
									0.96271
^{202}Pb		201.97216	5.3×10^4 y	EC/0.05		0+			Tl L x-ray
^{203m}Pb			6.2 s	I.T./0.8252		13/2+			Pb k x-ray
									0.8203
									0.8252
^{203}Pb		202.97339	2.163 d	EC/0.98		5/2-	+0.686	+0.10	Tl k x-ray
									0.279188
^{204m}Pb			1.13 h	I.T./2.185		9-			Pb k x-ray
									0.37481
									0.89922

Elem. or Isot.	Natural Abundance (Atom %)	Atomic Mass or Weight	Half-life/ Resonance Width (MeV)	Decay Mode/ Energy (/MeV)	Particle Energy/ Intensity (MeV/%)	Spin ($h/2\pi$)	Nuclear Magnetic Mom. (nm)	Elect. Quadr. Mom. (b)	γ-Energy / Intensity (MeV/%)
									0.91175
^{204}Pb	1.4(1)	203.973044				0+			
^{205}Pb		204.974482	1.51×10^7 y	EC/0.0512		5/2-	+0.712	+0.23	Tl L x-ray
^{206}Pb	24.1(1)	205.974465				0+			
^{207m}Pb			0.80 s	I.T./1.632		13/2+			Pb k x-ray
									0.56915
									1.06310
^{207}Pb	22.1(1)	206.975897				1/2-	+0.59258		
^{208}Pb	52.4(1)	207.976652	$> 2 \times 10^{19}$ y	sf		0+			
^{209}Pb		208.981090	3.25 h	β- /0.644	0.645/100	9/2+	-1.474	-0.3	
^{210}Pb		209.984189	22.6 y	β- /0.0635	0.017/81	0+			
					0.061/19				
				α	3.72				
^{211}Pb		210.988737	36.1 m	β- /1.37	0.57/5	(9/2+)	-1.404	+0.09	0.40486
					1.36/92				0.42700
									0.83186
									(0.09−1.27)
^{212}Pb		211.991898	10.64 h	β- /0.574	0.28/83	0+			Bi k x-ray
					0.57/12				0.23858
^{213}Pb		212.99658	10.2 m	β- /2.1					
^{214}Pb		213.999805	26.9 m	β- /1.0	0.67/48	0+			Bi k x-ray
					0.73/42				0.24192
									0.29509
									0.35187
^{215}Pb		215.0048	36 s						
$_{83}$Bi		208.98040(1)							
^{184m}Bi			0.007 s	α	(7.22-7.85)				0.449
^{184}Bi		184.0011	13. ms	α					0.124
^{185}Bi		184.9976	60. μs	p/90	1.55				
				α/10	8.03				
^{186m}Bi			15. ms	α	7.07-7.23				(0.087-0.520)
				p/<0.5					
^{186}Bi		185.9966	9.8 ms	α	7.26				0.1085
					7.37				
^{187m}Bi			~ 8. ms	α/12					
^{187}Bi		186.99316	32. ms	α/7	7.00/88.3				
					7.61/8.0				
					7.37/3.7				
^{188}Bi		187.99227	0.271 s	α	6.81				(0.071-0.320)
^{189m}Bi			7.0 ms	α	7.30				
^{189}Bi		188.9892	0.68 s	α					
^{190m}Bi			5.7 s	α/90	6.43				(0.105-0.314)
					(6.23-6.72)				
^{190}Bi		189.9883	~ 5.9 s	β+, EC/(30)/8.7	α/6.45				(0.089-0.374)
				α/70	(6.39-6.82)				
^{191m}Bi			0.12 ms	α/	6.87/100				
^{191}Bi		190.98579	12.4 s	β+, EC/(60)/7.3					
				α/(40)/	6.31				
^{192}Bi		191.98546	40. s	β+, EC/(80)/9.0					
				α/(20)/	6.06/				
^{193m}Bi			3.2 s	β+, EC/		1/2+			
				α/	6.48/				
^{193}Bi		192.98296	1.11 m	β+, EC/40/7.1		9/2+			
				α/(60)/	5.91/				
^{194}Bi		193.98283	1.8 m	β+, EC/99.9/8.2		(10-)			0.1661
				α/0.1/					0.1740
									0.2802
									0.421

Elem. or Isot.	Natural Abundance (Atom %)	Atomic Mass or Weight	Half-life/ Resonance Width (MeV)	Decay Mode/ Energy (/MeV)	Particle Energy/ Intensity (MeV/%)	Spin ($h/2\pi$)	Nuclear Magnetic Mom. (nm)	Elect. Quadr. Mom. (b)	γ-Energy / Intensity (MeV/%)
									0.5754
									0.9650
^{195m}Bi			1.45 m	β+, EC/(94)/					
				α/(6)/	6.11/				
^{195}Bi		194.98065	2.9 m	β+, EC/99.8/5.8		3/2-			
				α/(0.2)	5.45/				
^{196}Bi		195.98067	5. m	EC/~ 7.4					0.1376
									0.3720
									0.6880
									1.0486
^{197}Bi		196.97886	5. m	β+, EC/5.2		1/2+			
^{198m}Bi			7.7 s	I.T./0.2485		(10-)			0.2485
^{198}Bi		197.97921	11.8 m	β+, EC/6.6		(7+)			0.0900
									0.1976
									0.5624
									1.0635
^{199m}Bi			24.7 m	β+, EC/					ann.rad./
^{199}Bi		198.97767	27. m	β+, EC/4.3		9/2-	4.6		0.7203
									0.8374
									0.8417
									0.9460
									1.0528
									1.3056
									(0.12−3.2)
^{200m}Bi			31. m	β+, EC/		(2+)			0.2453
									0.4198
									0.4624
									1.0265
^{200}Bi		199.97813	36. m	EC/(90)/5.9		7+			ann.rad./
				β+ /(10)/					Pb k x-ray
									0.4198
									0.4623
									1.0265
^{201m}Bi			59.1 m	I.T./0.846		(1/2+)			Bi k x-ray
				β+, EC/					0.8464
^{201}Bi		200.97701	1.8 h	EC/3.84		9/2-	4.8		Pb k x-ray
									0.6288
									0.9357
									1.0138
									(0.13−2.4)
^{202}Bi		201.97774	1.72 h	β+ /(3)/5.16		5+	+4.26	-0.72	ann.rad./
				EC/(97)/					Pb k x-ray
									0.57860
									0.92734
									(0.08−3.5)
^{203}Bi		202.97688	11.8 h	EC/99.8/3.25		9/2-	+4.02	-0.69	Pb k x-ray
				β+ /(0.2)/	1.35/				0.1865
									0.8203
									0.8969
									1.8475
									(0.1−2.9)
^{204}Bi		203.97781	11.2 h	EC/4.44		6+	+4.32	-0.43	Pb k x-ray
									0.37481
									0.89922
									0.98409
^{205}Bi		204.97739	15.31 d	EC/2.71		9/2-	+4.07	-0.59	Pb k x-ray
									0.70347
									1.76435
^{206}Bi		205.97850	6.243 d	EC/3.76		6+	+4.36	-0.39	Pb k x-ray
									0.51619

Elem. or Isot.	Natural Abundance (Atom %)	Atomic Mass or Weight	Half-life/ Resonance Width (MeV)	Decay Mode/ Energy (/MeV)	Particle Energy/ Intensity (MeV/%)	Spin $(h/2\pi)$	Nuclear Magnetic Mom. (nm)	Elect. Quadr. Mom. (b)	γ-Energy / Intensity (MeV/%)
									0.80313
									0.88100
^{207}Bi		206.978471	31.55 y	EC/2.399		9/2-	4.08	-0.6	Pb k x-ray
									0.56915
									1.06310
^{208}Bi		207.979742	3.68×10^5 y	EC/2.880		5+	4.63	-0.64	Pb k x-ray
									2.61435
^{209}Bi	100.	208.980399	1.9×10^{19} y	α	3.13	9/2-	+4.111	-0.37	
^{210m}Bi			3.0×10^6 y	α/	4.420(3)/0.29	9-	+2.73	-0.47	Tl k x-ray
					4.569(3)/3.9				0.2661
					4.584(3)/1.4				0.3052
					4.908(4)/39				0.6502
					4.946(3)/55				
^{210}Bi		209.984120	5.01 d	β- /1.163	1.16/99	1-	-0.0445	+0.136	0.2661
									0.3.52
^{211}Bi		210.98727	2.14 m	α/(99.7)/	6.279/16	9/2-			Tl k x-ray
				β- /(0.3)/0.58	6.623/84				0.3501
^{212m2}Bi			7. m	β- /		(15-)			
^{212m1}Bi			25.0 m	α/(93)/	6.300/40	(9-)			0.120
				β- /(7)/	6.340/53				0.233
									0.275
									0.404
									0.727
^{212}Bi		211.991286	1.009 h	β- /(64)/2.254		(1-)	+0.32	+0.1	Tl k x-ray
				α/(36)/	6.051/25				Po k x-ray
					6.090/9.6				0.2881
									0.72725
									0.78551
									1.62066
^{213}Bi		212.994385	45.6 m	β- /(98)/1.43	1.02/31	9/2-	+3.72	-0.60	Po k x-ray
				α/(2)/	1.42/66				0.4404
					5.549/0.16				(0.15−1.328)
					5.869/2.0				
									1.10006
^{214}Bi		213.99871	19.7 m	β- /3.27					0.60931
									1.12027
									1.76449
									(0.19−3.2)
^{215m}Bi			37. s	β					(0.158-0.498)
^{215}Bi		215.00177	7.7 m	β- /2.3					0.2937/35.2
									(0.271−1.399)
^{216}Bi		216.00631	2.3 m	β-/4.0					0.5498
									0.4192
^{217}Bi		217.0095	98 s	β/					0.2646/100
									(0.254-1.017)
^{218}Bi		218.0143	33. s	β-					0.5097/134
									0.3857/100
									(0.174-0.703)

$_{84}$**Po**

Elem. or Isot.	Natural Abundance (Atom %)	Atomic Mass or Weight	Half-life/ Resonance Width (MeV)	Decay Mode/ Energy (/MeV)	Particle Energy/ Intensity (MeV/%)	Spin $(h/2\pi)$	Nuclear Magnetic Mom. (nm)	Elect. Quadr. Mom. (b)	γ-Energy / Intensity (MeV/%)
^{188}Po		187.99942	0.27 ms	α	7.91/80	0+			
					7.320				
^{189}Po		188.99848	5 ms	α	7.532/8				
					7.259/80				
					7.309/12				
^{190}Po		189.99510	2.4 ms	α/	7.53/96.4	0+			
					7.01/3.3				
^{191m}Po			93. ms	α	7.376/50				
					6.888/46				

Table of the Isotopes

Elem. or Isot.	Natural Abundance (Atom %)	Atomic Mass or Weight	Half-life/ Resonance Width (MeV)	Decay Mode/ Energy (/MeV)	Particle Energy/ Intensity (MeV/%)	Spin ($h/2\pi$)	Nuclear Magnetic Mom. (nm)	Elect. Quadr. Mom. (b)	γ-Energy / Intensity (MeV/%)
^{191}Po		190.99457	22 ms	α/	7.334/77				
					6.97/8				
^{192}Po		191.99134	32. ms	α/8.5	7.17/98.6	0+			
					6.59/1.4				
^{193m}Po			~ 0.07 s	α/	7.00				
^{193}Po		192.99103	0.45 s	α/	6.95				
^{194}Po		193.98819	0.2 s	α/	6.84/93	0+			
					6.19/0.22				
^{195m}Po			~ 2.8 s	α/	6.70/				
^{195}Po		194.98811	~ 3.9 s	α/	6.62/				
^{196}Po		195.98554	5. s	α/(95)/	6.53/94	0+			
				β+, EC/(5)/~4.6	5.77/0.02				
^{197m}Po			25.8 s	α/(84)/	6.385(3)/55	13/2+			
				β+, EC/(16)/					
^{197}Po		196.98566	53. s	α/(44)/	6.282(4)/76	(3/2-)			
				β+, EC/(56)/6.2					
^{198}Po		197.98339	1.76 m	α/(70)/	6.18/57	0+			
				β+, EC/(30)/4.0	5.27/7.6 × 10^{-4}				
^{199m}Po			4.2 m	β+, EC/(51)/		13/2+	0.99		ann.rad./
				α/(39)/	6.059/24				0.2745
									0.4998
									1.0020
^{199}Po		198.98367	5.2 m	β+, EC/(88)/7.		(3/2-)			Bi k x-ray
				α/(12)/	5.952/7.5				0.1877
									0.3616
									1.0214
									1.0344
^{200}Po		199.981780	11.5 m	β+, EC/85/3.4		0+			0.14748
				α/(15)/	5.863/11.1				0.32792
									0.6176
									0.6709
^{201m}Po			8.9 m	β+, EC/(57)/		13/2+	1.00		Bi k x-ray
				IT/40/0.418					Po k x-ray
				α/(3)/	5.786/~ 3.				0.2726
									0.4123
									0.4179
									0.9670
^{201}Po		200.98226	15.3 m	β+, EC/98/4.9		3/2-	0.94		Bi k x-ray
				α/(2)/	5.683(3)/1.1				0.2056
									0.2250
									0.8483
									0.9048
^{202}Po		201.98076	45. m	β+, EC/98/2.8		0+			0.0410
				α/(2)/	5.588/1.9				0.1656
									0.3158
									0.6884
^{203m}Po			1.2 m	IT/96/0.6414		13/2+			Bi k x-ray
				β-EC/(4)/					Po k x-ray
									0.6414
^{203}Po		202.98142	35. m	β+, EC/4.2		5/2-	+0.74		0.17516
									0.21477
									0.89350
									0.90863
									1.09095
^{204}Po		203.98032	3.53 h	EC/2.34		0+			Bi k x-ray
				α	5.377/0.66				0.2702
									0.8844
									1.0162
									(0.11−1.9)
^{205}Po		204.98120	1.7 h	β+, EC/3.53		5/2-	+0.76	+0.17	Bi k x-ray

Elem. or Isot.	Natural Abundance (Atom %)	Atomic Mass or Weight	Half-life/ Resonance Width (MeV)	Decay Mode/ Energy (/MeV)	Particle Energy/ Intensity (MeV/%)	Spin ($h/2\pi$)	Nuclear Magnetic Mom. (nm)	Elect. Quadr. Mom. (b)	γ-Energy / Intensity (MeV/%)
									0.83681
									0.84983
									0.87241
									1.00124
									(0.12−2.7)
^{206}Po		205.98048	8.8 d	EC/(95)/1.85		0+			Bi k x-ray
				α/(5)/	5.223/5.5				0.28644
									0.31156
									0.51134
									0.80737
									1.03228
									(0.11−1.5)
^{207m}Po			2.8 s	I.T./1.383		19/2−			Po k x-ray
									0.2682
									0.30074
									0.81448
^{207}Po		206.98159	5.80 h	EC, β+/2.91		5/2−	+0.79	+0.28	Bi k x-ray
									0.74263
									0.91176
									0.99225
^{208}Po		207.981246	2.898 y	α/5.213	4.233/0.0002	0+			
					5.1158/100				
^{209}Po		208.982430	102. y	α/4.976	4.624/0.56	1/2−	~ +0.77		0.26049
					4.879/99.2				0.8964
^{210}Po		209.982874	138.4 d	α/5.407	4.516/0.001	0+			0.80313
					5.304/100				
^{211m}Po			25.2 s	α/	7.273/91	25/2+			Pb k x-ray
					7.994/1.7				0.32808
					8.316/0.25				0.56915
					8.875/7.0				0.89723
									1.06310
^{211}Po		210.986653	0.516 s	α/7.594	6.570/0.54	9/2+			0.56915
					6.892/0.55				0.89723
					7.450/98.9				
^{212m}Po			45. s	α/	8.514/2.0	16+			
					9.086/1.0				
					11.650/97				
^{212}Po		211.988868	0.298 μs	α/8.953	8.784/100	0+			
^{213}Po		212.992857	3.7 μs	α/8.537	7.614/0.003	9/2+			
					8.375/100				
^{214}Po		213.995201	163.7 μs	α/7.833	6.904/0.01	0+			0.7995
					7.686/99.99				0.298
^{215}Po		214.999420	1.780 ms	α/7.526	6.950/0.02	(9/2+)			
					6.957/0.03				
					7.386/100				
^{216}Po		216.001915	0.145 s	α/6.906	5.895/0.002	0+			
					6.778/99.99				
^{217}Po		217.00634	1.53 s	α/6.662	6.539/				
^{218}Po		218.008973	3.04 m	α/6.114	6.003/99.999	0+			
					5.181/0.11				
^{219}Po		219.0137	~ 2 m						
^{220}Po		220.0166	> 0.3 μs			0+			

$_{85}$At

Elem. or Isot.	Natural Abundance (Atom %)	Atomic Mass or Weight	Half-life/ Resonance Width (MeV)	Decay Mode/ Energy (/MeV)	Particle Energy/ Intensity (MeV/%)	Spin ($h/2\pi$)	Nuclear Magnetic Mom. (nm)	Elect. Quadr. Mom. (b)	γ-Energy / Intensity (MeV/%)
^{191m}At			2.1 ms	α	7.65/98				
					7.72/2				
^{191}At			~ 1.7 ms	α	7.55/100				
^{193m}At			21 ms	α	7.33/98				
					7.42/2				

Elem. or Isot.	Natural Abundance (Atom %)	Atomic Mass or Weight	Half-life/ Resonance Width (MeV)	Decay Mode/ Energy (/MeV)	Particle Energy/ Intensity (MeV/%)	Spin ($h/2\pi$)	Nuclear Magnetic Mom. (nm)	Elect. Quadr. Mom. (b)	γ-Energy / Intensity (MeV/%)
^{193}At		192.9998	28 ms	α/	7.24/100				
^{194}At		193.9987	40 ms	α/					
^{195m}At			147 ms	α	7.07-7.22				
^{195}At		194.99627	0.33 s	α/	6.95				
^{196m}At			8 µs						0.158
^{196}At		195.9958	0.39 s	α/	7.05/				
^{197m}At			2.0 s	α	6.707	(1/2+)			
^{197}At		196.99319	0.39 s	β+, EC/7.8		(9/2-)			
				α/	6.960/				
^{198m}At			1.0 s	β+, EC/(75)/					
				α/(25)/	6.856/86				
^{198}At		197.99284	4.1 s	α/	6.755/94				
^{199}At		198.99053	6.9 s	β+, EC/8/5.6		9/2-			
				α/(92)/	6.643/				
^{200m2}At			47. s	α	6.411/				
^{200m}At			3.5 s	β+, EC/(80)		10-			
				α/(20)/	6.538/12				
^{200}At		199.99035	43. s	β+, EC/65/~ 8.0		5+			
				α/(35)/	6.412/44				
					6.465/57				
^{201}At		200.98842	1.48 m	β+, EC/29/5.9		9/2-			
				α/(71)/6.474	6.344/				
^{202m}At			0.46 s	I.T./0.391					
^{202}At		201.98863	3.02 m	β+, EC/88/7.2		5+			ann.rad./
				α/(12)/	6.135/7.7				0.4413
					6.225/4.3				0.5697
									0.6753
^{203}At		202.98694	7.4 m	β+, EC/69/5.1		9/2-			0.1458
				α/(31)/6.210	6.088/				0.2459
									0.6414
									1.0020
									1.0340
^{204}At		203.98725	9.1 m	β+, EC/95/6.5		(5+)			Po k x-ray
				α/(5)/	5.951/				0.3271
									0.4254
									0.5156
									0.6837
^{205}At		204.98607	26. m	β+, EC/90/4.54		(9/2-)			Po k x-ray
				α/(10)/6.020	5.902/				0.1543
									0.6696
									0.7194
^{206}At		205.98667	29.4 m	β+, EC/99/5.72		5+			Po k x-ray
				α/(1)/5.881	5.703/				0.20186
									0.39561
									0.47716
									0.70071
^{207}At		206.98578	1.81 h	β+, EC/90/3.91		9/2-			Po k x-ray
				α/(10)/5.873	5.758/				0.16801
									0.58842
									0.81448
^{208}At		207.98650	1.63 h	β+, EC/99/4.97		(6+)			Po k x-ray
				α/(1)/5.752	5.626/0.01				0.1770
					5.641/0.53				0.2060
									0.6601
									0.6852
									0.8450
									1.0281
^{209}At		208.98617	5.4 h	β+, EC/96/3.49		(6+)			Po k x-ray
				α/(4)/5.757	5.647/4.1				0.10422
									0.54503

Elem. or Isot.	Natural Abundance (Atom %)	Atomic Mass or Weight	Half-life/ Resonance Width (MeV)	Decay Mode/ Energy (/MeV)	Particle Energy/ Intensity (MeV/%)	Spin ($h/2\pi$)	Nuclear Magnetic Mom. (nm)	Elect. Quadr. Mom. (b)	γ-Energy / Intensity (MeV/%)
									0.78189
									0.79020
									(0.1−2.6)
²¹⁰At		209.98715	8.1 h	EC/99.8/3.98		5+			Po k x-ray
				α/(0.2)/5.632	5.361/0.05				0.24535
					5.442/0.05				0.52758
									1.18143
									1.43678
									1.48335
									(0.04−2.4)
²¹¹At		210.987496	7.21 h	EC/(58)/0.787		9/2-			Po k x-ray
				α/(42)/5.980	5.211/0.004				0.66956
					5.868/42				0.6870
									0.74263
²¹²ᵐAt			0.119 s	α/	7.837/65	(9-)			
					7.897/33				
²¹²At		211.99075	0.314 s	α/7.828	7.058/0.4	(1-)			
					7.088/0.6				
					7.618/15				
					7.681/84				
²¹³At		212.992937	0.11 μs	α/9.254	9.080/	9/2-			
²¹⁴ᵐAt			0.76 μs	α/8.762		(9-)			
²¹⁴At		213.996372	0.56 μs	α/8.987	8.819/100	(1-)			
²¹⁵At		214.99865	0.10 ms	α/8.178	7.626/0.045	(9/2-)			0.40486
					8.023/99.9				
²¹⁶At		216.002423	0.30 ms	α/7.947	7.595/0.2	(1-)			
					7.697/2.1				
					7.800/97				
²¹⁷At		217.004719	32. ms	α/7.202	6.812/0.06	(9/2-)			0.2595
					7.067/99.9				0.3345
									0.5940
²¹⁸At		218.00869	1.6 s	α/6.883	6.654/6				
					6.695/90				
					6.748/4				
²¹⁹At		219.011162	50. s	α/6.390	6.275/				
²²⁰At		220.0154	3.71 m	β- /3.7					(0.24−0.70)
²²¹At		221.0181	2.3 m	β					
²²²At		222.0223	0.9 m	β					
²²³At		223.0252	50. s	β					
₈₆Rn									
¹⁹⁵ᵐRn			5 ms	α	7.56				
¹⁹⁵Rn		195.00544	6 ms	α	7.54				
¹⁹⁶Rn		196.00212	4. ms	α/	7.46	0+			
¹⁹⁷ᵐRn			0.02 s	α	7.36				
¹⁹⁷Rn		197.0016	0.07 s	α/	7.26				
¹⁹⁸Rn		197.99868	64. ms	α	7.205	0+			
¹⁹⁹ᵐRn			0.32 s	α	7.060	(13/2+)			
¹⁹⁹Rn		198.9984	0.62 s	α/	6.989	3/2-			
²⁰⁰Rn		199.99570	1.06 s	α/(98)/	6.901/	0+			0.4329
				EC/(2)/5.					0.5043
²⁰¹ᵐRn			3.8 s	EC/(10)/		13/2+			
				α/(90)/	6.773/				
²⁰¹Rn		200.9956	7.0 s	α/(80)/	6.725/	(3/2-)			
				EC/(20)/	α/6.778				
²⁰²Rn		201.99326	9.9 s	α/(12)/	6.641/	0+			0.5695
				EC/(88)/					0.2876−0.6255
²⁰³ᵐRn			28. s	α/	6.551	13/2+	-0.96	+1.3	

Elem. or Isot.	Natural Abundance (Atom %)	Atomic Mass or Weight	Half-life/ Resonance Width (MeV)	Decay Mode/ Energy (/MeV)	Particle Energy/ Intensity (MeV/%)	Spin ($h/2\pi$)	Nuclear Magnetic Mom. (nm)	Elect. Quadr. Mom. (b)	γ-Energy / Intensity (MeV/%)
^{203}Rn		202.99339	45. s	α/(66)/6.629	6.499/	3/2-			
				EC/(34)/~ 7.4					
^{204}Rn		203.99143	1.24 m	α/(68)/	6.420/	0+			
				EC/(32)/3.8					
^{205}Rn		204.99172	2.8 m	α/(23)/6.390	6.123(3)/0.02	(5/2-)	+0.80	+0.06	0.2652
				EC/(77)/5.2	6.262(3)/23				0.3553
									0.4648
									0.6205
									0.6753
									0.7300
^{206}Rn		205.99021	5.7 m	α/(68)/6.384	6.258(3)/	0+			0.06170
				EC/(32)/3.3					0.0968
									0.3245
									0.3862
									0.4822
									0.4973
									0.7728
^{207}Rn		206.99073	9.3 m	β+, EC/77/4.6		5/2-	+0.82	+0.22	At k x-ray
				α/(23)/6.252	5.995(4)/0.02				0.32947
					6.068(3)/0.15				0.34455
					6.126(3)/22.8				0.36767
									0.40267
									0.74723
									(0.18−1.4)
^{208}Rn		207.98964	24.3 m	α/(60)/6.260	5.469(2)/0.003	0+			
				EC/(40)/2.85	6.140(2)/60				
^{209}Rn		208.99042	29. m	β+ /(83)/3.93	2.16/2.3	5/2-	+0.8388	+0.31	At k x-ray
				α/(17)/	5.887(3)/0.04				0.27933
					5.898(3)/0.02				0.33753
					6.039(2)/16.9				0.40841
									0.68942
									0.74594
									(0.18−3.2)
^{210}Rn		209.98970	2.4 h	α/(96)/6.157	5.351(2)/0.005	0+			At k x-ray
				EC/(4)/2.37	6.039(2)/96				0.19625
									0.45824
									0.57104
									0.64868
									(0.14−1.7)
^{211}Rn		210.99060	14.6 h	β+, EC/74/2.89		1/2-	+0.60		At k x-ray
				α/(26)/5.964	5.619(1)/0.7				0.16877
					5.784(1)/16.4				0.25022
					5.851(1)/8.8				0.37049
									0.67412
									0.67839
									1.36298
									(0.11−2.7)
^{212}Rn		211.990704	24. m	α/6.385	5.587(4)/0.05	0+			
					6.260(4)/99.95				
^{213}Rn		212.99388	19. ms	α/8.243	7.552(8)/1.0	9/2+			0.540
					8.087(8)/98.2				
					7.254/0.8				
^{214}Rn		213.99536	0.27 μs	α/9.209	9.037(9)/	0+			
^{215}Rn		214.99875	2.3 μs	α/8.840	8.674(8)/	(9/2+)			
^{216}Rn		216.00027	45. μs	α		0+			
^{217}Rn		217.003928	0.6 ms	α/7.885	7.500/0.1	9/2+			
					7.742(4)/100				
^{218}Rn		218.005601	35. ms	α/7.267	6.534(1)/0.16	0+			0.6093
					7.133(1)/99.8				0.6653
^{219}Rn		219.009480	3.96 s	α/6.946(1)	6.3130(5)/0.05	(5/2+)	-0.44	+0.93	Po k x-ray

Elem. or Isot.	Natural Abundance (Atom %)	Atomic Mass or Weight	Half-life/ Resonance Width (MeV)	Decay Mode/ Energy (/MeV)	Particle Energy/ Intensity (MeV/%)	Spin ($h/2\pi$)	Nuclear Magnetic Mom. (nm)	Elect. Quadr. Mom. (b)	γ-Energy / Intensity (MeV/%)
					6.425(3)/7.5				0.13057
					6.5309(4)/0.12				0.27113
					6.5531(3)/12.2				0.40170
					6.8193(3)/81				(0.1−1.05)
^{220}Rn		220.011394	55.6 s	α/6.404	5.7486(5)/0.07	0+			
					6.2883(1)/99.9				
^{221}Rn		221.01554	25. m	α/(22)/6.148	5.778(3)/1.8	7/2+	-0.020	-0.38	Fr L x-ray
				β- /(78)/1.2	5.788(3)/2.2				0.07384
					6.037(3)/18				0.08323
									0.0610
									0.18639
^{222}Rn		222.017578	3.823 d	α/5.590	4.987(1)/0.08	0+			0.510
					5.4897(3)/99.9				
^{223}Rn		223.0218	23. m	β- /			-0.78	+0.80	
^{224}Rn		224.0241	1.8 h	β- /		0+			0.1085
									0.2601
									0.2655
^{225}Rn		225.0284	4.5 m	β- /		7/2-	-0.70	+0.84	
^{226}Rn		226.0309	7.4 m	β- /		0+			
^{227}Rn		227.0354	2. s	β- /					
^{228}Rn		228.0380	65. s	β- /		0+			
$_{87}$Fr									
^{199}Fr		199.00726	12 s	α	7.66				
^{200}Fr		200.0066	49 ms	α	7.47				
^{201m}Fr			~ 0.02 s	α/	7.454				
^{201}Fr		201.0039	~ 60 ms	α/	7.36/	(9/2-)			
^{202m}Fr			0.29 s	α	7.236/				
^{202}Fr		202.00337	0.30 s	α/7.590	7.24/100				
^{203}Fr		203.00093	0.54 s	α/7.280	7.132(5)/	(9/2-)			
^{204m2}Fr			0.8 s	α	7.01				
^{204m1}Fr			2. s	α	6.97				
^{204}Fr		204.00065	1.8 s	α/	7.03/96				
					6.97/90				
					7.01/74				
^{205}Fr		204.99859	3.9 s	α/7.050	6.914(5)/	(9/2-)			
^{206m}Fr			0.7 s	α/	6.93				0.531(IT)
^{206}Fr		205.99867	16.0 s	α/7.416	6.792(5)/84				
^{207}Fr		206.99695	14.8 s	α/6.900	6.766(5)/	9/2-	+3.9	-0.16	
^{208}Fr		207.99714	59.1 s	α/(77)/6.770	6.636(5)/	7+	-4.8	+0.004	
				EC/(23)/6.99					
^{209}Fr		208.99595	50.0 s	α/(89)/5.1	6.646(3)/	9/2-	+3.9	-0.24	0.7978
				EC/(11)/5.16					(0.1103−1.384)
^{210}Fr		209.99641	3.2 m	α/6.670/71	6.543(5)/99.87	6+	+4.4	+0.19	0.2030
				EC/6.26	(5.90-6.42)				0.6438
									0.8175
									0.9008
^{211}Fr		210.99554	3.10 m	α/6.660/87	6.534(5)/99.94	9/2-	+4.0	-0.19	0.220
				EC/4.61	(5.87-6.20)				0.2799
									0.5389
									0.9169
^{212}Fr		211.99620	20. m	EC/(57)/5.12	6.261(1)/16	(5+)	+4.6	-0.10	Rn x-ray
				α/(43)/6.529	6.335(1)/4				0.08107
					6.335(1)/4				0.08378
					6.343(1)/1.3				0.2277
					6.383(1)/10				1.1856
					6.406(1)/9.5				1.2748
					6.08−6.18				0.014−1.178
^{213}Fr		212.99619	34.6 s	α/6.905	8.476(4)/51	9/2-	+4.0	-0.14	(0.408-0.577)

Elem. or Isot.	Natural Abundance (Atom %)	Atomic Mass or Weight	Half-life/ Resonance Width (MeV)	Decay Mode/ Energy (/MeV)	Particle Energy/ Intensity (MeV/%)	Spin ($h/2\pi$)	Nuclear Magnetic Mom. (nm)	Elect. Quadr. Mom. (b)	γ-Energy / Intensity (MeV/%)
^{214m}Fr			3.4 ms	α/	8.547(4)/46	9-			
					6.775−8.046				
^{214}Fr		213.99897	5.0 ms	α/8.587	7.409(3)/0.3	(1-)			(0.073−0.966)
					7.605(8)/1.0				
					7.940(3)/1.0				
					8.355(3)/4.7				
					8.427(3)/93				
^{215}Fr		215.00034	0.12 μs	α/9.537	9.360(8)/	(9/2-)			
^{216}Fr		216.00320	0.70 μs	α/9.175	9.005(10)/95				(0.045−0.160)
^{217}Fr		217.00463	0.016 ms	α/8.471	8.315(8)/	(9/2-)			
^{218m}Fr			22. ms	α					
^{218}Fr		218.007578	1. ms	α/8.014	7.384(10)/0.5	(1-)			
					7.542(15)/1.0				
					7.572(10)/5				
					7.732(10)/0.5				
					7.867(2)/93				
^{219}Fr		219.00925	21. ms	α/8.132	6.802(2)/0.25	(9/2-)			
					6.967(2)/0.6				
					7.146(2)/0.25				
					7.313(2)/99				
^{220}Fr		220.012327	27.4 s	α/6.800	6.582(1)/10	1+	-0.67	+0.47	0.0450
					6.630(2)/6				0.061
					6.641(1)/12				0.1060
					6.686(1)/61				0.1539
					6.39−6.58				0.1617
^{221}Fr		221.014255	4.8 m	α/6.457	5.9393(7)/0.17	(5/2-)	+1.58	-1.0	At k x-ray
					5.9797(7)/0.49				0.0995
					6.0751(7)/0.15				0.21798
					6.1270(7)/				0.4091
					6.2433(3)/1.3				
					6.3410(7)/83.4				
^{222}Fr		222.01755	14.3 m	β- /2.03	1.78/	2-	+0.63	+0.51	
				α/5.850					
^{223}Fr		223.019736	22.0 m	β- /1.149	α/5.291	(3/2+)	+1.17	+1.17	0.1509
				α//0.006	5.314				0.0589
					5.403				0.1453
^{224}Fr		224.02325	3.0 m	β- /2.82		1-	+0.40	+0.517	0.13150
									0.21575
									0.8367
									(0.1−2.21)
^{225}Fr		225.02557	3.9 m	β- /1.87		3/2	+1.07	+1.3	
^{226}Fr		226.0294	49. s	β- /3.6		1	+0.071	-1.35	0.18606
									0.25373
^{227}Fr		227.0318	2.48 m	β- /2.5		1/2	+1.50		
^{228}Fr		228.0357	39. s	β- /~ 3.5		2-	-0.76	+2.4	
^{229}Fr		229.03845	50. s	β- /					
^{230}Fr		230.0425	19. s	β- /		(3)			
^{231}Fr		231.0454	17. s	β- /					
^{232}Fr		232.050	5. s	β- /					(0.0545-0.721)

$_{88}$Ra

Elem. or Isot.	Natural Abundance (Atom %)	Atomic Mass or Weight	Half-life/ Resonance Width (MeV)	Decay Mode/ Energy (/MeV)	Particle Energy/ Intensity (MeV/%)	Spin ($h/2\pi$)	Nuclear Magnetic Mom. (nm)	Elect. Quadr. Mom. (b)	γ-Energy / Intensity (MeV/%)
^{201}Ra			~ 1.6 ms	α	7.91/				
^{202}Ra		202.0099	~ 0.02 ms	α	7.74	0+			
^{203m}Ra			24 ms	α	7.61				
^{203}Ra		203.0093	~ 31 ms	α	7.59				
^{204}Ra		204.0065	0.06 s	α	7.48	0+			
^{205m}Ra			~ 0.17 s						
^{205}Ra		205.0063	0.22 s	α	7.34				
^{206}Ra		206.00383	0.4 s	α/7.416	7.272(5)/	0+			

Elem. or Isot.	Natural Abundance (Atom %)	Atomic Mass or Weight	Half-life/ Resonance Width (MeV)	Decay Mode/ Energy (/MeV)	Particle Energy/ Intensity (MeV/%)	Spin ($h/2\pi$)	Nuclear Magnetic Mom. (nm)	Elect. Quadr. Mom. (b)	γ-Energy / Intensity (MeV/%)
^{207}Ra		207.0038	1.3 s	α/7.270	7.133(5)/				
^{208}Ra		208.00184	1.4 s	α/7.273	7.133(5)/	0+			
^{209}Ra		209.00199	4.6 s	α/7.150	(6.50-7.14)	5/2-	+0.87	+0.40	(0.387-0.634)
^{210m}Ra			2.4 µs						(0.0967-0.775)
^{210}Ra		210.00050	3.7 s	α/7.610	7.020(5)/	0+			574.9
^{211m}Ra			3.9 µs						(0.396-0.802)
^{211}Ra		211.00090	13. s	α/7.046	6.907/99.	(5/2-)	+0.878	+0.48	
				EC/5.0	(6.26-6.79)				(0.120-0.665)
^{212m}Ra			8.3 µs						(0.440-0.824)
^{212}Ra		211.99979	13.0 s	α/7.033	6.901(2)/	0+			
^{213m}Ra			2.1 ms	IT					
									(0.160-1.061)
^{213}Ra		213.00038	2.7 m	EC/(20)/3.88		(1/2-)	+0.613		0.1024
				α/(80)/6.860	6.521(3)/4.8				0.11010
					6.622(3)/39				0.2125
					6.730(3)/36				
^{214m}Ra			> 0.015 ms						(0.181-1.382)
^{214}Ra		214.00011	2.46 s	α/7.272	7.14/99.8/	0+			0.642
					6.51/0.2				
^{215m}Ra			7.6 µs						(0.196-1.048)
^{215}Ra		215.00272	1.64 ms	α/8.864	7.883(6)/2.8	(9/2+)			0.773/100
					8.171(3)/1.4				0.852/74
					8.700(3)/95.9				0.055-1.048
^{216}Ra		216.00353	0.18 µs	α/9.526	9.349(8)/	0+			
^{217}Ra		217.00632	1.6 µs	α/9.161	8.992(8)/	9/2-			
^{218}Ra		218.00714	26. µs	α/8.547	8.390(8)/	0+			
^{219}Ra		219.01009	0.010 s	α/8.132	7.680(10)/65				
					7.982(9)/35				
^{220}Ra		220.01103	18. ms	α/7.593	7.39/5	0+			0.465
					7.45/95				
^{221}Ra		221.013917	29. s	α/6.879	6.254(10)/0.7	5/2+	-0.180	+1.9	
					6.578(5)/3				
					6.585(3)/8				
					6.608(3)/35				
					6.669(3)/21				
					6.758(3)/31				
^{222}Ra		222.015375	36.2 s	α/5.590	6.237(2)/3.0	0+			0.324
					6.556(2)/97				0.1448-0.8402
^{223}Ra		223.018502	11.43 d	α/5.979	5.287(1)/0.15	(3/2+)	+0.271	+1.25	Rn k x-ray
					5.338(1)/0.13				0.12231
					5.365(1)/0.13				0.14418
					5.433(5)/2.3				0.15418
					5.502(1)/1.0				0.15859
					5.540(1)/9.2				0.26939
					5.607(3)/24				0.32388
					5.716(3)/52				0.33328
					5.747(1)/9				0.44494
					5.857(1)/0.32				(0.10-0.7)
					5.872(1)/0.85				
^{224}Ra		224.020212	3.66 d	α/5.789	5.034(10)/0.003	0+			Rn k x-ray
					5.047(1)/0.007				0.2407
					5.164(5)/0.007				0.4093
					5.449(2)/4.9				0.6501
					5.685(2)/95				
^{225}Ra		225.023612	14.9 d	β- /0.36	0.32/100	(3/2+)	-0.734		Ac k x-ray
				α	5.01×10^{-5}				0.0434
					4.98×10^{-6}				
^{226}Ra		226.025410	1599. y	α/4.870	4.194(1)/0.001	0+			Rn k x-ray
			> 4×10^{18} y	sf/4×10^{-14}	4.343(1)/0.006				0.1861/3.64
					4.601(1)/6.16				0.2624

Elem. or Isot.	Natural Abundance (Atom %)	Atomic Mass or Weight	Half-life/ Resonance Width (MeV)	Decay Mode/ Energy (/MeV)	Particle Energy/ Intensity (MeV/%)	Spin (h/2 π)	Nuclear Magnetic Mom. (nm)	Elect. Quadr. Mom. (b)	γ-Energy / Intensity (MeV/%)
					4.784(1)/93.8				0.053−2.448
²²⁷Ra		227.029178	42. m	β- /1.325	1.03/	(3/2+)	-0.404	+1.5	Ac L x-ray
					1.30/				Ac k x-ray
									0.02739
²²⁸Ra		228.031070	5.76 y	β- /0.046	0.039/50	0+			0.0135
					0.014/30				(0.006−0.0306)
					0.026/20				
²²⁹Ra		229.03496	4.0 m	β- /1.76	1.76/	(3/2+)	+0.503	+3.1	0.0145−0.1715
²³⁰Ra		230.03706	1.5 h	β- /1.0	0.7/	0+			0.0631
									0.0720
									0.2028
									0.4698
									0.4787
²³¹Ra		231.0412	1.7 m	β-					
²³²Ra		232.0436	4. m	β-		0+			
²³³Ra		233.0481	30. s	β-					
²³⁴Ra		234.051	~ 30. s	β-/		0+			
₈₉Ac									
²⁰⁶ᵐAc			0.04 s	α	7.79				
²⁰⁶Ac		206.0145	~ 26 ms	α	7.75				
²⁰⁷Ac		207.0120	27 ms	α/	7.69				
²⁰⁸ᵐAc			~ 25. ms	α/	7.72				
²⁰⁸Ac		208.0116	~ 0.1 s	α/	7.62				
²⁰⁹Ac		209.00949	~ 0.10 s	α/	7.58				
²¹⁰Ac		210.0094	0.34 s	α/7.610	7.462(8)/				
²¹¹Ac		211.0077	0.20 s	α/7.620	7.480(8)/				
²¹²Ac		212.0078	0.9 s	α/7.520	7.379(8)/				
²¹³Ac		213.0066	0.73 s	α/7.500	7.364(8)/	(9/2-)			
²¹⁴Ac		214.00690	8.2 s	α/(86)/7.350	7.215/54	(5+)			(0.0626-0.754)
				EC/(14)/6.34	7.081/42				
					(6.48-7.15)				
²¹⁵Ac		215.00645	0.17 s	α/7.750	7.60/99.57	(9/2-)			0.399
					7.21/0.46				0.582
					7.03/0.20				0.654
					6.96/0.14				
²¹⁶ᵐAc			0.44 ms	α/	8.198(8)/1.7	(9-)			(0.0826-1.375)
					8.283(8)/2.5				
					9.028(5)/49				
					9.106(5)/46				
²¹⁶Ac		216.00872	44. ms	α/9.241	8.990(2)/10	(1-)			
					9.070(8)/90				
²¹⁷ᵐAc			0.7 μs	α/	10.540/100				
²¹⁷Ac		217.00935	0.07 μs	α/9.832	9.650(10)/100	9/2-			
²¹⁸Ac		218.01164	1.1 μs	α/9.380	9.205(15)/				
²¹⁹Ac		219.01242	0.012 ms	α/8.830	8.664(10)/	(9/2-)			
²²⁰Ac		220.01476	26. ms	α/8.350	7.610(20)/23				
					4.680(20)/21				
					7.790(10)/13				
					7.850(10)/24				
					7.985(10)/4				
					8.005(10)/5				
					8.060(10)/6				
					8.195(10)/3				
²²¹Ac		221.01559	52. ms	α/7.790	7.170(10)/2				
					7.375(10)/10				
					7.440(15)/20				
					7.645(10)/70				
²²²ᵐAc			63. s	α/(>89)/	6.710(20)/7				

Elem. or Isot.	Natural Abundance (Atom %)	Atomic Mass or Weight	Half-life/ Resonance Width (MeV)	Decay Mode/ Energy (/MeV)	Particle Energy/ Intensity (MeV/%)	Spin ($h/2\pi$)	Nuclear Magnetic Mom. (nm)	Elect. Quadr. Mom. (b)	γ-Energy / Intensity (MeV/%)
				EC/(1)/	6.750(20)/13				
				I.T./(<10)/	6.810(20)/24				
					6.840(20)/9				
					6.890(20)/13				
					6.970(20)/7				
					7.000(20)/13				
^{222}Ac		222.01784	5. s	α/7.141	6.967(10)/6	1-			
					7.013(2)/94				
^{223}Ac		223.01914	2.1 m	α/(99)/6.783	6.131(2)/0.12	(5/2-)			0.0725
				EC/(1)/0.59	6.177(2)/0.94				0.0839
					6.293(1)/0.47				0.0927
					6.326(1)/0.3				0.0990
					6.332(2)/0.14				0.1917
					6.360(1)/0.22				0.2158
					6.397(1)/0.13				0.3588
					6.448(1)/0.2				0.4768
					6.473(1)/3.1				
					6.523(2)/0.6				
					6.528(1)/3.1				
					6.563(1)/13.6				
					6.582(3)/0.3				
					6.646(1)/44				
					6.661(1)/31				
^{224}Ac		224.021723	2.7 h	EC/(90)/1.403	5.841(1)/0.5	0-			Ra L x-ray
				α/(10)/6.323	5.860(1)/0.75				Ra k x-ray
					5.875(1)/1.7				0.08426
					5.941(1)/4.4				0.13150
					6.000(1)/6.7				0.1571
					6.013(1)/1.4				0.21575
					6.056(1)/22				0.2619
					6.138(1)/26				(0.03−0.3)
					6.154(1)/1.0				
					6.204(1)/12				
					6.210(1)/20				
^{225}Ac		225.023230	10.0 d	α/5.935	5.286(1)/0.2	3/2			Fr k x-ray
					5.444(3)/0.1				0.06296/0.48
					5.554(1)/0.1				0.09982/1.36
					5.608(1)/1.1				0.1084
					5.636(1)/4.5				0.1116
					5.681(1)/1.4				0.1451
					5.722(1)/2.9				0.150.02/0.691
					5.731(1)/10				0.15724
					5.791(1)/9				0.18795/0.54
					5.793(1)/18				0.0075−0.8085
^{226}Ac		226.026098	1.224 d	EC/(17)/0.640		(1-)			Ra k x-ray
				β- /(83)/1.116					Th k x-ray
				α/(0.006)/5.51	5.399(5)/0.006				0.07218
									0.15816
									0.23034
^{227}Ac		227.027752	21.77 y	β- /98.6/0.045	0.045/54	(3/2-)	+1.1	+1.7	0.0838/23.
				α/(1.4)/5.043	4.869(1)/0.09				0.0811/14.
					4.938(1)/0.52				0.2696/13.
					4.951(1)/0.65				(0.044−1.27)
^{228}Ac		228.031021	6.15 h	β- /2.127	1.11/32	(3+)			Th L x-ray
					1.85/12				Th k x-ray
					2.18/11				0.12903
									0.33842
									0.91116
									0.96897
									(0.2−1.96)

Elem. or Isot.	Natural Abundance (Atom %)	Atomic Mass or Weight	Half-life/ Resonance Width (MeV)	Decay Mode/ Energy (/MeV)	Particle Energy/ Intensity (MeV/%)	Spin ($h/2\pi$)	Nuclear Magnetic Mom. (nm)	Elect. Quadr. Mom. (b)	γ-Energy / Intensity (MeV/%)
^{229}Ac		229.03302	1.04 h	β- /1.10	1.1/	(3/2+)			0.09335/2.43
									0.16451/2.61
									0.56916/2.24
									0.0111−0.898
^{230}Ac		230.0363	2.03 m	β- /2.7	1.4/	1+			Th k x-ray
				β-, sf	/0.000119				0.45497
									0.50820
									(0.12−2.5)
^{231}Ac		231.0386	7.5 m	β- /2.1	2.1/100	(1/2+)			0.14379
									0.18574
									0.22140
									0.28250
									0.3070
^{232}Ac		232.0420	2.0 m	β- /3.7		(2-)			
^{233}Ac		233.0446	2.4 m	β- /		(1/2+)			
^{234}Ac		234.0484	40. s	β- /		(1+)			
$_{90}$Th		232.03806(2)							
^{209}Th		209.0177	~ 0.01 s	α	8.08				
^{210}Th		210.0158	~ 9 ms	α	7.90	0+			
^{211}Th		211.0149	0.04 s	α	7.79				
^{212}Th		212.01298	~ 30. ms	α/	7.80/	0+			
^{213}Th		213.0130	0.14 s	α/7.840	7.692(10)/				
^{214}Th		214.01150	0.10 s	α/7.825	7.677(10)/	0+			
^{215}Th		215.01173	1.2 s	α/7.660	7.33(10)/8	(1/2-)			0.134
					7.395(8)/52				0.192
					7.524(8)/40				(0.069-0.295)
^{216m}Th			0.14 ms	α	9.93/74				(0.0905-1.478)
					8.00, 9.31				
^{216}Th		216.01106	27. ms	α/8.071	7.92/99.46	0+			0.628
					7.30/0.54				
^{217}Th		217.01311	0.25 ms	α/9.424	9.27/94.6				(0.546-0.822)
					8.46/3.8				
					8.73/1.6				
^{218}Th		218.01328	0.11 µs	α/9.847	9.665(10)/	0+			
^{219}Th		219.01554	1.05 µs	α/9.510	9.340(20)/				
^{220}Th		220.01575	10. µs	α/8.953	8.790(20)/	0+			
^{221}Th		221.01818	2. ms	α/8.628	7.732/7				
					8.142/72				
					8.469/21				
^{222}Th		222.01847	2.24 ms	α/8.129	7.980/97.7	0+			
					7.599/2.3				
^{223}Th		223.02081	0.60 s	α/7.454	7.29(1)/41(5)				
					7.32(1)/29(5)				
					7.350(15)/20(5)				
					7.390(15)/10(4)				
^{224}Th		224.02147	1.05 s	α/7.305	6.768(5)/1.2	0+			
					6.997(5)/19				
					7.170(5)/7				
^{225}Th		225.023951	8.72 m	EC/(10)/0.68		(3/2+)			
				α/(90)/6.920	6.441(2)/15				
					6.479(2)/43				
					6.501(3)/14				
					6.627(3)/3				
					6.650(5)/3				
					6.700(5)/2				
					6.743(3)/7				
					6.796(2)/9				
^{226}Th		226.024903	30.83 m	α/6.454	6.026(1)/0.2	0+			Ra k x-ray

Elem. or Isot.	Natural Abundance (Atom %)	Atomic Mass or Weight	Half-life/ Resonance Width (MeV)	Decay Mode/ Energy (/MeV)	Particle Energy/ Intensity (MeV/%)	Spin ($h/2\pi$)	Nuclear Magnetic Mom. (nm)	Elect. Quadr. Mom. (b)	γ-Energy / Intensity (MeV/%)
					6.041(1)/0.19				0.1112
					6.098(1)/1.3				0.2421
					6.2283(4)/23				0.1310
					6.3375(4)/75				0.1733−0.9295
^{227}Th		227.027704	18.72 d	α/6.146		(3/2+)			Ra L x-ray
									Ra k x-ray
									0.05014
									0.23597
									0.25624
									(0.02−1.0)
^{228}Th		228.028741	1.913 y	α/5.520	5.1770(2)/0.18	0+			
					5.2114(1)/0.4				
					5.3405(1)/26.7				
					5.4233(1)/73				
^{229m}Th			13.9 h	α	4.83-5.08				
^{229}Th		229.031762	7.9×10^3 y	α/5.168	4.814/9.3	5/2+	+0.46	+4.	0.1935/4.3
					4.845(5)/56				0.21089/277
					4.9008(5)/10.2				0.13697/1.21
					4.689−5.077				0.0111−0.6036
^{230}Th		230.033134	7.54×10^4 y	α/4.771	4.4383(6)/0.03	0+			0.0677/0.46
					4.4798(6)/0.12				0.1439/0.078
			$> 2 \times 10^{18}$ y	sf/$< 4 \times 10^{-12}$	4.6211(6)/23.4				
					4.6876(6)/76.3				
^{231}Th		231.036304	1.063 d	β- /0.390	0.138/22	5/2+			Pa L x-ray
					0.218/20				Pa k x-ray
					0.305/52				0.02564
									0.084203/
									(0.02−0.3)
^{232}Th	100.	232.038055	1.40×10^{10} y	α/4.081	3.830(10)/0.2	0+			0.0590
			1.2×10^{21} y	sf/1.1×10^{-9}	3.952(5)/23				0.124
					4.010(5)/77				
^{233}Th		233.041582	22.3 m	β-/1.245	1.245/	½+			Pa L x-ray
									Pa k x-ray
									0.02938
									0.08653
									0.45930
									(0.02−1.2)
^{234}Th		234.043601	24.10 d	β- /0.273	0.102/20	0+			Pa L x-ray
					0.198/72				0.06329/4.1
									0.09235/2.4
									0.09278/2.4
^{235}Th		235.04751	7.2 m	β- /1.9					0.4162
									0.6594
									0.7272
									0.747
									0.9318
^{236}Th		236.0499	37.5 m	β- /~ 1.0		0+			Pa k x-ray
									0.1107
^{237}Th		237.0539	5.0 m	β-					
^{238}Th		238.0565	9.4 m			0+			0.0890
$_{91}$Pa		231.03588(2)							
^{212}Pa		212.0232	~ 5 ms	α	8.27				
^{213}Pa		213.0211	7 ms	α	8.24				
^{214}Pa		214.0209	17 ms	α	8.12				
^{215}Pa		215.0192	15. ms	α	8.08/100				
^{216}Pa		216.0191	0.19 s	α/	7.95/51				0.134
					7.82/45				
					7.79/4				

Table of the Isotopes

Elem. or Isot.	Natural Abundance (Atom %)	Atomic Mass or Weight	Half-life/ Resonance Width (MeV)	Decay Mode/ Energy (/MeV)	Particle Energy/ Intensity (MeV/%)	Spin ($h/2\pi$)	Nuclear Magnetic Mom. (nm)	Elect. Quadr. Mom. (b)	γ-Energy / Intensity (MeV/%)
^{217m}Pa			1.08 ms	α/	10.16/72				0.4504−0.8208
					8.306/11				
					9.55/6				
					9.69/2				
^{217}Pa		217.0183	3.8 ms	α/8.490	8.337/99				0.0466−0.634
					7.873/0.4				
					7.728/0.3				
					7.710/0.3				
^{218}Pa		218.02004	0.12 ms	α/	9.54/31				0.092
					9.61/69				
^{219}Pa		219.0199	0.05 µs	α					
^{220}Pa		220.0219	0.8 µs	α					
^{221}Pa		221.0219	6. µs	α	9.08(3)				
^{222}Pa		222.0237	~ 4.3 ms	α/8.700	8.180/50				
					8.330/20				
					8.540/30				
^{223}Pa		223.0240	~ 6.5 ms	α/8.340	8.006(10)/55				
					8.196(10)/45				
^{224}Pa		224.02563	0.84 s	α/7.630	7.555(10)/75(3)				0.1945
					7.46(1)/25(3)				(0.028−0.412)
^{225}Pa		225.0261	1.8 s	α/7.380	7.195(10)/30				
					7.245(10)/70				
^{226}Pa		226.02795	1.8 m	α/(74)/6.987	6.728(10)/0.7				
				EC/(26)/2.83	6.823(10)/35				
					6.863(10)/39				
^{227}Pa		227.02881	38.3 m	α/(85)/6.582	6.357(4)/7	(5/2-)			0.0649
				EC/(15)/1.02	6.376(10)/2.2				0.0669
					6.401(4)/8				0.1100
					6.416(4)/13				
					6.423(10)/10				
					6.465(4)/43				
^{228}Pa		228.031051	22. h	EC/(98)/2.111		(3+)	+3.5		Th k x-ray
				α/(2)	5.779/0.23				0.409/100
					5.805/0.15				0.4631/222
					6.078/0.4				0.91116/242
					6.105/0.25				0.96464/120
					6.118/0.22				0.96897/149
									0.058−1.96
^{229}Pa		229.032097	1.5 d	EC/(99.8)/0.32		(5/2+)			0.04244
				α/(0.2)/5.836	5.536(2)/0.02				(0.024−0.18)
					5.579(2)/0.09				
					5.668(2)/0.05				
^{230}Pa		230.034541	17.4 d	EC/(90)/1.310	0.51/	(2-)	2.0		Th L x-ray
				β-/(10)/0.563					Th k x-ray
									0.4437
									0.45477
									0.89876
									0.91856
									0.95199
									(0.053−1.07)
^{231}Pa		231.035884	3.25×10^4 y	α/5.148	4.6781(5)/1.5	3/2-	2.01	-1.7	Ac L x-ray
					4.7102(5)/1.0				Ac k x-ray
			$> 2 \times 10^{17}$ y	sf/$< 1.6 \times 10^{-15}$	4.7343(5)/8.4				0.01899
					4.8513(5)/1.4				0.027396
					4.9339(5)/3				0.03823
					4.9505(5)/22.8				0.04639
					4.9858(5)/1.4				0.25586
					5.0131(5)/25.4				0.26029
					5.0292(5)/20				0.28367
					5.0318(5)/2.5				0.30007

Elem. or Isot.	Natural Abundance (Atom %)	Atomic Mass or Weight	Half-life/ Resonance Width (MeV)	Decay Mode/ Energy (/MeV)	Particle Energy/ Intensity (MeV/%)	Spin ($h/2\pi$)	Nuclear Magnetic Mom. (nm)	Elect. Quadr. Mom. (b)	γ-Energy / Intensity (MeV/%)
					5.0587(5)/11				0.30264
									0.33007
									(0.02−0.61)
^{232}Pa		232.03859	1.31 d	β- /1.34		(2-)			U k x-ray
									0.10900
									0.15009
									0.89439
									0.96934
									(0.10−1.17)
^{233}Pa		233.040247	27.0 d	β-/0.571	0.15/40	3/2-	+4.0	-3.0	U L x-ray
					0.256/60				U k x-ray
									0.30017
									0.31201/38.4
									(0.0286-0.456)
^{234m}Pa			1.17 m	β- /99.9/2.29		(0-)			U k x-ray
				IT/0.13/					0.25818/0.07
									0.76641/0.32
									1.0009/0.86
									(0.06−1.96)
^{234}Pa		234.043308	6.69 h	β- /2.197	0.51/	(4+)			U L x-ray
									U k x-ray
									0.1312/0.03
									0.5695/0.02
									0.9256/0.02
									(0.02−1.99)
^{235}Pa		235.04544	24.4 m	β- /1.41	1.4/97	(3/2-)			0.0308−0.65893
^{236}Pa		236.0487	9.1 m	β- /2.9	1.1/40	(1-)			U k x-ray
					2.0/50				0.64235
					3.1/10				0.68759
									1.7630
									(0.04−2.18)
^{237}Pa		237.0512	8.7 m	β- /2.3	1.1/60	(1/2+)			0.4986
					1.6/30				0.5293
					2.3/10				0.5407
									0.8536
									0.8650
									(0.04−1.4)
^{238}Pa		238.0545	2.3 m	β- /3.5	1.2/	(3-)			0.10350
					1.7/				0.1785
									0.4484
									0.6350
									0.6800
									1.01446
									(0.04−2.5)
^{239}Pa		239.0573	1.8 h						
$_{92}$U		**238.02891(3)**							
^{217}U		217.0244	~ 0.2 ms	α	8.02				
^{218m}U			~ 0.56 ms	α	10.68				
^{218}U		218.02354	0.5 ms	α	8.61	0+			
^{219}U		219.0249	~ 0.08 ms	α	9.68(4)/				
^{222}U		222.0261	~ 1.μs	α		0+			
^{223}U		223.0277	0.02 s	α/	8.78(4)/				
^{224}U		224.02761	~ 1. ms	α/	8.46/100	0+			
^{225}U		225.02939	84. ms	α/	7.87/83				
					7.82/15				
					7.63/2				
^{226}U		226.02934	0.26 s	α/7.560	7.56/86	0+			
					7.38/14				

Table of the Isotopes

Elem. or Isot.	Natural Abundance (Atom %)	Atomic Mass or Weight	Half-life/ Resonance Width (MeV)	Decay Mode/ Energy (/MeV)	Particle Energy/ Intensity (MeV/%)	Spin ($h/2\pi$)	Nuclear Magnetic Mom. (nm)	Elect. Quadr. Mom. (b)	γ-Energy / Intensity (MeV/%)
^{227}U		227.03116	1.1 m	α/7.200	6.870/				
^{228}U		228.03137	9.1 m	α/6.803	6.404(6)/0.6	0+			0.095
					6.440(5)/0.7				0.152
					6.589(5)/29				0.187
					6.681(6)/70				0.246
^{229}U		229.03351	58. m	EC/(80)/1.31	6.223/3	(3/2+)			
				α/(20)/6.473	6.297(3)/11				
					6.332(3)/20				
					6.360(3)/64				
^{230}U		230.033940	20.8 d	α/5.992	5.5866(3)/0.01	0+			Th L x-ray
			> 4 × 10^{10} y	sf/< 10^{-10}	5.6624(3)/0.26				0.07218
					5.6663(3)/0.38				0.15421
					5.8178(3)/32				0.23034
					5.8887(3)/67				(0.081−0.8565)
^{231}U		231.036294	4.2 d	EC/0.36		(5/2−)			Pa L x-ray
				α/(10^{-3})	5.46/1.6 × 10^{-3}				Pa k x-ray
					5.47/1.4 × 10^{-3}				0.02564
					5.40/1. × 10^{-3}				0.08420
^{232}U		232.037156	70. y	α/5.414	4.9979(1)/0.003	0+			
			2.6 × 10^{15} y	sf/2.7 × 10^{-12}	5.1367(1)/0.3				
					5.2635(1)/31				
					5.3203(1)/69				
^{233}U		233.039635	1.592 × 10^{5} y	α/4.909	4.7830(8)/13.2	5/2+	+0.59	3.66	Th L x-ray
			> 2.7 × 10^{17} y	sf/6 × 10^{-11}	4.8247(8)/84.4				0.04244
					4.510−4.804				0.09714
									(0.0252−1.119)
^{234}U	0.0054(5)	234.040952	2.455 × 10^{5} y	α/4.856	4.604(1)/0.24	0+			0.05323/0.156
			1.5 × 10^{16} y	sf/1.6 × 10^{-9}	4.7231(1)/27.5				0.12091
					4.776(1)/72.5				
^{235m}U			26. m	IT/0.0007		1/2+			
^{235}U	0.7204(6)	235.043930	7.04 × 10^{8} y	α/4.6793	4.1525(9)/0.9	7/2−	−0.38	4.9	Th L x-ray
			1.0 × 10^{19} y	sf/7 × 10^{-9}	4.2157(9)/6.				Th k x-ray
					4.3237(9)/4.6				0.10917
					4.3641(9)/19.				0.14378
					4.370(4)/6				0.16338
					4.3952(9)/57.				0.18574
					4.4144(9)/2.1				0.20213
					4.5025(9)/1.7				0.20533
					4.5558(9)/4.2				0.22140
					4.5970(9)/4.8				(0.03−0.79)
^{236}U		236.045568	2.342 × 10^{7} y	α/4.569	4.332(8)/0.26	0+			Th L x-ray
			2.5 × 10^{16} y	sf/9 × 10^{-8}	4.445(5)/26				0.04946/100
					4.494(3)/74				0.11279/24.1
									0.17115/0.080
^{237}U		237.048730	6.75 d	β− /0.519	0.24/	1/2+			Np L x-ray
					0.25/				Np k x-ray
									0.05953
									0.20801
^{238}U	99.2742(10)	238.050788	4.47 × 10^{9} y	α	4.0395/0.23	0+			Th L x-ray
			8.2 × 10^{15}y	sf/5 × 10^{-5}	4.147(5)/23				0.04955/.06
					4.196(5)/77				0.1135/.01
^{239}U		239.054293	23.5 m	β− /1.265	1.2/	5/2+			(0.522−0.681)
					1.3/				
^{240}U		240.05659	14.1 h	β− /0.39	0.36/	0+			Np L x-ray
									0.04410
									0.05558
									0.06760
^{242}U		242.0629	16.8 m	β− /∼ 1.2		0+			

Elem. or Isot.	Natural Abundance (Atom %)	Atomic Mass or Weight	Half-life/ Resonance Width (MeV)	Decay Mode/ Energy (/MeV)	Particle Energy/ Intensity (MeV/%)	Spin ($h/2\pi$)	Nuclear Magnetic Mom. (nm)	Elect. Quadr. Mom. (b)	γ-Energy / Intensity (MeV/%)
$_{93}$Np									
^{225}Np		225.0339	> 2 µs						
^{226}Np		226.0352	0.03 s	α/	8.04(2)/				
^{227}Np		227.0350	0.51 s	α/	7.65(2)/				
					7.68(1)/				
^{228}Np		228.0362	61. s	EC/60(7)/					
				α/40(7)/, sf					
^{229}Np		229.0363	4.0 m	α/7.010	6.890(20)				
^{230}Np		230.0378	4.6 m	EC/97 /3.6					
				α/3	6.660(20)				
^{231}Np		231.03825	48.8 m	EC/98 /1.8		5/2			0.2629
				α/2 /6.368	6.280/2				0.3475
									0.3703
^{232}Np		232.0401	14.7 m	EC/99 /2.7		(4-)			U L x-ray
									U k x-ray
									0.3268
									0.81925
									0.86683
^{233}Np		233.04074	36.2 m	EC/1.2		(5/2+)			U L x-ray
									U k x-ray
									0.29887
									0.31201
^{234}Np		234.04290	4.4 d	β+, EC/1.81	0.79/	(0+)			U L x-ray
									U k x-ray
									1.5272
									1.5587
									1.6022
^{235}Np		235.044063	1.085 y	EC/99.9 /0.124		5/2+			U k x-ray
				α/0.001/5.191					
^{236m}Np			22.5 h	EC/52 /		(1-)			U L x-ray
				β- /48 /					Pu L x-ray
									U k x-ray
									0.64235
									0.68759
^{236}Np		236.04657	1.55×10^5 y	EC/91 /0.94		(6-)			U L x-ray
				β- /9 /0.49					U k x-ray
									0.10423
									0.16031
^{237}Np		237.048173	2.14×10^6 y	α/4.957	4.6395(5)/6.5	5/2+	+3.14	+3.89	Pa L x-ray
			1×10^{18} y	sf/2.1×10^{-10}	4.766(5)/9.7				Pa k x-ray
					4.7715(5)/22.7				0.029378/15
					4.7884(5)/47.8				0.08653/12
					4.558−4.873				(0.03−0.28)
^{238}Np		238.050946	2.117 d	β- /1.292	1.2/	2+			Pu L x-ray
									Pu k x-ray
									0.98447/25.2
									1.02855/18.3
									(.044−1.026)
^{239}Np		239.052939	2.355 d	β- /0.722	0.341/30	5/2+			Pu L x-ray
					0.438/48				Pu k x-ray
									0.10613
									0.228186/11
									0.27760/15
									(0.04−0.50)
^{240m}Np			7.22 m	β- /99.9 /	2.18/	(1+)			0.25143
				IT/0.1 /					0.26333
									0.55454
									0.59735

Table of the Isotopes

Elem. or Isot.	Natural Abundance (Atom %)	Atomic Mass or Weight	Half-life/ Resonance Width (MeV)	Decay Mode/ Energy (/MeV)	Particle Energy/ Intensity (MeV/%)	Spin ($h/2\pi$)	Nuclear Magnetic Mom. (nm)	Elect. Quadr. Mom. (b)	γ-Energy / Intensity (MeV/%)
^{240}Np		240.05616	1.032 h	β- /2.20	0.89/	5+			0.1471/
									0.5664
									0.6008
^{241}Np		241.0583	13.9 m	β- /1.3	1.3/	5/2+			0.1330/
									0.1740
									0.280
^{242m}Np			2.2 m	β- /		(1+)			0.15910
									0.2651/
									0.78570
									0.9448/
^{242}Np		242.0616	5.5 m	β- /2.7	2.7/	6+			0.6209
									0.73620
									0.78074
									1.47340
									(0.04–2.37)
^{243}Np		243.06428	1.9 m						
^{244}Np		244.0679	2.3 m						
$_{94}$Pu									
^{228}Pu		228.03874	~ 1.1 s	α/	7.81(2)/	0+			
^{229}Pu		229.0402	~ 1.5 m	α/	7.46/				
^{230}Pu		230.03965	1.7 m	α/	7.06/81	0+			
					7.00/19				
^{231}Pu		231.04110	8.6 m	EC/90					
				α/10	6.72				
^{232}Pu		232.04119	34. m	EC/>80/1.1		0+			
				α/<20/6.716	6.542(10)/38				
					6.600(10)/62				
^{233}Pu		233.04300	20.9 m	EC(99.9)/1.9					0.1503
				α/0.1 /6.416	6.300(20)/0.1				0.1804
									0.2353
									0.5002
									0.5346/
									1.0352/
^{234}Pu		234.04332	8.8 h	EC/94 /0.39		0+			
				α/6 /6.310	6.035(3)/0.024				
					6.149(3)/1.9				
					6.200(3)/4.				
^{235}Pu		235.04529	25.3 m	EC/99+ /1.2		(5/2+)			
				α/0.003/5.957	5.850(20)/0.003				
^{236m}Pu			1.2 µs						
^{236}Pu		236.046058	2.87 y	α/5.867	5.611/0.21	0+			0.0476/0.07
			1.5×10^9 y	sf/1.9×10^{-7}	5.7210/30.5				0.109/0.02
					5.7677(1)/69.3				(0.17–0.97)
^{237}Pu		237.048410	45.7 d	EC/99.9 /0.220		7/2-			Np L x-ray
				α/0.003 /5.747	5.334(4)/0.0015				Np k x-ray
					5.356(4)/0.0006				0.026344
					5.650(4)/0.0007				0.03319
									0.05954
									(0.03–0.5)
^{238}Pu		238.049560	87.7 y	α/5.593	5.3583(1)/0.10	0+			U k x-ray
			4.75×10^{10} y	sf/1.8×10^{-7}	5.465(1)/28.3				0.04347
					5.4992(1)/71.6				(0.04–1.1)
^{239}Pu		239.052163	2.410×10^4 y	α/5.244	5.055/0.047	1/2+	+0.203		U k x-ray
			$8. \times 10^{15}$ y	sf/3×10^{-10}	5.076/0.078				0.05162
					5.106/11.9				0.05682
					5.144/17.1				0.12928
					5.157/70.8				0.37502
					(4.74 –5.03)				0.41369

Elem. or Isot.	Natural Abundance (Atom %)	Atomic Mass or Weight	Half-life/ Resonance Width (MeV)	Decay Mode/ Energy (/MeV)	Particle Energy/ Intensity (MeV/%)	Spin ($h/2\pi$)	Nuclear Magnetic Mom. (nm)	Elect. Quadr. Mom. (b)	γ-Energy / Intensity (MeV/%)
^{240}Pu		240.053814	6.56×10^3 y	α/5.255	5.0212(1)/0.07	0+			U L x-ray
			1.14×10^{11} y	sf/5.7×10^{-6}	5.1237(1)/26.4				0.04524
					5.1681(1)/73.5				0.10423
									(0.04−0.97)
^{241}Pu		241.056852	14.3 y	β-/99+/0.0208	$4.853/3 \times 10^{-4}$	5/2+	-0.683	+6.	0.14854
				α/0.002 /5.139	4.897/0.002				0.1600
			$< 6 \times 10^{16}$ y	sf/$> 2.4 \times 10^{-14}$					
^{242}Pu		242.058743	3.75×10^5 y	α/4.983	4.7546(7)/0.098	0+			U L x-ray
			6.77×10^{10} y	sf/5.5×10^{-4}	4.8564(7)/22.4				0.04491
					4.9006(7)/78				0.10350
^{243}Pu		243.062003	4.956 h	β- /0.582	0.49/21	7/2+			Am L x-ray
					0.58/60				0.0417
									0.0839
^{244}Pu		244.064204	8.00×10^7 y	α/99.9/4.665	4.546(1)/19.4	0+			U L x-ray
			6.6×10^{10} y	sf/0.12	4.589(1)/80.5				0.0439
^{245}Pu		245.06775	10.5 h	β- /1.21	0.93/57	(9/2-)			Am L x-ray
					1.21/11				Am k x-ray
									0.2804 /
									0.30832
									0.32752
									0.56014
									(0.03−1.2)
^{246}Pu		246.07021	10.85 d	β- /0.40	0.150/85	0+			Am L x-ray
					0.35/10				Am k x-ray
									0.04379
									0.22371
^{247}Pu		247.0741	2.3 d						
$_{95}$Am									
^{232}Am		232.0466	0.9 m	EC/~ 5.0					
^{233}Am		233.0464	~ 3.2 m	α	6.78				
^{234}Am		234.0478	2.3 m	EC/4.2					
^{235}Am		235.0480	10.3 m	EC					Pu K x-ray
				α	6.46/0.4				0.291/100
									(0.170-0.828)
^{236m}Am			2.9 m			(1-)			(0.583-0.713)
^{236}Am		236.0496	3.6 m	EC		(5-)			(0.158-1.038)
^{237}Am		237.0500	1.22 h	EC/99.98 /1.7		(5/2-)			Pu k x-ray
				α/0.02 /6.20	6.042(5)/0.02				0.14559
									0.28026
									0.43845
^{238}Am		238.05198	1.63 h	EC/2.26		1+			Pu L x-ray
				α/0.0001 /6.04	5.940/0.0001				Pu k x-ray
									0.91870
									0.96278
^{239}Am		239.053025	11.9 h	EC/99.99/0.803		5/2-			Pu L x-ray
				α/0.01/5.924	5.734(2)/0.001				Pu k x-ray
					5.776(2)/0.008				0.18172
									0.22818
									0.27760
^{240}Am		240.05530	2.12 d	EC/1.38		(3-)			Pu L x-ray
				α/5.592	$5.378(1)/16 \times 10^{-4}$				Pu k x-ray
									0.88878
									0.98764
									(0.1−1.3)
^{241}Am		241.056829	432.7 y	α/5.637	5.2443(1)/0.002	5/2-	+1.58	+3.1	Np L x-ray
			1.2×10^{14} y	sf/3.6×10^{-10}	5.3221(1)/0.015				0.02634 /.024
					5.3884(1)/1.4				0.03319/.00126
					5.4431(1)/12.8				0.05954/0.359

Elem. or Isot.	Natural Abundance (Atom %)	Atomic Mass or Weight	Half-life/ Resonance Width (MeV)	Decay Mode/ Energy (/MeV)	Particle Energy/ Intensity (MeV/%)	Spin ($h/2\pi$)	Nuclear Magnetic Mom. (nm)	Elect. Quadr. Mom. (b)	γ-Energy / Intensity (MeV/%)
					5.4857(1)/85.2				(0.03−1.128)
					5.5116(1)/0.20				
					5.5442(1)/0.34				
^{242m}Am			141. y	IT/99.5/0.048		5-	+1.0	+6.5	Am L x-ray
				α/0.5/5.62	5.141(4)/0.026				0.04863
			> 3 × 10^{12} y	sf/< 4.7 × 10^{-9}	5.2070(2)/0.4				0.08648
									0.10944
									0.16304
^{242}Am		242.059549	16.02 h	β- /83 /0.665	0.63/46	1-	+0.388	-2.4	Pu L x-ray
				EC/17 /0.750	0.67/37				Cm L x-ray
									Pu k x-ray
									0.0422
									0.04453
^{243}Am		243.061381	7.37 × 10^3 y	α/5.438	5.1798(5)/1.1	5/2-	+1.5	+2.9	0.04354
			2. × 10^{14} y	sf/3.7 × 10^{-9}	5.2343(5)/11				0.07467
					5.2766(5)/88				0.08657
					5.394(5)/0.12				0.11770
					5.3500(5)/0.16				0.14197
^{244m}Am			~ 26. m	β- /1.498		(1-)			0.0429
^{244}Am		244.064285	10.1 h	β- /1.428					Am L x-ray
									Cm k x-ray
									0.7460
									0.9000
^{245}Am		245.066452	2.05 h	β- /0.894	0.65/19	(5/2+)			Cm L x-ray
					0.90/77				Cm k x-ray
									0.25299
^{246m}Am			25.0 m	β- /	1.3/79.	2-			Cm L x-ray
					1.60/14				Cm k x-ray
					2.1/7				0.27002
									0.79881
									1.06201
									1.07885
									(0.04−2.29)
^{246}Am		246.06978	39. m	β- /2.38	1.2/	(7-)			Cm L x-ray
									Cm k x-ray
									0.1529
									0.2046
									0.6786
^{247}Am		247.0721	22. m	β- /1.7					Cm L x-ray
									Cm k x-ray
									0.2267 /
									0.2853 /

$_{96}$Cm

Elem. or Isot.	Natural Abundance (Atom %)	Atomic Mass or Weight	Half-life/ Resonance Width (MeV)	Decay Mode/ Energy (/MeV)	Particle Energy/ Intensity (MeV/%)	Spin ($h/2\pi$)	Nuclear Magnetic Mom. (nm)	Elect. Quadr. Mom. (b)	γ-Energy / Intensity (MeV/%)
^{233}Cm		233.0508		α/	7.34/				
^{234}Cm		234.05016	~ 51. s	α	7.24/	0+			
^{235}Cm		235.0514							
^{236}Cm		236.0514		EC/1.7		0+			
^{237}Cm		237.0529		EC/2.5					
^{238}Cm		238.05303	2.4 h	EC/>90 /0.97		0+			
				α/<10 /6.632	6.520(50)/<10				
^{239}Cm		239.0550	~ 3. h	EC/1.7					
									0.0407
									0.1466
									0.1874
^{240}Cm		240.055530	27. d	α/6.397	5.989/0.014	0+			
					6.147/0.05				
			1.9 × 10^6 y	sf/3.9 × 10^{-6}	6.2478(6) /28.8				
					6.2906(6) /70.6				

Elem. or Isot.	Natural Abundance (Atom %)	Atomic Mass or Weight	Half-life/ Resonance Width (MeV)	Decay Mode/ Energy (/MeV)	Particle Energy/ Intensity (MeV/%)	Spin ($h/2\pi$)	Nuclear Magnetic Mom. (nm)	Elect. Quadr. Mom. (b)	γ-Energy / Intensity (MeV/%)
^{241}Cm		241.057653	32.8 d	EC/99 /0.768		1/2+			Am k x-ray
				α/1 /6.184	5.8842(4)/0.12				0.13241
					5.9291(4)/0.18				0.16505
					5.9389(4)/0.69				0.18028
									0.43063
									0.47181
^{242}Cm		242.058836	162.8 d	α/6.216	5.9694(1)/0.035	0+			Pu L x-ray
					6.069(1)/25				0.04408
			7.0×10^6 y	sf/6.4×10^{-6}	6.1129(1)/74				0.10189
									(0.04−1.2)
^{243}Cm		243.061389	29.1 y	α/6.167	5.6815(5) /0.2	5/2+	0.41		Pu L x-ray
					5.6856(5)/1.6				Pu k x-ray
			5.5×10^{11} y	sf/5.3×10^{-9}	5.7420(5)/10.6				0.10612
					5.7859(5)/73.3				0.20975
					5.9922(5)/6.5				0.22819
					6.0103(5)/1.0				0.27760
					6.0589(5)/5				0.28546
					6.0666(5)/1.5				0.33431
									(0.04−0.7)
^{244}Cm		244.062753	18.1 y	α/5.902	5.6656/0.02	0+			Pu L x-ray
					5.7528/23				0.04282
			1.32×10^7 y	sf/1.4×10^{-4}	5.8050/77				0.09885
					5.515/0.004				0.15262
^{245}Cm		245.065491	8.48×10^3 y	α/5.623	5.235(10)/0.3	7/2+	0.5		Pu L x-ray
					5.3038(10)/5.0				Pu k x-ray
			1.4×10^{12} y	sf/6.1×10^{-7}	5.3620(7)/93				0.04195
					5.4927(11)/0.8				0.13299
					5.5331(11)/0.6				0.13606
									0.17494
^{246}Cm		246.067224	4.76×10^3 y	α/5.476	5.343(3)/21	0+			Pu L x-ray
			1.8×10^7 y	sf/0.026	5.386(3)/79				0.04453
^{247}Cm		247.070354	1.56×10^7 y	α/5.352	4.818(4)/4.7	9/2-	0.37		Pu k x-ray
					4.8690(20)/71				0.2792
					4.941(4)/1.6				0.2886
					4.9820(20)/2.0				0.3471
					5.1436(20)/1.2				0.4035
					5.2104(20)/5.7				
					5.2659(20)/13.8				
^{248}Cm		248.072349	3.48×10^5 y	α/99.92 /5.162	4.931(5)/0.07	0+			
					5.0349(2)/16.5				
			4.15×10^6 y	sf/8.38	5.0784(2)/(75)/1				
^{249}Cm		249.075953	64.15 m	β- /0.900	0.9/	1/2+			Bk k x-ray
									0.56039/0.84
									0.63431/1.5
									(0.085-0.653)
^{250}Cm		250.07836	$\sim 9.7 \times 10^3$ y	sf/85.8		0+			
				α/5.27					
^{251}Cm		251.08229	16.8 m	β- /1.42	0.90/16	(1/2+)			0.3896 /
									0.5299
									0.5425
^{252}Cm		252.0849	< 2 d			0+			

$_{97}$**Bk**

^{238}Bk		238.0583	2.4 m	EC/5.0					
^{239}Bk		239.0583							
^{240}Bk		240.0598	~ 4.8 m						
^{241}Bk		241.0602	4.6 m	EC					(0.152-0.262)
^{242}Bk		242.0620	7.0 m	EC/3.0					
^{243}Bk		243.063008	4.5 h	EC/99.8 /1.508	6.542(4)/0.03	(3/2-)			0.1466

Elem. or Isot.	Natural Abundance (Atom %)	Atomic Mass or Weight	Half-life/ Resonance Width (MeV)	Decay Mode/ Energy (/MeV)	Particle Energy/ Intensity (MeV/%)	Spin ($h/2\,\pi$)	Nuclear Magnetic Mom. (nm)	Elect. Quadr. Mom. (b)	γ-Energy / Intensity (MeV/%)
				α/0.15 /6.871	6.5738(2)/0.04				0.1874
					6.7180(22)/0.02				0.755
					6.7581(20)/0.02				0.840
									0.946
^{244}Bk		244.06518	4.4 h	EC/99.99 /2.26		(4-)			0.1445
				α/0.01 /6.778	6.625(4)/0.003				0.1876
					6.667(4)/0.003				0.2176
									0.9815
									0.9215/
^{245}Bk		245.066362	4.94 d	EC/99.9 /0.810		3/2-			Cm L x-ray
				α/0.1 /6.453	5.8851(5)/0.03				Cm k x-ray
					6.1176(9)/0.01				0.25299
					6.1467(5)/0.02				0.3809
					6.3087(5)/0.014				0.3851
					6.3492(5)/0.018				
^{246}Bk		246.0687	1.80 d	EC/1.35		(2-)			Cm L x-ray
									Cm k x-ray
									0.79881
									1.08142
^{247}Bk		247.07031	1.4×10^3 y	α/5.889	5.465(5)/1.5	(3/2-)			0.04175
					5.501(5)/7				0.0839
					5.532(5)/45				0.268
					5.6535(20)/5.5				
					5.678(2)/13				
					5.712(2)/17				
					5.753(2)/4.3				
					5.794(2)/5.5				
^{248}Bk		248.07310	23.7 h	β- /70 /0.87	0.86/	(1-)			Cm L x-ray
				EC/30 /0.72					Cf L x-ray
									Cm k x-ray
									Cf k x-ray
									0.5507
^{249}Bk		249.074987	320. d	β- /0.125	0.125/100	7/2+	2.0		0.327/10^{-5}
				α/0.001 /5.525	5.390(1)/0.0002				0.308/10^{-6}
			1.8×10^9 y	sf/4.9 × 10^{-8}	5.4174(6)/0.001				
^{250}Bk		250.078317	3.217 h	β- /1.780	0.74/	2-			Cf L x-ray
									Cf k x-ray
									0.98912
									1.03184
									(0.04−1.6)
^{251}Bk		251.08076	56. m	β- /1.09		(3/2-)			0.02481
									0.1528
									0.1776
^{252}Bk		252.0843	1.8 m						

$_{98}$Cf

Elem. or Isot.	Natural Abundance (Atom %)	Atomic Mass or Weight	Half-life/ Resonance Width (MeV)	Decay Mode/ Energy (/MeV)	Particle Energy/ Intensity (MeV/%)	Spin ($h/2\,\pi$)	Nuclear Magnetic Mom. (nm)	Elect. Quadr. Mom. (b)	γ-Energy / Intensity (MeV/%)
^{237}Cf		237.062	2.1 s	α, sf/10					
^{238}Cf		238.0614	21 ms	sf/~ 100		0+			
				α/~ 0.2					
^{239}Cf		239.0624	~ 0.7 m	α					
^{240}Cf		240.0623	1.1 m	α/7.719	7.590(10)/	0+			
				sf/ ~ 2.1					
^{241}Cf		241.0637	4. m	EC/3.3					
				α/7.60	7.335(5)/				
^{242}Cf		242.06370	3.5 m	α/7.509	7.351(6)/20	0+			
				sf/<0.014	7.385(4)/80				
^{243}Cf		243.0654	11. m	EC/86 /2.2	7.060(6)/20	(1/2+)			
				α/14 /7.40	7.170/4				
^{244}Cf		244.066001	20. m	α/7.328	7.168(5)/25	0+			

Elem. or Isot.	Natural Abundance (Atom %)	Atomic Mass or Weight	Half-life/ Resonance Width (MeV)	Decay Mode/ Energy (/MeV)	Particle Energy/ Intensity (MeV/%)	Spin ($h/2\pi$)	Nuclear Magnetic Mom. (nm)	Elect. Quadr. Mom. (b)	γ-Energy / Intensity (MeV/%)
					7.210(5)/75				
^{245}Cf		245.068049	44. m	α/36 /7.255	7.14/91.7				Cm K x-ray
				EC/64 /1.569	6.983/0.31				0.5709
					7.09/7				0.6014
					7.065/0.68				0.6163
^{246}Cf		246.068805	1.49 d	α/6.869	6.6156(10)/0.18	0+			Cm L x-ray
					6.7086(7)/21.8				0.04221
			1.8×10^3 y	sf/2.3×10^{-4}	6.7501(7)/78.0				0.0945
									0.147
^{247}Cf		247.07100	3.11 h	EC/99.96 /0.65		7/2+			Bk k x-ray
				α/0.04 /6.55	6.301(5)/				0.2941
									0.4778
^{248}Cf		248.07219	334. d	α/6.369	6.220(5)/17	0+			
			3.2×10^4 y	sf/0.0029	6.262(5)/83				
^{249}Cf		249.074854	351. y	α/6.295	5.758/3.7	9/2-			Cm L x-ray
					5.812/85.7				Cm k x-ray
			$8. \times 10^{10}$ y	sf/4.4×10^{-7}	5.8488(2)/1.0				0.25299/2.5
					5.9029(2)/2.8				0.33351/13.6
					5.9451(2)/4.0				0.38832/63.6
					6.1401(2)/1.1				(0.0376−1.103)
					6.1940(2)/2.2				
^{250}Cf		250.076406	13.1 y	α/6.129	5.8913(4)/0.3	0+			Cm L x-ray
			1.7×10^4 y	sf/0.077	5.9889(4)/15				0.04285
					6.0310(4)/84.5				
^{251m}Cf			26.3 µs						
^{251}Cf		251.079587	9.0×10^2 y	α/6.172	5.56448(7)/1.5	1/2+			0.109/19.8
					5.632(1)/4.5				0.1775/17.3
					5.648(1)/3.5				(0.0385-0.354)
					5.6773(6)/35				
					5.762(3)/3.8				
					5.7937(7)/2.0				
					5.8124(8)/4.2				
					5.8514(6)/27				
					6.0140(7)/11.6				
					6.0744(7)/2.7				
^{252}Cf		252.081626	2.65 y	α/96.9 /6.217	5.7977(1)/0.23	0+			Cm L x-ray
			86. y	sf/3.1/	6.0756(4)/15.2				0.04339
					6.1184(4)/81.6				0.1002
^{253}Cf		253.08513	17.8 d	β- /99.7 /0.29	0.27/100	(7/2+)			
				α/0.3 /6.126	5.921(5)/0.02				
^{254}Cf		254.08732	60.5 d	sf/99.7/		0+			
				α/0.3/5.930	5.792(5)/0.05				
					5.834(5)/0.26				
^{255}Cf		255.0911	1.4 h	β- /0.7					
^{256}Cf		256.0934	12. m	sf		0+			
$_{99}$Es									
^{241}Es		241.0685	~ 8 s	α	8.11				
^{242}Es		242.0698	16 s	α	7.92				
^{243}Es		243.0696	21. s	α/>30 /	7.89/>30				
				EC/<70 /4.0					
^{244}Es		244.0709	37. s	EC/76 /4.6					
				α/4 /	7.57/4				
^{245}Es		245.0713	1.3 m	α/40 /7.858	7.74				
				EC/60 /3.1					
^{246}Es		246.0729	7.7 m	EC/90 /3.9					
				α/10 /	7.35				
^{247}Es		247.07366	4.8 m	EC/93 /2.48					
				α/7 /	7.32				

Elem. or Isot.	Natural Abundance (Atom %)	Atomic Mass or Weight	Half-life/ Resonance Width (MeV)	Decay Mode/ Energy (/MeV)	Particle Energy/ Intensity (MeV/%)	Spin ($h/2\pi$)	Nuclear Magnetic Mom. (nm)	Elect. Quadr. Mom. (b)	γ-Energy / Intensity (MeV/%)
[248]Es		248.0755	26. m	EC/99.7 /3.1					
				α/0.3 /	6.87				
[249]Es		249.07641	1.70 h	EC/99.4 /1.45		(7/2+)			0.3795
				α/0.6 /	6.77				0.8132
[250m]Es			2.2 h	EC/		(1-)			Cf L x-ray
				β+					Cf k x-ray
									0.9891
									1.0319
[250]Es		250.0786	8.6 h	EC/2.1		(6+)			Cf L x-ray
									Cf k x-ray
									0.30339
									0.34948
									0.82883
[251]Es		251.07999	1.38 d	EC/99.5 /0.38		(3/2-)			
				α/0.5 /	6.462/0.05				
					6.492/0.4				
[252]Es		252.08298	1.29 y	α/76 /	6.632/61.0	(5-)			
				EC/24 /1.26	6.562/10.3				
[253]Es		253.084825	20.47 d	α/	6.633/89.8	7/2+	+4.10	7.	0.04180/5.6
			6.3×10^5 y	sf/8.9×10^{-6}	6.5916/6.6				0.3892/2.7
									(0.0309-1.106)
[254m]Es			1.64 d	β- /99.6 /	0.475	2+	2.9	3.7	Fm L x-ray
				α/0.3 /6.67	6.382	2+			Fm k x-ray
			> 10. y	sf/0.045					0.6488
									0.6938
[254]Es		254.088022	276. d	α/	6.429	(7+)			0.064
			$> 2.5 \times 10^7$ y	sf/$< 3 \times 10^{-6}$					
[255]Es		255.09027	40. d	β- /92 /0.29		(7/2+)			
				α/8 /	6.26				
			2.6×10^3 y	sf/0.0042	6.300				
[256m]Es			7.6 h	β- /		(8+)			0.218
									0.232
									0.862
[256]Es		256.0936	25. m	β-/1.7		(1+)			
[257]Es		257.0960	7.7 d	β-					
$_{100}$Fm									
[242]Fm		242.0734	0.8 ms	sf/> 96		0+			
[243]Fm		243.0744	0.2 s	α/	8.55				
				sf/< 0.4					
[244]Fm		244.0741	3.3 ms	sf/> 97		0+			
[245]Fm		245.0754	4. s	α/	8.15/				
				sf/<0.1					
[246]Fm		246.07530	1.2 s	α/85/	8.24/	0+			
				sf/15/					
[247m]Fm			4.3 s	α/	8.17/				
[247]Fm		247.0769	29. s	α/8.20	7.87/70				
				EC/2.9	7.93/30				
[248]Fm		248.07720	33. s	α/99.9 /8.001	7.83/20	0+			
				sf/0.1/	7.87/80				
[249]Fm		249.0790	1.6 m	EC/2.4		(7/2+)			
				α/	7.57				
[250m]Fm			1.8 s	IT/					
				sf/$<8 \times 10^{-5}$					
[250]Fm		250.07952	30. m	α/	7.43/	0+			
				EC/0.8					
				sf/0.007					
[251]Fm		251.08158	5.3 h	EC/98 /1.47		(9/2-)			
				α/2 /	6.833				

Elem. or Isot.	Natural Abundance (Atom %)	Atomic Mass or Weight	Half-life/ Resonance Width (MeV)	Decay Mode/ Energy (/MeV)	Particle Energy/ Intensity (MeV/%)	Spin ($h/2\pi$)	Nuclear Magnetic Mom. (nm)	Elect. Quadr. Mom. (b)	γ-Energy / Intensity (MeV/%)
^{252}Fm		252.08247	1.058 d	α/7.154	6.998/15	0+			
				sf/0.0023	7.039/85				
^{253}Fm		253.085185	3.0 d	EC/88/0.333	6.676/	½+			Es k x-ray
				α/12 /	6.943/				0.2719
^{254}Fm		254.086854	3.240 h	α/	7.150	0+			
				sf/0.059	7.192				
^{255}Fm		255.089962	20.1 h	α/	6.9635(5)/5.0	7/2+			0.08148/1.
			1.0×10^4 y	sf/2.3×10^{-5}	7.0225(5)/93.4				(0.041–0.900)
^{256}Fm		256.09177	2.63 h	sf/91		0+			
				α/19	6.92/				
^{257}Fm		257.09511	100.5 d	α/99.79	6.519	(9/2+)			0.1794
				sf/0.21					0.2410
^{258}Fm		258.0971	0.37 ms	sf/		0+			
^{259}Fm		259.1006	1.5 s	sf/					
^{260}Fm		260.103	~ 4 ms	sf/		0+			
$_{101}$Md									
^{245m}Md			~ 0.4 s	α	8.64, 8.68				
^{245}Md		245.0808	0.9 ms	sf					
^{246}Md		246.0819	1.0 s	α	8.74				
					8.50–8.56				
^{247m}Md			~ 0.2 s	sf/					
^{247}Md		247.0816	3. s	α	8.43				
^{248}Md		248.0828	7. s	EC/80 /5.3	8.32/15				
				α/20 /	8.36/5				
				sf/<0.05					
^{249}Md		249.0830	24. s	EC>/<80 /3.7					
				α/>20 /8.46	8.030(20)/				
^{250}Md		250.0844	50. s	EC/94 /4.6	7.75/4				
				α/6 /8.25	7.83/2				
^{251}Md		251.0848	4.0 m	EC/>94 /3.1					
				α/<6 /	7.55/				
^{252}Md		252.0866	2. m	EC/>50 /3.9					
				α/<50 /	7.73/				
^{253}Md		253.0873	~ 6 m	EC/2.0					
^{254m}Md			30. m	EC/					
^{254}Md		254.0897	10. m	EC/2.7					
^{255}Md		255.09108	27. m	EC/92 /1.04	α/7.33/93	(7/2-)			0.121/100
				α/8 /	7.27/5				0.115/65
				sf/< 0.15	7.75/1				0.136/35
					7.71/1				0.141–0.453
^{256}Md		256.0941	1.30 h	EC/89 /2.13	7.21/71				Fm k x-ray
				α/11 /	7.14/22				0.121/409
				sf/< 2.6	7.68/2.5				0.115/266
					7.25/2.5				0.136/143
					7.64/2.1				0.634/119
									0.141–1.37
^{257}Md		257.095541	5.5 h	EC/85 /0.41	7.074	(7/2-)			Fm k x-ray
				α/15, sf/< 1	7.014				(0.181–0.389)
^{258m}Md			57. m	EC/		(1-)			Fm k x-ray
				sf/< 30					
^{258}Md		258.098431	51.5 d	α/7.40	6.718(2)/	(8-)			0.3678
				sf/< 0.003	6.763(4)/				0.057–0.448
^{259}Md		259.1005	1.64 h	sf/>98.7		7/2+			
				α/<1.3					
^{260}Md		260.1037	~ 27.8 d	sf/ 73–100					
$_{102}$No									
^{248}No		248.0866	< 1.0 μs	sf		0+			

Elem. or Isot.	Natural Abundance (Atom %)	Atomic Mass or Weight	Half-life/ Resonance Width (MeV)	Decay Mode/ Energy (/MeV)	Particle Energy/ Intensity (MeV/%)	Spin ($h/2\pi$)	Nuclear Magnetic Mom. (nm)	Elect. Quadr. Mom. (b)	γ-Energy / Intensity (MeV/%)
249No		249.0878	0.05 ms	sf					
				α/ < 20					
250No		250.0875	6. µs	sf/		0+			
				α/ < 10					
251mNo			0.9 s	α	8.67				
251No		251.0890	0.78 s	α/91	8.62/96				
				sf/0.26	8.58/4				
252No		252.08898	2.44 s	α/75/8.551	8.42	0+			
				sf/24/	8.37				
				EC, β+/<1.6					
253No		253.0907	1.7 m	α/	8.00	(9/2-)			0.222/100
				EC/3.2					(0.151-0.280)
254mNo			0.28 s	I.T./					
				sf/< .2					
254No		254.09096	49. s	α/	8.09	0+			0.102
				EC/1.1					0.152
				sf/0.17					
255No		255.09324	3.1 m	α/62 /	8.12/	½+			0.187
				EC/38/2.01	7.93				
					8.08				
256No		256.09428	2.9 s	α/	8.43	0+			
				sf/0.5					
257No		257.09688	24.5 s	α/	8.222/83	(7/2+)			0.0770
				sf/<1.5	8.27				0.1018
					8.323/17				0.1241
258No		258.0982	~ 1.2 ms	sf/		0+			
259No		259.1010	58. m	α/78 /7.794	7.52	(9/2+)			
				EC/22/0.5	7.55				
				sf/<9.7					
260No		260.1026	0.11 s	sf/		0+			
262No		262.1073	~ 8. ms	sf/		0+			
$_{103}$Lr									
251Lr		251.0944	39 m	sf					
252Lr		252.0954	~ 0.36 s	α	9.02/73				
				sf/<1	8.97/27				
253mLr			~ 0.57 s	α	8.79				
				sf/1.3					
253Lr		253.0952	1.5 s	α/	8.72				
				sf/8					
254Lr		254.0965	13. s	α/	8.45				
				EC/5.2					
				sf/<0.1					
255Lr		255.09669	22. s	α/	8.37/60				
				EC/3.2	8.43/40				
				sf/< 0.1					
256Lr		256.0986	27. s	α/99.7 /8.554	8.43/				
				EC/4.2	8.39				
				sf/< 0.03					
257Lr		257.0996	0.65 s	α/	8.80	7/2+			
				EC/2.5					
				sf/< 0.03					
258Lr		258.1018	3.9 s	α/	8.60/46				
				EC/3.4	8.62/25				
				sf/< 5	8.56/20				
					8.65/9				
259Lr		259.1029	6.1 s	α/80	8.44(1)				
				sf/20					
260Lr		260.1055	3. m	α/	8.03				

Elem. or Isot.	Natural Abundance (Atom %)	Atomic Mass or Weight	Half-life/ Resonance Width (MeV)	Decay Mode/ Energy (/MeV)	Particle Energy/ Intensity (MeV/%)	Spin ($h/2\pi$)	Nuclear Magnetic Mom. (nm)	Elect. Quadr. Mom. (b)	γ-Energy / Intensity (MeV/%)
^{261}Lr		261.1069	40. m	sf					
^{262}Lr		262.1096	3.6 h	EC/2.					
				sf/<10					
$_{104}$**Rf**									
^{253}Rf		253.1007	~ 48. μs	sf					
				α/<10					
^{254}Rf		254.1002	23. μs	sf/>98.5		0+			
				α/<1.5					
^{255}Rf		255.1013	1.6 s	α	8.72/<0.05				0.203
				sf/52	8.77/94				0.142
					8.67/<0.05				
					8.58/<0.05				
					8.92/<0.05				
^{256}Rf		256.10117	6.2 ms	sf/99.68		0+			
				α/0.32	8.81				
^{257}Rf		257.1030	4.7 s	α/9.22	8.77				0.117
				EC/11	9.01				
				sf/<1.4	8.95				
					8.62				
^{258}Rf		258.1035	12. ms	sf/87		0+			
				α/13					
^{259}Rf		259.1056	3.4 s	α/9.09/93	8.77(2)/				
				sf/7	8.86/				
^{260}Rf		260.1064	20. ms	sf/		0+			
^{261m}Rf			1.3 m	α	8.28				
^{261}Rf		261.10877	5. s	α/60, sf/40	8.52/				
^{262}Rf		262.1099	2.1 s	sf/>99.2		0+			
^{263}Rf		263.1126	10. m	sf, α					
^{265}Rf		265.1167	~ 13 h	α					
^{267}Rf		267.122	~ 0.1 d	sf					
$_{105}$**Db**									
^{255}Db		255.1074	~ 1.5 s	α,					
				sf/~ 20					
^{256}Db		256.1081	1.6 s	α/64	9.02/67				
				EC/35	8.89/11				
				sf/0.05	9.08/11				
					9.12/11				
^{257m}Db			0.8 s	α	9.16				
				sf/<13					
^{257}Db		257.1077	1.5 s	α/	8.97/33				
				sf/<6	9.07/38				
					9.12/5.5				
					8.94/9				
					9.02/9				
					8.89/5.5				
^{258}Db		258.1092	4.2 s	α/	9.30/				
				EC/5.3	9.17/				
				sf/<33	9.08/				
^{259}Db		259.1096	~ 0.51 s	sf/					
				α/	9.47/				
^{260m}Db			0.3 m						
^{260}Db		260.1113	1.5 s	α/	9.05/				
				sf/<9.6	9.08/				
					9.13/				
^{261}Db		261.1121	1.8 s	α/	8.93/				
				sf/<18					
^{262}Db		262.1141	0.5 m	sf/<33					

Elem. or Isot.	Natural Abundance (Atom %)	Atomic Mass or Weight	Half-life/ Resonance Width (MeV)	Decay Mode/ Energy (/MeV)	Particle Energy/ Intensity (MeV/%)	Spin ($h/2\pi$)	Nuclear Magnetic Mom. (nm)	Elect. Quadr. Mom. (b)	γ-Energy / Intensity (MeV/%)
				α/	8.45/				
					8.53/				
					8.67/				
^{263}Db		263.1150	~ 0.45 m	sf/57	8.36/				
				α/41	8.41/				
				EC/3					
^{267}Db		267.1224	1.2 h	sf					
^{268}Db		268.125	1.2 d	sf, EC					
$_{106}$Sg									
^{258}Sg		258.1132	~ 2.9 ms	sf		0+			
				α/<20					
^{259}Sg		259.1145	0.5 s	α/	9.62				
				sf/<20	9.35				
					9.03				
^{260}Sg		260.11442	4. ms	α/50	9.76	0+			
				sf/50	9.72				
					9.81				
^{261}Sg		261.1161	0.3 s	α, sf/<10	9.56				
^{262}Sg		262.1164	0.007 s	sf		0+			
				α/<22					
^{263m}Sg			0.3 s	α	9.2				
^{263}Sg		263.1183	0.8 s	α	9.06				
				sf/<30	9.25				
^{265}Sg		265.1211	8. s	α/>65	8.84/46				
				sf/<35	8.76/23				
					8.94/23				
					8.69/8				
^{266}Sg		266.1221	~ 21. s	α/	8.77/66	0+			
				sf/<82	8.52/33				
^{271}Sg		271.133	~ 0.04 h	α/50	8.53				
				sf/50					
$_{107}$Bh									
^{260}Bh		260.122		α					
^{261}Bh		261.1217	12. ms	α/, sf <10	10.40				
					10.10				
					10.03				
^{262m}Bh			8. ms	α/	10.37				
				sf/<12	10.24				
^{262}Bh		262.1229	0.10 s	α/	10.06				
				sf/<12	9.91				
					9.74				
^{264}Bh		264.1246	1.0 s	α/	9.3 – 9.8				
				sf/					
^{265}Bh		265.1252	0.9 s	α	9.24				
^{266}Bh		266.1269	~ 2 s	α	9.08				
^{267}Bh		267.1277	~ 17 s	α	8.83				
$_{108}$Hs									
^{263}Hs		263.1286		α/					
^{264}Hs		264.12839	~ 0.08 ms	α/, sf/~ 50	11.0	0+			
^{265m}Hs			~ 0.75 ms	α	10.57/63				
					10.73				
					10.52				
					10.34				
^{265}Hs		265.1301	2.0 ms	α/	10.30/90				
				sf/<1	10.43				
					10.37				

Elem. or Isot.	Natural Abundance (Atom %)	Atomic Mass or Weight	Half-life/ Resonance Width (MeV)	Decay Mode/ Energy (/MeV)	Particle Energy/ Intensity (MeV/%)	Spin ($h/2\pi$)	Nuclear Magnetic Mom. (nm)	Elect. Quadr. Mom. (b)	γ-Energy / Intensity (MeV/%)
					10.25				
^{266}Hs		266.1301	~ 2.3 ms	α	10.2	0+			
^{267m}Hs			~ 0.8 s						
^{267}Hs		267.1318	0.05 s	α/>88	9.88				
					9.83				
					9.75				
^{269}Hs		269.1341	~ 14 s	α	9.23				
					9.18				
^{270}Hs		270.1347	~ 3.6 s	α	9.16	0+			
^{275}Hs		275.146	~ 0.15 s	α	9.3				
^{277}Hs		277.150	~ 11 m	sf					
$_{109}$**Mt**									
^{266m}Mt			~ 1.2 ms	α	10.46–10.81				
^{266}Mt		266.1373	~ 0.7 ms	α	10.48–11.31				
^{267}Mt		267.137	19 ms	α					
^{268}Mt		268.1387	~ 0.03 s	α/>68	10.3 - 10.8				
^{270}Mt		270.141	5 ms	α	10.0				
^{275}Mt		275.149	0.01 s	α	10.3				
^{276}Mt		276.151	~ 0.7 s	α	9.7				
$_{110}$**Ds**									
^{267}Ds		267.1443	~ 3 μs	α/>32	11.6				
^{269}Ds		269.1451	0.17 ms	α/>75	11.11				
^{270m}Ds			~ 6 ms	α	10.95				
					11.15				
					12.15				
^{270}Ds		270.1447	0.1 ms	α	11.03	0+			
^{271m}Ds			0.07 s	α	9.9				
^{271}Ds		271.1461	1.6 ms	α	10.8				
^{273m}Ds			0.076 ms	α	11.8				
^{273}Ds		273.1489	118 ms	α/	9.73				
^{279}Ds		279.159	0.18 s	sf/90					
				α/10	9.7				
^{280}Ds		280.160	~ 7.6 s	sf/		0+			
^{281}Ds		281.162	10. s	sf					
^{282}Ds			0.5 ms	sf					
$_{111}$**Rg**									
^{272}Rg		272.1536	~ 2 ms	α/>68	10.82				
^{274}Rg		274.156	~ 65 ms	α	11.2				
^{279}Rg		279.162	~ 0.17 s	α	10.4				
^{280}Rg		280.164	~ 3.6 s	α	~ 9.75				
$_{112}$**112**									
^{277}112		277.1639	~ 0.24 ms	α	11.45				
					11.65				
^{283}112		283.172	~ 4. s	sf/< 10					
				α/~ 100	9.5				
^{284}112		284.172	0.10 s	sf					
^{285}112		285.174	~ 34. s	α	9.16				
$_{113}$**113**									
^{278}113			0.24 ms	α	11.7				
^{283}113		283.176	~ 0.1 s	α	10.1				
^{284}113		284.178	~ 0.5 s	α	10.0				

Elem. or Isot.	Natural Abundance (Atom %)	Atomic Mass or Weight	Half-life/ Resonance Width (MeV)	Decay Mode/ Energy (/MeV)	Particle Energy/ Intensity (MeV/%)	Spin ($h/2 \pi$)	Nuclear Magnetic Mom. (nm)	Elect. Quadr. Mom. (b)	γ-Energy / Intensity (MeV/%)
$_{114}$**114**									
^{286}114		286.184	0.16 s	α/40	10.2				
				sf/60					
^{287}114		287.186	0.5 s	α	10.0				
^{288}114		288.186	0.8 s	α	9.95				
^{289}114		289.187	~ 2.7 s	α	9.82				
$_{115}$**115**									
^{287}115		287.191	~ 0.03 s	α	10.6				
^{288}115		288.192	~ 0.09 s	α	10.5				
$_{116}$**116**									
^{290}116		290.199	~ 15 ms	α	10.9				
^{291}116		291.200	~ 6. ms	α	~ 10.74				
^{292}116		292.200	~ 18. ms	α	~ 10.66				
^{293}116			~ 0.05 s	α	10.5				
$_{118}$**118**									
^{294}118			~ 2.0 ms	α	11.7				

NEUTRON SCATTERING AND ABSORPTION PROPERTIES

Norman E. Holden

This table presents an evaluated set of values for experimental quantities that characterize the properties for scattering and absorption of neutrons. The neutron cross section is given for room temperature neutrons, 20.43°C, corresponding to a thermal neutron energy of 0.0253 electron volts (eV) or a neutron velocity 2200 meters/second. The neutron resonance integral is defined over the energy range from 0.5 eV to 0.1×10⁶ eV, or 0.1 MeV.

Bound neutron scattering lengths and neutron cross sections averaged over a Maxwellian spectrum at 30 keV for astrophysical applications are also presented. A list of the major references used is given below. The literature cutoff date is January 2003. Uncertainties are given in parentheses. Parentheses with two or more numbers indicate values to the excited state(s) and to the ground state of the product nucleus.

Table Layout

Column Number	Column Title	Description
1	Isotope/Element	For elements, atomic number and chemical symbol are listed. For nuclides, mass number and chemical symbol are listed. Isomers are indicated by the addition of m, m1, or m2.
2	Isotopic Abundance	in atom percent
3	Half-life	Half-life in decimal notation. μs = microsecond; ms = millisecond; s = second; m = minute; h = hour; d = day; y = year.
4	Thermal Neutron Cross Sections	Cross sections for neutron capture reactions in units of barns (10^{-24} cm^2) or millibarns (mb). Proton, alpha production and fission reactions are designated by σ_p, σ_α, σ_f, respectively. Separate values are listed for isomeric production.
5	Neutron Resonance Integrals	Resonance integrals for neutron capture reactions in barns (10^{-24} cm^2) or millibarns (mb). Proton, alpha production and fission reactions are designated by R.I.$_p$, R.I.$_\alpha$, R.I.$_f$, respectively. Separate values are listed for isomeric production.
6	Neutron Scattering Lengths	Bound coherent scattering lengths for neutron scattering reactions in units of femtometers (fm), which is equal to fermis (10^{-13} cm).
7	Maxwellian Averaged Cross Section	Astrophysical Cross Sections, averaged over a stellar neutron maxwellian spectrum characterized by a thermal energy of 30 keV, expressed in barns (10^{-24} cm^2), millibarns (mb) or microbarns (μb).

General Nuclear Data References

The following references represent the major sources of the nuclear data presented:

Mughabghab, S.F., Divadeenam, M., Holden, N.E.; Neutron Cross Sections, Vol. 1 Neutron Resonance Parameters and Thermal Cross Sections, Part A, Z = 1-60. Academic Press Inc., New York, New York (1981); Mughabghab, S.F.; Part B, Z = 61-100. Academic Press Inc., Orlando, Florida (1984).

Holden, N.E.; Fifty Years with Nuclear Fission Conference, Wash., D.C., Gaithersburg, Md. April 26-29, 1989, p. 946. American Nuclear Society, LaGrange Park, Illinois (1989).

Tuli, J.K.; Nuclear Wallet Cards, Brookhaven National Laboratory (Jan. 2000).

Holden, N.E.; Half-lives of Selected Nuclides, Pure & Applied Chemistry 62, 941 (1990).

Holden, N.E., Hoffman, D.C.; Spontaneous Fission Half-lives for Ground State Nuclides, Pure & Applied Chemistry 72, 1525 (2000).

Koester, L., Rauch, H., Seymann, E.; Neutron Scattering Lengths: A Survey of Experimental Data and Methods, Atomic Data Nuclear Data Tables 49, 65 (1991).

Sears, V.F.; Neutron Scattering Lengths and Cross Sections, Neutron News 3, (3), 26 (1992).

Bao, Z.Y., Beer, H., Käppeler, F., Voss, F., Wisshak, K., Raucher,T.; Neutron Cross Sections for Nucleo-synthesis Studies, Atomic Data Nuclear Data Tables 76, 70 (2000).

Elem. or Isot.	Natural Abundance (%)	Half-Life	Thermal Neut. Cross-Section (barns)	Resonance Integral (barns)	Coh. Scat. Length (fm)	σ (30 keV) Maxw. Avg. (barns)
$_1$H			0.332(2)	0.149(1)	-3.739(1)	
^{1}H	99.9885(70)	>2.8×10²³ y	0.332(2)	0.149(1)	- 3.741(1)	0.25(2) mb*
^{2}H	0.0115(70)		0.51(1)mb	0.23(2) mb	6.671(4)	2.1(4) μb
^{3}H		12.33 y	< 6. μb		4.79(3)	
$_2$He			< 0.05		3.26(3)	
^{3}He	0.000134(3)		σ_p = 5.33(1)×10³	RI$_p$ = 2.39(1)×10³	5.74(7)	
			0.05(1) mb			8.(1) μb*
^{4}He	99.999867(3)				3.26(3)	
$_3$Li			71.(2)	32.(1)	- 1.90(2)	
^{6}Li	7.59(4)		σ_t = 9.4(1)×10²	RI$_t$ = 422.(4)	2.0(1)	σ_t ≈ 1.

*Extrapolated value.

Elem. or Isot.	Natural Abundance (%)	Half-Life	Thermal Neut. Cross-Section (barns)	Resonance Integral (barns)	Coh. Scat. Length (fm)	σ (30 keV) Maxw. Avg. (barns)
			39.(5) mb	17.(2) mb		0.06(1) mb*
^{7}Li	92.41(4)		45.(5) mb	20.(2) mb	- 2.22(2)	42.(3) μb
^{8}Li		0.84 s				< ≈ 5.5 μb
$_{4}$Be			8.8(4) mb	3.9(2) mb	7.79(1)	
^{7}Be		53.28 d	$\sigma_p = 3.9(1)\times10^4$	$RI_p = 1.75(5)\times10^4$		$\sigma_p = 16(4)^*$
			$\sigma_\alpha \approx 0.1$			
^{9}Be	100.		8.8(4) mb	3.9(2) mb	7.79(1)	
^{10}Be		1.52×10^6 y	<1. mb			
$_{5}$B			$7.6(1)\times10^2$	$3.4(1)\times10^2$	5.30(4)	
^{10}B	19.9(7)		$\sigma_\alpha = 38.4(1)\times10^2$	$RI_\alpha = 17.3(1)\times10^2$	- 0.1(3)	
			0.3(1)	0.13(4)		
			$\sigma_p = 7.(1)$ mb			
			$\sigma_t = 8.(2)$ mb			
^{11}B	80.1(7)		5.(3) mb	2.(1) mb	6.65(4)	
$_{6}$C			3.5(1) mb	1.6(1) mb	6.646(1)	
^{12}C	98.93(8)		3.5(1) mb	1.6(1) mb	6.651(2)	16.(1) μb*
^{13}C	1.07(8)		1.4(1) mb	1.7(2) mb	6.19(9)	0.021(4) mb
^{14}C		5715. y	<1.4 μb			3.(1) μb*
$_{7}$N			2.00(6)	0.90(3)	9.36(2)	
^{14}N	99.636(20)		$\sigma_p = 1.93(5)$	$RI_p = 0.87(3)$	9.37(2)	$\sigma_p = 1.8(2)$ mb*
			0.080(1)	0.034(1)		0.04(1) mb
^{15}N	0.364(20)		0.04(1) mb	0.11(3) mb	6.44(3)	6.(1) μb*
$_{8}$O			0.29(1) mb	0.40(4) mb	5.805(4)	
^{16}O	99.757(16)		0.19(1) mb	0.36(4) mb	5.805(5)	34.(4) μb
^{17}O	0.038(1)		$\sigma_\alpha = 0.257(10)$	0.11(1)	5.8(2)	$\sigma_\alpha = 3.9(5)$ mb*
			0.54(7) mb	0.39(5) mb		
^{18}O	0.205(14)		0.16(1) mb	0.81(4) mb	5.84(7)	9.(1) μb*
$_{9}$F			9.5(1) mb	21.(3) mb	5.65(1)	6.(1) mb
^{19}F	100.		9.5(1) mb	21.(3) mb	5.65(1)	6.(1) mb
$_{10}$Ne			42.(5) mb	19.(3) mb	4.566(6)	
^{20}Ne	90.48(3)		39.(5) mb	18.(3) mb	4.631(6)	0.12(1) mb
^{21}Ne	0.27(1)		0.7(1)	0.31(5)	6.7(2)	≈ 1.5 mb
			$\sigma_\alpha = 0.18(9)$ mb			
^{22}Ne	9.25(3)		51.(5) mb	23.(3) mb	3.87(1)	58.(4) μb*
$_{11}$Na			0.53(2)	0.32(2)	3.63(2)	2.1(2) mb
^{22}Na		2.605 y	$\sigma_p = 2.8(3)\times10^4$	$RI_p< 2.\times10^5$		
			$\sigma_\alpha = 2.6(4)\times10^2$	$RI_\alpha = 1.2(2) \times 10^2$		
^{23}Na	100.		$\sigma_m = 0.43(3)$	$RI_m = 0.30(6)$	3.63(2)	2.1(2) mb
$_{12}$Mg			66.(6) mb	38.(5) mb	5.375(4)	
^{24}Mg	78.99(4)		0.053(6)	32.(4) mb	5.7(2)	3.3(4) mb
^{25}Mg	10.00(1)		0.20(1)	98.(15) mb	3.6(2)	6.4(4) mb
^{26}Mg	11.01(3)		0.038(1)	25.(2) mb	4.9(2)	0.13(1) mb*
^{27}Mg		9.45 m	0.07(2)	0.03(1)		
$_{13}$Al			0.230(2)	0.17(1)	3.45(1)	
^{26}Al		7.1×10^5 y	$\sigma_p = 1.97(10)$			0.14(2)
			$\sigma_\alpha = 0.34(1)$			
^{27}Al	100.		0.230(2)	0.17(1)	3.45(1)	2.9(3) mb
$_{14}$Si			0.166(9)	0.12(2)	4.15(1)	
^{28}Si	92.223(19)		0.17(1)	0.11(2)	4.11(1)	2.9(3) mb
^{29}Si	4.685(8)		0:12(1)	0.08(2)	4.7(1)	7.9(9) mb
^{30}Si	3.092(11)		0.107(3)	0.62(6)	4.61(1)	3.2(3) mb*
^{31}Si		2.62 h	73.(6) mb	33.(3) mb		
^{32}Si		1.6×10^2 y	< 0.5			
$_{15}$P			0.17(1)	0.08(1)	5.13(1)	
^{31}P	100.		0.17(1)	0.08(1)	5.13(1)	1.7(1) mb
$_{16}$S			0.54(2)	0.24(2)	2.847(1)	
^{32}S	94.93(31)		0.55(5)	0.25(2)	2.804(2)	4.1(2) mb
			$\sigma_\alpha < 0.5$ mb			

*Extrapolated value.

Elem. or Isot.	Natural Abundance (%)	Half-Life	Thermal Neut. Cross-Section (barns)	Resonance Integral (barns)	Coh. Scat. Length (fm)	σ (30 keV) Maxw. Avg. (barns)
^{33}S	0.76(2)		0.46(3)	0.21(2)	4.7(2)	7.4(15) mb
			$\sigma_\alpha = 0.12(1)$	$RI_\alpha = 0.05(1)$		$\sigma_\alpha = 0.18(1)$
			$\sigma_p = 2.$ mb			
^{34}S	4.29(28)		0.25(1)	0.13	3.48(3)	0.23(1) mb
^{36}S	0.02(1)		0.24(2)	0.26(3)		0.17(1) mb*
$_{17}$Cl			33.6(3)	15.(2)	9.58(1)	
^{35}Cl	75.78(4)		43.7(4)	20.(2)	11.7(1)	9.4(3) mb
			$\sigma_p = 0.44(1)$	$RI_p = 0.2$		$\sigma_p = 1.7(2)$ mb*
			$\sigma_\alpha \approx 0.08$ mb			
^{36}Cl		3.01×10^5 y	$\sigma_p = 46.(2)$ mb	$RI_p = 0.02$		$\sigma_p = 91.(8)$ mb
			<10.			
			$\sigma_\alpha = 0.59(7)$ mb			$\sigma_\alpha = 0.9(2)$ mb
^{37}Cl	24.22(4)		(0.05 + 0.38)	(0.04+0.26)	3.1(1)	2.0(2) mb
$_{18}$Ar			0.66(3)	0.42(5)	1.91(1)	
^{36}Ar	0.3365(30)		5.(1)	2.(1)	24.9(1)	
			$\sigma_\alpha = 5.4(3)$ mb			
			$\sigma_p < 1.5$ mb			
^{37}Ar		35.0 d	$\sigma_\alpha = 1.08(8)\times10^3$	$RI_\alpha = 900.$		$\sigma_\alpha \approx 1.3$
			$\sigma_p = 37.(4)$	$RI_p = 31.$		$\sigma_p \approx 0.04$
^{38}Ar	0.0632(5)		0.8(2)	0.4(1)	3.5(35)	
^{39}Ar		268. y	$6.(2)\times10^2$			
			$\sigma_\alpha <0.29$			
^{40}Ar	99.6003(30)		0.64(3)	0.41(5)	1.83(1)	2.5(3) mb
^{41}Ar		1.82 h	0.5(1)	0.2(1)		
$_{19}$K			2.1(1)	1.0(1)	3.67(2)	
^{39}K	93.2581(44)		2.1(2)	0.9(1)	3.74(2)	11.8(4) mb
			$\sigma_\alpha = 4.3(5)$ mb			
			$\sigma_p < 0.05$ mb			
^{40}K	0.0117(1)	1.26×10^9 y	30.(8)	13.(4)		$\sigma_p = 7.(1)$ mb
			$\sigma_p = 4.4(4)$	2.0(2)		$\sigma_\alpha = 40.(6)$ mb
			$\sigma_\alpha = 0.42(8)$			
^{41}K	6.7302(44)		1.46(3)	1.4(2)	2.69(8)	22.(1) mb
$_{20}$Ca			0.43(2)	0.23(2)	4.70(2)	
^{40}Ca	96.941(156)		0.41(3)	0.22(4)	4.80(2)	6.7(7) mb
			$\sigma_\alpha = 0.13(4)$ mb			
^{41}Ca		1.02×10^5 y	$\approx 4.$			
			$\sigma_\alpha = 0.18(3)$			
			$\sigma_p = 7.(2)$ mb			
^{42}Ca	0.647(23)		0.65(10)	0.39(4)	3.4(1)	16.(2) mb
^{43}Ca	0.135(10)		6.(1)	3.9(2)	- 1.56(9)	51.(6) mb
^{44}Ca	2.086(110)		0.8(2)	0.56(1)	1.42(6)	9.(1) mb
^{45}Ca		162.7 d	$\approx 15.$			
^{46}Ca	0.004(3)	$>4\times10^{15}$ y	0.70(3)	0.9(1)	3.6(2)	5.3(5) mb*
^{48}Ca	0.187(21)	4.3×10^{19} y	1.0(1)	0.5(1)	0.39(9)	0.8(1) mb*
$_{21}$Sc			27.2(2)	12.(1)	12.3(1)	
^{45}Sc	100.		(10.+17.)	(5.6+6.4)	12.3(1)	69.(5) mb
^{46}Sc		83.81 d	8.(1)	3.6(5)		
$_{22}$Ti			6.1(1)	2.8(2)	- 3.438(2)	
^{44}Ti		60 y	1.1(2)			
			$\sigma_p<0.2$			
^{46}Ti	8.25(3)		0.6(2)	0.4(1)	4.93(6)	27.(3) mb
^{47}Ti	7.44(2)		1.6(2)	1.6(2)	3.63(1)	64.(8) mb
^{48}Ti	73.72(3)		7.9(9)	3.6(2)	- 6.09(2)	32.(5) mb
^{49}Ti	5.41(2)		1.9(5)	1.2(2)	1.04(5)	22.(2) mb
^{50}Ti	5.18(2)		0.179(3)	0.12(2)	6.18(8)	3.6(4) mb
$_{23}$V			5.0(2)	2.8(1)	- 0.382(1)	
^{50}V	0.250(4)	1.4×10^{17} y	21.(4)	50.(20)	7.6(6)	
			$\sigma_p = 0.7(4)$ mb			

*Extrapolated value.

Elem. or Isot.	Natural Abundance (%)	Half-Life	Thermal Neut. Cross-Section (barns)	Resonance Integral (barns)	Coh. Scat. Length (fm)	σ (30 keV) Maxw. Avg. (barns)
^{51}V	99.750(4)		4.9(1)	2.7(2)	- 0.402(2)	38.(4) mb
$_{24}$Cr			3.0(2)	1.7(1)	3.635(7)	
^{50}Cr	4.345(13)	>1.8×10^{17} y	15.(1)	8.(1)	- 4.5(1)	0.05(1)
^{51}Cr		27.70 d	< 10.			
^{52}Cr	83.789(18)		0.8(1)	0.6(2)	4.91(2)	8.8(4) mb
^{53}Cr	9.501(17)		18.(2)	9.(1)	- 4.2(1)	0.06(1)
^{54}Cr	2.365(7)		0.36(4)	0.25(5)	4.6(1)	7.(2) mb
$_{25}$Mn			13.3(1)	14.0(3)	- 3.75(2)	
^{53}Mn		3.7×10^6 y	70.(10)	32.(5)		
^{54}Mn		312.1 d	< 10.			
^{55}Mn	100.		13.3(1)	14.0(3)	- 3.75(2)	40.(3) mb
$_{26}$Fe			2.7(1)	1.4(2)	9.45(2)	
^{54}Fe	5.845(35)		2.3(2) σ_α = 10. μb	1.3(2) RI$_\alpha$ = 1.1(1) mb	4.2(1)	29.(2) mb
^{55}Fe		2.73 y	13.(2) σ_α = 0.01	6.(1)		
^{56}Fe	91.754(36)		2.8(3)	1.4(2)	9.93(3)	11.7(5) mb
^{57}Fe	2.119(10)		1.4(2)	0.8(4)	2.3(1)	40.(4) mb
^{58}Fe	0.282(4)		1.3(1)	1.3(2)	15.(7)	12.(1) mb
^{59}Fe		44.51 d	13.(3)	6.(1)		
$_{27}$Co			37.19(8)	74.(2)	2.49(2)	
^{58m}Co		9.1 h	1.4(1)×10^5	2.5(10)×10^5		
^{58}Co		70.88 d	1.9(2)×10^3	7.(1)×10^3		
^{59}Co	100.		(20.7+16.5)	(39.+35.)	2.49(2)	38.(4) mb
^{60m}Co		10.47 m	58.(3)	230.(50)		
^{60}Co		5.271 y	2.0(2)	4.3(10)		
$_{28}$Ni			4.5(2)	2.3(2)	10.3(1)	
^{58}Ni	68.0769(89)	>4×10^{19} y	4.6(4) σ_α < 0.03 mb	2.3(2)	14.4(1)	41.(2) mb
^{59}Ni		≈ 7.6×10^4 y	σ_{abs} = 92.(4) σ_α = 14.(2) σ_p = 2.(1)	RI$_{abs}$ = 1.4(1)×10^2		
^{60}Ni	26.2231(77)		2.9(3)	1.5(2)	2.8(1)	25.(1) mb
^{61}Ni	1.1399(6)		2.5(5) σ_α = 0.03 mb	1.5(4)	7.60(6)	82.(8) mb
^{62}Ni	3.6345(17)		15.(1)	6.8(3)	- 8.7(2)	13.(4) mb
^{63}Ni		100. y	20.(5)	9.(2)		
^{64}Ni	0.9256(9)		1.6(1)	1.2(2)	- 0.37(7)	9.(1) mb
^{65}Ni		2.517 h	22.(2)	10.(1)		
$_{29}$Cu			3.8(1)	4.1(4)	7.718(4)	
^{63}Cu	69.15(15)		4.5(2)	5.(1)	6.43(15)	0.09(1)
^{64}Cu		12.701 h	≈ 270.			
^{65}Cu	30.85(15)		2.17(3)	2.2(1)	10.61(19)	41.(5) mb
^{66}Cu		5.09 m	1.4(1)×10^2	60.(20)		
$_{30}$Zn			1.1(2)	2.8(4)	5.680(5)	
^{64}Zn	48.27(32)	>2.3×10^{18} y	0.74(5) σ_p < 12. μb σ_α = 11.(3) μb	1.4(3)	5.23(4)	59.(5) mb
^{65}Zn		243.8 d	66.(8) σ_α = 2.0(2)	30.(4)		
^{66}Zn	27.977(77)		0.9(3) σ_α < 0.02 mb	1.8(2)	5.98(5)	35.(3) mb
^{67}Zn	4.102(21)		6.9(1.4) σ_α = 0.4 mb	25.(5)	7.58(8)	0.15(2)
^{68}Zn	19.02(12)		(0.072 + 0.8) σ_α <0.02 mb	(0.2 + 2.9)	6.04(3)	19.(2) mb σ_m = 3.(1) mb
^{70}Zn	0.631(9)		(8.1+83.) mb	0.9(2)		0.02(1)
$_{31}$Ga			2.9(1)	22.(3)	7.288(2)	

*Extrapolated value.

Elem. or Isot.	Natural Abundance (%)	Half-Life	Thermal Neut. Cross-Section (barns)	Resonance Integral (barns)	Coh. Scat. Length (fm)	σ (30 keV) Maxw. Avg. (barns)
^{69}Ga	60.108(9)		1.68(7)	16.(2)	7.88(4)	0.14(1)
^{71}Ga	39.892(9)	>2.4×10^{26} y	4.7(2)	31.(3)	6.40(3)	0.12(1)
			$\sigma_m = 0.15(5)$			
$_{32}$Ge			2.2(1)	6.(2)	8.19(2)	
^{68}Ge		270.8 d	1.0(5)			
^{70}Ge	20.370(89)		(0.3 + 2.7)	2.3(1)	10.0(1)	88.(5) mb
^{72}Ge	27.380(60)		0.9(2)	0.8(3)	8.5(1)	0.07(2)
^{73}Ge	7.759(78)	>1.8×10^{23} y	15.(1)	66.(20)	5.02(4)	0.3(1)
^{74}Ge	36.656(80)		(0.14 + 0.28)	(0.4+0.5)	7.6(1)	53.(7) mb
^{76}Ge	7.835(81)	1.6×10^{21} y	(0.09 + 0.06)	(1.3+0.6)	8.2(15)	0.03(2)
$_{33}$As			4.0(4)	61.(5)	6.58(1)	
^{75}As	100.		4.0(4)	61.(5)	6.58(1)	0.57(4)
$_{34}$Se			12.(1)	14.(3)	7.970(9)	
^{74}Se	0.89(4)		50.(2)	520(50)	0.8(3)	0.2(1)
^{75}Se		119.78 d	3.3(10)×10^2			
^{76}Se	9.37(29)		(22. + 63.)	(9.+31.)	12.2(1)	0.16(1)
^{77}Se	7.63(16)		42.(4)	30.(5)	8.25(8)	0.3(1)
			$\sigma_\alpha = 0.97(3)$ μb			
^{78}Se	23.77(28)		$\sigma_m = 0.38(2)$	$RI_m = 4.3(4)$	8.24(9)	0.1
^{80}Se	49.61(41)		(0.05+0.54)	(0.15+0.85)	7.48(3)	42.(3) mb
^{82}Se	8.73(22)	≈ 1×10^{20} y	(39.+ 5.2) mb	39.(4) mb	6.34(8)	0.04(2)
$_{35}$Br			6.8(2)	92.(8)	6.79(2)	
^{76}Br		16.0 h	224.(42)			
^{79}Br	50.69(7)		(2.5+8.3)	(36.+96.)	6.79(7)	0.63(4)
						$\sigma_m = 0.08(1)$
^{81}Br	49.31(7)		(2.4+0.24)	51.(5)	6.78(7)	0.31(2)
$_{36}$Kr			24.(1)	39.(6)	7.81(2)	
^{78}Kr	0.353(3)	>2.3×10^{20} y	(0.17+6.)	20.(1)		(0.11+0.19)
^{80}Kr	2.286(10)		(4.6+7.)	57.(6)		(0.09+0.18)
^{82}Kr	11.593(3)		(14.+7.)	130.(13)		90.(6) mb
^{83}Kr	11.500(19)		183.(30)	183.(20)		0.24(2)
^{84}Kr	56.987(15)		$(\sigma_m+ \sigma_g) = 0.11$	2.4(3)		(16.+33.) mb
			$\sigma_m = 0.09$			
^{85}Kr		10.73 y	1.7(2)	1.8(10)		0.07(2)
^{86}Kr	17.279(41)		3.(2) mb	≈ 1. mb	8.1(3)	3.2(4) mb
$_{37}$Rb			0.39(4)	6.(3)	7.08(2)	
^{84}Rb		32.9 d	$\sigma_p = 12.(2)$			
^{85}Rb	72.17(2)		(0.06+0.38)	(0.7+7.)	7.0(1)	0.24(1)
^{86}Rb		18.65 d	<20.			
^{87}Rb	27.83(2)	4.88×10^{10} y	0.10(1)	2.3(4)	7.3(1)	16.(1) mb
^{88}Rb		17.7 m	1.2(3)	0.5(1)		
$_{38}$Sr			1.2(1)	10.(1)	7.02(2)	
^{84}Sr	0.56(1)		(0.6+0.2)	(9.+1.)		0.4(1)
^{86}Sr	9.86(1)		$\sigma_m = 0.81(4)$	$RI_m = 4.(1)$	5.68(5)	(48.+22.) mb
^{87}Sr	7.00(1)		16.(3)	118.(30)	7.41(7)	97.(5) mb
^{88}Sr	82.58(1)		5.8(4) mb	0.07(3)	7.16(6)	6.0(2) mb
^{89}Sr		50.52 d	0.42(4)	0.2		
^{90}Sr		29.1 y	10.(1) mb	0.10(2)		
$_{39}$Y			1.25(5)	1.0(1)	7.75(2)	
^{89}Y	100.		(0.001+1.25)	(0.006+1.0)	7.75(2)	19.(1) mb
^{90}Y		2.67 d	<6.5			
^{91}Y		58.5 d	1.4(3)	0.6(1)		
$_{40}$Zr			0.19(1)	0.95(9)	7.16(3)	
			σ_α <0.1 mb			
^{90}Zr	51.45(40)		≈ 0.014	0.2(1)	6.4(1)	21.(2) mb
^{91}Zr	11.22(5)		1.2(3)	5.(2)	8.8(1)	60.(8) mb
^{92}Zr	17.15(8)		0.2(1)	0.6(2)	7.5(2)	33.(4) mb
^{93}Zr		1.5×10^6 y	<4.	16.(5)		0.10(1)

*Extrapolated value.

Elem. or Isot.	Natural Abundance (%)	Half-Life	Thermal Neut. Cross-Section (barns)	Resonance Integral (barns)	Coh. Scat. Length (fm)	σ (30 keV) Maxw. Avg. (barns)
^{94}Zr	17.38(28)	>10^{17} y	0.049(6)	0.25(3)	8.3(2)	26.(1) mb
^{96}Zr	2.80(9)	>1.7×10^{18} y	0.020(3)	5.0(5)	5.5(1)	11.(1) mb
$_{41}$Nb			1.11(1)	8.5(6)	7.14(3)	
			σ_α <0.1 mb			
^{93}Nb	100.		1.1	(6.3+2.2)	7.14(3)	266.(5) mb
			σ_m = 0.86			
^{94}Nb		2.4×10^4 y	$(\sigma_m+ \sigma_g)$ = 15.(1)	126.(13)		
			σ_m = 0.6(1)			
^{95}Nb		34.97 d	<7.	<200.		
$_{42}$Mo			2.5(1)	26.(5)	6.72(2)	
			σ_α <0.1 mb			
^{92}Mo	14.77(31)	>3×10^{17} y	0.06	≈ 0.8	6.93(8)	0.07(1)
			σ_m = 0.2 μb			
^{94}Mo	9.226(99)		0.02	≈ 0.8	6.82(7)	0.10(2)
^{95}Mo	15.900(85)		13.4(3)	109.(5)	6.93(6)	0.29(1)
			σ_α = 30.(4) μb			
^{96}Mo	16.674(12)		0.5	17.(3)	6.22(6)	0.11(1)
^{97}Mo	9.560(50)		2.5(2)	14.(3)	7.26(8)	0.34(1)
			σ_α = 0.4(2) μb			
^{98}Mo	24.20(25)		0.14(1)	7.2(7)	6.60(7)	0.10(1)
^{100}Mo	9.67(20)	≈ 1×10^{19} y	0.19(1)	3.6(3)	6.75(7)	0.11(1)
$_{43}$Tc						
^{98}Tc		≈ 6.6×10^6 y	σ_m = 0.9(2)			
^{99}Tc		2.13×10^5 y	23.(2)	4.0(4)×10^2	6.8(3)	0.93(5)
$_{44}$Ru			2.6 (1)	48.(5)	7.03(3)	
^{96}Ru	5.54(14)	>3.1×10^{16} y	0.23(4)	7.(2)		0.21(1)
^{98}Ru	1.87(3)		< 8.			0.3(1)
^{99}Ru	12.76(14)		4.(1)	195.(20)		1.2(3)
^{100}Ru	12.60(7)		5.8(6)	11.(2)		0.21(1)
^{101}Ru	17.06(2)		5.(1)	1.1(3)×10^2		1.00(4)
			σ_α <0.15 μb			
^{102}Ru	31.55(14)		1.2(1)	4.3(5)		0.15(1)
^{103}Ru		39.27 d	<20.	≈ 30.		
^{104}Ru	18.62(27)		0.49(2)	6.(2)		0.15(1)
^{105}Ru		4.44 h	0.29(3)	0.13(1)		
^{106}Ru		1.020 y	0.15(4)	2.0(6)		
$_{45}$Rh			145.(2)	1.2(1)×10^3	5.88(4)	
^{103}Rh	100.		(11.+ 134.)	(0.08+1.1)×10^3	5.88(4)	0.81(1)
^{104m}Rh		4.36 m	800.(100)			
^{104}Rh		42.3 s	40.(30)			
^{105}Rh		35.4 h	1.1(3)×10^4	1.7(4)×10^4		
$_{46}$Pd			7.(1)	82.(8)	5.91(6)	
^{102}Pd	1.02(1)		3.2(10)	10.(2)		0.3(1)
^{104}Pd	11.14(8)			16.(2)		0.29(3)
^{105}Pd	22.33(8)		22.(2)	60.(20)	5.5(3)	1.20(6)
			σ_α = 0.5(2) μb			
^{106}Pd	27.33(3)		(0.013+0.28)	(0.2+5.5)	6.4(4)	0.25(3)
^{107}Pd		6.5×10^6 y	1.8(2)	108.(4)		1.34(6)
^{108}Pd	26.46(9)		(0.19+8.5)	(2.+240.)	4.1(3)	0.20(2)
^{110}Pd	11.72(9)		(0.033+0.7)	(0.7+8.)		0.15(2)
$_{47}$Ag			62.(1)	767.(60)	5.922(7)	
^{107}Ag	51.839(8)		(1.+35.)	(3.+105.)	7.56(1)	0.80(3)
^{109}Ag	48.161(8)		(4.1 + 87.)	(0.7+14.1)×10^2	4.17(1)	0.79(3)
^{110m}Ag		249.8 d	82.(11)	20.(4)		
^{111}Ag		7.47 d	3.(2)	105.(20)		
$_{48}$Cd			2.52(5)×10^3	73.(8)	4.87(5)	
^{106}Cd	1.25(6)	>2.6×10^{17} y	0.20(3)	4.(1)		0.30(2)
^{108}Cd	0.89(3)	>4.1×10^{17} y	1.	14.(3)	5.4(1)	0.20(1)

*Extrapolated value.

Elem. or Isot.	Natural Abundance (%)	Half-Life	Thermal Neut. Cross-Section (barns)	Resonance Integral (barns)	Coh. Scat. Length (fm)	σ (30 keV) Maxw. Avg. (barns)
^{109}Cd		462.0 d	$\approx 180.$ $\sigma_\alpha < 0.05$	$6.7(12)\times10^3$		
^{110}Cd	12.49(18)		(0.06+11.)	(6.+34.)	5.9(1)	(0.01+0.22)
^{111}Cd	12.80(12)		3.5(20)	51.(6)	6.5(1)	0.75(1)
^{112}Cd	24.13(21)		(0.012+2.2)	15.	6.4(1)	0.19(1)
^{113}Cd	12.22(12)	7.7×10^{15} y	$2.06(4)\times10^4$ $\sigma_\alpha < 1.$ μb	390.(40)	- 8.0(2)	0.67(1)
^{114}Cd	28.73(42)		(0.04+0.29)	16.(7)	7.5(1)	(0.01+0.12)
^{116}Cd	7.49(18)	3.8×10^{19} y	(26.+52.) mb	1.2	6.3(1)	(12.+47.) mb
$_{49}$In			197.(4)	$3.3(2)\times10^3$	4.07(2)	
^{113}In	4.29(5)		(3.1+5.0+3.9)	(220.+90.)	5.39(6)	(0.48+0.31)
^{115}In	95.71(5)	4.4×10^{14} y	(88.+73.+44.)	$(1.5+1.2+0.7)\times10^3$	4.01(2)	(0.69+0.02)
$_{50}$Sn			0.61(3)	8.(2)	6.225(2)	
^{112}Sn	0.97(1)		(0.15+0.40)	(8.+19.)		0.21(1)
^{113}Sn		115.1 d	$\approx 9.$	210.(50)		
^{114}Sn	0.66(1)		≈ 0.12	5.(1)	6.2(3)	134.(3) mb
^{115}Sn	0.34(1)		$\sigma_\alpha = 0.06$ mb	29.(6)		0.34(1)
^{116}Sn	14.54(9)		(0.006+0.14)	(0.5+11.)	5.93(5)	91.(2) mb
^{117}Sn	7.68(7)		1.1(1)	16.(5)	6.48(5)	319.(7) mb
^{118}Sn	24.22(9)		$\sigma_m = 4.$ mb	4.7(5)	6.07(5)	62.(1) mb
^{119}Sn	8.59(4)		2.(1)	2.9(5)	6.12(5)	0.18(1)
^{120}Sn	32.58(9)		(0.001+0.13)	1.2(3)	6.49(5)	(0.5+36.) mb
^{122}Sn	4.63(3)		(0.15+0.001)	0.81(4)	5.74(5)	(18.+4.) mb
^{124}Sn	5.79(5)	$>2.2\times10^{18}$ y	(0.13+0.004)	(8.0+0.08)	5.97(5)	12.(2) mb
$_{51}$Sb			5.2(2)	169.(20)	5.57(3)	
^{121}Sb	57.21(5)		(0.4+5.8)	(13.+192.)	5.71(6)	0.53(2)
^{123}Sb	42.79(5)		(0.02+0.04+4.0)	(1.+119.)	5.38(7)	0.30(1)
^{124}Sb		60.20 d	17.(3)	$\approx 8.$		
$_{52}$Te			4.2(1)	47.(3)	5.80(3)	
^{120}Te	0.09(1)		(1.+5.)	$\approx 1.$	5.3(5)	0.4(1)
^{122}Te	2.55(12)		(0.4+3.)	(5.+75.)	3.8(2)	295.(3) mb
^{123}Te	0.89(3)	$>5.3\times10^{16}$ y	370.(40) $\sigma_\alpha = 0.05$ mb	$4.5(3)\times10^3$	- 0.05	0.83(1)
^{124}Te	4.74(14)		(1.+6.)	(1.4+4.)	8.0(1)	155.(2) mb
^{125}Te	7.07(15)		1.1(2)	21.(4)	5.02(8)	431.(4) mb
^{126}Te	18.84(25)		(0.12+0.8)	(0.6+7.4)	5.56(7)	(28.+53.) mb
^{128}Te	31.74(8)	2.2×10^{24} y	(0.03+0.2)	(0.2+1.6)	5.89(7)	(3.+41.) mb
^{130}Te	34.08(62)	$8.\times10^{20}$ y	(0.01+0.19)	(0.03+0.3)	6.02(7)	(4.+11.) mb
$_{53}$I			6.2(1)	$1.5(1)\times10^2$	5.28(2)	
^{125}I		59.4 d	900.(100)	$1.4(2)\times10^4$		
^{127}I	100.		6.2(1)	$1.5(1)\times10^2$	5.28(2)	0.64(3)
^{128}I		25.00 m	22.(4)	$\approx 10.$		
^{129}I		1.7×10^7 y	(20.7+10.3)	36.(4)		0.44(2)
^{130}I		12.36 h	18.(3)	$\approx 8.$		
^{131}I		8.021 d	≈ 0.7	8.(4)		
$_{54}$Xe			25.(1)	263.(50)	4.92(3)	
^{124}Xe	0.0953(27)	$>10^{17}$ y	(28.+137.)	$(0.6+3.0)\times10^3$		(0.13+0.51)
^{125}Xe		17.1 h	$\sigma_\alpha < 0.03$			
^{126}Xe	0.0890(14)		(0.45+3.)	(8.+52.)		(0.04+0.32)
^{127}Xe		36.34 d	$\sigma_\alpha \leq 0.01$			
^{128}Xe	1.910(22)		$\sigma_m = 0.48$	$RI_m = 38.(10)$		0.26(1)
^{129}Xe	26.40(18)		22.(5)	250.(50)		0.62(2)
^{130}Xe	4.071(53)		$\sigma_m = 0.45$	$RI_m = 16.(4)$		0.132(3)
^{131}Xe	21.233(62)		90.(10)	$9.(1)\times10^2$		0.45(8)
^{132}Xe	26.9087(680)		(0.05+0.4)	(0.9+3.7)		(5.+60.) mb
^{133}Xe		5.243 d	190.(90)			
^{134}Xe	10.436(29)	$>1.1\times10^{16}$ y	(0.003 + 0.26)	0.40(4)		20.(2) mb
^{135}Xe		9.10 h	$2.65(11)\times10^6$	$7.6(5)\times10^3$		

*Extrapolated value.

Elem. or Isot.	Natural Abundance (%)	Half-Life	Thermal Neut. Cross-Section (barns)	Resonance Integral (barns)	Coh. Scat. Length (fm)	σ (30 keV) Maxw. Avg. (barns)
^{136}Xe	8.858(33)	>8×10^{20} y	0.26(2)	0.7(2)		0.9(1) mb
$_{55}$Cs			30.4(8)	422.(50)	5.42(2)	
^{132}Cs		6.48 d	σ_α < 0.15			
^{133}Cs	100.		(2.7+27.3)	(32.+360.)	5.42(2)	(0.04+0.47)
^{134}Cs		2.065 y	140.(10)	54.(9)		
^{135}Cs		2.3×10^6 y	8.3(3)	38.(3)		
^{137}Cs		30.2 y	(0.20+0.07)	0.36(7)		
$_{56}$Ba			1.3(2)	10.(2)	5.07(3)	
^{130}Ba	0.106(1)	2.2×10^{21} y	(1.+8.)	(25.+200.)	- 3.6(6)	0.76(11)
^{132}Ba	0.101(1)	1.3×10^{21} y	(0.84+9.7)	(4.7+24.)	7.8(3)	0.6(1)
^{133}Ba		10.53 y	4.(1)	85.(30)		
^{134}Ba	2.417(18)		(0.1+1.3)	(5.6+18.)	5.7(1)	0.18(1)
^{135}Ba	6.592(12)		(0.014+5.8)	(0.47+131.)	4.7(1)	0.46(2)
^{136}Ba	7.854(24)		(0.010+0.44)	(0.1+1.5)	4.91(8)	61.(2) mb
^{137}Ba	11.232(24)		5.(1)	4.(1)	6.8(1)	76.(3) mb
^{138}Ba	71.698(42)		0.41(2)	0.4(1)	4.84(8)	4.0(2) mb
^{139}Ba		1.396 h	5.(1)	2.2(5)		
^{140}Ba		12.75 d	1.6(3)	14.(1)		
$_{57}$La			9.2(2)	12.(1)	8.24(4)	
^{138}La	0.090(1)	1.06×10^{11} y	57.(6)	4.1(9)×10^2		
^{139}La	99.910(1)		9.2(2)	12.(1)	8.24(4)	38.(3) mb
^{140}La		1.678 d	2.7(3)	69.(4)		
$_{58}$Ce			0.64(4)	0.71(6)	4.84(2)	
^{136}Ce	0.185(2)		(1.0+6.5)	58.(12)	5.80(9)	(0.028+0.3)
^{138}Ce	0.251(2)		(0.025+1.0)	(1.5+5.2)	6.70(9)	179.(5) mb
^{140}Ce	88.450(51)		0.58(4)	0.50(5)	4.84(9)	11.0(4) mb
^{141}Ce		32.50 d	29.(3)	13.(2)		
^{142}Ce	11.114(51)	>1.6×10^{17} y	0.97(3)	1.3(3)	4.75(9)	28.(1) mb
^{143}Ce		1.38 d	6.1(7)	2.7(3)		
^{144}Ce		284.6 d	1.0(1)	2.6(3)		
$_{59}$Pr			11.5(4)	14.(3)	4.58(5)	
^{141}Pr	100.		(4.+7.5)	14.(3)	4.58(5)	111.(2) mb
^{142}Pr		19.12 h	20.(3)	9.(1)		
^{143}Pr		13.57 d	90.(10)	190.(25)		
$_{60}$Nd			51.(2)	49.(5)	7.69(5)	
^{142}Nd	27.2(5)		19.(1)	34.(11)	7.7(3)	35.(1) mb
^{143}Nd	12.2(2)		330.(10)	128.(30)		0.24(1)
			σ_α = 17. mb			
^{144}Nd	23.8(3)	2.1×10^{15} y	3.6(3)	3.9(5)	2.8(3)	81.(2) mb
^{145}Nd	8.3(1)		47.(6)	260.(40)		0.42(1)
			σ_α = 12. μb			
^{146}Nd	17.2(3)		1.5(2)	3.0(4)	8.7(2)	91.(1) mb
^{147}Nd		10.98 d	440.(150)	200.		
^{148}Nd	5.7(1)		2.4(1)	13.(2)	5.7(3)	147.(2) mb
^{150}Nd	5.6(2)	≈ 1×10^{19} y	1.0(1)	14.(2)	5.3(2)	0.16(1)
$_{61}$Pm						
^{146}Pm		5.53 y	8.4(1.7)×10^3			
^{147}Pm		2.623 y	(84.+96.)	(1000.+1280.)	12.6(4)	2.(1)
^{148m}Pm		41.3 d	10600.(800)			
^{148}Pm		5.37 d	≈ 10^3	2.6(2.4)×10^3		
^{149}Pm		2.212 d	1400.(200)			
^{151}Pm		1.183 d	≈ 150.			
$_{62}$Sm			5.6(1)×10^3	1.4(2)×10^3		
^{144}Sm	3.07(7)		1.6(1)	2.4(3)		92.(6) mb
^{145}Sm		340. d	280.(20)	600.(90)		
^{147}Sm	14.99(18)	1.06×10^{11} y	56.(4), σ_α = 0.6 mb	710.(50)	14.(3)	0.97(1)
^{148}Sm	11.24(10)	7×10^{15} y	2.4(6)	27.(14)		241.(2) mb
^{149}Sm	13.82(7)	10^{16} y	4.01(6)×10^4, σ_α = 31. mb	3.1(5)×10^3		1.82(2)

*Extrapolated value.

Elem. or Isot.	Natural Abundance (%)	Half-Life	Thermal Neut. Cross-Section (barns)	Resonance Integral (barns)	Coh. Scat. Length (fm)	σ (30 keV) Maxw. Avg. (barns)
^{150}Sm	7.38(1)		102.(5)	290.(30)	14.(3)	422.(4) mb
^{151}Sm		90. y	$1.52(3)\times10^4$	3520.(60)		2.(1)
^{152}Sm	26.75(16)		206.(15)	$3.0(3)\times10^3$	- 5.0(6)	473.(4) mb
^{153}Sm		1.929 d	420.(180)			
^{154}Sm	22.75(29)		7.5(3)	32.(6)	9.(1)	0.21(1)
$_{63}$Eu			4570.(100)	$3.8(5)\times10^3$	5.3(3)	
^{151}Eu	47.81(6)		(4.+3150.+6000.)	$(2.+4.)\times10^3$		(1.6+2.2)
			$\sigma_\alpha = 8.7(3)$ μb			
^{152m1}Eu		9.30 h	$6.8(15)\times10^4$	$< 10^5$		
^{152}Eu		13.5 y	$1.1(2)\times10^4$	$1.6(2)\times10^3$		5.(2)
^{153}Eu	52.19(6)		300.(20), σ_α <1. μb	$1.8(4)\times10^3$	8.2(1)	2.8(1)
^{154}Eu		8.59 y	$1.5(3)\times10^3$	$1.6(2)\times10^3$		4.4(7)
^{155}Eu		4.76 y	$3.9(2)\times10^3$	$1.6(2)\times10^4$		1.3(1)
$_{64}$Gd			$48.8(6)\times10^3$	400.(10)	9.5(2)	
^{148}Gd		75. y	$1.40(14)\times10^4$			
^{152}Gd	0.20(1)	1.1×10^{14} y	700.(200), σ_α <7. mb	700.(200)		1.05(2)
^{153}Gd		240. d	$2.(1)\times10^4$, $\sigma_\alpha = 0.03$			
^{154}Gd	2.18(3)		(0.035+60.)	230.(50)		1.03(1)
^{155}Gd	14.80(12)		$61.(1)\times10^3$, $\sigma_\alpha = .08$ mb	1540.(100)		2.65(3)
^{156}Gd	20.47(9)		≈ 2.0	104.(15)	6.3(4)	615.(5) mb
^{157}Gd	15.65(2)		$2.54(3)\times10^5$, σ_α <0.05	800.(100)		1.37(2)
^{158}Gd	24.84(7)		2.3(3)	73.(7)	9.(2)	324.(3) mb
^{160}Gd	21.86(19)	$>1.9\times10^{19}$ y	1.5(7)	6.(1)	9.15(5)	0.15(2)
^{161}Gd		3.66 m	$2.0(6)\times10^4$			
$_{65}$Tb			23.2(5)	420.(50)	7.34(2)	
^{159}Tb	100.		23.2(5)	420.(50)	7.34(2)	1.6(2)
^{160}Tb		72.3 d	570.(110)			
$_{66}$Dy			$9.5(2)\times10^2$	$1.5(2)\times10^3$	16.9(3)	
^{156}Dy	0.056(3)		33.(3), σ_α < 9. mb	1000.(100)		1.6(2)
^{158}Dy	0.095(3)		43.(6), σ_α < 6. mb	120.(10)	6.1(5)	0.8(2)
^{159}Dy		144. d	$8.(2)\times10^3$			
^{160}Dy	2.39(18)		60.(10), σ_α < 0.3 mb	1100.(200)	6.7(4)	0.89(1)
^{161}Dy	18.889(42)		600.(50), σ_α < 1. μb	1100.(100)	10.3(4)	1.96(2)
^{162}Dy	25.475(36)		170.(20)	2755.(300)	- 1.4(5)	446.(4) mb
^{163}Dy	24.896(42)		120.(10), σ_α < 20. μb	1600.(400)	5.0(4)	1.11(1)
^{164}Dy	28.260(54)		$(1.7+1.0)\times10^3$	$(4.+2.)\times10^2$	49.4(2)	212.(3) mb
^{165m}Dy		1.26 m	$2.0(6)\times10^3$			
^{165}Dy		2.33 h	$3.5(3)\times10^3$	$2.2(3)\times10^4$		
$_{67}$Ho			61.(2)	670.(40)	8.01(8)	
^{163}Ho		4.57×10^3 y				(0.4+1.7)
^{165}Ho	100.		(3.1+58.), σ_α < 20. μb	(?+670.)	8.01(8)	(0.8+0.5)
^{166m}Ho		1.2×10^3 y	$3.1(8)\times10^3$	$10.(3)\times10^3$		
$_{68}$Er			$1.5(2)\times10^2$	730.(10)	7.79(2)	
^{162}Er	0.139(5)		19.(3), σ_α < 11. mb	480.(50)	8.8(2)	1.6(1)
^{164}Er	1.601(3)		13.(3), σ_α < 1.2 mb	105.(10)	8.2(2)	1.08(5)
^{166}Er	33.503(36)		(3.+14.), σ_α < 70. μb	96.(12)	10.6(2)	0.56(6)
^{167}Er	22.869(9)		$6.5(8)\times10^2$, $\sigma_\alpha = 3.$ μb	2970.(70)	3.0(3)	1.4(2)
^{168}Er	26.978(18)		2.3(3), $\sigma_\alpha = 0.09$ mb	37.(5)	7.4(4)	0.34(4)
^{170}Er	14.910(36)		8.(2)	26.(4)	9.6(5)	0.17(1)
^{171}Er		7.52 h	370.(40)	170.(20)		
$_{69}$Tm			108.(4)	$1.5(2)\times10^3$	7.07(3)	
^{169}Tm	100		(8.+100.)	$1.5(2)\times10^3$	7.07(3)	1.13(6)
^{170}Tm		128.6 d	100.(20)	460.(50)		
^{171}Tm		1.92 y	≈ 160.	118.(6)		
$_{70}$Yb			52.(10)	$1.7(2)\times10^2$	12.43(3)	
^{168}Yb	0.13(1)		$2.4(2)\times10^3$, σ_α < 0.1 mb	$2.0(5)\times10^4$	-4.07(2)	0.7(4)
^{169}Yb		32.02 d	$3.6(3)\times10^3$	5200.(500)		
^{170}Yb	3.04(15)		12.(2), σ_α < 10. μb	320.(30)	6.8(1)	0.77(1)

*Extrapolated value.

Elem. or Isot.	Natural Abundance (%)	Half-Life	Thermal Neut. Cross-Section (barns)	Resonance Integral (barns)	Coh. Scat. Length (fm)	σ (30 keV) Maxw. Avg. (barns)
^{171}Yb	14.28(57)		53.(5), σ_α < 1.5 μb	315.(30)	9.7(1)	1.21(1)
^{172}Yb	21.83(67)		≈ 1.3, σ_α < 1. μb	25.(3)	9.4(1)	0.34(1)
^{173}Yb	16.13(27)		16.(2), σ_α < 1. μb	380.(30)	9.56(7)	0.75(1)
^{174}Yb	31.83(92)		(46.+17.), σ_α < 0.02 mb	(13.+16.)	19.3(1)	151.(2) mb
^{176}Yb	12.76(41)		3.1(2), σ_α < 1. μb	8.(2)	8.7(1)	116.(2) mb
$_{71}$Lu			78.(7)	8.3(7)×10²	7.21(3)	
^{175}Lu	97.41(2)		(16.+8.)	(550.+270.)	7.24(3)	(1.04+0.11)
^{176}Lu	2.59(2)	3.73×10¹⁰ y	(2.+2100.)	(3.+930.)	6.1(2)	1.53(7)
^{177m}Lu		160.7 d	3.2(3)	1.4(2)		
^{177}Lu		6.65 d	1000.(300)			
$_{72}$Hf			106.(3)	19.7(5)×10²	7.8(1)	
^{174}Hf	0.16(1)	2.0×10¹⁵ y	600.(50)	400.(50)	11.(1)	0.8(2)
^{176}Hf	5.26(7)		23.(4)	700.(100)	6.6(2)	0.46(2)
^{177}Hf	18.60(9)		(1.+375.), σ_α < 20. μb	7170.(200)		1.5(1)
^{178m2}Hf		31. y	σ_{m2} = 45.(5)	RI_{m2} = 8(1)×10²		
^{178}Hf	27.28(7)		(54.+32.)	(0.9+1.0)×10³	5.9(2)	0.31(1)
^{179}Hf	13.62(2)		(0.43+46.)	(6.8+620.)	7.5(2)	(0.01+0.95)
^{180}Hf	35.08(16)		13.0(5), σ_α < 13. μb	32.(1)	13.2(3)	179.(5) mb
^{181}Hf		42.4 d	30.(25)			
$_{73}$Ta			20.(1)	650.(20.)	6.91(7)	
^{179}Ta		1.8 y	9.3(6)×10²	1.22(7)×10³		
^{180m}Ta	0.012(2)	> 1.2×10¹⁵ y	≈ 560.	1350.(100)		
^{181}Ta	99.988(2)		(0.012 + 20.), σ_α <1. μb	(0.4+650.)	6.91(7)	0.77(2)
^{182}Ta		114.43 d	8200.(600)	900.(90)		
$_{74}$W			18.(1)	3.6(3)×10²	4.86(2)	
^{180}W	0.12(1)	7.4×10¹⁶ y	≈ 4.	210.(30)		0.54(6)
^{182}W	26.50(16)	8.3×10¹⁸ y	20.(1)	600.(90)	6.97(4)	274.(8) mb
^{183}W	14.31(4)	1.9×10¹⁸ y	10.5(3)	340.(50)	6.53(4)	0.52(2)
^{184}W	30.64(2)	4.0×10¹⁸ y	(0.002 + 2.0)	15.(2)	7.48(6)	0.22(1)
^{185}W		74.8 d	≈ 3.3	300.(50)		
^{186}W	28.43(19)	6.5×10¹⁸ y	37.(2)	510.(50)	- 0.72(4)	176.(5) mb
^{187}W		23.9 h	70.(10)	2760.(550)		
^{188}W		69.78 h	12.(1)			
$_{75}$Re			90.(4)	8.4(2)×10²	9.2(3)	
^{185}Re	37.40(2)		(0.33+110.)	1700.(50)	9.0(3)	1.54(6)
^{187}Re	62.60(2)	4.2×10¹⁰ y	(2.+72.)	(9.+310.)	9.3(3)	1.16(6)
$_{76}$Os			17.(1)	1.5(1)×10²	10.7(2)	
^{184}Os	0.02(1)	>5.6×10¹³ y	3.3(3)×10³, σ_α <10. mb	1.4(1)×10³		0.4(2)
^{186}Os	1.59(3)	2.×10¹⁵ y	≈ 80., σ_α < 0.1 mb	3.8(9)×10²	12(2)	0.42(2)
^{187}Os	1.96(2)		2.(1)×10², σ_α < 0.1 mb	5.0(7)×10²		0.90(3)
^{188}Os	13.24(8)		≈ 5., σ_α < 30. μb	1.5(2)×10²	7.6(3)	0.40(2)
^{189}Os	16.15(5)		(0.00026+40.), σ_α<10. μb	(0.013+670.)	10.7(3)	1.17(5)
^{190}Os	26.26(2)		(9.+4.), σ_α < 20. μb	(22.+8.)	11.0(3)	0.30(5)
^{191}Os		15.4 d	3.8(6)×10²	1.7(3)×10²		
^{192}Os	40.78(19)		3.(1), σ_α < 10. μb	7.(1)	11.5(4)	0.31(5)
^{193}Os		30.5 h	2.5(5)×10²	1.1(2)×10²		
$_{77}$Ir			4.2(1)×10²	2.8(4)×10³	10.6(3)	
^{191}Ir	37.3(2)		(0.14+660.+260.)	(1.0+4.2)×10³		1.35(4)
^{192}Ir		73.83 d	1.4(3)×10³	4.8(7)×10³		
^{193}Ir	62.7(2)		(0.04+6.+109.)	1.4(2)×10³		0.99(7)
^{194}Ir		19.3 h	1.6(3)×10³	7.(2)×10²		
$_{78}$Pt			10.(1)	1.3(1)×10²	9.60(1)	
^{190}Pt	0.014(1)	4.5×10¹¹ y	1.5(1)×10², σ_α < 8. mb	70.(10)	9.(1)	0.7(2)
^{192}Pt	0.782(7)		(2.0+6.), σ_α < 0.2 mb	115.(20)	9.9(5)	0.6(1)
^{194}Pt	32.967(99)		(0.1+1.1), σ_α < 5. μb	(4.+?)	10.55(8)	(0.03+0.34)
^{195}Pt	33.832(10)		28.(1), σ_α < 5. μb	365.(50)	8.8(1)	0.9(2)
^{196}Pt	25.242(41)		(0.045+0.55)	7.(2)	9.89(8)	(0.01+0.19)
^{198}Pt	7.163(55)		(0.3+3.1)	(5.+53.)	7.8(1)	(3.+79.) mb

*Extrapolated value.

Elem. or Isot.	Natural Abundance (%)	Half-Life	Thermal Neut. Cross-Section (barns)	Resonance Integral (barns)	Coh. Scat. Length (fm)	σ (30 keV) Maxw. Avg. (barns)
^{199}Pt		30.8 m	≈ 15.	≈ 7.		
$_{79}$Au			98.7(1)	$1.55(3)\times10^3$	7.63(6)	
^{197}Au	100.		σ_{m+g} = 98.7(1)	RI_{m+g} = $1.55(3)\times10^3$	7.63(6)	582.(9) mb
			σ_m = 8.(2) mb	RI_m = 0.06(2)		
^{198}Au		2.695 d	$26.5(15)\times10^3$	≈ $4.\times10^4$		
^{199}Au		3.14 d	≈ 30.			
$_{80}$Hg			$3.7(1)\times10^2$	87.(5)	12.69(2)	
^{196}Hg	0.15(1)	$>2.5\times10^{18}$ y	(105.+3000.)	(53.+410.)	30.(1)	0.4(2)
^{198}Hg	9.97(8)		(0.017+2.)	(1.7+70.)		0.17(2)
^{199}Hg	16.87(10)		$2.1(2)\times10^3$	435(20)	16.9(4)	0.37(2)
^{200}Hg	23.10(16)		≈ 1.	2.1(5)		0.12(1)
^{201}Hg	13.18(8)		≈ 8.	30.(3)		0.26(1)
^{202}Hg	29.86(20)		4.9(5)	4.5(2)	11.(1)	74.(6) mb
^{204}Hg	6.87(4)		0.4(1)	0.8(2)		42.(4) mb
$_{81}$Tl			3.3(1)	12.5(8)	8.776(5)	
^{203}Tl	29.524(14)		11.(1), σ_α < 0.3 mb	41.(2)	7.0(2)	124.(8) mb
^{204}Tl		3.78 y	22.(2)	90.(20)		0.14(5)
^{205}Tl	70.476(14)		0.11(2)	0.6(2)	9.52(7)	54.(4) mb
$_{82}$Pb			0.172(2)	0.14(4)	9.402(2)	
^{204}Pb	1.4(1)		0.68(7)	2.0(2)	10.9(1)	90.(6) mb
^{205}Pb		1.51×10^7 y	≈ 5.	≈ 2.		0.06(1)
^{206}Pb	24.1(1)		0.027(1)	0.10(1)	9.23(5)	16.(1) mb
^{207}Pb	22.1(1)		0.61(3)	0.38(1)	9.28(2)	10.(1) mb
^{208}Pb	52.4(1)	$>2\times10^{19}$ y	0.23(1) mb, σ_α < 8. μb	2.0(2) mb	9.50(3)	0.36(4) mb
^{210}Pb		22.6 y	< 0.5			
$_{83}$Bi			0.034(1)	0.19(2)	8.532(2)	
^{209}Bi	100.		(11.+23.) mb, σ_α<0.3 μb	0.19(2)	8.532(2)	2.7(5) mb
^{210m}Bi		3.0×10^6 y	54.(4) mb	0.20(3)		
$_{84}$Po						
^{210}Po		138.4 d	σ_m<0.5 mb, σ_α < 2. mb			
			σ_g<30. mb, σ_f< 0.1			
$_{85}$At						
$_{86}$Rn						
^{220}Rn		55.6 s	<0.2			
^{222}Rn		3.823 d	0.74(5)			
$_{88}$Ra						
^{223}Ra		11.43 d	$1.3(2)\times10^2$, σ_f< 0.7			
^{224}Ra		3.66 d	12.0(5)			
^{226}Ra		1599. y	≈ 13., σ_f< 7. μb	280.(50)	10.(1)	
^{228}Ra		5.76 y	36.(5), σ_f< 2.			
$_{89}$Ac						
^{227}Ac		21.77 y	$8.8(7)\times10^2$, σ_f< 0.35 mb	$1.5(4)\times10^3$		
$_{90}$Th			7.4	85.(3)	10.31(3)	
^{227}Th		18.72 d	σ_f = $2.0(2)\times10^2$			
^{228}Th		1.913 y	$1.2(2)\times10^2$, σ_f<0.3	1014.(400)		
^{229}Th		7.9×10^3 y	≈ 60.	$1.0(2)\times10^3$		
			σ_f = 30.(3)	RI_f = 466.(75)		
^{230}Th		7.54×10^4 y	23.4(5)	$1.0(1)\times10^3$		
			σ_f < 0.5 mb			
^{232}Th	100.	1.40×10^{10} y	7.37(4)	85.(3)	10.31(3)	
			σ_f = 3.(1) μb			
			σ_α < 1. μb			
^{233}Th		22.3 m	$1.5(1)\times10^3$	$4.(1)\times10^2$		
			σ_f = 15.(2)			
^{234}Th		24.10 d	1.8(5)			
			σ_f < 0.01			

*Extrapolated value.

Elem. or Isot.	Natural Abundance (%)	Half-Life	Thermal Neut. Cross-Section (barns)	Resonance Integral (barns)	Coh. Scat. Length (fm)	σ (30 keV) Maxw. Avg. (barns)
$_{91}$Pa						
^{230}Pa		17.4 d	$1.5(3)\times10^3$			
^{231}Pa		3.25×10^4 y	$2.0(1)\times10^2$	750.(80)	9.1(3)	
			$\sigma_f = 20.(1)$ mb	$RI_f = 0.05(1)$		
^{232}Pa		1.31 d	$4.6(10)\times10^2$	300.(70)		
			$\sigma_f = 1.5(5)\times10^3$	$RI_f = 1.0(1)\times10^3$		
^{233}Pa		27.0 d	39.(2)	(460.+440.)		
			$\sigma_m = 20.(4)$			
			$\sigma_g = 19.(3)$			
			$\sigma_f < 0.1$			
$_{92}$U			$3.4(3)$; $\sigma_f = 4.2(1)$	$280.(20), RI_f = 2.0$	8.417(5)	
^{230}U		20.8 d	$\sigma_f \approx 25.$			
^{231}U		4.2 d	$\sigma_f \approx 250.$			
^{232}U		70. y	73.(2)	280.(15)		
			$\sigma_f = 74.(8)$	$RI_f = 350.(30)$		
^{233}U		1.592×10^5 y	47.(2)	137.(6)	10.1(2)	
			$\sigma_f = 5.3(1)\times10^2$	$RI_f = 760.(17)$		
			$\sigma_\alpha < 0.2$ mb			
^{234}U	0.0054(5)	2.455×10^5 y	96.(2)	660.(70)	12.(4)	
			$\sigma_f = 0.07(2)$	$RI_f = 6.5$		
^{235}U	0.7204(6)	7.04×10^8 y	95.(5)	144.(6)	10.47(4)	
			$\sigma_f = 586.(2)$	$RI_f = 275(5)$		
			$\sigma_\alpha < 0.1$ mb			
^{236}U		2.342×10^7 y	5.1(3)	360.(15)		
			$\sigma_f < 1.3$ mb	$RI_f = 4.38(50)$		
^{237}U		6.75 d	$\approx 10^2$	1200.(200)		
			$\sigma_f < 0.35$			
^{238}U	99.2742(10)	4.47×10^9 y	2.7(1)	277.(3)	8.402(5)	
			$\sigma_f \approx 3.$ μb	1.54(15) mb		
			$\sigma_\alpha = 1.4(5)$ μb			
^{239}U		23.5 m	22.(2)			
			$\sigma_f = 15.(3)$			
$_{93}$Np						
^{234}Np		4.4 d	$\sigma_f = 9.(3)\times10^2$			
^{235}Np		1.085 y	$1.6(1)\times10^2$			
^{236m}Np		22.5 h	$\sigma_f = 2.7(2)\times10^3$	$7.(4)\times10^2$		
^{236}Np		1.55×10^5 y	$\sigma_f = 3.0(2)\times10^3$	$1.35(30)\times10^3$		
^{237}Np		2.14×10^6 y	$1.7(1)\times10^2$	$6.5(3)\times10^2$	10.6(1)	
			$\sigma_f = 20.(1)$ mb	$RI_f = 4.7$		
^{238}Np		2.117 d	$\sigma_f = 2.6(3)\times10^3$	$1.4(3)\times10^3$		
^{239}Np		2.355 d	(32.+19.)			
			$\sigma_f < 1.$			
$_{94}$Pu						
^{236}Pu		2.87 y	$\sigma_f = 1.6(3)\times10^2$	1000.(60)		
^{237}Pu		45.7 d	$\sigma_f = 2.3(3)\times10^3$			
^{238}Pu		87.7 y	$5.1(2)\times10^2$	$1.6(2)\times10^2$	14.1(5)	
			$\sigma_f = 17.(1)$	$RI_f = 26.(2)$		
^{239}Pu		2.410 x 10^4 y	$2.7(1)\times10^2$	$2.0(2)\times10^2$	7.7(1)	
			$\sigma_f = 752.(3)$	$3.0(1)\times10^2$		
			$\sigma_\alpha \le 0.3$ mb			
^{240}Pu		6.56×10^3 y	$2.9(1)\times10^2$	$8.4(3)\times10^3$	3.5(1)	
			$\sigma_f \approx 59.$ mb	$RI_f = 3.2$		
^{241}Pu		14.4 y	$3.7(1)\times10^2$, $\sigma_\alpha < 0.2$ mb	$1.6(1)\times10^2$		
			$\sigma_f = 1.01(1)\times10^3$	$5.7(4)\times10^2$		
^{242}Pu		3.75 x 10^5 y	19.(1)	$1.1(1)\times10^3$	8.1(1)	
			$\sigma_f < 0.2$	$RI_f = 0.23$		
^{243}Pu		4.956 h	<100.			
			$\sigma_f = 2.0(2)\times10^2$			

*Extrapolated value.

Elem. or Isot.	Natural Abundance (%)	Half-Life	Thermal Neut. Cross-Section (barns)	Resonance Integral (barns)	Coh. Scat. Length (fm)	σ (30 keV) Maxw. Avg. (barns)
^{244}Pu		8.00×10^7 y	1.7(1)	41.(3)		
^{245}Pu		10.5 h	$1.5(3)\times10^2$	220.(40)		
$_{95}$Am						
^{241}Am		432.7 y	$(0.6+6.4)\times10^2$	$(1.+14.)\times10^2$		
			$\sigma_f = 3.15(10)$	14.(1)		
^{242m}Am		141. y	$1.7(4)\times10^3$	≈ 200.		
			$\sigma_f = 5.9(3)\times10^3$	$RI_f = 1.8(1)\times10^3$		
^{242}Am		16.02 h	$\sigma_f = 2.1(2)\times10^3$	$RI_f = < 300.$		
			$3.3(5)\times10^2$	≈ 1.5×10^2		
^{243}Am		7.37×10^3 y	(75.+5.)	$(17.1+1.0)\times10^2$	8.3(2)	
			$\sigma_f = 79.(2)$ mb	$RI_f = 0.056$		
^{244m}Am		≈ 26. m	$\sigma_f = 1.6(3)\times10^3$			
^{244}Am		10.1 h	$\sigma_f = 2.2(3)\times10^3$			
$_{96}$Cm						
^{242}Cm		162.8 d	≈ 20.	120.(50)		
			$\sigma_f ≈ 5.$			
^{243}Cm		29.1 y	$1.3(1)\times10^2$	214.(20)		
			$\sigma_f = 6.2(2)\times10^2$	$RI_f = 1.6(1)\times10^3$		
^{244}Cm		18.1 y	15.(1)	640.(50)	9.5(3)	
			$\sigma_f = 1.1(2)$	$RI_f = 10.8(8)$		
^{245}Cm		8.48×10^3 y	$3.5(2)\times10^2$	110.(10)		
			$\sigma_f = 2.1(1)\times10^3$	$RI_f = 8.(1)\times10^2$		
^{246}Cm		4.76×10^3 y	1.2(2)	120.(10)	9.3(2)	
			$\sigma_f = 0.16(7)$	13.(2)		
^{247}Cm		1.56×10^7 y	60.(30)	$5.(1)\times10^2$		
			$\sigma_f = 82.(5)$	$7.3(7)\times10^2$		
^{248}Cm		3.48×10^5 y	2.6(3)	270.(30)	7.7(2)	
			$\sigma_f = 0.36(7)$	13.(2)		
^{249}Cm		64.15 m	≈ 1.6			
^{250}Cm		≈ 9.7×10^3 y	≈ 80.			
$_{97}$Bk						
^{249}Bk		320. d	$7.(1)\times10^2$	$9.(1)\times10^2$		
			$\sigma_f ≈ 0.1$			
^{250}Bk		3.217 h	$\sigma_f = 1.0(2)\times10^3$			
$_{98}$Cf						
^{249}Cf		351. y	$5.0(3)\times10^2$	$7.7(4)\times10^2$		
			$\sigma_f = 1.7(1)\times10^3$	$RI_f = 2.1(3)\times10^3$		
^{250}Cf		13.1 y	$2.0(2)\times10^3$	$12.(2)\times10^3$		
			$\sigma_f = 110.(90)$	$RI_f = 160.(40)$		
^{251}Cf		9.0×10^2 y	$2.9(2)\times10^3$	$1.6(1)\times10^3$		
			$\sigma_f = 4.5(5)\times10^3$	$RI_f = 5.5(3)\times10^3$		
^{252}Cf		2.65 y	20.(2)	43.(3)		
			$\sigma_f = 32.(4)$	$RI_f = 1.1(3)\times10^2$		
^{253}Cf		17.8 d	18.(2)	8.(1)		
			$\sigma_f = 1.3(2)\times10^3$			
^{254}Cf		60.5 d	4.5(10)	2.		
$_{99}$Es						
^{253}Es		20.47 d	(180.+5.8)	$(37.5+1.1)\times10^2$		
^{254m}Es		1.64 d	$\sigma_f = 1.8(1)\times10^3$			
^{254}Es		276. d	28.(3)	18.(2)		
			$\sigma_f = 1.8(2)\times10^3$	$RI_f = 1.2(3)\times10^3$		
^{255}Es		40. d	≈ 55.			
$_{100}$Fm						
^{255}Fm		20.1 h	26.(3)	14.(2)		
			$\sigma_f = 3.3(2)\times10^3$			
^{257}Fm		100.5 d	$\sigma_f = 3.0(2)\times10^3$			

*Extrapolated value.

COSMIC RADIATION

A.G. Gregory and R.W. Clay

The Nature of Cosmic Rays

Primary cosmic radiation, in the form of high energy nuclear particles, electrons and photons from outside the solar system and from the Sun, continually bombards our atmosphere. Secondary radiation, resulting from the interaction of the primary cosmic rays with atmospheric gas, is present at sea-level and throughout the atmosphere.

The secondary radiation is collimated by absorption and scattering in the atmosphere and consists of a number of components associated with different particle species. High energy primary particles can produce large numbers of secondary particles forming an extensive air shower. Thus, a number of particles may then be detected simultaneously at sea-level.

Primary particle energies accessible in the vicinity of the earth range from $\sim 10^8$ eV to $\sim 10^{20}$ eV. At the lower energies, the limit is determined by the inability of charged particles to traverse the heliosphere to us through the outward-moving solar wind. The upper energy limit is set by the practicality of building detectors to record particles with the extremely low fluxes found at those energies (J.G. Wilson, 1976; O.C. Allkofer, 1975a).

Primary Cosmic Rays

Primary Particle Energy Spectrum

Figure 1 shows the spectrum of primary particle energies. This includes all particle species. In differential form it is roughly a power law of intensity versus energy with an index of ~ -3. There appears to be a knee (a steepening) at a little above 10^{15} eV and an ankle (a flattening) above $\sim 10^{18}$ eV. Figure 2 emphasizes the features in the spectrum at the highest energies through multiplying the flux with a strongly rising power law of energy. This figure should be used with caution as errors for the two axes are not now independent.

Data on the high energy cosmic ray spectrum are uncertain largely because of limited event statistics due to the very low flux which might best be measured in particles per square kilometer per century. The highest energy event recorded to 1995 had an energy of 3×10^{20} eV (D.J. Bird et al., 1993).

It is expected that the highest energy cosmic rays will interact with the 2.7 K cosmic microwave background through photoproduction or photodisintegration. These interactions will appreciably reduce the observed flux of cosmic rays with energies above 5×10^{19} eV if they travel further than ~ 150 million light years. This process is known as the Greisen-Zatsepin-Kuz'min (GZK) cut off (P. Sokolsky, 1989).

At energies below $\sim 10^{13}$ eV, solar system magnetic fields and plasma can modulate the primary component and Figure 3 shows the extent of this modulation between solar maximum and minimum (E. Juliusson, 1975; J. Linsley, 1981).

Primary Particle Energy Density

If the above spectrum is corrected for solar effects, the energy density above a particle energy of 10^9 eV outside the solar system is found to be $\sim 5 \times 10^5$ eV m^{-3}. As the threshold energy is increased, the energy density decreases rapidly, being 2×10^4 eV m^{-3} above 10^{12} eV and 10^2 eV m^{-3} above 10^{15} eV. The energy density at lower energies outside the heliosphere is unknown but may be substantially greater if the particle rest mass energy is included together with the kinetic energy (A. W. Wolfendale, 1979).

Primary Particle Isotropy

This is measured as an anisotropy $(I_{max} - I_{min})/(I_{max} + I_{min}) \times 100\%$, where I, the intensity (m^{-2}s^{-1}sr^{-1}), is usually measured with an angular resolution of a few degrees.

The measured anisotropy is small and energy dependent. It is roughly constant in amplitude at between 0.05 and 0.1% (with a phase of 0 to 6 hours in right ascension) for energies between 10^{11} eV and 10^{14} eV and appears to increase at higher energies roughly as $0.4 \times (\text{Energy(eV)}/10^{16})^{0.5}$ % up to $\sim 10^{18}$ eV. The latter rise may well be an artifact of the progressively more limited statistics as the flux drops rapidly with energy. It appears possible that a real anisotropy has been observed at the highest energies (above a few times 10^{19} eV) with a directional preference for the supergalactic plane (this plane reflects the directions of galaxies within about 100 million light years) (A.W. Wolfendale, 1979; R.W. Clay, 1987; T. Stanev et al., 1995).

Primary Particle Composition

The composition of low energy cosmic rays is close to universal nuclear abundances except where propagation effects are present. For example, Li, Be, and B which are spallation products, are over-abundant by about six orders of magnitude.

Composition at 10^{11} eV per nucleus

Charge	1	2	(3–5)	(6–8)	(10–14)	(16–24)	(26–28)	≥30
% Composition (10% uncertainty)	50	25	1	12	4	4	4	0.1

Measurements at higher energies indicate that there is an increase in the relative abundances of nuclei with charge greater than 6 at energies above 50 TeV/nucleus (K. Asakimori et al., 1993) (1 TeV = 10^{12} eV).

Cosmic ray composition at low energies is often quoted at a fixed energy per nucleon. When presented in this way, protons constitute roughly 90% of the flux, helium nuclei about 10% and the remainder sum to a total of about 1%.

Certain radioactive isotopic ratios show lifetime effects. The ratio of Be^{10}/B^9 abundances is used to measure an "age" of cosmic rays since Be^{10} is unstable with a half life of about 1.6×10^6 years. A ratio of 0.6 is expected in the absence of Be^{10} decay and a ratio of about 0.2 is found experimentally (E. Juliusson, 1975; P. Meyer, 1981).

At higher energies, composition determinations are indirect and are rather contradictory and controversial. Experiments aim to differentiate between broad composition models. The measurement technique is based on studies of cosmic ray shower development. A rather direct technique for such studies is to use fluorescence observations of the shower development to determine the atmospheric depth of maximum development of the shower. Such observations suggest a heavy composition (large atomic number) at energies $\sim 10^{17}$ eV which changes with increasing energy to a light composition (perhaps protonic) above $\sim 10^{19}$ eV (T. K. Gaisser et al., 1993).

FIGURE 1. The energy spectrum of cosmic ray particles. This spectrum is of a differential form and can be converted to an integral spectrum by integration over all energies above a required threshold (E). Insofar as the spectrum approximates a power law of index −3, a simple conversion to the integral at an energy E/1.8 is obtained by multiplying the differential flux by the energy and dividing by 0.62.

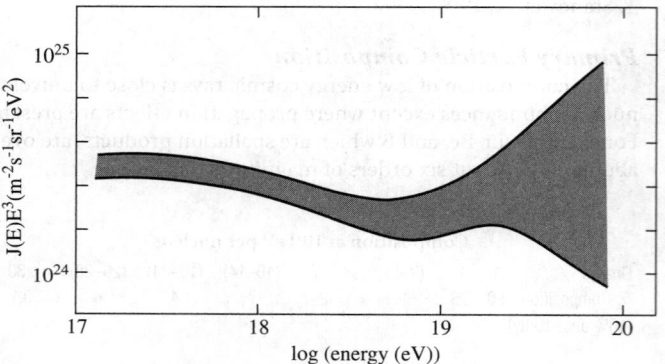

FIGURE 2. Energy spectrum at the highest energies. This spectrum (after Yoshida et al., 1995) has the differential spectrum multiplied by energy cubed. It is from a compilation of a number of measurements and indicates the good general agreement at the lower energies and a spread due to inadequate statistics at the highest energies.

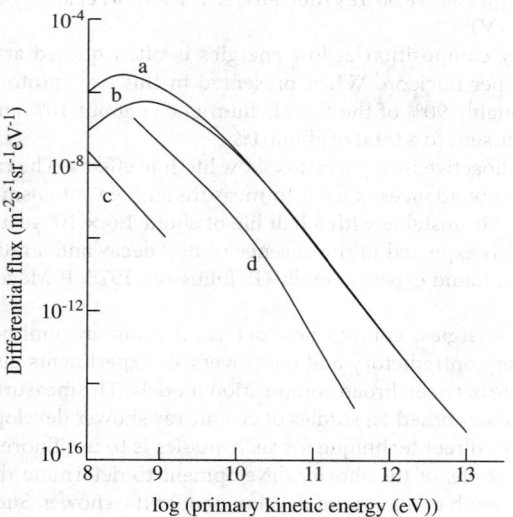

FIGURE 3. Energy spectrum of particles at lower energies. (a) Solar minimum proton energy spectrum. (b) Solar maximum proton energy spectrum. (c) Gamma-ray energy spectrum. (d) Local interstellar electron spectrum.

Primary Electrons

Primary electrons constitute about 1% of the cosmic ray beam. The positron to negative electron ratio is about 10% (J. M. Clem et al., 1995).

Antimatter in the Primary Beam

The ratio of antiprotons to protons in the primary cosmic ray beam (at about 400 MeV) is about 10^{-5}. At about 10 GeV the ratio is about 10^{-3}. At the highest measured energies (10 TeV), the upper limit to the ratio is about 20% (S. Orito et al., 1995; M. Amenomori et al., 1995).

Primary Gamma-Rays

The flux of primary gamma-rays is low at high energies. At 1 GeV the ratio of gamma-rays to protons is about 10^{-6}. The arrival directions of these gamma-rays are strongly concentrated in the plane of the Milky Way although there is a diffuse, near isotropic background flux and some point sources have been detected.

Since the absorption cross section for gamma-rays above 100 MeV is approximately 20 mbarn/electron, less than 10% of gamma-rays reach mountain altitudes (A. W. Wolfendale, 1979; P. F. Michelson, 1994).

Sea Level Cosmic Radiation

The sea level cosmic ray dose is 300 millirad·yr^{-1} and the sea level ionization is 2.2×10^6 ion pairs m^{-3}s^{-1}. The sea level flux has a soft component, which can be absorbed in about 100 mm of lead (about 100 g·cm^{-2} of absorber) and a more penetrating (largely muon) hard component. The sea level radiation is largely produced in the atmosphere and is a secondary component from interactions of the primary particles. The steep primary energy spectrum means that most secondaries at sea level are from rather low energy primaries. Thus the secondary flux is dependent on the solar cycle and the geomagnetic latitude of the observer.

Absolute Flux of the Hard Component

Vertical Integral Intensity $I(0) \sim 100$ m^{-2}s^{-1}sr^{-1}
Angular dependence $I(\theta) \sim I(0) \cos^2(\theta)$
Integrated Intensity ~ 200 m^{-2}s^{-1}
(O.C. Allkofer, 1975b).

Flux of the Soft Component

In free air, the soft component comprises about one third of the total cosmic ray flux.

Latitude Effect

The geomagnetic field influences the trajectories of lower energy cosmic rays approaching the Earth. As a result, the background flux is reduced by about 7% at the geomagnetic equator. The effect decreases towards the poles and is negligible at latitudes above about 40°.

Flux of Protons

The proton component is strongly attenuated by the atmosphere with an attenuation length (reduction by a factor of e) of about 120 g·cm^{-2}. It constitutes about 1% of the total vertical sea level flux.

Absorption

The soft component is absorbed in about 100 g·cm^{-2} of matter. The hard component is absorbed much more slowly:

Absorption in lead, 6% per 100 g·cm^{-2}
Absorption in rock, 8.5% per 100 g·cm^{-2}

Absorption in water, 10% per 100 g·cm^{-2}
(Absorption for depths less than 100 g·pd cm^{-2} is given by K. Greisen, 1943.)

Altitude Dependence

The cosmic ray background in the atmosphere has a maximum intensity of about 15 times that at sea level at a depth of about 150 g·cm^{-2} (15 km altitude). At maximum intensity, the soft and hard components contribute roughly equally but the hard component is then attenuated more slowly (S. Hayakawa, 1969).

Cosmic Ray Showers

High energy cosmic rays produce particle cascades in the atmosphere which can be detected at sea level provided that their energy exceeds about 100 GeV (such low energy cascades may be detected by using the most sensitive atmospheric Cerenkov detectors). The primary particle progressively loses energy which is transferred through the production of successive generations of secondary particles to a cascade of hadrons, an electromagnetic shower component (both positively and negatively charged electrons and gamma-rays) and muons. The secondary particles are relativistic and all travel effectively at the speed of light. As a result, they reach sea level at approximately the same time but, due to Coulomb scattering (for the electrons) and production angles (for the pions producing the muons), are spread laterally into a disk-like shower front with a characteristic lateral width of several tens of meters and thickness (near the central shower core) of 2 to 3 m. The number of particles at sea level is roughly proportional to the primary particle energy:

Number of particles at sea level $\sim 10^{-10} \times$ energy (eV).

At altitudes below a few kilometers, the number of particles in a shower attenuates with an *attenuation length* of about 200 g·cm^{-2}.

i.e., particle number = original number $\times$ exp($-$(depth increase)/200)

The above applies to an individual shower. The rate of observation of showers of a given size (particle number at the detector) at different depths of absorber attenuates with an *absorption length* of about 100 g·cm^{-2} (J.G. Wilson, 1976).

Atmospheric Background Light from Cosmic Rays

Cosmic ray particles produce Cerenkov light in the atmosphere and produce fluorescent light through the excitation of atmospheric molecules.

Cerenkov Light

High energy charged particles will cause the emission of Cerenkov light in air if their energies are above about 30 MeV (electrons). This threshold is pressure (and hence altitude) dependent. A typical Cerenkov light pulse (at sea level, 100 m from the central shower core) has a time spread of a few nanoseconds. Over this time, the photon flux between 430 and 530 nm would be $\sim 10^{14}$ m^{-2}s^{-1} for a primary particle energy of 10^{16} eV. For comparison, the night sky background flux is $\sim 6 \times 10^{11}$ photons m^{-2}s^{-1}sr^{-1} in the same wavelength band (J.V. Jelley, 1967).

Fluorescence Light

Cosmic ray particles in the atmosphere excite atmospheric molecules which then emit fluorescence light. This is weak compared to the highly collimated Cerenkov component when viewed in the direction of the incident cosmic ray particle but is emitted isotropically. Typical pulse widths are longer than 50 ns and may be up to several microseconds for the total pulse from distant large showers (R.M. Baltrusaitis et al., 1985).

Effects of Cosmic Rays

Cerenkov Effects in Transparent Media

Background cosmic ray particles will produce Cerenkov light in transparent material with a photon yield between wavelengths λ_1 and λ_2

$$\sim (2\pi/137)\sin^2(\theta_c)\int_{\lambda_1}^{\lambda_2} d\lambda/\lambda^2 \text{ photons(unit length)}^{-1}$$

where θ_c (the Cerenkov angle) = $\cos^{-1}$ (1/refractive index).

This background light is known to affect light detectors, e.g., photomultipliers, and can be a major source of background noise (R.W. Clay and A.G. Gregory, 1977).

Effects on Electronic Components

If background cosmic ray particles pass through electronic components, they may deposit sufficient energy to affect the state of, e.g., a transistor flip-flop. This effect may be significant where reliability is of great importance or the background flux is high. For instance, it has been estimated that, in communication satellite operation, an error rate of about 2×10^{-3} per transistor per year may be found. Permanent damage may also result. A significant error rate may be found even at sea level in large electronic memories. This error rate is dependent on the sensitivity of the component devices to the deposition of electrons in their sensitive volumes (J.F. Ziegler, 1981).

Biophysical Significance

When cosmic rays interact with living tissue, they produce radiation damage. The amount of the damage depends on the total dose of radiation. At sea level, this dose is small compared with doses from other sources but both the quantity and quality of the radiation change rapidly with altitude. Approximate dose rates under various conditions are:

Dose rates (mrem·yr^{-1})
Sea level cosmic rays, 30
Cosmic rays at 10 km (subsonic jets), 2000
Cosmic rays at 18 km (supersonic transports), 10,000
(c.f., mean total sea level dose, 300)

Astronauts would be subject to radiation from galactic (0.05 rads per day) and solar (a few hundred rads per solar flare) cosmic rays as well as large fluxes of low energy radiation when passing through the Van Allen belts (about 0.3 rads per traverse).

Both astronauts and SST travellers would be subject to a small flux of low energy heavy nuclei stopping in the body. Such particles are capable of destroying cell nuclei and could be particularly harmful in the early stages of the development of an embryo. The rates of heavy nuclei stopping in tissue in supersonic transports and spacecraft are approximately as follows:

Stopping nuclei ((cm^3 tissue)$^{-1}$ hr^{-1})
Supersonic transport (16 km), 0.0005
Supersonic transport (20 km), 0.005
Spacecraft, 0.15
(O. C. Allkofer, 1975a; O. C. Allkofer et al., 1974).

Carbon Dating

Radiocarbon is produced in the atmosphere due to the action of cosmic ray slow neutrons. Solar cycle modulation of the very low energy cosmic rays causes an anticorrelation of the atmospheric ^{14}C activity with sunspot number with a mean amplitude of about 0.5%. In the long term, modulation of cosmic rays by a varying magnetic field may be important (A.A. Burchuladze et al., 1979).

Practical Uses of Cosmic Rays

There are few direct practical uses of cosmic rays. Their attenuation in water and snow have, however, enabled automatic monitors of water and snow depth to be constructed. A search for hidden cavities in pyramids has been carried out using a muon "telescope".

Other Effects

Stellar X-rays have been observed to affect the transmission times of radio signals between distant stations by altering the depth of the ionospheric reflecting layer. It has also been suggested that variations in ionization of the atmosphere due to solar modulation may have observable effects on climatic conditions.

References

O.C. Allkofer, (1975a) *Introduction to Cosmic Radiation*, Verlag Karl Thiemig, Munchen, Germany.
O.O. Allkofer, (1975b) *J. Phys. G: Nucl. Phys.*, 1, L51.
O.C. Allkofer and W. Heinrich, (1974) *Health Phys.*, 27, 543.
M. Amenomori et al., (1995) Proc. 24th Int. Cosmic Ray Conf. Rome, 3, 85. Universita La Sapienza, Roma.
K. Asakimori et al., (1993) Proc. 23rd Int. Cosmic Ray Conf. Calgary, 2, 25, University of Calgary, Canada.
R.M. Baltrusaitis et al., (1985) *Nucl. Inst. Meth.*, A420, 410.
D.J. Bird et al., (1993) *Phys. Rev. Lett.*, 71, 3401.
A.A. Burchuladze, S.V. Pagava, P. Povinec, G. I. Togonidze, S. Usacev, (1979) Proc. 16th Int. Cosmic Ray Conf. Kyoto, 3, 201, Univ. of Tokyo, Japan.
R.W. Clay, (1987) *Aust. J. Phys.*, 40, 423.
R.W. Clay and A.G. Gregory, (1977) *J. Phys. A: Math. Gen.*, 10, 135.
J.M. Clem et al., (1995) Proc. 24th Int. Cosmic Ray Conf. Rome, 3, 5, Universita La Sapienza, Roma.
T.K. Gaisser et al., (1993) *Phys. Rev. D*, 47, 1919.
K. Greisen, (1943) *Phys. Rev.*, 63, 323.
S. Hayakawa, (1969) *Cosmic Ray Physics*, Wiley-Interscience, New York.
J.V. Jelley, (1967) *Prog. in Elementary Particle and Cosmic Ray Physics*, 9, 41.
E. Juliusson, (1975) Proc. 14th Int. Cosmic Ray Conf. Munich, 8, 2689, Max Planck Institute fur Extraterrestriche Physik, Munchen, Germany.
J. Linsley, (1981) *Origin of Cosmic Rays*, I.A.U. Symposium 94, 53, D. Reidel Publishing Co Dordrecht, Holland.
P. Meyer, (1981) *Origin of Cosmic Rays*, I.A.U. Symposium 94, 7, D. Reidel Publishing Co. Dordrecht, Holland.
P.F. Michelson (1994) in *Towards a Major Atmospheric Cerenkov Detector III*, 257, Ed. T. Kifune, Universal Academy Press Inc., Tokyo, Japan.
P. Sokolsky, (1989) *Introduction to Ultrahigh Energy Cosmic Ray Physics*, Addison Wesley Publishing Company.
T. Stanev et al., (1995) *Phys. Rev. Lett.*, 75, 3056.
S. Orito et al., (1995) Proc. 24th Int. Cosmic Ray Conf. Rome, 3, 76. Universita La Sapienza, Roma.
J.G. Wilson, (1976) *Cosmic Rays*, Wykeham Pub. (London) Lt., U.K.
A.W. Wolfendale, (1979) *Pramana*, 12, 631.
S. Yoshida et al., (1995) *Astroparticle Phys.*, 3, 105.
J.F. Ziegler, (1981) *IEEE Trans. Electron Devices*, ED-28, 560.

Section 12
Properties of Solids

TECHNIQUES FOR MATERIALS CHARACTERIZATION
Experimental Techniques Used to Determine the Composition, Structure, and Energy States of Solids and Liquids

H.P.R. Frederikse

The many experimental methods, originally designed to study the chemical and physical behavior of solids and liquids, have grown into a new field known as Materials Characterization (or Materials Analysis). During the past 30 years a host of techniques aimed at the study of surfaces and thin films has been added to the many tools for the analysis of bulk samples. The field has benefitted particularly from the development of computers and microprocessors, which have vastly increased the speed and accuracy of the measuring devices and the recording of their output. Materials characterization was and is a very important tool in the search for new physical and chemical phenomena. It plays an essential role in new applications of solids and liquids in industry, communications, and medicine. Many of its techniques are used in quality control, in safety regulations, and in the fight against pollution.

In most Materials Characterization experiments the sample is subjected to some kind of radiation: electromagnetic, acoustic, thermal, or particles (electrons, ions, neutrons, etc.). The surface analysis techniques usually require a high vacuum. As a result of interactions between the solid (or liquid) and the incoming radiation a beam of a similar (or a different) nature will emerge from the sample. Measurement of the physical and/or chemical attributes of this emerging radiation will yield qualitative, and often quantitative, information about the composition and the properties of the material being probed.

The modern tendency of describing practically everything in this world by a combination of a few letters (acronyms) has also penetrated the field of Materials Characterization. The table below gives the meaning of the acronym for every technique listed, the form and size of the required sample (bulk, surface, film, liquid, powder, etc.), the nature of the incoming and of the emerging radiation, the depth and the lateral spatial resolution that can be probed, and the information obtained from the experiment. The last column lists one or two major references to the technique described.

Technique	Sample	In	Out	Depth	Lateral resolution	Information obtained	Ref.
Optical and Mass Spectroscopies for Chemical Analysis							
1. AAS Atomic Absorption Spectroscopy	Atomize (flame, electro, thermal, etc.)	Light e.g., glow discharge	Absorption spectrum	–	–	Concentration of atomic species (quantitative, using standards)	1,2
2. ICP-AES Induct. Coupled Plasma – Atomic Emission Spectroscopy	Atomize (flame, electro, thermal, ICP, etc.)	–	Emission spectrum	–	–	Concentration of atomic species (quantitative, using standards)	3
3. Dynamic SIMS Dynamic Secondary Ion Mass Spectroscopy	Surface	Ion beam (1–20 keV)	Secondary ions; analysis with mass spectrometer	2 nm–1 μm (or deeper: ion milling)	0.50 nm	Elemental and isotopic analysis; depth profile (all elements); detection limits: ppb-ppm	4
4. Static SIMS Static Secondary Ion Mass Spectroscopy	Surface	Ion beam (0.5–20 keV)	Secondary ions, analysis with mass spectrometer	0.1–0.5 nm	10 μm	Elemental analysis of surface layers; molecular analysis; detection limits: ppb-ppm	4
5. SNMS Sputtered Neutral Mass Spectroscopy	Surface, bulk	Plasma discharge; noble gases: 0.5–20 keV	Sputtered atoms ionized by atoms or electrons; then mass analyzed	0.1–0.5 nm (or deeper: ion milling)	1 cm	Elemental analysis Z ≥ 3; depth profile; detection limit: ppm	4,6
6. SALI Surface Analysis by Laser Ionization	Surface	e-beam, ion-beam, or laser for sputtering	Sputtered atoms ionized by laser; then mass analyzed	0.1–0.5 nm up to 3 μm in milling mode	60 nm	Surface analysis; depth profiling	7
7. LIMS Laser Ionization Mass Spectroscopy	Surface, bulk	u.v. laser (ns pulses)	Ionized species; analyzed with mass spectrometer	50–150 nm	5 μm–1 mm	Elemental (micro)analysis; detection limits: 1–100 ppm	8
8. SSMS Spark Source Mass Spectroscopy	Sample in the form of two electrodes	High voltage R.F. spark produces ions	Ions – analyzed in mass spectrometer	1–5 μm	–	Survey of trace elements; detection limit: 0.01–0.05 ppm	9
9. GDMS Glow Discharge Mass Spectroscopy	Sample forms the cathode for a D.C. glow discharge	Sputtered atoms ionized in plasma	Ions – analyzed in mass spectrometer	0.1–100 μm	3–4 mm	(Bulk) trace element analysis; detection limit: sub-ppb	9,10
10. ICPMS Induct. Coupled Plasma Mass Spectroscopy	Liquid-dissolved sample carried by gas stream into R.F. induction coil	Ions produced in argon plasma	Ions – analyzed in quadrupole mass spectrometer	–	–	High sensitivity analysis of trace elements	11
Photons — Absorption, Reflection and Electron Emission							
11. IRS Infrared Spectroscopy	Thin crystal, glass, liquid	I.R. light (W-filament, globar, Hg-arc)	I.R. spectrum	–	–	Electronic transitions (mainly in semiconductors and superconductors); vibrational modes (in crystals and molecules)	12,13,14
12. FTIR Fourier Transform I.R. Spectroscopy	Solid, liquid; transmission or reflection	White light (all frequencies)	Fourier Transform of spectrum (interferometer)	–	–	Spectra obtained at higher speed and resolution	15
13. ATR Attenuated Total Reflection	Surface or thin crystal	–	–	μm's	–	Atomic or molecular spectra of surfaces and films	16
14. (μ)-RS (Micro-) Raman Spectroscopy	Solid, liquid (1 μm–1 cm)	Laser beam, e.g., Ar-line, YAG-line	Raman spectra	0.5 μm	0.5 μm	Molecular and crystal vibrations	12,14,17

	Technique	Sample	In	Out	Depth	Lateral resolution	Information obtained	Ref.
15.	CARS Coherent Anti-Stokes Raman Spectroscopy	Solid, liquid (50 μm–3 cm)	Pump beam (ω_0)+ probe beam (ω_s)	Anti-Stokes spectrum	–	–	High resolution Raman spectra	14
16.	Ellipsometry	Transparent films, crystals, adsorbed layers	Polarized light	Change in polarization	0.05 nm–5 μm	25 μm (or sample thickness)	Refractive index and absorption	18,19
17.	UPS Ultraviolet Photo-electron Spectroscopy	Surfaces, adsorbed layers	u.v. light, 10–100 eV; 200 eV (synchrotron)	Electrons	0.2–10 nm	0.1–10 nm	Energies of electronic states of surfaces and free molecules	20,21
18.	PSD Photon Stimulated Desorption	Surfaces with adsorbed species	Far u.v. light E > 10 eV	Ions – analyzed with mass spectrometer	0.1–2 nm		Structure and desorption kinetics of adsorbed atoms and molecules	22

X-Rays

	Technique	Sample	In	Out	Depth	Lateral resolution	Information obtained	Ref.
19.	XRD X-Ray Diffraction	Single crystals, powders films	X-rays: λ = 0.05–0.2 nm (6–17 keV)	Diffracted X-ray beam	1–1000 nm	0.1–10 nm	Identification of crystallographic structures; all elements (low Z difficult)	23,24
20.	XRF/EDS X-Ray Fluorescence/Energy Dispersive Spectroscopy	Thin films, single layer	Prim. X-ray beam λ = 0.02–0.1 nm 12–80 keV	Fluorescent X-rays	1–100 μm	10 mm	Elemental analysis; all elements except H, He, Li – (EDS also used in XRD, SEM, TEM and EPMA)	25,26
21.	EXAFS Extended X-Ray Absorption Fine Structure	Films, foils	High intensity X-rays (synchrotron)	Spectrum near absorption edge	nm–μm	–	Local atomic structure: order/ disorder in vicinity of absorbing atom	27
22.	XPS/ESCA X-Ray Photo-electron Spectroscopy/ Electron Spect. for Chemical Analysis	Surfaces, thin films (≈20 atomic layers)	Soft X-rays (1–20 keV)	Core electrons; valence electrons	0.5–10 nm	5 nm–50 μm	(Quantitative) identification of all elements in surface layer or film	28,29

Electrons

	Technique	Sample	In	Out	Depth	Lateral resolution	Information obtained	Ref.
23.	CL Cathode Luminescence	Insulators, semiconductors	Electrons 5–50 keV	Photons 0.1–5 eV	1 nm–2 μn	1 or 2 μm	Energy levels of impurities and point defects	30
24.	APS Appearance Potential Spectroscopy	Surface (≈20 atomic layers)	Electrons (energy scan) 50–2000 eV	X-rays to pinpoint electron energy threshold	–	–	Identification of surface species	21, see also C
25.	AES Auger Electron Spectroscopy	Thin films, surfaces	Electrons 3–10 keV	Auger electrons 20–2000 eV	0.3–3 nm	≈30 nm	Elemental composition of surface (except H, He); detection limit 0.1–1%	28,29
26.	EELS Electron Energy Loss Spectroscopy	Very thin samples (<200 nm)	Electrons (100–400 keV)	(Retarded) electrons; minus 1–1000 eV	<200 nm	1–100 nm	Local elemental concentration; electronic structure, chem. bonding; interatomic distances	31
27.	EXELFS Extended Electron Energy Loss Fine Structure	Thin films	Electrons (100–400 keV)	Electrons energies 0–30 eV above edge	<200 nm	1–100 nm	Density of states of valence electrons (above Fermi level)	27,32
28.	ESD Electron Stimulated Desorption	Adsorbed species	Electrons E > 10 eV	Ions – analyzed with mass spectrometer	–	–	Structure and desorption properties of adsorbed atoms and molecules	22
29.	ESDIAD ESD-Ion Angular Distribution	(See ESD)	(See ESD)	Directional dependence of emitted ions	–	–	Geometries of adsorbed species (atoms or molecules)	22
30.	EPMA Electron Probe (X-Ray) Micro Analysis	Solid conductors and insulators <1 cm thick	Electrons 5–30 keV	Characteristic X-ray 0.1–15 keV	100 nm–5 μm	1 μm	Elemental analysis, Z ≤ 4, major, minor and trace amounts	33,34
31.	LEED Low Energy Electron Diffraction	Surface	Mono-energetic electron beam 10–1000 eV	Diffracted electrons	0.4–2 nm	<5 μm	Crystallographic structure of surface; resolution: 0.01 nm	35
32.	RHEED Reflection High Energy Electron Diffraction	Surface	Electron beam at grazing angle 5–50 keV	Reflected electrons	0.2–10 nm	<5 μm	Surface symmetry	36,37
33.	SEM Scanning Electron Microscopy	Bulk, films (conducting)	High energy electrons usually ~30 keV	Secondary and backscattered electrons	1 nm–5 μm	1–20 nm	Surface image, defect structure; resolution 5–15 nm; magnification 300,000×	33,34
34.	(S)TEM (Scanning) Transmission Electron Microscopy	Thin specimen – <200 nm	High energy electrons typically 300 keV	Transmitted and diffracted electrons	(Sample thickness)	2–20 nm	(Defect) structure of cryst. solids; microchemistry; high resol.: 0.2 nm	33
35.	FEM Field Emission Microscopy	Metals, alloys (sharp point)	–	Electron emission (with appl. electric field – 50 kV)	≈0.5 nm	10–100 nm	Surface image, crystallographic structure	34
36.	STM Scanning Tunneling Microscopy	Polished or cleaved surface (conducting)	Tunneling current controls distance between sample and very sharp tip		1–5 nm	2–10 nm	Atomic-scale relief map of surface; resolution: vert. 0.002 nm, hor. 0.2 nm	39
37.	SPM Scanned Probe Microscopy	Very flat surface	Any field: e.g. mechan. vibration recorded with laser probe; same with magnetic, electric or thermal field		1–100 nm	1–100 nm	Surface-magnetic field, surface-thermal conductivity, etc.	39a
38.	AFM Atomic Force Microscopy	Very flat surface	Similar to STM; force measured with cantilever spring		0.5–5 nm	0.2–130 nm	Surface topography with atomic resolution; interatomic force	40

	Technique	Sample	In	Out	Depth	Lateral resolution	Information obtained	Ref.

Ions and Neutrons

	Technique	Sample	In	Out	Depth	Lateral resolution	Information obtained	Ref.
39.	ISS (or LEIS) Ion Scattering Spectroscopy (Low Energy Ion Scattering)	Surface	Ion beam He+ or Ne+ <3 keV	Sputtered ions (energy analysis)	0.1–0.5 nm	1–100 μm	Elemental analysis (better for low Z) detection limits: 0.01–1%	41
40.	FIM Field Ion Microscopy	Surface: metals, alloys; very sharp tip	(He gas above sample)	He ions + high electric field produce image	≈0.1 nm	0.1–2 nm	Atomic structure of surface	34,42
41.	RBS Rutherford Back Scattering	Solids, thin films	Mono-energetic ions (H+ or He++) 0.5–3 MeV	Backscattered ions	10 nm–1 μm	1 mm	Element identification (Li to U) detection limit: 0.01–1%	46
42.	NRA Nuclear Reaction Analysis	Solids, thin films	Mono-energetic ions (Li, Be, B, etc.) 200 keV–6 MeV	Protons, deuterons ^{3}He, α-particles, γ-rays	0.1–5 μm	10 μm–10 mm	Element identification (all) detection limit: 10^{-12}–10^{-2}	47
43.	PIXE Particle Induced X-ray Emission	Thin films, surface layers	High energy ions (H+ or He++)	Characteristic X-rays	<10 μm	1 μm–2 mm	Trace impurities: Z>3 detection limit: 0.1–100 ppm (depending on sample thickness)	48
44.	INS Ion Neutralization Spectroscopy	Surface	He-ions (≈5 eV)	Electrons	–		Energies of valence electrons	49
45.	NAA Neutron Activation Analysis	Bulk, >0.5 g	Thermal neutrons	Characteristic γ-rays, (≈1 MeV)	Bulk	–	Trace concentrations (of isotopes) of elements: trans. metals, Pt-group; detection limit: 10^8–10^{14} atoms/cm^3	43
46.	N(P)D Neutron (Powder) Diffraction	Crystalline solids	Thermal neutrons E ≈0.0025 eV	Diffracted neutrons	Bulk	–	Crystallographic structure; porosity, particle size	44
47.	SANS Small Angle Neutron Scattering	Inhomogeneous solids; powders; porous samples	Thermal neutrons 2 θ = 10^{-2}–10^{-4}	Scattered neutrons	1–25 mm	–	Average size of inhomogeneities; range: 1 nm–1 mm	45

Acoustic

	Technique	Sample	In	Out	Depth	Lateral resolution	Information obtained	Ref.
48.	SLAM Scanning Laser Acoustic Microscopy	Bulk, film	Acoustic wave produced by laser 1 MHz–1 GHz	Reflected acoustic wave	μm–cm	0.1–20 mm	Defect structure; thickness measurement	50

Thermal

	Technique	Sample	In	Out	Depth	Lateral resolution	Information obtained	Ref.
49.	DTA Differential Thermal Analysis	Specimen and reference sample	Uniform heating	Temperature difference	Bulk	–	Phase transitions, crystallization	51
50.	DSC Differential Scanning Calorimetry	Specimen and ref. sample	Controlled heating	Measure heat required for equal temperature	Bulk	–	Phase transitions, crystallization; activation energies	51
51.	TGA Thermo Gravimetric Analysis	Bulk, 1–100 g	Controlled heating	Weight as function of temperature (and time)	Bulk	–	Decomposition, non-stoichiometry, kinetics of reaction	52

Resonance

	Technique	Sample	In	Out	Depth	Lateral resolution	Information obtained	Ref.
52.	EPR (ESR) Electron Paramagnetic (Spin) Resonance	Paramagnetic solids or liquids	Microwave radiation in magnetic field 3–300 GHz; 1–100 kG	Microwave absorption (at resonance)	Bulk	–	Local environment of paramagnetic ion; concentration of paramagnetic, species; detection limit: 10^{11} spins/cm^3	53,54
53.	ECR Electron Cyclotron Resonance	Semiconductors, metals; free electrons (low temperature)	Microwave radiation in magnetic field 10–30 GHz; 5–10 kG	Microwave absorption (at resonance)	Bulk	–	Electronic energy bands, effective masses	55
54.	Mössbauer Effect	Source and absorber	Mono-energetic γ-rays: 5–100 keV	Mössbauer spectrum (Doppler shifted (lines)	50 m	1 cm	Interaction between nucleus and its environment (local electric, magnetic fields; bonds; valency; diffusion, etc.)	56
55.	NMR (MRI) Nuclear Magnetic Resonance (Magnetic Resonance Imaging)	Solids, liquids	R.F. radiation + magnetic field; e.g. for protons: 60 MHz, 14 kG	R.F. absorption	<1 cm	1 cm	Quant. analysis; local magnetic environment; diffusion; imaging	58
56.	ENDOR Electron Nuclear Double Resonance	Solids, liquids	R.F. + microwave radiation in magn. field.	Microwave absorption	–	–	Hyperfine interaction → local atomic structure	54
57.	NQR Nuclear Quadrupole Resonance	Solids	R.F. radiation 0.5–1000 MHz	R.F. absorption	–	–	Asymmetry of the charge distribution at the nucleus	55,59

Other

	Technique	Sample	In	Out	Depth	Lateral resolution	Information obtained	Ref.
58.	BET Brunauer-Emmett-Teller	(Large) surface area 1–20 m^2/g	Adsorbed gas (e.g., N$_2$ at low temp.) as function of pressure (monolayer coverage)		–	–	Surface area measurement	60

References

General References

A. Wachtman, J. B., *Characterization of Materials*, Butterworth-Heinemann, Boston, 1993.

B. Brundle, C. R., Evans, C. A., and Wilson, S., Eds., *Encyclopedia of Materials*, Butterworth-Heinemann, Boston, 1992.

C. Woodruff, D. P. and Delchar, T. A., *Modern Techniques of Surface Science*, Cambridge University Press, Cambridge, 1986.

D. *Metals Handbook*, 9th Edition, Vol. 10, Materials Characterization, Whan, R. E., Coordinator, American Society for Metals, Metals Park, OH, 1986.

Specific References

1. Slavin, M., *Atomic Absorption Spectroscopy*, 2nd Edition, John Wiley & Sons, New York, 1978.

2. Schrenk, W. G., *Analytical Atomic Spectroscopy*, Plenum Press, New York, 1975.

3. Dean, J. A. and Rains, T. E., *Flame Emission and Atomic Absorption Spectroscopy*, Vols. 1–3, Marcel Dekker, New York, 1969.

4. Benninghoven, A., Rudenauer, F. G., and Werner, H. W., *Secondary Ion Mass Spectroscopy*, John Wiley & Sons, New York, 1987.

5. Bird, J. R. and Williams, J. S., Eds., in *Ion Beams for Materials Analysis*, Academic Press, New York, 1989, pp. 515–537.

6. Smith, G. C., *Quantitative Surface Analysis for Materials Science*, The Institute of Metals, London, 1991.

7. Becker, E. H., in *Ion Spectroscopies for Surface Analysis*, Czanderna, A. W. and Hercules, D. M., Eds., Plenum Press, New York, 1991, p. 273.

8. Simons, D. S., *Int. J. Mass Spectrometry and Ion Processes*, 55, 15, 1983.

9. White, F. A. and Wood, G. M., *Mass Spectrometry: Applications in Science and Engineering*, John Wiley & Sons, New York, 1986.

10. Harrison, W. W. and Bentz, B. L., *Prog. Anal. Spectrometry*, 11, 53, 1988.

11. Bowmans, P. W. J. M., *Inductively Coupled Plasma Emission Spectroscopy*, Parts I and II, John Wiley & Sons, New York, 1987.

12. Brame, Jr., E. G. and Grasselli, J., *Infrared and Raman Spectroscopy*, Practical Spectroscopy Series, Vol. I, Marcel Dekker, New York, 1976.

13. Hollas, J. M., *Modern Spectroscopy*, John Wiley & Sons, New York, 1987.

14. Turrell, G., *Infrared and Raman Spectroscopy of Crystals*, Academic Press, New York and London, 1972.

15. Griffith, P. R. and Haseth, J. A., *Fourier Transform Infrared Spectroscopy*, John Wiley & Sons, New York, 1986.

16. Barnowski, M. K., *Fundamentals of Optical Fiber Communications*, Academic Press, New York, 1976.

17. Long, D. A., *Raman Spectroscopy*, McGraw-Hill, New York, 1977.

18. Azzam, R. M. A., *Ellipsometry and Polarized Light*, Elsevier-North Holland, Amsterdam, 1977.

19. Hecht, E., *Optics*, 2nd Edition, Addison-Wesley, Reading MA, 1987.

20. Brundle, C. R., in *Molecular Spectroscopy*, West, A. R., Ed., Heyden, London, 1976.

21. Park, R. L., in *Experimental Methods in Catalytic Research*, Vol. III, Anderson, R. B. and Dawson, P. T., Academic Press, New York, 1976, pp. 1–39.

22. Madey, T. E. and Stockbauer, R., in *Solid State Physics: Surfaces*, Vol. 22 of Methods of Experimental Physics, Park, R.L. and Lagally, M. G., Eds., Academic Press, New York, 1985.

23. Cullity, B. D., *Elements of X-Ray Diffraction*, 2nd Edition, Addison-Wesley, Reading, MA, 1978.

24. Schwartz, L. H. and Cohen, J. B., *Diffraction from Materials*, Springer Verlag, Berlin, 1987.

25. deBoer, D. K. G., in *Advances in X-Ray Analysis*, Vol. 34, Barrett, C. S. et. al., Eds., Plenum Press, New York, 1991.

26. Birks, L. S., *X-Ray Spectrochemical Analysis*, 2nd Edition, John Wiley & Sons, New York, 1969.

27. Bonnelle, C. and Mande, C., *Advances in X-Ray Spectroscopy*, Pergamon Press, Oxford, 1982.

28. *Practical Surface Analysis by Auger and X-Ray Photo-Electric Spectroscopy*, Briggs, D. and Seah, M. P., Eds., John Wiley & Sons, New York, 1983.

29. Powell, C. J. and Seah, M. P., *J. Vac. Sci. Technol. A*, Vol. 8, 735, 1990.

30. Yacobi, G. G. and Holt, D. B., *Cathodeluminescence Microscopy of Inorganic Solids*, Plenum Press, New York, 1990.

31. Egerton, R. F., *Electron Energy Loss Spectroscopy in the Electron Microscope*, Plenum Press, New York, 1986.

32. Disko, M. M., Krivanek, O. L., and Rez, P., *Phys. Rev.*, B25, 4252, 1982.

33. Goldstein, J. I., et. al., *Scanning Electron Microscopy and X-Ray Microanalysis*, 2nd Edition, Plenum Press, New York, 1986.

34. Murr, L. E., *Electron and Ion Microscopy and Microanalysis*, Marcel Dekker, New York, 1982.

35. Armstrong, R. A., in *Experimental Methods in Catalytic Research*, Vol. III, Anderson, R. B., and Dawson, P. T., Eds., Academic Press, New York, 1976.

36. Dobson, P. J. et. al., *Vacuum*, 33, 593, 1983.

37. Rymer, T. B., *Electron Diffraction*, Methuen, London, 1970.

38. Reimer, L., *Transmission Election Microscopy*, Springer-Verlag, Berlin, 1984.

39. *Scanning Tunneling Microscopy and Related Methods*, Behm, R. J., Garcia, N., and Rohrer, H., Eds., Kluwer Academic Publishers, Norwell, MA, 1990.

39a. Wikramasinghe, H.K., *Scientific American*, Vol. 261, No. 4, pp. 98–105, Oct. 1989.

40. Rugar, D. and Hansma, P., *Physics Today*, 43(10), pp. 23–30, 1990.

41. Feldman, C. C. and Mayer, J. W., *Fundamentals of Surface and Thin Film Analysis*, North-Holland, Amsterdam, 1986.

42. Muller, E. W. and Tsong, T. T., *Field Ion Microscopy*, Elsevier, Amsterdam, 1969.

43. Amiel, S., *Nondestructive Activation Analysis*, Elsevier, Amsterdam, 1981.

44. Bacon, G. E., *Neutron Diffraction*, 3rd Edition, Clarendon Press, Oxford, 1975.

45. Neutron Scattering, Part A., in *Methods of Experimental Physics*, Vol. 23, Skold, K. and Price, D. L., Eds., Academic Press, New York, 1986.

46. Chu, W. K., Mayer, J. W., and Nicolet, M. A., *Backscattering Spectroscopy*, Academic Press, New York, 1987.

47. Rickey, F. A., in *High Energy and Heavy Ion Beams in Materials Analysis*, Tesmer, J. R., et. al., Eds., MRS, 1990, pp. 3–26.

48. Johansson, S. A. E. and Campbell, J. L., *PIXE: A Novel Technique for Elemental Analysis*, John Wiley & Sons, New York, 1988.

49. Hagstrum, H. D., in *Inelastic Ion-Surface Collisions*, Tolk, N. H. et. al., Eds., Academic Press, New York, 1977, pp. 1–46.

50. Nikoonahad, M., in *Research Techniques in Nondestructive Testing*, Vol. VI, Sharpe, R.S., Ed., Academic Press, New York, 1984, pp. 217–257.

51. Gallagher, P. K., *Characterization of Materials by Thermoanalytical Techniques*, MRS - Bulletin, Vol. 13, No. 7, pp. 23–27, 1988.

52. Earnest, C. M., *Compositional Analysis by Thermogravimetry*, ASTM Special Technical Publication 997, 1988.

53. Poole, C. P., *Electron Spin Resonance – A Comprehensive Treatise on Experimental Techniques*, 2nd Edition, John Wiley & Sons, New York, 1983.

54. Atherton, N. M., *Principles of Electron Spin Resonance*, Ellis Horwood Ltd., Chichester, U.K., 1993.

55. Kittel, C., *Introduction to Solid State Physics*, 6th Edition, John Wiley & Sons, New York, 1986, p. 196.

56. Gibb, T. C., *Principles of Mössbauer Spectroscopy*, Chapman & Hall, London, 1976.

57. Slichter, C. P., *Principles of Magnetic Resonance*, 3rd Edition, Springer-Verlag, Berlin, 1990.

58. *NMR Spectroscopy Techniques*, Dybrowski, C. and Lichter, R. L., Eds., Marcel Dekker, New York, 1987.

59. Das, T. P. and Hahn, E. L., *Nuclear Quadrupole Resonance Spectroscopy*, Academic Press, New York, 1958.

60. Somorjai, G. A., *Principles of Surface Chemistry*, Prentice-Hall, Englewood Cliffs, NJ, 1972, p. 216

SYMMETRY OF CRYSTALS

L. I. Berger

The ability of a body to coincide with itself in its different positions regarding a coordinate system is called its symmetry. This property reveals itself in iteration of the parts of the body in space. The iteration may be done by reflection in mirror planes, rotation about certain axes, inversions and translations. These actions are called the symmetry operations. The planes, axes, points, etc., are known as symmetry elements. Essentially, mirror reflection is the only truly primitive symmetry operation. All other operations may be done by a sequence of reflections in certain mirror planes. Hence, the mirror plane is the only true basic symmetry element. But for clarity, it is convenient to use the other symmetry operations, and accordingly, the other aforementioned symmetry elements. The symmetry elements and operations are presented in Table 1.

The entire set of symmetry elements of a body is called its symmetry class. There are thirty-two symmetry classes that describe all crystals which have ever been noted in mineralogy or been synthesized (more than 150,000). The denominations and symbols of the symmetry classes are presented in Table 2.

There are several known approaches to classification of individual crystals in accordance with their symmetry and crystallochemistry. The particles which form a crystal are distributed in certain points in space. These points are separated by certain distances (translations) equal to each other in any chosen direction in the crystal. Crystal lattice is a diagram that describes the location of particles (individual or groups) in a crystal. The lattice parameters are three non-coplanar translations that form the crystal lattice. Three basic translations form the unit cell of a crystal. August Bravais (1848) has shown that all possible crystal lattice structures belong to one or another of fourteen lattice types (Bravais lattices). The Bravais lattices, both primitive and non-primitive, are the contents of Table 3.

Among the three-dimensional figures, there is a group of polyhedrons that are called regular, which have all faces of the same shape and all edges of the same size (regular polygons). It has been shown that there are only five regular polyhedrons. Because of their importance in crystallography and solid state physics, a brief description of these polyhedrons is included in Table 4.

The systematic description of crystal structures is presented primarily in the well known *Structurbericht*. The classification of crystals by the Structurbericht does not reflect their crystal class, the Bravais lattice, but is based on the crystallochemical type. This makes it inconvenient to use the Structurbericht categories for comparison of some individual crystals. Thus, there have been several attempts to provide a more convenient classification of crystals. Table 5 presents a compilation of different classifications which allows the reader to correlate the Structurbericht type with the international and Schoenflies point and space groups and with Pearson's symbols, based on the Bravais lattice and chemical composition of the class prototype. The information included in Table 5 has been chosen as an introduction to a more detailed crystallophysical and crystallochemical description of solids.

TABLE 1. Symmetry Operations and Elements

Symmetry operation	Name	Symmetry element Symbol International (Hermann-Mauguin)	Schoenflies	Presentation on the stereographic projection Parallel	Presentation on the stereographic projection Perpendicular
Reflection in a plane	Plane	m	C_s		
Rotation by angle $\alpha = 360°/n$ about an axis	Axis	n = 1, 2, 3, 4 or 6	C_n		
		n = 2	C_2		
		n = 3	C_3		
		n = 4	C_4		
		n = 6	C_6		
Rotation about an axis and inversion in a symmetry center lying on the axis	Inversion (improper) axis	$\bar{n} = \bar{3}, \bar{4}, \bar{6}$	C_{ni}		
		$\bar{n} = \bar{3}$	C_{3i}		
		$\bar{n} = \bar{4}$	C_{4i}		

TABLE 1. Symmetry Operations and Elements

Symmetry operation	Name	International (Hermann–Mauguin)	Schoenflies	Parallel	Perpendicular
		$\bar{n} = \bar{6}$	C_{6i}		
Inversion in a point	Center	$\bar{1}$	C_i		
Parallel translation	Translation vector $\vec{a}, \vec{b}, \vec{c}$				
Reflection in a plane and translation parallel to the plane	Glide–plane	a, b, c, n, d			
Rotation about an axis and translation parallel to the axis	Screw axis	n_m (m = 1, 2, .., n – 1)			
Rotation about an axis and reflection in a plane perpendicular to the axis	Rotatory-reflection axis	$\tilde{n}$ $\tilde{n} = \tilde{1}, \tilde{2}, \tilde{3}, \tilde{4}, \tilde{6}$	S_n		

Symmetry element — Symbol — Presentation on the stereographic projection

TABLE 2. The Thirty-Two Symmetry Classes

Class name[a] and its symbol – International (Int) and Schoenflies (Sch)

Crystal symbol	Primitive Int	Sch	Central Int	Sch	Planal Int	Sch	Axial Int	Sch	Plane-axial Int	Sch	Inversion primitive Int	Sch	Inversion-planal Int	Sch
Triclinic	1	C_1	$\bar{1}$	C_i										
Monoclinic					m	C_s	2	C_2	2/m	C_{2h}				
Ortho-rhombic					mm2	C_{2v}	222	D_2	mmm	D_{2h}				
Trigonal	3	C_3	$\bar{3}$	C_{3i}	3m	C_{3v}	32	D_3	$\bar{3}$m	C_{3d}				
Tetragonal	4	C_4	4/m	C_{4h}	4mm	C_{4v}	422	D_4	4/mmm	D_{4h}	$\bar{4}$	S_4	$\bar{4}$2m	D_{2d}
Hexagonal	6	C_6	6/m	C_{6h}	6mm	C_{6v}	622	D_6	6/mmm	D_{6h}	$\bar{6}$	C_{3h}	$\bar{6}$m2	D_{3h}
Cubic	23	T	m3	T_h	$\bar{4}$3m	T_d	432	O	m3m	O_h				

[a] Per Fedorov Institute of Crystallography, USSR Academy of Sciences, nomenclature.

TABLE 3. The Fourteen Possible Space Lattices (Bravais Lattices)

Crystal system	Metric category of the system	No. of different lattices in the system	Lattice type[a] (marked by +) P	C	I	F	R	No. of identi-points per unit cell	a	b	c	α	β	γ	Description a⊂X, b⊂Y, c⊂Z α≡(b,c), β≡(a,c), γ≡(a,b)	Symmetry Int	Sch
Triclinic	Trimetric	1	+					1	+	+	+	+	+	+	a ≠ b ≠ c, α ≠ β ≠ γ	1	C
Monoclinic	Trimetric	2	+	+				1 or 2	+	+	+		+		a ≠ b ≠ c, α = γ = 90° ≠ β	2/m	C_{2h}
Orthorhombic	Trimetric	4	+	+	+	+		1, 2 or 4	+	+	+				a ≠ b ≠ c, α = β = γ = 90°	mmm	D_{2h}
Trigonal (rhombohedral)	Dimetric	1					+	1	+			+			a = b = c, 120° > α = β = γ ≠ 90°	3m	D_{3d}
Tetragonal	Dimetric	2	+		+			1 or 2	+		+				a = b ≠ c, α = β = γ = 90°	4/mmm	D_{4h}
Hexagonal	Dimetric	1	+					1	+		+				a = b ≠ c, α = β = 90°, γ = 120°	6/mmm	D_{6h}
Isometric (cubic)	Monometric	3	+		+	+		1, 2 or 4	+						a = b = c, α = β = γ = 90°	m3m	O_h

[a] Designations of the space-lattice types: P – primitive, C – side-centered (base-centered), I – body-centered, F – face-centered, R – rhombohedral.

TABLE 4. The Five Possible Regular Polyhedrons

Polyhedron	Class	Elements	Form of faces	Faces (F)	Edges (E)	Vertices (V)
Tetrahedron	T	$4C_3 3C_2$	Equilateral triangle	4	6	4

Symmetry (Schoenflies) — Number of[a]

Cube (hexahedron)	O	$3C_4 4C_3 6C_2$	Square	6	12	8
Octahedron	O	$3C_4 4C_3 6C_2$	Equilateral triangle	8	12	6
Pentagonal dodecahedron	J	$6C_5 10C_3 15C_2$	Regular pentagon	12	30	20
Icosahedron	J	$6C_5 10C_3 15C_2$	Equilateral triangle	20	30	12

[a] Per formula by Leonhard Euler: $F + V - E = 2$

TABLE 5. Classification of Crystals

Strukturbericht symbol	Structure name	Symmetry group		Pearson symbol[a]	Standard ASTM E157-82a symbol[b]
		International	Schoenflies		
1	**2**	**3**	**4**	**5**	**6**
A1	Cu	$Fm3m$	O_h^4	cF4	F
A2	W	$Im3m$	O_h^9	cI2	B
A3	Mg	$P6_3/mmc$	D_{6h}^4	hP2	H
A4	C	$Fd3m$	O_h^7	cF8	F
A5	Sn	If_1/amd	D_{4h}^{19}	tI4	U
A6	In	$I4/mmm$	D_{4h}^{17}	tI2	U
A7	As	$R\bar{3}m$	D_{3d}^5	hR2	R
A8	Se	$P3_121$ or $P3_221$	D_3^4 (D_3^6)	hP3	H
A10	Hg	$R\bar{3}m$	D_{3d}^5	hR1	R
A11	Ga	$Cmca$	D_{2h}^{18}	oC8	Q
A12	α-Mn	$I\bar{4}3m$	T_d^3	cI58	B
A13	β-Mn	$P4_132$	O^7	cP20	C
A15	OW_3	$Pm3n$	O_h^3	cP8	C
A20	α-U	$Cmcm$	D_{2h}^{17}	oC4	Q
B1	ClNa	$Fm3m$	O_h^5	cF8	F
B2	ClCs	$Pm3m$	O_h^1	cP2	C
B3	SZn	$F\bar{4}3m$	T_d^2	cF8	F
B4	SZn	$P6_3mc$	C_{6v}^4	hP4	H
B8$_1$	AsNi	$P6_3/mmc$	D_{6h}^4	hP4	H
B8$_2$	$InNi_2$	$P6_3/mmc$	D_{6h}^4	hP6	H
B9	HgS	$P3_121$ or $P3_221$	D_3^4 or D_3^6	hP6	H
B10	OPb	$P4/nmm$	D_{4h}^7	tP4	T
B11	γ-CuTi	$P4/nmm$	D_{4h}^7	tP4	T
B13	NiS	$R\bar{3}m$	D_{3d}^5	hR6	R
B16	GeS	$Pnma$	D_{2h}^{16}	oP8	O
B17	PtS	$P4_2/mmc$	D_{4h}^9	tP4	T
B18	CuS	$P6_3/mmc$	D_{6h}^4	hP12	H
B19	AuCd	$Pmma$	D_{2h}^5	oP4	O
B20	FeSi	$P2_13$	T^4	cP8	C
B27	BFe	$Pnma$	D_{2h}^{16}	oP8	O
B31	MnP	$Pnma$	D_{2h}^{16}	oP8	O
B32	NaTl	$Fd3m$	O_h^7	cF16	F
B34	Pds	$P4_2/m$	C_{4h}^2	tP16	T
B35	CoSn	$P6/mmm$	D_{6h}^1	hP6	H
B37	SeTl	$I4/mcm$	D_{4h}^{18}	tI16	U
B$_e$	CdSb	$Pbca$	D_{2h}^{15}	oP16	O
B$_f$ (B33)	ξ-BCr	$Cmcm$	D_{2h}^{17}	oC8	Q
B$_g$	BMo	$I4_1/amd$	D_{4h}^{19}	tI4	U
B$_h$	CW	$P6m2$	D_{3h}^1	hP2	H
B$_i$	γ'CMo (AsTi)	$P6_3/mmc$	D_{6h}^4	hP8	H
C1	CaF_2	$Fm\bar{3}m$	O_h^5	cF12	F
C1$_b$	AgAsMg	$F\bar{4}3m$	T_d^2	cF12	F
C2	FeS_2	$Pa3$	T_h^6	cP12	C
C3	Cu_2O	$Pn3m$	O_h^4	cP6	C
C4	O_2Ti	$P4_2/mnm$	D_{4h}^{14}	tP6	T
C6	CdI_2	$P3m1$	D_{3d}^3	hP3	H
C7	MoS_2	$P6_3/mmc$	D_{6h}^4	hP6	H
C11$_a$	C_2Ca	$I4/mmm$	D_{4h}^{17}	tI6	U
C11$_b$	$MoSi_2$	$I4/mmm$	D_{4h}^{17}	tI6	U

TABLE 5. Classification of Crystals

Strukturbericht symbol	Structure name	Symmetry group International	Schoenflies	Pearson symbol[a]	Standard ASTM E157-82a symbol[b]
1	2	3	4	5	6
C12	$CaSi_2$	$R\bar{3}m$	D_{3d}^5	hR6	R
C14	$MgZn_2$	$P6_3/mmc$	D_{6h}^4	hP12	H
C15	Cu_2Mg	Fd3m	O_h^7	cF24	F
C15$_b$	$AuBe_5$	$F\bar{4}3m$ or F23	T_d^2 or T^2	cF24	F
C16	Al_2Cu	I4/mcm	D_{4h}^{18}	tI12	U
C18	FeS_2	Pnnm	D_{2h}^{12}	oP6	O
C19	$CdCl_2$	$R\bar{3}m$	D_{3d}^5	hR3	R
C22	Fe_2P	$P2\bar{6}m$	D_{3h}^1	hP9	H
C23	Cl_2Pb	Pnma	D_{2h}^{16}	oP12	O
C32	AlB_2	P6/mmm	D_{6h}^1	hP3	H
C33	Bi_2STe_2	$R\bar{3}m$	D_{3d}^5	hR5	R
C34	$AuTe_2$	C2/m (P2/m)	C_{2h}^3 (C_{2h}^1)	mC6	N
C36	$MgNi_2$	$P6_3/mmc$	D_{6h}^4	hP24	H
C38	Cu_2Sb	P4/nmm	D_{4h}^7	tP6	T
C40	$CrSi_2$	$P6_222$	D_6^4	hP9	H
C42	SiS_2	Ibam	D_{2h}^{26}	oI12	P
C44	GeS_2	Fdd2	C_{2v}^{19}	oF72	S
C46	$AuTe_2$	Pma2	C_{2v}^4	oP24	O
C49	Si_2Zr	Cmcm	D_{2h}^{17}	oC12	Q
C54	Si_2Ti	Fddd	D_{2h}^{24}	oF24	S
C$_c$	Si_2Th	$I4_1/amd$	D_{4h}^{19}	tI12	U
C$_e$	$CoGe_2$	Aba2	C_{2v}^{17}	oC23	Q
DO$_2$	As_3Co	Im3	T_h^5	cI32	B
DO$_3$	BiF_3	Fm3m	O_h^5	cF16	F
DO$_9$	O_3Re	Pm3m	O_h^1	cP4	C
DO$_{11}$	CFe_3	Pnma	D_{2h}^{16}	oP16	O
DO$_{18}$	$AsNa_3$	$P6_3/mmc$	D_{6h}^4	hP8	H
DO$_{19}$	Ni_3Sn	$P6_3/mmc$	D_{6h}^4	hP8	H
DO$_{20}$	Al_3Ni	Pnma	D_{2h}^{16}	oP16	O
DO$_{21}$	Cu_3P	$P\bar{3}c1$	D_{3d}^4	hP24	H
DO$_{22}$	Cu_3P	I4/mmm	D_{4h}^{17}	tI8	U
DO$_{23}$	Al_3Zr	I4/mmm	D_{4h}^{17}	tI16	U
DO$_{24}$	Ni_3Ti	$P6_3/mmc$	D_{6h}^4	hP16	H
DO$_c$	SiU_3	I4/mcm	D_{4h}^{18}	tI16	U
DO$_e$	Ni_3P	$I\bar{4}$	S_4^2	tI32	U
D1$_3$	Al_4Ba	I4/mmm	D_{4h}^{17}	tI10	U
D1$_a$	$MoNi_4$	I4/m	C_{4h}^5	tI10	U
D1$_b$	Al_4U	Imma	D_{2h}^{28}	oI20	P
D1$_c$	$PtSn_4$	Aba2	C_{2v}^{17}	oC20	Q
D1$_e$	B_4Th	P4/mbm	D_{4h}^5	tP20	T
D1$_f$	BMn_4	Fddd	D_{2h}^{24}	oF40	S
D2$_1$	B_6Ca	Pm3m	O_h^1	cP7	C
D2$_3$	$NaZn_{13}$	Fm3m	O_h^5	cF112	F
D2$_b$	$Mn_{12}Th$	I4/mmm	D_{4h}^{17}	tI26	U
D2$_c$	MnU_6	I4/mcm	D_{4h}^{18}	tI28	U
D2$_d$	$CaCu_5$	P6/mmm	D_{6h}^1	hP6	H
D2$_f$	$B_{12}U$	Fm3m	O_h^5	cF52	F
D2$_h$	Al_6Mn	Cmcm	D_{2h}^{17}	oC28	Q
D5$_1$	α-Al_2O_3	R3c	D_{3d}^6	hR10	R
D5$_2$	La_2O_3	$P\bar{3}m1$	D_{3d}^3	hP5	H
D5$_3$	Mn_2O_3	Ia3	T_h^7	cI80	B
D5$_8$	S_3Sb_2	Pnma	D_{2h}^{16}	oP20	O
D5$_9$	P_2Zn_3	$P4_2/mmc$	D_{4h}^9	tP40	T
D5$_{10}$	C_2C_3	Pnma	D_{2h}^{16}	oP20	O
D5$_{13}$	Al_3Ni_2	$P\bar{3}m1$	D_{3d}^3	hP5	H
D5$_a$	Si_2U_3	P4/mbm	D_{4h}^5	tP10	T
D5$_c$	C_3Pu_2	$I\bar{4}3d$	T_d^6	cI40	B

TABLE 5. Classification of Crystals

Strukturbericht symbol	Structure name	Symmetry group International	Schoenflies	Pearson symbol[a]	Standard ASTM E157-82a symbol[b]
1	2	3	4	5	6
$D7_1$	Al_4C_3	$R\bar{3}m$	D^5_{3d}	hR7	R
$D7_3$	P_4Th_3	$I\bar{4}3d$	T^6_d	cI28	B
$D7_b$	B_4Ta_3	Immm	D^{25}_{2h}	oI14	P
$D8_1$	Fe_3Zn_{10}	Im3m	O^9_h	cI52	B
$D8_2$	Cu_5Zn_8	$I\bar{4}3m$	T^3_d	cI52	B
$D8_3$	Al_4Cu_9	P43m	T^1_d	cP52	C
$D8_4$	C_6Cr23	Fm3m	O^5_h	cF116	F
$D8_5$	Fe_7W_6	$R\bar{3}m$	D^5_{3d}	hR13	R
$D8_6$	$Cu_{15}Si_4$	$I\bar{4}3m$	T^3_d	cI76	B
$D8_8$	Mn_5Si_3	$P6_3/mcm$	D^3_{6h}	hP16	H
$D8_9$	Co_9S_8	Fm3m	O^5_h	cF68	F
$D8_{10}$	Al_8Cr_5	R3m	C^5_{3v}	hR26	R
$D8_{11}$	Al_5Co_2	$P6_3/mcm$	D^3_{6h}	hP28	H
$D8_a$	$Mn_{23}Th_6$	Fm3m	O^5_h	cF116	F
$D8_b$	σ-phase of Cr-Fe	$p\bar{4}_2/mnm$	D^{14}_{4h}	tP30	T
$D8_e$	$(Al,Zn)_{49}Mg_{32}$	Im3	T^5_h	cI162	B
$D8_f$	Ge_7Ir_3	Im3m	O^9_h	cI40	B
$D8_h$	B_5W_2	$P6_3/mmc$	D^4_{6h}	hP14	H
$D8_i$	B_5Mo_2	$R\bar{3}m$	D^5_{3d}	hR7	R
$D8_l$	B_3Cr_5	I4/mcm	D^{18}_{4h}	tI32	U
$D8_m$	Si_3W_5	I4/mcm	D^{18}_{4h}	tI32	U
$D10_1$	C_3Cr_7	P31c	C^4_{3v}	hP80	H
$D10_2$	Fe_3Th_7	$P6_3mc$	C^4_{6v}	hP20	H
$E0_1$	ClFPb	P4/nmm	D^7_{4h}	tP6	T
$E1_1$	$CuFeS_2$	$I\bar{4}2d$	D^{12}_{2d}	tI16	U
$E2_1$	CaO_3Ti	Pm3m	O^1_h	cP5	C
$E2_4$	S_3Sn_2	Pnma	D^{16}_{2h}	oP20	O
E3	Al_2CdS_4	$I\bar{4}$	S^2_4	tI14	U
$E9_3$	$SiFe_3W_3$	Fd3m	O^7_h	cF112	F
$E9_a$	Al_7Cu_2Fe	P4/mnc	D^6_{4h}	tP40	T
$E9_b$	$AlLi_3N_2$	Ia3	T^7_h	cI96	B
$F0_1$	NiSSb	$P2_13$	T^4	cP12	C
$F5_1$	$CrNaS_2$	R3m or R32	D^5_{3d} or D^7_3	hR4	R
$F5_6$	CuS_2Sb	Pnma	D^{16}_{2h}	oP16	O
$H1_1$	Al_2MgO_4	Fd3m	O^7_h	cF56	F
$H2_4$	Cu_3S_4V	P43m	T^1_d	cP8	C
$H2_5$	$AsCu_3S_4$	$Pmn2_1$	C^7_{2v}	oP16	O
$L1_0$	AuCu	P4/mmm	D^1_{4h}	tP4	T
$L1_2$	$AlCu_3$	Pm3m	O^1_h	cP4	C
$L2_1$	$AlCu_2Mn$	Fm3m	O^5_h	cF16	F
$L2_2$	Sb_2Tl_7	Im3m	O^9_h	cI54	B
$L'2_b$	H_2Th	I4/mmm	D^{17}_{4h}	tI6	U
$L'3$	Fe_2N	$P6_3/mmc$	D^4_{6h}	hP3	H
$L6_0$	$CuTi_3$	P4/mmm	D^1_{4h}	tP4	T

[a] The first letter denotes the crystal system: triclinic (a), monoclinic (m), orthorhombic (o), tetragonal (t), hexagonal (h) and cubic (c). Trigonal (rhombohedral) system is denoted by combination hR. The second letter of Pearson's symbol denotes lattice type: primitive (P), edge-(base-) centered (C), body-centered (I) or face-centered (F). The following number denotes number of atoms in the crystal unit cell.

[b] Standard ASTM E157-82a has the Bravais lattices designations as following: C – primitive cubic; B – body-centered cubic; F – face-centered cubic; T – primitive tetragonal; U – body-centered tetragonal; R – rhombohedral; H – hexagonal; O – primitive orthorhombic; P – body-centered orthorhombic; Q – base-centered orthorhombic; S – face-centered orthorhombic; M – primitive monoclinic; N – centered monoclinic; A – triclinic.

References

1. A. Schoenflies, *Kristallsysteme und Kristallstructur*, Leipzig, 1891.
2. E. S. Fedorow, Zusammenstellung der kristallographischen Resultate, *Zs. Krist.*, 20, 1892.

3. P. Groth, *Elemente der physikalischen und chemischen Krystallographie*, R. Oldenbourg, München/Berlin, 1921.
4. N. V. Belov, *Class Method of Deriving Space Groups of Symmetry*, Trudy Instituta Kristallodraffi imeni Fedorova (Transactions of the Fedorov Inst. of Crystallography), 5, 25, 1951, in Russian.
5. W. B. Pearson, *Handbook of Lattice Spacings and Structures of Metals and Alloys*, Vol. 1, Pergamon Press, 1958; Vol. 2, 1967.
6. Ch. Kittel, *Introduction to Solid State Physics*, John Wiley & Sons, 1956.
7. G. S. Zhdanov, *Fizika Tverdogo Tela (Solid State Physics)*, Moscow University Press, 1962, in Russian.
8. M. J. Buerger, *Elementary Crystallography*, John Wiley & Sons, 1963.
9. F. D. Bloss, *Crystallography & Crystal Chemistry*, Holt, Rinehart & Winston, 1971.
10. T. Janssen, *Crystallographic Groups*, North-Holland/American Elsevier, 1973.
11. M. P. Shaskolskaya, *Kristallografiya (Crystallography)*, Vysshaya Shkola, Moscow, 1976, in Russian.
12. T. Hahn, Ed., Internat. *Tables for Crystallography*, Vol. A, D. Reidel Publishing, Boston, 1983.
13. Crystal Data. Determinative Tables, Volumes 1–6, 1966–1983, JCPDS-Intern Centre for Diffraction Data and U.S. Dept. of Commerce.
14. R. W. G. Wyckoff, *Crystal Structures*, 2nd ed., Volumes 1–6, Interscience, New York, 1963.
15. C. J. Bradley and A. P. Cracknell, *The Mathematical Theory of Symmetry in Solids*, Clarendon Press, Oxford, 1972.
16. International Tables for Crystallography. Volume A, *Space–Group Symmetry*, T. Hahn, Ed., 1989; Volume B, *Reciprocal Space*, U. Schmueli, Ed.; Volume C, *Mathematical, Physical and Chemical Tables*, A. J. C. Wilson, Ed., Kluwer Academic Publishers, Dordrecht, 1989.
17. G. R. Desiraju, *Crystal Engineering: The Design of Organic Solids*, Elsevier, Amsterdam, 1989.
18. M. Senechal, *Crystalline Symmetries: An Informal Mathematical Introduction*, Adam Hilger Publ., Bristol, 1990.
19. C. Hammond, *Introduction to Crystallography*, Oxford University Press, 1990.
20. N.W. Alcock, *Bonding and Structure: Structural Principles in Inorganic and Organic Chemistry*, Ellis Norwood Publ., 1990.
21. T. C. W. Mak and G. D. Zhou. *Crystallography in Modern Chemistry: A Resource Book of Crystal Structures*, Wiley–Interscience, New York, 1992.
22. S. C. Abrahams, K. Mirsky, and R. M. Nielson, *Acta Cryst*, B52, 806 (1996); B52, 1057 (1996).
23. C. Marcos, A. Panalague, D. B. Morciras, S. Garcia-Granda and M. R. Dias. *Acta Cryst*, B52, 899 (1996).
24. A. C. Larson, *Crystallographic Computing*, Manksgaard, Copenhagen, 1970.
25. G.M. Sheldrick, SHELXS86. Crystallographic Computing 3, Clarendon Press, Oxford, 1986; SHELXL93. Program for the Refinement of Crystal Structures, University of Göttingen Press, 1993.
26. Inorganic Crystal Structure Database, CD–ROM. Sci. Inf. Service. E-mail: SISI@Delphi.com.

IONIC RADII IN CRYSTALS

Ionic radii are a useful tool for predicting and visualizing crystal structures. This table lists a set of ionic radii R_i in Å units for the most common coordination numbers CN of positive and negative ions. The values are based on experimental crystal structure determinations, supplemented by empirical relationships, and theoretical calculations. The notation sq after the coordination number indicates a square configuration, while py indicates pyramidal.

The advice of Howard T. Evans and Marvin J. Weber in preparing this table is appreciated.

References

1. Shannon, R. D., *Acta Crystallogr.* A32, 751, 1976.
2. Jia, Y. Q., *J. Solid State Chem.* 95, 184, 1991.

Ion	CN	R_i/Å	Ion	CN	R_i/Å	Ion	CN	R_i/Å
Anions				8	1.12	Eu^{+3}	6	0.95
F^{-1}	6	1.33		10	1.23		8	1.07
Cl^{-1}	6	1.81		12	1.34	F^{+7}	6	0.08
Br^{-1}	6	1.96	Cd^{+2}	4	0.78	Fe^{+2}	4	0.63
I^{-1}	6	2.20		6	0.95		6	0.61
OH^{-1}	4	1.35		8	1.10		8	0.92
	6	1.37		12	1.31	Fe^{+3}	4	0.49
O^{-2}	2	1.21	Ce^{+3}	6	1.01		6	0.55
	6	1.40		8	1.14		8	0.78
	8	1.42		10	1.25	Fr^{+1}	6	1.80
S^{-2}	6	1.84		12	1.34	Ga^{+3}	4	0.47
Se^{-2}	6	1.98	Ce^{+4}	6	0.87		6	0.62
Te^{-2}	6	2.21		8	0.97	Gd^{+3}	6	0.94
Cations				10	1.07		8	1.05
Ac^{+3}	6	1.12		12	1.14	Ge^{+2}	6	0.73
Ag^{+1}	4	1.00	Cf^{+3}	6	0.95	Ge^{+4}	4	0.39
	6	1.15	Cf^{+4}	6	0.82		6	0.53
	8	1.28		8	0.92	Hf^{+4}	4	0.58
Ag^{+2}	4sq	0.79	Cl^{+5}	3py	0.12		6	0.71
	6	0.94	Cl^{+7}	4	0.08		8	0.83
Al^{+3}	4	0.39	Cm^{+3}	6	0.97	Hg^{+1}	6	1.19
	5	0.48	Cm^{+4}	6	0.85	Hg^{+2}	2	0.69
	6	0.54		8	0.95		4	0.96
Am^{+3}	6	0.98	Co^{+2}	4	0.56		6	1.02
	8	1.09		6	0.65		8	1.14
Am^{+4}	6	0.85		8	0.90	I^{+5}	3py	0.44
	8	0.95	Co^{+3}	6	0.55		6	0.95
As^{+3}	6	0.58	Cr^{+2}	6	0.73	I^{+7}	4	0.42
As^{+5}	4	0.34	Cr^{+3}	6	0.62		6	0.53
	6	0.46	Cr^{+4}	4	0.41	In^{+3}	4	0.62
Au^{+1}	6	1.37		6	0.55		6	0.80
Au^{+3}	4sq	0.64	Cr^{+6}	4	0.26	Ir^{+3}	6	0.68
	6	0.85		6	0.44	Ir^{+4}	6	0.63
Ba^{+2}	6	1.35	Cs^{+1}	6	1.67	Ir^{+5}	6	0.57
	8	1.42		8	1.74	K^{+1}	4	1.37
	12	1.61		10	1.81		6	1.38
Be^{+2}	4	0.27		12	1.88		8	1.51
	6	0.45	Cu^{+1}	2	0.46		12	1.64
Bi^{+3}	5	0.96		4	0.60	La^{+3}	6	1.03
	6	1.03		6	0.77		8	1.16
	8	1.17	Cu^{+2}	4sq	0.57		10	1.27
Bi^{+5}	6	0.76		6	0.73		12	1.36
Bk^{+3}	6	0.96	Dy^{+2}	6	1.07	Li^{+1}	4	0.59
Bk^{+4}	6	0.83		8	1.19		6	0.76
	8	0.93	Dy^{+3}	6	0.91		8	0.92
Br^{+5}	3py	0.31		8	1.03	Lu^{+3}	6	0.86
Br^{+7}	4	0.25	Er^{+3}	6	0.89		8	0.97
	6	0.39		8	1.00	Mg^{+2}	4	0.57
C^{+4}	4	0.15	Eu^{+2}	6	1.17		6	0.72
	6	0.16		8	1.25		8	0.89
Ca^{+2}	6	1.00		10	1.35	Mn^{+2}	4	0.66

Ion	CN	R_i/Å	Ion	CN	R_i/Å	Ion	CN	R_i/Å
	6	0.83	Pr^{+3}	6	0.99	Tc^{+4}	6	0.65
	8	0.96		8	1.13	Te^{+4}	4	0.66
Mn^{+3}	6	0.58	Pr^{+4}	6	0.85		6	0.97
Mn^{+4}	4	0.39		8	0.96	Te^{+6}	4	0.43
	6	0.53	Pt^{+2}	4sq	0.60		6	0.56
Mn^{+5}	4	0.33		6	0.80	Th^{+4}	6	0.94
Mn^{+6}	4	0.26	Pt^{+4}	6	0.63		8	1.05
Mn^{+7}	4	0.25	Pu^{+3}	6	1.00		10	1.13
Mo^{+3}	6	0.69	Pu^{+4}	6	0.86		12	1.21
Mo^{+4}	6	0.65	Pu^{+5}	6	0.74	Ti^{+2}	6	0.86
Mo^{+5}	4	0.46	Pu^{+6}	6	0.71	Ti^{+3}	6	0.67
	6	0.61	Ra^{+2}	8	1.48	Ti^{+4}	4	0.42
Mo^{+6}	4	0.41		12	1.70		6	0.61
	6	0.59	Rb^{+1}	6	1.52		8	0.74
	7	0.73		8	1.61	Tl^{+1}	6	1.50
N^{+3}	6	0.16		10	1.66		8	1.59
N^{+5}	6	0.13		12	1.72		12	1.70
Na^{+1}	4	0.99	Re^{+4}	6	0.63	Tl^{+3}	4	0.75
	6	1.02	Re^{+5}	6	0.58		6	0.89
	8	1.18	Re^{+6}	6	0.55		8	0.98
	9	1.24	Re^{+7}	4	0.38	Tm^{+2}	6	1.01
	12	1.39		6	0.53		7	1.09
Nb^{+3}	6	0.72	Rh^{+3}	6	0.67	Tm^{+3}	6	0.88
	8	0.79	Rh^{+4}	6	0.60		8	0.99
Nb^{+4}	6	0.68	Rh^{+5}	6	0.55	U^{+3}	6	1.03
Nb^{+5}	4	0.48	Ru^{+3}	6	0.68	U^{+4}	6	0.89
	6	0.64	Ru^{+4}	6	0.62		8	1.00
	8	0.74	Ru^{+5}	6	0.57		12	1.17
Nd^{+3}	6	0.98	Ru^{+7}	4	0.38	U^{+5}	6	0.76
	8	1.12	Ru^{+8}	4	0.36	U^{+6}	2	0.45
	9	1.16	S^{+4}	6	0.37		4	0.52
	12	1.27	S^{+6}	4	0.12		6	0.73
Ni^{+2}	4sq	0.49		6	0.29		8	0.86
	6	0.69	Sb^{+3}	4py	0.76	V^{+2}	6	0.79
Ni^{+3}	6	0.56		6	0.76	V^{+3}	6	0.64
Np^{+3}	6	1.01	Sb^{+5}	6	0.60	V^{+4}	5	0.53
Np^{+4}	6	0.87	Sc^{+3}	6	0.75		6	0.58
Np^{+5}	6	0.75		8	0.87		8	0.72
Np^{+6}	6	0.72	Se^{+4}	6	0.50	V^{+5}	4	0.36
Os^{+4}	6	0.63	Se^{+6}	4	0.28		5	0.46
Os^{+5}	6	0.58		6	0.42		6	0.54
Os^{+6}	6	0.55	Si^{+4}	4	0.26	W^{+4}	6	0.66
Os^{+8}	4	0.39		6	0.40	W^{+5}	6	0.62
P^{+5}	4	0.17	Sm^{+2}	6	1.19	W^{+6}	4	0.42
	6	0.38		8	1.27		5	0.51
Pa^{+3}	6	1.04	Sm^{+3}	6	0.96		6	0.60
Pa^{+4}	6	0.90		8	1.08	Y^{+3}	6	0.90
Pa^{+5}	6	0.78		12	1.24		8	1.02
Pb^{+2}	6	1.19	Sn^{+4}	4	0.55		9	1.08
	8	1.29		6	0.69	Yb^{+2}	6	1.02
	10	1.40		8	0.81		8	1.14
	12	1.49	Sr^{+2}	6	1.18	Yb^{+3}	8	0.99
Pb^{+4}	4	0.65		8	1.26		9	1.04
	6	0.78		10	1.36	Zn^{+2}	4	0.60
	8	0.94		12	1.44		6	0.74
Pd^{+2}	4sq	0.64	Ta^{+3}	6	0.72		8	0.90
	6	0.86	Ta^{+4}	6	0.68	Zr^{+4}	4	0.59
Pd^{+3}	6	0.76	Ta^{+5}	6	0.64		6	0.72
Pd^{+4}	6	0.62	Tb^{+3}	6	0.92		8	0.84
Pm^{+3}	6	0.97		8	1.04		9	0.89
	8	1.09	Tb^{+4}	6	0.76			
Po^{+4}	6	0.97		8	0.88			

POLARIZABILITIES OF ATOMS AND IONS IN SOLIDS

H. P. R. Frederikse

The polarization of a solid dielectric medium, P, is defined as the dipole moment per unit volume averaged over the volume of a crystal cell. A component of P can be expanded as a function of the electric field E:

$$P_i = \sum_j a_j E_j + \sum_{jk} b_{jk} E_j E_k$$

For relatively small electric fields in isotropic substances $P = \chi_e E$, where χ_e is the electric susceptibility. If the medium is made up of N atoms (or ions) per unit volume, the polarization is $P = N p_m$ where p_m is the average dipole moment per atom. The polarizability α can be defined as $p_m = \alpha E_0$, where E_0 is the local field at the position of the atom. Using the Lorentz method to calculate the local field one finds:

$$P = N\alpha(E + 4\pi P) = \chi_e E$$

Together with the definition of the dielectric constant (relative permittivity), $\varepsilon = 1 + 4\pi\chi_e$, this leads to:

$$\alpha = \frac{3}{4\pi N}\frac{\varepsilon - 1}{\varepsilon + 2}$$

This expression is known as the Clausius-Mossotti equation.

The total polarization associated with atoms, ions, or molecules is due to three different sources:

1. Electronic polarization arises because the center of the local electronic charge cloud around the nucleus is displaced under the action of the field: $P_e = N\alpha_e E_0$ where α_e is the *electronic polarizability*.

2. Ionic polarization occurs in ionic materials because the electric field displaces cations and anions in opposite directions: $P_i = N\alpha_i E_0$, where α_i is the *ionic polarizability*.

3. Orientational polarization can occur in substances composed of molecules that have permanent electric dipoles. The alignment of these dipoles depends on temperature and leads to an *orientational polarizability* per molecule: $\alpha_{or} = p^2/3kT$, where p is the permanent dipole moment per molecule, k is the Boltzmann constant, and T is the temperature.

Because of the different nature of these three polarization processes the response of a dielectric solid to an applied electric field will strongly depend on the frequency of the field. The resonance of the electronic excitation in insulators (dielectrics) takes place in the ultraviolet part of the spectrum; the characteristic frequency of the lattice vibrations is located in the infrared, while the orientation of dipoles requires fields of much lower frequencies (below 10^{10} Hz). This response to electric fields of different frequencies is shown in Figure 1. Values of the electronic polarizabilities for selected atoms and ions are given in Table 1.

References

1. Kittel, C., *Introduction to Solid State Physics*, Fourth Edition, John Wiley & Sons, New York, 1971.
2. Lerner, R.G., and Trigg, G.L., Eds., *Encyclopedia of Physics*, Second Edition, VCH Publishers, New York, 1990.
3. Ralls, K.M., Courtney, T.H., and Wulff, J., *An Introduction to Materials Science and Engineering*, John Wiley & Sons, New York, 1976.

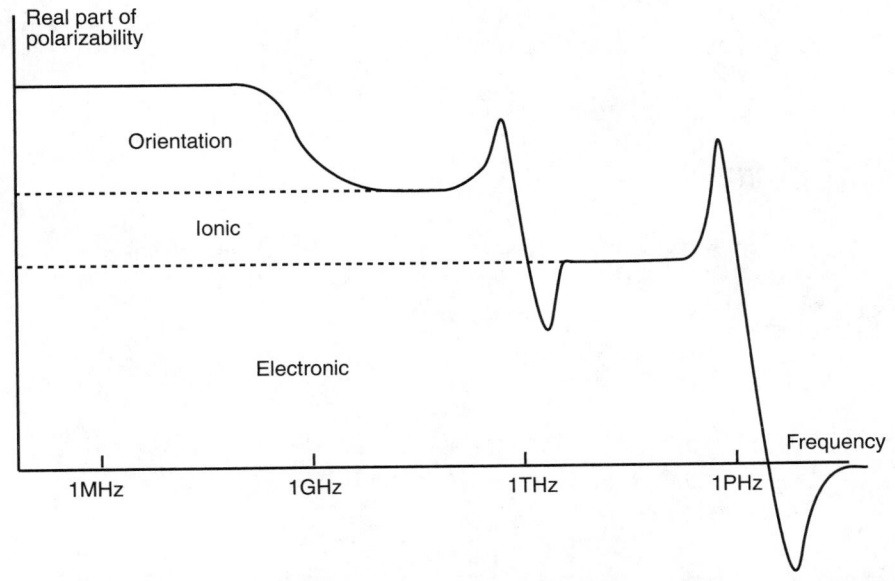

FIGURE 1. Schematic graph of the frequency dependence of the different contributions to polarizability.

TABLE 1. Electronic Polarizabilities in Units of 10^{-24} cm^3

						He
						0.201
Li$^+$	Be^{2+}	B^{3+}	C^{4+}	O^{2-}	F$^-$	Ne
0.029	0.008	0.003	0.0013	3.88	1.04	0.39
Na$^+$	Mg^{2+}	Al^{3+}	Si^{4+}	S^{2-}	Cl$^-$	Ar
0.179	0.094	0.052	0.0165	10.2	3.66	1.62
K$^+$	Ca^{2+}	Sc^{3+}	Ti^{4+}	Se^{2-}	Br$^-$	Kr
0.83	0.47	0.286	0.185	10.5	4.77	2.46
Rb$^+$	Sr^{2+}	Y^{3+}	Zr^{4+}	Te^{2-}	I$^-$	Xe
1.40	0.86	0.55	0.37	14.0	7.1	3.99
Cs$^+$	Ba^{2+}	La^{3+}	Ce^{4+}			
2.42	1.55	1.04	0.73			

Data from Pauling, L., *Proc. R. Soc.* London, A114, 181, 1927. See also Jaswal, S.S. and Sharma, T.P., *J. Phys. Chem. Solids*, 34, 509, 1973. Values are appropriate for cgs units. To convert to SI, use the relation $\alpha(\text{SI})/\text{C m}^2\text{V}^{-1} = 1.11265.10^{-16}\,\alpha(\text{cgs})/\text{cm}^3$

CRYSTAL STRUCTURES AND LATTICE PARAMETERS OF ALLOTROPES OF THE ELEMENTS

H. W. King

The crystal structures of the allotropic forms of the elements are presented in terms of the Pearson symbol, the Strukturbericht designation, and the prototype of the structure. The temperatures of the phase transformations are listed in degrees Celsius and the pressures are in the Gpa. A consistent nomenclature is used, whereby all allotropes are labeled by Greek letters. The lattice parameters of the units cells are given in nanometers (nm) and are considered to be accurate to ±2 in the last reported digit.

This compilation is restricted to changes in crystal structures that occur as a result of a change in temperature or pressure. Low-temperature structures are included for the diatomic and rare gases, which show many similarities with respect to the metallic elements. The elements identified with an asterisk (*) have polymorphic structures based on different molecular configurations. The crystal data given for these elements refer to the most stable structure at room temperature.

Reprinted with the permission of ASM International from T.B. Massalski, Ed., *Binary Alloy Phase Diagrams*, ASM International, Metals Park, Ohio, 1986; certain data on rare earth elements were provided by K.A. Gschneidner.

Element	Temperature, °C	Pressure, GPa	Pearson symbol	Space group	Strukturbericht designation	Prototype	Lattice parameters, nm a	b	c	Comment, c/a or α or β
Ac	25	atm	cF4	Fm3m	A1	Cu	0.5311	...	...	...
Ag	25	atm	cF4	Fm3m	A1	Cu	0.40857	...	...	...
αAl	25	atm	cF4	Fm3m	A1	Cu	0.40496	...	...	...
βAl	25	>20.5	hP2	P6₃/mmc	A3	Mg	0.2693	...	0.4398	1.6331
α′Am	25	atm	hP4	P6₃/mmc	A3′	αLa	0.34681	...	1.1241	2*1.621
αAm	>769	atm	cF4	Fm3m	A1	Cu	0.4894	...	...	...
βAm	>1074	atm	cI2	Im3m	A2	W	?	...	...	...
γAm	25	>15	oC4	Cmcm	A20	αU	0.3063	0.5968	0.5169	...
αAr	<-189.35	atm	cF4	Fm3m	A1	Cu	0.5316	...	...	...
(βAr)	<-189.40	atm	hP2	P6₃/mmc	A3	Mg	0.3760	...	0.6141	1.633
αAs	25	atm	hR2	R3̄m	A7	αAs	0.41319	...	......	α = 54.12°
εAs	>448	atm	oC8	Cmca	...	P(black)	0.362	1.085	0.448	...
Au	25	atm	cF4	Fm3m	A1	Cu	0.40782	...	...	...
βB	25	atm	hR105	R3m	...	βB	1.017	...	...	α = 65.12°
αBa	25	atm	cI2	Im3̄m	A2	W	0.50227	...	...	...
βBa	25	>5.33	hP2	P6₃/mmc	A3	Mg	0.3901	...	0.6154	1.5775
γBa	25	>23	?	?	...	...	...	...	...	...
αBe	25	atm	hP2	P6₃/mmc	A3	Mg	0.22859	...	0.35845	1.5681
βBe	>1270	atm	cI2	Im3̄m	A2	W	0.25515	...	...	...
γBe	25	>9.3	?	...	...	...	...	...	...	...
αBi	25	atm	hR2	r3̄m	A7	αAs	0.47460	...	...	α = 57.23°
βBi	25	>2.6	mC4	C2/m	...	βBi	0.6674	0.6117	0.3304	β = 110.33°
γBi	25	>3.0	mP3	?	...	...	0.605	0.42	0.465	β = 85.33°
σBi	25	>4.3	?	?	...	...	...	...	...	...
εBi	25	>6.5	?	?	...	...	...	...	...	...
ζBi	25	>9.0	cI2	Im3m	A2	W	0.3800	...	...	...
αBk	25	atm	hP4	P6₃/mmc	A3′	αLa	0.3416	...	1.1069	2*1.620
βBk	>977	atm	cF4	Fm3m	A1	Cu	0.4997	...	...	...
Br	<7.25	atm	oC8	Cmca	...	Cl	0.668	0.449	0.874	...
C (graphite)	25	atm	hP4	P6₃/mmc	A9	C (graphite)	0.24612	...	0.6709	2.7258
C (diamond)	25	>60	cF8	Fd3m	A4	C (diamond)	0.35669	...	...	...
C (hd)	25	HP	hP4	P6₃/mmc	...	C (hd)	0.2522	...	0.4119	1.633
αCa	25	atm	cF4	Fm3m	A1	Cu	0.55884	...	...	...
βCa	>443	atm	cI2	Im3m	A2	W	0.4480	...	...	...
γCa	25	>1.5	?	...	...	...	...	...	...	...
Cd	25	atm	hP2	P6₃/mmc	A3	Mg	0.29793	...	0.56196	1.8862
αCe	<-177	atm	cF4	Fm3m	A1	Cu	0.485	...	...	...
βCe	25	pressure	hP4	P6₃/mmc	A3′	αLa	0.36810	...	1.1857	2*1.611
γCe	25	atm	cF4	Fm3m	A1	Cu	0.51610	...	...	...
δ-Ce	>726	atm	cI2	Im3m	A2	W	0.412	...	...	...
α′Ce	25	>5.4	oC4	Cmcm	A20	αU	0.3049	0.5998	0.5215	...

Crystal Structures and Lattice Parameters of Allotropes of the Elements

Element	Temperature, °C	Pressure, GPa	Pearson symbol	Space group	Struktur-bericht designation	Prototype	Lattice parameters, nm			Comment, c/a or α or β
							a	b	c	
αCf	25	atm	hP4	P6₃/mmc	A3′	αLa	0.339	...	1.1015	2*1.625
βCf	>590	atm	cF4	Fm3m	A1	Cu	?	...	...	...
Cl	<-102	atm	oC8	Cmca	...	Cl	0.624	0.448	0.826	...
αCm	25	atm	hP4	P6₃/mmc	A3′	αLa	0.3496	...	1.1331	2*1.621
βCm	>1277	atm	cF4	Fm3m	A1	Cu	0.4382	...	...	...
eCo	25	atm	hP2	P6₃/mmc	A3	Mg	0.25071	...	0.40686	1.6228
αCo	>422	atm	cF4	Fm3m	A1	Cu	0.35447	...	...	...
αCr	25	atm	cI2	Im3m	A2	W	0.28848	...	...	...
α′Cr	25	HP	tI2	I4/mmm	...	α′Cr	0.2882	...	0.2887	1.002
aCs	25	atm	cI2	Im3m	A2	W	0.6141	...	...	...
βCs	25	>2.37	cF4	Fm3m	A1	Cu	0.6465	...	...	...
β′Cs	25	>4.22	cF4	Fm3m	A1	Cu	0.5800	...	...	...
γCs	25	>4.27	?	...	...	...	...	...	...	...
Cu	25	atm	cF4	Fm3m	A1	Cu	0.36146	...	...	...
α′Dy	<-187	atm	oC4	Cmcm	...	α′Dy	0.3595	0.6184	0.5678	...
αDy	25	atm	hP2	P6₃/mmc	A3	Mg	0.35915	...	0.56501	1.5732
βDy	>1381	atm	cI2	Im3m	A2	W	0.403	...	...	...
γDy	25	>7.5	hR3	R3̄m	...	αSm	0.3436	...	2.483	4.5*1.606
Er	25	atm	hP2	P6₃/mmc	A3	Mg	0.35592	...	0.55850	1.5692
αEs	25	atm	hP4	P6₃/mmc	A3′	αLa	?	...	...	...
βEs	?	atm	cF4	Fm3m	A1	Cu	?	...	...	...
Eu	25	atm	cI2	Im3m	A2	W	0.45827	...	...	...
αF	<-227.6	atm	mC8	C2/c	...	αF	0.550	0.338	0.728	β = 102.17°
βF	<-219.67	atm	cP16	Pm3n	...	γO	0.667	...	...	...
αFe	25	atm	cI2	Im3m	A2	W	0.28665	...	...	...
γFe	>912	atm	cF4	Fm3m	A1	Cu	0.36467	...	...	...
σFe	>1394	atm	cI2	Im3m	A2	W	0.29315	...	...	...
eFe	25	>13	hP2	P6₃/mmc	A3	Mg	0.2468	...	0.396	1.603
αGa	25	atm	oC8	Cmca	A11	αGa	0.45186	0.76570	0.45258	...
βGa	25	>1.2	tI2	I4/mmm	A6	In	0.2808	...	0.4458	1.588
γGa	-53	>3.0	oC40	Cmcm	...	γGa	1.0593	1.3523	0.5203	...
αGd	25	atm	hP2	P6₃/mmc	A3	Mg	0.36336	...	0.57810	0.5910
βGd	>1235	atm	cI2	Im3m	A2	W	0.406	...	...	...
γGd	25	>3.0	hR3	R3m	...	αSm	0.361	...	2.603	4*1.60
αGe	25	atm	cF8	Fd3m	A4	C (diamond)	0.56574	...	...	...
βGe	25	>12	tI4	I4₁/amd	A5	βSn	0.4884	...	0.2692	0.551
γGe	25	>12→atm	tP12	P4₁2₁2	...	σGe	0.593	...	0.698	1.18
σGe	LT	>12	cI16	Im3m	...	γSi	0.692	...	...	...
αH	<-271.9	atm	cF4	Fm3m	A1	Cu	0.5338	...	...	...
βH	<-259.34	atm	hP2	P6₃/mmc	A3	Mg	0.3776	...	0.6162	1.632
αHe	<-268.94	atm	hP2	P6₃/mmc	A3	Mg	0.3555	...	0.5798	1.631
βHe	>-258	0.125	cF4	Fm3m	A1	Cu	0.4240	...	...	...
γHe	<-271.47	0.03	cI2	Im3m	A2	W	0.4110	...	...	...
αHf	25	atm	hP2	P6₃/mmc	A3	Mg	0.31946	...	0.50510	1.5811
βHf	>1995	atm	cI2	Im3m	A2	W	0.3610	...	...	...
αHg	<-38.84	atm	hR1	R3̄m	A10	αHg	0.3005	...	...	α = 70.53°
βHg	<-194	HP	tI2	I4/mmm	...	βHg	0.3995	...	0.2825	0.707
γHg	<-194	c.w.	hR1	?	...	...	...	...	...	...
αHo	25	atm	hP2	P6₃/mmc	A3	Mg	0.35778	...	0.56178	1.5702
βHo	25	>7.5	hR3	R3̄m	...	αSm	0.334	...	2.45	4.5*1.63
I	25	atm	oC8	Cmca	...	Cl	0.72697	0.47903	0.97942	...
In	25	atm	tI2	I4/mmm	A6	In	0.3253	...	0.49470	1.5210
Ir	25	atm	cF4	Fm3m	A1	Cu	0.38392	...	...	...
K	25	atm	cI2	Im3m	A2	W	0.5321	...	...	...
Kr	<-157.39	atm	cF4	Fm3m	A1	Cu	0.5810	...	...	...
αLa	25	atm	hP4	P6₃/mmc	A3′	αLa	0.37740	...	1.2171	2*1.6125
βLa	>310	atm	cF4	Fm3m	A1	Cu	0.5303	...	...	...
γLa	>865	atm	cI2	Im3m	A2	W	0.426	...	...	...
β′La	25	>2.0	cF4	Fm3m	A1	Cu	0.517	...	...	...
αLi	<-193	atm	hP2	P6₃/mmc	A3	Mg	0.3111	...	0.5093	1.637

Crystal Structures and Lattice Parameters of Allotropes of the Elements

Element	Temperature, °C	Pressure, GPa	Pearson symbol	Space group	Strukturbericht designation	Prototype	Lattice parameters, nm			Comment, c/a or α or β
							a	b	c	
βLi	25	atm	cI2	Im3m	A2	W	0.35093	…	…	…
γLi	<-201	c.w.	cF4	Fm3m	A1	Cu	0.4388	…	…	…
Lu	25	atm	hP2	P6₃/mmc	A3	Mg	0.35052	…	0.55494	1.5832
Mg	25	atm	hP2	P6₃/mmc	A3	Mg	0.32094	…	0.52107	1.6236
αMn	25	atm	cI58	I4̄3m	A12	αMn	0.89126	…	…	…
βMn	>710	atm	cP20	P4₁32	A13	βMn	0.63152	…	…	…
γMn	>1079	atm	cF4	Fm3m	A1	Cu	0.3860	…	…	…
σMn	>1143	atm	cI2	Im3m	A2	W	0.3080	…	…	…
Mo	25	atm	cI2	Im3m	A2	W	0.31470	…	…	…
αN	<-237.6	atm	cP8	Pa3	…	αN	0.5661	…	…	…
βN	<-210.00	atm	hP4	P6₃/mmc	…	βN	0.4050	…	0.6604	1.631
γN	<-253	>3.3	tP4	P4₂/mnm	…	γN	0.3957	…	0.5109	1.291
αNa	<-233	atm	hP2	P6₃/mmc	A3	Mg	0.3767	…	0.6154	1.634
βNa	25	atm	cI2	Im3m	A2	W	0.42906	…	…	…
Nb	25	atm	cI2	Im3m	A2	W	0.33004	…	…	…
αNd	25	atm	hP4	P6₃/mmc	A3′	αLa	0.36582	…	1.17966	2*1.6124
βNd	>863	atm	cI2	Im3m	A2	W	0.413	…	…	…
γNd	25	>5.0	cF4	Fm3m	A1	Cu	0.480	…	…	…
Ne	<-243.59	atm	cF4	Fm3m	A1	Cu	0.4462	…	…	…
Ni	25	atm	cF4	Fm3m	A1	Cu	0.35240	…	…	…
αNp	25	atm	oP8	Pnma	A_c	αNp	0.6663	0.4723	0.4887	…
βNp	>280	atm	tP4	P42₁2	A_d	βNp	0.4883	…	0.3389	0.694
γNp	>576	atm	cI2	Im3m	A2	W	0.352	…	…	…
αO	<-243.3	atm	mC4	C2m	…	αO	0.5403	0.3429	0.5086	β = 132.53°
βO	<-229.6	atm	hR2	R3̄m	…	βO	0.4210	…	…	α = 46.27°
γO	<-218.79	atm	cP16	Pm3n	…	γO	0.683	…	…	…
Os	25	atm	hP2	P6₃/mmc	A3	Mg	0.27341	…	0.43918	1.6063
P (black)	25	atm	oC8	Cmca	…	P (black)	0.33136	1.0478	0.43763	…
αPa	25	atm	tI2	I4/mmm	A_a	αPa	0.3921	…	0.3235	0.825
βPa	>1170	atm	cI2	Im3m	A2	W	0.381	…	…	…
αPb	25	atm	cF4	Fm3m	A1	Cu	0.49502	…	…	…
βPb	25	>10.3	hP2	P6₃/mmc	A3	Mg	0.3265	…	0.5387	1.650
Pd	25	atm	cF4	Fm3m	A1	Cu	0.38903	…	…	…
αPm	25	atm	hP4	P6₃/mmc	A3′	αLa	0.365	…	1.165	2*1.60
βPm	>890	atm	cI2	Im3m	A2	W	(0.410)	…	…	…
αPo	25	atm	cP1	Pm3m	A_h	αPo	0.3366	…	…	…
βPo	>54	atm	hR1	R3̄m	…	βPo	0.3373	…	…	α = 98.08°
αPr	25	atm	hP4	P6₃/mmc	A3′	αLa	0.36721	…	1.18326	2*1.6111
βPr	>795	atm	cI2	Im3m	A2	W	0.413	…	…	…
γPr	25	>4.0	cF4	Fm3m	A1	Cu	0.488	…	…	…
Pt	25	atm	cF4	Fm3m	A1	Cu	0.39236	…	…	…
αPu	25	atm	mP16	P2₁/m	…	αPu	0.6183	0.4822	1.0963	β = 101.97°
βPu	>125	atm	mI34	I2/m	…	βPu	0.9284	1.0463	0.7859	β = 92.13°
γPu	>215	atm	oF8	Fddd	…	γPu	0.31587	0.57682	1.0162	…
σPu	>320	atm	cF4	Fm3m	A1	Cu	0.46371	…	…	…
σ′Pu	>463	atm	tI2	I4/mmm	A6	In	0.33261	…	0.44630	1.3418
εPu	>483	atm	cI2	Im3m	A2	W	0.36343	…	…	…
Ra	25	atm	cI2	Im3m	A2	W	0.5148	…	…	…
αRb	25	atm	cI2	Im3m	A2	W	0.5705	…	…	…
βRb	25	>1.08	?	…	…	…	…	…	…	…
γRb	25	>2.05	?	…	…	…	…	…	…	…
Re	25	atm	hP2	P6₃/mmc	A3	Mg	0.27609	…	0.4458	1.6145
Rh	25	atm	cF4	Fm3m	A1	Cu	0.38032	…	…	…
Ru	25	atm	hP2	P6₃/mmc	A3	Mg	0.27058	…	0.42816	1.5824
αS	25	atm	oF128	Fddd	A16	αS	1.0464	1.28660	2.44860	…
αSb	25	atm	hR2	R3̄m	A7	αAs	0.45067	…	…	α = 57.11°
βSb	25	>5.0	cP1	Pm3m	A_h	αPo	0.2992	…	…	…
γSb	25	>7.5	hP2	P6₃/mmc	A3	Mg	0.3376	…	0.5341	1.582
σSb	25	>14.0	mP3	?	…	…	0.556	0.404	0.422	β = 86.0°
αSc	25	atm	hP2	P6₃/mm	A3	Mg	0.33088	…	0.52680	1.5921

Crystal Structures and Lattice Parameters of Allotropes of the Elements

Element	Temperature, °C	Pressure, GPa	Pearson symbol	Space group	Strukturbericht designation	Prototype	Lattice parameters, nm			Comment, c/a or α or β
							a	b	c	
βSc	>1337	atm	cI2	Im3m	A2	W	(0.373)	...	...	...
γSe	25	atm	hP3	P3₁21	A8	γSe	0.43659	...	0.49537	1.1346
αSi	25	atm	cF8	Fd3m	A4	C (diamond)	0.54306	...	...	...
βSi	25	>9.5	tI4	I4₁/amd	A5	βSn	0.4686	...	0.2585	0.552
γSi	25	>16.0	cI16	Im3m	...	γSi	0.6636	...	...	...
σSi	25	>16→atm	hP4	P6₃/mmc	A3´	αLa	0.380	...	0.628	1.653
αSm	25	atm	hR3	R3̄m	...	αSm	0.36290	...	2.6207	4*1.6048
βSm	>734	atm	hP2	P6₃/mmc	A3	Mg	0.36630	...	0.58448	1.5956
γ´Sm	>922	atm	cI2	Im3m	A2	W	(0.410)	...	...	...
σSm	25	>4.0	hP4	P6₃/mmc	A3´	αLa	0.3618	...	1.166	2*1.611
αSn	<13	atm	cF8	Fd3m	A4	C (diamond)	0.64892	...	...	...
βSn	25	atm	tI4	I4₁/amd	A5	βSn	0.58318	...	0.31818	0.5456
γSn	25	>9.0	tI2	?	...	γSn	0.370	...	0.337	0.91
αSr	25	atm	cF4	Fm3m	A1	Cu	0.6084	...	...	...
βSr	>547	atm	cI2	Im3m	A2	W	0.487	...	...	...
β´Sr	25	>3.5	cI2	Im3m	A2	W	0.4437	...	...	...
Ta	25	atm	cI2	Im3m	A2	W	0.33030	...	...	...
α´Tb	<-53	atm	oC4	Cmcm	...	α´Dy	0.3605	0.6244	0.5706	...
aTb	25	atm	hP2	P6₃/mmc	A3	Mg	0.36055	...	0.56966	1.5800
βTb	>1289	atm	cI2	Im3m	A2	W	(0.407)	...	...	...
γTb	25	>6.0	hR3	R3̄m	...	αSm	0.341	...	2.45	4*1.60
Tc	25	atm	hP2	P6₃/mmc	A3	Mg	0.2738	...	0.4393	1.604
αTe	25	atm	hP3	P3₁21	A8	γSe	0.44566	...	0.59264	1.3298
βTe	25	>2.0	hR2	R3̄m	A7	αAs	0.469	...	...	α = 53.30°
γTe	25	>7.0	hR1	R3̄m	...	βPo	0.3002	...	...	α = 103.3°
αTh	25	atm	cF4	Fm3m	A1	Cu	0.50842	...	...	...
βTh	>1360	atm	cI2	Im3m	A2	W	0.411	...	...	...
αTi	25	atm	hP2	P6₃/mmc	A3	Mg	0.29506	...	0.46835	1.59873
βTi	>882	atm	cI2	Im3m	A2	W	0.33065	...	...	...
ωTi	25	HP→atm	hP3	P6/mmm	...	ωTi	0.4625	...	0.2813	0.6082
αTl	25	atm	hP2	P6₃/mmc	A3	Mg	0.34566	...	0.55248	1.5983
βTl	>230	atm	cI2	Im3m	A2	W	0.3879	...	...	...
γTl	25	HP	cF4	Fm3m	A1	Cu	?	...	...	...
Tm	25	atm	hP2	P6₃/mmc	A3	Mg	0.35375	...	0.55540	1.5700
αU	25	atm	oC4	Cmcm	A20	αU	0.28537	0.58695	0.49548	...
βU	>668	atm	tP30	P4₂/mnm	A_b	βU	1.0759	...	0.5656	0.526
γU	>776	atm	cI2	Im3m	A2	W	0.3524	...	...	...
V	25	atm	cI2	Im3m	A2	W	0.30240	...	...	...
W	25	atm	cI2	Im3m	A2	W	0.31652	...	...	...
Xe	<-111.76	atm	cF4	Fm3m	A1	Cu	0.6350	...	...	...
αY	25	atm	hP2	P6₃/mmc	A3	Mg	0.36482	...	0.57318	1.5711
βY	>1478	atm	cI2	Im3m	A2	W	(0.410)	...	...	...
αYb	<-3	atm	hP2	P6₃/mmc	A3	Mg	0.38799	...	0.63859	1.6459
βYb	25	atm	cF4	Fm3m	A1	Cu	0.54848	...	...	...
γYb	>795	atm	cI2	Im3m	A2	W	0.444	...	...	...
Zn	25	atm	hP2	P6₃/mmc	A3	Mg	0.26650	...	0.49470	1.8563
αZr	25	atm	hP2	P6₃/mmc	A3	Mg	0.32316	...	0.51475	1.5929
βZr	>863	atm	cI2	Im3m	A2	W	0.36090	...	...	...
ωZr	25	HP→atm	hP2	P6/mmm	...	ωTi	0.5036	...	0.3109	0.617

LATTICE ENERGIES

H. D. B. Jenkins and H. K. Roobottom

THERMOCHEMICAL CYCLE AND CALCULATED VALUES

Table 1 contains calculated values of the lattice energies (total lattice potential energies), U_{POT}, of crystalline salts, M_aX_b. U_{POT} is expressed in units of kilojoules per mole, kJ mol^{-1}. M and X can be either simple or complex ions. Substances are arranged by chemical class.

Also listed in the table is the lattice energy, U_{POT}^{BHFC}, obtained from the application of the Born - Fajans - Haber cycle (BHFC) described below, using the "Standard Thermochemical Properties of Chemical Substances" table in Section 5 of this *Handbook*, References 1 through 4, and certain other data which are given in Table 3 below.

The lattice enthalpy, ΔH_L, is given by the cycle:

where (ss) is the standard state of the element concerned.

The lattice enthalpy, ΔH_L, is obtained using the equation:

$$\Delta H_L = a\Delta_f H°(M^{b+}, g) + b\Delta_f H°(X^{a-}, g) - \Delta_f H°(M_aX_b, c)$$

and is futher related to the total lattice potential energy, U_{POT}, by the relationship:

$$\Delta H_L = U_{POT} + \left[a\left(\frac{n_M}{2} - 2\right) + b\left(\frac{n_X}{2} - 2\right) \right] RT$$

where n_M and n_X equal 3 for monatomic ions, 5 for linear polyatomic ions and 6 for polyatomic non-linear ions.

METHOD OF ESTIMATION OF VALUES NOT TABULATED

In cases where the lattice energy is not tabulated and we want to furnish an estimate, then the Kapustinskii equation[5] can be used to obtain a value (in kJ mol^{-1}):

$$U_{POT} = \frac{121.4 z_a z_b \nu}{(r_a + r_b)} \left[1 - \frac{0.0345}{(r_a + r_b)} \right]$$

where z_a and z_b are the moduli of the charges on the ν ions in the lattice and r_a and r_b (in nm) are the thermochemical radii given in Table 2. The r_a for metal ions is taken to be the Goldschmidt[6] radius.

To cite an example, if we wish to estimate the lattice energy of the salt $[NH_4^+][HF_2^-]$ using the above procedure, we see that Table 2 gives the thermochemical radius (r_a) for NH_4^+ to be 0.136 nm and that for HF_2^- (r_b) to be 0.172 nm. The lattice potential energy is then estimated to be 700 kJ mol^{-1} compared with the calculated value of 705 kJ mol^{-1} and the Born - Fajans - Haber cycle value of 658 kJ mol^{-1}.

References

1. Wagman, D. D., Evans, W. H., Parker, V. B., Schumm, R. H., Halow, I., Bailey, S. M., Churney, K. L., and Nuttall, R. L., The NBS Tables of Chemical Thermodynamic Properties, *J. Phys. Chem. Ref. Data*, Vol. 11, Suppl. 2, 1982.
2. Chase, M. W., Davies, C. A., Downey, J. R., Frurip, D. J., McDonald, R. A., and Syverud, A. N., JANAF Thermochemical Tables, Third Edition, *J. Phys. Chem. Ref. Data*, Vol. 14, Suppl. 1, 1985.
3. Lias, S. G., Bartmess, J. E., Liebman, J. F., Holmes, J. L., Levin, R. D., and Mallard, W. G., Gas-Phase Ion and Neutral Thermochemistry, *J. Phys. Chem. Ref. Data*, Vol. 17, Suppl. 1, 1988.
4. Jenkins, H. D. B., and Pratt, K. F., *Adv. Inorg. Chem. Radiochem.*, 22, 1, 1978.
5. Kapustinskii, A. F., *Quart. Rev.*, 10, 283-294, 1956.
6. Goldschmidt, V. M., *Skrifter Norske Videnskaps-Akad.* Oslo, I, Mat.-Naturn. Kl, 1926. See also Dasent, W. E., *Inorganic Energetics*, 2nd ed., Cambridge University Press, 1982.
7. Jenkins, H. D. B., Roobottom, H. K., Passmore, J., and Glasser, L., *J. Chem. Education*, in press.

Table 1. Lattice Energies (kJ mol^{-1})

Substance	Calc. U_{POT}	U_{POT}^{BHFC}
Acetates		
Li(CH$_3$COO)	–	843
Na(CH$_3$COO)	828	807
K(CH$_3$COO)	749	726
Rb(CH$_3$COO)	715	
Cs(CH$_3$COO)	682	
Acetylides		
CaC$_2$	2911	2902
SrC$_2$	2788	2782
BaC$_2$	2647	2652
Azides		
LiN$_3$	861	875
NaN$_3$	770	784
KN$_3$	697	–
RbN$_3$	674	691
CsN$_3$	665	674
AgN$_3$	854	910
TlN$_3$	689	742
Ca(N$_3$)$_2$	2186	2316
Sr(N$_3$)$_2$	2056	2187
Ba(N$_3$)$_2$	2021	–
Mn(N$_3$)$_2$	2408	2348
Cu(N$_3$)$_2$	2730	2738
Zn(N$_3$)$_2$	2840	2970
Cd(N$_3$)$_2$	2446	2576
Pb(N$_3$)$_2$	–	2300
Bihalide Salts		
LiHF$_2$	821	847
NaHF$_2$	755	748
KHF$_2$	657	660
RbHF$_2$	627	631
CsHF$_2$	607	–
NH$_4$HF$_2$	705	658
CsHCl$_2$	601	–
Me$_4$NHCl$_2$	427	–
Et$_4$NHCl$_2$	346	–
Bu$_4$NHCl$_2$	290	–
Bicarbonates		
NaHCO$_3$	820	656
KHCO$_3$	741	573
RbHCO$_3$	707	522
CsHCO$_3$	678	520
NH$_4$HCO$_3$	–	577
Borides		
CaB$_6$	5146	–
SrB$_6$	5104	–
BaB$_6$	5021	–
YB$_6$	7447	–
LaB$_6$	7406	–
CeB$_6$	10083	–
PrB$_6$	7447	–
NdB$_6$	7447	–
PmB$_6$	7406	–
SmB$_6$	7447	–
EuB$_6$	5104	–
GdB$_6$	7489	–
TbB$_6$	7489	–
DyB$_6$	7489	–
HoB$_6$	7489	–
ErB$_6$	7489	–
TmB$_6$	7489	–
YbB$_6$	5146	–
LuB$_6$	7489	–
ThB$_6$	10167	–
Borohydrides		
LiBH$_4$	778	–
NaBH$_4$	703	694
KBH$_4$	655	638
RbBH$_4$	648	–
CsBH$_4$	628	–
Borohalides		
LiBF$_4$	699	749
NaBF$_4$	657	674
KBF$_4$	611	616
RbBF$_4$	577	590
CsBF$_4$	556	565
NH$_4$BF$_4$	582	–
KBCl$_4$	506	497
RbBCl$_4$	489	486
CsBCl$_4$	473	–
Carbonates		
Li$_2$CO$_3$	2523	2254
Na$_2$CO$_3$	2301	2016
K$_2$CO$_3$	2084	1846
Rb$_2$CO$_3$	2000	1783
Cs$_2$CO$_3$	1920	1722
MgCO$_3$	3138	3122
CaCO$_3$	2804	2811
SrCO$_3$	2720	2688
BaCO$_3$	2615	2554
MnCO$_3$	3046	3092
FeCO$_3$	3121	3169
CoCO$_3$	3443	3235
CuCO$_3$	3494	–
ZnCO$_3$	3121	3273
CdCO$_3$	2929	3052
SnCO$_3$	2904	–
PbCO$_3$	2728	2750
Cyanates		
LiNCO	849	–
NaNCO	807	816
KNCO	726	734
RbNCO	692	–
CsNCO	661	–
NH$_4$NCO	724	–
Cyanides		
LiCN	874	–
NaCN	766	759
KCN	692	686
RbCN	638	–
CsCN	601	–
Ca(CN)$_2$	2268	2240
Sr(CN)$_2$	2138	–
Ba(CN)$_2$	2001	2009
NH$_4$CN	617	691
AgCN	(741)	935

Substance	Calc. U_{POT}	U_{POT}^{BHFC}	Substance	Calc. U_{POT}	U_{POT}^{BHFC}
Zn(CN)$_2$	2809	2817	AgI	881	892
Cd(CN)$_2$	2583	2591	AuCl	1013	1066
Formates			AuBr	1029	1059
Li(HCO$_2$)	865	–	AuI	1027	1070
Na(HCO$_2$)	791	804	InCl	–	764
K(HCO$_2$)	713	722	InBr	–	767
Rb(HCO$_2$)	685	–	InI	–	733
Cs(HCO$_2$)	651	–	TlF	–	850
NH$_4$(HCO$_2$)	715	–	TlCl	738	751
			TlBr	720	734
Germanates			TlI	692	710
Mg$_2$GeO$_4$	7991	–	Me$_4$NCl	566	–
Ca$_2$GeO$_4$	7301	7306	Me$_4$NBr	553	–
Sr$_2$GeO$_4$	6987	–	Me$_4$NI	544	–
Ba$_2$GeO$_4$	6653	6643	PH$_4$Br	616	–
			PH$_4$I	590	–
Halates			BeF$_2$	3464	3526
LiBrO$_3$	883	880	BeCl$_2$	3004	3033
NaBrO$_3$	803	791	BeBr$_2$	2950	2914
KBrO$_3$	740	722	BeI$_2$	2780	2813
RbBrO$_3$	720	705	MgF$_2$	2926	2978
CsBrO$_3$	694	681	MgCl$_2$	2477	2540
NaClO$_3$	770	785	MgBr$_2$	2406	2451
KClO$_3$	711	721	MgI$_2$	2293	2340
RbClO$_3$	690	703	CaF$_2$	2640	2651
CsClO$_3$	–	679	CaCl$_2$	2268	2271
LiIO$_3$	975	974	CaBr$_2$	2132	–
NaIO$_3$	883	876	CaI$_2$	1971	2087
KIO$_3$	820	780	SrF$_2$	2476	2513
RbIO$_3$	791	–	SrCl$_2$	2142	2170
CsIO$_3$	761	–	SrI$_2$	1984	1976
			BaF$_2$	2347	2373
Halides			BaCl$_2$	2046	2069
LiF	1030	1049	BaBr$_2$	1971	1995
LiCl	834	864	BaI$_2$	1862	1890
LiBr	788	820	RaF$_2$	2284	–
LiI	730	764	RaCl$_2$	2004	–
NaF	910	930	RaBr$_2$	1929	–
NaCl	769	790	RaI$_2$	1803	–
NaBr	732	754	ScCl$_2$	2380	–
NaI	682	705	ScBr$_2$	2291	–
KF	808	829	ScI$_2$	2201	–
KCl	701	720	TiF$_2$	2724	–
KBr	671	691	TiCl$_2$	2439	2514
KI	632	650	TiBr$_2$	2360	2430
RbF	774	795	TiI$_2$	2259	2342
RbCl	680	695	VCl$_2$	2607	2593
RbBr	651	668	VBr$_2$	–	2534
RbI	617	632	VI$_2$	–	2470
CsF	744	759	CrF$_2$	2778	2939
CsCl	657	670	CrCl$_2$	2540	2601
CsBr	632	647	CrBr$_2$	2377	2536
CsI	600	613	CrI$_2$	2269	2440
FrF	715	–	MoCl$_2$	2737	2746
FrCl	632	–	MoBr$_2$	2742	2753
FrBr	611	–	MoI$_2$	2630	–
FrI	582	–	MnF$_2$	2644	–
CuCl	992	996	MnCl$_2$	2510	2551
CuBr	969	978	MnBr$_2$	2448	2482
CuI	948	966	MnI$_2$	2212	–
AgF	953	974	FeF$_2$	2849	2967
AgCl	910	918			
AgBr	897	905			

Substance	Calc. U_{POT}	U_{POT}^{BHFC}	Substance	Calc. U_{POT}	U_{POT}^{BHFC}
$FeCl_2$	2569	2641	$CrCl_3$	5518	5529
$FeBr_2$	2515	2577	$CrBr_3$	5355	–
FeI_2	2439	2491	CrI_3	5275	5294
CoF_2	3004	3042	MoF_3	6459	–
$CoCl_2$	2707	2706	$MoCl_3$	5246	5253
$CoBr_2$	2640	2643	$MoBr_3$	5156	–
CoI_2	2569	2561	MoI_3	5073	–
NiF_2	3098	3089	MnF_3	6017	–
$NiCl_2$	2753	2786	$MnCl_3$	5544	–
$NiBr_2$	2729	2721	$MnBr_3$	5448	–
NiI_2	2607	2637	MnI_3	5330	–
$PdCl_2$	2778	2818	$TcCl_3$	5270	–
$PdBr_2$	2741	2751	$TcBr_3$	5215	–
PdI_2	2748	2760	TcI_3	5188	–
CuF_2	3046	3102	FeF_3	5870	–
$CuCl_2$	2774	2824	$FeCl_3$	5364	5436
$CuBr_2$	2715	2774	$FeBr_3$	5333	5347
CuI_2	2640	–	FeI_3	5117	–
AgF_2	2942	2967	$RuCl_3$	5245	5257
ZnF_2	3021	3053	$RuBr_3$	5223	5232
$ZnCl_2$	2703	2748	RuI_3	5222	5235
$ZnBr_2$	2648	2689	CoF_3	5991	–
ZnI_2	2581	2619	$RhCl_3$	5641	5665
CdF_2	2809	2830	IrF_3	(6112)	–
$CdCl_2$	2552	2565	$IrBr_3$	(4794)	–
$CdBr_2$	2507	2517	NiF_3	(6111)	–
CdI_2	2441	2455	AuF_3	(5777)	–
HgF_2	2757	–	$AuCl_3$	(4605)	–
$HgCl_2$	2657	2664	$ZnCl_3$	5832	–
$HgBr_2$	2628	2639	$ZnBr_3$	5732	–
HgI_2	2628	2624	ZnI_3	5636	–
SnF_2	2551	–	AlF_3	5924	6252
$SnCl_2$	2297	2310	$AlCl_3$	5376	5513
$SnBr_2$	2251	2256	$AlBr_3$	5247	5360
SnI_2	2193	2206	AlI_3	5070	5227
PbF_2	2535	2543	GaF_3	5829	6238
$PbCl_2$	2270	2282	$GaCl_3$	5217	5665
$PbBr_2$	2219	2230	$GaBr_3$	4966	5569
PbI_2	2163	2177	GaI_3	4611	5496
ScF_3	5492	5540	$InCl_3$	4736	5183
$ScCl_3$	4874	4901	$InBr_3$	4535	5117
$ScBr_3$	4729	4761	InI_3	4234	5001
ScI_3	4640	–	TlF_3	5493	–
YF_3	4983	–	$TlCl_3$	5258	5278
YCl_3	4506	4524	$TlBr_3$	5171	–
YI_3	4240	4258	TlI_3	5088	–
TiF_3	5644	–	$AsBr_3$	5497	5365
$TiCl_3$	5134	5153	AsI_3	4824	5295
$TiBr_3$	5012	5023	SbF_3	5295	5324
TiI_3	4845	–	$SbCl_3$	5032	4857
$ZrCl_3$	–	4791	$SbBr_3$	4954	4776
$ZrBr_3$	–	4758	SbI_3	4867	4692
ZrI_3	–	4591	$BiCl_3$	4689	4707
VF_3	5895	–	BiI_3	3774	–
VCl_3	5322	5329	LaF_3	4682	–
VBr_3	5214	5224	$LaCl_3$	4263	4242
VI_3	5121	5136	$LaBr_3$	4209	–
$NbCl_3$	5062	–	LaI_3	3916	3986
$NbBr_3$	4980	–	$CeCl_3$	4394	4348
NbI_3	4860	–	CeI_3	–	4061
CrF_3	6033	6065	$PrCl_3$	4322	4387

Substance	Calc. U_{POT}	U_{POT}^{BHFC}	Substance	Calc. U_{POT}	U_{POT}^{BHFC}
PrI_3	–	4101	RbH	686	684
$NdCl_3$	4343	4415	CsH	648	653
$SmCl_3$	4376	4450	VH	1184	(1344)
$EuCl_3$	4393	4490	NbH	1163	(1633)
$GdCl_3$	4406	4495	PdH	979	1368
$DyCl_3$	4481	4529	CuH	828	1254
$HoCl_3$	4501	4572	TiH	996	1407
$ErCl_3$	4527	4591	ZrH	916	1590
$TmCl_3$	4548	4608	HfH	904	–
TmI_3	–	4340	LaH	828	–
$YbCl_3$	–	4651	TaH	1021	–
$AcCl_3$	4096	–	CrH	1050	–
UCl_3	4243	–	NiH	929	–
$NpCl_3$	4268	–	PtH	937	–
$PuCl_3$	4289	–	AgH	941	–
$PuBr_3$	(3959)	–	AuH	1033	1108
$AmCl_3$	4293	–	TlH	745	–
TiF_4	10012	9908	GeH	950	–
$TiCl_4$	9431	–	PbH	778	–
$TiBr_4$	9288	9059	BeH_2	3205	3306
TiI_4	9108	8918	MgH_2	2791	2718
ZrF_4	8853	8971	CaH_2	2410	2406
$ZrCl_4$	8021	8144	SrH_2	2250	2265
$ZrBr_4$	7661	7984	BaH_2	2121	2133
ZrI_4	7155	7801	ScH_2	2711	2744
MoF_4	8795	–	YH_2	(2598)	2733
$MoCl_4$	8556	9603	LaH_2	2380	2522
$MoBr_4$	8510	9500	CeH_2	2414	2509
MoI_4	8427	–	PrH_2	2448	2405
$SnCl_4$	8355	8930	NdH_2	2464	2394
$SnBr_4$	7970	8852	PmH_2	2519	–
PbF_4	9519	–	SmH_2	2510	2389
CrF_2Cl	5795	–	GdH_2	2494	2651
CrF_2Br	5753	–	AcH_2	2372	–
CrF_2I	5669	–	ThH_2	2711	2738
$CrCl_2Br$	5448	–	PuH_2	2519	–
$CrCl_2I$	5381	5429	AmH_2	2544	–
$CrBr_2I$	5330	5370	TiH_2	2866	2864
$CuFCl$	2891	–	ZrH_2	2711	2999
$CuFBr$	2853	–	CuH_2	2941	–
$CuFI$	2803	–	ZnH_2	2870	–
$CuClBr$	2753	–	HgH_2	2707	–
$CuClI$	2694	–	AlH_3	5924	5969
$CuBrI$	2669	–	FeH_3	5724	–
FeF_2Cl	5711	–	ScH_3	5439	–
FeF_2Br	5653	–	YH_3	5063	4910
FeF_2I	5569	–	LaH_3	4895	4493
$FeCl_2Br$	5339	–	FeH_3	5724	–
$FeCl_2I$	5272	–	GaH_3	5690	–
$FeBr_2I$	5209	–	InH_3	5092	–
$LiIO_2F_2$	845	–	TlH_3	5092	–
$NaIO_2F_2$	766	756			
KIO_2F_2	699	689	**Hydroselenides**		
$RbIO_2F_2$	674	–	$NaHSe$	703	732
$CsIO_2F_2$	636	–	$KHSe$	644	712
$NH_4IO_2F_2$	678	–	$RbHSe$	623	689
$AgIO_2F_2$	736	685	$CsHse$	598	669
Hydrides			**Hydrosulphides**		
LiH	916	918	$LiHS$	768	862
NaH	807	807	$NaHS$	723	771
KH	711	713	$RbHS$	655	682

Substance	Calc. U_{POT}	U_{POT}^{BHFC}	Substance	Calc. U_{POT}	U_{POT}^{BHFC}
CsHS	628	657	$CsNO_3$	648	650
NH_4HS	661	718	$AgNO_3$	820	832
$Ca(HS)_2$	2184	(2171)	$TlNO_3$	690	707
$Sr(HS)_2$	2063	–	$Mg(NO_3)_2$	2481	2521
$Ba(HS)_2$	1979	(1956)	$Ca(NO_3)_2$	2268	2247
			$Sr(NO_3)_2$	2176	2151
Hydroxides			$Ba(NO_3)_2$	2062	2035
LiOH	1021	1028	$Mn(NO_3)_2$	2318	2478
NaOH	887	892	$Fe(NO_3)_2$	–	(2580)
KOH	789	796	$Co(NO_3)_2$	2560	2647
RbOH	766	765	$Ni(NO_3)_2$	–	2729
CsOH	721	732	$Cu(NO_3)_2$	–	2739
$Be(OH)_2$	3477	3620	$Zn(NO_3)_2$	2376	2649
$Mg(OH)_2$	2870	2998	$Cd(NO_3)_2$	2238	2462
$Ca(OH)_2$	2506	2637	$Sn(NO_3)_2$	2155	2254
$Sr(OH)_2$	2330	2474	$Pb(NO_3)_2$	2067	2208
$Ba(OH)_2$	2142	2330			
$Ti(OH)_2$	–	2953	*Nitrides*		
$Mn(OH)_2$	2909	3008	ScN	7547	7506
$Fe(OH)_2$	2653	3044	LaN	6876	6793
$Co(OH)_2$	2786	3109	TiN	8130	8033
$Ni(OH)_2$	2832	3186	ZrN	7633	7723
$Pd(OH)_2$	–	3189	VN	8283	8233
$Cu(OH)_2$	2870	3229	NbN	7939	8022
CuOH	1006	–	CrN	8269	8358
AgOH	918	845			
AuOH	1033	–	*Nitrites*		
TlOH	705	874	$NaNO_2$	774	772
$Zn(OH)_2$	2795	3151	KNO_2	699	687
$Cd(OH)_2$	2607	2909	$RbNO_2$	724	765
$Hg(OH)_2$	2669	–	$CsNO_2$	690	–
$Sn(OH)_2$	2489	2721			
$Pb(OH)_2$	2376	–	*Oxides*		
$Sc(OH)_3$	5063	5602	Li_2O	2799	2814
$Y(OH)_3$	4707	–	Na_2O	2481	2478
$La(OH)_3$	4443	–	K_2O	2238	2232
$Cr(OH)_3$	5556	6299	Rb_2O	2163	2161
$Mn(OH)_3$	6213	–	Cs_2O	2131	2063
$Al(OH)_3$	5627	–	Cu_2O	3273	3189
$Ga(OH)_3$	5732	6368	Ag_2O	3002	2910
$In(OH)_3$	5280	–	Tl_2O	2659	2575
$Tl(OH)_3$	5314	–	LiO_2	(878)	(872)
$Ti(OH)_4$	9456	–	NaO_2	799	821
$Zr(OH)_4$	8619	–	KO_2	741	751
$Mn(OH)_4$	10933	–	RbO_2	706	721
$Sn(OH)_4$	9188	9879	CsO_2	679	696
			Li_2O_2	2592	2557
Imides			Na_2O_2	2309	22717
CaNH	3293	–	K_2O_2	2114	2064
SrNH	3146	–	Rb_2O_2	2025	1994
BaNH	2975	–	Cs_2O_2	1948	1512
			MgO_2	3356	3526
Metavanadates			CaO_2	3144	3132
Li_3VO_4	3945	–	SrO_2	3037	2977
Na_3VO_4	3766	–	KO_3	697	707
K_3VO_4	3376	–	BeO	4514	4443
Rb_3VO_4	3243	–	MgO	3795	3791
Cs_3VO_4	3137	–	CaO	3414	3401
			SrO	3217	3223
Nitrates			BaO	3029	3054
$LiNO_3$	848	854	TiO	3832	3811
$NaNO_3$	755	763	VO	3932	3863
KNO_3	685	694	MnO	3724	3745
$RbNO_3$	662	671			

Lattice Energies

12-25

Substance	Calc. U_{POT}	U_{POT}^{BHFC}	Substance	Calc. U_{POT}	U_{POT}^{BHFC}
FeO	3795	3865	Ca(ClO$_4$)$_2$	1958	1971
CoO	3837	3910	Sr(ClO$_4$)$_2$	1862	1862
NiO	3908	4010	Ba(ClO$_4$)$_2$	1795	1769
PdO	3736	–			
CuO	4135	4050	**Permanganates**		
ZnO	4142	3971	NaMnO$_4$	661	–
CdO	3806	–	KMnO$_4$	607	–
HgO	3907	–	RbMnO$_4$	586	–
GeO	3919	–	CsMnO$_4$	565	–
SnO	3652	–	Ca(MnO$_4$)$_2$	1937	
PbO	3520	–	Sr(MnO$_4$)$_2$	1845	
Sc$_2$O$_3$	13557	13708	Ba(MnO$_4$)$_2$	1778	
Y$_2$O$_3$	12705	–			
La$_2$O$_3$	12452	–	**Phosphates**		
Ce$_2$O$_3$	12661	–	Mg$_3$(PO$_4$)$_2$	11632	11407
Pr$_2$O$_3$	12703	–	Ca$_3$(PO$_4$)$_2$	10602	10479
Nd$_2$O$_3$	12736	–	Sr$_3$(PO$_4$)$_2$	10125	10075
Pm$_2$O$_3$	12811	–	Ba$_3$(PO$_4$)$_2$	9652	9654
Sm$_2$O$_3$	12878	–	MnPO$_4$	7397	–
Eu$_2$O$_3$	12945	–	FePO$_4$	7251	7300
Gd$_2$O$_3$	12996	–	BPO$_4$	8201	–
Tb$_2$O$_3$	13071	–	AlPO$_4$	7427	7507
Dy$_2$O$_3$	13138	–	GaPO$_4$	7381	–
Ho$_2$O$_3$	13180	–			
Er$_2$O$_3$	13263	–	**Selenides**		
Tm$_2$O$_3$	13322	–	Li$_2$Se	2364	–
Yb$_2$O$_3$	13380	–	Na$_2$Se	2130	–
Lu$_2$O$_3$	13665	–	K$_2$Se	1933	–
Ac$_2$O$_3$	12573	–	Rb$_2$Se	1837	–
Ti$_2$O$_3$	–	14149	Cs$_2$Se	1745	–
V$_2$O$_3$	15096	14520	Ag$_2$Se	2686	–
Cr$_2$O$_3$	15276	14957	Tl$_2$Se	2209	–
Mn$_2$O$_3$	15146	15035	BeSe	3431	–
Fe$_2$O$_3$	14309	14774	MgSe	3071	–
Al$_2$O$_3$	15916	–	CaSe	2858	2862
Ga$_2$O$_3$	15590	15220	SrSe	2736	–
In$_2$O$_3$	13928	–	BaSe	2611	–
Pb$_2$O$_3$	(14841)	–	MnSe	3176	–
CeO$_2$	9627	–			
ThO$_2$	10397	–	**Selenites**		
PaO$_2$	10573	–	Li$_2$SeO$_3$	2171	–
VO$_2$(g)	10644	–	Na$_2$SeO$_3$	1950	1916
NpO$_2$	10707	–	K$_2$SeO$_3$	1774	1749
PuO$_2$	10786	–	Rb$_2$SeO$_3$	1715	1675
AmO$_2$	10799	–	Cs$_2$SeO$_3$	1640	–
CmO$_2$	10832	–	Tl$_2$SeO$_3$	1879	–
TiO$_2$	12150	–	Ag$_2$SeO$_3$	2113	2148
ZrO$_2$	11188	–	BeSeO$_3$	3322	–
MoO$_2$	11648	–	MgSeO$_3$	3012	2998
MnO$_2$	12970	–	CaSeO$_3$	2732	–
SiO$_2$	13125	–	SrSeO$_3$	2586	2588
GeO$_2$	12828	–	BaSeO$_3$	2460	2451
SnO$_2$	11807	–			
PbO$_2$	11217	–	**Selenates**		
			Li$_2$SeO$_4$	2054	–
Perchlorates			Na$_2$SeO$_4$	1879	–
LiClO$_4$	709	715	K$_2$SeO$_4$	1732	–
NaClO$_4$	643	641	Rb$_2$SeO$_4$	1686	–
KClO$_4$	599	595	Cs$_2$SeO$_4$	1615	–
RbClO$_4$	564	576	Cu$_2$SeO$_4$	2201	–
CsClO$_4$	636	550	Ag$_2$SeO$_4$	2033	–
NH$_4$ClO$_4$	583	580	Tl$_2$SeO$_4$	1766	–
			Hg$_2$SeO$_4$	2163	–
			BeSeO$_4$	3448	–

Substance	Calc. U_{POT}	U_{POT}^{BHFC}	Substance	Calc. U_{POT}	U_{POT}^{BHFC}
$MgSeO_4$	2895	–	Cs_2PdCl_6	1426	–
$CaSeO_4$	2632	–	Rb_2PbCl_6	1343	1343
$SrSeO_4$	2489	–	Cs_2PbCl_6	1344	–
			$(NH_4)_2PbCl_6$	1355	–
Sulphides			K_2PtCl_6	1468	1471
Li_2S	2464	2472	Rb_2PtCl_6	1464	–
Na_2S	2192	2203	Cs_2PtCl_6	1444	–
K_2S	1979	(2052)	$(NH_4)_2PtCl_6$	1468	–
Rb_2S	1929	1949	Tl_2PtCl_6	1546	–
Cs_2S	1892	1850	Ag_2PtCl_6	1773	1881
$(NH_4)_2S$	2008	(2026)	$BaPtCl_6$	2047	2070
Cu_2S	2786	2865	K_2PtBr_6	1423	1392
Ag_2S	2606	2677	Ag_2PtBr_6	1791	2276
Au_2S	2908	–	K_2PtI_6	1421	–
Tl_2S	2298	2258	K_2ReCl_6	1416	1442
			Rb_2ReCl_6	1414	–
Sulphates			Cs_2ReCl_6	1398	–
Li_2SO_4	2229	2142	K_2ReBr_6	1375	1375
Na_2SO_4	1827	1938	K_2SiF_6	1670	1765
K_2SO_4	1700	1796	Rb_2SiF_6	1639	1673
Rb_2SO_4	1636	1748	Cs_2SiF_6	1604	1498
Cs_2SO_4	1596	1658	Tl_2SiF_6	1675	–
$(NH_4)_2SO_4$	1766	1777	K_2SnCl_6	1363	1390
Cu_2SO_4	2276	2166	Rb_2SnCl_6	1361	1363
Ag_2SO_4	2104	1989	Cs_2SnCl_6	1358	–
Tl_2SO_4	1828	1722	Tl_2SnCl_6	1437	–
Hg_2SO_4	–	2127	$(NH_4)_2SnCl_6$	1370	1344
$CaSO_4$	2489	2480	Rb_2SnBr_6	1309	–
$SrSO_4$	2577	2484	Cs_2SnBr_6	1306	–
$BaSO_4$	2469	2374	Rb_2SnI_6	1226	–
$MnSO_4$	2920	2825	Cs_2SnBr_6	1243	–
			K_2TeCl_6	1318	1320
Ternary Salts			Rb_2TeCl_6	1321	–
Cs_2CuCl_4	1393	–	Cs_2TeCl_6	1323	–
Rb_2ZnCl_4	1529	–	Tl_2TeCl_6	1392	–
Cs_2ZnCl_4	1492	–	$(NH_4)_2TeCl_6$	1318	–
Rb_2ZnBr_4	1498	–	K_2RuCl_6	1451	–
Cs_2ZnBr_4	1454	–	Rb_2CoF_6	1688	–
Cs_2ZnI_4	1386	–	Cs_2CoF_6	1632	–
$CsGaCl_4$	494	–	K_2NiF_6	1721	–
$NaAlCl_4$	556	–	Rb_2NiF_6	1688	–
$CsAlCl_4$	486	–	Rb_2SbCl_6	1357	–
$NaFeCl_4$	492	–	Rb_2SeCl_6	1409	–
Rb_2CoCl_4	1447	–	Cs_2SeCl_6	1397	–
Cs_2CoCl_4	1391	–	$(NH_4)_2SeCl_6$	1420	–
K_2PtCl_4	1574	1550	$(NH_4)_2PoCl_6$	1338	–
Cs_2GeF_6	1573	–	Cs_2PoBr_6	1286	–
$(NH_4)_2GeF_6$	1657	–	Cs_2CrF_6	1603	–
Cs_2GeCl_6	1404	1419	Rb_2MnF_6	1688	–
K_2HfCl_6	1345	1461	Cs_2MnF_6	1620	–
K_2IrCl_6	1442	1440	K_2MnCl_6	1462	–
Na_2MoCl_6	1526	1504	Rb_2MnCl_6	1451	–
K_2MoCl_6	1418	1412	$(NH_4)_2MnCl_6$	1464	–
Rb_2MoCl_6	1399	1399	Cs_2TeBr_6	1306	–
Cs_2MoCl_6	1347	1347	Cs_2TeI_6	1246	–
K_2NbCl_6	1375	1398	K_2TiCl_6	1412	1447
Rb_2NbCl_6	1371	1385	Rb_2TiCl_6	1415	1416
Cs_2NbCl_6	1381	1344	Cs_2TiCl_6	1402	1384
K_2OsCl_6	1447	1447	Tl_2TiCl_6	1560	1553
Cs_2OsCl_6	1409	–	K_2TiBr_6	1379	1379
K_2OsBr_6	1396	–	Rb_2TiBr_6	1341	1331
K_2PdCl_6	1481	1493			
Rb_2PdCl_6	1449	–			

Substance	Calc. U_{POT}	U_{POT}^{BHFC}
Cs_2TiBr_6	1339	1306
Na_2UBr_6	1504	–
K_2UBr_6	1484	–
Rb_2UBr_6	1473	–
Cs_2UBr_6	1459	–
K_2WCl_6	1398	1423
Rb_2WCl_6	1397	1434
Cs_2WCl_6	1392	1366
K_2WBr_6	1408	1408
Rb_2WBr_6	1361	1391
Cs_2WBr_6	1362	1332
K_2ZrCl_6	1339	1371
Rb_2ZrCl_6	1341	–
Cs_2ZrCl_6	1339	1307
Tellurides		
Li_2Te	2212	–
Na_2Te	1997	2095
K_2Te	1830	–
Rb_2Te	1837	–
Cs_2Te	1745	–
Cu_2Te	2706	2683
Ag_2Te	2607	2600
Tl_2Te	2084	2172
BeTe	3319	–
MgTe	2878	3081
CaTe	2721	–

Substance	Calc. U_{POT}	U_{POT}^{BHFC}
Thiocyanates		
LiCNS	764	(765)
NaCNS	682	682
KCNS	623	616
RbCNS	623	619
CsCNS	623	568
NH4CNS	605	611
Ca(CNS)2	2184	2118
Sr(CNS)2	2063	1957
Ba(CNS)2	1979	1852
Mn(CNS)2	2280	2351
Zn(CNS)2	2335	2560
Cd(CNS)2	2201	2374
Hg(CNS)2	2146	2492
Sn(CNS)2	2117	2142
Pb(CNS)2	2058	–
Vanadates		
LiVO3	810	–
NaVO3	761	–
KVO3	686	–
RbVO3	657	–
CsVO3	628	–

TABLE 2. Thermochemical Radii (nm)

Ion	Radius		Ion	Radius	
Singly Charged Anions			$GaCl_4^-$	0.328	± 0.019
			H^-	0.148	± 0.019
AgF_4^-	0.231	± 0.019	$H_2AsO_4^-$	0.227	± 0.019
$AlBr_4^-$	0.321	± 0.023	$H_2PO_4^-$	0.213	± 0.019
$AlCl_4^-$	0.317	± 0.019	HCO_2^-	0.200	± 0.019
AlF_4^-	0.214	± 0.023	HCO_3^-	0.207	± 0.019
AlH_4^-	0.226	± 0.019	HF_2^-	0.172	± 0.019
AlI_4^-	0.374	± 0.019	HSO_4^-	0.221	± 0.019
AsF_6^-	0.243	± 0.019	I^-	0.211	± 0.019
AsO_2^-	0.211	± 0.019	I_2Br^-	0.261	± 0.019
$Au(CN)_2^-$	0.266	± 0.019	I_3^-	0.272	± 0.019
$AuCl_4^-$	0.288	± 0.019	I_4^-	0.300	± 0.019
AuF_4^-	0.240	± 0.019	IBr_2^-	0.251	± 0.019
AuF_6^-	0.235	± 0.038	ICl_2^-	0.235	± 0.019
$B(OH)_4^-$	0.229	± 0.019	ICl_4^-	0.307	± 0.019
BF_4^-	0.205	± 0.019	$IO_2F_2^-$	0.233	± 0.019
BH_4^-	0.205	± 0.019	IO_3^-	0.218	± 0.019
Br^-	0.190	± 0.019	IO_4^-	0.231	± 0.019
BrF_4^-	0.231	± 0.019	IrF_6^-	0.242	± 0.019
BrO_3^-	0.214	± 0.019	MnO_4^-	0.220	± 0.019
$CF_3SO_3^-$	0.230	± 0.049	MoF_6^-	0.241	± 0.019
$CH_3CO_2^-$	0.194	± 0.019	$MoOF_5^-$	0.241	± 0.019
Cl^-	0.168	± 0.019	N_3^-	0.180	± 0.019
ClO_2^-	0.195	± 0.019	NCO^-	0.193	± 0.019
ClO_3^-	0.208	± 0.019	$NbCl_6^-$	0.338	± 0.049
ClO_4^-	0.225	± 0.019	NbF_6^-	0.254	± 0.019
$ClS_2O_6^-$	0.260	± 0.049	$Nb_2F_{11}^-$	0.311	± 0.038
CN^-	0.187	± 0.023	NbO_3^-	0.194	± 0.019
$Cr_3O_8^-$	0.276	± 0.019	NH_2^-	0.168	± 0.019
$CuBr_4^-$	0.315	± 0.019	$NH_2CH_2COO^-$	0.252	± 0.019
F^-	0.126	± 0.019	NO_2^-	0.187	± 0.019
$FeCl_4^-$	0.317	± 0.019			

Ion	Radius		Ion	Radius	
NO_3^-	0.200	± 0.019	$MoCl_6^{2-}$	0.338	± 0.019
O_2^-	0.165	± 0.019	MoF_6^{2-}	0.274	± 0.019
O_3^-	0.199	± 0.034	MoO_4^{2-}	0.231	± 0.019
OH^-	0.152	± 0.019	$NbCl_6^{2-}$	0.343	± 0.019
OsF_6^-	0.252	± 0.020	NH^{2-}	0.128	± 0.019
PaF_6^-	0.249	± 0.019	$Ni(CN)_4^{2-}$	0.322	± 0.019
PdF_6^-	0.252	± 0.019	NiF_4^{2-}	0.211	± 0.019
PF_6^-	0.242	± 0.019	NiF_6^{2-}	0.249	± 0.019
PO_3^-	0.204	± 0.019	O^{2-}	0.141	± 0.019
PtF_6^-	0.247	± 0.019	O_2^{2-}	0.167	± 0.019
PuF_5^-	0.239	± 0.019	$OsBr_6^{2-}$	0.365	± 0.019
ReF_6^-	0.240	± 0.019	$OsCl_6^{2-}$	0.336	± 0.019
ReO_4^-	0.227	± 0.019	OsF_6^{2-}	0.276	± 0.019
RuF_6^-	0.242	± 0.019	$PbCl_4^{2-}$	0.279	± 0.019
S_6^-	0.305	± 0.019	$PbCl_6^{2-}$	0.347	± 0.019
SCN^-	0.209	± 0.019	PbF_6^{2-}	0.268	± 0.019
$SbCl_6^-$	0.320	± 0.019	$PdBr_6^{2-}$	0.354	± 0.019
SbF_6^-	0.252	± 0.019	$PdCl_4^{2-}$	0.313	± 0.019
$Sb_2F_{11}^-$	0.312	± 0.038	$PdCl_6^{2-}$	0.333	± 0.019
$Sb_3F_{14}^-$	0.374	± 0.038	PdF_6^{2-}	0.252	± 0.019
$SeCl_5^-$	0.258	± 0.038	$PoBr_6^{2-}$	0.380	± 0.019
$SeCN^-$	0.230	± 0.019	PoI_6^{2-}	0.428	± 0.019
SeH^-	0.195	± 0.019	$Pt(NO_2)_3Cl_3^{2-}$	0.364	± 0.019
SH^-	0.191	± 0.019	$Pt(NO_2)_4Cl_2^{2-}$	0.383	± 0.019
SO_3F^-	0.214	± 0.019	$Pt(OH)_2^{2-}$	0.333	± 0.019
$S_3N_3^-$	0.231	± 0.038	$Pt(SCN)_6^{2-}$	0.451	± 0.019
$S_3N_3O_4^-$	0.252	± 0.038	$PtBr_4^{2-}$	0.324	± 0.019
$TaCl_6^-$	0.352	± 0.019	$PtBr_6^{2-}$	0.363	± 0.019
TaF_6^-	0.250	± 0.019	$PtCl_4^{2-}$	0.307	± 0.019
TaO_3^-	0.192	± 0.019	$PtCl_6^{2-}$	0.333	± 0.019
UF_6^-	0.301	± 0.019	PtF_6^{2-}	0.245	± 0.019
VF_6^-	0.235	± 0.019	$PuCl_6^{2-}$	0.349	± 0.019
VO_3^-	0.201	± 0.019	$ReBr_6^{2-}$	0.371	± 0.019
WCl_6^-	0.337	± 0.019	$ReCl_6^{2-}$	0.337	± 0.019
WF_6^-	0.246	± 0.019	ReF_6^{2-}	0.256	± 0.019
WOF_5^-	0.241	± 0.019	ReF_8^{2-}	0.276	± 0.019
			ReH_9^{2-}	0.257	± 0.019
Doubly Charged Anions			ReI_6^{2-}	0.421	± 0.026
AmF_6^{2-}	0.255	± 0.019	RhF_6^{2-}	0.240	± 0.019
$Bi_2Br_8^{2-}$	0.392	± 0.055	$RuCl_6^{2-}$	0.336	± 0.019
$Bi_6Cl_{20}^{2-}$	0.501	± 0.073	RuF_6^{2-}	0.248	± 0.019
$CdCl_4^{2-}$	0.307	± 0.019	S^{2-}	0.189	± 0.019
$CeCl_6^{2-}$	0.352	± 0.019	$S_2O_3^{2-}$	0.251	± 0.019
CeF_6^{2-}	0.249	± 0.019	$S_2O_4^{2-}$	0.262	± 0.019
CO_3^{2-}	0.189	± 0.019	$S_2O_5^{2-}$	0.270	± 0.019
$CoCl_4^{2-}$	0.306	± 0.019	$S_2O_6^{2-}$	0.283	± 0.019
CoF_4^{2-}	0.209	± 0.019	$S_2O_7^{2-}$	0.275	± 0.019
CoF_6^{2-}	0.256	± 0.019	$S_2O_8^{2-}$	0.291	± 0.019
$Cr_2O_7^{2-}$	0.292	± 0.019	$S_3O_6^{2-}$	0.302	± 0.019
CrF_6^{2-}	0.253	± 0.019	$S_4O_6^{2-}$	0.325	± 0.019
CrO_4^{2-}	0.229	± 0.019	$S_6O_6^{2-}$	0.382	± 0.019
$CuCl_4^{2-}$	0.304	± 0.019	ScF_6^{2-}	0.276	± 0.019
CuF_4^{2-}	0.213	± 0.019	Se^{2-}	0.181	± 0.019
$GeCl_6^{2-}$	0.335	± 0.019	$SeBr_6^{2-}$	0.363	± 0.019
GeF_6^{2-}	0.244	± 0.019	$SeCl_6^{2-}$	0.336	± 0.019
HfF_6^{2-}	0.248	± 0.019	SeO_4^{2-}	0.229	± 0.019
HgI_4^{2-}	0.377	± 0.019	SiF_6^{2-}	0.248	± 0.019
$IrCl_6^{2-}$	0.332	± 0.019	SiO_3^{2-}	0.195	± 0.019
$MnCl_6^{2-}$	0.314	± 0.031	SmF_4^{2-}	0.218	± 0.019
MnF_4^{2-}	0.219	± 0.019	$Sn(OH)_6^{2-}$	0.279	± 0.020
MnF_6^{2-}	0.241	± 0.019	$SnBr_6^{2-}$	0.374	± 0.019
$MoBr_6^{2-}$	0.364	± 0.019			

Ion	Radius		Ion	Radius	
$SnCl_6^{2-}$	0.345	± 0.019	$Ni(NO_2)_6^{4-}$	0.383	± 0.038
SnF_6^{2-}	0.265	± 0.019	NiF_6^{3-}	0.250	± 0.042
SnI_6^{2-}	0.427	± 0.019	O^{3-}	0.288	± 0.038
SO_3^{2-}	0.204	± 0.019	P^{3-}	0.224	± 0.042
SO_4^{2-}	0.218	± 0.019	PaF_8^{3-}	0.299	± 0.042
$TcBr_6^{2-}$	0.363	± 0.019	PO_4^{3-}	0.230	± 0.042
$TcCl_6^{2-}$	0.337	± 0.019	PrF_6^{3-}	0.281	± 0.038
TcF_6^{2-}	0.244	± 0.019	$Rh(NO_2)_6^{3-}$	0.345	± 0.038
TcH_9^{2-}	0.260	± 0.019	$Rh(SCN)_6^{3-}$	0.428	± 0.042
TcI_6^{2-}	0.419	± 0.019	TaF_8^{3-}	0.284	± 0.042
Te^{2-}	0.220	± 0.019	TbF_7^{3-}	0.290	± 0.038
$TeBr_6^{2-}$	0.383	± 0.019	$Tc(CN)_6^{5-}$	0.410	± 0.042
$TeCl_6^{2-}$	0.353	± 0.019	ThF_7^{3-}	0.282	± 0.042
TeI_6^{2-}	0.430	± 0.019	$TiBr_6^{3-}$	0.315	± 0.038
TeO_4^{2-}	0.238	± 0.019	TlF_6^{3-}	0.271	± 0.038
$Th(NO_3)_6^{2-}$	0.424	± 0.019	UF_7^{3-}	0.285	± 0.042
$ThCl_6^{2-}$	0.360	± 0.019	YF_6^{3-}	0.275	± 0.038
ThF_6^{2-}	0.263	± 0.019	ZrF_7^{3-}	0.273	± 0.038
$TiBr_6^{2-}$	0.356	± 0.019			
$TiCl_6^{2-}$	0.335	± 0.019	**Singly Charged Cations**		
TiF_6^{2-}	0.252	± 0.019	$N(CH_3)_4^+$	0.234	± 0.019
UCl_6^{2-}	0.354	± 0.019	$N_2H_5^+$	0.158	± 0.019
UF_6^{2-}	0.256	± 0.019	$N_2H_6^{2+}$	0.158	± 0.029
VO_3^{2-}	0.204	± 0.019	$NH(C_2H_5)_3^+$	0.274	± 0.019
WBr_6^{2-}	0.363	± 0.019	$NH_3C_2H_5^+$	0.193	± 0.019
WCl_6^{2-}	0.339	± 0.019	$NH_3C_3H_7^+$	0.225	± 0.019
WO_4^{2-}	0.237	± 0.019	$NH_3CH_3^+$	0.177	± 0.019
$WOCl_5^{2-}$	0.334	± 0.019	NH_3OH^+	0.147	± 0.019
$ZnBr_4^{2-}$	0.335	± 0.019	NH_4^+	0.136	± 0.019
$ZnCl_4^{2-}$	0.306	± 0.019	$NH_3C_2H_4OH^+$	0.203	± 0.019
ZnF_4^{2-}	0.219	± 0.019	$As_3S_4^+$	0.244	± 0.027
ZnI_4^{2-}	0.384	± 0.019	$As_3Se_4^+$	0.253	± 0.027
$ZrBr_4^{2-}$	0.334	± 0.019	$AsCl_4^+$	0.221	± 0.027
$ZrCl_4^{2-}$	0.306	± 0.019	Br_2^+	0.155	± 0.027
$ZrCl_6^{2-}$	0.348	± 0.019	Br_3^+	0.204	± 0.027
ZrF_6^{2-}	0.258	± 0.019	Br_3^-	0.238	± 0.027
			Br_5^+	0.229	± 0.027
Multi-Charged Anions			$BrClCNH_2^+$	0.175	± 0.027
AlH_6^{3-}	0.256	± 0.042	BrF_2^+	0.183	± 0.027
AsO_4^{3-}	0.237	± 0.042	BrF_4^+	0.172	± 0.027
$CdBr_6^{4-}$	0.374	± 0.038	$C_{10}F_8^+$	0.265	± 0.027
$CdCl_6^{4-}$	0.352	± 0.038	$C_6F_6^+$	0.228	± 0.027
CeF_6^{3-}	0.278	± 0.038	$Cl(SNSCN)_2^+$	0.347	± 0.027
CeF_7^{3-}	0.282	± 0.038	$Cl_2C=NH_2^+$	0.173	± 0.027
$Co(CN)_6^{3-}$	0.349	± 0.038	Cl_2F^+	0.165	± 0.027
$Co(NO_2)_6^{3-}$	0.343	± 0.038	Cl_3^+	0.182	± 0.027
$CoCl_5^{3-}$	0.320	± 0.038	ClF_2^+	0.147	± 0.027
CoF_6^{3-}	0.258	± 0.042	ClO_2^+	0.118	± 0.027
$Cr(CN)_6^{3-}$	0.351	± 0.038	$GaBr_4^-$	0.317	± 0.038
CrF_6^{3-}	0.254	± 0.042	I_2^+	0.185	± 0.027
$Cu(CN)_4^{3-}$	0.312	± 0.038	I_3^+	0.225	± 0.027
$Fe(CN)_6^{3-}$	0.347	± 0.038	I_5^+	0.263	± 0.027
FeF_6^{3-}	0.298	± 0.042	IBr_2^+	0.196	± 0.027
HfF_7^{3-}	0.277	± 0.042	ICl_2^+	0.175	± 0.036
InF_6^{3-}	0.268	± 0.038	IF_6^+	0.209	± 0.027
$Ir(CN)_6^{3-}$	0.347	± 0.038	$N(S_3N_2)_2^+$	0.258	± 0.027
$Ir(NO_2)_6^{3-}$	0.338	± 0.038	$N(SCl)_2^+$	0.186	± 0.027
$Mn(CN)_6^{3-}$	0.350	± 0.038	$N(SeCl)_2^+$	0.246	± 0.027
$Mn(CN)_6^{5-}$	0.401	± 0.042	$N(SF_2)_2^+$	0.214	± 0.027
$MnCl_6^{4-}$	0.349	± 0.038	N_2F^+	0.156	± 0.027
N^{3-}	0.180	± 0.042	NO^+	0.145	± 0.027
$Ni(NO_2)_6^{3-}$	0.342	± 0.038	NO_2^+	0.153	± 0.027

Ion	Radius		Ion	Radius	
O_2^+	0.140	± 0.027	$(SNPMe_3)_3^+$	0.308	± 0.027
$O_2(SCCF_3Cl)_2^+$	0.275	± 0.027	$SNSC(CH_3)N^+$	0.225	± 0.027
$ONCH_3CF_3^+$	0.200	± 0.027	$SNSC(CN)CH^+$	0.209	± 0.027
$OsOF_5^-$	0.246	± 0.038	$SNSC(Ph)N^+$	0.251	± 0.027
$P(CH_3)_3Cl^+$	0.197	± 0.027	$SNSC(Ph)NS_3N_2^+$	0.327	± 0.027
$P(CH_3)_3D^+$	0.196	± 0.027	$SNSC(PhCH_3)N^+$	0.264	± 0.027
PCl_4^+	0.235	± 0.027	$(Te(N(SiMe_3)_2)_2^+$	0.371	± 0.027
$ReOF_5^-$	0.245	± 0.038	$Te(N_3)_3^+$	0.226	± 0.027
$S(CH_3)_2Cl^+$	0.207	± 0.027	$Te_4Nb_3OTe_2I_6^+$	0.407	± 0.027
$S(N(C_2H_5)_3)_3^+$	0.439	± 0.027	$TeBr_3^+$	0.235	± 0.027
$S_2(CH_3)_2Cl^+$	0.265	± 0.027	$TeCl_3^+$	0.216	± 0.027
$S_2(CH_3)_2CN^+$	0.223	± 0.027	$TeCl_3(15\text{-crown-}5)^+$	0.282	± 0.027
$S_2(CH_3)_3^+$	0.233	± 0.027	TeI_3^+	0.243	± 0.027
$S_2Br_5^+$	0.267	± 0.027	$Xe_2F_{11}^+$	0.266	± 0.027
S_2N^+	0.159	± 0.034	$Xe_2F_3^+$	0.221	± 0.027
$S_2N_2C_2H_3^+$	0.211	± 0.027	XeF^+	0.174	± 0.027
$S_2NC_2(PhCH_3)_2^+$	0.310	± 0.027	XeF_3^+	0.183	± 0.027
$S_2NC_3H_4^+$	0.218	± 0.027	XeF_5^+	0.186	± 0.027
$S_2NC_4H_8^+$	0.225	± 0.027	$XeOF_3^+$	0.186	± 0.027
$S_3(CH_3)_3^+$	0.239	± 0.027			
$S_3Br_3^+$	0.245	± 0.027	***Doubly Charged Cations***		
$S_3C_3H_7^+$	0.199	± 0.027	$Co_2S_2(CO)_6^{2+}$	0.263	± 0.035
$S_3C_4F_6^+$	0.261	± 0.027	$FeW(Se)_2(CO)^{2+}$	0.260	± 0.035
$S_3CF_3CN^+$	0.263	± 0.027	I_4^{2+}	0.207	± 0.035
$S_3Cl_3^+$	0.233	± 0.027	$Mo(Te_3)(CO)_4^{2+}$	0.234	± 0.035
$S_3N_2^+$	0.201	± 0.027	S_{19}^{2+}	0.292	± 0.035
$S_3N_2Cl^+$	0.232	± 0.027	$S_2(S(CH_3)_2)_2^{2+}$	0.230	± 0.035
$S_4N_3^+$	0.231	± 0.027	$S_2I_4^{2+}$	0.231	± 0.035
$S_4N_3(Ph)_2^+$	0.316	± 0.027	$S_3N_2^{2+}$	0.184	± 0.035
$S_4N_4H^+$	0.178	± 0.027	$S_3NCCNS_3^{2+}$	0.220	± 0.035
$S_5N_5^+$	0.257	± 0.027	S_3Se^{2+}	0.326	± 0.035
S_7I^+	0.262	± 0.027	$S_4N_4^{2+}$	0.186	± 0.035
$Sb(NPPh_3)_4^+$	0.518	± 0.027	$S_6N_4^{2+}$	0.232	± 0.035
SBr_3^+	0.220	± 0.027	S_8^{2+}	0.182	± 0.035
$SCH_3O_2^+$	0.183	± 0.027	Se_{10}^{2+}	0.253	± 0.035
$SCH_3P(CH_3)_3^+$	0.248	± 0.027	Se_{17}^{2+}	0.236	± 0.035
$SCH_3PCH_3Cl_2^+$	0.205	± 0.027	Se_{19}^{2+}	0.296	± 0.035
$SCl(C_2H_5)_2^+$	0.207	± 0.027	$Se_2I_4^{2+}$	0.218	± 0.035
$SCl_2CF_3^+$	0.207	± 0.027	$Se_3N_2^{2+}$	0.182	± 0.035
$SCl_2CH_3^+$	0.204	± 0.027	Se_4^{2+}	0.152	± 0.035
SCl_3^+	0.185	± 0.027	$Se_4S_2N_4^{2+}$	0.224	± 0.035
$Se_2Br_3^+$	0.253	± 0.027	Se_8^{2+}	0.186	± 0.035
$Se_2Cl_3^+$	0.245	± 0.027	$SeN_2S_2^{2+}$	0.182	± 0.035
$Se_3N_2^+$	0.288	± 0.042	$(SNP(C_2H_5)_3)_2^{2+}$	0.312	± 0.035
$Se_3NC_{l2}^+$	0.163	± 0.027	$TaBr_6^-$	0.351	± 0.049
Se_6I^+	0.260	± 0.027	$Te(trtu)_4^{2+}$	0.328	± 0.035
$SeBr_3^+$	0.182	± 0.027	$Te(tu)_4^{2+}$	0.296	± 0.035
$SeCl_3^+$	0.192	± 0.027	$Te_2(esu)_4Br_2^{2+}$	0.356	± 0.035
SeF_3^+	0.179	± 0.027	$Te_2(esu)_4Cl_2^{2+}$	0.361	± 0.035
SeI_3^+	0.238	± 0.027	$Te_2(esu)_4I_2^{2+}$	0.342	± 0.035
SeN_2Cl^+	0.196	± 0.027	$Te_2Se_2^{2+}$	0.192	± 0.035
$SeNCl_2^+$	0.157	± 0.027	$Te_2Se_4^{2+}$	0.222	± 0.035
$(SeNMe_3)_3^+$	0.406	± 0.027	$Te_2Se_8^{2+}$	0.252	± 0.035
$SeS_2N_2^+$	0.282	± 0.042	$Te_3S_3^{2+}$	0.217	± 0.035
$SF(C_6F_5)_2^+$	0.294	± 0.027	Te_3Se^{2+}	0.193	± 0.035
$SF_2CF_3^+$	0.198	± 0.027	Te_4^{2+}	0.169	± 0.035
$SF_2N(CH_3)_2^+$	0.210	± 0.027	Te_8^{2+}	0.187	± 0.035
SF_3^+	0.172	± 0.027	$W(CO)_4(h3\text{-}Te)^{2+}$	0.234	± 0.035
$SFS(C(CF_3)_2)_2^+$	0.275	± 0.027	$W_2(CO)_{10}Se_4^{2+}$	0.290	± 0.035
$SH_2C_3H_7^+$	0.210	± 0.027			
SN^+	0.158	± 0.027	***Multi-Charged Cations***		
$SNCl_5(CH_3CN)^-$	0.290	± 0.038	I_{15}^{3+}	0.442	± 0.051
			$Te_2(su)_6^{4+}$	0.453	± 0.034

Ligand abbreviations: su = selenourea; esu = ethyleneselenourea; tu = thiourea; ph = phenyl.

TABLE 3. Ancillary Thermochemical Data
(kJ mol^{-1})

Species	State	$\Delta_f H^o$
AsO_4^{3-}	g	(289)
BrO_3^-	g	−145
ClO_4^-	g	−344
CN^-	g	66
CO_3^{2-}	g	−321
$Fe(NO_3)_2$	c	(−448)
HF_2^-	g	−774
$HfCl_6^{2-}$	g	−1640
$IO_2F_2^-$	g	−693
IO_3^-	g	−208
$IrCl_6^{2-}$	g	−785
$LiCH_3O_2$	c	(−745)
$NbCl_6^{2-}$	g	−1224
$NH_2CH_2CO_2^-$	g	−564
O_2^{2-}	g	553
$PdCl_6^{2-}$	g	−749
PO_4^{3-}	g	291
$PtCl_6^{2-}$	g	−774
$ReBr_6^{2-}$	g	−689
$ReCl_6^{2-}$	g	−919
$Ti(OH)_2$	c	−778

THE MADELUNG CONSTANT AND CRYSTAL LATTICE ENERGY

If U is the crystal lattice energy and M is the Madelung constant, then[a]

$$U = \frac{N M z_i z_j e^2}{r}(1 - 1/n)$$

Substance	Ion type	Crystal form[b]	M
Sodium chloride, NaCl	M^+, X^-	FCC	1.74756
Cesium chloride, CsCl	M^+, X^-	BCC	1.76267
Calcium chloride, $CaCl_2$	M^{++}, $2X^-$	Cubic	2.365
Calcium fluoride (fluorite), CaF_2	M^{++}, $2X^-$	Cubic	2.51939
Cadmium chloride, $CdCl_2$	M^{++}, $2X^-$	Hexagonal	2.244[c]
Cadmium iodide (α), CdI_2	M^{++}, $2X^-$	Hexagonal	2.355[c]
Magnesium fluoride, MgF_2	M^{++}, $2X^-$	Tetragonal	2.381[c]
Cuprous oxide (cuprite), Cu_2O	$2M^+$, X^{--}	Cubic	2.22124
Zinc oxide, ZnO	M^{++}, X^{--}	Hexagonal	1.4985[c]
Sphalerite (zinc blende), ZnS	M^{++}, X^{--}	FCC	1.63806
Wurtzite, ZnS	M^{++}, X^{--}	Hexagonal	1.64132[c]
Titanium dioxide (anatase), TiO_2	M^{4+}, $2X^{--}$	Tetragonal	2.400[c]
Titanium dioxide (rutile), TiO_2	M^{4+}, $2X^{--}$	Tetragonal	2.408[c]
β-Quartz, SiO_2	M^{4+}, $2X^{--}$	Hexagonal	2.2197[c]
Corundum, Al_2O_3	$2M^{3+}$, $3X^{--}$	Rhombohedral	4.1719

[a] N is Avogadro's number, z_i and z_j are the integral charges on the ions (in units of e), and e is the charge on the electron in electrostatic units ($e = 4.803 \times 10^{-10}$ esu). r is the shortest distance between cation-anion pairs in centimeters. Then U is in ergs (1 erg = 10^{-7} J).

[b] FCC = face centered cubic; BCC = body centered cubic.

[c] For tetragonal and hexagonal crystals the value of M depends on the details of the lattice parameters.

The Born Exponent, n is:

Ion type	n
He, Li^+	5
Ne, Na^+, F^-	7
Ar, K^+, Cu^+, Cl^-	9
Kr, Rb^+, Ag^+, Br^-	10
Xe, Cs^+, Au^+, I^-	12

For a crystal with a mixed-ion type, an average of the values of n in this table is to be used (6 for LiF, for example).

ELASTIC CONSTANTS OF SINGLE CRYSTALS

H. P. R. Frederikse

This table gives selected values of elastic constants for single crystals. The values believed most reliable were selected from the original literature. The substances are arranged by crystal system and, within each system, alphabetically by name. A reference to the original literature is given for each value; a useful compilation of published values from many sources may be found in Reference 1 below.

Data are given for the single-crystal density and for the elastic constants c_{ij}, in units of 10^{11} N/m^2, which is equivalent to 10^{12} dyn/cm^2.

General References

1. Simmons, G., and Wang, H., *Single Crystal Elastic Constants and Calculated Aggregate Properties: A Handbook, Second Edition*, The MIT Press, Cambridge, MA, 1971.
2. Gray, D. E., Ed., *American Institute of Physics Handbook, Third Edition*, McGraw-Hill, New York, 1972.

CUBIC CRYSTALS

Name	Formula	ρ/g cm^{-3}	T/K	Ref.	C_{11}	C_{12}	C_{44}
Aluminum	Al	2.6970	298	1	1.0675	0.6041	0.2834
Aluminum antimonide	AlSb	4.3600	300	2	0.8939	0.4427	0.4155
Ammonium bromide	NH$_4$Br	2.4314	300	3	0.3414	0.0782	0.0722
Ammonium chloride	NH$_4$Cl	1.5279	290	4	0.3814	0.0866	0.0903
Argon	Ar	1.7710	4.2	5	0.0529	0.0135	0.0159
Barium fluoride	BaF$_2$	4.8860	298	6	0.9199	0.4157	0.2568
Barium nitrate	Ba(NO$_3$)$_2$	3.2560	293	7	0.2925	0.2065	0.1277
Calcium fluoride	CaF$_2$	3.810	298	8	1.6420	0.4398	0.8406
Calcium telluride	CaTe	5.8544	298	9	0.5351	0.3681	0.1994
Cesium	Cs	1.9800	78	10	0.0247	0.0206	0.0148
Cesium bromide	CsBr	4.4560	298	11	0.3063	0.0807	0.0750
Cesium chloride	CsCl	3.9880	298	11	0.3644	0.0882	0.0804
Cesium iodide	CsI	4.5250	298	11	0.2446	0.0661	0.0629
Chromite	FeCr$_2$O$_4$	4.4500	RT	12	3.2250	1.4370	1.1670
Chromium	Cr	7.20	298	13	3.398	0.586	0.990
Cobalt oxide	CoO	6.44	298	14	2.6123	1.4699	0.8300
Cobalt zinc ferrite	CoZnFeO$_2$	5.43	303	12	2.660	1.530	0.780
Copper	Cu	8.932	298	15	1.683	1.221	0.757
Gallium antimonide	GaSb	5.6137	298	16	0.8839	0.4033	0.4316
Gallium arsenide	GaAs	5.3169	298	17	1.1877	0.5372	0.5944
Gallium phosphide	GaP	4.1297	300	18	1.4120	0.6253	0.7047
Garnet (yttrium-iron)	Y$_3$Fe$_2$(FeO$_4$)$_3$	5.17	298	19	2.680	1.106	0.766
Germanium	Ge	5.313	298	20	1.2835	0.4823	0.6666
Gold	Au	19.283	296.5	21	1.9244	1.6298	0.4200
Indium antimonide	InSb	5.7890	298	22	0.6720	0.3670	0.3020
Indium arsenide	InAs	5.6720	293	23	0.8329	0.4526	0.3959
Indium phosphide	InP	4.78	RT	24	1.0220	0.5760	0.4600
Iridium	Ir	22.52	300	25	5.80	2.42	2.56
Iron	Fe	7.8672	298	26	2.26	1.40	1.16
Lead	Pb	11.34	296	27	0.4966	0.4231	0.1498
Lead fluoride	PbF$_2$	7.79	300	28	0.8880	0.4720	0.2454
Lead nitrate	Pb(NO$_3$)$_2$	4.547	293	29	0.3729	0.2765	0.1347
Lead telluride	PbTe	8.2379	303.2	30	1.0795	0.0764	0.1343
Lithium	Li	0.5326	298	31	0.1350	0.1144	0.0878
Lithium bromide	LiBr	3.47	RT	32	0.3940	0.1880	0.1910
Lithium chloride	LiCl	2.068	295	33	0.4927	0.2310	0.2495
Lithium fluoride	LiF	2.638	RT	34	1.1397	0.4767	0.6364
Lithium iodide	LiI	4.061	RT	32	0.2850	0.1400	0.1350
Magnesium oxide	MgO	3.579	298	20	2.9708	0.9536	1.5613
Magnetite	Fe$_3$O$_4$	5.18	RT	32	2.730	1.060	0.971
Manganese oxide	MnO	5.39	298	35	2.23	1.20	0.79
Mercury telluride	HgTe	8.079	290	36	0.548	0.381	0.204
Molybdenum	Mo	10.2284	273	37	4.637	1.578	1.092
Nickel	Ni	8.91	298	15	2.481	1.549	1.242
Niobium	Nb	8.578	300	38	2.4650	1.3450	0.2873

CUBIC CRYSTALS

Name	Formula	$\rho/\text{g cm}^{-3}$	T/K	Ref.	C_{11}	C_{12}	C_{44}
Palladium	Pd	12.038	300	39	2.2710	1.7604	0.7173
Platinum	Pt	21.50	300	40	3.4670	2.5070	0.7650
Potassium	K	0.851	295	41	0.0370	0.0314	0.0188
Potassium bromide	KBr	2.740	298	11	0.3468	0.0580	0.0507
Potassium chloride	KCl	1.984	298	11	0.4069	0.0711	0.0631
Potassium cyanide	KCN	1.553	RT	32	0.1940	0.1180	0.0150
Potassium fluoride	KF	2.480	295	33	0.6490	0.1520	0.1232
Potassium iodide	KI	3.128	300	42	0.2710	0.0450	0.0364
Pyrite	FeS_2	5.016	RT	43	3.818	0.310	1.094
Rubidium	Rb	1.58	170	44	0.0296	0.0250	0.0171
Rubidium bromide	RbBr	3.350	300	45	0.3152	0.0500	0.0380
Rubidium chloride	RbCl	2.797	300	45	0.3624	0.0612	0.0468
Rubidium iodide	RbI	3.551	300	45	0.2556	0.0382	0.0278
Silicon	Si	2.331	298	46	1.6578	0.6394	0.7962
Silver	Ag	10.50	300	47	1.2399	0.9367	0.4612
Silver bromide	AgBr	5.585	300	48	0.5920	0.3640	0.0616
Sodium	Na	0.971	299	49	0.0739	0.0622	0.0419
Sodium bromate	$NaBrO_3$	3.339	RT	32	0.5450	0.1910	0.1500
Sodium bromide	NaBr	3.202	300	33	0.3970	0.1001	0.0998
Sodium chlorate	$NaClO_3$	2.485	RT	50	0.4920	0.1420	0.1160
Sodium chloride	NaCl	2.163	298	11	0.4947	0.1288	0.1287
Sodium fluoride	NaF	2.804	300	51	0.9700	0.2380	0.2822
Sodium iodide	NaI	3.6689	300	52	0.3007	0.0912	0.0733
Spinel	$MgAl_2O_4$	3.6193	298	53	2.9857	1.5372	1.5758
Strontium fluoride	SrF_2	4.277	300	54	1.2350	0.4305	0.3128
Strontium nitrate	$Sr(NO_3)_2$	2.989	293	29	0.4255	0.2921	0.1590
Strontium oxide	SrO	4.99	300	55	1.601	0.435	0.590
Strontium titanate	$SrTiO_3$	5.123	RT	56	3.4817	1.0064	4.5455
Tantalum	Ta	16.626	298	57	2.6023	1.5446	0.8255
Tantalum carbide	TaC	14.65	RT	58	5.05	0.73	0.79
Thallium bromide	TlBr	7.4529	298	59	0.3760	0.1458	0.0757
Thorium	Th	11.694	300	60	0.7530	0.4890	0.4780
Thorium oxide	ThO_2	9.991	298	61	3.670	1.060	0.797
Tin telluride	SnTe	6.445	300	62	1.1250	0.0750	0.1172
Titanium carbide	TiC	4.940	RT	107	5.00	1.13	1.75
Tungsten	W	19.257	297	64	5.2239	2.0437	1.6083
Uranium carbide	UC	13.63	300	65	3.200	0.850	0.647
Uranium dioxide	UO_2	10.97	298	66	3.960	1.210	0.641
Vanadium	V	6.022	300	67	2.287	1.190	0.432
Zinc selenide	ZnSe	5.262	298	68	0.8096	0.4881	0.4405
Zinc sulfide	ZnS	4.088	298	68	1.0462	0.6534	0.4613
Zinc telluride	ZnTe	5.636	298	68	0.7134	0.4078	0.3115
Zirconium carbide	ZrC	6.606	298	63	4.720	0.987	1.593

TETRAGONAL CRYSTALS

Name	Formula	$\rho/\text{g cm}^{-3}$	T/K	Ref.	C_{11}	C_{12}	C_{13}	C_{16}	C_{33}	C_{44}	C_{66}
Ammonium dihydrogen arsenate (ADA)	$NH_4H_2AsO_4$	2.3110	298	69	0.6747	−0.106	0.1652		0.3022	0.0685	0.0639
Ammonium dihydrogen phosphate (ADP)	$NH_4H_2PO_4$	1.8030	293	69	0.6200	−0.050	0.1400		0.3000	0.0910	0.0610
Barium titanate	$BaTiO_3$	5.9988	298	70	2.7512	1.7897	1.5156		1.6486	0.5435	1.1312
Calcium molybdate	$CaMoO_4$	4.255	298	79	1.447	0.664	0.466	0.134	1.265	0.369	0.451
Indium	In	7.300	RT	71	0.4450	0.3950	0.4050		0.4440	0.0655	0.1220
Magnesium fluoride	MgF_2	3.177	RT	72	1.237	0.732	0.536		1.770	0.552	0.978
Nickel sulfate hexahydrate	$NiSO_4\cdot6H_2O$	2.070	RT	73	0.3209	0.2315	0.0209		0.2931	0.1156	0.1779
Potassium dihydrogen arsenate (KDA)	KH_2AsO_4	2.867	RT	12	0.530	−0.060	−0.020		0.370	0.120	0.070
Potassium dihydrogen phosphate (KDP)	KH_2PO_4	2.388	RT	71	0.7140	−0.049	0.1290		0.5620	0.1270	0.0628
Rubidium dihydrogen phosphate (RDP)	RbH_2PO_4	2.800	298	74	0.5562	−0.064	0.0279		0.4398	0.1142	0.0350
Rutile	TiO_2	4.260	298	75	2.7143	1.7796	1.4957		4.8395	1.2443	1.9477
Tellurium oxide	TeO_2	5.99	RT	76	0.5320	0.4860	0.2120		1.0850	0.2440	0.5520
Tin (white)	Sn	7.29	288	77	0.7529	0.6156	0.4400		0.9552	0.2193	0.2336
Zircon	$ZrSiO_4$	4.70	RT	78	2.585	1.791	1.542		3.805	0.733	1.113

ORTHORHOMBIC CRYSTALS

Name	Formula	$\rho/\text{g cm}^{-3}$	T/K	Ref.	C_{11}	C_{12}	C_{13}	C_{22}	C_{23}	C_{33}	C_{44}	C_{55}	C_{66}
Acenaphthene	$C_{12}H_{10}$	1.220	293	80	0.1380	0.0210	0.0410	0.1262	0.0460	0.1117	0.0265	0.0290	0.0185
Ammonium sulfate	$(NH_4)_2SO_4$	1.774	293	81	0.3607	0.1651	0.1580	0.2981	0.1456	0.3534	0.1025	0.0717	0.0974
Aragonite	$CaCO_3$	2.93	RT	82	1.5958	0.3663	0.0197	0.8697	0.1597	0.8503	0.4132	0.2564	0.4274
Barite	$BaSO_4$	4.40	RT	82	0.8941	0.4614	0.2691	0.7842	0.2676	1.0548	0.1190	0.2874	0.2778
Benzene	C_6H_6	1.061	250	83	0.0614	0.0352	0.0401	0.0656	0.0390	0.0583	0.0197	0.0378	0.0153
Benzophenone	$(C_6H_5)_2CO$	1.219	RT	32	0.1070	0.0550	0.0169	0.1000	0.0321	0.0710	0.0203	0.0155	0.0353
Bronzite	$(MgFe)SiO_3$	3.38	RT	78	1.876	0.686	0.605	1.578	0.561	2.085	0.700	0.592	0.544
Calcium sulfate	$CaSO_4$	2.962	RT	84	0.9382	0.1650	0.1520	1.845	0.3173	1.1180	0.3247	0.2653	0.0926
Celestite	$SrSO_3$	3.96	RT	12	1.044	0.773	0.605	1.061	0.619	1.286	0.135	0.279	0.266
Cesium sulfate	Cs_2SO_4	4.243	293	81	0.4490	0.1958	0.1815	0.4283	0.1800	0.3785	0.1326	0.1319	0.1323
Fosterite	Mg_2SiO_4	3.224	298	85	3.2848	0.6390	0.6880	1.9980	0.7380	2.3530	0.6515	0.8120	0.8088
Iodic acid	HIO_3	4.630	RT	73	0.3030	0.1194	0.1169	0.5448	0.0548	0.4359	0.1835	0.2193	0.1736
Lithium ammonium tartrate	$LiNH_4C_4H_4O_6\cdot4H_2O$	1.71	RT	12	0.3864	0.1655	0.0875	0.5393	0.2007	0.3624	0.1190	0.0667	0.2326
Magnesium sulfate heptahydrate	$MgSO_4\cdot7H_2O$	1.68	RT	86	0.325	0.174	0.182	0.288	0.182	0.315	0.078	0.156	0.090
Natrolite	$(Na,Al)SiO_3$	2.25	RT	78	0.716	0.261	0.297	0.632	0.297	1.378	0.196	0.248	0.423
Nickel sulfate heptahydrate	$NiSO_4\cdot7H_2O$	1.948	RT	86	0.353	0.198	0.201	0.311	0.201	0.335	0.091	0.172	0.099
Olivine	$(MgFe)SiO_4$	3.324	RT	87	3.240	0.590	0.790	1.980	0.780	2.490	0.667	0.810	0.793
Potassium pentaborate	$KB_5O_8\cdot4H_2O$	1.74	RT	71	0.582	0.229	0.174	0.359	0.231	0.255	0.164	0.046	0.057
Potassium sulfate	K_2SO_4	2.665	293	81	0.5357	0.1999	0.2095	0.5653	0.1990	0.5523	0.195	0.1879	0.1424
Rochelle salt	$NaK(C_4H_4O_6)\cdot4H_2O$	1.79	RT	71	0.255	0.141	0.116	0.381	0.146	0.371	0.134	0.032	0.098
Rubidium sulfate	Rb_2SO_4	3.621	293	81	0.5029	0.1965	0.1999	0.5098	0.1925	0.4761	0.1626	0.1589	0.1407
Sodium ammonium tartrate	$NaNH_4C_4H_4O_6\cdot4H_2O$	1.587	RT	12	0.3685	0.2725	0.3083	0.5092	0.3472	0.5541	0.1058	0.0303	0.0870
Sodium tartrate	$Na_2C_4H_4O_6\cdot2H_2O$	1.794	RT	12	0.461	0.286	0.320	0.547	0.352	0.665	0.124	0.031	0.098
Strontium formate dihydrate	$Sr(CHO_2)_2\cdot2H_2O$	2.25	RT	12	0.4391	0.1037	−0.149	0.3484	−0.014	0.3746	0.1538	0.1075	0.1724
Sulfur	S	2.07	RT	12	0.240	0.133	0.171	0.205	0.159	0.483	0.043	0.087	0.076
Thallium sulfate	$TlSO_4$	6.776	293	81	0.4106	0.2573	0.2288	0.3885	0.2174	0.4268	0.1125	0.1068	0.0751
Topaz	$Al_2SiO_3(OH,F)_2$	3.52	RT	82	2.8136	1.2582	0.8464	3.8495	0.8815	2.9452	1.0811	1.3298	1.3089
Uranium (alpha)	U	19.0453	293	88	2.1486	0.4622	0.2176	1.9983	1.0764	2.6763	1.2479	0.7379	0.7454
Zinc sulfate heptahydrate	$ZnSO_4\cdot7H_2O$	1.970	RT	86	0.3320	0.1720	0.2000	0.2930	0.1980	0.3200	0.0780	0.1530	0.0830

MONOCLINIC CRYSTALS

Name	Formula	$\rho/\mathrm{g\ cm^{-3}}$	T/K	Ref.	C_{11}	C_{12}	C_{13}	C_{15}	C_{22}
Aegirine	$(\mathrm{NaFe})\mathrm{Si_2O_6}$	3.50	RT	89	1.858	0.685	0.707	0.098	1.813
Anthracene	$\mathrm{C_{14}H_{10}}$	1.258	RT	90	0.0852	0.0672	0.0590	−0.0192	0.1170
Cobalt sulfate heptahydrate	$\mathrm{CoSO_4 \cdot 7H_2O}$	1.948	RT	86	0.335	0.205	0.158	0.016	0.378
Diopside	$(\mathrm{CaMg})\mathrm{Si_2O_6}$	3.31	RT	91	2.040	0.884	0.0883	−0.193	1.750
Dipotassium tartrate	$\mathrm{KHC_4H_4O_6}$	1.97	RT	12	0.4294	0.1399	0.3129	−0.0105	0.3460
Feldspar (microceine)	$\mathrm{KAlSi_3O_8}$	2.56	RT	92	0.664	0.438	0.259	−0.033	1.710
Ferrous sulfate heptahydrate	$\mathrm{FeSO_4 \cdot 7H_2O}$	1.898	RT	86	0.349	0.208	0.174	−0.020	0.376
Lithium sulfate monohydrate	$\mathrm{Li_2SO_4 \cdot H_2O}$	2.221	RT	32	0.5250	0.1715	0.1730	−0.0196	0.5060
Naphthalene	$\mathrm{C_{10}H_8}$	1.127	RT	93	0.0780	0.0445	0.0340	−0.006	0.0990
Potassium tartrate	$\mathrm{K_2C_4H_4O_6}$	1.987	RT	32	0.3110	0.1720	0.1690	0.0287	0.3900
Sodium thiosulfate	$\mathrm{Na_2S_2O_3}$	1.7499	RT	12	0.3323	0.1814	0.1875	0.0225	0.2953
Stilbene	$(\mathrm{C_6H_5CH})_2$	1.60	RT	94	0.0930	0.0570	0.0670	−0.003	0.0920
Triglycine sulfate (TGS)	$(\mathrm{NH_2CH_2COOH})_3 \cdot \mathrm{H_2SO_4}$	1.68	RT	32	0.4550	0.1720	0.1980	−0.030	0.3210

Name	C_{23}	C_{25}	C_{33}	C_{35}	C_{44}	C_{46}	C_{55}	C_{66}
Aegirine	0.626	0.094	2.344	0.214	0.692	0.077	0.510	0.474
Anthracene	0.0375	−0.0170	0.1522	−0.0187	0.0272	0.0138	0.0242	0.0399
Cobalt sulfate heptahydrate	0.158	−0.018	0.371	−0.047	0.060	0.016	0.058	0.101
Diopside	0.482	−0.196	2.380	−0.336	0.675	−0.113	0.588	0.705
Dipotassium tartrate	0.1173	0.0176	0.6816	0.0294	0.0961	−0.0044	0.1270	0.0841
Feldspar (microceine)	0.192	−0.148	1.215	−0.131	0.143	−0.015	0.238	0.361
Ferrous sulfate heptahydrate	0.172	−0.019	0.360	−0.014	0.064	0.001	0.056	0.096
Lithium sulfate monohydrate	0.0368	0.0571	0.5400	−0.0254	0.1400	−0.0054	0.1565	0.2770
Naphthalene	0.0230	−0.0270	0.1190	0.0290	0.0330	−0.0050	0.0210	0.0415
Potassium tartrate	0.1330	0.0182	0.5540	0.0710	0.0870	0.0072	0.1040	0.0826
Sodium thiosulfate	0.1713	0.0983	0.4590	−0.0678	0.0569	−0.0268	0.1070	0.0598
Stilbene	0.0485	−0.005	0.0790	−0.005	0.0325	0.0050	0.0640	0.0245
Triglycine sulfate (TGS)	0.2080	−0.0036	0.2630	−0.0500	0.0950	−0.0026	0.1110	0.0620

HEXAGONAL CRYSTALS

Name	Formula	ρ/g cm^{-3}	T/K	Ref.	C_{11}	C_{12}	C_{13}	C_{33}	C_{55}
Apatite	$Ca_5(PO_4)_3(OH,F,Cl)$	3.218	RT	12	1.667	0.131	0.655	1.396	0.663
Beryl	$Be_3Al_2Si_6O_{18}$	2.68	RT	12	2.800	0.990	0.670	2.480	0.658
Beryllium	Be	1.8477	300	95	2.923	0.267	0.140	3.364	1.625
Beryllium oxide	BeO	3.01	RT	96	4.70	1.68	1.19	4.94	1.53
Cadmium	Cd	8.652	300	97	1.1450	0.3950	0.3990	0.5085	0.1985
Cadmium selenide	CdSe	5.655	298	68	0.7046	0.4516	0.3930	0.8355	0.1317
Cadmium sulfide	CdS	4.824	298	98	0.8431	0.5208	0.4567	0.9183	0.1458
Cobalt	Co	8.836	298	99	3.071	1.650	1.027	3.581	0.755
Dysprosium	Dy	8.560	298	100	0.7466	0.2616	0.2233	0.7871	0.2427
Erbium	Er	9.064	298	100	0.8634	0.3050	0.2270	0.8554	0.2809
Gadolinium	Gd	7.888	298	101	0.6667	0.2499	0.2132	0.7191	0.2089
Hafnium	Hf	12.727	298	102	1.881	0.772	0.661	1.969	0.557
Ice	H_2O(solid)	0.920	250	103	0.1410	0.0660	0.0624	0.1515	0.0288
Indium	In	7.2788	300	104	0.4535	0.4006	0.4151	0.4515	0.0651
Magnesium	Mg	1.7364	298	105	0.5950	0.2612	0.2180	0.6155	0.1635
Rhenium	Re	21.024	298	100	6.1820	2.7530	2.0780	6.8350	1.6060
Ruthenium	Ru	12.3615	298	100	5.6260	1.8780	1.6820	6.2420	1.8060
Thallium	Tl	11.560	300	106	0.4080	0.3540	0.2900	0.5280	0.0726
Titanium	Ti	4.5063	298	102	1.6240	0.9200	0.6900	1.8070	0.4670
Titanium diboride	TiB_2	4.95	RT	107	6.90	4.10	3.20	4.40	2.50
Yttrium	Y	4.472	300	108	0.7790	0.2850	0.2100	0.7690	0.2431
Zinc	Zn	7.134	295	109	1.6368	0.3640	0.5300	0.6347	0.3879
Zinc oxide	ZnO	5.6760	298	110	2.0970	1.2110	1.0510	2.1090	0.4247
Zinc sulfide	ZnS	4.089	298	96	1.2420	0.6015	0.4554	1.4000	0.2864
Zirconium	Zr	6.505	298	102	1.434	0.728	0.653	1.648	0.320

TRIGONAL CRYSTALS

Name	Formula	ρ/g cm^{-3}	T/K	Ref.	C_{11}	C_{12}	C_{13}	C_{14}	C_{33}	C_{44}
Aluminum oxide	Al_2O_3	3.986	300	111	4.9735	1.6397	1.1220	−0.2358	4.9911	1.4739
Aluminum phosphate	$AlPO_4$	2.556	RT	73	1.0503	0.2934	0.6927	−0.1271	1.3353	0.2314
Antimony	Sb	6.70	295	112	1.0130	0.3450	0.2920	0.2090	0.4500	0.3930
Bismuth	Bi	9.80	295	112	0.6370	0.2490	0.2470	0.0717	0.3820	0.1123
Calcite	$CaCO_3$	2.712	300	113	1.4806	0.5578	0.5464	−0.2058	0.8557	0.3269
Hematite	Fe_2O_3	5.240	RT	82	2.4243	0.5464	0.1542	−0.1247	2.2734	0.8569
Lithium niobate	$LiNbO_3$	4.70	RT	114	2.030	0.530	0.750	0.090	2.450	0.600
Lithium tantalate	$LiTaO_3$	7.45	RT	114	2.330	0.470	0.800	−0.110	2.750	0.940
Quartz	SiO_2	2.6485	298	115	0.8680	0.0704	0.1191	−0.1804	1.0575	0.5820
Selenium	Se	4.838	300	116	0.1870	0.0710	0.2620	0.0620	0.7410	0.1490
Sodium nitrate	$NaNO_3$	2.27	RT	12	0.8670	0.1630	0.1600	0.0820	0.3740	0.2130
Tourmaline		3.05	RT	82	2.7066	0.6927	0.0872	−0.0774	1.6070	0.6682

References

1. Thomas, J. F., *Phys. Rev.*, 175, 955–962, 1968.
2. Bolef, D. I. and M. Menes, *J. Appl. Phys.*, 31, 1426–1427, 1960.
3. Garland, C. W. and C. F. Yarnell, *J. Chem. Phys.*, 44, 1112–1120, 1966.
4. Garland, C. W. and R. Renard, *J. Chem. Phys.*, 44, 1130–1139, 1966.
5. Gsänger, M., H. Egger and E. Lüscher, *Phys. Letters*, 27A, 695–696, 1968.
6. Wong, C. and D. E. Schuele, *J. Phys. Chem. Solids*, 29, 1309–1330, 1968.
7. Haussühl, S., *Phys. Stat. Sol.*, 3, 1072–1076, 1963.
8. Wong, C. and D. E. Schuele, *J. Phys. Chem. Solids*, 28, 1225–1231, 1967.
9. McSkimin, H. J. and D. G. Thomas, *J. Appl. Phys.*, 33, 56–59, 1962.
10. Kollarits, F. J. and J. Trivisonno, *J. Phys. Chem. Solids*, 29, 2133–2139, 1968.
11. Slagle, D. D. and H. A. McKinstry, *J. Appl. Phys.*, 38, 446–458, 1967.
12. Hearmon, R. F. S., *Adv. Phys.*, 5, 323–382, 1956.
13. Sumer, A. and J. F. Smith, *J. Appl. Phys.*, 34, 2691–2694, 1963.
14. Alexandrov, K. S. et. al., *Sov. Phys. Sol. State*, 10, 1316–1321, 1968.
15. Epstein, S. G. and O. N. Carlson, *Acta Metal.*, 13, 487–491, 1965.
16. McSkimin, H. J., et. al., *J. Appl. Phys.*, 39, 4127–4128, 1968.
17. McSkimin, H. J., et. al., *J. Appl. Phys.*, 38, 2362–2364, 1967.
18. Weil, R. and W. O. Groves, *J. Appl. Phys.*, 39, 4049–4051, 1968.
19. Bateman, T. B., *J. Appl. Phys.*, 37, 2194–2195, 1966.
20. Bogardus, E. H., *J. Appl. Phys.*, 36, 2504–2513, 1965.
21. Golding, B., S. C. Moss and B. L. Averbach, *Phys. Rev.*, 158, 637–645, 1967.
22. Bateman, T. B., H. J. McSkimin and J. M. Whelan, *J. Appl. Phys.*, 30, 544–545, 1959.
23. Gerlich, D., *J. Appl. Phys.*, 35, 3062, 1964.
24. Hickernell, F. S. and W. R. Gayton, *J. Appl. Phys.*, 37, 462, 1966.
25. MacFarlane, R. E., et. al., *Phys. Letters*, 20, 234–235, 1966.
26. Leese, J. and A. E. Lord Jr., *J. Appl. Phys.*, 39, 3986–3988, 1968.
27. Miller, R. A. and D. E. Schuele, *J. Phys. Chem. Solids*, 30, 589–600, 1969.
28. Wasilik, J. H. and M. L. Wheat, *J. Appl. Phys.*, 36, 791–793, 1965.
29. Haussühl, S., *Phys. Stat. Sol.*, 3, 1072–1076, 1963.
30. Houston, B., et. al., *J. Appl. Phys.*, 39, 3913–3916, 1968.
31. Trivisonno, J. and C. S. Smith, *Acta Metal.*, 9, 1064–1071, 1961.
32. Alexandrov, K. S. and T. V. Ryzhova, *Sov. Phys. Cryst.*, 6, 228–252, 1961.
33. Lewis, J. T., A. Lehoczky and C. V. Briscoe, *Phys. Rev.*, 161, 877–887, 1967.
34. Drabble, J. R. and R. E. B. Strathen, *Proc. Phys. Soc.*, 92, 1090–1095, 1967.
35. Oliver, D. W., *J. Appl. Phys.*, 40, 893, 1969.
36. Alper, T., and G. A. Saunders, *J. Phys. Chem. Solids*, 28, 1637–1642, 1967.
37. Dickinson, J. M. and P. E. Armstrong, *J. Appl. Phys.*, 38, 602–606, 1967.
38. Bolef, D. I., *J. Appl. Phys.*, 32, 100–105, 1961.
39. Rayne, J. A., *Phys. Rev.*, 112, 1125–1130, 1958.
40. MacFarlane, R. E., et. al., *Phys. Letters*, 18, 91–92, 1965.
41. Smith, P. A. and C. S. Smith, *J. Phys. Chem. Solids*, 26, 279–289, 1965.
42. Norwood, M. H. and C. V. Briscoe, *Phys. Rev.*, 112, 45–48, 1958.
43. Simmons, G. and F. Birch, *J. Appl. Phys.*, 34, 2736–2738, 1963.
44. Gutman, E. J. and J. Trivisonno, *J. Phys. Chem. Sol.*, 28, 805–809, 1967.
45. Ghafelehbashi, M., et. al., *J. Appl. Phys.*, 41, 652–666, 1970.
46. McSkimin, H. J. and P. Andreatch, Jr., *J. Appl. Phys.*, 35, 2161–2165, 1964.
47. Neighbours, J. R. and G. A. Alers, *Phys. Rev.*, 111, 707–712, 1958.
48. Hidshaw, W., J. T. Lewis, and C. V. Briscoe, *Phys. Rev.*, 163, 876–881, 1967.
49. Daniels, W. B., *Phys. Rev.*, 119, 1246–1252, 1960.
50. Viswanathan, R., *J. Appl. Phys.*, 37, 884–886, 1966.
51. Miller, R. A. and C. S. Smith, *J. Phys. Chem. Sol.*, 25, 1279–1292, 1964.
52. Claytor, R. N. and B. J. Marshall, *Phys. Rev.*, 120, 332–334, 1967.
53. Schreiber, E., *J. Appl. Phys.*, 38, 2508–2511, 1967.
54. Gerlich, D., *Phys. Rev.*, 136, A1366–A1368, 1964.
55. Johnston, D. L., P. H. Thrasher and R. J. Kearney, *J. Appl. Phys.*, 41, 427–428, 1970.
56. Poindexter, E. and A. A. Giardini, *Phys. Rev.*, 110, 1069, 1958.
57. Soga, N., *J. Appl. Phys.*, 37, 3416–3420, 1966.
58. Bartlett, R. W. and C. W. Smith, *J. Appl. Phys.*, 38, 5428–5429, 1967.
59. Morse, G. E. and A. W. Lawson, *J. Phys. Chem. Sol.*, 28, 939–950, 1967.
60. Armstrong, P. E., O. N. Carlson and J. F. Smith, *J. Appl. Phys.*, 30, 36–41, 1959.
61. Macedo, P. M., W. Capps and J. B. Wachtman, *J. Am. Cer. Soc.*, 47, 651, 1964.
62. Beattie, A. G., *J. Appl. Phys.*, 40, 4818–4821, 1969.
63. Chang, R. and L. J. Graham, *J. Appl Phys.*, 37, 3778–3783, 1966.
64. Lowrie, R. and A. M. Gonas, *J. Appl. Phys.*, 38, 4505–4509, 1967.
65. Graham, L. J., H. Nadler and R. Chang, *J. Appl. Phys.*, 34, 1572–1573, 1963.
66. Wachtman, J. B., Jr., et. al., *J. Nucl. Mat.*, 16, 39–41, 1965.
67. Bolef, D. I., *J. Appl. Phys.*, 32, 100–105, 1961.
68. Berlincourt, D., H. Jaffe and L. R. Shiozawa, *Phys. Rev.*, 129, 1009–1017, 1963.
69. Adhav. R. S. J. *Acoust. Soc. Am.*, 43, 835–838, 1968.
70. Berlincourt, D. and H. Jaffe, *Phys. Rev.*, 111, 143–148, 1958.
71. Huntington, H. B., in *Solid State Pysics*, *Vol. 7*, Seitz, F., and Turnbull, D., Ed., pp. 213–285, Academic Press, New York 1958.
72. Cutler, H. R., J. J. Gibson and K. A. McCarthy, *Sol. State Comm.*, 6, 431–433, 1968.
73. Mason, W. P., *Piezoelectric Crystals and Their Application to Ultrasonics*, D. Van Nostrand Co., Inc., New York, 1950.
74. Adhav, R. S., *J. Appl. Phys.*, 40, 2725–2727, 1969.
75. Manghnani, M. H., *J. Geophys. Res.*, 74, 4317–4328, 1969.
76. Uchida, N. and Y. Ohmachi, *J. Appl Phys.*, 40, 4692–4695, 1969.
77. House, D. G. and E. Y. Vernon, *Br. J. Appl. Phys.*, 11, 254–259, 1960.
78. Ryzhova, T. V., et. al., *Bull. Acad. Sci. USSR, Earth Phys. Ser.*, English Transl., no. 2, 111–113, 1966.
79. Alton, W. J. and A. J. Barlow, *J. Appl. Phys.*, 38, 3817–3820, 1967.
80. Michard, F., et. al., *C. R. Acad. Sci., Paris*, 265, 565–567, 1967.
81. Haussühl, S., *Acta Cryst.*, 18, 839–842, 1965.
82. Hearmon, R. F. S., *Rev. Mod. Phys.*, 18, 409–440, 1946.
83. Heseltine, J. C. W., D. W. Elliott and O. B. Wilson, *J. Chem. Phys.*, 40, 2584–2587, 1964.
84. Schwerdtner, W. M., et. al., *Canad. J. Earth Sci.*, 2, 673–683, 1965.
85. Kumazawa, M. and O. L. Anderson, *J. Geophys. Res.*, 74, 5961–5972, 1969.
86. Alexandrov, K. S., et. al., *Sov. Phys. Cryst.*, 7, 753–755, 1963.
87. Verma, R. K., *J. Geophys. Soc.*, 65, 757–766, 1960.
88. McSkimin, H. J. and E. S. Fisher, *J. Appl. Phys.*, 31, 1627–1639, 1960.
89. Alexandrov, K. S. and T.V. Ryzhova, *Bull. Acad. Sci. USSR, Geophys. Ser.*, English Transl., no.8, 871–875, 1961.
90. Afanaseva, G. K., et. al, *Phys. Stat. Sol.*, 24, K61–K63, 1967.
91. Alexandrov, K. S., et. al., *Sov. Phys. Cryst.*, 8, 589–591, 1964.
92. Alexandrov, K. S. and T. V Ryzhova, *Bull Acad. Sci. USSR, Geophys. Ser.*, English Transl., no.2, 129–131, 1962.
93. Alexandrov, K. S., et. al., *Sov. Phys. Cryst.*, 8, 164–166, 1963.
94. Teslenko, V. F., et. al., *Sov. Phys. Cryst.*, 10, 744–747, 1966.
95. Smith, J. F. and C. L. Arbogast, *J. Appl. Phys.*, 31, 99–102, 1960.
96. Cline, C. F., H. L. Dunegan and G. M. Henderson, *J. Appl. Phys.*, 38, 1944–1948, 1967.
97. Chang, Y. A. and L. Himmel, *J. Appl. Phys.*, 37, 3787–3790, 1966.
98. Gerlich, D., *J. Phys. Chem. Solids*, 28, 2575–2579, 1967.
99. McSkimin, H. J., *J. Appl. Phys.*, 26, 406–409, 1955.
100. Fisher, E. S. and D. Dever, *Trans. Met. Soc. AIME*, 239, 48–57, 1967.
101. Fisher, E. S. and D. Dever, *Proc. Conf. Rare Earth Res.*, 6th, Gatlinburg, Tenn., 522–533, 1967.
102. Fisher, E. S. and C. J. Renken, *Phys. Rev.*, 135, A482–A494, 1964.
103. Proctor, T. M., Jr., *J. Acoust. Soc. Am.*, 39, 972–977, 1966.
104. Chandrasekhar, B. S. and J. A. Rayne, *Phys. Rev.*, 124, 1011–1041, 1961.
105. Wazzan, A. R. and L. B. Robinson, *Phys. Rev.*, 155, 586–594, 1967.
106. Ferris, R. W., et. al., *J. Appl. Phys.*, 34, 768–770, 1963.
107. Gilman, J. J. and B. W. Roberts, *J. Appl. Phys.*, 32, 1405, 1961.
108. Smith, J. F. and J. A. Gjevre, *J. Appl. Phys.*, 31, 645–647, 1960.
109. Alers, G. A. and J. R. Neighbours, *J. Phys. Chem. Solids*, 7, 58–64, 1908.
110. Bateman, T. B., *J. Appl. Phys.*, 33, 3309–3312, 1962.
111. Tefft, W. E., *J. Res. Natl. Bur. Stand.*, 70A, 277–280, 1966.
112. DeBretteville, Jr., A. et. al., *Phys. Rev.*, 148, 575–579, 1966.
113. Dandekar, D. P. and A. L. Ruoff, *J. Appl. Phys.*, 39, 6004–6009, 1968.
114. Warner, A. W., M. Onoe and G. A. Coquin, *J. Acoust. Soc. Am.*, 42, 1223–1231, 1967.
115. McSkimin, H. J., P. Andreatch and R. N. Thurston, *J. Appl. Phys.*, 36, 1624–1632, 1965.
116. Mort, J., *J. Appl. Phys.*, 38, 3414–3415, 1967.

ELECTRICAL RESISTIVITY OF PURE METALS

The first part of this table gives the electrical resistivity, in units of 10^{-8} Ω m, for 28 common metallic elements as a function of temperature. The data refer to polycrystalline samples. The number of significant figures indicates the accuracy of the values. However, at low temperatures (especially below 50 K) the electrical resistivity is extremely sensitive to sample purity. Thus the low-temperature values refer to samples of specified purity and treatment. The references should be consulted for further information on this point, as well as for values at additional temperatures.

The second part of the table gives resistivity values in the neighborhood of room temperature for other metallic elements that have not been studied over an extended temperature range.

⟍ References

1. C. Y. Ho, et al., *J. Phys. Chem. Ref. Data*, 12, 183–322, 1983; 13, 1069–1096, 1984; 13, 1097–1130, 1984, 13, 1131–1172, 1984.
2. R. A. Matula, *J. Phys Chem. Ref. Data*, 8, 1147–1298, 1979.
3. T. C. Chi, *J. Phys. Chem. Ref. Data*, 8, 339–438, 1979; 8, 439–498, 1979.
4. K. H. Hellwege, Ed., *Landolt-Börnstein Numerical Data and Functional Relationships in Science and Technology*, Group III, Vol. 15, Subvolume a, Springer-Verlag, Heidelberg, 1982.
5. L. A. Hall, *Survey of Electrical Resistivity Measurements on 16 Pure Metals in the Temperature Range 0 to 273 K*, NBS Technical Note 365, U.S. Superintendent of Documents, 1968.

Electrical Resistivity in 10^{-8} Ω m

T/K	Aluminum	Barium	Beryllium	Calcium	Cesium	Chromium	Copper
1	0.000100	0.081	0.0332	0.045	0.0026		0.00200
10	0.000193	0.189	0.0332	0.047	0.243		0.00202
20	0.000755	0.94	0.0336	0.060	0.86		0.00280
40	0.0181	2.91	0.0367	0.175	1.99		0.0239
60	0.0959	4.86	0.067	0.40	3.07		0.0971
80	0.245	6.83	0.075	0.65	4.16		0.215
100	0.442	8.85	0.133	0.91	5.28	1.6	0.348
150	1.006	14.3	0.510	1.56	8.43	4.5	0.699
200	1.587	20.2	1.29	2.19	12.2	7.7	1.046
273	2.417	30.2	3.02	3.11	18.7	11.8	1.543
293	2.650	33.2	3.56	3.36	20.5	12.5	1.678
298	2.709	34.0	3.70	3.42	20.8	12.6	1.712
300	2.733	34.3	3.76	3.45	21.0	12.7	1.725
400	3.87	51.4	6.76	4.7		15.8	2.402
500	4.99	72.4	9.9	6.0		20.1	3.090
600	6.13	98.2	13.2	7.3		24.7	3.792
700	7.35	130	16.5	8.7		29.5	4.514
800	8.70	168	20.0	10.0		34.6	5.262
900	10.18	216	23.7	11.4		39.9	6.041

T/K	Gold	Hafnium	Iron	Lead	Lithium	Magnesium	Manganese
1	0.0220	1.00	0.0225		0.007	0.0062	7.02
10	0.0226	1.00	0.0238		0.008	0.0069	18.9
20	0.035	1.11	0.0287		0.012	0.0123	54
40	0.141	2.52	0.0758		0.074	0.074	116
60	0.308	4.53	0.271		0.345	0.261	131
80	0.481	6.75	0.693	4.9	1.00	0.557	132
100	0.650	9.12	1.28	6.4	1.73	0.91	132
150	1.061	15.0	3.15	9.9	3.72	1.84	136
200	1.462	21.0	5.20	13.6	5.71	2.75	139
273	2.051	30.4	8.57	19.2	8.53	4.05	143
293	2.214	33.1	9.61	20.8	9.28	4.39	144
298	2.255	33.7	9.87	21.1	9.47	4.48	144
300	2.271	34.0	9.98	21.3	9.55	4.51	144
400	3.107	48.1	16.1	29.6	13.4	6.19	147
500	3.97	63.1	23.7	38.3		7.86	149
600	4.87	78.5	32.9			9.52	151
700	5.82		44.0			11.2	152
800	6.81		57.1			12.8	
900	7.86					14.4	

T/K	Molybdenum	Nickel	Palladium	Platinum	Potassium	Rubidium	Silver
1	0.00070	0.0032	0.0200	0.002	0.0008	0.0131	0.00100
10	0.00089	0.0057	0.0242	0.0154	0.0160	0.109	0.00115
20	0.00261	0.0140	0.0563	0.0484	0.117	0.444	0.0042
40	0.0457	0.068	0.334	0.409	0.480	1.21	0.0539
60	0.206	0.242	0.938	1.107	0.90	1.94	0.162
80	0.482	0.545	1.75	1.922	1.34	2.65	0.289
100	0.858	0.96	2.62	2.755	1.79	3.36	0.418
150	1.99	2.21	4.80	4.76	2.99	5.27	0.726
200	3.13	3.67	6.88	6.77	4.26	7.49	1.029
273	4.85	6.16	9.78	9.6	6.49	11.5	1.467
293	5.34	6.93	10.54	10.5	7.20	12.8	1.587
298	5.47	7.12	10.73	10.7	7.39	13.1	1.617
300	5.52	7.20	10.80	10.8	7.47	13.3	1.629
400	8.02	11.8	14.48	14.6			2.241
500	10.6	17.7	17.94	18.3			2.87
600	13.1	25.5	21.2	21.9			3.53
700	15.8	32.1	24.2	25.4			4.21
800	18.4	35.5	27.1	28.7			4.91
900	21.2	38.6	29.4	32.0			5.64

T/K	Sodium	Strontium	Tantalum	Tungsten	Vanadium	Zinc	Zirconium
1	0.0009	0.80	0.10	0.000016		0.0100	0.250
10	0.0015	0.80	0.102	0.000137	0.0145	0.0112	0.253
20	0.016	0.92	0.146	0.00196	0.039	0.0387	0.357
40	0.172	1.70	0.751	0.0544	0.304	0.306	1.44
60	0.447	2.68	1.65	0.266	1.11	0.715	3.75
80	0.80	3.64	2.62	0.606	2.41	1.15	6.64
100	1.16	4.58	3.64	1.02	4.01	1.60	9.79
150	2.03	6.84	6.19	2.09	8.2	2.71	17.8
200	2.89	9.04	8.66	3.18	12.4	3.83	26.3
273	4.33	12.3	12.2	4.82	18.1	5.46	38.8
293	4.77	13.2	13.1	5.28	19.7	5.90	42.1
298	4.88	13.4	13.4	5.39	20.1	6.01	42.9
300	4.93	13.5	13.5	5.44	20.2	6.06	43.3
400		17.8	18.2	7.83	28.0	8.37	60.3
500		22.2	22.9	10.3	34.8	10.82	76.5
600		26.7	27.4	13.0	41.1	13.49	91.5
700		31.2	31.8	15.7	47.2		104.2
800		35.6	35.9	18.6	53.1		114.9
900			40.1	21.5	58.7		123.1

Element	T/K	Electrical resistivity 10^{-8} Ω m	Element	T/K	Electrical resistivity 10^{-8} Ω m
Antimony	273	39	Osmium	273	8.1
Bismuth	273	107	Polonium	273	40
Cadmium	273	6.8	Praseodymium	290–300	70.0
Cerium (β, hex)	290–300	82.8	Promethium	290–300	75 est.
Cerium (γ, cub)	298	74.4	Protactinium	273	17.7
Cobalt	273	5.6	Rhenium	273	17.2
Dysprosium	290–300	92.6	Rhodium	273	4.3
Erbium	290–300	86.0	Ruthenium	273	7.1
Europium	290–300	90.0	Samarium	290–300	94.0
Gadolinium	290–300	131	Scandium	290–300	56.2
Gallium	273	13.6	Terbium	290–300	115
Holmium	290–300	81.4	Thallium	273	15
Indium	273	8.0	Thorium	273	14.7
Iridium	273	4.7	Thulium	290–300	67.6
Lanthanum	290–300	61.5	Tin	273	11.5
Lutetium	290–300	58.2	Titanium	273	39
Mercury	298	96.1	Uranium	273	28
Neodymium	290–300	64.3	Ytterbium	290–300	25.0
Niobium	273	15.2	Yttrium	290–300	59.6

ELECTRICAL RESISTIVITY OF SELECTED ALLOYS

These values were obtained by fitting all available measurements to a theoretical formulation describing the temperature and composition dependence of the electrical resistivity of metals. Some of the values listed here fall in regions of temperature and composition where no actual measurements exist. Details of the procedure may be found in the reference.

Values of the resistivity are given in units of 10^{-8} Ω m. General comments in the preceding table for pure metals also apply here

Reference

C. Y. Ho, et al., *J. Phys. Chem. Ref. Data*, 12, 183–322, 1983.

Aluminum-Copper

Wt % Al	100 K	273 K	293 K	300 K	350 K	400 K
99[a]	0.531	2.51	2.74	2.82	3.38	3.95
95[a]	0.895	2.88	3.10	3.18	3.75	4.33
90[b]	1.38	3.36	3.59	3.67	4.25	4.86
85[b]	1.88	3.87	4.10	4.19	4.79	5.42
80[b]	2.34	4.33	4.58	4.67	5.31	5.99
70[b]	3.02	5.03	5.31	5.41	6.16	6.94
60[b]	3.49	5.56	5.88	5.99	6.77	7.63
50[b]	4.00	6.22	6.55	6.67	7.55	8.52
40[c]		7.57	7.96	8.10	9.12	10.2
30[c]		11.2	11.8	12.0	13.5	15.2
25[f]		16.3	17.2	17.6	19.8	22.2
15[h]			12.3			
10[g]	8.71	10.8	11.0	11.1	11.7	12.3
5[c]	7.92	9.43	9.61	9.68	10.2	10.7
1[b]	3.22	4.46	4.60	4.65	5.00	5.37

Aluminum-Magnesium

Wt % Al	100 K	273 K	293 K	300 K	350 K	400 K
99[c]	0.958	2.96	3.18	3.26	3.82	4.39
95[c]	3.01	5.05	5.28	5.36	5.93	6.51
90[c]	5.42	7.52	7.76	7.85	8.43	9.02
10[b]	14.0	17.1	17.4	17.6	18.4	19.2
5[b]	9.93	13.1	13.4	13.5	14.3	15.2
1[a]	2.78	5.92	6.25	6.37	7.20	8.03

Copper-Gold

Wt % Cu	100 K	273 K	293 K	300 K	350 K	400 K
99[c]	0.520	1.73	1.86	1.91	2.24	2.58
95[c]	1.21	2.41	2.54	2.59	2.92	3.26
90[c]	2.11	3.29	4.42	3.46	3.79	4.12
85[c]	3.01	4.20	4.33	4.38	4.71	5.05
80[c]	3.95	5.15	5.28	5.32	5.65	5.99
70[c]	5.91	7.12	7.25	7.30	7.64	7.99
60[c]	8.04	9.18	9.13	9.36	9.70	10.05
50[c]	9.88	11.07	11.20	11.25	11.60	11.94
40[c]	11.44	12.70	12.85	12.90	13.27	13.65
30[c]	12.43	13.77	13.93	13.99	14.38	14.78
25[c]	12.59	13.93	14.09	14.14	14.54	14.94
15[c]	11.38	12.75	12.91	12.96	13.36	13.77
10[c]	9.33	10.70	10.86	10.91	11.31	11.72
5[c]	5.91	7.25	7.41	7.46	7.87	8.28
1[c]	2.00	3.40	3.57	3.62	4.03	4.45

Copper-Nickel

Wt % Cu	100 K	273 K	293 K	300 K	350 K	400 K
99[c]	1.45	2.71	2.85	2.91	3.27	3.62
95[c]	6.19	7.60	7.71	7.82	8.22	8.62
90[c]	12.08	13.69	13.89	13.96	14.40	14.81
85[c]	18.01	19.63	19.83	19.90	20.32	20.70
80[c]	23.89	25.46	25.66	25.72	26.12	26.44
70[i]	35.73	36.67	36.72	36.76	36.85	36.89
60[i]	45.76	45.43	45.38	43.35	45.20	45.01
50[i]	50.22	50.19	50.05	50.01	49.73	49.50
40[c]	36.77	47.42	47.73	47.82	48.28	48.49
30[i]	26.73	40.19	41.79	42.34	44.51	45.40
25[c]	22.22	33.46	35.11	35.69	39.67	42.81
15[c]	13.49	22.00	23.35	23.85	27.60	31.38
10[c]	9.28	16.65	17.82	18.26	21.51	25.19
5[c]	5.20	11.49	12.50	12.90	15.69	18.78
1[c]	1.81	7.23	8.08	8.37	10.63	13.18

Copper-Palladium

Wt % Cu	100 K	273 K	293 K	300 K	350 K	400 K
99[c]	0.91	2.10	2.23	2.27	2.59	2.92
95[c]	2.99	4.21	4.35	4.40	4.74	5.08
90[c]	5.69	6.89	7.03	7.08	7.41	7.74
85[c]	8.30	9.48	9.61	9.66	10.01	10.36
80[c]	10.74	11.99	12.12	12.16	12.51	12.87
70[c]	15.67	16.87	17.01	17.06	17.41	17.78
60[c]	20.45	21.73	21.87	21.92	22.30	22.69
50[c]	26.07	27.62	27.79	27.86	28.25	28.64
40[c]	33.53	35.31	35.51	35.57	36.03	36.47
30[c]	45.03	46.50	46.66	46.71	47.11	47.47
25[c]	44.12	46.25	46.45	46.52	46.99	47.43
15[c]	31.79	36.52	36.99	37.16	38.28	39.35
10[c]	23.00	28.90	29.51	29.73	31.19	32.56
5[c]	13.09	20.00	20.75	21.02	22.84	24.54
1[c]	8.97	11.90	12.67	12.93	14.82	16.68

Copper-Zinc

Wt % Cu	100 K	273 K	293 K	300 K	350 K	400 K
99[b]	0.671	1.84	1.97	2.02	2.36	2.71
95[b]	1.54	2.78	2.92	2.97	3.33	3.69
90[b]	2.33	3.66	3.81	3.86	4.25	4.63
85[b]	2.93	4.37	4.54	4.60	5.02	5.44
80[b]	3.44	5.01	5.19	5.26	5.71	6.17
70[b]	4.08	5.87	6.08	6.15	6.67	7.19

Gold-Palladium

Wt % Au	100 K	273 K	293 K	300 K	350 K	400 K
99[c]	1.31	2.69	2.86	2.91	3.32	3.73
95[c]	3.88	5.21	5.35	5.41	5.79	6.17
90[i]	6.70	8.01	8.17	8.22	8.56	8.93
85[b]	9.14	10.50	10.66	10.72	11.10	11.48
80[b]	11.23	12.75	12.93	12.99	13.45	13.93
70[c]	16.44	18.23	18.46	18.54	19.10	19.67
60[b]	24.64	26.70	26.94	27.02	27.63	28.23
50[a]	23.09	27.23	27.63	27.76	28.64	29.42
40[a]	19.40	24.65	25.23	25.42	26.74	27.95
30[b]	14.94	20.82	21.49	21.72	23.35	24.92
25[b]	12.72	18.86	19.53	19.77	21.51	23.19
15[a]	8.54	15.08	15.77	16.01	17.80	19.61
10[a]	6.54	13.25	13.95	14.20	16.00	17.81
5[a]	4.58	11.49	12.21	12.46	14.26	16.07
1[a]	3.01	10.07	10.85	11.12	12.99	14.80

Gold-Silver

Wt % Au	100 K	273 K	293 K	300 K	350 K	400 K
99[b]	1.20	2.58	2.75	2.80	3.22	3.63
95[a]	3.16	4.58	4.74	4.79	5.19	5.59
90[j]	5.16	6.57	6.73	6.78	7.19	7.58
85[j]	6.75	8.14	8.30	8.36	8.75	9.15
80[j]	7.96	9.34	9.50	9.55	9.94	10.33
70[j]	9.36	10.70	10.86	10.91	11.29	11.68
60[j]	9.61	10.92	11.07	11.12	11.50	11.87
50[j]	8.96	10.23	10.37	10.42	10.78	11.14
40[j]	7.69	8.92	9.06	9.11	9.46	9.81
30[a]	6.15	7.34	7.47	7.52	7.85	8.19
25[a]	5.29	6.46	6.59	6.63	6.96	7.30
15[a]	3.42	4.55	4.67	4.72	5.03	5.34
10[a]	2.44	3.54	3.66	3.71	4.00	4.31
5[i]	1.44	2.52	2.64	2.68	2.96	3.25
1[b]	0.627	1.69	1.80	1.84	2.12	2.42

Iron-Nickel

Wt % Fe	100 K	273 K	293 K	300 K	400 K
99[a]	3.32	10.9	12.0	12.4	18.7
95[c]	10.0	18.7	19.9	20.2	26.8
90[c]	14.5	24.2	25.5	25.9	33.2
85[c]	17.5	27.8	29.2	29.7	37.3
80[c]	19.3	30.1	31.6	32.2	40.0
70[b]	20.9	32.3	33.9	34.4	42.4
60[c]	28.6	53.8	57.1	58.2	73.9
50[d]	12.3	28.4	30.6	31.4	43.7
40[d]	7.73	19.6	21.6	22.5	34.0
30[c]	5.97	15.3	17.1	17.7	27.4
25[b]	5.62	14.3	15.9	16.4	25.1
15[c]	4.97	12.6	13.8	14.2	21.1
10[c]	4.20	11.4	12.5	12.9	18.9
5[c]	3.34	9.66	10.6	10.9	16.1
1[b]	1.66	7.17	7.94	8.12	12.8

Silver-Palladium

Wt % Ag	100 K	273 K	293 K	300 K	350 K	400 K
99[b]	0.839	1.891	2.007	2.049	2.35	2.66
95[b]	2.528	3.58	3.70	3.74	4.04	4.34
90[b]	4.72	5.82	5.94	5.98	6.28	6.59
85[k]	6.82	7.92	8.04	8.08	8.38	8.68
80[k]	8.91	10.01	10.13	10.17	10.47	10.78
70[k]	13.43	14.53	14.65	14.69	14.99	15.30
60[i]	19.4	20.9	21.1	21.2	21.6	22.0
50[k]	29.3	31.2	31.4	31.5	32.0	32.4
40[m]	40.8	42.2	42.2	42.2	42.3	42.3
30[b]	37.1	40.4	40.6	40.7	41.3	41.7
25[k]	32.4	36.67	37.06	37.19	38.1	38.8
15[i]	21.0	27.08	26.68	27.89	29.3	30.6
10[i]	14.95	21.69	22.39	22.63	24.3	25.9
5[b]	8.91	15.98	16.72	16.98	18.8	20.5
1[a]	3.97	11.06	11.82	12.08	13.92	15.70

[a] Uncertainty in resistivity is ± 2%.
[b] Uncertainty in resistivity is ± 3%.
[c] Uncertainty in resistivity is ± 5%.
[d] Uncertainty in resistivity is ± 7% below 300 K and ± 5% at 300 and 400 K.
[e] Uncertainty in resistivity is ± 7%.
[f] Uncertainty in resistivity is ± 8%.
[g] Uncertainty in resistivity is ± 10%.
[h] Uncertainty in resistivity is ± 12%.
[i] Uncertainty in resistivity is ± 4%.
[j] Uncertainty in resistivity is ± 1%.
[k] Uncertainty in resistivity is ± 3% up to 300 K and ± 4% above 300 K.
[m] Uncertainty in resistivity is ± 2% up to 300 K and ± 4% above 300 K.

PERMITTIVITY (DIELECTRIC CONSTANT) OF INORGANIC SOLIDS

H. P. R. Frederikse

This table lists the permittivity ε, frequently called the dielectric constant, of a number of inorganic solids. When the material is not isotropic, the individual components of the permittivity are given. A superscript S indicates a measurement made under constant strain ("clamped" dielectric constant). If the constraint is removed, the measurement yields ε^T, the "unclamped" or free dielectric constant.

The temperature of the measurement is given when available; the symbol r.t. indicates a value at nominal room temperature. The frequency of the measurement is given in the last column (i.r. indicates a measurement in the infrared).

Substances are listed in alphabetical order by chemical formula.

Reference

Young, K. F. and Frederikse, H. P. R., *J. Phys. Chem. Ref. Data*, 2, 313, 1973.

Formula	Name	ε_{ijk}	T/K	ν/Hz
Ag_3AsS_3	Silver thioarsenate (Proustite)	$\varepsilon_{11}^T = 16.5, \varepsilon_{11}^S = 14.5$	r.t.	2×10^7
		$\varepsilon_{33}^T = 20.0, \varepsilon_{11}^S = 18.0$	r.t.	2×10^7
AgBr	Silver bromide	12.50	r.t.	
AgCN	Silver cyanide	5.6	r.t.	10^6
AgCl	Silver chloride	11.15	r.t.	
$AgNO_3$	Silver nitrate	9.0	293	5×10^5
$AgNa(NO_2)_2$	Silver sodium nitrite	4.5 ± 0.5	r.t.	9.4×10^9
Ag_2O	Silver oxide	8.8	r.t.	
$(AlF)_2SiO_4$	Aluminum fluosilicate (topaz)	$\varepsilon_{11} = 6.62$	297	7×10^3
		$\varepsilon_{22} = 6.58$	297	7×10^3
		$\varepsilon_{33} = 6.95$	297	7×10^3
Al_2O_3	Aluminum oxide (alumina)	$\varepsilon_{11} = \varepsilon_{22} = 9.34$	298	$10^2 - 8 \times 10^9$
		$\varepsilon_{33} = 11.54$	298	$10^2 - 8 \times 10^9$
$AlPO_4$	Aluminum phosphate	$\varepsilon_{11}^T = 6.05$	r.t.	10^3
AlSb	Aluminum antimonide	11.21	300	i.r.
AsF_3	Arsenic trifluoride	5.7	r.t.	
BN	Boron nitride	7.1	r.t.	i.r.
$BaCO_3$	Barium carbonate	8.53	291	2×10^5
$Ba(COOH)_2$	Barium formate	$\varepsilon_{11} = 7.9$	r.t.	10^3
		$\varepsilon_{22} = 5.9$	r.t.	10^3
		$\varepsilon_{33} = 7.5$	r.t.	10^3
$BaCl_2$	Barium chloride	9.81	r.t.	
$BaCl_2 \cdot 2H_2O$	Barium chloride dihydrate	9.00	r.t.	10^3
BaF_2	Barium fluoride	7.32	292	$5 \times 10^2 - 10^{11}$
$Ba(NO_3)_2$	Barium nitrate	4.95	292	2×10^5
$Ba_2NaNb_5O_{15}$	Barium sodium niobate ("Bananas")	$\varepsilon_{11}^S = 222, \varepsilon_{11}^T = 235$	296	10^4
		$\varepsilon_{22}^S = 227, \varepsilon_{22}^T = 247$	296	
		$\varepsilon_{33}^S = 32, \varepsilon_{33}^T = 51$	296	
BaO	Barium oxide (baria)	34 ± 1	248, 333	60×10^7
BaO_2	Barium peroxide	10.7	r.t.	2×10^6
BaS	Barium sulfide	19.23	r.t.	7.25×10^6
$BaSO_4$	Barium sulfate	11.4	288	10^8
$BaSnO_3$	Barium stannate	18	298	25×10^5
$BaTiO_3$	Barium titanate	$\varepsilon_{11}^T = 3600$	298	10^5
		$\varepsilon_{11}^S = 2300$	298	2.5×10^8
		$\varepsilon_{33}^T = 150$	298	10^5
		$\varepsilon_{33}^S = 80$	298	2.5×10^8
$Ba_6Ti_2Nb_8O_{30}$	Barium titanium niobate	$\varepsilon_{11} = \varepsilon_{22} \approx 190$	298	
		$\varepsilon_{33} \approx 220$	298	
$BaWO_4$	Barium tungstate	$\varepsilon_{11} = \varepsilon_{22} = 35.5 \pm 0.2$	297.5	1.6×10^3
		$\varepsilon_{33} = 37.2 \pm 0.2$	297.5	1.6×10^3

Formula	Name	ε_{ijk}	T/K	ν/Hz
$BaZrO_3$	Barium zirconate	43	r.t.	
$Be_3Al_2Si_6O_{18}$	Beryllium aluminum silicate (Beryl)	$\varepsilon_{33} = 5.95$	297	7×10^3
		$\varepsilon_{11} = \varepsilon_{22} = 6.86$	297	7×10^3
$BeCO_3$	Beryllium carbonate	9.7	291	2×10^5
BeO	Beryllium oxide (beryllia)	7.35 ± 0.2	293	2×10^6
$BiFeO_3$	Bismuth iron oxide	40 ± 3	300	9.4×10^9
$Bi_{12}GeO_{20}$	Bismuth germanite	$\varepsilon_{11}^S = 38$	r.t.	
$Bi(GeO_4)_3$	Bismuth germanate	16	293	
Bi_2O_3	Bismuth sesquioxide	18.2	r.t.	2×10^6
$Bi_4Ti_3O_{12}$	Bismuth titanate	112	r.t.	10^3
C	Diamond			
	Type I	5.87 ± 0.19	300	10^3
	Type IIa	5.66 ± 0.04	300	10^3
$C_4H_4O_6$	Tartaric acid	$\varepsilon_{11} = \varepsilon_{22} = 4.3$	298	
		$\varepsilon_{33} = 4.5$	298	
		$\varepsilon_{13} = 0.55$	298	
$C_6H_{14}N_2O_6$	Ethylene diamine tartrate (EDT)	$\varepsilon_{11}^T = 5.0$	293	
		$\varepsilon_{22}^T = 8.3$	293	
		$\varepsilon_{33}^T = 6.0$	293	
		$\varepsilon_{13}^T = 0.7$	293	
$C_6H_{12}O_6NaBr$	Dextrose sodium bromide	$\varepsilon_{11}^T = 4.0$	r.t.	10^3
$(CH_3NH_3)Al(SO_4)_2 \cdot 2H_2O$	Methyl ammonium alum (MASD)	19	197	
$Ca_2B_6O_{11} \cdot 5H_2O$	Colemanite	$\varepsilon_{11} = 20$	293	10^3
		$\varepsilon_{33} = 25$	293	10^3
$CaCO_3$	Calcium carbonate	$\varepsilon_{11} = 8.67$	r.t.	9.4×10^{10}
		$\varepsilon_{22} = 8.69$	r.t.	9.4×10^{10}
		$\varepsilon_{33} = 8.31$	r.t.	9.4×10^{10}
$CaCeO_3$	Calcium cerate	21	r.t.	
CaF_2	Calcium fluoride	6.81	300	5×10^2–10^{11}
$CaMoO_4$	Calcium molybdate	$\varepsilon_{11} = \varepsilon_{22} = 24.0 \pm 0.2$	297.5	<10
		$\varepsilon_{33} = 20.0 \pm 0.2$	297.5	<10
$Ca(NO_3)_2$	Calcium nitrate	6.54	292	2×10^5
$CaNb_2O_6$	Calcium niobate	$\varepsilon_{11} = 22.8 \pm 1.9$	r.t.	$(5–500) \times 10^3$
$Ca_2Nb_2O_7$	Calcium pyroniobate	~45	r.t.	5×10^7
CaO	Calcium oxide	11.8 ± 0.3	283	2×10^6
CaS	Calcium sulfide	6.699	r.t.	7.25×10^6
$CaSO_4 \cdot 2H_2O$	Calcium sulfate dihydrate	$\varepsilon_{11} = 5.10$	r.t.	
		$\varepsilon_{22} = 5.24$	r.t.	
		$\varepsilon_{33} = 10.30$	r.t.	
$CaTiO_3$	Calcium titanate	165	r.t.	
$CaWO_4$	Calcium tungstate	$\varepsilon_{11} = \varepsilon_{22} = 11.7 \pm 0.1$	297.5	1.59×10^3
		$\varepsilon_{33} = 9.5 \pm 0.2$	297.5	1.59×10^3
Cd_3As_2	Cadmium arsenide	$\varepsilon_{33} = 18.5$	4	
$CdBr_2$	Cadmium bromide	8.6	293	5×10^5
CdF_2	Cadmium fluoride	8.33 ± 0.08	300	10^5–10^7
CdS	Cadmium sulfide	$\varepsilon_{11} = \varepsilon_{22} = 8.7$	300	i.r.
		$\varepsilon_{33} = 9.25$	300	i.r.
		$\varepsilon_{11} = \varepsilon_{22} = 8.37$	8	i.r.
		$\varepsilon_{33} = 9.00$	8	i.r.
		$\varepsilon_{11}^T = 8.48$	77	10^4
		$\varepsilon_{33}^T = 9.48$	77	10^4
		$\varepsilon_{11}^S = 9.02, \varepsilon_{11}^T = 9.35$	298	10^4
		$\varepsilon_{33}^S = 9.53, \varepsilon_{33}^T = 10.33$	298	10^4
$CdSe$	Cadmium selenide	$\varepsilon_{11}^S = 9.53, \varepsilon_{11}^T = 9.70$	298	10^4

Formula	Name	ε_{ijk}	T/K	ν/Hz
		$\varepsilon_{33}^S = 10.2, \varepsilon_{33}^T = 10.65$	298	10^4
CdTe	Cadmium telluride	$\varepsilon_{11} = \varepsilon_{22} = 10.60 \pm 0.15$	297	i.r.
		$\varepsilon_{33} = 7.05 \pm 0.05$	297	i.r.
$Cd_2Nb_2O_7$	Cadmium pyroniobate	500–580	293	10^3
CeO_2	Cerium oxide	7.0	r.t.	2×10^6
$CoNb_2O_6$	Cobalt niobate	$\varepsilon_{11} = 18.4 \pm 1.1$	r.t.	$(5–500) \times 10^3$
		$\varepsilon_{22} = 21.4 \pm 1.1$	r.t.	$(5–500) \times 10^3$
		$\varepsilon_{33} = 33.0 \pm 0.7$	r.t.	$(5–500) \times 10^3$
CoO	Cobalt oxide	12.9	298	$10^2 – 10^{10}$
Cr_2O_3	Chromic sesquioxide	$\varepsilon_{11} = \varepsilon_{22} = 13.3$	298.5	10^3
		$\varepsilon_{33} = 11.9$	298.5	10^3
		8	315 (T_N)	6×10^{10}
$CsAl(SO_4)_2 \cdot 12H_2O$	Cesium alum	5.0	r.t.	$20–20 \times 10^3$
CsBr	Cesium bromide	6.38	298	1.6×10^3
Cs_2CO_3	Cesium carbonate	6.53	291	2×10^5
CsCl	Cesium chloride	7.2	298	
CsH_2AsO_4	Cesium dihydrogen arsenate (CDA)	4.8	273	9.5×10^9
CsH_2PO_4	Cesium dihydrogen phosphate (CDP)	6.15	285	9.5×10^9
$CsH_3(SeO_3)_2$	Cesium trihydrogen selenite	$\varepsilon_{11} = 80$	273	10^5
		$\varepsilon_{22} = 63$	273	10^5
		$\varepsilon_{33} = 12$	273	10^5
CsI	Cesium iodide	6.31	298	1.6×10^3
$CsNO_3$	Cesium nitrate	$\varepsilon_{11} = \varepsilon_{22} = 9.4$	r.t.	5×10^5
		$\varepsilon_{33} = 8.3$	r.t.	5×10^5
$CsPbCl_3$	Cesium lead chloride	14.37	300	$10^5 – 10^6$
CuBr	Cuprous bromide	8.0	293	5×10^5
CuCl	Cuprous chloride	9.8 ± 0.5	r.t.	10^3
CuO	Cupric oxide	18.1	r.t.	2×10^6
Cu_2O	Cuprous oxide (Cuprite)	7.60 ± 0.06	r.t.	10^5
$CuSO_4 \cdot 5H_2O$	Cupric sulfate pentahydrate	6.60	r.t.	
EuF_2	Europium fluoride	7.7 ± 0.2	298	$(1–300) \times 10^3$
$Eu_2(MoO_4)_3$	Europium molybdate	9.5	298	
EuS	Europium sulfide	13.10 ± 0.04	80	$5 \times 10^2 – 10^5$
FeO	Ferrous oxide	14.2	r.t.	2×10^6
Fe_2O_3	Ferric sesquioxide	4.5	r.t.	$10^5 – 10^7$
$Fe_2O_3\text{-}\alpha$	Ferric sesquioxide (hematite)	12		6×10^{10}
Fe_3O_4	Ferrosoferric oxide (magnetite)	20	r.t.	$10^5 – 10^7$
GaAs	Gallium arsenide	13.13	300	
		12.90	4	i.r.
GaP	Gallium phosphide	11.1	r.t.	
		10.75 ± 0.1	1.6	i.r.
GaSb	Gallium antimonide	15.69	r.t.	
		15.7	4	i.r.
$Gd_2(MoO_4)_3$	Gadolinium molybdate	$\varepsilon^T = 10$	298	
		$\varepsilon^S = 9.5$	298	10^3
Ge	Germanium	16.0 ± 0.3	4	9.2×10^9
		15.8 ± 0.2	r.t.	$500 – 3 \times 10^{10}$
GeO_2	Germanium dioxide	$\varepsilon_{11} = \varepsilon_{22} = 7.44$	r.t.	i.r.
HIO_3	Iodic acid	$\varepsilon_{11} = 7.5$	r.t.	10^3
		$\varepsilon_{22} = 12.4$	r.t.	10^3
		$\varepsilon_{33} = 8.1$	r.t.	10^3
$HNH_4(ClCH_2COO)_2$	Hydrogen ammonium dichloroacetate	$\varepsilon_{[102]} = 5.9$	r.t.	10^5
H_2O	Ice I (P = 0 kbar)	99	243	
	Ice III (P = 3 kbar)	117	243	
	Ice V (P = 5 kbar)	114	243	
	Ice VI (P = 8 kbar)	193	243	
HgCl	Mercurous chloride (Calumel)	$\varepsilon_{11} = \varepsilon_{22} = 14.0$	r.t.	10^{12}
$HgCl_2$	Mercuric chloride	6.5	r.t.	10^{12}
HgS	Mercurous sulfide (Cinnabar)	$\varepsilon_{11} = \varepsilon_{22} = 18.0$	r.t.	i.r.
		$\varepsilon_{33} = 32.5$	r.t.	i.r.

Formula	Name	ε_{ijk}	T/K	ν/Hz
HgSe	Mercurous selenide	25.6	r.t.	10^4–10^6
I_2	Iodine	$\varepsilon_{11} = 6$	r.t.	5×10^4–10^7
		$\varepsilon_{22} = 3$	r.t.	5×10^4–10^7
		$\varepsilon_{33} = 40$	r.t.	5×10^4–10^7
InAs	Indium arsenide	14.55 ± 0.3	r.t.	i.r.
		15.15	4	i.r.
InP	Indium phosphide	12.61	r.t.	i.r.
InSb	Indium antimonide	17.88	4	i.r.
$KAl(SO_4)_2 \cdot 12H_2O$	Potassium alum	6.5	r.t.	20–20×10^3
KBr	Potassium bromide	4.88	300	
		4.53	4.2	
$KBrO_3$	Potassium bromate	7.3	r.t.	2×10^6
KCN	Potassium cyanide	6.15	r.t.	2×10^6
K_2CO_3	Potassium carbonate	4.96	291	2×10^5
$K_2C_4H_4O_6 \cdot {}^1/_2 H_2O$	Dipotassium tartrate (DKT)	$\varepsilon_{11} = 6.44$	r.t.	
		$\varepsilon_{22} = 5.80$	r.t.	
		$\varepsilon_{33} = 6.49$	r.t.	
		$\varepsilon_{13} = 0.005$	r.t.	
KCl	Potassium chloride	4.86 ± 0.02	r.t.	5×10^3
		4.50	4.2	
$KClO_3$	Potassium chlorate	5.1	r.t.	2×10^6
$KClO_4$	Potassium perchlorate	5.9	r.t.	2×10^6
K_2CrO_4	Potassium chromate	7.3	r.t.	6×10^7
$KCr(SO_4)_2 \cdot 12H_2O$	Potassium chrome alum	6.5	100–240	175×10^3
KD_2AsO_4	Potassium dideuterium arsenate (KDDA)	$\varepsilon_{11} = 70$	298	
		$\varepsilon_{33} = 31$	298	
KD_2PO_4	Potassium dideuterium phosphate (KDDP)	50 ± 2	297	10^3
KF	Potassium fluoride	6.05		2×10^6
KH_2AsO_4	Potassium dihydrogen arsenate (KDA)	$\varepsilon_{11} = 60$	298	
		$\varepsilon_{33} = 24$	298	
KH_2PO_4	Potassium dihydrogen phosphate (KDP)	46	298	10^3
		$\varepsilon_{11} = 42$	r.t.	
		$\varepsilon_{33} = 21$	r.t.	
K_2HPO_4	Dipotassium monohydrogen orthophosphate	9.05	r.t.	2×10^6
KI	Potassium iodide	5.00	r.t.	9.4×10^{10}
KIO_3	Potassium iodate	170	255	10^5
		10	293	10^5
		$\varepsilon_{[101]} \approx 40,70$	r.t.	10^5
		16.85	r.t.	2×10^6
$(K,H)Al_3(SiO_4)_3$	Mica (muscovite)	5.4	299	10^2–3×10^9
$(K,H)Mg_3Al(SiO_4)_3$	Mica (Canadian)	$\varepsilon_{11} = \varepsilon_{22} = 6.9$	298	10^2–10^4
		$\varepsilon_{33} = 7.3$	298	10^4
KNO_2	Potassium nitrite	25	305	
KNO_3	Potassium nitrate	4.37	293	2×10^5
$KNbO_3$	Potassium niobate	700	r.t.	
K_3PO_4	Potassium orthophosphate	7.75	r.t.	2×10^6
KSCN	Potassium thiocyanate	7.9	r.t.	2×10^6
K_2SO_4	Potassium sulfate	6.4	r.t.	2×10^6
$K_2S_3O_6$	Potassium trithionate	5.7	293	1.8×10^6
$K_2S_4O_6$	Potassium tetrathionate	5.5	293	1.8×10^6
$K_2S_5O_6 \cdot H_2O$	Potassium pentathionate	7.8	293	1.8×10^6
$K_2S_6O_6$	Potassium hexathionate	7.8	293	1.8×10^6
K_2SeO_4	Potassium selenate	$\varepsilon_{11} = 5.9$	r.t.	10^3
		$\varepsilon_{22} = 7.7$	r.t.	10^3
$KSr_2Nb_5O_{15}$	Potassium strontium niobate	$\varepsilon_{11} = \varepsilon_{11} \approx 1200$	298	
		$\varepsilon_{33} \approx 800$	298	
$KTaNbO_3$	Potassium tantalate niobate (KTN)	34,000	273	10^4
		6,000	293	10^4
$KTaO_3$	Potassium tantalate	242	298	2×10^5
$LaScO_3$	Lanthanum scandate	30	r.t.	

Formula	Name	ε_{ijk}	T/K	ν/Hz
LiBr	Lithium bromide	12.1	r.t.	2×10^6
Li_2CO_3	Lithium carbonate	4.9	291	2×10^5
LiCl	Lithium chloride	11.05	r.t.	2×10^6
LiD	Lithium deuteride	14.0 ± 0.5	r.t.	i.r.
LiF	Lithium fluoride	9.00	298	$10^2–10^7$
		9.11	353	$10^2–10^7$
$LiGaO_2$	Lithium metagallate	$\varepsilon_{11}^T = 7.0, \varepsilon_{22}^T = 6.0$	r.t.	
		$\varepsilon_{33}^T = 9.5$	r.t.	
		$\varepsilon_{11}^S = 6.8, \varepsilon_{22}^S = 5.8$	r.t.	
Li^6H	Lithium-6 hydride	13.2 ± 0.5	r.t.	
Li^7H	Lithium-7 hydride	12.9 ± 0.5	r.t.	
$LiH_3(SeO_3)_2$	Lithium trihydrogen selenite	29	298	10^4
		$\varepsilon_{11} = 13.0$	r.t.	
		$\varepsilon_{22} = 12.9$	r.t.	
		$\varepsilon_{33} = 46$	r.t.	
LiI	Lithium iodide	11.03	r.t.	2×10^6
$LiIO_3$	Lithium iodate	$\varepsilon_{11} = \varepsilon_{22} = 65$	294.5	10^3
		$\varepsilon_{33} = 554$	298	
$LiNH_4C_4O_6 \cdot H_2O$	Lithium ammonium tartrate (LAT)	$\varepsilon_{11}^T = 7.2$	298	
		$\varepsilon_{22}^T = 8.0$	298	
		$\varepsilon_{33}^T = 6.9$	298	
$LiNa_3CrO_4 \cdot 6H_2O$	Lithium trisodium chromate	8.0	r.t.	10^3
$LiNa_3MoO_4 \cdot 6H_2O$	Lithium trisodium molybdate	$\varepsilon_{11} = 6.7$	r.t.	10^3
		$\varepsilon_{33} = 5.3$	r.t.	10^3
$LiNbO_3$	Lithium niobate	$\varepsilon_{11} = \varepsilon_{22} = 82$	298	10^5
		$\varepsilon_{33} = 30$	298	10^5
$Li_2SO_4 \cdot H_2O$	Lithium sulfate monohydrate	$\varepsilon_{11} = 5.6$	298	
		$\varepsilon_{22} = 10.3$	298	
		$\varepsilon_{33} = 6.5$	298	
		$\varepsilon_{13} = 0.07$	298	
$LiTaO_3$	Lithium tantalate	$\varepsilon_{11} = \varepsilon_{22} = 53$	r.t.	10^5
		$\varepsilon_{33} = 46$	r.t.	10^5
		$\varepsilon_{11}^S = \varepsilon_{22}^S = 41$	r.t.	
		$\varepsilon_{33}^S = 43$	r.t.	
		$\varepsilon_{11}^T = \varepsilon_{22}^T = 51$	r.t.	
		$\varepsilon_{33}^T = 45$	r.t.	
$LiTIC_4O_6 \cdot H_2O$	Lithium thallium tartrate (LTT)	$\varepsilon_{11} \approx 20$	80	
$Mg_3B_7O_{13}Cl$	Magnesium borate monochloride (boracite)	$\varepsilon_{11} = 14.1$	r.t.	5×10^5
$MgCO_3$	Magnesium carbonate	8.1	291	2×10^5
$MgNb_2O_6$	Magnesium niobate	$\varepsilon_{11} = 16.4 \pm 0.5$	r.t.	$(5–500) \times 10^3$
		$\varepsilon_{22} = 20.9 \pm 0.5$	r.t.	$(5–500) \times 10^3$
		$\varepsilon_{33} = 32.4 \pm 0.5$	r.t.	$(5–500) \times 10^3$
MgO	Magnesium oxide (Periclase)	9.65	298	$10^2–10^8$
$(MgO)_xAl_2O_3$	Spinel	8.6	r.t.	
$MgSO_4$	Magnesium sulfate	8.2	r.t.	
$MgSO_4 \cdot 7H_2O$	Magnesium sulfate septahydrate	5.46	r.t.	
$MgTiO_3$	Magnesium titanate	13.5	r.t.	
$MgWO_4$	Magnesium tungstate	$\varepsilon_{11} = 18.0 \pm 1$	r.t.	$(5–500) \times 10^3$
		$\varepsilon_{22} = 18.0 \pm 1$	r.t.	$(5–500) \times 10^3$
$MnNb_2O_6$	Manganese niobate	$\varepsilon_{11} = 17.4 \pm 2$	r.t.	$(5–500) \times 10^3$
		$\varepsilon_{22} = 16.1 \pm 0.5$	r.t.	$(5–500) \times 10^3$
		$\varepsilon_{33} = 30.7 \pm 1$	r.t.	$(5–500) \times 10^3$
MnO	Manganese oxide (Pyrolusite)	12.8	r.t.	6×10^{10}
MnO_2	Manganese dioxide	$\sim10^4$	298	10^4
Mn_2O_3	Manganese sesquioxide	8	r.t.	6×10^{10}
$MnWO_4$	Manganese tungstate	$\varepsilon_{11} = 19.3 \pm 1.3$	r.t.	$(5–500) \times 10^3$

Permittivity (Dielectric Constant) of Inorganic Solids

Formula	Name	ε_{ijk}	T/K	ν/Hz
		$\varepsilon_{22} = 14.3 \pm 0.5$	r.t.	$(5-500) \times 10^3$
		$\varepsilon_{33} = 16.5 \pm 1.1$	r.t.	$(5-500) \times 10^3$
$N(CH_3)_4HgBr_3$	Tetramethylammonium tribromomercurate (TTM)	~10	233–373	
$N(CH_3)_4HgI_3$	Tetramethylammonium triiodo mercurate (TTM)	~10	233–373	
$N_4(CH_2)_6$	Hexamethylene tetramine (HMTA)	2.6 ± 0.2	r.t.	10^9-10^{10}
$(ND_4)_2BeF_4$	Deuteroammonium fluoberyllate	$\varepsilon_{11} = 10$	r.t.	
		$\varepsilon_{22} = 9$	r.t.	
		$\varepsilon_{33} = 9$	r.t.	
$(ND_4)_2SO_4$	Deuteroammonium sulfate	$\varepsilon_{11} = 9$	r.t.	
		$\varepsilon_{22} = 10$	r.t.	
		$\varepsilon_{33} = 9$	r.t.	
$(NH_2 \cdot CH_2COOH)_3 \cdot H_2SO_4$	Triglycine sulfate (TGS)	$\varepsilon_{11} = 9$	273	10^4
		$\varepsilon_{22} = 30$	273	10^4
		$\varepsilon_{33} = 6.5$	273	10^4
$(NH_2 \cdot CH_2COOH)_3 \cdot H_2SeO_4$	Triglycine selenate (TGSe)	200	293	1.6×10^3
$(NH_2 \cdot CH_2COOH)_3 \cdot H_2BeF_4$	Triglycine fluorberyllate (TGFB)	$\varepsilon_{22} = 12$	273	10^4
$NH_4Al(SO_4)_2 \cdot 12H_2O$	Ammonium alum	6	r.t.	10^{12}
$(NH_4)_2BeF_4$	Ammonium fluorberyllate	$\varepsilon_{11} = \varepsilon_{22} = 7.8$	123	10^5
		$\varepsilon_{33} = 7.1$	123	10^5
		$\varepsilon_{11} = \varepsilon_{22} = 8.8$	293	10^5
		$\varepsilon_{33} = 9.2$	293	10^5
NH_4Br	Ammonium bromide	7.1	r.t.	7×10^5
NH_4I	Ammonium iodide	9.8	r.t.	
$(NH_4)_2C_2H_6O_6$	Ammonium tartrate	$\varepsilon_{11} = 6.45$	r.t.	10^3
		$\varepsilon_{22} = 6.8$	r.t.	10^3
		$\varepsilon_{33} = 6.0$	r.t.	10^3
$(NH_4)_2Cd_2(SO_4)_3$	Ammonium cadmium sulfate	10.0	r.t.	10^4
NH_4Cl	Ammonium chloride	6.9	r.t.	7×10^5
$NH_4(ClCH_2COO)$	Ammonium monochloroacetate	5	r.t.	2×10^6
$NH_4Cr(SO_4)_2 \cdot 12H_2O$	Ammonium chrome alum	6.5	r.t.	175×10^3
NH_4HSO_4	Ammonium bisulfate	165	273	5×10^4
$NH_4H_2AsO_4$	Ammonium dihydrogen arsenate (ADA)	5.1	265	9.5×10^9
		$\varepsilon_{11} = \varepsilon_{22} = 85$	298	10^3
		$\varepsilon_{33} = 22$	298	
$NH_4H_2PO_4$	Ammonium dihydrogen phosphate (ADP)	$\varepsilon_{11} = \varepsilon_{22} = 57.1 \pm 0.6$	294.5	$10^5-35 \times 10^9$
		$\varepsilon_{33} = 14.0 \pm 0.3$	294	$10^5-36 \times 10^9$
$ND_4D_2PO_4$	Ammonium dideuterium phosphate (ADDP)	$\varepsilon_{11} = \varepsilon_{22} = 74, \varepsilon_{33} = 24$	300	
NH_4NO_3	Ammonium nitrate	10.7	322	$(5-50) \times 10^3$
$(NH_4)_2SO_4$	Ammonium sulfate	$\varepsilon_{11} = \varepsilon_{22} = 8.0$	123	10^5
		$\varepsilon_{33} = 6.3$	123	10^5
		$\varepsilon_{11} = \varepsilon_{22} = 10.0$	293	10^5
		$\varepsilon_{33} = 9.3$	293	10^5
$(NH_4)_2UO_2(C_2O_4)_2$	Ammonium uranyl oxalate	8.03	r.t.	$10^4-3.3 \times 10^9$
$(NH_4)_2UO_2(C_2O_4)_2 \cdot 3H_2O$	Ammonium uranyl oxalate trihydrate	6.06	r.t.	$10^4-3.3 \times 10^9$
$NaBr$	Sodium bromide	6.44	298	1.6×10^3
$NaBrO_3$	Sodium bromate	$\varepsilon_{11}^T = 5.70$	298	10^3
$NaCN$	Sodium cyanide	7.55	293	10^5
$NaCO_3$	Sodium carbonate	8.75	291	2×10^5
$NaCO_3 \cdot 10H_2O$	Sodium carbonate decahydrate	5.3	r.t.	6×10^7
$NaCl$	Sodium chloride	5.9	298	10^2-10^7
		5.45	4.2	
$NaClO_3$	Sodium chlorate	$\varepsilon_{11}^T = 5.76$	301	10^3
		5.28	r.t.	10^3

Formula	Name	ε_{ijk}	T/K	ν/Hz
$NaClO_4$	Sodium perchlorate	5.76	r.t.	10^3
NaF	Sodium fluoride	5.08 ± 0.02	r.t.	5×10^3
$NaH_3(SeO_3)_2$	Sodium trihydrogen selenite	$\varepsilon_{11} \approx 75$	273	2×10^5
$NaD_3(SeO_3)_2$	Sodium trideuterium selenite	$\varepsilon_{11} \approx 220$	273	2×10^5
NaI	Sodium iodide	7.28 ± 0.03	r.t.	
$NaK(C_4H_2D_2O_6) \cdot 4D_2O$	Sodium potassium tartrate tetradeutrate	$\varepsilon_{11} = 70$	273	10^3
	(double deuterated Rochelle salt)	$\varepsilon_{22} = 8.9$	273	10^3
$NaK(C_4H_4O_6) \cdot 4H_2O$	Sodium potassium tartrate tetrahydrate	$\varepsilon_{11} = 170$	273	10^3
	(Rochelle salt)	$\varepsilon_{22} = 9.1$	273	10^3
$NaNH_4(C_4H_4O_6) \cdot 4H_2O$	Sodium ammonium tartrate	$\varepsilon_{11} = 8.4$	298	
	(Ammonium Rochelle salt)	$\varepsilon_{22} = 9.2$	298	
		$\varepsilon_{33} = 9.5$	298	
$NaNbO_3$	Sodium niobate	$\varepsilon_{33} = 670 \pm 13$	r.t.	
		$\varepsilon_{11} = \varepsilon_{22} = 76 \pm 2$	r.t.	
$NaNO_2$	Sodium nitrite	$\varepsilon_{11} = 7.4$	r.t.	5×10^5
		$\varepsilon_{22} = 5.5$	r.t.	5×10^5
		$\varepsilon_{33} = 5.0$	r.t.	5×10^5
$NaNO_3$	Sodium nitrate	6.85	292	2×10^5
$NaSO_4$	Sodium sulfate	7.90	r.t.	
$NaSO_4 \cdot 10H_2O$	Sodium sulfate decahydrate	5.0	r.t.	
$Na_2SO_4 \cdot 5H_2O$	Sodium sulfate pentahydrate	7	250–290	$300–10^4$
$Na_2UO_2(C_2O_4)_2$	Sodium uranyl oxalate	5.18	r.t.	
$NdAlO_3$	Neodymium aluminate	17.5	r.t.	
$NdScO_3$	Neodymium scandate	27	r.t.	
$Ni_3B_7O_{13}I$	Nickel iodine boracite	$\varepsilon_{11} = 14$	260	
$NiNb_2O_6$	Nickel niobate	$\varepsilon_{11} = 16.0 \pm 0.5$	r.t.	$(5–500) \times 10^3$
		$\varepsilon_{22} = 23.8 \pm 1.8$	r.t.	$(5–500) \times 10^3$
		$\varepsilon_{33} = 31.3 \pm 2.5$	r.t.	$(5–500) \times 10^3$
NiO	Nickel oxide	11.9	298	10^5
$NiSO_4 \cdot 6H_2O$	Nickel sulfate hexahydrate	$\varepsilon_{11} = 6.2$	r.t.	
		$\varepsilon_{33} = 6.8$	r.t.	
$NiWO_4$	Nickel tungstate	$\varepsilon_{11} = 17.4 \pm 2.4$	r.t.	$(5–500) \times 10^3$
		$\varepsilon_{22} = 13.6 \pm 1.0$	r.t.	$(5–500) \times 10^3$
		$\varepsilon_{33} = 19.7 \pm 0.6$	r.t.	$(5–500) \times 10^3$
P	Phosphorous (red)	4.1	r.t.	10^8
	Phosphorous (yellow)	3.6	r.t.	10^8
$[P(CH_3)_4]HgBr_3$	Tetramethylphosphonium tribromomercurate (TTM)	~10	233–373	
$PbBr_2$	Lead bromide	>30	293	$(0.5–3) \times 10^6$
$PbCO_3$	Lead carbonate	18.6	288	10.8
$Pb(C_2H_3O_2)_2$	Lead acetate	2.6	290–295	10^6
$PbCl_2$	Lead chloride	33.5	273	$(0.5–3) \times 10^6$
Pb_2CoWO_6	Lead cobalt tungstate	~250	r.t.	
PbF_2	Lead fluoride	26.3	r.t.	
$PbHfO_3$	Lead hafnate	390	300	10^5
		185	400	
PbI_2	Lead iodide	20.8	293	$(0.5–3) \times 10^6$
$Pb_3MgNb_2O_9$	Lead magnesium niobate	10,000	297	
$PbMoO_4$	Lead molybdate	$\varepsilon_{11} = 34.0 \pm 0.4$	297.5	1.6 ± 10^3
		$\varepsilon_{33} = 40.6 \pm 0.2$	297.5	1.6 ± 10^3
$Pb(NO_3)_2$	Lead nitrate	16.8	r.t.	$(0.5–3) \times 10^6$
$PbNb_2O_6$	Lead niobate	$\varepsilon_{33}^T = 180$	298	
PbO	Lead oxide	25.9	r.t.	2×10^6
PbS	Lead sulfide (Galena)	190	77	i.r.
		200 ± 35	r.t.	i.r.
$PbSO_4$	Lead sulfate	14.3	290—295	10^6
$PbSe$	Lead selenide	280	r.t.	i.r.
$PbTa_2O_6$	Lead metatantalate	$\varepsilon_{11} = \varepsilon_{22} \approx 300$	r.t.	10^4
		$\varepsilon_{33} = 150$	r.t.	10^4
$PbTe$	Lead telluride	450	r.t.	i.r.

Formula	Name	ε_{ijk}	T/K	ν/Hz
		40	77	$10^4 - 15 \times 10^4$
		430	4.2	$10^4 - 15 \times 10^4$
$PbTiO_3$	Lead titanate	~ 200	r.t.	10^3
$PbWO_4$	Lead tungstate	$\varepsilon_{11} = \varepsilon_{22} = 23.6 \pm 0.3$	297.5	1.59×10^3
		$\varepsilon_{33} = 31.0 \pm 0.4$	297.5	1.59×10^3
$Pb(Zn_{1/3}Nb_{2/3})O_3$	Lead zinc niobate	7	300	$10^3, 300 \times 10^3$
$PbZrO_3$	Lead zirconate	200	400	
$RbAl(SO_4)_2 \cdot 12H_2O$	Rubidium alum	5.1	r.t.	10^{12}
$RbBr$	Rubidium bromide	4.83	300	
Rb_2CO_3	Rubidium carbonate	4.87 ± 0.02	r.t.	5×10^3
$RbCl$	Rubidium chloride	4.91 ± 0.02	r.t.	5×10^3
$RbCr(SO_4)_2 \cdot 12H_2O$	Rubidium chrome alum	5.0	r.t.	10^{12}
RbF	Rubidium fluoride	5.91	r.t.	2×10^6
$RbHSO_4$	Rubidium bisulfate	$\varepsilon_{11} = 7$	r.t.	10^5
		$\varepsilon_{22} = 8$	r.t.	10^5
		$\varepsilon_{33} = 10$	r.t.	10^5
RbH_2AsO_4	Rubidium dihydrogen arsenate (RDA)	3.90	273	9.5×10^9
RbH_2PO_4	Rubidium dihydrogen phosphate (RDP)	6.15	285	9.5×10^9
RbI	Rubidium iodide	4.94 ± 0.02	r.t.	5×10^3
$RbInSO_4$	Rubidium indium sulfate	6.85	r.t.	
$RbNO_3$	Rubidium nitrate	20—380	433–488	10^6
		30	488–538	10^6
S	Sulfur	$\varepsilon_{11} = 3.75$	298	$10^2 - 10^3$
		$\varepsilon_{22} = 3.95$	298	$10^2 - 10^3$
		$\varepsilon_{33} = 4.44$	298	$10^2 - 10^3$
	sublimed	3.69	298	$10^2 - 10^3$
$SC(NH_2)_2$	Thiourea	$\varepsilon_{11} = \varepsilon_{22} \approx 3$	77–300	10^3
		$\varepsilon_{22} = 35$	300	10^3
Sb_2O_3	Antimonous sesquioxide	12.8	r.t.	$(1.5-2) \times 10^3$
Sb_2S_3	Antimonous sulfide (stibnite)	$\varepsilon_{11} = \varepsilon_{33} = 15$	r.t.	10^3
		$\varepsilon_{33} = 180$	r.t.	10^3
Sb_2Se_3	Antimonous selenide	~ 110	r.t.	$(10-16.5) \times 10^9$
$SbSI$	Antimonous sulfide iodide	2000	273	10^5
		$\varepsilon_{11} = \varepsilon_{22} \approx 25$	r.t.	$10^3 - 10^5$
		$\varepsilon_{33} \approx 5 \times 10^4$	295	$10^3 - 10^5$
Se	Selenium (monocrystal)	$\varepsilon_{11} = \varepsilon_{22} = 11$	300	24×10^9
		$\varepsilon_{33} = 21$	300	24×10^9
	(amorphous)	6.0	298	$10^2 - 10^{10}$
Si	Silicon	12.1	4.2	$10^7 - 10^9$
SiC	Silicon carbide			
	cubic	9.72	r.t.	i.r.
	6H	$\varepsilon_{11} = \varepsilon_{22} = 9.66$	r.t.	i.r.
		$\varepsilon_{33} = 10.03$	r.t.	i.r.
		9.7 ± 0.1	1.8	i.r.
Si_3N_4	Silicon nitride	4.2 (film)	r.t.	10^3
SiO	Silicon monoxide	5.8	r.t.	10^3
SiO_2	Silicon dioxide	$\varepsilon_{11} = 4.42$	r.t.	9.4×10^{10}
		$\varepsilon_{22} = 4.41$	r.t.	9.4×10^{10}
		$\varepsilon_{33} = 4.60$	r.t.	9.4×10^{10}
$Sm_2(MoO_4)_3$	Samarium molybdate	12	298	
SnO_2	Stannic dioxide	$\varepsilon_{11} = \varepsilon_{22} = 14 \pm 2$	r.t.	$10^4 - 10^{10}$
		$\varepsilon_{33} = 9.0 \pm 0.5$	r.t.	$10^4 - 10^{10}$
$SnSb$	Tin antimonide	147	r.t.	$10^4 - 10^6$
$SnTe$	Tin telluride	1770 ± 300	r.t.	i.r.
$Sr(COOH)_2 \cdot 2H_2O$	Strontium formate dihydrate	6.1	r.t.	10^3
$SrCO_3$	Strontium carbonate	8.85	298	2×10^5
$SrCl_2$	Strontium chloride	9.19	r.t.	
$SrCl_2 \cdot 6H_2O$	Strontium chloride hexahydrate	8.52	r.t.	
SrF_2	Strontium fluoride	6.50	300	$5 \times 10^2 - 10^{11}$
$SrMoO_4$	Strontium molybdate	$\varepsilon_{11} = \varepsilon_{22} = 31.7 \pm 0.2$	297.5	1.59×10^3

Formula	Name	ε_{ijk}	T/K	ν/Hz
		$\varepsilon_{33} = 41.7 \pm 0.2$	297.5	1.59×10^3
$Sr(NO_3)_2$	Strontium nitrate	5.33	292	2×10^5
$Sr_2Nb_2O_7$	Strontium niobate	$\varepsilon_{11} = 75$	r.t.	10^3
		$\varepsilon_{22} = 46$	r.t.	10^3
		$\varepsilon_{33} = 43$	r.t.	10^3
SrO	Strontium oxide	13.3 ± 0.3	273	2×10^6
SrS	Strontium sulfide	11.3	r.t.	7.25×10^6
$SrSO_4$	Strontium sulfate	11.5	r.t.	
$SrTiO_3$	Strontium titanate	332	298	10^3
		2080	78	10^3
$SrWO_4$	Strontium tungstate	$\varepsilon_{11} = \varepsilon_{22} = 25.7 \pm 0.2$	297.5	1.6×10^3
		$\varepsilon_{33} = 34.1 \pm 0.2$	297.5	1.6×10^3
Ta_2O_5	Tantalum pentoxide (tantala)			
	α phase	$\varepsilon_{11} = \varepsilon_{22} = 30$	77	10^3
		$\varepsilon_{33} = 65$	77	10^3
	β phase	24	292	10^3
$Tb(MoO_4)_3$	Terbium molybdate	11	298	
		$\varepsilon_{11} = \varepsilon_{22} = 33$	100–200	9.4×10^9
		$\varepsilon_{33} = 53$	100–200	9.4×10^9
Te	Tellurium	$\varepsilon_{11} = \varepsilon_{22} = 33$	r.t.	
		$\varepsilon_{33} = 54$	r.t.	
	polycrystalline	27.5	r.t.	i.r.
	monocrystalline	28.0	r.t.	i.r.
ThO_2	Thorium dioxide	18.9 ± 0.4	r.t.	3×10^5
TiO_2	Titanium dioxide (rutile)	$\varepsilon_{11} = \varepsilon_{22} = 86$	300	$10^4 – 10^6$
		$\varepsilon_{33} = 170$	300	$10^4 – 10^6$
Ti_2O_3	Titanium sesquioxide	30	77	6×10^{10}
$TlBr$	Thallium bromide	30	293	$10^3 – 10^7$
$TlCl$	Thallous chloride	32.2 ± 0.2	293	$10^3 – 10^5$
TlI	Thallous iodide (orthorhombic)	20.7 ± 0.2	293	10^4
		37.3	193	10^7
$TlNO_3$	Thallous nitrate	16.5	293	5×10^5
$TlSO_4$	Thallous sulfate	25.5	293	5×10^5
UO_2	Uranium dioxide	24	r.t.	3×10^5
WO_3	Tungsten trioxide	300		
$YMnO_3$	Yttrium manganate	20	r.t.	2×10^7
Y_2O_3	Yttrium sesquioxide	10	r.t.	10^6
$YbMnO_3$	Ytterbium manganate	20	r.t.	2×10^7
Yb_2O_3	Ytterbium sesquioxide	5.0 (film)	r.t.	10^3
ZnO	Zinc monoxide	$\varepsilon_{11}^S = 8.33$	r.t.	
		$\varepsilon_{33}^S = 8.84$	r.t.	
		$\varepsilon_{11}^T = 9.26$	r.t.	
		$\varepsilon_{33}^T = 11.0$	r.t.	
		$\varepsilon_{11} = 9.26$	r.t.	
		$\varepsilon_{33} = 8.2$	r.t.	
		8.15	r.t.	i.r.
ZnS	Zinc sulfide	$\varepsilon_{11}^S = 8.08 \pm 2\%$	77	10^4
		$\varepsilon_{11}^S = 8.32 \pm 2\%$	298	10^4
		$\varepsilon_{11}^T = 8.14 \pm 2\%$	77	10^4
		$\varepsilon_{11}^T = 8.37 \pm 2\%$	298	10^4
$ZnSe$	Zinc selenide	$\varepsilon_{11}^T = \varepsilon_{11}^S = 9.12 \pm 2\%$	298	10^4
$ZnTe$	Zinc telluride	$\varepsilon_{11}^T = \varepsilon_{11}^S = 10.10 \pm 2\%$	r.t.	
$ZnWO_4$	Zinc tungstate	$\varepsilon_{22} = 16.1 \pm 0.5$	r.t.	$(5–500) \times 10^3$
ZrO_2	Zirconium dioxide (zirconia)	12.5	r.t.	2×10^6

CURIE TEMPERATURE OF SELECTED FERROELECTRIC CRYSTALS

H. P. R. Frederikse

The following table lists the major ferroelectric crystals and their Curie temperatures, T_C.

Reference

Young, K. F. and Frederikse, H. P. R., *J. Phys. Chem. Ref. Data*, 2, 313, 1973.

Name or acronym	Formula	T_C/K
Potassium dihydrogen phosphate group		
KDP	KH_2PO_4	123
KDA	KH_2AsO_4	97
KDDP	KD_2PO_4	213
KDDA	KD_2AsO_4	162
RDP	RbH_2PO_4	146
RDA	RbH_2AsO_4	111
RDDP	RbD_2PO_4	218
RDDA	RbD_2AsO_4	178
CDP	CsH_2PO_4	159
CDA	CsH_2AsO_4	143
CDDA	CsD_2AsO_4	212
Rochelle salt group		
Rochelle salt	$NaKC_4H_4O_6 \cdot 4H_2O$	255–297
Deuterated Rochelle salt	$NaKC_4H_2D_2O_6 \cdot 4H_2O$	251–308
Ammonium Rochelle salt	$NaNH_4C_4H_4O_6 \cdot 4H_2O$	109
LAT	$LiNH_4C_4H_4O_6 \cdot H_2O$	106
Triglycine sulfate group		
TGS	$(NH_2CH_2COOH)_3 \cdot H_2SO_4$	322
TGSe	$(NH_2CH_2COOH)_3 \cdot H_2SeO_4$	295
TGFB	$(NH_2CH_2COOH)_3 \cdot H_2BeF_4$	346
AFB	$(NH_4)_2BeF_4$	176
HADA	$HNH_4(ClCH_2COO)_2$	128
Perovskites and related compounds		
Barium titanate	$BaTiO_3$	406, 278, 193
Lead titanate	$PbTiO_3$	765
Potassium niobate	$KNbO_3$	712
Potassium tantalate niobate	$KTa_{2/3}Nb_{1/3}O_3$	241, 220, 170
Lithium niobate	$LiNBO_3$	1483
Lithium tantalate	$LiTaO_3$	891
Barium titanium niobate	$Ba_6Ti_2Nb_8O_{30}$	521
Ba-Na niobate ("Bananas")	$Ba_2NaNb_5O_{15}$	833
Potassium iodate	KIO_3	485, 343, 257–263, 83
Lithium iodate	$LiIO_3$	529
Potassium nitrate	KNO_3	397
Sodium nitrate	$NaNO_3$	548
Rubidium nitrate	$RbNO_3$	437–487
Miscellaneous compounds		
Cesium trihydrogen selenite	$CsH_3(SeO_3)_2$	143
Lithium trihydrogen selenite	$LiH_3(SeO_3)_2$	$T_C > T_{mp}$
Potassium selenate	K_2SeO_4	93
Methyl ammonium alum (MASD)	$CH_3NH_3Al(SO_4)_2 \cdot 12H_2O$	177
Ammonium cadmium sulfate	$(NH_4)_2Cd_2(SO_4)_3$	95
Ammonium bisulfate	$(NH_4)HSO_4$	271
Ammonium sulfate	$(NH_4)_2SO_4$	224
Ammonium nitrate	NH_4NO_3	398, 357, 305, 255
Colemanite	$CaB_3O_4(OH)_3 \cdot H_2O$	266
Cadmium pyroniobite	$Cd_2Nb_2O_7$	185
Gadolinium molybdate	$Gd_2(MoO_4)_3$	432

PROPERTIES OF ANTIFERROELECTRIC CRYSTALS

H. P. R. Frederikse

Some important antiferroelectric crystals are listed here with their Curie Temperatures T_C. The last column gives the constant T_0 which appears in the Curie-Weiss law describing the dielectric constant of these materials above the Curie Temperature:

$$\varepsilon = const./(T-T_0)$$

Name or acronym	Formula	T_C/K	T_0/K
ADP	$NH_4H_2PO_4$	148	
ADA	$NH_4H_2AsO_4$	216	
ADDP	$NH_4D_2PO_4$	242, 245	
ADDA	$NH_4D_2AsO_4$	299	
A_dDDP	$ND_4D_2PO_4$	243	
A_dDDA	$ND_4D_2AsO_4$	304	
Sodium niobate	$NaNbO_3$	911, 793	
Lead hafnate	$PbHfO_3$	476	378
Lead zirconate	$PbZrO_3$	503	475
Lead metaniobate	$PbNb_2O_6$	843	530
Lead metatantalate	$PbTa_2O_6$	543	533
Tungsten trioxide	WO_3	1010	
Potassium strontium niobate	$KSr_2Nb_5O_{15}$	427	413
Sodium nitrite	$NaNO_2$	437	437
Sodium trihydrogen selenite	$NaH_3(SeO_3)_2$	193	192
Sodium trideuterium selenite	$NaD_3(SeO_3)_2$	271	245
Ammonium trihydrogen periodate	$(NH_4)_2H_3IO_6$	245	

DIELECTRIC CONSTANTS OF GLASSES

Type	Dielectric constant at 100 MHz (20°C)	Volume resistivity (In MΩ cm at 350°C)	Loss factor[a]
Corning 0010	6.32	10	0.015
Corning 0080	6.75	0.13	0.058
Corning 0120	6.65	100	0.012
Pyrex 1710	6.00	2,500	0.025
Pyrex 3320	4.71	–	0.019
Pyrex 7040	4.65	80	0.013
Pyrex 7050	4.77	16	0.017
Pyrex 7052	5.07	25	0.019
Pyrex 7060	4.70	13	0.018
Pyrex 7070	4.00	1,300	0.0048
Vycor 7230	3.83	–	0.0061
Pyrex 7720	4.50	16	0.014
Pyrex 7740	5.00	4	0.040
Pyrex 7750	4.28	50	0.011
Pyrex 7760	4.50	50	0.0081
Vycor 7900	3.9	130	0.0023
Vycor 7910	3.8	1,600	0.00091
Vycor 7911	3.8	4,000	0.00072
Corning 8870	9.5	5,000	0.0085
G. E. Clear (silica glass)	3.81	4,000–30,000	0.00038
Quartz (fused)	3.75 (4.1 at 1 MHz)	–	0.0002 (1 MHz)

[a] Power factor × dielectric constant equals loss factor.

PROPERTIES OF SUPERCONDUCTORS

L. I. Berger and B. W. Roberts

The following tables include superconductive properties of selected elements, compounds, and alloys. Individual tables are given for thin films, elements at high pressures, superconductors with high critical magnetic fields, and high critical temperature superconductors.

The historically first observed and most distinctive property of a superconductive body is the near total loss of resistance at a critical temperature (T_c) that is characteristic of each material. Figure 1(a) below illustrates schematically two types of possible transitions The sharp vertical discontinuity in resistance is indica-

tive of that found for a single crystal of a very pure element or one of a few well annealed alloy compositions. The broad transition, illustrated by broken lines, suggests the transition shape seen for materials that are not homogeneous and contain unusual strain distributions. Careful testing of the resistivity limit for superconductors shows that it is less than 4×10^{-23} ohm cm, while the lowest resistivity observed in metals is of the order of 10^{-13} ohm cm. If one compares the resistivity of a superconductive body to that of copper at room temperature, the superconductive body is at least 10^{17} times less resistive.

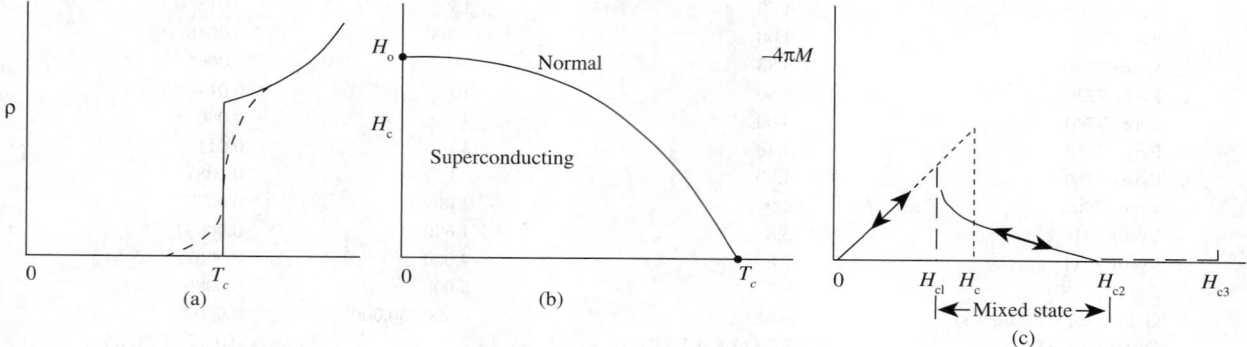

FIGURE 1. Physical properties of superconductors. (a) Resistivity vs. temperature for a pure and perfect lattice (solid line); impure and/or imperfect lattice (broken line). (b) Magnetic-field temperature dependence for Type-I or "soft" superconductors. (c) Schematic magnetization curve for "hard" or Type-II superconductors.

The temperature interval ΔT_c, over which the transition between the normal and superconductive states takes place, may be of the order of as little as 2×10^{-5} K or several K in width, depending on the material state. The narrow transition width was attained in 99.9999% pure gallium single crystals.

A Type-I superconductor below T_c, as exemplified by a pure metal, exhibits perfect diamagnetism and excludes a magnetic field up to some critical field H_c, whereupon it reverts to the normal state as shown in the H-T diagram of Figure 1(b).

The magnetization of a typical high-field superconductor is shown in Figure 1(c). The discovery of the large current-carrying capability of Nb_3Sn and other similar alloys has led to an extensive study of the physical properties of these alloys. In brief, a high-field superconductor, or Type-II superconductor, passes from the perfect diamagnetic state at low magnetic fields to a mixed state and finally to a sheathed state before attaining the normal resistive state of the metal. The magnetic field values separating the four stages are given as H_{c1}, H_{c2}, and H_{c3}. The superconductive state below H_{c1} is perfectly diamagnetic, identical to the state of most pure metals of the "soft" or Type-I superconductor. Between H_{c1} and H_{c2} a "mixed superconductive state" is found in which fluxons (a minimal unit of magnetic flux) create lines of normal flux in a superconductive matrix. The volume of the normal state is proportional to $-4\pi M$ in the "mixed state" region. Thus at H_{c2} the fluxon density has become so great as to drive the interior volume of the superconductive body completely normal. Between H_{c2} and H_{c3} the superconductor has a sheath of current-carrying superconductive material at the body surface, and above H_{c3} the normal state exists. With several types of careful measurement, it is possible to determine H_{c1}, H_{c2}, and H_{c3}. Table 6 contains some of the available data on high-field superconductive materials.

High-field superconductive phenomena are also related to specimen dimension and configuration. For example, the Type-I superconductor, Hg, has entirely different magnetization behavior in high magnetic fields when contained in the very fine sets of filamentary tunnels found in an unprocessed Vycor glass. The great majority of superconductive materials are Type-II. The elements in very pure form and a very few precisely stoichiometric and well annealed compounds are Type I with the possible exceptions of vanadium and niobium.

Metallurgical Aspects. The sensitivity of superconductive properties to the material state is most pronounced and has been used in a reverse sense to study and specify the detailed state of alloys. The mechanical state, the homogeneity, and the presence of impurity atoms and other electron-scattering centers are all capable of controlling the critical temperature and the current-carrying capabilities in high-magnetic fields. Well annealed specimens tend to show sharper transitions than those that are strained or inhomogeneous. This sensitivity to mechanical state underlines a general problem in the tabulation of properties for superconductive materials. The occasional divergent values of the critical temperature and of the critical fields quoted for a Type-II superconductor may lie in the variation in sample preparation. Critical temperatures of materials studied early in the history of superconductivity must be evaluated in light of the probable metallurgical state of the material, as well as the availability of less pure starting elements. It has been noted that recent work has given extended consideration to the metallurgical aspects of sample preparation.

Symbols in tables: T_c: Critical temperature; H_o: Critical magnetic field in the $T = 0$ limit; θ_D: Debye temperature; and γ: Electronic specific heat.

TABLE 1. Selective Properties of Superconductive Elements

Element	T_c(K)	H_o(oersted)	θ_D(K)	γ(mJ mol^{-1}K^{-1})
Al	1.175 ± 0.002	104.9 ± 0.3	420	1.35
Am* (α,?)	0.6			
Am* (β,?)	1.0			
Be	0.026			0.21
Cd	0.517 ± 0.002	28 ± 1	209	0.69
Ga	1.083 ± 0.001	58.3 ± 0.2	325	0.60
Ga (β)	5.9, 6.2	560		
Ga (γ)	7	950, HF[a]		
Ga (Δ)	7.85	815, HF		
Hf	0.128	12.7		2.21
Hg (α)	4.154 ± 0.001	411 ± 2	87, 71.9	1.81
Hg (β)	3.949	339	93	1.37
In	3.408 ± 0.001	281.5 ± 2	109	1.672
Ir	0.1125 ± 0.001	16 ± 0.05	425	3.19
La (α)	4.88 ± 0.02	800 ± 10	151	9.8
La (β)	6.00 ± 0.1	1096, 1600	139	11.3
Lu	0.1 ± 0.03	350 ± 50		
Mo	0.915 ± 0.005	96 ± 3	460	1.83
Nb	9.25 ± 0.02	2060 ± 50, HF	276	7.80
Os	0.66 ± 0.03	70	500	2.35
Pa	1.4			
Pb	7.196 ± 0.006	803 ± 1	96	3.1
Re	1.697 ± 0.006	200 ± 5	4.5	2.35
Ru	0.49 ± 0.015	69 ± 2	580	2.8
Sn	3.722 ± 0.001	305 ± 2	195	1.78
Ta	4.47 ± 0.04	829 ± 6	258	6.15
Tc	7.8 ± 0.1	1410, HF	411	6.28
Th	1.38 ± 0.02	1.60 ± 3	165	4.32
Ti	0.40 ± 0.04	56	415	3.3
Tl	2.38 ± 0.02	178 ± 2	78.5	1.47
U	0.2			
V	5.40 ± 0.05	1408	383	9.82
W	0.0154 ± 0.0005	1.15 ± 0.03	383	0.90
Zn	0.85 ± 0.01	54 ± 0.3	310	0.66
Zr	0.61 ± 0.15	47	290	2.77
Zr (ω)	0.65, 0.95			

TABLE 2. Range of Critical Temperatures Observed for Superconductive Elements in Thin Films Condensed Usually at Low Temperatures

Element	T_c Range (K)	Comments	Element	T_c Range (K)	Comments
Al	1.15–5.7	HF[a]	Nb	2.0–10.1	
Be	5–9.75	HF	Pb	1.8–7.5	
Bi	6.17–6.6		Re	1.7–7	
Cd			Sn	3.5–6	
(Disordered)	0.79–0.91		Ta	<1.7–4.51	HF[a]
(Ordered)	0.53–0.59		Tc	4.6–7.7	
Ga	2.5–8.5	HF	Ti	1.3 Max	
Hg	3.87–4.5		Tl	2.33–2.96	
In	2.2–5.6	HF	V	1.8–6.02	
La	3.55–6.74		W	<1.0–4.1	
Mo	3.3–8.0		Zn	0.77–1.9	

[a] HF denotes high magnetic field superconductive properties.

TABLE 3. Elements Exhibiting Superconductivity Under or After Application of High Pressure

Element	T_c Range (K)	Pressure (kbar)	Element	T_c Range (K)	Pressure (kbar)
Al	1.98–0.075	0–62	Pb II	3.55	160
As	0.31–0.5	220–140	Re II	2.3 Max.	"Plastic"
	0.2–0.25	140–100			compression
Ba II	1–1.8	55–85	Sb (prepared 120	2.6–2.7	
III	1.8–5	85–144	kbar, held below		
IV	4.5–5.4	144–190	77K)		
Bi II	3.9	25–27	Sb II	3.55–3.40	85–150
III	6.55–7.25	28–38	Se II	6.75, 6.95	130
IV	7.0, 8.7–6.0	43, 43–62	Si	6.7–7.1	120–130
V	6.7, 8.3	48–80	Sn II	5.2–4.85	125–160
VI	8.55	90, 92–101	III	5.30	113
VII(?)	8.2	30	Te II	2.4–5.1	38–55
Ce (α)	0.020–0.045	20–35		4.1–4.2	53–62
Ce (α′)	1.9–1.3	45–125	IV	4.72–4	63–80
Cs V	1.5	>125	( )	3.3–2.8	100–260
Ga II	6.38	≥35	Tl (cubic form)	1.45	35
II′	7.5	≥35 then P	(hexagonal form)	1.95	35
		removed	U	2.4–0.4	10–85
Ge	5.35	115	Y	1.7–2.5	110–160
Lu	0.022–1.0	45–190	Zr (omega form, metastable)	1–1.7	60–130
P	5.8	170			

TABLE 4. Superconductive Compounds and Alloys

All compositions are denoted on an atomic basis, i.e., AB, AB$_2$, or AB$_3$ for compounds, unless noted. Solid solutions or odd compositions may be denoted as A$_z$B$_{1-z}$ or A$_z$B. A series of three or more alloys is indicated as A$_x$B$_{1-x}$ or by actual indication of the atomic fraction range, such as A$_{0-0.6}$B$_{1-0.4}$. The critical temperature of such a series of alloys is denoted by a range of values or possibly the maximum value.

The selection of the critical temperature from a transition in the effective permeability, or the change in resistance, or possibly the incremental changes in frequency observed by certain techniques is not often obvious from the literature. Most authors choose the

mid-point of such curves as the probable critical temperature of the idealized material, while others will choose the highest temperature at which a deviation from the normal state property is observed. In view of the previous discussion concerning the variability of the superconductive properties as a function of purity and other metallurgical aspects, it is recommended that appropriate literature be checked to determine the most probable critical temperature or critical field of a given alloy.

A very limited amount of data on critical fields, H_o, is available for these compounds and alloys; these values are given at the end of the table.

A. Superconductors with $T_c < 10$ K

Substance	T_c, K	Crystal structure type	Substance	T_c, K	Crystal structure type
Ag$_{3.3}$Al	0.34	A12-cI58 (Mn)	Ag$_7$NO$_{11}$	1.04	Cubic
Ag$_x$Al$_y$Zn$_{1-x-y}$	0.15	Cubic	Ag$_x$Pb$_{1-x}$	7.2 max.	
AgBi$_2$	2.87–3.0		Ag$_4$Sn	0.1	h**
Ag$_7$F$_{0.25}$N$_{0.75}$O$_{10.25}$	0.85–0.90		Ag$_x$Sn$_{1-x}$	1.5–3.7	
Ag$_2$F	0.0.066		Ag$_x$Sn$_{1-x}$ (film)	2.0–3.8	
Ag$_7$FO$_8$	0.3	Cubic	AgTe$_3$	2.6	Cubic
Ag$_{0.8-0.3}$Ga$_{0.2-0.7}$	6.5–8		AgTh	2.2	C16-tI12 (Al$_2$Cu)
Ag$_4$Ge	0.85	Hex., c.p.	AgTh$_2$	2.26	C16
Ag$_{0.438}$Hg$_{0.562}$	0.64	D8$_2$	Ag$_{0.03}$Tl$_{0.97}$	2.67	
AgIn$_2$	~2.4	C16	Ag$_{0.94}$Tl$_{0.06}$	2.32	
Ag$_{0.1}$In$_{0.9}$Te (n = 1.4 × 10^{22})*	1.2–1.89	B1	AgY	0.33	B2-cP2 (CsCl)
Ag$_{0.2}$In$_{0.8}$Te (n = 1.07 × 10^{22})	0.77–1.00	B1	Ag$_x$Zn$_{1-x}$	0.5–0.845	
AgLa	0.94	B2-cP2 (CsCl)	AlAu$_4$	0.4–0.7	Like A13
AgLa (9.5 kbar)	1.2	B2	Al$_2$Au	0.1	C1-cF12 (CaF$_2$)
AgLu	0.33	B2-cP2	Al$_2$CMo$_3$	9.8–10.2	A13+trace 2nd. phase
AgMo$_6$S$_5$	9.1	hR15 (Mo$_6$PbS$_8$)	Al$_2$CaSi	5.8	
Ag$_{1.2}$Mo$_6$Se$_8$	5.9	Same	Al$_{0.131}$Cr$_{0.088}$V$_{0.781}$	1.46	Cubic
			AlGe$_2$	1.75	

Substance	T_c, K	Crystal structure type	Substance	T_c, K	Crystal structure type
Al_2Ge_2U	1.6	LI_2-cP4 (Cu_3Au)	$AuPb_2$	3.15	
$AlLa_3$	5.57	DO_{19}	$AuPb_2$ (film)	4.3	
Al_2La	3.23	C15	$AuPb_3$	4.40	
Al_2Lu	1.02	C15-cF24 (Cu_2Mg)	$AuPb_3$ (film)	4.25	
Al_3Mg_2	0.84	F.C.C.	Au_2Pb	1.18; 6–7	C15
$AlMo_3$	0.58	A15	$AuSb_2$	0.58	C2
$AlMo_6Pd$	2.1		$AuSn$	1.25	$B8_1$
AlN	1.55	B4	Au_xSn_{1-x} (film)	2.0–3.8	
Al_2NNb_3	1.3	A13	Au_5Sn	0.7–1.1	A3
Al_3Nb	0.64	tI8 (Al_3Ti)	$AuTa_{4.3}$	0.55	A15-cP8 (Cr_3Si)
$AlOs$	0.39	B2	Au_3Te_5	1.62	Cubic
Al_3Os	5.90		$AuTh_2$	3.08	C16
$AlPb$ (film)	1.2–7		$AuTl$	1.92	
Al_2Pt	0.48–0.55	C1	AuV_3	0.74	A15
Al_5Re_{24}	3.35	A12	Au_xZn_{1-x}	0.50–0.845	
$AlSb$	2.8	B4-tI4 (Sn)	$AuZn_3$	1.21	Cubic
Al_2Sc	1.02	C15-cF24 (Cu_2Mg)	Au_xZr_y	1.7–2.8	A3
Al_2Si_2U	1.34	LI_2-cP4 (Cu_3Au)	$AuZr_3$	0.92	A15
$AlTh_2$	0.1	C16-tI12 (Al_2Cu)	$B_2Ba_{0.67}Pt_3$	5.60	hP12 (B_2BaPt_3)
Al_3Th	0.75	DO_{19}	$BCMo_2$	5.4	Orthorhombic
$Al_xTi_yV_{1-x-y}$	2.05–3.62	Cubic	$BCMo_2$	5.3–7.0	Same
$Al_{0.108}V_{0.892}$	1.82	Cubic	$B_2Ca_{0.67}Pt_3$	1.57	hP12
Al_2Y	0.35	C15-cF24 (Cu_2Mg)	B_4ErIr_4	2.1	tP18 (B_4CeCo_4)
Al_3Yb	0.94	LI_2-cP4 (Cu_3Au)	B_4ErRh_4	4.3	oC108 (B_4LuRh_4)
Al_xZn_{1-x}	0.5–0.845		B_4ErRh_4	8.7	tP18 (B_4CeCo_4)
$AlZr_3$	0.73	LI_2	BHf	3.1	Cubic
$AsBiPb$	9.0		B_4HoIr_4	2.0	tP18
$AsBiPbSb$	9.0		B_4HoRh_4	1.4	oC108
$AsHfOs$	3.2	C22-hP9 (Fe_2P)	B_2Ir_3La	1.65	hP6 ($CaCu_5$)
$AsHfRu$	4.9	same	B_2Ir_3Th	2.09	Same
$As_{0.33}InTe_{0.67}$ ($n = 1.24 \times 10^{22}$)	0.85–1.15	B1	B_4Ir_4Tm	1.6	tP18
$As_{0.5}InTe_{0.5}$ ($n = 0.97 \times 10^{22}$)	0.44–0.62	B1	B_6La	5.7	
As_4La_3	0.6	cI28 (Th_3P_4)	B_2LaRh_3	2.82	hP6
$AsNb_3$	0.3	$L1_2$-tP32	$B_{12}Lu$	0.48	
$As_{0.50}Ni_{0.06}Pd_{0.44}$	1.39	C2	B_2LuOs	2.66	oP16 (B_2LuRu)
$AsNi_{0.25}Pd_{0.75}$	1.6	$B8_1$-hP4 (NiAs)	B_2LuOs_3	4.62	hP6
$AsOsZr$	8.0	C22-hP9 (Fe_2P)	B_4LuRh_4	6.2	oC108
$AsPb$	8.4		B_2LuRu	9.86	oP16
$AsPd_2$ (low-temp. phase)	0.60	Hexagonal	B_4LuRu_4	2.0	tI72 (B_4LuRu_4)
$AsPd_2$ (high-temp. phase)	1.70	C22	BMo	0.5	
$AsPd_5$	0.46	Complex		(extrapol.)	
As_3Pd_5	1.9		BMo_2	4.74	C16
$AsRh$	0.58	B31	BNb	8.25	B_f
$AsRh_{1.4-1.6}$	< 0.03–0.56	Hexagonal	B_4NdRh_4	5.3	tP18
$AsSn$	4.10		B_2OsSc	1.34	oP16
$AsSn$ ($n = 2.14 \times 10^{22}$)	3.41–3.65	B1	B_2OsY	2.22	oP16
	3.5–3.6;		$B_2Pt_3Sr_{0.67}$	2.78	hP12 (B_2BaPt_3)
$As_{-2}Sn_{-3}$	1.21–1.17		BRe_2	2.80; 4.6	
As_3Sn_4 ($n = 0.56 \times 10^{22}$)	1.16–1.19	Rhombohedral	$B_4Rh_{3.4}Ru_{0.6}$	8.38	tI72
AsV_3	0.20	A15-cP8 (Cr_3Si)	B_4Rh_4Sm	2.7	tP18
Au_5Ba	0.4–0.7	$D2_d$	B_4Rh_4Th	4.3	Same
$AuBe$	2.64	B20	B_4Rh_4Tm	9.8	Same
Au_2Bi	1.80	C15	B_4Rh_4Tm	5.4	oC108
Au_5Ca	0.34–0.38	$C15_b$	$B_{0.3}Ru_{0.7}$	2.58	$D10_2$
$AuGa_2$	1.6	C1-cF12 (CaF_2)	B_4Ru_4Sc	7.2	tI72
$AuGa$	1.2	B31	B_2Ru_3Th	1.79	hP6
$Au_{0.40-0.92}Ge_{0.60-0.08}$	<0.32–1.63	Complex	B_2Ru_3Y	2.85	Same
$AuIn_2$	0.2	C1-cF12	B_2RuY	7.80	oP16
$AuIn$	0.4–0.6	Complex	B_4Ru_4Y	1.4	tI72
$AuLu$	<0.35	B2	$B_{12}Sc$	0.39	
$AuNb_3$	1.2	A2	BTa	4.0	B_f

Substance	T_c, K	Crystal structure type
BTa_2	3.12	C16-tI12 (Al_2Cu)
B_6Th	0.74	
BW_2	3.1	C16
B_6Y	6.5–7.1	
$B_{12}Y$	4.7	
BZr	3.4	Cubic
$B_{12}Zr$	5.82	
$BaBi_3$	5.69	Tetragonal
$Ba_2Mo_{15}Se_{19}$	2.75	hP15 (Mo_6PbS_8)
$Ba_xO_3Sr_{1-x}Ti$ (n = 4.2 ×10^{19})	<0.1–0.55	
$Ba_{0.13}O_3W$	1.9	Tetragonal
$Ba_{0.14}O_3W$	<1.25–2.2	Hexagonal
$BaRh_2$	6.0	C15
$Be_{22}Mo$	2.51	Cubic ($Be_{22}Re$)
$Be_8Nb_5Zr_2$	5.2	
$Be_{0.98–0.92}Re_{0.02–0.08}$ (quenched)	9.5–9.75	Cubic
$Be_{0.957}Re_{0.043}$	9.62	Cubic ($Be_{22}Re$)
$BeTc$	5.21	Cubic
$Be_{22}W$	4.12	Cubic ($Be_{22}Re$)
$Be_{13}W$	4.1	Tetragonal
Bi_3Ca	2.0	
$Bi_{0.5}Cd_{0.13}Pb_{0.25}Sn_{0.12}$ (weight fractions)	8.2	
$BiCo$	0.42–0.49	
Bi_2Cs	4.75	C15
Bi_xCu_{1-x} (electrodeposited)	2.2	
$BiCu$	1.33–1.40	
Bi_3Fe	1.0	m**
$Bi_{0.019}In_{0.981}$	3.86	
$Bi_{0.05}In_{0.95}$	4.65	α-phase
$Bi_{0.10}In_{0.90}$	5.05	Same
$Bi_{0.15–0.30}In_{0.85–0.70}$	5.3–5.4	α- and β-phases
$Bi_{0.34–0.48}In_{0.66–0.52}$	4.0–4.1	
Bi_3In_5	4.1	
$BiIn_2$	5.65	β-phase
Bi_2Ir	1.7–2.3	
Bi_2Ir (quenched)	3.0–3.96	
BiK	3.6	
Bi_2K	3.58	C15
$BiLi$	2.47	$L1_o$, α-phase
$Bi_{4–9}Mg$	0.7–~1.0	
Bi_3Mo	3–3.7	
$BiNa$	2.25	$L1_o$
$BiNb_3$	4.5	A15-cP8 (Cr_3Si)
$BiNb_3$ (high pressure and temperature)	3.05	A15
$BiNi$	4.25	$B8_1$
Bi_3Ni	4.06	Orthorhombic
$BiNi_{0.5}Rh_{0.5}$	3.0	$B8_1$-hP4 (AsNi)
$Bi_{0.5}NiSb_{0.5}$	2.0	Same
$Bi_{1-0}Pb_{0-1}$	7.26–9.14	
$Bi_{1-0}Pb_{0-1}$ (film)	7.25–8.67	
$Bi_{0.05–0.40}Pb_{0.95–0.60}$	7.35–8.4	H.C.P. to ε-phase
Bi_2Pb	4.25	t**
$BiPbSb$	8.9	
$Bi_{0.5}Pb_{0.31}Sn_{0.19}$ (weight fractions)	8.5	
$Bi_{0.5}Pb_{0.25}Sn_{0.25}$	8.5	
$BiPd_2$	4.0	
$Bi_{0.4}Pd_{0.6}$	3.7–4	Hexagonal, ordered
$BiPd$	3.7	Orthorhombic
Bi_2Pd	1.70	Monoclinic, α-phase
Bi_2Pd	4.25	Tetragonal, β-phase
$BiPd_{0.45}Pt_{0.55}$	3.7	$B8_1$-hP4 (NiAs)
$BiPdSe$	1.0	C2
$BiPdTe$	1.2	C2
$BiPt$	1.21	$B8_1$
$Bi_{0.1}PtSb_{0.9}$	2.05; 1.5	$B8_1$-hP4 (NiAs)
$BiPtSe$	1.45	C2
$BiPtTe$	1.15	C2
Bi_2Pt	0.155	Hexagonal
Bi_2Rb	4.25	C15
$BiRe_2$	1.9–2.2	
$BiRh$	2.06	$B8_1$
Bi_3Rh	3.2	Orthorhombic (NiB_3)
Bi_4Rh	2.7	Hexagonal
$BiRu$	5.7	m**
Bi_3Sn	3.6–3.8	
$BiSn$	3.8	
Bi_3Sn_y	3.85–4.18	
Bi_2Sr	5.62	$L1_2$
Bi_2Te	0.75–1.0	
Bi_2Tl_3	6.4	
$Bi_{0.26}Tl_{0.74}$	4.4	Cubic, disordered
$Bi_{0.26}Tl_{0.74}$	4.15	$L1_2$, ordered (?)
Bi_2Y_3	2.25	
Bi_2Zn	0.8–0.9	
$Bi_{0.3}Zr_{0.7}$	1.51	
$BiZr_3$	2.4–2.8	
$BrMo_6Se_7$	7.1	hP15 (Mo_6PbS_8)
$Br_3Mo_6Se_5$	7.1	Same
CCs_x	0.020–0.135	Hexagonal
CFe_3	1.30	DO_{11}-oP16 (Fe_3C)
$CGaMo_2$	3.7–4.1	Hexagonal
$CHf_{0.5}Mo_{0.5}$	3.4	B1
$CHf_{0.3}Mo_{0.7}$	5.5	B1
$CHf_{0.25}Mo_{0.75}$	6.6	B1
$CHf_{0.7}Nb_{0.3}$	6.1	B1
$CHf_{0.6}Nb_{0.4}$	4.5	B1
$CHf_{0.5}Nb_{0.5}$	4.8	B1
$CHf_{0.4}Nb_{0.6}$	5.6	B1
$CHf_{0.25}Nb_{0.75}$	7.0	B1
$CHf_{0.2}Nb_{0.8}$	7.8	B1
$CHf_{0.9–0.1}Ta_{0.1–0.9}$	5.0–9.0	B1
CK (excess K)	0.55	Hexagonal
C_8K	0.39	Hexagonal
C_2La	1.66	tI6 (CaC_2)
C_2Lu	3.33	Same
$C_{0.40–0.44}Mo_{0.60–0.56}$	9–13	
C_3MoRe	3.8	B1-cF8
$C_{0.6}Mo_{4.8}Si_3$	7.6	$D8_8$
$CMo_{0.2}Ta_{0.8}$	7.5	B1
$CMo_{0.5}Ta_{0.5}$	7.7	B1
$CMo_{0.75}Ta_{0.25}$	8.5	B1
$CMo_{0.8}Ta_{0.2}$	8.7	B1
$CMo_{0.85}Ta_{0.15}$	8.9	B1
CMo_xV_{1-x}	2.9–9.3	B1
CMo_xZr_{1-x}	9.8	B1
$C_{0.984}Nb$	9.8	B1
CNb_2	9.1	
CNb_xTi_{1-x}	<4.2–8.8	B1
$CNb_{0.1–0.9}Zr_{0.9–0.1}$	4.2–8.4	B1

Substance	T_c, K	Crystal structure type	Substance	T_c, K	Crystal structure type
CRb_x (Au)	0.023–0.151	Hexagonal	$Co_4Sc_5Si_{10}$	5.0	tP38 ($Co_4Sc_5Si_{10}$)
$CRe_{0.06}W$	5.0		$CoSi_2$	1.40; 1.22	C1
CRu	2.00	hP2 (CW)	Co_xSn_yYb	2.5	cP40
$C_{0.98}7Ta$	9.7		Co_3Th_7	1.83	$D10_2$
$C_{0.848–0.987}$	2.04–9.7		Co_xTi_{1-x}	2.8 (max.)	Co in α-Ti
CTa (film)	5.09	B1	Co_xTi_{1-x}	3.8 (max.)	Co in β-Ti
CTa_2	3.26	L'_3	$CoTi_2$	3.44	$E9_3$
$CTa_{0.4}Ti_{0.6}$	4.8	B1	$CoTi$	0.71	A2
$Cta_{1-0.4}W_{0-0.6}$	8.5–10.5	B1	CoU	1.7	B2, distorted
$CTa_{0.2–0.9}Zr_{0.8–0.1}$	4.6–8.3	B1	CoU_6	2.29	$D2_c$
CTc (excess C)	3.85	Cubic	$Co_{0.28}Y_{0.72}$	0.34	
$CTi_{0.5–0.7}W_{0.5–0.3}$	6.7–2.1	B1	CoY_3	<0.34	
CW	1.0		$CoZr_2$	6.3	C16
CW_2	2.74	L'_3	$Co_{0.1}Zr_{0.9}$	3.9	A3
CW_2	5.2	F.C.C.	$Cr_{0.6}Ir_{0.4}$	0.4	H.C.P.
C_2Y	3.88	tI6 (CaC_2)	$Cr_{0.65}Ir_{0.35}$	0.59	H.C.P.
$Ca_3Co_4Sn_{13}$	5.9	cP40 ($Pr_3Rh_2Sn_{13}$)	$Cr_{0.7}Ir_{0.3}$	0.76	H.C.P.
$Ca_3Ge_{13}Rh_4$	2.1	Same	$Cr_{0.72}Ir_{0.28}$	0.83	
$CaHg$	1.6	B2-cP2 (CsCl)	Cr_3Ir	0.45	A15
$CaHg_3$	1.6	hP8 (Ni_3Sn)	$Cr_{0–0.1}Nb_{1–0.9}$	4.6–9.2	A2
$CaIr_2$	6.15	C15	$Cr_{0.80}Os_{0.20}$	2.5	Cubic
$Ca_3Ir_4Sn_{13}$	7.1	cP40	Cr_3Os	4.68	A15-cP8 (Cr_3Si)
$Ca_xO_3Sr_{1-x}Ti$ (n = 3.7–11 × 10^{19})	< 0.1–0.55		Cr_xRe_{1-x}	1.2–5.2	
$Ca_{0.1}O_3W$	1.4–3.4	Hexagonal	$Cr_{0.4}Re_{0.6}$	2.15	$D8_b$
$CaPb$	7.0		$Cr_{0.8–0.6}Rh_{0.2–0.4}$	0.5–1.10	A3
$CaRh_2$	6.40	C15	Cr_3Rh	0.3	A15-cP8
$CaRh_{1.2}Sn_{4.5}$	8.7	cP40	Cr_3Ru (annealed)	3.3	A15
$CaTl_3$	2.0	B2-cP2	Cr_2Ru	2.02	$D8_b$
$Cd_{0.3–0.5}Hg_{0.7–0.5}$	1.70–1.92		Cr_3Ru_2	2.10	$D8_b$-tP30 (CrFe)
$CdHg$	1.77; 2.15	Tetragonal	$Cr_{0.1–0.5}Ru_{0.9–0.5}$	0.34–1.65	A3
$Cd_{0.0075–0.05}In_{0.9925–0.95}$	3.24–3.36	Tetragonal	Cr_xTi_{1-x}	3.6 (max.)	Cr in α-Ti
$Cd_{0.97}Pb_{0.03}$	4.2		Cr_xTi_{1-x}	4.2 (max.)	Cr in β-Ti
$CdSn$	3.65		$Cr_{0.1}Ti_{0.3}V_{0.6}$	5.6	
$Cd_{0.17}Tl_{0.83}$	2.3		$Cr_{0.0175}U_{0.9825}$	0.75	β-phase
$Cd_{0.18}Tl_{0.82}$	2.54		$Cs_{0.32}O_3W$	1.12	Hexagonal
$CeCo_2$	0.84	C15	$Cu_{0.15}In_{0.85}$ (film)	3.75	
$CeCo_{1.67}Ni_{0.33}$	0.46	C15	$Cu_{0.04–0.08}In_{0.94–0.92}$	4.4	
$CeCo_{1.67}Rh_{0.33}$	0.47	C15	$CuLa$	5.85	
$Ce_xGd_{1-x}Ru_2$	3.2–5.2	C15	$Cu_2Mo_6O_2S_6$	9	hR15 (Mo_6PbS_8)
$CeIr_3$	3.34		$Cu_2Mo_6Se_8$	5.9	Same
$CeIr_5$	1.82		Cu_xPb_{1-x}	5.7–7.7	
$Ce_{0.005}La_{0.995}$	4.6		CuS	1.62	B18
Ce_xLa_{1-x}	1.3–6.3		CuS_2	1.48–1.53	C18
$Ce_xPr_{1-x}Ru_2$	1.4–5.3	C15	$CuSSe$	1.5–2.0	C18
Ce_xPt_{1-x}	0.7–1.55		$CuSe_2$	2.3–2.43	C18
$CeRu_2$	6.0	C15	$CuSeTe$	1.6–2.0	C18
$Ce_3Mo_6Se_5$	5.7	hR15 (Mo_6PbS_8)	Cu_xSn_{1-x}	3.2–3.7	
$Ce_2Mo_6Te_6$	1.7	Same	Cu_xSn_{1-x} (film, made at 10K)	3.6–7	
$Co_xFe_{1-x}Si_2$	1.4 (max.)	C1	Cu_xSn_{1-x} (film, made at 300K)	2.8–3.7	
$CoHf_2$	0.56	$E9_3$	$CuTe_2$	<1.25–1.3	C18
$CoLa_3$	4.28		$CuTh_2$	3.49	C16
$Co_4La_3Sn_{13}$	2.8	cP40	$Cu_{0–0.027}V$	3.9–5.3	A2
$CoLu_3$	~0.35		CuY	0.33	B2-cP2 (CsCl)
Co_xLuSn_y	1.5	cP40	Cu_xZn_{1-x}	0.5–0.845	
$Co_{0–0.01}Mo_{0.8}Re_{0.2}$	2–10		$DyMo_6S_8$	2.1	hR15
$Co_{0.02–0.10}Nb_3Rh_{0.98–0.90}$	2.28–1.90	A15	Er_xLa_{1-x}	1.4–6.3	
$Co_xNi_{1-x}Si_2$	1.4 (max.)	C1	$ErMo_6S_8$	2.2	hR15
$Co_{0.5}Rh_{0.5}Si_2$	2.5		$ErMo_6Se_8$	6.2	hR15
$Co_xRh_{1-x}Si_2$	3.65 (max.)		$Fe_3Lu_2Si_5$	6.1	tP40 ($Fe_3Sc_2Si_5$)
$Co_{-0.3}So_{~0.7}$	~0.35		$Fe_{0–0.04}Mo_{0.8}Re_{0.2}$	1–10	
			$Fe_{0.05}Ni_{0.05}Zr_{0.90}$	~3.9	

Substance	T_c, K	Crystal structure type	Substance	T_c, K	Crystal structure type
Fe_3Re_2	6.55	$D8_b$-tP30 (FeCr)	GeV_3	6.01	A15
$Fe_3Sc_2Si_5$	4.52	tP40	Ge_2Y	3.80	C_c
Fe_3Si_5Tm	1.3	Same	$Ge_{1.62}Y$	2.4	
$Fe_3Si_5Y_2$	2.4	Same	Ge_2Zr	0.30	oC12 ($ZrSi_2$)
Fe_3Th_7	1.86	D10	$GeZr_3$	0.4	$L1_2$-tP32 (Ti_3P)
Fe_xTi_{1-x}	3.2 (max.)	Fe in α-Ti	$H_{0.33}Nb_{0.67}$	7.28	B.C.C.
Fe_xTi_{1-x}	3.7 (max.)	Fe in β-Ti	$H_{0.1}Nb_{0.9}$	7.38	Same
$Fe_xTi_{0.6}V_{1-x}$	6.8 (max.)		$H_{0.05}Nb_{0.95}$	7.83	Same
FeU_6	3.86	$D2_c$	$H_{0.12}Ta_{0.88}$	2.81	B.C.C.
$Fe_{0.1}Zr_{0.9}$	1.0	A3	$H_{0.08}Ta_{0.92}$	3.26	Same
$Ga_{0.5}Ge_{0.5}Nb_3$	7.3	A15	$H_{0.04}Ta_{0.96}$	3.62	Same
Ga_2Ge_2U	0.87	B2-cP2	HfIrSi	3.50	C37-cP12 (Co_2Si)
$GaHf_2$	0.21	C16-tI12 (Al_2Cu)	$HfMo_2$	0.05	hP24 (Ni_2Mn)
$GaLa_3$	5.84		$HfN_{0.989}$	6.6	B1
Ga_3Lu	2.3	B2-cP2	$Hf_{0-0.5}Nb_{1-0.5}$	8.3–9.5	A2
Ga_2Mo	9.5		$Hf_{0.75}Nb_{0.25}$	> 4.2	
$GaMo_3$	0.76	A15	$HfOs_2$	2.69	C14
GaN (black)	5.85	B4	HfOsP	6.1	C22-hP9 (Fe_2P)
$Ga_{0.7}Pt_{0.3}$	2.9	C1	HfPRu	9.9	Same
GaPt	1.74	B20	$HfRe_2$	4.80	C14
GaSb (120kbar, 77K, annealed)	4.24	A5	$Hf_{0.14}Re_{0.86}$	5.86	A12
GaSb (unannealed)	~5.9		$Hf_{0.99-0.96}Rh_{0.01-0.04}$	0.85–1.51	
$Ga_{0-1}Sn_{1-0}$ (quenched)	3.47–4.18		$Hf_{0-0.55}Ta_{1-0.45}$	4.4–6.5	A2
$Ga_{0-1}Sn_{1-0}$ (annealed)	2.6–3.85		HfV_2	8.9–9.6	C15
GaTe	0.17	mC24 (GaTe)	Hg_xIn_{1-x}	3.14–4.55	
Ga_5V_2	3.55	Tetragonal (Mn_2Hg_5)	HgIn	3.81	
$GaV_{4.5}$	9.15		Hg_2K	1.20	Orthorhombic
Ga_3Zr	1.38		Hg_3K	3.18	
Ga_3Zr_5	3.8	$D8_b$-hP16 (Mn_5Si_3)	Hg_4K	3.27	
Gd_xLa_{1-x}	< 1.0–5.5		Hg_8K	3.42	
$GdMo_6S_8$	3.5	hR15	Hg_3Li	1.7	Hexagonal
$GdMo_6Se_8$	5.6	hR15	$HgMg_3$	0.17	hP8 (Na_3As)
$Gd_xOs_2Y_{1-x}$	1.4–4.7		Hg_2Mg	4.0	tI6 ($MoSi_2$)
$Gd_xRu_2Th_{1-x}$	3.6 (max.)	C15	Hg_3Mg_5	0.48	$D8_b$-hP16 (Mn_5Si_3)
$Ge_{10}As_4Y_5$	9.06	tP38 ($C0_4Sc_5Si_{10}$)	Hg_2Na	1.62	Hexagonal
GeIr	4.7	B31	Hg_4Na	3.05	
GeIrLa	1.64	tI12 (LaPtSi)	Hg_xPb_{1-x}	4.14–7.26	
$Ge_{10}Ir_4Lu_5$	2.60	tP38	HgSn	4.2	
$Ge_{10}Ir_4Y_5$	2.62	tP38	Hg_xTl_{1-x}	2.30–4.19	
Ge_2La	1.49; 2.2	Orthorhombic, distorted (Mn_2Hg_5)	Hg_xTl_2	3.86	
			Ho_xLa_{1-x}	1.3–6.3	
GeLaPt	3.53	tI12	$Ho_{1.2}Mo_6Se_8$	6.1	$D10_2$-hR12 (Be_3Nb)
$Ge_{13}Lu_3Os_4$	3.6	cP40 ($Pr_3Rh_2Sn_{13}$)	$In_{1-0.86}Mg_{0-0.14}$	3.395–3.363	
$Ge_{10}Lu_5Rh_4$	2.79	tP38	$In_2Mo_6Te_6$	2.6	hR15 (Mo_6PbS_8)
$Ge_{13}Lu_3Ru_4$	2.3	cP40	$InNb_3$ (high pressure and temp.)	4–8; 9.2	A15
$GeMo_3$	1.43	A15			
$GeNb_2$	1.9		$In_{0.5}Nb_3Zr_{0.5}$	6.4	
$Ge_{0.29}Nb_{0.71}$	6	A15	$In_{0.11}O_3W$	< 1.25–2.8	Hexagonal
GePt	0.40	B31	$In_{0.95-0.85}Pb_{0.05-0.15}$	3.6–5.05	
Ge_3Rh_5	2.12	Orthorhombic, related to $InNi_2$	$In_{0.98-0.91}Pb_{0.02-0.09}$	3.45–4.2	
			InPb	6.65	
GeRh	0.96	B31-oP8 (MnP)	InPd	0.7	B2
$Ge_{13}Rh_4Sc_3$	1.9	c P40	InSb (quenched from 170 kbar into liquid N_2)	4.8	Like A5
$Ge_{10}Rh_4Y_5$	1.35	tP38			
$Ge_{13}Ru_4Y_3$	1.7	cP40	InSb	2.1	B4
Ge_2So	1.3		$(InSb)_{0.95-0.10}Sn_{0.05-0.90}$ (various heat treatments)	3.8–5.1	
$GeTa_3$	8.0	A15-cP8 (Cr_3Si)			
Ge_3Te_4 (n = 1.06×10^{22})	1.55–1.80	Rhombohedral	$(InSb)_{0-0.07}Sn_{1-0.93}$	3.67–3.74	
Ge_xTe_{1-x} (n = $8.5–64 \times 10^{20}$)	0.07–0.41	R1	In_3Sn	~5.5	
			In_xSn_{1-x}	3.4–7.3	

Properties of Superconductors

Substance	T_c, K	Crystal structure type	Substance	T_c, K	Crystal structure type
$In_{0.82-1}Te$ (n = 0.83–1.71 × 10^{22})	1.02–3.45	B1	Ir_2Y_3	1.61	
$In_{1.000}Te_{1.002}$	3.5–3.7	B1	Ir_3Y	3.50	$D10_2$-hR13 (Be_3Nb)
In_3Te_4 (n = 4.7 × 10^{21})	1.15–1.25	Rhombohedral	Ir_xY_{1-x}	0.3–3.7	
In_xTl_{1-x}	2.7–3.374		Ir_2Zr	4.10	C15
$In_{0.8}Tl_{0.2}$	3.223		$Ir_{0.1}Zr_{0.9}$	5.5	A3
$In_{0.62}Tl_{0.38}$	2.760		$K_2Mo_{15}S_{19}$	3.32	hR15
$In_{0.78-0.69}Tl_{0.22-0.31}$	3.18–3.32	Tetragonal	$K_{0.27-0.31}O_3W$	0.50	Hexagonal
$In_{0.69-0.62}Tl_{0.31-0.38}$	2.98–3.3	F.C.C.	$K_{0.40-0.57}O_3W$	1.5	Tetragonal
Ir_2La	0.48	C15	$La_{0.55}Lu_{0.45}$	2.2	Hexagonal, La type
Ir_3La	2.32	$D10_2$	$La_{0.8}Lu_{0.2}$	3.4	Same
Ir_3La_7	2.24	$D10_2$	$LaMg_2$	1.05	C15
Ir_5La	2.13		$LaMo_6S_8$	7.1	hR15
$IrLaSi_2$	2.03	oC16 ($CeNiSi_2$)	LaN	1.35	
$IrLaSi_3$	2.7	tI10 ($BaNiSn_3$)	$LaOs_2$	6.5	C15
Ir_2Lu	2.47	C15	$LaPt_2$	0.46	C15
Ir_3Lu	2.89	C15	$La_{0.28}Pt_{0.72}$	0.54	C15
$Ir_4Lu_5Si_{10}$	3.9	tP38 ($Co_4Sc_5Si_{10}$)	LaPtSi	3.48	tI12
IrMo	< 1.0	A3	$LaRh_3$	2.60	
$IrMo_3$	9.6	A15	$LaRh_5$	1.62	
$IrMo_3$	6.8	$D8_b$	La_7Rh_3	2.58	$D10_2$
$IrNb_3$	1.9	A15	$LaRhSi_2$	3.42	oC16 ($CeNiSi_2$)
$Ir_{0.4}Nb_{0.6}$	9.8	$D8_b$	$La_2Rh_3Si_5$	4.45	oI40 ($Co_3Si_5U_2$)
$Ir_{0.37}Nb_{0.63}$	2.32	$D8_b$	$LaRhSi_3$	2.7	tI10 ($BaNiSn_3$)
IrNb	7.9	$D8_b$	$LaRh_2Si_2$	3.90	tI10 (Al_4Ba)
$Ir_{1.15}Nb_{0.85}$	4.6	oP12 (IrTa)	$LaRu_2$	1.63	C15
$Ir_{0.02}Nb_3Rh_{0.98}$	2.43	A15	La_3S_4	6.5	$D7_3$
$Ir_{0.05}Nb_3Rh_{0.95}$	2.38	A15	La_3Se_4	8.6	$D7_3$
$Ir_{0.287}O_{0.14}Ti_{0.573}$	5.5	$E9_3$	$LaSi_2$	2.3	C_c
$Ir_{0.265}O_{0.035}Ti_{0.65}$	2.30	$E9_3$	La_xY_{1-x}	1.7–5.4	
Ir_xOs_{1-x}	0.3–0.98		LaZn	1.04	B2
$Ir_{1.5}Os_{0.5}$	2.4	C14	$Li_2Mo_6S_8$	4.2	hR15
IrOsY	2.6	C15	LiPb	7.2	
IrSiY	2.70	C37-oP12 (Co_2Si)	$LuOs_2$	3.49	C14
IrSiZr	2.04	Same	$Lu_{0.275}Rh_{0.725}$	1.27	C15
Ir_2Sc	2.07	C15	$LuRh_5$	0.49	
$Ir_{2.5}Sc$	2.46	C15	$Lu_5Rh_4Si_{10}$	3.95	tP38 ($Co_4So_5Si_{10}$)
$Ir_4Sc_5Si_{10}$	8.46	tP38	$LuRu_2$	0.86	C14
Ir_2Si_2Th	2.14	tI10	$Mg_{1.14}Mo_{6.6}S_8$	3.5	hR15
$IrSi_3Th$	1.75	tI10	Mg2Nb	5.6	
IrSiTh	6.50	tI12 (LaPtSi)	$Mg_{~0.47}Tl_{~0.53}$	2.75	B2
Ir_2Si_2Y	2.60	tI10 (Al4Ba)	MgZn	0.9	A3-oP4 (AuCd)
$Ir_4Si_{10}Y_5$	3.10	tP38	Mn_xTi_{1-x}	2.3 (max.)	Mn in -Ti
$Ir_3Si_5Y_2$	2.83	oI40	Mn_xTi_{1-x}	1.1–3.0	Mn in -Ti
$IrSn_2$	0.65–0.78	C1	MnU_6	2.32	$D2_c$
Ir_2Sr	5.70	C15	Mo_2N	5.0	F.C.C.
Ir_7Ta_{13}	1.2	$D8_b$-tP30 (FeCr)	$Mo_6Na_2S_8$	8.6	hR15
$Ir_{0.5}Te_{0.5}$	~3		Mo_xNb_{1-x}	0.016–9.2	
$IrTe_3$	1.18	C2	$Mo_{5.25}Nb_{0.75}Se_8$	6.2	hR15
IrTh	< 0.37	B_f	Mo_6NdSa_8	8.2	hR15
Ir_2Th	6.50	C15	Mo_3Os	7.2	A15
Ir_3Th	4.71		$Mo_{0.62}Cs_{0.38}$	5.65	$D8_b$
Ir_3Th_7	1.52	$D10_2$	Mo_3P	5.31	DO_e
Ir_5Th	3.93	$D2_d$	$Mo_6Pb_{1.2}Se_8$	6.75	hR15
$IrTi_3$	5.40	A15	$Mo_{0.5}Pd_{0.5}$	3.52	A3
IrV_2	1.39	A15	Mo_6PrSe_8	9.2	hR15
IrW_3	3.82		MoRe	7.8	$D8_b$-tP30
$Ir_{0.28}W_{0.72}$	4.49		$MoRe_3$	9.25; 9.89	A12
Ir_2Y	2.18; 1.38	C15	Mo_xRe_{1-x}	1.2–12.2	
$Ir_{0.69}Y_{0.31}$	1.98; 1.44	C15	$Mo_{0.42}Re_{0.58}$	6.35	$D8_b$
$Ir_{0.70}Y_{0.30}$	2.16	C15	MoRh	1.97	A3
			Mo_xRh_{1-x}	1.5–8.2	B.C.C.

Substance	T_c, K	Crystal structure type
MoRu	9.5–10.5	A3
$Mo_{0.61}Ru_{0.39}$	7.18	$D8_b$
$Mo_{0.2}Ru_{0.8}$	1.66	A3
Mo_3Ru_2	7.0	$D8_b$-tP30
$Mo_4Ru_2Te_8$	1.7	hR15
Mo_6S_8	1.85	hR15
Mo_6S_8Sc	3.6	hR15
$Mo_6S_8Sm_{1.2}$	2.9	hR15
Mo_6S_8Tb	2.0	hR15
Mo_6S_8Tl	8.7	hR15
$Mo_6S_8Tm_{1.2}$	2.1	hR15
$Mo_6S_8Y_{1.2}$	3.0	hR15
Mo_6S_8Yb	9.2	hR15
$Mo_{6.6}S_8Zn_{11}$	3.6	hR15
Mo_3Sb_4	2.1	
Mo_6Se_8	6.3	hR15
$Mo_6Se_8Sm_{1.2}$	6.8	hR15
$Mo_6Se_8Sn_{1.2}$	6.8	hR15
Mo_6Se_8Tb	5.7	hR15
Mo_3Se_3Tl	4.0	hP14
$Mo_6Se_8Tm_{1.2}$	6.3	hR15
Mo_6Se_8Yb	6.2	hR15
Mo_3Si	1.30	A15
$MoSi_{0.7}$	1.34	
Mo_xSiV_{3-x}	4.54–16.0	A15
$Mo_{5.25}Ta_{0.75}Te_8$	1.7	hR15
Mo_6Te_8	1.7	hR15
$Mo_{0.16}Ti_{0.84}$	4.18; 4.25	
$Mo_{0.913}Ti_{0.087}$	2.95	
$Mo_{0.04}Ti_{0.96}$	2.0	Cubic
$Mo_{0.025}Ti_{0.975}$	1.8	
Mo_xU_{1-x}	0.7–2.1	
Mo_xV_{1-x}	0–~5.3	
Mo_2Zr	4.25–4.75	C15
NNb (film)	6–9	B1
$N_xO_yTi_z$	2.9–5.6	Cubic
$N_xO_yV_z$	5.8–8.2	Cubic
$N_{0.34}Re$	4–5	F.C.C.
NTa (film)	4.84	B1
$N_{0.6-0.987}Ti$	< 1.17–5.8	B1
$N_{0.82-0.99}V$	2.9–7.9	B1
NZr	9.8	B1
$N_{0.906-0.984}Zr$	3.0–9.5	B1
$Na_{0.28-0.35}O_3W$	0.56	Tetragonal
$Na_{0.28}Pb_{0.72}$	7.2	
NbO	1.25	
$NbOs_2$	2.52	A12
Nb_3Os	1.05	A15
$Nb_{0.6}Os_{0.4}$	1.89; 1.78	$D8_b$
$Nb_3Os_{0.02-0.10}Rh_{0.98-0.90}$	2.42–2.30	A15
Nb_3P	1.8	$L1_2$-tP32 (Ti_3P)
NbPRh	4.08	C37-oP12 (Co_2Si)
$Nb_{0.6}Pd_{0.4}$	1.60	$D8_f$ plus cubic
$Nb_3Pd_{0.02-0.10}Rh_{0.92-0.90}$	2.49–2.55	A15
$Nb_{0.62}Pt_{0.38}$	4.21	$D8_b$
Nb_5Pt_3	3.73	$D8_b$
$Nb_3Pt_{0.02-0.98}Rh_{0.98-0.02}$	2.52–9.6	A15
$NbRe_3$	5.27	$D8_b$-tP30 (FeCr)
$Nb_{0.38-0.18}Re_{0.62-0.82}$	2.43–9.70	A15
NbRe	3.8	$D8_b$-tP30
NbReSi	5.1	oI36 (FeTiSi)

Substance	T_c, K	Crystal structure type
Nb_3Rh	2.64	A15
$Nb_{0.6}Rh_{0.40}$	4.21	$D8_b$ plus other
$Nb_{0.9}Rh_{1.1}$	3.07	A3-oP4 (AuCd)
$Nb_3Rh_{0.98-0.90}Ru_{0.02-0.10}$	2.42–2.44	A15
Nb_xRu_{1-x}	1.2–4.8	
NbRuSi	2.65	oI36
NbS_2	6.1–6.3	Hexagonal, $NbSe_2$ type
NbS_2	5.0–5.5	Hexagonal, three-layer type
Nb_3Sb	0.2	$L1_2$-tP32 (Ti_3P)
$Nb_3Sb_{0-0.7}Sn_{1-0.3}$	6.8–18	A15
$NbSe_2$	5.15–5.62	Hexagonal
$Nb_{1-1.05}Se_2$	2.2–7.0	Same
Nb_3Se_4	2.0	hP14
Nb_3Si	1.5	$L1_2$
Nb_3SiSnV_3	4.0	
$NbSn_2$	2.60	Orthorhombic
Nb_6Sn_5	2.8	oI44 (Sn_5Ti_6)
NbSnTaV	6.2	A15
$NbSnV_2$	5.5	A15
Nb_3SnV	9.8	A15
Nb_xTa_{1-x}	4.4–9.2	A2
Nb_3Te_4	1.8	hP14
Nb_xTi_{1-x}	0.6–9.8	
$Nb_{0.6}Ti_{0.4}$	9.8	
Nb_xU_{1-x}	1.95 (max.)	
$Nb_{0.88}V_{0.12}$	5.7	A2
$Nb_{0.5}V_{1.5}Zr$	4.3	C15-hP12 ($MgZn_2$)
$Ni_{0.3}Th_{0.7}$	1.98	$D10_2$
$NiZr_2$	1.52	
$Ni_{0.1}Zr_{0.9}$	1.5	A3
$O_3Rb_{0.27-0.29}W$	1.98	Hexagonal
OSn	3.81	tP4 (PbO)
O_3SrTi ($n = 1.7–12.0 \times 10^{19}$)	0.12–0.37	
O_3SrTi ($n = 10^{18}–10^{21}$)	0.05–0.47	
O_3SrTi ($n = 10^{20}$)	0.47	
$O_3Sr_{0.08}W$	2–4	Hexagonal
OTi	0.58	
$O_3Tl_{0.30}W$	2.0–2.14	Hexagonal
OV_3Zr_3	7.5	$E9_3$
OW_3 (film)	3.35; 1.1	A15
OsPti	1.2	C22-hP9 (Fe_2P)
OsPZr	7.4	Same
OsReY	2.0	C14
Os_2Sc	4.6	C14
OsTa	1.95	A12
Os_3Th_7	1.51	$D10_2$
Os_xW_{1-x}	0.9–4.1	
OsW_3	~3	
Os_2Y	4.7	C14
Os_2Zr	3.0	C14
Os_xZr_{1-x}	1.5–5.6	
PPb	7.8	
OsW_2	3.81	$D8_b$-tP30 (FeCr)
$PPd_{3.0-3.2}$	<0.35–0.7	DO_{11}
P_3Pd_7 (high temperature)	1.0	Rhombohedral
P_3Pd_7 (low temperature)	0.70	Complex
PRh	1.22	
PRh_2	1.3	C1
P_4Rh_5	1.22	oP28 ($CaFe_2O_4$)
PRhTa	4.41	C37-oP12 (Co_2Si)

Substance	T_c, K	Crystal structure type	Substance	T_c, K	Crystal structure type
PRhZr	1.55	Same	$PtV_{3.5}$	1.26	A15
PRuTi	1.3	C22-hP9 (Fe_2P)	$Pt_{0.5}W_{0.5}$	1.45	A1
PRuZr	3.46	C37-oP12	Pt_xW_{1-x}	0.4–2.7	
PW_3	2.26	DO_e	Pt_2Y_3	0.90	
Pb_2Pd	2.95	C16	Pt_2Y	1.57; 1.70	C15
Pb_4Pt	2.80	Related to C16	Pt_3Y_7	0.82	$D10_2$
Pb_2Rh	2.66	C16	PtZr	3.0	A3
PbSb	6.6		Re_2Sc	4.2	C15-hP12 ($MgZn_2$)
PbTe (plus 0.1 w/o Pb)	5.19		$Re_{24}Sc_5$	2.2	A12-cI58 (Mg)
PbTe (plus 0.1 w/o Te)	5.24–5.27		ReSiTa	4.4	oI36 (FeTiSi)
$PbTl_{0.27}$	6.43		$Re_2Si_5Y_2$	1.76	tP40 ($Fe_3Sc_2Si_5$)
$PbTl_{0.17}$	6.73		Re_3Ta_2	1.4	$D8_b$-tP30 (FeCr)
$PbTl_{0.12}$	6.88		$Re_{0.64}Ta_{0.36}$	1.46	A12
$PbTl_{0.075}$	6.98		Re_3Ta	6.78	A12-cI58 (Mn)
$PbTl_{0.04}$	7.06		$Re_{24}Ti_5$	6.60	A12
$Pb_{1-0.26}Tl_{0-0.74}$	7.20–3.68		Re_xTi_{1-x}	6.6 (max.)	
$PbTl_2$	3.75–4.1		$Re_{0.76}V_{0.24}$	4.52	$D8_b$
Pb_3Zr_5	4.60	$D8_8$	Re_3V	6.26	$D8_b$-tP30
$PbZr_3$	0.76	A15	$Re_{0.92}V_{0.08}$	6.8	A3
$Pd_{0.9}Pt_{0.1}Te_2$	1.65	C6	$Re_{0.6}W_{0.4}$	6.0	
$Pd_{0.05}Ru_{0.05}Zr_{0.9}$	~9		$Re_{0.5}W_{0.5}$	5.12	$D8_b$
$Pd_{2.2}S$ (quenched)	1.63	Cubic	$Re_{13}W_{12}$	5.2	$D8_b$-tP30
$PdSb_2$	1.25	C2	Re_3W	9.0	A12-cI58
PdSb	1.5	$B8_1$	Re_2Y	1.83	C14
PdSbSe	1.0	C2	Re_2Zr	5.9	C14
PdSbTe	1.2	C2	Re_3Zr	7.40	A12-cI58
Pd_4Se	0.42	Tetragonal	Re_6Zr	7.40	Same
$Pd_{6-7}Se$	0.66	Like Pd_4Te	$Rh_{17}S_{15}$	5.8	Cubic
$Pd_{2.8}Se$	2.3		$Rh_{~0.24}Sc_{~0.76}$	0.88; 0.92	
Pd_xSe_{1-x}	2.5 (max.)		$Rh_4Sc_5Si_{10}$	8.54	tP38
PdSi	0.93	B31	$Rh_4Sc_3Sn_{13}$	4.5	cP40
PdSn	0.41	B31	Rh_xSe_{1-x}	6.0 (max.)	
$PdSn_2$	3.34		$RhSi_3Th$	1.76	tI10
Pd_2Sn	0.41	C37	$Rh_{0.86}Sc_{1.04}Th$	6.45	tI12
Pd_3Sn	0.47–0.64	$B8_2$	Rh_2Si_2Y	3.11	tI10
Pd_2SnTm	1.77	DO_3-cF16 (BiF_3)	$Rh_3Si_5Y_2$	2.70	oI40
Pd_2SnY	4.92	Same	$Rh_4Sn_{13}Sr_3$	4.3	cP40
Pd_2SnYb	1.79	Same	Rh_xSn_yTh	1.9	cI2 (W)
PdTe	2.3; 3.85	$B8_1$	Rh_xSn_yTm	2.3	cP40
$PdTe_{1.02-1.08}$	2.56–1.88	$B8_1$	$Rh_4Sn_{13}Y_3$	3.2	cP40
$PdTe_2$	1.69	C6	Rh_2Sr	6.2	C15
$PdTe_{2.1}$	1.89	C6	$Rh_{0.4}Ta_{0.6}$	2.35	$D8_b$
$PdTe_{2.3}$	1.85	C6	$RhTe_2$	1.51	C2
$Pd_{1.1}Te$	4.07	$B8_1$	$Rh_{0.67}Te_{0.33}$	0.49	
Pd_3Te	0.76	cI2 (W)	Rh_xTe_{1-x}	1.51 (max.)	
$PdTh_2$	0.85	C16	RhTh	0.36	B_f
$Pd_{0.1}Zr_{0.9}$	7.5	A3	Rh_3Th_7	2.15	$D10_2$
PtSb	2.1	$B8_1$	Rh_5Th	1.07	
PtSi	0.88	B31	Rh_xTi_{1-x}	2.25–3.95	
PtSn	0.37	$B8_1$	$Rh_{0.02}U_{0.98}$	0.96	
$PtSn_4$	2.38	C16-oC20 ($PdSn_4$)	RhV_3	0.38	A15
Pt_3Ta_7	1.5	$D8_b$-tP30	RhW	~3.4	A3
$PtTa_3$	0.4	A15-cP8 (Cr_3Si)	RhY_3	0.65	
PtTe	0.59	Orthorhombic	Rh_2Y_3	1.48	
PtTh	0.44	B_f	Rh_3Y	1.07	C15
Pt_3Th_7	0.98	$D10_2$	Rh_5Y	0.56	
Pt_5Th	3.13		Rh_3Y_7	0.32	hP20 (Fe_3Th_7)
$PtTi_3$	0.58	A15	$Rh_{0.005}Zr$ (annealed)	5.8	
$Pt_{0.02}U_{0.98}$	0.87	β-phase	$Rh_{0-0.45}Zr_{1-0.55}$	2.1–10.8	
$PtV_{2.5}$	1.36	A15	$Rh_{0.1}Zr_{0.9}$	9.0	H.C.P.
PtV_3	2.87–3.20	A15	Ru_2Sc	1.67	C14
			RuSiTa	3.15	oI36

Substance	T_c, K	Crystal structure type
Ru_3Si_2Th	3.98	hP12
Ru_3Si_2Y	3.51	hP12
$Ru_{1.1}Sn_{3.1}Y$	1.3	cP40
Ru_2Th	3.56	C15
RuTi	1.07	B2
$Ru_{0.05}Ti_{0.95}$	2.5	
$Ru_{0.1}Ti_{0.9}$	3.5	
$Ru_xTi_{0.6}V_y$	6.6 (max.)	
Ru_3U	0.15	$L1_2$-cP4
$Ru_{0.45}V_{0.55}$	4.0	B2
RuW	7.5	A3
Ru_2Y	1.52	C14
Ru_2Zr	1.84	C14
$Ru_{0.1}Zr_{0.9}$	5.7	A3
STh	0.5	B1-cF8 (NaCl)
SbSn	1.30–1.42	B1 or distorted
$SbTa_3$	0.72	A15-cP8 (Cr_3Si)
$SbTi_3$	5.8	Same
Sb_2Ti_7	5.2	
$Sb_{0.01-0.03}V_{0.99-0.97}$	3.76–2.63	A2
SbV_3	0.80	A15
SeTh	1.7	B1-cF8
$SiMo_3$	1.4	A15-cP8
Si_2Th	3.2	C_c, α-phase
Si_2Th	2.4	C32, β-phase
$SiV_{2.7}Ru_{0.3}$	2.9	A15
Si_2W_3	2.8; 2.84	
$SiZr_3$	0.5	$L1_2$-tP32 (Ti_3P)
$Sn_{0.174-0.104}Ta_{0.826-0.896}$	6.5–< 4.2	A15
$SnTa_3$	8.35	A15, highly ordered
$SnTa_3$	6.2	A15, partially ordered
$SnTaV_2$	2.8	A15
$SnTa_2V$	3.7	A15
Sn_xTe_{1-x} ($n = 10.5-20 \times 10^{20}$)	0.07–0.22	B1
Sn_3Th	3.33	$L1_2$-cP4
$SnTi_3$	5.80	A15-cP8
Sn_xTl_{1-x}	2.37–5.2	
SnV_3	3.8	A15
$Sn_{0.02-0.057}V_{0.98-0.943}$	2.87–~1.6	A2
$SnZr_3$	0.92	A15-cP8
$Ta_{0.025}Ti_{0.975}$	1.3	Hexagonal
$Ta_{0.05}Ti_{0.95}$	2.9	Hexagonal
$Ta_{0.05-0.75}V_{0.95-0.25}$	4.30–2.65	A2
$Ta_{0.8-1}W_{0.2-0}$	1.2–4.4	A2
$Tc_{0.1-0.4}W_{0.9-0.6}$	1.25–7.18	Cubic
$Tc_{0.50}W_{0.50}$	7.52	α plus
$Tc_{0.60}W_{0.40}$	7.88	plus α
Tc_6Zr	9.7	A12
TeY	1.02	B1-cF8
$ThTl_3$	0.87	$L1_2$-cP4
$Th_{0-0.55}Y_{1-0.45}$	1.2–1.8	
$Ti_{0.70}V_{0.30}$	6.14	Cubic
Ti_xV_{1-x}	0.2–7.5	
$Ti_{0.5}Zr_{0.5}$ (annealed)	1.23	
$Ti_{0.5}Zr_{0.5}$ (quenched)	2.0	
Tl_3Y	1.52	$L1_2$-cP4
V_2Zr	8.80	C15
$V_{0.26}Zr_{0.74}$	5.9	
W_2Zr	2.16	C15
YZn	0.33	B2-cP2 (CsCl)

* n denotes current carriers concentration in cm^{-3}.

B. Superconductors with $T_c > 10$K

Substance	T_c,K	Crystal structure type	
Al_2CMo_3	10.0	A13	
$Al_{0.5}Ge_{0.5}Nb$	12.6	A15	
$Al_{-0.8}Ge_{-0.2}Nb_3$	20.7	A15	
$AlNb_3$	18.0	A15	(Cr_3Si)
$AlNb_3$	12.0		(FeCr)
Al_xNb_{1-x}	<4.2–13.5	$D8_b$	
Al_xNb_{1-x}	12–17.5	A15	
$Al_{0.27}Nb_{0.73-0.48}V_{0-0.25}$	14.5–17.5	A15	
$Al\ Nb_xV_{1-x}$	4.4–13.5		
$Al_{0.1}Si_{0.9}V_3$	14.05		
AlV_3	11.8	A15	(Cr_3Si)
$AuNb_3$	11.5	A15	
$Au_{0-0.3}Nb_{1-0.7}$	1.1–11.0		
$Au_{0.02-0.98}Nb_3Rh_{0.98-0.02}$	2.53–10.9	A15	
$AuNb_{3(1-x)}V_{3x}$	1.5–11.0	A15	
$B_{0.03}C_{0.51}Mo_{0.47}$	12.5		
B_4LuRh_4	11.7		(B_4CeCo_4)
B_2LuRu	10		
B_4Rh_4Y	11.3		(B_4CeCo_4)
$B_{0.1}Si_{0.9}V_3$	15.8	A15	
$BaBi_{0.2}O_3Pb_{0.8}$	13.2		
$Ba_2CaCu_2O_8Tl_2$	120		
$Ba_2Cu_3LaO_6$	80		
$Ba_2Cu_3O_7Tm$	101		
$Ba_2Cu_3O_7Y$	90		
$(Ba,La)_2CuO_4$	36	A15	(K_2NiF_4)
$Bi_2CaCu_2O_8Sr_2$	110		
$Br_2Mo_6S_6$	13.8		(Mo_6PbS_8)
C_3La	11.0		(C_3Pu_2)
CMo	14.3	B1	(NaCl)
CMo_2	12.2	o**	
$C_{0.5}Mo_xNb_{1-x}$	10.8–12.5	B1	
CMo_xTi_{1-x}	10.2(max)	B1	
$CMo_{0.83}Ti_{0.17}$	10.2	B1	
$C_{0-0.38}N_{1-0.62}Ta$	10.0–11.3		
CNb (whiskers)	7.5–10.5		
CNb	11.5	B1	
$C_{0.7-1.0}Nb_{0.3-0}$	6–11	B1	
CNb_xTa_{1-x}	8.2–13.9		
$CNb_{0.6-0.9}W_{0.4-0.1}$	12.5–11.6	B1	
$C_{0.1}Si_{0.9}V_3$	16.4	A15	
CTa	10.3	B1	
$CTa_{1-0.4}W_{0-0.6}$	8.5–10.5	B1	
$C_{0.66}Th_{0.13}Y_{0.21}$	17		(C_3Pu_2)
C_3Y_2	11.5		(C_3Pu_2)
CW	10	B1	
$(Ca,La)_2CuO_4$	18		(K_2NiF_4)
$Cu(La,Sr)_2O_4$	39		
$Cu_{1.8}Mo_6S_8$	10.8		(Mo_6PbS_8)
$Cr_{0.3}SiV_{2.7}$	11.3	A15	
$GaNb_3$	14.5	A15	(Cr_3Si)
$Ga_xNb_3Sn_{1-x}$	14–18.37	A15	
GaV_3	16.8	A15	
$GaV_{2.1-3.5}$	6.3–14.45	A15	
$GeNb_3$	23.2	A15	
$GeNb_3$ (quenched)	6–17	A15	
$Ge_xNb_3Sn_{1-x}$	17.6–18.0	A15	
$Ge_{0.5}Nb_3Sn_{0.5}$	11.3	A15	
$Ge_{0.1}Si_{0.9}V_3$	14.0	A15	
GeV_3	11	A15	
$InLa_3$	9.83; 10.4	$L1_2$	($AuCu_3$)

Substance	T_c,K	Crystal structure type	Substance	T_c,K	Crystal structure type
$In_{0-0.3}Nb_3Sn_{1-0.7}$	18.0–18.19	A15	$N_{100-42w/o}Nb_{0-58w/o}Ti$	15–16.8	
InV_3	13.9	A15	$N_{100-75w/o}Nb_{0-25w/o}Zr$	12.5–16.35	
$Ir_{0.4}Nb_{0.6}$	10	(FeCr)	NNb_xZr_{1-x}	9.8–13.8	B1
$LaMo_6Se_8$	11.4	(Mo_6PbS_8)	$N_{0.93}Nb_{0.85}Zr_{0.15}$	13.8	B1
LiO_4Ti_2	13.7	(Al_2MgO_4)	NTa	12–14	B1
MgB_2	39.0±0.5	C32	NZr	10.7	B1
MoN	12; 14.8	h*	Nb_3Pt	10.9	A15
Mo_3Os	12.7	A15	$Nb_{0.18}Re_{0.82}$	10	(Mn)
$Mo_6Pb_{0.9}S_{7.5}$	15.2	(Mo_6PbS_8)	Nb_3Si	19	A15
Mo_3Re	10.0; 15	A15	$Nb_{0.3}SiV_{2.7}$	12.8	A15
Mo_xRe_{1-x}	1.2–12.2		Nb_3Sn	18.05	A15
$Mo_{0.52}Re_{0.48}$	11.1		$Nb_{0.8}Sn_{0.2}$	18.18; 18.5	A15
$Mo_{0.57}Re_{0.43}$	14.0		Nb_3Sn_{1-x} (film)	2.6–18.5	o*
$Mo_{~0.60}Re_{0.395}$	10.6		Nb_3Sn_2	16.6	t*
$MoRu$	9.5–10.5	A3	$NbSnTa_2$	10.8	A15
Mo_3Ru	10.6	A15	Nb_2SnTa	16.4	A15
Mo_6Se_8T1	12.2	(Mo_6PbS_8)	$Nb_{2.5}SnTa_{0.5}$	17.6	A15
$Mo_{0.3}SiV_{2.7}$	11.7	A15	$Nb_{2.75}SnTa_{0.25}$	17.8	A15
Mn_3Si	12.5	A15	$Nb_{3x}SnTa_{3(1-x)}$	6.0–18.0	
Mo_3Tc	15	A15	$Nb_2SnTa_{0.5}V_{0.5}$	12.2	A15
$Mo_{0.3}Tc_{0.7}$	12.0	A15	$NbTc_3$	10.5	A12
Mo_xTc_{1-x}	10.8–15.8		$Nb_{0.75}Zr_{0.25}$	10.8	
$MoTc_3$	15.8		$Nb_{0.66}Zr_{0.33}$	10.8	
NNb (whiskers)	10–14.5		$PbTa_3$	17	A15
NNb (diffusion wires)	16.10		$RhTa_3$	10	A15
$N_{0.988}Nb$	14.9; 17.3	B1	$RhZr_2$	10.8; 11.3	C16 (Al$_2$Cu)
$N_{0.824-0.988}Nb$	14.4–15.3	B1	$Rh_{0-0.45}Zr_{1-0.55}$	2.1–10.8	
$N_{0.7-0.795}Nb$	11.3–12.9		$SiTi_{0.3}V_{2.7}$	10.9	A15
NNb_xO_y	13.5–17.0	B1	SiV_3	17.1	A15
NNb_xO_y	6.0–11		$SiV_{2.7}Zr_{0.3}$	13.2	A15

TABLE 5. Critical Field Data

Substance	H_o (oersteds)	Substance	H_o (oersteds)
Ag_2F	2.5	$InSb$	1100
Ag_7NO_{11}	57	In_xTl_{1-x}	252–284
Al_2CMo_3	1700	$In_{0.8}Tl_{0.2}$	252
$BaBi_3$	740	$Mg_{0.47}Tl_{0.53}$	220
Bi_2Pt	10	$Mo_{0.16}Ti_{0.84}$	<985
Bi_3Sr	530	$NbSn_2$	620
Bi_5Tl_3	>400	$PbTl_{0.27}$	756
$CdSn$	>266	$PbTl_{0.17}$	796
$CoSi_2$	105	$PbTl_{0.12}$	849
$Cr_{0.1}Ti_{0.3}V_{0.6}$	1360	$PbTl_{0.075}$	880
$In_{1-0.86}Mg_{0-0.14}$	272.4–259.2	$PbTl_{0.04}$	864

TABLE 6. High Critical Magnetic-Field Superconductive Compounds and Alloys

Substance	T_c, K	H_{c1}, kOe	H_{c2}, kOe	H_{c3}, kOe	T_{obs}, K[a]
Al_2CMo_3	9.8–10.2	0.091	156		1.2
$AlNb_3$		0.375			
$Ba_xO_3Sr_{1-x}Ti$	<0.1–0.55	0.0039 max.			
$Bi_{0.5}Cd_{0.1}Pb_{0.27}Sn_{0.13}$			>24		3.06
Bi_xPb_{1-x}	7.35–8.4	0.122 max.	30 max.		4.2
$Bi_{0.56}Pb_{0.44}$	8.8		15		4.2
$Bi_{7.5w/o}Pb_{92.5w/o}$ [b]			2.32		
$Bi_{0.099}Pb_{0.901}$		0.29	2.8		
$Bi_{0.02}Pb_{0.98}$		0.46	0.73		
$Bi_{0.53}Pb_{0.32}Sn_{0.16}$			>25		3.06
$Bi_{1-0.93}Sn_{0-0.07}$			0–0.032		3.7
Bi_5Tl_3	6.4		>5.6		3.35
C_8K (excess K)	0.55		0.160 (H⊥c)		0.32
			0.730 (H|c)		0.32
C_8K	0.39		0.025 (H⊥c)		0.32
			0.250 (H|c)		0.32
$C_{0.44}Mo_{0.56}$	12.5–13.5	0.087	98.5		1.2
CNb	8–10	0.12	16.9		4.2
$CNb_{0.4}Ta_{0.6}$	10–13.6	0.19	14.1		1.2
CTa	9–11.4	0.22	4.6		1.2
$Ca_xO_3Sr_{1-x}Ti$	<0.1–0.55	0.002–0.004			
$Cd_{0.1}Hg_{0.9}$ (by weight)		0.23	0.34		2.04
$Cd_{0.05}Hg_{0.95}$		0.28	0.31		2.16
$Cr_{0.10}Ti_{0.30}V_{0.60}$	5.6	0.071	84.4		0
GaN	5.85	0.725			4.2
Ga_xNb_{1-x}			>28		4.2
$GaSb$ (annealed)	4.24		2.64		3.5
$GaV_{1.95}$	5.3		73[e]		
$GaV_{2.1-3.5}$	6.3–14.45		230–300[d]		0
GaV_3		0.4	350[e]		0
			500[d]		
$GaV_{4.5}$	9.15		121[c]		0
Hf_xNb_y			>52–>102		1.2
Hf_xTa_y			>28–>86		1.2
$Hg_{0.05}Pb_{0.95}$		0.235	2.3		
$Hg_{0.101}Pb_{0.899}$		0.23	4.3		4.2
$Hg_{0.15}Pb_{0.85}$	6.75		>13		2.93
$In_{0.98}Pb_{0.02}$	3.45	0.1		0.12	2.76
$In_{0.96}Pb_{0.04}$	3.68	0.1	0.12	0.25	2.94
$In_{0.94}Pb_{0.06}$	3.90	0.095	0.18	0.35	3.12
$In_{0.913}Pb_{0.087}$	4.2	~10.17	0.55	2.65	
$In_{0.316}Pb_{0.684}$		0.155	3.7		4.2
$In_{0.17}Pb_{0.83}$			2.8	5.5	4.2
$In_{1.000}Te_{1.002}$	3.5–3.7		1.2[c]		0
$In_{0.95}Tl_{0.05}$		0.263	0.263		3.3
$In_{0.90}Tl_{0.10}$		0.257	0.257		3.25
$In_{0.83}Tl_{0.17}$		0.242	0.39		3.21
$In_{0.75}Tl_{0.25}$		0.216	0.50		3.16
LaN	1.35	0.45			0.76
La_3S_4	6.5	≈0.15	>25		1.3
La_3Se_4	8.6	≈0.2	>25		1.25
$Mo_{0.52}Re_{0.48}$	11.1		14–21	22–33	4.2
			18–28	37–43	1.3
$Mo_{0.6}Re_{0.395}$	10.6		14–20	20–37	4.2
			19–26	26–37	1.3
$Mo_{0.5}Ti_{0.5}$			75[c]		0
$Mo_{0.16}Ti_{0.84}$	4.18	0.028	98.7[c]		0
			36–38		3.0
$Mo_{0.913}Ti_{0.087}$	2.95	0.060	15		4.2
$Mo_{0.1-0.3}U_{0.9-0.7}$	1.85–2.06		>25		
$Mo_{0.17}Zr_{0.83}$			30		

Substance	T_c, K	H_{c1}, kOe	H_{c2}, kOe	H_{c3}, kOe	T_{obs}, K[a]
$N_{(12.8 w/o)}Nb$	15.2		>9.5		13.2
NNb (wires)	16.1		153[c]		0
			132		4.2
			95		8
			53		12
NNb_xO_{1-x}	13.5–17.0		38		
NNb_xZr_{1-x}	9.8–13.8		4- >130		4.2
$N_{0.93}Nb_{0.85}Zr_{0.15}$	13.8		>130		4.2
$Na_{0.086}Pb_{0.914}$		0.19	6.0		
$Na_{0.016}Pb_{0.984}$		0.28	2.05		
Nb	9.15		2.020		1.4
			1.710		4.2
Nb		0.4–1.1	3–5.5		4.2
Nb (unstrained)		1.1–1.8	3.40	6–9.1	4.2
Nb (strained)		1.25–1.92	3.44	6.0–8.7	4.2
Nb (cold-drawn wire)		2.48	4.10	≈10	4.2
Nb (film)			>25		4.2
NbSc			>30		
Nb_3Sn		0.170	221		4.2
			70		14.15
			54		15
			34		16
			17		17
$Nb_{0.1}Ta_{0.9}$		0.084	0.154		4.195
$Nb_{0.2}Ta_{0.8}$			10		4.2
$Nb_{0.65-0.73}Ta_{0.02-0.10}Zr_{0.25}$			>70->90		4.2
Nb_xTi_{1-x}			148 max.		1.2
			120 max.		4.2
$Nb_{0.222}U_{0.778}$		1.98	23		1.2
Nb_xZr_{1-x}			127 max.		1.2
			94 max.		4.2
O_3SrTi	0.43	0.0049[c]	0.504[c]		0
O_3SrTi	0.33	0.00195[c]	0.420[c]		0
$PbSb_{1 w/o}$(quenched)			>1.5		4.2
$PbSb_{1 w/o}$(annealed)			>0.7		4.2
$PbSb_{2.8 w/o}$(quenched)			>2.3		4.2
$PbSb_{2.8 w/o}$(annealed)			>0.7		4.2
$Pb_{0.871}Sn_{0.129}$		0.45	1.1		
$Pb_{0.965}Sn_{0.035}$		0.53	0.56		
$Pb_{1-0.26}Tl_{0-0.74}$	7.20–3.68		2–6.9[c]		0
$PbTl_{0.17}$	6.73		4.5[c]		0
$Re_{0.26}W_{0.74}$			>30		
$Sb_{0.93}Sn_{0.07}$			0.12		3.7
SiV_3	17.0	0.55	156[e]		
Sn_xTe_{1-x}		0.00043–0.00236	0.005–0.0775		0.012–0.079
Ta (99.95%)		0.425	1.850		1.3
		0.325	1.425		2.27
		0.275	1.175		2.66
		0.090	0.375		3.72
$Ta_{0.5}Nb_{0.5}$			3.55		4.2
$Ta_{0.65-0}Ti_{0.35-1}$	4.4–7.8		>14–138		1.2
$Ta_{0.5}Ti_{0.5}$			138		1.2
Te	3.3	0.25[c]			0
Tc_xW_{1-x}	5.75–7.88		8–44		4.2
Ti				2.7	4.2
$Ti_{0.75}V_{0.25}$	5.3	0.029[c]	199[c]		0
$Ti_{0.775}V_{0.225}$	4.7	0.024[c]	172[c]		0
$Ti_{0.615}V_{0.385}$	7.07	0.050	34		4.2
$Ti_{0.516}V_{0.484}$	7.20	0.062	28		4.2
$Ti_{0.415}V_{0.585}$	7.49	0.078	25		4.2
$Ti_{0.12}V_{0.88}$			17.3	28.1	4.2
$Ti_{0.09}V_{0.91}$			14.3	16.4	4.2
$Ti_{0.06}V_{0.94}$			8.2	12.7	4.2

Substance	T_c, K	H_{c1}, kOe	H_{c2}, kOe	H_{c3}, kOe	T_{obs}, K[a]
$Ti_{0.03}V_{0.97}$			3.8	6.8	4.2
Ti_xV_{1-x}			108 max.		1.2
V	5.31	0.8	3.4		1.79
		0.75	3.15		2
		0.45	2.2		3
		0.30	1.2		4
$V_{0.26}Zr_{0.74}$	≈5.9	0.238			1.05
		0.227			1.78
		0.185			3.04
		0.165			3.5
W (film)	1.7–4.1		>34		1

[a] Temperature of critical field measurement.
[b] w/o denotes weight percent.
[c] Extrapolated.
[d] Linear extrapolation.
[e] Parabolic extrapolation

References

1. B. W. Roberts, in *Superconductive Materials and Some of Their Properties. Progress in Cryogenics*, Vol. IV, 1964, pp. 160–231.
2. B. W. Roberts, Superconductive Materials and Some of Their Properties, NBS Technical Notes 408 and 482, U.S. Government Printing Office, 1966 and 1969; B. W. Roberts, *J. Phys. Chem. Ref. Data*, 5, 581, 1976.
3. B. W. Roberts, Properties of Selected Superconductive Materials, 1978 Supplement, NBS Technical Note 983, 1978.
4. T. Claeson, *Phys. Rev.*, 147, 340, 1966.
5. C. J. Raub, W. H. Zachariasen, T. H. Geballe, and B. T. Matthias, *J. Phys. Chem. Solids*, 24, 1093, 1963.
6. T. H. Geballe, B. T. Matthias, V. B. Compton, E. Corenzwit, G. W. Hull, Jr., and L. D. Longinotti, *Phys. Rev.*, 1A, 119, 1965.
7. C. J. Raub, V. B. Compton, T. H. Geballe, B. T. Matthias, J. P. Maita, and G. W. Hull, Jr., *J. Phys. Chem. Solids*, 26, 2051, 1965.
8. R. D. Blaugher, J. K. Hulm, and P. N. Yocom, *J. Phys. Chem. Solids*, 26, 2037, 1965.
9. T. Claeson and H. L. Luo, *J. Phys. Chem. Solids*, 27, 1081, 1966.
10. S. C. Ng and B. N. Brockhouse, *Solid State Comm.*, 5, 79, 1967.
11. O. I. Shulishova and I. A. Shcherbak, *Izv. AN SSSR, Neorg. Materials*, 3, 1495, 1967.
12. T. F. Smith and H. L. Luo, *J. Phys. Chem. Solids*, 28, 569, 1967.
13. A. C. Lawson, *J. Less-Common Metals*, 23, 103, 1971.
14. R. Chevrel, M. Sergent, and J. Prigent, *J. Solid State Chem.*, 3, 515, 1971.
15. M. Marezio, P. D. Dernier, J. P. Remeika, and B. T. Matthias, *Mat. Res. Bull.*, 8, 657, 1973.
16. J. K. Hulm and R. D. Blaugher in *Superconductivity in d- and f-Band Metals*, D. H. Douglass, Ed., American Institute of Physics, 4, 1, 1972.
17. R. N. Shelton, A. C. Lawson, and D. C. Johnston, *Mat. Res. Bull.*, 10, 297, 1975.
18. H. D. Wiesinger, *Phys. Status Sol.*, 41A, 465, 1977.
19. O. Fisher, *Applied Phys.*, 16, 1, 1978.
20. D. C. Johnston, *Solid State Comm.*, 24, 699, 1977.
21. H. C. Ku and R. H. Shelton, *Mat. Res. Bull.*, 15, 1441, 1980.
22. H. Barz, *Mat. Res. Bull.*, 15, 1489, 1980.
23. G. P. Espinosa, A. S. Cooper, H. Barz, and J. P. Remeika, *Mat. Res. Bull.*, 15, 1635, 1980.
24. E. M. Savitskii, V. V. Baron, Yu. V. Efimov, M. I. Bychkova, and L. F. Myzenkova, in *Superconducting Materials*, Plenum Press, 1981, p. 107.
25. R. Fluckiger and R. Baillif, in Topics in *Current Physics*, O. Fischer and M. B. Maple, Eds., Springer Verlag, 34, 113, 1982.
26. R. N. Shelton, in *Superconductivity in d- and f-Band Metals*, W. Buckel and W. Weber, Eds., Kernforschungszentrum, Karlsruhe, 1982, p. 123.
27. D. C. Johnston and H. F. Braun, *Topics in Current Phys.*, 32, 11, 1982.
28. R. Chevrel and M. Sergent, *Topics in Current Phys.*, 32, 25, 1982.
29. G. P. Espinosa, A. S. Cooper, and H. Barz, *Mat. Res. Bull.*, 17, 963, 1982.
30. R. Muller, R. N. Shelton, J. W. Richardson, Jr., and R. A. Jacobson, *J. Less-Comm. Met.*, 92, 177, 1983.
31. You-Xian Zhao and Shou-An He, in *High Pressure in Science and Technology*, North Holland, 22, 51, 1983.
32. You-Xian Zhao and Shou-An He, *Solid State Comm.*, 24, 699, 1983.
33. G. P. Meisner and H. C. Ku, *Appl. Phys.*, A31, 201, 1983.
34. R. J. Cava, D. W. Murphy, and S. M. Zahurak, *J. Electrochem. Soc.*, 130, 2345, 1983.
35. R. N. Shelton, *J. Less-Comm. Met.*, 94, 69, 1983.
36. B. Chevalier, P. Lejay, B. Lloret, Wang Xian–Zhong, J. Etourneau, and P. Hagenmuller, *Annales de Chemie*, 9, 191, 1984.
37. G. Venturini, M. Meot-Meyer, E. McRae, J. F. Mareche, and B. Rogues, *Mat. Res. Bull.*, 19, 1647, 1984.
38. J. M. Tarascon, F. G. DiSalvo, D. W. Murphy, G. Hull, and J. V. Waszczak, *Phys. Rev.*, 29B, 172, 1984.
39. G. V. Subba and G. Balakrishnan, *Bull. Mat. Sci.*, 6, 283, 1984.
40. B. Batlog, *Physica*, 126B, 275, 1984.
41. M. J. Johnson, Ames Lab (USA) Report IS-T-1140, 1984.
42. I. M. Chapnik, *J. Mat. Sci. Lett.*, 4, 370, 1985.
43. W. Rong-Yao, L. Q-Guang, and Z. Xiao, *Phys. Status Sol.*, 90A, 763, 1985.
44. W. Xian-Zhong, B. Chevalier, J. Etourneau, and P. Hagenmuller, *Mat. Res. Bull.*, 20, 517, 1985.
45. H. R. Ott, F. Hulliger, H. Rudigier, and Z. Fisk, *Phys. Rev.*, 31B, 1329, 1985.
46. P. Villars and L. D. Calver, *Pearson's Handbook of Crystallographic Data for Intermetallic Phases*, Vol. 1–3, ASM, 1985.
47. G. V. Subba Rao, K. Wagner, G. Balakrishnan, J. Jakani, W. Paulus, and R. Scollhorn, *Bull. Mat. Sci.*, 7, 215, 1985.
48. J. G. Bednorz and K. A. Muller, *Zs. Physik*, B64, 189, 1986.
49. W. Rong-Yao, *Phys. Status Sol.*, 94A, 445, 1986.
50. H. D. Yang, R. N. Shelton, and H. F. Braun, *Phys. Rev.*, 33B, 5062, 1986.
51. G. Venturini, M. Kanta, E. McRae, J. F. Mareche, B. Malaman, and B. Roques, *Mat. Res. Bull.*, 21, 1203, 1986.
52. W. Rong-Yao, *J. Mat. Sci. Lett.*, 5, 87, 1986.
53. M. K. Wu, J. R. Ashburn, C. J. Torng, P. H. Hor, R. L. Meng, L. Gao, Z. J. Huang, Y. Q. Wang, and C. W. Chu, *Phys. Rev. Lett.*, 58, 908, 1987.
54. R. J. Cava, R. B. Van Dover, B. Batlog, and E. A. Rietman, *Phys. Rev. Lett.*, 58, 408, 1987.
55. L. C. Porter, T. J. Thorn, U. Geiser, A. Umezawa, H. H. Wang, W. K. Kwok, H-C. I. Kao, M. R. Monaghan, G. W. Crabtree, K. D. Carlson, and J. M. Williams, *Inorg. Chem.*, 26, 1645, 1987.
56. A. M. Kini, U. Geiser, H-C. I. Kao, K. D. Carlson, H. H. Wang, M. R. Monaghan, and K. M. Williams, *Inorg. Chem.*, 26, 1834, 1987.

57. T. Penney, S. von Molnar, D. Kaiser, F. Holtzberg, and A. W. Kleinsasser, *Phys. Rev.*, B38, 2918, 1988.

58. Y. K. Tao, J. S. Swinnea, A. Manthiram, J. S. Kim, J. B. Goodenoug, and H. Steinfink, *J. Mat. Res.*, 3, 248, 1988.

59. G. G. Peterson, B. R. Weinberger, L. Lynds, and H. A. Krasinski, *J. Mat. Res.*, 3, 605, 1988.

60. J. B. Torrance, Y. Tokura, A. Nazzai, and S. S. P. Parkin, *Phys. Rev. Lett.*, 60, 542, 1988.

61. K. Kourtakis, M. Robbins, P. K. Gallagher, and T. Teifel, *J. Mat. Res.*, 4, 1289, 1989.

62. J. C. Phillips, *Physics of High-T_c Superconductors*, Academic Press, 1989, p. 336.

63. Shui Wai Lin and L. I. Berger, *Rev. Sci. Instrum.*, 60, 507, 1989.

64. M. Tinkham, *Introduction to Superconductivity*, McGraw–Hill, New York, 1975.

65. O. Fischer and M.B. Maple, Eds., *Topics in Current Physics*, Volume 32: Superconductivity in Ternary Compounds I; Volume 34: Superconductivity in Ternary Compounds II, Springer–Verlag, Berlin, 1982.

66. K. J. Dunn and F. P. Bundy, *Phys. Rev.*, B25, 194, 1982.

67. A. Barone and G. Paterno, *Physics and Applications of the Josephson Effect*, Wiley, New York, 1982.

68. D. H. Douglass, Ed., *Superconductivity in d- and f-Band Metals*, Plenum Press, New York, 1976.

69. D. M. Ginsberg, Ed., *Physical Properties of High Temperature Superconductors*, (Volume II, 1990; Volume III, 1992; Volume V, 1996), World Scientific, Singapore.

70. T. Ishiguro and K. Yamji, *Organic Superconductors*, Springer-Verlag, Berlin, 1990.

71. Sh. Okada, K. Shimizu, T. C. Kobayashi, K. Amaya, and Sh. Endo., *J. Phys. Soc. Jpn.*, 65, 1924, 1996.

72. A. Bourdillon and N. X. Tan Bourdillon, *High Temperature Superconductors: Processing and Science*, Academic Press, 1994.

73. J. M. Williams, J. R. Ferraro, R. J. Thorn, K. Carlson, U. Geiser, H. H. Wang, A. M. Kini, and M.-H. Whangbo, *Organic Superconductors (Including Fullerenes): Synthesis, Structure, Properties, and Theory*, Prentice–Hall, 1992.

74. J. Nagamatsu, N. Nakagawa, T. Muranaka, Y. Zenitani, and J. Akimitsu, *Nature (London)*, 410, 63, 2001.

75. Y. Boguslavsky, G. K. Perkins, X. Qi, L. F. Cohen, and A. D. Caplin, *Nature (London)*, 410, 563, 2001.

76. B. Q. Fu, Y. Feng, G. Yan, Y. Zhao, A. K. Pradhan, C. H. Cheng, P. Ji, X. H. Liu, C. F. Liu, L. Zhou, and K. F. Yau, *J. Appl. Phys.*, 92, 7341, 2002.

HIGH TEMPERATURE SUPERCONDUCTORS

C. N. R. Rao and A. K. Raychaudhuri

The following tables give properties of a number of high temperature superconductors. Table 1 lists the crystal structure (space group and lattice constants) and the critical transition temperature T_c for the more important high temperature superconductors so far studied. Table 2 gives energy gap, critical current density, and penetration depth in the superconducting state. Table 3 gives electrical and thermal properties of some of these materials in the normal state. The tables were prepared in November 1992 and updated in November 1994.

References

1. Ginsburg, D. M., Ed., *Physical Properties of High-Temperature Superconductors*, Vols. I–III, World Scientific, Singapore, 1989–1992.

2. Rao, C. N .R., Ed., *Chemistry of High-Temperature Superconductors*, World Scientific, Singapore, 1991.

3. Shackelford, J. F., *The CRC Materials Science and Engineering Handbook*, CRC Press, Boca Raton, 1992, 98–99 and 122–123.

4. Kaldis, E., Ed., *Materials and Crystallographic Aspects of HT_c-Superconductivity*, Kluwer Academic Publ., Dordrecht, The Netherlands, 1992.

5. Malik, S. K. and Shah, S. S., Ed., *Physical and Material Properties of High Temperature Superconductors*, Nova Science Publ., Commack, N.Y., 1994.

6. Chmaissem, O. et. al., *Physica*, C230, 231–238, 1994.

7. Antipov, E. V. et. al., *Physica*, C215, 1–10, 1993.

TABLE 1. Structural Parameters and Approximate T_c Values of High-Temperature Superconductors

Material	Structure	T_c/K (maximum value)
$La_2CuO_{4+\delta}$	Bmab; $a = 5.355$, $b = 5.401$, c = 13.15 Å	39
$La_{2-x}Sr_x(Ba_x)CuO_4$	I4/mmm; $a = 3.779$, $c = 13.23$ Å	35
$La_2Ca_{1-x}Sr_xCu_2O_6$	I4/mmm; $a = 3.825$, $c = 19.42$ Å	60
$YBa_2Cu_3O_7$	Pmmm; $a = 3.821$, $b = 3.885$, $c = 11.676$ Å	93
$YBa_2Cu_4O_8$	Ammm; $a = 3.84$, $b = 3.87$, $c = 27.24$ Å	80
$Y_2Ba_4Cu_7O_{15}$	Ammm; $a = 3.851$, $b = 3.869$, $c = 50.29$ Å	93
$Bi_2Sr_2CuO_6$	Amaa; $a = 5.362$, $b = 5.374$, $c = 24.622$ Å	10
$Bi_2CaSr_2Cu_2O_8$	A_2aa; $a = 5.409$, $b = 5.420$, $c = 30.93$ Å	92
$Bi_2Ca_2Sr_2Cu_3O_{10}$	A_2aa; $a = 5.39$, $b = 5.40$, $c = 37$ Å	110
$Bi_2Sr_2(Ln_{1-x}Ce_x)_2Cu_2O_{10}$	P4/mmm; $a = 3.888$, $c = 17.28$ Å	25
$Tl_2Ba_2CuO_6$	A_2aa; $a = 5.468$, $b = 5.472$, $c = 23.238$ Å; I4/mmm; $a = 3.866$, $c = 23.239$ Å	92
$Tl_2CaBa_2Cu_2O_8$	I4/mmm; $a = 3.855$, $c = 29.318$ Å	119
$Tl_2Ca_2Ba_2Cu_3O_{10}$	I4/mmm; $a = 3.85$, $c = 35.9$ Å	128
$Tl(BaLa)CuO_5$	P4/mmm; $a = 3.83$, $c = 9.55$ Å	40
$Tl(SrLa)CuO_5$	P4/mmm; $a = 3.7$, $c = 9$ Å	40
$(Tl_{0.5}Pb_{0.5})Sr_2CuO_5$	P4/mmm; $a = 3.738$, $c = 9.01$ Å	40
$TlCaBa_2Cu_2O_7$	P4/mmm; $a = 3.856$, $c = 12.754$ Å	103
$(Tl_{0.5}Pb_{0.5})CaSr_2Cu_2O_7$	P4/mmm; $a = 3.80$, $c = 12.05$ Å	90
$TlSr_2Y_{0.5}Ca_{0.5}Cu_2O_7$	P4/mmm; $a = 3.80$, $c = 12.10$ Å	90
$TlCa_2Ba_2Cu_3O_8$	P4/mmm; $a = 3.853$, $c = 15.913$ Å	110
$(Tl_{0.5}Pb_{0.5})Sr_2Ca_2Cu_3O_9$	P4/mmm; $a = 3.81$, $c = 15.23$ Å	120
$TlBa_2(La_{1-x}Ce_x)_2Cu_2O_9$	I4/mmm; $a = 3.8$, $c = 29.5$ Å	40
$Pb_2Sr_2La_{0.5}Ca_{0.5}Cu_3O_8$	Cmmm; $a = 5.435$, $b = 5.463$, $c = 15.817$ Å	70
$Pb_2(Sr,La)_2Cu_2O_6$	$P2_12_12$; $a = 5.333$, $b = 5.421$, $c = 12.609$ Å	32
$(Pb,Cu)Sr_2(La,Ca)Cu_2O_7$	P4/mmm; $a = 3.820$, $c = 11.826$ Å	50
$(Pb,Cu)(Sr,Eu)(Eu,Ce)Cu_2O_x$	I4/mmm; $a = 3.837$, $c = 29.01$ Å	25
$Nd_{2-x}Ce_xCuO_4$	I4/mmm; $a = 3.95$, $c = 12.07$ Å	30
$Ca_{1-x}Sr_xCuO_2$	P4/mmm; $a = 3.902$, $c = 3.35$ Å	110
$Sr_{1-x}Nd_xCuO_2$	P4/mmm; $a = 3.942$, $c = 3.393$ Å	40
$Ba_{0.6}K_{0.4}BiO_3$	Pm3m; $a = 4.287$ Å	31
Rb_2CsC_{60}	$a = 14.493$ Å	31
$NdBa_2Cu_3O_7$	Pmmm; $a = 3.878$, $b = 3.913$, $c = 11.753$	58
$SmBaSrCu_3O_7$	I4/mmm; $a = 3.854$, $c = 11.62$	84
$EuBaSrCu_3O_7$	I4/mmm; $a = 3.845$, $c = 11.59$	88
$GdBaSrCu_3O_7$	I4/mmm; $a = 3.849$, $c = 11.53$	86
$DyBaSrCu_3O_7$	Pmmm; $a = 3.802$, $b = 3.850$, $c = 11.56$	90
$HoBaSrCu_3O_7$	Pmmm; $a = 3.794$, $b = 3.849$, $c = 11.55$	87
$ErBaSrCu_3O_7$ (multiphase)	Pmmm; $a = 3.787$, $b = 3.846$, $c = 11.54$	82
$TmBaSrCu_3O_7$ (multiphase)	Pmmm; $a = 3.784$, $b = 3.849$, $c = 11.55$	88
$YBaSrCu_3O_7$	Pmmm; $a = 3.803$, $b = 3.842$, $c = 11.54$	84
$HgBa_2CuO_4$	I4/mmm; $a = 3.878$, $c = 9.507$	94
$HgBa_2CaCu_2O_6$ (annealed in O_2)	I4/mmm; $a = 3.862$, $c = 12.705$	127
$HgBa_2Ca_2Cu_3O_8$	Pmmm; $a = 3.85$, $c = 15.85$	133
HgBa2Ca3Cu4O10	Pmmm; a = 3.854, c = 19.008	126

TABLE 2. Superconducting Properties

J_c (0): Critical current density extrapolated to 0 K
λ_{ab}: Penetration depth in a-b plane
k_B: Boltzmann constant

		Energy gap (Δ)			
Material	Form	$2\Delta_{pp}/k_B T_c$ *	$2\Delta_{fit}/k_B T_c$ †	$10^{-6} \times J_c$ (0)/A cm^{-2}	λ_{ab}/Å
Y Ba$_2$Cu$_3$O$_7$	Single Crystal	5–6	4–5	30 (film)	1400
Bi$_2$Sr$_2$CaCu$_2$O$_8$	Single Crystal	8–9	5.5–6.5	2	2700
Tl$_2$Ba$_2$CaCu$_2$O$_8$	Ceramic	6–7	4–6	10 (film, 80 K)	2000
La$_{2-x}$Sr$_x$CuO$_4$, x = 0.15	Ceramic	7–9	4–6		
Nd$_{2-x}$Ce$_x$CuO$_4$	Ceramic	8	4–5	0.2 (film)	

* Obtained from peak to peak value.
† Obtained from fit to BCS-type relation.

TABLE 3. Normal State Properties

ρ_{ab}: Resistivity in the a-b plane
ρ_c: Resistivity along the c axis
+ve: ρ_c has positive temperature coefficient of resistivity
−ve: ρ_c has negative temperature coefficient of resistivity
n_H: Hall density
k: Thermal conductivity
in plane: Along a-b plane
out of plane: Perpendicular to a-b plane

		ρ_{ab}/$\mu\Omega$ cm		ρ_c/mΩ cm		$10^{-21} \times n_H$/cm^{-3}		k/(mW/cm K) at 300 K	
Material	Form	300 K	100 K	300 K	dρ_c/dT	300 K	100 K	in plane	out of plane
YBa$_2$Cu$_3$O$_7$	Single crystal	110	35	5	+ve	11–16	4–6	120	3
	Film	200–300	60–100			5–9	2–3		
YBa$_2$Cu$_4$O$_8$	Single crystal	75	20	10	−ve	14			
	Film	100–200	20–50			22	17		
Bi$_2$Sr$_2$CuO$_6$	Single crystal	300	150	5000	−ve	6	5		
Bi$_2$Sr$_2$CaCu$_2$O$_8$	Single crystal	150	50	>1000	−ve	4	3	60	8
Tl$_2$Ba$_2$CuO$_6$	Single crystal	300–400	50–75	200–300	+ve	3.1	2.5		
Tl$_2$Ba$_2$Ca$_2$Cu$_3$O$_{10}$	Ceramic	***	**				≈ 2*		
La$_{2-x}$Sr$_x$CuO$_4$, x = 0.12	Single crystal	900	350	200	+ve for T >225 K	2.5			
La$_{2-x}$Sr$_x$CuO$_4$, x = 0.20	Single crystal	400	200	80	+ve for T >150 K	10		50 (for x = 0.04)	20
	Film	400	160			8.4	6.3		
Nd$_{2-x}$Ce$_x$CuO$_4$, x = 0.17	Single crystal	500	275			53	17		
x = 0.15	Film	140–180	35			32	11	250 (for x = 0.15)	

* At 200 K
** ρ ~0.4 mΩ cm at 120 K
*** ρ ~1.5 mΩ cm at 300 K

ORGANIC SUPERCONDUCTORS

H.P.R. Frederikse

Although the vast majority of organic compounds are insulators, a small number of organic solids show considerable electrical conductivity. Some of these materials appear to be superconductors. The superconducting organics fall primarily into two groups: those containing fulvalenes (pentagonal rings containing sulfur or selenium) and those based on fullerenes, involving the nearly spherical cluster C_{60}.

The transition temperatures T_c of the fulvalene derivatives are shown in Table 1. The abbreviations of the various molecular groups are listed in Table 2 and their chemical structures are depicted in Figure 1. Most of the T_c's are between 1 and 12 K. Several of the compounds only show superconductivity under pressure.

The fullerenes are A_3C_{60} compounds, where A represents a single or a combination of alkali atoms. The C_{60} cluster is shown in Figure 2a, while Figure 2b illustrates how the alkali atoms fit into the A_3C_{60} molecule to form the A15 crystallographic structure. Their superconducting transition temperatures range from 8 to 31.3 K (see Table 3).

References

1. Ishigura, T. and Yamaji, K., *Organic Superconductors*, Springer-Verlag, Berlin, 1990.
2. Williams, Jack M. et al., *Organic Superconductors (Including Fullerenes)*, Prentice Hall, Englewood Cliffs, N.J., 1992.
3. *The Fullerenes*, Ed.: Krato, H. W., Fisher, J. E., and Cox, D. E., Pergammon Press, Oxford, 1993.
4. Schluter, M. et al., in *The Fullerenes* (Ref. 3), p. 303.

TABLE 1. Critical Pressure and Maximum Critical Temperature of Organic Superconductors

Material	P_c/kbar	T_c/K	Material	P_c/kbar	T_c/K
$(TMTSF)_2PF_6$	6.5	1.2	β-$(ET)_2IBr_2$	0	2.8
$(TMTSF)_2AsF_6$	9	1.3	β-$(ET)_2AuI_2$	0	4.8
$(TMTSF)_2SbF_6$	11	0.4	$(ET)_4Hg_{2.89}Cl_8$	0	4.2
$(TMTSF)_2TaF_6$	12	1.4	$(ET)_4Hg_{2.89}Br_8$	12	1.8
$(TMTSF)_2ClO_4$	0	1.4	$(ET)_3Cl_2(H_2O)_2$	16	2
$(TMTSF)_2ReO_4$	9.5	1.3	κ-$(ET)_2Cu(NCS)_2$	0	10.4
$(TMTSF)_2FSO_3$	5	3	κ-$(d$-$ET)_2Cu(NCS)_2$	0	11.4
$(ET)_4(ReO_4)_2$	4.5	2	$(DMET)_2Au(CN)_2$	1.5	0.9
β_L-$(ET)_2I_3$	0	1.4	$(DMET)_2AuI_2$	5	0.6
β_H-$(ET)_2I_3$	0	8.1	$(DMET)_2AuBr_2$	0	1.9
γ-$(ET)_3I_{2.5}$	0	2.5	$(DMET)_2AuCl_2$	0	0.9
ε-$(ET)_2I_3(I_8)_{0.5}$	0	2.5	$(DMET)_2I_3$	0	0.6
α-$(ET)_2I_3I_2$-doped	0	3.3	$(DMET)_2IBr_2$	0	0.7
α_t-$(ET)_2I_3$	0	8	$(MDT$-$TTF)_2AuI_2$	0	3.5
$\varepsilon \rightarrow \beta$-$(ET)_2I_3$ [a]	0	6	$TTF[Ni(dmit)_2]_2$	2	1.6 [b]
θ-$(ET)_2I_3$	0	3.6	$TTF[Pd(dmit)_2]_2$	20	6.5
κ-$(ET)_2I_3$	0	3.6	$(CH_3)_4N[Ni(dmit)_2]_2$	7	5

[a] Converted form ε-type to β-type by thermal treatment.
[b] For 7 kbar.

From Ishigura, T. and Yamaji, K., *Organic Superconductors*, Springer-Verlag, Berlin, 1990. With permission.

TABLE 2. List of Symbols and Abbreviations

TTF	tetrathiafulvalene
TMTSF	tetramethyltetraselenafulvalene
BEDT-TTF or "ET"	bis(ethylenedithio)tetrathiafulvalene
MDT-TTF	methylenedithiotetrathiafulvalene
DMET	[dimethyl(ethylenedithio)diselenadithiafulvalene]
dmit	4,5-dimercapto-1,3-dithiole-2-thione
T_c	transition temperature to superconducting state
P_c	minimum pressure required for superconducting transition

TMTSF

Tetramethyltetraselenafulvalene

TTF

Tetrathiafulvalene

BEDT – TTF or ET

Bis(ethylenedithio)tetrathiafulvalene

DMET

Dimethyl(ethylenedithio)diselenadithiafulvalene

MDT – TTF

Methylenedithiotetrathiafulvalene

M=Ni, Pd, Pt

M(dmit)$_2^{2-}$

Ligand is 4,5-dimercapto-1.3-dithiole-2-thione

FIGURE 1. Structures of various donor molecules and acceptor species.

a

b

FIGURE 2. (a) C_{60} cluster placed in a fcc lattice. Each crystal axis crosses a double bond shared by two hexagons. (b) A hypothetical A_3C_{60} with the A15 structure. The structure can be seen to be an ordered defect structure of A_6C_{60}.

TABLE 3. Unit Cell and T_c for FCC-A_3 C_{60}

	Lattice parameter(s) (Å)	T_c/K
$Na_2Rb_{0.5}Cs_{0.5}C_{60}$	14.148(3)	8.0
Na_2CsC_{60} No. 1[a]	14.132(2)	10.5
Na_2CsC_{60} No. 2[a]	14.176(9)	14.0
K_3C_{60}	14.253(3)	19.3
K_2RbC_{60}	14.299(2)	21.8
Rb_2KC_{60} No. 1[a]	14.336(1)	24.4
Rb_2KC_{60} No. 2[a]	14.364(5)	26.4
Rb_3C_{60}	14.436(2)	29.4
Rb_2CsC_{60}	14.493(2)	31.3

[a] Samples labeled No. 1 and No. 2 have the same nominal composition.

From Schluter, M et al., *The Fullerenes*, Ed.: Krato, H.W., Fisher, J.E., and Cox, D.E., Pergamon Press, Oxford, 1993. With permission.

PROPERTIES OF SEMICONDUCTORS

L.I. Berger

The term *semiconductor* is applied to a material in which electric current is carried by electrons or holes and whose electrical conductivity, when extremely pure, rises exponentially with temperature and may be increased from its low "intrinsic" value by many orders of magnitude by "doping" with electrically active impurities.

Semiconductors are characterized by an energy gap in the allowed energies of electrons in the material which separates the normally filled energy levels of the *valence band* (where "missing" electrons behave like positively charged current carriers "holes") and the *conduction band* (where electrons behave rather like a gas of free negatively charged carriers with an effective mass dependent on the material and the direction of the electrons' motion). This energy gap depends on the nature of the material and varies with direction in anisotropic crystals. It is slightly dependent on temperature and pressure, and this dependence is usually almost linear at normal temperatures and pressures.

Data are presented in five tables. Table 1 lists the main crystallographic and semiconducting properties of a large number of semiconducting materials in three main categories: "Tetrahedral Semiconductors" in which every atom is tetrahedrally coordinated to four nearest neighbor atoms (or atomic sites) as for example in the diamond structure; "Octahedral Semiconductors" in which every atom is octahedrally coordinated to six nearest neighbor atoms—as for example the halite structure; and "Other Semiconductors."

Table 2 gives electrical, magnetic, and optical properties, while Tables 3 and 4 give more details on the semiconducting properties and band structure of the most common semiconductors. Table 5 lists semiconducting minerals with typical resistivity ranges.

TABLE 1. Physico-Chemical Properties of Semiconductors (Listed by Crystal Structure)

Substance	Molecular weight	Average atomic weight	Lattice parameters (Å, room temp.)	Density (g/cm³)	Melting point (K)	Microhardness, N/mm² (M-Mohs Scale)	Specific heat, J/kg·K (300 K)	Debye temp. (K)	Coefficient of thermal linear expansion [10⁻⁶ K⁻¹ (300K)]	Thermal conductivity [mW/cm·K (300K)]
1.1. Adamantine Semiconductors										
1.1.1. Diamond Structure Elements (Strukturbericht symbol A4, Space Group Fd3m-O$_h^7$)										
C (Diamond)	12.01	3.56683	3.513	≈4713 (12.4 GPa) Transition to graphite > 980	10 (M)	471.5	2340	1.18	99000 23200(II 13600(II.
Si	28.09	5.43072	2.329	1687	11270	702	645	2.6	12
Ge	72.64	5.65754	5.323	1211.35	7644	321.9	374	5.8	6
α-Sn	118.71	6.4912	5.769	505.1 (Tr. 286.4)		213	230	5.4 (220 K)	
1.1.2. Sphalerite (Zinc Blende) Structure Compounds (Strukturbericht symbol B3 Space Group F$\bar{4}$ 3m-T$_d^2$)										
I-VII Compounds										
CuF	82.54	41.27	4.255		1181					
CuCl	98.99	49.49	5.4057	3.53	695	2.3 (M)	490	240	12.1	8
CuBr	143.45	71.73	5.6905	4.98	770	2.5 (M)	381	207	15.4	12
CuI	190.45	95.23	6.60427	5.63	878	192	276	181	19.2	16
AgBr	187.77	93.89		6.473	>1570 (Tr. 410)	2.5 (M)	270			
AgI	234.77	117.39	6.502	5.67	831	2.5 (M)	232	134	−2.5	4
II-VI Compounds										
BeS	41.08	20.54	4.865	2.36	dec.					
BeSe	87.97	43.99	5.139	4.315						
BeTe	136.61	68.31	5.626	5.090						
BePo	(2318)	(109)	5.838	7.3						
ZnO	81.39	40.69	4.63	5.675	2248	5.0 (M)	494	416	2.9	2
ZnS	97.46	48.72	5.4093	4.079	2100 (Tr. 1295)	1780	472	530	6.36	2

Substance	Molecular weight	Average atomic weight	Lattice parameters (Å, room temp.)	Density (g/cm³)	Melting point (K)	Microhardness, N/mm² (M-Mohs Scale)	Specific heat, J/kg·K (300 K)	Debye temp. (K)	Coefficient of thermal linear expansion [10^{-6} K^{-1} (300K)]	Thermal conductivity [mW/cm·K (300K)]
ZnSe	144.34	72.17	5.6676	5.42	1790	1350	339	400	7.2	140
ZnTe	192.99	96.5	6.101	6.34	1568	900	264	223	8.19	108
ZnPo	(274)	(137)	6.309							
CdS	144.48	72.24	5.832	4.826	1750	1250	330	219	4.7	200
CdSe	191.37	95.68	6.05	5.674	1512	1300	255	181	3.8	90
CdTe	240.01	120.00	6.477	5.86	1365	600	205	200	4.9	58.5
CdPo	(321)	(161)	6.665							
HgS	232.66	116.33	5.8517	7.73	1820	3 (M)	210			
HgSe	279.55	139.78	6.084	8.25	1070	2.5 (M)	178	151	5.46	10
HgTe	328.19	164.10	6.4623	8.17	943	300	164	242	4.6	20

III-V Compounds

Substance	Molecular weight	Average atomic weight	Lattice parameters (Å, room temp.)	Density (g/cm³)	Melting point (K)	Microhardness, N/mm² (M-Mohs Scale)	Specific heat, J/kg·K (300 K)	Debye temp. (K)	Coefficient of thermal linear expansion [10^{-6} K^{-1} (300K)]	Thermal conductivity [mW/cm·K (300K)]
BN	24.82	12.41	3.615	3.49	3239	10 (M)	793	≈1900		200
BP(L.T.)	41.78	20.87	4.538	2.9	1398 (dec)	37000		≈980		
BAs	85.73	42.87	4.777		≈2300	19000		≈625		
AlP	57.95	28.98	5.451	2.42	≈2100	5.5 (M)		588		920
AlAs	101.90	50.95	5.6622	3.81	2013	5000		417	3.5	840
AlSb	148.74	74.37	6.1355	4.218	1330	4000		292	4.2	600
GaP	100.70	50.35	5.4905	4.13	1750	9450		446	5.3	752
GaAs	144.64	72.32	5.65315	5.316	1510	7500		344	5.4	560
GaSb	191.48	95.74	6.0954	5.619	980	4480	320	265	6.1	270
InP	145.79	72.90	5.86875	4.787	1330	4100		321	4.6	800
InAs	189.74	94.87	6.05838	5.66	1215	3300	268	249	4.7	290
InSb	236.58	118.29	6.47877	5.775	798	2200	144	202	4.7	160

Other sphalerite structure compounds

Substance	Molecular weight	Average atomic weight	Lattice parameters (Å, room temp.)	Density (g/cm³)	Melting point (K)	Microhardness, N/mm² (M-Mohs Scale)	Specific heat, J/kg·K (300 K)	Debye temp. (K)	Coefficient of thermal linear expansion [10^{-6} K^{-1} (300K)]	Thermal conductivity [mW/cm·K (300K)]
MnS	87.00	43.5	5.011							
MnSe	133.90	66.95	5.82							
β-SiC (3-C SiC)	40.10	20.1	4.348	3.21	3070				2.9	4.9
Ga₂Se₃	376.32	75.26	5.429	4.92	1020	3160			8.9	50
Ga₂Te₃	522.24	104.45	5.899	5.75	1063	2370				47
In₂Te₃(H.T.)	608.44	121.7	6.173	5.8	940	1660				69
MgGeP₂	158.84	39.71	5.652							
ZnSnP₂	246.00	61.5	5.65		1200					
ZnSnAs₂(H.T.)	333.90	82.38	5.851	5.53	1050					76
ZnSnSb₂	427.56	106.89	6.281	5.67	870	2500				76

1.1.3. Wurtzite (Zincite) Structure Compounds (Strukturbericht symbol B4, Space Group P 6₃mc-C⁴₆ᵥ)

I-VII Compounds

Substance	Molecular weight	Average atomic weight	Lattice parameters (Å, room temp.)	Density (g/cm³)	Melting point (K)	Microhardness, N/mm² (M-Mohs Scale)	Specific heat, J/kg·K (300 K)	Debye temp. (K)	Coefficient of thermal linear expansion [10^{-6} K^{-1} (300K)]	Thermal conductivity [mW/cm·K (300K)]
CuCl	99.0	49.5	3.91	6.42		703				
CuBr	143.45	71.73	4.06	6.66		770				
CuI	190.45	95.23	4.31	7.09						
AgI	234.77	117.40	4.580	7.494						

II-VI Compounds

Substance	Molecular weight	Average atomic weight	Lattice parameters (Å, room temp.)	Density (g/cm³)	Melting point (K)	Microhardness, N/mm² (M-Mohs Scale)	Specific heat, J/kg·K (300 K)	Debye temp. (K)	Coefficient of thermal linear expansion [10^{-6} K^{-1} (300K)]	Thermal conductivity [mW/cm·K (300K)]
BeO	25.01	12.51	2.698	4.380		2800				
MgTe	151.9	76.0	4.54	7.39	3.85	≈2800				
ZnO	81.37	40.69	3.24950	5.2069	5.66	2250				600
ZnS	97.43	48.72	3.8140	6.2576	4.1	2100				460

Substance	Molecular weight	Average atomic weight	Lattice parameters (Å, room temp.)		Density (g/cm^3)	Melting point (K)	Microhardness, N/mm^2 (M-Mohs Scale)	Specific heat, J/kg·K (300 K)	Debye temp. (K)	Coefficient of thermal linear expansion [10^{-6} K^{-1} (300K)]	Thermal conductivity [mW/cm·K (300K)]
ZnTe	192.99	46.50	4.27	6.99		1568					
CdS	144.48	72.23	4.1348	6.7490	4.82	1748					401
CdSe	191.37	95.68	4.299	7.010	5.66	1512					316
CdTe	240.01	120.00	4.57	7.47							

III-V Compounds

Substance	Molecular weight	Average atomic weight	Lattice parameters (Å, room temp.)		Density (g/cm^3)	Melting point (K)	Microhardness, N/mm^2 (M-Mohs Scale)	Specific heat, J/kg·K (300 K)	Debye temp. (K)	Coefficient of thermal linear expansion [10^{-6} K^{-1} (300K)]	Thermal conductivity [mW/cm·K (300K)]
BP(H.T.)	41.79	20.90	3.562	5.900							
AlN	40.99	20.50	3.111	4.978	3.26	≈2500					823
GaN	83.73	41.87	3.190	5.189	6.10	1500					656
InN	128.83	64.42	3.533	5.693	6.88	1200					556

Other wurtzite structure compounds

Substance	Molecular weight	Average atomic weight	Lattice parameters (Å, room temp.)		Density (g/cm^3)	Melting point (K)	Microhardness, N/mm^2 (M-Mohs Scale)	Specific heat, J/kg·K (300 K)	Debye temp. (K)	Coefficient of thermal linear expansion [10^{-6} K^{-1} (300K)]	Thermal conductivity [mW/cm·K (300K)]
MnS	87.00	43.5	3.985	6.45	3.248						
MnSe	133.90	66.95	4.12	6.72							
SiC	40.10	20.1	3.076	5.048							
MnTe	182.54	91.27	4.078	6.701							
Al$_2$S$_3$	150.14	30.03	3.579	5.829	2.55	1400					
Al$_2$Se$_3$	290.84	58.17	3.890	6.30	3.91	1250					

1.1.4. Chalcopyrite Structure Compounds (Strukturbericht symbol E1$_1$, Space Group I $\bar{4}$ 2d-D$_{24}^{12}$)

I-III-VI$_2$ Compounds

Substance	Molecular weight	Average atomic weight	Lattice parameters (Å, room temp.)		Density (g/cm^3)	Melting point (K)	Microhardness, N/mm^2 (M-Mohs Scale)	Specific heat, J/kg·K (300 K)	Debye temp. (K)	Coefficient of thermal linear expansion [10^{-6} K^{-1} (300K)]	Thermal conductivity [mW/cm·K (300K)]
CuAlS$_2$	154.65	38.66	5.323	10.44	3.47	2500					
CuAlSe$_2$	248.45	62.11	5.617	10.92	4.70	2260					
CuAlTe$_2$	345.73	86.43	5.976	11.80	5.50	2550					
CuGaS$_2$	197.39	49.53	5.360	10.49	4.35	2300					
CuGaSe$_2$	291.19	72.80	5.618	11.01	5.56	1970	4200		275	5.4	42
CuGaTe$_2$	388.47	97.12	6.013	11.93	5.99	2400	3500			6.9	27
CuInS$_2$	242.49	60.62	5.528	11.08	4.75	1400	2550				
CuInSe$_2$	336.29	84.07	5.785	11.56	5.77	1600	2050			6.6	37
CuInTe$_2$	433.57	108.39	6.179	12.365	6.10	1660	400		195	7.1	49
CuTlS$_2$	322.05	83.01	5.580	11.17	6.32						
CuTlSe$_2$(L.T.)	425.85	106.46	5.844	11.65	7.11	900					
CuFeS$_2$	183.51	45.88	5.29	10.32	4.088	1135					
CuFeSe$_2$	277.31	69.33				850					
CuLaS$_2$	266.58	66.65	5.65	10.86							
AgAlS$_2$	198.97	49.74	5.707	10.28	3.94						
AgAlSe$_2$	292.77	73.19	5.968	10.77	5.07	1220					
AgAlTe$_2$	390.05	97.51	6.309	11.85	6.18	1000					
AgGaS$_2$	241.71	60.43	5.755	10.28	4.72						
AgGaSe$_2$	335.51	83.88	5.985	10.90	5.84	1120	4400				
AgGaTe$_2$	432.79	108.2	6.301	11.96	6.05	990	1800		212		10
AgInS$_2$(L.T.)	286.87	71.70	5.828	11.19	5.00		2250				
AgInSe$_2$	380.61	95.15	6.102	11.69	5.81	1053	1850				30
AgInTe$_2$	477.89	119.47	6.42	12.59	6.12	965				9.49, 0.69	
AgFeS$_2$	227.83	56.96	5.66	10.30	4.53						

II-IV-V$_2$ Compounds

Substance	Molecular weight	Average atomic weight	Lattice parameters (Å, room temp.)		Density (g/cm^3)	Melting point (K)	Microhardness, N/mm^2 (M-Mohs Scale)	Specific heat, J/kg·K (300 K)	Debye temp. (K)	Coefficient of thermal linear expansion [10^{-6} K^{-1} (300K)]	Thermal conductivity [mW/cm·K (300K)]
ZnSiP$_2$	155.40	38.85	5.400	10.441	3.39	1640	1100				
ZnGeP$_2$	199.90	49.98	5.465	10.771	4.17	1295	8100				180
ZnSnP$_2$	246.00	61.5					6500				
CdSiP$_2$	202.43	50.61	5.678	10.431	4.00	≈1470	10500		282		

Substance	Molecular weight	Average atomic weight	Lattice parameters (Å, room temp.)		Density (g/cm³)	Melting point (K)	Microhardness, N/mm² (M-Mohs Scale)	Specific heat, J/kg·K (300 K)	Debye temp. (K)	Coefficient of thermal linear expansion [10⁻⁶ K⁻¹ (300K)]	Thermal conductivity [mW/cm·K (300K)]
CdGeP$_2$	246.94	61.74	5.741	10.775	4.48	1049	5650				110
CdSnP$_2$	243.03	73.26	5.900	11.518		840	5000		195		140
ZnSiAs$_2$	242.20	60.55	5.61	10.88	4.70	1311	9200				
ZnGeAs$_2$	287.80	71.95	5.672	11.153	5.32	1150	6800		263		110
ZnSnAs$_2$	333.90	83.48	5.8515	11.704	5.53	1048	4550		271		150
CdSiAs$_2$	290.34	72.58	5.884	10.882		>1120	6850				
CdGeAs$_2$	334.83	83.71	5.9427	11.2172	5.60	938	4700				48
CdSnAs$_2$	380.93	95.23	6.0944	11.9182	5.72	880	3450				40

1.1.5. Other Ternary Semiconductors with Tetrahedral Coordination

I$_2$-IV-VI$_3$ Compounds

Substance	Molecular weight	Average atomic weight	Lattice parameters (Å, room temp.)		Density (g/cm³)	Melting point (K)	Microhardness, N/mm² (M-Mohs Scale)	Specific heat, J/kg·K (300 K)	Debye temp. (K)	Coefficient of thermal linear expansion [10⁻⁶ K⁻¹ (300K)]	Thermal conductivity [mW/cm·K (300K)]
Cu$_2$SiS$_3$(H.T.)	251.36	41.89	3.684	6.004	3.81	1200					23
Cu$_2$SiS$_3$(L.T.)			5.290	10.156	3.63						
Cu$_2$SiTe$_3$	537.98	89.66	5.93		5.47						
Cu$_2$GeS$_3$(H.T.)	295.88	49.31	5.317		4.45	1210	4550	510	254	7.2	12
Cu$_2$GeS$_3$(L.T.)			5.327	5.215	4.46						
Cu$_2$GeSe$_3$	436.56	72.76	5.589	5.485	5.57	1030	3840	340	168	8.4	24
Cu$_2$GeTe$_3$	582.51	97.09	5.958	5.935	5.92		2890				130
Cu$_2$SnS$_3$	341.98	57.00	5.436		5.02	1110	2770	440	214	7.8	28
CuSnSe$_3$	482.66	80.44	5.687		5.94	960	2510	310	148	8.9	35
Cu$_2$SnTe$_3$	628.61	104.77	6.048		6.51	680	1970				144
Ag$_2$GeSe$_3$	525.21	87.54				810					
Ag$_2$SnSe$_3$	571.31	95.22									
Ag$_2$GeTe$_3$	671.13	111.86				600					
Ag$_2$SnTe$_3$	717.23	119.54									

I$_3$-V -VI$_4$-Compounds

Substance	Molecular weight	Average atomic weight	Lattice parameters (Å, room temp.)		Density (g/cm³)	Melting point (K)	Microhardness, N/mm² (M-Mohs Scale)	Specific heat, J/kg·K (300 K)	Debye temp. (K)	Coefficient of thermal linear expansion [10⁻⁶ K⁻¹ (300K)]	Thermal conductivity [mW/cm·K (300K)]
Cu$_3$PS$_4$	349.85	40.73	7.44	6.19							
Cu$_3$AsS$_4$	393.79	49.22	6.43	6.14	4.37	931				3.2	30.2
Cu$_3$AsSe$_4$	581.37	72.67	5.570	10.957	5.61	733			169	9.5	19
Cu$_3$SbS$_4$	440.64	55.08	5.38	16.76	4.90	830					
Cu$_3$SbSe$_4$	628.22	78.53	5.654	11.256	6.0	700			131	12.4	14.6

I-IV$_2$-V$_3$ Compounds

Substance	Molecular weight	Average atomic weight	Lattice parameters (Å, room temp.)		Density (g/cm³)	Melting point (K)	Microhardness, N/mm² (M-Mohs Scale)	Specific heat, J/kg·K (300 K)	Debye temp. (K)	Coefficient of thermal linear expansion [10⁻⁶ K⁻¹ (300K)]	Thermal conductivity [mW/cm·K (300K)]
CuSi$_2$P$_3$	212.64	35.44	5.25								
CuGe$_2$P$_3$	301.65	50.28	5.375		4.318	1113	8500	429		8.21	37.6
AgGe$_2$P$_3$	345.97	57.66				1015	6150				

1.1.6. "Defect Chalcopyrite" Structure Compounds (Strukturbericht symbol E3, Space Group I $\bar{4}$-S$_4^2$)

Substance	Molecular weight	Average atomic weight	Lattice parameters (Å, room temp.)		Density (g/cm³)	Melting point (K)	Microhardness, N/mm² (M-Mohs Scale)	Specific heat, J/kg·K (300 K)	Debye temp. (K)	Coefficient of thermal linear expansion [10⁻⁶ K⁻¹ (300K)]	Thermal conductivity [mW/cm·K (300K)]
ZnAl$_2$Se$_4$	435.18	62.17	5.503	10.90	4.37						
ZnAl$_2$Te$_4$(?)	629.74	84.96	5.904	12.05	4.95						
ZnGa$_2$S$_4$(?)	333.06	47.58	5.274	10.44	3.80						
ZnGa$_2$Se$_4$(?)	520.66	74.38	5.496	10.99	5.21						
ZnGa$_2$Te$_4$(?)	715.22	102.17	5.937	11.87	5.67						
ZnIn$_2$Se$_4$	610.86	87.27	5.711	11.42	5.44	1250					
ZnIn$_2$Te$_4$	805.42	115.06	6.122	12.24	5.83	1075					
CdAl$_2$S$_4$	294.61	42.09	5.564	10.32	3.06						
CdAl$_2$Se$_4$	482.21	68.89	5.747	10.68	4.54						
CdAl$_2$Te$_4$(?)	676.77	97.68	6.011	12.21	5.10						

Substance	Molecular weight	Average atomic weight	Lattice parameters (Å, room temp.)		Density (g/cm³)	Melting point (K)	Microhardness, N/mm² (M-Mohs Scale)	Specific heat, J/kg·K (300 K)	Debye temp. (K)	Coefficient of thermal linear expansion [10⁻⁶ K⁻¹ (300K)]	Thermal conductivity [mW/cm·K] (300K)]
CdGa$_2$S$_4$	380.09	54.30	5.577	10.08	4.03						
CdGa$_2$Se$_4$	567.69	81.10	5.743	10.73	5.32						
CdGa$_2$Te$_4$	762.25	108.89	6.093	11.81	5.77						
CdIn$_2$Te$_4$	852.45	121.78	6.205	12.41	5.9	1060					
HgAl$_2$S$_4$	382.79	54.68	5.488	10.26	4.11						
HgAl$_2$Se$_4$	570.39	82.48	5.708	10.74	5.05						
HgAl$_2$Te$_4$(?)	764.48	109.28	6.004	12.11	5.81						
HgGa$_2$S$_4$	468.27	66.90	5.507	10.23	5.00						
HgGa$_2$Se$_4$	655.87	93.70	5.715	10.78	6.18						
HgIn$_2$Se$_4$	746.07	106.58	5.764	11.80	6.3	1100					
HgIn$_2$Te$_4$(?)	940.63	134.38	6.186	12.37	6.3	980					

1.1.7. Other Adamantine Compounds

Substance	Molecular weight	Average atomic weight	Lattice parameters (Å, room temp.)		Density (g/cm³)	Melting point (K)					
α–SiC	40.10	20.10	3.0817	15.12	3.21	3070					
Hg$_5$Ga$_2$Te$_8$	2163.19	144.21	6.235								
Hg$_5$In$_2$Te$_8$	2253.39	150.23	6.328								
CdIn$_2$Se$_4$	657.89	93.98	a = c = 5.823								

1.2. Octahedral Semiconductors

1.2.1. Halite Structure Semiconductors (Strukturbericht symbol B1, Space Group Fm3m-O$_h^5$)

Substance	Molecular weight	Average atomic weight	Lattice parameters (Å, room temp.)	Density (g/cm³)	Melting point (K)					Thermal conductivity [mW/cm·K] (300K)]
GeTe	200.21	100.10	5.98	6.14						
SnSe	197.67	98.83	6.020		1133					
SnTe	246.31	123.15	6.313	6.45	1080 (max)					91
PbS	239.3	119.63	5.9362	7.61	1390					23
PbSe	286.2	143.08	6.1243	8.15	1340					17
PbTe	334.8	167.4	6.454	8.16	1180					23

1.2.2. Selected Other Binary Halites

Substance	Molecular weight	Average atomic weight	Lattice parameters (Å, room temp.)	Density (g/cm³)	Melting point (K)					Thermal conductivity [mW/cm·K] (300K)]
BiSe	287.94	143.97	5.99	7.98	880					
BiTe	336.58	168.29	6.47							
EuSe	230.92	115.46	6.191		2300					2.4
GdSe	236.21	118.11	5.771		2400					
NiO	74.69	37.35	4.1684	6.6	2260					
CdO	128.41	64.21	4.6953		1700					7
SrS	119.69	59.84	6.0199	3.643	3000					

1.3. Other Semiconductors

1.3.1. Antifluorite Structure Compounds (Fm3m–O$_h^5$)

Substance	Molecular weight	Average atomic weight	Lattice parameters (Å, room temp.)	Density (g/cm³)	Melting point (K)				Coefficient of thermal linear expansion [10⁻⁶ K⁻¹ (300K)]	Thermal conductivity [mW/cm·K] (300K)]
Mg$_2$Si	76.70	25.57	6.338	1.88	1375				11.5	
Mg$_2$Ge	121.22	40.4	6.380	3.08	1388				15.0	
Mg$_2$Sn	167.32	55.77	6.765	3.53	1051				9.9	92
Mg$_2$Pb	225.81	85.27	6.836	5.1	823				10.0	

1.3.2. Tetradymite Structure Compounds ($\bar{R}3m$–D$_{3d}^5$)

Substance	Molecular weight	Average atomic weight	Lattice parameters (Å, room temp.)		Density (g/cm³)	Melting point (K)	Microhardness, N/mm² (M-Mohs Scale)	Specific heat, J/kg·K (300 K)			Thermal conductivity [mW/cm·K] (300K)]
Sb$_2$Te$_3$	626.3	125.26	4.25	30.3	6.44	895					24
Bi$_2$Se$_3$	654.84	130.97	4.14	28.7	7.51	979	167				24
Bi$_2$Te$_3$	800.76	160.15	4.38	30.45	7.73	858	155	16			30

Substance	Molecular weight	Average atomic weight	Lattice parameters (Å, room temp.)	Density (g/cm³)	Melting point (K)	Microhardness, N/mm² (M-Mohs Scale)	Specific heat, J/kg·K (300 K)	Debye temp. (K)	Coefficient of thermal linear expansion [10⁻⁶ K⁻¹ (300K)]	Thermal conductivity [mW/cm·K (300K)]

1.3.3. Skutterudite Structure Compounds (Im3–T_h^5)

Substance	MW	AAW	Lattice	Density	MP	Micro	Spec	Debye	Coeff	Thermal
CoP₃	151.85	37.96	7.7073		>1270					
CoAs₃	286.70	71.65	8.2060	6.73	1230					
CoSb₃	424.18	106.05	9.0385		1123			307		50
NiAs₃	283.45	70.86	8.330	6.43						
RhP₃	195.83	48.96	7.9951		>1470					
RhAs₃	327.67	81.92	8.4427		>1270					100
RhSb₃	468.16	117.04	9.2322		1170					
IrP₃	285.14	71.29	8.0151	7.36	>1470					
IrAs₃	416.98	104.25	8.4673	9.12	>1470					90
IrSb₃	557.47	139.37	9.2533	9.35	1170			303		

1.3.4. Selected Multinary Compounds

Substance	MW	AAW	Lattice	Density	MP	Micro	Spec	Debye	Coeff	Thermal
AgSbSe₂	387.54	96.88	5.786	6.60	910					10.5
AgSbTe₂ (or Ag₁₉Sb₂₉Te₅₂)	484.82	121.2	6.078	7.12	830					86
AgBiS₂(H.T.)	380.97	95.24	5.648							
AgBiSe₂(H.T.)	474.77	118.69	5.82							
AgBiTe₂(H.T.)	572.05	143.01	6.155							
Cu₂CdSnS₄	486.43	60.80	5.586	10.83						

1.3.5. Some Elemental Semiconductors

Substance	MW	AAW	Lattice	Density	MP	Micro	Spec	Debye	Coeff	Thermal
B		10.81	4.91	12.6	2.34	2348	9.5 (M)	1277	1370	8.3	600
Se(gray)		78.96	4.36	4.95	4.81	493	350	292.6		(‖C) 17.89 (⊥C) 74.09	(‖C) 45.2 (⊥C) 13.1
Te		127.60	4.45	5.91	6.23	723		196.5		16.8	(‖C) 33.8 (⊥C) 19.7

Note: In the B, Se, Te rows the columns are: Average atomic weight | Lattice parameters | Density | Melting point | Microhardness | Specific heat | Debye temp | Coeff thermal linear expansion | Thermal conductivity.

TABLE 2. Basic Thermodynamic, Electrical, and Magnetic Properties of Semiconductors (Listed by Crystal Structure)

Substance	Heat of formation [kJ/mol (300K)]	Volume compressibility (10⁻¹⁰m²/N)	Static dielectric constant	Atomic magnetic susceptibility (10⁻⁶ cgs)	Index of refraction	Miniumum room temperature energy gap (eV)	Mobility (room temp.) (cm²/V·s) Electrons	Holes	Optical transition	Breakdown voltage kV/mm	Remarks

2.1. Adamantine Semiconductors

2.1.1. Diamond Structure Elements (Strukturbericht symbol A4, Space Group Fd 3m–O_h^7)

Substance	Heat of formation	Vol compr	Static diel	Atomic mag susc	Index of refraction	Min energy gap (eV)	Electrons	Holes	Optical transition	Breakdown voltage	Remarks
C	714.4	18	5.7	−5.88	2.419 (589 nm)	5.4	1800	1400	i*	500	
Si	324	0.306	11.9	−3.9	3.49 (589 nm)	1.12	1900	500	i	30	
Ge	291	0.768	16	−0.12	3.99 (589 nm)	0.67	3800	1820	i		
α-Sn	267.5		24		2.75 (589 nm)	0.0; 0.8	2500	2400			

2.1.2. Sphalerite (Zinc Blende) Structure Compounds (Strukturbericht symbol B3 Space Group F $\bar{4}$ 3m–T_d^2)

I-VII Compounds

Substance	Heat of formation	Vol compr	Static diel	Atomic mag susc	Index of refraction	Min energy gap (eV)	Electrons	Holes	Optical transition	Breakdown voltage	Remarks
CuF											
CuCl	481	0.26	7.9		1.93	3.17			d		Nantokite

Substance	Heat of formation [kJ/mol (300K)]	Volume compressibility (10^{-10} m²/N)	Static dielectric constant	Atomic magnetic susceptibility (10^{-6} cgs)	Index of refraction	Miniumum room temperature energy gap (eV)	Mobility (room temp.) (cm²/V·s)		Optical transition	Breakdown voltage kV/mm	Remarks
							Electrons	Holes			
CuBr	481	0.26	7.9		2.12	2.91			d		
CuI	439	0.27	6.5		2.346	2.95			d		Marshite
AgBr	486		12.4		2.253	2.50	4000		i		Bromirite
AgI	389	0.41	10		2.22	2.22	30		d		Miersite
II-VI Compounds											
BeS						4.17			i		
BeSe						3.61			i		
BeTe						1.45	20		d		
BePo											
ZnO											See 2.1.3.
ZnS	477		8.9	−9.9	2.356	3.54	180	5(400°C)	d		See also 2.1.3.
ZnSe	422		9.2		2.89	2.58	540	28	d		
ZnTe	376		10.4		3.56	2.26	340	100	d		
ZnP											
CdS											See 2.1.3.
CdSe											See 2.1.3.
CdTe	339		7.2		2.50	1.44	1200	50	d		
CdPo											
HgS					2.85		250		d		Metacinnabarite
HgSe	247					2.10 (α)	20000	≈1.5	s		Tiemannite
HgTe	242					−0.06	25000	350	s		Coloradoite
III-V Compounds											
BN	815					4.6					Borazone
BP(L.T.)						≈2.1	500	70			Ignites 470K
BAs						≈1.5					
AlP						2.45	80		i		
AlAs	627		10.9			2.16	1200	420	i		
AlSb	585	0.571	11		3.2	1.60	200–400	550	i		
GaP	635	0.110	11.1	−13.8	3.2	2.24	300	150	i		
GaAs	535	0.771	13.2	−16.2	3.30	1.35	8800	400	d		
GaSb	493	0.457	15.7	−14.2	3.8	0.67	4000	1400	d		
InP	560	0.735	12.4	−22.8	3.1	1.27	4600	150	d		
InAs	477	0.549	14.6	−27.7	3.5	0.36	33000	460	d		
InSb	447	0.442	17.7	−32.9	3.96	0.163	78000	750	d		

* i = indirect, d = direct, s = semimetal.

Other sphalerite structure compounds

Substance	Heat of formation [kJ/mol (300K)]	Volume compressibility (10^{-10} m²/N)	Static dielectric constant	Atomic magnetic susceptibility (10^{-6} cgs)	Index of refraction	Miniumum room temperature energy gap (eV)	Electrons	Holes	Optical transition	Breakdown voltage kV/mm	Remarks
MnS											See also 2.1.3.
MnSe											See also 2.1.3.
β-SiC					2.697	2.3	4000				
Ga_2Te_3	271			−13.5		1.35	50				
In_2Te_3(H.T.)	198			−13.6		1.04	50				
$MgGeP_2$											El–T^{d12}
$ZnSnP_2$						2.1					Same
$ZnSnAs_2$(H. T.)						≈0.7					Same
$ZnSnSb_2$						0.4					Same

2.1.3. Wurtzite (Zincite) Structure Compounds (Strukturbericht symbol B4, Space Group $P 6_3 mc$-C_{6v}^4)
I-VII Compounds

Substance						Energy gap (eV)	Electrons				Remarks
CuCl											
CuBr											
CuI											
AgI						2.63					Iodargirite

II-VI Compounds

Substance						Energy gap (eV)	Electrons				
BeO											
MgTe											
ZnO	−350					3.2	180				
ZnS	−206					3.67					
ZnTe	−163										

Substance	Heat of formation [kJ/mol (300K)]	Volume compressibility (10^{-10} m²/N)	Static dielectric constant	Atomic magnetic susceptibility (10^{-6} cgs)	Index of refraction	Minimum room temperature energy gap (eV)	Mobility (room temp.) (cm²/V·s)		Optical transition	Breakdown voltage kV/mm	Remarks
							Electrons	Holes			
CdS				8.45; 9.12	2.32	2.42	350	40	d		Greenockide
CdSe						1.74	900	50	d		Cadmoselite
CdTe						1.50	650				
III-V Compounds											
BP(H.T.)											
AlN						6.02					
GaN						3.34					
InN						2.0					
Other wurtzite structure compounds											
MnS											
MnSe											
SiC					2.654						
MnTe						≈1.0					
Al₂S₃	426					4.1					
Al₂Se₃	367					3.1					

2.1.4 Chalcopyrite Structure Compounds (Strukturbericht symbol E1$_1$, Space Group I $\bar{4}$ 2d-D$_{2d}^{12}$)

I-III-VI₂ Compounds

Substance	Heat of formation [kJ/mol (300K)]	Volume compressibility (10^{-10} m²/N)	Static dielectric constant	Atomic magnetic susceptibility (10^{-6} cgs)	Index of refraction	Minimum room temperature energy gap (eV)	Electrons	Holes	Optical transition	Breakdown voltage kV/mm	Remarks
CuAlS₂		0.106				2.5					
CuAlSe₂						2.67					
CuAlTe₂						0.88					
CuCaS₂		0.106				2.38					
CuGaSe₂		0.141				0.96, 1.63					
CuGaTe₂		0.227				0.82, 1.0					
CuInS₂		0.141				1.2					
CuInSe₂		0.187				0.86, 0.92					
CuInTe₂		0.278				0.95					
CuTlS₂											
CuTlSe₂ (L.T.)						1.07					
CuFeS₂						0.53					Chalcopyrite
CuFeSe₂						0.16					
CuLaS₂											
AgAlS₂											
AgAlSe₂						0.7					
AgAlTe₂						0.56					
AgGaS₂		0.150				1.66					
AgGaSe₂		0.182				1.1					
AgGaTe₂		0.280				1.9					
AgInS₂ (L.T.)		0.185				1.18					
AgInSe₂		0.238				0.96, 0.52					
AgInTe₂		0.338									
AgFeS₂											

II-IV-V₂ Compounds

Substance	Heat of formation [kJ/mol (300K)]	Volume compressibility (10^{-10} m²/N)	Static dielectric constant	Atomic magnetic susceptibility (10^{-6} cgs)	Index of refraction	Minimum room temperature energy gap (eV)	Electrons	Holes	Optical transition	Breakdown voltage kV/mm	Remarks
ZnSiP₂	312					2.3	1000				
ZnGeP₂	293					2.2					
ZnSnP₂	275					1.45					
CdSiP₂		0.103				2.2	1000				
CdGeP₂	289					1.8					
CdSnP₂	270					1.5					
ZnSiAs₂	290					1.7		50			
ZnGeAs₂	271			−14.4		0.85					
ZnSnAs₂	252			−18.4		0.65		300			Disorders at 910 K
CdSiAs₂		0.143				1.6					
CdGeAs₂	266			−23.4		0.53	70	25			Disorders at 903 K
CdSnAs₂	247		13.7	−21.5		0.26	22000	250			

Substance	Heat of formation [kJ/mol] (300K)	Volume compressibility (10^{-10}m²/N)	Static dielectric constant	Atomic magnetic susceptibility (10^{-6} cgs)	Index of refraction	Miniumum room temperature energy gap (eV)	Mobility (room temp.) (cm²/V·s)		Optical transition	Breakdown voltage kV/mm	Remarks
							Electrons	Holes			
2.1.5. Other Ternary Semiconductors with Tetrahedral Coordination											
II$_2$-IV-VI$_3$ Compounds											
Cu$_2$SiS$_3$ (H.T.)											Wurtzite
Cu$_2$SiS$_3$ (L.T.)											Tetragonal
Cu$_2$SiTe$_3$											Cubic
Cu$_2$GeS$_3$ (H.T.)				−18.7							Cubic
Cu$_2$GeS$_3$ (L.T.)							360				Tetragonal
Cu$_2$GeSe$_3$	211.5			−21.3		0.94	238				Same
Cu$_2$GeTe$_3$	190.2			−23.4							Same
Cu$_2$SnS$_3$				−18.2		0.91	405				Cubic
CuSnSe$_3$				−21.0		0.66	870				Cubic
Cu$_2$SnTe$_3$				−28.4							Cubic
Ag$_2$GeSe$_3$				−29.6		0.91 (77K)					
Ag$_2$SnSe$_3$				−29.5		0.81					
Ag$_2$GeTe$_3$				−31.4		0.25					
Ag$_2$SnTe$_3$				−31.0		0.08					
II$_3$-V-VI$_4$ Compounds											
Cu$_3$PS$_4$											Enargite
Cu$_3$AsS$_4$	269.6			−15.8		1.24					
Cu$_3$AsSe$_4$	161.3			−13.1		0.88					Famatinite
Cu$_3$SbS$_4$				−8.3		0.74					Famatinite
Cu$_3$SbSe$_4$	127.1			−20.5		0.31					
II-IV$_2$-V$_3$ Compounds											
CuSi$_2$P$_3$											El
CuGe$_2$P$_3$		0.12				0.90					El
AgGe$_2$P$_3$											
2.1.6. "Defect Chalcopyrite" Structure Compounds (Strukturbericht symbol E3, Space Group I $\overline{4}$-S$_4^2$)											
ZnAl$_2$Se$_4$											
ZnAl$_2$Te$_4$(?)											
ZnGa$_2$S$_4$ (?)						≈3.4					
ZnGa$_2$Se$_4$(?)						≈2.2					
ZnGa$_2$Te$_4$(?)						1.35					
ZnIn$_2$Se$_4$	206					1.82	35				
ZnIn$_2$Te$_4$	198					1.2					
CdAl$_2$S$_4$											
CdAl$_2$Se$_4$											
CdAl$_2$Te$_4$(?)											
CdGa$_2$S$_4$	256					3.44	60				
CdGa$_2$Se$_4$	216					2.43	33				
CdGa$_2$Te$_4$											
CdIn$_2$Te$_4$	195					(1.26 or 0.9)	4000				
HgAl$_2$S$_4$											
HgAl$_2$Se$_4$											
HgAl$_2$Te$_4$(?)											
HgGa$_2$S$_4$	249					2.84					
HgGa$_2$Se$_4$	204					1.95	400				
HgIn$_2$Se$_4$	196					0.6	290				
HgIn$_2$Te$_4$(?)	188					0.86	200				
2.1.7. Other Adamantine Compounds											
α–SiC		10.2		−6.4	2.67	2.86	400				6H structure
Hg$_5$Ga$_2$Te$_8$											B3 with superlattice
Hg$_5$In$_2$Te$_8$						0.7	2000				B3 with superlattice
CdIn$_2$Se$_4$						1.55					

Substance	Heat of formation [kJ/mol (300K)]	Volume compressibility (10^{-10} m²/N)	Static dielectric constant	Atomic magnetic susceptibility (10^{-6} cgs)	Index of refraction	Miniumum room temperature energy gap (eV)	Mobility (room temp.) (cm²/V·s) Electrons	Holes	Optical transition	Breakdown voltage kV/mm	Remarks
2.2. Octahedral Semiconductors											
2.2.1. Halite Structure Semiconductors (Strukturbericht symbol B1, Space Group $Fm3m$-O_h^5)											
GeTe											
SnSe											
SnTe											
PbS	435					0.5	600	600			
PbSe	393		161			0.37	1000	900			
PbTe	393		280			0.26	1600	600			Altaite
			360			0.25					
2.2.2. Selected Other Binary Halites											
BiSe											
BiTe						0.4					
EuSe											
GdSe						1.8	4				
NiO						2.0 or 3.7	100				
CdO	531					2.5					
SrS						4.1					
2.3. Other Semiconductors											
2.3.1. Antifluorite Structure Compounds ($Fm3m$-O_h^5)											
Mg_2Si	79.08					0.77	405	70			
Mg_2Ge						0.74	520	110			
Mg_2Sn	76.57					0.36	320	260			
Mg_2Pb	52.72					0.1					
2.3.2. Tetradymite Structure Compounds ($R\bar{3}m$-D_{3d}^5)											
Sb_2Te_3						0.3		360			
Bi_2Se_3						0.35	600				
Bi_2Te_3						0.21	1140	680			R3m (166)
2.3.3. Skutterudite Structure Compounds ($Im3$-T_h^5)											
CoP_3						0.43					
$CoAs_3$						0.69		~4000			
$CoSb_3$						0.63	70	~3000			
RhP_3								700			
$RhAs_3$						0.85		~3000			
$RhSb_3$						0.80		~7000			
$IrSb_3$						1.18		1500			
2.3.4. Selected Multinary Compounds											
$AgSbSe_2$						0.58					
$AgSbTe_2$						0.7, 0.27					
(or$Ag_{19}Sb_{29}$ Te_{52})											
$AgBiS_2$ (H.T.)											
$AgBiSe_2$ (H.T.)											
$AgBiTe_2$ (H.T.)											
Cu_2CdSnS_4						1.16	<2				
2.3.5. Some Elemental Semiconductors											
B	397.1			−6.7	3.4	1.55	10				
Se(gray)		6.6 (0.1 GHz)		−22.1	2.5	1.5			5		$P3_121$(152)
Te				−39.5	3.3	0.33	1700	1200			Same

TABLE 3. Semiconducting Properties of Selected Materials

Substance	Minimum energy gap (eV) R.T.	0 K	$dE_g/dT \times 10^4$ eV/°C	$dE_g/dP \times 10^6$ eV·cm²/kg	Density of states electron effective mass m_{da} (m_o)	Electron mobility μ_n (cm²/V·s)	$-x$	Density of states hole effective Mass m_{dp} (m_n)	Hole mobility μ_p (cm²/V·s)	$-x$
Elements										
Si	1.107	1.153	−2.3	−2.0	1.1	1900	2.6	0.56	500	2.3
Ge	0.67	0.744	−3.7	+7.3	0.55	3800	1.66	0.3	1820	2.33
α–Sn	0.08	0.094	−0.5		0.02	2500	1.65	0.3	2400	2.0
Te	0.33				0.08	1100		0.19	560	
III-V Compounds										
AlAs	2.2	2.3				1200			420	
AlSb	1.6	1.7	−3.5	−1.6	0.09	200	1.5	0.4	500	1.8
GaP	2.24	2.40	−5.4	−1.7	0.35	300	1.5	0.5	150	1.5
GaAs	1.35	1.53	−5.0	+9.4	0.068	9000	1.0	0.5	500	2.1
GaSb	0.67	0.78	−3.5	+12	0.050	5000	2.0	0.23	1400	0.9
InP	1.27	1.41	−4.6	+4.6	0.067	5000	2.0		200	2.4
InAs	0.36	0.43	−2.8	+8	0.022	33,000	1.2	0.41	460	2.3
InSb	0.165	0.23	−2.8	+15	0.014	78,000	1.6	0.4	750	2.1
II-VI Compounds										
ZnO	3.2		−9.5	+0.6	0.38	180	1.5			
ZnS	3.54		−5.3	+5.7		180			5(400°C)	
ZnSe	2.58	2.80	−7.2	+6		540			28	
ZnTe	2.26			+6		340			100	
CdO	2.5 ± .1		−6		0.1	120				
CdS	2.4		−5	+3.3	0.165	400		0.8		
CdSe	1.74	1.85	−4.6		0.13	650	1.0	0.6		
GdTe	1.44	1.56	−4.1	+8	0.14	1200		0.35	50	
HgSe	0.30				0.030	20,000	2.0			
HgTe	0.15		−1		0.017	25,000		0.5	350	
Halite Structure Compounds										
PbS	0.37	0.28	+4		0.16	800		0.1	1000	2.2
PbSe	0.26	0.16	+4		0.3	1500		0.34	1500	2.2
PbTe	0.25	0.19	+4	−7	0.21	1600		0.14	750	2.2
Others										
ZnSb	0.50	0.56			0.15	10				1.5
CdSb	0.45	0.57	−5.4		0.15	300			2000	1.5
Bi₂S₃	1.3					200			1100	
Bi₂Se₃	0.27					600			675	
Bi₂Te₃	0.13		−0.95		0.58	1200	1.68	1.07	510	1.95
Mg₂Si		0.77	−6.4		0.46	400	2.5		70	
Mg₂Ge		0.74	−9			280	2		110	
Mg₂Sn	0.21	0.33	−3.5		0.37	320			260	
Mg₃Sb₂		0.32				20			82	
Zn₃As₂	0.93					10	1.1		10	
Cd₃As₂	0.55				0.046	100,000	0.88			
GaSe	2.05		−3.8						20	
GaTe	1.66	1.80	−3.6			14				
InSe	1.8					900				
TlSe	0.57		−3.9		0.3	30		0.6	20	1.5
CdSnAs₂	0.23				0.05	25,000	1.7			
Ga₂Te₂	1.1	1.55	−4.8							
α-In₂Te₂	1.1	1.2			0.7				50	1.1
β-In₂Te₂	1.0								5	
Hg₅In₂Te₈	0.5								11,000	
SnO₂									78	

TABLE 4. Band Properties of Semiconductors

4.1. Data on Valence Bands of Semiconductors (Room Temperature)

Substance	Band curvature effective mass (Expressed as fraction of free electron mass)			Energy separation of "split-off" band (eV)	Measured (light) hole mobility (cm²/V·s)
	Heavy holes	Light holes	"Split-off" band holes		

4.1.1. Semiconductors with Valence Band Maximum at the Center of the Brillouin Zone ("Γ")

Substance	Heavy holes	Light holes	"Split-off" band holes	Energy separation of "split-off" band (eV)	Measured (light) hole mobility (cm²/V·s)
Si	0.52	0.16	0.25	0.044	500
Ge	0.34	0.043	0.08	0.3	1820
Sn	0.3				2400
AlAs					
AlSb	0.4			0.7	550
GaP				0.13	100
GaAs	0.8	0.12	0.20	0.34	400
GaSb	0.23	0.06		0.7	1400
InP				0.21	150
InAs	0.41	0.025	0.083	0.43	460
InSb	0.4	0.015		0.85	750
CdTe	0.35				50
HgTe	0.5				350

4.1.2. Semiconductors with Multiple Band Maxima

Substance	Number of equivalent valleys and direction	Band curvature effective masses		Anistrophy $K = m_L/m_T$	Measured (light) hole mobility (cm²/V·s)
		Longitudinal m_L	Transverse m_T		
PbSe	4 "L" [111]	0.095	0.047	2.0	1500
PbTe	4 "L" [111]	0.27	0.02	10	750
Bi_2Te_3	6	0.207	~0.045	4.5	515

4.2. Data on Conduction Bands of Semiconductors (Room Temperature Data)

4.2.1. Single Valley Semiconductors

Substance	Energy gap (eV)	Effective mass (m_o)	Mobility (cm²/V·s)	Comments
GaAs	1.35	0.067	8500	3(or 6?) equivalent [100] valleys 0.36 eV above this maximum with a mobility of ~50.
InP	1.27	0.067	5000	3(or 6?) equivalent [100] valleys 0.4 eV above this minimum.
InAs	0.36	0.022	33,000	Equivalent valleys ~1.0 eV above this minimum.
InSb	0.165	0.014	78,000	
CdTe	1.44	0.11	1000	4(or 8?) equivalent [111] valleys 0.51 eV above this minimum.

4.2.2. Multivalley Semiconductors

Substance	Energy gap	Number of equivalent valleys and direction	Band curvature effective mass		Anisotropy $K = m_L/m_T$
			Longitudinal m_L	Transverse m_T	
Si	1.107	6 in [100] "Δ"	0.00	0.192	4.7
Ge	0.67	4 in [111] at "L"	1.588	0.0815	19.5
GaSb	0.67	as Ge (?)	~1.0	~0.2	~5
PbSe	0.26	4 in [111] at "L"	0.085	0.05	1.7
PbTe	0.25	4 in [111] at "L"	0.21	0.029	5.5
Bi_2Te_2	0.13	6			~0.05

TABLE 5. Resistivity of Semiconducting Minerals

Mineral	ρ (ohm · m)	Mineral	ρ (ohm · m)
Diamond (C)	2.7	Pentlandite, $(Fe, Ni)_4S_4$	1 to 11 × 10⁻⁶
Sulfides		Pyrrhotite, Fe_7S_4	2 to 160 × 10⁻⁶
Argentite, Ag_2S	1.5 to 2.0 × 10⁻³	Pyrite, FeS_2	1.2 to 600 × 10⁻³
Bismuthinite, Bi_2S_3	3 to 570	Sphalerite, ZnS	2.7 × 10⁻³ to 1.2 × 10⁴
Bornite, $Fe_2S_3 \cdot nCu_2S$	1.6 to 6000 × 10⁻⁶	Antimony-sulfur compounds	
Chalcocite, Cu_2S	80 to 100 × 10⁻⁶	Berthierite, $FeSb_2S_4$	0.0083 to 2.0
Chalcopyrite, $Fe_2S_3 \cdot Cu_2S$	150 to 9000 × 10⁻⁶	Boulangerite, $Pb_5Sb_3S_{11}$	2 × 10³ to 4 × 10⁴
Covellite, CuS	0.30 to 83 × 10⁻⁶	Cylindrite, $Pb_3Sn_4Sb_2S_{14}$	2.5 to 60
Galena, PbS	6.8 × 10⁻⁶ to 9.0 × 10⁻²	Franckeite, $Pb_5Sn_3Sb_2S_{14}$	1.2 to 4
Haverite, MnS_2	10 to 20	Hauchecornite, $Ni_4(Bi, Sb)_2S_{14}$	1 to 83 × 10⁻⁶
Marcasite, FeS_2	1 to 150 × 10⁻³	Jamesonite, $Pb_4FeSb_6S_{14}$	0.020 to 0.15
Metacinnabarite, HgS	2 × 10⁻⁶ to 1 × 10⁻³	Tetrahedrite, Cu_3SbS_3	0.30 to 30,000
Millerite, NiS	2 to 4 × 10⁻⁷	Arsenic-sulfur compounds	
Molybdenite, MoS_2	0.12 to 7.5	Arsenopyrite, FeAsS	20 to 300 × 10⁻⁶

Cobaltite, CoAsS	6.5 to 130×10^{-3}	Hessite, Ag_2Te	4 to 100×10^{-6}
Enargite, Cu_3AsS_4	0.2 to 40×10^{-3}	Nagyagite, $Pb_6Au(S,Te)_{14}$	20 to 80×10^{-6}
Gersdorfite, NiAsS	1 to 160×10^{-6}	Sylvanite, $AgAuTe_4$	4 to 20×10^{-6}
Glaucodote, (Co, Fe)AsS	5 to 100×10^{-6}	Oxides	
Antimonide		Braunite, Mn_2O_3	0.16 to 1.0
Dyscrasite, Ag_3Sb	0.12 to 1.2×10^{-6}	Cassiterite, SnO_2	4.5×10^{-4} to 10,000
Arsenides		Cuprite, Cu_2O	10 to 50
Allemonite, $SbAs_3$	70 to 60,000	Hollandite, (Ba, Na, K) Mn_8O_{16}	2 to 100×10^{-3}
Lollingite, $FeAs_2$	2 to 270×10^{-6}	Ilmenite, $FeTiO_3$	0.001 to 4
Nicollite, NiAs	0.1 to 2×10^{-6}	Magnetite, Fe_3O_4	52×10^{-6}
Skutterudite, $CoAs_3$	1 to 400×10^{-6}	Manganite, $MnO \cdot OH$	0.018 to 0.5
Smaltite, $CoAs_2$	1 to 12×10^{-6}	Melaconite, CuO	6000
Tellurides		Psilomelane, $BaMn_9O_{18} \cdot 2H_2O$	0.04 to 6000
Altaite, PbTe	20 to 200×10^{-6}	Pyrolusite, MnO_2	0.007 to 30
Calavarite, $AuTe_2$	6 to 12×10^{-6}	Rutile, TiO_2	29 to 910
Coloradoite, HgTe	4 to 100×10^{-6}	Uraninite, UO_2	1.5 to 200

References

1. Beer, A. C., *Galvanomagnetic Effects in Semiconductors*, Academic Press, 1963.
2. Goryunova, N. A., *The Chemistry of Diamond-Like Semiconductors*, The MIT Press, 1965.
3. Abrikosov, N. Kh., Bankina, V. F., Poretskaya, L. E., Shelimova, L. E., and Skudnova, E.V., *Semiconducting II-VI, IV-VI, and V-VI Compounds*, Plenum Press, 1969.
4. Berger, L. I. and Prochukhan, V. D., *Ternary Diamond-Like Semiconductors*, Cons. Bureau/Plenum Press, 1969.
5. Shay, J. L. and Wernick, J. H., *Ternary Chalcopyrite Semiconductors: Growth, Electronic Properties, and Applications*, Pergammon Press, 1975.
6. Bergman, R., *Thermal Conductivity in Solids*, Clarendon, Oxford, 1976.
7. Handbook of Semiconductors, Vol. 1, Moss, T.S. and Paul, W., Eds., *Band Theory and Transport Properties*; Vol. 2, Moss, T.S. and Balkanski, M., Eds., *Optical Properties of Solids*; Vol. 3, Moss, T.S. and Keller, S.P., Eds., *Materials Properties and Preparation*, North Holland Publ. Co., 1980.
8. Böer, K. W., *Survey of Semiconductor Physics*, Van Nostrand Reinhold, 1990.
9. Rowe, D. M., Ed., *CRC Handbook of Thermoelectrics*, CRC Press, Boca Raton, FL, 1995.
10. Berger, L. I., *Semiconductor Materials*, CRC Press, Boca Raton, FL, 1997.
11. Glazov, V. M., Chizhevskaya, S.N., and Glagoleva, N.N., *Liquid Semiconductors*, Plenum Press, New York, 1969.
12. Phillips, J. C., *Bonds and Bands in Semiconductors*, Academic Press, New York, 1973.
13. Harrison, W. A., *Electronic Structure and the Properties of Solids*, Freeman Publ. House, San Francisco, 1980.
14. Balkanski, M., Ed., *Optical Properties of Solids*, North-Holland, Amsterdam, 1980.
15. *Landolt-Börnstein. Numerical Data and Functional Relationships in Science and Technology, New Series, Group III: Crystal and Solid State Physics*, Hellwege, K.-H. and Madelung, O., Eds., Volumes 17 and 22, Springer Verlag, Berlin, 1984 (and further).
16. Shklovskii, B. L. and Efros, A.L., *Electronic Processes in Doped Semiconductors*, Springer Verlag, Berlin, 1984.
17. Cohen, M. L. and Chelikowsky, J. R., *Electronic Structure and Optical Properties of Semiconductors*, Springer Verlag, New York, 1988.
18. Glass, J.T., Messier, R.F., and Fujimori, N., Eds., *Diamond, Silicon Carbide, and Related Wide Bandgap Semiconductors*, MRS Symposia Proc. 1652, Mater. Res. Soc., Pittsburgh, 1990.
19. Palik, E., Ed., *Handbook of Optical Constants of Solids II*, Academic Press, New York, 1991.
20. Reed, M., Ed., *Semiconductors and Semimetals*, Volume 35, Academic Press, Boston, 1992.
21. Haug, H. and Koch, S. W., *Quantum Theory of the Optical and Electronic Properties of Semiconductors*, 2nd Edition, World Scientific, Singapore, 1993.
22. Lockwood, D. J., Ed., *Proc. 22nd Intl. Conf. on the Physics of Semiconductors, Vancouver, 1994*, World Scientific, Singapore, 1994.
23. Morelli, D. T., Caillat, T., Fleurial, J.-P., Borschchevsky, A., Vandersande, J., Chen, B., and Uher, C., *Phys. Rev.*, B51, 9622, 1995.
24. Caillat, T., Borshchevsky, A., and Fleurial, J.-P., *J. Appl. Phys.*, 80, 4442, 1996.
25. Fleurial, J.-P.,Caillat, T., and Borshchevsky, A., *Proc. XVI Intl. Conf. Thermoelectrics*, Dresden, Germany, August 26–29, 1997 (in print).
26. Borshchevsky, A. et al., U.S. Patents 5,610,366 (March 1977) and 5,831,286 (March 1998)
27. Jarrendahl, K. and Davis, R. F., *Semiconductors and Semimetals*, Vol. 52, Y.S. Park, Ed., 1998, pp. 1-20.
28. Bettini, M., *Solid State Comm.*,13, 599, 1973.
29. Chen, A. and Sher, A., *Semiconductor Alloys, Physics and Material Engineering*, Plenum Press, New York, 1995.
30. Holloway, P. H. and MgGuire, G. E., Eds., *Handbook of Compound Semiconductors*, Noyes Publ., 1995.

SELECTED PROPERTIES OF SEMICONDUCTOR SOLID SOLUTIONS

L.I. Berger

Alloy system	Limits of solubility	Energy gap in eV (300 K)	Remarks, references
Adamantine Semiconductors IV-IV			
Si_xGe_{1-x}	$0 \leq x \leq 1$	$0.8941+0.0421x+0.1691x^2$	Transition Γ - X [Ref.1]
		$0.7596+1.0860x+0.3306x^2$	Trans. Γ - L [Ref. 1]
Adamantine Semiconductors III-V/III-V			
Common Anion			
$Al_xGa_{1-x}N$	$0 \leq x \leq 1$		Wurtzite Structure [Ref. 2 & 3]
$Al_xGa_{1-x}P$	$0 \leq x \leq 0.5$	$2.28+0.16x$	[Ref. 2]
$Al_xIn_{1-x}P$	$0 \leq x \leq 0.44$	at Γ: 134+2.23x; at X: 2.24+0.18x	[Ref. 2]
$Al_xGa_{1-x}As$	$0 \leq x \leq 0.5$	$1.42=0.75x$ [Ref.3]; $1.424+1.429x-0.14x^2$ [Ref.4]	
$Al_xIn_{1-x}As$	$0 \leq x \leq 1$	at Γ: $0.37+1.91x+0.74x^2$; at X: 1.8+0.4x	[Ref. 2 and 6]
$Al_xGa_{1-x}Sb$	$0 \leq x < 1$	$0.73+1.10x+0.47x^2$	Trans. Γ_{8v}- Γ_{6c} [Ref. 2]
$Al_xIn_{1-x}Sb$	$0 \leq x \leq 1$		[Ref. 6]
$Ga_xIn_{1-x}N$	$0 \leq x \leq 1$	$1.950+1.487x-1.000x(1-x)$	Wurtzite [Ref. 8 and 10]
$Ga_xIn_{1-x}P$	$0 \leq x \leq 1$		[Ref. 2]
$Ga_xIn_{1-x}As$	$0 \leq x \leq 1$	$0.360+0.629x+0.436x^2$	[Ref. 5]
$Ga_xIn_{1-x}Sb$	$0 < x < 1$	$0.235+0.1653x+0.413x^2$	[Ref. 2, see also Ref. 9]
Common Cation			
GaN_xAs_{1-x}	$0 \leq x \leq 0.05$	$1.42-9.9x$	[Ref. 2]
GaP_xAs_{1-x}	$0 < x < 1$	$2.270-0.846x$	[Ref. 2]
GaP_xAs_{1-x}	$0 \leq x \leq 0.05$	$1.515+1.172x+0.186x^2$	(at 2K, $\Gamma-\Gamma$) [Ref. 7]
		$1.9715+0.144x+0.211x^2$	[Ref. 2]
$GaAs_xSb_{1-x}$	$0 \leq x \leq 0.45, 0.6 \leq x \leq 1$	$1.43-1.9x+1.2x^2$	[Ref. 5]
InP_xAs_{1-x}	$0 < x < 1$	$0.356+0.675x+0.32x^2$	[Ref. 2]
Adamantine Binary Semiconductors II-VI/II-VI [Ref. 3 and 6]			
Common Anion			
$Zn_xCd_{1-x}S$	$0 \leq x \leq 1$		Wurtzite Structure
$Zn_xHg_{1-x}S$	$0 \leq x \leq 1$		
$Cd_xHg_{1-x}S$	$0 \leq x \leq 1$		Wurtzite Structure at x<0.6
$Zn_xCd_{1-x}Se$	$0.7 \leq x \leq 1$		
$Zn_xHg_{1-x}Se$	$0 \leq x \leq 1$		
$Cd_xHg_{1-x}Se$	$0 \leq x \leq 0.7$ and $0.75 \leq x^* \leq 1$		x^*- Wurtzite Structure
$Zn_xCd_{1-x}Te$	$0 \leq x \leq 1$		
$Zn_xHg_{1-x}Te$	$0 \leq x \leq 1$		
$Cd_xHg_{1-x}Te$	$0 \leq x \leq 1$		
Common Cation			
ZnS_xSe_{1-x}	$0 \leq x \leq 1$		
ZnS_xTe_{1-x}	$0 \leq x \leq 0.1$ and $0.9 \leq x^* \leq 1$		x^*- Wurtzite Structure
$ZnSe_xTe_{1-x}$	$0 \leq x \leq 1$		
CdS_xSe_{1-x}	$0 \leq x \leq 1$		Wurtzite Structure
CdS_xTe_{1-x}	$0 \leq x \leq 0.25$ and $0.8 \leq x^* \leq 1$		x^*- Wurtzite Structure
$CdSe_xTe_{1-x}$	$0 \leq x \leq 0.4$ and $0.6 \leq x^* \leq 1$		x^*- Wurtzite Structure
HgS_xSe_{1-x}	$0 \leq x \leq 1$		
HgS_xTe_{1-x}	$0 \leq x \leq 1$		
$HgSe_xTe_{1-x}$	$0 \leq x \leq 1$		
Quaternary Adamantine Semiconductors II-VI/III-V [Ref. 6]			
$(ZnS)_x(AlP)_{1-x}$	$0.99 \leq x \leq 1$		
$(ZnSe)_x(GaAs)_{1-x}$	$0 \leq x < 1$		
$(CdTe)_x(InAs)_{1-x}$	$0 < x \leq 0.2$ and $0.7 \leq x \leq 1$		
$(CdTe)_x(AlSb)_{1-x}$	$0 \leq x \leq 1$		
$(HgTe)_x(InAs)_{1-x}$	$0 \leq x \leq 1$		

Alloy system	Limits of solubility	Energy gap in eV (300 K)	Remarks, references
	Quaternary Adamantine Semiconductors III_x-III_{1-x}-V_y - V_{1-y}		
$Ga_xIn_{1-x}As_yP_{1-y}$	$0 \leq x \leq 1$, $0 \leq x \leq 1$	$1.35+0.668x-1.068y+0.758x^2+0.078y^2-0.069xy-0.322x^2y+0.03xy^2$	[Ref. 2 and 6]
	Quaternary Adamantine Semiconductors III_{1-x-y}-III_x-III_y-V		
$Al_xGa_yIn_{1-x-y}Sb$	$0 \leq x \leq 1$, $0 \leq y \leq 1$	$0.095+1.76x+0.28y+0.345(x^2+y^2)+0.085(1-x-y)^2+xy\ (1-x-y)(23-28y)$	[Ref. 2 and 6]

References

1. Krishnamurti, S., Sher, A., and Chen, A. *Appl. Phys. Lett.* 47, 160, 1985.
2. Madelung, O., Ed., *Semiconductors Group IV Elements and III-V Compounds*, Springer, 1991; *Semiconductors Other than Group IV Elements and III-V Compounds*, Springer, 1992.
3. Goryunova, N. A., Kesamanly, F.P., and Nasledov, D.N., *Semiconductors and Semimetals*, Vol. 4, 1968, p. 413.
4. El Allali, M., Sorensen, C. B., Veje, E., and Tideman-Peterson, P., *Phys. Rev. B*, 48, 4398, 1993.
5. Nahorny, R. E., Pollack, M. A., Johnson, W. D., and Barns, R. L. *Appl. Phys. Lett.* 33, 695, 1978.
6. Goryunova, N. A. *Multicomponent Diamond-Like Semiconductors*, Sov. Radio, Moscow, 1968 (in Russian).
7. Capizzi, M., Modesti, S., Martelli, F., and Frova, A., *Solid State Comm.* 39, 333, 1981.
8. Nakamura, S., Pearton, S., and Fasol, G., *The Blue Laser Diode*, 2nd ed., Springer, 2000.
9. Roth, A. P., Keeler, W. J., and Fortin, E. *Canad. J. Phys.* 58, 560, 1980.
10. Wu, J., Walukiewicz, Yu, K. M., Ager, J. W., Haller, E. E., Lu, H., and Schaff, W.J., *Appl. Phys. Lett.* 80, 4741, 2002.

DIFFUSION DATA FOR SEMICONDUCTORS

B. L. Sharma

The diffusion coefficient D in many semiconductors may be expressed by an Arrhenius-type relation

$$D = D_o \exp(-Q/kT)$$

where D_o is a frequency factor, Q is the activation energy for diffusion, k is the Boltzmann constant, and T is the absolute temperature. This table lists D_o and Q for various diffusants in common semiconductors.

Abbreviations used in the table are

AES – Auger Electron Spectroscopy
DLTS – Deep Level Transient Spectroscopy
SEM – Scanning Electron Microscopy
SIMS – Secondary Ion Mass Spectrometry
$D(c)$ – Concentration Dependent Diffusion Coefficient
D_{max} – Maximum Diffusion Coefficient
(f) – Fast Diffusion Component
(i) – Interstitial Diffusion Component
(s) – Slow Diffusion Component
($\parallel$) – Parallel to c Direction
($\perp$) – Perpendicular to c Direction

Semiconductor	Diffusant	Frequency factor, D_o (cm²/s)	Activation energy, Q (eV)	Temperature range (°C)	Method of measurement	Ref.
Si	H	6×10^{-1}	1.03	120–1207	Electrical and SIMS	1
	Li	2.5×10^{-3}	0.65	25–1350	Electrical	2
	Na	1.65×10^{-3}	0.72	530–800	Electrical and flame photometry	3
	K	1.1×10^{-3}	0.76	740–800	Electrical and flame photometry	3
	Cu	4×10^{-2}	1.0	800–1100	Radioactive	4
		4.7×10^{-3}	0.43 (i)	300–700	Radioactive	5
	Ag	2×10^{-3}	1.6	1100–1350	Radioactive	6
	Au	2.4×10^{-4}	0.39 (i)	700–1300	Radioactive	7
		2.75×10^{-3}	2.05 (s)			
	Be	$(D \sim 10^{-7})$	–	1050	Electrical	8
	Ca	$(D \sim 6 \times 10^{-14})$	–	1100	Electrical and SIMS	1
	Zn	1×10^{-1}	1.4	980–1270	Electrical	9
	B	2.46	3.59	1100–1250	Electrical	10
		2.4×10^{1}	3.87	840–1250	Electrical	11
	Al	1.38	3.41	1119–1390	Electrical	12
		1.8	3.2	1025–1175	Electrical	13
	Ga	3.74×10^{-1}	3.39	1143–1393	Electrical	12
		6×10^{1}	3.89	900–1050	Radioactive	14
	In	7.85×10^{-1}	3.63	1180–1389	Electrical	12
		1.94×10^{1}	3.86	1150–1242	Radioactive	15
	Tl	1.37	3.7	1244–1338	Electrical	12
		1.65×10^{1}	3.9	1105–1360	Electrical	16
	Sc	8×10^{-2}	3.2	1100–1250	Radioactive	1
	Ce	$(D \sim 3.9 \times 10^{-13})$	–	1050	SIMS	1
	Pr	2.5×10^{-7}	1.74	1100–1280	Electrical	1
	Pm	7.5×10^{-9}	1.2 (s)	730–1270	Radioactive	1
		4.2×10^{-12}	0.13 (f)			
	Er	2×10^{-3}	2.9	1100–1250	Radioactive	1
	Tm	8×10^{-3}	3.0	1100–1280	Radioactive	1
	Yb	2.8×10^{-5}	0.95	947–1097	Neutron activation	1
	Ti	1.45×10^{-2}	1.79	950–1200	DLTS	17
	C	3.3×10^{-1}	2.92	1070–1400	Radioactive	18
	Si (self)	1.54×10^{2}	4.65	855–1175	SIMS	19
		1.6×10^{3}	4.77	1200–1400	Radioactive	20
	Ge	3.5×10^{-1}	3.92	855–1000	Radioactive	21
		2.5×10^{3}	4.97	1030–1302	Radioactive	21
		7.55×10^{3}	5.08	1100–1300	SIMS	22
	Sn	3.2×10^{1}	4.25	1050–1294	Neutron activation	23
	N	2.7×10^{-3}	2.8	800–1200	Out Diffusion; SIMS	1

Semiconductor	Diffusant	Frequency factor, D_o (cm²/s)	Activation energy, Q(eV)	Temperature range (°C)	Method of measurement	Ref.
	P	2.02×10^1	3.87	1100–1250	Electrical	10
		1.1	3.4	900–1200	Radioactive	24
		7.4×10^{-2}	3.3	1130–1405	Electrical	25
	As	6.0×10^1	4.2	950–1350	Radioactive	26
		6.55×10^{-2}	3.44	1167–1394	Electrical	27
		2.29×10^1	4.1	900–1250	Electrical	28
	Sb	1.29×10^1	3.98	1190–1398	Radioactive	29
		2.14×10^{-1}	3.65	1190–1405	Electrical	27
	Bi	1.03×10^3	4.64	1220–1380	Electrical	16
		1.08	3.85	1190–1394	Electrical	27
	Cr	1×10^{-2}	1	1100–1250	Radioactive	30
	Mo	$(D \sim 2 \times 10^{-10})$	–	1000	DLTS	1
	W	$(D \sim 10^{-12})$	–	1100	DLTS	1
	O	7×10^{-2}	2.44	700–1250	SIMS	31
		1.4×10^{-1}	2.53	700–1160	SIMS	32
	S	5.95×10^{-3}	1.83	975–1200	Radioactive	33
	Se	9.5×10^{-1}	2.6	1050–1250	Electrical	34
	Te	5×10^{-1}	3.34	900–1250	SIMS	1
	Mn	6.9×10^{-4}	0.63	900–1200	Radioactive	35
	Fe	1.3×10^{-3}	0.68	30–1250	Radioactive	36
	Co	2×10^{-3}	0.69	700–1300	Radioactive	37
	Ni	2×10^{-3}	0.47	800–1300	Radioactive	38
	Ru	$(D \sim 5 \times 10^{-7}$ $- 5 \times 10^{-6})$	–	1000–1280	Electrical	1
	Rh	$(D \sim 10^{-6}\text{–}10^{-4})$	–	1000–1200	Electrical	39
	Pd	2.95×10^{-4}	0.22 (i)	702–1320	Nuclear Activation	1
	Pt	1.5×10^2	2.22	800–1000	Electrical	1
	Os	$(D \sim 2 \times 10^{-6})$	–	1280	Electrical	40
	Ir	4.2×10^{-2}	1.3	950–1250	Electrical	41
Ge	Li	1.3×10^{-3}	0.46	350–800	Electrical	42
		9.1×10^{-3}	0.57	800–500	Electrical	43
	Na	3.95×10^{-1}	2.03	700–850	Radioactive	44
	Cu	1.9×10^{-4}	0.18 (i)	750–900	Radioactive	45
		4×10^{-2}	0.99 (s)	600–700		
		4×10^{-3}	0.33 (i)	350–750	Radioactive	5
	Ag	4.4×10^{-2}	1.0 (i)	700–900	Radioactive	46, 47
		4×10^{-2}	2.23 (s)	800–900	Radioactive	48
	Au	2.25×10^2	2.5	600–900	Radioactive	49
	Be	5×10^{-1}	2.5	720–900	Electrical	50
	Mg	$(D \sim 8 \times 10^{-9})$	–	900	Electrical	1
	Zn	5	2.7	600–900	Radioactive and electrical	51
	Cd	1.75×10^9	4.4	760–915	Radioactive	52
	B	1.8×10^9	4.55	600–900	Electrical	51
	Al	1.0×10^3	3.45	554–905	SIMS	53
		$\sim 1.6 \times 10^2$	~ 3.24	750–850	Electrical	54
	Ga	1.4×10^2	3.35	554–916	SIMS	55
		3.4×10^1	3.1	600–900	Electrical	51
	In	1.8×10^4	3.67	554–919	SIMS	56
		3.3×10^1	3.02	700–855	Radioactive	57
	Tl	1.7×10^3	3.4	800–930	Radioactive	58
	Si	2.4×10^{-1}	2.9	650–900	(γ) resonance	59
	Ge (self)	2.48×10^1	3.14	549–891	Radioactive	60
		7.8	2.95	766–928	Radioactive	61
	Sn	1.7×10^{-2}	1.9	–	Radioactive	45
	P	3.3	2.5	600–900	Electrical	51
	As	2.1	2.39	700–900	Electrical	62
	Sb	3.2	2.41	700–855	Radioactive	57
		1.0×10^1	2.5	600–900	Radioactive and electrical	51

Semiconductor	Diffusant	Frequency factor, D_o (cm²/s)	Activation energy, Q(eV)	Temperature range (°C)	Method of measurement	Ref.
	Bi	3.3	2.57	650–850	–	63
	O	4×10^{-1}	2.08	–	Optical	64
	S	$(D \sim 10^{-9})$	–	920	–	65
	Se	$(D \sim 10^{-10})$	–	920	–	65
	Te	5.6	2.43	750–900	Radioactive	66
	Fe	1.3×10^{-1}	1.08	750–900	Radioactive	67
	Co	1.6×10^{-1}	1.12	750–850	Radioactive	47
	Ni	8×10^{-1}	0.9	670–900	Electrical	68
GaAs	Li	5.3×10^{-1}	1.0	250–500	Electrical and chemical	69
	Cu	3×10^{-2}	0.53	100–500	Radioactive	69
		6×10^{-2}	0.98	450–750	Ultrasonic	69
		1.5×10^{-3}	0.6	800–1000	Radioactive	69
	Ag	4×10^{-4}	0.8	500–1150	Radioactive	69
	Au	1×10^{-3}	1.0	740–1025	Radioactive	69
	Be	7.3×10^{-6}	1.2	800–990	Electrical	69
	Mg	4×10^{-5}	1.22	800–1200	Electrical	69
	Zn	1.5×10^{1}	2.49	600–980	Radioactive	69
		2.5×10^{-1}	3.0	750–1000	Radioactive	69
	Cd	1.3×10^{-3}	2.2	800–1100	Radioactive	69
		5×10^{-2}	2.43	868–1149	Radioactive	69
	Hg	$(D \sim 5 \times 10^{-14})$	–	1100	Radioactive	69
	Al	$(D \sim 4 \times 10^{-18}\text{–}10^{-14})$	4.3	850–1100	AES	70
	Ga (self)	4×10^{-5}	2.6	1025–1100	Radioactive	69
		1×10^{7}	5.6	1125–1230	Radioactive	69
	In	$(D \sim 7 \times 10^{-11})$	–	1000	Radioactive	69
	C	$(D \sim 1.04 \times 10^{-16})$	–	825	SIMS	69
	Si	1.1×10^{-1}	2.5	850–1050	SIMS	69
	Ge	1.6×10^{-5}	2.06	650–850	SIMS	69
	Sn	6×10^{-4}	2.5	1060–1200	Radioactive	69
		1×10^{-5}	2	800–1000	Radioactive	69
	P	$(D \sim 10^{-12}\text{–}10^{-10})$	2.9	800–1150	Reflectance measurements	69
	As (self)	7×10^{-1}	3.2	–	Radioactive	69
	Cr	2.04×10^{-6}	0.83 (f)	750–1000	SIMS	69
			1.7 (s)	700–900		
		7.9×10^{-3}	2.2	800–1100	Chemical analysis	69
	O	2×10^{-3}	1.1	700–900	Mass spectroscopy	69
	S	1.85×10^{-2}	2.6	1000–1300	Radioactive	69
		1.1×10^{1}	2.95	750–900	Electrical	69
	Se	3×10^{3}	4.16	1025–1200	Radioactive	69
	Te	1.5×10^{-1}	3.5	1000–1150	Radioactive	69
	Mn	6.5×10^{-1}	2.49	850–1100	Radioactive	69
	Fe	4.2×10^{-2}	1.8	850–1150	Radioactive	69
		2.2×10^{-3}	2.32	750–1050	Radioactive	69
	Co	5×10^{2}	2.5	800–1000	Radioactive	69
		1.2×10^{-1}	2.64	750–1050	Radioactive	69
	Tm	2.3×10^{-16}	1.0	800–1000	Radioactive	69
GaSb	Li	2.3×10^{-4}	1.9 (s)	527–657	Electrical and flame photometry	69
		1.2×10^{-1}	0.7 (f)	277–657		
	Cu	4.7×10^{-3}	0.9	470–650	Radioactive	69
	Zn	$(D \sim 2 \times 10^{-13} - 1 \times 10^{-11})$	2	510–600	Radioactive	69
	Cd	1.5×10^{-6}	0.72	640–800	Electrical	69
	Ga (self)	3.2×10^{3}	3.15	658–700	Radioactive	69
	In	1.2×10^{-7}	0.53	320–650	Radioactive	69
	Sn	2.4×10^{-5}	0.8	320–650	Radioactive	69
		1.3×10^{-5}	1.1	500–650	Radioactive	69
	Sb (self)	3.4×10^{4}	3.45	658–700	Radioactive	69
	Se	$(D \sim 2.4 \times 10^{-13} - 1.37 \times 10^{-11})$	–	400–500	Radioactive	69

Semiconductor	Diffusant	Frequency factor, D_o (cm²/s)	Activation energy, Q(eV)	Temperature range (°C)	Method of measurement	Ref.
	Te	3.8×10^{-4}	1.20	320–650	Radioactive	69
	Fe	5×10^{-2}	1.9 (I)	500–650	Radioactive	69
		5×10^{2}	2.3 (II)	500–650		
GaP	Ag	–	–	1000–1300	Radioactive	69
	Au	8	2.5 (I)	1050–1250	Radioactive	69
		20	2.4 (II)	1100–1250	Diffusion (I) A face and (II) B face	
	Be	($D_{max} \sim 2.4 \times 10^{-9} - 8.5 \times 10^{-8}$)	–	900–1000	Atomic absorption analysis	69
	Mg	5×10^{-5}	1.4	700–1050	Electrical	69
	Zn	1.0	2.1	700–1300	Radioactive	69
	Ge	–	–	900–1000	Radioactive	69
	Cr	6.2×10^{-4}	1.2	900–1130	Radioactive; ESR	69
	S	3.2×10^{3}	4.7	1120–1305	Radioactive	69
	Mn	2.1×10^{9}	4.7	T < 950	Radioactive; ESR	69
		1.1×10^{-6}	0.9	950–1130		
	Fe	1.6×10^{-1}	2.3	980–1180	Radioactive	69
	Co	2.8×10^{-3}	2.9	850–1100	Radioactive	69
InP	Cu	3.8×10^{-3}	0.69	600–900	Radioactive	69
	Ag	3.6×10^{-4}	0.59	500–900	Radioactive	69
	Au	1.32×10^{-5}	0.48	600–820	Radioactive	69
		1.37×10^{-4}	0.73	600–900	Radioactive	69
	Zn	1.6×10^{-8}	0.3	750–900	Electrical	69
		($D \sim 2 \times 10^{-9} - 4 \times 10^{-8}$)	–	700–900	Radioactive	69
	Cd	1.8	1.9	700–900	Radioactive	69
		1.1×10^{-7}	0.72	700–900	Electrical	69
		($D \sim 7 \times 10^{-13} - 2 \times 10^{-10}$)	–	450–650	Electrical	69
	In (self)	1×10^{5}	3.85	830–990	Radioactive	69
	Sn	($D \sim 3 \times 10^{-8}$)	–	550	Etching and cathodo-luminescence	69
	P (self)	7×10^{10}	5.65	900–1000	Radioactive	69
	Cr	–	–	600–900	Radioactive	69
	S	3.6×10^{-4}	1.94	585–708	Electrical	69
	Se	($D \sim 2 \times 10^{-8}$)	–	550	Cathodoluminescence	69
	Mn	–	2.9	650–750	SIMS	69
	Fe	3	2	600–950	Radioactive	69
		6.8×10^{5}	3.4	600–700	SIMS	69
	Co	9×10^{-1}	1.8	600–950	Radioactive	69
InAs	Cu	3.6×10^{-3}	0.52	342–875	Radioactive	69
		2.2×10^{-2}	0.54	525–890	Radioactive	69
	Ag	7.3×10^{-4}	0.26	450–900	Radioactive	69
	Au	5.8×10^{-3}	0.65	600–900	Radioactive	69
	Mg	1.98×10^{-6}	1.17	600–900	Electrical	69
	Zn	4.2×10^{-3}	0.96	600–900	Radioactive	69
		3.11×10^{-3}	1.17	600–900	Electrical	69
	Cd	7.4×10^{-4}	1.15	650–900	Radioactive	69
	Hg	1.45×10^{-5}	1.32	650–850	Radioactive	69
	In (self)	6×10^{5}	4.0	740–900	Radioactive	69
	Ge	3.74×10^{-6}	1.17	600–900	Electrical	69
	Sn	1.49×10^{-6}	1.17	600–900	Electrical	69
	As (self)	3×10^{7}	4.45	740–900	Radioactive	69
	S	6.78	2.2	600–900	Electrical	69
	Se	12.6	2.2	600–900	Electrical	69
	Te	3.43×10^{-5}	1.28	600–900	Electrical	69
InSb	Li	7×10^{-4}	0.28	0–210	Electrical	69
	Cu	9×10^{-4}	1.08	200–500	Radioactive	69
		3×10^{-5}	0.37	230–490	Radioactive	69
	Ag	1×10^{-7}	0.25	440–510	Radioactive	69
	Au	7×10^{-4}	0.32	140–510	Radioactive	69
	Zn	5×10^{-1}	1.35	362–508	Radioactive	69

Semiconductor	Diffusant	Frequency factor, D_o (cm²/s)	Activation energy, Q(eV)	Temperature range (°C)	Method of measurement	Ref.
		–	1.5	355–455	SIMS	69
	Cd	1×10^{-5}	1.1	250–500	Radioactive	69
		1.3×10^{-4}	1.2	360–500	Electrical	69
	Hg	4×10^{-6}	1.17	425–500	Radioactive	69
	In (self)	6×10^{-7}	1.45	400–500	Radioactive	69
		1.8×10^{13}	4.3	475–517	Radioactive	69
	Sn	5.5×10^{-8}	0.75	390–512	Radioactive	69
	Pb	$(D \sim 2.7 \times 10^{-15})$	–	500	Radioactive	71
	Sb (self)	5.35×10^{-4}	1.91	400–500	Radioactive	69
		3.1×10^{13}	4.3	475–517	Radioactive	69
	S	9×10^{-2}	1.4	360–500	Electrical	69
	Se	1.6	1.87	380–500	Electrical	69
	Te	1.7×10^{-7}	0.57	300–500	Radioactive	69
	Fe	1×10^{-7}	0.25	440–510	Radioactive	69
	Co	2.7×10^{-11}	0.39	420–500	Radioactive	69
AlAs	Ga	$(D \sim 2 \times 10^{-18} - 10^{-15})$	3.6	850–1100	AES	70
	Zn	$(D \sim 9 \times 10^{-11})$	–	557	SEM	69
AlSb	Cu	3.5×10^{-3}	0.36	150–500	Radioactive	69
	Zn	3.3×10^{-1}	1.93	660–860	Radioactive	69
	Cd	$D(c) \sim 4 \times 10^{-12} - 3 \times 10^{-10}$	–	900	Radioactive	69
	Al (self)	2	1.88	570–620	X-ray	69
	Sb (self)	1	1.7	570–620	X-ray	69
ZnS	Cu	2.6×10^{-3}	0.79	470–750	Radioactive	69
		4.3×10^{-4}	0.64	250–1200	Electroluminescence	69
		9.75×10^{-3}	1.04	400–800	Luminescence	69
	Au	1.75×10^{-4}	1.16	500–800	Radioactive	69
	Zn (self)	3×10^{-4}	1.5	$925 < T < 940$	Radioactive	69
		1.5×10^{4}	3.26	$940 < T < 1030$		
		1×10^{16}	6.5	$1030 < T < 1075$		
	Cd	$(D \sim 10^{-10})$	–	1100	Luminescence	72
	Al	5.69×10^{-4}	1.28	800–1000	Luminescence	69
	In	3×10^{1}	2.2	750–1000	Radioactive	69
	S (self)	2.16×10^{4}	3.15	600–800	Radioactive	69
		8×10^{-5}	2.2	740–1100	Radioactive	69
	Se	$(D \sim 5 \times 10^{-13})$	–	1070	X-ray microprobe	69
	Mn	2.3×10^{3}	2.46	500–800	Radioactive	69
ZnSe	Li	2.66×10^{-6}	0.49	950–980	Electrical	69
	Cu	1×10^{-4}	0.66	400–800	Luminescence	69
		1.7×10^{-5}	0.56	200–570	Radioactive	69
	Ag	2.2×10^{-2}	1.18	400–800	Luminescence	69
	Zn (self)	9.8	3.0	760–1150	Radioactive	69
	Cd	6.39×10^{-4}	1.87	700–950	Photoluminescence	69
	Al	2.3×10^{-2}	1.8	800–1100	Luminescence	69
	Ga	1.81×10^{2}	3.0	900–1100	Luminescence	69
		–	1.3	700–850	Electron probe	69
	In	$(D \sim 2 \times 10^{-12})$	–	940	–	69
	S	$(D \sim 8 \times 10^{-12})$	–	1060	X-ray microprobe	69
	Se (self)	1.3×10^{1}	2.5	860–1020	Radioactive	69
		2.3×10^{-1}	2.7	1000–1050	Radioactive	69
	Ni	$(D \sim 1.5 \times 10^{-8} - 1.7 \times 10^{-7})$	–	740–910	Luminescence	69
ZnTe	Li	2.9×10^{-2}	1.22 (s)	400–700	Nuclear and chemical analysis	69
		1.7×10^{-4}	0.78 (f)			
	Zn (self)	2.34	2.56	760–860	Radioactive	69
		1.4×10^{1}	2.69	667–1077	Radioactive	69
	Al	–	2.0	700–1000	Electrical and optical	69
	In	4	1.96	1100–1300	Radioactive	69
	Te (self)	2×10^{4}	3.8	727–977	Radioactive	69
CdS	Li	3×10^{-6}	0.68	610–960	Microhardness	69
	Na	$(D \sim 3 \times 10^{-7})$	–	800	Radioactive	69

Semiconductor	Diffusant	Frequency factor, D_o (cm²/s)	Activation energy, Q(eV)	Temperature range (°C)	Method of measurement	Ref.
	Cu	1.5×10^{-3}	0.76	400–700	Radioactive	69
		1.2×10^{-2}	1.05	300–700	Ultrasonic	69
		8×10^{-5}	0.72	20–200	Electrical	69
	Ag	2.5×10^{1}	1.2 (s)	300–500	Radioactive	69
		2.4×10^{-1}	0.8 (f)			
	Au	2×10^{2}	1.8	500–800	Radioactive	69
	Zn	1.27×10^{-9}	0.86 (s)	720–1000	Radioactive	69
		1.22×10^{-8}	0.66 (f)			
	Cd (self)	3.4	2.0	700–1100	Radioactive	69
	Ga	–	–	667–967	Optical and microprobe	69
	In	6×10^{1}	2.3 (∥)	650–930	Radioactive, optical and microprobe	69
		1×10^{1}	2.03 (⊥)			
	P	6.5×10^{-4}	1.6	800–1100	Radioactive	69
	S (self)	1.6×10^{-2}	2.05	800–900	Radioactive	69
		–	2.4	750–1050	Radioactive	69
	Se	$(D \sim 1.2 \times 10^{-9})$	–	900	Radioactive	69
	Te	1.3×10^{-7}	10.4	700–1000	Radioactive	69
	Cl	$(D \sim 3 \times 10^{-10})$	–	800	Electrical	69
	I	$(D \sim 5 \times 10^{-12})$	–	1000	Radioactive	69
	Ni	6.75×10^{-3}	10.9	570–900	Luminescence	69
	Yb	$(D \sim 1.3 \times 10^{-9})$	–	960	Photoluminescence	69
CdSe	Ag	2×10^{-4}	0.53	22–400	Ultrasonic	69
	Cd (self)	1.6×10^{-3}	1.5	700–1000	Radioactive	69
		6.3×10^{-2}	1.25 (I)	600–900	Radioactive;	69
		4.12×10^{-2}	2.18 (II)	600–900	(I) saturated Cd and (II) saturated Se pressure	
	P	$(D \sim 5.3 \times 10^{-12} - 6 \times 10^{-11})$	–	900–1000	Radioactive	69
	Se (self)	2.6×10^{3}	1.55	700–1000	Radioactive; saturated Se pressure	69
CdTe	Li	$(D \sim 1.5 \times 10^{-10})$	–	300	Ion microprobe	69
	Cu	3.7×10^{-4}	0.67	97–300	Radioactive	69
		8.2×10^{-8}	0.64	290–350	Ion backscattering	69
	Ag	–	–	700–800	Electrical and photo-luminescence	69
	Au	6.7×10^{1}	2.0	600–1000	Radioactive	69
	Cd (self)	1.26	2.07	700–1000	Radioactive	69
		3.26×10^{2}	2.67 (I)	650–900	Radioactive;	69
		1.58×10^{1}	2.44 (II)		(I) saturated Cd and (II) saturated Te pressure	
	In	8×10^{-2}	1.61	650–1000	Radioactive	69
		1.17×10^{2}	2.21 (I)	500–850	Radioactive; (I) saturated Cd and (II) saturated Te pressure	69
		6.48×10^{-4}	1.15 (II)			
	Sn	8.3×10^{-2}	2.2	700–925	Radioactive	69
	P	$(D \sim 1.2 \times 10^{-10})$	–	900	Radioactive	69
	As	–	–	850	–	69
	O	5.6×10^{-9}	1.22	200–650	Mass spectrometry	69
		6.0×10^{-10}	0.29	650–900		
	Se	1.7×10^{-4}	1.35	700–1000	Radioactive	69
	Te (self)	8.54×10^{-7}	1.42 (I)	600–900	Radioactive; (I) saturated Cd and (II) saturated Te pressure	69
		1.66×10^{-4}	1.38 (II)	500–800		
	Cl	7.1×10^{-2}	1.6	520–800	Radioactive	69
	Fe	$(D \sim 4 \times 10^{-8})$	0.77	900	Radioactive	69

Semiconductor	Diffusant	Frequency factor, D_o (cm²/s)	Activation energy, Q(eV)	Temperature range (°C)	Method of measurement	Ref.
HgSe	Sb	6.3×10^{-5}	0.85	540–630	Radioactive	69
	Se (self)	–	–	200–400	Radioactive	69
HgTe	Ag	6×10^{-4}	0.8	250–350	Radioactive	69
	Zn	5×10^{-8}	0.6	250–350	Radioactive	69
	Cd	3.1×10^{-4}	0.66	250–350	Radioactive	69
	Hg (self)	2×10^{-8}	0.6	200–350	Radioactive	69
	In	6×10^{-6}	0.9	200–300	Radioactive	69
	Sn	1.72×10^{-6}	0.66 (s)	200–300	Radioactive	69
		1.8×10^{-3}	0.80 (f)			
	Te (self)	10^{-6}	1.4	200–400	Radioactive	69
	Mn	1.5×10^{-4}	1.3	250–350	Radioactive	69
PbS	Cu	4.6×10^{-4}	0.36	150–450	Electrical	69
		5×10^{-3}	0.31	100–400	Electrical	69
	Pb (self)	8.6×10^{-5}	1.52	500–800	Radioactive	69
	S (self)	6.8×10^{-5}	1.38	500–750	Radioactive	69
	Ni	1.78×10^{1}	0.95	200–500	Electrical	69
PbSe	Na	1.5×10^{1}	1.74 (s)	400–850	Radioactive	69
		5.6×10^{-6}	0.4 (f)			
	Cu	2×10^{-5}	0.31	93–520	Radioactive	69
	Ag	7.4×10^{-4}	0.35	400–850	Radioactive	69
	Pb (self)	4.98×10^{-6}	0.83	400–800	Radioactive	69
	Sb	3.4×10^{-1}	2.0	650–850	Radioactive	69
	Se (self)	2.1×10^{-5}	1.2	650–850	Radioactive	69
	Cl	1.6×10^{-8}	0.45	400–850	Radioactive	69
	Ni	($D \sim 1 \times 10^{-10}$)	–	700	Radioactive	69
PbTe	Na	1.7×10^{-1}	1.91	600–850	Radioactive	69
	Sn	3.1×10^{-2}	1.56	500–800	Radioactive	69
	Pb (self)	2.9×10^{-5}	0.6	250–500	Radioactive	69
	Sb	4.9×10^{-2}	1.54	500–800	Radioactive	69
	Te	2.7×10^{-6}	0.75	500–800	Radioactive	69
	Cl	($D > 2.3 \times 10^{-10}$)	–	700	Radioactive	69
	Ni	($D > 1 \times 10^{-6}$)	–	700	Radioactive	69

References

1. N. A. Stolwijk and H. Bracht, in *Diffusion in Semiconductors and Non-Metallic Solids*, D. L. Beke, Ed., Springer-Verlag, Berlin, 1998, 2-1.
2. E. M. Pell, *Phys. Rev.*, 119, 1960; 119, 1014, 1960.
3. L. Svob, *Solid State Electron*, 10, 991, 1967.
4. B. I. Boltaks and I. I. Sosinov, *Zh. Tekh. Fiz.*, 28, 3, 1958.
5. R. N. Hall and J. N. Racette, *J. Appl. Phys.*, 35, 379, 1964.
6. B. I. Boltaks and Hsueh Shih-Yin, *Sov. Phys. Solid State*, 2, 2383, 1961.
7. W. R. Wilcox and T. J. LaChapelle, *J. Appl. Phys.*, 35, 240, 1964.
8. E. A. Taft and R. O. Carlson, *J. Electrochem. Soc.*, 117, 711, 1970.
9. R. Sh. Malkovich and N. A. Alimbarashvili, *Sov. Phys. Solid State*, 4, 1725, 1963.
10. R. N. Ghoshtagore, *Solid State Electron*, 15, 1113, 1972.
11. C. Hill, *Semiconductor Silicon 1981*, H. R. Huff, R. J. Kreiger, and Y. Takeishi, Eds., p. 988, *Electrochem. Soc.*, 1981.
12. R. N. Ghoshtagore, *Phys. Rev. B*, 3, 2507, 1971.
13. W. Rosnowski, *J. Electrochem. Soc.*, 125, 957, 1978.
14. J. S. Makris and B. J. Masters, *J. Appl. Phys.*, 42, 3750, 1971.
15. M. F. Millea, *J. Phys. Chem. Solids*, 27, 315, 1965 (refer Reference 2).
16. C. S. Fuller and J. A. Ditzenberger, *J. Appl. Phys.*, 27, 544, 1956.
17. S. Hocine and D. Mathiot, *Appl. Phys. Lett.*, 53, 1269, 1988.
18. R. C. Newman and J. Wakefield, *J. Phys. Chem. Solids*, 19, 230, 1961.
19. L. Kalinowski and R. Seguin, *Appl. Phys. Lett.*, 35, 211, 1979; *Appl. Phys. Lett.*, 36, 171, 1980.
20. R. F. Peart, *Phys. Stat. Sol.*, 15, K 119, 1966.
21. G. Hettich, H. Mehrer and K. Maler, *Inst. Phys. Conf. Ser.*, 46, 500, 1979.
22. M. Ogina, Y. Oana and M. Watanabe, *Phys. Stat. Sol. (a)*, 72, 535, 1982.
23. T. H. Yeh, S. M. Hu, and R. H. Kastl, *Appl. Phys.*, 39, 4266, 1968.
24. I. Franz and W. Langheinrich, *Solid State Electron*, 14, 835, 1971.
25. R. N. Ghoshtagore, *Hys. Rev. B*, 3, 389, 1971.
26. B. J. Masters and J. M. Fairfield, *J. Appl. Phys.*, 40, 2390, 1969.
27. R. N. Goshtagore, *Phys. Rev. B*, 3, 397, 1971.
28. R. S. Fair and J. C. C. Tsai, *J. Electrochem. Soc.*, 122, 1689, 1975.
29. J. J. Rohan, N. E. Pickering, and J. Kennedy, *J. Electrochem. Soc.*, 106, 705, 1969.
30. W. Wuerker, K. Roy, and J. Hesse, *Matsr. Res. Bull.*, 9, 971, 1974.
31. J. C. Mikkelsen, Jr., *Appl. Phys. Lett.*, 40, 336, 1982.
32. S. Tang Lee and D. Nicols, *Appl. Phys. Lett.*, 47, 1001, 1985.
33. P. L. Gruzin, S. V. Zemskii, A. D. Bullkin, and N. M. Makarov, *Sov. Phys. Sem.*, 7, 1241, 1974.
34. N. S. Zhdanovich and Yu. I. Kozlov, *Svoistva Legir, Poluprovodn.*, V. S. Zemskov, Ed., Nauka, Moscow, 1977, 115–120; *Fiz Tekh. Poluprovod.*, 9, 1594, 1975.
35. D. Gilles, W. Bergholze, and W. Schroeter, *J. Appl. Phys.*, 59, 3590, 1986.
36. E. R. Weber, *Appl. Phys. A*, 30, 1, 1983.
37. E. R. Weber, Properties of Silicon, EMIS Datareviews Ser. No. 4, INSPEC Publications, 1988, 409–451.
38. M. K. Bakhadyrkhanov, S. Zainabidinov, and A. Khamidov, *Sov. Phys. Sem.*, 14, 243, 1980.
39. S. A. Azimov, M. S. Yunosov, F. K. Khatamkulov, and G. Nasyrov, *Poluprovod.*, N. Kh. Abrikosov and V. S. Zemskov, Eds., Nauka, Moscow, 1975, 21–23.

40. S. A. Azimov, M. S. Yunosov, G. Nurkuziev, and F. R. Karimov, *Sov. Phys. Sem.*, 12, 981, 1978.

41. S. A. Azimov, B. V. Umarov, and M. S. Yunusov, *Sov. Phys. Sem.*, 10, 842, 1976.

42. C. S. Fuller and J. A. Ditzenberger, *Phys. Rev.*, 91, 193, 1953.

43. B. Pratt and F. Friedman, *J. Appl. Phys.*, 37, 1893, 1966.

44. M. Stojic, V. Spiric, and D. Kostoski, *Inst. Phys. Conf. Ser.*, 31, 304, 1976.

45. B. I. Boltaks, *Diffusion in Semiconductors*, Inforsearch, London, 1963, 162.

46. A. A. Bugai, V. E. Kosenko, and E. G. Miselyuk, *Zh. Tekh. Fiz.*, 27, 67, 1957.

47. L. Y. Wei, *J. Phys. Chem. Solids*, 18, 162, 1961.

48. V. E. Kosenko, *Sov. Phys. Solid State*, 4, 42, 1962.

49. W. C. Dunlap, Jr., *Phys. Rev.*, 97, 614, 1955

50. Yu. I. Belyaev and V. A. Zhidkov, *Sov. Phys. Solid State*, 3, 133, 1961.

51. W. C. Dunlap, Jr. *Phys. Rev.*, 94, 1531, 1954.

52. V. E. Kosenko, *Sov. Phys. Solid State*, 1, 1481, 1960.

53. P. Dorner, W. Gust, A. Lodding, H. Odelius, B. Predel, and U. Roll, *Acta Metall.*, 30, 941, 1982.

54. W. Meer and D. Pommerrening, *Z. Agnew. Phys.*, 23, 369, 1967.

55. U. Sodervall, H. Odelius, A. Lodding, U. Roll, B. Predel, W. Gust, and P. Dorner, *Phil. Mag. A*, 54, 539, 1986.

56. P. Dorner, W. Gust, A. Lodding, H. Odelius, B. Predel, and U. Roll, *Z. Metalkd.*, 73, 325, 1982.

57. P. V. Pavlov, *Sov. Phys. Solid State*, 8, 2377, 1967.

58. V. I. Tagirov and A. A. Kuliev, *Sov. Phys. Solid State*, 4, 196, 1962.

59. J. Raisanen, J. Hirvonen, and A. Anttila, *Solid State Electron.*, 24, 333, 1981.

60. C. Vogel, G. Hettich, and H. Mehrer, *J. Phys. C.*, 16, 6197, 1983.

61. H. Letaw, Jr., W. M. Portnoy, and L. Slifkin, *Phys. Rev.*, 102, 363, 1956.

62. W. Bosenberg, *Z. Naturforsch.*, 10a, 285, 1955.

63. V. M. Glazov and V. S. Zemskov, Physicochemical Principles of Semiconductor Doping, Israel Program for Scientific Translation, Jerusalem, 1968.

64. J. W. Corbett, R. S. McDonald, and G. D. Watkins, *J. Phys. Chem. Solids*, 25, 873, 1964.

65. W. W. Tyler, *J. Phys. Chem. Solids*, 8, 59, 1959.

66. V. D. Ignatkov and V. E. Kosenko, *Sov. Phys. Solid State*, 4, 1193, 1962.

67. A. A. Bugal, V. E. Kosenko, and E. G. Miseluk, *Zh. Tekh. Fiz.*, 27, 210, 1957.

68. F. van der Maesen and J. A. Brenkman, *Phillips Res. Rep.*, 9, 255, 1954.

69. M. B. Dutt and B. L. Sharma, in *Diffusion in Semiconductors and Non-Metallic Solids*, D. L. Beke, Ed., Springer-Verlag, Berlin, 1998, 3-1.

70. L. L. Chang and A. Koma, *Appl. Physics Lett.*, 29, 138, 1976.

71. D. L. Kendall, *Semiconductors and Semimetals*, Vol. 4, R. K. Willardson and A. C. Beer, Eds., Academic, 1968, 255.

72. H. J. Biter and F. Williams, *J. Luminescence*, 3, 395, 1971.

PROPERTIES OF MAGNETIC MATERIALS

H. P. R. Frederikse

Glossary of Symbols

Quantity	Symbol	Units SI	emu
Magnetic field	H	A m^{-1}	Oe (oersted)
Magnetic induction	B	T (tesla)	G (gauss)
Magnetization	M	A m^{-1}	emu cm^{-3}
Spontaneous magnetization	M_s	A m^{-1}	emu cm^{-3}
Saturation magnetization	M_0	A m^{-1}	emu cm^{-3}
Magnetic flux	Φ	Wb (weber)	maxwell
Magnetic moment	m, μ	A m^2	erg/G
Coercive field	H_c	A m^{-1}	Oe
Remanence	B_r	T	G
Saturation magnetic polarization	J_s	T	G
Magnetic susceptibility	χ		
Magnetic permeability	μ	H m^{-1} (henry/meter)	
Magnetic permeability of free space	μ_0	H m^{-1}	
Saturation magnetostriction	λ ($\Delta l/l$)		
Curie temperature	T_C	K	K
Néel temperature	T_N	K	K

Magnetic moment $\mu = \gamma \hbar J = g \mu_B J$
where

γ = gyromagnetic ratio; J = angular momentum; g = spectroscopic splitting factor ($\sim$2)
μ_B = Bohr magneton = $9.2741 \cdot 10^{-24}$ J/T = $9.2741 \cdot 10^{-21}$ erg/G

Earth's magnetic field H = 56 A m^{-1} = 0.7 Oe
For iron: $M_0 = 1.7 \cdot 10^6$ A m^{-1}; $B_r = 0.8 \cdot 10^6$ A m^{-1}
1 Oe = (1000/4π) A m^{-1}; 1 G = 10^{-4} T; 1 emu cm^{-3} = 10^3 A m^{-1}
1 Maxwell = 10^{-8} Wb
$\mu_0 = 4\pi\ 10^{-7}$ H m^{-1}

Relation between Magnetic Induction and Magnetic Field

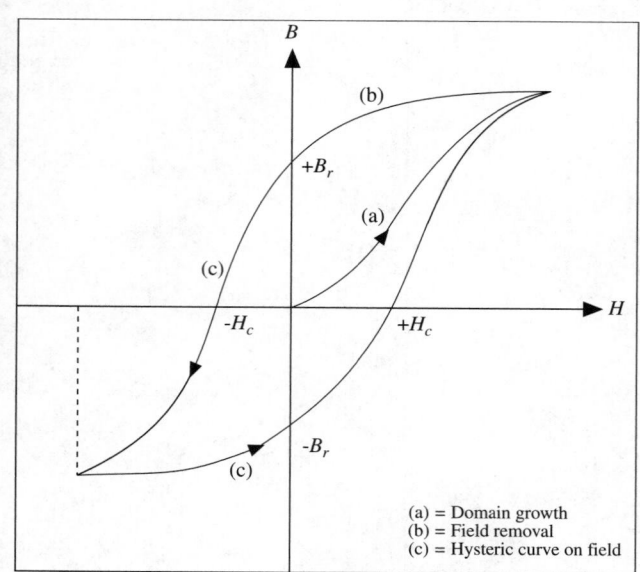

(a) = Domain growth
(b) = Field removal
(c) = Hysteric curve on field

FIGURE 1. Typical curve representing the dependence of magnetic induction B on magnetic field H for a ferromagnetic material. When H is first applied, B follows curve **a** as the favorably oriented magnetic domains grow. This curve flattens as saturation is approached. When H is then reduced, B follows curve **b**, but retains a finite value (the remanence B_r) at H = 0. In order to demagnetize the material, a negative field $-H_c$ (where H_c is called the coercive field or coercivity) must be applied. As H is further decreased and then increased to complete the cycle (curve **c**), a hysteresis loop is obtained. The area within this loop is a measure of the energy loss per cycle for a unit volume of the material.

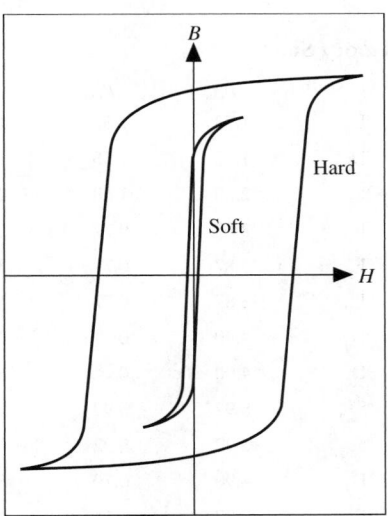

FIGURE 2. Schematic curve illustrating the *B* vs. *H* dependence for hard and soft magnetic materials. Hard materials have a larger remanence and coercive field, and a correspondingly large hysteresis loss.

Reference

Ralls, K. M., Courtney, T. H., and Wulff, J., *Introduction to Materials Science and Engineering*, J. Wiley & Sons, New York, 1976, p. 577, 582. With permission.

Magnetic Susceptibility of the Elements

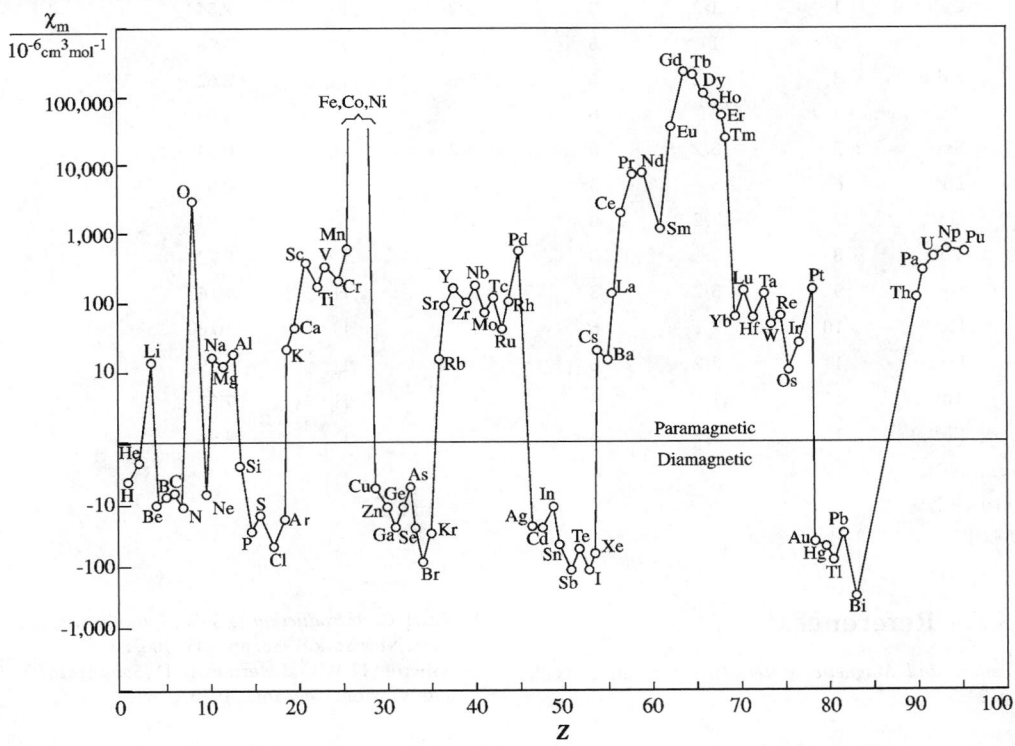

FIGURE 3. Molar susceptibility of the elements at room temperature (cgs units of 10^{-6} cm³/mol). Values are not available for Z = 9, 61, and 84–89; Fe, Co, and Ni (Z = 26–28) are ferromagnetic. Data taken from the table "Magnetic Susceptibility of the Elements and Inorganic Compounds" in Section 4.

Reference

Gray, D. E., Ed., *American Institute of Physics Handbook, Third Edition*, McGraw Hill, New York, 1972, p. 5–224. With permission.

Ground State of Ions with Partly Filled d or f Shells

Z	Element	n	S	L	J	Gr. state	p_{calc} [a]	p_{calc} [b]	p_{meas}
22	Ti^{3+}	1	1/2	2	3/2	$^2D_{3/2}$	1.73	1.55	1.8
23	V^{4+}	1	1/2	2	3/2	$^2D_{3/2}$	1.73	1.55	1.8
23	V^{3+}	2	1	3	2	$^3F_2$	2.83	1.63	2.8
23	V^{2+}	3	3/2	3	3/2	$^4F_{3/2}$	3.87	0.77	3.8
24	Cr^{3+}	3	3/2	3	3/2	$^4F_{3/2}$	3.87	0.77	3.7
25	Mn^{4+}	3	3/2	3	3/2	$^4F_{3/2}$	3.87	0.77	4.0
24	Cr^{2+}	4	2	2	0	$^5D_0$	4.90	0	4.9
25	Mn^{3+}	4	2	2	0	$^5D_0$	4.90	0	5.0
25	Mn^{2+}	5	5/2	0	5/2	$^6S_{5/2}$	5.92	5.92	5.9
26	Fe^{3+}	5	5/2	0	5/2	$^6S_{5/2}$	5.92	5.92	5.9
26	Fe^{2+}	6	2	2	4	$^5D_4$	4.90	6.70	5.4
27	Co^{2+}	7	3/2	3	9/2	$^4F_{9/2}$	3.87	6.54	4.8
28	Ni^{2+}	8	1	3	4	$^3F_4$	2.83	5.59	3.2
29	Cu^{2+}	9	1/2	2	5/2	$^2D_{5/2}$	1.73	3.55	1.9

Z	Element	n	S	L	J	Gr. state	p_{calc} [c]		p_{meas}
58	Ce^{3+}	1	1/2	3	5/2	$^2F_{5/2}$	2.54		2.4
59	Pr^{3+}	2	1	5	4	$^3H_4$	3.58		3.5
60	Nd^{3+}	3	3/2	6	9/2	$^4I_{9/2}$	3.62		3.5
61	Pm^{3+}	4	2	6	4	$^5I_4$	2.68		
62	Sm^{3+}	5	5/2	5	5/2	$^6H_{5/2}$	0.84		1.5
63	Eu^{3+}	6	3	3	0	$^7F_0$	0.0		3.4
64	Gd^{3+}	7	7/2	0	7/2	$^8S_{7/2}$	7.94		8.0
65	Tb^{3+}	8	3	3	6	$^7F_6$	9.72		9.5
66	Dy^{3+}	9	5/2	5	15/2	$^6H_{15/2}$	10.63		10.6
67	Ho^{3+}	10	2	6	8	$^5I_8$	10.60		10.4
68	Er^{3+}	11	3/2	6	15/2	$^4I_{15/2}$	9.59		9.5
69	Tm^{3+}	12	1	5	6	$^3H_6$	7.57		7.3
70	Yb^{3+}	13	1/2	3	7/2	$^2F_{7/2}$	4.54		4.5

[a] $p_{calc} = 2[S(S + 1)]^{1/2}$
[b] $p_{calc} = 2[J(J + 1)]^{1/2}$
[c] $p_{calc} = g[J(J + 1)]^{1/2}$

References

1. Jiles, D., *Magnetism and Magnetic Materials*, Chapman & Hall, London, 1991, p. 243.

2. Kittel, C., *Introduction to Solid State Physics, 6th Edition*, J. Wiley & Sons, New York, 1986, pp. 405–406.
3. Ashcroft, N. W. and Mermin, N. D., *Solid State Physics*, Holt, Rinehart, and Winston, New York, 1976, p. 652.

Ferro- and Antiferromagnetic Elements

M_0 is the saturation magnetization at $T = 0$ K

n_B is the number of Bohr magnetons per atom

T_C is the Curie temperature

T_N is the Néel temperature

	M_0/gauss	n_B	T_C/K	T_N/K	Comments
Fe	22020	2.22	1043		
Co	18170	1.72	1388		
Ni	6410	0.62	627		
Cr				311	
Mn				100	
Ce				12.5	c-Axis antiferromagnetic
Nd				19.2	Basal plane modulation on hexagonal sites
				7.8	Cubic sites order (periodicity different from high-T phase)
Sm				106	Ordering on hexagonal sites
				13.8	Cubic site order
Eu				90.5	Spiral along cube axis
Gd	24880	7	293		
Tb		9	220		Basal plane ferromagnet
				230.2	Basal plane spiral
Dy		10	87		Basal plane ferromagnet
				176	Basal plane spiral
Ho		10	20		Bunched cone structure
				133	Basal plane spiral
Er		9	32		c-Axis ferrimagnetic cone structure
				80	c-Axis modulated structure
Tm		7	32		c-Axis ferrimagnetic cone structure
				56	c-Axis modulated structure

References

1. Ashcroft, N. W., and Mermin, N. D., *Solid State Physics*, Holt, Rinehart, and Winston, New York, 1976, p.652.

2. Gschneidner, K. A., and Eyring, L., *Handbook on the Physics and Chemistry of Rare Earths*, North Holland Publishing Co., Amsterdam, 1978.

Selected Ferromagnetic Compounds

M_0 is the saturation magnetization at $T = 293$ K

T_C is the Curie temperature

Compound	M_0/gauss	T_C/K	Crystal system
MnB	152	578	orthorh(FeB)
MnAs	670	318	hex(FeB)
MnBi	620	630	hex(FeB)
MnSb	710	587	hex(FeB)
Mn_4N	183	743	
MnSi		34	cub(FeSi)
CrTe	247	339	hex(NiAs)
$CrBr_3$	270	37	hex(BiI_3)
CrI_3		68	hex(BiI_3)
CrO_2	515	386	tetr(TiO_2)
EuO	1910*	77	cub
EuS	1184*	16.5	cub
$GdCl_3$	550*	2.2	orthorh
FeB		598	orthorh
Fe_2B		1043	tetr ($CuAl_2$)
$FeBe_5$		75	cub($MgCu_2$)
Fe_3C		483	orthorh
FeP		215	orthorh (MnP)

* At $T = 0$ K

References

1. Kittel, C., *Introduction to Solid State Physics, 6th Edition*, J. Wiley & Sons, New York, 1986.

2. Ashcroft, N. W., and Mermin, N. D., *Solid State Physics*, Holt, Rinehart, and Winston, New York, 1976.

Magnetic Properties of High-Permeability Metals and Alloys (Soft)

μ_i is the initial permeability

μ_m is the maximum permeability

H_c is the coercive force

J_s is the saturation polarization

W_H is the hysteresis loss per cycle

T_C is the Curie temperature

Material	Composition (mass %)	μ_i/μ_0	μ_m/μ_0	H_c/A m^{-1}	J_s/T	W_H/J m^{-3}	T_C/K
Iron	Commercial 99Fe	200	6000	70	2.16	500	1043
Iron	Pure 99.9Fe	25000	350000	0.8	2.16	60	1043
Silicon-iron	96Fe-4Si	500	7000	40	1.95	50–150	1008
Silicon-iron (110) [001]	97Fe-3Si	9000	40000	12	2.01	35–140	1015
Silicon-iron {100} <100>	97Fe-3Si		100000	6	2.01		1015
Mild steel	Fe-0.1C-0.1Si-0.4Mn	800	1100	200			
Hypernik	50Fe-50Ni	4000	70000	4	1.60	22	753
Deltamax {100} <100>	50Fe-50Ni	500	200000	16	1.55		773
Isoperm {100} <100>	50Fe-50Ni	90	100	480	1.60		
78 Permalloy	78Ni-22Fe	4000	100000	4	1.05	50	651
Supermalloy	79Ni-16Fe-5Mo	100000	1000000	0.15	0.79	2	673
Mumetal	77Ni-16Fe-5Cu-2Cr	20000	100000	4	0.75	20	673
Hyperco	64Fe-35Co-0.5Cr	650	10000	80	2.42	300	1243
Permendur	50Fe-50Co	500	6000	160	2.46	1200	1253
2V-Permendur	49Fe-49Co-2V	800	4000	160	2.45	600	1253
Supermendur	49Fe-49Co-2V		60000	16	2.40	1150	1253
25Perminvar	45Ni-30Fe-25Co	400	2000	100	1.55		
7Perminvar	70Ni-23Fe-7Co	850	4000	50	1.25		
Perminvar (magnet. annealed)	43Ni-34Fe-23Co		400000	2.4	1.50		
Alfenol (or Alperm)	84Fe-16Al	3000	55000	3.2	0.8		723
Alfer	87Fe-13Al	700	3700	53	1.20		673
Aluminum-Iron	96.5Fe-3.5Al	500	19000	24	1.90		
Sendust	85Fe-10Si-5Al	36000	120000	1.6	0.89		753

References

1. McCurrie, R. A., *Structure and Properties of Ferromagnetic Materials*, Academic Press, London, 1994, p. 42.

2. Gray, D. E., Ed., *American Institute of Physics Handbook, Third Edition*, McGraw Hill, New York, 1972, p. 5–224.

Applications of High-Permeability Materials

Applications	Requirements
Power applications	
Distribution and power transformers	Low core losses, high permeability, high saturation magnetic polarization
High-quality motors and generators, stators and armatures, switched-mode power supplies	
Instrument transformers	
Audiofrequency transformers	Low core losses, high permeability, high magnetic polarization
Pulse transformers	High permeability
Cores for inductor coils	
Audiofrequency	Low hysteresis, high permeability
Carrier frequency	Very low hysteresis and eddy current loss
Radiofrequency	High permeability at low fields
Miscellaneous	
Relays, switches Earth leakage circuit	High permeability, low remanence, low coercivity
Magnetic shielding	Low core loss for AC applications

Applications of High-Permeability Materials

Applications	Requirements
Magnetic recording heads	High initial permeability, low or zero remanence
Magnetic amplifiers Saturable reactors Saturable transformers Transformer cores	Rectangular hysteresis loops, low hysteresis loss
Magnetic shunts for temperature compensation in magnetic circuits	Low Curie temperature, appropriate decrease in permeability with increase in temperature
Electromagnets in indicating instruments, fire detection, quartz watches, electromechanical devices	High permeability, high saturation magnetic polarization
Magnetic yokes in permanent magnet devices, such as lifting and holding magnets, loudspeakers	High permeability, high saturation magnetic polarization

Reference

McCurrie, R. A., *Structure and Properties of Ferromagnetic Materials*, Academic Press, London, 1994. With permission.

Saturation Magnetostriction of Selected Materials

The tabulated parameter λ_s is related to the fractional change in length $\Delta l/l$ by $\Delta l/l = (3/2)\lambda_s(\cos^2\theta - 1/3)$, where θ is the angle of rotation.

Material	$\lambda_s \times 10^6$
Iron	−7
Fe - 3.2% Si	+9
Nickel	−33
Cobalt	−62
45 Permalloy, 45% Ni - 55% Fe	+27
Permalloy, 82% Ni - 18% Fe	0
Permendur, 49% Co - 49% Fe - 2% V	+70
Alfer, 87% Fe - 13% Al	+30
Magnetite, Fe_3O_4	+40
Cobalt ferrite, $CoFe_2O_4$	−110
$SmFe_2$	−1560
$TbFe_2$	+1753
$Tb_{0.3}Dy_{0.7}Fe_{1.93}$ (Terfenol D)	+2000
$Fe_{66}Co_{18}B_{15}Si$ (amorphous)	+35
$Co_{72}Fe_3B_6A_{13}$ (amorphous)	0

Reference

McCurrie, R.A., *Structure and Properties of Ferromagnetic Materials*, Academic Press, London, 1994, p. 91; additional data provided by A.E. Clark, Adelphi, MD.

Properties of Various Permanent Magnetic Materials (Hard)

B_r is the remanence
$_BH_c$ is the flux coercivity
$_iH_c$ is the intrinsic coercivity

$(BH)_{max}$ is the maximum energy product
T_C is the Curie temperature
T_{max} is the maximum operating temperature

Composition	B_r/T	$_BH_c/10^3$ A m^{-1}	$_iH_c/10^3$ A m^{-1}	$(BH)_{max}$/kJ m^{-3}	T_C/°C	T_{max}/°C
Alnico1 20Ni;12Al;5Co	0.72		35	25		
Alnico2 17Ni;10Al;12.5Co;6Cu	0.72		40−50	13−14		
Alnico3 24-30Ni;12-14Al;0-3Cu	0.5−0.6		40−54	10		
Alnico4 21-28Ni;11-13Al;3-5Co;2-4Cu	0.55−0.75		36−56	11−12		
Alnico5 14Ni;8Al;24Co;3Cu	1.25	53	54	40	850	520
Alnico6 16Ni;8Al;24Co;3Cu;2Ti	1.05		75	52		
Alnico8 15Ni;7Al;35Co;4Cu;5Ti	0.83	1.6	160	45		
Alnico9 15Ni;7Al;35Co;4Cu;5Ti	1.10	1.45	1.45	75	850	520
Alnico12 13.5Ni;8Al;24.5Co;2Nb	1.20		64	76.8		

Composition	B_r/T	$_BH_c$/10^3 A m^{-1}	$_iH_c$/10^3 A m^{-1}	$(BH)_{max}$/kJ m^{-3}	T_C/°C	T_{max}/°C
BaFe$_{12}$O$_{19}$ (Ferroxdur)	0.4	1.6	192	29	450	400
SrFe$_{12}$O$_{19}$	0.4	2.95	3.3	30	450	400
LaCo$_5$	0.91			164	567	
CeCo$_5$	0.77			117	380	
PrCo$_5$	1.20			286	620	
NdCo$_5$	1.22			295	637	
SmCo$_5$	1.00	7.9	696	196	700	250
Sm(Co$_{0.76}$Fe$_{0.10}$Cu$_{0.14}$)$_{6.8}$	1.04	4.8	5	212	800	300
Sm(Co$_{0.65}$Fe$_{0.28}$Cu$_{0.05}$Zr$_{0.02}$)$_{7.7}$	1.2	10	16	264	800	300
Nd$_2$Fe$_{14}$B sintered	1.22	8.4	1120	280	300	100
Fe;52Co;14V (Vicalloy II)	1.0	42		28	700	500
Fe;24Cr;15Co;3Mo (anisotropic)	1.54	67		76	630	500
Fe;28Cr;10.5Co (Chromindur II)	0.98	32		16	630	500
Fe;23Cr;15Co;3V;2Ti	1.35	4		44	630	500
Cu;20Ni;20Fe (Cunife)	0.55	4		12	410	350
Cu;21Ni;29Fe (Cunico)	0.34	0.5		8		
Pt;23Co	0.64	4		76	480	350
Mn;29.5Al;0.5C (anisotropic)	0.61	2.16	2.4	56	300	120

References

1. McCurrie, R. A., *Structure and Properties of Ferromagnetic Materials*, Academic Press, London, 1994, p. 204.

2. Gray, D. E., Ed., *American Institute of Physics Handbook, Third Edition*, McGraw Hill, New York, 1972, p. 5–165.

3. Jiles, D., *Magnetism and Magnetic Materials*, Chapman & Hall, London, 1991.

Selected Ferrites

J_s is the saturation magnetic polarization
T_C is the Curie temperature
ΔH is the line width

Material	J_s/T	T_C/°C	ΔH/kA m^{-1}	Applications
Spinels				
γ-Fe$_2$O$_3$	0.52	575		
Fe$_3$O$_4$	0.60	585		
NiFe$_2$O$_4$	0.34	575	350	Microwave devices
MgFe$_2$O$_4$	0.14	440	70	
NiZnFe$_2$O$_4$	0.50	375	120	Transformer cores
MnFe$_2$O$_4$	0.50	300	50	Microwave devices
NiCoFe$_2$O$_4$	0.31	590	140	Microwave devices
NiCoAlFe$_2$O$_4$	0.15	450	330	Microwave devices
NiAl$_{0.35}$Fe$_{1.65}$O$_4$	0.12	430	67	Microwave devices
NiAlFe$_2$O$_4$	0.05	1860	32	Microwave devices
Mg$_{0.9}$Mn$_{0.1}$Fe$_2$O$_4$	0.25	290	56	Microwave devices
Ni$_{0.5}$Zn$_{0.5}$Al$_{0.8}$Fe$_{1.2}$O$_4$	0.14		17	Microwave devices
CuFe$_2$O$_4$	0.17	455		Electromechanical transducers
CoFe$_2$O$_4$	0.53	520		
LiFe$_5$O$_8$	0.39	670		Microwave devices
Garnets				
Y$_3$Fe$_5$O$_{12}$	0.178	280	55	Microwave devices
Y$_3$Fe$_5$O$_{12}$ (single crys.)	0.178	292	0.5	Microwave devices
(Y,Al)$_3$Fe$_5$O$_{12}$	0.12	250	80	Microwave devices
(Y,Gd)$_3$Fe$_5$O$_{12}$	0.06	250	150	Microwave devices
Sm$_3$Fe$_5$O$_{12}$	0.170	305		Microwave devices
Eu$_3$Fe$_5$O$_{12}$	0.116	293		Microwave devices
GdFe$_5$O$_{12}$	0.017	291		Microwave devices
Hexagonal crystals				
BaFe$_{12}$O$_{19}$	0.45	430	1.5	Permanent magnets
Ba$_3$Co$_2$Fe$_{24}$O$_{41}$	0.34	470	12	Microwave devices
Ba$_2$Zn$_2$Fe$_{12}$O$_{22}$	0.28	130	25	Microwave devices
Ba$_3$Co$_{1.35}$Zn$_{0.65}$Fe$_{24}$O$_{41}$		390	16	Microwave devices
Ba$_2$Ni$_2$Fe$_{12}$O$_{22}$	0.16	500	8	Microwave devices
SrFe$_{12}$O$_{19}$	0.4	450		Permanent magnets

Reference

McCurrie, R. A., *Structure and Properties of Ferromagnetic Materials*, Academic Press, London, 1994.

Spinel Structure (AB$_2$O$_4$)

FIGURE 4. Arrangement of metal ions in the two octants A and B, showing tetrahedrally (A) and octahedrally (B) coordinated sites. (Reprinted from McCurrie, R.A., *Ferromagnetic Materials*, Academic Press, London, 1994. With permission.)

- A
- B
- O^{2-}

Selected Antiferromagnetic Solids
T_N is the Néel temperature

Material	Structure	T_N/K	Material	Structure	T_N/K
			ZnCr$_2$O$_4$	cub	15
Binary oxides			ZnFe$_2$O$_4$	cub	9
MnO	cub(fcc)	122	GeFe$_2$O$_4$	cub	10
FeO	cub(fcc)	198	MgV$_2$O$_4$	cub	45
CoO	cub(fcc)	291	MnGa$_2$O$_4$	cub	33
NiO	cub(fcc)	525			
α-Mn$_2$O$_3$	cub	90	***NiAs and related structures***		
CuO	monocl	230	CrAs	orth	300
UO$_2$	cub	30.8	CrSb	hex	705–723
Er$_2$O$_3$	cub	3.4	CrSe	hex	300
Gd$_2$O$_3$	cub	1.6	MnTe	hex	320–323
			NiS	hex	263
Perovskites			CrS	monocl	460
LaCrO$_3$	orth	282			
LaMnO$_3$	orth	100	***Rutile and related structures***		
LaFeO$_3$	orth	750	CoF$_2$	tetr	38
NdCrO$_3$	orth	224	CrF$_2$	monocl	53
NdFeO$_3$	orth	760	FeF$_2$	tetr	79
YbCrO$_3$	orth	118	MnF$_2$	tetr	67
CaMnO$_3$	cub	110	NiF$_2$	tetr	83
EuTiO$_3$	cub	5.3	CrCl$_2$	orth	20
YCrO$_3$	orth	141	MnO$_2$	tetr	84
BiFeO$_3$	cub*	673	FeOF	tetr	315
KCoF$_3$	cub	125			
KMnF$_3$	cub*	88.3	***Corundum and related structures***		
KFeF$_3$	cub	115	Cr$_2$O$_3$	rhomb	318
KNiF$_3$	cub	275	α-Fe$_2$O$_3$	rhomb	948
NaMnF$_3$	cub*	60	FeTiO$_3$	rhomb	68
NaNiF$_3$	orth	149	MnTiO$_3$	rhomb	41
RbMnF$_3$	cub	82	CoTiO$_3$	rhomb	38
Spinels					
Co$_3$O$_4$	cub	40	***VF$_3$ and related structures***		
NiCr$_2$O$_4$	tetr	65	CoF$_3$	rhomb	460
			CrF$_3$	rhomb	80

Material	Structure	T_N/K
FeF_3	rhomb	394
MnF_3	monocl	43
MoF_3	rhomb	185
Miscellaneous		
K_2NiF_4	tetr	97
MnI_2	hex	3.4
$CoUO_4$	orth	12
$CaMn_2O_4$	orth	225
CrN	cub*	273
CeC_2	tetr	33
$FeSn$	hex	373
Mn_2P	hex	103

* Distorted.

References

1. Gray, D. E., Ed., *American Institute of Physics Handbook, Third Edition*, McGraw Hill, New York, 1972, p. 5–168 to 5–183.
2. Kittel, C., *Introduction to Solid State Physics, 6th Edition*, J. Wiley & Sons, New York, 1986.
3. Ashcroft, N. W., and Mermin, N. D., *Solid State Physics*, Holt, Rinehart, and Winston, New York, 1976, p. 697.

ORGANIC MAGNETS

J.S. Miller

Magnetic ordering, e.g., ferromagnetism, like superconductivity, is a property of a solid, not of an individual molecule or ion, and very rarely occurs for organic compounds. In contrast to superconductivity, where all electron spins pair to form a perfect diamagnetic material, magnetic ordering requires unpaired electron spins; hence, superconductivity and ferromagnetism are mutually exclusive.

The vast majority of organic compounds are diamagnetic (i.e., all electron spins are paired), and a relative few possess unpaired electrons (designated by an arrow, ↑) and are paramagnetic (PM), i.e., they are oriented in random directions. A few organic solids, however, exhibit strong magnetic behavior and magnetically order as ferromagnets (FO) with all spins aligned in the same direc-

tion. In some cases the spins align in the opposite direction and compensate to form an antiferromagnet (AF). In some cases these spins are not opposed to each other and do not compensate and lead to a canted antiferromagnet or weak ferromagnet (WF). If the number of spins that align in one direction differs from the number of spins that align in the opposite direction, the spins cannot compensate and a ferrimagnet (FI) results. Metamagnets (MM) are antiferromagnets in which all the spins become aligned like a ferromagnet in an applied magnetic field. Above the ordering or critical temperature, T_c, all magnets are paramagnets (PM). Organic magnets all possess electron spins in p-orbitals, but these may be in conjunction with metal ion-based spins.

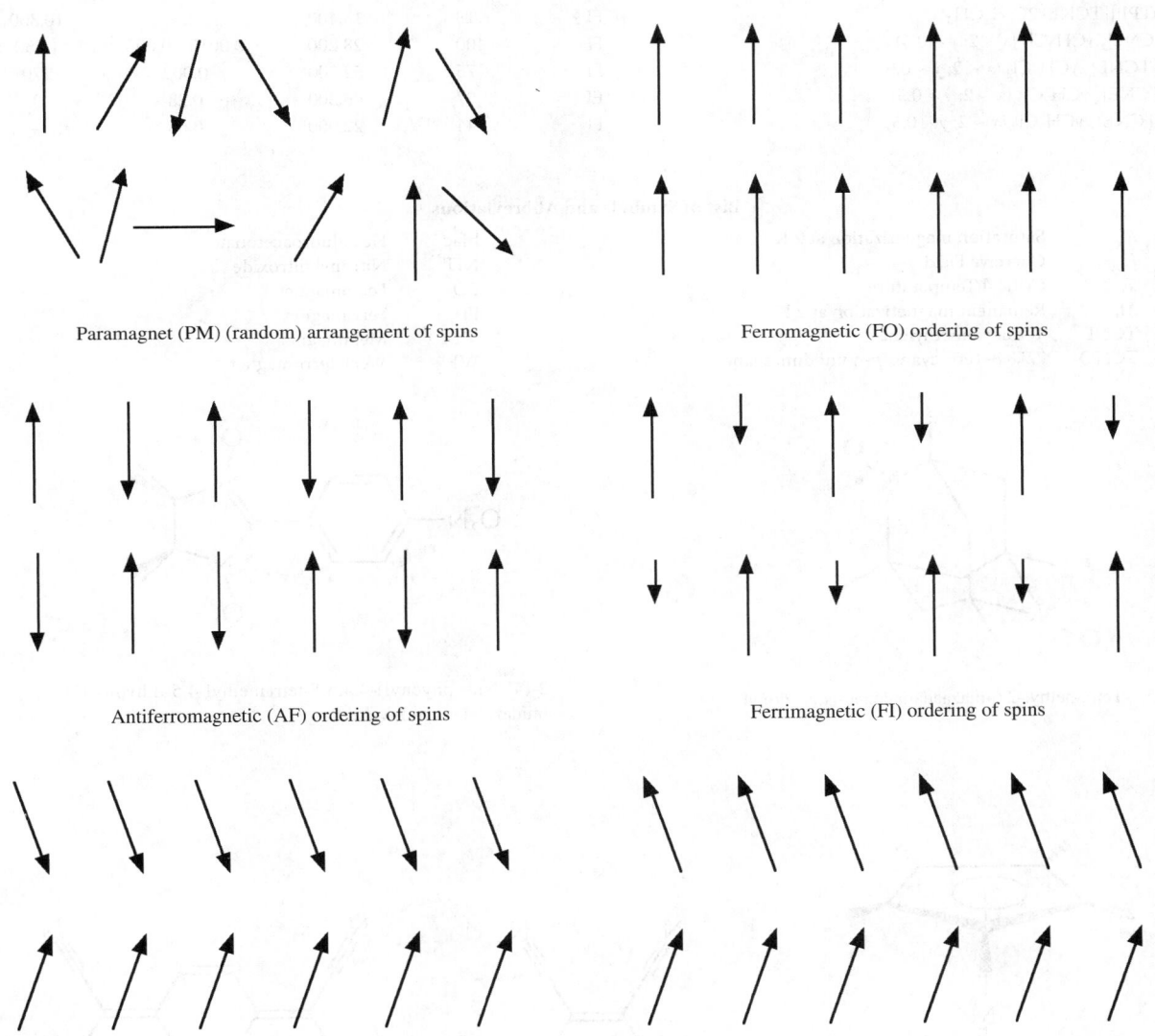

Paramagnet (PM) (random) arrangement of spins

Ferromagnetic (FO) ordering of spins

Antiferromagnetic (AF) ordering of spins

Ferrimagnetic (FI) ordering of spins

Canted antiferromagnet or weak ferromagnet (WF) ordering of spins

FIGURE 1.   Schematic illustration of the different types of magnetic behavior.

Summary of the Critical Temperature, T_c, Saturation Magnetization, M_s, Coercive Field, H_{cr}, and Remanent Magnetization, M_r, for Selected Organic-Based Magnets

Magnet	Type	T_c/K	M_s/A m^{-1}	H_{cr}/T	M_r/A m^{-1}
α-1,3,5,7-Tetramethyl-2,6-diazaadamantane-N,N'-doxyl	FO	1.48	48,300	<0.00001	—
β-2-(4'-Nitrophenyl)-4,4,5,5-tetramethyl-4,5-dihydro-1H-imidazol-1-oxyl-3-N-oxide	FO	0.6	22,300	0.00008	<200
{Fe^{III}[C$_5$(CH$_3$)$_5$]$_2$}[TCNE]	FO	4.8	37,600	0.10	2,300
{Mn^{III}[C$_5$(CH$_3$)$_5$]$_2$}[TCNE]	FO	8.8	58,200	0.12	3,700
{Cr^{III}[C$_5$(CH$_3$)$_5$]$_2$}[TCNE]	FO	3.65	46,300	—	—
α-{Fe^{III}[C$_5$(CH$_3$)$_5$]$_2$}[TCNQ]	MM	2.55	34,200	—	—
β-{Fe^{III}[C$_5$(CH$_3$)$_5$]$_2$}[TCNQ]	FO	3.0	21,600	—	—
Tanol subarate	MM	0.38	20,700	—	—
NCC$_6$F$_4$CN$_2$S$_2$	WF	35.5	45	0.00009	—
Mn^{II}(hfac)$_2$NITC$_2$H$_5$	FI	7.8	39,400	0.03	27,600
Mn^{II}(hfac)$_2$NIT(i-C$_3$H$_8$)	FI	7.6	42,400	<0.0005	<420
[Mn(hfac)$_2$]$_3$[{ON[C$_6$H$_3$(t-C(CH$_3$)$_3$]$_2$NO]$_2$}	FI	46	24,400	—	—
[MnTPP][TCNE]·2C$_6$H$_5$CH$_3$	FI	13	18,400	2.4	10,300
V[TCNE]$_x$·yCH$_2$Cl$_2$ ($x \sim 2$; $y \sim 0.5$)	FI	~400	28,200	0.0015 - 0.006	1,650
Mn[TCNE]$_x$·yCH$_2$Cl$_2$ ($x \sim 2$; $y \sim 0.5$)	FI	75	52,000	0.002	270
Fe[TCNE]$_x$·yCH$_2$Cl$_2$ ($x \sim 2$; $y \sim 0.5$)	FI	97	46,300	0.23	3
Co[TCNE]$_x$·yCH$_2$Cl$_2$ ($x \sim 2$; $y \sim 0.5$)	FI	44	22,000	0.65	—

List of Symbols and Abbreviations

M_s	Saturation magnetization at 2 K	hfac	Hexafluoroacetonate
H_{cr}	Coercive Field	NIT	Nitronyl nitroxide
T_c	Critical Temperature	FO	Ferromagnet
M_r	Remanent magnetization at 2 K	FI	Ferrimagnet
TCNE	Tetracyanoethylene	MM	Metamagnet
TCNQ	7,7,8,8-Tetracyano-p-quinodimethane	WF	Weak ferromagnet

1,3,5,7-Tetramethyl-2,6-diazaadamantane-N,N'-doxyl

2-(4'-Nitrophyenyl)-4,4,5,5-tetramethyl-4,5-dihydro-1H-imidazol-1-oxyl-3-N-oxide

M[C$_5$(CH$_3$)$_5$]$_2$(M = Cr, Mn, Fe)

TCNE

TCNQ

Mn(hfac)₂

Tanol subarate

NITR (R = C₂H₅, i-C₃H₈, n-C₃H₈)

{ON[C₆H₃(t-C(CH₃)₃)₂NO]₂}

MnTPP

NCC₆F₄CN₂S₂

References

1. Miller, J. S. and Epstein, A. J., *Angew. Chem. Internat. Ed.*, 33, 385, 1994.
2. Chiarelli, R., Rassat, A., Dromzee, Y., Jeannin, Y., Novak, M. A., and Tholence, J. L., *Phys. Scrip.*, T49, 706, 1993.
3. Kinoshita, M., *Jap. J. Appl. Phys.*, 33, 5718, 1994.
4. Gatteschi, D., *Adv. Mat.*, 6, 635, 1994.
5. Miller, J. S. and Epstein, A. J., *J. Chem. Soc., Chem. Commun.*, 1319, 1998.
6. Broderick, W. E., Eichorn, D. M., Lu, X., Toscano, P. J., Owens, S. M. and Hoffman, B. M., *J. Am. Chem. Soc.*, 117, 3641, 1995.
7. Banister, A. J., Bricklebank, N., Lavander, I., Rawson, J., Gregory, C. I., Tanner, B. K., Clegg, W. J., Elsegood, M. R., and Palacio, F., *Angew. Chem. Internat. Ed.*, 35, 2533, 1996.

ELECTRON INELASTIC MEAN FREE PATHS

Cedric J. Powell

The inelastic mean free path (IMFP) of electrons is defined as the average of distances, measured along the trajectories, that electrons with a given energy travel between inelastic collisions in the substance. It is an important parameter in analyzing results from surface-characterization techniques such as Auger electron spectroscopy, x-ray photoelectron spectroscopy, low-energy electron diffraction, and others. IMFPs can be measured by the elastic-peak electron spectroscopy technique and other methods, and they can be calculated from optical data. A detailed analysis of the experimental and theoretical considerations in obtaining reliable IMPF values can be found in reference 4.

The table below gives recommended IMFP values for several elements, simple inorganic compounds, and organic materials. Values are given in Ångström units (1 Å = 10^{-10} m) for a range of electron energies. Substances are given in alphabetical order by name, with elements and inorganic compounds listed before the organic materials.

References

1. Tanuma, S., Powell, C. J., and Penn, D. R., *Surf. Interface Anal.* 17, 911, 1991.
2. Tanuma, S., Powell, C. J., and Penn, D. R., *Surf. Interface Anal.* 17, 927, 1991.
3. Tanuma, S., Powell, C. J., and Penn, D. R., *Surf. Interface Anal.* 21, 165, 1994.
4. Powell, C. J., and Jablonski, A., *J. Phys. Chem. Ref. Data* 28, 19, 1999.
5. Tanuma, S., Powell, C. J., and Penn, D. R., *Surf. Interface Anal.*, 37, 1, 2005.

Electron Inelastic Mean Free Path in Å

Electron Energy in eV

Substance	Formula	50	100	150	200	300	400	600	800	1000	1200	1400	1600	1800	2000	Ref.
Aluminum	Al	3.5	4.6	5.7	6.8	8.9	10.9	14.5	18.0	21.3	24.6	27.8	31.0	34.1	37.2	5
Bismuth	Bi	4.9	5.5	6.3	7.2	8.8	10.6	14.0	17.2	20.2	23.2	26.1	28.9	31.6	34.4	1
Carbon	C	5.9	6.4	7.5	8.8	11.2	13.7	18.4	22.8	27.0	31.1	35.2	39.1	43.0	46.8	1
Chromium	Cr	4.4	4.3	5.0	5.7	7.2	8.6	11.4	14.0	16.6	19.1	21.6	24.0	26.3	28.6	1
Copper	Cu	5.2	5.0	5.4	6.0	7.3	8.6	11.3	13.9	16.5	18.9	21.3	23.7	26.0	28.3	4
Gallium phosphide	GaP	5.6	6.5	7.8	9.0	11.4	13.7	18.1	22.3	26.3	30.2	34.1	37.8	41.5	45.2	2
Gold	Au	6.3	4.7	4.7	5.1	6.1	7.2	9.5	11.7	13.8	15.8	17.8	19.7	21.6	23.4	4
Hafnium	Hf	5.8	6.2	7.1	8.0	10.2	12.0	15.6	19.0	22.2	25.3	28.4	31.4	34.4	37.3	1
Indium phosphide	InP	4.8	4.9	5.6	6.4	8.1	9.7	12.8	15.7	18.7	21.4	24.2	26.8	29.4	32.0	2
Iridium	Ir	5.3	4.3	4.7	5.2	6.4	7.5	9.7	11.8	13.8	15.7	17.6	19.4	21.2	22.9	1
Iron	Fe	4.3	4.4	5.1	5.8	7.2	8.5	11.2	13.7	16.2	18.6	20.9	23.2	25.5	27.7	1
Lead(II) sulfide	PbS	4.8	5.6	6.7	7.8	10.0	12.1	16.1	19.8	23.6	27.1	30.6	33.9	37.2	40.5	2
Lead(II) telluride	PbTe	4.3	5.5	6.6	7.7	9.8	11.9	15.8	19.6	23.4	26.9	30.3	33.7	37.0	40.2	2
Magnesium	Mg	4.0	5.4	6.8	8.2	10.7	13.0	17.5	21.7	25.9	29.9	33.9	37.7	41.6	45.3	1
Molybdenum	Mo	5.1	4.5	5.0	5.6	7.1	8.5	11.3	14.0	16.5	18.9	21.2	23.5	25.8	28.0	1
Nickel	Ni	4.9	4.5	4.9	5.4	6.5	7.7	10.1	12.4	14.6	16.7	18.8	20.8	22.8	24.7	4
Niobium	Nb	6.0	6.0	6.7	7.7	9.7	11.7	15.6	19.2	22.6	25.9	29.1	32.3	35.3	38.4	1
Osmium	Os	5.5	4.3	4.5	5.0	6.0	7.1	9.1	11.1	12.9	14.7	16.4	18.1	19.8	21.5	1
Palladium	Pd	4.8	4.8	5.4	6.2	7.8	9.4	12.5	15.4	18.2	20.9	23.5	26.0	28.4	30.9	1
Platinum	Pt	5.0	4.2	4.5	4.9	6.0	7.1	9.2	11.2	13.1	14.9	16.7	18.5	20.2	21.9	1
Potassium chloride	KCl	7.5	7.8	9.3	10.9	14.2	17.3	23.2	28.7	34.0	39.2	44.2	49.1	54.0	58.8	2
Rhenium	Re	5.2	3.8	3.9	4.3	5.1	6.0	7.7	9.4	10.9	12.5	14.0	15.4	16.9	18.3	1
Rhodium	Rh	4.8	4.1	4.5	5.0	6.1	7.3	9.7	12.0	14.1	16.2	18.2	20.1	22.0	23.9	1
Ruthenium	Ru	4.9	4.2	4.6	5.2	6.5	7.8	10.4	12.8	15.1	17.4	19.5	21.6	23.7	25.7	1
Silicon	Si	4.1	5.3	6.5	7.8	10.3	12.5	16.6	20.6	24.4	28.2	31.8	35.4	39.0	42.5	1
Silicon carbide	SiC	4.7	4.9	5.8	6.8	8.7	10.5	13.9	17.1	20.3	23.3	26.3	29.2	32.1	35.0	2
Silicon dioxide (vitreous)	SiO_2	8.0	7.7	8.8	10.0	12.6	15.2	20.0	24.7	29.3	33.7	38.0	42.2	46.4	50.5	2
Silver	Ag	6.4	4.7	4.8	5.2	6.4	7.7	10.2	12.6	14.9	17.2	19.3	21.4	23.5	25.5	4
Tantalum	Ta	4.8	4.5	5.0	5.5	6.8	8.0	10.4	12.7	14.8	16.9	19.0	21.0	22.9	24.9	1
Titanium	Ti	4.5	5.1	6.2	7.3	9.5	11.6	15.6	19.5	23.2	26.8	30.2	33.6	36.9	40.2	1
Tungsten	W	5.0	4.1	4.5	5.0	6.1	7.3	9.4	11.4	13.4	15.2	17.1	18.9	20.6	22.4	1
Vanadium	V	4.2	4.9	5.9	6.8	8.8	10.7	14.3	17.7	21.0	24.3	27.4	30.5	33.5	36.4	1
Yttrium	Y	5.0	5.5	6.4	7.5	9.8	11.9	16.0	19.8	23.4	27.0	30.4	33.8	37.1	40.4	1
Zinc sulfide	ZnS	5.8	6.5	7.7	8.9	11.3	13.6	17.9	22.0	26.0	29.8	33.6	37.3	40.9	44.5	2
Zirconium	Zr	4.4	4.8	5.7	6.6	8.6	10.5	14.1	17.5	20.7	23.8	26.9	29.9	32.8	35.7	1
Adenine	$C_5H_5N_5$	6.4	6.6	7.8	9.2	11.8	14.4	19.2	24.1	28.6	33.1	37.4	41.6	45.8	49.9	3
Bovine plasma albumin		7.3	7.2	8.5	9.9	12.7	15.4	20.7	25.8	30.8	35.6	40.2	44.8	49.4	53.8	3
β-Carotene	$C_{40}H_{56}$	6.4	7.0	8.5	10.0	13.0	15.9	21.4	26.9	32.0	37.0	41.9	46.6	51.3	56.0	3

Substance	Formula	50	100	150	200	300	400	600	800	1000	1200	1400	1600	1800	2000	Ref.
Deoxyribonucleic acid (DNA)		7.3	7.3	8.5	9.8	12.6	15.4	20.7	25.9	30.8	35.6	40.3	44.9	49.4	53.8	3
1,6-Diphenyl-1,3,5-hexatriene	$C_{18}H_{16}$	6.4	7.0	8.4	9.9	12.9	15.8	21.3	26.7	31.8	36.8	41.7	46.4	51.1	55.7	3
Guanine	$C_5H_5N_5O$	6.2	6.2	7.2	8.4	10.8	13.1	17.5	21.8	25.9	29.9	33.8	37.6	41.4	45.1	3
Hexacosane	$C_{26}H_{54}$	7.0	7.6	9.2	10.9	14.1	17.2	23.2	29.2	34.7	40.1	45.4	50.6	55.7	60.7	3
Kapton		7.0	6.8	7.9	9.2	11.7	14.2	19.0	23.7	28.2	32.5	36.7	40.9	44.9	49.0	3
Polyacetylene		5.3	5.7	6.8	7.9	10.2	12.5	16.9	21.1	25.1	29.0	32.8	36.5	40.2	43.8	3
Poly(butene-1-sulfone)		7.1	7.2	8.5	9.9	12.7	15.4	20.7	25.8	30.6	35.3	39.9	44.4	48.8	53.2	3
Polyethylene		6.9	7.2	8.6	10.1	13.0	15.9	21.4	26.8	31.8	36.8	41.6	46.3	51.0	55.6	3
Poly(methyl methacrylate)		7.8	7.9	9.3	10.8	13.9	16.9	22.7	28.3	33.7	38.8	43.9	48.9	53.8	58.6	3
Polystyrene		6.9	7.3	8.7	10.2	13.2	16.1	21.6	27.1	32.2	37.2	42.1	46.9	51.6	56.2	3
Poly(2-vinylpyridine)		6.9	7.3	8.7	10.3	13.3	16.2	21.8	27.3	32.5	37.5	42.4	47.3	52.0	56.7	3

ELECTRON WORK FUNCTION OF THE ELEMENTS

The electron work function Φ is a measure of the minimum energy required to extract an electron from the surface of a solid. It is defined more precisely as the energy difference between the state in which an electron has been removed to a distance from the surface of a single crystal face that is large enough that the image force is negligible but small compared to the distance to any other face (typically about 10^{-4} cm) and the state in which the electron is in the bulk solid. In general, Φ differs for each face of a monocrystalline sample.

Since Φ is dependent on the cleanliness of the surface, measurements reported in the literature often cover a considerable range. This table contains selected values for the electron work function of the elements which may be regarded as typical values for a reasonably clean surface. The method of measurement is indicated for each value. The following abbreviations appear:

TE – Thermionic emission
PE – Photoelectric effect
FE – Field emission
CPD – Contact potential difference
polycr – Polycrystalline sample
amorp – Amorphous sample

Values in parentheses are only approximate.

References

1. Hölzl, J., and Schulte, F. K., Work Functions of Metals, in *Solid Surface Physics*, Höhler, G., Ed., Springer-Verlag, Berlin, 1979.
2. Riviere, J. C., Work Function: Measurements and Results, in *Solid State Surface Science, Vol.1*, Green, M., Ed., Decker, New York, 1969.
3. Michaelson, H. B., *J. Appl. Phys.*, 48, 4729, 1977.

Element	Plane	Φ/eV	Method
Ag	100	4.64	PE
	110	4.52	PE
	111	4.74	PE
Al	100	4.20	PE
	110	4.06	PE
	111	4.26	PE
As	polycr	(3.75)	PE
Au	100	5.47	PE
	110	5.37	PE
	111	5.31	PE
B	polycr	(4.45)	TH
Ba	polycr	2.52	TH
Be	polycr	4.98	PE
Bi	polycr	4.34	PE
C	polycr	(5.0)	CPD
Ca	polycr	2.87	PE
Cd	polycr	4.08	CPD
Ce	polycr	2.9	PE
Co	polycr	5.0	PE
Cr	polycr	4.5	PE
Cs	polycr	1.95	PE
Cu	100	5.10	FE
	110	4.48	PE
	111	4.94	PE
	112	4.53	PE
Eu	polycr	2.5	PE
Fe	100	4.67	PE
	111	4.81	PE
Ga	polycr	4.32	PE
Gd	polycr	2.90	CPD
Ge	polycr	5.0	CPD
Hf	polycr	3.9	PE
Hg	liquid	4.475	PE
In	polycr	4.09	PE
Ir	100	5.67	PE
	110	5.42	PE
	111	5.76	PE
	210	5.00	PE
K	polycr	2.29	PE
La	polycr	3.5	PE
Li	polycr	2.93	FE
Lu	polycr	(3.3)	CPD
Mg	polycr	3.66	PE
Mn	polycr	4.1	PE
Mo	100	4.53	PE
	110	4.95	PE
	111	4.55	PE
	112	4.36	PE
	114	4.50	PE
	332	4.55	PE
Na	polycr	2.36	PE
Nb	001	4.02	TH
	110	4.87	TH
	111	4.36	TH
	112	4.63	TH
	113	4.29	TH
	116	3.95	TH
	310	4.18	TH
Nd	polycr	3.2	PE
Ni	100	5.22	PE
	110	5.04	PE
	111	5.35	PE
Os	polycr	5.93	PE
Pb	polycr	4.25	PE
Pd	polycr	5.22	PE
	111	5.6	PE
Pt	polycr	5.64	PE
	110	5.84	FE
	111	5.93	FE
	320	5.22	FE
	331	5.12	FE
Rb	polycr	2.261	PE
Re	polycr	4.72	TE
Rh	polycr	4.98	PE
Ru	polycr	4.71	PE
Sb	amorp	4.55	
	100	4.7	
Sc	polycr	3.5	PE
Se	polycr	5.9	PE
Si	n	4.85	CPD
	p 100	(4.91)	CPD
	p 111	4.60	PE
Sm	polycr	2.7	PE
Sn	polycr	4.42	CPD
Sr	polycr	(2.59)	TH
Ta	polycr	4.25	TH
	100	4.15	TH
	110	4.80	TH
	111	4.00	TH
Tb	polycr	3.0	PE
Te	polycr	4.95	PE
Th	polycr	3.4	TH
Ti	polycr	4.33	PE
Tl	polycr	(3.84)	CPD
U	polycr	3.63	PE
	100	3.73	PE
	110	3.90	PE
	113	3.67	PE
V	polycr	4.3	PE
W	polycr	4.55	CPD
	100	4.63	FE
	110	5.22	FE
	111	4.45	FE
	113	4.46	FE
	116	4.32	TH
Y	polycr	3.1	PE
Zn	polycr	3.63	PE
Zr	polycr	(4.9)	CPD
	polycr	4.05	PE

SECONDARY ELECTRON EMISSION

The secondary emission yield, or secondary emission ratio, δ, is the average number of secondary electrons emitted from a bombarded material for every incident primary electron. It is a function of the primary electron energy E_p. The maximum yield δ_{max} corresponds to a primary electron energy E_{pmax} (see figure). The two primary electron energies corresponding to a yield of unity are denoted the first and second crossovers (E_I and E_{II}). An insulating target, or a conducting target that is electrically floating, will charge positively or negatively depending on the primary electron energy. For $E_I < E_p < E_{II}$, $\delta > 1$ and the surface charges positively provided there is a collector present that is positive with respect to the target. For $E_p < E_I$ or $E_p > E_{II}$, $\delta < 1$, and the surface charges negatively with respect to the potential of the source of primary electrons.

Primary Electron Energy (Ep)

Element	δ_{max}	E_{pmax} (eV)	E_I (eV)	E_{II} (eV)
Ag	1.5	800	200	>2000
Al	1.0	300	300	300
Au	1.4	800	150	>2000
B	1.2	150	50	600
Ba	0.8	400	None	None
Bi	1.2	550	None	None
Be	0.5	200	None	None
C (diamond)	2.8	750	None	>5000
C (graphite)	1.0	300	300	300
C (soot)	0.45	500	None	None
Cd	1.1	450	300	700
Co	1.2	600	200	None
Cs	0.7	400	None	None
Cu	1.3	600	200	1500
Fe	1.3	400	120	1400
Ga	1.55	500	75	None
Ge	1.15	500	150	900
Hg	1.3	600	350	>1200
K	0.7	200	None	None

Element	δ_{max}	E_{pmax} (eV)	E_I (eV)	E_{II} (eV)
Li	0.5	85	None	None
Mg	0.95	300	None	None
Mo	1.25	375	150	1200
Na	0.82	300	None	None
Nb	1.2	375	150	1050
Ni	1.3	550	150	>1500
Pb	1.1	500	250	1000
Pd	>1.3	>250	120	None
Pt	1.8	700	350	3000
Rb	0.9	350	None	None
Sb	1.3	600	250	2000
Si	1.1	250	125	500
Sn	1.35	500	None	None
Ta	1.3	600	250	>2000
Th	1.1	800	None	None
Ti	0.9	280	None	None
Tl	1.7	650	70	>1500
W	1.4	650	250	>1500
Zr	1.1	350	None	None

Compound	δ_{max}	E_{pmax} (eV)
Alkali halides		
CsCl	6.5	
KBr (crystal)	14	1800
KCl (crystal)	12	1600
KCl (layer)	7.5	1200
KI (crystal)	10	1600
KI (layer)	5.6	
LiF (crystal)	8.5	
LiF (layer)	5.6	700
NaBr (crystal)	24	1800
NaBr (layer)	6.3	
NaCl (crystal)	14	1200
NaCl (layer)	6.8	600
NaF (crystal)	14	1200
NaF (layer)	5.7	
NaI (crystal)	19	1300
NaI (layer)	5.5	
RbCl (layer)	5.8	
Oxides		
Ag_2O	1.0	
Al_2O_3 (layer)	2–9	
BaO (layer)	2.3–4.8	400

Compound	δ_{max}	E_{pmax} (eV)
BeO	3.4	2000
CaO	2.2	500
Cu_2O	1.2	400
MgO (crystal)	20–25	1500
MgO (layer)	3–15	400–1500
MoO_2	1.2	
SiO_2 (quartz)	2.1–4	400
SnO_2	3.2	640
Sulfides		
MoS_2	1.1	
PbS	1.2	500
WS_2	1.0	
ZnS	1.8	350
Others		
BaF_2 (layer)	4.5	
CaF_2 (layer)	3.2	
$BiCs_3$	6	1000
BiCs	1.9	1000
GeCs	7	700
Rb_3Sb	7.1	450
$SbCs_3$	6	700
Mica	2.4	350
Glasses	2–3	300–450

OPTICAL PROPERTIES OF SELECTED ELEMENTS

J. H. Weaver and H. P. R. Frederikse

These tables list the index of refraction n, the extinction coefficient k, and the normal incidence reflection R ($\phi = 0$) as a function of photon energy E, which is expressed in electron volts (eV). To convert the energy in eV to wavelength in μm, use $\lambda = 1.2398/E$. To compute the dielectric function $\tilde{\varepsilon} = \varepsilon_1 + i\varepsilon_2$ from the complex index of refraction $\tilde{N} = n + ik$, use $\varepsilon_1 = n^2 - k^2$ and $\varepsilon_2 = 2nk$.

The optical constants in these tables are abridged from three more extensive tabulations:

- *Optical Properties of Metals* (OPM), Volumes I and II, *Physics Data*, Nr. 18-1 and 18-2, J. H. Weaver, C. Krafka, D. W. Lynch, and E. E. Koch, Fachinformationzentrum, Karlsruhe, Germany.
- *Handbook of Optical Constants* (HOC), Vol. I, 1985, and Vol. II, 1991. E. D. Palik, Ed., Academic Press, Inc.
- *American Institute of Physics Handbook* (AIPH), 3rd Edition, D. E. Gray, Ed., McGraw-Hill, New York, 1972.

The first two of these major sources provide detailed comparisons of all optical data available in the literature at the time of the compilation. For critical applications the reader should refer to the original work. References for individual metals and semiconductors are listed at the end of the tables. Generally, tabulated values for the optical properties are accurate to better than 10%. Data in parentheses are extrapolated or interpolated values. For most elements the spectral range covered is from the far infrared (0.010 or 0.10 eV) to the far ultraviolet (10, 30 or 300 eV). The intervals between successive energies in the tables are chosen in such a way that the major spectral features are preserved.

Very small values of k are expressed in exponential notation, e.g., 1.23E-5 means 1.23×10^{-5}.

The following table is convenient for identifying the energy entries in these tables with the corresponding wavelengths:

λ	E/eV	λ	E/eV
1 mm	0.00124	6000 Å	2.066
500 μm	0.00248	5000 Å	2.480
100 μm	0.01240	4000 Å	3.100
50 μm	0.02480	3000 Å	4.133
10 μm	0.12398	2000 Å	6.199
5 μm	0.24797	1000 Å	12.398
1 μm	1.240	400 Å	30.996

Energy (eV)	n	k	R(φ = 0)	Energy (eV)	n	k	R(φ = 0)	Energy (eV)	n	k	R(φ = 0)
Aluminium[1]				2.200	1.018	6.846	0.9200	14.400	0.058	0.327	0.8102
0.040	98.595	203.701	0.9923	2.400	0.826	6.283	0.9228	14.600	0.067	0.273	0.7802
0.050	74.997	172.199	0.9915	2.600	0.695	5.800	0.9238	14.800	0.086	0.211	0.7202
0.060	62.852	150.799	0.9906	2.800	0.598	5.385	0.9242	15.000	0.125	0.153	0.6119
0.070	53.790	135.500	0.9899	3.000	0.523	5.024	0.9241	15.200	0.178	0.108	0.4903
0.080	45.784	123.734	0.9895	3.200	0.460	4.708	0.9243	15.400	0.234	0.184	0.3881
0.090	39.651	114.102	0.9892	3.400	0.407	4.426	0.9245	15.600	0.280	0.073	0.3182
0.100	34.464	105.600	0.9889	3.600	0.363	4.174	0.9246	15.800	0.318	0.065	0.2694
0.125	24.965	89.250	0.9884	3.800	0.326	3.946	0.9247	16.000	0.351	0.060	0.2326
0.150	18.572	76.960	0.9882	4.000	0.294	3.740	0.9248	16.200	0.380	0.055	0.2031
0.175	14.274	66.930	0.9879	4.200	0.267	3.552	0.9248	16.400	0.407	0.050	0.1789
0.200	11.733	59.370	0.9873	4.400	0.244	3.380	0.9249	16.750	0.448	0.045	0.1460
0.250	8.586	48.235	0.9858	4.600	0.223	3.222	0.9249	17.000	0.474	0.042	0.1278
0.300	6.759	40.960	0.9844	4.800	0.205	3.076	0.9249	17.250	0.498	0.040	0.1129
0.350	5.438	35.599	0.9834	5.000	0.190	2.942	0.9244	17.500	0.520	0.038	0.1005
0.400	4.454	31.485	0.9826	6.000	0.130	2.391	0.9257	17.750	0.540	0.036	0.0899
0.500	3.072	25.581	0.9817	6.500	0.110	2.173	0.9260	18.000	0.558	0.035	0.0809
0.600	2.273	21.403	0.9806	7.000	0.095	1.983	0.9262	18.500	0.591	0.032	0.0664
0.700	1.770	18.328	0.9794	7.500	0.082	1.814	0.9265	19.000	0.620	0.030	0.0554
0.800	1.444	15.955	0.9778	8.000	0.072	1.663	0.9269	19.500	0.646	0.028	0.0467
0.900	1.264	14.021	0.9749	8.500	0.063	1.527	0.9272	20.000	0.668	0.027	0.0398
1.000	1.212	12.464	0.9697	9.000	0.056	1.402	0.9277	20.500	0.689	0.025	0.0342
1.100	1.201	11.181	0.9630	9.500	0.049	1.286	0.9282	21.000	0.707	0.024	0.0296
1.200	1.260	10.010	0.9521	10.000	0.044	1.178	0.9286	21.500	0.724	0.023	0.0258
1.300	1.468	8.949	0.9318	10.500	0.040	1.076	0.9293	22.000	0.739	0.022	0.0226
1.400	2.237	8.212	0.8852	11.000	0.036	0.979	0.9298	22.500	0.753	0.021	0.0199
1.500	2.745	8.309	0.8678	11.500	0.033	0.883	0.9283	23.000	0.766	0.021	0.0177
1.600	2.625	8.597	0.8794	12.000	0.033	0.791	0.9224	23.500	0.778	0.020	0.0157
1.700	2.143	8.573	0.8972	12.500	0.034	0.700	0.9118	24.000	0.789	0.019	0.0140
1.800	1.741	8.205	0.9069	13.000	0.038	0.609	0.8960	24.500	0.799	0.018	0.0126
1.900	1.488	7.821	0.9116	13.500	0.041	0.517	0.8789	25.000	0.809	0.018	0.0113
2.000	1.304	7.479	0.9148	14.000	0.048	0.417	0.8486	25.500	0.817	0.017	0.0102
				14.200	0.053	0.373	0.8312	26.000	0.826	0.016	0.0092

Energy (eV)	n	k	$R(\phi = 0)$	Energy (eV)	n	k	$R(\phi = 0)$	Energy (eV)	n	k	$R(\phi = 0)$
27.000	0.840	0.015	0.0076	0.36		2.47E-05		7.30	3.376	0.473	0.303
28.000	0.854	0.014	0.0063	0.37		2.80E-05		7.40	3.335	0.515	0.300
29.000	0.865	0.014	0.0053	0.38		3.11E-05		7.50	3.321	0.533	0.299
30.000	0.876	0.013	0.0044	0.39		3.67E-05		7.60	3.306	0.592	0.300
35.000	0.915	0.010	0.0020	0.40		3.58E-05		7.80	3.276	0.659	0.300
40.000	0.940	0.008	0.0010	0.41		3.25E-05		8.00	3.251	0.712	0.300
45.000	0.957	0.007	0.0005	0.4133	2.3795		0.167	8.25	3.232	0.765	0.301
50.000	0.969	0.006	0.0003	0.42		2.94E-05		8.50	3.228	0.806	0.303
55.000	0.979	0.005	0.0001	0.43		2.87E-05		8.75	3.247	0.855	0.308
60.000	0.987	0.004	0.0000	0.44		3.14E-05		9.00	3.272	0.910	0.314
65.000	0.995	0.004	0.0000	0.45		3.62E-05		9.25	3.308	0.978	0.322
70.000	1.006	0.004	0.0000	0.46		3.22E-05		9.50	3.348	1.055	0.331
72.500	1.025	0.004	0.0002	0.47		1.57E-05		9.75	3.398	1.147	0.342
75.000	1.011	0.024	0.0002	0.48		6.17E-06		10.00	3.453	1.258	0.355
77.500	1.008	0.025	0.0002	0.4959	2.3801		0.167	10.25	3.514	1.403	0.371
80.000	1.007	0.024	0.0002	0.6199	2.3813		0.167	10.50	3.565	1.581	0.389
85.000	1.007	0.028	0.0002	0.8266	2.3837		0.167	10.75	3.600	1.813	0.411
90.000	1.005	0.031	0.0002	1.240	2.3905		0.168	11.00	3.582	2.078	0.434
95.000	0.999	0.036	0.0003	1.378	2.3934		0.169	11.25	3.507	2.380	0.460
100.000	0.991	0.030	0.0002	1.459	2.3953		0.169	11.50	3.346	2.693	0.488
110.000	0.994	0.025	0.0002	1.550	2.3975		0.169	11.75	3.090	2.986	0.518
120.000	0.991	0.024	0.0002	1.653	2.4003		0.170	12.00	2.736	3.228	0.551
130.000	0.987	0.021	0.0001	1.771	2.4036		0.170	12.20	2.383	3.354	0.580
140.000	0.989	0.016	0.0001	1.889	2.4073		0.171	12.40	1.983	3.382	0.610
150.000	0.990	0.015	0.0001	1.926	2.4084		0.171	12.60	1.532	3.265	0.641
160.000	0.989	0.014	0.0001	2.066	2.4133		0.171	12.80	1.312	2.953	0.627
170.000	0.989	0.011	0.0001	2.105	2.4147		0.172	13.00	1.223	2.722	0.604
180.000	0.990	0.010	0.0000	2.271	2.4210		0.173	13.50	1.129	2.379	0.557
190.000	0.990	0.009	0.0000	2.480	2.4299		0.174	14.00	1.070	2.178	0.526
200.000	0.991	0.007	0.0000	2.650	2.4380		0.175	14.50	1.018	2.034	0.504
220.000	0.992	0.006	0.0000	2.845		3.82E-07		15.00	0.972	1.929	0.489
240.000	0.993	0.005	0.0000	3.100	2.4627		0.178	15.50	0.917	1.845	0.482
260.000	0.993	0.004	0.0000	3.434	2.4849		0.182	16.00	0.861	1.767	0.477
280.000	0.994	0.003	0.0000	3.576	2.4955		0.183	16.50	0.805	1.692	0.474
300.000	0.995	0.002	0.0000	3.961		8.97E-07		17.00	0.753	1.619	0.471
				4.160	2.5465		0.190	17.50	0.707	1.546	0.467
Carbon (diamond)[2]				4.511		1.29E-06		18.00	0.665	1.476	0.463
0.06199	2.3741		0.166	4.8187	2.6205	1.47E-06	0.200	18.50	0.626	1.408	0.459
0.06888	2.3741		0.166	5.00	2.6383		0.203	19.00	0.589	1.341	0.455
0.07749	2.3745		0.166	5.30		2.98E-06		19.50	0.557	1.273	0.449
0.08856	2.3750		0.166	5.35		6.45E-06		20.00	0.527	1.203	0.442
0.1033	2.3757		0.166	5.40		1.04E-05		21.00	0.487	1.052	0.413
0.1240	2.3765		0.166	5.50		3.41E-05		22.00	0.518	0.888	0.330
0.1550	2.3772		0.166	5.55		5.48E-04		23.00	0.597	0.850	0.270
0.1907		3.1 E-05		5.60	2.740	1.48E-03	0.216	24.00	0.586	0.829	0.268
0.2066	2.3779	5.7 E-05	0.166	5.80	2.780	5.02E-03	0.222	25.00	0.562	0.787	0.265
0.22		1.21E-04		6.00	2.826	7.99E-03	0.228	26.00	0.538	0.736	0.260
0.23		2.36E-04		6.10	2.852	8.62E-03	0.231	27.00	0.516	0.679	0.252
0.24		3.82E-04		6.20	2.879	9.30E-03	0.235	28.00	0.501	0.616	0.239
0.25		5.21E-04		6.30	2.910	9.74E-03	0.239	29.00	0.494	0.552	0.221
0.26		2.96E-04		6.40	2.944	9.87E-03	0.243	30.00	0.493	0.490	0.201
0.27		4.39E-04		6.50	2.985	1.10E-02	0.248				
0.28		2.75E-04		6.60	3.031	1.47E-02	0.254	**Cesium (evaporated)[3]**			
0.29		7.82E-05		6.70	3.085	2.20E-02	0.261	2.145	0.264	1.123	0.631
0.30		1.32E-04		6.80	3.146	3.44E-02	0.268	2.271	0.278	0.950	0.561
0.31	2.3787	1.30E-04	0.167	6.90	3.220	5.24E-02	0.277	2.845	0.425	0.438	0.235
0.32		1.11E-04		7.00	3.322	9.35E-02	0.289	3.064	0.540	0.320	0.127
0.33		2.99E-05		7.10	3.444	0.210	0.304	3.397	0.671	0.233	0.057
0.34		1.89E-05		7.15	3.464	0.307	0.308	3.966	0.827	0.174	0.018
0.35		2.11E-05		7.20	3.437	0.388	0.307	4.889	0.916	0.143	0.007

Chromium[4]

Energy (eV)	n	k	R(φ = 0)
0.06	21.19	42.00	0.962
0.10	11.81	29.76	0.955
0.14	15.31	26.36	0.936
0.18	8.73	25.37	0.53
0.22	5.30	20.62	0.954
0.26	3.91	17.12	0.951
0.30	3.15	14.28	0.943
0.42	3.47	8.97	0.862
0.54	3.92	7.06	0.788
0.66	3.96	5.95	0.736
0.78	4.13	5.03	0.680
0.90	4.43	4.60	0.650
1.00	4.47	4.43	0.639
1.12	4.53	4.31	0.631
1.24	4.50	4.28	0.629
1.36	4.42	4.30	0.631
1.46	4.31	4.32	0.632
1.77	3.84	4.37	0.639
2.00	3.48	4.36	0.644
2.20	3.18	4.41	0.656
2.40	2.75	4.46	0.677
2.60	2.22	4.36	0.698
2.80	1.80	4.06	0.703
3.00	1.54	3.71	0.695
3.20	1.44	3.40	0.670
3.40	1.39	3.24	0.657
3.60	1.26	3.12	0.661
3.80	1.12	2.95	0.660
4.00	1.02	2.76	0.651
4.20	0.94	2.58	0.639
4.40	0.90	2.42	0.620
4.50	0.89	2.35	0.607
4.60	0.88	2.28	0.598
4.70	0.86	2.21	0.586
4.80	0.86	2.13	0.572
4.90	0.86	2.07	0.557
5.00	0.85	2.01	0.542
5.10	0.86	1.94	0.523
5.20	0.87	1.87	0.503
5.40	0.93	1.80	0.466
5.60	0.95	1.74	0.443
5.80	0.97	1.74	0.437
6.00	0.94	1.73	0.444
6.20	0.89	1.69	0.446
6.40	0.85	1.66	0.447
6.60	0.80	1.59	0.444
6.80	0.75	1.51	0.439
7.00	0.74	1.45	0.425
7.20	0.71	1.39	0.414
7.40	0.69	1.33	0.404
7.60	0.66	1.23	0.378
7.80	0.67	1.15	0.347
8.00	0.68	1.07	0.315
8.20	0.71	1.00	0.278
8.50	0.74	0.92	0.235
9.0	0.83	0.81	0.170
9.50	0.92	0.74	0.132
10.00	0.98	0.73	0.120
10.50	1.01	0.72	0.112

Energy (eV)	n	k	R(φ = 0)
11.00	1.05	0.69	0.103
11.50	1.09	0.69	0.100
12.00	1.13	0.70	0.101
12.50	1.15	0.73	0.108
13.00	1.15	0.77	0.119
13.50	1.12	0.80	0.128
14.00	1.09	0.82	0.135
14.50	1.03	0.82	0.142
15.00	1.00	0.82	0.143
15.50	0.96	0.80	0.141
16.00	0.92	0.77	0.139
16.50	0.31	0.75	0.134
17.00	0.90	0.73	0.132
17.50	0.88	0.72	0.130
18.00	0.87	0.70	0.129
18.50	0.84	0.69	0.130
19.00	0.82	0.68	0.131
20.00	0.77	0.64	0.130
20.5	0.76	0.63	0.129
21.0	0.74	0.58	0.121
21.5	0.72	0.55	0.116
22.0	0.71	0.52	0.112
22.5	0.70	0.50	0.109
23.0	0.69	0.48	0.105
23.5	0.68	0.45	0.101
24.0	0.68	0.43	0.096
24.5	0.67	0.39	0.089
25.0	0.68	0.36	0.080
25.5	0.68	0.33	0.072
26.0	0.70	0.31	0.063
26.5	0.71	0.28	0.055
27.0	0.72	0.26	0.048
27.5	0.73	0.25	0.043
28.0	0.75	0.23	0.037
29.0	0.77	0.22	0.032
30.0	0.78	0.21	0.030

Cobalt, single crystal, $\vec{E} \parallel \hat{c}$[5]

Energy (eV)	n	k	R(φ = 0)
0.10	6.71	37.87	0.982
0.15	4.66	25.47	0.973
0.20	3.55	18.78	0.962
0.25	3.98	14.59	0.933
0.30	4.04	12.16	0.907
0.40	4.24	9.13	0.847
0.50	4.41	7.19	0.782
0.60	4.91	6.13	0.729
0.70	5.24	5.85	0.713
0.80	5.17	5.89	0.716
0.90	4.94	5.95	0.720
1.00	4.46	5.86	0.722
1.10	4.07	5.61	0.715
1.20	3.81	5.36	0.706
1.30	3.60	5.20	0.701
1.40	3.37	5.09	0.701
1.50	3.10	4.96	0.701
1.60	2.84	4.77	0.697
1.70	2.66	4.57	0.690
1.80	2.45	4.41	0.687
1.90	2.31	4.18	0.675
2.00	2.21	4.00	0.664
2.10	2.13	3.85	0.654

Energy (eV)	n	k	R(φ = 0)
2.20	2.07	3.70	0.642
2.30	2.01	3.59	0.634
2.40	1.95	3.49	0.627
2.50	1.88	3.40	0.622
2.60	1.81	3.32	0.618
2.70	1.73	3.24	0.615
2.80	1.66	3.13	0.607
2.90	1.61	3.05	0.600
3.00	1.55	2.96	0.594
3.20	1.46	2.80	0.579
3.40	1.38	2.64	0.563
3.60	1.31	2.48	0.544
3.80	1.28	2.33	0.519
4.00	1.26	2.20	0.495
4.20	1.25	2.10	0.471
4.40	1.24	2.01	0.452
4.60	1.24	1.94	0.435
4.80	1.23	1.88	0.423
5.00	1.22	1.83	0.411
5.20	1.21	1.79	0.403
5.40	1.19	1.77	0.399
5.60	1.16	1.75	0.400
5.80	1.10	1.73	0.406
6.00	1.03	1.68	0.407
6.20	0.97	1.62	0.401
6.40	0.94	1.53	0.386
6.60	0.91	1.46	0.368
6.80	0.91	1.38	0.345
7.00	0.91	1.32	0.326
7.00	0.91	1.26	0.305
7.40	0.92	1.21	0.286
7.60	0.93	1.17	0.269
7.80	0.94	1.13	0.253
8.00	0.95	1.09	0.239

Cobalt, single crystal, $\vec{E} \perp \hat{c}$[5]

Energy (eV)	n	k	R(φ = 0)
0.10	5.83	32.36	0.979
0.15	4.24	21.37	0.965
0.20	3.87	15.53	0.042
0.30	4.34	10.01	0.865
0.40	4.66	7.39	0.785
0.50	5.17	5.75	0.709
0.60	5.77	5.17	0.682
0.70	6.15	5.20	0.685
0.80	6.08	5.61	0.702
0.90	5.57	5.93	0.715
1.00	4.83	5.94	0.721
1.10	4.31	5.60	0.711
1.20	4.02	5.34	0.701
1.30	3.78	5.16	0.694
1.40	3.55	5.05	0.692
1.50	3.26	4.93	0.692
1.60	3.03	4.74	0.687
1.70	2.83	4.60	0.684
1.80	2.61	4.45	0.683
1.90	2.41	4.27	0.677
2.00	2.25	4.09	0.670
2.10	2.13	3.89	0.659
2.20	2.04	3.72	0.646
2.30	1.99	3.56	0.632
2.40	1.95	3.44	0.620

Energy (eV)	n	k	$R(\varphi = 0)$
2.50	1.90	3.34	0.611
2.60	1.86	3.26	0.605
2.70	1.79	3.19	0.602
2.80	1.72	3.11	0.596
2.90	1.66	3.03	0.591
3.00	1.60	2.94	0.586
3.20	1.50	2.78	0.571
3.40	1.42	2.62	0.553
3.60	1.36	2.47	0.533
3.80	1.33	2.33	0.511
4.00	1.31	2.21	0.488
4.20	1.28	2.12	0.471
4.40	1.27	2.03	0.452
4.60	1.26	1.95	0.435
4.80	1.25	1.90	0.423
5.00	1.24	1.84	0.411
5.20	1.22	1.80	0.403
5.40	1.21	1.78	0.399
5.60	1.17	1.76	0.400
5.80	1.11	1.74	0.406
6.00	1.04	1.69	0.407
6.20	0.98	1.62	0.401
6.40	0.94	1.54	0.386
6.60	0.92	1.46	0.368
6.80	0.91	1.38	0.345
7.00	0.91	1.32	0.326
7.20	0.91	1.26	0.305
7.40	0.92	1.21	0.285
7.60	0.93	1.17	0.269
7.80	0.94	1.13	0.253

Copper[6]

Energy (eV)	n	k	$R(\varphi = 0)$
0.10	29.69	71.57	0.980
0.50	1.71	17.63	0.979
1.00	0.44	8.48	0.976
1.50	0.26	5.26	0.965
1.70	0.22	4.43	0.958
1.75	0.21	4.25	0.956
1.80	0.21	4.04	0.952
1.85	0.22	3.85	0.947
1.90	0.21	3.67	0.943
2.00	0.27	3.24	0.910
2.10	0.47	2.81	0.814
2.20	0.83	2.60	0.673
2.30	1.04	2.59	0.618
2.40	1.12	2.60	0.602
2.60	1.15	2.50	0.577
2.80	1.17	2.36	0.545
3.00	1.18	2.21	0.509
3.20	1.23	2.07	0.468
3.40	1.27	1.95	0.434
3.60	1.31	1.87	0.407
3.80	1.34	1.81	0.387
4.00	1.34	1.72	0.364
4.20	1.42	1.64	0.336
4.40	1.49	1.64	0.329
4.60	1.52	1.67	0.334
4.80	1.53	1.71	0.345
5.00	1.47	1.78	0.366
5.20	1.38	1.80	0.380
5.40	1.28	1.78	0.389

Energy (eV)	n	k	$R(\varphi = 0)$
5.60	1.18	1.74	0.391
5.80	1.10	1.67	0.389
6.00	1.04	1.59	0.380
6.50	0.96	1.37	0.329
7.00	0.97	1.20	0.271
7.50	1.00	1.09	0.230
8.00	1.03	1.03	0.206
8.50	1.03	0.98	0.189
9.00	1.03	0.92	0.171
9.50	1.03	0.87	0.154
10.00	1.04	0.82	0.139
11.00	1.07	0.75	0.118
12.00	1.09	0.73	0.111
13.00	1.08	0.72	0.109
14.00	1.06	0.72	0.111
14.50	1.03	0.72	0.111
15.00	1.01	0.71	0.111
15.50	0.98	0.69	0.109
16.00	0.95	0.67	0.106
17.00	0.91	0.62	0.097
18.00	0.89	0.56	0.084
19.00	0.88	0.51	0.071
20.00	0.88	0.45	0.059
21.00	0.90	0.41	0.048
22.00	0.92	0.38	0.040
23.00	0.94	0.37	0.035
24.00	0.96	0.37	0.035
25.00	0.96	0.40	0.040
26.00	0.92	0.40	0.044
27.00	0.88	0.38	0.043
28.00	0.86	0.35	0.039
29.00	0.85	0.30	0.032
30.00	0.86	0.26	0.025
31.00	0.88	0.24	0.020
32.00	0.89	0.22	0.017
33.00	0.90	0.21	0.015
34.00	0.91	0.20	0.014
35.00	0.92	0.20	0.013
36.00	0.92	0.19	0.012
37.00	0.92	0.19	0.011
38.00	0.93	0.18	0.010
39.00	0.93	0.17	0.009
40.00	0.93	0.17	0.009
41.00	0.94	0.16	0.008
42.00	0.94	0.16	0.007
43.00	0.94	0.15	0.007
44.00	0.95	0.15	0.007
45.00	0.95	0.15	0.006
46.00	0.95	0.15	0.006
47.00	0.95	0.14	0.006
48.00	0.95	0.14	0.006
49.00	0.95	0.14	0.005
50.00	0.95	0.13	0.005
51.00	0.95	0.13	0.005
52.00	0.95	0.13	0.005
53.00	0.96	0.12	0.004
54.00	0.96	0.12	0.004
55.00	0.96	0.12	0.004
56.00	0.96	0.11	0.004
57.00	0.96	0.11	0.004

Energy (eV)	n	k	$R(\varphi = 0)$
58.00	0.96	0.11	0.004
59.00	0.97	0.11	0.003
60.00	0.97	0.11	0.003
61.00	0.97	0.11	0.003
62.00	0.97	0.11	0.003
63.00	0.96	0.10	0.003
64.00	0.96	0.10	0.003
65.00	0.97	0.10	0.003
66.00	0.97	0.10	0.003
67.00	0.97	0.09	0.003
68.00	0.97	0.09	0.002
69.00	0.97	0.09	0.002
70.00	0.97	0.09	0.002
75.00	0.98	0.09	0.002
80.00	0.98	0.09	0.002
85.00	0.97	0.09	0.002
90.00	0.96	0.08	0.002

Gallium (liquid)[7]

Energy (eV)	n	k	$R(\varphi = 0)$
1.425	2.40	9.20	0.900
1.550	2.09	8.50	0.898
1.771	1.65	7.60	0.898
2.066	1.25	6.60	0.897
2.480	0.89	5.60	0.898
3.100	0.59	4.50	0.896

Germanium, single crystal[8]

Energy (eV)	n	k	$R(\varphi = 0)$
0.01240	(4.0065)	3.00E-03	0.361
0.01364	4.0063	2.40E-03	0.361
0.01488	(4.0060)	1.70E-03	0.361
0.01612	(4.0060)	1.55E-03	0.361
0.01736	(4.0060)	1.50E-03	0.361
0.01860		1.50E-03	
0.01984		1.60E-03	
0.02108		1.60E-03	
0.02232		1.55E-03	
0.02356		1.53E-03	
0.02480		1.50E-03	
0.02604		1.25E-03	
0.02728		8.50E-04	
0.02852		6.50E-04	
0.02976		7.00E-04	
0.03100	3.9827	8.50E-04	0.358
0.03224		1.55E-03	
0.03348		2.75E-03	
0.03472		3.55E-03	
0.03596	(3.9900)	3.05E-03	0.359
0.03720		2.75E-03	
0.03844		2.70E-03	
0.03968	(3.9930)	2.90E-03	0.359
0.04092		2.95E-03	
0.04215		3.20E-03	
0.04339		6.30E-03	
0.04463		3.40E-03	
0.04587	(3.9955)	2.50E-03	0.360
0.04711		2.10E-03	
0.04835		2.00E-03	
0.04959		8.00E-04	
0.05083		1.40E-03	
0.05207		1.35E-03	
0.05331		1.10E-03	
0.05455		8.00E-04	

Energy (eV)	n	k	$R(\phi = 0)$
0.05579		6.00E-04	
0.05703		9.0 E-04	
0.05827		6.5 E-04	
0.05951		4.6 E-04	
0.06075		4.0 E-04	
0.06199	3.9992	3.98E-04	0.360
0.06323		4.0 E-04	
0.06447		4.3 E-04	
0.06571		4.4 E-04	
0.06695	(4.0000)	4.3 E-04	0.360
0.06819		3.1 E-04	
0.06943		3.3 E-04	
0.07067		3.8 E-04	
0.07191		3.3 E-04	
0.07315		2.5 E-04	
0.07439		1.9 E-04	
0.07514		1.58E-04	
0.07749	4.0009	9.55E-05	0.360
0.07999	4.0011	1.71E-04	0.360
0.08266	4.0013	9.78E-05	0.360
0.08551	4.0015	5.77E-05	0.360
0.08920		3.98E-05	
0.09460		4.59E-05	
0.09840		3.51E-05	
0.1	4.0063	3.70E-05	0.361
0.2	4.0108		0.361
0.3	4.0246		0.362
0.4	4.0429		0.364
0.5	(4.074)		0.367
0.6	(4.104)	6.58E-07	0.370
0.7	4.180	1.27E-04	0.377
0.8	4.275	5.67E-03	0.385
0.9	4.285	7.45E-02	0.386
1.0	4.325	8.09E-02	0.390
1.1	4.385	0.103	0.395
1.2	4.420	0.123	0.398
1.3	4.495	0.167	0.405
1.4	4.560	0.190	0.411
1.5	4.635	0.298	0.418
1.6	4.763	0.345	0.428
1.7	4.897	0.401	0.439
1.8	5.067	0.500	0.453
1.9	5.380	0.540	0.475
2.0	5.588	0.933	0.495
2.1	5.748	1.634	0.523
2.2	5.283	2.049	0.516
2.3	5.062	2.318	0.519
2.4	4.610	2.455	0.508
2.5	4.340	2.384	0.492
2.6	4.180	2.309	0.480
2.7	4.082	2.240	0.471
2.8	4.035	2.181	0.464
2.9	4.037	2.140	0.461
3.0	4.082	2.145	0.463
3.1	4.141	2.215	0.471
3.2	4.157	2.340	0.482
3.3	4.128	2.469	0.490
3.4	4.070	2.579	0.497
3.5	4.020	2.667	0.502
3.6	3.985	2.759	0.509

Energy (eV)	n	k	$R(\phi = 0)$
3.7	3.958	2.863	0.517
3.8	3.936	2.986	0.527
3.9	3.920	3.137	0.539
4.0	3.905	3.336	0.556
4.1	3.869	3.614	0.579
4.2	3.745	4.009	0.612
4.3	3.338	4.507	0.659
4.4	2.516	4.669	0.705
4.5	1.953	4.297	0.713
4.6	1.720	3.960	0.702
4.7	1.586	3.709	0.690
4.8	1.498	3.509	0.677
4.9	1.435	3.342	0.664
5.0	1.394	3.197	0.650
5.1	1.370	3.073	0.636
5.2	1.364	2.973	0.622
5.3	1.371	2.897	0.609
5.4	1.383	2.854	0.600
5.5	1.380	2.842	0.598
5.6	1.360	2.846	0.602
5.7	1.293	2.163	0.479
5.8	1.209	2.873	0.632
5.9	1.108	2.813	0.641
6.0	1.30	2.34	0.517
6.5	1.10	2.05	0.489
7.0	1.00	1.80	0.448
7.5		1.60	
8.0	0.92	1.40	0.348
8.5	0.92	1.20	0.282
9.0	0.92	1.14	0.262
9.5		1.00	
10.0	0.93	0.86	0.167
20.0		0.237	
22.0		0.179	
24.0		0.144	
26.0		0.110	
28.0		0.0747	
30.0		0.1020	
32.0		0.0999	
34.0		0.0856	
36.0		0.0740	
38.0		0.0651	
40.0		0.0604	

Gold, electropolished, Au (110)[9]

Energy (eV)	n	k	$R(\phi = 0)$
0.10	8.17	82.83	0.995
0.20	2.13	41.73	0.995
0.30	0.99	27.82	0.995
0.40	0.59	20.83	0.995
0.50	0.39	16.61	0.994
0.60	0.28	13.78	0.994
0.70	0.22	11.75	0.994
0.80	0.18	10.21	0.993
0.90	0.15	9.01	0.993
1.00	0.13	8.03	0.992
1.20	0.10	6.54	0.991
1.40	0.08	5.44	0.989
1.60	0.08	4.56	0.986
1.80	0.09	3.82	0.979
2.00	0.13	3.16	0.953
2.10	0.18	2.84	0.925

Energy (eV)	n	k	$R(\phi = 0)$
2.20	0.24	2.54	0.880
2.40	0.50	1.86	0.647
2.50	0.82	1.59	0.438
2.60	1.24	1.54	0.331
2.70	1.43	1.72	0.356
2.80	1.46	1.77	0.368
2.90	1.50	1.79	0.368
3.00	1.54	1.80	0.369
3.10	1.54	1.81	0.371
3.20	1.54	1.80	0.368
3.30	1.55	1.78	0.362
3.40	1.56	1.76	0.356
3.50	1.58	1.73	0.349
3.60	1.62	1.73	0.346
3.70	1.64	1.75	0.351
3.80	1.63	1.79	0.360
3.90	1.59	1.81	0.366
4.00	1.55	1.81	0.369
4.10	1.51	1.79	0.368
4.20	1.48	1.78	0.367
4.30	1.45	1.77	0.368
4.40	1.41	1.76	0.370
4.50	1.35	1.74	0.370
4.60	1.30	1.69	0.364
4.70	1.27	1.64	0.354
4.80	1.25	1.59	0.344
4.90	1.23	1.54	0.332
5.00	1.22	1.49	0.319
5.20	1.21	1.40	0.295
5.40	1.21	1.33	0.275
5.60	1.21	1.27	0.256
5.80	1.21	1.20	0.236
6.00	1.22	1.14	0.218
6.20	1.24	1.09	0.203
6.40	1.25	1.05	0.190
6.60	1.27	1.01	0.177
6.80	1.30	0.97	0.167
7.00	1.34	0.95	0.162
7.20	1.36	0.95	0.161
7.40	1.38	0.96	0.164
7.60	1.38	0.98	0.169
7.80	1.35	0.99	0.171
8.00	1.31	0.96	0.165
8.20	1.30	0.92	0.155
8.40	1.30	0.89	0.147
8.60	1.31	0.88	0.144
8.80	1.31	0.86	0.140
9.00	1.30	0.83	0.133
9.20	1.31	0.81	0.126
9.40	1.33	0.78	0.122
9.60	1.36	0.78	0.121
9.80	1.37	0.79	0.124
10.00	1.37	0.80	0.126
10.20	1.36	0.80	0.127
10.40	1.35	0.80	0.125
10.60	1.34	0.79	0.123
10.80	1.34	0.77	0.120
11.00	1.34	0.76	0.116
11.20	1.34	0.74	0.113
11.40	1.35	0.73	0.111

Energy (eV)	n	k	$R(\phi = 0)$
11.60	1.36	0.72	0.109
11.80	1.38	0.71	0.108
12.00	1.39	0.71	0.109
12.40	1.44	0.73	0.115
12.80	1.45	0.79	0.127
13.20	1.42	0.84	0.137
13.60	1.37	0.86	0.140
14.00	1.33	0.86	0.140
14.40	1.29	0.86	0.139
14.80	1.26	0.84	0.135
15.20	1.24	0.83	0.132
15.60	1.22	0.81	0.127
16.00	1.21	0.79	0.123
16.40	1.20	0.78	0.119
16.80	1.19	0.76	0.116
17.20	1.19	0.75	0.114
17.60	1.19	0.74	0.111
18.00	1.19	0.74	0.109
18.40	1.19	0.73	0.109
18.80	1.20	0.74	0.110
19.20	1.21	0.76	0.116
19.60	1.21	0.80	0.125
20.00	1.18	0.83	0.133
20.40	1.14	0.85	0.141
20.80	1.10	0.87	0.149
21.20	1.05	0.88	0.156
21.60	1.00	0.88	0.162
22.00	0.94	0.86	0.164
22.40	0.89	0.83	0.163
22.80	0.85	0.79	0.157
23.20	0.82	0.75	0.149
23.60	0.80	0.70	0.138
24.00	0.80	0.66	0.125
24.40	0.80	0.62	0.113
24.80	0.80	0.58	0.101
25.20	0.82	0.56	0.090
25.60	0.83	0.54	0.084
26.00	0.84	0.52	0.079
26.40	0.85	0.51	0.074
26.80	0.85	0.50	0.071
27.20	0.86	0.49	0.068
27.60	0.86	0.49	0.065
28.00	0.87	0.48	0.063
28.40	0.88	0.48	0.062
28.80	0.88	0.48	0.062
29.20	0.88	0.48	0.062
29.60	0.87	0.48	0.064
30.00	0.86	0.48	0.064

Hafnium, single crystal, $\vec{\mathrm{E}} \parallel \hat{c}$[10]

Energy (eV)	n	k	$R(\phi = 0)$
0.52	1.48	4.11	0.747
0.56	1.84	3.29	0.615
0.60	2.34	2.62	0.486
0.66	3.21	2.13	0.428
0.70	3.70	2.03	0.441
0.76	4.31	2.10	0.476
0.80	4.61	2.31	0.504
0.86	4.71	2.70	0.533
0.90	4.64	2.85	0.541
0.95	4.54	2.96	0.545
1.00	4.45	3.00	0.545
1.10	4.28	3.08	0.547
1.20	4.08	3.10	0.544
1.30	3.87	3.04	0.536
1.40	3.72	2.95	0.525
1.50	3.60	2.85	0.514
1.60	3.52	2.73	0.500
1.70	3.52	2.61	0.488
1.80	3.57	2.56	0.485
1.90	3.63	2.59	0.489
2.00	3.65	2.67	0.498
2.10	3.64	2.81	0.511
2.20	3.53	2.99	0.526
2.30	3.34	3.09	0.534
2.40	3.15	3.11	0.537
2.50	2.99	3.13	0.540
2.60	2.83	3.12	0.542
2.70	2.68	3.10	0.542
2.80	2.54	3.08	0.543
2.90	2.40	3.04	0.544
3.00	2.27	3.00	0.544
3.10	2.14	2.95	0.544
3.20	2.00	2.89	0.544
3.30	1.87	2.79	0.538
3.40	1.78	2.68	0.528
3.50	1.71	2.58	0.517
3.60	1.66	2.48	0.503
3.70	1.63	2.40	0.491
3.80	1.60	2.33	0.481
3.90	1.56	2.27	0.473
4.00	1.52	2.21	0.466
4.10	1.48	2.14	0.455
4.20	1.45	2.07	0.442
4.30	1.43	2.01	0.431
4.40	1.41	1.95	0.420
4.50	1.39	1.89	0.407
4.60	1.39	1.83	0.394
4.70	1.39	1.79	0.382
4.80	1.38	1.75	0.373
4.90	1.38	1.71	0.364
5.00	1.37	1.68	0.356
5.20	1.36	1.61	0.341
5.40	1.35	1.55	0.324
5.60	1.35	1.51	0.314
5.80	1.32	1.48	0.308
6.00	1.28	1.41	0.295
6.20	1.26	1.35	0.278
6.40	1.26	1.28	0.258
6.60	1.27	1.22	0.240
6.80	1.28	1.16	0.224
7.00	1.31	1.13	0.212
7.20	1.33	1.10	0.204
7.40	1.34	1.07	0.197
7.60	1.36	1.05	0.191
7.80	1.37	1.02	0.183
8.00	1.40	1.01	0.179
8.20	1.43	1.01	0.178
8.40	1.45	1.01	0.180
8.60	1.47	1.02	0.183
8.80	1.48	1.04	0.186
9.00	1.49	1.07	0.193
9.20	1.50	1.10	0.201
9.40	1.48	1.14	0.211
9.60	1.46	1.18	0.222
9.80	1.41	1.21	0.230
10.00	1.36	1.22	0.235
10.20	1.32	1.22	0.238
10.40	1.28	1.22	0.240
10.60	1.24	1.21	0.241
10.80	1.20	1.20	0.242
11.00	1.16	1.19	0.242
11.20	1.13	1.17	0.241
11.40	1.10	1.16	0.241
11.60	1.07	1.14	0.239
11.80	1.04	1.12	0.238
12.00	1.02	1.10	0.236
12.40	0.96	1.06	0.232
12.80	0.92	1.01	0.225
13.20	0.88	0.96	0.218
13.60	0.84	0.90	0.205
14.00	0.83	0.83	0.186
14.40	0.83	0.80	0.172
14.80	0.81	0.76	0.167
15.20	0.79	0.70	0.153
15.60	0.79	0.64	0.132
16.00	0.83	0.60	0.111
16.40	0.81	0.60	0.114
16.80	0.79	0.55	0.105
17.20	0.79	0.50	0.089
17.60	0.80	0.46	0.077
18.00	0.81	0.42	0.064
18.40	0.84	0.38	0.051
18.80	0.87	0.34	0.040
19.00	0.89	0.33	0.036
19.60	0.93	0.32	0.030
20.00	0.94	0.31	0.027
20.60	0.97	0.30	0.023
21.00	0.99	0.29	0.022
21.60	1.01	0.28	0.020
22.00	1.03	0.28	0.020
22.60	1.06	0.28	0.020
23.00	1.07	0.28	0.021
23.60	1.09	0.29	0.022
24.00	1.09	0.30	0.023
24.60	1.10	0.31	0.024

Hafnium, single crystal, $\vec{\mathrm{E}} \perp \hat{c}$[10]

Energy (eV)	n	k	$R(\phi = 0)$
0.52	2.25	4.65	0.723
0.56	2.34	3.66	0.623
0.60	2.84	2.89	0.512
0.66	3.71	2.35	0.469
0.70	4.26	2.21	0.482
0.76	4.97	2.33	0.521
0.80	5.41	2.62	0.554
0.86	5.46	3.36	0.593
0.90	5.22	3.62	0.601
0.95	4.95	3.72	0.602
1.00	4.76	3.76	0.602
1.10	4.43	3.80	0.601
1.20	4.07	3.74	0.594
1.30	3.79	3.55	0.578
1.40	3.61	3.36	0.561

Energy (eV)	n	k	$R(\phi = 0)$	Energy (eV)	n	k	$R(\phi = 0)$	Energy (eV)	n	k	$R(\phi = 0)$
1.50	3.55	3.13	0.540	10.00	1.32	1.21	0.230	1.60	2.68	5.08	0.728
1.60	3.58	3.01	0.529	10.20	1.27	1.21	0.234	1.70	2.69	4.92	0.716
1.70	3.63	2.98	0.526	10.40	1.23	1.20	0.235	1.80	2.64	4.81	0.710
1.80	3.66	3.02	0.530	10.60	1.19	1.20	0.237	1.90	2.57	4.68	0.704
1.90	3.63	3.14	0.541	10.80	1.15	1.19	0.237	2.00	2.50	4.57	0.699
2.00	3.51	3.26	0.551	11.00	1.12	1.17	0.237	2.10	2.40	4.48	0.697
2.10	3.35	3.33	0.558	11.20	1.08	1.16	0.237	2.20	2.29	4.38	0.695
2.20	3.18	3.36	0.563	11.40	1.05	1.14	0.236	2.30	2.18	4.26	0.692
2.30	2.99	3.39	0.568	11.60	1.03	1.12	0.235	2.40	2.07	4.14	0.689
2.40	2.78	3.35	0.569	11.80	1.00	1.10	0.233	2.50	1.98	4.00	0.682
2.50	2.65	3.26	0.562	12.00	0.97	1.08	0.231	2.60	1.91	3.86	0.673
2.60	2.54	3.22	0.560	12.40	0.92	1.04	0.226	2.70	1.85	3.73	0.665
2.70	2.42	3.17	0.559	12.80	0.88	0.99	0.219	2.80	1.81	3.61	0.655
2.80	2.31	3.13	0.558	13.20	0.83	0.94	0.211	2.90	1.77	3.51	0.646
2.90	2.20	3.08	0.558	13.60	0.80	0.88	0.196	3.00	1.73	3.43	0.640
3.00	2.08	3.05	0.561	14.00	0.79	0.81	0.177	3.20	1.62	3.26	0.629
3.10	1.94	2.98	0.560	14.40	0.80	0.77	0.160	3.40	1.53	3.05	0.610
3.20	1.83	2.88	0.555	14.80	0.77	0.73	0.154	3.60	1.52	2.81	0.573
3.30	1.74	2.78	0.547	15.20	0.76	0.68	0.140	3.80	1.61	2.69	0.541
3.40	1.68	2.69	0.538	15.60	0.76	0.61	0.119	4.00	1.64	2.68	0.535
3.50	1.62	2.61	0.529	16.00	0.81	0.58	0.099	4.20	1.58	2.71	0.549
3.60	1.57	2.52	0.519	16.40	0.78	0.57	0.102	4.40	1.45	2.68	0.561
3.70	1.53	2.45	0.510	16.80	0.77	0.53	0.092	4.60	1.31	2.60	0.567
3.80	1.49	2.38	0.501	17.20	0.77	0.48	0.077	4.80	1.18	2.49	0.570
3.90	1.45	2.32	0.493	17.60	0.79	0.44	0.065	5.00	1.10	2.35	0.559
4.00	1.41	2.25	0.484	18.00	0.80	0.39	0.053	5.20	1.04	2.22	0.543
4.10	1.38	2.18	0.474	18.40	0.82	0.36	0.041	5.40	1.00	2.09	0.522
4.20	1.35	2.11	0.462	18.80	0.86	0.33	0.032	5.60	0.98	1.98	0.499
4.30	1.33	2.05	0.451	19.00	0.88	0.32	0.030	5.80	0.96	1.86	0.474
4.40	1.31	1.99	0.438	19.60	0.91	0.31	0.025	6.00	0.95	1.78	0.454
4.50	1.30	1.93	0.427	20.00	0.93	0.30	0.023	6.20	0.94	1.68	0.427
4.60	1.29	1.88	0.415	20.60	0.96	0.29	0.021	6.40	0.94	1.59	0.401
4.70	1.28	1.82	0.402	21.00	0.97	0.29	0.020	6.60	0.94	1.50	0.375
4.80	1.28	1.77	0.389	21.60	1.00	0.28	0.019	6.80	0.95	1.42	0.345
4.90	1.27	1.73	0.379	22.00	1.01	0.28	0.019	7.00	0.97	1.34	0.318
5.00	1.27	1.69	0.367	22.60	1.03	0.27	0.018	7.20	0.99	1.27	0.290
5.20	1.27	1.62	0.349	23.00	1.05	0.28	0.019	7.40	1.02	1.20	0.262
5.40	1.27	1.57	0.335	23.60	1.06	0.28	0.020	7.60	1.03	1.14	0.241
5.60	1.26	1.52	0.322	24.00	1.07	0.29	0.021	7.80	1.08	1.06	0.208
5.80	1.24	1.48	0.313	24.60	1.09	0.30	0.022	8.00	1.13	1.03	0.191
6.00	1.21	1.42	0.302					8.20	1.18	1.00	0.179
6.20	1.19	1.36	0.285	*Iridium*[11]				8.40	1.22	0.98	0.171
6.40	1.18	1.29	0.265	0.10	28.49	60.62	0.975	8.60	1.26	0.96	0.164
6.60	1.19	1.22	0.244	0.15	15.32	45.15	0.973	8.80	1.29	0.95	0.160
6.80	1.21	1.18	0.230	0.20	9.69	35.34	0.972	9.00	1.33	0.94	0.157
7.00	1.22	1.14	0.217	0.25	6.86	28.84	0.969	9.20	1.36	0.95	0.159
7.20	1.23	1.10	0.206	0.30	5.16	24.25	0.967	9.40	1.39	0.95	0.161
7.40	1.26	1.06	0.194	0.35	4.11	20.79	0.964	9.60	1.42	0.97	0.163
7.60	1.28	1.04	0.187	0.40	3.42	18.06	0.960	9.80	1.44	0.99	0.169
7.80	1.30	1.02	0.180	0.45	3.05	15.82	0.954	10.00	1.45	1.01	0.175
8.00	1.33	1.00	0.174	0.50	2.98	14.06	0.944	10.20	1.45	1.04	0.182
8.20	1.35	0.99	0.173	0.60	2.79	11.58	0.925	10.40	1.44	1.07	0.187
8.40	1.38	0.99	0.173	0.70	2.93	9.78	0.895	10.60	1.43	1.09	0.193
8.60	1.40	1.00	0.174	0.80	3.14	8.61	0.862	10.80	1.41	1.12	0.200
8.80	1.42	1.02	0.178	0.90	3.19	7.88	0.840	11.00	1.38	1.13	0.206
9.00	1.43	1.04	0.184	1.00	3.15	7.31	0.822	11.20	1.34	1.14	0.208
9.20	1.45	1.08	0.193	1.10	3.04	6.84	0.808	11.40	1.31	1.13	0.208
9.40	1.43	1.12	0.204	1.20	2.96	6.41	0.791	11.60	1.28	1.12	0.206
9.60	1.40	1.16	0.214	1.30	2.85	6.07	0.779	11.80	1.25	1.10	0.203
9.80	1.37	1.19	0.223	1.40	2.72	5.74	0.767	12.00	1.24	1.08	0.199
				1.50	2.65	5.39	0.750				

Optical Properties of Selected Elements

Energy (eV)	n	k	$R(\phi = 0)$
12.40	1.21	1.05	0.191
12.80	1.19	1.01	0.181
13.20	1.18	0.98	0.173
13.60	1.17	0.95	0.165
14.00	1.16	0.91	0.155
14.40	1.17	0.88	0.147
14.80	1.18	0.87	0.142
15.20	1.19	0.84	0.136
15.60	1.20	0.83	0.133
16.00	1.21	0.83	0.131
16.40	1.23	0.82	0.129
16.80	1.25	0.82	0.127
17.20	1.28	0.83	0.131
17.60	1.30	0.87	0.140
18.00	1.30	0.93	0.154
18.40	1.27	0.97	0.166
18.80	1.24	1.00	0.176
19.20	1.20	1.03	0.187
19.60	1.15	1.05	0.197
20.00	1.10	1.06	0.205
20.50	1.04	1.05	0.210
21.00	0.99	1.04	0.215
21.50	0.94	1.02	0.220
22.00	0.89	1.00	0.222
22.50	0.84	0.99	0.228
23.00	0.79	0.96	0.232
23.50	0.76	0.92	0.228
24.00	0.73	0.87	0.223
24.50	0.70	0.83	0.218
25.00	0.69	0.79	0.209
25.50	0.68	0.76	0.200
26.00	0.67	0.72	0.192
26.50	0.67	0.69	0.181
27.00	0.66	0.66	0.174
27.50	0.66	0.63	0.166
28.00	0.66	0.61	0.158
28.50	0.66	0.59	0.151
29.00	0.65	0.57	0.148
29.50	0.64	0.55	0.145
30.00	0.64	0.53	0.140
32.00	0.62	0.44	0.119
34.00	0.64	0.35	0.091
36.00	0.69	0.27	0.059
38.00	0.73	0.24	0.044
40.00	0.76	0.22	0.034

Iron[5]

Energy (eV)	n	k	$R(\phi = 0)$
0.10	6.41	33.07	0.978
0.15	6.26	22.82	0.956
0.20	3.68	18.23	0.958
0.26	4.98	13.68	0.911
0.30	4.87	12.05	0.892
0.36	4.68	10.44	0.867
0.40	4.42	9.75	0.858
0.50	4.14	8.02	0.817
0.60	3.93	6.95	0.783
0.70	3.78	6.17	0.752
0.80	3.65	5.60	0.725
0.90	3.52	5.16	0.700
1.00	3.43	4.79	0.678
1.10	3.33	4.52	0.660

Energy (eV)	n	k	$R(\phi = 0)$
1.20	3.24	4.26	0.641
1.30	3.16	4.07	0.626
1.40	3.12	3.87	0.609
1.50	3.05	3.77	0.601
1.60	3.00	3.60	0.585
1.70	2.98	3.52	0.577
1.80	2.92	3.46	0.573
1.90	2.89	3.37	0.563
2.00	2.85	3.36	0.563
2.10	2.80	3.34	0.562
2.20	2.74	3.33	0.563
2.30	2.65	3.34	0.567
2.40	2.56	3.31	0.567
2.50	2.46	3.31	0.570
2.60	2.34	3.30	0.576
2.70	2.23	3.25	0.575
2.80	2.12	3.23	0.580
2.90	2.01	3.17	0.580
3.00	1.88	3.12	0.583
3.10	1.78	3.04	0.580
3.20	1.70	2.96	0.576
3.30	1.62	2.87	0.572
3.40	1.55	2.79	0.565
3.50	1.50	2.70	0.556
3.60	1.47	2.63	0.548
3.70	1.43	2.56	0.542
3.83	1.38	2.49	0.534
4.00	1.30	2.39	0.527
4.17	1.26	2.27	0.510
4.33	1.23	2.18	0.494
4.50	1.20	2.10	0.482
4.67	1.16	2.02	0.470
4.83	1.14	1.93	0.451
5.00	1.14	1.87	0.435
5.17	1.12	1.81	0.425
5.33	1.11	1.75	0.408
5.50	1.09	1.17	0.401
5.67	1.09	1.65	0.383
5.83	1.10	1.61	0.373
6.00	1.09	1.59	0.366
6.17	1.08	1.57	0.365
6.33	1.04	1.55	0.365
6.50	1.02	1.51	0.358
6.67	1.00	1.47	0.351
6.83	0.97	1.43	0.346
7.00	0.96	1.39	0.333
7.17	0.94	1.35	0.327
7.33	0.94	1.30	0.311
7.50	0.94	1.26	0.298
7.67	0.94	1.23	0.288
7.83	0.94	1.21	0.279
8.00	0.94	1.18	0.272
8.17	0.94	1.16	0.265
8.33	0.94	1.14	0.258
8.50	0.94	1.12	0.251
8.67	0.94	1.10	0.246
8.83	0.92	1.08	0.240
9.00	0.93	1.07	0.236
9.17	0.92	1.06	0.233
9.33	0.91	1.04	0.231

Energy (eV)	n	k	$R(\phi = 0)$
9.50	0.90	1.02	0.226
9.67	0.90	1.00	0.221
9.83	0.89	0.99	0.218
10.00	0.88	0.97	0.213
10.17	0.87	0.94	0.203
10.33	0.87	0.91	0.196
10.50	0.87	0.89	0.189
10.67	0.88	0.87	0.179
10.83	0.89	0.85	0.170
11.00	0.91	0.83	0.162
11.17	0.92	0.83	0.159
11.33	0.93	0.84	0.159
11.50	0.93	0.84	0.160
11.67	0.93	0.84	0.162
11.83	0.92	0.84	0.163
12.00	0.91	0.84	0.163
12.17	0.90	0.84	0.165
12.33	0.89	0.83	0.164
12.50	0.98	0.83	0.165
12.67	0.87	0.82	0.166
12.83	0.86	0.81	0.166
13.00	0.85	0.80	0.162
13.17	0.84	0.79	0.161
13.33	0.84	0.78	0.160
13.50	0.83	0.77	0.159
13.67	0.82	0.76	0.157
13.83	0.81	0.75	0.154
14.00	0.81	0.73	0.151
14.17	0.80	0.72	0.149
14.33	0.80	0.71	0.146
14.50	0.79	0.79	0.144
14.67	0.79	0.69	0.141
14.83	0.78	0.67	0.138
15.00	0.78	0.66	0.135
15.17	0.78	0.65	0.131
15.33	0.78	0.64	0.238
15.50	0.77	0.63	0.126
15.67	0.77	0.62	0.123
15.83	0.77	0.61	0.119
16.00	0.77	0.60	0.116
16.17	0.78	0.58	0.112
16.33	0.78	0.58	0.110
16.50	0.78	0.57	0.107
16.67	0.77	0.56	0.106
16.83	0.78	0.55	0.103
17.00	0.78	0.55	0.102
17.17	0.78	0.54	0.100
17.33	0.78	0.54	0.098
17.50	0.77	0.53	0.097
17.67	0.77	0.52	0.095
17.83	0.78	0.51	0.092
18.00	0.78	0.51	0.091
18.17	0.78	0.51	0.090
18.33	0.78	0.50	0.089
18.50	0.77	0.50	0.089
18.67	0.77	0.50	0.088
18.83	0.77	0.49	0.087
19.00	0.77	0.49	0.087
19.17	0.76	0.49	0.088
19.33	0.76	0.48	0.087

Energy (eV)	n	k	R(φ = 0)
19.50	0.75	0.47	0.086
19.67	0.75	0.47	0.085
19.83	0.75	0.46	0.084
20.00	0.74	0.45	0.083
20.17	0.74	0.44	0.081
20.33	0.74	0.44	0.081
20.50	0.74	0.42	0.080
20.67	0.73	0.43	0.079
20.83	0.73	0.42	0.078
21.00	0.73	0.41	0.077
21.17	0.72	0.40	0.076
21.33	0.72	0.39	0.074
21.50	0.72	0.38	0.073
21.67	0.72	0.38	0.071
21.83	0.72	0.37	0.070
22.00	0.72	0.36	0.068
22.17	0.71	0.35	0.067
22.33	0.72	0.34	0.064
22.50	0.72	0.34	0.063
22.67	0.72	0.33	0.062
22.83	0.72	0.32	0.059
23.00	0.72	0.31	0.058
23.17	0.72	0.30	0.056
23.33	0.72	0.29	0.054
23.50	0.73	0.28	0.050
23.67	0.73	0.28	0.049
23.83	0.74	0.27	0.047
24.00	0.74	0.27	0.045
24.17	0.74	0.26	0.044
24.33	0.74	0.26	0.043
24.50	0.74	0.25	0.042
24.67	0.75	0.25	0.040
24.83	0.75	0.24	0.039
25.00	0.75	0.24	0.038
26.00	0.76	0.21	0.031
27.00	0.78	0.18	0.026
28.00	0.79	0.16	0.021
29.00	0.81	0.14	0.017
30.00	0.82	0.13	0.014

Lithium[12]

Energy (eV)	n	k	R(φ = 0)
0.14	0.659	38.0	0.998
0.54	0.661	12.6	0.984
0.75	0.561	7.68	0.963
1.05	0.448	5.58	0.946
1.35	0.338	4.36	0.935
1.65	0.265	3.55	0.925
1.95	0.221	2.94	0.913
2.25	0.206	2.48	0.892
2.55	0.217	2.11	0.854
2.85	0.247	1.82	0.797
3.15	0.304	1.60	0.715
3.45	0.334	1.45	0.656
3.75	0.345	1.32	0.611
4.05	0.346	1.21	0.578
4.35	0.333	1.11	0.557
4.65	0.317	1.01	0.540
4.95	0.302	0.906	0.520
5.25	0.299	0.795	0.484
5.55	0.310	0.688	0.434
5.85	0.342	0.594	0.365

Energy (eV)	n	k	R(φ = 0)
6.15	0.376	0.522	0.306
6.45	0.408	0.460	0.256
6.75	0.440	0.407	0.214
7.05	0.466	0.364	0.183
7.35	0.492	0.320	0.155
7.65	0.517	0.282	0.131
7.95	0.545	0.246	0.109
8.25	0.572	0.214	0.091
8.55	0.601	0.189	0.075
8.85	0.624	0.163	0.063
9.15	0.657	0.144	0.050
9.45	0.680	0.130	0.042
9.75	0.708	0.119	0.034
10.1	0.726	0.108	0.029
10.4	0.743	0.102	0.025
10.6	0.753	0.080	0.022

Magnesium (evaporated)[13]

Energy (eV)	n	k	R(φ = 0)
2.145	0.48	3.71	0.880
2.270	0.57	3.47	0.843
2.522	0.53	2.92	0.805
2.845	0.52	2.65	0.777
3.064	0.52	2.05	0.681
5.167	0.10	1.60	0.894
5.636	0.15	1.50	0.832
6.200	0.20	1.40	0.765
6.889	0.25	1.30	0.693
7.750	0.20	1.20	0.722
8.857	0.15	0.95	0.730
10.335	0.25	0.40	0.419

Manganese[14]

Energy (eV)	n	k	R(φ = 0)
0.64	3.89	5.95	0.738
0.77	3.78	5.41	0.710
0.89	3.65	5.02	0.688
1.02	3.48	4.74	0.673
1.14	3.30	4.53	0.662
1.26	3.10	4.35	0.653
1.39	2.97	4.18	0.643
1.51	2.83	4.03	0.634
1.64	2.70	3.91	0.627
1.76	2.62	3.78	0.617
1.88	2.56	3.65	0.606
2.01	2.51	3.54	0.596
2.13	2.47	3.43	0.585
2.26	2.39	3.33	0.577
2.38	2.32	3.23	0.567
2.50	2.25	3.14	0.559
2.63	2.19	3.06	0.552
2.75	2.11	2.98	0.545
2.88	2.06	2.90	0.536
3.00	2.00	2.82	0.528
3.12	1.96	2.74	0.518
3.25	1.92	2.67	0.509
3.37	1.89	2.59	0.498
3.50	1.89	2.51	0.484
3.62	1.87	2.45	0.475
3.74	1.86	2.38	0.463
3.87	1.86	2.32	0.451
3.99	1.86	2.25	0.438
4.12	1.86	2.19	0.427

Energy (eV)	n	k	R(φ = 0)
4.24	1.85	2.14	0.417
4.36	1.85	2.08	0.406
4.49	1.86	2.03	0.395
4.61	1.85	1.99	0.388
4.74	1.84	1.94	0.378
4.86	1.83	1.91	0.372
4.98	1.82	1.86	0.362
5.11	1.82	1.82	0.354
5.23	1.81	1.79	0.348
5.36	1.78	1.76	0.342
5.48	1.74	1.73	0.337
5.60	1.73	1.70	0.331
5.73	1.72	1.67	0.325
5.85	1.70	1.64	0.319
5.98	1.67	1.61	0.313
6.10	1.63	1.58	0.307
6.22	1.62	1.55	0.301
6.35	1.59	1.52	0.295
6.47	1.55	1.50	0.292
6.60	1.48	1.47	0.288

Mercury (liquid)[15]

Energy (eV)	n	k	R(φ = 0)
0.2	13.99	14.27	0.869
0.3	11.37	11.95	0.846
0.4	9.741	10.65	0.830
0.5	8.528	9.805	0.818
0.6	7.574	9.195	0.808
0.8	6.086	8.312	0.796
1.0	4.962	7.643	0.789
1.2	4.050	7.082	0.786
1.4	3.324	6.558	0.785
1.6	2.746	6.054	0.783
1.8	2.284	5.582	0.782
2.0	1.910	5.150	0.782
2.2	1.620	4.751	0.780
2.4	1.384	4.407	0.779
2.6	1.186	4.090	0.779
2.8	1.027	3.802	0.779
3.0	0.898	3.538	0.777
3.2	0.798	3.294	0.773
3.4	0.713	3.074	0.770
3.6	0.644	2.860	0.763
3.8	0.589	2.665	0.755
4.0	0.542	2.502	0.749
4.2	0.507	2.341	0.738
4.4	0.477	2.195	0.727
4.6	0.452	2.058	0.715
4.8	0.431	1.929	0.701
5.0	0.414	1.806	0.685
5.2	0.401	1.687	0.666
5.4	0.394	1.569	0.642
5.6	0.386	1.454	0.617
5.7	0.386	1.396	0.601
5.8	0.386	1.341	0.585
5.9	0.385	1.287	0.569
6.0	0.386	1.232	0.551
6.1	0.388	1.176	0.531
6.2	0.390	1.118	0.510
6.3	0.399	1.058	0.481
6.4	0.412	1.002	0.450
6.5	0.428	0.949	0.418

Energy (eV)	n	k	$R(\phi = 0)$
6.6	0.436	0.898	0.392
6.7	0.438	0.836	0.367
6.8	0.459	0.756	0.320
6.9	0.510	0.676	0.255
7.0	0.585	0.617	0.191
7.1	0.663	0.589	0.148
7.2	0.717	0.584	0.128
7.3	0.769	0.575	0.111
7.4	0.817	0.574	0.100
7.5	0.860	0.580	0.094
7.6	0.893	0.597	0.093
7.8	0.929	0.623	0.096
8.0	0.946	0.639	0.098
8.2	0.952	0.645	0.099
8.4	0.953	0.638	0.097
8.6	0.956	0.624	0.093
8.8	0.965	0.607	0.087
9.0	0.975	0.588	0.082
9.2	0.988	0.568	0.076
9.4	1.009	0.548	0.069
9.6	1.044	0.541	0.066
9.8	1.061	0.557	0.069
10.0	1.062	0.567	0.071
10.2	1.054	0.569	0.072
10.4	1.045	0.561	0.070
10.6	1.041	0.550	0.068
10.8	1.039	0.537	0.065
11.0	1.039	0.523	0.062
11.5	1.050	0.491	0.055
12.0	1.064	0.467	0.050
12.5	1.078	0.445	0.045
13.0	1.092	0.430	0.042
13.5	1.104	0.416	0.040
14.0	1.115	0.404	0.038
14.5	1.125	0.394	0.037
15.0	1.135	0.383	0.035
15.5	1.146	0.374	0.034
16.0	1.159	0.368	0.034
16.5	1.170	0.367	0.034
17.0	1.177	0.367	0.034
17.5	1.184	0.366	0.034
18.0	1.191	0.367	0.035
18.5	1.195	0.367	0.035
19.0	1.200	0.366	0.035
19.5	1.208	0.364	0.035

Molybdenum[16]

Energy (eV)	n	k	$R(\phi = 0)$
0.10	18.53	68.51	0.985
0.15	8.78	47.54	0.985
0.20	5.10	35.99	0.985
0.25	3.36	28.75	0.984
0.30	2.44	23.80	0.983
0.34	2.00	20.84	0.982
0.38	1.70	18.44	0.980
0.42	1.57	16.50	0.978
0.46	1.46	14.91	0.975
0.50	1.37	13.55	0.971
0.54	1.35	12.36	0.966
0.58	1.34	11.34	0.960
0.62	1.38	10.44	0.952
0.66	1.43	9.67	0.942
0.70	1.48	8.99	0.932
0.74	1.51	8.38	0.921
0.78	1.60	7.83	0.906
0.82	1.64	7.35	0.892
0.86	1.70	6.89	0.876
0.90	1.74	6.48	0.859
1.00	1.94	5.58	0.805
1.10	2.15	4.85	0.743
1.20	2.44	4.22	0.671
1.30	2.77	3.74	0.608
1.40	3.15	3.40	0.562
1.50	3.53	3.30	0.550
1.60	3.77	3.41	0.562
1.70	3.84	3.51	0.570
1.80	3.81	3.58	0.576
1.90	3.74	3.58	0.576
2.00	3.68	3.52	0.571
2.10	3.68	3.45	0.565
2.20	3.76	3.41	0.562
2.30	3.79	3.61	0.578
2.40	3.59	3.78	0.594
2.50	3.36	3.73	0.591
2.60	3.22	3.61	0.582
2.70	3.13	3.51	0.573
2.80	3.08	3.42	0.565
2.90	3.05	3.33	0.566
3.00	3.04	3.27	0.550
3.10	3.03	3.21	0.544
3.20	3.05	3.18	0.540
3.30	3.06	3.18	0.540
3.40	3.06	3.19	0.541
3.50	3.06	3.21	0.543
3.60	3.05	3.23	0.546
3.70	3.04	3.27	0.550
3.80	3.04	3.31	0.554
3.90	3.04	3.40	0.564
4.00	3.01	3.51	0.576
4.20	2.77	3.77	0.610
4.40	2.39	3.88	0.640
4.60	2.06	3.84	0.658
4.80	1.75	3.76	0.678
5.00	1.46	3.62	0.695
5.20	1.22	3.42	0.706
5.40	1.07	3.20	0.706
5.60	0.96	2.99	0.700
5.80	0.89	2.80	0.688
6.00	0.85	2.64	0.674
6.20	0.81	2.50	0.660
6.40	0.79	2.36	0.641
6.60	0.78	2.24	0.619
6.80	0.78	2.13	0.592
7.00	0.80	2.04	0.568
7.20	0.81	1.98	0.548
7.40	0.81	1.95	0.542
7.60	0.75	1.90	0.552
7.80	0.71	1.81	0.542
8.00	0.69	1.73	0.530
8.20	0.67	1.65	0.512
8.40	0.66	1.57	0.495
8.60	0.65	1.49	0.475
8.80	0.65	1.41	0.450
9.00	0.65	1.33	0.420
9.20	0.67	1.25	0.385
9.40	0.69	1.19	0.355
9.60	0.71	1.12	0.320
9.80	0.74	1.05	0.285
10.00	0.77	0.99	0.250
10.20	0.81	0.93	0.217
10.40	0.86	0.88	0.188
10.60	0.91	0.83	0.162
10.80	0.98	0.79	0.138
11.00	1.05	0.77	0.125
11.20	1.12	0.78	0.123
11.40	1.18	0.80	0.125
11.60	1.23	0.85	0.135
11.80	1.25	0.89	0.145
12.00	1.26	0.92	0.154
12.40	1.25	0.98	0.168
12.80	1.23	1.00	0.178
13.20	1.20	1.02	0.185
13.60	1.17	1.02	0.187
14.00	1.15	1.01	0.185
14.40	1.13	1.00	0.182
14.80	1.13	0.99	0.179
15.00	1.14	0.99	0.179
15.60	1.15	1.01	0.184
16.00	1.14	1.04	0.194
16.60	1.10	1.10	0.216
17.00	1.04	1.12	0.233
17.60	0.94	1.14	0.257
18.00	0.87	1.12	0.270
18.60	0.77	1.08	0.283
19.00	0.71	1.02	0.284
19.60	0.66	0.94	0.275
20.00	0.64	0.89	0.264
20.60	0.62	0.81	0.245
21.00	0.61	0.77	0.234
21.60	0.61	0.71	0.215
22.00	0.60	0.69	0.207
22.60	0.59	0.63	0.195
23.00	0.58	0.60	0.185
23.60	0.58	0.53	0.166
24.00	0.58	0.49	0.151
24.60	0.60	0.43	0.124
25.00	0.62	0.39	0.106
25.60	0.66	0.35	0.085
26.00	0.68	0.33	0.072
26.50	0.71	0.31	0.060
27.00	0.73	0.29	0.050
27.50	0.76	0.28	0.041
28.00	0.79	0.27	0.036
28.50	0.81	0.26	0.031
29.00	0.83	0.26	0.028
29.50	0.86	0.26	0.025
30.00	0.88	0.26	0.023
31.00	0.92	0.29	0.024
32.00	0.92	0.32	0.030
33.00	0.90	0.33	0.032
34.00	0.91	0.34	0.034
35.00	0.87	0.37	0.043

Energy (eV)	n	k	$R(\phi = 0)$	Energy (eV)	n	k	$R(\phi = 0)$	Energy (eV)	n	k	$R(\phi = 0)$
36.00	0.82	0.34	0.043	6.60	1.01	1.40	0.325	25.00	0.89	0.42	0.050
37.00	0.81	0.30	0.038	6.80	1.02	1.35	0.308	26.00	0.88	0.39	0.046
38.00	0.81	0.27	0.033	7.00	1.03	1.30	0.291	27.00	0.87	0.37	0.042
39.00	0.82	0.25	0.029	7.20	1.03	1.27	0.282	28.00	0.87	0.35	0.040
40.00	0.83	0.23	0.025	7.40	1.03	1.24	0.273	29.00	0.86	0.34	0.037
				7.60	1.02	1.22	0.265	30.00	0.86	0.32	0.034
Nickel[17]				7.80	1.01	1.18	0.256	35.00	0.86	0.24	0.022
0.10	9.54	45.82	0.983	8.00	1.01	1.15	0.248	40.00	0.87	0.18	0.014
0.15	5.45	30.56	0.978	8.20	1.00	1.13	0.242	45.00	0.88	0.13	0.008
0.20	4.12	22.48	0.969	8.40	0.99	1.11	0.235	50.00	0.92	0.10	0.004
0.25	4.25	17.68	0.950	8.60	0.98	1.08	0.228	60.00	0.96	0.08	0.002
0.30	4.19	15.05	0.934	8.80	0.97	1.05	0.220	65.00	0.98	0.09	0.002
0.35	4.03	13.05	0.918	9.00	0.97	1.01	0.211	68.00	0.96	0.12	0.004
0.40	3.84	11.43	0.900	9.20	0.96	0.99	0.203	70.00	0.94	0.11	0.004
0.50	4.03	9.64	0.864	9.40	0.95	0.96	0.194	75.00	0.94	0.09	0.003
0.60	3.84	8.35	0.835	9.60	0.95	0.93	0.185	80.00	0.94	0.07	0.002
0.70	3.59	7.48	0.813	9.80	0.95	0.89	0.175	90.00	0.94	0.06	0.002
0.80	3.38	6.82	0.794	10.00	0.95	0.87	0.166				
0.90	3.18	6.23	0.774	10.20	0.95	0.83	0.155	**Niobium[18]**			
1.00	3.06	5.74	0.753	10.40	0.95	0.80	0.145	0.12	15.99	53.20	0.979
1.10	2.97	5.38	0.734	10.60	0.97	0.76	0.129	0.20	7.25	34.14	0.976
1.20	2.85	5.10	0.721	10.80	0.99	0.75	0.123	0.24	5.47	28.88	0.975
1.30	2.74	4.85	0.708	11.00	1.01	0.73	0.115	0.28	4.26	24.95	0.974
1.40	2.65	4.63	0.695	11.25	1.04	0.72	0.111	0.35	3.11	20.03	0.970
1.50	2.53	4.47	0.688	11.50	1.05	0.71	0.109	0.45	2.28	15.58	0.964
1.60	2.43	4.31	0.679	11.75	1.07	0.71	0.108	0.55	1.83	12.67	0.956
1.70	2.28	4.18	0.677	12.00	1.07	0.71	0.108	0.65	1.57	10.59	0.947
1.80	2.14	4.01	0.670	12.25	1.07	0.71	0.107	0.75	1.41	9.00	0.935
1.90	2.02	3.82	0.659	12.50	1.08	0.71	0.106	0.85	1.35	7.74	0.918
2.00	1.92	3.65	0.649	12.75	1.08	0.71	0.106	0.95	1.35	6.70	0.893
2.10	1.85	3.48	0.634	13.00	1.08	0.71	0.105	1.05	1.44	5.86	0.857
2.20	1.80	3.33	0.620	13.25	1.08	0.71	0.105	1.15	1.55	5.18	0.814
2.30	1.75	3.19	0.605	13.50	1.07	0.70	0.105	1.25	1.65	4.63	0.768
2.40	1.71	3.06	0.590	13.75	1.07	0.70	0.105	1.35	1.76	4.13	0.715
2.50	1.67	2.93	0.575	14.00	1.07	0.71	0.106	1.45	1.95	3.68	0.650
2.60	1.65	2.81	0.557	14.25	1.06	0.70	0.106	1.55	2.15	3.37	0.595
2.70	1.64	2.71	0.542	14.50	1.05	0.70	0.106	1.65	2.36	3.13	0.552
2.80	1.63	2.61	0.525	14.75	1.04	0.70	0.107	1.75	2.54	2.99	0.527
2.90	1.62	2.52	0.509	15.00	1.03	0.70	0.107	1.85	2.69	2.89	0.510
3.00	1.61	2.44	0.495	15.25	1.02	0.69	0.106	1.95	2.82	2.86	0.505
3.10	1.61	2.36	0.480	15.50	1.01	0.69	0.105	2.05	2.89	2.87	0.505
3.20	1.61	2.30	0.467	15.75	1.00	0.68	0.104	2.15	2.92	2.87	0.505
3.30	1.61	2.23	0.454	16.00	0.99	0.67	0.103	2.25	2.93	2.87	0.505
3.40	1.62	2.17	0.441	16.50	0.98	0.66	0.101	2.35	2.92	2.88	0.506
3.50	1.63	2.11	0.428	17.00	0.96	0.64	0.098	2.45	2.89	2.90	0.509
3.60	1.64	2.07	0.416	17.50	0.94	0.63	0.096	2.55	2.83	2.92	0.512
3.70	1.66	2.02	0.405	18.00	0.92	0.61	0.092	2.65	2.74	2.90	0.511
3.80	1.69	1.99	0.397	18.50	0.91	0.58	0.087	2.75	2.66	2.86	0.507
3.90	1.72	1.98	0.393	19.00	0.90	0.56	0.082	2.85	2.58	2.80	0.500
4.00	1.73	1.98	0.392	19.50	0.90	0.54	0.077	3.00	2.51	2.68	0.485
4.20	1.74	2.01	0.396	20.00	0.89	0.51	0.071	3.10	2.48	2.60	0.475
4.40	1.71	2.06	0.409	20.50	0.89	0.49	0.066	3.20	2.45	2.53	0.465
4.60	1.63	2.09	0.421	21.00	0.90	0.47	0.061	3.30	2.44	2.45	0.453
4.80	1.53	2.11	0.435	21.50	0.91	0.46	0.057	3.40	2.46	2.38	0.442
5.00	1.40	2.10	0.449	22.00	0.91	0.45	0.055	3.50	2.48	2.33	0.435
5.20	1.27	2.04	0.454	22.50	0.91	0.44	0.053	3.60	2.52	2.29	0.428
5.40	1.16	1.94	0.449	23.00	0.92	0.44	0.051	3.70	2.56	2.27	0.426
5.60	1.09	1.83	0.435	23.50	0.91	0.44	0.052	3.80	2.59	2.28	0.427
5.80	1.04	1.73	0.417	24.00	0.90	0.43	0.051	3.90	2.62	2.29	0.429
6.20	1.00	1.54	0.371	24.50	0.90	0.43	0.051	4.00	2.64	2.33	0.434
6.40	1.01	1.46	0.345					4.20	2.64	2.42	0.447

Energy (eV)	n	k	$R(\phi = 0)$
4.40	2.53	2.56	0.467
4.60	2.39	2.56	0.470
4.80	2.32	2.52	0.465
5.00	2.26	2.57	0.475
5.20	2.16	2.62	0.487
5.40	2.00	2.68	0.505
5.60	1.81	2.67	0.518
5.80	1.63	2.60	0.522
6.00	1.49	2.49	0.520
6.20	1.38	2.38	0.512
6.40	1.31	2.25	0.496
6.60	1.26	2.14	0.480
6.80	1.24	2.04	0.460
7.00	1.23	1.96	0.441
7.20	1.22	1.91	0.430
7.40	1.20	1.88	0.427
7.60	1.14	1.85	0.430
7.80	1.07	1.78	0.428
8.00	1.02	1.69	0.412
8.20	1.00	1.60	0.390
8.40	0.99	1.51	0.365
8.60	0.99	1.43	0.340
8.70	0.99	1.39	0.328
8.80	1.00	1.36	0.315
9.00	1.01	1.29	0.290
9.20	1.04	1.22	0.265
9.40	1.07	1.18	0.245
9.60	1.10	1.13	0.227
9.80	1.13	1.09	0.209
10.00	1.18	1.05	0.194
10.20	1.23	1.04	0.187
10.40	1.27	1.04	0.185
10.60	1.30	1.06	0.190
10.80	1.32	1.08	0.195
11.00	1.32	1.10	0.200
11.20	1.31	1.12	0.204
11.40	1.30	1.13	0.207
11.60	1.28	1.13	0.209
11.80	1.27	1.13	0.210
12.00	1.25	1.12	0.209
12.40	1.24	1.10	0.204
12.80	1.24	1.09	0.200
13.20	1.24	1.09	0.201
13.60	1.23	1.12	0.208
14.00	1.20	1.13	0.216
14.40	1.16	1.15	0.225
14.80	1.11	1.16	0.234
15.00	1.08	1.16	0.238
15.60	0.99	1.14	0.247
16.00	0.92	1.11	0.250
16.60	0.85	1.04	0.245
17.00	0.80	0.99	0.240
17.20	0.79	0.96	0.236
17.40	0.77	0.93	0.230
17.80	0.75	0.87	0.217
18.00	0.74	0.85	0.209
18.60	0.73	0.77	0.185
19.00	0.72	0.72	0.170
19.60	0.72	0.66	0.150
20.00	0.72	0.62	0.137

Energy (eV)	n	k	$R(\phi = 0)$
20.60	0.71	0.55	0.119
21.00	0.72	0.50	0.100
21.60	0.75	0.43	0.075
22.00	0.78	0.40	0.063
22.60	0.82	0.35	0.045
23.00	0.85	0.33	0.038
23.60	0.88	0.30	0.029
24.00	0.91	0.29	0.025
24.60	0.94	0.28	0.022
25.00	0.96	0.27	0.020
25.60	0.99	0.26	0.018
26.00	1.00	0.26	0.017
26.60	1.03	0.25	0.016
27.00	1.04	0.25	0.015
27.60	1.06	0.25	0.015
28.00	1.08	0.24	0.015
28.60	1.11	0.24	0.016
29.00	1.13	0.25	0.017
29.60	1.16	0.26	0.020
30.00	1.18	0.28	0.023
31.00	1.18	0.31	0.026
32.00	1.20	0.34	0.031
33.00	1.21	0.38	0.038
34.00	1.20	0.42	0.044
35.20	1.17	0.47	0.051
36.00	1.15	0.50	0.056
37.50	1.07	0.53	0.064
39.50	0.95	0.50	0.063
40.50	0.92	0.47	0.059

Osmium (Polycrystalline)[9]

Energy (eV)	n	k	$R(\phi = 0)$
0.10	4.08	50.23	0.994
0.15	2.90	33.60	0.990
0.20	2.44	25.11	0.985
0.25	2.35	19.99	0.977
0.30	2.23	16.54	0.969
0.35	2.33	14.06	0.955
0.40	2.45	12.32	0.940
0.45	2.43	11.02	0.927
0.50	2.41	9.97	0.913
0.55	2.33	9.12	0.901
0.60	2.21	8.37	0.890
0.65	2.11	7.68	0.877
0.70	2.02	7.04	0.862
0.75	2.00	6.46	0.842
0.80	2.00	5.95	0.820
0.85	2.01	5.51	0.796
0.90	2.03	5.10	0.769
0.95	2.05	4.74	0.742
1.00	2.09	4.41	0.712
1.10	2.15	3.84	0.651
1.20	2.16	3.35	0.592
1.30	2.25	2.77	0.506
1.40	2.49	2.23	0.419
1.50	2.84	1.80	0.369
1.60	3.36	1.62	0.379
1.70	3.70	1.75	0.411
1.80	3.78	1.83	0.423
1.90	3.81	1.75	0.418
2.00	3.98	1.60	0.418
2.10	4.26	1.54	0.432
2.20	4.58	1.62	0.457
2.30	4.84	1.76	0.479
2.40	5.10	2.01	0.506
2.50	5.28	2.38	0.532
2.60	5.36	2.82	0.557
2.70	5.30	3.29	0.580
2.80	5.07	3.78	0.603
2.90	4.65	4.18	0.624
3.00	4.05	4.40	0.639
3.20	3.29	3.96	0.614
3.40	2.93	3.79	0.607
3.60	2.75	3.45	0.577
3.80	2.73	3.32	0.562
4.00	2.71	3.34	0.565
4.20	2.53	3.44	0.584
4.40	2.24	3.44	0.599
4.60	2.01	3.31	0.598
4.80	1.88	3.19	0.592
5.00	1.74	3.12	0.596
5.20	1.58	3.00	0.597
5.40	1.46	2.88	0.593
5.60	1.36	2.77	0.589
5.80	1.27	2.65	0.582
6.00	1.20	2.54	0.575
6.20	1.13	2.44	0.571
6.40	1.06	2.33	0.562
6.60	1.01	2.21	0.548
6.80	0.97	2.11	0.532
7.00	0.95	2.00	0.514
7.20	0.92	1.91	0.497
7.40	0.91	1.81	0.476
7.60	0.90	1.72	0.451
7.80	0.90	1.63	0.426
8.00	0.91	1.55	0.400
8.20	0.91	1.48	0.375
8.40	0.94	1.40	0.344
8.60	0.96	1.34	0.319
8.80	0.98	1.29	0.296
9.00	1.01	1.24	0.274
9.20	1.04	1.19	0.255
9.40	1.08	1.16	0.238
9.60	1.10	1.14	0.229
9.80	1.13	1.11	0.217
10.00	1.16	1.10	0.209
10.20	1.19	1.08	0.203
10.30	1.20	1.08	0.201
10.40	1.22	1.08	0.200
10.50	1.23	1.09	0.201
10.60	1.24	1.10	0.203
10.80	1.25	1.11	0.206
11.00	1.24	1.13	0.213
11.20	1.23	1.14	0.217
11.40	1.19	1.15	0.223
11.60	1.17	1.12	0.216
11.80	1.16	1.10	0.211
12.00	1.15	1.08	0.205
12.40	1.14	1.03	0.191
12.80	1.15	1.01	0.183
13.20	1.16	0.98	0.174
13.60	1.17	0.97	0.170

Energy (eV)	n	k	$R(\phi = 0)$
14.00	1.17	0.96	0.169
14.40	1.16	0.94	0.165
14.80	1.16	0.91	0.156
15.20	1.17	0.89	0.148
15.60	1.20	0.86	0.140
16.00	1.25	0.87	0.140
16.40	1.28	0.90	0.147
16.80	1.28	0.94	0.157
17.20	1.27	0.97	0.167
17.60	1.26	1.01	0.178
18.00	1.23	1.04	0.189
18.40	1.19	1.08	0.200
18.80	1.14	1.10	0.210
19.20	1.10	1.10	0.219
19.60	1.05	1.11	0.227
20.00	0.96	1.10	0.239
20.40	0.93	1.09	0.240
20.80	0.89	1.05	0.240
21.20	0.86	1.02	0.237
21.60	0.83	0.99	0.235
22.00	0.80	0.96	0.230
22.40	0.78	0.93	0.226
22.80	0.77	0.90	0.220
23.20	0.75	0.88	0.217
23.60	0.75	0.86	0.211
24.00	0.73	0.84	0.209
24.40	0.72	0.82	0.207
24.80	0.70	0.80	0.205
25.20	0.69	0.77	0.202
25.60	0.67	0.75	0.199
26.00	0.66	0.72	0.195
26.40	0.65	0.69	0.189
26.80	0.63	0.66	0.183
27.20	0.65	0.62	0.165
28.00	0.64	0.59	0.156
28.40	0.64	0.57	0.148
28.80	0.65	0.55	0.140
29.20	0.65	0.53	0.134
29.60	0.65	0.51	0.128
30.00	0.65	0.49	0.121
31.00	0.65	0.45	0.111
32.00	0.66	0.41	0.095
33.00	0.68	0.37	0.079
34.00	0.70	0.34	0.068
35.00	0.72	0.31	0.057
36.00	0.74	0.29	0.048
37.00	0.77	0.27	0.040
38.00	0.79	0.26	0.035
39.00	0.81	0.26	0.031
40.00	0.84	0.26	0.026

Palladium[19]

Energy (eV)	n	k	$R(\phi = 0)$
0.10	4.13	54.15	0.994
0.15	3.13	35.82	0.990
0.20	3.07	26.59	0.983
0.26	3.11	20.15	0.971
0.30	3.56	17.27	0.955
0.36	3.98	14.41	0.932
0.40	4.27	13.27	0.916
0.46	4.27	12.11	0.902
0.50	4.10	11.44	0.896

Energy (eV)	n	k	$R(\phi = 0)$
0.56	3.92	10.49	0.883
0.60	3.80	9.96	0.876
0.72	3.51	8.70	0.854
0.80	3.35	8.06	0.840
1.00	2.99	6.89	0.811
1.10	2.81	6.46	0.800
1.20	2.65	6.10	0.790
1.30	2.50	5.78	0.781
1.40	2.34	5.50	0.774
1.50	2.17	5.22	0.767
1.60	2.08	4.95	0.755
1.70	2.00	4.72	0.745
1.80	1.92	4.54	0.737
1.90	1.82	4.35	0.729
2.00	1.75	4.18	0.721
2.10	1.67	4.03	0.714
2.20	1.60	3.88	0.707
2.30	1.53	3.75	0.700
2.40	1.47	3.61	0.693
2.50	1.41	3.48	0.685
2.60	1.37	3.36	0.676
2.70	1.32	3.25	0.668
2.80	1.29	3.13	0.658
2.90	1.26	3.03	0.648
3.00	1.23	2.94	0.639
3.10	1.20	2.85	0.630
3.20	1.17	2.77	0.622
3.30	1.14	2.68	0.613
3.40	1.12	2.60	0.602
3.50	1.10	2.52	0.591
3.60	1.08	2.45	0.581
3.70	1.07	2.38	0.570
3.80	1.06	2.31	0.558
3.90	1.05	2.25	0.547
4.00	1.03	2.19	0.537
4.20	1.04	2.09	0.510
4.40	1.03	2.01	0.493
4.60	1.03	1.94	0.476
4.80	1.01	1.90	0.470
5.00	0.96	1.86	0.472
5.20	0.90	1.79	0.474
5.40	0.85	1.70	0.463
5.60	0.81	1.62	0.449
5.80	0.78	1.54	0.437
6.00	0.76	1.45	0.418
6.20	0.74	1.37	0.397
6.40	0.73	1.29	0.375
6.60	0.72	1.21	0.350
6.80	0.73	1.13	0.316
7.00	0.73	1.05	0.287
7.20	0.75	0.98	0.255
7.40	0.77	0.91	0.223
7.60	0.79	0.85	0.195
7.80	0.83	0.78	0.163
8.00	0.88	0.73	0.133
8.20	0.94	0.70	0.117
8.40	0.96	0.70	0.114
8.60	1.00	0.65	0.097
8.80	1.04	0.65	0.094
9.00	1.07	0.64	0.090

Energy (eV)	n	k	$R(\phi = 0)$
9.50	1.12	0.65	0.089
10.00	1.14	0.65	0.088
10.50	1.16	0.65	0.087
11.00	1.18	0.64	0.086
11.50	1.19	0.65	0.087
12.00	1.20	0.66	0.089
12.50	1.19	0.67	0.091
13.00	1.18	0.67	0.091
13.50	1.18	0.67	0.092
14.00	1.17	0.67	0.093
14.50	1.15	0.68	0.095
15.00	1.13	0.69	0.098
15.50	1.10	0.68	0.096
16.00	1.08	0.66	0.092
16.50	1.06	0.63	0.086
17.00	1.07	0.61	0.081
17.50	1.06	0.61	0.080
18.00	1.07	0.59	0.077
18.50	1.07	0.59	0.077
19.00	1.08	0.59	0.077
19.50	1.08	0.61	0.080
20.00	1.07	0.65	0.090
20.50	1.03	0.67	0.098
21.00	0.99	0.67	0.103
21.50	0.95	0.66	0.103
22.00	0.91	0.64	0.103
22.50	0.88	0.62	0.101
23.00	0.86	0.59	0.097
23.50	0.85	0.56	0.091
24.00	0.84	0.54	0.086
25.00	0.81	0.51	0.084
26.40	0.80	0.43	0.066
27.80	0.81	0.38	0.052
29.20	0.82	0.35	0.046

Platinum[20]

Energy (eV)	n	k	$R(\phi = 0)$
0.10	13.21	44.72	0.976
0.15	8.18	31.16	0.969
0.20	5.90	23.95	0.962
0.25	4.70	19.40	0.954
0.30	3.92	16.16	0.945
0.35	3.28	13.66	0.936
0.40	2.81	11.38	0.922
0.45	3.03	9.31	0.882
0.50	3.91	7.71	0.813
0.55	4.58	7.14	0.777
0.60	5.13	6.75	0.753
0.65	5.52	6.66	0.746
0.70	5.71	6.83	0.751
0.75	5.57	7.02	0.759
0.80	5.31	7.04	0.762
0.85	5.05	6.98	0.763
0.90	4.77	6.91	0.765
0.95	4.50	6.77	0.763
1.00	4.25	6.62	0.762
1.10	3.86	6.24	0.753
1.20	3.55	5.92	0.746
1.30	3.29	5.61	0.736
1.40	3.10	5.32	0.725
1.50	2.92	5.07	0.716
1.60	2.76	4.84	0.706

Energy (eV)	n	k	$R(\phi = 0)$	Energy (eV)	n	k	$R(\phi = 0)$	Energy (eV)	n	k	$R(\phi = 0)$
1.70	2.63	4.64	0.697	12.80	1.29	0.94	0.158	4.00	0.34	0.08	0.245
1.80	2.51	4.43	0.686	13.20	1.31	0.93	0.155	4.065	0.38	0.07	0.204
1.90	2.38	4.26	0.678	13.60	1.31	0.93	0.155	4.133	0.41	0.07	0.177
2.00	2.30	4.07	0.664	14.00	1.31	0.93	0.155	4.203	0.45	0.06	0.145
2.10	2.23	3.92	0.654	14.40	1.30	0.93	0.156	4.275	0.48	0.06	0.125
2.20	2.17	3.77	0.642	14.80	1.27	0.93	0.157	4.350	0.52	0.05	0.101
2.30	2.10	3.67	0.636	15.20	1.27	0.93	0.155	4.428	0.55	0.05	0.085
2.40	2.03	3.54	0.626	15.60	1.25	0.92	0.151	4.509	0.58	0.05	0.072
2.50	1.96	3.42	0.616	16.00	1.24	0.89	0.146	4.592	0.61	0.05	0.060
2.60	1.91	3.30	0.605	16.50	1.24	0.87	0.142	4.679	0.64	0.04	0.049
2.70	1.87	3.20	0.595	17.00	1.25	0.86	0.138	4.769	0.66	0.04	0.043
2.80	1.83	3.10	0.585	17.50	1.27	0.85	0.135	4.862	0.68	0.04	0.037
2.90	1.79	3.01	0.575	18.00	1.31	0.88	0.142	4.959	0.70	0.04	0.032
3.00	1.75	2.92	0.565	18.50	1.30	0.94	0.157	5.061	0.72	0.04	0.027
3.20	1.68	2.76	0.546	19.00	1.28	0.99	0.171	5.166	0.74	0.04	0.023
3.40	1.63	2.62	0.527	19.50	1.23	1.03	0.184	5.276	0.76	0.04	0.019
3.60	1.58	2.48	0.507	20.00	1.18	1.06	0.197	5.391	0.78	0.04	0.016
3.80	1.53	2.37	0.491	20.50	1.11	1.09	0.212	5.510	0.79	0.05	0.015
4.00	1.49	2.25	0.472	21.00	1.03	1.10	0.226	5.637	0.81	0.05	0.012
4.20	1.45	2.14	0.452	21.50	0.94	1.08	0.238	5.767	0.83	0.05	0.009
4.40	1.43	2.04	0.432	22.00	0.87	1.04	0.240	6.048	0.85	0.05	0.007
4.60	1.39	1.95	0.415	22.50	0.81	0.98	0.235	6.199	0.87	0.05	0.006
4.80	1.38	1.85	0.392	23.00	0.77	0.92	0.226	6.358	0.88	0.05	0.005
5.00	1.36	1.76	0.372	23.50	0.75	0.87	0.213	6.526	0.90	0.06	0.004
5.20	1.36	1.67	0.350	24.00	0.74	0.82	0.201	6.702	0.91	0.06	0.003
5.40	1.36	1.61	0.332	24.50	0.73	0.77	0.187	6.888	0.92	0.06	0.003
5.60	1.36	1.54	0.315	25.00	0.73	0.73	0.174	7.085	0.92	0.06	0.003
5.80	1.36	1.47	0.295	25.50	0.73	0.70	0.162	7.293	0.93	0.06	0.002
6.00	1.38	1.40	0.276	26.00	0.74	0.67	0.150	7.514	0.93	0.06	0.002
6.20	1.39	1.35	0.261	26.50	0.74	0.65	0.142	7.749	0.94	0.06	0.002
6.40	1.42	1.29	0.246	27.00	0.74	0.63	0.136	7.999	0.94	0.06	0.002
6.60	1.45	1.26	0.236	27.50	0.74	0.62	0.130	8.260	0.94	0.06	0.002
6.80	1.48	1.24	0.231	28.00	0.75	0.60	0.125	8.551	0.94	0.06	0.002
7.00	1.50	1.24	0.230	28.50	0.75	0.59	0.121	8.856	0.94	0.05	0.002
7.20	1.50	1.25	0.231	29.00	0.75	0.58	0.118	9.184	0.94	0.05	0.002
7.40	1.49	1.23	0.228	29.50	0.74	0.58	0.120	9.537	0.94	0.04	0.001
7.60	1.48	1.22	0.225	30.00	0.73	0.58	0.124	9.919	0.94	0.04	0.001
7.80	1.48	1.20	0.221					10.33	0.94	0.03	0.001
8.00	1.47	1.18	0.216	*Potassium*[21]				11.0		0.03	
8.20	1.47	1.17	0.212	0.55	0.139	7.10	0.989	12.0		0.028	
8.40	1.47	1.15	0.209	0.58	0.119	6.72	0.990				
8.60	1.47	1.14	0.205	0.63	0.106	6.32	0.990	*Rhenium, single crystal*, $\vec{E} \parallel \hat{c}$[9]			
8.80	1.47	1.13	0.202	0.67	0.091	5.79	0.990	0.10	6.06	51.03	0.991
9.00	1.48	1.12	0.200	0.73	0.079	5.30	0.989	0.15	4.66	33.96	0.984
9.20	1.49	1.11	0.198	0.81	0.066	4.75	0.989	0.20	4.16	25.36	0.975
9.40	1.49	1.12	0.200	0.92	0.056	4.19	0.988	0.25	4.03	20.10	0.962
9.60	1.49	1.13	0.203	1.05	0.044	3.58	0.987	0.30	4.37	16.69	0.943
9.80	1.48	1.15	0.207	1.23	0.040	3.04	0.985	0.35	4.50	14.53	0.925
10.00	1.46	1.15	0.209	1.44	0.040	2.56	0.979	0.40	4.53	12.96	0.909
10.20	1.43	1.16	0.211	1.65	0.044	2.19	0.970	0.45	4.53	11.78	0.893
10.40	1.40	1.15	0.210	1.87	0.050	1.84	0.955	0.50	4.53	10.88	0.878
10.60	1.37	1.14	0.207	2.07	0.053	1.62	0.943	0.55	4.50	10.26	0.867
10.80	1.35	1.12	0.203	2.27	0.049	1.43	0.938	0.60	4.29	9.75	0.861
11.00	1.33	1.10	0.199	2.45	0.046	1.28	0.933	0.65	4.07	9.35	0.856
11.20	1.31	1.08	0.194	2.64	0.043	1.14	0.928	0.70	3.80	8.94	0.853
11.40	1.30	1.06	0.188	2.82	0.043	1.02	0.919	0.75	3.48	8.55	0.850
11.60	1.29	1.04	0.183	2.95	0.041	0.898	0.913	0.80	3.21	8.10	0.846
11.80	1.29	1.01	0.177	3.06	0.041	0.799	0.905	0.85	2.96	7.68	0.841
12.00	1.29	1.00	0.173	3.40	0.052	0.549	0.852	0.90	2.73	7.24	0.835
12.40	1.29	0.97	0.165	3.71	0.089	0.288	0.719	0.95	2.56	6.79	0.826
				3.97	0.287	0.091	0.310	1.00	2.45	6.36	0.813

Energy (eV)	n	k	R(φ = 0)
1.10	2.38	5.61	0.778
1.20	2.35	5.02	0.742
1.30	2.39	4.54	0.702
1.40	2.44	4.13	0.662
1.50	2.50	3.79	0.624
1.60	2.59	3.49	0.587
1.70	2.70	3.27	0.557
1.80	2.82	3.10	0.535
1.90	2.90	3.00	0.520
2.00	2.97	2.91	0.510
2.10	3.03	2.86	0.504
2.20	3.06	2.84	0.501
2.30	3.07	2.82	0.499
2.40	3.06	2.81	0.498
2.50	3.02	2.80	0.497
2.60	2.96	2.77	0.493
2.70	2.89	2.68	0.482
2.80	2.89	2.57	0.468
2.90	2.99	2.47	0.457
3.00	3.11	2.57	0.470
3.20	2.90	2.68	0.482
3.40	2.83	2.50	0.459
3.60	2.93	2.48	0.457
3.80	2.86	2.56	0.467
4.00	2.81	2.51	0.460
4.20	2.86	2.55	0.466
4.40	2.81	2.74	0.489
4.60	2.56	2.83	0.504
4.80	2.41	2.71	0.493
5.00	2.39	2.68	0.488
5.20	2.34	2.75	0.500
5.40	2.20	2.81	0.515
5.60	2.02	2.84	0.530
5.80	1.83	2.80	0.538
6.00	1.65	2.71	0.541
6.20	1.54	2.59	0.532
6.40	1.45	2.50	0.526
6.80	1.32	2.31	0.508
7.00	1.26	2.23	0.500
7.20	1.20	2.15	0.493
7.40	1.16	2.06	0.480
7.60	1.12	1.99	0.470
7.80	1.08	1.89	0.454
8.00	1.05	1.80	0.435
8.20	1.05	1.71	0.411
8.40	1.05	1.62	0.386
8.60	1.06	1.55	0.360
8.80	1.09	1.48	0.336
9.00	1.11	1.43	0.317
9.20	1.13	1.39	0.301
9.40	1.16	1.34	0.281
9.60	1.18	1.32	0.274
9.80	1.20	1.29	0.264
10.00	1.23	1.26	0.252
10.20	1.25	1.25	0.246
10.40	1.28	1.25	0.242
10.60	1.29	1.25	0.242
10.80	1.30	1.26	0.244
11.00	1.30	1.27	0.247
11.20	1.29	1.28	0.249

Energy (eV)	n	k	R(φ = 0)
11.40	1.28	1.28	0.252
11.60	1.26	1.28	0.252
11.80	1.24	1.26	0.249
12.00	1.23	1.24	0.244
12.40	1.22	1.21	0.237
12.80	1.21	1.18	0.230
13.20	1.22	1.16	0.222
13.60	1.22	1.13	0.215
14.00	1.24	1.12	0.209
14.40	1.27	1.11	0.204
14.80	1.29	1.15	0.213
15.20	1.29	1.19	0.225
15.60	1.26	1.22	0.236
16.00	1.23	1.25	0.248
16.40	1.19	1.27	0.259
16.80	1.14	1.29	0.269
17.00	1.12	1.30	0.275
17.40	1.07	1.30	0.286
18.00	0.99	1.30	0.300
18.40	0.93	1.29	0.311
18.80	0.87	1.28	0.321
19.20	0.81	1.25	0.330
19.60	0.77	1.21	0.332
20.00	0.73	1.18	0.333
20.40	0.70	1.14	0.332
20.80	0.67	1.11	0.332
21.20	0.64	1.08	0.334
21.60	0.61	1.04	0.335
22.00	0.58	1.01	0.340
22.40	0.55	0.97	0.341
22.80	0.53	0.93	0.338
23.20	0.51	0.89	0.334
23.60	0.50	0.85	0.329
24.00	0.48	0.80	0.319
24.40	0.48	0.76	0.207
24.80	0.47	0.72	0.296
25.20	0.47	0.68	0.282
25.60	0.47	0.65	0.270
26.00	0.47	0.61	0.255
26.40	0.48	0.57	0.240
26.80	0.48	0.54	0.225
27.20	0.49	0.51	0.208
27.60	0.50	0.48	0.193
28.00	0.51	0.45	0.176
29.00	0.54	0.39	0.145
30.00	0.57	0.33	0.114
31.00	0.62	0.29	0.086
32.00	0.66	0.26	0.065
33.00	0.68	0.24	0.054
34.00	0.72	0.21	0.041
35.00	0.76	0.20	0.031
36.00	0.79	0.20	0.025
37.00	0.82	0.19	0.021
38.00	0.85	0.20	0.018
39.00	0.89	0.21	0.016
40.00	0.88	0.26	0.022
42.00	0.88	0.26	0.022
44.00	0.89	0.29	0.026
46.00	0.85	0.32	0.035
48.00	0.82	0.30	0.036

Energy (eV)	n	k	R(φ = 0)
50.00	0.80	0.30	0.038
52.00	0.78	0.30	0.044
54.00	0.72	0.30	0.055
56.00	0.66	0.24	0.061
58.00	0.65	0.16	0.055

Rhenium, single crystal, $\vec{E} \perp \vec{c}$[9]

Energy (eV)	n	k	R(φ = 0)
0.10	4.25	42.83	0.991
0.15	3.28	28.08	0.984
0.20	3.28	20.66	0.971
0.25	3.47	16.27	0.951
0.30	3.73	13.44	0.926
0.35	3.93	11.54	0.900
0.40	3.99	10.15	0.875
0.45	4.17	9.03	0.846
0.50	4.34	8.26	0.821
0.55	4.45	7.73	0.801
0.60	4.53	7.40	0.788
0.65	4.44	7.26	0.784
0.70	4.13	7.09	0.784
0.75	3.77	6.75	0.779
0.80	3.55	6.32	0.766
0.85	3.39	5.95	0.752
0.90	3.26	5.61	0.737
0.95	3.17	5.27	0.719
1.00	3.09	4.96	0.701
1.10	3.05	4.39	0.658
1.20	3.08	3.89	0.613
1.30	3.20	3.56	0.578
1.40	3.23	3.38	0.559
1.50	3.23	3.12	0.532
1.60	3.29	2.88	0.507
1.70	3.38	2.72	0.491
1.80	3.47	2.59	0.480
1.90	3.54	2.50	0.473
2.00	3.63	2.43	0.469
2.10	3.74	2.40	0.470
2.20	3.83	2.38	0.472
2.30	3.93	2.44	0.481
2.40	4.00	2.55	0.492
2.50	4.01	2.70	0.505
2.60	3.90	2.84	0.514
2.70	3.74	2.92	0.517
2.80	3.57	2.88	0.511
2.90	3.49	2.75	0.497
3.00	3.53	2.71	0.493
3.20	3.55	2.84	0.506
3.40	3.34	2.88	0.508
3.60	3.25	2.83	0.501
3.80	3.24	2.84	0.502
4.00	3.19	2.94	0.513
4.20	3.05	3.06	0.526
4.40	2.88	3.15	0.539
4.60	2.67	3.18	0.548
4.80	2.44	3.17	0.554
5.00	2.25	3.12	0.556
5.20	2.10	3.04	0.555
5.40	1.96	2.96	0.553
5.60	1.84	2.88	0.551
5.80	1.73	2.81	0.549
6.00	1.61	2.74	0.549

Energy (eV)	n	k	R(φ = 0)	Energy (eV)	n	k	R(φ = 0)	Energy (eV)	n	k	R(φ = 0)
6.20	1.51	2.64	0.545	24.80	0.48	0.75	0.303	3.10	1.41	4.20	0.760
6.40	1.42	2.56	0.541	25.20	0.47	0.72	0.295	3.20	1.30	4.09	0.764
6.80	1.28	2.37	0.526	25.60	0.47	0.68	0.286	3.30	1.20	3.97	0.767
7.00	1.22	2.28	0.517	26.00	0.46	0.64	0.276	3.40	1.11	3.84	0.769
7.20	1.16	2.19	0.508	26.40	0.46	0.61	0.263	3.50	1.04	3.71	0.768
7.40	1.12	2.08	0.493	26.80	0.46	0.57	0.249	3.60	0.99	3.58	0.764
7.60	1.12	1.98	0.468	27.20	0.47	0.53	0.231	3.70	0.95	3.45	0.759
7.80	1.08	1.93	0.463	27.60	0.48	0.50	0.216	3.80	0.91	3.34	0.753
8.00	1.05	1.83	0.443	28.00	0.49	0.47	0.198	3.90	0.88	3.23	0.747
8.20	1.05	1.74	0.418	29.00	0.51	0.41	0.164	4.00	0.86	3.12	0.739
8.40	1.05	1.66	0.397	30.00	0.55	0.34	0.129	4.20	0.83	2.94	0.722
8.60	1.06	1.58	0.372	31.00	0.59	0.29	0.097	4.40	0.80	2.76	0.706
8.80	1.07	1.52	0.351	32.00	0.64	0.26	0.072	4.60	0.78	2.60	0.684
9.00	1.09	1.46	0.327	33.00	0.67	0.24	0.060	4.80	0.79	2.46	0.659
9.20	1.11	1.41	0.309	34.00	0.70	0.22	0.047	5.00	0.79	2.34	0.635
9.40	1.14	1.36	0.290	35.00	0.74	0.20	0.036	5.20	0.79	2.23	0.613
9.60	1.17	1.31	0.273	36.00	0.77	0.19	0.029	5.40	0.80	2.14	0.591
9.80	1.20	1.27	0.258	37.00	0.80	0.19	0.023	5.60	0.80	2.06	0.573
10.00	1.24	1.24	0.244	38.00	0.84	0.19	0.018	5.80	0.79	2.00	0.561
10.20	1.29	1.22	0.234	39.00	0.88	0.21	0.016	6.00	0.76	1.93	0.556
10.40	1.33	1.23	0.233	40.00	0.87	0.25	0.023	6.20	0.73	1.85	0.544
10.60	1.36	1.25	0.238	42.00	0.87	0.25	0.023	6.40	0.70	1.77	0.534
10.80	1.38	1.28	0.245	44.00	0.88	0.28	0.026	6.60	0.68	1.69	0.518
11.00	1.37	1.31	0.253	46.00	0.84	0.31	0.035	6.80	0.67	1.60	0.498
11.20	1.36	1.33	0.259	48.00	0.82	0.30	0.036	7.00	0.66	1.52	0.476
11.40	1.33	1.34	0.264	50.00	0.80	0.30	0.039	7.20	0.66	1.43	0.452
11.60	1.31	1.34	0.266	52.00	0.77	0.30	0.044	7.40	0.66	1.35	0.423
11.80	1.28	1.33	0.266	54.00	0.71	0.29	0.055	7.60	0.67	1.27	0.394
12.00	1.26	1.32	0.264	56.00	0.66	0.23	0.061	7.80	0.68	1.20	0.363
12.40	1.23	1.29	0.257	58.00	0.64	0.16	0.055	8.00	0.69	1.12	0.329
12.80	1.22	1.26	0.251					8.20	0.71	1.04	0.288
13.20	1.20	1.23	0.245	*Rhodium*[11]				8.40	0.74	0.97	0.252
13.60	1.19	1.20	0.236	0.10	18.48	69.43	0.986	8.60	0.78	0.89	0.212
14.00	1.20	1.16	0.225	0.20	8.66	37.46	0.977	8.80	0.83	0.83	0.179
14.40	1.22	1.13	0.214	0.30	5.85	25.94	0.967	9.00	0.88	0.77	0.148
14.80	1.27	1.12	0.207	0.40	4.74	19.80	0.955	9.20	0.95	0.73	0.125
15.20	1.31	1.17	0.218	0.50	4.20	16.07	0.941	9.40	1.01	0.71	0.110
15.60	1.31	1.23	0.234	0.60	3.87	13.51	0.925	9.60	1.07	0.69	0.102
16.00	1.28	1.28	0.251	0.70	3.67	11.72	0.908	9.80	1.12	0.69	0.098
16.40	1.24	1.33	0.270	0.80	3.63	10.34	0.887	10.00	1.17	0.69	0.098
16.80	1.17	1.37	0.288	0.90	3.62	9.36	0.867	10.60	1.26	0.73	0.106
17.00	1.14	1.38	0.297	1.00	3.71	8.67	0.848	11.00	1.29	0.76	0.113
17.40	1.06	1.39	0.314	1.10	3.67	8.26	0.837	11.60	1.32	0.80	0.124
18.00	0.95	1.38	0.334	1.20	3.51	7.94	0.832	12.00	1.32	0.82	0.127
18.40	0.88	1.36	0.346	1.30	3.26	7.63	0.829	12.60	1.32	0.82	0.129
18.80	0.82	1.33	0.355	1.40	3.01	7.31	0.827	13.00	1.32	0.83	0.131
19.20	0.76	1.29	0.360	1.50	2.78	6.97	0.823	13.60	1.32	0.85	0.134
19.60	0.72	1.25	0.363	1.60	2.60	6.64	0.818	14.00	1.32	0.86	0.138
20.00	0.67	1.21	0.369	1.70	2.42	6.33	0.813	14.60	1.30	0.89	0.144
20.40	0.64	1.15	0.364	1.80	2.30	6.02	0.805	15.00	1.28	0.90	0.147
20.80	0.61	1.10	0.357	1.90	2.20	5.76	0.798	15.60	1.25	0.90	0.147
21.20	0.60	1.06	0.349	2.00	2.12	5.51	0.789	16.00	1.24	0.89	0.147
21.60	0.58	1.02	0.342	2.10	2.05	5.30	0.780	16.50	1.23	0.88	0.145
22.00	0.57	0.98	0.336	2.20	2.00	5.11	0.772	17.00	1.22	0.88	0.144
22.40	0.56	0.95	0.328	2.30	1.94	4.94	0.765	17.50	1.22	0.87	0.143
22.80	0.55	0.92	0.325	2.40	1.90	4.78	0.756	18.00	1.23	0.88	0.145
23.20	0.53	0.89	0.322	2.50	1.88	4.65	0.748	18.50	1.25	0.92	0.155
23.60	0.52	0.85	0.317	2.60	1.85	4.55	0.743	19.00	1.24	0.98	0.172
24.00	0.50	0.82	0.314	2.70	1.80	4.49	0.742	19.50	1.18	1.05	0.193
24.40	0.49	0.79	0.309	2.90	1.63	4.36	0.748	20.00	1.10	1.09	0.213
				3.00	1.53	4.29	0.753				

Energy (eV)	n	k	$R(\phi = 0)$
20.50	1.00	1.09	0.230
21.00	0.91	1.05	0.234
21.50	0.86	1.00	0.228
22.00	0.83	0.95	0.219
22.50	0.81	0.92	0.214
23.00	0.79	0.90	0.213
23.50	0.75	0.87	0.214
24.00	0.73	0.84	0.210
24.50	0.70	0.81	0.208
25.00	0.69	0.77	0.202
25.50	0.67	0.74	0.195
26.00	0.66	0.70	0.188
26.50	0.65	0.66	0.176
27.00	0.65	0.64	0.168
27.50	0.65	0.61	0.159
28.00	0.65	0.59	0.152
29.00	0.65	0.54	0.137
30.00	0.66	0.51	0.127
31.00	0.64	0.49	0.127
32.00	0.61	0.44	0.126
33.00	0.60	0.37	0.110
34.00	0.65	0.30	0.074
35.00	0.69	0.28	0.058
36.00	0.73	0.27	0.049
37.00	0.74	0.28	0.047
38.00	0.74	0.27	0.045
39.00	0.75	0.25	0.041

Ruthenium, single crystal, $\vec{E} \parallel \hat{c}^9$

Energy (eV)	n	k	$R(\phi = 0)$
0.10	11.50	51.38	0.984
0.20	5.93	27.14	0.970
0.30	4.33	18.50	0.953
0.40	3.60	13.97	0.933
0.50	3.18	11.04	0.909
0.60	3.28	8.89	0.865
0.70	3.62	7.73	0.822
0.80	3.42	7.02	0.801
0.90	3.25	6.12	0.766
1.00	3.39	5.33	0.715
1.10	3.66	4.83	0.675
1.20	3.84	4.57	0.654
1.30	3.94	4.38	0.638
1.40	4.02	4.19	0.624
1.50	4.16	4.07	0.614
1.60	4.33	4.08	0.615
1.70	4.42	4.21	0.624
1.80	4.40	4.38	0.636
1.90	4.29	4.61	0.651
2.00	4.04	4.81	0.667
2.10	3.69	4.90	0.679
2.20	3.35	4.82	0.683
2.30	3.09	4.70	0.681
2.40	2.89	4.55	0.677
2.50	2.74	4.40	0.671
2.60	2.64	4.25	0.663
2.70	2.58	4.14	0.656
2.80	2.54	4.05	0.650
2.90	2.48	4.03	0.650
3.00	2.38	4.03	0.656
3.10	2.26	4.00	0.661
3.20	2.13	3.96	0.666

Energy (eV)	n	k	$R(\phi = 0)$
3.30	2.00	3.91	0.671
3.40	1.87	3.83	0.673
3.50	1.76	3.74	0.674
3.60	1.66	3.65	0.675
3.70	1.57	3.55	0.673
3.80	1.49	3.45	0.672
3.90	1.42	3.35	0.668
4.00	1.37	3.24	0.661
4.20	1.29	3.08	0.649
4.40	1.22	2.93	0.639
4.60	1.16	2.79	0.628
4.80	1.11	2.67	0.617
5.00	1.06	2.56	0.607
5.20	1.01	2.46	0.600
5.40	0.95	2.35	0.593
5.60	0.92	2.23	0.576
5.80	0.90	2.14	0.559
6.00	0.88	2.05	0.545
6.20	0.87	1.98	0.531
6.40	0.84	1.91	0.521
6.60	0.82	1.84	0.510
6.80	0.79	1.77	0.500
7.00	0.76	1.69	0.489
7.20	0.75	1.61	0.472
7.40	0.73	1.54	0.455
7.60	0.73	1.46	0.433
7.80	0.73	1.39	0.411
8.00	0.72	1.33	0.391
8.20	0.72	1.26	0.366
8.40	0.73	1.20	0.342
8.60	0.74	1.14	0.318
8.80	0.74	1.08	0.295
9.00	0.75	1.02	0.267
9.20	0.77	0.97	0.243
9.40	0.79	0.91	0.217
9.60	0.82	0.86	0.190
9.80	0.85	0.81	0.167
10.00	0.88	0.76	0.144
10.20	0.92	0.72	0.125
10.40	0.96	0.69	0.110
10.60	1.01	0.67	0.100
10.80	1.05	0.66	0.094
11.00	1.09	0.65	0.090
11.20	1.12	0.65	0.088
11.40	1.15	0.65	0.087
11.60	1.18	0.65	0.088
11.80	1.21	0.66	0.090
12.00	1.23	0.67	0.092
12.40	1.26	0.69	0.098
12.80	1.27	0.72	0.104
13.20	1.28	0.74	0.108
13.60	1.28	0.75	0.111
14.00	1.28	0.76	0.114
14.40	1.27	0.76	0.114
14.80	1.27	0.76	0.114
15.00	1.27	0.76	0.114
15.60	1.28	0.77	0.115
16.00	1.30	0.78	0.118
16.50	1.32	0.80	0.123
17.00	1.34	0.85	0.136

Energy (eV)	n	k	$R(\phi = 0)$
17.50	1.32	0.93	0.155
18.00	1.26	0.99	0.173
18.50	1.18	1.02	0.185
19.00	1.11	1.02	0.192
19.50	1.05	1.02	0.199
20.00	0.99	1.02	0.208
20.50	0.92	0.99	0.212
21.00	0.86	0.94	0.209
21.50	0.83	0.90	0.203
22.00	0.81	0.86	0.193
23.00	0.77	0.79	0.182
24.00	0.74	0.74	0.171
25.00	0.71	0.69	0.163
26.00	0.68	0.63	0.154
27.00	0.67	0.57	0.140
28.00	0.66	0.51	0.124
29.00	0.67	0.46	0.107
30.00	0.67	0.43	0.097
31.00	0.67	0.37	0.084
32.00	0.69	0.33	0.070
33.00	0.71	0.30	0.058
34.00	0.73	0.27	0.048
35.00	0.75	0.25	0.039
36.00	0.77	0.24	0.035
37.00	0.79	0.23	0.039
38.00	0.80	0.22	0.027
39.00	0.82	0.22	0.024
40.00	0.83	0.22	0.022

Ruthenium, single crystal, $\vec{E} \perp \hat{c}^5$

Energy (eV)	n	k	$R(\phi = 0)$
0.10	11.85	50.81	0.983
0.20	6.68	27.18	0.966
0.30	4.94	18.92	0.950
0.40	3.90	14.51	0.933
0.50	3.27	11.63	0.915
0.60	2.98	9.54	0.888
0.70	2.82	7.99	0.856
0.80	2.73	6.71	0.815
0.90	2.82	5.54	0.751
1.00	3.17	4.59	0.670
1.10	3.69	3.91	0.604
1.20	4.28	3.66	0.585
1.30	4.66	3.72	0.593
1.40	4.86	3.79	0.601
1.50	4.99	3.89	0.609
1.60	5.08	4.03	0.618
1.70	5.12	4.22	0.629
1.80	5.10	4.45	0.642
1.90	4.96	4.78	0.660
2.00	4.61	5.06	0.677
2.10	4.21	5.09	0.682
2.20	3.94	5.00	0.681
2.30	3.69	4.97	0.684
2.40	3.44	4.88	0.684
2.50	3.27	4.77	0.681
2.60	3.14	4.66	0.677
2.70	3.06	4.59	0.674
2.80	2.99	4.59	0.676
2.90	2.87	4.64	0.686
3.00	2.64	4.69	0.701
3.10	2.40	4.64	0.710

Energy (eV)	n	k	R(φ = 0)
3.20	2.18	4.55	0.717
3.30	2.00	4.43	0.721
3.40	1.84	4.30	0.723
3.50	1.71	4.16	0.723
3.60	1.60	4.03	0.722
3.70	1.50	3.90	0.721
3.80	1.41	3.77	0.718
3.90	1.35	3.64	0.713
4.00	1.29	3.53	0.707
4.20	1.21	3.31	0.694
4.40	1.16	3.13	0.679
4.60	1.13	2.97	0.662
4.80	1.09	2.86	0.652
5.00	1.03	2.75	0.648
5.20	0.97	2.64	0.643
5.40	0.91	2.52	0.635
5.60	0.88	2.40	0.622
5.80	0.86	2.29	0.605
6.00	0.84	2.20	0.591
6.20	0.82	2.11	0.576
6.40	0.81	2.04	0.564
6.60	0.78	1.97	0.556
6.80	0.76	1.89	0.545
7.00	0.73	1.82	0.538
7.20	0.70	1.75	0.527
7.40	0.68	1.67	0.513
7.60	0.67	1.59	0.496
7.80	0.66	1.51	0.476
8.00	0.66	1.44	0.454
8.20	0.65	1.36	0.430
8.40	0.66	1.29	0.403
8.60	0.66	1.22	0.378
8.80	0.68	1.15	0.346
9.00	0.69	1.09	0.317
9.20	0.70	1.02	0.286
9.40	0.73	0.95	0.251
9.60	0.77	0.89	0.216
9.80	0.82	0.84	0.185
10.00	0.86	0.81	0.163
10.20	0.90	0.77	0.143
10.40	0.94	0.74	0.127
10.60	0.99	0.72	0.115
10.80	1.04	0.71	0.108
11.00	1.08	0.70	0.104
11.20	1.11	0.70	0.102
11.40	1.14	0.70	0.101
11.60	1.17	0.71	0.102
11.80	1.20	0.72	0.104
12.00	1.22	0.73	0.107
12.40	1.25	0.76	0.113
12.80	1.26	0.78	0.118
13.20	1.27	0.81	0.124
13.60	1.27	0.83	0.129
14.00	1.26	0.84	0.132
14.40	1.25	0.84	0.132
14.80	1.25	0.84	0.133
15.00	1.25	0.84	0.133
15.60	1.25	0.85	0.134
16.00	1.27	0.85	0.134
16.50	1.28	0.89	0.145

Energy (eV)	n	k	R(φ = 0)
17.00	1.28	0.94	0.158
17.50	1.25	1.00	0.175
18.00	1.19	1.04	0.190
18.50	1.12	1.05	0.200
19.00	1.07	1.05	0.205
19.50	1.02	1.04	0.212
20.00	0.97	1.04	0.219
20.50	0.91	1.03	0.228
21.00	0.85	1.01	0.234
21.50	0.80	0.97	0.234
22.00	0.77	0.94	0.233
23.00	0.71	0.87	0.229
24.00	0.67	0.79	0.218
25.00	0.64	0.73	0.205
26.00	0.61	0.66	0.194
27.00	0.60	0.59	0.177
28.00	0.60	0.53	0.155
29.00	0.61	0.48	0.134
30.00	0.62	0.45	0.123
31.00	0.61	0.40	0.114
32.00	0.63	0.34	0.093
33.00	0.65	0.31	0.077
34.00	0.67	0.28	0.065
35.00	0.70	0.26	0.054
36.00	0.72	0.25	0.047
37.00	0.73	0.23	0.041
38.00	0.75	0.22	0.035
39.00	0.77	0.22	0.031
40.00	0.79	0.22	0.028

Selenium, single crystal, $\vec{E} \parallel \hat{c}$[22]

Energy (eV)	n	k	R(φ = 0)
0.01364	2.914	0.248	0.242
0.01488	3.175	9.95E-02	0.272
0.01612	3.263	2.13E-03	0.282
0.01736	3.306	3.81E-02	0.287
0.01860	3.330	7.04E-03	0.290
0.01984	3.346	4.23E-02	0.291
0.02108	3.358	3.40E-03	0.293
0.02232	3.366	5.31E-02	0.294
0.02356	3.372	1.96E-03	0.294
0.02480	3.377	2.39E-02	0.295
0.02604	3.380		0.295
0.02728		1.16E-02	
0.02976		7.96E-03	
0.03224		8.57E-03	
0.03472		2.70E-02	
0.03720	3.397	1.72E-02	0.297
0.04463		1.13E-02	
0.04959	3.403	2.79E-03	0.298
0.05703		1.56E-03	
0.06199	3.405	1.35E-03	0.298
0.06819		5.79E-04	
0.07439	3.407	4.44E-04	0.298
0.08059		4.41E-04	
0.08679	3.408	4.32E-04	0.298
0.09299		2.44E-04	
0.09919	3.409	3.23E-04	0.299
0.1116	3.409	2.87E-04	0.299
0.1240	3.410	2.71E-04	0.299
0.2480	3.417	2.67E-04	0.299
0.3720	3.427	1.90E-04	0.301

Energy (eV)	n	k	R(φ = 0)
0.4959	3.442	1.41E-04	0.302
0.6199	3.462	1.12E-04	0.304
0.7439	3.486	9.42E-05	0.307
0.8679	3.516	8.07E-05	0.310
0.9919	3.551	7.11E-05	0.314
1.116	3.592	6.37E-05	0.319
1.240	3.640	5.81E-05	0.324
1.50		1.33E-04	
1.60		1.59E-04	
1.70		6.27E-04	
1.80	4.46	2.20E-02	0.402
2.0	4.79	0.76	0.438
2.2	4.49	1.19	0.431
2.4	4.28	1.21	0.417
2.6	4.40	1.32	0.430
2.8	4.59	1.70	0.462
3.0	4.44	2.29	0.490
3.2	3.92	2.59	0.493
3.4	3.69	2.76	0.502
3.6	3.39	3.01	0.521
3.8	(3.00)		
4.0	(2.65)		
4.2	(2.30)		
4.5	1.92	2.78	0.528
5.0	1.50	2.31	0.482
6.0	1.57	1.49	0.288
7.0	1.84	1.45	0.276
8.0	1.35	1.68	0.353
9.0	1.35	1.64	0.342
10.0	0.92	1.07	0.238
12.0	1.00	1.10	0.232
14.0	0.81	0.91	0.211
16.0	0.65	0.61	0.160
18.0	0.65	0.48	0.120
20.0	0.69	0.36	0.076
22.0	0.81	0.25	0.030
24.0	0.91	0.18	0.011
26.0	0.86	0.15	0.012
28.0	0.85	0.13	0.011
30.0	0.87	0.11	0.008

Selenium, single crystal, $\vec{E} \perp \hat{c}$[22]

Energy (eV)	n	k	R(φ = 0)
0.01364	2.854	0.0239	0.231
0.01488	2.932	0.0325	0.241
0.01612	3.140	0.1750	0.269
0.01736	2.959	1.3300	0.321
0.01860	2.111	0.2550	0.133
0.01984	2.356	0.0746	0.164
0.02108	2.462	0.0276	0.178
0.02232	2.502	0.0442	0.184
0.02356	2.543	0.0097	0.190
0.02480	2.550	0.0239	0.191
0.02604	2.582		0.195
0.02728	2.600	0.0101	0.198
0.02976	2.576	9.95E-03	0.194
0.03224	2.598	1.16E-02	0.197
0.03472	2.607	1.68E-02	0.199
0.03720	2.613	1.54E-02	0.199
0.04463		1.17E-02	
0.04959	2.627	3.58E-03	0.201
0.05703		8.65E-04	

Energy (eV)	n	k	$R(\phi = 0)$
0.06199	2.632	2.07E-03	0.202
0.06819		2.89E-04	
0.07439	2.635	1.59E-04	0.202
0.08059		1.35E-04	
0.08679	2.636	1.42E-04	0.202
0.09299		1.04E-04	
0.09919	2.637	8.95E-05	0.203
0.1116	2.638	8.84E-05	0.203
0.1240	2.639	8.51E-05	0.203
0.2480	2.645	5.97E-05	0.204
0.3720	2.652	5.44E-05	0.205
0.4959	2.654	4.58E-05	0.205
0.6199	2.675	3.82E-05	0.208
0.7439	2.692	3.32E-05	0.210
0.8679	2.713	2.96E-05	0.213
0.9919	2.739	2.69E-05	0.216
1.116	2.772	2.48E-05	0.221
1.240	2.816	2.31E-05	0.226
1.50		7.37E-05	
1.60		8.63E-05	
1.70		3.60E-04	
1.80	3.32	0.11	0.289
2.00	3.38	0.65	0.310
2.20	3.07	0.73	0.282
2.40	2.93	0.61	0.259
2.60	3.00	0.53	0.263
2.80	3.12	0.58	0.279
3.00	3.30	0.70	0.305
3.20	3.35	1.01	0.328
3.40	3.22	1.24	0.334
3.60	3.06	1.47	0.344
3.80	2.84	1.66	0.351
4.00	2.51	1.81	0.356
4.20	2.18	1.83	0.352
4.50	1.75	1.94	0.382
5.00	1.25	1.50	0.316
6.00	1.32	0.73	0.107
7.00	1.62	0.61	0.105
8.00	1.81	0.69	0.135
9.00	1.66	1.02	0.182
10.00	1.72	0.95	0.171
12.00	1.25	1.02	0.181
14.00	0.98	0.92	0.178
16.00	0.68	0.96	0.274
18.00	0.61	0.65	0.191
20.00	0.73	0.48	0.094
22.00	0.78	0.39	0.060
24.00	0.78	0.32	0.046
26.00	0.78	0.26	0.036
28.00	0.80	0.19	0.023
30.00	0.79	0.14	0.020

Silicon, single crystal[23]

Energy (eV)	n	k	$R(\phi = 0)$
0.01240	3.4185	2.90E-04	0.300
0.01488	3.4190	2.30E-04	0.300
0.01736	3.4192	1.90E-04	0.300
0.01984	3.4195	1.70E-04	0.300
0.02480	3.4197		0.300
0.03100	3.4199		0.300
0.04092	3.4200		0.300
0.04463		1.08E-04	
0.04959	3.4201	9.15E-05	0.300
0.05703		1.56E-04	
0.06199	3.4204	2.86E-04	0.300
0.06943		3.84E-04	
0.07439		7.16E-04	
0.08059	(3.4207)	1.52E-04	0.300
0.08679		1.02E-04	
0.09299		2.59E-04	
0.09919		1.77E-04	
0.1054		1.53E-04	
0.1116		2.02E-04	
0.1178		1.22E-04	
0.1240	3.4215	6.76E-05	0.300
0.1364		5.49E-05	
0.1488		2.41E-05	
0.1612		2.49E-05	
0.1736	(3.4230)	1.68E-05	0.300
0.1798		2.45E-05	
0.1860		2.66E-06	
0.1922		1.74E-06	
0.1984		8.46E-07	
0.2046		5.64E-07	
0.2108	(3.4244)	4.17E-07	0.300
0.2170		4.05E-07	
0.2232		3.94E-07	
0.2294		3.26E-07	
0.2356		2.97E-07	
0.2418		2.82E-07	
0.2480	3.4261	1.99E-07	0.300
0.3100	3.4294		0.301
0.3626	3.4327		0.301
0.4568	3.4393	2.50E-09	0.302
0.6199	3.4490		0.303
0.8093	3.4784		0.306
1.033	3.5193		0.311
1.1	(3.5341)	1.30E-05	0.312
1.2		1.80E-04	
1.3		2.26E-03	
1.4		7.75E-03	
1.5	3.673	5.00E-03	0.327
1.6	3.714	8.00E-03	0.331
1.7	3.752	1.00E-02	0.335
1.8	3.796	0.013	0.340
1.9	3.847	0.016	0.345
2.0	3.906	0.022	0.351
2.1	3.969	0.030	0.357
2.2	4.042	0.032	0.364
2.3	4.123	0.048	0.372
2.4	4.215	0.060	0.380
2.5	4.320	0.073	0.390
2.6	4.442	0.090	0.400
2.7	4.583	0.130	0.412
2.8	4.753	0.163	0.426
2.9	4.961	0.203	0.442
3.0	5.222	0.269	0.461
3.1	5.570	0.387	0.486
3.2	6.062	0.630	0.518
3.3	6.709	1.321	0.561
3.4	6.522	2.705	0.592
3.5	5.610	3.014	0.575
3.6	5.296	2.987	0.564
3.7	5.156	3.058	0.563
3.8	5.065	3.182	0.568
3.9	5.016	3.346	0.577
4.0	5.010	3.587	0.591
4.1	5.020	3.979	0.614
4.2	4.888	4.639	0.652
4.3	4.086	5.395	0.703
4.4	3.120	5.344	0.726
4.5	2.451	5.082	0.740
4.6	1.988	4.678	0.742
4.7	1.764	4.278	0.728
4.8	1.658	3.979	0.710
4.9	1.597	3.749	0.693
5.0	1.570	3.565	0.675
5.1	1.571	3.429	0.658
5.2	1.589	3.354	0.646
5.3	1.579	3.353	0.647
5.4	1.471	3.366	0.663
5.5	1.340	3.302	0.673
5.6	1.247	3.206	0.675
5.7	1.180	3.112	0.673
5.8	1.133	3.045	0.672
5.9	1.083	2.982	0.673
6.0	1.010	2.909	0.677
6.5	0.847	2.73	0.688
7.0	0.682	2.45	0.691
7.5	0.563	2.21	0.693
8.0	0.478	2.00	0.691
8.5	0.414	1.82	0.688
9.0	0.367	1.66	0.683
9.5	0.332	1.51	0.672
10.0	0.306	1.38	0.661
12.0	0.257	0.963	0.590
14.0	0.275	0.641	0.460
16.0	0.345	0.394	0.297
18.0	0.455	0.219	0.159
20.0	0.567	0.0835	0.079
22.14	0.675	0.0405	0.038
24.31	0.752	0.0243	0.020
26.38	0.803	0.0178	0.012
28.18	0.834	0.0152	0.008
30.24	0.860	0.0138	0.006
31.79	0.877	0.0132	0.004
34.44	0.899	0.0121	0.003
36.47	0.913	0.0113	0.002
38.75	0.925	0.0104	0.002
40.00	0.930	0.0100	0.001

Silver[6]

Energy (eV)	n	k	$R(\phi = 0)$
0.10	9.91	90.27	0.995
0.20	2.84	45.70	0.995
0.30	1.41	30.51	0.994
0.40	0.91	22.89	0.993
0.50	0.67	18.32	0.992
1.00	0.28	9.03	0.987
1.50	0.27	5.79	0.969
2.00	0.27	4.18	0.944
2.50	0.24	3.09	0.914
3.00	0.23	2.27	0.864
3.25	0.23	1.86	0.816

Energy (eV)	n	k	R(φ = 0)
3.50	0.21	1.42	0.756
3.60	0.23	1.13	0.671
3.70	0.30	0.77	0.475
3.77	0.53	0.40	0.154
3.80	0.73	0.30	0.053
3.90	1.30	0.36	0.040
4.00	1.61	0.60	0.103
4.10	1.73	0.85	0.153
4.20	1.75	1.06	0.194
4.30	1.73	1.13	0.208
4.50	1.69	1.28	0.238
4.75	1.61	1.34	0.252
5.00	1.55	1.36	0.257
5.50	1.45	1.34	0.257
6.00	1.34	1.28	0.246
6.50	1.25	1.18	0.225
7.00	1.18	1.06	0.196
7.50	1.14	0.91	0.157
8.00	1.16	0.75	0.114
9.00	1.33	0.56	0.074
10.00	1.46	0.56	0.082
11.00	1.52	0.56	0.088
12.00	1.61	0.59	0.100
13.00	1.66	0.64	0.112
14.00	1.72	0.78	0.141
14.50	1.64	0.88	0.152
15.00	1.56	0.92	0.156
16.00	1.42	0.91	0.151
17.00	1.33	0.86	0.139
18.00	1.28	0.80	0.124
19.00	1.27	0.75	0.111
20.00	1.29	0.71	0.103
21.00	1.35	0.75	0.112
21.50	1.37	0.80	0.124
22.00	1.34	0.87	0.141
22.50	1.26	0.93	0.157
23.00	1.17	0.94	0.163
23.50	1.10	0.93	0.165
24.00	1.04	0.90	0.165
24.50	0.99	0.87	0.160
25.00	0.95	0.83	0.154
25.50	0.91	0.78	0.144
26.00	0.90	0.74	0.133
26.50	0.89	0.69	0.121
27.00	0.89	0.65	0.109
27.50	0.89	0.62	0.099
28.00	0.90	0.59	0.090
28.50	0.91	0.57	0.084
29.00	0.92	0.56	0.079
30.00	0.93	0.54	0.074
31.00	0.93	0.53	0.072
32.00	0.92	0.53	0.072
33.00	0.90	0.51	0.071
34.00	0.88	0.49	0.067
35.00	0.86	0.45	0.061
36.00	0.89	0.44	0.055
38.00	0.89	0.39	0.043
40.00	0.90	0.37	0.039
42.00	0.90	0.35	0.036
44.00	0.90	0.33	0.033

Energy (eV)	n	k	R(φ = 0)
46.00	0.90	0.32	0.031
48.00	0.89	0.31	0.030
50.00	0.88	0.29	0.027
52.00	0.89	0.28	0.024
54.00	0.88	0.17	0.024
56.00	0.87	0.26	0.024
58.00	0.87	0.24	0.021
60.00	0.87	0.22	0.018
62.00	0.88	0.21	0.016
64.00	0.88	0.21	0.016
66.00	0.88	0.21	0.016
68.00	0.87	0.21	0.017
70.00	0.83	0.20	0.021
72.00	0.85	0.18	0.016
74.00	0.85	0.17	0.014
76.00	0.85	0.16	0.013
78.00	0.85	0.15	0.013
80.00	0.85	0.14	0.012
85.00	0.85	0.11	0.011
90.00	0.85	0.08	0.009
95.00	0.86	0.06	0.007
100.00	0.87	0.04	0.005

Sodium[24]

Energy (eV)	n	k	R(φ = 0)
0.55	0.262	9.97	0.990
0.58	0.241	9.45	0.989
0.63	0.207	8.80	0.990
0.67	0.175	8.09	0.990
0.73	0.147	7.42	0.990
0.81	0.123	6.67	0.989
0.92	0.099	5.82	0.989
1.05	0.078	5.11	0.989
1.23	0.064	4.35	0.987
1.44	0.053	3.72	0.986
1.65	0.050	3.22	0.983
1.87	0.049	2.76	0.978
2.07	0.053	2.48	0.971
2.27	0.059	2.23	0.961
2.45	0.063	2.07	0.953
2.64	0.066	1.88	0.943
2.82	0.068	1.76	0.936
2.95	0.068	1.63	0.928
3.06	0.069	1.54	0.921
3.20	0.065	1.47	0.921
3.40	0.061	1.33	0.916
3.71	0.055	1.13	0.908
3.97	0.049	1.01	0.908
6.199	0.390		0.193
6.358	0.454		0.141
6.526	0.485		0.120
6.702	0.533		0.093
6.888	0.574		0.073
7.130	0.616		0.056
7.328	0.641		0.048
7.583	0.674		0.038
7.847	0.700		0.031
8.015	0.710		0.029
8.634	0.762		0.018
9.143	0.800		0.012
9.709	0.819		0.010
10.20	0.843		0.007

Energy (eV)	n	k	R(φ = 0)
11.08	0.870		0.005
11.83	0.887		0.004
12.73	0.907		0.002
13.05	0.913		0.002
13.42	0.914		0.002
13.73	0.917		0.002
14.07	0.922		0.002
14.83	0.934		0.001
15.05	0.936		0.001
15.46	0.942		0.001
16.21	0.948		0.001
18.10	0.964		0.000
21.12	0.979		0.000
25.51	0.993		0.000
26.95	1.00		0.000
27.68	1.01		0.000
28.37	1.01		0.000
29.52	1.02		0.000

Tantalum[16]

Energy (eV)	n	k	R(φ = 0)
0.10	10.14	66.39	0.984
0.15	9.45	46.41	0.9834
0.20	5.77	35.46	0.982
0.26	3.67	27.53	0.981
0.30	2.87	23.90	0.980
0.38	2.03	18.87	0.978
0.50	1.37	14.26	0.974
0.58	1.15	12.19	0.970
0.70	0.96	9.92	0.962
0.78	0.89	8.77	0.956
0.90	0.84	7.38	0.942
1.00	0.89	6.47	0.992
1.10	0.93	5.75	0.899
1.20	0.98	5.14	0.872
1.30	1.00	4.62	0.842
1.40	1.04	4.15	0.805
1.50	1.09	3.73	0.762
1.60	1.15	3.33	0.707
1.70	1.24	2.95	0.640
1.80	1.35	2.60	0.560
1.90	1.57	2.24	0.460
2.00	1.83	1.99	0.388
2.10	2.10	1.84	0.354
2.20	2.36	1.81	0.351
2.30	2.56	1.86	0.365
2.40	2.68	1.92	0.378
2.50	2.75	1.98	0.388
2.60	2.80	2.02	0.395
2.70	2.84	2.08	0.405
2.80	2.85	2.14	0.412
2.90	2.84	2.20	0.420
3.00	2.81	2.24	0.425
3.20	2.73	2.31	0.432
3.40	2.61	2.33	0.435
3.60	2.49	2.30	0.430
3.80	2.40	2.22	0.418
4.00	2.36	2.14	0.406
4.20	2.35	2.06	0.392
4.40	2.39	2.01	0.384
4.60	2.45	2.00	0.384
4.80	2.53	2.06	0.394

Energy (eV)	n	k	R(φ = 0)
5.00	2.58	2.20	0.416
5.20	2.52	2.44	0.450
5.40	2.31	2.61	0.480
5.60	2.06	2.67	0.501
5.80	1.83	2.63	0.510
6.00	1.63	2.56	0.515
6.20	1.48	2.45	0.512
6.40	1.37	2.33	0.504
6.60	1.29	2.22	0.492
6.80	1.23	2.11	0.478
7.00	1.18	2.01	0.462
7.20	1.15	1.91	0.445
7.40	1.13	1.82	0.425
7.60	1.12	1.75	0.406
7.80	1.11	1.68	0.390
8.00	1.11	1.61	0.370
8.20	1.12	1.55	0.350
8.40	1.13	1.50	0.332
8.60	1.14	1.45	0.317
8.80	1.17	1.41	0.301
9.00	1.19	1.40	0.294
9.20	1.21	1.38	0.289
9.40	1.21	1.38	0.287
9.60	1.21	1.38	0.285
9.80	1.21	1.37	0.285
10.00	1.20	1.37	0.286
10.20	1.19	1.37	0.286
10.40	1.18	1.37	0.287
10.60	1.16	1.36	0.288
10.80	1.15	1.36	0.289
11.00	1.13	1.35	0.290
11.20	1.11	1.35	0.292
11.40	1.09	1.34	0.293
11.60	1.07	1.33	0.294
11.80	1.05	1.32	0.295
12.00	1.02	1.31	0.296
12.20	1.00	1.29	0.295
12.40	0.98	1.28	0.294
12.60	0.96	1.26	0.292
12.80	0.94	1.24	0.289
13.00	0.93	1.22	0.286
13.60	0.91	1.16	0.272
14.00	0.90	1.15	0.272
14.60	0.85	1.15	0.285
15.00	0.80	1.13	0.293
15.60	0.72	1.08	0.301
16.00	0.68	1.04	0.304
16.60	0.63	0.97	0.301
17.00	0.60	0.92	0.296
17.60	0.60	0.92	0.296
18.00	0.55	0.79	0.274
18.60	0.53	0.71	0.254
19.00	0.53	0.65	0.236
19.60	0.53	0.57	0.207
20.00	0.54	0.52	0.185
20.60	0.55	0.44	0.153
21.00	0.57	0.39	0.127
21.60	0.64	0.34	0.089
22.00	0.64	0.32	0.081
22.60	0.69	0.27	0.058
23.00	0.73	0.24	0.043
23.60	0.80	0.26	0.033
24.00	0.80	0.26	0.034
24.60	0.82	0.25	0.029
25.00	0.83	0.25	0.026
25.60	0.86	0.24	0.022
26.00	0.88	0.25	0.022
26.60	0.87	0.26	0.023
27.00	0.87	0.25	0.022
27.60	0.89	0.23	0.019
28.00	0.90	0.23	0.017
28.60	0.91	0.22	0.015
29.00	0.92	0.22	0.014
29.60	0.94	0.22	0.014
30.00	0.95	0.22	0.014
31.00	0.97	0.23	0.014
32.00	0.98	0.24	0.015
33.00	0.98	0.25	0.015
34.00	0.99	0.25	0.016
35.00	0.99	0.26	0.017
36.00	0.99	0.27	0.018
37.00	0.99	0.28	0.019
38.00	0.98	0.28	0.021
39.00	0.97	0.29	0.022
40.00	0.95	0.29	0.023

Tellurium, $\vec{E} \parallel \hat{c}$[25]

Energy (eV)	n	k	R(φ = 0)
0.01364	4.82	0.118	0.431
0.01488	5.26	0.0505	0.463
0.01612	5.47	0.0278	0.477
0.01736	5.59	0.0174	0.485
0.01860		0.0796	
0.01984		0.0696	
0.02108		0.0749	
0.02232		0.1900	
0.02356		0.2220	
0.02480		0.0716	
0.02604		0.0682	
0.02728		0.0832	
0.02976		0.0149	
0.03224		2.14E-03	
0.03472		1.71E-02	
0.03720	5.94	3.71E-03	0.507
0.03968		2.44E-03	
0.04339	5.96	1.59E-03	0.508
0.04711		7.85E-04	
0.05083		7.38E-04	
0.05579		3.89E-04	
0.06199	5.98	3.09E-04	0.509
0.07439		2.52E-04	
0.08679		2.96E-04	
0.09919		3.68E-04	
0.12400	6.246	3.34E-04	0.524
0.15500	6.253		0.525
0.20660	6.286		0.526
0.24800	6.316	7.48E-05	0.528
0.31	6.372	1.18E-05	0.531
0.35		4.93E-04	
0.41		6.74E-03	
0.5	6.53	2.30E-02	0.539
0.6	6.71	7.50E-02	0.549
0.7	7.00	0.24	0.563
0.8	7.23	0.48	0.574
0.9	7.48	0.94	0.589
1.0	7.70	1.56	0.606
1.2	6.99	2.22	0.593
1.4	7.11	2.46	0.604
1.6	6.75	2.91	0.606
1.8	6.89	3.70	0.637
2.0	4.67	4.67	0.654
2.2	4.94	5.16	0.681
2.4	3.94	5.08	0.686
2.6	3.25	4.77	0.681
2.8	2.73	4.42	0.674
3.0	2.30	4.16	0.674
3.5	1.69	3.44	0.646
4.0	1.33	2.64	0.571
4.5	1.32	1.96	0.428
5.0	1.63	1.60	0.312
5.5	1.72	1.57	0.302
6.0	1.73	1.45	0.276
6.5	1.78	1.36	0.257
7.0	1.83	1.36	0.257
7.5	1.72	1.51	0.289
8.0	1.54	1.37	0.260
8.5	1.55	1.23	0.226
9.0	0.99	0.93	0.179
9.5	1.47	1.25	0.233
10.0	0.86	0.86	0.181
11.0	0.80	0.77	0.165
12.0	0.79	0.76	0.164
14.0	0.67	0.59	0.146
16.0	0.59	0.49	0.147
18.0	0.48	0.31	0.160
20.0	0.74	0.20	0.035
22.0	0.83	0.18	0.018
24.0	0.85	0.15	0.013
26.0	0.87	0.12	0.009
28.0	0.89	0.090	0.006
30.0	0.90	0.045	0.003

Tellurium, $\vec{E} \perp \hat{c}$[25]

Energy (eV)	n	k	R(φ = 0)
0.01364	2.61	0.2980	0.204
0.01488	3.65	0.0894	0.325
0.01612	4.10	0.0535	0.370
0.01736	4.63	0.4990	0.420
0.01860		0.1170	
0.01984		0.0343	
0.02108	(4.42)	0.0421	0.398
0.02232		0.1060	
0.02356		0.0880	
0.02480		0.0458	
0.02604		0.0928	
0.02728		0.0886	
0.02976		0.0232	
0.03224		3.06E-03	
0.03472		1.25E-02	
0.03720	4.71	2.65E-03	0.422
0.03968		1.89E-03	
0.04339	4.74	1.41E-03	0.425
0.04711		8.38E-04	
0.05083		6.79E-04	

Energy (eV)	n	k	R(φ = 0)	Energy (eV)	n	k	R(φ = 0)	Energy (eV)	n	k	R(φ = 0)
0.05579		1.59E-04		0.40	2.49	4.68	0.708	8.40	1.08	0.95	0.175
0.06199	4.77	1.16E-04	0.427	0.45	3.35	3.25	0.545	8.60	1.04	0.96	0.181
0.07439		7.23E-05		0.50	4.43	3.22	0.555	8.80	1.02	0.95	0.181
0.08679		5.34E-05		0.60	4.71	3.77	0.597	9.00	1.00	0.94	0.182
0.09919		4.28E-05		0.70	4.38	3.89	0.603	9.20	0.97	0.93	0.182
0.1240	4.796	3.18E-05	0.429	0.80	4.04	3.82	0.596	9.40	0.95	0.91	0.181
0.1550	4.809		0.430	0.90	3.80	3.65	0.582	9.60	0.94	0.90	0.179
0.2066	4.838		0.432	1.00	3.62	3.52	0.570	9.80	0.91	0.88	0.179
0.2480	4.864	2.19E-05	0.434	1.10	3.47	3.40	0.560	10.00	0.89	0.88	0.180
0.31	4.929	3.18E-05	0.439	1.20	3.35	3.30	0.550	10.20	0.86	0.85	0.178
0.35		7.89E-02		1.30	3.28	3.25	0.546	10.40	0.85	0.83	0.175
0.41		0.149		1.40	3.17	3.28	0.549	10.60	0.81	0.79	0.167
0.5	4.90		0.437	1.50	2.98	3.32	0.557	10.80	0.80	0.76	0.162
0.6	4.93		0.439	1.60	2.74	3.30	0.559	11.00	0.79	0.72	0.152
0.7	4.95	0.11	0.441	1.70	2.54	3.23	0.557	11.20	0.81	0.69	0.139
0.8	5.10	0.13	0.452	1.80	2.36	3.11	0.550	11.40	0.81	0.69	0.139
0.9	5.22	0.22	0.461	1.90	2.22	2.99	0.540	11.60	0.79	0.68	0.139
1.0	5.35	0.45	0.472	2.00	2.11	2.88	0.530	11.80	0.78	0.67	0.137
1.2	5.17	0.63	0.462	2.10	2.01	2.77	0.520	12.00	0.77	0.65	0.132
1.4	5.56	0.63	0.488	2.20	1.92	2.67	0.509	12.80	0.76	0.55	0.106
1.6	5.88	1.15	0.517	2.30	1.86	2.56	0.495	13.20	0.76	0.52	0.097
1.8	6.10	1.80	0.545	2.40	1.81	2.47	0.483	13.60	0.76	0.48	0.087
2.0	5.94	2.69	0.571	2.50	1.78	2.39	0.471	14.00	0.77	0.45	0.077
2.2	5.10	3.61	0.594	2.60	1.75	2.34	0.462	14.40	0.77	0.42	0.069
2.4	4.24	3.77	0.593	2.70	1.71	2.29	0.456	14.80	0.79	0.38	0.058
2.6	3.57	3.75	0.591	2.80	1.68	2.25	0.451	15.20	0.79	0.36	0.052
2.8	3.03	3.63	0.588	2.90	1.63	2.21	0.447	15.60	0.79	0.32	0.045
3.0	2.51	3.39	0.578	3.00	1.59	2.17	0.444	16.00	0.83	0.31	0.037
3.5	1.72	2.70	0.532	3.10	1.55	2.15	0.442	16.40	0.84	0.28	0.030
4.0	1.32	2.01	0.440	3.20	1.50	2.12	0.442	16.80	0.87	0.27	0.025
4.5	1.28	1.28	0.251	3.30	1.44	2.09	0.442	17.20	0.90	0.25	0.020
5.0	1.47	0.82	0.132	3.40	1.37	2.06	0.443	17.60	0.93	0.25	0.017
5.5	1.74	0.51	0.104	3.50	1.30	2.01	0.443	18.00	0.94	0.24	0.165
6.0	1.94	0.39	0.118	3.60	1.24	1.96	0.441	18.40	0.94	0.23	0.017
6.5	2.19	0.32	0.148	3.70	1.17	1.90	0.436	18.80	0.95	0.24	0.016
7.0	2.48	0.40	0.192	3.80	1.11	1.83	0.430	19.20	0.96	0.25	0.016
7.5	2.60	0.69	0.226	3.85	1.08	1.78	0.423	19.60	0.97	0.25	0.017
8.0	2.59	0.91	0.245	3.90	1.06	1.73	0.413	20.00	0.98	0.27	0.018
8.5	2.39	1.00	0.235	4.00	1.04	1.62	0.389	20.40	0.98	0.27	0.019
9.0	1.11	1.24	0.259	4.20	1.05	1.45	0.333	20.60	1.00	0.29	0.020
9.5	2.08	1.11	0.224	4.40	1.13	1.33	0.284	21.20	0.99	0.31	0.023
10.0	0.99	1.04	0.215	4.60	1.17	1.29	0.265	21.60	0.99	0.31	0.024
11.0	0.84	1.01	0.237	4.80	1.21	1.23	0.244	22.00	0.98	0.32	0.025
12.0	0.87	0.87	0.182	5.00	1.24	1.21	0.236	22.40	0.98	0.33	0.027
14.0	0.59	0.87	0.282	5.20	1.27	1.20	0.228	22.80	0.97	0.33	0.028
16.0	0.64	0.55	0.144	5.40	1.17	1.16	0.228	23.20	0.96	0.34	0.030
18.0	0.52	0.41	0.161	5.60	1.24	1.21	0.234	23.60	0.95	0.35	0.031
20.0	0.50	0.38	0.165	5.80	1.21	1.22	0.241	24.00	0.92	0.35	0.033
22.0	0.56	0.29	0.110	6.00	1.15	1.21	0.244	24.5	0.91	0.34	0.032
24.0	0.54	0.25	0.113	6.20	1.11	1.18	0.240	25.0	0.91	0.33	0.032
26.0	0.50	0.20	0.127	6.40	1.08	1.14	0.232	25.5	0.89	0.33	0.032
28.0	0.48	0.17	0.135	6.60	1.04	1.06	0.212	26.0	0.89	0.33	0.032
30.0	0.46	0.088	0.140	6.80	1.05	1.02	0.198	26.5	0.88	0.32	0.032
				7.00	1.06	0.97	0.182	27.0	0.86	0.31	0.032
Titanium (Polycrystalline)[14]				7.20	1.07	0.95	0.175	27.5	0.85	0.30	0.033
0.10	5.03	23.38	0.965	7.40	1.11	0.94	0.167	28.0	0.84	0.29	0.033
0.15	3.00	15.72	0.954	7.60	1.09	0.92	0.165	28.5	0.82	0.26	0.029
0.20	2.12	11.34	0.939	7.80	1.11	0.93	0.165	29.0	0.83	0.25	0.027
0.25	2.05	8.10	0.890	8.00	1.10	0.94	0.169	30.0	0.84	0.22	0.022
0.30	6.39	9.94	0.833	8.20	1.10	0.95	0.171				
0.35	2.74	6.21	0.792								

Tungsten[27]

Energy (eV)	n	k	R(φ = 0)
0.10	14.06	54.71	0.983
0.20	3.87	28.30	0.981
0.25	2.56	22.44	0.980
0.30	1.83	18.32	0.979
0.34	1.71	15.71	0.973
0.38	1.86	13.88	0.963
0.42	1.92	12.63	0.954
0.46	1.69	11.59	0.952
0.50	1.40	10.52	0.952
0.54	1.23	9.45	0.948
0.58	1.17	8.44	0.938
0.62	1.28	7.52	0.917
0.66	1.45	6.78	0.888
0.70	1.59	6.13	0.856
0.74	1.83	5.52	0.810
0.78	2.12	5.00	0.759
0.82	2.36	4.61	0.710
0.86	2.92	4.37	0.661
0.90	3.11	4.44	0.660
0.94	3.15	4.43	0.658
0.98	3.15	4.36	0.653
1.00	3.14	4.32	0.649
1.10	3.05	4.04	0.627
1.20	3.00	3.64	0.590
1.30	3.12	3.24	0.545
1.40	3.29	2.96	0.515
1.50	3.48	2.79	0.500
1.60	3.67	2.68	0.494
1.70	3.84	2.79	0.507
1.80	3.82	2.91	0.518
1.90	3.70	2.94	0.518
2.00	3.60	2.89	0.512
2.10	3.54	2.84	0.506
2.20	3.49	2.76	0.497
2.30	3.49	2.72	0.494
2.40	3.45	2.72	0.493
2.50	3.38	2.68	0.487
2.60	3.34	2.62	0.480
2.70	3.31	2.55	0.472
2.80	3.31	2.49	0.466
2.90	3.32	2.45	0.461
3.00	3.35	2.42	0.459
3.10	3.39	2.41	0.460
3.20	3.43	2.45	0.465
3.30	3.45	2.55	0.476
3.40	3.39	2.66	0.485
3.50	3.24	2.70	0.488
3.60	3.13	2.67	0.482
3.70	3.05	2.62	0.476
3.80	2.99	2.56	0.468
3.90	2.96	2.50	0.460
4.00	2.95	2.43	0.451
4.20	3.02	2.33	0.440
4.40	3.13	2.32	0.442
4.60	3.24	2.41	0.455
4.80	3.33	2.57	0.475
5.00	3.40	2.85	0.505
5.20	3.27	3.27	0.548
5.40	2.92	3.58	0.586
5.60	2.43	3.70	0.618
5.80	2.00	3.61	0.637
6.00	1.70	3.42	0.643
6.20	1.47	3.24	0.646
6.40	1.32	3.04	0.640
6.60	1.21	2.87	0.631
6.80	1.12	2.70	0.619
7.00	1.06	2.56	0.607
7.20	1.01	2.43	0.593
7.40	0.98	2.30	0.573
7.60	0.95	2.18	0.556
7.80	0.93	2.06	0.533
8.00	0.94	1.95	0.505
8.20	0.94	1.86	0.481
8.40	0.96	1.76	0.449
8.60	0.99	1.70	0.422
8.80	1.01	1.65	0.401
9.00	1.01	1.60	0.388
9.20	1.02	1.55	0.369
9.40	1.03	1.50	0.352
9.60	1.05	1.44	0.329
9.80	1.09	1.38	0.307
10.00	1.13	1.34	0.287
10.20	1.19	1.33	0.274
10.40	1.24	1.34	0.270
10.60	1.27	1.36	0.274
10.80	1.29	1.39	0.282
11.00	1.28	1.42	0.290
11.20	1.27	1.44	0.297
11.40	1.25	1.46	0.305
11.60	1.22	1.48	0.313
11.80	1.20	1.48	0.318
12.00	1.16	1.48	0.323
12.40	1.10	1.47	0.329
12.80	1.04	1.44	0.333
13.20	0.98	1.40	0.332
13.60	0.94	1.35	0.325
14.00	0.91	1.28	0.312
14.40	0.90	1.23	0.296
14.80	0.90	1.17	0.276
15.20	0.93	1.13	0.255
15.60	0.97	1.12	0.246
16.00	0.98	1.14	0.249
16.40	0.97	1.17	0.260
16.80	0.94	1.19	0.273
17.20	0.90	1.21	0.289
17.60	0.85	1.21	0.304
18.00	0.80	1.20	0.317
18.40	0.74	1.18	0.330
18.80	0.69	1.15	0.340
19.20	0.64	1.11	0.347
19.60	0.60	1.07	0.353
20.00	0.56	1.02	0.354
20.40	0.54	0.97	0.350
20.80	0.52	0.92	0.342
21.20	0.50	0.87	0.331
21.60	0.50	0.82	0.318
22.00	0.49	0.77	0.303
22.40	0.49	0.73	0.287
22.80	0.49	0.69	0.272
23.20	0.49	0.66	0.263
23.60	0.48	0.62	0.252
24.00	0.49	0.57	0.234
24.40	0.50	0.53	0.213
24.80	0.51	0.49	0.191
25.20	0.53	0.46	0.171
25.60	0.55	0.43	0.150
26.00	0.57	0.40	0.132
26.40	0.59	0.38	0.117
26.80	0.61	0.37	0.105
27.00	0.62	0.36	0.099
27.50	0.64	0.34	0.085
28.00	0.67	0.32	0.073
28.50	0.69	0.31	0.065
29.00	0.71	0.30	0.057
29.50	0.73	0.30	0.052
30.00	0.75	0.29	0.047
31.00	0.78	0.29	0.042
32.00	0.79	0.29	0.040
33.00	0.82	0.28	0.033
34.00	0.84	0.29	0.032
35.00	0.85	0.31	0.033
36.00	0.85	0.32	0.036
37.00	0.84	0.33	0.039
38.00	0.83	0.33	0.040
39.00	0.81	0.33	0.042
40.00	0.80	0.33	0.045

Vanadium[9]

Energy (eV)	n	k	R(φ = 0)
0.10	12.83	45.89	0.978
0.20	3.90	24.30	0.975
0.28	2.13	17.35	0.973
0.36	1.54	13.32	0.966
0.44	1.28	10.74	0.957
0.52	1.16	8.93	0.945
0.60	1.10	7.59	0.929
0.68	1.07	6.54	0.909
0.76	1.08	5.67	0.882
0.80	1.10	5.30	0.864
0.90	1.18	4.50	0.811
1.00	1.34	3.80	0.730
1.10	1.60	3.26	0.632
1.20	1.93	2.88	0.543
1.30	2.25	2.71	0.498
1.40	2.48	2.72	0.491
1.50	2.57	2.79	0.499
1.60	2.57	2.84	0.507
1.70	2.52	2.88	0.512
1.80	2.45	2.88	0.515
1.90	2.36	2.85	0.514
2.00	2.34	2.81	0.509
2.10	2.31	2.78	0.506
2.20	2.28	2.80	0.510
2.30	2.23	2.83	0.516
2.40	2.15	2.88	0.528
2.50	2.02	2.91	0.540
2.60	1.89	2.92	0.552
2.70	1.74	2.89	0.561
2.80	1.61	2.85	0.569
2.90	1.48	2.80	0.577

Energy (eV)	n	k	$R(\phi = 0)$
3.00	1.36	2.73	0.582
3.20	1.16	2.55	0.585
3.40	0.99	2.37	0.586
3.60	0.87	2.17	0.575
3.80	0.80	1.96	0.547
4.00	0.78	1.76	0.503
4.20	0.80	1.60	0.449
4.40	0.83	1.47	0.400
4.60	0.87	1.38	0.355
4.80	0.90	1.31	0.326
5.00	0.91	1.26	0.304
5.25	0.93	1.18	0.271
5.50	0.94	1.14	0.258
5.75	0.96	1.09	0.235
6.00	0.98	1.06	0.223
6.25	0.97	1.02	0.212
6.50	0.97	0.98	0.199
6.75	0.97	0.94	0.185
7.00	0.98	0.91	0.175
7.33	0.97	0.89	0.170
7.66	0.98	0.87	0.162
8.00	0.98	0.85	0.155
8.33	0.98	0.81	0.146
8.66	0.98	0.81	0.145
9.00	0.96	0.79	0.142
9.50	0.94	0.77	0.136
10.00	0.91	0.74	0.133
10.50	0.89	0.71	0.126
11.00	0.87	0.65	0.112
11.50	0.88	0.58	0.091
12.00	0.90	0.58	0.089
12.50	0.89	0.57	0.086
13.00	0.88	0.55	0.082
13.50	0.87	0.53	0.079
14.00	0.86	0.51	0.075
14.50	0.86	0.49	0.070
15.00	0.86	0.47	0.065
15.50	0.86	0.46	0.062
16.00	0.85	0.45	0.061
16.50	0.84	0.43	0.059
17.00	0.84	0.41	0.056
17.50	0.83	0.40	0.054
18.00	0.82	0.38	0.051
18.50	0.82	0.37	0.048
19.00	0.82	0.35	0.045
19.50	0.82	0.34	0.043
20.00	0.81	0.32	0.041
20.50	0.81	0.31	0.038
21.00	0.81	0.29	0.036
21.50	0.81	0.28	0.033
22.00	0.81	0.27	0.032
22.50	0.81	0.25	0.029
23.00	0.82	0.24	0.027
23.50	0.82	0.23	0.025
24.00	0.82	0.22	0.024
24.50	0.83	0.21	0.022
25.00	0.83	0.20	0.020
25.50	0.83	0.19	0.019
26.00	0.83	0.18	0.018
26.50	0.84	0.17	0.016

Energy (eV)	n	k	$R(\phi = 0)$
27.00	0.84	0.16	0.015
27.50	0.85	0.16	0.014
28.00	0.85	0.15	0.013
28.50	0.86	0.14	0.012
29.00	0.86	0.14	0.011
29.50	0.86	0.13	0.010
30.00	0.87	0.13	0.009
31.00	0.88	0.12	0.008
32.00	0.90	0.11	0.007
33.00	0.90	0.10	0.005
34.00	0.91	0.10	0.005
35.00	0.92	0.09	0.004
36.00	0.94	0.10	0.004
37.00	0.94	0.10	0.004
38.00	0.95	0.11	0.004
39.00	0.95	0.12	0.004
40.00	0.95	0.13	0.005

Zinc, $\vec{E} \parallel \hat{c}$[28]

Energy (eV)	n	k	$R(\phi = 0)$
0.7514	1.9241	7.5619	0.883
0.827	1.7921	6.9973	0.874
0.866	1.5571	6.7753	0.881
0.952	1.4824	6.2296	0.868
0.992	1.5762	5.8843	0.847
1.033	1.5407	5.3192	0.823
1.078	1.5853	4.9013	0.793
1.127	1.7768	4.5307	0.748
1.181	1.9808	4.2004	0.701
1.240	2.8821	3.4766	0.575
1.305	3.2039	3.0042	0.520
1.377	2.9459	3.5761	0.584
1.459	3.2523	4.2447	0.640
1.550	3.8086	4.6212	0.657
1.653	3.7577	4.6239	0.659
1.722	3.5908	4.4614	0.650
1.823	3.4234	4.3232	0.642
1.937	3.0132	3.9974	0.624
1.984	1.8562	3.9706	0.690
2.066	1.4856	4.0555	0.737
2.094	1.2525	3.9961	0.762
2.119	1.0017	3.8683	0.789
2.275	0.7737	3.9129	0.832
2.445	0.6395	3.4013	0.821
2.666	0.4430	3.1379	0.851
2.917	0.3589	2.8140	0.853
3.220	0.3069	2.5088	0.847
3.594	0.2737	2.1737	0.828
4.065	0.2510	1.8528	0.799
4.678	0.2354	1.6357	0.776

Zinc, $\vec{E} \perp \hat{c}$[28]

Energy (eV)	n	k	$R(\phi = 0)$
0.751	1.4469	7.4158	0.905
0.827	1.4744	6.9688	0.892
0.866	1.3628	6.6886	0.892
0.952	1.3165	6.2212	0.881
0.992	1.3835	5.8910	0.863
1.033	1.2889	5.4001	0.850
1.078	1.3095	4.9025	0.822
1.127	1.6897	4.4062	0.746
1.181	1.9701	4.0176	0.684
1.240	2.8717	3.2873	0.555

Energy (eV)	n	k	$R(\phi = 0)$
1.305	3.3991	2.7684	0.497
1.377	3.1807	3.4709	0.569
1.459	3.5064	4.1994	0.630
1.550	4.1241	4.7768	0.664
1.653	4.0269	4.8027	0.667
1.722	3.9369	4.6356	0.657
1.823	3.7549	4.3042	0.635
1.937	3.4512	4.1942	0.631
1.984	3.2515	4.2980	0.644
2.066	2.0802	4.7231	0.738
2.094	1.7084	4.7923	0.774
2.119	1.3329	4.4751	0.791
2.275	0.9725	4.2879	0.825
2.455	0.7568	3.7627	0.824
2.666	0.5470	3.4277	0.845
2.917	0.4774	3.0476	0.834
3.220	0.3911	2.7463	0.835
3.594	0.3147	2.3041	0.821
4.065	0.3013	2.0077	0.789
4.678	0.2806	1.7997	0.770

Zirconium (Polycrystalline)[28]

Energy (eV)	n	k	$R(\phi = 0)$
0.10	6.18	1.76	0.300
0.15	3.37	1.30	0.123
0.20	2.34	1.08	0.058
0.26	2.24	1.06	0.052
0.30	2.59	1.14	0.073
0.36	3.17	1.26	0.110
0.40	3.09	1.24	0.105
0.46	3.36	1.30	0.123
0.50	4.13	1.44	0.175
0.56	5.01	1.58	0.231
0.60	5.18	1.61	0.242
0.70	4.54	1.51	0.202
0.80	4.03	1.42	0.168
0.90	3.74	1.37	0.149
0.96	3.69	1.36	0.145
1.00	3.66	1.35	0.143
1.10	3.65	1.35	0.142
1.20	3.53	1.33	0.134
1.30	3.25	1.27	0.116
1.40	3.10	1.25	0.106
1.50	3.02	1.23	0.100
1.60	2.88	1.20	0.091
1.70	2.68	1.16	0.078
1.80	2.49	1.12	0.067
2.00	2.14	1.03	0.047
2.10	1.99	1.00	0.040
2.20	1.87	0.97	0.034
2.30	1.78	0.94	0.030
2.40	1.71	0.92	0.027
2.50	1.62	0.90	0.024
2.60	1.54	0.88	0.022
2.70	1.46	0.86	0.019
2.80	1.40	0.84	0.018
2.90	1.34	0.82	0.016
3.00	0.30	0.81	0.016
3.10	1.26	0.80	0.015
3.30	1.19	0.77	0.014
3.40	1.16	0.76	0.013
3.50	1.13	0.75	0.013

Energy (eV)	n	k	$R(\phi = 0)$	Energy (eV)	n	k	$R(\phi = 0)$	Energy (eV)	n	k	$R(\phi = 0)$
3.60	1.10	0.74	0.013	9.00	1.65	0.91	0.025	17.20	1.09	0.74	0.013
3.70	1.07	0.73	0.013	9.20	1.63	0.90	0.025	17.60	1.13	0.75	0.013
3.80	1.04	0.72	0.012	9.40	1.60	0.89	0.024	18.00	1.17	0.76	0.014
3.90	1.01	0.71	0.012	9.60	1.57	0.89	0.023	18.40	1.21	0.78	0.014
4.00	0.98	0.70	0.012	9.80	1.52	0.87	0.021	18.80	1.24	0.79	0.014
4.20	0.94	0.68	0.013	10.00	1.47	0.86	0.020	19.20	1.27	0.80	0.015
4.40	0.89	0.67	0.013	10.20	1.42	0.84	0.018	19.60	1.29	0.80	0.015
4.60	0.85	0.65	0.014	10.40	1.35	0.82	0.016	20.00	1.30	0.81	0.015
4.80	0.81	0.64	0.014	10.50	1.32	0.81	0.016	20.60	1.29	0.80	0.015
5.00	0.78	0.63	0.015	10.60	1.28	0.80	0.015	21.00	1.27	0.80	0.015
5.20	0.77	0.62	0.016	10.80	1.23	0.78	0.014	21.60	1.23	0.78	0.014
5.40	0.77	0.62	0.016	11.00	1.19	0.77	0.014	22.00	1.20	0.77	0.014
5.60	0.80	0.63	0.014	11.20	1.16	0.76	0.013	22.60	1.15	0.76	0.013
5.80	0.87	0.66	0.013	11.40	1.13	0.75	0.013	23.00	1.12	0.75	0.013
6.00	1.00	0.71	0.012	11.60	1.11	0.74	0.013	23.60	1.08	0.73	0.013
6.20	1.11	0.75	0.013	11.80	1.09	0.74	0.013	24.00	1.05	0.73	0.013
6.40	1.23	0.78	0.014	12.00	1.08	0.73	0.013	24.60	1.02	0.71	0.012
6.60	1.33	0.81	0.016	12.40	1.05	0.72	0.012	25.00	1.00	0.71	0.012
6.80	1.42	0.84	0.018	12.80	1.01	0.71	0.012	25.60	0.97	0.69	0.012
7.00	1.49	0.86	0.020	13.20	0.98	0.70	0.012	26.00	0.95	0.69	0.013
7.20	1.54	0.88	0.022	13.60	0.95	0.69	0.013	26.60	0.91	0.67	0.013
7.40	1.58	0.89	0.023	14.00	0.92	0.68	0.013	27.00	0.88	0.66	0.013
7.60	1.61	0.90	0.024	14.40	0.89	0.67	0.013	27.60	0.84	0.65	0.014
7.80	1.63	0.90	0.025	14.80	0.90	0.67	0.013	28.00	0.83	0.64	0.014
8.00	1.66	0.91	0.026	15.20	0.92	0.68	0.013	28.60	0.82	0.64	0.014
8.20	0.67	0.91	0.026	15.60	0.95	0.69	0.013	29.00	0.81	0.64	0.014
8.40	1.68	0.92	0.026	16.00	0.98	0.70	0.012	29.60	0.82	0.64	0.014
8.60	1.68	0.92	0.026	16.40	1.01	0.71	0.012	30.00	0.82	0.64	0.014
8.80	1.66	0.91	0.026	16.80	1.04	0.72	0.012				

References

1. Shiles, E., Sasaki, T., Inokuti, M., and Smith, D. Y., *Phys. Rev. Sect. B*, 22, 1612, 1980.
2. Edwards, D. F., and Philipp, H. R., in *HOC-I*, p. 665.
3. Ives, H. E., and Briggs, N. B., *J. Opt. Soc. Am.*, 27, 395, 1937.
4. Bos, L. W., and Lynch, D. W., *Phys. Rev. Sect. B*, 2, 4567, 1970.
5. Weaver, J. H., Colavita, E., Lynch, D. W., and Rosei, R., *Phys. Rev. Sect. B*, 19, 3850, 1979.
6. Hagemann, H. J., Gudat, W., and Kunz, C., *J. Opt. Soc. Am.*, 65, 742, 1975.
7. Schulz, L. G., *J. Opt. Soc. Am.*, 47, 64, 1957.
8. Potter, R. F., in *HOC-I*, p. 465.
9. Olson, C. G., Lynch, D. W., and Weaver, J. H., unpublished.
10. Lynch, D. W., Olson, C. G., and Weaver, J. H., unpublished.
11. Weaver, J. H., Olson, C. G., and Lynch, D. W., *Phys. Rev. Sect. B*, 15, 4115, 1977.
12. Lynch, D. W., and Hunter, W. R., in *HOC-II*, p. 345.
13. Priol, M. A., Daudé, A., and Robin, S., *Compt. Rend.*, 264, 935, 1967.
14. Johnson, P. B., and Christy, R. W., *Phys. Rev. Sect. B*, 9, 5056, 1974.
15. Arakawn, E. T., and Inagaki, T., in *HOC-II*, p. 461.
16. Weaver, J. H., Lynch, D. W., and Olson, D. G., *Phys. Rev. Sect. B*, 10, 501, 1973.
17. Lynch, D. W., Rosei, R., and Weaver, J. H., *Solid State Commun.*, 9, 2195, 1971.
18. Weaver, J. H., Lynch, D. W., and Olson, C. G., *Phys. Rev. Sect. B*, 7, 4311, 1973.
19. Weaver, J. H., and Benbow, R. L., *Phys. Rev. Sect. B*, 12, 3509, 1975.
20. Weaver, J. H., *Phys. Rev., Sect. B*, 11, 1416, 1975.
21. Lynch, D. W., and Hunter, W. R., in *HOC-II*, p. 364.
22. Palik, E. D., in *HOC-II*, p. 691.
23. Edwards, D. F., in *HOC-I*, p. 547.
24. Lynch, D. W., and Hunter, W. R., in *HOC-II*, p. 354.
25. Palik, E. D., in *HOC-II*, p. 709.
26. Lynch, D. W., Olson, C. G., and Weaver, J. H., *Phys. Rev. Sect. B*, 11, 3671, 1975.
27. Weaver, J. H., Lynch, D. W., and Olson, C. G., *Phys. Rev. Sect. B*, 12, 1293, 1975.
28. Lanham, A. P., and Terherne, D. M., *Proc. Phys. Soc.*, 83, 1059, 1964.

OPTICAL PROPERTIES OF SELECTED INORGANIC AND ORGANIC SOLIDS

L. I. Berger

Optical properties of materials are closely related to their dielectric properties. The complex dielectric function (relative permittivity) of a material is equal to

$$\varepsilon(\omega) = \varepsilon'(\omega) - j\varepsilon''(\omega),$$

where $\varepsilon'(\omega)$ and $\varepsilon''(\omega)$ are its real and imaginary parts, respectively, and ω is the angular frequency of the applied electric field. For a non-absorbing medium, the index of refraction is $n = (\varepsilon\mu)^{1/2}$, where μ is the relative magnetic permeability of the medium (material); in the majority of dielectrics, $\mu \cong 1$.

For many applications, the most important optical properties of materials are the index of refraction, the extinction coefficient, k, and the reflectivity, R. The common index of refraction of a material is equal to the ratio of the phase velocity of propagation of an electromagnetic wave of a given frequency in vacuum to that in the material. Hence, $n \gtrsim 1$. The optical properties of highly conductive materials like metals and semiconductors (at photon energy range above the energy gap) differ from those of optically transparent media. Free electrons absorb the incident electromagnetic wave in a thin surface layer (a few hundred nanometers thick) and then release the absorbed energy in the form of secondary waves reflected from the surface. Thus, the light reflection becomes very strong; for example, highly conductive sodium reflects 99.8% of the incident wave (at 589 nm). Introduction of the effective index of refraction, $n_{eff} = (\varepsilon')^{1/2} = n - jk$, where $\varepsilon' = \varepsilon - j\delta/\omega\,\varepsilon_o$, δ is the electrical conductivity of the material in S/m, and $\varepsilon_o = 8.8542\cdot10^{-12}$ F/m is the permittivity of vacuum, allows one to apply the expressions of the optics of transparent media to the conductive materials. It is clear that the effective index of refraction may be smaller than 1. For example, $n = 0.05$ for pure sodium and $n = 0.18$ for pure silver (at 589.3 nm). At very high photon energies, the quantum effects, such as the internal photoeffect, start playing a greater role, and the optical properties of these materials become similar to those of insulators (low reflectance, existence of Brewster's angle, etc.).

The extinction coefficient characterizes absorption of the electromagnetic wave energy in the process of propagation of a wave through a material. The wave intensity, I, after it passes a distance x in an isotropic medium is equal to

$$I = I_0 \exp(-\alpha x),$$

where I_0 is the intensity at $x = 0$ and α is called the absorption coefficient. For many applications, the extinction coefficient, k, which is equal to

$$k = \alpha\frac{\lambda}{4\pi},$$

where λ is the wavelength of the wave in the medium, is more commonly used for characterization of the electromagnetic losses in materials.

Reflection of an electromagnetic wave from the interface between two media depends on the media indices of refraction and on the angle of incidence. It is characterized by the reflectivity, which is equal to the ratio of the intensity of the wave reflected back into the first medium to the intensity of the wave approaching the interface. For polarized light and two non-absorbing media,

$$R = \frac{(N_1 - N_2)^2}{(N_1 + N_2)^2},$$

where $N_1 = n_1/\cos\theta_1$ and $N_2 = n_2/\cos\theta_2$ for the wave polarized in the plane of incidence, and $N_1 = n_1\cos\theta_1$ and $N_2 = n_2\cos\theta_2$ for the wave polarized normal to the plane of incidence; θ_1 and θ_2 are the angles between the normal to the interface in the point of incidence and the directions of the beams in the first and second medium, respectively. The reflectivity at normal incidence in this case is

$$R = [(n_1 - n_2)/(n_1 + n_2)]^2$$

For any two opaque (absorbing) media, the normal incidence reflectivity is

$$R = \frac{(n_1 - n_2)^2 + k_2^2}{(n_1 + n_2)^2 + k_2^2}.$$

In the majority of experiments, the first medium is air ($n \approx 1$), and hence,

$$R = \frac{(1 - n)^2 + k^2}{(1 + n)^2 + k^2}.$$

The data on n and k in the following table are abridged from the sources listed in the references. The reflectivity at normal incidence, R, has been calculated from the last equation. For convenience, the energy E, wavenumber $\bar{v}$, and wavelength λ are given for the incidence radiation.

E/eV	$\bar{\nu}$ /cm⁻¹	λ/μm	n	n_a	n_c	k	k_a	k_c	R	R_a	R_c
						Crystalline Arsenic Selenide (As₂Se₃) [Ref. 1]*					
2.194	17700	0.565					0.30				
2.168	17480	0.572					0.25				
2.141	17270	0.579					0.20				
2.123	17120	0.584					0.17				
2.098	16920	0.591					0.13				
2.094	16890	0.592						0.26			
2.091	16860	0.593						0.26			
2.073	16720	0.598					0.10	0.23			
2.060	16610	0.602						0.20			
2.049	16530	0.605					0.079	0.17			
2.036	16420	0.609						0.15			
2.023	16310	0.613						0.12			
2.013	16230	0.616					0.050				
2.009	16210	0.617						0.097			
2.000	16130	0.620						0.082			
1.987	16030	0.624						0.063			
1.977	15940	0.627					0.031				
1.974	15920	0.628						0.051			
1.962	15820	0.632						0.038			
1.953	15750	0.635						0.030			
1.949	15720	0.636					0.020				
1.937	15630	0.640						0.022			
1.925	15530	0.644						0.017			
1.922	15500	0.645					0.012				
1.905	15360	0.651					8.6·10⁻³				
1.893	15270	0.655					6.4				
1.881	15170	0.659					5.2				
1.859	14990	0.667					3.1				
1.848	14900	0.671						1.7·10⁻³			
1.845	14880	0.672					2.0				
1.842	14860	0.673						1.2·10⁻³			
1.831	14770	0.677					1.3·10⁻³	9.0·10⁻⁴			
1.826	14730	0.679						6.4			
1.821	14680	0.681						4.7			
1.818	14660	0.682					8.6·10⁻⁴				
1.815	14640	0.683						3.4			
1.807	14580	0.686					5.5				
1.802	14530	0.688					4.1				
0.06199	500.0	20.0		3.2	2.9		1.7·10⁻³	1.8·10⁻³		0.27	0.24
0.05904	476.2	21.0		3.1	2.9		2.1·10⁻³	2.2·10⁻³		0.26	0.24
0.05636	454.5	22.0		3.1	2.9		2.5·10⁻³	2.6·10⁻³		0.26	0.24
0.05391	434.8	23.0		3.1	2.9		3.0·10⁻³	3.1·10⁻³			
0.04592	370.4	27.0		3.0	2.8		6.3·10⁻³	6.4·10⁻³		0.25	0.22
0.04428	357.1	28.0		3.0	2.8		7.6·10⁻³	7.7·10⁻³		0.25	0.22
0.04275	344.8	29.0		3.0	2.8		0.0092	0.0093		0.25	0.22
0.04133	333.3	30.0		3.0	2.7		0.011	0.011		0.25	0.21
0.03542	285.7	35.0		2.7	2.5		0.037	0.034		0.21	0.18
0.03100	250.0	40.0		1.9	1.7		0.38	1.0		0.19	0.18
0.03061	247.0	40.5		2.0	2.6		0.33	0.95		0.12	0.25
0.03024	244.0	41.0		1.7	2.4		0.41	0.46		0.088	0.18
0.02883	232.6	43.0		1.2	1.3		2.2	0.94		0.50	0.16
0.02850	229.9	43.5		1.6	1.2		2.8	1.4		0.56	0.29
0.02818	227.3	44.0		2.3	1.2		3.3	2.0		0.58	0.48
0.02755	222.2	45.0		4.2	2.0		2.5	3.3		0.50	0.60
0.02480	200.0	50.0		6.5	4.0		3.6	0.26		0.62	0.36
0.02254	181.8	55.0		4.5	3.5		0.17	0.10		0.40	0.31
0.02066	166.7	60.0		4.0	3.2		0.089	0.10		0.36	0.27
0.01907	153.8	65.0		3.8	3.1		0.097	0.16		0.34	0.26
0.01771	142.9	70.0		3.6	3.0		0.19	0.30		0.32	0.25

E/eV	$\bar{\nu}$ /cm^{-1}	λ/μm	n	n_a	n_c	k	k_a	k_c	R	R_a	R_c
0.01653	133.3	75.0		3.7	3.0		0.41	0.44		0.34	0.26
0.01550	125.0	80.0		3.8	3.1		0.29	0.40		0.34	0.27
0.01459	117.6	85.0		3.6	2.9		0.20	0.34		0.32	0.24
0.01378	111.1	90.0		3.2	2.6		0.43	0.49		0.28	0.21
0.01305	105.3	95.0		4.7	3.0		1.5	1.5		0.46	0.34
0.01240	100.0	100.0		4.4	2.7		0.22	0.81		0.40	0.25
0.01181	95.24	105.0		4.2	3.0		0.094	3.9		0.38	0.62
0.01127	90.91	110.0		4.1	5.3		0.059	0.70		0.37	0.47
0.01033	83.33	120.0		3.9	4.2		0.034	0.13		0.35	0.38
0.009537	76.92	130.0		3.9	4.0		0.024	0.069		0.35	0.36
0.008856	71.43	140.0		3.9	3.8		0.019	0.048		0.35	0.34
0.007749	63.50	160.0		3.8	3.7		0.014	0.032		0.34	0.33
0.006888	55.55	180.0		3.8	3.7		0.011	0.024		0.34	0.33
0.006199	50.0	200.0		3.8	3.6		0.0091	0.019		0.34	0.32

*Indices a and c relate to the radiation electric field parallel to the a and c axes of the crystal, respectively.

Vitreous Arsenic Selenide (As$_2$Se$_3$) [Ref. 1]

E/eV	$\bar{\nu}$ /cm^{-1}	λ/μm	n	k	R
2.056	16580	0.603		0.12	
2.026	16340	0.612		0.11	
2.006	16180	0.618		0.099	
1.990	16050	0.623		9.0	
1.925	15530	0.644		5.6	
1.826	14730	0.679		1.4	
1.810	14600	0.685		0.012	
1.794	14470	0.691		0.0089	
1.771	14290	0.700		6.2	
1.715	13830	0.723		2.6	
1.701	13720	0.729		0.0022	
1.647	13280	0.753		0.00046	
1.629	13140	0.761	3.07	4.0	0.62
1.596	12870	0.777	3.06	2.7	0.49
1.579	12740	0.785	3.05	1.9	0.39
1.562	12590	0.794	3.05	0.00013	0.26
1.544	12450	0.803	3.04	0.000094	0.25
1.529	12330	0.811	3.03	6.3	0.78
1.512	12200	0.820	3.03	4.2	0.64
1.494	12050	0.830	3.02	2.8	0.50
1.476	11910	0.840	3.01	1.8	0.38
1.378	11110	0.90	2.98		
1.240	10000	1.00	2.93		
1.127	9091	1.10	2.90		
1.051	8475	1.18	2.89		
1.033	8333	1.20	2.88		
0.2555	1980	5.05		$1.6 \cdot 10^{-7}$	
0.2380	1919	5.21		$9.9 \cdot 10^{-8}$	
0.2344	1890	5.29		$1.1 \cdot 10^{-7}$	
0.1345	1085	9.22		4.4	
0.1339	1080	9.26		3.7	
0.1333	1075	9.30		4.4	
0.1308	1055	9.48		4.5	
0.1215	980	10.20		8.9	
0.1203	970	10.31		$9.9 \cdot 10^{-7}$	
0.1196	965	10.36		$1.0 \cdot 10^{-6}$	
0.1178	950	10.53		1.1	
0.1116	900	11.11		1.8	
0.1004	810	12.35		4.9	
0.09919	800	12.50		$7.0 \cdot 10^{-6}$	
0.09795	790	12.66		$1.0 \cdot 10^{-5}$	
0.09671	780	12.82		1.5	
0.09299	750	13.33		3.7	
0.08555	690	14.49		6.9	

E/eV	$\bar{v}$ /cm^{-1}	λ/μm	n	n_a	n_c	k	k_a	k_c	R	R_a	R_c
0.08431	680	14.71				5.9					
0.08059	650	15.38				6.1					
0.07811	630	15.87				6.3					
0.07687	620	16.13				7.7					
0.07563	610	16.39				7.8					
0.07439	600	16.67				$9.3 \cdot 10^{-5}$					
0.07315	590	16.95	2.8			$1.2 \cdot 10^{-4}$			0.22		
0.07191	580	17.24	2.8			1.4			0.32		
0.07067	570	17.54	2.8			1.8			0.37		
0.06943	560	17.86	2.8			2.8			0.50		
0.06633	535	18.69	2.8			5.2			0.73		
0.06571	530	18.87	2.8			$7.2 \cdot 10^{-4}$			0.22		
0.06509	525	19.05	2.8			$1.2 \cdot 10^{-3}$			0.22		
0.06447	520	19.23	2.8			1.7			0.35		
0.06075	490	20.41	2.7			4.9			0.71		
0.06024	485.9	20.58	2.7			5.2			0.73		
0.05331	430	23.26	2.7			1.4			0.31		
0.05269	425	23.53	2.7			$1.1 \cdot 10^{-3}$			0.21		
0.05207	420	23.81	2.7			$8.5 \cdot 10^{-4}$			0.21		
0.05145	415	24.10	2.7			7.3			0.84		
0.05083	410	24.39	2.7			8.3			0.87		
0.05021	405	24.69	2.7			$9.4 \cdot 10^{-4}$			0.21		
0.04959	400	25.0	2.7			$1.2 \cdot 10^{-3}$			0.21		
0.04862	392.2	25.5	2.6			1.6			0.33		
0.04679	377.4	26.5	2.6			5.0			0.73		
0.04592	370.4	27.0	2.6			$8.0 \cdot 10^{-3}$			0.20		
0.04509	363.6	27.5	2.6			$1.2 \cdot 10^{-2}$			0.20		
0.04428	357.1	28.0	2.6			1.7			0.34		
0.03875	312.5	32.0	2.5			8.2			0.87		
0.03815	307.7	32.5	2.5			$9.3 \cdot 10^{-3}$			0.18		
0.03757	303.0	33.0	2.4			0.11			0.17		
0.02988	241.0	41.5	2.2			0.89			0.20		
0.02952	238.1	42.0	2.2			1.0			0.22		
0.02725	219.8	45.5	3.2			1.8			0.39		
0.02362	190.5	52.5	3.6			0.30			0.32		
0.01937	156.2	64.0	3.2			0.10			0.27		
0.01922	155.0	64.5	3.2			$9.6 \cdot 10^{-2}$			0.27		
0.01907	153.8	65.0	3.2			9.4			0.88		
0.01734	139.9	71.5	3.1			8.7			0.87		
0.01653	133.3	75.0	3.1			9.4			0.88		
0.01642	132.5	75.5	3.1			0.096			0.26		
0.01494	120.5	83.0	3.0			0.15			0.25		
0.01246	100.5	99.5	3.2			0.60			0.26		
0.007606	61.35	163.0	3.3			0.12			0.29		
0.006199	50.00	200.0	3.2								
0.004592	37.04	270.0	3.1			0.072			0.26		
0.002799	22.57	443.0	3.0			4.5			0.67		
0.001826	14.73	679.0	3.0			2.8			0.50		
0.001273	10.27	974.0	3.0			2.1			0.41		
0.0006491	5.236	1910.0	3.0			$1.1 \cdot 10^{-2}$			0.25		
0.0004376	3.530	2833.0	3.0			$7.5 \cdot 10^{-3}$			0.25		
0.0002903	2.341	4271.0	3.0			5.0			0.71		
0.0001716	1.384	7224.0	3.0			3.1			0.53		
0.00009047	0.7297	13704	3.0			$1.6 \cdot 10^{-3}$			0.25		
0.00005621	0.4534	22056	3.0			$9.9 \cdot 10^{-4}$			0.25		
0.00002774	0.2237	44699	3.0			5.2			0.72		
0.00001439	0.1161	86153	3.0			2.6			0.47		

Vitreous Arsenic Sulfide (As$_2$S$_3$) - [Ref. 2]

E/eV	$\bar{v}$ /cm^{-1}	λ/μm	n	n_a	n_c	k	k_a	k_c	R	R_a	R_c
4.959	40000	0.2500	2.48			1.21			0.27		
3.100	25000	0.40	3.09			0.34			0.27		

E/eV	$\bar{\nu}$ /cm^{-1}	λ/μm	n	n_a	n_c	k	k_a	k_c	R	R_a	R_c
2.48	20000	0.4999	2.83			0.013			0.23		
1.879	15150	0.66	2.59			$1.7 \cdot 10^{-6}$			0.20		
1.240	10000	1.0	2.48			$2.4 \cdot 10^{-7}$			0.18		
0.6199	5000	2.0	2.43						0.17		
0.3100	2500	4.0	2.41						0.17		
0.2480	2000	5.0	2.41						0.17		
0.1736	1400	7.143	2.40			$7.4 \cdot 10^{-7}$			0.17		
0.1240	1000	10.00	2.38			$1.3 \cdot 10^{-4}$			0.17		
0.09299	750	13.33	2.35			$3.0 \cdot 10^{-3}$			0.16		
0.07439	600	16.67	2.31			$4.6 \cdot 10^{-4}$			0.16		
0.04959	400.0	25.0	1.79			0.2			0.085		
0.03757	303.0	33.0	3.59			1.4			0.38		
0.03100	250.0	40.0	2.98			0.15			0.25		
0.02480	200.0	50	2.66			0.11			0.21		
0.02066	166.7	60	2.64			0.57			0.22		
0.01771	142.9	70	2.99			0.17			0.25		
0.01550	125.0	80	2.89			0.14			0.24		
0.01378	111.1	90	2.84			0.12			0.23		
0.01240	100	100	2.81			0.10			0.23		
0.008183	66	152	2.76			0.072			0.22		
0.004029	32.5	308	2.74			0.044			0.22		
0.002418	19.5	513	2.74			0.031			0.22		
0.001984	16	625	2.74			0.025			0.22		
0.001048	8.45	1180	2.73			$8.8 \cdot 10^{-3}$			0.22		
0.0001033	0.833	12000	2.73			$1.3 \cdot 10^{-3}$			0.22		
$4.129 \cdot 10^{-12}$	$3.33 \cdot 10^{-8}$	$3 \cdot 10^{11}$	2.73						0.22		

Cadmium Telluride (CdTe) - [Ref. 3]

E/eV	$\bar{\nu}$ /cm^{-1}	λ/μm	n	n_a	n_c	k	k_a	k_c	R	R_a	R_c
4.9	39520	0.2530	2.48			2.04			0.39		
4.1	33070	0.3024	2.33			1.59			0.32		
3.9	31460	0.3179	2.57			1.90			0.37		
3.5	28230	0.3542	2.89			1.52			0.34		
3.1	25000	0.4000	3.43			1.02			0.34		
3.0	24200	0.4133	3.37			0.861			0.32		
2.755	22220	0.45	3.080			0.485			0.27		
2.75	22180	0.4509	3.23			0.636			0.29		
2.610	21050	0.475	3.045								
2.5	20160	0.4959	3.14			0.525			0.28		
2.25	18150	0.5510	3.05			0.411			0.26		
1.771	14290	0.70	2.861			0.210			0.23		
1.512	12200	0.82	2.880			0.040			0.23		
1.50	12100	0.8266	2.98			0.319			0.25		
1.475	11900	0.840	2.905			0.00134			0.24		
1.47	11860	0.8434				0.000671					
1.465	11820	0.8463				3.37					
1.46	11780	0.8492				1.89					
1.459	11760	0.850	2.948						0.24		
1.455	11740	0.8521				$1.08 \cdot 10^{-4}$					
1.45	11690	0.8551	2.9565			$5.10 \cdot 10^{-5}$			0.24		
1.445	11650	0.8580				2.73					
1.442	11630	0.860	2.952						0.24		
1.44	11610	0.8610	2.9479			1.37			0.32		
1.43	11530	0.8670	2.9402						0.24		
1.30	10490	0.9537	2.8720						0.23		
1.24	10000	1.0	2.840						0.23		
1.20	9679	1.033	2.8353						0.23		
1.10	8872	1.127	2.8050						0.23		
1.00	8065	1.240	2.7793						0.22		
0.90	7259	1.378	2.7537						0.22		
0.80	6452	1.550	2.7384						0.22		
0.70	5646	1.771	2.7223						0.21		

E/eV	$\bar{\nu}$ /cm⁻¹	λ/μm	n	n_a	n_c	k	k_a	k_c	R	R_a	R_c
0.60	4839	2.066	2.7086						0.21		
0.50	4033	2.480	2.6972						0.21		
0.40	3226	3.100	2.6878						0.21		
0.30	2420	4.133	2.6800						0.21		
0.20	1613	6.199	2.6722						0.21		
0.10	806.5	12.40	2.6535						0.20		
0.09	725.9	13.78	2.6482						0.20		
0.06819	550	18.18	2.623						0.20		
0.0573	462	21.6				$3.8 \cdot 10^{-6}$					
0.05	403.3	24.80	2.5801						0.19		
0.0469	378	26.5				$8.0 \cdot 10^{-5}$					
0.04592	370.3	27				$9.88 \cdot 10^{-5}$					
0.04133	333.3	30	2.55916			$2.86 \cdot 10^{-4}$			0.19		
0.04092	330	30.30	2.531			3.34			0.57		
0.03720	300	33.33	2.494			4.97			0.73		
0.03647	294.1	34.00				8.93					
0.03596	290	34.48	2.478			$5.77 \cdot 10^{-3}$			0.18		
0.03493	281.7	35.5				7.91					
0.03472	280	35.71	2.459			6.76			0.83		
0.03100	250	40	2.378			$1.18 \cdot 10^{-2}$			0.17		
0.02917	235.3	42.5				6.93					
0.02852	230	43.48	2.289			1.87			0.36		
0.02728	220	45.45	2.224			$2.47 \cdot 10^{-2}$			0.14		
0.02604	210	47.62	2.137			$3.4 \cdot 10^{-2}$			0.13		
0.02480	200	50.00	2.013			$4.97 \cdot 10^{-2}$			0.11		
0.02384	192.3	52.0				6.21					
0.01798	145	68.97	1.8			5.2			0.79		
0.01736	140	71.43	6.778			4.50			0.66		
0.01550	125	80.0	4.598			0.294			0.41		
0.01364	110	90.91	3.868			$9.47 \cdot 10^{-2}$			0.35		
0.01240	100	100	3.649			$5.68 \cdot 10^{-2}$			0.32		
0.009919	80	125	3.415			0.0262			0.30		
0.008679	70	142.9	3.348			0.0189			0.29		
0.007439	60	166.7	3.299			1.39			0.35		
0.006199	50	200	3.263			1.03			0.32		
0.004959	40	250	3.236			$7.52 \cdot 10^{-3}$			0.28		
0.003720	30	333.3	3.217						0.28		
0.023015	18.563		538.71			3.2096			0.28		
0.001550	12.50	800				6.18					

Gallium Arsenide (GaAs) - [Ref. 4]

E/eV	$\bar{\nu}$ /cm⁻¹	λ/μm	n	n_a	n_c	k	k_a	k_c	R	R_a	R_c
155		0.007999				0.0181					
145		0.008551				0.0203					
130		0.009537				0.0224					
110		0.01127				0.0278					
90		0.01378				0.0323					
70		0.01771				0.0376					
40		0.03100				0.0426					
23		0.05391	1.037			0.228					
7.0		0.1771	1.063			1.838					
6.0	48390	0.2066	1.264			2.472			0.61		
5.00	40330	0.2480	2.273			4.084			0.67		
4.00	32260	0.3100	3.601			1.920			0.42		
3.00	24200	0.4133	4.509			1.948			0.47		

E/eV	$\bar{\nu}$ /cm^{-1}	λ/μm	n	n_a	n_c	k	k_a	k_c	R	R_a	R_c
2.50	20160	0.4959	4.333			0.441			0.39		
2.00	16130	0.6199	3.878			0.211			0.35		
1.80	14520	0.8888	3.785			0.151			0.34		
1.60	12900	0.7749	3.700			0.091			0.33		
1.50	12100	0.8266	3.666			0.080			0.33		
1.40	11290	0.8856	3.6140			$1.69 \cdot 10^{-3}$			0.32		
1.20	9679	1.033	3.4920						0.31		
1.00	8065	1.240	3.4232						0.30		
0.80	6452	1.550	3.3737						0.29		
0.50	4033	2.480	3.3240						0.29		
0.25	2016	4.959	3.2978						0.29		
0.15	1210	8.266	3.2831						0.28		
0.100	806.5	12.40	3.2597			$4.93 \cdot 10^{-6}$			0.28		
0.090	725.9	13.78	3.2493			$1.64 \cdot 10^{-5}$			0.28		
0.070	564.6	17.71	3.2081			$2.32 \cdot 10^{-4}$			0.28		
0.060	483.9	20.66	3.1609			$3.45 \cdot 10^{-3}$			0.27		
0.0495	399.2	25.05	3.058			$2.07 \cdot 10^{-3}$			0.26		
0.03968	320	31.25	2.495			$2.43 \cdot 10^{-2}$			0.18		
0.03496	282	35.46	0.307			$294 \cdot 10^{-2}$					
0.02976	240	41.67	4.57			$4.26 \cdot 10^{-2}$			0.41		
0.02066	166.7	60	3.77			$3.89 \cdot 10^{-3}$			0.34		
0.01550	125	80	3.681			$1.84 \cdot 10^{-3}$			0.33		
0.008266	66.67	150	3.62			$2.14 \cdot 10^{-3}$			0.32		
0.002480	20	500	3.607			$1.3 \cdot 10^{-3}$			0.32		
0.001240	10	1000	3.606						0.32		

Gallium Phosphide (GaP) - [Ref. 5]

E/eV	$\bar{\nu}$ /cm^{-1}	λ/μm	n	n_a	n_c	k	k_a	k_c	R	R_a	R_c
154.0		0.00805				$1.7 \cdot 10^{-2}$					
110.0		0.0113				$2.15 \cdot 10^{-2}$					
100.0		0.0124				$215 \cdot 10^{-2}$					
80.0		0.0155				$3.0 \cdot 10^{-2}$					
50.0		0.0248				$4.7 \cdot 10^{-2}$					
27.0		0.0459				$9.3 \cdot 10^{-2}$					
25.0		0.0496				0.122					
20.0		0.0620				0.180					
15.0		0.0826	0.748			0.628					
5.5	44360	0.2254	1.543			3.556			0.68		
4.68	37750	0.2649	4.181			2.634			0.50		
3.50	28230	0.3542	5.050			0.819			0.46		
3.00	24200	0.4133	4.081			0.224			0.37		
2.78	22420	0.4460	3.904			0.103			0.35		
2.621	21140	0.473	3.73			$6.37 \cdot 10^{-3}$			0.33		
2.480	20000	0.500	3.590			$2.47 \cdot 10^{-3}$			0.32		
2.18	17580	0.5687	3.411			$2.8 \cdot 10^{-7}$			0.30		
2.000	16130	0.62	3.3254						0.29		
1.6	12900	0.7749	3.209						0.28		
1.240	10000	1.0	3.1192						0.26		
0.6888	5556	1.8	3.0439						0.26		
0.4769	3846	2.6	3.0271						0.25		
0.1907	1538	6.5	2.995			$4.29 \cdot 10^{-4}$			0.25		
0.1550	1250	8.0	2.984						0.25		
0.1240	1000	10	2.964						0.25		
0.06199	500	20	2.615			$7.16 \cdot 10^{-3}$			0.20		
0.03100	250	40	3.594			$1.81 \cdot 10^{-2}$			0.32		

E/eV	$\bar{\nu}$ /cm^{-1}	λ/μm	n	n_a	n_c	k	k_a	k_c	R	R_a	R_c
0.02480	200	50	3.461			$5.77 \cdot 10^{-3}$			0.30		
0.01727	139.27	71.80	3.3922			$4.34 \cdot 10^{-3}$			0.30		
0.01168	94.21	106.1	3.3621			$4.26 \cdot 10^{-3}$			0.29		
0.006199	50.00	200	3.3447			$1.3 \cdot 10^{-4}$			0.29		
0.004133	33.33	300	3.3413						0.29		
0.001240	10.00	1000	3.3319						0.29		

Indium Antimonide (InSb) - [Ref. 6]

E/eV	$\bar{\nu}$ /cm^{-1}	λ/μm	n	n_a	n_c	k	k_a	k_c	R	R_a	R_c
155		0.007999				$4.77 \cdot 10^{-3}$					
60		0.02066				$7.30 \cdot 10^{-2}$					
25		0.04959	1.15			.015					
24		0.05166	1.15			0.18					
15		0.08266	0.97			0.230					
10		0.1240	0.74			0.88					
5.00	40330	0.2480	1.307			2.441			0.53		
4.50	36290	0.2755	1.443			2.894			0.60		
4.00	32260	0.3100	2.632			3.694			0.61		
3.34	26940	0.3712	3.528			2.280			0.45		
2.84	22910	0.4366	3.340			2.021			0.45		
1.80	14520	0.6888	4.909			1.396			0.47		
1.50	12100	0.8266	4.418			0.643			0.41		
0.6	4839	2.066	4.03						0.36		
0.2480	2000	5.0	4.14			$9.1 \cdot 10^{-2}$			0.37		
0.1907	1538	6.5	4.30			$6.3 \cdot 10^{-2}$			0.39		
0.1653	1333	7.5	4.18			$2.7 \cdot 10^{-2}$			0.38		
0.06199	500	20.00	3.869			$2.0 \cdot 10^{-3}$			0.35		
0.03100	250	40.00	2.98			$2.6 \cdot 10^{-3}$			0.25		
0.02480	200	50.00	2.22			0.165			0.14		
0.02244	181	55.25	3.05			7.59			0.84		
0.02207	178	56.18	9.61			4.20			0.70		
0.02033	164	60.98	4.94			0.140			0.44		
0.01054	85	117.6	2.12			0.423			0.14		
0.005579	45	222.2	1.02			5.59			0.88		
0.001860	15	666.7	6.03			17.9			0.93		
0.001240	10	1000	10.7			24.0			0.94		

Indium Arsenide (InAs) - [Ref. 7]

E/eV	$\bar{\nu}$ /cm^{-1}	λ/μm	n	n_a	n_c	k	k_a	k_c	R	R_a	R_c
25		0.04959				1.139			0.168		
20		0.06199				1.125			0.225		
15		0.08266				0.894			0.336		
10		0.1240				0.835			1.071		
6	48390	0.2066	1.434			2.112			0.45		
5.0	40330	0.2480	1.524			2.871			0.58		
4.0	32260	0.3100	3.313			1.799			0.39		
3.5	28230	0.3542	3.008			1.754			0.37		
3.0	24200	0.4133	3.197			2.034			0.41		
2.5	20160	0.4959	4.364			1.786			0.45		
2.44	19680	0.5081	4.489			1.446			0.44		
1.86	15000	0.6666	3.889			0.554			0.36		
1.8	14520	0.6888	3.851			0.530			0.35		
1.7	13710	0.7293	3.798			0.493			0.35		
1.6	12900	0.7749	3.755			0.463			0.34		
1.5	12100	0.8266	3.714			0.432			0.34		
1.2	9679	1.033	3.613						0.32		
1.0	8065	1.240	3.548						0.31		
0.6	4839	2.066				0.161					
0.35	2823	3.542	3.608			$9.58 \cdot 10^{-3}$			0.32		
0.32	2581	3.875	3.512			$1.23 \cdot 10^{-4}$			0.31		
0.20	1613	6.199	3.427						0.30		
0.1240	1000	10.00	3.402						0.30		
0.06199	500	20.00	3.334						0.29		
0.04959	400	25.00	3.264						0.28		

E/eV	$\bar{\nu}$ /cm^{-1}	λ/µm	n	n_a	n_c	k	k_a	k_c	R	R_a	R_c
0.04339	350	28.57	3.182			$5.46 \cdot 10^{-3}$			0.27		
0.03720	300	33.33	2.988						0.25		
0.03100	250	40.00	1.970			$6.37 \cdot 10^{-2}$			0.11		
0.02765	222	44.84	5.90			6.53			0.74		
0.02480	200	50.00	6.91			0.30			0.56		
0.01984	160	62.50	5.27			0.41			0.47		
0.01860	150	66.67	5.27			0.51			0.47		
0.01736	140	71.43	3.99			$1.1 \cdot 10^{-2}$			0.36		
0.01488	120	83.33	3.91			$6.6 \cdot 10^{-3}$			0.35		
0.01240	100	100.0	3.85			$4.3 \cdot 10^{-3}$			0.35		
0.009919	80	125.0	3.817						0.34		
0.007439	60	166.7	3.793						0.34		
0.004959	40	250.0	3.778						0.34		
0.002480	20	500	3.769						0.37		
0.001240	10	1000	3.766						0.34		

Indium Phosphide (InP) - [Ref. 8]

E/eV	$\bar{\nu}$ /cm^{-1}	λ/µm	n	n_a	n_c	k	k_a	k_c	R	R_a	R_c
20		0.06199	0.793			0.494					
15		0.08266	0.695			0.574					
10		0.1240	0.806			1.154					
5.5	44360	0.2254	1.426			2.562			0.79		
5.0	40330	0.2480	2.131			3.495			0.61		
4.0	32260	0.3100	3.141			1.730			0.38		
3.0	24200	0.4133	4.395			1.247			0.43		
2.0	16130	0.6199	3.549			0.317			0.32		
1.5	12100	0.8266	3.456			0.203			0.31		
1.25	10085	0.9915	3.324						0.29		
1.00	8068	1.239	3.220						0.28		
0.50	4034	2.479	3.114						0.26		
0.30	2420	4.131	3.089						0.26		
0.10	806.8	12.39	3.012						0.25		
0.075	605.1	16.53	2.932						0.24		
0.060	484.1	20.66	2.780			$1.46 \cdot 10^{-2}$			0.22		
0.050	403.4	24.79	2.429			$3.35 \cdot 10^{-2}$			0.17		
0.03992	322	31.06	0.307			3.57					
0.03496	282	35.46	3.89			0.282			0.35		
0.03100	250	40.00	4.27			$3.0 \cdot 10^{-2}$			0.39		
0.02728	220	45.45	3.93			$1.3 \cdot 10^{-2}$			0.35		
0.02480	200	50.0	3.81			$8.7 \cdot 10^{-3}$			0.34		
0.02418	195	51.28	3.19						0.27		
0.02232	180	55.56	3.19						0.27		
0.01860	150	66.67	3.65						0.32		
0.01240	100	100	3.57						0.32		
0.009919	80	125.0	3.551						0.31		
0.007439	60	166.7	3.538						0.31		
0.004959	40	250.0	3.529						0.31		
0.002480	20	500	3.523						0.31		
0.001240	10	1000.0	3.522						0.31		

Lead Selenide (PbSe) - [Ref. 9]

E/eV	$\bar{\nu}$ /cm^{-1}	λ/µm	n	n_a	n_c	k	k_a	k_c	R	R_a	R_c
14.5		0.08551	0.72			0.20					
10		0.1240	0.68			0.50					
5	40330	0.2480	0.54			1.2					
2.0	16130	0.6199	3.65			2.9			0.51		
1.65	13310	0.7514	4.51			1.73			0.46		
1.5	12100	0.8266	4.64			2.64			0.52		
1.0	8065	1.240	4.65			1.1			0.44		
0.75	6049	1.653				0.269					
0.62	5001	2.000	4.59			0.770			0.42		
0.48	3871	2.583	4.90						0.44		
0.40	3226	3.100	4.91						0.44		
0.32	2581	3.875	4.98			0.173			0.44		

E/eV	$\bar{\nu}$ /cm^{-1}	λ/μm	n	n_a	n_c	k	k_a	k_c	R	R_a	R_c
0.20	1613	6.199	4.82						0.43		
0.1190	960	10.42	4.74			$1.20 \cdot 10^{-3}$			0.42		
0.09919	800	12.50	4.72			$2.09 \cdot 10^{-3}$			0.42		
0.07935	640	15.63	4.68			$4.12 \cdot 10^{-3}$			0.42		
0.05951	480	20.83	4.59			$1.00 \cdot 10^{-2}$			0.41		
0.04959	400	25.00	4.49			$1.77 \cdot 10^{-2}$			0.40		
0.03968	320	31.25	4.31			$3.62 \cdot 10^{-2}$			0.39		
0.02976	240	41.67	3.89			$9.61 \cdot 10^{-2}$			0.24		
0.01984	160	62.50	2.34			0.56			0.18		
0.009919	80	125.0	1.73			7.38			0.88		
0.007935	64	156.3	2.91			10.1			0.90		
0.004959	40	250.0	11.2			14.6			0.88		
0.002480	20	500.0	12.6			12.2					
0.001736	14	714.3	14.1			16.6					
0.001240	10	1000	17.4			21.1					

Lead Sulfide (PbS) - [Ref. 10]

E/eV	$\bar{\nu}$ /cm^{-1}	λ/μm	n	n_a	n_c	k	k_a	k_c	R	R_a	R_c
150		0.008266				$3.86 \cdot 10^{-3}$					
125		0.009919				$5.59 \cdot 10^{-3}$					
100		0.01240				$1.54 \cdot 10^{-2}$					
80		0.01550				$2.88 \cdot 10^{-2}$					
60		0.02066				$6.17 \cdot 10^{-2}$					
25		0.04959	0.845			0.171					
18.0		0.06888	0.846			0.294					
14.0		0.08856	0.651			0.665					
10.0		0.1240	0.879			1.050					
4.95	39920	0.2505	1.52			2.10			0.43		
4.0	32260	0.3100	1.73			2.83			0.55		
3.00	24200	0.4133	3.88			3.00			0.53		
2.90	23390	0.4275	4.12			2.70			0.51		
2.75	22180	0.4509	4.25			2.33			0.48		
2.55	20570	0.4862	4.35			2.00			0.47		
2.00	16130	0.6199	4.29			1.48			0.43		
1.60	12910	0.7749	4.62			0.94			0.43		
1.24	10000	1.00	4.43			0.597			0.41		
1.03	8333	1.2	4.30			0.458			0.39		
0.650	5263	1.9	4.24			0.318			0.39		
0.496	4000	2.5	4.30			0.235			0.39		
0.400	3226	3.1	4.30			$2.27 \cdot 10^{-2}$			0.39		
0.3100	2500	4.0	4.16			$6.38 \cdot 10^{-4}$			0.38		
0.2480	2000	5	4.115			$9.25 \cdot 10^{-4}$			0.37		
0.1240	1000	10	4.01			$6.32 \cdot 10^{-3}$			0.36		
0.1033	833.3	12	3.90			$1.14 \cdot 10^{-2}$			0.35		
0.08059	650	15.38	3.90						0.35		
0.06819	550	18.18	3.81						0.34		
0.04959	400	25.00	3.53						0.31		
0.03720	300	33.33	2.99						0.25		
0.02480	200.0	50	0.514			1.59					
0.01378	111.1	90	1.175			8.48			0.94		
0.01240	100.0	100	1.79			10.51			0.94		
0.008856	71.43	140	17.41			17.94			0.89		
0.006199	50.0	200	16.27			2.20			0.79		
0.003100	25.00	400	12.96			0.495			0.73		
0.001653	13.33	750	12.44			0.228			0.72		
0.001240	10.00	1000	12.35			0.167			0.72		
0.0006199	5.000	2000	12.27			0.0815			0.72		

Lead Telluride (PbTe) - [Ref. 11]

E/eV	$\bar{\nu}$ /cm^{-1}	λ/μm	n	n_a	n_c	k	k_a	k_c	R	R_a	R_c
150		0.008266				$2.37 \cdot 10^{-3}$					
125		0.009919				$9.71 \cdot 10^{-3}$					
100		0.01240				$4.39 \cdot 10^{-2}$					
75		0.01653				$6.43 \cdot 10^{-2}$					

E/eV	$\bar{\nu}$ /cm^{-1}	λ/μm	n	n_a	n_c	k	k_a	k_c	R	R_a	R_c
50		0.02480				$6.87 \cdot 10^{-2}$					
30		0.04133				$7.77 \cdot 10^{-2}$					
15		0.08266	0.72			0.17					
10		0.1240	0.66			0.60					
7.5		0.1653	0.8			0.92					
5.0	40330	0.2480	0.72			1.0					
3.0	24200	0.4133	1.0			2.2					
2.5	20160	0.4959	1.35			2.86			0.61		
1.5	12100	0.8266	3.8			3.1			0.53		
1.0	8065	1.240	4.55			2.2			0.49		
0.80	6452	1.550	6.25			0.71			0.53		
0.60	4839	2.066	6.10			0.521			0.52		
0.40	3226	3.100	6.075			0.331			0.52		
0.30	2420	4.133	5.95			$3.55 \cdot 10^{-2}$			0.51		
0.20	1613	6.199	5.77						0.50		
0.15	1210	8.266	5.76						0.50		
0.1017	820	12.20	5.47			$9.16 \cdot 10^{-3}$			0.48		
0.08927	720	13.89	5.38			$1.37 \cdot 10^{-2}$			0.47		
0.06943	560	17.86	5.13			$3.06 \cdot 10^{-2}$			0.45		
0.04959	400	25.00	4.50			$9.6 \cdot 10^{-2}$			0.40		
0.03968	320	31.25	3.58			0.23			0.32		
0.02976	240	41.67	1.01			1.9					
0.009919	80	125.0	2.95			16.6			0.96		
0.007439	60	166.7	4.9			22.5			0.96		
0.006199	50	200.0	6.9			27.2			0.97		
0.004959	40	250.0	11.6			34.8			0.97		
0.003720	30	333.3	27.7			35.7			0.95		
0.002480	20	500.0	27.6			39.1			0.95		
0.001240	10	1000	45.1			57.8			0.97		

Lithium Fluoride (LiF) - [Ref. 12]

E/eV	$\bar{\nu}$ /cm^{-1}	λ/μm	n	n_a	n_c	k	k_a	k_c	R	R_a	R_c
2000		$6.199 \cdot 10^{-4}$	0.9999347			$4.33 \cdot 10^{-6}$					
1496		$8.287 \cdot 10^{-4}$	0.999883			$1.28 \cdot 10^{-5}$					
1016		$1.220 \cdot 10^{-3}$	0.999757			$5.18 \cdot 10^{-5}$					
725		$1.710 \cdot 10^{-3}$	0.999643			$1.62 \cdot 10^{-4}$					
504		$2.460 \cdot 10^{-3}$	0.999162			$4.96 \cdot 10^{-5}$					
303		$4.092 \cdot 10^{-3}$	0.99752			$3.12 \cdot 10^{-4}$					
250		$4.959 \cdot 10^{-3}$	0.99632			$6.17 \cdot 10^{-5}$					
200		$6.199 \cdot 10^{-3}$				$2.12 \cdot 10^{-3}$					
150		$8.265 \cdot 10^{-3}$	0.9899			$3.54 \cdot 10^{-3}$					
100		$1.240 \cdot 10^{-2}$	0.9801			$1.32 \cdot 10^{-2}$					
75		$1.653 \cdot 10^{-2}$				$2.63 \cdot 10^{-2}$					
50		$2.480 \cdot 10^{-2}$				$7.89 \cdot 10^{-2}$					
25		$4.959 \cdot 10^{-2}$	0.558			0.521					
20		$6.199 \cdot 10^{-2}$	1.20			0.58			0.10		
15.1		$8.211 \cdot 10^{-2}$	1.08			0.68			0.10		
13		$9.537 \cdot 10^{-2}$	1.04			1.64					
12.0		0.1033	2.28			0.11			0.15		
11.0		0.1127	1.77			$8.07 \cdot 10^{-7}$			0.08		
10.00		0.12398	1.606			$7.70 \cdot 10^{-7}$			0.05		
9		0.1375	1.53						0.04		
7		0.1771	1.46								
4.959	40000	0.250	1.4189						0.03		
4.000	32260	0.31	1.4073						0.03		
2.952	23810	0.42	1.3978						0.03		
2.000	16130	0.62	1.3915						0.03		
0.9919	8000	1.25	1.3851								
0.7999	6452	1.55	1.3858						0.03		
0.4959	4000	2.5	1.3731						0.02		
0.4000	3226	3.1	1.3650								
0.3100	2500	4.0	1.3493								

E/eV	$\bar{v}$ /cm^{-1}	λ/μm	n	n_a	n_c	k	k_a	k_c	R	R_a	R_c
0.2480	2000	5.0	1.3266			$1.8 \cdot 10^{-6}$			0.02		
0.2000	1613	6.2	1.2912								
0.1698	1370	7.3	1.2499								
0.1494	1205	8.3	1.2036								
0.1240	1000	10.0	1.1005			$2.6 \cdot 10^{-3}$					
0.1127	909.1	11.0	1.0208			$8.0 \cdot 10^{-3}$					
0.1033	833.3	12.0				$1.9 \cdot 10^{-2}$					
0.09537	769.2	13.0				$3.7 \cdot 10^{-2}$					
0.08679	700	14.29	0.508			$7.74 \cdot 10^{-2}$					
0.07439	600	16.67	0.124			0.804					
0.06199	500	20.00	0.306			1.47			0.68		
0.05579	450	22.22	0.191			1.88			0.85		
0.04959	400	25.00	0.208			2.71			0.91		
0.03720	300	33.33	8.76			3.91			0.68		
0.03100	250	40.00	4.64			0.287			0.42		
0.02480	200	50.00	3.69			0.102			0.33		
0.01240	100.0	100	3.067			0.106			0.26		
0.06199	50.0	200	3.067			$4.0 \cdot 10^{-2}$			0.26		
0.04959	40.00	250	3.067			$2.2 \cdot 10^{-2}$			0.26		
0.02480	20.00	500	3.067			$6.3 \cdot 10^{-3}$					
0.01378	11.11	900				$3.1 \cdot 10^{-3}$					
$4.798 \cdot 10^{-4}$	3.870	2584	3.023			$1.19 \cdot 10^{-3}$			0.25		
$1.464 \cdot 10^{-4}$	1.181	8469	3.023			$6.20 \cdot 10^{-4}$			0.25		
$4.053 \cdot 10^{-5}$	0.3269	30590	3.023			$2.63 \cdot 10^{-4}$			0.25		
$1.861 \cdot 10^{-7}$	$1.501 \cdot 10^{-3}$	$6.662 \cdot 10^{6}$	3.018			$1.6 \cdot 10^{-5}$					
$3.718 \cdot 10^{-8}$	$2.999 \cdot 10^{-4}$	$3.335 \cdot 10^{7}$	3.018			$1.6 \cdot 10^{-5}$					

Potassium Chloride (KCl) - [Ref. 13]

E/eV	$\bar{v}$ /cm^{-1}	λ/μm	n	n_a	n_c	k	k_a	k_c	R	R_a	R_c
2860.3		$4.3347 \cdot 10^{-4}$				$3.93 \cdot 10^{-6}$					
2855.3		$4.3423 \cdot 10^{-4}$				$3.39 \cdot 10^{-6}$					
2849.3		$4.3514 \cdot 10^{-4}$				$4.61 \cdot 10^{-6}$					
2835.8		$4.3721 \cdot 10^{-4}$				$5.85 \cdot 10^{-6}$					
2832.3		$4.3775 \cdot 10^{-4}$				$5.85 \cdot 10^{-6}$					
2829.8		$4.3814 \cdot 10^{-4}$				$1.57 \cdot 10^{-6}$					
2828.3		$4.3837 \cdot 10^{-4}$				$4.19 \cdot 10^{-7}$					
219		$5.661 \cdot 10^{-3}$				$1.82 \cdot 10^{-3}$					
215		$5.767 \cdot 10^{-3}$				$1.84 \cdot 10^{-3}$					
212.5		$5.834 \cdot 10^{-3}$				$2.19 \cdot 10^{-3}$					
211		$5.876 \cdot 10^{-3}$				$1.82 \cdot 10^{-3}$					
185.1		$6.7 \cdot 10^{-3}$	0.99874						$1.01 \cdot 10^{-3}$		
109.7		$1.13 \cdot 10^{-2}$	0.99578						$4.22 \cdot 10^{-3}$		
43		0.02883	0.96			$3.0 \cdot 10^{-2}$					
40		0.03179	0.925			$1.8 \cdot 10^{-2}$					
29.9		0.04147	0.756			0.145					
20.1		0.06168	0.910			0.495					
15.1		0.08211	0.965			0.344					
10.0		0.1240	1.16			0.38			0.035		
9.0		0.1378	1.99			0.50			0.13		
8.0		0.1550	1.15			0.46			0.048		
7.0		0.1771	2.0			$8.46 \cdot 10^{-7}$			0.11		
6.199	50000	0.20	1.71739						0.070		
4.959	40000	0.25	1.58972								
3.999	32260	0.31	1.54005								
2.952	23810	0.42	1.50701								
2.695	21740	0.46	1.50115						0.040		
2.616	21100	0.474				$7.6 \cdot 10^{-11}$					
2.384	19230	0.52	1.49501								
2.066	16670	0.60	1.48969						0.039		
1.550	12500	0.80	1.48291						0.038		
1.033	8333	1.2	1.47813						0.037		
0.5166	4167	2.4	1.47464						0.037		

E/eV	$\bar{\nu}/\mathrm{cm^{-1}}$	$\lambda/\mathrm{\mu m}$	n	n_a	n_c	k	k_a	k_c	R	R_a	R_c
0.2480	2000	5.0	1.47048						0.036		
0.2000	1.613	6.2	1.46796						0.036		
0.1512	1220	8.2	1.46260						0.035		
0.09999	806.5	12.4	1.44611						0.033		
0.07560	609.8	16.4	1.42295						0.030		
0.04959	400.0	25.0	1.34059			$6.57\cdot10^{-4}$			0.021		
0.03999	322.6	31.0	1.2431						0.012		
0.02976	240	41.67	0.85			0.16					
0.02728	220	45.45	0.53			0.35					
0.02232	180	55.56	0.31			1.05					
0.01860	150	66.67	0.44			4.0					
0.01612	130	76.92	4.1			0.32			0.37		
0.01240	100	100.0	2.7			0.11			0.21		
0.008679	70	142.9	2.4			$9.2\cdot10^{-2}$			0.17		
0.006199	50	200.0	2.2						0.14		
0.001240	10.00	1000				$9.0\cdot10^{-3}$					
0.0006199	5.000	2000				$3.7\cdot10^{-3}$					
0.0004133	3.333	3000				$2.0\cdot10^{-3}$					

Silicon Dioxide (Glass) - [Ref. 14]

E/eV	$\bar{\nu}/\mathrm{cm^{-1}}$	$\lambda/\mathrm{\mu m}$	n	n_a	n_c	k	k_a	k_c	R	R_a	R_c
2000		$6.199\cdot10^{-4}$	0.99993			$1.503\cdot10^{-5}$					
1860		$6.665\cdot10^{-4}$	0.99991			$1.936\cdot10^{-5}$					
1609		$7.705\cdot10^{-4}$	0.99989			$9.941\cdot10^{-6}$					
1496		$8.287\cdot10^{-4}$	0.99987			$1.308\cdot10^{-5}$					
1204		$1.030\cdot10^{-3}$	0.99980			$2.916\cdot10^{-5}$					
1093		$1.134\cdot10^{-3}$	0.99975			$4.155\cdot10^{-5}$					
1016		$1.220\cdot10^{-3}$	0.99971			$5.423\cdot10^{-5}$					
798		$1.554\cdot10^{-3}$	0.99954			$1.289\cdot10^{-4}$					
597		$2.077\cdot10^{-3}$	0.99917			$3.560\cdot10^{-4}$					
396		$3.131\cdot10^{-3}$	0.99812			$4.04\cdot10^{-4}$					
303		$4.092\cdot10^{-3}$	0.99678			$9.91\cdot10^{-4}$					
201		$6.168\cdot10^{-3}$	0.99269			$3.63\cdot10^{-3}$					
151.2		$8.2\cdot10^{-3}$	0.9871			$7.3\cdot10^{-3}$					
99.99		$1.24\cdot10^{-2}$	0.9813			$7.0\cdot10^{-3}$					
49.59		$2.50\cdot10^{-2}$	0.9164			$6.5\cdot10^{-2}$					
40.00		$3.10\cdot10^{-2}$	0.907			$9.2\cdot10^{-2}$					
31.00		$4.00\cdot10^{-2}$	0.851			0.156					
25.00		0.04959	0.733			0.325					
20.00		0.06199	0.859			0.585					
15.00		0.08266	1.168			0.711			0.10		
13.00		0.09537	1.368			0.747			0.11		
11.00		0.1127	1.739			0.569			0.11		
10.00		0.1240	2.330			0.323			0.17		
9.00		0.1378	1.904			$1.89\cdot10^{-2}$			0.097		
7.00		0.1771	1.600						0.053		
6.00	48390	0.2066	1.543						0.046		
4.9939	40278.4	0.248272	1.50841						0.041		
4.1034	33096.1	0.302150	1.48719						0.038		
3.0640	24712.3	0.404656	1.46961						0.036		
2.5504	20570.5	0.486133	1.46313						0.035		
2.4379	19662.5	0.508582	1.46187						0.035		
2.2705	18312.5	0.546074	1.46008						0.035		
2.1489	17332.3	0.576959	1.45885						0.035		
2.1411	17269.2	0.579065	1.45877						0.035		
2.1102	17019.5	0.587561	1.45847						0.035		
2.1041	16970.4	0.589262	1.45841						0.035		
1.9257	15531.6	0.643847	1.45671						0.035		
1.8892	15237.6	0.656272	1.45637						0.035		
1.8566	14974.2	0.667815	1.45608						0.034		
1.7549	14153.9	0.706519	1.45515						0.034		
1.4550	11735.6	0.852111	1.45248						0.034		

E/eV	$\bar{\nu}$ /cm^{-1}	λ/μm	n	n_a	n_c	k	k_a	k_c	R	R_a	R_c
1.0985	8860.06	1.12866	1.44888						0.034		
0.60243	4858.9	2.0581	1.43722						0.032		
0.35354	2851.4	3.5070	1.40568						0.028		
0.2976	2400	4.176	1.383			$1.07 \cdot 10^{-4}$			0.026		
0.2728	2200	4.545	1.365			$2.56 \cdot 10^{-4}$			0.024		
0.2480	2000	5.000	1.342			$3.98 \cdot 10^{-3}$			0.021		
0.2232	1800	5.556	1.306			$5.63 \cdot 10^{-3}$					
0.1984	1600	6.250	1.239			$6.52 \cdot 10^{-3}$					
0.1736	1400	7.143	1.053			$1.06 \cdot 10^{-2}$					
0.1674	1350	7.407	0.9488			$1.48 \cdot 10^{-2}$					
0.1612	1300	7.692	0.7719			$3.72 \cdot 10^{-2}$					
0.1500	1210	8.265	0.4530			0.704			0.30		
0.1401	1130	8.850	0.3563			1.53			0.66		
0.1302	1050	9.524	2.760			1.65			0.35		
0.1209	975	10.26	2.448			0.231			0.18		
0.1091	880	11.36	1.784			$7.75 \cdot 10^{-2}$			0.079		
0.09919	800	12.50	1.753			0.343			0.089		
0.08989	725	13.79	1.698			0.175			0.071		
0.06943	560	17.86	1.337			0.298			0.036		
0.06199	500	20.00	0.6616						0.882		
0.04959	400	25.0	2.739			0.397			0.23		
0.03720	300	33.33	2.210			$6.7 \cdot 10^{-2}$			0.14		
0.01240	100	100.0	1.967			$1.59 \cdot 10^{-2}$			0.11		
0.007439	60	166.7	1.959			$8.62 \cdot 10^{-3}$			0.11		
0.002480	20	500.0	1.955			$7.96 \cdot 10^{-3}$			0.10		

Silicon Monoxide (Noncrystalline) - [Ref. 15]

E/eV	$\bar{\nu}$ /cm^{-1}	λ/μm	n	n_a	n_c	k	k_a	k_c	R	R_a	R_c
25		0.04959	0.8690			0.2717					
20		0.06199	0.8853			0.4919					
17.5		0.07085	0.9825			0.5961					
15		0.08266	1.132			0.6651			0.092		
12.5		0.09919	1.283			0.6523			0.090		
10		0.1240	1.378			0.6843			0.10		
7.5		0.1653	1.593			0.7473			0.12		
5	40330	0.2480	2.001			0.6052			0.15		
4	32260	0.3100	2.141			0.4006			0.15		
3	24200	0.4133	2.116			0.1211			0.13		
2.8	22580	0.4428	2.085			0.08374			0.12		
2.6	20970	0.4769	2.053			0.05544			0.12		
2.4	19360	0.5166	2.021			0.03533			0.11		
2.2	17740	0.5636	1.994			0.02153			0.11		
2	16130	0.6199	1.969			0.01175			0.11		
1.8	14520	0.6888	1.948			0.00523			0.10		
1.6	12900	0.7749	1.929			0.00151			0.10		
1.240	10000	1.000	1.87						0.092		
0.6199	5000	2.000	1.84						0.087		
0.3100	2500	4.000	1.80						0.082		
0.2480	2000	5.000	1.75						0.074		
0.2066	1667	6.000	1.70						0.067		
0.1771	1492	7.000	1.60						0.053		
0.1653	1333	7.500	1.42								
0.1459	1176	8.500	0.90			0.18					
0.1305	1053	9.500	1.20			1.20			0.024		
0.1240	1000	10.00	2.00			1.38			0.27		
0.1181	952.4	10.50	2.85			0.90			0.27		
0.1153	930.2	10.75	2.86			0.58			0.25		
0.1127	909.1	11.00	2.82			0.40			0.24		
0.1078	869.6	11.50	2.50			0.20			0.19		
0.1033	833.3	12.00	2.13			0.14			0.13		
0.09537	769.2	13.00	2.04			0.20			0.12		
0.08856	714.3	14.00	2.01			0.30			0.12		

E/eV	$\bar{v}\,/\text{cm}^{-1}$	$\lambda/\mu\text{m}$	n	n_a	n_c	k	k_a	k_c	R	R_a	R_c
					Noncrystalline Silicon Nitride (Si_3N_4) - [Ref. 16]						
24		0.05166	0.655			0.420			0.28		
23		0.05391	0.625			0.481			0.22		
22		0.05636	0.611			0.560			0.16		
21		0.05904	0.617			0.647			0.19		
20		0.06199	0.635			0.743			0.21		
19		0.06526	0.676			0.841			0.23		
18		0.06888	0.735			0.936			0.26		
17		0.07293	0.810			1.03			0.25		
16		0.07749	0.902			1.11			0.26		
15		0.08266	1.001			1.18			0.26		
14		0.08856	1.111			1.26			0.26		
13		0.09537	1.247			1.35			0.27		
12	96790	0.1033	1.417			1.43			0.28		
11	88720	0.1127	1.657			1.52			0.29		
10.5	84690	0.1181	1.827			1.53			0.29		
10	80650	0.1240	2.000			1.49			0.29		
9.5	76620	0.1305	2.162			1.44			0.28		
9	72590	0.1378	2.326			1.32			0.27		
8	64520	0.1550	2.651			0.962			0.26		
7	56460	0.1771	2.752			0.493			0.23		
6	48390	0.2066	2.541			0.102			0.19		
5	40330	0.2480	2.278			$4.9 \cdot 10^{-3}$			0.15		
4.75	38310	0.2610	2.234			$1.2 \cdot 10^{-3}$			0.15		
4.5	36290	0.2755	2.198			$2.2 \cdot 10^{-4}$			0.14		
4	32260	0.3100	2.141						0.13		
3.5	28230	0.3542	2.099						0.13		
3	24200	0.4133	2.066						0.12		
2.5	20160	0.4959	2.041						0.12		
2	16130	0.6199	2.022						0.11		
1.5	12100	0.8266	2.008						0.11		
1	8065	1.240	1.998						0.11		
					Sodium Chloride (NaCl) - [Ref. 17]						
209.5		$5.918 \cdot 10^{-3}$				$2.54 \cdot 10^{-3}$					
206		$6.019 \cdot 10^{-3}$				$2.62 \cdot 10^{-3}$					
203		$6.107 \cdot 10^{-3}$				$2.08 \cdot 10^{-3}$					
200		$6.199 \cdot 10^{-3}$				$1.92 \cdot 10^{-3}$					
26.0		0.04769	0.83			0.15			0.015		
25.0		0.04959	0.83			0.18			0.018		
22.0		0.05636	0.83			0.31			0.057		
20.0		0.06199	0.88			0.34			0.036		
18.0		0.06888	0.89			0.33			0.033		
16.1		0.07700	0.74			0.45			0.084		
14.0		0.08856	0.98			0.89			0.17		
12.0		0.1033	1.22			0.79			0.12		
10.0		0.1240	1.55			0.71			0.12		
8.00		0.1550	1.38			1.10			0.20		
6.00	48390	0.2066	1.75						0.074		
5.00	40330	0.2480	1.65						0.060		
2.952	23810	0.42	1.56324						0.048		
2.480	20000	0.50	1.55157						0.047		
2.214	17860	0.56	1.54613						0.046		
2.000	16130	0.62	1.54228						0.045		
1.771	14290	0.70	1.53865						0.045		
1.675	13510	0.74	1.53728						0.045		
1.550	12500	0.80	1.53560						0.045		
1.240	10000	1.00	1.53200						0.044		
1.033	8333	1.2	1.53000						0.044		
0.6888	5556	1.8	1.52712						0.043		
0.4959	4000	2.5	1.52531						0.043		

E/eV	$\bar{\nu}$ /cm^{-1}	λ/μm	n	n_a	n_c	k	k_a	k_c	R	R_a	R_c
0.4000	3226	3.1	1.52395						0.043		
0.3263	2632	3.8	1.52226			(1.8 ± 0.2) $\cdot10^{-9}$			0.043		
0.2952	2381	4.2	1.52121						0.043		
0.2755	2222	4.5	1.52036						0.043		
0.2480	2000	5.0	1.51883						0.042		
0.1240	1000	10.0	1.49473						0.039		
0.1033	833.3	12.0	1.48000						0.037		
0.08856	714.3	14.0	1.46188						0.035		
0.07749	625.0	16.0	1.4399						0.033		
0.06888	555.5	18.0	1.41364						0.029		
0.06199	500.0	20.0	1.3822						0.026		
0.04959	400	25.0	1.27			$3.5\cdot10^{-3}$			0.014		
0.04215	340	29.41	1.12			$1.7\cdot10^{-2}$			0.0032		
0.03720	300	33.33	0.85			0.85			0.18		
0.03410	275	36.36	0.59			0.22			0.084		
0.03286	265	37.74	0.42			0.50			0.26		
0.03224	260	38.46	0.45			0.45			0.22		
0.02480	200	50.00	0.14			1.99			0.89		
0.02108	170	58.82	1.35			6.03			0.87		
0.01984	160	62.50	6.92			2.14			0.59		
0.01922	155	64.52	5.50			0.87			0.49		
0.01860	150	66.67	4.52			0.380			0.41		
0.01736	140	71.43	3.72			0.219			0.33		
0.01612	130	76.92	3.31			0.135			0.29		
0.01488	120	83.33	3.02			0.110			0.25		
0.01240	100	100.0	2.74			0.087			0.22		
0.009919	80	125.0	2.57			0.077			0.19		
0.07439	60	166.7	2.48			0.055			0.18		
0.04959	40	250.00	2.44			0.041			0.18		
0.002480	20	500.0	2.43			0.024			0.17		
0.001240	10	1000	2.43			0.006			0.17		
0.001033	8.333	1200				$8.8\cdot10^{-3}$					
0.0006888	5.556	1800				$5.4\cdot10^{-3}$					
0.0006199	5.000	2000	2.43						0.17		
0.0004959	4.000	2500				$4.4\cdot10^{-3}$					
0.0004797	3.869	2584	2.43			$2.1\cdot10^{-3}$			0.17		
0.0003875	3.125	3200				$3.3\cdot10^{-3}$					
0.0001464	1.181	8469	2.43			$5.8\cdot10^{-4}$			0.17		
0.00004053	0.3269	30590	2.43			$2.5\cdot10^{-4}$					

Cubic Zinc Sulfide (ZnS) - [Ref. 18]

E/eV	$\bar{\nu}$ /cm^{-1}	λ/μm	n	n_a	n_c	k	k_a	k_c	R	R_a	R_c
2000		$6.199\cdot10^{-4}$	0.999904			$1.76\cdot10^{-5}$					
1204		$1.030\cdot10^{-3}$	0.999777			$1.00\cdot10^{-4}$					
1016		$1.220\cdot10^{-3}$	0.999838			$3.61\cdot10^{-5}$					
901		$1.376\cdot10^{-3}$	0.999647			$5.42\cdot10^{-5}$					
798		$1.554\cdot10^{-3}$	0.999520			$8.28\cdot10^{-5}$					
707		$1.754\cdot10^{-3}$	0.999372			$1.25\cdot10^{-4}$					
597		$2.077\cdot10^{-3}$	0.999160			$2.19\cdot10^{-4}$					
377		$9.50\cdot10^{-3}$	0.99789			$9.50\cdot10^{-4}$					
201		$6.168\cdot10^{-3}$	0.99553			$4.82\cdot10^{-3}$					
100		$1.240\cdot10^{-2}$	0.99061			$1.17\cdot10^{-2}$					
61.99		$2.000\cdot10^{-2}$	0.964			$3.32\cdot10^{-2}$			$6.2\cdot10^{-4}$		
41.33		$3.000\cdot10^{-2}$	0.941			$5.10\cdot10^{-2}$					
31.00		$4.000\cdot10^{-2}$	0.847			$9.95\cdot10^{-2}$					
24.80		$5.000\cdot10^{-2}$	0.796			0.171			$2.2\cdot10^{-2}$		
17.71		$7.000\cdot10^{-2}$	0.747			0.431			$7.7\cdot10^{-2}$		
13.78		$9.000\cdot10^{-2}$	0.758			0.824			0.20		
12.40		0.1000	0.862			0.876			0.19		
9.919		0.125	1.02			1.36			0.31		
8.266		0.150	1.41			1.47			0.29		

E/eV	$\bar{\nu}$ /cm^{-1}	λ/μm	n	n_a	n_c	k	k_a	k_c	R	R_a	R_c
6.199		0.200	2.32			1.62			0.32		
6.00	48390	0.2066	2.24			1.65			0.59		
4.00	32260	0.3100	2.70			0.44			0.22		
3.00	24200	0.4133	2.54			$4 \cdot 10^{-2}$			0.19		
2.50	20160	0.4959	2.42			$3 \cdot 10^{-2}$			0.17		
2.30	18550	0.5391	2.3950						0.17		
2.00	16130	0.6199	2.3576						0.16		
1.75	14110	0.7085	2.3319						0.16		
1.55	12500	0.7999	2.3146			$3.50 \cdot 10^{-6}$			0.16		
1.40	11290	0.8856	2.3033						0.16		
1.240	10000	1.000	2.2907			$3.02 \cdot 10^{-6}$			0.15		
1.00	8065	1.240	2.2795						0.15		
0.80	6452	1.550	2.2706						0.15		
0.6199	5000	2.000	2.2631			$6.2 \cdot 10^{-6}$			0.15		
0.45	3629	2.755	2.2587						0.15		
0.30	2420	4.133	2.2529						0.15		
0.20	1613	6.199	2.2443						0.15		
0.1550	1250	8.0	2.2213			$4.5 \cdot 10^{-6}$			0.14		
0.1240	1000	10.00	2.1986			$8.8 \cdot 10^{-6}$			0.14		
0.100	806.5	12.4	2.1969						0.14		
0.09	725.9	13.78	2.1793						0.14		
0.07999	645.2	15.5	2.1518			$3.82 \cdot 10^{-3}$			0.14		
0.07	564.6	17.71	2.1040						0.13		
0.06075	490	20.41	2.03			$8.0 \cdot 10^{-3}$			0.12		
0.05	403.3	24.80	1.6866						0.065		
0.03546	286	34.97	3.29			$8.3 \cdot 10^{-2}$			0.28		
0.03472	280	35.71	9.54			$5.2 \cdot 10^{-2}$			0.66		
0.02480	200	50.00	3.48			$3.1 \cdot 10^{-2}$			0.31		
0.01240	100	100.0	3.06			$5.8 \cdot 10^{-3}$			0.26		
0.004955	40	250.0	2.903			$6.2 \cdot 10^{-3}$			0.24		
0.004339	35	285.7	2.899			$7.0 \cdot 10^{-3}$			0.24		
0.003720	30	333.3	2.896						0.24		
0.003100	25	400.0	2.894						0.24		
0.002480	20	500.0	2.892						0.24		
0.001860	15	666.7	2.890						0.24		

Polytetrafluoroethylene (Teflon) - [Ref. 19]

E/eV	$\bar{\nu}$ /cm^{-1}	λ/μm	n	n_a	n_c	k	k_a	k_c	R	R_a	R_c
4.960	40000	0.250							0.970		
4.769	38462	0.260							0.972		
4.593	37037	0.270							0.975		
4.426	35714	0.280							0.978		
4.276	34483	0.290							0.980		
4.133	33333	0.300							0.983		
4.000	32258	0.310							0.986		
3.875	31250	0.320							0.988		
3.758	30303	0.330							0.990		
3.647	29412	0.340							0.991		
3.543	28571	0.350							0.992		
3.444	27778	0.360							0.992		
3.351	27027	0.370							0.993		
2.255	18182	0.550							0.993		
2.067	16667	0.600							0.992		
1.378	11111	0.900							0.992		
1.305	10526	0.950							0.991		
1.078	8696	1.150							0.991		
1.033	8333	1.200							0.990		
0.9920	8000	1.250							0.990		
0.9538	7692	1.300							0.989		
0.9185	7407	1.350							0.988		
0.8857	7143	1.400							0.988		
0.8552	6897	1.450							0.989		

E/eV	$\bar{v}$ /cm^{-1}	λ/μm	n	n_a	n_c	k	k_a	k_c	R	R_a	R_c
0.8267	6667	1.500							0.989		
0.8000	6452	1.550							0.988		
0.7750	6250	1.600							0.988		
0.7515	6061	1.650							0.987		
0.7294	5882	1.700							0.986		
0.7086	5714	1.750							0.986		
0.6889	5556	1.800							0.985		
0.6703	5405	1.850							0.980		
0.6526	5263	1.900							0.978		
0.6359	51282	1.950							0.978		
0.6200	5000	2.000							0.970		
0.6049	4878	2.050							0.959		
0.5905	4762	2.100							0.951		
0.5767	4651	2.150							0.946		
0.5636	4545	2.200							0.966		
0.5511	44444	2.250							0.965		
0.5487	44247	2.260							0.964		
0.5439	4386	2.280							0.963		
0.5415	4367	2.290							0.961		
0.5368	4329	2.310							0.959		
0.5345	4310	2.320							0.957		
0.5322	4292	2.330							0.956		
0.5299	4274	2.340							0.954		
0.5277	4255	2.350							0.951		
0.5232	4219	2.370							0.950		
0.5188	4184	2.390							0.949		
0.5167	4167	2.400							0.947		
0.5061	4082	2.450							0.946		
0.4960	4000	2.500							0.945		

References

1. Arsenic Selenide
D. J. Treacy in *Handbook of Optical Constants of Solids*, E. D. Palik, Editor, Academic Press, 1985, p. 623. (Hereafter abbreviated as *HOCS*.)
R. Zallen, R. E. Drews, R. L. Emerald, and M. L. Slade, *Phys. Rev. Lett.* 26, 1564 (1971)
R. Zallen, M. L. Slade, and A. T. Ward, *Phys. Rev. B* 3, 4257 (1971).
U. Strom and P. C. Taylor, *Phys. Rev. B* 16, 5512 (1977).
G. Lucovsky, *Phys. Rev. B* 6, 1480 (1972).
C. T. Moynihan, P. B. Macedo, M. S. Maklad, R. K. Mohr, and R. E. Howard, *J. Non-Cryst. Solids*, 17, 369 (1975).
Y. Ohmachi, *J. Opt. Soc. Am.* 63, 630 (1973).
2. Arsenic Sulfide
D. J. Treacy in *HOCS*, 1985, p. 641.
P. A. Young, *J. Phys. C* 4, 93 (1971).
W. S. Rodny, I. H. Malitson, and T. A. King, *J. Opt. Soc. Am.* 48, 633 (1958).
R. Zallen, R.E. Drew, R. L. Emerald, and M.L. Slade, *Phys. Rev. Lett.* 26, 1564 (1971).
M. S. Maklad, R. K. Mohr, R. E. Howard, P. B. Macedo, and C. T. Moynihan, *Solid State Commun.* 15, 855 (1974).
P. B. Klein, P. C. Taylor, and D. J. Treacy, *Phys. Rev. B16*, 4511 (1977).
G. Lucovsky, *Phys. Rev. B* 6, 1480 (1972).
3. Cadmium Telluride
E. D. Palik in *HOCS*, 1985, p. 409.
D. T. F. Marple and H. Ehrenreich, *Phys. Lett.* 8, 87 (1962).
T. H. Myers, S. W. Edwards, and J. F. Schetzina, *J. Appl. Phys.* 52, 4231 (1981).
D. T. F. Marple, *Phys. Rev.* 150, 728 (1966).
A. N. Pikhtin and A. D. Yas'kov, *Sov. Phys. Semicond.* 12, 622 (1978).
L. S. Ladd, *Infrared Phys.* 6, 145 (1966).
J. E. Harvey and W. L. Wolfe, *J. Opt. Soc. Am.* 65, 1267 (1975).

A. Manabe, A. Mitsuishi, and H. Yoshinaga, *Jpn. J. Appl. Phys.* 6, 593 (1967).
A. Manabe, A. Mitsuishi, H. Oshinaga, Y. Ueda, and H. Sei, *Technol. Rep. Osaka Univ. Jpn.* 17, 263 (1967).
J. R. Birch and D. K. Murrey, *Infrared Phys.* 18, 283 (1978).
4. Gallium Arsenide
E. D. Palik in *HOCS*, 1985, p. 429.
M. Cardona, W. Gudat, B. Sonntag, and P. Y. Yu, in *Proc. Intl. Conf. Phys. Semicond.*, 10th. Cambridge, 1970, p. 208. US Atom. Energy Commission, Oak Ridge, TN, 1970.
H. R. Philipp and H. Ehrenreich, *Phys. Rev.* 129, 1550 (1963).
J. B. Theeten, D. E. Aspnes, and R. P. H. Chang, *J. Appl. Phys.* 49, 6097 (1978).
H. C. Casey, D. D. Sell, and K. W. Wecht, *J. Appl. Phys.* 46, 250 (1975).
A. H. Kachare, W. G. Spitzer, F. K. Euler, and A. Kahan, *J. Appl. Phys.* 45, 2938 (1974).
R. T. Holm, J. W. Gibson, and E. D. Palik, *J. Appl. Phys.* 48, 212 (1977).
W. Cochran, S. J. Fray, F. A. Johnson, J. E. Quarrington, and N. Williams, *J. Appl. Phys. Suppl.* 32, 2102 (1961).
C. P. Christensen, R. Joiner, S. K. T. Nieh, and W. H. Steier, *J. Appl. Phys.* 45, 4957 (1974).
R. H. Stolen, *Phys. Rev. B* 11, 767 (1975); *Appl. Phys. Lett.* 15, 74 (1969).
5. Gallium Phosphide
A. Borghesi and G. Guizzetti in *HOCS*, 1985, p. 445.
M. Cardona, W. Gudat, B. Sonntag, and P. Y. Yu, *Proc. Intl. Conf. Phys. Semicond.* Cambridge, 1970, p. 208. US Atom. Energy Commission, Oak Ridge, TN, 1970.
M. Cardona, W. Gudat, E. E. Koch, M. Skibowski, B. Sonntag, and P. Yu. *Phys. Rev. Lett.* 25, 659 (1970).
S. E. Stokowski and D. D. Sell, *Phys. Rev. B* 5, 1636 (1972).
S. A. Abagyan, G. A. Ivanov, Y. E. Shanurin, and V. I. Amosov, *Sov. Phys. Semicond.* 5, 889 (1971).
P. G. Dean, G. Kaminsky, and R. B. Zetterstorm, *J. Appl. Phys.* 38, 3551 (1967).

D. E. Aspnes and A. A. Studna, *Phys. Rev.* B 27, 985 (1983).

6. Indium Antimonide

R. T. Holm in *HOCS*, 1985, p. 491.

M. Cardona, W. Gudat, B. Sonntag, and P. Y. Yu, Proc. *Int. Conf. Phys. Semicond.*, 10th. Cambridge, 1970, p. 208. US Atom. Comm., Oak Ridge, TN, 1970.

H. R. Philipp and H. Ehrenreich, *Phys. Rev.* 129, 1550 (1963).

D. E. Aspnes and A. A. Studna, *Phys. Rev.* B 27, 985 (1983).

T. S. Moss, S. D. Smith, and T. D. F. Hawkins, *Proc. Phys. Soc. London* 70B, 776 (1957).

H. Yoshinaga and R. A. Oetjen, *Phys. Rev.* 101, 526 (1956).

R. B. Sanderson, *J. Phys. Chem. Solids* 26, 803 (1965).

7. Indium Arsenide

E. D. Palick and R. T. Holm in *HOCS*, 1985, p. 479.

H. R. Philipp and H. Ehrenreich, *Phys. Rev.* 129, 1550 (1963).

B. O. Seraphin and H. E. Bennett in *Semiconductors and Semimetals* (R. K. Willardson and A. C. Beer, Eds.), vol. 3, Academic, 1967, p. 499.

D. E. Aspnes and A. A. Studna, *Phys. Rev.* B 27, 985 (1983).

J. R. Dixon and J. M. Ellis, *Phys. Rev.* 123, 1560 (1961).

A. Memon, T. J. Parker, and J. R. Birch, *Proc. SPIE*, 289, 20 (1981).

8. Indium Phosphide

O. J. Glembocki and H. Piller in *HOCS*, 1985, p. 503.

M. Cardona, *J. Appl. Phys.* 32, 958 (1961); 36, 2181 (1965).

D. E. Aspnes and A. A. Studna, *Phys. Rev.* B 27, 985 (1983).

G. D. Pettit and W. J. Turner, *J. Appl. Phys.* 36, 2081 (1965).

R. Newman, *Phys. Rev.* 111, 1518 (1958).

W. N. Reynolds, M. T. Lilburne, and R. M. Dell, *Proc. Phys. Soc. London* 71, 416 (1958).

H. Jamshidi and T. J. Parker, Int. Meet. Infrared Mm. Waves, 7th., Marseilles, 1983.

9. Lead Selenide

G. Bauer and H. Krenn in *HOCS*, 1985, p. 517.

M. Cardona and D. L. Greenaway, *Phys. Rev.* A 133, 1685 (1964).

T. S. Moss, *Optical Properties of Semiconductors*, Butterworth, 1959, p. 189.

J. N. Zemel, J. D. Jensen, and R. B. Schoolar, *Phys. Rev.* A 140, 330 (1965).

W. W. Scanlon, *J. Phys. Chem. Solids*, 8, 423 (1959).

K. V. Vyatkin and A. P. Shotov, *Sov. Phys. Semicond.* 14, 785 (1980); Fiz. Tekh. Poluprovodn. 14, 1331 (1980).

10. Lead Sulfide

G. Guizzetti and A. Borghesi in *HOCS*, 1985, p. 525.

M. Cardona and R. Haensel, *Phys. Rev.* B 1, 2605 (1970).

M. Cardona and D. L. Greenaway, *Phys. Rev.* A 133, 1685 (1964).

M. Cardona, C. M. Penchina, E. E. Koch, and P. Y. Yu, *Phys. Status Solidi* B 53, 327 (1972).

P. R. Wessel, *Phys. Rev.* 153, 836 (1967).

C. E. Rossi and W. Paul, *J. Appl. Phys.* 38, 1803 (1967).

J. N. Zemel, J. D. Jensen, and R. B. Schoolar, *Phys. Rev.* A 140, 330 (1965).

11. Lead Telluride

G. Bauer and H. Krenn in *HOCS*, 1985, p. 535.

M. Cardona and R. Haensel, *Phys. Rev.* B 1, 2605 (1970).

M. Cardona and D. L. Greenaway, *Phys. Rev.* 133, A1685 (1964).

D. M. Korn and R. Braunstein, *Phys. Rev.* B 5, 4837 (1972).

W. W. Scanlon, *J. Phys. Chem. Solids* 8, 423 (1959).

J. N. Zemel, J. D. Jensen, and R. B. Schoolar, *Phys. Rev.* 140, A330 (1965).

12. Lithium Fluoride

E. D. Palik and W. R. Hunter in *HOCS*, 1985, p. 675.

B. L. Henke, P. Lee, T. J. Tanaka, R. L. Shimabukuro, and B. K. Fujikawa, *Low Energy X-ray Diagnostics-1981* (D. T. Attwood and B. L. Henke, Eds.), AIP Conf. Proc. No. 75, 1981.

A. P. Lukirskii, E. P. Savinov, O. A. Ershov, and Y. F. Shepelev, *Opt. Spektrosk.* 16, 168 (1964); 16, 310 (1964).

F. C. Brown, C. Gahwiller, A. B. Kunz, and N. O. Lipari, *Phys. Rev. Lett.* 25, 927 (1970).

A. Milgram and M. P. Givens, *Phys. Rev.* 125, 1506 (1962).

T. Tomiki and T. Miyata, *J. Phys. Soc. Jpn.* 27, 658 (1969).

A. Kachare, G. Andermann, and L. R. Brantley, *J. Phys. Chem. Solids* 33, 467 (1972).

13. Potassium Chloride

E. D. Palik in *HOCS*, 1985, p. 703.

O. Aita, I. Nagakura, and T. Sagawa, *J. Phys. Soc. Jpn.* 30, 1414 (1971).

A. P. Lukirskii, E. P. Savinov, O. A. Ershov, and Y. F. Shepelev, *Opt. Spectrosc.* 16, 168 (1964); Opt. Spektrosk. 16, 310 (1964).

T. Tomika, *J. Phys. Soc. Jpn.* 22, 463 (1967).

M. Antinori, A. Balzarotti, and M. Piacentini, *Phys. Rev.* B 7, 1541 (1973).

H. H. Li, *J. Phys. Chem. Ref. Data* 5, 329 (1976).

S. D. Allen and J. A. Harrington, *Appl. Opt.* 17, 1679 (1978).

K. W. Johnson and E. E. Bell, *Phys. Rev.* 139A, 1295 (1965).

14. Silicon Dioxide

H. R. Philipp in *HOCS*, 1985, p. 749.

J. Rife and J. Osantowski, *J. Opt. Soc. Am.* 70, 1513 (1980).

B. L. Henke, P. Lee, T. J. Tanaka, R. L. Shimabukuro, and B. K. Fujikawa, *Low Energy X-ray Diagnostics-1981* (D. T. Attwood and B. L. Henke, Eds.), AIP Conf. Proc. No. 75, 1981.

H. R. Philipp, *Solid State Commun.* 4, 73 (1966); *J. Phys. Chem. Solids*, 32, 1935 (1971).

P. L. Lamy, *Appl. Opt.* 16, 2212 (1977).

H. R. Philipp, *J. Appl. Phys.* 50 1053 (1979).

D. G. Drummond, *Proc. Roy. Soc. London*, 153, 328 (1935).

15. Silicon Monoxide

H. R. Philipp in *HOCS*, 1985, p. 765.

H. R. Philipp, *J. Phys. Chem. Solids*, 32, 1935 (1971).

G. Hass and C. D. Salzberg, *J. Opt. Soc. Am.* 44, 181 (1954).

E. Cremer, T. Kraus, and E. Ritter, *Zs. Electrochem.* 62, 939 (1958).

A. P. Bradford, G. Hass, M. McFarland, and E. Ritter, *Appl. Opt.* 4, 971 (1965).

16. Silicon Nitride

H. R. Philipp in *HOCS*, 1985, p. 771.

H. R. Philipp, *J. Electrochem. Soc.* 120, 295 (1973).

J. B. Theeten, D. E. Aspnes, F. Simondet, M. Errman, and P. C. Mürau, *J. Appl. Phys.* 52, 6788 (1981).

J. Bauer, *Phys. Status Solidi*, A 39, 411 (1977).

17. Sodium Chloride

J. E. Eldridge and E. D. Palik in *HOCS*, p. 775.

J. A. Harrington, C. J. Duthler, F. W. Patten, and M. Hass, *Solid State Commun.* 18, 1043 (1976).

T. Miyata and T. Tomiki, *J. Phys. Soc. Jpn.* 24, 1286 (1968); ibid., 22, 209 (1967).

D. M. Roessler and W. C. Walker, *J. Opt. Soc. Am.* 58, 279 (1968).

D. M. Roessler and W. C. Walker, *Phys. Rev.* 166, 599 (1968).

S. Allen and J. A. Harrington, *Appl. Opt.* 17, 1679 (1978).

O. Aita, I. Nagakura, and T. Sagawa, *J. Phys. Soc. Jpn.* 30, 1414 (1971).

18. Zinc Sulfide

E. D. Palik and A. Addamiano in *HOCS*, 1985, p. 597.

B. L. Henke, P. L. Lee, T. J. Tanaka, R. L. Shimabukuro, and B. F. Fujikawa, *Low Energy X-ray Diagnostics-1981* (D. T. Attwood and B. L. Henke, Eds.), *AIP Conf. Proc.* No. 75, 1981.

M. Cardona and G. Harbeke, *Phys. Rev.* 137, A1467 (1965).

Eastman Kodak, Publ. No. U-72, Rochester, New York (1981).

C. A. Klein and R. N. Donadio, *J. Appl. Phys.* 51, 797 (1980).

T. Deutsch, *Proc. Int. Conf. Phys. Semicond.*, 6th Exeter 1962, p. 505. The Inst. of Physics and the Physical Soc., London, 1962.

A. Manabe, A. Mitsuishi, and H. Yoshinaga, *Jpn. J. Appl. Phys.* 6, 593 (1967).

W. W. Piper, D. T. F. Marple, and P. D. Johnson, *Phys. Rev.* 110, 323 (1958).

19. Polytetrafluoroethylene

J. W. L. Thomas (NIST), Private communication. NIST Certificate, STM 2044.

P. Y. Barnes, E. A. Early, and A. C. Parr, *NIST Special Publ. 250-48*, NIST Measurement Services: Spectral Reflectance.

Diffuse Reflectance Coatings and Materials Sections, Labsphere Catalog, 1996.

A. Arecchi and C. Ryder (Labsphere, North Sutten, NJ), private communication.

ELASTO-OPTIC, ELECTRO-OPTIC, AND MAGNETO-OPTIC CONSTANTS

When a crystal is subjected to a stress field, an electric field, or a magnetic field, the resulting optical effects are in general dependent on the orientation of these fields with respect to the crystal axes. It is useful, therefore, to express the optical properties in terms of the refractive index ellipsoid (or indicatrix):

$$\frac{x^2}{n_x^2} + \frac{y^2}{n_y^2} + \frac{z^2}{n_z^2} = 1$$

or

$$\sum_{ij} B_{ij} x_i y_j = 1 \, (i, j = 1, 2, 3)$$

where

$$B_{ij} = \frac{1}{\varepsilon}\bigg|_{ij} = \frac{1}{n^2}\bigg|_{ij}$$

ε is the dielectric constant or permeability; the quantity B_{ij} has the name impermeability.

A crystal exposed to a *stress* **S** will show a change of its impermeability. The photo-elastic (or elasto-optic) constants, P_{ijkl}, are defined by

$$\frac{1}{\varepsilon}\bigg|_{ij} = \frac{1}{n^2}\bigg|_{ij} = \sum_{kl} P_{ijkl} S_{kl}$$

where n is the refractive index and S_{kl} are the strain tensor elements; the P_{ijkl} are the elements of a 4th rank tensor.

When a crystal is subjected to an *electric field* **E** two possible changes of the refractive index may occur depending on the symmetry of the crystal.

1. All materials, including isotropic solids and polar liquids, show an electro-optic birefringence (Kerr effect) which is proportional to the square of the electric field, *E*:

$$\frac{1}{n^2}\bigg|_{ij} = \sum_k K_{ijkl} E_k E_l = \sum_{k,l=1,2,3} g_{ijkl} p_k p_l$$

where E_k and E_l are the components of the electric field and P_k and P_l the electric polarizations. The coefficients, K_{ijkl}, are the quadratic electro-optic coefficients, while the constants g_{ijkl} are known as the Kerr constants.

2. The other electro-optic effect only occurs in the 20 piezo-electric crystal classes (no center of symmetry). This effect is known as the Pockels effect. The optical impermeability changes linearly with the static field

$$\frac{1}{n^2}\bigg|_{ij} = \sum_k r_{ij,k} E_k$$

The coefficients $r_{ij,k}$ have the name (linear) electro-optic coefficients.

The values of the electro-optic coefficients depend on the boundary conditions. If the superscripts T and S denote, respectively, the conditions of zero stress (free) and zero strain (clamped) one finds:

$$r_{ij}^T = r_{ij}^S + q_{ik}^E e_{jk} = r_{ij}^S + P_{ik}^E d_{jk}$$

where $e_{jk} = (\partial T_k / \partial E_j)_S$ and $d_{jk} = (\partial S_k / \partial E_j)_T$ are the appropriate piezo-electric coefficients.

The interaction between a *magnetic field* and a light wave propagating in a solid or in a liquid gives rise to a rotation of the plane of polarization. This effect is known as *Faraday rotation*. It results from a difference in propagation velocity for left and right circular polarized light.

The Faraday rotation, θ_F, is linearly proportional to the magnetic field H:

$$\theta_F = VlH$$

where l is the light path length and V is the *Verdet* constant (minutes/oersted·cm).

For ferromagnetic, ferrimagnetic, and antiferromagnetic materials the magnetic field in the above expression is replaced by the magnetization M and the magneto-optic coefficient in this case is known as the Kund constant K:

$$\text{Specific Faraday rotation } F = KM$$

In the tables below the *Faraday rotation* is listed at the saturation magnetization per unit length, together with the absorption coefficient α, the temperature T, the critical temperature T_C (or T_N), and the wavelength of the measurement.

In the tables that follow, the properties are presented in groups:

- Elasto-optic coefficients (photoelastic constants)
- Linear electro-optic coefficients (Pockels constants)
- Quadratic electro-optic coefficients (Kerr constants)
- Magneto-optic coefficients:
  - Verdet constants
  - Faraday rotation parameters

Within each group, materials are classified by crystal system or physical state. References are given at the end of each group of tables.

ELASTO-OPTIC COEFFICIENTS (PHOTOELASTIC CONSTANTS)

Name Cubic (43m, 432, m3m)	Formula	$\lambda/\mu m$	p_{11}	p_{12}	p_{44}	$p_{11}-p_{12}$	Ref.
Sodium fluoride	NaF	0.633	0.08	0.20	−0.03	−0.12	1
Sodium chloride	NaCl	0.589	0.115	0.159	−0.011	−0.042	2
Sodium bromide	NaBr	0.589	0.148	0.184	−0.0036	−0.035	1
Sodium iodide	NaI	0.589	−	−	0.0048	−0.0141	3
Potassium fluoride	KF	0.546	0.26	0.20	−0.029	0.06	1
Potassium chloride	KCl	0.633	0.22	0.16	−0.025	0.06	4
Potassium bromide	KBr	0.589	0.212	0.165	−0.022	0.047	5
Potassium iodide	KI	0.590	0.212	0.171	−	0.041	6
Rubidium chloride	RbCl	0.589	0.288	0.172	−0.041	0.116	7,8
Rubidium bromide	RbBr	0.589	0.293	0.185	−0.034	0.108	7,8
Rubidium iodide	RbI	0.589	0.262	0.167	−0.023	0.095	7,8
Lithium fluoride	LiF	0.589	0.02	0.13	−0.045	−0.11	5
Lithium chloride	LiCl	0.589	−	−	−0.0177	−0.0407	3
Ammonium chloride	NH_4Cl	0.589	0.142	0.245	0.042	−0.103	9
Cadmium telluride	CdTe	1.06	−0.152	−0.017	−0.057	−0.135	10
Calcium fluoride	CaF_2	0.55–0.65	0.038	0.226	0.0254	−0.183	11
Copper chloride	CuCl	0.633	0.120	0.250	−0.082	−0.130	12
Copper bromide	CuBr	0.633	0.072	0.195	−0.083	−0.123	12
Copper iodide	CuI	0.633	0.032	0.151	−0.068	−0.119	12
Diamond	C	0.540–0.589	−0.278	0.123	−0.161	−0.385	13
Germanium	Ge	3.39	−0.151	−0.128	−0.072	−0.023	14
Gallium arsenide	GaAs	1.15	−0.165	−0.140	−0.072	−0.025	15
Gallium phosphide	GaP	0.633	−0.151	−0.082	−0.074	−0.069	15
Strontium fluoride	SrF_2	0.633	0.080	0.269	0.0185	−0.189	16
Strontium titanate	$SrTiO_3$	0.633	0.15	0.095	0.072	−	17
KRS-5	Tl(Br,I)	0.633	−0.140	0.149	−0.0725	−0.289	18,20
KRS-6	Tl(Br,Cl)	0.633	−0.451	−0.337	−0.164	−0.114	19,20
Zinc sulfide	Zn	0.633	0.091	−0.01	0.075	0.101	15

Rare Gases	Formula	$\lambda/\mu m$	p_{11}	p_{12}	p_{44}	$p_{11}-p_{12}$	Ref.
Neon (T = 24.3 K)	Ne	0.488	0.157	0.168	0.004	−0.011	21
Argon (T = 82.3 K)	Ar	0.488	0.256	0.302	0.015	−0.046	22
Krypton (T = 115.6 K)	Kr	0.488	0.34	0.34	0.037	0	21
Xenon (T = 160.5 K)	Xe	0.488	0.284	0.370	0.029	−0.086	22

Garnets	Formula	$\lambda/\mu m$	p_{11}	p_{12}	p_{44}	$p_{11}-p_{12}$	Ref.
GGG	$Gd_3Ga_5O_{12}$	0.514	−0.086	−0.027	−0.078	−0.059	23
YIG	$Y_3Fe_5O_{12}$	1.15	0.025	0.073	0.041	−	15
YGG	$Y_3Ga_5O_{12}$	0.633	0.091	0.019	0.079	−	17
YAG	$Y_3Al_5O_{12}$	0.633	−0.029	0.0091	−0.0615	−0.038	15

Cubic (23, m3)	Formula	$\lambda/\mu m$	p_{11}	p_{12}	p_{44}	p_{13}	Ref.
Barium nitrate	$Ba(NO_3)_2$	0.589	−	$p_{11}-p_{22}=$ 0.992	−0.0205	$p_{11}-p_{13}=$ 0.713	13
Lead nitrate	$Pb(NO_3)_2$	0.589	0.162	0.24	−0.0198	0.20	24,25
Sodium bromate	$NaBrO_3$	0.589	0.185	0.218	−0.0139	0.213	26
Sodium chlorate	$NaClO_3$	0.589	0.162	0.24	−0.0198	0.20	26
Strontium nitrate	$Sr(NO_3)_2$	0.41	0.178	0.362	−0.014	0.316	27

Hexagonal (mmc, 6mm)	Formula	$\lambda/\mu m$	p_{11}	p_{12}	p_{13}	p_{31}	p_{33}	p_{44}	Ref.
Beryl	$Be_3Al_2Si_6O_{18}$	0.589	0.0099	0.175	0.191	0.313	0.023	−0.152	28
Cadmium sulfide	CdS	0.633	−0.142	−0.066	−0.057	−0.041	−0.20	−0.099	15,2
Zinc oxide	ZnO	0.633	±0.222	±0.099	−0.111	±0.088	−0.235	0.0585	30
Zinc sulfide	ZnS	0.633	−0.115	0.017	0.025	0.0271	−0.13	−0.0627	31

ELASTO-OPTIC COEFFICIENTS (PHOTOELASTIC CONSTANTS)

Trigonal (3m, 32, 3̄m)

	Formula	λ/μm	p_{11}	p_{12}	p_{13}	p_{14}	p_{31}	p_{33}	p_{41}	p_{44}	Ref.
Sapphire	Al_2O_3	0.644	−0.23	−0.03	0.02	0.00	−0.04	−0.20	0.01	−0.10	15,32
Calcite	$CaCO_3$	0.514	0.062	0.147	0.186	−0.011	0.241	0.139	−0.036	−0.058	33
Lithium niobate	$LiNbO_3$	0.633	±0.034	±0.072	±0.139	±0.066	±0.178	±0.060	±0.154	±0.300	15,34
Lithium tantalate	$LiTaO_3$	0.633	−0.081	0.081	0.093	−0.026	0.089	−0.044	−0.085	0.028	15,35
Cinnabar	HgS	0.633			±0.445			±0.115	–	–	36
Quartz	SiO_2	0.589	0.16	0.27	0.27	−0.030	0.29	0.10	−0.047	−0.079	37
Proustite	Ag_3AsS_3	0.633	±0.10	±0.19	±0.22		±0.24	±0.20	–	–	38
Sodium nitrite	$NaNO_2$	0.633		±0.21	±0.215	±0.027	±0.25		0.055	−0.06	39
Tellurium	Te	10.6	0.155	0.130	–	–	–		–	–	15

Tetragonal (4/mmm, 4̄2m, 422)

	Formula	λ/μm	p_{11}	p_{12}	p_{13}	p_{31}	p_{33}	p_{44}	p_{66}	Ref.
Ammonium dihydrogen phosphate	ADP	0.589	0.319	0.277	0.169	0.197	0.167	−0.058	−0.091	40
Barium titanate	$BaTiO_3$	0.633	0.425	–	–	–	–	–	–	41
Cesium dihydrogen arsenate	CDA	0.633	0.267	0.225	0.200	0.195	0.227	–	–	42
Magnesium fluoride	MgF_2	0.546	–	–	–	–	–	±0.0776	±0.0488	43
Calomel	Hg_2Cl_2	0.633	±0.551	±0.440	±0.256	±0.137	±0.010		±0.047	44
Potassium dihydrogen phosphate	KDP	0.589	0.287	0.282	0.174	0.241	0.122	−0.019	−0.064	45
Rubidium dihydrogen arsenate	RDA	0.633	0.227	0.239	0.200	0.205	0.182	–	–	41
Rubidium dihydrogen phosphate	RDP	0.633	0.273	0.240	0.218	0.210	0.208	–	–	41
Strontium barium niobate	$Sr_{0.75}Ba_{0.25}Nb_2O_6$	0.633	0.16	0.10	0.08	0.11	0.47			46
Strontium barium niobate	$Sr_{0.5}Ba_{0.5}Nb_2O_6$	0.633	0.06	0.08	0.17	0.09	0.23			46
Tellurium oxide	TeO_2	0.633	0.0074	0.187	0.340	0.090	0.240	−0.17	−0.046	47
Rutile	TiO_2	0.633	0.017	0.143	−0.139	−0.080	−0.057	−0.009	−0.060	48

Tetragonal (4, 4̄, 4/m)

	Formula	λ/μm	p_{11}	p_{12}	p_{13}	p_{16}	p_{31}	p_{33}	p_{44}	p_{45}	p_{61}	p_{66}	Ref.
Cadmium molybdate	$CdMoO_4$	0.633	0.12	0.10	0.13	–	0.11	0.18	–	–	–	–	49
Lead molybdate	$PbMoO_4$	0.633	0.24	0.24	0.255	0.017	0.175	0.300	0.067	−0.01	0.013	0.05	52
Sodium bismuth molybdate	$NaBi(MoO_4)_2$	0.633	0.243	0.205	0.25	–	0.21	0.29	–	–	–	–	

Orthorhombic (222, m22, mmm)

	Formula	λ/μm	p_{11}	p_{12}	p_{13}	p_{21}	p_{22}	p_{23}	p_{31}	p_{32}	p_{33}	p_{44}	p_{55}	p_{66}	Ref.
Ammonium chlorate	NH_4ClO_3	0.633	–	0.24	0.18	0.23	–	0.20	0.19	0.18	±0.02	<±0.02	–	±0.04	51
Ammonium sulfate	$(NH_4)_2SO_4$	0.633	0.26	0.19	±0.260	±0.230	±0.27	±0.254	0.20	±0.26	0.26	0.015	±0.0015	0.012	52
Rochelle salt	$NaKC_4H_4O_6$	0.589	0.35	0.41	0.42	0.37	0.28	0.34	0.36	0.35	0.36	−0.030	0.0046	−0.025	53
Iodic acid (α)	HIO_3	0.633	0.302	0.496	0.339	0.263	0.412	0.304	0.251	0.345	0.336	0.084	−0.030	0.098	54
Sulfur (α)	S	0.633	0.324	0.307	0.268	0.272	0.301	0.310	0.203	0.232	0.270	0.143	0.019	0.118	54
Barite	$BaSO_4$	0.589	0.21	0.25	0.16	0.34	0.24	0.19	0.28	0.22	0.31	0.002	−0.012	0.037	55
Topaz	$Al_2SiO_4(OH,F)_2$	–	−0.085	0.069	0.052	0.095	−0.120	0.065	0.095	0.085	−0.083	−0.095	−0.031	0.098	28

Monoclinic (2, m, 2/m)

	Formula	λ/μm
Taurine	$C_2H_7NO_3S$	0.589

$$p_{11} = 0.313 \quad p_{25} = -0.0025 \quad p_{51} = -0.014$$
$$p_{12} = 0.251 \quad p_{31} = 0.362 \quad p_{52} = 0.006$$
$$p_{13} = 0.270 \quad p_{32} = 0.275 \quad p_{53} = 0.0048$$
$$p_{15} = -0.10 \quad p_{33} = 0.308 \quad p_{55} = 0.047$$
$$p_{21} = 0.281 \quad p_{35} = -0.003 \quad p_{64} = 0.0024$$
$$p_{22} = 0.252 \quad p_{44} = 0.0025 \quad p_{66} = 0.0028$$
$$p_{23} = 0.272 \quad p_{46} = -0.0056$$

Isotropic	Formula	λ/μm	p_{11}	p_{12}	p_{44}	Ref.
Fused silica	SiO_2	0.633	0.121	0.270	−0.075	15
Water	H_2O	0.633	±0.31	±0.31		15
Polystyrene		0.633	±0.30	±0.31		25
Lucite		0.633	±0.30	0.28		25
Orpiment	As_2S_3-glass	1.15	0.308	0.299	0.0045	15
Tellurium oxide	TeO_2-glass	0.633	0.257	0.241	0.0079	56
Laser glasses	LGS-247-2	0.488	±0.168	±0.230		57
	LGS-250-3		±0.135	±0.198		
	LGS-1		±0.214	±0.250		
	KGSS-1621		±0.205	±0.239		
Dense flint glasses	LaSF	0.633	0.088	0.147	−0.030	58
(examples)	SF_4		0.215	0.243	−0.014	
	U10502		0.172	0.179	−0.004	
	$TaFd_7$		0.099	0.138	−0.020	

References

A. Narasimhamurty, T. S., *Photoelastic and Electro-Optic Properties of Crystals*, Plenum Press, New York, 1981, pp. 290–293.

B. Weber, M. J., Ed., *CRC Handbook of Laser Science and Technology*, Volume IV, Part 2, CRC Press, Boca Raton, FL, 1986, pp. 324–331.

1. Petterson, H. E., *J. Opt. Soc. Am.*, 63, 1243, 1973.
2. Burstein, E. and Smith, P. L., *Phys. Rev.*, 74, 229, 1948.
3. Pakhnev, A. V., et al., *Sov. Phys. J. (transl.)*, 18, 1662, 1975.
4. Feldman, A., Horovitz, D., and Waxler, R. M., *Appl. Opt.*, 16, 2925, 1977.
5. Iyengar, K. S., *Nature (London)*, 176, 1119, 1955.
6. Bansigir, K. G. and Iyengar, K. S., *Acta Crystallogr.*, 14, 727, 1961.
7. Pakhev, A. V., et al., *Sov. Phys. J. (transl.)*, 20, 648, 1975.
8. Bansigir, K. G., *Acta Crystallogr.*, 23, 505, 1967.
9. Krishna Rao, K. V. and Krishna Murty, V. G., *Ind. J. Phys.*, 41, 150, 1967.
10. Weil, R. and Sun, M. J., *Proc. Int. Symp. CdTe (Detectors)*, Strasbourg Centre de Rech. Nucl., 1971, XIX-1 to 6, 1972.
11. Schmidt, E. D. D. and Vedam, K., *J. Phys. Chem. Solids*, 27, 1563, 1966.
12. Biegelsen, D. K., et al., *Phys. Rev. B*, 14, 3578, 1976.
13. Helwege, K. H., *Landolt-Börnstein, New Series Group III*, Vol. II, Springer-Verlag, Berlin, 1979.
14. Feldman, A., Waxler, R. M., and Horovitz, D., *J. Appl. Phys.*, 49, 2589, 1978.
15. Dixon, R. W., *J. Appl. Phys.*, 38, 5149, 1967.
16. Shabin, O. V., et al., *Sov. Phys. Sol. State (transl.)*, 13, 3141, 1972.
17. Reintjes, J. and Schultz, M. B., *J. Appl. Phys.*, 39, 5254, 1968.
18. Rivoallan, L. and Favre, F., *Opt. Commun.*, 8, 404, 1973.
19. Rivoallan, L. and Favre, F., *Opt. Commun.*, 11, 296, 1974.
20. Afanasev, I. I., et al., *Sov. J. Opt. Technol.*, 46, 663, 1979.
21. Rand, S. C., et al., *Phys. Rev. B*, 19, 4205, 1979.
22. Sipe, J. E., *Can J. Phys.*, 56, 199, 1978.
23. Christyi, I. L., et al., *Sov. Phys. Sol. State (transl.)*, 17, 922, 1975.
24. Narasimhamurty, T. S., *Curr. Sci. (India)*, 23, 149, 1954.
25. Smith, T. M. and Korpel, A., *IEEE J. Quant. Electron.*, QE-1, 283, 1965.

26. Narasimhamurty, T. S., *Proc. Ind. Acad. Sci.*, A40, 164, 1954.
27. Rabman, A., *Bhagarantam Commem. Vol.*, Bangalore Print. and Publ., 173, 1969.
28. Eppendahl, R., *Ann. Phys. (IV)*, 61, 591, 1920.
29. Laurenti, J. P. and Rouzeyre, M., *J. Appl. Phys.*, 52, 6484, 1981.
30. Sasaki, H., et al., *J. Appl. Phys.*, 47, 2046, 1976.
31. Uchida, N. and Saito, S., *J. Appl. Phys.*, 43, 971, 1972.
32. Waxler, R. M. and Farabaugh, E. M., *J. Res. Natl. Bur. Stand.*, A74, 215, 1970.
33. Nelson, D. F., Lazay, P. D., and Lax, M., *Phys. Rev.*, B6, 3109, 1972.
34. O'Brien, R. J., Rosasco, G. J., and Weber, A., *J. Opt. Soc. Am.*, 60, 716, 1970.
35. Avakyants, L. P., et al., *Sov. Phys.*, 18, 1242, 1976.
36. Sapriel, J., *Appl. Phys. Litt.*, 19, 533, 1971.
37. Narasimhamurty, T. S., *J. Opt. Soc. Am.*, 59, 682, 1969.
38. Zubrinov, I. I., et al., *Sov. Phys. Sol. State (transl.)*, 15, 1921, 1974.
39. Kachalov, O. V. and Shpilko, I. O., *Sov. Phys. JETP (transl.)*, 35, 957, 1972.
40. Narasimhamurty, T. S., et al., *J. Mater. Sci.*, 8, 577, 1973.
41. Tada, K. and Kikuchi, K., *Jpn. J. Appl. Phys.*, 19, 1311, 1980.
42. Aleksandrov, K. S., et al., *Sov. Phys. Sol. State (transl.)*, 19, 1090, 1977.
43. Afanasev, I. I., et al., *Sov. Phys. Sol. State (transl.)*, 17, 2006, 1975.
44. Silvestrova, I. M., et al., *Sov. Phys. Cryst. (transl.)*, 20, 649, 1975.
45. Veerabhadra Rao, K. and Narasimhamurty, T. S., *J. Mater. Sci.*, 10, 1019, 1975.
46. Venturini, E. L., et al., *J. Appl. Phys.*, 40, 1622, 1969.
47. Vehida, N. and Ohmachi, Y., *J. Appl. Phys.*, 40, 4692, 1969.
48. Grimsditch, M. H. and Ramdus, A. K., *Phys. Rev. B*, 22, 4094, 1980.
49. Schinke, D. P. and Viehman,W., unpublished data.
50. Coquin, G. A., et al., *J. Appl. Phys.*, 42, 2162, 1971.
51. Vasquez, F., et al., *J. Phys. Chem. Solids*, 37, 451, 1976.
52. Luspin, Y. and Hauret, G., *C.R.Ac. Sci. Paris*, B274, 995 1972.
53. Narasimhamurty, T. S., *Phys. Rev.*, 186, 945, 1969.
54. Haussühl, S. and Weber, H. J., *Z. Kristall.*, 132, 266, 1970.
55. Vedam, K., *Proc. Ind. Ac. Sci.*, A34, 161, 1951.
56. Yano, T., Fukumoto, A., and Watanabe, A., *J. Appl. Phys.*, 42, 3674, 1971.
57. Manenkov, A. A. and Ritus, A. I., *Sov. J. Quant. Electr.*, 8, 78, 1978.
58. Eschler, H. and Weidinger, F., *J. Appl. Phys.*, 46, 65, 1975.

LINEAR ELECTRO-OPTIC COEFFICIENTS

Name

Cubic ($\bar{4}$3m)	Formula	$\lambda/\mu m$	r_{41} pm/V
Cuprous bromide	CuBr	0.525	0.85
Cuprous chloride	CuCl	0.633	3.6
Cuprous iodide	CuI	0.55	−5.0
Eulytite (BSO)	$Bi_4Si_3O_{12}$	0.63	0.54
Germanium eulytite (BGO)	$Bi_4Ge_3O_{12}$	0.63	1.0
Gallium arsenide	GaAs	10.6	1.6
Gallium phosphide	GaP	0.56	−1.07
Hexamethylenetetramine	$C_6H_{12}N_4$	0.633	0.78
Sphalerite	ZnS	0.65	2.1
Zinc selenide	ZnSe	0.546	2.0
Zinc telluride	ZnTe	3.41	4.2
Cadmium telluride	CdTe	3.39	6.8

Cubic (23)	Formula	$\lambda/\mu m$	r_{41} pm/V
Ammonium chloride (77 K)	NH_4Cl	–	1.5
Ammonium cadmium langbeinite	$(NH_4)_2Cd_2(SO_4)_3$	0.546	0.70
Ammonium manganese langbeinite	$(NH_4)_2Mn_2(SO_4)_3$	0.546	0.53
Thallium cadmium langbeinite	$Tl_2Cd_2(SO_4)_3$	0.546	0.37
Potassium magnesium langbeinite	$K_2Mg_2(SO_4)_3$	0.546	0.40
Bismuth monogermanate	$Bi_{12}GeO_{20}$	–	3.3
Bismuth monosilicate	$Bi_{12}SiO_{20}$	–	3.3
Sodium chlorate	$NaClO_3$	0.589	0.4
Sodium uranyl acetate	$NaUO_2(CH_3COO)_3$	0.546	0.87
Trenhydrobromide	$N(CH_2CH_2NH_2)_3 \cdot 3HBr$	–	1.5
Trenhydrochloride	$N(CH_2CH_2NH_2)_3 \cdot 3HCl$	–	1.7

Tetragonal ($\bar{4}$2m)	Formula	T_{tran} K	r_{41} pm/V	r_{63} pm/V
Ammonium dihydrogen phosphate (ADP)	$NH_4H_2PO_4$	148	24.5	−8.5
Ammonium dideuterium phosphate (AD*P)	$NH_4D_2PO_4$	242	–	11.9
Ammonium dihydrogen arsenate (ADA)	$NH_4H_2AsO_4$	–	–	9.2
Cesium dihydrogen arsenate (CsDA)	CsH_2AsO_4	143	–	18.6
Cesium dideuterium arsenate (CsD*A)	CsD_2AsO_4	212	–	36.6
Potassium dihydrogen phosphate (KDP)	KH_2PO_4	123	8.6	−10.5
Potassium dideuterium phosphate (KD*P)	KD_2PO_4	222	8.8	23.8
Potassium dihydrogen arsenate (KDA)	KH_2AsO_4	97	12.5	10.9
Potassium dideuterium arsenate (KD*A)	KD_2AsO_4	162	–	18.2
Rubidium dihydrogen phosphate (RDP)	RbH_2PO_4	147	–	15.5
Rubidium dihydrogen arsenate (RDA)	RbH_2AsO_4	110	–	13.0
Rubidium dideuterium arsenate (RD*A)	RbD_2AsO_4	178	–	21.4

Tetragonal (4mm)	Formula	T_{tran} K	r_{13} pm/V	r_{33} pm/V	r_{51} pm/V
Barium titanate	$BaTiO_3$	406	8	28	–
Potassium lithium niobate	$K_3Li_2Nb_5O_{15}$	693	8.9	5.9	–
Lead titanate	$PbTiO_3$	765	13.8	5.9	–
Strontium barium niobate (SBN75)	$Sr_{0.75}Ba_{0.25}Nb_2O_6$	330	6.7	1340	42
Strontium barium niobate (SBN46)	$Sr_{0.46}Ba_{0.54}Nb_2O_6$	602	~180	35	–

Hexagonal (6mm)	Formula	r_{13} pm/V	r_{33} pm/V	r_{42} pm/V	r_{51} pm/V
Greenockite	CdS	3.1	2.9	2.0	3.7
Greenockite (const. strain)	CdS	1.1	2.4	–	–
Wurzite	ZnS	0.9	1.8	–	–
Zincite	ZnO	−1.4	+2.6	–	–

Hexagonal (6)	Formula	r_{13} pm/V	r_{33} pm/V	r_{42} pm/V	r_{51} pm/V
Lithium iodate	$LiIO_3$	4.1	6.4	1.4	3.3
Lithium potassium sulfate	$LiKSO_4$	$r_{13}-r_{33} = 1.6$	–	–	–

Trigonal (3m)	Formula	T_{tran} K	r_{13} pm/V	r_{22} pm/V	r_{33} pm/V	r_{42} pm/V
Cesium nitrate	$CsNO_3$	425	–	0.43	–	–
Lithium niobate	$LiNbO_3$	1483	8.6	7.0	30.8	28
Lithium tantalate	$LiTaO_3$	890	8.4	–	30.5	–
Lithium sodium sulfate	$LiNaSO_4$	–	–	<0.02	–	–
Tourmaline	–	–	–	0.3	–	–

Trigonal (32)	Formula	T_{tran} K	r_{11} pm/V	r_{41} pm/V
Cesium tartrate	$Cs_2C_4H_4O_6$	–	1.0	–
Cinnabar	HgS	659	3.1	1.5
Potassium dithionate	$K_2S_2O_6$	–	0.26	–
Strontium dithionate	$SrS_2O_6 \cdot 4H_2O$	–	0.1	–
Quartz	SiO_2	1140	−0.47	0.2
Selenium	Se	398	2.5	

Orthorhombic (222)	Formula	T_{tran} K	r_{41} pm/V	r_{52} pm/V	r_{63} pm/V
Ammonium oxalate	$(NH_4)_2C_2O_4 \cdot 4H_2O$	–	230	330	250
Rochelle salt	$KNaC_4H_4O_6 \cdot 4H_2O$	$T_u = 297$ $T_1 = 255$	−2.0	−1.7	+0.32

Orthorhombic (mm2)	Formula	Ttrans K	r_{13} pm/V	r_{23} pm/V	r_{33} pm/V	r_{42} pm/V	r_{51} pm/V
Barium sodium niobate (BSN)	Ba_2NaNbO_{15}	833	15	13	48	92	90
Potassium niobate	$KNbO_3$	476	28	1.3	64	380	105

Monoclinic (2)	Formula	T_{trans} K	r_{22} pm/V	r_{32} pm/V
Calcium pyroniobate	$Ca_2Nb_2O_7$	–	0.33	13.7
Triglycine sulfate (TGS)	$(NH_2CH_2COOH)_3 \cdot H_2SO_4$	322	7.2	13.6

References

1. Narasimhamurty, T. S., *Photoelastic and Electro–Optic Properties of Crystals*, Plenum Press, New York, 1981, pp. 405–407.
2. Weber, M. J., Ed., *CRC Handbook of Laser Science and Technology*, Vol. IV, CRC Press, Boca Raton, FL, 1986, pp. 258–278.

QUADRATIC ELECTRO–OPTIC COEFFICIENTS

Kerr Constants of Ferroelectric Crystals[1,2]

Name	Formula	T_{tran} K	λ μm	g_{11} 10^{10} esu	g_{12} 10^{10} esu	$g_{11}-g_{12}$ 10^{10} esu	g_{44} 10^{10} esu
Barium titanate	$BaTiO_3$	406	0.633	1.33	−0.11	1.44	
Strontium titanate	$SrTiO_3$	–	0.633	–	–	1.56	–
Potassium tantalate niobate	$KTa_{0.65}Nb_{0.35}O_3$	330	0.633	1.50	−0.42	1.92	1.63
Potassium tantalate	$KTaO_3$	13	0.633	–	–	1.77	1.33
Lithium niobate	$LiNbO_3$	1483	–	0.94	0.25	0.7	0.6
Lithium tantalate	$LiTaO_3$	938	–	1.0	0.17	0.8	0.7
Barium sodium niobate (BSN)	$Ba_{0.8}Na_{0.4}Nb_2O_6$	833	–	1.55	0.44	1.11	

Kerr Constants of Selected Liquids[2]

K is the Kerr constant at a wavelength of 589 nm and at room temperature; ε is the static dielectric constant; t_m is the melting point; and t_b is the normal boiling point

Name	Molecular formula	K 10^{-7} esu	ε	t_m °C	t_b °C
Carbon disulfide	CS_2	+3.23	2.63	−111.5	+46.3
Acetone	C_3H_6O	+16.3	21.0	−94.8	+56.1
Methyl ethyl ketone	C_4H_8O	+13.6	18.56	−86.67	+79.6
Pyridine	C_5H_5N	+20.4	13.26	−42	+115.23
Ethyl cyanoacetate	$C_5H_7NO_2$	+38.8	31.6	−22.5	205
o-Dichlorobenzene	$C_6H_4Cl_2$	+42.6	10.12	−16.7	180
Benzenesulfonyl chloride	$C_6H_5ClO_2S$	+89.9	28.90	+14.5	247
Nitrobenzene	$C_6H_5NO_2$	+326	35.6	+5.7	210.8
Ethyl 3-aminocrotonate	$C_6H_{11}NO_2$	+31.0	–	+33.9	210
Paraldehyde	$C_6H_{12}O_3$	−23.0	14.7 12.0[a]	+12.6	124
Benzaldehyde	C_7H_6O	+80.8	17.85 14.1[a]	−26	179.05
p-Chlorotoluene	C_7H_7Cl	+23.0	6.25	+7.5	162.4
o-Nitrotoluene	$C_7H_7NO_2$	+174	26.26	−10	222.3
m-Nitrotoluene	$C_7H_7NO_2$	+177	24.95	+15.5	232
p-Nitrotoluene	$C_7H_7NO_2$	+222	22.2	+51.6	238.3
Benzyl alcohol	C_7H_8O	−15.4	11.92 10.8[a]	−15.3	205.8
m-Cresol	C_7H_8O	+21.2	12.44 5.0[a]	+11.8	202.27
m-Chloroacetophenone	C_8H_7ClO	+69.1			
Acetophenone	C_8H_8O	+66.6	17.44 15.8[a]	+19.7	202.3
Quinoline	C_9H_7N	+15.0	9.16	−14.78	237.16
Ethyl salicylate	$C_9H_{10}O_3$	+19.6	8.48	+1.3	231.5
Carvone	$C_{10}H_{14}O$	+23.6	11.2	<0	230
Ethyl benzoylacetate	$C_{11}H_{12}O_3$	+16.0	13.50	<0	270
Water	H_2O	+4.0	80.10	0.00	100.0

[a] Dielectric constant at radio frequencies (108–109 Hz).

References

1. Narasimhamurty, T. S., *Photoelastic and Electro-Optic Properties of Crystals*, Plenum Press, New York, 1981, p. 408.
2. Gray, D. E., Ed., *AIP Handbook of Physics*, McGraw Hill, New York, 1972, p. 6–241.

MAGNETO–OPTIC CONSTANTS

Verdet Constants of Non-Magnetic Crystals[1]

V is the Verdet constant; n is the refractive index; and λ is the wavelength.

Material	T K	λ nm	n	V min/Oe cm
Al_2O_3	300	546.1	1.771	0.0240
	300	589.3	1.768	0.0210
$BaTaO_3$	403	427		0.95
	403	496		0.38
	403	620		0.18
	403	826		0.072
$Bi_4Ge_3O_{12}$	300	442	2.077	0.289
	300	632.8	2.048	0.099
	300	1064	2.031	0.026
C (diamond)	300	589.3	2.417	0.0233
$CaCO_3$	300	589.3	1.658	0.019
CaF_2	300	589.3	1.434	0.0088
$Cd_{0.55}Mn_{0.45}Te$	300	632.8		6.87
CuCl	300	546.1	1.93	0.20
GaSe	298	632.8		0.80
$KAl(SO_4)_2 \cdot 12H_2O$	300	589.3	1.456	0.0124
KBr	300	546.1	1.564	0.0500
	300	589.3	1.560	0.0425
KCl	300	589.3	1.490	0.0275
KI	300	546.1	1.673	0.083
	300	589.3	1.666	0.070
$KTaO_3$	296	352		0.44
	296	413		0.19
	296	496		0.096
	296	620		0.051
	296	826		0.022
LaF_3	300	325	1.639	0.054
(H∥c)	300	442	1.615	0.028
	300	632.8	1.601	0.012
	300	1064	1.592	0.006
$MgAl_2O_4$	300	589.3	1.718	0.021
$NH_4AlSO_4 \cdot 12H_2O$	300	589.3	1.459	0.0128
NH_4Br	300	589.3	1.711	0.0504
NH_4Cl	300	546.1		0.0410
	300	589.3	1.643	0.0362
NaBr	300	546.1		0.0621
NaCl	300	546.1		0.0410
	300	589.3	1.544	0.0345
$NaClO_3$	300	546.1		0.0105
	300	589.3	1.515	0.0081
$NiSO_4 \cdot 6H_2O$	297	546.1		0.0256
	297	589.3	1.511	0.0221
SiO_2	300	546.1	1.546	0.0195
	300	589.3	1.544	0.0166
$SrTiO_3$	298	413	2.627	0.78
	298	496		0.31
	298	620		0.14
	298	826		0.066
ZnS	300	546.1		0.287
	300	589.3	2.368	0.226
ZnSe	300	476	2.826	1.50
	300	496	2.759	1.04
	300	514	2.721	0.839
	300	587	2.627	0.529
	300	632.8	2.592	0.406

Verdet Constants of Rare-Earth Aluminum Garnets at Various Wavelengths[1]

The absorption coefficient α for these materials ranges from 0.2 to 0.6 cm^{-1} at 300 K.

Material	T/K	λ = 405 nm	450 nm	480 nm	520 nm	546 nm	578 nm	635 nm	670 nm
$Tb_2Al_5O_{12}$	300	−2.266	−1.565	−1.290	−1.039	−0.912	−0.787	−0.620	−0.542
	77		−102.16	−83.45	−3.425	−3.051	−2.603	−2.008	−1.815
	4.2				−64.80	−58.35	−53.77	48.39	−45.15
	1.45		−200.95	−172.52	−139.28	−125.07	−111.27	97.47	−93.42
$Dy_3Al_5O_{12}$	300	−1.241	−0.942	−0.803	−0.667	−0.592	−0.518	−0.411	−0.359
$Ho_3Al_5O_{12}$	300	−0.709	−0.320	−0.260	−0.335	−0.304	−0.299		−0.206
$Er_3Al_5O_{12}$	300	−0.189	−0.240	−0.154	−0.162	−0.157	−0.145	−0.105	−0.089
$Tm_3Al_5O_{12}$	300	+0.151	+0.103	+0.093	0.076	0.069	+0.059	+0.048	
$Yb_3Al_5O_{12}$	298	0.287	0.215	0.186	0.140	0.133	0.116	0.094	
	77	0.718	0.540	0.481	0.393	0.342	0.302	0.239	

The header spanning label: *V* in min/Oe cm

Verdet Constants for KDP-Type Crystals[1]

Measurements refer to T = 298 K and λ = 632.8 nm, with $k \parallel$ [001].

Material	V min/Oe cm
KH_2PO_4 (KDP)	0.0124
$KH_{0.3}D_{1.7}PO_4$ (KD*P)	0.145
$NH_4H_2PO_4$ (ADP)	0.138
KH_2AsO_4 (KDA)	0.238
$KH_{0.1}D_{1.9}AsO_4$ (KD*A)	0.245
$NH_4H_2AsO_4$ (ADH)	0.244

Verdet Constants of Gases[2]

Values refer to T = 0°C and P = 101.325 kPa (760 mmHg); n_D is the refractive index at a wavelength of 589 nm.

Gas	$(n_D - 1) \times 10^3$	$10^6 \times V$ min/Oe cm
He	0.036	+0.40
Ar	2.81	+9.36
H_2		+6.29
N_2	0.297	+6.46
O_2	0.272	+5.69
Air	0.293	+6.27
Cl_2	0.773	+31.9
HCl	0.447	+21.5
H_2S	0.63	+41.5
NH_3	0.376	+19.0
CO	0.34	+11.0
CO_2	0.45	+9.39
NO	0.297	−58
CH4	0.444	+17.4
$n\text{-}C_4H_{10}$		+44.0

Verdet Constants of Liquids[2]

n_D is the refractive index at a wavelength of 589 nm and a temperature of 20°C, unless otherwise indicated. V is the Verdet constant.

Liquid	λ/nm	t/°C	$10^2 \times V$ min/Oe cm	n_D
P	589	33	+13.3	
S	589	114	+8.1	1.929 (110°C)
H_2O	589	20	+1.309	1.3328
D_2O	589	19.7	+1.257	1.3384
H_3PO_4	578	97.4	+1.35	
CS_2	589	20	+4.255	1.6255
CCl_4	578–589	25.1	+1.60	1.463 (15°C)
$SbCl_5$	578	18	+7.45	1.601 (14°C)
$TiCl_4$	578	17	−1.65	1.61
$TiBr_4$	578	46	−5.3	
Methanol	589	18.7	+0.958	1.3289
Acetone	578–589	20.0	+1.116	1.3585
Toluene	578–589	15.0	+2.71	1.4950
Benzene	578–589	15.0	+3.00	1.5005
Chlorobenzene	589	15	+2.92	1.5246
Nitrobenzene	589	15	+2.17	1.5523
Bromoform	589	17.9	+3.13	1.5960

Verdet Constants of Rare Earth Paramagnetic Crystals[1]

n is the refractive index, and V is the Verdet constant at the wavelength and temperature indicated.

Rare Earth	Host	T/K	λ/nm	n	V min/Oe cm
Ce^{3+}(30%)	CaF_2	300	325	1.516	−0.956
		300	442	1.502	−0.297
		300	633	1.494	−0.111
		300	1064	1.489	−0.035
Ce^{3+}	CeF_3	300	442	1.613	−1.05
		300	633	1.598	−0.406
		77	633		−1.418
		300	1064	1.589	−0.113
Pr^{3+}(5%)	CaF_2	300	266	1.471	−0.172
		300	325	1.461	−0.0818
		300	442	1.451	−0.0089
		300	633	1.445	−0.0168
		300	1064	1.441	−0.0045
Nd^{3+}(2.9%)	CaF_2	4.2	426		−0.19
Nd^{3+}	NdF_3	300	442	1.60	−0.553
		290	633	1.59	−0.209
		77	633		−0.755
		300	1064	1.58	−0.097
Eu^{3+}(3%)	CaF_2	4.2	430		29
		4.2	440		22
Eu^{2+}	EuF_2	300	450		−4.5
		300	500		−2.6
		300	550		−1.6
		300	600		−1.1
		300	650		−0.8
		300	1064		−0.19
Tb^{3+}	KTb_3F_{10}	300	325	1.531	−2.174
		300	442	1.518	−0.933
		300	633	1.510	−0.386
		77	633		−1.94
		300	1064	1.505	−0.114
Tb^{3+}	$LiTbF_4$	300	325	1.493	−1.9
		300	442	1.481	−0.98
		300	633	1.473	−0.44
		300	1064	1.469	−0.13
Tb^{3+}	$Tb_3Ga_5O_{12}$	300	500	1.989	−0.749
		300	570	1.981	−0.581
		300	633	1.976	−0.461
		300	830	1.967	−0.21
		300	1060	1.954	−0.12

Verdet Constants of Paramagnetic Glasses[1]

The Verdet constant V is given at room temperature for the wavelengths indicated.

Rare earth phosphate glasses of composition $R_2O_3 \cdot xP_2O_5$, where x is given in the second column

R	x	$\lambda = 405$ nm	$\lambda = 436$ nm	$\lambda = 480$ nm	$\lambda = 500$ nm	$\lambda = 520$ nm	$\lambda = 546$ nm	$\lambda = 578$ nm	$\lambda = 600$ nm	$\lambda = 635$ nm	$\lambda = 670$ nm
						Verdet constant V in min/Oe cm					
La		0.037	0.030	0.024	0.022	0.020	0.018	0.015	−0.014	0.013	–
Ce	2.67	−0.672	0.510	−0.366	−0.326	−0.287	−0.253	−0.217	−0.197	−0.173	−0.150
Pr	3.09	−0.447	−0.332	−0.283	−0.261	−0.236	−0.208	−0.182	−0.170	−0.150	−0.132
Nd	2.92	−0.250	−0.209	−0.167	−0.155	−0.136	−0.134	−0.094	−0.080	−0.080	−0.071
Sm	2.87	0.026	0.024	0.020	0.020	0.017	0.015	0.014	0.012	0.011	0.010
Eu	2.93	−0.025	−0.017	−0.010	−0.006	−0.006	−0.005	−0.004	−0.003	−0.002	−0.002
Gd	3.01	0.018	0.015	0.014	0.012	0.012	0.011	0.011	0.010	0.009	0.009
Tb	2.94	−0.560	−0.458	−0.357	−0.323	−0.295	−0.261	−0.226	−0.206	−0.190	−0.164
Dy	2.51	−0.540	−0.453	−0.359	−0.331	−0.301	0.268	−0.237	−0.217	−0.197	−0.173
Ho	2.94	−0.299	−0.313	−0.156	−0.153	−0.138	−0.138	−0.119	−0.110	−0.098	−0.084
Er	3.01	−0.139	−0.121	−0.100	−0.111	−0.095	−0.062	−0.060	−0.057	−0.051	−0.044
Tm	2.79	0.019	0.013	0.012	0.009	0.008	0.006	0.005	0.004	0.004	0.007
Yb	3.01	0.087	0.072	0.056	0.050	0.045	0.041	0.036	0.032	0.029	0.024

The following are rare earth borate glasses with composition:

for La and Pr: $R_2O_3 \cdot xP_2O_5$; for Tb–Pr and Dy–Pr: $R_2O_3 \cdot xB_2O_3$; and for other elements: $R_2O_3 \cdot 0.85La_2O_3 \cdot xB_2O_3$.

R	x	$\lambda = 405$ nm	$\lambda = 436$ nm	$\lambda = 480$ nm	$\lambda = 500$ nm	$\lambda = 520$ nm	$\lambda = 546$ nm	$\lambda = 578$ nm	$\lambda = 600$ nm	$\lambda = 635$ nm	$\lambda = 670$ nm
La	3.04	0.043	0.036	0.029	0.026	0.023	0.022	0.019	0.018	0.016	0.014
Pr-La	5.44	−0.380	−0.307	−0.230	−0.220	−0.201	−0.178	−0.153	−0.146	−0.128	−0.110
Nd-La	5.41	−0.180	−0.147	−0.120	−0.111	−0.096	−0.094	−0.100	−0.059	−0.056	−0.046
Sm-La	4.97	0.032	0.030	0.025	0.024	0.022	0.019	0.017	0.016	0.014	0.012
Eu-La	4.69	−0.081	−0.060	−0.038	−0.033	−0.029	−0.024	0.019	−0.016	0.014	−0.012
Gd-La	4.71	0.032	0.026	0.024	0.022	0.021	0.020	0.018	0.017	0.015	0.013
Tb-La	4.73	−0.512	−0.419	−0.319	−0.288	−0.262	−0.234	−0.205	−0.186	−0.167	−0.142
Dy-La	4.88	−0.436	−0.361	−0.299	−0.273	−0.246	−0.220	−0.193	−0.177	−0.159	−0.138
Ho-La	4.36	−0.269	−0.252	−0.123	−0.131	−0.112	−0.128	−0.104	−0.096	–	−0.074
Er-La	4.50	−0.093	−0.078	−0.068	−0.082	–	−0.045	−0.042	−0.040	−0.035	−0.034
Tm-La	4.75	0.060	0.046	0.039	0.034	0.031	0.026	0.023	0.021	0.018	0.016
Yb-La	8.58	0.115	0.094	0.073	0.066	0.060	0.054	0.046	0.043	0.037	0.033
Tb-Pr	4.99	−0.940	−0.786	−0.560	−0.536	−0.489	−0.436	−0.380	−0.348	−0.306	−0.265
Dy-Pr	4.63	−0.850	–	–	−0.497	−0.465	−0.413	−0.358	−0.332	−0.290	−0.252
Pr	2.56	−0.843	−0.646	−0.471	−0.480	−0.432	−0.390	−0.334	−0.317	−0.271	−0.243

Verdet Constants of Diamagnetic Glasses[1]

The Verdet constant V is given at room temperature for the wavelengths indicated.

Glass type	Composition (wt. %)	Verdet constant V in min/Oe cm			
		$\lambda = 325$ nm	$\lambda = 442$ nm	$\lambda = 633$ nm	$\lambda = 1064$ nm
SiO_2	100% SiO_2			0.013	
B_2O_3	100% B_2O_3			0.010	
CdO	47.5% CdO, 52.5% P_2O_5	0.079	0.033	0.022	
ZnO	36.4% ZnO, 63.6% P_2O_5	0.072	0.044	0.020	
TeO_2	88.9% TeO_2, 11.1% P_2O_5		0.196	0.076	0.022
ZrF_4	63.1% ZrF_4, 14.9% BaF_2, 7.2% LaF_3, 1.9% AlF_3, 9.1% PbF_2, 3.8% LiF			0.011	

		$\lambda = 700$ nm	$\lambda = 853$ nm	$\lambda = 1060$ nm
Bi_2O_3	95% Bi2O3, 5% B2O3	0.086	0.051	0.033
PbO	95% PbO, 5% B2O3	0.093	0.061	0.031
	82% PbO, 18% SiO2	0.077	0.045	0.027
	50% PbO, 15% K2O, 35% SiO2	0.032	0.020	0.011
Tl_2O	95% Tl2O, 5% B2O3	0.092	0.061	0.032
	82% Tl2O, 18% SiO2	0.100	0.067	0.043
	50% Tl2O, 15% K2O, 35% SiO2	0.036	0.022	0.012
SnO	76% SnO, 13% B2O3, 11% SiO2	0.071	0.046	0.026
TeO_3	75% TeO2, 25% Sb2O3	0.076	0.052	0.032
	80% TeO2, 20% ZnCl2	0.073	0.046	0.025
	84% TeO2, 16% BaO	0.056	0.041	0.029
	70% TeO2, 30% WO3	0.052	0.035	0.022
	20% TeO2, 80% PbO	0.128	0.075	0.048
Sb_2O_3	25% Sb2O3, 75% TeO2	0.076	0.050	0.032
	75% Sb2O3, 75% Cs2O, 5% Al2O3	0.074	0.044	0.025
	75% Sb2O3, 10% Cs2O, 10% Rb2O, 5% Al2O3	0.078	0.052	0.030

Verdet Constants of Commercial Glasses[1]

This table gives the density, ρ, refractive index at 589 nm, n_D, and Verdet constant, V, for the wavelengths indicated; the data refer to room temperature.

Glass type	ρ g/cm^3	n_D	V in min/Oe cm				
			$\lambda = 365.0$ nm	$\lambda = 404.7$ nm	$\lambda = 435.8$ nm	$\lambda = 546.1$ nm	$\lambda = 578.0$ nm
BSC	2.49	1.5096	0.0499	0.0392	0.0333	0.02034	0.01798
HC	2.53	1.5189	0.0561	0.0440	0.0372	0.0225	0.01995
LBC	2.87	1.5406	0.0609	0.0477	0.0403	0.0245	0.0216
LF	3.23	1.5785	0.1143	0.0850	0.0693	0.0394	0.0344
BLF	3.48	1.6047	0.1112	0.0832	0.0685	0.0393	0.0344
DBC	3.56	1.6122	0.0662	0.0517	0.0435	0.0261	0.0231
DF	3.63	1.6203	0.1473	0.1076	0.0872	0.0485	0.0423
EDF	3.9	1.6533	0.1725	0.1248	0.1007	0.0556	0.0483

The composition of the glasses in weight percent is:

Glass type	SiO_2	B_2O_3	K_2O	CaO	Al_2O_3	As_2O_3	Na_2O	BaO	ZnO	PbO
BSC	69.6	6.7	20.5	2.9	0.3	0.1	–	–	–	–
HC	72.0	–	10.1	11.4	0.3	0.2	6.1	–	–	–
LBC	57.1	1.8	13.7	0.3	0.2	0.1	–	26.9	–	–
LF	52.5	–	9.5	0.3	0.2	0.1	–	–	–	37.6
BLF	45.2	–	7.8	–	–	0.4	–	16.0	8.3	22.2
DBC	36.2	7.7	0.2	0.2	3.5	0.7	–	44.6	6.7	–
DF	46.3	–	1.1	0.3	0.2	0.1	5.0	–	–	47.0
EDF	40.6	–	7.5	0.2	0.2	0.2	0.1	–	–	51.5

References

1. Weber, M. J., *CRC Handbook of Laser Science and Technology*, Vol. IV, Part 2, CRC Press, Boca Raton, FL, 1988, pp. 299–310.
2. Gray, D. E., Ed., *American Institute of Physics Handbook*, Third edition, McGraw Hill, New York, 1972, p. 6–230.

FARADAY ROTATION

Ferro-, Ferri-, and Antiferromagnetic Solids

Material	T_c K	$4\pi M_s$ gauss	F deg/cm	α cm^{-1}	$2F/\alpha$ deg	T K	λ nm
Fe	1043	21,800	4.4×10^5	6.5×10^5	1.4	300	500
			6.5×10^5	5.0×10^5	2.6	300	1000
			7×10^5	4.2×10^5	3.3	300	1500
			7×10^5	3.5×10^5	4.0	300	2000
Co	1390	18,200	2.9×10^5	–	–	300	500
			5.5×10^5	6.1×10^5	1.8	300	1000
			5.5×10^5	4.5×10^5	2.4	300	1500
			5.5×10^5	3.6×10^5	2.7	300	2000
Ni	633	6,400	0.8×10^5	–	–	300	500
			2.6×10^5	5.8×10^5	0.9	300	1000
			1.5×10^5	4.8×10^5	0.6	300	1500
			1×10^5	4.1×10^5	0.25	300	2000
Permalloy (Ni/Fe = 82/18)	803	10,700	1.2×10^5	6×10^5	0.4	300	500
Ni/Fe = 100/0		6,000	1.2×10^5	7.05×10^5	0.34	300	632.8
Ni/Fe = 80/20		10,800	2.2×10^5	7.10×10^5	0.62	300	632.8
Ni/Fe = 60/40		14,900	2.9×10^5	7.54×10^5	0.77	300	632.8
Ni/Fe = 40/60		14,400	2.2×10^5	8.17×10^5	0.54	300	632.8
Ni/Fe = 20/80		19,400	3.3×10^5	8.10×10^5	0.81	300	632.8
Ni/Fe = 0/100	639	21,600	3.5×10^5	8.13×10^5	0.86	300	632.8
MnBi		7,700	4.2×10^5	6.1×10^5	1.4	300	450
			7.5×10^5	4.2×10^5	3.6	300	900
MnAs	313	–	0.44×10^5	5.0×10^5	0.174	300	500
			0.62×10^5	4.4×10^5	0.28	300	900
CrTe	334	1015	0.5×10^5	2.0×10^5	0.5	300	550
			0.4×10^5	1.2×10^5	0.7	300	900
FeRh	333	–	0.9×10^5	3.3×10^5	0.56	348	700
$Y_3Fe_5O_{12}$ (YIG)	560	2500	2400	1500	3.2	300	555
			1250	1400	1.8	300	625
			750	450	3.3	300	770
			175	<0.06	$>3 \times 10^3$	300	5000 to 1500
$Gd_3Fe_5O_{12}$ (GdIG)	$T_n = 564$ $T = 286$	7300	−2000	6000	0.6	300	500
			−1050	900	2.3	300	600
			−300	100	6.0	300	800
			−80	70	2.3	300	1000
$NiFe_2O_4$	858	3350	2.0×10^4	5.9×10^4	0.7	300	286
			-1.0×10^4	10×10^4	0.2	300	500
			−120	38	6	300	1500
			+75	15	10	300	3000
			+110	32	7	300	5000
$CoFe_2O_4$	793	4930	2.75×10^4	12×10^4	0.5	300	286
			3.6×10^4	17×10^4	0.4	300	400
			-2.5×10^4	6×10^4	0.8	300	660
$MgFe_2O_4$	593-713[e]	1450[e]	−60	100	1	300	2500
			0	12	0	300	4000
			+35	6	11	300	6000
$Li_{0.5}Fe_{2.5}O_4$	863–953[e]	3240[e] to 3900	−440	150	6	300	1500
			+10	85	0.2	300	3000
			+110	44	5	300	5000
			+135	80	3	300	7000
$BaFe_{12}O_{19}$	723	–	−50	−38	3	300	2000
			+75	20	7.5	300	3000
			+150	20	15	300	5000
			+165	22	15	300	7000
$Ba_2Zn_2Fe_{12}O_{19}$	–	–	90	120	1.5	300	5000

Material	T_c K	$4\pi M_s$ gauss	F deg/cm	α cm^{-1}	$2F/\alpha$ deg	T K	λ nm
			75	65	2.0	300	7000
RbNiF$_3$	220	1250	360	35	20	77	450[a]
			70	10	14	77	600[a]
			310	70	9	77	800[a]
			75	25	6	77	1000[a]
RbNi$_{0.75}$Co$_{0.25}$F$_3$	109	–	180	9	40	77	600[b]
RbFeF$_3$	102	–	3400	7	900	82	300[c]
			1600	3	1100	82	400[c]
			620	1.5	830	82	600[c]
			300	2.5	240	82	800[c]
FeF$_3$	365	40 at 300 K	670	14	95	300	349[d]
			180	4.4	82	300	522.5[d]
CrCl$_3$	16.8	3880	2000	200	20	1.5	410
			–500	300	3	1.5	450
			–1000	70	30	1.5	590
CrBr$_3$	32.5	3390	3×10^5	3×10^3	200	1.5	478
			1.6×10^5	1.4×10^4	23	1.5	500
CrI$_3$	68	2690	1.1×10^5	6.3×10^3	35	1.5	970
			0.8×10^5	3×10^3	53	1.5	1000
FeBO$_3$	348	115 at 300 K	3200	140	45	300	500
			450	38	24	300	700
EuO	69	23700	-1.0×10^5	0.5×10^4	40	5	1100
			5×10^5	9.7×10^4	10	5	700
			0.5×10^5	7.8×10^4	1.3	5	500
			3×10^4	>0.5	~105	20	2500
			660	>1.0	1300	20	10600
EuS	16.3	–	-1.6×10^5	0	–	6	825
			-9.6×10^5	3.3×10^4	58	6	690
			$+5.5 \times 10^5$	1.2×10^5	9.2	6	563
EuSe	7.0	13,200	1.45×10^5	80	3600	4.2	750
			0.95×10^5	60	3170	4.2	800

[a] Measured along the C-axis (magnetic hard axis).

[b] Measured along the C-axis (magnetic easy axis).

[c] Measured along the C-axis ([100]-direction at room temperature).

[d] Strong natural birefringence interferes with the Faraday effect.

[e] Depends on heat treatment.

Reference

1. Weber, M. J., Ed., *CRC Handbook of Laser Science and Technology,* Vol. IV, Part 2, CRC Press, Boca Raton, FL, 1988, pp. 288–296.

NONLINEAR OPTICAL CONSTANTS

H. P. R. Frederikse

The relation between the polarization density P of a dielectric medium and the electric field E is linear when E is small, but becomes nonlinear as E acquires values comparable with interatomic electric fields (10^5 to 10^8 V/cm). Under these conditions the relation between P and E can be expanded in a Taylor's series

$$P = \varepsilon_0 \chi^{(1)} E + 2\chi^{(2)} E^2 + 4\chi^{(3)} E^3 + \cdots \tag{1}$$

where ε_0 is the permittivity of free space, while $\chi^{(1)}$ is the linear and $\chi^{(2)}$, $\chi^{(3)}$ etc. the nonlinear optical susceptibilities.

If we consider two optical fields, the first $E_j^{\omega_1}$ (along the j-direction at frequency ω_1) and the second $E_k^{\omega_2}$ (along the k-direction at frequency ω_2) one can write the second term of the Taylor's series as follows

$$P_i(\omega_1 \omega_2) = 2\chi_{ijk}^{\omega_3 = \omega_1 \pm \omega_2} E_j^{\omega_1} E_k^{\omega_2}$$

When $\omega_1 \neq \omega_2$ the (parametric) mixing of the two fields gives rise to two new polarizations at the frequencies $\omega_3 = \omega_1 + \omega_2$ and $\omega_3' = \omega_1 - \omega_2$. When the two frequencies are equal, $\omega_1 = \omega_2 = \omega$, the result is Second Harmonic Generation (SHG): $\chi_{ijk}(2\omega, \omega, \omega)$, while equal and opposite frequencies, $\omega_1 = \omega$ and $\omega_2 = -\omega$ leads to Optical Rectification (OR): $\chi_{ijk}(0, \omega, -\omega)$. In the SHG case the following convention is adopted: the second order nonlinear coefficient d is equal to one half of the second order nonlinear susceptibility

$$d_{ijk} = 1/2\chi^{(2)}$$

Because of the symmetry of the indices j and k one can replace these two by a single index (subscript) m. Consequently the notation for the SHG nonlinear coefficient in reduced form is d_{im} where m takes the values 1 to 6. Only noncentrosymmetric crystals can possess a nonvanishing d_{ijk} tensor (third rank). The unit of the SHG coefficients is m/V (in the MKSQ/SI system).

In centrosymmetric media the dominant nonlinearity is of the third order. This effect is represented by the third term in the Taylor's series (Equation 1); it is the result of the interaction of a number of optical fields (one to three) producing a new frequency $\omega_4 = \omega_1 + \omega_2 + \omega_3$. The third order polarization is given by

$$P_j(\omega_1 \omega_2 \omega_3) = g_4 \chi_{jklm} E_k^{\omega_1} E_l^{\omega_2} E_m^{\omega_3}$$

Third Harmonic Generation (THG) is achieved when $\omega_1 = \omega_2 = \omega_3 = \omega$. In this case the constant $g_4 = 1/4$. The third order nonlinear coefficient C is related to the third order susceptibility as follows:

$$C_{jklm} = 1/4\chi_{jklm}$$

This coefficient is a fourth rank tensor. In the THG case the matrices must be invariant under permutation of the indices k, l, and m; as a result the notation for the third order nonlinear coefficient can be simplified to C_{jn}. The unit of C_{jn} is m$^2\cdot$V^{-2} (in the MKSQ/SI system).

Applications of second order nonlinear optical materials include the generation of higher (up to sixth) optical harmonics, the mixing of monochromatic waves to generate sum or difference frequencies (frequency conversion), the use of two monochromatic waves to amplify a third wave (parametric amplification) and the addition of feedback to such an amplifier to create an oscillation (parametric oscillation).

Third order nonlinear optical materials are used for THG, self-focusing, four wave mixing, optical amplification, and optical conjugation. Many of these effects – as well as the variation and modulation of optical propagation caused by mechanical, electric, and magnetic fields (see the preceeding table on "Elasto-Optic, Electro-Optic, and Magneto-Optic Constants") – are used in the areas of optical communication, optical computing, and optical imaging.

References

1. *Handbook of Laser Science and Technology*, Vol. 111, Part 1; Weber, M. J. Ed., CRC Press, Boca Raton, FL, 1986.
2. Dmitriev, V.G., Gurzadyan, G.G., and Nikogosyan, D., *Handbook of Nonlinear Optical Crystals*, Springer-Verlag, Berlin, 1991.
3. Shen, Y.R., *The Principles of Nonlinear Optics*, John Wiley, New York, 1984.
4. Yariv, A., *Quantum Electronics*, 3rd edition, John Wiley, New York, 1988.
5. Bloembergen, N., *Nonlinear Optics*, W.A. Benjamin, New York, 1965.
6. Zernike F. and Midwinter, J.E., *Applied Nonlinear Optics*, John Wiley, New York, 1973.
7. Hopf, F.A. and Stegeman, G.I., *Applied Classical Electrodynamics*, Volume 2: Nonlinear Optics, John Wiley, New York, 1986.
8. *Nonlinear Optical Properties of Organic Molecules and Crystals*, Chemla, D. S., and Zyss, J., Eds., Academic Press, Orlando, FL, 1987.
9. *Optical Phase Conjugation*, Fisher, R. A., Ed., Academic Press, New York, 1983.
10. Zyss, J., *Molecular Nonlinear Optics: Materials, Devices and Physics*, Academic Press, Boston, 1994.
11. Nonlinear Optics, 5 articles in *Physics Today, (Am. Inst. of Phys.)*, Vol. 47, No. 5, May, 1994.

Selected SHG Coefficients of NLO Crystals*

Material	Symmetry class	$d_{im} \times 10^{12}$ m/V	λ μm	Material	Symmetry class	$d_{im} \times 10^{12}$ m/V	λ μm
GaAs	$\bar{4}3\,m$	$d_{14} = 134.1 \pm 42$	10.6	AgGaSe$_2$	$\bar{4}2\,m$	$d_{36} = 37.4 \pm 6.0$	10.6
GaP	$\bar{4}3\,m$	$d_{14} = 71.8 \pm 12.3$	1.058	(NH$_2$)$_2$CO (urea)	$\bar{4}2\,m$	$d_{36} = 1.3$	1.06
InAs	$\bar{4}3\,m$	$d_{14} = 364 \pm 47$	1.058	AlPO$_4$	32	$d_{11} = 0.35 \pm 0.03$	1.058
		$d_{14} = 210$	10.6	Se	32	$d_{11} = 97 \pm 25$	10.6
ZnSe	$\bar{4}3\,m$	$d_{14} = 78.4 \pm 29.3$	10.6	Te	32	$d_{11} = 650 \pm 30$	10.6
		$d_{36} = 26.6 \pm 1.7$	1.058	SiO$_2$ (quartz)	32	$d_{11} = 0.335$	1.064
β-ZnS	$\bar{4}3\,m$	$d_{14} = 30.6 \pm 8.4$	10.6	HgS	32	$d_{11} = 50.3 \pm 17$	10.6
		$d_{36} = 20.7 \pm 1.3$	1.058	(C$_6$H$_5$CO)$_2$ [benzil]	32	$d_{11} = 3.6 \pm 0.5$	1.064
ZnTe	$\bar{4}3\,m$	$d_{14} = 92.2 \pm 33.5$	10.6	β-BaB$_2$O$_4$ [BBO]	3 m	$d_{22} = 2.22 \pm 0.09$	1.06
		$d_{14} = 83.2 \pm 8.4$	1.058			$d_{31} = 0.16 \pm 0.08$	1.06
		$d_{36} = 89.6 \pm 5.7$	1.058	LiNbO$_3$	3 m	$d_{33} = 34.4$	1.06
CdTe	$\bar{4}3\,m$	$d_{14} = 167.6 \pm 63$	10.6			$d_{31} = -5.95$	1.06
Bi$_4$GeO$_{12}$	$\bar{4}3\,m$	$d_{14} = 1.28$	1.064			$d_{22} = 2.76$	1.06
N$_4$(CH$_2$)$_6$ (hexamine)	$\bar{4}3\,m$	$d_{14} = 4.1$	1.06	LiTaO$_3$	3 m	$d_{33} = -16.4 \pm 2$	1.058
LiIO$_3$	6	$d_{33} = -7.02$	1.06			$d_{31} = -1.07 \pm 0.2$	1.058
		$d_{31} = -5.53 \pm 0.3$	1.064			$d_{22} = +1.76 \pm 0.2$	1.058
ZnO	6 mm	$d_{33} = -5.86 \pm 0.16$	1.058	Ag$_3$AsS$_3$ [proustite]	3 m	$d_{31} = 11.3 \pm 2.5$	10.6
		$d_{31} = 1.76 \pm 0.16$	1.058			$d_{22} = 18.0 \pm 2.5$	10.6
		$d_{15} = 1.93 \pm 0.16$	1.058	Ag$_3$SbS$_3$ [pyrargerite]	3m	$d_{31} = 12.6 \pm 4$	10.6
α-ZnS	6 mm	$d_{33} = 11.37 \pm 0.07$	1.058			$d_{22} = 13.4 \pm 4$	10.6
		$d_{33} = 37.3 \pm 12.6$	10.6	α-HIO$_3$	222	$d_{36} = 5.15 \pm 0.16$	1.064
		$d_{31} = -18.9 \pm 6.3$	10.6	NO$_2 \cdot$ CH$_3$NOC$_5$H$_4 \cdot$ (POM)	222	$d_{36} = 6.4 \pm 1.0$	1.064
		$d_{15} = 21.37 \pm 8.4$	10.6	Ba$_2$NaNb$_5$O$_{15}$ [Banana]	mm 2	$d_{33} = -17.6 \pm 1.28$	1.064
CdS	6 mm	$d_{33} = 25.8 \pm 1.6$	1.058			$d_{31} = -12.8 \pm 1.28$	1.064
		$d_{31} = -13.1 \pm 0.8$	1.058				
		$d_{15} = 14.4 \pm 0.8$	1.058	C$_6$H$_4$(NO$_2$)$_2$ [MDB]	mm 2	$d_{33} = 0.74$	1.064
CdSe	6 mm	$d_{33} = 54.5 \pm 12.6$	10.6			$d_{32} = 2.7$	1.064
		$d_{31} = -26.8 \pm 2.7$	10.6			$d_{31} = 1.78$	1.064
BaTiO$_3$	4 mm	$d_{33} = 6.8 \pm 1.0$	1.064	Gd$_2$(MoO$_4$)$_3$	mm 2	$d_{33} = -0.044 \pm 0.008$	1.064
		$d_{31} = 15.7 \pm 1.8$	1.064			$d_{32} = +2.42 \pm 0.36$	1.064
		$d_{15} = 17.0 \pm 1.8$	1.064			$d_{31} = -2.49 \pm 0.37$	1.064
PbTiO$_3$	4 mm	$d_{33} = 7.5 \pm 1.2$	1.064	KNbO$_3$	mm 2	$d_{33} = -19.58 \pm 1.03$	1.064
		$d_{31} = 37.6 \pm 5.6$	1.064			$d_{32} = +11.34 \pm 1.03$	1.064
		$d_{15} = 33.3 \pm 5$	1.064			$d_{31} = -12.88 \pm 1.03$	1.064
K$_3$Li$_2$Nb$_5$O$_{15}$	4 mm	$d_{33} = 11.2 \pm 1.6$	1.064				
		$d_{31} = 6.18 \pm 1.28$	1.064	KTiOPO$_4$ [KTP]	mm 2	$d_{33} = 13.7$	1.06
		$d_{15} = 5.45 \pm 0.54$	1.064			$d_{32} = \pm 5.0$	1.06
K$_{0.8}$Na$_{0.2}$Ba$_2$Nb$_5$O$_{15}$	4 mm	$d_{31} = 13.6 \pm 1.6$	1.064			$d_{31} = \pm 6.5$	1.06
SrBaNb$_5$O$_{15}$	4 mm	$d_{33} = 11.3 \pm 3.3$	1.064	NO$_2$C$_6$H$_4 \cdot$ NH$_2$ [mNA]	mm 2	$d_{33} = 13.12 \pm 1.28$	1.064
		$d_{31} = 4.31 \pm 1.32$	1.064			$d_{32} = 1.02 \pm 0.22$	1.064
		$d_{15} = 5.98 \pm 2$	1.064			$d_{31} = 12.48 \pm 1.28$	1.064
NH$_4$H$_2$PO$_4$ (ADP)	$\bar{4}2\,m$	$d_{36} = 0.53$	1.064	C$_{10}$H$_{12}$N$_3$O$_6$ [MAP]	2	$d_{23} = 10.67 \pm 1.3$	1.064
		$d_{36} = 0.85$	0.694			$d_{22} = 11.7 \pm 1.3$	1.064
KH$_2$PO$_4$ (KDP)	$\bar{4}2\,m$	$d_{36} = 0.44$	1.064			$d_{21} = 2.35 \pm 0.5$	1.064
		$d_{36} = 0.47 \pm 0.07$	0.694			$d_{25} = -0.35 \pm 0.3$	1.064
KD$_2$PO$_4$ (KD*P)	$\bar{4}2\,m$	$d_{36} = 0.38 \pm 0.016$	1.058	(NH$_2$CH$_2$COOH)$_3$H$_2$SO$_4$ [TGS]	2	$d_{23} = 0.32$	0.694
		$d_{36} = 0.34 \pm 0.06$	0.694				
		$d_{14} = 0.37$	1.058				
KH$_2$AsO$_4$ (KDA)	$\bar{4}2\,m$	$d_{36} = 0.43 \pm 0.025$	1.06				
		$d_{36} = 0.39 \pm 0.4$	0.694				
CdGeAs$_2$	$\bar{4}2\,m$	$d_{36} = 351 \pm 105$	10.6				
AgGaS$_2$	$\bar{4}2\,m$	$d_{36} = 18 \pm 2.7$	10.6				

* These data are taken from References 1 and 2.

Selected THG Coefficients of Some NLO Materials*

Material	NLO process	$C_{jn} \times 10^{20}$ m^2/V^{-2}	λ μm
$NH_4H_2PO_4$ [ADP]	$(-3\omega,\omega,\omega,\omega)$	$C_{11} = 0.0104$	1.06
		$C_{18} = 0.0098$	1.06
C_6H_6 [benzene]	$(-3\omega,\omega,\omega,\omega)$	$C_{11} = 0.0184 \pm 0.0042$	1.89
$CdGeAs_2$	$(-3\omega,\omega,\omega,\omega)$	$C_{11} = 182 \pm 84$	10.6
p-type: 5×10^{16} cm^{-3}		$C_{16} = 175$	10.6
		$C_{18} = -35$	10.6
$C_{40}H_{56}$ [β-carotene]	$(-3\omega,\omega,\omega,\omega)$	C_{11} 0.263 ± 0.08	1.89
GaAs high-resistivity	$(-3\omega,\omega,\omega,-\omega)$	$C_{11} = 62 \pm 31$	1.06
Ge	$(-3\omega,\omega,\omega,-\omega)$	$C_{11} = 23.5 \pm 12$	1.06
$LiIO_3$	$(-3\omega,\omega,\omega,-\omega)$	$C_{12} = 0.2285$	1.06
		$C_{35} = 6.66 \pm 1$	1.06
KBr	$(-3\omega,\omega,\omega,-\omega)$	$C_{11} = 0.0392$	1.06
		$C_{18}/C_{11} = 0.3667$	1.06
KCl	$(-3\omega,\omega,\omega,-\omega)$	$C_{11} = 0.0168$	1.06
		$C_{18}/C_{11} = 0.28$	1.06
KH_2PO_4 [KDP]	$(-3\omega,\omega,\omega,-\omega)$	$C_{11}-3C_{18} = 0.04$	1.06
Si p-type: 10^{14} cm^{-3}	$(-3\omega,\omega,\omega,-\omega)$	$C_{11} = 82.8 \pm 25$	1.06
NaCl	$(-3,\omega,\omega,\omega,-\omega)$	$C_{11} = 0.0168$	1.06
		$C_{18}/C_{11} = 0.4133$	1.06
NaF	$(-3\omega,\omega,\omega,-\omega)$	$C_{11} = 0.0035$	1.06

* These data are taken from Reference 1.

PHASE DIAGRAMS

H. P. R. Frederikse

A phase is a structurally homogeneous portion of matter. Regardless of the number of chemical constituents of a gas, there is only one vapor phase. This is true also for the liquid form of a pure substance, although a mixture of several liquid substances may exist as one or several phases, depending on the interactions among the substances. On the other hand a pure solid may exist in several phases at different temperatures and pressures because of differences in crystal structure (Reference 1). At the phase transition temperature, T_{tr}, the chemical composition of the solid remains the same, but often a change in the physical properties will take place. Such changes are found in ferroelectric crystals (example $BaTiO_3$) which develop a spontaneous polarization below T_{tr}, in superconductors (example Pb) which loose all electrical resistance below the transition point, and in many other classes of solids.

In quite a few cases it is difficult to bring about the phase transition, and the high- (or low-) temperature phase persists in its metastable form. Many liquids remain in the liquid state for shorter or longer periods of time when cooled below the melting point (supercooling). However, often the slightest disturbance will cause solidification. Persistence of the high temperature phase in solid-solid transitions is usually of much longer duration. An example of this behavior is found in white tin; although gray tin is the thermodynamically stable form below T_{tr} (286.4 K), the metal remains in its undercooled, white tin state all the way to $T = 0$ K, and crystals of gray tin are very difficult to produce.

A *phase diagram* is a map which indicates the areas of stability of the various phases as a function of external conditions (temperature and pressure). Pure materials, such as mercury, helium, water, and methyl alcohol are considered one-component systems and they have *unary* phase diagrams. The equilibrium phases in two-component systems are presented in *binary* phase diagrams. Because many important materials consist of three, four, and more components, many attempts have been made to deduce their multicomponent phase diagrams. However, the vast majority of systems with three or more components are very complex, and no overall maps of the phase relationships have been worked out.

It has been shown during the last 20 to 25 years that very useful partial phase diagrams of complex systems can be obtained by means of thermodynamic modeling (References 2, 3). Especially for complicated, multicomponent alloy systems the CALPHAD method has proved to be a successful approach for producing valuable portions of very intricate phase diagrams (Reference 4). With this method thermodynamic descriptions of the free energy functions of various phases are obtained which are consistent with existing (binary) phase diagram information and other thermodynamic data. Extrapolation methods are then used to extend the thermodynamic functions into a ternary system. Comparison of the results of this procedure with available experimental data is then used to fine-tune the phase diagram and add ternary interaction functions if necessary. In principle this approximation strategy can be extended to four, five, and more component systems.

The nearly two dozen phase diagrams shown below present the reader with examples of some important types of single and multicomponent systems, especially for ceramics and metal alloys. This makes it possible to draw attention to certain features like the kinetic aspects of phase transitions (see Figure 22, which presents a time-temperature-transformation, or TTT, diagram for the precipitation of α-phase particles from the β-phase in a Ti-Mo alloy; Reference 1, pp. 358–360). The general references listed below and the references to individual figures contain phase diagrams for many additional systems.

General References

1. Ralls, K. M., Courtney, T. H., and Wulff, J., *Introduction to Materials Science and Engineering*, Chapters 16 and 17, John Wiley & Sons, New York, 1976.
2. Kaufman, L., and Bernstein, H., *Computer Calculation of Phase Diagrams*, Academic Press, New York, 1970.
3. Kattner, U. R., Boettinger, W. J. B., and Coriell, S. R., *Z. Metallkd.*, 87, 9, 1996.
4. Dinsdale, A. T., Ed., *CALPHAD*, Vol. 1–20, Pergamon Press, Oxford, 1977–1996 and continuing.
5. Baker, H., Ed., *ASM Handbook, Volume 3: Alloy Phase Diagrams*, ASM International, Materials Park, OH, 1992.
6. Massalski, T. B., Ed., *Binary Alloy Phase Diagrams, Second Edition*, ASM International, Materials Park, OH, 1990.
7. Roth. R. S., Ed., *Phase Diagrams for Ceramists*, Vol. I (1964) to Volume XI (1995), American Ceramic Society, Waterville, OH.

References to Individual Phase Diagrams

Figure 1. Carbon: Reference 7, Vol. X (1994), Figure 8930. Reprinted with permission.

Figure 2. Si-Ge : Ref. 5, p. 2.231. Reprinted with permission.

Figure 3. H_2O (ice): See figure.

Figure 4. SiO_2: Reference 7, Vol. XI (1995), Figure 9174. Reprinted with permission.

Figure 5. Fe-O: Darken, L.S., and Gurry, R.W., *J. Am. Chem. Soc.*, 68, 798, 1946. Reprinted with permission.

Figure 6. Ti-O: Reference 5, p. 2.324. Reprinted with permission.

Figure 7. $BaO-TiO_2$: Reference 7, Vol. III (1975), Figure 4302. Reprinted with permission.

Figure 8. $MgO-Al_2O_3$: Reference 7, Vol. XI (1995), Figure 9239. Reprinted with permission.

Figure 9. $Y_2O_3-ZrO_2$: Reference 7, Vol. XI (1995), Figure 9348. Reprinted with permission.

Figure 10. Si-N-Al-O (Sialon): Reference 7, Vol. X (1994), Figure 8759. Reprinted with permission.

Figure 11. $PbO-ZrO_2-TiO_2$ (PZT): Reference 7, Vol. III (1975), Figure 4587. Reprinted with permission.

Figure 12. Al-Si-Ca-O: Reference 7 (1964), Vol. I, Figure 630. Reprinted with permission.

Figure 13. Y-Ba-Cu-O: Whitler, J.D., and Roth, R.S., *Phase Diagrams for High T_c Superconductors*, Figure S-082, American Ceramic Society, Waterville, OH, 1990. Reprinted with permission.

Figure 14. Al-Cu: Reference 5, p. 2.44. Reprinted with permission.

Figure 15. Fe-C: Ralls, K.M., Courtney, T.H., and Wulff, J., *Introduction to Materials Science and Engineering*, Figure 16.13 , John Wiley & Sons, New York, 1976. Reprinted with permission.

Figure 16. Fe-Cr: Reference 5, p. 2.152. Reprinted with permission.

Figure 17. Cu-Sn: Reference 5, p. 2.178. Reprinted with permission.

Figure 18. Cu-Ni: Reference 5, p. 2.173. Reprinted with permission.

Figure 19. Pb-Sn (solder): Reference 5, p. 2.335. Reprinted with permission.

Figure 20. Cu-Zn (brass): Subramanian, P.R., Chakrabarti, D.J., and Laughlin, D.E., Editors, *Phase Diagrams of Binary Copper Alloys*, p. 487, ASM International, Materials Park, OH, 1994. Reprinted with permission.

Figure 21. Co-Sm: Reference 5, p. 2.148. Reprinted with permission.

Figure 22. Ti-Mo: Reference 5, p. 2.296; Reference 1, p. 359. Reprinted with permission.

Figure 23: Fe-Cr-Ni: Reference 5, Figure 48. Reprinted with permission.

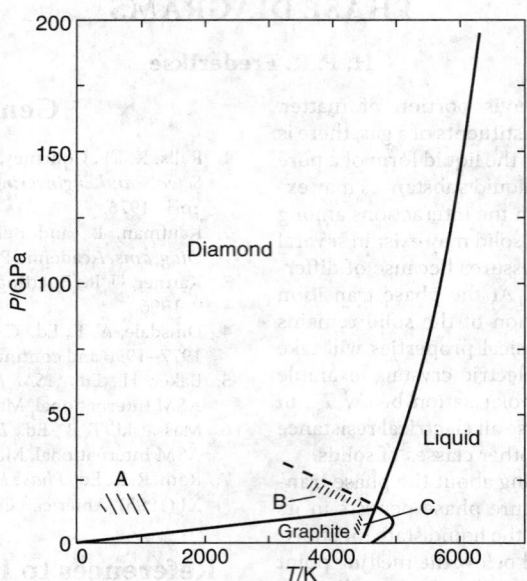

FIGURE 1.   Phase diagram of carbon. (A) Martensitic transition: hex graphite → hex diamond. (B) Fast graphite-to-diamond transition. (C) Fast diamond-to-graphite transition.

FIGURE 2.   Si-Ge system.

Phase	Composition, mass % Si	Pearson symbol	Space group
(Ge,Si)	0 to 100	cF8	Fd$\bar{3}$m
High-pressure phases			
GeII	—	tI4	I4$_1$/amd
SiII	—	tI4	I4$_1$/amd

FIGURE 3.   Diagram of the principal phases of ice. Solid lines are measured boundaries between stable phases; dotted lines are extrapolated. Ice IV is a metastable phase which exists in the region of ice V. Ice IX exists in the region below −100°C and pressures in the range 200–400 MPa. Ice X exists at pressures above 44 GPa. See Table 1 for the coordinates of the triple points, where liquid water is in equilibrium with two adjacent solid phases.

TABLE 1. Crystal Structure, Density, and Transition Temperatures for the Phases of Ice

Phase	Crystal system	Cell parameters	Z	n	ρ/g cm^{-3}	Triple ponts
Ih	Hexagonal	$a = 4.513$; $c = 7352$	4	4	0.93	I-III: −21.99°C, 209.9 MPa
Ic	Cubic	$a = 6.35$	8	4	0.94	
II	Rhombohedral	$a = 7.78$; $\alpha = 113.1°$	12	4	1.18	
III	Tetragonal	$a = 6.73$; $c = 6.83$	12	4	1.15	III-V: −16.99°C, 350.1 MPa
IV	Rhombohedral	$a = 7.60$; $\alpha = 70.1°$	16	4	1.27	
V	Monoclinic	$a = 9.22$; $b = 7.54$, $c = 10.35$; $\beta = 109.2°$	28	4	1.24	V-VI: 0.16°C, 632.4 MPa
VI	Tetragonal	$a = 6.27$; $c = 5.79$	10	4	1.31	VI-VII: 82°C, 2216 MPa
VII	Cubic	$a = 3.41$	2	8	1.56	
VIII	Tetragonal	$a = 4.80$; $c = 6.99$	8	8	1.56	
IX	Tetragonal	$a = 6.73$; $c = 6.83$	12	4	1.16	
X	Cubic	$a = 2.83$	2	8	2.51	

References

1. Wagner, W., Saul, A., and Pruss, A., *J. Phys. Chem. Ref. Data*, 23, 515, 1994.
2. Lerner, R.G. and Trigg, G.L., Eds., *Encyclopedia of Physics*, VCH Publishers, New York, 1990.
3. Donnay, J.D.H. and Ondik, H.M, *Crystal Data Determinative Tables, Third Edition, Volume 2, Inorganic Compounds*, Joint Committee on Powder Diffraction Standards, Swarthmore, PA, 1973.
4. Hobbs, P.V., *Ice Physics*, Oxford University Press, Oxford, 1974.
5. Glasser, L., *J. Chem. Edu.*, 81, 414, 2004.

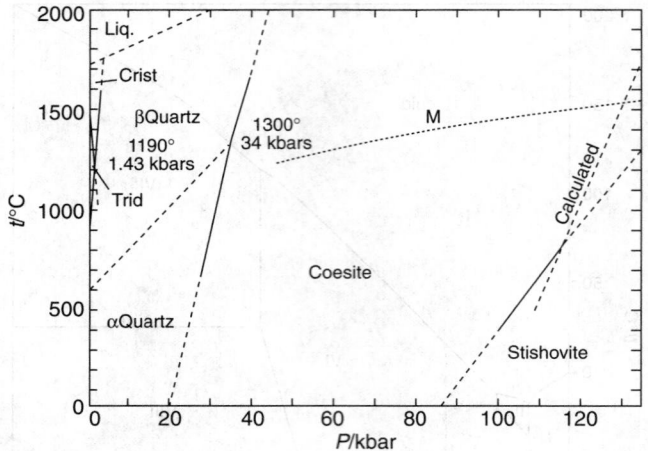

FIGURE 4.   SiO₂ system. Crist = cristobalite; Trid = tridymite.

FIGURE 5.   Fe-O system.

Point	$t/°C$	% O	p_{CO_2}/p_{CO}	Point	$t/°C$	% O	p_{CO_2}/p_{CO}	p_{O_2}/atm
A	1539			Q	560	23.26	1.05	
B	1528	0.16	0.209	R	1583	28.30		1
C	1528	22.60	0.209	R'	1583	28.07		1
G	1400[a]	22.84	0.263	S	1424	27.64	16.2	
H	1424	25.60	16.2	V	1597	27.64		0.0575
I	1424	25.31	16.2	Y	1457	28.36		1
J	1371	23.16	0.282	Z	1457	30.04		1
L	911[a]	23.10	0.447	Z'		30.6		
N	1371	22.91	0.282					

[a] Values for pure iron.

FIGURE 6.   Ti-O system.

Phase	Composition, mass % O	Pearson symbol	Space group
(βTi)	0 to 3	cI2	$Im\bar{3}m$
(αTi)	0 to 13.5	hP2	$P6_3/mmc$
Ti_3O	~8 to ~13	hP~16	$P\bar{3}c$
Ti_2O	~10 to 14.4	hP3	$P\bar{3}m1$
γTiO	15.2 to 29.4	cF8	$Fm\bar{3}m$
Ti_3O_2	~18	hP~5	$P6/mmm$
βTiO	~24 to 29.4	c**	–
αTiO	~25.0	mC16	$A2/m$ or $B*/*$
$βTi_{1-x}O$	~29.5	oI12	$I222$
$αTi_{1-x}O$	~29.5	tI18	$I4/m$
$βTi_2O_3$	33.2 to 33.6	hR30	$R\bar{3}c$
$αTi_2O_3$	33.2 to 33.6	hR30	$R\bar{3}c$
$βTi_3O_5$	35.8	m**	–
$αTi_3O_5$	35.8	mC32	$C2/m$
$α'Ti_3O_5$	35.8	mC32	Cc
$γTi_4O_7$	36.9	aP44	$P\bar{1}$
$βTi_4O_7$	36.9	aP44	$P\bar{1}$
$αTi4O_7$	36.9	aP44	$P\bar{1}$
$γTi_5O_9$	37.6	aP28	$P\bar{1}$
$βTi_6O_{11}$	38.0	aC68	$A\bar{1}$
Ti_7O_{13}	38.3	aP40	$P\bar{1}$
Ti_8O_{15}	38.5	aC92	$A\bar{1}$
Ti_9O_{17}	38.7	aP52	$P\bar{1}$
Rutile TiO_2	40.1	tP6	$P4_2/mnm$
Metastable phases			
Anatase	–	tI12	$I4_1/amd$
Brookite	–	oP24	$Pbca$
High-pressure phases			
TiO_2-II	–	oP12	$Pbcn$
TiO_2-III	–	hP~48	–

FIGURE 7.   BaO-TiO₂ system.

FIGURE 8.   MgO-Al₂O₃ system.

FIGURE 9. Y_2O_3-ZrO_2 system. C_{ss} = cubic ZrO_2 ss (fluorite-type ss); Y_{ss} = cubic Y_2O_3 ss; Tet_{ss} = tetragonal ZrO_2 ss; Mon_{ss} = monoclinic ZrO_2 ss; H_{ss} = hexagonal Y_2O_3 ss; 3:4 = $Zr_3Y_4O_{12}$; 1:6 = ZrY_6O_{11} ss.

FIGURE 10. 3(SiO₂)-Si₃N₄-4(AlN)-2(Al₂O₃) system. "Behavior" diagram at 1700°C. The labels 8H, 15R, 12H, 21R, 27R, 2Hᵟ indicate defect AlN polytypes. β′ = 3-sialon (Si₆₋ₓAlₓOₓN₈₋ₓ); O′ = sialon of Si₂ON₂ type; X = SiAlO₂N ("nitrogen mullite"). ALON ss = aluminum oxynitride ss extending from approximately Al₇O₉N to Al₃O₃N.

FIGURE 11. PbO-ZrO₂-TiO₂ (PZT) system, subsolidus at 1100°C. P = PbO; T = TiO₂; Z = ZrO₂.

FIGURE 12. CaO-Al₂O₃-SiO₂ system (temperatures in °C).

Crystalline Phases

Notation	Oxide formula
Cristobalite }	SiO₂
Tridymite	
Pseudowollastonite	CaO·SiO₂
Rankinite	3CaO·2SiO₂
Lime	CaO
Corundum	Al₂O₃
Mullite	3Al₂O₃·2SiO₂
Anorthite	CaO·Al₂O₃·2SiO₂
Gehlenite	2CaO·Al₂O₃·SiO₂

Temperatures up to approximately 1550°C are on the Geophysical Laboratory Scale; those above 1550°C are on the 1948 International Scale.

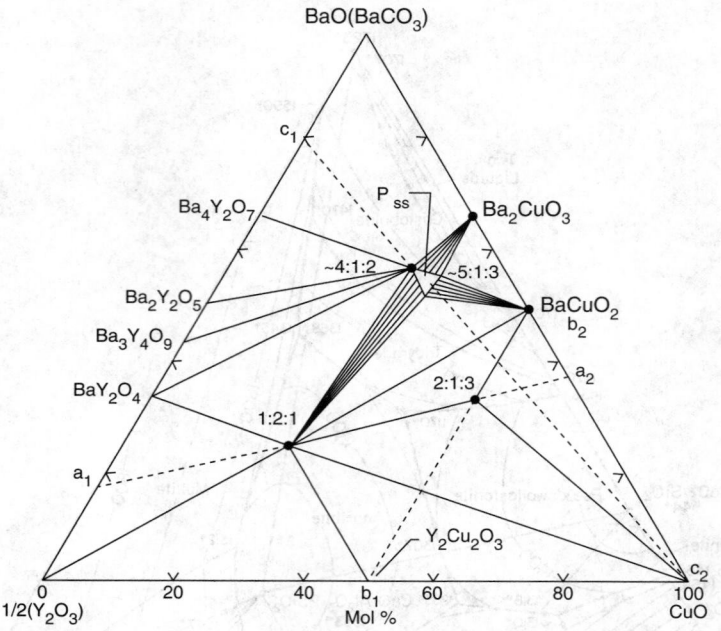

FIGURE 13.   BaO-Y$_2$O$_3$-CuO system. 2:1:3 = Ba$_2$YCu$_3$O$_{7-x}$; 1:2:1 = BaY$_2$CuO$_5$; 4:1:2 = Ba$_4$YCu$_2$O$_{7.5+x}$; and 5:1:3 = Ba$_5$YCu$_3$O$_{9.5+x}$. The superconducting 2:1:3 phase was prepared using barium peroxide.

FIGURE 14.   Al-Cu system.

Phase	Composition, wt % Cu	Pearson symbol	Space group
(Al)	0 to 5.65	cF4	$Fm\bar{3}m$
θ	52.5 to 53.7	tI12	I4/mcm
η₁	70.0 to 72.2	oP16 or oC16	Pban or Cmmm
η₂	70.0 to 72.1	mC20	C2/m
ζ₁	74.4 to 77.8	hP42	P6/mmm
ζ₂	74.4 to 75.2	(a)	–
ε₁	77.5 to 79.4	(b)	–
ε₂	72.2 to 78.7	hP4	P63/mmc
δ	77.4 to 78.3	(c)	$R\bar{3}m$
γ₀	77.8 to 84	(d)	–
γ₁	79.7 to 84	cP52	$P\bar{4}3m$
β₀	83.1 to 84.7	(d)	–
β	85.0 to 91.5	cI2	$Im\bar{3}m$
α₂	88.5 to 89	(e)	–
(Cu)	90.6 to 100	cF4	$Fm\bar{3}m$
Metastable phases			
θ´	–	tP6	–
β´	–	cF16	$Fm\bar{3}m$
Al₃Cu₂	61 to 70	hp5	$P\bar{3}m1$

(a) Monoclinic? (b) Cubic? (c) Rhombohedral. (d) Unknown. (e) D0₂₂-type long-period superlattice.

FIGURE 15. Fe-C system.

Phase	Composition, mass % C	Pearson symbol	Space group
(δFe)	0 to 0.09	cI2	$Im\bar{3}m$
(γFe)	0 to 2.1	cF4	$Fm\bar{3}m$
(αFe)	0 to 0.021	cI2	$Im\bar{3}m$
(C)	100	hP4	P6₃/mmc
Metastable/high-pressure phases			
(εFe)	0	hP2	P6₃/mmc
Martensite	< 2.1	tI4	I4/mmm
Fe₄C	5.1	cP5	$P\bar{4}3m$
Fe₃C (θ)	6.7	oP16	Pnma
Fe₅C₂ (χ)	7.9	mC28	C2/c
Fe₇C₃	8.4	hP20	P6₃mc
Fe₇C₃	8.4	oP40	Pnma
Fe₂C (η)	9.7	oP6	Pnnm
Fe₂C (ε)	9.7	hP*	P6₃22
Fe₂C	9.7	hP*	$P\bar{3}m1$
(C)	100	cF8	$Fd\bar{3}m$

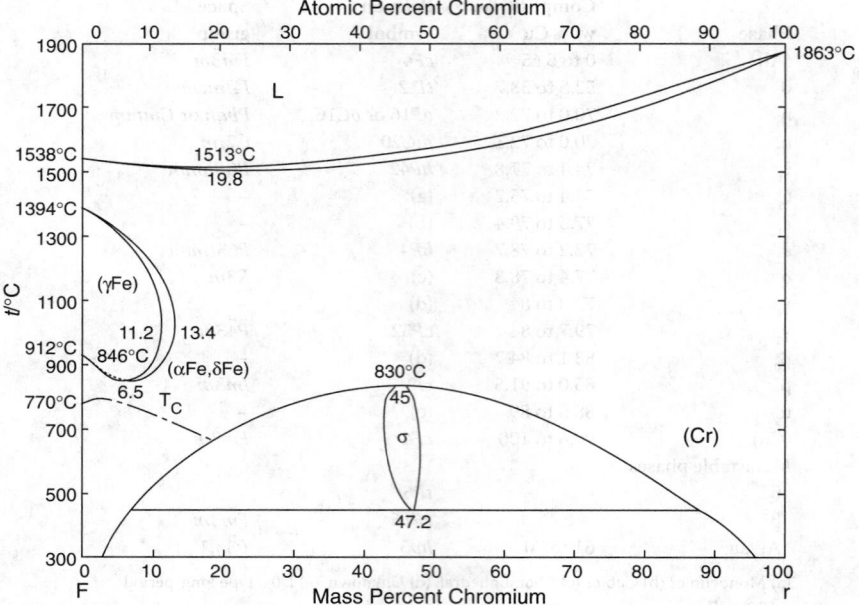

FIGURE 16. Fe-Cr system.

Phase	Composition, mass % Cr	Pearson symbol	Space group
(aFe, Cr)	0 to 100	cI2	Im3̄m
(γFe)	0 to 11.2	cF4	Fm3̄m
σ	42.7 to 48.2	tP30	P4₂/mnm

FIGURE 17. Cu-Sn system.

Phase	Composition, mass % Sn	Pearson symbol	Space group
α	0 to 15.8	cF4	Fm$\bar{3}$m
β	22.0 to 27.0	cI2	Im$\bar{3}$m
γ	25.5 to 41.5	cF16	Fm$\bar{3}$m
δ	32 to 33	cF416	F$\bar{4}$3m
ζ	32.2 to 35.2	hP26	P6$_3$
ε	27.7 to 39.5	oC80	Cmcm
η	59.0 to 60.9	hP4	P6$_3$/mmc
η´	44.8 to 60.9	(a)	–
(βSn)	~100	tI4	I4$_1$/amd
(αSn)	100	cF8	Fd$\bar{3}$m

(a) Hexagonal; superlattice based on NiAs-type structure.

FIGURE 18.   Cu-Ni system.

Phase	Composition, mass % Ni	Pearson symbol	Space group
(Cu, Ni) (above 354.5°C)	0 to 100	cF4	Fm$\bar{3}$ m

FIGURE 19.   Pb-Sn system.

Phase	Composition, mass % Sn	Pearson symbol	Space group
(Pb)	0 to 18.3	cF4	Fm$\bar{3}$m
(βSn)	97.8 to 100	tI4	I4$_1$/amd
(αSn)	100	cF8	Fd$\bar{3}$m
High-pressure phases			
ε(a)	52 to 74	hP1	P6/mmm
ε'(b)	52	hP2	P6$_3$/mmc

(a) From phase diagram calculated at 2500 MPa. (b) This phase was claimed for alloys at 350°C and 5500 MPa.

FIGURE 20.   Cu-Zn system.

Phase	Composition, mass % Zn	Pearson symbol	Space group
α or (Cu)	0 to 38.95	cF4	$Fm\bar{3}m$
β	36.8 to 56.5	cI2	$Im\bar{3}m$
β′	45.5 to 50.7	cP2	$Pm\bar{3}m$
γ	57.7 to 70.6	cI52	$I\bar{4}3m$
δ	73.02 to 76.5	hP3	$P\bar{6}$
ε	78.5 to 88.3	hP2	$P6_3/mmc$
η or (Zn)	97.25 to 100	hP2	$P6_3/mmc$

FIGURE 21.   Co-Sm system.

Phase	Composition, mass % Sm	Pearson symbol	Space group
(αCo)	0 to ~3.7	cF4	$Fm\bar{3}m$
(εCo)	~0	hP2	$P6_3/mmc$
βCo₁₇Sm₂	~23.0	hP38	$P6_3/mmc$
αCo₁₇Sm₂	~23.0	hR19	$R\bar{3}m$
		hP8	$P6/mmm$
Co₅₊ₓSm	~33 to 34	–	–
Co₅₋ₓSm	~34 to 35	–	–
Co₁₉Sm₅	~40.1	hR24	$R\bar{3}m$
		hP48	$P6_3/mmc$
αCo₇Sm₂	~42.1	hR18	$R\bar{3}m$
βCo₇Sm₂	~42.1	hP36	$P6_3/mmc$
Co₃Sm	46	hR12	$R\bar{3}m$
Co₂Sm	56.0	hR4	$R\bar{3}m$
		cF24	$Fd\bar{3}m$
Co₄Sm₉	~85.1	o**	–
CoSm₃	88	oP16	$Pnma$
(γSm)	~100	cI2	$Im\bar{3}m$
(βSm)	~100	hP2	$P6_3/mmc$
(αSm)	~100	hR3	$R\bar{3}m$

Other reported phases

Co₅Sm	~33.8	hP6	$P6/mmm$
Co₂Sm₅	~86.4	mC28	$C2/c$

FIGURE 22.   Ti-Mo system.

Phase	Composition, mass % Mo	Pearson symbol	Space group
(βTi, Mo)	0 to 100	$cI2$	$Im\overline{3}m$
(αTi)	0 to 0.8	$hP2$	$P6_3/mmc$
α′	(a)	$hP2$	$P6_3/mmc$
α″	(a)	$oC4$	$Cmcm$
ω	(a)	$hP3$	$P6/mmm$

(a) Metastable.

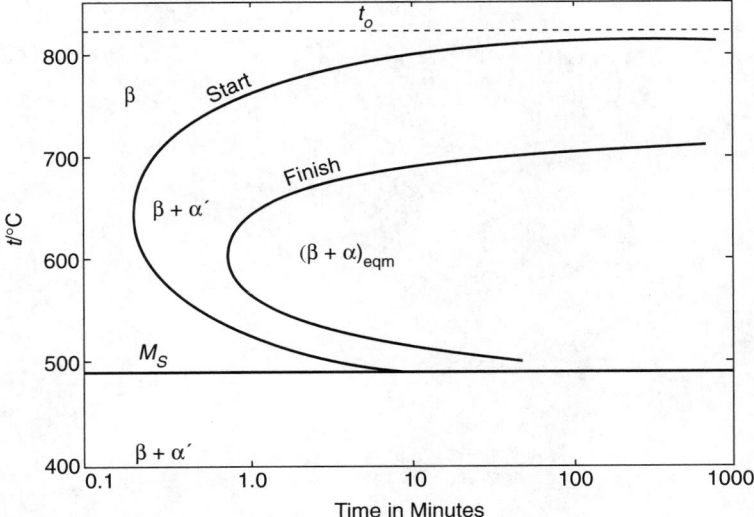

Experimental time-temperature-transformation (TTT) diagram for Ti-Mo. The start and finish times of the isothermal precipitation reaction vary with temperature as a result of the temperature dependence of the nucleation and growth processes. Precipitation is complete, at any temperature, when the equilibrium fraction of α is established in accordance with the lever rule. The solid horizontal line represents the athermal (or nonthermally activated) martensitic transformation that occurs when the β phase is quenched.

FIGURE 23. The isothermal section at 900°C (1652°F) of the iron-chromium-nickel ternary phase diagram, showing the nominal composition of 18-8 stainless steel.

HEAT CAPACITY OF SELECTED SOLIDS

This table gives the molar heat capacity at constant pressure of representative metals, semiconductors, and other crystalline solids as a function of temperature in the range 200 to 600 K.

References

1. Chase, M. W., et al., *JANAF Thermochemical Tables*, 3rd ed., *J. Phys. Chem. Ref. Data*, 14, Suppl. 1, 1985.
2. Garvin, D., Parker, V. B., and White, H. J., *CODATA Thermodynamic Tables*, Hemisphere Press, New York, 1987.
3. DIPPR Database of Pure Compound Properties, Design Institute for Physical Properties Data, American Institute of Chemical Engineers, New York, 1987.

Name	C_p in J/mol K						
	200 K	250 K	300 K	350 K	400 K	500 K	600 K
Aluminum	21.33	23.08	24.25	25.11	25.78	26.84	27.89
Aluminum oxide	51.12	67.05	79.45	88.91	96.14	106.17	112.55
Anthracene	138.6	173.9	210.7	248.8	288.4		
Benzoic acid	102.7	123.5	147.4	172.0			
Beryllium	9.98	13.58	16.46	18.53	19.95	21.94	23.34
Biphenyl	131.0	162.5	197.2				
Boron	5.99	8.82	11.40	13.65	15.69	18.72	20.78
Calcium	24.54	25.41	25.94	26.32	26.87	28.49	30.38
Calcium carbonate	66.50	75.66	83.82	91.51	96.97	104.52	109.86
Calcium oxide	33.64	38.59	42.18	45.07	46.98	49.33	50.72
Cesium chloride	50.13	51.34	52.48	53.58	54.68	56.90	59.10
Chromium	19.86	22.30	23.47	24.39	25.23	26.63	27.72
Cobalt	22.23	23.98	24.83	25.68	26.53	28.20	29.66
Copper	22.63	23.77	24.48	24.95	25.33	25.91	26.48
Copper oxide	34.80		42.41	44.95	46.78	49.19	50.83
Copper sulfate	77.01	89.25	99.25	107.65	114.93	127.19	136.31
Germanium			23.25	23.85	24.31	24.96	25.45
Gold			25.41	25.37	25.51	26.06	26.65
Graphite	5.01	6.82	8.58	10.24	11.81	14.62	16.84
Hexachlorobenzene	162.7	183.6	202.4				
Iodine	51.57	53.24	54.51	58.60			
Iron	21.59	23.74	25.15	26.28	27.39	29.70	32.05
Lead	25.87	26.36	26.85	27.30	27.72	28.55	29.40
Lithium	21.57	23.42	24.64	25.96	27.60	29.28	
Lithium chloride	43.35	46.08	48.10	49.66	50.97	53.34	55.59
Magnesium	22.72	24.02	24.90	25.57	26.14	27.17	28.18
Magnesium oxide			37.38	40.59	42.77	45.56	47.30
Manganese	23.05	24.95	26.35	27.52	28.53	30.29	31.90
Naphthalene	105.8	134.1	167.8	204.1			
Potassium	27.00	28.01	29.60				
Potassium chloride	48.44	50.10	51.37	52.31	53.08	54.71	56.35
Silicon	15.64	18.22	20.04	21.28	22.14	23.33	24.15
Silicon dioxide	32.64	39.21	44.77	49.47	53.43	59.64	64.42
Silver			25.36	25.55	25.79	26.36	26.99
Sodium	22.45	27.01	28.20	30.14			
Sodium chloride	46.89	48.85	50.21	51.25	52.14	53.96	55.81
Tantalum	24.08	24.86	25.31	25.60	25.84	26.35	26.84
Titanium	22.37	24.07	25.28	26.17	26.86	27.88	28.60
Tungsten	22.49	23.69	24.30	24.65	24.92	25.36	25.79
Vanadium	21.88	23.70	24.93	25.68	26.23	26.94	27.49
Zinc	24.05	25.02	25.45	25.88	26.35	27.39	28.59
Zirconium	23.87	24.69	25.22	25.61	25.93	26.56	27.28

THERMAL AND PHYSICAL PROPERTIES OF PURE METALS

This table gives the following properties for the metallic elements:

t_m: Melting point in °C

t_b: Normal boiling point in °C, at a pressure of 101.325 kPa (760 Torr)

$\Delta_{fus}H$: Enthalpy of fusion at the melting point in J/g

ρ_{25}:   Density at 25°C in g/cm³

α: Coefficient of linear expansion at 25°C in K⁻¹ (the quantity listed is $10^6 \times \alpha$)

c_p: Specific heat capacity at constant pressure at 25°C in J/g K

λ: Thermal conductivity at 27°C in W/cm K

References

1. Dinsdale, A. T., *CALPHAD*, 15, 317, 1991 (melting points, enthalpy of fusion).
2. Touloukian, Y. S., *Thermophysical Properties of Matter*, Vol. 12, Thermal Expansion, IFI/Plenum, New York, 1975 (coefficient of expansion, density).
3. Ho, C. Y., Powell, R. W., and Liley, P. E., *J. Phys. Chem. Ref. Data*, 3, Suppl. 1, 1974 (thermal conductivity).
4. Cox, J. D., Wagman, D. D., and Medvedev, V. A., *CODATA Key Values for Thermodynamics*, Hemisphere Publishing Corp., New York, 1989 (heat capacity).
5. Glushko, V. P., Ed., *Thermal Constants of Substances*, VINITI, Moscow, (enthalpy of fusion, heat capacity).
6. Wagman, D. D., et. al., *The NBS Tables of Chemical Thermodynamic Properties*, J. Phys. Chem. Ref. Data, 11, Suppl. 2, 1982 (heat capacity).
7. Chase, M. W., et. al., *JANAF Thermochemical Tables*, 3rd ed., *J. Phys. Chem. Ref. Data*, 14, Suppl. 1, 1985 (heat capacity, enthalpy of fusion).
8. Gschneidner, K. A., *Bull. Alloy Phase Diagrams*, 11, 216–224, 1990 (various properties of the rare earth metals).
9. Hellwege, K. H., Ed., *Landolt Börnstein, Numerical Values and Functions in Physics, Chemistry, Astronomy, Geophysics, and Technology*, Vol. 2, Part 1, Mechanical-Thermal Properties of State, 1971 (density).
10. *Physical Encyclopedic Dictionary*, Vol. 1–5, Encyclopedy Publishing House, Moscow, 1960–66.

Metal (symbol)	Atomic weight	t_m °C	t_b °C	$\Delta_{fus}H$ J/g	ρ_{25} g/cm³	$\alpha \times 10^6$ K⁻¹	c_p J/g K	λ W/cm K
Actinium (Ac)		1050	3198		10		0.12	
Aluminum (Al)	26.98	660.32	2519	399.9	2.70	23.1	0.897	2.37
Antimony (Sb)	121.76	630.628	1587	162.5	6.68	11.0	0.207	0.243
Barium (Ba)	137.33	727	1897	51.8	3.62	20.6	0.205	0.184
Beryllium (Be)	9.01	1287	2471	876.0	1.85	11.3	1.82	2.00
Bismuth (Bi)	208.98	271.406	1564	53.3	9.79	13.4	0.122	0.0787
Cadmium (Cd)	112.41	321.069	767	55.2	8.69	30.8	0.231	0.968
Calcium (Ca)	40.08	842	1484	213.1	1.54	22.3	0.646	2.00
Cerium (Ce)	140.11	799	3443	39.0	6.77	6.3	0.192	0.113
Cesium (Cs)	132.91	28.44	671	15.7	1.93	97	0.242	0.359
Chromium (Cr)	52.00	1907	2671	404	7.15	4.9	0.450	0.937
Cobalt (Co)	58.93	1495	2927	272.5	8.86	13.0	0.421	1.00
Copper (Cu)	63.55	1084.62	2562	203.5	8.96	16.5	0.384	4.01
Dysprosium (Dy)	162.50	1412	2567	68.1	8.55	9.9	0.170	0.107
Erbium (Er)	167.26	1529	2868	119	9.07	12.2	0.168	0.145
Europium (Eu)	151.96	822	1529	60.6	5.24	35.0	0.182	0.139[a]
Gadolinium (Gd)	157.25	1313	3273	63.6	7.90	9.4[b]	0.235	0.105
Gallium (Ga)	69.72	29.7666	2204	80.0	5.91	18	0.374	0.406
Gold (Au)	196.97	1064.18	2856	64.6	19.3	14.2	0.129	3.17
Hafnium (Hf)	178.49	2233	4603	152.4	13.3	5.9	0.144	0.230
Holmium (Ho)	164.93	1472	2700	103[a]	8.80	11.2	0.165	0.162
Indium (In)	114.82	156.60	2072	28.6	7.31	32.1	0.233	0.816
Iridium (Ir)	192.22	2446	4428	213.9	22.5	6.4	0.131	1.47
Iron (Fe)	55.85	1538	2861	247.3	7.87	11.8	0.449	0.802
Lanthanum (La)	138.91	920	3464	44.6	6.15	12.1	0.195	0.134
Lead (Pb)	207.20	327.462	1749	23.1	11.3	28.9	0.127	0.353
Lithium (Li)	6.94	180.5	1342	432	0.534	46	3.57	0.847
Lutetium (Lu)	174.97	1663	3402	126[a]	9.84	9.9	0.154	0.164
Magnesium (Mg)	24.30	650	1090	348.9	1.74	24.8	1.024	1.56
Manganese (Mn)	54.94	1246	2061	235.0	7.3	21.7	0.479	0.0782
Mercury (Hg)	200.59	−38.8290	356.62	11.4	13.5336	60.4	0.139	0.0834
Molybdenum (Mo)	95.94	2623	4639	390.7	10.2	4.8	0.251	1.38
Neodymium (Nd)	144.24	1016	3074	49.5	7.01	9.6	0.191	0.165
Neptunium (Np)		644		13.5	20.2			0.063
Nickel (Ni)	58.69	1455	2913	290.3	8.90	13.4	0.445	0.907

Metal (symbol)	Atomic weight	t_m °C	t_b °C	$\Delta_{fus} H$ J/g	ρ_{25} g/cm³	$\alpha \times 10^6$ K⁻¹	c_p J/g K	λ W/cm K
Niobium (Nb)	92.91	2477	4744	323	8.57	7.3	0.265	0.537
Osmium (Os)	190.23	3033	5012	304.1	22.59	5.1	0.130	0.876
Palladium (Pd)	106.42	1554.8	2963	157.3	12.0	11.8	0.244	0.718
Platinum (Pt)	195.08	1768.2	3825	113.6	21.5	8.8	0.133	0.716
Plutonium (Pu)		640	3228	11.6	19.7	46.7		0.0674
Polonium (Po)		254	962		9.20	23.5		0.20
Potassium (K)	39.10	63.38	759	59.6	0.89	83.3	0.757	1.024
Praseodymium (Pr)	140.91	931	3520	48.9	6.77	6.7	0.193	0.125
Promethium (Pm)		1042	3000ᵃ		7.26	11ᵃ	0.19ᵃ	0.15ᵃ
Protactinium (Pa)	231.04	1572		53.4	15.4			
Radium (Ra)		696			5			
Rhenium (Re)	186.21	3185	5596	324.5	20.8	6.2	0.137	0.479
Rhodium (Rh)	102.91	1964	3695	258.4	12.4	8.2	0.243	1.50
Rubidium (Rb)	85.47	39.30	688	25.6	1.53		0.364	0.582
Ruthenium (Ru)	101.07	2334	4150	381.8	12.1	6.4	0.238	1.17
Samarium (Sm)	150.36	1072	1794	57.3	7.52	12.7	0.196	0.133
Scandium (Sc)	44.96	1541	2836	314	2.99	10.2	0.567	0.158
Silver (Ag)	107.87	961.78	2162	104.6	10.5	18.9	0.235	4.29
Sodium (Na)	22.99	97.794	882.94	113.1	0.97	71	1.225	1.41
Strontium (Sr)	87.62	777	1382	84.8	2.64	22.5	0.306	0.353
Tantalum (Ta)	180.95	3017	5458	202.1	16.4	6.3	0.140	0.575
Technetium (Tc)		2157	4265	339.7	11			0.506
Terbium (Tb)	158.93	1359	3230	67.9	8.23	10.3	0.182	0.111
Thallium (Tl)	204.38	304	1473	20.3	11.8	29.9	0.129	0.461
Thorium (Th)	232.04	1750	4788	59.5	11.7	11.0	0.118	0.540
Thulium (Tm)	168.93	1545	1950	99.7	9.32	13.3	0.160	0.169
Tin (Sn)	118.71	231.93	2602	60.4	7.26	22.0	0.227	0.666
Titanium (Ti)	47.88	1668	3287	295.6	4.51	8.6	0.522	0.219
Tungsten (W)	183.84	3422	5555	284.5	19.3	4.5	0.132	1.74
Uranium (U)	238.03	1135	4131	38.4	19.1	13.9	0.116	0.276
Vanadium (V)	50.94	1910	3407	422	6.0	8.4	0.489	0.307
Ytterbium (Yb)	173.04	824	1196	44.3	6.90	26.3	0.154	0.385
Yttrium (Y)	88.91	1522	3345	128	4.47	10.6	0.298	0.172
Zinc (Zn)	65.39	419.53	907	108.1	7.14	30.2	0.388	1.16
Zirconium (Zr)	91.22	1854.7	4409	230.2	6.52	5.7	0.278	0.227

ᵃ Estimated.

ᵇ At 100°C.

THERMAL CONDUCTIVITY OF METALS AND SEMICONDUCTORS AS A FUNCTION OF TEMPERATURE

This table gives the temperature dependence of the thermal conductivity of several metals and of carbon, germanium, and silicon. For graphite, separate entries are given for the thermal conductivity parallel ($\parallel$) and perpendicular ($\perp$) to the layer planes. The thermal conductivity of all these materials is very sensitive to impurities at low temperatures, especially below 100 K. Therefore, the values given here should be regarded as typical values for a highly purified specimen; the thermal conductivity of different specimens can vary by more than an order of magnitude in the low-temperature range. See Reference 2 for details.

References

1. Ho, C. Y., Powell, R. W., and Liley, P. E., *J. Phys. Chem. Ref. Data*, 1, 279, 1972.
2. White, G. K., and Minges, M. L., *Thermophysical Properties of Some Key Solids*, CODATA Bulletin No. 59, 1985.

Thermal Conductivity in W/cm K

T/K	Ag	Al	Au	Diamond (type) I	Diamond (type) IIa	Diamond (type) IIb	Pyrolytic graphite $\parallel$	Pyrolytic graphite $\perp$	Cr	Cu
1	39.4	41.1	5.46						0.402*	42.2
2	78.3	81.8	10.9	0.0138*	0.033*	0.0200*			0.803	84.0
3	115	121	16.1	0.0461	0.111	0.0676			1.20	125
4	147	157	20.9	0.108	0.261	0.160			1.60	162
5	172	188	25.2	0.206	0.494	0.307			2.00	195
6	187	213	28.5	0.344	0.820	0.510			2.39	222
7	193	229	30.9	0.523	1.24	0.778			2.27	239
8	190	237	32.3	0.762	1.77	1.12			3.14	248
9	181	239	32.7	1.05	2.41	1.53			3.50	249
10	168	235	32.4	1.40	3.17	2.03	0.811	0.0116	3.85	243
15	96.0	176	24.6	3.96	8.65	5.66			5.24	171
20	51.0	117	15.8	7.87	16.8	11.2	4.20	0.0397	5.93	108
30	19.3	49.5	7.55	18.8	38.9	26.5	9.86	0.0786	5.49	44.5
40	10.5	24.0	5.15	29.4	65.9	44.0	16.4	0.120	4.25	21.7
50	7.0	13.5	4.21	35.3	92.1	59.1	23.1	0.152	3.17	12.5
60	5.5	8.5	3.74	37.4	112	67.5	29.8	0.173	2.48	8.29
70	4.97	5.85	3.48	36.9	119	69.1	36.6	0.181	2.07	6.47
80	4.71	4.32	3.32	35.1	117	65.7	42.8	0.181	1.84	5.57
90	4.60	3.42	3.28	32.7	109	60.0	47.5	0.176	1.69	5.08
100	4.50	3.02	3.27	30.0	100	54.2	49.7	0.168	1.59	4.82
150	4.32	2.48	3.25	19.5	60.2	32.5	45.1	0.125	1.29	4.29
200	4.30	2.37	3.23	14.1	40.3	22.6	32.3	0.0923	1.11	4.13
250	4.29	2.35	3.21	11.0	29.7	17.0	24.4	0.0711	1.00	4.06
300	4.29	2.37	3.17	8.95	23.0	13.5	19.5	0.0570	0.937	4.01
350	4.27	2.40	3.14	7.55*	18.5*	11.1*	16.2	0.0477	0.929	3.96
400	4.25	2.40	3.11	6.5*	15.4*	9.32*	13.9	0.0409	0.909	3.93
500	4.19	2.36	3.04				10.8	0.0322	0.860	3.86
600	4.12	2.31	2.98				8.92	0.0268	0.807	3.79
800	3.96	2.18	2.84				6.67	0.0201	0.713	3.66
1000	3.79		2.70				5.34	0.0160	0.654	3.52
1200	3.61*		2.55				4.48	0.0134	0.619	3.39
1400							3.84	0.0116	0.588	
1600							3.33	0.0100	0.556	
1800							2.93	0.00895	0.526*	
2000							2.62	0.00807	0.494*	

T/K	Fe	Ge[a]	Mg	Ni	Pb	Pt	Si[a]	Sn	Ti	W
1	1.71	0.274	9.86	2.17	27.9	2.31	0.0693*	183	0.0144*	14.4
2	3.42	2.06	19.6	4.34	44.6	4.60	0.454	323	0.0288*	28.7
3	5.11	5.35	29.0	6.49	35.8	6.79	1.38	297	0.0432	42.8
4	6.77	8.77	37.6	8.59	22.2	8.8	2.97	181	0.0575	56.3
5	8.39	11.6	45.0	10.6	13.8	10.5	5.27	117	0.0719	68.7
6	9.93	13.9	50.8	12.5	8.10	11.8	8.23	76	0.0863	79.5
7	11.4	15.5	54.7	14.2	4.86	12.6	11.7	52	0.101	88.0
8	12.7	16.6	56.7	15.8	3.20	12.9	15.5	36	0.115	93.8
9	13.9	17.3	57.0	17.1	2.30	12.8	19.5	26	0.129	96.8
10	14.8	17.7	55.8	18.1	1.78	12.3	23.3	19.3	0.143	97.1
15	17.0	17.3	41.1	19.5	0.845	8.41	41.6	6.3	0.212	72.0
20	15.4	14.9	27.2	16.5	0.591	4.95	49.8	3.2	0.275	40.5
30	10.0	10.8	12.9	9.56	0.477	2.15	48.1	1.79	0.365	14.4
40	6.23	7.98	7.19	5.82	0.451	1.39	35.3	1.33	0.390	6.92
50	4.05	6.15	4.65	4.00	0.436	1.09	26.8	1.15	0.374	4.27
60	2.85	4.87	3.27	3.08	0.425	0.947	21.1	1.04	0.355	3.14
70	2.16	3.93	2.49	2.50	0.416	0.862	16.8	0.96	0.340	2.58
80	1.75	3.25	2.02	2.10	0.409	0.815	13.4	0.915	0.326	2.29
90	1.50	2.70	1.78	1.83	0.403	0.789	10.8	0.880	0.315	2.17
100	1.34	2.32	1.69	1.64	0.397	0.775	8.84	0.853	0.305	2.08
150	1.04	1.32	1.61	1.22	0.379	0.740	4.09	0.779	0.270	1.92
200	0.94	0.968	1.59	1.07	0.367	0.726	2.64	0.733	0.245	1.85
250	0.865	0.749	1.57	0.975	0.360	0.718	1.91	0.696	0.229	1.80
300	0.802	0.599	1.56	0.907	0.353	0.716	1.48	0.666	0.219	1.74
350	0.744	0.495	1.55	0.850	0.347	0.717	1.19	0.642	0.210	1.67
400	0.695	0.432	1.53	0.802	0.340	0.718	0.989	0.622	0.204	1.59
500	0.613	0.338	1.51	0.722	0.328	0.723	0.762	0.596	0.197	1.46
600	0.547	0.273	1.49	0.656	0.314	0.732	0.619		0.194	1.37
800	0.433	0.198	1.46*	0.676		0.756	0.422		0.197	1.25
1000	0.323	0.174		0.718		0.787	0.312		0.207	1.18
1200	0.283	0.174		0.762		0.826	0.257		0.220	1.12
1400	0.312			0.804		0.871	0.235		0.236	1.08
1600	0.330					0.919	0.221		0.253	1.04
1800	0.345*					0.961			0.270*	1.01
2000						0.994*				0.98

[a] Values below 300 K are typical values.

* Extrapolated.

THERMAL CONDUCTIVITY OF ALLOYS AS A FUNCTION OF TEMPERATURE

This table lists the thermal conductivity of selected alloys at various temperatures. The indicated compositions refer to weight percent. Since the thermal conductivity is sensitive to exact composition and processing history, especially at low temperatures, these values should be considered approximate.

References

1. Powell, R. L., and Childs, G. E., in *American Institute of Physics Handbook, 3rd Edition*, Gray, D. E., Ed., McGraw-Hill, New York, 1972.
2. Ho, C. Y., et al., *J. Phys. Chem. Ref. Data*, 7, 959, 1978.

Thermal conductivity in W/m K

	Alloy	4 K	20 K	77 K	194 K	273 K	373 K	573 K	973 K
Aluminum:	1100	50	240	270	220	220			
	2024	3.2	17	56	95	130			
	3003	11	58	140	150	160			
	5052	4.8	25	77	120	140			
	5083, 5086	3	17	55	95	120			
	Duralumin	5.5	30	91	140	160	180		
Bismuth:	Rose metal		5.5	8.3	14	16			
	Wood's metal	4	17	23					
Copper:	electrolytic tough pitch	330	1300	550	400	390	380	370	350
	free cutting, leaded	200	800	460	380	380			
	phosphorus, deoxidized	7.5	42	120	190	220			
	brass, leaded	2.3	12	39	70	120			
	bronze, 68% Cu; 32% Zn	2.3	16	48	92	110			
	beryllium	2	17	36	70	90	113	172	
	german silver	0.75	7.5	17	20	23	25	30	40
	silicon bronze A		3.4	11	23	30			
	manganin	0.48	3.2	14	17	22			
	constantan	0.9	8.6	17	19	22			
Ferrous:	commercial pure iron	15	72	106	82	76	66	54	34
	plain carbon steel(AISI 1020)	13	20	58	65	65			
	plain carbon steel(AISI 1095)		8.5	31	41	45			
	3% Ni; 0.7% Cr; 0.6% Mo		6	22		33	35	36	30
	4% Si					20	24	28	26
	stainless steel	0.3	2	8	13	14	16	19	25
	27% Ni; 15% Cr		1.7	55		11	12	16	21
Gold:	colbalt thermocouple	1.2	8.6	20					
	65% Au; 35% Ag		12	24		61	89		
Indium:	85.5% In; 14.5% Pb	1.9	7.8	24	41				
Lead:	60% Pb; 40% Sn (soft solder)		28	44					
	64.35% Pb; 35.65% In	0.8	3.26	9.1		20.2			
Nickel:	80% Ni; 20% Cr					12	14	17	23
	contracid	0.2	2	7.3	9.5	13			
	inconel	0.5	4.2	12.5	13	15	16	19	26
	monel	0.9	7.1	15	20	21	24	30	43
Platinum:	90% Pt; 10% Ir					31	31.4		
	90% Pt; 10% Rh					30.1	30.5		
Silver:	silver solder		12	34	58				
	normal Ag thermocouple	48	230	310					
Tin:	60% Sn; 40% Pb	16	55	51					
Titanium:	5.5% Al; 2.5% Sn;0.2% Fe		1.8	4.3	6.4	7.8	8.4	10.8	
	4.7% Mn; 3.99% Al; 0.14% C		1.7	4.5	6.5	8.5			

THERMAL CONDUCTIVITY OF CRYSTALLINE DIELECTRICS

This table lists the thermal conductivity of a number of crystalline dielectrics, including some which find use as optical materials. Values are given at temperatures for which data are available.

Reference

Powell, R. L., and Childs, G. E., in *American Institute of Physics Handbook, 3rd Edition*, Gray, D. E., Ed., McGraw-Hill, New York, 1972.

Material	T/K	Ther. cond. W/m K	Material	T/K	Ther. cond. W/m K
AgCl	223	1.3		20	16
	273	1.2		77	270
	323	1.1		373	210
	373	1.1		573	120
Al,B silicate (tourmaline)	398	2.9		1273	29
‖ to c axis	540	3.2	Bi_2Te_3	80	6.4
	723	3.5		204	2.8
Al,Be silicate (beryl)	315	6.4		303	3.6
Al,F silicate (topaz)	315	17.7		370	4.6
‖ to c axis	358	15.6	C (diamond)	4.2	13
	417	13.3	type I	20	800
Al,Fe silicate (garnet)	315	35.8		77	3550
	358	35.4		194	1450
	377	35.6		273	1000
Al_2O_3 (sapphire):			$CaCO_3$		
36° to c axis	4.2	110	‖ to c axis	83	25
	20	3500		273	5.5
	35	6000	⊥ to c axis	83	17
	77	1100		194	6.5
⊥ to c axis	373	2.6		273	4.6
	523	3.9		373	3.6
	773	5.8	CaF_2	83	39
Al_2O_3 (sintered)	4.2	0.5		223	18
	20	23		273	10
	77	150		323	9.2
	194	48		373	9
	273	35	$CaWO_4$ (scheelite)	422	11.3
	373	26	CdTe	160	7.0
	973	8		297	3.6
Ar	8	6.0		422	2.9
	10	3.7	CsBr	223	1.2
	20	1.4		273	0.94
	77	0.31		323	0.81
As_2S_3 (glass)	283	0.16		373	0.77
	323	0.21	CsI	223	1.4
	373	0.27		273	1.2
BN	1047	36.2		323	1
	1475	22.7		373	0.95
	1928	21.9	Cu_2O (cuprite)	102	3.74
	2111	18.5		163	7.76
BaF_2	225	20		299	5.58
	260	13.4		360	4.86
	305	10.9	Fe_3O_4 (magnetite)	4.5	27.4
	370	10.5		20.5	293.0
$BaTiO_3$	5	4.2		126.5	7.4
	30	24.0		304	7.0
	40	25.0	Glass:		
	100	12.0	phoenix	4.2	0.095
	250	4.8		20	0.13
	300	6.2		77	0.37
BeO	4.2	0.3	plastic perspex	4.2	0.058

Material	T/K	Ther. cond. W/m K	Material	T/K	Ther. cond. W/m K
pyrex	20	0.074	NaCl	4.2	440
	77	0.44		20	300
	194	0.88		77	30
	273	1		273	6.4
H_2 (para + 0.5% ortho)	2.5	100		323	5.6
	3	150		373	5.4
	4	200	NaF	5	1100
	6	30		50	250
	10	3		100	90
H_2O (ice)	173	3.5	Ne	2	3.0
	223	2.8		3	4.6
	273	2.2		4.2	4.2
He^3 (high pressure)	0.6	25		10	0.8
	1	2		20	0.3
	1.5	0.57	NH_4Cl	77	17
	2	0.21		194	23
He^4 (high pressure)	0.5	42		230	38
	0.8	120		273	27
	1	24	$NH_4H_2PO_4$		
	2	0.18	|| to optic axis	315	0.71
I_2	300	0.45		339	0.71
	325	0.42	$\perp$ to optic axis	313	1.26
	350	0.4		342	1.34
KBr	2	150	NiO	4.2	5.9
	4.2	360		40	400
	100	12		194	82
	273	5	SiO_2 (quartz)		
	323	4.8	|| to c axis	20	720
	373	4.8		194	20
KCl	4.2	500		273	12
	25	140	$\perp$ to c axis	20	370
	80	35		194	10
	194	10		273	6.8
	273	7.0	SiO_2 (fused silica)	4.2	0.25
	323	6.5		20	0.7
	373	6.3		77	0.8
KI	4.2	700		194	1.2
	80	13		273	1.4
	194	4.6		373	1.6
	273	3.1		673	1.8
Kr	4.2	0.48	$SrTiO_3$	5	2.4
	10	1.7		30	21.0
	20	1.2		40	19.2
	77	0.36		100	18.5
LaF_3	78	7.8		250	12.5
	197	5.0		300	11.2
	274	5.4	TlBr	316	0.59
LiF	4.2	620	TlCl	311	0.75
	20	1800	TiO_2 (rutile)		
	77	150	|| to optic axis	4.2	200
$MgO \cdot Al_2O_3$ (spinel)	373	13		20	1000
	773	8.5		273	13
MnO	4.2	0.25	$\perp$ to optic axis	4.2	160
	40	55		20	690
	120	8		273	9
	573	3.5			

THERMAL CONDUCTIVITY OF CERAMICS AND OTHER INSULATING MATERIALS

Thermal conductivity values for ceramics, refractory oxides, and miscellaneous insulating materials are given here. The thermal conductivity refers to samples with density indicated in the second column. Since most of these materials are highly variable, the values should only be considered as a rough guide.

References

1. Powell, R. L., and Childs, G. E., in *American Institute of Physics Handbook, 3rd Edition*, Gray, D. E., Ed., McGraw-Hill, New York, 1972.
2. Perry, R. H., and Green, D., *Perry's Chemical Engineers' Handbook, Sixth Edition*, McGraw-Hill, New York, 1984.

Material	Dens. g/cm³	t °C	Ther. cond. W/m K
Alumina (Al_2O_3)	3.8	100	30
		400	13
		1300	6
		1800	7.4
	3.5	100	17
		800	7.6
Al_2O_3 + MgO		100	15
		400	10
		1000	5.6
Asbestos	0.4	−100	0.07
		0	0.09
		100	0.10
Asbestos + 85% MgO	0.3	30	0.08
Asphalt	2.1	20	0.06
Beryllia (BeO)	2.8	100	210
		400	90
		1000	20
		1800	15
	1.85	50	64
		200	40
		600	23
Brick, dry	1.54	0	0.04
Brick, refractory:			
alosite		1000	1.3
aluminous	1.99	400	1.2
		1000	1.3
diatomaceous	0.77	100	0.2
		500	0.24
	0.4	100	0.08
		500	0.1
fireclay	2	400	1
		1000	1.2
silicon carbide	2	200	2
		600	2.4
vermiculite	0.77	200	0.26
		600	0.31
Calcium oxide		100	16
		400	9
		1000	7.5
Cement mortar	2	90	0.55
Charcoal	0.2	20	0.055
Coal	1.35	20	0.26
Concrete	1.6	0	0.8
Cork	0.05	0	0.03
		100	0.04
	0.35	0	0.06
		100	0.08
Cotton wool	0.08	30	0.04
Diatomite	0.2	0	0.05

Material	Dens. g/cm³	t °C	Ther. cond. W/m K
		400	0.09
	0.5	0	0.09
		400	0.16
Ebonite	1.2	0	0.16
Felt, flax	0.2	30	0.05
	0.3	30	0.04
Fuller's earth	0.53	30	0.1
Glass wool	0.2	−200 to 20	0.005
		50	0.04
		100	0.05
		300	0.08
Graphite			
100 mesh	0.48	40	0.18
20-40 mesh	0.7	40	1.29
Linoleum cork	0.54	20	0.08
Magnesia (MgO)		100	36
		400	18
		1200	5.8
		1700	9.2
MgO + SiO_2		100	5.3
		400	3.5
		1500	2.3
Mica:			
muscovite		100	0.72
		300	0.65
		600	0.69
phlogopite		100	0.66
Canadian		300	0.19
		600	0.2
Micanite		30	0.3
Mineral wool	0.15	30	0.04
Perlite, expanded	0.1	−200 to 20	0.002
Plastics:			
bakelite	1.3	20	1.4
celluloid	1.4	30	0.02
polystyrene foam	0.05	−200 to 20	0.033
mylar foil	0.05	−200 to 20	0.0001
nylon		−253	0.10
		−193	0.23
		25	0.30
polytetrafluoroethylene		−253	0.13
		−193	0.16
		25	0.26
		230	2.5
urethane foam	0.07	20	0.06
Porcelain		90	1
Rock:			
basalt		20	2
chalk		20	0.92

Material	Dens. g/cm^3	t °C	Ther. cond. W/m K	Material	Dens. g/cm^3	t °C	Ther. cond. W/m K
granite	2.8	20	2.2	Uranium dioxide		100	9.8
limestone	2	20	1			400	5.5
sandstone	2.2	20	1.3			1000	3.4
slate, ⊥		95	1.4	Wood:			
slate, ‖		95	2.5	balsa, ⊥	0.11	30	0.04
Rubber:				fir, ⊥	0.54	20	0.14
sponge	0.2	20	0.05	fir, ‖	0.54	20	0.35
92 percent		25	0.16	oak		20	0.16
Sand, dry	1.5	20	0.33	plywood		20	0.11
Sawdust	0.2	30	0.06	pine, ⊥	0.45	60	0.11
Shellac		20	0.23	pine, ‖	0.45	60	0.26
Silica aerogel	0.1	−200 to 20	0.003	walnut, ⊥	0.65	20	0.14
Snow	0.25	0	0.16	Wool	0.09	30	0.04
Steel wool	0.1	55	0.09	Zinc oxide		200	17
Thoria (ThO$_2$)		100	10			800	5.3
		400	5.8	Zirconia (ZrO$_2$)		100	2
		1500	2.4			400	2
Titanium dioxide		100	6.5			1500	2.5
		400	3.8	Zirconia + silica		200	5.6
		1200	3.3			600	4.6
						1500	3.7

THERMAL CONDUCTIVITY OF GLASSES

This table gives the composition of various types of glasses and the thermal conductivity k as a function of temperature. Because of the variability of glasses, the data should be regarded as only approximate.

Type of glass	SiO$_2$ (wt%)	Other oxides	(wt%)	t °C	k W/m K
		Composition			
Vitreous silica	100			−150	0.85
				−100	1.05
				−50	1.20
				0	1.30
				50	1.40
				100	1.50
Vycor glass	96	B$_2$O$_3$	3	−100	1.00
				0	1.25
				100	1.40
Pyrex type chemically-resistant borosilicate glasses	80–81	B$_2$O$_3$	12–13	−100	0.90
		Na$_2$O	4	0	1.10
		Al	2	100	1.25
Borosilicate crown glasses	60–65	B$_2$O$_3$	15–20	−100	0.65–0.75
				0	0.90–0.95
				100	1.00–1.05
	65–70	B$_2$O$_3$	10–15	−100	0.75–0.80
				0	0.95–1.00
				100	1.05–1.15
	70–75	B$_3$O$_3$	5–10	−100	0.80–0.85
				0	1.05–1.10
				100	1.15–1.20
Zinc crown glasses (i)	55–65	ZnO Remainder: B$_2$O$_3$, Al$_2$O$_3$	5–15	−100	0.88–0.92
				0	1.10–1.15
				100	1.15–1.25
		ZnO Remainder: Na$_2$O, K$_2$O	5–15	−100	0.60–0.70
				0	0.70–0.90
				100	0.85–0.95
		ZnO Remainder: B$_2$O$_3$, Al$_2$O$_3$	15–25	−100	0.88–0.92
				0	1.10–1.15
				100	1.15–1.20
		ZnO Remainder: Na$_2$O, K$_2$O	15–25	−100	0.65–0.80
				0	0.85–0.95
				100	0.90–1.05
Zinc crown glasses (ii)	65–75	ZnO Remainder: B$_2$O$_3$, Al$_2$O$_3$	5–15	−100	0.88–0.92
				0	1.15–1.15
				100	1.20–1.30
		ZnO Remainder: Na$_2$O, K$_2$O	5–15	−100	0.70–0.85
				0	0.90–1.05
				100	1.00–1.15
		ZnO	15–25	−100	0.90–0.95

Type of glass	SiO$_2$ (wt%)	Other oxides (wt%)		t °C	k W/m K
		Remainder:		0	1.15–1.15
		B$_2$O$_3$, Al$_2$O$_3$		100	1.20–1.25
		ZnO	15–25	−100	0.65–0.85
		Remainder:		0	0.85–1.00
		Na$_2$O, K$_2$O		100	1.05–1.20
Barium crown glasses	31	B$_2$O$_3$	12	−100	0.55
		Al$_2$O$_3$	8	0	0.70
		BaO	48	100	0.80
	41	B$_2$O$_3$	6	−100	0.60
		Al$_2$O$_3$	2	0	0.75
		ZnO	8	100	0.85
		BaO	43		
	47	B$_2$O$_3$	4	−100	0.65
		Na$_2$O	1	0	0.75
		K$_2$O	7	100	0.90
		ZnO	8		
		BaO	32		
	65	B$_2$O$_3$	2	−100	0.70
		Na$_2$O	5	0	0.90
		K$_2$O	15	100	1.00
		ZnO	2		
		BaO	10		
Borate glasses					
Borate flint glass	9	B$_2$O$_3$	36	−100	0.55
		Na$_2$O	1	0	0.65
		K$_2$O	2	100	0.80
		PbO	36		
		Al$_2$O$_3$	10		
		ZnO	6		
Borate flint glass	0	B$_2$O$_3$	56	−100	0.50
		Al$_2$O$_3$	12	0	0.65
		PbO	32	100	0.85
Borate flint glass	0	B$_2$O$_3$	43	−100	0.40
		Al$_2$O$_3$	5	0	0.55
		PbO	52	100	0.70
Borate glass	4	B$_2$O$_3$	55	−100	0.65
		Al$_2$O$_3$	14	0	0.80
		PbO	11	100	0.90
		K$_2$O	4		
		ZnO	12		
Borate crown glass	0	B$_2$O$_3$	64	−100	0.50
		Na$_2$O	8	0	0.65
		K$_2$O	3	100	0.85
		BaO	4		
		PbO	3		
		Al$_2$O$_3$	18		
Light borate crown glass	0	B$_2$O$_3$	69	−100	0.55
		Na$_2$O	8	0	0.70

Type of glass	SiO₂ (wt%)	Other oxides (wt%)		t °C	k W/m K
		SiO_2 (wt%)	Other oxides (wt%)	t °C	k W/m K
		BaO	5	100	0.90
		Al_2O_3	18		
Zinc borate glass	0	B_2O_3	40	−100	0.65
		ZnO	60	0	0.75
				100	0.85
Phosphate crown glasses					
Potash phosphate glass	0	P_2O_5	70	0	0.75
		B_2O_3	3	100	0.85
		K_2O	12		
		Al_2O_3	10		
		MgO	4		
Baryta phosphate glass	0	P_2O_5	60	45	0.75
		B_2O_3	3		
		Al_2O_3	8		
		BaO	28		
Soda-lime glasses	75	Na_2O	17	−100	0.75
		CaO	8	0	0.95
				100	1.10
	75	Na_2O	12	−100	0.90
		CaO	13	0	1.10
				100	1.15
	72	Na_2O	15	−100	0.80
		CaO	11	0	1.00
		Al_2O_3	2	100	1.15
	65	Na_2O	25	−100	0.65
		CaO	10	0	0.85
				100	0.95
	65	Na_2O	15	−100	0.85
		CaO	20	0	1.00
				100	1.10
	60	Na_2O	20	−100	0.75
		CaO	20	0	0.90
				100	1.00
Other crown glasses					
Crown glass	75	Na_2O	9	−100	0.80
		K_2O	11	0	1.00
		CaO	5	100	1.10
High dispersion crown glass	68	Na_2O	16	−100	0.65
		ZnO	3	0	0.85
		PbO	13	100	1.00
Miscellaneous flint glasses					
(i) Silicate flint glasses					
Light flint glasses	65	PbO	25	−100	0.65−0.70
		Others	10	0	0.88−0.92
				100	1.00−1.05
	55	PbO	35	−100	0.60−0.65

Type of glass	Composition			t	k
	SiO$_2$ (wt%)	Other oxides (wt%)		°C	W/m K
		Others	10	0	0.75–0.85
				100	0.88–0.92
Ordinary flint glass	45	PbO	45	−100	0.50–0.60
		Others	10	0	0.65–0.75
				100	0.80–0.85
Heavy flint glass	35	PbO	60	−100	0.45–0.50
		Others	5	0	0.60–0.65
				100	0.70–0.75
Very heavy flint glasses	25	PbO	73	−100	0.40–0.45
		Others	2	0	0.55–0.60
				100	0.63–0.67
	20	PbO	80	−100	0.40
				0	0.50
				100	0.60
(ii) Borosilicate flint glass	33	B$_2$O$_3$	31	−100	0.65
		PbO	25	0	0.85
		Al$_2$O$_3$	7	100	0.95
		K$_2$O	3		
		Na$_2$O	1		
(iii) Barium flint glass	50	BaO	24	−100	0.60
		PbO	6	0	0.70
		K$_2$O	8	100	0.85
		Na$_2$O	3		
		ZnO	8		
		Sb$_2$O$_3$	1		
Other glasses					
Potassium glass	59	K$_2$O	33	50	0.88–0.92
		CaO	8		
Iron glasses	63	Fe$_2$O$_3$	10	−100	0.80
		Na$_2$O	17	0	0.95
		MgO	4	100	1.05
		CaO	3		
		Al$_2$O$_3$	2		
	67	Fe$_2$O$_3$	15	0	0.88–0.92
		Na$_2$O$_3$	18	100	1.00–1.05
	62	Fe$_2$O$_3$	20	0	0.85–0.90
		Na$_2$O	18	100	0.95–1.00
Rock glasses					
Obsidian				0	1.35
Artificial diabase				100	1.25

FERMI ENERGY AND RELATED PROPERTIES OF METALS

Lev. I. Berger

In the classical Drude theory of metals, the Maxwell-Boltzmann velocity distribution of electrons is used. It states that the number of electrons per unit volume with velocities in the range of $d\vec{v}$ about any magnitude $\vec{v}$ at temperature T is

$$f_B(\vec{v})d\vec{v} = n\left(\frac{m}{2\pi k_B T}\right) \exp\left(-\frac{mv^2}{2k_B T}\right) d\vec{v}$$

where n is the total number of conduction electrons in a unit volume of a metal, m is the free electron mass, and k_B is the Boltzmann constant. In an attempt to explain a substantial discrepancy between the experimental data on the specific heat of metals and the values calculated on the basis of the Drude model, Sommerfeld suggested a model of the metal in which the Pauli exclusion principle is applied to free electrons. In this case, the Maxwell-Boltzmann distribution is replaced by the Fermi-Dirac distribution:

$$f(\vec{v})d\vec{v} = 2\left(\frac{m}{h}\right)^3 d\vec{v}\left\{\exp\left[\left(\frac{mv^2}{2} - k_B T_0\right)\Big/k_B T\right]+1\right\}^{-1}$$

Here h is the Planck constant and T_0 is a characteristic temperature which is determined by the normalization condition

$$n = \int d\vec{v} \cdot f(\vec{v})$$

The magnitude of T_0 is quite high; usually, $T_0 > 10^4$ K. So, at common temperatures ($T < 10^3$ K), the free electron density of a metal is much smaller than in the case of the Maxwell-Boltzmann distribution. This allows us to explain why the experimental data on specific heat for metals are close to those for insulators.

The maximum kinetic energy the electrons of a metal may possess at $T = 0$ K is called the Fermi energy, e.g.,

$$E_F = \frac{\hbar^2 k_F^2}{2m} = \left(\frac{e^2}{2k_B}\right)(k_F r_B)^2$$

where k_F is the Fermi momentum or the Fermi wave vector

$$k_F = (3\pi^2 n)^{1/3}$$

e is the electron charge, and r_B is the Bohr radius

$$r_B = \hbar^2/me^2 = 0.529 \cdot 10^{-10} \text{ m}$$

Another, more common expression for the Fermi energy is

$$E_F = \tfrac{1}{2}mv_F^2$$

where $v_F = \hbar k_F/m$ is the Fermi velocity which can be expressed using the concept of the electron radius, r_s. It is equal to radius of a sphere occupied by one free electron. If the total volume of a metal sample is V and the number of conduction electrons in this volume is N, then the volume per electron is equal to

$$\frac{V}{N} = \frac{1}{n} = \frac{4}{3}\pi r_s^3$$

and

$$r_s = \left(\frac{3}{4\pi n}\right)^{1/3}$$

The following table contains information pertinent to the Sommerfeld model for some metals. The magnitudes of T_0 are calculated using the expression

$$T_0 = \frac{E_F}{k_B} = \frac{58.2 \cdot 10^4}{(r_s/r_B)^2} \text{ K}$$

Ground State Properties of the Electron Gas in Some Metals

Metal	Valency	$n/10^{28}$ m^{-3}	r_S/pm	r_S/r_B	E_F/eV	$T_0/10^4$ K	$k_F/10^{10}$ m^{-1}	$v_F/10^6$ m s^{-1}
Li[a]	1	4.70	172	3.25	4.74	5.51	1.12	1.29
Na[b]	1	2.65	208	3.93	3.24	3.77	0.92	1.07
K[b]	1	1.40	257	4.86	2.12	2.46	0.75	0.86
Rb[b]	1	1.15	275	5.20	1.85	2.15	0.70	0.81
Cs[b]	1	0.91	298	5.62	1.59	1.84	0.65	0.75
Cu	1	8.47	141	2.67	7.00	8.16	1.36	1.57
Ag	1	5.86	160	3.02	5.49	6.38	1.20	1.39
Au	1	5.90	159	3.01	5.53	6.42	1.21	1.40
Be	2	24.7	99	1.87	14.3	16.6	1.94	2.25
Mg	2	8.61	141	2.66	7.08	8.23	1.36	1.58
Ca	2	4.61	173	3.27	4.69	5.44	1.11	1.28
Sr	2	3.55	189	3.57	3.93	4.57	1.02	1.18
Ba	2	3.15	196	3.71	3.64	4.23	0.98	1.13
Nb	1	5.56	163	3.07	5.32	6.18	1.18	1.37
Fe	2	17.0	112	2.12	11.1	13.0	1.71	1.98
Mn[c]	2	16.5	113	2.14	10.9	12.7	1.70	1.96
Zn	2	13.2	122	2.30	9.47	11.0	1.58	1.83
Cd	2	9.27	137	2.59	7.47	8.68	1.40	1.62

Metal	Valency	$n/10^{28}$ m^{-3}	r_S/pm	r_S/r_B	E_F/eV	$T_0/10^4$ K	$k_F/10^{10}$ m^{-1}	$v_F/10^6$ m s^{-1}
Hg[a]	2	8.65	140	2.65	7.13	8.29	1.37	1.58
Al	3	18.1	110	2.07	11.7	13.6	1.75	2.03
Ga	3	15.4	116	2.19	10.4	12.1	1.66	1.92
In	3	11.5	127	2.41	8.63	10.0	1.51	1.74
Tl	3	10.5	131	2.48	8.15	9.46	1.46	1.69
Sn	4	14.8	117	2.22	10.2	11.8	1.64	1.90
Pb	4	13.2	122	2.30	9.47	11.0	1.58	1.83
Bi	5	14.1	119	2.25	9.90	11.5	1.61	1.87
Sb	5	16.5	113	2.14	10.9	12.7	1.70	1.96

[a] At 78 K.
[b] At 5 K.
[c] α-phase.
The data in the table are for atmospheric pressure and room temperature unless otherwise noted.

References

1. Drude, P., *Ann. Physik*, 1, 566, 1900; *ibid.*, 3, 369, 1900.
2. Sommerfeld, A. and Bethe, H., *Handbuch der Physik*, Chapter 3, Springer, 1933.
3. Wyckoff, R. W. G., *Crystal Structures*, 2nd. ed., Interscience, 1963.
4. Ashcroft, N. W. and Mermin, N. D., *Solid State Physics*, Holt, Rinehart and Winston, 1976.

COMMERCIAL METALS AND ALLOYS

This table gives typical values of mechanical, thermal, and electrical properties of several common commercial metals and alloys. Values refer to ambient temperature (0 to 25°C). All values should be regarded as typical, since these properties are dependent on the particular type of alloy, heat treatment, and other factors. Values for individual specimens can vary widely.

References

1. *ASM Metals Reference Book, Second Edition*, American Society for Metals, Metals Park, OH, 1983.
2. Lynch, C. T., *CRC Practical Handbook of Materials Science*, CRC Press, Boca Raton, FL, 1989.
3. Shackelford, J. F., and Alexander, W., *CRC Materials Science and Engineering Handbook*, CRC Press, Boca Raton, FL, 1991.

Common name	Thermal conductivity W/cm K	Density g/cm³	Coeff. of linear expansion 10⁻⁶/°C	Electrical resistivity μΩ cm	Modulus of elasticity GPa	Tensile strength MPa	Approx. melting point °C
Ingot iron	0.7	7.86	11.7	9.7	205	–	1540
Plain carbon steel AISI-SAE 1020	0.52	7.86	11.7	18	205	450	1515
Stainless steel type 304	0.15	7.9	17.3	72	195	550	1425
Cast gray iron	0.47	7.2	10.5	67	90	180	1175
Malleable iron		7.3	12	30	170	345	1230
Hastelloy C	0.12	8.94	11.3	125	200	780	1350
Inconel	0.15	8.25	11.5	103	200	800	1370
Aluminum alloy 3003, rolled	1.9	2.73	23.2	3.7	70	110	650
Aluminum alloy 2014, annealed	1.9	2.8	23.0	3.4	70	185	650
Aluminum alloy 360	1.5	2.64	21.0	7.5	70	325	565
Copper, electrolytic (ETP)	3.9	8.94	16.5	1.7	120	300	1080
Yellow brass (high brass)	1.2	8.47	20.3	6.4	100	300-800	930
Aluminum bronze	0.7	7.8	16.4	12	120	400-600	1050
Beryllium copper 25	0.8	8.23	17.8	7	130	500-1400	925
Cupronickel 30%	0.3	8.94	16.2		150	400-600	1200
Red brass, 85%	1.6	8.75	18.7	11	90	300-700	1000
Chemical lead	0.35	11.34	29.3	21	13	17	327
Antimonial lead (hard lead)	0.3	10.9	26.5	23	20	47	290
Solder 50-50	0.5	8.89	23.4	15	–	42	215
Magnesium alloy AZ31B	1.0	1.77	26	9	45	260	620
Monel	0.3	8.84	14.0	58	180	545	1330
Nickel (commercial)	0.9	8.89	13.3	10	200	460	1440
Cupronickel 55-45 (constantan)	0.2	8.9	18.8	49	160	–	1260
Titanium (commercial)	1.8	4.5	8.5	43	110	330-500	1670
Zinc (commercial)	1.1	7.14	32.5	6	–	130	419
Zirconium (commercial)	0.2	6.5	5.85	41	95	450	1855

HARDNESS OF MINERALS AND CERAMICS

There are several hardness scales for describing the resistance of a material to indentation or scratching. This table lists a number of common materials in order of increasing hardness. Values are given, when available, on three different hardness scales: the original Mohs Scale (range 1 to 10); the modified Mohs Scale (range 1 to 15), and the Knoop Hardness Scale. In the last case, a load of 100 g is assumed.

Reference

Shackelford, J. F. and Alexander, W., *CRC Materials Science and Engineering Handbook*, CRC Press, Boca Raton, FL, 1991.

Material	Formula	Mohs	Modified mohs	Knoop
Graphite	C	0.5		
Talc	$3MgO \cdot 4SiO_2 \cdot H_2O$	1	1	
Alabaster	$CaSO_4 \cdot 2H_2O$	1.7		
Gypsum	$CaSO_4 \cdot 2H_2O$	2	2	32
Halite (rock salt)	NaCl	2		
Stibnite (antimonite)	Sb_2S_3	2.0		
Galena	PbS	2.5		
Mica		2.8		
Calcite	$CaCO_3$	3	3	135
Barite	$BaSO_4$	3.3		
Marble		3.5		
Aragonite	$CaCO_3$	3.5		
Dolomite	$CaMg(CO_3)_2$	3.5		
Fluorite	CaF_2	4	4	163
Magnesia	MgO	5		370
Apatite	$CaF_2 \cdot 3Ca_3(PO_4)_2$	5	5	430
Opal		5		
Feldspar (orthoclase)	$K_2O \cdot Al_2O \cdot 6SiO_2$	6	6	560
Augite		6		
Hematite	Fe_2O_3	6		750
Magnetite	Fe_3O_4	6		
Rutile	TiO_2	6.2		
Pyrite	FeS_2	6.3		
Agate	SiO_2	6.5		
Uranium dioxide	UO_2	6.7		600
Silica (fused)	SiO_2		7	
Quartz	SiO_2	7	8	820
Flint		7		
Silicon	Si	7		
Andalusite	Al_2OSiO_4	7.5		
Zircon	$ZrSiO_4$	7.5		
Zirconia	ZrO_2			1200
Aluminum nitride	AlN			1225
Beryl	$Be_3Al_2Si_6O_{18}$	7.8		
Beryllia	BeO			1300
Topaz	$Al_2SiO_4(OH,F)_2$	8	9	1340
Garnet	$Al_2O_3 \cdot 3FeO \cdot 3SiO_2$		10	1360
Emery	Al_2O_3 (impure)	8		
Zirconium nitride	ZrN	8+		1510
Zirconium boride	ZrB_2			1560
Titanium nitride	TiN	9		1770
Zirconia (fused)	ZrO_2		11	
Tantalum carbide	TaC			1800
Tungsten carbide	WC			1880
Corundum (alumina)	Al_2O_3	9		2025
Zirconium carbide	ZrC			2150
Alumina (fused)	Al_2O_3		12	
Beryllium carbide	Be_2C			2400
Titanium carbide	TiC			2470
Carborundum (silicon carbide)	SiC	9.3	13	2500
Aluminum boride	AlB			2500
Tantalum boride	TaB_2			2600
Boron carbide	B_4C		14	2800
Boron	B	9.5		
Titanium boride	TiB_2			2850
Diamond	C	10	15	7000

Section 13
Polymer Properties

NOMENCLATURE FOR ORGANIC POLYMERS

Robert B. Fox and Edward S. Wilks

Organic polymers have traditionally been named on the basis of the monomer used, a hypothetical monomer or a semi-systematic structure. Alternatively, they may be named in the same way as organic compounds, i.e., on the basis of a structure as drawn. The former method, often called "source-based nomenclature" or "monomer-based nomenclature," sometimes results in ambiguity and multiple names for a single material. The latter method, termed "structure-based nomenclature," generates a sometimes cumbersome unique name for a given polymer, independent of its source. Within their limitations, both types of names are acceptable and well-documented.[1] The use of stereochemical descriptors with both types of polymer nomenclature has been published.[2]

Traditional Polymer Names

Monomer-Based Names

"Polystyrene" is the name of a homopolymer made from the single monomer styrene. When the name of a monomer comprises two or more words, the name should be enclosed in parentheses, as in "poly(methyl methacrylate)" or "poly(4-bromostyrene)" to identify the monomer more clearly. This method can result in several names for a given polymer: thus, "poly(ethylene glycol)," "poly(ethylene oxide)," and "poly(oxirane)" describe the same polymer. Sometimes, the name of a hypothetical monomer is used, as in "poly(vinyl alcohol)." Even though a name like "polyethylene" covers a multitude of materials, the system does provide understandable names when a single monomer is involved in the synthesis of a single polymer. When one monomer can yield more than one polymer, e.g., 1,3-butadiene or acrolein, some sort of structural notation must be used to identify the product, and one is not far from a formal structure-based name.

Copolymers, Block Polymers, and Graft Polymers. When more than one monomer is involved, monomer-based names are more complex. Some common polymers have been given names based on an apparent structure, as with "poly(ethylene terephthalate)." A better system has been approved by the IUPAC.[1] With this method, the arrangement of the monomeric units is introduced through use of an italicized connective placed between the names of the monomers. For monomer names represented by A, B, and C, the various types of arrangements are shown in Table 1.

Table 2 contains examples of common or semi-systematic names of copolymers. The systematic names of comonomers may also be used; thus, the polyacrylonitrile-*block*-polybutadiene-*block*-polystyrene polymer in Table 2 may also be named poly(prop-2-enenitrile)-*block*-polybuta-1,3-diene-*block*-poly(ethenylbenzene). IUPAC does not require alphabetized names of comonomers within a polymer name; many names are thus possible for some copolymers.

These connectives may be used in combination and with small, non-repeating (i.e. non-polymeric) junction units; see, for example, Table 2, line 8. A long dash may be used in place of the connective -*block*-; thus, in Table 2, the polymers of lines 7 and 8 may also be written as shown on lines 9 and 10.

IUPAC also recommends an alternative scheme for naming copolymers that comprises use of "copoly" as a prefix followed by the names of the comonomers, a solidus (an oblique stroke) to separate comonomer names, and addition before "copoly" of any applicable connectives listed in Table 2 except -*co*-.

Table 3 gives the same examples shown in Table 2 but with the alternative format. Comonomer names need not be parenthesized.

TABLE 1. IUPAC Source-Based Copolymer Classification

No.	Copolymer type	Connective	Example
1	Unspecified or unknown	-*co*-	poly(A-*co*-B)
2	Random (obeys Bernoullian distribution)	-*ran*-	poly(A-*ran*-B)
3	Statistical (obeys known statistical laws)	-*stat*-	poly(A-*stat*-B)
4	Alternating (for two monomeric units)	-*alt*-	poly(A-*alt*-B)
5	Periodic (ordered sequence for 2 or more monomeric units)	-*per*-	poly(A-*per*-B-*per*-C)
6	Block (linear block arrangement)	-*block*-	polyA-*block*-polyB
7	Graft (side chains connected to main chains)	-*graft*-	polyA-*graft*-polyB

TABLE 2. Examples of Source-Based Copolymer Nomenclature

No.	Copolymer name
1	poly(propene-*co*-methacrylonitrile)
2	poly[(acrylic acid)-*ran*-(ethyl acrylate)]
3	poly(butene-*stat*-ethylene-*stat*-styrene)
4	poly[(sebacic acid)-*alt*-butanediol]
5	poly[(ethylene oxide)-*per*-(ethylene oxide)-*per*-tetrahydrofuran]
6	polyisoprene-*graft*-poly(methacrylic acid)
7	polyacrylonitrile-*block*-polybutadiene-*block*-polystyrene
8	polystyrene-*block*-dimethylsilylene-*block*-polybutadiene
9	polyacrylonitrile—polybutadiene—polystyrene
10	polystyrene—dimethylsilylene—polybutadiene

TABLE 3. Examples of Source-Based Copolymer Nomenclature (Alternative Format)

No.	Polymer name
1	copoly(propene/methacrylonitrile)
2	*ran*-copoly(acrylic acid/ethyl acrylate)
3	*stat*-copoly(butene/ethylene/styrene)
4	*alt*-copoly(sebacic acid/butanediol)
5	*block*-copoly(acrylonitrile/butadiene/styrene)
6	*per*-copoly(ethylene oxide/ethylene oxide/tetrahydrofuran)
7	*graft*-copoly(isoprene/methacrylic acid)

Source-based nomenclature for non-linear macromolecules and macromolecular assemblies is covered by a 1997 IUPAC document.[11] The types of polymers in these classes, together with their connectives, are given in Table 4; the terms shown may be used as connectives, prefixes, or both to designate the features present.

TABLE 4. Connectives for Non-Linear Macromolecules and Macromolecular Assemblies

No.	Type	Connective
	Branched (type unspecified)	branch
	Branched with branch point of functionality f	f-branch
	Comb	comb
	Cross-link	ι (Greek iota)
	Cyclic	cyclo
	Interpenetrating polymer network	ipn
	Long-chain branched	l-branch
	Network	net
	Polymer blend	blend
	Polymer-polymer complex	compl
	Semi-interpenetrating polymer network	sipn
	Short-chain branched	sh-branch
	Star	star
	Star with f arms	f-star

Non-linear polymers are named by using the italicized connective as a *prefix* to the source-based name of the polymer component or components to which the prefix applies; some examples are listed in Table 5.

TABLE 5. Non-Linear Macromolecules

Polymer name	Polymer structural features
poly(methacrylic acid)-*comb*-polyacrylonitrile	Comb polymer with a poly(methacrylic acid) backbone and polyacrylonitrile side chains
comb-poly[ethylene-*stat*-(vinyl chloride)]	Comb polymer with unspecified backbone composition and statistical ethylene/vinyl chloride copolymer side chains
polybutadiene-*comb*-(polyethylene; polypropene)	Comb polymer with butadiene backbone and side chains of polyethylene and polypropene
star-(polyA; polyB; polyC; polyD; polyE)	Star polymer with arms derived from monomers A, B, C, D, and E, respectively
star-(polyA-*block*-polyB-*block*-polyC)	Star polymer with every arm comprising a tri-block segment derived from comonomers A, B, and C
star-poly(propylene oxide)	A star polymer prepared from propylene oxide
5-*star*-poly(propylene oxide)	A 5-arm star polymer prepared from propylene oxide
star-(polyacrylonitrile; polypropylene) (M_r 10000: 25000)	A star polymer containing polyacrylonitrile arms of MW 10000 and polypropylene arms of MW 25000

Macromolecular assemblies held together by forces other than covalent bonds are named by inserting the appropriate italicized connective between names of individual components; Table 6 gives examples.

TABLE 6. Examples of Polymer Blends and Nets

No.	Polymer name
1	polyethylene-*blend*-polypropene
2	poly(methacrylic acid)-*blend*-poly(ethyl acrylate)
3	*net*-poly(4-methylstyrene-ι-divinylbenzene)
4	*net*-poly[styrene-*alt*-(maleic anhydride)]-ι-(polyethylene glycol; polypropylene glycol)
5	*net*-poly(ethyl methacrylate)-*sipn*-polyethylene
6	[*net*-poly(butadiene-*stat*-styrene)]-*ipn*-[*net*-poly(4-methylstyrene-ι-divinylbenzene)]

Structure-Based Polymer Nomenclature
Regular Single-Strand Polymers

Structure-based nomenclature has been approved by the IUPAC[4] and is currently being updated; it is used by *Chemical Abstracts*.[5] Monomer names are not used. To the extent that a polymer chain can be described by a repeating unit in the chain, it can be named "poly(repeating unit)". For regular single-strand polymers, "repeating unit" is a bivalent group; for regular double-strand (ladder and spiro) polymers, "repeating unit" is usually a tetravalent group.[9]

Since there are usually many possible repeating units in a given chain, it is necessary to select one, called the "constitutional repeating unit" (CRU) to provide a unique and unambiguous name, "poly(CRU)," where "CRU" is a recitation of the names of successive units as one proceeds through the CRU from left to right. For this purpose, a portion of the main chain structure that includes at least two repeating sequences is written out. These sequences will typically be composed of bivalent subunits such as -CH_2-, -O-, and groups from ring systems, each of which can be named by the usual nomenclature rules.[6,7]

Where a chain is simply one long sequence comprising repetition of a single subunit, that subunit is itself the CRU, as in "poly(methylene)" or "poly(1,4-phenylene)." In chains having more than one kind of subunit, a seniority system is used to determine the beginning of the CRU and the direction in which to move along the main chain atoms (following the shortest path in rings) to complete the CRU. Determination of the first, most senior, subunit, is based on a descending order of seniority: (1) heterocyclic rings, (2) hetero atoms, (3) carbocyclic rings, and, lowest, (4) acyclic carbon chains.

Within each of these classes, there is a further order of seniority that follows the usual rules of nomenclature.

Heterocycles: A nitrogen-containing ring system is senior to a ring system not containing nitrogen.[4,9] Further descending order of seniority is determined by:

(i)   the highest number of rings in the ring system
(ii) the largest individual ring in the ring system
(iii) the largest number of hetero atoms
(iv) the greatest variety of hetero atoms

Hetero atoms: The senior bivalent subunit is the one nearest the top right-hand corner of the Periodic Table; the order of seniority is: O, S, Se, Te, N, P, As, Sb, Bi, Si, Ge, Sn, Pb, B, Hg.

Carbocycles: Seniority[4] is determined by:

(i)   the highest number of rings in the ring system
(ii) the largest individual ring in the ring system
(iii) degree of ring saturation; an unsaturated ring is senior to a saturated ring of the same size

Carbon chains: Descending order of seniority is determined by:

(i)　chain length (longer is senior to shorter)

(ii)　highest degree of unsaturation

(iii)　number of substituents (higher number is senior to lower number)

(iv)　ascending order of locants

(v)　alphabetical order of names of substituent groups

Among equivalent ring systems, preference is given to the one ving lowest locants for the free valences in the subunit, and ong otherwise identical ring systems, the one having least drogenation is senior. Lowest locants in unsaturated chains are o given preference. Lowest locants for substituents are the final terminant of seniority.

Direction within the repeating unit depends upon the shortest th, which is determined by counting main chain atoms, both clic and acyclic, from the most senior subunit to another subit of the same kind or to a subunit next lower in seniority. hen identification and orientation of the CRU have been complished, the CRU is named by writing, in sequence, the mes of the largest possible subunits within the CRU from left to ht. For example, the main chain of the polymer traditionally med "poly(ethylene terephthalate)" has the structure shown in gure 1.

gure 1. Structure-based name: poly(oxyethyleneoxyterephthaloyl); traional name: poly(ethylene terephthalate).

The CRU in Figure 1 is enclosed in brackets and read from left right. It is selected because (1) either backbone oxygen atom alifies as the "most senior subunit," (2) the shortest path length om either -O- to the other -O- is via the ethylene subunit. Oritation of the CRU is thus defined by (1) beginning at the -O-arked with an asterisk, and (2) reading in the direction of the row. The structure-based name of this polymer is therefore oly(oxyethyleneoxyterephthaloyl)," not much longer than the aditional name and much more adaptable to the complexities of bstitution. As organic nomenclature evolves, more systematic mes may be used for subunits, e.g. "ethane-1,2-diyl" instead of thylene." IUPAC still prefers "ethylene" for the -CH_2-CH_2- unit, wever, but also accepts "ethane-1,2-diyl."

Structure-based nomenclature can also be used when the CRU ckbone has no carbon atoms. An example is the polymer tradionally named "poly(dimethylsiloxane)," which on the basis of ructure would be named "poly(oxydimethylsilylene)" or oly(oxydimethylsilanediyl)." This nomenclature method has also een applied to inorganic and coordination polymers[8] and to doue-strand (ladder and spiro) organic polymers.[9]

Irregular Single-Strand Polymers

Polymers that cannot be described by the repetition of a single RU or comprise units not all connected identically in a direcional sense can also be named on a structure basis.[10] These clude copolymers, block and graft polymers, and star polymers. hey are given names of the type "poly(A/B/C...)," where A, B, C, c. are the names of the component constitutional units, the

number of which are minimized. The constitutional units may include regular or irregular blocks as well as atoms or atomic groupings, and each is named by the method described above or by the rules of organic nomenclature.

The solidus denotes an unspecified arrangement of the units within the main chain.[10] For example, a statistical copolymer derived from styrene and vinyl chloride with the monomeric units joined head-to-tail is named "poly(l-chloroethylene/l-phenylethylene)." A polymer obtained by 1,4-polymerization and both head-to-head and head-to-tail 1,2- polymerization of 1,3-butadiene would be named "poly(but-1-ene-l,4-diyl/l-vinylethylene/2-vinylethylene)."[12] In graphic representations of these polymers, shown in Figure 2, the hyphens or dashes at each end of each CRU depiction are shown *completely within* the enclosing parentheses; this indicates that they are not necessarily the terminal bonds of the macromolecule.

Figure 2. Graphic representations of copolymers.

A long hyphen is used to separate components in names of block polymers, as in "poly(A)—poly(B)—poly(C)," or "poly(A) —X—poly(B)" in which X is a non-polymeric junction unit, e.g., dimethylsilylene.

In graphic representations of these polymers, the blocks are shown connected when the bonding is known (Figure 3, for example); when the bonding between the blocks is unknown, the blocks are separated by solidi and are shown *completely within* the outer set of enclosing parentheses (Figure 4, for example).[10,13]

Figure 3. polystyrene—polyethylene—polystyrene.

Figure 4. poly[poly(methyl methacrylate)—polystyrene—poly(methyl acrylate)].

Graft polymers are named in the same way as a substituted polymer but without the ending "yl" for the grafted chain; the name of a regular polymer, comprising Z units in which some have grafts of "poly(A)," is "poly[Z/poly(A)Z]." Star polymers are treated as a central unit with substituent blocks, as in "tetrakis(polymethylene)silane."[10,13]

Other Nomenclature Articles and Publications

In addition to the *Chemical Abstracts* and IUPAC documents cited above and listed below, other articles on polymer nomenclature are available. A 1999 article lists significant documents on polymer nomenclature published during the last 50 years in books, encyclopedias, and journals by *Chemical Abstracts*,

JPAC, and individual authors.[14] A comprehensive review of ~urce-based and structure-based nomenclature for all of the ~ajor classes of polymers,[15] and a short tutorial on the correct entification, orientation, and naming of most commonly ~countered constitutional repeating units were both published 2000.[16]

References and Notes

. International Union of Pure and Applied Chemistry, *Compendium of Macromolecular Nomenclature,* Blackwell Scientific Publications, Oxford, 1991.

?. International Union of Pure and Applied Chemistry, Stereochemical Definitions and Notations Relating to Polymers (Recommendations 1980), *Pure Appl. Chem.,* 53, 733-752 (1981).

\. International Union of Pure and Applied Chemistry, Source-Based Nomenclature for Copolymers (Recommendations 1985), *Pure Appl. Chem.,* 57, 1427-1440 (1985).

. International Union of Pure and Applied Chemistry, Nomenclature of Regular Single-Strand Organic Polymers (Recommendations 1975, *Pure Appl. Chem.,* 48, 373-385 (1976).

. Chemical Abstracts Service, Naming and Indexing of Chemical Substances for Chemical Abstracts, Appendix IV, *Chemical Abstracts 1999 Index Guide.*

. International Union of Pure and Applied Chemistry, *A Guide to IUPAC Nomenclature of Organic Compounds* (1993), Blackwell Scientific Publications, Oxford, 1993.

. International Union of Pure and Applied Chemistry, *Nomenclature of Organic Chemistry, Sections A, B, C, D, E, F, and H,* Pergamon Press, Oxford, 1979.

. International Union of Pure and Applied Chemistry, Nomenclature of Regular Double-Strand and Quasi-Single-Strand Inorganic and Coordination Polymers (Recommendations 1984), *Pure Appl. Chem.,* 57, 149-168 (1985).

9. International Union of Pure and Applied Chemistry, Nomenclature of Regular Double-Strand (Ladder and Spiro) Organic Polymers (Recommendations 1993), *Pure Appl. Chem.,* 65, 1561-1580 (1993).

10. International Union of Pure and Applied Chemistry, Structure-Based Nomenclature for Irregular Single-Strand Organic Polymers (Recommendations 1994), *Pure Appl. Chem.,* 66, 873-889 (1994).

11. International Union of Pure and Applied Chemistry, "Source-Based Nomenclature for Non-Linear Macromolecules and Macromolecular Assemblies (Recommendations 1997)." *Pure Appl. Chem.,* 69, 2511-2521 (1997).

12. Poly(1,3-butadiene) obtained by polymerization of 1,3-butadiene in the so-called 1,4- mode is frequently drawn incorrectly in publications as $-(CH_2-CH=CH-CH_2)_n-$; the double bond should be assigned the lowest locant possible, i.e., the structure should be drawn as $-(CH=CH-CH_2-CH_2)_n-$.

13. International Union of Pure and Applied Chemistry, "Graphic Representations (Chemical Formulae) of Macromolecules (Recommendations 1994)." *Pure Appl. Chem.,* 66, 2469-2482 (1994).

14. Wilks, E. S. Macromolecular Nomenclature Note No. 17: "Whither Nomenclature?" *Polym. Prepr.* 40(2), 6-11 (1999); also available at www.chem.umr.edu/~poly/nomenclature.html.

15. Wilks, E. S. "Polymer Nomenclature: The Controversy Between Source-Based and Structure-Based Representations (A Personal Perspective)." *Prog. Polym. Sci.* 25, 9-100 (2000).

16. Wilks, E. S. Macromolecular Nomenclature Note No. 18: "SRUs: Using the Rules." *Polym. Prepr.* 41(1), 6a-11a (2000); also available at www.chem.umr.edu/~poly/nomenclature.html; a .pdf format version is also available.

SOLVENTS FOR COMMON POLYMERS

Abbreviations:
HC: hydrocarbons
MEK: methyl ethyl ketone

THF: tetrahydrofuran
DMF: dimethylformamide
DMSO: dimethylsulfoxide

Polyethylene (HDPE)	HC and halogenated HC
Polypropylene (atactic)	HC and halogenated HC
Polybutadiene	HC, THF, ketones
Polystyrene	ethylbenzene, CHCl$_3$, CCl$_4$, THF, MEK
Polyacrylates	aromatic HC, chlorinated HC, THF, esters, ketones
Polymethacrylates	aromatic HC, chlorinated HC, THF, esters, MEK
Polyacrylamide	water
Poly(vinyl ethers)	halogenated HC, MEK, butanol
Poly(vinyl alcohol)	glycols (hot), DMF
Poly(vinyl acetate)	aromatic HC, chlorinated HC, THF, esters, DMF
Poly(vinyl chloride)	THF, DMF, DMSO
Poly(vinylidene chloride)	THF (hot), dioxane, DMF
Poly(vinyl fluoride)	DMF, DMSO (hot)
Polyacrylonitrile	DMF, DMSO
Poly(oxyethylene)	aromatic HC, CHCl$_3$, alcohols, esters, DMF
Poly(2,6-dimethylphenylene oxide)	aromatic HC, halogenated HC
Poly(ethylene terephthalate)	phenol, DMSO (hot)
Polyurethanes (linear)	aromatic HC, THF, DMF
Polyureas	phenol, formic acid
Polysiloxanes	HC, THF, DMF
Poly[bis(2,2,2-trifluoroethoxy)-phosphazene]	THF, ketones, ethyl acetate

GLASS TRANSITION TEMPERATURE FOR SELECTED POLYMERS

Robert B. Fox

Polymer names are based on the IUPAC structure-based nomenclature system described in the table "Naming Organic Polymers." Within each category, names are listed in alphabetical order. Source-based and trivial names are also given (in italics) for the most common polymers. The table does not include polymers for which T_g is not clearly defined because of variability of structure or because of reactions taking place near the glass transition.

All values of T_g cited in this table have been determined by differential scanning calorimetry (DSC) except those values indicated by:

(D)	dynamic method
(Dil)	dilatometry
(M)	mechanical method

Polymer name	Glass transition temperature (T_g/K)
ACYCLIC CARBON CHAINS	
Polyalkadienes	
Poly(alkenylene) *Polyalkadiene* $-[CH=CHCH_2CH_2]-$	
Poly(*cis*-1-butenylene)	171
cis-1,3-polybutadiene [PBD]	
Poly(*trans*-1-butenylene)	215
trans-1,3-polybutadiene [PBD]	
Poly(1-chloro-*cis*-1-butenylene)	253
cis-1,3-polychloroprene	
Poly(1-chloro-*trans*-1-butenylene)	233
trans-1,3-polychloroprene	
Poly(1-methyl-*cis*-1-butenylene)	200
cis-1,3-polyisoprene	
Poly(1-methyl-*trans*-1-butenylene)	207
trans-1,3-polyisoprene	
Poly(1,4,4-trifluoro-1-butenylene)	238
Polyalkenes	
Poly(alkylethylene) *Poly(alkylethylene)* $-[RCHCH_2]-$	
Poly(1-benzylethylene)	333
Poly(1-butylethylene)	223
Poly(1-cyclohexylethylene) (atactic)	393
Poly(1-cyclohexylethylene) (isotactic)	406 (D)
Poly(1,1-dimethylethylene)	200
Polyisobutylene [PIB]	
Poly(ethylene)	148
Poly(methylene)	155
Poly(1-phenethylethylene)	283
Poly(propylene) (isotactic)	272
Poly(propylene) (syndiotactic)	ca. 265
Poly[1-(2-pyridyl)ethylene]	377
Poly[1-(4-pyridyl)ethylene]	415
Poly(1-vinylethylene)	273
Polyacrylics	
Poly[1-(alkoxycarbonyl)ethylene] *Poly(alkyl acrylate)* $-[(ROCO)CHCH_2]-$	
Poly[1-(benzyloxycarbonyl)ethylene]	279
Poly[1-(butoxycarbonyl)ethylene]	219 (M)
Poly(butyl acrylate) [PBA]	
Poly[1-(*sec*-butoxycarbonyl)ethylene]	251
Poly[1-(butoxycarbonyl)-1-cyanoethylene]	358
Poly[1-(butylcarbamoyl)ethylene]	319 (M)
Poly(1-carbamoylethylene)	438
Polyacrylamide [PAM]	
Poly(1-carboxyethylene)	379

Polymer name	Glass transition temperature (T_g/K)
Poly(acrylic acid) [PAA]	
Poly[1-(2-chlorophenoxycarbonyl)ethylene]	326
Poly[1-(4-chlorophenoxycarbonyl)ethylene]	331
Poly[1-(4-cyanobenzyloxycarbonyl)ethylene]	317
Poly[1-(2-cyanoethoxycarbonyl)ethylene]	277
Poly[1-(cyanomethoxycarbonyl)ethylene)]	433 Dil
Poly[1-(4-cyanophenoxycarbonyl)ethylene]	363
Poly[1-(cyclohexyloxycarbonyl)ethylene]	292
Poly[1-(2,4-dichlorophenoxycarbonyl)ethylene]	333
Poly[1-(dimethylcarbamoyl)ethylene]	362
Poly[1-(ethoxycarbonyl)ethylene]	249
Poly(ethyl acrylate) [PEA]	
Poly[1-(ethoxycarbonyl)-1-fluoroethylene]	316
Poly[1-(2-ethoxycarbonylphenoxycarbonyl)ethylene]	303
Poly[1-(3-ethoxycarbonylphenoxycarbonyl)ethylene]	297
Poly[1-(4-ethoxycarbonylphenoxycarbonyl)ethylene]	310
Poly[1-(2-ethoxyethoxycarbonyl)ethylene]	223
Poly[1-(3-ethoxypropoxycarbonyl)ethylene]	218
Poly[1-(isopropoxycarbonyl)ethylene]	267-270
Poly[1-(methoxycarbonyl)ethylene]	283
Poly(methyl acrylate) [PMA]	
Poly[1-(2-methoxycarbonylphenoxycarbonyl)ethylene]	319
Poly[1-(3-methoxycarbonylphenoxycarbonyl)ethylene]	311
Poly[1-(4-methoxycarbonylphenoxycarbonyl)ethylene]	340
Poly[1-(2-methoxyethoxycarbonyl)ethylene]	223
Poly[1-(4-methoxyphenoxycarbonyl)ethylene]	324
Poly[1-(3-methoxypropoxycarbonyl)ethylene]	198
Poly[1-(2-naphthyloxycarbonyl)ethylene]	358
Poly[1-(pentachlorophenoxycarbonyl)ethylene]	420
Poly[1-(phenethoxycarbonyl)ethylene]	270
Poly[1-(phenoxycarbonyl)ethylene]	330
Poly[1-(*m*-tolyloxycarbonyl)ethylene]	298
Poly[1-(*o*-tolyloxycarbonyl)ethylene]	325
Poly[1-(*p*-tolyloxycarbonyl)ethylene]	316
Poly[1-(2,2,2-trifluorethoxycarbonyl)ethylene]	263

Polymethacrylics

Poly[1-(alkoxycarbonyl)-1-methylethylene] *Poly(alkyl methacrylate)* $-[(ROCO)(Me)CCH_2]-$

Polymer name	Glass transition temperature (T_g/K)
Poly[1-(benzyloxycarbonyl)-1-methylethylene]	327
Poly[1-(2-bromoethoxycarbonyl)-1-methylethylene]	325
Poly[(1-(butoxycarbonyl)-1-methylethylene]	293
Poly(butyl methacrylate) [PBMA]	
Poly[1-(*sec*-butoxycarbonyl)-1-methylethylene]	333
Poly[1-(*tert*-butoxycarbonyl)-1-methylethylene)]	391
Poly[1-(2-chloroethoxycarbonyl)-1-methylethylene]	ca 315
Poly[1-(2-cyanoethoxycarbonyl)-1-methylethylene]	364
Poly[1-(4-cyanophenoxycarbonyl)-1-methylethylene]	428
Poly[1-(cyclohexyloxycarbonyl)-1-methylethylene] (atactic)	356
Poly[1-(cyclohexyloxycarbonyl)-1-methylethylene)] (isotactic)	324
Poly[1-(dimethylaminoethoxycarbonyl)-1-methylethylene]	292
Poly[1-(ethoxycarbonyl)-1-ethylethylene]	300
Poly[1-(ethoxycarbonyl)-1-methylethylene] (atactic) *Poly(ethyl methacrylate)* [PEMA]	338
Poly[1-(ethoxycarbonyl)-1-methylethylene] (isotactic)	285
Poly[1-(ethoxycarbonyl)-1-methylethylene)] (syndiotactic)	339
Poly[1-(hexyloxycarbonyl)-1-methylethylene]	268
Poly[1-(isobutoxycarbonyl)-1-methylethylene]	326
Poly[1-(isopropoxycarbonyl)-1-methylethylene]	354
Poly[1-(methoxycarbonyl)-1-methylethylene] (atactic) *Poly(methyl methacrylate)* [PMMA]	378
Poly[1-(methoxycarbonyl)-1-methylethylene)] (isotactic)	311
Poly[1-(methoxycarbonyl)-1-methylethylene)] (syndiotactic)	378
Poly[1-(4-methoxycarbonylphenoxy)-1-methylethylene]	379

Polymer name	Glass transition temperature (T_g/K)
Poly[1-(methoxycarbonyl)-1-phenylethylene)] (atactic)	391
Poly[1-(methoxycarbonyl)-1-phenylethylene)] (isotactic)	397
Poly[1-methyl-1-(phenethoxycarbonyl)ethylene]	299
Poly[1-methyl-1-(phenoxycarbonyl)ethylene]	383

Polyvinyl ethers, alcohols, and ketones

Poly(1-alkoxyethylene) *Poly(alkyl vinyl ether)* –[ROCHCH$_2$]–
Poly(1-hydroxyethylene) *Poly(vinyl alcohol)* –[HOCHCH$_2$]–
Poly(1-alkanoylethylene) *Poly(alkyl vinyl ketone)* –[RCOCHCH$_2$]–

Poly(1-butoxyethylene)	218
Poly(1-*sec*-butoxyethylene)	253
Poly(1-*tert*-butoxyethylene)	361
Poly[1-(butylthio)ethylene]	253
Poly(1-ethoxyethylene)	230
Poly[1-(4-ethylbenzoyl)ethylene]	325
Poly(1-hydroxyethylene)	358 (D)
Poly(vinyl alcohol) [PVA]	
Poly(hydroxymethylene)	407
Poly(1-isopropoxyethylene)	270
Poly[1-(4-methoxybenzoyl)ethylene]	319 (M)
Poly(1-methoxyethylene)	242
Poly(methyl vinyl ether) [PMVE]	
Poly[1-(methylthio)ethylene]	272
Poly(1-propoxyethylene)	224
Poly[1-(trifluoromethoxy)trifluoroethylene]	268

Polyvinyl halides and nitriles

Poly(1-haloethylene) *Poly(vinyl halide)* –[XCHCH$_2$]–
Poly(1-cyanoethylene) *Poly(acrylonitrile)* –[NCCHCH$_2$]–

Poly(1-chloroethylene)	354
Poly(vinyl chloride) [PVC]	
Poly(chlorotrifluoroethylene)	373
Poly(1-cyanoethylene)	370
Polyacrylonitrile [PAN]	
Poly(1-cyano-1-methylethylene)	393
Polymethacrylonitrile	
Poly(1,1-dichloroethylene)	255
Poly(vinylidene chloride)	
Poly(1,1-difluoroethylene)	ca 233
Poly(vinylidene fluoride)	
Poly(1-fluoroethylene)	314 (M)
Poly(vinyl fluoride)	
Poly(1-hexafluoropropylene)	425
Poly[1-(2-iodoethyl)ethylene]	343
Poly(tetrafluoroethylene)	(160)
Poly[1-(trifluoromethyl)ethylene]	300

Polyvinyl esters

Poly[1-(alkanoyloxy)ethylene] *Poly(vinyl alkanoate)* –[RCOOCHCH$_2$]–

Poly(1-acetoxyethylene)	305
Poly(vinyl acetate) [PVAc]	
Poly[1-(benzoyloxy)ethylene]	344
Poly[1-(4-bromobenzoyloxy)ethylene]	365
Poly[1-(2-chlorobenzoyloxy)ethylene]	335
Poly[1-(3-chlorobenzoyloxy)ethylene]	338
Poly[1-(4-chlorobenzoyloxy)ethylene]	357
Poly[1-(cyclohexanoyloxy)ethylene]	349 (M)
Poly[1-(4-ethoxybenzoyloxy)ethylene]	343
Poly[1-(4-ethylbenzoyloxy)ethylene]	326

Polymer name	Glass transition temperature (T_g/K)
Poly[1-(4-isopropylbenzoyloxy)ethylene]	342
Poly[1-(2-methoxybenzoyloxy)ethylene]	338
Poly[1-(3-methoxybenzoyloxy)ethylene]	ca 317
Poly[1-(4-methoxybenzoyloxy)ethylene]	360
Poly[1-(4-methylbenzoyloxy)ethylene]	343
Poly[1-(4-nitrobenzoyloxy)ethylene]	395
Poly[1-(propionoyloxy)ethylene]	283 (M)

Polystyrenes

Poly(1-phenylethylene) *Polystyrene* $-[C_6H_5CHCH_2]-$

Poly[1-(4-acetylphenyl)ethylene]	389 (M)
Poly[1-(4-benzoylphenyl)ethylene]	371 (M)
Poly[1-(4-bromophenyl)ethylene]	391
Poly[1-(4-butoxyphenyl)ethylene]	ca 320 (M)
Poly[1-(4-butoxycarbonylphenyl)ethylene]	349 (M)
Pol[(1-(4-butylphenyl)ethylene]	279
Poly[1-(4-carboxyphenyl)ethylene]	386 (M)
Poly[1-(2-chlorophenyl)ethylene]	392
Poly[1-(3-chlorophenyl)ethylene]	363
Poly[1-(4-chlorophenyl)ethylene]	383
Poly[1-(2,4-dichlorophenyl)ethylene]	406
Poly[1-(2,5-dichlorophenyl)ethylene]	379
Poly[1-(2,6-dichlorophenyl)ethylene]	440
Poly[1-(3,4-dichlorophenyl)ethylene]	401
Poly[1-(2,4-dimethylphenyl)ethylene]	385
Poly[1-(4-(dimethylamino)phenyl)ethylene]	398 (M)
Poly[1-(4-ethoxyphenyl)ethylene]	ca 359 (M)
Poly[1-(4-ethoxycarbonylphenyl)ethylene]	367 (M)
Poly[1-(4-fluorophenyl)ethylene]	368
Poly[1-(4-iodophenyl)ethylene]	429
Poly[1-(4-methoxyphenyl)ethylene]	386
Poly[1-(4-methoxycarbonylphenyl)ethylene]	386 (M)
Poly(1-methyl-1-phenylethylene)	373
Poly(α-methylstyrene)	
Poly[1-(2-(methylamino)phenyl)ethylene]	462 (M)
Poly(1-phenylethylene)	373
Polystyrene [PS]	
Poly[1-(4-propoxyphenyl)ethylene]	343 (M)
Poly[1-(4-propoxycarbonylphenyl)ethylene]	365 (M)
Poly(1-*o*-tolylethylene)	409

CHAINS WITH CARBOCYCLIC UNITS

Poly(arylenealkylene) $-[-Ar-(CH_2)_n]-$

Poly[1-(2-bromo-1,4-phenylene)ethylene]	353 (M)
Poly[1-(2-chloro-1,4-phenylene)ethylene]	343 (M)
Poly[1-(2-cyano-1,4-phenylene)ethylene]	363 (M)
Poly[1-(2,5-dimethyl-1,4-phenylene)ethylene]	373 (M)
Poly[1-(2-ethyl-1,4-phenylene)ethylene]	298 (M)
Poly[1-(1,4-naphthylene)ethylene]	433 (M)
Poly[1-(1,4-phenylene)ethylene]	ca 353 (M)

CHAINS WITH HETEROATOM UNITS

Main chain oxide units

Poly(oxyalkylene) *Poly(alkylene oxide)* $-[O(CH_2)_n]-$

Poly[oxy(1,1-bis(chloromethyl)trimethylene)]	265
Poly[oxy(1-(bromomethyl)ethylene)]	259
Poly[oxy(1-(butoxymethyl)ethylene)]	194
Poly[oxy(1-butylethylene)]	203
Poly[oxy(1-*tert*-butylethylene)]	308
Poly[oxy(1-(chloromethyl)ethylene)]	251

Polymer name	Glass transition temperature (T_g/K)
Poly(epichlorohydrin)	
Poly[oxy(2,6-dimethoxy-1,4-phenylene)]	440
Poly[oxy(1,1-dimethylethylene)]	264
Poly[oxy(2,6-dimethyl-1,4-phenylene)]	482
Poly[oxy(2,6-diphenyl-1,4-phenylene)]	493
Poly[oxy(1-ethylethylene)]	203
Poly(oxyethylidene)	243
Polyacetaldehyde	
Poly[oxy(1-(methoxymethyl)ethylene)]	211
Poly[oxy(2-methyl-6-phenyl-1,4-phenylene)]	428
Poly[oxy(1-methyltrimethylene)]	223 (D)
Poly[oxy(2-methyltrimethylene)]	218
Poly(oxy-1,4-phenylene)	358
Poly(phenylene oxide) [PPO]	
Poly[oxy(1-phenylethylene)]	313
Poly(oxytetramethylene)	189
Poly(tetrahydrofuran) [PTMO]	
Poly(oxytrimethylene)	195

Main-chain ester or anhydride units

Poly(oxyalkyleneoxyalkanedioyl) *Poly(alkylene alkanedioate)*--[O(CH$_2$)$_m$OCO(CH$_2$)$_n$CO]–

Poly(oxyadipoyloxydecamethylene)	217
Poly(oxyadipoyloxy-1,4-phenyleneisopropylidene-1,4-phenylene)	341
Poly(oxycarbonyloxy-1,4-phenylene-isopropylidene-1,4-phenylene)	422
Bisphenol A polycarbonate	
Poly(oxycarbonylpentamethylene)	213
Poly(oxycarbonyl-1,4-phenylenemethylene-1,4-phenylene)	395
Poly(oxycarbonyl-1,4-phenyleneisopropylidene-1,4-phenylene)	333
Poly[oxy(2,6-dimethyl-1,4-phenyleneisopropylidene-3,5-dimethyl-1,4-phenylene)oxysebacoyl]	318
Poly(oxyethylenecarbonyl-1,4-cyclohexylenecarbonyl) (trans)	291
Poly(oxyethyleneoxycarbonyl-1,4-naphthylenecarbonyl)	337
Poly(oxyethyleneoxycarbonyl-1,5-naphthylenecarbonyl)	344
Poly(oxyethyleneoxycarbonyl-2,6-naphthylenecarbonyl)	386
Poly(oxyethyleneoxycarbonyl-2,7-naphthylenecarbonyl)	392
Poly(oxyethyleneoxyterephthaloyl)	342
Poly(ethylene terephthalate) [PET]	
Poly(oxyisophthaloyl)	403 (D)
Poly(oxy(1-oxo-2,2-dimethyltrimethylene))	263
Poly(pivalolactone)	
Poly(oxy-1,4-phenyleneisopropylidene-1,4-phenyleneoxysebacoyl)	280
Poly(oxy-1,4-phenyleneoxy-1,4-phenyleneoxy-carbonyl-1,4-phenylene) [PEEK]	416
Poly(oxypropyleneoxyterephthaloyl)	341
Poly[oxyterephthaloyloxy(2,6-dimethyl-1,4-phenyleneisopropylidene-3,5-dimethyl-1,4-(D)phenylene)]	498
Poly(oxyterephthaloyloxyoctamethylene)	318 (D)
Poly(oxyterephthaloyloxy-1,4-phenyleneisopropylidene-1,4-phenylene)	478
Poly(bisphenol A terephthalate)	
Poly(oxytetramethyleneoxyterephthaloyl)	323
Poly(butylene terephthalate) [PBT]	

Main-chain amide units

Poly(iminoalkyleneiminoalkanedioyl) *Poly(alkylene alkanediamide)*–[NH(CH$_2$)$_m$NHCO(CH$_2$)$_n$CO]–

Poly(iminoadipoyliminodecamethylene)	313
Nylon 10,6	
Poly(iminoadipoyliminohexamethylene)	ca 323
Nylon 6,6	
Poly(iminoadipoyliminooctamethylene)	318
Nylon 8,6	
Poly[iminoadipoyliminotrimethylene(methylimino)trimethylene]	278
Poly(iminocarbonyl-1,4-cyclohexylenemethylene)	466
Poly[iminocarbonyl-1,4-phenylene(2-oxoethylene)iminohexamethylene]	377
Poly(iminoethylene-1,4-phenyleneethyleneiminosebacoyl)	378 (D)

Polymer name	Glass transition temperature (T_g/K)
Poly(iminohexamethyleneiminoazelaoyl)	331
Nylon 6,9	
Poly(iminohexamethyleneiminododecanedioyl)	319
Nylon 6, 12	
Poly(iminohexamethyleneiminopimeloyl)	331
Nylon 6,7	
Poly(iminohexamethyleneiminosebacoyl)	323
Nylon 6,10	
Poly(iminohexamethyleneiminosuberoyl)	330
Nylon 6,8	
Poly(iminoisophthaloylimino-4,4′-biphenylylene)	558
Poly(iminoisophthaloyliminohexamethylene)	390
Poly(iminoisophthaloyliminomethylene-1,4-cyclohexylenemethylene)	481
Poly(iminoisophthaloyliminomethylene-1,3-phenylenemethylene)	438 (M)
Poly[iminomethylene(2,5-dimethyl-1,4-phenylene)methyleneiminosuberoyl]	351
Poly(imino-1,5-naphthyleneiminoisophthaloyl)	598
Poly(imino-1,5-naphthyleneiminoterephthaloyl)	578
Poly(iminooctamethyleneiminodecanedioyl)	333
Nylon 8,10	
Poly(iminooxalyliminohexamethylene)	430
Nylon 6,2	
Poly[imino(1-oxohexamethylene)]	326
Nylon 6	
Poly[imino(1-oxodecamethylene)]	315
Nylon 10	
Poly[imino(1-oxoheptamethylene)]	325
Nylon 7	
Poly[imino(1-oxo-3-methyltrimethylene]	369
Poly[imino(1-oxononamethylene)]	319
Nylon 9	
Poly[imino(1-oxooctamethylene)]	323
Nylon 8	
Poly[imino(1-oxotrimethylene)]	384
Nylon 3	
Poly(iminopentamethyleneiminoadipoyl)	318
Nylon 5,6	
Poly[iminopentamethyleneiminocarbonyl-1,4-phenylene(2-oxoethylene)]	376
Poly(imino-1,3-phenyleneiminoisophthaloyl)	553 (M)
Poly(imino-1,4-phenyleneiminoterephthaloyl)	618
Poly(iminopimeloyliminoheptamethylene)	328
Nylon 7,7	
Poly(iminoterephthaloylimino-4,4′-biphenylylene)	613
Poly(iminotetramethyleneiminoadipoyl)	316
Nylon 4,6	
Poly[iminotetramethyleneiminocarbonyl-1,4-phenylene(2-oxoethylene)]	357
Poly(iminotrimethyleneiminoadipoyliminotrimethylene)	307
Poly[iminotrimethyleneiminocarbonyl-1,4-phenylene(2-oxoethylene)]	382
Poly(oxy-1,4-phenyleneiminoterephthaloyl-imino-1,4-phenylene)	613
Poly(sulfonylimino-1,4-phenyleneiminoadipoylimino-1,4-phenylene)	467

Main-chain urethane units

Poly(oxyalkyleneoxycarbonyliminoalkyleneiminocarbonyl)–[O(CH$_2$)$_m$OCONH(CH$_2$)$_n$NHCO]–

Poly(oxyethyleneoxycarbonyliminohexamethyleneiminocarbonyl)	329
Poly[oxyethyleneoxycarbonylimino(6-methyl-1,3-phenylene)iminocarbonyl]	325
Poly(oxyethyleneoxycarbonylimino-1,4-phenylenemethylene-1,4-phenyleneiminocarbonyl)	412
Poly(oxyhexamethyleneoxycarbonyliminohexamethyleneiminocarbonyl)	332
Poly[oxyhexamethyleneoxycarbonylimino(6-methyl-1,3-phenylene)iminocarbonyl]	305
Poly(oxyhexamethyleneoxycarbonylimino-1,4-phenylenemethylene-1,4-phenyleneiminocarbonyl)	364
Poly(oxyoctamethyleneoxycarbonyliminohexamethyleneiminocarbonyl)	331
Poly[oxyoctamethyleneoxycarbonylimino(6-methyl-1,3-phenylene)iminocarbonyl]	337
Poly(oxyoctamethyleneoxycarbonylimino-1,4-phenylenemethylene-1,4-phenyleneiminocarbonyl)	352

Polymer name	Glass transition temperature (T_g/K)
Poly(oxytetramethyleneoxycarbonyliminohexamethyleneiminocarbonyl)	332
Poly[oxytetramethyleneoxycarbonylimino(6-methyl-1,3-phenylene)iminocarbonyl]	315
Poly(oxytetramethyleneoxycarbonylimino-1,4-phenylenemethylene-1,4-phenyleneiminocarbonyl)	382

Main-chain siloxanes

Poly[oxy(dialkylsilylene)] *Poly(dialkylsiloxane)* $-[O(R_2Si)]-$	
Poly[oxy(dimethylsilylene)]	148
Poly(dimethylsiloxane) [PDMS]	
Poly[oxy(dimethylsilylene)oxy-1,4-phenylene]	363 (M)
Poly[oxy(dimethylsilylene)oxy-1,4-phenyleneisopropylidene-1,4-phenylene]	318 (M)
Poly[oxy(diphenylsilylene)]	238
Poly(diphenylsiloxane)	
Poly[oxy(diphenylsilylene)-1,3-phenylene]	ca 331
Poly[oxy((methyl)phenylsilylene)]	187
Poly[oxy((methyl)-3,3,3-trifluoropropylsilylene)]	<193

Main-chain sulfur-containing units

Poly(dithioethylene)	223
Poly(dithiomethylene-1,4-phenylenemethylene)	296
Poly(oxy-4,4′-biphenylylene-1,4-phenylenesulfonyl-1,4-phenylene)	503 (M)
Poly(oxycarbonyloxy-1,4-phenylenethio-1,4-phenylene)	ca 383
Poly(oxyethylenedithioethylene)	220 (M)
Poly[oxy(2-hydroxytrimethylene)oxy-1,4-phenylenesulfonyl-1,4-phenylene]	428
Poly(oxymethyleneoxyethylenedithioethylene)	214
Poly(oxy-1,4-phenylenesulfinyl-1,4-phenyleneoxy-1,4-phenylenecarbonyl-1,4-phenylene)	478 (M)
Poly(oxy-1,4-phenylenesulfinyl-1,4-phenyleneoxy-1,4-phenyleneisopropylidene-1,4-phenylene)	438 (M)
Poly(oxy-1,4-phenylenesulfonyl-1,4-phenylene)	487
Poly(oxy-1,4-phenylenesulfonyl-4,4′-biphenylylenesulfonyl-1,4-phenylene)	533
Poly[oxy-1,4-phenylenesulfonyl-1,4-phenyleneoxy(2,6-dimethyl-1,4-phenylene)isopropylidene (3,5-dimethyl-1,4-phenylene)]	508 (M)
Poly(oxy-1,4-phenylenesulfonyl-1,4-phenyleneoxy-1,4-phenylenecarbonyl-1,4-phenylene)	478 (M)
Poly[oxy-1,4-phenylenesulfonyl-1,4-phenyleneoxy-1,4-phenylene(hexafluoroisopropylidene)1,4-phenylene]	478 (M)
Poly(oxy-1,4-phenylenesulfonyl-1,4-phenyleneoxy-1,4-phenyleneisopropylidene-1,4-phenylene)	449
Poly(oxy-1,4-phenylenesulfonyl-1,4-phenyleneoxy-1.4-phenylenemethylene-1,4-phenylene)	453 (M)
Poly(oxy-1,4-phenylenesulfonyl-1,4-phenyleneoxy-1.4-phenylenethio-1,4-phenylene)	448 (M)
Poly(oxy-1,4-phenylenesulfonyl-1,4-phenyleneoxyterephthaloyl)	522
Poly(oxytetramethylenedithiotetramethylene)	197
Poly(sulfonyl-1,2-cyclohexylene)	401
Poly(sulfonyl-1,3-cyclohexylene)	381
Poly(sulfonyl-1,4-phenylenemethylene-1,4-phenylene)	497
Poly(thio-1,3-cyclohexylene)	221
Poly[thio(difluoromethylene)]	155
Poly(thioethylene)	223
Poly[thio(1-ethylethylene]	218
Poly[thio(1-methyl-3-oxotrimethylene)]	285
Poly[thio(1-methyltrimethylene)]	214
Pol[(thio(1-oxohexamethylene)]	292
Poly(thio-1,4-phenylene)	370
Poly(thiopropylene)	226

Main-chain heterocyclic units

Poly(1,3-dioxa-4,6-cyclohexylenemethylene)	378
Poly(vinyl formal)	
Poly[(2,6-dioxopiperidine-1,4-diyl)trimethylene]	363
Poly[(2-methyl-1,3-dioxa-4,6-cyclohexylene)methylene]	355
Poly(vinyl acetal)	
Poly(1,4-piperazinediylcarbonyloxyethyleneoxycarbonyl)	333
Poly(1,4-piperazinediylisophthaloyl)	465 (M)
Poly[(2-propyl-1,3-dioxa-4,6-cyclohexylene)methylene]	322
Poly(vinyl butyral)	
Poly(3,6-pyridazinediyloxy-1,4-phenyleneisopropylidene-1,4-phenyleneoxy)	453 (M)
Poly(2,5-pyridinediylcarbonyliminohexamethyleneiminocarbonyl)	322

DIELECTRIC CONSTANT OF SELECTED POLYMERS

This table lists typical values of the dielectric constant (more properly called relative permittivity) of some important polymers. Values are given for frequencies of 1 kHz, 1 MHz, and 1 GHz; in most cases the dielectric constant at frequencies below 1 kHz does not differ significantly from the value at 1 kHz. Since the dielectric constant of a polymeric material can vary with density, degree of crystallinity, and other details of a particular sample, the values given here should be regarded as only typical or average values.

References

1. Gray, D. E., Ed., *American Institute of Physics Handbook, Third Edition*, p. 5-132, McGraw Hill, New York, 1972.
2. Anderson, H. L., Ed., *A Physicist's Desk Reference*, American Institute of Physics, New York, 1989.
3. Brandrup, J., and Immergut, E. H., *Polymer Handbook, Third Edition*, John Wiley & Sons, New York, 1989.

Name	t/°C	1 kHz	1 MHz	1 GHz
Polyacrylonitrile	25	5.5	4.2	
Polyamides (nylons)	25	3.50	3.14	2.8
	84	11	4.4	2.8
Polybutadiene	25	2.5		
Polycarbonate	23	2.92	2.8	
Polychloroprene (neoprene)	25	6.6	6.3	4.2
Polychlorotrifluoroethylene	23	2.65	2.46	2.39
Polyethylene	23	2.3		
Poly(ethylene terephthalate) (Mylar)	23	3.25	3.0	2.8
Polyisoprene (natural rubber)	27	2.6	2.5	2.4
Poly(methyl methacrylate)	27	3.12	2.76	2.6
	80	3.80	2.7	2.6
Polyoxymethylene (polyformaldehyde)	25	3.8		
Poly(phenylene oxide)	23	2.59	2.59	
Polypropylene	25	2.3	2.3	2.3
Polystyrene	25	2.6	2.6	2.6
Polysulfones	25	3.13	2.10	
Polytetrafluoroethylene (teflon)	25	2.1	2.1	2.1
Poly(vinyl acetate)	50		3.5	
	150		8.3	
Poly(vinyl chloride)	25	3.39	2.9	2.8
	100	5.3	3.3	2.7
Poly(vinylidene chloride)	23	4.6	3.2	2.7
Poly(vinylidene fluoride)	23	12.2	8.9	4.7

PRESSURE–VOLUME–TEMPERATURE RELATIONSHIP FOR POLYMER MELTS

Christian Wohlfarth

Numerous theoretical equations of state for polymer liquids have been developed. These, at the minimum, have to provide accurate fitting functions to experimental data. However, for the purpose of this table, the empirical Tait equation along with a polynomial expression for the zero pressure isobar is used. This equation is able to represent the experimental data for the melt state within the limits of experimental errors, i.e., the maximum deviations between measured and calculated specific volumes are about 0.001-0.002 cm³/g.

The general form of the Tait equation is:

$$V(P,T) = V(0,T)\{1 - C \ln[1 + P/B(T)]\} \qquad (1)$$

where the coefficient C is usually taken to be a universal constant equal to 0.0894. T is the absolute temperature in K and P the pressure in MPa. The volume V is the specific volume in cm³/g. The Tait parameter $B(T)$ has the very simple meaning that it is inversely proportional to the compressibility κ at constant temperature and zero pressure:

$$\kappa(0,T) = -[1/V(0,T)](dV/dP) = C/B(T) \qquad (2)$$

The $B(T)$ function is usually given by:

$$B(T) = B_0 \exp[-B_1(T\text{-}273.15)] \qquad (3)$$

but, sometimes a polynomial expression is used:

$$B(T) = b_0 + b_1(T\text{-}273.15) + b_2(T\text{-}273.15)^2 \qquad (4)$$

The zero-pressure isobar $V(0,T)$ is usually given by:

$$V(0,T) = A_0 + A_1(T\text{-}273.15) + A_2(T\text{-}273.15)^2 \qquad (5)$$

where A_0, A_1, A_2 are specific constants for a given polymer (the expression T-273.15 is used because fitting to the zero-pressure isobar is usually done in terms of Celsius temperature). Other forms for $V(0,T)$ are also found in the literature, such as

$$V(0,T) = A_3 \exp[A_4(T\text{-}273.15)] \qquad (6)$$

or

$$V(0,T) = A_5 \exp(A_6 T^{1.5}) \qquad (7)$$

where A_3 and A_4 or A_5 and A_6 are again specific constants for a given polymer.

The Tait equation is particularly useful to calculate derivative quantities, such as the isothermal compressibility and the thermal expansivity coefficients. The isothermal compressibility $\kappa(P,T)$ is derived from equation (1) as:

$$\kappa(P,T) = -(1/V)(dV/dP) = 1/\{[P + B(T)][1/C - \ln(1 + P/B(T))]\} \qquad (8)$$

and the thermal expansivity $\alpha(P,T)$ as:

$$\alpha(P,T) = (1/V)(dV/dT) = \alpha(0,T) - PB_1\kappa(P,T) \qquad (9)$$

where $\alpha(0,T)$ represents the thermal expansivity at zero (atmospheric) pressure and is calculated from any suitable fit for the zero-pressure volume, such as equations (5) through (7) above.

Because polymer melt PVT-behavior depends only slightly on polymer molar mass above the oligomeric region, usually no information is given in the original literature for the average molar mass of the polymers.

Table 1 summarizes the polymers or copolymers considered here and the experimental ranges of pressure and temperature over which data are available. In Table 2 the Tait-equation functions, with parameters obtained from the fit, are given for 90 polymer or copolymer melts.

References

1. Zoller, P., *J. Appl. Polym. Sci.*, 23, 1051-1056, 1979.
2. Starkweather, H. W., Jones, G. A., and Zoller, P., *J. Polym. Sci., Pt. B Polym. Phys.*, 26, 257-266,1988.
3. Fakhreddine, Y. A., and Zoller, P., *J. Polym. Sci., Pt. B Polym. Phys.*, 29, 1141-1146, 1991.
4. Rodgers, P. A., *J. Appl. Polym. Sci.*, 48, 1061-1080, 1993.
5. Rodgers, P. A., *J. Appl. Polym. Sci.*, 48, 2075-2083, 1993.
6. Yi, Y. X., and Zoller, P., *J. Polym. Sci., Pt. B Polym. Phys.*, 31, 779-788, 1993.
7. Callaghan, T. A., and Paul, D. R., *Macromolecules*, 26, 2439–2450, 1993.
8. Wang, Y. Z., Hsieh, K. H., Chen, L. W.,and Tseng, H. C., *J. Appl. Polym. Sci.*, 53, 1191-1201, 1994.
9. Privalko, V. P., Arbuzova, A. P., Korskanov, V. V., and Zagdanskaya, N. E., *Polym. Intern.*, 35, 161-169, 1994.
10. Sachdev, V. K., Yashi, U., and Jain, R. K., *J. Polym. Sci., Pt. B Polym. Phys.*, 36, 841-850, 1998.

TABLE 1. Names of the Polymers, Abbreviation Used, and Range of Experimental Data Applied in the Determination of the Equation Constants

Polymer	Symbol	T/K	P/MPa	Ref.
Ethylene/propylene copolymer (50 wt%)	EP50	413-523	0.1-63	4
Ethylene/vinyl acetate copolymer				
18 wt% vinyl acetate	EVA18	385-491	0.1-177	4
25 wt% vinyl acetate	EVA25	367-506	0.1-177	4
28 wt% vinyl acetate	EVA28	367-508	0.1-177	4
40 wt% vinyl acetate	EVA40	348-508	0.1-177	4
Polyamide-6	PA6	509-569	0.1-196	4
Polyamide-11	PA11	478-542	0.1-200	5
Polyamide-66	PA66	519-571	0.1-196	4
cis-1,4-Polybutadiene	cPBD	277-328	0.1-284	4
Polybutadiene, 8% 1,2-content	PBD-8	298-473	0.1-200	6
Polybutadiene, 24% 1,2-content	PBD-24	298-473	0.1-200	6
Polybutadiene, 40% 1,2-content	PBD-40	298-473	0.1-200	6
Polybutadiene, 50% 1,2-content	PBD-50	298-473	0.1-200	6
Polybutadiene, 87% 1,2-content	PBD-87	298-473	0.1-200	6
Poly(1-butene), isotactic	iPB	406-519	0.1-196	4
Poly(butyl methacrylate)	PnBMA	307-473	0.1-200	4
Poly(butylene terephthalate)	PBT	508-576	0.1-200	3
Poly(ε-caprolactone)	PCL	373-421	0.1-200	4
Polycarbonate-bisphenol-A	PC	424-613	0.1-177	4
Polycarbonate-bisphenol-chloral	BCPC	428-557	0.1-200	4
Polycarbonate-hexafluorobisphenol-A	HFPC	432-553	0.1-200	4
Polycarbonate-tetramethylbisphenol-A	TMPC	491-563	0.1-160	4
Poly(cyclohexyl methacrylate)	PcHMA	396-471	0.1-200	4
Poly(2,5-dimethylphenylene oxide)	PPO	473-593	0.1-177	4
Poly(dimethyl siloxane)	PDMS	298-343	0.1-100	4
Poly(dimethyl siloxane) M_n = 1000	PDMS-10	304-420	0.1-250	10
Poly(dimethyl siloxane) M_n = 4000	PDMS-40	298-418	0.1-250	10
Poly(dimethyl siloxane) M_n = 6000	PDMS-60	291-423	0.1-250	10
Poly(epichlorohydrin)	PECH	333-413	0.1-200	4
Poly(ether ether ketone)	PEEK	619-671	0.1-200	4
Poly(ethyl acrylate)	PEA	310-490	0.1-196	4
Poly(ethyl methacrylate)	PEMA	386-434	0.1-196	4
Polyethylene, high density	HDPE	413-476	0.1-196	4
Polyethylene, linear	LPE	415-473	0.1-200	4
Polyethylene, linear, high MW	HMLPE	410-473	0.1-200	4
Polyethylene, branched	BPE	398-471	0.1-200	4
Polyethylene, low density	LDPE	394-448	0.1-196	4
Polyethylene, low density, type A	LDPE-A	385-498	0.1-196	1
Polyethylene, low density, type B	LDPE-B	385-498	0.1-196	1
Polyethylene, low density, type C	LDPE-C	385-498	0.1-196	1
Poly(ethylene oxide)	PEO	361-497	0.1-68	4
Poly(ethylene terephthalate)	PET	547-615	0.1-196	4
Poly(4-hexylstyrene)	P4HS	303-403	30-100	4
Polyisobutylene	PIB	326-383	0.1-100	4
Polyisoprene, 8% 3,4-content	PI-8	298-473	0.1-200	6
Polyisoprene, 14% 3,4-content	PI-14	298-473	0.1-200	6
Polyisoprene, 41% 3,4-content	PI-41	298-473	0.1-200	6
Polyisoprene, 56% 3,4-content	PI-56	298-473	0.1-200	6
Poly(methyl acrylate)	PMA	310-493	0.1-196	4
Poly(methyl methacrylate)	PMMA	387-432	0.1-200	4
Poly(4-methyl-1-pentene)	P4MP	514-592	0.1-196	4
Poly(α-methylstyrene)	PαMS	473-533	0.1-170	7
Poly(o-methylstyrene)	PoMS	412-471	0.1-180	4
Polyoxymethylene	POM	463-493	0.1-196	2
Phenoxy[a]	PH	341-573	0.1-177	4
Polysulfone[b]	PSF	475-644	0.1-196	4
Polyarylate[c]	PAr	450-583	0.1-177	4
Polypropylene, atactic	aPP	353-393	0.1-100	4

Polymer	Symbol	T/K	P/MPa	Ref.
Polypropylene, isotactic	iPP	443-570	0.1-196	4
Polystyrene	PS	388-469	0.1-200	4
Poly(tetrafluoroethylene)	PTFE	603-645	0.1- 39	4
Poly(tetrahydrofuran)	PTHF	335-439	0.1- 78	4
Poly(vinyl acetate)	PVAc	308-373	0.1- 80	4
Poly(vinyl chloride)	PVC	373-423	0.1-200	4
Poly(vinyl methyl ether)	PVME	303-471	0.1-200	4
Poly(vinylidene fluoride)	PVdF	451-521	0.1-200	5
Styrene/acrylonitrile copolymer				
2.7 wt% acrylonitrile	SAN3	378-539	0.1-200	4
5.7 wt% acrylonitrile	SAN6	370-540	0.1-200	4
15.3 wt% acrylonitrile	SAN15	405-531	0.1-200	4
18.0 wt% acrylonitrile	SAN18	377-528	0.1-200	4
40 wt% acrylonitrile	SAN40	373-543	0.1-200	4
70 wt% acrylonitrile	SAN70	373-544	0.1-200	4
Styrene/butadiene copolymer				
10 wt% styrene	SBR10	393-533	0.1-196	8
23.5 wt% styrene	SBR23	393-533	0.1-196	8
60 wt% styrene	SBR60	393-533	0.1-196	8
85 wt% styrene	SBR85	393-533	0.1-196	8
Styrene/methyl methacrylate copolymer				
20 wt% methyl methacrylate	SMMA20	383-543	0.1-200	4
60 wt% methyl methacrylate	SMMA60	383-543	0.1-200	4
N-Vinylcarbazole/4-ethylstyrene copolymer				
50 mol% ethylstyrene	VCES50	393-443	30-100	9
N-Vinylcarbazole/4-hexylstyrene copolymer				
80 mol% hexylstyrene	VCHS80	313-423	30-100	9
67 mol% hexylstyrene	VCHS67	333-423	30-100	9
60 mol% hexylstyrene	VCHS60	383-453	30-100	9
50 mol% hexylstyrene	VCHS50	373-443	30-100	9
40 mol% hexylstyrene	VCHS40	423-493	30-100	9
33 mol% hexylstyrene	VCHS33	463-523	30-100	9
20 mol% hexylstyrene	VCHS20	473-523	30-100	9
N-Vinylcarbazole/4-octylstyrene copolymer				
50 mol% octylstyrene	VCOS50	403-453	30-100	9
N-Vinylcarbazole/4-pentylstyrene copolymer				
50 mol% pentylstyrene	VCPS50	383-443	30-100	9

[a] Phenoxy = Poly(oxy-2-hydroxytrimethyleneoxy-1,4-phenyleneisopropylidene-1,4-phenylene)

[b] Polysulfone = Poly(oxy-1,4-phenylenesulfonyl-1,4-phenyleneoxy-1,4-phenyleneisopropylidene-1,4-phenylene)

[c] Polyarylate = Poly(oxyterephthaloyl/isophthaloyl T/I=50/50)oxy-1,4-phenyleneisopropylidene-1,4-phenylene

TABLE 2. Tait Equation Parameter Functions for Polymer Melts

Polymer	$V(0,T)/\text{cm}^3\text{g}^{-1}$	$B(T)/\text{MPa}$
EP50	$1.2291 + 5.799 \cdot 10^{-5}(T-273.15) + 1.964 \cdot 10^{-6}(T-273.15)^2$	$487.0 \exp[-8.103 \cdot 10^{-3}(T-273.15)]$
EVA18	$1.02391 \exp(2.173 \cdot 10^{-5}T^{1.5})$	$188.2 \exp[-4.537 \cdot 10^{-3}(T-273.15)]$
EVA25	$1.00416 \exp(2.244 \cdot 10^{-5}T^{1.5})$	$184.4 \exp[-4.734 \cdot 10^{-3}(T-273.15)]$
EVA28	$1.00832 \exp(2.241 \cdot 10^{-5}T^{1.5})$	$183.5 \exp[-4.457 \cdot 10^{-3}(T-273.15)]$
EVA40	$1.06332 \exp(2.288 \cdot 10^{-5}T^{1.5})$	$205.1 \exp[-4.989 \cdot 10^{-3}(T-273.15)]$
PA6	$0.7597 \exp[4.701 \cdot 10^{-4}(T-273.15)]$	$376.7 \exp[-4.660 \cdot 10^{-3}(T-273.15)]$
PA11	$0.9581 \exp[6.664 \cdot 10^{-4}(T-273.15)]$	$254.7 \exp[-4.178 \cdot 10^{-3}(T-273.15)]$
PA66	$0.7657 \exp[6.600 \cdot 10^{-4}(T-273.15)]$	$316.4 \exp[-5.040 \cdot 10^{-3}(T-273.15)]$
cPBD	$1.0970 \exp[6.600 \cdot 10^{-4}(T-273.15)]$	$177.7 \exp[-3.593 \cdot 10^{-3}(T-273.15)]$
PBD-8	$1.1004 + 6.718 \cdot 10^{-4}(T-273.15) + 6.584 \cdot 10^{-7}(T-273.15)^2$	$200.0 \exp[-4.606 \cdot 10^{-3}(T-273.15)]$
PBD-24	$1.1049 + 6.489 \cdot 10^{-4}(T-273.15) + 7.099 \cdot 10^{-7}(T-273.15)^2$	$193.0 \exp[-4.519 \cdot 10^{-3}(T-273.15)]$
PBD-40	$1.1013 + 6.593 \cdot 10^{-4}(T-273.15) + 5.776 \cdot 10^{-7}(T-273.15)^2$	$188.0 \exp[-4.437 \cdot 10^{-3}(T-273.15)]$
PBD-50	$1.1037 + 5.955 \cdot 10^{-4}(T-273.15) + 7.789 \cdot 10^{-7}(T-273.15)^2$	$183.0 \exp[-4.425 \cdot 10^{-3}(T-273.15)]$
PBD-87	$1.1094 + 6.729 \cdot 10^{-4}(T-273.15) + 4.470 \cdot 10^{-7}(T-273.15)^2$	$175.0 \exp[-4.538 \cdot 10^{-3}(T-273.15)]$
iPB	$1.1417 \exp[6.751 \cdot 10^{-4}(T-273.15)]$	$167.5 \exp[-4.533 \cdot 10^{-3}(T-273.15)]$
PnBMA	$0.9341 + 5.5254 \cdot 10^{-4}(T-273.15) + 6.5803 \cdot 10^{-6}(T-273.15)^2 + 1.5691 \cdot 10^{-10}(T-273.15)^3$	$226.7 \exp[-5.344 \cdot 10^{-3}(T-273.15)]$
PBT	$0.9640 - 1.017 \cdot 10^{-3}(T-273.15) + 3.065 \cdot 10^{-6}(T-273.15)^2$	$263.0 \exp[-3.444 \cdot 10^{-3}(T-273.15)]$
PCL	$0.9049 \exp[6.392 \cdot 10^{-4}(T-273.15)]$	$189.0 \exp[-3.931 \cdot 10^{-3}(T-273.15)]$
PC	$0.73565 \exp(1.859 \cdot 10^{-5}T^{1.5})$	$310.0 \exp[-4.078 \cdot 10^{-3}(T-273.15)]$
BCPC	$0.6737 + 3.634 \cdot 10^{-4}(T-273.15) + 2.370 \cdot 10^{-7}(T-273.15)^2$	$363.4 \exp[-4.921 \cdot 10^{-3}(T-273.15)]$
HFPC	$0.6111 + 4.898 \cdot 10^{-4}(T-273.15) + 1.730 \cdot 10^{-7}(T-273.15)^2$	$236.6 \exp[-5.156 \cdot 10^{-3}(T-273.15)]$
TMPC	$0.8497 + 5.073 \cdot 10^{-4}(T-273.15) + 3.832 \cdot 10^{-7}(T-273.15)^2$	$231.4 \exp[-4.242 \cdot 10^{-3}(T-273.15)]$
PcHMA	$0.8793 + 4.0504 \cdot 10^{-4}(T-273.15) + 7.774 \cdot 10^{-7}(T-273.15)^2 - 7.7534 \cdot 10^{-10}(T-273.15)^3$	$295.2 \exp[-5.220 \cdot 10^{-3}(T-273.15)]$
PPO	$0.78075 \exp(2.151 \cdot 10^{-5}T^{1.5})$	$227.8 \exp[-4.290 \cdot 10^{-3}(T-273.15)]$
PDMS	$1.0079 \exp[9.121 \cdot 10^{-4}(T-273.15)]$	$89.4 \exp[-5.701 \cdot 10^{-3}(T-273.15)]$
PDMS-10	$0.8343 + 5.991 \cdot 10^{-4}(T-273.15) + 5.734 \cdot 10^{-7}(T-273.15)^2$	$542.63 \exp[-6.69 \cdot 10^{-3}(T-273.15)]$
PDMS-40	$0.8018 + 7.072 \cdot 10^{-4}(T-273.15) + 3.635 \cdot 10^{-7}(T-273.15)^2$	$482.73 \exp[-6.09 \cdot 10^{-3}(T-273.15)]$
PDMS-60	$0.8146 + 5.578 \cdot 10^{-4}(T-273.15) + 5.774 \cdot 10^{-7}(T-273.15)^2$	$482.73 \exp[-6.09 \cdot 10^{-3}(T-273.15)]$
PECH	$0.7216 \exp[5.825 \cdot 10^{-4}(T-273.15)]$	$238.3 \exp[-4.171 \cdot 10^{-3}(T-273.15)]$
PEEK	$0.7158 \exp[6.690 \cdot 10^{-4}(T-273.15)]$	$388.0 \exp[-4.124 \cdot 10^{-3}(T-273.15)]$
PEA	$0.8756 \exp[7.241 \cdot 10^{-4}(T-273.15)]$	$193.2 \exp[-4.839 \cdot 10^{-3}(T-273.15)]$
PEMA	$0.8614 \exp[7.468 \cdot 10^{-4}(T-273.15)]$	$260.9 \exp[-5.356 \cdot 10^{-3}(T-273.15)]$
HDPE	$1.1595 + 8.0394 \cdot 10^{-4}(T-273.15)$	$179.9 \exp[-4.739 \cdot 10^{-3}(T-273.15)]$
LPE	$0.9172 \exp[7.806 \cdot 10^{-4}(T-273.15)]$	$176.7 \exp[-4.661 \cdot 10^{-3}(T-273.15)]$
HMLPE	$0.8992 \exp[8.502 \cdot 10^{-4}(T-273.15)]$	$168.3 \exp[-4.292 \cdot 10^{-3}(T-273.15)]$
BPE	$0.9399 \exp[7.341 \cdot 10^{-4}(T-273.15)]$	$177.1 \exp[-4.699 \cdot 10^{-3}(T-273.15)]$
LDPE	$1.1944 + 2.841 \cdot 10^{-4}(T-273.15) + 1.872 \cdot 10^{-6}(T-273.15)^2$	$202.2 \exp[-5.243 \cdot 10^{-3}(T-273.15)]$
LDPE-A	$1.1484 \exp[6.950 \cdot 10^{-4}(T-273.15)]$	$192.9 \exp[-4.701 \cdot 10^{-3}(T-273.15)]$
LDPE-B	$1.1524 \exp[6.700 \cdot 10^{-4}(T-273.15)]$	$196.6 \exp[-4.601 \cdot 10^{-3}(T-273.15)]$
LDPE-C	$1.1516 \exp[6.730 \cdot 10^{-4}(T-273.15)]$	$186.7 \exp[-4.391 \cdot 10^{-3}(T-273.15)]$
PEO	$0.8766 \exp[7.087 \cdot 10^{-4}(T-273.15)]$	$207.7 \exp[-3.947 \cdot 10^{-3}(T-273.15)]$
PET	$0.6883 + 5.90 \cdot 10^{-4}(T-273.15)$	$369.7 \exp[-4.150 \cdot 10^{-3}(T-273.15)]$
P4HS	$0.8251 + 6.77 \cdot 10^{-4}T$	$103.1 \exp[-2.417 \cdot 10^{-3}(T-273.15)]$
PIB	$1.0750 \exp[5.651 \cdot 10^{-4}(T-273.15)]$	$200.3 \exp[-4.329 \cdot 10^{-3}(T-273.15)]$
PI-8	$1.1030 + 6.488 \cdot 10^{-4}(T-273.15) + 5.125 \cdot 10^{-7}(T-273.15)^2$	$188.0 \exp[-4.541 \cdot 10^{-3}(T-273.15)]$
PI-14	$1.0943 + 6.293 \cdot 10^{-4}(T-273.15) + 6.231 \cdot 10^{-7}(T-273.15)^2$	$202.0 \exp[-4.653 \cdot 10^{-3}(T-273.15)]$
PI-41	$1.0951 + 6.188 \cdot 10^{-4}(T-273.15) + 6.629 \cdot 10^{-7}(T-273.15)^2$	$199.0 \exp[-4.622 \cdot 10^{-3}(T-273.15)]$
PI-56	$1.0957 + 6.655 \cdot 10^{-4}(T-273.15) + 5.661 \cdot 10^{-7}(T-273.15)^2$	$200.0 \exp[-4.644 \cdot 10^{-3}(T-273.15)]$
PMA	$0.8365 \exp[6.795 \cdot 10^{-4}(T-273.15)]$	$235.8 \exp[-4.493 \cdot 10^{-3}(T-273.15)]$
PMMA	$0.8254 + 2.8383 \cdot 10^{-4}(T-273.15) + 7.792 \cdot 10^{-7}(T-273.15)^2$	$287.5 \exp[-4.146 \cdot 10^{-3}(T-273.15)]$
P4MP	$1.4075 - 9.095 \cdot 10^{-4}(T-273.15) + 3.497 \cdot 10^{-6}(T-273.15)^2$	$37.67 + 0.2134(T-273.15)] - 7.0445 \cdot 10^{-4}(T-273.15)^2$
PαMS	$0.89365 + 3.4864 \cdot 10^{-4}(T-273.15) + 5.0184 \cdot 10^{-7}(T-273.15)^2$	$297.7 \exp[-4.074 \cdot 10^{-3}(T-273.15)]$
PoMS	$0.9396 \exp[5.306 \cdot 10^{-4}(T-273.15)]$	$261.9 \exp[-4.114 \cdot 10^{-3}(T-273.15)]$
POM	$0.7484 \exp[6.770 \cdot 10^{-4}(T-273.15)]$	$305.6 \exp[-4.326 \cdot 10^{-3}(T-273.15)]$
PH	$0.76644 \exp(1.921 \cdot 10^{-5}T^{1.5})$	$359.9 \exp[-4.378 \cdot 10^{-3}(T-273.15)]$
PSF	$0.7644 + 3.419 \cdot 10^{-4}(T-273.15) + 3.126 \cdot 10^{-7}(T-273.15)^2$	$365.9 \exp[-3.757 \cdot 10^{-3}(T-273.15)]$
PAr	$0.73381 \exp(1.626 \cdot 10^{-5}T^{1.5})$	$296.9 \exp[-3.375 \cdot 10^{-3}(T-273.15)]$
aPP	$1.1841 - 1.091 \cdot 10^{-4}(T-273.15) + 5.286 \cdot 10^{-6}(T-273.15)^2$	$162.1 \exp[-6.604 \cdot 10^{-3}(T-273.15)]$
iPP	$1.1606 \exp[6.700 \cdot 10^{-4}(T-273.15)]$	$149.1 \exp[-4.177 \cdot 10^{-3}(T-273.15)]$

Polymer	$V(0,T)/\mathrm{cm^3 g^{-1}}$	$B(T)/\mathrm{MPa}$
PS	$0.9287 \exp[5.131 \cdot 10^{-4}(T-273.15)]$	$216.9 \exp[-3.319 \cdot 10^{-3}(T-273.15)]$
PTFE	$0.3200 + 9.5862 \cdot 10^{-4}(T-273.15)$	$425.2 \exp[-9.380 \cdot 10^{-3}(T-273.15)]$
PTHF	$1.0043 \exp[6.691 \cdot 10^{-4}(T-273.15)]$	$178.6 \exp[-4.223 \cdot 10^{-3}(T-273.15)]$
PVAc	$0.82496 + 5.820 \cdot 10^{-4}(T-273.15) + 2.940 \cdot 10^{-7}(T-273.15)^2$	$204.9 \exp[-4.346 \cdot 10^{-3}(T-273.15)]$
PVC	$0.7196 + 5.581 \cdot 10^{-5}(T-273.15) + 1.468 \cdot 10^{-6}(T-273.15)^2$	$294.2 \exp[-5.321 \cdot 10^{-3}(T-273.15)]$
PVME	$0.9585 \exp[6.653 \cdot 10^{-4}(T-273.15)]$	$215.8 \exp[-4.588 \cdot 10^{-3}(T-273.15)]$
PVdF	$0.5790 \exp[8.051 \cdot 10^{-4}(T-273.15)]$	$244.0 \exp[-5.210 \cdot 10^{-3}(T-273.15)]$
SAN3	$0.9233 + 3.936 \cdot 10^{-4}(T-273.15) + 5.685 \cdot 10^{-7}(T-273.15)^2$	$239.8 \exp[-4.376 \cdot 10^{-3}(T-273.15)]$
SAN6	$0.9211 + 4.370 \cdot 10^{-4}(T-273.15) + 5.846 \cdot 10^{-7}(T-273.15)^2$	$226.9 \exp[-4.286 \cdot 10^{-3}(T-273.15)]$
SAN15	$0.9044 + 4.207 \cdot 10^{-4}(T-273.15) + 4.077 \cdot 10^{-7}(T-273.15)^2$	$238.4 \exp[-3.943 \cdot 10^{-3}(T-273.15)]$
SAN18	$0.9016 + 4.036 \cdot 10^{-4}(T-273.15) + 4.206 \cdot 10^{-7}(T-273.15)^2$	$240.4 \exp[-3.858 \cdot 10^{-3}(T-273.15)]$
SAN40	$0.8871 + 3.406 \cdot 10^{-4}(T-273.15) + 4.938 \cdot 10^{-7}(T-273.15)^2$	$289.3 \exp[-4.431 \cdot 10^{-3}(T-273.15)]$
SAN70	$0.8528 + 3.616 \cdot 10^{-4}(T-273.15) + 2.634 \cdot 10^{-7}(T-273.15)^2$	$335.4 \exp[-3.923 \cdot 10^{-3}(T-273.15)]$
SBR10	$0.9053 \exp(2.437 \cdot 10^{-5} T^{1.5})$	$530.3 \exp[-3.99 \cdot 10^{-3}(T-273.15)]$
SBR23	$0.8986 \exp(2.317 \cdot 10^{-5} T^{1.5})$	$551.6 \exp[-4.17 \cdot 10^{-3}(T-273.15)]$
SBR60	$0.8812 \exp(2.031 \cdot 10^{-5} T^{1.5})$	$486.0 \exp[-4.34 \cdot 10^{-3}(T-273.15)]$
SBR85	$0.8704 \exp(1.846 \cdot 10^{-5} T^{1.5})$	$356.7 \exp[-4.24 \cdot 10^{-3}(T-273.15)]$
SMMA20	$0.9063 + 3.570 \cdot 10^{-4}(T-273.15) + 6.532 \cdot 10^{-7}(T-273.15)^2$	$232.0 \exp[-4.143 \cdot 10^{-3}(T-273.15)]$
SMMA60	$0.8610 + 3.350 \cdot 10^{-4}(T-273.15) + 6.980 \cdot 10^{-7}(T-273.15)^2$	$261.0 \exp[-4.611 \cdot 10^{-3}(T-273.15)]$
VCES50	$0.6676 + 6.63 \cdot 10^{-4} T$	$5281.7 \exp[-9.264 \cdot 10^{-3}(T-273.15)]$
VCHS80	$0.7753 + 6.17 \cdot 10^{-4} T$	$247.6 \exp[-2.604 \cdot 10^{-3}(T-273.15)]$
VCHS67	$0.8028 + 6.50 \cdot 10^{-4} T$	$581.7 \exp[-4.553 \cdot 10^{-3}(T-273.15)]$
VCHS60	$0.8213 + 6.23 \cdot 10^{-4} T$	$229.1 \exp[-2.133 \cdot 10^{-3}(T-273.15)]$
VCHS50	$0.7827 + 5.05 \cdot 10^{-4} T$	$136.0 \exp[-1.083 \cdot 10^{-3}(T-273.15)]$
VCHS40	$0.7805 + 4.92 \cdot 10^{-4} T$	$155.0 \exp[-1.605 \cdot 10^{-3}(T-273.15)]$
VCHS33	$0.7710 + 4.86 \cdot 10^{-4} T$	$460.4 \exp[-3.453 \cdot 10^{-3}(T-273.15)]$
VCHS20	$0.6416 + 5.42 \cdot 10^{-4} T$	$489.8 \exp[-3.193 \cdot 10^{-3}(T-273.15)]$
VCOS50	$0.7081 + 7.40 \cdot 10^{-4} T$	$666.5 \exp[-4.503 \cdot 10^{-3}(T-273.15)]$
VCPS50	$0.7814 + 4.36 \cdot 10^{-4} T$	$880.1 \exp[-4.393 \cdot 10^{-3}(T-273.15)]$

UPPER CRITICAL (UCST) AND LOWER CRITICAL (LCST) SOLUTION TEMPERATURES OF BINARY POLYMER SOLUTIONS

Christian Wohlfarth

Liquid–liquid demixing in solutions of polymers in low molar mass solvents is not a rare phenomenon. Demixing depends on concentration, temperature, pressure, molar mass and molar mass distribution function of the polymer, chain branching and end groups of the polymer, the chemical nature of the solvent, isotope substitution in solvents or polymers, chemical composition of copolymers and its distributions, and other variables. Phase diagrams of polymer solutions can therefore show a quite complicated behavior when they have to be considered in detail (see Ref. 1a).

Polymer solutions can undergo demixing when cooling a homogeneous solution as well as when heating such a solution. The corresponding cloud-point curves show a maximum (UCST behavior) or a minimum (LCST behavior). For common polymer solutions, the LCST region is at higher temperatures (in many cases near the critical temperature of the solvent) than the UCST region. The temperature range between both extrema provides the essential information where the one-phase region of a polymer solution can be found. In the case of monodisperse polymers the extrema are equal to the critical points. However, in the case of polydisperse polymers with distribution functions, these extrema are threshold temperatures whereas the critical point shifts to higher concentrations on the shoulder of the cloud-point curve. Usually, the critical concentration is much more strongly influenced than the critical temperature. Thus, the table below does not distinguish between threshold and critical temperatures.

UCST and LCST values depend somewhat on pressure. LCST values in the table are usually given at the vapor pressure of the solvent at this temperature. UCST values are measured in most cases at normal pressure; data at higher pressures are neglected here. The interested reader can find such information, for example, in Refs. 76, 84, 104, 157, 165, 177, 185–187, or 192.

However, UCST and LCST values of a given polymer/solvent pair depend strongly on the molar mass of the polymer. In the case of monodisperse polymers, this dependency can be described in good approximation by the so-called Shultz-Flory plot (see Refs. 6 and 8):

$$\frac{1}{T_{crit}} = \frac{1}{\theta}\left[1 + const.\left(\frac{1}{\sqrt{r}} - \frac{1}{2r}\right)\right] \quad (1)$$

where r denotes the number of segments of a polymer (being proportional to the degree of polymerization or to the molar mass or molar volume of the polymer). Extrapolation to $r \to \infty$, i.e., to infinite molar mass, leads to the value of the θ-temperature. This θ-temperature is the highest temperature for UCST behavior or the lowest temperature for LCST behavior and a given polymer/solvent pair. In the case of polydisperse polymers, the segment number in equation (1) is to be replaced by its weight average, r_w (related to M_w). The constant in equation (1) reflects further thermodynamic properties of the given polymer/solvent pair, but should not depend on molar mass. A detailed discussion can be found in Ref. 1b.

The printed table in the *Handbook* provides only one data line for a given polymer/solvent pair and does not show the molar mass dependence of UCST or LCST data. The entire table with all data at different molar masses for many of the systems is given in the electronic version, however. Nevertheless, the necessary molar mass information for a system is always provided in the table by the corresponding number average, M_n, mass average, M_w, or viscosity average, M_η, values of the polymer as given in the original sources.

Polymer	M_n/g mol⁻¹	M_w/g mol⁻¹	M_η/g mol⁻¹	Solvent	UCST/K	LCST/K	Ref.
Acrylonitrile/butadiene copolymer							
(18% Acrylonitrile)			840000	Ethyl acetate		427	220
(26% Acrylonitrile)			1000000	Ethyl acetate		412	220
Butadiene/α-methylstyrene copolymer							
(10% α-Methylstyrene)			100000	Ethyl acetate	387	393	220
Carbon monoxide/ethylene copolymer (1:1, alternating)							
		1000000		1,1,1,3,3,3-Hexafluoro-2-propanol		453	159
Cellulose diacetate			120000	Benzyl alcohol	372		86
	59900	75500		2-Butanone	279.7	471.5	111
	59300			2-Propanone	216.2	438.2	42
Cellulose diacetate/styrene graft copolymer (77.4 wt% grafted polystyrene)							
		750000		*N,N*-Dimethylformamide	262	399	106
		750000		Tetrahydrofuran		363	106
Cellulose nitrate (13.3 wt% N)							
	unknown			2-Propanone	328	182	148

Polymer	M_n/g mol^{-1}	M_w/g mol^{-1}	M_η/g mol^{-1}	Solvent	UCST/K	LCST/K	Ref.
Cellulose triacetate							
			20000	Benzyl alcohol	322		86
		100500		2-Propanone	290.0	472.0	42
Cellulose tricaprylate							
	infinite			N,N-Dimethylformamide	413		5
	infinite			3-Phenyl-1-propanol	321		5
Decamethyltetrasiloxane							
	310.69			Tetradecafluorohexane	332.59		195
N,N-Dimethylacrylamide/2-butoxyethyl acrylate copolymer (50 wt% 2-butoxyethyl acrylate)							
				Water		<273.2	164
N,N-Dimethylacrylamide/butyl acrylate copolymer							
(15 wt% Butyl acrylate)				Water		346.2	164
(20 wt% Butyl acrylate)				Water		323.2	164
(30 wt% Butyl acrylate)				Water		294.2	164
(35 wt% Butyl acrylate)				Water		281.2	164
N,N-Dimethylacrylamide/2-ethoxyethyl acrylate copolymer							
(50 wt% 2-Ethoxyethyl acrylate)				Water		319.2	164
(75 wt% 2-Ethoxyethyl acrylate)				Water		285.2	164
N,N-Dimethylacrylamide/ethyl acrylate copolymer							
(25 wt% Ethyl acrylate)				Water		347.2	164
(30 wt% Ethyl acrylate)				Water		334.2	164
(50 wt% Ethyl acrylate)				Water		287.2	164
(55 wt% Ethyl acrylate)				Water		<273.2	164
N,N-Dimethylacrylamide/2-methoxyethyl acrylate copolymer							
(38 mol% 2-Methoxyethyl acrylate)				Water		353	184
(45 mol% 2-Methoxyethyl acrylate)				Water		333	184
(55 mol% 2-Methoxyethyl acrylate)				Water		315	184
(68 mol% 2-Methoxyethyl acrylate)				Water		305	184
(82 mol% 2-Methoxyethyl acrylate)				Water		288	184
(92 mol% 2-Methoxyethyl acrylate)				Water		283	184
N,N-Dimethylacrylamide/methyl acrylate copolymer							
(30 wt% Methyl acrylate)				Water		371.2	164
(40 wt% Methyl acrylate)				Water		338.2	164
(50 wt% Methyl acrylate)				Water		314.2	164
(55 wt% Methyl acrylate)				Water		294.2	164
(60 wt% Methyl acrylate)				Water		279.2	164
(70 wt% Methyl acrylate)				Water		<273.2	164
N,N-Dimethylacrylamide/propyl acrylate copolymer							
(20 wt% Propyl acrylate)				Water		353.2	164
(30 wt% Propyl acrylate)				Water		337.2	164
(40 wt% Propyl acrylate)				Water		294.2	164
(50 wt% Propyl acrylate)				Water		281.2	164
Dimethylsiloxane/methylphenylsiloxane copolymer (15 wt% methylphenylsiloxane)							
	9100	41200		Anisole	291.45		198

Polymer	M_n/g mol^{-1}	M_w/g mol^{-1}	M_η/g mol^{-1}	Solvent	UCST/K	LCST/K	Ref.
	9100	41200		2-Propanone	282.45		198
Ethylene/propylene copolymer (33 mol% ethylene)							
			145000	Cyclohexane		534	101
			145000	Cyclopentane		490	101
			145000	2,2-Dimethylbutane		428	101
			145000	2,3-Dimethylbutane		452	101
			145000	3,4-Dimethylhexane		541	101
			145000	2,2-Dimethylpentane		472	101
			145000	2,3-Dimethylpentane		500	101
			145000	2,4-Dimethylpentane		464	101
			145000	3-Ethylpentane		511	101
			145000	Heptane		502	101
			145000	Hexane		455	101
			145000	2-Methylbutane		396	101
			145000	Methylcyclohexane		558	101
			145000	Methylcyclopentane		512	101
			145000	2-Methylhexane		486	101
			145000	Nonane		558	101
			145000	Octane		528	101
			145000	Pentane		409	101
			145000	2,2,4,4-Tetramethylpentane		539	101
			145000	2,2,3-Trimethylbutane		500	101
			145000	2,2,4-Trimethylpentane		503	101
Ethylene/propylene copolymer (43 mol% ethylene)							
	70000	140000		Hexane		436	127
	70000	140000		2-Methylpentane		474	127
	70000	140000		Pentane		441	127
Ethylene/propylene copolymer (53 mol% ethylene)							
			154000	2,2-Dimethylbutane		407	101
			154000	2,3-Dimethylbutane		437	101
			154000	2,2-Dimethylpentane		453	101
			154000	2,3-Dimethylpentane		488	101
			154000	2,4-Dimethylpentane		445	101
			154000	3-Ethylpentane		500	101
			154000	Heptane		493	101
			154000	Hexane		443	101
			154000	Pentane		395	101
			154000	2,2,3-Trimethylbutane		488	101
			154000	2,3,4-Trimethylhexane		565	101
			154000	2,2,4-Trimethylpentane		484	101
Ethylene/propylene copolymer (63 mol% ethylene)							
			236000	Cyclohexane		526	101
			236000	Cyclopentane		481	101
			236000	2,3-Dimethylbutane		429	101
			236000	3,4-Dimethylhexane		530	101
			236000	2,2-Dimethylpentane		444	101
			236000	2,3-Dimethylpentane		482	101
			236000	2,4-Dimethylpentane		434	101

Polymer	M_n/g mol^{-1}	M_w/g mol^{-1}	M_η/g mol^{-1}	Solvent	UCST/K	LCST/K	Ref.
			236000	3-Ethylpentane		492	101
			236000	Heptane		485	101
			236000	Hexane		436	101
			236000	2-Methylbutane		348	101
			236000	Methylcyclopentane		498	101
			236000	Nonane		547	101
			236000	Octane		512	101
			236000	Pentane		387	101
			236000	2,2,4,4-Tetramethylpentane		528	101
			236000	2,2,3-Trimethylbutane		479	101
			236000	2,2,4-Trimethylpentane		479	101
Ethylene/propylene copolymer (75 mol% ethylene)							
			109000	2,2-Dimethylpentane		431	101
			109000	2,4-Dimethylpentane		425	101
			109000	Heptane		475	101
			109000	Hexane		427	101
			109000	Nonane		542	101
			109000	Octane		509	101
			109000	Pentane		378	101
			109000	2,2,4,4-Tetramethylpentane		523	101
			109000	2,2,4-Trimethylpentane		469	101
Ethylene/propylene copolymer (81 mol% ethylene)							
			195000	Cyclohexane		522	101
			195000	Cyclopentane		474	101
			195000	2,2-Dimethylbutane		381	101
			195000	2,3-Dimethylbutane		413	101
			195000	2,4-Dimethylhexane		478	101
			195000	2,5-Dimethylhexane		466	101
			195000	3,4-Dimethylhexane		522	101
			195000	2,2-Dimethylpentane		425	101
			195000	2,3-Dimethylpentane		471	101
			195000	2,4-Dimethylpentane		420	101
			195000	3-Ethylpentane		478	101
			195000	Heptane		468	101
			195000	Hexane		425	101
			195000	2-Methylbutane		327	101
			195000	Methylcyclohexane		541	101
			195000	Methylcyclopentane		493	101
			195000	2-Methylhexane		453	101
			195000	3-Methylhexane		459	101
			195000	Nonane		540	101
			195000	Octane		506	101
			195000	Pentane		370	101
			195000	2,2,4,4-Tetramethylpentane		519	101
			195000	2,2,3-Trimethylbutane		461	101
			195000	2,2,4-Trimethylpentane		460	101
Ethylene/vinyl acetate copolymer							
(2.3 wt% vinyl acetate)	52000	465000		Diphenyl ether	404.2		143
(4.0 wt% vinyl acetate)	47000	280000		Diphenyl ether	392.5		143

Polymer	M_n/g mol⁻¹	M_w/g mol⁻¹	M_η/g mol⁻¹	Solvent	UCST/K	LCST/K	Ref.
(7.1 wt% Vinyl acetate)	34000	460000		Diphenyl ether	378.2		143
(9.5 wt% Vinyl acetate)	53000	350000		Diphenyl ether	367.3		143
(9.7 wt% Vinyl acetate)	55000	490000		Diphenyl ether	370.8		143
(12.1 wt% Vinyl acetate)	66000	300000		Diphenyl ether	360.4		143
(42.6 mol% Vinyl acetate)	14800	41500		Methyl acetate	307.0		130
Ethylene/vinyl alcohol copolymer							
(87.2 mol% Vinyl alcohol)			infinite	Water	463.55	285.65	44
(88.9 mol% Vinyl alcohol)			infinite	Water	449.15	290.75	44
(91.0 mol% Vinyl alcohol)			infinite	Water	428.45	302.95	44
(94.1 mol% Vinyl alcohol)			infinite	Water	389.25	324.45	44
Ethylene oxide/propylene oxide copolymer							
(20.0 mol% Ethylene oxide)	3400			Water		303	211
(27.0 mol% Ethylene oxide)	3000			Water		309	210
(30.0 mol% Ethylene oxide)	5400			Water		313	211
(38.5 mol% Ethylene oxide)	5000			Water		309	210
(50.0 mol% Ethylene oxide)	3900			Water		323	211
(58.8 mol% Ethylene oxide)	3000			Water		326.65	210
(72.4 mol% Ethylene oxide)		36000		Water		333	153
(79.5 mol% Ethylene oxide)		30800		Water		345	153
(86.6 mol% Ethylene oxide)		30100		Water		355.5	153
Gutta Percha							
			194000	Propyl acetate	318.95		7
Hydroxypropylcellulose							
		75000		Water		318.45	43
		300000		Water		331.25	43
N-Isopropylacrylamide/acrylamide copolymer							
(15 mol% Acrylamide)		3100000		Water		315.15	172
(30 mol% Acrylamide)		4500000		Water		326.15	172
(45 mol% Acrylamide)		3900000		Water		347.15	172
N-Isopropylacrylamide/1-deoxy-1-methacrylamido-D-glucitol							
(12.9 mol% Glucitol)	78000	170000		Water		311.3	218
(13.7 mol% Glucitol)	51600	110000		Water		314.9	218
(14.0 mol% Glucitol)	145000	432000		Water		307.5	218
N-Isopropylacrylamide/N-isopropylmethacrylamide copolymer							
(10.56 mol% N-Isopropylmethacrylamide)	55300	177000		Water		307.15	212
(30.00 mol% N-Isopropylmethacrylamide)	28800	92000		Water		309.75	212
(39.99 mol% N-Isopropylmethacrylamide)	23100	74000		Water		311.05	212
(59.89 mol% N-Isopropylmethacrylamide)	23100	74000		Water		314.65	212
(79.81 mol% N-Isopropylmethacrylamide)	16600	53000		Water		317.35	212
(89.99 mol% N-Isopropylmethacrylamide)	14700	47000		Water		318.75	212
Methylcellulose (about 30 mol% methyl subsitution)							
		70000		Water		324.75	47
Methylcellulose/hydroxypropylcellulose copolymer (25 mol% methyl, 8 mol% hydroxypropyl subsitution)		80000		Water		340.15	63
Natural rubber							
		300000		Pentane		403	10
			74500	2-Pentanone	274.45		7

Polymer	M_n/g mol^{-1}	M_w/g mol^{-1}	M_η/g mol^{-1}	Solvent	UCST/K	LCST/K	Ref.
Phenol-formaldehyde resin (acetylated)				2-Ethoxyethanol	378.2		200
Poly(acrylic acid)		120000		Tetrahydrofuran		268.3	189
Poly[bis(2,3-dimethoxypropanoxy) phosphazene]	1070000	1500000		Water		317.15	183
Poly[bis(2-(2'-methoxyethoxy)ethoxy) phosphazene]	667000	1000000		Water		338.15	183
Poly[bis(2,3-bis(2-methoxyethoxy) propanoxy)phosphazene]	714000	1000000		Water		311.15	183
Poly[bis(2,3-bis(2-(2'-methoxyethoxy) ethoxy)propanoxy)phosphazene]	1420000	1700000		Water		322.65	183
Poly[bis(2,3-bis(2-(2'-(2"-dimethoxyethoxy) ethoxy)ethoxy)propanoxy)phosphazene]	857000	1200000		Water		334.65	183
Poly(1-butene) (atactic)	infinite			Anisole	359.4		11
	infinite			Toluene	356.2		28
Poly(1-butene) (isotactic)	infinite			Anisole	362.3		11
			530000	Cyclopentane		498	102
			530000	2,2-Dimethylbutane		444	102
			530000	2,5-Dimethylhexane		519	102
			530000	3,4-Dimethylhexane		559	102
			530000	2,3-Dimethylpentane		517	102
			530000	2,4-Dimethylpentane		480	102
			530000	3-Ethylpentane		523	102
			530000	Heptane		509	102
	infinite			Hexane		464	102
			530000	2-Methylbutane		416	102
	infinite			Nonane		564	102
			530000	Octane		540	102
	infinite			Pentane		421	102
			530000	2,2,3-Trimethylbutane		507	102
Poly(butyl methacrylate)	278000	470000		1-Butanol	287.15		132
	278000	470000		Decane	357.25		132
	278000	470000		Ethanol	315.25		132
	278000	470000		Heptane	342.55		132
	278000	470000		Octane	345.80		132
	278000	470000		1-Pentanol	286.30		132
	278000	470000		2-Propanol	294.90		132
	278000	470000		2,2,4-Trimethylpentane	347.50		132
Poly(2-chlorostyrene)	infinite			Benzene		298	40
Poly(4-chlorostyrene)	infinite			Benzene	274.0		22
	infinite			2-(Butoxyethoxy)ethanol		323.25	46
	infinite			Butyl acetate		502.4	22
	infinite			*tert*-Butyl acetate		338.55	46

Polymer	M_n/g mol⁻¹	M_w/g mol⁻¹	M_η/g mol⁻¹	Solvent	UCST/K	LCST/K	Ref.
	infinite			Chlorobenzene	128.8		22
	infinite			2-(Ethoxyethoxy)ethanol		300.95	46
	infinite			Ethyl acetate		613.2	22
	infinite			Ethylbenzene	283.2		22
	infinite			Ethylbenzene	258.45		46
	infinite			Ethyl chloroacetate	271.35		46
	infinite			Isopropyl acetate		348.65	46
	infinite			Isopropylbenzene	332.15		46
	infinite			Isopropyl chloroacetate	264.95		46
	infinite			Methyl chloroacetate	337.75		46
	infinite			Propyl acetate		908.7	22
	infinite			Tetrachloroethene	317.55		46
	infinite			Tetrachloromethane	323.85		46
	infinite			Toluene	236.8		22
Poly(decyl methacrylate)							
	390000	468000		1-Butanol	304.85		113
	390000	468000		1-Pentanol	278.40		113
	220000	252000		2-Propanol	346.85		132
Polydimethylsiloxane (cyclic)							
	9810	10300		2,2-Dimethylpropane		433	133
	9810	10300		Tetramethylsilane		448	133, 171
Polydimethylsiloxane							
			626000	Butane		392.95	53
	infinite			Decane		603	30
	14750	16370		2,2-Dimethylpropane		428	133
	infinite			Dodecane		643	30
			626000	Ethane		259.65	53
			100000	Ethoxybenzene	341.99		108
	infinite			Heptane		528	30
	infinite			Hexadecane		708	30
	infinite			Hexane		493	30
	infinite			Octane		553	30
	infinite			Pentane		453	30
			203000	Propane		340.15	53
	14750	16370		Tetramethylsilane		443	133, 171
Poly(ethyl acrylate)							
			48000	1-Butanol	310.05		27
			48000	Ethanol	301.15		27
			380000	Methanol	287.25		27
			48000	1-Propanol	305.15		27
Polyethylene (branched)							
	8400	32000		Diphenyl ether	384.7		95, 98
	24000	123000		Diphenyl ether	396.7		95, 98
	65000	425000		Diphenyl ether	415.3		95, 98
Polyethylene (linear)							
			20000	Anisole	368.15		24
			20000	Benzyl acetate	459.65		24
			20000	Benzyl phenyl ether	437.15		24
			20000	Benzyl propionate	436.15		24
			50900	Biphenyl	383.55		25
			61100	Butyl acetate	448	497	70

Polymer	M_n/g mol^{-1}	M_w/g mol^{-1}	M_η/g mol^{-1}	Solvent	UCST/K	LCST/K	Ref.
			20000	4-*tert*-Butylphenol	466.15		24
			134000	Cyclohexane		518	101
			20000	Cyclohexanone	389.65		24
			134000	Cyclopentane		472	101
	36700	49300		Decane		563.75	91
			20000	1-Decanol	400.15		24
			20000	Dibenzyl ether	448.65		24
			134000	3,4-Dimethylhexane		515	101
			134000	2,2-Dimethylpentane		399	101
			134000	2,3-Dimethylpentane		463	101
			134000	2,4-Dimethylpentane		395	101
	12000	150000		Diphenyl ether	416.2		95, 98
			97200	Diphenylmethane	400.25		25
	60400	82600		Dodecane		610.85	91
		218000		1-Dodecanol	405.15		141
			134000	3-Ethylpentane		471	101
	36700	49300		Heptane		464.70	91
			20000	1-Heptanol	440.15		24
	36700	49300		Hexane		414.65	91
	7900	92000		1-Hexanol	458.15		154
			20000	2-Methoxynaphthalene	427.65		24
			20000	3-Methylbutyl acetate	407.15		24
			134000	Methylcyclohexane		537	101
			134000	Methylcyclopentane		488	101
	60400	82600		Nonane		531.90	91
			20000	1-Nonanol	431.15		24
			20000	4-Nonylphenol	410.15		24
	36700	49300		Octane		502.40	91
	7900	92000		1-Octanol	426.65		154
			20000	4-Octylphenol	424.65		24
			134000	Pentane		353	101
			20000	1-Pentanol	445.15		24
			175000	Pentyl acetate	421	528	70
			20000	4-*tert*-Pentylphenol	443.65		24
			20000	Phenetole	366.65		24
			134000	2,2,4,4-Tetramethylpentane		513	101
	60400	82600		Tridecane		639.30	91
			134000	2,2,3-Trimethylbutane		444	101
			134000	2,3,4-Trimethylhexane		545	101
			134000	2,2,4-Trimethylpentane		495	101
	97700	135900		Undecane		583.95	91
Poly(ethylene glycol)							
			8000	*tert*-Butyl acetate	321.2	464.2	83
			21200	*tert*-Butyl acetate	353.2	431.2	83
	6100	6200		Water		404.79	185
	10457	11615		Water		394.33	205
	40800	151000		Water		378.25	205
Poly(ethylene oxide)-b-poly[bis(methoxyethoxyethoxy)-phosphazene] block copolymer (about 67 mol% Ethylene oxide)							
	22000	31500		Water		338	222

Polymer	M_n/g mol^{-1}	M_w/g mol^{-1}	M_η/g mol^{-1}	Solvent	UCST/K	LCST/K	Ref.
Poly(ethylene oxide)-b-poly(propylene oxide)-b-poly(ethylene oxide) triblock copolymer (about 30 mol% Ethylene oxide)							
		4400		Water		286.65	209
Polyethylethylene							
	48000	52000		Diphenyl ether	411.2		95, 98
Poly(*p*-hexylstyrene)							
	infinite			2-Butanone	302.6		135
Poly(2-hydroxyethyl methacrylate)							
			77400	1-Butanol	337.25		35
			233600	2-Butanol	287		35
			233600	2-Metyl-1-propanol	342		35
			77400	1,2,3-Propanetriol	345		35
			77400	1-Propanol	311		35
Polyisobutylene							
	infinite			Anisole	377		3
			72000	Benzene		540.5	39
			703000	Butane		264.75	53
	infinite			Cycloheptane		572	34
			1500000	Cyclohexane		412	10
	infinite			Cyclooctane		637	34
			1500000	Cyclopentane		344	10
	infinite			Decane		535	34
			1500000	2,2-Dimethylbutane		376	10
			1500000	2,3-Dimethylbutane		404	10
	infinite			2,2-Dimethylhexane		454	34
	infinite			2,4-Dimethylhexane		458	34
	infinite			2,5-Dimethylhexane		446	34
	infinite			3,4-Dimethylhexane		497	34
	infinite			2,2-Dimethylpentane		404	34
	infinite			2,3-Dimethylpentane		451	34
	infinite			2,4-Dimethylpentane		403	34
	infinite			3,3-Dimethylpentane		451	34
	infinite			Diphenyl ether	306		3
	infinite			Decane		585	30
	infinite			Dodecane		582	34
	infinite			Ethylbenzene	249		3
	infinite			Ethylcyclopentane		524	34
	infinite			Ethyl heptanoate	306		3
	infinite			Ethyl hexanoate	330		3
	infinite			3-Ethylpentane		458	34
	infinite			Heptane		442	34
			72000	Hexane		428.5	39
			6030	2-Methylbutane		357.85	53
	infinite			Methylcyclohexane		526	34
	infinite			Methylcyclopentane		478	34
	infinite			2-Methylheptane		466	34
	infinite			3-Methylheptane		478	34
	infinite			2-Methylhexane		426	34
	infinite			3-Methylhexane		446	34
	infinite			2-Methylpentane		376	34

Polymer	M_n/g mol^{-1}	M_w/g mol^{-1}	M_η/g mol^{-1}	Solvent	UCST/K	LCST/K	Ref.
	infinite			3-Methylpentane		405	34
			470	2-Methylpropane		387	10
			72000	Octane		506.0	39
			6030	Pentane		403.55	53
			72000	Pentane		373.5	39
	infinite			Phenetole	357		3
			470	Propane		358	10
	infinite			Propylcyclopentane		547	34
	infinite			Toluene	260		3
	infinite			2,2,3-Trimethylbutane		445	34
	infinite			2,2,4-Trimethylpentane		435	34
1,4-*cis*-Polyisoprene							
			780000	2,5-Dimethylhexane		474.15	140
			780000	3,4-Dimethylhexane		520.15	140
			780000	2,2-Dimethylpentane		445.15	140
			780000	2,3-Dimethylpentane		484.15	140
			780000	2,4-Dimethylpentane		442.15	140
			780000	3-Methylpentane		483.15	140
			780000	Heptane		488.15	140
			780000	Hexane		434.15	140
			780000	Nonane		541.15	140
			780000	Octane		509.15	140
			780000	2,2,4,4-Tetramethylpentane		518.15	140
			780000	2,3,4-Trimethylhexane		548.15	140
			780000	2,2,4-Trimethylpentane		471.15	140
1,4-*trans*-Polyisoprene							
			180000	2,5-Dimethylhexane		451.15	140
			180000	3,4-Dimethylhexane		521.15	140
			180000	2,2-Dimethylpentane		405.15	140
			180000	2,3-Dimethylpentane		460.15	140
			180000	2,4-Dimethylpentane		404.15	140
			180000	3-Methylpentane		473.15	140
			180000	Heptane		467.15	140
			180000	Hexane		407.15	140
			180000	Nonane		540.15	140
			180000	Octane		503.15	140
			180000	2,2,4,4-Tetramethylpentane		519.15	140
			180000	2,3,4-Trimethylhexane		548.15	140
Poly(*N*-isopropylacrylamide)							
	5400	14000		Water		307.45	146
	146000	530000		Water		305.85	146
Poly(*N*-isopropylacrylamide)-poly[(*N*-acetylimino)ethylene] block copolymer (80 wt% *N*-Isopropylacrylamide)							
	5500			Water		306.2	223
Poly(*N*-isopropylacrylamide)-poly[(*N*-acetylimino)ethylene] graft copolymer (75 wt% *N*-Isopropylacrylamide)							
	6030			Water		306.2	223
Poly(*N*-isopropylmethacrylamide)							
	6250	20000		Water		319.95	212
Poly(methyl methacrylate)							

Polymer	M_n/g mol^{-1}	M_w/g mol^{-1}	M_η/g mol^{-1}	Solvent	UCST/K	LCST/K	Ref.
			127000	Acetonitrile	267.15		16
			970000	Acetonitrile	303.15		16
			50000	1-Butanol	353.25		2
	infinite			2-Butanone		482	80
	infinite			1-Chlorobutane	320	463	80
			970000	2,2-Dimethyl-3-pentanone	301.55		16
			127000	2,4-Dimethyl-3-pentanone	280.15		16
	200000	264000		2-Ethoxyethanol	312.15		196
		77000		Ethyl acetate	290	533	190
			127000	2-Ethylbutanal	264.65		16
	infinite			3-Heptanone	307.7		126
			970000	4-Heptanone	299.95		16
	infinite			3-Hexanone		522	80
	infinite			Methyl acetate		451	80
			50000	1-Methyl-4-isopropylbenzene	400.15		2
			1400000	2-Octanone	321.15		16
	572400	595300		3-Octanone	329.88		166
	infinite			3-Pentanone		506	80
			50000	1-Propanol	349.95		2
	infinite			2-Propanone		439	80
	200000	264000		Tetra(ethylene glycol)	390.15		196
			400000	Toluene	225.35		2
			50000	Trichloromethane	231.15		2
	200000	264000		Tri(ethylene glycol)	407.15		196
Poly(methyl methacrylate) (isotactic)							
	infinite			Acetonitrile	301	461	80
	infinite			2-Butanone		464	80
	infinite			1-Chlorobutane	309	454	80
	infinite			4-Heptanone	319	522	80
	infinite			3-Hexanone	279	511	80
	infinite			Methyl acetate		441	80
	infinite			3-Pentanone		497	80
	infinite			2-Propanone		428	80
Poly(4-methyl-1-pentene) (isotactic)							
			152000	Butane		388	102
			152000	Cyclopentane		505	102
			152000	2,2-Dimethylbutane		462	102
			152000	2,2-Dimethylpentane		499	102
			152000	2,4-Dimethylpentane		499	102
	infinite			Diphenyl	467.8		62
	infinite			Diphenyl ether	483.2		62
	infinite			Diphenylmethane	449.8		62
			152000	3-Ethylpentane		532	102
			152000	Heptane		522	102
			152000	Hexane		487	102
			152000	2-Methylbutane		431	102
			152000	Nonane		579	102
			152000	Octane		553	102
			152000	Pentane		441	102
			152000	2,2,3-Trimethylbutane		521	102

Polymer	M_n/g mol^{-1}	M_w/g mol^{-1}	M_η/g mol^{-1}	Solvent	UCST/K	LCST/K	Ref.
Poly(α-methylstyrene)							
	58500	61400		Butyl acetate	262.05	457.15	181
	99100	113000		Cyclohexane	293.55		152
	26000	31200		Cyclopentane	276.7	435.95	181
		289000		trans-Decahydronaphthalene	273		181
	69500	76500		Hexyl acetate	285.05	508.15	181
	72000	75600		Methylcyclohexane	328.9		203
	58500	61400		Pentyl acetate	287.1	484.6	181
Poly(2-methyl-5-vinylpyridine)							
			600000	Butyl acetate	287.95		20
			263000	Ethyl butyrate	319.05		20
			335000	Ethyl propionate	293.55		20
			275000	3-Methylbutyl acetate	314.75		20
			335000	4-Methyl-2-pentanone	299.95		20
			170000	2-Methylpropyl acetate	312.35		20
			165000	Pentyl acetate	316.95		20
			284000	Propionitrile	262.35		20
			152000	Propyl acetate	282.65		20
			181000	Propyl propionate	312.15		20
			233000	Tetrahydronaphthalene	316.95		20
Poly(1-pentene) (isotactic)							
			4500000	Cyclopentane		502	102
			4500000	2,2-Dimethylbutane		457	102
			4500000	3,4-Dimethylhexane		>569	102
			4500000	2,2-Dimethylpentane		502	102
			4500000	2,3-Dimethylpentane		529	102
			4500000	2,4-Dimethylpentane		493	102
			4500000	3-Ethylpentane		537	102
			4500000	Heptane		522	102
			4500000	Hexane		482	102
			4500000	2-Methylbutane		422	102
			4500000	Octane		556	102
			4500000	Pentane		433	102
			4500000	2,2,4-Trimethylpentane		527	102
Polypropylene (atactic)							
	infinite			Diphenyl ether	426.5		9
	infinite			Diethyl ether		383	68
			242000	Heptane		511	101
	infinite			Hexane		441	68
			242000	2-Methylbutane		413	101
			242000	Methylcyclohexane		564	101
	infinite			Pentane		397	68
Polypropylene (isotactic)							
			28000	Benzyl phenyl ether	429.2		31
			28000	Benzyl propionate	405.2		31
			28000	1-Butanol	395.2		31
			28000	4-tert-Butylphenol	413.2		31
			242000	Cyclohexane		540	101
			242000	Cyclopentane		495	101
			28000	Dibenzyl ether	433.2		31

Polymer	M_n/g mol^{-1}	M_w/g mol^{-1}	M_η/g mol^{-1}	Solvent	UCST/K	LCST/K	Ref.
			242000	2,2-Dimethylbutane		441	101
			242000	2,3-Dimethylbutane		465	101
			242000	3,4-Dimethylhexane		553	101
			242000	2,2-Dimethylpentane		489	101
			242000	2,3-Dimethylpentane		513	101
			242000	2,4-Dimethylpentane		481	101
			28000	Diphenyl	388.2		31
			28000	Diphenyl ether	395.2		31
			28000	Diphenylmethane	389.7		31
			242000	3-Ethylpentane		520	101
			28000	4-Ethylphenol	457.2		31
			242000	Heptane		511	101
			242000	Hexane		470	101
			242000	2-Methylbutane		413	101
			28000	3-Methylbutyl benzyl ether	384.2		31
			242000	Methylcyclohexane		564	101
			242000	Methylcyclopentane		518	101
			28000	4-Methylphenol	479.2		31
			28000	2-Methyl-1-propanol	395.2		31
			242000	Nonane		571	101
			242000	Octane		542	101
			28000	4-Octylphenol	379.2		31
			28000	4-Isooctylphenol	383.2		31
			242000	Pentane		422	101
			242000	2,2,4,4-Tetramethylpentane		548	101
			242000	2,2,3-Trimethylbutane		511	101
			242000	2,3,4-Trimethylhexane		585	101
			242000	2,2,4-Trimethylpentane		510	101
Poly(propylene glycol)							
	1000			Hexane	288.15		88
	575			Water		318.2	65
Polystyrene							
	34900	37000		Benzene		538.7	61
			62600	Butanedioic acid dimethyl ester	335.15		2
	3700	4000		1-Butanol	383.45		154
	91700	97200		2-Butanone		448.8	61
	545500	600000		Butyl acetate		489	181
	104000	110000		tert-Butyl acetate	250.0	417.9	74
			62600	Butyl stearate	387.15		2
	18400	19200		1-Chlorododecane	274.65		154
	18400	19200		1-Chlorohexadecane	337.05		154
	18400	19200		1-Chlorooctadecane	365.55		154
	18400	19200		1-Chlorotetradecane	309.35		154
	46400	51000		Cyclodecane	278.9		128
	46400	51000		Cycloheptane	276.2		128
	34900	37000		Cyclohexane	285.6	510.9	60
			236000	Cyclohexanol	353.5		8
	46400	51000		Cyclooctane	275.2		128
	91700	97200		Cyclopentane	275.2	445.5	61

Polymer	M_n/g mol⁻¹	M_w/g mol⁻¹	M_η/g mol⁻¹	Solvent	UCST/K	LCST/K	Ref.
	91500	97000		*trans*-Decahydronaphthalene	281.95		81
		4800		Decane	360.95		154
	3700	4000		1-Decanol	375.15		154
			570000	Decyl acetate		650	64
	18700	19800		Diethyl ether	235.6	314.5	51
	187000	200000		Diethyl malonate	285.8	589.6	74
	47200	50000		Diethyl oxalate	280.05		131
	151000	160000		Dimethoxymethane		401.2	51
		240000		1,4-Dimethylcyclohexane	387	482	116
			62600	Dimethyl malonate	409.15		2
			62600	Dimethyl oxalate	453.15		2
	116000	123000		Dodecadeuterocyclohexane	298.10		224
		25000		Dodecadeuteromethyl-cyclopentane	310.07		180
		4800		Dodecane	368.65		154
	3700	4000		1-Dodecanol	379.75		154
	infinite			Dodecyl acetate	285.2		206
	104000	110000		Ethyl acetate	213.9	435.4	72
	104000	110000		Ethyl butanoate		490.8	74
	221000	239000		Ethylcyclohexane	330.52		18
	9440	10000		Ethyl formate	272	451	74
		900000		Bis(2-ethylhexyl) phthalate	283.05		136
	4530	4800		Heptane	359	477	112
	3700	4000		1-Dexadecanol	386.25		154
	5500	5770		1,1,1,3,3,3-Hexadeutero-2-propanone	270	436	157
	1920	2030		Hexane	318	470	112
			62600	Hexanoic acid	448.15		2
	3700	4000		1-Hexanol	372.15		154
			62600	3-Hexanol	396.65		2
			90000	Hexyl acetate		578	64
	104000	110000		Methyl acetate	284.2	415.7	72
	104000	110000		3-Methyl-1-butyl acetate	210.1	510.1	72
	91700	97200		Methylcyclohexane	321.8	505.9	60
	10750	11500		Methylcyclopentane	295	480	157
	104000	110000		2-Methyl-1-propyl acetate	210.4	468.5	72
		48000		Nitroethane	303.1		151
		4800		Octadecane	403.55		154
	3700	4000		1-Octadecanol	390.55		154
	4530	4800		Octane	353	527	112
	3700	4000		1-Octanol	372.35		154
			62600	1-Octene	355.15		2
		4800		Pentadecane	385.25		154
		1100		Pentane	292		137
	3700	4000		1-Pentanol	375.05		154
	219800	233000		Pentyl acetate		519	181
		100000		1-Phenyldecane	283.60		105
	5500	5770		2-Propanone	251	452	157
	12750	13500		Propionitrile	312		187
	104000	110000		Propyl acetate	183.7	469.0	72
	104000	110000		2-Propyl acetate	220.9	414.2	72

Polymer	M_n/g mol^{-1}	M_w/g mol^{-1}	M_η/g mol^{-1}	Solvent	UCST/K	LCST/K	Ref.
	3700	4000		1-Tetradecanol	383.25		154
	34900	37000		Toluene		567.2	60
			62600	Vinyl acetate	384.15		2
Polystyrene (three-arm star)							
		230000		Cyclohexane	297.1	496.8	93
Polystyrene (four-arm star)							
		155000		Cyclohexane	294.13		199
Poly(trimethylene oxide)							
		infinite		Cyclohexane	300		79
Poly(vinyl alcohol)							
		40000		Water		514	45
Poly(N-vinyl caprolactam)							
		150000		Water		306.45	217
Poly(vinyl chloride)							
	55000			Dibutyl phthalate	353		114
	55000			Tricresyl phosphate	383		114
			85000	Dimethyl phthalate	355		219
Poly(N-vinylisobutyramide)							
	66000	105600		Water		313.25	208
Poly(vinyl methyl ether)							
	46500	98600		Deuterium oxide		307.2	173
	83000	155000		Water		306.95	146
Poly(N-vinyl-N-propylacetamide)							
			30000	Water		313.5	176
Styrene/acrylonitrile copolymer							
(21.1 wt% acrylonitrile)	infinite			Toluene	325.4		52
(23.2 wt% Acrylonitrile)	infinite			Toluene	355.1		52
(25.0 wt% Acrylonitrile)	90000	147000		Toluene	313.15		198
(51.0 wt% Acrylonitrile)		347000		Ethyl acetate		344.15	107
Styrene/methyl methacrylate copolymer							
(52.0 mol% Styrene)							
		infinite		Cyclohexanol	334.65		38
Styrene/α-methylstyrene copolymer							
(20.0 mol% Styrene)							
	100000	114000		Butyl acetate	288.85	453.05	181
	100000	114000		Cyclohexane	285.85	484.85	181
	100000	114000		Cyclopentane	290.95	421.05	181
	100000	114000		trans-Decahydronaphthalene	264.15		181
	100000	114000		Hexyl acetate	288.55	514.15	181
	100000	114000		Pentyl acetate	303.15	480.65	181
Trifluoronitrosomethane/ tetrafluoroethylene copolymer (1:1) alternating							
	infinite			1,1,2-Trichloro-1,2,2-trifluoroethane	301.6		12
N-Vinylacetamide/vinyl acetate copolymer							
(58 mol% Vinyl acetate)	30000	57000		Water		340.15	225
(63 mol% Vinyl acetate)	27000	48600		Water		323.15	225
(78 mol% Vinyl acetate)	26000	46800		Water		282.15	225
Vinyl alcohol/vinyl butyrate copolymer							
(7.5 mol% Butyralized PVA)	infinite			Water	408.0	298.25	121

Polymer	M_n/g mol^{-1}	M_w/g mol^{-1}	M_η/g mol^{-1}	Solvent	UCST/K	LCST/K	Ref.
N-Vinylcaprolactam/N-vinylamine copolymer (3.8 mol% Vinyl amine)							
			160000	Water		308.8	176
N-Vinylformamide/vinyl acetate copolymer							
(60 mol% Vinyl acetate)	24000	45600		Water		310.15	225
(66 mol% Vinyl acetate)	25000	47500		Water		291.15	225
(73 mol% Vinyl acetate)	23000	50600		Water		277.15	225

References

1a. Koningsveld, R., Stockmayer, W.H., and Nies, E., *Polymer Phase Diagrams*, Oxford University Press, Oxford, 2001.

1b. Kamide, K., *Thermodynamics of Polymer Solutions*, Elsevier, Amsterdam, 1990.

2. Jenckel, E. and Gorke, K., *Z. Naturforsch.*, 5a, 317, 556, 1950.

3. Fox, T.G. and Flory, P.J., *J. Amer. Chem. Soc.*, 73, 1909, 1951.

4. Fox, T.G. and Flory, P.J., *J. Amer. Chem. Soc.*, 73, 1915, 1951.

5. Mandelkern, L. and Flory, P.J., *J. Amer. Chem. Soc.*, 74, 2517, 1952.

6. Shultz, A.R. and Flory, P.J., *J. Amer. Chem. Soc.*, 74, 4760, 1952.

7. Wagner, H.L. and Flory, P.J., *J. Amer. Chem. Soc.*, 74, 195, 1952.

8. Shultz, A.R.and Flory, P.J., *J. Amer. Chem. Soc.*, 75, 3888, 1953.

9. Kinsinger, J.B. and Wessling, R.A., *J. Amer. Chem. Soc.*, 81, 2908, 1959.

10. Freeman, P.I. and Rowlinson, J.S., *Polymer*, 1, 20, 1960.

11. Krigbaum, W.R., Kurz, J.E., and Smith, P., *J. Phys. Chem.*, 65, 1984, 1961.

12. Morneau, G.A., Roth, P.I., and Shultz, A.R., *J. Polym. Sci.*, 55, 609, 1961.

13. Debye, P., Coll, H., and Woermann, D., *J. Chem. Phys.*, 32, 939, 1960.

14. Debye, P., Coll, H., and Woermann, D., *J. Chem. Phys.*, 33, 1746, 1960.

15. Debye, P., Chu, B., and Woermann, D., *J. Chem. Phys.*, 36, 1803, 1962.

16. Fox, T.G., *Polymer*, 3, 111, 1962.

17. Ham, J.S., Bolen, M.C., and Hughes, J.K., *J. Polym. Sci.*, 57, 25, 1962.

18. Debye, P., Woermann, D., and Chu, B., *J. Polym. Sci.: Part A*, 1, 255, 1963.

19. Allen, G. and Baker, C.H., *Polymer*, 6, 181, 1965.

20. Gechele, G.B., Crescentini, L., *J. Polym. Sci.: Part A*, 3, 3599, 1965.

21. Myrat, C.D. and Rowlinson, J.S., *Polymer*, 6, 645, 1965.

22. Kubo, K. and Ogino, K., *Sci. Pap. Coll. Art. Sci. Univ. Tokyo*, 16, 193, 1966.

23. Rehage, G., Moeller, D., and Ernst, O., *Makromol.Chem.*, 88, 232, 1965.

24. Nakajima, A., Fujiwara, H., and Hamada, F., *J. Polym. Sci.: Part A-2*, 4, 507, 1966.

25. Nakajima, A., Hamada, F., and Hayashi, S., *J. Polym. Sci.: Part C*, 15, 285, 1966.

26. Koningsveld, R., *Proefschrift Univ. Leiden*, Heerlen, 1967.

27. Llopis, J., Albert, A., and Usobinaga P., *Eur. Polym. J.*, 3, 259, 1967.

28. Moraglio, G., Gianotti, G., and Danusso, F., *Eur. Polym. J.*, 3, 251, 1967.

29. Orwoll, R.A. and Flory, P.J., *J. Amer. Chem. Soc.*, 89, 6822, 1967.

30. Patterson, D., Delmas, G., and Somcynsky, T., *Polymer*, 8, 503, 1967.

31. Nakajima, A. and Fujiwara, H., *J. Polym. Sci.: Part A-2*, 6, 723, 1968.

32. Rehage, G. and Koningsveld, R., *J. Polym. Sci.: Polym. Lett.*, 6, 421, 1968.

33. Andreeva, V. M., et al., *Vysokomol. Soedin., Ser. B*, 11, 555, 1969.

34. Bardin, J.-M. and Patterson, D., *Polymer*, 10, 247, 1969.

35. Dusek, K., *Coll. Czech. Chem. Commun.*, 34, 3309, 1969.

36. Delmas, G. and Patterson, D., *J. Polym. Sci.: Part C*, 30, 1, 1970.

37. Koningsveld, R., Kleintjens, L.A., and Shultz, A.R., *J. Polym. Sci.: Part A-2*, 8, 1261, 1970.

38. Kotaka, T., et al., *Polym. J.*, 1, 245, 1970.

39. Liddell, A.H. and Swinton, F.L., *Discuss. Faraday Soc.*, 49, 115, 1970.

40. Matsumura, K., *Polym. J.*, 1, 322, 1970.

41. Nakayama, H., *Bull. Chem. Soc. Japan*, 43, 1683, 1970.

42. Cowie, J.M.G., Maconnachie, A., and Ranson, R.J., *Macromolecules*, 4, 57, 1971.

43. Kagemoto, A. and Baba, Y., *Kobunshi Kagaku*, 28, 784, 1971.

44. Shibatani, K. and Oyanagi, Y., *Kobunshi Kagaku*, 28 (1971) 361-367

45. Tager, A.A., et al., *Vysokomol. Soedin., Ser. A*, 13, 659, 1971.

46. Izumi, Y. and Miyake, Y., *Polym. J.*, 3, 647, 1972.

47. Kagemoto, A., Baba, Y., and Fujishiro, R., *Makromol. Chem.*, 154, 105, 1972.

48. Kennedy, J.W., Gordon, M., and Koningsveld, R., *J. Polym. Sci.: Part C*, 39, 43, 1972.

49. Lirova, B.I., et al., *Vysokomol. Soedin., Ser. B*, 14,265, 1972.

50. Nakayama, H., *Bull. Chem. Soc. Japan*, 45, 1371, 1972.

51. Siow, K.S., Delmas, G., and Patterson, D., *Macromolecules*, 5, 29, 1972.

52. Teramachi, S. and Fujikawa, T., *J. Macromol. Sci.-Chem. A*, 6, 1393, 1972.

53. Zeman, L., Biros, J., Delmas, G., and Patterson, D., *J. Phys. Chem.*, 76, 1206, 1972.

54. Zeman, L. and Patterson, D., *J. Phys. Chem.*, 76, 1214, 1972.

55. Baba, Y., Fujita, Y., and Kagemoto, A., *Makromol. Chem.*, 164, 349, 1973.

56. Candau, F., Strazielle, C., and Benoit, H., *Makromol. Chem.*, 170, 165, 1973.

57. Hamada, F., Fujisawa, K., and Nakajima, A., *Polym. J.*, 4, 316, 1973.

58. Kuwahara, N., Nakata, M., and Kaneko, M., *Polymer*, 14, 415, 1973.

59. Kuwahara, N., Kojima, J., and Kaneko, M., *J. Polym. Sci.: Polym. Phys. Ed.*, 11, 2307, 1973.

60. Saeki, S., Kuwahara, N., Konno, S., and Kaneko, M., *Macromolecules*, 6, 246, 1973.

61. Saeki, S., Kuwahara, N., Konno, S., and Kaneko, M., *Macromolecules*, 6, 589, 1973.

62. Tani, S., Hamada, F., and Nakajima, A., *Polym.J.*, 5, 86, 1973.

63. Baba, Y. and Kagemoto, A., *Kobunshi Ronbunshu*, 31, 446, 1974.

64. Bataille, P., *J. Chem. Eng. Data*, 19, 224, 1974.

65. Bessonov, Yu.S. and Tager, A.A., *Trud. Khim. Khim. Tekhnol.*, 1, 150, 1974.

66. Cowie, J.M.G. and McEwen, L.J., *J. Chem. Soc., Faraday Trans. I*, 70, 171, 1974.

67. Cowie, J.M.G. and McEwen, I.J., *Macromolecules*, 7, 291, 1974.

68. Cowie, J.M.G. and McEwen, I.J., *J. Polym. Sci.: Polym. Phys. Ed.*, 12, 441, 1974.

69. Derham, K.W., Goldsbrough, J., and Gordon, M., *Pure Appl. Chem.*, 38, 97, 1974.

70. Kuwahara, N., Saeki, S., Chiba, T., and Kaneko, M., *Polymer*, 15, 777, 1974.

71. Nakajima, A., et al., *Makromol. Chem.*, 175, 197, 1974.

72. Saeki, S., Konno, S., Kuwahara, N., Nakata, M., and Kaneko, M., *Macromolecules*, 7, 521, 1974.

73. Ver Strate, G. and Philippoff, W., *J. Polym. Sci.: Polym. Lett. Ed.*, 12, 267, 1974.

74. Konno, S., et al., *Macromolecules*, 8, 799, 1975.

75. Nakata, M., Kuwahara, N., and Kaneko, M., *J. Chem. Phys.*, 62, 4278, 1975.

76. Saeki, S., Kuwahara, N., Nakata, M., and Kaneko, M., *Polymer*, 16, 445, 1975.

77. Strazielle, C. and Benoit, H., *Macromolecules*, 8, 203, 1975.
78. Tager, A. A., et al., *Vysokomol. Soedin., Ser. B*, 17, 61, 1975.
79. Chiu, D.S., Takahashi, Y., and Mark, J.E., *Polymer*, 17, 670, 1976.
80. Cowie, J.M.G. and McEwen, I.J., *J. Chem. Soc., Faraday Trans. I*, 72, 526, 1976.
81. Nakata, M., et al., *J. Chem. Phys.*, 64, 1022, 1976.
82. Nose, T. and Tan, T.V., *J. Polym. Sci.: Polym. Lett. Ed.*, 14, 705, 1976.
83. Saeki, S., Kuwahara, N., Nakata, M., and Kaneko, M., *Polymer*, 17, 685, 1976.
84. Saeki, S., Kuwahara, N., and Kaneko, M., *Macromolecules*, 101, 1976.
85. Slagowski, E., Tsai, B., and McIntyre, D., *Macromolecules*, 9, 687, 1976.
86. Panina, N.I., Lozgacheva, V.P., and Aver'yanova, V.M., *Vysokomol. Soedin., Ser. B*, 19, 786,.1977.
87. Rigler, J.K., Wolf, B.A., and Breitenbach, J.W., *Angew. Makromol. Chem.*, 57, 15, 1977.
88. Vshivkov, S.A., et al., *Prots. Studneobras. Polimern. Sistem.*, (2), 3, 1977.
89. Wolf, B.A. and Jend, R., *Makromol. Chem.*, 178, 1811, 1977.
90. Wolf, B.A. and Sezen, M.C., *Macromolecules*, 10, 1010, 1977.
91. Kodama, Y. and Swinton, F.L., *Brit. Polym. J.*, 10, 191, 1978.
92. Nakata, M., Dobashi, T., Kuwahara, N., Kaneko, M., and Chu, B., *Phys. Rev. A*, 18, 2683, 1978.
93. Cowie, J.M.G., Horta, A., McEwen, I.J., and Prochazka, K., *Polym. Bull.*, 1, 329, 1979.
94. Hamano, K., Kuwahara, N., and Kaneko, M., *Phys.Rev. A*, 20, 1135, 1979.
95. Kleintjens, L.A.L., *Ph.D. Thesis*, Univ. Essex, U.K., 1979.
96. Dobashi, T., Nakata, M., and Kaneko, M., *J. Chem. Phys.*, 72, 6685, 1980.
97. Irvine, P. and Gordon, M., *Macromolecules*, 13, 761, 1980.
98. Kleintjens, L.A., Koningsveld, R., and Gordon, M., *Macromolecules*, 13, 303, 1980.
99. Lang, J.C. and Morgan, R.D., *J. Chem. Phys.*, 73, 5849, 1980.
100. Richards, R.W., *Polymer*, 21, 715, 1980.
101. Charlet, G. and Delmas, G., *Polymer*, 22, 1181, 1981.
102. Charlet, G., Ducasse, R., and Delmas, G., *Polymer*, 22, 1190, 1981.
103. Hashizume, J., Teramoto, A., and Fujita, H., *J. Polym. Sci.: Polym. Phys. Ed.*, 19, 1405, 1981.
104. Wolf, B.A. and Geerissen, H., *Colloid Polym. Sci.*, 259, 1214, 1981.
105. Geerissen, H. and Wolf, B.A., *Makromol. Chem., Rapid Commun.*, 3, 17, 1982.
106. Goloborod'ko, V.I., Valatin, S.M., and Tashmukhamedov, I.P., *Uzb. Khim. Zh.*, (3), 33, 1982.
107. Mangalam, P. V. and Kalpagam, V., *J. Polym. Sci.: Polym. Phys. Ed.*, 20, 773, 1982.
108. Shinozaki, K., Abe, M., and Nose, T., *Polymer*, 23, 722, 1982.
109. Shinozaki, K., Van Tan, T., Saito, Y., and Nose, T., *Polymer*, 23, 728, 1982.
110. Suzuki, H., Kamide, K., and Saitoh, M., *Eur. Polym. J.*, 18, 123, 1982.
111. Suzuki, H., Muraoka, Y., Saitoh, M., and Kamide, K., *Brit. Polym. J.*, 14, 23, 1982.
112. Cowie, J.M.G. and McEwen, I.J., *Polymer*, 24, 1445, 1983.
113. Herold, F.K., Schulz, G.V., and Wolf, B.A., *Materials Chem. Phys.*, 8, 243, 1983.
114. Tager, A.A., et al., *Vysokomol. Soedin., Ser. A*, 25, 1444, 1983.
115. Corti, M., Minero, C., and Degiorgio, V., *J. Phys. Chem.*, 88, 309, 1984.
116. Cowie, J.M.G. and McEwen, I.J., *Polymer*, 25, 1107, 1984.
117. Dobashi, T., Nakata, M., and Kaneko, M., *J. Chem. Phys.*, 80, 948, 1984.
118. Florin, E., Kjellander, R., and Eriksson, J.C., *J. Chem. Soc., Faraday Trans. I*, 80, 2889, 1984.
119. Gilluck, M., *Dissertation*, TH Leuna-Merseburg, 1984.
120. Rangel-Nafaile, C., Metzner, A.B., and Wissbrun, K.F., *Macromolecules*, 17, 1187, 1984.
121. Shiomi, T., et al., *J. Polym. Sci.: Polym. Phys. Ed.*, 22, 1305, 1984.
122. Tsuyumoto, M., Einaga, Y., and Fujita, H., *Polym. J.*, 16, 229, 1984.

123. Varennes, S., Charlet, G., and Delmas, G., *Polym. Eng. Sci.*, 24, 98, 1984.
124. Hamano, K., Kuwahara, N., Koyama, T., and Harada, S., *Phys. Rev. A*, 32, 3168, 1985.
125. Kraemer, H. and Wolf, B.A., *Makromol. Chem., Rapid Commun.*, 6, 21, 1985.
126. Herold, F.K. and Wolf, B.A., *Mater. Chem. Phys.*, 14, 311, 1986.
127. Irani, C.A. and Cozewith, C., *J. Appl. Polym. Sci.*, 31, 1879, 1986.
128. Cowie, J.M.G. and McEwen, I.J., *Brit. Polym. J.*, 18, 387, 1986.
129. Krüger, B., *Dissertation*, TH Leuna-Merseburg, 1986.
130. Rätzsch, M.T., et al., *J. Macromol. Sci.-Chem. A*, 23, 1349, 1986.
131. Saeki, S., et al., *Macromolecules*, 19, 2353, 1986.
132. Sander, U. and Wolf, B.A., *Angew. Makromol. Chem.*, 139, 149, 1986.
133. Barbarin-Castillo, J.-M., et al., *Polym.Commun.*, 28, 212, 1987.
134. Gruner, K. and Greer, S.C., *Macromolecules*, 20, 2238, 1987.
135. Magarik, S.Ya., Filippov, A.P., and D'yakonova, N.V., *Vysokomol. Soedin., Ser. A*, 29, 698, 1987.
136. Rangel-Nafaile, C. and Munoz-Lara, J.J., *Chem. Eng. Commun.*, 53, 177, 1987.
137. Kiepen, F. and Borchard, W., *Macromolecules*, 21, 1784, 1988.
138. Schuster, R., *Diploma Paper*, TH Leuna-Merseburg, 1988.
139. Tveekrem, J.L., Greer, S.C., and Jacobs, D.T., *Macromolecules*, 21, 147, 1988.
140. Bohossian, T., Charlet, G., and Delmas, G., *Polymer*, 30, 1695, 1989.
141. Chiu, G. and Mandelkern, L., *Macromolecules*, 23, 5356, 1990.
142. Goedel, W.A., et al., *Ber. Bunsenges. Phys. Chem.*, 94, 17, 1990.
143. Van der Haegen, R. and Van Opstal, L., *Makromol. Chem.*, 191, 1871, 1990.
144. Iwai, Y., et al., *Sekiyu Gakkaishi*, 33, 117, 1990.
145. Raetzsch, M.T., Krueger, B., and Kehlen, H., *J. Macromol. Sci.-Chem. A*, 27, 683, 1990.
146. Schild, H.G. and Tirrell, D.A., *J. Phys. Chem.*, 94, 4352, 1990.
147. Stafford, S.G., Ploplis, A.C., and Jacobs, D.T., *Macromolecules*, 23, 470, 1990.
148. Akhmadeev, I.R., et al., *Vysokomol. Soedin., Ser. B*, 33, 543, 1991.
149. Bae, Y.C., Lambert, S.M., Soane, D.S., and Prausnitz, J.M., *Macromolecules*, 24, 4403, 1991.
150. Chu, B., Linliu, K., Xie, P., Ying, Q., Wang, Z., and Shook, J.W., *Rev. Sci. Instr.*, 62, 2252, 1991.
151. Kawate, K., Imagawa, I., and Nakata, M., *Polym. J.*, 23, 233, 1991.
152. Lee, K.D. and Lee, D.C., *Pollimo*, 15, 274, 1991.
153. Louai, A., Sarazin, D., Pollet, G., Francois, J., and Moreaux, F., *Polymer*, 32, 703, 1991.
154. Van Opstal, L., Koningsveld, R., and Kleintjens, L.A., *Macromolecules*, 24, 161, 1991.
155. Schubert, K.-V., Strey, R., and Kahlweit, M., *J. Colloid Interface Sci.*, 141, 21, 1991.
156. Shen, W., Smith, G.R., Knobler, C.M., and Scott, R.L., *J. Phys. Chem.*, 95, 3376, 1991.
157. Szydlowski, J. and Van Hook, W.A., *Macromolecules*, 24, 4883, 1991.
158. Tager, A.A., et al., *Vysokomol. Soedin., Ser. B*, 33, 572, 1991.
159. Wakker, A., *Polymer*, 32, 279, 1991.
160. Yokoyama, H., Takano, A., Okada, M., and Nose, T., *Polymer*, 32, 3218, 1991.
161. Heinrich, M. and Wolf, B.A., *Polymer*, 33, 1926, 1992.
162. Heinrich, M. and Wolf, B.A., *Macromolecules*, 25, 3817, 1992.
163. Lecointe, J.P., Pascault, J.P., Suspene, L., and Yang, Y.S., *Polymer*, 33, 3226, 1992.
164. Mueller, K.F., *Polymer*, 33, 3470, 1992.
165. Szydlowski, J., Rebelo, L., and Van HooK, W.A., *Rev. Sci. Instrum.*, 63, 1717, 1992.
166. Xia, K.-Q., Franck, C., and Widom, B., *J. Chem. Phys.*, 97, 1446, 1992.
167. Arnauts, J., Berghmans, H., and Koningsveld, R., *Makromol. Chem.*, 194, 77, 1993.
168. Iwai, Y., Shigematsu, Y., Furuya, T., Fukuda, H., Arai, Y., *Polym. Eng. Sci.*, 33, 480, 1993.
169. Wakker, A., Van Dijk, F., and Van Dijk, M.A., *Macromolecules*, 26, 5088, 1993.

170. Wells, P.A., de Loos, Th.W., and Kleintjens, L.A., *Fluid Phase Equil.*, 83, 383, 1993.
171. Barbarin-Castillo, J.-M. and McLure, I.A., *Polymer*, 35, 3075, 1994.
172. Mumick, P.S. and McCormick, C.L., *Polym. Eng. Sci.*, 34, 1419, 1994.
173. Okano, K., Takada, M., Kurita, K., and Furusaka, M., *Polymer*, 35, 2284, 1994.
174. Sato, H., Kuwahara, N., and Kubota, K., *Phys. Rev. E*, 50, 1752, 1994.
175. Song, S.-W. and Torkelson, J.M., *Macromolecules*, 27, 6389, 1994.
176. Tager, A.A., et al., *Colloid Polym. Sci.*, 272, 1234, 1994.
177. Vanhee, S., et al., *Makromol. Chem. Phys.*, 195, 759, 1994.
178. Haas, C.K. and Torkelson, J.M., *Phys. Rev. Lett.*, 75, 3134, 1995.
179. Ikier, C. and Klein, H., *Macromolecules*, 28, 1003, 1995.
180. Luszczyk, M., Rebelo, L.P.N., and Van Hook, W.A., *Macromolecules*, 28, 745, 1995.
181. Pfohl, O., Hino, T., and Prausnitz, J.M., Polymer, 36, 2065, 1995.
182. Vshivkov, S.A. and Safronov, A.P., *Vysokomol. Soedin., Ser. B*, 37, 1779, 1995.
183. Allcock, H.R. and Dudley, G.K., *Macromolecules*, 29, 1313, 1996.
184. El-Ejmi, A.A.S. and Huglin, M.B., *Polym.Int.* 39, 113, 1996.
185. Fischer, V., Borchard, W., and Karas, M., *J. Phys. Chem.*, 100, 15992, 1996.
186. Imre, A. and Van Hook, W.A., *J. Polym. Sci.: Part B: Polym. Sci.*, 34, 751, 1996.
187. Luszczyk, M. and Van Hook, W.A., *Macromolecules*, 29, 6612, 1996.
188. Rong, Z., Wang, H., Ying, X., and Hu, Y., *J. East China Univ. Sci. Technol.*, 22, 754, 1996.
189. Safronov, A.P., Tager, A.A., and Koroleva, E.V., *Vysokomol. Soedin., Ser. B*, 38, 900, 1996.
190. Vshivkov, S.A. and Rusinova, E.V., *Vysokomol. Soedin., Ser. A*, 38, 1746, 1996.
191. Xia, K.-Q., An, X.-Q., and Shen, W.-G., *J.Chem.Phys.*, 105, 6018, 1996.
192. Imre, A. and Van Hook, W.A., *J. Polym. Sci.: Part B: Polym. Phys.*, 35, 1251, 1997.
193. Kita, R., Dobashi, T., Yamamoto, T., Nakata, M., and Kamide, K., *Phys. Rev. E*, 55, 3159, 1997.
194. Li, M., Zhu, Z.-Q., and Mei, L.-H., *Biotechnol. Progr.*, 13, 105, 1997.
195. McLure, I.A., Mokhtari, A., and Bowers, J., *J. Chem. Soc., Faraday Trans.*, 93, 249, 1997.
196. Chalykh, A.E., Dement'eva, O.V., and Gerasimov, V.K., *Vysokomol. Soedin., Ser. A*, 40, 815,.1998.
197. Kubota, K., Kita, R., and Dobashi, T., *J. Chem. Phys.*, 109, 711, 1998.

198. Schneider, A., *Dissertation*, Johannes Gutenberg Universität Mainz. 1998.
199. Terao, K., et al., *Macromolecules*, 31, 6885, 1998.
200. Yamagishi, T.-A., et al., *Macromol. Chem. Phys.*, 199, 423, 1998.
201. Lau, A.C.W. and Wu, C., *Macromolecules*, 32, 581, 1999.
202. Nakata, M., Dobashi, T., Inakuma, Y.-I., and Yamamura, K., *J. Chem. Phys.*, 111, 6617, 1999.
203. Pruessner, M.D., Retzer, M.E., and Greer, S.C., *J. Chem. Eng. Data*, 44, 1419, 1999.
204. Shimofure, S., Kubota, K., Kita, R., and Dobashi, T., *J.Chem.Phys.*, 111, 4199, 1999.
205. Fischer, V. and Borchard, W., *J. Phys. Chem. B*, 104, 4463, 2000.
206. Imre, A., and Van Hook, W.A., *Macromolecules*, 33, 5308, 2000.
207. Koizumi, J., et al., *J. Phys. Soc. Japan*, 69, 2543, 2000.
208. Kunugi, S., Tada, T., Yamazaki, Y., Yamamoto, K., and Akashi, M., *Langmuir*, 16, 2042, 2000.
209. La Mesa, C., *J. Therm. Anal. Calorim.*, 61, 493, 2000.
210. Persson, J., et al., *Bioseparation*, 9, 105, 2000.
211. Persson, J., Kaul, A., and Tjerneld, F., *J. Chromatogr. B*, 743, 115, 2000.
212. Djokpe, E. and Vogt, W., *Macromol. Chem. Phys.*, 202, 750, 2001.
213. Kujawa, P. and Winnik, F.M., *Macromolecules*, 34, 4130, 2001.
214. Pendyala, K.S., Greer, S.C., and Jacobs, D.T., *J. Chem. Phys.*, 115, 9995, 2001.
215. Berlinova, I. V., Nedelcheva, A. N., Samchikov, V., and Ivanov, Ya., *Polymer*, 43, 7243, 2002.
216. Freitag, R. and Garret-Flaudy, F., *Langmuir*, 18, 3434, 2002.
217. Maeda, Y., Nakamura, T., and Ikeda, I., *Macromolecules*, 35, 217, 2002.
218. Rebelo, L.P.N., et al., J., *Macromolecules*, 35, 1887, 2002.
219. Safronov, A.P. and Somova, T.V., *Vysokomol. Soedin., Ser. A*, 44, 2014, 2002.
220. Vshvikov, S.A., Rusinova, E.V., and Gur'ev, A.A., *Vysokomol. Soedin., Ser. B*, 44, 504, 2002.
221. Zhou, C.-S., An, X.-Q., Xia, K.-Q., Yin, X.-L., and Shen, W.-G., *J. Chem. Phys.*, 117, 4557, 2002.
222. Chang, Y., Powell, E.S., Allcock, H.R., Park, S.M., and Kim, C., *Macromolecules*, 36, 2568, 2003.
223. David, G., et al., *Eur. Polym. J.*, 39, 1209, 2003.
224. Siporska, A., Szydlowski, J., and Rebelo, L.P.N., *Phys. Chem. Chem. Phys.*, 5, 2996, 2003.
225. Yamamoto, K., Serizawa, T., and Akashi, M., *Macromol. Chem. Phys.*, 204, 1027, 2003.

VAPOR PRESSURES (SOLVENT ACTIVITIES) FOR BINARY POLYMER SOLUTIONS

Christian Wohlfarth

The vapor pressure of a binary polymer solution is given by the activity of the solvent A, a_A. Solvent activities in polymer solutions are measured either by the isopiestic method applying a reference system whose solvent activity is precisely known or by determining the solvent partial pressure, P_A, and calculating the activity of the solvent by equation (1):

$$a_A = \left(P_A/P_A^s\right)\exp\left[\frac{\left(B_{AA}-V_A^L\right)\left(P-P_A^s\right)}{RT}\right] \tag{1}$$

where B_{AA} is the second virial coefficient, P_A^s is the saturation vapor pressure, and V_A^L is the molar volume of the pure solvent A at the measuring temperature T. The exponential term is neglected in quite a lot of original papers, however, and only the reduced vapor pressures are given (such data are indicated by an asterisk in the table below). Vapor pressures of polymer solutions have been measured since the 1940s, but the amount of experimental data for polymer solutions is still relatively small in comparison to low-molecular mixtures and solutions. The data scatter with respect to temperature, concentration, molar mass, and other polymer characterization variables. Furthermore, the concentration range for measuring vapor pressures in good thermodynamic quality is often limited to the polymer mass fraction range between 0.4 and 0.85. A recent review on methods for the measurement of vapor pressures/solvent activities of polymer solutions and on related problems is given in Ref. [1]. Experimental data have been collected in several books [2-6].

The table in this *Handbook* provides data for a number of polymer solutions as smoothed values over the complete range of solvent activities between 0 (polymer mass fraction = 1) and 1 (polymer mass fraction = 0). For this purpose, the data were selected from data books [4–6] as well as from a number of original sources [7–22] which are not included in these books. The appropriate data were smoothed. The final table provides then the polymer mass fractions at given fixed solvent activities between 0.1 and 0.9. Of course, the user must keep in mind that the activity vs. concentration range of the experimental data is sometimes smaller than the below given complete range, thus the smoothed data should be used with sufficient care.

Generally, vapor pressures or solvent activities of binary polymer solutions depend on molar mass. However, for high molecular weight polymers (well above the oligomer region), this molar-mass dependence can be neglected in many cases. Therefore, the table below presents only data for polymer solutions where the number average molar mass, M_n, is in the order of 10^5 g/mol or even higher, therefore, the molar mass is not specified. The temperature is stated, even though the temperature dependence of a_A is relatively small for the temperature ranges where most of the experimental data exist.

References

1. Wohlfarth, C., Methods for the measurement of solvent activity of polymer solutions, in *Handbook of Solvents*, Wypych, G., Ed., ChemTec Publishing, Toronto, 2000, 146.
2. Wen, H., Elbro, H. S., and Alessi, P., *Polymer Solution Data Collection.* I. Vapor-liquid equilibrium; II. Solvent activity coefficients at infinite dilution; III. Liquid-liquid equilibrium, Chemistry Data Series, Vol. 15, DECHEMA, Frankfurt am Main, 1992.
3. Danner, R. P. and High, M. S., *Handbook of Polymer Solution Thermodynamics*, American Institute of Chemical Engineers, New York, 1993.
4. Wohlfarth, C., *Vapour-Liquid Equilibrium Data of Binary Polymer Solutions: Physical Science Data*, 44, Elsevier, Amsterdam, 1994.
5. Wohlfarth, C., *CRC Handbook of Thermodynamic Data of Copolymer Solutions*, CRC Press, Boca Raton, FL, 2001.
6. Wohlfarth, C., *CRC Handbook of Thermodynamic Data of Aqueous Polymer Solutions*, CRC Press, Boca Raton, FL, 2003.
7. Wang, K., Hu, Y., and Wu, D. T., *J. Chem. Eng. Data*, 39, 916, 1994.
8. Choi, J. S., Tochigi, K., and Kojima, K., *Fluid Phase Equil.*, 111, 143, 1995.
9. Tochigi, K., Kurita, S., Ohashi, M., and Kojima, K., *Kagaku Kogaku Ronbunshu*, 23, 720, 1997.
10. Wong, H. C., Campbell, S. W., and Bhethanabotla, V. R., *Fluid Phase Equil.*, 139, 371, 1997.
11. Kim, J., Joung, K. C., Hwang, S., Huh, W., Lee, C. S., and Yoo, K.-P., *Korean J. Chem. Eng.*, 15, 199, 1998.
12. Kim, N. H., Kim, S.J., Won, Y. S., and Choi, J. S., *Korean J. Chem. Eng.*, 15, 141, 1998.
13. Feng, W., Wang, W., and Feng, Z., *J. Chem. Ind. Eng. (China)*, 49, 271, 1998.
14. French, R. N. and Koplos, G. J., *Fluid Phase Equil.*, 160, 879, 1999.
15. Striolo, A. and Praunsitz, J. M., *Polymer*, 41, 1109, 2000.
16. Fornasiero, F., Halim, M., and Prausnitz, J. M., *Macromolecules*, 33, 8435, 2000.
17. Wong, H. C., Campbell, S. W., and Bhethanabotla, V. R., *Fluid Phase Equil.*, 179, 181, 2001.
18. Wibawa, G., Takahashi, M., Sato, Y., Takishima, S., and Masuoka, H., *J. Chem. Eng. Data*, 47, 518, 2002.
19. Wibawa, G., Hatano, R., Sato, Y., Takishima, S., and Masuoka, H., *J. Chem. Eng. Data*, 47, 1022, 2002.
20. Pfohl, O., Riebesell, C., and Dohrn, R., *Fluid Phase Equil.*, 202, 289, 2002.
21. Jung, J. K., Joung, S. N., Shin, H.Y., Kim, S. Y., Yoo, K.-P., Huh, W., Lee, C. S., *Korean J. Chem. Eng.*, 19, 296, 2002.
22. Kang, S., Huang, Y., Fu, J., Liu, H., and Hu, Y., *J. Chem. Eng. Data*, 47, 788, 2002.

Solvent Activity a_A as Function of Temperature and Mass Fraction

Polymer/ solvents	a_A: T/K	0.1	0.2	0.3	0.4	0.5	0.6	0.7	0.8	0.9
					Mass Fraction of the Polymer					
Acrylonitrile/Styrene Copolymer (28 wt% Acrylonitrile)										
Benzene*⁾	343.15	0.982	0.962	0.940	0.915	0.886	0.851	0.809	0.753	0.670
1,2-Dimethylbenzene*⁾	398.15	0.983	0.964	0.942	0.918	0.890	0.857	0.817	0.764	0.685
1,3-Dimethylbenzene*⁾	398.15	0.983	0.965	0.944	0.921	0.893	0.861	0.821	0.769	0.690
1,4-Dimethylbenzene*⁾	398.15	0.983	0.964	0.942	0.918	0.890	0.857	0.817	0.763	0.684
Propylbenzene*⁾	398.15	0.987	0.972	0.955	0.935	0.913	0.885	0.851	0.804	0.732
Toluene*⁾	343.15	0.982	0.962	0.940	0.915	0.886	0.851	0.809	0.753	0.669

Polymer/ solvents	a_A: T/K	0.1	0.2	0.3	0.4	0.5	0.6	0.7	0.8	0.9
						Mass Fraction of the Polymer				
Butadiene/Styrene Copolymer (41 wt% Styrene)										
Benzene*)	343.15	0.968	0.934	0.896	0.853	0.805	0.748	0.680	0.591	0.461
Cyclohexane*)	343.15	0.978	0.953	0.925	0.893	0.856	0.811	0.754	0.678	0.556
Ethylbenzene*)	398.15	0.974	0.945	0.912	0.875	0.831	0.779	0.713	0.625	0.491
Mesitylene*)	398.15	0.977	0.950	0.921	0.887	0.847	0.799	0.738	0.656	0.526
Toluene*)	343.15	0.970	0.936	0.899	0.857	0.808	0.751	0.682	0.591	0.456
Cellulose Triacetate										
Dichloromethane	298.15	0.979	0.956	0.930	0.899	0.863	0.819	0.762	0.683	0.554
Trichloromethane	298.15	0.978	0.953	0.924	0.892	0.853	0.806	0.747	0.665	0.533
Dextran										
Water	313.15	0.988	0.975	0.960	0.942	0.921	0.894	0.860	0.810	0.725
Hydroxyethylcellulose										
Water	368.15	0.988	0.974	0.958	0.939	0.915	0.884	0.841	0.775	0.650
Hydroxypropylstarch										
Water	293.15	0.989	0.977	0.963	0.947	0.927	0.903	0.872	0.827	0.749
Nitrocellulose										
Ethyl acetate	293.15	0.938	0.885	0.835	0.786	0.737	0.685	0.627	0.560	0.471
Ethyl formate	293.15	0.958	0.916	0.873	0.828	0.780	0.728	0.668	0.595	0.494
Ethyl propionate	293.15	0.941	0.889	0.839	0.789	0.739	0.685	0.625	0.555	0.460
Methyl acetate	293.15	0.890	0.820	0.763	0.711	0.660	0.609	0.554	0.490	0.406
2-Propanone	293.15	0.922	0.861	0.807	0.756	0.706	0.653	0.596	0.530	0.443
Propyl acetate	293.15	0.937	0.881	0.827	0.775	0.722	0.665	0.602	0.528	0.426
Polybutadiene (random cis-trans-vinyl)										
Benzene	298.15	0.964	0.925	0.884	0.839	0.788	0.731	0.663	0.578	0.455
Cyclohexane	298.15	0.974	0.945	0.913	0.876	0.833	0.782	0.719	0.635	0.507
Dichloromethane	298.15	0.951	0.902	0.852	0.800	0.745	0.684	0.616	0.532	0.415
Hexane	298.15	0.984	0.965	0.943	0.916	0.881	0.837	0.775	0.683	0.534
Tetrachloromethane	298.15	0.932	0.865	0.799	0.731	0.660	0.585	0.503	0.409	0.288
Toluene	298.15	0.969	0.935	0.898	0.856	0.809	0.754	0.688	0.603	0.476
Trichloromethane	298.15	0.925	0.855	0.788	0.720	0.650	0.578	0.498	0.406	0.289
1,4-cis-Polybutadiene										
Benzene	298.15	0.966	0.930	0.890	0.846	0.796	0.738	0.668	0.580	0.450
Cyclohexane	298.15	0.977	0.951	0.922	0.888	0.849	0.803	0.747	0.677	0.581
Dichloromethane	298.15	0.948	0.898	0.848	0.796	0.742	0.683	0.616	0.536	0.424
Hexane	298.15	0.983	0.963	0.941	0.916	0.886	0.850	0.804	0.741	0.639
Tetrachloromethane	298.15	0.936	0.871	0.805	0.736	0.665	0.588	0.505	0.409	0.287
Toluene	298.15	0.969	0.936	0.900	0.860	0.815	0.763	0.701	0.622	0.506
Trichloromethane	298.15	0.915	0.840	0.770	0.702	0.634	0.562	0.485	0.396	0.283
Poly(butyl acrylate)										
Benzene	298.15	0.964	0.926	0.887	0.845	0.799	0.749	0.691	0.619	0.519
Dichloromethane	298.15	0.868	0.801	0.744	0.690	0.636	0.577	0.511	0.430	0.318
Tetrachloromethane	298.15	0.932	0.868	0.805	0.742	0.677	0.607	0.529	0.438	0.317
Toluene	298.15	0.967	0.932	0.893	0.849	0.801	0.744	0.676	0.590	0.463
Trichloromethane	298.15	0.901	0.811	0.733	0.662	0.595	0.529	0.459	0.381	0.282
Poly(butyl methacrylate)										
Benzene	313.15	0.971	0.939	0.902	0.861	0.813	0.756	0.685	0.592	0.453
1-Butanol	313.15	0.991	0.980	0.968	0.953	0.936	0.914	0.885	0.842	0.762
2-Butanol	313.15	0.992	0.982	0.969	0.953	0.933	0.906	0.869	0.815	0.719
2-Butanone	313.15	0.982	0.963	0.940	0.914	0.884	0.846	0.799	0.732	0.623
Butyl acetate*)	308.15	0.982	0.961	0.936	0.908	0.875	0.836	0.789	0.730	0.652
Cyclohexane	313.15	0.985	0.968	0.948	0.925	0.899	0.866	0.823	0.764	0.666
Cyclopentane	313.15	0.984	0.965	0.944	0.918	0.886	0.846	0.792	0.714	0.579
Diethyl ether*)	298.15	0.987	0.973	0.956	0.937	0.914	0.885	0.848	0.795	0.703
1,4-Dimethylbenzene	333.15	0.971	0.940	0.905	0.866	0.822	0.770	0.706	0.622	0.497
Ethylbenzene	333.15	0.969	0.935	0.899	0.859	0.815	0.764	0.704	0.627	0.517
Methyl acetate	313.15	0.984	0.965	0.944	0.920	0.891	0.856	0.811	0.748	0.645

Polymer/ solvents	a_A: T/K	0.1	0.2	0.3	0.4	0.5	0.6	0.7	0.8	0.9
					Mass Fraction of the Polymer					
2-Methyl-1-propanol	333.15	0.988	0.974	0.958	0.940	0.919	0.893	0.860	0.815	0.744
Octane	313.15	0.988	0.974	0.959	0.942	0.921	0.896	0.865	0.823	0.758
1-Propanol	333.15	0.990	0.980	0.967	0.952	0.934	0.911	0.881	0.834	0.746
2-Propanol	313.15	0.991	0.981	0.970	0.956	0.939	0.918	0.889	0.845	0.755
2-Propanone	313.15	0.989	0.976	0.961	0.944	0.921	0.892	0.850	0.783	0.647
Propyl acetate	313.15	0.980	0.957	0.932	0.903	0.870	0.830	0.780	0.714	0.612
Toluene	313.15	0.971	0.939	0.903	0.863	0.818	0.764	0.698	0.613	0.485
Poly(ε-caprolacton)										
Tetrachloromethane[a)]	338.15	0.956	0.910	0.864	0.815	0.762	0.704	0.637	0.554	0.438
Poly(dimethylsiloxane)										
Chlorodifluoromethane	298.15	0.976	0.950	0.921	0.888	0.850	0.805	0.750	0.677	0.565
Cyclohexane	303.15	0.979	0.955	0.928	0.898	0.863	0.822	0.770	0.702	0.596
Hexane	303.15	0.982	0.962	0.939	0.912	0.880	0.842	0.793	0.724	0.611
Pentane	308.15	0.982	0.962	0.940	0.913	0.881	0.842	0.791	0.720	0.600
Pentane	423.15	0.984	0.966	0.946	0.922	0.893	0.858	0.813	0.749	0.641
Poly(ethyl acrylate)										
Benzene	298.15	0.970	0.939	0.904	0.866	0.823	0.774	0.716	0.641	0.533
Dichloromethane	298.15	0.900	0.830	0.768	0.709	0.648	0.584	0.512	0.427	0.313
Tetrachloromethane	298.15	0.950	0.900	0.848	0.794	0.736	0.672	0.598	0.509	0.385
Toluene	298.15	0.972	0.942	0.910	0.874	0.833	0.786	0.730	0.659	0.555
Trichloromethane	298.15	0.866	0.776	0.701	0.632	0.566	0.499	0.428	0.349	0.248
Poly(ethylene oxide)										
Benzene	323.15	0.972	0.942	0.908	0.869	0.824	0.771	0.706	0.620	0.490
2-Butanone	353.15	0.981	0.959	0.934	0.902	0.863	0.813	0.746	0.651	0.503
Cyclohexane	353.15	0.989	0.976	0.960	0.943	0.921	0.893	0.855	0.798	0.688
Methanol	303.15	0.964	0.927	0.887	0.844	0.797	0.744	0.682	0.604	0.494
2-Propanone	353.15	0.979	0.947	0.896	0.815	0.719	0.625	0.532	0.434	0.315
Water	293.15	0.977	0.951	0.923	0.890	0.852	0.806	0.748	0.671	0.550
Poly(ethylenimine)										
Water	353.15	0.975	0.947	0.917	0.883	0.845	0.801	0.748	0.680	0.581
Poly(ethyl methacrylate)										
Benzene	298.15	0.970	0.938	0.903	0.864	0.821	0.771	0.712	0.637	0.529
Dichloromethane	298.15	0.912	0.838	0.769	0.703	0.636	0.567	0.491	0.404	0.292
Tetrachloromethane	298.15	0.935	0.873	0.812	0.750	0.686	0.616	0.540	0.449	0.328
Toluene	298.15	0.974	0.945	0.913	0.877	0.836	0.787	0.727	0.647	0.527
Trichloromethane	298.15	0.859	0.760	0.678	0.604	0.533	0.464	0.392	0.313	0.217
Polyisobutylene										
Benzene	313.15	0.984	0.965	0.945	0.921	0.892	0.858	0.813	0.751	0.645
Cyclohexane	313.15	0.976	0.950	0.921	0.888	0.850	0.805	0.749	0.676	0.563
Cyclopentane	313.15	0.977	0.952	0.924	0.892	0.855	0.812	0.758	0.687	0.579
1,4-Dimethylbenzene	313.15	0.979	0.955	0.929	0.899	0.863	0.821	0.767	0.694	0.579
2,2-Dimethylbutane	298.15	0.983	0.964	0.942	0.917	0.887	0.852	0.806	0.743	0.640
Ethylbenzene	313.15	0.979	0.955	0.927	0.895	0.857	0.810	0.750	0.668	0.535
Heptane	298.15	0.983	0.964	0.942	0.917	0.887	0.851	0.804	0.741	0.637
Hexane	298.15	0.980	0.959	0.934	0.906	0.873	0.834	0.784	0.715	0.606
Octane	298.15	0.983	0.963	0.940	0.914	0.883	0.845	0.797	0.729	0.617
Tetrachloromethane	298.15	0.962	0.921	0.877	0.829	0.776	0.715	0.643	0.552	0.423
Toluene	313.15	0.984	0.966	0.944	0.918	0.884	0.840	0.779	0.688	0.537
Trichloromethane	298.15	0.969	0.935	0.899	0.858	0.813	0.761	0.698	0.619	0.503
2,4,4-Trimethylpentane	298.15	0.981	0.961	0.937	0.911	0.879	0.842	0.794	0.730	0.628
1,4-cis-Polyisoprene										
Benzene	313.15	0.982	0.962	0.937	0.908	0.873	0.827	0.766	0.679	0.537
2-Butanone	353.15	0.986	0.970	0.953	0.933	0.910	0.883	0.850	0.808	0.746
Cyclohexane	313.15	0.978	0.954	0.928	0.899	0.865	0.825	0.778	0.716	0.625
Dichloromethane	298.15	0.969	0.935	0.898	0.857	0.811	0.757	0.693	0.610	0.488
1,4-Dimethylbenzene	313.15	0.977	0.951	0.923	0.892	0.857	0.816	0.767	0.704	0.613

Polymer/ solvents	a_A: T/K	0.1	0.2	0.3	0.4	0.5	0.6	0.7	0.8	0.9
						Mass Fraction of the Polymer				
Ethylbenzene	313.15	0.978	0.954	0.928	0.898	0.864	0.823	0.774	0.709	0.612
Methyl acetate	313.15	0.968	0.935	0.900	0.862	0.820	0.773	0.717	0.649	0.554
Octane	313.15	0.984	0.967	0.948	0.926	0.901	0.871	0.834	0.785	0.711
Propyl acetate	333.15	0.983	0.964	0.942	0.916	0.886	0.850	0.803	0.738	0.633
Tetrachloromethane	298.15	0.929	0.864	0.800	0.737	0.672	0.602	0.526	0.435	0.316
Toluene	313.15	0.978	0.954	0.927	0.898	0.865	0.827	0.782	0.725	0.645
Trichloromethane	298.15	0.930	0.867	0.807	0.747	0.685	0.620	0.547	0.462	0.346
Poly(methyl acrylate)										
Benzene	298.15	0.979	0.956	0.930	0.901	0.867	0.826	0.776	0.710	0.608
Dichloromethane	298.15	0.917	0.851	0.791	0.732	0.671	0.605	0.532	0.444	0.326
Tetrachloromethane	298.15	0.963	0.924	0.882	0.838	0.788	0.733	0.668	0.586	0.470
Toluene	298.15	0.981	0.960	0.936	0.909	0.878	0.840	0.792	0.727	0.626
Trichloromethane	298.15	0.912	0.830	0.753	0.678	0.603	0.527	0.446	0.357	0.248
Poly(methyl methacrylate)										
Benzene	298.15	0.982	0.961	0.938	0.912	0.881	0.843	0.795	0.729	0.622
2-Butanone*)	308.15	0.989	0.976	0.961	0.945	0.925	0.900	0.869	0.825	0.751
Cyclohexanone*)	323.15	0.978	0.954	0.928	0.899	0.866	0.827	0.781	0.723	0.640
Dichloromethane	298.15	0.939	0.882	0.825	0.766	0.704	0.637	0.560	0.468	0.343
Ethyl acetate*)	308.15	0.986	0.969	0.950	0.928	0.902	0.869	0.826	0.763	0.649
Toluene	298.15	0.981	0.959	0.935	0.908	0.877	0.841	0.795	0.736	0.646
Trichloromethane	298.15	0.924	0.848	0.771	0.694	0.616	0.536	0.451	0.358	0.246
Poly(α-methylstyrene)										
Cumene	338.15	0.984	0.965	0.944	0.918	0.887	0.848	0.796	0.721	0.593
α-Methylstyrene	338.15	0.978	0.954	0.927	0.896	0.859	0.816	0.761	0.687	0.570
Poly(propylene oxide)										
Benzene	333.15	0.967	0.932	0.893	0.850	0.801	0.744	0.675	0.588	0.460
Metvhanol	298.15	0.992	0.982	0.970	0.955	0.936	0.910	0.872	0.812	0.689
Polystyrene										
Benzene	333.15	0.978	0.953	0.924	0.891	0.852	0.804	0.742	0.657	0.521
2-Butanone*)	298.15	0.986	0.971	0.954	0.935	0.912	0.885	0.851	0.804	0.724
Cyclohexane	313.15	0.990	0.978	0.965	0.949	0.931	0.908	0.877	0.833	0.754
Cyclohexanone*)	313.15	0.970	0.937	0.900	0.858	0.810	0.753	0.684	0.593	0.459
Dichloromethane	298.15	0.949	0.899	0.849	0.797	0.743	0.684	0.617	0.536	0.423
1,3-Dimethylbenzene*)	323.15	0.980	0.956	0.926	0.891	0.846	0.791	0.723	0.638	0.524
1,4-Dimethylbenzene	423.15	0.974	0.944	0.911	0.872	0.826	0.770	0.698	0.601	0.452
Ethyl acetate*)	313.15	0.976	0.948	0.918	0.882	0.841	0.791	0.728	0.642	0.507
Hexane	423.15	0.980	0.958	0.933	0.904	0.869	0.827	0.772	0.697	0.574
2-Propanone	323.15	0.991	0.980	0.969	0.955	0.938	0.918	0.892	0.854	0.788
Propyl acetate	343.15	0.983	0.965	0.943	0.919	0.891	0.858	0.815	0.758	0.667
Tetrachloromethane	298.15	0.961	0.917	0.869	0.814	0.751	0.678	0.592	0.486	0.344
Toluene	313.15	0.981	0.959	0.933	0.901	0.861	0.809	0.738	0.638	0.481
Trichloromethane	298.15	0.949	0.898	0.847	0.793	0.736	0.675	0.604	0.519	0.400
Poly(tetramethylene glycol)										
Methanol	303.15	0.981	0.961	0.938	0.913	0.883	0.849	0.806	0.751	0.671
Poly(vinyl acetate)										
Benzene	313.15	0.985	0.967	0.945	0.919	0.886	0.844	0.784	0.696	0.548
1-Butanol	313.15	0.992	0.982	0.971	0.958	0.942	0.923	0.896	0.856	0.779
2-Butanol	313.15	0.987	0.972	0.956	0.937	0.915	0.889	0.856	0.813	0.747
2-Butanone	313.15	0.980	0.958	0.934	0.906	0.873	0.835	0.787	0.724	0.626
1,2-Dichloroethane*)	300.15	0.955	0.906	0.851	0.790	0.722	0.644	0.556	0.450	0.315
1,4-Dimethylbenzene	313.15	0.990	0.978	0.964	0.948	0.928	0.903	0.868	0.814	0.705
Ethylbenzene	313.15	0.990	0.979	0.966	0.950	0.932	0.910	0.880	0.836	0.759
Methanol	333.15	0.990	0.978	0.965	0.949	0.931	0.908	0.877	0.834	0.757
Methyl acetate	313.15	0.976	0.949	0.919	0.886	0.849	0.805	0.752	0.684	0.583
2-Methyl-1-propanol	353.15	0.984	0.966	0.946	0.924	0.899	0.868	0.832	0.784	0.715
1-Propanol	353.15	0.987	0.972	0.955	0.936	0.914	0.888	0.856	0.815	0.753
2-Propanol	353.15	0.988	0.974	0.958	0.940	0.919	0.894	0.863	0.820	0.754

Polymer/ solvents	a_A: T/K	0.1	0.2	0.3	0.4	0.5	0.6	0.7	0.8	0.9
					Mass Fraction of the Polymer					
2-Propanone	333.15	0.983	0.963	0.940	0.913	0.880	0.838	0.784	0.707	0.578
Propyl acetate	333.15	0.979	0.955	0.930	0.901	0.869	0.831	0.786	0.728	0.645
Tetrahydrofuran	323.15	0.973	0.943	0.911	0.874	0.831	0.781	0.720	0.640	0.519
Toluene	333.15	0.983	0.965	0.944	0.920	0.891	0.857	0.815	0.756	0.664
Poly(vinyl chloride)										
2-Butanone[*]	313.15	0.976	0.949	0.920	0.887	0.849	0.804	0.749	0.676	0.566
Cyclohexanone[*]	333.15	0.971	0.934	0.889	0.839	0.781	0.714	0.635	0.536	0.397
Poly(vinyl methyl ether)										
Benzene[*]	298.15	0.969	0.935	0.897	0.855	0.807	0.751	0.683	0.596	0.466
Chlorobenzene[*]	343.15	0.972	0.941	0.906	0.867	0.822	0.769	0.705	0.620	0.494
1,2-Dimethylbenzene[*]	363.15	0.973	0.943	0.910	0.871	0.826	0.772	0.705	0.616	0.478
Ethylbenzene[*]	343.15	0.978	0.954	0.927	0.895	0.857	0.811	0.753	0.672	0.542
Propylbenzene[*]	373.15	0.977	0.951	0.923	0.890	0.852	0.808	0.752	0.678	0.563
Poly(4-vinylpyridine)										
Methanol	343.15	0.986	0.971	0.953	0.931	0.905	0.871	0.825	0.756	0.627
2-Propanol	343.15	0.989	0.977	0.964	0.948	0.928	0.904	0.872	0.826	0.743
Poly(1-vinyl-2-pyrrolidinone)										
Water	368.15	0.984	0.966	0.946	0.924	0.899	0.870	0.835	0.790	0.727
Starch (amorphous)										
Water	383.15	0.991	0.981	0.970	0.956	0.939	0.918	0.889	0.845	0.754
Styrene/Methyl methacrylate Copolymer (41.45 wt% Styrene)										
Benzene[*]	308.15	0.982	0.963	0.940	0.913	0.881	0.841	0.789	0.716	0.590
Vinyl acetate/Vinyl chloride Copolymer (12 wt% Vinyl acetate)										
Benzene	398.15	0.976	0.949	0.918	0.883	0.841	0.791	0.728	0.643	0.509
Chlorobenzene	398.15	0.984	0.965	0.944	0.920	0.891	0.856	0.810	0.746	0.638
1,4-Dimethylbenzene	398.15	0.989	0.977	0.963	0.946	0.926	0.899	0.863	0.807	0.692
Ethylbenzene	398.15	0.989	0.976	0.961	0.944	0.924	0.899	0.866	0.818	0.735
Octane	398.15	0.992	0.982	0.971	0.958	0.942	0.922	0.893	0.847	0.739

[*] $a_A = P_A / P_A^S$

SPECIFIC ENTHALPIES OF SOLUTION OF POLYMERS AND COPOLYMERS

Christian Wohlfarth

Enthalpies of solution or mixing, expressed as the enthalpy change per unit mass of polymer, are given in the table at infinite dilution, i.e., a very small amount of polymer and a large excess of solvent were mixed isothermally to form a homogeneous solution. By thermodynamics, $\Delta_{sol}H_B^\infty$ or $\Delta_M H_B^\infty$ are obtained from the following derivatives:

$$\Delta_{sol}H_B^\infty = \lim_{m_B \to 0}\left(\partial \Delta_{sol}h / \partial m_B\right)_{P,T,m_{j \neq B}} \qquad (1)$$

$$\Delta_M H_B^\infty = \lim_{m_B \to 0}\left(\partial \Delta_M h / \partial m_B\right)_{P,T,m_{j \neq B}} \qquad (2)$$

with a unit of J/g. Thus, they are the partial specific enthalpies of solution or mixing of the polymer B at infinite dilution where $\Delta_{sol}h$ or $\Delta_M h$ is the extensive enthalpy of the solution or mixing process.

The state of the polymer before dissolution can significantly affect the enthalpy of solution. The dissolving of a semicrystalline polymer requires an additional amount of heat associated with the disordering of crystalline regions. Consequently, its enthalpy of solution is usually positive and depends on the degree of crystallinity of the given polymer sample. An amorphous polymer below its glass transition temperature, T_g (see the T_g-table of this Section), often dissolves with the release of heat. The enthalpy of solution of a glassy polymer is additionally dependent to some extent on the thermal history of the glass-forming process. An amorphous polymer above T_g can show endothermic or exothermic dissolution behavior depending on the nature of the solvent and the interaction energies involved as is the case for any enthalpy of mixing. This enthalpy of mixing is then independent of any crystalline or glassy aspects of the polymer. It can be obtained without difficulties for liquid/molten polymers mixed with a solvent. Therefore, the enthalpies given in the table are either enthalpies of solution or enthalpies of mixing, depending on the state of the polymer.

The enthalpies depend on temperature and molar mass. The necessary molar mass information for a system is provided in the table (if available) by the corresponding number average, M_n, mass average, M_w, or viscosity average, M_η, values of the polymer as given in the original sources. Outside the oligomer range, specific enthalpies of solution or mixing do not remarkably depend on molar mass, however. More enthalpy data of polymer-solvent systems can be found in Ref. 106.

Polymer	$M_n/$ g/mol	$M_w/$ g/mol	$M_\eta/$ g/mol	Solvent	T/K	$\Delta H_B^\infty/$ J/g	Ref.
Acrylonitrile/butadiene copolymer							
(18 wt% Acrylonitrile)				Benzene	298.15	0.0	18
(26 wt% Acrylonitrile)				Benzene	298.15	−1.9	18
(40 wt% Acrylonitrile)				Benzene	298.15	−2.9	18
Acrylonitrile/isoprene copolymer							
(15 mol% Isoprene)				N,N-Dimethylformamide	323.15	−32	66
Acrylonitrile/vinyl chloride copolymer							
(13 wt% Acrylonitrile)				N,N-Dimethylformamide	293.15	−38	35
(13 wt% Acrylonitrile)				N,N-Dimethylformamide	308.15	−22	35
(13 wt% Acrylonitrile)				N,N-Dimethylformamide	323.15	−18	35
(13 wt% Acrylonitrile)				N,N-Dimethylformamide	338.15	−15	35
(13 wt% Acrylonitrile)				N,N-Dimethylformamide	353.15	−12	35
(29 wt% Acrylonitrile)				N,N-Dimethylformamide	295.15	−42	35
(29 wt% Acrylonitrile)				N,N-Dimethylformamide	308.15	−27	35
(29 wt% Acrylonitrile)				N,N-Dimethylformamide	323.15	−21	35
(29 wt% Acrylonitrile)				N,N-Dimethylformamide	338.15	−19	35
(29 wt% Acrylonitrile)				N,N-Dimethylformamide	353.15	−16	35
(40 wt% Acrylonitrile)				N,N-Dimethylformamide	295.15	−47	35
(40 wt% Acrylonitrile)				N,N-Dimethylformamide	308.15	−30	35
(40 wt% Acrylonitrile)				N,N-Dimethylformamide	323.15	−28	35
(40 wt% Acrylonitrile)				N,N-Dimethylformamide	338.15	−18	35
(40 wt% Acrylonitrile)				N,N-Dimethylformamide	353.15	−17	35
Benzylcellulose							
				Benzene	298.15	−11	8
				Cyclohexanone	298.15	−15	25
				Trichloromethane	298.15	−38	25

Polymer	$M_n/$ g/mol	$M_w/$ g/mol	$M_\eta/$ g/mol	Solvent	T/K	$\Delta H_B^\infty/$ J/g	Ref.
Bisphenol A-isophthaloyl chloride/terephthaloyl chloride							
(50/50 Iso/terephtaloyl chloride)				N,N-Dimethylacetamide	298.15	−56	69
(50/50 Iso/terephtaloyl chloride)				1,1,2,2-Tetrachloroethane	298.15	+72	69
Butadiene/styrene copolymer							
(10 wt% Styrene)				Benzene	293.65	4.9	17
(30 wt% Styrene)				Benzene	293.65	3.0	17
(30 wt% Styrene)				Benzene	298.15	3.0	25
(50 wt% Styrene)				Benzene	293.65	1.8	17
(60 wt% Styrene)				Benzene	293.65	0.0	17
(70 wt% Styrene)				Benzene	293.65	0.0	17
(75 wt% Styrene)				Benzene	298.15	1.5	7
(80 wt% Styrene)				Benzene	293.65	−0.6	17
(90 wt% Styrene)				Benzene	293.65	−4.9	17
Butyl methacrylate/isobutyl methacrylate copolymer (50 wt%/50 wt%)							
Glass		150000		Cyclohexanone	303.15	5.9	98
Liquid		150000		Cyclohexanone	303.15	14.0	98
Butyl methacrylate/methyl methacrylate copolymer (45 wt%/55 wt%)							
Glass	107000	250000		Cyclohexanone	304.15	−5.4	99
Liquid	107000	250000		Cyclohexanone	304.15	+9.1	99
Cellulose acetate							
(52.2 wt% Acetate)				Formic acid	298.15	−30	10
(55.8 wt% Acetate)				Formic acid	298.15	−44	10
(52.5 wt% Acetate)				Methyl acetate	298.15	−80	1
(48 wt% Acetate)				2-Propanone	298.15	−35	25
(52.2 wt% Acetate)				2-Propanone	298.15	−30	10
(55.8 wt% Acetate)				2-Propanone	298.15	−26	10
(56 wt% Acetate)				2-Propanone	298.15	−45	25
(56 wt% Acetate)				2-Propanone	298.15	−30	4
Cellulose triacetate							
				2-Propanone	298.15	−29	4
				Trichloromethane	298.15	−47	4
Dextran							
	8200	10400		Dimethylsulfoxide	298.15	−185	75
	75900	101000		Dimethylsulfoxide	298.15	−187	70
	75900	101000		1,2-Ethanediol	298.15	−98	70
	75900	101000		Formamide	298.15	−228	70
	8200	10400		Water	298.15	−140	75
	75900	101000		Water	298.15	−150	75
(amorph)				Water	298.15	−123	65
Ethylene/propylene copolymer							
(33 mol% Ethylene)				Cyclohexane	298.15	1.4	74
(63 mol% Ethylene)				Cyclohexane	298.15	8.1	74
(75 mol% Ethylene)				Cyclohexane	298.15	11.8	74
(33 mol% Ethylene)				Cyclooctane	298.15	1.2	74

Polymer	$M_n/$ g/mol	$M_w/$ g/mol	$M_\eta/$ g/mol	Solvent	T/K	$\Delta H_B^\infty/$ J/g	Ref.
(63 mol% Ethylene)				Cyclooctane	298.15	6.9	74
(75 mol% Ethylene)				Cyclooctane	298.15	8.6	74
(33 mol% Ethylene)				Cyclopentane	298.15	−3.5	74
(63 mol% Ethylene)				Cyclopentane	298.15	1.1	74
(33 mol% Ethylene)				cis-Decahydronaphthalene	298.15	−2.4	74
(63 mol% Ethylene)				cis-Decahydronaphthalene	298.15	2.4	74
(75 mol% Ethylene)				cis-Decahydronaphthalene	298.15	3.9	74
(33 mol% Ethylene)				trans-Decahydronaphthalene	298.15	−4.8	74
(63 mol% Ethylene)				trans-Decahydronaphthalene	298.15	−1.3	74
(75 mol% Ethylene)				trans-Decahydronaphthalene	298.15	−0.3	74
(63 mol% Ethylene)				3,3-Diethylpentane	298.15	−1.4	74
(75 mol% Ethylene)				3,3-Diethylpentane	298.15	<0.1	74
(63 mol% Ethylene)				2,2-Dimethylpentane	298.15	5.3	74
(75 mol% Ethylene)				2,2-Dimethylpentane	298.15	2.3	74
(63 mol% Ethylene)				2,3-Dimethylpentane	298.15	0.7	74
(75 mol% Ethylene)				2,3-Dimethylpentane	298.15	0.4	74
(33 mol% Ethylene)				2,4-Dimethylpentane	298.15	−1.2	74
(63 mol% Ethylene)				2,4-Dimethylpentane	298.15	3.0	74
(75 mol% Ethylene)				2,4-Dimethylpentane	298.15	0.2	74
(33 mol% Ethylene)				3,3-Dimethylpentane	298.15	−2.7	74
(63 mol% Ethylene)				3,3-Dimethylpentane	298.15	0.3	74
(33 mol% Ethylene)				Dodecane	298.15	−0.1	73
(63 mol% Ethylene)				Dodecane	298.15	0.8	73
(75 mol% Ethylene)				Dodecane	298.15	−4.0	73
(63 mol% Ethylene)				3-Ethylpentane	298.15	2.6	74
(75 mol% Ethylene)				3-Ethylpentane	298.15	−0.6	74
(33 mol% Ethylene)				2,2,4,4,6,8,8-Heptamethylnonane	298.15	−0.5	73
(63 mol% Ethylene)				2,2,4,4,6,8,8-Heptamethylnonane	298.15	2.2	73
(75 mol% Ethylene)				2,2,4,4,6,8,8-Heptamethylnonane	298.15	−0.9	73
(33 mol% Ethylene)				Hexadecane	298.15	0.7	73
(63 mol% Ethylene)				Hexadecane	298.15	−1.1	73
(75 mol% Ethylene)				Hexadecane	298.15	−4.6	73
(63 mol% Ethylene)				3-Methylhexane	298.15	0.7	74
(75 mol% Ethylene)				3-Methylhexane	298.15	1.7	74
(33 mol% Ethylene)				Octane	298.15	−1.6	73
(63 mol% Ethylene)				Octane	298.15	3.6	73
(75 mol% Ethylene)				Octane	298.15	0.3	73
(33 mol% Ethylene)				2,2,4,6,6-Pentamethylheptane	298.15	−0.3	73
(63 mol% Ethylene)				2,2,4,6,6-Pentamethylheptane	298.15	3.6	73
(75 mol% Ethylene)				2,2,4,6,6-Pentamethylheptane	298.15	0.0	73
(63 mol% Ethylene)				2,2,4,4-Tetramethylpentane	298.15	2.7	74
(75 mol% Ethylene)				2,2,4,4-Tetramethylpentane	298.15	3.1	74
(33 mol% Ethylene)				2,2,4-Trimethylpentane	298.15	−0.2	73
(63 mol% Ethylene)				2,2,4-Trimethylpentane	298.15	1.9	73
(75 mol% Ethylene)				2,2,4-Trimethylpentane	298.15	3.5	73

Polymer	$M_n/$ g/mol	$M_w/$ g/mol	$M_\eta/$ g/mol	Solvent	T/K	$\Delta H_B^\infty/$ J/g	Ref.
Ethylene/vinylacetate copolymer							
(85 wt% Vinyl acetate)				Cyclopentanone	298.15	−0.5	104
(70 wt% Vinyl acetate)		220000		Tetrahydrofuran	304.65	−1.3	93
Gelatine							
				Water	293.15	−92	29
				Water	323.15	−63	29
Guttapercha							
				Trichloromethane	303.15	47	22
Isobutyl methacrylate/methyl methacrylate copolymer (51 wt%/49 wt%)							
Glass		150000		Cyclohexanone	303.15	−11	98
Liquid		150000		Cyclohexanone	303.15	15	98
Natural rubber							
				Benzene	298.15	10	25
				Benzene	298.15	12	20
Nitrocellulose							
			16600	2-Butanone	298.15	−80	4
			23000	2-Butanone	298.15	−81	4
			40000	2-Butanone	298.15	−81	4
				Butyl acetate	293.15	−75	23
				Butyl acetate	298.15	−75	23
				Butyl acetate	298.15	−73	26
				Butyl acetate	303.15	−75	23
				Butyl acetate	308.15	−71	23
				Butyl acetate	313.15	−65	23
				Butyl acetate	313.15	−67	26
				Butyl acetate	318.15	−59	23
				Butyl acetate	323.15	−54	23
				Butyl acetate	328.15	−50	23
				Butyl acetate	333.15	−59	26
				Butyl acetate	343.15	−55	26
				Butyl acetate	353.15	−47	26
				Diethyl ether	295.15	−62	3
				Dibutyl phthalate	273.15	−45	26
				Dibutyl phthalate	298.15	−46	26
				Dibutyl phthalate	313.15	−46	26
				Dibutyl phthalate	333.15	−42	26
				Ethanol	295.15	−46	3
				Ethyl acetate	293.15	−76	23
				Ethyl acetate	298.15	−75	23
				Ethyl acetate	303.15	−69	23
				Ethyl acetate	308.15	−61	23
				Ethyl acetate	313.15	−54	23
				Ethyl acetate	318.15	−50	23
				Ethyl acetate	323.15	−50	23
				Ethyl acetate	328.15	−50	23
				Methanol	293.15	−69	23

Polymer	$M_n/$ g/mol	$M_w/$ g/mol	$M_\eta/$ g/mol	Solvent	T/K	$\Delta H_B^\infty/$ J/g	Ref.
				Methanol	298.15	−56	23
				Methanol	303.15	−50	23
				Methanol	308.15	−50	23
				Methanol	313.15	−50	23
				Methanol	318.15	−50	23
				Methanol	323.15	−50	23
				Methanol	328.15	−50	23
				2,4-Pentanedione	298.15	−74	4
				2-Pentanone	298.15	−64	4
				2-Propanone	273.15	−75	26
				2-Propanone	293.15	−75	23
				2-Propanone	298.15	−83	2
				2-Propanone	298.15	−68	4
				2-Propanone	298.15	−71	8
				2-Propanone	298.15	−74	23
				2-Propanone	298.15	−79	25
				2-Propanone	298.15	−75	26
				2-Propanone	303.15	−60	23
				2-Propanone	308.15	−51	23
				2-Propanone	313.15	−50	23
				2-Propanone	313.15	−65	26
				2-Propanone	318.15	−50	23
				2-Propanone	323.15	−50	23
				2-Propanone	323.15	−50	26
				2-Propanone	328.15	−50	23
				Pyridine	298.15	−106	2
				Tri(4-methylphenyl) phosphate	298.15	−16	26
				Tri(4-methylphenyl) phosphate	313.15	−28	26
				Tri(4-methylphenyl) phosphate	333.15	−41	26
				Tri(4-methylphenyl) phosphate	343.15	−44	26
				Tri(4-methylphenyl) phosphate	353.15	−47	26
Nylon-6 (unoriented)							
				Formic acid	295.15	−53	24
				Tricresol	323.55	−66	22
				Tricresol	345.55	−66	22
Poly(acrylonitrile)							
				Benzene	298.15	0.0	18
				N,N-Dimethylformamide	295.15	−23	35
				N,N-Dimethylformamide	298.15	−21	18
				N,N-Dimethylformamide	298.15	−43	42
				N,N-Dimethylformamide	308.15	−17	35
				N,N-Dimethylformamide	323.15	−13	35

Polymer	$M_n/$ g/mol	$M_w/$ g/mol	$M_\eta/$ g/mol	Solvent	T/K	$\Delta H_B^\infty/$ J/g	Ref.
				N,N-Dimethylformamide	323.15	−15	66
				N,N-Dimethylformamide	338.15	−10	35
				Dimethylsulfoxide	298.15	−70	42
Poly(γ-benzyl-L-glutamate)							
		160000		Dichloroacetic acid	303.15	−35	46
		160000		1,2-Dichloroethane	303.15	−1.6	46
Polybutadiene							
				Benzene	298.15	6.1	7
				Benzene	298.15	7.1	25
				Benzene	298.15	10.5	32
				2,2,4-Trimethylpentane	298.15	1.1	32
1,4-cis-Polybutadiene							
	low			Cyclohexane	298.15	5.4	74
	low			Cyclooctane	298.15	5.8	74
	low			Cyclopentane	298.15	<0.1	74
	low			cis-Decahydronaphthalene	298.15	4.2	74
	low			trans-Decahydronaphthalene	298.15	2.6	74
	low			3,3-Diethylpentane	298.15	5.2	74
	low			2,2-Dimethylpentane	298.15	4.1	74
	low			2,3-Dimethylpentane	298.15	4.5	74
	low			2,4-Dimethylpentane	298.15	3.2	74
	low			3,3-Dimethylpentane	298.15	3.2	74
	low			Dodecane	298.15	4.2	74
	low			3-Ethylpentane	298.15	3.7	74
	low			2,2,4,4,6,8,8-Heptamethylnonane	298.15	4.8	74
	low			Hexadecane	298.15	4.9	74
	low			3-Methylhexane	298.15	3.6	74
	low			Octane	298.15	4.3	74
	low			2,2,4,6,6-Pentamethylheptane	298.15	5.0	74
	low			2,2,4,4-Tetramethylpentane	298.15	5.8	74
	low			2,3,3,4-Tetramethylpentane	298.15	5.1	74
Poly(1-butene)							
			20000	Cyclohexane	298.15	1.0	74
			20000	Cyclooctane	298.15	1.8	74
			20000	Cyclopentane	298.15	−2.9	74
			20000	cis-Decahydronaphthalene	298.15	<0.1	74
			20000	trans-Decahydronaphthalene	298.15	−2.0	74
			20000	Decane	298.15	1.2	62
			20000	3,3-Diethylpentane	298.15	−2.6	74
			20000	2,2-Dimethylpentane	298.15	−4.0	74
			20000	2,3-Dimethylpentane	298.15	−2.8	74
			20000	2,4-Dimethylpentane	298.15	−2.3	74
			20000	3,3-Dimethylpentane	298.15	−2.2	74
			20000	Dodecane	298.15	2.1	74

Polymer	$M_n/$ g/mol	$M_w/$ g/mol	$M_\eta/$ g/mol	Solvent	T/K	$\Delta H_B^\infty/$ J/g	Ref.
			20000	3-Ethylpentane	298.15	−2.8	74
			20000	Heptane	298.15	0.0	73
			20000	Hexadecane	298.15	3.0	62
			20000	Hexane	298.15	−1.2	62
			20000	3-Methylhexane	298.15	−2.1	74
			20000	Nonane	298.15	0.9	73
			20000	Octane	298.15	0.4	73
			20000	2,2,4,6,6-Pentamethylheptane	298.15	0.6	73
			20000	Pentane	298.15	−2.6	62
			20000	Tetradecane	298.15	2.7	62
			20000	2,2,4,4-Tetramethylpentane	298.15	−1.4	74
			20000	2,3,3,4-Tetramethylpentane	298.15	−2.2	74
			20000	2,2,4-Trimethylpentane	298.15	−0.5	73
Poly(butyl acrylate)							
				2-Propanone	298.15	0.8	25
Poly(butyl methacrylate)							
Glass	91300	210000		Cyclohexanone	304.15	7.7	99
Liquid	91300	210000		Cyclohexanone	304.15	8.2	99
				2-Propanone	298.15	19.5	25
Polychloroprene							
				Benzene	298.15	0.5	7
Poly(2,6-dimethyl phenylene oxide)							
	17000	46400		1,2-Dichlorobenzene	303.05	55	89
Poly(dimethylsiloxane)							
	13000			Benzene	298.15	11.2	50
			20000	Benzene	298.15	13.5	61
			100000	Benzene	298.15	14.2	40
			170000	Bromocyclohexane	303.15	10.2	51
			80000	2-Butanone	293.15	14.4	77
	30900			2-Butanone	303.15	14.2	44
			170000	2-Butanone	303.15	14.7	51
			80000	2-Butanone	308.15	14.3	77
			80000	2-Butanone	323.15	14.3	77
			80000	Butyl acetate	298.15	6.1	41
			80000	Butyl propanoate	298.15	4.9	41
			100000	Chlorobenzene	298.15	7.5	40
	13000			Cyclohexane	298.15	3.0	50
			20000	Cyclohexane	298.15	5.2	74
			100000	Cyclohexane	298.15	5.2	40
			20000	Cyclooctane	298.15	6.8	74
			20000	Cyclopentane	298.15	1.0	74
			20000	*cis*-Decahydronapthalene	298.15	7.1	74
			20000	*trans*-Decahydronapthalene	298.15	4.3	74
			20000	Decamethyltetrasiloxane	297.65	0.45	37
			20000	Decane	298.15	3.8	37
			20000	Decane	298.15	3.9	61

Polymer	M_n/ g/mol	M_w/ g/mol	M_η/ g/mol	Solvent	T/K	ΔH_B^∞/ J/g	Ref.
			80000	Decane	298.15	3.8	41
			80000	Decyl acetate	298.15	4.5	41
			80000	Dibutyl ether	298.15	0.6	41
			80000	Diethoxymethane	298.15	1.8	41
			80000	Diethyl ether	298.15	−1.3	41
			20000	3,3-Diethylpentane	298.15	1.9	74
			80000	Dihexyl ether	298.15	3.0	41
			80000	1,2-Dimethoxyethane	298.15	12.2	41
			80000	Dimethoxymethane	298.15	7.4	41
	13000			1,2-Dimethylbenzene	298.15	4.3	50
	13000			1,3-Dimethylbenzene	298.15	3.0	50
	13000			1,4-Dimethylbenzene	298.15	3.2	50
			20000	1,4-Dimethylbenzene	298.15	4.2	61
			80000	2,6-Dimethyl-4-heptanone	293.15	6.1	77
			20000	2,2-Dimethylpentane	298.15	0.8	74
			20000	2,3-Dimethylpentane	298.15	1.4	74
			20000	2,4-Dimethylpentane	298.15	1.6	74
			20000	3,3-Dimethylpentane	298.15	0.5	74
			80000	Dipentyl ether	298.15	2.1	41
			80000	Dipropyl ether	298.15	−1.2	41
			20000	Dodecamethylpentasiloxane	297.65	−0.3	37
			20000	Dodecane	297.65	4.5	37
			20000	Dodecane	298.15	4.4	73
			80000	Dodecane	298.15	4.5	41
			80000	Ethyl acetate	298.15	12.7	41
			170000	Ethyl acetate	303.15	13.7	51
	13000			Ethylbenzene	298.15	6.4	50
			20000	Ethylbenzene	298.15	6.2	61
			80000	Ethyl butanoate	298.15	6.0	41
			80000	Ethyl decanoate	298.15	3.8	41
			80000	Ethyl dodecanoate	298.15	3.8	41
			80000	Ethyl heptanoate	298.15	4.1	41
			80000	Ethyl hexanoate	298.15	4.3	41
			80000	Ethyl nonanoate	298.15	3.7	41
			80000	Ethyl octanoate	298.15	3.8	41
			20000	3-Ethylpentane	298.15	0.6	74
			80000	Ethyl propanoate	298.15	8.0	41
			20000	2,2,4,4,6,8,8- Heptamethylnonane	298.15	3.5	74
	13000			Heptane	298.15	1.8	50
			20000	Heptane	298.15	1.9	73
			20000	Heptane	297.65	2.0	37
			20000	Heptane	298.15	2.0	61
			80000	Heptane	298.15	2.0	41
			100000	Heptane	298.15	2.1	40
			170000	Heptane	303.15	1.9	51
			80000	3-Heptanone	308.15	8.8	77
			80000	3-Heptanone	323.15	8.8	77

Polymer	$M_n/$ g/mol	$M_w/$ g/mol	$M_\eta/$ g/mol	Solvent	T/K	$\Delta H_B^\infty/$ J/g	Ref.
			20000	Hexadecane	297.65	5.5	37
			20000	Hexadecane	298.15	5.5	73
			20000	Hexamethyldisiloxane	298.15	−1.2	37
			20000	Hexamethyldisiloxane	298.15	−1.6	61
			170000	Hexamethyldisiloxane	303.15	−1.5	51
			20000	Hexane	297.65	0.7	37
			20000	Hexane	298.15	0.7	61
			80000	Hexane	298.15	0.7	41
			170000	Hexane	303.15	0.3	51
			80000	Hexyl acetate	298.15	5.0	41
	13000			Isopropylbenzene	298.15	4.1	50
			80000	Methyl butanoate	298.15	8.6	41
	13000			Methylcyclohexane	298.15	2.9	50
			100000	Methylcyclohexane	298.15	1.9	40
			80000	Methyl decanoate	298.15	4.8	41
			20000	3-Methylhexane	298.15	1.3	74
			80000	Methyl hexanoate	298.15	5.3	41
			80000	Methyl octanoate	298.15	5.0	41
			80000	4-Methyl-2-pentanone	293.15	9.9	77
			80000	4-Methyl-2-pentanone	308.15	9.0	77
			80000	Methyl propanoate	298.15	12.1	41
			20000	Nonane	297.65	3.4	37
			20000	Nonane	298.15	3.3	73
			80000	Nonane	298.15	3.4	41
			80000	Octamethylcyclotetrasiloxane	293.15	−0.4	78
			20000	Octamethyltrisiloxane	297.65	−0.6	37
			20000	Octamethyltrisiloxane	298.15	−0.8	61
			170000	Octamethyltrisiloxane	303.15	−1.0	51
			20000	Octane	297.65	2.6	37
			20000	Octane	298.15	2.4	73
			20000	Octane	298.15	2.6	61
			80000	Octane	298.15	2.6	41
			20000	2,2,4,6,6-Pentamethylheptane	298.15	2.7	73
			20000	Pentane	298.15	−0.9	37
			20000	Pentane	298.15	−0.9	61
			80000	Pentane	298.15	−1.0	41
			80000	Pentyl acetate	298.15	5.8	41
			80000	Pentyl propanoate	298.15	3.8	41
			80000	Propyl acetate	298.15	8.6	41
			170000	Propyl acetate	303.15	9.9	51
			80000	Propyl propanoate	298.15	5.6	41
			100000	Tetrachloromethane	298.15	2.4	40
			20000	Tetradecane	297.65	5.1	37
			20000	Tetradecane	298.15	5.1	61
			80000	Tetradecane	298.15	5.1	41
			20000	2,2,4,4-Tetramethylpentane	298.15	2.1	73
			20000	2,2,4,4-Tetramethylpentane	298.15	2.3	74

Polymer	$M_n/$ g/mol	$M_w/$ g/mol	$M_\eta/$ g/mol	Solvent	T/K	$\Delta H_B^\infty/$ J/g	Ref.
			20000	2,3,3,4-Tetramethylpentane	298.15	1.9	74
	13000			Toluene	298.15	5.5	50
			20000	Toluene	298.15	6.7	61
			20000	Tridecane	297.65	4.8	37
			80000	Tridecane	298.15	4.8	41
	13000			1,3,5-Trimethylbenzene	298.15	3.7	50
			20000	2,2,4-Trimethylpentane	298.15	1.4	73
			20000	Undecane	297.65	4.2	37
			80000	Undecane	298.15	4.3	41
Polyethylene							
Semicrystalline		65000		1-Chloronaphthalene	373.15	780	47
Semicrystalline		65000		1-Chloronaphthalene	383.15	980	47
Semicrystalline		65000		1-Chloronaphthalene	393.15	800	47
Liquid		65000		1-Chloronaphthalene	403.15	49	47
Semicrystalline		144000		1-Chloronaphthalene	383.15	920	47
Semicrystalline		144000		1-Chloronaphthalene	393.15	990	47
Semicrystalline		144000		1-Chloronaphthalene	403.15	690	47
Liquid		144000		1-Chloronaphthalene	413.15	67	47
Liquid		144000		1-Chloronaphthalene	423.15	85	47
Semicrystalline		670000		1-Chloronaphthalene	363.15	380	47
Semicrystalline		670000		1-Chloronaphthalene	373.15	430	47
Semicrystalline		670000		1-Chloronaphthalene	383.15	165	47
Liquid		670000		1-Chloronaphthalene	393.15	39	47
Liquid		670000		1-Chloronaphthalene	403.15	36	47
Semicrystalline		900000		1-Chloronaphthalene	391.80	245	105
Semicrystalline		900000		Cyclohexane	379.50	205	105
Semicrystalline		900000		Cyclopentane	380.00	190	105
Alkathene				Decahydronaphthalene	349.85	142	56
Rigidex-3				Decahydronaphthalene	366.65	180	56
Rigidex-50				Decahydronaphthalene	374.05	233	56
Semicrystalline		900000		Decahydronaphthalene	384.00	260	105
Semicrystalline				1,2-Dichloroethane	333.15	30	27
Semicrystalline				1,2-Dichloroethane	338.15	38	27
Semicrystalline				1,2-Dichloroethane	343.15	54	27
Semicrystalline				1,2-Dichloroethane	348.15	65	27
Semicrystalline			10000	1,4-Dimethylbenzene	354.15	139	5
Semicrystalline			11800	1,4-Dimethylbenzene	352.15	139	5
Semicrystalline			15600	1,4-Dimethylbenzene	353.65	154	5
Semicrystalline			15600	1,4-Dimethylbenzene	363.65	113	5
Semicrystalline			15600	1,4-Dimethylbenzene	368.65	104	5
Semicrystalline		900000		2,4-Dimethylpentane	393.00	230	105
Semicrystalline		900000		2,2,4,6,8,8-Heptamethylnonane	399.50	170	105
Semicrystalline		900000		Hexadecane	399.50	262	105
Semicrystalline		900000		2-Methylbutane	394.20	165	105
Semicrystalline		65000		1,2,3,4-Tetrahydronaphthalene	373.15	940	47

Polymer	M_n/ g/mol	M_w/ g/mol	M_η/ g/mol	Solvent	T/K	ΔH_B^∞/ J/g	Ref.
Semicrystalline		65000		1,2,3,4-Tetrahydronaphthalene	383.15	990	47
Semicrystalline		65000		1,2,3,4-Tetrahydronaphthalene	393.15	790	47
Liquid		65000		1,2,3,4-Tetrahydronaphthalene	403.15	58	47
Semicrystalline		84000		1,2,3,4-Tetrahydronaphthalene	373.15	835	47
Semicrystalline		130000		1,2,3,4-Tetrahydronaphthalene	353.15	630	47
Semicrystalline		130000		1,2,3,4-Tetrahydronaphthalene	373.15	520	47
Liquid		130000		1,2,3,4-Tetrahydronaphthalene	393.15	69	47
Semicrystalline		144000		1,2,3,4-Tetrahydronaphthalene	373.15	610	47
Semicrystalline		144000		1,2,3,4-Tetrahydronaphthalene	383.15	1200	47
Semicrystalline		144000		1,2,3,4-Tetrahydronaphthalene	393.15	1130	47
Semicrystalline		144000		1,2,3,4-Tetrahydronaphthalene	403.15	800	47
Liquid		144000		1,2,3,4-Tetrahydronaphthalene	413.15	136	47
Liquid		144000		1,2,3,4-Tetrahydronaphthalene	423.15	88	47
Semicrystalline		310000		1,2,3,4-Tetrahydronaphthalene	343.15	485	47
Semicrystalline		670000		1,2,3,4-Tetrahydronaphthalene	353.15	560	47
Semicrystalline		670000		1,2,3,4-Tetrahydronaphthalene	363.15	560	47
Semicrystalline		670000		1,2,3,4-Tetrahydronaphthalene	373.15	460	47
Semicrystalline		670000		1,2,3,4-Tetrahydronaphthalene	383.15	155	47
Liquid		670000		1,2,3,4-Tetrahydronaphthalene	393.15	67	47
Liquid		670000		1,2,3,4-Tetrahydronaphthalene	403.15	39	47
Semicrystalline			16000	Toluene	353.15	110	22
Semicrystalline			22000	Toluene	358.35	118	22
Semicrystalline			22000	Toluene	367.35	106	22
Semicrystalline		900000		1,2,4-Trichlorobenzene	386.50	255	105
Poly(ethylene glycol)							
	180			Benzene	303.15	110	71
	385			Benzene	303.15	60	71
	560			Benzene	303.15	40	71
	1050			Benzene	303.15	90	71
	1610			Benzene	303.15	140	71
	1940			Benzene	303.15	215	71
	3200			Benzene	303.15	195	71
	4330			Benzene	303.15	195	71

Polymer	$M_n/$ g/mol	$M_w/$ g/mol	$M_\eta/$ g/mol	Solvent	T/K	$\Delta H_B^\infty/$ J/g	Ref.
	5850			Benzene	303.15	190	71
	9950			Benzene	303.15	195	71
			43400	Benzene	303.15	190	71
	400	420		Trichloromethane	303.15	−79	95
	590	615		Trichloromethane	303.15	−88	95
	180			Water	303.15	−136	71
	200			Water	321.35	−125	83
	355			Water	303.15	−159	71
	400			Water	321.35	−150	83
	560			Water	303.15	−150	71
	990			Water	321.35	−101	83
	1050			Water	303.15	−106	71
	1460			Water	321.35	−137	83
	1610			Water	303.15	−6	71
	1940			Water	303.15	57	71
	3200			Water	303.15	58	71
	4330			Water	303.15	28	71
	5850			Water	303.15	39	71
	9950			Water	303.15	30	71
			14000	Water	303.15	7	55
			14000	Water	313.15	27	55
			20300	Water	303.15	45	71
			34500	Water	303.15	34	71
			43300	Water	303.15	+40	71
Poly(ethylene glycol) dimethyl ether							
	250			Tetrachloromethane	303.15	−12	95
	250			Tetrachloromethane	318.15	−7.6	95
	400			Tetrachloromethane	303.15	−12	95
	520	550		Tetrachloromethane	303.15	−12	95
	520	550		Tetrachloromethane	303.15	−7.6	95
	250			Trichloromethane	303.15	−184	95
	520	550		Trichloromethane	303.15	−135	95
Poly(ethylene glycol) monododecyl ether							
	230			Dodecane	302.15	42	68
	274			Dodecane	302.15	23	68
	318			Dodecane	302.15	34	68
	362			Dodecane	302.15	37	68
	406			Dodecane	302.15	42	68
Poly(ethylene glycol) monomethyl ether							
	353	377		Trichloromethane	303.15	−125	95
	550	580		Trichloromethane	303.15	−117	95
Poly(ethylene oxide)							
Semicrystalline	6000			Dichloromethane	303.15	+84	58
Liquid	6000			Dichloromethane	303.15	−160	58

Polymer	$M_n/$ g/mol	$M_w/$ g/mol	$M_\eta/$ g/mol	Solvent	T/K	$\Delta H_B^\infty/$ J/g	Ref.
Semicrystalline	6000			Trichloromethane	303.15	+52	58
Liquid	6000			Trichloromethane	303.15	−186	58
Quenched	1520	1720		Water	293.15	−403	64
Annealed	1520	1720		Water	293.15	−392	64
Quenched	1520	1720		Water	298.15	−180	64
Annealed	1520	1720		Water	298.15	−150	64
Quenched	1520	1720		Water	303.15	+68	64
Annealed	1520	1720		Water	303.15	+109	64
Liquid	6000			Water	303.15	−50	58
Quenched	6840	7525		Water	293.15	−28	64
Annealed	6840	7525		Water	293.15	+209	64
Quenched	6840	7525		Water	298.15	+241	64
Annealed	6840	7525		Water	298.15	+540	64
Quenched	16600	19600		Water	293.15	−160	64
Annealed	16600	19600		Water	293.15	−143	64
Quenched	16600	19600		Water	298.15	+59	64
Annealed	16600	19600		Water	298.15	+155	64
Quenched	16600	19600		Water	303.15	+353	64
Annealed	16600	19600		Water	303.15	+490	64
Semicrystalline		20000		Water	298.15	+10	84
Polyindene							
	765	1023		Anisole	299.15	2.1	102
	765	1023		Benzene	299.15	−0.04	102
	765	1023		Benzonitrile	299.15	−4.4	102
	765	1023		Bromobenzene	299.15	−3.9	102
	765	1023		2-Butanone	299.15	1.9	102
	765	1023		Chlorobenzene	299.15	−3.9	102
	765	1023		1-Chlorobutane	299.15	−4.0	102
	765	1023		1-Chloroheptane	299.15	1.5	102
	765	1023		Cyclohexane	299.15	15	102
	765	1023		N,N-dimethylaniline	299.15	−8.2	102
	765	1023		Ethyl acetate	299.15	4.2	102
	765	1023		Ethylbenzene	299.15	−1.4	102
	765	1023		Ethyl benzoate	299.15	−0.7	102
	765	1023		Nitrobenzene	299.15	4.6	102
	765	1023		1-Nitropropane	299.15	8.4	102
	765	1023		Pyridine	299.15	−6.7	102
	765	1023		1,1,2,2-Tetrachloroethane	299.15	−19	102
	765	1023		Tetrachloromethane	299.15	−2.5	102
	765	1023		1,1,1-Trichloroethane	299.15	−1.8	102
	765	1023		Trichloromethane	299.15	−20	102
Polyisobutylene							
	360		700	Benzene	298.15	30	67
	1000		2000	Benzene	298.15	25	67
	1300		2500	Benzene	298.15	23	67
			30000	Benzene	297.65	19	38
			30000	Benzene	298.15	19	40

Polymer	$M_n/$ g/mol	$M_w/$ g/mol	$M_\eta/$ g/mol	Solvent	T/K	$\Delta H_B^\infty/$ J/g	Ref.
	44700			Benzene	303.15	16	44
			48000	Benzene	303.4	19	72
			50000	Benzene	303.15	16	45
			72000	Benzene	300.15	19	53
			72000	Benzene	323.15	18	53
			72000	Benzene	343.15	16	53
			72000	Benzene	375.15	13	53
			72000	Benzene	394.15	9.2	53
			72000	Benzene	423.15	3.5	53
			72000	Benzene	437.15	−0.5	53
			72000	Benzene	453.15	−4.7	53
			90000	Benzene	298.15	6.7	32
			160000	Benzene	303.15	16	51
			560000	Benzene	298.15	18	34
				Benzene	298.15	6.8	7
				Benzene	298.15	6.8	10
			30000	Chlorobenzene	297.65	12	38
			30000	Chlorobenzene	298.15	13	40
			50000	Chlorobenzene	303.15	12	45
			160000	Chlorobenzene	303.15	12	51
			560000	Chlorobenzene	298.15	12	34
	360		700	Cyclohexane	298.15	3.8	67
	1000		2000	Cyclohexane	298.15	1.2	67
	1300		2500	Cyclohexane	298.15	1.1	67
			4500	Cyclohexane	298.15	−0.6	74
			30000	Cyclohexane	297.65	−0.7	38
			30000	Cyclohexane	298.15	−0.6	40
			50000	Cyclohexane	303.15	−0.7	45
			160000	Cyclohexane	303.15	−0.6	51
			1990000	Cyclohexane	298.15	−0.7	39
			4500	Cyclooctane	298.15	+0.3	74
			4500	Cyclopentane	298.15	−5.9	74
			4500	cis-Decahydronaphthalene	298.15	0.2	74
			4500	trans-Decahydronaphthalene	298.15	−0.8	74
			30000	Decane	297.65	−0.5	38
			50000	Decane	303.15	−0.5	45
			30000	Dibutyl ether	297.65	1.2	37
			30000	Diethyl ether	297.65	2.8	37
			30000	Diethyl ether	297.65	2.8	38
			4500	3,3-Diethylpentane	298.15	−1.4	74
			30000	Dihexyl ether	297.65	0.9	37
			4500	2,2-Dimethylpentane	298.15	−1.1	74
			4500	2,3-Dimethylpentane	298.15	−1.9	74
			4500	2,4-Dimethylpentane	298.15	−1.1	74
			4500	3,3-Dimethylpentane	298.15	−1.7	74
			30000	Dipentyl ether	297.65	1.0	37
			30000	Dipropyl ether	297.65	1.8	37

Polymer	$M_n/$ g/mol	$M_w/$ g/mol	$M_\eta/$ g/mol	Solvent	T/K	$\Delta H_B^\infty/$ J/g	Ref.
	360		700	Dodecane	298.15	1.9	67
	1000		2000	Dodecane	298.15	0.7	67
	1300		2500	Dodecane	298.15	0.5	67
			4500	Dodecane	298.15	0.2	73
			30000	Dodecane	297.65	−0.1	38
			30000	Dodecane	298.15	−0.1	40
			48000	Ethylbenzene	291.15	9.5	72
			48000	Ethylbenzene	343.15	3.6	72
			30000	Ethyl decanoate	297.65	3.0	37
			30000	Ethyl heptanoate	297.65	5.6	37
			30000	Ethyl hexadecanoate	297.65	1.3	37
			30000	Ethyl hexanoate	297.65	6.7	37
			30000	Ethyl nonanoate	297.65	3.7	37
			30000	Ethyl octanoate	297.65	4.6	37
			4500	3-Ethylpentane	298.15	−2.0	74
			30000	Ethyl tetradecanoate	297.65	1.8	37
			4500	2,2,4,4,6,8,8-Heptamethylnonane	298.15	−0.5	74
	360		700	Heptane	298.15	−0.5	67
	1000		2000	Heptane	298.15	−1.0	67
	1300		2500	Heptane	298.15	−1.4	67
			4500	Heptane	298.15	−1.7	73
			30000	Heptane	297.65	−1.8	38
			30000	Heptane	298.15	−2.0	40
			50000	Heptane	303.15	−1.8	45
			160000	Heptane	303.15	−1.6	51
				Heptane	298.15	−1.4	7
				Heptane	298.15	−1.4	10
	360		700	Hexadecane	298.15	4.5	67
	1000		2000	Hexadecane	298.15	2.1	67
	1300		2500	Hexadecane	298.15	1.0	67
			4500	Hexadecane	298.15	0.9	73
			30000	Hexadecane	297.65	0.04	38
			30000	Hexane	297.65	−2.5	38
			30000	Hexane	298.15	−2.6	40
			50000	Hexane	303.15	−2.5	45
			72000	Hexane	303.15	−1.8	53
			72000	Hexane	324.15	−2.3	53
			72000	Hexane	348.15	−2.9	53
			72000	Hexane	373.15	−3.7	53
			72000	Hexane	393.15	−5.3	53
			72000	Hexane	408.15	−6.7	53
			72000	Hexane	423.15	−9.0	53
			72000	Hexane	433.15	−9.9	53
			160000	Hexane	303.15	−2.5	51
			30000	2-Methylbutane	297.65	−3.1	38
			30000	Methylcyclohexane	297.65	−1.2	38
			50000	Methylcyclohexane	303.15	−1.2	45

Polymer	M_n/ g/mol	M_w/ g/mol	M_η/ g/mol	Solvent	T/K	ΔH_B^∞/ J/g	Ref.
			160000	Methylcyclohexane	303.15	−1.2	51
			4500	3-Methylhexane	298.15	−1.0	74
			30000	3-Methylpentane	297.65	−2.8	38
			30000	Nonane	297.65	−0.8	38
			4500	Nonane	298.15	−0.8	73
			4500	Octane	298.15	−1.1	73
			30000	Octane	297.65	−1.2	38
			72000	Octane	303.15	−0.3	53
			72000	Octane	324.15	−0.8	53
			72000	Octane	348.15	−0.9	53
			72000	Octane	373.15	−1.1	53
			72000	Octane	393.15	−1.3	53
			72000	Octane	423.15	−3.6	53
			4500	2,2,4,6,6-Pentamethylheptane	298.15	−0.1	73
	360		700	Pentane	298.15	−1.9	67
	1000		2000	Pentane	298.15	−2.9	67
	1300		2500	Pentane	298.15	−3.2	67
			30000	Pentane	297.65	−3.6	38
			72000	Pentane	303.15	−2.8	53
			72000	Pentane	333.15	−3.4	53
			72000	Pentane	352.15	−4.5	53
			72000	Pentane	365.15	−5.5	53
	360		700	Tetrachloromethane	298.15	5.9	67
	1000		2000	Tetrachloromethane	298.15	5.8	67
	1300		2500	Tetrachloromethane	298.15	5.0	67
			1990000	Tetrachloromethane	298.15	4.1	39
			30000	Tetradecane	297.65	0.0	38
			4500	2,2,4,4-Tetramethylpentane	298.15	−0.6	74
			4500	2,3,3,4-Tetramethylpentane	298.15	−2.3	74
			50000	Toluene	303.15	7.4	45
			160000	Toluene	303.15	7.4	51
			1990000	Toluene	298.15	8.8	39
				Toluene	298.15	1.8	7
				Toluene	298.15	1.8	10
			30000	Tridecane	297.65	−0.04	38
			4500	2,2,4-Trimethylpentane	298.15	−0.4	73
			30000	2,2,4-Trimethylpentane	297.65	−0.6	38
			1990000	2,2,4-Trimethylpentane	298.15	0.0	39
				2,2,4-Trimethylpentane	298.15	0.0	7
				2,2,4-Trimethylpentane	298.15	0.0	10
			30000	Undecane	297.65	−0.4	38
Poly(isobutyl methacrylate)							
Glass	260000			Cyclohexanone	303.15	−5.2	98
Liquid	260000			Cyclohexanone	303.15	13	98
Poly(methyl acrylate)							
				2-Propanone	298.15	0.0	25

Polymer	$M_n/$ g/mol	$M_w/$ g/mol	$M_\eta/$ g/mol	Solvent	T/K	$\Delta H_B^\infty/$ J/g	Ref.
Poly(methyl methacrylate)							
Glass	73900	170000		Cyclohexanone	304.15	−14	98
Liquid	73900	170000		Cyclohexanone	304.15	17	98
	1930			1,2-Dichloroethane	298.15	−20	32
	240000			1,2-Dichloroethane	298.15	−27	32
			53000	Ethylbenzene	298.15	−31	28
			180000	Ethylbenzene	298.15	−29	28
	28900	35900		4-Methyl-2-pentanone	303.15	−21	76
	93940	101000		4-Methyl-2-pentanone	303.15	−24	76
	137000	215000		4-Methyl-2-pentanone	303.15	−28	76
				2-Propanone	298.15	−30	25
	93940	101000		Toluene	303.15	−22	76
	689000	782000		Toluene	303.15	−24	76
			12000	Trichloromethane	298.15	−65	52
			54000	Trichloromethane	298.15	−80	52
			80000	Trichloromethane	298.15	−81	52
			100000	Trichloromethane	298.15	−84	52
			320000	Trichloromethane	298.15	−83	52
	93940	101000		Trichloromethane	303.15	−71	76
	689000	782000		Trichloromethane	303.15	−72	76
		2320000		Trichloromethane	303.15	−73	76
Poly(4-methyl-1-pentene)							
Semicrystalline			350000	Cyclohexane	303.15	30	79
Poly(α-methylstyrene)							
	1030	1180		Toluene	298.15	−7.1	43
		1430		Toluene	298.15	−30	43
	1820	2230		Toluene	298.15	−34	43
	1920			Toluene	298.15	−37	43
	2700	3300		Toluene	298.15	−39	43
	3280			Toluene	298.15	−46	43
	5260			Toluene	298.15	−46	43
	8600			Toluene	298.15	−45	43
	12200			Toluene	298.15	−46	43
		10500		Toluene	310.15	−8.4	92
		53000		Toluene	310.15	−13	92
		55000		Toluene	333.15	−16	96
		87000		Toluene	298.15	−17	90
		87000		Toluene	310.15	−16	92
		87000		Toluene	333.15	−11	90
Poly(2-methyl-5-vinyltetrazole)							
				Acetic acid	298.15	47	100
				Acetonitrile	298.15	14	100
				1,2-Dichloroethane	298.15	17	100
				N,N-Diethylacetamide	298.15	17	100
				N,N-Dimethylformamide	298.15	33	100
				Dimethylsulfoxide	298.15	10	100

Polymer	M_n/ g/mol	M_w/ g/mol	M_η/ g/mol	Solvent	T/K	ΔH_B^∞/ J/g	Ref.
				Formamide	298.15	12	100
				Formic acid	298.15	110	100
				Nitromethane	298.15	10	100
				Pyridine	298.15	16	100
Poly(octamethylene oxide)							
	7000			Benzene	298.15	20	64
	7000			Benzene	303.15	22	64
	7000			Benzene	308.15	25	64
Polypentenamer							
			50000	Cyclohexane	298.15	4.6	74
			50000	Cyclooctane	298.15	5.1	74
			50000	Cyclopentane	298.15	−2.3	74
			50000	*cis*-Decahydronaphthalene	298.15	2.6	74
			50000	*trans*-Decahydronaphthalene	298.15	<0.1	74
			50000	3,3-Diethylpentane	298.15	2.4	74
			50000	2,2-Dimethylpentane	298.15	3.3	74
			50000	2,3-Dimethylpentane	298.15	2.2	74
			50000	2,4-Dimethylpentane	298.15	3.3	74
			50000	3,3-Dimethylpentane	298.15	2.7	74
			50000	Dodecane	298.15	2.9	74
			50000	3-Ethylpentane	298.15	2.1	74
			50000	2,2,4,4,6,8,8-Heptamethylnonane	298.15	3.2	74
			50000	Hexadecane	298.15	2.6	74
			50000	3-Methylhexane	298.15	2.4	74
			50000	Octane	298.15	2.2	74
			50000	2,2,4,6,6-Pentamethylheptane	298.15	3.8	74
			50000	2,2,4,4-Tetramethylpentane	298.15	4.5	74
			50000	2,3,3,4-Tetramethylpentane	298.15	2.4	74
			50000	2,2,4-Trimethylpentane	298.15	4.3	74
Poly(m-phenyleneisophthalamide)							
Glass				*N,N*-Dimethylacetamide	298.15	−171	60
Semicrystalline				*N,N*-Dimethylacetamide	298.15	−128	60
Glass				*N,N*-Dimethylformamide	298.15	−149	60
Semicrystalline				*N,N*-Dimethylformamide	298.15	−125	60
Glass				1-Methyl-2-pyrrolidone	298.15	−177	60
Semicrystalline				1-Methyl-2-pyrrolidone	298.15	−118	60
Polypropylene (atactic)							
	18000			Benzene	298.15	31	80
	6000			Cyclohexane	298.15	2.3	74
	18000			Cyclohexane	298.15	3.9	80
	6000			Cyclooctane	298.15	3.0	74
	6000			Cyclopentane	298.15	−2.3	74
	6000			*cis*-Decahydronaphthalene	298.15	0.5	74

Polymer	$M_n/$ g/mol	$M_w/$ g/mol	$M_\eta/$ g/mol	Solvent	T/K	$\Delta H_B^\infty/$ J/g	Ref.
			6000	trans-Decahydronaphthalene	298.15	−2.4	74
			18000	Decane	298.15	3.1	80
			6000	3,3-Diethylpentane	298.15	−3.9	74
			18000	1,2-Dimethylbenzene	298.15	13	80
			18000	1,3-Dimethylbenzene	298.15	12	80
			18000	1,4-Dimethylbenzene	298.15	10	80
			6000	2,2-Dimethylpentane	298.15	−2.2	74
			6000	2,3-Dimethylpentane	298.15	−2.5	74
			6000	2,4-Dimethylpentane	298.15	−1.8	74
			6000	3,3-Dimethylpentane	298.15	−3.0	74
			6000	Dodecane	298.15	1.7	73
			18000	Ethylbenzene	298.15	14	80
			6000	3-Ethylpentane	298.15	−2.5	74
			6000	2,2,4,4,6,8,8-Heptamethylnonane	298.15	−0.7	73
			6000	Heptane	298.15	−1.6	73
			18000	Heptane	298.15	0.5	80
			6000	Hexadecane	298.15	2.3	73
			18000	Hexane	298.15	−1.4	80
			6000	3-Methylhexane	298.15	−1.8	74
			6000	Nonane	298.15	0.8	73
			18000	Nonane	298.15	2.4	80
			6000	Octane	298.15	−1.2	73
			18000	Octane	298.15	1.0	80
			6000	2,2,4,6,6-Pentamethylheptane	298.15	−0.2	73
			18000	Pentane	298.15	−4.7	80
			18000	Tetrachloromethane	298.15	6.6	80
			6000	2,2,4,4-Tetramethylpentane	298.15	−0.8	74
			6000	2,3,3,4-Tetramethylpentane	298.15	−3.1	74
			18000	Toluene	298.15	17	80
			18000	Trichloromethane	298.15	17	80
			6000	2,2,4-Trimethylpentane	298.15	−1.0	73
Polypropylene (isotactic)							
				1-Chloronaphthalene	383.15	26	47
				1-Chloronaphthalene	393.15	170	47
				1-Chloronaphthalene	403.15	245	47
				1-Chloronaphthalene	423.15	275	47
				1,2,3,4-Tetrahydronaphthalene	373.15	140	47
				1,2,3,4-Tetrahydronaphthalene	383.15	215	47
				1,2,3,4-Tetrahydronaphthalene	393.15	330	47
				1,2,3,4-Tetrahydronaphthalene	403.15	330	47
				1,2,3,4-Tetrahydronaphthalene	413.15	335	47
				1,2,3,4-Tetrahydronaphthalene	423.15	290	47

Polymer	M_n/ g/mol	M_w/ g/mol	M_η/ g/mol	Solvent	T/K	ΔH_B^∞/ J/g	Ref.
Poly(propylene glycol)							
	150			Benzene	321.35	200	103
	425			Benzene	321.35	80	103
	2025			Benzene	321.35	45	103
	150			Ethanol	321.35	40	103
	425			Ethanol	321.35	60	103
	2025			Ethanol	321.35	65	103
	396	412		Tetrachloromethane	303.15	4.7	95
	396	412		Tetrachloromethane	318.15	5.2	95
	1900			Tetrachloromethane	303.15	−8.2	95
	1900			Tetrachloromethane	318.15	11	95
	1900			Trichloromethane	303.15	−81	95
		400		Water	298.15	−165	97
	150			Water	321.35	−90	103
	425			Water	321.35	−95	103
Polystyrene							
		600		Benzene	298.15	−1.3	54
		600		Benzene	313.15	−2.5	54
		900		Benzene	291.15	−10	54
		900		Benzene	318.15	−5.8	54
		2000		Benzene	291.15	−16	54
		2000		Benzene	318.15	−6.8	54
		5000		Benzene	291.15	−23	54
		5000		Benzene	318.15	−12	54
		10300		Benzene	291.15	−26	54
		10300		Benzene	318.15	−18	54
			18000	Benzene	298.15	−4.1	19
	20000			Benzene	296.15	−15	9
			29000	Benzene	298.15	−5.0	19
			30000	Benzene	298.15	−7.5	19
			59000	Benzene	298.15	−13	19
			91000	Benzene	298.15	−15	19
		97200		Benzene	318.15	−21	54
			142000	Benzene	298.15	−17	19
			190000	Benzene	303.15	−18	51
	214000			Benzene	300.15	−16	85
			216000	Benzene	298.15	−18	19
			272000	Benzene	298.15	−21	19
			300000	Benzene	298.15	−21	12
				Benzene	298.15	−27	25
				Benzene	298.15	−10	7
	20000			Butyl acetate	296.15	−13	9
	20000			2-Butanone	296.15	−15	9
			142000	2-Butanone	296.15	−17	30
			190000	Butylbenzene	303.15	−14	51
	150000			Chlorobenzene	293.15	−32	49
			266000	Chlorobenzene	298.15	5.4	34

Polymer	M_n/ g/mol	M_w/ g/mol	M_η/ g/mol	Solvent	T/K	ΔH_B^∞/ J/g	Ref.
				Chlorobenzene	293.15	−39	21
	1260			Cyclohexane	298.15	10	16
	1910			Cyclohexane	298.15	5.4	16
	3160			Cyclohexane	298.15	−5.4	16
	3980			Cyclohexane	298.15	−6.9	16
	5630			Cyclohexane	298.15	−9.3	16
	9070			Cyclohexane	298.15	−11	16
	20000			Cyclohexane	296.15	2.5	9
			190000	Cyclohexane	303.15	−2.1	51
				Cyclohexane	293.15	−14	21
	22400			Cyclohexanone	298.15	−29	25
	20000			Cyclohexene	296.15	−9.4	9
			190000	Decahydronaphthalene	303.15	3.8	51
	110000	115000		1,2-Dichlorobenzene	303.05	26	89
	20000			1,2-Dimethylbenzene	296.15	−13	9
	20000			1,3-Dimethylbenzene	296.15	−12	9
			190000	1,3-Dimethylbenzene	303.15	−15	51
			190000	1,4-Dioxane	303.15	−12	51
	20000			Ethyl acetate	296.15	−11	9
			142000	Ethyl acetate	296.15	−13	30
			785	Ethylbenzene	298.15	0.0	14
			18000	Ethylbenzene	298.15	−3.8	14
			18000	Ethylbenzene	298.15	−3.9	19
			30000	Ethylbenzene	298.15	−5.7	19
			35000	Ethylbenzene	298.15	−6.5	19
			45000	Ethylbenzene	298.15	−8.4	19
			91000	Ethylbenzene	298.15	−11	19
			142000	Ethylbenzene	298.15	−13	19
			216000	Ethylbenzene	298.15	−17	19
	60000			Ethylbenzene	303.15	−22	57
	113000	122000		Ethylbenzene	306.65	−24	63
	113000	122000		Ethylbenzene	317.15	−19	63
	113000	122000		Ethylbenzene	337.15	−11	63
	113000	122000		Ethylbenzene	347.15	−6.4	63
	113000	122000		Ethylbenzene	350.65	−4.9	63
	113000	122000		Ethylbenzene	366.65	−2.3	63
	113000	122000		Ethylbenzene	367.15	−2.1	63
	113000	122000		Ethylbenzene	368.65	−2.6	63
	113000	122000		Ethylbenzene	372.15	−4.2	63
	113000	122000		Ethylbenzene	378.15	−4.6	63
	113000	122000		Ethylbenzene	385.15	−5.7	63
	150000			Ethylbenzene	293.15	−34	49
			190000	Ethylbenzene	303.15	−15	51
			272000	Ethylbenzene	298.15	−18	14
			272000	Ethylbenzene	298.15	−18	19
			413000	Ethylbenzene	298.15	−24	39
				Ethylbenzene	293.15	−34	21

Polymer	$M_n/$ g/mol	$M_w/$ g/mol	$M_\eta/$ g/mol	Solvent	T/K	$\Delta H_B^\infty/$ J/g	Ref.
				Ethylbenzene	298.15	−17	11
				Ethylbenzene	298.15	−30	28
	1260			2-Propanone	298.15	−0.6	16
	1910			2-Propanone	298.15	−7.7	16
	3160			2-Propanone	298.15	−16	16
	3980			2-Propanone	298.15	−17	16
	5630			2-Propanone	298.15	−19	16
	9070			2-Propanone	298.15	−21	16
	20000			2-Propanone	296.15	−11	9
			190000	Propylbenzene	303.15	−14	51
	20000			Styrene	296.15	−18	9
				Styrene	296.15	−35	6
			413000	Tetrachloromethane	298.15	−22	39
	600			Toluene	296.15	−2.1	16
	600			Toluene	309.15	−1.8	16
	600			Toluene	318.15	−1.5	16
	1260			Toluene	296.15	−11	16
	1260			Toluene	303.15	−8.0	16
	1260			Toluene	309.15	−5.9	16
	1260			Toluene	318.15	−3.4	16
	1260			Toluene	328.15	−2.3	16
	1260			Toluene	338.15	−1.9	16
	1260			Toluene	346.65	−1.3	16
	1910			Toluene	298.15	−16	16
	1910			Toluene	318.15	−6.7	16
	1910			Toluene	338.15	−3.4	16
	1910			Toluene	348.15	−2.5	16
	3160			Toluene	298.15	−23	16
	3980			Toluene	298.15	−24	16
	3980			Toluene	318.15	−17	16
	3980			Toluene	338.15	−10	16
	5630			Toluene	298.15	−26	16
	9070			Toluene	298.15	−28	16
	270000			Toluene	298.15	−33	16
		600		Toluene	298.15	−1.4	54
		600		Toluene	313.15	−3.2	54
		900		Toluene	291.15	−7.3	54
		900		Toluene	318.15	−6.6	54
		2000		Toluene	291.15	−11	54
		2000		Toluene	318.15	−7.2	54
		5000		Toluene	291.15	−21	54
		5000		Toluene	318.15	−11	54
		10300		Toluene	291.15	−24	54
		10300		Toluene	318.15	−15	54
		97200		Toluene	318.15	−17	54
		9000		Toluene	310.15	−9.2	92
	20000			Toluene	296.15	−17	9

Polymer	$M_n/$ g/mol	$M_w/$ g/mol	$M_\eta/$ g/mol	Solvent	T/K	$\Delta H_B^\infty/$ J/g	Ref.
	20400			Toluene	298.15	−8.2	90
	20400			Toluene	310.15	−8.4	92
	20400			Toluene	333.15	−6.4	90
	47000			Toluene	310.15	−5.0	92
	50000			Toluene	333.15	−6.8	96
	60000			Toluene	303.15	−21	57
	113000	122000		Toluene	304.15	−29	63
	113000	122000		Toluene	306.15	−27	63
	113000	122000		Toluene	306.65	−26	63
	113000	122000		Toluene	316.15	−23	63
	113000	122000		Toluene	333.15	−15	63
	113000	122000		Toluene	337.15	−12	63
	113000	122000		Toluene	346.15	−8.3	63
	113000	122000		Toluene	347.15	−7.8	63
	113000	122000		Toluene	348.15	−8.2	63
	113000	122000		Toluene	350.65	−6.5	63
	113000	122000		Toluene	359.15	−4.3	63
	113000	122000		Toluene	362.15	−2.7	63
	113000	122000		Toluene	369.15	−3.3	63
	113000	122000		Toluene	372.15	−2.8	63
	115000			Toluene	310.15	−5.0	92
	150000			Toluene	293.15	−34	49
			190000	Toluene	303.15	−18	51
		214000		Toluene	300.15	−19	85
			250000	Toluene	303.15	−18	51
				Toluene	293.15	−34	21
				Toluene	298.65	−39	27
				Toluene	308.15	−34	27
				Toluene	318.15	−30	27
				Toluene	333.15	−23	27
				Toluene	343.15	−13	27
				Toluene	353.15	−13	27
		600		Trichloromethane	298.15	−13	54
		600		Trichloromethane	313.15	−9.9	54
		900		Trichloromethane	291.15	−22	54
		900		Trichloromethane	313.15	−15	54
		2000		Trichloromethane	291.15	−28	54
		2000		Trichloromethane	313.15	−16	54
		5000		Trichloromethane	291.15	−30	54
		5000		Trichloromethane	313.15	−18	54
		10300		Trichloromethane	291.15	−33	54
		10300		Trichloromethane	313.15	−23	54
	22400			Trichloromethane	298.15	−17	25
		97200		Trichloromethane	313.15	−25	54
	20000			1,3,5-Trimethylbenzene	296.15	−11	9
			190000	1,3,5-Trimethylbenzene	303.15	−13	51

Poly(tetramethylene oxide)

Polymer	$M_n/$ g/mol	$M_w/$ g/mol	$M_\eta/$ g/mol	Solvent	T/K	$\Delta H_B^\infty/$ J/g	Ref.
	650			Benzene	313.15	4.0	86
	1000			Benzene	313.15	2.0	86
	2000			Benzene	313.15	1.1	86
	650			1,2-Dichloroethane	313.35	3.1	88
	2000			1,2-Dichloroethane	313.35	0.3	88
	650			1,2-Dimethylbenzene	313.15	5.9	87
	1000			1,2-Dimethylbenzene	313.15	1.8	87
	2000			1,2-Dimethylbenzene	313.15	0.9	87
	650			1,3-Dimethylbenzene	313.15	6.4	87
	1000			1,3-Dimethylbenzene	313.15	0.6	87
	2000			1,3-Dimethylbenzene	313.15	0.8	87
	650			1,4-Dimethylbenzene	313.15	4.3	87
	1000			1,4-Dimethylbenzene	313.15	1.8	87
	2000			1,4-Dimethylbenzene	313.15	0.7	87
	650			1,4-Dioxane	321.35	4.0	81
	1000			1,4-Dioxane	321.35	2.3	81
	2000			1,4-Dioxane	321.35	1.0	81
	650			Ethylbenzene	313.15	6.9	86
	1000			Ethylbenzene	313.15	3.4	86
	2000			Ethylbenzene	313.15	−0.05	86
	650			Propylbenzene	313.15	5.8	86
	1000			Propylbenzene	313.15	1.3	86
	2000			Propylbenzene	313.15	0.9	86
	650			Tetrachloromethane	313.15	3.3	88
	1000			Tetrachloromethane	313.15	1.4	88
	2000			Tetrachloromethane	321.35	0.7	82
	650			Toluene	313.15	4.3	86
	1000			Toluene	313.15	2.0	86
	2000			Toluene	313.15	0.9	86
	650			1,3,5-Trimethylbenzene	313.15	6.1	87
	1000			1,3,5-Trimethylbenzene	313.15	2.7	87
	2000			1,3,5-Trimethylbenzene	313.15	0.6	87
Poly(vinyl acetate)							
			140000	Benzene	298.15	2.3	13
			350000	2-Butanone	303.15	−1.7	51
			350000	Butyl acetate	303.15	1.0	51
			135000	Chlorobenzene	298.15	5.0	34
			350000	Ethyl acetate	303.15	−6.7	51
				Ethyl acetate	303.15	0.0	11
			26000	3-Heptanone	303.15	7.0	44
			350000	3-Heptanone	303.15	4.9	51
			140000	Methanol	298.15	−45	13
			350000	Methyl acetate	303.15	−9.7	51
			350000	2-Pentanone	303.15	0.0	51
			93000	2-Propanone	303.15	−0.4	32
			350000	2-Propanone	303.15	−3.9	51
				2-Propanone	303.15	−2.9	25

Polymer	$M_n/$ g/mol	$M_w/$ g/mol	$M_\eta/$ g/mol	Solvent	T/K	$\Delta H_B^\infty/$ J/g	Ref.
			350000	Propyl acetate	303.15	−2.7	51
		150000		Tetrahydrofuran	304.65	4.5	93
			140000	Trichloromethane	298.15	28	13
Poly(vinyl alcohol)							
				Ethanol	298.15	3.8	11
				Ethanol	298.15	9.6	31
	7260			Water	303.15	−34	15
	17000			Water	303.15	−18	32
	61600			Water	303.15	−41	15
				Water	303.15	−8.4	11
Poly(vinyl chloride)							
Glass	23200			Chlorobenzene	298.15	−17	36
Liquid	23200			Cyclohexanone	303.15	−27	59
Glass	38700			Cyclohexanone	303.15	−7.5	59
Liquid	38700			Cyclohexanone	303.15	−29	59
Glass	53500			Cyclohexanone	303.15	−6.6	59
Liquid	53500			Cyclohexanone	303.15	−28	59
Glass	66700			Cyclohexanone	303.15	−6.3	59
Liquid	66700			Cyclohexanone	303.15	−29	59
Glass	136000			Cyclohexanone	303.15	−6.1	59
Liquid	136000			Cyclohexanone	303.15	−31	59
Glass	155400			Cyclohexanone	303.15	−5.8	59
Liquid	155400			Cyclohexanone	303.15	−32	59
	48000			Cyclohexanone	303.15	−5.8	59
				Cyclopentanone	298.15	−28	104
				1,2-Dichloroethane	323.65	24	27
				1,2-Dichloroethane	328.15	34	27
				1,2-Dichloroethane	333.15	38	27
				1,2-Dichloroethane	368.15	44	27
				1,2-Dichloroethane	373.15	46	27
				1,2-Dichloroethane	378.15	46	27
				N,N-Dimethylformamide	293.15	−28	35
				N,N-Dimethylformamide	308.15	−19	35
				N,N-Dimethylformamide	323.15	−14	35
				N,N-Dimethylformamide	338.15	−7.5	35
				N,N-Dimethylformamide	353.15	2.4	35
Glass	23200			Tetrahydrofuran	303.15	−34	59
Liquid	23200			Tetrahydrofuran	303.15	−14	59
Glass	38700			Tetrahydrofuran	303.15	−35	59
Liquid	38700			Tetrahydrofuran	303.15	−14	59
Glass	53500			Tetrahydrofuran	303.15	−39	59
Liquid	53500			Tetrahydrofuran	303.15	−14	59
Glass	66700			Tetrahydrofuran	303.15	−36	59
Liquid	66700			Tetrahydrofuran	303.15	−14	59
Glass	136000			Tetrahydrofuran	303.15	−39	59
Liquid	136000			Tetrahydrofuran	303.15	−14	59
Glass	155400			Tetrahydrofuran	303.15	−39	59

Polymer	$M_n/$ g/mol	$M_w/$ g/mol	$M_\eta/$ g/mol	Solvent	T/K	$\Delta H_B^\infty/$ J/g	Ref.
Liquid	155400			Tetrahydrofuran	303.15	−14	59
Poly(1-vinyl-3,5-dimethyl-1,2,4-triazole)							
				N,N-Dimethylformamide	298.15	−28	94
				Water	298.15	−139	94
Poly(1-vinylimidazole)							
	20700			Acetic acid	298.15	−393	94
	20700			Butanoic acid	298.15	−322	94
	20700			N,N-Dimethylacetamide	298.15	−48	94
	20700			N,N-Dimethylformamide	298.15	−48	94
	20700			1-Methyl-2-pyrrolidinone	298.15	−54	91
	20700			Pentanoic acid	298.15	−325	94
	20700			Propanoic acid	298.15	−278	94
	20700			Water	298.15	−119	91
Poly(1-vinylpyrazole)							
	18900			Acetic acid	298.15	−88	94
	18900			Butanoic acid	298.15	−60	94
	18900			N,N-Dimethylacetamide	298.15	−26	94
	18900			N,N-Dimethylformamide	298.15	−28	94
	18900			Pentanoic acid	298.15	−52	94
	18900			Propanoic acid	298.15	−36	94
Poly(1-vinyl-2-pyrrolidone)							
			32000	Trichloromethane	298.15	−75	48
			32000	Water	298.15	−150	48
Poly(1-vinyl-1,2,4-triazole)							
	69500			Acetic acid	298.15	−85	94
	69500			Butanoic acid	298.15	−74	94
	69500			N,N-Dimethylacetamide	298.15	−49	94
	69500			N,N-Dimethylformamide	298.15	−47	94
	69500			1-Methyl-2-pyrrolidinone	298.15	−55	91
	69500			Pentanoic acid	298.15	−75	94
	69500			Propanoic acid	298.15	−72	94
	69500			Water	298.15	−68	91
Vinyl acetate/vinyl alcohol copolymer							
(9 wt% Vinyl acetate)				2-Propanone	298.15	6.3	33
(44 wt% Vinyl acetate)				2-Propanone	298.15	4.6	33
(57 wt% Vinyl acetate)				2-Propanone	298.15	0.0	33
(67 wt% Vinyl acetate)				2-Propanone	298.15	−1.3	33
(4.2 mol% Vinyl acetate) 7560				Water	303.15	−41	15
(4.3 mol% Vinyl acetate) 64300				Water	303.15	−49	15
(9.0 mol% Vinyl acetate) 66900				Water	303.15	−55	15
(10.3 mol% Vinyl acetate) 7970				Water	303.15	−41	15
(15.3 mol% Vinyl acetate) 8300				Water	303.15	−60	15

Polymer	$M_n/$ g/mol	$M_w/$ g/mol	$M_\eta/$ g/mol	Solvent	T/K	$\Delta H_B^\infty/$ J/g	Ref.
(15.4 mol% Vinyl acetate)							
70700				Water	303.15	−65	15
(19.5 mol% Vinyl acetate)							
73100				Water	303.15	−66	15
(22.1 mol% Vinyl acetate)							
8800				Water	303.15	−60	15
(26.2 mol% Vinyl acetate)							
77000				Water	303.15	−64	15
(30.6 mol% Vinyl acetate)							
9370				Water	303.15	−53	15
(34.0 mol% Vinyl acetate)							
81600				Water	303.15	−60	15
(34.7 mol% Vinyl acetate)							
9670				Water	303.15	−44	15
Vinyl acetate/vinyl chloride copolymer (90 wt% Vinyl chloride)							
Glass	12400	26000		Cyclohexanone	304.15	−37	101
Liquid	12400	26000		Cyclohexanone	304.15	−16	101
Vinyl chloride/vinylidene chloride copolymer							
				Trichloromethane	297.15	−17	2

References

1. Liepatoff, S. and Preobagenskaja, S., *Kolloid Z. Z. Polym.*, 68, 324, 1934.
2. Kargin, V. and Papkov, S., *Acta Physicochim. URSS*, 3, 839, 1935.
3. Papkov, S. and Kargin, V., *Acta Physicochim. URSS*, 7, 667, 1937.
4. Tager, A. and Kargin, V., *Acta Physicochim. URSS*, 14, 713, 1941.
5. Raine, H.C., Richards, R.B., and Ryder, H., *Trans Faraday Soc.*, 41, 56, 1945.
6. Roberts, D.E., Walton, W.W., and Jessup, R.S., *J. Polym. Sci.*, 2, 420, 1947.
7. Tager, A. and Sanatina, V., *Kolloidn. Zhur.*, 12, 474, 1950.
8. Glikman, S.A. and Root, L.A., *Zh. Obshch. Khim.*, 21, 58, 1951.
9. Hellfritz, H., *Makromol. Chem.*, 7, 191, 1951.
10. Tager, A. and Vershkain, R., *Kolloidn. Zhur.*, 13, 123, 1951.
11. Tager, A.A., and Kargin, V.A., *Kolloidn. Zhur.*, 14, 367, 1952.
12. Tager, A.A. and Dombek, Zh.S., *Kolloidn. Zhur.*, 15, 69, 1953.
13. Daoust, H. and Rinfret, M., *Can. J. Chem.*, 32, 492, 1954.
14. Gatovskaya, T.V., Kargin, V.A., and Tager, A.A., *Zh. Fiz. Khim.*, 29, 883, 1955.
15. Oya, S., *Chem. High Polym. Japan*, 12, 122, 1955.
16. Schulz, G.V., Guenner, K. von, and Gerrens, H., *Z. Phys. Chem., N. F.*, 4, 192, 1955.
17. Tager, A.A., Kosova, L.K., Karlinskaya, D.Yu., and Yurina, I.A., *Kolloid. Zhur.*, 17, 315, 1955.
18. Tager, A.A. and Kosova, L.K., *Kolloid. Zhur.*, 17, 391, 1955.
19. Tager, A.A., Krivokorytova, R.V., and Khodorov, P.M., *Dokl. Akad. Nauk SSSR*, 100, 741, 1955.
20. Glikman, S.A. and Root, L.A., *Kolloidn. Zhur.*, 18, 523, 1956.
21. Jenckel, E. and Gorke, K., *Z. Elektrochem.*, 60, 579, 1956.
22. Lipatov, Yu.S., Kargin, V.A., and Slonimskii, G.L., *Zh. Fiz. Khim.*, 30, 1202, 1956
23. Meerson, S.I. and Lipatov, S.M., *Kolloidn.,Zh.*, 18, 447, 1956.
24. Mikhailov, N.V. and Fainberg, E.Z., *Kolloidn. Zhur.*, 18, 44, 1956.
25. Struminskii, G.V. and Slonimskii, G.L., *Zh. Fiz. Khim.*, 30, 1941, 1956.
26. Gal'perin, D.I. and Moseev, L.I., *Kolloidn. Zhur.*, 19, 167, 1957.
27. Akhmedov, K.S., *Uzb. Khim. Zh.* (1), 19, 1958.
28. Kargin, V.A. and Lipatov, Yu.S., *Zh. Fiz. Khim.*, 32, 326, 1958.
29. Meerson, S.I. and Lipatov, S.M., *Kolloidn. Zhur.*, 20, 353, 1958.
30. Tager, A.A. and Galkina, L.A., *Nauchn. Dokl. Vyssh. Shkol., Khim. Khim. Tekhnol.*, (2), 357, 1958.
31. Tager, A. A. and Kargin, V. A., *Zh. Fiz. Khim.*, 32, 1362, 1958.
32. Tager, A. A. and Kargin, V. A., *Zh. Fiz. Khim.*, 32, 2694, 1958.
33. Tager, A.A. and Iovleva, M., *Zh. Fiz. Khim.*, 32, 1774, 1958.
34. Horth, A., Patterson, D., and Rinfret, M., *J. Polym. Sci.*, 39. 189, 1959.

35. Zelikman, S.G. and Mikhailov, N.V., *Vysokomol. Soedin.*, 1, 1077, 1959.
36. Mueller, F.H. and Engelter, A., *Kolloid Z.*, 171, 152, 1960.
37. Delmas, G., Patterson, D., and Boehme, A., *Trans. Faraday Soc.*, 58, 2116, 1962.
38. Delmas, G., Patterson, D., and Somcynsky, T., *J. Polym. Sci.*, 57, 79, 1962.
39. Tager, A.A. and Podlesnyak, A.I., *Vysokomol. Soedin., Ser. A*, 5, 87, 1963.
40. Delmas, G., Patterson, D., and Bhattacharyya, S.N., *J. Phys. Chem.*, 68, 1468, 1964.
41. Patterson, D., *J. Polym. Sci.: Part A*, 2, 5177, 1964.
42. Zverev, M.P., Barash, A.N., and Zubov, P.I., *Vysokomol. Soedin., Ser. A*, 6, 1012, 1964.
43. Cottam, B.J., Cowie, J.M.G., and Bywater, S., *Makromol. Chem.*, 86, 116, 1965.
44. Bianchi, U., Pedemonte, E., and Rossi, C., *Makromol. Chem.*, 92, 114, 1966.
45. Cuniberti, C. and Bianchi, U., *Polymer*, 7, 151, 1966.
46. Giacommeti, G. and Turolla, A., *Z. Phys. Chem., N.F.*, 51, 108, 1966.
47. Schreiber, H.P. and Waldman, M.H., *J. Polym. Sci.: Part A-2*, 5, 555, 1967.
48. Goldfarb, J. and Rodriguez, S., *Makromol. Chem.*, 116, 96, 1968.
49. Maron, S.H. and Daniels, C.A., *J. Macromol. Sci.-Phys. B*, 2, 769, 1968.
50. Morimoto, S., *J. Polym. Sci.: Part A-1*, 6, 1547, 1968.
51. Bianchi, U., Cuniberti, C., Pedemonte, E., and Rossi, C., *J. Polym. Sci.: Part A-2*, 7, 855, 1969.
52. Gerth, Ch. and Mueller, F.H., *Kolloid-Z. Z. Polym.*, 241, 1071, 1970.
53. Liddell, A.H. and Swinton, F.L., *Discuss. Faraday Soc.*, 49, 115, 1970.
54. Morimoto, S., *Nippon Kagaku Zasshi*, 91, 31, 1970.
55. Nakayama, H., *Bull. Chem. Soc. Japan*, 43, 1683, 1970.

56. Blackadder, D.A. and Roberts, T.L., *Angew. Makromol. Chem.*, 27, 165, 1972.
57. Maron, S.H. and Filisko, F.E., *J. Macromol. Sci.-Phys. B*, 6, 57, 1972.
58. Maron, S.H. and Filisko, F.E., *J. Macromol. Sci.-Phys. B*, 6, 79, 1972.
59. Maron, S.H. and Filisko, F.E., *J. Macromol. Sci.-Phys. B*, 6, 413, 1972.
60. Sokolova, D.F., Sokolov, L.B., and Gerasimov, V.D., *Vysokomol. Soedin., Ser. B*, 14, 580, 1972.
61. Chahal, R.S., Kao, W.-P., and Patterson, D., *J. Chem. Soc., Faraday Trans. I*, 69, 1834, 1973.
62. Delmas, G. and Tancrede, P., *Eur. Polym. J.*, 9, 199, 1973.
63. Filisko, F.E., Raghava, R.S., and Yeh, G.S.Y., *J. Macromol. Sci.-Phys. B*, 10, 371, 1974.
64. Ikeda, M., Suga, H., and Seki, S., *Polymer*, 16, 634, 1975.
65. Kiselev, V.P., Shakhova, E.M., Fainberg, E.Z., Virnik, A.D, and Rogovin, Z.A., *Vysokomol. Soedin., Ser. B*, 18, 847, 1976.
66. Petrosyan, V.A., Gabrielyan, G.A., and Rogovin, Z.A., *Arm. Khim. Zhur.*, 29, 516, 1976.
67. Deshpande, D.D. and Prabhu, C.S., *Macromolecules*, 10, 433, 1977.
68. Miura, T. and Nakamura, M., *Bull. Chem. Soc. Japan*, 50, 2528, 1977.
69. Sokolova, D.F., Kudim, T.V., Sokolov, L.B., Zhegalova, N.I., and Zhuravlev, N.D., *Vysokomol. Soedin., Ser. B*, 20, 596, 1978.
70 Basedow, A.M. and Ebert, K.H., *J. Polym. Sci.: Polym. Symp.*, 66, 101, 1979.
71 Koller, J., *Dissertation*, TU München, 1979.
72. Lee, J.-O., Ono, M., Hamada, F., and Nakajima, A., *Polym. Bull.*, 1, 763, 1979.
73. Phuong-Nguyen, H. and Delmas, G., *Macromolecules*, 12, 740, 1979.
74. Phuong-Nguyen, H. and Delmas, G., *Macromolecules*, 12, 746, 1979.
75. Basedow, A.M., Ebert, K.H., and Feigenbutz, W., *Makromol. Chem.*, 181, 1071, 1980.
76. Graun, K., *Dissertation*, TU München, 1980.
77. Shiomi, T., Izumi, Z., Hamada, F., and Nakajima, A., *Macromolecules*, 13, 1149, 1980.
78. Shiomi, T., Kohra, Y., Hamada, F., and Nakajima, A., *Macromolecules*, 13, 1154, 1980.
79. Aharoni, S.M., Charlet, G., and Delmas, G., *Macromolecules*, 14, 1390, 1981.
80. Ochiai, H., Ohashi, T., Tadokoro, Y., and Murakami, I., *Polym. J.*, 14, 457, 1982.
81. Sharma, S.C., Mahajan, R., Sharma, V.K., and Lakhanpal, M.L., *Indian J. Chem.*, 21A, 682, 1982.
82. Sharma, S.C., Mahajan, R., Sharma, V.K., and Lakhanpal, M.L., *Indian J. Chem.*, 21A, 685, 1982.
83. Lakhanpal, M.L. and Parashar, R.N., *Indian J. Chem.*, 22A, 48, 1983.
84. Daoust, H. and St-Cyr, D., *Macromolecules*, 17, 596, 1984.
85. Aeleni, N., *Mater. Plast. (Bucharest)*, 22, 92, 1985.
86. Sharma, S.C. and Sharma, V.K., *Indian J. Chem.*, 24A, 292, 1985.

87. Sharma, S.C., Bhalla, S., and Sharma, V.K., *Indian J. Chem.*, 25A, 131, 1986.
88. Sharma, S.C., Syngal, M., and Sharma, V.K., *Indian J. Chem.*, 26A, 285, 1987.
89. Aukett, P.N. and Brown, C.S., *J. Therm. Anal.*, 33, 1079, 1988.
90. Lanzavecchia, L. and Pedemonte, E., *Thermochim. Acta*, 137, 123, 1988.
91. Tager, A.A., Safronov, A.P., Voit, V.V., Lopyrev, V.A., Ermakova, T.G., Tatarova, L.A., and Shagelaeva, N.S., *Vysokomol. Soedin., Ser. A*, 30, 2360, 1988.
92. Pedemonte, E. and Lanzavecchia, L., *Thermochim. Acta*, 162, 223, 1990.
93. Shiomi, T., Ishimatsu, H., Eguchi, T., and Imai, K., *Macromolecules*, 23, 4970, 1990.
94. Tager, A.A. and Safronov, A.P., *Vysokomol. Soedin., Ser. A*, 33, 67, 1991.
95. Zellner, H., *Dissertation*, TU München, 1993.
96. Brunacci, A., Pedemonte, E., Cowie, J.M.G., and McEwen, I. *J., Polymer*, 35, 2893, 1994.
97. Carlsson, M., Hallen, D., and Linse, P., *J. Chem. Soc. Faraday Trans.*, 91, 2081, 1995.
98. Sato, T., Tohyama, M., Suzuki, M., Shiomi, T., and Imai, K., *Macromolecules*, 29, 8231, 1996.
99. Shiomi, T., Tohyama, M., Endo, M., Sato, T., and Imai, K., *J. Polym. Sci.: Part B: Polym. Phys.*, 34, 2599, 1996.
100. Kizhnyaev, V.N., Gorkovenko, O.P., Bazhenov, D.N., and Smirnov, A.I., Vysokomol. Soedin., Ser. A, 39, 856, 1997.
101. Sato, T., Suzuki, M., Tohyama, M., Endo, M., Shiomi, T., and Imai, K., *Polym. J.*, 29, 417, 1997.
102. Vanderryn, J. and Zettlemoyer, A.C., Ind. Eng. Chem., *Chem. Eng. Data Ser.*, 2, 56, 1957.
103. Parashar, R. and Sharma, S.C., Indian *J. Chem.*, 27A, 1092, 1988.
104. Righetti, M.C., Cardelli, C., Scalari, M., Tombari, E., and Conti, G., *Polymer*, 43, 5035, 2002.
105. Phuong-Nguyen, H. and Delmas, G., *J. Solution Chem.*, 23, 249, 1994.
106. Wohlfarth, C., *CRC Handbook of Enthalpy Data of Polymer-Solvent Systems*, CRC Press, Boca Raton, 2006.

Section 14
Geophysics, Astronomy, and Acoustics

ASTRONOMICAL CONSTANTS

Victor Abalakin

The constants in this table are based primarily on the set of constants adopted by the International Astronomical Union (IAU) in 1976. Updates have been made when new data were available. All values are given in SI Units; thus masses are expressed in kilograms and distances in meters.

The astronomical unit of time is a time interval of one day (1 d) equal to 86400 s. An interval of 36525 d is one Julian century (1 cy).

References

1. Seidelmann, P. K., *Explanatory Supplement to the Astronomical Almanac*, University Science Books, Mill Valley, CA, 1990.
2. Lang, K. R., *Astrophysical Data: Planets and Stars*, Springer-Verlag, New York, 1992.
3. *The Astronomical Almanac for the Year 2005*, U.S. Government Printing Office, Washington, and Her Majesty's Stationary Office, London (2003).

Defining constants

Gaussian gravitational constant	$k = 0.01720209895$ m^3 kg^{-1} s^{-2}
Speed of light	$c = 299792458$ m s^{-1}

Primary constants

Light-time for unit distance (1 ua)	$\tau_A = 499.004782$ s
Equatorial radius of earth	$a_e = 6378140$ m
Equatorial radius of earth (IUGG value)	$a_e = 6378136$ m
Dynamical form-factor for earth	$J_2 = 0.001082626$
Geocentric gravitational constant	$GE = 3.986005 \times 10^{14}$ m^{3}s^{-2}
Constant of gravitation	$G = 6.6742 \times 10^{-11}$ m^{3}kg^{-1}s^{-2}
Ratio of mass of moon to that of earth	$\mu = 0.01230002$
	$1/\mu = 81.300587$
General precession in longitude, per Julian century, at standard epoch J2000	$\rho = 5029''.0966$
Obliquity of the ecliptic at standard epoch J2000	$\varepsilon = 23°26'21''.448$

Derived constants

Constant of nutation at standard epoch J2000	$N = 9''.2025$
Unit distance (ua = $c\tau_A$)	ua $= 1.49597870660 \times 10^{11}$ m
Solar parallax ($\pi_0 = \arcsin(a_e/$ua$)$)	$\pi_0 = 8''.794148$
Constant of aberration for standard epoch J2000	$\kappa = 20''.49552$
Flattening factor for the earth	$f = 1/298.257 = 0.00335281$
Heliocentric gravitational constant ($GS = A^3k^2/D^2$)	$GS = 1.32712438 \times 10^{20}$ m^3 s^{-2}
Ratio of mass of sun to that of the earth (S/E) = (GS)/(GE))	$S/E = 332946.0$
Ratio of mass of sun to that of earth + moon	$(S/E)/(1 + \mu) = 328900.5$
Mass of the sun ($S = (GS)/G$)	$S = 1.98844 \times 10^{30}$ kg

Ratios of mass of sun to masses of the planets

Mercury	6023600
Venus	408523.5
Earth + moon	328900.5
Mars	3098710
Jupiter	1047.355
Saturn	3498.5
Uranus	22869
Neptune	19314
Pluto	3000000

PROPERTIES OF THE SOLAR SYSTEM

The following tables give various properties of the planets and characteristics of their orbits in the solar system. Certain properties of the sun and of the earth's moon are also included.

Explanations of the column headings:

- *Den.*: mean density in g/cm^3
- *Radius*: radius at the equator in km
- *Flattening*: degree of oblateness, defined as $(r_e - r_p)/r_e$, where r_e and r_p are the equatorial and polar radii, respectively
- *Potential coefficients*: coefficients in the spherical harmonic representation of the gravitational potential U by the equation

$$U(r,\phi) = (GM/r)\,[1 - \Sigma\, J_n(a/r)^n\, P_n(\sin \phi)]$$

where G is the gravitational constant, r the distance from the center of the planet, a the radius of the planet, M the mass, ϕ the latitude, and P_n the Legendre polynomial of degree n.

- *Gravity*: acceleration due to gravity at the surface
- *Escape velocity*: velocity needed at the surface of the planet to escape the gravitational pull
- *Dist. to sun*: semi-major axis of the elliptical orbit (1 ua = 1.496×10^8 km)
- *ε*: eccentricity of the orbit
- *Ecliptic angle*: angle between the planetary orbit and the plane of the earth's orbit around the sun
- *Inclin.*: angle between the equatorial plane and the plane of the planetary orbit
- *Rot. period*: period of rotation of the planet measured in earth days
- *Albedo*: ratio of the light reflected from the planet to the light incident on it
- T_{sur}: mean temperature at the surface
- P_{sur}: pressure of the atmosphere at the surface

The following general information on the solar system is of interest:

Mass of the earth = M_e = 5.9742×10^{24} kg
Total mass of planetary system = 2.669×10^{27} kg = $447\ M_e$
Total angular momentum of planetary system = 3.148×10^{43} kg m^2/s
Total kinetic energy of the planets = 1.99×10^{35} J
Total rotational energy of planets = 0.7×10^{35} J

Properties of the sun:

Mass = 1.9891×10^{30} kg = $332946.0\ M_e$
Radius = 6.9599×10^8 m
Surface area = 6.087×10^{18} m^2
Volume = 1.412×10^{27} m^3
Mean density = 1.409 g/cm^3
Gravity at surface = 27398 cm/s^2
Escape velocity at surface = 6.177×10^5 m/s
Effective temperature = 5780 K
Total radiant power emitted (luminosity) = 3.86×10^{26} W
Surface flux of radiant energy = 6.340×10^7 W/m^2
Flux of radiant energy at the earth (Solar Constant) = 1373 W/m^2

References

1. Seidelmann, P. K., Ed., *Explanatory Supplement to the Astronomical Almanac*, University Science Books, Mill Valley, CA, 1992.
2. Lang, K. R., *Astrophysical Data: Planets and Stars*, Springer-Verlag, New York, 1992.
3. Allen, C. W., *Astrophysical Quantities, Third Edition*, Athlone Press, London, 1977.

Planet	Mass 10^{24} kg	Den. g/cm^3	Radius km	Flattening	Potential coeffients $10^3\, J_2$	$10^6\, J_3$	$10^6\, J_4$	Gravity cm/s^2	Escape vel. km/s
Mercury	0.33022	5.43	2439.7	0				370	4.25
Venus	4.8690	5.24	6051.9	0	0.027			887	10.4
Earth	5.9742	5.515	6378.140	0.00335364	1.08263	−2.54	−1.61	980	11.2
(Moon)	0.073483	3.34	1738	0	0.2027			162	2.37
Mars	0.64191	3.94	3397	0.00647630	1.964	36		371	5.02
Jupiter	1898.8	1.33	71492	0.0648744	14.75	−580		2312	59.6
Saturn	568.50	0.70	60268	0.0979624	16.45	−1000		896	35.5
Uranus	86.625	1.30	25559	0.0229273	12			777	21.3
Neptune	102.78	1.76	24764	0.0171	4			1100	23.3
Pluto	0.015	1.1	1151	0				72	1.1

Planet	Dist. to sun ua	ε	Ecliptic angle	Inclin.	Rot. period d	Albedo	No. of satellites
Mercury	0.38710	0.2056	7.00°	0°	58.6462	0.106	0
Venus	0.72333	0.0068	3.39°	177.3°	−243.01	0.65	0
Earth	1.00000	0.0167		23.45°	0.99726968	0.367	1
(Moon)				6.68°	27.321661	0.12	
Mars	1.52369	0.0933	1.85°	25.19°	1.02595675	0.150	2
Jupiter	5.20283	0.048	1.31°	3.12°	0.41354	0.52	16
Saturn	9.53876	0.056	2.49°	26.73°	0.4375	0.47	18
Uranus	19.19139	0.046	0.77°	97.86°	−0.65	0.51	15
Neptune	30.06107	0.010	1.77°	29.56°	0.768	0.41	8
Pluto	39.52940	0.248	17.15°	118°	−6.3867	0.3	1

Planet	T_{sur} K	P_{sur} bar	Atmospheric composition								
			CO_2	N_2	O_2	H_2O	H_2	He	Ar	Ne	CO
Mercury	440	2×10^{-15}					2%	98%			
Venus	730	90	96.4%	3.4%	69 ppm	0.1%			4 ppm		20 ppm
Earth	288	1	0.03%	78.08%	20.95%	0 to 3%			0.93%	18 ppm	1 ppm
Mars	218	0.007	95.32%	2.7%	0.13%	0.03%			1.6%	3 ppm	0.07%
Jupiter	129						86.1%	13.8%			
Saturn	97						92.4%	7.4%			
Uranus	58						89%	11%			
Neptune	56						89%	11%			
Pluto	50	1×10^{-5}									

SATELLITES OF THE PLANETS

This table gives characteristics of the known satellites of the planets. The parameters covered are:

- Orbital period in units of earth days. An R following the value indicates a retrograde motion.
- Distance from the planet, as measured by the semi-major axis of the orbit.
- Eccentricity of the orbit.
- Inclination of the satellite orbit with respect to the equator of the planet.
- Mass of the satellite relative to the planet.
- Radius of the satellite in km.
- Mean density of the satellite.
- Geometric albedo, which is a measure of the fraction of incident sunlight reflected by the satellite.

References

1. Seidelmann, P. K., Ed., *Explanatory Supplement to the Astronomical Almanac*, University Science Books, Mill Valley, CA, 1992.
2. Lang, K. R., *Astrophysical Data: Planets and Stars*, Springer-Verlag, New York, 1992.
3. Burns, J. A., and Matthews, M. S., Eds., *Satellites*, University of Arizona Press, Tucson, 1986.

Planet		Satellite	Orb. Period d	Distance 10^3 km	Eccentricity	Inclination	Rel. mass	Radius km	Den. g/cm³	Albedo
Earth		Moon	27.321661	384.400	0.054900489	18.28–28.58°	0.01230002	1738	3.34	0.12
Mars	I	Phobos	0.31891023	9.378	0.015	1.0°	1.5×10^{-8}	13.5 × 10.8 × 9.4	<2	0.06
	II	Deimos	1.2624407	23.459	0.0005	0.9–2.7°	3×10^{-9}	7.5 × 6.1 × 5.5	<2	0.07
Jupiter	I	Io	1.769137786	422	0.004	0.04°	4.68×10^{-5}	1815	3.55	0.61
	II	Europa	3.551181041	671	0.009	0.47°	2.52×10^{-5}	1569	3.04	0.64
	III	Ganymede	7.15455296	1070	0.002	0.21°	7.80×10^{-5}	2631	1.93	0.42
	IV	Callisto	16.6890184	1883	0.007	0.51°	5.66×10^{-5}	2400	1.83	0.20
	V	Amalthea	0.49817905	181	0.003	0.40°	3.8×10^{-9}	135 × 83 × 75		0.05
	VI	Himalia	250.5662	11480	0.15798	27.63°	5.0×10^{-9}	93		0.03
	VII	Elara	259.6528	11737	0.20719	24.77°	4×10^{-10}	38		0.03
	VIII	Pasiphae	735 R	23500	0.378	145°	1×10^{-10}	25		
	IX	Sinope	758 R	23700	0.275	153°	0.4×10^{-10}	18		
	X	Lysithea	259.22	11720	0.107	29.02°	0.4×10^{-10}	18		
	XI	Carme	692 R	22600	0.20678	164°	0.5×10^{-10}	20		
	XII	Ananke	631 R	21200	0.16870	147°	0.2×10^{-10}	15		
	XIII	Leda	238.72	11094	0.14762	26.07°	0.03×10^{-10}	8		
	XIV	Thebe	0.6745	222	0.015	0.8°	4×10^{-10}	55 × 45		0.05
	XV	Adrastea	0.29826	129			0.1×10^{-10}	12.5 × 10 × 7.5		0.05
	XVI	Metis	0.294780	128			0.5×10^{-10}	20		0.05
Saturn	I	Mimas	0.942421813	185.52	0.0202	1.53°	8.0×10^{-8}	196	1.44	0.5
	II	Enceladus	1.370217855	238.02	0.00452	1.86°	1.3×10^{-7}	250	1.13	1.0
	III	Tethys	1.887802160	294.66	0.00000	1.86°	1.3×10^{-6}	530	1.20	0.9
	IV	Dione	2.736914742	377.40	0.002230	0.02°	1.85×10^{-6}	560	1.41	0.7
	V	Rhea	4.517500436	527.04	0.00100	0.35°	4.4×10^{-6}	765	1.33	0.7
	VI	Titan	15.94542068	1221.83	0.029192	0.33°	2.38×10^{-4}	2575	1.88	0.21
	VII	Hyperion	21.2766088	1481.1	0.104	0.43°	3×10^{-8}	205 × 130 × 110		0.3
	VIII	Iapetus	79.3301825	3561.3	0.02828	14.72°	3.3×10^{-6}	730	1.15	0.2
	IX	Phoebe	550.48 R	12952	0.16326	177°	7×10^{-10}	110		0.06
	X	Janus	0.6945	151.472	0.007	0.14°		110 × 100 × 80		0.8
	XI	Epimetheus	0.6942	151.422	0.009	0.34°		70 × 60 × 50		0.8
	XII	Helene	2.7369	377.40	0.005	0.0°		18 × 16 × 15		0.7
	XIII	Telesto	1.8878	294.66				17 × 14 × 13		0.5
	XIV	Calypso	1.8878	294.66				17 × 11 × 11		0.6
	XV	Atlas	0.6019	137.670	0.000	0.3°		20 × 10		0.9
	XVI	Prometheus	0.6130	139.353	0.003	0.0°		70 × 50 × 40		0.6
	XVII	Pandora	0.6285	141.700	0.004	0.0°		55 × 45 × 35		0.9
	XVIII	Pan	0.5750	133.583				10		0.5
Uranus	I	Ariel	2.52037935	191.02	0.0034	0.3°	1.56×10^{-5}	579	1.55	0.34
	II	Umbriel	4.1441772	266.30	0.0050	0.36°	1.35×10^{-5}	586	1.58	0.18
	III	Titania	8.7058717	435.91	0.0022	0.14°	4.06×10^{-5}	790	1.69	0.27
	IV	Oberon	13.4632389	583.52	0.0008	0.10°	3.47×10^{-5}	762	1.64	0.24

Planet		Satellite	Orb. Period d	Distance 10³ km	Eccentricity	Inclination	Rel. mass	Radius km	Den. g/cm³	Albedo
	V	Miranda	1.41347925	129.39	0.0027	4.2°	0.08×10^{-5}	240	1.25	0.27
	VI	Cordelia	0.335033	49.77	<0.001	0.1°		13		0.07
	VII	Ophelia	0.376409	53.79	0.010	0.1°		15		0.07
	VIII	Bianca	0.434577	59.17	<0.001	0.2°		21		0.07
	IX	Cressida	0.463570	61.78	<0.001	0.0°		31		0.07
	X	Desdemona	0.473651	62.68	<0.001	0.2°		27		0.07
	XI	Juliet	0.493066	64.35	<0.001	0.1°		42		0.07
	XII	Portia	0.513196	66.09	<0.001	0.1°		54		0.07
	XIII	Rosalind	0.558459	69.94	<0.001	0.3°		27		0.07
	XIV	Belinda	0.623525	75.26	<0.001	0.0°		33		0.07
	XV	Puck	0.761832	86.01	<0.001	0.31°		77		0.07
Neptune	I	Triton	5.8768541 R	354.76	0.000016	157.345°	2.09×10^{-4}	1353	2.05	0.7
	II	Nereid	360.13619	5513.4	0.7512	27.6°	2×10^{-7}	170		0.4
	III	Naiad	0.294396	117.6	<0.001	4.74°		29		0.06
	IV	Thalassa	0.311485	73.6	<0.001	0.21°		40		0.06
	V	Despina	0.334655	52.6	<0.001	0.07°		74		0.06
	VI	Galatea	0.428745	62.0	<0.001	0.05°		79		0.06
	VII	Larissa	0.554654	50.0	<0.0014	0.20°		104×89		0.06
	VIII	Proteus	1.122315	48.2	<0.001	0.55°		$218 \times 208 \times 201$		0.06
Pluto	I	Charon	6.38725	19.6	<0.001	99°	0.22	593		0.5

INTERSTELLAR MOLECULES

Frank J. Lovas and Lewis E. Snyder

A number of molecules have been detected in the interstellar medium, in circumstellar envelopes around evolved stars, and comae and tails of comets through observation of their microwave, infrared, or optical spectra. The following list gives the molecules and the particular isotopic species that have been reported thus far. Molecules are listed by molecular formula in the Hill order. All species not footnoted otherwise are observed in interstellar clouds, while some are also found in comets and circumstellar clouds. The list was last updated in November 2005 and lists 147 molecules (263 isotopic forms).

References

1. Lovas, F. J., "Recommended Rest Frequencies for Observed Interstellar Molecule Microwave Transitions - 1991 Revision," *J. Phys. Chem. Ref. Data*, 21, 181–272, 1992.
2. Snyder, L. E., "Cometary Molecules," Internat. Astron. Union Symposium No. 150, *Astrochemistry of Cosmic Phenomena*, Ed. P.D. Singh, Kluwer Academic Publishers, Dordrecht, The Netherlands, pp. 427–434 (1992).

Molecular formula	Name	Isotopic species
AlCl	Aluminum monochloride	AlCl[a]
		Al^{37}Cl[a]
AlF	Aluminum monofluoride	AlF[a]
CAlN	Aluminum isocyanide	AlNC[a]
CF$^+$	Flouromethylidynium ion	CF$^+$
CH	Methylidyne	CH
CH$^+$	Methyliumylidene	CH$^+$
CHN	Hydrogen cyanide	HCN
		H^{13}CN
		HC^{15}N
		DCN
CHN	Hydrogen isocyanide	HNC
		H^{15}NC
		HN^{13}C
		DNC
		D^{15}NC
CHNO	Isocyanic acid	HNCO
		DNCO
CHNS	Isothiocyanic acid	HNCS
CHO	Oxomethyl	HCO
CHO$^+$	Oxomethylium	HCO$^+$
		H^{13}CO$^+$
		HC^{17}O$^+$
		HC^{18}O$^+$
		DCO$^+$
		D^{13}CO$^+$
CHO$^+$	Hydroxymethylidyne	HOC$^+$
CHO$_2$$^+$	Hydroxyoxomethylium	HOCO$^+$
CHS$^+$	Thiooxomethylium	HCS$^+$
CH$_2$	Methylene	CH$_2$
CH$_2$N$^+$	Iminomethylium	HCNH$^+$
CH$_2$N	Methylene amidogen	CH$_2$N
CH$_2$N$_2$	Cyanamide	NH$_2$CN
CH$_2$O	Formaldehyde	H$_2$CO
		H$_2$^{13}CO
		H$_2$C^{18}O
		HDCO
		D$_2$CO
CH$_2$O$_2$	Formic acid	HCOOH
		H^{13}COOH
		HCOOD
		DCOOH
CH$_2$S	Thioformaldehyde	H$_2$CS
		H$_2$^{13}CS

Molecular formula	Name	Isotopic species
		H$_2$C^{34}S
		HDCS
		D$_2$CS
CH$_3$	Methyl	CH$_3$ [a]
CH$_3$N	Methanimine	CH$_2$NH
		^{13}CH$_2$NH
CH$_3$NO	Formamide	NH$_2$CHO
		NH$_2$^{13}CHO
CH$_3$O$^+$	Hydroxy methylium ion	H$_2$COH$^+$
CH$_4$	Methane	CH$_4$
CH$_4$O	Methanol	CH$_3$OH
		^{13}CH$_3$OH
		CH$_3$^{18}OH
		CH$_2$DOH
		CH$_3$OD
		CHD$_2$OH
		CD$_3$OH
CH$_4$S	Methanethiol	CH$_3$SH
CH$_5$N	Methylamine	CH$_3$NH$_2$
CMgN	Magnesium cyanide	MgCN[a]
CMgN	Magnesium isocyanide	^{24}MgNC[a]
		^{25}MgNC[a]
		^{26}MgNC[a]
CN	Cyanide radical	CN
		^{13}CN
		C^{15}N
CN$^+$	Cyanide radical ion	CN$^{+ b}$
CNNa	Sodium cyanide	NaCN[a]
CNSi	Silicon cyanide	SiNC[a]
CNSi	Silicon isocyanide	SiNC[a]
CN$_2$	Cyanoimidogen	NCN[b]
CO	Carbon monoxide	CO
		^{13}CO
		C^{17}O
		C^{18}O
		^{13}C^{18}O
CO$^+$	Carbon monoxide ion	CO$^+$
COS	Carbon oxysulfide	OCS
		OC^{34}S
		O^{13}CS
		^{18}OCS
CO$_2$	Carbon dioxide	CO$_2$
CO$_2$$^+$	Carbon dioxide ion	CO$_2$$^{+ b}$
CP	Carbon phosphide	CP[a]

Molecular formula	Name	Isotopic species	Molecular formula	Name	Isotopic species
CS	Carbon monosulfide	CS	C_3H_3N	Acrylonitrile (vinyl cyanide)	CH_2CHCN
		$C^{33}S$	C_3H_4	Propyne	CH_3CCH
		$C^{34}S$			$CH_3C^{13}CH$
		$C^{36}S$			$^{13}CH_3CCH$
		^{13}CS			CH_2DCCH
		$^{13}C^{34}S$			CH_3CCD
CSi	Silicon carbide	SiC^a	C_3H_4O	Propenal	CH_2CHCHO
C_2	Dicarbon	C_2	C_3H_5N	Propanenitrile (ethyl cyanide)	CH_3CH_2CN
C_2H	Ethynyl	C_2H			
		^{13}CCH	C_3H_6O	Acetone	$(CH_3)_2CO$
		$C^{13}CH$	C_3H_6O	Propanal	CH_3CH_2CHO
		C_2D	C_3N	Cyanoethynyl	$CCCN$
C_2HN	Cyanomethylene	$HCCN$	C_3O	1,2-Propadienylidene, 3-oxo	$CCCO$
C_2H_2	Acetylene	$HCCH$	C_3S	1,2-Propadienylidene, 3-thioxo	$CCCS$
C_2H_2N	Cyanomethyl	CH_2CN			
C_2H_2O	Ketene	H_2CCO	C_3Si	Silicon tricarbon	SiC_3
C_2H_3N	Acetonitrile	CH_3CN	C_4H	1,3-Butadiynyl radical	$HCCCC$
		$^{13}CH_3CN$			$H^{13}CCCC$
		$CH_3^{13}CN$			$HC^{13}CCC$
		$CH_3C^{15}N$			$HCC^{13}CC$
		CH_2DCN			$HCCC^{13}C$
C_2H_3N	Isocyanomethane	CH_3NC			$DCCCC$
C_2H_4	Ethylene	H_2CCH_2	C_4HN	3-Cyano-1,2-propadienylidene	$HCCCN$
C_2H_4O	Acetaldehyde	CH_3CHO			
C_2H_4O	Ethylene oxide	$c\text{-}C_2H_4O^c$	C_4H_2	Butatrienylidene	H_2CCCC
C_2H_4O	Ethenol	CH_2CHOH	C_4H_2	1,3-Butadiyne	$HCCCCH^a$
$C_2H_4O_2$	Methyl formate	CH_3OCHO	C_4H_3N	2-Butynenitrile	CH_3CCCN
$C_2H_4O_2$	Acetic acid	CH_3COOH	C_4H_3N	Cyanoallene	CH_2CCHCN
$C_2H_4O_2$	Glycolaldehyde	CH_2OHCHO	C_4Si	Silicon tetracarbide	SiC_4^a
C_2H_6	Ethane	$CH_3CH_3^b$	C_5	Pentacarbon	C_5^a
C_2H_6O	*trans*-Ethanol	$t\text{-}CH_3CH_2OH$	C_5H	2,4-Pentadiynylidyne	$HCCCCC$
C_2H_6O	*gauche*-Ethanol	$g\text{-}CH_3CH_2OH$	C_5HN	2,4-Pentadiynenitrile	$HCCCCCN$
C_2H_6O	Dimethyl ether	CH_3OCH_3			$H^{13}CCCCCN$
$C_2H_6O_2$	Ethylene glycol	$HOCH_2CH_2OH$			$HC^{13}CCCCN$
C_2O	Oxoethenylidene	CCO			$HCC^{13}CCCN$
C_2S	Thioxoethenylidene	CCS			$HCCC^{13}CCN$
		$CC^{34}S$			$HCCCC^{13}CN$
C_2Si	Silicon dicarbide	$c\text{-}SiC_2$			$DCCCCCN$
		$c\text{-}^{29}SiC_2$	C_5H_4	1,3-Pentadiyne	CH_3C_4H
		$c\text{-}^{30}SiC_2$	C_5N	1,3-Butadiynylium, 4-cyano	C_5N
		$c\text{-}Si^{13}CC$	C_6H	1,3,5-Hexatriynyl	$HCCCCCC$
C_3	Tricarbon	C_3	C_6H_2	1,3,5-Hexatriyne	$HCCCCCCH^a$
C_3H	Cyclopropenylidyne	$c\text{-}C_3H$	C_6H_2	1,2,3,4,5-Hexapentaenylidene	$H_2CCCCCC$
C_3H	Propenylidyne	$l\text{-}C_3H$			
C_3HN	Cyanoacetylene	$HCCCN$	C_6H_3N	Methylcyanodiacetylene	CH_3C_4CN
		$H^{13}CCCN$	C_6H_6	Benzene	C_6H_6
		$HC^{13}CCN$	C_7H	2,4,6-Heptatriynylidyne	$HCCCCCCC$
		$HCC^{13}CN$	C_7HN	2,4,6-Heptatriynenitrile	HC_7N
		$HCCC^{15}N$	C_8H	1,3,5,7-Octatetraynyl	HC_8
		$DCCCN$	C_9HN	2,4,6,8-Nonatetraynenitrile	HC_9N
C_3HN	Isocyanoacetylene	$HCCNC$	$C_{11}HN$	2,4,6,8,10-Undecapentaynenitrile	$HC_{11}N$
C_3HN	1,2-Propadienylidene, 3-imino	$HNCCC$			
			ClH	Hydrogen chloride	$H^{35}Cl$
C_3H_2	Cyclopropenylidene	$c\text{-}C_3H_2$			$H^{37}Cl$
		$c\text{-}H^{13}CCCH$	ClK	Potassium chloride	$K^{35}Cl^a$
		$c\text{-}HC^{13}CCH$			$K^{37}Cl^a$
		$c\text{-}C_3HD$	$ClNa$	Sodium chloride	$Na^{35}Cl^a$
C_3H_2	Propadienylidene	$l\text{-}H_2CCC$			$Na^{37}Cl^a$
$C_3H_2N^+$	Protonated cyanoacetylene	$HCCCNH^+$	FH	Hydrogen fluoride	HF
C_3H_2O	2-Propynal	$HCCCHO$	FeO	Iron monoxide	FeO
C_3H_2O	Cyclopropenone	$c\text{-}C_3H_2O$	HLi	Lithium hydride	^7LiH

Molecular formula	Name	Isotopic species	Molecular formula	Name	Isotopic species
HN	Imidogen	HN	H_4Si	Silane	SiH_4 [a]
HNO	Nitrosyl hydride	HNO	NO	Nitric oxide	NO
HN_2^+	Hydrodinitrogen(1+)	N_2H^+	NP	Phosphorus nitride	NP
		$^{15}NNH^+$	NS	Nitrogen sulfide	NS
		$N^{15}NH^+$			$N^{34}S$
		N_2D^+	NSi	Silicon nitride	SiN
HO	Hydroxyl	OH	N_2	Nitrogen	N_2
		^{17}OH	N_2^+	Nitrogen ion	N_2^+ [b]
		^{18}OH	N_2O	Nitrous oxide	N_2O
HO^+	Oxoniumylidene	HO^+ [b]	OS	Sulfur monoxide	SO
HS	Mercapto	SH			^{34}SO
H_2	Hydrogen	H_2			^{33}SO
H_2N	Amidogen	NH_2			$S^{18}O$
H_2O	Water	H_2O	OS^+	Sulfur monoxide ion	SO^+
		$H_2^{18}O$	OSi	Silicon monoxide	SiO
		HDO			$Si^{18}O$
H_2O^+	Oxoniumyl	H_2O^+ [b]			^{29}SiO
H_2S	Hydrogen sulfide	H_2S			^{30}SiO
		$H_2^{34}S$	O_2S	Sulfur dioxide	SO_2
		HDS			$^{33}SO_2$
		D_2S			$^{34}SO_2$
H_3^+	Trihydrogen ion	H_3^+			$OS^{18}O$
		H_2D^+	SSi	Silicon monosulfide	SiS
		D_2H^+			$Si^{33}S$
H_3N	Ammonia	NH_3			$Si^{34}S$
		$^{15}NH_3$			^{29}SiS
		NH_2D			^{30}SiS
		NHD_2			$Si^{36}S$
		ND_3	S_2	Disulfur	S_2 [b]
H_3O^+	Oxonium hydride	H_3O^+			

l- before the isotopic species indicates a linear configuration, while *c-* indicates a cyclic molecule.

[a] Reported only in circumstellar clouds.

[b] Reported only in comets.

MASS, DIMENSIONS, AND OTHER PARAMETERS OF THE EARTH

This table is a collection of data on various properties of the Earth. Most of the values are given in SI units. Note that 1 ua (astronomical unit) = 149,597,870 km.

References

1. Seidelmann, P. K., Ed., *Explanatory Supplement to the Astronomical Almanac*, University Science Books, Mill Valley, CA, 1992.
2. Lang, K. R., *Astrophysical Data: Planets and Stars*, Springer-Verlag, New York, 1992.

Quantity	Symbol	Value	Unit
Mass	M	$5.9723 \cdot 10^{27}$	g
Major orbital semi-axis	a_{orb}	1.000000	ua
		$1.4959787 \cdot 10^{8}$	km
Distance from sun at perihelion	r_{π}	0.9833	ua
Distance from sun at aphelion	r_{α}	1.0167	ua
Moment of perihelion passage	T_{π}	Jan. 2, 4 h 52 min	
Moment of aphelion passage	T_{α}	July 4, 5 h 05 min	
Siderial rotation period around sun	P_{orb}	$31.5581 \cdot 10^{6}$	s
		365.25636	d
Mean rotational velocity	U_{orb}	29.78	km/s
Mean equatorial radius	$\bar{a}$	6378.140	km
Mean polar compression (flattening factor)	α	1/298.257	
Difference in equatorial and polar semi-axes	$a - c$	21.385	km
Compression of meridian of major equatorial axis	α_{a}	1/295.2	
Compression of meridian of minor equatorial axis	α_{b}	1/298.0	
Equatorial compression	ε	1/30 000	
Difference in equatorial semi-axes	$a - b$	213	m
Difference in polar semi-axes	$c_{N} - c_{S}$	~70	m
Polar asymmetry	η	$\sim 1 \cdot 10^{-5}$	
Mean acceleration of gravity at equator	g_{e}	9.78036	m/s²
Mean acceleration of gravity at poles	g_{p}	9.83208	m/s²
Difference in acceleration of gravity at pole and at equator	$g_{p} - g_{e}$	5.172	cm/s²
Mean acceleration of gravity for entire surface of terrestrial ellipsoid	g	9.7978	m/s²
Mean radius	R	6371.0	km
Area of surface	S	$5.10 \cdot 10^{8}$	km²
Volume	V	$1.0832 \cdot 10^{12}$	km³
Mean density	ρ	5.515	g/cm³
Siderial rotational period	P	86,164.09	s
Rotational angular velocity	ω	$7.292116 \cdot 10^{-5}$	rad/s
Mean equatorial rotational velocity	v	0.46512	km/s
Rotational angular momentum	L	$5.861 \cdot 10^{33}$	J s
Rotational energy	E	$2.137 \cdot 10^{29}$	J
Ratio of centrifugal force to force of gravity at equator	q_{c}	0.0034677 = 1/288	
Moment of inertia	I	$8.070 \cdot 10^{37}$	kg m²
Relative braking of earth's rotation due to tidal friction	$\Delta\omega_{e}/\omega$	$-4.2 \cdot 10^{-8}$	century⁻¹
Relative secular acceleration of earth's rotation	$\Delta\omega_{i}/\omega$	$+1.4 \cdot 10^{-8}$	century⁻¹
Not secular braking of earth's rotation	$\Delta\omega/\omega$	$-2.8 \cdot 10^{-8}$	century⁻¹
Probable value of total energy of tectonic deformation of earth	E_{t}	$\sim 1 \cdot 10^{23}$	J/century
Secular loss of heat of earth through radiation into space	$\Delta'E_{k}$	$1 \cdot 10^{23}$	J/century
Portion of earth's kinetic energy transformed into heat as a result of lunar and solar tides in the hydrosphere	$\Delta''E_{k}$	$1.3 \cdot 10^{23}$	J/century
Differences in duration of days in March and August	ΔP	0.0025 (March-August)	s
Corresponding relative annual variation in earth's rotational velocity	$\Delta^{*}\omega/\omega$	$2.9 \cdot 10^{-8}$ (Aug.-March)	
Presumed variation in earth's radius between August and March	$\Delta^{*}R$	−9.2 (Aug.-March)	cm
Annual variation in level of world ocean	Δh_{o}	~10 (Sept.-March)	cm
Area of continents	S_{C}	$1.49 \cdot 10^{8}$	km²
		29.2	% of surface

Quantity	Symbol	Value	Unit
Area of world ocean	S_o	$3.61 \cdot 10^8$	km²
		70.8	% of surface
Mean height of continents above sea level	h_c	875	m
Mean depth of world ocean	h_o	3794	m
Mean thickness of lithosphere within the limits of the continents	$h_{c.l.}$	35	km
Mean thickness of lithosphere within the limits of the ocean	$h_{o.l.}$	4.7	km
Mean rate of thickening of continental lithosphere	$\Delta h / \Delta t$	$10 - 40$	m/10⁶ y
Mean rate of horizontal extension of continental lithosphere	$\Delta l / \Delta t$	$0.75 - 20$	km/10⁶ y
Mass of crust	m_l	$2.36 \cdot 10^{22}$	kg
Mass of mantle		$4.05 \cdot 10^{24}$	kg
Amount of water released from the mantle and core in the course of geological time		$3.40 \cdot 10^{21}$	kg
Total reserve of water in the mantle		$2 \cdot 10^{23}$	kg
Present content of free and bound water in the earth's lithosphere		$2.4 \cdot 10^{21}$	kg
Mass of hydrosphere	m_h	$1.664 \cdot 10^{21}$	kg
Amount of oxygen bound in the earth's crust		$1.300 \cdot 10^{21}$	kg
Amount of free oxygen		$1.5 \cdot 10^{18}$	kg
Mass of atmosphere	m_a	$5.136 \cdot 10^{18}$	kg
Mass of biosphere	m_b	$1.148 \cdot 10^{16}$	kg
Mass of living matter in the biosphere		$3.6 \cdot 10^{14}$	kg
Density of living matter on dry land		0.1	g/cm²
Density of living matter in ocean		$15 \cdot 10^{-8}$	g/cm³
Age of the earth		$4.55 \cdot 10^9$	y
Age of oldest rocks		$4.0 \cdot 10^9$	y
Age of most ancient fossils		$3.4 \cdot 10^9$	y

GEOLOGICAL TIME SCALE

Period or epoch	Beginning and end, in 10^6 years	Key events
Cenozoic era		
Quaternian		
Contemporary	0–10,000 y ± 2,000 y	
Pleistocene	10,000–1,000,000 y ± 50,000 y	Homo Erectus breakout
Tertiary		
Pliocene	1.8–5.3	Ape man fossils
Miocene	5–25	Origin of grass
Oligocene	25–37	Rise of cats, dogs, pigs
Eocene	37–55	Debut of hoofed mammals
Paleocene	55–67	Earliest primates
Mesozoic era		
Cretaceous	67–138	Demise of dinosaurs
Jurassic	138–208	First birds
Triassic	208–245	Appearance of dinosaurs
Paleozoic era		
Permian	245–290	Flowers, insect pollination
Carboniferous	290–360	First conifers
Devonian	360–410	First vertebrates ashore
Silurian	410–435	Spore-bearing plants
Ordovician	435–520	First animals ashore
Cambrian	520–570	Vertebrates appear
Pre-Cambrian		
Pre-Cambrian III (Proterozoic)	570–2500	First plants, jellyfish
Pre-Cambrian II (Archean)	2500–3800	Photosynthetic bacteria
Pre-Cambrian I (Hadean)	3800–4450	Earth formed 4600 million years ago

Reference: Calder, N., *Timescale - An Atlas of the Fourth Dimension*, Viking Press, New York, 1983.

ACCELERATION DUE TO GRAVITY

The acceleration due to gravity is tabulated here as a function of latitude and height above the Earth's surface. Values were calculated from the expression

$$g/(\text{m/s}^2) = 9.780356 \, (1 + 0.0052885 \sin^2 \phi - 0.0000059 \sin^2 2\phi) - 0.003086 \, H$$

where ϕ is the latitude and H is the height in kilometers.

Reference

Jursa, A. S., Ed., *Handbook of Geophysics and the Space Environment*, 4th ed., Air Force Geophysics Laboratory, 1985, p. 14–17.

ϕ	$H = 0$	$H = 1$ km	$H = 5$ km	$H = 10$ km
0	9.78036	9.77727	9.76493	9.74950
5	9.78075	9.77766	9.76532	9.74989
10	9.78191	9.77882	9.76648	9.75105
15	9.78381	9.78072	9.76838	9.75295
20	9.78638	9.78330	9.77095	9.75552
25	9.78956	9.78647	9.77413	9.75870
30	9.79324	9.79016	9.77781	9.76238
35	9.79732	9.79424	9.78189	9.76646
40	9.80167	9.79858	9.78624	9.77081
45	9.80616	9.80307	9.79073	9.77530
50	9.81065	9.80757	9.79522	9.77979
55	9.81501	9.81193	9.79958	9.78415
60	9.81911	9.81602	9.80368	9.78825
65	9.82281	9.81972	9.80738	9.79195
70	9.82601	9.82292	9.81058	9.79515
75	9.82860	9.82551	9.81317	9.79774
80	9.83051	9.82743	9.81508	9.79965
85	9.83168	9.82860	9.81625	9.80082
90	9.83208	9.82899	9.81665	9.80122

DENSITY, PRESSURE, AND GRAVITY AS A FUNCTION OF DEPTH WITHIN THE EARTH

This table gives the density ρ, pressure p, and acceleration due to gravity g as a function of depth below the Earth's surface, as calculated from the model of the structure of the Earth in Reference 1. The model assumes a radius of 6371 km for the Earth. The boundary between the crust and mantle (the Mohorovicic discontinuity) is taken as 21 km, while in reality it varies considerable with location.

References

1. Anderson, D. L., and Hart, R. S., *J. Geophys. Res.*, 81, 1461, 1976.
2. Carmichael, R. S., *CRC Practical Handbook of Physical Properties of Rocks and Minerals*, p. 467, CRC Press, Boca Raton, FL, 1989.

Depth km	ρ g/cm^3	p kbar	g cm/s^2	Depth km	ρ g/cm^3	p kbar	g cm/s^2
				1771	4.96	752	994
				2071	5.12	903	1002
Crust				2371	5.31	1061	1017
0	1.02	0	981	2671	5.45	1227	1042
3	1.02	3	982	2886	5.53	1352	1069
3	2.80	3	982				
21	2.80	5	983	**Outer core (liquid)**			
				2886	9.96	1352	1069
Mantle (solid)				2971	10.09	1442	1050
21	3.49	5	983	3371	10.63	1858	953
41	3.51	12	983	3671	11.00	2154	874
61	3.52	19	984	4071	11.36	2520	760
81	3.48	26	984	4471	11.69	2844	641
101	3.44	33	984	4871	11.99	3116	517
121	3.40	39	985	5156	12.12	3281	427
171	3.37	56	987				
221	3.34	73	989	**Inner core (solid)**			
271	3.37	89	991	5156	12.30	3281	427
321	3.47	106	993	5371	12.48	3385	355
371	3.59	124	994	5771	12.52	3529	218
571	3.95	199	999	6071	12.53	3592	122
871	4.54	328	997	6371	12.58	3617	0
1171	4.67	466	992				
1471	4.81	607	991				

OCEAN PRESSURE AS A FUNCTION OF DEPTH AND LATITUDE

The following table is based upon an ocean model which takes into account the equation of state of standard seawater and the dependence on latitude of the acceleration of gravity. The tabulated pressure value is the excess pressure over the ambient atmospheric pressure at the surface.

References

1. *International Oceanographic Tables, Volume 4*, Unesco Technical Papers in Marine Science No. 40, Unesco, Paris, 1987.
2. Saunders, P. M., and Fofonoff, N. P., *Deep-Sea Res.* 23, 109–111, 1976.

Pressure in MPa at the Specified Latitude

Depth (meters)	0°	15°	30°	45°	60°	75°	90°
0	0.0000	0.0000	0.0000	0.0000	0.0000	0.0000	0.0000
500	5.0338	5.0355	5.0404	5.0471	5.0537	5.0586	5.0605
1000	10.0796	10.0832	10.0930	10.1064	10.1198	10.1296	10.1333
1500	15.1376	15.1431	15.1577	15.1778	15.1980	15.2127	15.2182
2000	20.2076	20.2148	20.2344	20.2613	20.2882	20.3080	20.3153
2500	25.2895	25.2985	25.3231	25.3568	25.3905	25.4153	25.4244
3000	30.3831	30.3940	30.4236	30.4641	30.5047	30.5345	30.5453
3500	35.4886	35.5012	35.5358	35.5832	35.6307	35.6654	35.6782
4000	40.6056	40.6201	40.6598	40.7140	40.7683	40.8082	40.8229
4500	45.7342	45.7505	45.7952	45.8564	45.9176	45.9626	45.9791
5000	50.8742	50.8924	50.9421	51.0102	51.0785	51.1285	51.1469
5500	56.0255	56.0456	56.1004	56.1755	56.2508	56.3059	56.3262
6000	61.1882	61.2100	61.2700	61.3521	61.4344	61.4947	61.5168
6500	66.3619	66.3857	66.4508	66.5399	66.6292	66.6947	66.7187
7000	71.5467	71.5724	71.6427	71.7388	71.8352	71.9059	71.9318
7500	76.7426	76.7701	76.8456	76.9488	77.0523	77.1282	77.1560
8000	81.9493	81.9788	82.0594	82.1697	82.2804	82.3614	82.3911
8500	87.1669	87.1983	87.2841	87.4016	87.5193	87.6057	87.6373
9000	92.3950	92.4284	92.5194	92.6440	92.7689	92.8606	92.8941
9500	97.6346	97.6698	97.7661	97.8978	98.0300	98.1269	98.1624
10000	102.8800	102.9170	103.0185	103.1572	103.2961	103.3981	103.4355

PROPERTIES OF SEAWATER

In addition to the dependence on temperature and pressure, the physical properties of seawater vary with the concentration of the dissolved constituents. A convenient parameter for describing the composition is the salinity, S, which is defined in terms of the electrical conductivity of the seawater sample. The defining equation for the practical salinity is:

$$S = a_0 + a_1 K^{1/2} + a_2 K + a_3 K^{3/2} + a_4 K^2 + a_5 K^{5/2},$$

where K is the ratio of the conductivity of the seawater sample at 15°C and atmospheric pressure to the conductivity of a potassium chloride solution in which the mass fraction of KCl is 0.0324356, at the same temperature and pressure. The values of the coefficients are:

$$a_0 = 0.0080 \qquad a_3 = 14.0941$$
$$a_1 = -0.1692 \qquad a_4 = -7.0261$$
$$a_2 = 25.3851 \qquad a_5 = 2.7081$$
$$\Sigma\, a_i = 35.0000$$

Thus when $K = 1$, $S = 35$ exactly (S is normally quoted in units of ‰, i.e., parts per thousand). The value of S can be roughly equated with the mass of dissolved material in grams per kilogram of seawater. Salinity values in the open oceans at mid latitudes typically fall between 34 and 36.

It is customary in oceanography to define the pressure at a given point as the pressure due to the column of water between that point and the surface. Thus by convention $P = 0$ at the sea surface. To a good approximation the pressure in decibars (dbar) can be equated to the depth in meters. Thus at 45° latitude the pressure is 5000 dbar at 4902 m, 10000 dbar at 9700 m.

The first table below gives several properties of seawater as a function of temperature for a salinity of 35. The second and third give density and electrical conductivity as a function of salinity at several temperatures, and the fourth lists typical concentrations of the main constituents of seawater as a function of salinity. The final table gives the freezing point as a function of salinity and pressure.

References

1. *The Practical Salinity Scale 1978 and the International Equation of State of Seawater 1980*, Unesco Technical Papers in Marine Science No. 36, Unesco, Paris, 1981; sections No. 37, 38, 39, and 40 in this series give background papers and detailed tables.
2. Kennish, M. J., *CRC Practical Handbook of Marine Science*, CRC Press, Boca Raton, FL, 1989.
3. Poisson, A. *IEEE J. Ocean. Eng.* OE-5, 50, 1981.
4. Webster, F., in *AIP Physics Desk Reference*, E. R. Cohen, D. R. Lide and G. L. Trigg, Eds., Springer-Verlag, New York, 2003.

Properties of Seawater as a Function of Temperature at Salinity $S = 35$ and Normal Atmospheric Pressure

ρ = density in g/cm^3

β = $(1/\rho)\,(d\rho/dS)$ = fractional change in density per unit change in salinity

α = $-(1/\rho)\,(d\rho/dt)$ = fractional change in density per unit change in temperature (°C^{-1})

κ = electrical conductivity in S/cm

η = viscosity in mPa s (equal to cP)

c_p = specific heat in J/kg °C

v = speed of sound in m/s

t/°C	ρ/g cm^{-3}	$10^7\beta$	$10^7\,\alpha$/°C^{-1}	κ/S cm^{-1}	η/mPa s	c_p/J kg^{-1}°C^{-1}	v/m s^{-1}
0	1.028106	7854	526	0.029048	1.892	3986.5	1449.1
5	1.027675	7717	1136	0.033468	1.610		
10	1.026952	7606	1668	0.038103	1.388	3986.3	1489.8
15	1.025973	7516	2141	0.042933	1.221		
20	1.024763	7444	2572	0.047934	1.085	3993.9	1521.5
25	1.023343	7385	2970	0.053088	0.966		
30	1.021729	7338	3341	0.058373	0.871	4000.7	1545.6
35	1.019934	7300	3687				
40		7270	4004			4003.5	1563.2

Density of Surface Seawater in g/cm^3 as a Function of Temperature and Salinity

t/°C	$S = 0$	$S = 5$	$S = 10$	$S = 15$	$S = 20$	$S = 25$	$S = 30$	$S = 35$	$S = 40$
0	0.999843	1.003913	1.007955	1.011986	1.016014	1.020041	1.024072	1.028106	1.032147
5	0.999967	1.003949	1.007907	1.011858	1.015807	1.019758	1.023714	1.027675	1.031645
10	0.999702	1.003612	1.007501	1.011385	1.015269	1.019157	1.023051	1.026952	1.030862
15	0.999102	1.002952	1.006784	1.010613	1.014443	1.018279	1.022122	1.025973	1.029834
20	0.998206	1.002008	1.005793	1.009576	1.013362	1.017154	1.020954	1.024763	1.028583
25	0.997048	1.000809	1.004556	1.008301	1.012050	1.015806	1.019569	1.023343	1.027128
30	0.995651	0.999380	1.003095	1.006809	1.010527	1.014252	1.017985	1.021729	1.025483
35	0.994036	0.997740	1.001429	1.005118	1.008810	1.012509	1.016217	1.019934	1.023662
40	0.992220	0.995906	0.999575	1.003244	1.006915	1.010593	1.014278	1.017973	1.021679

Electrical Conductivity of Seawater in S/cm as a Function of Temperature and Salinity

$t/°C$	$S = 5$	$S = 10$	$S = 15$	$S = 20$	$S = 25$	$S = 30$	$S = 35$	$S = 40$
0	0.004808	0.009171	0.013357	0.017421	0.021385	0.025257	0.029048	0.032775
5	0.005570	0.010616	0.015441	0.020118	0.024674	0.029120	0.033468	0.037734
10	0.006370	0.012131	0.017627	0.022947	0.028123	0.033171	0.038103	0.042935
15	0.007204	0.013709	0.019905	0.025894	0.031716	0.037391	0.042933	0.048355
20	0.008068	0.015346	0.022267	0.028948	0.035438	0.041762	0.047934	0.053968
25	0.008960	0.017035	0.024703	0.032097	0.039276	0.046267	0.053088	0.059751
30	0.009877	0.018771	0.027204	0.035330	0.043213	0.050888	0.058373	0.065683

Composition of Seawater and Ionic Strength at Various Salinities (Ref. 2)

	Expressed as molality			As grams per kilogram of seawater		
Constituent	$S = 30$	$S = 35$	$S = 40$	$S = 30$	$S = 35$	$S = 40$
Cl^-	0.482	0.562	0.650	16.58	19.33	22.36
Br^-	0.00074	0.00087	0.00100	0.057	0.067	0.078
F^-		0.00007			0.001	
SO_4^{2-}	0.0104	0.0114	0.0122	0.97	1.06	1.14
HCO_3^-	0.00131	0.00143	0.00100	0.078	0.085	0.059
$NaSO_4^-$	0.0085	0.0108	0.0139	0.98	1.25	1.60
KSO_4^-	0.00010	0.00012	0.00015	0.013	0.016	0.020
Na^+	0.405	0.472	0.544	9.03	10.53	12.13
K^+	0.00892	0.01039	0.01200	0.338	0.394	0.455
Mg^{2+}	0.0413	0.0483	0.0561	0.974	1.139	1.323
Ca^{2+}	0.00131	0.00143	0.00154	0.051	0.056	0.060
Sr^{2+}	0.00008	0.00009	0.00011	0.007	0.008	0.009
$MgHCO_3^+$	0.00028	0.00036	0.00045	0.023	0.030	0.037
$MgSO_4$	0.00498	0.00561	0.00614	0.582	0.655	0.717
$CaSO_4$	0.00102	0.00115	0.00126	0.135	0.152	0.166
$NaHCO_3$	0.00015	0.00020	0.00024	0.012	0.016	0.020
H_3BO_3	0.00032	0.00037	0.00042	0.019	0.022	0.025
Ionic strength	0.5736	0.6675	0.7701			

Freezing Point of Seawater in °C as a Function of Salinity and Pressure

$P/dbar$	$S = 0$	5	10	15	20	25	30	35	40
0	0.000	−0.274	−0.542	−0.812	−1.083	−1.358	−1.638	−1.922	−2.212
50	−0.038	−0.311	−0.580	−0.849	−1.121	−1.396	−1.676	−1.960	−2.250
100	−0.075	−0.349	−0.618	−0.887	−1.159	−1.434	−1.713	−1.998	−2.287
500	−0.377	−0.650	−0.919	−1.188	−1.460	−1.735	−2.014	−2.299	−2.589

ABUNDANCE OF ELEMENTS IN THE EARTH'S CRUST AND IN THE SEA

This table gives the estimated abundance of the elements in the continental crust (in mg/kg, equivalent to parts per million by mass) and in seawater near the surface (in mg/L). Values represent the median of reported measurements. The concentrations of the less abundant elements may vary with location by several orders of magnitude.

References

1. Carmichael, R. S., Ed., *CRC Practical Handbook of Physical Properties of Rocks and Minerals*, CRC Press, Boca Raton, FL, 1989.
2. Bodek, I., et al., *Environmental Inorganic Chemistry*, Pergamon Press, New York, 1988.
3. Ronov, A. B., and Yaroshevsky, A. A., "Earth's Crust Geochemistry," in *Encyclopedia of Geochemistry and Environmental Sciences*, Fairbridge, R. W., Ed., Van Nostrand, New York, 1969.

	Abundance	
	Crust	Sea
Element	mg/kg	mg/L
Ac	5.5×10^{-10}	
Ag	7.5×10^{-2}	4×10^{-5}
Al	8.23×10^{4}	2×10^{-3}
Ar	3.5	4.5×10^{-1}
As	1.8	3.7×10^{-3}
Au	4×10^{-3}	4×10^{-6}
B	1.0×10^{1}	4.44
Ba	4.25×10^{2}	1.3×10^{-2}
Be	2.8	5.6×10^{-6}
Bi	8.5×10^{-3}	2×10^{-5}
Br	2.4	6.73×10^{1}
C	2.00×10^{2}	2.8×10^{1}
Ca	4.15×10^{4}	4.12×10^{2}
Cd	1.5×10^{-1}	1.1×10^{-4}
Ce	6.65×10^{1}	1.2×10^{-6}
Cl	1.45×10^{2}	1.94×10^{4}
Co	2.5×10^{1}	2×10^{-5}
Cr	1.02×10^{2}	3×10^{-4}
Cs	3	3×10^{-4}
Cu	6.0×10^{1}	2.5×10^{-4}
Dy	5.2	9.1×10^{-7}
Er	3.5	8.7×10^{-7}
Eu	2.0	1.3×10^{-7}
F	5.85×10^{2}	1.3
Fe	5.63×10^{4}	2×10^{-3}
Ga	1.9×10^{1}	3×10^{-5}
Gd	6.2	7×10^{-7}
Ge	1.5	5×10^{-5}
H	1.40×10^{3}	1.08×10^{5}
He	8×10^{-3}	7×10^{-6}
Hf	3.0	7×10^{-6}
Hg	8.5×10^{-2}	3×10^{-5}
Ho	1.3	2.2×10^{-7}
I	4.5×10^{-1}	6×10^{-2}
In	2.5×10^{-1}	2×10^{-2}
Ir	1×10^{-3}	
K	2.09×10^{4}	3.99×10^{2}
Kr	1×10^{-4}	2.1×10^{-4}
La	3.9×10^{1}	3.4×10^{-6}
Li	2.0×10^{1}	1.8×10^{-1}
Lu	8×10^{-1}	1.5×10^{-7}
Mg	2.33×10^{4}	1.29×10^{3}
Mn	9.50×10^{2}	2×10^{-4}
Mo	1.2	1×10^{-2}
N	1.9×10^{1}	5×10^{-1}
Na	2.36×10^{4}	1.08×10^{4}
Nb	2.0×10^{1}	1×10^{-5}
Nd	4.15×10^{1}	2.8×10^{-6}
Ne	5×10^{-3}	1.2×10^{-4}
Ni	8.4×10^{1}	5.6×10^{-4}
O	4.61×10^{5}	8.57×10^{5}
Os	1.5×10^{-3}	
P	1.05×10^{3}	6×10^{-2}
Pa	1.4×10^{-6}	5×10^{-11}
Pb	1.4×10^{1}	3×10^{-5}
Pd	1.5×10^{-2}	
Po	2×10^{-10}	1.5×10^{-14}
Pr	9.2	6.4×10^{-7}
Pt	5×10^{-3}	
Ra	9×10^{-7}	8.9×10^{-11}
Rb	9.0×10^{1}	1.2×10^{-1}
Re	7×10^{-4}	4×10^{-6}
Rh	1×10^{-3}	
Rn	4×10^{-13}	6×10^{-16}
Ru	1×10^{-3}	7×10^{-7}
S	3.50×10^{2}	9.05×10^{2}
Sb	2×10^{-1}	2.4×10^{-4}
Sc	2.2×10^{1}	6×10^{-7}
Se	5×10^{-2}	2×10^{-4}
Si	2.82×10^{5}	2.2
Sm	7.05	4.5×10^{-7}
Sn	2.3	4×10^{-6}
Sr	3.70×10^{2}	7.9
Ta	2.0	2×10^{-6}
Tb	1.2	1.4×10^{-7}
Te	1×10^{-3}	
Th	9.6	1×10^{-6}
Ti	5.65×10^{3}	1×10^{-3}
Tl	8.5×10^{-1}	1.9×10^{-5}
Tm	5.2×10^{-1}	1.7×10^{-7}
U	2.7	3.2×10^{-3}
V	1.20×10^{2}	2.5×10^{-3}
W	1.25	1×10^{-4}
Xe	3×10^{-5}	5×10^{-5}
Y	3.3×10^{1}	1.3×10^{-5}
Yb	3.2	8.2×10^{-7}
Zn	7.0×10^{1}	4.9×10^{-3}
Zr	1.65×10^{2}	3×10^{-5}

SOLAR SPECTRAL IRRADIANCE

The solar luminosity (total radiant power emitted) is $3.86 \cdot 10^{26}$ W, of which 1373 W/m² reaches the top of the earth's atmosphere. To a zeroth approximation the sun can be considered a black body with an effective temperature of 5780 K, which implies a peak in the radiation at around 0.520 μm (5200 Å). The actual solar spectral emission is more complex, especially at ultraviolet and shorter wavelengths. The graph below, which was taken from Reference 1, summarizes the solar irradiance at the top of the atmosphere in the range 0.3 to 10 μm.

References

1. Jursa, A. S., Ed., *Handbook of Geophysics and the Space Environment*, Air Force Geophysics Laboratory, 1985.
2. Pierce, A. K., and Allen, R. G., "The Solar Spectrum between 0.3 and 10 μm", in *The Solar Output and Its Variation*, White, O. R., Ed., Colorado Associated University Press, Boulder, CO, 1977.
3. Lang, K. R., *Astrophysical Data. Planets and Stars*, Springer-Verlag, New York, 1992.

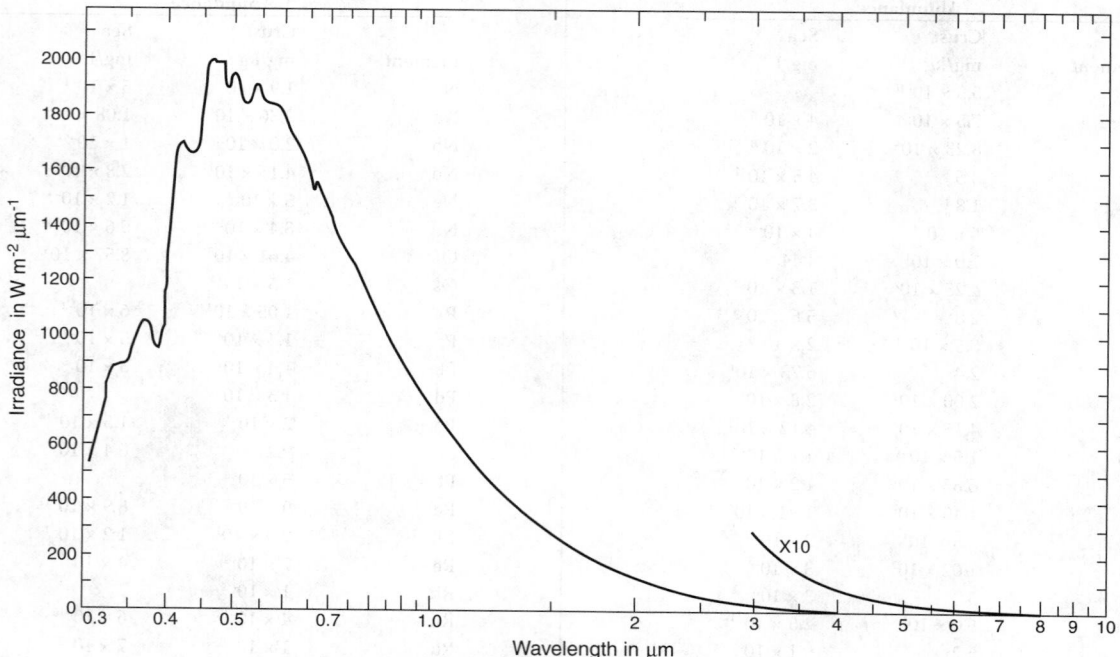

U.S. STANDARD ATMOSPHERE (1976)

A Standard Atmosphere is a hypothetical vertical distribution of atmospheric temperature, pressure, and density which is roughly representative of year-round, midlatitude conditions. Typical uses are to serve as a basis for pressure altimeter calibrations, aircraft performance calculations, aircraft and rocket design, ballistic tables, meteorological diagrams, and various types of atmospheric modeling. The air is assumed to be dry and to obey the perfect gas law and the hydrostatic equation which, taken together, relate temperature, pressure, and density with vertical position. The atmosphere is considered to rotate with the earth and to be an average over the diurnal cycle, the semiannual variation, and the range from active to quiet geomagnetic and sunspot conditions.

The U.S. Standard Atmosphere, (1976) is an idealized, steady-state representation of mean annual conditions of the Earth's atmosphere from the surface to 1000 km at latitude 45°N, as it is assumed to exist during a period with moderate solar activity. The defining meteorological elements are sea-level temperature and pressure and a temperature-height profile to 1000 km. The 1976 Standard Atmosphere uses the following sea-level values which have been standard for many decades:

Temperature — 288.15 K (15°C)
Pressure — 101325 Pa (1013.25 mbar, 760 mm of Hg, or 29.92 in. of Hg)

Density — 1225 g/m^3 (1.225 g/L)
Mean molar mass — 28.964 g/mol

The parameters included in this condensed version of the U.S. Standard Atmosphere are:

Z — Height (geometric) above mean sea level in meters
T — Temperature in kelvins
P — Pressure in pascals (1 Pa = 0.01 millibars)
ρ — Density in kilograms per cubic meter (1 kg/m^3 = 1 g/L)
n — Number density in molecules per cubic meter
ν — Mean collision frequency in collisions per second
l — Mean free path in meters
η — Absolute viscosity in pascal seconds (1 Pa s = 1000 cP)
k — Thermal conductivity in joules per meter second kilogram (W/m K)
v_s — Speed of sound in meters per second
g — Acceleration of gravity in meters per second square

The sea-level composition (percent by volume) is taken to be:

N_2 — 78.084%	He — 0.000524
O_2 — 20.9476	Kr — 0.000114
Ar — 0.934	Xe — 0.0000087
CO_2 — 0.0314	CH_4 — 0.0002
Ne — 0.001818	H_2 — 0.00005

The T and P columns for the troposphere and lower stratosphere were generated from the following formulas:

	T/K	P/Pa
$H \leq 11000$ m	$288.15 - 0.0065\,H$	$101325(288.15/T)^{-5.25577}$
11000 m $< H \leq 20000$ m	216.65	$22632\,e^{-0.00015768832(H-11000)}$
20000 m $< H \leq 32000$ m	$216.65 + 0.0010(H-20000)$	$5474.87(216.65/T)^{34.16319}$

where $H = rZ/(r + Z)$ is the geopotential height in meters and r is the mean Earth radius at 45° N latitude, taken as 6356766 m. For altitudes up to 32 km, $\rho = 0.003483677(P/T)$ in the units used here. Formulas for the other quantities may be found in the references.

References

1. COESA, *U.S. Standard Atmosphere*, 1976, U.S. Government Printing Office, Washington, D.C., 1976.
2. Jursa, A. S., Ed., *Handbook of Geophysics and the Space Environment*, Air Force Geophysics Laboratory, 1985.

Z/m	T/K	P/Pa	ρ/kg m⁻³	n/m⁻³	v/s⁻¹	l/m	η/Pa s	k/J m⁻¹s⁻¹K⁻¹	v_s/m s⁻¹	g/m s⁻²
−5000	320.68	1.778E+05	1.931	4.015E+25	1.151E+10	4.208E−08	1.942E−05	0.02788	359.0	9.822
−4500	317.42	1.685E+05	1.849	3.845E+25	1.096E+10	4.395E−08	1.927E−05	0.02763	357.2	9.830
−4000	314.17	1.596E+05	1.770	3.680E+25	1.044E+10	4.592E−08	1.912E−05	0.02738	355.3	9.819
−3500	310.91	1.511E+05	1.693	3.520E+25	9.933E+09	4.800E−08	1.897E−05	0.02713	353.5	9.818
−3000	307.66	1.430E+05	1.619	3.366E+25	9.448E+09	5.019E−08	1.882E−05	0.02688	351.6	9.816
−2500	304.41	1.352E+05	1.547	3.217E+25	8.982E+09	5.252E−08	1.867E−05	0.02663	349.8	9.814
−2000	301.15	1.278E+05	1.478	3.102E+25	8.623E+09	5.447E−08	1.852E−05	0.02638	347.9	9.813
−1500	297.90	1.207E+05	1.411	2.935E+25	8.106E+09	5.757E−08	1.836E−05	0.02613	346.0	9.811
−1000	294.65	1.139E+05	1.347	2.801E+25	7.693E+09	6.032E−08	1.821E−05	0.02587	344.1	9.810
−500	291.40	1.075E+05	1.285	2.672E+25	7.298E+09	6.324E−08	1.805E−05	0.02562	342.2	9.808
0	288.15	1.013E+05	1.225	2.547E+25	6.919E+09	6.633E−08	1.789E−05	0.02533	340.3	9.807
500	284.90	9.546E+04	1.167	2.427E+25	6.556E+09	6.961E−08	1.774E−05	0.02511	338.4	9.805
1000	281.65	8.988E+04	1.112	2.311E+25	6.208E+09	7.310E−08	1.758E−05	0.02485	336.4	9.804
1500	278.40	8.456E+04	1.058	2.200E+25	5.874E+09	7.680E−08	1.742E−05	0.02459	334.5	9.802
2000	275.15	7.950E+04	1.007	2.093E+25	5.555E+09	8.073E−08	1.726E−05	0.02433	332.5	9.801
2500	271.91	7.469E+04	0.957	1.990E+25	5.250E+09	8.491E−08	1.710E−05	0.02407	330.6	9.799
3000	268.66	7.012E+04	0.909	1.891E+25	4.959E+09	8.937E−08	1.694E−05	0.02381	328.6	9.797
3500	265.41	6.579E+04	0.863	1.795E+25	4.680E+09	9.411E−08	1.678E−05	0.02355	326.6	9.796
4000	262.17	6.166E+04	0.819	1.704E+25	4.414E+09	9.917E−08	1.661E−05	0.02329	324.6	9.794
4500	258.92	5.775E+04	0.777	1.616E+25	4.160E+09	1.046E−07	1.645E−05	0.02303	322.6	9.793
5000	255.68	5.405E+04	0.736	1.531E+25	3.918E+09	1.103E−07	1.628E−05	0.02277	320.6	9.791
5500	252.43	5.054E+04	0.697	1.450E+25	3.687E+09	1.165E−07	1.612E−05	0.02250	318.5	9.790
6000	249.19	4.722E+04	0.660	1.373E+25	3.467E+09	1.231E−07	1.595E−05	0.02224	316.5	9.788
6500	245.94	4.408E+04	0.664	1.299E+25	3.258E+09	1.302E−07	1.578E−05	0.02197	314.4	9.787
7000	242.70	4.111E+04	0.590	1.227E+25	3.058E+09	1.377E−07	1.561E−05	0.02170	312.3	9.785
7500	239.46	3.830E+04	0.557	1.159E+25	2.869E+09	1.458E−07	1.544E−05	0.02144	310.2	9.784
8000	236.22	3.565E+04	0.526	1.093E+25	2.689E+09	1.545E−07	1.527E−05	0.02117	308.1	9.782
8500	232.97	3.315E+04	0.496	1.031E+25	2.518E+09	1.639E−07	1.510E−05	0.02090	306.0	9.781
9000	229.73	3.080E+04	0.467	9.711E+24	2.356E+09	1.740E−07	1.493E−05	0.02063	303.9	9.779
9500	226.49	2.858E+04	0.440	9.141E+24	2.202E+09	1.848E−07	1.475E−05	0.02036	301.7	9.777
10000	223.25	2.650E+04	0.414	8.598E+24	2.056E+09	1.965E−07	1.458E−05	0.02009	299.5	9.776
10500	220.01	2.454E+04	0.389	8.079E+24	1.918E+09	2.091E−07	1.440E−05	0.01982	297.4	9.774
11000	216.77	2.270E+04	0.365	7.585E+24	1.787E+09	2.227E−07	1.422E−05	0.01954	295.2	9.773
11500	216.65	2.098E+04	0.337	7.016E+24	1.653E+09	2.408E−07	1.422E−05	0.01953	295.1	9.771
12000	216.65	1.940E+04	0.312	6.486E+24	1.528E+09	2.605E−07	1.422E−05	0.01953	295.1	9.770
12500	216.65	1.793E+04	0.288	5.996E+24	1.412E+09	2.818E−07	1.422E−05	0.01953	295.1	9.768
13000	216.65	1.658E+04	0.267	5.543E+24	1.306E+09	3.048E−07	1.422E−05	0.01953	295.1	9.767
13500	216.65	1.533E+04	0.246	5.124E+24	1.207E+09	3.297E−07	1.422E−05	0.01953	295.1	9.765
14000	216.65	1.417E+04	0.228	4.738E+24	1.116E+09	3.566E−07	1.422E−05	0.01953	295.1	9.764
14500	216.65	1.310E+04	0.211	4.380E+24	1.032E+09	3.857E−07	1.422E−05	0.01953	295.1	9.762
15000	216.65	1.211E+04	0.195	4.049E+24	9.538E+08	4.172E−07	1.422E−05	0.01953	295.1	9.761
16000	216.65	1.035E+04	0.166	3.461E+24	8.153E+08	4.881E−07	1.422E−05	0.01953	295.1	9.758
17000	216.65	8.850E+03	0.142	2.959E+24	6.969E+08	5.710E−07	1.422E−05	0.01953	295.1	9.754
18000	216.65	7.565E+03	0.122	2.529E+24	5.958E+08	6.680E−07	1.422E−05	0.01953	295.1	9.751
19000	216.65	6.467E+03	0.104	2.162E+24	5.093E+08	7.814E−07	1.422E−05	0.01953	295.1	9.748
20000	216.65	5.529E+03	8.891E−02	1.849E+24	4.354E+08	9.139E−07	1.422E−05	0.01953	295.1	9.745
21000	217.58	4.729E+03	7.572E−02	1.574E+24	3.716E+08	1.073E−06	1.427E−05	0.01961	295.1	9.742
22000	218.57	4.048E+03	6.451E−02	1.341E+24	3.173E+08	1.260E−06	1.432E−05	0.01970	296.4	9.739
23000	219.57	3.467E+03	5.501E−02	1.144E+24	2.712E+08	1.477E−06	1.438E−05	0.01978	297.1	9.736
24000	220.56	2.972E+03	4.694E−02	9.759E+23	2.319E+08	1.731E−06	1.443E−05	0.01986	297.7	9.733
25000	221.55	2.549E+03	4.008E−02	8.334E+23	1.985E+08	2.027E−06	1.448E−05	0.01995	298.4	9.730
26000	222.54	2.188E+03	3.426E−02	7.123E+23	1.700E+08	2.372E−06	1.454E−05	0.02003	299.1	9.727
27000	223.54	1.880E+03	2.930E−02	6.092E+23	1.458E+08	2.773E−06	1.459E−05	0.02011	299.7	9.724
28000	224.53	1.610E+03	2.508E−02	5.214E+23	1.250E+08	3.240E−06	1.465E−05	0.02020	300.4	9.721
29000	225.52	1.390E+03	2.148E−02	4.466E+23	1.073E+08	3.783E−06	1.470E−05	0.02028	301.1	9.718
30000	226.51	1.197E+03	1.841E−02	3.828E+23	9.219E+07	4.414E−06	1.475E−05	0.02036	301.7	9.715
31000	227.50	1.031E+03	1.579E−02	3.283E+23	7.925E+07	5.146E−06	1.481E−05	0.02044	302.4	9.712
32000	228.49	8.891E+02	1.356E−02	2.813E+23	6.818E+07	5.995E−06	1.486E−05	0.02053	303.0	9.709
33000	230.97	7.673E+02	1.157E−02	2.406E+23	5.852E+07	7.021E−06	1.499E−05	0.02073	304.7	9.706
34000	233.74	6.634E+02	9.887E−03	2.056E+23	5.030E+07	8.218E−06	1.514E−05	0.02096	306.5	9.703
35000	236.51	5.746E+02	8.463E−03	1.760E+23	4.331E+07	9.601E−06	1.529E−05	0.02119	308.3	9.700

U.S. Standard Atmosphere (1976)

Z/m	T/K	P/Pa	$\rho/\text{kg m}^{-3}$	n/m^{-3}	ν/s^{-1}	l/m	$\eta/\text{Pa s}$	$k/\text{J m}^{-1}\text{s}^{-1}\text{K}^{-1}$	$\nu_s/\text{m s}^{-1}$	$g/\text{m s}^{-2}$
36000	239.28	4.985E+02	7.258E−03	1.509E+23	3.736E+07	1.120E−05	1.543E−05	0.02142	310.1	9.697
38000	244.82	3.771E+02	5.367E−03	1.116E+23	2.794E+07	1.514E−05	1.572E−05	0.02188	313.7	9.690
40000	250.35	2.871E+02	3.996E−03	8.308E+22	2.104E+07	2.034E−05	1.601E−05	0.02233	317.2	9.684
42000	255.88	2.200E+02	2.995E−03	6.227E+22	1.594E+07	2.713E−05	1.629E−05	0.02278	320.7	9.678
44000	261.40	1.695E+02	2.259E−03	4.697E+22	1.215E+07	3.597E−05	1.657E−05	0.02323	324.1	9.672
46000	266.93	1.313E+02	1.714E−03	3.564E+22	9.318E+06	4.740E−05	1.685E−05	0.02376	327.5	9.666
48000	270.65	1.023E+02	1.317E−03	2.738E+22	7.208E+06	6.171E−05	1.704E−05	0.02397	329.8	9.660
50000	270.65	7.978E+01	1.027E−03	2.135E+22	5.620E+06	7.913E−05	1.703E−05	0.02397	329.8	9.654
52000	269.03	6.221E+01	8.056E−04	1.675E+22	4.397E+06	1.009E−04	1.696E−05	0.02384	328.8	9.648
54000	263.52	4.834E+01	6.390E−04	1.329E+22	3.452E+06	1.272E−04	1.660E−05	0.02340	325.4	9.642
56000	258.02	3.736E+01	5.045E−04	1.049E+22	2.696E+06	1.611E−04	1.640E−05	0.02296	322.0	9.636
58000	252.52	2.872E+01	3.963E−04	8.239E+21	2.095E+06	2.051E−04	1.612E−05	0.02251	318.6	9.632
60000	247.02	2.196E+01	3.097E−04	6.439E+21	1.620E+06	2.624E−04	1.584E−05	0.02206	315.1	9.624
65000	233.29	1.093E+01	1.632E−04	3.393E+21	8.294E+05	4.979E−04	1.512E−05	0.02093	306.2	9.609
70000	219.59	5.221	8.283E−05	1.722E+21	4.084E+05	9.810E−04	1.438E−05	0.01978	297.1	9.594
75000	208.40	2.388	3.992E−05	8.300E+20	1.918E+05	2.035E−03	1.376E−05	0.01883	289.4	9.579
80000	198.64	1.052	1.846E−05	3.838E+20	8.656E+04	4.402E−03	1.321E−05	0.01800	282.5	9.564
85000	188.89	4.457E−01	8.220E−06	1.709E+20	3.766E+04	9.886E−03	1.265E−05	0.01716	275.5	9.550
90000	186.87	1.836E−01	3.416E−06	7.116E+19	1.560E+04	2.370E−02				9.535
95000	188.42	7.597E−02	1.393E−06	2.920E+19	6.440E+03	5.790E−02				9.520
100000	195.08	3.201E−02	5.604E−07	1.189E+19	2.680E+03	1.420E−01				9.505
110000	240.00	7.104E−03	9.708E−08	2.144E+18	5.480E+02	7.880E−01				9.476
120000	360.00	2.538E−03	2.222E−08	5.107E+17	1.630E+02	3.310				9.447
130000	469.27	1.251E−03	8.152E−09	1.930E+17	7.100E+01	8.800				9.418
140000	559.63	7.203E−04	3.831E−09	9.322E+16	3.800E+01	1.800E+01				9.389
150000	634.39	4.542E−04	2.076E−09	5.186E+16	2.300E+01	3.300E+01				9.360
160000	696.29	3.040E−04	1.233E−09	3.162E+16	1.500E+01	5.300E+01				9.331
170000	747.57	2.121E−04	7.815E−10	2.055E+16	1.000E+01	8.200E+01				9.302
180000	790.07	1.527E−04	5.194E−10	1.400E+16	7.200	1.200E+02				9.274
190000	825.16	1.127E−04	3.581E−10	9.887E+15	5.200	1.700E+02				9.246
200000	854.56	8.474E−05	2.541E−10	7.182E+15	3.900	2.400E+02				9.218
220000	899.01	5.015E−05	1.367E−10	4.040E+15	2.300	4.200E+02				9.162
240000	929.73	3.106E−05	7.858E−11	2.420E+15	1.400	7.000E+02				9.106
260000	950.99	1.989E−05	4.742E−11	1.515E+15	9.300E−01	1.100E+03				9.051
280000	965.75	1.308E−05	2.971E−11	9.807E+14	6.100E−01	1.700E+03				8.997
300000	976.01	8.770E−06	1.916E−11	6.509E+14	4.200E−01	2.600E+03				8.943
320000	983.16	5.980E−06	1.264E−11	4.405E+14	2.900E−01	3.800E+03				8.889
340000	988.15	4.132E−06	8.503E−12	3.029E+14	2.000E−01	5.600E+03				8.836
360000	991.65	2.888E−06	5.805E−12	2.109E+14	1.400E−01	8.000E+03				8.784
380000	994.10	2.038E−06	4.013E−12	1.485E+14	1.000E−01	1.100E+04				8.732
400000	995.83	1.452E−06	2.803E−12	1.056E+14	7.200E−02	1.600E+04				8.680
450000	998.22	6.447E−07	1.184E−12	4.678E+13	3.300E−02	3.600E+04				8.553
500000	999.24	3.024E−07	5.215E−13	2.192E+13	1.600E−02	7.700E+04				8.429
550000	999.67	1.514E−07	2.384E−13	1.097E+13	8.400E−03	1.500E+05				8.307
600000	999.85	8.213E−08	1.137E−13	5.950E+12	4.800E−03	2.800E+05				8.188
650000	999.93	4.887E−08	5.712E−14	3.540E+12	3.100E−03	4.800E+05				8.072
700000	999.97	3.191E−08	3.070E−14	2.311E+12	2.200E−03	7.300E+05				7.958
750000	999.98	2.260E−08	1.788E−14	1.637E+12	1.700E−03	1.000E+06				7.846
800000	999.99	1.704E−08	1.136E−14	1.234E+12	1.400E−03	1.400E+06				7.737
850000	1000.00	1.342E−08	7.824E−15	9.717E+11	1.200E−03	1.700E+06				7.630
900000	1000.00	1.087E−08	5.759E−15	7.876E+11	1.000E−03	2.100E+06				7.525
950000	1000.00	8.982E−09	4.453E−15	6.505E+11	8.700E−04	2.600E+06				7.422
1000000	1000.00	7.514E−09	3.561E−15	5.442E+11	7.500E−04	3.100E+06				7.322

FIGURE 1. Temperature-height profile for U.S. Standard Atmosphere.

FIGURE 3. Mean molecular weight as a function of geometric altitude.

FIGURE 2. Total pressure and mass density as a function of geometric altitude.

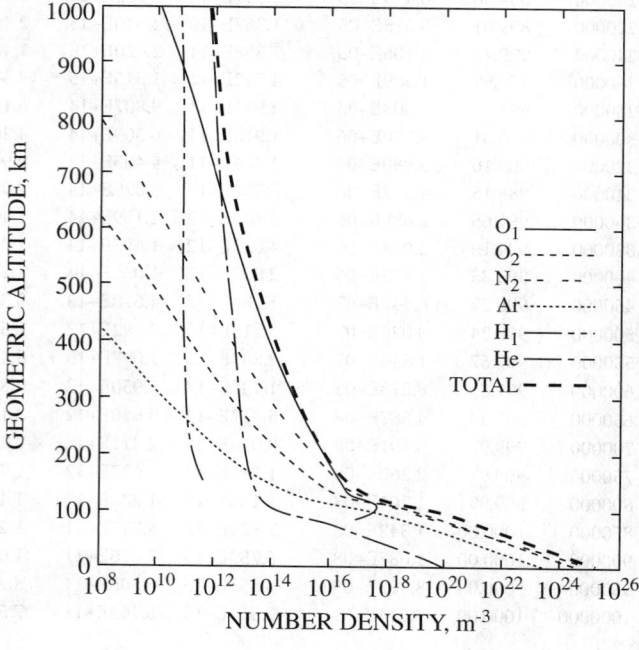

FIGURE 4. Number density of individual species and total number density as a function of geometric altitude.

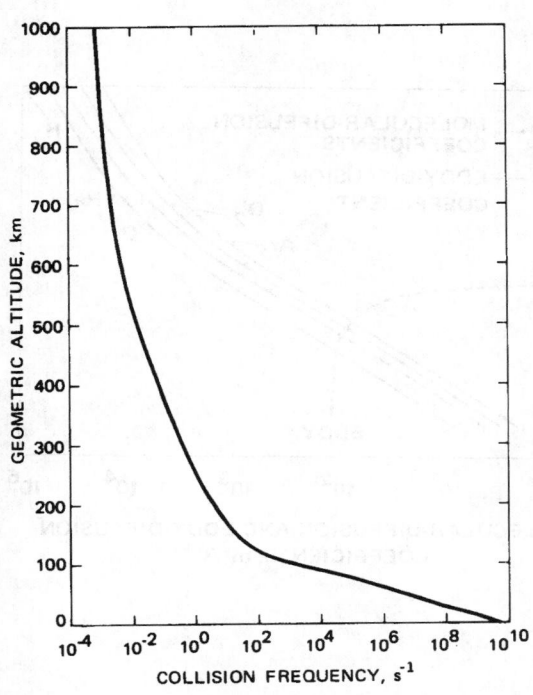

FIGURE 5. Collision frequency as a function of geometric altitude.

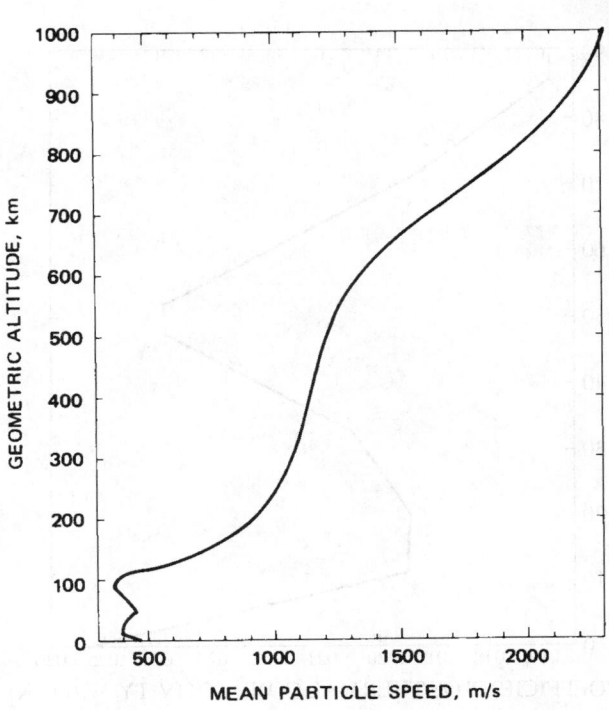

FIGURE 7. Mean air-particle speed as a function of geometric altitude.

FIGURE 6. Mean free path as a function of geometric altitude.

FIGURE 8. Dynamic viscosity as a function of geometric altitude.

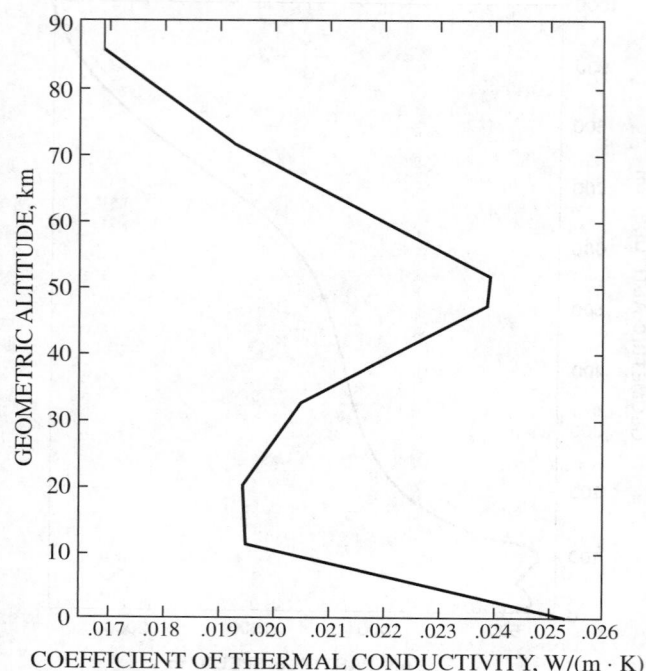

FIGURE 9. Coefficient of thermal conductivity as a function of geometric altitude.

FIGURE 11. Molecular-diffusion and eddy-diffusion coefficients as a function of geometric altitude.

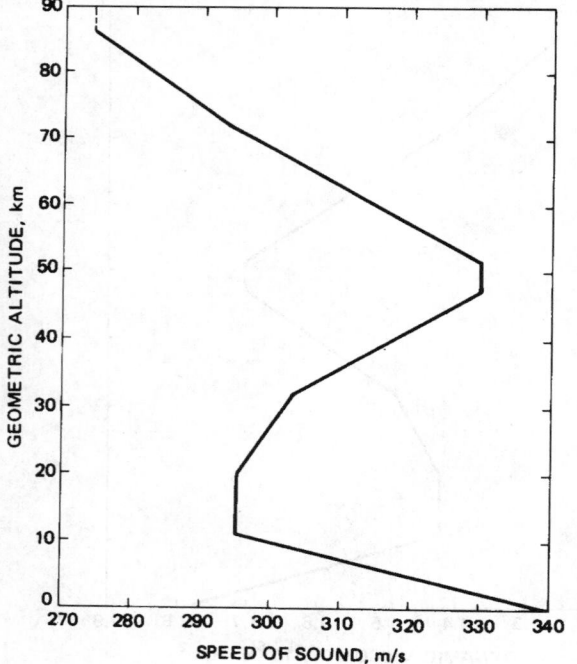

FIGURE 10. Speed of sound as a function of geometric altitude.

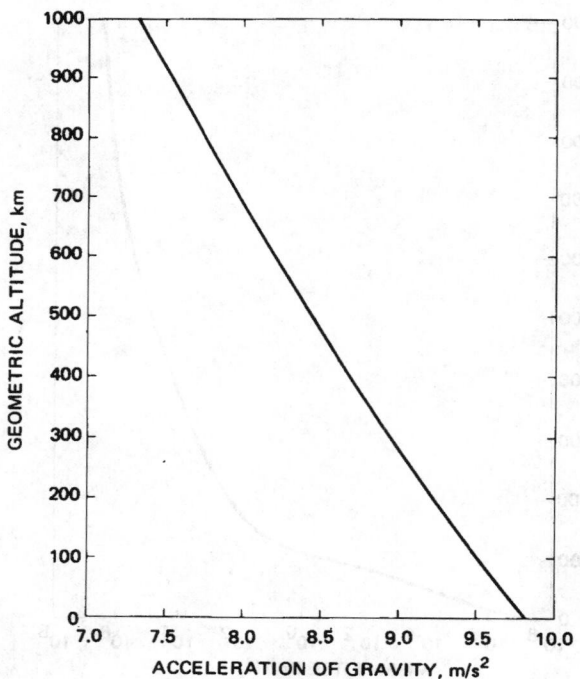

FIGURE 12. Acceleration of gravity as a function of geometric altitude.

GEOGRAPHICAL AND SEASONAL VARIATION IN SOLAR RADIATION

This table gives the amount of solar radiation reaching a unit area at the top of the Earth's atmosphere per day as a function of latitude and approximate date. It is based upon a solar constant (total energy per unit area at the Earth's average orbital distance) of 1373 W/m². Absorption of radiation by the atmosphere is not taken into consideration.

Reference

List, R. J., *Smithsonian Meteorological Tables, Seventh Edition*, Smithsonian Institution Press, Washington, D.C., 1962.

Daily Solar Radiation in MJ/m²

Lat.	Mar. 21	Apr. 13	May 6	May 29	Jun. 2	Jul. 15	Aug. 8	Aug. 31
90°		18.0	32.8	42.4	45.7	42.2	32.5	17.7
80	6.6	18.0	32.3	41.8	45.0	41.6	32.0	17.7
70	13.0	22.3	31.8	39.9	43.0	39.7	31.5	22.0
60	19.0	27.0	34.4	39.7	41.6	39.4	34.0	26.7
50	24.4	31.1	36.8	40.7	42.0	40.5	36.5	30.8
40	29.1	34.3	38.6	41.3	42.1	41.1	38.3	33.9
30	32.9	36.7	39.4	41.1	41.4	40.8	39.1	36.3
20	35.7	38.0	39.2	39.7	39.7	39.5	38.9	37.5
10	37.4	38.1	37.9	37.4	37.1	37.2	37.6	37.7
0	38.0	37.1	35.5	34.1	33.5	34.0	35.2	36.6
−10	37.4	35.0	32.3	30.0	29.2	29.9	32.0	34.6
−20	35.7	31.8	28.0	25.2	24.1	25.1	27.8	31.5
−30	32.9	27.8	23.1	19.7	18.5	19.7	22.8	27.4
−40	29.1	22.8	17.5	14.0	12.6	13.9	17.4	22.6
−50	24.4	17.3	11.7	8.2	7.0	8.2	11.6	17.2
−60	19.0	11.4	5.9	2.9	2.0	2.9	5.9	11.3
−70	13.0	5.4	1.0				1.0	5.3
−80	6.6	0.3						0.3
−90								

Lat.	Sep. 23	Oct. 16	Nov. 8	Nov. 30	Dec. 22	Jan. 13	Feb. 4	Feb. 26
90°								
80	6.5	0.3						0.3
70	12.9	5.5	1.0				1.0	5.6
60	18.8	11.6	6.2	3.1	2.1	3.1	6.2	11.7
50	24.1	17.6	12.1	8.7	7.5	8.7	12.3	17.8
40	28.7	23.1	18.2	14.8	13.5	14.9	18.4	23.5
30	32.5	28.2	23.9	20.9	19.8	21.0	24.1	28.4
20	35.3	32.3	29.1	26.6	25.7	26.7	29.3	32.7
10	37.0	35.5	33.5	31.8	31.1	31.9	33.8	35.9
0	37.6	37.6	36.9	36.1	35.8	36.3	37.3	38.0
−10	37.0	38.6	39.4	39.5	39.6	39.7	39.7	39.1
−20	35.3	38.5	40.7	42.0	42.4	42.2	41.1	39.0
−30	32.5	37.2	40.9	43.3	44.2	43.5	41.3	37.7
−40	28.7	34.8	40.1	43.6	45.0	43.8	40.5	35.2
−50	24.1	31.5	38.3	43.1	44.8	43.2	38.6	31.9
−60	18.8	27.3	35.7	41.9	44.4	42.1	36.0	27.7
−70	12.9	22.6	33.0	42.2	45.9	42.4	33.3	22.9
−80	6.5	18.2	33.5	44.2	48.1	44.4	33.8	18.4
−90		18.2	34.0	44.8	48.8	45.1	34.4	18.4

INFRARED ABSORPTION BY THE EARTH'S ATMOSPHERE

Several constituents of the Earth's atmosphere absorb infrared radiation. At ground level the strongest absorbers are H_2O and CO_2, but 30 to 40 other compounds can make significant contributions. The centers of the most important absorption bands are listed below:

Molecule	Vibrational mode	Band center in cm^{-1}
H_2O	Bend	1595
H_2O	Symmetric O-H stretch	3657
H_2O	Antisymmetric O-H stretch	3756
CO_2	Bend	667
CO_2	Antisymmetric C-O stretch	2349
O_3	Bend	701
O_3	Antisymmetric O-O stretch	1042
O_3	Symmetric O-O stretch	1103
N_2O	Bend	589
N_2O	N-O stretch	1285
N_2O	N-N stretch	2224
CO	C-O stretch	2143
CH_4	Degenerate deformation	1306
CH_4	Degenerate stretch	3019

The HITRAN Molecular Spectroscopy Database (References 1 and 2) is a compilation of wavenumbers and intensities of more than 1.7 million spectral lines of atmospheric constituents. It is a valuable resource for calculating transmission of the atmosphere, radiative energy transfer, and other phenomena. The graph below, which was supplied by Walter J. Lafferty (Reference 3), gives the transmittance of the atmosphere for one set of conditions.

References

1. Rothman, L. S., et al., *J. Quant. Spectros. Radiat. Trans*fer 82, 5, 2003; *ibid.*, to be published, 2005.
2. HITRAN Molecular Spectroscopy Database, <http://cfa-www.Harvard.edu/HITRAN/hitrandata04/>.
3. Lafferty, W. J., Some Aspects of High Resolution Molecular Spectroscopy, in *Lectures on Molecular Physics*, Institute for the Structure of Matter, Centro de Fisica Miguel A. Catalan, Madrid, 1997.

Transmittance of U.S. Standard Atmosphere at Ground Level for a Path of 1 km at 296 K

ATMOSPHERIC CONCENTRATION OF CARBON DIOXIDE, 1958–2004

The data in this table were taken at the Mauna Loa Observatory in Hawaii and represent averages adjusted to the 15th of each month. The last column gives the average over the year. The concentration of CO_2 is given in parts per million by volume. Data from other measurement sites may be found in Reference 1.

The first graph illustrates the seasonal variation of CO_2 concentration and the steady increase over the last 45 years. The second graph summarizes the growth in the emissions of CO_2 into the atmosphere as a result of burning of fossil fuels (Reference 2).

References

1. Keeling, C.D., and Whorf, T.P., Atmospheric carbon dioxide record from Mauna Loa, Carbon Dioxide Information Analysis Center, Oak Ridge National Laboratory, U.S. Department of Energy, Oak Ridge, TN, 2005; <cdiac.esd.ornl.gov/trends/co2/sio-mlo.htm>.
2. Marland, G., Boden, T. A., and Andres, R. J., Global, Regional, and National CO_2 Emissions. In *Trends: A Compendium of Data on Global Change, 2001*. Carbon Dioxide Information Analysis Center, Oak Ridge National Laboratory, U.S. Department of Energy, Oak Ridge, TN; <cdiac.esd.ornl.gov/trends/emis/tre_glob.htm>.

CO_2 Concentration in ppm at Mauna Loa

Year	Jan.	Feb.	March	April	May	June	July	Aug.	Sept.	Oct.	Nov.	Dec.	Annual
1958			315.71	317.45	317.50		315.85	314.93	313.19		313.34	314.67	
1959	315.58	316.47	316.65	317.72	318.29	318.16	316.55	314.80	313.84	313.34	314.82	315.59	315.98
1960	316.43	316.97	317.58	319.03	320.03	319.59	318.18	315.91	314.16	313.84	315.00	316.19	316.91
1961	316.89	317.70	318.54	319.48	320.58	319.77	318.58	316.79	314.99	315.31	316.10	317.01	317.65
1962	317.94	318.56	319.69	320.58	321.01	320.61	319.61	317.40	316.26	315.42	316.69	317.69	318.45
1963	318.74	319.08	319.86	321.39	322.24	321.47	319.74	317.77	316.21	315.99	317.06	318.36	318.99
1964	319.57				322.24	321.89	320.44	318.70	316.70	316.87	317.68	318.71	
1965	319.44	320.44	320.89	322.13	322.16	321.87	321.21	318.87	317.81	317.30	318.87	319.42	320.03
1966	320.62	321.59	322.39	323.70	324.07	323.75	322.41	320.37	318.64	318.10	319.79	321.03	321.37
1967	322.33	322.50	323.04	324.42	325.00	324.09	322.55	320.92	319.26	319.39	320.72	321.96	322.18
1968	322.57	323.15	323.89	325.03	325.57	325.36	324.14	322.11	320.33	320.25	321.33	322.90	323.05
1969	324.00	324.42	325.64	326.66	327.38	326.70	325.89	323.67	322.38	321.78	322.85	324.12	324.62
1970	325.06	325.98	326.93	328.14	328.07	327.66	326.35	324.69	323.10	323.07	324.01	325.13	325.68
1971	326.17	326.68	327.18	327.78	328.92	328.57	327.37	325.43	323.36	323.57	324.80	326.01	326.32
1972	326.77	327.63	327.75	329.72	330.07	329.09	328.05	326.32	324.84	325.20	326.50	327.55	327.46
1973	328.54	329.56	330.30	331.50	332.48	332.07	330.87	329.31	327.51	327.18	328.16	328.64	329.68
1974	329.35	330.71	331.48	332.65	333.08	332.25	331.18	329.40	327.44	327.37	328.46	329.58	330.25
1975	330.40	331.41	332.04	333.31	333.96	333.59	331.91	330.06	328.56	328.34	329.49	330.76	331.15
1976	331.74	332.56	333.50	334.58	334.87	334.34	333.05	330.94	329.30	328.94	330.31	331.68	332.15
1977	332.92	333.41	334.70	336.07	336.74	336.27	334.93	332.75	331.58	331.16	332.40	333.85	333.90
1978	334.97	335.39	336.64	337.76	338.01	337.89	336.54	334.68	332.76	332.54	333.92	334.95	335.50
1979	336.23	336.76	337.96	338.89	339.47	339.29	337.73	336.09	333.91	333.86	335.29	336.73	336.85
1980	338.01	338.36	340.08	340.77	341.46	341.17	339.56	337.60	335.88	336.02	337.10	338.21	338.69
1981	339.23	340.47	341.38	342.51	342.91	342.25	340.49	338.43	336.69	336.85	338.36	339.61	339.93
1982	340.75	341.61	342.70	343.57	344.13	343.35	342.06	339.82	337.97	337.86	339.26	340.49	341.13
1983	341.37	342.52	343.10	344.94	345.75	345.32	343.99	342.39	339.86	339.99	341.16	342.99	342.78
1984	343.70	344.50	345.29	347.08	347.43	346.79	345.40	343.28	341.07	341.35	342.98	344.22	344.42
1985	344.97	346.00	347.43	348.35	348.93	348.25	346.56	344.69	343.09	342.80	344.24	345.56	345.91
1986	346.29	346.96	347.86	349.55	350.21	349.54	347.94	345.91	344.86	344.17	345.66	346.90	347.15
1987	348.02	348.47	349.42	350.99	351.84	351.25	349.52	348.11	346.44	346.36	347.81	348.96	348.93
1988	350.43	351.72	352.22	353.59	354.22	353.79	352.39	350.44	348.72	348.88	350.07	351.34	351.48
1989	352.76	353.07	353.68	355.42	355.67	355.13	353.90	351.67	349.80	349.99	351.30	352.53	352.91
1990	353.66	354.70	355.39	356.20	357.16	356.22	354.82	352.91	350.96	351.18	352.83	354.21	354.19
1991	354.72	355.75	357.16	358.60	359.33	358.24	356.18	354.03	352.16	352.21	353.75	354.99	355.59
1992	355.98	356.72	357.81	359.15	359.66	359.25	357.03	355.00	353.01	353.31	354.16	355.40	356.37
1993	356.70	357.16	358.38	359.46	360.28	359.59	357.58	355.52	353.70	353.98	355.33	356.80	357.04
1994	358.36	358.91	359.97	361.27	361.68	360.94	359.55	357.49	355.84	355.99	357.58	359.04	358.89
1995	359.96	361.00	361.64	363.45	363.79	363.26	361.90	359.46	358.06	357.75	359.56	360.70	360.88
1996	362.05	363.25	364.03	364.72	365.41	364.97	363.65	361.49	359.46	359.60	360.76	362.33	362.64
1997	363.18	364.00	364.57	366.35	366.80	365.62	364.47	362.51	360.19	360.77	362.43	364.28	363.76
1998	365.32	366.15	367.31	368.61	369.29	368.87	367.64	365.77	363.90	364.23	365.46	366.97	366.63
1999	368.15	368.87	369.59	371.14	371.00	370.35	369.27	366.94	364.63	365.12	366.67	368.01	368.31
2000	369.14	369.46	370.52	371.66	371.82	371.70	370.12	368.12	366.62	366.73	368.29	369.53	369.48
2001	370.28	371.50	372.12	372.87	374.02	373.30	371.62	369.55	367.96	368.09	369.68	371.24	371.02

Year	Jan.	Feb.	March	April	May	June	July	Aug.	Sept.	Oct.	Nov.	Dec.	Annual
2002	372.43	373.09	373.52	374.86	375.55	375.40	374.02	371.49	370.71	370.24	372.08	373.78	373.10
2003	374.68	375.63	376.11	377.65	378.35	378.13	376.62	374.50	372.99	373.00	374.35	375.70	375.64
2004	376.79	377.37	378.41	380.52	380.63	379.57	377.79	375.86	374.06	374.24	375.86	377.48	377.38

CO_2 Concentration at Mauna Loa

CO_2 Emmissions from Burning of Fossil Fuels

MEAN TEMPERATURES IN THE UNITED STATES, 1900–1992

Historical records of atmospheric temperatures have been analyzed to obtain mean temperatures in °C for 23 climatically distinct regions of the United States. The table below gives the average over these 23 regions, which cover completely the contiguous 48 states. Data for the individual regions and for other parts of the world may be found in the references.

The data are presented as temperature anomalies, i.e., as deviations (in °C) from the average temperature at each individual recording station over a 1961–1990 reference period. The trend in the temperature anomaly thus gives an indication of the long-term variation in average temperatures.

CY Mean: Calendar year mean (January-December)
Winter: December-February
Spring: March-May
Summer: June-August
Fall: September-November

References

1. Karl, T. R., Easterling, D. R., Knight, R. W., and Hughes, P. Y., in *Trends '93: A Compendium of Data on Global Change*, p. 686, Boden, T. A., Kaiser, D. P., Sepanski, R. J., and Stoss, F. W., Eds., ORNL/CDIAC-65, Oak Ridge National Laboratory, Oak Ridge, TN, 1994.
2. Carbon Dioxide Information Analysis Center, WWW site <http://cdiac.esd.ornl.gov/ftp/trends93>.

Year	CY Mean	Winter	Spring	Summer	Fall
1900	0.46		0.21	0.27	0.85
1901	−0.21	0.07	−0.46	0.48	−0.28
1902	−0.14	−0.69	0.48	−0.58	0.12
1903	−0.77	−0.84	0.10	−0.83	−1.02
1904	−0.72	−1.86	−0.39	−0.92	−0.21
1905	−0.45	−1.86	0.67	−0.32	−0.25
1906	−0.04	0.23	−0.69	−0.42	−0.10
1907	−0.23	1.09	−0.61	−0.85	−0.39
1908	−0.11	0.73	0.36	−0.64	−0.59
1909	−0.36	0.82	−1.06	0.06	0.01
1910	−0.14	−2.08	0.95	−0.55	0.13
1911	0.02	0.52	0.20	−0.19	−0.62
1912	−0.88	−1.50	−0.75	−0.76	−0.51
1913	−0.23	−0.74	−0.65	−0.07	0.22
1914	−0.05	0.48	0.04	0.14	0.32
1915	−0.11	−0.37	−0.18	−1.16	0.31
1916	−0.77	−0.29	−0.36	−0.30	−1.23
1917	−1.34	−1.93	−1.75	−0.73	−1.04
1918	−0.14	−2.02	0.30	0.03	−0.09
1919	−0.16	0.69	0.00	0.12	−0.13
1920	−0.37	−0.83	−0.96	−0.67	0.02
1921	0.87	1.56	0.83	0.52	0.32
1922	0.01	−0.44	0.15	0.16	0.41
1923	−0.10	0.23	−1.02	−0.07	−0.09
1924	−1.01	0.13	−1.27	−0.64	−0.53
1925	0.20	−0.44	0.57	0.05	−0.41
1926	−0.01	0.97	−0.58	−0.27	0.02
1927	0.20	1.11	0.41	−0.83	0.83
1928	−0.08	−0.40	−0.28	−0.43	−0.08
1929	−0.68	−1.94	0.30	−0.39	−0.78
1930	−0.12	0.07	0.09	−0.07	−0.25
1931	0.81	1.16	−0.71	0.51	1.41
1932	−0.14	1.75	−0.58	0.26	−0.78
1933	0.35	−0.60	−0.08	0.45	0.54
1934	0.86	1.45	0.77	0.86	0.82
1935	0.12	0.84	0.12	0.31	−0.47
1936	−0.10	−2.23	0.48	1.00	−0.32
1937	−0.13	−0.65	−0.24	0.66	−0.03
1938	0.71	1.31	0.98	0.39	0.08
1939	0.38	0.36	0.34	0.18	0.15
1940	0.06	0.03	−0.10	0.18	0.12
1941	0.79	1.56	0.44	0.28	0.85

Year	CY Mean	Winter	Spring	Summer	Fall
1942	0.01	0.35	0.22	0.08	0.06
1943	−0.17	0.20	−0.46	0.54	−0.79
1944	0.04	0.61	−0.38	−0.20	0.43
1945	−0.02	0.30	0.25	−0.47	0.08
1946	0.53	−0.26	1.21	−0.21	0.18
1947	0.10	0.47	−0.42	0.03	0.85
1948	−0.22	−0.67	−0.01	0.02	−0.15
1949	0.05	−0.77	0.39	0.33	0.23
1950	−0.37	0.39	−1.02	−0.86	0.13
1951	−0.45	0.05	−0.69	−0.30	−0.70
1952	0.03	0.68	−0.40	0.49	−1.22
1953	0.61	1.91	0.03	0.31	0.31
1954	0.57	1.47	−0.22	0.36	0.70
1955	−0.25	−0.33	0.12	0.19	−0.60
1956	−0.05	−0.16	−0.42	0.01	−0.37
1957	0.45	1.02	0.44	0.25	−0.13
1958	0.14	0.93	0.06	0.11	0.45
1959	0.12	−0.60	0.16	0.50	−0.52
1960	−0.27	0.43	−0.98	0.01	0.56
1961	−0.01	−0.02	−0.17	0.18	−0.29
1962	−0.04	−0.52	−0.09	−0.41	0.65
1963	−0.06	−1.35	0.61	0.12	1.38
1964	−0.27	−1.30	−0.24	−0.08	−0.40
1965	−0.03	−0.07	−0.56	−0.42	0.28
1966	−0.34	−0.31	−0.25	−0.07	−0.13
1967	−0.13	0.23	−0.05	−0.37	−0.32
1968	−0.28	−0.31	−0.16	−0.12	0.01
1969	−0.06	−0.36	−0.54	0.12	−0.16
1970	−0.12	−0.17	−0.44	0.32	−0.07
1971	−0.04	−0.08	−0.99	0.01	0.56
1972	−0.13	0.20	0.26	−0.19	0.05
1973	0.48	−0.23	0.47	0.22	0.90
1974	0.16	0.52	0.74	−0.32	−0.39
1975	−0.09	0.63	−0.79	0.03	−0.20
1976	−0.62	0.88	−0.09	−0.54	−1.72
1977	0.32	−1.95	1.07	0.60	0.68
1978	−0.37	−1.31	0.23	0.09	0.21
1979	−0.53	−2.92	0.09	−0.21	−0.07
1980	0.18	0.72	−0.25	0.43	−0.04
1981	0.64	0.90	0.80	0.57	0.36
1982	−0.08	−0.86	0.03	−0.11	0.06
1983	0.40	2.33	−0.36	0.58	0.94
1984	0.21	−0.78	−0.30	0.31	0.07
1985	−0.26	−0.78	1.24	−0.23	0.05
1986	0.93	0.22	1.22	0.45	0.49
1987	0.67	1.52	0.97	0.33	−0.17
1988	−0.07	−0.26	−0.06	0.57	0.04
1989	−0.30	−0.28	0.36	0.12	−0.27
1990	0.72	0.41	0.72	0.41	0.66
1991	0.77	0.32	1.36	0.56	−0.31
1992		2.48	0.82	−0.70	

GLOBAL TEMPERATURE TREND, 1856–2004

This table and graph summarize the trend in annual mean global surface temperature from 1856 to 2004. The values were calculated from mean temperature anomalies by assuming an absolute global mean of 14.00°C, which is the best estimate for the 1961–1990 period. The 95% confidence interval for the annual mean temperature values since 1951 is ± 0.12°C; prior to 1900 this interval is ± 0.18°C.

Reference

Jones, P. D., Parker, D. E., Osborn, T. J., and Briffa, K. R., Global and hemispheric temperature anomalies--land and marine instrumental records, Carbon Dioxide Information Analysis Center, Oak Ridge National Laboratory, U.S. Department of Energy, Oak Ridge, TN; <cdiac.esd.ornl.gov/trends/temp/jonescru/jones.html>.

Year	t/°C	Year	t/°C	Year	t/°C	Year	t/°C	Year	t/°C
1856	13.621	1887	13.657	1918	13.590	1949	13.903	1980	14.102
1857	13.538	1888	13.689	1919	13.708	1950	13.790	1981	14.133
1858	13.584	1889	13.800	1920	13.707	1951	13.907	1982	14.019
1859	13.775	1890	13.587	1921	13.783	1952	13.975	1983	14.226
1860	13.628	1891	13.647	1922	13.677	1953	14.045	1984	14.031
1861	13.596	1892	13.593	1923	13.701	1954	13.830	1985	14.014
1862	13.473	1893	13.551	1924	13.653	1955	13.811	1986	14.096
1863	13.727	1894	13.586	1925	13.754	1956	13.728	1987	14.253
1864	13.526	1895	13.638	1926	13.884	1957	13.994	1988	14.240
1865	13.733	1896	13.801	1927	13.782	1958	14.061	1989	14.163
1866	13.777	1897	13.814	1928	13.773	1959	14.014	1990	14.310
1867	13.709	1898	13.661	1929	13.639	1960	13.972	1991	14.251
1868	13.775	1899	13.750	1930	13.848	1961	14.015	1992	14.116
1869	13.695	1900	13.808	1931	13.901	1962	14.008	1993	14.179
1870	13.704	1901	13.744	1932	13.863	1963	14.039	1994	14.232
1871	13.649	1902	13.651	1933	13.761	1964	13.767	1995	14.373
1872	13.734	1903	13.555	1934	13.861	1965	13.834	1996	14.227
1873	13.673	1904	13.557	1935	13.828	1966	13.916	1997	14.411
1874	13.624	1905	13.629	1936	13.879	1967	13.907	1998	14.579
1875	13.576	1906	13.708	1937	13.976	1968	13.895	1999	14.340
1876	13.548	1907	13.495	1938	14.076	1969	14.040	2000	14.290
1877	13.789	1908	13.523	1939	13.962	1970	13.969	2001	14.422
1878	13.943	1909	13.553	1940	13.917	1971	13.813	2002	14.475
1879	13.713	1910	13.558	1941	14.028	1972	13.959	2003	14.475
1880	13.704	1911	13.535	1942	13.980	1973	14.093	2004	14.455
1881	13.754	1912	13.594	1943	14.001	1974	13.829		
1882	13.738	1913	13.607	1944	14.158	1975	13.877		
1883	13.682	1914	13.752	1945	14.039	1976	13.800		
1884	13.651	1915	13.841	1946	13.880	1977	14.057		
1885	13.650	1916	13.627	1947	13.895	1978	13.964		
1886	13.746	1917	13.505	1948	13.907	1979	14.067		

Annual Mean Global Temperature in °C

ATMOSPHERIC ELECTRICITY

Hans Dolezalek, Hannes Tammet, John Latham, and Martin A. Uman

I. SURVEY AND GLOBAL CIRCUIT

Hans Dolezalek

The science of atmospheric electricity originated in 1752 by an experimental proof of a related earlier hypothesis (that lightning is an electrical event). In spite of a large effort, in part by such eminent physicists as Coulomb, Lord Kelvin, and many others, an overall, proven theory able to generate models with sufficient resolution is not yet available. Generally accepted and encompassing text books are now more than 20 years old. The voluminous proceedings of the, so far, nine international atmospheric electricity conferences (1954 to 1992) give much valuable detail and demonstrate impressive progress, as do a number of less comprehensive textbooks published in the last 20 years, but a general theory as indicated above is not yet created. Only now, certain related measuring techniques and mathematical possibilities are emerging.

Applications to practical purposes do exist in the field of lightning research (including the electromagnetic radiation emanating from lightning) by the establishment of lightning-location networks and by the now developing possibility to detect electrified clouds which pose hazards to aircraft. Application of atmospheric electricity to other parts of meteorology seems to be promising but so far has seldom been instituted. Because some atmospheric electric signals propagate around the earth and because of the existence of a global circuit, applications for the monitoring of global change processes and conditions are now being proposed. Significant secular changes in the global circuit would indicate a change in the global climate; the availability of many old data (about a span of 100 years) could help detect a long-term trend.

The concept of the "global circuit" is based on the theory of the global spherical capacitor: both, the solid (and liquid) earth as one electrode, and the high atmospheric layers (about the ionosphere) as the other, are by orders of magnitude more electrically conductive than the atmosphere between them. According to the "classical picture of atmospheric electricity", this capacitor is continuously charged by the common action of all thunderstorms to a d.c. voltage difference of several hundred kilovolts, the earth being negative. The much smaller but still existing conductivity of the atmosphere allows a current flowing from the ionosphere to the ground, integrated for all sink areas of the whole earth, of the order of 1.5 kA. In this way, a global circuit is created with many generators and sink-areas both interspaced and distributed over the whole globe, all connected to two nodes: ionosphere and ground. Within the scope of the global circuit, for each location, the current density (order of several pA/m²) is determined by the voltage difference between ionosphere and ground (which is the same for all locations but varying in time) and the columnar resistance reaching from the ground up to the ionosphere (in the order of 10^{17} Ωm^2).

Natural processes, especially meteorological processes and some human activity, which produce or move electric charges ("space charges") or affect the ion distribution, constitute local generators and thereby "local circuits", horziontally and/or in parallel or antiparallel to the local part of the global circuit. In many cases, the local currents are much stronger than the global ones, making the measurement of the global current at a given location and/or during a period of time very difficult or, often, impossible. The strongest local circuits usually occur with certain weather conditions (precipitation, fog, high wind, blown-up dust or snow, heavy cloudiness) which make measurement of the global circuit impossible everywhere: but even in their absence local generators exist in varying magnitudes and of different characters. The separation of the local and global shares in the measured values of current density is a central problem of the science of atmospheric electricity. Aerological measurements are of high value in this regard.

The above description is within the "classical picture" of atmospheric electricity, a group of hypotheses to explain the electrification of the atmosphere. It is probably fundamentally correct but certainly not complete; it has not yet been confirmed by systems of measurements resulting in no inner contradictions. In particular, extraterrestrial influences must be permitted; their general significance is still under debate.

Within this "classical picture" a kind of electric standard atmosphere may be constructed as shown in Table 1.

Values with a star, *, are rough average values from measurement. A star in parentheses, (*), points to a typical value from one or a few measurements. All other values have been calculated from starred values, under the assumption that at 2 km 50% and at 12 km 90% of the columnar resistance is reached. Voltage drop along one of the partial columns can be calculated by subtracting the value for the lower column from that of the upper one. Columnar resistances, conductances, and capacitances are valid for that particular part of the column which is indicated at the left. Capacitances are calculated with the formula for plate capacitors, and this fact must be considered also for the time constants for columns.

According to measurements, U, the potential difference between 0 m and 65 km may vary by a factor of approximately 2. The total columnar resistance, R_c, is estimated to vary up to a factor of 3, the variation being due to either reduction of conductivity in the exchange layer (about lowest 2 km of this table) or to the presence of high mountains; in both cases the variation is caused in the troposphere. Smaller variations in the stratosphere and mesosphere are being discussed because of aerosols there. The air-earth current density in fair weather varies by a factor of 3 to 6 accordingly. Conductivity near the ground varies by a factor of about 3 but only decreasing; increase of conductivity due to extraordinary radioactivity is a singular event. The field strength near the ground varies as a consequence of variations of air-earth current density and conductivity from about 1/3 to about 10 times of the value quoted in the table. Conductivity near the ground shows a diurnal and an annual variation which depends strongly on the locality: air-earth current density shows a diurnal and annual variation because the earth-ionosphere potential difference undergoes such variations, and also because the columnar resistance is supposed to have a diurnal and probably an annual variation.

Conductivities and air-earth current densities on high mountains are greater than at sea level by factors of up to 10. Conductivity decreases when atmospheric humidity increases. Values for space charges are not quoted because measurements are too few to allow calculation of average values. Values of parameters over the oceans are still rather uncertain.

Theoretically, in fair-weather conditions, Ohm's law must be fulfilled for the electric field, the conduction current density, and the electrical conductivity of the atmosphere. Deviations point to shortcomings in the applied measuring techniques. Data which are representative for a large area (in the extreme, "globally representative data", i.e. data on the global circuit), can on the ground be obtained only by stations on an open plane and only if local generators are either small or constant or are independently measured. Certain measurements with instrumented aircraft provide globally representative information valid for the period of the actual measurement.

TABLE 1. Electrical Parameters of the Clear (Fair-Weather) Atmosphere, Pertinent to the Classical Picture of Atm. Electricity (Electric Standard Atmosphere)

Part of atmosphere for which the values are calculated (elements are in free, cloudless atmosphere)	Currents, I, in A; and current densities, i, in A/m²	Potential differences, U, in V; field strength E in V/m; $U = 0$ at sea level	Resistances, R, in Ω; columnar resistances, R_c, in Ω m² and resistivities, ρ, in Ω m	Conductances, G, in Ω^{-1}; columnar conductances G_c, in Ω^{-1}m²; total conductivities, γ, in Ω^{-1}m⁻¹	Capacitances, C, in F; columnar capacitances, C_c, in F m⁻² and capacitivities, ε, in F m⁻¹	Time constants τ, in seconds
Volume element at about sea level, 1 m³	$i = 3 \times 10^{-12*}$	$E_0 = 1.2 \times 10^{2*}$	$\rho_0 = 4 \times 10^{13}$	$\gamma_0 = 2.5 \times 10^{-14}$	$\varepsilon_0 = 8.9 \times 10^{-12*}$	$\tau_0 = 3.6 \times 10^2$
Lower column of 1 m² cross section from sea level to 2 km height	$i = 3 \times 10^{-12}$	At upper end: $U_1 = 1.8 \times 10^5$	$R_{c1} = 6 \times 10^{16}$	$G_{c1} = 1.7 \times 10^{-17}$	$C_{c1} = 4.4 \times 10^{-15}$	$\tau_{c1} = 2.6 \times 10^2$
Volume element at about 2 km height, 1 m³	$i = 3 \times 10^{-12}$	$E_2 = 6.6 \times 10^1$	$\rho_2 = 2.2 \times 10^{13(*)}$	$\gamma_2 = 4.5 \times 10^{-14}$	$\varepsilon_2 = 8.9 \times 10^{-12*}$	$\tau_2 = 2 \times 10^2$
Center column of 1 m² cross section from 2 to 12 km	$i = 3 \times 10^{-12}$	At upper end: $U_m = 3.15 \times 10^5$	$R_{cm} = 4.5 \times 10^{16}$	$G_{cm} = 5 \times 10^{-17}$	$C_{cm} = 8.8 \times 10^{-16}$	$\tau_{cm} = 1.8 \times 10^1$
Volume element at about 12 km height, 1 m³	$i = 3 \times 10^{-12}$	$E_{12} = 3.9 \times 10^0$	$\rho_{12} = 1.3 \times 10^{12(*)}$	$\gamma_{12} = 7.7 \times 10^{-13}$	$\varepsilon_{12} = 8.9 \times 10^{-12}$	$\tau_{12} = 1.2 \times 10^1$
Upper column of 1 m² cross section from 12 to 65 km height	$i = 3 \times 10^{-12}$	At upper end: $U_u = 3.5 \times 10^5$	$R_{cu} = 1.5 \times 10^{16}$	$G_{cu} = 2.5 \times 10^{-17}$	$C_{cu} = 1.67 \times 10^{-16}$	$\tau_{cm} = 6.7 \times 10^0$
Whole column of 1 m² cross section from 0 to 65 km height	$i = 3 \times 10^{-12}$	At upper end: $U = 3.5 \times 10^5$	$R_c = 1.2 \times 10^{17}$	$G_c = 8.3 \times 10^{-18}$	$C_c = 1.36 \times 10^{-16}$	$\tau_c = 1.64 \times 10^1$
Total spherical capacitor area: 5×10^{14} m²	$i = 1.5 \times 10^3$	$U = 3.5 \times 10^{5*}$	$R = 2.4 \times 10^2$	$G = 4.2 \times 10^{-3}$	$C = 6.8 \times 10^{-2}$	$\tau = 1.64 \times 10^1$

Note: All currents and fields listed are part of the global circuit, i.e., circuits of local generators are not included. Values are subject to variations due to latitude and altitude of the point of observation above sea level, locality with respect to sources of disturbances, meteorological and climatological factors, and man-made changes. For more explanations, see text.

II. AIR IONS

Hannes Tammet

The term "air ions" signifies all airborne particles which are the carriers of the electrical current in the air and have drift velocities determined by the electric field.

The probability of electrical dissociation of molecules in the atmospheric air under thermodynamic equilibrium is near to zero. The average ionization at the ground level over the ocean is $2 \cdot 10^6$ ion pairs m⁻³s⁻¹. This ionization is produced mainly by cosmic rays. Over the continents the ionizing radiation from soil and from radioactive substances in the air each add about $4 \cdot 10^6$ m⁻³s⁻¹. The total average ionization rate of 10^7 m⁻³s⁻¹ is equivalent to 17 μR/h which is a customary expression of the background level of the ionizing radiations. The ionization rate over the ground varies in space due to the radioactivity of soil, and in time depending on the exchange of air between the atmosphere and radon-containing soil. Radioactive pollution increases the ionization rate. A temporary increase of about 10 times was registered in Sweden after the Chernobyl accident in 1986. The emission of Kr⁸⁵ from nuclear power plants can noticeably increase the global ionization rate in the next century. The ionization rate decreases with altitude near the ground and increases at higher altitudes up to 15 km, where it has a maximum of about $5 \cdot 10^7$ m⁻³s⁻¹. Solar X-ray and extreme UV radiation cause a new increase at altitudes over 60 km.

Local sources of air ions are point discharges in strong electric fields, fluidization of charged drops from waves, etc.

The enhanced chemical activity of an ion results in a chain of ion-molecule reactions with the colliding neutrals, and, in the first microsecond of the life of an air ion, a charged molecular cluster called the *cluster ion* is formed. According to theoretical calcula-

tions in the air free from exotic trace gases the following cluster ions should be dominant:

$NO_3^- \cdot (HNO_3) \cdot H_2O$, $NO_2^- \cdot (H_2O)_2$, $NO_3^- \cdot H_2O$, $O_2^- \cdot (H_2O)_4$, $O_2^- \cdot (H_2O)_5$,

$H_3O^+ \cdot (H_2O)_6$, $NH_4^+ \cdot (H_2O)_2$, $NH_4^+ \cdot (H_2O)$, $H_3O^+ \cdot (H_2O)_5$, $NH_4^+ \cdot NH_3$

A measurable parameter of air ions is the electrical mobility k, characterizing the drift velocity in the unit electric field. The mobility is inversely proportional to the density of air, and the results of measurements are as a rule reduced to normal conditions. According to mobility the air ions are called: fast or small or light ions with mobility $k > 5 \cdot 10^{-5}$ m²V⁻¹s¹, intermediate ions, and slow or large or heavy ions with mobility $k > 10^{-6}$ m²V⁻¹s¹. The boundary between intermediate and slow ions is conventional.

Cluster ions are fast ions. The masses of cluster ions may be measured with mass spectrometers, but the possible ion-molecule reactions during the passage of the air through nozzles to the vacuum chamber complicate the measurement. Mass and mobility of cluster ions are highly correlated. The experimental results[5] can be expressed by the empirical formula

$$m \approx \frac{850 \text{ u}}{[0.3 + k / (10^{-4} \text{ m}^2 \text{V}^{-1} \text{s}^{-1})]^3}$$

where u is the unified atomic mass unit.

The value of the transport cross-section of a cluster ion is needed to calculate its mobility according to the kinetic theory of Chapman and Enskog. The theoretical estimation of transport cross-sections is rough and cannot be used to identify the chemical structures of cluster ions. Mass spectrometry is the main technique of identification of cluster ions.[2]

Märk and Castleman[4] presented an overview of over 1000 publications on the experimental studies of cluster ions. Most of them present information about ions of millisecond age range. The low concentration makes it difficult to get detailed information about masses and mobilities of the natural atmospheric ions at ground level. The results of a 1-year continuous measurement[6] are as follows:

	+ ions	− ions	unit
Average mobility	1.36	1.56	$10^{-4} m^2 V^{-1} s^{-1}$
The corresponding mass	190	130	u
The corresponding diameter	0.69	0.61	nm
The average concentration	400	360	$10^6 m^{-3}$
The corresponding conductivity	8.7	9.0	fS

The distribution of tropospheric cluster ions according to the mobility and estimated mass is depicted in Figure 1.

The problems and results of direct mass spectrometry of natural cluster ions are analyzed by Eisele[2] for ground level and by Meyerott, Reagan and Joiner[5] for stratospheric measurements. Air ions in the high atmosphere are a subject of ionospheric physics.

During its lifetime (about 1 min), a cluster ion at ground level collides with nearly 10^{12} molecules. Thus the cluster ions are able to concentrate trace gases of very low concentration if they have an extra high electron or proton affinity. For example, Eisele[2] demon-

strated that a considerable fraction of positive atmospheric cluster ions in the unpolluted atmosphere at ground level probably consist of a molecule derived from pyridine. The concentration of these constituents is estimated to be about 10^{-12}. Therefore, air- ion mass and mobility spectrometry is considered as a promising technique for trace analysis in the air. Mass and mobility spectrometry of millisecond-age air ions has been developed as a technique of chemical analysis known as "plasma chromatography".[1] The sensitivity of the detection grows with the age of the cluster ions measured.

The mechanisms of annihilation of cluster ions are ion-ion recombination (on the average 3%) and sedimentation on aerosol particles (on the average 97% of cluster ions at ground level). The result of the combination of a cluster ion and neutral particle is a charged particle called an *aerosol ion*. In conditions of detailed thermodynamic equilibrium the probability that a spherical particle of diameter d carries q elementary charges is calculated from the Boltzmann distribution:

$$p_q(d) = (2\pi d/d_0)^{1/2} \exp(-q^2 d_0 / 2d)$$

where $d_0 = 115$ nm (at 18°C). The supposition about the detailed equilibrium is an approximation and the formula is not valid for particles less than d_0. On the basis of numerical calculations by Hoppel and Frick[3] the following charge probabilities can be derived:

d	3	10	30	100	300	1000	3000	nm
P_0	98	90	70	42	24	14	8	%
$p_{-1} + p_1$	2	10	30	48	41	25	15	%
$p_{-2} + p_2$	0	0	0	10	23	21	14	%
$P_{q>2}$	0	0	0	0	12	40	63	%
k_1	15000	1900	250	28	5.1	1.11	0.33	$10^{-9} m^2 V^{-1} s^{-1}$

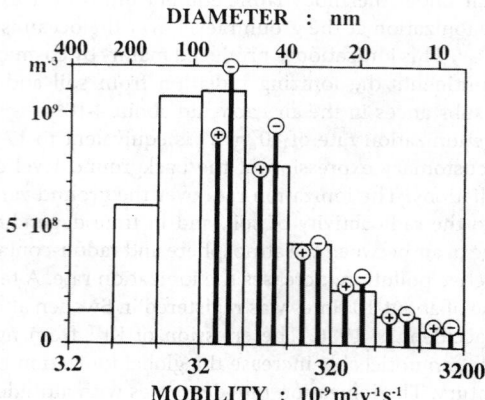

FIGURE 1. Average mobility and mass spectra of natural tropospheric cluster ions. Concentrations of the mobility fractions were measured in a rural site every 5 min over 1 year.[6] Ion mass is estimated according to the above empirical formula.

FIGURE 2. Mobility and size spectra of tropospheric aerosol ions.[6] The wide bars mark the fraction concentrations theoretically estimated on the basis of the standard size distribution of tropospheric aerosol. The pin bars with head + and − mark average values of positive and negative aerosol ion fraction concentrations measured in a rural site every 5 min during 4 months.

The last line of the table presents the mobility of a particle carrying one elementary charge. The distribution of the atmospheric aerosol ions over mobility is demonstrated in Figure 2.

Although the concentration of aerosol in continental air at ground level is an order of magnitude higher than the concentration of cluster ions, the mobilities of aerosol ions are so small that their percentage in air conductivity is less than 1%.

A specific class of aerosol ions are condensed aerosol ions produced as a result of the condensation of gaseous matter on the cluster ions. In aerosol physics the process is called ion-induced nucleation; it is considered as one among the processes of gas-to-particle conversion. The condensed aerosol ions have an inherent charge. Their sizes and mobilities are between the sizes and mobilities of cluster ions and of ordinary aerosol ions. Water and standard constituents of atmospheric air are not able to condense on the cluster ions in the real atmosphere. Thus the concentration of condensed aerosol ions depends on the trace constituents in the air and is very low in unpolluted air. Knowledge about condensed aerosol ions is poor because of measurement difficulties.

References

1. Carr, T. W., Ed., *Plasma Chromatography*, Plenum Press, New York and London, XII + 259 pp., 1984.
2. Eisele, F. L., Identification of Tropospheric Ions, *J. Geophys. Res.*, vol. 91, no. D7, pp. 7897–7906, 1986.
3. Hoppel, W. A., and Frick, G. M., The Nonequilibrium Character of the Aerosol Charge Distributions Produced by Neutralizers, *Aerosol Sci. Technol.*, vol. 12, no. 3, pp. 471–496, 1990.
4. Mark, T. D., and Castleman, A. W., Experimental Studies on Cluster Ions, in *Advances in Atomic and Molecular Physics*, vol. 20, pp. 65–172, Academic Press, 1985.
5. Meyerott, R. E., Reagan, J. B., and Joiner, R. G., The Mobility and Concentration of Ions and the Ionic Conductivity in the Lower Stratosphere, *J. Geophys. Res.*, vol. 85, no. A3, pp. 1273–1278, 1980.
6. Salm, J., Tammet, H., Iher, H., and Hörrak, U., Atmospheric Electrical Measurements in Tahkuse, Estonia (in Russian), in *Voprosy Atmosfernogo Elektrichestva*, pp. 168–175, Gidrometeoizdat, Leningrad, 1990.

III. THUNDERSTORM ELECTRICITY

John Latham

The development of improved radar techniques and instruments for in-cloud electrical and physical measurements, coupled with a much clearer recognition by the research community that establishment of the mechanism or mechanisms responsible for electric field development in thunderclouds, culminating in lightning, is inextricably linked to the concomitant dynamical and microphysical evolution of the clouds, has led to significant progress over the past decade.

Field studies indicate that in most thunderclouds the electrical development is associated with the process of glaciation, which can occur in a variety of incompletely understood ways. In the absence of ice, field growth is slow, individual hydrometeor charges are low, and lightning is produced only rarely. Precipitation — in the solid form, as graupel — also appears to be a necessary ingredient for significant electrification, as does significant convective activity and mixing between the clouds and their environments, via entrainment.

Increasingly, the view is being accepted that charge transfer leading to field-growth is largely a consequence of rebounding collisions between graupel pellets and smaller vapor-grown ice crystals, followed by the separation under gravity of these two types of hydrometeor. These collisions occur predominantly within the temperature range -15 to $-30°C$, and for significant charge transfer need to occur in the presence of supercooled cloud droplets.

The field evidence is inconsistent with an inductive mechanism, and extensive laboratory studies indicate that the principal charging mechanism is non-inductive and associated — in ways yet to be identified — with differences in surface characteristics of the interacting hydrometeors.

Laboratory studies indicate that the two most favored sites for corona emission leading to the lightning discharge are the tips of ephemeral liquid filaments, produced during the glancing collisions of supercooled raindrops, and protuberances on large ice crystals or graupel pellets. The relative importance of these alternatives will depend on the hydrometeor characteristics and the temperature in the regions of strongest fields; these features are themselves dependent on air-mass characteristics and climatological considerations.

A recently identified but unresolved question is why, in continental Northern Hemisphere thunderclouds at least, the sign of the charge brought to ground by lightning is predominantly negative in summer but more evenly balanced in winter.

IV. LIGHTNING

Martin A. Uman

From both ground-based weather-station data and satellite measurements, it has been estimated that there are about 100 lightning discharges, both cloud and ground flashes, over the whole earth each second; representing an average global lightning flash density of about 6 km^{-2}yr^{-1}. Most of this lightning occurs over the earth's land masses. For example, in central Florida, where thunderstorms occur about 90 days/yr, the flash density for discharges to earth is about 15 km^{-2}yr^{-1}. Some tropical areas of the earth have thunderstorms up to 300 days/yr.

Lightning can be defined as a transient, high-current electric discharge whose path length is measured in kilometers and whose most common source is the electric charge separated in the ordinary thunderstorm or cumulonimbus cloud. Well over half of all lightning discharges occur totally within individual thunderstorm clouds and are referred to as intracloud discharges. Cloud-to-ground lightning, however, has been studied more extensively than any other lightning form because of its visibility and its more practical interest. Cloud-to-cloud and cloud-to-air discharges are less common than intracloud or cloud-to-ground lightning.

Lightning between the cloud and earth can be categorized in terms of the direction of motion, upward or downward, and the sign of the charge, positive or negative, of the developing discharge (called a *leader*) which initiates the overall event. Over 90% of the worldwide cloud-to-ground discharges is initiated in the thundercloud by downward-moving negatively charged leaders and subsequently results in the lowering of negative charge to earth. Cloud-to-ground lightning can also be initiated by downward-moving positive leaders, less than 10% of the worldwide cloud-to-ground lightning being of this type although the exact percentage is a function of season and latitude. Lightning between cloud and ground can also be initiated by leaders which develop upward from the earth. These upward-initiated discharges are relatively rare, may be of either polarity, and generally occur from mountaintops and tall man-made structures.

We discuss next the most common type of cloud-to-ground lightning. A negative cloud-to-ground discharge or *flash* has an overall duration of some tenths of a second and is made up of various components, among which are typically three or four high-current pulses called *strokes*. Each stroke lasts about a millisecond, the separation time between strokes being typically several tens of milliseconds. Such lightning often appears to "flicker" because the human eye can just resolve the individual light pulse associated with each stroke. A drawing of the components of a negative cloud-to-ground flash is found in Figure 3. Some values for salient parameters are found in Table 1. The negatively charged *stepped leader* initiates the first stroke in a flash by propagating from cloud to ground through virgin air in a series of discrete steps. Photographically observed leader steps in clear air are typically 1 µs in duration and tens of meters in length, with a pause time between steps of about 50 µs. A fully developed stepped leader lowers up to 10 or more coulombs of negative cloud charge toward ground in tens of milliseconds with an average downward speed of about 2×10^5 m/s. The average leader current is in the 100 to 1000 A range. The steps have pulse currents of at least 1 kA. Associated with these currents are electric- and magnetic-field pulses with widths of about 1 µs or less and risetimes of about 0.1 µs or less. The stepped leader, during its trip toward ground, branches in a downward direction, resulting in the characteristic downward-branched geometrical structure commonly observed. The electric potential of the bottom of the negatively charged leader channel with respect to ground has a magnitude in excess of 10^7 V. As the leader tip nears ground, the electric field at sharp objects on the ground or at irregularities of the ground itself exceeds the breakdown value of air, and one or more upward-moving discharges (often called upward leaders) are initiated from those points, thus beginning the *attachment process*. An understanding of the physics of the attachment process is central to an understanding of the operation of lightning protection of ground-based objects and the effects of lightning on humans and animals, since it is the attachment process that determines where the lightning connects to objects on the ground and the value of the early currents which flow. When one of the upward-moving discharges from the ground (or from a lightning rod or an individual) contacts the tip of the downward-moving stepped leader, typically some tens of meters above the ground, the leader tip is effectively connected to ground potential. The negatively charged leader channel is then discharged to earth when a ground potential wave, referred to as the first *return stroke*, propagates continuously up the leader path. The upward speed of a return stroke near the ground is typically near one third the speed of light, and the speed decreases with height. The first return stroke produces a peak current near ground of typically 30 kA, with a time from zero to peak of a few microseconds. Currents measured at the ground fall to half of the peak value in about 50 µs, and currents of the order of hundreds of amperes may flow for times of a few milliseconds up to several hundred milliseconds. The longer-lasting currents are known as *continuing currents*. The rapid release of return stroke energy heats the leader channel to a temperature near 30,000 K and creates a high-pressure channel which expands and generates the shock waves that eventually become thunder, as further discussed later. The return stroke effectively lowers to ground the charge originally deposited onto the stepped-leader channel and additionally initiates the lowering of other charges which may be available to the top of its channel. First return-stroke electric fields exhibit a microsecond scale rise to peak with a typical peak value of 5 V/m, normalized to a distance of 100 km by an inverse distance relationship. Roughly half of the field rise to peak, the so-called "fast transition", takes place in tenths of a microsecond, an observation that can only be made if the field propagation is over a highly conducting surface such as salt water.

After the first return-stroke current has ceased to flow, the flash, including charge motion in the cloud, may end. The lightning is then called a single-stroke flash. On the other hand, if additional charge is made available to the top of the channel, a continuous or *dart leader* may propagate down the residual first-stroke channel at a typical speed of about 1×10^7 m/s. The dart leader lowers a charge of the order of 1 C by virtue of a current of about 1 kA. The dart leader then initiates the second (or any subsequent) return stroke. Subsequent return-stroke currents generally have faster zero-to-peak rise times than do first-stroke currents, but similar maximum rates of change, about 100 kA/µs. Some leaders begin as dart leaders, but toward the end of their trip toward ground become stepped leaders. These leaders are known as *dart-stepped leaders* and may have different ground termination points (and separate upward leaders) from the first stroke. Most often the dart-stepped leaders are associated with the second stroke of the flash. Nearly half of all flashes exhibit more than one termination point on ground with the distance between separate terminations being up to several kilometers. Subsequent return-stroke radiated electric and magnetic fields are similar to, but usually a factor of two or so smaller, than first return-stroke fields. About one third of all multiple-stroke flashes has at least one subsequent stroke which is larger than the first stroke.

Cloud-to-ground flashes that lower positive charge, though not common, are of considerable practical interest because their peak currents and total charge transfer can be much larger than for the more common negative ground flash. The largest recorded peak currents, those in the 200- to 300-kA range, are due to the return strokes of positive lightning. Such positive flashes to ground are initiated by downward-moving leaders which do not exhibit the distinct steps of their negative counterparts. Rather, they show a luminosity which is more or less continuous but modulated in intensity. Positive flashes are generally composed of a single stroke followed by a period of continuing current. Positive flashes are probably initiated from the upper positive charge in the thundercloud charge dipole when that cloud charge is horizontally separated from the negative charge beneath it, the source of the usual negative cloud-to-ground lightning. Positive flashes are relatively common in winter thunderstorms (snow storms), which produce few flashes overall, and are relatively uncommon in summer thunderstorms. The fraction of positive lightning in summer thunderstorms apparently increases with increasing latitude and with increasing height of the ground above sea level.

Distant lightning return stroke fields are often referred to as sferics (called "atmospherics" in the older literature). The peak in the sferics frequency spectrum is near 5 kHz due to the bipolar or ringing nature of the distant return-stroke electromagnetic signal and to the effects of propagation.

Thunder, the acoustic radiation associated with lightning, is sometimes divided into the categories "audible", sounds that one can hear, and "infrasonic", below a few tens of hertz, a frequency range that is inaudible. This division is made because it is thought that the mechanisms that produce audible and infrasonic thunder are different. Audible thunder is thought to be due to the expansion of a rapidly heated return stroke channel, as noted earlier, whereas infrasonic thunder is thought to be associated with the conversion to sound of the energy stored in the electrostatic field of the thundercloud when lightning rapidly reduces that cloud field.

The technology of artificially initiating lightning by firing upward small rocket trailing grounded wire of a few hundred meters length has been well-developed during the past decade. Such "triggered" flashes are similar to natural upward-initiated discharges from tall structure. They often contain subsequent strokes which,

FIGURE 3. Sequence of steps in cloud-to-ground lightning.

when they occur, are similar to the subsequent strokes in natural lightning. These triggered subsequent strokes have been the subject of considerable recent research.

Also in the past 10 years or so sophisticated lightning locating equipment has been installed throughout the world. For example, all ground flashes in the U.S. are now centrally monitored for research, for better overall weather prediction, and for hazard warning for aviation, electric utilities and other lightning-sensitive facilities.

Information on lightning physics can be found in M. A. Uman, *The Lightning Discharge*, Academic Press, San Diego, 1987; on lightning death and injury in *Medical Aspects of Lightning Injury*, C. Andrews, M. A. Cooper, M. Darveniza, and D. Mackerras, Eds.,

CRC Press, 1992. Ground flash location information for the U.S., in real time or archived, is available from Geomet Data Service of Tucson, AZ, which is also a source of the names of providers of those data in other countries.

Table 2 has data for cloud-to-ground lightning discharges bringing negative charge to earth. The values listed are intended to convey a rough feeling for the various physical parameters of lightning. No great accuracy is claimed since the results of different investigators are often not in good agreement. These values may, in fact, depend on the particular environment in which the lightning discharge is generated. The choice of some of the entries in the table is arbitrary.

TABLE 2. Data for Cloud-to-Ground Lightning Discharges

	Minimum[a]	Representative values	Maximum[a]
Stepped leader			
Length of step, m	3	50	200
Time interval between steps, μs	30	50	125
Average speed of propagation of stepped leader, m/s[b]	1.0×10^5	2.0×10^5	3.0×10^6
Charged deposited on stepped-leader channel, coulombs	3	5	20
Dart leader			
Speed of propagation, m/s[b]	1.0×10^6	1.0×10^7	2.4×10^7
Charged deposited on dart-leader channel, coulombs	0.2	1	6
Return stroke[c]			
Speed of propagation, m/s[b]	2.0×10^7	1.0×10^8	2.0×10^8
Maximum current rate of increase, kA/μs	<1	100	400
Time to peak current, μs	<1	2	30
Peak current, kA	2	30	200
Time to half of peak current, μs	10	50	250
Charge transferred excluding continuing current, coulombs	0.02	3	20
Channel length, km	2	5	15
Lightning flash			
Number of strokes per flash	1	4	26
Time interval between strokes in absence of continuing current, ms	3	60	100
Time duration of flash, s	10^{-2}	0.5	2
Charge transferred including continuing current, coulombs	3	30	200

[a] The words maximum and minimum are used in the sense that most measured values fall between these limits.

[b] Speeds of propagation are generally determined from photographic data and are "two-dimensional." Since many lightning flashes are not vertical, values stated are probably slight underestimates of actual values.

[c] First return strokes have longer times to current peak and generally larger charge transfer than do subsequent return strokes

Adapted from Uman, M.A., *Lightning*, Dover Paperbook, New York, 1986, and Uman, M.A., *The Lightning Discharge*, Academic Press, San Diego, 1987.

SPEED OF SOUND IN VARIOUS MEDIA

The speed of sound in various solids, liquids, and gases is given in these tables. While only a single parameter v is needed for liquids and gases, sound propagation in isotropic solids is characterized by three velocity parameters. For a solid of infinite extent (or of finite extent if all dimensions are much larger than a wavelength), there are two relevant quantities,

v_l: velocity of longitudinal waves
v_s: velocity of shear waves.

For a cylindrical rod with diameter much smaller than a wavelength,

v_{ext}: velocity of extensional waves along the rod. (Torsional waves in the rod are propagated at the same speed as sheer waves in an infinite solid.)

Table 1 lists values for a variety of solid materials. Table 2 covers gases liquids and gases; values for cryogenic liquids are given at the normal boiling point. Table 3 gives the speed of sound in pure water and in seawater of salinity $S = 3.5\%$ as a function of temperature. All values are in meters per second and are given for normal atmospheric pressure.

References

1. Gray, D. E., Ed., *American Institute of Physics Handbook, Third Edition*, McGraw Hill, New York, 1972.
2. Anderson, H.L., Ed., *A Physicist's Desk Reference*, American Institute of Physics, New York, 1989.
3. Younglove, B. A., Thermophysical Proeprties of Fluids. Part I, *J. Phys. Chem. Ref. Data*, 11, Suppl. 1, 1982.
4. Younglove, B. A., and Ely, J. F., Thermophysical Properties of Fluids. Part II, *J. Phys. Chem. Ref. Data*, 16, 577, 1987.
5. Haar, L., Gallagher, J. S., and Kell, G. S., *NBS/NRC Steam Tables*, Hemisphere Publishing Corp., New York, 1984.
6. Mason, W. P., *Physical Acoustics and the Properties of Solids*, D. Van Nostrand Co., Princeton, N.J., 1958.
7. *Landolt-Börnstein, Numerical Data and Functional Relationships in Science and Technology, New Series, II/5, Molecular Acoustics*, Springer-Verlag, Heidelberg, 1967.

TABLE 1. Speed of Sound in Solids at Room Temperature

Name	v_l/m s^{-1}	v_s/m s^{-1}	v_{ext}/m s^{-1}	Name	v_l/m s^{-1}	v_s/m s^{-1}	v_{ext}/m s^{-1}
				Steel, K9	5940	3250	5250
Metals				Tin, rolled	3320	1670	2730
Aluminum, rolled	6420	3040	5000	Titanium	6070	3125	5090
Beryllium	12890	8880	12870	Tungsten, annealed	5220	2890	4620
Brass (70 Cu, 30 Zn)	4700	2110	3480	Tungsten, drawn	5410	2640	4320
Constantan	5177	2625	4270	Zinc, rolled	4210	2440	3850
Copper, annealed	4760	2325	3810				
Copper, rolled	5010	2270	3750	*Other materials*			
Duralumin 17S	6320	3130	5150	Fused silica	5968	3764	5760
Gold, hard-drawn	3240	1200	2030	Glass, heavy silicate			
Iron, cast	4994	2809	4480	flint	3980	2380	3720
Iron, electrolytic	5950	3240	5120	Glass, light borate			
Iron, Armco	5960	3240	5200	crown	5100	2840	4540
Lead, annealed	2160	700	1190	Glass, pyrex	5640	3280	5170
Lead, rolled	1960	690	1210	Lucite	2680	1100	1840
Magnesium, annealed	5770	3050	4940	Nylon 6-6	2620	1070	1800
Molybdenum	6250	3350	5400	Polyethylene	1950	540	920
Monel metal	5350	2720	4400	Polystyrene	2350	1120	1840
Nickel	6040	3000	4900	Rubber, butyl	1830		
Platinum	3260	1730	2800	Rubber, gum	1550		
Silver	3650	1610	2680	Rubber, neoprene	1600		
Steel (1% C)	5940	3220	5180	Tungsten carbide	6655	3980	6220
Steel, 347 Stainless	5790	3100	5000				

TABLE 2. Speed of Sound in Liquids and Gases

Name	$t/°C$	$v/\text{m s}^{-1}$	Name	$t/°C$	$v/\text{m s}^{-1}$
			1-Pentadecene	20	1351
Liquids			Pentane	20	1008
Acetone	20	1203	Propane	−42.1	1158
Argon	−185.9	813	1-Propanol	20	1223
Benzene	25	1310	Tetrachloromethane	25	930
Bromobenzene	20	1169	Trichloromethane	25	987
Butane	−0.5	1034	1-Undecene	20	1275
1-Butanol	20	1258	Water	25	1497
Carbon disulphide	25	1140	Water (sea, $S = 3.5\%$)	25	1535
Chlorobenzene	20	1311			
Cyclohexane	19	1280	*Gases at 1 atm*		
1-Decene	20	1250	Air, dry	25	346
Diethyl ether	25	976	Ammonia	0	415
Ethane	−88.6	1326	Argon	27	323
Ethanol	20	1162	Carbon monoxide	0	338
Ethylene	−103.8	1309	Carbon dioxide	0	259
Ethylene glycol	25	1658	Chlorine	0	206
Fluorobenzene	20	1183	Deuterium	0	890
Glycerol	25	1904	Ethane	27	312
Helium	−268.9	180	Ethylene	27	331
Heptane	20	1162	Helium	0	965
1-Heptene	20	1128	Hydrogen	27	1310
Hexane	20	1083	Hydrogen bromide	0	200
Hydrogen	−252.9	1101	Hydrogen chloride	0	296
Iodobenzene	20	1114	Hydrogen iodide	0	157
Mercury	25	1450	Hydrogen sulfide	0	289
Methane	−161.5	1337	Methane	27	450
Methanol	20	1121	Neon	0	435
Nitrobenzene	25	1463	Nitric oxide	10	325
Nitrogen	−195.8	939	Nitrogen	27	353
1-Nonene	20	1218	Nitrous oxide	0	263
Octane	20	1197	Oxygen	27	330
1-Octene	20	1184	Sulfur dioxide	0	213
Oxygen	−183.0	906	Water (steam)	100	473

TABLE 3. Speed of Sound in Water and Seawater (S = 3.5%) at Different Temperatures

$t/°C$	$v/\text{m s}^{-1}$ Water	Seawater
0	1401.0	1449.4
10	1447.8	1490.4
20	1483.2	1522.2
25	1497.4	1535.1
30	1509.5	1546.2
40	1528.4	
50	1541.4	
60	1549.5	
70	1553.2	
80	1552.8	

ATTENUATION AND SPEED OF SOUND IN AIR AS A FUNCTION OF HUMIDITY AND FREQUENCY

This table gives the attenuation and speed of sound as a function of frequency at various values of relative humidity. All values refer to still air at 20°C.

References

1. Tables of Absorption and Velocity of Sound in Still Air at 68°F (20°C), AD-738576, National Technical Information Service, Springfield, VA.
2. Evans, L. B., Bass, H. E., and Sutherland, L. C., *J. Acoust. Soc. Am.*, 51, 1565, 1972.

Frequency (Hz)	Attenuation (dB/km)	Speed (m/s)
Relative humidity 0%		
20	0.51	343.477
40	1.07	343.514
50	1.26	343.525
63	1.43	343.536
100	1.67	343.550
200	1.84	343.559
400	1.96	343.561
630	2.11	343.562
800	2.27	343.562
1250	2.82	343.562
2000	4.14	343.562
4000	8.84	343.564
6300	14.89	343.565
10000	26.28	343.566
12500	35.81	343.566
16000	52.15	343.567
20000	75.37	343.567
40000	267.01	343.567
63000	644.66	343.567
80000	1032.14	343.567
Relative humidity 30%		
20	0.03	343.807
40	0.11	343.808
50	0.17	343.810
63	0.25	343.810
100	0.50	343.814
200	1.01	343.821
400	1.59	343.826
630	2.24	343.827
800	2.85	343.828
1250	5.09	343.828
2000	10.93	343.829
4000	38.89	343.831
6300	90.61	343.836
10000	204.98	343.846
12500	294.08	343.854
16000	422.51	343.865
20000	563.66	343.877
40000	1110.97	343.911
63000	1639.47	343.924
80000	2083.08	343.929

Frequency (Hz)	Attenuation (dB/km)	Speed (m/s)
Relative humidity 60%		
20	0.02	344.182
40	0.06	344.183
50	0.09	344.183
63	0.15	344.184
100	0.34	344.185
200	0.99	344.190
400	1.94	344.197
630	2.57	344.200
800	2.94	344.201
1250	4.01	344.202
2000	6.55	344.203
4000	18.73	344.204
6300	42.51	344.204
10000	101.84	344.206
12500	155.67	344.208
16000	247.78	344.211
20000	373.78	344.215
40000	1195.37	344.238
63000	2220.64	344.262
80000	2951.71	344.274
Relative humidity 100%		
20	0.01	344.685
40	0.04	344.685
50	0.06	344.685
63	0.09	344.685
100	0.22	344.686
200	0.77	344.689
400	2.02	344.695
630	3.05	344.699
800	3.57	344.701
1250	4.59	344.704
2000	6.29	344.705
4000	13.58	344.706
6300	27.72	344.706
10000	63.49	344.706
12500	96.63	344.707
16000	154.90	344.708
20000	237.93	344.709
40000	884.28	344.718
63000	1973.62	344.731
80000	2913.01	344.742

SPEED OF SOUND IN DRY AIR

The values in this table were calculated from the equation of state for dry air (average molecular weight 28.96) treated as a real gas. Values refer to standard atmospheric pressure. The speed of sound varies only slightly with pressure; at two atmospheres and −100°C the value decreases by 0.13%, while at two atmospheres and 80°C the speed increases by 0.04%.

Reference

Sytchev, V. V., Vasserman, A. A., Kozlov, A. D., Spiridonov, G. A., and Tsymarny, V. A., *Thermodynamic Properties of Air*, Hemisphere Publishing Corp., New York, 1987.

$t/°C$	$v_s/\text{m s}^{-1}$	$t/°C$	$v_s/\text{m s}^{-1}$	$t/°C$	$v_s/\text{m s}^{-1}$
−100	263.5	−35	309.5	30	349.1
−95	267.3	−30	312.7	35	352.0
−90	271.1	−25	315.9	40	354.8
−85	274.8	−20	319.1	45	357.6
−80	278.5	−15	322.3	50	360.4
−75	282.1	−10	325.4	55	363.2
−70	285.7	−5	328.4	60	365.9
−65	289.2	0	331.5	65	368.6
−60	292.7	5	334.5	70	371.3
−55	296.1	10	337.5	75	374.0
−50	299.5	15	340.4	80	376.7
−45	302.9	20	343.4		
−40	306.2	25	346.3		

MUSICAL SCALES

Equal Tempered Chromatic Scale
$A_4 = 440$ Hz
International Concert Pitch

Note	Frequency	Note	Frequency	Note	Frequency	Note	Frequency
C_0	16.35	C_2	65.41	C_4	261.63	C_6	1046.50
$C\#_0$	17.32	$C\#_2$	69.30	$C\#_4$	277.18	$C\#_6$	1108.73
D_0	18.35	D_2	73.42	D_4	293.66	D_6	1174.66
$D\#_0$	19.45	$D\#_2$	77.78	$D\#_4$	311.13	$D\#_6$	1244.51
E_0	20.60	E_2	82.41	E_4	329.63	E_6	1318.51
F_0	21.83	F_2	87.31	F_4	349.23	F_6	1396.91
$F\#_0$	23.12	$F\#_2$	92.50	$F\#_4$	369.99	$F\#_6$	1479.98
G_0	24.50	G_2	98.00	G_4	392.00	G_6	1567.98
$G\#_0$	25.96	$G\#_2$	103.83	$G\#_4$	415.30	$G\#_6$	1661.22
A_0	27.50	A_2	110.00	A_4	440.00	A_6	1760.00
$A\#_0$	29.14	$A\#_2$	116.54	$A\#_4$	466.16	$A\#_6$	1864.66
B_0	30.87	B_2	123.47	B_4	493.88	B_6	1975.53
C_1	32.70	C_3	130.81	C_5	523.25	C_7	2093.00
$C\#_1$	34.65	$C\#_3$	138.59	$C\#_5$	554.37	$C\#_7$	2217.46
D_1	36.71	D_3	146.83	D_5	587.33	D_7	2349.32
$D\#_1$	38.89	$D\#_3$	155.56	$D\#_5$	622.25	$D\#_7$	2489.02
E_1	41.20	E_3	164.81	E_5	659.26	E_7	2637.02
F_1	43.65	F_3	174.61	F_5	698.46	F_7	2793.83
$F\#_1$	46.25	$F\#_3$	185.00	$F\#_5$	739.99	$F\#_7$	2959.96
G_1	49.00	G_3	196.00	G_5	783.99	G_7	3135.96
$G\#_1$	51.91	$G\#_3$	207.65	$G\#_5$	830.61	$G\#_7$	3322.44
A_1	55.00	A_3	220.00	A_5	880.00	A_7	3520.00
$A\#_1$	58.27	$A\#_3$	233.08	$A\#_5$	932.33	$A\#_7$	3729.31
B_1	61.74	B_3	246.94	B_5	987.77	B_7	3951.07
						C_8	4186.01

Scientific or Just Scale
$C_4 = 256$ Hz

Note	Frequency	Note	Frequency	Note	Frequency	Note	Frequency
C_0	16	C_2	64	C_4	256	C_6	1024
D_0	18	D_2	72	D_4	288	D_6	1152
E_0	20	E_2	80	E_4	320	E_6	1280
F_0	21.33	F_2	85.33	F_4	341.33	F_6	1365.33
G_0	24	G_2	96	G_4	384	G_6	1536
A_0	26.67	A_2	106.67	A_4	426.67	A_6	1706.67
B_0	30	B_2	120	B_4	480	B_6	1920
C_1	32	C_3	128	C_5	512	C_7	2048
D_1	36	D_3	144	D_5	576	D_7	2304
E_1	40	E_3	160	E_5	640	E_7	2560
F_1	42.67	F_3	170.67	F_5	682.67	F_7	2730.67
G_1	48	G_3	192	G_5	768	G_7	3072
A_1	53.33	A_3	213.33	A_5	853.33	A_7	3413.33
B_1	60	B_3	240	B_5	960	B_7	3840
						C_8	4096

CHARACTERISTICS OF HUMAN HEARING

The human ear is sensitive to sound waves with frequencies in the range from a few hertz to almost 20 kHz. Auditory response is usually expressed in terms of the *loudness level* of a sound, which is a measure of the sound pressure. The reference level, which is given in the unit *phon,* is a pure tone of frequency 1000 Hz with sound pressure of 20 μPa (in cgs units, $2 \cdot 10^{-4}$ dyn/cm^2); loudness level is usually expressed in decibels (dB) relative to this reference level. If a normal observer perceives an arbitrary sound to be equally loud as this reference sound, the sound is said to have the loudness level of the reference. The sensitivity of the typical human ear ranges from about 0 dB, the threshold loudness level, to about 140 dB, the level at which pain sets in. The minimum detectable level thus represents a sound wave of pressure 20 μPa and intensity (power density) 10^{-16} W/cm^2.

The following figure illustrates the frequency dependence of the threshold for an average young adult.

The relation between loudness level and frequency for a typical person is expressed by the following table:

Sound pressure level in dB relative to 20 μPa	Frequency in Hz					
	125	500	1000	4000	8000	10000
10			10	18		
20		16	20	28	11	
30	4	27	30	37	21	17
40	17	39	40	45	30	26
50	34	52	50	54	38	35
60	52	65	60	64	47	44
70	70	76	70	73	56	54
80	86	86	80	83	66	64
90	98	96	90	94	77	74
100	108	105	100	106	88	86

Thus, a 10,000 Hz tone at a pressure level of 50 dB seems equally loud as a 1000 Hz tone at a pressure of 35 dB.

The term *noise* refers to any unwanted sound, either a pure tone or a mixture of frequencies. Since the sensitivity of the ear is frequency dependent, as illustrated by the above table, noise level is expressed in a frequency-weighted scale, known as A-weighting. Decibel readings on this scale are designated as dBa. Typical noise levels from various sources are illustrated in this table:

Source	Noise level in dBa
Rocket engine	200
Jet aircraft engine	160
Light aircraft, cruising	140
Tractor, 150 hp	115
Electric motor, 100 hp at 2600 rpm	105
Pneumatic drill	100
Subway train	90
Vacuum cleaner	85
Heavy automobile traffic	75
Conversational speech	65
Whispered speech	40
Background noise, recording studio	25-30

Recommended noise thresholds in the workplace have been established by the American Conference of Government Industrial Hygenists. Some examples of the maximum safe levels for different daily exposure times are given below.

Duration of exposure	Max. level in dBa
24 h	80
8 h	85
4 h	88
1 h	94
30 min	97
15 min	100
2 min	109
28 s	115
0.11 s	139

No exposure greater than 140 dBa is permitted. Further details may be found in Reference 3.

References

1. Anderson, H. L., Ed., *A Physicist's Desk Reference*, American Institute of Physics, New York, 1989, chap. 2.
2. Gray, D. E., Ed., *American Institute of Physics Handbook, Third Edition*, McGraw Hill, New York, 1972, chap. 3.
3. *Threshold Limit Values for Chemical Substances and Physical Agents; Biological Exposure Indices*, 1999 Edition, American Conference of Governmental Industrial Hygienists, 1330 Kemper Meadow Drive, Cincinnati, OH 45240-1634.

Section 15
Practical Laboratory Data

STANDARD ITS-90 THERMOCOUPLE TABLES

The Instrument Society of America (ISA) has assigned standard letter designations to a number of thermocouple types having specified emf-temperature relations. These designations and the approximate metal compositions which meet the required relations, as well as the useful temperature ranges, are given below:

Type B	(Pt + 30% Rh) vs. (Pt + 6% Rh)	0 to 1820°C
Type E	(Ni + 10% Cr) vs. (Cu + 43% Ni)	−270 to 1000°C
Type J	Fe vs. (Cu + 43% Ni)	−210 to 1200°C
Type K	(Ni + 10% Cr) vs. (Ni + 2% Al + 2% Mn + 1% Si)	−270 to 1372°C
Type N	(Ni + 14% Cr + 1.5% Si) vs. (Ni + 4.5% Si + 0.1% Mg)	−270 to 1300°C
Type R	(Pt + 13% Rh) vs. Pt	−50 to 1768°C
Type S	(Pt + 10% Rh) vs. Pt	−50 to 1768°C
Type T	Cu vs. (Cu + 43% Ni)	−270 to 400°C

The compositions are given in weight percent, and the positive leg is listed first. It should be emphasized that the standard letter designations do not imply a precise composition but rather that the specified emf-temperature relation is satisfied.

The first set of tables below lists, for each thermocouple type, the emf as a function of temperature on the International Temperature Scale of 1990 (ITS-90). The coefficients in the equation used to generate the table are also given. The second set of tables gives the inverse relationships, i.e., the coefficients in the polynomial equation which expresses the temperature as a function of thermocouple emf. The accuracy of these equations is also stated.

Further details and tables at closer intervals may be found in Reference 1.

References

1. Burns, G. W., Seroger, M. G., Strouse, G. F., Croarkin, M. C., and Guthrie, W. F., *Temperature-Electromotive Force Reference Functions and Tables for the Letter-Designated Thermocouple Types Based on the ITS-90*, Natl. Inst. Stand. Tech. (U.S.) Monogr. 175, 1993.
2. Schooley, J. F., *Thermometry*, CRC Press, Boca Raton, FL, 1986.

Type B Thermocouples: emf-Temperature (°C) Reference Table and Equations
Thermocouple emf as a Function of Temperature in Degrees Celsius (ITS-90)

°C	emf in Millivolts								Reference junctions at 0 °C		
	0	10	20	30	40	50	60	70	80	90	100
0	0.000	−0.002	−0.003	−0.002	−0.000	0.002	0.006	0.011	0.017	0.025	0.033
100	0.033	0.043	0.053	0.065	0.078	0.092	0.107	0.123	0.141	0.159	0.178
200	0.178	0.199	0.220	0.243	0.267	0.291	0.317	0.344	0.372	0.401	0.431
300	0.431	0.462	0.494	0.527	0.561	0.596	0.632	0.669	0.707	0.746	0.787
400	0.787	0.828	0.870	0.913	0.957	1.002	1.048	1.095	1.143	1.192	1.242
500	1.242	1.293	1.344	1.397	1.451	1.505	1.561	1.617	1.675	1.733	1.792
600	1.792	1.852	1.913	1.975	2.037	2.101	2.165	2.230	2.296	2.363	2.431
700	2.431	2.499	2.569	2.639	2.710	2.782	2.854	2.928	3.002	3.078	3.154
800	3.154	3.230	3.308	3.386	3.466	3.546	3.626	3.708	3.790	3.873	3.957
900	3.957	4.041	4.127	4.213	4.299	4.387	4.475	4.564	4.653	4.743	4.834
1000	4.834	4.926	5.018	5.111	5.205	5.299	5.394	5.489	5.585	5.682	5.780
1100	5.780	5.878	5.976	6.075	6.175	6.276	6.377	6.478	6.580	6.683	6.786
1200	6.786	6.890	6.995	7.100	7.205	7.311	7.417	7.524	7.632	7.740	7.848
1300	7.848	7.957	8.066	8.176	8.286	8.397	8.508	8.620	8.731	8.844	8.956
1400	8.956	9.069	9.182	9.296	9.410	9.524	9.639	9.753	9.868	9.984	10.099
1500	10.099	10.215	10.331	10.447	10.563	10.679	10.796	10.913	11.029	11.146	11.263
1600	11.263	11.380	11.497	11.614	11.731	11.848	11.965	12.082	12.199	12.316	12.433
1700	12.433	12.549	12.666	12.782	12.898	13.014	13.130	13.246	13.361	13.476	13.591
1800	13.591	13.706	13.820								

Temperature ranges and coefficients of equations used to compute the above table: The equations are of the form: $E = c_0 + c_1 t + c_2 t^2 + c_3 t^3 + \ldots c_n t^n$, where E is the emf in millivolts, t is the temperature in degrees Celsius (ITS-90), and c_0, c_1, c_2, c_3, etc. are the coefficients. These coefficients are extracted from Reference 1.

		0 °C to 630.615 °C	630.615 °C to 1820 °C
c_0	=	0.000 000 000 0	$-3.893\ 816\ 862\ 1\ \ldots$
c_1	=	$-2.465\ 081\ 834\ 6 \times 10^{-4}$	$2.857\ 174\ 747\ 0 \times 10^{-2}$
c_2	=	$5.904\ 042\ 117\ 1 \times 10^{-6}$	$-8.488\ 510\ 478\ 5 \times 10^{-5}$
c_3	=	$-1.325\ 793\ 163\ 6 \times 10^{-9}$	$1.578\ 528\ 016\ 4 \times 10^{-7}$
c_4	=	$1.566\ 829\ 190\ 1 \times 10^{-12}$	$-1.683\ 534\ 486\ 4 \times 10^{-10}$
c_5	=	$-1.694\ 452\ 924\ 0 \times 10^{-15}$	$1.110\ 979\ 401\ 3 \times 10^{-13}$
c_6	=	$6.299\ 034\ 709\ 4 \times 10^{-19}$	$-4.451\ 543\ 103\ 3 \times 10^{-17}$
c_7	=	....	$9.897\ 564\ 082\ 1 \times 10^{-21}$
c_8	=	....	$-9.379\ 133\ 028\ 9 \times 10^{-25}$

Type E Thermocouples: emf-Temperature (°C) Reference Table and Equations
Thermocouple emf as a Function of Temperature in Degrees Celsius (ITS-90)

emf in Millivolts   Reference junctions at 0 °C

°C	0	-10	-20	-30	-40	-50	-60	-70	-80	-90	-100
-200	-8.825	-9.063	-9.274	-9.455	-9.604	-9.718	-9.797	-9.835			
-100	-5.237	-5.681	-6.107	-6.516	-6.907	-7.279	-7.632	-7.963	-8.273	-8.561	-8.825
0	0.000	-0.582	-1.152	-1.709	-2.255	-2.787	-3.306	-3.811	-4.302	-4.777	-5.237
°C	0	10	20	30	40	50	60	70	80	90	100
0	0.000	0.591	1.192	1.801	2.420	3.048	3.685	4.330	4.985	5.648	6.319
100	6.319	6.998	7.685	8.379	9.081	9.789	10.503	11.224	11.951	12.684	13.421
200	13.421	14.164	14.912	15.664	16.420	17.181	17.945	18.713	19.484	20.259	21.036
300	21.036	21.817	22.600	23.386	24.174	24.964	25.757	26.552	27.348	28.146	28.946
400	28.946	29.747	30.550	31.354	32.159	32.965	33.772	34.579	35.387	36.196	37.005
500	37.005	37.815	38.624	39.434	40.243	41.053	41.862	42.671	43.479	44.286	45.093
600	45.093	45.900	46.705	47.509	48.313	49.116	49.917	50.718	51.517	52.315	53.112
700	53.112	53.908	54.703	55.497	56.289	57.080	57.870	58.659	59.446	60.232	61.017
800	61.017	61.801	62.583	63.364	64.144	64.922	65.698	66.473	67.246	68.017	68.787
900	68.787	69.554	70.319	71.082	71.844	72.603	73.360	74.115	74.869	75.621	76.373
1000	76.373										

Temperature ranges and coefficients of equations used to compute the above table: The equations are of the form: $E = c_0 + c_1 t + c_2 t^2 + c_3 t^3 + \ldots c_n t^n$, where E is the emf in millivolts, t is the temperature in degrees Celsius (ITS- 90), and c_0, c_1, c_2, c_3, etc. are the coefficients. These coefficients are extracted from Reference 1.

		−270 to 0°C	0°C to 1000°C
c_0	=	$0.000\ 000\ 000\ 0\ \ldots$	$0.000\ 000\ 000\ 0\ \ldots$
c_1	=	$5.866\ 550\ 870\ 8 \times 10^{-2}$	$5.866\ 550\ 871\ 0 \times 10^{-2}$
c_2	=	$4.541\ 097\ 712\ 4 \times 10^{-5}$	$4.503\ 227\ 558\ 2 \times 10^{-5}$
c_3	=	$-7.799\ 804\ 868\ 6 \times 10^{-7}$	$2.890\ 840\ 721\ 2 \times 10^{-8}$
c_4	=	$-2.580\ 016\ 084\ 3 \times 10^{-8}$	$-3.305\ 689\ 665\ 2 \times 10^{-10}$
c_5	=	$-5.945\ 258\ 305\ 7 \times 10^{-10}$	$6.502\ 440\ 327\ 0 \times 10^{-13}$
c_6	=	$-9.321\ 405\ 866\ 7 \times 10^{-12}$	$-1.919\ 749\ 550\ 4 \times 10^{-16}$
c_7	=	$-1.028\ 760\ 553\ 4 \times 10^{-13}$	$-1.253\ 660\ 049\ 7 \times 10^{-18}$
c_8	=	$-8.037\ 012\ 362\ 1 \times 10^{-16}$	$2.148\ 921\ 756\ 9 \times 10^{-21}$
c_9	=	$-4.397\ 949\ 739\ 1 \times 10^{-18}$	$-1.438\ 804\ 178\ 2 \times 10^{-24}$
c_{10}	=	$-1.641\ 477\ 635\ 5 \times 10^{-20}$	$3.596\ 089\ 948\ 1 \times 10^{-28}$
c_{11}	=	$-3.967\ 361\ 951\ 6 \times 10^{-23}$	....
c_{12}	=	$-5.582\ 732\ 872\ 1 \times 10^{-26}$	....
c_{13}	=	$-3.465\ 784\ 201\ 3 \times 10^{-29}$	....

Type J Thermocouples: emf-Temperature (°C) Reference Table and Equations
Thermocouple emf as a Function of Temperature in Degrees Celsius (ITS-90)

emf in Millivolts   Reference junctions at 0 °C

°C	0	-10	-20	-30	-40	-50	-60	-70	-80	-90	-100
-200	-7.890	-8.095									
-100	-4.633	-5.037	-5.426	-5.801	-6.159	-6.500	-6.821	-7.123	-7.403	-7.659	-7.890
0	0.000	-0.501	-0.995	-1.482	-1.961	-2.431	-2.893	-3.344	-3.786	-4.215	-4.633

°C	emf in Millivolts 0	10	20	30	40	50	60	70	Reference junctions at 0 °C 80	90	100
0	0.000	0.507	1.019	1.537	2.059	2.585	3.116	3.650	4.187	4.726	5.269
100	5.269	5.814	6.360	6.909	7.459	8.010	8.562	9.115	9.669	10.224	10.779
200	10.779	11.334	11.889	12.445	13.000	13.555	14.110	14.665	15.219	15.773	16.327
300	16.327	16.881	17.434	17.986	18.538	19.090	19.642	20.194	20.745	21.297	21.848
400	21.848	22.400	22.952	23.504	24.057	24.610	25.164	25.720	26.276	26.834	27.393
500	27.393	27.953	28.516	29.080	29.647	30.216	30.788	31.362	31.939	32.519	33.102
600	33.102	33.689	34.279	34.873	35.470	36.071	36.675	37.284	37.896	38.512	39.132
700	39.132	39.755	40.382	41.012	41.645	42.281	42.919	43.559	44.203	44.848	45.494
800	45.494	46.141	46.786	47.431	48.074	48.715	49.353	49.989	50.622	51.251	51.877
900	51.877	52.500	53.119	53.735	54.347	54.956	55.561	56.164	56.763	57.360	57.953
1000	57.953	58.545	59.134	59.721	60.307	60.890	61.473	62.054	62.634	63.214	63.792
1100	63.792	64.370	64.948	65.525	66.102	66.679	67.255	67.831	68.406	68.980	69.553
1200	69.553										

Temperature ranges and coefficients of equations used to compute the above table: The equations are of the form: $E = c_0 + c_1 t + c_2 t^2 + c_3 t^3 + \ldots c_n t^n$, where E is the emf in millivolts, t is the temperature in degrees Celsius (ITS-90), and c_0, c_1, c_2, c_3, etc. are the coefficients. These coefficients are extracted from Reference 1.

		−260°C to 760°C	760°C to 1200°C
c_0	=	0.000 000 000 0 …	$2.964\ 562\ 568\ 1 \times 10^2$
c_1	=	$5.038\ 118\ 781\ 5 \times 10^{-2}$	$-1.497\ 612\ 778\ 6 \ldots$
c_2	=	$3.047\ 583\ 693\ 0 \times 10^{-5}$	$3.178\ 710\ 392\ 4 \times 10^{-3}$
c_3	=	$-8.568\ 106\ 572\ 0 \times 10^{-8}$	$-3.184\ 768\ 670\ 1 \times 10^{-6}$
c_4	=	$1.322\ 819\ 529\ 5 \times 10^{-10}$	$1.572\ 081\ 900\ 4 \times 10^{-9}$
c_5	=	$-1.705\ 295\ 833\ 7 \times 10^{-13}$	$-3.069\ 136\ 905\ 6 \times 10^{-13}$
c_6	=	$2.094\ 809\ 069\ 7 \times 10^{-16}$	…………
c_7	=	$-1.253\ 839\ 533\ 6 \times 10^{-19}$	…………
c_8	=	$1.563\ 172\ 569\ 7 \times 10^{-23}$	…………

Type K Thermocouples: emf-Temperature (°C) Reference Table and Equations
Thermocouple emf as a Function of Temperature in Degrees Celsius (ITS-90)

°C	emf in Millivolts 0	-10	-20	-30	-40	-50	-60	-70	Reference junctions at 0 °C -80	-90	-100
-200	-5.891	-6.035	-6.158	-6.262	-6.344	-6.404	-6.441	-6.458			
-100	-3.554	-3.852	-4.138	-4.411	-4.669	-4.913	-5.141	-5.354	-5.550	-5.730	-5.891
0	0.000	-0.392	-0.778	-1.156	-1.527	-1.889	-2.243	-2.587	-2.920	-3.243	-3.554

°C	0	10	20	30	40	50	60	70	80	90	100
0	0.000	0.397	0.798	1.203	1.612	2.023	2.436	2.851	3.267	3.682	4.096
100	4.096	4.509	4.920	5.328	5.735	6.138	6.540	6.941	7.340	7.739	8.138
200	8.138	8.539	8.940	9.343	9.747	10.153	10.561	10.971	11.382	11.795	12.209
300	12.209	12.624	13.040	13.457	13.874	14.293	14.713	15.133	15.554	15.975	16.397
400	16.397	16.820	17.243	17.667	18.091	18.516	18.941	19.366	19.792	20.218	20.644
500	20.644	21.071	21.497	21.924	22.350	22.776	23.203	23.629	24.055	24.480	24.905
600	24.905	25.330	25.755	26.179	26.602	27.025	27.447	27.869	28.289	28.710	29.129
700	29.129	29.548	29.965	30.382	30.798	31.213	31.628	32.041	32.453	32.865	33.275
800	33.275	33.685	34.093	34.501	34.908	35.313	35.718	36.121	36.524	36.925	37.326
900	37.326	37.725	38.124	38.522	38.918	39.314	39.708	40.101	40.494	40.885	41.276
1000	41.276	41.665	42.053	42.440	42.826	43.211	43.595	43.978	44.359	44.740	45.119
1100	45.119	45.497	45.873	46.249	46.623	46.995	47.367	47.737	48.105	48.473	48.838
1200	48.838	49.202	49.565	49.926	50.286	50.644	51.000	51.355	51.708	52.060	52.410
1300	52.410	52.759	53.106	53.451	53.795	54.138	54.479	54.819			

Temperature ranges and coefficients of equations used to compute the above table: The equations are of the form: $E = c_0 + c_1 t + c_2 t^2 + c_3 t^3 + \ldots c_n t^n$, where E is the emf in millivolts, t is the temperature in degrees Celsius (ITS-90), and c_0, c_1, c_2, c_3, etc. are the coefficients. In the 0°C to 1372°C range there is also an exponential term that must be evaluated and added to the equation. The exponential term is of the form: $c_0 e^{c_1(t-126.9686)^2}$, where t is the temperature in °C, e is the natural logarithm base, and c_0 and c_1 are the coefficients. These coefficients are extracted from Reference 1.

		−270°C to 0°C	0°C to 1372°C	0°C to 1372°C (Exponential term)
c_0	=	0.000 000 000 0	$-1.760\,041\,368\,6 \times 10^{-2}$	$1.185\,976 \times 10^{-1}$
c_1	=	$3.945\,012\,802\,5 \times 10^{-2}$	$3.892\,120\,497\,5 \times 10^{-2}$	$-1.183\,432 \times 10^{-4}$
c_2	=	$2.362\,237\,359\,8 \times 10^{-5}$	$1.855\,877\,003\,2 \times 10^{-5}$	....
c_3	=	$-3.285\,890\,678\,4 \times 10^{-7}$	$-9.945\,759\,287\,4 \times 10^{-8}$	....
c_4	=	$-4.990\,482\,877\,7 \times 10^{-9}$	$3.184\,094\,571\,9 \times 10^{-10}$	....
c_5	=	$-6.750\,905\,917\,3 \times 10^{-11}$	$-5.607\,284\,488\,9 \times 10^{-13}$	....
c_6	=	$-5.741\,032\,742\,8 \times 10^{-13}$	$5.607\,505\,905\,9 \times 10^{-16}$	....
c_7	=	$-3.108\,887\,289\,4 \times 10^{-15}$	$-3.202\,072\,000\,3 \times 10^{-19}$	....
c_8	=	$-1.045\,160\,936\,5 \times 10^{-17}$	$9.715\,114\,715\,2 \times 10^{-23}$	....
c_9	=	$-1.988\,926\,687\,8 \times 10^{-20}$	$-1.210\,472\,127\,5 \times 10^{-26}$	....
c_{10}	=	$-1.632\,269\,748\,6 \times 10^{-23}$	......	....

Type N Thermocouples: emf-Temperature (°C) Reference Table and Equations
Thermocouple emf as a Function of Temperature in Degrees Celsius (ITS-90)

emf in Millivolts — Reference junctions at 0 °C

°C	0	-10	-20	-30	-40	-50	-60	-70	-80	-90	-100
-200	3.990	-4.083	-4.162	-4.226	-4.277	-4.313	-4.336	-4.345			
-100	-2.407	-2.612	2.808	2.994	3.171	3.336	-3.491	3.634	-3.766	-3.884	-3.990
0	0.000	-0.260	-0.518	-0.772	-1.023	-1.269	-1.509	1.744	-1.972	-2.193	-2.407

°C	0	10	20	30	40	50	60	70	80	90	100
0	0.000	0.261	0.525	0.793	1.065	1.340	1.619	1.902	2.189	2.480	2.774
100	2.774	3.072	3.374	3.680	3.989	4.302	4.618	4.937	5.259	5.585	5.913
200	5.913	6.245	6.579	6.916	7.255	7.597	7.941	8.288	8.637	8.988	9.341
300	9.341	9.696	10.054	10.413	10.774	11.136	11.501	11.867	12.234	12.603	12.974
400	12.974	13.346	13.719	14.094	14.469	14.846	15.225	15.604	15.984	16.366	16.748
500	16.748	17.131	17.515	17.900	18.286	18.672	19.059	19.447	19.835	20.224	20.613
600	20.613	21.003	21.393	21.784	22.175	22.566	22.958	23.350	23.742	24.134	24.527
700	24.527	24.919	25.312	25.705	26.098	26.491	26.883	27.276	27.669	28.062	28.455
800	28.455	28.847	29.239	29.632	30.024	30.416	30.807	31.199	31.590	31.981	32.371
900	32.371	32.761	33.151	33.541	33.930	34.319	34.707	35.095	35.482	35.869	36.256
1000	36.256	36.641	37.027	37.411	37.795	38.179	38.562	38.944	39.326	39.706	40.087
1100	40.087	40.466	40.845	41.223	41.600	41.976	42.352	42.727	43.101	43.474	43.846
1200	43.846	44.218	44.588	44.958	45.326	45.694	46.060	46.425	46.789	47.152	47.513
1300	47.513										

Temperature ranges and coefficients of equations used to compute the above table: The equations are of the form: $E = c_0 + c_1 t + c_2 t^2 + c_3 t^3 + \ldots c_n t^n$, where E is the emf in millivolts, t is the temperature in degrees Celsius (ITS-90), and c_0, c_1, c_2, c_3, etc. are the coefficients. These coefficients are extracted from Reference 1.

		−270°C to 0°C	0°C to 1300°C
c_0	=	0.000 000 000 0 ...	0.000 000 000 0...
c_1	=	$2.615\,910\,596\,2 \times 10^{-2}$	$2.592\,939\,460\,1 \times 10^{-2}$
c_2	=	$1.095\,748\,422\,8 \times 10^{-5}$	$1.571\,014\,188\,0 \times 10^{-5}$
c_3	=	$-9.384\,111\,155\,4 \times 10^{-8}$	$4.382\,562\,723\,7 \times 10^{-8}$
c_4	=	$-4.641\,203\,975\,9 \times 10^{-11}$	$-2.526\,116\,979\,4 \times 10^{-10}$
c_5	=	$-2.630\,335\,771\,6 \times 10^{-12}$	$6.431\,181\,933\,9 \times 10^{-13}$
c_6	=	$-2.265\,343\,800\,3 \times 10^{-14}$	$-1.006\,347\,151\,9 \times 10^{-15}$
c_7	=	$-7.608\,930\,079\,1 \times 10^{-17}$	$9.974\,533\,899\,2 \times 10^{-19}$
c_8	=	$-9.341\,966\,783\,5 \times 10^{-20}$	$-6.086\,324\,560\,7 \times 10^{-22}$
c_9	=	....	$2.084\,922\,933\,9 \times 10^{-25}$
c_{10}	=	....	$-3.068\,219\,615\,1 \times 10^{-29}$

Type R Thermocouples: emf-Temperature (°C) Reference Table and Equations
Thermocouple emf as a Function of Temperature in Degrees Celsius (ITS-90)

emf in Millivolts — Reference junctions at 0 °C

°C	0	-10	-20	-30	-40	-50	-60	-70	-80	-90	-100
0	0.000	-0.051	-0.100	-0.145	-0.188	-0.226					

°C	0	10	20	30	40	50	60	70	80	90	100
0	0.000	0.054	0.111	0.171	0.232	0.296	0.363	0.431	0.501	0.573	0.647
100	0.647	0.723	0.800	0.879	0.959	1.041	1.124	1.208	1.294	1.381	1.469
200	1.469	1.558	1.648	1.739	1.831	1.923	2.017	2.112	2.207	2.304	2.401
300	2.401	2.498	2.597	2.696	2.796	2.896	2.997	3.099	3.201	3.304	3.408
400	3.408	3.512	3.616	3.721	3.827	3.933	4.040	4.147	4.255	4.363	4.471
500	4.471	4.580	4.690	4.800	4.910	5.021	5.133	5.245	5.357	5.470	5.583
600	5.583	5.697	5.812	5.926	6.041	6.157	6.273	6.390	6.507	6.625	6.743
700	6.743	6.861	6.980	7.100	7.220	7.340	7.461	7.583	7.705	7.827	7.950
800	7.950	8.073	8.197	8.321	8.446	8.571	8.697	8.823	8.950	9.077	9.205
900	9.205	9.333	9.461	9.590	9.720	9.850	9.980	10.111	10.242	10.374	10.506
1000	10.506	10.638	10.771	10.905	11.039	11.173	11.307	11.442	11.578	11.714	11.850
1100	11.850	11.986	12.123	12.260	12.397	12.535	12.673	12.812	12.950	13.089	13.228
1200	13.228	13.367	13.507	13.646	13.786	13.926	14.066	14.207	14.347	14.488	14.629
1300	14.629	14.770	14.911	15.052	15.193	15.334	15.475	15.616	15.758	15.899	16.040
1400	16.040	16.181	16.323	16.464	16.605	16.746	16.887	17.028	17.169	17.310	17.451
1500	17.451	17.591	17.732	17.872	18.012	18.152	18.292	18.431	18.571	18.710	18.849
1600	18.849	18.988	19.126	19.264	19.402	19.540	19.677	19.814	19.951	20.087	20.222
1700	20.222	20.356	20.488	20.620	20.749	20.877	21.003				

Temperature ranges and coefficients of equations used to compute the above table: The equations are of the form: $E = c_0 + c_1 t + c_2 t^2 + c_3 t^3 + \dots + c_n t^n$, where E is the emf in millivolts, t is the temperature in degrees Celsius (ITS-90), and c_0, c_1, c_2, c_3, etc. are the coefficients. These coefficients are extracted from Reference 1.

		−50°C to 1064.18°C	1064.18°C to 1664.5°C	1664.5°C to 1768.1°C
c_0	=	0.000 000 000 00 ...	2.951 579 253 16 ...	$1.522\,321\,182\,09 \times 10^2$
c_1	=	$5.289\,617\,297\,65 \times 10^{-3}$	$-2.520\,612\,513\,32 \times 10^{-3}$	$-2.688\,198\,885\,45 \times 10^{-1}$
c_2	=	$1.391\,665\,897\,82 \times 10^{-5}$	$1.595\,645\,018\,65 \times 10^{-5}$	$1.712\,802\,804\,71 \times 10^{-4}$
c_3	=	$-2.388\,556\,930\,17 \times 10^{-8}$	$-7.640\,859\,475\,76 \times 10^{-9}$	$-3.458\,957\,064\,53 \times 10^{-8}$
c_4	=	$3.569\,160\,010\,63 \times 10^{-11}$	$2.053\,052\,910\,24 \times 10^{-12}$	$-9.346\,339\,710\,46 \times 10^{-15}$
c_5	=	$-4.623\,476\,662\,98 \times 10^{-14}$	$-2.933\,596\,681\,73 \times 10^{-16}$	....
c_6	=	$5.007\,774\,410\,34 \times 10^{-17}$	....	....
c_7	=	$-3.731\,058\,861\,91 \times 10^{-20}$	....	....
c_8	=	$1.577\,164\,823\,67 \times 10^{-23}$	....	....
c_9	=	$-2.810\,386\,252\,51 \times 10^{-27}$	....	....

Type S Thermocouples: emf-Temperature (°C) Reference Table and Equations.
Thermocouple emf as a Function of Temperature in Degrees Celsius (ITS-90)

emf in Millivolts — Reference junctions at 0 °C

°C	0	-10	-20	-30	-40	-50	-60	-70	-80	-90	-100
0	0.000	-0.053	-0.103	-0.150	-0.194	-0.236					

°C	0	10	20	30	40	50	60	70	80	90	100
0	0.000	0.055	0.113	0.173	0.235	0.299	0.365	0.433	0.502	0.573	0.646
100	0.646	0.720	0.795	0.872	0.950	1.029	1.110	1.191	1.273	1.357	1.441
200	1.441	1.526	1.612	1.698	1.786	1.874	1.962	2.052	2.141	2.232	2.323
300	2.323	2.415	2.507	2.599	2.692	2.786	2.880	2.974	3.069	3.164	3.259
400	3.259	3.355	3.451	3.548	3.645	3.742	3.840	3.938	4.036	4.134	4.233
500	4.233	4.332	4.432	4.532	4.632	4.732	4.833	4.934	5.035	5.137	5.239
600	5.239	5.341	5.443	5.546	5.649	5.753	5.857	5.961	6.065	6.170	6.275
700	6.275	6.381	6.486	6.593	6.699	6.806	6.913	7.020	7.128	7.236	7.345
800	7.345	7.454	7.563	7.673	7.783	7.893	8.003	8.114	8.226	8.337	8.449
900	8.449	8.562	8.674	8.787	8.900	9.014	9.128	9.242	9.357	9.472	9.587
1000	9.587	9.703	9.819	9.935	10.051	10.168	10.285	10.403	10.520	10.638	10.757
1100	10.757	10.875	10.994	11.113	11.232	11.351	11.471	11.590	11.710	11.830	11.951
1200	11.951	12.071	12.191	12.312	12.433	12.554	12.675	12.796	12.917	13.038	13.159

Type S Thermocouples: emf-Temperature (°C) Reference Table and Equations.
Thermocouple emf as a Function of Temperature in Degrees Celsius (ITS-90)

emf in Millivolts

Reference junctions at 0 °C

°C	0	-10	-20	-30	-40-	-50	-60	-70	-80	-90	-100
1300	13.159	13.280	13.402	13.523	13.644	13.766	13.887	14.009	14.130	14.251	14.373
1400	14.373	14.494	14.615	14.736	14.857	14.978	15.099	15.220	15.341	15.461	15.582
1500	15.582	15.702	15.822	15.942	16.062	16.182	16.301	16.420	16.539	16.658	16.777
1600	16.777	16.895	17.013	17.131	17.249	17.366	17.483	17.600	17.717	17.832	17.947
1700	17.947	18.061	18.174	18.285	18.395	18.503	18.609				

Temperature ranges and coefficients of equations used to compute the above table: The equations are of the form: $E = c_0 + c_1 t + c_2 t^2 + c_3 t^3 + \ldots c_n t^n$, where E is the emf in millivolts, t is the temperature in degrees Celsius (ITS-90), and c_0, c_1, c_2, c_3, etc. are the coefficients. These coefficients are extracted from Reference 1.

		−50°C to 1064.18°C	1064.18°C to 1664.5°C	1664.5°C to 1768.1°C
c_0	=	0.000 000 000 00 ...	1.329 004 440 85 ...	$1.466\,282\,326\,36 \times 10^2$
c_1	=	$5.403\,133\,086\,31 \times 10^{-3}$	$3.345\,093\,113\,44 \times 10^{-3}$	$-2.584\,305\,167\,52 \times 10^{-1}$
c_2	=	$1.259\,342\,897\,40 \times 10^{-5}$	$6.548\,051\,928\,18 \times 10^{-6}$	$1.636\,935\,746\,41 \times 10^{-4}$
c_3	=	$-2.324\,779\,686\,89 \times 10^{-8}$	$-1.648\,562\,592\,09 \times 10^{-9}$	$-3.304\,390\,469\,87 \times 10^{-8}$
c_4	=	$3.220\,288\,230\,36 \times 10^{-11}$	$1.299\,896\,051\,74 \times 10^{-14}$	$-9.432\,236\,906\,12 \times 10^{-15}$
c_5	=	$-3.314\,651\,963\,89 \times 10^{-14}$	....	....
c_6	=	$2.557\,442\,517\,86 \times 10^{-17}$	....	....
c_7	=	$-1.250\,688\,713\,93 \times 10^{-20}$	....	....
c_8	=	$2.714\,431\,761\,45 \times 10^{-24}$	....	....

Type T Thermocouples: emf-Temperature (°C) Reference Table and Equations.
Thermocouple emf as a Function of Temperature in Degrees Celsius (ITS-90)

emf in Millivolts

Reference junctions at 0 °C

°C	0	-10	-20	-30	-40	-50	-60	-70	-80	-90	-100
-200	-5.603	-5.753	-5.888	-6.007	-6.105	-6.180	-6.232	-6.258			
-100	-3.379	-3.657	-3.923	-4.177	-4.419	-4.648	-4.865	-5.070	-5.261	-5.439	-5.603
0	0.000	-0.383	-0.757	-1.121	-1.475	-1.819	-2.153	-2.476	-2.788	-3.089	-3.379

°C	0	10	20	30	40	50	60	70	80	90	100
0	0.000	0.391	0.790	1.196	1.612	2.036	2.468	2.909	3.358	3.814	4.279
100	4.279	4.750	5.228	5.714	6.206	6.704	7.209	7.720	8.237	8.759	9.288
200	9.288	9.822	10.362	10.907	11.458	12.013	12.574	13.139	13.709	14.283	14.862
300	14.862	15.445	16.032	16.624	17.219	17.819	18.422	19.030	19.641	20.255	20.872
400	20.872										

Temperature ranges and coefficients of equations used to compute the above table: The equations are of the form: $E = c_0 + c_1 t + c_2 t^2 + c_3 t^3 + \ldots c_n t^n$, where E is the emf in millivolts, t is the temperature in degrees Celsius (ITS-90), and c_0, c_1, c_2, c_3, etc. are the coefficients. These coefficients are extracted from Reference 1.

		270 °C to 0 °C	0 °C to 400 °C
c_0	=	0.000 000 000 0 ...	0.000 000 000 0 ...
c_1	=	$3.874\,810\,636\,4 \times 10^{-2}$	$3.874\,810\,636\,4 \times 10^{-2}$
c_2	=	$4.419\,443\,434\,7 \times 10^{-5}$	$3.329\,222\,788\,0 \times 10^{-5}$
c_3	=	$1.184\,432\,310\,5 \times 10^{-7}$	$2.061\,824\,340\,4 \times 10^{-7}$
c_4	=	$2.003\,297\,355\,4 \times 10^{-8}$	$-2.188\,225\,684\,6 \times 10^{-9}$
c_5	=	$9.013\,801\,955\,9 \times 10^{-10}$	$1.099\,688\,092\,8 \times 10^{-11}$
c_6	=	$2.265\,115\,659\,3 \times 10^{-11}$	$-3.081\,575\,877\,2 \times 10^{-14}$
c_7	=	$3.607\,115\,420\,5 \times 10^{-13}$	$4.547\,913\,529\,0 \times 10^{-17}$
c_8	=	$3.849\,393\,988\,3 \times 10^{-15}$	$-2.751\,290\,167\,3 \times 10^{-20}$
c_9	=	$2.821\,352\,192\,5 \times 10^{-17}$	....
c_{10}	=	$1.425\,159\,477\,9 \times 10^{-19}$	....
c_{11}	=	$4.876\,866\,228\,6 \times 10^{-22}$	....
c_{12}	=	$1.079\,553\,927\,0 \times 10^{-24}$	....
c_{13}	=	$1.394\,502\,706\,2 \times 10^{-27}$	....
c_{14}	=	$7.979\,515\,392\,7 \times 10^{-31}$	....

Type B Thermocouples: Coefficients (c_i) of Polynomials for the Computation of Temperatures in °C as a Function of the Thermocouple emf in Various Temperature and emf Ranges

Temperature range:	250 °C to 700 °C	700 °C to 1820 °C
emf range:	0.291 mV to 2.431 mV	2.431 mV to 13.820 mV
$c_0 =$	$9.842\ 332\ 1 \times 10^1$	$2.131\ 507\ 1 \times 10^2$
$c_1 =$	$6.997\ 150\ 0 \times 10^2$	$2.851\ 050\ 4 \times 10^2$
$c_2 =$	$-8.476\ 530\ 4 \times 10^2$	$-5.274\ 288\ 7 \times 10^1$
$c_3 =$	$1.005\ 264\ 4 \times 10^3$	$9.916\ 080\ 4\ ...$
$c_4 =$	$-8.334\ 595\ 2 \times 10^2$	$-1.296\ 530\ 3\ ...$
$c_5 =$	$4.550\ 854\ 2 \times 10^2$	$1.119\ 587\ 0 \times 10^{-1}$
$c_6 =$	$-1.552\ 303\ 7 \times 10^2$	$-6.062\ 519\ 9 \times 10^{-3}$
$c_7 =$	$2.988\ 675\ 0 \times 10^1$	$1.866\ 169\ 6 \times 10^{-4}$
$c_8 =$	$-2.474\ 286\ 0\ ...$	$-2.487\ 858\ 5 \times 10^{-6}$

NOTE— The above coefficients are extracted from Reference 1 and are for an expression of the form shown in Section 10.3.2. They yield approximate values of temperature that agree within ±0.03 °C with the values given in Table 10.2.

Type E Thermocouples: Coefficients (c_i) of Polynomials for the Computation of Temperatures in °C as a Function of the Thermocouple emf in Various Temperature and emf Ranges

Temperature range:	-200 °C TO 0 °C	0 °C to 1000 °C
emf range:	-8.825 mV to 0.0 mV	0.0 mV to 76.373 mV
$c_0 =$	$0.000\ 000\ 0\ ...$	$0.000\ 000\ 0\ ...$
$c_1 =$	$1.697\ 728\ 8 \times 10^1$	$1.705\ 703\ 5 \times 10^1$
$c_2 =$	$-4.351\ 497\ 0 \times 10^{-1}$	$-2.330\ 175\ 9 \times 10^{-1}$
$c_3 =$	$-1.585\ 969\ 7 \times 10^{-1}$	$6.543\ 558\ 5 \times 10^{-3}$
$c_4 =$	$-9.250\ 287\ 1 \times 10^{-2}$	$-7.356\ 274\ 9 \times 10^{-5}$
$c_5 =$	$-2.608\ 431\ 4 \times 10^{-2}$	$-1.789\ 600\ \ \times 10^{-6}$
$c_6 =$	$-4.136\ 019\ 9 \times 10^{-3}$	$8.403\ 616\ 5 \times 10^{-8}$
$c_7 =$	$-3.403\ 403\ 0 \times 10^{-4}$	$-1.373\ 587\ 9 \times 10^{-9}$
$c_8 =$	$-1.156\ 489\ 0 \times 10^{-5}$	$1.062\ 982\ 3 \times 10^{-11}$
$c_9 =$	....	$-3.244\ 708\ 7 \times 10^{-14}$

NOTE— The above coefficients are extracted from Reference 1 and are for an expression of the form shown in Section 10.3.2. They yield approximate values of temperature that agree within ±0.02 °C with the values given in Table 10.4

Type J Thermocouples: Coefficients (c_i) of Polynomials for the Computation of Temperatures in °C as a Function of the Thermocouple emf in Various Temperature and emf Ranges

Temperature range:	-210 °C to 0 °C	0 °C to 760 °C	760 °C to 1200 °C
emf Range:	-8.095 mV to 0.0 mV	0.0 mV to 42.919 mV	42.919 mV to 69.553 mV
$c_0 =$	$0.000\ 000\ 0...$	$0.000\ 000\ ...$	$-3.113\ 581\ 87 \times 10^3$
$c_1 =$	$1.952\ 826\ 8 \times 10^1$	$1.978\ 425 \times 10^1$	$3.005\ 436\ 84 \times 10^2$
$c_2 =$	$-1.228\ 618\ 5\ ...$	$-2.001\ 204 \times 10^{-1}$	$-9.947\ 732\ 30\ ...$
$c_3 =$	$-1.075\ 217\ 8\ ...$	$1.036\ 969 \times 10^{-2}$	$1.702\ 766\ 30 \times 10^{-1}$
$c_4 =$	$-5.908\ 693\ 3 \times 10^{-1}$	$-2.549\ 687 \times 10^{-4}$	$1.430\ 334\ 68 \times 10^{-3}$
$c_5 =$	$-1.725\ 671\ 3 \times 10^{-1}$	$3.585\ 153 \times 10^{-6}$	$4.438\ 860\ 84 \times 10^{-6}$
$c_6 =$	$-2.813\ 151\ 3 \times 10^{-2}$	$-5.344\ 285 \times 10^{-8}$	....
$c_7 =$	$-2.396\ 337\ 0 \times 10^{-3}$	$5.099\ 890 \times 10^{-10}$	....
$c_8 =$	$-8.382\ 332\ 1 \times 10^{-5}$	....	....

NOTE— The above coefficients are extracted from Reference 1 and are for an expression of the form shown in Section 10.3.2. They yield approximate values of temperature that agree within ± 0.5 °C with the values given in Table 10.6.

Type K Thermocouples: Coefficients (c_i) of Polynomials for the Computation of Temperatures in °C as a Function of the Thermocouple emf in Various Temperature and emf Ranges

Temperature range:	−200 °C to 0 °C	0 °C to 500 °C	500 °C to 1372 °C
emf Range:	−5.891 mV to 0.0 mV	0.0 mV to 20.644 mV	20.644 mV to 54.886 mV
$c_0 =$	0.000 000 0 ...	0.000 000 ...	$-1.318\ 058 \times 10^2$
$c_1 =$	$2.517\ 346\ 2 \times 10^1$	$2.508\ 355 \times 10^1$	$4.830\ 222 \times 10^1$
$c_2 =$	−1.166 287 8 ...	$7.860\ 106 \times 10^{-2}$	−1.646 031 ...
$c_3 =$	−1.083 363 8 ...	$-2.503\ 131 \times 10^{-1}$	$5.464\ 731 \times 10^{-2}$
$c_4 =$	$-8.977\ 354\ 0 \times 10^{-1}$	$8.315\ 270 \times 10^{-2}$	$-9.650\ 715 \times 10^{-4}$
$c_5 =$	$-3.734\ 237\ 7 \times 10^{-1}$	$-1.228\ 034 \times 10^{-2}$	$8.802\ 193 \times 10^{-6}$
$c_6 =$	$-8.663\ 264\ 3 \times 10^{-2}$	$9.804\ 036 \times 10^{-4}$	$3.110\ 810 \times 10^{-8}$
$c_7 =$	$-1.045\ 059\ 8 \times 10^{-2}$	$-4.413\ 030 \times 10^{-5}$	....
$c_8 =$	$-5.192\ 057\ 7 \times 10^{-4}$	$1.057\ 734 \times 10^{-6}$	....
$c_9 =$	....	$-1.052\ 755 \times 10^{-8}$	....

NOTE—The above coefficients are extracted from Reference 1 and are for an expression of the form shown in Section 10.3.2. They yield approximate values of temperature that agree within ±0.05 °C with the values given in Table 10.8.

Type N Thermocouples: Coefficients (c_i) of Polynomials for the Computation of Temperatures in °C as a Function of the Thermocouple emf in Various Temperature and emf Ranges

Temperature range:	−200 °C to 0 °C	0 °C to 600 °C	600 °C to 1300 °C
emf Range:	−3.990 mV to 0.0 mV	0.0 mV to 20.613 mV	20.613 mV to 47.513 mV
$c_0 =$	0.000 000 0 ...	0.000 00 ...	$1.972\ 485 \times 10^1$
$c_1 =$	$3.843\ 684\ 7 \times 10^1$	$3.868\ 96 \times 10^1$	$3.300\ 943 \times 10^1$
$c_2 =$	1.101 048 5 ...	−1.082 67 ...	$-3.915\ 159 \times 10^{-1}$
$c_3 =$	5.222 931 2 ...	$4.702\ 05 \times 10^{-2}$	$9.855\ 391 \times 10^{-3}$
$c_4 =$	7.206 052 5 ...	$-2.121\ 69 \times 10^{-6}$	$-1.274\ 371 \times 10^{-4}$
$c_5 =$	5.848 858 6 ...	$-1.172\ 72 \times 10^{-4}$	$7.767\ 022 \times 10^{-7}$
$c_6 =$	2.775 491 6 ...	$5.392\ 80 \times 10^{-6}$	....
$c_7 =$	$7.707\ 516\ 6 \times 10^{-1}$	$-7.981\ 56 \times 10^{-8}$	....
$c_8 =$	$1.158\ 266\ 5 \times 10^{-1}$	....	....
$c_9 =$	$7.313\ 886\ 8 \times 10^{-3}$	....	....

NOTE—The above coefficients are extracted from Reference 1 and are for an expression of the form shown in Section 10.3.2. They yield approximate values of temperature that agree within ± 0.04 °C with the values given in Table 10.10.

Type R Thermocouples: Coefficients (c_i) of Polynomials for the Computation of Temperatures in °C as a Function of the Thermocouple emf in Various Temperature and emf Ranges

Temperature range:	−50 °C to 250 °C	250 °C to 1200 °C	1064 °C to 1664.5 °C	1664.5 °C to 1768.1 °C
emf Range:	−0.226 mV to 1.923 mV	1.923 mV to 13.228 mV	11.361 mV to 19.739 mV	19.739 mV to 21.103 mV
$c_0 =$	0.000 000 0 ...	$1.334\ 584\ 505 \times 10^1$	$-8.199\ 599\ 416 \times 10^1$	$3.406\ 177\ 836 \times 10^4$
$c_1 =$	$1.889\ 138\ 0 \times 10^2$	$1.472\ 644\ 573 \times 10^2$	$1.553\ 962\ 042 \times 10^2$	$-7.023\ 729\ 171 \times 10^3$
$c_2 =$	$-9.383\ 529\ 0 \times 10^1$	$-1.844\ 024\ 844 \times 10^1$	$-8.342\ 197\ 663$	$5.582\ 903\ 813 \times 10^2$
$c_3 =$	$1.306\ 861\ 9 \times 10^2$	4.031 129 726 ...	$4.279\ 433\ 549 \times 10^{-1}$	$-1.952\ 394\ 635 \times 10^1$
$c_4 =$	$-2.270\ 358\ 0 \times 10^2$	$-6.249\ 428\ 360 \times 10^{-1}$	$-1.191\ 577\ 910 \times 10^{-2}$	$2.560\ 740\ 231 \times 10^{-1}$
$c_5 =$	$3.514\ 565\ 9 \times 10^2$	$6.468\ 412\ 046 \times 10^{-2}$	$1.492\ 290\ 091 \times 10^{-4}$	....
$c_6 =$	$-3.895\ 390\ 0 \times 10^2$	$-4.458\ 750\ 426 \times 10^{-3}$	....	....
$c_7 =$	$2.823\ 947\ 1 \times 10^2$	$1.994\ 710\ 149 \times 10^{-4}$	....	....
$c_8 =$	$-1.260\ 728\ 1 \times 10^2$	$-5.313\ 401\ 790 \times 10^{-6}$	....	....
$c_9 =$	$3.135\ 361\ 1 \times 10^1$	$6.481\ 976\ 217 \times 10^{-8}$	....	....
$c_{10} =$	−3.318 776 9 ...	....	....	....

NOTE—The above coefficients are extracted from Reference 1 and are for an expression of the form shown in Section 10.3.2. They yield approximate values of temperature that agree within ±0.02 °C with the values given in Table 10.12.

Type S Thermocouples: Coefficients (c_i) of Polynomials for the Computation of Temperatures in °C as a Function of the Thermocouple emf in Various Temperature and emf Ranges

Temperature range:	−50 °C to 250 °C	250 °C to 1200 °C	1064 °C to 1664.5 °C	1664.5 °C to 1768.1 °C
emf Range:	−0.235 mV to 1.874 mV	1.874 mV to 11.950 mV	10.332 mV to 17.536 mV	17.536 mV to 18.693 mV
$c_0 =$	$0.000\,000\,00\ldots$	$1.291\,507\,177 \times 10^1$	$-8.087\,801\,117 \times 10^1$	$5.333\,875\,126 \times 10^4$
$c_1 =$	$1.849\,494\,60 \times 10^2$	$1.466\,298\,863 \times 10^2$	$1.621\,573\,104 \times 10^2$	$-1.235\,892\,298 \times 10^4$
$c_2 =$	$-8.005\,040\,62 \times 10^1$	$-1.534\,713\,402 \times 10^1$	$-8.536\,869\,453\ldots$	$1.092\,657\,613 \times 10^3$
$c_3 =$	$1.022\,374\,30 \times 10^2$	$3.145\,945\,973\ldots$	$4.719\,686\,976 \times 10^{-1}$	$-4.265\,693\,686 \times 10^1$
$c_4 =$	$-1.522\,485\,92 \times 10^2$	$-4.163\,257\,839 \times 10^{-1}$	$-1.441\,693\,666 \times 10^{-2}$	$6.247\,205\,420 \times 10^{-1}$
$c_5 =$	$1.888\,213\,43 \times 10^2$	$3.187\,963\,771 \times 10^{-2}$	$2.081\,618\,890 \times 10^{-4}$	....
$c_6 =$	$-1.590\,859\,41 \times 10^2$	$-1.291\,637\,500 \times 10^{-3}$	....	....
$c_7 =$	$8.230\,278\,80 \times 10^1$	$2.183\,475\,087 \times 10^{-5}$	....	....
$c_8 =$	$-2.341\,819\,44 \times 10^1$	$-1.447\,379\,511 \times 10^{-7}$	....	....
$c_9 =$	$2.797\,862\,60\ldots$	$8.211\,272\,125 \times 10^{-9}$	....	....

NOTE—The above coefficients are extracted from Reference 1 and are for an expression of the form shown in Section 10.3.2. They yield approximate values of temperature that agree within ± 0.02 °C with the values given in Table 10.14.

Type T Thermocouples: Coefficients (c_i) of Polynomials for the Computation of temperatures in °C as a Function of the Thermocouple emf in Various Temperature and emf Ranges

Temperature range:	−200 °C to 0 °C	0 °C to 400 °C
emf Range:	−5.603 mV to 0.0 mV	0.0 mV to 20.872 mV
$c_0 =$	$0.000\,000\,0\ldots$	$0.000\,000\ldots$
$c_1 =$	$2.594\,919\,2 \times 10^1$	$2.592\,800 \times 10^1$
$c_2 =$	$-2.131\,696\,7 \times 10^{-1}$	$-7.602\,961 \times 10^{-1}$
$c_3 =$	$7.901\,869\,2 \times 10^{-1}$	$4.637\,791 \times 10^{-2}$
$c_4 =$	$4.252\,777\,7 \times 10^{-1}$	$-2.165\,394 \times 10^{-3}$
$c_5 =$	$1.330\,447\,3 \times 10^{-1}$	$6.048\,144 \times 10^{-5}$
$c_6 =$	$2.024\,144\,6 \times 10^{-2}$	$-7.293\,422 \times 10^{-7}$
$c_7 =$	$1.266\,817\,1 \times 10^{-3}$	....

NOTE—The above coefficients are extracted from Reference 1 and are for an expression of the form shown in Section 10.3.2. They yield approximate values of temperature that agree within ± 0.04 °C with the values given in Table 10.16.

SECONDARY REFERENCE POINTS ON THE ITS-90 TEMPERATURE SCALE

The International Temperature Scale of 1990 is described in
ction 1 of this *Handbook*, where the defining fixed points are
ted. The Consultative Committee on Thermometry (CCT) of
e International Committee on Weights and Measures
IPM), which oversees the temperature scale, has recom-
ended a number of secondary reference points whose values
ve been accurately determined with respect to the primary
ed points. The most accurate of these, referred to as "first
ality points", satisfy several criteria involving purity of the
aterial, reproducibility, and documentation of the measure-
ents. The CCT also lists "second quality points" that do not
t satisfy all the criteria but are still useful. Taken together,

these secondary reference points, help fill in the gaps between
the primary fixed points.

The table below describes these secondary reference points.
The best values resulting from the CCT evaluation are listed on
both the Kelvin and Celsius scales, along with an estimate of
uncertainty. Full details are given in the reference.

The entries within each quality group are listed in order of
increasing temperature.

Reference

Bedford, R. E., Bonnier, G., Maas, H., and Pavese, F., *Metrologia* 33, 133, 1996.

Substance	Type of Transition	T_{90}/K	t_{90}/°C	Uncert.
First quality points				
Zinc	Superconductive transition	0.8500	-272.300	0.0030
Aluminum	Superconductive transition	1.1810	-271.9690	0.0025
Helium (^{4}He)	Superfluid transition	2.1768	-270.9732	0.0001
Indium	Superconductive transition	3.4145	-269.7355	0.0025
Lead	Superconductive transition	7.1997	-265.9503	0.0025
Niobium	Superconductive transition	9.2880	-263.8620	0.0025
Deuterium (^{2}H$_2$)	Triple point (equilibrium D$_2$)	18.689	-254.461	0.001
Deuterium (^{2}H$_2$)	Triple point (normal D$_2$)	18.724	-254.426	0.001
Neon (^{20}Ne)	Triple point	24.541	-248.609	0.001
Neon	Boiling point	27.097	-246.053	0.001
Nitrogen	Triple point	63.151	-209.999	0.001
Nitrogen	Boiling point	77.352	-195.798	0.002
Argon	Boiling point	87.303	-185.847	0.001
Oxygen	Condensation point	90.197	-182.953	0.001
Methane	Triple point	90.694	-182.456	0.001
Xenon	Triple point	161.405	-111.745	0.001
Carbon dioxide	Triple point	216.592	-56.558	0.001
Mercury	Freezing point	234.3210	-38.8290	0.0005
Water	Ice point	273.15	0	
Gallium	Triple point	302.9166	29.7666	0.0001
Water	Boiling point	373.124	99.974	0.001
Indium	Triple point	429.7436	156.5936	0.0002
Bismuth	Freezing point	544.552	271.402	0.001
Cadmium	Freezing point	594.219	321.069	0.001
Lead	Freezing point	600.612	327.462	0.001
Antimony	Freezing point	903.778	630.628	0.001
Copper/71.9% silver	Eutectic melting point	1052.78	779.63	0.05
Palladium	Freezing point	1828.0	1554.8	0.1
Platinum	Freezing point	2041.3	1768.2	0.4
Rhodium	Freezing point	2236	1963	3
Iridium	Freezing point	2719	2446	6
Molybdenum	Melting point	2895	2622	4
Tungsten	Melting point	3687	3414	7
Second quality points				
Hydrogen	Triple point (normal H$_2$)	13.952	-259.198	0.002
Hydrogen	Boiling point (normal H$_2$)	20.388	-252.762	0.002
Oxygen	α-β transition	23.868	-249.282	0.005
Nitrogen	α-β transition	35.614	-237.536	0.006
Oxygen	β-γ transition	43.796	-229.354	0.001
Krypton	Triple point	115.775	-157.375	0.001
Carbon dioxide	Sublimation point	194.686	-78.464	0.003

Substance	Type of Transition	T_{90}/K	t_{90}/°C	Uncert.
Sulfur hexafluoride	Triple point	223.554	-49.596	0.005
Gallium/20% indium	Eutectic melting point	288.800	15.650	0.001
Gallium/8% tin	Eutectic melting point	293.626	20.476	0.002
Diphenyl ether	Triple point	300.014	26.864	0.001
Ethylene carbonate	Triple point	309.465	36.315	0.001
Succinonitrile	Triple point	331.215	58.065	0.002
Sodium	Freezing point	370.944	97.794	0.005
Benzoic acid	Triple point	395.486	122.336	0.002
Benzoic acid	Freezing point	395.502	122.352	0.007
Mercury	Boiling point	629.769	356.619	0.004
Sulfur	Boiling point	717.764	444.614	0.002
Copper/66.9% aluminum	Eutectic melting point	840.957	567.807	0.010
Silver/30% aluminum	Eutectic melting point	840.957	567.807	0.002
Sodium chloride	Freezing point	1075.168	802.018	0.011
Sodium	Boiling point	1156.090	882.940	0.005
Nickel	Freezing point	1728	1455	1
Cobalt	Freezing point	1768	1495	3
Iron	Freezing point	1811	1538	3
Titanium	Melting point	1943	1670	2
Zirconium	Melting point	2127	1854	8
Aluminum oxide	Melting point	2326	2053	2
Ruthenium	Melting point	2606	2333	10

RELATIVE SENSITIVITY OF BAYARD-ALPERT IONIZATION GAUGES TO VARIOUS GASES

Paul Redhead

The ion current I_+ in a hot-cathode ionization gauge is given by $I_+ = KI_eP$. The gauge constant is $K = (I_+/I_e)(1/P)$, where I_e is the electron current, and P the pressure. The sensitivity is given by $S = KI_e = I_+/P$. The constant K is independent of pressure below about 10^{-3} Pa.

Relative sensitivities for different Bayard-Alpert ionization gauges may differ by as much as ± 15% as a result of differences in applied voltages, electron current, and electrode structure. The table below presents the average of the measurements of 12 experimenters on Bayard-Alpert ionization gauges in various gases. The sensitivity relative to nitrogen is tabulated.

Gas		Relative sensitivity $S/S(N_2)$
Helium	He	0.18
Neon	Ne	0.31
Argon	Ar	1.4
Krypton	Kr	1.9
Xenon	Xe	2.7
Nitrogen	N_2	1.00
Hydrogen	H_2	0.43
Oxygen	O_2	0.96
Carbon monoxide	CO	1.0
Carbon dioxide	CO_2	1.4
Water	H_2O	0.93
Sulfur hexafluoride	SF_6	2.3
Mercury	Hg	3.5
Methane	CH_4	1.6
Ethane	C_2H_6	2.6
Propane	C_3H_8	3.5
Butane	C_4H_{10}	4.3
Ethene	C_2H_4	1.3
Propene	C_3H_6	1.8
Acetylene	C_2H_2	0.61
Allene	C_3H_4	1.3
1-Propyne (Methyl acetylene)	C_3H_4	1.4
Benzene	C_6H_6	3.8

References

1. Hollanda, R., *J. Vac. Sci. Technol.*, **10**, 1133, 1973.
2. Nakayama, K., and Hojo, H., *Jap. J. Appl. Phys.*, Suppl. 2, part 1, p. 113, 1974.
3. Tilford, C. R., *J. Vac. Sci. Technol. A*, **1**, 152, 1983.
4. Tilford, C. R., in *Physical Methods of Chemistry, vol.6, Determination of Thermodynamic Properties*, B. W. Rossiter and R. C. Baetzoid, Eds., pp. 101-173, John Wiley, New York, 1992.

LABORATORY SOLVENTS AND OTHER LIQUID REAGENTS

This table summarizes the properties of 575 liquids that are commonly used in the laboratory as solvents or chemical reagents.

The properties tabulated are:

M_r : Molecular weight

t_m : Melting point in °C

t_b : Normal boiling point in °C

ρ : Density in g/mL at the temperature in °C indicated by the superscript

η : Viscosity in mPa s (1 mPa s = 1 centipoise)

ε : Dielectric constant at ambient temperature (15 to 30°C)

μ : Dipole moment in D

c_p : Specific heat capacity of the liquid at constant pressure at 25°C in J/g K

vp: Vapor pressure at 25°C in kPa (1 kPa = 7.50 mmHg)

FP: Flash point in °C

Fl.Lim: Flammable (explosive) limit in air in percent by volume

IT Autoignition temperature in °C

TLV Threshold limit for allowable airborne concentration in parts per million by volume at 25°C and atmospheric pressure

Data on the temperature dependence of viscosity, dielectric constant, and vapor pressure can be found in the pertinent tables in this *Handbook*.

References

1. Lide, D. R., *Handbook of Organic Solvents*, CRC Press, Boca Raton, FL, 1994.
2. Lide, D. R., and Kehiaian, H. V., *Handbook of Thermophysical and Thermochemical Data*, CRC Press, Boca Raton, FL, 1994.
3. Riddick, J. A., Bunger, W. B., and Sakano, T. K., *Organic Solvents, Fourth Edition*, John Wiley & Sons, New York, 1986.
4. *Fire Protection Guide to Hazardous Materials, 11th Edition*, National Fire Protection Association, Quincy, MA, 1994.
5. Urben, P. G., Ed., *Bretherick's Handbook of Reactive Chemical Hazards, 5th Edition*, Butterworth-Heinemann, Oxford, 1995.
6. *2004 TLV's and BEI's*, American Conference of Governmental Industrial Hygienists, 1330 Kemper Meadow Drive, Cincinnati, OH 45240-1634, 2004.

Name	Mol. form.	M_r	t_m/°C	t_b/°C	ρ/g mL^{-1}	η/mPa s	ε	μ/D	c_p/J g^{-1}K^{-1}	vp/kPa	FP/°C	Fl. lim.	IT/°C	TLV/ppm
Acetaldehyde	C$_2$H$_4$O	44.052	-123.37	20.1	0.7834^{18}		21.0	2.750	2.020	120	-39	4-60%	175	25
Acetic acid	C$_2$H$_4$O$_2$	60.052	16.64	117.9	1.0446^{25}	1.056	6.20	1.70	2.053	2.07	39	4-20%	463	10
Acetic anhydride	C$_4$H$_6$O$_3$	102.089	-74.1	139.5	1.082^{20}	0.843	22.45	≈ 2.8	1.648	0.680	49	2.7-10.3%	316	5
Acetone	C$_3$H$_6$O	58.079	-94.7	56.05	0.7845^{25}	0.306	21.01	2.88	2.175	30.8	-20	3-13%	465	500
Acetone cyanohydrin	C$_4$H$_7$NO	85.105	-19	95	0.932^{19}						74	2.2-12%	688	4.6
Acetonitrile	C$_2$H$_3$N	41.052	-43.82	81.65	0.7857^{20}	0.369	36.64	3.92	2.229	11.9	6	3-16%	524	20
Acetophenone	C$_8$H$_8$O	120.149	20.5	202	1.0281^{20}	1.681	17.44	3.02	1.703	0.049	77		570	10
Acetyl bromide	C$_2$H$_3$BrO	122.948	-96	76	1.6625^{16}					16.2				
Acetyl chloride	C$_2$H$_3$ClO	78.497	-112.8	50.7	1.1051^{20}	0.368	15.8	2.72	1.491	38.4	4		390	
Acrolein	C$_3$H$_4$O	56.063	-87.7	52.6	0.840^{20}			3.1		36.2	-26	2.8-31%	220	0.1
Acrylic acid	C$_3$H$_4$O$_2$	72.063	12.5	141	1.0511^{20}				2.022	0.53	50	2.4-8%	438	2
Acrylonitrile	C$_3$H$_3$N	53.063	-83.48	77.3	0.8007^{25}		33.0	3.87	2.05	14.1	0	3-17%	481	2
Allyl alcohol	C$_3$H$_6$O	58.079	-129	97.0	0.8540^{20}	1.218	19.7	1.60	2.392	3.14	21	3-18%	378	0.5
Allylamine	C$_3$H$_7$N	57.095	-88.2	53.3	0.758^{20}			1.2		33.1	-29	2-22%	374	
2-Amino-2-methyl-1-propanol	C$_4$H$_{11}$NO	89.136	25.5	165.5	0.934^{20}						67			
3-Amino-1-propanol	C$_3$H$_9$NO	75.109	12.4	187.5	0.9824^{26}						80			
Aniline	C$_6$H$_7$N	93.127	-6.02	184.17	1.0217^{20}	3.85	7.06	1.13	2.061	0.090	70	1.3-11%	615	2
Anisole	C$_7$H$_8$O	108.138	-37.13	153.7	0.9940^{20}	1.056	4.30	1.38	1.840	0.472	52		475	
Antimony(V) chloride	Cl$_5$Sb	299.024	4	140 dec	2.34		3.222							
Antimony(V) fluoride	F$_5$Sb	216.752	8.3	141	3.10									
Arsenic(III) chloride	AsCl$_3$	181.280	-16	130	2.150			1.59		5.38				
Benzaldehyde	C$_7$H$_6$O	106.122	-57.1	178.8	1.0401^{25}		17.85	3.0	1.621	0.169	63		192	
Benzene	C$_6$H$_6$	78.112	5.49	80.09	0.8765^{20}	0.604	2.2825	0	1.741	12.7	-11	1-8%	498	0.5
Benzeneacetonitrile	C$_8$H$_7$N	117.149	-23.8	233.5	1.0205^{15}		17.87	3.5		0.012	113			
Benzeneethanamine	C$_8$H$_{11}$N	121.180	<0	195	0.9640^{25}									
Benzeneethanol	C$_8$H$_{10}$O	122.164	-27	218.2	1.0202^{20}		12.31		2.068	0.01	96			
Benzenemethanethiol	C$_7$H$_8$S	124.204	-30	194.5	1.058^{20}		4.705							
Benzenesulfonyl chloride	C$_6$H$_5$ClO$_2$S	176.621	14.5	251 dec	1.3470^{15}		28.90			0.008				
Benzenethiol	C$_6$H$_6$S	110.177	-14.93	169.1	1.0775^{20}		4.26	1.23	1.572	0.26				0.1
Benzonitrile	C$_7$H$_5$N	103.122	-13.99	191.1	1.0093^{15}	1.267	25.9	4.18	1.602	0.11				
Benzoyl chloride	C$_7$H$_5$ClO	140.567	-0.4	197.2	1.2120^{20}		23.0			0.084	72			0.5
Benzyl acetate	C$_9$H$_{10}$O$_2$	150.174	-51.3	213	1.0550^{20}		5.34	1.22	0.989	0.022	90		460	10
Benzyl alcohol	C$_7$H$_8$O	108.138	-15.4	205.31	1.0419^{24}	5.47	11.916	1.71	2.015	0.015	93		436	
Benzylamine	C$_7$H$_9$N	107.153		185	0.9813^{20}	1.624	5.18			0.096				
2,2'-Bioxirane	C$_4$H$_6$O$_2$	86.090	2.0	144	1.113^{20}									

Name	Mol. form.	M_r	$t_m/°C$	$t_b/°C$	$\rho/\text{g mL}^{-1}$	$\eta/\text{mPa s}$	ε	μ/D	$c_p/\text{J g}^{-1}\text{K}^{-1}$	vp/kPa	FP/°C	Fl. lim.	IT/°C	TLV/ppm
Bis(2-aminoethyl)amine	$C_4H_{13}N_3$	103.166	-39	207	0.9569[20]		12.62	1.9	2.462	0.03	98	2-7%	358	1
N,N'-Bis(2-aminoethyl)-1,2-ethanediamine	$C_6H_{18}N_4$	146.234	12	266.5			10.76							
Bis(2-chloroethyl) ether	$C_4H_8Cl_2O$	143.012	-51.9	178.5	1.22[20]		21.20	2.6	1.545	0.143	55	3%-	369	5
Bis(chloromethyl) ether	$C_2H_4Cl_2O$	114.958	-41.5	106	1.323[15]		3.51							0.001
Bis(2-ethylhexyl) phthalate	$C_{24}H_{38}O_4$	390.557	-55	384	0.981[25]		5.3	2.84	1.804	0.00000005	218			0.3
Bis(2-hydroxyethyl) sulfide	$C_4H_{10}O_2S$	122.186	-10.2	282	1.1793[25]		28.61			0.08	160		298	
Boron tribromide	BBr_3	250.523	-45	91	2.6			0						1
Boron trichloride	BCl_3	117.169	-107	12.65				0	0.911	156				
Bromine	Br_2	159.808	-7.2	58.8	3.1028	0.944	3.1484	0	0.474	28.2				0.1
Bromobenzene	C_6H_5Br	157.008	-30.72	156.06	1.4950[20]	1.074	5.45	1.70	0.983	0.556	51		565	
1-Bromobutane	C_4H_9Br	137.018	-112.6	101.6	1.2758[20]	0.606	7.315	2.08	0.798	5.26	18	2.6-6.6%	265	
2-Bromobutane, (±)-	C_4H_9Br	137.018	-112.65	91.3	1.2585[20]		8.64	2.23		9.32	21			
Bromochloromethane	CH_2BrCl	129.384	-87.9	68.0	1.9344[20]			1.7	0.41	19.5				200
Bromodichloromethane	$CHBrCl_2$	163.829	-57	90	1.980[20]									
Bromoethane	C_2H_5Br	108.965	-118.6	38.5	1.4604[20]	0.374	9.01	2.03	0.925	62.5		7-8%	511	5
Bromoethene	C_2H_3Br	106.949	-139.54	15.8	1.4933[20]		5.63	1.42	1.007	141		9-15%	530	0.5
2-Bromo-2-methylpropane	C_4H_9Br	137.018	-16.2	73.3	1.4278[20]		10.98	2.17	1.102	17.7				
1-Bromopentane	$C_5H_{11}Br$	151.045	-88.0	129.8	1.2182[20]		6.31	2.20	0.875	1.68	32			
1-Bromopropane	C_3H_7Br	122.992	-110.3	71.1	1.3537[20]	0.489	8.09	2.18	0.702	18.6			490	
2-Bromopropane	C_3H_7Br	122.992	-89.0	59.5	1.3140[20]	0.458	9.46	2.21	1.075	28.9				
3-Bromopropene	C_3H_5Br	120.976	-119	70.1	1.398[20]	0.471	7.0	≈ 1.9		18.6	-1	4.4-7.3%	295	
2-Bromotoluene	C_7H_7Br	171.035	-27.8	181.7	1.4232[20]		4.641			0.17	79			
Bromotrichloromethane	$CBrCl_3$	198.274	-5.65	105	2.012[25]		2.405			5.35				
Butanal	C_4H_8O	72.106	-96.86	74.8	0.8016[20]		13.45	2.72	2.270	15.7	-22	2-12.5%	218	
1,3-Butanediol	$C_4H_{10}O_2$	90.121	-77	207.5	1.0053[20]		28.8		2.521	0.008	121		395	
1,4-Butanediol	$C_4H_{10}O_2$	90.121	20.4	235	1.0171[20]		31.9	2.58	2.220	0.002	121			
2,3-Butanediol	$C_4H_{10}O_2$	90.121	7.6	182.5	1.0033[20]				2.363	0.02			402	
2,3-Butanedione	$C_4H_6O_2$	86.090	-1.2	88	0.9808[18]		4.04			7.45	27			
Butanenitrile	C_4H_7N	69.106	-111.9	117.6	0.7936[20]	0.553	24.83	3.9	2.301	2.55	24	>1.6%	501	
1-Butanethiol	$C_4H_{10}S$	90.187	-115.7	98.5	0.8416[20]		5.204	1.53	1.898	6.07	2			0.5
2-Butanethiol	$C_4H_{10}S$	90.187	-165	85.0	0.8295[20]		5.645			10.8	-23			
Butanoic acid	$C_4H_8O_2$	88.106	-5.1	163.75	0.9528[25]	1.426	2.98	1.65	2.027	0.221	72	2-10%	443	
Butanoic anhydride	$C_8H_{14}O_3$	158.195	-75	200	0.9668[20]		12.8		1.793	0.07	54	0.9-5.8%	279	
1-Butanol	$C_4H_{10}O$	74.121	-88.6	117.73	0.8095[20]	2.54	17.84	1.66	2.391	0.86	37	1-11%	343	20
2-Butanol	$C_4H_{10}O$	74.121	-88.5	99.51	0.8063[20]	3.10	17.26	1.8	2.656	2.32	24	2-10%	405	100
2-Butanone	C_4H_8O	72.106	-86.64	79.59	0.7999[25]	0.405	18.56	2.78	2.201	12.6	-9	1-11%	404	200
trans-2-Butenal	C_4H_6O	70.090	-76	102.2	0.8516[20]			3.67	1.361	4.92	13	2.1-15.5%	232	0.3
cis-2-Butenoic acid	$C_4H_6O_2$	86.090	15	169	1.0267[20]					0.06				
2-Butoxyethanol	$C_6H_{14}O_2$	118.174	-74.8	168.4	0.9015[20]		9.30	2.1	2.378	0.15	69	4-13%	238	20
Butyl acetate	$C_6H_{12}O_2$	116.158	-78	126.1	0.8825[20]	0.685	5.07	1.9	1.961	1.66	22	2-8%	425	150
sec-Butyl acetate	$C_6H_{12}O_2$	116.158	-98.9	112	0.8748[20]		5.135	1.87			31	1.7-9.8%		200
Butyl acrylate	$C_7H_{12}O_2$	128.169	-64.6	145	0.8898[20]		5.25		1.958	0.731	29	1.7-9.9%	292	2
Butylamine	$C_4H_{11}N$	73.137	-49.1	77.00	0.7414[20]	0.574	4.71	1.0	2.450	12.2	-12	2-10%	312	5
sec-Butylamine	$C_4H_{11}N$	73.137	<-72	62.73	0.7246[20]			1.28			-9			
tert-Butylamine	$C_4H_{11}N$	73.137	-66.94	44.04	0.6958[20]		58.5	1.3	2.627	48.4	-9	2-9%	380	
Butylbenzene	$C_{10}H_{14}$	134.218	-87.85	183.31	0.8601[20]	0.950	2.359	≈ 0	1.813	0.150	71	0.8-5.8%	410	
tert-Butylbenzene	$C_{10}H_{14}$	134.218	-57.8	169.1	0.8665[20]		2.359	≈ 0.83	1.773	0.280	60	0.7-5.7%	450	
Butyl benzoate	$C_{11}H_{14}O_2$	178.228	-22.4	250.3	1.000[20]		5.52			0.005	107			
tert-Butyl ethyl ether	$C_6H_{14}O$	102.174	-94	72.6	0.736[25]				2.13	16.5				5
tert-Butyl hydroperoxide	$C_4H_{10}O_2$	90.121	6	89 dec	0.8960[20]						27			
1-tert-Butyl-4-methylbenzene	$C_{11}H_{16}$	148.245	-52	190	0.8612[20]			≈ 0		0.09	68			1
Butyl vinyl ether	$C_6H_{12}O$	100.158	-92	94	0.7888[20]			1.25	2.316	6.65	-9		255	
γ-Butyrolactone	$C_4H_6O_2$	86.090	-43.61	204	1.1296[20]		39.0	4.27	1.642	0.43	98			
Carbon disulfide	CS_2	76.141	-112.1	46	1.2632[20]	0.352	2.6320	0	1.003	48.2	-30	1-50%	90	10
Chloroacetaldehyde	C_2H_3ClO	78.497	-16.3	85.5	1.19									1
Chloroacetone	C_3H_5ClO	92.524	-44.5	119	1.15[20]					2				1
Chloroacetyl chloride	$C_2H_2Cl_2O$	112.942	-22	106	1.4202[20]			2.23		3.33				0.05
2-Chloroaniline	C_6H_6ClN	127.572	-1.9	208.8		3.32	13.40	1.77		0.034				
3-Chloroaniline	C_6H_6ClN	127.572	-10.28	230.5	1.2161[20]		13.3		1.558	0.0156			705	
Chlorobenzene	C_6H_5Cl	112.557	-45.31	131.72	1.1058[20]	0.753	5.6895	1.69	1.334	1.6	28	1-10%	593	10
2-Chloro-1,3-butadiene	C_4H_5Cl	88.536	-130	59.4	0.956[20]		4.914			29.5	-20	4-20%		10
1-Chlorobutane	C_4H_9Cl	92.567	-123.1	78.4	0.8857[20]	0.422	7.276	2.05	1.891	13.7	-12	2-10%	240	
2-Chlorobutane	C_4H_9Cl	92.567	-131.3	68.2	0.8732[20]		8.564	2.04		21.0	-10			
Chlorocyclohexane	$C_6H_{11}Cl$	118.604	-43.81	142	1.000[20]		7.9505	2.1		1.0	32			
Chlorodibromomethane	$CHBr_2Cl$	208.280	-20	120	2.451[20]									
Chloroethane	C_2H_5Cl	64.514	-138.4	12.3	0.9239[0]		9.45	2.05	1.617	160	-50	4-15%	519	100

Name	Mol. form.	M_r	t_m/°C	t_b/°C	ρ/g mL⁻¹	η/mPa s	ε	μ/D	c_p/J g⁻¹K⁻¹	vp/kPa	FP/°C	Fl. lim.	IT/°C	TLV/ppm
2-Chloroethanol	C_2H_5ClO	80.513	-67.5	128.6	1.2019[20]		25.80	1.78		1.2	60	5-16%	425	1
2-Chloroethyl vinyl ether	C_4H_7ClO	106.551	-70	108	1.0495[20]						27			
(Chloromethyl)benzene	C_7H_7Cl	126.584	-45	179	1.1004[20]		6.854	1.8	1.44	0.164	67	1%-	585	1
Chloromethyl methyl ether	C_2H_5ClO	80.513	-103.5	59.5	1.063[10]					24.9				
1-Chloro-2-methylpropane	C_4H_9Cl	92.567	-130.3	68.5	0.8773[20]		7.027	2.00	1.713	19.9	-6	2-8.7%		
2-Chloro-2-methylpropane	C_4H_9Cl	92.567	-25.60	50.9	0.8420[20]		9.663	2.13	1.867	42.7	0			
1-Chloronaphthalene	$C_{10}H_7Cl$	162.616	-2.5	259	1.1880[25]		5.04	1.57	1.307	0.003	121		>558	
1-Chlorooctane	$C_8H_{17}Cl$	148.674	-57.8	183.5	0.8734[20]		5.05	2.00	1.335	0.11	70			
1-Chloropentane	$C_5H_{11}Cl$	106.594	-99.0	108.4	0.8820[20]		6.654	2.16		4.36	13	1.6-8.6%	260	
2-Chlorophenol	C_6H_5ClO	128.556	9.4	174.9	1.2634[20]	3.59	7.40		1.468	0.308	64			
1-Chloropropane	C_3H_7Cl	78.541	-122.9	46.5	0.8899[20]	0.334	8.588	2.05	1.683	45.8	<-18	2.6-11%	520	
2-Chloropropane	C_3H_7Cl	78.541	-117.18	35.7	0.8617[20]	0.303		2.17		68.9	-32	2.8-11%	593	
3-Chloro-1,2-propanediol	$C_3H_7ClO_2$	110.540		213 dec	1.325[18]		31.0							
3-Chloropropanenitrile	C_3H_4ClN	89.524	-51	175.5	1.1573[20]						76			
2-Chloropropene	C_3H_5Cl	76.525	-137.4	22.6	0.9017[20]		8.92	1.647		110	-37	4.5-16%		
3-Chloropropene	C_3H_5Cl	76.525	-134.5	45.1	0.9376[20]	0.314	8.2	1.94	1.635	48.9	-32	2.9-11%	485	1
Chlorosulfonic acid	$ClHO_3S$	116.525	-80	152	1.75					0.42				
2-Chlorotoluene	C_7H_7Cl	126.584	-35.8	159.0	1.0825[20]	0.964	4.721	1.56	1.318	0.482				50
4-Chlorotoluene	C_7H_7Cl	126.584	7.5	162.4	1.0697[20]	0.837	6.25	2.21		0.4				
Chromyl chloride	Cl_2CrO_2	154.900	-96.5	117	1.91									0.025
trans-Cinnamaldehyde	C_9H_8O	132.159	-7.5	246	1.0497[20]		17.72			0.005				
o-Cresol	C_7H_8O	108.138	31.03	191.04	1.0327[35]		6.76	1.45	2.160	0.041	81	>1.4%	599	5
m-Cresol	C_7H_8O	108.138	12.24	202.27	1.0339[20]	12.91	12.44	1.48	2.080	0.019	86	>1.1%	558	5
p-Cresol	C_7H_8O	108.138	34.77	201.98	1.0185[40]		13.05	1.48	2.044	0.017	86	>1.1%	558	5
Cyanogen chloride	$CClN$	61.471	-6.5	13	1.186[20]			2.8331						0.3
Cyclobutane	C_4H_8	56.107	-90.7	12.6	0.7038[0]			0		157	<10	>1.8%		
Cyclohexane	C_6H_{12}	84.159	6.59	80.73	0.7739[25]	0.894	2.0243	≈ 0	1.841	13.0	-20	1-8%	245	100
Cyclohexanol	$C_6H_{12}O$	100.158	25.93	160.84	0.9624[20]	57.5	16.40		2.079	0.10	68	1-9%	300	50
Cyclohexanone	$C_6H_{10}O$	98.142	-27.9	155.43	0.9478[20]	2.02	16.1	2.87	1.856	0.53	44	1-9%	420	20
Cyclohexene	C_6H_{10}	82.143	-103.5	82.98	0.8110[20]	0.625	2.2176	0.33	1.805	11.8	-12	>1.2%	310	300
Cyclohexylamine	$C_6H_{13}N$	99.174	-17.8	134	0.8191[20]	1.944	4.547	1.3		1.20	31	1-9%	293	10
1,3-Cyclopentadiene	C_5H_6	66.102	-85	41	0.8021[20]			0.419		58.5				75
Cyclopentane	C_5H_{10}	70.133	-93.4	49.3	0.7457[20]	0.413	1.9687	≈ 0	1.837	42.3	-25	2%-	361	600
Cyclopentanol	$C_5H_{10}O$	86.132	-17.5	140.42	0.9488[20]		18.5		2.119	0.294	51			
Cyclopentanone	C_5H_8O	84.117	-51.90	130.57	0.9487[20]		13.58	3.3	1.84	1.55	26			
cis-Decahydronaphthalene	$C_{10}H_{18}$	138.250	-42.9	195.8	0.8965[20]	3.04	2.219	≈ 0	1.678	0.10				
trans-Decahydronaphthalene	$C_{10}H_{18}$	138.250	-30.4	187.3	0.8659[25]	1.948	2.184	≈ 0	1.653	0.164	54	1-5%	255	
Decamethylcyclopenta- siloxane	$C_{10}H_{30}O_5Si_5$	370.770	-38	210	0.9593[20]		2.50			0.02				
Decanal	$C_{10}H_{20}O$	156.265	-4.0	208.5	0.830[15]					0.02				
Decane	$C_{10}H_{22}$	142.282	-29.6	174.15	0.7266[25]	0.838	1.9853	≈ 0	2.210	0.170	51	0.8-5.4%	210	
Decanoic acid	$C_{10}H_{20}O_2$	172.265	31.4	268.7	0.8858[40]				2.761		82		288	
1-Decanol	$C_{10}H_{22}O$	158.281	6.9	231.1	0.8297[20]	10.91	7.93		2.341	0.009	<55		235	
1-Decene	$C_{10}H_{20}$	140.266	-66.3	170.5	0.7408[20]	0.756	2.136	≈ 0	2.144	0.210	58	2-7%	643	50
Diacetone alcohol	$C_6H_{12}O_2$	116.158	-44	167.9	0.9387[20]	2.80	18.2	3.2	1.905	0.224	135			
Dibenzyl ether	$C_{14}H_{14}O$	198.260	1.8	298	1.0428[20]		3.821							
Dibromodifluoromethane	CBr_2F_2	209.816	-110.1	22.76				0.66		110				100
1,2-Dibromoethane	$C_2H_4Br_2$	187.861	9.84	131.6	2.1683[25]	1.595	4.9612	1.2	0.724	1.55				
Dibromomethane	CH_2Br_2	173.835	-52.5	97	2.4969[20]	0.980	7.77	1.43	0.61	6.12				
1,2-Dibromotetrafluoroethane	$C_2Br_2F_4$	259.823	-110.32	47.35	2.149[25]		2.34		0.69	43.4				
Dibutylamine	$C_8H_{19}N$	129.244	-62	159.6	0.7670[20]	0.918	2.765	1.0	2.266	0.34	47	1-6%		
Dibutyl ether	$C_8H_{18}O$	130.228	-95.2	140.28	0.7684[20]	0.637	3.0830	1.17	2.136	0.898	25	1.5-7.6%	194	
Di-tert-butyl peroxide	$C_8H_{18}O_2$	146.228	-40	111	0.704[20]					3.43	18			
Dibutyl phthalate	$C_{16}H_{22}O_4$	278.344	-35	340	1.0465[20]	16.63	6.58	2.82	1.789		157	>0.5%	402	0.4
Dibutyl sebacate	$C_{18}H_{34}O_4$	314.461	-10	344.5	0.9405[15]		4.54	2.48	1.968		178	>0.4%	365	
Dibutyl sulfide	$C_8H_{18}S$	146.294	-79.7	185	0.8386[20]		4.29	1.61	1.943	0.09	76			
Dichloroacetic acid	$C_2H_2Cl_2O_2$	128.942	10	194	1.5634[20]		8.33			0.03				
o-Dichlorobenzene	$C_6H_4Cl_2$	147.002	-17.0	180	1.3059[20]	1.324	10.12	2.50	1.105	0.18	66	2-9%	648	25
m-Dichlorobenzene	$C_6H_4Cl_2$	147.002	-24.8	173	1.2884[20]	1.044	5.02	1.72	1.163	0.252	72			
trans-1,4-Dichloro-2-butene	$C_4H_6Cl_2$	124.997	1.0	155.4	1.183[25]									0.005
Dichlorodimethylsilane	$C_2H_6Cl_2Si$	129.061	-16	70.3	1.064[25]					18.9	<21	3.4-9.5%		
1,1-Dichloroethane	$C_2H_4Cl_2$	98.959	-96.9	57.3	1.1757[20]	0.464	10.10	2.06	1.276	30.5	-17	5-11%	458	100
1,2-Dichloroethane	$C_2H_4Cl_2$	98.959	-35.7	83.5	1.2454[25]	0.779	10.42	1.8	1.298	10.6	13	6-16%	413	10
1,1-Dichloroethene	$C_2H_2Cl_2$	96.943	-122.56	31.6	1.213[20]		4.60	1.34	1.148	80.0	-28	7-16%	570	5
cis-1,2-Dichloroethene	$C_2H_2Cl_2$	96.943	-80.0	60.1	1.2837[20]	0.445	9.20	1.90	1.201	26.8	6	3-15%	460	200
trans-1,2-Dichloroethene	$C_2H_2Cl_2$	96.943	-49.8	48.7	1.2565[20]	0.317	2.14	0	1.205	44.2	2	6-13%	460	200
Dichloromethane	CH_2Cl_2	84.933	-97.2	40	1.3266[20]	0.413	8.93	1.60	1.192	58.2		13-23%	556	50

Name	Mol. form.	M_r	t_m/°C	t_b/°C	ρ/g mL⁻¹	η/mPa s	ε	μ/D	c_p/J g⁻¹K⁻¹	vp/kPa	FP/°C	Fl. lim.	IT/°C	TLV/ppm
(Dichloromethyl)benzene	C₇H₆Cl₂	161.029	-17	205	1.26²⁵		6.9	2.1		0.06				
1,1-Dichloropropane	C₃H₆Cl₂	112.986		88.1	1.1321²⁰					9.09				
1,2-Dichloropropane, (±)-	C₃H₆Cl₂	112.986	-100.53	96.4	1.1560²⁰		8.37	1.8	1.320	6.62	21	3-15%	557	75
1,3-Dichloropropane	C₃H₆Cl₂	112.986	-99.5	120.9	1.1785²⁵		10.27	2.08		2.44				
2,3-Dichloropropene	C₃H₄Cl₂	110.970	10	94	1.211²⁰						15	2.6-7.8%		
2,4-Dichlorotoluene	C₇H₆Cl₂	161.029	-13.5	201	1.2476²⁰		5.68	1.70		0.055				
Dicyclohexylamine	C₁₂H₂₃N	181.318	-0.1	256 dec	0.9123²⁰					0.003	>99			
Diethanolamine	C₄H₁₁NO₂	105.136	28	268.8	1.0966²⁰		25.75	2.8	2.22	<0.01	172	2-13%	662	0.5
1,1-Diethoxyethane	C₆H₁₄O₂	118.174	-100	102.25	0.8254²⁰		3.80	1.4	2.01	3.68	-21	2-10%	230	
1,2-Diethoxyethane	C₆H₁₄O₂	118.174	-74.0	121.2	0.8351²⁵		3.90		2.195	4.33	27		205	
Diethylamine	C₄H₁₁N	73.137	-49.8	55.5	0.7056²⁰	0.319	3.680	0.92	2.313	30.1	-23	2-10%	312	5
N,N-Diethylaniline	C₁₀H₁₅N	149.233	-38.8	216.3	0.9307²⁰		5.15			0.025	85		630	
o-Diethylbenzene	C₁₀H₁₄	134.218	-31.2	184	0.8800²⁰		2.594			0.13	57		395	
m-Diethylbenzene	C₁₀H₁₄	134.218	-83.9	181.1	0.8602²⁰		2.369			0.14	56		450	
p-Diethylbenzene	C₁₀H₁₄	134.218	-42.83	183.7	0.8620²⁰		2.259			0.13	55	0.7-6%	430	
Diethyl carbonate	C₅H₁₀O₃	118.131	-43	126	0.9692²⁵		2.820	1.10	1.80	1.63	25			
Diethylene glycol	C₄H₁₀O₃	106.120	-10.4	245.8	1.1197¹⁵	30.2	31.82	2.3	2.307	0.001	124	2-17%	224	
Diethylene glycol diethyl ether	C₈H₁₈O₃	162.227	-45	188	0.9063²⁰		5.70		2.104	0.10	82			
Diethylene glycol dimethyl ether	C₆H₁₄O₃	134.173	-68	162	0.9434²⁰	0.989	7.23	2.0	2.043	0.315	67			
Diethylene glycol monobutyl ether	C₈H₁₈O₃	162.227	-68	231	0.9553²⁰				2.188	0.0032				
Diethylene glycol monoethyl ether	C₆H₁₄O₃	134.173		196	0.9885²⁰			1.6	2.243	0.017	96			
Diethylene glycol monoethyl ether acetate	C₈H₁₆O₄	176.211	-25	218.5	1.0096²⁰			1.8		0.029	110		425	
Diethylene glycol monomethyl ether	C₅H₁₂O₃	120.147		193	1.035²⁰			1.6	2.256	0.024	96	1-23%	240	
Diethyl ether	C₄H₁₀O	74.121	-116.2	34.5	0.7138²⁰	0.224	4.2666	1.15	2.369	71.7	-45	2-36%	180	400
Diethyl maleate	C₈H₁₂O₄	172.179	-8.8	223	1.0662²⁰		7.560			0.015	121		350	
Diethyl malonate	C₇H₁₂O₄	160.168	-50	200	1.0551²⁰		7.550	2.54	1.779	0.048	93			
Diethyl oxalate	C₆H₁₀O₄	146.141	-40.6	185.7	1.0785²⁰		8.266	2.49	1.784	0.030	76			
Diethyl phthalate	C₁₂H₁₄O₄	222.237	-40.5	295	1.232¹⁴		7.86		1.647	0.002	161	>0.7%	457	0.6
Diethyl succinate	C₈H₁₄O₄	174.195	-21	217.7	1.0402²⁰		6.098			0.15	90			
Diethyl sulfate	C₄H₁₀O₄S	154.185	-24	208	1.172²⁵		29.2			0.05	104		436	
Diethyl sulfide	C₄H₁₀S	90.187	-103.91	92.1	0.8362²⁰	0.422	5.723	1.54	1.900	7.78				
Diiodomethane	CH₂I₂	267.836	6.1	182	3.3211²⁰		5.32	1.08	0.500	0.172				
Diiodosilane	H₂I₂Si	283.911	-1	150										
Diisobutylamine	C₈H₁₉N	129.244	-73.5	139.6		0.723				0.972	29			
Diisopentyl ether	C₁₀H₂₂O	158.281		172.5	0.7777²⁰		2.817	1.23	2.394	0.210				
Diisopropylamine	C₆H₁₅N	101.190	-61	83.9	0.7153²⁰	0.393		1.15		10.7	-1	1.1-7.1%	316	5
Diisopropyl ether	C₆H₁₄O	102.174	-85.4	68.4	0.7192²⁵	0.379	3.805	1.13	2.122	19.9	-28	1-8%	443	250
1,2-Dimethoxyethane	C₄H₁₀O₂	90.121	-69.20	84.5	0.8637²⁵	0.455	7.30		2.145	9.93	-2		202	
Dimethoxymethane	C₃H₈O₂	76.095	-105.1	42	0.8593²⁰		2.644	0.7	2.129	53.1	-32	2-14%	237	1000
Dimethylacetal	C₄H₁₀O₂	90.121	-113.2	64.5	0.8501²⁰					22.9				
N,N-Dimethylacetamide	C₄H₉NO	87.120	-18.59	165	0.9372²⁵	1.927	38.85	3.7	2.016	0.075	70	2-12%	490	10
2,3-Dimethylaniline	C₈H₁₁N	121.180	<-15	221.5	0.9931²⁰						97	>1%		
2,6-Dimethylaniline	C₈H₁₁N	121.180	11.2	215	0.9842²⁰			1.63	1.971	0.45	96			
N,N-Dimethylaniline	C₈H₁₁N	121.180	2.42	194.15	0.9557²⁰	1.300	4.90	1.68	1.771	0.107	63		371	5
2,2-Dimethylbutane	C₆H₁₄	86.175	-98.8	49.73	0.6444²⁵	0.351	1.869	≈ 0	2.227	42.5	-48	1.2-7%	405	500
2,3-Dimethylbutane	C₆H₁₄	86.175	-128.10	57.93	0.6616²⁰	0.361	1.889	≈ 0	2.201	31.3	-29	1.2-7%	405	500
3,3-Dimethyl-2-butanone	C₆H₁₂O	100.158	-52.5	106.1	0.7229²⁵		12.73			4.27				
Dimethylcarbamic chloride	C₃H₆ClNO	107.539	-33	167	1.168²⁵									
Dimethyl disulfide	C₂H₆S₂	94.199	-84.67	109.74	1.0625²⁰		9.6	1.8	1.551	3.82	24			
N,N-Dimethylethanolamine	C₄H₁₁NO	89.136	-59	134	0.8866²⁰					0.9				
N,N-Dimethylformamide	C₃H₇NO	73.094	-60.48	153	0.9445²⁵	0.794	38.25	3.82	2.060	0.439	58	2-15%	445	10
2,6-Dimethyl-4-heptanone	C₉H₁₈O	142.238	-41.5	169.4	0.8062²⁰		9.91	2.7	2.090	0.23	49	1-7%	396	25
1,1-Dimethylhydrazine	C₂H₈N₂	60.098	-57.20	63.9	0.791²²				2.731	20.9	-15	2-95%	249	0.01
Dimethyl phthalate	C₁₀H₁₀O₄	194.184	5.5	283.7	1.1905²⁰	14.36	8.66		1.561	0.001	146	>0.9%	490	0.6
2,6-Dimethylpyridine	C₇H₉N	107.153	-6.1	144.01	0.9226²⁰		7.33	1.7	1.728	0.746				
Dimethyl sulfate	C₂H₆O₄S	126.132	-31.7	188 dec	1.3322²⁰		55.0			0.13	83		188	0.1
Dimethyl sulfide	C₂H₆S	62.134	-98.24	37.33	0.8483²⁰	0.284	6.70	1.554	1.901	64.4	-37	2.2-20%	206	10
Dimethyl sulfoxide	C₂H₆OS	78.133	17.89	189	1.1010²⁵	1.987	47.24	3.96	1.958	0.084	95	3-42%	215	
1,4-Dioxane	C₄H₈O₂	88.106	11.85	101.5	1.0337²⁰	1.177	2.2189	0	1.726	4.95	12	2-22%	180	20
1,3-Dioxolane	C₃H₆O₂	74.079	-97.22	78	1.060²⁰			1.19	1.593	14.6	2			20
Dipentyl ether	C₁₀H₂₂O	158.281	-69	190	0.7833²⁰		2.798	1.20	1.579	0.13	57		170	
Dipropylamine	C₆H₁₅N	101.190	-63	109.3	0.7400²⁰	0.517	2.923	1.03	2.500	3.21	17		299	

Name	Mol. form.	M_r	t_m/°C	t_b/°C	ρ/g mL^{-1}	η/mPa s	ε	μ/D	c_p/J g^{-1}K^{-1}	vp/kPa	FP/°C	Fl. lim.	IT/°C	TLV/ppm
Dipropylene glycol monomethyl ether	$C_7H_{16}O_3$	148.200	-80	188.3	0.95									
Dipropyl ether	$C_6H_{14}O$	102.174	-114.8	90.08	0.7466[20]	0.396	3.38	1.21	2.169	8.35	21	1.3-7%	188	
Dodecane	$C_{12}H_{26}$	170.334	-9.57	216.32	0.7495[20]	1.383	2.0120	≈ 0	2.206	0.016	74	>0.6%	203	
1-Dodecanol	$C_{12}H_{26}O$	186.333	23.9	260	0.8309[24]		5.82		2.351	0.000016	127		275	
1-Dodecene	$C_{12}H_{24}$	168.319	-35.2	213.8	0.7584[20]	1.20	2.152	≈ 0	2.143	0.019	79			
Epichlorohydrin	C_3H_5ClO	92.524	-26	118	1.1812[20]	1.073	22.6	1.8	1.422	2.2	31	4-21%	411	0.5
1,2-Epoxybutane	C_4H_8O	72.106	-150	63.4	0.8297[20]			1.891	2.039	31.7	-22	1.7-19%	439	
1,2-Epoxy-4-(epoxyethyl)cyclohexane	$C_8H_{12}O_2$	140.180	<-55	227	1.0966[20]									0.1
1,2-Ethanediamine	$C_2H_8N_2$	60.098	11.14	117	0.8979[20]		13.82	1.99	2.872	1.62	40	3-12%	385	10
1,2-Ethanediol	$C_2H_6O_2$	62.068	-12.69	197.3	1.1135[20]	16.06	41.4	2.28	2.394	0.01	111	3-22%	398	40
1,2-Ethanediol, diacetate	$C_6H_{10}O_4$	146.141	-31	190	1.1043[20]		7.7	2.34	2.121	0.030	88	1.6-8.4%	482	
1,2-Ethanediol, dinitrate	$C_2H_4N_2O_6$	152.062	-22.3	198.5	1.4918[20]		28.26			0.009				0.05
1,2-Ethanedithiol	$C_2H_6S_2$	94.199	-41.2	146.1	1.234[20]		7.26	2.03						
Ethanethiol	C_2H_6S	62.134	-147.88	35.0	0.8315[25]	0.287	6.667	1.60	1.898	70.3	-17	2.8-18%	300	0.5
Ethanol	C_2H_6O	46.068	-114.14	78.29	0.7893[20]	1.074	25.3	1.69	2.438	7.87	13	3-19%	363	1000
Ethanolamine	C_2H_7NO	61.083	10.5	171	1.0180[20]	21.1	31.94	2.3	3.201	0.05	86	3-24%	410	3
4-Ethoxyaniline	$C_8H_{11}NO$	137.179	4.6	254	1.0652[16]		7.43			0.0007	116			
Ethoxybenzene	$C_8H_{10}O$	122.164	-29.43	169.81	0.9651[20]	1.197	4.216	1.45	1.870	0.204	63			
2-Ethoxyethanol	$C_4H_{10}O_2$	90.121	-70	135	0.9253[25]		13.38	2.1	2.339	0.71	43	3-18%	235	5
2-Ethoxyethyl acetate	$C_6H_{12}O_3$	132.157	-61.7	156.4	0.9740[20]		7.567	2.2	2.845	0.24	56	2-8%	379	5
Ethyl acetate	$C_4H_8O_2$	88.106	-83.8	77.11	0.9003[20]	0.423	6.0814	1.78	1.937	12.6	-4	2-12%	426	400
Ethyl acetoacetate	$C_6H_{10}O_3$	130.141	-45	180.8	1.0368[10]		14.0		1.906	0.095	57	1-10%	295	
Ethyl acrylate	$C_5H_8O_2$	100.117	-71.2	99.4	0.9234[20]		6.05	1.96		5.14	10	1.4-14%	372	5
Ethylamine	C_2H_7N	45.084	-80.5	16.5	0.689[15]		8.7	1.22	2.884	141	-16	4-14%	385	5
N-Ethylaniline	$C_8H_{11}N$	121.180	-63.5	203.0	0.9625[20]	2.05	5.87			0.039	85			
Ethylbenzene	C_8H_{10}	106.165	-94.96	136.19	0.8626[25]	0.631	2.4463	0.59	1.726	1.28	21	1-7%	432	100
Ethyl benzoate	$C_9H_{10}O_2$	150.174	-34	212	1.0415[25]		6.20	2.00	1.638	0.04	88		490	
Ethyl butanoate	$C_6H_{12}O_2$	116.158	-98	121.3	0.8735[25]	0.639	5.18	1.74	1.963	2.01	24		463	
2-Ethyl-1-butanol	$C_6H_{14}O$	102.174	<-15	147	0.8326[20]		6.19			0.206	57			
Ethyl chloroacetate	$C_4H_7ClO_2$	122.551	-21	144.3	1.1585[20]					0.640	64			
Ethyl chloroformate	$C_3H_5ClO_2$	108.524	-80.6	95	1.1352[20]		9.736				16		500	
Ethyl cyanoacetate	$C_5H_7NO_2$	113.116	-22.5	205	1.0654[20]		31.62	2.17	1.947	0.003	110			
Ethyleneimine	C_2H_5N	43.068	-77.9	56	0.832[25]		18.3	1.90		28.9	-11	3.3-55%	320	0.5
Ethyl formate	$C_3H_6O_2$	74.079	-79.6	54.4	0.9208[20]	0.380	8.57	1.9	2.015	32.3	-20	3-16%	455	100
2-Ethylhexanal	$C_8H_{16}O$	128.212	<-100	163	0.8540[20]						44	0.9-7.2%	190	
2-Ethyl-1,3-hexanediol	$C_8H_{18}O_2$	146.228	-40	244	0.9325[22]		18.73				127		360	
2-Ethyl-1-hexanol	$C_8H_{18}O$	130.228	-70	184.6	0.8319[25]	6.27	7.58	1.74	2.438	0.019	73	0.8-9.7%	231	
2-Ethylhexyl acetate	$C_{10}H_{20}O_2$	172.265	-80	199	0.8718[20]			1.8		0.09	71	1-8%	268	
Ethyl lactate	$C_5H_{10}O_3$	118.131	-26	154.5	1.0328[20]		15.4	2.4	2.150		46	>1.5%		400
Ethyl 3-methylbutanoate	$C_7H_{14}O_2$	130.185	-99.3	135	0.8656[20]		4.71			1.07				
Ethyl 2-methylpropanoate	$C_6H_{12}O_2$	116.158	-88.2	110.1	0.868[20]					3.25	13			
Ethyl nitrite	$C_2H_5NO_2$	75.067		18	0.899[15]					135	-35	4-50%	90	
Ethyl propanoate	$C_5H_{10}O_2$	102.132	-73.9	99.1	0.8843[25]	0.501	5.76	1.74	1.920	4.97	12	1.9-11%	440	
Ethyl silicate	$C_8H_{20}O_4Si$	208.329	-82.5	168.8	0.9320[20]		2.50		1.749	1.17	52			10
Eucalyptol	$C_{10}H_{18}O$	154.249	0.8	176.4	0.9267[20]		4.57			0.260	48			
Fluorobenzene	C_6H_5F	96.102	-42.18	84.73	1.0225[20]	0.550	5.465	1.60	1.523	10.4	-15			
Fluorosulfonic acid	FHO_3S	100.070	-89	163	1.726					0.08				
Formamide	CH_3NO	45.041	2.49	220	1.1334[20]	3.34	111.0	3.73	2.389	0.01	154			10
Formic acid	CH_2O_2	46.026	8.3	101	1.220[20]	1.607	51.1	1.425	2.151	5.75	50	18-57%	434	5
Furan	C_4H_4O	68.074	-85.61	31.5	0.9514[20]	0.361	2.94	0.66	1.686	80.0	-36	2-14%		
Furfural	$C_5H_4O_2$	96.085	-38.1	161.7	1.1594[20]	1.587	42.1	3.5	1.698	0.29	60	2-19%	316	2
Furfuryl alcohol	$C_5H_6O_2$	98.101	-14.6	171	1.1296[20]		16.85	1.9	2.079	0.097	75	2-16%	491	10
Germanium(IV) chloride	Cl_4Ge	214.42	-51.50	86.55	1.88			0						
Glycerol	$C_3H_8O_3$	92.094	18.1	290	1.2613[20]	934	46.53	2.6	2.377	<0.01	199	3-19%	370	2.7
Glycerol triacetate	$C_9H_{14}O_6$	218.203	-78	259	1.1583[20]		7.11		1.763	<0.01	138	1%-	433	
Glycerol trioleate	$C_{57}H_{104}O_6$	885.432	-4		0.915[15]		3.109							
Heptanal	$C_7H_{14}O$	114.185	-43.4	152.8	0.8132[25]		9.07		2.015	0.46				
Heptane	C_7H_{16}	100.202	-90.55	98.4	0.6795[25]	0.387	1.9209	≈ 0	2.242	6.09	-4	1-7%	204	400
Heptanoic acid	$C_7H_{14}O_2$	130.185	-7.17	222.2	0.9124[25]	3.84	3.04		2.039	0.001			275	
1-Heptanol	$C_7H_{16}O$	116.201	-33.2	176.45	0.8219[20]	5.81	11.75		2.342	0.0044				
2-Heptanone	$C_7H_{14}O$	114.185	-35	151.05	0.8111[20]	0.714	11.95	2.6	2.037	0.49	39	1-8%	393	50
3-Heptanone	$C_7H_{14}O$	114.185	-39	147	0.8183[20]		12.7	2.78		0.5	46			50
4-Heptanone	$C_7H_{14}O$	114.185	-33	144	0.8174[20]		12.60			0.164	49			50
1-Heptene	C_7H_{14}	98.186	-118.9	93.64	0.6970[20]	0.340	2.092	≈ 0	2.157	7.52	-1		260	
Hexachloro-1,3-butadiene	C_4Cl_6	260.761	-21	215	1.556[25]		2.55			0.13			610	0.02

Name	Mol. form.	M_r	t_m/°C	t_b/°C	ρ/g mL⁻¹	η/mPa s	ε	μ/D	c_p/J g⁻¹K⁻¹	vp/kPa	FP/°C	Fl. lim.	IT/°C	TLV/ppm
Hexachloro-1,3-cyclopentadiene	C_5Cl_6	272.772	-9	239	1.7019[25]									0.01
Hexafluorobenzene	C_6F_6	186.054	5.03	80.26	1.6184[20]	2.79	2.029	0	1.191	11.3				
Hexamethyldisiloxane	$C_6H_{18}OSi_2$	162.377	-66	99	0.7638[20]		2.179		1.918	5.57				
Hexamethylphosphoric triamide	$C_6H_{18}N_3OP$	179.200	7.2	232.5	1.03[20]		31.3	5.5	1.791					
Hexanal	$C_6H_{12}O$	100.158	-56	131	0.8335[20]				2.101	1.48	32			
Hexane	C_6H_{14}	86.175	-95.35	68.73	0.6606[25]	0.300	1.8865	≈ 0	2.270	20.2	-22	1-8%	225	50
Hexanedinitrile	$C_6H_8N_2$	108.141	1	295	0.9676[20]				1.190	<0.01	93	2-5%	550	2
Hexanoic acid	$C_6H_{12}O_2$	116.158	-3	205.2	0.9212[25]		2.600	1.13	1.937	0.005	102		380	
1-Hexanol	$C_6H_{14}O$	102.174	-47.4	157.6	0.8136[20]	4.58	13.03		2.353	0.11	63		290	
2-Hexanone	$C_6H_{12}O$	100.158	-55.5	127.6	0.8113[20]	0.583	14.56	2.7	2.130	1.54	25	1-8%	423	5
1-Hexene	C_6H_{12}	84.159	-139.76	63.48	0.6685[25]	0.252	2.077	≈ 0	2.178	24.8	-26	1.2-6.9%	253	50
Hexyl acetate	$C_8H_{16}O_2$	144.212	-80.9	171.5	0.8779[15]		4.42		1.961	0.185	45			
Hydrazine	H_4N_2	32.045	1.4	113.55	1.0036	0.876	51.7	1.75	3.086	1.91	38	5-100%		0.01
Hydrazoic acid	HN_3	43.028	-80	35.7				1.70		68.2				0.11
Hydrogen cyanide	CHN	27.026	-13.29	26	0.6876[20]	0.183	114.9	2.985	2.612	98.8	-18	6-40%	538	4.7
Hydrogen peroxide	H_2O_2	34.015	-0.43	150.2	1.44		74.6	1.573	2.619	0.26				1
3-Hydroxypropanenitrile	C_3H_5NO	71.078	-46	221	1.0404[25]			3.2		0.010	129			
Indan	C_9H_{10}	118.175	-51.38	177.97	0.9639[20]	1.357			1.609	0.2				
Indene	C_9H_8	116.160	-1.5	182	0.9960[25]				1.609	0.220				10
Iodine bromide	BrI	206.808	40	116 dec	4.3			0.726						
Iodine chloride	ClI	162.357	27.39	100 dec	3.24			1.24		3.59				
Iodobenzene	C_6H_5I	204.008	-31.3	188.4	1.8308[20]	1.554	4.59	1.70	0.778	0.133				
1-Iodobutane	C_4H_9I	184.018	-103	130.5	1.6154[20]		6.27	1.93		1.85				
Iodoethane	C_2H_5I	155.965	-111.1	72.3	1.9357[20]	0.556	7.82	1.976	0.738	18.2				
Iodomethane	CH_3I	141.939	-66.4	42.43	2.2789[20]	0.469	6.97	1.62	0.888	53.9				2
1-Iodopropane	C_3H_7I	169.992	-101.3	102.5	1.7489[20]	0.703	7.07	2.04	0.746	5.75				
2-Iodopropane	C_3H_7I	169.992	-90	89.5	1.7042[20]	0.653	8.19	1.95	0.535	9.36				
Iron pentacarbonyl	C_5FeO_5	195.896	-20	103	1.5[20]		2.602		1.228	4				0.1
Isobutanal	C_4H_8O	72.106	-65.9	64.5	0.7891[20]			2.75		23.0	-18	1.6-10.6%	196	
Isobutyl acetate	$C_6H_{12}O_2$	116.158	-98.8	116.5	0.8712[20]	0.676	5.068	1.9	2.013	2.39	18	1-11%	421	150
Isobutyl acrylate	$C_7H_{12}O_2$	128.169	-61	132	0.8896[20]						30		427	
Isobutylamine	$C_4H_{11}N$	73.137	-86.7	67.75	0.724[25]	0.571	4.43	1.3	2.505	19.0	-9	2-12%	378	
Isobutylbenzene	$C_{10}H_{14}$	134.218	-51.4	172.79	0.8532[20]		2.318	≈ 0	1.793	0.257	55	0.8-6%	427	
Isobutyl formate	$C_5H_{10}O_2$	102.132	-95.8	98.2	0.8776[20]		6.41	1.88		5.34	5	2-9%	320	
Isobutyl isobutanoate	$C_8H_{16}O_2$	144.212	-80.7	148.6	0.8542[20]			1.9		0.552	38	1-8%	432	
Isopentane	C_5H_{12}	72.149	-159.77	27.88	0.6201[20]	0.214	1.845	0.13	2.284	91.7	-51	1.4-7.6%	420	600
Isopentyl acetate	$C_7H_{14}O_2$	130.185	-78.5	142.5	0.876[15]		4.72	1.9	1.909	0.728	25	1-8%	360	50
Isophorone	$C_9H_{14}O$	138.206	-8.1	215.2	0.9255[20]	2.33			1.834	0.06	84	1-4%	460	5
Isopropenyl acetate	$C_5H_8O_2$	100.117	-92.9	94	0.9090[20]					6.02	26		432	
Isopropenylbenzene	C_9H_{10}	118.175	-23.2	165.4	0.9106[20]		2.28		1.711	0.40	54	1.9-6.1%	574	50
Isopropyl acetate	$C_5H_{10}O_2$	102.132	-73.4	88.6	0.8718[20]				1.952	7.88	2	2-8%	460	100
Isopropylamine	C_3H_9N	59.110	-95.13	31.76	0.6891[20]	0.325	5.6268	1.19	2.771	78.0	-37		402	5
Isopropylbenzene	C_9H_{12}	120.191	-96.02	152.41	0.8640[25]	0.737	2.381	0.79	1.753	0.61	36	1-7%	424	50
Isopropylbenzene hydroperoxide	$C_9H_{12}O_2$	152.190		153	1.03[20]					0.004				
1-Isopropyl-2-methylbenzene	$C_{10}H_{14}$	134.218	-71.5	178.1	0.8766[20]					0.2				
1-Isopropyl-3-methylbenzene	$C_{10}H_{14}$	134.218	-63.7	175.1	0.8610[20]					0.22				
1-Isopropyl-4-methylbenzene	$C_{10}H_{14}$	134.218	-67.94	177.1	0.8573[20]		2.2322	≈0	1.761	0.19	47	1-6%	436	
Isoquinoline	C_9H_7N	129.159	26.47	243.22	1.0910[30]		11.0	2.73	1.519	0.007				
d-Limonene	$C_{10}H_{16}$	136.234	-74.0	178	0.8411[20]	1.47	2.3746		1.828	0.277	45	0.7-6.1%	237	
l-Limonene	$C_{10}H_{16}$	136.234		178	0.843[20]		2.3738			0.254				
Mesityl oxide	$C_6H_{10}O$	98.142	-59	130	0.8653[20]	0.602	15.6	2.8	2.165	1.47	31	1-7%	344	15
Methacrylic acid	$C_4H_6O_2$	86.090	16	162.5	1.0153[20]			1.65	1.871	0.12	77	1.6-8.8%	68	20
Methanol	CH_4O	32.042	-97.53	64.6	0.7914[20]	0.544	33.0	1.70	2.531	16.9	11	6-36%	464	200
2-Methoxyaniline	C_7H_9NO	123.152	6.2	224	1.0923[20]		5.230			0.013	118			0.1
4-Methoxybenzaldehyde	$C_8H_8O_2$	136.149	0	248	1.119[15]		22.0			0.004				
2-Methoxyethanol	$C_3H_8O_2$	76.095	-85.1	124.1	0.9647[20]		17.2	2.36	2.249	1.31	39	2-14%	285	5
2-Methoxyethyl acetate	$C_5H_{10}O_3$	118.131	-70	143	1.0074[19]		8.25	2.1	2.624	0.67	49	2-12%	392	5
Methyl acetate	$C_3H_6O_2$	74.079	-98.25	56.87	0.9342[20]	0.364	7.07	1.72	1.916	28.8	-10	3-16%	454	200
Methyl acrylate	$C_4H_6O_2$	86.090	<-75	80.7	0.9535[20]		7.03	1.77	1.845	11.0	-3	2.8-25%	468	2
2-Methylacrylonitrile	C_4H_5N	67.090	-35.8	90.3	0.8001[20]			3.69	1.883	8.26	1	2-6.8%		1
2-Methylaniline	C_7H_9N	107.153	-14.41	200.3	0.9984[20]	3.82	6.138	1.6	1.96	0.043	85		482	2
3-Methylaniline	C_7H_9N	107.153	-31.3	203.3	0.9889[20]	3.31	5.816	1.45	2.118	0.036				2
N-Methylaniline	C_7H_9N	107.153	-57	196.2	0.9891[20]	2.04	5.96		1.933	0.05				2
Methyl benzoate	$C_8H_8O_2$	136.149	-12.4	199	1.0837[25]	1.857	6.642	1.9	1.625	0.052	83			0.5

Name	Mol. form.	M_r	$t_m/°C$	$t_b/°C$	ρ/g mL^{-1}	η/mPa s	ε	μ/D	c_p/J g^{-1}K^{-1}	vp/kPa	FP/°C	Fl. lim.	IT/°C	TLV/ppm
2-Methyl-1,3-butadiene	C_5H_8	68.118	-145.9	34.0	0.679[20]		2.098	0.25	2.240	73.4	-54	1.5-8.9%	395	
Methyl butanoate	$C_5H_{10}O_2$	102.132	-85.8	102.8	0.8984[20]	0.541	5.48		1.941	4.30	14			
3-Methylbutanoic acid	$C_5H_{10}O_2$	102.132	-29.3	176.5	0.931[20]			0.63	1.930	0.067			416	
3-Methyl-1-butanol	$C_5H_{12}O$	88.148	-117.2	131.1	0.8104[20]	3.69	15.63		2.382	0.315	43	1.2-9%	350	100
2-Methyl-2-butanol	$C_5H_{12}O$	88.148	-9.1	102.4	0.8096[20]	3.55	5.78	1.82	2.803	2.19	19	1.2-9%	437	
3-Methyl-2-butanol, (±)-	$C_5H_{12}O$	88.148		112.9	0.8180[20]		12.1			1.20	38			
3-Methyl-2-butanone	$C_5H_{10}O$	86.132	-93.1	94.33	0.8051[20]		10.37		2.089	6.99				200
2-Methyl-1-butene	C_5H_{10}	70.133	-137.53	31.2	0.6504[20]		2.180		2.241	81.4	-20			
2-Methyl-2-butene	C_5H_{10}	70.133	-133.72	38.56	0.6623[20]	0.203	1.979		2.179	62.1	-20			
Methyl tert-butyl ether	$C_5H_{12}O$	88.148	-108.6	55.0	0.7353[25]				2.127	33.6				50
Methyl chloroacetate	$C_3H_5ClO_2$	108.524	-32.1	129.5	1.236[20]		12.0			1.0	57	7.5-18.5%		
Methylcyclohexane	C_7H_{14}	98.186	-126.6	100.93	0.7694[20]	0.679	2.024	≈ 0	1.882	6.18	-4	1-7%	250	400
Methylcyclopentane	C_6H_{12}	84.159	-142.42	71.8	0.7486[20]	0.479	1.9853	≈ 0	1.886	18.3	-29	1-8%	258	
N-Methylformamide	C_2H_5NO	59.067	-3.8	199.51	1.011[19]	1.678	189.0	3.83	2.096	0.03				
Methyl formate	$C_2H_4O_2$	60.052	-99	31.7	0.9713[20]	0.325	9.20	1.77	1.983	78.1	-19	5-23%	449	100
5-Methyl-2-hexanone	$C_7H_{14}O$	114.185		144	0.888[20]		13.53			0.691	36	1-8%	191	50
Methylhydrazine	CH_6N_2	46.072	-52.36	87.5					2.928	6.61	-8	2.5-92%	194	0.01
Methyl isocyanate	C_2H_3NO	57.051	-45	39.5	0.9230[27]		21.75	≈ 2.8		57.7	-7	5.3-26%	534	0.02
Methyl lactate, (±)-	$C_4H_8O_3$	104.105		144.8	1.0928[20]					0.62	49	>2.2%	385	
Methyl methacrylate	$C_5H_8O_2$	100.117	-47.55	100.5	0.9377[25]		6.32	1.67	1.910	5.10	10	1.7-8.2%		50
1-Methylnaphthalene	$C_{11}H_{10}$	142.197	-30.43	244.7	1.0202[20]		2.915	≈ 0	1.578	0.009			529	
Methyloxirane	C_3H_6O	58.079	-111.9	35	0.859[0]			2.01	2.073	71.7	-37	3.1-27.5%	449	2
2-Methylpentane	C_6H_{14}	86.175	-153.6	60.26	0.650[25]	0.286	1.886	≈ 0	2.248	28.2	<-29	1-7%	264	500
3-Methylpentane	C_6H_{14}	86.175	-162.90	63.27	0.6598[25]	0.306	1.886	≈ 0	2.213	25.3	-7	1.2-7%	278	500
2-Methyl-2,4-pentanediol	$C_6H_{14}O_2$	118.174	-50	197.1	0.923[15]		23.4	2.9	2.843	<0.01	102	1-9%	306	25
2-Methyl-1-pentanol	$C_6H_{14}O$	102.174		149	0.8263[20]				2.427	0.236	54	1.1-9.65%	310	
4-Methyl-2-pentanol	$C_6H_{14}O$	102.174	-90	131.6	0.8075[20]	4.07			2.672	0.698	41	1-6%		25
4-Methyl-2-pentanone	$C_6H_{12}O$	100.158	-84	116.5	0.7965[25]	0.545	13.11		2.130	2.64	18	1-8%	448	50
2-Methylpropanenitrile	C_4H_7N	69.106	-71.5	103.9	0.7704[20]		24.42	4.29			8		482	
2-Methyl-2-propanethiol	$C_4H_{10}S$	90.187	-0.5	64.2	0.7943[25]		5.475	1.66		24.2	<-29			
Methyl propanoate	$C_4H_8O_2$	88.106	-87.5	79.8	0.9150[20]	0.431	6.200		1.943	11.5	-2	2.5-13%	469	
2-Methylpropanoic acid	$C_4H_8O_2$	88.106	-46	154.45	0.9681[20]	1.226	2.58	1.08	1.964	0.17	56	2-9.2%	481	
2-Methyl-1-propanol	$C_4H_{10}O$	74.121	-101.9	107.89	0.8018[20]	3.33	17.93	1.64	2.449	1.39	28	2-11%	415	50
2-Methyl-2-propanol	$C_4H_{10}O$	74.121	25.69	82.4	0.7887[20]	4.31	12.47	1.7	2.949	5.52	11	2-8%	478	100
2-Methylpyridine	C_6H_7N	93.127	-66.68	129.38	0.9443[20]		10.18	1.85	1.703	1.5	39		538	
3-Methylpyridine	C_6H_7N	93.127	-18.14	144.14	0.9566[20]		11.10	2.40	1.704	0.795				
4-Methylpyridine	C_6H_7N	93.127	3.67	145.36	0.9548[20]		12.2	2.70	1.707	0.759	57			
N-Methyl-2-pyrrolidone	C_5H_9NO	99.131	-23.09	202	1.0230[25]		32.55	4.1	3.105	0.04	96	1-10%	346	
Methyl salicylate	$C_8H_8O_3$	152.148	-8	222.9	1.181[25]		8.80	2.47	1.637	0.015	96		454	
4-Methylstyrene	C_9H_{10}	118.175	-34.1	172.8	0.9173[25]					0.245	53	0.8-11%	538	50
Morpholine	C_4H_9NO	87.120	-4.8	128	1.0005[20]	2.02	7.42	1.55	1.892	1.34	37	1-11%	290	20
β-Myrcene	$C_{10}H_{16}$	136.234		167	0.8013[15]		2.3			0.280				
Nickel carbonyl	C_4NiO_4	170.734	-19.3	43 (exp 60)	1.31[25]				1.198					0.05
L-Nicotine	$C_{10}H_{14}N_2$	162.231	-79	247	1.0097[20]		8.937							0.1
Nitric acid	HNO_3	63.013	-41.6	83	1.55			2.17	1.744	8.34				2
2-Nitroanisole	$C_7H_7NO_3$	153.136	10.5	272	1.2540[20]		45.75	5.0		0.002				
Nitrobenzene	$C_6H_5NO_2$	123.110	5.7	210.8	1.2037[20]	1.863	35.6	4.22	1.509	0.03	88	2-9%	482	1
Nitroethane	$C_2H_5NO_2$	75.067	-89.5	114.0	1.0448[25]	0.688	29.11	3.23	1.790	2.79	28	3-17%	414	100
Nitromethane	CH_3NO_2	61.041	-28.38	101.19	1.1371[20]	0.630	37.27	3.46	1.746	4.79	35	7-22%	418	20
1-Nitropropane	$C_3H_7NO_2$	89.094	-108	131.1	0.9961[25]	0.798	24.70	3.66	1.97	1.36	36	2%-	421	25
2-Nitropropane	$C_3H_7NO_2$	89.094	-91.3	120.2	0.9821[25]		26.74	3.73	1.911	2.3	24	3-11%	428	10
N-Nitrosodiethylamine	$C_4H_{10}N_2O$	102.134		176.9	0.9422[20]									
N-Nitrosodimethylamine	$C_2H_6N_2O$	74.081		152	1.0048[20]					0.73				
2-Nitrotoluene	$C_7H_7NO_2$	137.137	-10.4	222	1.1611[19]		26.26		1.474	0.0014	106			2
3-Nitrotoluene	$C_7H_7NO_2$	137.137	15.5	232	1.1581[20]		24.95		1.474	0.03	106			2
Nonane	C_9H_{20}	128.255	-53.46	150.82	0.7192[20]	0.665	1.9722	≈ 0	2.217	0.570	31	0.8-2.9%	205	200
Nonanoic acid	$C_9H_{18}O_2$	158.238	12.4	254.5	0.9052[20]	7.01	2.475	0.79	2.290	0.00005				
1-Nonanol	$C_9H_{20}O$	144.254	-5	213.37	0.8280[20]	9.12	8.83		2.470	0.00050			260	
1-Nonene	C_9H_{18}	126.239	-81.3	146.9	0.7253[25]	0.586	2.180	≈ 0	2.142	0.714	26			
4-Nonylphenol	$C_{15}H_{24}O$	220.351	42	≈ 295	0.950[20]									
cis,cis-9,12-Octadecadienoic acid	$C_{18}H_{32}O_2$	280.446	-7		0.9022[20]		2.754							
cis-9-Octadecenoic acid	$C_{18}H_{34}O_2$	282.462	13.4	360	0.8935[20]		2.336	1.18	2.043	0.000001	189		363	
Octane	C_8H_{18}	114.229	-56.82	125.67	0.6986[25]	0.508	1.948	≈0	2.229	1.86	13	1-7%	206	300
Octanoic acid	$C_8H_{16}O_2$	144.212	16.5	239	0.9073[25]	5.02	2.85	1.15	2.066	0.0002				
1-Octanol	$C_8H_{18}O$	130.228	-14.8	195.16	0.8262[25]	7.29	10.30	1.8	2.344	0.01	81		270	
2-Octanol	$C_8H_{18}O$	130.228	-31.6	179.3	0.8193[20]	6.49	8.13	1.71	2.535		88		265	

Name	Mol. form.	M_r	t_m/°C	t_b/°C	ρ/g mL⁻¹	η/mPa s	ε	μ/D	c_p/J g⁻¹K⁻¹	vp/kPa	FP/°C	Fl. lim.	IT/°C	TLV/ppm
2-Octanone	C₈H₁₆O	128.212	-16	172.5	0.820²⁰		9.51	2.7	2.132	0.12	52			
1-Octene	C₈H₁₆	112.213	-101.7	121.29	0.7149²⁰	0.447	2.113	≈ 0	2.148	2.30	21		230	
Oxetane	C₃H₆O	58.079	-97	47.6	0.8930²⁵			1.94						
2-Oxetanone	C₃H₄O₂	72.063	-33.4	162	1.1460²⁰		4.18		1.694	0.3	74	>2.9%		0.5
Oxirane	C₂H₄O	44.052	-112.5	10.6	0.8821¹⁰		12.42	1.89	1.998	175	-20	3-100%	429	1
Oxiranemethanol, (±)-	C₃H₆O₂	74.079	-45	167 dec	1.1143²⁵									2
Paraldehyde	C₆H₁₂O₃	132.157	12.6	124.3	0.9943²⁰	1.079		1.43		1.6	36	>1.3%	238	
Parathion	C₁₀H₁₄NO₅PS	291.261	6.1	375	1.2681²⁰									0.01
Pentachloroethane	C₂HCl₅	202.294	-28.78	162.0	1.6796²⁰	2.25	3.716	0.92	0.859	0.478				
cis-1,3-Pentadiene	C₅H₈	68.118	-140.8	44.1	0.6910²⁰		2.319	0.500		50.6				
trans-1,3-Pentadiene	C₅H₈	68.118	-87.4	42	0.6710²⁵			0.585		54.7				
Pentanal	C₅H₁₀O	86.132	-91.5	103	0.8095²⁰		10.00			4.58	12		222	50
Pentane	C₅H₁₂	72.149	-129.67	36.06	0.6262²⁰	0.224	1.8371	≈ 0	2.317	68.3	-40	2-8%	260	600
Pentanedial	C₅H₈O₂	100.117	-14	188 dec										0.05
1,5-Pentanediol	C₅H₁₂O₂	104.148	-18	239	0.9914²⁰		26.2	2.5	3.08	0.001	129		335	
2,4-Pentanedione	C₅H₈O₂	100.117	-23	138	0.9721²⁵		26.524	2.8	2.08	1.02	34		340	
1-Pentanethiol	C₅H₁₂S	104.214	-75.65	126.6	0.850²⁰		4.847			1.83	18			
Pentanoic acid	C₅H₁₀O₂	102.132	-33.6	186.1	0.9339²⁵		2.661	1.61	2.059	0.024	96		400	
1-Pentanol	C₅H₁₂O	88.148	-77.6	137.98	0.8144²⁰	3.62	15.13	1.7	2.361	0.259	33	1-10%	300	
2-Pentanol	C₅H₁₂O	88.148	-73	119.3	0.8094²⁰	3.47	13.71	1.66	2.716	0.804	34	1.2-9%	343	
3-Pentanol	C₅H₁₂O	88.148	-69	116.25	0.8203²⁰	4.15	13.35	1.64	2.719	1.10	41	1.2-9%	435	
2-Pentanone	C₅H₁₀O	86.132	-76.8	102.26	0.809²⁰	0.470	15.45	2.7	2.137	4.97	7	2-8%	452	200
3-Pentanone	C₅H₁₀O	86.132	-39	101.7	0.8098²⁵	0.444	17.00	2.82	2.216	4.72	13	>1.6%	450	200
1-Pentene	C₅H₁₀	70.133	-165.12	29.96	0.6405²⁰	0.195	2.011	≈ 0.5	2.196	85.0	-18	1.5-8.7%	275	
cis-2-Pentene	C₅H₁₀	70.133	-151.36	36.93	0.6556²⁰			≈ 0	2.163	66.0	<-20			
trans-2-Pentene	C₅H₁₀	70.133	-140.21	36.34	0.6431²⁵			≈ 0	2.239	67.4	<-20			
Pentyl acetate	C₇H₁₄O₂	130.185	-70.8	149.2	0.8756²⁰		4.79	1.75	2.005	0.60	16	1-8%	360	50
Pentylamine	C₅H₁₃N	87.164	-55	104.3	0.7544²⁰	0.702	4.27		2.501	4.00	-1	2.2-22%		
Perchloric acid	ClHO₄	100.459	-112	≈ 90 dec	1.77									
Peroxyacetic acid	C₂H₄O₃	76.051	-0.2	110	1.226¹⁵					1.93	41			
Phenol	C₆H₆O	94.111	40.89	181.87	1.0545⁴⁵		12.40	1.224	2.123	0.055	79	1.8-8.6%	715	5
2-Phenoxyethanol	C₈H₁₀O₂	138.164	14	245	1.102²²					0.001	121			
Phenylhydrazine	C₆H₈N₂	108.141	20.6	243.5	1.0986²⁰	13.03	7.15		2.007	0.003	88			0.1
1-Phenyl-2-propylamine, (±)-	C₉H₁₃N	135.206		203	0.9306²⁵					0.06	<100			
Phosphinic acid	H₃O₂P	65.997	26.5	130	1.49									
Phosphoric acid	H₃O₄P	97.995	42.4	407					1.480					0.25
Phosphorothioc trichloride	Cl₃PS	169.398	-36.2	125	1.635		4.94							
Phosphorus(III) bromide	Br₃P	270.686	-41.5	173.2	2.8					0.38				
Phosphorus(III) chloride	Cl₃P	137.332	-93.6	76.1	1.574	0.529	3.498	0.56		16.1				0.2
Phosphoryl chloride	Cl₃OP	153.331	1.18	105.5	1.645		14.1	2.54	0.905	4.97				0.1
α-Pinene	C₁₀H₁₆	136.234	-64	156.2	0.8539²⁵		2.1787			0.64	33		255	
β-Pinene	C₁₀H₁₆	136.234	-61.5	166	0.860²⁵		2.4970			0.61	38		275	
Piperidine	C₅H₁₁N	85.148	-11.02	106.22	0.8606²⁰	1.573	4.33	1.2	2.113	4.28	16	1-10%		
Propanal	C₃H₆O	58.079	-80	48	0.8657²⁵	0.321	18.5	2.72	2.362	42.2	-30	2.6-17%	207	20
1,2-Propanediol	C₃H₈O₂	76.095	-60	187.6	1.0361²⁰	40.4	27.5	2.2	2.507	0.02	99	3-13%	371	
1,3-Propanediol	C₃H₈O₂	76.095	-27.7	214.4	1.0538²⁰		35.1	2.5		0.007			400	
Propanenitrile	C₃H₅N	55.079	-92.78	97.14	0.7818²⁰	0.294	29.7	4.05	2.166	6.14	2	3-14%	512	
Propanoic acid	C₃H₆O₂	74.079	-20.5	141.15	0.9882²⁵	1.030	3.44	1.75	2.063	0.553	52	2.9-12.1%	465	10
Propanoic anhydride	C₆H₁₀O₃	130.141	-45	170	1.0110²⁰		18.30		1.806	0.45	63	1.3-9.5%	285	
1-Propanol	C₃H₈O	60.095	-124.39	97.2	0.7997²⁵	1.945	20.8	1.55	2.395	2.76	23	2-14%	412	200
2-Propanol	C₃H₈O	60.095	-87.9	82.3	0.7809²⁵	2.04	20.18	1.56	2.604	6.02	12	2-13%	399	200
Propargyl alcohol	C₃H₄O	56.063	-51.8	113.6	0.9478²⁰		20.8	1.13			36			1
Propyl acetate	C₅H₁₀O₂	102.132	-93	101.54	0.8878²⁰	0.544	5.62	1.8	1.921	4.49	13	2-8%	450	200
Propylamine	C₃H₉N	59.110	-84.75	47.22	0.7173²⁰	0.376	5.08	1.17	2.776	42.1	-37	2-10%	318	
Propylbenzene	C₉H₁₂	120.191	-99.6	159.24	0.8593²⁵		2.370	≈ 0	1.786	0.45	30	1-6%	450	
Propyl butanoate	C₇H₁₄O₂	130.185	-95.2	143.0	0.8730²⁰		4.3			0.618	37			
Propylene carbonate	C₄H₆O₃	102.089	-48.8	242	1.2047²⁰		66.14	4.9	2.141	0.05	135			
Propyl formate	C₄H₈O₂	88.106	-92.9	80.9	0.9073²⁰	0.485	6.92	1.89	1.945	10.9	-3		455	
Propyl propanoate	C₆H₁₂O₂	116.158	-75.9	122.5	0.8809²⁰		5.249			1.88	79			
Pyridine	C₅H₅N	79.101	-41.70	115.23	0.9819²⁰	0.879	13.260	2.21	1.678	2.76	20	2-12%	482	1
Pyrrole	C₄H₅N	67.090	-23.39	129.79	0.9698²⁰	1.225	8.00	1.74	1.903	1.10	39			
Pyrrolidine	C₄H₉N	71.121	-57.79	86.56	0.8586²⁰	0.704	8.30	1.6	2.202	8.40	3			
2-Pyrrolidone	C₄H₇NO	85.105	25	251	1.120²⁰		28.18	3.5	1.99		129			
Quinoline	C₉H₇N	129.159	-14.78	237.16	1.0977¹⁵	3.34	9.16	2.29	1.51	0.011			480	
Safrole	C₁₀H₁₀O₂	162.185	11.2	234.5	1.1000²⁰					0.01	100			
Salicylaldehyde	C₇H₆O₂	122.122	-7	197	1.1674²⁰		18.35	2.86	1.818	0.075	78			

Name	Mol. form.	M_r	t_m/°C	t_b/°C	ρ/g mL⁻¹	η/mPa s	ε	μ/D	c_p/J g⁻¹K⁻¹	vp/kPa	FP/°C	Fl. lim.	IT/°C	TLV/ppm
Selenium chloride	Cl₂Se₂	228.83	-85	130 dec	2.774					0.02				
Selenium oxychloride	Cl₂OSe	165.86	8.5	177	2.44		46.2			0.02				
Selenium oxyfluoride	F₂OSe	132.96	15	125	2.8					0.56				
Styrene	C₈H₈	104.150	-30.65	145	0.9016²⁵	0.695	2.4737	0.123	1.747	0.81	31	1-7%	490	20
Sulfolane	C₄H₈O₂S	120.171	27.6	287.3	1.2723¹⁸		43.26	4.8	1.498	<0.01	177			
Sulfur chloride	Cl₂S₂	135.037	-77	137	1.69		4.79			1.27				1
Sulfur dichloride	Cl₂S	102.971	-122	59.6	1.62		2.915	0.36		17.9				
Sulfuric acid	H₂O₄S	98.080	10.31	337	1.8				1.416					0.05
Sulfuryl chloride	Cl₂O₂S	134.970	-51	69.4	1.680		9.1	1.81	0.993	18.7				
α-Terpinene	C₁₀H₁₆	136.234		174	0.8375¹⁹		2.4526							
1,1,2,2-Tetrabromoethane	C₂H₂Br₄	345.653	0	243.5	2.9655²⁰		6.72	1.38	0.479	0.003			335	1
Tetrabromosilane	Br₄Si	347.702	5.39	154	2.8			0						
1,1,2,2-Tetrachloro-1,2-difluoroethane	C₂Cl₄F₂	203.830	24.8	92.8	1.5951⁵⁰		2.52		0.852	7.51				500
1,1,1,2-Tetrachloroethane	C₂H₂Cl₄	167.849	-70.2	130.2	1.5406²⁰	1.437			0.92	1.6	47	5-12%		
1,1,2,2-Tetrachloroethane	C₂H₂Cl₄	167.849	-42.4	145.2	1.5953²⁰		8.50	1.32	0.967	0.622	62	20-54%		1
Tetrachloroethene	C₂Cl₄	165.833	-22.3	121.3	1.6230²⁰	0.844	2.268	0	0.865	2.42	45			25
Tetrachloromethane	CCl₄	153.823	-22.62	76.8	1.5940²⁰	0.908	2.2379	0	0.850	15.2				5
Tetrachlorosilane	Cl₄Si	169.897	-68.74	57.65	1.5	99.4		0	0.855	31.3				
Tetradecane	C₁₄H₃₀	198.388	5.82	253.58	0.7596²⁰	2.13	2.0343	≈ 0		0.002	112	>0.5%	200	
Tetraethylene glycol	C₈H₁₈O₅	194.226	-6.2	328	1.1285¹⁵		20.44		2.208	0.000001	182			
Tetrafluoroboric acid	BF₄H	87.813		130 dec	~1.8									
Tetrahydrofuran	C₄H₈O	72.106	-108.44	65	0.8833²⁵	0.456	7.52	1.75	1.720	21.6	-14	2-12%	321	200
Tetrahydrofurfuryl alcohol	C₅H₁₀O₂	102.132	<-80	178	1.0524²⁰		13.48	2.1	1.774	0.100	75	1.5-9.7%	282	
1,2,3,4-Tetrahydronaphthalene	C₁₀H₁₂	132.202	-35.7	207.6	0.9645²⁵	2.14	2.771	≈ 0	1.645	0.05	71	1-5%	385	
Tetrahydropyran	C₅H₁₀O	86.132	-49.1	88	0.8814²⁰		5.66	1.74	1.82	9.54	-20			
Tetrahydrothiophene	C₄H₈S	88.172	-96.2	121.1	0.9987²⁰	0.973		1.90		2.45				
Tetramethylsilane	C₄H₁₂Si	88.224	-99.06	26.6	0.648¹⁹		1.921	0	2.313	94.2				
Tetramethylurea	C₅H₁₂N₂O	116.161	-0.6	176.5	0.9687²⁰		23.10	3.5		0.138	77			
Tetranitromethane	CN₄O₈	196.033	13.8	126.1	1.6380²⁰		2.317	0		1.13				0.005
Thionyl bromide	Br₂OS	207.873	-50	140			9.06			0.84				
Thionyl chloride	Cl₂OS	118.970	-101	75.6	1.631		8.675	1.45	1.017	16.0				1
Thiophene	C₄H₄S	84.140	-38.21	84.0	1.0649²⁰		2.739	0.55	1.471	10.6	-1			
Tin(IV) chloride	Cl₄Sn	260.521	-34.07	114.15	2.234			0	0.634					
Titanium(IV) chloride	Cl₄Ti	189.678	-24.12	136.45	1.73				0.766					
Toluene	C₇H₈	92.139	-94.95	110.63	0.8668²⁰	0.560	2.379	0.37	1.707	3.79	4	1-7%	480	50
Toluene-2,4-diisocyanate	C₉H₆N₂O₂	174.156	20.5	251	1.2244²⁰		8.433		1.653	0.003	127	0.9-9.5%		0.005
Tribromomethane	CHBr₃	252.731	8.69	149.1	2.8788²⁵	1.857	4.404	0.99	0.517	0.726	83			0.5
Tributylamine	C₁₂H₂₇N	185.349	-70	216.5	0.7770²⁰		2.340	0.8		0.01	63	1-5%		
Tributyl borate	C₁₂H₂₇BO₃	230.151	<-70	234	0.8567²⁰		2.23	0.77			93			
Tributyrin	C₁₅H₂₆O₆	302.363	-75	307.5	1.0350²⁰		5.72		1.837		180	>0.5%	407	
Trichloroacetaldehyde	C₂HCl₃O	147.387	-57.5	97.8	1.512²⁰		6.8		1.025	6.66				
1,2,4-Trichlorobenzene	C₆H₃Cl₃	181.447	16.92	213.5	1.459²⁵					0.057	105	2.5-6.6%	571	5
1,1,1-Trichloroethane	C₂H₃Cl₃	133.404	-30.01	74.09	1.3390²⁰	0.793	7.243	1.76	1.082	16.5	-1	8-13%	500	350
1,1,2-Trichloroethane	C₂H₃Cl₃	133.404	-36.3	113.8	1.4397²⁰		7.1937	1.4	1.131	3.1	32	6-28%	460	10
Trichloroethene	C₂HCl₃	131.388	-84.7	87.21	1.4642²⁰	0.545	3.390	0.8	0.947	9.91	32	8-11%	420	50
Trichloroethylsilane	C₂H₅Cl₃Si	163.506	-105.6	100.5	1.2373²⁰			2.04		6.29	22			
Trichlorofluoromethane	CCl₃F	137.368	-110.44	23.7	1.4879²⁰	0.421	3.00	0.46	0.885	106				1000
Trichloromethane	CHCl₃	119.378	-63.41	61.17	1.4788²⁵	0.537	4.8069	1.04	0.957	26.2				10
(Trichloromethyl)benzene	C₇H₅Cl₃	195.474	-4.42	221	1.3723²⁰		6.9	2.03		0.35	127		211	0.1
Trichloromethylsilane	CH₃Cl₃Si	149.480	-90	65.6	1.273²⁰			1.91	1.091	22.5	-9	7.6->20%	>404	
Trichloronitromethane	CCl₃NO₂	164.376	-64	112	1.6558²⁰		7.319			3.18				0.1
1,2,3-Trichloropropane	C₃H₅Cl₃	147.431	-14.7	157	1.3889²⁰		7.5		1.245	0.492	71	3.2-12.6%		10
Trichlorosilane	Cl₃HSi	135.452	-128.2	33	1.331	0.326		0.86			-50		104	
1,1,2-Trichloro-1,2,2-trifluoroethane	C₂Cl₃F₃	187.375	-36.22	47.7	1.5635²⁵	0.656	2.41		0.908	44.8				1000
Tri-o-cresyl phosphate	C₂₁H₂₁O₄P	368.363	11	410	1.1955²⁰		6.7	2.87	1.57	0.0000002	225		385	0.01
Tridecane	C₁₃H₂₈	184.361	-5.4	235.47	0.7564²⁰	1.724	2.0213	≈ 0	2.206	0.005	79			
1-Tridecene	C₁₃H₂₆	182.345	-13	232.8	0.7658²⁰	1.50	2.139	≈ 0	2.149	0.0047	79			
Triethanolamine	C₆H₁₅NO₃	149.188	20.5	335.4	1.1242²⁰	609	29.36	3.6	2.61	<0.01	179	1-10%		0.8
Triethylamine	C₆H₁₅N	101.190	-114.7	89	0.7275²⁰	0.347	2.418	0.66	2.173	7.70	-7	1-8%	249	1
Triethylene glycol	C₆H₁₄O₄	150.173	-7	285	1.1274¹⁵		23.69		2.18	0.0002	177	1-9%	371	
Triethylene glycol dimethyl ether	C₈H₁₈O₄	178.227	-45	216	0.986²⁰		7.62				111			
Triethyl phosphate	C₆H₁₅O₄P	182.154	-56.4	215.5	1.0695²⁰		13.20	3.1			115		454	
Trifluoroacetic acid	C₂HF₃O₂	114.023	-15.2	73	1.5351²⁵	0.808	8.42	2.28		15.1				
(Trifluoromethyl)benzene	C₇H₅F₃	146.110	-28.95	102.1	1.1884²⁰		9.22	2.86	1.289	5.14	12			

Name	Mol. form.	M_r	t_m/°C	t_b/°C	ρ/g mL⁻¹	η/mPa s	ε	μ/D	c_p/J g⁻¹K⁻¹	vp/kPa	FP/°C	Fl. lim.	IT/°C	TLV/ppm
1,2,3-Trimethylbenzene	C_9H_{12}	120.191	-25.4	176.12	0.8944[20]		2.656	≈ 0	1.800	0.20	44	0.8-6.6%	470	25
1,2,4-Trimethylbenzene	C_9H_{12}	120.191	-43.77	169.38	0.8758[20]		2.377	≈ 0	1.789	0.30	44	1-6%	500	25
1,3,5-Trimethylbenzene	C_9H_{12}	120.191	-44.72	164.74	0.8615[25]		2.279	0	1.741	0.33	50	1-5%	559	25
Trimethyl borate	$C_3H_9BO_3$	103.912	-29.3	67.5	0.915[25]		2.2762		1.828	17.2	-8			
Trimethylchlorosilane	C_3H_9ClSi	108.642	-40	60	0.856[25]					30.7	-28		395	
2,2,4-Trimethylpentane	C_8H_{18}	114.229	-107.3	99.22	0.6878[25]		1.943	≈ 0	2.093	6.50	-12		418	300
2,3,3-Trimethylpentane	C_8H_{18}	114.229	-100.9	114.8	0.7262[20]		1.9780	≈ 0	2.150	3.60	<21		425	300
Trimethyl phosphate	$C_3H_9O_4P$	140.074	-46	197.2	1.2144[20]		20.6	3.2		0.11	107			
2,4,6-Trimethylpyridine	$C_8H_{11}N$	121.180	-46	170.6	0.9166[22]		7.807	2.05		4.1				
Trinitroglycerol	$C_3H_5N_3O_9$	227.087	13.5	exp 218	1.5931[20]		19.25			0.00005			270	0.05
Undecane	$C_{11}H_{24}$	156.309	-25.5	195.9	0.7402[20]	1.098	1.9972	≈ 0	2.207	0.05	69			
Vanadium(IV) chloride	Cl_4V	192.753	-25.7	148	1.816		3.05							
Vanadyl trichloride	Cl_3OV	173.299	-79	127	1.829		3.4							
Vinyl acetate	$C_4H_6O_2$	86.090	-93.2	72.8	0.9256[25]			1.79	1.969	15.4	-8	2.6-13.4%	402	10
4-Vinylcyclohexene	C_8H_{12}	108.181	-108.9	128	0.8299[20]					1.87	16		269	0.1
Water	H_2O	18.015	0.00	100.0	0.9970	0.890	80.100	1.8546	4.180	3.17				
o-Xylene	C_8H_{10}	106.165	-25.2	144.5	0.8802[10]	0.760	2.562	0.64	1.753	0.88	32	1-7%	463	100
m-Xylene	C_8H_{10}	106.165	-47.8	139.12	0.8596[25]	0.581	2.359	≈ 0	1.724	1.13	27	1-7%	527	100
p-Xylene	C_8H_{10}	106.165	13.25	138.37	0.8566[25]	0.603	2.2735	0	1.710	1.19	27	1-7%	528	100
2,4-Xylenol	$C_8H_{10}O$	122.164	24.5	210.98	0.9650[20]		5.060	1.4		0.022				

MISCIBILITY OF ORGANIC SOLVENTS

The chart below gives qualitative information on the miscibility of pairs of organic liquids. Two liquids are considered miscible (indicated by **M** in the chart) if mixing equal volumes produces a single liquid phase. If two phases separate, they are considered immiscible (**I**). An entry of **P** indicates two phases whose volumes differ appreciably, suggesting a partial miscibility of the components. The symbol **R** indicates a reaction between the components. All data refer to room temperature.

The codes for the columns are:

A Acetone	**J** Diethyl ether	**S** Methyl isopropyl ketone
B Benzaldehyde	**K** *N,N*-Dimethylaniline	**T** Nitromethane
C Benzene	**L** Dipentylamine	**U** 1-Octanol
D Butyl acetate	**M** Ethyl alcohol	**V** 1,3-Propanediol
E Butyl alcohol	**N** Ethylene glycol	**W** Pyridine
F Carbon tetrachloride	**O** Ethylene glycol monoethyl ether	**X** Triethylenetetramine
G 2-Chloroethanol	**P** Formamide	**Y** Triethyl phosphate
H Chloroform	**Q** Furfuryl alcohol	
I *o*-Cresol	**R** Glycerol	

References

1. Drury, J. S., *Ind. Eng. Chem.* 44, 2744, 1959.
2. Jackson, W. M., and Drury, J. S., *Ind. Eng. Chem.* 51, 1491, 1959.

	A	B	C	D	E	F	G	H	I	J	K	L	M	N	O	P	Q	R	S	T	U	V	W	X	Y
Acetone	-	M	M	M	M	M	M	M	M	M	M	M	M	M	M	M	M	I	M	M	M	M	M	M	M
Adiponitrile	M		M		M	I			I	M		M	I		M	M	I					M	M		
2-Amino-2-methyl-1-propanol	M	M	M	M	M	M	M						M	M		M	M	M			M	M	M	M	
p-Anisaldehyde							M				I		M			I					I				M
Benzaldehyde	M	-	M	M	M	M				M	M		M	P		M	M	P			M	M	M	M	
Benzene	M	M	-	M	M	M	M	M	M	M	M	M	I	M	I	M	I	M	I	M	I	M		M	M
Benzonitrile	M		M		M	M				M	M		M	I		I	M	I						I	M
Benzothiazole	M		M		M	M				M	M		M	I		I	M	I							
Benzyl alcohol	M	M	M	M	M	M				M	M		M	M		M	M	M			M	M	M	M	
Benzyl mercaptan	M		M		M	M				M	M		M	I		I	M	I						I	M
2-Bromoethyl acetate	M		P				M					R	M											R	
1,3-Butanediol	M		I				M	M	P				M	M			M							M	M
2,3-Butanediol	M		P				M	M	M				M	M			M							M	M
Butyl acetate	M	M	M	-	M	M				M	M		M	P		I	M	I			M	M	P	M	
Butyl alcohol	M	M	M	M	-	M				M	M		M	M		M	M	M			M	M	M	M	
Carbon tetrachloride	M	M	M	M	M	-				M	M		M	I		I	M	I			M	M	I	M	
2-Chloroethanol	M		M				-	M	M				M				M						M	M	M
Chloroform	M						M	-		M	M		M	M	M		I	M					M	M	M
3-Chloro-1,2-propanediol	M		I				M	M	M			R	M				M							M	R
Cinnamaldehyde	M		M				M	M		M	M	I					I	M			I			R	M
o-Cresol							M		-				M	M			M				M		M		
Diacetone alcohol	M	M	P	M	M	P				M	M		M	M		M	M	M	I		M	M	M	M	
Dibenzyl ether	M		M				M	M				M				I		M	I					M	M
Dibutylamine							R						M	M		P									M
Dibutyl carbonate	M		M					M				M	M					M					I		
Dibutyl ether	M	M	M	M	M	M				M	M		M	I		I	M	I	I	M	I	I	M		
Diethanolamine	M	I	I	I	M	I				I	P		M	M	M	M	M	M			I	M	M	M	
Diethylacetic acid	M		M				M	M				R	M	M	M		I	M			M			R	M
Diethylene glycol dibutyl ether	M		M					M	M	M		R	M								M				
Diethylene glycol diethyl ether	M		M					M	M	M		M	M								M			M	M
Diethylene glycol monobutyl ether	M		M					M	M	M		M	M								M			M	M
Diethylene glycol monoethyl ether	M		M					M	M	M		M	M								M			M	M
Diethylene glycol monomethyl ether	M		M					M				M		M	M									M	M
Diethylenetriamine	M		M				R	M				I	M	M	M		M	R						M	M
Diethyl ether	M	M	M	M	M	M	M	M	M	-	M	M	M	M	I	M	I	M	I	M	M	M	M	I	M
Diethylformamide	M		M				M	M				R	M	M	M		M	M				M		R	M
Dihexyl ether	M		M				M	M				M	M	I	M		I	M			I				I

	A	B	C	D	E	F	G	H	I	J	K	L	M	N	O	P	Q	R	S	T	U	V	W	X	Y
Diisobutyl ketone	M		M				M	M			M	M	I	M				I	M			I		M	M
Diisopropylamine	M		M				R	M			M	M	M	M				M	M			M		M	M
N,N-Dimethylaniline	M	M	M	M	M	M				M	-		M	I		I	M	I		M	M	I	M		
Dipentylamine	M		M					M			M	-	M	P	M		P	M			M			I	M
N,N-Dipropylaniline	M		M	M	M	M	M			M	M	M	M	I	M	I	M	I		M		I	M	M	M
Dipropylene glycol	M		M				M	M	M	M			M						M					M	M
Ethyl alcohol	M	M	M	M	M	M				M	M	M	M	-	M	M	M	M	M	M	M	M	M	M	M
Ethyl benzoate	M	M	M	M	M	M				M	M	M	M	M		I		I	M	I	M	M	P	M	M
Ethyl chloroacetate	M		M				M	M			M	M	I	M				I	M			I		R	M
Ethyl cinnamate	M		M					M			M	M	I	M				I	M			I		M	M
Ethylene glycol	M	P	I	P	M	I			P	M	I	I	P	M	-	M	M	M	I	I	M	M	M	M	M
Ethylene glycol monobutyl ether	M		M				M	M	M		M	M							M					M	M
Ethylene glycol monoethyl ether	M		M				M	M	M		M	M		-					M					M	M
Ethylene glycol monomethyl ether	M		M					M	M		M	M												M	M
2-Ethyl-1-hexanol	M	M	M	M	M	M					M	M	M	M		I	M	I		I	M			M	M
Ethyl phenylacetate	M		M				M	M			M		I	M				I	M			I		M	M
Ethyl thiocyanate	M		M		M	M					M	M	M	I		I	M	I				I	M		
Formamide	M	M	I	I	M	I			I	I			M	M		-	M	M		M	I	M	M		
Furfuryl alcohol	M	M	M	M	M						M	M	M	M	M	-		M		M	M	M	M		
Glycerol	I	P	I	I	M	I			I		M	I	I	P	M	M	M	M	-	I	I	I	M	M	M
1-Heptadecanol	M		M					M			M	M	M						M						M
3-Heptanol	M		M				M	M			M	M	M	M				I				M		M	M
Heptyl acetate	M		M				M	M			M		I	M				I	M			I		R	M
Hexanenitrile	M		M				M	M			M	M	I	M				I	M			I		M	M
Isobutyl mercaptan	M		M		M	M					M	M	M	M		I		I	M	I				R	M
Isopentyl acetate	M		M				M	M			M	M	I	M				I	M			I		M	M
Isopentyl alcohol	M	M	M	M	M	M					M	M	M	M		M	M	I		M	M	M	M		
Isopentyl sulfide	M		M		M	M					M	M	M	M		I		I	I	I				I	M
Methyl disulfide	M		M		M	M					M	M	M	M		I		I	M	I				R	M
Methyl isobutyl ketone	M	M	M	M	M		M	H			M	M	I			P	M	I		M	M	I		M	
Methyl isopropyl ketone	M		M				M	M			M	M	M	I	M			I	-			M		R	M
4-Methylpentanoic acid	M		M					M			M	M	M	M				I	M			M		R	M
Nitromethane	M	M	I	M	M	M					M	M	M	I		M	M	I	-	P	I	M			
1-Octanol	M	M	M	M	M	M					M	M	M	M		I	M	I		P	-	M		M	M
o-Phenetidine	M		M				M	M			M	M	M	M					M			M		M	
1,2-Propanediol	M		I				M	M	P	M	M	M							M					M	M
1,3-Propanediol	M	M	I	P	M	I				M	M	I	I	M	M	M	M	M	M	I	M	-		M	M
Pyridine	M	M	M	M	M	M	M			M	M	M	M	M	M	M	M	M	M	M	M	-			M
Tetradecanol	M		M				M	M			M	M	I	M				I	M			P		M	M
Tributyl phosphate	M		M				M	M			M	M	P	M				I	M			M		M	M
Triethylene glycol	M	P					M	M	I		P	M							M					M	M
Triethylenetetramine	M		M				M	M			M	I	M	M				M	R			M		-	M
Triethyl phosphate	M		M				M	M	M		M	M							M				M	M	-
2,6,8-Trimethyl-4-nonanone	M		M				M	M			M	M	I	M				I	M			I		I	M

DENSITY OF SOLVENTS AS A FUNCTION OF TEMPERATURE

The table below lists the density of several common solvents in the temperature range from 0°C to 100°C. The values have been calculated from the Rackett Equation using parameters in the reference. Density values refer to the liquid at its saturation vapor pressure; thus entries for temperatures above the normal boiling point are for pressures greater than atmospheric.

Reference

Lide, D. R., and Kehiaian, H. V., *Handbook of Thermophysical and Thermochemical Data*, CRC Press, Boca Raton, FL, 1994.

					Density in g/mL						
Solvent	0°C	10°C	20°C	30°C	40°C	50°C	60°C	70°C	80°C	90°C	100°C
Acetic acid			1.051	1.038	1.025	1.012	0.9993	0.9861	0.9728	0.9592	0.9454
Acetone	0.8129	0.8016	0.7902	0.7785	0.7666	0.7545	0.7421	0.7293	0.7163	0.7029	0.6890
Acetonitrile			0.7825	0.7707	0.7591	0.7473	0.7353	0.7231	0.7106	0.6980	0.6851
Aniline	1.041	1.033	1.025	1.016	1.008	1.000	0.9909	0.9823	0.9735	0.9646	0.9557
Benzene		0.8884	0.8786	0.8686	0.8584	0.8481	0.8376	0.8269	0.8160	0.8049	0.7935
1-Butanol	0.8293	0.8200	0.8105	0.8009	0.7912	0.7812	0.7712	0.7609	0.7504	0.7398	0.7289
Butylamine	0.7606	0.7512	0.7417	0.7320	0.7221	0.7120	0.7017	0.6911	0.6803	0.6693	0.6579
Carbon disulfide	1.290	1.277	1.263	1.248	1.234						
Chlorobenzene	1.127	1.116	1.106	1.096	1.085	1.074	1.064	1.053	1.042	1.030	1.019
Cyclohexane		0.7872	0.7784	0.7694	0.7602	0.7509	0.7414	0.7317	0.7218	0.7117	0.7013
Decane	0.7447	0.7374	0.7301	0.7226	0.7151	0.7074	0.6997	0.6919	0.6839	0.6758	0.6676
1-Decanol			0.8294	0.8229	0.8162	0.8093	0.8024	0.7955	0.7884	0.7813	0.7740
Dichloromethane	1.362	1.344	1.326	1.307	1.289	1.269	1.250	1.229	1.208	1.187	1.165
Diethyl ether	0.7368	0.7254	0.7137	0.7018	0.6896	0.6770	0.6639	0.6505	0.6366	0.6220	0.6068
N,N-Dimethylaniline		0.9638	0.9562	0.9483	0.9401	0.9318	0.9234	0.9150	0.9064	0.8978	0.8890
Ethanol	0.8121	0.8014	0.7905	0.7793	0.7680	0.7564	0.7446	0.7324	0.7200	0.7073	0.6942
Ethyl acetate	0.9245	0.9126	0.9006	0.8884	0.8759	0.8632	0.8503	0.8370	0.8234	0.8095	0.7952
Ethylbenzene	0.8836	0.8753	0.8668	0.8582	0.8495	0.8407	0.8318	0.8228	0.8136	0.8043	0.7948
Ethyl formate	0.9472	0.9346	0.9218	0.9087	0.8954	0.8818	0.8678	0.8535	0.8389	0.8238	0.8082
Ethyl propanoate	0.9113	0.9005	0.8895	0.8784	0.8671	0.8556	0.8439	0.8319	0.8197	0.8072	0.7944
Heptane	0.7004	0.6921	0.6837	0.6751	0.6664	0.6575	0.6485	0.6393	0.6298	0.6202	0.6102
Hexane	0.6774	0.6685	0.6594	0.6502	0.6407	0.6311	0.6212	0.6111	0.6006	0.5899	0.5789
1-Hexanol	0.8359	0.8278	0.8195	0.8111	0.8027	0.7941	0.7854	0.7766	0.7676	0.7585	0.7492
Isopropylbenzene	0.8769	0.8696	0.8615	0.8533	0.8450	0.8366	0.8280	0.8194	0.8106	0.8017	0.7927
Methanol	0.8157	0.8042	0.7925	0.7807	0.7685	0.7562	0.7435	0.7306	0.7174	0.7038	0.6898
Methyl acetate	0.9606	0.9478	0.9346	0.9211	0.9074	0.8933	0.8790	0.8643	0.8491	0.8336	0.8176
N-Methylaniline	1.0010	0.9933	0.9859	0.9785	0.9709	0.9633	0.9556	0.9478	0.9399	0.9319	0.9239
Methylcyclohexane	0.7858	0.7776	0.7693	0.7608	0.7522	0.7435	0.7346	0.7255	0.7163	0.7069	0.6973
Methyl formate	1.003	0.9887	0.9739	0.9588	0.9433	0.9275	0.9112	0.8945	0.8772	0.8594	0.8409
Methyl propanoate	0.9383	0.9268	0.9150	0.9030	0.8907	0.8783	0.8656	0.8526	0.8393	0.8257	0.8117
Nitromethane			1.139	1.125	1.111	1.097	1.083	1.069	1.055	1.040	1.026
Nonane	0.7327	0.7252	0.7176	0.7099	0.7021	0.6941	0.6861	0.6779	0.6696	0.6611	0.6525
Octane	0.7185	0.7106	0.7027	0.6945	0.6863	0.6779	0.6694	0.6608	0.6520	0.6430	0.6338
Pentanoic acid	0.9563	0.9476	0.9389	0.9301	0.9211	0.9121	0.9029	0.8937	0.8843	0.8748	0.8652
1-Propanol	0.8252	0.8151	0.8048	0.7943	0.7837	0.7729	0.7619	0.7506	0.7391	0.7273	0.7152
2-Propanol	0.8092	0.7982	0.7869	0.7755	0.7638	0.7519	0.7397	0.7272	0.7143	0.7011	0.6876
Propyl acetate	0.9101	0.8994	0.8885	0.8775	0.8662	0.8548	0.8432	0.8313	0.8192	0.8069	0.7942
Propylbenzene	0.8779	0.8700	0.8619	0.8538	0.8456	0.8373	0.8289	0.8204	0.8117	0.8030	0.7943
Propyl formate	0.9275	0.9166	0.9053	0.8938	0.8821	0.8702	0.8581	0.8457	0.8330	0.8201	0.8068
Tetrachloromethane	1.629	1.611	1.593	1.575	1.557	1.538	1.518	1.499	1.479	1.458	1.437
Toluene	0.8846	0.8757	0.8667	0.8576	0.8483	0.8389	0.8294	0.8197	0.8098	0.7998	0.7896
Trichloromethane	1.524	1.507	1.489	1.471	1.452	1.433	1.414	1.394			
2,2,4-Trimethylpentane			0.6921	0.6836	0.6750	0.6663	0.6574	0.6484	0.6391	0.6296	0.6199
o-Xylene			0.8801	0.8717	0.8633	0.8547	0.8460	0.8372	0.8282	0.8191	0.8099
m-Xylene	0.8813	0.8729	0.8644	0.8558	0.8470	0.8382	0.8292	0.8201	0.8109	0.8015	0.7920
p-Xylene			0.8609	0.8523	0.8436	0.8347	0.8258	0.8167	0.8075	0.7981	0.7886

DEPENDENCE OF BOILING POINT ON PRESSURE

The normal boiling point of a liquid is defined as the temperature at which the vapor pressure reaches standard atmospheric pressure, 101.325 kPa. The change in boiling point with pressure may be calculated from the representation of the vapor pressure by the Antoine Equation,

$$\ln p = A_1 - A_2/(T + A_3)$$

where p is the vapor pressure, T the absolute temperature, and A_1, A_2, and A_3 are constants. This table, which has been calculated using the Antoine constants in Reference 1, gives values of $\Delta t/\Delta p$ for a number of liquids, in units of both °C/kPa and °C/mmHg. The correction to the boiling point is generally accurate to 0.1 to 0.2 °C as long as the pressure is within 10% of standard atmospheric pressure.

A slightly less accurate estimate of $\Delta t/\Delta p$ may be obtained from the Clausius-Clapeyron equation, with the assumption that the change in volume upon vaporization equals the ideal-gas volume of the vapor. This leads to the equation

$$\Delta t/\Delta p = RT_b^2/p_0 \Delta_{vap} H(T_b)$$

where R is the molar gas constant, p_0 is 101.325 kPa, T_b is the normal boiling point temperature (absolute), and $\Delta_{vap} H(T_b)$ is the molar enthalpy of vaporization at the normal boiling point. Values of the last quantity may be obtained from the table "Enthalpy of Vaporization" in Section 6.

Reference

1. Lide, D. R., and Kehiaian, H. V., *CRC Handbook of Thermophysical and Thermochemical Data*, CRC Press, Boca Raton, FL, 1994, pp. 49-59.

Compound	t_b °C	$\Delta t/\Delta p$ °C/kPa	$\Delta t/\Delta p$ °C/mmHg
Acetaldehyde	20.1	0.261	0.0348
Acetic acid	117.9	0.324	0.0432
Acetone	56.0	0.289	0.0385
Acetonitrile	81.6	0.316	0.0421
Ammonia	-33.33	0.198	0.0264
Aniline	184.1	0.378	0.0504
Anisole	153.7	0.367	0.0489
Benzaldehyde	179.0	0.392	0.0523
Benzene	80.0	0.321	0.0428
Bromine	58.8	0.300	0.0400
Butane	-0.5	0.267	0.0356
1-Butanol	117.7	0.278	0.0371
Carbon disulfide	46.2	0.304	0.0405
Chlorine	-34.04	0.224	0.0299
Chlorobenzene	131.7	0.365	0.0487
1-Chlorobutane	78.6	0.321	0.0428
Chloroethane	12.3	0.262	0.0349
Chloroethylene	-13.3	0.241	0.0321
Cyclohexane	80.7	0.328	0.0437
Cyclohexanol	160.8	0.344	0.0459
Cyclohexanone	155.4	0.382	0.0509
Decane	174.1	0.388	0.0517
Dibutyl ether	140.2	0.363	0.0484
Dichloromethane	39.6	0.276	0.0368
Diethyl ether	34.5	0.278	0.0371
Dimethyl sulfoxide	189.0	0.379	0.0505
1,4-Dioxane	101.5	0.321	0.0428
Dipropyl ether	90.0	0.326	0.0435
Ethanol	78.2	0.249	0.0332
Ethyl acetate	77.1	0.300	0.0400
Ethylene glycol	197.3	0.331	0.0441
Heptane	98.5	0.336	0.0448
Hexafluorobenzene	80.2	0.305	0.0407
Hexane	68.7	0.314	0.0419

Compound	t_b °C	$\Delta t/\Delta p$ °C/kPa	$\Delta t/\Delta p$ °C/mmHg
1-Hexanol	157.6	0.318	0.0424
Hydrogen fluoride	20.1	0.276	0.0368
Iodomethane	42.5	0.291	0.0388
Isobutane	-11.7	0.254	0.0339
Methanol	64.6	0.251	0.0335
Methyl acetate	56.8	0.282	0.0376
Methyl formate	31.7	0.582	0.0776
N-Methylaniline	196.2	0.396	0.0528
N-Methylformamide	199.5	0.371	0.0495
Nitrobenzene	210.8	0.418	0.0557
Nitromethane	101.1	0.320	0.0427
1-Octanol	195.1	0.360	0.0480
Pentane	36.0	0.289	0.0385
1-Pentanol	137.9	0.296	0.0395
Phenol	181.8	0.349	0.0465
Propane	-42.1	0.224	0.0299
1-Propanol	97.2	0.261	0.0348
2-Propanol	82.3	0.247	0.0329
Pyridine	115.2	0.340	0.0453
Pyrrole	129.7	0.330	0.0440
Pyrrolidine	86.5	0.309	0.0412
Styrene	145.1	0.369	0.0492
Sulfur dioxide	-10.05	0.221	0.0295
Tetrachloroethylene	121.3	0.354	0.0472
Tetrachloromethane	76.8	0.325	0.0433
Toluene	110.6	0.353	0.0471
Trichloroethylene	87.2	0.330	0.0440
Trichloromethane	61.1	0.302	0.0403
Trimethylamine	2.8	0.248	0.0331
Water	100.0	0.276	0.0368
o-Xylene	144.5	0.373	0.0497
m-Xylene	139.1	0.368	0.0491
p-Xylene	138.3	0.369	0.0492

EBULLIOSCOPIC CONSTANTS FOR CALCULATION OF BOILING POINT ELEVATION

The boiling point T_b of a dilute solution of a non-volatile, non-dissociating solute is elevated relative to that of the pure solvent. If the solution is ideal (i.e., follows Raoult's Law), the amount of elevation depends only on the number of particles of solute present. Hence the change in boiling point ΔT_b can be expressed as

$$\Delta T_b = E_b\, m_2$$

where m_2 is the molality (moles of solute per kilogram of solvent) and E_b is the Ebullioscopic Constant, a characteristic property of the solvent. The Ebullioscopic Constant may be calculated from the relation

$$E_b = R\, T_b^2\, M / \Delta_{vap} H$$

where R is the molar gas constant, T_b is the normal boiling point temperature (absolute) of the solvent, M the molar mass of the solvent, and $\Delta_{vap} H$ the molar enthalpy (heat) of vaporization of the solvent at its normal boiling point.

This table lists E_b values for some common solvents, as calculated from data in the table "Enthalpy of Vaporization" in Section 6.

Compound	E_b/K kg mol^{-1}	Compound	E_b/K kg mol^{-1}
Acetic acid	3.22	Hexane	2.90
Acetone	1.80	Iodomethane	4.31
Acetonitrile	1.44	Methanol	0.86
Aniline	3.82	Methyl acetate	2.21
Anisole	4.20	N-Methylaniline	4.3
Benzaldehyde	4.24	N-Methylformamide	2.2
Benzene	2.64	Nitrobenzene	5.2
1-Butanol	2.17	Nitromethane	2.09
Carbon disulfide	2.42	1-Octanol	5.06
Chlorobenzene	4.36	Phenol	3.54
1-Chlorobutane	3.13	1-Propanol	1.66
Cyclohexane	2.92	2-Propanol	1.58
Cyclohexanol	3.5	Pyridine	2.83
Decane	6.10	Pyrrole	2.33
Dichloromethane	2.42	Pyrrolidine	2.32
Diethyl ether	2.20	Tetrachloroethylene	6.18
Dimethyl sulfoxide	3.22	Tetrachloromethane	5.26
1,4-Dioxane	3.01	Toluene	3.40
Ethanol	1.23	Trichloroethylene	4.52
Ethyl acetate	2.82	Trichloromethane	3.80
Ethylene glycol	2.26	Water	0.513
Heptane	3.62	o-Xylene	4.25

CRYOSCOPIC CONSTANTS FOR CALCULATION OF FREEZING POINT DEPRESSION

The freezing point T_f of a dilute solution of a non-volatile, non-dissociating solute is depressed relative to that of the pure solvent. If the solution is ideal (i.e., follows Raoult's Law), this lowering is a function only of the number of particles of solute present. Thus the absolute value of the lowering of freezing point ΔT_f can be expressed as

$$\Delta T_f = E_f m_2$$

where m_2 is the molality (moles of solute per kilogram of solvent) and E_f is the Cryoscopic Constant, a characteristic property of the solvent. The Cryoscopic Constant may be calculated from the relation

$$E_f = R\, T_f^2\, M/\Delta_{fus}H$$

where R is the molar gas constant, T_b is the freezing point temperature (absolute) of the solvent, M the molar mass of the solvent, and $\Delta_{fus}H$ the molar enthalpy (heat) of fusion of the solvent.

This table lists cryscopic constants for selected substances, as calculated from data in the table "Enthalpy of Fusion" in Section 6.

Compound	E_f/K kg mol^{-1}
Acetamide	3.92
Acetic acid	3.63
Acetophenone	5.16
Aniline	5.23
Benzene	5.07
Benzonitrile	5.35
Benzophenone	8.58
(+)-Camphor	37.8
1-Chloronaphthalene	7.68
o-Cresol	5.92
m-Cresol	7.76
p-Cresol	7.20
Cyclohexane	20.8
Cyclohexanol	42.2
cis-Decahydronaphthalene	6.42
trans-Decahydronaphthalene	4.70
Dibenzyl ether	6.17
p-Dichlorobenzene	7.57
Diethanolamine	3.16
Dimethyl sulfoxide	3.85

Compound	E_f/K kg mol^{-1}
1,4-Dioxane	4.63
Diphenylamine	8.38
Ethylene glycol	3.11
Formamide	4.25
Formic acid	2.38
Glycerol	3.56
Methylcyclohexane	2.60
Naphthalene	7.45
Nitrobenzene	6.87
Phenol	6.84
Pyridine	4.26
Quinoline	6.73
Succinonitrile	19.3
1,1,2,2-Tetrabromoethane	21.4
1,1,2,2-Tetrachloro-1,2-difluoroethane	41.0
Toluene	3.55
p-Toluidine	4.91
Tribromomethane	15.0
Water	1.86
p-Xylene	4.31

FREEZING POINT LOWERING BY ELECTROLYTES IN AQUEOUS SOLUTION

Reference

Forsythe, W. E., *Smithsonian Physical Tables, Ninth Edition*, Smithsonian Institution, Washington, 1956.

Compound	Lowering of freezing point of water (in °C) as function of molality (mol/kg)									
	0.05	0.10	0.25	0.50	0.75	1.00	1.50	2.00	2.50	3.00
$CaCl_2$	0.25	0.49	1.27	2.66	4.28	6.35	10.78	15.27	20.42	28.08
$CuSO_4$	0.13	0.23	0.47	0.96						
HCl	0.18	0.36	0.90	1.86	2.90	4.02	6.63	9.94		
HNO_3	0.18	0.35	0.88	1.80	2.78	3.80	5.98	8.34	10.95	13.92
H_2SO_4	0.20	0.39	0.96	1.95	3.04	4.28	7.35	11.35	16.32	
KBr	0.18	0.36	0.92	1.78						
KCl	0.17	0.35	0.86	1.68	2.49	3.29	4.88	6.50	8.14	9.77
KNO_3	0.17	0.33	0.78	1.47	2.11	2.66				
K_2SO_4	0.23	0.43	1.01	1.87						
LiCl	0.18	0.35	0.88	1.80	2.78					
$MgSO_4$	0.13	0.24	0.55	1.01	1.50	2.08	3.41			
NH_4Cl	0.17	0.34	0.85	1.70	2.55					
NaCl	0.18	0.35	0.85	1.68	2.60					
$NaNO_3$	0.18	0.36	0.80	1.62	2.63	3.10				

CORRECTION OF BAROMETER READINGS TO 0°C TEMPERATURE

The following corrections are used to reduce the reading of a mercury barometer with a brass scale to 0°C. The number in the table should be subtracted from the observed height of the mercury column to give the true pressure in mmHg (1mmHg = 133.322 Pa). The table is calculated from the formula

$$\Delta h = -0.0001634\, ht/(1+0.0001818\, t),$$

where h is the observed column height in mm and t the Celsius temperature. This relation is based on thermal expansion coefficients of $181.8 \cdot 10^{-6}\ °C^{-1}$ for mercury and $18.4 \cdot 10^{-6}\ °C^{-1}$ for brass.

									Observed Height in mm										
$t/°C$	620	630	640	650	660	670	680	690	700	710	720	730	740	750	760	770	780	790	800
0	0.00	0.00	0.00	0.00	0.00	0.00	0.00	0.00	0.00	0.00	0.00	0.00	0.00	0.00	0.00	0.00	0.00	0.00	0.00
1	0.10	0.10	0.10	0.11	0.11	0.11	0.11	0.11	0.11	0.12	0.12	0.12	0.12	0.12	0.12	0.13	0.13	0.13	0.13
2	0.20	0.21	0.21	0.21	0.22	0.22	0.22	0.23	0.23	0.23	0.24	0.24	0.24	0.25	0.25	0.25	0.25	0.26	0.26
3	0.30	0.31	0.31	0.32	0.32	0.33	0.33	0.34	0.34	0.35	0.35	0.36	0.36	0.37	0.37	0.38	0.38	0.39	0.39
4	0.40	0.41	0.42	0.42	0.43	0.44	0.44	0.45	0.46	0.46	0.47	0.48	0.48	0.49	0.50	0.50	0.51	0.52	0.52
5	0.51	0.51	0.52	0.53	0.54	0.55	0.56	0.56	0.57	0.58	0.59	0.60	0.60	0.61	0.62	0.63	0.64	0.64	0.65
6	0.61	0.62	0.63	0.64	0.65	0.66	0.67	0.68	0.69	0.70	0.71	0.71	0.72	0.73	0.74	0.75	0.76	0.77	0.78
7	0.71	0.72	0.73	0.74	0.75	0.77	0.78	0.79	0.80	0.81	0.82	0.83	0.85	0.86	0.87	0.88	0.89	0.90	0.91
8	0.81	0.82	0.84	0.85	0.86	0.87	0.89	0.90	0.91	0.93	0.94	0.95	0.97	0.98	0.99	1.01	1.02	1.03	1.04
9	0.91	0.92	0.94	0.95	0.97	0.98	1.00	1.01	1.03	1.04	1.06	1.07	1.09	1.10	1.12	1.13	1.15	1.16	1.17
10	1.01	1.03	1.04	1.06	1.08	1.09	1.11	1.13	1.14	1.16	1.17	1.19	1.21	1.22	1.24	1.26	1.27	1.29	1.30
11	1.11	1.13	1.15	1.17	1.18	1.20	1.22	1.24	1.26	1.27	1.29	1.31	1.33	1.35	1.36	1.38	1.40	1.42	1.44
12	1.21	1.23	1.25	1.27	1.29	1.31	1.33	1.35	1.37	1.39	1.41	1.43	1.45	1.47	1.49	1.51	1.53	1.55	1.57
13	1.31	1.34	1.36	1.38	1.40	1.42	1.44	1.46	1.48	1.50	1.53	1.55	1.57	1.59	1.61	1.63	1.65	1.67	1.70
14	1.41	1.44	1.46	1.48	1.51	1.53	1.55	1.57	1.60	1.62	1.64	1.67	1.69	1.71	1.73	1.76	1.78	1.80	1.83
15	1.52	1.54	1.56	1.59	1.61	1.64	1.66	1.69	1.71	1.74	1.76	1.78	1.81	1.83	1.86	1.88	1.91	1.93	1.96
16	1.62	1.64	1.67	1.69	1.72	1.75	1.77	1.80	1.82	1.85	1.88	1.90	1.93	1.96	1.98	2.01	2.03	2.06	2.09
17	1.72	1.74	1.77	1.80	1.83	1.86	1.88	1.91	1.94	1.97	1.99	2.02	2.05	2.08	2.10	2.13	2.16	2.19	2.22
18	1.82	1.85	1.88	1.91	1.93	1.96	1.99	2.02	2.05	2.08	2.11	2.14	2.17	2.20	2.23	2.26	2.29	2.32	2.35
19	1.92	1.95	1.98	2.01	2.04	2.07	2.10	2.13	2.17	2.20	2.23	2.26	2.29	2.32	2.35	2.38	2.41	2.44	2.48
20	2.02	2.05	2.08	2.12	2.15	2.18	2.21	2.25	2.28	2.31	2.34	2.38	2.41	2.44	2.47	2.51	2.54	2.57	2.60
21	2.12	2.15	2.19	2.22	2.26	2.29	2.32	2.36	2.39	2.43	2.46	2.50	2.53	2.56	2.60	2.63	2.67	2.70	2.73
22	2.22	2.26	2.29	2.33	2.36	2.40	2.43	2.47	2.51	2.54	2.58	2.61	2.65	2.69	2.72	2.76	2.79	2.83	2.86
23	2.32	2.36	2.40	2.43	2.47	2.51	2.54	2.58	2.62	2.66	2.69	2.73	2.77	2.81	2.84	2.88	2.92	2.96	2.99
24	2.42	2.46	2.50	2.54	2.58	2.62	2.66	2.69	2.73	2.77	2.81	2.85	2.89	2.93	2.97	3.01	3.05	3.08	3.12
25	2.52	2.56	2.60	2.64	2.68	2.72	2.77	2.81	2.85	2.89	2.93	2.97	3.01	3.05	3.09	3.13	3.17	3.21	3.25
26	2.62	2.66	2.71	2.75	2.79	2.83	2.88	2.92	2.96	3.00	3.04	3.09	3.13	3.17	3.21	3.26	3.30	3.34	3.38
27	2.72	2.77	2.81	2.85	2.90	2.94	2.99	3.03	3.07	3.12	3.16	3.20	3.25	3.29	3.34	3.38	3.42	3.47	3.51
28	2.82	2.87	2.91	2.96	3.00	3.05	3.10	3.14	3.19	3.23	3.28	3.32	3.37	3.41	3.46	3.51	3.55	3.60	3.64
29	2.92	2.97	3.02	3.06	3.11	3.16	3.21	3.25	3.30	3.35	3.39	3.44	3.49	3.54	3.58	3.63	3.68	3.72	3.77
30	3.02	3.07	3.12	3.17	3.22	3.27	3.32	3.36	3.41	3.46	3.51	3.56	3.61	3.66	3.71	3.75	3.80	3.85	3.90
31	3.12	3.17	3.22	3.27	3.32	3.37	3.43	3.48	3.53	3.58	3.63	3.68	3.73	3.78	3.83	3.88	3.93	3.98	4.03
32	3.22	3.28	3.33	3.38	3.43	3.48	3.54	3.59	3.64	3.69	3.74	3.79	3.85	3.90	3.95	4.00	4.05	4.11	4.16
33	3.32	3.38	3.43	3.48	3.54	3.59	3.64	3.70	3.75	3.81	3.86	3.91	3.97	4.02	4.07	4.13	4.18	4.23	4.29
34	3.42	3.48	3.53	3.59	3.64	3.70	3.75	3.81	3.87	3.92	3.98	4.03	4.09	4.14	4.20	4.25	4.31	4.36	4.42
35	3.52	3.58	3.64	3.69	3.75	3.81	3.86	3.92	3.98	4.03	4.09	4.15	4.21	4.26	4.32	4.38	4.43	4.49	4.55
36	3.62	3.68	3.74	3.80	3.86	3.92	3.97	4.03	4.09	4.15	4.21	4.27	4.32	4.38	4.44	4.50	4.56	4.62	4.68
37	3.72	3.78	3.84	3.90	3.96	4.02	4.08	4.14	4.20	4.26	4.32	4.38	4.44	4.50	4.56	4.62	4.68	4.74	4.80
38	3.82	3.88	3.95	4.01	4.07	4.13	4.19	4.25	4.32	4.38	4.44	4.50	4.56	4.62	4.69	4.75	4.81	4.87	4.93
39	3.92	3.99	4.05	4.11	4.18	4.24	4.30	4.37	4.43	4.49	4.56	4.62	4.68	4.75	4.81	4.87	4.94	5.00	5.06
40	4.02	4.09	4.15	4.22	4.28	4.35	4.41	4.48	4.54	4.61	4.67	4.74	4.80	4.87	4.93	5.00	5.06	5.13	5.19

DETERMINATION OF RELATIVE HUMIDITY FROM DEW POINT

The relative humidity of a water vapor-air mixture is defined as 100 times the partial pressure of water divided by the saturation vapor pressure of water at the same temperature. The relative humidity may be determined from the dew point t_{dew}, which is the temperature at which liquid water first condenses when the mixture is cooled from an initial temperature t. This table gives relative humidity as a function of the dew point depression $t - t_{dew}$ for several values of the dew point. Values are calculated from the vapor pressure table in Section 6.

$t - t_{dew}$	t_{dew}/°C					$t - t_{dew}$	t_{dew}/°C				
	−10	0	10	20	30		−10	0	10	20	30
0.0	100	100	100	100	100	8.2	54	56	59	61	63
0.2	99	99	99	99	99	8.4	53	56	58	60	63
0.4	97	97	97	98	98	8.6	52	55	57	60	62
0.6	95	96	96	96	97	8.8	51	54	57	59	61
0.9	94	94	95	95	96	9.0	51	53	56	58	61
1.0	92	93	94	94	94	9.2	50	53	55	58	60
1.2	91	92	92	93	93	9.4	49	52	55	57	59
1.4	90	90	91	92	92	9.6	48	51	54	56	59
1.6	88	89	90	91	91	9.8	48	51	53	56	58
1.8	87	88	89	90	90	10.0	47	50	53	55	57
2.0	86	87	88	88	89	10.5	45	48	51	54	56
2.2	84	85	86	87	89	11.0	44	47	49	52	55
2.4	83	84	85	86	87	11.5	42	45	48	51	53
2.6	82	83	84	85	86	12.0	41	44	47	49	52
2.8	80	82	83	84	85	12.5	39	42	45	48	50
3.0	79	81	82	83	84	13.0	38	41	44	46	49
3.2	78	80	81	82	83	13.5	37	40	43	45	48
3.4	77	79	80	81	82	14.0	35	38	41	44	47
3.6	76	77	79	80	82	14.5	34	37	40	43	45
3.8	75	76	78	79	81	15.0	33	36	39	42	44
4.0	73	75	77	78	80	15.5	32	35	38	40	
4.2	72	74	76	77	79	16.0	31	34	37	39	
4.4	71	73	75	77	78	16.5	30	33	36	38	
4.6	70	72	74	76	77	17.0	29	32	35	37	
4.8	69	71	73	75	76	17.5	28	31	34	36	
5.0	69	70	72	74	75	18.0	27	30	33	35	
5.2	67	69	71	73	75	18.5	26	29	32	34	
5.4	66	68	70	72	74	19.0	25	28	31	33	
5.6	65	67	69	71	73	19.5	24	27	30	33	
5.9	64	66	69	70	72	20.0	24	26	29	32	
6.0	63	66	68	70	71	21.0	22	25	27	30	
6.2	62	65	67	69	71	22.0	21	23	26	29	
6.4	61	64	66	68	70	23.0	19	22	24	27	
6.6	60	63	65	67	69	24.0	18	21	23	26	
6.8	60	62	64	66	68	25.0	17	19	22	24	
7.0	59	61	63	66	68	26.0	16	18	21	23	
7.2	58	60	63	65	67	27.0	15	17	20	22	
7.4	57	60	62	64	66	28.0	14	16	19	21	
7.6	56	59	61	63	65	29.0	13	15	18	20	
7.8	55	58	60	63	65	30.0	12	14	17	19	
8.0	54	57	60	62	64						

DETERMINATION OF RELATIVE HUMIDITY FROM WET AND DRY BULB TEMPERATURES

Relative humidity may be determined by comparing temperature readings of wet and dry bulb thermometers. The following table, extracted from more extensive U.S. National Weather Service tables, gives the relative humidity as a function of air temperature t_d (dry bulb) and the difference $t_d - t_w$ between dry and wet bulb temperatures. The data assume a pressure near normal atmospheric pressure and an instrumental configuration with forced ventilation.

t_d/°C	\multicolumn{12}{c}{$(t_d - t_w)$/°C}
	0.5	1.0	1.5	2.0	2.5	3.0	3.5	4.0	4.5	5.0	5.5	6.0
−10	83	67	51	35	19							
−8	86	71	57	43	29	15						
−6	88	74	61	49	37	25	8					
−4	89	77	66	55	44	33	23	12				
−2	90	79	69	60	50	40	31	22	12			
0	91	81	72	64	55	46	38	29	21	13	5	
2	91	84	76	68	60	52	44	37	29	22	14	7
4	92	85	78	71	63	57	49	43	36	29	22	16
6	93	86	79	73	66	60	54	48	41	35	29	24
8	93	87	81	75	69	63	57	51	46	40	35	29
10	94	88	82	77	71	66	60	55	50	44	39	34
12	94	89	83	78	73	68	63	58	53	48	43	39
14	95	90	85	79	75	70	65	60	56	51	47	42
16	95	90	85	81	76	71	67	63	58	54	50	46
18	95	91	86	82	77	73	69	65	61	57	53	49
20	96	91	87	83	78	74	70	66	63	59	55	51
22	96	92	87	83	80	76	72	68	64	61	57	54
24	96	92	88	84	80	77	73	69	66	62	59	56
26	96	92	88	85	81	78	74	71	67	64	61	58
28	96	93	89	85	82	78	75	72	69	65	62	59
30	96	93	89	86	83	79	76	73	70	67	64	61
35	97	94	90	87	84	81	78	75	72	69	67	64
40	97	94	91	88	85	82	80	77	74	72	69	67

t_d/°C	\multicolumn{12}{c}{$(t_d - t_w)$/°C}
	6.5	7.0	7.5	8.0	8.5	9.0	10.0	11.0	12.0	13.0	14.0	15.0
4	9											
6	17	11	5									
8	24	19	14	8								
10	29	24	20	15	10	6						
12	34	29	25	21	16	12	5					
14	38	34	30	26	22	18	10					
16	42	38	34	30	26	23	15	8				
18	45	41	38	34	30	27	20	14	7			
20	48	44	41	37	34	31	24	18	12	6		
22	50	47	44	40	37	34	28	22	17	11	6	
24	53	49	46	43	40	37	31	26	20	15	10	5
26	54	51	49	46	43	40	34	29	24	19	14	10
28	56	53	51	48	45	42	37	32	27	22	18	13
30	58	55	52	50	47	44	39	35	30	25	21	17
32	60	57	54	51	49	46	41	37	32	28	24	20
34	61	58	56	53	51	48	43	39	35	30	26	23
36	62	59	57	54	52	50	45	41	37	33	29	25
38	63	61	58	56	54	51	47	43	39	35	31	27
40	64	62	59	57	54	53	48	44	40	36	33	29

CONSTANT HUMIDITY SOLUTIONS

Anthony Wexler

An excess of a water soluble salt in contact with its saturated solution and contained within an enclosed space produces a constant relative humidity and water vapor pressure according to

$$RH = A \exp(B/T)$$

where RH is the percent relative humidity (generally accurate to ±2 %), T is the temperature in kelvin, and the constants A and B and the range of valid temperatures are given in the table below. The vapor pressure, p, can be calculated from

$$p = (RH/100) \times p_0$$

where p_0 is the vapor pressure of pure water at temperature T as given in the table in Section 6 titled "Vapor Pressure of Water from 0 to 370°C".

References

1. Wexler, A. S. and Seinfeld, J. H., *Atmospheric Environment*, 25A, 2731, 1991.
2. Greenspan, L., *J. Res. National Bureau of Standards*, 81A, 89, 1977.
3. Broul, et al., *Solubility of Inorganic Two-Component Systems*, Elsevier, New York, 1981.
4. Wagman, D. D. et al., *J. Phys. Chem. Ref. Data*, Vol. 11, Suppl. 2, 1982.

Compound	Temperature range (°C)	RH (25°C)	A	B
$NaOH \cdot H_2O$	15—60	6	5.48	27
$LiBr \cdot 2H_2O$	10—30	6	0.23	996
$ZnBr_2 \cdot 2H_2O$	5—30	8	1.69	455
$KOH \cdot 2H_2O$	5—30	9	0.014	1924
$LiCl \cdot H_2O$	20—65	11	14.53	−75
$CaBr_2 \cdot 6H_2O$	11—22	16	0.17	1360
$LiI \cdot 3H_2O$	15—65	18	0.15	1424
$CaCl_2 \cdot 6H_2O$	15—25	29	0.11	1653
$MgCl_2 \cdot 6H_2O$	5—45	33	29.26	34
$NaI \cdot 2H_2O$	5—45	38	3.62	702
$Ca(NO_3)_2 \cdot 4H_2O$	10—30	51	1.89	981
$Mg(NO_3)_2 \cdot 6H_2O$	5—35	53	25.28	220
$NaBr \cdot 2H_2O$	0—35	58	20.49	308
NH_4NO_3	10—40	62	3.54	853
KI	5—30	69	29.35	254
$SrCl_2 \cdot 6H_2O$	5—30	71	31.58	241
$NaNO_3$	10—40	74	26.94	302
$NaCl$	10—40	75	69.20	25
NH_4Cl	10—40	79	35.67	235
KBr	5—25	81	40.98	203
$(NH_4)_2SO_4$	10—40	81	62.06	79
KCl	5—25	84	49.38	159
$Sr(NO_3)_2 \cdot 4H_2O$	5—25	85	28.34	328
$BaCl_2 \cdot 2H_2O$	5—25	90	69.99	75
CsI	5—25	91	70.77	75
KNO_3	0—50	92	43.22	225
K_2SO_4	10—50	97	86.75	34

STANDARD SALT SOLUTIONS FOR HUMIDITY CALIBRATION

Saturated aqueous solutions of inorganic salts are convenient secondary standards for calibration of instruments for measurement of relative humidity. The International Union of Pure and Applied Chemistry has recommended salt solutions for calibrations in the range of 10% to 90% relative humidity, and the American Society for Testing and Materials has published similar standards. The data in this table are taken from the IUPAC recommendations, except for K_2CO_3 and K_2SO_4, which are ASTM recommendations.

Details on the preparation and use of these standards may be found in References 1 and 2. Data for other salts are given in Reference 3.

References

1. Marsh, K. N., Editor, *Recommended Reference Materials for the Realization of Physicochemical Properties*, Blackwell Scientific Publications, Oxford, 1987, pp.157-162.
2. *Standard Practice for Maintaining Constant Relative Humidity by Means of Aqueous Solutions*, ASTM Standard E 104-85, Reapproved 1991.
3. Greenspan, L., *J. Res. Nat. Bur. Stand.*, 81A, 89, 1977.

			Relative Humidity in %				
$t/°C$	LiCl	$MgCl_2$	K_2CO_3	$Mg(NO_3)_2$	NaCl	KCl	K_2SO_4
0		33.66 ± 0.33	43.1 ± 0.7	60.35 ± 0.55	75.51 ± 0.34	88.61 ± 0.53	98.8 ± 2.1
5		33.60 ± 0.28	43.1 ± 0.5	58.86 ± 0.43	75.65 ± 0.27	87.67 ± 0.45	98.5 ± 0.9
10		33.47 ± 0.24	43.1 ± 0.4	57.36 ± 0.33	75.67 ± 0.22	86.77 ± 0.39	98.2 ± 0.8
15		33.30 ± 0.21	43.2 ± 0.3	55.87 ± 0.27	75.61 ± 0.18	85.92 ± 0.33	97.9 ± 0.6
20	11.31 ± 0.31	33.07 ± 0.18	43.2 ± 0.3	54.38 ± 0.23	75.47 ± 0.14	85.11 ± 0.29	97.6 ± 0.5
25	11.30 ± 0.27	32.78 ± 0.16	43.2 ± 0.4	52.89 ± 0.22	75.29 ± 0.12	84.34 ± 0.26	97.3 ± 0.5
30	11.28 ± 0.24	32.44 ± 0.14	43.2 ± 0.5	51.40 ± 0.24	75.09 ± 0.11	83.62 ± 0.25	97.0 ± 0.4
35	11.25 ± 0.22	32.05 ± 0.13		49.91 ± 0.29	74.87 ± 0.12	82.95 ± 0.25	96.7 ± 0.4
40	11.21 ± 0.21	31.60 ± 0.13		48.42 ± 0.37		82.32 ± 0.25	96.4 ± 0.4
45	11.16 ± 0.21	31.10 ± 0.13		46.93 ± 0.47		81.74 ± 0.28	96.1 ± 0.4
50	11.10 ± 0.22	30.54 ± 0.14		45.44 ± 0.60		81.20 ± 0.31	95.8 ± 0.5
55	11.03 ± 0.23	29.93 ± 0.16				80.70 ± 0.35	
60	10.95 ± 0.26	29.26 ± 0.18				80.25 ± 0.41	
65	10.86 ± 0.29	28.54 ± 0.21				79.85 ± 0.48	
70	10.75 ± 0.33	27.77 ± 0.25				79.49 ± 0.57	
75	10.64 ± 0.38	26.94 ± 0.29				79.17 ± 0.66	
80	10.51 ± 0.44	26.05 ± 0.34				78.90 ± 0.77	

LOW TEMPERATURE BATHS FOR MAINTAINING CONSTANT TEMPERATURE

A liquid-solid slurry is a convenient means of maintaining a constant temperature environment below room temperature. The following is a list of readily available organic liquids suitable for this purpose, arranged in order of their melting (freezing) points t_m. The normal boiling points t_b are also given.

Compound	t_m/°C	t_b/°C
Isopentane (2-Methylbutane)	−159.9	27.8
Methylcyclopentane	−142.5	71.8
3-Chloropropene (Allyl chloride)	−134.5	45.1
Pentane	−129.7	36.0
Allyl alcohol	−129	97.0
Ethanol	−114.1	78.2
Carbon disulfide	−111.5	46
Isobutyl alcohol	−108	107.8
Toluene	−94.9	110.6
Acetone	−94.8	56.0
Ethyl acetate	−83.6	77.1
Dry ice + acetone	−78	
p-Cymene	−68.9	177.1
Trichloromethane (Chloroform)	−63.6	61.1
N-Methylaniline	−57	196.2
Chlorobenzene	−45.2	131.7
Anisole	−37.5	153.7
Bromobenzene	−30.6	156.0
Tetrachloromethane (Carbon tetrachloride)	−23	76.8
Benzonitrile	−12.7	191.1

METALS AND ALLOYS WITH LOW MELTING TEMPERATURE

L. I. Berger

Metal or alloy system	Composition, % * Weight	Atomic	Melting temperature (°C)	Comments	Ref.
Hg	100	100	−38.84		
Cs–K	77.0–23.0	50.0–50.0	−37.5	Eutectic (?)	1
Cs–Na	94.5–5.5	75.0–25.0	−30.0	Eutectic	2
K–Na	76.7–23.3	65.9–34.1	−12.65	Eutectic	3
Na–Rb	8.0–92.0	24.4–75.6	−5	Eutectic	4
Ga–In–Sn	62.5–21.5–16.0	73.6–15.3–11.1	11	Eutectic	5
Ga–Sn–Zn	82.0–12.0–6.0	86.0–7.3–6.7	17	Eutectic	5
Cs	100	100	28.44		
Ga	100	100	29.77		
K–Rb	32.0–68.0	50–50	33	Eutectic	4
Bi–Cd–In–Pb–Sn	44.7–5.3–19.1–22.6–8.3	35.1–8.2–27.3–17.9–11.5	46.7	Eutectic	6
Bi–In–Pb–Sn	49.5–21.3–17.6–11.6	39.2–30.7–14.0–16.2	58.2	Eutectic	6
Bi–In–Sn	32.5–51.0–16.5	21.1–60.1–18.8	60.5	Eutectic	7
K	100	100	63.38		
Bi–Cd–Pb–Sn	50.0–12.5–25.0–12.5	41.5–19.3–21.0–18.2	70	Wood's alloy	6
Bi–In	33.0–67.0	21.3–78.7	72	Eutectic	8
Bi–Cd–Pb	51.6–8.2–40.2	48.1–14.2–37.7	91.5	Eutectic	6
Bi–Pb–Sn	52.5–32.0–15.5	46.8–28.7–24.5	95	Eutectic	6
Na	100	100	97.8		
Bi–Cd–Sn	54.0–20.0–26.0	39.4–27.2–33.4	102.5	Eutectic	6
In–Sn	51.8–48.2	52.6–47.4	119	Eutectic	9
Cd–In	25.3–74.7	25.7–74.3	120	Eutectic	10
Bi–Pb	55.5–44.5	55.3–44.7	124	Eutectic	11
Bi–Sn–Zn	56.0–40.0–4.0	40.2–50.6–9.2	130	Eutectic	6, 7
Bi–Sn	70–30	57.0–43.0	138.5	Eutectic	6, 12
Bi–Cd	60.3–39.7	45.0–55.0	145.5	Eutectic	13, 14
In	100	100	156.6		
Li	100	100	180.5		
Pb–Sn	38.1–61.9	26.1–73.9	183	Eutectic	6,15
Bi–Tl	48.0–52.0	47.5–52.5	185	Eutectic	13
Sn–Zn	91.0–9.0	85.0–15.0	198	Eutectic	14
Sb–Sn	8.0–92.0	7.8–92.2	199	White Metal	16
Au–Pb	14.6–85.4	15.2–84.8	212	Eutectic	17
Ag–Sn	3.5–96.5	3.8–96.2	221	Eutectic	13,18
Bi–Pb–Sb–Sn	48.0–28.5–9.0–14.5	40.8–24.5–13.1–21.6	226	Matrix Alloy	6
Cu–Sn	0.75–99.25	1.3–98.7	227	Eutectic	13, 19
Sn	100	100	231.9		

* The useful expression for correlations between the atomic and weight concentrations of an alloy components are:

$$f(\mathrm{a}, A_k) = \frac{f(\mathrm{w}, A_k)}{M_k \sum_{i=1}^{N} \frac{f(\mathrm{w}, A_i)}{M_i}} \quad \text{and} \quad f(\mathrm{w}, A_k) = \frac{M_k \cdot f(\mathrm{a}, A_k)}{\sum_{i=1}^{N} M_i \cdot f(\mathrm{a}, A_i)} \quad (i = 1, \quad, k, \quad, N)$$

where $f(\mathrm{a}, A_i)$ and $f(\mathrm{w}, A_i)$ are the atomic and weight concentrations of component A_i, respectively, and M_i is the atomic weight of this component.

References

1. Zintle, E. and Hauke, W., *Z. Electrochem.*, 44, 104, 1938.
2. Rinck, E., *Compt. Rend.*, 199, 1217, 1934.
3. Krier, C. A., Craign, R. S., and Wallace, W. E., *J. Phys. Chem.*, 61, 522, 1957.
4. Goria, C., *Gazz. Chim. Ital.*, 65, 865, 1935.
5. Baker, H., Ed., *ASM Handbook, Volume 3: Alloy Phase Diagrams*, ASM Intl., Materials Park, OH, 1992.
6. Sedlacek, V., *Non–Ferrous Metals and Alloys*, Elsevier, 1986.
7. Villars, P., Prince, A., Okamoto, H., Eds., *Handbook of Ternary Alloy Phase Diagrams*, ASM Intl., 1994.
8. Palatnik, L. S., Kosevich, V. M., and Tyrina, L. V., *Phys. Metals Metallog. (USSR)*, 11, 75, 1961.
9. Neumann, T. and Alpout, O., *J. Less–Common Metals*, 6, 108, 1964.
10. Neumann, T. and Predel, B., *Z. Metallk.*, 50, 309, 1959.
11. Roy, P., Orr, R. L., and Hultgren, R., *J. Phys. Chem.*, 64, 1034, 1960.
12. Dobovicek, B. and Smajic, N., *Rudarsko–Met. Zbornik*, 4, 353, 1962.
13. Massalski, T. B., Okamoto, H., Subramanian, P. R., and Kacprzak, L., Eds., *Binary Alloy Phase Diagrams*, 2nd ed., ASM Intl., 1990.
14. Dobovicek, B. and Straus, B., *Rudarsko–Met. Zbornik*, 3, 273, 1960.
15. Schurmann, E. and Gilhaus, F. J., *Arch. Eisenhuettenw.*, 32, 867, 1961.
16. Rosenblatt, G. M. and Birchenall, C. E., *Trans. AIME*, 224, 481, 1962.
17. Evans, D. S. and Prince, A., in *Alloy Phase Diagrams*, MRS Simposia Proc., Vol. 19, North–Holland, 1983, p. 383.
18. Umanskiy, M. M., *Zh. Fiz. Khim.*, 14, 846, 1940.
19. Homer, C. E. and Plummer, H., *J. Inst. Met.*, 64, 169, 1939.

WIRE TABLES

The resistance per unit length of wires of various metals is tabulated here. Values were calculated from resistivity values in the tables "Electrical Resistivity of Pure Metals" and "Electrical Resistivity of Selected Alloys", which appear in Section 12. In practice, resistance may vary because of differing heat treatments and metal composition. The values in the table refer to 20°C, but values at other temperatures may be calculated from the following resistivity data:

Resistivity in 10^{-8} Ω m at temperature

Metal	0°C	20°C	25°C	100°C
Aluminum	2.417	2.650	2.709	3.56
Brass (70% Cu, 30% Zn)	5.87	6.08	6.13	6.91
Constantan (60% Cu, 40% Ni)	45.43	45.38	45.35	45.11
Copper	1.543	1.678	1.712	2.22
Nichrome (79% Ni, 21% Cr)	107.3	107.5	107.6	108.3
Platinum	9.6	10.5	10.7	13.6
Silver	1.467	1.587	1.617	2.07
Tungsten	4.82	5.28	5.39	7.18

Resistance per unit length at 20°C in Ω/m

B & S Gauge	Diameter (mm)	Aluminum	Brass	Constantan	Copper	Nichrome	Platinum	Silver	Tungsten
0	8.252	0.000495	0.00114	0.00848	0.000314	0.0201	0.00196	0.000297	0.00099
2	6.543	0.000788	0.00181	0.0135	0.000499	0.0320	0.00312	0.000472	0.00157
4	5.189	0.00125	0.00287	0.0214	0.000793	0.0508	0.00496	0.000750	0.00250
6	4.115	0.00199	0.00457	0.0341	0.00126	0.0808	0.00789	0.00119	0.00397
8	3.264	0.00317	0.00727	0.0542	0.00200	0.128	0.0125	0.00190	0.00631
10	2.588	0.00504	0.0115	0.0863	0.00319	0.204	0.0200	0.00302	0.0100
12	2.053	0.00800	0.0184	0.137	0.00507	0.325	0.0317	0.00479	0.0159
14	1.628	0.0127	0.0292	0.218	0.00806	0.516	0.0504	0.00762	0.0254
16	1.291	0.0202	0.0464	0.347	0.0128	0.821	0.0802	0.0121	0.0403
18	1.024	0.0322	0.0738	0.551	0.0204	1.30	0.127	0.0193	0.0641
20	0.8118	0.0512	0.117	0.877	0.0324	2.08	0.203	0.0307	0.102
22	0.6439	0.0814	0.187	1.39	0.0515	3.30	0.322	0.0487	0.162
24	0.5105	0.129	0.297	2.22	0.0820	5.25	0.513	0.0775	0.258
26	0.4049	0.206	0.472	3.52	0.130	8.35	0.815	0.123	0.410
28	0.3211	0.327	0.751	5.60	0.207	13.3	1.30	0.196	0.652
30	0.2548	0.520	1.19	8.90	0.329	21.1	2.06	0.311	1.03
32	0.2019	0.828	1.90	14.2	0.524	33.6	3.28	0.496	1.65
34	0.1601	1.32	3.02	22.5	0.833	53.4	5.22	0.788	2.62
36	0.1270	2.09	4.80	35.8	1.32	84.9	8.29	1.25	4.17
38	0.1007	3.33	7.63	57.0	2.11	135	13.2	1.99	6.63
40	0.07988	5.29	12.1	90.5	3.35	214	20.9	3.17	10.5

CHARACTERISTICS OF PARTICLES AND PARTICLE DISPERSOIDS

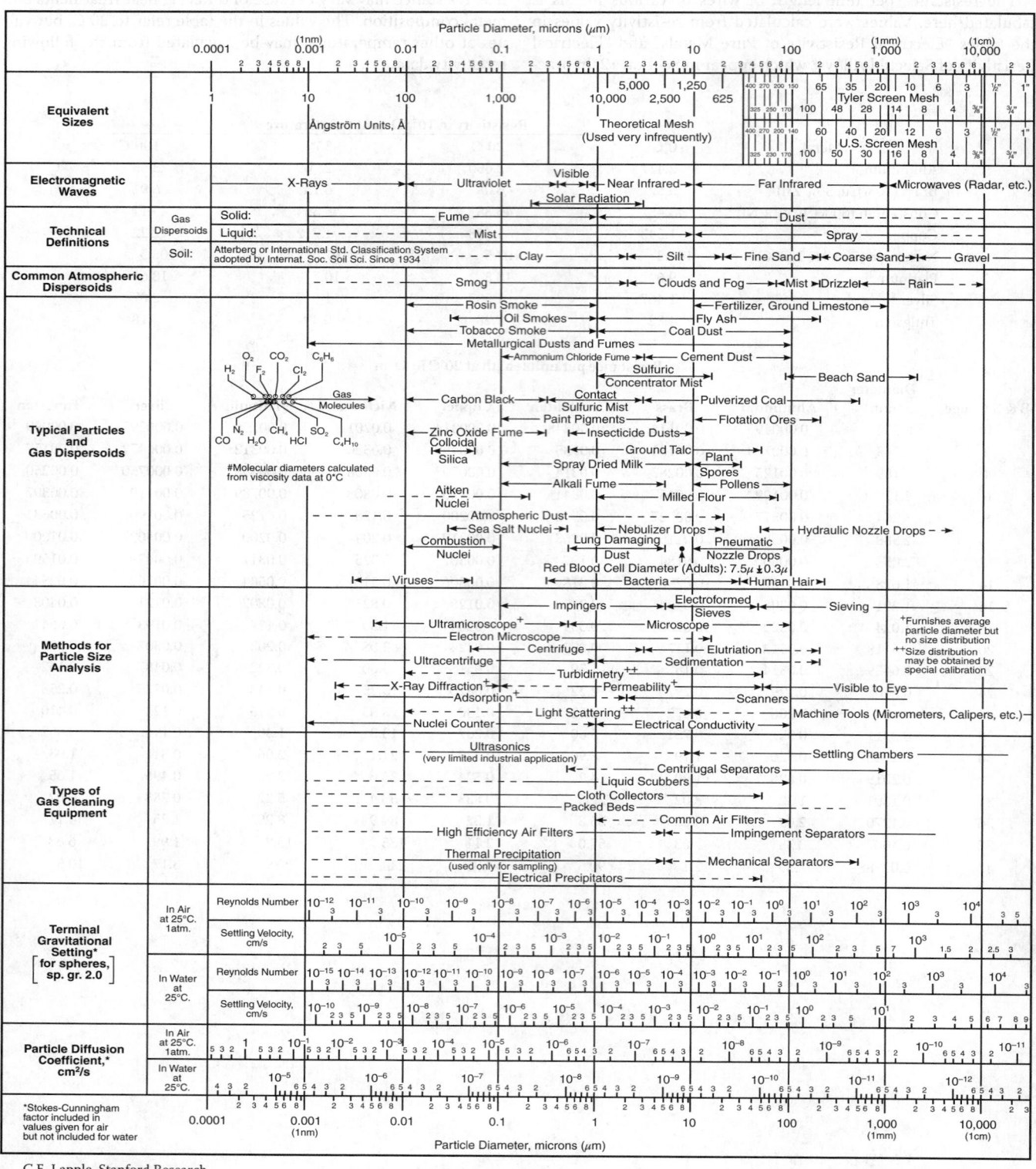

C.E. Lapple, Stanford Research
Institute Journal, Vol. 5, p.95
(Third Quarter, 1961)

DENSITY OF VARIOUS SOLIDS

This table gives the range of density for miscellaneous solid materials whose characteristics depend on the source or method of preparation.

References

1. Forsythe, W. E., *Smithsonian Physical Tables, Ninth Edition*, Smithsonian Institution, Washington, D.C., 1956.

2. Kaye, G. W. C., and Laby, T. H., *Tables of Physical and Chemical Constants, 16th Edition*, Longman, London, 1995.

3. Brandrup, J., and Immergut, E. H., *Polymer Handbook, Third Edition*, John Wiley & Sons, New York, 1989.

Material	$\rho/\ \mathrm{g\ cm^{-3}}$
Agate	2.5-2.7
Alabaster,	
carbonate	2.69-2.78
sulfate	2.26-2.32
Albite	2.62-2.65
Amber	1.06-1.11
Amphiboles	2.9-3.2
Anorthite	2.74-2.76
Asbestos	2.0-2.8
Asbestos slate	1.8
Asphalt	1.1-1.5
Basalt	2.4-3.1
Beeswax	0.96-0.97
Beryl	2.69-2.70
Biotite	2.7-3.1
Bone	1.7-2.0
Brasses	8.44-8.75
Brick	1.4-2.2
Bronzes	8.74-8.89
Butter	0.86-0.87
Calamine	4.1-4.5
Calcspar	2.6-2.8
Camphor	0.99
Cardboard	0.69
Celluloid	1.4
Cement, set	2.7-3.0
Chalk	1.9-2.8
Charcoal,	
oak	0.57
pine	0.28-0.44
Cinnabar	8.12
Clay	1.8-2.6
Coal,	
anthracite	1.4-1.8
bituminous	1.2-1.5
Coke	1.0-1.7
Copal	1.04-1.14
Cork	0.22-0.26
Corundum	3.9-4.0
Diamond	3.51
Dolomite	2.84
Ebonite	1.15
Emery	4.0
Epidote	3.25-3.50
Feldspar	2.55-2.75
Flint	2.63
Fluorite	3.18
Galena	7.3-7.6
Garnet	3.15-4.3
Gelatin	1.27
Glass,	
common	2.4-2.8
lead	3-4

Material	$\rho/\ \mathrm{g\ cm^{-3}}$
Pyrex	2.23
Granite	2.64-2.76
Graphite	2.30-2.72
Gum arabic	1.3-1.4
Gypsum	2.31-2.33
Hematite	4.9-5.3
Hornblende	3.0
Ice	0.917
Iron, cast	7.0-7.4
Ivory	1.83-1.92
Kaolin	2.6
Leather, dry	0.86
Lime, slaked	1.3-1.4
Limestone	2.68-2.76
Linoleum	1.18
Magnetite	4.9-5.2
Malachite	3.7-4.1
Marble	2.6-2.84
Meerschaum	0.99-1.28
Mica	2.6-3.2
Muscovite	2.76-3.00
Ochre	3.5
Opal	2.2
Paper	0.7-1.15
Paraffin	0.87-0.91
Peat blocks	0.84
Pitch	1.07
Polyamides	1.15-1.25
Polyethylene	0.92-0.97
Poly(methyl methacrylate)	1.19
Polypropylene	0.91-0.94
Polystyrene	1.06-1.12
Polytetrafluoroethylene	2.28-2.30
Poly(vinyl acetate)	1.19
Poly(vinyl chloride)	1.39-1.42
Porcelain	2.3-2.5
Porphyry	2.6-2.9
Pyrite	4.95-5.10
Quartz (α)	2.65
Resin	1.07
Rock salt	2.18
Rubber,	
hard	1.19
soft	1.1
pure gum	0.91-0.93
Neoprene	1.23-1.25
Sandstone	2.14-2.36
Serpentine	2.50-2.65
Silica, fused,	2.21
Silicon carbide	3.16
Slag	2.0-3.9
Slate	2.6-3.3
Soapstone	2.6-2.8

Material	$\rho/\ \mathrm{g\ cm^{-3}}$
Solder	8.7-9.4
Starch	1.53
Steel, stainless	7.8
Sugar	1.59
Talc	2.7-2.8
Tallow, beef	0.94
Tar	1.02
Topaz	3.5-3.6
Tourmaline	3.0-3.2
Tungsten carbide	14.0-15.0
Wax, sealing	1.8
Wood (seasoned)	
alder	0.42-0.68
apple	0.66-0.84
ash	0.65-0.85
balsa	0.11-0.14
bamboo	0.31-0.40
basswood	0.32-0.59
beech	0.70-0.90
birch	0.51-0.77
blue gum	1.00
box	0.95-1.16
butternut	0.38
cedar	0.49-0.57
cherry	0.70-0.90
dogwood	0.76
ebony	1.11-1.33
elm	0.54-0.60
hickory	0.60-0.93
holly	0.76
juniper	0.56
larch	0.50-0.56
locust	0.67-0.71
logwood	0.91
mahogany	0.66-0.85
maple	0.62-0.75
oak	0.60-0.90
pear	0.61-0.73
pine, pitch	0.83-0.85
white	0.35-0.50
yellow	0.37-0.60
plum	0.66-0.78
poplar	0.35-0.50
satinwood	0.95
spruce	0.48-0.70
sycamore	0.40-0.60
teak, Indian	0.66-0.98
walnut	0.64-0.70
water gum	1.00
willow	0.40-0.60
Wood's metal	9.70

DENSITY OF SULFURIC ACID

This table gives the density of aqueous sulfuric acid solutions as a function of concentration (in mass percent of H_2SO_4) and temperature.

Density in g/mL

Mass %	0°C	10°C	15°C	20°C	25°C	30°C	40°C	50°C	60°C	80°C	100°C
1	1.0074	1.0068	1.0060	1.0051	1.0038	1.0022	0.9986	0.9944	0.9895	0.9779	0.9645
2	1.0147	1.0138	1.0129	1.0118	1.0104	1.0087	1.0050	1.0006	0.9956	0.9839	0.9705
3	1.0219	1.0206	1.0197	1.0184	1.0169	1.0152	1.0113	1.0067	1.0017	0.9900	0.9766
4	1.0291	1.0275	1.0264	1.0250	1.0234	1.0216	1.0176	1.0129	1.0078	0.9961	0.9827
5	1.0364	1.0344	1.0332	1.0317	1.0300	1.0281	1.0240	1.0192	1.0140	1.0022	0.9888
6	1.0437	1.0414	1.0400	1.0385	1.0367	1.0347	1.0305	1.0256	1.0203	1.0084	0.9950
7	1.0511	1.0485	1.0469	1.0453	1.0434	1.0414	1.0371	1.0321	1.0266	1.0146	1.0013
8	1.0585	1.0556	1.0539	1.0522	1.0502	1.0481	1.0437	1.0386	1.0330	1.0209	1.0076
9	1.0660	1.0628	1.0610	1.0591	1.0571	1.0549	1.0503	1.0451	1.0395	1.0273	1.0140
10	1.0735	1.0700	1.0681	1.0661	1.0640	1.0617	1.0570	1.0517	1.0460	1.0338	1.0204
12	1.0886	1.0846	1.0825	1.0802	1.0780	1.0756	1.0705	1.0651	1.0593	1.0469	1.0335
14	1.1039	1.0994	1.0971	1.0947	1.0922	1.0897	1.0844	1.0788	1.0729	1.0603	1.0469
16	1.1194	1.1145	1.1120	1.1094	1.1067	1.1040	1.0985	1.0927	1.0868	1.0740	1.0605
18	1.1351	1.1298	1.1271	1.1243	1.1215	1.1187	1.1129	1.1070	1.1009	1.0879	1.0744
20	1.1510	1.1453	1.1424	1.1394	1.1365	1.1335	1.1275	1.1215	1.1153	1.1021	1.0885
22	1.1670	1.1609	1.1579	1.1548	1.1517	1.1486	1.1424	1.1362	1.1299	1.1166	1.1029
24	1.1832	1.1768	1.1736	1.1704	1.1672	1.1640	1.1576	1.1512	1.1448	1.1313	1.1176
26	1.1996	1.1929	1.1896	1.1862	1.1829	1.1796	1.1730	1.1665	1.1599	1.1463	1.1325
28	1.2160	1.2091	1.2057	1.2023	1.1989	1.1955	1.1887	1.1820	1.1753	1.1616	1.1476
30	1.2326	1.2255	1.2220	1.2185	1.2150	1.2115	1.2046	1.1977	1.1909	1.1771	1.1630
32	1.2493	1.2421	1.2385	1.2349	1.2314	1.2278	1.2207	1.2137	1.2068	1.1928	1.1787
34	1.2661	1.2588	1.2552	1.2515	1.2479	1.2443	1.2371	1.2300	1.2229	1.2088	1.1946
36	1.2831	1.2757	1.2720	1.2684	1.2647	1.2610	1.2538	1.2466	1.2394	1.2251	1.2109
38	1.3004	1.2929	1.2891	1.2855	1.2818	1.2780	1.2707	1.2635	1.2561	1.2418	1.2276
40	1.3179	1.3103	1.3065	1.3028	1.2991	1.2953	1.2880	1.2806	1.2732	1.2589	1.2446
42	1.3357	1.3280	1.3242	1.3205	1.3167	1.3129	1.3055	1.2981	1.2907	1.2762	1.2619
44	1.3538	1.3461	1.3423	1.3384	1.3346	1.3308	1.3234	1.3160	1.3086	1.2939	1.2796
46	1.3724	1.3646	1.3608	1.3569	1.3530	1.3492	1.3417	1.3343	1.3269	1.3120	1.2976
48	1.3915	1.3835	1.3797	1.3758	1.3719	1.3680	1.3604	1.3528	1.3455	1.3305	1.3159
50	1.4110	1.4029	1.3990	1.3951	1.3911	1.3872	1.3795	1.3719	1.3644	1.3494	1.3348
52	1.4310	1.4228	1.4188	1.4148	1.4109	1.4069	1.3991	1.3914	1.3837	1.3687	1.3540
54	1.4515	1.4431	1.4391	1.4350	1.4310	1.4270	1.4191	1.4113	1.4036	1.3884	1.3735
56	1.4724	1.4640	1.4598	1.4557	1.4516	1.4475	1.4396	1.4317	1.4239	1.4085	1.3934
58	1.4937	1.4852	1.4809	1.4768	1.4726	1.4685	1.4604	1.4524	1.4446	1.4290	1.4137
60	1.5154	1.5067	1.5024	1.4983	1.4940	1.4898	1.4816	1.4735	1.4656	1.4497	1.4344
62	1.5375	1.5287	1.5243	1.5200	1.5157	1.5115	1.5031	1.4950	1.4869	1.4708	1.4554
64	1.5600	1.5510	1.5465	1.5421	1.5378	1.5335	1.5250	1.5167	1.5086	1.4923	1.4766
66	1.5828	1.5736	1.5691	1.5646	1.5602	1.5558	1.5472	1.5388	1.5305	1.5140	1.4981
68	1.6059	1.5965	1.5920	1.5874	1.5829	1.5785	1.5697	1.5611	1.5528	1.5359	1.5198
70	1.6293	1.6198	1.6151	1.6105	1.6059	1.6014	1.5925	1.5838	1.5753	1.5582	1.5417
72	1.6529	1.6433	1.6385	1.6338	1.6292	1.6246	1.6155	1.6067	1.5981	1.5806	1.5637
74	1.6768	1.6670	1.6622	1.6574	1.6526	1.6480	1.6387	1.6297	1.6209	1.6031	1.5857
76	1.7008	1.6908	1.6858	1.6810	1.6761	1.6713	1.6619	1.6526	1.6435	1.6252	1.6074
78	1.7247	1.7144	1.7093	1.7043	1.6994	1.6944	1.6847	1.6751	1.6657	1.6469	1.6286
80	1.7482	1.7376	1.7323	1.7272	1.7221	1.7170	1.7069	1.6971	1.6873	1.6680	1.6493
82	1.7709	1.7599	1.7544	1.7491	1.7437	1.7385	1.7281	1.7180	1.7080	1.6882	1.6692
84	1.7916	1.7804	1.7748	1.7693	1.7639	1.7585	1.7479	1.7375	1.7274	1.7072	1.6878
86	1.8095	1.7983	1.7927	1.7872	1.7818	1.7763	1.7657	1.7552	1.7449	1.7245	1.7050
88	1.8243	1.8132	1.8077	1.8022	1.7968	1.7914	1.7809	1.7705	1.7602	1.7397	1.7202
90	1.8361	1.8252	1.8198	1.8144	1.8091	1.8038	1.7933	1.7829	1.7729	1.7525	1.7331
91	1.8410	1.8302	1.8248	1.8195	1.8142	1.8090	1.7986	1.7883	1.7783	1.7581	1.7388
92	1.8453	1.8346	1.8293	1.8240	1.8188	1.8136	1.8033	1.7932	1.7832	1.7633	1.7439
93	1.8490	1.8384	1.8331	1.8279	1.8227	1.8176	1.8074	1.7974	1.7876	1.7681	1.7485
94	1.8520	1.8415	1.8363	1.8312	1.8260	1.8210	1.8109	1.8011	1.7914		
95	1.8544	1.8439	1.8388	1.8337	1.8286	1.8236	1.8137	1.8040	1.7944		
96	1.8560	1.8457	1.8406	1.8355	1.8305	1.8255	1.8157	1.8060	1.7965		
97	1.8569	1.8466	1.8414	1.8364	1.8314	1.8264	1.8166	1.8071	1.7977		
98	1.8567	1.8463	1.8411	1.8361	1.8310	1.8261	1.8163	1.8068	1.7976		
99	1.8551	1.8445	1.8393	1.8342	1.8292	1.8242	1.8145	1.8050	1.7958		
100	1.8517	1.8409	1.8357	1.8305	1.8255	1.8205	1.8107	1.8013	1.7922		

DENSITY OF ETHANOL–WATER MIXTURES

This table gives the density of mixtures of ethanol and water as a function of composition and temperature. The composition is specified in weight percent of ethanol, i.e., mass of ethanol per 100 g of solution. Values from the reference have been converted to true densities.

Reference

Washburn, E. W., Ed., *International Critical Tables of Numerical Data of Physics, Chemistry, and Technology*, Vol. 3, McGraw-Hill, New York, 1926–1932.

Weight % Ethanol	Density in g/cm³						
	10 °C	15 °C	20 °C	25 °C	30 °C	35 °C	40 °C
0	0.99970	0.99910	0.99820	0.99705	0.99565	0.99403	0.99222
5	0.99095	0.99029	0.98935	0.98814	0.98667	0.98498	0.98308
10	0.98390	0.98301	0.98184	0.98040	0.97872	0.97682	0.97472
15	0.97797	0.97666	0.97511	0.97331	0.97130	0.96908	0.96667
20	0.97249	0.97065	0.96861	0.96636	0.96392	0.96131	0.95853
25	0.96662	0.96421	0.96165	0.95892	0.95604	0.95303	0.94988
30	0.95974	0.95683	0.95379	0.95064	0.94738	0.94400	0.94052
35	0.95159	0.94829	0.94491	0.94143	0.93787	0.93422	0.93048
40	0.94235	0.93879	0.93515	0.93145	0.92767	0.92382	0.91989
45	0.93223	0.92849	0.92469	0.92082	0.91689	0.91288	0.90881
50	0.92159	0.91773	0.91381	0.90982	0.90577	0.90165	0.89747
55	0.91052	0.90656	0.90255	0.89847	0.89434	0.89013	0.88586
60	0.89924	0.89520	0.89110	0.88696	0.88275	0.87848	0.87414
65	0.88771	0.88361	0.87945	0.87524	0.87097	0.86664	0.86224
70	0.87599	0.87184	0.86763	0.86337	0.85905	0.85467	0.85022
75	0.86405	0.85985	0.85561	0.85131	0.84695	0.84254	0.83806
80	0.85194	0.84769	0.84341	0.83908	0.83470	0.83027	0.82576
85	0.83948	0.83522	0.83093	0.82658	0.82218	0.81772	0.81320
90	0.82652	0.82225	0.81795	0.81360	0.80920	0.80476	0.80026
95	0.81276	0.80850	0.80422	0.79989	0.79553	0.79112	0.78668
100	0.79782	0.79358	0.78932	0.78504	0.78073	0.77639	0.77201

DIELECTRIC STRENGTH OF INSULATING MATERIALS

L. I. Berger

The loss of the dielectric properties by a sample of a gaseous, liquid, or solid insulator as a result of application to the sample of an electric field* greater than a certain critical magnitude is called *dielectric breakdown*. The critical magnitude of electric field at which the breakdown of a material takes place is called the *dielectric strength* of the material (or *breakdown voltage*). The dielectric strength of a material depends on the specimen thickness (as a rule, thin films have greater dielectric strength than that of thicker samples of a material), the electrode shape**, the rate of the applied voltage increase, the shape of the voltage vs. time curve, and the medium surrounding the sample, e.g., air or other gas (or a liquid — for solid materials only).

Breakdown in Gases

The current carriers in gases are free electrons and ions generated by external radiation. The equilibrium concentration of these particles at normal pressure is about 10^3 cm^{-3}, and hence the electrical conductivity is very small, of the order of $10^{-16} - 10^{-15}$ S/cm. But in a strong electric field, these particles acquire kinetic energy along their free path, large enough to ionize the gas molecules. The new charged particles ionize more molecules; this avalanche-like process leads to formation between the electrodes of channels of conducting plasma (streamers), and the electrical resistance of the space between the electrodes decreases virtually to zero.

Because the dielectric strength (breakdown voltage) of gases strongly depends on the electrode geometry and surface condition and the gas pressure, it is generally accepted to present the data for a particular gas as a fraction of the dielectric strength of either nitrogen or sulfur hexafluoride measured at the same conditions. In Table 1, the data are presented in comparison with the dielectric strength of nitrogen, which is considered equal to 1.00. For convenience to the reader, a few average magnitudes of the dielectric strength of some gases are expressed in kilovolts per millimeter. The data in the table relate to the standard conditions, unless indicated otherwise.

Breakdown in Liquids

If a liquid is pure, the breakdown mechanism in it is similar to that in gases. If a liquid contains liquid impurities in the form of small drops with greater dielectric constant than that of the main liquid, the breakdown is the result of formation of ellipsoids from these drops by the electric field. In a strong enough electric field, these ellipsoids merge and form a high-conductivity channel between the electrodes. The current increases the temperature in the channel, liquid boils, and the current along the steam canal leads to breakdown. Formation of a conductive channel (bridge) between the electrodes is observed also in liquids with solid impurities. If a liquid contains gas impurities in the form of small bubbles, breakdown is the result of heating of the liquid in strong electric fields. In the locations with the highest current density, the liquid boils, the size of the gas bubbles increases, they merge and form gaseous channels between the electrodes, and the breakdown medium is again the gas plasma.

Breakdown in Solids

It is known that the current in solid insulators does not obey Ohm's law in strong electric fields. The current density increases almost exponentially with the electric field, and at a certain field magnitude it jumps to very high magnitudes at which a specimen of a material is destroyed. The two known kinds of electric breakdown are thermal and electrical breakdowns. The former is the result of material heating by the electric current. Destruction of a sample of a material happens when, at a certain voltage, the amount of heat produced by the current exceeds the heat release through the sample surface; the breakdown voltage in this case is proportional to the square root of the ratio of the thermal conductivity and electrical conductivity of the material. A semi-empirical expression for dependence of the breakdown voltage, V_B, on the physical properties and geometry of a sample of a solid material for the one-dimensional case is

$$V_B = [A\rho\kappa / a\varphi(d)]^{1/2}$$

where A is a numerical constant related to the system of units used, ρ and κ are the volume resistivity and thermal conductivity of the sample material, a is a constant related to the chemical bond nature and crystal structure of the sample material, and $\varphi(d)$ is a function of the sample geometry, first of all, thickness, d (see, e.g., Ref. R6). In the majority of materials, $\varphi(d)$ increases with d, hence, the magnitude of V_B is greater in the thinner samples of a particular material.

The electrical breakdown results from the tunneling of the charge carriers from electrodes or from the valence band or from the impurity levels into the conduction band, or by the impact ionization. The tunnel effect breakdown happens mainly in thin layers, e.g., in thin p-n junctions. Otherwise, the impact ionization mechanism dominates. For this mechanism, the dielectric strength of an insulator can be estimated using Boltzmann's kinetic equation for electrons in a crystal.

In the following tables, the dielectric strength values are for room temperature and normal atmospheric pressure, unless indicated otherwise.

---

* The unit of electric field in the SI system is newton per coulomb or volt per meter.
** For example, the U.S. standard ASTM D149 is based on use of symmetrical electrodes, while per U.K. standard BS2918 one electrode is a plane and the other is a rod with the axis normal to the plane.

TABLE 1. Dielectric Strength of Gases

Material	Dielectric* strength	Ref.
Nitrogen, N_2	1.00	
Hydrogen, H_2	0.50	1,2
Helium, He	0.15	1
Oxygen, O_2	0.92	2
Air	0.97	6
Air (flat electrodes), kV/mm	3.0	3
Air, kV/mm	0.4-0.7	4
Air, kV/mm	1.40	5
Neon, Ne	0.25	1
	0.16	2
Argon, Ar	0.18	2
Chlorine, Cl_2	1.55	1
Carbon monoxide, CO	1.02	1
	1.05	2
Carbon dioxide, CO_2	0.88	1
	0.82	2
	0.84	6
Nitrous oxide, N_2O	1.24	2
Sulfur dioxide, SO_2	2.63	2
	2.68	6
Sulfur monochloride, S_2Cl_2 (at 12.5 Torr)	1.02	1
Thionyl fluoride, SOF_2	2.50	1
Sulfur hexafluoride, SF_6	2.50	1
	2.63	2
Sulfur hexafluoride, SF_6, kV/mm	8.50	7
	9.8	8
Perchloryl fluoride, ClO_3F	2.73	1
Tetrachloromethane, CCl_4	6.33	1
	6.21	2
Tetrafluoromethane, CF_4	1.01	1
Methane, CH_4	1.00	1
	1.13	2
Bromotrifluoromethane, CF_3Br	1.35	1
	1.97	2
Bromomethane, CH_3Br	0.71	2
Chloromethane, CH_3Cl	1.29	2
Iodomethane, CH_3I	3.02	2
Iodomethane, CH_3I, at 370 Torr	2.20	7
Dichloromethane, CH_2Cl_2	1.92	2
Dichlorodifluoromethane, CCl_2F_2	2.42	1
	2.63	2,6
Chlorotrifluoromethane, $CClF_3$	1.43	1
	1.53	2

Material	Dielectric* strength	Ref.
Trichlorofluoromethane, CCl_3F	3.50	1
	4.53	2
Trichloromethane, $CHCl_3$	4.2	1
	4.39	2
Methylamine, CH_3NH_2	0.81	1
Difluoromethane, CH_2F_2	0.79	2
Trifluoromethane, CHF_3	0.71	2
Bromochlorodifluoromethane, CF_2ClBr	3.84	2
Chlorodifluoromethane, $CHClF_2$	1.40	1
	1.11	2
Dichlorofluoromethane, $CHCl_2F$	1.33	1
	2.61	2
Chlorofluoromethane, CH_2ClF	1.03	1
Hexafluoroethane, C_2F_6	1.82	1
	2.55	2
Ethyne (Acetylene), C_2H_2	1.10	1
	1.11	2
Chloropentafluoroethane, C_2ClF_5	2.3	1
	3.0	6
Dichlorotetrafluoroethane, $C_2Cl_2F_4$	2.52	1
Chlorotrifluoroethylene, C_2ClF_3	1.82	2
1,1,1-Trichloro-2,2,2-trifluoroethane	6.55	2
1,1,2-Trichloro-1,2,2-trifluoroethane	6.05	2
Chloroethane, C_2H_5Cl	1.00	1
1,1-Dichloroethane	2.66	2
Trifluoroacetonitrile, CF_3CN	3.5	1
Acetonitrile, CH_3CN	2.11	2
Dimethylamine, $(CH_3)_2NH$	1.04	1
Ethylamine, $C_2H_5NH_2$	1.01	1
Ethylene oxide (oxirane), CH_3CHO	1.01	1
Perfluoropropene, C_3F_6	2.55	2
Octafluoropropane, C_3F_8	2.19	1
	2.47	2
3,3,3-Trifluoro-1-propene, CH_2CHCF_3	2.11	2
Pentafluoroisocyanoethane, C_2F_5NC	4.5	1
1,1,1,4,4,4-Hexafluoro-2-butyne, CF_3CCCF_3	5.84	2
Octafluorocyclobutane, C_4F_8	3.34	2
1,1,1,2,3,4,4-Octafluoro-2-butene	2.8	1
Decafluorobutane, C_4F_{10}	3.08	1
Perfluorobutanenitrile, C_3F_7CN	5.5	1
Perfluoro-2-methyl-1,3-butadiene, C_5F_8	5.5	1
Hexafluorobenzene, C_6F_6	2.11	2
Perfluorocyclohexane, C_6F_{12}, (saturated vapor)	6.18	2

* Relative to nitrogen, unless units of kV/mm are indicated.

TABLE 2. Dielectric Strength of Liquids

Material	Dielectric strength kV/mm	Ref.	Material	Dielectric strength kV/mm	Ref.
Helium, He, liquid, 4.2 K	10	9		20.4	15
Static	10	11		179	17,18
Dynamic	5	11	Ethylbenzene, C_8H_{10}	226	17,18
	23	12	Propylbenzene, C_9H_{12}	250	17,18
Nitrogen, N_2, liquid, 77K			Isopropylbenzene, C_9H_{12}	238	17,18
Coaxial cylinder electrodes	20	10	Decane, $C_{10}H_{22}$	192	17,18
Sphere to plane electrodes	60	10	Synthetic Paraffin Mixture		
Water, H_2O, distilled	65-70	13	Synfluid 2cSt PAO	29.5	37
Carbon tetrachloride, CCl_4	5.5	14	Butylbenzene, $C_{10}H_{14}$	275	17,18
	16.0	15	Isobutylbenzene, $C_{10}H_{14}$	222	17,18
Hexane, C_6H_{14}	42.0	16	Silicone oils—polydimethylsiloxanes,		
Two 2.54 cm diameter spherical			$(CH_3)_3Si-O-[Si(CH_3)_2]_x-O-Si(CH_3)_3$		
electrodes, 50.8 µm space	156	17,18	Polydimethylsiloxane silicone fluid	15.4	20
Cyclohexane, C_6H_{12}	42-48	16	Dimethyl silicone	24.0	21,22
2-Methylpentane, C_6H_{14}	149	17,18	Phenylmethyl silicone	23.2	22
2,2-Dimethylbutane, C_6H_{14}	133	17,18	Silicone oil, Basilone M50	10-15	23
2,3-Dimethylbutane, C_6H_{14}	138	17,18	Mineral insulating oils	11.8	6
Benzene, C_6H_6	163	17,18	Polybutene oil for capacitors	13.8	6
Chlorobenzene, C_6H_5Cl	7.1	14	Transformer dielectric liquid	28-30	6
	18.8	15	Isopropylbiphenyl capacitor oil	23.6	6
2,2,4-Trimethylpentane, C_8H_{18}	140	17,18	Transformer oil	110.7	24
Phenylxylylethane	23.6	19	Transformer oil Agip ITE 360	9-12.6	23
Heptane, C_7H_{16}	166	17,18	Perfluorinated hydrocarbons		
2,4-Dimethylpentane, C_7H_{16}	133	17,18	Fluorinert FC 6001	8.0	23
Toluene, $C_6H_5CH_3$	199	17,18	Fluorinert FC 77	10.7	23
	46	16	Perfluorinated polyethers		
	12.0	14	Galden XAD (Mol. wt. 800)	10.5	23
	20.4	15	Galden D40 (Mol. wt. 2000)	10.2	23
Octane, C_8H_{18}	16.6	14	Castor oil	65	25

TABLE 3. Dielectric Strength of Solids

Material	Dielectric strength kV/mm	Ref	Material	Dielectric strength kV/mm	Ref
Sodium chloride, NaCl, crystalline	150	26	Phlogopite, amber, natural	118	6
Potassium bromide, KBr, crystalline	80	26	Fluorophlogopite, synthetic	118	6
Ceramics			Glass-bonded mica	14.0-15.7	6
Alumina (99.9% Al_2O_3)	13.4	6,27a	Thermoplastic Polymers		
Aluminum silicate, Al_2SiO_5	5.9	6	Polypropylene	23.6	6
Berillia (99% BeO)	13.8	6,27b	Amide polymer nylon 6/6, dry	23.6	6
Boron nitride, BN	37.4	6	Polyamide-imide copolymer	22.8	6
Cordierite, $Mg_2Al_4Si_5O_{18}$	7.9	6,27c	Modified polyphenylene oxide	21.7	6
Forsterite, Mg_2SiO_4	9.8	28	Polystyrene	19.7	6
Porcelain	35-160	26	Polymethyl methacrylate	19.7	6
Steatite, $Mg_3Si_4O_{11} \cdot H_2O$	9.1-15.4	6	Polyetherimide	18.9	6
Titanates of Mg, Ca, Sr, Ba, and Pb	20-120	3	Amide polymer nylon 11(dry)	16.7	6
Barium titanate, glass bonded	>30	36	Polysulfone	16.7	6
Zirconia, ZrO_2	11.4	29	Styrene-acrylonitrile copolymer	16.7	6
Glasses			Acrylonitrile-butadiene-styrene	16.7	6
Fused silica, SiO_2	470-670	26	Polyethersulfone	15.7	6
Alkali-silicate glass	200	26	Polybutylene terephthalate	15.7	6
Standard window glass	9.8-13.8	28	Polystyrene-butadiene copolymer	15.7	6
Micas			Acetal homopolymer	15.0	6
Muscovite, ruby, natural	118	6	Acetal copolymer	15.0	6
			Polyphenylene sulfide	15.0	6

Material	Dielectric strength kV/mm	Ref
Polycarbonate	15.0	6
Acetal homopolymer resin (molding resin)	15.0	6
Acetal copolymer resin	15.0	6
Thermosetting Molding Compounds		
Glass-filled allyl (Type GDI-30 per MIL-M-14G)	15.7	6
Glass-filled epoxy, electrical grade	15.4	6
Glass-filled phenolic (Type GPI-100 per MIL-M-14G)	15.0	6
Glass-filled alkyd/polyester (Type MAI-60 per MIL-M-14G)	14.8	6
Glass-filled melamine (Type MMI-30 per MIL-M-14G)	13.4	6
Extrusion Compounds for High-Temperature Insulation		
Polytetrafluoroethylene	19.7	6
Perfluoroalkoxy polymer	21.7	6
Fluorinated ethylene-propylene copolymer	19.7	6
Ethylene-tetrafluoroethylene copolymer	15.7	6
Polyvinylidene fluoride	10.2	6
Ethylene-chlorotrifluoroethylene copolymer	19.3	6
Polychlorotrifluoroethylene	19.7	6
Extrusion Compounds for Low-Temperature Insulation		
Polyvinyl chloride		
Flexible	11.8-15.7	30
Rigid	13.8-19.7	30
Polyethylene	18.9	28
Polyethylene, low-density	21.7	6
	300	31
Polyethylene, high-density	19.7	6
Polypropylene/polyethylene copolymer	23.6	6
Embedding Compounds		
Basic epoxy resin: bisphenol-A/epichlorohydrin polycondensate	19.7	6
Cycloaliphatic epoxy: alicyclic diepoxy carboxylate	19.7	6
Polyetherketone	18.9	30
Polyurethanes		
Two-component, polyol-cured	25.4	6
Two-part solventless, polybutylene-based	24.0	6
Silicones		
Clear two-part heat curing eletrical grade silicone embedding resin	21.7	6
Red insulating enamel (MIL-E-22118)		
Dry	47.2	6
Wet	11.8	6
Enamels		
Red enamel, fast cure		
Standard conditions	78.7	6
Immersion conditions	47.2	6
Black enamel		
Standard conditions	70.9	6
Immersion conditions	47.2	6
Varnishes		
Vacuum-pressure impregnated baking type solventless polyester varnish		

Material	Dielectric strength kV/mm	Ref
Rigid, two-part	70.9	6
Semiflexible high-bond thixotropic	78.7	6
Rigid high-bond high-flash freon-resistant	68.9	6
Baking type epoxy varnish		
Solventless, rigid, low viscosity, one-part	90.6	6
Solventless, semiflexible, one-part	82.7	6
Solventless, semirigid, chemical resistant, low dielectric constant	106.3	6
Solvable, for hermetic electric motors	181.1	6
Polyurethane coating		
Clear conformal, fast cure		
Standard conditions	78.7	6
Immersion conditions	47.2	6
Insulating Films and Tapes		
Low-density polyethylene film (40 μm thick)	300	31
Poly-p-xylylene film	410-590	32
Aromatic polymer films		
Kapton H (Du Pont)	389-430	33
Ultem (GE Plastic and Roem AG)	437-565	33
Hostaphan (Hoechst AG)	338-447	33
Amorphous Stabar K2000 (ICI film)	404-422	33
Stabar S100 (ICI film)	353-452	33
Polyetherimide film (26 μm)	486	34
Parylene N/D (poly-p-xylylene/poly-dichloro-p-xylylene) 25 μm film	275	6
Cellulose acetate film	157	6
Cellulose triacetate film	157	6
Polytetrafluoroethylene film	87-173	6
Perfluoroalkoxy film	157-197	6
Fluorinated ethylene-propylene copolymer film	197	6
Ethylene-tetrafluoroethylene film	197	6
Ethylene-chlorotrifluoroethylene copolymer film	197	6
Polychlorotrifluoroethylene film	118-153.5	6
High-voltage rubber insulating tape	28	6
Composites		
Isophthalic polyester (vinyl toluene monomer) filled with		
Calcium carbonate, $CaCO_3$	15.0	38
Gypsum, $CaSO_4$	14.4	38
Alumina trihydrate	15.4	38
Clay	14.4	38
BPA fumarate polyester (vinyl toluene monomer) filled with		
Calcium carbonate	6.1	38
Gypsum	5.9	38
Alumina trihydrate	11.8	38
Clay	12.6	38
Polysulfone resin—30% glass fiber	16.5-18.7	38
Polyamid resin (Nylon 66)—		
30% carbon fiber	13.0	38
Polyimide thermoset resin, glass reinforced	12.0	39
Polyester resin (thermoplastic)—		

Material	Dielectric strength kV/mm	Ref
40% glass fiber	20.0	38
Epoxy resin (diglycidyl ether of bisphenol A), glass reinforced	16.0	40
Various Insulators		
Rubber, natural	100-215	26
Butyl rubber	23.6	6
Neoprene	15.7-27.6	6
Silicone rubber	26-36	6

Material	Dielectric strength kV/mm	Ref
Room-temperature vulcanized silicone rubber	9.2-10.9	35
Ureas (from carbamide to tetraphenylurea)	11.8-15.7	28
Dielectric papers		
Aramid paper, calendered	28.7	6
Aramid paper, uncalendered	12.2	6
Aramid with Mica	39.4	6

References

1. Vijh, A. K. *IEEE Trans.*, EI-12, 313, 1997.
2. Brand, K. P., *IEEE Trans.*, EI-17, 451, 1982.
3. *Encyclopedic Dictionary in Physics*, Vedensky, B. A. and Vul, B. M., Eds., Vol. 4, Soviet Encyclopedia Publishing House, Moscow, 1965.
4. Kubuki, M., Yoshimoto, R., Yoshizumi, K., Tsuru, S., and Hara, M., *IEEE Trans.*, DEI-4, 92, 1997.
5. Al-Arainy, A. A. Malik, N. H., and Cureshi, M. I., *IEEE Trans.*, DEI-1, 305, 1994.
6. Shugg, W. T., *Handbook of Electrical and Electronic Insulating Materials*, Van Nostrand Reinhold, New York, 1986.
7. Devins, J. C., *IEEE Trans.*, EI-15, 81, 1980.
8. Xu, X., Jayaram, S., and Boggs, S. A., *IEEE Trans.*, DEI-3, 836, 1996.
9. Okubo, H., Wakita, M., Chigusa, S., Nayakawa, N., and Hikita, M., *IEEE Trans.*, DEI-4, 120, 1997.
10. Hayakawa, H., Sakakibara, H., Goshima, H., Hikita, M., and Okubo, H., *IEEE Trans.*, DEI-4, 127, 1997.
11. Okubo, H., Wakita, M., Chigusa, S., Hayakawa, N., and Hikita, M., *IEEE Trans.*, DEI-4, 220, 1997.
12. Von Hippel, A. R., *Dielectric Materials and Applications*, MIT Press, Cambridge, MA, 1954.
13. Jones, H. M. and Kunhards, E. E., *IEEE Trans.*, DEI-1, 1016, 1994.
14. Nitta, Y. and Ayhara, Y., *IEEE Trans.*, EI-11, 91, 1976.
15. Gallagher, T. J., *IEEE Trans.*, EI-12, 249, 1977.
16. Wong, P. P. and Forster, E. O., in *Dielectric Materials. Measurements and Applications*, IEE Conf. Publ. 177, 1, 1979.
17. Kao, K. C. *IEEE Trans.*, EI-11, 121, 1976.
18. Sharbaugh, A. H., Crowe, R. W., and Cox, E. B., *J. Appl. Phys.*, 27, 806, 1956.
19. Miller, R. L., Mandelcorn, L., and Mercier, G. E., in *Proc. Intl. Conf. on Properties and Applications of Dielectric Materials*, Xian, China, June 24-28, 1985; cited in Ref. 6, p. 492.
20. Hakim, R. M., Oliver, R. G., and St-Onge, H., *IEEE Trans.*, EI-12, 360, 1977.
21. Hosticka, C., *IEEE Trans.*, 389, 1977.
22. Yasufuku, S., Umemura, T., and Ishioka, Y., *IEEE Trans.*, EI-12, 402, 1977.
23. Forster, E. O., Yamashita, H., Mazzetti, C., Pompini, M., Caroli, L., and Patrissi, S., *IEEE Trans.*, DEI-1, 440, 1994.
24. Bell, W. R., *IEEE Trans.*, 281, 1977.
25. Ramu, T. C. and Narayana Rao, Y., in *Dielectric Materials. Measurements and Applications*, IEE Conf. Publ. 177, 37.
26. Skanavi, G. I., *Fizika Dielektrikov; Oblast Silnykh Polei* (Physics of Dielectrics; Strong Fields). Gos. Izd. Fiz. Mat. Nauk (State Publ. House for Phys. and Math. Scis.), Moscow, 1958.
27. Kleiner, R. N., in *Practical Handbook of Materials Science*, Lynch, C. T., Ed., CRC Press, 1989; 27a: p. 304; 27b: p.300; 27c: p. 316.
28. *Materials Selector Guide. Materials and Methods*, Reinhold Publ., New York, 1973.
29. Flinn, R. A. and Trojan, P. K., *Engineering Materials and Their Applications*, 2nd ed., Houghton Mifflin, 1981, p. 614.
30. Lynch, C. T., Ed., *Practical Handbook of Materials Science*, CRC Press, Boca Raton, FL, 1989.
31. Suzuki, H., Mukai, S., Ohki, Y., Nakamichi, Y., and Ajiki, K., *IEEE Trans.*, DEI-4, 238, 1997.
32. Mori, T., Matsuoka, T., and Muzitani, T., *IEEE Trans.*, DEI-1, 71, 1994.
33. Bjellheim, P. and Helgee, B., *IEEE Trans.*, DEI-1, 89, 1994.
34. Zheng, J. P., Cygan, P. J., and Jow, T. R., *IEEE Trans.*, DEI-3, 144, 1996.
35. Danukas, M. G., *IEEE Trans.*, DEI-1, 1196, 1994.
36. Burn, I. and Smithe, D. H., *J. Mater. Sci.*, 7, 339, 1972.
37. Hope, K.D., Chevron Chemical, Private Communication.
38. *Engineering Materials Handbook*, Vol. 1, Composites, C.A. Dostal, Ed., ASM Intl., 1987.
39. 1985 Materials Selector, *Mater. Eng.*, (12) 1984.
40. *Modern Plastics Encyclopedia*, McGraw-Hill, v. 62 (No. 10A) 1985–1986.

Review Literature on the Subject
R1. Kuffel, E. and Zaengl, W. S., *HV Engineering Fundamentals*, Pergamon, 1989.
R2. Kok, J. A., *Electrical Breakdown of Insulating Liquids*, Phillips Tech. Library, Cleaver-Hum, London, 1961.
R3. Gallagher, T. J., *Simple Dielectric Liquids*, Clarendon, Oxford, 1975.
R4. Meek, J. M. and Craggs, J. D., Eds., *Electric Breakdown in Gases*, John Wiley & Sons, 1976.
R5. Von Hippel, A. R., *Dielectric Materials and Applications*, MIT Press, Cambridge, MA, 1954.
R6. O'Dwyer, J. J. *The Theory of Dielectric Breakdown of Solids*, Clarendon Press, 1964.

COEFFICIENT OF FRICTION

The coefficient of friction between two surfaces is the ratio of the force required to move one over the other to the force pressing the two together. Thus if F is the minimum force needed to move one surface over the other, and W is the force pressing the surfaces together, the coefficient of friction μ is given by $\mu = F/W$. A greater force is generally needed to initiate movement from rest than to continue the motion once sliding has started. Thus the static coefficient of friction μ(static) is usually larger that the sliding or kinetic coefficient μ(sliding).

This table gives characteristic values of both the static and sliding coefficients of friction for a number of material combinations. In each case Material 1 is moving over the surface of Material 2.

The type of lubrication or any other special condition is indicated in the third column. All values refer to room temperature unless otherwise indicated. It should be emphasized that the coefficient of friction is very sensitive to the condition of the surface, so that these values represent only a rough guide.

References

1. Minshall, H., in *CRC Handbook of Chemistry and Physics, 73rd Edition*, Lide, D. R., Ed., CRC Press, Boca Raton, FL, 1992.
2. Fuller, D. D., in *American Institute of Physics Handbook, 3rd Edition*, Gray, D. E., Ed., McGraw-Hill, New York, 1972.

Material 1	Material 2	Conditions	μ(static)	μ(sliding)
Metals				
Hard steel	Hard steel	Dry	0.78	0.42
		Castor oil	0.15	0.081
		Steric acid	0.005	0.029
		Lard	0.11	0.084
		Light mineral oil	0.23	
		Graphite		0.058
Hard steel	Graphite	Dry	0.21	
Mild steel	Mild steel	Dry	0.74	0.57
		Oleic acid		0.09
Mild steel	Phosphor bronze	Dry		0.34
Mild steel	Cast iron	Dry		0.23
Mild steel	Lead	Dry	0.95	0.95
		Mineral oil	0.5	0.3
Mild steel	Brass	Dry	0.35	
Cast iron	Cast iron	Dry	1.10	0.15
Aluminum	Aluminum	Dry	1.05	1.4
Aluminum	Mild steel	Dry	0.61	0.47
Brass	Mild steel	Dry	0.51	0.44
		Castor oil	0.11	
Brass	Cast iron	Dry		0.30
Bronze	Cast iron	Dry		0.22
Cadmium	Mild steel	Dry		0.46
Copper	Copper	Dry	1.6	
Copper	Mild steel	Dry	0.53	0.36
		Oleic acid		0.18
Copper	Cast iron	Dry	1.05	0.29
Copper	Glass	Dry	0.68	0.53
Lead	Cast iron	Dry		0.43
Magnesium	Magnesium	Dry	0.6	
Magnesium	Mild steel	Dry		0.42
Magnesium	Cast iron	Dry		0.25
Nickel	Nickel	Dry	1.10	0.53
Nickel	Mild steel	Dry		0.64
Tin	Cast iron	Dry		0.32
Zinc	Cast iron	Dry	0.85	0.21
Nonmetals				
Diamond	Diamond	Dry	0.1	
Diamond	Metals	Dry	0.12	
Garnet	Mild steel	Dry		0.39
Glass	Glass	Dry	0.94	0.4
Glass	Nickel	Dry	0.78	0.56

Material 1	Material 2	Conditions	μ(static)	μ(sliding)
Graphite	Graphite	Dry	0.1	
Mica	Mica	Freshly cleaved	1.0	
Nylon	Nylon	Dry	0.2	
Nylon	Steel	Dry	0.40	
Polyethylene	Polyethylene	Dry	0.2	
Polyethylene	Steel	Dry	0.2	
Polystyrene	Polystyrene	Dry	0.5	
Polystyrene	Steel	Dry	0.3	
Sapphire	Sapphire	Dry	0.2	
Teflon	Teflon	Dry	0.04	0.04
Teflon	Steel	Dry	0.04	0.04
Tungsten carbide	Tungsten carbide	Dry, room temp.	0.17	
		Dry, 1000°C	0.45	
		Dry, 1600°C	1.8	
		Oleic acid	0.12	
Tungsten carbide	Graphite	Dry	0.15	
Tungsten carbide	Steel	Dry	0.5	
		Oleic acid	0.08	

Miscellaneous materials

Cotton	Cotton	Threads	0.3	
Leather	Cast iron	Dry	0.6	0.56
Leather	Oak	Parallel to grain	0.61	0.52
Oak	Oak	Parallel to grain	0.62	0.48
		Perpendicular to grain	0.54	0.32
Silk	Silk	Clean	0.25	
Wood	Wood	Dry	0.35	
		Wet	0.2	
Wood	Brick	Dry	0.6	
Wood	Leather	Dry	0.35	

Various materials on ice and snow

Ice	Ice	Clean, 0°C	0.1	0.02
		Clean, -12°C	0.3	0.035
		Clean, -80°C	0.5	0.09
Aluminum	Snow	Wet, 0°C	0.4	
		Dry, 0°C	0.35	
Brass	Ice	Clean, 0°C		0.02
		Clean, -80°C		0.15
Nylon	Snow	Wet, 0°C	0.4	
		Dry, -10°C	0.3	
Teflon	Snow	Wet, 0°C	0.05	
		Dry, 0°C	0.02	
Wax, ski	Snow	Wet, 0°C	0.1	
		Dry, 0°C	0.04	
		Dry, -10°C	0.2	

FLAME TEMPERATURES

This table gives the adiabatic flame temperature for stoichemetric mixtures of various fuels and oxidizers. The temperatures are calculated from thermodynamic and transport properties under ideal adiabatic conditions, using methods described in the reference.

Reference

Fristrom, R. M., *Flame Structures and Processes*, Oxford University Press, New York, 1995.

Adiabatic Flame Temperature in K for Various Fuel-Oxidizer Combinations

Fuel	Oxidizer					
	Air	O_2	F_2	Cl_2	N_2O	NO
Organic liquids and gases						
Acetaldehyde	2288					
Acetone	2253					
Acetylene	2607					
Benzene	2363					
Butane	2248					
Carbon disulfide	2257					
Cyanogen	2596	4855				
Cyclohexane	2250					
Cyclopropane	2370					
Decane	2286					
Ethane	2244					
Ethanol	2238					
Ethylene	2375					
Hexane	2238					
Methane	2236					
Methanol	2222					
Oxirane	2177					
Pentane	2250					
Propane	2250					
Toluene	2344					
Solids						
Aluminum		4005				
Lithium		2711				
Phosphorus (white)		3242				
Zirconium		4278				
Other						
Ammonia		2845				
Carbon monoxide	1388					
Diborane		3350				
Hydrazine		3037				
Hydrogen	2169	3000	4006	2493	2965	3127
Hydrogen sulfide	2091	3414				
Phosphine		3139				
Silane		3043				

ALLOCATION OF FREQUENCIES IN THE RADIO SPECTRUM

In the United States the National Telecommunications and Information Administration (NTIA) has responsibility for assigning each portion of the radio spectrum (9 kHz to 300 GHz) for different uses. These assignments must be compatible with the rules of the International Telecommunications Union (ITU), to which the United States is bound by treaty. The current assignments are given in a wall chart (Reference 1) and may also be found on the NTIA web site (Reference 2). The list below summarizes the broad features of the spectrum allocation, with particular attention to those sections of scientific interest. The references should be consulted for details of the allocations in the frequency bands listed here, which in some cases are quite complex.

References

1. *United States Frequency Allocations*, 1996 Spectrum Wall Chart, Stock No. 003-000-00652-2, U. S. Government Printing Office, P. O. Box 371954, Pittsburgh, PA 15250-7954.
2. http://www.ntia.doc.gov/osmhome/allochrt.html

Frequency range	Allocation
9 - 19.95 kHz	Maritime communication, navigation
19.95 - 20.05 kHz	Standard frequency and time signal (also at 60 kHz and 2.5, 5, 10, 15, 20, 25 MHz)
20.05 - 535 kHz	Maritime and aeronautical communication, navigation
535 - 1605 kHz	AM radio broadcasting
1605 - 3500 kHz	Mobile communication and navigation, amateur radio (1800-1900 kHz)
3.5 - 4.0 MHz	Amateur radio
4.0 - 5.95 MHz	Mobile communication
5.95 - 13.36 MHz	Mobile communication, amateur, short-wave broadcasting
13.36 - 13.41 MHz	Radioastronomy
13.41 - 25.55 MHz	Mobile communication, amateur, short-wave broadcasting
25.55 - 25.67 MHz	Radioastronomy
25.67 - 37.5 MHz	Mobile communication, amateur, short-wave broadcasting
37.5 -38.25 MHz	Radioastronomy
38.25 - 50.0 MHz	Mobile communication
50.0 - 54.0 MHz	Amateur
54.0 - 72.0 MHz	TV channels 2-4
72.0 - 73.0 MHz	Mobile communication
73.0 - 74.6 MHz	Radioastronomy
74.6 - 76.0 MHz	Mobile communication
76.0 - 88.0 MHz	TV channels 5-6
88.0 - 108.0 MHz	FM radio broadcasting
108.0 - 118.0 MHz	Aeronautical navigation
118.0 - 174.0 MHz	Mobile communication, space research, meteorological satellites
174.0 - 216.0 MHz	TV channels 7-13
216.0 - 400.05 MHz	Mobile communication
400.05 - 400.15 MHz	Standard frequency and time satellite (also 20 and 25 GHz)
400.15 - 406.1 MHz	Meteorological aids (radiosonde)
406.1 - 410.0 MHz	Radioastronomy
410.0 - 470.0 MHz	Mobile communication, amateur
470.0 - 512.0 MHz	TV channels 14-20
512.0 - 608.0 MHz	TV channels 21-36
608.0 - 614.0 MHz	Radioastronomy
614.0 - 806.0 MHz	TV channels 38-69
806 -1400 MHz	Mobile communication, navigation
1400 - 1427 MHz	Radioastronomy, space research
1427 - 1660 MHz	Various navigation and satellite applications
1660 - 1710 MHz	Radioastronomy, space research, meteorology
1710 - 2655 MHz	Various navigation and satellite applications
2655 - 2700 MHz	Radioastronomy, space research
2.7 - 4.99 GHz	Various navigation and satellite applications
4.99 - 5.0 GHz	Radioastronomy, space research
5.0 - 10.6 GHz	Various navigation and satellite applications
10.6 - 10.7 GHz	Radioastronomy, space research
10.7 - 15.35 GHz	Various navigation and satellite applications
15.35 - 15.4 GHz	Radioastronomy, space research
15.4 - 22.21 GHz	Various navigation and satellite applications

Frequency range	Allocation
22.21 - 22.5 GHz	Radioastronomy, space research
22.25 - 23.6 GHz	Various navigation and satellite applications
23.6 - 24.0 GHz	Radioastronomy, space research
24.0 - 31.3 GHz	Various navigation and satellite applications
31.3 - 31.8 GHz	Radioastronomy, space research
31.8 - 42.5 GHz	Various navigation and satellite applications
42.5 - 43.5 GHz	Radioastronomy
43.5 - 51.4 GHz	Various navigation and satellite applications
51.4 - 54.25 GHz	Radioastronomy, space research
54.25 - 58.2 GHz	Space research
58.2 - 59.0 GHz	Radioastronomy, space research
59.0 - 64.0 GHz	Satellite applications
64.0 - 65.0 GHz	Radioastronomy, space research
65.0 - 72.77 GHz	Various navigation and satellite applications
72.77 - 72.91 GHz	Radioastronomy, space research
72.91 - 86.0 GHz	Various navigation and satellite applications
86.0 - 92.0 GHz	Radioastronomy, space research
92.0 - 105.0 GHz	Various navigation and satellite applications
105.0 - 116.0 GHz	Radioastronomy, space research
116.0 - 164.0 GHz	Various navigation and satellite applications
164.0 - 168.0 GHz	Radioastronomy, space research
168.0 - 182.0 GHz	Various navigation and satellite applications
182.0 - 185.0 GHz	Radioastronomy, space research
185.0 - 217.0 GHz	Various navigation and satellite applications
217.0 - 231.0 GHz	Radioastronomy, space research
231.0 - 265.0 GHz	Various navigation and satellite applications
265.0 - 275.0 GHz	Radioastronomy
275.0 - 300.0 GHz	Mobile communications

Section 16
Health and Safety Information

HANDLING AND DISPOSAL OF CHEMICALS IN LABORATORIES

Robert Joyce and Blaine C. McKusick

The following material has been extracted from two books prepared under the auspices of the Committee on Hazardous Substances in the Laboratory of the National Academy of Sciences – National Research Council. Readers are referred to these books for full details:

Prudent Practices for Handling Hazardous Chemicals in Laboratories, National Academy Press, Washington, 1981.

Prudent Practices for Disposal of Chemicals from Laboratories, National Academy Press, Washington, 1983.

The permission of the National Academy Press to use these extracts is gratefully acknowledged.

INCOMPATIBLE CHEMICALS

The term "incompatible chemicals" refers to chemicals that can react with each other

- Violently
- With evolution of substantial heat
- To produce flammable products
- To produce toxic products

Good laboratory safety practice requires that incompatible chemicals be stored, transported, and disposed of in ways that will prevent their coming together in the event of an accident. Tables 1 and 2 give some basic guidelines for the safe handling of acids, bases, reactive metals, and other chemicals. Neither of these tables is exhaustive, and additional information on incompatible chemicals can be found in the following references.

1. Urben, P. G., Ed., *Bretherick's Handbook of Reactive Chemical Hazards,* 5th ed., Butterworth-Heinemann, Oxford, 1995.
2. Luxon, S. G., Ed., *Hazards in the Chemical Laboratory,* 5th ed., Royal Society of Chemistry, Cambridge, 1992.
3. *Fire Protection Guide to Hazardous Materials,* 11th ed., National Fire Protection Association, Quincy, MA, 1994.

TABLE 1. General Classes of Incompatible Chemicals

A	B
Acids	Bases, reactive metals
Oxidizing agents[a]	Reducing agents[a]
Chlorates	Ammonia, anhydrous and aqueous
Chromates	Carbon
Chromium trioxide	Metals
Dichromates	Metal hydrides
Halogens	Nitrites
Halogenating agents	Organic compounds
Hydrogen peroxide	Phosphorus
Nitric acid	Silicon
Nitrates	Sulfur
Perchlorates	
Peroxides	
Permanganates	
Persulfates	

[a] The examples of oxidizing and reducing agents are illustrative of common laboratory chemicals; they are not intended to be exhaustive.

TABLE 2. Examples of Incompatible Chemicals

Chemical	Is incompatible with
Acetic acid	Chromic acid, nitric acid, hydroxyl compounds, ethylene glycol, perchloric acid, peroxides, permanaganates
Acetylene	Chlorine, bromine, copper, fluorine, silver, mercury
Acetone	Concentrated nitric and sulfuric acid mixtures
Alkali and alkaline earth metals (such as powdered aluminum or magnesium, calcium, lithium, sodium, potassium)	Water, carbon tetrachloride or other chlorinated hydrocarbons, carbon dioxide, halogens
Ammonia (anhydrous)	Mercury (in manometers, for example), chlorine, calcium hypochlorite, iodine, bromine, hydrofluoric acid (anhydrous)
Ammonium nitrate	Acids, powdered metals, flammable liquids, chlorates, nitrites, sulfur, finely divided organic or combustible materials
Aniline	Nitric acid, hydrogen peroxide
Arsenical materials	Any reducing agent
Azides	Acids

Chemical	Is incompatible with
Bromine	See Chlorine
Calcium oxide	Water
Carbon (activated)	Calcium hypochlorite, all oxidizing agents
Carbon tetrachloride	Sodium
Chlorates	Ammonium salts, acids, powdered metals, sulfur, finely divided organic or combustible materials
Chromic acid and chromium troixide	Acetic acid, naphthalene, camphor, glycerol, alcohol, flammable liquids in general
Chlorine	Ammonia, acetylene, butadiene, butane, methane, propane (or other petroleum gases), hydrogen, sodium carbide, benzene, finely divided metals, turpentine
Chlorine dioxide	Ammonia, methane, phosphine, hydrogen sulfide
Copper	Acetylene, hydrogen peroxide
Cumene hydroperoxide	Acids (organic or inorganic)
Cyanides	Acids
Flammable liquids	Ammonium nitrate, chromic acid, hydrogen peroxide, nitric acid, sodium peroxide, halogens
Fluorine	Everything
Hydrocarbons (such as butane, propane, benzene)	Fluorine, chlorine, bromine, chromic acid, sodium peroxide
Hydrocyanic acid	Nitric acid, alkali
Hydrofluoric acid (anhydrous)	Ammonia (aqueous or anhydrous)
Hydrogen peroxide	Copper, chromium, iron, most metals or their salts, alcohols, acetone, organic materials, aniline, nitro-methane, combustible materials
Hydrogen sulfide	Fuming nitric acid, oxidizing gases
Hypochlorites	Acids, activated carbon
Iodine	Acetylene, ammonia (aqueous or anhydrous), hydrogen
Mercury	Acetylene, fulminic acid, ammonia
Nitrates	Sulfuric acid
Nitric acid (concentrated)	Acetic acid, aniline, chromic acid, hydrocyanic acid, hydrogen sulfide, flammable liquids, flammable gases, copper, brass, any heavy metals
Nitrites	Acids
Nitroparaffins	Inorganic bases, amines
Oxalic acid	Silver, mercury
Oxygen	Oils, grease, hydrogen, flammable liquids, solids, or gases
Perchloric acid	Acetic anhydride, bismuth and its alloys, alcohol, paper, wood, grease, oils
Peroxides, organic	Acids (organic or mineral), avoid friction, store cold
Phosphorus (white)	Air, oxygen, alkalis, reducing agents
Potassium	Carbon tetrachloride, carbon dioxide, water
Potassium chlorate	Sulfuric and other acids
Potassium perchlorate (see also chlorates)	Sulfuric and other acids
Potassium permanganate	Glycerol, ethylene glycol, benzaldehyde, surfuric acid
Selenides	Reducing agents
Silver	Acetylene, oxalic acid, tartartic acid, ammonium compounds, fulminic acid
Sodium	Carbon tetrachloride, carbon dioxide, water
Sodium nitrite	Ammonium nitrate and other ammonium salts
Sodium peroxide	Ethyl or methyl alcohol, glacial acetic acid, acetic anhydride, benzaldehyde, carbon disulfide, glycerin, ethylene glycol, ethyl acetate, methyl acetate, furfural
Sulfides	Acids
Sulfuric acid	Potassium chlorate, potassium perchlorate, potassium permanganate (similar compounds of light metals, such as sodium, lithium)
Tellurides	Reducing agents

EXPLOSION HAZARDS

Table 3 lists some common classes of laboratory chemicals that have potential for producing a violent explosion when subjected to shock or friction. These chemicals should never be disposed of as such, but should be handled by procedures given in *Prudent Practices for Disposal of Chemicals from Laboratories*, National Academy Press, 1983, chapters 6 and 7. Additional information on these, as well as on some less common classes of explosives, can be found in L. Bretherick, *Handbook of Reactive Chemical Hazards*, 3rd ed., Butterworths, London–Boston, 1985.

Table 4 lists some illustrative combinations of common laboratory reagents that can produce explosions when they are brought together or that form reaction products that can explode without any apparent external initiating action. This list is not exhaustive, and additional information on potentially explosive reagent combinations can be found in *Manual of Hazardous Chemical Reactions, A Compilation of Chemical Reactions Reported to be Potentially Hazardous*, National Fire Protection Association, NFPA 491M, 1975, NFPA, 470 Atlantic Avenue, Boston, MA 02210.

WATER-REACTIVE CHEMICALS

Table 5 lists some common laboratory chemicals that react violently with water and that should always be stored and handled so that they do not come into contact with liquid water or water vapor.

Procedures for decomposing laboratory quantities are given in *Prudent Practices for Disposal of Chemicals from Laboratories*, chapter 6; the pertinent section of that chapter is given in parentheses.

PYROPHORIC CHEMICALS

Many members of the classes of readily oxidized, common laboratory chemicals listed in Table 6 ignite spontaneously in air. A more extensive list can be found in L. Bretherick, *Handbook of Reactive Chemical Hazards*, 3rd ed., Butterworths, London-Boston, 1985. Pyrophoric chemicals should be stored in tightly closed containers under an inert atmosphere (or, for some, an inert liquid),

and all transfers and manipulations of them must be carried out under an inert atmosphere or liquid. Suggested procedures for decomposing them are given in *Prudent Practices for Disposal of Chemicals from Laboratories*, chapter 6; the pertinent section of that chapter is given in parentheses.

TABLE 3. Shock-Sensitive Compounds

Acetylenic compounds, especially polyacetylenes, haloacetylenes, and heavy metal salts of acetylenes (copper, silver, and mercury salts are particularly sensitive)

Acyl nitrates

Alkyl nitrates, particularly polyol nitrates such as nitrocellulose and nitroglycerine

Alkyl and acyl nitrites

Alkyl perchlorates

Amminemetal oxosalts: metal compounds with coordinated ammonia, hydrazine, or similar nitrogenous donors and ionic perchlorate, nitrate, permanganate, or other oxidizing group

Azides, including metal, nonmetal, and organic azides

Chlorite salts of metals, such as $AgClO_2$ and $Hg(ClO_2)_2$

Diazo compounds such as CH_2N_2

Diazonium slats, when dry

Fulminates (silver fulminate, AgCNO, can form in the reaction mixture from the Tollens' test for aldehydes if it is allowed to stand for some time; this can be prevented by adding dilute nitric acid to the test mixture as soon as the test has been completed)

Hydrogen peroxide becomes increasingly treacherous as the concentration rises above 30%, forming explosive mixtures with organic materials and decomposing violently in the presence of traces of transition metals

N–Halogen compounds such as difluoroamino compounds and halogen azides

N–Nitro compounds such as N–nitromethylamine, nitrourea, nitroguanidine, and nitric amide

Oxo salts of nitrogenous bases: perchlorates, dichromates, nitrates, iodates, chlorites, chlorates, and permanganates of ammonia, amines, hydroxylamine, guanidine, etc.

Perchlorate salts. Most metal, nonmetal, and amine perchlorates can be detonated and may undergo violent reaction in contact with combustible materials

Peroxides and hydroperoxides, organic (see Chapter 6, Section II.P)

Peroxides (solid) that crystallize from or are left from evaporation of peroxidizable solvents (see Chapter 6 and Appendix I)

Peroxides, transition–metal salts

Picrates, especially salts of transition and heavy metals, such as Ni, Pb, Hg, Cu, and Zn; picric acid is explosive but is less sensitive to shock or friction than its metal salts and is relatively safe as a water–wet paste (see Chapter 7)

Polynitroalkyl compounds such as tetranitromethane and dinitroacetonitrile

Polynitroaromatic compounds, especially polynitro hydrocarbons, phenols, and amines

TABLE 4. Potentially Explosive Combinations of Some Common Reagents

Acetone + chloroform in the presence of base
Acetylene + copper, silver, mercury, or their salts
Ammonia (including aqueous solutions) + Cl_2, Br_2, or I_2
Carbon disulfide + sodium azide
Chlorine + an alcohol
Chloroform or carbon tetrachloride + powdered Al or Mg
Decolorizing carbon + an oxidizing agent
Diethyl ether + chlorine (including a chlorine atmosphere)
Dimethyl sulfoxide + an acyl halide, $SOCl_2$ or $POCl_3$
Dimethyl sulfoxide + CrO_3
Ethanol + calcium hypochlorite
Ethanol + silver nitrate
Nitric acid + acetic anhydride or acetic acid
Picric acid + a heavy–metal salt, such as of Pb, Hg, or Ag
Silver oxide + ammonia + ethanol
Sodium + a chlorinated hydrocarbon
Sodium hypochlorite + an amine

TABLE 5. Water-Reactive Chemicals

Alkali metals (III.D)
Alkali metal hydrides (III.C.2)
Alkali metal amides (III.C.7)
Metal alkyls, such as lithium alkyls and aluminum alkyls (IV.A)
Grignard reagents (IV.A)
Halides of nonmetals, such as BCl_3, BF_3, PCl_3, PCl_5, $SiCl_4$, S_2Cl_2 (III.F)
Inorganic acid halides, such as $POCl_3$, $SOCl_2$, SO_2Cl_2 (III.F)
Anhydrous metal halides, such as $AlCl_3$, $TiCl_4$, $ZrCl_4$, $SnCl_4$ (III.E)
Phosphorus pentoxide (III.I)
Calcium carbide (IV.E)
Organic acid halides and anhydrides of low molecular weight (II.J)

TABLE 6. Classes of Pyrophoric Chemicals

Grignard reagents, RMgX (IV.A)
Metal alkyls and aryls, such as RLi, RNa, R_3Al, R_2Zn (IV.A)
Metal carbonyls, such as Ni $(CO)_4$, $Fe(CO)_5$, $Co_2(CO)_8$ (IV.B)
Alkali metals such as Na, K (III.D.1)
Metal powders, such as Al, Co, Fe, Mg, Mn, Pd, Pt, Ti, Sn, Zn, Zr (III.D.2)
Metal hydrides, such as NaH, $LiAlH_4$ (IV.C.2)
Nonmetal hydrides, such as B_2H_6 and other boranes, PH_3, AsH_3 (III.G)
Nonmetal alkyls, such as R_3B, R_3P, R_3As (IV.C)
Phosphorus (white) (III.H)

HAZARDS FROM PEROXIDE FORMATION

Many common laboratory chemicals can form peroxides when allowed access to air over a period of time. A single opening of a container to remove some of the contents can introduce enough air for peroxide formation to occur. Some types of compounds form peroxides that are treacherously and violently explosive in concentrated solution or as solids. Accordingly, peroxide–containing liquids should never be evaporated near to or to dryness. Peroxide formation can also occur in many polymerizable unsaturated compounds, and these peroxides can initiate a runaway, sometimes explosive, polymerization reaction. Procedures for testing for peroxides and for removing small amounts from laboratory chemicals are given in *Prudent Practices for Disposal of Chemicals from Laboratories*, chapter 6, Section II.P.

Table 7 provides a list of structural characteristics in organic compounds that can peroxidize. These structures are listed in approximate order of decreasing hazard. Reports of serious incidents involving the last five structural types are extremely rare, but these structures are listed because laboratory workers should be aware that they can form peroxides that can influence the course of experiments in which they are used.

Table 8 gives examples of common laboratory chemicals that are prone to form peroxides on exposure to air. The lists are not exhaustive, and analogous organic compounds that have any of the structural features given in Table 7 should be tested for peroxides before being used as solvents or reagents, or before being distilled. The recommended retention times begin with the date of synthesis or of opening the original container.

DISPOSAL OF TOXIC CHEMICALS

It is often desirable to precipitate toxic cations or hazardous anions from solution to facilitate recovery or disposal. Table 9 lists precipitants for many common cations, and Table 10 gives precipitants for some hazardous anions. Many cations can be precipitated as sulfides by adding sodium sulfide solution (preferable to the highly toxic hydrogen sulfide) to a neutral solution of the cation (Table 11). Control of pH is important because some sulfides will redissolve in excess sulfide ion. After precipitation, excess sulfide can be destroyed by addition of hypochlorite.

Most metal cations are precipitated as hydroxides or oxides at high pH. Since many of these precipitates will redissolve in excess

base, it is often necessary to control pH. Table 12 shows the recommended pH range for precipitating many cations in their most common oxidation state. The notation "1 N" in the right–hand column indicates that the precipitate will not dissolve in 1 N sodium hydroxide (pH 14).

The distinctions between high and low toxicity or hazard are based on toxicological and other data, and are relative. There is no implication of a sharp distinction between high and low, or that any cations or anions are totally without hazard.

TABLE 7. Types of Chemicals That Are Prone to Form Peroxides

A. Organic structures (in approximate order of decreasing hazard)

	Structure	Description
1.	$\overset{H}{\underset{}{C}}-O-$	Ethers and acetals with α hydrogen atoms
2.	$C=C-\overset{H}{\underset{}{C}}$	Olefins with allylic hydrogen atoms
3.	$C=\overset{X}{\underset{}{C}}-$	Chloroolefins and fluoroolefins
4.	$CH_2=C$	Vinyl halides, esters, and ethers
5.	$C=C-C=C$	Dienes
6.	$C=\overset{H}{\underset{}{C}}-C\equiv CH$	Vinylacetylenes with α hydrogen atoms
7.	$\overset{H}{\underset{}{C}}-C\equiv CH$	Alkylacetylenes with α hydrogen atoms
8.	$\overset{H}{\underset{}{C}}-Ar$	Alkylarenes that contain tertiary hydrogen atoms
9.	$-C-H$	Alkanes and cycloalkanes that contain tertiary hydrogen atoms

10.   C=C—CO₂R   Acrylates and methacrylates

11.   C—OH   Secondary alcohols

12.   —C—C   Ketones that contain α hydrogen atoms

13.   —C=O   Aldehydes

14.   —C—N—C   Ureas, amides, and lactams that have a hydrogen atom on a carbon atom attached to nitrogen

B. Inorganic substances

1. Alkali metals, especially potassium, rubidium, and cesium (see Chapter 6, Section III.D)
2. Metal amides (see Chapter 6, Section III.C.7)
3. Organometallic compounds with a metal atom bonded to carbon (see Chapter 6, Section IV)
4. Metal alkoxides

TABLE 8. Common Peroxide-Forming Chemicals
LIST A
Severe Peroxide Hazard on Storage with Exposure to Air
Discard within 3 months

- Diisopropyl ether (isopropyl ether)
- Divinylacetylene (DVA)[a]
- Potassium metal
- Potassium amide
- Sodium amide (sodamide)
- Vinylidene chloride (1,1–dichloroethylene)[a]

LIST B
Peroxide Hazard on Concentration; Do Not Distill or Evaporate without First Testing for the Presence of Peroxides
Discard or Test for Peroxides after 6 Months

- Acetaldehyde diethyl acetal (acetal)
- Cumene (isopropylbenzene)
- Cyclohexene
- Cyclopentene
- Decalin (decahydronaphthalene)
- Diacetylene
- Dicyclopentadiene
- Diethyl ether (ether)
- Diethylene glycol dimethyl ether (diglyme)
- Dioxane
- Ethylene glycol dimethyl ether (glyme)
- Ethylene glycol ether acetates
- Ethylene glycol monoethers (cellosolves)
- Furan
- Methylacetylene
- Methylcyclopentane
- Methyl isobutyl ketone
- Tetrahydrofuran (THF)
- Tetralin (tetrahydronaphthalene)
- Vinyl ethers[a]

LIST C
Hazard of Rapid Polymerization Initiated by Internally Formed Peroxides[a]

a. Normal Liquids; discard or test for peroxides after 6 months[b]
- Chloroprene (2–chloro–1,3–butadiene)[c]
- Styrene
- Vinyl acetate
- Vinylpyridine

b. Normal Gases; discard after 12 months[d]
- Butadiene[c]
- Tetrafluoroethylene (TFE)[c]
- Vinylacetylene (MVA)[c]
- Vinyl chloride

[a] Polymerizable monomers should be stored with a polymerization inhibitor from which the monomer can be separated by distillation just before use.
[b] Although common acrylic monomers such as acrylonitrile, acrylic acid, ethyl acrylate, and methyl methacrylate can form peroxides, they have not been reported to develop hazardous levels in normal use and storage.
[c] The hazard from peroxides in these compounds is substantially greater when they are stored in the liquid phase, and if so stored without an inhibitor they should be considered as in LIST A.
[d] Although air will not enter a gas cylinder in which gases are stored under pressure, these gases are sometimes transferred from the original cylinder to another in the laboratory, and it is difficult to be sure that there is no residual air in the receiving cylinder. An inhibitor should be put into any such secondary cylinder before one of these gases is transferred into it; the supplier can suggest inhibitors to be used. The hazard posed by these gases is much greater if there is a liquid phase in such a secondary container, and even inhibited gases that have been put into a secondary container under conditions that create a liquid phase should be discarded within 12 months.

Note: Laboratory workers should label all containers of peroxidizable solvents or reagents with one of the following:

[LIST A]

	Peroxidizable compound	
	Received	Opened
Date	———	———

Discard 3 months after opening

[LISTS B AND C]

	Peroxidizable compound	
	Received	Opened
Date	———	———

Discard or test for peroxides
6 months after opening

TABLE 9. Relative Toxicity of Cations

High toxic hazard	Precipitant[a]	Low toxic hazard	Precipitant[a]
Antimony	OH^-, S^{2-}	Aluminum	OH^-
Arsenic	S^{2-}	Bismuth	OH^-, S^{2-}
Barium	SO_4^{2-}, CO_3^{2-}	Calcium	SO_4^{2-}, CO_3^{2-}
Beryllium	OH^-	Cerium	OH^-
Cadmium	OH^-, S^{2-}	Cesium	
Chromium (III)[b]	OH^-	Copper[c]	OH^-, S^{2-}
Cobalt (II)[b]	OH^-, S^{2-}	Gold	OH^-, S^{2-}
Gallium	OH^-	Iron[c]	OH^-, S^{2-}
Germanium	OH^-, S^{2-}	Lanthanides	OH^-
Hafnium	OH^-	Lithium	
Indium	OH^-, S^{2-}	Magnesium	OH^-
Iridium	OH^-, S^{2-}	Molybdenum (VI)[b,d]	
Lead	OH^-, S^{2-}	Niobium (V)	OH^-
Manganese (II)[b]	OH^-, S^{2-}	Palladium	OH^-, S^{2-}
Mercury	OH^-, S^{2-}	Potassium	
Nickel	OH^-, S^{2-}	Rubidium	
Osmium (IV)[b,e]	OH^-, S^{2-}	Scandium	OH^-
Platinum (II)[b]	OH^-, S^{2-}	Sodium	
Rhenium (VII)[b]	S^{2-}	Strontium	SO_4^{2-} CO_3^{2-}
Rhodium (III)[b]	OH^-, S^{2-}	Tantalum	OH^-
Ruthenium (III)[b]	OH^-, S^{2-}	Tin	OH^-, S^{2-}
Selenium	S^{2-}	Titanium	OH^-
Silver	Cl^-, OH^-, S^{2-}	Yttrium	OH^-
Tellurium	S^{2-}	Zinc[c]	OH^-, S^{2-}
Thallium	OH^-, S^{2-}	Zirconium	OH^-
Tungsten (VI)[b,d]			
Vanadium	OH^-, S^{2-}		

[a] Precipitants are listed in order of preference:
  OH^- = base (sodium hydroxide or sodium carbonate)
  S^{2-} = sulfide
  Cl^- = chloride
  SO_4^{2-} = sulfate
  CO_3^{2-} = carbonate
[b] The precipitant is for the indicated valence state.
[c] Maximum tolerance levels have been set for these low–toxicity ions by the U.S. Public Health Service, and large amounts should not be put into public sewer systems. The small amounts typically used in laboratories will not normally affect water supplies.
[d] These ions are best precipitated as calcium molybdate or calcium tungstate.
[e] CAUTION: OsO_4, a volatile, extremely poisonous substance, is formed from almost any osmium compound under acid conditions in the presence of air.

TABLE 10. Relative Hazard of Anions

High-hazard anions			Low-hazard anions
Ion	Hazard type[a]	Precipitant	
Aluminum hydride, AlH_4^-	F	—	Bisulfite, HSO_3^-
Amide, NH_2^-	F,E[b]	—	Borate, BO_3^{3-}, $B_4O_7^{2-}$
Arsenate, AsO_3^-, AsO_4^{3-}	T	Cu^{2+}, Fe^{2+}	Bromide, Br^-
Arsenite, AsO_2^-, AsO_3^{3-}	T	Pb^{2+}	Carbonate, CO_3^{2-}
Azide, N_3^-	E, T	—	Chloride, Cl^-
Borohydride, BH_4^-	F	—	Cyanate, OCN^-
Bromate, BrO_3^-	O, E	—	Hydroxide, OH^-
Chlorate, ClO_3^-	O, E	—	Iodide, I^-
Chromate, CrO_4^{2-}, $Cr_2O_7^{2-}$	T, O	[c]	Oxide, O^{2-}
Cyanide, CN^-	T	—	Phosphate, PO_4^{3-}
Ferricyanide, $Fe(CN)_6^{3-}$	T	Fe^{2+}	Sulfate, SO_4^2
Ferrocyanide, $Fe(CN)_6^{4-}$	T	Fe^{3+}	Sulfite, SO_3^{2-}
Fluoride, F^-	T	Ca^{2+}	Thiocyanate, SCN^-
Hydride, H^-	F	—	
Hydroperoxide, O_2H^-	O, E	—	
Hydrosulfide, SH^-	T	—	
Hypochlorite, OCl^-	O	—	
Iodate, IO_3^-	O, E	—	
Nitrate, NO_3^-	O	—	
Nitrite, NO_2^-	T, O	—	
Perchlorate, ClO_4^-	O, E	—	
Permanganate, MnO_4^-	T, O	[d]	
Peroxide, O_2^{2-}	O, E	—	
Persulfate, $S_2O_8^{2-}$	O	—	
Selenate, SeO_4^{2-}	T	Pb^{2+}	
Selenide, Se^{2-}	T	Cu^{2+}	
Sulfide, S^{2-}	T	[e]	

[a] Toxic, T: oxidant, O; flammable, F; explosive, E.
[b] Metal amides readily form explosive peroxides on exposure to air.
[c] Reduce and precipitate as Cr(III); see Table 9.
[d] Reduce and precipitate as Mn(II); see Table 9.
[e] See Table 11.

TABLE 11. Precipitation of Sulfides

Precipitated at pH 7	Not precipitated at low pH	Forms a soluble complex at high pH
Ag^+		
As^{3+a}		X
Au^{+a}		X
Bi^{3+}		
Cd^{2+}		
Co^{2+}	X	
Cr^{3+a}		
Cu^{2+}		
Fe^{2+a}	X	
Ge^{2+}		X
Hg^{2+}		X
In^{3+}	X	
Ir^{4+}		X
Mn^{2+a}	X	
Mo^{3+}		X
Ni^{2+a}	X	
Os^{4+}		
Pb^{2+}		
Pd^{2+a}		
Pt^{2+a}		X
Re^{4+}		
Rh^{2+a}		
Ru^{4+}		

TABLE 11. Precipitation of Sulfides

Precipitated at pH 7	Not precipitated at low pH	Forms a soluble complex at high pH
Sb^{3+a}		X
Se^{2+}		X
Sn^{2+}		X
Te^{4+}		X
Tl^{+a}	X	
V^{4+a}	X	
Zn^{2+}	X	

ᵃ Higher oxidation states of this ion are reduced by sulfide ion and precipitated as this sulfide.

TABLE 12. pH Range for Precipitation of Metal Hydroxides and Oxides

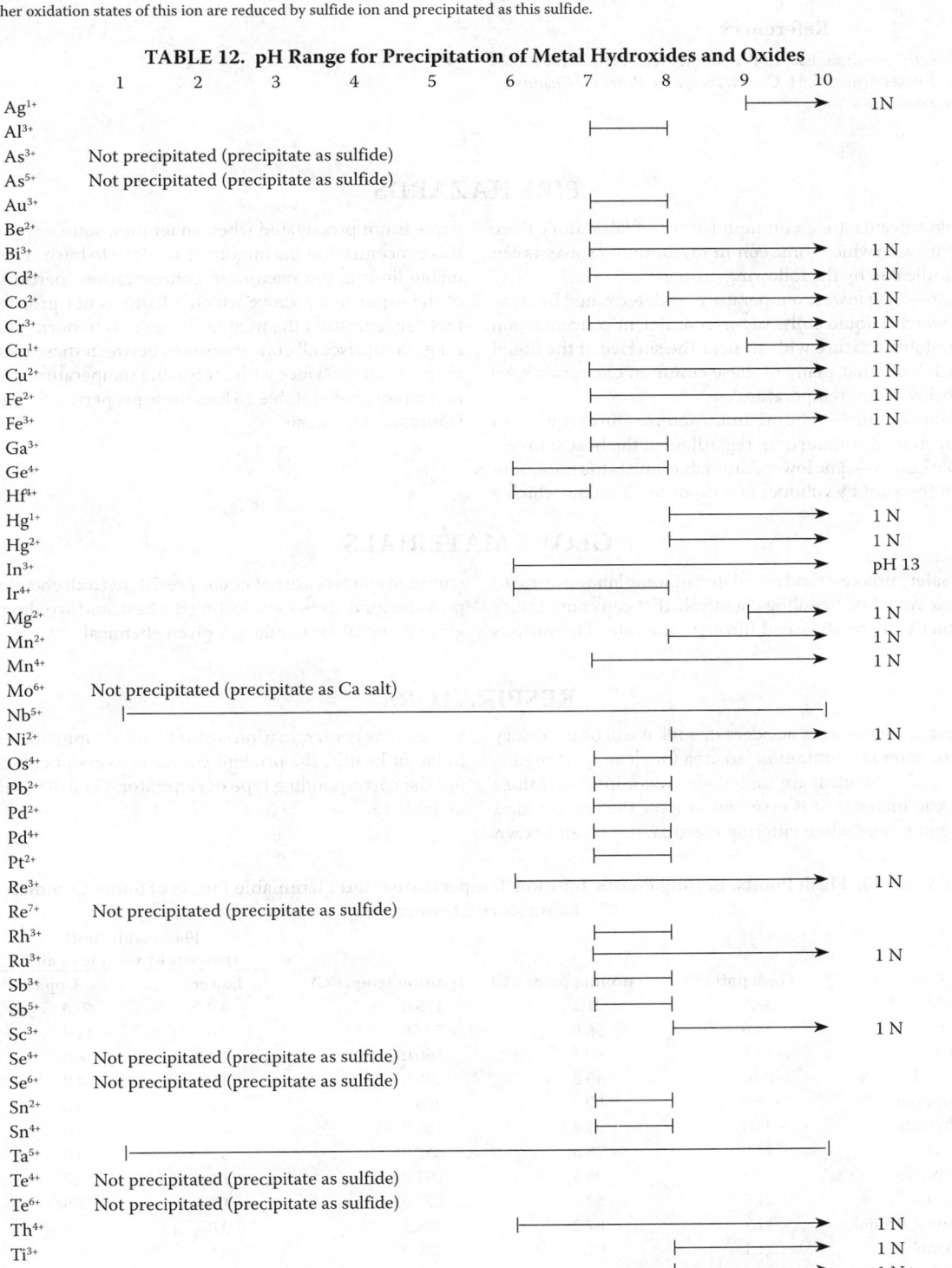

Ion	pH range for precipitation
Ag^{1+}	9 → 1 N
Al^{3+}	7–8
As^{3+}	Not precipitated (precipitate as sulfide)
As^{5+}	Not precipitated (precipitate as sulfide)
Au^{3+}	7–8
Be^{2+}	7 → 1 N
Bi^{3+}	7 → 1 N
Cd^{2+}	7 → 1 N
Co^{2+}	8 → 1 N
Cr^{3+}	7 → 1 N
Cu^{1+}	7 → 1 N
Cu^{2+}	7 → 1 N
Fe^{2+}	7 → 1 N
Fe^{3+}	7 → 1 N
Ga^{3+}	7–8
Ge^{4+}	6–8
Hf^{4+}	6–7
Hg^{1+}	8 → 1 N
Hg^{2+}	8 → 1 N
In^{3+}	6 → pH 13
Ir^{4+}	6–8
Mg^{2+}	9 → 1 N
Mn^{2+}	8 → 1 N
Mn^{4+}	7 → 1 N
Mo^{6+}	Not precipitated (precipitate as Ca salt)
Nb^{5+}	1–10
Ni^{2+}	8 → 1 N
Os^{4+}	7–8
Pb^{2+}	7–8
Pd^{2+}	7–8
Pd^{4+}	7–8
Pt^{2+}	7–8
Re^{3+}	6 → 1 N
Re^{7+}	Not precipitated (precipitate as sulfide)
Rh^{3+}	7–8
Ru^{3+}	7 → 1 N
Sb^{3+}	7–8
Sb^{5+}	7–8
Sc^{3+}	8 → 1 N
Se^{4+}	Not precipitated (precipitate as sulfide)
Se^{6+}	Not precipitated (precipitate as sulfide)
Sn^{2+}	7–8
Sn^{4+}	7–8
Ta^{5+}	1–10
Te^{4+}	Not precipitated (precipitate as sulfide)
Te^{6+}	Not precipitated (precipitate as sulfide)
Th^{4+}	6 → 1 N
Ti^{3+}	8 → 1 N
Ti^{4+}	8 → 1 N

TABLE 12. pH Range for Precipitation of Metal Hydroxides and Oxides

	1	2	3	4	5	6	7	8	9	10	
Tl^{3+}									→→→		1 N
V^{4+}							⊢——⊣				
V^{5+}							⊢——⊣				
W^{6+}	Not precipitated (precipitate as Ca salt)										
Zn^{2+}							⊢——⊣				
Zr^{4+}						⊢——⊣					

References

L. Erdey, *Gravimetric Analysis*, Part II, Pergamon Press, New York, 1965.
D. T. Burns, A. Towsend, and A. H. Carter, *Inorganic Reaction Chemistry*,
Vol. 2, Ellis Horwood, New York, 1981.

FIRE HAZARDS

Flammable solvents are a common source of laboratory fires. The relative ease with which some common laboratory solvents can be ignited is indicated by the following properties.

Flash Point — The lowest temperature, as determined by standard tests, at which a liquid emits vapor in sufficient concentration to form an ignitable mixture with air near the surface of the liquid in a test vessel. Note that many of these common chemicals have flash points below room temperature.

Ignition Temperature — The minimum temperature required to initiate self–sustained combustion, regardless of the heat source.

Flammable Limits — The lower flammable limit is the minimum concentration (percent by volume) of a vapor in air below which a flame is not propagated when an ignition source is present. Below this concentration the mixture is too lean to burn. The upper flammable limit is the maximum concentration (percent by volume) of the vapor in air above which a flame is not propagated. Above this concentration the mixture is too rich to burn. The flammable range comprises all concentrations between these two limits. This range becomes wider with increasing temperature and in oxygen-rich atmospheres. Table 13 lists these properties for a few common laboratory chemicals.

GLOVE MATERIALS

It is good safety practice (and mandated in some laboratories) to wear rubber gloves while handling chemicals that can cause injury when in contact with, or absorbed through, the skin. The various common rubbers are not equally resistant to all chemicals. Table 14 provides guidelines for selecting the best, and avoiding the poorest, glove material for handling a given chemical.

RESPIRATORS

In the event of a laboratory accident or spill, it will be necessary for someone to enter the contaminated area for cleanup. If significant quantities of a chemical are spilled, or even minor quantities of a known toxic material, it is essential to wear the correct kind of respirator equipment when entering the area. If it is not known whether the contamination is of a chemical "immediately dangerous to life or health", the prudent course is to assume that it is, and to use the corresponding type of respirator. Guidelines are presented in Table 15.

TABLE 13. Flash Points, Boiling Points, Ignition Temperatures, and Flammable Limits of Some Common Laboratory Chemicals

Chemical	Flash point (°C)	Boiling point (°C)	Ignition temp. (°C)	Flammable limit (percent by volume in air)	
				Lower	Upper
Acetaldehyde	−37.8	21.1	175.0	4.0	60.0
Acetone	−19.0	56.0	538.0	2.6	12.8
Benzene	−11.1	80.1	560.0	1.4	8.0
Carbon disulfide	−30.0	45.8	90.0	1.0	44.0
Cyclohexane	−18.0	80.7	260.0	1.3	8.0
Diethyl ether	−45.0	34.4	160.0	1.8	48.0
Ethanol	12.0	78.3	363.0	3.3	19.0
n–Heptane	−3.9	98.4	204.0	1.0	6.7
n–Hexane	−21.7	68.7	223.0	1.2	7.5
Isopropyl alcohol	11.7	82.2	398.9	2.0	12.0
Methanol	11.1	64.5	385.0	6.0	36.5
Methyl ethyl ketone	−6.1	79.6	515.6	1.9	11.0

TABLE 13. Flash Points, Boiling Points, Ignition Temperatures, and Flammable Limits of Some Common Laboratory Chemicals

Chemical	Flash point (°C)	Boiling point (°C)	Ignition temp. (°C)	Flammable limit (percent by volume in air)	
				Lower	Upper
Pentane	−40.0	36.1	260.0	1.4	7.8
Styrene	31.0	145.0	490.0	1.1	6.1
Toluene	4.4	110.6	530.0	1.3	7.0
p–Xylene	25.0	132.4	529.0	1.1	7.0

Note: For a more extensive listing, see the table "Properties of Common Solvents" in Section 15.

TABLE 14. Resistance to Chemicals of Common Glove Materials (E = Excellent, G = Good, F = Fair, P = Poor)

Chemical	Natural rubber	Neoprene	Nitrile	Vinyl
Acetaldehyde	G	G	E	G
Acetic acid	E	E	E	E
Acetone	G	G	G	F
Acrylonitrile	P	G	—	F
Ammonium hydroxide (sat)	G	E	E	E
Aniline	F	G	E	G
Benzaldehyde	F	F	E	G
Benzene[a]	P	F	G	F
Benzyl chloride[a]	F	P	G	P
Bromine	G	G	—	G
Butane	P	E	—	P
Butyraldehyde	P	G	—	G
Calcium hypochlorite	P	G	G	G
Carbon disulfide	P	P	G	F
Carbon tetrachloride[a]	P	F	G	F
Chlorine	G	G	—	G
Chloroacetone	F	E	—	P
Chloroform[a]	P	F	G	P
Chromic acid	P	F	F	E
Cyclohexane	F	E	—	P
Dibenzyl ether	F	G	—	P
Dibutyl phthalate	F	G	—	P
Diethanolamine	F	E	—	E
Diethyl ether	F	G	E	P
Dimethyl sulfoxide[b]	—	—	—	—
Ethyl acetate	F	G	G	F
Ethylene dichloride[a]	P	F	G	
Ethylene glycol	G	G	E	E
Ethylene trichloride[a]	P	P	—	P
Fluorine	G	G	—	G
Formaldehyde	G	E	E	E
Formic acid	G	E	E	E
Glycerol	G	G	E	E
Hexane	P	E	—	P
Hydrobromic acid (40%)	G	E	—	E
Hydrochloric acid (conc)	G	G	G	E
Hydrofluoric acid (30%)	G	G	G	E
Hydrogen peroxide	G	G	G	E
Iodine	G	G	—	G
Methylamine	G	G	E	E
Methyl cellosolve	F	E	—	P
Methyl chloride[a]	P	E	—	P
Methyl ethyl ketone	F	G	G	P
Methylene chloride[a]	F	F	G	F
Monoethanolamine	F	E	—	E
Morpholine	F	E	—	E
Naphthalene[a]	G	G	E	G
Nitric acid (conc)	P	P	P	G
Perchloric acid	F	G	F	E

TABLE 14. Resistance to Chemicals of Common Glove Materials (E = Excellent, G = Good, F = Fair, P = Poor)

Chemical	Natural rubber	Neoprene	Nitrile	Vinyl
Phenol	G	E	—	E
Phosphoric acid	G	E	—	E
Potassium hydroxide (sat)	G	G	G	E
Propylene dichloride[a]	P	F	—	P
Sodium hydroxide	G	G	G	E
Sodium hypochlorite	G	P	F	G
Sulfuric acid (conc)	G	G	F	G
Toluene[a]	P	F	G	F
Trichloroethylene[a]	P	F	G	F
Tricresyl phosphate	P	F	—	F
Triethanolamine	F	E	E	E
Trinitrotoluene	P	E	—	P

[a] Aromatic and halogenated hydrocarbons will attack all types of natural and synthetic glove materials. Should swelling occur, the user should change to fresh gloves and allow the swollen gloves to dry and return to normal.

[b] No data on the resistance to dimethyl sulfoxide of natural rubber, neoprene, nitrile rubber, or vinyl materials are available; the manufacturer of the substance recommends the use of butyl rubber gloves.

TABLE 15. Guide for Selection of Respirators

Type of hazard	Type of respirator
Oxygen deficiency	Self-contained breathing apparatus
	Hose mask with blower
	Combination of air-line respirator and auxiliary self-contained air supply or air-storage receiver with alarm
Gas and vapor contaminants	
Immediately dangerous to life or health	Self-contained breathing apparatus
	Hose mask with blower
	Air-purifying full-facepiece respirator with chemical canister (gas mask)
	Self-rescue mouthpiece respirator (for escape only)
	Combination of air-line respirator and auxiliary self-contained air supply or air-storage receiver with alarm
Not immediately dangerous to life or health	Air-line respirator
	Hose mask with blower
	Air-purifying half-mask or mouthpiece respirator with chemical cartridge
Particulate Contaminants	
Immediately dangerous to life or health	Self-contained breathing apparatus
	Hose mask with blower
	Air-purifying full-facepiece respirator with appropriate filter
	Self-rescue mouthpiece respirator (for escape only)
	Combination of air-line respirator and auxiliary self-contained air supply or air-storage receiver with alarm
Not immediately dangerous to life or health	Air-purifying half-mask or mouthpiece respirator with filter pad or cartridge
	Air-line respirator
	Air-line abrasive-blasting respirator
	Hose mask with blower
Combination of gas, vapor, and particulate contaminants	
Immediately dangerous to life or health	Self-contained breathing apparatus
	Hose mask with blower
	Air-purifying full-facepiece respirator with chemical canister and appropriate filter (gas mask with filter)
	Self-rescue mouthpiece respirator (for escape only)
	Combination of air-line respirator and auxiliary self-contained air supply or air-storage receiver with alarm
Not immediately dangerous to life or health	Air-line respirator
	Hose mask without blower
	Air-purifying half-mask or mouthpiece respirator with chemical cartridge and appropriate filter

Source: ANSI Standard Z88.2 (1969).

FLAMMABILITY OF CHEMICAL SUBSTANCES

This table gives properties related to the flammability of about 900 chemical substances. The properties listed are:

t_B: Normal boiling point in °C (at 101.325 kPa pressure).

FP: Flash point, which is the minimum temperature at which the vapor pressure of a liquid is sufficient to form an ignitable mixture with air near the surface of the liquid. Flash point is not an intrinsic physical property but depends on the conditions of measurement (see Reference 1).

Fl. Limits: Flammable limits (often called explosive limits), which specify the range of concentration of the vapor in air (in percent by volume) for which a flame can propagate. Below the lower flammable limit, the gas mixture is too lean to burn; above the upper flammable limit, the mixture is too rich. Values refer to ambient temperature and pressure and are dependent on the precise test conditions. A ? indicates that one of the limits is not known.

IT: Ignition temperature (sometimes called autoignition temperature), which is the minimum temperature required for self-sustained combustion in the absence of an external ignition source. As in the case of flash point, the value depends on specified test conditions.

Even in cases where very careful measurements of flash point have been replicated in several laboratories, observed values can differ by 3 to 6°C (Reference 4). For more typical measurements, larger uncertainties should be assumed in both flash points and autoignition temperatures. The absence of a flash point entry in this table does not mean that the substance is nonflammable, but only that no reliable value is available.

Compounds are listed by molecular formula following the Hill convention. Substances not containing carbon are listed first, followed by those that contain carbon. To locate an organic compound by name or CAS Registry Number when the molecular formula is not known, use the table "Physical Constants of Organic Compounds" in Section 3 and its indexes to determine the molecular formula.

References

1. *Fire Protection Guide to Hazardous Materials, 11th Edition*, National Fire Protection Association, Quincy, MA, 1994.
2. Urben, P. G., Ed., *Bretherick's Handbook of Reactive Chemical Hazards, 5th Edition*, Butterworth-Heinemann, Oxford, 1995.
3. Daubert, T. E., Danner, R. P., Sibul, H. M., and Stebbins, C. C., *Physical and Thermodynamic Properties of Pure Compounds: Data Compilation*, extant 1994 (core with 4 supplements), Taylor & Francis, Bristol, PA.
4. *Report of Investigation: Flash Point Reference Materials*, National Institute of Standards and Technology, Standard Reference Materials Program, Gaithersburg, MD, 1995.

Mol. form.	Name	t_B/°C	FP/°C	Fl. limits	IT/°C
		Compounds not containing carbon			
B_2H_6	Diborane	−92.4	−90	1–98%	≈40
B_5H_9	Pentaborane(9)	60	30	0.4–?	35
BrH_3Si	Bromosilane	1.9	<0		≈20
Br_3HSi	Tribromosilane	109			≈20
Cl_2H_2Si	Dichlorosilane	8.3		4.1–99%	36
Cl_3HSi	Trichlorosilane	33	−50		104
GeH_4	Germane	−88.1			≈20
Ge_2H_6	Digermane	29			≈50
H_2	Hydrogen	−252.8		4–74%	
H_2S	Hydrogen sulfide	−59.55		4–44%	260
H_2S_2	Hydrogen disulfide	70.7	<22		
H_2Te	Hydrogen telluride	−2			−50
H_3N	Ammonia	−33.33		16–25%	
H_3P	Phosphine	−87.75		1.8–?	
H_4N_2	Hydrazine	113.55	38	5–100%	
H_4P_2	Diphosphine	63.5			≈20
H_4Si	Silane	−111.9	−112	1.4–?	≈20
H_6Si_2	Disilane	−14.3	−14		≈20
H_8Si_3	Trisilane	52.9	<0		≈20
P	Phosphorus (white)	280.5			38
		Compounds containing carbon			
CHN	Hydrogen cyanide	26	−18	6–40%	538
CH_2Cl_2	Dichloromethane	40		13–23%	556
CH_2N_2	Cyanamide		141		
CH_2O	Formaldehyde	−19.1	85	7.0–73%	424
$(CH_2O)_x$	Paraformaldehyde		70	7.0–73%	300
CH_2O_2	Formic acid	101	50	18–57%	434
CH_3Br	Bromomethane	3.5		10–16%	537
CH_3Cl	Chloromethane	−24.0		8.1–17.4%	632

Mol. form.	Name	t_B/°C	FP/°C	Fl. limits	IT/°C
CH$_3$Cl$_3$Si	Methyltrichlorosilane	65.6	−9	7.6−>20%	>404
CH$_3$NO	Formamide	220	154		
CH$_3$NO$_2$	Nitromethane	101.1	35	7.3−?	418
CH$_4$	Methane	−161.5		5.0−15.0%	537
CH$_4$Cl$_2$Si	Dichloromethylsilane	41	−9	6.0−55%	316
CH$_4$O	Methanol	64.6	11	6.0−36%	464
CH$_4$S	Methanethiol	5.9	−18	3.9−21.8%	
CH$_5$N	Methylamine	−6.3	0	4.9−20.7%	430
CH$_6$N$_2$	Methylhydrazine	87.5	−8	2.5−92%	194
CO	Carbon monoxide	−191.5		12.5−74%	609
COS	Carbon oxysulfide	−50		12−29%	
CS$_2$	Carbon disulfide	46	−30	1.3−50.0%	90
C$_2$ClF$_3$	Chlorotrifluoroethylene	−27.8		8.4−16.0%	
C$_2$F$_4$	Tetrafluoroethylene	−75.9		10.0−50.0%	200
C$_2$HCl$_3$	Trichloroethylene	87.2		8−10.5%	420
C$_2$HCl$_3$O	Dichloroacetyl chloride	108	66		
C$_2$H$_2$	Acetylene	−84.7		2.5−100%	305
C$_2$H$_2$Cl$_2$	1,1-Dichloroethylene	31.6	−28	6.5−15.5%	570
C$_2$H$_2$Cl$_2$	cis-1,2-Dichloroethylene	60.1	6	3−15%	460
C$_2$H$_2$Cl$_2$	trans-1,2-Dichloroethylene	48.7	2	6−13%	460
C$_2$H$_2$F$_2$	1,1-Difluoroethylene	−85.7		5.5−21.3%	
C$_2$H$_3$Br	Bromoethylene	15.8		9−15%	530
C$_2$H$_3$Cl	Chloroethylene	−13.3	−78	3.6−33.0%	472
C$_2$H$_3$ClF$_2$	1-Chloro-1,1-difluoroethane	−9.7		6−18%	632
C$_2$H$_3$ClO	Acetyl chloride	50.7	4		390
C$_2$H$_3$Cl$_2$NO$_2$	1,1-Dichloro-1-nitroethane	123.5	76		
C$_2$H$_3$Cl$_3$	1,1,1-Trichloroethane	74.0		8−10.5%	500
C$_2$H$_3$Cl$_3$	1,1,2-Trichloroethane	113.8	32	6−28%	460
C$_2$H$_3$Cl$_3$Si	Trichlorovinylsilane	91.5	21		
C$_2$H$_3$F	Fluoroethylene	−72		2.6−21.7%	
C$_2$H$_3$N	Acetonitrile	81.6	6	3.0−16.0%	524
C$_2$H$_3$NO	Methyl isocyanate	39.5	−7	5.3−26%	534
C$_2$H$_4$	Ethylene	−103.7		2.7−36%	450
C$_2$H$_4$ClNO$_2$	1-Chloro-1-nitroethane	124.5	56		
C$_2$H$_4$Cl$_2$	1,1-Dichloroethane	57.4	−17	5.4−11.4%	458
C$_2$H$_4$Cl$_2$	1,2-Dichloroethane	83.5	13	6.2−16%	413
C$_2$H$_4$O	Acetaldehyde	20.1	−39	4.0−60%	175
C$_2$H$_4$O	Ethylene oxide	10.6	−20	3.0−100%	429
C$_2$H$_4$O$_2$	Acetic acid	117.9	39	4.0−19.9%	463
C$_2$H$_4$O$_2$	Methyl formate	31.7	−19	4.5−23%	449
C$_2$H$_4$O$_3$	Ethaneperoxoic acid	110	41		
C$_2$H$_5$Br	Bromoethane	38.5		6.8−8.0%	511
C$_2$H$_5$Cl	Chloroethane	12.3	−50	3.8−15.4%	519
C$_2$H$_5$ClO	Ethylene chlorohydrin	128.6	60	4.9−15.9%	425
C$_2$H$_5$Cl$_3$Si	Trichloroethylsilane	100.5	22		
C$_2$H$_5$N	Ethyleneimine	56	−11	3.3−54.8%	320
C$_2$H$_5$NO$_2$	Nitroethane	114.0	28	3.4−17%	414
C$_2$H$_5$NO$_2$	Ethyl nitrite	18	−35	4.0−50%	90
C$_2$H$_5$NO$_3$	Ethyl nitrate	87.2	10	4−?	
C$_2$H$_6$	Ethane	−88.6		3.0−12.5%	472
C$_2$H$_6$Cl$_2$Si	Dichlorodimethylsilane	70.3	<21	3.4−9.5%	
C$_2$H$_6$O	Ethanol	78.2	13	3.3−19%	363
C$_2$H$_6$O	Dimethyl ether	−24.8	−41	3.4−27.0%	350
C$_2$H$_6$OS	2-Mercaptoethanol	158	74		
C$_2$H$_6$OS	Dimethyl sulfoxide	189	95	2.6−42%	215
C$_2$H$_6$O$_2$	Ethylene glycol	197.3	111	3.2−22%	398
C$_2$H$_6$O$_4$S	Dimethyl sulfate		83		188
C$_2$H$_6$S	Ethanethiol	35.1	−17	2.8−18.0%	300
C$_2$H$_6$S	Dimethyl sulfide	37.3	−37	2.2−19.7%	206
C$_2$H$_6$S$_2$	Dimethyl disulfide	109.8	24		
C$_2$H$_7$N	Ethylamine	16.5	−16	3.5−14%	385

Mol. form.	Name	$t_B/°C$	FP/°C	Fl. limits	IT/°C
C_2H_7N	Dimethylamine	6.8	20	2.8–14.4%	400
C_2H_7NO	Ethanolamine	171	86	3.0–23.5%	410
$C_2H_8N_2$	1,2-Ethanediamine	117	40	2.5–12.0%	385
$C_2H_8N_2$	1,1-Dimethylhydrazine	63.9	−15	2–95%	249
C_2N_2	Cyanogen	−21.1		6.6–32%	
C_3H_3Br	3-Bromo-1-propyne	89	10	3.0–?	324
C_3H_3N	2-Propenenitrile	77.3	0	3.0–17.0%	481
C_3H_4	Propyne	−23.2		2.1–12.5%	
C_3H_4ClN	3-Chloropropanenitrile	175.5	76		
$C_3H_4Cl_2$	2,3-Dichloropropene	94	15	2.6–7.8%	
C_3H_4O	Propargyl alcohol	113.6	36		
C_3H_4O	Acrolein	52.6	−26	2.8–31%	220
$C_3H_4O_2$	Propenoic acid	141	50	2.4–8.0%	438
$C_3H_4O_2$	2-Oxetanone	162	74	2.9–?	
$C_3H_4O_3$	Ethylene carbonate	248	143		
C_3H_5Br	3-Bromopropene	70.1	−1	4.4–7.3%	295
C_3H_5Cl	2-Chloropropene	22.6	−37	4.5–16%	
C_3H_5Cl	3-Chloropropene	45.1	−32	2.9–11.1%	485
C_3H_5ClO	Epichlorohydrin	118	31	3.8–21.0%	411
C_3H_5ClO	Propanoyl chloride	80	12		
$C_3H_5ClO_2$	2-Chloropropanoic acid	185	107		500
$C_3H_5ClO_2$	Ethyl chloroformate	95	16		500
$C_3H_5ClO_2$	Methyl chloroacetate	129.5	57	7.5–18.5%	
$C_3H_5Cl_2NO_2$	1,1-Dichloro-1-nitropropane	145	66		
$C_3H_5Cl_3$	1,2,3-Trichloropropane	157	71	3.2–12.6%	
$C_3H_5Cl_3Si$	Trichloro-2-propenylsilane	117.5	35		
C_3H_5N	Propanenitrile	97.1	2	3.1–14%	512
C_3H_5NO	3-Hydroxypropanenitrile	221	129		
$C_3H_5N_3O_9$	Trinitroglycerol				270
C_3H_6	Propene	−47.6		2.0–11.1%	455
C_3H_6	Cyclopropane	−32.8		2.4–10.4%	498
$C_3H_6ClNO_2$	1-Chloro-1-nitropropane	142	62		
$C_3H_6ClNO_2$	2-Chloro-2-nitropropane		57		
$C_3H_6Cl_2$	1,2-Dichloropropane	96.4	21	3.4–14.5%	557
$C_3H_6Cl_2O$	1,3-Dichloro-2-propanol	176	74		
$C_3H_6N_2$	Dimethylcyanamide	163.5	71		
C_3H_6O	Allyl alcohol	97.0	21	2.5–18.0%	378
C_3H_6O	Methyl vinyl ether	5.5			287
C_3H_6O	Propanal	48	−30	2.6–17%	207
C_3H_6O	Acetone	56.0	−20	2.5–12.8%	465
C_3H_6O	Methyloxirane	35	−37	3.1–27.5%	449
$C_3H_6O_2$	Propanoic acid	141.1	52	2.9–12.1%	465
$C_3H_6O_2$	Ethyl formate	54.4	−20	2.8–16.0%	455
$C_3H_6O_2$	Methyl acetate	56.8	−10	3.1–16%	454
$C_3H_6O_2$	1,3-Dioxolane	78	2		
$C_3H_6O_3$	Dimethyl carbonate	90.5	19		
$C_3H_6O_3$	1,3,5-Trioxane	114.5	45	3.6–29%	414
C_3H_7Br	1-Bromopropane	71.1			490
C_3H_7Cl	1-Chloropropane	46.5	<−18	2.6–11.1%	520
C_3H_7Cl	2-Chloropropane	35.7	−32	2.8–10.7%	593
C_3H_7ClO	2-Chloro-1-propanol	133.5	52		
C_3H_7ClO	1-Chloro-2-propanol	127	52		
$C_3H_7Cl_3Si$	Trichloropropylsilane	123.5	37		
C_3H_7N	Allylamine	53.3	−29	2.2–22%	374
C_3H_7NO	*N,N*-Dimethylformamide	153	58	2.2–15.2%	445
$C_3H_7NO_2$	1-Nitropropane	131.1	36	2.2–?	421
$C_3H_7NO_2$	2-Nitropropane	120.2	24	2.6–11.0%	428
$C_3H_7NO_3$	Propyl nitrate	110	20	2–100%	175
C_3H_8	Propane	−42.1	−104	2.1–9.5%	450
C_3H_8O	1-Propanol	97.2	23	2.2–13.7%	412
C_3H_8O	2-Propanol	82.3	12	2.0–12.7%	399

Mol. form.	Name	$t_B/°C$	FP/°C	Fl. limits	IT/°C
C₃H₈O	Ethyl methyl ether	7.4	−37	2.0–10.1%	190
C₃H₈O₂	1,2-Propylene glycol	187.6	99	2.6–12.5%	371
C₃H₈O₂	1,3-Propylene glycol	214.4			400
C₃H₈O₂	Ethylene glycol monomethyl ether	124.1	39	1.8–14%	285
C₃H₈O₂	Dimethoxymethane	42	−32	2.2–13.8%	237
C₃H₈O₃	Glycerol	290	199	3–19%	370
C₃H₉BO₃	Trimethyl borate	67.5	−8		
C₃H₉ClSi	Trimethylchlorosilane	60	−28		395
C₃H₉N	Propylamine	47.2	−37	2.0–10.4%	318
C₃H₉N	Isopropylamine	31.7	−37		402
C₃H₉N	Trimethylamine	2.8	−5	2.0–11.6%	190
C₃H₉NO	3-Amino-1-propanol	187.5	80		
C₃H₉NO	1-Amino-2-propanol	159.4	77		374
C₃H₉NO	N-Methyl-2-ethanolamine	158	74		
C₃H₉O₃P	Trimethyl phosphite	111.5	54		
C₃H₉O₄P	Trimethyl phosphate	197.2	107		
C₃H₁₀N₂	1,3-Propanediamine	139.8	24		
C₄Cl₆	Hexachloro-1,3-butadiene	215			610
C₄H₂O₃	Maleic anhydride	202	102	1.4–7.1%	477
C₄H₄	1-Buten-3-yne	5.1		21–100%	
C₄H₄N₂	Succinonitrile	266	132		
C₄H₄O	Furan	31.5	−36	2.3–14.3%	
C₄H₄O₂	Diketene	126.1	34		
C₄H₄S	Thiophene	84.0	−1		
C₄H₅Cl	2-Chloro-1,3-butadiene	59.4	−20	4.0–20.0%	
C₄H₅N	2-Butenenitrile	120.5	16		
C₄H₅N	Methylacrylonitrile	90.3	1	2–6.8%	
C₄H₅N	Pyrrole	129.7	39		
C₄H₆	1,3-Butadiene	−4.4		2.0–12.0%	420
C₄H₆	2-Butyne	26.9	−31	1.4–?	
C₄H₆O	Divinyl ether	28.3	<−30	1.7–27%	360
C₄H₆O	Ethoxyacetylene	50	<−7		
C₄H₆O	trans-2-Butenal	102.2	13	2.1–15.5%	232
C₄H₆O	3-Buten-2-one	81.4	−7	2.1–15.6%	491
C₄H₆O	Vinyloxirane	68	<−50		
C₄H₆O₂	Methacrylic acid	162.5	77	1.6–8.8%	68
C₄H₆O₂	Vinyl acetate	72.5	−8	2.6–13.4%	402
C₄H₆O₂	Methyl acrylate	80.7	−3	2.8–25%	468
C₄H₆O₂	2,3-Butanedione	88	27		
C₄H₆O₂	γ-Butyrolactone	204	98		
C₄H₆O₃	Acetic anhydride	139.5	49	2.7–10.3%	316
C₄H₆O₃	Propylene carbonate	242	135		
C₄H₆O₆	L-Tartaric acid		210		425
C₄H₇Br	1-Bromo-2-butene	104.5		4.6–12.0%	
C₄H₇BrO₂	Ethyl bromoacetate	168.5	48		
C₄H₇Cl	2-Chloro-1-butene	58.5	−19	2.3–9.3%	
C₄H₇Cl	3-Chloro-2-methylpropene	71.5	−12	3.2–8.1%	
C₄H₇ClO	2-Chloroethyl vinyl ether	108	27		
C₄H₇ClO₂	Ethyl chloroacetate	144.3	64		
C₄H₇N	Butanenitrile	117.6	24	1.6–?	501
C₄H₇N	2-Methylpropanenitrile	103.9	8		482
C₄H₇NO	Acetone cyanohydrin		74	2.2–12.0%	688
C₄H₇NO	2-Pyrrolidone	251	129		
C₄H₈	1-Butene	−6.2		1.6–10.0%	385
C₄H₈	cis-2-Butene	3.7		1.7–9.0%	325
C₄H₈	trans-2-Butene	0.8		1.8–9.7%	324
C₄H₈	Isobutene	−6.9		1.8–9.6%	465
C₄H₈	Cyclobutane	12.6	<10	1.8–?	
C₄H₈Cl₂	1,2-Dichlorobutane	124.1			275
C₄H₈Cl₂	1,4-Dichlorobutane	161	52		
C₄H₈Cl₂O	Bis(2-chloroethyl) ether	178.5	55	2.7–?	369

Flammability of Chemical Substances

Mol. form.	Name	t_B/°C	FP/°C	Fl. limits	IT/°C
C_4H_8O	2-Buten-1-ol	121.5	27	4.2–35.3%	349
C_4H_8O	2-Methyl-2-propenol	114.5	33		
C_4H_8O	Ethyl vinyl ether	35.5	<−46	1.7–28%	202
C_4H_8O	1,2-Epoxybutane	63.4	−22	1.7–19%	439
C_4H_8O	Butanal	74.8	−22	1.9–12.5%	218
C_4H_8O	Isobutanal	64.5	−18	1.6–10.6%	196
C_4H_8O	2-Butanone	79.5	−9	1.4–11.4%	404
C_4H_8O	Tetrahydrofuran	65	−14	2–11.8%	321
C_4H_8OS	1,4-Oxathiane	147	42		
$C_4H_8O_2$	Butanoic acid	163.7	72	2.0–10.0%	443
$C_4H_8O_2$	2-Methylpropanoic acid	154.4	56	2.0–9.2%	481
$C_4H_8O_2$	Propyl formate	80.9	−3		455
$C_4H_8O_2$	Isopropyl formate	68.2	−6		485
$C_4H_8O_2$	Ethyl acetate	77.1	−4	2.0–11.5%	426
$C_4H_8O_2$	Methyl propanoate	79.8	−2	2.5–13%	469
$C_4H_8O_2$	3-Hydroxybutanal		66		250
$C_4H_8O_2$	1,4-Dioxane	101.5	12	2.0–22%	180
$C_4H_8O_2S$	Sulfolane	287.3	177		
$C_4H_8O_3$	Methyl lactate	144.8	49	2.2–?	385
$C_4H_8O_3$	Ethylene glycol monoacetate	188	102		
C_4H_9Br	1-Bromobutane	101.6	18	2.6–6.6%	265
C_4H_9Br	2-Bromobutane	91.2	21		
C_4H_9Cl	1-Chlorobutane	78.6	−12	1.9–10.1%	240
C_4H_9Cl	2-Chlorobutane	68.2	−10		
C_4H_9Cl	1-Chloro-2-methylpropane	68.5	−6	2.0–8.7%	
C_4H_9Cl	2-Chloro-2-methylpropane	50.9	0		
$C_4H_9Cl_3Si$	Butyltrichlorosilane	148.5	54		
C_4H_9N	Pyrrolidine	86.5	3		
C_4H_9NO	N-Ethylacetamide	205	110		
C_4H_9NO	N,N-Dimethylacetamide	165	70	1.8–11.5%	490
C_4H_9NO	Butanal oxime	154	58		
C_4H_9NO	2-Butanone oxime	152.5	≈70		
C_4H_9NO	Morpholine	128	37	1.4–11.2%	290
$C_4H_9NO_2$	N-Acetylethanolamine		179		460
$C_4H_9NO_3$	Butyl nitrate	133	36		
C_4H_{10}	Butane	−0.5	−60	1.9–8.5%	287
C_4H_{10}	Isobutane	−11.7	−87	1.8–8.4%	460
$C_4H_{10}N_2$	Piperazine	146	81		
$C_4H_{10}O$	1-Butanol	117.7	37	1.4–11.2%	343
$C_4H_{10}O$	2-Butanol	99.5	24	1.7–9.8%	405
$C_4H_{10}O$	2-Methyl-1-propanol	107.8	28	1.7–10.6%	415
$C_4H_{10}O$	2-Methyl-2-propanol	82.4	11	2.4–8.0%	478
$C_4H_{10}O$	Diethyl ether	34.5	−45	1.9–36.0%	180
$C_4H_{10}O$	Methyl propyl ether	39.1	−20	2.0–14.8%	
$C_4H_{10}O_2$	1,2-Butanediol	190.5	40		
$C_4H_{10}O_2$	1,3-Butanediol	207.5	121		395
$C_4H_{10}O_2$	1,4-Butanediol	235	121		
$C_4H_{10}O_2$	2,3-Butanediol	182.5			402
$C_4H_{10}O_2$	Ethylene glycol monoethyl ether	135	43	3–18%	235
$C_4H_{10}O_2$	Ethylene glycol dimethyl ether	85	−2		202
$C_4H_{10}O_2$	tert-Butyl hydroperoxide		27		
$C_4H_{10}O_2S$	2,2'-Thiodiethanol	282	160		298
$C_4H_{10}O_3$	Diethylene glycol	245.8	124	2–17%	224
$C_4H_{10}O_4S$	Diethyl sulfate	208	104		436
$C_4H_{10}S$	1-Butanethiol	98.5	2		
$C_4H_{10}S$	2-Butanethiol	85	−23		
$C_4H_{10}S$	2-Methyl-1-propanethiol	88.5	2		
$C_4H_{10}S$	2-Methyl-2-propanethiol	64.3	<−29		
$C_4H_{10}Se$	Diethyl selenide	108		2.5–?	
$C_4H_{11}N$	Butylamine	77.0	−12	1.7–9.8%	312
$C_4H_{11}N$	sec-Butylamine	63.5	−9		

Mol. form.	Name	t_B/°C	FP/°C	Fl. limits	IT/°C
$C_4H_{11}N$	tert-Butylamine	44.0	−9	1.7–8.9%	380
$C_4H_{11}N$	Isobutylamine	67.7	−9	2–12%	378
$C_4H_{11}N$	Diethylamine	55.5	−23	1.8–10.1%	312
$C_4H_{11}NO$	2-Amino-1-butanol	178	74		
$C_4H_{11}NO$	2-Amino-2-methyl-1-propanol	165.5	67		
$C_4H_{11}NO_2$	Diethanolamine	268.8	172	2–13%	662
$C_4H_{12}Sn$	Tetramethylstannane	78	−12	1.9–?	
$C_4H_{13}N_3$	Diethylenetriamine	207	98	2–6.7%	358
$C_5H_4O_2$	Furfural	161.7	60	2.1–19.3%	316
C_5H_5N	Pyridine	115.2	20	1.8–12.4%	482
C_5H_6	2-Methyl-1-buten-3-yne	32	<−7		
$C_5H_6N_2$	2-Methylpyrazine	137	50		
C_5H_6O	3-Methylfuran	66	−30		
$C_5H_6O_2$	Furfuryl alcohol	171	75	1.8–16.3%	491
C_5H_7N	1-Methylpyrrole	115	16		
C_5H_7NO	2-Furanmethanamine	145.5	37		
$C_5H_7NO_2$	Ethyl cyanoacetate	205	110		
C_5H_8	2-Methyl-1,3-butadiene	34.0	−54	1.5–8.9%	395
C_5H_8	1-Pentyne	40.1	<−20		
C_5H_8	Cyclopentene	44.2	−29		395
C_5H_8O	3-Methyl-3-buten-2-one	98		1.8–9.0%	
C_5H_8O	Cyclopentanone	130.5	26		
C_5H_8O	3,4-Dihydro-2H-pyran	86	−18		
$C_5H_8O_2$	Allyl acetate	103.5	22		374
$C_5H_8O_2$	Isopropenyl acetate	94	26		432
$C_5H_8O_2$	Vinyl propanoate	91.2	1		
$C_5H_8O_2$	Ethyl acrylate	99.4	10	1.4–14%	372
$C_5H_8O_2$	Methyl methacrylate	100.5	10	1.7–8.2%	
$C_5H_8O_2$	2,4-Pentanedione	138	34		340
$C_5H_8O_3$	Methyl acetoacetate	171.7	77		280
C_5H_9NO	N-Methyl-2-pyrrolidone	202	96	1–10%	346
C_5H_{10}	1-Pentene	29.9	−18	1.5–8.7%	275
C_5H_{10}	cis-2-Pentene	36.9	<−20		
C_5H_{10}	trans-2-Pentene	36.3	<−20		
C_5H_{10}	2-Methyl-1-butene	31.2	−20		
C_5H_{10}	3-Methyl-1-butene	20.1	−7	1.5–9.1%	365
C_5H_{10}	2-Methyl-2-butene	38.5	−20		
C_5H_{10}	Cyclopentane	49.3	−25	1.5–?	361
$C_5H_{10}Cl_2$	1,5-Dichloropentane	179	>27		
$C_5H_{10}N_2$	3-(Dimethylamino)propanenitrile	173	65		
$C_5H_{10}O$	Cyclopentanol	140.4	51		
$C_5H_{10}O$	Pentanal	103	12		222
$C_5H_{10}O$	2-Pentanone	102.2	7	1.5–8.2%	452
$C_5H_{10}O$	3-Pentanone	101.9	13	1.6–?	450
$C_5H_{10}O$	Tetrahydropyran	88	−20		
$C_5H_{10}O$	2-Methyltetrahydrofuran	78	−11		
$C_5H_{10}O_2$	Pentanoic acid	186.1	96		400
$C_5H_{10}O_2$	3-Methylbutanoic acid	176.5			416
$C_5H_{10}O_2$	Butyl formate	106.1	18	1.7–8.2%	322
$C_5H_{10}O_2$	Isobutyl formate	98.2	5	2–9%	320
$C_5H_{10}O_2$	Propyl acetate	101.5	13	1.7–8%	450
$C_5H_{10}O_2$	Isopropyl acetate	88.6	2	1.8–8%	460
$C_5H_{10}O_2$	Ethyl propanoate	99.1	12	1.9–11%	440
$C_5H_{10}O_2$	Methyl butanoate	102.8	14		
$C_5H_{10}O_2$	3-Ethoxypropanal	135.2	38		
$C_5H_{10}O_2$	Tetrahydrofurfuryl alcohol	178	75	1.5–9.7%	282
$C_5H_{10}O_3$	Diethyl carbonate	126	25		
$C_5H_{10}O_3$	Ethylene glycol monomethyl ether acetate	143	49	1.5–12.3%	392
$C_5H_{10}O_3$	Ethyl lactate	154.5	46	1.5–?	400
$C_5H_{11}Br$	1-Bromopentane	129.8	32		

Mol. form.	Name	t_B/°C	FP/°C	Fl. limits	IT/°C
$C_5H_{11}Cl$	1-Chloropentane	107.8	13	1.6–8.6%	260
$C_5H_{11}Cl$	2-Chloro-2-methylbutane	85.6		1.5–7.4%	345
$C_5H_{11}Cl$	1-Chloro-3-methylbutane	98.9	<21	1.5–7.4%	
$C_5H_{11}Cl_3Si$	Trichloropentylsilane	172	63		
$C_5H_{11}N$	Piperidine	106.2	16		
$C_5H_{11}N$	N-Methylpyrrolidine	81	−14		
$C_5H_{11}NO$	4-Methylmorpholine	116	24		
$C_5H_{11}NO_2$	Isopentyl nitrite	99.2			210
C_5H_{12}	Pentane	36.0	−40	1.4–8.0%	260
C_5H_{12}	Isopentane	27.8	−51	1.4–7.6%	420
C_5H_{12}	Neopentane	9.4	−65	1.4–7.5%	450
$C_5H_{12}N_2$	1-Methylpiperazine	138	42		
$C_5H_{12}N_2O$	Tetramethylurea	176.5	77		
$C_5H_{12}O$	1-Pentanol	137.9	33	1.2–10.0%	300
$C_5H_{12}O$	2-Pentanol	119.3	34	1.2–9.0%	343
$C_5H_{12}O$	3-Pentanol	116.2	41	1.2–9.0%	435
$C_5H_{12}O$	2-Methyl-1-butanol	128	50		385
$C_5H_{12}O$	3-Methyl-1-butanol	131.1	43	1.2–9.0%	350
$C_5H_{12}O$	2-Methyl-2-butanol	102.4	19	1.2–9.0%	437
$C_5H_{12}O$	3-Methyl-2-butanol	112.9	38		
$C_5H_{12}O$	2,2-Dimethyl-1-propanol	113.5	37		
$C_5H_{12}O$	Ethyl propyl ether	63.2	<−20	1.7–9.0%	
$C_5H_{12}O_2$	1,5-Pentanediol	239	129		335
$C_5H_{12}O_2$	2-Isopropoxyethanol	145	33		
$C_5H_{12}O_2$	2,2-Dimethyl-1,3-propanediol	208	129		399
$C_5H_{12}O_3$	Diethylene glycol monomethyl ether	193	96	1.38–22.7%	240
$C_5H_{12}S$	1-Pentanethiol	126.6	18		
$C_5H_{12}S$	3-Methyl-2-butanethiol		3		
$C_5H_{13}N$	Pentylamine	104.3	−1	2.2–22%	
$C_5H_{13}N$	Butylmethylamine	91	13		
$C_6H_2Cl_4$	1,2,4,5-Tetrachlorobenzene	244.5	155		
$C_6H_3ClN_2O_4$	1-Chloro-2,4-dinitrobenzene	315	194	2.0–22%	
$C_6H_3Cl_3$	1,2,4-Trichlorobenzene	213.5	105	2.5–6.6%	571
$C_6H_4ClNO_2$	1-Chloro-4-nitrobenzene	242	127		
$C_6H_4Cl_2$	o-Dichlorobenzene	180	66	2.2–9.2%	648
$C_6H_4Cl_2$	m-Dichlorobenzene	173	72		
$C_6H_4Cl_2$	p-Dichlorobenzene	174	66		
$C_6H_4Cl_2O$	2,4-Dichlorophenol	210	114		
C_6H_5Br	Bromobenzene	156.0	51		565
C_6H_5Cl	Chlorobenzene	131.7	28	1.3–9.6%	593
C_6H_5ClO	o-Chlorophenol	174.9	64		
C_6H_5ClO	p-Chlorophenol	220	121		
$C_6H_5Cl_2N$	3,4-Dichloroaniline	272	166		
$C_6H_5Cl_3Si$	Trichlorophenylsilane	201	91		
C_6H_5F	Fluorobenzene	84.7	−15		
$C_6H_5NO_2$	Nitrobenzene	210.8	88	1.8–?	482
$C_6H_5N_3O_4$	2,4-Dinitroaniline		224		
C_6H_6	1,5-Hexadien-3-yne	85	<−20	1.5–?	
C_6H_6	Benzene	80.0	−11	1.2–7.8%	498
$C_6H_6N_2O_2$	p-Nitroaniline	332	199		
C_6H_6O	Phenol	181.8	79	1.8–8.6%	715
$C_6H_6O_2$	1,2-Benzenediol	245	127		
$C_6H_6O_2$	Resorcinol		127	1.4–?	608
$C_6H_6O_2$	p-Hydroquinone	287	165		516
C_6H_7N	Aniline	184.1	70	1.3–11%	615
C_6H_7N	2-Methylpyridine	129.3	39		538
C_6H_7N	4-Methylpyridine	145.3	57		
C_6H_8ClN	Aniline, hydrochloride		193		
$C_6H_8Cl_2O_2$	Hexanedioyl dichloride		72		
$C_6H_8N_2$	Adiponitrile	295	93	1.0–?	550
$C_6H_8N_2$	o-Phenylenediamine	257	156	1.5–?	

Mol. form.	Name	$t_B/°C$	FP/°C	Fl. limits	IT/°C
$C_6H_8N_2$	Phenylhydrazine	243.5	88		
$C_6H_8N_2$	2,5-Dimethylpyrazine	155	64		
C_6H_8O	2,5-Dimethylfuran	93.5	7		
$C_6H_8O_4$	Dimethyl maleate	202	113		
C_6H_{10}	1,4-Hexadiene	65	−21	2.0–6.1%	
C_6H_{10}	2-Methyl-1,3-pentadiene	75.8	−12		
C_6H_{10}	4-Methyl-1,3-pentadiene	76.5	−34		
C_6H_{10}	2-Hexyne	84.5	−10		
C_6H_{10}	Cyclohexene	82.9	−12	1.2–?	310
$C_6H_{10}O$	Diallyl ether	94	−7		
$C_6H_{10}O$	Cyclohexanone	155.4	44	1.1–9.4%	420
$C_6H_{10}O$	Mesityl oxide	130	31	1.4–7.2%	344
$C_6H_{10}O_2$	Vinyl butanoate	116.7	20	1.4–8.8%	
$C_6H_{10}O_2$	Ethyl 2-butenoate	136.5	2		
$C_6H_{10}O_2$	Ethyl methacrylate	117	20		
$C_6H_{10}O_2$	2,5-Hexanedione	194	79		499
$C_6H_{10}O_3$	Ethyl acetoacetate	180.8	57	1.4–9.5%	295
$C_6H_{10}O_3$	Propanoic anhydride	170	63	1.3–9.5%	285
$C_6H_{10}O_4$	Adipic acid	337.5	196		420
$C_6H_{10}O_4$	Diethyl oxalate	185.7	76		
$C_6H_{10}O_4$	Ethylene glycol diacetate	190	88	1.6–8.4%	482
$C_6H_{11}Cl$	Chlorocyclohexane	142	32		
$C_6H_{11}NO$	Caprolactam	270	125		
$C_6H_{11}NO_2$	Nitrocyclohexane	205	88		
$C_6H_{11}NO_2$	4-Acetylmorpholine		113		
C_6H_{12}	1-Hexene	63.4	−26	1.2–6.9%	253
C_6H_{12}	cis-2-Hexene	68.8	−21		
C_6H_{12}	2-Methyl-1-pentene	62.1	−28		300
C_6H_{12}	4-Methyl-1-pentene	53.9	−7		300
C_6H_{12}	4-Methyl-cis-2-pentene	56.3	−32		
C_6H_{12}	4-Methyl-trans-2-pentene	58.6	−29		
C_6H_{12}	2-Ethyl-1-butene	64.7	<−20		315
C_6H_{12}	2,3-Dimethyl-1-butene	55.6	<−20		360
C_6H_{12}	2,3-Dimethyl-2-butene	73.3	<−20		401
C_6H_{12}	Cyclohexane	80.7	−20	1.3–8%	245
C_6H_{12}	Methylcyclopentane	71.8	−29	1.0–8.35%	258
C_6H_{12}	Ethylcyclobutane	70.8	−15	1.2–7.7%	210
C_6H_{12}	2-Methyl-2-pentene	67.3	<−7		
$C_6H_{12}Cl_2O_2$	1,2-Bis(2-chloroethoxy)ethane	232	121		
$C_6H_{12}O$	cis-3-Hexen-1-ol	156.5	54		
$C_6H_{12}O$	Butyl vinyl ether	94	−9		255
$C_6H_{12}O$	Isobutyl vinyl ether	83	−9		
$C_6H_{12}O$	Hexanal	131	32		
$C_6H_{12}O$	2-Ethylbutanal		21	1.2–7.7%	
$C_6H_{12}O$	2-Methylpentanal	117	17		199
$C_6H_{12}O$	2-Hexanone	127.6	25	1–8%	423
$C_6H_{12}O$	3-Hexanone	123.5	35	1–8%	
$C_6H_{12}O$	4-Methyl-2-pentanone	116.5	18	1.2–8.0%	448
$C_6H_{12}O$	Cyclohexanol	160.8	68	1–9%	300
$C_6H_{12}O_2$	Hexanoic acid	205.2	102		380
$C_6H_{12}O_2$	2-Methylpentanoic acid	195.6	107		378
$C_6H_{12}O_2$	Diethylacetic acid	194	99		400
$C_6H_{12}O_2$	Pentyl formate	130.4	26		
$C_6H_{12}O_2$	Butyl acetate	126.1	22	1.7–7.6%	425
$C_6H_{12}O_2$	sec-Butyl acetate	112	31	1.7–9.8%	
$C_6H_{12}O_2$	Isobutyl acetate	116.5	18	1.3–10.5%	421
$C_6H_{12}O_2$	Propyl propanoate	122.5	79		
$C_6H_{12}O_2$	Ethyl butanoate	121.5	24		463
$C_6H_{12}O_2$	Ethyl 2-methylpropanoate	110.1	13		
$C_6H_{12}O_2$	Diacetone alcohol	167.9	58	1.8–6.9%	643

Mol. form.	Name	t_B/°C	FP/°C	Fl. limits	IT/°C
$C_6H_{12}O_3$	Ethylene glycol monoethyl ether acetate	156.4	56	2–8%	379
$C_6H_{12}O_3$	Paraldehyde	124.3	36	1.3–?	238
$C_6H_{12}S$	Cyclohexanethiol	158.9	43		
$C_6H_{13}Cl$	1-Chlorohexane	135	35		
$C_6H_{13}N$	Cyclohexylamine	134	31	1.9–9.4%	293
$C_6H_{13}NO$	N-Butylacetamide	229	116		
$C_6H_{13}NO$	2,6-Dimethylmorpholine	146.6	44		
$C_6H_{13}NO$	N-Ethylmorpholine	138.5	32		
$C_6H_{13}NO_2$	4-Morpholineethanol	227	99		
C_6H_{14}	Hexane	68.7	−22	1.1–7.5%	225
C_6H_{14}	2-Methylpentane	60.2	<−29	1.0–7.0%	264
C_6H_{14}	3-Methylpentane	63.2	−7	1.2–7.0%	278
C_6H_{14}	2,2-Dimethylbutane	49.7	−48	1.2–7.0%	405
C_6H_{14}	2,3-Dimethylbutane	57.9	−29	1.2–7.0%	405
$C_6H_{14}N_2O$	1-Piperazineethanol	246	124		
$C_6H_{14}O$	1-Hexanol	157.6	63		
$C_6H_{14}O$	2-Methyl-1-pentanol	149	54	1.1–9.65%	310
$C_6H_{14}O$	4-Methyl-2-pentanol	131.6	41	1.0–5.5%	
$C_6H_{14}O$	2-Ethyl-1-butanol	147	57		
$C_6H_{14}O$	Dipropyl ether	90.0	21	1.3–7.0%	188
$C_6H_{14}O$	Diisopropyl ether	68.5	−28	1.4–7.9%	443
$C_6H_{14}O$	Butyl ethyl ether	92.3	4		
$C_6H_{14}O_2$	2,5-Hexanediol	218	110		
$C_6H_{14}O_2$	2-Methyl-2,4-pentanediol	197.1	102	1–9%	306
$C_6H_{14}O_2$	Ethylene glycol monobutyl ether	168.4	69	4–13%	238
$C_6H_{14}O_2$	1,1-Diethoxyethane	102.2	−21	1.6–10.4%	230
$C_6H_{14}O_2$	Ethylene glycol diethyl ether	119.4	27		205
$C_6H_{14}O_3$	1,2,6-Hexanetriol		191		
$C_6H_{14}O_3$	Diethylene glycol monoethyl ether	196	96		
$C_6H_{14}O_3$	Diethylene glycol dimethyl ether	162	67		
$C_6H_{14}O_3$	Trimethylolpropane		149		
$C_6H_{14}O_4$	Triethylene glycol	285	177	0.9–9.2%	371
$C_6H_{15}N$	Hexylamine	132.8	29		
$C_6H_{15}N$	Butylethylamine	107.5	18		
$C_6H_{15}N$	Dipropylamine	109.3	17		299
$C_6H_{15}N$	Diisopropylamine	83.9	−1	1.1–7.1%	316
$C_6H_{15}N$	Triethylamine	89	−7	1.2–8.0%	249
$C_6H_{15}NO_2$	Diisopropanolamine	250	127		374
$C_6H_{15}NO_3$	Triethanolamine	335.4	179	1–10%	
$C_6H_{15}N_3$	1-Piperazineethanamine	220	93		
$C_6H_{15}O_4P$	Triethyl phosphate	215.5	115		454
$C_6H_{16}N_2$	N,N-Diethylethylenediamine	144	46		
$C_7H_3ClF_3NO_2$	1-Chloro-4-nitro-2-(trifluoromethyl)benzene	232	135		
$C_7H_4ClF_3$	1-Chloro-2-(trifluoromethyl)benzene	152.2	59		
$C_7H_4F_3NO_2$	1-Nitro-3-(trifluoromethyl)benzene	202.8	103		
C_7H_5ClO	Benzoyl chloride	197.2	72		
C_7H_5ClO	4-Chlorobenzaldehyde	213.5	88		
$C_7H_5Cl_3$	(Trichloromethyl)benzene	221	127		211
$C_7H_5F_3$	(Trifluoromethyl)benzene	102.1	12		
$C_7H_6N_2O_4$	1-Methyl-2,4-dinitrobenzene		207		
C_7H_6O	Benzaldehyde	179.0	63		192
$C_7H_6O_2$	Benzoic acid	249.2	121		570
$C_7H_6O_2$	Salicylaldehyde	197	78		
$C_7H_6O_3$	Salicylic acid		157	1.1–?	540
C_7H_7Br	o-Bromotoluene	181.7	79		
C_7H_7Br	p-Bromotoluene	184.3	85		
C_7H_7Cl	(Chloromethyl)benzene	179	67	1.1–?	585
$C_7H_7NO_2$	o-Nitrotoluene	222	106		
$C_7H_7NO_2$	m-Nitrotoluene	232	106		
$C_7H_7NO_2$	p-Nitrotoluene	238.3	106		

Mol. form.	Name	t_B/°C	FP/°C	Fl. limits	IT/°C
C$_7$H$_8$	Toluene	110.6	4	1.1–7.1%	480
C$_7$H$_8$	Bicyclo[2.2.1]hepta-2,5-diene	89.5	−21		
C$_7$H$_8$O	o-Cresol	191.0	81	1.4–?	599
C$_7$H$_8$O	m-Cresol	202.2	86	1.1–?	558
C$_7$H$_8$O	p-Cresol	201.9	86	1.1–?	558
C$_7$H$_8$O	Benzyl alcohol	205.3	93		436
C$_7$H$_8$O	Anisole	153.7	52		475
C$_7$H$_8$O$_2$	4-Methoxyphenol	243	132		421
C$_7$H$_8$O$_3$S	p-Toluenesulfonic acid		184		
C$_7$H$_9$N	o-Methylaniline	200.3	85		482
C$_7$H$_9$N	p-Methylaniline	200.4	87		482
C$_7$H$_9$NO	o-Anisidine	224	118		
C$_7$H$_{10}$O	3-Cyclohexene-1-carboxaldehyde	105	57		
C$_7$H$_{10}$O$_4$	3,3-Diacetoxy-1-propene	180	82		
C$_7$H$_{12}$	4-Methylcyclohexene	102.7	−1		
C$_7$H$_{12}$O$_2$	Butyl acrylate	145	29	1.7–9.9%	292
C$_7$H$_{12}$O$_2$	Isobutyl acrylate	132	30		427
C$_7$H$_{12}$O$_2$	Cyclohexyl formate	162	51		
C$_7$H$_{12}$O$_4$	Diethyl malonate	200	93		
C$_7$H$_{14}$	1-Heptene	93.6	−1		260
C$_7$H$_{14}$	trans-2-Heptene	98	<0		
C$_7$H$_{14}$	Cycloheptane	118.4	<21	1.1–6.7%	
C$_7$H$_{14}$	Methylcyclohexane	100.9	−4	1.2–6.7%	250
C$_7$H$_{14}$	Ethylcyclopentane	103.5	<21	1.1–6.7%	260
C$_7$H$_{14}$O	2-Heptanone	151.0	39	1.1–7.9%	393
C$_7$H$_{14}$O	3-Heptanone	147	46		
C$_7$H$_{14}$O	4-Heptanone	144	49		
C$_7$H$_{14}$O	5-Methyl-2-hexanone	144	36	1.0–8.2%	191
C$_7$H$_{14}$O	cis-2-Methylcyclohexanol	165	65		296
C$_7$H$_{14}$O	trans-2-Methylcyclohexanol	167.5	65		296
C$_7$H$_{14}$O	cis-3-Methylcyclohexanol	174.5	70		295
C$_7$H$_{14}$O	trans-3-Methylcyclohexanol	174.5	70		295
C$_7$H$_{14}$O	cis-4-Methylcyclohexanol	173	70		295
C$_7$H$_{14}$O	trans-4-Methylcyclohexanol	174	70		295
C$_7$H$_{14}$O$_2$	Pentyl acetate	149.2	16	1.1–7.5%	360
C$_7$H$_{14}$O$_2$	Isopentyl acetate	142.5	25	1.0–7.5%	360
C$_7$H$_{14}$O$_2$	sec-Pentyl acetate	130.5	32		
C$_7$H$_{14}$O$_2$	Butyl propanoate	146.8	32		426
C$_7$H$_{14}$O$_2$	Propyl butanoate	143.0	37		
C$_7$H$_{15}$NO$_2$	Ethyl N-butylcarbamate	202	92		
C$_7$H$_{16}$	Heptane	98.5	−4	1.05–6.7%	204
C$_7$H$_{16}$	2-Methylhexane	90.0	−1	1.0–6.0%	280
C$_7$H$_{16}$	3-Methylhexane	92	−4		280
C$_7$H$_{16}$	2,3-Dimethylpentane	89.7	−56	1.1–6.7%	335
C$_7$H$_{16}$	2,4-Dimethylpentane	80.4	−12		
C$_7$H$_{16}$	2,2,3-Trimethylbutane	80.8	<0		412
C$_7$H$_{16}$N$_2$O	4-Morpholinepropanamine	220	104		
C$_7$H$_{16}$O	2-Heptanol	159	71		
C$_7$H$_{16}$O	3-Heptanol	157	60		
C$_7$H$_{16}$O	2,4-Dimethyl-3-pentanol	138.7	49		
C$_7$H$_{16}$O	2,3,3-Trimethyl-2-butanol	131	<0		375
C$_7$H$_{17}$N	Heptylamine	156	54		
C$_7$H$_{18}$N$_2$	N,N-Diethyl-1,3-propanediamine	168.5	59		
C$_8$H$_4$O$_3$	Phthalic anhydride	295	152	1.7–10.5%	570
C$_8$H$_6$O$_4$	Phthalic acid		168		
C$_8$H$_6$O$_4$	Terephthalic acid		260		496
C$_8$H$_7$ClO	α-Chloroacetophenone	247	118		
C$_8$H$_7$N	Benzeneacetonitrile	233.5	113		
C$_8$H$_8$	Styrene	145	31	0.9–6.8%	490
C$_8$H$_8$O	Phenyloxirane	194.1	74		498
C$_8$H$_8$O	Benzeneacetaldehyde	195	71		

Mol. form.	Name	t_B/°C	FP/°C	Fl. limits	IT/°C
C_8H_8O	Acetophenone	202	77		570
$C_8H_8O_2$	Benzeneacetic acid	265.5	>100		
$C_8H_8O_2$	Phenyl acetate	196	80		
$C_8H_8O_2$	Methyl benzoate	199	83		
$C_8H_8O_2$	2-Methoxybenzaldehyde	243.5	118		
$C_8H_8O_3$	Methyl salicylate	222.9	96		454
C_8H_9Cl	1-Chloro-4-ethylbenzene	184.4	64		
C_8H_9NO	Acetanilide	304	169		530
$C_8H_9NO_2$	Methyl 2-aminobenzoate	256	>100		
C_8H_{10}	Ethylbenzene	136.1	21	0.8–6.7%	432
C_8H_{10}	o-Xylene	144.5	32	0.9–6.7%	463
C_8H_{10}	m-Xylene	139.1	27	1.1–7.0%	527
C_8H_{10}	p-Xylene	138.3	27	1.1–7.0%	528
$C_8H_{10}O$	p-Ethylphenol	217.9	104		
$C_8H_{10}O$	Benzeneethanol	218.2	96		
$C_8H_{10}O$	α-Methylbenzyl alcohol	205	93		
$C_8H_{10}O$	Phenetole	169.8	63		
$C_8H_{10}O$	Benzyl methyl ether	170	135		
$C_8H_{10}O$	4-Methylanisole	175.5	60		
$C_8H_{10}O_2$	2-Phenoxyethanol	245	121		
$C_8H_{11}N$	N-Ethylaniline	203.0	85		
$C_8H_{11}N$	N,N-Dimethylaniline	194.1	63		371
$C_8H_{11}N$	2,3-Xylidine	221.5	97	1.0–?	
$C_8H_{11}N$	2,6-Xylidine	215	96		
$C_8H_{11}N$	α-Methylbenzylamine	187	79		
$C_8H_{11}N$	5-Ethyl-2-picoline	178.3	68	1.1–6.6%	
$C_8H_{11}NO$	N-Phenylethanolamine	279.5	152		
$C_8H_{11}NO$	o-Phenetidine	232.5	115		
$C_8H_{11}NO$	p-Phenetidine	254	116		
C_8H_{12}	1,5-Cyclooctadiene	150.8	35		
C_8H_{12}	4-Vinylcyclohexene	128	16		269
$C_8H_{12}O_4$	Diethyl maleate	223	121		350
$C_8H_{12}O_4$	Diethyl fumarate	214	104		
$C_8H_{14}O_2$	Cyclohexyl acetate	173	58		335
$C_8H_{14}O_2$	Butyl methacrylate	160	52		
$C_8H_{14}O_3$	Butanoic anhydride	200	54	0.9–5.8%	279
$C_8H_{14}O_3$	2-Methylpropanoic anhydride	183	59	1.0–6.2%	329
$C_8H_{14}O_3$	Butyl acetoacetate		85		
$C_8H_{14}O_4$	Ethyl succinate	217.7	90		
$C_8H_{14}O_5$	Diethylene glycol diacetate	200	135		
$C_8H_{14}O_6$	Diethyl tartrate	281	93		
$C_8H_{15}ClO$	Octanoyl chloride	195.6	82		
C_8H_{16}	1-Octene	121.2	21		230
C_8H_{16}	2,4,4-Trimethyl-1-pentene	101.4	−5	0.8–4.8%	391
C_8H_{16}	2,4,4-Trimethyl-2-pentene	104.9	2		305
C_8H_{16}	Ethylcyclohexane	131.9	35	0.9–6.6%	238
C_8H_{16}	cis-1,2-Dimethylcyclohexane	129.8	16		304
C_8H_{16}	trans-1,2-Dimethylcyclohexane	123.5	11		304
C_8H_{16}	cis-1,4-Dimethylcyclohexane	124.4	16		
C_8H_{16}	Propylcyclopentane	131			269
$C_8H_{16}O$	Octanal	171	52		
$C_8H_{16}O$	2-Ethylhexanal	163	44	0.85–7.2%	190
$C_8H_{16}O$	2-Octanone	172.5	52		
$C_8H_{16}O_2$	Hexyl acetate	171.5	45		
$C_8H_{16}O_2$	sec-Hexyl acetate	147.5	45		
$C_8H_{16}O_2$	2-Ethylbutyl acetate	162.5	54		
$C_8H_{16}O_2$	Pentyl propanoate	168.6	41		378
$C_8H_{16}O_2$	Butyl butanoate	166	53		
$C_8H_{16}O_2$	Isobutyl butanoate	156.9	50		
$C_8H_{16}O_2$	Isobutyl isobutanoate	148.6	38	0.96–7.59%	432
$C_8H_{16}O_2$	Ethyl hexanoate	167	49		

Mol. form.	Name	t_B/°C	FP/°C	Fl. limits	IT/°C
$C_8H_{16}O_2$	1,4-Cyclohexanedimethanol	283	167		316
$C_8H_{16}O_3$	Pentyl lactate		79		
$C_8H_{16}O_4$	Diethylene glycol monoethyl ether acetate	218.5	110		425
$C_8H_{17}Cl$	1-Chlorooctane	181.5	70		
$C_8H_{17}Cl$	3-(Chloromethyl)heptane	172	60		
C_8H_{18}	Octane	125.6	13	1.0–6.5%	206
C_8H_{18}	2,3-Dimethylhexane	115.6	7		438
C_8H_{18}	2,4-Dimethylhexane	109.5	10		
C_8H_{18}	3-Ethyl-2-methylpentane	115.6	<21		460
C_8H_{18}	2,2,3-Trimethylpentane	110	<21		346
C_8H_{18}	2,2,4-Trimethylpentane	99.2	−12		418
C_8H_{18}	2,3,3-Trimethylpentane	114.8	<21		425
$C_8H_{18}O$	1-Octanol	195.1	81		
$C_8H_{18}O$	2-Octanol	180	88		
$C_8H_{18}O$	2-Ethyl-1-hexanol	184.6	73	0.88–9.7%	231
$C_8H_{18}O$	Dibutyl ether	140.2	25	1.5–7.6%	194
$C_8H_{18}O_2$	2-Ethyl-1,3-hexanediol	244	127		360
$C_8H_{18}O_2$	2,2,4-Trimethyl-1,3-pentanediol	235	113		346
$C_8H_{18}O_2$	Di-tert-butyl peroxide	111	18		
$C_8H_{18}O_3$	Diethylene glycol diethyl ether	188	82		
$C_8H_{18}O_4$	2,5,8,11-Tetraoxadodecane	216	111		
$C_8H_{18}O_5$	Tetraethylene glycol	328	182		
$C_8H_{18}S$	1-Octanethiol	199.1	69		
$C_8H_{18}S$	Dibutyl sulfide	185	76		
$C_8H_{19}N$	Octylamine	179.6	60		
$C_8H_{19}N$	Dibutylamine	159.6	47	1.1–6%	
$C_8H_{19}N$	Diisobutylamine	139.6	29		
$C_8H_{19}N$	2-Ethylhexylamine	169.2	60		
$C_8H_{20}O_4Si$	Ethyl silicate	168.8	52		
$C_8H_{23}N_5$	Tetraethylenepentamine	341.5	163		321
$C_9H_6N_2O_2$	Toluene-2,4-diisocyanate	251	127	0.9–9.5%	
C_9H_7N	Quinoline	237.1			480
C_9H_{10}	o-Methylstyrene	169.8	53	0.8–11.0%	538
C_9H_{10}	m-Methylstyrene	164	53	0.8–11.0%	538
C_9H_{10}	p-Methylstyrene	172.8	53	0.8–11.0%	538
C_9H_{10}	Isopropenylbenzene	165.4	54	1.9–6.1%	574
$C_9H_{10}O$	1-Phenyl-1-propanone	217.5	99		
$C_9H_{10}O$	4-Methylacetophenone	226	96		
$C_9H_{10}O_2$	Ethyl benzoate	212	88		490
$C_9H_{10}O_2$	Benzyl acetate	213	90		460
$C_9H_{10}O_2$	Methyl 2-phenylacetate	216.5	91		
$C_9H_{11}NO$	4-Methylacetanilide	307	168		
C_9H_{12}	Propylbenzene	159.2	30	0.8–6.0%	450
C_9H_{12}	Isopropylbenzene	152.4	36	0.9–6.5%	424
C_9H_{12}	o-Ethyltoluene	165.2			440
C_9H_{12}	m-Ethyltoluene	161.3			480
C_9H_{12}	p-Ethyltoluene	162			475
C_9H_{12}	1,2,3-Trimethylbenzene	176.1	44	0.8–6.6%	470
C_9H_{12}	1,2,4-Trimethylbenzene	169.3	44	0.9–6.4%	500
C_9H_{12}	1,3,5-Trimethylbenzene	164.7	50	1–5%	559
$C_9H_{12}O$	α–Ethylbenzyl alcohol	219	100		
$C_9H_{12}O_2$	Ethylene glycol monobenzyl ether	256	129		352
$C_9H_{12}O_3S$	Ethyl p-toluenesulfonate		158		
$C_9H_{13}N$	Amphetamine	203	<100		
$C_9H_{14}O$	Phorone	197.5	85		
$C_9H_{14}O$	Isophorone	215.2	84	0.8–3.8%	460
$C_9H_{14}O_6$	Triacetin	259	138	1.0–?	433
C_9H_{16}	Octahydroindene	167			296
$C_9H_{16}O_2$	Allyl hexanoate	186	66		
C_9H_{18}	1-Nonene	146.9	26		

Mol. form.	Name	t_B/°C	FP/°C	Fl. limits	IT/°C
C_9H_{18}	Propylcyclohexane	156.7			248
C_9H_{18}	Isopropylcyclohexane	154.8			283
C_9H_{18}	Butylcyclopentane	156.6			250
$C_9H_{18}O$	2-Nonanone	195.3	60	0.9–5.9%	360
$C_9H_{18}O$	Diisobutyl ketone	169.4	49	0.8–7.1%	396
$C_9H_{18}O_2$	Pentyl butanoate	186.4	57		
$C_9H_{18}O_2$	Isopentyl butanoate	179	59		
$C_9H_{18}O_2$	Butyl 3-methylbutanoate		53		
C_9H_{20}	Nonane	150.8	31	0.8–2.9%	205
C_9H_{20}	3-Ethyl-4-methylhexane	140	24		
C_9H_{20}	4-Ethyl-2-methylhexane	133.8	<21	0.7–?	280
C_9H_{20}	2,2,5-Trimethylhexane	124.0	13		
C_9H_{20}	3,3-Diethylpentane	146.3		0.7–5.7%	290
C_9H_{20}	3-Ethyl-2,4-dimethylpentane	136.7	390		
C_9H_{20}	2,2,3,3-Tetramethylpentane	140.2	<21	0.8–4.9%	430
C_9H_{20}	2,2,3,4-Tetramethylpentane	133.0	<21		
$C_9H_{21}BO_3$	Triisopropyl borate	140	28		
$C_9H_{21}N$	Tripropylamine	156	41		
$C_9H_{21}NO_3$	Triisopropanolamine		160		320
$C_{10}H_7Cl$	1-Chloronaphthalene	259	121		>558
$C_{10}H_8$	Naphthalene	217.9	79	0.9–5.9%	526
$C_{10}H_8O$	2-Naphthol	285	153		
$C_{10}H_9N$	1-Naphthalenamine	300.8	157		
$C_{10}H_{10}O_2$	Safrole	234.5	100		
$C_{10}H_{10}O_4$	Dimethyl phthalate	283.7	146	0.9–?	490
$C_{10}H_{10}O_4$	Dimethyl isophthalate	282	138		
$C_{10}H_{10}O_4$	Dimethyl terephthalate	288	153		518
$C_{10}H_{11}NO_2$	Acetoacetanilide		185		
$C_{10}H_{12}$	1,2,3,4-Tetrahydronaphthalene	207.6	71	0.8–5.0%	385
$C_{10}H_{12}O_2$	Isopropyl benzoate	216	99		
$C_{10}H_{12}O_2$	Ethyl phenylacetate	227	99		
$C_{10}H_{14}$	Butylbenzene	183.3	71	0.8–5.8%	410
$C_{10}H_{14}$	sec-Butylbenzene	173.3	52	0.8–6.9%	418
$C_{10}H_{14}$	tert-Butylbenzene	169.1	60	0.7–5.7%	450
$C_{10}H_{14}$	Isobutylbenzene	172.7	55	0.8–6.0%	427
$C_{10}H_{14}$	p-Cymene	177.1	47	0.7–5.6%	436
$C_{10}H_{14}$	1,2,3,4-Tetramethylbenzene	205	74		427
$C_{10}H_{14}$	1,2,3,5-Tetramethylbenzene	198	71		427
$C_{10}H_{14}$	1,2,4,5-Tetramethylbenzene	196.8	54		
$C_{10}H_{14}$	o-Diethylbenzene	184	57		395
$C_{10}H_{14}$	m-Diethylbenzene	181.1	56		450
$C_{10}H_{14}$	p-Diethylbenzene	183.7	55	0.7–6.0%	430
$C_{10}H_{14}O$	Butyl phenyl ether	210	82		
$C_{10}H_{14}O_2$	4-tert-Butyl-1,2-benzenediol	285	130		
$C_{10}H_{15}N$	N-Butylaniline	243.5	107		
$C_{10}H_{15}N$	N,N-Diethylaniline	216.3	85		630
$C_{10}H_{15}NO_2$	N-Phenyl-N,N-diethanolamine		196	0.7–?	387
$C_{10}H_{16}$	Dipentene	178	45		237
$C_{10}H_{16}$	d-Limonene	178	45	0.7–6.1%	237
$C_{10}H_{16}$	α-Pinene	156.2	33		255
$C_{10}H_{16}$	β-Pinene	166	38		275
$C_{10}H_{16}$	β-Phellandrene	171.5	49		
$C_{10}H_{16}O$	Camphor	207.4	66	0.6–3.5%	466
$C_{10}H_{18}$	trans-Decahydronaphthalene	187.3	54	0.7–5.4%	255
$C_{10}H_{18}O$	Borneol		66		
$C_{10}H_{18}O$	Linalol	198	71		
$C_{10}H_{18}O$	α-Terpineol	220	90		
$C_{10}H_{18}O$	Cineole	176.4	48		
$C_{10}H_{18}O$	trans-Geraniol	230	>100		
$C_{10}H_{18}O_4$	Dibutyl oxalate	241	104		
$C_{10}H_{19}NO_2$	N-tert-Butylaminoethyl methacrylate		96		

Mol. form.	Name	t_B/°C	FP/°C	Fl. limits	IT/°C
$C_{10}H_{20}$	1-Decene	170.5	<55		235
$C_{10}H_{20}$	Butylcyclohexane	180.9			246
$C_{10}H_{20}$	Isobutylcyclohexane	171.3			274
$C_{10}H_{20}$	*tert*-Butylcyclohexane	171.5			342
$C_{10}H_{20}O$	Citronellol	224	96		
$C_{10}H_{20}O_2$	2-Ethylhexyl acetate	199	71	0.76–8.14%	268
$C_{10}H_{20}O_2$	Ethyl octanoate	208.5	79		
$C_{10}H_{21}N$	*N*-Butylcyclohexanamine		93		
$C_{10}H_{22}$	Decane	174.1	51	0.8–5.4%	210
$C_{10}H_{22}$	2-Methylnonane	167.1			210
$C_{10}H_{22}$	3-Ethyloctane	166.5			230
$C_{10}H_{22}$	4-Ethyloctane	163.7			229
$C_{10}H_{22}O$	1-Decanol	231.1	82		288
$C_{10}H_{22}O$	Dipentyl ether	190	57		170
$C_{10}H_{22}O_2$	Ethylene glycol dibutyl ether	203.3	85		
$C_{10}H_{22}O_5$	Tetraethylene glycol dimethyl ether	275.3	141		
$C_{10}H_{22}S$	Dipentyl sulfide		85		
$C_{10}H_{23}N$	Decylamine	220.5	99		
$C_{10}H_{23}N$	Dipentylamine	202.5	51		
$C_{11}H_{10}$	1-Methylnaphthalene	244.7			529
$C_{11}H_{12}O_3$	Ethyl benzoylacetate		141		
$C_{11}H_{14}O_2$	Butyl benzoate	250.3	107		
$C_{11}H_{16}$	*p-tert*-Butyltoluene	190	68		
$C_{11}H_{16}$	Pentylbenzene	205.4	66		
$C_{11}H_{16}$	1,3-Diethyl-5-methylbenzene	205			455
$C_{11}H_{16}$	Pentamethylbenzene	232	93		427
$C_{11}H_{16}O$	4-*tert*-Butyl-2-methylphenol	237	118		
$C_{11}H_{17}N$	*p-tert*-Pentylaniline	260.5	102		
$C_{11}H_{20}O_2$	2-Ethylhexyl acrylate		82		252
$C_{11}H_{22}$	Pentylcyclohexane	203.7			239
$C_{11}H_{22}O$	2-Undecanone	231.5	89		
$C_{11}H_{22}O_2$	Nonyl acetate	210	68		
$C_{11}H_{24}$	Undecane	195.9	69		
$C_{11}H_{24}$	2-Methyldecane	189.3			225
$C_{11}H_{24}O$	2-Undecanol	228	113		
$C_{12}H_9Br$	4-Bromo-1,1'-biphenyl	310	144		
$C_{12}H_{10}$	Biphenyl	256.1	113	0.6–5.8%	540
$C_{12}H_{10}Cl_2Si$	Dichlorodiphenylsilane	305	142		
$C_{12}H_{10}O$	*o*-Phenylphenol	286	124		530
$C_{12}H_{10}O$	Diphenyl ether	258.0	112	0.8–1.5%	618
$C_{12}H_{11}N$	2-Aminobiphenyl	299			450
$C_{12}H_{11}N$	Diphenylamine	302	153		634
$C_{12}H_{12}$	1-Ethylnaphthalene	258.6			480
$C_{12}H_{14}O_4$	Diethyl phthalate	295	161	0.7–?	457
$C_{12}H_{14}O_4$	Diethyl terephthalate	302	117		
$C_{12}H_{16}$	Cyclohexylbenzene	240.1	99		
$C_{12}H_{16}O_3$	Pentyl salicylate	270	132		
$C_{12}H_{17}NO$	*N*-Butyl-*N*-phenylacetamide	281	141		
$C_{12}H_{18}$	1,5,9-Cyclododecatriene	240	71		
$C_{12}H_{20}O_4$	Dibutyl maleate	280	141		
$C_{12}H_{22}O_4$	Dimethyl sebacate		145		
$C_{12}H_{22}O_6$	Dibutyl tartrate	320	91		284
$C_{12}H_{23}N$	Dicyclohexylamine		>99		
$C_{12}H_{24}$	1-Dodecene	213.8	79		
$C_{12}H_{24}O_2$	Ethyl decanoate	241.5	>100		
$C_{12}H_{25}Br$	1-Bromododecane	276	144		
$C_{12}H_{26}$	Dodecane	216.3	74	0.6–?	203
$C_{12}H_{26}O$	1-Dodecanol	259	127		275
$C_{12}H_{26}O$	2-Butyl-1-octanol	246.5	110		
$C_{12}H_{26}O_3$	Diethylene glycol dibutyl ether	256	118		310
$C_{12}H_{26}S$	1-Dodecanethiol	277	128		

Mol. form.	Name	t_B/°C	FP/°C	Fl. limits	IT/°C
$C_{12}H_{27}BO_3$	Tributyl borate	234	93		
$C_{12}H_{27}N$	Tributylamine	216.5	63		
$C_{12}H_{27}O_4P$	Tributyl phosphate	289	146		
$C_{13}H_{12}$	2-Methylbiphenyl	255.5	137		502
$C_{13}H_{12}$	Diphenylmethane	265.0	130		485
$C_{13}H_{14}N_2$	p,p'-Diaminodiphenylmethane	398	220		
$C_{13}H_{26}$	1-Tridecene	232.8	79		
$C_{13}H_{26}O$	2-Tridecanone	263	107		
$C_{13}H_{28}$	Tridecane	235.4	79		
$C_{13}H_{28}O$	1-Tridecanol		121		
$C_{14}H_8O_2$	9,10-Anthracenedione	377	185		
$C_{14}H_{10}$	Anthracene	339.9	121	0.6–?	540
$C_{14}H_{10}$	Phenanthrene	340	171		
$C_{14}H_{12}O_2$	Benzyl benzoate	323.5	148		480
$C_{14}H_{12}O_3$	Benzyl salicylate	320	>100		
$C_{14}H_{14}$	1,1-Diphenylethane	272.6	>100		440
$C_{14}H_{14}O$	Dibenzyl ether	298	135		
$C_{14}H_{16}$	1-Butylnaphthalene	289.3	360		
$C_{14}H_{16}N_2O_2$	o-Dianisidine		206		
$C_{14}H_{23}N$	N,N-Dibutylaniline	274.8	110		
$C_{14}H_{28}$	1-Tetradecene	233	110		235
$C_{14}H_{30}$	Tetradecane	253.5	112	0.5–?	200
$C_{14}H_{30}O$	1-Tetradecanol	289	141		
$C_{15}H_{18}$	1-Pentylnaphthalene	307	124		
$C_{15}H_{24}$	Nonylbenzene	280.5	99		
$C_{15}H_{24}O$	2,6-Di-$tert$-butyl-4-methylphenol	265	127		
$C_{15}H_{26}O_6$	Tributyrin	307.5	180	0.5–?	407
$C_{15}H_{33}N$	Tripentylamine	242.5	102		
$C_{16}H_{14}O$	1,3-Diphenyl-2-buten-1-one	342.5	177		
$C_{16}H_{18}$	2-Butyl-1,1'-biphenyl		>100		430
$C_{16}H_{22}O_4$	Dibutyl phthalate	340	157	0.5–?	402
$C_{16}H_{26}$	Decylbenzene	298	107		
$C_{16}H_{34}$	Hexadecane	286.8	136		202
$C_{16}H_{34}O$	Dioctyl ether	283	>100		205
$C_{16}H_{35}N$	Bis(2-ethylhexyl)amine		132		
$C_{17}H_{20}N_2O$	N,N'-Diethylcarbanilide		150		
$C_{17}H_{34}O$	2-Heptadecanone	320	120		
$C_{17}H_{36}O$	1-Heptadecanol	333	154		
$C_{18}H_{14}$	o-Terphenyl	332	163		
$C_{18}H_{14}$	m-Terphenyl	363	191		
$C_{18}H_{15}O_3P$	Triphenyl phosphite	360	218		
$C_{18}H_{15}O_4P$	Triphenyl phosphate		220		
$C_{18}H_{15}P$	Triphenylphosphine		180		
$C_{18}H_{30}$	Dodecylbenzene	328	140		
$C_{18}H_{32}O_7$	Butyl citrate		157		368
$C_{18}H_{34}O_2$	Oleic acid	360	189		363
$C_{18}H_{34}O_4$	Dibutyl sebacate	344.5	178	0.4–?	365
$C_{18}H_{36}O_2$	Stearic acid		196		395
$C_{18}H_{37}Cl_3Si$	Trichlorooctadecylsilane		89		
$C_{18}H_{38}$	Octadecane	316.3	>100		227
$C_{18}H_{38}O$	1-Octadecanol				450
$C_{19}H_{16}$	Triphenylmethane	359	>100		
$C_{19}H_{38}O$	2-Nonadecanone		124		
$C_{19}H_{38}O_2$	Methyl stearate	443	153		
$C_{19}H_{40}$	Nonadecane	329.9	>100		230
$C_{20}H_{14}O_4$	Diphenyl phthalate		224		
$C_{20}H_{28}$	1-Decylnaphthalene	379	177		
$C_{20}H_{42}$	Eicosane	343	>100		232
$C_{21}H_{21}O_4P$	Tri-o-cresyl phosphate	410	225		385
$C_{21}H_{26}O_3$	4-Octylphenyl salicylate		216		416
$C_{21}H_{32}O_2$	Methyl abietate		180		

Mol. form.	Name	t_B/°C	FP/°C	Fl. limits	IT/°C
$C_{22}H_{42}O_2$	Butyl oleate		180		
$C_{22}H_{42}O_4$	Bis(2-ethylhexyl) adipate		206	0.4–?	377
$C_{22}H_{44}O_2$	Butyl stearate	343	160		355
$C_{23}H_{46}O_2$	Pentyl stearate		185		
$C_{24}H_{20}Sn$	Tetraphenylstannane	420	232		
$C_{24}H_{38}O_4$	Bis(2-ethylhexyl) phthalate	384	218		
$C_{25}H_{48}O_4$	Bis(2-ethylhexyl) azelate		227	0.3–?	374

THRESHOLD LIMITS FOR AIRBORNE CONTAMINANTS

Several organizations recommend limits of exposure to airborne contaminants in the workplace. These include the Occupational Safety and Health Administration (OSHA), the National Institute for Occupational Safety and Health (NIOSH), and the non-governmental organization, American Conference of Governmental Industrial Hygienists (ACGIH). The threshold limit value (TLV) for a substance is defined as the concentration level under which the majority of workers may be repeatedly exposed, day after day, without adverse effects. The TLV recommendations are given in two forms:

- Time-weighted average (TWA) concentration for a normal 8-hr workday and 40-hr workweek.
- Short-term exposure limit (STEL), which should not be exceeded for more than 15 min.

Both kinds of limits are specified for some substances.

The following table gives threshold limit values for a number of substances that may be encountered in the atmosphere of a chemical laboratory or industrial facility. All values refer to the concentration in air at 25°C and normal atmospheric pressure. Data for gases are given in parts per million by volume (ppm). Values for liquids refer to mists or aerosols, and those for solids to dusts or fumes; both are stated in mass concentration (mg/m3). A "C" preceding a value indicates a ceiling limit, which should not be exceeded even for very brief periods because of acute toxic effects of the substance.

Substances are listed by systematic name; molecular formula in the Hill format and Chemical Abstracts Service Registry Number are also given. The Notes column gives further information on the form of the substance and the basis on which the TLV is reported. This column also includes common synonyms and acronyms in brackets (e.g., [MTBE]).

References

1. *2004 TLV's and BEI's*, American Conference of Governmental Industrial Hygienists, 1330 Kemper Meadow Drive, Cincinnati, OH 45240-1634, 2004.
2. *NIOSH Pocket Guide to Chemical Hazards*, U.S. Department of Health and Human Services, National Institute for Occupational Health and Safety, U.S. Government Printing Office, Washington, DC, 1994.
3. Chemical Information Manual, U.S. Department of Labor, Occupational Safety and Health Administration, Washington, DC, 1991.

Substance	Notes	Molecular formula	CAS Reg. No.	Time-weighted average	Short-term exposure limit
Abate	[Temephos]	$C_{16}H_{20}O_6P_2S_3$	3383-96-8	10 mg/m^3	
Acetaldehyde		C_2H_4O	75-07-0		C 25 ppm
Acetic acid		$C_2H_4O_2$	64-19-7	10 ppm	15 ppm
Acetic anhydride		$C_4H_6O_3$	108-24-7	5 ppm	
Acetone	[2-Propanone]	C_3H_6O	67-64-1	500 ppm	750 ppm
Acetone cyanohydrin	as CN	C_4H_7NO	75-86-5		C 5 mg/m^3
Acetonitrile	[Methyl cyanide]	C_2H_3N	75-05-8	20 ppm	
Acetophenone	[Methyl phenyl ketone]	C_8H_8O	98-86-2	10 ppm	
2-(Acetyloxy)benzoic acid	[Aspirin]	$C_9H_8O_4$	50-78-2	5 mg/m^3	
Acrolein	[2-Propenal]	C_3H_4O	107-02-8		C 0.1 ppm
Acrylamide	[2-Propenamide]	C_3H_5NO	79-06-1	0.03 mg/m^3	
Acrylic acid	[2-Propenoic acid]	$C_3H_4O_2$	79-10-7	2 ppm	
Acrylonitrile	[Propenenitrile]	C_3H_3N	107-13-1	2 ppm	
Aldrin		$C_{12}H_8Cl_6$	309-00-2	0.25 mg/m^3	
Allyl alcohol	[2-Propen-1-ol]	C_3H_6O	107-18-6	0.5 ppm	
Allyl glycidyl ether	[AGE]	$C_6H_{10}O_2$	106-92-3	1 ppm	
Allyl propyl disulfide		$C_6H_{12}S_2$	2179-59-1	0.5 ppm	
Aluminum	metal dust	Al	7429-90-5	10 mg/m^3	
Aluminum oxide		Al_2O_3	1344-28-1	10 mg/m^3	
4-Amino-3,5,6-trichloro-2-pyridinecarboxlic acid	[Picloram]	$C_6H_3Cl_3N_2O_2$	1918-02-1	10 mg/m^3	
Ammonia		H_3N	7664-41-7	25 ppm	35 ppm
Ammonium chloride		ClH_4N	12125-02-9	10 mg/m^3	20 mg/m^3
Ammonium perfluorooctanoate		$C_8H_4F_{15}NO_2$	3825-26-1	0.01 mg/m^3	
Ammonium sulfamate		$H_6N_2O_3S$	7773-06-0	10 mg/m^3	
Aniline		C_6H_7N	62-53-3	2 ppm	
Antimony		Sb	7440-36-0	0.5 mg/m^3	
Arsenic		As	7440-38-2	0.01 mg/m^3	
Arsine		AsH_3	7784-42-1	0.05 ppm	
Atrazine		$C_8H_{14}ClN_5$	1912-24-9	5 mg/m^3	
Azinphos-methyl		$C_{10}H_{12}N_3O_3PS_2$	86-50-0	0.2 mg/m^3	
Barium	soluble compounds, as Ba	Ba	7440-39-3	0.5 mg/m^3	
Barium sulfate		BaO_4S	7727-43-7	10 mg/m^3	
Benomyl		$C_{14}H_{18}N_4O_3$	17804-35-2	10 mg/m^3	

Substance	Notes	Molecular formula	CAS Reg. No.	Time-weighted average	Short-term exposure limit
Benzene		C_6H_6	71-43-2	0.5 ppm	2.5 ppm
1,3-Benzenedimethanamine	[*m*-Xylene diamine]	$C_8H_{12}N_2$	1477-55-0		C 0.1 mg/m³
Benzenethiol	[Phenyl mercaptan]	C_6H_6S	108-98-5	0.1 ppm	
p-Benzoquinone	[Quinone]	$C_6H_4O_2$	106-51-4	0.1 ppm	
Benzoyl chloride		C_7H_5ClO	98-88-4		C 0.5 ppm
Benzoyl peroxide		$C_{14}H_{10}O_4$	94-36-0	5 mg/m³	
Benzyl acetate		$C_9H_{10}O_2$	140-11-4	10 ppm	
Beryllium	and compounds, as Be	Be	7440-41-7	0.002 mg/m³	0.01 mg/m³
Biphenyl		$C_{12}H_{10}$	92-52-4	0.2 ppm	
Bis(2-aminoethyl)amine	[Diethylenetriamine]	$C_4H_{13}N_3$	111-40-0	1 ppm	
Bis(2-chloroethyl) ether	[2,2'-Dichlorethyl ether]	$C_4H_8Cl_2O$	111-44-4	5 ppm	10 ppm
Bis(chloromethyl) ether		$C_2H_4Cl_2O$	542-88-1	0.001 ppm	
Bis(2-dimethylaminoethyl) ether	[DMAEE]	$C_8H_{20}N_2O$	3033-62-3	0.05 ppm	0.15 ppm
Bis(2-ethylhexyl) phthalate	[Di-*sec*-octyl phthalate; DEHP]	$C_{24}H_{38}O_4$	117-81-7	5 mg/m³	
Bismuth telluride		Bi_2Te_3	1304-82-1	10 mg/m³	
Boron oxide		B_2O_3	1303-86-2	10 mg/m³	
Boron tribromide		BBr_3	10294-33-4		C 1 ppm
Boron trifluoride		BF_3	7637-07-2		C 1 ppm
Bromacil	[5-Bromo-3-*sec*-butyl-6-methyluracil]	$C_9H_{13}BrN_2O_2$	314-40-9	10 mg/m³	
Bromine		Br_2	7726-95-6	0.1 ppm	0.2 ppm
Bromine pentafluoride		BrF_5	7789-30-2	0.1 ppm	
Bromochloromethane	[Halon 1011]	CH_2BrCl	74-97-5	200 ppm	
2-Bromo-2-chloro-1,1,1-trifluoroethane	[Halothane]	$C_2HBrClF_3$	151-67-7	50 ppm	
Bromoethane	[Ethyl bromide]	C_2H_5Br	74-96-4	5 ppm	
Bromoethene	[Vinyl bromide]	C_2H_3Br	593-60-2	0.5 ppm	
Bromomethane	[Methyl bromide]	CH_3Br	74-83-9	1 ppm	
Bromotrifluoromethane		$CBrF_3$	75-63-8	1000 ppm	
1,3-Butadiene		C_4H_6	106-99-0	2 ppm	
Butane		C_4H_{10}	106-97-8	1000 ppm	
1-Butanethiol	[Butyl mercaptan]	$C_4H_{10}S$	109-79-5	0.5 ppm	
1-Butanol	[Butyl alcohol]	$C_4H_{10}O$	71-36-3	20 ppm	
2-Butanol	[*sec*-Butyl alcohol]	$C_4H_{10}O$	78-92-2	100 ppm	
2-Butanone	[Methyl ethyl ketone; MEK]	C_4H_8O	78-93-3	200 ppm	300 ppm
2-Butanone peroxide	[Methyl ethyl ketone peroxide]	$C_8H_{16}O_4$	1338-23-4		C 0.2 ppm
trans-2-Butenal	[Crotonaldehyde]	C_4H_6O	4170-30-3		C 0.3 ppm
3-Buten-2-one	[Methyl vinyl ketone]	C_4H_6O	78-94-4		C 0.2 ppm
2-Butoxyethanol	[EGBE]	$C_6H_{14}O_2$	111-76-2	20 ppm	
2-Butoxyethyl acetate	[EGBEA]	$C_8H_{16}O_3$	112-07-2	20 ppm	
Butyl acetate		$C_6H_{12}O_2$	123-86-4	150 ppm	200 ppm
sec-Butyl acetate		$C_6H_{12}O_2$	105-46-4	200 ppm	
tert-Butyl acetate		$C_6H_{12}O_2$	540-88-5	200 ppm	
Butyl acrylate		$C_7H_{12}O_2$	141-32-2	2 ppm	
Butylamine		$C_4H_{11}N$	109-73-9		C 5 ppm
tert-Butyl chromate	as CrO_3	$C_8H_{18}CrO_4$	1189-85-1		C 0.1 mg/m³
tert-Butyl ethyl ether	[ETBE]	$C_6H_{14}O$	637-92-3	5 ppm	
Butyl glycidyl ether	[BGE]	$C_7H_{14}O_2$	2426-08-6	25 ppm	
Butyl lactate		$C_7H_{14}O_3$	138-22-7	5 ppm	
1-*tert*-Butyl-4-methylbenzene	[*p-tert*-Butyltoluene]	$C_{11}H_{16}$	98-51-1	1 ppm	
2-*sec*-Butylphenol		$C_{10}H_{14}O$	89-72-5	5 ppm	
Cadmium	metal	Cd	7440-43-9	0.01 mg/m³	
Cadmium	compounds, as Cd	Cd	7440-43-9	0.002 mg/m³	
Calcium carbonate		$CCaO_3$	1317-65-3	10 mg/m³	
Calcium chromate	as Cr	$CaCrO_4$	13765-19-0	0.001 mg/m³	
Calcium cyanamide		$CCaN_2$	156-62-7	0.5 mg/m³	
Calcium hydroxide		CaH_2O_2	1305-62-0	5 mg/m³	
Calcium metasilicate		CaO_3Si	1344-95-2	10 mg/m³	
Calcium oxide		CaO	1305-78-8	2 mg/m³	
Calcium sulfate		CaO_4S	7778-18-9	10 mg/m³	

Substance	Notes	Molecular formula	CAS Reg. No.	Time-weighted average	Short-term exposure limit
Camphor		$C_{10}H_{16}O$	76-22-2	2 ppm	3 ppm
Caprolactam		$C_6H_{11}NO$	105-60-2	5 mg/m³	
Captafol		$C_{10}H_9Cl_4NO_2S$	2425-06-1	0.1 mg/m³	
Captan		$C_9H_8Cl_3NO_2S$	133-06-2	5 mg/m³	
Carbaryl		$C_{12}H_{11}NO_2$	63-25-2	5 mg/m³	
Carbofuran		$C_{12}H_{15}NO_3$	1563-66-2	0.1 mg/m³	
Carbon black		C	1333-86-4	3.5 mg/m³	
Carbon dioxide		CO_2	124-38-9	5000 ppm	30,000 ppm
Carbon disulfide		CS_2	75-15-0	10 ppm	
Carbon monoxide		CO	630-08-0	25 ppm	
Carbonyl chloride	[Phosgene]	CCl_2O	75-44-5	0.1 ppm	
Carbonyl fluoride		CF_2O	353-50-4	2 ppm	5 ppm
Cesium hydroxide		CsHO	21351-79-1	2 mg/m³	
Chlordane		$C_{10}H_6Cl_8$	57-74-9	0.5 mg/m³	
Chlorine		Cl_2	7782-50-5	0.5 ppm	1 ppm
Chlorine dioxide		ClO_2	10049-04-4	0.1 ppm	0.3 ppm
Chlorine trifluoride		ClF_3	7790-91-2		C 0.1 ppm
Chloroacetaldehyde		C_2H_3ClO	107-20-0		C 1 ppm
Chloroacetone		C_3H_5ClO	78-95-5		C 1 ppm
α-Chloroacetophenone		C_8H_7ClO	532-27-4	0.05 ppm	
Chloroacetyl chloride		$C_2H_2Cl_2O$	79-04-9	0.05 ppm	0.15 ppm
Chlorobenzene		C_6H_5Cl	108-90-7	10 ppm	
o-Chlorobenzylidene malononitrile		$C_{10}H_5ClN_2$	2698-41-1		C 0.05 ppm
2-Chloro-1,3-butadiene	[Chloroprene]	C_4H_5Cl	126-99-8	10 ppm	
Chlorodifluoromethane		$CHClF_2$	75-45-6	1000 ppm	
Chloroethane	[Ethyl chloride]	C_2H_5Cl	75-00-3	100 ppm	
2-Chloroethanol	[Ethylene chlorohydrin]	C_2H_5ClO	107-07-3		C 1 ppm
Chloroethene	[Vinyl chloride]	C_2H_3Cl	75-01-4	1 ppm	
Chloromethane	[Methyl chloride]	CH_3Cl	74-87-3	50 ppm	100 ppm
(Chloromethyl)benzene	[Benzyl chloride]	C_7H_7Cl	100-44-7	1 ppm	
1-Chloro-4-nitrobenzene		$C_6H_4ClNO_2$	100-00-5	0.1 ppm	
1-Chloro-1-nitropropane		$C_3H_6ClNO_2$	600-25-9	2 ppm	
Chloropentafluoroethane		C_2ClF_5	76-15-3	1000 ppm	
2-Chloropropanoic acid		$C_3H_5ClO_2$	598-78-7	0.1 ppm	
2-Chloro-1-propanol		C_3H_7ClO	78-89-7	1 ppm	
1-Chloro-2-propanol		C_3H_7ClO	127-00-4	1 ppm	
3-Chloropropene	[Allyl chloride]	C_3H_5Cl	107-05-1	1 ppm	2 ppm
2-Chlorostyrene		C_8H_7Cl	2039-87-4	50 ppm	75 ppm
2-Chlorotoluene		C_7H_7Cl	95-49-8	50 ppm	
Chlorpyrifos		$C_9H_{11}Cl_3NO_3PS$	2921-88-2	0.1 mg/m³	
Chromium	metal	Cr	7440-47-3	0.5 mg/m³	
Chromium	Cr(III) compounds, as Cr	Cr	7440-47-3	0.5 mg/m³	
Chromium	soluble Cr(VI) compounds, as Cr	Cr	7440-47-3	0.05 mg/m³	
Chromium	insoluble Cr(VI) compounds, as Cr	Cr	7440-47-3	0.01 mg/m³	
Chromyl chloride		Cl_2CrO_2	14977-61-8	0.025 ppm	
Clopidol		$C_7H_7Cl_2NO$	2971-90-6	10 mg/m³	
Cobalt	metal and inorganic compounds, as Co	Co	7440-48-4	0.02 mg/m³	
Cobalt carbonyl	as Co	$C_8Co_2O_8$	10210-68-1	0.1 mg/m³	
Cobalt hydrocarbonyl	as Co	C_4HCoO_4	16842-03-8	0.1 mg/m³	
Copper	fume	Cu	7440-50-8	0.2 mg/m³	
Cresol	all isomers	C_7H_8O	1319-77-3	5 ppm	
Crufomate		$C_{12}H_{19}ClNO_3P$	299-86-5	5 mg/m³	
Cyanamide		CH_2N_2	420-04-2	2 mg/m³	
Cyanide ion [CN⁻]	cyanide salts, as CN	CN	57-12-5		C 5 mg/m³
Cyanogen		C_2N_2	460-19-5	10 ppm	
Cyanogen chloride		CCIN	506-77-4		C 0.3 ppm
Cyclohexane		C_6H_{12}	110-82-7	100 ppm	
Cyclohexanol		$C_6H_{12}O$	108-93-0	50 ppm	
Cyclohexanone		$C_6H_{10}O$	108-94-1	20 ppm	50 ppm

Substance	Notes	Molecular formula	CAS Reg. No.	Time-weighted average	Short-term exposure limit
Cyclohexene		C_6H_{10}	110-83-8	300 ppm	
Cyclohexylamine		$C_6H_{13}N$	108-91-8	10 ppm	
Cyclonite	[Hexahydro-1,3,5-trinitro-1,3,5-triazine]	$C_3H_6N_6O_6$	121-82-4	0.5 mg/m³	
1,3-Cyclopentadiene		C_5H_6	542-92-7	75 ppm	
Cyclopentane		C_5H_{10}	287-92-3	600 ppm	
Cyhexatin		$C_{18}H_{34}OSn$	13121-70-5	5 mg/m³	
Decaborane(14)		$B_{10}H_{14}$	17702-41-9	0.05 ppm	0.15 ppm
Demeton-S-methyl		$C_6H_{15}O_3PS_2$	919-86-8	0.05 mg/m³	
Diacetone alcohol		$C_6H_{12}O_2$	123-42-2	50 ppm	
4,4'-Diaminodiphenylmethane	[4,4-Methylene dianiline]	$C_{13}H_{14}N_2$	101-77-9	0.1 ppm	
Diazinon		$C_{12}H_{21}N_2O_3PS$	333-41-5	0.01 mg/m³	
Diazomethane		CH_2N_2	334-88-3	0.2 ppm	
Diborane		B_2H_6	19287-45-7	0.1 ppm	
Dibromodifluoromethane		CBr_2F_2	75-61-6	100 ppm	
2-Dibutylaminoethanol		$C_{10}H_{23}NO$	102-81-8	0.5 ppm	
2,6-Di-tert-butyl-4-methylphenol	[Butylated hydroxytoluene; BHT]	$C_{15}H_{24}O$	128-37-0	2 mg/m³	
Dibutylphenyl phosphate		$C_{14}H_{23}O_4P$	2528-36-1	0.3 ppm	
Dibutyl phosphate		$C_8H_{19}O_4P$	107-66-4	1 ppm	2 ppm
Dibutyl phthalate		$C_{16}H_{22}O_4$	84-74-2	5 mg/m³	
Dichloroacetylene		C_2Cl_2	7572-29-4		C 0.1 ppm
o-Dichlorobenzene		$C_6H_4Cl_2$	95-50-1	25 ppm	50 ppm
p-Dichlorobenzene		$C_6H_4Cl_2$	106-46-7	10 ppm	
1,4-Dichloro-2-butene	both isomers	$C_4H_6Cl_2$	764-41-0	0.005 ppm	
Dichlorodifluoromethane		CCl_2F_2	75-71-8	1000 ppm	
1,3-Dichloro-5,5-dimethyl hydantoin		$C_5H_6Cl_2N_2O_2$	118-52-5	0.2 mg/m³	0.4 mg/m³
Dichlorodiphenyltrichloro-ethane	[DDT]	$C_{14}H_9Cl_5$	50-29-3	1 mg/m³	
1,1-Dichloroethane	[Ethylidene dichloride]	$C_2H_4Cl_2$	75-34-3	100 ppm	
1,2-Dichloroethane	[Ethylene dichloride]	$C_2H_4Cl_2$	107-06-2	10 ppm	
1,1-Dichloroethene	[Vinylidene chloride]	$C_2H_2Cl_2$	75-35-4	5 ppm	
1,2-Dichloroethene	both isomers	$C_2H_2Cl_2$	540-59-0	200 ppm	
Dichlorofluoromethane		$CHCl_2F$	75-43-4	10 ppm	
Dichloromethane	[Methylene chloride]	CH_2Cl_2	75-09-2	50 ppm	
1,1-Dichloro-1-nitroethane		$C_2H_3Cl_2NO_2$	594-72-9	2 ppm	
(2,4-Dichlorophenoxy)acetic acid	[2,4-D]	$C_8H_6Cl_2O_3$	94-75-7	10 mg/m³	
1,2-Dichloropropane		$C_3H_6Cl_2$	78-87-5	75 ppm	110 ppm
2,2-Dichloropropanoic acid		$C_3H_4Cl_2O_2$	75-99-0	5 mg/m³	
1,3-Dichloropropene	both isomers	$C_3H_4Cl_2$	542-75-6	1 ppm	
1,2-Dichloro-1,1,2,2-tetrafluoroethane		$C_2Cl_2F_4$	76-14-2	1000 ppm	
Dichlorvos		$C_4H_7Cl_2O_4P$	62-73-7	0.1 mg/m³	
Dicrotophos		$C_8H_{16}NO_5P$	141-66-2	0.05 mg/m³	
m-Dicyanobenzene	[m-Phthalodinitrile]	$C_8H_4N_2$	626-17-5	5 mg/m³	
Dicyclopentadiene		$C_{10}H_{12}$	77-73-6	5 ppm	
Dieldrin		$C_{12}H_8Cl_6O$	60-57-1	0.25 mg/m³	
Diethanolamine	[Bis(2-hydroxyethyl)amine]	$C_4H_{11}NO_2$	111-42-2	2 mg/m³	
Diethylamine		$C_4H_{11}N$	109-89-7	5 ppm	15 ppm
2-Diethylaminoethanol		$C_6H_{15}NO$	100-37-8	2 ppm	
Diethyl ether	[Ethyl ether]	$C_4H_{10}O$	60-29-7	400 ppm	500 ppm
Diethyl phthalate		$C_{12}H_{14}O_4$	84-66-2	5 mg/m³	
1,1-Difluoroethene		$C_2H_2F_2$	75-38-7	500 ppm	
Diglycidyl ether		$C_6H_{10}O_3$	2238-07-5	0.1 ppm	
Diisopropylamine		$C_6H_{15}N$	108-18-9	5 ppm	
Diisopropyl ether		$C_6H_{14}O$	108-20-3	250 ppm	310 ppm
Dimethoxymethane	[Methylal]	$C_3H_8O_2$	109-87-5	1000 ppm	
N,N-Dimethylacetamide		C_4H_9NO	127-19-5	10 ppm	

Substance	Notes	Molecular formula	CAS Reg. No.	Time-weighted average	Short-term exposure limit
Dimethylamine		C_2H_7N	124-40-3	5 ppm	15 ppm
N,N-Dimethylaniline		$C_8H_{11}N$	121-69-7	5 ppm	10 ppm
2,2-Dimethylbutane	[Neohexane]	C_6H_{14}	75-83-2	500 ppm	1000 ppm
2,3-Dimethylbutane		C_6H_{14}	79-29-8	500 ppm	1000 ppm
N,N-Dimethylformamide	[DMF]	C_3H_7NO	68-12-2	10 ppm	
2,6-Dimethyl-4-heptanone	[Diisobutyl ketone]	$C_9H_{18}O$	108-83-8	25 ppm	
1,1-Dimethylhydrazine		$C_2H_8N_2$	57-14-7	0.01 ppm	
Dimethyl mercury		C_2H_6Hg	593-74-8	0.01 mg/m³	
Dimethyl phthalate		$C_{10}H_{10}O_4$	131-11-3	5 mg/m³	
2,2-Dimethyl-1-propanol acetate		$C_7H_{14}O_2$	926-41-0	50 ppm	100 ppm
Dimethyl sulfate		$C_2H_6O_4S$	77-78-1	0.1 ppm	
Dimethyl sulfide		C_2H_6S	75-18-3	10 ppm	
Dinitrobenzene	all isomers	$C_6H_4N_2O_4$	25154-54-5	0.15 ppm	
Dinitrotoluene	all isomers	$C_7H_6N_2O_4$	25321-14-6	0.2 mg/m³	
1,4-Dioxane		$C_4H_8O_2$	123-91-1	20 ppm	
Dioxathion		$C_{12}H_{26}O_6P_2S_4$	78-34-2	0.1 mg/m³	
1,3-Dioxolane		$C_3H_6O_2$	646-06-0	20 ppm	
Diphenylamine		$C_{12}H_{11}N$	122-39-4	10 mg/m³	
Diphenyl ether		$C_{12}H_{10}O$	101-84-8	1 ppm	2 ppm
4,4′-Diphenylmethane diisocyanate	[Methylene diphenyl isocyanate; MDI]	$C_{15}H_{10}N_2O_2$	101-68-8	0.005 ppm	
Diquat		$C_{12}H_{12}N_2$	231-36-7	0.5 mg/m³	
Disulfiram		$C_{10}H_{20}N_2S_4$	97-77-8	2 mg/m³	
Disulfoton		$C_8H_{19}O_2PS_3$	298-04-4	0.05 mg/m³	
Diuron		$C_9H_{10}Cl_2N_2O$	330-54-1	10 mg/m³	
Divinyl benzene	all isomers	$C_{10}H_{10}$	1321-74-0	10 ppm	
1-Dodecanethiol		$C_{12}H_{26}S$	112-55-0	0.1 ppm	
Endosulfan		$C_9H_6Cl_6O_3S$	115-29-7	0.1 mg/m³	
Endrin		$C_{12}H_8Cl_6O$	72-20-8	0.1 mg/m³	
Enflurane		$C_3H_2ClF_5O$	13838-16-9	75 ppm	
Epichlorohydrin	[(Chloromethyl)oxirane]	C_3H_5ClO	106-89-8	0.5 ppm	
1,2-Epoxy-4-(epoxyethyl)cyclohexane	[Vinylcyclohexene dioxide]	$C_8H_{12}O_2$	106-87-6	0.1 ppm	
Ethane		C_2H_6	74-84-0	1000 ppm	
1,2-Ethanediamine	[Ethylenediamine]	$C_2H_8N_2$	107-15-3	10 ppm	
1,2-Ethanediol	[Ethylene glycol]	$C_2H_6O_2$	107-21-1		C 100 mg/m³
1,2-Ethanediol, dinitrate	[Ethylene glycol dinitrate; EGDN]	$C_2H_4N_2O_6$	628-96-6	0.05 ppm	
Ethanethiol	[Ethyl mercaptan]	C_2H_6S	75-08-1	0.5 ppm	
Ethanol	[Ethyl alcohol]	C_2H_6O	64-17-5	1000 ppm	
Ethanolamine		C_2H_7NO	141-43-5	3 ppm	6 ppm
Ethion		$C_9H_{22}O_4P_2S_4$	563-12-2	0.05 mg/m³	
Ethoxydimethylsilane		$C_4H_{12}OSi$	14857-34-2	0.5 ppm	1.5 ppm
2-Ethoxyethanol	[Ethylene glycol monoethyl ether; EGEE]	$C_4H_{10}O_2$	110-80-5	5 ppm	
2-Ethoxyethyl acetate	[Ethylene glycol monoethyl ether acetate; EGEEA]	$C_6H_{12}O_3$	111-15-9	5 ppm	
Ethyl acetate		$C_4H_8O_2$	141-78-6	400 ppm	
Ethyl acrylate	[Ethyl propenoate]	$C_5H_8O_2$	140-88-5	5 ppm	15 ppm
Ethylamine		C_2H_7N	75-04-7	5 ppm	15 ppm
Ethylbenzene		C_8H_{10}	100-41-4	100 ppm	125 ppm
Ethyl 2-cyanoacrylate	[Ethyl 2-cyano-2-propenoate]	$C_6H_7NO_2$	7085-85-0	0.2 ppm	
Ethyleneimine	[Aziridine]	C_2H_5N	151-56-4	0.5 ppm	
Ethyl formate		$C_3H_6O_2$	109-94-4	100 ppm	
2-Ethylhexanoic acid		$C_8H_{16}O_2$	149-57-5	5 mg/m³	
5-Ethylidene-2-norbornene		C_9H_{12}	16219-75-3		C 5 ppm
N-Ethylmorpholine		$C_6H_{13}NO$	100-74-3	5 ppm	
Ethyl p-nitrophenyl benzenethiophosphate	[EPN]	$C_{14}H_{14}NO_4PS$	2104-64-5	0.1 mg/m³	
Ethyl silicate		$C_8H_{20}O_4Si$	78-10-4	10 ppm	
Fenamiphos		$C_{13}H_{22}NO_3PS$	22224-92-6	0.1 mg/m³	

Substance	Notes	Molecular formula	CAS Reg. No.	Time-weighted average	Short-term exposure limit
Fensulfothion		$C_{11}H_{17}O_4PS_2$	115-90-2	0.1 mg/m³	
Fenthion		$C_{10}H_{15}O_3PS_2$	55-38-9	0.2 mg/m³	
Ferbam		$C_9H_{18}FeN_3S_6$	14484-64-1	10 mg/m³	
Ferrocene	[Dicyclopentadienyl iron]	$C_{10}H_{10}Fe$	102-54-5	10 mg/m³	
Fluoride ion [F⁻]	fluoride salts, as F	F	16984-48-8	2.5 mg/m³	
Fluorine		F_2	7782-41-4	1 ppm	2 ppm
Fluorine monoxide	[Oxygen difluoride]	F_2O	7783-41-7		C 0.05 ppm
Fluoroethene	[Vinyl fluoride]	C_2H_3F	75-02-5	1 ppm	
Fonofos		$C_{10}H_{15}OPS_2$	944-22-9	0.1 mg/m³	
Formaldehyde		CH_2O	50-00-0		C 0.3 ppm
Formamide		CH_3NO	75-12-7	10 ppm	
Formic acid		CH_2O_2	64-18-6	5 ppm	10 ppm
Furfural	[2-Furaldehyde]	$C_5H_4O_2$	98-01-1	2 ppm	
Furfuryl alcohol	[2-Furanmethanol]	$C_5H_6O_2$	98-00-0	10 ppm	15 ppm
Germane	[Germanium tetrahydride]	GeH_4	7782-65-2	0.2 ppm	
Glycerol	[1,2,3-Propanetriol]	$C_3H_8O_3$	56-81-5	10 mg/m³	
Glyoxal		$C_2H_2O_2$	107-22-2	0.1 mg/m³	
Graphite	except fibers	C	7440-44-0	2 mg/m³	
Hafnium	metal and compounds, as Hf	Hf	7440-58-6	0.5 mg/m³	
Heptachlor		$C_{10}H_5Cl_7$	76-44-8	0.05 mg/m³	
Heptachlor epoxide		$C_{10}H_5Cl_7O$	1024-57-3	0.05 mg/m³	
Heptane		C_7H_{16}	142-82-5	400 ppm	500 ppm
2-Heptanone	[Methyl pentyl ketone]	$C_7H_{14}O$	110-43-0	50 ppm	
3-Heptanone	[Ethyl butyl ketone]	$C_7H_{14}O$	106-35-4	50 ppm	75 ppm
4-Heptanone	[Dipropyl ketone]	$C_7H_{14}O$	123-19-3	50 ppm	
Hexachlorobenzene		C_6Cl_6	118-74-1	0.002 mg/m³	
Hexachloro-1,3-butadiene		C_4Cl_6	87-68-3	0.02 ppm	
1,2,3,4,5,6-Hexachlorocyclohexane	[Lindane]	$C_6H_6Cl_6$	58-89-9	0.5 mg/m³	
Hexachloro-1,3-cyclopentadiene		C_5Cl_6	77-47-4	0.01 ppm	
Hexachloroethane	[Perchloroethane]	C_2Cl_6	67-72-1	1 ppm	
Hexachloronaphthalene	all isomers	$C_{10}H_2Cl_6$	1335-87-1	0.2 mg/m³	
Hexahydro-1,3-isobenzofurandione	[Hexahydrophthalic anhydride]	$C_8H_{10}O_3$	85-42-7		C 0.005 mg/m³
Hexamethylene diisocyanate		$C_8H_{12}N_2O_2$	822-06-0	0.005 ppm	
Hexane		C_6H_{14}	110-54-3	50 ppm	
1,6-Hexanediamine	[Hexamethylenediamine]	$C_6H_{16}N_2$	124-09-4	0.5 ppm	
Hexanedinitrile	[Adiponitrile]	$C_6H_8N_2$	111-69-3	2 ppm	
1,6-Hexanedioic acid	[Adipic acid]	$C_6H_{10}O_4$	124-04-9	5 mg/m³	
2-Hexanone	[Butyl methyl ketone]	$C_6H_{12}O$	591-78-6	5 ppm	10 ppm
1-Hexene		C_6H_{12}	592-41-6	50 ppm	
sec-Hexyl acetate		$C_8H_{16}O_2$	108-84-9	50 ppm	
Hydrazine		H_4N_2	302-01-2	0.01 ppm	
Hydrazoic acid		HN_3	7782-79-8		C 0.11 ppm
Hydrogen bromide		BrH	10035-10-6		C 2 ppm
Hydrogen chloride		ClH	7647-01-0		C 2 ppm
Hydrogen cyanide		CHN	74-90-8		C 4.7 ppm
Hydrogen fluoride		FH	7664-39-3		C 3 ppm
Hydrogen peroxide		H_2O_2	7722-84-1	1 ppm	
Hydrogen selenide		H_2Se	7783-07-5	0.05 ppm	
Hydrogen sulfide		H_2S	7783-06-4	10 ppm	15 ppm
p-Hydroquinone	[1,4-Benzenediol]	$C_6H_6O_2$	123-31-9	2 mg/m³	
2-Hydroxypropyl acrylate		$C_6H_{10}O_3$	999-61-1	0.5 ppm	
Indene		C_9H_8	95-13-6	10 ppm	
Indium	metal and compounds, as In	In	7440-74-6	0.1 mg/m³	
Iodine		I_2	7553-56-2		C 0.1 ppm
Iodomethane	[Methyl iodide]	CH_3I	74-88-4	2 ppm	
Iron ion [Fe⁺²]	soluble ferrous salts, as Fe	Fe	15438-31-0	1 mg/m³	
Iron ion [Fe⁺³]	soluble ferric salts, as Fe	Fe	20074-52-6	1 mg/m³	

Substance	Notes	Molecular formula	CAS Reg. No.	Time-weighted average	Short-term exposure limit
Iron(III) oxide	dust and fume, as Fe	Fe_2O_3	1309-37-1	5 mg/m^3	
Iron pentacarbonyl		C_5FeO_5	13463-40-6	0.1 ppm	0.2 ppm
Isobutane	[2-Methylpropane]	C_4H_{10}	75-28-5	1000 ppm	
Isobutyl acetate		$C_6H_{12}O_2$	110-19-0	150 ppm	
Isobutyl nitrite		$C_4H_9NO_2$	542-56-3		C 1 ppm
Isopentane	[2-Methylbutane]	C_5H_{12}	78-78-4	600 ppm	
Isopentyl acetate	[Isoamyl acetate]	$C_7H_{14}O_2$	123-92-2	50 ppm	100 ppm
Isophorone		$C_9H_{14}O$	78-59-1		C 5 ppm
Isophorone diisocyanate		$C_{12}H_{18}N_2O_2$	4098-71-9	0.005 ppm	
Isopropenylbenzene	[α-Methyl styrene]	C_9H_{10}	98-83-9	50 ppm	100 ppm
2-Isopropoxyethanol		$C_5H_{12}O_2$	109-59-1	25 ppm	
Isopropyl acetate		$C_5H_{10}O_2$	108-21-4	100 ppm	200 ppm
Isopropylamine		C_3H_9N	75-31-0	5 ppm	10 ppm
N-Isopropylaniline		$C_9H_{13}N$	768-52-5	2 ppm	
Isopropylbenzene	[Cumene]	C_9H_{12}	98-82-8	50 ppm	
Isopropyl glycidyl ether	[IGE]	$C_6H_{12}O_2$	4016-14-2	50 ppm	75 ppm
Kaolin			1332-58-7	2 mg/m^3	
Ketene		C_2H_2O	463-51-4	0.5 ppm	1.5 ppm
Lead	metal and compounds, as Pb	Pb	7439-92-1	0.05 mg/m^3	
Lead(II) arsenate		$As_2O_8Pb_3$	7784-40-9	0.15 mg/m^3	
Lead(II) chromate	as Pb	CrO_4Pb	7758-97-6	0.05 mg/m^3	
Lithium hydride		HLi	7580-67-8	0.025 mg/m^3	
Magnesium carbonate	[Magnesite]	$CMgO_3$	546-93-0	10 mg/m^3	
Magnesium oxide		MgO	1309-48-4	10 mg/m^3	
Malathion		$C_{10}H_{19}O_6PS_2$	121-75-5	1 mg/m^3	
Maleic anhydride		$C_4H_2O_3$	108-31-6	0.1 ppm	
Manganese	metal and inorganic compounds, as Mn	Mn	7439-96-5	0.2 mg/m^3	
Manganese cyclopentadienyl tricarbonyl	as Mn	$C_8H_5MnO_3$	12079-65-1	0.1 mg/m^3	
Manganese 2-methylcyclopentadienyl tricarbonyl	as Mn	$C_9H_7MnO_3$	12108-13-3	0.2 mg/m^3	
Mercury	metal and inorganic compounds, as Hg	Hg	7439-97-6	0.025 mg/m^3	
Mercury	alkyl compounds, as Hg	Hg	7439-97-6	0.01 mg/m^3	0.03 mg/m^3
Mercury	aryl compounds, as Hg	Hg	7439-97-6	0.1 mg/m^3	
Mesityl oxide	[Isobutenyl methyl ketone]	$C_6H_{10}O$	141-79-7	15 ppm	25 ppm
Methacrylic acid	[2-Methylpropenoic acid]	$C_4H_6O_2$	79-41-4	20 ppm	
Methane		CH_4	74-82-8	1000 ppm	
Methanethiol	[Methyl mercaptan]	CH_4S	74-93-1	0.5 ppm	
Methanol	[Methyl alcohol]	CH_4O	67-56-1	200 ppm	250 ppm
Methomyl		$C_5H_{10}N_2O_2S$	16752-77-5	2.5 mg/m^3	
2-Methoxyaniline	[o-Anisidine]	C_7H_9NO	90-04-0	0.5 mg/m^3	
4-Methoxyaniline	[p-Anisidine]	C_7H_9NO	104-94-9	0.5 mg/m^3	
Methoxychlor		$C_{16}H_{15}Cl_3O_2$	72-43-5	10 mg/m^3	
2-Methoxyethanol	[Ethylene glycol monomethyl ether; EGME]	$C_3H_8O_2$	109-86-4	5 ppm	
2-Methoxyethyl acetate	[Ethylene glycol monomethyl ether acetate; EGMEA]	$C_5H_{10}O_3$	110-49-6	5 ppm	
2-Methoxy-2-methylbutane	[Methyl tert-pentyl ether; TAME]	$C_6H_{14}O$	994-05-8	20 ppm	
4-Methoxyphenol		$C_7H_8O_2$	150-76-5	5 mg/m^3	
1-Methoxy-2-propanol	[1,2-Propylene glycol monomethyl ether; PGME]	$C_4H_{10}O_2$	107-98-2	100 ppm	150 ppm
Methyl acetate		$C_3H_6O_2$	79-20-9	200 ppm	250 ppm
Methyl acrylate	[Methyl propenoate]	$C_4H_6O_2$	96-33-3	2 ppm	
2-Methylacrylonitrile	[2-Methylpropenenitrile]	C_4H_5N	126-98-7	1 ppm	
Methylamine		CH_5N	74-89-5	5 ppm	15 ppm
2-Methylaniline	[o-Toluidine]	C_7H_9N	95-53-4	2 ppm	
3-Methylaniline	[m-Toluidine]	C_7H_9N	108-44-1	2 ppm	
4-Methylaniline	[p-Toluidine]	C_7H_9N	106-49-0	2 ppm	
N-Methylaniline		C_7H_9N	100-61-8	0.5 ppm	

Substance	Notes	Molecular formula	CAS Reg. No.	Time-weighted average	Short-term exposure limit
3-Methyl-1-butanol	[Isoamyl alcohol]	$C_5H_{12}O$	123-51-3	100 ppm	125 ppm
2-Methyl-1-butanol acetate		$C_7H_{14}O_2$	624-41-9	50 ppm	100 ppm
3-Methyl-2-butanol acetate		$C_7H_{14}O_2$	5343-96-4	50 ppm	100 ppm
3-Methyl-2-butanone	[Methyl isopropyl ketone]	$C_5H_{10}O$	563-80-4	200 ppm	
Methyl tert-butyl ether	[MTBE]	$C_5H_{12}O$	1634-04-4	50 ppm	
Methyl 2-cyanoacrylate	[Mecrylate]	$C_5H_5NO_2$	137-05-3	0.2 ppm	
Methylcyclohexane		C_7H_{14}	108-87-2	400 ppm	
Methylcyclohexanol	all isomers	$C_7H_{14}O$	25639-42-3	50 ppm	
2-Methylcyclohexanone		$C_7H_{12}O$	583-60-8	50 ppm	75 ppm
Methyl demeton		$C_6H_{15}O_3PS_2$	8022-00-2	0.5 mg/m³	
2-Methyl-3,5-dinitrobenzamide	[Dinitolmide]	$C_8H_7N_3O_5$	148-01-6	5 mg/m³	
2-Methyl-4,6-dinitrophenol	[Dinitro-o-cresol]	$C_7H_6N_2O_5$	534-52-1	0.2 mg/m³	
4,4'-Methylenebis[2-chloroaniline]	[MBOCA]	$C_{13}H_{12}Cl_2N_2$	101-14-4	0.01 ppm	
Methylenebis(4-cyclohexylisocyanate)		$C_{15}H_{22}N_2O_2$	5124-30-1	0.005 ppm	
Methyl formate		$C_2H_4O_2$	107-31-3	100 ppm	150 ppm
6-Methyl-1-heptanol	[Isooctyl alcohol]	$C_8H_{18}O$	26952-21-6	50 ppm	
5-Methyl-3-heptanone		$C_8H_{16}O$	541-85-5	25 ppm	
5-Methyl-2-hexanone	[Methyl isopentyl ketone]	$C_7H_{14}O$	110-12-3	50 ppm	
Methylhydrazine		CH_6N_2	60-34-4	0.01 ppm	
Methyl isocyanate		C_2H_3NO	624-83-9	0.02 ppm	
Methyl methacrylate	[Methyl 2-methyl-2-propenoate]	$C_5H_8O_2$	80-62-6	50 ppm	100 ppm
Methyloxirane	[1,2-Propylene oxide]	C_3H_6O	75-56-9	2 ppm	
Methyl parathion		$C_8H_{10}NO_5PS$	298-00-0	0.2 mg/m³	
2-Methylpentane	[Isohexane]	C_6H_{14}	107-83-5	500 ppm	1000 ppm
3-Methylpentane		C_6H_{14}	96-14-0	500 ppm	1000 ppm
2-Methyl-2,4-pentanediol	[Hexylene glycol]	$C_6H_{14}O_2$	107-41-5		C 25 ppm
4-Methyl-2-pentanol	[Methyl isobutyl carbinol]	$C_6H_{14}O$	108-11-2	25 ppm	40 ppm
4-Methyl-2-pentanone	[Isobutyl methyl ketone]	$C_6H_{12}O$	108-10-1	50 ppm	75 ppm
2-Methyl-1-propanol	[Isobutyl alcohol]	$C_4H_{10}O$	78-83-1	50 ppm	
2-Methyl-2-propanol	[tert-Butyl alcohol]	$C_4H_{10}O$	75-65-0	100 ppm	
Methylstyrene	all isomers	C_9H_{10}	25013-15-4	50 ppm	100 ppm
N-Methyl-N,2,4,6-tetranitroaniline	[Tetryl]	$C_7H_5N_5O_8$	479-45-8	1.5 mg/m³	
Metribuzin		$C_8H_{14}N_4OS$	21087-64-9	5 mg/m³	
Mevinphos		$C_7H_{13}O_6P$	7786-34-7	0.01 ppm	
Mica			12001-26-2	3 mg/m³	
Molybdenum	metal and insoluble compounds, as Mo	Mo	7439-98-7	10 mg/m³	
Molybdenum	soluble compounds, as Mo	Mo	7439-98-7	0.5 mg/m³	
Monocrotophos		$C_7H_{14}NO_5P$	6923-22-4	0.05 mg/m³	
Morpholine		C_4H_9NO	110-91-8	20 ppm	
Naled		$C_4H_7Br_2Cl_2O_4P$	300-76-5	0.1 mg/m³	
Naphthalene		$C_{10}H_8$	91-20-3	10 ppm	15 ppm
1-Naphthalenylthiourea	[ANTU]	$C_{11}H_{10}N_2S$	86-88-4	0.3 mg/m³	
Neopentane	[2,2-Dimethylpropane]	C_5H_{12}	463-82-1	600 ppm	
Nickel	metal	Ni	7440-02-0	1.5 mg/m³	
Nickel	soluble compounds, as Ni	Ni	7440-02-0	0.1 mg/m³	
Nickel	insoluble compounds, as Ni	NI	7440-02-0	0.2 mg/m³	
Nickel carbonyl	as Ni	C_4NiO_4	13463-39-3	0.05 ppm	
Nickel(III) sulfide	as Ni	Ni_3S_2	12035-72-2	0.1 mg/m³	
Nicotine		$C_{10}H_{14}N_2$	54-11-5	0.5 mg/m³	
Nitrapyrin		$C_6H_3Cl_4N$	1929-82-4	10 mg/m³	
Nitric acid		HNO_3	7697-37-2	2 ppm	4 ppm
Nitric oxide		NO	10102-43-9	25 ppm	
4-Nitroaniline		$C_6H_6N_2O_2$	100-01-6	3 mg/m³	
Nitrobenzene		$C_6H_5NO_2$	98-95-3	1 ppm	
Nitroethane		$C_2H_5NO_2$	79-24-3	100 ppm	
Nitrogen dioxide		NO_2	10102-44-0	3 ppm	5 ppm

Substance	Notes	Molecular formula	CAS Reg. No.	Time-weighted average	Short-term exposure limit
Nitrogen trifluoride		F_3N	7783-54-2	10 ppm	
Nitromethane		CH_3NO_2	75-52-5	20 ppm	
1-Nitropropane		$C_3H_7NO_2$	108-03-2	25 ppm	
2-Nitropropane		$C_3H_7NO_2$	79-46-9	10 ppm	
Nitrotoluene	all isomers	$C_7H_7NO_2$	1321-12-6	2 ppm	
Nitrous oxide		N_2O	10024-97-2	50 ppm	
Nonane	all isomers	C_9H_{20}	111-84-2	200 ppm	
Octachloronaphthalene		$C_{10}Cl_8$	2234-13-1	0.1 mg/m^3	
Octane	all isomers	C_8H_{18}	111-65-9	300 ppm	375 ppm
Osmium(VIII) oxide	[Osmium tetroxide]	O_4Os	20816-12-0	0.0002 ppm	0.0006 ppm
Oxalic acid		$C_2H_2O_4$	144-62-7	1 mg/m^3	2 mg/m^3
2-Oxetanone	[β-Propiolactone]	$C_3H_4O_2$	57-57-8	0.5 ppm	
Oxirane	[Ethylene oxide]	C_2H_4O	75-21-8	1 ppm	
Oxiranemethanol	[Glycidol]	$C_3H_6O_2$	556-52-5	2 ppm	
4,4'-Oxybis(benzenesulfonyl hydrazide)		$C_{12}H_{14}N_4O_5S_2$	80-51-3	0.1 mg/m^3	
Ozone	depends on workload	O_3	10028-15-6	0.1 ppm	
Paraquat		$C_{12}H_{14}N_2$	4685-14-7	0.5 mg/m^3	
Parathion		$C_{10}H_{14}NO_5PS$	56-38-2	0.05 mg/m^3	
Pentaborane(9)		B_5H_9	19624-22-7	0.005 ppm	0.015 ppm
Pentachloronaphthalene	all isomers	$C_{10}H_3Cl_5$	1321-64-8	0.5 mg/m^3	
Pentachloronitrobenzene		$C_6Cl_5NO_2$	82-68-8	0.5 mg/m^3	
Pentachlorophenol		C_6HCl_5O	87-86-5	0.5 mg/m^3	
Pentaerythritol		$C_5H_{12}O_4$	115-77-5	10 mg/m^3	
Pentanal	[Valeraldehyde]	$C_5H_{10}O$	110-62-3	50 ppm	
Pentane		C_5H_{12}	109-66-0	600 ppm	750 ppm
Pentanedial	[Glutaraldehyde]	$C_5H_8O_2$	111-30-8		C 0.05 ppm
3-Pentanol acetate		$C_7H_{14}O_2$	620-11-1	50 ppm	100 ppm
2-Pentanone	[Methyl propyl ketone]	$C_5H_{10}O$	107-87-9	200 ppm	250 ppm
3-Pentanone	[Diethyl ketone]	$C_5H_{10}O$	96-22-0	200 ppm	300 ppm
Pentyl acetate	[Amyl acetate]	$C_7H_{14}O_2$	628-63-7	50 ppm	100 ppm
sec-Pentyl acetate		$C_7H_{14}O_2$	626-38-0	50 ppm	100 ppm
Perchloryl fluoride		$ClFO_3$	7616-94-6	3 ppm	6 ppm
Perfluoroacetone	[Hexafluoroacetone]	C_3F_6O	684-16-2	0.1 ppm	
Perfluoroisobutene		C_4F_8	382-21-8		C 0.01 ppm
Phenol		C_6H_6O	108-95-2	5 ppm	
10H-Phenothiazine	[Thiodiphenylamine]	$C_{12}H_9NS$	92-84-2	5 mg/m^3	
Phenylenediamine	all isomers	$C_6H_8N_2$	25265-76-3	0.1 mg/m^3	
Phenyl glycidyl ether	[PGE]	$C_9H_{10}O_2$	122-60-1	0.1 ppm	
Phenylhydrazine		$C_6H_8N_2$	100-63-0	0.1 ppm	
Phenylphosphine		C_6H_7P	638-21-1		C 0.05 ppm
Phorate		$C_7H_{17}O_2PS_3$	298-02-2	0.05 mg/m^3	0.2 mg/m^3
Phosphine		H_3P	7803-51-2	0.3 ppm	1 ppm
Phosphoric acid		H_3O_4P	7664-38-2	1 mg/m^3	3 mg/m^3
Phosphorus (yellow)	[White phosphorus]	P	7723-14-0	0.1 mg/m^3	
Phosphorus(III) chloride	[Phosphorus trichloride]	Cl_3P	7719-12-2	0.2 ppm	0.5 ppm
Phosphorus(V) chloride	[Phosphorus pentachloride]	Cl_5P	10026-13-8	0.1 ppm	
Phosphorus(V) sulfide		P_2S_5	1314-80-3	1 mg/m^3	3 mg/m^3
Phosphoryl chloride	[Phosphorus(V) oxychloride]	Cl_3OP	10025-87-3	0.1 ppm	
Phthalic anhydride		$C_8H_4O_3$	85-44-9	1 ppm	
Piperazine dihydrochloride		$C_4H_{12}Cl_2N_2$	142-64-3	5 mg/m^3	
2-Pivaloyl-1,3-indandione	[Pindone]	$C_{14}H_{14}O_3$	83-26-1	0.1 mg/m^3	
Platinum		Pt	7440-06-4	1 mg/m^3	
Platinum	soluble salts, as Pt	Pt	7440-06-4	0.002 mg/m^3	
Potassium hydroxide		HKO	1310-58-3		C 2 mg/m^3
Propanal	[Propionaldehyde]	C_3H_6O	123-38-6	20 ppm	
Propane		C_3H_8	74-98-6	1000 ppm	
Propanoic acid		$C_3H_6O_2$	79-09-4	10 ppm	
1-Propanol	[Propyl alcohol]	C_3H_8O	71-23-8	200 ppm	400 ppm
2-Propanol	[Isopropyl alcohol]	C_3H_8O	67-63-0	200 ppm	400 ppm

Substance	Notes	Molecular formula	CAS Reg. No.	Time-weighted average	Short-term exposure limit
Propargyl alcohol	[2-Propyn-1-ol]	C_3H_4O	107-19-7	1 ppm	
Propoxur		$C_{11}H_{15}NO_3$	114-26-1	0.5 mg/m^3	
Propyl acetate		$C_5H_{10}O_2$	109-60-4	200 ppm	250 ppm
1,2-Propylene glycol dinitrate		$C_3H_6N_2O_6$	6423-43-4	0.05 ppm	
Propyleneimine		C_3H_7N	75-55-8	2 ppm	
Propyl nitrate		$C_3H_7NO_3$	627-13-4	25 ppm	40 ppm
Propyne	[Methylacetylene]	C_3H_4	74-99-7	1000 ppm	
2-Pyridinamine	[2-Aminopyridine]	$C_5H_6N_2$	504-29-0	0.5 ppm	
Pyridine		C_5H_5N	110-86-1	1 ppm	
Pyrocatechol	[Catechol]	$C_6H_6O_2$	120-80-9	5 ppm	
Resorcinol	[1,3-Benzenediol]	$C_6H_6O_2$	108-46-3	10 ppm	20 ppm
Rhodium	metal and insoluble compounds, as Rh	Rh	7440-16-6	1 mg/m^3	
Rhodium	soluble compounds, as Rh	Rh	7440-16-6	0.01 mg/m^3	
Ronnel		$C_8H_8Cl_3O_3PS$	299-84-3	10 mg/m^3	
Rotenone		$C_{23}H_{22}O_6$	83-79-4	5 mg/m^3	
Selenium	element and compounds, as Se	Se	7782-49-2	0.2 mg/m^3	
Selenium hexafluoride		F_6Se	7783-79-1	0.05 ppm	
Sesone		$C_8H_7Cl_2NaO_5S$	136-78-7	10 mg/m^3	
Silane		H_4Si	7803-62-5	5 ppm	
Silicon		Si	7440-21-3	10 mg/m^3	
Silicon carbide		CSi	409-21-2	10 mg/m^3	
Silicon dioxide (α-quartz)		O_2Si	14808-60-7	0.05 mg/m^3	
Silicon dioxide (tridymite)		O_2Si	15468-32-3	0.05 mg/m^3	
Silicon dioxide (cristobalite)		O_2Si	14464-46-1	0.05 mg/m^3	
Silicon dioxide (vitreous)	[Fused silica]	O_2Si	60676-86-0	0.1 mg/m^3	
Silver		Ag	7440-22-4	0.1 mg/m^3	
Silver	soluble compounds, as Ag	Ag	7440-22-4	0.01 mg/m^3	
Sodium azide		N_3Na	26628-22-8		C 0.29 mg/m^3
Sodium fluoroacetate		$C_2H_2FNaO_2$	62-74-8	0.05 mg/m^3	
Sodium hydrogen sulfite		$HNaO_3S$	7631-90-5	5 mg/m^3	
Sodium hydroxide		HNaO	1310-73-2		C 2 mg/m^3
Sodium metabisulfite		$Na_2O_5S_2$	7681-57-4	5 mg/m^3	
Sodium pyrophosphate		$Na_4O_7P_2$	7722-88-5	5 mg/m^3	
Sodium tetraborate decahydrate	[Borax]	$B_4H_{20}Na_2O_{17}$	1303-96-4	5 mg/m^3	
Stibine		H_3Sb	7803-52-3	0.1 ppm	
Strontium chromate	as Cr	CrO_4Sr	7789-06-2	0.0005 mg/m^3	
Strychnine		$C_{21}H_{22}N_2O_2$	57-24-9	0.15 mg/m^3	
Styrene	[Vinylbenzene]	C_8H_8	100-42-5	20 ppm	40 ppm
Sucrose		$C_{12}H_{22}O_{11}$	57-50-1	10 mg/m^3	
Sulfotep	[Tetraethyl thiodiphosphate; TEDP]	$C_8H_{20}O_5P_2S_2$	3689-24-5	0.2 mg/m^3	
Sulfur chloride		Cl_2S_2	10025-67-9		C 1 ppm
Sulfur decafluoride		$F_{10}S_2$	5714-22-7		C 0.01 ppm
Sulfur dioxide		O_2S	7446-09-5	2 ppm	5 ppm
Sulfur hexafluoride		F_6S	2551-62-4	1000 ppm	
Sulfuric acid		H_2O_4S	7664-93-9	0.2 mg/m^3	
Sulfur tetrafluoride		F_4S	7783-60-0		C 0.1 ppm
Sulfuryl fluoride		F_2O_2S	2699-79-8	5 ppm	10 ppm
Sulprofos		$C_{12}H_{19}O_2PS_3$	35400-43-2	1 mg/m^3	
Talc			14807-96-6	2 mg/m^3	
Tantalum	dust	Ta	7440-25-7	5 mg/m^3	
Tantalum(V) oxide	dust, as Ta	O_5Ta_2	1314-61-0	5 mg/m^3	
Tellurium	and compounds, as Te (except H_2Te)	Te	13494-80-9	0.1 mg/m^3	
Tellurium hexafluoride		F_6Te	7783-80-4	0.02 ppm	
Terbufos		$C_9H_{21}O_2PS_3$	13071-79-9	0.01 mg/m^3	
Terephthalic acid	[1,4-Benzenedicarboxylic acid]	$C_8H_6O_4$	100-21-0	10 mg/m^3	
Terphenyl	all isomers	$C_{18}H_{14}$	26140-60-3		5 mg/m^3
1,1,2,2-Tetrabromoethane	[Acetylene tetrabromide]	$C_2H_2Br_4$	79-27-6	1 ppm	
Tetrabromomethane	[Carbon tetrabromide]	CBr_4	558-13-4	0.1 ppm	0.3 ppm

Substance	Notes	Molecular formula	CAS Reg. No.	Time-weighted average	Short-term exposure limit
1,1,1,2-Tetrachloro-2,2-difluoroethane		$C_2Cl_4F_2$	76-11-9	500 ppm	
1,1,2,2-Tetrachloro-1,2-difluoroethane		$C_2Cl_4F_2$	76-12-0	500 ppm	
1,1,2,2-Tetrachloroethane		$C_2H_2Cl_4$	79-34-5	1 ppm	
Tetrachloroethene	[Perchloroethylene]	C_2Cl_4	127-18-4	25 ppm	100 ppm
Tetrachloromethane	[Carbon tetrachloride]	CCl_4	56-23-5	5 ppm	10 ppm
Tetrachloronaphthalene	all isomers	$C_{10}H_4Cl_4$	1335-88-2	2 mg/m³	
Tetraethyl lead	as Pb	$C_8H_{20}Pb$	78-00-2	0.1 mg/m³	
Tetraethyl pyrophosphate	[TEPP]	$C_8H_{20}O_7P_2$	107-49-3	0.05 mg/m³	
Tetrafluoroethene		C_2F_4	116-14-3	2 ppm	
Tetrahydrofuran	[Oxolane]	C_4H_8O	109-99-9	200 ppm	250 ppm
Tetramethyl lead	as Pb	$C_4H_{12}Pb$	75-74-1	0.15 mg/m³	
Tetramethyl silicate		$C_4H_{12}O_4Si$	681-84-5	1 ppm	
Tetramethylsuccinonitrile		$C_8H_{12}N_2$	3333-52-6	0.5 ppm	
Tetranitromethane		CN_4O_8	509-14-8	0.005 ppm	
Thallium	and soluble compounds, as Tl	Tl	7440-28-0	0.1 mg/m³	
4,4'-Thiobis(6-*tert*-butyl-*m*-cresol)		$C_{22}H_{30}O_2S$	96-69-5	10 mg/m³	
Thioglycolic acid		$C_2H_4O_2S$	68-11-1	1 ppm	
Thionyl chloride		Cl_2OS	7719-09-7		C 1 ppm
Thiram		$C_6H_{12}N_2S_4$	137-26-8	1 mg/m³	
Tin		Sn	7440-31-5	2 mg/m³	
Tin	inorganic compounds, as Sn	Sn	7440-31-5	2 mg/m³	
Tin	organic compounds, as Sn	Sn	7440-31-5	0.1 mg/m³	
Titanium(IV) oxide	[Titanium dioxide]	O_2Ti	13463-67-7	10 mg/m³	
Toluene		C_7H_8	108-88-3	50 ppm	
Toluene-2,4-diisocyanate		$C_9H_6N_2O_2$	584-84-9	0.005 ppm	0.02 ppm
Toluene-2,6-diisocyanate		$C_9H_6N_2O_2$	91-08-7	0.005 mg/m³	0.02 ppm
1*H*-1,2,4-Triazol-3-amine	[Amitrole]	$C_2H_4N_4$	61-82-5	0.2 mg/m³	
Tribromomethane	[Bromoform]	$CHBr_3$	75-25-2	0.5 ppm	
Tributyl phosphate		$C_{12}H_{27}O_4P$	126-73-8	0.2 ppm	
Trichlorfon		$C_4H_8Cl_3O_4P$	52-68-6	1 mg/m³	
Trichloroacetic acid		$C_2HCl_3O_2$	76-03-9	1 ppm	
1,2,4-Trichlorobenzene		$C_6H_3Cl_3$	120-82-1		C 5 ppm
1,1,1-Trichloroethane	[Methyl chloroform]	$C_2H_3Cl_3$	71-55-6	350 ppm	450 ppm
1,1,2-Trichloroethane		$C_2H_3Cl_3$	79-00-5	10 ppm	
Trichloroethene		C_2HCl_3	79-01-6	50 ppm	100 ppm
Trichlorofluoromethane		CCl_3F	75-69-4		C 1000 ppm
Trichloromethane	[Chloroform]	$CHCl_3$	67-66-3	10 ppm	
Trichloromethanesulfenyl chloride	[Perchloromethyl mercaptan]	CCl_4S	594-42-3	0.1 ppm	
(Trichloromethyl)benzene	[Benzotrichloride]	$C_7H_5Cl_3$	98-07-7		C 0.1 ppm
Trichloronaphthalene	all isomers	$C_{10}H_5Cl_3$	1321-65-9	5 mg/m³	
Trichloronitromethane	[Chloropicrin]	CCl_3NO_2	76-06-2	0.1 ppm	
2,4,5-Trichlorophenoxyacetic acid	[2,4,5-T]	$C_8H_5Cl_3O_3$	93-76-5	10 mg/m³	
1,2,3-Trichloropropane		$C_3H_5Cl_3$	96-18-4	10 ppm	
1,1,2-Trichloro-1,2,2-trifluoroethane		$C_2Cl_3F_3$	76-13-1	1000 ppm	1250 ppm
Tri-*o*-cresyl phosphate		$C_{21}H_{21}O_4P$	78-30-8	0.1 mg/m³	
Triethanolamine	[Tris(2-hydroxyethyl)amine]	$C_6H_{15}NO_3$	102-71-6	5 mg/m³	
Triethylamine		$C_6H_{15}N$	121-44-8	1 ppm	3 ppm
1,3,5-Triglycidyl-*s*-triazinetrione		$C_{12}H_{15}N_3O_6$	2451-62-9	0.05 mg/m³	
Triiodomethane	[Iodoform]	CHI_3	75-47-8	0.6 ppm	
Trimellitic anhydride	[1,2,4-Benzenetricarboxylic anhydride]	$C_9H_4O_5$	552-30-7		C 0.04 mg/m³
Trimethylamine		C_3H_9N	75-50-3	5 ppm	15 ppm
Trimethylbenzene	all isomers	C_9H_{12}	25551-13-7	25 ppm	
Trimethyl phosphite		$C_3H_9O_3P$	121-45-9	2 ppm	

Substance	Notes	Molecular formula	CAS Reg. No.	Time-weighted average	Short-term exposure limit
Trinitroglycerol	[Nitroglycerin; NG]	$C_3H_5N_3O_9$	55-63-0	0.05 ppm	
2,4,6-Trinitrophenol	[Picric acid]	$C_6H_3N_3O_7$	88-89-1	0.1 mg/m³	
2,4,6-Trinitrotoluene	[TNT]	$C_7H_5N_3O_6$	118-96-7	0.1 mg/m³	
Triphenylamine		$C_{18}H_{15}N$	603-34-9	5 mg/m³	
Triphenyl phosphate		$C_{18}H_{15}O_4P$	115-86-6	3 mg/m³	
Tungsten	metal and insoluble compounds, as W	W	7440-33-7	5 mg/m³	10 mg/m³
Tungsten	soluble compounds, as W	W	7440-33-7	1 mg/m³	3 mg/m³
Uranium	metal and compounds, as U	U	7440-61-1	0.2 mg/m³	
Vanadium(V) oxide	dust or fume; [Vanadium pentoxide]	O_5V_2	1314-62-1	0.05 mg/m³	
Vinyl acetate		$C_4H_6O_2$	108-05-4	10 ppm	15 ppm
4-Vinylcyclohexene		C_8H_{12}	100-40-3	0.1 ppm	
1-Vinyl-2-pyrrolidinone		C_6H_9NO	88-12-0	0.05 ppm	
Warfarin	[Coumadin]	$C_{19}H_{16}O_4$	81-81-2	0.1 mg/m³	
Xylene	all isomers	C_8H_{10}	1330-20-7	100 ppm	150 ppm
Xylidine	all isomers	$C_8H_{11}N$	1300-73-8	0.5 ppm	
Yttrium	metal and compounds, as Y	Y	7440-65-5	1 mg/m³	
Zinc chloride		Cl_2Zn	7646-85-7	1 mg/m³	2 mg/m³
Zinc chromate hydroxide	as Cr	$CrH_2O_6Zn_2$	13530-65-9	0.01 mg/m³	
Zinc oxide		OZn	1314-13-2	2 mg/m³	10 mg/m³
Zirconium	metal and compounds, as Zr	Zr	7440-67-7	5 mg/m³	10 mg/m³

OCTANOL–WATER PARTITION COEFFICIENTS

The octanol–water partition coefficient, P, is a widely used parameter for correlating biological effects of organic substances. It is a property of the two-phase system in which water and 1-octanol are in equilibrium at a fixed temperature and the substance is distributed between the water-rich and octanol-rich phases. P is defined as the ratio of the equilibrium concentration of the substance in the octanol-rich phase to that in the water-rich phase, in the limit of zero concentration. In general, P tends to be large for compounds with extended non-polar structures (such as long chain or multi-ring hydrocarbons) and small for compounds with highly polar groups. Thus P (or, in its more common form of expression, log P) provides a measure of the lipophilic vs. hydrophilic nature of a compound, which is an important consideration in assessing the potential toxicity. A discussion of methods of measurement and accuracy considerations for log P may be found in Reference 1.

This table gives selected values of log P for about 450 organic compounds, including many of environmental importance. All values refer to a nominal temperature of 25°C. The source of each value is indicated in the last column. These references contain data on many more compounds than are included here.

Compounds are listed by molecular formula following the Hill convention. To locate a compound by name or CAS Registry Number when the molecular formula is not known, use the table "Physical Constants of Organic Compounds" in Section 3 and its indexes to determine the molecular formula.

References

1. Sangster, J., *J. Phys. Chem. Ref. Data*, 18, 1111, 1989.
2. Mackay, D., Shiu, W. Y., and Ma, K. C., *Illustrated Handbook of Physical-Chemical Properties and Environmental Fate for Organic Chemicals*, Lewis Publishers/CRC Press, Boca Raton, FL, 1992.
3. Shiu, W. Y., and Mackay, D., *J. Phys. Chem. Ref. Data*, 15, 911, 1986.
4. Pinsuwan, S., Li, L., and Yalkowsky, S. H., *J. Chem. Eng. Data*, 40, 623, 1995.
5. *Solubility Data Series, International Union of Pure and Applied Chemistry, Vol. 20*, Pergamon Press, Oxford, 1985.
6. *Solubility Data Series, International Union of Pure and Applied Chemistry, Vol. 38*, Pergamon Press, Oxford, 1985.
7. Miller, M. M., Ghodbane, S., Wasik, S. P., Tewari, Y. B., and Martire, D. E., *J. Chem. Eng. Data*, 29, 184, 1984.

Mol. form.	Name	log P	Ref.
CCl_2F_2	Dichlorodifluoromethane	2.16	2
CCl_3F	Trichlorofluoromethane	2.53	2
CCl_4	Tetrachloromethane	2.64	2
$CHBr_3$	Tribromomethane	2.38	2
$CHCl_3$	Trichloromethane	1.97	2
CH_2BrCl	Bromochloromethane	1.41	2
CH_2Br_2	Dibromomethane	2.3	2
CH_2Cl_2	Dichloromethane	1.25	2
CH_2F_2	Difluoromethane	0.20	1
CH_2I_2	Diiodomethane	2.5	2
CH_2O	Formaldehyde	0.35	1
CH_2O_2	Formic acid	−0.54	1
CH_3Br	Bromomethane	1.19	2
CH_3Cl	Chloromethane	0.91	2
CH_3F	Fluoromethane	0.51	1
CH_3I	Iodomethane	1.5	2
CH_3NO	Formamide	−1.51	1
CH_3NO_2	Nitromethane	−0.33	1
CH_4O	Methanol	−0.74	1
CH_5N	Methylamine	−0.57	1
$C_2Cl_3F_3$	1,1,2-Trichlorotrifluoroethane	3.16	2
C_2Cl_4	Tetrachloroethylene	2.88	2
C_2Cl_6	Hexachloroethane	4.00	4
C_2HCl_3	Trichloroethylene	2.53	2
C_2HCl_5	Pentachloroethane	2.89	2
$C_2H_2Cl_2$	1,1-Dichloroethylene	2.13	2
$C_2H_2Cl_2$	*cis*-1,2-Dichloroethylene	1.86	2
$C_2H_2Cl_2$	*trans*-1,2-Dichloroethylene	1.93	2
$C_2H_2Cl_4$	1,1,2,2-Tetrachloroethane	2.39	2
C_2H_3Cl	Chloroethylene	1.38	2
$C_2H_3Cl_3$	1,1,1-Trichloroethane	2.49	2
$C_2H_3Cl_3$	1,1,2-Trichloroethane	2.38	2
C_2H_3N	Acetonitrile	−0.34	1
$C_2H_4Cl_2$	1,1-Dichloroethane	1.79	2
$C_2H_4Cl_2$	1,2-Dichloroethane	1.48	2
C_2H_4O	Acetaldehyde	0.45	1
C_2H_4O	Ethylene oxide	−0.30	1
$C_2H_4O_2$	Acetic acid	−0.17	1
C_2H_5Br	Bromoethane	1.6	2
C_2H_5Cl	Chloroethane	1.43	2
C_2H_5I	Iodoethane	2	2
C_2H_5NO	Acetamide	−1.26	1
$C_2H_5NO_2$	Nitroethane	0.18	1
C_2H_6O	Ethanol	−0.30	1
C_2H_6O	Dimethyl ether	0.10	1
C_2H_6OS	Dimethyl sulfoxide	−1.35	1
$C_2H_6O_2S$	Dimethyl sulfone	−1.41	1
C_2H_7N	Ethylamine	−0.13	1
C_2H_7N	Dimethylamine	−0.38	1
C_3H_3N	2-Propenenitrile	0.25	1
$C_3H_4Cl_2$	*cis*-1,3-Dichloropropene	2.03	2
C_3H_4O	Propargyl alcohol	−0.38	1
C_3H_4O	Acrolein	−0.01	1
C_3H_5Br	3-Bromopropene	1.79	1
C_3H_5ClO	Epichlorohydrin	0.30	2
$C_3H_5Cl_3$	1,2,3-Trichloropropane	2.63	2
C_3H_5N	Propanenitrile	0.16	1
C_3H_5NO	Acrylamide	−0.78	1
$C_3H_6Cl_2$	1,2-Dichloropropane	2.0	2
C_3H_6O	Allyl alcohol	0.17	1
C_3H_6O	Propanal	0.59	1
C_3H_6O	Acetone	−0.24	1
C_3H_6O	Methyloxirane	0.03	1
$C_3H_6O_2$	Propanoic acid	0.33	1
$C_3H_6O_2$	Methyl acetate	0.18	1
C_3H_7Br	1-Bromopropane	2.1	2
C_3H_7Br	2-Bromopropane	1.9	2
C_3H_7Cl	1-Chloropropane	2.04	1
C_3H_7Cl	2-Chloropropane	1.90	1
C_3H_7I	1-Iodopropane	2.5	2
C_3H_7N	Allylamine	0.03	1
C_3H_7NO	N,N-Dimethylformamide	−1.01	1
C_3H_7NO	N-Methylacetamide	−1.05	1
$C_3H_7NO_2$	1-Nitropropane	0.87	1

Mol. form.	Name	log P	Ref.	Mol. form.	Name	log P	Ref.
C_3H_8O	1-Propanol	0.25	1	$C_5H_{10}O$	2-Methyltetrahydrofuran	1.85	2
C_3H_8O	2-Propanol	0.05	1	$C_5H_{10}O_2$	Pentanoic acid	1.39	1
C_3H_8S	1-Propanethiol	1.81	1	$C_5H_{10}O_2$	Propyl acetate	1.24	1
C_3H_9N	Propylamine	0.48	1	$C_5H_{10}O_2$	Ethyl propanoate	1.21	1
C_3H_9N	Isopropylamine	0.26	1	$C_5H_{10}O_3$	Diethyl carbonate	1.21	1
C_3H_9N	Ethylmethylamine	0.15	1	$C_5H_{11}Br$	1-Bromopentane	3.37	1
C_3H_9N	Trimethylamine	0.16	1	$C_5H_{11}F$	1-Fluoropentane	2.33	1
C_4H_4O	Furan	1.34	1	$C_5H_{11}N$	Piperidine	0.84	1
C_4H_4S	Thiophene	1.81	1	$C_5H_{11}NO_2$	1-Nitropentane	2.01	1
C_4H_5N	Pyrrole	0.75	1	C_5H_{12}	Pentane	3.45	1
C_4H_6	1,3-Butadiene	1.99	1	C_5H_{12}	Neopentane	3.11	1
C_4H_6	2-Butyne	1.46	1	$C_5H_{12}O$	1-Pentanol	1.51	1
C_4H_6O	2,5-Dihydrofuran	0.46	1	$C_5H_{12}O$	2-Pentanol	1.25	1
$C_4H_6O_2$	Methacrylic acid	0.93	1	$C_5H_{12}O$	3-Pentanol	1.21	1
$C_4H_6O_2$	Vinyl acetate	0.73	1	$C_5H_{12}O$	3-Methyl-1-butanol	1.28	1
$C_4H_6O_2$	Methyl acrylate	0.80	1	$C_5H_{12}O$	2-Methyl-2-butanol	0.89	1
C_4H_7N	Butanenitrile	0.60	1	$C_5H_{12}O$	3-Methyl-2-butanol	1.28	1
C_4H_8	cis-2-Butene	2.33	1	$C_5H_{12}O$	2,2-Dimethyl-1-propanol	1.31	1
C_4H_8	trans-2-Butene	2.31	1	$C_5H_{12}O$	Methyl tert-butyl ether	0.94	1
C_4H_8	Isobutene	2.35	1	$C_5H_{13}N$	Pentylamine	1.49	1
$C_4H_8Cl_2O$	Bis(2-chloroethyl) ether	1.12	2	C_6Cl_6	Hexachlorobenzene	5.47	5
C_4H_8O	Ethyl vinyl ether	1.04	1	C_6HCl_5	Pentachlorobenzene	5.03	5
C_4H_8O	Butanal	0.88	1	C_6HCl_5O	Pentachlorophenol	5.07	4
C_4H_8O	2-Butanone	0.29	1	$C_6H_2Cl_4$	1,2,3,4-Tetrachlorobenzene	4.55	5
C_4H_8O	Tetrahydrofuran	0.46	1	$C_6H_2Cl_4$	1,2,3,5-Tetrachlorobenzene	4.65	5
$C_4H_8O_2$	Butanoic acid	0.79	1	$C_6H_2Cl_4$	1,2,4,5-Tetrachlorobenzene	4.51	5
$C_4H_8O_2$	Propyl formate	0.83	1	$C_6H_3Cl_3$	1,2,3-Trichlorobenzene	4.04	5
$C_4H_8O_2$	Ethyl acetate	0.73	1	$C_6H_3Cl_3$	1,2,4-Trichlorobenzene	3.98	5
C_4H_9Br	1-Bromobutane	2.75	1	$C_6H_3Cl_3$	1,3,5-Trichlorobenzene	4.02	5
C_4H_9Cl	1-Chlorobutane	2.64	2	$C_6H_4Cl_2$	o-Dichlorobenzene	3.38	5
C_4H_9F	1-Fluorobutane	2.58	1	$C_6H_4Cl_2$	m-Dichlorobenzene	3.48	5
C_4H_9I	1-Iodobutane	3	2	$C_6H_4Cl_2$	p-Dichlorobenzene	3.38	5
C_4H_9N	Pyrrolidine	0.46	1	$C_6H_4Cl_2O$	2,4-Dichlorophenol	3.23	4
C_4H_9NO	Butanamide	−0.21	1	C_6H_5Br	Bromobenzene	2.99	2
C_4H_9NO	N,N-Dimethylacetamide	−0.77	1	C_6H_5Cl	Chlorobenzene	2.84	1
$C_4H_9NO_2$	1-Nitrobutane	1.47	1	C_6H_5F	Fluorobenzene	2.27	2
C_4H_{10}	Isobutane	2.8	2	C_6H_5I	Iodobenzene	3.28	2
$C_4H_{10}O$	1-Butanol	0.84	1	$C_6H_5NO_2$	Nitrobenzene	1.85	1
$C_4H_{10}O$	2-Butanol	0.65	1	C_6H_6	Benzene	2.13	1
$C_4H_{10}O$	2-Methyl-1-propanol	0.76	1	C_6H_6O	Phenol	1.48	4
$C_4H_{10}O$	2-Methyl-2-propanol	0.35	1	C_6H_6S	Benzenethiol	2.52	1
$C_4H_{10}O$	Diethyl ether	0.89	1	C_6H_7N	Aniline	0.90	1
$C_4H_{10}S$	1-Butanethiol	2.28	1	C_6H_7N	2-Methylpyridine	1.11	1
$C_4H_{10}S$	Diethyl sulfide	1.95	1	C_6H_7N	3-Methylpyridine	1.20	1
$C_4H_{11}N$	Butylamine	0.86	1	C_6H_7N	4-Methylpyridine	1.22	1
$C_4H_{11}N$	tert-Butylamine	0.40	1	C_6H_8	1,4-Cyclohexadiene	2.3	2
$C_4H_{11}N$	Diethylamine	0.58	1	C_6H_8O	5-Hexyn-2-one	0.58	1
C_5H_5N	Pyridine	0.65	1	C_6H_8O	2-Cyclohexen-1-one	0.61	1
C_5H_6O	2-Methylfuran	1.85	1	C_6H_8O	2-Ethylfuran	2.40	1
C_5H_7N	1-Methylpyrrole	1.21	1	C_6H_{10}	1,5-Hexadiene	2.8	2
C_5H_8	1,4-Pentadiene	2.48	1	C_6H_{10}	1-Hexyne	2.73	2
C_5H_8	1-Pentyne	1.98	1	C_6H_{10}	Cyclohexene	2.86	1
$C_5H_8O_2$	Methyl methacrylate	1.38	1	$C_6H_{10}O$	5-Hexen-2-one	1.02	1
$C_5H_8O_2$	Ethyl acrylate	1.32	1	$C_6H_{10}O$	Cyclohexanone	0.81	1
C_5H_9N	Pentanenitrile	0.94	1	$C_6H_{10}O_2$	Ethyl methacrylate	1.94	1
C_5H_{10}	1-Pentene	2.2	2	$C_6H_{11}Br$	Bromocyclohexane	3.20	1
C_5H_{10}	Cyclopentane	3.00	1	$C_6H_{11}N$	Hexanenitrile	1.66	1
$C_5H_{10}O$	2-Pentanone	0.84	1	C_6H_{12}	1-Hexene	3.40	1
$C_5H_{10}O$	3-Pentanone	0.82	1	C_6H_{12}	4-Methyl-1-pentene	2.5	2
$C_5H_{10}O$	3-Methyl-2-butanone	0.56	1	C_6H_{12}	Cyclohexane	3.44	1
$C_5H_{10}O$	Tetrahydropyran	0.82	1	C_6H_{12}	Methylcyclopentane	3.37	2

Mol. form.	Name	log P	Ref.
$C_6H_{12}O$	Cyclohexanol	1.23	1
$C_6H_{12}O$	Hexanal	1.78	1
$C_6H_{12}O$	2-Hexanone	1.38	1
$C_6H_{12}O$	4-Methyl-2-pentanone	1.31	1
$C_6H_{12}O_2$	Hexanoic acid	1.92	1
$C_6H_{12}O_2$	Butyl acetate	1.82	1
$C_6H_{13}Br$	1-Bromohexane	3.80	1
$C_6H_{13}N$	Cyclohexylamine	1.49	1
C_6H_{14}	Hexane	4.00	1
C_6H_{14}	3-Methylpentane	3.60	2
C_6H_{14}	2,2-Dimethylbutane	3.82	1
C_6H_{14}	2,3-Dimethylbutane	3.85	2
$C_6H_{14}O$	1-Hexanol	2.03	1
$C_6H_{14}O$	2-Hexanol	1.76	1
$C_6H_{14}O$	3-Hexanol	1.65	1
$C_6H_{14}O$	3,3-Dimethyl-2-butanol	1.48	1
$C_6H_{14}O$	Dipropyl ether	2.03	1
$C_6H_{14}O$	Diisopropyl ether	1.52	1
$C_6H_{15}N$	Hexylamine	2.06	1
$C_6H_{15}N$	Dipropylamine	1.67	1
$C_6H_{15}N$	Triethylamine	1.45	1
$C_7H_5BrO_2$	2-Bromobenzoic acid	2.20	4
$C_7H_5BrO_2$	3-Bromobenzoic acid	2.87	4
$C_7H_5BrO_2$	4-Bromobenzoic acid	2.86	4
C_7H_5N	Benzonitrile	1.56	1
C_7H_6O	Benzaldehyde	1.48	1
$C_7H_6O_2$	Benzoic acid	1.88	4
$C_7H_6O_2$	Phenyl formate	1.26	1
$C_7H_6O_3$	Salicylic acid	2.20	4
C_7H_7Br	(Bromomethyl)benzene	2.92	1
C_7H_7Cl	o-Chlorotoluene	3.42	1
C_7H_7Cl	m-Chlorotoluene	3.28	1
C_7H_7Cl	p-Chlorotoluene	3.33	1
C_7H_7Cl	(Chloromethyl)benzene	2.30	1
$C_7H_7NO_2$	p-Nitrotoluene	2.42	1
C_7H_8	Toluene	2.73	1
C_7H_8	1,3,5-Cycloheptatriene	2.63	2
C_7H_8O	o-Cresol	1.98	1
C_7H_8O	m-Cresol	1.98	1
C_7H_8O	p-Cresol	1.97	1
C_7H_8O	Benzyl alcohol	1.05	1
C_7H_8O	Anisole	2.11	1
C_7H_9N	Benzylamine	1.09	1
C_7H_9N	o-Methylaniline	1.32	1
C_7H_9N	m-Methylaniline	1.40	1
C_7H_9N	p-Methylaniline	1.39	1
C_7H_9N	N-Methylaniline	1.66	1
C_7H_{14}	1-Heptene	3.99	1
C_7H_{14}	Methylcyclohexane	3.88	1
$C_7H_{14}O$	2-Heptanone	1.98	1
$C_7H_{14}O$	5-Methyl-2-hexanone	1.88	1
$C_7H_{15}Br$	1-Bromoheptane	4.36	1
$C_7H_{15}Cl$	1-Chloroheptane	4.15	1
$C_7H_{15}I$	1-Iodoheptane	4.70	1
C_7H_{16}	Heptane	4.50	1
$C_7H_{16}O$	1-Heptanol	2.62	1
$C_7H_{16}O$	2-Heptanol	2.31	1
$C_7H_{16}O$	3-Heptanol	2.24	1
$C_7H_{16}O$	4-Heptanol	2.22	1
$C_7H_{17}N$	Heptylamine	2.57	1
C_8H_6	Phenylacetylene	2.40	1
C_8H_6O	Benzofuran	2.67	1
C_8H_6S	Benzo[b]thiophene	3.12	1
C_8H_7N	Benzeneacetonitrile	1.56	1
C_8H_7N	Indole	2.14	1
C_8H_8	Styrene	3.05	1
C_8H_8O	Acetophenone	1.63	1
C_8H_8O	2-Methylbenzaldehyde	2.26	1
C_8H_8O	Benzeneacetaldehyde	1.78	1
C_8H_8O	2,3-Dihydrobenzofuran	2.14	1
C_8H_8O	Phenyloxirane	1.61	1
$C_8H_8O_2$	o-Toluic acid	2.32	4
$C_8H_8O_2$	m-Toluic acid	2.37	1
$C_8H_8O_2$	p-Toluic acid	2.34	1
$C_8H_8O_2$	Benzeneacetic acid	1.41	1
$C_8H_8O_2$	Phenyl acetate	1.49	1
$C_8H_8O_2$	Methyl benzoate	2.20	1
C_8H_{10}	Ethylbenzene	3.15	1
C_8H_{10}	o-Xylene	3.12	1
C_8H_{10}	m-Xylene	3.20	1
C_8H_{10}	p-Xylene	3.15	1
$C_8H_{10}O$	o-Ethylphenol	2.47	1
$C_8H_{10}O$	m-Ethylphenol	2.50	1
$C_8H_{10}O$	p-Ethylphenol	2.50	1
$C_8H_{10}O$	2,4-Xylenol	2.35	1
$C_8H_{10}O$	2,5-Xylenol	2.34	1
$C_8H_{10}O$	2,6-Xylenol	2.36	1
$C_8H_{10}O$	3,4-Xylenol	3.23	1
$C_8H_{10}O$	3,5-Xylenol	2.35	1
$C_8H_{10}O$	Benzeneethanol	1.36	1
$C_8H_{10}O$	α-Methylbenzyl alcohol	1.42	1
$C_8H_{10}O$	3-Methylbenzenemethanol	1.60	1
$C_8H_{10}O$	4-Methylbenzenemethanol	1.58	1
$C_8H_{10}O$	Phenetole	2.51	1
$C_8H_{10}O$	Benzyl methyl ether	1.35	1
$C_8H_{10}O$	2-Methylanisole	2.74	1
$C_8H_{10}O$	3-Methylanisole	2.66	1
$C_8H_{10}O$	4-Methylanisole	2.81	1
$C_8H_{11}N$	p-Ethylaniline	1.96	1
$C_8H_{11}N$	N,N-Dimethylaniline	2.31	1
$C_8H_{11}N$	Benzeneethanamine	1.41	1
$C_8H_{14}O_2$	Butyl methacrylate	2.88	1
$C_8H_{15}N$	Octanenitrile	2.75	1
C_8H_{16}	1-Octene	4.57	1
C_8H_{16}	Cyclooctane	4.45	2
$C_8H_{16}O$	2-Octanone	2.37	1
$C_8H_{16}O_2$	Octanoic acid	3.05	1
$C_8H_{17}Br$	1-Bromooctane	4.89	1
C_8H_{18}	Octane	5.15	1
$C_8H_{18}O$	1-Octanol	3.07	1
$C_8H_{18}O$	2-Octanol	2.90	1
$C_8H_{18}O$	4-Octanol	2.68	1
$C_8H_{18}O$	Dibutyl ether	3.21	1
C_9H_7N	Quinoline	2.03	1
C_9H_7N	Isoquinoline	2.08	1
C_9H_8	Indene	2.92	1
$C_9H_8O_2$	trans-Cinnamic acid	2.13	1
C_9H_9N	Benzenepropanenitrile	1.72	1
C_9H_{10}	Indan	3.33	1
$C_9H_{10}O$	1-Phenyl-1-propanone	2.19	1
$C_9H_{10}O$	1-Phenyl-2-propanone	1.44	1
$C_9H_{10}O$	4-Methylacetophenone	2.19	1

Mol. form.	Name	log P	Ref.	Mol. form.	Name	log P	Ref.
$C_9H_{10}O_2$	2-Phenylpropanoic acid	1.80	1	$C_{12}H_6Cl_4$	2,3,4,5-Tetrachlorobiphenyl	5.72	3
$C_9H_{10}O_2$	Benzyl acetate	1.96	1	$C_{12}H_6Cl_4$	2,2′,4′,5-Tetrachlorobiphenyl	5.73	7
$C_9H_{10}O_2$	4-Methylphenyl acetate	2.11	1	$C_{12}H_7Cl_3$	2,4,5-Trichlorobiphenyl	5.60	3
$C_9H_{10}O_2$	Ethyl benzoate	2.64	1	$C_{12}H_7Cl_3$	2,4,6-Trichlorobiphenyl	5.47	3
C_9H_{12}	Propylbenzene	3.69	1	$C_{12}H_8Cl_2$	2,5-Dichlorobiphenyl	5.10	3
C_9H_{12}	Isopropylbenzene	3.66	1	$C_{12}H_8Cl_2$	2,6-Dichlorobiphenyl	5.00	3
C_9H_{12}	o-Ethyltoluene	3.53	1	$C_{12}H_8O$	Dibenzofuran	4.12	1
C_9H_{12}	p-Ethyltoluene	3.63	2	$C_{12}H_9Cl$	2-Chlorobiphenyl	4.52	1
C_9H_{12}	1,2,3-Trimethylbenzene	3.60	1	$C_{12}H_9Cl$	3-Chlorobiphenyl	4.58	1
C_9H_{12}	1,2,4-Trimethylbenzene	3.63	1	$C_{12}H_9Cl$	4-Chlorobiphenyl	4.61	1
C_9H_{12}	1,3,5-Trimethylbenzene	3.42	1	$C_{12}H_9N$	Carbazole	3.72	1
$C_9H_{12}O$	2-Propylphenol	2.93	1	$C_{12}H_{10}$	Acenaphthene	3.96	4
$C_9H_{12}O$	4-Propylphenol	3.20	1	$C_{12}H_{10}$	Biphenyl	3.76	6
$C_9H_{12}O$	2,3,6-Trimethylphenol	2.67	1	$C_{12}H_{10}N_2$	Azobenzene	3.82	1
$C_9H_{12}O$	2,4,6-Trimethylphenol	2.46	1	$C_{12}H_{10}O$	Diphenyl ether	4.21	1
$C_9H_{12}O$	Benzenepropanol	1.88	1	$C_{12}H_{10}S$	Diphenyl sulfide	4.45	1
$C_9H_{13}N$	N,N-Dimethylbenzylamine	1.98	1	$C_{12}H_{11}N$	Diphenylamine	3.44	4
$C_9H_{13}N$	Amphetamine	1.76	1	$C_{12}H_{12}$	1-Ethylnaphthalene	4.40	1
C_9H_{18}	1-Nonene	5.15	1	$C_{12}H_{12}$	1,2-Dimethylnaphthalene	4.31	1
$C_9H_{18}O$	2-Nonanone	3.16	1	$C_{12}H_{12}$	1,4-Dimethylnaphthalene	4.37	1
$C_9H_{18}O$	5-Methyl-2-octanone	2.92	1	$C_{12}H_{14}O$	4-Phenylcyclohexanone	2.45	1
C_9H_{20}	Nonane	5.65	1	$C_{12}H_{18}$	Hexylbenzene	5.52	1
$C_9H_{20}O$	1-Nonanol	4.02	1	$C_{12}H_{18}$	Hexamethylbenzene	4.69	4
$C_9H_{21}N$	Tripropylamine	2.79	1	$C_{12}H_{22}O$	Cyclododecanone	4.10	1
$C_{10}H_7Cl$	1-Chloronaphthalene	3.90	1	$C_{12}H_{24}O_2$	Dodecanoic acid	4.6	1
$C_{10}H_7Cl$	2-Chloronaphthalene	3.98	1	$C_{12}H_{26}O$	1-Dodecanol	5.13	1
$C_{10}H_8$	Naphthalene	3.34	4	$C_{13}H_8O$	9H-Fluoren-9-one	3.58	1
$C_{10}H_8$	Azulene	3.22	1	$C_{13}H_9N$	Acridine	3.40	1
$C_{10}H_8O$	1-Naphthol	2.84	1	$C_{13}H_{10}$	9H-Fluorene	4.20	4
$C_{10}H_8O$	2-Naphthol	2.70	1	$C_{13}H_{10}O$	Benzophenone	3.18	1
$C_{10}H_{12}O_2$	Isopropyl benzoate	3.18	1	$C_{13}H_{10}O_2$	Phenyl benzoate	3.59	1
$C_{10}H_{14}$	Butylbenzene	4.26	1	$C_{13}H_{11}NO$	N-Phenylbenzamide	2.62	1
$C_{10}H_{14}$	tert-Butylbenzene	4.11	1	$C_{13}H_{12}$	Diphenylmethane	4.14	1
$C_{10}H_{14}$	Isobutylbenzene	4.01	2	$C_{13}H_{12}$	4-Methylbiphenyl	4.63	1
$C_{10}H_{14}$	p-Cymene	4.10	1	$C_{13}H_{12}O$	Diphenylmethanol	2.67	1
$C_{10}H_{14}$	1,2,4,5-Tetramethylbenzene	4.10	2	$C_{13}H_{12}O$	Benzyl phenyl ether	3.79	1
$C_{10}H_{14}$	1,2,3,4-Tetramethylbenzene	4.00	1	$C_{14}H_{10}$	Anthracene	4.56	4
$C_{10}H_{14}$	1,2,3,5-Tetramethylbenzene	4.10	1	$C_{14}H_{10}$	Phenanthrene	4.52	4
$C_{10}H_{14}O$	4-Butylphenol	3.65	1	$C_{14}H_{12}$	trans-Stilbene	4.81	1
$C_{10}H_{20}O$	2-Decanone	3.77	1	$C_{14}H_{12}$	1-Methylfluorene	4.97	1
$C_{10}H_{20}O_2$	Decanoic acid	4.09	1	$C_{14}H_{12}O$	2-Phenylacetophenone	3.18	1
$C_{10}H_{22}$	Decane	6.25	1	$C_{14}H_{12}O_2$	Benzyl benzoate	3.97	1
$C_{10}H_{22}O$	1-Decanol	4.57	1	$C_{14}H_{14}$	1,2-Diphenylethane	4.70	1
$C_{11}H_9N$	4-Phenylpyridine	2.59	1	$C_{14}H_{14}$	4,4′-Dimethylbiphenyl	5.09	1
$C_{11}H_{10}$	1-Methylnaphthalene	3.87	1	$C_{14}H_{22}$	Octylbenzene	6.30	1
$C_{11}H_{10}$	2-Methylnaphthalene	4.00	1	$C_{14}H_{28}O_2$	Tetradecanoic acid	6.1	1
$C_{11}H_{16}$	Pentylbenzene	4.90	1	$C_{15}H_{12}$	2-Methylanthracene	5.15	2
$C_{11}H_{16}$	Pentamethylbenzene	4.56	1	$C_{15}H_{12}$	9-Methylanthracene	5.07	1
$C_{11}H_{22}O$	2-Undecanone	4.09	1	$C_{15}H_{12}$	1-Methylphenanthrene	5.14	2
$C_{11}H_{22}O_2$	Methyl decanoate	4.41	1	$C_{16}H_{10}$	Fluoranthene	5.07	4
$C_{12}Cl_{10}$	Decachlorobiphenyl	8.26	3	$C_{16}H_{10}$	Pyrene	5.08	4
$C_{12}HCl_9$	2,2′,3,3′,4,5,5′,6,6′-Nonachlorobiphenyl	8.16	3	$C_{16}H_{14}$	9,10-Dimethylanthracene	5.69	1
$C_{12}H_2Cl_8$	2,2′,3,3′,5,5′,6,6′-Octachlorobiphenyl	7.10	3	$C_{16}H_{32}O_2$	Hexadecanoic acid	7.17	1
$C_{12}H_3Cl_7$	2,2′,3,3′,4,4′,6-Heptachlorobiphenyl	6.70	3	$C_{17}H_{12}$	11H-Benzo[a]fluorene	5.40	1
$C_{12}H_4Cl_6$	2,2′,3,3′,4,4′-Hexachlorobiphenyl	7.00	3	$C_{17}H_{12}$	11H-Benzo[b]fluorene	5.75	1
$C_{12}H_4Cl_6$	2,2′,4,4′,6,6′-Hexachlorobiphenyl	7.00	3	$C_{18}H_{12}$	Benz[a]anthracene	5.91	1
$C_{12}H_4Cl_6$	2,2′,3,3′,6,6′-Hexachlorobiphenyl	6.70	3	$C_{18}H_{12}$	Chrysene	5.73	4
$C_{12}H_5Cl_5$	2,3,4,5,6-Pentachlorobiphenyl	6.30	3	$C_{18}H_{12}$	Naphthacene	5.76	1
$C_{12}H_5Cl_5$	2,2′,4,5,5′-Pentachlorobiphenyl	6.40	3	$C_{18}H_{12}$	Triphenylene	5.49	4
				$C_{18}H_{15}N$	Triphenylamine	5.74	1
				$C_{18}H_{30}O_2$	Linolenic acid	6.46	1

Mol. form.	Name	log P	Ref.	Mol. form.	Name	log P	Ref.
$C_{18}H_{32}O_2$	Linoleic acid	7.05	1	$C_{20}H_{40}O_2$	Arachidic acid	9.29	1
$C_{18}H_{34}O_2$	Oleic acid	7.64	1	$C_{21}H_{16}$	1,2-Dihydro-3-methylbenz[j]	6.75	1
$C_{18}H_{36}O_2$	Stearic acid	8.23	1		aceanthrylene		
$C_{19}H_{16}O$	Triphenylmethanol	3.68	1	$C_{22}H_{12}$	Benzo[ghi]perylene	6.90	1
$C_{20}H_{12}$	Perylene	6.25	1	$C_{24}H_{12}$	Coronene	6.05	4
$C_{20}H_{12}$	Benzo[a]pyrene	6.20	4				
$C_{20}H_{32}O_2$	Arachidonic acid	6.98	1				

PROTECTION AGAINST IONIZING RADIATION

The following data and rules of thumb are helpful in estimating the penetrating capability of and danger of exposure to various types of ionizing radiation. More precise data should be used for critical applications.

Alpha Particles

Alpha particles of at least 7.5 MeV are required to penetrate the epidermis, the protective layer of skin, 0.07 mm thick.

Electrons

Electrons of at least 70 keV are required to penetrate the epidermis, the protective layer of skin, 0.07 mm thick.

The range of electrons in g/cm^2 is approximately equal to the maximum energy (E) in MeV divided by 2.

The range of electrons in air is about 3.65 m per MeV; for example, a 3 MeV electron has a range of about 11 m in air.

A chamber wall thickness of 30 mg/cm^2 will transmit 70% of the initial fluence of 1 MeV electrons and 20% of that of 0.4 MeV electrons.

When electrons of 1 to 2 MeV pass through light materials such as water, aluminum, or glass, less than 1% of their energy is dissipated as bremsstrahlung.

The bremsstrahlung from 1 Ci of ^{32}P aqueous solution in a glass bottle is about 1 mR/h at 1 meter distance.

When electrons from a 1 Ci source of $^{90}Sr - ^{90}Y$ are absorbed, the bremsstrahlung hazard is approximately equal to that presented by the gamma radiation from 12 mg of radium. The average energy of the bremsstrahlung is about 300 keV.

Gamma Rays

The air-scattered radiation (sky-shine) from a 100 Ci ^{60}Co source placed 1 ft behind a 4 ft high shield is about 100 mrad/h at 6 ft from the outside of the shield.

Within ±20% for point source gamma emitters with energies between 0.07 and 4 MeV, the exposure rate (R/h) at 1 ft is $6C \cdot E \cdot n$ where C is the activity in curies, E is the energy in MeV, and n is the number of gammas per disintegration.

Neutrons

An approximate HVL (thickness of absorber for which the neutron flux falls to half its initial value) for 1 MeV neutrons is 3.2 cm of paraffin; that for 5 MeV neutrons is 6.9 cm of paraffin.

Miscellaneous

The activity of any radionuclide is reduced to less than 1% after 7 half-lives (i.e., $2^{-7} = 0.8\%$).

For nuclides with a half-life greater than 6 days, the change in activity in 24 hours will be less than 10%.

10 HVL (half-value layers) attenuates approximately by 10^{-3}.

There is 0.64 mm^3 of radon gas at STP in transient equilibrium with 1 Ci of radium.

The natural background from all sources in most parts of the world leads to an equivalent dose rate of about 0.04 to 4 mSv per year for the average person. About 84% of this comes from terrestrial sources, the remainder from cosmic rays. The U. S. average is about 3.6 mSv/yr but can range up to 50 mSv/yr in some areas. A passenger in a plane flying at 12,000 meters receives 5 μSv/hr from cosmic rays (as compared to about 0.03 μSv/hr at sea level).

The ICRP recommended exposure limit to man-made sources of ionizing radiation (Reference 2) is 20 mSv/yr averaged over 5 years, with the dose in any one year not to exceed 50 mSv.

A whole-body dose of about 3 Gy over a short time interval will typically lead to 50% mortality in 30 days assuming no medical treatment.

Units

The gray (Gy) is the SI unit of absorbed dose; it is a measure of the mean energy imparted to a sample of irradiated matter, divided by the mass of the sample. Gy is a special name for the SI unit J/kg.

The sievert (Sv) is the SI unit of equivalent dose, which is defined as the absorbed dose multiplied by a weighting factor that expresses the long-term biological risk from low-level chronic exposure to a specified type of radiation. The Sv is another special name for J/kg.

1 curie (Ci) = $3.7 \cdot 10^{10}$ becquerel (Bq); i.e., $3.7 \cdot 10^{10}$ disintegrations per second.

1 roentgen (R) = $2.58 \cdot 10^{-4}$ coulomb per kilogram (C/kg); a measure of the charge (positive or negative) liberated by x-ray or gamma radiation in air, divided by the mass of air.

1 rad = 0.01 Gy

1 rem = 0.01 Sv

References

1. Padikal, T. N., and Fivozinsky, S. P., *Medical Physics Data Book, National Bureau of Standards Handbook 138*, U.S. Government Printing Office, Washington, D.C., 1981.
2. *1990 Recommendations of the International Commission on Radiological Protection*, ICRP Publication 60, *Annals of the ICRP*, Pergamon Press, Oxford, 1991.
3. *Radiation: Doses, Effects, Risks*, United Nations Sales No. E.86.III.D.4, 1985.
4. Eidelman, S., et al., *Physics Letters*, B592, 1, 2004.

ANNUAL LIMITS ON INTAKES OF RADIONUCLIDES

K. F. Eckerman

The following table lists, for workers, the annual limits on oral and inhalation intakes (ALI) for selected radionuclides based on the occupational radiation protection guidance of the International Commission on Radiological Protection (References 1 and 2). An intake of one ALI corresponds to an annual whole body dose of 0.02 Sv (2 rem).

The ALI is expressed in the SI unit of activity, the becquerel (Bq), and in the conventional unit, the microcurie (µCi); 1 µCi = $3.7 \cdot 10^4$ Bq. The chemical form of inhaled radionuclides is, in most instances, stated in terms of the rate of absorption to blood from the lungs and the fractional absorption from the small intestine. Type F, M, and S denote chemical forms which are absorbed from the lungs at rates characterized as fast, moderate, and slow, respectively. The time to absorb 90% of the deposited radionuclide, in the absence of radioactive decay, corresponds to about 10 minutes, 150 days, and 7000 days for Type F, M, and S compounds, respec-

tively. Type F compounds can be considered to be more soluble than M or S, S being the most insoluble. Chemical form consideration for ingestion is specified by the fractional absorption from the small intestine, denoted as f_1. The f_1 values range from 10^{-5} to 1. Higher fractional absorption is associated with greater solubility of the compound.

References

1. *1990 Recommendations of the International Commission on Radiological Protection, ICRP Publication 60, Annals of the ICRP 21, (1–3)*, Pergamon Press, Oxford, 1991.
2. *Dose Coefficients for Intakes of Radionuclides by Workers, ICRP Publication 68, Annals of the ICRP, 24(4)*, Pergamon Press, Oxford, 1995.

	Physical half-life	Inhalation intakes			Oral intakes		
		Chemical form Type/f_1	ALI Bq	ALI µCi	Chemical form f_1	ALI Bq	ALI µCi
^{3}H	12.3 y	HT gas	1.1E+13	3.0E+08	1.000	1.1E+13	3.0E+08
		HTO vapor	1.1E+09	3.0E+04			
^{11}C	0.340 h	CO	1.7E+10	4.5E+05	1.000	8.3E+08	2.3E+04
		CO_2	9.1E+09	2.5E+05			
		Organic compounds	6.2E+09	1.7E+05			
^{14}C	5730 y	CO	2.5E+10	6.8E+05	1.000	3.4E+07	9.3E+02
		CO_2	3.1E+09	8.3E+04			
		Organic compounds	3.4E+07	9.3E+02			
^{18}F	1.83 h	F 1.000	3.7E+08	1.0E+04	1.000	4.1E+08	1.1E+04
		M 1.000	2.2E+08	6.1E+03			
		S 1.000	2.2E+08	5.8E+03			
^{22}Na	2.60 y	F 1.000	1.0E+07	2.7E+02	1.000	6.3E+06	1.7E+02
^{24}Na	15.0 h	F 1.000	3.8E+07	1.0E+03	1.000	4.7E+07	1.3E+03
^{32}P	14.3 d	F 0.800	1.8E+07	4.9E+02	0.800	8.3E+06	2.3E+02
		M 0.800	6.9E+06	1.9E+02			
^{35}S	87.4 d	Inorganic compounds					
		F 0.800	2.5E+08	6.8E+03	0.800	1.4E+08	3.9E+03
		M 0.800	1.8E+07	4.9E+02	0.100	1.1E+08	2.8E+03
		Vapor	1.7E+08	4.5E+03			
		Organic compounds			1.000	2.6E+07	7.0E+02
^{42}K	12.4 h	F 1.000	1.0E+08	2.7E+03	1.000	4.7E+07	1.3E+03
^{43}K	22.6 h	F 1.000	7.7E+07	2.1E+03	1.000	8.0E+07	2.2E+03
^{45}Ca	163 d	M 0.300	8.7E+06	2.4E+02	0.300	2.6E+07	7.1E+02
^{47}Ca	4.53 d	M 0.300	9.5E+06	2.6E+02	0.300	1.3E+07	3.4E+02
^{51}Cr	27.7 d	F 0.100	6.7E+08	1.8E+04	0.100	5.3E+08	1.4E+04
		M 0.100	5.9E+08	1.6E+04	0.010	5.4E+08	1.5E+04
		S 0.100	5.6E+08	1.5E+04			
^{54}Mn	312 d	F 0.100	1.8E+07	4.9E+02	0.100	2.8E+07	7.6E+02
		M 0.100	1.7E+07	4.5E+02			
^{52}Fe	8.28 h	F 0.100	2.9E+07	7.8E+02	0.100	1.4E+07	3.9E+02
		M 0.100	2.1E+07	5.7E+02			
^{55}Fe	2.70 y	F 0.100	2.2E+07	5.9E+02	0.100	6.1E+07	1.6E+03
		M 0.100	6.1E+07	1.6E+03			

	Physical half-life	Inhalation intakes Chemical form Type/f_1	Inhalation ALI Bq	Inhalation ALI µCi	Oral intakes Chemical form f_1	Oral ALI Bq	Oral ALI µCi
^{59}Fe	44.5 d	F 0.100	6.7E+06	1.8E+02	0.100	1.1E+07	3.0E+02
		M 0.100	6.3E+06	1.7E+02			
^{57}Co	271 d	M 0.100	5.1E+07	1.4E+03	0.100	9.5E+07	2.6E+03
		S 0.050	3.3E+07	9.0E+02	0.050	1.1E+08	2.8E+03
^{58}Co	70.8 d	M 0.100	1.4E+07	3.9E+02	0.100	2.7E+07	7.3E+02
		S 0.050	1.2E+07	3.2E+02	0.050	2.9E+07	7.7E+02
^{60}Co	5.27 y	M 0.100	2.8E+06	7.6E+01	0.100	5.9E+06	1.6E+02
		S 0.050	1.2E+06	3.2E+01	0.050	8.0E+06	2.2E+02
^{64}Cu	12.7 h	F 0.500	2.9E+08	7.9E+03	0.500	1.7E+08	4.5E+03
		M 0.500	1.3E+08	3.6E+03			
		S 0.500	1.3E+08	3.6E+03			
^{59}Ni	75000 y	F 0.050	9.1E+07	2.5E+03	0.050	3.2E+08	8.6E+03
		M 0.050	2.1E+08	5.8E+03			
		Vapor	2.4E+07	6.5E+02			
^{63}Ni	96.0 y	F 0.050	3.8E+07	1.0E+03	0.050	1.3E+08	3.6E+03
		M 0.050	6.5E+07	1.7E+03			
		Vapor	1.0E+07	2.7E+02			
^{65}Zn	244 d	S 0.500	7.1E+06	1.9E+02	0.500	5.1E+06	1.4E+02
^{67}Ga	3.26 d	F 0.001	1.8E+08	4.9E+03	0.001	1.1E+08	2.8E+03
		M 0.001	7.1E+07	1.9E+03			
^{68}Ga	1.13 h	F 0.001	4.1E+08	1.1E+04	0.001	2.0E+08	5.4E+03
		M 0.001	2.5E+08	6.7E+03			
^{68}Ge	288 d	F 1.000	2.4E+07	6.5E+02	1.000	1.5E+07	4.2E+02
		M 1.000	2.5E+06	6.8E+01			
^{75}Se	120 d	F 0.800	1.4E+07	3.9E+02	0.800	7.7E+06	2.1E+02
		M 0.800	1.2E+07	3.2E+02	0.050	4.9E+07	1.3E+03
^{79}Se	65000 y	F 0.800	1.3E+07	3.4E+02	0.800	6.9E+06	1.9E+02
		M 0.800	6.5E+06	1.7E+02	0.050	5.1E+07	1.4E+03
^{86}Rb	18.6 d	F 1.000	1.5E+07	4.2E+02	1.000	7.1E+06	1.9E+02
^{85}Sr	64.8 d	F 0.300	3.6E+07	9.7E+02	0.300	3.6E+07	9.7E+02
		S 0.010	3.1E+07	8.4E+02	0.010	6.1E+07	1.6E+03
^{87m}Sr	2.80 h	F 0.300	9.1E+08	2.5E+04	0.300	6.7E+08	1.8E+04
		S 0.010	5.7E+08	1.5E+04	0.010	6.1E+08	1.6E+04
^{89}Sr	50.5 d	F 0.300	1.4E+07	3.9E+02	0.300	7.7E+06	2.1E+02
		S 0.010	3.6E+06	9.7E+01	0.010	8.7E+06	2.4E+02
^{90}Sr	29.1 y	F 0.300	6.7E+05	1.8E+01	0.300	7.1E+05	1.9E+01
		S 0.010	2.6E+05	7.0E+00	0.010	7.4E+06	2.0E+02
^{99}Mo	2.75 d	F 0.800	5.6E+07	1.5E+03	0.800	2.7E+07	7.3E+02
		S 0.050	1.8E+07	4.9E+02	0.050	1.7E+07	4.5E+02
^{99m}Tc	6.02 h	F 0.800	1.0E+09	2.7E+04	0.800	9.1E+08	2.5E+04
		M 0.800	6.9E+08	1.9E+04			
^{99}Tc	213000 y	F 0.800	5.0E+07	1.4E+03	0.800	2.6E+07	6.9E+02
		M 0.800	6.3E+06	1.7E+02			
^{106}Ru	1.01 y	F 0.050	2.0E+06	5.5E+01	0.050	2.9E+06	7.7E+01
		M 0.050	1.2E+06	3.2E+01			
		S 0.050	5.7E+05	1.5E+01			
^{111}In	2.83 d	F 0.020	9.1E+07	2.5E+03	0.020	6.9E+07	1.9E+03
		M 0.020	6.5E+07	1.7E+03			
^{113m}In	1.66 h	F 0.020	1.1E+09	2.8E+04	0.020	7.1E+08	1.9E+04
		M 0.020	6.3E+08	1.7E+04			
^{113}Sn	115 d	F 0.020	2.5E+07	6.8E+02	0.020	2.7E+07	7.4E+02
		M 0.020	1.1E+07	2.8E+02			
^{123}I	13.2 h	F 1.000	1.8E+08	4.9E+03	1.000	9.5E+07	2.6E+03
		Vapor	9.5E+07	2.6E+03			
^{125}I	60.1 d	F 1.000	2.7E+06	7.4E+01	1.000	1.3E+06	3.6E+01
		Vapor	1.4E+06	3.9E+01			
^{129}I	$1.57 \cdot 10^7$ y	F 1.000	3.9E+05	1.1E+01	1.000	1.8E+05	4.9E+00
		Vapor	2.1E+05	5.6E+00			

	Physical half-life	Inhalation intakes			Oral intakes		
		Chemical form	ALI		Chemical form	ALI	
		Type/f_1	Bq	µCi	f_1	Bq	µCi
^{131}I	8.04 d	F 1.000	1.8E+06	4.9E+01	1.000	9.1E+05	2.5E+01
		Vapor	1.0E+06	2.7E+01			
^{129}Cs	1.34 d	F 1.000	2.5E+08	6.7E+03	1.000	3.3E+08	9.0E+03
^{134}Cs	2.06 y	F 1.000	2.1E+06	5.6E+01	1.000	1.1E+06	2.8E+01
^{136}Cs	13.1 d	F 1.000	1.1E+07	2.8E+02	1.000	6.7E+06	1.8E+02
^{137}Cs	30.0 y	F 1.000	3.0E+06	8.1E+01	1.000	1.5E+06	4.2E+01
^{141}Ce	32.5 d	M 5.0E-04	7.4E+06	2.0E+02	5.0E-04	2.8E+07	7.6E+02
		S 5.0E-04	6.5E+06	1.7E+02			
^{144}Ce	284 d	M 5.0E-04	8.7E+05	2.4E+01	5.0E-04	3.8E+06	1.0E+02
		S 5.0E-04	6.9E+05	1.9E+01			
^{133}Ba	10.7 y	F 0.100	1.1E+07	3.0E+02	0.100	2.0E+07	5.4E+02
^{140}Ba	12.7 d	F 0.100	1.3E+07	3.4E+02	0.100	8.0E+06	2.2E+02
^{169}Yb	32.0 d	M 5.0E-04	9.5E+06	2.6E+02	5.0E-04	2.8E+07	7.6E+02
		S 5.0E-04	8.3E+06	2.3E+02			
^{198}Au	2.69 d	F 0.100	5.1E+07	1.4E+03	0.100	2.0E+07	5.4E+02
		M 0.100	2.0E+07	5.5E+02			
		S 0.100	1.8E+07	4.9E+02			
^{198m}Au	2.30 d	F 0.100	3.4E+07	9.2E+02	0.100	1.5E+07	4.2E+02
		M 0.100	1.0E+07	2.7E+02			
		S 0.100	1.1E+07	2.8E+02			
^{197}Hg	2.67 d	Inorganic compounds					
		F 0.400	2.4E+08	6.4E+03	1.000	2.0E+08	5.5E+03
					0.400	1.2E+08	3.2E+03
		Vapor	4.5E+06	1.2E+02			
		Organic compounds					
		F 0.020	2.0E+08	5.4E+03	0.020	8.7E+07	2.4E+03
		M 0.020	7.1E+07	1.9E+03			
^{203}Hg	46.6 d	Inorganic compounds					
		F 0.400	2.7E+07	7.2E+02	1.000	1.1E+07	2.8E+02
					0.400	1.8E+07	4.9E+02
		Vapor	2.9E+06	7.7E+01			
		Organic compounds					
		F 0.020	3.4E+07	9.2E+02	0.020	3.7E+07	1.0E+03
		M 0.020	1.1E+07	2.8E+02			
^{201}Tl	3.04 d	F 1.000	2.6E+08	7.1E+03	1.000	2.1E+08	5.7E+03
^{210}Pb	22.3 y	F 0.200	1.8E+04	4.9E-01	0.200	2.9E+04	7.9E-01
^{207}Bi	38.0 y	F 0.050	2.4E+07	6.4E+02	0.050	1.5E+07	4.2E+02
		M 0.050	6.3E+06	1.7E+02			
^{210}Po	138 d	F 0.100	2.8E+04	7.6E-01	0.100	8.3E+04	2.3E+00
		M 0.100	9.1E+03	2.5E-01			
^{224}Ra	3.66 d	M 0.200	8.3E+03	2.3E-01	0.200	3.1E+05	8.3E+00
^{226}Ra	1600 y	M 0.200	1.7E+03	4.5E-02	0.200	7.1E+04	1.9E+00
^{228}Ra	5.75 y	M 0.200	1.2E+04	3.2E-01	0.200	3.0E+04	8.1E-01
^{228}Th	1.91 y	M 5.0E-04	8.7E+02	2.4E-02	5.0E-04	2.9E+05	7.7E+00
		S 2.0E-04	6.3E+02	1.7E-02	2.0E-04	5.7E+05	1.5E+01
^{230}Th	77000 y	M 5.0E-04	7.1E+02	1.9E-02	5.0E-04	9.5E+04	2.6E+00
		S 2.0E-04	2.8E+03	7.5E-02	2.0E-04	2.3E+05	6.2E+00
^{232}Th	1.40·10^{10} y	M 5.0E-04	6.9E+02	1.9E-02	5.0E-04	9.1E+04	2.5E+00
		S 2.0E-04	1.7E+03	4.5E-02	2.0E-04	2.2E+05	5.9E+00
^{234}U	2.44·10^5 y	F 0.020	3.1E+04	8.4E-01	0.020	4.1E+05	1.1E+01
		M 0.020	9.5E+03	2.6E-01	0.002	2.4E+06	6.5E+01
		S 0.002	2.9E+03	7.9E-02			
^{235}U	7.04·10^8 y	F 0.020	3.3E+04	9.0E-01	0.020	4.3E+05	1.2E+01
		M 0.020	1.1E+04	3.0E-01	0.002	2.4E+06	6.5E+01

		Inhalation intakes			Oral intakes		
	Physical half-life	Chemical form Type/f_1	ALI Bq	ALI µCi	Chemical form f_1	ALI Bq	ALI µCi
		S 0.002	3.3E+03	8.9E-02			
^{238}U	4.47·10^9 y	F 0.020	3.4E+04	9.3E-01	0.020	4.5E+05	1.2E+01
		M 0.020	1.3E+04	3.4E-01	0.002	2.6E+06	7.1E+01
		S 0.002	3.5E+03	9.5E-02			
^{237}Np	2.14·10^6 y	M 5.0E-04	1.3E+03	3.6E-02	5.0E-04	1.8E+05	4.9E+00
^{239}Np	2.36 d	M 5.0E-04	1.8E+07	4.9E+02	5.0E-04	2.5E+07	6.8E+02
^{238}Pu	87.7 y	M 5.0E-04	6.7E+02	1.8E-02	5.0E-04	8.7E+04	2.4E+00
		S 1.0E-05	1.8E+03	4.9E-02	1.0E-05	2.3E+06	6.1E+01
					1.0E-04	4.1E+05	1.1E+01
^{239}Pu	24100 y	M 5.0E-04	6.3E+02	1.7E-02	5.0E-04	8.0E+04	2.2E+00
		S 1.0E-05	2.4E+03	6.5E-02	1.0E-05	2.2E+06	6.0E+01
					1.0E-04	3.8E+05	1.0E+01
^{241}Pu	14.4 y	M 5.0E-04	3.4E+04	9.3E-01	5.0E-04	4.3E+06	1.2E+02
		S 1.0E-05	2.4E+05	6.4E+00	1.0E-05	1.8E+08	4.9E+03
					1.0E-04	2.1E+07	5.6E+02
^{241}Am	432 y	M 5.0E-04	7.4E+02	2.0E-02	5.0E-04	1.0E+05	2.7E+00
^{244}Cm	18.1 y	M 5.0E-04	1.2E+03	3.2E-02	5.0E-04	1.7E+05	4.5E+00
^{252}Cf	2.64 y	M 5.0E-04	1.5E+03	4.2E-02	5.0E-04	2.2E+05	6.0E+00

CHEMICAL CARCINOGENS

The following substances are listed in the 11th Report on Carcinogens 2004, released by the National Institute of Environmental Health Sciences (NIEHS) under the National Toxicology Program (NTP). Substances are grouped in two classes:

- **Known to be human carcinogens**: There is sufficient evidence of carcinogenicity from studies in humans which indicates a causal relationship between exposure to the substance and human cancer.

- **Reasonably anticipated to be human carcinogens**: There is limited evidence of carcinogenicity from studies in humans which indicates that causal interpretation is credible, but that alternative explanations, such as chance, bias, or confounding factors, could not be adequately excluded; or there is sufficient evidence of carcinogenicity from studies in experimental animals.

The NTP report also lists many poorly defined materials such as soots, tars, mineral oils, coke oven emissions, etc. These materials are not included here.

The table lists the name normally used in the *Handbook of Chemistry and Physics*, followed by additional names and acronyms by which the substance is known. In many cases the primary name given here is different from that used in the NTP report; however, names used in the NTP report appear in the Other Names column. The Chemical Abstracts Service Registry Number (CAS RN), given in the last column, is taken from the NTP report. Extensive details on each substance are given in the reference.

Reference

Public Health Service, National Toxicology Program, 11th Report on Carcinogens, available on the Internet at <http://ntp.niehs.nih.gov/>

Substance	Other Names	CAS RN
Known to be Human Carcinogens		
Aflatoxins		1402-68-2
4-Aminobiphenyl	*p*-Biphenylamine	92-67-1
Arsenic compounds, inorganic		
Asbestos		1332-21-4
Azathioprine	6-[(1-Methyl-4-nitro-1*H*-imidazol-5-yl)thio]-1*H*-purine	446-86-6
Benzene		71-43-2
p-Benzidine (includes dyes metabolized to benzidene)	[1,1'-Biphenyl]-4,4'-diamine	92-87-5
Beryllium and beryllium compounds		7440-41-7
Bis(2-chloroethyl) sulfide	Mustard gas	505-60-2
Bis(chloromethyl) ether		542-88-1
1,3-Butadiene		106-99-0
1,4-Butanediol dimethylsulfonate	Myleran; Busulfan	55-98-1
Cadmium and cadmium compounds		7440-43-9
Chlorambucil		305-03-3
Chloroethene	Vinyl chloride; Chloroethylene	75-01-4
1-(2-Chloroethyl)-3-(4-methylcyclohexyl)-1-nitrosourea	MeCCNU	13909-09-6
Chloromethyl methyl ether		107-30-2
Chromium hexavalent compounds		
Cyclophosphamide	Cyclophosphane; 2*H*-1,3,2-Oxazaphosphorin-2-amine, *N,N*-bis(2-chloroethyl)tetrahydro-, 2-oxide	50-18-0
Cyclosporin A	Cyclosporine	59865-13-3
Diethylstilbestrol		56-53-1
Erionite		66733-21-9
Estrogens, steroidal		
N-(4-Ethoxyphenyl)acetamide	Phenacetin	62-44-2
Melphalan	*L*-Phenylalanine, 4-[bis(2-chloroethyl)amino]-	148-82-3
Methoxsalen (with UV therapy)	PUVA; 9-Methoxy-7*H*-furo[3,2-g][1]benzopyran-7-one	298-81-7
2-Naphthylamine	2-Aminonaphthalene; β-Naphthylamine	91-59-8
Nickel compounds		
Oxirane	Ethylene oxide	75-21-8

Substance	Other Names	CAS RN
Radon		10043-92-2
Silicon dioxide (respirable size)	Quartz; Silica	14808-60-7
Silicon dioxide (respirable size)	Cristobalite; Silica	14464-46-1
Silicon dioxide (respirable size)	Tridymite; Silica	15468-32-3
Sulfuric acid (strong acid mists)	Oil of vitriol	7664-93-9
Tamoxifen		10540-29-1
2,3,7,8-Tetrachlorodibenzo-p-dioxin	TCDD; Dioxin	1746-01-6
Thorium(IV) oxide	Thorium dioxide	1314-20-1
Triethylenethiophosphoramide	Thiotepa; Tris(1-aziridinyl)phosphine, sulfide	52-24-4

Reasonably Anticipated to be Human Carcinogens

Substance	Other Names	CAS RN
Acetaldehyde	Ethanal	75-07-0
2-(Acetylamino)fluorene		53-96-3
Acrylamide	2-Propenamide	79-06-1
Acrylonitrile	Propenenitrile	107-13-1
4-Allyl-1,2-dimethoxybenzene	Methyleugenol; 1,2-Dimethoxy-4-allylbenzene	93-15-2
2-Amino-9,10-anthracenedione	2-Aminoanthraquinone	117-79-3
1-Amino-2,4-dibromo-9,10-anthracenedione	1-Amino-2,4-dibromoanthraquinone	81-49-2
2-Amino-3,4-dimethylimidazo[4,5-f]quinoline	MeIQ	77094-11-2
2-Amino-3,8-dimethylimidazo[4,5-f]quinoxaline	MeIQx	77500-04-0
1-Amino-2-methyl-9,10-anthracenedione	1-Amino-2-methylanthraquinone	82-28-0
2-Amino-3-methyl-3H-imidazo[4,5-f]quinoline	IQ	76180-96-6
2-Amino-1-methyl-6-phenylimidazo[4,5-b]pyridine	PhIP	105650-23-5
Azacitidine	5-Azacytidine; 4-Amino-1-β-D-ribofuranosyl-1,3,5-triazine-2(1H)-one	320-67-2
Benz[a]anthracene		56-55-3
Benzo[b]fluoranthene	Benz[e]acephenanthrylene	205-99-2
Benzo[j]fluoranthene		205-82-3
Benzo[k]fluoranthene	2,3,1',8'-Binaphthylene	207-08-9
Benzo[a]pyrene		50-32-8
2,2'-Bioxirane	Diepoxybutane	1464-53-5
Bis(4-amino-3-chlorophenyl)methane	4,4-Methylene-bis(2-chloraniline); MBOCA	101-14-4
2,2-Bis(bromomethyl)-1,3-propanediol	BBMP; Pentaerythritol dibromide	3296-90-0
Bis(2-chloroethyl)methylamine	Nitrogen mustard hydrochloride	55-86-7
N,N'-Bis(2-chloroethyl)-N-nitrosourea	BCNU; Carmustine	154-93-8
Bis[4-(dimethylamino)phenyl]methane	Michler's Base; 4,4-Methylenebis(N,N-dimethylbenzenamine)	101-61-1
1,3-Bis(2,3-epoxypropoxy)benzene	Diglycidyl resorcinol ether	101-90-6
Bis(2-ethylhexyl) phthalate	DEHP; Di(2-ethylhexyl) phthalate	117-81-7
Bromodichloromethane		75-27-4
Bromoethene	Vinyl bromide	593-60-2
tert-Butyl-4-hydroxyanisole	BHA; Butylated hydroxyanisole	25013-16-5
Chloramphenicol		56-75-7
Chlorendic acid	1,4,5,6,7,7-Hexachloro-5-norbornene-2,3-dicarboxylic acid	115-28-6
Chlorinated paraffins (C_{12}, 60% Cl)		108171-26-2
4-Chloro-1,2-benzenediamine	4-Chloro-o-phenylenediamine	95-83-0
2-Chloro-1,3-butadiene	Chloroprene	126-99-8
1-(2-Chloroethyl)-3-cyclohexyl-1-nitrosourea	CCNU; Lomustine; Belustine	13010-47-4
4-Chloro-2-methylaniline	p-Chloro-o-toluidine	95-69-2
4-Chloro-2-methylaniline hydrochloride	p-Chloro-o-toluidine hydrochloride	3165-93-3

Substance	Other Names	CAS RN
1-Chloro-2-methylpropene	Dimethylvinyl chloride	513-37-1
3-Chloro-2-methylpropene		563-47-3
Chlorozotocin	2-[[[(2-Chloroethyl)nitrosoamino]carbonyl]amino]-2-deoxy-D-glucose	54749-90-5
Fuchsin	C.I. Basic Red 9, monohydrochloride	569-61-9
Cobalt(II) sulfate	Cobaltous sulfate	10124-43-3
Cupferron		135-20-6
Dacarbazine	5-(3,3-Dimethyl-1-triazenyl)-1H-imidazole-4-carboxamide	4342-03-4
Decabromobiphenyl		13654-09-6
cis-Diaminedichloroplatinum	Cisplatin	15663-27-1
2,4-Diaminoanisole sulfate	1,3-Benzenediamine, 4-methoxy, sulfate	39156-41-7
4,4'-Diaminodiphenyl ether	4,4-Oxydianiline	101-80-4
4,4'-Diaminodiphenylmethane	4,4'-Methylenedianiline	101-77-9
4,4'-Diaminodiphenylmethane dihydrochloride	4,4'-Methylenedianiline dihydrochloride	13552-44-8
4,4'-Diaminodiphenyl sulfide	4,4'-Thiodianiline	139-65-1
Dibenz[a,h]acridine		226-36-8
Dibenz[a,j]acridine	7-Azadibenz[a,j]anthracene	224-42-0
Dibenz[a,h]anthracene	1,2:5,6-Dibenzanthracene	53-70-3
7H-Dibenzo[c,g]carbazole		194-59-2
Dibenzo[a,e]pyrene	Naphtho[1,2,3,4-def]chrysene	192-65-4
Dibenzo[a,h]pyrene	Dibenzo[b,def]chrysene	189-64-0
Dibenzo[a,i]pyrene	Benzo[rst]pentaphene	189-55-9
Dibenzo[a,l]pyrene	Dibenzo[def,p]chrysene	191-30-0
1,2-Dibromo-3-chloropropane		96-12-8
1,2-Dibromoethane	Ethylene dibromide; EDB	106-93-4
2,3-Dibromo-1-propanol	DBP	96-13-9
2,3-Dibromo-1-propanol, phosphate (3:1)	Tris(2,3-dibromopropyl) phosphate	126-72-7
p-Dichlorobenzene	1,4-Dichlorobenzene	106-46-7
3,3'-Dichloro-p-benzidine	3,3'-Dichloro[1,1'-biphenyl]-4,4'-diamine	91-94-1
3,3'-Dichloro-p-benzidine dihydrochloride	3,3'-Dichloro-[1,1'-biphenyl]-4,4'-diamine dihydrochloride	612-83-9
1,2-Dichloroethane	Ethylene dichloride	107-06-2
Dichloromethane	Methylene chloride	75-09-2
1,3-Dichloropropene (unspecified isomer)		542-75-6
Diethyl sulfate		64-67-5
2,3-Dihydro-6-propyl-2-thioxo-4(1H)-pyrimidinone	Propylthiouracil	51-52-5
1,8-Dihydroxy-9,10-anthracenedione	Danthron; 1,8-Dihydroxyanthraquinone	117-10-2
3,3'-Dimethoxybenzidine (and dyes metabolized to 3,3'-Dimethoxybenzidine)	Dianisidine	119-90-4
p-(Dimethylamino)azobenzene		60-11-7
2',3-Dimethyl-4-aminoazobenzene	o-Aminoazotoluene; 4-o-Tolylazo-o-toluidine	97-56-3
Dimethylcarbamic chloride	Dimethylcarbamoyl chloride	79-44-7
1,1-Dimethylhydrazine	UDMH	57-14-7
Dimethyl sulfate		77-78-1
1,6-Dinitropyrene		42397-64-8
1,8-Dinitropyrene		42397-65-9
1,4-Dioxane		123-91-1
1,2-Diphenylhydrazine	Hydrazobenzene	122-66-7
1,3-Diphenyl-1-triazene	Diazoaminobenzene	136-35-6

Substance	Other Names	CAS RN
Disperse Blue No. 1	1,4,5,8-Tetraamino-9,10-anthracenedione	2475-45-8
Doxorubicin hydrochloride	Adriamycin	23214-92-8
Epichlorohydrin	(Chloromethyl)oxirane	106-89-8
1,2-Epoxy-4-(epoxyethyl)cyclohexane	4-Vinyl-1-cyclohexene dioxide	106-87-6
Ethyl carbamate	Urethane	51-79-6
Ethyl methanesulfonate		62-50-0
N-Ethyl-N-nitrosourea	ENU; N-Nitroso-N-ethylurea	759-73-9
Fluoroethene	Vinyl fluoride	75-02-5
Formaldehyde (gas)	Methanal	50-00-0
Furan		110-00-9
Hexabromobiphenyl isomers	Firemaster FF-1	67774-32-7
Hexachlorobenzene	Perchlorobenzene	118-74-1
Hexachlorocyclohexane isomers		608-73-1
1,2,3,4,5,6-Hexachlorocyclohexane, (1α,2α,3β,4α,5β,6β)	α-Hexachlorocyclohexane	319-84-6
1,2,3,4,5,6-Hexachlorocyclohexane, (1α,2β,3α,4β,5α,6β)	β-Hexachlorocyclohexane	319-85-7
1,2,3,4,5,6-Hexachlorocyclohexane, (1α,2α,3β,4β,5α,6β)	Lindane; γ-Hexachlorocyclohexane	58-89-9
Hexachloroethane	Perchloroethane	67-72-1
Hexamethylphosphoric triamide	Hexamethylphosphoramide; Tris(dimethylamino)phosphine oxide	680-31-9
Hydrazine		302-01-2
Hydrazine sulfate		10034-93-2
2-Imidazolidinethione	Ethylene thiourea	96-45-7
Indeno[1,2,3-cd]pyrene	1,10-(1,2-Phenylene)pyrene	193-39-5
Kepone	Chlordecone	143-50-0
Lead and lead compounds	Includes all lead compounds	7439-92-1
o-Methoxyaniline hydrochloride	o-Anisidine hydrochloride	134-29-2
2-Methoxy-5-methylaniline	p-Cresidine; 5-Methyl-o-anisidine	120-71-8
o-Methylaniline	o-Toluidine	95-53-4
o-Methylaniline hydrochloride	o-Toluidine hydrochloride	636-21-5
2-Methyl-1,3-butadiene	Isoprene	78-79-5
5-Methylchrysene		3697-24-3
4,4-Methylenedianiline dihydrochloride	Benzenamine, 4,4'-methylenedi-, dihydrochloride	13552-44-8
Methyl methanesulfonate		66-27-3
N-Methyl-N′-nitro-N-nitrosoguanidine		70-25-7
N-Methyl-N-nitrosourea	N-Nitroso-N-methylurea	684-93-5
Methyloxirane	1,2-Propylene oxide	75-56-9
Metronidazole	2-Methyl-5-nitro-1H-imidazole-1-ethanol	443-48-1
Mirex	Hexachloropentadiene dimer	2385-85-5
Naphthalene		91-20-3
Nickel (metallic)		7440-02-0
Nitrilotriacetic acid	N,N-Bis(carboxymethyl)glycine	139-13-9
2-Nitroanisole	1-Methoxy-2-nitrobenzene	91-23-6
Nitrobenzene		98-95-3
6-Nitrochrysene		7496-02-8
Nitrofen	2,4-Dichloro-1-(4-nitrophenoxy)benzene	1836-75-5
Nitromethane		75-52-5
2-Nitropropane		79-46-9
1-Nitropyrene		5522-43-0

Substance	Other Names	CAS RN
4-Nitropyrene		57835-92-4
N-Nitrosodibutylamine	Dibutylnitrosamine	924-16-3
N-Nitrosodiethanolamine	2,2′-(Nitrosoimino)ethanol	1116-54-7
N-Nitrosodiethylamine	DEN; Diethylnitrosamine	55-18-5
N-Nitrosodimethylamine	DMN; Dimethylnitrosamine	62-75-9
4-(*N*-Nitrosomethylamino)-1-(3-pyridyl)-1-butanone	NNK; Ketone, 3-pyridyl-3-(*N*-methyl-*N*-nitrosamino)propyl	64091-91-4
N-Nitroso-*N*-methylvinylamine	*N*-Methyl-*N*-nitrosoethenamine	4549-40-0
4-Nitrosomorpholine	*N*-Nitrosomorpholine	59-89-2
N-Nitrosonornicotine	*N*′-Nitroso-3-(2-pyrrolidinyl)pyridine	16543-55-8
N-Nitrosopiperidine	1-Nitrosopiperidine	100-75-4
N-Nitroso-*N*-propyl-1-propanamine	*N*-Nitrosodipropylamine	621-64-7
N-Nitrosopyrrolidine		930-55-2
N-Nitrososarcosine	*N*-Methyl-*N*-nitrosoglycine	13256-22-9
Norethisterone	19-Norpregn-4-en-20-yn-3-one, 17-hydroxy-, (17 α)-	68-22-4
Ochratoxin A		303-47-9
Octabromobiphenyl isomers		61288-13-9
2-Oxetanone	β-Propiolactone	57-57-8
Oxiranemethanol	Glycidol	556-52-5
Oxymetholone	Androstan-3-one, 17-hydroxy-2-(hydroxymethylene)-17-methyl-	434-07-1
Phenazopyridine hydrochloride	3-(Phenylazo)-2,6-pyridinediamine, monohydrochloride	136-40-3
Phenolphthalein	3,3-Bis(4-hydroxyphenyl)-1(3*H*)-isobenzofuranone	77-09-8
Phenoxybenzamine hydrochloride		63-92-3
Phenyloxirane	Styrene-7,8-oxide	96-09-3
Phenytoin	5,5-Diphenyl-2,4-imidazolidinedione	57-41-0
Polybrominated biphenyls	PBBs	
Polychlorinated biphenyls	PCBs	1336-36-3
Procarbazine hydrochloride		366-70-1
Progesterone	Pregn-4-ene-3,20-dione	57-83-0
1,3-Propane sultone	1,2-Oxathiolane, 2,2-dioxide	1120-71-4
Propyleneimine	2-Methylaziridine	75-55-8
Reserpine		50-55-5
Safrole	5-(2-Propenyl)-1,3-benzodioxole	94-59-7
Selenium sulfide		7446-34-6
Streptozotocin	*D*-Glucopyranose, 2-deoxy-2-[[(methylnitrosoamino)carbonyl]amino]-	18883-66-4
Sulfallate	*N,N*-Diethyldithiocarbamic acid, 2-chloroallyl ester	95-06-7
Tetrachloroethene	Tetrachloroethylene; Perchloroethylene	127-18-4
Tetrachloromethane	Carbon tetrachloride	56-23-5
Tetrafluoroethene	Tetrafluoroethylene	116-14-3
N,N,N′,N′-Tetramethyl-4,4′-diaminobenzophenone	Bis(dimethylamino)benzophenone; Michler's Ketone	90-94-8
Tetranitromethane		509-14-8
Thioacetamide	Ethanethioamide	62-55-5
Thiourea	Thiocarbamide	62-56-6
o-Tolidine	3,3′-Dimethylbenzidine	119-93-7
Toluene-2,4-diamine	2,4-Diaminotoluene	95-80-7
Toluene diisocyanate (unspecified isomer)		26471-62-5
Toxaphene	Polychlorocamphene	8001-35-2

Substance	Other Names	CAS RN
1*H*-1,2,4-Triazol-3-amine	Amitrole	61-82-5
1,1,1-Trichloro-2,2-bis(4-chlorophenyl)ethane	DDT; Dichlorodiphenyltrichloroethane	50-29-3
Trichloroethene	Trichloroethylene	79-01-6
Trichloromethane	Chloroform	67-66-3
(Trichloromethyl)benzene	Benzotrichloride	98-07-7
2,4,6-Trichlorophenol		88-06-2
1,2,3-Trichloropropane		96-18-4

Appendix A
Mathematical Tables

MISCELLANEOUS MATHEMATICAL CONSTANTS

π CONSTANTS

$$\pi = 3.14159\ 26535\ 89793\ 23846\ 26433\ 83279\ 50288\ 41971\ 69399\ 37511$$
$$1/\pi = 0.31830\ 98861\ 83790\ 67153\ 77675\ 26745\ 02872\ 40689\ 19291\ 48091$$
$$\pi^2 = 9.86960\ 44010\ 89358\ 61883\ 44909\ 99876\ 15113\ 53136\ 99407\ 24079$$
$$\log_e \pi = 1.14472\ 98858\ 49400\ 17414\ 34273\ 51353\ 05871\ 16472\ 94812\ 91531$$
$$\log_{10} \pi = 0.49714\ 98726\ 94133\ 85435\ 12682\ 88290\ 89887\ 36516\ 78324\ 38044$$
$$\log_{10} \sqrt{2\pi} = 0.39908\ 99341\ 79057\ 52478\ 25035\ 91507\ 69595\ 02099\ 34102\ 92128$$

CONSTANTS INVOLVING e

$$e = 2.71828\ 18284\ 59045\ 23536\ 02874\ 71352\ 66249\ 77572\ 47093\ 69996$$
$$1/e = 0.36787\ 94411\ 71442\ 32159\ 55237\ 70161\ 46086\ 74458\ 11131\ 03177$$
$$e^2 = 7.38905\ 60989\ 30650\ 22723\ 04274\ 60575\ 00781\ 31803\ 15570\ 55185$$
$$M = \log_{10} e = 0.43429\ 44819\ 03251\ 82765\ 11289\ 18916\ 60508\ 22943\ 97005\ 80367$$
$$1/M = \log_e 10 = 2.30258\ 50929\ 94045\ 68401\ 79914\ 54684\ 36420\ 76011\ 01488\ 62877$$
$$\log_{10} M = 9.63778\ 43113\ 00536\ 78912\ 29674\ 98645\ -10$$

π^e AND e^π CONSTANTS

$$\pi^e = 22.45915\ 77183\ 61045\ 47342\ 71522$$
$$e^\pi = 23.14069\ 26327\ 79269\ 00572\ 90864$$
$$e^{-\pi} = 0.04321\ 39182\ 63772\ 24977\ 44177$$
$$e^{\pi/2} = 4.81047\ 73809\ 65351\ 65547\ 30357$$
$$i^i = e^{-\pi/2} = 0.20787\ 95763\ 50761\ 90854\ 69556$$

NUMERICAL CONSTANTS

$$\sqrt{2} = 1.41421\ 35623\ 73095\ 04880\ 16887\ 24209\ 69807\ 85696\ 71875\ 37695$$
$$\sqrt[3]{2} = 1.25992\ 10498\ 94873\ 16476\ 72106\ 07278\ 22835\ 05702\ 51464\ 70151$$
$$\log_e 2 = 0.69314\ 71805\ 59945\ 30941\ 72321\ 21458\ 17656\ 80755\ 00134\ 36026$$
$$\log_{10} 2 = 0.30102\ 99956\ 63981\ 19521\ 37388\ 94724\ 49302\ 67881\ 89881\ 46211$$
$$\sqrt{3} = 1.73205\ 08075\ 68877\ 29352\ 74463\ 41505\ 87236\ 69428\ 05253\ 81039$$
$$\sqrt[3]{3} = 1.44224\ 95703\ 07408\ 38232\ 16383\ 10780\ 10958\ 83918\ 69253\ 49935$$
$$\log_e 3 = 1.09861\ 22886\ 68109\ 69139\ 52452\ 36922\ 52570\ 46474\ 90557\ 82275$$
$$\log_{10} 3 = 0.47712\ 12547\ 19662\ 43729\ 50279\ 03255\ 11530\ 92001\ 28864\ 19070$$

OTHER CONSTANTS

$$\text{Euler's Constant } \gamma = 0.57721\ 56649\ 01532\ 86061$$
$$\log_e \gamma = -0.54953\ 93129\ 81644\ 82234$$
$$\text{Golden Ratio } \phi = 1.61803\ 39887\ 49894\ 84820\ 45868\ 34365\ 63811\ 77203\ 09180$$

DECIMAL EQUIVALENTS OF COMMON FRACTIONS

		1/64	0.015625			33/64	0.515625
	1/32	2/64	0.03125		17/32	34/64	0.53125
		3/64	0.046875			35/64	0.546875
1/16	2/32	4/64	0.0625	9/16	18/32	36/64	0.5625
		5/64	0.078125			37/64	0.578125
	3/32	6/64	0.09375		19/32	38/64	0.59375
		7/64	0.109375			39/64	0.609375
1/8	4/32	8/64	0.125	5/8	20/32	40/64	0.625
		9/64	0.140625			41/64	0.640625
	5/32	10/64	0.15625		21/32	42/64	0.65625
		11/64	0.171875			43/64	0.671875
3/16	6/32	12/64	0.1875	11/16	22/32	44/64	0.6875
		13/64	0.203125			45/64	0.703125
	7/32	14/64	0.21875		23/32	46/64	0.71875
		15/64	0.234375			47/64	0.734375
1/4	8/32	16/64	0.25	3/4	24/32	48/64	0.75
		17/64	0.265625			49/64	0.765625
	9/32	18/64	0.28125		25/32	50/64	0.78125
		19/64	0.296875			51/64	0.796875
5/16	10/32	20/64	0.3125	13/16	26/32	52/64	0.8125
		21/64	0.328125			53/64	0.828125
	11/32	22/64	0.34375		27/32	54/64	0.84375
		23/64	0.359375			55/64	0.859375
3/8	12/32	24/64	0.375	7/8	28/32	56/64	0.875
		25/64	0.390625			57/64	0.890625
	13/32	26/64	0.40625		29/32	58/64	0.90625
		27/64	0.421875			59/64	0.921875
7/16	14/32	28/64	0.4375	15/16	30/32	60/64	0.9375
		29/64	0.453125			61/64	0.953125
	15/32	30/64	0.46875		31/32	62/64	0.96875
		31/64	0.484375			63/64	0.984375
1/2	16/32	32/64	0.5	1/1	32/32	64/64	1

QUADRATIC FORMULA

The solution of the equation $ax^2 + bx + c = 0$, where $a \neq 0$, is given by

$$x = \frac{-b \pm \sqrt{b^2 - 4ac}}{2a}.$$

EXPONENTIAL AND HYPERBOLIC FUNCTIONS AND THEIR COMMON LOGARITHMS

x	e^x Value	e^x log$_{10}$	e^{-x} Value	sinh x Value	sinh x log$_{10}$	cosh x Value	cosh x log$_{10}$	tanh x Value
0.00	1.0000	0.00000	1.00000	0.0000	-∞	1.0000	0.00000	0.00000
0.01	1.0101	.00434	0.99005	.0100	−2.00001	1.0001	.00002	.01000
0.02	1.0202	.00869	.98020	.0200	−2.30106	1.0002	.00009	.02000
0.03	1.0305	.01303	.97045	.0300	−2.47719	1.0005	.00020	.02999
0.04	1.0408	.01737	.96079	.0400	−2.60218	1.0008	.00035	.03998
0.05	1.0513	.02171	.95123	.0500	−2.69915	1.0013	.00054	.04996
0.06	1.0618	.02606	.94176	.0600	−2.77841	1.0018	.00078	.05993
0.07	1.0725	.03040	.93239	.0701	−2.84545	1.0025	.00106	.06989
0.08	1.0833	.03474	.92312	.0801	−2.90355	1.0032	.00139	.07983
0.09	1.0942	.03909	.91393	.0901	−2.95483	1.0041	.00176	.08976
0.10	1.1052	.04343	.90484	.1002	−1.00072	1.0050	.00217	.09967
0.11	1.1163	.04777	.89583	.1102	−1.04227	1.0061	.00262	.10956
0.12	1.1275	.05212	.88692	.1203	−1.08022	1.0072	.00312	.11943
0.13	1.1388	.05646	.87809	.1304	−1.11517	1.0085	.00366	.12927
0.14	1.1503	.06080	.86936	.1405	−1.14755	1.0098	.00424	.13909
0.15	1.1618	.06514	.86071	.1506	−1.17772	1.0113	.00487	.14889
0.16	1.1735	.06949	.85214	.1607	−1.20597	1.0128	.00554	.15865
0.17	1.1853	.07383	.84366	.1708	−1.23254	1.0145	.00625	.16838
0.18	1.1972	.07817	.83527	.1810	−1.25762	1.0162	.00700	.17808
0.19	1.2092	.08252	.82696	.1911	−1.28136	1.0181	.00779	.18775
0.20	1.2214	.08686	.81873	.2013	−1.30392	1.0201	.00863	.19738
0.21	1.2337	.09120	.81058	.2115	−1.32541	1.0221	.00951	.20697
0.22	1.2461	.09554	.80252	.2218	−1.34592	1.0243	.01043	.21652
0.23	1.2586	.09989	.79453	.2320	−1.36555	1.0266	.01139	.22603
0.24	1.2712	.10423	.78663	.2423	−1.38437	1.0289	.01239	.23550
0.25	1.2840	.10857	.77880	.2526	−1.40245	1.0314	.01343	.24492
0.26	1.2969	.11292	.77105	.2629	−1.41986	1.0340	.01452	.25430
0.27	1.3100	.11726	.76338	.2733	−1.43663	1.0367	.01564	.26362
0.28	1.3231	.12160	.75578	.2837	−1.45282	1.0395	.01681	.27291
0.29	1.3364	.12595	.74826	.2941	−1.46847	1.0423	.01801	.28213
0.30	1.3499	.13029	.74082	.3045	−1.48362	1.0453	.01926	.29131
0.31	1.3634	.13463	.73345	.3150	−1.49830	1.0484	.02054	.30044
0.32	1.3771	.13897	.72615	.3255	−1.51254	1.0516	.02187	.30951
0.33	1.3910	.14332	.71892	.3360	−1.52637	1.0549	.02323	.31852
0.34	1.4049	.14766	.71177	.3466	−1.53981	1.0584	.02463	.32748
0.35	1.4191	.15200	.70469	.3572	−1.55290	1.0619	.02607	.33638
0.36	1.4333	.15635	.69768	.3678	−1.56564	1.0655	.02755	.34521
0.37	1.4477	.16069	.69073	.3785	−1.57807	1.0692	.02907	.35399
0.38	1.4623	.16503	.68386	.3892	−1.59019	1.0731	.03063	.36271
0.39	1.4770	.16937	.67706	.4000	−1.60202	1.0770	.03222	.37136
0.40	1.4918	.17372	.67032	.4108	−1.61358	1.0811	.03385	.37995
0.41	1.5063	.17806	.66365	.4216	−1.62488	1.0852	.03552	.33847
0.42	1.5220	.18240	.65705	.4325	−1.63594	1.0895	.03723	.39693
0.43	1.5373	.18675	.65051	.4434	−1.64677	1.0939	.03897	.40532
0.44	1.5527	.19109	.64404	.4543	−1.65738	1.0984	.04075	.41364
0.45	1.5683	.19543	.63763	.4653	−1.66777	1.1030	.04256	.42190
0.46	1.5841	.19978	.63128	.4764	−1.67797	1.1077	.04441	.43008
0.47	1.6000	.20412	.62500	.4875	−1.68797	1.1125	.04630	.43820
0.48	1.6161	.20846	.61878	.4986	−1.69779	1.1174	.04822	.44624
0.49	1.6323	.21280	.61263	.5098	−1.70744	1.1225	.05018	.45422
0.50	1.6487	.21715	.60653	.5211	−1.71692	1.1276	.05217	.46212
0.51	1.6653	.22149	.60050	.5324	−1.72624	1.1329	.05419	.46995
0.52	1.6820	.22583	.59452	.5438	−1.73540	1.1383	.05625	.47770
0.53	1.6989	.23018	.58860	.5552	−1.74442	1.1438	.05834	.48538

x	e^x Value	e^x $\log_{10}$	e^{-x} Value	$\sinh x$ Value	$\sinh x$ $\log_{10}$	$\cosh x$ Value	$\cosh x$ $\log_{10}$	$\tanh x$ Value
0.54	1.7160	.23452	.58275	.5666	−1.75330	1.1494	.06046	.49299
0.55	1.7333	.23886	.57695	.5782	−1.76204	1.1551	.06262	.50052
0.56	1.7507	.24320	.57121	.5897	−1.77065	1.1609	.06481	.50798
0.57	1.7683	.24755	.56553	.6014	−1.77914	1.1669	.06703	.51536
0.58	1.7860	.25189	.55990	.6131	−1.78751	1.1730	.06929	.52267
0.59	1.8040	.25623	.55433	.6248	−1.79576	1.1792	.07157	.52990
0.60	1.8221	.26058	.54881	.6367	−1.80390	1.1855	.07389	.53705
0.61	1.8404	.26492	.54335	.6485	−1.81194	1.1919	.07624	.54413
0.62	1.8589	.26926	.53794	.6605	−1.81987	1.1984	.07861	.55113
0.63	1.8776	.27361	.53259	.6725	−1.82770	1.2051	.08102	.55805
0.64	1.8965	.27795	.52729	.6846	−1.83543	1.2119	.08346	.56490
0.65	1.9155	.28229	.52205	.6967	−1.84308	1.2188	.08593	.57167
0.66	1.9348	.28664	.51685	.7090	−1.85063	1.2258	.08843	.57836
0.67	1.9542	.29098	.51171	.7213	−1.85809	1.2330	.09095	.58498
0.68	1.9739	.29532	.50662	.7336	−1.86548	1.2402	.09351	.59152
0.69	1.9937	.29966	.50158	.7461	−1.87278	1.2476	.09609	.59798
0.70	2.0138	.30401	.49659	.7586	−1.88000	1.2552	.09870	.60437
0.71	2.0340	.30835	.49164	.7712	−1.88715	1.2628	.10134	.61068
0.72	2.0544	.31269	.48675	.7838	−1.89423	1.2706	.10401	.61691
0.73	2.0751	.31703	.48191	.7966	−1.90123	1.2785	.10670	.62307
0.74	2.0959	.32138	.47711	.8094	−1.90817	1.2865	.10942	.62915
0.75	2.1170	.32572	.47237	.8223	−1.91504	1.2947	.11216	.63515
0.76	2.1383	.33006	.46767	.8353	−1.92185	1.3030	.11493	.64108
0.77	2.1598	.33441	.46301	.8484	−1.92859	1.3114	.11773	.64693
0.78	2.1815	.33875	.45841	.8615	−1.93527	1.3199	.12055	.65721
0.79	2.2034	.34309	.45384	.8748	−1.94190	1.3286	.12340	.65841
0.80	2.2255	.34744	.44933	.8881	−1.94846	1.3374	.12627	.66404
0.81	2.2479	.35178	.44486	.9015	−1.95498	1.3464	.12917	.66959
0.82	2.2705	.35612	.44043	.9150	−1.96144	1.3555	.13209	.67507
0.83	2.2933	.36046	.43605	.9286	−1.96784	1.3647	.13503	.68048
0.84	2.3164	.36481	.43171	.9423	−1.97420	1.3740	.13800	.68581
0.85	2.3396	.36915	.42741	.9561	−1.98051	1.3835	.14099	.69107
0.86	2.3632	.37349	.42316	.9700	−1.98677	1.3932	.14400	.69626
0.87	2.3869	.37784	.41895	.9840	−1.99299	1.4029	.14704	.70137
0.88	2.4100	.38218	.41478	.9981	−1.99916	1.4128	.15009	.70642
0.89	2.4351	.38652	.41066	1.0122	0.00528	1.4229	.15317	.71139
0.90	2.4596	.39087	.40657	1.0265	.01137	1.4331	.15627	.21630
0.91	2.4843	.39521	.40242	1.0409	.01741	1.4434	.15939	.72113
0.92	2.5093	.39955	.39852	1.0554	.02341	1.4539	.16254	.72590
0.93	2.5345	.40389	.39455	1.0700	.02937	1.4645	.16570	.73059
0.94	2.5600	.40824	.39063	1.0847	.03530	1.4753	.16888	.73522
0.95	2.5857	.41258	.38674	1.0995	.04119	1.4862	.17208	.73978
0.96	2.6117	.41692	.38289	1.1144	.04704	1.4973	.17531	.74428
0.97	2.6379	.42127	.37908	1.1294	.05286	1.5085	.17855	.74870
0.98	2.6645	.42561	.37531	1.1446	.05864	1.5199	.18181	.75307
0.99	2.6912	.42995	.37158	1.1598	.06439	1.5314	.18509	.75736
1.00	2.7183	.43429	.36788	1.1752	.07011	1.5431	.18839	.76159
1.10	3.0042	.47772	.33287	1.3356	.12569	1.6685	.22233	.80050
1.20	3.3201	.52115	.30119	1.5095	.17882	1.8107	.25784	.83365
1.30	3.6693	.56458	.27253	1.6984	.23004	1.9709	.29467	.86172
1.40	4.0552	.60801	.24660	1.9043	.27974	2.1509	.33262	.88535
1.50	4.4817	.65144	.22313	2.1293	.32823	2.3524	.37151	.90515
1.60	4.9530	.69487	.20190	2.3756	.37577	2.5775	.41119	.92167
1.70	5.4739	.73830	.18268	2.6456	.42253	2.8283	.45153	.93541
1.80	6.0496	.78173	.16530	2.9422	.46867	3.1075	.49241	.94681
1.90	6.6859	.82516	.14957	3.2682	.51430	3.4177	.53374	.95624
2.00	7.3891	.86859	.13534	3.6269	.55953	3.7622	.57544	.96403

x	e^x Value	e^x $\log_{10}$	e^{-x} Value	$\sinh x$ Value	$\sinh x$ $\log_{10}$	$\cosh x$ Value	$\cosh x$ $\log_{10}$	$\tanh x$ Value
2.10	8.1662	.91202	.12246	4.0219	.60443	4.1443	.61745	.97045
2.20	9.0250	.95545	.11080	4.4571	.64905	4.5679	.65972	.97574
2.30	9.9742	.99888	.10026	4.9370	.69346	5.0372	.70219	.98010
2.40	11.023	1.04231	.09072	5.4662	.73769	5.5569	.74484	.98367
2.50	12.182	1.08574	.08208	6.0502	.78177	6.1323	.78762	.98661
2.60	13.464	1.12917	.07427	6.6947	.82573	6.7690	.83052	.98903
2.70	14.880	1.17260	.06721	7.4063	.86960	7.4735	.87352	.99101
2.80	16.445	1.21602	.06081	8.1919	.91339	8.2527	.91660	.99263
2.90	18.174	1.25945	.05502	9.0596	.95711	9.1146	.95974	.99396
3.00	20.086	1.30288	.04979	10.018	1.00078	10.068	1.00293	0.99505
3.50	33.115	1.52003	.03020	16.543	1.21860	16.573	1.21940	0.99818
4.00	54.598	1.73718	.01832	27.290	1.43600	27.308	1.43629	0.99933
4.50	90.017	1.95433	.01111	45.003	1.65324	45.014	1.65335	0.99975
5.00	148.41	2.17147	.00674	74.203	1.87042	74.210	1.87046	0.99991
5.50	244.69	2.38862	.00409	122.34	2.08758	122.35	2.08760	0.99997
6.00	403.43	2.60577	.00248	201.71	2.30473	201.72	2.30474	0.99999
6.50	665.14	2.82291	.00150	332.57	2.52188	332.57	2.52189	1.00000
7.00	1096.6	3.04006	.00091	548.32	2.73904	548.32	2.73903	1.00000
7.50	1808.0	3.25721	.00055	904.02	2.95618	904.02	2.95618	1.00000
8.00	2981.0	3.47436	.00034	1490.5	3.17333	1490.5	3.17333	1.00000
8.50	4914.8	3.69150	.00020	2457.4	3.39047	2457.4	3.39047	1.00000
9.00	8103.1	3.90865	.00012	4051.5	3.60762	4051.5	3.60762	1.00000
9.50	13360.	4.12580	.00007	6679.9	3.82477	6679.9	3.82477	1.00000
10.00	22026.	4.34294	.00005	11013.	4.04191	11013.	4.04191	1.00000

NATURAL TRIGONOMETRIC FUNCTIONS TO FOUR PLACES

x radians	x degrees	$\sin x$	$\cos x$	$\tan x$	$\cot x$	$\sec x$	$\csc x$		
.0000	0° 00′	.000	1.0000	.0000	–	1.000	–	90° 00′	1.5708
.0029	10	.0029	1.0000	.0029	343.8	1.000	343.8	50	1.5679
.0058	20	.0058	1.0000	.0058	171.9	1.000	171.9	40	1.5650
.0087	30	.0087	1.0000	.0087	114.6	1.000	114.6	30	1.5621
.0116	40	.0116	.9999	.0116	85.94	1.000	85.95	20	1.5592
.0145	50	.0145	.9999	.0145	68.75	1.000	68.76	10	1.5563
.0175	1° 00′	.0175	.9998	.0175	57.29	1.000	57.30	89° 00′	1.5533
.0262	30	.0262	.9997	.0262	38.19	1.000	38.20	30	1.5446
.0349	2° 00′	.0349	.9994	.0349	28.64	1.001	28.65	88° 00′	1.5359
.0436	30	.0436	.9990	.0437	22.90	1.001	22.93	30	1.5272
.0524	3° 00′	.0523	.9986	.0524	19.08	1.001	19.11	87° 00′	1.5184
.0611	30	.0610	.9981	.0612	16.35	1.002	16.38	30	1.5097
.0698	4° 00′	.0698	.9976	.0699	14.30	1.002	14.34	86° 00′	1.5010
.0785	30	.0785	.9969	.0787	12.71	1.003	12.75	30	1.4923
.0873	5° 00′	.0872	.9962	.0875	11.43	1.004	11.47	85° 00′	1.4835
.0960	30	.0958	.9954	.0963	10.39	1.005	10.43	30	1.4748
.1047	6° 00′	.1045	.9945	.1051	9.514	1.006	9.597	84° 00′	1.4661
.1134	30	.1132	.9936	.1139	8.777	1.006	8.834	30	1.4573
.1222	7° 00′	.1219	.9925	.1228	8.144	1.008	8.206	83° 00′	1.4486
.1309	30	.1305	.9914	.1317	7.596	1.009	7.661	30	1.4399
.1396	8° 00′	.1392	.9903	.1405	7.115	1.010	7.185	82° 00′	1.4312
.1484	30	.1478	.9890	.1495	6.691	1.011	6.765	30	1.4224
.1571	9° 00′	.1564	.9877	.1584	6.314	1.012	6.392	81° 00′	1.4137
.1658	30	.1650	.9863	.1673	5.976	1.014	6.059	30	1.4050
.1745	10° 00′	.1736	.9848	.1763	5.671	1.015	5.759	80° 00′	1.3963
.1833	30	.1822	.9833	.1853	5.396	1.017	5.487	30	1.3875
.1920	11° 00′	.1908	.9816	.1944	5.145	1.019	5.241	79° 00′	1.3788
.2007	30	.1994	.9799	.2035	4.915	1.020	5.016	30	1.3701
.2094	12° 00′	.2079	.9781	.2126	4.705	1.022	4.810	78° 00′	1.3614
.2182	30	.2164	.9763	.2217	4.511	1.025	4.620	30	1.3526
.2269	13° 00′	.2250	.9744	.2309	4.331	1.026	4.445	77° 00′	1.3439
.2356	30	.2334	.9724	.2401	4.165	1.028	4.284	30	1.3352
.2443	14° 00′	.2419	.9703	.2493	4.011	1.031	4.134	76° 00′	1.3265
.2531	30	.2404	.9681	.2586	3.867	1.033	3.994	30	1.3177
.2618	15° 00′	.2588	.9659	.2679	3.732	1.035	3.864	75° 00′	1.3090
.2705	30	.2672	.9636	.2773	3.606	1.038	3.742	30	1.3003
.2793	16° 00′	.2756	.9613	.2867	3.487	1.040	3.628	74° 00′	1.2915
.2880	30	.2840	.9588	.2962	3.376	1.043	3.521	30	1.2828
.2967	17° 00′	.2924	.9563	.3057	3.271	1.046	3.420	73° 00′	1.2741
.3054	30	.3007	.9537	.3153	3.172	1.049	3.326	30	1.2654
.3142	18° 00′	.3090	.9511	.3249	3.078	1.051	3.236	72° 00′	1.2566
.3229	30	.3173	.9483	.3346	2.989	1.054	3.152	30	1.2479
.3316	19° 00′	.3256	.9455	.3443	2.904	1.058	3.072	71° 00′	1.2392
.3403	30	.3338	.9426	.3541	2.824	1.061	2.996	30	1.2305
.3491	20° 00′	.3420	.9397	.3640	2.747	1.064	2.924	70° 00′	1.2217
.3578	30	.3502	.9367	.3739	2.675	1.068	2.855	30	1.2130
.3665	21° 00′	.3584	.9336	.3839	2.605	1.071	2.790	69° 00′	1.2043
.3752	30	.3665	.9304	.3939	2.539	1.075	2.729	30	1.1956
.3840	22° 00′	.3746	.9272	.4040	2.475	1.079	2.669	68° 00′	1.1868
.3927	30	.3827	.9239	.4142	2.414	1.082	2.613	30	1.1781
.4014	23° 00′	.3907	.9205	.4245	2.356	1.086	2.559	67° 00′	1.1694
.4102	30	.3987	.9171	.4348	2.300	1.090	2.508	30	1.1606
.4189	24° 00′	.4067	.9135	.4452	2.246	1.095	2.459	66° 00′	1.1519
		$\cos y$	$\sin y$	$\cot y$	$\tan y$	$\csc y$	$\sec y$	y degree	y radians

x radians	x degrees	$\sin x$	$\cos x$	$\tan x$	$\cot x$	$\sec x$	$\csc x$		
.4276	30	.4147	.9100	.4557	2.194	1.099	2.411	30	1.1432
.4363	25° 00′	.4226	.9063	.4663	2.145	1.103	2.366	65° 00′	1.1345
.4451	30	.4305	.9026	.4770	2.097	1.108	2.323	30	1.1257
.4538	26° 00′	.4384	.8988	.4877	2.050	1.113	2.281	64° 00′	1.1170
.4625	30	.4462	.8949	.4986	2.006	1.117	2.241	30	1.1083
.4712	27° 00′	.4540	.8910	.5095	1.963	1.122	2.203	63° 00′	1.0996
.4800	30	.4617	.8870	.5206	1.921	1.127	2.166	30	1.0908
.4887	28° 00′	.4695	.8829	.5317	1.881	1.133	2.130	62° 00′	1.0821
.4974	30	.4772	.8788	.5430	1.842	1.138	2.096	30	1.0734
.5061	29° 00′	.4848	.8746	.5543	1.804	1.143	2.063	61° 00′	1.0647
.5149	30	.4924	.8704	.5658	1.767	1.149	2.031	30	1.0559
.5236	30° 00′	.5000	.8660	.5774	1.732	1.155	2.000	60° 00′	1.0472
.5323	30	.5075	.8616	.5890	1.698	1.161	1.970	30	1.0385
.5411	31° 00′	.5150	.8572	.6009	1.664	1.167	1.942	59° 00′	1.0297
.5498	30	.5225	.8526	.6128	1.632	1.173	1.914	30	1.0210
.5585	32° 00′	.5299	.8480	.6249	1.600	1.179	1.887	58° 00′	1.0123
.5672	30	.5373	.8434	.6371	1.570	1.186	1.861	30	1.0036
.5760	33° 00′	.5446	.8397	.6494	1.540	1.192	1.836	57° 00′	.9948
.5847	30	.5519	.8339	.6619	1.511	1.199	1.812	30	.9861
.5934	34° 00′	.5592	.8290	.6745	1.483	1.206	1.788	56° 00′	.9774
.6021	30	.5664	.8241	.6873	1.455	1.213	1.766	30	.9687
.6109	35° 00′	.5736	.8192	.7002	1.428	1.221	1.743	55° 00′	.9599
.6196	30	.5807	.8141	.7133	1.402	1.228	1.722	30	.9512
.6283	36° 00′	.5878	.8090	.7265	1.376	1.236	1.701	54° 00′	.9425
.6370	30	.5948	.8039	.7400	1.351	1.244	1.681	30	.9338
.6458	37° 00′	.6018	.7986	.7536	1.327	1.252	1.662	53° 00′	.9250
.6545	30	.6088	.7934	.7673	1.303	1.260	1.643	30	.9163
.6632	38° 00′	.6157	.7880	.7813	1.280	1.269	1.624	52° 00′	.9076
.6720	30	.6225	.7826	.7954	1.257	1.278	1.606	30	.8988
.6807	39° 00′	.6293	.7771	.8098	1.235	1.287	1.589	51° 00′	.8901
.6894	30	.6361	.7716	.8243	1.213	1.296	1.572	30	.8814
.6981	40° 00′	.6428	.7660	.8391	1.192	1.305	1.556	50° 00′	.8727
.7069	30	.6494	.7604	.8541	1.171	1.315	1.540	30	.8639
.7156	41° 00′	.6561	.7547	.8693	1.150	1.325	1.524	49° 00′	.8552
.7243	30	.6626	.7490	.8847	1.130	1.335	1.509	30	.8465
.7330	42° 00′	.6691	.7431	.9004	1.111	1.346	1.494	48° 00′	.8378
.7418	30	.6756	.7373	.9163	1.091	1.356	1.480	30	.8290
.7505	43° 00′	.6820	.7314	.9325	1.072	1.367	1.466	47° 00′	.8203
.7592	30	.6884	.7254	.9490	1.054	1.379	1.453	30	.8116
.7679	44° 00′	.6947	.7193	.9657	1.036	1.390	1.440	46° 00′	.8029
.7767	30	.7009	.7133	.9827	1.018	1.402	1.427	30	.7941
.7854	45° 00′	.7071	.7071	1.0000	1.0000	1.414	1.414	45° 00′	.7854
		$\cos y$	$\sin y$	$\cot y$	$\tan y$	$\csc y$	$\sec y$	y degree	y radians

RELATION OF ANGULAR FUNCTIONS IN TERMS OF ONE ANOTHER

Trigonometric Functions

Function	$\sin\alpha$	$\cos\alpha$	$\tan\alpha$	$\cot\alpha$	$\sec\alpha$	$\csc\alpha$
$\sin\alpha$	$\sin\alpha$	$\pm\sqrt{1-\cos^2\alpha}$	$\dfrac{\tan\alpha}{\pm\sqrt{1+\tan^2\alpha}}$	$\dfrac{1}{\pm\sqrt{1+\cot^2\alpha}}$	$\dfrac{\pm\sqrt{\sec^2\alpha-1}}{\sec\alpha}$	$\dfrac{1}{\csc\alpha}$
$\cos\alpha$	$\pm\sqrt{1-\sin^2\alpha}$	$\cos\alpha$	$\dfrac{1}{\pm\sqrt{1+\tan^2\alpha}}$	$\dfrac{\cot\alpha}{\pm\sqrt{1+\cot^2\alpha}}$	$\dfrac{1}{\sec\alpha}$	$\dfrac{\pm\sqrt{\csc^2\alpha-1}}{\csc\alpha}$
$\tan\alpha$	$\dfrac{\sin\alpha}{\pm\sqrt{1-\sin^2\alpha}}$	$\dfrac{\pm\sqrt{1-\cos^2\alpha}}{\cos\alpha}$	$\tan\alpha$	$\dfrac{1}{\cot\alpha}$	$\pm\sqrt{\sec^2\alpha-1}$	$\dfrac{1}{\pm\sqrt{\csc^2\alpha-1}}$
$\cot\alpha$	$\dfrac{\pm\sqrt{1-\sin^2\alpha}}{\sin\alpha}$	$\dfrac{\cos\alpha}{\pm\sqrt{1-\cos^2\alpha}}$	$\dfrac{1}{\tan\alpha}$	$\cot\alpha$	$\dfrac{1}{\pm\sqrt{\sec^2\alpha-1}}$	$\pm\sqrt{\csc^2\alpha-1}$
$\sec\alpha$	$\dfrac{1}{\pm\sqrt{1-\sin^2\alpha}}$	$\dfrac{1}{\cos\alpha}$	$\pm\sqrt{1+\tan^2\alpha}$	$\dfrac{\pm\sqrt{1+\cot^2\alpha}}{\cot\alpha}$	$\sec\alpha$	$\dfrac{\csc\alpha}{\pm\sqrt{\csc^2\alpha-1}}$
$\csc\alpha$	$\dfrac{1}{\sin\alpha}$	$\dfrac{1}{\pm\sqrt{1-\cos^2\alpha}}$	$\dfrac{\pm\sqrt{1+\tan^2\alpha}}{\tan\alpha}$	$\pm\sqrt{1+\cot^2\alpha}$	$\dfrac{\sec\alpha}{\pm\sqrt{\sec^2\alpha-1}}$	$\csc\alpha$

Note: The choice of sign depends upon the quadrant in which the angle terminates.

Hyperbolic Functions

Function	$\sinh x$	$\cosh x$	$\tanh x$
$\sinh x =$	$\sinh x$	$+\sqrt{\cosh^2 x-1}$	$\dfrac{\tanh x}{\sqrt{1-\tanh^2 x}}$
$\cosh x =$	$\sqrt{1+\sinh^2 x}$	$\cosh x$	$\dfrac{1}{\sqrt{1-\tanh^2 x}}$
$\tanh x =$	$\dfrac{\sinh x}{\sqrt{1+\sinh^2 x}}$	$\pm\dfrac{\sqrt{\cosh^2 x-1}}{\cosh x}$	$\tanh x$
$\operatorname{cosech} x =$	$\dfrac{1}{\sinh x}$	$\pm\dfrac{1}{\sqrt{\cosh^2 x-1}}$	$\dfrac{\sqrt{1-\tanh^2 x}}{\tanh x}$
$\operatorname{sech} x =$	$\dfrac{1}{\sqrt{1+\sinh^2 x}}$	$\dfrac{1}{\cosh x}$	$\sqrt{1-\tanh^2 x}$
$\coth x =$	$\dfrac{\sqrt{1+\sinh^2 x}}{\sinh x}$	$\dfrac{\pm\cosh x}{\sqrt{\cosh^2 x-1}}$	$\dfrac{1}{\tanh x}$

Function	$\operatorname{cosech} x$	$\operatorname{sech} x$	$\coth x$
$\sinh x =$	$\dfrac{1}{\operatorname{cosech} x}$	$\pm\dfrac{\sqrt{1-\operatorname{sech}^2 x}}{\operatorname{sech} x}$	$\dfrac{\pm 1}{\sqrt{\coth^2 x-1}}$
$\cosh x =$	$\pm\dfrac{\sqrt{\operatorname{cosech}^2 x+1}}{\operatorname{cosech} x}$	$\dfrac{1}{\operatorname{sech} x}$	$\pm\dfrac{\coth x}{\sqrt{\coth^2 x-1}}$
$\tanh x =$	$\dfrac{1}{\sqrt{\operatorname{cosech}^2 x+1}}$	$\pm\sqrt{1+\operatorname{sech}^2 x}$	$\dfrac{1}{\coth x}$
$\operatorname{cosech} x =$	$\operatorname{cosech} x$	$\pm\dfrac{\operatorname{sech} x}{\sqrt{1-\operatorname{sech}^2 x}}$	$\pm\dfrac{\sqrt{\coth^2 x-1}}{1}$
$\operatorname{sech} x =$	$\pm\dfrac{\operatorname{cosech} x}{\sqrt{\operatorname{cosech}^2 x+1}}$	$\operatorname{sech} x$	$\pm\dfrac{\sqrt{\coth^2 x-1}}{\coth x}$
$\coth x =$	$\sqrt{\operatorname{cosech}^2 x+1}$	$\pm\dfrac{1}{\sqrt{1-\operatorname{sech}^2 x}}$	$\coth x$

Whenever two signs are shown, choose $+$ sign if x is positive, $-$ sign if x is negative.

DERIVATIVES

In the following formulas u, v, w represent functions of x, while a, c, n represent fixed real numbers. All arguments in the trigonometric functions are measured in radians, and all inverse trigonometric and hyperbolic functions represent principal values

1. $\dfrac{d}{dx}(a) = 0$

2. $\dfrac{d}{dx}(x) = 1$

3. $\dfrac{d}{dx}(au) = a\dfrac{du}{dx}$

4. $\dfrac{d}{dx}(u + v - w) = \dfrac{du}{dx} + \dfrac{dv}{dx} - \dfrac{dw}{dx}$

5. $\dfrac{d}{dx}(uv) = u\dfrac{dv}{dx} + v\dfrac{du}{dx}$

6. $\dfrac{d}{dx}(uvw) = uv\dfrac{dw}{dx} + vw\dfrac{du}{dx} + uw\dfrac{dv}{dx}$

7. $\dfrac{d}{dx}\left(\dfrac{u}{v}\right) = \dfrac{v\frac{du}{dx} - u\frac{dv}{dx}}{v^2} = \dfrac{1}{v}\dfrac{du}{dx} - \dfrac{u}{v^2}\dfrac{dv}{dx}$

8. $\dfrac{d}{dx}(u^n) = nu^{n-1}\dfrac{du}{dx}$

9. $\dfrac{d}{dx}(\sqrt{u}) = \dfrac{1}{2\sqrt{u}}\dfrac{du}{dx}$

10. $\dfrac{d}{dx}\left(\dfrac{1}{u}\right) = -\dfrac{1}{u^2}\dfrac{du}{dx}$

11. $\dfrac{d}{dx}\left(\dfrac{1}{u^n}\right) = -\dfrac{n}{u^{n+1}}\dfrac{du}{dx}$

12. $\dfrac{d}{dx}\left(\dfrac{u^n}{v^m}\right) = \dfrac{u^{n-1}}{v^{m+1}}\left(nv\dfrac{du}{dx} - mu\dfrac{dv}{dx}\right)$

13. $\dfrac{d}{dx}(u^n v^m) = u^{n-1}v^{m-1}\left(nv\dfrac{du}{dx} + mu\dfrac{dv}{dx}\right)$

14. $\dfrac{d}{dx}[f(u)] = \dfrac{d}{du}[f(u)] \cdot \dfrac{du}{dx}$

*Let $y = f(x)$ and $\frac{dy}{dx} = \frac{d[f(x)]}{dx} = f'(x)$ define respectively a function and its derivative for any value x in their common domain. The differential for the function at such a value x is accordingly defined as

$$dy = d[f(x)] = \frac{dy}{dx}dx = \frac{d[f(x)]}{dx}dx = f'(x)dx$$

Each derivative formula has an associated differential formula. For example, formula 6 above has the differential formula

$$d(uvw) = uv\,dw + vw\,du + uw\,dv$$

15. $\dfrac{d^2}{dx^2}[f(u)] = \dfrac{df(u)}{du} \cdot \dfrac{d^2u}{dx^2} + \dfrac{d^2 f(u)}{du^2} \cdot \left(\dfrac{du}{dx}\right)^2$

16. $\dfrac{d^n}{dx^n}[uv] = \binom{n}{0}v\dfrac{d^n u}{dx^n} + \binom{n}{1}\dfrac{dv}{dx}\dfrac{d^{n-1}u}{dx^{n-1}} + \binom{n}{2}\dfrac{d^2v}{dx^2}\dfrac{d^{n-2}u}{dx^{n-2}} + \cdots + \binom{n}{k}\dfrac{d^k v}{dx^k}\dfrac{d^{n-k}u}{dx^{n-k}} + \cdots + \binom{n}{n}u\dfrac{d^n v}{dx^n}$

where $\binom{n}{r} = \dfrac{n!}{r!(n-r)!}$ is the binomial coefficient, n non-negative integer, and $\binom{n}{0} = 1$.

17. $\dfrac{du}{dx} = \dfrac{1}{\frac{dx}{du}}$ if $\dfrac{dx}{du} \neq 0$

18. $\dfrac{d}{dx}(\log_a u) = (\log_a e)\dfrac{1}{u}\dfrac{du}{dx}$

19. $\dfrac{d}{dx}(\log_e u) = \dfrac{1}{u}\dfrac{du}{dx}$

20. $\dfrac{d}{dx}(a^u) = a^u(\log_e a)\dfrac{du}{dx}$

21. $\dfrac{d}{dx}(e^u) = e^u\dfrac{du}{dx}$

22. $\dfrac{d}{dx}(u^v) = vu^{v-1}\dfrac{du}{dx} + (\log_e u)\,u^v\dfrac{dv}{dx}$

23. $\dfrac{d}{dx}(\sin u) = (\cos u)\dfrac{du}{dx}$

24. $\dfrac{d}{dx}(\cos u) = -(\sin u)\dfrac{du}{dx}$

25. $\dfrac{d}{dx}(\tan u) = (\sec^2 u)\dfrac{du}{dx}$

26. $\dfrac{d}{dx}(\cot u) = -(\csc^2 u)\dfrac{du}{dx}$

27. $\dfrac{d}{dx}(\sec u) = \sec u \cdot \tan u\dfrac{du}{dx}$

28. $\dfrac{d}{dx}(\csc u) = -\csc u \cdot \cot u\dfrac{du}{dx}$

29. $\dfrac{d}{dx}(\operatorname{vers} u) = \sin u\dfrac{du}{dx}$

30. $\dfrac{d}{dx}(\arcsin u) = \dfrac{1}{\sqrt{1 - u^2}}\dfrac{du}{dx}, \quad \left(-\dfrac{\pi}{2} \leq \arcsin u \leq \dfrac{\pi}{2}\right)$

31. $\dfrac{d}{dx}(\arccos u) = -\dfrac{1}{\sqrt{1 - u^2}}\dfrac{du}{dx}, \quad (0 \leq \arccos u \leq \pi)$

32. $\dfrac{d}{dx}(\arctan u) = \dfrac{1}{1 + u^2}\dfrac{du}{dx}, \quad \left(-\dfrac{\pi}{2} < \arctan u < \dfrac{\pi}{2}\right)$

33. $\dfrac{d}{dx}(\operatorname{arc\,cot} u) = -\dfrac{1}{1 + u^2}\dfrac{du}{dx}, \quad (0 \leq \operatorname{arc\,cot} u \leq \pi)$

34. $\dfrac{d}{dx}(\operatorname{arc\,sec} u) = \dfrac{1}{u\sqrt{u^2 - 1}}\dfrac{du}{dx}, \quad \left(0 \leq \operatorname{arc\,sec} u < \dfrac{\pi}{2}, \; -\pi \leq \operatorname{arc\,sec} u < -\dfrac{\pi}{2}\right)$

35. $\dfrac{d}{dx}(\operatorname{arc\,csc} u) = -\dfrac{1}{u\sqrt{u^2 - 1}}\dfrac{du}{dx}, \quad \left(0 < \operatorname{arc\,csc} u \leq \dfrac{\pi}{2}, \; -\pi < \operatorname{arc\,csc} u \leq -\dfrac{\pi}{2}\right)$

36. $\dfrac{d}{dx}(\operatorname{arc\,vers} u) = \dfrac{1}{\sqrt{2u - u^2}}\dfrac{du}{dx}, \quad (0 \leq \operatorname{arc\,vers} u \leq \pi)$

37. $\dfrac{d}{dx}(\sinh u) = (\cosh u)\dfrac{du}{dx}$

38. $\dfrac{d}{dx}(\cosh u) = (\sinh u)\dfrac{du}{dx}$

39. $\dfrac{d}{dx}(\tanh u) = (\operatorname{sech}^2 u)\dfrac{du}{dx}$

40. $\dfrac{d}{dx}(\coth u) = -(\operatorname{csch}^2 u)\dfrac{du}{dx}$

41. $\dfrac{d}{dx}(\operatorname{sech} u) = -(\operatorname{sech} u \cdot \tanh u)\dfrac{du}{dx}$

42. $\dfrac{d}{dx}(\operatorname{csch} u) = -(\operatorname{csch} u \cdot \coth u)\dfrac{du}{dx}$

43. $\dfrac{d}{dx}(\sinh^{-1} u) = \dfrac{d}{dx}[\log(u + \sqrt{u^2 + 1})] = \dfrac{1}{\sqrt{u^2 + 1}}\dfrac{du}{dx}$

44. $\dfrac{d}{dx}(\cosh^{-1} u) = \dfrac{d}{dx}[\log(u + \sqrt{u^2 - 1})] = \dfrac{1}{\sqrt{u^2 - 1}}\dfrac{du}{dx}, \quad (u > 1, \cosh^{-1} u > 0)$

45. $\dfrac{d}{dx}(\tanh^{-1} u) = \dfrac{d}{dx}\left[\dfrac{1}{2}\log\dfrac{1+u}{1-u}\right] = \dfrac{1}{1-u^2}\dfrac{du}{dx}, \quad (u^2 < 1)$

46. $\dfrac{d}{dx}(\coth^{-1} u) = \dfrac{d}{dx}\left[\dfrac{1}{2}\log\dfrac{u+1}{u-1}\right] = \dfrac{1}{1-u^2}\dfrac{du}{dx}, \quad (u^2 > 1)$

47. $\dfrac{d}{dx}(\operatorname{sech}^{-1} u) = \dfrac{d}{dx}\left[\log\dfrac{1+\sqrt{1-u^2}}{u}\right] = -\dfrac{1}{u\sqrt{1-u^2}}\dfrac{du}{dx}, \quad (0 < u < 1,\ \operatorname{sech}^{-1} u > 0)$

48. $\dfrac{d}{dx}(\operatorname{csch}^{-1} u) = \dfrac{d}{dx}\left[\log\dfrac{1+\sqrt{1+u^2}}{u}\right] = -\dfrac{1}{|u|\sqrt{1+u^2}}\dfrac{du}{dx}$

49. $\dfrac{d}{dq}\displaystyle\int_p^q f(x)\,dx = f(q), \quad [p\text{ constant}]$

50. $\dfrac{d}{dp}\displaystyle\int_p^q f(x)\,dx = -f(p), \quad [q\text{ constant}]$

51. $\dfrac{d}{da}\displaystyle\int_p^q f(x,a)\,dx = \int_p^q \dfrac{\partial}{\partial a}[f(x,a)]\,dx + f(q,a)\dfrac{dq}{da} - f(p,a)\dfrac{dp}{da}$

INTEGRATION

The following is a brief discussion of some integration techniques. A more complete discussion can be found in a number of good text books. However, the purpose of this introduction is simply to discuss a few of the important techniques which may be used, in conjunction with the integral table which follows, to integrate particular functions.

No matter how extensive the integral table, it is a fairly uncommon occurrence to find in the table the exact integral desired. Usually some form of transformation will have to be made. The simplest type of transformation, and yet the most general, is substitution. Simple forms of substitution, such as $y = ax$, are employed almost unconsciously by experienced users of integral tables. Other substitutions may require more thought. In some sections of the tables, appropriate substitutions are suggested for integrals which are similar to, but not exactly like, integrals in the table. Finding the right substitution is largely a matter of intuition and experience.

Several precautions must be observed when using substitutions:

1. Be sure to make the substitution in the dx term, as well as everywhere else in the integral.

2. Be sure that the function substituted is one-to-one and continuous. If this is not the case, the integral must be restricted in such a way as to make it true. See the example following.

3. With definite integrals, the limits should also be expressed in terms of the new dependent variable. With indefinite integrals, it is necessary to perform the reverse substitution to obtain the answer in terms of the original independent variable. This may also be done for definite integrals, but it is usually easier to change the limits.

Example:

$$\int \frac{x^4}{\sqrt{a^2 - x^2}}\,dx$$

Here we make the substitution $x = |a|\sin\theta$. Then $dx = |a|\cos\theta\,d\theta$, and

$$\sqrt{a^2 - x^2} = \sqrt{a^2 - a^2\sin^2\theta} = |a|\sqrt{1 - \sin^2\theta} = |a\cos\theta|$$

Notice the absolute value signs. It is very important to keep in mind that a square root radical always denotes the positive square root, and to assure the sign is always kept positive. Thus $\sqrt{x^2} = |x|$. Failure to observe this is a common cause of errors in integration.

Notice also that the indicated substitution is not a one-to-one function, that is, it does not have a unique inverse. Thus we must restrict the range of θ in such a way as to make the function one-to-one. Fortunately, this is easily done by solving for θ

$$\theta = \sin^{-1}\frac{x}{|a|}$$

and restricting the inverse sine to the principal values, $-\frac{\pi}{2} \le \theta \le \frac{\pi}{2}$.

Thus the integral becomes

$$\int \frac{a^4 \sin^4 \theta |a| \cos \theta \, d\theta}{|a| |\cos \theta|}$$

Now, however, in the range of values chosen for θ, $\cos \theta$ is always positive. Thus we may remove the absolute value signs from $\cos \theta$ in the denominator. (This is one of the reasons that the principal values of the inverse trigonometric functions are defined as they are.) Then the $\cos \theta$ terms cancel, and the integral becomes

$$a^4 \int \sin^4 \theta \, d\theta$$

By application of integral formulas 299 and 296, we integrate this to

$$-a^4 \frac{\sin^3 \theta \cos \theta}{4} - \frac{3a^4}{8} \cos \theta \sin \theta + \frac{3a^4}{8} \theta + C$$

We now must perform the inverse substitution to get the result in terms of x. We have

$$\theta = \sin^{-1} \frac{x}{|a|}$$

$$\sin \theta = \frac{x}{|a|}$$

Then

$$\cos \theta = \pm \sqrt{1 - \sin^2 \theta} = \pm \sqrt{1 - \frac{x^2}{a^2}} = \pm \frac{\sqrt{a^2 - x^2}}{|a|}.$$

Because of the previously mentioned fact that $\cos \theta$ is positive, we may omit the $\pm$ sign. The reverse substitution then produces the final answer

$$\int \frac{x^4}{\sqrt{a^2 - x^2}} \, dx = -\frac{1}{4} x^3 \sqrt{a^2 - x^2} - \frac{3}{8} a^2 x \sqrt{a^2 - x^2} + \frac{3}{8} a^4 \sin^{-1} \frac{x}{|a|} + C.$$

Any rational function of x may be integrated, if the denominator is factored into linear and irreducible quadratic factors. The function may then be broken into partial fractions, and the individual partial fractions integrated by use of the appropriate formula from the integral table. See the section on partial fractions for further information.

Many integrals may be reduced to rational functions by proper substitutions. For example,

$$z = \tan \frac{x}{2}$$

will reduce any rational function of the six trigonometric functions of x to a rational function of z. (Frequently there are other substitutions which are simpler to use, but this one will always work. See integral formula number 484.)

Any rational function of x and $\sqrt{ax + b}$ may be reduced to a rational function of z by making the substitution

$$z = \sqrt{ax + b}.$$

Other likely substitutions will be suggested by looking at the form of the integrand.

The other main method of transforming integrals is integration by parts. This involves applying formula number 5 or 6 in the accompanying integral table. The critical factor in this method is the choice of the functions u and v. In order for the method to be successful, $v = \int dv$ and $\int v \, du$ must be easier to integrate than the original integral. Again, this choice is largely a matter of intuition and experience.

Example:

$$\int x \sin x \, dx$$

Two obvious choices are $u = x$, $dv = \sin x \, dx$, or $u = \sin x$, $dv = x \, dx$. Since a preliminary mental calculation indicates that $\int v \, du$ in the second choice would be more, rather than less, complicated than the original integral (it would contain x^2), we use the first choice.

$$u = x \qquad\qquad du = dx$$
$$dv = \sin x \, dx \qquad\qquad v = -\cos x$$
$$\int x \sin x \, dx = \int u \, dv = uv - \int v \, du = -x \cos x + \int \cos x \, dx$$
$$= \sin x - x \cos x$$

Of course, this result could have been obtained directly from the integral table, but it provides a simple example of the method. In more complicated examples the choice of u and v may not be so obvious, and several different choices may have to be tried. Of course, there is no guarantee that any of them will work.

Integration by parts may be applied more than once, or combined with substitution. A fairly common case is illustrated by the following example.

Example:

$$\int e^x \sin x \, dx$$

Let

$$u = e^x \qquad \text{Then } du = e^x dx$$
$$dv = \sin x \, dx \qquad v = -\cos x$$
$$\int e^x \sin x \, dx = \int u \, dv = uv - \int v \, du = -e^x \cos x + \int e^x \cos x \, dx$$

In this latter integral,

$$\text{Let } u = e^x \qquad \text{Then } du = e^x dx$$
$$dv = \cos x \, dx \qquad v = \sin x$$

$$\int e^x \sin x \, dx = -e^x \cos x + \int e^x \cos x \, dx \;\; = \;\; -e^x \cos x + \int u \, dv$$

$$= \;\; -e^x \cos x + uv - \int v \, du$$

$$= \;\; -e^x \cos x + e^x \sin x - \int e^x \sin x \, dx$$

This looks as if a circular transformation has taken place, since we are back at the same integral we started from. However, the above equation can be solved algebraically for the required integral:

$$\int e^x \sin x \, dx = \frac{1}{2} e^x \sin x - \frac{1}{2} e^x \cos x$$

In the second integration by parts, if the parts had been chosen as $u = \cos x$, $dv = e^x dx$, we would indeed have made a circular transformation, and returned to the starting place.

In general, when doing repeated integration by parts, one should never choose the function u at any stage to be the same as the function v at the previous stage, or a constant times the previous v

The following rule is called the extended rule for integration by parts. It is the result of $n+1$ successive applications of integration by parts. If

$$g_1(x) \;\; = \;\; \int g(x) \, dx, \qquad g_2(x) = \int g_1(x) \, dx,$$

$$g_3(x) \;\; = \;\; \int g_2(x) \, dx, \dots, g_m(x) = \int g_{m-1}(x) \, dx, \dots,$$

then

$$\int f(x) \cdot g(x) \, dx \;\; = \;\; f(x) \cdot g_1(x) - f'(x) \cdot g_2(x) + f''(x) \cdot g_3(x) - + \cdots$$

$$+ (-1)^n f^{(n)}(x) g_{n+1}(x) + (-1)^{n+1} \int f^{(n+1)}(x) g_{n+1}(x) \, dx.$$

A useful special case of the above rule is when $f(x)$ is a polynomial of degree n. Then $f^{(n+1)}(x) = 0$, and

$$\int f(x) \cdot g(x) \, dx = f(x) \cdot g_1(x) - f'(x) \cdot g_2(x) + f''(x) \cdot g_3(x) - + \cdots + (-1)^n f^{(n)}(x) g_{n+1}(x) + C$$

Example: If $f(x) = x^2$, $g(x) = \sin x$

$$\int x^2 \sin x \, dx = -x^2 \cos x + 2x \sin x + 2 \cos x + C$$

Another application of this formula occurs if

$$f''(x) = af(x) \qquad \text{and} \qquad g''(x) = bg(x),$$

where a and b are unequal constants. In this case, by a process similar to that used in the above example for $\int e^x \sin x \, dx$, we get the formula

$$\int f(x)g(x) \, dx = \frac{f(x) \cdot g'(x) - f'(x) \cdot g(x)}{b - a} + C$$

This formula could have been used in the example mentioned. Here is another example.

Example: If $f(x) = e^{2x}$, $g(x) = \sin 3x$, then $a = 4$, $b = -9$, and

$$\int e^{2x} \sin 3x \, dx = \frac{3 e^{2x} \cos 3x - 2 e^{2x} \sin 3x}{-9 - 4} + C = \frac{e^{2x}}{13}(2 \sin 3x - 3 \cos 3x) + C$$

The following additional points should be observed when using this table.

1. A constant of integration is to be supplied with the answers for indefinite integrals.

2. Logarithmic expressions are to base $e = 2.71828\ldots$, unless otherwise specified, and are to be evaluated for the absolute value of the arguments involved therein.

3. All angles are measured in radians, and inverse trigonometric and hyperbolic functions represent principal values, unless otherwise indicated.

4. If the application of a formula produces either a zero denominator or the square root of a negative number in the result, there is usually available another form of the answer which avoids this difficulty. In many of the results, the excluded values are specified, but when such are omitted it is presumed that one can tell what these should be, especially when difficulties of the type herein mentioned are obtained.

5. When inverse trigonometric functions occur in the integrals, be sure that any replacements made for them are strictly in accordance with the rules for such functions. This causes little difficulty when the argument of the inverse trigonometric function is positive, since then all angles involved are in the first quadrant. However, if the argument is negative, special care must be used. Thus if $u > 0$,

$$\sin^{-1} u = \cos^{-1} \sqrt{1 - u^2} = \csc^{-1} \frac{1}{u}, \text{ etc.}$$

However, if $u < 0$,

$$\sin^{-1} u = -\cos^{-1} \sqrt{1 - u^2} = -\pi - \csc^{-1} \frac{1}{u}, \text{ etc.}$$

See the section on inverse trigonometric functions for a full treatment of the allowable substitutions.

6. In integrals 340–345 and some others, the right side includes expressions of the form

$$A \tan^{-1} [B + C \tan f(x)].$$

In these formulas, the $\tan^{-1}$ does not necessarily represent the principal value. Instead of always employing the principal branch of the inverse tangent function, one must instead use that branch of the inverse tangent function upon which $f(x)$ lies for any particular choice of x.

Example:

$$\int_0^{4\pi} \frac{dx}{2 + \sin x} = \frac{2}{\sqrt{3}} \tan^{-1} \frac{2 \tan(x/2 + 1)}{\sqrt{3}} \Bigg]_0^{4\pi}$$

$$= \frac{2}{\sqrt{3}} \left[\tan^{-1} \frac{2 \tan 2\pi + 1}{\sqrt{3}} - \tan^{-1} \frac{2 \tan 0 + 1}{\sqrt{3}} \right]$$

$$= \frac{2}{\sqrt{3}} \left[\frac{13\pi}{6} - \frac{\pi}{6} \right] = \frac{4\pi}{\sqrt{3}} = \frac{4\sqrt{3}\pi}{3}$$

Here

$$\tan^{-1}\frac{2\tan 2\pi + 1}{\sqrt{3}} = \tan^{-1}\frac{1}{\sqrt{3}} = \frac{13\pi}{6},$$

since $f(x) = 2\pi$; and

$$\tan^{-1}\frac{2\tan 0 + 1}{\sqrt{3}} = \tan^{-1}\frac{1}{\sqrt{3}} = \frac{\pi}{6},$$

since $f(x) = 0$.

7. B_n and E_n where used in Integrals represents the Bernoulli and Euler numbers as defined in tables of Bernoulli and Euler polynomials contained in certain mathematics reference and hand-books.

INTEGRALS

ELEMENTARY FORMS

1. $\displaystyle\int a\,dx = ax$

2. $\displaystyle\int a\cdot f(x)\,dx = a\int f(x)\,dx$

3. $\displaystyle\int \phi(y)\,dx = \int \frac{\phi(y)}{y'}\,dy,\quad$ where $y' = \dfrac{dy}{dx}$

4. $\displaystyle\int (u+v)\,dx = \int u\,dx + \int v\,dx$, where u and v are any functions of x

5. $\displaystyle\int u\,dv = u\int dv - \int v\,du = uv - \int v\,du$

6. $\displaystyle\int u\frac{dv}{dx}\,dx = uv - \int v\frac{du}{dx}\,dx$

7. $\displaystyle\int x^n\,dx = \frac{x^{n+1}}{n+1},\quad$ except $n = -1$

8. $\displaystyle\int \frac{f'(x)\,dx}{f(x)} = \log f(x),\quad (df(x) = f'(x)\,dx)$

9. $\displaystyle\int \frac{dx}{x} = \log x$

10. $\displaystyle\int \frac{f'(x)\,dx}{2\sqrt{f(x)}} = \sqrt{f(x)},\quad (df(x) = f'(x)\,dx)$

11. $\displaystyle\int e^x\,dx = e^x$

12. $\displaystyle\int e^{ax}\,dx = e^{ax}/a$

13. $\displaystyle\int b^{ax}\,dx = \frac{b^{ax}}{a\log b},\quad (b > 0)$

14. $\displaystyle\int \log x\,dx = x\log x - x$

15. $\displaystyle\int a^x\log a\,dx = a^x,\quad (a > 0)$

16. $\displaystyle\int \frac{dx}{a^2 + x^2} = \frac{1}{a}\tan^{-1}\frac{x}{a}$

17. $\displaystyle\int \frac{dx}{a^2 - x^2} = \begin{cases} \frac{1}{a}\tanh^{-1}\frac{x}{a} \\ \text{or} \\ \frac{1}{2a}\log\frac{a+x}{a-x}, \quad (a^2 > x^2) \end{cases}$

18. $\displaystyle\int \frac{dx}{x^2 - a^2} = \begin{cases} -\frac{1}{a}\coth^{-1}\frac{x}{a} \\ \text{or} \\ \frac{1}{2a}\log\frac{x-a}{x+a}, \quad (x^2 > a^2) \end{cases}$

19. $\displaystyle\int \frac{dx}{\sqrt{a^2 - x^2}} = \begin{cases} \sin^{-1}\frac{x}{|a|} \\ \text{or} \\ -\cos^{-1}\frac{x}{|a|}, \quad (a^2 > x^2) \end{cases}$

20. $\displaystyle\int \frac{dx}{\sqrt{x^2 \pm a^2}} = \log(x + \sqrt{x^2 \pm a^2})$

21. $\displaystyle\int \frac{dx}{x\sqrt{x^2 - a^2}} = \frac{1}{|a|}\sec^{-1}\frac{x}{a}$

22. $\displaystyle\int \frac{dx}{x\sqrt{a^2 \pm x^2}} = -\frac{1}{a}\log\left(\frac{a + \sqrt{a^2 \pm x^2}}{x}\right)$

FORMS CONTAINING $(a + bx)$

For forms containing $a + bx$, but not listed in the table, the substitution $u = \frac{a+bx}{x}$ may prove helpful.

23. $\displaystyle\int (a + bx)^n \, dx = \frac{(a + bx)^{n+1}}{(n+1)b}, \quad (n \neq -1)$

24. $\displaystyle\int x(a + bx)^n \, dx = \frac{1}{b^2(n+2)}(a + bx)^{n+2} - \frac{a}{b^2(n+1)}(a + bx)^{n+1}, \quad (n \neq -1, -2)$

25. $\displaystyle\int x^2(a + bx)^n \, dx = \frac{1}{b^3}\left[\frac{(a + bx)^{n+3}}{n+3} - 2a\frac{(a + bx)^{n+2}}{n+2} + a^2\frac{(a + bx)^{n+1}}{n+1}\right]$

26. $\displaystyle\int x^m(a + bx)^n \, dx = \begin{cases} \frac{x^{m+1}(a+bx)^n}{m+n+1} + \frac{an}{m+n+1}\int x^m(a + bx)^{n-1} \, dx \\ \text{or} \\ \frac{1}{a(n+1)}\left[-x^{m+1}(a + bx)^{n+1} + (m + n + 2)\int x^m(a + bx)^{n+1} \, dx\right] \\ \text{or} \\ \frac{1}{b(m+n+1)}\left[x^m(a + bx)^{n+1} - ma\int x^{m-1}(a + bx)^n \, dx\right] \end{cases}$

27. $\displaystyle\int \frac{dx}{a + bx} = \frac{1}{b}\log(a + bx)$

28. $\displaystyle\int \frac{dx}{(a + bx)^2} = -\frac{1}{b(a + bx)}$

29. $\displaystyle\int \frac{dx}{(a + bx)^3} = -\frac{1}{2b(a + bx)^2}$

30. $\displaystyle\int \frac{x\,dx}{a + bx} = \begin{cases} \frac{1}{b^2}[a + bx - a\log(a + bx)] \\ \text{or} \\ \frac{x}{b} - \frac{a}{b^2}\log(a + bx) \end{cases}$

31. $\displaystyle\int \frac{x\,dx}{(a + bx)^2} = \frac{1}{b^2}\left[\log(a + bx) + \frac{a}{a + bx}\right]$

32. $\displaystyle\int \frac{x\,dx}{(a + bx)^n} = \frac{1}{b^2}\left[\frac{-1}{(n-2)(a + bx)^{n-2}} + \frac{a}{(n-1)(a + bx)^{n-1}}\right], \quad n \neq 1, 2$

33. $\displaystyle\int \frac{x^2\,dx}{a + bx} = \frac{1}{b^3}\left[\frac{1}{2}(a + bx)^2 - 2a(a + bx) + a^2\log(a + bx)\right]$

34. $\displaystyle\int \frac{x^2\,dx}{(a + bx)^2} = \frac{1}{b^3}\left[a + bx - 2a\log(a + bx) - \frac{a^2}{a + bx}\right]$

35. $\displaystyle\int \frac{x^2\,dx}{(a + bx)^3} = \frac{1}{b^3}\left[\log(a + bx) + \frac{2a}{a + bx} - \frac{a^2}{2(a + bx)^2}\right]$

36. $\displaystyle\int \frac{x^2\,dx}{(a + bx)^n} = \frac{1}{b^3}\left[\frac{-1}{(n-3)(a + bx)^{n-3}} + \frac{2a}{(n-2)(a + bx)^{n-2}} - \frac{a^2}{(n-1)(a + bx)^{n-1}}\right], \quad n \neq 1, 2, 3$

37. $\displaystyle\int \frac{dx}{x(a + bx)} = -\frac{1}{a}\log\frac{a + bx}{x}$

38. $\displaystyle\int \frac{dx}{x(a + bx)^2} = \frac{1}{a(a + bx)} - \frac{1}{a^2}\log\frac{a + bx}{x}$

39. $\displaystyle\int \frac{dx}{x(a+bx)^3} = \frac{1}{a^3}\left[\frac{1}{2}\left(\frac{2a+bx}{a+bx}\right)^2 + \log\frac{x}{a+bx}\right]$

40. $\displaystyle\int \frac{dx}{x^2(a+bx)} = -\frac{1}{ax} + \frac{b}{a^2}\log\frac{a+bx}{x}$

41. $\displaystyle\int \frac{dx}{x^3(a+bx)} = \frac{2bx-a}{2a^2x^2} + \frac{b^2}{a^3}\log\frac{x}{a+bx}$

42. $\displaystyle\int \frac{dx}{x^2(a+bx)^2} = -\frac{a+2bx}{a^2x(a+bx)} + \frac{2b}{a^3}\log\frac{a+bx}{x}$

FORMS CONTAINING $c^2 \pm x^2$ or $x^2 - c^2$

43. $\displaystyle\int \frac{dx}{c^2+x^2} = \frac{1}{c}\tan^{-1}\frac{x}{c}$

44. $\displaystyle\int \frac{dx}{c^2-x^2} = \frac{1}{2c}\log\frac{c+x}{c-x}, \quad (c^2 > x^2)$

45. $\displaystyle\int \frac{dx}{x^2-c^2} = \frac{1}{2c}\log\frac{x-c}{x+c}, \quad (x^2 > c^2)$

46. $\displaystyle\int \frac{x\,dx}{c^2\pm x^2} = \pm\frac{1}{2}\log(c^2\pm x^2)$

47. $\displaystyle\int \frac{x\,dx}{(c^2\pm x^2)^{n+1}} = \mp\frac{1}{2n(c^2\pm x^2)^n}$

48. $\displaystyle\int \frac{dx}{(c^2\pm x^2)^n} = \frac{1}{2c^2(n-1)}\left[\frac{x}{(c^2\pm x^2)^{n-1}} + (2n-3)\int\frac{dx}{(c^2\pm x^2)^{n-1}}\right]$

49. $\displaystyle\int \frac{dx}{(x^2-c^2)^n} = \frac{1}{2c^2(n-1)}\left[-\frac{x}{(x^2-c^2)^{n-1}} - (2n-3)\int\frac{dx}{(x^2-c^2)^{n-1}}\right]$

50. $\displaystyle\int \frac{x\,dx}{x^2-c^2} = \frac{1}{2}\log(x^2-c^2)$

51. $\displaystyle\int \frac{x\,dx}{(x^2-c^2)^{n+1}} = -\frac{1}{2n(x^2-c^2)^n}$

FORMS CONTAINING $a+bx$ AND $c+dx$

$$u = a+bx \quad v = c+dx \quad k = ad-bc$$

If $k = 0$, then $v = \frac{c}{a}u$

52. $\displaystyle\int \frac{dx}{u\cdot v} = \frac{1}{k}\cdot\log\left(\frac{v}{u}\right)$

53. $\displaystyle\int \frac{x\,dx}{u\cdot v} = \frac{1}{k}\left[\frac{a}{b}\log(u) - \frac{c}{d}\log(v)\right]$

54. $\displaystyle\int \frac{dx}{u^2\cdot v} = \frac{1}{k}\left(\frac{1}{u} + \frac{d}{k}\log\frac{v}{u}\right)$

55. $\displaystyle\int \frac{x\,dx}{u^2\cdot v} = \frac{-a}{bku} - \frac{c}{k^2}\log\frac{v}{u}$

56. $\displaystyle\int \frac{x^2\,dx}{u^2\cdot v} = \frac{a^2}{b^2ku} + \frac{1}{k^2}\left[\frac{c^2}{d}\log(v) + \frac{a(k-bc)}{b^2}\log(u)\right]$

57. $\displaystyle\int \frac{dx}{u^n\cdot v^m} = \frac{1}{k(m-1)}\left[\frac{-1}{u^{n-1}\cdot v^{m-1}} - (m+n-2)b\int\frac{dx}{u^n\cdot v^{m-1}}\right]$

58. $\displaystyle\int \frac{u}{v}\,dx = \frac{bx}{d} + \frac{k}{d^2}\log(v)$

59. $\displaystyle\int \frac{u^m\,dx}{v^n} = \begin{cases} \dfrac{-1}{k(n-1)}\left[\dfrac{u^{m+1}}{v^{n-1}} + b(n-m-2)\displaystyle\int\dfrac{u^m}{v^{n-1}}\,dx\right] \\ \qquad\text{or} \\ \dfrac{-1}{d(n-m-1)}\left[\dfrac{u^m}{v^{n-1}} + mk\displaystyle\int\dfrac{u^{m-1}}{v^n}\,dx\right] \\ \qquad\text{or} \\ \dfrac{-1}{d(n-1)}\left[\dfrac{u^m}{v^{n-1}} - mb\displaystyle\int\dfrac{u^{m-1}}{v^{n-1}}\,dx\right] \end{cases}$

FORMS CONTAINING $(a + bx^n)$

60. $\displaystyle\int \frac{dx}{a + bx^2} = \frac{1}{\sqrt{ab}} \tan^{-1} \frac{x\sqrt{ab}}{a}, \quad (ab > 0)$

61. $\displaystyle\int \frac{dx}{a + bx^2} = \begin{cases} \frac{1}{2\sqrt{-ab}} \log \frac{a+x\sqrt{-ab}}{a-x\sqrt{-ab}}, & (ab < 0) \\ \text{or} \\ \frac{1}{\sqrt{-ab}} \tanh^{-1} \frac{x\sqrt{-ab}}{a}, & (ab < 0) \end{cases}$

62. $\displaystyle\int \frac{dx}{a^2 + b^2 x^2} = \frac{1}{ab} \tan^{-1} \frac{bx}{a}$

63. $\displaystyle\int \frac{x\,dx}{a + bx^2} = \frac{1}{2b} \log(a + bx^2)$

64. $\displaystyle\int \frac{x^2\,dx}{a + bx^2} = \frac{x}{b} - \frac{a}{b} \int \frac{dx}{a + bx^2}$

65. $\displaystyle\int \frac{dx}{(a + bx^2)^2} = \frac{x}{2a(a + bx^2)} + \frac{1}{2a} \int \frac{dx}{a + bx^2}$

66. $\displaystyle\int \frac{dx}{a^2 - b^2 x^2} = \frac{1}{2ab} \log \frac{a + bx}{a - bx}$

67. $\displaystyle\int \frac{dx}{(a + bx^2)^{m+1}} = \begin{cases} \frac{1}{2ma} \frac{x}{(a+bx^2)^m} + \frac{2m-1}{2ma} \int \frac{dx}{(a+bx^2)^m} \\ \text{or} \\ \frac{(2m)!}{(m!)^2} \left[\frac{x}{2a} \sum_{r=1}^{m} \frac{r!(r-1)!}{(4a)^{m-r}(2r)!(a+bx^2)^r} + \frac{1}{(4a)^m} \int \frac{dx}{a+bx^2} \right] \end{cases}$

68. $\displaystyle\int \frac{x\,dx}{(a + bx^2)^{m+1}} = -\frac{1}{2bm(a + bx^2)^m}$

69. $\displaystyle\int \frac{x^2\,dx}{(a + bx^2)^{m+1}} = \frac{-x}{2mb(a + bx^2)^m} + \frac{1}{2mb} \int \frac{dx}{(a + bx^2)^m}$

70. $\displaystyle\int \frac{dx}{x(a + bx^2)} = \frac{1}{2a} \log \frac{x^2}{a + bx^2}$

71. $\displaystyle\int \frac{dx}{x^2(a + bx^2)} = -\frac{1}{ax} - \frac{b}{a} \int \frac{dx}{a + bx^2}$

72. $\displaystyle\int \frac{dx}{x(a + bx^2)^{m+1}} = \begin{cases} \frac{1}{2am(a+bx^2)^m} + \frac{1}{a} \int \frac{dx}{x(a+bx^2)^m} \\ \text{or} \\ \frac{1}{2a^{m+1}} \left[\sum_{r=1}^{m} \frac{a^r}{r(a+bx^2)^r} + \log \frac{x^2}{a+bx^2} \right] \end{cases}$

73. $\displaystyle\int \frac{dx}{x^2(a + bx^2)^{m+1}} = \frac{1}{a} \int \frac{dx}{x^2(a + bx^2)^m} - \frac{b}{a} \int \frac{dx}{(a + bx^2)^{m+1}}$

74. $\displaystyle\int \frac{dx}{a + bx^3} = \frac{k}{3a} \left[\frac{1}{2} \log \frac{(k + x)^3}{a + bx^3} + \sqrt{3} \tan^{-1} \frac{2x - k}{k\sqrt{3}} \right], \quad \left(k = \sqrt[3]{\frac{a}{b}} \right)$

75. $\displaystyle\int \frac{x\,dx}{a + bx^3} = \frac{1}{3bk} \left[\frac{1}{2} \log \frac{a + bx^3}{(k + x)^3} + \sqrt{3} \tan^{-1} \frac{2x - k}{k\sqrt{3}} \right], \quad \left(k = \sqrt[3]{\frac{a}{b}} \right)$

76. $\displaystyle\int \frac{x^2\,dx}{a + bx^3} = \frac{1}{3b} \log(a + bx^3)$

77. $\displaystyle\int \frac{dx}{a + bx^4} = \frac{k}{2a} \left[\frac{1}{2} \log \frac{x^2 + 2kx + 2k^2}{x^2 - 2kx + 2k^2} + \tan^{-1} \frac{2kx}{2k^2 - x^2} \right], \quad \left(ab > 0, k = \sqrt[4]{\frac{a}{4b}} \right)$

78. $\displaystyle\int \frac{dx}{a + bx^4} = \frac{k}{2a} \left[\frac{1}{2} \log \frac{x + k}{x - k} + \tan^{-1} \frac{x}{k} \right], \quad \left(ab < 0, k = \sqrt[4]{-\frac{a}{b}} \right)$

79. $\displaystyle\int \frac{x\,dx}{a + bx^4} = \frac{1}{2bk} \tan^{-1} \frac{x^2}{k}, \quad \left(ab > 0, k = \sqrt{\frac{a}{b}} \right)$

80. $\displaystyle\int \frac{x\,dx}{a + bx^4} = \frac{1}{4bk} \log \frac{x^2 - k}{x^2 + k}, \quad \left(ab < 0, k = \sqrt{-\frac{a}{b}} \right)$

81. $\displaystyle\int \frac{x^2\,dx}{a + bx^4} = \frac{1}{4bk} \left[\frac{1}{2} \log \frac{x^2 - 2kx + 2k^2}{x^2 + 2kx + 2k^2} + \tan^{-1} \frac{2kx}{2k^2 - x^2} \right], \quad \left(ab > 0, k = \sqrt[4]{\frac{a}{4b}} \right)$

82. $\displaystyle\int \frac{x^2\,dx}{a + bx^4} = \frac{1}{4bk} \left[\log \frac{x - k}{x + k} + 2\tan^{-1} \frac{x}{k} \right], \quad \left(ab < 0, k = \sqrt[4]{-\frac{a}{b}} \right)$

83. $\displaystyle\int \frac{x^3\,dx}{a + bx^4} = \frac{1}{4b} \log(a + bx^4)$

84. $\displaystyle\int \frac{dx}{x(a + bx^n)} = \frac{1}{an} \log \frac{x^n}{a + bx^n}$

85. $\displaystyle\int \frac{dx}{(a + bx^n)^{m+1}} = \frac{1}{a} \int \frac{dx}{(a + bx^n)^m} - \frac{b}{a} \int \frac{x^n\,dx}{(a + bx^n)^{m+1}}$

86. $$\int \frac{x^m\,dx}{(a+bx^n)^{p+1}} = \frac{1}{b}\int \frac{x^{m-n}\,dx}{(a+bx^n)^p} - \frac{a}{b}\int \frac{x^{m-n}\,dx}{(a+bx^n)^{p+1}}$$

87. $$\int \frac{dx}{x^m(a+bx^n)^{p+1}} = \frac{1}{a}\int \frac{dx}{x^m(a+bx^n)^p} - \frac{b}{a}\int \frac{dx}{x^{m-n}(a+bx^n)^{p+1}}$$

88. $$\int x^m(a+bx^n)^p\,dx = \begin{cases} \frac{1}{b(np+m+1)}\left[x^{m-n+1}(a+bx^n)^{p+1} - a(m-n+1)\int x^{m-n}(a+bx^n)^p\,dx\right] \\[4pt] \text{or} \\[4pt] \frac{1}{np+m+1}\left[x^{m+1}(a+bx^n)^p + anp\int x^m(a+bx^n)^{p-1}\,dx\right] \\[4pt] \text{or} \\[4pt] \frac{1}{a(m+1)}\left[x^{m+1}(a+bx^n)^{p+1} - (m+1+np+n)b\int x^{m+n}(a+bx^n)^p\,dx\right] \\[4pt] \text{or} \\[4pt] \frac{1}{an(p+1)}\left[-x^{m+1}(a+bx^n)^{p+1} + (m+1+np+n)\int x^m(a+bx^n)^{p+1}\,dx\right] \end{cases}$$

FORMS CONTAINING $c^3 \pm x^3$

89. $$\int \frac{dx}{c^3 \pm x^3} = \pm\frac{1}{6c^2}\log\frac{(c \pm x)^3}{c^3 \pm x^3} + \frac{1}{c^2\sqrt{3}}\tan^{-1}\frac{2x \mp c}{c\sqrt{3}}$$

90. $$\int \frac{dx}{(c^3 \pm x^3)^2} = \frac{x}{3c^3(c^3 \pm x^3)} + \frac{2}{3c^3}\int \frac{dx}{c^3 \pm x^3}$$

91. $$\int \frac{dx}{(c^3 \pm x^3)^{n+1}} = \frac{1}{3nc^3}\left[\frac{x}{(c^3 \pm x^3)^n} + (3n-1)\int \frac{dx}{(c^3 \pm x^3)^n}\right]$$

92. $$\int \frac{x\,dx}{c^3 \pm x^3} = \frac{1}{6c}\log\frac{c^3 \pm x^3}{(c \pm x)^3} \pm \frac{1}{c\sqrt{3}}\tan^{-1}\frac{2x \mp c}{c\sqrt{3}}$$

93. $$\int \frac{x\,dx}{(c^3 \pm x^3)^2} = \frac{x^2}{3c^3(c^3 \pm x^3)} + \frac{1}{3c^3}\int \frac{x\,dx}{c^3 \pm x^3}$$

94. $$\int \frac{x\,dx}{(c^3 \pm x^3)^{n+1}} = \frac{1}{3nc^3}\left[\frac{x^2}{(c^3 \pm x^3)^n} + (3n-2)\int \frac{x\,dx}{(c^3 \pm x^3)^n}\right]$$

95. $$\int \frac{x^2\,dx}{c^3 \pm x^3} = \pm\frac{1}{3}\log(c^3 \pm x^3)$$

96. $$\int \frac{x^2\,dx}{(c^3 \pm x^3)^{n+1}} = \mp\frac{1}{3n(c^3 \pm x^3)^n}$$

97. $$\int \frac{dx}{x(c^3 \pm x^3)} = \frac{1}{3c^3}\log\frac{x^3}{c^3 \pm x^3}$$

98. $$\int \frac{dx}{x(c^3 \pm x^3)^2} = \frac{1}{3c^3(c^3 \pm x^3)} + \frac{1}{3c^6}\log\frac{x^3}{c^3 \pm x^3}$$

99. $$\int \frac{dx}{x(c^3 \pm x^3)^{n+1}} = \frac{1}{3nc^3(c^3 \pm x^3)^n} + \frac{1}{c^3}\int \frac{dx}{x(c^3 \pm x^3)^n}$$

100. $$\int \frac{dx}{x^2(c^3 \pm x^3)} = -\frac{1}{c^3 x} \mp \frac{1}{c^3}\int \frac{x\,dx}{c^3 \pm x^3}$$

101. $$\int \frac{dx}{x^2(c^3 \pm x^3)^{n+1}} = \frac{1}{c^3}\int \frac{dx}{x^2(c^3 \pm x^3)^n} \mp \frac{1}{c^3}\int \frac{x\,dx}{(c^3 \pm x^3)^{n+1}}$$

FORMS CONTAINING $c^4 \pm x^4$

102. $$\int \frac{dx}{c^4 + x^4} = \frac{1}{2c^3\sqrt{2}}\left[\frac{1}{2}\log\frac{x^2 + cx\sqrt{2} + c^2}{x^2 - cx\sqrt{2} + c^2} + \tan^{-1}\frac{cx\sqrt{2}}{c^2 - x^2}\right]$$

103. $$\int \frac{dx}{c^4 - x^4} = \frac{1}{2c^3}\left[\frac{1}{2}\log\frac{c+x}{c-x} + \tan^{-1}\frac{x}{c}\right]$$

104. $$\int \frac{x\,dx}{c^4 + x^4} = \frac{1}{2c^2}\tan^{-1}\frac{x^2}{c^2}$$

105. $$\int \frac{x\,dx}{c^4 - x^4} = \frac{1}{4c^2}\log\frac{c^2 + x^2}{c^2 - x^2}$$

106. $$\int \frac{x^2\,dx}{c^4 + x^4} = \frac{1}{2c\sqrt{2}}\left[\frac{1}{2}\log\frac{x^2 - cx\sqrt{2} + c^2}{x^2 + cx\sqrt{2} + c^2} + \tan^{-1}\frac{cx\sqrt{2}}{c^2 - x^2}\right]$$

107. $$\int \frac{x^2\,dx}{c^4 - x^4} = \frac{1}{2c}\left[\frac{1}{2}\log\frac{c+x}{c-x} - \tan^{-1}\frac{x}{c}\right]$$

108. $\displaystyle \int \frac{x^3\,dx}{c^4 \pm x^4} = \pm \frac{1}{4}\log(c^4 \pm x^4)$

FORMS CONTAINING $(a + bx + cx^2)$

$$X = a + bx + cx^2 \quad \text{and} \quad q = 4ac - b^2$$

If $q = 0$, then $X = c\left(x + \frac{b}{2c}\right)^2$, and formulas starting with 23 should be used in place of these.

109. $\displaystyle \int \frac{dx}{X} = \frac{2}{\sqrt{q}}\tan^{-1}\frac{2cx+b}{\sqrt{q}}, \quad (q > 0)$

110. $\displaystyle \int \frac{dx}{X} = \begin{cases} \frac{-2}{\sqrt{-q}}\tanh^{-1}\frac{2cx+b}{\sqrt{-q}} \\ \text{or} \\ \frac{1}{\sqrt{-q}}\log\frac{2cx+b-\sqrt{-q}}{2cx+b+\sqrt{-q}}, \quad (q < 0) \end{cases}$

111. $\displaystyle \int \frac{dx}{X^2} = \frac{2cx+b}{qX} + \frac{2c}{q}\int \frac{dx}{X}$

112. $\displaystyle \int \frac{dx}{X^3} = \frac{2cx+b}{q}\left(\frac{1}{2X^2} + \frac{3c}{qX}\right) + \frac{6c^2}{q^2}\int \frac{dx}{X}$

113. $\displaystyle \int \frac{dx}{X^{n+1}} = \begin{cases} \dfrac{2cx+b}{nqX^n} + \dfrac{2(2n-1)c}{qn}\displaystyle\int \frac{dx}{X^n} \\ \text{or} \\ \dfrac{(2n)!}{(n!)^2}\left(\dfrac{c}{q}\right)^n\left[\dfrac{2cx+b}{q}\displaystyle\sum_{r=1}^{n}\left(\dfrac{q}{cX}\right)^r\left(\dfrac{(r-1)!r!}{(2r)!}\right) + \displaystyle\int \frac{dx}{X}\right] \end{cases}$

114. $\displaystyle \int \frac{x\,dx}{X} = \frac{1}{2c}\log X - \frac{b}{2c}\int \frac{dx}{X}$

115. $\displaystyle \int \frac{x\,dx}{X^2} = \frac{bx+2a}{qX} - \frac{b}{q}\int \frac{dx}{X}$

116. $\displaystyle \int \frac{x\,dx}{X^{n+1}} = -\frac{2a+bx}{nqX^n} - \frac{b(2n-1)}{nq}\int \frac{dx}{X^n}$

117. $\displaystyle \int \frac{x^2}{X}\,dx = \frac{x}{c} - \frac{b}{2c^2}\log X + \frac{b^2-2ac}{2c^2}\int \frac{dx}{X}$

118. $\displaystyle \int \frac{x^2}{X^2}\,dx = \frac{(b^2-2ac)x+ab}{cqX} + \frac{2a}{q}\int \frac{dx}{X}$

119. $\displaystyle \int \frac{x^m\,dx}{X^{n+1}} = -\frac{x^{m-1}}{(2n-m+1)cX^n} - \frac{n-m+1}{2n-m+1}\cdot\frac{b}{c}\int \frac{x^{m-1}\,dx}{X^{n+1}} + \frac{m-1}{2n-m+1}\cdot\frac{a}{c}\int \frac{x^{m-2}\,dx}{X^{n+1}}$

120. $\displaystyle \int \frac{dx}{xX} = \frac{1}{2a}\log\frac{x^2}{X} - \frac{b}{2a}\int \frac{dx}{X}$

121. $\displaystyle \int \frac{dx}{x^2 X} = \frac{b}{2a^2}\log\frac{X}{x^2} - \frac{1}{ax} + \left(\frac{b^2}{2a^2} - \frac{c}{a}\right)\int \frac{dx}{X}$

122. $\displaystyle \int \frac{dx}{xX^n} = \frac{1}{2a(n-1)X^{n-1}} - \frac{b}{2a}\int \frac{dx}{X^n} + \frac{1}{a}\int \frac{dx}{xX^{n-1}}$

123. $\displaystyle \int \frac{dx}{x^m X^{n+1}} = -\frac{1}{(m-1)ax^{m-1}X^n} - \frac{n+m-1}{m-1}\cdot\frac{b}{a}\int \frac{dx}{x^{m-1}X^{n+1}} - \frac{2n+m-1}{m-1}\cdot\frac{c}{a}\int \frac{dx}{x^{m-2}X^{n+1}}$

FORMS CONTAINING $\sqrt{a + bx}$

124. $\displaystyle \int \sqrt{a+bx}\,dx = \frac{2}{3b}\sqrt{(a+bx)^3}$

125. $\displaystyle \int x\sqrt{a+bx}\,dx = -\frac{2(2a-3bx)\sqrt{(a+bx)^3}}{15b^2}$

126. $\displaystyle \int x^2\sqrt{a+bx}\,dx = \frac{2(8a^2-12abx+15b^2x^2)\sqrt{(a+bx)^3}}{105b^3}$

127. $\displaystyle \int x^m\sqrt{a+bx}\,dx = \begin{cases} \dfrac{2}{b(2m+3)}\left[x^m\sqrt{(a+bx)^3} - ma\displaystyle\int x^{m-1}\sqrt{a+bx}\,dx\right] \\ \text{or} \\ \dfrac{2}{b^{m+1}}\sqrt{a+bx}\displaystyle\sum_{r=0}^{m}\frac{m!(-a)^{m-r}}{r!(m-r)!(2r+3)}(a+bx)^{r+1} \end{cases}$

128. $\displaystyle \int \frac{\sqrt{a+bx}}{x}dx = 2\sqrt{a+bx} + a\int \frac{dx}{x\sqrt{a+bx}}$

129. $\displaystyle \int \frac{\sqrt{a+bx}}{x^2}dx = \frac{\sqrt{a+bx}}{x} + \frac{b}{2}\int \frac{dx}{x\sqrt{a+bx}}$

130. $\displaystyle \int \frac{\sqrt{a+bx}}{x^m}dx = -\frac{1}{(m-1)a}\left[\frac{\sqrt{(a+bx)^3}}{x^{m-1}} + \frac{(2m-5)b}{2}\int \frac{\sqrt{a+bx}}{x^{m-1}}dx\right]$

131. $\displaystyle \int \frac{dx}{\sqrt{a+bx}} = \frac{2\sqrt{a+bx}}{b}$

132. $\displaystyle \int \frac{x\,dx}{\sqrt{a+bx}} = -\frac{2(2a-bx)}{3b^2}\sqrt{a+bx}$

133. $\displaystyle \int \frac{x^2\,dx}{\sqrt{a+bx}} = \frac{2(8a^2-4abx-3b^2x^2)}{15b^3}\sqrt{a+bx}$

134. $\displaystyle \int \frac{x^m\,dx}{\sqrt{a+bx}} = \begin{cases} \dfrac{2}{(2m+1)b}\left[x^m\sqrt{a+bx} - ma\int \dfrac{x^{m-1}dx}{\sqrt{a+bx}}\right] \\ \text{or} \\ \dfrac{2(-a)^m\sqrt{a+bx}}{b^{m+1}}\sum\limits_{r=0}^{m}\dfrac{(-1)^r m!(a+bx)^r}{(2r+1)r!(m-r)!a^r} \end{cases}$

135. $\displaystyle \int \frac{dx}{x\sqrt{a+bx}} = \frac{1}{\sqrt{a}}\log\left(\frac{\sqrt{a+bx}-\sqrt{a}}{\sqrt{a+bx}+\sqrt{a}}\right), \quad (a>0)$

136. $\displaystyle \int \frac{dx}{x\sqrt{a+bx}} = \frac{2}{\sqrt{-a}}\tan^{-1}\sqrt{\frac{a+bx}{-a}}, \quad (a<0)$

137. $\displaystyle \int \frac{dx}{x^2\sqrt{a+bx}} = -\frac{\sqrt{a+bx}}{ax} - \frac{b}{2a}\int \frac{dx}{x\sqrt{a+bx}}$

138. $\displaystyle \int \frac{dx}{x^n\sqrt{a+bx}} = \begin{cases} -\dfrac{\sqrt{a+bx}}{(n-1)ax^{n-1}} - \dfrac{(2n-3)b}{(2n-2)a}\int \dfrac{dx}{x^{n-1}\sqrt{a+bx}} \\ \text{or} \\ \dfrac{(2n-2)!}{[(n-1)!]^2}\left[-\dfrac{\sqrt{a+bx}}{a}\sum\limits_{r=1}^{n-1}\dfrac{r!(r-1)!}{x^r 2(r)!}\left(-\dfrac{b}{4a}\right)^{n-r-1} + \left(-\dfrac{b}{4a}\right)^{n-1}\int \dfrac{dx}{x\sqrt{a+bx}}\right] \end{cases}$

139. $\displaystyle \int (a+bx)^{\pm\frac{n}{2}}dx = \frac{2(a+bx)^{\frac{2\pm n}{2}}}{b(2\pm n)}$

140. $\displaystyle \int x(a+bx)^{\pm\frac{n}{2}}dx = \frac{2}{b^2}\left[\frac{(a+bx)^{\frac{4\pm n}{2}}}{4\pm n} - \frac{a(a+bx)^{\frac{2\pm n}{2}}}{2\pm n}\right]$

141. $\displaystyle \int \frac{dx}{x(a+bx)^{\frac{m}{2}}} = \frac{1}{a}\int \frac{dx}{x(a+bx)^{\frac{m-2}{2}}} - \frac{b}{a}\int \frac{dx}{(a+bx)^{\frac{m}{2}}}$

142. $\displaystyle \int \frac{(a+bx)^{n/2}dx}{x} = b\int (a+bx)^{(n-2)/2}dx + a\int \frac{(a+bx)^{(n-2)/2}}{x}dx$

143. $\displaystyle \int f(x,\sqrt{a+bx})\,dx = \frac{2}{b}\int f\left(\frac{z^2-a}{b},z\right)z\,dz, \quad (z=\sqrt{a+bx})$

FORMS CONTAINING $\sqrt{a+bx}$ and $\sqrt{c+dx}$

$$u = a + bx \qquad v = c + dx \qquad k = ad - bc$$

If $k=0$, then, $v = (\frac{c}{a})u$, and formulas starting with 124 should be used in place of these.

144. $\displaystyle \int \frac{dx}{\sqrt{uv}} = \begin{cases} \dfrac{2}{\sqrt{bd}}\tanh^{-1}\dfrac{\sqrt{bduv}}{bv}, & bd>0, \quad k<0 \\ \text{or} \\ \dfrac{2}{\sqrt{bd}}\tanh^{-1}\dfrac{\sqrt{bduv}}{du}, & bd>0, \quad k>0. \\ \text{or} \\ \dfrac{1}{\sqrt{bd}}\log\dfrac{(bv+\sqrt{bduv})^2}{v}, & (bd>0) \end{cases}$

145.
$$\int \frac{dx}{\sqrt{uv}} = \begin{cases} \frac{2}{\sqrt{-bd}} \tan^{-1} \frac{\sqrt{-bduv}}{bv} \\ \text{or} \\ -\frac{1}{\sqrt{-bd}} \sin^{-1}\left(\frac{2bdx+ad+bc}{|k|}\right), \quad (bd < 0) \end{cases}$$

146.
$$\int \sqrt{uv}\, dx = \frac{k+2bv}{4bd}\sqrt{uv} - \frac{k^2}{8bd}\int \frac{dx}{\sqrt{uv}}$$

147.
$$\int \frac{dx}{v\sqrt{u}} = \begin{cases} \frac{1}{\sqrt{kd}} \log \frac{d\sqrt{u}-\sqrt{kd}}{d\sqrt{u}+\sqrt{kd}} \\ \text{or} \\ \frac{1}{\sqrt{kd}} \log \frac{(d\sqrt{u}-\sqrt{kd})^2}{v}, \quad (kd > 0) \end{cases}$$

148.
$$\int \frac{dx}{v\sqrt{u}} = \frac{2}{\sqrt{-kd}} \tan^{-1} \frac{d\sqrt{u}}{\sqrt{-kd}}, \quad (kd < 0)$$

149.
$$\int \frac{x\,dx}{\sqrt{uv}} = \frac{\sqrt{uv}}{bd} - \frac{ad+bc}{2bd}\int \frac{dx}{\sqrt{uv}}$$

150.
$$\int \frac{dx}{v\sqrt{uv}} = \frac{-2\sqrt{uv}}{kv}$$

151.
$$\int \frac{v\,dx}{\sqrt{uv}} = \frac{\sqrt{uv}}{b} - \frac{k}{2b}\int \frac{dx}{\sqrt{uv}}$$

152.
$$\int \sqrt{\frac{v}{u}}\,dx = \frac{v}{|v|}\int \frac{v\,dx}{\sqrt{uv}}$$

153.
$$\int v^m \sqrt{u}\, dx = \frac{1}{(2m+3)d}\left(2v^{m+1}\sqrt{u} + k\int \frac{v^m dx}{\sqrt{u}}\right)$$

154.
$$\int \frac{dx}{v^m \sqrt{u}} = -\frac{1}{(m-1)k}\left(\frac{\sqrt{u}}{v^{m-1}} + \left(m-\frac{3}{2}\right)b\int \frac{dx}{v^{m-1}\sqrt{u}}\right)$$

155.
$$\int \frac{v^m dx}{\sqrt{u}} = \begin{cases} \frac{2}{b(2m+1)}\left[v^m\sqrt{u} - mk\int \frac{v^{m-1}}{\sqrt{u}}dx\right] \\ \text{or} \\ \frac{2(m!)^2\sqrt{u}}{b(2m+1)!}\sum_{r=0}^{m}\left(-\frac{4k}{b}\right)^{m-r}\frac{(2r)!}{(r!)^2}v^r \end{cases}$$

FORMS CONTAINING $\sqrt{x^2 \pm a^2}$

156.
$$\int \sqrt{x^2 \pm a^2}\,dx = \frac{1}{2}\left[x\sqrt{x^2 \pm a^2} \pm a^2 \log\left(x + \sqrt{x^2 \pm a^2}\right)\right]$$

157.
$$\int \frac{dx}{\sqrt{x^2 \pm a^2}} = \log\left(x + \sqrt{x^2 \pm a^2}\right)$$

158.
$$\int \frac{dx}{x\sqrt{x^2 - a^2}} = \frac{1}{|a|}\sec^{-1}\frac{x}{a}$$

159.
$$\int \frac{dx}{x\sqrt{x^2 + a^2}} = -\frac{1}{a}\log\left(\frac{a + \sqrt{x^2 + a^2}}{x}\right)$$

160.
$$\int \frac{\sqrt{x^2 + a^2}}{x}\,dx = \sqrt{x^2 + a^2} - a\log\left(\frac{a + \sqrt{x^2 + a^2}}{x}\right)$$

161.
$$\int \frac{\sqrt{x^2 - a^2}}{x}\,dx = \sqrt{x^2 - a^2} - |a|\sec^{-1}\frac{x}{a}$$

162.
$$\int \frac{x\,dx}{\sqrt{x^2 \pm a^2}} = \sqrt{x^2 \pm a^2}$$

163.
$$\int x\sqrt{x^2 \pm a^2}\,dx = \frac{1}{3}\sqrt{(x^2 \pm a^2)^3}$$

164.
$$\int \sqrt{(x^2 \pm a^2)^3}\,dx = \frac{1}{4}\left[x\sqrt{(x^2 \pm a^2)^3} \pm \frac{3a^2 x}{2}\sqrt{x^2 \pm a^2} + \frac{3a^4}{2}\log(x + \sqrt{x^2 \pm a^2})\right]$$

165.
$$\int \frac{dx}{\sqrt{(x^2 \pm a^2)^3}} = \frac{\pm x}{a^2\sqrt{x^2 \pm a^2}}$$

166.
$$\int \frac{x\,dx}{\sqrt{(x^2 \pm a^2)^3}} = \frac{-1}{\sqrt{x^2 \pm a^2}}$$

167. $\displaystyle\int x\sqrt{(x^2 \pm a^2)^3}\,dx = \frac{1}{5}\sqrt{(x^2 \pm a^2)^5}$

168. $\displaystyle\int x^2\sqrt{x^2 \pm a^2}\,dx = \frac{x}{4}\sqrt{(x^2 \pm a^2)^3} \mp \frac{a^2}{8}x\sqrt{x^2 \pm a^2} - \frac{a^4}{8}\log(x + \sqrt{x^2 \pm a^2})$

169. $\displaystyle\int x^3\sqrt{x^2 + a^2}\,dx = \left(\frac{1}{5}x^2 - \frac{2}{15}a^2\right)\sqrt{(a^2 + x^2)^3}$

170. $\displaystyle\int x^3\sqrt{x^2 - a^2}\,dx = \frac{1}{5}\sqrt{(x^2 - a^2)^5} + \frac{a^2}{3}\sqrt{(x^2 - a^2)^3}$

171. $\displaystyle\int \frac{x^2\,dx}{\sqrt{x^2 \pm a^2}} = \frac{x}{2}\sqrt{x^2 \pm a^2} \mp \frac{a^2}{2}\log(x + \sqrt{x^2 \pm a^2})$

172. $\displaystyle\int \frac{x^3\,dx}{\sqrt{x^2 \pm a^2}} = \frac{1}{3}\sqrt{(x^2 \pm a^2)^3} \mp a^2\sqrt{x^2 \pm a^2}$

173. $\displaystyle\int \frac{dx}{x^2\sqrt{x^2 \pm a^2}} = \mp \frac{\sqrt{x^2 \pm a^2}}{a^2 x}$

174. $\displaystyle\int \frac{dx}{x^3\sqrt{x^2 + a^2}} = \frac{\sqrt{x^2 + a^2}}{2a^2 x^2} + \frac{1}{2a^3}\log\frac{a + \sqrt{x^2 + a^2}}{x}$

175. $\displaystyle\int \frac{dx}{x^3\sqrt{x^2 - a^2}} = \frac{\sqrt{x^2 - a^2}}{2a^2 x^2} + \frac{1}{2|a^3|}\sec^{-1}\frac{x}{a}$

176. $\displaystyle\int x^2\sqrt{(x^2 \pm a^2)^3}\,dx = \frac{x}{6}\sqrt{(x^2 \pm a^2)^5} \mp \frac{a^2 x}{24}\sqrt{(x^2 \pm a^2)^3} - \frac{a^4 x}{16}\sqrt{x^2 \pm a^2} \pm \frac{a^6}{16}\log(x + \sqrt{x^2 \pm a^2})$

177. $\displaystyle\int x^3\sqrt{(x^2 \pm a^2)^3}\,dx = \frac{1}{7}\sqrt{(x^2 \pm a^2)^7} \mp \frac{a^2}{5}\sqrt{(x^2 \pm a^2)^5}$

178. $\displaystyle\int \frac{\sqrt{x^2 \pm a^2}\,dx}{x^2} = -\frac{\sqrt{x^2 \pm a^2}}{x} + \log(x + \sqrt{x^2 \pm a^2})$

179. $\displaystyle\int \frac{\sqrt{x^2 + a^2}}{x^3}\,dx = -\frac{\sqrt{x^2 + a^2}}{2x^2} - \frac{1}{2a}\log\frac{a + \sqrt{x^2 + a^2}}{x}$

180. $\displaystyle\int \frac{\sqrt{x^2 - a^2}}{x^3}\,dx = -\frac{\sqrt{x^2 - a^2}}{2x^2} + \frac{1}{2|a|}\sec^{-1}\frac{x}{a}$

181. $\displaystyle\int \frac{\sqrt{x^2 \pm a^2}}{x^4}\,dx = \mp\frac{\sqrt{(x^2 \pm a^2)^3}}{3a^2 x^3}$

182. $\displaystyle\int \frac{x^2\,dx}{\sqrt{(x^2 \pm a^2)^3}} = \frac{-x}{\sqrt{x^2 \pm a^2}} + \log(x + \sqrt{x^2 \pm a^2})$

183. $\displaystyle\int \frac{x^3\,dx}{\sqrt{(x^2 \pm a^2)^3}} = \sqrt{x^2 \pm a^2} \pm \frac{a^2}{\sqrt{x^2 \pm a^2}}$

184. $\displaystyle\int \frac{dx}{x\sqrt{(x^2 + a^2)^3}} = \frac{1}{a^2\sqrt{x^2 + a^2}} - \frac{1}{a^3}\log\frac{a + \sqrt{x^2 + a^2}}{x}$

185. $\displaystyle\int \frac{dx}{x\sqrt{(x^2 - a^2)^3}} = -\frac{1}{a^2\sqrt{x^2 - a^2}} - \frac{1}{|a^3|}\sec^{-1}\frac{x}{a}$

186. $\displaystyle\int \frac{dx}{x^2\sqrt{(x^2 \pm a^2)^3}} = -\frac{1}{a^4}\left[\frac{\sqrt{x^2 \pm a^2}}{x} + \frac{x}{\sqrt{x^2 \pm a^2}}\right]$

187. $\displaystyle\int \frac{dx}{x^3\sqrt{(x^2 + a^2)^3}} = -\frac{1}{2a^2 x^2\sqrt{x^2 + a^2}} - \frac{3}{2a^4\sqrt{x^2 + a^2}} + \frac{3}{2a^5}\log\frac{a + \sqrt{x^2 + a^2}}{x}$

188. $\displaystyle\int \frac{dx}{x^3\sqrt{(x^2 - a^2)^3}} = \frac{1}{2a^2 x^2\sqrt{x^2 - a^2}} - \frac{3}{2a^4\sqrt{x^2 - a^2}} - \frac{3}{2|a^5|}\sec^{-1}\frac{x}{a}$

189. $\displaystyle\int \frac{x^m}{\sqrt{x^2 \pm a^2}}\,dx = \frac{1}{m}x^{m-1}\sqrt{x^2 \pm a^2} \mp \frac{m-1}{m}a^2\int \frac{x^{m-2}}{\sqrt{x^2 \pm a^2}}\,dx$

190. $\displaystyle\int \frac{x^{2m}}{\sqrt{x^2 \pm a^2}}\,dx = \frac{(2m)!}{2^{2m}(m!)^2}\left[\sqrt{x^2 \pm a^2}\sum_{r=1}^{m}\frac{r!(r-1)!}{(2r)!}(\mp a^2)^{m-r}(2x)^{2r-1} + (\mp a^2)^m\log(x + \sqrt{x^2 \pm a^2})\right]$

191. $\displaystyle\int \frac{x^{2m+1}}{\sqrt{x^2 \pm a^2}}\,dx = \sqrt{x^2 \pm a^2}\sum_{r=0}^{m}\frac{(2r)!(m!)^2}{(2m+1)!(r!)^2}(\mp 4a^2)^{m-r}x^{2r}$

192. $\displaystyle\int \frac{dx}{x^m\sqrt{x^2 \pm a^2}} = \mp\frac{\sqrt{x^2 \pm a^2}}{(m-1)a^2 x^{m-1}} \mp \frac{(m-2)}{(m-1)a^2}\int \frac{dx}{x^{m-2}\sqrt{x^2 \pm a^2}}$

193. $\displaystyle\int \frac{dx}{x^{2m}\sqrt{x^2 \pm a^2}} = \sqrt{x^2 \pm a^2}\sum_{r=0}^{m-1}\frac{(m-1)!m!(2r)!2^{2m-2r-1}}{(r!)^2(2m)!(\mp a^2)^{m-r}x^{2r+1}}$

194. $\displaystyle\int \frac{dx}{x^{2m+1}\sqrt{x^2+a^2}} = \frac{(2m)!}{(m!)^2}\left[\frac{\sqrt{x^2+a^2}}{a^2}\sum_{r=1}^{m}(-1)^{m-r+1}\frac{r!(r-1)!}{2(2r)!(4a^2)^{m-r}x^{2r}} + \frac{(-1)^{m+1}}{2^{2m}a^{2m+1}}\log\frac{\sqrt{x^2+a^2}+a}{x}\right]$

195. $\displaystyle\int \frac{dx}{x^{2m+1}\sqrt{x^2-a^2}} = \frac{(2m)!}{(m!)^2}\left[\frac{\sqrt{x^2-a^2}}{a^2}\sum_{r=1}^{m}\frac{r!(r-1)!}{2(2r)!(4a^2)^{m-r}x^{2r}} + \frac{1}{2^{2m}|a|^{2m+1}}\sec^{-1}\frac{x}{a}\right]$

196. $\displaystyle\int \frac{dx}{(x-a)\sqrt{x^2-a^2}} = -\frac{\sqrt{x^2-a^2}}{a(x-a)}$

197. $\displaystyle\int \frac{dx}{(x+a)\sqrt{x^2-a^2}} = \frac{\sqrt{x^2-a^2}}{a(x+a)}$

198. $\displaystyle\int f(x,\sqrt{x^2+a^2})\,dx = a\int f(a\tan u, a\sec u)\sec^2 u\,du, \quad \left(u=\tan^{-1}\frac{x}{a}, a>0\right)$

199. $\displaystyle\int f(x,\sqrt{x^2-a^2})\,dx = a\int f(a\sec u, a\tan u)\sec u\tan u\,du, \quad \left(u=\sec^{-1}\frac{x}{a}, a>0\right)$

FORMS CONTAINING $\sqrt{a^2-x^2}$

200. $\displaystyle\int \sqrt{a^2-x^2}\,dx = \frac{1}{2}\left[x\sqrt{a^2-x^2} + a^2\sin^{-1}\frac{x}{|a|}\right]$

201. $\displaystyle\int \frac{dx}{\sqrt{a^2-x^2}} = \begin{cases}\sin^{-1}\frac{x}{|a|} \\ \text{or} \\ -\cos^{-1}\frac{x}{|a|}\end{cases}$

202. $\displaystyle\int \frac{dx}{x\sqrt{a^2-x^2}} = -\frac{1}{a}\log\left(\frac{a+\sqrt{a^2-x^2}}{x}\right)$

203. $\displaystyle\int \frac{\sqrt{a^2-x^2}}{x}\,dx = \sqrt{a^2-x^2} - a\log\left(\frac{a+\sqrt{a^2-x^2}}{x}\right)$

204. $\displaystyle\int \frac{x\,dx}{\sqrt{a^2-x^2}} = -\sqrt{a^2-x^2}$

205. $\displaystyle\int x\sqrt{a^2-x^2}\,dx = -\frac{1}{3}\sqrt{(a^2-x^2)^3}$

206. $\displaystyle\int \sqrt{(a^2-x^2)^3}\,dx = \frac{1}{4}\left[x\sqrt{(a^2-x^2)^3} + \frac{3a^2x}{2}\sqrt{a^2-x^2} + \frac{3a^4}{2}\sin^{-1}\frac{x}{|a|}\right]$

207. $\displaystyle\int \frac{dx}{\sqrt{(a^2-x^2)^3}} = \frac{x}{a^2\sqrt{a^2-x^2}}$

208. $\displaystyle\int \frac{x\,dx}{\sqrt{(a^2-x^2)^3}} = \frac{1}{\sqrt{a^2-x^2}}$

209. $\displaystyle\int x\sqrt{(a^2-x^2)^3}\,dx = -\frac{1}{5}\sqrt{(a^2-x^2)^5}$

210. $\displaystyle\int x^2\sqrt{a^2-x^2}\,dx = -\frac{x}{4}\sqrt{(a^2-x^2)^3} + \frac{a^2}{8}\left(x\sqrt{a^2-x^2} + a^2\sin^{-1}\frac{x}{|a|}\right)$

211. $\displaystyle\int x^3\sqrt{a^2-x^2}\,dx = \left(-\frac{1}{5}x^2 - \frac{2}{15}a^2\right)\sqrt{(a^2-x^2)^3}$

212. $\displaystyle\int x^2\sqrt{(a^2-x^2)^3}\,dx = -\frac{1}{6}x\sqrt{(a^2-x^2)^5} + \frac{a^2x}{24}\sqrt{(a^2-x^2)^3} + \frac{a^4x}{16}\sqrt{a^2-x^2} + \frac{a^6}{16}\sin^{-1}\frac{x}{|a|}$

213. $\displaystyle\int x^3\sqrt{(a^2-x^2)^3}\,dx = \frac{1}{7}\sqrt{(a^2-x^2)^7} - \frac{a^2}{5}\sqrt{(a^2-x^2)^5}$

214. $\displaystyle\int \frac{x^2\,dx}{\sqrt{a^2-x^2}} = -\frac{x}{2}\sqrt{a^2-x^2} + \frac{a^2}{2}\sin^{-1}\frac{x}{|a|}$

215. $\displaystyle\int \frac{dx}{x^2\sqrt{a^2-x^2}} = -\frac{\sqrt{a^2-x^2}}{a^2x}$

216. $\displaystyle\int \frac{\sqrt{a^2-x^2}}{x^2}\,dx = -\frac{\sqrt{a^2-x^2}}{x} - \sin^{-1}\frac{x}{|a|}$

217. $\displaystyle\int \frac{\sqrt{a^2-x^2}}{x^3}\,dx = -\frac{\sqrt{a^2-x^2}}{2x^2} + \frac{1}{2a}\log\frac{a+\sqrt{a^2-x^2}}{x}$

218. $\displaystyle\int \frac{\sqrt{a^2-x^2}}{x^4}\,dx = -\frac{\sqrt{(a^2-x^2)^3}}{3a^2x^3}$

219. $\displaystyle\int \frac{x^2 dx}{\sqrt{(a^2 - x^2)^3}} = \frac{x}{\sqrt{a^2 - x^2}} - \sin^{-1}\frac{x}{|a|}$

220. $\displaystyle\int \frac{x^3 dx}{\sqrt{a^2 - x^2}} = -\frac{2}{3}(a^2 - x^2)^{3/2} - x^2(a^2 - x^2)^{1/2} = -\frac{1}{3}\sqrt{a^2 - x^2}(x^2 + 2a^2)$

221. $\displaystyle\int \frac{x^3 dx}{\sqrt{(a^2 - x^2)^3}} = 2(a^2 - x^2)^{1/2} + \frac{x^2}{(a^2 - x^2)^{1/2}} = -\frac{a^2}{\sqrt{a^2 - x^2}} + \sqrt{a^2 - x^2}$

222. $\displaystyle\int \frac{dx}{x^3\sqrt{a^2 - x^2}} = -\frac{\sqrt{a^2 - x^2}}{2a^2 x^2} - \frac{1}{2a^3}\log\frac{a + \sqrt{a^2 - x^2}}{x}$

223. $\displaystyle\int \frac{dx}{x\sqrt{(a^2 - x^2)^3}} = \frac{1}{a^2\sqrt{a^2 - x^2}} - \frac{1}{a^3}\log\frac{a + \sqrt{a^2 - x^2}}{x}$

224. $\displaystyle\int \frac{dx}{x^2\sqrt{(a^2 - x^2)^3}} = \frac{1}{a^4}\left[-\frac{\sqrt{a^2 - x^2}}{x} + \frac{x}{\sqrt{a^2 - x^2}}\right]$

225. $\displaystyle\int \frac{dx}{x^3\sqrt{(a^2 - x^2)^3}} = -\frac{1}{2a^2 x^2\sqrt{a^2 - x^2}} + \frac{3}{2a^4\sqrt{a^2 - x^2}} - \frac{3}{2a^5}\log\frac{a + \sqrt{a^2 - x^2}}{x}$

226. $\displaystyle\int \frac{x^m}{\sqrt{a^2 - x^2}}dx = -\frac{x^{m-1}\sqrt{a^2 - x^2}}{m} + \frac{(m-1)a^2}{m}\int \frac{x^{m-2}}{\sqrt{a^2 - x^2}}dx$

227. $\displaystyle\int \frac{x^{2m}}{\sqrt{a^2 - x^2}}dx = \frac{(2m)!}{(m!)^2}\left[-\sqrt{a^2 - x^2}\sum_{r=1}^{m}\frac{r!(r-1)!}{2^{2m-2r+1}(2r)!}a^{2m-2r}x^{2r-1} + \frac{a^{2m}}{2^{2m}}\sin^{-1}\frac{x}{|a|}\right]$

228. $\displaystyle\int \frac{x^{2m+1}}{\sqrt{a^2 - x^2}}dx = -\sqrt{a^2 - x^2}\sum_{r=0}^{m}\frac{(2r)!(m!)^2}{(2m+1)!(r!)^2}(4a^2)^{m-r}x^{2r}$

229. $\displaystyle\int \frac{dx}{x^m\sqrt{a^2 - x^2}} = -\frac{\sqrt{a^2 - x^2}}{(m-1)a^2 x^{m-1}} + \frac{m-2}{(m-1)a^2}\int \frac{dx}{x^{m-2}\sqrt{a^2 - x^2}}$

230. $\displaystyle\int \frac{ax}{x^{2m}\sqrt{a^2 - x^2}} = -\sqrt{a^2 - x^2}\sum_{r=0}^{m-1}\frac{(m-1)!m!(2r)!2^{2m-2r-1}}{(r!)^2(2m)!a^{2m-2r}x^{2r+1}}$

231. $\displaystyle\int \frac{dx}{x^{2m+1}\sqrt{a^2 - x^2}} = \frac{(2m)!}{(m!)^2}\left[-\frac{\sqrt{a^2 - x^2}}{a^2}\sum_{r=1}^{m}\frac{r!(r-1)!}{2(2r)!(4a^2)^{m-r}x^{2r}} + \frac{1}{2^{2m}a^{2m+1}}\log\frac{a - \sqrt{a^2 - x^2}}{x}\right]$

232. $\displaystyle\int \frac{dx}{(b^2 - x^2)\sqrt{a^2 - x^2}} = \frac{1}{2b\sqrt{a^2 - b^2}}\log\frac{(b\sqrt{a^2 - x^2} + x\sqrt{a^2 - b^2})^2}{b^2 - x^2}, \quad (a^2 > b^2)$

233. $\displaystyle\int \frac{dx}{(b^2 - x^2)\sqrt{a^2 - x^2}} = \frac{1}{b\sqrt{b^2 - a^2}}\tan^{-1}\frac{x\sqrt{b^2 - a^2}}{b\sqrt{a^2 - x^2}}, \quad (b^2 > a^2)$

234. $\displaystyle\int \frac{dx}{(b^2 + x^2)\sqrt{a^2 - x^2}} = \frac{1}{b\sqrt{a^2 + b^2}}\tan^{-1}\frac{x\sqrt{a^2 + b^2}}{b\sqrt{a^2 - x^2}}$

235. $\displaystyle\int \frac{\sqrt{a^2 - x^2}}{b^2 + x^2}dx = \frac{\sqrt{a^2 + b^2}}{|b|}\sin^{-1}\frac{x\sqrt{a^2 + b^2}}{|a|\sqrt{x^2 + b^2}} - \sin^{-1}\frac{x}{|a|}$

236. $\displaystyle\int f(x, \sqrt{a^2 - x^2})\,dx = a\int f(a\sin u, a\cos u)\cos u\,du, \quad \left(u = \sin^{-1}\frac{x}{a}, a > 0\right)$

FORMS CONTAINING $\sqrt{a + bx + cx^2}$

$$X = a + bx + cx^2, \quad q = 4ac - b^2, \quad \text{and } k = \frac{4c}{q}$$

If $q = 0$, then $\sqrt{X} = \sqrt{c}\,\left|x + \frac{b}{2c}\right|$

237. $\displaystyle\int \frac{dx}{\sqrt{X}} = \begin{cases} \frac{1}{\sqrt{c}}\log(2\sqrt{cX} + 2cx + b) \\ \text{or} \\ \frac{1}{\sqrt{c}}\sinh^{-1}\frac{2cx+b}{\sqrt{q}}, \quad (c > 0) \end{cases}$

238. $\displaystyle\int \frac{dx}{\sqrt{X}} = -\frac{1}{\sqrt{-c}}\sin^{-1}\frac{2cx + b}{\sqrt{-q}}, \quad (c < 0)$

239. $\displaystyle\int \frac{dx}{X\sqrt{X}} = \frac{2(2cx + b)}{q\sqrt{X}}$

240. $\displaystyle\int \frac{dx}{X^2\sqrt{X}} = \frac{2(2cx + b)}{3q\sqrt{X}}\left(\frac{1}{X} + 2k\right)$

241. $\displaystyle \int \frac{dx}{X^n \sqrt{X}} = \begin{cases} \frac{2(2cx+b)\sqrt{X}}{(2n-1)q X^n} + \frac{2k(n-1)}{2n-1} \int \frac{dx}{X^{n-1}\sqrt{X}} \\ \quad\text{or} \\ \frac{(2cx+b)(n!)(n-1)!4^n k^{n-1}}{q[(2n)!]\sqrt{X}} \sum_{r=0}^{n-1} \frac{(2r)!}{(4kX)^r (r!)^2} \end{cases}$

242. $\displaystyle \int \sqrt{X}\,dx = \frac{(2cx+b)\sqrt{X}}{4c} + \frac{1}{2k} \int \frac{dx}{\sqrt{X}}$

243. $\displaystyle \int X\sqrt{X}\,dx = \frac{(2cx+b)\sqrt{X}}{8c}\left(X + \frac{3}{2k}\right) + \frac{3}{8k^2} \int \frac{dx}{\sqrt{X}}$

244. $\displaystyle \int X^2\sqrt{X}\,dx = \frac{(2cx+b)\sqrt{X}}{12c}\left(X^2 + \frac{5X}{4k} + \frac{15}{8k^2}\right) + \frac{5}{16k^3} \int \frac{dx}{\sqrt{X}}$

245. $\displaystyle \int X^n\sqrt{X}\,dx = \begin{cases} \frac{(2cx+b)X^n\sqrt{X}}{4(n+1)c} + \frac{2n+1}{2(n+1)k} \int X^{n-1}\sqrt{X}\,dx \\ \quad\text{or} \\ \frac{(2n+2)!}{[(n+1)!]^2(4k)^{n+1}}\left[\frac{k(2cx+b)\sqrt{X}}{c} \sum_{r=0}^{n} \frac{r!(r+1)!(4kX)^r}{(2r+2)!} + \int \frac{dx}{\sqrt{X}}\right] \end{cases}$

246. $\displaystyle \int \frac{x\,dx}{\sqrt{X}} = \frac{\sqrt{X}}{c} - \frac{b}{2c} \int \frac{dx}{\sqrt{X}}$

247. $\displaystyle \int \frac{x\,dx}{X\sqrt{X}} = -\frac{2(bx+2a)}{q\sqrt{X}}$

248. $\displaystyle \int \frac{x\,dx}{X^n\sqrt{X}} = -\frac{\sqrt{X}}{(2n-1)c X^n} - \frac{b}{2c} \int \frac{dx}{X^n\sqrt{X}}$

249. $\displaystyle \int \frac{x^2\,dx}{\sqrt{X}} = \left(\frac{x}{2c} - \frac{3b}{4c^2}\right)\sqrt{X} + \frac{3b^2 - 4ac}{8c^2} \int \frac{dx}{\sqrt{X}}$

250. $\displaystyle \int \frac{x^2\,dx}{X\sqrt{X}} = \frac{(2b^2 - 4ac)x + 2ab}{cq\sqrt{X}} + \frac{1}{c} \int \frac{dx}{\sqrt{X}}$

251. $\displaystyle \int \frac{x^2\,dx}{X^n\sqrt{X}} = \frac{(2b^2 - 4ac)x + 2ab}{(2n-1)cq\, X^{n-1}\sqrt{X}} + \frac{4ac + (2n-3)b^2}{(2n-1)cq} \int \frac{dx}{X^{n-1}\sqrt{X}}$

252. $\displaystyle \int \frac{x^3\,dx}{\sqrt{X}} = \left(\frac{x^2}{3c} - \frac{5bx}{12c^2} + \frac{5b^2}{8c^3} - \frac{2a}{3c^2}\right)\sqrt{X} + \left(\frac{3ab}{4c^2} - \frac{5b^3}{16c^3}\right) \int \frac{dx}{\sqrt{X}}$

253. $\displaystyle \int \frac{x^n\,dx}{\sqrt{X}} = \frac{1}{nc}x^{n-1}\sqrt{X} - \frac{(2n-1)b}{2nc} \int \frac{x^{n-1}\,dx}{\sqrt{X}} - \frac{(n-1)a}{nc} \int \frac{x^{n-2}\,dx}{\sqrt{X}}$

254. $\displaystyle \int x\sqrt{X}\,dx = \frac{X\sqrt{X}}{3c} - \frac{b(2cx+b)}{8c^2}\sqrt{X} - \frac{b}{4ck} \int \frac{dx}{\sqrt{X}}$

255. $\displaystyle \int xX\sqrt{X}\,dx = \frac{X^2\sqrt{X}}{5c} - \frac{b}{2c} \int X\sqrt{X}\,dx$

256. $\displaystyle \int xX^n\sqrt{X}\,dx = \frac{X^{n+1}\sqrt{X}}{(2n+3)c} - \frac{b}{2c} \int X^n\sqrt{X}\,dx$

257. $\displaystyle \int x^2\sqrt{X}\,dx = \left(x - \frac{5b}{6c}\right)\frac{X\sqrt{X}}{4c} + \frac{5b^2 - 4ac}{16c^2} \int \sqrt{X}\,dx$

258. $\displaystyle \int \frac{dx}{x\sqrt{X}} = -\frac{1}{\sqrt{a}} \log \frac{2\sqrt{aX} + bx + 2a}{x}, \quad (a > 0)$

259. $\displaystyle \int \frac{dx}{x\sqrt{X}} = \frac{1}{\sqrt{-a}} \sin^{-1}\left(\frac{bx + 2a}{|x|\sqrt{-q}}\right), \quad (a < 0)$

260. $\displaystyle \int \frac{dx}{x\sqrt{X}} = -\frac{2\sqrt{X}}{bx}, \quad (a = 0)$

261. $\displaystyle \int \frac{dx}{x^2\sqrt{X}} = -\frac{\sqrt{X}}{ax} - \frac{b}{2a} \int \frac{dx}{x\sqrt{X}}$

262. $\displaystyle \int \frac{\sqrt{X}\,dx}{x} = \sqrt{X} + \frac{b}{2} \int \frac{dx}{\sqrt{X}} + a \int \frac{dx}{x\sqrt{X}}$

263. $\displaystyle \int \frac{\sqrt{X}\,dx}{x^2} = -\frac{\sqrt{X}}{x} + \frac{b}{2} \int \frac{dx}{x\sqrt{X}} + c \int \frac{dx}{\sqrt{X}}$

FORMS INVOLVING $\sqrt{2ax - x^2}$

264. $\displaystyle \int \sqrt{2ax - x^2}\,dx = \frac{1}{2}\left[(x-a)\sqrt{2ax - x^2} + a^2 \sin^{-1}\frac{x-a}{|a|}\right]$

265. $\displaystyle \int \frac{dx}{\sqrt{2ax - x^2}} = \begin{cases} \cos^{-1} \frac{a-x}{|a|} \\ \quad\text{or} \\ \sin^{-1} \frac{x-a}{|a|} \end{cases}$

266. $\displaystyle \int x^n \sqrt{2ax - x^2}\, dx = \begin{cases} -\frac{x^{n-1}(2ax - x^2)^{3/2}}{n+2} + \frac{(2n+1)a}{n+2} \int x^{n-1} \sqrt{2ax - x^2}\, dx \\ \quad\text{or} \\ \sqrt{2ax - x^2}\left[\frac{x^{n+1}}{n+2} - \sum_{r=0}^{n} \frac{(2n+1)!(r!)^2 a^{n-r+1}}{2^{n-r}(2r+1)!(n+2)!n!} x^r \right] \\ \qquad + \frac{(2n+1)!a^{n+2}}{2^n n!(n+2)!} \sin^{-1} \frac{x-a}{|a|} \end{cases}$

267. $\displaystyle \int \frac{\sqrt{2ax - x^2}}{x^n}\, dx = \frac{(2ax - x^2)^{1/2}}{(3 - 2n)ax^n} + \frac{n-3}{(2n-3)a} \int \frac{\sqrt{2ax - x^2}}{x^{n-1}}\, dx$

268. $\displaystyle \int \frac{x^n dx}{\sqrt{2ax - x^2}} = \begin{cases} \frac{-x^{n-1}\sqrt{2ax - x^2}}{n} + \frac{a(2n-1)}{n} \int \frac{x^{n-1}}{\sqrt{2ax - x^2}}\, dx \\ \quad\text{or} \\ -\sqrt{2ax - x^2} \sum_{r=1}^{n} \frac{(2n)!r!(r-1)!a^{n-r}}{2^{n-r}(2r)!(n!)^2} x^{r-1} + \frac{(2n)!a^n}{2^n(n!)^2} \sin^{-1} \frac{x-a}{|a|} \end{cases}$

269. $\displaystyle \int \frac{dx}{x^n \sqrt{2ax - x^2}} = \begin{cases} \frac{\sqrt{2ax - x^2}}{a(1-2n)x^n} + \frac{n-1}{(2n-1)a} \int \frac{dx}{x^{n-1}\sqrt{2ax - x^2}} \\ \quad\text{or} \\ -\sqrt{2ax - x^2} \sum_{r=0}^{n-1} \frac{2^{n-r}(n-1)!n!(2r)!}{(2n)!(r!)^2 a^{n-r} x^{r+1}} \end{cases}$

270. $\displaystyle \int \frac{dx}{(2ax - x^2)^{3/2}} = \frac{x - a}{a^2 \sqrt{2ax - x^2}}$

271. $\displaystyle \int \frac{x\, dx}{(2ax - x^2)^{3/2}} = \frac{x}{a\sqrt{2ax - x^2}}$

MISCELLANEOUS ALGEBRAIC FORMS

272. $\displaystyle \int \frac{dx}{\sqrt{2ax + x^2}} = \log(x + a + \sqrt{2ax + x^2})$

273. $\displaystyle \int \sqrt{ax^2 + c}\, dx = \frac{x}{2} \sqrt{ax^2 + c} + \frac{c}{2\sqrt{a}} \log\left(x\sqrt{a} + \sqrt{ax^2 + c} \right), \quad (a > 0)$

274. $\displaystyle \int \sqrt{ax^2 + c}\, dx = \frac{x}{2} \sqrt{ax^2 + c} + \frac{c}{2\sqrt{-a}} \sin^{-1}\left(x\sqrt{-\frac{a}{c}} \right), \quad (a < 0)$

275. $\displaystyle \int \sqrt{\frac{1 + x}{1 - x}}\, dx = \sin^{-1} x - \sqrt{1 - x^2}$

276. $\displaystyle \int \frac{dx}{x\sqrt{ax^n + c}} = \begin{cases} \frac{1}{n\sqrt{c}} \log \frac{\sqrt{ax^n + c} - \sqrt{c}}{\sqrt{ax^n + c} + \sqrt{c}} \\ \quad\text{or} \\ \frac{2}{n\sqrt{c}} \log \frac{\sqrt{ax^n + c} - \sqrt{c}}{\sqrt{x^n}}, \quad (c > 0) \end{cases}$

277. $\displaystyle \int \frac{dx}{x\sqrt{ax^n + c}} = \frac{2}{n\sqrt{-c}} \sec^{-1} \sqrt{-\frac{ax^n}{c}}, \quad (c < 0)$

278. $\displaystyle \int \frac{dx}{\sqrt{ax^2 + c}} = \frac{1}{\sqrt{a}} \log(x\sqrt{a} + \sqrt{ax^2 + c}), \quad (a > 0)$

279. $\displaystyle \int \frac{dx}{\sqrt{ax^2 + c}} = \frac{1}{\sqrt{-a}} \sin^{-1}\left(x\sqrt{-\frac{a}{c}} \right), \quad (a < 0)$

280. $\displaystyle \int (ax^2 + c)^{m+1/2}\, dx = \begin{cases} \frac{x(ax^2 + c)^{m+1/2}}{2(m+1)} + \frac{(2m+1)c}{2(m+1)} \int (ax^2 + c)^{m-1/2}\, dx \\ \quad\text{or} \\ x\sqrt{ax^2 + c} \sum_{r=0}^{m} \frac{(2m+1)!(r!)^2 c^{m-r}}{2^{2m-2r+1}m!(m+1)!(2r+1)!} (ax^2 + c)^r + \frac{(2m+1)!c^{m+1}}{2^{2m+1}m!(m+1)!} \int \frac{dx}{\sqrt{ax^2 + c}} \end{cases}$

281. $\displaystyle \int x(ax^2 + c)^{m+\frac{1}{2}}\, dx = \frac{(ax^2 + c)^{m+\frac{3}{2}}}{(2m+3)a}$

282. $\displaystyle \int \frac{(ax^2 + c)^{m+1/2}}{x}\, dx = \begin{cases} \frac{(ax^2 + c)^{m+1/2}}{2m + 1} + c \int \frac{(ax^2 + c)^{m-1/2}}{x}\, dx \\ \quad\text{or} \\ \sqrt{ax^2 + c} \sum_{r=0}^{m} \frac{c^{m-r}(ax^2 + c)^r}{2r + 1} + c^{m+1} \int \frac{dx}{x\sqrt{ax^2 + c}} \end{cases}$

283. $\displaystyle \int \frac{dx}{(ax^2+c)^{m+1/2}} = \begin{cases} \dfrac{x}{(2m-1)c(ax^2+c)^{m-1/2}} + \dfrac{2m-2}{(2m-1)c} \displaystyle\int \dfrac{dx}{(ax^2+c)^{m-1/2}} \\ \qquad\qquad \text{or} \\ \dfrac{x}{\sqrt{ax^2+c}} \displaystyle\sum_{r=0}^{m-1} \dfrac{2^{2m-2r-1}(m-1)!\,m!\,(2r)!}{(2m)!(r!)^2 c^{m-r}(ax^2+c)^r} \end{cases}$

284. $\displaystyle \int \frac{dx}{x^m\sqrt{ax^2+c}} = -\frac{\sqrt{ax^2+c}}{(m-1)cx^{m-1}} - \frac{(m-2)a}{(m-1)c}\int \frac{dx}{x^{m-2}\sqrt{ax^2+c}}$

285. $\displaystyle \int \frac{1+x^2}{(1-x^2)\sqrt{1+x^4}}dx = \frac{1}{\sqrt{2}}\log\frac{x\sqrt{2}+\sqrt{1+x^4}}{1-x^2}$

286. $\displaystyle \int \frac{1-x^2}{(1+x^2)\sqrt{1+x^4}}dx = \frac{1}{\sqrt{2}}\tan^{-1}\frac{x\sqrt{2}}{\sqrt{1+x^4}}$

287. $\displaystyle \int \frac{dx}{x\sqrt{x^n+a^2}} = -\frac{2}{na}\log\frac{a+\sqrt{x^n+a^2}}{\sqrt{x^n}}$

288. $\displaystyle \int \frac{dx}{x\sqrt{x^n-a^2}} = -\frac{2}{na}\sin^{-1}\frac{a}{\sqrt{x^n}}$

289. $\displaystyle \int \sqrt{\frac{x}{a^3-x^3}}\,dx = \frac{2}{3}\sin^{-1}\left(\frac{x}{a}\right)^{3/2}$

FORMS INVOLVING TRIGONOMETRIC FUNCTIONS

290. $\displaystyle \int (\sin ax)\,dx = -\frac{1}{a}\cos ax$

291. $\displaystyle \int (\cos ax)\,dx = \frac{1}{a}\sin ax$

292. $\displaystyle \int (\tan ax)\,dx = -\frac{1}{a}\log\cos ax = \frac{1}{a}\log\sec ax$

293. $\displaystyle \int (\cot ax)\,dx = \frac{1}{a}\log\sin ax = -\frac{1}{a}\log\csc ax$

294. $\displaystyle \int (\sec ax)\,dx = \frac{1}{a}\log(\sec ax + \tan ax) = \frac{1}{a}\log\tan\left(\frac{\pi}{4}+\frac{ax}{2}\right)$

295. $\displaystyle \int (\csc ax)\,dx = \frac{1}{a}\log(\csc ax - \cot ax) = \frac{1}{a}\log\tan\frac{ax}{2}$

296. $\displaystyle \int (\sin^2 ax)\,dx = -\frac{1}{2a}\cos ax \sin ax + \frac{1}{2}x = \frac{1}{2}x - \frac{1}{4a}\sin 2ax$

297. $\displaystyle \int (\sin^3 ax)\,dx = -\frac{1}{3a}(\cos ax)(\sin^2 ax + 2)$

298. $\displaystyle \int (\sin^4 ax)\,dx = \frac{3x}{8} - \frac{\sin 2ax}{4a} + \frac{\sin 4ax}{32a}$

299. $\displaystyle \int (\sin^n ax)\,dx = -\frac{\sin^{n-1} ax \cos ax}{na} + \frac{n-1}{n}\int (\sin^{n-2} ax)\,dx$

300. $\displaystyle \int (\sin^{2m} ax)\,dx = -\frac{\cos ax}{a}\sum_{r=0}^{m-1}\frac{(2m)!(r!)^2}{2^{2m-2r}(2r+1)!(m!)^2}\sin^{2r+1} ax + \frac{(2m)!}{2^{2m}(m!)^2}x$

301. $\displaystyle \int (\sin^{2m+1} ax)\,dx = -\frac{\cos ax}{a}\sum_{r=0}^{m}\frac{2^{2m-2r}(m!)^2(2r)!}{(2m+1)!(r!)^2}\sin^{2r} ax$

302. $\displaystyle \int (\cos^2 ax)\,dx = \frac{1}{2a}\sin ax \cos ax + \frac{1}{2}x = \frac{1}{2}x + \frac{1}{4a}\sin 2ax$

303. $\displaystyle \int (\cos^3 ax)\,dx = \frac{1}{3a}(\sin ax)(\cos^2 ax + 2)$

304. $\displaystyle \int (\cos^4 ax)\,dx = \frac{3x}{8} + \frac{\sin 2ax}{4a} + \frac{\sin 4ax}{32a}$

305. $\displaystyle \int (\cos^n ax)\,dx = \frac{1}{na}\cos^{n-1} ax \sin ax + \frac{n-1}{n}\int (\cos^{n-2} ax)\,dx$

306. $\displaystyle \int (\cos^{2m} ax)\,dx = \frac{\sin ax}{a}\sum_{r=0}^{m-1}\frac{(2m)!(r!)^2}{2^{2m-2r}(2r+1)!(m!)^2}\cos^{2r+1} ax + \frac{(2m)!}{2^{2m}(m!)^2}x$

307. $\displaystyle \int (\cos^{2m+1} ax)\,dx = \frac{\sin ax}{a}\sum_{r=0}^{m}\frac{2^{2m-2r}(m!)^2(2r)!}{(2m+1)!(r!)^2}\cos^{2r} ax$

308. $\displaystyle\int \frac{dx}{\sin^2 ax} = \int (\csc^2 ax)\,dx = -\frac{1}{a}\cot ax$

309. $\displaystyle\int \frac{dx}{\sin^m ax} = \int (\csc^m ax)\,dx = -\frac{1}{(m-1)a}\cdot\frac{\cos ax}{\sin^{m-1} ax} + \frac{m-2}{m-1}\int \frac{dx}{\sin^{m-2} ax}$

310. $\displaystyle\int \frac{dx}{\sin^{2m} ax} = \int (\csc^{2m} ax)\,dx = -\frac{1}{a}\cos ax\sum_{r=0}^{m-1}\frac{2^{2m-2r-1}(m-1)!\,m!\,(2r)!}{(2m)!\,(r!)^2\sin^{2r+1} ax}$

311. $\displaystyle\int \frac{dx}{\sin^{2m+1} ax} = \int (\csc^{2m+1} ax)\,dx = -\frac{1}{a}\cos ax\sum_{r=0}^{m-1}\frac{(2m)!\,(r!)^2}{2^{2m-2r}(m!)^2(2r+1)!\sin^{2r+2} ax} + \frac{1}{a}\cdot\frac{(2m)!}{2^{2m}(m!)^2}\log\tan\frac{ax}{2}$

312. $\displaystyle\int \frac{dx}{\cos^2 ax} = \int (\sec^2 ax)\,dx = \frac{1}{a}\tan ax$

313. $\displaystyle\int \frac{dx}{\cos^n ax} = \int (\sec^n ax)\,dx = \frac{1}{(n-1)a}\cdot\frac{\sin ax}{\cos^{n-1} ax} + \frac{n-2}{n-1}\int \frac{dx}{\cos^{n-2} ax}$

314. $\displaystyle\int \frac{dx}{\cos^{2m} ax} = \int (\sec^{2m} ax)\,dx = \frac{1}{a}\sin ax\sum_{r=0}^{m-1}\frac{2^{2m-2r-1}(m-1)!\,m!\,(2r)!}{(2m)!\,(r!)^2\cos^{2r+1} ax}$

315. $\displaystyle\int \frac{dx}{\cos^{2m+1} ax} = \int (\sec^{2m+1} ax)\,dx = \frac{1}{a}\sin ax\sum_{r=0}^{m-1}\frac{(2m)!\,(r!)^2}{2^{2m-2r}(m!)^2(2r+1)!\cos^{2r+2} ax} + \frac{1}{a}\cdot\frac{(2m)!}{2^{2m}(m!)^2}\log(\sec ax + \tan ax)$

316. $\displaystyle\int (\sin mx)(\sin nx)\,dx = \frac{\sin(m-n)x}{2(m-n)} - \frac{\sin(m+n)x}{2(m+n)}, \quad (m^2 \neq n^2)$

317. $\displaystyle\int (\cos mx)(\cos nx)\,dx = \frac{\sin(m-n)x}{2(m-n)} + \frac{\sin(m+n)x}{2(m+n)}, \quad (m^2 \neq n^2)$

318. $\displaystyle\int (\sin ax)(\cos ax)\,dx = \frac{1}{2a}\sin^2 ax$

319. $\displaystyle\int (\sin mx)(\cos nx)\,dx = -\frac{\cos(m-n)x}{2(m-n)} - \frac{\cos(m+n)x}{2(m+n)}, \quad (m^2 \neq n^2)$

320. $\displaystyle\int (\sin^2 ax)(\cos^2 ax)\,dx = -\frac{1}{32a}\sin 4ax + \frac{x}{8}$

321. $\displaystyle\int (\sin ax)(\cos^m ax)\,dx = -\frac{\cos^{m+1} ax}{(m+1)a}$

322. $\displaystyle\int (\sin^m ax)(\cos ax)\,dx = \frac{\sin^{m+1} ax}{(m+1)a}$

323. $\displaystyle\int (\cos^m ax)(\sin^n ax)\,dx = \begin{cases} \frac{\cos^{m-1} ax\,\sin^{n+1} ax}{(m+n)a} + \frac{m-1}{m+n}\int (\cos^{m-2} ax)(\sin^n ax)\,dx \\ \text{or} \\ -\frac{\sin^{n-1} ax\,\cos^{m+1} ax}{(m+n)a} + \frac{n-1}{m+n}\int (\cos^m ax)(\sin^{n-2} ax)\,dx \end{cases}$

324. $\displaystyle\int \frac{\cos^m ax}{\sin^n ax}\,dx = \begin{cases} -\frac{\cos^{m+1} ax}{(n-1)a\,\sin^{n-1} ax} - \frac{m-n+2}{n-1}\int \frac{\cos^m ax}{\sin^{n-2} ax}\,dx \\ \text{or} \\ \frac{\cos^{m-1} ax}{a(m-n)\sin^{n-1} ax} + \frac{m-1}{m-n}\int \frac{\cos^{m-2} ax}{\sin^n ax}\,dx \end{cases}$

325. $\displaystyle\int \frac{\sin^m ax}{\cos^n ax}\,dx = \begin{cases} \frac{\sin^{m+1} ax}{a(n-1)\cos^{n-1} ax} - \frac{m-n+2}{n-1}\int \frac{\sin^m ax}{\cos^{n-2} ax}\,dx \\ \text{or} \\ -\frac{\sin^{m-1} ax}{a(m-n)\cos^{n-1} ax} + \frac{m-1}{m-n}\int \frac{\sin^{m-2} ax}{\cos^n ax}\,dx \end{cases}$

326. $\displaystyle\int \frac{\sin ax}{\cos^2 ax}\,dx = \frac{1}{a\cos ax} = \frac{\sec ax}{a}$

327. $\displaystyle\int \frac{\sin^2 ax}{\cos ax}\,dx = -\frac{1}{a}\sin ax + \frac{1}{a}\log\tan\left(\frac{\pi}{4} + \frac{ax}{2}\right)$

328. $\displaystyle\int \frac{\cos ax}{\sin^2 ax}\,dx = -\frac{1}{a\sin ax} = -\frac{\csc ax}{a}$

329. $\displaystyle\int \frac{dx}{(\sin ax)(\cos ax)} = \frac{1}{a}\log\tan ax$

330. $\displaystyle\int \frac{dx}{(\sin ax)(\cos^2 ax)} = \frac{1}{a}\left(\sec ax + \log\tan\frac{ax}{2}\right)$

331. $\displaystyle\int \frac{dx}{(\sin ax)(\cos^n ax)} = \frac{1}{a(n-1)\cos^{n-1} ax} + \int \frac{dx}{(\sin ax)(\cos^{n-2} ax)}$

332. $\displaystyle\int \frac{dx}{(\sin^2 ax)(\cos ax)} = -\frac{1}{a}\csc ax + \frac{1}{a}\log\tan\left(\frac{\pi}{4}+\frac{ax}{2}\right)$

333. $\displaystyle\int \frac{dx}{(\sin^2 ax)(\cos^2 ax)} = -\frac{2}{a}\cot 2ax$

334. $\displaystyle\int \frac{dx}{\sin^m ax \cos^n ax} = \begin{cases} \dfrac{-\frac{1}{a(m-1)(\sin^{m-1}ax)(\cos^{n-1}ax)}}{} \\ +\frac{m+n-2}{m-1}\displaystyle\int \frac{dx}{(\sin^{m-2}ax)(\cos^n ax)} \\ \text{or} \\ \frac{1}{a(n-1)\sin^{m-1}ax\cos^{n-1}ax} + \frac{m+n-2}{n-1}\displaystyle\int \frac{dx}{\sin^m ax\cos^{n-2}ax} \end{cases}$

335. $\displaystyle\int \sin(a+bx)\,dx = -\frac{1}{b}\cos(a+bx)$

336. $\displaystyle\int \cos(a+bx)\,dx = \frac{1}{b}\sin(a+bx)$

337. $\displaystyle\int \frac{dx}{1\pm\sin ax} = \mp\frac{1}{a}\tan\left(\frac{\pi}{4}\mp\frac{ax}{2}\right)$

338. $\displaystyle\int \frac{dx}{1+\cos ax} = \frac{1}{a}\tan\frac{ax}{2}$

339. $\displaystyle\int \frac{dx}{1-\cos ax} = -\frac{1}{a}\cot\frac{ax}{2}$

340. $\displaystyle\int \frac{dx}{a+b\sin x} = \begin{cases} \frac{2}{\sqrt{a^2-b^2}}\tan^{-1}\frac{a\tan\frac{x}{2}+b}{\sqrt{a^2-b^2}} \\ \text{or} \\ \frac{1}{\sqrt{b^2-a^2}}\log\frac{a\tan\frac{x}{2}+b-\sqrt{b^2-a^2}}{a\tan\frac{x}{2}+b+\sqrt{b^2-a^2}} \end{cases}$

341. $\displaystyle\int \frac{dx}{a+b\cos x} = \begin{cases} \frac{2}{\sqrt{a^2-b^2}}\tan^{-1}\frac{\sqrt{a^2-b^2}\tan\frac{x}{2}}{a+b} \\ \text{or} \\ \frac{1}{\sqrt{b^2-a^2}}\log\left(\frac{\sqrt{b^2-a^2}\tan\frac{x}{2}+a+b}{\sqrt{b^2-a^2}\tan\frac{x}{2}-a-b}\right) \end{cases}$

342. $\displaystyle\int \frac{dx}{a+b\sin x+c\cos x}$

$= \begin{cases} \frac{1}{\sqrt{b^2+c^2-a^2}}\log\frac{b-\sqrt{b^2+c^2-a^2}+(a-c)\tan\frac{x}{2}}{b+\sqrt{b^2+c^2-a^2}+(a-c)\tan\frac{x}{2}}, & \text{if } a^2 < b^2+c^2,\, a\neq c \\ \text{or} \\ \frac{2}{\sqrt{a^2-b^2-c^2}}\tan^{-1}\frac{b+(a-c)\tan\frac{x}{2}}{\sqrt{a^2-b^2-c^2}}, & \text{if } a^2 > b^2+c^2 \\ \text{or} \\ \frac{1}{a}\left[\frac{a-(b+c)\cos x-(b-c)\sin x}{a-(b-c)\cos x+(b+c)\sin x}\right], & \text{if } a^2 = b^2+c^2,\, a\neq c. \end{cases}$

343. $\displaystyle\int \frac{\sin^2 x\,dx}{a+b\cos^2 x} = \frac{1}{b}\sqrt{\frac{a+b}{a}}\tan^{-1}\left(\sqrt{\frac{a}{a+b}}\tan x\right) - \frac{x}{b}, \quad (ab>0, \text{ or } |a| > |b|)$

344. $\displaystyle\int \frac{dx}{a^2\cos^2 x+b^2\sin^2 x} = \frac{1}{ab}\tan^{-1}\left(\frac{b\tan x}{a}\right)$

345. $\displaystyle\int \frac{\cos^2 cx}{a^2+b^2\sin^2 cx}\,dx = \frac{\sqrt{a^2+b^2}}{ab^2 c}\tan^{-1}\frac{\sqrt{a^2+b^2}\tan cx}{a} - \frac{x}{b^2}$

346. $\displaystyle\int \frac{\sin cx\cos cx}{a\cos^2 cx+b\sin^2 cx}\,dx = \frac{1}{2c(b-a)}\log(a\cos^2 cx+b\sin^2 cx)$

347. $\displaystyle\int \frac{\cos cx}{a\cos cx+b\sin cx}\,dx = \int \frac{dx}{a+b\tan cx}$

$= \frac{1}{c(a^2+b^2)}[acx+b\log(a\cos cx+b\sin cx)]$

348. $\displaystyle\int \frac{\sin cx}{a\sin cx+b\cos cx}\,dx = \int \frac{dx}{a+b\cot cx} = \frac{1}{c(a^2+b^2)}[acx-b\log(a\sin cx+b\cos cx)]$

349. $\displaystyle\int \frac{dx}{a\cos^2 x+2b\cos x\sin x+c\sin^2 x} = \begin{cases} \frac{1}{2\sqrt{b^2-ac}}\log\frac{c\tan x+b-\sqrt{b^2-ac}}{c\tan x+b+\sqrt{b^2-ac}}, & (b^2>ac) \\ \text{or} \\ \frac{1}{\sqrt{ac-b^2}}\tan^{-1}\frac{c\tan x+b}{\sqrt{ac-b^2}}, & (b^2<ac) \\ \text{or} \\ -\frac{1}{c\tan x+b}, & (b^2=ac) \end{cases}$

350. $\int \dfrac{\sin ax}{1 \pm \sin ax}\,dx = \pm x + \dfrac{1}{a}\tan\left(\dfrac{\pi}{4} \mp \dfrac{ax}{2}\right)$

351. $\int \dfrac{dx}{(\sin ax)\,(1 \pm \sin ax)} = \dfrac{1}{a}\tan\left(\dfrac{\pi}{4} \mp \dfrac{ax}{2}\right) + \dfrac{1}{a}\log\tan\dfrac{ax}{2}$

352. $\int \dfrac{dx}{(1 + \sin ax)^2} = -\dfrac{1}{2a}\tan\left(\dfrac{\pi}{4} - \dfrac{ax}{2}\right) - \dfrac{1}{6a}\tan^3\left(\dfrac{\pi}{4} - \dfrac{ax}{2}\right)$

353. $\int \dfrac{dx}{(1 - \sin ax)^2} = \dfrac{1}{2a}\cot\left(\dfrac{\pi}{4} - \dfrac{ax}{2}\right) + \dfrac{1}{6a}\cot^3\left(\dfrac{\pi}{4} - \dfrac{ax}{2}\right)$

354. $\int \dfrac{\sin ax}{(1 + \sin ax)^2}\,dx = -\dfrac{1}{2a}\tan\left(\dfrac{\pi}{4} - \dfrac{ax}{2}\right) + \dfrac{1}{6a}\tan^3\left(\dfrac{\pi}{4} - \dfrac{ax}{2}\right)$

355. $\int \dfrac{\sin ax}{(1 - \sin ax)^2}\,dx = -\dfrac{1}{2a}\cot\left(\dfrac{\pi}{4} - \dfrac{ax}{2}\right) + \dfrac{1}{6a}\cot^3\left(\dfrac{\pi}{4} - \dfrac{ax}{2}\right)$

356. $\int \dfrac{\sin x\,dx}{a + b\sin x} = \dfrac{x}{b} - \dfrac{a}{b}\int \dfrac{dx}{a + b\sin x}$

357. $\int \dfrac{dx}{(\sin x)\,(a + b\sin x)} = \dfrac{1}{a}\log\tan\dfrac{x}{2} - \dfrac{b}{a}\int \dfrac{dx}{a + b\sin x}$

358. $\int \dfrac{dx}{(a + b\sin x)^2} = \dfrac{b\cos x}{(a^2 - b^2)\,(a + b\sin x)} + \dfrac{a}{a^2 - b^2}\int \dfrac{dx}{a + b\sin x}$

359. $\int \dfrac{\sin x\,dx}{(a + b\sin x)^2} = \dfrac{a\cos x}{(b^2 - a^2)(a + b\sin x)} + \dfrac{b}{b^2 - a^2}\int \dfrac{dx}{a + b\sin x}$

360. $\int \dfrac{dx}{a^2 + b^2\sin^2 cx} = \dfrac{1}{ac\sqrt{a^2 + b^2}}\tan^{-1}\dfrac{\sqrt{a^2 + b^2}\,\tan cx}{a}$

361. $\int \dfrac{dx}{a^2 - b^2\sin^2 cx} = \begin{cases} \dfrac{1}{ac\sqrt{a^2-b^2}}\tan^{-1}\dfrac{\sqrt{a^2-b^2}\,\tan cx}{a}, & (a^2 > b^2) \\ \text{or} \\ \dfrac{1}{2ac\sqrt{b^2-a^2}}\log\dfrac{\sqrt{b^2-a^2}\,\tan cx+a}{\sqrt{b^2-a^2}\,\tan cx-a}, & (a^2 < b^2) \end{cases}$

362. $\int \dfrac{\cos ax}{1 + \cos ax}\,dx = x - \dfrac{1}{a}\tan\dfrac{ax}{2}$

363. $\int \dfrac{\cos ax}{1 - \cos ax}\,dx = -x - \dfrac{1}{a}\cot\dfrac{ax}{2}$

364. $\int \dfrac{dx}{(\cos ax)(1 + \cos ax)} = \dfrac{1}{a}\log\tan\left(\dfrac{\pi}{4} + \dfrac{ax}{2}\right) - \dfrac{1}{a}\tan\dfrac{ax}{2}$

365. $\int \dfrac{dx}{(\cos ax)(1 - \cos ax)} = \dfrac{1}{a}\log\tan\left(\dfrac{\pi}{4} + \dfrac{ax}{2}\right) - \dfrac{1}{a}\cot\dfrac{ax}{2}$

366. $\int \dfrac{dx}{(1 + \cos ax)^2} = \dfrac{1}{2a}\tan\dfrac{ax}{2} + \dfrac{1}{6a}\tan^3\dfrac{ax}{2}$

367. $\int \dfrac{dx}{(1 - \cos ax)^2} = -\dfrac{1}{2a}\cot\dfrac{ax}{2} - \dfrac{1}{6a}\cot^3\dfrac{ax}{2}$

368. $\int \dfrac{\cos ax}{(1 + \cos ax)^2}\,dx = \dfrac{1}{2a}\tan\dfrac{ax}{2} - \dfrac{1}{6a}\tan^3\dfrac{ax}{2}$

369. $\int \dfrac{\cos ax}{(1 - \cos ax)^2}\,dx = \dfrac{1}{2a}\cot\dfrac{ax}{2} - \dfrac{1}{6a}\cot^3\dfrac{ax}{2}$

370. $\int \dfrac{\cos x\,dx}{a + b\cos x} = \dfrac{x}{b} - \dfrac{a}{b}\int \dfrac{dx}{a + b\cos x}$

371. $\int \dfrac{dx}{(\cos x)(a + b\cos x)} = \dfrac{1}{a}\log\tan\left(\dfrac{x}{2} + \dfrac{\pi}{4}\right) - \dfrac{b}{a}\int \dfrac{dx}{a + b\cos x}$

372. $\int \dfrac{dx}{(a + b\cos x)^2} = \dfrac{b\sin x}{(b^2 - a^2)(a + b\cos x)} - \dfrac{a}{b^2 - a^2}\int \dfrac{dx}{a + b\cos x}$

373. $\int \dfrac{\cos x}{(a + b\cos x)^2}\,dx = \dfrac{a\sin x}{(a^2 - b^2)(a + b\cos x)} - \dfrac{b}{a^2 - b^2}\int \dfrac{dx}{a + b\cos x}$

374. $\int \dfrac{dx}{a^2 + b^2 - 2ab\cos cx} = \dfrac{2}{c(a^2 - b^2)}\tan^{-1}\left(\dfrac{a + b}{a - b}\tan\dfrac{cx}{2}\right)$

375. $\int \dfrac{dx}{a^2 + b^2\cos^2 cx} = \dfrac{1}{ac\sqrt{a^2 + b^2}}\tan^{-1}\dfrac{a\tan cx}{\sqrt{a^2 + b^2}}$

376. $\int \dfrac{dx}{a^2 - b^2\cos^2 cx} = \begin{cases} \dfrac{1}{ac\sqrt{a^2-b^2}}\tan^{-1}\dfrac{a\tan cx}{\sqrt{a^2-b^2}}, & (a^2 > b^2) \\ \text{or} \\ \dfrac{1}{2ac\sqrt{b^2-a^2}}\log\dfrac{a\tan cx-\sqrt{b^2-a^2}}{a\tan cx+\sqrt{b^2-a^2}}, & (b^2 > a^2) \end{cases}$

377. $\displaystyle\int \frac{\sin ax}{1 \pm \cos ax}\,dx = \mp\frac{1}{a}\log(1 \pm \cos ax)$

378. $\displaystyle\int \frac{\cos ax}{1 \pm \sin ax}\,dx = \pm\frac{1}{a}\log(1 \pm \sin ax)$

379. $\displaystyle\int \frac{dx}{(\sin ax)(1 \pm \cos ax)} = \pm\frac{1}{2a(1 \pm \cos ax)} + \frac{1}{2a}\log\tan\frac{ax}{2}$

380. $\displaystyle\int \frac{dx}{(\cos ax)(1 \pm \sin ax)} = \mp\frac{1}{2a(1 \pm \sin ax)} + \frac{1}{2a}\log\tan\left(\frac{\pi}{4} + \frac{ax}{2}\right)$

381. $\displaystyle\int \frac{\sin ax}{(\cos ax)(1 \pm \cos ax)}\,dx = \frac{1}{a}\log(\sec ax \pm 1)$

382. $\displaystyle\int \frac{\cos ax}{(\sin ax)(1 \pm \sin ax)}\,dx = -\frac{1}{a}\log(\csc ax \pm 1)$

383. $\displaystyle\int \frac{\sin ax}{(\cos ax)(1 \pm \sin ax)}\,dx = \frac{1}{2a(1 \pm \sin ax)} \pm \frac{1}{2a}\log\tan\left(\frac{\pi}{4} + \frac{ax}{2}\right)$

384. $\displaystyle\int \frac{\cos ax}{(\sin ax)(1 \pm \cos ax)}\,dx = -\frac{1}{2a(1 \pm \cos ax)} \pm \frac{1}{2a}\log\tan\frac{ax}{2}$

385. $\displaystyle\int \frac{dx}{\sin ax \pm \cos ax} = \frac{1}{a\sqrt{2}}\log\tan\left(\frac{ax}{2} \pm \frac{\pi}{8}\right)$

386. $\displaystyle\int \frac{dx}{(\sin ax \pm \cos ax)^2} = \frac{1}{2a}\tan\left(ax \mp \frac{\pi}{4}\right)$

387. $\displaystyle\int \frac{dx}{1 + \cos ax \pm \sin ax} = \pm\frac{1}{a}\log\left(1 \pm \tan\frac{ax}{2}\right)$

388. $\displaystyle\int \frac{dx}{a^2 \cos^2 cx - b^2 \sin^2 cx} = \frac{1}{2abc}\log\frac{b\tan cx + a}{b\tan cx - a}$

389. $\displaystyle\int x(\sin ax)\,dx = \frac{1}{a^2}\sin ax - \frac{x}{a}\cos ax$

390. $\displaystyle\int x^2(\sin ax)\,dx = \frac{2x}{a^2}\sin ax - \frac{a^2x^2 - 2}{a^3}\cos ax$

391. $\displaystyle\int x^3(\sin ax)\,dx = \frac{3a^2x^2 - 6}{a^4}\sin ax - \frac{a^2x^3 - 6x}{a^3}\cos ax$

392. $\displaystyle\int x^m \sin ax\,dx = \begin{cases} -\frac{1}{a}x^m \cos ax + \frac{m}{a}\int x^{m-1}\cos ax\,dx \\ \text{or} \\ \cos ax \sum_{r=0}^{\left[\frac{m}{2}\right]} (-1)^{r+1}\frac{m!}{(m-2r)!} \cdot \frac{x^{m-2r}}{a^{2r+1}} \\ \quad + \sin ax \sum_{r=0}^{\left[\frac{m-1}{2}\right]} (-1)^r \frac{m!}{(m-2r-1)!} \cdot \frac{x^{m-2r-1}}{a^{2r+2}} \end{cases}$

Note: $[s]$ means greatest integer $\leq s$; Thus $[3.5]$ means 3; $[5] = 5$, $\left[\frac{1}{2}\right] = 0$.

393. $\displaystyle\int x(\cos ax)\,dx = \frac{1}{a^2}\cos ax + \frac{x}{a}\sin ax$

394. $\displaystyle\int x^2(\cos ax)\,dx = \frac{2x\cos ax}{a^2} + \frac{a^2x^2 - 2}{a^3}\sin ax$

395. $\displaystyle\int x^3(\cos ax)\,dx = \frac{3a^2x^2 - 6}{a^4}\cos ax + \frac{a^2x^3 - 6x}{a^3}\sin ax$

396. $\displaystyle\int x^m(\cos ax)\,dx = \begin{cases} \frac{x^m \sin ax}{a} - \frac{m}{a}\int x^{m-1}\sin ax\,dx \\ \text{or} \\ \sin ax \sum_{r=0}^{|m/2|} (-1)^r \frac{m!}{(m-2r)!} \cdot \frac{x^{m-2r}}{a^{2r+1}} \\ \quad + \cos ax \sum_{r=0}^{|(m-1)/2|} (-1)^r \frac{m!}{(m-2r-1)!} \cdot \frac{x^{m-2r-1}}{a^{2r+2}} \end{cases}$

Note: $[s]$ means greatest integer $\leq s$; Thus $[3.5]$ means 3; $[5] = 5$, $\left[\frac{1}{2}\right] = 0$.

397. $\displaystyle\int \frac{\sin ax}{x}\,dx = \sum_{n=0}^{r} (-1)^n \frac{(ax)^{2n+1}}{(2n+1)(2n+1)!}$

398. $\displaystyle\int \frac{\cos ax}{x}\,dx = \log x + \sum_{n=1}^{r} (-1)^n \frac{(ax)^{2n}}{2n(2n)!}$

399. $\displaystyle\int x(\sin^2 ax)\,dx = \frac{x^2}{4} - \frac{x\sin 2ax}{4a} - \frac{\cos 2ax}{8a^2}$

400. $\displaystyle\int x^2(\sin^2 ax)\,dx = \frac{x^3}{6} - \left(\frac{x^2}{4a} - \frac{1}{8a^3}\right)\sin 2ax - \frac{x\cos 2ax}{4a^2}$

401. $\displaystyle\int x(\sin^3 ax)\,dx = \frac{x\cos 3ax}{12a} - \frac{\sin 3ax}{36a^2} - \frac{3x\cos ax}{4a} + \frac{3\sin ax}{4a^2}$

402. $\displaystyle\int x(\cos^2 ax)\,dx = \frac{x^2}{4} + \frac{x\sin 2ax}{4a} + \frac{\cos 2ax}{8a^2}$

403. $\displaystyle\int x^2(\cos^2 ax)\,dx = \frac{x^3}{6} + \left(\frac{x^2}{4a} - \frac{1}{8a^3}\right)\sin 2ax + \frac{x\cos 2ax}{4a^2}$

404. $\displaystyle\int x(\cos^3 ax)\,dx = \frac{x\sin 3ax}{12a} + \frac{\cos 3ax}{36a^2} + \frac{3x\sin ax}{4a} + \frac{3\cos ax}{4a^2}$

405. $\displaystyle\int \frac{\sin ax}{x^m}\,dx = -\frac{\sin ax}{(m-1)x^{m-1}} + \frac{a}{m-1}\int \frac{\cos ax}{x^{m-1}}\,dx$

406. $\displaystyle\int \frac{\cos ax}{x^m}\,dx = -\frac{\cos ax}{(m-1)x^{m-1}} - \frac{a}{m-1}\int \frac{\sin ax}{x^{m-1}}\,dx$

407. $\displaystyle\int \frac{x}{1 \pm \sin ax}\,dx = \mp\frac{x\cos ax}{a(1 \pm \sin ax)} + \frac{1}{a^2}\log(1 \pm \sin ax)$

408. $\displaystyle\int \frac{x}{1 + \cos ax}\,dx = \frac{x}{a}\tan\frac{ax}{2} + \frac{2}{a^2}\log\cos\frac{ax}{2}$

409. $\displaystyle\int \frac{x}{1 - \cos ax}\,dx = -\frac{x}{a}\cot\frac{ax}{2} + \frac{2}{a^2}\log\sin\frac{ax}{2}$

410. $\displaystyle\int \frac{x + \sin x}{1 + \cos x}\,dx = x\tan\frac{x}{2}$

411. $\displaystyle\int \frac{x - \sin x}{1 - \cos x}\,dx = -x\cot\frac{x}{2}$

412. $\displaystyle\int \sqrt{1 - \cos ax}\,dx = -\frac{2\sin ax}{a\sqrt{1 - \cos ax}} = -\frac{2\sqrt{2}}{a}\cos\left(\frac{ax}{2}\right)$

413. $\displaystyle\int \sqrt{1 + \cos ax}\,dx = \frac{2\sin ax}{a\sqrt{1 + \cos ax}} = \frac{2\sqrt{2}}{a}\sin\left(\frac{ax}{2}\right)$

414. $\displaystyle\int \sqrt{1 + \sin x}\,dx = \pm 2\left(\sin\frac{x}{2} - \cos\frac{x}{2}\right),$
[use $+$ if $(8k - 1)\frac{\pi}{2} < x \le (8k + 3)\frac{\pi}{2}$, otherwise $-$; k an integer]

415. $\displaystyle\int \sqrt{1 - \sin x}\,dx = \pm 2\left(\sin\frac{x}{2} + \cos\frac{x}{2}\right),$
[use $+$ if $(8k - 3)\frac{\pi}{2} < x \le (8k + 1)\frac{\pi}{2}$, otherwise $-$; k an integer]

416. $\displaystyle\int \frac{dx}{\sqrt{1 - \cos x}} = \pm\sqrt{2}\log\tan\frac{x}{4},$
[use $+$ if $4k\pi < x < (4k + 2)\pi$, otherwise $-$; k an integer]

417. $\displaystyle\int \frac{dx}{\sqrt{1 + \cos x}} = \pm\sqrt{2}\log\tan\left(\frac{x + \pi}{4}\right),$
[use $+$ if $(4k - 1)\pi < x < (4k + 1)\pi$, otherwise $-$; k an integer]

418. $\displaystyle\int \frac{dx}{\sqrt{1 - \sin x}} = \pm\sqrt{2}\log\tan\left(\frac{x}{4} - \frac{\pi}{8}\right),$
[use $+$ if $(8k + 1)\frac{\pi}{2} < x < (8k + 5)\frac{\pi}{2}$, otherwise $-$; k an integer]

419. $\displaystyle\int \frac{dx}{\sqrt{1 + \sin x}} = \pm\sqrt{2}\log\tan\left(\frac{x}{4} + \frac{\pi}{8}\right),$
[use $+$ if $(8k - 1)\frac{\pi}{2} < x < (8k + 3)\frac{\pi}{2}$, otherwise $-$; k an integer]

420. $\displaystyle\int (\tan^2 ax)\,dx = \frac{1}{a}\tan ax - x$

421. $\displaystyle\int (\tan^3 ax)\,dx = \frac{1}{2a}\tan^2 ax + \frac{1}{a}\log\cos ax$

422. $\displaystyle\int (\tan^4 ax)\,dx = \frac{\tan^3 ax}{3a} - \frac{1}{a}\tan ax + x$

423. $\displaystyle\int (\tan^n ax)\,dx = \frac{\tan^{n-1} ax}{a(n-1)} - \int (\tan^{n-2} ax)\,dx$

424. $\displaystyle\int (\cot^2 ax)\,dx = -\frac{1}{a}\cot ax - x$

425. $\displaystyle\int (\cot^3 ax)\,dx = -\frac{1}{2a}\cot^2 ax - \frac{1}{a}\log\sin ax$

426. $\displaystyle\int (\cot^4 ax)\,dx = -\frac{1}{3a}\cot^3 ax + \frac{1}{a}\cot ax + x$

427. $\displaystyle\int (\cot^n ax)\,dx = -\frac{\cot^{n-1} ax}{a(n-1)} - \int (\cot^{n-2} ax)\,dx$

428. $\displaystyle\int \frac{x}{\sin^2 ax}\,dx = \int x(\csc^2 ax)\,dx = -\frac{x\cot ax}{a} + \frac{1}{a^2}\log\sin ax$

429. $\displaystyle\int \frac{x}{\sin^n ax}\,dx = \int x(\csc^n ax)\,dx = -\frac{x\cos ax}{a(n-1)\sin^{n-1} ax} - \frac{1}{a^2(n-1)(n-2)\sin^{n-2} ax} + \frac{(n-2)}{(n-1)}\int \frac{x}{\sin^{n-2} ax}\,dx$

430. $\displaystyle\int \frac{x}{\cos^2 ax}\,dx = \int x(\sec^2 ax)\,dx = \frac{1}{a}x\tan ax + \frac{1}{a^2}\log\cos ax$

431. $\displaystyle\int \frac{x}{\cos^n ax}\,dx = \int x(\sec^n ax)\,dx = \frac{x\sin ax}{a(n-1)\cos^{n-1} ax} - \frac{1}{a^2(n-1)(n-2)\cos^{n-2} ax} + \frac{n-2}{n-1}\int \frac{x}{\cos^{n-2} ax}\,dx$

432. $\displaystyle\int \frac{\sin ax}{\sqrt{1+b^2\sin^2 ax}}\,dx = -\frac{1}{ab}\sin^{-1}\frac{b\cos ax}{\sqrt{1+b^2}}$

433. $\displaystyle\int \frac{\sin ax}{\sqrt{1-b^2\sin^2 ax}}\,dx = -\frac{1}{ab}\log(b\cos ax + \sqrt{1-b^2\sin^2 ax})$

434. $\displaystyle\int (\sin ax)\sqrt{1+b^2\sin^2 ax}\,dx = -\frac{\cos ax}{2a}\sqrt{1+b^2\sin^2 ax} - \frac{1+b^2}{2ab}\sin^{-1}\frac{b\cos ax}{\sqrt{1+b^2}}$

435. $\displaystyle\int (\sin ax)\sqrt{1-b^2\sin^2 ax}\,dx = -\frac{\cos ax}{2a}\sqrt{1-b^2\sin^2 ax} - \frac{1-b^2}{2ab}\log(b\cos ax + \sqrt{1-b^2\sin^2 ax})$

436. $\displaystyle\int \frac{\cos ax}{\sqrt{1+b^2\sin^2 ax}}\,dx = \frac{1}{ab}\log(b\sin ax + \sqrt{1+b^2\sin^2 ax})$

437. $\displaystyle\int \frac{\cos ax}{\sqrt{1-b^2\sin^2 ax}}\,dx = \frac{1}{ab}\sin^{-1}(b\sin ax)$

438. $\displaystyle\int (\cos ax)\sqrt{1+b^2\sin^2 ax}\,dx = \frac{\sin ax}{2a}\sqrt{1+b^2\sin^2 ax} + \frac{1}{2ab}\log(b\sin ax + \sqrt{1+b^2\sin^2 ax})$

439. $\displaystyle\int (\cos ax)\sqrt{1-b^2\sin^2 ax}\,dx = \frac{\sin ax}{2a}\sqrt{1-b^2\sin^2 ax} + \frac{1}{2ab}\sin^{-1}(b\sin ax)$

440. $\displaystyle\int \frac{dx}{\sqrt{a+b\tan^2 cx}} = \frac{\pm 1}{c\sqrt{a-b}}\sin^{-1}\left(\sqrt{\frac{a-b}{a}}\sin cx\right), \quad (a > |b|)$

[use + if $(2k-1)\frac{\pi}{2} < x \le (2k+1)\frac{\pi}{2}$, otherwise −; k an integer]

FORMS INVOLVING INVERSE TRIGONOMETRIC FUNCTIONS

441. $\displaystyle\int (\sin^{-1} ax)\,dx = x\sin^{-1} ax + \frac{\sqrt{1-a^2x^2}}{a}$

442. $\displaystyle\int (\cos^{-1} ax)\,dx = x\cos^{-1} ax - \frac{\sqrt{1-a^2x^2}}{a}$

443. $\displaystyle\int (\tan^{-1} ax)\,dx = x\tan^{-1} ax - \frac{1}{2a}\log(1+a^2x^2)$

444. $\displaystyle\int (\cot^{-1} ax)\,dx = x\cot^{-1} ax + \frac{1}{2a}\log(1+a^2x^2)$

445. $\displaystyle\int (\sec^{-1} ax)\,dx = x\sec^{-1} ax - \frac{1}{a}\log(ax + \sqrt{a^2x^2 - 1})$

446. $\displaystyle\int (\csc^{-1} ax)\,dx = x\csc^{-1} ax + \frac{1}{a}\log(ax + \sqrt{a^2x^2 - 1})$

447. $\displaystyle\int \left(\sin^{-1}\frac{x}{a}\right)dx = x\sin^{-1}\frac{x}{a} + \sqrt{a^2 - x^2}, \quad (a > 0)$

448. $\displaystyle\int \left(\cos^{-1}\frac{x}{a}\right)dx = x\cos^{-1}\frac{x}{a} - \sqrt{a^2 - x^2}, \quad (a > 0)$

449. $\displaystyle\int \left(\tan^{-1}\frac{x}{a}\right)dx = x\tan^{-1}\frac{x}{a} - \frac{a}{2}\log(a^2 + x^2)$

450. $\displaystyle\int \left(\cot^{-1}\frac{x}{a}\right)dx = x\cot^{-1}\frac{x}{a} + \frac{a}{2}\log(a^2 + x^2)$

451. $\displaystyle\int x[\sin^{-1}(ax)]\,dx = \frac{1}{4a^2}[(2a^2x^2 - 1)\sin^{-1}(ax) + ax\sqrt{1-a^2x^2}]$

452. $\displaystyle\int x[\cos^{-1}(ax)]\,dx = \frac{1}{4a^2}[(2a^2x^2 - 1)\cos^{-1}(ax) - ax\sqrt{1-a^2x^2}]$

453. $\displaystyle\int x^n[\sin^{-1}(ax)]\,dx = \frac{x^{n+1}}{n+1}\sin^{-1}(ax) - \frac{a}{n+1}\int \frac{x^{n+1}dx}{\sqrt{1-a^2x^2}}, \quad (n \ne -1)$

454. $\displaystyle\int x^n[\cos^{-1}(ax)]\,dx = \frac{x^{n+1}}{n+1}\cos^{-1}(ax) + \frac{a}{n+1}\int \frac{x^{n+1}dx}{\sqrt{1-a^2x^2}}, \quad (n \ne -1)$

455. $\displaystyle\int x(\tan^{-1} ax)\,dx = \frac{1 + a^2 x^2}{2a^2}\tan^{-1} ax - \frac{x}{2a}$

456. $\displaystyle\int x^n(\tan^{-1} ax)\,dx = \frac{x^{n+1}}{n+1}\tan^{-1} ax - \frac{a}{n+1}\int \frac{x^{n+1}}{1 + a^2 x^2}\,dx$

457. $\displaystyle\int x(\cot^{-1} ax)\,dx = \frac{1 + a^2 x^2}{2a^2}\cot^{-1} ax + \frac{x}{2a}$

458. $\displaystyle\int x^n(\cot^{-1} ax)\,dx = \frac{x^{n+1}}{n+1}\cot^{-1} ax + \frac{a}{n+1}\int \frac{x^{n+1}}{1 + a^2 x^2}\,dx$

459. $\displaystyle\int \frac{\sin^{-1}(ax)}{x^2}\,dx = a\log\left(\frac{1 - \sqrt{1 - a^2 x^2}}{x}\right) - \frac{\sin^{-1}(ax)}{x}$

460. $\displaystyle\int \frac{\cos^{-1}(ax)\,dx}{x^2} = -\frac{1}{x}\cos^{-1}(ax) + a\log\frac{1 + \sqrt{1 - a^2 x^2}}{x}$

461. $\displaystyle\int \frac{\tan^{-1}(ax)\,dx}{x^2} = -\frac{1}{x}\tan^{-1}(ax) - \frac{a}{2}\log\frac{1 + a^2 x^2}{x^2}$

462. $\displaystyle\int \frac{\cot^{-1} ax}{x^2}\,dx = -\frac{1}{x}\cot^{-1} ax - \frac{a}{2}\log\frac{x^2}{a^2 x^2 + 1}$

463. $\displaystyle\int (\sin^{-1} ax)^2\,dx = x(\sin^{-1} ax)^2 - 2x + \frac{2\sqrt{1 - a^2 x^2}}{a}\sin^{-1} ax$

464. $\displaystyle\int (\cos^{-1} ax)^2\,dx = x(\cos^{-1} ax)^2 - 2x - \frac{2\sqrt{1 - a^2 x^2}}{a}\cos^{-1} ax$

465. $\displaystyle\int (\sin^{-1} ax)^n\,dx = \begin{cases} x(\sin^{-1} ax)^n + \dfrac{n\sqrt{1 - a^2 x^2}}{a}(\sin^{-1} ax)^{n-1} \\[2mm] \qquad - n(n-1)\displaystyle\int (\sin^{-1} ax)^{n-2}\,dx \\[4mm] \text{or} \\[2mm] \displaystyle\sum_{r=0}^{[n/2]}(-1)^r\frac{n!}{(n - 2r)!}x(\sin^{-1} ax)^{n-2r} \\[4mm] \qquad + \displaystyle\sum_{r=0}^{[n-1/2]}(-1)^r\frac{n!\sqrt{1 - a^2 x^2}}{(n - 2r - 1)!a}(\sin^{-1} ax)^{n-2r-1} \end{cases}$

Note: [s] means greatest integer $\le s$. Thus [3.5] means 3; [5] = 5, $\left[\tfrac{1}{2}\right] = 0$.

466. $\displaystyle\int (\cos^{-1} ax)^n\,dx = \begin{cases} x(\cos^{-1} ax)^n - \dfrac{n\sqrt{1 - a^2 x^2}}{a}(\cos^{-1} ax)^{n-1} \\[2mm] \qquad - n(n-1)\displaystyle\int (\cos^{-1} ax)^{n-2}\,dx \\[4mm] \text{or} \\[2mm] \displaystyle\sum_{r=0}^{[n/2]}(-1)^r\frac{n!}{(n - 2r)!}x(\cos^{-1} ax)^{n-2r} \\[4mm] \qquad \times \displaystyle\sum_{r=0}^{[n-1/2]}(-1)^r\frac{n!\sqrt{1 - a^2 x^2}}{(n - 2r - 1)!a}(\cos^{-1} ax)^{n-2r-1} \end{cases}$

467. $\displaystyle\int \frac{1}{\sqrt{1 - a^2 x^2}}(\sin^{-1} ax)\,dx = \frac{1}{2a}(\sin^{-1} ax)^2$

468. $\displaystyle\int \frac{x^n}{\sqrt{1 - a^2 x^2}}(\sin^{-1} ax)\,dx = -\frac{x^{n-1}}{na^2}\sqrt{1 - a^2 x^2}\sin^{-1} ax + \frac{x^n}{n^2 a} + \frac{n-1}{na^2}\int \frac{x^{n-2}}{\sqrt{1 - a^2 x^2}}\sin^{-1} ax\,dx$

469. $\displaystyle\int \frac{1}{\sqrt{1 - a^2 x^2}}(\cos^{-1} ax)\,dx = -\frac{1}{2a}(\cos^{-1} ax)^2$

470. $\displaystyle\int \frac{x^n}{\sqrt{1 - a^2 x^2}}(\cos^{-1} ax)\,dx = -\frac{x^{n-1}}{na^2}\sqrt{1 - a^2 x^2}\cos^{-1} ax - \frac{x^n}{n^2 a} + \frac{n-1}{na^2}\int \frac{x^{n-2}}{\sqrt{1 - a^2 x^2}}\cos^{-1} ax\,dx$

471. $\displaystyle\int \frac{\tan^{-1} ax}{a^2 x^2 + 1}\,dx = \frac{1}{2a}(\tan^{-1} ax)^2$

472. $\displaystyle\int \frac{\cot^{-1} ax}{a^2 x^2 + 1}\,dx = -\frac{1}{2a}(\cot^{-1} ax)^2$

473. $\displaystyle\int x\sec^{-1} ax\,dx = \frac{x^2}{2}\sec^{-1} ax - \frac{1}{2a^2}\sqrt{a^2 x^2 - 1}$

474. $\int x^n \sec^{-1} ax\, dx = \dfrac{x^{n+1}}{n+1} \sec^{-1} ax - \dfrac{1}{n+1} \int \dfrac{x^n dx}{\sqrt{a^2 x^2 - 1}}$

475. $\int \dfrac{\sec^{-1} ax}{x^2}\, dx = -\dfrac{\sec^{-1} ax}{x} + \dfrac{\sqrt{a^2 x^2 - 1}}{x}$

476. $\int x \csc^{-1} ax\, dx = \dfrac{x^2}{2} \csc^{-1} ax + \dfrac{1}{2a^2} \sqrt{a^2 x^2 - 1}$

477. $\int x^n \csc^{-1} ax\, dx = \dfrac{x^{n+1}}{n+1} \csc^{-1} ax + \dfrac{1}{n+1} \int \dfrac{x^n dx}{\sqrt{a^2 x^2 - 1}}$

478. $\int \dfrac{\csc^{-1} ax}{x^2}\, dx = -\dfrac{\csc^{-1} ax}{x} - \dfrac{\sqrt{a^2 x^2 - 1}}{x}$

FORMS INVOLVING TRIGONOMETRIC SUBSTITUTIONS

479. $\int f(\sin x)\, dx = 2 \int f\left(\dfrac{2z}{1+z^2}\right) \dfrac{dz}{1+z^2}, \quad \left(z = \tan \dfrac{x}{2}\right)$

480. $\int f(\cos x)\, dx = 2 \int f\left(\dfrac{1-z^2}{1+z^2}\right) \dfrac{dz}{1+z^2}, \quad \left(z = \tan \dfrac{x}{2}\right)$

481. $\int f(\sin x)\, dx = \int f(u) \dfrac{du}{\sqrt{1-u^2}}, \quad (u = \sin x)$

482. $\int f(\cos x)\, dx = - \int f(u) \dfrac{du}{\sqrt{1-u^2}}, \quad (u = \cos x)$

483. $\int f(\sin x, \cos x)\, dx = \int f\left(u, \sqrt{1-u^2}\right) \dfrac{du}{\sqrt{1-u^2}}, \quad (u = \sin x)$

484. $\int f(\sin x, \cos x)\, dx = 2 \int f\left(\dfrac{2z}{1+z^2}, \dfrac{1-z^2}{1+z^2}\right) \dfrac{dz}{1+z^2}, \quad \left(z = \tan \dfrac{x}{2}\right)$

LOGARITHMIC FORMS

485. $\int (\log x)\, dx = x \log x - x$

486. $\int x(\log x)\, dx = \dfrac{x^2}{2} \log x - \dfrac{x^2}{4}$

487. $\int x^2 (\log x)\, dx = \dfrac{x^3}{3} \log x - \dfrac{x^3}{9}$

488. $\int x^n (\log ax)\, dx = \dfrac{x^{n+1}}{n+1} \log ax - \dfrac{x^{n+1}}{(n+1)^2}$

489. $\int (\log x)^2 dx = x(\log x)^2 - 2x \log x + 2x$

490. $\int (\log x)^n dx = \begin{cases} x(\log x)^n - n \int (\log x)^{n-1} dx, \quad (n \neq -1) \\ \text{or} \\ (-1)^n n! x \sum_{r=0}^{n} \dfrac{(-\log x)^r}{r!} \end{cases}$

491. $\int \dfrac{(\log x)^n}{x}\, dx = \dfrac{1}{n+1} (\log x)^{n+1}$

492. $\int \dfrac{dx}{\log x} = \log(\log x) + \log x + \dfrac{(\log x)^2}{2 \cdot 2!} + \dfrac{(\log x)^3}{3 \cdot 3!} + \cdots$

493. $\int \dfrac{dx}{x \log x} = \log(\log x)$

494. $\int \dfrac{dx}{x(\log x)^n} = -\dfrac{1}{(n-1)(\log x)^{n-1}}$

495. $\int \dfrac{x^m dx}{(\log x)^n} = -\dfrac{x^{m+1}}{(n-1)(\log x)^{n-1}} + \dfrac{m+1}{n-1} \int \dfrac{x^m dx}{(\log x)^{n-1}}$

496. $\int x^m (\log x)^n dx = \begin{cases} \dfrac{x^{m+1}(\log x)^n}{m+1} - \dfrac{n}{m+1} \int x^m (\log x)^{n-1} dx \\ \text{or} \\ (-1)^n \dfrac{n!}{m+1} x^{m+1} \sum_{r=0}^{n} \dfrac{(-\log x)^r}{r!(m+1)^{n-r}} \end{cases}$

497. $\int x^p \cos(b \ln x)\, dx = \dfrac{x^{p+1}}{(p+1)^2 + b^2} [b \sin(b \ln x) + (p+1) \cos(b \ln x)] + c$

498. $\int x^p \sin(b \ln x)\, dx = \dfrac{x^{p+1}}{(p+1)^2 + b^2}\,[(p+1)\sin(b\ln x) - b\cos(b\ln x)] + c$

499. $\int [\log(ax+b)]\, dx = \dfrac{ax+b}{a}\log(ax+b) - x$

500. $\int \dfrac{\log(ax+b)}{x^2}\, dx = \dfrac{a}{b}\log x - \dfrac{ax+b}{bx}\log(ax+b)$

501. $\int x^m[\log(ax+b)]\, dx = \dfrac{1}{m+1}\left[x^{m+1} - \left(-\dfrac{b}{a}\right)^{m+1}\right]\log(ax+b) - \dfrac{1}{m+1}\left(-\dfrac{b}{a}\right)^{m+1}\sum_{r=1}^{m+1}\dfrac{1}{r}\left(-\dfrac{ax}{b}\right)^r$

502. $\int \dfrac{\log(ax+b)}{x^m}\, dx = -\dfrac{1}{m-1}\dfrac{\log(ax+b)}{x^{m-1}} + \dfrac{1}{m-1}\left(-\dfrac{a}{b}\right)^{m-1}\log\dfrac{ax+b}{x}$

$$+ \dfrac{1}{m-1}\left(-\dfrac{a}{b}\right)^{m-1}\sum_{r=1}^{m-2}\dfrac{1}{r}\left(-\dfrac{b}{ax}\right)^r,\ (m > 2)$$

503. $\int \left[\log\dfrac{x+a}{x-a}\right]\, dx = (x+a)\log(x+a) - (x-a)\log(x-a)$

504. $\int x^m\left[\log\dfrac{x+a}{x-a}\right]\, dx = \dfrac{x^{m+1} - (-a)^{m+1}}{m+1}\log(x+a) - \dfrac{x^{m+1} - a^{m+1}}{m+1}\log(x-a) + \dfrac{2a^{m+1}}{m+1}\sum_{r=1}^{\left[\frac{m+1}{2}\right]}\dfrac{1}{m-2r+2}\left(\dfrac{x}{a}\right)^{m-2r+2}$

Note: [s] means greatest integer $\le s$; Thus [3.5] means 3; [5] = 5, $\left[\frac{1}{2}\right] = 0$.

505. $\int \dfrac{1}{x^2}\left[\log\dfrac{x+a}{x-a}\right]\, dx = \dfrac{1}{x}\log\dfrac{x-a}{x+a} - \dfrac{1}{a}\log\dfrac{x^2-a^2}{x^2}$

506. $\int (\log X)\, dx = \begin{cases} \left(x + \dfrac{b}{2c}\right)\log X - 2x + \dfrac{\sqrt{4ac-b^2}}{c}\tan^{-1}\dfrac{2cx+b}{\sqrt{4ac-b^2}}, & (b^2 - 4ac < 0) \\[1em] \text{or} \\[0.5em] \left(x + \dfrac{b}{2c}\right)\log X - 2x + \dfrac{\sqrt{b^2-4ac}}{c}\tanh^{-1}\dfrac{2cx+b}{\sqrt{b^2-4ac}}, & (b^2 - 4ac > 0) \\[1em] \text{where} \\ X = a + bx + cx^2 \end{cases}$

507. $\int x^n(\log X)\, dx = \dfrac{x^{n+1}}{n+1}\log X - \dfrac{2c}{n+1}\int\dfrac{x^{n+2}}{X}\, dx - \dfrac{b}{n+1}\int\dfrac{x^{n+1}}{X}\, dx$

where $X = a + bx + cx^2$

508. $\int [\log(x^2 + a^2)]\, dx = x\log(x^2 + a^2) - 2x + 2a\tan^{-1}\dfrac{x}{a}$

509. $\int [\log(x^2 - a^2)]\, dx = x\log(x^2 - a^2) - 2x + a\log\dfrac{x+a}{x-a}$

510. $\int x[\log(x^2 \pm a^2)]\, dx = \dfrac{1}{2}(x^2 \pm a^2)\log(x^2 \pm a^2) - \dfrac{1}{2}x^2$

511. $\int [\log(x + \sqrt{x^2 \pm a^2})]\, dx = x\log(x + \sqrt{x^2 \pm a^2}) - \sqrt{x^2 \pm a^2}$

512. $\int x[\log(x + \sqrt{x^2 \pm a^2})]\, dx = \left(\dfrac{x^2}{2} \pm \dfrac{a^2}{4}\right)\log(x + \sqrt{x^2 \pm a^2}) - \dfrac{x\sqrt{x^2 \pm a^2}}{4}$

513. $\int x^m[\log(x + \sqrt{x^2 \pm a^2})]\, dx = \dfrac{x^{m+1}}{m+1}\log(x + \sqrt{x^2 \pm a^2}) - \dfrac{1}{m+1}\int\dfrac{x^{m+1}}{\sqrt{x^2 \pm a^2}}\, dx$

514. $\int \dfrac{\log(x + \sqrt{x^2 + a^2})}{x^2}\, dx = -\dfrac{\log(x + \sqrt{x^2 + a^2})}{x} - \dfrac{1}{a}\log\dfrac{a + \sqrt{x^2 + a^2}}{x}$

515. $\int \dfrac{\log(x + \sqrt{x^2 - a^2})}{x^2}\, dx = -\dfrac{\log(x + \sqrt{x^2 - a^2})}{x} + \dfrac{1}{|a|}\sec^{-1}\dfrac{x}{a}$

516. $\int x^n\log(x^2 - a^2)\, dx = \dfrac{1}{n+1}\left[x^{n+1}\log(x^2-a^2) - a^{n+1}\log(x-a) - (-a)^{n+1}\log(x+a) - 2\sum_{r=0}^{[n/2]}\dfrac{a^{2r}x^{n-2r+1}}{n-2r+1}\right]$

Note: [s] means greatest integer $\le s$; Thus [3.5] means 3; [5] = 5, $\left[\frac{1}{2}\right] = 0$.

EXPONENTIAL FORMS

517. $\int e^x\, dx = e^x$

518. $\int e^{-x}\, dx = -e^{-x}$

519. $\int e^{ax}\, dx = \dfrac{e^{ax}}{a}$

520. $\int x e^{ax} dx = \dfrac{e^{ax}}{a^2}(ax-1)$

521. $\int x^m e^{ax} dx = \begin{cases} \dfrac{x^m e^{ax}}{a} - \dfrac{m}{a}\int x^{m-1} e^{ax} dx \\ \quad\text{or} \\ e^{ax}\sum_{r=0}^{m}(-1)^r \dfrac{m!\,x^{m-r}}{(m-r)!\,a^{r+1}} \end{cases}$

522. $\int \dfrac{e^{ax} dx}{x} = \log x + \dfrac{ax}{1!} + \dfrac{a^2 x^2}{2\cdot 2!} + \dfrac{a^3 + x^3}{3\cdot 3!} + \cdots$

523. $\int \dfrac{e^{ax}}{x^m} dx = -\dfrac{1}{m-1}\dfrac{e^{ax}}{x^{m-1}} + \dfrac{a}{m-1}\int \dfrac{e^{ax}}{x^{m-1}} dx$

524. $\int e^{ax} \log x\, dx = \dfrac{e^{ax}\log x}{a} - \dfrac{1}{a}\int \dfrac{e^{ax}}{x} dx$

525. $\int \dfrac{dx}{1+e^x} = x - \log(1+e^x) = \log\dfrac{e^x}{1+e^x}$

526. $\int \dfrac{dx}{a+be^{px}} = \dfrac{x}{a} - \dfrac{1}{ap}\log(a+be^{px})$

527. $\int \dfrac{dx}{ae^{mx}+be^{-mx}} = \dfrac{1}{m\sqrt{ab}}\tan^{-1}\left(e^{mx}\sqrt{\dfrac{a}{b}}\right), \quad (a>0, b>0)$

528. $\int \dfrac{dx}{ae^{mx}-be^{-mx}} = \begin{cases} \dfrac{1}{2m\sqrt{ab}}\log\dfrac{\sqrt{a}\,e^{mx}-\sqrt{b}}{\sqrt{a}\,e^{mx}+\sqrt{b}} \\ \quad\text{or} \\ \dfrac{-1}{m\sqrt{ab}}\tanh^{-1}\left(\sqrt{\tfrac{a}{b}}e^{mx}\right), \quad (a>0, b>0) \end{cases}$

529. $\int (a^x - a^{-x})\, dx = \dfrac{a^x + a^{-x}}{\log a}$

530. $\int \dfrac{e^{ax}}{b+ce^{ax}} dx = \dfrac{1}{ac}\log(b+ce^{ax})$

531. $\int \dfrac{x e^{ax}}{(1+ax)^2} dx = \dfrac{e^{ax}}{a^2(1+ax)}$

532. $\int x e^{-x^2} dx = -\dfrac{1}{2}e^{-x^2}$

533. $\int e^{ax}[\sin(bx)]\, dx = \dfrac{e^{ax}[a\sin(bx)-b\cos(bx)]}{a^2+b^2}$

534. $\int e^{ax}[\sin(bx)][\sin(cx)]\, dx = \dfrac{e^{ax}[(b-c)\sin(b-c)x + a\cos(b-c)x]}{2[a^2+(b-c)^2]} - \dfrac{e^{ax}[(b+c)\sin(b+c)x + a\cos(b+c)x]}{2[a^2+(b+c)^2]}$

535. $\int e^{ax}[\sin(bx)][\cos(cx)]\, dx = \begin{cases} \dfrac{e^{ax}[a\sin(b-c)x-(b-c)\cos(b-c)x]}{2[a^2+(b-c)^2]} \\ \quad + \dfrac{e^{ax}[a\sin(b+c)x-(b+c)\cos(b+c)x]}{2[a^2+(b+c)^2]} \\ \quad\text{or} \\ \dfrac{e^{ax}}{\rho}[(a\sin bx - b\cos bx)[\cos(cx-\alpha)] \\ \quad -c(\sin bx)\sin(cx-\alpha)] \\ \text{where} \\ \rho = \sqrt{(a^2+b^2-c^2)^2 + 4a^2c^2}, \\ \rho\cos\alpha = a^2+b^2-c^2, \quad \rho\sin\alpha = 2ac \end{cases}$

536. $\int e^{ax}[\sin(bx)][\sin(bx+c)]\, dx = \dfrac{e^{ax}\cos c}{2a} - \dfrac{e^{ax}[a\cos(2bx+c)+2b\sin(2bx+c)]}{2(a^2+4b^2)}$

537. $\int e^{ax}[\sin(bx)][\cos(bx+c)]\, dx = -\dfrac{e^{ax}\sin c}{2a} + \dfrac{e^{ax}[a\sin(2bx+c)-2b\cos(2bx+c)]}{2(a^2+4b^2)}$

538. $\int e^{ax}[\cos(bx)]\, dx = \dfrac{e^{ax}}{a^2+b^2}[a\cos(bx)+b\sin(bx)]$

539. $\int e^{ax}[\cos(bx)][\cos(cx)]\, dx = \dfrac{e^{ax}[(b-c)\sin(b-c)x + a\cos(b-c)x]}{2[a^2+(b-c)^2]} + \dfrac{e^{ax}[(b+c)\sin(b+c)x + a\cos(b+c)x]}{2[a^2+(b+c)^2]}$

540. $\int e^{ax}[\cos(bx)][\cos(bx+c)]\, dx = \dfrac{e^{ax}\cos c}{2a} + \dfrac{e^{ax}[a\cos(2bx+c)+2b\sin(2bx+c)]}{2(a^2+4b^2)}$

541. $\int e^{ax}[\cos(bx)][\sin(bx+c)]\, dx = \dfrac{e^{ax}\sin c}{2a} + \dfrac{e^{ax}[a\sin(2bx+c)-2b\cos(2bx+c)]}{2(a^2+4b^2)}$

542. $\int e^{ax}[\sin^n bx]\, dx = \dfrac{1}{a^2+n^2 b^2}\left[(a\sin bx - nb\cos bx)e^{ax}\sin^{n-1}bx + n(n-1)b^2\int e^{ax}[\sin^{n-2}bx]\, dx\right]$

543. $\int e^{ax}[\cos^n bx]\, dx = \dfrac{1}{a^2+n^2 b^2}\left[(a\cos bx + nb\sin bx)e^{ax}\cos^{n-1}bx + n(n-1)b^2\int e^{ax}[\cos^{n-2}bx]\, dx\right]$

544. $\int x^m e^x \sin x\, dx = \dfrac{1}{2}x^m e^x(\sin x - \cos x) - \dfrac{m}{2}\int x^{m-1}e^x\sin x\, dx + \dfrac{m}{2}\int x^{m-1}e^x\cos x\, dx$

545. $\displaystyle\int x^m e^{ax}[\sin bx]\,dx = \begin{cases} x^m e^{ax}\dfrac{a\sin bx - b\cos bx}{a^2+b^2} \\[2mm] \quad -\dfrac{m}{a^2+b^2}\displaystyle\int x^{m-1}e^{ax}(a\sin bx - b\cos bx)\,dx \\[2mm] \text{or} \\[2mm] e^{ax}\sum_{r=0}^{m}\dfrac{(-1)^r m!\,x^{m-r}}{\rho^{r+1}(m-r)!}\,\sin[bx-(r+1)\alpha] \\[2mm] \quad\text{where} \\[1mm] \rho = \sqrt{a^2+b^2},\quad \rho\cos\alpha = a,\quad \rho\sin\alpha = b \end{cases}$

546. $\displaystyle\int x^m e^x \cos x\,dx = \dfrac{1}{2}x^m e^x(\sin x + \cos x) - \dfrac{m}{2}\int x^{m-1}e^x \sin x\,dx - \dfrac{m}{2}\int x^{m-1}e^x \cos x\,dx$

547. $\displaystyle\int x^m e^{ax}\cos bx\,dx = \begin{cases} x^m e^{ax}\dfrac{a\cos bx + b\sin bx}{a^2+b^2} \\[2mm] \quad -\dfrac{m}{a^2+b^2}\displaystyle\int x^{m-1}e^{ax}(a\cos bx + b\sin bx)\,dx \\[2mm] \text{or} \\[2mm] e^{ax}\sum_{r=0}^{m}\dfrac{(-1)^r m!\,x^{m-r}}{\rho^{r+1}(m-r)!}\,\cos[bx-(r+1)\alpha] \\[2mm] \rho = \sqrt{a^2+b^2},\quad \rho\cos\alpha = a,\quad \rho\sin\alpha = b \end{cases}$

548. $\displaystyle\int e^{ax}(\cos^m x)(\sin^n x)\,dx = \begin{cases} \dfrac{e^{ax}\cos^{m-1}x\,\sin^n x[a\cos x + (m+n)\sin x]}{(m+n)^2+a^2} \\[2mm] \quad -\dfrac{na}{(m+n)^2+a^2}\displaystyle\int e^{ax}(\cos^{m-1}x)(\sin^{n-1}x)\,dx \\[2mm] \quad +\dfrac{(m-1)(m+n)}{(m+n)^2+a^2}\displaystyle\int e^{ax}(\cos^{m-2}x)(\sin^n x)\,dx \\[3mm] \text{or} \\[2mm] \dfrac{e^{ax}\cos^m x\,\sin^{n-1}x[a\sin x - (m+n)\cos x]}{(m+n)^2+a^2} \\[2mm] \quad +\dfrac{ma}{(m+n)^2+a^2}\displaystyle\int e^{ax}(\cos^{m-1}x)(\sin^{n-1}x)\,dx \\[2mm] \quad +\dfrac{(n-1)(m+n)}{(m+n)^2+a^2}\displaystyle\int e^{ax}(\cos^m x)(\sin^{n-2}x)\,dx \\[3mm] \text{or} \\[2mm] \dfrac{e^{ax}(\cos^{m-1}x)(\sin^{n-1}x)(a\sin x\cos x + m\sin^2 x - n\cos^2 x)}{(m+n)^2+a^2} \\[2mm] \quad +\dfrac{m(m-1)}{(m+n)^2+a^2}\displaystyle\int e^{ax}(\cos^{m-2}x)(\sin^n x)\,dx \\[2mm] \quad +\dfrac{n(n-1)}{(m+n)^2+a^2}\displaystyle\int e^{ax}(\cos^m x)(\sin^{n-2}x)\,dx \\[3mm] \text{or} \\[2mm] \dfrac{e^{ax}(\cos^{m-1}x)(\sin^{n-1}x)(a\cos x\sin x + m\sin^2 x - n\cos^2 x)}{(m+n)^2+a^2} \\[2mm] \quad +\dfrac{m(m-1)}{(m+n)^2+a^2}\displaystyle\int e^{ax}(\cos^{m-2}x)(\sin^{n-2}x)\,dx \\[2mm] \quad +\dfrac{(n-m)(n+m-1)}{(m+n)^2+a^2}\displaystyle\int e^{ax}(\cos^m x)(\sin^{n-2}x)\,dx \end{cases}$

549. $\displaystyle\int xe^{ax}(\sin bx)\,dx = \dfrac{xe^{ax}}{a^2+b^2}(a\sin bx - b\cos bx) - \dfrac{e^{ax}}{(a^2+b^2)^2}[(a^2-b^2)\sin bx - 2ab\cos bx]$

550. $\displaystyle\int xe^{ax}(\cos bx)\,dx = \dfrac{xe^{ax}}{a^2+b^2}(a\cos bx - b\sin bx) - \dfrac{e^{ax}}{(a^2+b^2)^2}[(a^2-b^2)\cos bx - 2ab\sin bx]$

551. $\displaystyle\int \dfrac{e^{ax}}{\sin^n x}\,dx = -\dfrac{e^{ax}[a\sin x + (n-2)\cos x]}{(n-1)(n-2)\sin^{n-1}x} + \dfrac{a^2+(n-2)^2}{(n-1)(n-2)}\int\dfrac{e^{ax}}{\sin^{n-2}x}\,dx$

552. $\displaystyle\int \dfrac{e^{ax}}{\cos^n x}\,dx = -\dfrac{e^{ax}[a\cos x - (n-2)\sin x]}{(n-1)(n-2)\cos^{n-1}x} + \dfrac{a^2+(n-2)^2}{(n-1)(n-2)}\int\dfrac{e^{ax}}{\cos^{n-2}x}\,dx$

553. $\displaystyle\int e^{ax}\tan^n x\,dx = e^{ax}\dfrac{\tan^{n-1}x}{n-1} - \dfrac{a}{n-1}\int e^{ax}\tan^{n-1}x\,dx - \int e^{ax}\tan^{n-2}x\,dx$

HYPERBOLIC FORMS

554. $\displaystyle\int (\sinh x)\,dx = \cosh x$

555. $\displaystyle\int (\cosh x)\,dx = \sinh x$

556. $\displaystyle\int (\tanh x)\,dx = \log\cosh x$

557. $\displaystyle\int (\coth x)\,dx = \log\sinh x$

558. $\displaystyle\int (\operatorname{sech} x)\, dx = \tan^{-1}(\sinh x)$

559. $\displaystyle\int (\operatorname{csch} x)\, dx = \log \tanh\left(\frac{x}{2}\right)$

560. $\displaystyle\int x(\sinh x)\, dx = x\cosh x - \sinh x$

561. $\displaystyle\int x^n(\sinh x)\, dx = x^n \cosh x - n\int x^{n-1}(\cosh x)\,dx$

562. $\displaystyle\int x(\cosh x)\, dx = x\sinh x - \cosh x$

563. $\displaystyle\int x^n(\cosh x)\, dx - x^n \sinh x - n\int x^{n-1}(\sinh x)\, dx$

564. $\displaystyle\int (\operatorname{sech} x)(\tanh x)\, dx = -\operatorname{sech} x$

565. $\displaystyle\int (\operatorname{csch} x)(\coth x)\, dx = -\operatorname{csch} x$

566. $\displaystyle\int (\sinh^2 x)\,dx = \frac{\sinh 2x}{4} - \frac{x}{2}$

567. $\displaystyle\int (\sinh^m x)(\cosh^n x)\, dx = \begin{cases} \frac{1}{m+n}(\sinh^{m+1} x)(\cosh^{n-1} x) \\ \quad + \frac{n-1}{m+n}\int (\sinh^m x)(\cosh^{n-2} x)\, dx \\ \text{or} \\ \frac{1}{m+n}\sinh^{m-1} x\cosh^{n+1} x \\ \quad - \frac{m-1}{m+n}\int (\sinh^{m-2} x)(\cosh^n x)\,dx, \quad (m+n \neq 0) \end{cases}$

568. $\displaystyle\int \frac{dx}{(\sinh^m x)(\cosh^n x)} \begin{cases} -\frac{1}{(m-n)(\sinh^{m-1} x)(\cosh^{n-1} x)} \\ \quad -\frac{m+n-2}{m-1}\int \frac{dx}{(\sinh^{m-2} x)(\cosh^n x)}, \quad (m \neq 1) \\ \text{or} \\ \frac{1}{(n-1)\sinh^{m-1} x\cosh^{n-1} x} \\ \quad + \frac{m+n-2}{n-1}\int \frac{dx}{(\sinh^m x)(\cosh^{n-2} x)}, \quad (n \neq 1) \end{cases}$

569. $\displaystyle\int (\tanh^2 x)\, dx = x - \tanh x$

570. $\displaystyle\int (\tanh^n x)\, dx = -\frac{\tanh^{n-1} x}{n-1} + \int (\tanh^{n-2} x)\,dx, \quad (n \neq 1)$

571. $\displaystyle\int (\operatorname{sech}^2 x)\, dx = \tanh x$

572. $\displaystyle\int (\cosh^2 x)\, dx = \frac{\sinh 2x}{4} + \frac{x}{2}$

573. $\displaystyle\int (\coth^2 x)\, dx = x - \coth x$

574. $\displaystyle\int (\coth^n x)\, dx = -\frac{\coth^{n-1} x}{n-1} + \int \coth^{n-2} x\, dx, \quad (n \neq 1)$

575. $\displaystyle\int (\operatorname{csch}^2 x)\, dx = -\operatorname{ctnh} x$

576. $\displaystyle\int (\sinh mx)(\sinh nx)\, dx = \frac{\sinh(m+n)x}{2(m+n)} - \frac{\sinh(m-n)x}{2(m-n)}, \quad (m^2 \neq n^2)$

577. $\displaystyle\int (\cosh mx)(\cosh nx)\, dx = \frac{\sinh(m+n)x}{2(m+n)} + \frac{\sinh(m-n)x}{2(m-n)}, \quad (m^2 \neq n^2)$

578. $\displaystyle\int (\sinh mx)(\cosh nx)\, dx = \frac{\cosh(m+n)x}{2(m+n)} + \frac{\cosh(m-n)x}{2(m-n)}, \quad (m^2 \neq n^2)$

579. $\displaystyle\int \left(\sinh^{-1}\frac{x}{a}\right) dx = x\sinh^{-1}\frac{x}{a} - \sqrt{x^2 + a^2}, \quad (a > 0)$

580. $\displaystyle\int x\left(\sinh^{-1}\frac{x}{a}\right) dx = \left(\frac{x^2}{2} + \frac{a^2}{4}\right)\sinh^{-1}\frac{x}{a} - \frac{x}{4}\sqrt{x^2 + a^2}, \quad (a > 0)$

581. $\displaystyle\int x^n\left(\sinh^{-1} x\right) dx = \left(\frac{x^{n+1}}{n+1}\right)\sinh^{-1} x - \frac{1}{n+1}\int \frac{x^{n+1}}{(1+x^2)^{\frac{1}{2}}}dx, \quad (n \neq -1)$

582. $\displaystyle \int \left(\cosh^{-1} \frac{x}{a} \right) dx = \begin{cases} x \cosh^{-1} \frac{x}{a} - \sqrt{x^2 - a^2}, \ \left(\cosh^{-1} \frac{x}{a} > 0 \right) \\ \qquad \text{or} \\ x \cosh^{-1} \frac{x}{a} + \sqrt{x^2 - a^2}, \ \left(\cosh^{-1} \frac{x}{a} < 0 \right), \quad (a > 0) \end{cases}$

583. $\displaystyle \int x \left(\cosh^{-1} \frac{x}{a} \right) dx = \frac{2x^2 - a^2}{4} \cosh^{-1} \frac{x}{a} - \frac{x}{4}(x^2 - a^2)^{\frac{1}{2}}$

584. $\displaystyle \int x^n (\cosh^{-1} x) \, dx = \frac{x^{n+1}}{n+1} \cosh^{-1} x - \frac{1}{n+1} \int \frac{x^{n+1}}{(x^2 - 1)^{\frac{1}{2}}} dx, \quad (n \neq -1)$

585. $\displaystyle \int \left(\tanh^{-1} \frac{x}{a} \right) dx = x \tanh^{-1} \frac{x}{a} + \frac{a}{2} \log(a^2 - x^2), \quad \left(\left| \frac{x}{a} \right| < 1 \right)$

586. $\displaystyle \int \left(\coth^{-1} \frac{x}{a} \right) dx = x \coth^{-1} \frac{x}{a} + \frac{a}{2} \log(x^2 - a^2), \quad \left(\left| \frac{x}{a} \right| > 1 \right)$

587. $\displaystyle \int x \left(\tanh^{-1} \frac{x}{a} \right) dx = \frac{x^2 - a^2}{2} \tanh^{-1} \frac{x}{a} + \frac{ax}{2}, \quad \left(\left| \frac{x}{a} \right| < 1 \right)$

588. $\displaystyle \int x^n (\tanh^{-1} x) \, dx = \frac{x^{n+1}}{n+1} \tanh^{-1} x - \frac{1}{n+1} \int \frac{x^{n+1}}{1 - x^2} dx, \quad (n \neq -1)$

589. $\displaystyle \int x \left(\coth^{-1} \frac{x}{a} \right) dx = \frac{x^2 - a^2}{2} \coth^{-1} \frac{x}{a} + \frac{ax}{2}, \quad \left(\left| \frac{x}{a} \right| > 1 \right)$

590. $\displaystyle \int x^n \left(\coth^{-1} x \right) dx = \frac{x^{n+1}}{n+1} \coth^{-1} x + \frac{1}{n+1} \int \frac{x^{n+1}}{x^2 - 1} dx, \quad (n \neq -1)$

591. $\displaystyle \int (\text{sech}^{-1} x) \, dx = x \, \text{sech}^{-1} x + \sin^{-1} x$

592. $\displaystyle \int x \, \text{sech}^{-1} x \, dx = \frac{x^2}{2} \text{sech}^{-1} x - \frac{1}{2} \sqrt{1 - x^2}$

593. $\displaystyle \int x^n \, \text{sech}^{-1} x \, dx = \frac{x^{n+1}}{n+1} \text{sech}^{-1} x + \frac{1}{n+1} \int \frac{x^n}{(1 - x^2)^{\frac{1}{2}}} dx, \quad (n \neq -1)$

594. $\displaystyle \int \text{csch}^{-1} x \, dx = x \, \text{csch}^{-1} x + \frac{x}{|x|} \sinh^{-1} x$

595. $\displaystyle \int x \, \text{csch}^{-1} x \, dx = \frac{x^2}{2} \text{csch}^{-1} x + \frac{1}{2} \frac{x}{|x|} \sqrt{1 + x^2}$

596. $\displaystyle \int x^n \text{csch}^{-1} x \, dx = \frac{x^{n+1}}{n+1} \text{csch}^{-1} x + \frac{1}{n+1} \frac{x}{|x|} \int \frac{x^n}{(x^2 + 1)^{\frac{1}{2}}} dx, \quad (n \neq -1)$

DEFINITE INTEGRALS

597. $\displaystyle \int_0^\infty x^{n-1} e^{-x} dx = \int_0^1 \left(\log \frac{1}{x} \right)^{n-1} dx = \frac{1}{n} \prod_{m=1}^\infty \frac{\left(1 + \frac{1}{m} \right)^n}{1 + \frac{n}{m}} = \Gamma(n)$

for $n \neq 0, -1, -2, -3, \ldots$ (This is the Gamma function)

598. $\displaystyle \int_0^\infty t^n p^{-t} dt = \frac{n!}{(\log p)^{n+1}}, \quad (n = 0, 1, 2, 3, \ldots \text{ and } p > 0)$

599. $\displaystyle \int_0^\infty t^{n-1} e^{-(a+1)t} dt = \frac{\Gamma(n)}{(a+1)^n}, \quad (n > 0, a > -1)$

600. $\displaystyle \int_0^1 x^m \left(\log \frac{1}{x} \right)^n dx = \frac{\Gamma(n+1)}{(m+1)^{n+1}}, \quad (m > -1, n > -1)$

601. $\Gamma(n)$ is finite if $n > 0$; $\Gamma(n+1) = n\Gamma(n)$

602. $\Gamma(n) \cdot \Gamma(1-n) = \frac{\pi}{\sin n\pi}$

603. $\Gamma(n) = (n-1)!$ if $n = \text{integer} > 0$

604. $\displaystyle \Gamma\left(\frac{1}{2} \right) = 2 \int_0^\infty e^{-t^2} dt = \sqrt{\pi} = 1.7724538509 \cdots = \left(-\frac{1}{2} \right)!$

605. $\Gamma(n + \frac{1}{2}) = \frac{1 \cdot 3 \cdot 5 \ldots (2n-1)}{2^n} \sqrt{\pi} \quad n = 1, 2, 3, \ldots$

606. $\Gamma(-n + \frac{1}{2}) = \frac{(-1)^n 2^n \sqrt{\pi}}{1 \cdot 3 \cdot 5 \ldots (2n-1)} \quad n = 1, 2, 3, \ldots$

607. $\displaystyle \int_0^1 x^{m-1} (1-x)^{n-1} dx = \int_0^\infty \frac{x^{m-1}}{(1+x)^{m+n}} dx = \frac{\Gamma(m)\Gamma(n)}{\Gamma(m+n)} = B(m, n)$

(This is the Beta function)

608. $B(m, n) = B(n, m) = \frac{\Gamma(m)\Gamma(n)}{\Gamma(m+n)}$, where m and n are any positive real numbers.

609. $\displaystyle\int_a^b (x-a)^m (b-x)^n \, dx = (b-a)^{m+n+1} \frac{\Gamma(m+1) \cdot \Gamma(n+1)}{\Gamma(m+n+2)}, \quad (m > -1, n > -1, b > a)$

610. $\displaystyle\int_1^\infty \frac{dx}{x^m} = \frac{1}{m-1}, \quad [m > 1]$

611. $\displaystyle\int_0^\infty \frac{dx}{(1+x)x^p} = \pi \csc p\pi, \quad [0 < p < 1]$

612. $\displaystyle\int_0^\infty \frac{dx}{(1-x)x^p} = -\pi \cot p\pi, \quad [0 < p < 1]$

613. $\displaystyle\int_0^\infty \frac{x^{p-1} dx}{(1+x)} = \frac{\pi}{\sin p\pi} = B(p, 1-p) = \Gamma(p)\Gamma(1-p), \quad [0 < p < 1]$

614. $\displaystyle\int_0^\infty \frac{x^{m-1} dx}{1+x^n} = \frac{\pi}{n \sin \frac{m\pi}{n}}, \quad [0 < m < n]$

615. $\displaystyle\int_0^\infty \frac{x^a dx}{(m + x^b)^c} = \frac{m^{\frac{a+1-bc}{b}}}{b} \left[\frac{\Gamma\left(\frac{a+1}{b}\right) \Gamma\left(c - \frac{a+1}{b}\right)}{\Gamma(c)} \right]$

$(a > -1, \, b > 0, \, m > 0, \, c > \frac{a+1}{b})$

616. $\displaystyle\int_0^\infty \frac{dx}{(1+x)\sqrt{x}} = \pi$

617. $\displaystyle\int_0^\infty \frac{a \, dx}{a^2 + x^2} = \frac{\pi}{2}, \quad \text{if } a > 0; \, 0, \text{ if } a = 0; \, -\frac{\pi}{2}, \text{ if } a < 0$

618. $\displaystyle\int_0^a (a^2 - x^2)^{n/2} \, dx = \frac{1}{2} \int_{-a}^a (a^2 - x^2)^{n/2} \, dx = \frac{1 \cdot 3 \cdot 5 \dots n}{2 \cdot 4 \cdot 6 \dots (n+1)} \cdot \frac{\pi}{2} \cdot a^{n+1} \quad (n \text{ odd, } a > 0)$

619. $\displaystyle\int_0^a x^m (a^2 - x^2)^{n/2} \, dx = \begin{cases} \frac{1}{2} a^{m+n+1} B\left(\frac{m+1}{2}, \frac{n+2}{2}\right) & (a > 0, \, m > -1, \, n > -2) \\ \quad \text{or} \\ \frac{1}{2} a^{m+n+1} \frac{\Gamma\left(\frac{m+1}{2}\right)\Gamma\left(\frac{n+2}{2}\right)}{\Gamma\left(\frac{m+n+3}{2}\right)} & (a > 0, \, m > -1, \, n > -2) \end{cases}$

620. $\displaystyle\int_0^{\pi/2} (\sin^n x) \, dx = \begin{cases} \int_0^{\pi/2} (\cos^n x) \, dx \\ \quad \text{or} \\ \frac{1 \cdot 3 \cdot 5 \cdot 7 \dots (n-1)}{2 \cdot 4 \cdot 6 \cdot 8 \dots (n)} \frac{\pi}{2}, & (n \text{ an even integer, } n \neq 0) \\ \quad \text{or} \\ \frac{1 \cdot 3 \cdot 5 \cdot 7 \dots (n-1)}{2 \cdot 4 \cdot 6 \cdot 8 \dots (n)}, & (n \text{ an odd integer, } n \neq 0) \\ \frac{\sqrt{\pi}}{2} \frac{\Gamma\left(\frac{n+1}{2}\right)}{\Gamma\left(\frac{n}{2}+1\right)}, & (n > -1) \end{cases}$

621. $\displaystyle\int_0^\infty \frac{\sin mx \, dx}{x} = \frac{\pi}{2}; \quad \text{if } m > 0; \, 0, \text{ if } m = 0; \, -\frac{\pi}{2}, \text{ if } m < 0$

622. $\displaystyle\int_0^\infty \frac{\cos x \, dx}{x} = \infty$

623. $\displaystyle\int_0^\infty \frac{\tan x \, dx}{x} = \frac{\pi}{2}$

624. $\displaystyle\int_0^\pi \sin ax \cdot \sin bx \, dx = \int_0^\pi \cos ax \cdot \cos bx \, dx = 0, \quad (a \neq b; \, a, \, b \text{ integers})$

625. $\displaystyle\int_0^{\pi/a} [\sin(ax)][\cos(ax)] \, dx = \int_0^\pi [\sin(ax)][\cos(ax)] dx = 0$

626. $\displaystyle\int_0^\pi [\sin(ax)][\cos(bx)] \, dx = \frac{2a}{a^2 - b^2}, \text{ if } a - b \text{ is odd, or } 0 \quad \text{if } a - b \text{ is even}$

627. $\displaystyle\int_0^\infty \frac{\sin x \cos mx \, dx}{x} = 0, \quad \text{if } m < -1 \text{ or } m > 1; \, \frac{\pi}{4}, \text{ if } m = \pm 1; \, \frac{\pi}{2}, \text{ if } m^2 < 1$

628. $\displaystyle\int_0^\infty \frac{\sin ax \sin bx}{x^2} \, dx = \frac{\pi a}{2}, \quad (a \leq b)$

629. $\displaystyle\int_0^\pi \sin^2 mx \, dx = \int_0^\pi \cos^2 mx \, dx = \frac{\pi}{2} \quad (m \text{ is a non-zero integer})$

630. $\displaystyle\int_0^\infty \frac{\sin^2(px)}{x^2} \, dx = \frac{\pi |p|}{2}$

631. $\displaystyle\int_0^\infty \frac{\sin x}{x^p} \, dx = \frac{\pi}{2\Gamma(p) \sin(p\pi/2)}, \quad 0 < p < 1$

632. $\displaystyle\int_0^\infty \frac{\cos x}{x^p} \, dx = \frac{\pi}{2\Gamma(p) \cos(p\pi/2)}, \quad 0 < p < 1$

633. $\displaystyle\int_0^\infty \frac{1 - \cos px}{x^2} \, dx = \frac{\pi |p|}{2}$

634. $\int_0^\infty \dfrac{\sin px \cos qx}{x}\, dx = \left\{ 0, q > p > 0; \quad \dfrac{\pi}{2}, p > q > 0; \dfrac{\pi}{4}, p = q > 0 \right\}$

635. $\int_0^\infty \dfrac{\cos(mx)}{x^2 + a^2}\, dx = \dfrac{\pi}{2\,|a|} e^{-|ma|}$

636. $\int_0^\infty \cos(x^2)\, dx = \int_0^\infty \sin(x^2)\, dx = \dfrac{1}{2}\sqrt{\dfrac{\pi}{2}}$

637. $\int_0^\infty \sin ax^n dx = \dfrac{1}{na^{1/n}}\Gamma(1/n)\sin\dfrac{\pi}{2n}, \quad n > 1$

638. $\int_0^\infty \cos ax^n dx = \dfrac{1}{na^{1/n}}\Gamma(1/n)\cos\dfrac{\pi}{2n}, \quad n > 1$

639. $\int_0^\infty \dfrac{\sin x}{\sqrt{x}}\, dx = \int_0^\infty \dfrac{\cos x}{\sqrt{x}}\, dx = \sqrt{\dfrac{\pi}{2}}$

640. (a) $\int_0^\infty \dfrac{\sin^3 x}{x}\, dx = \dfrac{\pi}{4}$ (b) $\int_0^\infty \dfrac{\sin^3 x}{x^2}\, dx = \dfrac{3}{4}\log 3$

641. $\int_0^\infty \dfrac{\sin^3 x}{x^3}\, dx = \dfrac{3\pi}{8}$

642. $\int_0^\infty \dfrac{\sin^4 x}{x^4}\, dx = \dfrac{\pi}{3}$

643. $\int_0^{\pi/2} \dfrac{dx}{1 + a\cos x} = \dfrac{\cos^{-1} a}{\sqrt{1 - a^2}}, \quad (|a| < 1)$

644. $\int_0^\pi \dfrac{dx}{a + b\cos x} = \dfrac{\pi}{\sqrt{a^2 - b^2}}, \quad (a > b \geq 0)$

645. $\int_0^{2\pi} \dfrac{dx}{1 + a\cos x} = \dfrac{2\pi}{\sqrt{1 - a^2}}, \quad (a^2 < 1)$

646. $\int_0^\infty \dfrac{\cos ax - \cos bx}{x}\, dx = \log\left|\dfrac{b}{a}\right|$

647. $\int_0^{\pi/2} \dfrac{dx}{a^2\sin^2 x + b^2\cos^2 x} = \dfrac{\pi}{2|ab|}$

648. $\int_0^{\pi/2} \dfrac{dx}{(a^2\sin^2 x + b^2\cos^2 x)^2} = \dfrac{\pi(a^2 + b^2)}{4a^3 b^3}, \quad (a, b > 0)$

649. $\int_0^{\pi/2} \sin^{n-1} x \cos^{m-1} x\, dx = \dfrac{1}{2}\mathrm{B}\left(\dfrac{n}{2}, \dfrac{m}{2}\right), \quad m \text{ and } n \text{ positive integers}$

650. $\int_0^{\pi/2} (\sin^{2n+1}\theta)\, d\theta = \dfrac{2 \cdot 4 \cdot 6 \dots (2n)}{1 \cdot 3 \cdot 5 \dots (2n+1)}, \quad (n = 1, 2, 3, \dots)$

651. $\int_0^{\pi/2} (\sin^{2n}\theta)\, d\theta = \dfrac{1 \cdot 3 \cdot 5 \dots (2n-1)}{2 \cdot 4 \dots (2n)}\left(\dfrac{\pi}{2}\right), \quad (n = 1, 2, 3, \dots)$

652. $\int_0^{\pi/2} \dfrac{x}{\sin x}\, dx = 2\left\{ \dfrac{1}{1^2} - \dfrac{1}{3^2} + \dfrac{1}{5^2} - \dfrac{1}{7^2} + \dots \right\}$

653. $\int_0^{\pi/2} \dfrac{dx}{1 + \tan^m x} = \dfrac{\pi}{4}$

654. $\int_0^{\pi/2} \sqrt{\cos\theta}\, d\theta = \dfrac{(2\pi)^{\frac{3}{2}}}{\left[\Gamma(\frac{1}{4})\right]^2}$

655. $\int_0^{\pi/2} (\tan^b\theta)\, d\theta = \dfrac{\pi}{2\cos\left(\frac{b\pi}{2}\right)}, \quad (0 < b < 1)$

656. $\int_0^\infty \dfrac{\tan^{-1}(ax) - \tan^{-1}(bx)}{x}\, dx = \dfrac{\pi}{2}\log\dfrac{a}{b}, \quad (a, b > 0)$

657. The area enclosed by a curve defined through the equation $x^{\frac{b}{c}} + y^{\frac{b}{c}} = a^{\frac{b}{c}}$ where $a > 0$, c a positive odd integer and b a positive even integer is given by $\dfrac{[\Gamma(\frac{c}{b})]^2}{\Gamma(\frac{2c}{b})}\left(\dfrac{2ca^2}{b}\right)$

658. $I = \displaystyle\iiint_R x^{b-1} y^{m-1} z^{n-1}\, dv$, where R denotes the region of space bounded by the co-ordinate planes and that portion of

the surface $\left(\frac{x}{a}\right)^p + \left(\frac{y}{b}\right)^q + \left(\frac{z}{c}\right)^k = 1$, which lies in the first octant and where $h, m, n, p, q, k, a, b, c$, denote positive real numbers is given by

$$\int_0^a x^{b-1} dx \int_0^b \left[1 - \left(\dfrac{x}{a}\right)^p\right]^{\frac{1}{e}} y^m dy \int_0^c \left[1 - \left(\dfrac{x}{a}\right)^p - \left(\dfrac{y}{b}\right)^q\right]^{\frac{1}{e}} z^{n-1} dz = \dfrac{a^b b^m c^n}{pqk} \dfrac{\Gamma\left(\frac{h}{p}\right)\Gamma\left(\frac{m}{q}\right)\Gamma\left(\frac{n}{k}\right)}{\Gamma\left(\frac{h}{p} + \frac{m}{q} + \frac{n}{k} + 1\right)}$$

659. $\displaystyle\int_0^\infty e^{-ax}\,dx = \frac{1}{a}, \quad (a > 0)$

660. $\displaystyle\int_0^\infty \frac{e^{-ax} - e^{-bx}}{x}\,dx = \log\frac{b}{a}, \quad (a, b > 0)$

661. $\displaystyle\int_0^\infty x^n e^{-ax}\,dx = \begin{cases} \frac{\Gamma(n+1)}{a^{n+1}} & n > -1,\ a > 0 \\ \text{or} \\ \frac{n!}{a^{n+1}} & (a > 0,\ n\text{ positive integer}) \end{cases}$

662. $\displaystyle\int_0^\infty x^n \exp(-ax^p)\,dx = \frac{\Gamma(k)}{pa^k}, \quad \left(n > -1,\ p > 0,\ a > 0,\ k = \frac{n+1}{p}\right)$

663. $\displaystyle\int_0^\infty e^{-a^2 x^2}\,dx = \frac{1}{2a}\sqrt{\pi} = \frac{1}{2a}\Gamma\left(\frac{1}{2}\right), \quad (a > 0)$

664. $\displaystyle\int_0^\infty x e^{-x^2}\,dx = \frac{1}{2}$

665. $\displaystyle\int_0^\infty x^2 e^{-x^2}\,dx = \frac{\sqrt{\pi}}{4}$

666. $\displaystyle\int_0^\infty x^{2n} e^{-ax^2}\,dx = \frac{1 \cdot 3 \cdot 5 \ldots (2n-1)}{2^{n+1}a^n}\sqrt{\frac{\pi}{a}} \quad (a > 0,\ n > -\tfrac{1}{2})$

667. $\displaystyle\int_0^\infty x^{2n+1} e^{-ax^2}\,dx = \frac{n!}{2a^{n+1}}, \quad (a > 0,\ n > -1)$

668. $\displaystyle\int_0^1 x^m e^{-ax}\,dx = \frac{m!}{a^{m+1}}\left[1 - e^{-a}\sum_{r=0}^m \frac{a^r}{r!}\right]$

669. $\displaystyle\int_0^\infty e^{\left(-x^2 - \frac{a^2}{x^2}\right)}\,dx = \frac{e^{-2a}\sqrt{\pi}}{2}, \quad (a \geq 0)$

670. $\displaystyle\int_0^\infty e^{-nx}\sqrt{x}\,dx = \frac{1}{2n}\sqrt{\frac{\pi}{n}} \quad (n > 0)$

671. $\displaystyle\int_0^\infty \frac{e^{-nx}}{\sqrt{x}}\,dx = \sqrt{\frac{\pi}{n}} \quad (n > 0)$

672. $\displaystyle\int_0^\infty e^{-ax}(\cos mx)\,dx = \frac{a}{a^2 + m^2}, \quad (a > 0)$

673. $\displaystyle\int_0^\infty e^{-ax}(\sin mx)\,dx = \frac{m}{a^2 + m^2}, \quad (a > 0)$

674. $\displaystyle\int_0^\infty x e^{-ax}[\sin(bx)]\,dx = \frac{2ab}{(a^2 + b^2)^2}, \quad (a > 0)$

675. $\displaystyle\int_0^\infty x e^{-ax}[\cos(bx)]\,dx = \frac{a^2 - b^2}{(a^2 + b^2)^2}, \quad (a > 0)$

676. $\displaystyle\int_0^\infty x^n e^{-ax}[\sin(bx)]\,dx = \frac{n![(a+ib)^{n+1} - (a-ib)^{n+1}]}{2i(a^2 + b^2)^{n+1}}, \quad (i^2 = -1,\ a > 0)$

677. $\displaystyle\int_0^\infty x^n e^{-ax}[\cos(bx)]\,dx = \frac{n![(a-ib)^{n+1} + (a+ib)^{n+1}]}{2(a^2 + b^2)^{n+1}}, \quad (i^2 = -1,\ a > 0,\ n > -1)$

678. $\displaystyle\int_0^\infty \frac{e^{-ax}\sin x}{x}\,dx = \cot^{-1}a, \quad (a > 0)$

679. $\displaystyle\int_0^\infty e^{-a^2 x^2}\cos bx\,dx = \frac{\sqrt{\pi}}{2|a|}\exp\left(-\frac{b^2}{4a^2}\right), \quad (ab \neq 0)$

680. $\displaystyle\int_0^\infty e^{-t\cos\phi}t^{b-1}[\sin(t\sin\phi)]\,dt - [\Gamma(b)]\sin(b\phi), \quad \left(b > 0,\ -\frac{\pi}{2} < \phi < \frac{\pi}{2}\right)$

681. $\displaystyle\int_0^\infty e^{-t\cos\phi}t^{b-1}[\cos(t\sin\phi)]\,dt - [\Gamma(b)]\cos(b\phi), \quad \left(b > 0,\ -\frac{\pi}{2} < \phi < \frac{\pi}{2}\right)$

682. $\displaystyle\int_0^\infty t^{b-1}\cos t\,dt = [\Gamma(b)]\cos\left(\frac{b\pi}{2}\right), \quad (0 < b < 1)$

683. $\displaystyle\int_0^\infty t^{b-1}(\sin t)\,dt = [\Gamma(b)]\sin\left(\frac{b\pi}{2}\right), \quad (0 < b < 1)$

684. $\displaystyle\int_0^1 (\log x)^n\,dx = (-1)^n \cdot n! \quad (n > -1)$

685. $\displaystyle\int_0^1 \left(\log\frac{1}{x}\right)^{\frac{1}{2}}\,dx = \frac{\sqrt{\pi}}{2}$

686. $\displaystyle\int_0^1 \left(\log\frac{1}{x}\right)^{-\frac{1}{2}}\,dx = \sqrt{\pi}$

687. $\displaystyle\int_0^1 \left(\log\frac{1}{x}\right)^n dx = n!$

688. $\displaystyle\int_0^1 x\log(1-x)\, dx = -\frac{3}{4}$

689. $\displaystyle\int_0^1 x\log(1+x)\, dx = \frac{1}{4}$

690. $\displaystyle\int_0^1 x^m(\log x)^n\, dx = \frac{(-1)^n n!}{(m+1)^{n+1}}, \quad m > -1,\ n = 0, 1, 2, \ldots$
If $n \neq 0, 1, 2, \ldots$ replace $n!$ by $\Gamma(n+1)$.

691. $\displaystyle\int_0^1 \frac{\log x}{1+x}\, dx = -\frac{\pi^2}{12}$

692. $\displaystyle\int_0^1 \frac{\log x}{1-x}\, dx = -\frac{\pi^2}{6}$

693. $\displaystyle\int_0^1 \frac{\log(1+x)}{x}\, dx = \frac{\pi^2}{12}$

694. $\displaystyle\int_0^1 \frac{\log(1-x)}{x}\, dx = -\frac{\pi^2}{6}$

695. $\displaystyle\int_0^1 (\log x)[\log(1+x)]\, dx = 2 - 2\log 2 - \frac{\pi^2}{12}$

696. $\displaystyle\int_0^1 (\log x)[\log(1-x)]\, dx = 2 - \frac{\pi^2}{6}$

697. $\displaystyle\int_0^1 \frac{\log x}{1-x^2}\, dx = -\frac{\pi^2}{8}$

698. $\displaystyle\int_0^1 \log\left(\frac{1+x}{1-x}\right)\cdot\frac{dx}{x} = \frac{\pi^2}{4}$

699. $\displaystyle\int_0^1 \frac{\log x\, dx}{\sqrt{1-x^2}} = -\frac{\pi}{2}\log 2$

700. $\displaystyle\int_0^1 x^m \left[\log\left(\frac{1}{x}\right)\right]^n dx = \frac{\Gamma(n+1)}{(m+1)^{n+1}}, \quad \text{if } m+1 > 0,\ n+1 > 0$

701. $\displaystyle\int_0^1 \frac{(x^p - x^q)\, dx}{\log x} = \log\left(\frac{p+1}{q+1}\right), \quad (p+1 > 0,\ q+1 > 0)$

702. $\displaystyle\int_0^1 \frac{dx}{\sqrt{\log\left(\frac{1}{x}\right)}} = \sqrt{\pi}, \text{ (same as integral 686)}$

703. $\displaystyle\int_0^\infty \log\left(\frac{e^x + 1}{e^x - 1}\right) dx = \frac{\pi^2}{4}$

704. $\displaystyle\int_0^{\pi/2} (\log\sin x)\, dx = \int_0^{\pi/2} \log\cos x\, dx = -\frac{\pi}{2}\log 2$

705. $\displaystyle\int_0^{\pi/2} (\log\sec x)\, dx = \int_0^{\pi/2} \log\csc x\, dx = \frac{\pi}{2}\log 2$

706. $\displaystyle\int_0^\pi x(\log\sin x)\, dx = -\frac{\pi^2}{2}\log 2$

707. $\displaystyle\int_0^{\pi/2} (\sin x)(\log\sin x)\, dx = \log 2 - 1$

708. $\displaystyle\int_0^{\pi/2} (\log\tan x)\, dx = 0$

709. $\displaystyle\int_0^\pi \log(a \pm b\cos x)\, dx = \pi\log\left(\frac{a + \sqrt{a^2 - b^2}}{2}\right), \quad (a \geq b)$

710. $\displaystyle\int_0^\pi \log(a^2 - 2ab\cos x + b^2)\, dx = \begin{cases} 2\pi\log a & a \geq b > 0 \\ 2\pi\log b & b \geq a > 0 \end{cases}$

711. $\displaystyle\int_0^\infty \frac{\sin ax}{\sinh bx}\, dx = \frac{\pi}{2|b|}\tanh\frac{a\pi}{2b}$

712. $\displaystyle\int_0^\infty \frac{\cos ax}{\cosh bx}\, dx = \frac{\pi}{2|b|}\operatorname{sech}\frac{a\pi}{2b}$

713. $\displaystyle\int_0^\infty \frac{dx}{\cosh ax} = \frac{\pi}{2|a|}$

714. $\displaystyle\int_0^\infty \frac{x\,dx}{\sinh ax} = \frac{\pi^2}{4a^2}$ $(a>0)$

715. $\displaystyle\int_0^\infty e^{-ax}(\cosh bx)\,dx = \frac{a}{a^2-b^2},$ $(0 \le |b| < a)$

716. $\displaystyle\int_0^\infty e^{-ax}(\sinh bx)\,dx = \frac{b}{a^2-b^2},$ $(0 \le |b| < a)$

717. $\displaystyle\int_0^\infty \frac{\sinh ax}{e^{bx}+1}\,dx = \frac{\pi}{2b}\csc\frac{a\pi}{b} - \frac{1}{2a}$ $(b>0)$

718. $\displaystyle\int_0^\infty \frac{\sinh ax}{e^{bx}-1}\,dx = \frac{1}{2a} - \frac{\pi}{2b}\cot\frac{a\pi}{b}$ $(b>0)$

719. $\displaystyle\int_0^{\pi/2} \frac{dx}{\sqrt{1-k^2\sin^2 x}} = \frac{\pi}{2}\left[1+\left(\frac{1}{2}\right)^2 k^2 + \left(\frac{1\cdot3}{2\cdot4}\right)^2 k^4 + \left(\frac{1\cdot3\cdot5}{2\cdot4\cdot6}\right)^2 k^6 + \cdots\right],$

if $k^2 < 1$

720. $\displaystyle\int_0^{\pi/2} \sqrt{1-k^2\sin^2 x}\,dx = \frac{\pi}{2}\left[1-\left(\frac{1}{2}\right)^2 k^2 - \left(\frac{1\cdot3}{2\cdot4}\right)^2 \frac{k^4}{3} - \left(\frac{1\cdot3\cdot5}{2\cdot4\cdot6}\right)2\frac{k^6}{5} - \cdots\right],$

if $k^2 < 1$

721. $\displaystyle\int_0^\infty e^{-x}\log x\,dx = -\gamma = -0.5772157\ldots$

722. $\displaystyle\int_0^\infty e^{-x^2}\log x\,dx = -\frac{\sqrt{\pi}}{4}(\gamma + 2\log 2)$

723. $\displaystyle\int_0^\infty \left(\frac{1}{1-e^{-x}} - \frac{1}{x}\right)e^{-x}\,dx = \gamma = 0.5772157\ldots$ [Euler's Constant]

724. $\displaystyle\int_0^\infty \frac{1}{x}\left(\frac{1}{1+x} - e^{-x}\right)dx = \gamma = 0.5772157\ldots$

For n even :

725. $\displaystyle\int \cos^n x\,dx = \frac{1}{2^{n-1}}\sum_{k=0}^{n/2-1}\binom{n}{k}\frac{\sin(n-2k)x}{(n-2k)} + \frac{1}{2^n}\binom{n}{n/2}x$

726. $\displaystyle\int \sin^n x\,dx = \frac{1}{2^{n-1}}\sum_{k=0}^{n/2-1}\binom{n}{k}\frac{\sin[(n-2k)(\frac{\pi}{2}-x)]}{2k-n} + \frac{1}{2^n}\binom{n}{n/2}x$

For n odd:

727. $\displaystyle\int \cos^n x\,dx = \frac{1}{2^{n-1}}\sum_{k=0}^{(n-1)/2}\binom{n}{k}\frac{\sin(n-2k)x}{n-2k}$

728. $\displaystyle\int \sin^n x\,dx = \frac{1}{2^{n-1}}\sum_{k=0}^{(n-1)/2}\binom{n}{k}\frac{\sin\left[(n-2k)\left(\frac{\pi}{2}-x\right)\right]}{2k-n}$

DIFFERENTIAL EQUATIONS

Special Formulas

Certain types of differential equations occur sufficiently often to justify the use of formulas for the corresponding particular solutions. The following set of tables I to XIV covers all first, second, and nth order ordinary linear differential equations with constant coefficients for which the right members are of the form $P(x)e^{rx}\sin sx$ or $P(x)e^{rx}\cos sx$, where r and s are constants and $P(x)$, is a polynomial of degree n.

When the right member of a reducible linear partial differential equation with constant coefficients is not zero, particular solutions for certain types of right members are contained in tables XV to XXI. In these tables both F and P are used to denote polynomials, and it is assumed that no denominator is zero. In any formula the roles of x and y may be reversed throughout, changing a formula in which x dominates to one in which y dominates. Tables XIX, XX, XXI are applicable whether the equations are reducible or not. The symbol $\binom{m}{n}$ stands for $\frac{m!}{(m-n)!n!}$ and is the $(n+1)^{\text{st}}$ coefficient in the expansion of $(a+b)^m$. Also $0! = 1$ by definition.

SOLUTION OF LINEAR DIFFERENTIAL EQUATIONS WITH CONSTANT COEFFICIENTS

Any linear differential equation with constant coefficients may be written in the form

$$p(D)y = R(x)$$

where

- D is the differential operation: $Dy = \frac{dy}{dx}$

- $p(D)$ is a polynomial in D,

- y is the dependent variable,

- x is the independent variable,

- $R(x)$ is an arbitrary function of x.

A power of D represents repeated differentiation, that is

$$D^n y = \frac{d^n y}{dx^n}$$

For such an equation, the general solution may be written in the form

$$y = y_c + y_p$$

where y_p is any particular solution, and y_c is called the *complementary function*. This complementary function is defined as the general solution of the *homogeneous equation*, which is the original differential equation with the right side replaced by zero, i.e.

$$p(D)y = 0$$

The complementary function y_c may be determined as follows:

1. Factor the polynomial $p(D)$ into real and complex linear factors, just as if D were a variable instead of an operator.

2. For each nonrepeated linear factor of the form $(D - a)$, where a is real, write down a term of the form

$$ce^{ax}$$

   where c is an arbitrary constant.

3. For each repeated real linear factor of the form $(D - a)^n$, write down n terms of the form

$$c_1 e^{ax} + c_2 x e^{ax} + c_3 x^2 e^{ax} + \cdots + c_n x^{n-1} e^{ax}$$

   where the c_i's are arbitrary constants.

4. For each non-repeated conjugate complex pair of factors of the form $(D - a + ib)(D - a - ib)$, write down 2 terms of the form

$$c_1 e^{ax} \cos bx + c_2 e^{ax} \sin bx$$

5. For each repeated conjugate complex pair of factors of the form $(D - a + ib)^n (D - a - ib)^n$, write down $2n$ terms of the form

$$c_1 e^{ax} \cos bx + c_2 e^{ax} \sin bx + c_3 x e^{ax} \cos bx + c_4 x e^{ax} \sin bx$$
$$+ \cdots + c_{2n-1} x^{n-1} e^{ax} \cos bx + c_{2n} x^{n-1} e^{ax} \sin bx$$

6. The sum of all the terms thus written down is the complementary function y_c.

To find the particular solution y_p, use the following tables, as shown in the examples. For cases not shown in the tables, there are various methods of finding y_p. The most general method is called *variation of parameters*. The following example illustrates the method:

Example: Find y_p for $(D^2 - 4)y = e^x$.

This example can be solved most easily by use of equation 63 in the tables following. However it is given here as an example of the method of variation of parameters.

The complementary function is

$$y_c = c_1 e^{2x} + c_2 e^{-2x}$$

To find y_p, replace the constants in the complementary function with unknown functions,

$$y_p = u e^{2x} + v e^{-2x}$$

We now prepare to substitute this assumed solution into the original equation. We begin by taking all the necessary derivatives:

$$y_p = u e^{2x} + v e^{-2x}$$
$$y'_p = 2u e^{2x} - 2v e^{-2x} + u' e^{2x} + v' e^{-2x}$$

For each derivative of y_p except the highest, we set the sum of all the terms containing u' and v' to 0. Thus the above equation becomes

$$u' e^{2x} + v' e^{-2x} = 0 \quad \text{and} \quad y'_p = 2u e^{2x} - 2v e^{-2x}$$

Continuing to differentiate, we have

$$y''_p = 4u e^{2x} + 4v e^{-2x} + 2u' e^{2x} - 2v' e^{-2x}$$

When we substitute into the original equation, all the terms not containing u' or v' cancel out. This is a consequence of the method by which y_p was set up.

Thus all that is necessary is to write down the terms containing u' or v' in the highest order derivative of y_p, multiply by the constant coefficient of the highest power of D in $p(D)$, and set it equal to $R(x)$. Together with the previous terms in u' and v' which were set equal to 0, this gives us as many linear equations in the first derivatives of the unknown functions as there are unknown functions. The first derivatives may then be solved for by algebra, and the unknown functions found by integration. In the present example, this becomes

$$u' e^{2x} + v' e^{-2x} = 0$$
$$2u' e^{2x} - 2v' e^{-2x} = e^x$$

We eliminate v' and u' separately, getting

$$4u' e^{2x} = e^x$$
$$4v' e^{-2x} = -e^x$$

Thus

$$u' = \tfrac{1}{4} e^{-x}$$
$$v' = -\tfrac{1}{4} e^{3x}$$

Therefore, by integrating

$$u = -\tfrac{1}{4} e^{-x}$$
$$v = -\tfrac{1}{12} e^{3x}$$

A constant of integration is not needed, since we need only one particular solution. Thus

$$y_p = u e^{2x} + v e^{-2x} = -\frac{1}{4} e^{-x} e^{2x} - \frac{1}{12} e^{3x} e^{-2x}$$
$$= -\frac{1}{4} e^x - \frac{1}{12} e^x = -\frac{1}{3} e^x$$

and the general solution is

$$y = y_c + y_p = c_1 e^{2x} + c_2 e^{-2x} - \frac{1}{3} e^x$$

The following samples illustrate the use of the tables.

Example 1: Solve $(D^2 - 4)y = \sin 3x$. Substitution of $q = -4$, $s = 3$ in formula 24 gives

$$y_p = \frac{\sin 3x}{-9 - 4}$$

wherefore the general solution is

$$y = c_1 e^{2x} + c_2 e^{-2x} - \frac{\sin 3x}{13}$$

Example 2: Obtain a particular solution of $(D^2 - 4D + 5)y = x^2 e^{3x} \sin x$.
  Applying formula 40 with $a = 2$, $b = 1$, $r = 3$, $s = 1$, $P(x) = x^2$, $s + b = 2$, $s - b = 0$, $a - r = -1$, $(a - r)^2 + (s + b)^2 = 5$, $(a - r)^2 + (s - b)^2 = 1$, we have

$$
\begin{aligned}
y_p &= \frac{e^{3x} \sin x}{2}\left[\left(\frac{2}{5} - \frac{0}{1}\right)x^2 + \left(\frac{2(-1)2}{25} - \frac{2(-1)0}{1}\right)2x + \left(\frac{3\cdot1\cdot2 - 2^3}{125} - \frac{3\cdot1\cdot0 - 0}{1}\right)2\right] \\
&\quad - \frac{e^{3x} \cos x}{2}\left[\left(\frac{-1}{5} - \frac{-1}{1}\right)x^2 + \left(\frac{1-4}{25} - \frac{1-0}{1}\right)2x + \left(\frac{-1 - 3(-1)4}{125} - \frac{-1 - 3(-1)0}{1}\right)2\right] \\
&= \left(\frac{1}{5}x^2 - \frac{4}{25}x - \frac{2}{125}\right)e^{3x}\sin x + \left(-\frac{2}{5}x^2 + \frac{28}{25} - \frac{136}{125}\right)e^{3x}\cos x
\end{aligned}
$$

The special formulas effect a very considerable saving of time in problems of this type.

Example 3: Obtain a particular solution of $(D^2 - 4D + 5)y = x^2 e^{2x} \cos x$. (Compare with Example 2.)
  Formula 40 is not applicable here since for this equation $r = a$, $s = b$, wherefore the denominator $(a - r)^2 + (s - b)^2 = 0$. We turn instead to formula 44. Substituting $a = 2$, $b = 1$, $P(x) = x^2$ and replacing sin by cos, cos by -sin, we obtain

$$
\begin{aligned}
y_p &= \frac{e^{2x} \cos x}{4}\left(x^2 - \frac{2}{4}\right) + \frac{e^{2x} \sin x}{2}\int\left(x^2 - \frac{1}{2}\right)dx \\
&= \left(\frac{x^2}{4} - \frac{1}{8}\right)e^{2x}\cos x + \left(\frac{x^3}{6} - \frac{x}{4}\right)e^{2x}\sin x
\end{aligned}
$$

which is the required solution.

Example 4: Find z_p for $(D_x - 3D_y)z = \ln(y + 3x)$. Referring to Table XV we note that formula 69 (not 68) is applicable. This gives

$$z_p = x\ln(y + 3x)$$

It is easily seen that $-y/3\,\ln(y + 3x)$ would serve equally well.

Example 5: Solve $(D_x + 2D_y - 4)z = y\cos(y - 2x)$.
  Since R in formula 76 contains a polynomial in x, not y, we rewrite the given equation in the form $(D_y + \frac{1}{2}D_x - 2)z = \frac{1}{2}y\cos(y - 2x)$. Then

$$z_c = e^{2y}F\left(x - \frac{1}{2}y\right) = e^{2x}f(2x - y)$$

and by the formula

$$z_p = -\frac{1}{2}\cos(y - 2x)\cdot\left(\frac{y}{2} + \frac{\frac{1}{2}}{2}\right) = -\frac{1}{8}(2y + 1)\cos(y - 2x)$$

Example 6: Find z_p for $(D_x + 4D_y)^3 z = (2x - y)^2$.
  Using formula 79, we obtain

$$z_p = \frac{\int\int\int u^2\,du^3}{[2 + 4(-1)]^3} = \frac{u^5}{5\cdot4\cdot3\cdot(-8)} = -\frac{(2x - y)^5}{480}$$

Example 7:   Find z_p for $(D_x^3 + 5D_x^2 D_y - 7D_x + 4)z = e^{2x+3y}$. By formula 87

$$z_p = \frac{e^{2x+3y}}{2^3 + 5 \cdot 2^2 \cdot 3 - 7 \cdot 2 + 4} = \frac{e^{2x+3y}}{58}$$

Example 8:   Find z_p for

$$(D_x^4 + 6D_x^3 D_y + D_x D_y + D_y^2 + 9)z = \sin(3x + 4y)$$

Since every term in the left number is of even degree in the two operators D_x and D_y, formula 90 is applicable. It gives

$$z_p = \frac{\sin(3x + 4y)}{(-9)^2 + 6(-9)(-12) + (-12) + (-16) + 9}$$

$$= \frac{\sin(3x + 4y)}{710}$$

Table I: $(D - a)y = R$

R	y_p
1. e^{rx}	$\dfrac{e^{rx}}{r-a}$
2. $\sin sx*$	$-\dfrac{a \sin sx + s \cos sx}{a^2 + s^2} = \dfrac{1}{\sqrt{a^2 + s^2}} \sin\left(sx + \tan^{-1}\dfrac{s}{a}\right)$
3. $P(x)$	$-\dfrac{1}{a}\left[P(x) + \dfrac{P'(x)}{a} + \dfrac{P''(x)}{a^2} + \cdots + \dfrac{P^{(n)}(x)}{a^n}\right]$
4. $e^{rx} \sin sx*$	Replace a by $a - r$ in formula 2 and multiply by e^{rx}.
5. $P(x)\ \ e^{rx}$	Replace a by $a - r$ in formula 3 and multiply by e^{rx}.

6. $P(x) \sin sx*$ $-\sin sx\left[\dfrac{a}{a^2+s^2}P(x) + \dfrac{a^2-s^2}{(a^2+s^2)^2}P'(x) + \dfrac{a^3-3as^2}{(a^2+s^2)^3}P''(x) + \cdots + \dfrac{a^k - \binom{k}{2}a^{k-2}s^2 + \binom{k}{4}a^{k-4}s^4 - \cdots}{(a^2+s^2)^k}P^{(k-1)}(x) + \cdots\right]$

$-\cos sx\left[\dfrac{s}{a^2+s^2}P(x) + \dfrac{2as}{(a^2+s^2)^2}P'(x) + \dfrac{3a^2s-s^3}{(a^2+s^2)^3}P''(x) + \cdots + \dfrac{\binom{k}{1}a^{k-1}s - \binom{k}{3}a^{k-3}s^3 + \cdots}{(a^2+s^2)^k}P^{(k-1)}(x) + \cdots\right]$

7. $P(x)e^{rx}\sin sx*$	Replace a by $a - r$ in formula 6 and multiply by e^{rx}.
8. e^{ax}	xe^{ax}
9. $e^{ax}\sin sx*$	$-\dfrac{e^{ax}\cos sx}{s}$
10. $P(x)e^{ax}$	$e^{ax}\int P(x)\,dx$

11. $P(x)e^{ax}\sin sx$ $\dfrac{e^{ax}\sin sx}{s}\left[\dfrac{P'(x)}{s^3} - \dfrac{P'''(x)}{s^3} + \dfrac{P^v(x)}{s^5} - \cdots\right] - \dfrac{e^{ax}\cos sx}{s}\left[P(x) - \dfrac{P''(x)}{s^2} + \dfrac{P^{iv}(x)}{s^4} - \cdots\right]$

*For $\cos sx$ in R replace "sin" by "cos" and "cos" by "$-$sin" in y_p.

$$D^n = \frac{d^n}{dx^n} \qquad \binom{m}{n} = \frac{m!}{(m-n)!n!} \qquad 0! = 1$$

Table II: $(D - a)^2 y = R$

R	y_p
12. e^{rx}	$\dfrac{e^{rx}}{(r-a)^2}$
13. $\sin sx*$	$\dfrac{1}{(a^2+s^2)}[(a^2-s^2)\sin sx + 2as\cos sx] = \dfrac{1}{a^2+s^2}\sin\left(sx + \tan^{-1}\dfrac{2as}{a^2-s^2}\right)$
14. $P(x)$	$\dfrac{1}{a^2}\left[P(x) + \dfrac{2P'(x)}{a} + \dfrac{3P''(x)}{a^2} + \cdots + \dfrac{(n+1)P^{(n)}(x)}{a^n}\right]$
15. $e^{rx}\sin sx*$	Replace a by $a - r$ in formula 13 and multiply by e^{rx}.
16. $P(x)e^{rx}$	Replace a by $a - r$ in formula 14 and multiply by e^{rx}.

17. $P(x)\sin sx*$ $\sin sx\left[\dfrac{a^2-s^2}{(a^2+s^2)^2}P(x) + 2\dfrac{a^3-3as^2}{(a^2+s^2)^3}P'(x) + 3\dfrac{a^4-6a^2s^2+s^4}{(a^2+s^2)^4}P''(x) + \cdots\right.$

$\left. +(k-1)\dfrac{a^k - \binom{k}{2}a^{k-2}s^2 + \binom{k}{4}a^{k-4}s^4 - \cdots}{(a^2+s^2)^k}P^{(k-2)}(x) + \cdots\right]$

$+\cos sx\left[\dfrac{2as}{(a^2+s^2)^2}P(x) + 2\dfrac{3a^2s-s^3}{(a^2+s^2)^3}P'(x) + 3\dfrac{4a^3s-4as^3}{(a^2+s^2)^4}P''(x) + \cdots\right.$

$\left. +(k-1)\dfrac{\binom{k}{1}a^{k-1}s - \binom{k}{3}a^{k-3}s^3 + \cdots}{(a^2+s^2)^k}P^{(k-2)}(x) + \cdots\right]$

18. $P(x)e^{rx}\sin sx*$ Replace a by $a - r$ in formula 17 and multiply by e^{rx}.

19. e^{ax} $\qquad \dfrac{1}{2}x^2 e^{ax}$

20. $e^{ax}\sin sx*$ $\qquad -\dfrac{e^{ax}\sin sx}{s^2}$

21. $P(x)e^{ax}$ $\qquad e^{ax}\displaystyle\int^{s^2}\!\!\int P(x)\,dx\,dx$

22. $P(x)e^{ax}\sin sx*$
$$-\frac{e^{ax}\sin sx}{s^2}\left[P(x) - \frac{3P''(x)}{s^2} + \frac{5P^{iv}(x)}{s^4} - \frac{7P^{vi}(x)}{s^6} + \cdots\right]$$
$$-\frac{e^{ax}\cos sx}{s^2}\left[\frac{2P'(x)}{s} + \frac{4P'''(x)}{s^3} - \frac{6P^{v}(x)}{s^5} - \cdots\right]$$

* For $\cos sx$ in R replace "sin" by "cos" and "cos" by "-sin" in y_p.

Table III: $(D^2 + q)y = R$

R	y_p
23. e^{rx}	$\dfrac{e^{rx}}{r^2 + q}$
24. $\sin sx*$	$\dfrac{\sin sx}{-s^2 + q}$
25. $P(x)$	$\dfrac{1}{q}\left[P(x) - \dfrac{P''(x)}{q} + \dfrac{P^{iv}(x)}{q^2} - \cdots + (-1)^k\dfrac{P^{(2k)}(x)}{qk} \cdots\right]$
26. $e^{rx}\sin sx$	$\dfrac{(r^2 - s^2 + q)e^{rx}\sin sx - 2rse^{rx}\cos sx}{(r^2 - s^2 + q)^2 + (2rs)^2} = \dfrac{e^{rx}}{\sqrt{(r^2 - s^2 + q)^2 + (2rs)^2}}\sin\left[sx - \tan^{-1}\dfrac{2rs}{r^2 - s^2 + q}\right]$
27. $P(x)e^{rx}$	$\dfrac{e^{rx}}{r^2 + q}\left[P(x) - \dfrac{2r}{r^2 + q}P'(x) + \dfrac{3r^2 - q}{(r^2 + q)^2}P''(x) - \dfrac{4r^3 - 4qr}{(r^2 + q)^3}P'''(x) + \cdots \right.$ $\left. + \cdots + (-1)^{k-1}\dfrac{\binom{k}{1}r^{k-1} - \binom{k}{3}r^{k-3}q + \binom{k}{5}r^{k-5}q^2 - \cdots}{(r^2 + q)^{k-1}}P^{(k-1)}(x) + \cdots\right]$
28. $P(x)\sin sx*$	$\dfrac{\sin sx}{(-s^2 + q)}\left[P(x) - \dfrac{3s^2 + q}{(-s^2 + q)^2}P''(x) + \dfrac{5s^4 + 10s^2q + q^2}{(-s^2 + q)^4}P^{iv}(x) + \cdots \right.$ $\left. + (-1)^k\dfrac{\binom{2k+1}{1}s^{2k} + \binom{2k+1}{3}s^{2k-2}q + \binom{2k+1}{5}s^{2k-4}q^2 + \cdots}{(-s^2 + q)^{2k}}P^{(2k)}(x) + \cdots\right]$ $-\dfrac{s\cos sx}{(-s^2 + q)}\left[\dfrac{2P'(x)}{(-s^2 + q)} - \dfrac{4s^2 + 4q}{(-s^2 + q)^3}P'''(x) + \cdots + (-1)^{k+1}\dfrac{\binom{2k}{1}s^{2k-2} + \binom{2k}{3}s^{2k-4}q + \cdots}{(-s^2 + q)^{2k-1}}P^{(2k-1)}(x) + \cdots\right]$

Table IV: $(D^2 + b^2)y = R$

29. $\sin bx*$	$-\dfrac{x\cos bx}{2b}$
30. $P(x)\sin bx*$	$\dfrac{\sin bx}{(2b)^2}\left[P(x) - \dfrac{P''(x)}{(2b)^2} + \dfrac{P^{iv}(x)}{(2b)^4} - \cdots\right] - \dfrac{\cos bx}{2b}\displaystyle\int\left[P(x) - \dfrac{P''(x)}{(2b)^2} + \cdots\right]dx$

* For $\cos sx$ in R replace "sin" by "cos" and "cos" by "$-\sin$" in y_p.

Table V: $(D^2 + pD + q)y = R$

R	y_p
31. e^{rx}	$\dfrac{e^{rx}}{r^2 + pr + q}$
32. $\sin sx*$	$\dfrac{(q - s^2)\sin sx - ps\cos sx}{(q - s^2)^2 + (ps)^2} = \dfrac{1}{\sqrt{(q - s^2)^2 + (ps)^2}}\sin\left(sx - \tan^{-1}\dfrac{ps}{q - s^2}\right)$
33. $P(x)$	$\dfrac{1}{q}\left[P(x) - \dfrac{p}{q}P'(x) + \dfrac{p^2 - q}{q^2}P''(x) - \dfrac{p^3 - 2pq}{q^3}P'''(x) + \cdots + (-1)^n\dfrac{p^n - \binom{n-1}{1}p^{n-2}q + \binom{n-2}{2}p^{n-4}q^2 - \cdots}{q^n}P^{(n)}(x)\right]$
34. $e^{rx}\sin sx*$	Replace p by $p + 2r$, q by $q + pr + r^2$ in formula 32 and multiply by e^{rx}.
35. $P(x)e^{rx}$	Replace p by $p + 2r$, q by $q + pr + r^2$ in formula 33 and multiply by e^{rx}.

Table VI: $(D - b)(D - a)y = R$

36. $P(x) \sin sx*$

$$\frac{\sin sx}{b-a}\left[\left(\frac{a}{a^2+s^2}-\frac{b}{b^2+s^2}\right)P(x)+\left(\frac{a^2-s^2}{(a^2+s^2)^2}-\frac{b^2-s^2}{(b^2+s^2)^2}\right)P'(x)\right.$$
$$\left.+\left(\frac{a^3-3as^2}{(a^2+s^2)^3}-\frac{b^3-3bs^2}{(b^2+s^2)^3}\right)P''(x)+\cdots\right]$$
$$+\frac{\cos sx}{b-a}\left(\frac{s}{a^2+s^2}-\frac{s}{b^2+s^2}\right)P(x)+\left(\frac{2as}{(a^2+s^2)^2}-\frac{2bs}{(b^2+s^2)^2}\right)P'(x)$$
$$+\left(\frac{3a^2s-s^2}{(a^2+s^2)^3}-\frac{3b^2s-s^3}{(b^2+s^2)^3}\right)P''(x)+\cdots\right]$$

37. $P(x)e^{rx}\sin sx*$ Replace a by $a-r$, b by $b-r$ in formula 36 and multiply by e^{rx}.

38. $P(x)e^{ax}$

$$\frac{e^{ax}}{a-b}\left[\int P(x)\,dx+\frac{P(x)}{(b-a)}+\frac{P'(x)}{(b-a)^2}+\frac{P''(x)}{(b-a)^3}+\cdots+\frac{P^{(n)}(x)}{(b-a)^{n+1}}\right]$$

* For $\cos sx$ in R replace "sin" by "cos" and "cos" by "-sin" in y_p.

† For additional terms, com e with formula 6.

Table VII: $(D^2 - 2aD + a^2 + b^2)y = R$

R	y_p

39. $P(x)\sin sx*$

$$\frac{\sin sx}{2b}\left[\left(\frac{s+b}{a^2+(s+b)^2}-\frac{s-b}{a^2+(s-b)^2}\right)P(x)+\left(\frac{2a(s+b)}{\left[a^2+(s+b)^2\right]^2}-\frac{2a(s-b)}{\left[a^2+(s-b)^2\right]^2}\right)P'(x)\right.$$
$$\left.+\left(\frac{3a^2(s+b)-(s+b)^3}{\left[a^2+(s+b)^2\right]^3}-\frac{3a^2(s-b)-(s-b)^3}{\left[a^2+(s-b)^2\right]^3}\right)P''(x)+\cdots\right]$$
$$-\frac{\cos sx}{2b}\left[\left(\frac{a}{a^2+(s+b)^2}-\frac{a}{a^2+(s-b)^2}\right)P(x)+\left(\frac{a^2-(s+b)^2}{\left[a^2+(s+b)^2\right]^2}-\frac{a^2-(s-b)^2}{\left[a^2+(s-b)^2\right]^2}\right)P'(x)\right.$$
$$\left.+\left(\frac{a^2-3a(s+b)^2}{\left[a^2+(s+b)^2\right]^3}-\frac{a^3-3a(s-b)^2}{\left[a^2+(s-b)^2\right]^3}\right)P''(x)+\cdots\right]^{\dagger}$$

40. $P(x)e^{rx}\sin sx*$ Replace a by $a-r$ in formula 39 and multiply by e^{rx}.

41. $P(x)e^{ax}$

$$\frac{e^{ax}}{b^2}\left[P(x)-\frac{P''(x)}{b^2}+\frac{P^{iv}(x)}{b^4}-\cdots\right]$$

42. $e^{ax}\sin sx*$

$$\frac{e^{ax}\sin sx}{-s^2+b^2}$$

43. $e^{ax}\sin bx*$

$$-\frac{xe^{ax}\cos bx}{2b}$$

44. $P(x)e^{ax}\sin bx*$

$$\frac{e^{ax}\sin bx}{(2b)^2}\left[P(x)-\frac{P''(x)}{(2b)^2}+\frac{P^{iv}(x)}{(2b)^4}-\cdots\right]$$
$$-\frac{e^{ax}\cos bx}{2b}\int\left[P(x)-\frac{P''(x)}{(2b)^2}+\frac{P^{iv}(x)}{(2b)^4}-\cdots\right]dx$$

* For $\cos sx$ in R replace "sin' by "cos' and "cos" by "-sin" in y_p.

For additional terms, com e with formula 6.

Table VIII: $f(D)y = [D^n + a_{n-1}D^{n-1} + \cdots + a_1 D + a_0]y = R$

R	y_p

45. e^{rx}

$$\frac{e^{rx}}{f(r)}$$

46. $\sin sx*$

$$\frac{[a_0-a_2s^2+a_4s^4-\cdots]\sin sx-[a_1s-a_3s^3+a_5s^5+\cdots]\cos sx}{[a_0-a_2s^2+a_4s^4-\cdots]^2+[a_1s-a_3s^3+a_5s^5-\cdots]^2}$$

Table IX: $f(D^2)y = R$

47. $\sin sx*$

$$\frac{\sin sx}{f(-s^2)}=\frac{\sin sx}{a_0-a_2s^2+\cdots\pm s^{2n}}$$

Table X: $(D - a)^n y = R$

48. e^{rx} $\quad\dfrac{e^{rx}}{(r - a)^n}$

49. $\sin sx*$ $\quad\dfrac{(-1)^n}{(a^2 + s^2)^2}\{[a^n - \binom{n}{2}a^{n-2}s^2 + \binom{4}{n}a^{n-4}s^4 - \cdots]\sin sx$

$\qquad\qquad\qquad + [\binom{n}{1}a^{n-1}s - \binom{n}{3}a^{n-3}s^3 + \cdots]\cos sx\}$

50. $P(x)$ $\quad\dfrac{(-1)^n}{a^n}\left[P(x) + \binom{n}{1}\dfrac{P'(x)}{a} + \binom{n+1}{2}\dfrac{P'(x)}{a^2} + \binom{n+2}{3}\dfrac{P'''(x)}{a^2} + \cdots\right]$

51. $e^{rx}\sin sx*$ $\quad$ Replace a by $a - r$ in formula 49 and multiply by e^{rx}.

52. $e^{rx}P(x)$ $\quad$ Replace a by $a - r$ in formula 50 and multiply by e^{rx}.

53. $P(x)\sin sx*$ $\quad(-1)^n\sin sx[A_n P(x) + \binom{n}{1}A_{n+1}P'(x) + \binom{n+1}{2}A_{n+2}P''(x) + \binom{n+2}{3}A_{n+3}P'''(x) + \cdots]$

$\qquad\qquad + (-1)^n\cos sx[B_n P(x) + \binom{n}{1}B_{n+1}P'(x) + \binom{n+1}{2}B_{n+2}P''(x) + \binom{n+2}{3}B_{n+3}P'''(x) + \cdots]$

$$A_1 = \frac{a}{a^2 + s^2},\quad A_2 = \frac{a^2 - s^2}{(a^2 + s^2)^2},\quad\dots,\quad A_k = \frac{a^k - \binom{k}{2}a^{k-2}s^2 + \binom{k}{4}a^{k-4}s^4 - \cdots}{(a^2 + s^2)^k}$$

$$B_1 = \frac{a}{a^2 + s^2},\quad B_2 = \frac{2as}{(a^2 + s^2)^2},\quad\dots,\quad B_k = \frac{\binom{k}{1}a^{k-1}s - \binom{k}{3}a^{k-3}s^3 + \cdots}{(a^2 + s^2)^k}$$

54. $e^{rx}\sin sx*$ $\quad$ Replace a by $a - r$ in formula 53 and multiply by e^{rx}.

55. $e^{ax}P(x)$ $\quad e^{ax}\underbrace{\displaystyle\int\int\cdots\int}_{n-1}P(x)\,dx^n$

56. $P(x)e^{ax}\sin sx*$ $\quad\dfrac{(-1)^{\frac{n-1}{2}}e^{ax}\sin sx}{s^n}\left[\binom{n}{n-1}\dfrac{P'(x)}{s} - \binom{n+2}{n-1}\dfrac{P'''(x)}{s^3} + \binom{n+4}{n-1}\dfrac{P^{v}(x)}{s^5} - \cdots\right]$

$\qquad + \dfrac{(-1)^{\frac{n+1}{2}}e^{ax}\cos sx}{s^n}\left[\binom{n-1}{n-1}P(x) - \binom{n+1}{n-1}\dfrac{P''(x)}{s^2} + \binom{n+3}{n-1}\dfrac{P^{iv}(x)}{s^4} - \cdots\right]\quad(n\text{ odd})$

$\qquad\dfrac{(-1)^{\frac{n}{2}}e^{ax}\sin sx}{s^n}\left[\binom{n-1}{n-1}P(x) - \binom{n+1}{n-1}\dfrac{P''(x)}{s^2} + \binom{n+3}{n-1}\dfrac{P^{iv}(x)}{s^4} - \cdots\right]$

$\qquad + \dfrac{(-1)^{\frac{n}{2}}e^{ax}\cos sx}{s^n}\left[\binom{n}{n-1}\dfrac{P'(x)}{s} - \binom{n+2}{n-1}\dfrac{P'''(x)}{s^3} + \binom{n+4}{n-1}\dfrac{P^{v}(x)}{s^5} - \cdots\right]\quad(n\text{ even})$

$\quad$ * For $\cos sx$ in R replace "sin" by "cos" and "cos" by "$-\sin$" in y_p.

Table XI: $(D - a)^n f(D)y = R$

57. e^{ax} $\quad\dfrac{x^n}{n!}\cdot\dfrac{e^{ax}}{f(a)}$

$\quad$ *For $\cos sx$ in R replace "sin" by "cos" and "cos" by "$-\sin$" in y_p.

Table XII: $(D^2 + q)^n y = R$

R	y_p
58. e^{rx}	$e^{rx}/(r^2 + q)^n$
59. $\sin sx*$	$\sin sx/(q - s^2)^n$

60. $P(x)$ $\quad\dfrac{1}{q^n}\left[P(x) - \binom{n}{1}\dfrac{P''(x)}{q^2} + \binom{n+1}{2}\dfrac{P^{iv}(x)}{q^2} - \binom{n+2}{3}\dfrac{P^{vi}(x)}{q^3} + \cdots\right]$

61. $e^{rx}\sin sx*$ $\quad\dfrac{e^{rx}}{(A^2 + B^2)^n}\left\{\left[A^n - \binom{n}{2}A^{n-2}B^2 + \binom{n}{4}A^{n-4}B^4 - \cdots\right]\sin sx - \left[\binom{n}{1}A^{n-1}B - \binom{n}{3}A^{n-3}B^3 + \cdots\right]\cos sx\right\}$

$\qquad\qquad A = r^2 - s^2 + q,\quad B = 2rs$

Table XIII: $(D^2 + b^2)^n y = R$

62. $\sin bx*$ $\qquad (-1)^{n+1/2}\dfrac{x^n \cos bx}{n!(2b)^n}$ $\quad$ (n odd), $(-1)^{n/2}\dfrac{x^n \sin bx}{n!(2b)^n}$ $\quad$ (n even)

Table XIV: $(D^n - q)y = R$

63. e^{rx} $\qquad e^{rx}/(r^n - q)$

64. $P(x)$ $\qquad -\dfrac{1}{q}\left[P(x)\dfrac{P^{(n)}(x)}{q} + \dfrac{P^{(2n)}(x)}{q^2} + \cdots \right]$

65. $\sin sx*$ $\qquad -\dfrac{q\sin sx + (-1)^{\frac{n-1}{2}} s^n \cos sx}{q^2 + s^{2n}}$ $\quad$ (n odd), $\qquad \dfrac{\sin sx}{(-s^2)^{n/2} - q}$ $\quad$ (n even)

66. $e^{rx}\sin sx*$ $\qquad \dfrac{Ae^{rx}\sin sr - Be^{rx}\cos sx}{A^2 + B^2} = \dfrac{e^{rx}}{\sqrt{A^2 + B^2}}\sin\left(sx - \tan^{-1}\dfrac{B}{A}\right)$

$A = \left[r^n - \binom{n}{2}r^{n-2}s^2 + \binom{n}{4}r^{n-4}s^4 - \cdots \right] - q,$

$B = \left[\binom{n}{1}r^{n-1}s - \binom{n}{3}r^{n-3}s^3 + \cdots \right]$

*For $\cos sx$ in R replace "sin" by "cos" and cos by "$-\sin$" in y_p.

Table XV: $(D_x + mD_y)z = R$

R	z_p
67. e^{ax+by}	$\dfrac{e^{ax+by}}{a+mb}$
68. $f(ax + by)$	$\dfrac{\int f(u)du}{a+mb}, \quad u = ax + by$
69. $f(y - mx)$	$xf(y - mx)$
70. $\phi(x, y)f(y - mx)$	$f(y - mx)\int \phi(x, a + mx)dx \quad (a = y - mx$ after integration)

Table XVI: $(D_x + mD_y - k)z = R$

71. e^{ax+by} $\qquad \dfrac{e^{ax+by}}{a + mb - k}$

72. $\sin(ax + by)*$ $\qquad -\dfrac{(a+bm)\cos(ax + by) + k\sin(ax + by)}{(a + bm)^2 + k^2}$

73. $e^{\alpha x + \beta y}\sin(ax + by)*$ $\qquad$ Replace k in 72 by $k - \alpha - m\beta$ and multiply by $e^{\alpha x + \beta y}$

74. $e^{xk}f(ax + by)$ $\qquad \dfrac{e^{kx}\int f(u)du}{a + mb}, \quad u = ax + by$

75. $f(y - mx)$ $\qquad -\dfrac{f(y - mx)}{k}$

76. $p(x)f(y - mx)$ $\qquad -\dfrac{1}{k}f(y - mx)\left[p(x) + \dfrac{p'(x)}{k} + \dfrac{p''(x)}{k^2} + \cdots + \dfrac{p^{(n)}(x)}{k^n} \right]$

77. $e^{kx}f(y - mx)$ $\qquad xe^{kx}f(y - mx)$

*For $\cos(ax + by)$ replace "sin" by "cos" and "cos" by "-sin" in z_p.

$$D_x = \frac{\partial}{\partial x}; \quad D_y = \frac{\partial}{\partial y}; \quad D_{x^k}D_{y^r} = \frac{\partial^{k+r}}{\partial_{x^k}\partial_{y^r}}$$

Table XVII: $(D_z + mD_y)^n z = R$

R	z_p
78. e^{ax+by}	$\dfrac{e^{ax+by}}{(a + mb)^n}$
79. $f(ax + by)$	$\dfrac{\int\int\cdots\int f(u)du^n}{(a + mb)^n}, \quad u = ax + by$
80. $f(y - mx)$	$\dfrac{x^n}{n!}f(y - mx)$
81. $\phi(x, y)f(y + mx)$	$f(y - mx)\int\int\cdots\int \phi(x, a + mx)dx^n (a = y - mx$ after integration)

Table XVIII: $(D_x + mD_y - k)^n z = R$

82. e^{ax+by}

$$\dfrac{e^{ax+by}}{(a+mb-k)^n}$$

83. $f(y - mx)$

$$\dfrac{(-1)^n f(y-mx)}{k^n}$$

84. $P(x) f(y - mx)$

$$\dfrac{(-1)^n}{k^n} f(y-mx)\left[p(x) + \binom{n}{1}\dfrac{p'(x)}{k} + \binom{n+1}{2}\dfrac{p''(x)}{k^3} + \binom{n+2}{3}\dfrac{p'''(x)}{k^3} + \cdots\right]$$

85. $e^{kz} f(ax + by)$

$$\dfrac{e^{kx}\int\int\cdots\int f(u)du^n}{(a+mb)^n}, \quad u = ax + by$$

86. $e^{kx} f(y - mx)$

$$\dfrac{x^n}{n!} e^{kx} f(y-mx)$$

Table XIX: $[D_x^n + a_1 D_x^{n-1} D_y + a_2 D_x^{n-2} D_y^2 + \cdots + a^n D_y^n]z = R$

87. e^{ax+by}

$$\dfrac{e^{ax+by}}{a + a_1 a^{n-1}b + a_2 a^{n-2}b^2 + \cdots + a_n b^n}$$

88. $f(ax + by)$

$$\dfrac{\int\int\cdots\int\int f(u)du^n}{a^n + a_1 a^{n-1}b + a_2 a^{n-2}b^2 + \cdots + a^n b^n}, \quad (u = ax + by)$$

Table XX: $F(D_x{}^c D_y)z = R$

89. e^{ax+by}

$$\dfrac{e^{ax+by}}{F(a,b)}$$

Table XXI: $F(D_x^2{}^c D_x D_y{}^c D_y^2)z = R$

90. $\sin(ax + by)*$

$$\dfrac{\sin(ax+by)}{F(-a^2, -ab, -b^2)}$$

*For $\cos(ax + by)$ replace "sin " by "cos", and "cos" by "-sin" in z_p.

DIFFERENTIAL EQUATIONS

$yF(xy)\,dx + x\,G(xy)\,dy = 0$	$\ln x = \displaystyle\int \frac{G(v)\,dv}{v\{G(v) - F(v)\}} + c$ where $v = xy$. If $G(v) = F(v)$, the solution is $xy = c$.
Linear, homogeneous, second order equation $\dfrac{d^2 y}{dx^2} + b\dfrac{dy}{dx} + cy = 0$ b, c are real constants	Let m_1, m_2 be the roots of $m^2 + bm + c = 0$. Then there are 3 cases: Case 1.　m_1, m_2 real and distinct: $$y = c_1 e^{m_1 x} + c_2 e^{m_2 x}$$ Case 2.　m_1, m_2 real and equal: $$y = c_1 e^{m_1 x} + c_2 x e^{m_1 x}$$ Case 3.　$m_1 = p + qi,\ m_2 = p - qi$: $$y = e^{px}(c_1 \cos qx + c_2 \sin qx)$$ where $p = -b/2,\ q = \sqrt{4c - b^2}/2$
Linear, nonhomogeneous, second order equation $\dfrac{d^2 y}{dx^2} + b\dfrac{dy}{dx} + cy = R(x)$ b, c are real constants	There are 3 cases corresponding to those immediately above: Case 1. $$y = c_1 e^{m_1 x} + c_2 e^{m_2 x}$$ $$+ \frac{e^{m_1 x}}{m_1 - m_2} \int e^{-m_1 x} R(x)\,dx$$ $$+ \frac{e^{m_2 x}}{m_2 - m_1} \int e^{-m_2 x} R(x)\,dx$$ Case 2. $$y = c_1 e^{m_1 x} + c_2 x e^{m_1 x}$$ $$+ x e^{m_1 x} \int e^{-m_1 x} R(x)\,dx$$ $$- e^{m_1 x} \int x e^{-m_1 x} R(x)\,dx$$ Case 3. $$y = e^{px}(c_1 \cos qx + c_2 \sin qx)$$ $$+ \frac{e^{px} \sin qx}{q} \int e^{-px} R(x) \cos qx\,dx$$ $$- \frac{e^{px} \cos qx}{q} \int e^{-px} R(x) \sin qx\,dx$$

Euler or Cauchy equation $$x^2\frac{d^2y}{dx^2} + bx\frac{dy}{dx} + cy = S(x)$$	Putting $x = e^t$, the equation becomes $$\frac{d^2y}{dt^2} + (b-1)\frac{dy}{dt} + cy = S(e^t)$$ and can then be solved as a linear second order equation.
Bessel's equation $$x^2\frac{d^2y}{dx^2} + x\frac{dy}{dx} + (\lambda^2 x^2 - n^2)y = 0$$	$y = c_1 J_n(\lambda x) + c_2 Y_n(\lambda x)$
Transformed Bessel's equation $$x^2\frac{d^2y}{dx^2} + (2p+1)x\frac{dy}{dx} + (\alpha^2 x^{2r} + \beta^2)y = 0$$	$y = x^{-p}\left\{c_1 J_{q/r}\left(\frac{\alpha}{r}x^r\right) + c_2 Y_{q/r}\left(\frac{\alpha}{r}x^r\right)\right\}$ where $q = \sqrt{p^2 - \beta^2}$.
Legendre's equation $$(1-x^2)\frac{d^2y}{dx^2} - 2x\frac{dy}{dx} + n(n+1)y = 0$$	$y = c_1 P_n(x) + c_2 Q_n(x)$

Differential equation	Method of solution
Separation of variables $f_1(x)g_1(y)\,dx + f_2(x)g_2(y)\,dy = 0$	$\int \frac{f_1(x)}{f_2(x)}dx + \int \frac{g_2(y)}{g_1(y)}dy = c$
Exact equation $M(x,y)\,dx + N(x,y)\,dy = 0$ where $\partial M/\partial y = \partial N/\partial x$	$\int M\partial x + \int \left(n - \frac{\partial}{\partial y}\int M\partial x\right)dy = c$ where ∂x indicates that the integration is to be performed with respect to x keeping y constant.
Linear first order equation $\frac{dy}{dx} + P(x)y = Q(x)$	$ye^{\int P\,dx} = \int Qe^{\int P\,dx}dx + c$
Bernoulli's equation $\frac{dy}{dx} + P(x)y = Q(x)y^n$	$ve^{(1-n)\int P\,dx} = \int Qe^{(1-n)\int P\,dx}dx + c$ where $v = y^{1-n}$ if $n = 1$, the solution is $\ln y = \int (Q - P)\,dx + c$
Homogeneous equation $\frac{dy}{dx} = F\left(\frac{y}{x}\right)$	$\ln x = \int \frac{dv}{F(v)-v} + c$ where $v = y/x$. If $F(v) = v$, the solution is $y = cx$
Reducible to homogeneous $(a_1 x + b_1 y + c_1)dx$ $+(a_2 x + b_2 y + c_2)\,dy = 0$ $\frac{a_1}{a_2} \neq \frac{b_1}{b_2}$	Set $u = a_1 x + b_1 y + c_1$ $v = a_2 x + b_2 y + c_2$ Eliminate x and y and the equation becomes homogenous
Reducible to separable $(a_1 x + b_1 y + c_1)\,dx$ $+(a_2 x + b_2 y + c_2)\,dy = 0$ $\frac{a_1}{a_2} = \frac{b_1}{b_2}$	Set $u = a_1 x + b_1 y$ Eliminate x or y and equation becomes separable

FOURIER SERIES

1. If $f(x)$ is a bounded periodic function of period 2L (i.e. $f(x + 2L) = f(x)$, and satisfies the *Dirichlet conditions*:

   (a) In any period $f(x)$ is continuous, except possibly for a finite number of jump discontinuities.

   (b) In any period $f(x)$ has only a finite number of maxima and minima.

   Then $f(x)$ may be represented by the *Fourier series*

   $$\frac{a_0}{2} + \sum_{n=1}^{\infty} \left(a_n \cos \frac{n\pi x}{L} + b_n \sin \frac{n\pi x}{L} \right)$$

   where a_n and b_n are as determined below. This series will converge to $f(x)$ at every point where $f(x)$ is continuous, and to

   $$\frac{f(x^+) + f(x^-)}{2}$$

   (i.e., the average of the left-hand and right-hand limits) at every point where $f(x)$ has a jump discontinuity.

   $$a_n = \frac{1}{L} \int_{-L}^{L} f(x) \cos \frac{n\pi x}{L} \, dx, \quad n = 0, 1, 2, 3, \dots,$$

   $$b_n = \frac{1}{L} \int_{-L}^{L} f(x) \sin \frac{n\pi x}{L} \, dx, \quad n = 1, 2, 3, \dots$$

   we may also write

   $$a_n = \frac{1}{L} \int_{\alpha}^{\alpha+2L} f(x) \cos \frac{n\pi x}{L} dx \text{ and } b_n = \frac{1}{L} \int_{\alpha}^{\alpha+2L} f(x) \sin \frac{n\pi x}{L} dx$$

   where α is any real number. Thus if $\alpha = 0$,

   $$a_n = \frac{1}{L} \int_{0}^{2L} f(x) \cos \frac{n\pi x}{L} dx, \quad n = 0, 1, 2, 3, \dots,$$

   $$b_n = \frac{1}{L} \int_{0}^{2L} f(x) \sin \frac{n\pi x}{L} dx, \quad n = 1, 2, 3, \dots$$

2. If in addition to the restrictions in (1), $f(x)$ is an even function (i.e., $f(-x) = f(x)$), then the Fourier series reduces to

   $$\frac{a_0}{2} + \sum_{n=1}^{\infty} a_n \cos \frac{n\pi x}{L}$$

   That is, $b_n = 0$. In this case, a simpler formula for a_n is

   $$a_n = \frac{2}{L} \int_{0}^{L} f(x) \cos \frac{n\pi x}{L} \, dx, \quad n = 0, 1, 2, 3, \dots$$

3. If in addition to the restrictions in (1), $f(x)$ is an odd function (i.e., $f(-x) = -f(x)$), then the Fourier series reduces to

   $$\sum_{n=1}^{\infty} b_n \sin \frac{n\pi x}{L}$$

   That is, $a_n = 0$. In this case, a simpler formula for the b_n is

   $$b_n = \frac{2}{L} \int_{0}^{L} f(x) \sin \frac{n\pi x}{L} \, dx, \quad n = 1, 2, 3, \dots$$

4. If in addition to the restrictions in (2) above, $f(x) = -f(L - x)$, then a_n will be 0 for all even values of n, including $n = 0$. Thus in this case, the expansion reduces to

   $$\sum_{m=1}^{\infty} a_{2m-1} \cos \frac{(2m-1)\pi x}{L}$$

5. If in addition to the restrictions in (3) above, $f(x) = f(L - x)$, then b_n will be 0 for all even values of n. Thus in this case, the expansion reduces to

$$\sum_{m=1}^{\infty} b_{2m-1} \sin \frac{(2m-1)\pi x}{L}$$

(The series in (4) and (5) are known as *odd-harmonic series*, since only the odd harmonics appear. Similar rules may be stated for even-harmonic series, but when a series appears in the even-harmonic form, it means that $2L$ has not been taken as the smallest period of $f(x)$. Since any integral multiple of a period is also a period, series obtained in this way will also work, but in general computation is simplified if $2L$ is taken to be the smallest period.)

6. If we write the Euler definitions for $\cos \theta$ and $\sin \theta$, we obtain the complex form of the Fourier Series known either as the "Complex Fourier Series" or the "Exponential Fourier Series" of $f(x)$. It is represented as

$$f(x) = \frac{1}{2} \sum_{n=-\infty}^{n=+\infty} c_n e^{i\omega_n x}$$

where

$$c_n = \frac{1}{L} \int_{-L}^{L} f(x) e^{-i\omega_n x} \, dx, \quad n = 0, \pm 1, \pm 2, \pm 3, \dots$$

with $\omega_n = \frac{n\pi}{L}$ for $n = 0, \pm 1, \pm 2, \dots$ The set of coefficients c_n is often referred to as the Fourier spectrum.

7. If both sine and cosine terms are present and if $f(x)$ is of period $2L$ and expandable by a Fourier series, it can be represented as

$$f(x) = \frac{a_0}{2} + \sum_{n=1}^{\infty} c_n \sin \left(\frac{n\pi x}{L} + \phi_n \right),$$

$$\text{where } a_n = c_n \sin \phi_n, \quad b_n = c_n \cos \phi_n, \quad c_n = \sqrt{a_n^2 + b_n^2}, \quad \phi_n = \arctan \left(\frac{a_n}{b_n} \right)$$

It can also be represented as

$$f(x) = \frac{a_0}{2} + \sum_{n=1}^{\infty} c_n \cos \left(\frac{n\pi x}{L} + \phi_n \right), \quad \text{where} \quad a_n = c_n \cos \phi_n, \quad b_n = -c_n \sin \phi_n, \quad c_n = \sqrt{a_n^2 + b_n^2}, \quad \phi_n = \arctan \left(-\frac{b_n}{a_n} \right)$$

where ϕ_n is chosen so as to make a_n, b_n, and c_n hold.

8. The following table of trigonometric identities should be helpful for developing Fourier series.

	n	n even	n odd	$n/2$ odd	$n/2$ even
$\sin n\pi$	0	0	0	0	0
$\cos n\pi$	$(-1)^n$	$+1$	-1	$+1$	$+1$
$* \sin \frac{n\pi}{2}$		0	$(-1)^{(n-1)/2}$	0	0
$* \cos \frac{n\pi}{2}$		$(-1)^{n/2}$	0	-1	$+1$
$\sin \frac{n\pi}{4}$			$\frac{\sqrt{2}}{2}(-1)^{(n^2+4n+11)/8}$	$(-1)^{(n-2)/4}$	0

*A useful formula for $\sin \frac{n\pi}{2}$ and $\cos \frac{n\pi}{2}$ is given by

$$\sin \frac{n\pi}{2} = \frac{(i)^{n+1}}{2}[(-1)^n - 1] \quad \text{and} \quad \cos \frac{n\pi}{2} = \frac{(i)^n}{2}[(-1)^n + 1], \quad \text{where} \quad i^2 = -1.$$

Auxiliary Formulas for Fourier Series

$$1 = \frac{4}{\pi} \left[\sin \frac{\pi x}{k} + \frac{1}{3} \sin \frac{3\pi x}{k} + \frac{1}{5} \sin \frac{5\pi x}{k} + \cdots \right] \quad [0 < x < k]$$

$$x = \frac{2k}{\pi} \left[\sin \frac{\pi x}{k} - \frac{1}{2} \sin \frac{2\pi x}{k} + \frac{1}{3} \sin \frac{3\pi x}{k} - \cdots \right] \quad [-k < x < k]$$

$$x = \frac{k}{2} - \frac{4k}{\pi^2}\left[\cos\frac{\pi x}{k} + \frac{1}{3^2}\cos\frac{3\pi x}{k} + \frac{1}{5^2}\cos\frac{5\pi x}{k} + \cdots\right] \qquad [0 < x < k]$$

$$x^2 = \frac{2k^2}{\pi^3}\left[\left(\frac{\pi^2}{1} - \frac{4}{1}\right)\sin\frac{\pi x}{k} - \frac{\pi^2}{2}\sin\frac{2\pi x}{k} + \left(\frac{\pi^2}{3} - \frac{4}{3^3}\right)\sin\frac{3\pi x}{k}\right.$$
$$\left. - \frac{\pi^2}{4}\sin\frac{4\pi x}{k} + \left(\frac{\pi^2}{5} - \frac{4}{5^3}\right)\sin\frac{5\pi x}{k} + \cdots\right] \qquad [0 < x < k]$$

$$x^2 = \frac{k^2}{3} - \frac{4k^2}{\pi^2}\left[\cos\frac{\pi x}{k} - \frac{1}{2^2}\cos\frac{2\pi x}{k} + \frac{1}{3^2}\cos\frac{3\pi x}{k} - \frac{1}{4^2}\cos\frac{4\pi x}{k} + \cdots\right]$$
$$[-k < x < k]$$

$$1 - \frac{1}{3} + \frac{1}{5} - \frac{1}{7} + \cdots = \frac{\pi}{4}$$

$$1 - \frac{1}{2^2} + \frac{1}{3^2} + \frac{1}{4^2} + \cdots = \frac{\pi^2}{6}$$

$$1 - \frac{1}{2^2} + \frac{1}{3^2} - \frac{1}{4^2} + \cdots = \frac{\pi^2}{12}$$

$$1 + \frac{1}{3^2} + \frac{1}{5^2} - \frac{1}{7^2} + \cdots = \frac{\pi^2}{8}$$

$$\frac{1}{2^2} + \frac{1}{4^2} + \frac{1}{6^2} + \frac{1}{8^2} + \cdots = \frac{\pi^2}{24}$$

FOURIER EXPANSIONS FOR BASIC PERIODIC FUNCTIONS

$$f(x) = \frac{4}{\pi}\sum_{n=1,3,5\ldots}\frac{1}{n}\sin\frac{n\pi x}{L}$$

$$f(x) = \frac{2}{\pi}\sum_{n=1}^{\infty}\frac{(-1)^n}{n}\left(\cos\frac{n\pi c}{L} - 1\right)\sin\frac{n\pi x}{L}$$

$$f(x) = \frac{c}{L} + \frac{2}{\pi}\sum_{n=1}^{\infty}\frac{(-1)^n}{n}\sin\frac{n\pi c}{L}\cos\frac{n\pi x}{L}$$

$$f(x) = \frac{2}{L}\sum_{n=1}^{\infty}\sin\frac{n\pi}{2}\frac{\sin(\frac{1}{2}n\pi c/L)}{\frac{1}{2}n\pi c/L}\sin\frac{n\pi x}{L}$$

$$f(x) = \frac{2}{\pi} \sum_{n=1}^{\infty} \frac{(-1)^{n+1}}{n} \sin \frac{n\pi x}{L}$$

$$f(x) = \frac{1}{2} - \frac{4}{\pi^2} \sum_{n=1,3,5,\ldots} \frac{1}{n^2} \cos \frac{n\pi x}{L}$$

$$f(x) = \frac{8}{\pi^2} \sum_{n=1,3,5,\ldots} \frac{(-1)^{(n-1)/2}}{n^2} \sin \frac{n\pi x}{L}$$

$$f(x) = \frac{1}{2} - \frac{1}{\pi} \sum_{n=1}^{\infty} \frac{1}{n} \sin \frac{n\pi x}{L}$$

$$f(x) = \frac{1}{2}(1+a) + \frac{2}{\pi^2(1-a)} \sum_{n=1}^{\infty} \frac{1}{n^2}[(-1)^n \cos n\pi a - 1] \cos \frac{n\pi x}{L};$$
$$\left(a = \frac{c}{2L}\right)$$

$$f(x) = \frac{2}{\pi} \sum_{n=1}^{\infty} \frac{(-1)^{n-1}}{n} \left[1 + \frac{\sin n\pi a}{n\pi(1-a)}\right] \sin \frac{n\pi x}{L}; \left(a = \frac{c}{2L}\right)$$

$$f(x) = \frac{1}{2} - \frac{4}{\pi^2(1-2a)} \sum_{n=1,3,5,\ldots} \frac{1}{n^2} \cos n\pi a \cos \frac{n\pi x}{L}; \left(a = \frac{c}{2L}\right)$$

$$f(x) = \frac{2}{\pi} \sum_{n=1}^{\infty} \frac{(-1)^n}{n} \left[1 + \frac{1+(-1)^n}{n\pi(1-2a)} \sin n\pi a\right] \sin \frac{n\pi x}{L}; \left(a = \frac{c}{L}\right)$$

$$f(x) \frac{4}{\pi} \sum_{n=1}^{\infty} \frac{1}{n} \sin \frac{n\pi}{4} \sin n\pi a \sin \frac{n\pi x}{L}; \left(a = \frac{c}{2L}\right)$$

$$f(x) = \frac{9}{\pi^2} \sum_{n=1}^{\infty} \frac{1}{n^2} \sin \frac{n\pi}{3} \sin \frac{n\pi x}{L}; \left(a = \frac{c}{2L}\right)$$

$$f(x) = \frac{32}{3\pi^2} \sum_{n=1}^{\infty} \frac{1}{n^2} \sin \frac{n\pi}{4} \sin \frac{n\pi x}{L}; \left(a = \frac{c}{2L}\right)$$

$$f(x) = \frac{1}{\pi} + \frac{1}{2} \sin \omega t - \frac{2}{\pi} \sum_{n=2,4,6,\ldots} \frac{1}{n^2-1} \cos n\omega t$$

Extracted from graphs and formulas, pages 372, 373, Differential Equations in Engineering Problems, Salvadori and Schwarz, published by Prentice-Hall, Inc.,1954.

THE FOURIER TRANSFORMS

For a piecewise continuous function $F(x)$ over a finite interval $0 \leq x \leq \pi$; the *finite Fourier cosine transform* of $F(x)$ is

$$f_c(n) = \int_0^\pi F(x) \cos nx \, dx \quad (n = 0, 1, 2, \ldots)$$

If x ranges over the interval $0 \leq x \leq L$, the substitution $x' = \pi x/L$ allows the use of this definition, also. The inverse transform is written.

$$\overline{F}(x) = \frac{1}{\pi} f_c(0) - \frac{2}{\pi} \sum_{n=1}^{x} f_c(n) \cos nx \quad (0 < x < \pi)$$

where $F(x) = \frac{F(x+\epsilon)+F(x-\epsilon)}{2}$. We observe that $F(x+) = F(x-) = F(x)$ at points of continuity. The formula

$$f_c^{(2)}(n) = \int_0^\pi F''(x) \cos nx \, dx$$
$$= -n^2 f_c(n) - F'(0) + (-1)^n F'(\pi)$$

(1)

makes the finite Fourier cosine transform useful in certain boundary value problems. Analogously, the *finite Fourier sine transform* of $F(x)$ is

$$f_s(n) = \int_0^\pi F(x) \sin nx \, dx \quad (n = 1, 2, 3, \ldots)$$

and

$$\overline{F}(x) = \frac{2}{\pi} \sum_{n=1}^{\infty} f_s(n) \sin nx \quad (0 < x < \pi)$$

Corresponding to (1) we have

$$f_s^{(2)}(n) = \int_0^\pi F''(x) \sin nx \, dx$$

(2)

$$= -n^2 f_s(n) - n F(0) - n(-1)^n F(\pi)$$

If $F(x)$ is defined for $x \leq 0$ and is piecewise continuous over any finite interval, and if $\int_0^x F(x) \, dx$ is absolutely convergent, then

$$f_c(\alpha) = \sqrt{\frac{2}{\pi}} \int_0^x F(x) \cos(\alpha x) \, dx$$

A-62

is the *Fourier cosine transform* of $F(x)$. Furthermore,

$$\overline{F}(x) = \sqrt{\frac{2}{\pi}} \int_0^x f_c(\alpha) \cos(\alpha x) \, d\alpha.$$

If $\lim_{x \to \infty} d^n F/dx^n = 0$, then an important property of the Fourier cosine transform is

$$f_c^{(2r)}(\alpha) = \sqrt{\frac{2}{\pi}} \int_0^x \left(\frac{d^{2r} F}{dx^{2r}} \right) \cos(\alpha x) \, dx$$

$$= -\sqrt{\frac{2}{\pi}} \sum_{n=0}^{r-1} (-1)^n a_{2r-2n-1} \alpha^{2n} + (-1)^r \alpha^{2r} f_c(\alpha)$$

(3)

where $\lim_{x \to \infty} d^r F/dx^r = a_r$, makes it useful in the solution of many problems.
    Under the same conditions.

$$f_s(\alpha) = \sqrt{\frac{2}{\pi}} \int_0^x F(x) \sin(\alpha x) \, dx$$

defines the *Fourier sine transform* of $F(x)$, and

$$\overline{F}(x) = \sqrt{\frac{2}{\pi}} \int_0^x f_s(\alpha) \sin(\alpha x) \, d\alpha$$

Corresponding to (3) we have

$$f_s^{(2r)}(\alpha) = \sqrt{\frac{2}{\pi}} \int_0^\infty \frac{d^{2r} F}{dx^{2r}} \sin(\alpha x) \, dx$$

(4)

$$= -\sqrt{\frac{2}{\pi}} \sum_{n=1}^{r} (-1)^n \alpha^{2n-1} a_{2r-2n} + (-1)^{r-1} \alpha^{2r} f_s(\alpha)$$

Similarly, if $F(x)$ is defined for $-\infty < x < \infty$, and if $\int_{-\infty}^\infty F(x) dx$ is absolutely convergent, then

$$f(\alpha) = \frac{1}{\sqrt{2\pi}} \int_{-\infty}^\infty F(x) e^{i\alpha x} \, dx$$

is the *Fourier transform* of $F(x)$, and

$$\overline{F}(x) = \frac{1}{\sqrt{2\pi}} \int_{-\infty}^\infty f(\alpha) e^{-i\alpha x} \, d\alpha$$

Also, if

$$\lim_{|x| \to \infty} \left| \frac{d^n F}{dx^n} \right| = 0 \quad (n = 1, 2, \ldots, r - 1)$$

then

$$f^{(r)}(\alpha) = \frac{1}{\sqrt{2\pi}} \int_{-\infty}^\infty F^{(r)}(x) e^{i\alpha x} dx = (-i\alpha)^r f(\alpha)$$

Finite Sine Transforms

	$f_s(n)$	$F(x)$
1	$f_s(n) = \int_0^\pi F(x) \sin nx\, dx\ (n = 1, 2, \ldots)$	$F(x)$
2	$(-1)^{n+1} f_s(n)$	$F(\pi - x)$
3	$\frac{1}{n}$	$\frac{\pi - x}{\pi}$
4	$\frac{(-1)^{n+1}}{n}$	$\frac{x}{\pi}$
5	$\frac{1-(-1)^n}{n}$	1
6	$\frac{2}{n^2} \sin \frac{n\pi}{2}$	$\begin{cases} x & \text{when } 0 < x < \pi/2 \\ \pi - x & \text{when } \pi/2 < x < \pi \end{cases}$
7	$\frac{(-1)^{n+1}}{n^3}$	$\frac{x(\pi^2 - x^2)}{6\pi}$
8	$\frac{1-(-1)^n}{n^3}$	$\frac{x(\pi - x)}{2}$
9	$\frac{\pi^2 (-1)^{n-1}}{n} - \frac{2[1-(-1)^n]}{n^3}$	x^2
10	$\pi(-1)^n \left(\frac{6}{n^3} - \frac{\pi^2}{n} \right)$	x^3
11	$\frac{n}{n^2+c^2}[1 - (-1)^n e^{c\pi}]$	e^{cx}
12	$\frac{n}{n^2+c^2}$	$\frac{\sinh c(\pi - x)}{\sinh c\pi}$

	$f_s(n)$	$F(x)$
13	$\frac{n}{n^2-k^2}\quad (k \neq 0, 1, 2, \ldots)$	$\frac{\sin k(\pi - x)}{\sin k\pi}$
14	$\begin{cases} \frac{\pi}{2} & \text{when } n = m \\ 0 & \text{when } n \neq m \end{cases}\quad (m = 1, 2, \ldots)$	$\sin mx$
15	$\frac{n}{n^2 - k^2}[1 - (-1)^n \cos k\pi]$ $(k \neq 1, 2, \ldots)$	$\cos kx$
16	$\begin{cases} \frac{n}{n^2-m^2}[1 - (-1)^{n+m}] \\ \qquad \text{when } n \neq m = 1, 2, \ldots \\ 0 \qquad \text{when } n = m \end{cases}$	$\cos mx$
17	$\frac{n}{(n^2-k^2)^2}\ (k \neq 0, 1, 2, \ldots)$	$\frac{\pi \sin kx}{2k \sin^2 k\pi} - \frac{x \cos k(\pi - x)}{2k \sin k\pi}$
18	$\frac{b^n}{n}\ (|b| \leq 1)$	$\frac{2}{\pi} \arctan \frac{b \sin x}{1 - b \cos x}$
19	$\frac{1-(-1)^n}{n} b^n\quad (|b| \leq 1)$	$\frac{2}{\pi} \arctan \frac{2b \sin x}{1 - b^2}$

Finite Cosine Transforms

	$f_c(n)$	$F(x)$
1	$f_c(n) = \int_0^\pi F(x)\cos nx\,dx \quad (n = 0, 1, 2, \ldots)$	$F(x)$
2	$(-1)^n f_c(n)$	$F(\pi - x)$
3	0 when $n = 1, 2, \cdots;\quad f_c(0) = \pi$	1
4	$\frac{2}{n}\sin\frac{n\pi}{2};\quad f_c(0) = 0$	$\begin{cases} 1 & \text{when } 0 < x < \pi/2 \\ -1 & \text{when } \pi/2 < x < \pi \end{cases}$
5	$-\frac{1-(-1)^n}{n^2};\quad f_c(0) = \frac{\pi^2}{2}$	x
6	$\frac{(-1)^n}{n^2};\quad f_c(0) = \frac{\pi^2}{6}$	$\frac{x^2}{2\pi}$
7	$\frac{1}{n^2};\quad f_c(0) = 0$	$\frac{(\pi-x)^2}{2\pi} - \frac{\pi}{6}$
8	$3\pi^2\frac{(-1)^n}{n^2} - 6\frac{1-(-1)^n}{n^4};\quad f_c(0) = \frac{\pi^4}{4}$	x^3
9	$\frac{(-1)^n e^{c\pi}-1}{n^2+c^2}$	$\frac{1}{c}e^{cx}$
10	$\frac{1}{n^2+c^2}$	$\frac{\cosh c(\pi-x)}{c\sinh c\pi}$
11	$\frac{k}{n^2-k^2}[(-1)^n\cos\pi k - 1](k \neq 0, 1, 2, \cdots)$	$\sin kx$
12	$\frac{(-1)^{n+m}-1}{n^2-m^2};\ f_c(m) = 0 \quad (m = 1, 2, \cdots)$	$\frac{1}{m}\sin mx$
13	$\frac{1}{n^2-k^2} \quad (k \neq 0, 1, 2, \ldots)$	$-\frac{\cos k(\pi-x)}{k\sin k\pi}$
14	$\begin{cases} 0 & \text{for } n = 1, 2, \cdots;\ n \neq m \\ \frac{\pi}{2} & \text{for } n = m \end{cases}$	$\cos mx \quad \text{for } m = 1, 2, 3, \ldots$

Fourier Sine Transforms

	$F(x)$	$f_s(\alpha)$
1	$\begin{cases} 1 & (0 < x < a) \\ 0 & (x > a) \end{cases}$	$\sqrt{\frac{2}{\pi}}\left[\frac{1-\cos\alpha}{\alpha}\right]$
2	$x^{p-1}(0 < p < 1)$	$\sqrt{\frac{2}{\pi}}\frac{\Gamma(p)}{\alpha^p}\sin\frac{p\pi}{2}$
3	$\begin{cases} \sin x & (0 < x < a) \\ 0 & (x > a) \end{cases}$	$\frac{1}{\sqrt{2\pi}}\left[\frac{\sin[a(1-\alpha)]}{1-\alpha} - \frac{\sin[a(1+\alpha)]}{1+\alpha}\right]$
4	e^{-x}	$\sqrt{\frac{2}{\pi}}\left[\frac{\alpha}{1+\alpha^2}\right]$
5	$xe^{-x^2/2}$	$\alpha e^{-\alpha^2/2}$
6	$\cos\frac{x^2}{2}$	$\sqrt{2}\left[\sin\frac{\alpha^2}{2}C\left(\frac{\alpha^2}{2}\right) - \cos\frac{\alpha^2}{2}S\left(\frac{\alpha^2}{2}\right)\right]^*$
7	$\sin\frac{x^2}{2}$	$\sqrt{2}\left[\cos\frac{\alpha^2}{2}C\left(\frac{\alpha^2}{2}\right) + \sin\frac{\alpha^2}{2}S\left(\frac{\alpha^2}{2}\right)\right]^*$

Here $C(y)$ and $S(y)$ are the Fresnel integrals:

$$C(y) = \frac{1}{\sqrt{2\pi}}\int_0^y \frac{1}{\sqrt{t}}\cos t\,dt, \qquad S(y) = \frac{1}{\sqrt{2\pi}}\int_0^y \frac{1}{\sqrt{t}}\sin t\,dt$$

*More extensive tables of the Fourier sine and cosine transforms can be found in Fritz Oberhettinger, *Tabellen zur-Fourier Transformation*, Springer, 1957.

Fourier Cosine Transforms

	$F(x)$	$f_c(\alpha)$
1	$\begin{cases} 1 & (0 < x < a) \\ 0 & (x > a) \end{cases}$	$\sqrt{\frac{2}{\pi}}\frac{\sin a\alpha}{\alpha}$
2	$x^{p-1} \quad (0 < p < 1)$	$\sqrt{\frac{2}{\pi}}\frac{\Gamma(p)}{\alpha^p}\cos\frac{p\pi}{2}$
3	$\begin{cases} \cos x & (0 < x < a) \\ 0 & (x > a) \end{cases}$	$\frac{1}{\sqrt{2\pi}}\left[\frac{\sin[a(1-\alpha)]}{1-\alpha} + \frac{\sin[a(1+\alpha)]}{1+\alpha}\right]$
4	e^{-x}	$\sqrt{\frac{2}{\pi}}\left(\frac{1}{1+\alpha^2}\right)$
5	$e^{-x^2/2}$	$e^{-\alpha^1/2}$
6	$\cos\frac{x^2}{2}$	$\cos\left(\frac{\alpha^2}{2} - \frac{\pi}{4}\right)$
7	$\sin\frac{x^2}{2}$	$\cos\left(\frac{\alpha^2}{2} + \frac{\pi}{4}\right)$

Fourier Transforms

	$F(x)$	$f(\alpha)$
1	$\frac{\sin ax}{x}$	$\begin{cases} \sqrt{\frac{\pi}{2}} & |\alpha| < a \\ 0 & |\alpha| > a \end{cases}$
2	$\begin{cases} e^{iwx} & (p < x < q) \\ 0 & (x < p,\ x > q) \end{cases}$	$\frac{i}{\sqrt{2\pi}}\,\frac{e^{ip(w+\alpha)} - e^{iq(w+\alpha)}}{(w+\alpha)}$
3	$\begin{cases} e^{-cx+iwx} & (x > 0) \\ 0 & (x < 0) \end{cases} \quad (c > 0)$	$\frac{i}{\sqrt{2\pi}(w+\alpha+ic)}$
4	$e^{-px^2}\ R(p) > 0$	$\frac{1}{\sqrt{2p}} e^{-\alpha^2/4p}$
5	$\cos px^2$	$\frac{1}{\sqrt{2p}} \cos\left[\frac{\alpha^2}{4p} - \frac{\pi}{4}\right]$
6	$\sin px^2$	$\frac{1}{\sqrt{2p}} \cos\left[\frac{\alpha^2}{4p} + \frac{\pi}{4}\right]$
7	$|x|^{-p} \quad (0 < p < 1)$	$\sqrt{\frac{2}{\pi}}\,\frac{\Gamma(1-p)\sin\frac{p\pi}{2}}{|\alpha|^{(1-p)}}$
8	$\frac{e^{-a|x|}}{\sqrt{|x|}}$	$\frac{\sqrt{\sqrt{(a^2+\alpha^2)}+a}}{\sqrt{a^2+\alpha^2}}$
9	$\frac{\cosh ax}{\cosh \pi x} \quad (-\pi < a < \pi)$	$\sqrt{\frac{2}{\pi}}\,\frac{\cos\frac{a}{2}\cosh\frac{\alpha}{2}}{\cosh\alpha + \cos a}$
10	$\frac{\sinh ax}{\sinh \pi x} \quad (-\pi < a < \pi)$	$\frac{1}{\sqrt{2\pi}}\,\frac{\sin a}{\cosh\alpha + \cos a}$
11	$\begin{cases} \frac{1}{\sqrt{a^2-x^2}} & (|x| < a) \\ 0 & (|x| > a) \end{cases}$	$\sqrt{\frac{\pi}{2}} J_0(a\alpha)$
12	$\frac{\sin[b\sqrt{a^2+x^2}]}{\sqrt{a^2+x^2}}$	$\begin{cases} 0 & (|\alpha| > b) \\ \sqrt{\frac{\pi}{2}} J_0(a\sqrt{b^2-\alpha^2}) & (|\alpha| < b) \end{cases}$
13	$\begin{cases} P_{n(x)} & (|x| < 1) \\ 0 & (|x| > 1) \end{cases}$	$\frac{i^n}{\sqrt{\alpha}} J_{n+\frac{1}{2}}(\alpha)$
14	$\begin{cases} \frac{\cos[b\sqrt{a^2-x^2}]}{\sqrt{a^2-x^2}} & (|x| < a) \\ 0 & (|x| > a) \end{cases}$	$\sqrt{\frac{\pi}{2}} J_0(a\sqrt{a^2+b^2})$
15	$\begin{cases} \frac{\cosh[b\sqrt{a^2-x^2}]}{\sqrt{a^2-x^2}} & (|x| < a) \\ 0 & (|x| > a) \end{cases}$	$\sqrt{\frac{\pi}{2}} J_0(a\sqrt{\alpha^2-b^2})$

*More extensive tables of Fourier transforms can be found in W. Magnus and F. Oberhettinger, *Formulas and Theorems of the Special Functions of Mathematical Physics*. Chelsea, 1949, 116–120.

SERIES EXPANSION

The expression in parentheses following certain of the series indicates the region of convergence. If not otherwise indicated it is to be understood that the series converges for all finite values of x.

Binomial Series

$$(x + y)^n = x^n + nx^{n-1}y + \frac{n(n-1)}{2!}x^{n-2}y^2 + \frac{n(n-1)(n-2)}{3!}x^{n-3}y^3 + \cdots (y^2 < x^2)$$

$$(1 \pm x)^n = 1 \pm nx + \frac{n(n-1)x^2}{2!} \pm \frac{n(n-1)(n-2)x^3}{3!} + \cdots (x^2 < 1)$$

$$(1 \pm x)^{-n} = 1 \mp nx + \frac{n(n+1)x^2}{2!} \mp \frac{n(n+1)(n+2)x^3}{3!} + \cdots (x^2 < 1)$$

$$(1 \pm x)^{-1} = 1 \mp x + x^2 \mp x^3 + x^4 \mp x^5 + \cdots \quad (x^2 < 1)$$

$$(1 \pm x)^{-2} = 1 \mp 2x + 3x^2 \mp 4x^3 + 5x^4 \mp 6x^5 + \cdots \quad (x^2 < 1)$$

Reversion of Series

Let a series be represented by

$$y = a_1 x + a_2 x^2 + a_3 x^3 + a_4 x^4 + a_5 x^5 + a_6 x^6 + \cdots$$

with $a_1 \neq 0$. The coefficients of the series

$$x = A_1 y + A_2 y^2 + A_3 y^3 + A_4 y^4 + \cdots$$

are

$$A_1 = \frac{1}{a_1} \qquad A_2 = -\frac{a_2}{a_1^3} \qquad A_3 = \frac{1}{a_1^5}(2a_2^2 - a_1 a_3)$$

$$A_4 = \frac{1}{a_1^7}(5a_1 a_2 a_3 - a_1^2 a_4 - 5a_2^3)$$

$$A_5 = \frac{1}{a_1^9}(6a_1^2 a_2 a_4 + 3a_1^2 a_3^2 + 14a_2^4 - a_1^3 a_5 - 21a_1 a_2^2 a_3)$$

$$A_6 = \frac{1}{a_1^{11}}(7a_1^3 a_2 a_5 + 7a_1^3 a_3 a_4 + 84a_1 a_2^3 a_3 - a_1^4 a_6 - 28a_1^2 a_2^2 a_4 - 28a_1^2 a_2 a_3^2 - 42a_2^5)$$

$$A_7 = \frac{1}{a_1^{13}}(8a_1^4 a_2 a_6 + 8a_1^4 a_3 a_5 + 4a_1^4 a_4^2 + 120a_1^2 a_2^3 a_4 + 180a_1^2 a_2^2 a_3^2 + 132a_2^6 - a_1^5 a_7$$
$$- 36a_1^3 a_2^2 a_5 - 72a_1^3 a_2 a_3 a_4 - 12a_1^3 a_3^3 - 330a_1 a_2^4 a_3)$$

Taylor Series

1. $$f(x) = f(a) + (x-a)f'(a) + \frac{(x-a)^2}{2!}f''(a) + \frac{(x-a)^3}{3!}f'''(a)$$
    $$+ \cdots + \frac{(x-a)^n}{n!}f^{(n)}(a) + \cdots \qquad \text{(Taylor's Series)}$$

(Increment form)

2. $$f(x+h) = f(x) + hf'(x) + \frac{h^2}{2!}f''(x) + \frac{h^3}{3!}f'''(x) + \cdots$$
    $$= f(h) + xf'(h) + \frac{x^2}{2!}f''(h) + \frac{x^3}{3!}f'''(h) + \cdots$$

3. If $f(x)$ is a function possessing derivatives of all orders throughout the interval $a \le x \le b$, then there is a value X, with $a < X < b$, such that

$$f(b) = f(a) + (b-a)f'(a) + \frac{(b-a)^2}{2!}f''(a) + \cdots + \frac{(b-a)^{n-1}}{(n-1)!}f^{(n-1)}(a) + \frac{(b-a)^n}{n!}f^{(n)}(X)$$

$$f(a+h) = f(a) + hf'(a) + \frac{h^2}{2!}f''(a) + \cdots + \frac{h^{n-1}}{(n-1)!}f^{(n-1)}(a) + \frac{h^n}{n!}f^{(n)}(a+\theta h)$$

where $b = a + h$ and $0 < \theta < 1$. Or

$$f(x) = f(a) + (x-a)f'(a) + \frac{(x-a)^2}{2!}f''(a) + \cdots + (x-a)^{n-1}\frac{f^{(n-1)}(a)}{(n-1)!} + R_n,$$

where

$$R_n = \frac{f^{(n)}[a + \theta \cdot (x-a)]}{n!}(x-a)^n, \; 0 < \theta < 1.$$

The above forms are known as Taylor's series with the remainder term.

4. Taylor's series for a function of two variables

If $\left(h\frac{\partial}{\partial x} + k\frac{\partial}{\partial y}\right) f(x, y) = h\frac{\partial f(x, y)}{\partial x} + k\frac{\partial f(x, y)}{\partial y}$;

$$\left(h\frac{\partial}{\partial x} + k\frac{\partial}{\partial y}\right)^2 f(x, y) = h^2\frac{\partial^2 f(x, y)}{\partial x^2} + 2hk\frac{\partial^2 f(x, y)}{\partial x \partial y} + k^2\frac{\partial^2 f(x, y)}{\partial y^2}$$

etc., and if $\left(h\frac{\partial}{\partial x} + k\frac{\partial}{\partial y}\right)^n f(x, y)\Big|_{x=a}^{y=b}$ where the bar and subscripts means that after differentiation we are to replace x by a and y by b, then

$$f(a+h, b+k) = f(a, b) + \left(h\frac{\partial}{\partial x} + k\frac{\partial}{\partial y}\right) f(x, y)\Big|_{x=a}^{y=b} + \cdots + \frac{1}{n!}\left(h\frac{\partial}{\partial x} + k\frac{\partial}{\partial y}\right)^n f(x, y)\Big|_{x=a}^{y=b} + \cdots$$

Maclaurin Series

$$f(x) = f(0) + xf'(0) + \frac{x^2}{2!}f''(0) + \frac{x^3}{3!}f'''(0) + \cdots + x^{n-1}\frac{f^{(n-1)}(0)}{(n-1)!} + R_n,$$

where

$$R_n = \frac{x^n f^{(n)}(\theta x)}{n!}, \quad 0 < \theta < 1.$$

Exponential Series

$$e = 1 + \frac{1}{1!} + \frac{1}{2!} + \frac{1}{3!} + \frac{1}{4!} + \cdots$$

$$e^x = 1 + x + \frac{x^2}{2!} + \frac{x^3}{3!} + \frac{x^4}{4!} + \cdots$$

$$a^x = 1 + x\log_e a + \frac{(x\log_e a)^2}{2!} + \frac{(x\log_e a)^3}{3!} + \cdots$$

$$e^x = e^a\left[1 + (x-a) + \frac{(x-a)^2}{2!} + \frac{(x-a)^3}{3!} + \cdots\right]$$

Logarithmic Series

$$\log_e x = \frac{x-1}{x} + \frac{1}{2}\left(\frac{x-1}{x}\right)^2 + \frac{1}{3}\left(\frac{x-1}{x}\right)^3 + \cdots \qquad (x > \tfrac{1}{2})$$

$$\log_e x = (x-1) - \frac{1}{2}(x-1)^2 + \frac{1}{3}(x-1)^3 - \cdots \qquad (2 \geq x > 0)$$

$$\log_e x = 2\left[\frac{x-1}{x+1} + \frac{1}{3}\left(\frac{x-1}{x+1}\right)^3 + \frac{1}{5}\left(\frac{x-1}{x+1}\right)^5 + \cdots\right] \qquad (x > 0)$$

$$\log_e(1+x) = x - \frac{1}{2}x^2 + \frac{1}{3}x^3 - \frac{1}{4}x^4 + \cdots \qquad (-1 < x \leq 1)$$

$$\log_e(n+1) - \log_e(n-1) = 2\left[\frac{1}{n} + \frac{1}{3n^3} + \frac{1}{5n^5} + \cdots\right]$$

$$\log_e(a+x) = \log_e a + 2\left[\frac{x}{2a+x} + \frac{1}{3}\left(\frac{x}{2a+x}\right)^3\right.$$
$$\left. + \frac{1}{5}\left(\frac{x}{2a+x}\right)^5 + \cdots\right] \qquad (a > 0, -a < x < +\infty)$$

$$\log_e\frac{1+x}{1-x} = 2\left[x + \frac{x^3}{3} + \frac{x^5}{5} + \cdots + \frac{x^{2n-1}}{2n-1} + \cdots\right] \qquad -1 < x < 1$$

$$\log_e x = \log_e a + \frac{(x-a)}{a} - \frac{(x-a)^2}{2a^2} + \frac{(x-a)^3}{3a^3} - + \cdots \qquad 0 < x \leq 2a$$

Trigonometric Series

$$\sin x = x - \frac{x^3}{3!} + \frac{x^5}{5!} - \frac{x^7}{7!} + \cdots \quad \text{(all real values of } x\text{)}$$

$$\cos x = 1 - \frac{x^2}{2!} + \frac{x^4}{4!} - \frac{x^6}{6!} + \cdots \quad \text{(all real values of } x\text{)}$$

$$\tan x = x + \frac{x^3}{3} + \frac{2x^5}{15} + \frac{17x^7}{315} + \frac{62x^9}{2835} + \cdots + \frac{(-1)^{n-1}2^{2n}(2^{2n}-1)B_{2n}}{(2n)!}x^{2n-1} + \cdots,$$
$$\left[x^2 < \frac{\pi^2}{4} \quad \text{and } B_n \text{ represents the } n^{\text{th}} \text{ Bernoulli number}\right]$$

$$\cot x = \frac{1}{x} - \frac{x}{3} - \frac{x^3}{45} - \frac{2x^5}{945} - \frac{x^7}{4725} - \cdots - \frac{(-1)^{n+1}2^{2n}}{(2n)!}B_{2n}x^{2n-1} - \cdots,$$
$$\left[x^2 < \pi^2 \quad \text{and } B_n \text{ represents the } n^{\text{th}} \text{ Bernoulli number}\right]$$

$$\sec x = 1 + \frac{x^2}{2} + \frac{5}{24}x^4 + \frac{61}{720}x^6 + \frac{277}{8064}x^8 + \cdots + \frac{(-1)^n}{(2n)!}E_{2n}x^{2n} + \cdots,$$
$$\left[x^2 < \frac{\pi^2}{4} \quad \text{and } E_n \text{ represents the } n^{\text{th}} \text{ Euler number}\right]$$

$$\csc x = \frac{1}{x} + \frac{x}{6} + \frac{7}{360}x^3 + \frac{31}{15,120}x^5 + \frac{127}{604,800}x^7 + \cdots$$
$$+ \frac{(-1)^{n+1}2(2^{2n-1}-1)}{(2n)!}B_{2n}x^{2n-1} + \cdots,$$
$$\left[x^2 < \pi^2 \quad \text{and } B_n \text{ represents the } n^{\text{th}} \text{ Bernoulli number}\right]$$

$$\sin x = x\left(1 - \frac{x^2}{\pi^2}\right)\left(1 - \frac{x^2}{2^2\pi^2}\right)\left(1 - \frac{x^2}{3^2\pi^2}\right)\cdots \qquad (x^2 < \infty)$$

$$\cos x = \left(1 - \frac{4x^2}{\pi^2}\right)\left(1 - \frac{4x^2}{3^2\pi^2}\right)\left(1 - \frac{4x^2}{5^2\pi^2}\right)\cdots \qquad (x^2 < \infty)$$

$$\sin^{-1} x = x + \frac{x^3}{2\cdot3} + \frac{1\cdot3}{2\cdot4\cdot5}x^5 + \frac{1\cdot3\cdot5}{2\cdot4\cdot6\cdot7}x^7 + \cdots \qquad \left(x^2 < 1,\ -\frac{\pi}{2} < \sin^{-1} x < \frac{\pi}{2}\right)$$

$$\cos^{-1} x = \frac{\pi}{2} - \left(x + \frac{x^3}{2\cdot3} + \frac{1\cdot3}{2\cdot4\cdot5}x^5 + \frac{1\cdot3\cdot5x^7}{2\cdot4\cdot6\cdot7} + \cdots\right) \qquad (x^2 < 1,\ 0 < \cos^{-1} x < \pi)$$

$$\tan^{-1} x = x - \frac{x^3}{3} + \frac{x^5}{5} - \frac{x^7}{7} + \cdots \qquad (x^2 < 1)$$

$$\tan^{-1} x = \frac{\pi}{2} - \frac{1}{x} + \frac{1}{3x^3} - \frac{1}{5x^5} + \frac{1}{7x^7} - \cdots \qquad (x > 1)$$

$$\tan^{-1} x = -\frac{\pi}{2} - \frac{1}{x} + \frac{1}{3x^3} - \frac{1}{5x^5} + \frac{1}{7x^7} - \cdots \qquad (x < -1)$$

$$\cot^{-1} x = \frac{\pi}{2} - x + \frac{x^3}{3} - \frac{x^5}{5} + \frac{x^7}{7} - \cdots \qquad (x^2 < 1)$$

$$\log_e \sin x = \log_e x - \frac{x^2}{6} - \frac{x^4}{180} - \frac{x^6}{2835} - \cdots \qquad (x^2 < \pi^2)$$

$$\log_e \cos x = -\frac{x^2}{2} - \frac{x^4}{12} - \frac{x^6}{45} - \frac{17x^8}{2520} - \cdots \qquad \left(x^2 < \frac{\pi^2}{4}\right)$$

$$\log_e \tan x = \log_e x + \frac{x^2}{3} + \frac{7x^4}{90} + \frac{62x^6}{2835} + \cdots \qquad \left(x^2 < \frac{\pi^2}{4}\right)$$

$$e^{\sin x} = 1 + x + \frac{x^2}{2!} - \frac{3x^4}{4!} - \frac{8x^5}{5!} - \frac{3x^6}{6!} + \frac{56x^7}{7!} + \cdots$$

$$e^{\cos x} = e\left(1 - \frac{x^2}{2!} + \frac{4x^4}{4!} - \frac{31x^6}{6!} + \cdots\right)$$

$$e^{\tan x} = 1 + x + \frac{x^2}{2!} + \frac{3x^3}{3!} + \frac{9x^4}{4!} + \frac{37x^5}{5!} + \cdots \qquad \left(x^2 < \frac{\pi^2}{4}\right)$$

$$\sin x = \sin a + (x - a)\cos a - \frac{(x-a)^2}{2!}\sin a$$
$$- \frac{(x-a)^3}{3!}\cos a + \frac{(x-a)^4}{4!}\sin a + \cdots$$

VECTOR ANALYSIS

Definitions

Any quantity which is completely determined by its magnitude is called a *scalar*. Examples of such are mass, density, temperature, etc. Any quantity which is completely determined by its magnitude and direction is called a *vector*. Examples of such are velocity, acceleration, force, etc. A vector quantity is represented by a directed line segment, the length of which represents the magnitude of the vector. A vector quantity is usually represented by a boldfaced letter such as **V**. Two vectors **V**$_1$ and **V**$_2$ are equal to one another if they have equal magnitudes and are acting in the same directions. A negative vector, written as -**V**, is one which acts in the opposite direction to **V**, but is of equal magnitude to it. If we represent the magnitude of **V** by v, we write $|\mathbf{V}| = v$. A vector parallel to **V**, but equal to the reciprocal of its magnitude is written as $\mathbf{V}^{-1}$ or as $1/\mathbf{V}$.

The *unit vector* $\mathbf{V}/v$ (when $v \neq 0$) is that vector which has the same direction as **V**, but has a magnitude of unity (sometimes represented as $\mathbf{V}_0$ or $\hat{\mathbf{v}}$).

Vector Algebra

The vector sum of **V**$_1$ and **V**$_2$ is represented by **V**$_1$+**V**$_2$. The vector sum of **V**$_1$ and -**V**$_2$, or the difference of the vector **V**$_2$ from **V**$_1$ is represented by $\mathbf{V}_1 - \mathbf{V}_2$.

If r is a scalar, then $r\mathbf{V}=\mathbf{V}r$, and represents a vector r times the magnitude of **V**, in the same direction as **V** if r is positive, and in the opposite direction if r is negative. If r and s are scalars, **V**$_1$, **V**$_2$, **V**$_3$, vectors, then the following rules of scalars and vectors hold:

$$\mathbf{V}_1 + \mathbf{V}_2 = \mathbf{V}_2 + \mathbf{V}_1$$
$$(r + s)\mathbf{V}_1 = r\mathbf{V}_1 + s\mathbf{V}_1; \quad r(\mathbf{V}_1 + \mathbf{V}_2) = r\mathbf{V}_1 + r\mathbf{V}_2$$
$$\mathbf{V}_1 + (\mathbf{V}_2 + \mathbf{V}_3) = (\mathbf{V}_1 + \mathbf{V}_2) + \mathbf{V}_3 = \mathbf{V}_1 + \mathbf{V}_2 + \mathbf{V}_3$$

Vectors in Space

A plane is described by two distinct vectors **V**$_1$ and **V**$_2$. Should these vectors not intersect each other, then one is displaced parallel to itself until they do (Figure 1). Any other vector **V** lying in this plane is given by

$$\mathbf{V} = r\mathbf{V}_1 + s\mathbf{V}_2$$

A *position vector* specifies the position in space of a point relative to a fixed origin. If therefore **V**$_1$ and **V**$_2$ are the position vectors of the points A and B, relative to the origin O, then any point P on the line AB has a position vector **V** given by

$$\mathbf{V} = r\mathbf{V}_1 + (1 - r)\mathbf{V}_2$$

The scalar "r" can be taken as the metric representation of P since $r = 0$ implies $P = B$ and $r = 1$ implies $P = A$ (Figure 2). If P divides the line AB in the ratio $r:s$ then

$$\mathbf{V} = \left(\frac{r}{r + s}\right)\mathbf{V}_1 + \left(\frac{s}{r + s}\right)\mathbf{V}_2$$

Figure 1.   Figure 2.

The vectors $V_1, V_2, V_3, \ldots, V_n$ are said to be *linearly dependent* if there exist scalars $r_1, r_2, r_3, \ldots, r_n$, not all zero, such that

$$r_1 V_1 + r_2 V_2 + \cdots + r_n V_n = 0$$

A vector V is linearly dependent upon the set of vectors $V_1, V_2, V_3, \ldots, V_n$ if

$$V = r_1 V_1 + r_2 V_2 + r_3 V_3 + \cdots + r_n V_n$$

Three vectors are linearly dependent if and only if they are co-planar.

All points in space can be uniquely determined by linear dependence upon three *base vectors* i.e., three vectors any one of which is linearly independent of the other two. The simplest set of base vectors are the unit vectors along the coordinate Ox, Oy and Oz axes. These are usually designated by i, j and k respectively.

If V is a vector in space, and a, b and c are the respective magnitudes of the projections of the vector along the axes then

$$V = ai + bj + ck$$

and

$$v = \sqrt{a^2 + b^2 + c^2}$$

and the direction cosines of V are

$$\cos \alpha = a/v, \quad \cos \beta = b/v, \quad \cos \gamma = c/v.$$

The law of addition yields

$$V_1 + V_2 = (a_1 + a_2)i + (b_1 + b_2)j + (c_1 + c_2)k$$

The Scalar, Dot, or Inner Product of Two Vectors

This product is represented as $V_1 \cdot V_2$ and is defined to be equal to $v_1 v_2 \cos \theta$, where θ is the angle from V_1 to V_2, i.e.,

$$V_1 \cdot V_2 = v_1 v_2 \cos \theta$$

The following rules apply for this product:

$$V_1 \cdot V_2 = a_1 a_2 + b_1 b_2 + c_1 c_2 = V_2 \cdot V_1$$

It should be noted that this verifies that scalar multiplication is commutative.

$$(V_1 + V_2) \cdot V_3 = V_1 \cdot V_3 + V_2 \cdot V_3$$
$$V_1 \cdot (V_2 + V_3) = V_1 \cdot V_2 + V_1 \cdot V_3$$

If V_1 is perpendicular to V_2 then $V_1 \cdot V_2 = 0$, and if V_1 is parallel to V_2 then $V_1 \cdot V_2 = v_1 v_2 = rw_1^2$ In particular

$$i \cdot i = j \cdot j = k \cdot k = 1,$$

and

$$i \cdot j = j \cdot k = k \cdot i = 0$$

The Vector or Cross Product of Two Vectors

This product is represented as $V_1 \times V_2$ and is defined to be equal to $v_1 v_2 (\sin \theta) 1$, where θ is the angle from V_1 to V_2 and 1 is a unit vector perpendicular to the plane of V_1 and V_2 and so directed that a right-handed screw driven in the direction of 1 would carry V_1 into V_2, i.e.,

$$V_1 \times V_2 = v_1 v_2 (\sin \theta) 1$$

and $\tan \theta = \dfrac{|V_1 \times V_2|}{V_1 \cdot V_2}$

The following rules apply for vector products:

$$\mathbf{V}_1 \times \mathbf{V}_2 = -\mathbf{V}_2 \times \mathbf{V}_1$$
$$\mathbf{V}_1 \times (\mathbf{V}_2 + \mathbf{V}_3) = \mathbf{V}_1 \times \mathbf{V}_2 + \mathbf{V}_1 \times \mathbf{V}_3$$
$$(\mathbf{V}_1 + \mathbf{V}_2) \times \mathbf{V}_3 = \mathbf{V}_1 \times \mathbf{V}_3 + \mathbf{V}_2 \times \mathbf{V}_3$$
$$\mathbf{V}_1 \times (\mathbf{V}_2 \times \mathbf{V}_3) = \mathbf{V}_2(\mathbf{V}_3 \cdot \mathbf{V}_1) - \mathbf{V}_3(\mathbf{V}_1 \cdot \mathbf{V}_2)$$
$$\mathbf{i} \times \mathbf{i} = \mathbf{j} \times \mathbf{j} = \mathbf{k} \times \mathbf{k} = 0 \quad \text{(the zero vector)}$$
$$\mathbf{i} \times \mathbf{j} = \mathbf{k}, \qquad \mathbf{j} \times \mathbf{k} = \mathbf{i}, \qquad \mathbf{k} \times \mathbf{i} = \mathbf{j}$$

If $\mathbf{V}_1 = a_1\mathbf{i} + b_1\mathbf{j} + c_1\mathbf{k}$, $\mathbf{V}_2 = a_2\mathbf{i} + b_2\mathbf{j} + c_2\mathbf{k}$, and $\mathbf{V}_3 = a_3\mathbf{i} + b_3\mathbf{j} + c_3\mathbf{k}$, then

$$\mathbf{V}_1 \times \mathbf{V}_2 = \begin{vmatrix} \mathbf{i} & \mathbf{j} & \mathbf{k} \\ a_1 & b_1 & c_1 \\ a_2 & b_2 & c_2 \end{vmatrix} = (b_1c_2 - b_2c_1)\mathbf{i} + (c_1a_2 - c_2a_1)\mathbf{j} + (a_1b_2 - a_2b_1)\mathbf{k}$$

It should be noted that, since $\mathbf{V}_1 \times \mathbf{V}_2 = -\mathbf{V}_2 \times \mathbf{V}_1$, the vector product is not commutative.

Scalar Triple Product

There is only one possible interpretation of the expression $\mathbf{V}_1 \cdot \mathbf{V}_2 \times \mathbf{V}_3$ and that is $\mathbf{V}_1 \cdot (\mathbf{V}_2 \times \mathbf{V}_3)$ which is obviously a scalar. Further

$$\mathbf{V}_1 \cdot (\mathbf{V}_2 \times \mathbf{V}_3) = (\mathbf{V}_1 \times \mathbf{V}_2) \cdot \mathbf{V}_3 = \mathbf{V}_2 \cdot (\mathbf{V}_3 \times \mathbf{V}_1)$$
$$= \begin{vmatrix} a_1 & b_1 & c_1 \\ a_2 & b_2 & c_2 \\ a_3 & b_3 & c_3 \end{vmatrix}$$
$$= r_1 r_2 r_3 \cos\phi \sin\theta,$$

Where θ is the angle between $\mathbf{V}_2$ and $\mathbf{V}_3$ and ϕ is the angle between $\mathbf{V}_1$ and the normal to the plane of $\mathbf{V}_2$ and $\mathbf{V}_3$.

This product is called the *scalar triple product* and is written as $[\mathbf{V}_1\mathbf{V}_2\mathbf{V}_3]$.

The determinant indicates that it can be considered as the volume of the parallelepiped whose three determining edges are $\mathbf{V}_1$, $\mathbf{V}_2$ and $\mathbf{V}_3$.

It also follows that cyclic permutation of the subscripts does not change the value of the scalar triple product so that

$$[\mathbf{V}_1\mathbf{V}_2\mathbf{V}_3] = [\mathbf{V}_2\mathbf{V}_3\mathbf{V}_1] = [\mathbf{V}_3\mathbf{V}_1\mathbf{V}_2]$$
$$\text{but} \quad [\mathbf{V}_1\mathbf{V}_2\mathbf{V}_3] = -[\mathbf{V}_2\mathbf{V}_1\mathbf{V}_3] \quad \text{etc.} \qquad \text{and} \qquad [\mathbf{V}_1\mathbf{V}_1\mathbf{V}_2] \equiv 0 \quad \text{etc.}$$

Given three non-coplanar reference vectors $\mathbf{V}_1$, $\mathbf{V}_2$ and $\mathbf{V}_3$, the *reciprocal system* is given by $\mathbf{V}_1^*$, $\mathbf{V}_2^*$ and $\mathbf{V}_3^*$, where

$$1 = v_1 v_1^* = v_2 v_2^* = v_3 v_3^*$$
$$0 = v_1 v_2^* = v_1 v_3^* = v_2 v_1^* \quad \text{etc.}$$
$$\mathbf{V}_1^* = \frac{\mathbf{V}_2 \times \mathbf{V}_3}{[\mathbf{V}_1\mathbf{V}_2\mathbf{V}_3]}, \quad \mathbf{V}_2^* = \frac{\mathbf{V}_3 \times \mathbf{V}_1}{[\mathbf{V}_1\mathbf{V}_2\mathbf{V}_3]}, \quad \mathbf{V}_3^* = \frac{\mathbf{V}_1 \times \mathbf{V}_2}{[\mathbf{V}_1\mathbf{V}_2\mathbf{V}_3]}$$

The system $\mathbf{i}, \mathbf{j}, \mathbf{k}$ is its own reciprocal.

Vector Triple Product

The product $\mathbf{V}_1 \times (\mathbf{V}_2 \times \mathbf{V}_3)$ defines the *vector triple product*. Obviously, in this case, the brackets are vital to the definition.

$$\mathbf{V}_1 \times (\mathbf{V}_2 \times \mathbf{V}_3) = (\mathbf{V}_1 \cdot \mathbf{V}_3)\mathbf{V}_2 - (\mathbf{V}_1 \cdot \mathbf{V}_2)\mathbf{V}_3$$
$$= \begin{vmatrix} \mathbf{i} & \mathbf{j} & \mathbf{k} \\ a_1 & b_1 & c_1 \\ \begin{vmatrix} b_2 & c_2 \\ b_3 & c_3 \end{vmatrix} & \begin{vmatrix} c_2 & a_2 \\ c_3 & a_3 \end{vmatrix} & \begin{vmatrix} a_2 & b_2 \\ a_3 & b_3 \end{vmatrix} \end{vmatrix}$$

i.e. it is a vector, perpendicular to V_1, lying in the plane of V_2, V_3. Similarly

$$(V_1 \times V_2) \times V_3 = \begin{vmatrix} i & j & k \\ \begin{vmatrix} b_1 & c_1 \\ b_2 & c_2 \end{vmatrix} & \begin{vmatrix} c_1 & a_1 \\ c_2 & a_2 \end{vmatrix} & \begin{vmatrix} a_1 & b_1 \\ a_2 & b_2 \end{vmatrix} \\ a_3 & b_3 & c_3 \end{vmatrix}$$

$$V_1 \times (V_2 \times V_3) + V_2 \times (V_3 \times V_1) + V_3 \times (V_1 \times V_2) \equiv 0$$

If $V_1 \times (V_2 \times V_3) = (V_1 \times V_2) \times V_3$ then V_1, V_2, V_3 form an *orthogonal set*. Thus **i**, **j**, **k** form an orthogonal set.

Geometry of the Plane, Straight Line and Sphere

The position vectors of the fixed points A, B, C, D relative to O are V_1, V_2, V_3, V_4 and the position vector of the variable point P is V.

The vector form of the equation of the straight line through A parallel to V_2 is

$$V = V_1 + rV_2$$
$$\text{or} \quad (V - V_1) = rV_2$$
$$\text{or} \quad (V - V_1) \times V_2 = 0$$

while that of the plane through A perpendicular to V_2 is

$$(V - V_1) \cdot V_2 = 0$$

The equation of the line AB is

$$V = rV_1 + (1 - r)V_2$$

and those of the bisectors of the angles between V_1 and V_2 are

$$V = r\left(\frac{V_1}{v_1} \pm \frac{V_2}{v_2}\right) \quad \text{or}$$
$$V = r(\hat{v}_1 \pm \hat{v}_2)$$

The perpendicular from C to the line through A parallel to V_2 has as its equation

$$V = V_1 - V_3 - \hat{v}_2 \cdot (V_1 - V_3)\hat{v}_2.$$

The condition for the intersection of the two lines, $V = V_1 + rV_3$ and $V = V_2 + sV_4$ is

$$[(V_1 - V_2)V_3V_4] = 0.$$

The common perpendicular to the above two lines is the line of intersection of the two planes

$$[(V - V_1)V_3(V_3 \times V_4)] = 0 \quad \text{and} \quad [(V - V_2)V_4(V_3 \times V_4)] = 0$$

and the length of this perpendicular is

$$\frac{[(V_1 - V_2)V_3V_4]}{|V_3 \times V_4|}.$$

The equation of the line perpendicular to the plane ABC is

$$V = V_1 \times V_2 + V_2 \times V_3 + V_3 \times V_1$$

and the distance of the plane from the origin is

$$\frac{[V_1V_2V_3]}{|(V_2 - V_1) \times (V_3 - V_1)|}.$$

In general the vector equation

$$V \cdot V_2 = r$$

defines the plane which is perpendicular to V_2, and the perpendicular distance from A to this plane is

$$\frac{r - V_1 \cdot V_2}{v_2}$$

The distance from A, measured along a line parallel to V_3, is

$$\frac{r - V_1 \cdot V_2}{V_2 \cdot \hat{v}_3} \quad \text{or} \quad \frac{r - V_1 \cdot V_2}{v_2 \cos \theta}$$

where θ is the angle between V_2 and V_3.

(If this plane contains the point C then $r = \mathbf{V}_3 \cdot \mathbf{V}_2$ and if it passes through the origin then $r = 0$.) Given two planes

$$\mathbf{V} \cdot \mathbf{V}_1 = r$$
$$\mathbf{V} \cdot \mathbf{V}_2 = s$$

then any plane through the line of intersection of these two planes is given by

$$\mathbf{V} \cdot (\mathbf{V}_1 + \lambda \mathbf{V}_2) = r + \lambda s$$

where λ is a scalar parameter. In particular $\lambda = \pm v_1/v_2$ yields the equation of the two planes bisecting the angle between the given planes.

The plane through A parallel to the plane of $\mathbf{V}_2$, $\mathbf{V}_3$ is

$$\mathbf{V} = \mathbf{V}_1 + r\mathbf{V}_2 + s\mathbf{V}_3$$
$$\text{or} \quad (\mathbf{V} - \mathbf{V}_1) \cdot \mathbf{V}_2 \times \mathbf{V}_3 = 0$$
$$\text{or} \quad [\mathbf{V}\mathbf{V}_2\mathbf{V}_3] - [\mathbf{V}_1\mathbf{V}_2\mathbf{V}_3] = 0$$

so that the expansion in rectangular Cartesian coordinates yields (where $\mathbf{V} \equiv x\mathbf{i} + y\mathbf{j} + z\mathbf{k}$):

$$\begin{vmatrix} (x - a_1) & (y - b_1) & (z - c_1) \\ a_2 & b_2 & c_2 \\ a_3 & b_3 & c_3 \end{vmatrix} = 0$$

which is obviously the usual linear equation in x, y, and z.

The plane through AB parallel to $\mathbf{V}_3$ is given by

$$[(\mathbf{V} - \mathbf{V}_1)(\mathbf{V}_1 - \mathbf{V}_2)\mathbf{V}_3] = 0$$
$$\text{or} \quad [\mathbf{V}\mathbf{V}_2\mathbf{V}_3] - [\mathbf{V}\mathbf{V}_1\mathbf{V}_3] - [\mathbf{V}_1\mathbf{V}_2\mathbf{V}_3] = 0$$

The plane through the three points A, B and C is

$$\mathbf{V} = \mathbf{V}_1 + s(\mathbf{V}_2 - \mathbf{V}_1) + t(\mathbf{V}_3 - \mathbf{V}_1)$$
$$\text{or} \quad \mathbf{V} = r\mathbf{V}_1 + s\mathbf{V}_2 + t\mathbf{V}_3 \qquad (r + s + t \equiv 1)$$
$$\text{or} \quad [(\mathbf{V} - \mathbf{V}_1)(\mathbf{V}_1 - \mathbf{V}_2)(\mathbf{V}_2 - \mathbf{V}_3)] = 0$$
$$\text{or} \quad [\mathbf{V}\mathbf{V}_1\mathbf{V}_2] + [\mathbf{V}\mathbf{V}_2\mathbf{V}_3] + [\mathbf{V}\mathbf{V}_3\mathbf{V}_1] - [\mathbf{V}_1\mathbf{V}_2\mathbf{V}_3] = 0$$

For four points A, B, C, D to be coplanar, then

$$r\mathbf{V}_1 + s\mathbf{V}_2 + t\mathbf{V}_3 + u\mathbf{V}_4 \equiv 0 \equiv r + s + t + u$$

The following formulae relate to a sphere when the vectors are taken to lie in three dimensional space and to a circle when the space is two dimensional. For a circle in three dimensions take the intersection of the sphere with a plane.

The equation of a sphere with center O and radius $\mathbf{OA}$ is

$$\mathbf{V} \cdot \mathbf{V} = v_1^2 \quad (\text{not} \mathbf{V} = \mathbf{V}_1)$$
$$\text{or} \quad (\mathbf{V} - \mathbf{V}_1) \cdot (\mathbf{V} + \mathbf{V}_1) = 0$$

while that of a sphere with center B radius v_1 is

$$(\mathbf{V} - \mathbf{V}_2) \cdot (\mathbf{V} - \mathbf{V}_2) = v_1^2$$
$$\text{or} \quad \mathbf{V} \cdot (\mathbf{V} - 2\mathbf{V}_2) = v_1^2 - v_2^2$$

If the above sphere passes through the origin then

$$\mathbf{V} \cdot (\mathbf{V} - 2\mathbf{V}_2) = 0$$

(note that in two dimensional polar coordinates this is simply)

$$r = 2a \cdot \cos \theta$$

while in three dimensional Cartesian coordinates it is

$$x^2 + y^2 + z^2 - 2(a_2x + b_2y + c_2x) = 0.$$

The equation of a sphere having the points A and B as the extremities of a diameter is

$$(\mathbf{V} - \mathbf{V}_1) \cdot (\mathbf{V} - \mathbf{V}_2) = 0.$$

The square of the length of the tangent from C to the sphere with center B and radius v_1 is given by

$$(\mathbf{V}_3 - \mathbf{V}_2) \cdot (\mathbf{V}_3 - \mathbf{V}_2) = v_1^2$$

The condition that the plane $\mathbf{V} \cdot \mathbf{V}_3 = s$ is tangential to the sphere $(\mathbf{V} - \mathbf{V}_2) \cdot (\mathbf{V} - \mathbf{V}_2) = v_1^2$ is

$$(s - \mathbf{V}_3 \cdot \mathbf{V}_2) \cdot (s - \mathbf{V}_3 \cdot \mathbf{V}_2) = v_1^2 v_3^2.$$

The equation of the tangent plane at D, on the surface of sphere $(\mathbf{V} - \mathbf{V}_2) \cdot (\mathbf{V} - \mathbf{V}_2) = v_1^2$, is

$$(\mathbf{V} - \mathbf{V}_4) \cdot (\mathbf{V}_4 - \mathbf{V}_2) = 0$$

$$\text{or} \quad \mathbf{V} \cdot \mathbf{V}_4 - \mathbf{V}_2 \cdot (\mathbf{V} + \mathbf{V}_4) = v_1^2 - v_2^2$$

The condition that the two circles $(\mathbf{V} - \mathbf{V}_2) \cdot (\mathbf{V} - \mathbf{V}_2) = v_1^2$ and $(\mathbf{V} - \mathbf{V}_4) \cdot (\mathbf{V} - \mathbf{V}_4) = v_3^2$ intersect orthogonally is clearly

$$(\mathbf{V}_2 - \mathbf{V}_4) \cdot (\mathbf{V}_2 - \mathbf{V}_4) = v_1^2 + v_3^2$$

The polar plane of D with respect to the circle

$$(\mathbf{V} - \mathbf{V}_2) \cdot (\mathbf{V} - \mathbf{V}_2) = v_1^2 \quad \text{is}$$
$$\mathbf{V} \cdot \mathbf{V}_4 - \mathbf{V}_2 \cdot (\mathbf{V} + \mathbf{V}_4) = v_1^2 - v_2^2$$

Any sphere through the intersection of the two spheres $(\mathbf{V} - \mathbf{V}_2) \cdot (\mathbf{V} - \mathbf{V}_2) = v_1^2$ and $(\mathbf{V} - \mathbf{V}_4) \cdot (\mathbf{V} - \mathbf{V}_4) = v_3^2$ is given by

$$(\mathbf{V} - \mathbf{V}_2) \cdot (\mathbf{V} - \mathbf{V}_2) + \lambda(\mathbf{V} - \mathbf{V}_4) \cdot (\mathbf{V} - \mathbf{V}_4) = v_1^2 + \lambda v_3^2$$

while the radical plane of two such spheres is

$$\mathbf{V} \cdot (\mathbf{V}_2 - \mathbf{V}_4) = -\frac{1}{2}(v_1^2 - v_2^2 - v_3^2 + v_4^2)$$

Differentiation of Vectors

If $\mathbf{V}_1 = a_1\mathbf{i} + b_1\mathbf{j} + c_1\mathbf{k}$, and $\mathbf{V}_2 = a_2\mathbf{i} + b_2\mathbf{j} + c_2\mathbf{k}$, and if $\mathbf{V}_1$ and $\mathbf{V}_2$ are functions of the scalar t, then

$$\frac{d}{dt}(\mathbf{V}_1 + \mathbf{V}_2 + \cdots) = \frac{d\mathbf{V}_1}{dt} + \frac{d\mathbf{V}_2}{dt} + \cdots$$

$$\frac{d\mathbf{V}_1}{dt} = \frac{da_1}{dt}\mathbf{i} + \frac{db_1}{dt}\mathbf{j} + \frac{dc_1}{dt}\mathbf{k}, \quad \text{etc}$$

$$\frac{d}{dt}(\mathbf{V}_1 \cdot \mathbf{V}_2) = \frac{d\mathbf{V}_1}{dt} \cdot \mathbf{V}_2 + \mathbf{V}_1 \cdot \frac{d\mathbf{V}_2}{dt}$$

$$\frac{d}{dt}(\mathbf{V}_1 \times \mathbf{V}_2) = \frac{d\mathbf{V}_1}{dt} \times \mathbf{V}_2 + \mathbf{V}_1 \times \frac{d\mathbf{V}_2}{dt}$$

$$\mathbf{V} \cdot \frac{d\mathbf{V}}{dt} = v \cdot \frac{dv}{dt}$$

In particular, if $\mathbf{V}$ is a vector of constant length then the right hand side of the last equation is identically zero showing that $\mathbf{V}$ is perpendicular to its derivative.

The derivatives of the triple products are

$$\frac{d}{dt}[\mathbf{V}_1\mathbf{V}_2\mathbf{V}_3] = \left[\left(\frac{d\mathbf{V}_1}{dt}\right)\mathbf{V}_2\mathbf{V}_3\right] + \left[\mathbf{V}_1\left(\frac{d\mathbf{V}_2}{dt}\right)\mathbf{V}_3\right] + \left[\mathbf{V}_1\mathbf{V}_2\left(\frac{d\mathbf{V}_3}{dt}\right)\right] \quad \text{and}$$

$$\frac{d}{dt}\{\mathbf{V}_1 \times (\mathbf{V}_2 \times \mathbf{V}_3)\} = \left(\frac{d\mathbf{V}_1}{dt}\right) \times (\mathbf{V}_2 \times \mathbf{V}_3) + \mathbf{V}_1 \times \left(\left(\frac{d\mathbf{V}_2}{dt}\right) \times \mathbf{V}_3\right) + \mathbf{V}_1 \times \left(\mathbf{V}_2 \times \left(\frac{d\mathbf{V}_3}{dt}\right)\right)$$

Geometry of Curves in Space

s = the *length of arc*, measured from some fixed point on the curve (Figure 3).

$\mathbf{V}_1$ = the position vector of the point A on the curve

$\mathbf{V}_1 + \delta\mathbf{V}_1$ = the position vector of the point P in the neighborhood of A

$\hat{\mathbf{t}}$ = the *unit tangent* to the curve at the point A, measured in the direction of s increasing.

The *normal plane* is that plane which is perpendicular to the unit tangent. The principal normal is defined as the intersection of the normal plane with the plane defined by $\mathbf{V}_1$ and $\mathbf{V}_1 + \delta\mathbf{V}_1$ in the limit as $\delta\mathbf{V}_1 - 0$.

$\hat{\mathbf{n}}$ = the *unit normal* (principal) at the point A. The plane defined by $\hat{\mathbf{t}}$ and $\hat{\mathbf{n}}$ is called the *osculating plane* (alternatively plane of curvature or local plane).

ρ = the radius of curvature at A.

$\delta\theta$ = the angle subtended at the origin by $\delta\mathbf{V}_1$.

$$\kappa = \frac{d\theta}{ds} = \frac{1}{\rho}$$

$\hat{\mathbf{b}}$ = the *unit binormal* i.e. the unit vector which is parallel to $\hat{\mathbf{t}} \times \hat{\mathbf{n}}$ at the point A:

λ = the *torsion* of the curve at A

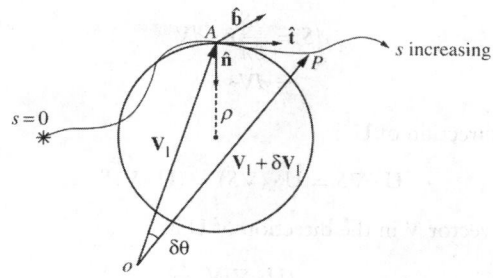

Figure 3.

Frenet's Formulae:

$$\frac{d\hat{\mathbf{t}}}{ds} = \kappa\hat{\mathbf{n}}$$

$$\frac{d\hat{\mathbf{n}}}{ds} = -\kappa\hat{\mathbf{t}} + \lambda\hat{\mathbf{b}}$$

$$\frac{d\hat{\mathbf{b}}}{ds} = -\lambda\hat{\mathbf{n}}$$

The following formulae are also applicable:

Unit tangent	$\hat{\mathbf{t}} = \frac{d\mathbf{V}_1}{ds}$
Equation of the tangent	$(\mathbf{V} - \mathbf{V}_1) \times \hat{\mathbf{t}} = 0$ or $\mathbf{V} = \mathbf{V}_1 + q\hat{\mathbf{t}}$
Unit normal	$\hat{\mathbf{n}} = \frac{1}{\kappa}\frac{d^2\mathbf{V}_1}{ds^2}$
Equation of the normal plane	$(\mathbf{V} - \mathbf{V}_1) \cdot \hat{\mathbf{t}} = 0$
Equation of the normal	$(\mathbf{V} - \mathbf{V}_1) \times \hat{\mathbf{n}} = 0$ or $\mathbf{V} = \mathbf{V}_1 + r\hat{\mathbf{n}}$
Unit binormal	$\hat{\mathbf{b}} = \hat{\mathbf{t}} \times \hat{\mathbf{n}}$
Equation of the binormal	$(\mathbf{V} - \mathbf{V}_1) \times \hat{\mathbf{b}} = 0$
	or $\mathbf{V} = \mathbf{V}_1 + u\hat{\mathbf{b}}$
	or $\mathbf{V} = \mathbf{V}_1 + w\frac{d\mathbf{V}_1}{ds} \times \frac{d^2\mathbf{V}_1}{ds^2}$
Equation of the osculating plane:	$[(\mathbf{V} - \mathbf{V}_1)\hat{\mathbf{t}}\hat{\mathbf{n}}] = 0$
	or $\left[(\mathbf{V} - \mathbf{V}_1)\left(\frac{d\mathbf{V}_1}{ds}\right)\left(\frac{d^2\mathbf{V}_1}{ds^2}\right)\right] = 0$

Differential Operators—Rectangular Coordinates

$$dS = \frac{\partial S}{\partial x} \cdot dx + \frac{\partial S}{\partial y} \cdot dy + \frac{\partial S}{\partial z} \cdot dz$$

By definition

$$\nabla \equiv \text{del} \equiv \mathbf{i}\frac{\partial}{\partial x} + \mathbf{j}\frac{\partial}{\partial y} + \mathbf{k}\frac{\partial}{\partial z}$$

$$\nabla^2 \equiv \text{Laplacian} \equiv \frac{\partial^2}{\partial x^2} + \frac{\partial^2}{\partial y^2} + \frac{\partial^2}{\partial z^2}$$

If S is a scalar function, then $\nabla S \equiv \text{grad } S \equiv \frac{\partial S}{\partial x}\mathbf{i} + \frac{\partial S}{\partial y}\mathbf{j} + \frac{\partial S}{\partial z}\mathbf{k}$

Grad S defines both the direction and magnitude of the maximum rate of increase of S at any point. Hence the name *gradient* and also its vectorial nature. ∇S is independent of the choice of rectangular coordinates.

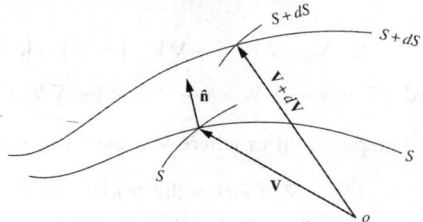

Figure 4.

$$\nabla S = \frac{\partial S}{\partial n}\hat{\mathbf{n}}$$

(5)

where $\hat{\mathbf{n}}$ is the unit normal to the surface S =constant, in the direction of S increasing. The total derivative of S at a point having the position vector $\mathbf{V}$ is given by (Figure 4)

$$dS = \frac{\partial S}{\partial n}\hat{\mathbf{n}} \cdot d\mathbf{V}$$
$$= d\mathbf{V} \cdot \nabla S$$

and the directional derivative of S in the direction of $\mathbf{U}$ is

$$\mathbf{U} \cdot \nabla S = \mathbf{U} \cdot (\nabla S) = (\mathbf{U} \cdot \nabla)S$$

Similarly the directional derivative of the vector $\mathbf{V}$ in the direction of $\mathbf{U}$ is

$$(\mathbf{U} \cdot \nabla)\mathbf{V}$$

The *distributive* law holds for finding a gradient. Thus if S and T are scalar functions

$$\nabla(S + T) = \nabla S + \nabla T$$

The *associative* law becomes the rule for differentiating a product:

$$\nabla(ST) = S\nabla T + T\nabla S$$

If $\mathbf{V}$ is a vector function with the magnitudes of the components parallel to the three coordinate axes V_x, V_y, V_z, then

$$\nabla \cdot \mathbf{V} \equiv \operatorname{div}\mathbf{V} \equiv \frac{\partial V_x}{\partial x} + \frac{\partial V_y}{\partial y} + \frac{\partial V_z}{\partial z}$$

The divergence obeys the distributive law. Thus, if $\mathbf{V}$ and $\mathbf{U}$ are vector functions, then

$$\nabla \cdot (\mathbf{V} + \mathbf{U}) = \nabla \cdot \mathbf{V} + \nabla \cdot \mathbf{U}$$
$$\nabla \cdot (S\mathbf{V}) = (\nabla S) \cdot \mathbf{V} + S(\nabla \cdot \mathbf{V})$$
$$\nabla \cdot (\mathbf{U} \times \mathbf{V}) = \mathbf{V} \cdot (\nabla \times \mathbf{U}) - \mathbf{U} \cdot (\nabla \times \mathbf{V})$$

As with the gradient of a scalar, the divergence of a vector is invariant under a transformation from one set of rectangular coordinates to another.

$$\nabla \times \mathbf{V} \equiv \operatorname{curl}\mathbf{V} \qquad (\text{sometimes } \nabla \wedge \mathbf{V} \text{ or } \operatorname{rot}\mathbf{V})$$
$$\equiv \left(\frac{\partial V_x}{\partial y} - \frac{\partial V_y}{\partial z}\right)\mathbf{i} + \left(\frac{\partial V_x}{\partial z} - \frac{\partial V_z}{\partial x}\right)\mathbf{j} + \left(\frac{\partial V_y}{\partial x} - \frac{\partial V_x}{\partial y}\right)\mathbf{k}$$
$$= \begin{vmatrix} \mathbf{i} & \mathbf{j} & \mathbf{k} \\ \frac{\partial}{\partial x} & \frac{\partial}{\partial y} & \frac{\partial}{\partial z} \\ V_x & V_y & V_z \end{vmatrix}$$

The *curl* (or *rotation*) of a vector is a vector which is invariant under a transformation from one set of rectangular coordinates to another.

$$\nabla \times (\mathbf{U} + \mathbf{V}) = \nabla \times \mathbf{U} + \nabla \times \mathbf{V}$$
$$\nabla \times (S\mathbf{V}) = (\nabla S) \times \mathbf{V} + S(\nabla \times \mathbf{V})$$
$$\nabla \times (\mathbf{U} \times \mathbf{V}) = (\mathbf{V} \cdot \nabla)\mathbf{U} - (\mathbf{U} \cdot \nabla)\mathbf{V} + \mathbf{U}(\nabla \cdot \mathbf{V}) - \mathbf{V}(\nabla \cdot \mathbf{U})$$

If $\mathbf{V} = V_x\mathbf{i} + V_y\mathbf{j} + V_z\mathbf{k}$ then

$$\nabla \cdot \mathbf{V} = \nabla V_x \cdot \mathbf{i} + \nabla V_y \cdot \mathbf{j} + \nabla V_z \cdot \mathbf{k}$$
$$\text{and} \quad \nabla \times \mathbf{V} = \nabla V_x \times \mathbf{i} + \nabla V_y \times \mathbf{j} + \nabla V_z \times \mathbf{k}$$

The operator ∇ can be used more than once. The possibilities where ∇ is used twice are:

$$\nabla \cdot (\nabla\theta) \equiv \operatorname{div}\operatorname{grad}\theta$$
$$\nabla \times (\nabla\theta) \equiv \operatorname{curl}\operatorname{grad}\theta$$
$$\nabla(\nabla \cdot \mathbf{V}) \equiv \operatorname{grad}\operatorname{div}\mathbf{V}$$
$$\nabla \cdot (\nabla \times \mathbf{V}) \equiv \operatorname{div}\operatorname{curl}\mathbf{V}$$
$$\nabla \times (\nabla \times \mathbf{V}) \equiv \operatorname{curl}\operatorname{curl}\mathbf{V}$$

Thus, if S is a scalar and $\mathbf{V}$ is a vector:

$$\text{div grad } S \equiv \nabla \cdot (\nabla S) \equiv \text{Laplacian } S \equiv \nabla^2 S \equiv \frac{\partial^2 S}{\partial x^2} + \frac{\partial^2 S}{\partial y^2} + \frac{\partial^2 S}{\partial z^2}$$

$$\text{curl grad } S \equiv 0$$

$$\text{curl curl } \mathbf{V} \equiv \text{grad div } \mathbf{V} - \nabla^2 \mathbf{V};$$

$$\text{div curl } \mathbf{V} \equiv 0$$

Taylor's expansion in three dimensions can be written

$$f(\mathbf{V} + \varepsilon) = e^{\varepsilon \cdot \nabla} f(\mathbf{V}) \qquad \text{where} \quad \mathbf{V} = x\mathbf{i} + y\mathbf{j} + z\mathbf{k}$$

$$\text{and} \quad \varepsilon = h\mathbf{i} + l\mathbf{j} + m\mathbf{k}$$

ORTHOGONAL CURVILINEAR COORDINATES

If at a point P there exist three uniform point functions u, v and w so that the surfaces $u = $const., $v = $ const., and $w = $ const., intersect in three distinct curves through P then the surfaces are called the *coordinate surfaces* through P. The three lines of intersection are referred to as the *coordinate lines* and their tangents a, b, and c as the *coordinate axes*. When the coordinate axes form an orthogonal set the system is said to define *orthogonal curvilinear coordinates* at P.

Consider an infinitesimal volume enclosed by the surfaces u, v, w, $u + du$, $v + dv$, and $w + dw$ (Figure 5).

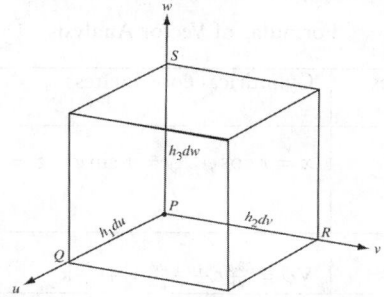

Figure 5.

The surface $P\,RS \equiv u = $ constant, and the face of the curvilinear figure immediately opposite this is $u + du = $constant, etc. In terms of these surface constants

$$P = P(u, v, w)$$
$$Q = Q(u + du, v, w) \quad \text{and} \quad PQ = h_1\, du$$
$$R = R(u, v + dv, w) \quad \text{and} \quad PR = h_2\, dv$$
$$S = S(u, v, w + dw) \quad \text{and} \quad PS = h_3\, dw$$

where h_1, h_2, and h_3 are functions of u, v, and w.

- In rectangular Cartesians $\mathbf{i}, \mathbf{j}, \mathbf{k}$

$$h_1 = 1, \qquad h_2 = 1, \qquad h_3 = 1.$$

$$\frac{\hat{\mathbf{a}}}{h_1}\frac{\partial}{\partial u} = \mathbf{i}\frac{\partial}{\partial x}, \qquad \frac{\hat{\mathbf{b}}}{h_2}\frac{\partial}{\partial v} = \frac{\hat{\Phi}}{r}\frac{\partial}{\partial \phi}, \qquad \frac{\hat{\mathbf{c}}}{h_3}\frac{\partial}{\partial w} = \hat{\mathbf{k}}\frac{\partial}{\partial z}.$$

- In cylindrical Cartesians $\hat{\mathbf{r}}, \hat{\theta}, \hat{\Phi}$

$$h_1 = 1, \qquad h_2 = 1, \qquad h_3 = 1.$$

$$\frac{\hat{\mathbf{a}}}{h_1}\frac{\partial}{\partial u} = \hat{r}\frac{\partial}{\partial r}, \qquad \frac{\hat{\mathbf{b}}}{h_2}\frac{\partial}{\partial v} = \frac{\hat{\Phi}}{r}\frac{\partial}{\partial \phi}, \qquad \frac{\hat{\mathbf{c}}}{h_3}\frac{\partial}{\partial w} = \hat{\mathbf{k}}\frac{\partial}{\partial z}.$$

- In spherical coordinates $\hat{\mathbf{r}}, \hat{\theta}, \hat{\Phi}$

$$h_1 = 1, \qquad h_2 = r, \qquad h_3 = r\sin\theta$$

$$\frac{\hat{\mathbf{a}}}{h_1}\frac{\partial}{\partial u} = \hat{\mathbf{r}}\frac{\partial}{\partial r}, \qquad \frac{\mathbf{b}}{h_2}\frac{\partial}{\partial v} = \frac{\hat{\Phi}}{r}\frac{\partial}{\partial \theta}, \qquad \frac{\hat{\mathbf{c}}}{h_3}\frac{\partial}{\partial w} = \frac{\hat{\Phi}}{r\sin\theta}\frac{\partial}{\partial \phi}$$

The general expressions for grad, div and curl together with those for ∇^2 and the directional derivative are, in orthogonal curvilinear coordinates, given by:

$$\nabla S = \frac{\hat{\mathbf{a}}}{h_1}\frac{\partial S}{\partial u} + \frac{\hat{\mathbf{b}}}{h_2}\frac{\partial S}{\partial v} + \frac{\hat{\mathbf{c}}}{h_3}\frac{\partial S}{\partial w}$$

$$(\mathbf{V}\cdot\nabla)S = \frac{V_1}{h_1}\frac{\partial S}{\partial u} + \frac{V_2}{h_2}\frac{\partial S}{\partial v} + \frac{V_3}{h_3}\frac{\partial S}{\partial w}$$

$$\nabla\cdot\mathbf{V} = \frac{1}{h_1 h_2 h_3}\left\{\frac{\partial}{\partial u}(h_2 h_3 V_1) + \frac{\partial}{\partial v}(h_3 h_1 V_2) + \frac{\partial}{\partial w}(h_1 h_2 V_3)\right\}.$$

$$\nabla\times\mathbf{V} = \frac{\hat{\mathbf{a}}}{h_2 h_3}\left\{\frac{\partial}{\partial v}(h_3 V_3) - \frac{\partial}{\partial w}(h_2 V_2)\right\} + \frac{\hat{\mathbf{b}}}{h_3 h_1}\left\{\frac{\partial}{\partial w}(h_1 V_1) - \frac{\partial}{\partial u}(h_3 V_3)\right\}$$

$$+ \frac{\hat{\mathbf{c}}}{h_1 h_2}\left\{\frac{\partial}{\partial u}(h_2 V_2) - \frac{\partial}{\partial v}(h_1 V_1)\right\}$$

$$\nabla^2 S = \frac{1}{h_1 h_2 h_3}\left\{\frac{\partial}{\partial u}\left(\frac{h_2 h_3}{h_1}\frac{\partial S}{\partial u}\right) + \frac{\partial}{\partial v}\left(\frac{h_3 h_1}{h_2}\frac{\partial S}{\partial v}\right) + \frac{\partial}{\partial w}\left(\frac{h_1 h_2}{h_3}\frac{\partial S}{\partial w}\right)\right\}$$

Formulas of Vector Analysis

	Rectangular coordinates	Cylindrical coordinates	Spherical coordinates
Conversion to rectangular coordinates		$x = r\cos\varphi \quad y = r\sin\varphi \quad z = z$	$x = r\cos\varphi\sin\theta \quad y = r\sin\varphi\sin\theta$ $z = r\cos\theta$
Gradient ...	$\nabla\phi = \frac{\partial\phi}{\partial x}\mathbf{i} + \frac{\partial\phi}{\partial y}\mathbf{j} + \frac{\partial\phi}{\partial z}\mathbf{k}$	$\nabla\phi = \frac{\partial\phi}{\partial r}\mathbf{r} + \frac{1}{r}\frac{\partial\phi}{\partial\varphi}\Phi + \frac{\partial\phi}{\partial z}\mathbf{k}$	$\nabla\phi = \frac{\partial\phi}{\partial r}\mathbf{r} + \frac{1}{r}\frac{\partial\phi}{\partial\theta}\theta + \frac{1}{r\sin\theta}\frac{\partial\phi}{\partial\varphi}\Phi$
Divergence ...	$\nabla\cdot\mathbf{A} = \frac{\partial A_x}{\partial x} + \frac{\partial A_y}{\partial y} + \frac{\partial A_z}{\partial z}$	$\nabla\cdot\mathbf{A} = \frac{1}{r}\frac{\partial(r A_r)}{\partial r} + \frac{1}{r}\frac{\partial A_\varphi}{\partial\varphi}$ $+ \frac{\partial A_z}{\partial z}$	$\nabla\cdot\mathbf{A} = \frac{1}{r^2}\frac{\partial(r^2 A_r)}{\partial r} + \frac{1}{r\sin\theta}\frac{\partial(A_\theta\sin\theta)}{\partial\theta}$ $+ \frac{1}{r\sin\theta}\frac{\partial A_\varphi}{\partial\varphi}$
Curl ...	$\nabla\times\mathbf{A} = \begin{vmatrix} \mathbf{i} & \mathbf{j} & \mathbf{k} \\ \frac{\partial}{\partial x} & \frac{\partial}{\partial y} & \frac{\partial}{\partial z} \\ A_x & A_y & A_z \end{vmatrix}$	$\nabla\times\mathbf{A} = \begin{vmatrix} \frac{1}{r}\mathbf{r} & \Phi & \frac{1}{r}\mathbf{k} \\ \frac{\partial}{\partial r} & \frac{\partial}{\partial\varphi} & \frac{\partial}{\partial z} \\ A_r & r A_\varphi & A_z \end{vmatrix}$	$\nabla\times\mathbf{A} = \begin{vmatrix} \frac{\mathbf{r}}{r^2\sin\theta} & \frac{\theta}{r\sin\theta} & \frac{\Phi}{r} \\ \frac{\partial}{\partial r} & \frac{\partial}{\partial\theta} & \frac{\partial}{\partial\varphi} \\ A_r & r A_\theta & r A_\varphi\sin\theta \end{vmatrix}$
Laplacian ...	$\nabla^2\phi = \frac{\partial^2\phi}{\partial x^2} + \frac{\partial^2\phi}{\partial y^2} + \frac{\partial^2\phi}{\partial z^2}$	$\nabla^2\phi = \frac{1}{r}\frac{\partial}{\partial r}\left(r\frac{\partial\phi}{\partial r}\right) + \frac{1}{r^2}\frac{\partial^2\phi}{\partial\varphi^2}$ $+ \frac{\partial^2\phi}{\partial z^2}$	$\nabla^2\phi = \frac{1}{r^2}\frac{\partial}{\partial r}\left(r^2\frac{\partial\phi}{\partial r}\right) + \frac{1}{r^2\sin\theta}\frac{\partial}{\partial\theta}\left(\sin\theta\frac{\partial\phi}{\partial\theta}\right)$ $+ \frac{1}{r^2\sin^2\theta}\frac{\partial^2\phi}{\partial\varphi^2}$

TRANSFORMATION OF INTEGRALS

If

1. s is the distance along a curve "C" in space and is measured from some fixed point.

2. S is a surface area

3. V is a volume contained by a specified surface

4. $\hat{\mathbf{t}} =$ the unit tangent to C at the point

5. $P\,\hat{\mathbf{n}} =$ the unit outward pointing normal

6. F is some vector function

7. ds is the vector element of curve $(= \hat{\mathbf{t}}\,ds\,)$

8. dS is the vector element of surface $(= \hat{\mathbf{n}}\,dS\,)$

then

$$\int_{(c)} \mathbf{F} \cdot \hat{\mathbf{t}}\, ds = \int_{(c)} \mathbf{F}$$

and when $\mathbf{F} = \nabla\phi$

$$\int_{(c)} (\nabla\phi) \cdot \hat{\mathbf{t}}\, ds = \int_{(c)} d\phi$$

Gauss' Theorem
   When S defines a closed region having a volume V:

$$\iiint_{(v)} (\nabla \cdot \mathbf{F})\, dV = \iint_{(s)} (\mathbf{F} \cdot \hat{\mathbf{n}})\, dS = \iint_{(s)} \mathbf{F} \cdot dS$$

also

$$\iiint_{(v)} (\nabla\phi)\, dV = \iint_{(s)} \phi\hat{\mathbf{n}}\, dS$$

and

$$\iiint_{(v)} (\nabla \times \mathbf{F})\, dV = \iint_{(s)} (\hat{\mathbf{n}} \times \mathbf{F})\, dS$$

Stokes' Theorem
   When C is closed and bounds the open surface S:

$$\iint_{(s)} \hat{\mathbf{n}} \cdot (\nabla \times \mathbf{F})\, dS = \int_{(c)} \mathbf{F} \cdot d\mathbf{s}$$

also

$$\iint_{(s)} (\hat{\mathbf{n}} \times \nabla\phi)\, dS = \int_{(c)} \phi\, d\mathbf{s}$$

Green's Theorem

$$\iint_{(s)} (\nabla\phi \cdot \nabla\theta)\, dS = \iint_{(s)} \phi\hat{\mathbf{n}} \cdot (\nabla\theta)\, dS = \iiint_{(v)} \phi(\nabla^2\theta)\, dV$$

$$= \iint_{(s)} \theta \cdot \hat{\mathbf{n}}(\nabla\phi)\, dS = \iiint_{(v)} \phi(\nabla^2\theta)\, dV$$

BESSEL FUNCTIONS

1. Bessel's differential equation for a real variable x is

$$x^2 \frac{d^2 y}{dx^2} + x\frac{dy}{dx} + (x^2 - n^2)y = 0$$

2. When n is not an integer, two independent solutions of the equation are $J_n(x)$, $J_{-n}(x)$, where

$$J_n(x) = \sum_{k=0}^{\infty} \frac{(-1)^k}{k!\,\Gamma(n+k+1)} \left(\frac{x}{2}\right)^{n+2k}$$

3. If n is an integer $J_n(x) = (-1)^n J_n(x)$, where

$$J_n(x) = \frac{x^n}{2^n n!} \left\{ 1 - \frac{x^2}{2^2 \cdot 1!(n+1)} + \frac{x^4}{2^4 \cdot 2!(n+1)(n+2)} + \frac{x^6}{2^6 \cdot 3!(n+1)(n+2)(n+3)} + \cdots \right\}$$

4. For $n = 0$ and $n = 1$, this formula becomes

$$J_0(x) = 1 - \frac{x^2}{2^2(1!)^2} + \frac{x^4}{2^4(2!)^2} - \frac{x^6}{2^6(3!)^2} + \frac{x^8}{2^8(4!)^2} - \cdots$$

$$J_1(x) = \frac{x}{2} - \frac{x^3}{2^3 \cdot 1!2!} + \frac{x^5}{2^5 \cdot 2!3!} - \frac{x^7}{2^7 \cdot 3!4!} + \frac{x^9}{2^9 \cdot 4!5!} - \cdots$$

5. When x is large and positive, the following asymptotic series may be used

$$J_0(x) = \left(\frac{2}{\pi x}\right)^{\frac{1}{2}} \left\{ P_0(x) \cos\left(x - \frac{\pi}{4}\right) - Q_0(x) \sin\left(x - \frac{\pi}{4}\right) \right\}$$

$$J_1(x) = \left(\frac{2}{\pi x}\right)^{\frac{1}{2}} \left\{ P_1(x) \cos\left(x - \frac{3\pi}{4}\right) - Q_1(x) \sin\left(x - \frac{3\pi}{4}\right) \right\}$$

where

$$P_0(x) \sim 1 - \frac{1^2 \cdot 3^2}{2!(8x)^2} + \frac{1^2 \cdot 3^2 \cdot 5^2 \cdot 7^2}{4!(8x)^4} - \frac{1^2 \cdot 3^2 \cdot 5^2 \cdot 7^2 \cdot 9^2 \cdot 11^2}{6!(8x)^6} + \cdots$$

$$Q_0(x) \sim -\frac{1^2}{1!8x} + \frac{1^2 \cdot 3^2 \cdot 5^2}{3!(8x)^3} - \frac{1^2 \cdot 3^2 \cdot 5^2 \cdot 7^2 \cdot 9^2}{5!(8x)^5} + - \cdots$$

$$P_1(x) \sim 1 + \frac{1^2 \cdot 3 \cdot 5}{2!(8x)^2} - \frac{1^2 \cdot 3^2 \cdot 5^2 \cdot 7 \cdot 9}{4!(8x)^4} + \frac{1^2 \cdot 3^2 \cdot 5^2 \cdot 7^2 \cdot 9^2 \cdot 11 \cdot 13}{6!(8x)^6} - + \cdots$$

$$Q_1(x) \sim \frac{1 \cdot 3}{1!8x} - \frac{1^2 \cdot 3^2 \cdot 5 \cdot 7}{3!(8x)^3} + \frac{1^2 \cdot 3^2 \cdot 5^2 \cdot 7^2 \cdot 9 \cdot 11}{5!(8x)^5} - \cdots$$

[In $P_1(x)$ the signs alternate from $+$ to $-$ after the first term]

6. The zeros of $J_0(x)$ and $J_1(x)$.

If j_{0s} and j_{1s} are the sth zeros of $J_0(x)$ and $J_1(x)$ respectively, and if $a = 4_s - 1$, $b = 4_s + 1$

$$j_{0,s} \sim \frac{1}{4}\pi a \left\{ 1 + \frac{2}{\pi^2 a^2} - \frac{62}{3\pi^4 a^4} + \frac{15,116}{15\pi^6 a^6} - \frac{12,554,474}{105\pi^8 a^8} + \frac{8,368,654,292}{315\pi^{10}a^{10}} - + \cdots \right\}$$

$$j_{1,s} \sim \frac{1}{4}\pi b \left\{ 1 - \frac{6}{\pi^2 b^2} + \frac{6}{\pi^4 b^4} - \frac{4716}{5\pi^6 b^6} + \frac{3,902,418}{35\pi^8 b^8} - \frac{895,167,324}{35\pi^{10}b^{10}} + \cdots \right\}$$

$$J_1(j_{0,s}) \sim \frac{(-1)^{s+1}2^{\frac{3}{2}}}{\pi a^{\frac{1}{2}}} \left\{ 1 - \frac{56}{3\pi^4 a^4} + \frac{9664}{5\pi^6 a^6} - \frac{7,381,280}{21\pi^8 a^8} + \cdots \right\}$$

$$J_0(j_{1,s}) \sim \frac{(-1)^s 2^{\frac{3}{2}}}{\pi b^{\frac{1}{2}}} \left\{ 1 + \frac{24}{\pi^4 b^4} - \frac{19,584}{10\pi^6 b^6} + \frac{2,466,720}{7\pi^8 b^8} - \cdots \right\}$$

7. Table of zeros for $J_0(x)$ and $J_1(x)$

Define $\{\alpha_n, \beta_n\}$ by $J_1(\alpha_n) = 0$ and $J_0(\beta_n) = 0$.

Roots α_n	$J_1(\alpha_n)$	Roots β_n	$J_0(\beta_n)$
2.4048	0.5191	0.0000	1.0000
5.5201	-0.3403	3.8317	-0.4028
8.6537	0.2715	7.0156	0.3001
11.7915	-0.2325	10.1735	-0.2497
14.9309	0.2065	13.3237	0.2184
18.0711	-0.1877	16.4706	-0.1965
21.2116	0.1733	19.6159	0.1801

8. Recurrence formulas

$$J_{n-1}(x) + J_{n+1}(x) = \frac{2n}{x}J_n(x) \qquad nJ_n(x) + xJ_n'(x) = xJ_{n-1}(x)$$

$$J_{n-1}(x) - J_{n+1}(x) = 2J_n'(x) \qquad nJ_n(x) - xJ_n'(x) = xJ_{n+1}(x)$$

9. If J_n is written for $J_n(x)$ and $J_n^{(k)}$ is written for $\frac{d^k}{dx^k}\{J_n(x)\}$, then the following derivative relationships are important

$$J_0^{(r)} = -J_1^{(r-1)}$$
$$J_0^{(2)} = -J_0 + \frac{1}{x}J_1 = \frac{1}{2}(J_2 - J_0)$$
$$J_0^{(3)} = \frac{1}{x}J_0 + \left(1 - \frac{2}{x^2}\right)J_1 = \frac{1}{4}(-J_3 + 3J_1)$$
$$J_0^{(4)} = \left(1 - \frac{3}{x^2}\right)J_0 - \left(\frac{2}{x} - \frac{6}{x^3}\right)J_1 = \frac{1}{8}(J_4 - 4J_2 + 3J_0), \text{ etc.}$$

10. Half order Bessel functions

$$J_{\frac{1}{2}}(x) = \sqrt{\tfrac{2}{\pi x}} \sin x$$
$$J_{-\frac{1}{2}}(x) = \sqrt{\tfrac{2}{\pi x}} \cos x$$
$$J_{n+\frac{3}{2}}(x) = -x^{n+\frac{1}{2}} \tfrac{d}{dx}\{x^{-(n+\frac{1}{2})} J_{n+\frac{1}{2}}(x)\}$$
$$J_{n-\frac{1}{2}}(x) = x^{-(n+\frac{1}{2})} \tfrac{d}{dx}\{x^{n+\frac{1}{2}} J_{n+\frac{1}{2}}(x)\}$$

n	$\left(\frac{\pi x}{2}\right)^{\frac{1}{2}} J_{n+\frac{1}{2}}(x)$	$\left(\frac{\pi x}{2}\right)^{\frac{1}{2}} J_{-(n+\frac{1}{2})}(x)$
0	$\sin x$	$\cos x$
1	$\frac{\sin x}{x} - \cos x$	$-\frac{\cos x}{x} - \sin x$
2	$\left(\frac{3}{x^2} - 1\right)\sin x - \frac{3}{x}\cos x$	$\left(\frac{3}{x^2} - 1\right)\cos x + \frac{3}{x}\sin x$
3	$\left(\frac{15}{x^3} - \frac{6}{x}\right)\sin x - \left(\frac{15}{x^2} - 1\right)\cos x$ etc.	$-\left(\frac{15}{x^3} - \frac{6}{x}\right)\cos x - \left(\frac{15}{x^2} - 1\right)\sin x$

11. Additional solutions to Bessel's equation are

$Y_n(x)$ (also called Weber's function, and sometimes denoted by $N_n(x)$)
$H_n^{(1)}(x)$ and $H_n^{(2)}(x)$ (also called Hankel functions)

These solutions are defined as follows

$$Y_n(x) = \begin{cases} \dfrac{J_n(x)\cos(n\pi) - J_{-n}(x)}{\sin(n\pi)} & n \quad \text{not an integer} \\ \lim\limits_{\nu \to n} \dfrac{J_\nu(x)\cos(\nu\pi) - J_{-\nu}(x)}{\sin(\nu\pi)} & n \quad \text{an integer} \end{cases}$$

$$H_n^{(1)}(x) = J_n(x) + i\, Y_n(x)$$
$$H_n^{(2)}(x) = J_n(x) - i\, Y_n(x)$$

The additional properties of these functions may all be derived from the above relations and the known properties of $J_n(x)$.

12. Complete solutions to Bessel's equation may be written as

$$c_1 J_n(x) + c_2 J_{-n}(x) \qquad \text{if } n \text{ is not an integer}$$

or, for any value of n,

$$c_1 J_n(x) + c_2 Y_n(x) \qquad \text{or} \qquad c_1 H_n^{(1)} x + c_2 H_n^{(2)}(x)$$

13. The modified (or hyperbolic) Bessel's differential equation is

$$x^2 \frac{d^2 y}{dx^2} + x\frac{dy}{dx} - (x^2 + n^2)y = 0$$

14. When n is not an integer, two independent solutions of the equation are $I_n(x)$ and $I_{-n}(x)$, where

$$I_n(x) = \sum_{k=0}^{\infty} \frac{1}{k!\,\Gamma(n+k+1)} \left(\frac{x}{2}\right)^{n+2k}$$

15. If n is an integer,

$$I_n(x) = I_{-n}(x) = \frac{x^n}{2^n n!}\left\{1 + \frac{x^2}{2^2 \cdot 1!(n+1)} + \frac{x^4}{2^4 \cdot 2!(n+1)(n+2)} \right.$$
$$\left. + \frac{x^6}{2^6 \cdot 3!(n+1)(n+2)(n+3)} + \cdots\right\}$$

16. For $n = 0$ and $n = 1$, this formula becomes

$$I_0(x) = 1 + \frac{x^2}{2^2(1!)^2} + \frac{x^4}{2^4(2!)^2} + \frac{x^6}{2^6(3!)^2} + \frac{x^8}{2^8(4!)^2} + \cdots$$
$$I_1(x) = \frac{x}{2} + \frac{x^3}{2^3 \cdot 1!2!} + \frac{x^5}{2^5 \cdot 2!3!} + \frac{x^7}{2^7 \cdot 3!4!} + \frac{x^9}{2^9 \cdot 4!5!} + \cdots$$

17. Another solution to the modified Bessel's equation is

$$K_n(x) = \begin{cases} \frac{1}{2}\pi \frac{I_{-n}(x)-I_n(x)}{\sin(n\pi)} & n \text{ not an integer} \\ \lim_{v\to n} \frac{1}{2}\pi \frac{I_{-v}(x)-I_v(x)}{\sin(v\pi)} & n \text{ an integer} \end{cases}$$

This function is linearly independent of $I_n(x)$ for all values of n. Thus the complete solution to the modified Bessel's equation may be written as

$$c_1 I_n(x) + c_2 I_{-n}(x) \quad n \text{ not an integer}$$

or

$$c_1 I_n(x) + c_2 K_n(x) \quad \text{any } n$$

18. The following relations hold among the various Bessel functions:

$$I_n(z) = i^{-m} J_m(iz)$$
$$Y_n(iz) = (i)^{n+1} I_n(z) - \frac{2}{\pi} i^{-n} K_n(z)$$

Most of the properties of the modified Bessel function may be deduced from the known properties of $J_n(x)$ by use of these relations and those previously given.

19. Recurrence formulas

$$I_{n-1}(x) - I_{n+1}(x) = \frac{2n}{x} I_n(x) \quad I_{n-1}(x) + I_{n+1}(x) = 2I'_n(x)$$
$$I_{n-1}(x) - \frac{n}{x} I_n(x) = I'_n(x) \quad I'_n(x) = I_{n+1}(x) + \frac{n}{x} I_n(z)$$

THE FACTORIAL FUNCTION

For non-negative integers n, the factorial of n, denoted $n!$, is the product of all positive integers less than or equal to n; $n! = n \cdot (n-1) \cdot (n-2) \cdots 2 \cdot 1$. If n is a negative integer ($n = -1, -2, \ldots$) then $n! = \pm\infty$.

Approximations to $n!$ for large n include Stirling's formula

$$n! \approx \sqrt{2\pi e} \left(\frac{n}{e}\right)^{n+\frac{1}{2}},$$

and Burnsides's formula

$$n! \approx \sqrt{2\pi} \left(\frac{n+\frac{1}{2}}{e}\right)^{n+\frac{1}{2}}.$$

n	$n!$	$\log_{10} n!$	n	$n!$	$\log_{10} n!$
0	1	0.00000	1	1	0.00000
2	2	0.30103	3	6	0.77815
4	24	1.38021	5	120	2.07918
6	720	2.85733	7	5040	3.70243
8	40320	4.60552	9	3.6288×10^5	5.55976
10	3.6288×10^6	6.55976	11	3.9917×10^7	7.60116
12	4.7900×10^8	8.68034	13	6.2270×10^9	9.79428
14	8.7178×10^{10}	10.94041	15	1.3077×10^{12}	12.11650
16	2.0923×10^{13}	13.32062	17	3.5569×10^{14}	14.55107
18	6.4024×10^{15}	15.80634	19	1.2165×10^{17}	17.08509
20	2.4329×10^{18}	18.38612	25	1.5511×10^{25}	25.19065
30	2.6525×10^{32}	32.42366	40	8.1592×10^{47}	47.91165
50	3.0414×10^{64}	64.48307	60	8.3210×10^{81}	81.92017
70	1.1979×10^{100}	100.07841	80	7.1569×10^{118}	118.85473
90	1.4857×10^{138}	138.17194	100	9.3326×10^{157}	157.97000
110	1.5882×10^{178}	178.20092	120	6.6895×10^{198}	198.82539
130	6.4669×10^{219}	219.81069	150	5.7134×10^{262}	262.75689
500	1.2201×10^{1134}	1134.0864	1000	4.0239×10^{2567}	2567.6046

THE GAMMA FUNCTION

Definition: $\Gamma(n) = \int_0^\infty t^{n-1} e^{-t}\, dt \quad n > 0$

Recursion Formula: $\Gamma(n+1) = n\Gamma(n)$

$\Gamma(n+1) = n!$ if $n = 0, 1, 2, \ldots$ where $0! = 1$

For $n < 0$ the gamma function can be defined by using

$\Gamma(n) = \frac{\Gamma(n+1)}{n}$

Graph:

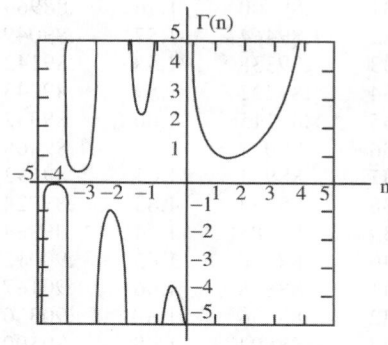

Special Values: $\Gamma(1/2) = \sqrt{\pi}$

$$\Gamma(m + 1/2) = \frac{1 \cdot 3 \cdot 5 \cdots (2m-1)}{2^m} \sqrt{\pi} \quad m = 1, 2, 3, \ldots$$

$$\Gamma(-m + 1/2) = \frac{(-1)^m 2^m \sqrt{\pi}}{1 \cdot 3 \cdot 5 \cdots (2m-1)} \quad m = 1, 2, 3, \ldots$$

Definition:

$$\Gamma(x+1) = \lim_{k \to \infty} \frac{1 \cdot 2 \cdot 3 \cdots k}{(x+1)(x+2) \cdots (x+k)} k^x$$

$$\frac{1}{\Gamma(x)} = x e^{\gamma x} \prod_{m=1}^{\infty} \left\{ \left(1 + \frac{x}{m}\right) e^{-x/m} \right\}$$

This is an infinite product representation for the gamma function where γ is Euler's constant.

Properties:

$$\Gamma'(1) = \int_0^\infty e^{\gamma x} \ln x\, dx = -\gamma$$

$$\frac{\Gamma'(x)}{\Gamma(x)} = -\gamma + \left(\frac{1}{1} - \frac{1}{x}\right) + \left(\frac{1}{2} - \frac{1}{x+1}\right) + \ldots + \left(\frac{1}{n} - \frac{1}{x+n-1}\right) + \cdots$$

$$\Gamma(x+1) = \sqrt{2\pi x}\, x^x e^{-x} \left\{ 1 + \frac{1}{12x} + \frac{1}{288x^2} - \frac{139}{51,840x^3} + \cdots \right\}$$

This is called *Stirling's asymptotic series.*

$$\text{Values of } \Gamma(n) = \int_0^\infty e^{-x}x^{n-1}dx; \quad \Gamma(n+1) = n\Gamma(n)$$

n	$\Gamma(n)$	n	$\Gamma(n)$	n	$\Gamma(n)$	n	$\Gamma(n)$
1.00	1.00000	1.25	.90640	1.50	.88623	1.75	.91906
1.01	.99433	1.26	.90440	1.51	.88659	1.76	.92137
1.02	.98884	1.27	.90250	1.52	.88704	1.77	.92376
1.03	.98355	1.28	.90072	1.53	.88757	1.78	.92623
1.04	.97844	1.29	.89904	1.54	.88818	1.79	.92877
1.05	.97350	1.30	.89747	1.55	.88887	1.80	.93138
1.06	.96874	1.31	.89600	1.56	.88964	1.81	.93408
1.07	.96415	1.32	.89464	1.57	.89049	1.82	.93685
1.08	.95973	1.33	.89338	1.58	.89142	1.83	.93969
1.09	.95546	1.34	.89222	1.59	.89243	1.84	.94261
1.10	.95135	1.35	.89115	1.60	.89352	1.85	.94561
1.11	.94740	1.36	.89018	1.61	.89468	1.86	.94869
1.12	.94359	1.37	.88931	1.62	.89592	1.87	.95184
1.13	.93993	1.38	.88854	1.63	.89724	1.88	.95507
1.14	.93642	1.39	.88785	1.64	.89864	1.89	.95838
1.15	.93304	1.40	.88726	1.65	.90012	1.90	.96177
1.16	.92980	1.41	.88676	1.66	.90167	1.91	.96523
1.17	.92670	1.42	.88636	1.67	.90330	1.92	.96877
1.18	.92373	1.43	.88604	1.68	.90500	1.93	.97240
1.19	.92089	1.44	.88581	1.69	.90678	1.94	.97610
1.20	.91817	1.45	.88566	1.70	.90864	1.95	.97988
1.21	.91558	1.46	.88560	1.71	.91057	1.96	.98374
1.22	.91311	1.47	.88563	1.72	.91258	1.97	.98768
1.23	.91075	1.48	.88575	1.73	.91466	1.98	.99171
1.24	.90852	1.49	.88595	1.74	.91683	1.99	.99581
						2.00	1.00000

THE BETA FUNCTION

Definition: $B(m, n) = \int_0^1 t^{m-1}(1 - t)^{m-1}dt \quad m > 0, n > 0$

Relationship with Gamma function: $B(m, n) = \dfrac{\Gamma(m)\Gamma(n)}{\Gamma(m + n)}$

Properties:

$$B(m, n) = B(n, m)$$
$$B(m, n) = 2 \int_0^{\pi/2} \sin^{2m-1}\theta \cos^{2n-1}\theta\, d\theta$$
$$B(m, n) = \int_0^\infty \frac{t^{m-1}}{(1+t)^{m+n}}dt$$
$$B(m, n) = r^n(r + 1)^m \int_0^1 \frac{t^{m-1}(1-t)^{n-1}}{(r+t)^{m+n}}\, dt$$

THE ERROR FUNCTION

Definition: $\text{erf}(x) = \dfrac{2}{\sqrt{\pi}} \int_0^x e^{-t^2}dt$

Series: $erf(x) = \dfrac{2}{\sqrt{\pi}}\left(x - \dfrac{x^3}{3} + \dfrac{1}{2!}\dfrac{x^5}{5} - \dfrac{1}{3!}\dfrac{x^7}{7} + \cdots\right)$

Property: $\text{erf}(x) = -\text{erf}(-x)$

Relationship with Normal Probability Function $f(t)$: $\int_0^x f(t)\, dt = \dfrac{1}{2}\,\text{erf}\left(\dfrac{x}{\sqrt{2}}\right)$ To evaluate erf(2.3), one proceeds as follows:

For $\frac{x}{\sqrt{2}} = 2.3$, one finds $x = (2.3)\,(\sqrt{2}) = 3.25$. In the normal probability function table (page A-104), one finds the entry 0.4994 opposite the value 3.25. Thus erf(2.3) = 2(0.4994) = 0.9988.

$$\text{erfc}(z) = 1 - \text{erf}(z) = \frac{2}{\sqrt{\pi}} \int_z^\infty e^{-t^2}dt$$

is known as the complementary error function.

ORTHOGONAL POLYNOMIALS

I: Legendre

Name: Legendre   *Symbol:* $P_n(x)$ *Interval:* $[-1, 1]$

Differential Equation: $(1 - x^2)y'' - 2xy' + n(n + 1)y = 0$

Explicit Expression: $P_n(x) = \dfrac{1}{2^n} \displaystyle\sum_{m=0}^{[n/2]} (-1)^m \binom{n}{m} \binom{2n-2m}{n} x^{n-2m}$

Recurrence Relation: $(n+1)P_{n+1}(x) = (2n+1)xP_n(x) - nP_{n-1}(x)$

Weight: 1

Standardization: $P_n(1) = 1$

Norm: $\displaystyle\int_{-1}^{+1} [P_n(x)]^2 dx = \dfrac{2}{2n+1}$

Rodrigues' Formula: $P_n(x) = \dfrac{(-1)^n}{2^n n!} \dfrac{d^n}{dx^n}\{(1-x^2)^n\}$

Generating Function: $R^{-1} = \displaystyle\sum_{n=0}^{\infty} P_n(x)z^n; \ -1 < x < 1, \quad |z| < 1,$

$$R = \sqrt{1 - 2xz + z^2}$$

Inequality: $|P_n(x)| \le 1, \ -1 \le x \le 1.$

II: Tschebysheff, First Kind

Name: Tschebysheff, First Kind   *Symbol:* $T_n(x)$ *Interval:* $[-1, 1]$

Differential Equation: $(1-x^2)y - xy' + n^2 y = 0$

Explicit Expression: $\dfrac{n}{2} \displaystyle\sum_{m=0}^{[n/2]} (-1)^m \dfrac{(n-m-1)!}{m!(n-2m)!} (2x)^{n-2m} = \cos(n \arccos x) = T_n(x)$

Recurrence Relation: $T_{n+1}(x) = 2xT_n(x) - T_{n-1}(x)$

Weight: $(1-x^2)^{-1/2}$

Standardization: $T_n(1) = 1$

Norm: $\int_{-1}^{+1} (1-x^2)^{-1/2}[T_n(x)]^2 dx = \begin{cases} \pi/2, & n \ne 0 \\ \pi, & n = 0 \end{cases}$

Rodrigues' Formula: $\dfrac{(-1)^n(1-x^2)^{1/2}\sqrt{\pi}}{2^{n+1}\Gamma(n+\frac{1}{2})} \dfrac{d^n}{dx^n}\{(1-x^2)^{n-(1/2)}\} = T_n(x)$

Generating Function: $\dfrac{1-xz}{1-2xz-z^2} = \displaystyle\sum_{n=0}^{\infty} T_n(x)z^n, \ -1 < x < 1, \quad |z| < 1$

Inequality: $|T_n(x)| \le 1, \ -1 \le x \le 1.$

III: Tschebysheff, Second Kind

Name: Tschebysheff, Second Kind   *Symbol* $U_n(x)$ *Interval:* $[-1, 1]$

Differential Equation: $(1-x^2)y'' - 3xy' + n(n+2)y = 0$

Explicit Expression: $U_n(x) = \displaystyle\sum_{m=0}^{[n/2]} (-1)^m \dfrac{(m-n)!}{m!(n-2m)!} (2x)^{n-2m}$

$$U_n(\cos\theta) = \dfrac{\sin[(n+1)\theta]}{\sin\theta}$$

Recurrence Relation: $U_{n+1}(x) = 2xU_n(x) - U_{n-1}(x)$

Weight: $(1-x^2)^{1/2}$ *Standardization:* $U_n(1) = n+1$

Norm: $\displaystyle\int_{-1}^{+1} (1-x^2)^{1/2}[U_n(x)]^2 dx = \dfrac{\pi}{2}$

Rodrigues' Formula: $U_n(x) = \dfrac{(-1)^n(n+1)\sqrt{\pi}}{(1-x^2)^{1/2}2^{n+1}\Gamma(n+\frac{3}{2})} \dfrac{d^n}{dx^n}\{(1-x^2)^{n+(1/2)}\}$

Generating Function: $\dfrac{1}{1-2xz+z^2} = \displaystyle\sum_{n=0}^{\infty} U_n(x)z^n, \ -1 < x < 1, \quad |z| < 1$

Inequality: $|U_n(x)| \le n+1, \ -1 \le x \le 1.$

IV: Jacobi

Name: Jacobi   *Symbol:* $P_n^{(\alpha,\beta)}(x)$ *Interval:* $[-1, 1]$

Differential Equation: $(1-x^2)y'' + [\beta - \alpha - (\alpha+\beta+2)x]y' + n(n+\alpha+\beta+1)y = 0$

Explicit Expression: $P_n^{(\alpha,\beta)}(x) = \dfrac{1}{2^n} \sum\limits_{m=0}^{n} \binom{n+\alpha}{m}\binom{n+\beta}{n-m}(x-1)^{n-m}(x+1)^m$

Recurrence Relation:

$$2(n+1)(n+\alpha+\beta+1)(2n+\alpha+\beta)P_{n+1}^{(\alpha,\beta)}(x)$$
$$= (2n+\alpha+\beta+1)[(\alpha^2-\beta^2)+(2n+\alpha+\beta+2)(2n+\alpha+\beta)x]P_n^{(\alpha,\beta)}(x)$$
$$- 2(n+\alpha)(n+\beta)(2n+\alpha+\beta+2)P_{n-1}^{(\alpha,\beta)}(x)$$

Weight: $(1-x)^\alpha(1+x)^\beta; \alpha, \beta > 1$
Standardization: $P_n^{(\alpha,\beta)}(x) = \binom{n+\alpha}{n}$

Norm: $\displaystyle\int_{-1}^{+1}(1-x)^\alpha(1+x)^\beta[P_n^{(\alpha,\beta)}(x)]^2\,dx = \dfrac{2^{\alpha+\beta+1}\Gamma(n+\alpha+1)\Gamma(n+\beta+1)}{(2n+\alpha+\beta+1)n!\Gamma(n+\alpha+\beta+1)}$

Rodrigues' Formula: $P_n^{(\alpha,\beta)}(x) = \dfrac{(-1)^n}{2^n n!(1-x)^\alpha(1+x)^\beta}\dfrac{d^n}{dx^n}\{(1-x)^{n+\alpha}(1+x)^{n+\beta}\}$

Generating Function: $R^{-1}(1-z+R)^{-\alpha}(1+z+R)^{-\beta} = \sum\limits_{n=0}^{\infty}2^{-\alpha-\beta}P_n^{(\alpha,\beta)}(x)z^n,$

$$R = \sqrt{1-2xz+z^2}, \quad |z| < 1$$

Inequality: $\max\limits_{-1\leq x\leq 1}|P_n^{(\alpha,\beta)}(x)| = \begin{cases} \binom{n+q}{n} \sim n^q \text{if } q = \max(\alpha,\beta) \geq -\dfrac{1}{2} \\ |P_n^{(\alpha,\beta)}(x')| \sim n^{-1/2} \text{if } q < -\dfrac{1}{2} \\ x' \text{ is one of the two maximum points nearest} \\ \dfrac{\beta-\alpha}{\alpha+\beta+1} \end{cases}$

V: Generalized Laguerre

Name: Generalized Laguerre   *Symbol:* $L_n^{(\alpha)}(x)$ *Interval:* $[0,\infty]$
Differential Equation: $xy'' + (\alpha+1-x)y' + ny = 0$

Explicit Expression: $L_n^{(\alpha)}(x) = \sum\limits_{m=0}^{n}(-1)^m\binom{n+\alpha}{n-m}\dfrac{1}{m!}x^m$

Recurrence Relation: $(n+1)L_n^{(\alpha)} + 1(x) = [(2n+\alpha+1)-x]L_n^{(\alpha)}(x) - (n+\alpha)L_n^{(\alpha)} - 1(x)$
Weight: $x^\alpha e^{-x}, \alpha > -1$ *Standardization:* $L_n^{(\alpha)}(x) = \dfrac{(-1)^n}{n!}x^n + \cdots$

Norm: $\displaystyle\int_0^\infty x^\alpha e^{-x}[L_n^{(\alpha)}(x)]^2\,dx = \dfrac{\Gamma(n+\alpha+1)}{n!}$

Rodrigues' Formula: $L_n^{(\alpha)}(x) = \dfrac{1}{n!x^\alpha e^{-x}}\dfrac{d^n}{dx^n}\{x^{n+\alpha}e^{-x}\}$

Generating Function: $(1-z)^{-\alpha-1}\exp\left(\dfrac{xz}{z-1}\right) = \sum\limits_{n=0}^{\infty}L_n^{(\alpha)}(x)z^n$

Inequality: $|L_n^{(\alpha)}(x) \leq \dfrac{\Gamma(n+\alpha+1)}{n!\Gamma(\alpha+1)}e^{x/2}; \quad \begin{matrix} x \geq 0 \\ \alpha > 0 \end{matrix}$

$|L_n^{(\alpha)}(x)| \leq \left[2 - \dfrac{\Gamma(\alpha+n+1)}{n!\Gamma(\alpha+1)}\right]e^{x/2}; \quad \begin{matrix} x \geq 0 \\ -1 < \alpha < 0 \end{matrix}$

VI: Hermite

Name: Hermite   *Symbol:* $H_n(x)$ *Interval:* $[-\infty,\infty]$
Differential Equation: $y'' - 2xy' + 2ny = 0$

Explicit Expression: $H_n(x) = \sum\limits_{m=0}^{[n/2]}\dfrac{(-1)^m n!(2x)^{n-2m}}{m!(n-2m)!}$

Recurrence Relation: $H_{n+1}(x) = 2xH_n(x) - 2nH_{n-1}(x)$
Weight: e^{-x^2} *Standardization:* $H_n(1) = 2^n x^n + \cdots$

Norm: $\displaystyle\int_{-\infty}^{\infty}e^{-x^2}[H_n(x)]^2\,dx = 2^n n!\sqrt{\pi}$

Rodrigues' Formula: $H_n(x) = (-1)^n e^{x^2}\dfrac{d^n}{dx^n}(e^{-x^2})$

Generating Function: $e^{-x^2+2zx} = \sum\limits_{n=0}^{\infty}H_n(x)\dfrac{z^n}{n!}$

Inequality: $|H_n(x)|e^{x^2/2}k2^{n/2}\sqrt{n!}\, k \approx 1.086435$

TABLES OF ORTHOGONAL POLYNOMIALS

$H_0 = 1$
$H_1 = 2x$
$H_2 = 4x^2 - 2$
$H_3 = 8x^3 - 12x$
$H_4 = 16x^4 - 48x^2 + 12$
$H_5 = 32x^5 - 160x^3 + 120x$
$H_6 = 64x^6 - 480x^4 + 720x^2 - 120$
$H_7 = 128x^7 - 1344x^5 + 3360x^3 - 1680x$
$H_8 = 256x^8 - 3584x^6 + 13440x^4 - 13440x^2 + 1680$
$H_9 = 512x^9 - 9216x^7 + 48384x^5 - 80640x^3 + 30240x$
$H_{10} = 1024x^{10} - 23040x^8 + 161280x^6 - 403200x^4 + 302400x^2 - 30240$

$x^{10} = (30240H_0 + 75600H_2 + 25200H_4 + 2520H_6 + 90H_8 + H_{10})/1024$
$x^9 = (15120H_1 + 10080H_3 + 1512H_5 + 72H_7 + H_9)/512$
$x^8 = (1680H_0 + 3360H_2 + 840H_4 + 56H_6 + H_8)/256$
$x^7 = (840H_1 + 420H_3 + 42H_5 + H_7)/128$
$x^6 = (120H_0 + 180H_2 + 30H_4 + H_6)/64$
$x^5 = (60H_1 + 20H_3 + H_5)/32$
$x^4 = (12H_0 + 12H_2 + H_4)/16$
$x^3 = (6H_1 + H_3)/8$
$x^2 = (2H_0 + H_2)/4$
$x = (H_1)/2$
$1 = H_0$

$L_0 = 1$
$L_1 = -x + 1$
$L_2 = (x^2 - 4x + 2)/2$
$L_3 = (-x^3 + 9x^2 - 18x + 6)/6$
$L_4 = (x^4 - 16x^3 + 72x^2 - 96x + 24)/24$
$L_5 = (-x^5 + 25x^4 - 200x^3 + 600x^2 - 600x + 120)/120$
$L_6 = (x^6 - 36x^5 + 450x^4 - 2400x^3 + 5400x^2 - 4320x + 720)/720$

$x^6 = 720L_0 - 4320L_1 + 10800L_2 - 14400L_3 + 10800L_4 - 4320L_5 + 720L_6$
$x^5 = 120L_0 - 600L_1 + 1200L_2 - 1200L_3 + 600L_4 - 120L_5$
$x^4 = 24L_0 - 96L_1 + 144L_2 - 96L_3 + 24L_4$
$x^3 = 6L_0 - 18L_1 + 18L_2 - 6L_3$
$x^2 = 2L_0 - 4L_1 + 2L_2$
$x = L_0 - L_1$
$1 = L_0$

$P_0 = 1$
$P_1 = x$
$P_2 = (3x^2 - 1)/2$
$P_3 = (5x^3 - 3x)/2$
$P_4 = (35x^4 - 30x^2 + 3)/8$
$P_5 = (63x^5 - 70x^3 + 15x)/8$
$P_6 = (231x^6 - 315x^4 + 105x^2 - 5)/16$
$P_7 = (429x^7 - 693x^5 + 315x^3 - 35x)/16$
$P_8 = (6435x^8 - 12012x^6 + 6930x^4 - 1260x^2 + 35)/128$
$P_9 = (12155x^9 - 25740x^7 + 18018x^5 - 4620x^3 + 315x)/128$
$P_{10} = (46189x^{10} - 109395x^8 + 90090x^6 - 30030x^4 + 3465x^2 - 63)/256$

$x^{10} = (4199P_0 + 16150P_2 + 15504P_4 + 7904P_6 + 2176P_8 + 256P_{10})/46189$
$x^9 = (3315P_1 + 4760P_3 + 2992P_5 + 960P_7 + 128P_9)/12155$
$x^8 = (715P_0 + 2600P_2 + 2160P_4 + 832P_6 + 128P_8)/6435$
$x^7 = (143P_1 + 182P_3 + 88P_5 + 16P_7)/429$
$x^6 = (33P_0 + 110P_2 + 72P_4 + 16P_6)/231$
$x^5 = (27P_1 + 28P_3 + 8P_5)/63$
$x^4 = (7P_0 + 20P_2 + 8P_4)/35$
$x^3 = (3P_1 + 2P_3)/5$
$x^2 = (P_0 + 2P_2)/3$
$x = P_1$
$1 = P_0$

$T_0 = 1$
$T_1 = x$
$T_2 = 2x^2 - 1$
$T_3 = 4x^3 - 3x$
$T_4 = 8x^4 - 8x^2 + 1$
$T_5 = 16x^5 - 20x^3 + 5x$
$T_6 = 32x^6 - 48x^4 + 18x^2 - 1$
$T_7 = 64x^7 - 112x^5 + 56x^3 - 7x$
$T_8 = 128x^8 - 256x^6 + 160x^4 - 32x^2 + 1$
$T_9 = 256x^9 - 576x^7 + 432x^5 - 120x^3 + 9x$
$T_{10} = 512x^{10} - 1280x^8 + 1120x^6 - 400x^4 + 50x^2 - 1$

$x^{10} = (126T_0 + 210T_2 + 120T_4 + 45T_6 + 10T_8 + T_{10})/512$
$x^9 = (126T_1 + 84T_3 + 36T_5 + 9T_7 + T_9)/256$
$x^8 = (35T_0 + 56T_2 + 28T_4 + 8T_6 + T_8)/128$
$x^7 = (35T_1 + 21T_3 + 7T_5 + T_7)/64$
$x^6 = (10T_0 + 15T_2 + 6T_4 + T_6)/32$
$x^5 = (10T_1 + 5T_3 + T_5)/16$
$x^4 = (3T_0 + 4T_2 + T_4)/8$
$x^3 = (3T_1 + T_3)/4$
$x^2 = (T_0 + T_2)/2$
$x = T_1$
$1 = T_0$

$U_0 = 1$
$U_1 = 2x$
$U_2 = 4x^2 - 1$
$U_3 = 8x^3 - 4x$
$U_4 = 16x^4 - 12x^2 + 1$
$U_5 = 32x^5 - 32x^3 + 6x$
$U_6 = 64x^6 - 80x^4 + 24x^2 - 1$
$U_7 = 128x^7 - 192x^5 + 80x^3 - 8x$
$U_8 = 256x^8 - 448x^6 + 240x^4 - 40x^2 + 1$
$U_9 = 512x^9 - 1024x^7 + 672x^5 - 160x^3 + 10x$
$U_{10} = 1024x^{10} - 2304x^8 + 1792x^6 - 560x^4 + 60x^2 - 1$

$x^{10} = (42U_0 + 90U_2 + 75U_4 + 35U_6 + 9U_8 + U_{10})/1024$
$x^9 = (42U_1 + 48U_3 + 27U_5 + 8U_7 + U_9)/512$
$x^8 = (14U_0 + 28U_2 + 20U_4 + 7U_6 + U_8)/256$
$x^7 = (14U_1 + 14U_3 + 6U_5 + U_7)/128$
$x^6 = (5U_0 + 9U_2 + 5U_4 + U_6)/64$
$x^5 = (5U_1 + 4U_3 + U_5)/32$
$x^4 = (2U_0 + 3U_2 + U_4)/16$
$x^3 = (2U_1 + U_3)/8$
$x^2 = (U_0 + U_2)/4$
$x = (U_1)/2$
$1 = U_0$

CLEBSCH–GORDAN COEFFICIENTS

$$\begin{pmatrix} j_1 & j_2 & j \\ m_1 & m_2 & m \end{pmatrix} = \delta_{m,m_1+m_2} \sqrt{\frac{(j_1 + j_2 - j)!(j + j_1 - j_2)!(j + j_2 - j_1)!(2j + 1)}{(j + j_1 + j_2 + 1)!}}$$

$$\times \sum_k \frac{(-1)^k \sqrt{(j_1 + m_1)!(j_1 - m_1)!(j_2 + m_2)!(j_2 - m_2)!(j + m)!(j - m)!}}{k!(j_1 + j_2 - j - k)!(j_1 - m_1 - k)!(j_2 + m_2 - k)!(j - j_2 + m_1 + k)!(j - j_1 - m_2 + k)!}.$$

1. Conditions:

   (a) Each of $\{j_1, j_2, j, m_1, m_2, m\}$ may be an integer, or half an integer. Additionally: $j > 0$, $j_1 > 0$, $j_2 > 0$ and $j + j_1 + j_2$ is an integer.

   (b) $j_1 + j_2 - j \geq 0$.

   (c) $j_1 - j_2 + j \geq 0$.

   (d) $-j_1 + j_2 + j \geq 0$.

   (e) $|m_1| \leq j_1$, $|m_2| \leq j_2$, $|m| \leq j$.

2. Special values:

   (a) $\begin{pmatrix} j_1 & j_2 & j \\ m_1 & m_2 & m \end{pmatrix} = 0$ if $m_1 + m_2 \neq m$.

   (b) $\begin{pmatrix} j_1 & 0 & j \\ m_1 & 0 & m \end{pmatrix} = \delta_{j_1,j}\delta_{m_1,m}$.

   (c) $\begin{pmatrix} j_1 & j_2 & j \\ 0 & 0 & 0 \end{pmatrix} = 0$ when $j_1 + j_2 + j$ is an odd integer.

   (d) $\begin{pmatrix} j_1 & j_1 & j \\ m_1 & m_1 & m \end{pmatrix} = 0$ when $2j_1 + j$ is an odd integer.

3. Symmetry relations: all of the following are equal to $\begin{pmatrix} j_1 & j_2 & j \\ m_1 & m_2 & m \end{pmatrix}$:

   (a) $\begin{pmatrix} j_2 & j_1 & j \\ -m_2 & -m_1 & -m \end{pmatrix}$,

   (b) $(-1)^{j_1+j_2-j} \begin{pmatrix} j_2 & j_1 & j \\ m_1 & m_2 & m \end{pmatrix}$,

   (c) $(-1)^{j_1+j_2-j} \begin{pmatrix} j_1 & j_2 & j \\ -m_1 & -m_2 & -m \end{pmatrix}$,

   (d) $\sqrt{\frac{2j+1}{2j_1+1}}(-1)^{j_2+m_2} \begin{pmatrix} j & j_2 & j_1 \\ -m & m_2 & -m_1 \end{pmatrix}$,

   (e) $\sqrt{\frac{2j+1}{2j_1+1}}(-1)^{j_1-m_1+j-m} \begin{pmatrix} j & j_2 & j_1 \\ m & -m_2 & m_1 \end{pmatrix}$,

   (f) $\sqrt{\frac{2j+1}{2j_1+1}}(-1)^{j-m+j_1-m_1} \begin{pmatrix} j_2 & j & j_1 \\ m_2 & -m & -m_1 \end{pmatrix}$,

   (g) $\sqrt{\frac{2j+1}{2j_2+1}}(-1)^{j_1-m_1} \begin{pmatrix} j_1 & j & j_2 \\ m_1 & -m & -m_2 \end{pmatrix}$,

   (h) $\sqrt{\frac{2j+1}{2j_2+1}}(-1)^{j_1-m_1} \begin{pmatrix} j & j_1 & j_2 \\ m & -m_1 & m_2 \end{pmatrix}$.

By use of the symmetry relations, Clebsch–Gordan coefficients may be put in the standard form $j_1 \leq j_2 \leq j$ and $m \geq 0$.

m_2	m	j_1	j	$\begin{pmatrix} j_1 & \frac{1}{2} & j \\ m_1 & m_2 & m \end{pmatrix}$	
$-\frac{1}{2}$	0	$\frac{1}{2}$	1	$\frac{\sqrt{2}}{2}$	≈ 0.707107
0	$\frac{1}{2}$	$\frac{1}{2}$	1	$\frac{\sqrt{3}}{2}$	≈ 0.866025
$\frac{1}{2}$	0	$\frac{1}{2}$	1	$\frac{\sqrt{2}}{2}$	≈ 0.707107
$\frac{1}{2}$	$\frac{1}{2}$	$\frac{1}{2}$	1	$\frac{\sqrt{3}}{2}$	≈ 0.866025
$\frac{1}{2}$	1	$\frac{1}{2}$	1	1	≈ 1.000000

m_2	m	j_1	j	$\begin{pmatrix} j_1 & 1 & j \\ m_1 & m_2 & m \end{pmatrix}$	
-1	0	1	1	$\frac{\sqrt{2}}{2}$	≈ 0.707107
-1	0	1	2	$\frac{\sqrt{6}}{6}$	≈ 0.408248
$-\frac{1}{2}$	0	$\frac{1}{2}$	$\frac{3}{2}$	$\frac{\sqrt{2}}{2}$	≈ 0.707107
$-\frac{1}{2}$	$\frac{1}{2}$	1	1	$\frac{3}{4}$	≈ 0.750000
$-\frac{1}{2}$	$\frac{1}{2}$	1	2	$\frac{\sqrt{5}}{4}$	≈ 0.559017
0	0	1	2	$\frac{\sqrt{6}}{3}$	≈ 0.816496
0	0	$\frac{1}{2}$	$\frac{3}{2}$	$\frac{\sqrt{3}}{2}$	≈ 0.866025
0	$\frac{1}{2}$	$\frac{1}{2}$	$\frac{3}{2}$	$\frac{\sqrt{6}}{3}$	≈ 0.8164967
0	$\frac{1}{2}$	1	1	$\frac{\sqrt{2}}{4}$	≈ 0.353553
0	$\frac{1}{2}$	1	2	$\frac{\sqrt{10}}{4}$	≈ 0.790569
0	1	1	1	$\frac{\sqrt{2}}{2}$	≈ 0.707107

m_2	m	j_1	j	$\begin{pmatrix} j_1 & 1 & j \\ m_1 & m_2 & m \end{pmatrix}$	
0	1	1	2	$\frac{\sqrt{2}}{2}$	≈ 0.707107
$\frac{1}{2}$	0	$\frac{1}{2}$	$\frac{3}{2}$	$\frac{\sqrt{2}}{2}$	≈ 0.707107
$\frac{1}{2}$	$\frac{1}{2}$	1	1	$-\frac{\sqrt{2}}{4}$	≈ -0.353553
$\frac{1}{2}$	$\frac{1}{2}$	1	2	$\frac{\sqrt{10}}{4}$	≈ 0.790569
$\frac{1}{2}$	1	$\frac{1}{2}$	$\frac{3}{2}$	$\frac{\sqrt{30}}{6}$	≈ 0.912871
$\frac{1}{2}$	$\frac{3}{2}$	1	2	$\frac{\sqrt{105}}{12}$	≈ 0.853913
1	0	1	1	$-\frac{\sqrt{2}}{2}$	≈ -0.707107
1	0	1	2	$\frac{\sqrt{6}}{6}$	≈ 0.408248
1	$\frac{1}{2}$	$\frac{1}{2}$	$\frac{3}{2}$	$\frac{\sqrt{3}}{3}$	≈ 0.577350
1	$\frac{1}{2}$	1	1	$-\frac{3}{4}$	≈ -0.750000
1	$\frac{1}{2}$	1	2	$\frac{\sqrt{5}}{4}$	≈ 0.559017
1	1	$\frac{1}{2}$	$\frac{3}{2}$	$\frac{\sqrt{10}}{4}$	≈ 0.790569
1	1	1	1	$-\frac{\sqrt{2}}{2}$	≈ -0.707107
1	1	1	2	$\frac{\sqrt{2}}{2}$	≈ 0.707107
1	$\frac{3}{2}$	$\frac{1}{2}$	$\frac{3}{2}$	1	≈ 1.000000
1	$\frac{3}{2}$	1	2	$\frac{\sqrt{105}}{12}$	≈ 0.853913
1	2	1	2	1	≈ 1.000000

NORMAL PROBABILITY FUNCTION

Table of the normal distribution

For a standard normal random variable ($\Phi(z)$ is the area under the Standard Normal Curve from $-\infty$ to z).

Limits		Proportion of the total area	Remaining area
$\mu - \lambda\sigma$	$\mu + \lambda\sigma$	(%)	(%)
$\mu - \sigma$	$\mu + \sigma$	68.27	31.73
$\mu - 1.65\sigma$	$\mu + 1.65\sigma$	90	10
$\mu - 1.96\sigma$	$\mu + 1.96\sigma$	95	5
$\mu - 2\sigma$	$\mu + 2\sigma$	95.45	4.55
$\mu - 2.58\sigma$	$\mu + 2.58\sigma$	99.0	0.99
$\mu - 3\sigma$	$\mu + 3\sigma$	99.73	0.27
$\mu - 3.09\sigma$	$\mu + 3.09\sigma$	99.8	0.2
$\mu - 3.29\sigma$	$\mu + 3.29\sigma$	99.9	0.1

x	1.282	1.645	1.960	2.326	2.576	3.090
$\Phi(x)$	0.90	0.95	0.975	0.99	0.995	0.999
$2[1 - \Phi(x)]$	0.20	0.10	0.05	0.02	0.01	0.002

x	3.09	3.72	4.26	4.75	5.20	5.61	6.00	6.36
$1 - \Phi(x)$	10^{-3}	10^{-4}	10^{-5}	10^{-6}	10^{-7}	10^{-8}	10^{-9}	10^{-10}

Areas under the Standard Normal Curve from 0 to z

z	0	1	2	3	4	5	6	7	8	9
0.0	.0000	.0040	.0080	.0120	.0160	.0199	.0239	.0279	.0319	.0359
0.1	.0398	.0438	.0478	.0517	.0557	.0596	.0636	.0675	.0714	.0754
0.2	.0793	.0832	.0871	.0910	.0948	.0987	.1026	.1064	.1103	.1141
0.3	.1179	.1217	.1255	.1293	.1331	.1368	.1406	.1443	.1480	.1517
0.4	.1554	.1591	.1628	.1664	.1700	.1736	.1772	.1808	.1844	.1879
0.5	.1915	.1950	.1985	.2019	.2054	.2088	.2123	.2157	.2190	.2224
0.6	.2258	.2291	.2324	.2357	.2389	.2422	.2454	.2486	.2518	.2549
0.7	.2580	.2612	.2652	.2673	.2704	.2734	.2764	.2794	.2823	.2852
0.8	.2881	.2910	.2939	.2967	.2996	.3023	.3051	.3078	.3106	.3133
0.9	.3159	.3186	.3212	.3238	.3264	.3289	.3315	.3340	.3365	.3389
1.0	.3413	.3438	.3461	.3485	.3508	.3531	.3554	.3577	.3599	.3621
1.1	.3643	.3665	.3686	.3708	.3729	.3749	.3770	.3790	.3810	.3830
1.2	.3849	.3869	.3888	.3907	.3925	.3944	.3962	.3980	.3997	.4015
1.3	.4032	.4049	.4066	.4082	.4099	.4115	.4131	.4147	.4162	.4177
1.4	.4192	.4207	.4222	.4236	.4251	.4265	.4279	.4292	.4306	.4319
1.5	.4332	.4345	.4357	.4370	.4382	.4394	.4406	.4418	.4429	.4441
1.6	.4452	.4463	.4474	.4484	.4495	.4505	.4515	.4525	.4535	.4545
1.7	.4554	.4564	.4573	.4582	.4591	.4599	.4608	.4616	.4625	.4633
1.8	.4641	.4649	.4656	.4664	.4671	.4678	.4686	.4693	.4699	.4706
1.9	.4713	.4719	.4726	.4732	.4738	.4744	.4750	.4756	.4761	.4767
2.0	.4772	.4778	.4783	.4788	.4793	.4798	.4803	.4808	.4812	.4817
2.1	.4821	.4826	.4830	.4834	.4838	.4842	.4846	.4850	.4854	.4857
2.2	.4861	.4864	.4868	.4871	.4875	.4878	.4881	.4884	.4887	.4890
2.3	.4893	.4896	.4898	.4901	.4904	.4906	.4909	.4911	.4913	.4916
2.4	.4918	.4920	.4922	.4925	.4927	.4929	.4931	.4932	.4934	.4936
2.5	.4938	.4940	.4941	.4943	.4945	.4946	.4948	.4949	.4951	.4952
2.6	.4953	.4955	.4956	.4957	.4959	.4960	.4961	.4962	.4963	.4964
2.7	.4965	.4966	.4967	.4968	.4969	.4970	.4971	.4972	.4973	.4974
2.8	.4974	.4975	.4976	.4977	.4977	.4978	.4979	.4979	.4980	.4981
2.9	.4981	.4982	.4982	.4983	.4984	.4984	.4985	.4985	.4986	.4986
3.0	.4987	.4987	.4987	.4988	.4988	.4989	.4989	.4989	.4990	.4990
3.1	4990	.4991	.4991	.4991	.4992	.4992	.4992	.4992	.4993	.4993
3.2	4993	.4993	.4994	.4994	.4994	.4994	.4994	.4995	.4995	.4995
3.3	4995	.4995	.4995	.4996	.4996	.4996	.4996	.4996	.4996	.4997
3.4	4997	.4997	.4997	.4997	.4997	.4997	.4997	.4997	.4997	.4998
3.5	4998	.4998	.4998	.4998	.4998	.4998	.4998	.4998	.4998	.4998
3.6	4998	.4998	.4999	.4999	.4999	.4999	.4999	.4999	.4999	.4999
3.7	4999	.4999	.4999	.4999	.4999	.4999	.4999	.4999	.4999	.4999
3.8	4999	.4999	.4999	.4999	.4999	.4999	.4999	.4999	.4999	.4999
3.9	5000	.5000	.5000	.5000	.5000	.5000	.5000	.5000	.5000	.5000

COMMON SAMPLE SIZE CALCULATIONS

Parameter	Estimate	Sample size
μ	$\bar{x}$	$n = \left(\dfrac{z_{\alpha/2} \cdot \sigma}{E} \right)^2$
p	$\hat{p}$	$n = \dfrac{(z_{\alpha/2})^2 \cdot pq}{E^2}$
$\mu_2 - \mu_2$	$\bar{x}_1 - \bar{x}_2$	$n_1 = n_2 = \dfrac{(z_{\alpha/2})^2(\sigma_1^2 + \sigma_2^2)}{E^2}$
$p_1 - p_2$	$\hat{p}_1 - \hat{p}_2$	$n_1 = n_2 = \dfrac{(z_{\alpha/2})^2(p_1 q_1 + p_2 q_2)}{E^2}$

COMMON ONE SAMPLE CONFIDENCE INTERVALS

Parameter	Assumptions	$100(1 - \alpha)\%$ Confidence interval
μ	n large, σ^2 known, or normality, σ^2 known	$\bar{x} \pm z_{\alpha/2} \cdot \dfrac{\sigma}{\sqrt{n}}$
μ	normality, σ^2 unknown	$\bar{x} \pm t_{\alpha/2, n-1} \cdot \dfrac{s}{\sqrt{n}}$
σ^2	normality	$\left(\dfrac{(n-1)s^2}{\chi_{\alpha/2, n-1}^2}, \dfrac{(n-1)s^2}{\chi_{1-\alpha/2, n-1}^2} \right)$
p	binomial experiment, n large	$\hat{p} \pm z_{\alpha/2} \cdot \sqrt{\dfrac{\hat{p}(1 - \hat{p})}{n}}$

COMMON TWO SAMPLE CONFIDENCE INTERVALS

Parameter	Assumptions	$100(1 - \alpha)\%$ Confidence interval
$\mu_1 - \mu_2$	normality, independence, σ_1^2, σ_2^2 known or n_1, n_2 large, independence, σ_1^2, σ_2^2 known	$(\bar{x}_1 - \bar{x}_2) \pm z_{\alpha/2} \cdot \sqrt{\dfrac{\sigma_1^2}{n_1} + \dfrac{\sigma_2^2}{n_2}}$
$\mu_1 - \mu_2$	normality, independence, $\sigma_1^2 = \sigma_2^2$ unknown	$(\bar{x}_1 - \bar{x}_2) \pm$ $t_{\frac{\alpha}{2}, n_1 + n_2 - 2} \cdot s_p \sqrt{\dfrac{1}{n_1} + \dfrac{1}{n_2}}$ $s_p^2 = \dfrac{(n_1 - 1)s_1^2 + (n_2 - 1)s_2^2}{n_1 + n_2 - 2}$
$\mu_1 - \mu_2$	normality, independence, $\sigma_1^2 \neq \sigma_2^2$ unknown	$(\bar{x}_1 - \bar{x}_2) \pm t_{\alpha/2, \nu} \cdot \sqrt{\dfrac{s_1^2}{n_1} + \dfrac{s_2^2}{n_2}}$ $\nu \approx \dfrac{\left(\frac{s_1^2}{n_1} + \frac{s_2^2}{n_2} \right)^2}{\frac{(s_1^2/n_1)^2}{n_1 - 1} + \frac{(s_2^2/n_2)^2}{n_2 - 1}}$
$\mu_1 - \mu_2$	normality, n pairs, dependence	$\bar{d} \pm t_{\alpha/2, n-1} \cdot \dfrac{s_d}{\sqrt{n}}$
$p_1 - p_2$	binomial experiments, n_1, n_2 large, independence	$(\hat{p}_1 - \hat{p}_2) \pm$ $z_{\alpha/2} \cdot \sqrt{\dfrac{\hat{p}_1(1 - \hat{p}_1)}{n_1} + \dfrac{\hat{p}_2(1 - \hat{p}_2)}{n_2}}$

PERCENTAGE POINTS, STUDENT'S t-DISTRIBUTION

This table gives values of t such that

$$F(t) = \int_{-\infty}^{t} \frac{\Gamma\left(\frac{n+1}{2}\right)}{\sqrt{n\pi}\,\Gamma\left(\frac{n}{2}\right)} \left(1 + \frac{x^2}{n}\right) - \frac{n+1}{2}\, dx$$

for n, the number of degrees of freedom, equal to $1, 2, \ldots, 30, 40, 60, 120, \infty$; and for $F(t) = 0.60, 0.75, 0.90, 0.95, 0.975, 0.99,$ $0.995,$ and 0.9995. The t-distribution is symmetrical, so that $F(-t) = 1 - F(t)$

n/F	.60	.75	.90	.95	.975	.99	.995	.9995
1	.325	1.000	3.078	6.314	12.706	31.821	63.657	636.619
2	.289	.816	1.886	2.920	4.303	6.965	9.925	31.598
3	.277	.765	1.638	2.353	3.182	4.541	5.841	12.924
4	.271	.741	1.533	2.132	2.776	3.747	4.604	8.610
5	.267	.727	1.476	2.015	2.571	3.365	4.032	6.869
6	.265	.718	1.440	1.943	2.447	3.143	3.707	5.959
7	.263	.711	1.415	1.895	2.365	2.998	3.499	5.408
8	.262	.706	1.397	1.860	2.306	2.896	3.355	5.041
9	.261	.703	1.383	1.833	2.262	2.821	3.250	4.781
10	.260	.700	1.372	1.812	2.228	2.764	3.169	4.587
11	.260	.697	1.363	1.796	2.201	2.718	3.106	4.437
12	.259	.695	1.356	1.782	2.179	2.681	3.055	4.318
13	.259	.694	1.350	1.771	2.160	2.650	3.012	4.221
14	.258	.692	1.345	1.761	2.145	2.624	2.977	4.140
15	.258	.691	1.341	1.753	2.131	2.602	2.947	4.073
16	.258	.690	1.337	1.746	2.120	2.583	2.921	4.015
17	.257	.689	1.333	1.740	2.110	2.567	2.898	3.965
18	.257	.688	1.330	1.734	2.101	2.552	2.878	3.922
19	.257	.688	1.328	1.729	2.093	2.539	2.861	3.883
20	.257	.687	1.325	1.725	2.086	2.528	2.845	3.850
21	.257	.686	1.323	1.721	2.080	2.518	2.831	3.819
22	.256	.686	1.321	1.717	2.074	2.508	2.819	3.792
23	.256	.685	1.319	1.714	2.069	2.500	2.807	3.767
24	.256	.685	1.318	1.711	2.064	2.492	2.797	3.745
25	.256	.684	1.316	1.708	2.060	2.485	2.787	3.725
26	.256	.684	1.315	1.706	2.056	2.479	2.779	3.707
27	.256	.684	1.314	1.703	2.052	2.473	2.771	3.690
28	.256	.683	1.313	1.701	2.048	2.467	2.763	3.674
29	.256	.683	1.311	1.699	2.045	2.462	2.756	3.659
30	.256	.683	1.310	1.697	2.042	2.457	2.750	3.646
40	.255	.681	1.303	1.684	2.021	2.423	2.704	3.551
60	.254	.679	1.296	1.671	2.000	2.390	2.660	3.460
120	.254	.677	1.289	1.658	1.980	2.358	2.617	3.373
∞	.253	.674	1.282	1.645	1.960	2.326	2.576	3.291

*This table is abridged from the "Statistical Tables" of R. A. Fisher and Frank Yates published by Oliver & Boyd. Ltd., Edinburgh and London, 1938. It is here published with the kind permission of the authors and their publishers.

PERCENTAGE POINTS, CHI-SQUARE DISTRIBUTION

This table gives values of χ^2 such that

$$F(\chi)^2 = \int_0^{\chi^2} \frac{1}{2^{n/2}\Gamma\left(\frac{n}{2}\right)} x^{(n-2)/2} e^{-x/2}\, dx$$

for n, the number of degrees of freedom, equal to $1, 2, \ldots, 30$. For $n > 30$, a normal approximation is quite accurate. The expression $\sqrt{2x^2} - \sqrt{2n-1}$ is approximately normally distributed as the standard normal distribution. Thus χ_α^2, the α-point of the distribution, may be computed by the formula

$$\chi_\alpha^2 = \frac{1}{2}[x_\alpha + \sqrt{2n-1}]^2,$$

where x_α is the α-point of the cumulative normal distribution. For even values of n, $F(\chi^2)$ can be written as

$$1 - F(\chi^2) = \sum_{x=0}^{x'-1} \frac{e^{-\lambda}\lambda^x}{x!}$$

with $\lambda = \frac{1}{2}\chi^2$ and $x' = \frac{1}{2}n$. Thus the cumulative Chi-Square distribution is related to the cumulative Poisson distribution.
Another approximate formula for large n

$$\chi_\alpha^2 = n\left(1 - \frac{2}{9n} + z_\alpha\sqrt{\frac{2}{9n}}\right)^3$$

n = degrees of freedom
z_α = the normal deviate (the value of x for which $F(x)$ = the desired percentile).

x	1.282	1.645	1.960	2.326	2.576	3.090
$F(x)$	.90	.95	.975	.99	.995	.999

$\chi_{.99}^2 = 60[1 - 0.00370 + 2.326(0.06086)]^3 = 88.4$ is the 99th percentile for 60 degrees of freedom.

$$F(\chi^2) = \int_0^{\chi^2} \frac{1}{2^{n/2}\Gamma\left(\frac{n}{2}\right)} x^{n-2/2} e^{-x/2}\, dx$$

$n\backslash F$	.005	.010	.025	.050	.100	.250	.500	.750	.900	.950	.975	.990	.995
1	.0000393	.000157	.000982	.00393	.0158	.102	.455	1.32	2.71	3.84	5.02	6.63	7.88
2	.0100	.0201	.0506	.103	.211	.575	1.39	2.77	4.61	5.99	7.38	9.21	10.6
3	.0717	.115	.216	.352	.584	1.21	2.37	4.11	6.25	7.81	9.35	11.3	12.8
4	.207	.297	.484	.711	1.06	1.92	3.36	5.39	7.78	9.49	11.1	13.3	14.9
5	.412	.554	.831	1.15	1.61	2.67	4.35	6.63	9.24	11.1	12.8	15.1	16.7
6	.676	.872	1.24	1.64	2.20	3.45	5.35	7.84	10.6	12.6	14.4	16.8	18.5
7	.989	1.24	1.69	2.17	2.83	4.25	6.35	9.04	12.0	14.1	16.0	18.5	20.3
8	1.34	1.65	2.18	2.73	3.49	5.07	7.34	10.2	13.4	15.5	17.5	20.1	22.0
9	1.73	2.09	2.70	3.33	4.17	5.90	8.34	11.4	14.7	16.9	19.0	21.7	23.6
10	2.16	2.56	3.25	3.94	4.87	6.74	9.34	12.5	16.0	18.3	20.5	23.2	25.2
11	2.60	3.05	3.82	4.57	5.58	7.58	10.3	13.7	17.3	19.7	21.9	24.7	26.8
12	3.07	3.57	4.40	5.23	6.30	8.44	11.3	14.8	18.5	21.0	23.3	26.2	28.3
13	3.57	4.11	5.01	5.89	7.04	9.30	12.3	16.0	19.8	22.4	24.7	27.7	29.8
14	4.07	4.66	5.63	6.57	7.79	10.2	13.3	17.1	21.1	23.7	26.1	29.1	31.3
15	4.60	5.23	6.26	7.26	8.55	11.0	14.3	18.2	22.3	25.0	27.5	30.6	32.8
16	5.14	5.81	6.91	7.96	9.31	11.9	15.3	19.4	23.5	26.3	28.8	32.0	34.3
17	5.70	6.41	7.56	8.67	10.1	12.8	16.3	20.5	24.8	27.6	30.2	33.4	35.7
18	6.26	7.01	8.23	9.39	10.9	13.7	17.3	21.6	26.0	28.9	31.5	34.8	37.2
19	6.84	7.63	8.91	10.1	11.7	14.6	18.3	22.7	27.2	30.1	32.9	36.2	38.6
20	7.43	8.26	9.59	10.9	12.4	15.5	19.3	23.8	28.4	31.4	34.2	37.6	40.0
21	8.03	8.90	10.3	11.6	13.2	16.3	20.3	24.9	29.6	32.7	35.5	38.9	41.4
22	8.64	9.54	11.0	12.3	14.0	17.2	21.3	26.0	30.8	33.9	36.8	40.3	42.8
23	9.26	10.2	11.7	13.1	14.8	18.1	22.3	27.1	32.0	35.2	38.1	41.6	44.2
24	9.89	10.9	12.4	13.8	15.7	19.0	23.3	28.2	33.2	36.4	39.4	43.0	45.6
25	10.5	11.5	13.1	14.6	16.5	19.9	24.3	29.3	34.4	37.7	40.6	44.3	46.9
26	11.2	12.2	13.8	15.4	17.3	20.8	25.3	30.4	35.6	38.9	41.9	45.6	48.3
27	11.8	12.9	14.6	16.2	18.1	21.7	26.3	31.5	36.7	40.1	43.2	47.0	49.6
28	12.5	12.6	15.3	16.9	18.9	22.7	27.3	32.6	37.9	41.3	44.5	48.3	51.0
29	13.1	14.3	16.0	17.7	19.8	23.6	28.3	33.7	39.1	42.6	45.7	49.6	52.3
30	13.8	15.0	16.8	18.5	20.6	24.5	29.3	34.8	40.3	43.8	47.0	50.9	53.7

PERCENTAGE POINTS, F-DISTRIBUTION

This table gives values of F such that

$$F(F) = \int_0^F \frac{\Gamma\left(\frac{m+n}{2}\right)}{\Gamma\left(\frac{m}{2}\right)\Gamma\left(\frac{n}{2}\right)} m^{m/2} n^{n/2} x^{m-2/2}(n+mx)^{-(m+n)/2}\, dx$$

for selected values of m, the number of degrees of freedom of the numerator of F; and for selected values of n, the number of degrees freedom of the denominator of F. The table also provides values corresponding to $F(F) = .10, .05, .025, .01, .005, .001$ since $F_{1-\alpha}$ for m and n degrees of freedom is the reciprocal of F_α for n and m degrees of freedom. Thus

$$F_{.05}(4, 7) = \frac{1}{F_{.95}(7, 4)} = \frac{1}{6.09} = .164$$

$$F(F) = \int_0^F \frac{\Gamma\left(\frac{m+n}{2}\right)}{\Gamma\left(\frac{m}{2}\right)\Gamma\left(\frac{n}{2}\right)} m^{m/2} n^{n/2} x^{(m/2)-1}(n+mx)^{-(m+n)/2}\, dx = .90$$

$n\backslash m$	1	2	3	4	5	6	7	8	9	10	12	15	20	24	30	40	60	120	∞
1	39.86	49.50	53.59	55.83	57.24	58.20	58.91	59.44	59.86	60.19	60.71	61.22	61.74	62.00	62.26	62.53	62.79	63.06	63.33
2	8.53	9.00	9.16	9.24	9.29	9.33	9.35	9.37	9.38	9.39	9.41	9.42	9.44	9.45	9.46	9.47	9.47	9.48	9.49
3	5.54	5.46	5.39	5.34	5.31	5.28	5.27	5.25	5.24	5.23	5.22	5.20	5.18	5.18	5.17	5.16	5.15	5.14	5.13
4	4.54	4.32	4.19	4.11	4.05	4.01	3.98	3.95	3.94	3.92	3.90	3.87	3.84	3.83	3.82	3.80	3.79	3.78	3.76
5	4.06	3.78	3.62	3.52	3.45	3.40	3.37	3.34	3.32	3.30	3.27	3.24	3.21	3.19	3.17	3.16	3.14	3.12	3.10
6	3.78	3.46	3.29	3.18	3.11	3.05	3.01	2.98	2.96	2.94	2.90	2.87	2.84	2.82	2.80	2.78	2.76	2.74	2.72
7	3.59	3.26	3.07	2.96	2.88	2.83	2.78	2.75	2.72	2.70	2.67	2.63	2.59	2.58	2.56	2.54	2.51	2.49	2.47
8	3.46	3.11	2.92	2.81	2.73	2.67	2.62	2.59	2.56	2.54	2.50	2.46	2.42	2.40	2.38	2.36	2.34	2.32	2.29
9	3.36	3.01	2.81	2.69	2.61	2.55	2.51	2.47	2.44	2.42	2.38	2.34	2.30	2.28	2.25	2.23	2.21	2.18	2.16
10	3.29	2.92	2.73	2.61	2.52	2.46	2.41	2.38	2.35	2.32	2.28	2.24	2.20	2.18	2.16	2.13	2.11	2.08	2.06
11	3.23	2.86	2.66	2.54	2.45	2.39	2.34	2.30	2.27	2.25	2.21	2.17	2.12	2.10	2.08	2.05	2.03	2.00	1.97
12	3.18	2.81	2.61	2.48	2.39	2.33	2.28	2.24	2.21	2.19	2.15	2.10	2.06	2.04	2.01	1.99	1.96	1.93	1.90
13	3.14	2.76	2.56	2.43	2.35	2.28	2.23	2.20	2.16	2.14	2.10	2.05	2.01	1.98	1.96	1.93	1.90	1.88	1.85
14	3.10	2.73	2.52	2.39	2.31	2.24	2.19	2.15	2.12	2.10	2.05	2.01	1.96	1.94	1.91	1.89	1.86	1.83	1.80
15	3.07	2.70	2.49	2.36	2.27	2.21	2.16	2.12	2.09	2.06	2.02	1.97	1.92	1.90	1.87	1.85	1.82	1.79	1.76
16	3.05	2.67	2.46	2.33	2.24	2.18	2.13	2.09	2.06	2.03	1.99	1.94	1.89	1.87	1.84	1.81	1.78	1.75	1.72
17	3.03	2.64	2.44	2.31	2.22	2.15	2.10	2.06	2.03	2.00	1.96	1.91	1.86	1.84	1.81	1.78	1.75	1.72	1.69
18	3.01	2.62	2.42	2.29	2.20	2.13	2.08	2.04	2.00	1.98	1.93	1.89	1.84	1.81	1.78	1.75	1.72	1.69	1.66
19	2.99	2.61	2.40	2.27	2.18	2.11	2.06	2.02	1.98	1.96	1.91	1.86	1.81	1.79	1.76	1.73	1.70	1.67	1.63
20	2.97	2.59	2.38	2.25	2.16	2.09	2.04	2.00	1.96	1.94	1.89	1.84	1.79	1.77	1.74	1.71	1.68	1.64	1.61
21	2.96	2.57	2.36	2.23	2.14	2.08	2.02	1.98	1.95	1.92	1.87	1.83	1.78	1.75	1.72	1.69	1.66	1.62	1.59
22	2.95	2.56	2.35	2.22	2.13	2.06	2.01	1.97	1.93	1.90	1.86	1.81	1.76	1.73	1.70	1.67	1.64	1.60	1.57
23	2.94	2.55	2.34	2.21	2.11	2.05	1.99	1.95	1.92	1.89	1.84	1.80	1.74	1.72	1.69	1.66	1.62	1.59	1.55
24	2.93	2.54	2.33	2.19	2.10	2.04	1.98	1.94	1.91	1.88	1.83	1.78	1.73	1.70	1.67	1.64	1.61	1.57	1.53
25	2.92	2.53	2.32	2.18	2.09	2.02	1.97	1.93	1.89	1.87	1.82	1.77	1.72	1.69	1.66	1.63	1.59	1.56	1.52
26	2.91	2.52	2.31	2.17	2.08	2.01	1.96	1.92	1.88	1.86	1.81	1.76	1.71	1.68	1.65	1.61	1.58	1.54	1.50
27	2.90	2.51	2.30	2.17	2.07	2.00	1.95	1.91	1.87	1.85	1.80	1.75	1.70	1.67	1.64	1.60	1.57	1.53	1.49
28	2.89	2.50	2.29	2.16	2.06	2.00	1.94	1.90	1.87	1.84	1.79	1.74	1.69	1.66	1.63	1.59	1.56	1.52	1.48
29	2.89	2.50	2.28	2.15	2.06	1.99	1.93	1.89	1.86	1.83	1.78	1.73	1.68	1.65	1.62	1.58	1.55	1.51	1.47
30	2.88	2.49	2.28	2.14	2.05	1.98	1.93	1.88	1.85	1.82	1.77	1.72	1.67	1.64	1.61	1.57	1.54	1.50	1.46
40	2.84	2.44	2.23	2.09	2.00	1.93	1.87	1.83	1.79	1.76	1.71	1.66	1.61	1.57	1.54	1.51	1.47	1.42	1.38
60	2.79	2.39	2.18	2.04	1.95	1.87	1.82	1.77	1.74	1.71	1.66	1.60	1.54	1.51	1.48	1.44	1.40	1.35	1.29
120	2.75	2.35	2.13	1.99	1.90	1.82	1.77	1.72	1.68	1.65	1.60	1.55	1.48	1.45	1.41	1.37	1.32	1.26	1.19
∞	2.71	2.30	2.08	1.94	1.85	1.77	1.72	1.67	1.63	1.60	1.55	1.49	1.42	1.38	1.34	1.30	1.24	1.17	1.00

$F = \frac{s_1^2}{s_2^2} = \frac{S_1}{m} / \frac{S_2}{n}$, where $s_1^2 = S_1/m$ and $s_2^2 = S_2/n$ are independent mean squares estimating a common variance σ^2 and based on m and n degrees of freedom, respectively.

$$F(F) = \int_0^F \frac{\Gamma\left(\frac{m+n}{2}\right)}{\Gamma\left(\frac{m}{2}\right)\Gamma\left(\frac{n}{2}\right)} m^{m/2} n^{n/2} x^{(m/2)-1}(n+mx)^{-(m+n)/2}\, dx = .95$$

$n\backslash m$	1	2	3	4	5	6	7	8	9	10	12	15	20	24	30	40	60	120	∞
1	161.4	199.5	215.7	224.6	230.2	234.0	236.8	238.9	240.5	241.9	243.9	245.9	248.0	249.1	250.1	251.1	252.2	253.3	254.3
2	18.51	19.00	19.16	19.25	19.30	19.33	19.35	19.37	19.38	19.40	19.41	19.43	19.45	19.45	19.46	19.47	19.48	19.49	19.50
3	10.13	9.55	9.28	9.12	9.01	8.94	8.89	8.85	8.81	8.79	8.74	8.70	8.66	8.64	8.62	8.59	8.57	8.55	8.53
4	7.71	6.94	6.59	6.39	6.26	6.16	6.09	6.04	6.00	5.96	5.91	5.86	5.80	5.77	5.75	5.72	5.69	5.66	5.63
5	6.61	5.79	5.41	5.19	5.05	4.95	4.88	4.82	4.77	4.74	4.68	4.62	4.56	4.53	4.50	4.46	4.43	4.40	4.36
6	5.99	5.14	4.76	4.53	4.39	4.28	4.21	4.15	4.10	4.06	4.00	3.94	3.87	3.84	3.81	3.77	3.74	3.70	3.67
7	5.59	4.74	4.35	4.12	3.97	3.87	3.79	3.73	3.68	3.64	3.57	3.51	3.44	3.41	3.38	3.34	3.30	3.27	3.23
8	5.32	4.46	4.07	3.84	3.69	3.58	3.50	3.44	3.39	3.35	3.28	3.22	3.15	3.12	3.08	3.04	3.01	2.97	2.93
9	5.12	4.26	3.86	3.63	3.48	3.37	3.29	3.23	3.18	3.14	3.07	3.01	2.94	2.90	2.86	2.83	2.79	2.75	2.71
10	4.96	4.10	3.71	3.48	3.33	3.22	3.14	3.07	3.02	2.98	2.91	2.85	2.77	2.74	2.70	2.66	2.62	2.58	2.54
11	4.84	3.98	3.59	3.36	3.20	3.09	3.01	2.95	2.90	2.85	2.79	2.72	2.65	2.61	2.57	2.53	2.49	2.45	2.40
12	4.75	3.89	3.49	3.26	3.11	3.00	2.91	2.85	2.80	2.75	2.69	2.62	2.54	2.51	2.47	2.43	2.38	2.34	2.30
13	4.67	3.81	3.41	3.18	3.03	2.92	2.83	2.77	2.71	2.67	2.60	2.53	2.46	2.42	2.38	2.34	2.30	2.25	2.21
14	4.60	3.74	3.34	3.11	2.96	2.85	2.76	2.70	2.65	2.60	2.53	2.46	2.39	2.35	2.31	2.27	2.22	2.18	2.13
15	4.54	3.68	3.29	3.06	2.90	2.79	2.71	2.64	2.59	2.54	2.48	2.40	2.33	2.29	2.25	2.20	2.16	2.11	2.07
16	4.49	3.63	3.24	3.01	2.85	2.74	2.66	2.59	2.54	2.49	2.42	2.35	2.28	2.24	2.19	2.15	2.11	2.06	2.01
17	4.45	3.59	3.20	2.96	2.81	2.70	2.61	2.55	2.49	2.45	2.38	2.31	2.23	2.19	2.15	2.10	2.06	2.01	1.96
18	4.41	3.55	3.16	2.93	2.77	2.66	2.58	2.51	2.46	2.41	2.34	2.27	2.19	2.15	2.11	2.06	2.02	1.97	1.92
19	4.38	3.52	3.13	2.90	2.74	2.63	2.54	2.48	2.42	2.38	2.31	2.23	2.16	2.11	2.07	2.03	1.98	1.93	1.88
20	4.35	3.49	3.10	2.87	2.71	2.60	2.51	2.45	2.39	2.35	2.28	2.20	2.12	2.08	2.04	1.99	1.95	1.90	1.84
21	4.32	3.47	3.07	2.84	2.68	2.57	2.49	2.42	2.37	2.32	2.25	2.18	2.10	2.05	2.01	1.96	1.92	1.87	1.81
22	4.30	3.44	3.05	2.82	2.66	2.55	2.46	2.40	2.34	2.30	2.23	2.15	2.07	2.03	1.98	1.94	1.89	1.84	1.78
23	4.28	3.42	3.03	2.80	2.64	2.53	2.44	2.37	2.32	2.27	2.20	2.13	2.05	2.01	1.96	1.91	1.86	1.81	1.76
24	4.26	3.40	3.01	2.78	2.62	2.51	2.42	2.36	2.30	2.25	2.18	2.11	2.03	1.98	1.94	1.89	1.84	1.79	1.73
25	4.24	3.39	2.99	2.76	2.60	2.49	2.40	2.34	2.28	2.24	2.16	2.09	2.01	1.96	1.92	1.87	1.82	1.77	1.71
26	4.23	3.37	2.98	2.74	2.59	2.47	2.39	2.32	2.27	2.22	2.15	2.07	1.99	1.95	1.90	1.85	1.80	1.75	1.69
27	4.21	3.35	2.96	2.73	2.57	2.46	2.37	2.31	2.25	2.20	2.13	2.06	1.97	1.93	1.88	1.84	1.79	1.73	1.67
28	4.20	3.34	2.95	2.71	2.56	2.45	2.36	2.29	2.24	2.19	2.12	2.04	1.96	1.91	1.87	1.82	1.77	1.71	1.65
29	4.18	3.33	2.93	2.70	2.55	2.43	2.35	2.28	2.22	2.18	2.10	2.03	1.94	1.90	1.85	1.81	1.75	1.70	1.64
30	4.17	3.32	2.92	2.69	2.53	2.42	2.33	2.27	2.21	2.16	2.09	2.01	1.93	1.89	1.84	1.79	1.74	1.68	1.62
40	4.08	3.23	2.84	2.61	2.45	2.34	2.25	2.18	2.12	2.08	2.00	1.92	1.84	1.79	1.74	1.69	1.64	1.58	1.51
60	4.00	3.15	2.76	2.53	2.37	2.25	2.17	2.10	2.04	1.99	1.92	1.84	1.75	1.70	1.65	1.59	1.53	1.47	139
120	3.92	3.07	2.68	2.45	2.29	2.17	2.09	2.02.	1.96	1.91	1.83	1.75	1.66	1.61	1.55	1.50	1.43	1.35	1.25
∞	3.84	3.00	2.60	2.37	2.21	2.10	2.01	1.94	1.88	1.83	1.75	1.67	1.57	1.52	1.46	1.39	1.32	1.22	1.00

$F = \frac{s_1^2}{s_2^2} = \frac{S_1}{m} / \frac{S_2}{n}$, where $s_1^2 = S_1/m$ and $s_2^2 = S_2/n$ are independent mean squares estimating a common variance σ^2 and based on m and n degrees of freedom, respectively.

$$F(F) = \int_0^F \frac{\Gamma\left(\frac{m+n}{2}\right)}{\Gamma\left(\frac{m}{2}\right)\Gamma\left(\frac{n}{2}\right)} m^{m/2} n^{n/2} x^{(m/2)-1} (n+mx)^{-(m+n)/2} \, dx = .975$$

$n\backslash m$	1	2	3	4	5	6	7	8	9	10	12	15	20	24	30	40	60	120	∞
1	647.8	799.5	864.2	899.6	921.8	937.1	948.2	956.7	963.3	968.6	976.7	984.9	993.1	997.2	1001	1006	1010	1014	1018
2	38.51	39.00	39.17	39.25	39.30	39.33	39.36	39.37	39.39	39.40	39.41	39.43	39.45	39.46	39.46	39.47	39.48	39.49	39.50
3	17.44	16.04	15.44	15.10	14.88	14.73	14.62	14.54	14.47	14.42	14.34	14.25	14.17	14.12	14.08	14.04	13.99	13.95	13.90
4	12.22	10.65	9.98	9.60	9.36	9.20	9.07	8.98	8.90	8.84	8.75	8.66	8.56	8.51	8.46	8.41	8.36	8.31	8.26
5	10.01	8.43	7.76	7.39	7.15	6.98	6.85	6.76	6.68	6.62	6.52	6.43	6.33	6.28	6.23	6.18	6.12	6.07	6.02
6	8.81	7.26	6.60	6.23	5.99	5.82	5.70	5.60	5.52	5.46	5.37	5.27	5.17	5.12	5.07	5.01	4.96	4.90	4.85
7	8.07	6.54	5.89	5.52	5.29	5.12	4.99	4.90	4.82	4.76	4.67	4.57	4.47	4.42	4.36	4.31	4.25	4.20	4.14
8	7.57	6.06	5.42	5.05	4.82	4.65	4.53	4.43	4.36	4.30	4.20	4.10	4.00	3.95	3.89	3.84	3.78	3.73	3.67
9	7.21	5.71	5.08	4.72	4.48	4.32	4.20	4.10	4.03	3.96	3.87	3.77	3.67	3.61	3.56	3.51	3.45	3.39	3.33
10	6.94	5.46	4.83	4.47	4.24	4.07	3.95	3.85	3.78	3.72	3.62	3.52	3.42	3.37	3.31	3.26	3.20	3.14	3.08
11	6.72	5.26	4.63	4.28	4.04	3.88	3.76	3.66	3.59	3.53	3.43	3.33	3.23	3.17	3.12	3.06	3.00	2.94	2.88
12	6.55	5.10	4.47	4.12	3.89	3.73	3.61	3.51	3.44	3.37	3.28	3.18	3.07	3.02	2.96	2.91	2.85	2.79	2.72
13	6.41	4.97	4.35	4.00	3.77	3.60	3.48	3.39	3.31	3.25	3.15	3.05	2.95	2.89	2.84	2.78	2.72	2.66	2.60
14	6.30	4.86	4.24	3.89	3.66	3.50	3.38	3.29	3.21	3.15	3.05	2.95	2.84	2.79	2.73	2.67	2.61	2.55	2.49
15	6.20	4.77	4.15	3.80	3.58	3.41	3.29	3.20	3.12	3.06	2.96	2.86	2.76	2.70	2.64	2.59	2.52	2.46	2.40
16	6.12	4.69	4.08	3.73	3.50	3.34	3.22	3.12	3.05	2.99	2.89	2.79	2.68	2.63	2.57	2.51	2.45	2.38	2.32
17	6.04	4.62	4.01	3.66	3.44	3.28	3.16	3.06	2.98	2.92	2.82	2.72	2.62	2.56	2.50	2.44	2.38	2.32	2.25
18	5.98	4.56	3.95	3.61	3.38	3.22	3.10	3.01	2.93	2.87	2.77	2.67	2.56	2.50	2.44	2.38	2.32	2.26	2.19
19	5.92	4.51	3.90	3.56	3.33	3.17	3.05	2.96	2.88	2.82	2.72	2.62	2.51	2.45	2.39	2.33	2.27	2.20	2.13
20	5.87	4.46	3.86	3.51	3.29	3.13	3.01	2.91	2.84	2.77	2.68	2.57	2.46	2.41	2.35	2.29	2.22	2.16	2.09
21	5.83	4.42	3.82	3.48	3.25	3.09	2.97	2.87	2.80	2.73	2.64	2.53	2.42	2.37	2.31	2.25	2.18	2.11	2.04
22	5.79	4.38	3.78	3.44	3.22	3.05	2.93	2.84	2.76	2.70	2.60	2.50	2.39	2.33	2.27	2.21	2.14	2.08	2.00
23	5.75	4.35	3.75	3.41	3.18	3.02	2.90	2.81	2.73	2.67	2.57	2.47	2.36	2.30	2.24	2.18	2.11	2.04	1.97
24	5.72	4.32	3.72	3.38	3.15	2.99	2.87	2.78	2.70	2.64	2.54	2.44	2.33	2.27	2.21	2.15	2.08	2.01	1.94
25	5.69	4.29	3.69	3.35	3.13	2.97	2.85	2.75	2.68	2.61	2.51	2.41	2.30	2.24	2.18	2.12	2.05	1.98	1.91
26	5.66	4.27	3.67	3.33	3.10	2.94	2.82	2.73	2.65	2.59	2.49	2.39	2.28	2.22	2.16	2.09	2.03	1.95	1.88
27	5.63	4.24	3.65	3.31	3.08	2.92	2.80	2.71	2.63	2.57	2.47	2.36	2.25	2.19	2.13	2.07	2.00	1.93	1.85
28	5.61	4.22	3.63	3.29	3.06	2.90	2.78	2.69	2.61	2.55	2.45	2.34	2.23	2.17	2.11	2.05	1.98	1.91	1.83
29	5.59	4.20	3.61	3.27	3.04	2.88	2.76	2.67	2.59	2.53	2.43	2.32	2.21	2.15	2.09	2.03	1.96	1.89	1.81
30	5.57	4.18	3.59	3.25	3.03	2.87	2.75	2.65	2.57	2.51	2.41	2.31	2.20	2.14	2.07	2.01	1.94	1.87	1.79
40	5.42	4.05	3.46	3.13	2.90	2.74	2.62	2.53	2.45	2.39	2.29	2.18	2.07	2.01	1.94	1.88	1.80	1.72	1.64
60	5.29	3.93	3.34	3.01	2.79	2.63	2.51	2.41	2.33	2.27	2.17	2.06	1.94	1.88	1.82	1.74	1.67	1.58	1.48
120	5.15	3.80	3.23	2.89	2.67	2.52	2.39	2.30	2.22	2.16	2.05	1.94	1.82	1.76	1.69	1.61	1.53	1.43	1.31
∞	5.02	3.69	3.12	2.79	2.57	2.41	2.29	2.19	2.11	2.05	1.94	1.83	1.71	1.64	1.57	1.48	1.39	1.27	1.00

$F = \frac{s_1^2}{s_2^2} = \frac{S_1}{m} / \frac{S_2}{n}$, where $s_1^2 = S_1/m$ and $s_2^2 = S_2/n$ are independent mean squares estimating a common variance σ^2 and based on m and n degrees of freedom, respectively.

$$F(F) = \int_0^F \frac{\Gamma\left(\frac{m+n}{2}\right)}{\Gamma\left(\frac{m}{2}\right)\Gamma\left(\frac{n}{2}\right)} m^{m/2} n^{n/2} x^{(m/2)-1} (n+mx)^{-(m+n)/2}\, dx = .99$$

$n\backslash m$	1	2	3	4	5	6	7	8	9	10	12	15	20	24	30	40	60	120	∞
1	4052	4999.5	5403	5625	5764	5859	5928	5982	6022	6056	6106	6157	6209	6235	6261	6287	6313	6339	6366
2	98.50	99.00	99.17	99.25	99.30	99.33	99.36	99.37	99.39	99.40	99.42	99.43	99.45	99.46	99.47	99.47	99.48	99.49	99.50
3	34.12	30.82	29.46	28.71	28.24	27.91	27.67	27.49	27.35	27.23	27.05	26.87	26.69	26.60	26.50	26.41	26.32	26.22	26.13
4	21.20	18.00	16.69	15.98	15.52	15.21	14.98	14.80	14.66	14.55	14.37	14.20	14.02	13.93	13.84	13.75	13.65	13.56	13.46
5	16.26	13.27	12.06	11.39	10.97	10.67	10.46	10.29	10.16	10.05	9.89	9.72	9.55	9.47	9.38	9.29	9.20	9.11	9.02
6	13.75	10.92	9.78	9.15	8.75	8.47	8.26	8.10	7.98	7.87	7.72	7.56	7.40	7.31	7.23	7.14	7.06	6.97	6.88
7	12.25	9.55	8.45	7.85	7.46	7.19	6.99	6.84	6.72	6.62	6.47	6.31	6.16	6.07	5.99	5.91	5.82	5.74	5.65
8	11.26	8.65	7.59	7.01	6.63	6.37	6.18	6.03	5.91	5.81	5.67	5.52	5.36	5.28	5.20	5.12	5.03	4.95	4.86
9	10.56	8.02	6.99	6.42	6.06	5.80	5.61	5.47	5.35	5.26	5.11	4.96	4.81	4.73	4.65	4.57	4.48	4.40	4.31
10	10.04	7.56	6.55	5.99	5.64	5.39	5.20	5.06	4.94	4.85	4.71	4.56	4.41	4.33	4.25	4.17	4.08	4.00	3.91
11	9.65	7.21	6.22	5.67	5.32	5.07	4.89	4.74	4.63	4.54	4.40	4.25	4.10	4.02	3.94	3.86	3.78	3.69	3.60
12	9.33	6.93	5.95	5.41	5.06	4.82	4.64	4.50	4.39	4.30	4.16	4.01	3.86	3.78	3.70	3.62	3.54	3.45	3.36
13	9.07	6.70	5.74	5.21	4.86	4.62	4.44	4.30	4.19	4.10	3.96	3.82	3.66	3.59	3.51	3.43	3.34	3.25	3.17
14	8.86	6.51	5.56	5.04	4.69	4.46	4.28	4.14	4.03	3.94	3.80	3.66	3.51	3.43	3.35	3.27	3.18	3.09	3.00
15	8.68	6.36	5.42	4.89	4.56	4.32	4.14	4.00	3.89	3.80	3.67	3.52	3.37	3.29	3.21	3.13	3.05	2.96	2.87
16	8.53	6.23	5.29	4.77	4.44	4.20	4.03	3.89	3.78	3.69	3.55	3.41	3.26	3.18	3.10	3.02	2.93	2.84	2.75
17	8.40	6.11	5.18	4.67	4.34	4.10	3.93	3.79	3.68	3.59	3.46	3.31	3.16	3.08	3.00	2.92	2.83	2.75	2.65
18	8.29	6.01	5.09	4.58	4.25	4.01	3.84	3.71	3.60	3.51	3.37	3.23	3.08	3.00	2.92	2.84	2.75	2.66	2.57
19	8.18	5.93	5.01	4.50	4.17	3.94	3.77	3.63	3.52	3.43	3.30	3.15	3.00	2.92	2.84	2.76	2.67	2.58	2.49
20	8.10	5.85	4.94	4.43	4.10	3.87	3.70	3.56	3.46	3.37	3.23	3.09	2.94	2.86	2.78	2.69	2.61	2.52	2.42
21	8.02	5.78	4.87	4.37	4.04	3.81	3.64	3.51	3.40	3.31	3.17	3.03	2.88	2.80	2.72	2.64	2.55	2.46	2.36
22	7.95	5.72	4.82	4.31	3.99	3.76	3.59	3.45	3.35	3.26	3.12	2.98	2.83	2.75	2.67	2.58	2.50	2.40	2.31
23	7.88	5.66	4.76	4.26	3.94	3.71	3.54	3.41	3.30	3.21	3.07	2.93	2.78	2.70	2.62	2.54	2.45	2.35	2.26
24	7.82	5.61	4.72	4.22	3.90	3.67	3.50	3.36	3.26	3.17	3.03	2.89	2.74	2.66	2.58	2.49	2.40	2.31	2.21
25	7.77	5.57	4.68	4.18	3.85	3.63	3.46	3.32	3.22	3.13	2.99	2.85	2.70	2.62	2.54	2.45	2.36	2.27	2.17
26	7.72	5.53	4.64	4.14	3.82	3.59	3.42	3.29	3.18	3.09	2.96	2.81	2.66	2.58	2.50	2.42	2.33	2.23	2.13
27	7.68	5.49	4.60	4.11	3.78	3.56	3.39	3.26	3.15	3.06	2.93	2.78	2.63	2.55	2.47	2.38	2.29	2.20	2.10
28	7.64	5.45	4.57	4.07	3.75	3.53	3.36	3.23	3.12	3.03	2.90	2.75	2.60	2.52	2.44	2.35	2.26	2.17	2.06
29	7.60	5.42	4.54	4.04	3.73	3.50	3.33	3.20	3.09	3.00	2.87	2.73	2.57	2.49	2.41	2.33	2.23	2.14	2.03
30	7.56	5.39	4.51	4.02	3.70	3.47	3.30	3.17	3.07	2.98	2.84	2.70	2.55	2.47	2.39	2.30	2.21	2.11	2.01
40	7.31	5.18	4.31	3.83	3.51	3.29	3.12	2.99	2.89	2.80	2.66	2.52	2.37	2.29	2.20	2.11	2.02	1.92	1.80
60	7.08	4.98	4.13	3.65	3.34	3.12	2.95	2.82	2.72	2.63	2.50	2.35	2.20	2.12	2.03	1.94	1.84	1.73	1.60
120	6.85	4.79	3.95	3.48	3.17	2.96	2.79	2.66	2.56	2.47	2.34	2.19	2.03	1.95	1.86	1.76	1.66	1.53	1.38
∞	6.63	4.61	3.78	3.32	3.02	2.80	2.64	2.51	2.41	2.32	2.18	2.04	1.88	1.79	1.70	1.59	1.47	1.32	1.00

MOMENT OF INERTIA FOR VARIOUS BODIES OF MASS

The mass of the body is indicated by m

Body	Axis	Moment of inertia
Uniform thin rod of length l	Normal to the length, at one end	$m\frac{1}{3}l^2$
Uniform thin rod of length l	Normal to the length, at the center	$m\frac{1}{12}l^2$
Thin rectangular sheet, sides a and b	Through the center parallel to b	$m\frac{1}{12}a^2$
Thin rectangular sheet, sides a and b	Through the center perpendicular to the sheet	$m\frac{1}{12}(a^2 + b^2)$
Thin circular sheet of radius r	Normal to the plate through the center	$m\frac{1}{2}r^2$
Thin circular sheet of radius r	Along any diameter	$m\frac{1}{4}r^2$
Thin circular ring. Radii r_1 and r_2	Through center normal to plane of ring	$m\frac{1}{2}(r_1^2 + r_2^2)$
Thin circular ring. Radii r_1 and r_2	Any diameter	$m\frac{1}{4}(r_1^2 + r_2^2)$
Rectangular parallelepiped, edges a, b, and c	Through center perpendicular to face ab, (parallel to edge c)	$m\frac{1}{12}(a^2 + b^2)$
Sphere, radius r	Any diameter	$m\frac{2}{5}r^2$
Spherical shell, external radius, r_1, internal radius r_2	Any diameter	$m\frac{2}{5}\frac{(r_1^5 - r_2^5)}{(r_1^3 - r_2^3)}$
Spherical shell, very thin, mean radius, r	Any diameter	$m\frac{2}{3}r^2$
Right circular cylinder of radius r, length l	The longitudinal axis of the solid	$m\frac{1}{2}r^2$
Right circular cylinder of radius r, length l	Transverse diameter	$m\left(\frac{r^2}{4} + \frac{l^2}{12}\right)$
Hollow circular cylinder, length l, radii r_1 and r_2	The longitudinal axis of the figure	$m\frac{1}{2}(r_1^2 + r_2^2)$
Thin cylindrical shell, length l, mean radius, r	The longitudinal axis of the figure	mr^2
Hollow circular cylinder, length l, radii r_1 and r_2	Transverse diameter	$m\left(\frac{r_1^2 + r_2^2}{4} + \frac{l^2}{12}\right)$
Hollow circular cylinder, length l, very thin, mean radius r	Transverse diameter	$m\left(\frac{r^2}{2} + \frac{l^2}{12}\right)$
Elliptic cylinder, length l, transverse semiaxes a and b	Longitudinal axis	$m\frac{1}{4}(a^2 + b^2)$
Right cone, altitude h, radius of base r	Axis of the figure	$m\frac{3}{10}r^2$
Spheroid of revolution, equatorial radius r	Polar axis	$m\frac{2}{5}r^2$
Ellipsoid, axes $2a$, $2b$, $2c$	Axis $2a$	$m\frac{1}{5}(b^2 + c^2)$

Appendix B
Sources of Physical and Chemical Data

SOURCES OF PHYSICAL AND CHEMICAL DATA

In addition to the primary research journals, there are many useful sources of property data of the type contained in the *CRC Handbook of Chemistry and Physics*. A selected list of these is presented here, with emphasis on print and electronic sources whose contents have been subject to a reasonable level of quality control.

A. Data Journals

1. *Journal of Physical and Chemical Reference Data* – Published jointly by the National Institute of Standards and Technology and the American Institute of Physics, this quarterly journal contains compilations of evaluated data in chemistry, physics, and materials science. It is available in print and on the Internet. [ojps.aip.org/jpcrd/]
2. *Journal of Chemical and Engineering Data* – This bi-monthly journal of the American Chemical Society publishes articles reporting original experimental measurements carried out under carefully controlled conditions. The main emphasis is on thermochemical and thermophysical properties. Review articles with evaluated data from the literature are also published. [pubs.acs.org/journals/jceaax/index.html]
3. *Journal of Chemical Thermodynamics* – This journal publishes original research papers that include highly accurate measurements of thermodynamic and thermophysical properties. [www.sciencedirect.com/science/journal/00219614]
4. *Atomic Data and Nuclear Data Tables* – This is a bi-monthly journal containing compilations of data in atomic physics, nuclear physics, and related fields. [www.sciencedirect.com/science/journal/0092640x]
5. *Journal of Phase Equilibria and Diffusion* – This journal presents critically evaluated phase diagrams and related data on alloy systems. It is published by ASM International and is the successor to the previous ASM periodical *Bulletin Of Alloy Phase Diagrams*. [www.asm-intl.org]

B. Data Centers

This section lists selected organizations that perform a continuing function of compiling and critically evaluating data in specific fields of science.

1. **National Institute of Standards and Technology** – Under its Standard Reference Data program, NIST supports a number of data centers in chemistry, physics, and materials science. Topics covered include thermodynamics, fluid properties, chemical kinetics, mass spectroscopy, atomic spectroscopy, fundamental physical constants, ceramics, and crystallography. Address: Office of Standard Reference Data, National Institute of Standards and Technology, Gaithersburg, MD 20899 [www.nist.gov/srd/].
2. **Thermodynamics Research Center** – Now located at the National Institute of Standards and Technology, TRC maintains an extensive archive of data covering thermodynamic, thermochemical, and transport properties of organic compounds and mixtures. Data are distributed in both print and electronic form. Address: Mail code 838.00, 325 Broadway, Boulder, CO 80305-3328 [www.trc.nist.gov] .
3. **Design Institute for Physical Property Data** – Under the auspices of the American Institute of Chemical Engineers [www.aiche.org/dippr/], DIPPR offers evaluated data on industrially-important chemical compounds. The largest project deals with physical, thermodynamic, and transport properties of pure compounds. Address: Brigham Young University, Provo, UT 84602 [dippr.byu.edu] .
4. **Dortmund Data Bank** – Maintains extensive databases on thermodynamic and transport properties of pure compounds and mixtures of industrial interest. The data are distributed through DECHEMA, FIZ CHEMIE, and other outlets. An abbreviated database system is also available for educational use. Address: DDBST GmbH, Industriestr. 1, 26121 Oldenburg, Germany [www.ddbst.de].
5. **Cambridge Crystallographic Data Centre** – Maintains the Cambridge Structural Database of over 330,000 organic compounds. The data files and manipulation software are distributed in several ways. Address: 12 Union Rd., Cambridge CB2 1EZ, UK [www.ccdc.cam.ac.uk].
6. **FIZ Karlsruhe** – In addition to many bibliographic databases, FIZ Karlsruhe maintains the Inorganic Crystal Structure Database in collaboration with the National Institute of Standards and Technology. The ICSD contains the atomic coordinates and related data on over 50,000 inorganic crystals. Address: Fachinformationszentrum (FIZ) Karlsruhe, Hermann-von-Helmholtz-Platz 1, D-76344 Eggenstein-Leopoldshafen, Germany [www.fiz-karlsruhe.de].
7. **International Centre for Diffraction Data** – Maintains and distributes the Powder Diffraction File (PDF), a file of x-ray powder diffraction patterns used for identification of crystalline materials. The ICDD also distributes the NIST Crystal Data file, which contains lattice parameters for over 235,000 inorganic, organic, metal, and mineral crystalline materials. Address: 12 Campus Blvd., Newton Square, PA 19073-3273 [www.icdd.com].
8. **Research Collaboratory for Structural Bioinformatics** – Maintains the Protein Data Bank (PDB), a file of 3-dimensional structures of proteins and other biological macromolecules. Address: Department of Chemistry and Chemical Biology, Rutgers University, 610 Taylor Road, Piscataway, NJ 08854-8087 [www.rcsb.org].
9. **Toth Information Systems** – Maintains the Metals Crystallographic Data File (CRYSTMET). Address: 2045 Quincy Ave., Gloucester, ON, Canada K1J 6B2 [www.toth-canada.com].
10. **Atomic Mass Data Center** – Collects and evaluates high-precision data on masses of individual isotopes and maintains a comprehensive database. Address: C.S.N.S.M (IN2P3-CNRS), Batiment 108, F-91405 Orsay Campus, France [www.nndc.bnl.gov/amdc].
11. **Particle Data Group** – International center for data of high-energy physics; maintains database of properties of fundamental particles, which is published in both print and electronic form. Address: MS 50-308, Lawrence Berkeley National Laboratory, Berkeley, CA 94720 [pdg.lbl.gov].
12. **National Nuclear Data Center** – Maintains databases on nuclear structure and reactions, including neutron cross

sections. The NNDC is the U. S. node in an international network of nuclear data centers. Address: Brookhaven National Laboratory, Upton, NY 11973-5000 [www.nndc.bnl.gov].

13. **International Union of Pure and Applied Chemistry** – Address: PO Box 13757, Research Triangle Park, NC 27709-3757 [www.iupac.org]. IUPAC supports a number of long-term data projects, including these examples:

   a. **Solubility Data Project** – Carries out evaluation of all types of solubility data. The results are published in the Solubility Data Series, whose current outlet is the *Journal of Physical and Chemical Reference Data.* [www.iupac.org/divisions/V/cp5.html].

   b. **Kinetic Data for Atmospheric Chemistry** – Maintains a comprehensive database on the kinetics of reactions important in the chemistry of the atmosphere. [www.iupac-kinetic.ch.cam.ac.uk/]

   c. **International Thermodynamic Tables for the Fluid State** – Prepares definitive tables of the thermodynamic properties of industrially important fluids. Thirteen volumes have been published by IUPAC. [www.iupac.org/publications/books/seriestitles/]

   d. **Stability Constants Database** – Collection of metal-ligand stability constants and associated software. [www.acadsoft.co.uk]

C. Major Multi-Volume Handbook Series

1. *Chapman & Hall/CRC Chemical Dictionaries* – These originally appeared in print form as the *Dictionary of Organic Compounds, Dictionary of Natural Products,* etc. They are now published in electronic form and are available in CDROM format [www.crcpress.com] and on the Internet [www.chemnetbase.com]. The consolidated version, called the *Combined Chemical Dictionary,* has data on more than 450,000 compounds spanning all branches of chemistry. The coverage includes physical properties, biological sources, hazard information, uses, and literature references.

2. *Properties of Organic Compounds* – Originally published in three editions as the *Handbook of Data on Organic Compounds,* it is now in electronic form as *Properties of Organic Compounds.* The database includes about 30,000 compounds; physical properties and spectral data (mass, infrared, Raman, ultraviolet, and NMR) are covered. It is offered as CDROM [www.crcpress.com] and web access [www.chemnetbase.com].

3. *Beilstein Handbook of Organic Chemistry* – The classic source of data on organic compounds, dating from the 18th century, *Beilstein* was converted to electronic form in the last decade of the 20th century. Over 8 million compounds and 5 million chemical reactions are now covered, with a broad range of physical properties as well as synthetic methods and ecological data. The database is accessed by the CrossFire software [www.mdli.com].

4. *Gmelin Handbook of Inorganic and Organometallic Chemistry* – A subset of the information in the print series has been converted to electronic form and is now distributed in the same manner as *Beilstein.* In addition to the standard physical properties, the coverage includes a wide range of optical, magnetic, spectroscopic, thermal, and transport properties for about 1.4 million compounds [www.mdli.com].

5. *DECHEMA Chemical Data Series* – DECHEMA distributes the DETHERM database, which emphasizes data used in process design in the chemical industry, including thermodynamic and transport properties of about 20,000 pure compounds and 90,000 mixtures. Access is available through in-house databases and via the Internet. [www.dechema.de].

6. *Landolt-Börnstein Numerical Data and Functional Relationships in Science and Technology* - *Landolt-Börnstein* covers a very broad range of data in physics, chemistry, crystallography, materials science, biophysics, astronomy, and geophysics. Hard-copy volumes in the New Series (started in 1961) are still being published, and the entire New Series is now accessible on the Internet [www.landolt-boernstein.com].

D. Selected Single-Volume Handbooks

The following handbooks offer broad coverage of high-quality data in a single volume. This list is only representative; an extensive listing of handbooks in all fields of science may be found in *Handbooks and Tables in Science and Technology, Third Edition* (Russell H. Powell, ed., Oryx Press, Westport, CT, 1994).

1. *American Institute of Physics Handbook* – Although an old book, it contains much data that is still useful, especially in acoustics, mechanics, optics, and solid state physics. (Dwight E. Gray, ed., McGraw-Hill, New York, 1972)

2. *Constants of Inorganic Substances* - This book presents physical constants, thermodynamic data, solubility, reactivity, and other information on over 3000 inorganic compounds. Since it draws heavily on Russian literature, it contains a great deal of data that does not make its way into most U. S. handbooks. (R. A. Lidin, L. L. Andreeva, and V. A. Molochko, Begell House, New York, 1995)

3. *Handbook of Chemistry and Physics* – Now in the 84th Edition, the *CRC Handbook* covers data from most branches of chemistry and physics. The annual revisions permit regular updating of the information. Also available on CDROM [www.crcpress.com] and the web [hbcpnetbase.com]. (David R. Lide, ed., CRC Press, Boca Raton, FL, 2002)

4. *Handbook of Inorganic Compounds* – This book covers physical constants and solubility for about 3300 inorganic compounds. Also available on CDROM [www.crcpress.com]. (Dale L. Perry and Sidney L. Phillips, eds., CRC Press, Boca Raton, FL, 1995)

5. *Handbook of Physical Properties of Liquids and Gases* – This is a valuable source of data on all types of fluids, ranging from liquid and gaseous hydrocarbons to molten metals and ionized gases. Detailed tables of physical, thermodynamic, and transport properties are given for temperatures from the cryogenic region to 6000 K. Both Western and Russian literature is covered. (N. B. Vargaftik, Y. K. Vinogradov, and V. S. Yargin, Begell House, New York, 1996)

6. *Handbook of Physical Quantities* – The range of coverage is somewhat similar to the *CRC Handbook of Chemistry and Physics,* but with a stronger emphasis on physics than on chemistry. Solid state physics, lasers, nuclear physics, geophysics, and astronomy receive considerable attention. (Igor S. Grigoriev and Evgenii Z. Meilikhov, eds., CRC Press, Boca Raton, FL, 1997)

7. *Kaye & Laby Tables of Physical and Chemical Constants* – *Kaye & Laby* dates from 1911, and the 16th Edition was

prepared in 1995 by a committee of experts. The coverage extends to almost every field of physics and chemistry; data on a limited number of representative substances or materials are given for each topic. (Longman Group Limited, Harlow, Essex, UK, 1995)

8. *Lange's Handbook of Chemistry* –Provides broad coverage of chemical data; last updated in 1998. Also available on the web [www.knovel.com]. (John A. Dean, ed., McGraw-Hill, New York, 1998)

9. *Recommended Reference Materials for the Realization of Physicochemical Properties* – This IUPAC book emphasizes highly accurate data on substances and materials that can be used as calibration standards. It covers physical, thermal, optical, and electrical properties. (K. N. Marsh, ed., Blackwell Scientific Publications, Oxford, 1987)

10. *The Merck Index* – Now in its 13th Edition (published in 2001), The Merck Index is a widely used source of data on over 10,000 compounds, chosen particularly for their importance in biology, medicine, and ecology. A short monograph on each compound gives information on the synthesis and uses as well as physical and toxicological properties. Also available on CDROM [www.camsoft.com]. (Maryadele J. O'Neil, ed., Merck & Co., Whitehouse Station, NJ, 2001)

E. Summary of Useful Web Sites for Physical and Chemical Properties

Most of the web sites in the following list provide direct access to factual data on physical and chemical properties. However, the list also includes portals that link to different property databases or describe the procedure for gaining access to electronic sources of property data. There are also a few chemical directory sites, which are useful for obtaining formulas, synonyms, and registry numbers for substances of interest.

Web Site	Address	Comments
Acronyms and Symbols	www3.interscience.wiley.com/stasa/	Free service; useful for identifying acronyms for chemicals
Advanced Chemistry Development	www.acdlabs.com	Chemical directory, with programs for estimating physical and spectral properties
Alloy Center	products.asminternational.org/alloycenter/	Physical, electrical, thermal, and mechanical properties of alloys
American Mineralogist Crystal Structure Database	www.geo.arizona.edu/AMS/amcsd.php	Lattice constants of minerals
Atomic Mass Data Center	www.nndc.bnl.gov/amdc/	See B.10
Beilstein	www.mdli.com	See C.3
Biocatalysis/Biodegradation Database	umbbd.ahc.umn.edu/	Biocatalytic reactions, biodegradation of chemical compounds
BioCyc	biocyc.org/	Metabolic pathways of microorganisms
BioInfo Bank	gibk26.bse.kyutech.ac.jp/jouhou/jouhoubank.html	Portal to ProTherm (protein thermodynamics), ProNit (protein-nucleic acid interactions), biomolecule structures, etc.
Biological Macromolecule Crystallization Database	www.cstl.nist.gov/biotech/carb/gilliland-group/database/database/html	Crystal data and crystallization conditions for proteins, nucleic acids, and complexes
BRENDA	www.brenda.uni-koeln.de	Enzyme nomenclature and properties
Cambridge Structural Database	www.ccdc.cam.ac.uk	See B.5
Ceramic Properties Databases	www.ceramics.org/cic/propertiesdb.asp	Mechanical, thermal, and other properties of ceramic materials
Chapman & Hall/CRC Combined Chemical Dictionary	www.chemnetbase.com/scripts/ccdweb.exe	See C.1
ChemExper	www.chemexper.com/	Consolidated chemical catalogs from various suppliers; provides physical properties and safety data; links to molfiles and MSDS
Chemfinder	www.chemfinder.com	Chemical directory, with links to several property databases
Chemical Acronyms Database	www.oscar.chem.indiana.edu/cfdocs/libchem/acronyms/ acronymsearch.html	Useful for associating chemical names and acronyms
ChemIDplus	chem.sis.nlm.nih.gov/chemidplus/	Chemical directory
ChemIndustry	www.chemindustry.com/chemicals/	Chemical directory
CHEMnetBASE	www.chemnetbase.com/content/databases	Portal to C&H/CRC Chemical Dictionaries, Handbook of Chemistry and Physics, Properties of Organic Compounds, etc.
ChemWeb Databases	www.chemweb.com/databases/	Portal to many databases
CODATA Home Page	www.codata.org	Thermodynamic key values and fundamental constants
Crystallography Open Database (COD)	www.crystallography.net	Crystal data on 12,000 compounds

Web Site	Address	Comments
DECHEMA (DETHERM)	www.dechema.de	See C.5
DIPPR Pure Compound Database	dippr.byu.edu	See B.3
Dortmund Data Bank	www.ddbst.de	See B.4
Enzyme Nomenclature Database	www.expasy.ch/enzyme/	IUBMB nomenclature for enzymes
European Bioinformatics Institute	www.ebi.ac.uk/Databases/	Nucleotide and protein sequences, protein structures, enzyme nomenclature and reactions
FDM Reference Spectra Databases	www.fdmspectra.com/	Infrared, Raman, and mass spectra
FIZ Chemie Berlin	www.fiz-chemie.de	Portal to DETHERM (C.5), Dortmund Data Bank (B.4), Infotherm
FIZ Karlsruhe - ICSD	www.fiz-karlsruhe.de	See B.6
Fundamental Physical Constants	physics.nist.gov/constants	CODATA fundamental constants
Gmelin	www.mdli.com	See C.4
Handbook of Chemistry and Physics	hbcpnetbase.com	Web version of CRC Handbook
Hazardous Substances Data Bank	toxnet.nlm.nih.gov/cgi-bin/sis/htmlgen?HSDB	Physical and toxicological properties of chemicals of health or environmental importance
HITRAN Database	cfa-www.harvard.edu/hitran/	High resolution spectroscopic data for constituents of the atmosphere; parameters for calculating atmospheric transmission
Infotherm	www.fiz-chemie.de/infotherm	Physical and thermal properties of pure compounds and mixtures
International Centre for Diffraction Data	icdd.com	See B.7
International Spectroscopic Data Bank	www.is-db.org	All types of spectra, deposited by users. Access is free.
IUBMB	www.chem.qmw.ac.uk/iubmb/	Enzyme and nucleic acid nomenclature
IUPAC Home Page	www.iupac.org	See B.13
IUPAC Kinetics Data	www.iupac-kinetic.ch.cam.ac.uk/	See B.13.b
IUPAC Nomenclature Rules	www.chem.qmul.ac.uk/iupac/	Useful site for organic and biochemical nomenclature
Klotho Biochemical Compounds Declarative Database	www.biocheminfo.org/klotho/	Structure diagrams of biochemical molecules
Kovel.com	www.knovel.com	Portal to Lange's Handbook, Perry's Chemical Engineers' Handbook, etc.
Landolt-Börnstein	www.landolt-boernstein.com	See C.6
Lipidat	www.lipidat.chemistry.ohio-state.edu/	Structures and thermodynamic properties of lipids; crystal polymorphic transitions
MatWeb	www.matweb.com	Thermal, electrical, and mechanical properties of engineering materials
Metals Crystallographic Data File	www.tothcanada.com	See B.9
NASA Chemical Kinetics Data	jpldataeval.jpl.nasa.gov	Kinetic and photochemical data for stratospheric modeling
National Center for Biotechnology Information	www.ncbi.nlm.nih.gov	Portal to GenBank and other sequence databases
National Nuclear Data Center	www.nndc.bnl.gov	See B.12
National Toxicology Program	ntp-server.niehs.nih.gov	Chemical health and safety data
NIST Atomic Spectra Database	physics.nist.gov/PhysRefData/contents-atomic.html	Energy levels, wavelengths, and transition probabilities of atoms and atomic ions
NIST Ceramics Webbook	www.ceramics.nist.gov/	See B.1
NIST Chemistry Webbook	webbook.nist.gov	Broad range of physical, thermal, and spectral properties
NIST Data Gateway	srdata.nist.gov/gateway/	Portal to all NIST data systems; see B.1
NIST Physical Reference Data	physics.nist.gov/PhysRefData/	Atomic and molecular spectra, cross sections, x-ray attenuation, and dosimetry data

Web Site	Address	Comments
NLM Gateway	gateway.nlm.nih.gov/gw/Cmd	Portal to all National Library of Medicine databases
NMR Shift DB	www.nmrshiftdb.org	NMR data submitted by users
Nucleic Acid Database	ndbserver.rutgers.edu/	Crystal structures of nucleic acids
Particle Data Group	pdg.lbl.gov	See B.11
Polymers - A Property Database	www.polymersdatabase.com/	Properties of commercial polymers
Powder Diffraction File	www.icdd.com	See B.7
Properties of Organic Compounds	www.chemnetbase.com/scripts/pocweb.exe	See C.2
Protein Data Bank	www.rcsb.org	See B.8
Sigma-Aldrich	www.sigmaaldrich.com/	Chemical catalogs; includes some physical property data
SpecInfo	www3.interscience.wiley.com/cgi-bin/ mrwhome/109609148/HOME	IR, NMR, and mass spectra
Spectra Online	www.ftirsearch.com	FTIR and Raman spectra
SPRESI-web	www.spresi.de/	Structures, reactions, and some physical properties
STN Easy	stneasy.cas.org	Chemical directory (and access to Chemical Abstracts)
STN Easy-Europe	stneasy.fiz-karlsruhe.de	European node of STN Easy
STN Easy-Japan	stneasy-japan.cas.org	Japanese node of STN Easy
Swissprot	bo.expasy.org/enzyme/	Enzyme nomenclature and related information
Syracuse Research Corporation	www.syrres.com/esc/databases.htm	Properties of environmental interest
Table of Isotopes	ie.lbl.gov/education/isotopes.htm	Nuclear energy levels, moments, and other properties
Thermodynamics of Enzyme-Catalyzed Reactions	xpdb.nist.gov/enzyme_thermodynamics/	Equilibrium constants of biochemical reactions
Thermodynamics Research Center	www.trc.nist.gov	See B.2
TOXNET	toxnet.nlm.nih.gov	Portal to HSDB and other databases on hazardous chemicals
Wiley Interscience	www3.interscience.wiley.com/reference.html	Portal to Kirk-Othmer Encyclopedia of Chemical Technology, Ullmann's Encyclopedia of Industrial Chemistry, Encyclopedia of Reagents for Organic Synthesis, SpecInfo Database, etc.

Index

The most efficient way to use this index is to look for the pertinent *property* (e.g., vapor pressure, entropy), *process* (e.g., disposal of chemicals, calibration), or *general concept* (e.g., units, radiation). Most primary entries are subdivided into several secondary entries, e.g., under heat capacity there are 17 secondary entries such as air, metals, water, etc. Primary entries will be found for certain *classes of substances*, such as alloys, elements, organic compounds, refrigerants, semiconductors, etc. Primary entries are also given for the individual chemical elements and for a few compounds such as water and carbon dioxide. However, only the most important tables are listed under these substances. Therefore, the user will find in most cases that it is best to look first for the property of interest, then examine the table or tables that are referenced.

The reference given for each index term is the inclusive pages of the pertinent table (e.g., **8**-45 to 55). The introduction to each table describes the method of ordering the substances within that table.

The editor would be grateful for comments and suggestions on this index.

INDEX